CRC Handbook
of
Chemistry and Physics

A Ready-Reference Book of Chemical and Physical Data

HANDBOOK OF CHEMISTRY AND PHYSICS

2012-2013

93rd

EDITION

CRC PRESS

Editor-in-Chief

W. M. Haynes, Ph.D.
Scientist Emeritus
National Institute of Standards and Technology

Associate Editors

David R. Lide, Ph.D.
Former Director, Standard Reference Data
National Institute of Standards and Technology

Thomas J. Bruno, Ph.D.
Group Leader
National Institute of Standards and Technology

CRC Press
Taylor & Francis Group
Boca Raton London New York

CRC Press is an imprint of the
Taylor & Francis Group, an **informa** business

CRC Press
Taylor & Francis Group
6000 Broken Sound Parkway NW, Suite 300
Boca Raton, FL 33487-2742

© 2012 by Taylor & Francis Group, LLC
CRC Press is an imprint of Taylor & Francis Group, an Informa business

No claim to original U.S. Government works

Printed in the United States of America on acid-free paper
Version Date: 20120329

International Standard Book Number: 978-1-4398-8049-4 (Hardback)

Visit the Taylor & Francis Web site at
http://www.taylorandfrancis.com

and the CRC Press Web site at
http://www.crcpress.com

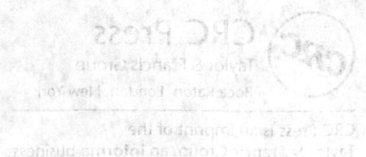

PREFACE

The 93rd Edition of the *Handbook* includes new tables, major updates and expansions, and a reorganization of two sections. A major effort was carried out to strengthen the section on analytical chemistry. As a result, the following new tables were added:

Section 8: Analytical Chemistry
- Introduction
- Analytical Standardization and Calibration
- Mass- and Volume-Based Concentration Units
- Properties of Common Cross-Linked Silicone Stationary Phases
- Detectors for Gas Chromatography
- Solid-Phase Microextraction Sorbents
- Eluotropic Values of Solvents on Octadecylsilane and Octylsilane
- Wavelength–Wavenumber Conversion Table
- Middle Range Infrared Absorption Correlation Charts
- Common Spurious Infrared Absorption Bands
- Properties of Important NMR Nuclei
- Proton NMR Absorption of Major Chemical Families
- ^{15}N NMR Chemical Shifts of Major Chemical Families
- Natural Abundance of Important Isotopes
- Common Mass Spectral Fragmentation Patterns of Organic Compound Families
- Common Mass Spectral Fragments Lost
- Major Reference Masses in the Spectrum of Heptacosafluorotributylamine (Perfluorotributylamine)
- Common Spurious Signals Observed in Mass Spectrometers
- Standards for Laboratory Weights
- Organic Analytical Reagents for the Determination of Inorganic Cations

In addition to adding the new tables to the section on Analytical Chemistry, several tables were moved from Section 9 on Molecular Structure and Spectroscopy to Section 8 on Analytical Chemistry for which the fit is better based on the subject matter.

The results of the IUPAC 2009 biennial review of atomic-weight determinations were released just as the 92nd edition of the *Handbook* was going into production. At that time, the new values were added only to the table of standard atomic weights in Section 1 and to the table on the inside back cover. However, these changes have now been made to the atomic weights in the Periodic Table at the front of the *Handbook* and to The Elements in Section 4.

Other significant updates and expansions of tables for the 93rd Edition include the following:

Section 1: Basic Constants, Units, and Conversion Factors
- Major update of CODATA Recommended Values of the Fundamental Physical Constants: 2010
- Update of Atomic Masses and Abundances

Section 8: Analytical Chemistry
- Major expansion of Abbreviations and Symbols Used in Analytical Chemistry
- Major update of ^{13}C NMR Absorptions of Major Functional Groups
- Major update of Indicators for Acids and Bases
- Major update of Preparation of Special Analytical Reagents

Section 9: Molecular Structure and Spectroscopy
- Update of Bond Dissociation Energies

Section 10: Atomic, Molecular, and Optical Physics
- Update of Electronic Affinities
- Update of Atomic and Molecular Polarizabilities

Section 14: Geophysics, Astronomy, and Acoustics
- Update of Solar Irradiance at the Earth
- Update of the Global Temperature Trend table to include 2011 data from NASA

Section 16: Health and Safety Information
- Major update of Chemical Carcinogens to include data from the 2011 National Toxicology Program report

Again this year, in order to maintain a manageable number of pages and allow space for growth of the *Handbook*, the indexes for molecular formulae and CAS registry numbers for the Physical Constants of Organic Compounds Table and the index for CAS registry numbers for the Physical Constants of Inorganic Compounds Table have been omitted from the hard-copy edition of the *Handbook*. However, they are available in the electronic versions of the *Handbook* and by email request to the Editor-in-Chief: william.haynes@taylorandfrancis.com

The success of the *Handbook* is very dependent on feedback from its users. The Editor-in-Chief will appreciate any suggestions from readers on proposed new topics for the *Handbook* or comments on how the usefulness of the *Handbook* may be improved in future editions. Please send your comments to the Editor-in-Chief: william.haynes@taylorandfrancis.com

Numerous international experts make key contributions to the *Handbook*. These contributors are listed on pages immediately following the Preface. Their efforts play a key role in the quality and diversity of the subject matter covered in the *Handbook*. I also acknowledge the sound advice and guidance of the Editorial Advisory Board members of the *Handbook*, who are listed in the front matter. Fiona Macdonald, Publisher – Chemical & Life Sciences, CRC Press/Taylor & Francis Group has been of great assistance and support in providing oversight to ensure that we meet our goals. Thanks also to Glen Butler, Pam Morrell, Theresa Delforn, and James Yanchak for their detailed, cooperative work and extreme care in the production of the *Handbook*.

W. M. Haynes
March 2012

The 93rd Edition of the *CRC Handbook of Chemistry and Physics* is dedicated in memory of my grandparents,
Willie Cameron and Elsie Craig Haynes
Charles William and Martha Ann Juliana Frances Young

Note on the Ordering of Chemical Compounds: Several different ordering schemes for lists of chemical compounds are used in this book. The long tables, Physical Constants of Organic Compounds and Physical Constants of Inorganic Compound, are ordered by name (generally the systematic name), but indexes to synonyms, formulas, and CAS Registry Numbers are available. If the table is very short and includes only familiar substances, the listing is usually alphabetical by name or common formula. Many tables of intermediate length are ordered by molecular formula using a modification of the Hill convention. In this convention the molecular formula is written with C first, H second, and then all other elements in alphabetical order of their chemical symbols. For tables with organic compounds only, the sequence of entries is determined by the alphabetical order of elements in the molecular formula and the number of atoms of each element, in ascending order, e.g., C_3H_7Cl, C_3H_7N, C_3H_7NO, $C_3H_7NO_2$, etc. (For organic compounds, a quick way to look up the molecular formula is to use the Physical Constants of Organic Compounds table, which starts on Page **3-1**, and its synonym index on Page **3-554**.) In tables containing non-carbon compounds, those are usually listed first, followed by a separate listing of compounds that do contain carbon. This is a departure from the strict Hill convention as followed by Chemical Abstracts Service, where the molecular formulas beginning with A and B precede the formulas for carbon-containing compounds, while those beginning with D... Z follow. For tabular displays, as opposed to an index, it appears more convenient to the user if the non-carbon compounds are listed as a block, rather than being split by the longer list of carbon compounds.

FOREWORD

It's an honor for me to be asked to write the Foreword for this 93rd Edition of the *CRC Handbook of Chemistry and Physics*, following in the footsteps of some of my scientific heroes such as Linus Pauling. I wish I had something as powerful to say as Pauling, who in his Foreword to the 74th Edition revealed that he had "spent much of [his] time during five months in the summer of 1919 pouring over the [*CRC Handbook*] tables and thinking about the properties of substances [when] working as a paving engineer in the mountains of southern Oregon."

Not having such a great story to tell, I pulled out my green 48th Edition (1967–1968), hoping to find some pages more dog-eared than others. This helped spark my memory of the sections that I'd found most valuable as an undergraduate chemistry major at Grinnell College. Not surprisingly, given the paucity of electronic calculators at the time, the tables of logarithms and antilogarithms are well used, along with the "Natural Trigonometric Functions." (I do wonder, however, what an unnatural trig function would look like.) The compilation of derivatives and integrals is so clearly organized that it remains a handy reference as I teach in Boulder's new Interdisciplinary Quantitative Biology (IQ Biology) graduate curriculum. And certainly the 554-page table of "Physical Constants of Organic Compounds" brings back memories of identifying unknowns in organic chem lab, where this compilation was much used.

One feature of the *Handbook* that has gone extinct is the blank rectangle on the cover of my copy, on which—using the sheet of gold leaf included with the book—I inscribed my name. At the risk of revealing my "nerdiness" as a college student, I recall that just possessing the *Handbook* gave me a sense of power over all constants, physical and chemical. I shared some of these memories with my wife, Carol, also a PhD chemist, and she quickly one-upped me by pulling out her first copy of the *Handbook* — a tan 47th Edition (1966–1967), one year senior to mine. She had won Grinnell's Sophomore Book Award for her stellar freshman academic year and had used the gift certificate to buy the *Handbook*. I guess one good nerd deserves another.

Well, time goes on, and I'm delighted to see that this 93rd Edition is now available as an e-book. This is most appropriate, as we live in a transitional period where for many applications electronic resources are most convenient; yet the print version allows one to see the scope of a topic and the organization of data in a way that's hard for many of us to grasp from electronic resources. Some of you will have a strong preference for the printed copy, others the electronic. Some of you, like Linus Pauling, may be pouring over the tables, page after page, while most will be incisively pulling out individual bits of information as needed. Whatever your particular need, I trust that you'll find your new copy of the *CRC Handbook of Chemistry and Physics* useful, clearly organized… and perhaps even inspiring.

Thomas R. Cech, PhD
Nobel Laureate (Chemistry, 1989)
Director, University of Colorado Biofrontiers Institute
March 2012

EDITORIAL ADVISORY BOARD

CURRENT CONTRIBUTORS

Lev I. Berger
California Institute of Electronics and
  Materials Science
2115 Flame Tree Way
Hemet, California 92545

Thomas J. Bruno
Thermophysical Properties Division
National Institute of Standards and
  Technology
Boulder, Colorado 80305

Charles E. Carraher
Department of Chemistry and
  Biochemistry
Florida Atlantic University
Boca Raton, Florida 33431

Robert D. Chirico
Thermodynamics Research Center
Thermophysical Properties Division
National Institute of Standards and
  Technology
Boulder, Colorado 80305

Ivan Cibulka
Department of Physical Chemistry
Institute of Chemical Technology
CZ-166 28 Prague, Czech Republic

Arthur K. Covington
Department of Chemistry
University of Newcastle
Newcastle upon Tyne NE1 7RU
England

Christopher J. Cramer
Department of Chemistry
University of Minnesota
Minneapolis, Minnesota 55455

Vladimir Diky
Thermodynamics Research Center
Thermophysical Properties Division
National Institute of Standards and
  Technology
Boulder, Colorado 80305

Michael Frenkel
Thermodynamics Research Center
Thermophysical Properties Division
National Institute of Standards and
  Technology
Boulder, Colorado 80305

Jeffrey R. Fuhr
Quantum Measurement Division
National Institute of Standards and
  Technology
Gaithersburg, Maryland 20899

Jürgen Gmehling
Universität Oldenburg
Falkutät V, Technische Chemie
D-26111 Oldenburg, Germany

Robert N. Goldberg
Biochemical Science Division
National Institute of Standards and
  Technology
Gaithersburg, Maryland 20899

Allan H. Harvey
Thermophysical Properties Division
National Institute of Standards and
  Technology
Boulder, Colorado 80305

Steven R. Heller
Chemical and Biochemical Reference Data
  Division
National Institute of Standards and
  Technology
Gaithersburg, Maryland 20899

Norman E. Holden
National Nuclear Data Center
Brookhaven National Laboratory
Upton, New York 11973

Marcia L. Huber
Thermophysical Properties Division
National Institute of Standards and
  Technology
Boulder, Colorado 80305

Andrei Kazakov
Thermodynamics Research Center
Thermophysical Properties Division
National Institute of Standards and
  Technology
Boulder, Colorado 80305

Daniel E. Kelleher
Quantum Measurement Division
National Institute of Standards and
  Technology
Gaithersburg, Maryland 20899

Carolyn A. Koh
Center for Hydrate Research
Colorado School of Mines
1600 Illinois Street
Golden, Colorado 80401

Willem H. Koppenol
Dept. CHAB
Lab. f. Anorg. Chemie, HC1 H211
Wolfgang-Pauli-Strasse 10
ETH Hönggerberg
CH-8093 Zürich, Switzerland

Eric W. Lemmon
Thermophysical Properties Division
National Institute of Standards and
  Technology
Boulder, Colorado 80305

Frank J. Lovas
8616 Melwood Rd.
Bethesda, Maryland 20817

Yu-Ran Luo
College of Chemistry and Chemical
  Engineering
Chongqing University
Chongqing 400044, China

Serguei N. Lvov
Department of Energy and Mineral
  Engineering
Pennsylvania State University
University Park, Pennsylvania 16802

Manjeera Mantina
Department of Chemistry
University of Minnesota
Minneapolis, Minnesota 55455

William C. Martin
Quantum Measurement Division
National Institute of Standards and
  Technology
Gaithersburg, Maryland 20899

Alan D. McNaught
8 Cavendish Avenue
Cambridge CB1 7US
England

Thomas M. Miller
Air Force Research Laboratory/VSBP
29 Randolph Rd.
Hanscom AFB, Massachusetts 01731-3010

Nasser Moazzen-Ahmadi
Department of Physics and Astronomy
University of Calgary
2500 University Drive NW
Calgary, Alberta T2N 1N4, Canada

Peter J. Mohr
Quantum Measurement Division
National Institute of Standards and
  Technology
Gaithersburg, Maryland 20899

Chris D. Muzny
Thermodynamics Research Center
Thermophysical Properties Division
National Institute of Standards and
  Technology
Boulder, Colorado 80305

David B. Newell
Quantum Measurement Division
National Institute of Standards and
　Technology
Gaithersburg, Maryland 20899

Irving Ozier
Department of Physics and Astronomy
University of British Columbia
6224 Agricultural Road
Vancouver, British Columbia V6T 1Z1,
　Canada

Larissa I. Podobedova
Quantum Measurement Division
National Institute of Standards and
　Technology
Gaithersburg, Maryland 20899

Cedric J. Powell
Surface and Microanalysis Science
　Division
National Institute of Standards and
　Technology
Gaithersburg, Maryland 20899

Joseph Reader
Quantum Measurement Division
National Institute of Standards and
　Technology
Gaithersburg, Maryland 20899

E. Dendy Sloan
Center for Hydrate Research
Colorado School of Mines
1600 Illinois Street
Golden, Colorado 80401

Lewis E. Snyder
Astronomy Department
University of Illinois
Urbana, Illinois 61801

Paris D. N. Svoronos
Queensborough Community College
City University of New York
Bayside, NY 11364

Barry N. Taylor
Quantum Measurement Division
National Institute of Standards and
　Technology
Gaithersburg, Maryland 20899

Donald G. Truhlar
Department of Chemistry
University of Minnesota
Minneapolis, Minnesota 55455

Rosendo Valero
Chemistry Department
University of Coimbra
Coimbra, Portugal

Wolfgang L. Wiese
Quantum Measurement Division
National Institute of Standards and
　Technology
Gaithersburg, Maryland 20899

Christian Wohlfarth
Martin Luther University
Institute of Physical Chemistry
Mühlpforte 1
06108 Halle (Saale), Germany

Daniel Zwillinger
Mathematics Department
Rensselaer Polytechnic Institute
Troy, New York 12180

TABLE OF CONTENTS

SECTION 6: FLUID PROPERTIES

SECTION 9: MOLECULAR STRUCTURE AND SPECTROSCOPY

SECTION 10: ATOMIC, MOLECULAR, AND OPTICAL PHYSICS

SECTION 11: NUCLEAR AND PARTICLE PHYSICS

SECTION 12: PROPERTIES OF SOLIDS

SECTION 13: POLYMER PROPERTIES

SECTION 14: GEOPHYSICS, ASTRONOMY, AND ACOUSTICS

SECTION 15: PRACTICAL LABORATORY DATA

SECTION 16: HEALTH AND SAFETY INFORMATION

Section 1
Basic Constants, Units, and Conversion Factors

CODATA RECOMMENDED VALUES OF THE FUNDAMENTAL PHYSICAL CONSTANTS: 2010*

Peter J. Mohr,[†] Barry N. Taylor,[‡] and David B. Newell[§]

National Institute of Standards and Technology, Gaithersburg, Maryland 20899-8420, USA

This report gives the 2010 self-consistent set of values of the basic constants and conversion factors of physics and chemistry recommended by the Committee on Data for Science and Technology (CODATA) for international use. The 2010 adjustment takes into account the data considered in the 2006 adjustment as well as the data that became available from 1 January 2007, after the closing date of that adjustment, until 31 December 2010, the closing date of the new adjustment. The 2010 set replaces the previously recommended 2006 CODATA set and may also be found on the World Wide Web at physics.nist.gov/constants.

Reference

1. Nakamura, K., K . Hagiwara, K . Hikasa, H. Murayama, M. Tanabashi, T. Watari, C. Amsler, M. Antonelli, D. M. Asner, H. Baer, and e. al, 2010, J. Phys. G 37, 075021.

TABLE I: An abbreviated list of the CODATA recommended values of the fundamental constants of physics and chemistry based on the 2010 adjustment.

Quantity	Symbol	Numerical value	Unit	Relative std. uncert. u_r
speed of light in vacuum	c, c_0	299 792 458	m s^{-1}	exact
magnetic constant	μ_0	$4\pi \times 10^{-7}$	N A^{-2}	
		$= 12.566\,370\,614... \times 10^{-7}$	N A^{-2}	exact
electric constant $1/\mu_0 c^2$	ϵ_0	$8.854\,187\,817... \times 10^{-12}$	F m^{-1}	exact
Newtonian constant of gravitation	G	$6.673\,84(80) \times 10^{-11}$	m^3 kg^{-1} s^{-2}	1.2×10^{-4}
Planck constant	h	$6.626\,069\,57(29) \times 10^{-34}$	J s	4.4×10^{-8}
$h/2\pi$	$\hbar$	$1.054\,571\,726(47) \times 10^{-34}$	J s	4.4×10^{-8}
elementary charge	e	$1.602\,176\,565(35) \times 10^{-19}$	C	2.2×10^{-8}
magnetic flux quantum $h/2e$	Φ_0	$2.067\,833\,758(46) \times 10^{-15}$	Wb	2.2×10^{-8}
conductance quantum $2e^2/h$	G_0	$7.748\,091\,7346(25) \times 10^{-5}$	S	3.2×10^{-10}
electron mass	m_e	$9.109\,382\,91(40) \times 10^{-31}$	kg	4.4×10^{-8}
proton mass	m_p	$1.672\,621\,777(74) \times 10^{-27}$	kg	4.4×10^{-8}
proton-electron mass ratio	m_p/m_e	1836.152 672 45(75)		4.1×10^{-10}
fine-structure constant $e^2/4\pi\epsilon_0 \hbar c$	α	$7.297\,352\,5698(24) \times 10^{-3}$		3.2×10^{-10}
inverse fine-structure constant	α^{-1}	137.035 999 074(44)		3.2×10^{-10}
Rydberg constant $\alpha^2 m_e c/2h$	R_∞	10 973 731.568 539(55)	m^{-1}	5.0×10^{-12}

*This report was prepared by the authors under the auspices of the CODATA Task Group on Fundamental Constants. The members of the task group are:

F. Cabiati, Istituto Nazionale di Ricerca Metrologica, Italy

J. Fischer, Physikalisch-Technische Bundesanstalt, Germany

J. Flowers, National Physical Laboratory, United Kingdom

K. Fujii, National Metrology Institute of Japan, Japan

S. G. Karshenboim, Pulkovo Observatory, Russian Federation

P. J. Mohr, National Institute of Standards and Technology, United States of America

D. B. Newell, National Institute of Standards and Technology, United States of America

F. Nez, Laboratoire Kastler-Brossel, France

K. Pachucki, University of Warsaw, Poland

T. J. Quinn, Bureau international des poids et mesures

B. N. Taylor, National Institute of Standards and Technology, United States of America

B. M. Wood, National Research Council, Canada

Z. Zhang, National Institute of Metrology, China (People's Republic of)

[†]Electronic address: mohr@nist.gov

[‡]Electronic address: barry.taylor@nist.gov

[§]Electronic address: dnewell@nist.gov

TABLE I: (Continued.)

Quantity	Symbol	Numerical value	Unit	Relative std. uncert. u_r
Avogadro constant	N_A, L	$6.022\,141\,29(27) \times 10^{23}$	mol^{-1}	4.4×10^{-8}
Faraday constant $N_A e$	F	$96\,485.3365(21)$	C mol^{-1}	2.2×10^{-8}
molar gas constant	R	$8.314\,4621(75)$	$\text{J mol}^{-1}\,\text{K}^{-1}$	9.1×10^{-7}
Boltzmann constant R/N_A	k	$1.380\,6488(13) \times 10^{-23}$	J K^{-1}	9.1×10^{-7}
Stefan-Boltzmann constant $(\pi^2/60)k^4/\hbar^3 c^2$	σ	$5.670\,373(21) \times 10^{-8}$	$\text{W m}^{-2}\,\text{K}^{-4}$	3.6×10^{-6}

Non-SI units accepted for use with the SI

Quantity	Symbol	Numerical value	Unit	Relative std. uncert. u_r
electron volt (e/C) J	eV	$1.602\,176\,565(35) \times 10^{-19}$	J	2.2×10^{-8}
(unified) atomic mass unit $\frac{1}{12}m(^{12}\text{C})$	u	$1.660\,538\,921(73) \times 10^{-27}$	kg	4.4×10^{-8}

TABLE II: The CODATA recommended values of the fundamental constants of physics and chemistry based on the 2010 adjustment.

Quantity	Symbol	Numerical value	Unit	Relative std. uncert. u_r
UNIVERSAL				
speed of light in vacuum	c, c_0	$299\,792\,458$	m s^{-1}	exact
magnetic constant	μ_0	$4\pi \times 10^{-7}$	N A^{-2}	
		$= 12.566\,370\,614... \times 10^{-7}$	N A^{-2}	exact
electric constant $1/\mu_0 c^2$	ϵ_0	$8.854\,187\,817... \times 10^{-12}$	F m^{-1}	exact
characteristic impedance of vacuum $\mu_0 c$	Z_0	$376.730\,313\,461...$	Ω	exact
Newtonian constant of gravitation	G	$6.673\,84(80) \times 10^{-11}$	$\text{m}^3\,\text{kg}^{-1}\,\text{s}^{-2}$	1.2×10^{-4}
	$G/\hbar c$	$6.708\,37(80) \times 10^{-39}$	$(\text{GeV}/c^2)^{-2}$	1.2×10^{-4}
Planck constant	h	$6.626\,069\,57(29) \times 10^{-34}$	J s	4.4×10^{-8}
		$4.135\,667\,516(91) \times 10^{-15}$	eV s	2.2×10^{-8}
$h/2\pi$	$\hbar$	$1.054\,571\,726(47) \times 10^{-34}$	J s	4.4×10^{-8}
		$6.582\,119\,28(15) \times 10^{-16}$	eV s	2.2×10^{-8}
	$\hbar c$	$197.326\,9718(44)$	MeV fm	2.2×10^{-8}
Planck mass $(\hbar c/G)^{1/2}$	m_P	$2.176\,51(13) \times 10^{-8}$	kg	6.0×10^{-5}
energy equivalent	$m_P c^2$	$1.220\,932(73) \times 10^{19}$	GeV	6.0×10^{-5}
Planck temperature $(\hbar c^5/G)^{1/2}/k$	T_P	$1.416\,833(85) \times 10^{32}$	K	6.0×10^{-5}
Planck length $\hbar/m_P c = (\hbar G/c^3)^{1/2}$	l_P	$1.616\,199(97) \times 10^{-35}$	m	6.0×10^{-5}
Planck time $l_P/c = (\hbar G/c^5)^{1/2}$	t_P	$5.391\,06(32) \times 10^{-44}$	s	6.0×10^{-5}
ELECTROMAGNETIC				
elementary charge	e	$1.602\,176\,565(35) \times 10^{-19}$	C	2.2×10^{-8}
	e/h	$2.417\,989\,348(53) \times 10^{14}$	A J^{-1}	2.2×10^{-8}
magnetic flux quantum $h/2e$	Φ_0	$2.067\,833\,758(46) \times 10^{-15}$	Wb	2.2×10^{-8}
conductance quantum $2e^2/h$	G_0	$7.748\,091\,7346(25) \times 10^{-5}$	S	3.2×10^{-10}
inverse of conductance quantum	G_0^{-1}	$12\,906.403\,7217(42)$	Ω	3.2×10^{-10}
Josephson constant[1] $2e/h$	K_J	$483\,597.870(11) \times 10^9$	Hz V^{-1}	2.2×10^{-8}
von Klitzing constant[2] $h/e^2 = \mu_0 c/2\alpha$	R_K	$25\,812.807\,4434(84)$	Ω	3.2×10^{-10}
Bohr magneton $e\hbar/2m_e$	μ_B	$927.400\,968(20) \times 10^{-26}$	J T^{-1}	2.2×10^{-8}
		$5.788\,381\,8066(38) \times 10^{-5}$	eV T^{-1}	6.5×10^{-10}
	μ_B/h	$13.996\,245\,55(31) \times 10^9$	Hz T^{-1}	2.2×10^{-8}
	μ_B/hc	$46.686\,4498(10)$	$\text{m}^{-1}\,\text{T}^{-1}$	2.2×10^{-8}
	μ_B/k	$0.671\,713\,88(61)$	K T^{-1}	9.1×10^{-7}
nuclear magneton $e\hbar/2m_p$	μ_N	$5.050\,783\,53(11) \times 10^{-27}$	J T^{-1}	2.2×10^{-8}
		$3.152\,451\,2605(22) \times 10^{-8}$	eV T^{-1}	7.1×10^{-10}
	μ_N/h	$7.622\,593\,57(17)$	MHz T^{-1}	2.2×10^{-8}
	μ_N/hc	$2.542\,623\,527(56) \times 10^{-2}$	$\text{m}^{-1}\,\text{T}^{-1}$	2.2×10^{-8}
	μ_N/k	$3.658\,2682(33) \times 10^{-4}$	K T^{-1}	9.1×10^{-7}

---

[1]See Table IV for the conventional value adopted internationally for realizing representations of the volt using the Josephson effect.
[2]See Table IV for the conventional value adopted internationally for realizing representations of the ohm using the quantum Hall effect.

TABLE II: *(Continued)*.

Quantity	Symbol	Numerical value	Unit	Relative std. uncert. u_r
ATOMIC AND NUCLEAR				
General				
fine-structure constant $e^2/4\pi\epsilon_0\hbar c$	α	$7.297\,352\,5698(24) \times 10^{-3}$		3.2×10^{-10}
inverse fine-structure constant	α^{-1}	$137.035\,999\,074(44)$		3.2×10^{-10}
Rydberg constant $\alpha^2 m_e c/2h$	R_∞	$10\,973\,731.568\,539(55)$	m^{-1}	5.0×10^{-12}
	$R_\infty c$	$3.289\,841\,960\,364(17) \times 10^{15}$	Hz	5.0×10^{-12}
	$R_\infty hc$	$2.179\,872\,171(96) \times 10^{-18}$	J	4.4×10^{-8}
		$13.605\,692\,53(30)$	eV	2.2×10^{-8}
Bohr radius $\alpha/4\pi R_\infty = 4\pi\epsilon_0\hbar^2/m_e e^2$	a_0	$0.529\,177\,210\,92(17) \times 10^{-10}$	m	3.2×10^{-10}
Hartree energy $e^2/4\pi\epsilon_0 a_0 = 2R_\infty hc = \alpha^2 m_e c^2$	E_h	$4.359\,744\,34(19) \times 10^{-18}$	J	4.4×10^{-8}
		$27.211\,385\,05(60)$	eV	2.2×10^{-8}
quantum of circulation	$h/2m_e$	$3.636\,947\,5520(24) \times 10^{-4}$	$m^2\ s^{-1}$	6.5×10^{-10}
	h/m_e	$7.273\,895\,1040(47) \times 10^{-4}$	$m^2\ s^{-1}$	6.5×10^{-10}
Electroweak				
Fermi coupling constant[3]	$G_F/(\hbar c)^3$	$1.166\,364(5) \times 10^{-5}$	GeV^{-2}	4.3×10^{-6}
weak mixing angle[4] θ_W (on-shell scheme) $\sin^2\theta_W = s_W^2 \equiv 1 - (m_W/m_Z)^2$	$\sin^2\theta_W$	$0.2223(21)$		9.5×10^{-3}
Electron, e^-				
electron mass	m_e	$9.109\,382\,91(40) \times 10^{-31}$	kg	4.4×10^{-8}
		$5.485\,799\,0946(22) \times 10^{-4}$	u	4.0×10^{-10}
energy equivalent	$m_e c^2$	$8.187\,105\,06(36) \times 10^{-14}$	J	4.4×10^{-8}
		$0.510\,998\,928(11)$	MeV	2.2×10^{-8}
electron-muon mass ratio	m_e/m_μ	$4.836\,331\,66(12) \times 10^{-3}$		2.5×10^{-8}
electron-tau mass ratio	m_e/m_τ	$2.875\,92(26) \times 10^{-4}$		9.0×10^{-5}
electron-proton mass ratio	m_e/m_p	$5.446\,170\,2178(22) \times 10^{-4}$		4.1×10^{-10}
electron-neutron mass ratio	m_e/m_n	$5.438\,673\,4461(32) \times 10^{-4}$		5.8×10^{-10}
electron-deuteron mass ratio	m_e/m_d	$2.724\,437\,1095(11) \times 10^{-4}$		4.0×10^{-10}
electron-triton mass ratio	m_e/m_t	$1.819\,200\,0653(17) \times 10^{-4}$		9.1×10^{-10}
electron-helion mass ratio	m_e/m_h	$1.819\,543\,0761(17) \times 10^{-4}$		9.2×10^{-10}
electron to alpha particle mass ratio	m_e/m_α	$1.370\,933\,555\,78(55) \times 10^{-4}$		4.0×10^{-10}
electron charge to mass quotient	$-e/m_e$	$-1.758\,820\,088(39) \times 10^{11}$	$C\ kg^{-1}$	2.2×10^{-8}
electron molar mass $N_A m_e$	$M(e), M_e$	$5.485\,799\,0946(22) \times 10^{-7}$	$kg\ mol^{-1}$	4.0×10^{-10}
Compton wavelength $h/m_e c$	λ_C	$2.426\,310\,2389(16) \times 10^{-12}$	m	6.5×10^{-10}
$\lambda_C/2\pi = \alpha a_0 = \alpha^2/4\pi R_\infty$	$\lambdabar_C$	$386.159\,268\,00(25) \times 10^{-15}$	m	6.5×10^{-10}
classical electron radius $\alpha^2 a_0$	r_e	$2.817\,940\,3267(27) \times 10^{-15}$	m	9.7×10^{-10}
Thomson cross section $(8\pi/3)r_e^2$	σ_e	$0.665\,245\,8734(13) \times 10^{-28}$	m^2	1.9×10^{-9}
electron magnetic moment	μ_e	$-928.476\,430(21) \times 10^{-26}$	$J\ T^{-1}$	2.2×10^{-8}
to Bohr magneton ratio	μ_e/μ_B	$-1.001\,159\,652\,180\,76(27)$		2.6×10^{-13}
to nuclear magneton ratio	μ_e/μ_N	$-1838.281\,970\,90(75)$		4.1×10^{-10}
electron magnetic moment anomaly $\|\mu_e\|/\mu_B - 1$	a_e	$1.159\,652\,180\,76(27) \times 10^{-3}$		2.3×10^{-10}
electron g-factor $-2(1 + a_e)$	g_e	$-2.002\,319\,304\,361\,53(53)$		2.6×10^{-13}
electron-muon magnetic moment ratio	μ_e/μ_μ	$206.766\,9896(52)$		2.5×10^{-8}
electron-proton magnetic moment ratio	μ_e/μ_p	$-658.210\,6848(54)$		8.1×10^{-9}
electron to shielded proton magnetic moment ratio (H_2O, sphere, 25 °C)	μ_e/μ_p'	$-658.227\,5971(72)$		1.1×10^{-8}
electron-neutron magnetic moment ratio	μ_e/μ_n	$960.920\,50(23)$		2.4×10^{-7}
electron-deuteron magnetic moment ratio	μ_e/μ_d	$-2143.923\,498(18)$		8.4×10^{-9}
electron to shielded helion magnetic moment ratio (gas, sphere, 25 °C)	μ_e/μ_h'	$864.058\,257(10)$		1.2×10^{-8}

---

[3] Value recommended by the Particle Data Group (Nakamura et al., 2010).

[4] Based on the ratio of the masses of the W and Z bosons m_W/m_Z recommended by the Particle Data Group (Nakamura et al., 2010). The value for $\sin^2\theta_W$ they recommend, which is based on a particular variant of the modified minimal subtraction ($\overline{MS}$) scheme, is $\sin^2\hat{\theta}_W(M_Z) = 0.231\,16(13)$.

TABLE II: *(Continued)*.

Quantity	Symbol	Numerical value	Unit	Relative std. uncert. u_r
electron gyromagnetic ratio $2\|\mu_e\|/\hbar$	γ_e	$1.760\,859\,708(39) \times 10^{11}$	$\mathrm{s^{-1}\,T^{-1}}$	2.2×10^{-8}
	$\gamma_e/2\pi$	$28\,024.952\,66(62)$	$\mathrm{MHz\,T^{-1}}$	2.2×10^{-8}
Muon, μ^-				
muon mass	m_μ	$1.883\,531\,475(96) \times 10^{-28}$	kg	5.1×10^{-8}
		$0.113\,428\,9267(29)$	u	2.5×10^{-8}
energy equivalent	$m_\mu c^2$	$1.692\,833\,667(86) \times 10^{-11}$	J	5.1×10^{-8}
		$105.658\,3715(35)$	MeV	3.4×10^{-8}
muon-electron mass ratio	m_μ/m_e	$206.768\,2843(52)$		2.5×10^{-8}
muon-tau mass ratio	m_μ/m_τ	$5.946\,49(54) \times 10^{-2}$		9.0×10^{-5}
muon-proton mass ratio	m_μ/m_p	$0.112\,609\,5272(28)$		2.5×10^{-8}
muon-neutron mass ratio	m_μ/m_n	$0.112\,454\,5177(28)$		2.5×10^{-8}
muon molar mass $N_A m_\mu$	$M(\mu)$, M_μ	$0.113\,428\,9267(29) \times 10^{-3}$	$\mathrm{kg\,mol^{-1}}$	2.5×10^{-8}
muon Compton wavelength $h/m_\mu c$	$\lambda_{C,\mu}$	$11.734\,441\,03(30) \times 10^{-15}$	m	2.5×10^{-8}
$\lambda_{C,\mu}/2\pi$	$\lambdabar_{C,\mu}$	$1.867\,594\,294(47) \times 10^{-15}$	m	2.5×10^{-8}
muon magnetic moment	μ_μ	$-4.490\,448\,07(15) \times 10^{-26}$	$\mathrm{J\,T^{-1}}$	3.4×10^{-8}
to Bohr magneton ratio	μ_μ/μ_B	$-4.841\,970\,44(12) \times 10^{-3}$		2.5×10^{-8}
to nuclear magneton ratio	μ_μ/μ_N	$-8.890\,596\,97(22)$		2.5×10^{-8}
muon magnetic moment anomaly $\|\mu_\mu\|/(e\hbar/2m_\mu) - 1$	a_μ	$1.165\,920\,91(63) \times 10^{-3}$		5.4×10^{-7}
muon g-factor $-2(1 + a_\mu)$	g_μ	$-2.002\,331\,8418(13)$		6.3×10^{-10}
muon-proton magnetic moment ratio	μ_μ/μ_p	$-3.183\,345\,107(84)$		2.6×10^{-8}
Tau, τ^-				
tau mass[5]	m_τ	$3.167\,47(29) \times 10^{-27}$	kg	9.0×10^{-5}
		$1.907\,49(17)$	u	9.0×10^{-5}
energy equivalent	$m_\tau c^2$	$2.846\,78(26) \times 10^{-10}$	J	9.0×10^{-5}
		$1776.82(16)$	MeV	9.0×10^{-5}
tau-electron mass ratio	m_τ/m_e	$3477.15(31)$		9.0×10^{-5}
tau-muon mass ratio	m_τ/m_μ	$16.8167(15)$		9.0×10^{-5}
tau-proton mass ratio	m_τ/m_p	$1.893\,72(17)$		9.0×10^{-5}
tau-neutron mass ratio	m_τ/m_n	$1.891\,11(17)$		9.0×10^{-5}
tau molar mass $N_A m_\tau$	$M(\tau)$, M_τ	$1.907\,49(17) \times 10^{-3}$	$\mathrm{kg\,mol^{-1}}$	9.0×10^{-5}
tau Compton wavelength $h/m_\tau c$	$\lambda_{C,\tau}$	$0.697\,787(63) \times 10^{-15}$	m	9.0×10^{-5}
$\lambda_{C,\tau}/2\pi$	$\lambdabar_{C,\tau}$	$0.111\,056(10) \times 10^{-15}$	m	9.0×10^{-5}
Proton, p				
proton mass	m_p	$1.672\,621\,777(74) \times 10^{-27}$	kg	4.4×10^{-8}
		$1.007\,276\,466\,812(90)$	u	8.9×10^{-11}
energy equivalent	$m_p c^2$	$1.503\,277\,484(66) \times 10^{-10}$	J	4.4×10^{-8}
		$938.272\,046(21)$	MeV	2.2×10^{-8}
proton-electron mass ratio	m_p/m_e	$1836.152\,672\,45(75)$		4.1×10^{-10}
proton-muon mass ratio	m_p/m_μ	$8.880\,243\,31(22)$		2.5×10^{-8}
proton-tau mass ratio	m_p/m_τ	$0.528\,063(48)$		9.0×10^{-5}
proton-neutron mass ratio	m_p/m_n	$0.998\,623\,478\,26(45)$		4.5×10^{-10}
proton charge to mass quotient	e/m_p	$9.578\,833\,58(21) \times 10^{7}$	$\mathrm{C\,kg^{-1}}$	2.2×10^{-8}
proton molar mass $N_A m_p$	$M(p)$, M_p	$1.007\,276\,466\,812(90) \times 10^{-3}$	$\mathrm{kg\,mol^{-1}}$	8.9×10^{-11}
proton Compton wavelength $h/m_p c$	$\lambda_{C,p}$	$1.321\,409\,856\,23(94) \times 10^{-15}$	m	7.1×10^{-10}
$\lambda_{C,p}/2\pi$	$\lambdabar_{C,p}$	$0.210\,308\,910\,47(15) \times 10^{-15}$	m	7.1×10^{-10}
proton rms charge radius	r_p	$0.8775(51) \times 10^{-15}$	m	5.9×10^{-3}
proton magnetic moment	μ_p	$1.410\,606\,743(33) \times 10^{-26}$	$\mathrm{J\,T^{-1}}$	2.4×10^{-8}
to Bohr magneton ratio	μ_p/μ_B	$1.521\,032\,210(12) \times 10^{-3}$		8.1×10^{-9}
to nuclear magneton ratio	μ_p/μ_N	$2.792\,847\,356(23)$		8.2×10^{-9}
proton g-factor $2\mu_p/\mu_N$	g_p	$5.585\,694\,713(46)$		8.2×10^{-9}

---

[5]This and all other values involving m_τ are based on the value of $m_\tau c^2$ in MeV recommended by the Particle Data Group (Nakamura et al., 2010).

TABLE II: (Continued).

Quantity	Symbol	Numerical value	Unit	Relative std. uncert. u_r
proton-neutron magnetic moment ratio	μ_p/μ_n	$-1.459\,898\,06(34)$		2.4×10^{-7}
shielded proton magnetic moment (H_2O, sphere, 25 °C)	μ'_p	$1.410\,570\,499(35) \times 10^{-26}$	$J\,T^{-1}$	2.5×10^{-8}
to Bohr magneton ratio	μ'_p/μ_B	$1.520\,993\,128(17) \times 10^{-3}$		1.1×10^{-8}
to nuclear magneton ratio	μ'_p/μ_N	$2.792\,775\,598(30)$		1.1×10^{-8}
proton magnetic shielding correction $1 - \mu'_p/\mu_p$ (H_2O, sphere, 25 °C)	σ'_p	$25.694(14) \times 10^{-6}$		5.3×10^{-4}
proton gyromagnetic ratio $2\mu_p/\hbar$	γ_p	$2.675\,222\,005(63) \times 10^8$	$s^{-1}\,T^{-1}$	2.4×10^{-8}
	$\gamma_p/2\pi$	$42.577\,4806(10)$	$MHz\,T^{-1}$	2.4×10^{-8}
shielded proton gyromagnetic ratio $2\mu'_p/\hbar$ (H_2O, sphere, 25 °C)	γ'_p	$2.675\,153\,268(66) \times 10^8$	$s^{-1}\,T^{-1}$	2.5×10^{-8}
	$\gamma'_p/2\pi$	$42.576\,3866(10)$	$MHz\,T^{-1}$	2.5×10^{-8}

Neutron, n

Quantity	Symbol	Numerical value	Unit	Relative std. uncert. u_r
neutron mass	m_n	$1.674\,927\,351(74) \times 10^{-27}$	kg	4.4×10^{-8}
		$1.008\,664\,916\,00(43)$	u	4.2×10^{-10}
energy equivalent	$m_n c^2$	$1.505\,349\,631(66) \times 10^{-10}$	J	4.4×10^{-8}
		$939.565\,379(21)$	MeV	2.2×10^{-8}
neutron-electron mass ratio	m_n/m_e	$1838.683\,6605(11)$		5.8×10^{-10}
neutron-muon mass ratio	m_n/m_μ	$8.892\,484\,00(22)$		2.5×10^{-8}
neutron-tau mass ratio	m_n/m_τ	$0.528\,790(48)$		9.0×10^{-5}
neutron-proton mass ratio	m_n/m_p	$1.001\,378\,419\,17(45)$		4.5×10^{-10}
neutron-proton mass difference	$m_n - m_p$	$2.305\,573\,92(76) \times 10^{-30}$	kg	3.3×10^{-7}
		$0.001\,388\,449\,19(45)$	u	3.3×10^{-7}
energy equivalent	$(m_n - m_p)c^2$	$2.072\,146\,50(68) \times 10^{-13}$	J	3.3×10^{-7}
		$1.293\,332\,17(42)$	MeV	3.3×10^{-7}
neutron molar mass $N_A m_n$	$M(n), M_n$	$1.008\,664\,916\,00(43) \times 10^{-3}$	$kg\,mol^{-1}$	4.2×10^{-10}
neutron Compton wavelength $h/m_n c$	$\lambda_{C,n}$	$1.319\,590\,9068(11) \times 10^{-15}$	m	8.2×10^{-10}
$\lambda_{C,n}/2\pi$	$\lambdabar_{C,n}$	$0.210\,019\,415\,68(17) \times 10^{-15}$	m	8.2×10^{-10}
neutron magnetic moment	μ_n	$-0.966\,236\,47(23) \times 10^{-26}$	$J\,T^{-1}$	2.4×10^{-7}
to Bohr magneton ratio	μ_n/μ_B	$-1.041\,875\,63(25) \times 10^{-3}$		2.4×10^{-7}
to nuclear magneton ratio	μ_n/μ_N	$-1.913\,042\,72(45)$		2.4×10^{-7}
neutron g-factor $2\mu_n/\mu_N$	g_n	$-3.826\,085\,45(90)$		2.4×10^{-7}
neutron-electron magnetic moment ratio	μ_n/μ_e	$1.040\,668\,82(25) \times 10^{-3}$		2.4×10^{-7}
neutron-proton magnetic moment ratio	μ_n/μ_p	$-0.684\,979\,34(16)$		2.4×10^{-7}
neutron to shielded proton magnetic moment ratio (H_2O, sphere, 25 °C)	μ_n/μ'_p	$-0.684\,996\,94(16)$		2.4×10^{-7}
neutron gyromagnetic ratio $2|\mu_n|/\hbar$	γ_n	$1.832\,471\,79(43) \times 10^8$	$s^{-1}\,T^{-1}$	2.4×10^{-7}
	$\gamma_n/2\pi$	$29.164\,6943(69)$	$MHz\,T^{-1}$	2.4×10^{-7}

Deuteron, d

Quantity	Symbol	Numerical value	Unit	Relative std. uncert. u_r
deuteron mass	m_d	$3.343\,583\,48(15) \times 10^{-27}$	kg	4.4×10^{-8}
		$2.013\,553\,212\,712(77)$	u	3.8×10^{-11}
energy equivalent	$m_d c^2$	$3.005\,062\,97(13) \times 10^{-10}$	J	4.4×10^{-8}
		$1875.612\,859(41)$	MeV	2.2×10^{-8}
deuteron-electron mass ratio	m_d/m_e	$3670.482\,9652(15)$		4.0×10^{-10}
deuteron-proton mass ratio	m_d/m_p	$1.999\,007\,500\,97(18)$		9.2×10^{-11}
deuteron molar mass $N_A m_d$	$M(d), M_d$	$2.013\,553\,212\,712(77) \times 10^{-3}$	$kg\,mol^{-1}$	3.8×10^{-11}
deuteron rms charge radius	r_d	$2.1424(21) \times 10^{-15}$	m	9.8×10^{-4}
deuteron magnetic moment	μ_d	$0.433\,073\,489(10) \times 10^{-26}$	$J\,T^{-1}$	2.4×10^{-8}
to Bohr magneton ratio	μ_d/μ_B	$0.466\,975\,4556(39) \times 10^{-3}$		8.4×10^{-9}
to nuclear magneton ratio	μ_d/μ_N	$0.857\,438\,2308(72)$		8.4×10^{-9}
deuteron g-factor μ_d/μ_N	g_d	$0.857\,438\,2308(72)$		8.4×10^{-9}
deuteron-electron magnetic moment ratio	μ_d/μ_e	$-4.664\,345\,537(39) \times 10^{-4}$		8.4×10^{-9}
deuteron-proton magnetic moment ratio	μ_d/μ_p	$0.307\,012\,2070(24)$		7.7×10^{-9}
deuteron-neutron magnetic moment ratio	μ_d/μ_n	$-0.448\,206\,52(11)$		2.4×10^{-7}

TABLE II: *(Continued).*

Quantity	Symbol	Numerical value	Unit	Relative std. uncert. u_r
Triton, t				
triton mass	m_t	$5.007\,356\,30(22) \times 10^{-27}$	kg	4.4×10^{-8}
		$3.015\,500\,7134(25)$	u	8.2×10^{-10}
energy equivalent	$m_t c^2$	$4.500\,387\,41(20) \times 10^{-10}$	J	4.4×10^{-8}
		$2808.921\,005(62)$	MeV	2.2×10^{-8}
triton-electron mass ratio	m_t/m_e	$5496.921\,5267(50)$		9.1×10^{-10}
triton-proton mass ratio	m_t/m_p	$2.993\,717\,0308(25)$		8.2×10^{-10}
triton molar mass $N_A m_t$	$M(t)$, M_t	$3.015\,500\,7134(25) \times 10^{-3}$	kg mol^{-1}	8.2×10^{-10}
triton magnetic moment	μ_t	$1.504\,609\,447(38) \times 10^{-26}$	J T^{-1}	2.6×10^{-8}
to Bohr magneton ratio	μ_t/μ_B	$1.622\,393\,657(21) \times 10^{-3}$		1.3×10^{-8}
to nuclear magneton ratio	μ_t/μ_N	$2.978\,962\,448(38)$		1.3×10^{-8}
triton g-factor $2\mu_t/\mu_N$	g_t	$5.957\,924\,896(76)$		1.3×10^{-8}
Helion, h				
helion mass	m_h	$5.006\,412\,34(22) \times 10^{-27}$	kg	4.4×10^{-8}
		$3.014\,932\,2468(25)$	u	8.3×10^{-10}
energy equivalent	$m_h c^2$	$4.499\,539\,02(20) \times 10^{-10}$	J	4.4×10^{-8}
		$2808.391\,482(62)$	MeV	2.2×10^{-8}
helion-electron mass ratio	m_h/m_e	$5495.885\,2754(50)$		9.2×10^{-10}
helion-proton mass ratio	m_h/m_p	$2.993\,152\,6707(25)$		8.2×10^{-10}
helion molar mass $N_A m_h$	$M(h)$, M_h	$3.014\,932\,2468(25) \times 10^{-3}$	kg mol^{-1}	8.3×10^{-10}
helion magnetic moment	μ_h	$-1.074\,617\,486(27) \times 10^{-26}$	J T^{-1}	2.5×10^{-8}
to Bohr magneton ratio	μ_h/μ_B	$-1.158\,740\,958(14) \times 10^{-3}$		1.2×10^{-8}
to nuclear magneton ratio	μ_h/μ_N	$-2.127\,625\,306(25)$		1.2×10^{-8}
helion g-factor $2\mu_h/\mu_N$	g_h	$-4.255\,250\,613(50)$		1.2×10^{-8}
shielded helion magnetic moment (gas, sphere, 25 °C)	μ'_h	$-1.074\,553\,044(27) \times 10^{-26}$	J T^{-1}	2.5×10^{-8}
to Bohr magneton ratio	μ'_h/μ_B	$-1.158\,671\,471(14) \times 10^{-3}$		1.2×10^{-8}
to nuclear magneton ratio	μ'_h/μ_N	$-2.127\,497\,718(25)$		1.2×10^{-8}
shielded helion to proton magnetic moment ratio (gas, sphere, 25 °C)	μ'_h/μ_p	$-0.761\,766\,558(11)$		1.4×10^{-8}
shielded helion to shielded proton magnetic moment ratio (gas/H$_2$O, spheres, 25 °C)	μ'_h/μ'_p	$-0.761\,786\,1313(33)$		4.3×10^{-9}
shielded helion gyromagnetic ratio $2\lvert\mu'_h\rvert/\hbar$ (gas, sphere, 25 °C)	γ'_h	$2.037\,894\,659(51) \times 10^8$	s^{-1} T^{-1}	2.5×10^{-8}
	$\gamma'_h/2\pi$	$32.434\,100\,84(81)$	MHz T^{-1}	2.5×10^{-8}
Alpha particle, α				
alpha particle mass	m_α	$6.644\,656\,75(29) \times 10^{-27}$	kg	4.4×10^{-8}
		$4.001\,506\,179\,125(62)$	u	1.5×10^{-11}
energy equivalent	$m_\alpha c^2$	$5.971\,919\,67(26) \times 10^{-10}$	J	4.4×10^{-8}
		$3727.379\,240(82)$	MeV	2.2×10^{-8}
alpha particle to electron mass ratio	m_α/m_e	$7294.299\,5361(29)$		4.0×10^{-10}
alpha particle to proton mass ratio	m_α/m_p	$3.972\,599\,689\,33(36)$		9.0×10^{-11}
alpha particle molar mass $N_A m_\alpha$	$M(\alpha)$, M_α	$4.001\,506\,179\,125(62) \times 10^{-3}$	kg mol^{-1}	1.5×10^{-11}
PHYSICOCHEMICAL				
Avogadro constant	N_A, L	$6.022\,141\,29(27) \times 10^{23}$	mol^{-1}	4.4×10^{-8}
atomic mass constant $m_u = \frac{1}{12}m(^{12}C) = 1$ u	m_u	$1.660\,538\,921(73) \times 10^{-27}$	kg	4.4×10^{-8}
energy equivalent	$m_u c^2$	$1.492\,417\,954(66) \times 10^{-10}$	J	4.4×10^{-8}
		$931.494\,061(21)$	MeV	2.2×10^{-8}
Faraday constant[6] $N_A e$	F	$96\,485.3365(21)$	C mol^{-1}	2.2×10^{-8}

---

[6]The numerical value of F to be used in coulometric chemical measurements is $96\,485.3321(43)$ $[4.4 \times 10^{-8}]$ when the relevant current is measured in terms of representations of the volt and ohm based on the Josephson and quantum Hall effects and the internationally adopted conventional values of the Josephson and von Klitzing constants K_{J-90} and R_{K-90} given in Table IV.

TABLE II: *(Continued).*

Quantity	Symbol	Numerical value	Unit	Relative std. uncert. u_r
molar Planck constant	$N_A h$	$3.990\,312\,7176(28) \times 10^{-10}$	J s mol^{-1}	7.0×10^{-10}
	$N_A hc$	$0.119\,626\,565\,779(84)$	J m mol^{-1}	7.0×10^{-10}
molar gas constant	R	$8.314\,4621(75)$	J mol^{-1} K^{-1}	9.1×10^{-7}
Boltzmann constant R/N_A	k	$1.380\,6488(13) \times 10^{-23}$	J K^{-1}	9.1×10^{-7}
		$8.617\,3324(78) \times 10^{-5}$	eV K^{-1}	9.1×10^{-7}
	k/h	$2.083\,6618(19) \times 10^{10}$	Hz K^{-1}	9.1×10^{-7}
	k/hc	$69.503\,476(63)$	m^{-1} K^{-1}	9.1×10^{-7}
molar volume of ideal gas RT/p				
$\quad T = 273.15$ K, $p = 100$ kPa	V_m	$22.710\,953(21) \times 10^{-3}$	m^3 mol^{-1}	9.1×10^{-7}
$\quad$ Loschmidt constant N_A/V_m	n_0	$2.651\,6462(24) \times 10^{25}$	m^{-3}	9.1×10^{-7}
molar volume of ideal gas RT/p				
$\quad T = 273.15$ K, $p = 101.325$ kPa	V_m	$22.413\,968(20) \times 10^{-3}$	m^3 mol^{-1}	9.1×10^{-7}
$\quad$ Loschmidt constant N_A/V_m	n_0	$2.686\,7805(24) \times 10^{25}$	m^{-3}	9.1×10^{-7}
Sackur-Tetrode (absolute entropy) constant[7]				
$\frac{5}{2} + \ln[(2\pi m_u kT_1/h^2)^{3/2} kT_1/p_0]$				
$\quad T_1 = 1$ K, $p_0 = 100$ kPa	S_0/R	$-1.151\,7078(23)$		2.0×10^{-6}
$\quad T_1 = 1$ K, $p_0 = 101.325$ kPa		$-1.164\,8708(23)$		1.9×10^{-6}
Stefan-Boltzmann constant				
$(\pi^2/60)k^4/\hbar^3 c^2$	σ	$5.670\,373(21) \times 10^{-8}$	W m^{-2} K^{-4}	3.6×10^{-6}
first radiation constant $2\pi hc^2$	c_1	$3.741\,771\,53(17) \times 10^{-16}$	W m^2	4.4×10^{-8}
first radiation constant for spectral radiance $2hc^2$	c_{1L}	$1.191\,042\,869(53) \times 10^{-16}$	W m^2 sr^{-1}	4.4×10^{-8}
second radiation constant hc/k	c_2	$1.438\,7770(13) \times 10^{-2}$	m K	9.1×10^{-7}
Wien displacement law constants				
$b = \lambda_{max}T = c_2/4.965\,114\,231...$	b	$2.897\,7721(26) \times 10^{-3}$	m K	9.1×10^{-7}
$b' = \nu_{max}/T = 2.821\,439\,372...c/c_2$	b'	$5.878\,9254(53) \times 10^{10}$	Hz K^{-1}	9.1×10^{-7}

TABLE III: The variances, covariances, and correlation coefficients of the values of a selected group of constants based on the 2010 CODATA adjustment. The numbers in bold above the main diagonal are 10^{16} times the numerical values of the relative covariances; the numbers in bold on the main diagonal are 10^{16} times the numerical values of the relative variances; and the numbers in italics below the main diagonal are the correlation coefficients.[1]

	α	h	e	m_e	N_A	m_e/m_μ	F
α	**0.0010**	**0.0010**	**0.0010**	**−0.0011**	**0.0009**	**−0.0021**	**0.0019**
h	*0.0072*	**19.4939**	**9.7475**	**19.4918**	**−19.4912**	**−0.0020**	**−9.7437**
e	*0.0145*	*1.0000*	**4.8742**	**9.7454**	**−9.7452**	**−0.0020**	**−4.8709**
m_e	*−0.0075*	*0.9999*	*0.9998*	**19.4940**	**−19.4929**	**0.0021**	**−9.7475**
N_A	*0.0060*	*−0.9999*	*−0.9997*	*−1.0000*	**19.4934**	**−0.0017**	**9.7483**
m_e/m_μ	*−0.0251*	*−0.0002*	*−0.0004*	*0.0002*	*−0.0002*	**6.3872**	**−0.0037**
F	*0.0265*	*−0.9993*	*−0.9990*	*−0.9997*	*0.9997*	*−0.0007*	**4.8774**

[1] The relative covariance is $u_r(x_i, x_j) = u(x_i, x_j)/(x_i x_j)$, where $u(x_i, x_j)$ is the covariance of x_i and x_j; the relative variance is $u_r^2(x_i) = u_r(x_i, x_i)$: and the correlation coefficient is $r(x_i, x_j) = u(x_i, x_j)/[u(x_i)u(x_j)]$.

---

[7] The entropy of an ideal monoatomic gas of relative atomic mass A_r is given by $S = S_0 + \frac{3}{2} R \ln A_r - R \ln(p/p_0) + \frac{5}{2} R \ln(T/K)$.

TABLE IV: Internationally adopted values of various quantities.

Quantity	Symbol	Numerical value	Unit	Relative std. uncert. u_r
relative atomic mass[1] of ^{12}C	$A_r(^{12}C)$	12		exact
molar mass constant	M_u	1×10^{-3}	kg mol^{-1}	exact
molar mass of ^{12}C	$M(^{12}C)$	12×10^{-3}	kg mol^{-1}	exact
conventional value of Josephson constant[2]	K_{J-90}	483 597.9	GHz V^{-1}	exact
conventional value of von Klitzing constant[3]	R_{K-90}	25 812.807	Ω	exact
standard-state pressure		100	kPa	exact
standard atmosphere		101.325	kPa	exact

[1] The relative atomic mass $A_r(X)$ of particle X with mass $m(X)$ is defined by $A_r(X) = m(X)/m_u$, where $m_u = m(^{12}C)/12 = M_u/N_A = 1$ u is the atomic mass constant, M_u is the molar mass constant, N_A is the Avogadro constant, and u is the unified atomic mass unit. Thus the mass of particle X is $m(X) = A_r(X)$ u and the molar mass of X is $M(X) = A_r(X) M_u$.

[2] This is the value adopted internationally for realizing representations of the volt using the Josephson effect.

[3] This is the value adopted internationally for realizing representations of the ohm using the quantum Hall effect.

TABLE V: Values of some x-ray-related quantities based on the 2010 CODATA adjustment of the values of the constants.

Quantity	Symbol	Numerical value	Unit	Relative std. uncert. u_r
Cu x unit: $\lambda(CuK\alpha_1)/1\,537.400$	$xu(CuK\alpha_1)$	$1.002\,076\,97(28) \times 10^{-13}$	m	2.8×10^{-7}
Mo x unit: $\lambda(MoK\alpha_1)/707.831$	$xu(MoK\alpha_1)$	$1.002\,099\,52(53) \times 10^{-13}$	m	5.3×10^{-7}
ångstrom star: $\lambda(WK\alpha_1)/0.209\,010\,0$	Å*	$1.000\,014\,95(90) \times 10^{-10}$	m	9.0×10^{-7}
lattice parameter[1] of Si (in vacuum, 22.5 °C)	a	$543.102\,0504(89) \times 10^{-12}$	m	1.6×10^{-8}
{220} lattice spacing of Si $a/\sqrt{8}$ (in vacuum, 22.5 °C)	d_{220}	$192.015\,5714(32) \times 10^{-12}$	m	1.6×10^{-8}
molar volume of Si $M(Si)/\rho(Si) = N_A a^3/8$ (in vacuum, 22.5 °C)	$V_m(Si)$	$12.058\,833\,01(80) \times 10^{-6}$	m^3 mol^{-1}	6.6×10^{-8}

[1] This is the lattice parameter (unit cell edge length) of an ideal single crystal of naturally occurring Si free of impurities and imperfections, and is deduced from measurements on extremely pure and nearly perfect single crystals of Si by correcting for the effects of impurities.

TABLE VI: The values in SI units of some non-SI units based on the 2010 CODATA adjustment of the values of the constants.

Quantity	Symbol	Numerical value	Unit	Relative std. uncert. u_r
Non-SI units accepted for use with the SI				
electron volt: (e/C) J	eV	$1.602\,176\,565(35) \times 10^{-19}$	J	2.2×10^{-8}
(unified) atomic mass unit: $\frac{1}{12}m(^{12}C)$	u	$1.660\,538\,921(73) \times 10^{-27}$	kg	4.4×10^{-8}
Natural units (n.u.)				
n.u. of velocity	c, c_0	299 792 458	m s^{-1}	exact
n.u. of action: $h/2\pi$	$\hbar$	$1.054\,571\,726(47) \times 10^{-34}$	J s	4.4×10^{-8}
		$6.582\,119\,28(15) \times 10^{-16}$	eV s	2.2×10^{-8}
	$\hbar c$	$197.326\,9718(44)$	MeV fm	2.2×10^{-8}
n.u. of mass	m_e	$9.109\,382\,91(40) \times 10^{-31}$	kg	4.4×10^{-8}
n.u. of energy	$m_e c^2$	$8.187\,105\,06(36) \times 10^{-14}$	J	4.4×10^{-8}
		$0.510\,998\,928(11)$	MeV	2.2×10^{-8}
n.u. of momentum	$m_e c$	$2.730\,924\,29(12) \times 10^{-22}$	kg m s^{-1}	4.4×10^{-8}
		$0.510\,998\,928(11)$	MeV/c	2.2×10^{-8}
n.u. of length: $\hbar/m_e c$	$\lambdabar_C$	$386.159\,268\,00(25) \times 10^{-15}$	m	6.5×10^{-10}
n.u. of time	$\hbar/m_e c^2$	$1.288\,088\,668\,33(83) \times 10^{-21}$	s	6.5×10^{-10}

TABLE VI: (Continued.)

Quantity	Symbol	Numerical value	Unit	Relative std. uncert. u_r
		Atomic units (a.u.)		
a.u. of charge	e	$1.602\,176\,565(35) \times 10^{-19}$	C	2.2×10^{-8}
a.u. of mass	m_e	$9.109\,382\,91(40) \times 10^{-31}$	kg	4.4×10^{-8}
a.u. of action: $h/2\pi$	$\hbar$	$1.054\,571\,726(47) \times 10^{-34}$	J s	4.4×10^{-8}
a.u. of length: Bohr radius (bohr)				
$\alpha/4\pi R_\infty$	a_0	$0.529\,177\,210\,92(17) \times 10^{-10}$	m	3.2×10^{-10}
a.u. of energy: Hartree energy (hartree)				
$e^2/4\pi\epsilon_0 a_0 = 2R_\infty hc = \alpha^2 m_e c^2$	E_h	$4.359\,744\,34(19) \times 10^{-18}$	J	4.4×10^{-8}
a.u. of time	$\hbar/E_h$	$2.418\,884\,326\,502(12) \times 10^{-17}$	s	5.0×10^{-12}
a.u. of force	E_h/a_0	$8.238\,722\,78(36) \times 10^{-8}$	N	4.4×10^{-8}
a.u. of velocity: αc	$a_0 E_h/\hbar$	$2.187\,691\,263\,79(71) \times 10^{6}$	m s^{-1}	3.2×10^{-10}
a.u. of momentum	$\hbar/a_0$	$1.992\,851\,740(88) \times 10^{-24}$	kg m s^{-1}	4.4×10^{-8}
a.u. of current	$eE_h/\hbar$	$6.623\,617\,95(15) \times 10^{-3}$	A	2.2×10^{-8}
a.u. of charge density	e/a_0^3	$1.081\,202\,338(24) \times 10^{12}$	C m^{-3}	2.2×10^{-8}
a.u. of electric potential	E_h/e	$27.211\,385\,05(60)$	V	2.2×10^{-8}
a.u. of electric field	E_h/ea_0	$5.142\,206\,52(11) \times 10^{11}$	V m^{-1}	2.2×10^{-8}
a.u. of electric field gradient	E_h/ea_0^2	$9.717\,362\,00(21) \times 10^{21}$	V m^{-2}	2.2×10^{-8}
a.u. of electric dipole moment	ea_0	$8.478\,353\,26(19) \times 10^{-30}$	C m	2.2×10^{-8}
a.u. of electric quadrupole moment	ea_0^2	$4.486\,551\,331(99) \times 10^{-40}$	C m^2	2.2×10^{-8}
a.u. of electric polarizability	$e^2 a_0^2/E_h$	$1.648\,777\,2754(16) \times 10^{-41}$	C^2 m^2 J^{-1}	9.7×10^{-10}
a.u. of 1st hyperpolarizability	$e^3 a_0^3/E_h^2$	$3.206\,361\,449(71) \times 10^{-53}$	C^3 m^3 J^{-2}	2.2×10^{-8}
a.u. of 2nd hyperpolarizability	$e^4 a_0^4/E_h^3$	$6.235\,380\,54(28) \times 10^{-65}$	C^4 m^4 J^{-3}	4.4×10^{-8}
a.u. of magnetic flux density	$\hbar/ea_0^2$	$2.350\,517\,464(52) \times 10^{5}$	T	2.2×10^{-8}
a.u. of magnetic dipole moment: $2\mu_B$	$\hbar e/m_e$	$1.854\,801\,936(41) \times 10^{-23}$	J T^{-1}	2.2×10^{-8}
a.u. of magnetizability	$e^2 a_0^2/m_e$	$7.891\,036\,607(13) \times 10^{-29}$	J T^{-2}	1.6×10^{-9}
a.u. of permittivity: $10^7/c^2$	$e^2/a_0 E_h$	$1.112\,650\,056\ldots \times 10^{-10}$	F m^{-1}	exact

TABLE VII: The values of some energy equivalents derived from the relations $E = mc^2 = hc/\lambda = h\nu = kT$, and based on the 2010 CODATA adjustment of the values of the constants; $1\text{ eV} = (e/\text{C})$ J, $1\text{ u} = m_u = \frac{1}{12}m(^{12}\text{C}) = 10^{-3}$ kg mol$^{-1}/N_A$, and $E_h = 2R_\infty hc = \alpha^2 m_e c^2$ is the Hartree energy (hartree).

Relevant unit

	J	kg	m^{-1}	Hz
1 J	$(1\text{ J}) =$ 1 J	$(1\text{ J})/c^2 =$ $1.112\,650\,056\ldots \times 10^{-17}$ kg	$(1\text{ J})/hc =$ $5.034\,117\,01(22) \times 10^{24}$ m^{-1}	$(1\text{ J})/h =$ $1.509\,190\,311(67) \times 10^{33}$ Hz
1 kg	$(1\text{ kg})c^2 =$ $8.987\,551\,787\ldots \times 10^{16}$ J	$(1\text{ kg}) =$ 1 kg	$(1\text{ kg})c/h =$ $4.524\,438\,73(20) \times 10^{41}$ m^{-1}	$(1\text{ kg})c^2/h =$ $1.356\,392\,608(60) \times 10^{50}$ Hz
1 m^{-1}	$(1\text{ m}^{-1})hc =$ $1.986\,445\,684(88) \times 10^{-25}$ J	$(1\text{ m}^{-1})h/c =$ $2.210\,218\,902(98) \times 10^{-42}$ kg	$(1\text{ m}^{-1}) =$ 1 m^{-1}	$(1\text{ m}^{-1})c =$ $299\,792\,458$ Hz
1 Hz	$(1\text{ Hz})h =$ $6.626\,069\,57(29) \times 10^{-34}$ J	$(1\text{ Hz})h/c^2 =$ $7.372\,496\,68(33) \times 10^{-51}$ kg	$(1\text{ Hz})/c =$ $3.335\,640\,951\ldots \times 10^{-9}$ m^{-1}	$(1\text{ Hz}) =$ 1 Hz
1 K	$(1\text{ K})k =$ $1.380\,6488(13) \times 10^{-23}$ J	$(1\text{ K})k/c^2 =$ $1.536\,1790(14) \times 10^{-40}$ kg	$(1\text{ K})k/hc =$ $69.503\,476(63)$ m^{-1}	$(1\text{ K})k/h =$ $2.083\,6618(19) \times 10^{10}$ Hz
1 eV	$(1\text{ eV}) =$ $1.602\,176\,565(35) \times 10^{-19}$ J	$(1\text{ eV})/c^2 =$ $1.782\,661\,845(39) \times 10^{-36}$ kg	$(1\text{ eV})/hc =$ $8.065\,544\,29(18) \times 10^{5}$ m^{-1}	$(1\text{ eV})/h =$ $2.417\,989\,348(53) \times 10^{14}$ Hz
1 u	$(1\text{ u})c^2 =$ $1.492\,417\,954(66) \times 10^{-10}$ J	$(1\text{ u}) =$ $1.660\,538\,921(73) \times 10^{-27}$ kg	$(1\text{ u})c/h =$ $7.513\,006\,6042(53) \times 10^{14}$ m^{-1}	$(1\text{ u})c^2/h =$ $2.252\,342\,7168(16) \times 10^{23}$ Hz
1 E_h	$(1\,E_h) =$ $4.359\,744\,34(19) \times 10^{-18}$ J	$(1\,E_h)/c^2 =$ $4.850\,869\,79(21) \times 10^{-35}$ kg	$(1\,E_h)/hc =$ $2.194\,746\,313\,708(11) \times 10^{7}$ m^{-1}	$(1\,E_h)/h =$ $6.579\,683\,920\,729(33) \times 10^{15}$ Hz

TABLE VIII: The values of some energy equivalents derived from the relations $E = mc^2 = hc/\lambda = h\nu = kT$, and based on the 2010 CODATA adjustment of the values of the constants; $1\ \mathrm{eV} = (e/\mathrm{C})$ J, $1\ \mathrm{u} = m_\mathrm{u} = \frac{1}{12}m(^{12}\mathrm{C}) = 10^{-3}$ kg mol^{-1}/N_A, and $E_\mathrm{h} = 2R_\infty hc = \alpha^2 m_\mathrm{e}c^2$ is the Hartree energy (hartree).

Relevant unit

	K	eV	u	E_h
1 J	$(1\ \mathrm{J})/k =$ $7.242\,9716(66) \times 10^{22}$ K	$(1\ \mathrm{J}) =$ $6.241\,509\,34(14) \times 10^{18}$ eV	$(1\ \mathrm{J})/c^2 =$ $6.700\,535\,85(30) \times 10^{9}$ u	$(1\ \mathrm{J}) =$ $2.293\,712\,48(10) \times 10^{17}\ E_\mathrm{h}$
1 kg	$(1\ \mathrm{kg})c^2/k =$ $6.509\,6582(59) \times 10^{39}$ K	$(1\ \mathrm{kg})c^2 =$ $5.609\,588\,85(12) \times 10^{35}$ eV	$(1\ \mathrm{kg}) =$ $6.022\,141\,29(27) \times 10^{26}$ u	$(1\ \mathrm{kg})c^2 =$ $2.061\,485\,968(91) \times 10^{34}\ E_\mathrm{h}$
1 m^{-1}	$(1\ \mathrm{m}^{-1})hc/k =$ $1.438\,7770(13) \times 10^{-2}$ K	$(1\ \mathrm{m}^{-1})hc =$ $1.239\,841\,930(27) \times 10^{-6}$ eV	$(1\ \mathrm{m}^{-1})h/c =$ $1.331\,025\,051\,20(94) \times 10^{-15}$ u	$(1\ \mathrm{m}^{-1})hc =$ $4.556\,335\,252\,755(23) \times 10^{-8}\ E_\mathrm{h}$
1 Hz	$(1\ \mathrm{Hz})h/k =$ $4.799\,2434(44) \times 10^{-11}$ K	$(1\ \mathrm{Hz})h =$ $4.135\,667\,516(91) \times 10^{-15}$ eV	$(1\ \mathrm{Hz})h/c^2 =$ $4.439\,821\,6689(31) \times 10^{-24}$ u	$(1\ \mathrm{Hz})h =$ $1.519\,829\,846\,0045(76) \times 10^{-16}\ E_\mathrm{h}$
1 K	$(1\ \mathrm{K}) =$ 1 K	$(1\ \mathrm{K})k =$ $8.617\,3324(78) \times 10^{-5}$ eV	$(1\ \mathrm{K})k/c^2 =$ $9.251\,0868(84) \times 10^{-14}$ u	$(1\ \mathrm{K})k =$ $3.166\,8114(29) \times 10^{-6}\ E_\mathrm{h}$
1 eV	$(1\ \mathrm{eV})/k =$ $1.160\,4519(11) \times 10^{4}$ K	$(1\ \mathrm{eV}) =$ 1 eV	$(1\ \mathrm{eV})/c^2 =$ $1.073\,544\,150(24) \times 10^{-9}$ u	$(1\ \mathrm{eV}) =$ $3.674\,932\,379(81) \times 10^{-2}\ E_\mathrm{h}$
1 u	$(1\ \mathrm{u})c^2/k =$ $1.080\,954\,08(98) \times 10^{13}$ K	$(1\ \mathrm{u})c^2 =$ $931.494\,061(21) \times 10^{6}$ eV	$(1\ \mathrm{u}) =$ 1 u	$(1\ \mathrm{u})c^2 =$ $3.423\,177\,6845(24) \times 10^{7}\ E_\mathrm{h}$
1 E_h	$(1\ E_\mathrm{h})/k =$ $3.157\,7504(29) \times 10^{5}$ K	$(1\ E_\mathrm{h}) =$ $27.211\,385\,05(60)$ eV	$(1\ E_\mathrm{h})/c^2 =$ $2.921\,262\,3246(21) \times 10^{-8}$ u	$(1\ E_\mathrm{h}) =$ $1\ E_\mathrm{h}$

STANDARD ATOMIC WEIGHTS (2009)

This table of atomic weights includes the changes made in 2009 by the International Union of Pure and Applied Chemistry (IUPAC) Commission on Isotopic Abundances and Atomic Weights. Those changes affected the following 11 elements: boron, carbon, chlorine, germanium, hydrogen, lithium, nitrogen, oxygen, silicon, sulfur, and thallium.

IUPAC made a significant policy change in its 2009 report (Refs. 1, 3). Each atomic weight had previously been given as a single value with an uncertainty that took into account both the measurement uncertainty and the variation in isotopic abundance in samples of the element from different terrestrial sources. For a variety of reasons (Ref. 2), this fails to give complete information on the natural variability in isotopic abundance of several elements. Therefore, the 2009 recommendations express the atomic weights of 10 elements as intervals rather than single numbers plus uncertainties. The symbol for these intervals is [a; b], where a is the lower bound of values found in normal materials, and b the upper bound. For the other elements in the table, a single recommended atomic weight value is given; the number in parentheses following the value gives the uncertainty in the last digit.

Table 1 gives the 2009 atomic weights of the elements listed in alphabetical order by name. Table 2 gives reference atomic weights for the 10 elements whose entries in Table 1 are intervals rather than single numbers. These conventional values are suggested for use on samples of unspecified origin and for calculation of molecular weights in tables intended to be broadly applicable. They have been selected such that most or all natural terrestrial atomic-weight variation is covered in an interval of plus or minus one in the last digit. It should be emphasized that the conventional values are not simply midpoints of the intervals, but rather represent the best judgment of the data evaluators.

References

1. Wieser, M. E., and Coplen, T. D., *Pure Appl. Chem.* 83, 359, 2011.
2. Coplen, T. B., and Holden, N. E., *Chemistry International*, Vol. 33, No. 2, p. 10, 2011.
3. Berglund, M., and Wieser, M. E., *Pure Appl. Chem.* 83, 397, 2011.

TABLE 1. STANDARD ATOMIC WEIGHTS 2009

Element	Symbol	Atomic Number	Atomic Weight	Footnotes	Element	Symbol	Atomic Number	Atomic Weight	Footnotes
Actinium*	Ac	89			Francium*	Fr	87		
Aluminum	Al	13	26.9815386(8)		Gadolinium	Gd	64	157.25(3)	g
Americium*	Am	95			Gallium	Ga	31	69.723(1)	
Antimony	Sb	51	121.760(1)	g	Germanium	Ge	32	72.63(1)	
Argon	Ar	18	39.948(1)	g r	Gold	Au	79	196.966569(4)	
Arsenic	As	33	74.92160(2)		Hafnium	Hf	72	178.49(2)	
Astatine*	At	85			Hassium*	Hs	108		
Barium	Ba	56	137.327(7)		Helium	He	2	4.002602(2)	g r
Berkelium*	Bk	97			Holmium	Ho	67	164.93032(2)	
Beryllium	Be	4	9.012182(3)		Hydrogen	H	1	[1.00784; 1.00811]	m
Bismuth	Bi	83	208.98040(1)		Indium	In	49	114.818(3)	
Bohrium*	Bh	107			Iodine	I	53	126.90447(3)	
Boron	B	5	[10.806; 10.821]	m	Iridium	Ir	77	192.217(3)	
Bromine	Br	35	79.904(1)		Iron	Fe	26	55.845(2)	
Cadmium	Cd	48	112.411(8)	g	Krypton	Kr	36	83.798(2)	g m
Calcium	Ca	20	40.078(4)	g	Lanthanum	La	57	138.90547(7)	g
Californium*	Cf	98			Lawrencium*	Lr	103		
Carbon	C	6	[12.0096; 12.0116]		Lead	Pb	82	207.2(1)	g r
Cerium	Ce	58	140.116(1)	g	Lithium	Li	3	[6.938; 6.997]	m
Cesium	Cs	55	132.9054519(2)		Lutetium	Lu	71	174.9668(1)	g
Chlorine	Cl	17	[35.446; 35.457]	m	Magnesium	Mg	12	24.3050(6)	
Chromium	Cr	24	51.9961(6)		Manganese	Mn	25	54.938045(5)	
Cobalt	Co	27	58.933195(5)		Meitnerium*	Mt	109		
Copernicium*	Cn	112			Mendelevium*	Md	101		
Copper	Cu	29	63.546(3)	r	Mercury	Hg	80	200.59(2)	
Curium*	Cm	96			Molybdenum	Mo	42	95.96(2)	g
Darmstadtium*	Ds	110			Neodymium	Nd	60	144.242(3)	g
Dubnium*	Db	105			Neon	Ne	10	20.1797(6)	g m
Dysprosium	Dy	66	162.500(1)	g	Neptunium*	Np	93		
Einsteinium*	Es	99			Nickel	Ni	28	58.6934(4)	r
Erbium	Er	68	167.259(3)	g	Niobium	Nb	41	92.90638(2)	
Europium	Eu	63	151.964(1)	g	Nitrogen	N	7	[14.00643; 14.00728]	
Fermium*	Fm	100			Nobelium*	No	102		
Fluorine	F	9	18.9984032(5)		Osmium	Os	76	190.23(3)	g

Element	Symbol	Atomic Number	Atomic Weight	Footnotes	Element	Symbol	Atomic Number	Atomic Weight	Footnotes
Oxygen	O	8	[15.99903; 15.99977]		Strontium	Sr	38	87.62(1)	g r
Palladium	Pd	46	106.42(1)	g	Sulfur	S	16	[32.059; 32.076]	
Phosphorus	P	15	30.973762(2)		Tantalum	Ta	73	180.94788(2)	
Platinum	Pt	78	195.084(9)		Technetium*	Tc	43		
Plutonium*	Pu	94			Tellurium	Te	52	127.60(3)	g
Polonium*	Po	84			Terbium	Tb	65	158.92535(2)	
Potassium	K	19	39.0983(1)		Thallium	Tl	81	[204.382; 204.385]	
Praseodymium	Pr	59	140.90765(2)		Thorium**	Th	90	232.03806(2)	g
Promethium*	Pm	61			Thulium	Tm	69	168.93421(2)	
Protactinium**	Pa	91	231.03588(2)		Tin	Sn	50	118.710(7)	g
Radium*	Ra	88			Titanium	Ti	22	47.867(1)	
Radon*	Rn	86			Tungsten	W	74	183.84(1)	
Rhenium	Re	75	186.207(1)		Ununhexium*	Uuh	116		
Rhodium	Rh	45	102.90550(2)		Ununoctium*	Uuo	118		
Roentgenium*	Rg	111			Ununpentium*	Uup	115		
Rubidium	Rb	37	85.4678(3)	g	Ununquadium*	Uuq	114		
Ruthenium	Ru	44	101.07(2)	g	Ununseptium	Uus	117		
Rutherfordium*	Rf	104			Ununtrium*	Uut	113		
Samarium	Sm	62	150.36(2)	g	Uranium**	U	92	238.02891(3)	g m
Scandium	Sc	21	44.955912(6)		Vanadium	V	23	50.9415(1)	
Seaborgium*	Sg	106			Xenon	Xe	54	131.293(6)	g m
Selenium	Se	34	78.96(3)	r	Ytterbium	Yb	70	173.054(5)	g
Silicon	Si	14	[28.084; 28.086]		Yttrium	Y	39	88.90585(2)	
Silver	Ag	47	107.8682(2)	g	Zinc	Zn	30	65.38(2)	r
Sodium	Na	11	22.98976928(2)		Zirconium	Zr	40	91.224(2)	g

* Element has no stable isotopes, and no characteristic terrestrial isotopic abundance can be established. See "Table of the Isotopes" in Sec.11 for individual isotopic masses.

** Element has no stable isotopes but does have a characteristic terrestrial isotopic abundance.

g Geological specimens are known in which the element has an isotopic composition outside the limits for the normal material. The difference between the atomic weight of the element in such specimens and that given in the table may exceed the stated uncertainty.

m Modified isotopic compositions may be found in commercially available material because the material has been subjected to an undisclosed or inadvertent isotopic fractionation. Substantial deviations in atomic weight of the element from that given in the table can occur.

r Range in isotopic composition of normal terrestrial material prevents a more precise atomic weight being given; the tabulated value and uncertainty should be applicable to any normal material.

TABLE 2. CONVENTIONAL ATOMIC WEIGHTS 2009

Element	Symbol	Atomic Number	Reference Atomic Weight[a]
Boron	B	5	10.81
Carbon	C	6	12.011
Chlorine	Cl	17	35.45
Hydrogen	H	1	1.008
Lithium	Li	3	6.94
Nitrogen	N	7	14.007
Oxygen	O	8	15.999
Silicon	Si	14	28.085
Sulfur	S	16	32.06
Thallium	Tl	81	204.38

a For users needing an atomic-weight value for an unspecified sample, such as for trade or commerce. See text.

ATOMIC MASSES AND ABUNDANCES

This table lists the mass (in atomic mass units, symbol u) and the natural abundance (in percent) of the stable nuclides and a few important radioactive nuclides. A complete table of all nuclides may be found in Section 11, "Table of the Isotopes" (Reference 1).

The majority of the atomic masses were taken from the 2003 evaluation of Audi, Wapstra, and Thibault (References 2, 3). The number in parentheses following the mass value is the uncertainty in the last digit(s) given. The mass values for elements with $Z = 102$ and higher were derived from a combination of experimental data and systematic trends.

A comprehensive reevaluation of the 2003 mass data by the Atomic Mass Data Center is in progress, and a preliminary report was released in April 2011 (Reference 4). In the table below several updated mass values, indicated by an asterisk *, have been taken from that preliminary report.

Natural abundance values were taken from the IUPAC Technical Report "Atomic Weight of the Elements: Review 2000" (Reference 5); these entries are also followed by uncertainties in the last digit(s) of the stated values. This uncertainty includes both the es-

timated measurement uncertainty and the reported range of variation in different terrestrial sources of the element (see Reference 5 for full details and caveats regarding elements whose abundance is variable). The absence of an entry in the Abundance column indicates a radioactive nuclide not present in nature or an element whose isotopic composition varies so widely that a meaningful natural abundance cannot be defined.

References

1. Holden, N. E., "Table of the Isotopes", in Haynes, W. M., Ed., *CRC Handbook of Chemistry and Physics*, *93rd Ed.*, CRC Press, Boca Raton, FL, 2012.
2. Audi, G., Wapstra, A. H., and Thibault, C., *Nucl. Phys.* A729, 337, 2003.
3. Audi, G., and Wapstra, A. H., Atomic Mass Data Center, <www.nndc.bnl.gov/amdc/index.html>
4. Audi, G., and Wang, M., <amdc.in2p3.fr/masstables/Ame2011int/mass.mas114>.
5. de Laeter, J. R., Böhlke, J. K., De Bièvre, P., Hidaka, H., Peiser, H. S., Rosman, K. J. R., and Taylor, P. D. P., *Pure Appl. Chem.* 75, 683, 2003.

Z	Isotope	Mass in u	Abundance in %
1	^{1}H	1.00782503207(10)	99.9885(70)
	^{2}H	2.0141017778(4)	0.0115(70)
	^{3}H	3.0160492777(25)	
2	^{3}He	3.0160293191(26)	0.000134(3)
	^{4}He	4.00260325415(6)	99.999866(3)
3	^{6}Li	6.015122795(16)	7.59(4)
	^{7}Li	7.016003427(5)*	92.41(4)
4	^{9}Be	9.01218305(8)*	100
5	^{10}B	10.0129370(4)	19.9(7)
	^{11}B	11.0093054(4)	80.1(7)
6	^{11}C	11.0114336(10)	
	^{12}C	12.0000000(0)	98.93(8)
	^{13}C	13.0033548378(10)	1.07(8)
	^{14}C	14.003241989(4)	
7	^{14}N	14.0030740048(6)	99.636(7)
	^{15}N	15.0001088982(7)	0.364(7)
8	^{16}O	15.99491461956(16)	99.757(16)
	^{17}O	16.99913170(12)	0.038(1)
	^{18}O	17.9991596129(8)*	0.205(14)
9	^{18}F	18.0009380(6)	
	^{19}F	18.99840322(7)	100
10	^{20}Ne	19.9924401754(19)	90.48(3)
	^{21}Ne	20.99384668(4)	0.27(1)
	^{22}Ne	21.991385114(19)	9.25(3)
11	^{22}Na	21.9944364(4)	
	^{23}Na	22.9897692809(29)	100
	^{24}Na	23.99096278(8)	
12	^{24}Mg	23.985041700(14)	78.99(4)
	^{25}Mg	24.98583692(3)	10.00(1)
	^{26}Mg	25.982592929(30)	11.01(3)
13	^{27}Al	26.98153863(12)	100
14	^{28}Si	27.9769265325(19)	92.223(19)
	^{29}Si	28.976494700(22)	4.685(8)
	^{30}Si	29.97377017(3)	3.092(11)
15	^{31}P	30.97376163(20)	100
	^{32}P	31.97390727(20)	
16	^{32}S	31.97207100(15)	94.99(26)

Z	Isotope	Mass in u	Abundance in %
	^{33}S	32.97145876(15)	0.75(2)
	^{34}S	33.96786690(12)	4.25(24)
	^{35}S	34.96903216(11)	
	^{36}S	35.96708076(20)	0.01(1)
17	^{35}Cl	34.96885268(4)	75.76(10)
	^{37}Cl	36.96590259(5)	24.24(10)
18	^{36}Ar	35.967545106(29)	0.3365(30)
	^{38}Ar	37.9627324(4)	0.0632(5)
	^{40}Ar	39.9623831225(29)	99.6003(30)
19	^{39}K	38.96370668(20)	93.2581(44)
	^{40}K	39.96399848(21)	0.0117(1)
	^{41}K	40.96182576(21)	6.7302(44)
	^{42}K	41.96240281(24)	
	^{43}K	42.9607347(5)*	
20	^{40}Ca	39.96259098(22)	96.941(156)
	^{42}Ca	41.95861801(27)	0.647(23)
	^{43}Ca	42.9587666(3)	0.135(10)
	^{44}Ca	43.9554818(4)	2.086(110)
	^{45}Ca	44.9561866(4)	
	^{46}Ca	45.9536873(24)*	0.004(3)
	^{47}Ca	46.9545407(24)*	
	^{48}Ca	47.9525241(23)*	0.187(21)
21	^{45}Sc	44.9559091(7)*	100
22	^{46}Ti	45.9526277(3)*	8.25(3)
	^{47}Ti	46.9517588(4)*	7.44(2)
	^{48}Ti	47.9479419(4)*	73.72(3)
	^{49}Ti	48.9478656(4)*	5.41(2)
	^{50}Ti	49.9447868(4)*	5.18(2)
23	^{50}V	49.9471585(11)	0.250(4)
	^{51}V	50.9439595(11)	99.750(4)
24	^{50}Cr	49.9460442(11)	4.345(13)
	^{51}Cr	50.9447674(11)	
	^{52}Cr	51.9405075(8)	83.789(18)
	^{53}Cr	52.9406494(8)	9.501(17)
	^{54}Cr	53.9388804(8)	2.365(7)
25	^{54}Mn	53.9403589(14)	
	^{55}Mn	54.9380451(7)	100

Z	Isotope	Mass in u	Abundance in %		Z	Isotope	Mass in u	Abundance in %
26	^{52}Fe	51.948114(7)				^{89}Sr	88.9074507(12)	
	^{54}Fe	53.9396105(7)	5.845(35)			^{90}Sr	89.9077295(30)*	
	^{55}Fe	54.9382934(7)			39	^{89}Y	88.9058398(26)*	100
	^{56}Fe	55.9349375(7)	91.754(36)		40	^{90}Zr	89.9046969(24)*	51.45(40)
	^{57}Fe	56.9353940(7)	2.119(10)			^{91}Zr	90.9056383(24)*	11.22(5)
	^{58}Fe	57.9332756(8)	0.282(4)			^{92}Zr	91.9050334(24)*	17.15(8)
	^{59}Fe	58.9348755(8)				^{94}Zr	93.9063090(24)*	17.38(28)
27	^{57}Co	56.9362914(8)				^{96}Zr	95.9082734(30)	2.80(9)
	^{58}Co	57.9357528(13)			41	^{93}Nb	92.9063717(24)*	100
	^{59}Co	58.9331950(7)	100		42	^{92}Mo	91.906811(4)	14.77(31)
	^{60}Co	59.9338171(7)				^{94}Mo	93.9050883(21)	9.23(10)
28	^{58}Ni	57.9353429(7)	68.0769(89)			^{95}Mo	94.9058421(21)	15.90(9)
	^{59}Ni	58.9343467(7)				^{96}Mo	95.9046795(21)	16.68(1)
	^{60}Ni	59.9307864(7)	26.2231(77)			^{97}Mo	96.9060215(21)	9.56(5)
	^{61}Ni	60.9310560(7)	1.1399(6)			^{98}Mo	97.9054082(21)	24.19(26)
	^{62}Ni	61.9283451(6)	3.6345(17)			^{99}Mo	98.9077119(21)	
	^{63}Ni	62.9296694(6)				^{100}Mo	99.907477(6)	9.67(20)
	^{64}Ni	63.9279660(7)	0.9256(9)		43	^{97}Tc	96.906365(5)	
29	^{63}Cu	62.9295975(6)	69.15(3)			^{98}Tc	97.907216(4)	
	^{64}Cu	63.9297642(6)				^{99}Tc	98.9062547(21)	
	^{65}Cu	64.9277895(7)	30.85(3)		44	^{96}Ru	95.9075889(16)*	5.54(14)
30	^{64}Zn	63.9291422(7)	48.268(321)			^{98}Ru	97.905287(7)	1.87(3)
	^{65}Zn	64.9292410(7)				^{99}Ru	98.9059393(22)	12.76(14)
	^{66}Zn	65.9260334(10)	27.975(77)			^{100}Ru	99.9042195(22)	12.60(7)
	^{67}Zn	66.9271273(10)	4.102(21)			^{101}Ru	100.9055821(22)	17.06(2)
	^{68}Zn	67.9248442(10)	19.024(123)			^{102}Ru	101.9043493(22)	31.55(14)
	^{70}Zn	69.9253193(21)	0.631(9)			^{104}Ru	103.905433(3)	18.62(27)
31	^{67}Ga	66.9282017(14)				^{106}Ru	105.907329(8)	
	^{68}Ga	67.9279801(16)			45	^{103}Rh	102.905504(3)	100
	^{69}Ga	68.9255736(13)	60.108(9)		46	^{102}Pd	101.905609(3)	1.02(1)
	^{71}Ga	70.9247013(11)	39.892(9)			^{104}Pd	103.904036(4)	11.14(8)
32	^{68}Ge	67.928094(7)				^{105}Pd	104.905085(4)	22.33(8)
	^{70}Ge	69.9242474(11)	20.38(18)			^{106}Pd	105.903486(4)	27.33(3)
	^{72}Ge	71.9220758(18)	27.31(26)			^{108}Pd	107.903892(4)	26.46(9)
	^{73}Ge	72.9234589(18)	7.76(8)			^{110}Pd	109.905153(12)	11.72(9)
	^{74}Ge	73.9211778(18)	36.72(15)		47	^{107}Ag	106.905097(5)	51.839(8)
	^{76}Ge	75.9214026(18)	7.83(7)			^{109}Ag	108.904752(3)	48.161(8)
33	^{75}As	74.9215965(20)	100		48	^{106}Cd	105.906459(6)	1.25(6)
34	^{74}Se	73.9224764(18)	0.89(4)			^{108}Cd	107.904184(6)	0.89(3)
	^{75}Se	74.9225234(18)				^{110}Cd	109.9030021(29)	12.49(18)
	^{76}Se	75.9192136(18)	9.37(29)			^{111}Cd	110.9041781(29)	12.80(12)
	^{77}Se	76.9199140(18)	7.63(16)			^{112}Cd	111.9027578(29)	24.13(21)
	^{78}Se	77.9173091(18)	23.77(28)			^{113}Cd	112.9044017(29)	12.22(12)
	^{79}Se	78.9184991(18)				^{114}Cd	113.9033585(29)	28.73(42)
	^{80}Se	79.9165213(21)	49.61(41)			^{116}Cd	115.904756(3)	7.49(18)
	^{82}Se	81.9166994(22)	8.73(22)		49	^{111}In	110.905103(5)	
35	^{79}Br	78.9183371(22)	50.69(7)			^{113}In	112.904058(3)	4.29(5)
	^{81}Br	80.9162906(21)	49.31(7)			^{115}In	114.903878(5)	95.71(5)
36	^{78}Kr	77.9203648(12)	0.355(3)		50	^{112}Sn	111.904818(5)	0.97(1)
	^{80}Kr	79.9163790(16)	2.286(10)			^{113}Sn	112.905171(4)	
	^{82}Kr	81.9134836(19)	11.593(31)			^{114}Sn	113.902779(3)	0.66(1)
	^{83}Kr	82.9141271(3)*	11.500(19)			^{115}Sn	114.903342(3)	0.34(1)
	^{84}Kr	83.911497728(4)*	56.987(15)			^{116}Sn	115.901741(3)	14.54(9)
	^{86}Kr	85.91061073(11)	17.279(41)			^{117}Sn	116.902952(3)	7.68(7)
37	^{85}Rb	84.911789738(12)	72.17(2)			^{118}Sn	117.901603(3)	24.22(9)
	^{86}Rb	85.91116742(21)				^{119}Sn	118.903308(3)	8.59(4)
	^{87}Rb	86.909180527(13)	27.83(2)			^{120}Sn	119.9022021(24)*	32.58(9)
38	^{84}Sr	83.9134194(13)*	0.56(1)			^{122}Sn	121.9034390(29)	4.63(3)
	^{85}Sr	84.912933(3)				^{124}Sn	123.9052739(15)	5.79(5)
	^{86}Sr	85.9092602(12)	9.86(1)		51	^{121}Sb	120.9038157(24)	57.21(5)
	^{87}Sr	86.9088771(12)	7.00(1)			^{123}Sb	122.9042140(22)	42.79(5)
	^{88}Sr	87.9056121(12)	82.58(1)		52	^{120}Te	119.9040577(34)*	0.09(1)

Z	Isotope	Mass in u	Abundance in %	Z	Isotope	Mass in u	Abundance in %
	^{122}Te	121.9030439(16)	2.55(12)	64	^{152}Gd	151.9197910(27)	0.20(1)
	^{123}Te	122.9042700(16)	0.89(3)		^{154}Gd	153.9208656(27)	2.18(3)
	^{124}Te	123.9028179(16)	4.74(14)		^{155}Gd	154.9226220(27)	14.80(12)
	^{125}Te	124.9044307(16)	7.07(15)		^{156}Gd	155.9221227(27)	20.47(9)
	^{126}Te	125.9033117(16)	18.84(25)		^{157}Gd	156.9239601(27)	15.65(2)
	^{128}Te	127.9044631(19)	31.74(8)		^{158}Gd	157.9241039(27)	24.84(7)
	^{130}Te	129.9062244(21)	34.08(62)		^{160}Gd	159.9270541(27)	21.86(19)
53	^{123}I	122.905589(4)		65	^{159}Tb	158.9253468(27)	100
	^{125}I	124.9046302(16)		66	^{156}Dy	155.924283(7)	0.056(3)
	^{127}I	126.904473(4)	100		^{158}Dy	157.924409(4)	0.095(3)
	^{129}I	128.904988(3)			^{160}Dy	159.9251975(27)	2.329(18)
	^{131}I	130.9061246(12)			^{161}Dy	160.9269334(27)	18.889(42)
54	^{124}Xe	123.9058930(20)	0.0952(3)		^{162}Dy	161.9267984(27)	25.475(36)
	^{126}Xe	125.9042976(39)*	0.0890(2)		^{163}Dy	162.9287312(27)	24.896(42)
	^{128}Xe	127.9035313(15)	1.9102(8)		^{164}Dy	163.9291748(27)	28.260(54)
	^{129}Xe	128.9047794(8)	26.4006(82)	67	^{165}Ho	164.9303221(27)	100
	^{130}Xe	129.9035080(8)	4.0710(13)	68	^{162}Er	161.9287887(20)*	0.139(5)
	^{131}Xe	130.9050824(10)	21.2324(30)		^{164}Er	163.929200(3)	1.601(3)
	^{132}Xe	131.9041535(10)	26.9086(33)		^{166}Er	165.9302931(27)	33.503(36)
	^{134}Xe	133.9053945(9)	10.4357(21)		^{167}Er	166.9320482(27)	22.869(9)
	^{136}Xe	135.907219(8)	8.8573(44)		^{168}Er	167.9323702(27)	26.978(18)
55	^{129}Cs	128.906064(5)			^{170}Er	169.9354643(30)	14.910(36)
	^{133}Cs	132.905451933(24)	100	69	^{169}Tm	168.9342133(27)	100
	^{134}Cs	133.906718475(28)		70	^{168}Yb	167.933897(5)	0.13(1)
	^{136}Cs	135.9073116(20)			^{169}Yb	168.935190(5)	
	^{137}Cs	136.9070895(5)			^{170}Yb	169.9347618(26)	3.04(15)
56	^{130}Ba	129.9063208(30)	0.106(1)		^{171}Yb	170.9363258(26)	14.28(57)
	^{132}Ba	131.9050613(11)	0.101(1)		^{172}Yb	171.9363815(26)	21.83(67)
	^{133}Ba	132.9060075(11)			^{173}Yb	172.9382108(26)	16.13(27)
	^{134}Ba	133.9045084(4)	2.417(18)		^{174}Yb	173.9388621(26)	31.83(92)
	^{135}Ba	134.9056886(4)	6.592(12)		^{176}Yb	175.9425717(28)	12.76(41)
	^{136}Ba	135.9045759(4)	7.854(24)	71	^{175}Lu	174.9407718(23)	97.41(2)
	^{137}Ba	136.9058274(5)	11.232(24)		^{176}Lu	175.9426863(23)	2.59(2)
	^{138}Ba	137.9052472(5)	71.698(42)	72	^{174}Hf	173.940046(3)	0.16(1)
	^{140}Ba	139.910605(9)			^{176}Hf	175.9414086(24)	5.26(7)
57	^{138}La	137.907112(4)	0.090(1)		^{177}Hf	176.9432207(23)	18.60(9)
	^{139}La	138.9063533(26)	99.910(1)		^{178}Hf	177.9436988(23)	27.28(7)
58	^{136}Ce	135.907172(14)	0.185(2)		^{179}Hf	178.9458161(23)	13.62(2)
	^{138}Ce	137.905991(11)	0.251(2)		^{180}Hf	179.9465500(23)	35.08(16)
	^{140}Ce	139.9054387(26)	88.450(51)	73	^{180}Ta	179.9474648(24)	0.012(2)
	^{141}Ce	140.9082763(26)			^{181}Ta	180.9479958(19)	99.988(2)
	^{142}Ce	141.909244(3)	11.114(51)	74	^{180}W	179.946704(4)	0.12(1)
	^{144}Ce	143.913647(4)			^{182}W	181.9482042(9)	26.50(16)
59	^{141}Pr	140.9076528(26)	100		^{183}W	182.9502230(9)	14.31(4)
60	^{142}Nd	141.9077233(25)	27.2(5)		^{184}W	183.9509312(9)	30.64(2)
	^{143}Nd	142.9098143(25)	12.2(2)		^{186}W	185.9543641(19)	28.43(19)
	^{144}Nd	143.9100873(25)	23.8(3)	75	^{185}Re	184.9529550(13)	37.40(2)
	^{145}Nd	144.9125736(25)	8.3(1)		^{187}Re	186.9557531(15)	62.60(2)
	^{146}Nd	145.9131169(25)	17.2(3)	76	^{184}Os	183.9524891(14)	0.02(1)
	^{148}Nd	147.916893(3)	5.7(1)		^{186}Os	185.9538382(15)	1.59(3)
	^{150}Nd	149.920891(3)	5.6(2)		^{187}Os	186.9557505(15)	1.96(2)
61	^{145}Pm	144.912749(3)			^{188}Os	187.9558382(15)	13.24(8)
	^{147}Pm	146.9151385(26)			^{189}Os	188.9581475(16)	16.15(5)
62	^{144}Sm	143.911999(3)	3.07(7)		^{190}Os	189.9584470(16)	26.26(2)
	^{147}Sm	146.9148979(26)	14.99(18)		^{192}Os	191.9614807(27)	40.78(19)
	^{148}Sm	147.9148227(26)	11.24(10)	77	^{191}Ir	190.9605940(18)	37.3(2)
	^{149}Sm	148.9171847(26)	13.82(7)		^{193}Ir	192.9629264(18)	62.7(2)
	^{150}Sm	149.9172755(26)	7.38(1)	78	^{190}Pt	189.959932(6)	0.014(1)
	^{152}Sm	151.9197324(27)	26.75(16)		^{192}Pt	191.9610380(27)	0.782(7)
	^{154}Sm	153.9222093(27)	22.75(29)		^{194}Pt	193.9626803(9)	32.967(99)
63	^{151}Eu	150.9198502(26)	47.81(6)		^{195}Pt	194.9647911(9)	33.832(10)
	^{153}Eu	152.9212303(26)	52.19(6)		^{196}Pt	195.9649515(9)	25.242(41)

Z	Isotope	Mass in u	Abundance in %	Z	Isotope	Mass in u	Abundance in %
	^{198}Pt	197.967893(3)	7.163(55)		^{234}U	234.0409521(20)	0.0054(5)
79	^{197}Au	196.9665687(6)	100		^{235}U	235.0439299(20)	0.7204(6)
	^{198}Au	197.9682423(6)			^{236}U	236.0455680(20)	
80	^{196}Hg	195.965833(3)	0.15(1)		^{238}U	238.0507882(20)	99.2742(10)
	^{197}Hg	196.967213(3)		93	^{237}Np	237.0481734(20)	
	^{198}Hg	197.9667690(4)	9.97(20)		^{239}Np	239.0529390(22)	
	^{199}Hg	198.9682799(4)	16.87(22)	94	^{238}Pu	238.0495599(20)	
	^{200}Hg	199.9683260(4)	23.10(19)		^{239}Pu	239.0521634(20)	
	^{201}Hg	200.9703023(6)	13.18(9)		^{240}Pu	240.0538135(20)	
	^{202}Hg	201.9706430(6)	29.86(26)		^{241}Pu	241.0568515(20)	
	^{203}Hg	202.9728725(18)			^{242}Pu	242.0587426(20)	
	^{204}Hg	203.9734939(4)	6.87(15)		^{244}Pu	244.064204(5)	
81	^{201}Tl	200.970819(16)		95	^{241}Am	241.0568291(20)	
	^{203}Tl	202.9723442(14)	29.52(1)		^{243}Am	243.0613811(25)	
	^{205}Tl	204.9744275(14)	70.48(1)	96	^{243}Cm	243.0613891(22)	
82	^{204}Pb	203.9730436(13)	1.4(1)		^{244}Cm	244.0627526(20)	
	^{206}Pb	205.9744653(13)	24.1(1)		^{245}Cm	245.0654912(22)	
	^{207}Pb	206.9758969(13)	22.1(1)		^{246}Cm	246.0672237(22)	
	^{208}Pb	207.9766521(13)	52.4(1)		^{247}Cm	247.070354(5)	
	^{210}Pb	209.9841885(16)			^{248}Cm	248.072349(5)	
83	^{207}Bi	206.9784707(26)		97	^{247}Bk	247.070307(6)	
	^{209}Bi	208.9803987(16)	100		^{249}Bk	249.0749867(28)	
84	^{209}Po	208.9824304(20)		98	^{249}Cf	249.0748535(24)	
	^{210}Po	209.9828737(13)			^{250}Cf	250.0764061(22)	
85	^{210}At	209.987148(8)			^{251}Cf	251.079587(5)	
	^{211}At	210.9874963(30)			^{252}Cf	252.081626(5)	
86	^{211}Rn	210.990601(7)		99	^{252}Es	252.082980(50)	
	^{220}Rn	220.0113940(24)		100	^{257}Fm	257.095105(7)	
	^{222}Rn	222.0175777(25)		101	^{256}Md	256.094060(60)	
87	^{223}Fr	223.0197359(26)			^{258}Md	258.098431(5)	
88	^{223}Ra	223.0185022(27)		102	^{259}No	259.10103(11)*	
	^{224}Ra	224.0202118(24)		103	^{262}Lr	262.10946(22)*	
	^{226}Ra	226.0254098(25)		104	^{261}Rf	261.108770(30)*	
	^{228}Ra	228.0310703(26)		105	^{262}Db	262.11408(20)*	
89	^{227}Ac	227.0277521(26)		106	^{263}Sg	263.11832(13)*	
90	^{228}Th	228.0287411(24)		107	^{264}Bh	264.12426(19)*	
	^{230}Th	230.0331338(19)		108	^{265}Hs	265.13009(15)*	
	^{232}Th	232.0380553(21)	100	109	^{268}Mt	268.13839(25)*	
91	^{231}Pa	231.0358840(24)	100	110	^{281}Ds	281.16206(78)*	
92	^{233}U	233.0396352(29)		111	^{272}Rg	273.15362(36)*	

ELECTRON CONFIGURATION AND IONIZATION ENERGY OF NEUTRAL ATOMS IN THE GROUND STATE

William C. Martin

The ground state electron configuration, ground level, and ionization energy of the elements hydrogen through rutherfordium are listed in this table. The electron configurations of elements heavier than neon are shortened by using rare-gas element symbols in brackets to represent the corresponding electrons. See the references for details of the notation for Pa, U, and Np. Ionization energies to higher states (and more precise values of the first ionization energy for certain elements) may be found in the table "Ionization Energies of Atoms and Atomic Ions" in Section 10 of this *Handbook*.

References

1. Martin, W. C., Musgrove, A., Kotochigova, S., and Sansonetti, J. E., NIST Physical Reference Data Web Site, <http://physics.nist.gov/PhysRefData/IonEnergy/ionEnergy.html>, October 2004.
2. Martin, W. C., and Wiese, W. L., "Atomic Spectroscopy", in *Atomic, Molecular, & Optical Physics Handbook*, ed. by G.W.F. Drake (AIP, Woodbury, NY, 1996) Chapter 10, pp. 135-153.

Z	Element		Ground-state configuration	Ground level	Ionization energy (eV)
1	H	Hydrogen	$1s$	$^2S_{1/2}$	13.5984
2	He	Helium	$1s^2$	$^1S_0$	24.5874
3	Li	Lithium	$1s^2\,2s$	$^2S_{1/2}$	5.3917
4	Be	Beryllium	$1s^2\,2s^2$	$^1S_0$	9.3227
5	B	Boron	$1s^2\,2s^2\,2p$	$^2P^o_{1/2}$	8.2980
6	C	Carbon	$1s^2\,2s^2\,2p^2$	$^3P_0$	11.2603
7	N	Nitrogen	$1s^2\,2s^2\,2p^3$	$^4S^o_{3/2}$	14.5341
8	O	Oxygen	$1s^2\,2s^2\,2p^4$	$^3P_2$	13.6181
9	F	Fluorine	$1s^2\,2s^2\,2p^5$	$^2P^o_{3/2}$	17.4228
10	Ne	Neon	$1s^2\,2s^2\,2p^6$	$^1S_0$	21.5645
11	Na	Sodium	[Ne] $3s$	$^2S_{1/2}$	5.1391
12	Mg	Magnesium	[Ne] $3s^2$	$^1S_0$	7.6462
13	Al	Aluminum	[Ne] $3s^2\,3p$	$^2P^o_{1/2}$	5.9858
14	Si	Silicon	[Ne] $3s^2\,3p^2$	$^3P_0$	8.1517
15	P	Phosphorus	[Ne] $3s^2\,3p^3$	$^4S^o_{3/2}$	10.4867
16	S	Sulfur	[Ne] $3s^2\,3p^4$	$^3P_2$	10.3600
17	Cl	Chlorine	[Ne] $3s^2\,3p^5$	$^2P^o_{3/2}$	12.9676
18	Ar	Argon	[Ne] $3s^2\,3p^6$	$^1S_0$	15.7596
19	K	Potassium	[Ar] $4s$	$^2S_{1/2}$	4.3407
20	Ca	Calcium	[Ar] $4s^2$	$^1S_0$	6.1132
21	Sc	Scandium	[Ar] $3d\,4s^2$	$^2D_{3/2}$	6.5615
22	Ti	Titanium	[Ar] $3d^2\,4s^2$	$^3F_2$	6.8281
23	V	Vanadium	[Ar] $3d^3\,4s^2$	$^4F_{3/2}$	6.7462
24	Cr	Chromium	[Ar] $3d^5\,4s$	$^7S_3$	6.7665
25	Mn	Manganese	[Ar] $3d^5\,4s^2$	$^6S_{5/2}$	7.4340
26	Fe	Iron	[Ar] $3d^6\,4s^2$	$^5D_4$	7.9024
27	Co	Cobalt	[Ar] $3d^7\,4s^2$	$^4F_{9/2}$	7.8810
28	Ni	Nickel	[Ar] $3d^8\,4s^2$	$^3F_4$	7.6398
29	Cu	Copper	[Ar] $3d^{10}\,4s$	$^2S_{1/2}$	7.7264
30	Zn	Zinc	[Ar] $3d^{10}\,4s^2$	$^1S_0$	9.3942
31	Ga	Gallium	[Ar] $3d^{10}\,4s^2\,4p$	$^2P^o_{1/2}$	5.9993
32	Ge	Germanium	[Ar] $3d^{10}\,4s^2\,4p^2$	$^3P_0$	7.8994
33	As	Arsenic	[Ar] $3d^{10}\,4s^2\,4p^3$	$^4S^o_{3/2}$	9.7886
34	Se	Selenium	[Ar] $3d^{10}\,4s^2\,4p^4$	$^3P_2$	9.7524
35	Br	Bromine	[Ar] $3d^{10}\,4s^2\,4p^5$	$^2P^o_{3/2}$	11.8138
36	Kr	Krypton	[Ar] $3d^{10}\,4s^2\,4p^6$	$^1S_0$	13.9996
37	Rb	Rubidium	[Kr] $5s$	$^2S_{1/2}$	4.1771
38	Sr	Strontium	[Kr] $5s^2$	$^1S_0$	5.6949
39	Y	Yttrium	[Kr] $4d\,5s^2$	$^2D_{3/2}$	6.2173
40	Zr	Zirconium	[Kr] $4d^2\,5s^2$	$^3F_2$	6.6339
41	Nb	Niobium	[Kr] $4d^4\,5s$	$^6D_{1/2}$	6.7589
42	Mo	Molybdenum	[Kr] $4d^5\,5s$	$^7S_3$	7.0924
43	Tc	Technetium	[Kr] $4d^5\,5s^2$	$^6S_{5/2}$	7.28
44	Ru	Ruthenium	[Kr] $4d^7\,5s$	$^5F_5$	7.3605

Z		Element	Ground-state configuration	Ground level	Ionization energy (eV)
45	Rh	Rhodium	[Kr] $4d^8\,5s$	$^4F_{9/2}$	7.4589
46	Pd	Palladium	[Kr] $4d^{10}$	$^1S_0$	8.3369
47	Ag	Silver	[Kr] $4d^{10}\,5s$	$^2S_{1/2}$	7.5762
48	Cd	Cadmium	[Kr] $4d^{10}\,5s^2$	$^1S_0$	8.9938
49	In	Indium	[Kr] $4d^{10}\,5s^2\,5p$	$^2P^o_{1/2}$	5.7864
50	Sn	Tin	[Kr] $4d^{10}\,5s^2\,5p^2$	$^3P_0$	7.3439
51	Sb	Antimony	[Kr] $4d^{10}\,5s^2\,5p^3$	$^4S^o_{3/2}$	8.6084
52	Te	Tellurium	[Kr] $4d^{10}\,5s^2\,5p^4$	$^3P_2$	9.0096
53	I	Iodine	[Kr] $4d^{10}\,5s^2\,5p^5$	$^2P^o_{3/2}$	10.4513
54	Xe	Xenon	[Kr] $4d^{10}\,5s^2\,5p^6$	$^1S_0$	12.1298
55	Cs	Cesium	[Xe] $6s$	$^2S_{1/2}$	3.8939
56	Ba	Barium	[Xe] $6s^2$	$^1S_0$	5.2117
57	La	Lanthanum	[Xe] $5d\,6s^2$	$^2D_{3/2}$	5.5769
58	Ce	Cerium	[Xe] $4f\,5d\,6s^2$	$^1G^o_4$	5.5387
59	Pr	Praseodymium	[Xe] $4f^3\,6s^2$	$^4I^o_{9/2}$	5.473
60	Nd	Neodymium	[Xe] $4f^4\,6s^2$	$^5I_4$	5.5250
61	Pm	Promethium	[Xe] $4f^5\,6s^2$	$^6H^o_{5/2}$	5.582
62	Sm	Samarium	[Xe] $4f^6\,6s^2$	$^7F_0$	5.6437
63	Eu	Europium	[Xe] $4f^7\,6s^2$	$^8S^o_{7/2}$	5.6704
64	Gd	Gadolinium	[Xe] $4f^7\,5d\,6s^2$	$^9D^o_2$	6.1498
65	Tb	Terbium	[Xe] $4f^9\,6s^2$	$^6H^o_{15/2}$	5.8638
66	Dy	Dysprosium	[Xe] $4f^{10}\,6s^2$	$^5I_8$	5.9389
67	Ho	Holmium	[Xe] $4f^{11}\,6s^2$	$^4I^o_{15/2}$	6.0215
68	Er	Erbium	[Xe] $4f^{12}\,6s^2$	$^3H_6$	6.1077
69	Tm	Thulium	[Xe] $4f^{13}\,6s^2$	$^2F^o_{7/2}$	6.1843
70	Yb	Ytterbium	[Xe] $4f^{14}\,6s^2$	$^1S_0$	6.2542
71	Lu	Lutetium	[Xe] $4f^{14}\,5d\,6s^2$	$^2D_{3/2}$	5.4259
72	Hf	Hafnium	[Xe] $4f^{14}\,5d^2\,6s^2$	$^3F_2$	6.8251
73	Ta	Tantalum	[Xe] $4f^{14}\,5d^3\,6s^2$	$^4F_{3/2}$	7.5496
74	W	Tungsten	[Xe] $4f^{14}\,5d^4\,6s^2$	$^5D_0$	7.8640
75	Re	Rhenium	[Xe] $4f^{14}\,5d^5\,6s^2$	$^6S_{5/2}$	7.8335
76	Os	Osmium	[Xe] $4f^{14}\,5d^6\,6s^2$	$^5D_4$	8.4382
77	Ir	Iridium	[Xe] $4f^{14}\,5d^7\,6s^2$	$^4F_{9/2}$	8.9670
78	Pt	Platinum	[Xe] $4f^{14}\,5d^9\,6s$	$^3D_3$	8.9588
79	Au	Gold	[Xe] $4f^{14}\,5d^{10}\,6s$	$^2S_{1/2}$	9.2255
80	Hg	Mercury	[Xe] $4f^{14}\,5d^{10}\,6s^2$	$^1S_0$	10.4375
81	Tl	Thallium	[Xe] $4f^{14}\,5d^{10}\,6s^2\,6p$	$^2P^o_{1/2}$	6.1082
82	Pb	Lead	[Xe] $4f^{14}\,5d^{10}\,6s^2\,6p^2$	$^3P_0$	7.4167
83	Bi	Bismuth	[Xe] $4f^{14}\,5d^{10}\,6s^2\,6p^3$	$^4S^o_{3/2}$	7.2855
84	Po	Polonium	[Xe] $4f^{14}\,5d^{10}\,6s^2\,6p^4$	$^3P_2$	8.414
85	At	Astatine	[Xe] $4f^{14}\,5d^{10}\,6s^2\,6p^5$	$^2P^o_{3/2}$	
86	Rn	Radon	[Xe] $4f^{14}\,5d^{10}\,6s^2\,6p^6$	$^1S_0$	10.7485
87	Fr	Francium	[Rn] $7s$	$^2S_{1/2}$	4.0727
88	Ra	Radium	[Rn] $7s^2$	$^1S_0$	5.2784
89	Ac	Actinium	[Rn] $6d\,7s^2$	$^2D_{3/2}$	5.17
90	Th	Thorium	[Rn] $6d^2\,7s^2$	$^3F_2$	6.3067
91	Pa	Protactinium	[Rn] $5f^2(^3H_4)\,6d\,7s^2$	$(4,3/2)_{11/2}$	5.89
92	U	Uranium	[Rn] $5f^3(^4I^o_{9/2})\,6d\,7s^2$	$(9/2,3/2)^o_6$	6.1941
93	Np	Neptunium	[Rn] $5f^4(^5I_4)\,6d\,7s^2$	$(4,3/2)_{11/2}$	6.2657
94	Pu	Plutonium	[Rn] $5f^6\,7s^2$	$^7F_0$	6.0260
95	Am	Americium	[Rn] $5f^7\,7s^2$	$^8S^o_{7/2}$	5.9738
96	Cm	Curium	[Rn] $5f^7\,6d\,7s^2$	$^9D^o_2$	5.9914
97	Bk	Berkelium	[Rn] $5f^9\,7s^2$	$^6H^o_{15/2}$	6.1979
98	Cf	Californium	[Rn] $5f^{10}\,7s^2$	$^5I_8$	6.2817
99	Es	Einsteinium	[Rn] $5f^{11}\,7s^2$	$^4I^o_{15/2}$	6.42
100	Fm	Fermium	[Rn] $5f^{12}\,7s^2$	$^3H_6$	6.50
101	Md	Mendelevium	[Rn] $5f^{13}\,7s^2$	$^2F^o_{7/2}$	6.58
102	No	Nobelium	[Rn] $5f^{14}\,7s^2$	$^1S_0$	6.65
103	Lr	Lawrencium	[Rn] $5f^{14}\,7s^2\,7p$?	$^2P^o_{1/2}$?	4.9?
104	Rf	Rutherfordium	[Rn] $5f^{14}\,6d^2\,7s^2$?	$^3F_2$?	6.0?

INTERNATIONAL TEMPERATURE SCALE OF 1990 (ITS-90)

B. W. Mangum

A new temperature scale, the International Temperature Scale of 1990 (ITS-90), was officially adopted by the Comité International des Poids et Mesures (CIPM), meeting 26—28 September 1989 at the Bureau International des Poids et Mesures (BIPM). The ITS-90 was recommended to the CIPM for its adoption following the completion of the final details of the new scale by the Comité Consultatif de Thermométrie (CCT), meeting 12—14 September 1989 at the BIPM in its 17th Session. The ITS-90 became the official international temperature scale on 1 January 1990. The ITS-90 supersedes the present scales, the International Practical Temperature Scale of 1968 (IPTS-68) and the 1976 Provisional 0.5 to 30 K Temperature Scale (EPT-76).

The ITS-90 extends upward from 0.65 K, and temperatures on this scale are in much better agreement with thermodynamic values that are those on the IPTS-68 and the EPT-76. The new scale has subranges and alternative definitions in certain ranges that greatly facilitate its use. Furthermore, its continuity, precision, and reproducibility throughout its ranges are much improved over that of the present scales. The replacement of the thermocouple with the platinum resistance thermometer at temperatures below 961.78 °C resulted in the biggest improvement in reproducibility.

The ITS-90 is divided into four primary ranges:

1. Between 0.65 and 3.2 K, the ITS-90 is defined by the vapor pressure-temperature relation of ^{3}He, and between 1.25 and 2.1768 K (the λ point) and between 2.1768 and 5.0 K by the vapor pressure–temperature relations of ^{4}He. T_{90} is defined by the vapor pressure equations of the form:

$$T_{90} / \mathrm{K} = A_0 + \sum_{i=1}^{9} A_i \left[(\ln(p / \mathrm{Pa}) - B) / C \right]^i$$

The values of the coefficients A_i, and of the constants A_o, B, and C of the equations are given below.

2. Between 3.0 and 24.5561 K, the ITS-90 is defined in terms of a ^{3}He or ^{4}He constant volume gas thermometer (CVGT). The thermometer is calibrated at three temperatures — at the triple point of neon (24.5561 K), at the triple point of equilibrium hydrogen (13.8033 K), and at a temperature between 3.0 and 5.0 K, the value of which is determined by using either ^{3}He or ^{4}He vapor pressure thermometry.

3. Between 13.8033 K (−259.3467 °C) and 1234.93 K (961.78 °C), the ITS-90 is defined in terms of the specified fixed points given below, by resistance ratios of platinum resistance thermometers obtained by calibration at specified sets of the fixed points, and by reference functions and deviation functions of resistance ratios which relate to T_{90} between the fixed points.

4. Above 1234.93 K, the ITS-90 is defined in terms of Planck's radiation law, using the freezing-point temperature of either silver, gold, or copper as the reference temperature.

Full details of the calibration procedures and reference functions for various subranges are given in:

The International Temperature Scale of 1990, *Metrologia*, 27, 3, 1990; errata in *Metrologia*, 27, 107, 1990.

Defining Fixed Points of the ITS-90

Material[a]	Equilibrium state[b]	Temperature T_{90} (K)	t_{90} (°C)
He	VP	3 to 5	−270.15 to −268.15
e-H$_2$	TP	13.8033	−259.3467
e-H$_2$ (or He)	VP (or CVGT)	≈17	≈ −256.15
e-H$_2$ (or He)	VP (or CVGT)	≈20.3	≈ −252.85
Ne[c]	TP	24.5561	−248.5939
O$_2$	TP	54.3584	−218.7916
Ar	TP	83.8058	−189.3442
Hg[c]	TP	234.3156	−38.8344
H$_2$O	TP	273.16	0.01
Ga[c]	MP	302.9146	29.7646
In[c]	FP	429.7485	156.5985
Sn	FP	505.078	231.928
Zn	FP	692.677	419.527
Al[c]	FP	933.473	660.323
Ag	FP	1234.93	961.78
Au	FP	1337.33	1064.18
Cu[c]	FP	1357.77	1084.62

[a] e-H$_2$ indicates equilibrium hydrogen, that is, hydrogen with the equilibrium distribution of its ortho and para states. Normal hydrogen at room temperature contains 25% para hydrogen and 75% ortho hydrogen.

[b] VP indicates vapor pressure point; CVGT indicates constant volume gas thermometer point; TP indicates triple point (equilibrium temperature at which the solid, liquid, and vapor phases coexist); FP indicates freezing point, and MP indicates melting point (the equilibrium temperatures at which the solid and liquid phases coexist under a pressure of 101 325 Pa, one standard atmosphere). The isotopic composition is that naturally occurring.

[c] Previously, these were secondary fixed points.

Values of Coefficients in the Vapor Pressure Equations for Helium

Coef. or constant	^{3}He 0.65—3.2 K	^{4}He 1.25—2.1768 K	^{4}He 2.1768—5.0 K
A_0	1.053 447	1.392 408	3.146 631
A_1	0.980 106	0.527 153	1.357 655
A_2	0.676 380	0.166 756	0.413 923
A_3	0.372 692	0.050 988	0.091 159
A_4	0.151 656	0.026 514	0.016 349
A_5	−0.002 263	0.001 975	0.001 826
A_6	0.006 596	−0.017 976	−0.004 325
A_7	0.088 966	0.005 409	−0.004 973
A_8	−0.004 770	0.013 259	0
A_9	−0.054 943	0	0
B	7.3	5.6	10.3
C	4.3	2.9	1.9

CONVERSION OF TEMPERATURES FROM THE 1948 AND 1968 SCALES TO ITS-90

This table gives temperature corrections from older scales to the current International Temperature Scale of 1990 (see the preceding table for details on ITS-90). The first part of the table may be used for converting Celsius temperatures in the range −180 to 4000 °C from IPTS-68 or IPTS-48 to ITS-90. Within the accuracy of the corrections, the temperature in the first column may be identified with either t_{68}, t_{48}, or t_{90}. The second part of the table is designed for use at lower temperatures to convert values expressed in kelvins from EPT-76 or IPTS-68 to ITS-90.

The references give analytical equations for expressing these relations. Note that Reference 1 supersedes Reference 2 with respect to corrections in the 630 to 1064 °C range.

References

1. Burns, G. W. et al., in *Temperature: Its Measurement and Control in Science and Industry*, Vol. 6, Schooley, J. F., Ed., American Institute of Physics, New York, 1993.
2. Goldberg, R. N. and Weir, R. D., *Pure and Appl. Chem.*, 64, 1545, 1992.

$t/°C$	$t_{90}-t_{68}$	$t_{90}-t_{48}$
−180	0.008	0.020
−170	0.010	0.017
−160	0.012	0.007
−150	0.013	0.000
−140	0.014	0.001
−130	0.014	0.008
−120	0.014	0.017
−110	0.013	0.026
−100	0.013	0.035
−90	0.012	0.041
−80	0.012	0.045
−70	0.011	0.045
−60	0.010	0.042
−50	0.009	0.038
−40	0.008	0.032
−30	0.006	0.024
−20	0.004	0.016
−10	0.002	0.008
0	0.000	0.000
10	−0.002	−0.006
20	−0.005	−0.012
30	−0.007	−0.016
40	−0.010	−0.020
50	−0.013	−0.023
60	−0.016	−0.026
70	−0.018	−0.026
80	−0.021	−0.027
90	−0.024	−0.027
100	−0.026	−0.026
110	−0.028	−0.024
120	−0.030	−0.023
130	−0.032	−0.020
140	−0.034	−0.018
150	−0.036	−0.016
160	−0.037	−0.012
170	−0.038	−0.009
180	−0.039	−0.005
190	−0.039	−0.001
200	−0.040	0.003
210	−0.040	0.007
220	−0.040	0.011
230	−0.040	0.014
240	−0.040	0.018
250	−0.040	0.021
260	−0.040	0.024
270	−0.039	0.028
280	−0.039	0.030
290	−0.039	0.032
300	−0.039	0.034
310	−0.039	0.035
320	−0.039	0.036
330	−0.040	0.036
340	−0.040	0.037
350	−0.041	0.036
360	−0.042	0.035
370	−0.043	0.034
380	−0.045	0.032
390	−0.046	0.030
400	−0.048	0.028
410	−0.051	0.024
420	−0.053	0.022
430	−0.056	0.019
440	−0.059	0.015
450	−0.062	0.012
460	−0.065	0.009
470	−0.068	0.007
480	−0.072	0.004
490	−0.075	0.002
500	−0.079	0.000
510	−0.083	−0.001
520	−0.087	−0.002
530	−0.090	−0.001
540	−0.094	0.000
550	−0.098	0.002
560	−0.101	0.007
570	−0.105	0.011
580	−0.108	0.018
590	−0.112	0.025
600	−0.115	0.035
610	−0.118	0.047
620	−0.122	0.060
630	−0.125	0.075
640	−0.11	0.12
650	−0.10	0.15
660	−0.09	0.19
670	−0.07	0.24
680	−0.05	0.29
690	−0.04	0.32
700	−0.02	0.37
710	−0.01	0.41
720	0.00	0.45
730	0.02	0.49
740	0.03	0.53
750	0.03	0.56
760	0.04	0.60
770	0.05	0.63
780	0.05	0.66
790	0.05	0.69
800	0.05	0.72
810	0.05	0.75
820	0.04	0.76
830	0.04	0.79
840	0.03	0.81
850	0.02	0.83
860	0.01	0.85
870	0.00	0.87
880	−0.02	0.87
890	−0.03	0.89
900	−0.05	0.90
910	−0.06	0.92
920	−0.08	0.93
930	−0.10	0.94
940	−0.11	0.96
950	−0.13	0.97
960	−0.15	0.97
970	−0.16	0.99
980	−0.18	1.00
990	−0.19	1.02
1000	−0.20	1.04
1010	−0.22	1.05
1020	−0.23	1.07
1030	−0.23	1.10
1040	−0.24	1.12
1050	−0.25	1.14
1060	−0.25	1.17
1070	−0.25	1.19
1080	−0.26	1.20
1090	−0.26	1.20
1100	−0.26	1.2
1200	−0.30	1.4
1300	−0.35	1.5
1400	−0.39	1.6
1500	−0.44	1.8
1600	−0.49	1.9
1700	−0.54	2.1
1800	−0.60	2.2
1900	−0.66	2.3
2000	−0.72	2.5
2100	−0.79	2.7
2200	−0.85	2.9
2300	−0.93	3.1
2400	−1.00	3.2
2500	−1.07	3.4
2600	−1.15	3.7
2700	−1.24	3.8
2800	−1.32	4.0
2900	−1.41	4.2
3000	−1.50	4.4
3100	−1.59	4.6
3200	−1.69	4.8
3300	−1.78	5.1
3400	−1.89	5.3
3500	−1.99	5.5
3600	−2.10	5.8
3700	−2.21	6.0
3800	−2.32	6.3
3900	−2.43	6.6
4000	−2.55	6.8

T/K	$T_{90}-T_{76}$	$T_{90}-T_{68}$
5	−0.0001	
6	−0.0002	
7	−0.0003	
8	−0.0004	
9	−0.0005	
10	−0.0006	
11	−0.0007	
12	−0.0008	
13	−0.0010	
14	−0.0011	−0.006
15	−0.0013	−0.003
16	−0.0014	−0.004
17	−0.0016	−0.006
18	−0.0018	−0.008
19	−0.0020	−0.009
20	−0.0022	−0.009
21	−0.0025	−0.008
22	−0.0027	−0.007
23	−0.0030	−0.007
24	−0.0032	−0.006
25	−0.0035	−0.005
26	−0.0038	−0.004
27	−0.0041	−0.004
28		−0.005
29		−0.006
30		−0.006
31		−0.007
32		−0.008

T/K	$T_{90}-T_{76}$	$T_{90}-T_{68}$	T/K	$T_{90}-T_{76}$	$T_{90}-T_{68}$	T/K	$T_{90}-T_{76}$	$T_{90}-T_{68}$	T/K	$T_{90}-T_{76}$	$T_{90}-T_{68}$
33		−0.008	57		0.000	81		0.008	150		0.014
34		−0.008	58		0.001	82		0.008	160		0.014
35		−0.007	59		0.002	83		0.008	170		0.013
36		−0.007	60		0.003	84		0.008	180		0.012
37		−0.007	61		0.003	85		0.008	190		0.012
38		−0.006	62		0.004	86		0.008	200		0.011
39		−0.006	63		0.004	87		0.008	210		0.010
40		−0.006	64		0.005	88		0.008	220		0.009
41		−0.006	65		0.005	89		0.008	230		0.008
42		−0.006	66		0.006	90		0.008	240		0.007
43		−0.006	67		0.006	91		0.008	250		0.005
44		−0.006	68		0.007	92		0.008	260		0.003
45		−0.007	69		0.007	93		0.008	270		0.001
46		−0.007	70		0.007	94		0.008	273.16		0.000
47		−0.007	71		0.007	95		0.008	300		−0.006
48		−0.006	72		0.007	96		0.008	400		−0.031
49		−0.006	73		0.007	97		0.009	500		−0.040
50		−0.006	74		0.007	98		0.009	600		−0.040
51		−0.005	75		0.008	99		0.009	700		−0.055
52		−0.005	76		0.008	100		0.009	800		−0.089
53		−0.004	77		0.008	110		0.011	900		−0.124
54		−0.003	78		0.008	120		0.013			
55		−0.002	79		0.008	130		0.014			
56		−0.001	80		0.008	140		0.014			

INTERNATIONAL SYSTEM OF UNITS (SI)

The International System of Units, abbreviated as SI (from the French name *Le Système International d'Unités***),** was established in 1960 by the 11th General Conference on Weights and Measures (CGPM) as the modern metric system of measurement. The core of the SI is the seven base units for the physical quantities length, mass, time, electric current, thermodynamic temperature, amount of substance, and luminous intensity. These base units are:

	SI base unit	
Base quantity	Name	Symbol
length	meter	m
mass	kilogram	kg
time	second	s
electric current	ampere	A
thermodynamic temperature	kelvin	K
amount of substance	mole	mol
luminous intensity	candela	cd

The SI base units are defined as follows:

ampere: The ampere is that constant current which, if maintained in two straight parallel conductors of infinite length, of negligible circular cross-section, and placed 1 meter apart in vacuum, would produce between these conductors a force equal to $2 \cdot 10^{-7}$ newton per meter of length.

candela: The candela is the luminous intensity, in a given direction, of a source that emits monochromatic radiation of frequency $540 \cdot 10^{12}$ hertz and that has a radiant intensity in that direction of 1/683 watt per steradian.

kelvin: The kelvin, unit of thermodynamic temperature, is the fraction 1/273.16 of the thermodynamic temperature of the triple point of water.

kilogram: The kilogram is the unit of mass; it is equal to the mass of the international prototype of the kilogram.

meter: The meter is the length of the path travelled by light in vacuum during a time interval of 1/299 792 458 of a second.

mole: The mole is the amount of substance of a system which contains as many elementary entities as there are atoms in 0.012 kilogram of carbon 12. When the mole is used, the elementary entities must be specified and may be atoms, molecules, ions, electrons, other particles, or specified groups of such particles.

second: The second is the duration of 9 192 631 770 periods of the radiation corresponding to the transition between the two hyperfine levels of the ground state of the cesium 133 atom.

SI derived units

Derived units are units which may be expressed in terms of base units by means of the mathematical symbols of multiplication and division (and, in the case of °C, subtraction). Certain derived units have been given special names and symbols, and these special names and symbols may themselves be used in combination with those for base and other derived units to express the units of other quantities. The next table lists some examples of derived units expressed directly in terms of base units:

	SI derived unit	
Physical quantity	Name	Symbol
area	square meter	m^2
volume	cubic meter	m^3
speed, velocity	meter per second	m/s
acceleration	meter per second squared	m/s^2
wave number	reciprocal meter	m^{-1}
density, mass density	kilogram per cubic meter	kg/m^3
specific volume	cubic meter per kilogram	m^3/kg
current density	ampere per square meter	A/m^2
magnetic field strength	ampere per meter	A/m
concentration (of amount of substance)	mole per cubic meter	mol/m^3
luminance	candela per square meter	cd/m^2
refractive index	(the number) one	$1^{(a)}$

(a) The symbol "1" is generally omitted in combination with a numerical value.

For convenience, certain derived units, which are listed in the next table, have been given special names and symbols. These names and symbols may themselves be used to express other derived units. The special names and symbols are a compact form for the expression of units that are used frequently. The final column shows how the SI units concerned may be expressed in terms of SI base units. In this column, factors such as $m^0, kg^0 \ldots$, which are all equal to 1, are not shown explicitly.

Physical quantity	Name	Symbol	SI derived unit expressed in terms of:	
			Other SI units	SI base units
plane angle	radian[a]	rad	$m \cdot m^{-1} = 1^{(b)}$	
solid angle	steradian[a]	sr[c]	$m^2 \cdot m^{-2} = 1^{(b)}$	
frequency	hertz	Hz	s^{-1}	
force	newton	N	$m \cdot kg \cdot s^{-2}$	
pressure, stress	pascal	Pa	N/m^2	$m^{-1} \cdot kg \cdot s^{-2}$
energy, work, quantity of heat	joule	J	$N \cdot m$	$m^2 \cdot kg \cdot s^{-2}$
power, radiant flux	watt	W	J/s	$m^2 \cdot kg \cdot s^{-3}$
electric charge, quantity of electricity	coulomb	C	$s \cdot A$	
electric potential difference, electromotive force	volt	V	W/A	$m^2 \cdot kg \cdot s^{-3} \cdot A^{-1}$
capacitance	farad	F	C/V	$m^{-2} \cdot kg^{-1} \cdot s^4 \cdot A^2$
electric resistance	ohm	Ω	V/A	$m^2 \cdot kg \cdot s^{-3} \cdot A^{-2}$
electric conductance	siemens	S	A/V	$m^{-2} \cdot kg^{-1} \cdot s^3 \cdot A^2$
magnetic flux	weber	Wb	$V \cdot s$	$m^2 \cdot kg \cdot s^{-2} \cdot A^{-1}$

| Physical quantity | Name | Symbol | SI derived unit expressed in terms of: | |
			Other SI units	SI base units
magnetic flux density	tesla	T	Wb/m^2	kg $\cdot$ s^{-2} $\cdot$ A^{-1}
inductance	henry	H	Wb/A	m^2 $\cdot$ kg $\cdot$ s^{-2} $\cdot$ A^{-2}
Celsius temperature	degree Celsius[d]	°C		K
luminous flux	lumen	lm	cd $\cdot$ sr[c]	m^2 $\cdot$ m^{-2} $\cdot$ cd = cd
illuminance	lux	lx	lm/m^2	m^2 $\cdot$ m^{-4} $\cdot$ cd = m^{-2} $\cdot$ cd
activity (of a radionuclide)	becquerel	Bq		s^{-1}
absorbed dose, specific energy (imparted), kerma	gray	Gy	J/kg	m^2 $\cdot$ s^{-2}
dose equivalent, ambient dose equivalent, directional dose equivalent, personal dose equivalent, organ equivalent dose	sievert	Sv	J/kg	m^2 $\cdot$ s^{-2}
catalytic activity	katal	kat		s^{-1} $\cdot$ mol

[a] The radian and steradian may be used with advantage in expressions for derived units to distinguish between quantities of different nature but the same dimension. Some examples of their use in forming derived units are given in the next table.

[b] In practice, the symbols rad and sr are used where appropriate, but the derived unit "1" is generally omitted in combination with a numerical value.

[c] In photometry, the name steradian and the symbol sr are usually retained in expressions for units.

[d] It is common practice to express a thermodynamic temperature, symbol T, in terms of its difference from the reference temperature T_0 = 273.15 K. The numerical value of a Celsius temperature t expressed in degrees Celsius is given by t/°C = T/K-273.15. The unit °C may be used in combination with SI prefixes, e.g., millidegree Celsius, m°C. Note that there should never be a space between the ° sign and the letter C, and that the symbol for kelvin is K, not °K.

The SI derived units with special names may be used in combinations to provide a convenient way to express more complex physical quantities. Examples are given in the next table:

| Physical Quantity | SI derived unit | | |
	Name	Symbol	As SI base units
dynamic viscosity	pascal second	Pa $\cdot$ s	m^{-1} $\cdot$ kg $\cdot$ s^{-1}
moment of force	newton meter	N $\cdot$ m	m^2 $\cdot$ kg $\cdot$ s^{-2}
surface tension	newton per meter	N/m	kg $\cdot$ s^{-2}
angular velocity	radian per second	rad/s	m $\cdot$ m^{-1} $\cdot$ s^{-1} = s^{-1}
angular acceleration	radian per second squared	rad/s^2	m $\cdot$ m^{-1} $\cdot$ s^{-2} = s^{-2}
heat flux density, irradiance	watt per square meter	W/m^2	kg $\cdot$ s^{-3}
heat capacity, entropy	joule per kelvin	J/K	m^{-3} $\cdot$ kg $\cdot$ s^{-2} $\cdot$ K^{-1}
specific heat capacity, specific entropy	joule per kilogram kelvin	J/(kg $\cdot$ K)	m^2 $\cdot$ s^{-2} $\cdot$ K^{-1}
specific energy	joule per kilogram	J/kg	m^2 $\cdot$ s^{-2}
thermal conductivity	watt per meter kelvin	W/(m $\cdot$ K)	m $\cdot$ kg $\cdot$ s^{-3} $\cdot$ K^{-1}
energy density	joule per cubic meter	J/m^3	m^{-1} $\cdot$ kg $\cdot$ s^{-2}
electric field strength	volt per meter	V/m	m $\cdot$ kg $\cdot$ s^{-3} $\cdot$ A^{-1}
electric charge density	coulomb per cubic meter	C/m^3	m^{-3} $\cdot$ s $\cdot$ A
electric flux density	coulomb per square meter	C/m^2	m^{-2} $\cdot$ s $\cdot$ A
permittivity	farad per meter	F/m	m^{-3} $\cdot$ kg^{-1} $\cdot$ s^4 $\cdot$ A^2
permeability	henry per meter	H/m	m $\cdot$ kg $\cdot$ s^{-2} $\cdot$ A^{-2}
molar energy	joule per mole	J/mol	m^2 $\cdot$ kg $\cdot$ s^{-2} $\cdot$ mol^{-1}
molar entropy, molar heat capacity	joule per mole kelvin	J/(mol $\cdot$ K)	m^2 $\cdot$ kg $\cdot$ s^{-2} $\cdot$ K^{-1} $\cdot$ mol^{-1}
exposure (x and γ rays)	coulomb per kilogram	C/kg	kg^{-1} $\cdot$ s $\cdot$ A
absorbed dose rate	gray per second	Gy/s	m^2 $\cdot$ s^{-3}
radiant intensity	watt per steradian	W/sr	m^4 $\cdot$ m^{-2} $\cdot$ kg $\cdot$ s^{-3} = m^2 $\cdot$ kg $\cdot$ s^{-3}
radiance	watt per square meter steradian	W/(m^2 $\cdot$ sr)	m^2 $\cdot$ m^{-2} $\cdot$ kg $\cdot$ s^{-3} = kg $\cdot$ s^{-3}
catalytic (activity) concentration	katal per cubic meter	kat/m^3	m^{-3} $\cdot$ s^{-1} $\cdot$ mol

In practice, with certain quantities preference is given to the use of certain special unit names, or combinations of unit names, in order to facilitate the distinction between different quantities having the same dimension. For example, the SI unit of frequency is designated the hertz, rather than the reciprocal second, and the SI unit of angular velocity is designated the radian per second rather than the reciprocal second (in this case retaining the word radian emphasizes that angular velocity is equal to 2π times the rotational frequency). Similarly the SI unit of moment of force is designated the newton meter rather than the joule.

In the field of ionizing radiation, the SI unit of activity is designated the becquerel rather than the reciprocal second, and the SI units of absorbed dose and dose equivalent the gray and sievert, respectively, rather than the joule per kilogram. In the field of catalysis, the SI unit of catalytic activity is designated the katal rather than the mole per second. The special names becquerel, gray, sievert, and katal were specifically introduced because of the dangers to human health which might arise from mistakes involving the units reciprocal second, joule per kilogram and mole per second.

Units for dimensionless quantities, quantities of dimension one

Certain quantities are defined as the ratios of two quantities of the same kind, and thus have a dimension which may be expressed by the number one. The unit of such quantities is necessarily a derived unit coherent with the other units of the SI and, since it is formed as the ratio of two identical SI units, the unit also may be expressed by the number one. Thus the SI unit of all quantities having the dimensional product one is the number one. Examples of such quantities are refractive index, relative permeability, and friction factor. Other quantities having the unit 1 include "characteristic numbers" like the Prandtl number and numbers which represent a count, such as a number of molecules, degeneracy (number of energy levels), and partition function in statistical thermodynamics. All of these quantities are described as being dimensionless, or of dimension one, and have the coherent SI unit 1. Their values are simply expressed as numbers and, in general, the unit 1 is not explicitly shown. In a few cases, however, a special name is given to this unit, mainly to avoid confusion between some compound derived units. This is the case for the radian, steradian and neper.

SI prefixes

The following prefixes have been approved by the CGPM for use with SI units. Only one prefix may be used before a unit. Thus 10^{-12} farad should be designated pF, not μμF.

Factor	Name	Symbol	Factor	Name	Symbol
10^{24}	yotta	Y	10^{-1}	deci	d
10^{21}	zetta	Z	10^{-2}	centi	c
10^{18}	exa	E	10^{-3}	milli	m
10^{15}	peta	P	10^{-6}	micro	μ
10^{12}	tera	T	10^{-9}	nano	n
10^{9}	giga	G	10^{-12}	pico	p
10^{6}	mega	M	10^{-15}	femto	f
10^{3}	kilo	k	10^{-18}	atto	a
10^{2}	hecto	h	10^{-21}	zepto	z
10^{1}	deka	da	10^{-24}	yocto	y

The kilogram

Among the base units of the International System, the unit of mass is the only one whose name, for historical reasons, contains a prefix. Names and symbols for decimal multiples and submultiples of the unit of mass are formed by attaching prefix names to the unit name "gram" and prefix symbols to the unit symbol "g".

Example : 10^{-6} kg = 1 mg (1 milligram) *but not* 1 μkg (1 microkilogram).

Units used with the SI

Many units that are not part of the SI are important and widely used in everyday life. The CGPM has adopted a classification of non-SI units: (1) units accepted for use with the SI (such as the traditional units of time and of angle); (2) units accepted for use with the SI whose values are obtained experimentally; and (3) other units currently accepted for use with the SI to satisfy the needs of special interests.

(1) Non-SI units accepted for use with the International System

Name	Symbol	Value in SI units
minute	min	1 min = 60 s
hour	h	1 h= 60 min = 3600 s
day	d	1 d = 24 h = 86 400 s
degree	°	1° = $(\pi/180)$ rad
minute	'	1' = $(1/60)°$ = $(\pi/10\ 800)$ rad
second	"	1" = $(1/60)'$ = $(\pi/648\ 000)$ rad
liter	l, L	1L= 1 dm³ = 10^{-3} m³
metric ton	t	1 t = 10^3 kg
neper[a]	Np	1 Np = 1
bel[b]	B	1 B = $(1/2)$ ln 10 Np

[a] The neper is used to express values of such logarithmic quantities as field level, power level, sound pressure level, and logarithmic decrement. Natural logarithms are used to obtain the numerical values of quantities expressed in nepers. The neper is coherent with the SI, but is not yet adopted by the CGPM as an SI unit. In using the neper, it is important to specify the quantity.

[b] The bel is used to express values of such logarithmic quantities as field level, power level, sound-pressure level, and attenuation. Logarithms to base ten are used to obtain the numerical values of quantities expressed in bels. The submultiple decibel, dB, is commonly used.

(2) Non-SI units accepted for use with the International system, whose values in SI units are obtained experimentally

Name	Symbol	Value in SI Units
electronvolt[b]	eV	1 eV = 1.602 176 53(14) $\cdot 10^{-19}$ J[a]
dalton[c]	Da	1 Da = 1.660 538 86(28) $\cdot 10^{-27}$ kg[a]
unified atomic mass unit[c]	u	1 u = 1 Da
astronomical unit[d]	ua	1 ua = 1.495 978 706 91(06) $\cdot 10^{11}$ m[a]

[a] For the electronvolt and the dalton (unified atomic mass unit), values are quoted from the 2002 CODATA set of the Fundamental Physical Constants (p. 1-1 of this Handbook). The value given for the astronomical unit is quoted from the IERS Conventions 2003 (D.D. McCarthy and G. Petit, eds., IERS Technical Note 32, Frankfurt am Main: Verlag des Bundesamts für Kartographie und Geodäsie, 200). The value of ua in meters comes from the JPL ephemerides DE403 (Standish E.M. 1995, "Report of the IAU WGAS Sub-Group on Numerical Standards", in "Highlights of Astronomy", Appenlzer ed., pp 180-184, Kluwer Academic Publishers, Dordrecht). It has been determined in "TDB" units using Barycentric Dynamical Time TDB as a time coordinate for the barycentric system.

[b] The electronvolt is the kinetic energy acquired by an electron in passing through a potential difference of 1 V in vacuum.

[c] The Dalton and unified atomic mass unit are alternative names for the same unit, equal to 1/12 of the mass of an unbound atom of the nuclide ^{12}C, at rest and in its ground state. The dalton may be combined with SI prefixes to express the masses of large molecules in kilodalton, kDa, or megadalton, MDa.

[d] The astronomical unit is a unit of length approximately equal to the mean Earth-Sun distance. It is the radius of an unperturbed circular Newtonian orbit about the Sun of a particle having infinitesimal mass, moving with a mean motion of 0.017 202 098 95 radians/day (known as the Gaussian constant).

(3) Other non-SI units currently accepted for use with the International System

Name	Symbol	Value in SI Units
nautical mile		1 nautical mile = 1852 m
knot		1 nautical mile per hour = $(1852/3600)$ m/s
are		1 a = 1 dam² = 10^2 m²
hectare	ha	1 ha = 1 hm² = 10^4 m²
bar	bar	1 bar = 0.1 MPa = 100 kPa = 10^5 Pa
ångström	Å	1 Å = 0.1 nm = 10^{-10} m
barn	b	1 b = 100 fm² = 10^{-28} m²

Other non-SI units

The SI does not encourage the use of cgs units, but these are frequently found in old scientific texts. The following table gives the relation of some common cgs units to SI units.

Name	Symbol	Value in SI units
erg	erg	1 erg = 10^{-7} J
dyne	dyn	1 dyn = 10^{-5} N
poise	P	1P = 1 dyn $\cdot$ s/cm² = 0.1 Pa $\cdot$ s
stokes	St	1 St = 1 cm²/s = 10^{-4} m²/s
gauss	G	1G $\triangleq 10^{-4}$ T
oersted	Oe	1 Oe $\triangleq (1000/4\pi)$ A/m
maxwell	Mx	1Mx $\triangleq 10^{-8}$ Wb
stilb	sb	1 sb = 1 cd/cm² = 10^4 cd/m²
phot	ph	1 ph = 10^4 lx
gal	Gal	1 Gal = 1 cm/s² = 10^{-2} m/s²

Note: The symbol $\triangleq$ should be read as "corresponds to"; these units cannot strictly be equated because of the different dimensions of the electromagnetic cgs and the SI.

Examples of other non-SI units found in the older literature and their relation to the SI are given below. Use of these units in current texts is discouraged.

Name	Symbol	Value in SI units
curie	Ci	1 Ci = 3.7 · 10^{10} Bq
roentgen	R	1 R = 2.58 · 10^{-4} C/kg
rad	rad	1 rad = 1 cGy = 10^{-2} Gy
rem	rem	1 rem = 1 cSv = 10^{-2} Sv
X unit		1 X unit ≈ 1.002 · 10^{-4} nm
gamma	γ	1 γ = 1 nT = 10^{-9} T
jansky	Jy	1 Jy = 10^{-26} W · m^{-2} · Hz^{-1}
fermi		1 fermi = 1 fm = 10^{-15} m
metric carat		1 metric carat = 200 mg = 2 · 10^{-4} kg
torr	Torr	1 Torr = (101325/760) Pa
standard atmosphere	atm	1 atm = 101325 Pa
calorie[a]	cal	1 cal = 4.184 J
micron	μ	1 μ = 1 μm = 10^{-6} m

[a] Several types of calorie have been used; the value given here is the so-called "thermochemical calorie".

Prefixes for binary multiples

In December 1998, the International Electrotechnical Commission (IEC), the leading international organization for worldwide standardization in electrotechnology, approved as an IEC International Standard names and symbols for prefixes for binary multiples for use in the fields of data processing and data transmission. The prefixes are as follows:

Prefixes for binary multiples

Factor	Name	Symbol	Origin	Derivation
2^{10}	kibi	Ki	kilobinary: (2^{10})^1	kilo: (10^3)^1
2^{20}	mebi	Mi	megabinary: (2^{10})^2	mega: (10^3)^2
2^{30}	gibi	Gi	gigabinary: (2^{10})^3	giga: (10^3)^3
2^{40}	tebi	Ti	terabinary: (2^{10})^4	tera: (10^3)^4
2^{50}	pebi	Pi	petabinary: (2^{10})^5	peta: (10^3)^5
2^{60}	exbi	Ei	exabinary: (2^{10})^6	exa: (10^3)^6

Examples and comparisons with SI prefixes

one **kibibit**	1 Kibit = 2^{10} bit =	**1024 bit**
one **kilobit**	1 kbit = 10^3 bit =	**1000 bit**
one **mebibyte**	1 MiB = 2^{20} B =	**1 048 576 B**
one **megabyte**	1 MB = 10^6 B =	**1 000 000 B**
one **gibibyte**	1 GiB = 2^{30} B =	**1 073 741 824 B**
one **gigabyte**	1 GB = 10^9 B =	**1 000 000 000 B**

It is suggested that in English, the first syllable of the name of the binary-multiple prefix should be pronounced in the same way as the first syllable of the name of the corresponding SI prefix, and that the second syllable should be pronounced as "bee."

It is important to recognize that the new prefixes for binary multiples are not part of the International System of Units (SI), the modern metric system. However, for ease of understanding and recall, they were derived from the SI prefixes for positive powers of ten. As can be seen from the above table, the name of each new prefix is derived from the name of the corresponding SI prefix by retaining the first two letters of the name of the SI prefix and adding the letters "bi," which recalls the word "binary." Similarly, the symbol of each new prefix is derived from the symbol of the corresponding SI prefix by adding the letter "i," which again recalls the word "binary." (For consistency with the other prefixes for binary multiples, the symbol Ki is used for 2^{10} rather than ki.)

References

1. Taylor, B. N., and Thompson, A., *The International System of Unit (SI)*, NIST Special Publication 330, National Institute of Standards and Technology, Gaithersburg, MD, 2008.
2. Bureau International des Poids et Mesures, *Le Système International d'Unités (SI)*, 8th French and English Edition, BIPM, Sèvres, France, 2006.
3. Thompson, A., and Taylor, B. N., *Guide for the Use of the International System of Unit (SI)*, NIST Special Publication 811, National Institute of Standards and Technology, Gaithersburg, MD, 2008.
4. NIST Physical Reference Data web site, http://physics.nist.gov/cuu/Units/index.html, October 2004.
5. Amendment 2 to IEC International Standard IEC 60027-2, 1999-01, Letter symbols to be used in electrical technology – Part 2: Telecommunications and electronics.
6. IEC 60027-2, Second edition, 2000-11, Letter symbols to be used in electrical technology - Part 2: Telecommunications and electronics.
7. Barrow, B., "A Lesson in Megabytes," *IEEE Stand. Bearer*, January 1997, p. 5.

UNITS FOR MAGNETIC PROPERTIES

Quantity	Symbol	Gaussian & cgs emu [a]	Conversion factor, C [b]	SI & rationalized mks [c]
Magnetic flux density, magnetic induction	B	gauss (G) [d]	10^{-4}	tesla (T), Wb/m^2
Magnetic flux	Φ	maxwell (Mx), $G \cdot cm^2$	10^{-8}	weber (Wb), volt second (V · s)
Magnetic potential difference, magnetomotive force	U, F	gilbert (Gb)	$10/4\pi$	ampere (A)
Magnetic field strength, magnetizing force	H	oersted (Oe), [e] Gb/cm	$10^3/4\pi$	A/m [f]
(Volume) magnetization [g]	M	emu/cm^3 [h]	10^3	A/m
(Volume) magnetization	$4\pi M$	G	$10^3/4\pi$	A/m
Magnetic polarization, intensity of magnetization	J, I	emu/cm^3	$4\pi \times 10^{-4}$	T, Wb/m^2 [i]
(Mass) magnetization	σ, M	emu/g	1	$A \cdot m^2/kg$
			$4\pi \times 10^{-7}$	$Wb \cdot m/kg$
Magnetic moment	m	emu, erg/G	10^{-3}	$A \cdot m^2$, joule per tesla (J/T)
Magnetic dipole moment	j	emu, erg/G	$4\pi \times 10^{-10}$	$Wb \cdot m$ [i]
(Volume) susceptibility	χ, κ	dimensionless, emu/cm^3	4π	dimensionless
			$(4\pi)^2 \times 10^{-7}$	henry per meter (H/m), $Wb/(A \cdot m)$
(Mass) susceptibility	χ_ρ, κ_ρ	cm^3/g, emu/g	$4\pi \times 10^{-3}$	m^3/kg
			$(4\pi)^2 \times 10^{-10}$	$H \cdot m^2/kg$
(Molar) susceptibility	χ_{mol}, κ_{mol}	cm^3/mol, emu/mol	$4\pi \times 10^{-6}$	m^3/mol
			$(4\pi)^2 \times 10^{-13}$	$H \cdot m^2/mol$
Permeability	μ	dimensionless	$4\pi \times 10^{-7}$	H/m, $Wb/(A \cdot m)$
Relative permeability [j]	μ_r	not defined		dimensionless
(Volume) energy density, energy product [k]	W	erg/cm^3	10^{-1}	J/m^3
Demagnetization factor	D, N	dimensionless	$1/4\pi$	dimensionless

[a.] Gaussian units and cgs emu are the same for magnetic properties. The defining relation is $B = H + 4\pi M$.

[b.] Multiply a number in Gaussian units by C to convert it to SI (e.g., $1 \text{ G} \times 10^{-4} \text{ T/G} = 10^{-4} \text{ T}$).

[c.] SI (*Système International d'Unités*) has been adopted by the National Bureau of Standards. Where two conversion factors are given, the upper one is recognized under, or consistent with, SI and is based on the definition $B = \mu_0(H + M)$, where $\mu_0 = 4\pi \times 10^{-7}$ H/m. The lower one is not recognized under SI and is based on the definition $B = \mu_0 H + J$, where the symbol I is often used in place of J.

[d.] 1 gauss = 10^5 gamma (γ).

[e.] Both oersted and gauss are expressed as $cm^{-1/2} \cdot g^{1/2} \cdot s^{-1}$ in terms of base units.

[f.] A/m was often expressed as "ampere–turn per meter" when used for magnetic field strength.

[g.] Magnetic moment per unit volume.

[h.] The designation "emu" is not a unit.

[i.] Recognized under SI, even though based on the definition $B = \mu_0 H + J$. See footnote c.

[j.] $\mu_r = \mu/\mu_0 = 1 + \chi$, all in SI. μ_r is equal to Gaussian μ.

[k.] $B \cdot H$ and $\mu_0 M \cdot H$ have SI units J/m^3; $M \cdot H$ and $B \cdot H/4\pi$ have Gaussian units erg/cm^3.

Reference

R. B. Goldfarb and F. R. Fickett, U.S. Department of Commerce, National Bureau of Standards, Boulder, Colorado 80303, March 1985, NBS Special Publication 696. Superintendent of Documents, U.S. Government Printing Office, Washington, DC 20402, 1985.

CONVERSION FACTORS

The following table gives conversion factors from various units of measure to SI units. It is reproduced from NIST Special Publication 811, *Guide for the Use of the International System of Units (SI)*. The table gives the factor by which a quantity expressed in a non-SI unit should be multiplied in order to calculate its value in the SI. The SI values are expressed in terms of the base, supplementary, and derived units of SI in order to provide a coherent presentation of the conversion factors and facilitate computations (see the table "International System of Units" in this section). If desired, powers of ten can be avoided by using SI prefixes and shifting the decimal point if necessary.

Conversion from a non-SI unit to a different non-SI unit may be carried out by using this table in two stages, e.g.,

$$1 \text{ cal}_{th} = 4.184 \text{ J}$$

$$1 \text{ Btu}_{IT} = 1.055056 \text{ E}+03 \text{ J}$$

Thus,

$$1 \text{ Btu}_{IT} = (1.055056 \text{ E}+03 \div 4.184) \text{ cal}_{th} = 252.164 \text{ cal}_{th}$$

Conversion factors are presented for ready adaptation to computer readout and electronic data transmission. The factors are written as a number equal to or greater than one and less than ten with six or fewer decimal places. This number is followed by the letter E (for exponent), a plus or a minus sign, and two digits that indicate the power of 10 by which the number must be multiplied to obtain the correct value. For example:

$$3.523\,907 \text{ E}-02 \text{ is } 3.523\,907 \times 10^{-2}$$

or

$$0.035\,239\,07$$

Similarly:

$$3.386\,389 \text{ E}+03 \text{ is } 3.386\,389 \times 10^{3}$$

or

$$3\,386.389$$

A factor in boldface is exact; i.e., all subsequent digits are zero. All other conversion factors have been rounded to the figures given in accordance with accepted practice. Where less than six digits after the decimal point are shown, more precision is not warranted.

It is often desirable to round a number obtained from a conversion of units in order to retain information on the precision of the value. The following rounding rules may be followed:

1. If the digits to be discarded begin with a digit less than 5, the digit preceding the first discarded digit is not changed.

Example: 6.974 951 5 rounded to 3 digits is 6.97

2. If the digits to be discarded begin with a digit greater than 5, the digit preceding the first discarded digit is increased by one.

Example: 6.974 951 5 rounded to 4 digits is 6.975

3. If the digits to be discarded begin with a 5 and at least one of the following digits is greater than 0, the digit preceding the 5 is increased by 1.

Example: 6.974 851 rounded to 5 digits is 6.974 9

4. If the digits to be discarded begin with a 5 and all of the following digits are 0, the digit preceding the 5 is unchanged if it is even and increased by one if it is odd. (Note that this means that the final digit is always even.)

Examples:
6.974 951 5 rounded to 7 digits is 6.974 952
6.974 950 5 rounded to 7 digits is 6.974 950

Reference

Thompson, A., and Taylor, B. N., *Guide for the Use of the International System of Units (SI)*, NIST Special Publication 811, 2008 Edition, Superintendent of Documents, U.S. Government Printing Office, Washington, DC 20402, 2008.

Factors in **boldface** are exact

To convert from	to	Multiply by	
abampere	ampere (A)	**1.0**	**E+01**
abcoulomb	coulomb (C)	**1.0**	**E+01**
abfarad	farad (F)	**1.0**	**E+09**
abhenry	henry (H)	**1.0**	**E−09**
abmho	siemens (S)	**1.0**	**E+09**
abohm	ohm (Ω)	**1.0**	**E−09**
abvolt	volt (V)	**1.0**	**E−08**
acceleration of free fall, standard (g_n)	meter per second squared (m/s²)	**9.806 65**	**E+00**
acre (based on U.S. survey foot)[a]	square meter (m²)	4.046 873	E+03
acre foot (based on U.S. survey foot)[a]	cubic meter (m³)	1.233 489	E+03
ampere hour (A · h)	coulomb (C)	**3.6**	**E+03**
ångström (Å)	meter (m)	**1.0**	**E−10**
ångström (Å)	nanometer (nm)	**1.0**	**E−01**
apostilb (asb)	candela per meter squared (cd/m²)	3.183 098	E−01
are (a)	square meter (m²)	**1.0**	**E+02**
astronomical unit (ua or AU)	meter (m)	1.495 979	E+11
atmosphere, standard (atm)	pascal (Pa)	**1.013 25**	**E+05**
atmosphere, standard (atm)	kilopascal (kPa)	**1.013 25**	**E+02**
atmosphere, technical (at)[b]	pascal (Pa)	**9.806 65**	**E+04**
atmosphere, technical (at)[b]	kilopascal (kPa)	**9.806 65**	**E+01**

[a] The U.S. survey foot equals (1200/3937) m. 1 international foot = 0.999998 survey foot.

[b] One technical atmosphere equals one kilogram-force per square centimeter (1 at = 1 kgf/cm²).

To convert from	to	Multiply by	
bar (bar)	pascal (Pa)	**1.0**	**E+05**
bar (bar)	kilopascal (kPa)	**1.0**	**E+02**
barn (b)	square meter (m^2)	**1.0**	**E−28**
barrel [for petroleum, 42 gallons (U.S.)](bbl)	cubic meter (m^3)	1.589 873	E−01
barrel [for petroleum, 42 gallons (U.S.)](bbl)	liter (L)	1.589 873	E+02
biot (Bi)	ampere (A)	**1.0**	**E+01**
British thermal unit$_{IT}$ (Btu$_{IT}$)[c]	joule (J)	1.055 056	E+03
British thermal unit$_{th}$ (Btu$_{th}$)[c]	joule (J)	1.054 350	E+03
British thermal unit (mean) (Btu)	joule (J)	1.055 87	E+03
British thermal unit (39 °F) (Btu)	joule (J)	1.059 67	E+03
British thermal unit (59 °F) (Btu)	joule (J)	1.054 80	E+03
British thermal unit (60 °F) (Btu)	joule (J)	1.054 68	E+03
British thermal unit$_{IT}$ foot per hour square foot degree Fahrenheit [Btu$_{IT}$ · ft/(h · ft^2 · °F)]	watt per meter kelvin [W/(m · K)]	1.730 735	E+00
British thermal unit$_{th}$ foot per hour square foot degree Fahrenheit [Btu$_{th}$ · ft/(h · ft^2 · °F)]	watt per meter kelvin [W/(m · K)]	1.729 577	E+00
British thermal unit$_{IT}$ inch per hour square foot degree Fahrenheit [Btu$_{IT}$ · in/(h · ft^2 · °F)]	watt per meter kelvin [W/(m · K)]	1.442 279	E−01
British thermal unit$_{th}$ inch per hour square foot degree Fahrenheit [Btu$_{th}$ · in/(h · ft^2 · °F)]	watt per meter kelvin [W/(m · K)]	1.441 314	E−01
British thermal unit$_{IT}$ inch per second square foot degree Fahrenheit [Btu$_{IT}$ · in/(s · ft^2 · °F)]	watt per meter kelvin [W/(m · K)]	5.192 204	E+02
British thermal unit$_{th}$ inch per second square foot degree Fahrenheit [Btu$_{th}$ · in/(s · ft^2 · °F)]	watt per meter kelvin [W/(m · K)]	5.188 732	E+02
British thermal unit$_{IT}$ per cubic foot (Btu$_{IT}$/ft^3)	joule per cubic meter (J/m^3)	3.725 895	E+04
British thermal unit$_{th}$ per cubic foot (Btu$_{th}$/ft^3)	joule per cubic meter (J/m^3)	3.723 403	E+04
British thermal unit$_{IT}$ per degree Fahrenheit (Btu$_{IT}$/°F)	joule per kelvin (J/k)	1.899 101	E+03
British thermal unit$_{th}$ per degree Fahrenheit (Btu$_{th}$/°F)	joule per kelvin (J/k)	1.897 830	E+03
British thermal unit$_{IT}$ per degree Rankine (Btu$_{IT}$/°R)	joule per kelvin (J/k)	1.899 101	E+03
British thermal unit$_{th}$ per degree Rankine (Btu$_{th}$/°R)	joule per kelvin (J/k)	1.897 830	E+03
British thermal unit$_{IT}$ per hour (Btu$_{IT}$/h)	watt (W)	2.930 711	E−01
British thermal unit$_{th}$ per hour (Btu$_{th}$/h)	watt (W)	2.928 751	E−01
British thermal unit$_{IT}$ per hour square foot degree Fahrenheit [Btu$_{IT}$/(h · ft^2 · °F)]	watt per square meter kelvin [W/(m^2 · K)]	5.678 263	E+00
British thermal unit$_{th}$ per hour square foot degree Fahrenheit [Btu$_{th}$/(h · ft^2 · °F)]	watt per square meter kelvin [W/(m^2 · K)]	5.674 466	E+00
British thermal unit$_{th}$ per minute (Btu$_{th}$/min)	watt (W)	1.757 250	E+01
British thermal unit$_{IT}$ per pound (Btu$_{IT}$/lb)	joule per kilogram (J/kg)	**2.326**	**E+03**
British thermal unit$_{th}$ per pound (Btu$_{th}$/lb)	joule per kilogram (J/kg)	2.324 444	E+03
British thermal unit$_{IT}$ per pound degree Fahrenheit [Btu$_{IT}$/(lb · °F)]	joule per kilogram kelvin (J/(kg · K))	**4.1868**	**E+03**
British thermal unit$_{th}$ per pound degree Fahrenheit [Btu$_{th}$/(lb · °F)]	joule per kilogram kelvin [J/(kg · K)]	**4.184**	**E+03**
British thermal unit$_{IT}$ per pound degree Rankine [Btu$_{IT}$/(lb · °R)]	joule per kilogram kelvin [J/(kg · K)]	**4.1868**	**E+03**
British thermal unit$_{th}$ per pound degree Rankine [Btu$_{th}$/(lb · °R)]	joule per kilogram kelvin [J/(kg · K)]	**4.184**	**E+03**
British thermal unit$_{IT}$ per second (Btu$_{IT}$/s)	watt (W)	1.055 056	E+03
British thermal unit$_{th}$ per second (Btu$_{th}$/s)	watt (W)	1.054 350	E+03

[c] The Fifth International Conference on the Properties of Steam (London, July 1956) defined the International Table calorie as 4.1868 J. Therefore the exact conversion factor for the International Table Btu is 1.055 055 852 62 kJ. Note that the notation for the International Table used in this listing is subscript "IT". Similarly, the notation for thermochemical is subscript "th." Further, the thermochemical Btu, Btu$_{th}$, is based on the thermochemical calorie, cal$_{th}$, where cal$_{th}$ = 4.184 J exactly.

To convert from	to		Multiply by	
British thermal unit$_{IT}$ per second square foot degree Fahrenheit				
[Btu$_{IT}$/(s · ft² · °F)]	watt per square meter kelvin			
	[W/(m² · K)]		2.044 175	E+04
British thermal unit$_{th}$ per second square foot degree Fahrenheit				
[Btu$_{th}$/(s · ft² · °F)]	watt per square meter kelvin			
	[W/(m² · K)]		2.042 808	E+04
British thermal unit$_{IT}$ per square foot				
(Btu$_{IT}$/ft²)	joule per square meter (J/m²)		1.135 653	E+04
British thermal unit$_{th}$ per square foot				
(Btu$_{th}$/ft²)	joule per square meter (J/m²)		1.134 893	E+04
British thermal unit$_{IT}$ per square foot hour				
[(Btu$_{IT}$/(ft² · h)]	watt per square meter (W/m²)		3.154 591	E+00
British thermal unit$_{th}$ per square foot hour				
[Btu$_{th}$/(ft² · h)]	watt per square meter (W/m²)		3.152 481	E+00
British thermal unit$_{th}$ per square foot minute				
[Btu$_{th}$/(ft² · min)]	watt per square meter (W/m²)		1.891 489	E+02
British thermal unit$_{IT}$ per square foot second				
[(Btu$_{IT}$/(ft² · s)]	watt per square meter (W/m²)		1.135 653	E+04
British thermal unit$_{th}$ per square foot second				
[Btu$_{th}$/(ft² · s)]	watt per square meter (W/m²)		1.134 893	E+04
British thermal unit$_{th}$ per square inch second				
[Btu$_{th}$/(in² · s)]	watt per square meter (W/m²)		1.634 246	E+06
bushel (U.S.) (bu)	cubic meter (m³)		3.523 907	E−02
bushel (U.S.) (bu)	liter (L)		3.523 907	E+01
calorie$_{IT}$ (cal$_{IT}$)^c	joule (J)		**4.1868**	**E+00**
calorie$_{th}$ (cal$_{th}$)^c	joule (J)		**4.184**	**E+00**
calorie (cal) (mean)	joule (J)		4.190 02	E+00
calorie (15 °C) (cal$_{15}$)	joule (J)		4.185 80	E+00
calorie (20 °C) (cal$_{20}$)	joule (J)		4.181 90	E+00
calorie$_{IT}$, kilogram (nutrition)^d	joule (J)		**4.1868**	**E+03**
calorie$_{th}$, kilogram (nutrition)^d	joule (J)		**4.184**	**E+03**
calorie (mean), kilogram (nutrition)^d ...	joule (J)		4.190 02	E+03
calorie$_{th}$ per centimeter second degree Celsius				
[cal$_{th}$/(cm · s · °C)]	watt per meter kelvin [W/(m · K)]		**4.184**	**E+02**
calorie$_{IT}$ per gram (cal$_{IT}$/g)	joule per kilogram (J/kg)		**4.1868**	**E+03**
calorie$_{th}$ per gram (cal$_{th}$/g)	joule per kilogram (J/kg)		**4.184**	**E+03**
calorie$_{IT}$ per gram degree Celsius				
[cal$_{IT}$/(g · °C)]	joule per kilogram kelvin [J/(kg · K)]		**4.1868**	**E+03**
calorie$_{th}$ per gram degree Celsius				
[cal$_{th}$/(g · °C)]	joule per kilogram kelvin [J/(kg · K)]		**4.184**	**E+03**
calorie$_{IT}$ per gram kelvin [cal$_{IT}$/(g · K)] ...	joule per kilogram kelvin [J/(kg · K)]		**4.1868**	**E+03**
calorie$_{th}$ per gram kelvin [cal$_{th}$/(g · K)] ...	joule per kilogram kelvin [J/(kg · K)]		**4.184**	**E+03**
calorie$_{th}$ per minute (cal$_{th}$/min)	watt (W)		6.973 333	E−02
calorie$_{th}$ per second (cal$_{th}$/s)	watt (W)		**4.184**	**E+00**
calorie$_{th}$ per square centimeter (cal$_{th}$/cm²)	joule per square meter (J/m²)		**4.184**	**E+04**
calorie$_{th}$ per square centimeter minute				
[cal$_{th}$/(cm² · min)]	watt per square meter (W/m²)		6.973 333	E+02
calorie$_{th}$ per square centimeter second				
[cal$_{th}$/(cm² · s)]	watt per square meter (W/m²)		**4.184**	**E+04**
candela per square inch (cd/in²)	candela per square meter (cd/m²)		1.550 003	E+03
carat, metric	kilogram (kg)		**2.0**	**E−04**
carat, metric	gram (g)		**2.0**	**E−01**
centimeter of mercury (0 °C)^e	pascal (Pa)		1.333 22	E+03
centimeter of mercury (0 °C)^e	kilopascal (kPa)		1.333 22	E+00
centimeter of mercury, conventional (cmHg)^e ...	pascal (Pa)		1.333 224	E+03

^{d} The kilogram calorie or "large calorie" is an obsolete term used for the kilocalorie, which is the calorie used to express the energy content of foods. However, in practice, the prefix "kilo" is usually omitted.

^{e} Conversion factors for mercury manometer pressure units are calculated using the standard value for the acceleration of gravity and the density of mercury at the stated temperature. Additional digits are not justified because the definitions of the units do not take into account the compressibility of mercury or the change in density caused by the revised practical temperature scale, ITS-90. Similar comments also apply to water manometer pressure units. Conversion factors for conventional mercury and water manometer pressure factors are based on ISO 31-3.

To convert from	to	Multiply by	
centimeter of mercury, conventional (cmHg)[e]	kilopascal (kPa)	1.333 224	E+00
centimeter of water (4 °C)[e]	pascal (Pa)	9.806 38	E+01
centimeter of water, conventional (cmH₂O)[e]	pascal (Pa)	**9.806 65**	**E+01**
centipoise (cP)	pascal second (Pa · s)	**1.0**	**E−03**
centistokes (cSt)	meter squared per second (m²/s)	**1.0**	**E−06**
chain (based on U.S. survey foot) (ch)[a]	meter (m)	2.011 684	E+01
circular mil	square meter (m²)	5.067 075	E−10
circular mil	square millimeter (mm²)	5.067 075	E−04
clo	square meter kelvin per watt (m² · K/W)	1.55	E−01
cord (128 ft³)	cubic meter (m³)	3.624 556	E+00
cubic foot (ft³)	cubic meter (m³)	2.831 685	E−02
cubic foot per minute (ft³/min)	cubic meter per second (m³/s)	4.719 474	E−04
cubic foot per minute (ft³/min)	liter per second (L/s)	4.719 474	E−01
cubic foot per second (ft³/s)	cubic meter per second (m³/s)	2.831 685	E−02
cubic inch (in³)[f]	cubic meter (m³)	1.638 706	E−05
cubic inch per minute (in³/min)	cubic meter per second (m³/s)	2.731 177	E−07
cubic mile (mi³)	cubic meter (m³)	4.168 182	E+09
cubic yard (yd³)	cubic meter (m³)	7.645 549	E−01
cubic yard per minute (yd³/min)	cubic meter per second (m³/s)	1.274 258	E−02
cup (U.S.)	cubic meter (m³)	2.365 882	E−04
cup (U.S.)	liter (L)	2.365 882	E−01
cup (U.S.)	milliliter (mL)	2.365 882	E+02
curie (Ci)	becquerel (Bq)	**3.7**	**E+10**
darcy[g]	meter squared (m²)	9.869 233	E−13
day (d)	second (s)	**8.64**	**E+04**
day (sidereal)	second (s)	8.616 409	E+04
debye (D)	coulomb meter (C · m)	3.335 641	E−30
degree (angle) (°)	radian (rad)	1.745 329	E−02
degree Celsius (temperature) (°C)	kelvin (K)	$T/K = t/°C + 273.15$	
degree Celsius (temperature interval) (°C)	kelvin (K)	**1.0**	**E+00**
degree centigrade (temperature)[h]	degree Celsius (°C)	$t/°C \approx t/\text{deg.cent.}$	
degree centigrade (temperature interval)[h]	degree Celsius (°C)	1.0	E+00
degree Fahrenheit (temperature) (°F)	degree Celsius (°C)	$t/°C = (t/°F - 32)/1.8$	
degree Fahrenheit (temperature) (°F)	kelvin (K)	$T/K = (t/°F + 459.67)/1.8$	
degree Fahrenheit (temperature interval)(°F)	degree Celsius (°C)	5.555 556	E−01
degree Fahrenheit (temperature interval) (°F)	kelvin (K)	5.555 556	E−01
degree Fahrenheit hour per British thermal unit$_{IT}$ (°F · h/Btu$_{IT}$)	kelvin per watt (K/W)	1.895 634	E+00
degree Fahrenheit hour per British thermal unit$_{th}$ (°F · h/Btu$_{th}$)	kelvin per watt (K/W)	1.896 903	E+00
degree Fahrenheit hour square foot per British thermal unit$_{IT}$ (°F · h · ft²/Btu$_{IT}$)	square meter kelvin per watt (m² · K/W)	1.761 102	E−01
degree Fahrenheit hour square foot per British thermal unit$_{th}$ (°F · h · ft²/Btu$_{th}$)	square meter kelvin per watt (m² · K/W)	1.762 280	E−01
degree Fahrenheit hour square foot per British thermal unit$_{IT}$ inch [°F · h · ft²/(Btu$_{IT}$ · in)]	meter kelvin per watt (m · K/W)	6.933 472	E+00
degree Fahrenheit hour square foot per British thermal unit$_{th}$ inch [°F · h · ft²/(Btu$_{th}$ · in)]	meter kelvin per watt (m · K/W)	6.938 112	E+00
degree Fahrenheit second per British thermal unit$_{IT}$ (°F · s/Btu$_{IT}$)	kelvin per watt (K/W)	5.265 651	E−04
degree Fahrenheit second per British thermal unit$_{th}$ (°F · s/Btu$_{th}$)	kelvin per watt (K/W)	5.269 175	E−04
degree Rankine (°R)	kelvin (K)	$T/K = (T/°R)/1.8$	
degree Rankine (temperature interval) (°R)	kelvin (K)	5.555 556	E−01
denier	kilogram per meter (kg/m)	1.111 111	E−07
denier	gram per meter (g/m)	1.111 111	E−04
dyne (dyn)	newton (N)	**1.0**	**E−05**
dyne centimeter (dyn · cm)	newton meter (N · m)	**1.0**	**E−07**
dyne per square centimeter (dyn/cm²)	pascal (Pa)	**1.0**	**E−01**

[f] The exact conversion factor is 1.638 706 4 E−05.

[g] The darcy is a unit for expressing the permeability of porous solids, not area.

[h] The centigrade temperature scale is obsolete; the degree centigrade is only approximately equal to the degree Celsius.

To convert from	to	Multiply by	
electronvolt (eV)	joule (J)	1.602 177	E−19
EMU of capacitance (abfarad)	farad (F)	**1.0**	**E+09**
EMU of current (abampere)	ampere (A)	**1.0**	**E+01**
EMU of electric potential (abvolt)	volt (V)	**1.0**	**E−08**
EMU of inductance (abhenry)	henry (H)	**1.0**	**E−09**
EMU of resistance (abohm)	ohm (Ω)	**1.0**	**E−09**
erg (erg)	joule (J)	**1.0**	**E−07**
erg per second (erg/s)	watt (W)	**1.0**	**E−07**
erg per square centimeter second [erg/(cm² · s)]	watt per square meter (W/m²)	**1.0**	**E−03**
ESU of capacitance (statfarad)	farad (F)	1.112 650	E−12
ESU of current (statampere)	ampere (A)	3.335 641	E−10
ESU of electric potential (statvolt)	volt (V)	2.997 925	E+02
ESU of inductance (stathenry)	henry (H)	8.987 552	E+11
ESU of resistance (statohm)	ohm (Ω)	8.987 552	E+11
faraday (based on carbon 12)	coulomb (C)	9.648 531	E+04
fathom (based on U.S survey foot)[a]	meter (m)	1.828 804	E+00
fermi	meter (m)	**1.0**	**E−15**
fermi	femtometer (fm)	**1.0**	**E+00**
fluid ounce (U.S.) (fl oz)	cubic meter (m³)	2.957 353	E−05
fluid ounce (U.S.) (fl oz)	milliliter (mL)	2.957 353	E+01
foot (ft)	meter (m)	**3.048**	**E−01**
foot (U.S. survey ft)[a]	meter (m)	3.048 006	E−01
footcandle	lux (lx)	1.076 391	E+01
footlambert	candela per square meter (cd/m²)	3.426 259	E+00
foot of mercury, conventional (ftHg)[e]	pascal (Pa)	4.063 666	E+04
foot of mercury, conventional (ftHg)[e]	kilopascal (kPa)	4.063 666	E+01
foot of water (39.2 °F)[e]	pascal (Pa)	2.988 98	E+03
foot of water (39.2 °F)[e]	kilopascal (kPa)	2.988 98	E+00
foot of water, conventional (ftH₂O)[e]	pascal (Pa)	2.989 067	E+03
foot of water, conventional (ftH₂O)[e]	kilopascal (kPa)	2.989 067	E+00
foot per hour (ft/h)	meter per second (m/s)	8.466 667	E−05
foot per minute (ft/min)	meter per second (m/s)	**5.08**	**E−03**
foot per second (ft/s)	meter per second (m/s)	**3.048**	**E−01**
foot per second squared (ft/s²)	meter per second squared (m/s²)	**3.048**	**E−01**
foot poundal	joule (J)	4.214 011	E−02
foot pound-force (ft · lbf)	joule (J)	1.355 818	E+00
foot pound-force per hour (ft · lbf/h)	watt (W)	3.766 161	E−04
foot pound-force per minute (ft · lbf/min)	watt (W)	2.259 697	E−02
foot pound-force per second (ft · lbf/s)	watt (W)	1.355 818	E+00
foot to the fourth power (ft⁴)[i]	meter to the fourth power (m⁴)	8.630 975	E−03
franklin (Fr)	coulomb (C)	3.335 641	E−10
gal (Gal)	meter per second squared (m/s²)	**1.0**	**E−02**
gallon [Canadian and U.K. (Imperial)] (gal)	cubic meter (m³)	**4.546 09**	**E−03**
gallon [Canadian and U.K. (Imperial)] (gal)	liter (L)	**4.546 09**	**E+00**
gallon (U.S.) (gal)	cubic meter (m³)	3.785 412	E−03
gallon (U.S.) (gal)	liter (L)	3.785 412	E+00
gallon (U.S.) per day (gal/d)	cubic meter per second (m³/s)	4.381 264	E−08
gallon (U.S.) per day (gal/d)	liter per second (L/s)	4.381 264	E−05
gallon (U.S.) per horsepower hour [gal/(hp · h)]	cubic meter per joule (m³/J)	1.410 089	E−09
gallon (U.S.) per horsepower hour [gal/(hp · h)]	liter per joule (L/J)	1.410 089	E−06
gallon (U.S.) per minute (gpm)(gal/min)	cubic meter per second (m³/s)	6.309 020	E−05
gallon (U.S.) per minute (gpm)(gal/min)	liter per second (L/s)	6.309 020	E−02
gamma (γ)	tesla (T)	**1.0**	**E−09**
gauss (Gs, G)	tesla (T)	**1.0**	**E−04**
gilbert (Gi)	ampere (A)	7.957 747	E−01

[i] This is a unit for the quantity second moment of area, which is sometimes called the "moment of section" or "area moment of inertia" of a plane section about a specified axis.

To convert from	to	Multiply by	
gill [Canadian and U.K. (Imperial)] (gi)	cubic meter (m^3)	1.420 653	E−04
gill [Canadian and U.K. (Imperial)] (gi)	liter (L)	1.420 653	E−01
gill (U.S.) (gi)	cubic meter (m^3)	1.182 941	E−04
gill (U.S.) (gi)	liter (L)	1.182 941	E−01
gon (also called grade) (gon)	radian (rad)	1.570 796	E−02
gon (also called grade) (gon)	degree (angle) (°)	**9.0**	**E−01**
grain (gr)	kilogram (kg)	**6.479 891**	**E−05**
grain (gr)	milligram (mg)	**6.479 891**	**E+01**
grain per gallon (U.S.) (gr/gal)	kilogram per cubic meter (kg/m^3)	1.711 806	E−02
grain per gallon (U.S.) (gr/gal)	milligram per liter (mg/L)	1.711 806	E+01
gram-force per square centimeter (gf/cm^2)	pascal (Pa)	**9.806 65**	**E+01**
gram per cubic centimeter (g/cm^3)	kilogram per cubic meter (kg/m^3)	**1.0**	**E+03**
hectare (ha)	square meter (m^2)	**1.0**	**E+04**
horsepower (550 ft · lbf/s) (hp)	watt (W)	7.456 999	E+02
horsepower (boiler)	watt (W)	9.809 50	E+03
horsepower (electric)	watt (W)	**7.46**	**E+02**
horsepower (metric)	watt (W)	7.354 988	E+02
horsepower (U.K.)	watt (W)	7.4570	E+02
horsepower (water)	watt (W)	7.460 43	E+02
hour (h)	second (s)	**3.6**	**E+03**
hour (sidereal)	second (s)	3.590 170	E+03
hundredweight (long, 112 lb)	kilogram (kg)	5.080 235	E+01
hundredweight (short, 100 lb)	kilogram (kg)	4.535 924	E+01
inch (in)	meter (m)	**2.54**	**E−02**
inch (in)	centimeter (cm)	**2.54**	**E+00**
inch of mercury (32 °F)[e]	pascal (Pa)	3.386 38	E+03
inch of mercury (32 °F)[e]	kilopascal (kPa)	3.386 38	E+00
inch of mercury (60 °F)[e]	pascal (Pa)	3.376 85	E+03
inch of mercury (60 °F)[e]	kilopascal (kPa)	3.376 85	E+00
inch of mercury, conventional (inHg)[e]	pascal (Pa)	3.386 389	E+03
inch of mercury, conventional (inHg)[e]	kilopascal (kPa)	3.386 389	E+00
inch of water (39.2 °F)[e]	pascal (Pa)	2.490 82	E+02
inch of water (60 °F)[e]	pascal (Pa)	2.4884	E+02
inch of water, conventional (inH_2O)[e]	pascal (Pa)	2.490 889	E+02
inch per second (in/s)	meter per second (m/s)	**2.54**	**E−02**
inch per second squared (in/s^2)	meter per second squared (m/s^2)	**2.54**	**E−02**
inch to the fourth power (in^4)[i]	meter to the fourth power (m^4)	4.162 314	E−07
kayser (K)	reciprocal meter (m^{-1})	**1.0**	**E+02**
kelvin (K)	degree Celsius (°C)	$t/°C = T/K − 273.15$	
kilocalorie$_{IT}$ (kcal$_{IT}$)	joule (J)	**4.1868**	**E+03**
kilocalorie$_{th}$ (kcal$_{th}$)	joule (J)	**4.184**	**E+03**
kilocalorie (mean) (kcal)	joule (J)	4.190 02	E+03
kilocalorie$_{th}$ per minute (kcal$_{th}$/min)	watt (W)	6.973 333	E+01
kilocalorie$_{th}$ per second (kcal$_{th}$/s)	watt (W)	**4.184**	**E+03**
kilogram-force (kgf)	newton (N)	**9.806 65**	**E+00**
kilogram-force meter (kgf · m)	newton meter (N · m)	**9.806 65**	**E+00**
kilogram-force per square centimeter (kgf/cm^2)	pascal (Pa)	**9.806 65**	**E+04**
kilogram-force per square centimeter (kgf/cm^2)	kilopascal (kPa)	**9.806 65**	**E+01**
kilogram-force per square meter (kgf/m^2)	pascal (Pa)	**9.806 65**	**E+00**
kilogram-force per square millimeter (kgf/mm^2)	pascal (Pa)	**9.806 65**	**E+06**
kilogram-force per square millimeter (kgf/mm^2)	megapascal (MPa)	**9.806 65**	**E+00**
kilogram-force second squared per meter ($kgf · s^2/m$)	kilogram (kg)	**9.806 65**	**E+00**
kilometer per hour (km/h)	meter per second (m/s)	2.777 778	E−01
kilopond (kilogram-force) (kp)	newton (N)	**9.806 65**	**E+00**
kilowatt hour (kW · h)	joule (J)	**3.6**	**E+06**
kilowatt hour (kW · h)	megajoule (MJ)	**3.6**	**E+00**

To convert from	to	Multiply by	
kip (1 kip=1000 lbf)	newton (N)	4.448 222	E+03
kip (1 kip=1000 lbf)	kilonewton (kN)	4.448 222	E+00
kip per square inch (ksi) (kip/in²)	pascal (Pa)	6.894 757	E+06
kip per square inch (ksi) (kip/in²)	kilopascal (kPa)	6.894 757	E+03
knot (nautical mile per hour)	meter per second (m/s)	5.144 444	E−01
lambertʲ	candela per square meter (cd/m²) ...	3.183 099	E+03
langley (cal$_{th}$/cm²)	joule per square meter (J/m²)	**4.184**	**E+04**
light year (l.y.)ᵏ	meter (m)	9.460 73	E+15
liter (L)ˡ	cubic meter (m³)	**1.0**	**E−03**
lumen per square foot (lm/ft²)	lux (lx)	1.076 391	E+01
maxwell (Mx)	weber (Wb)	**1.0**	**E−08**
mho	siemens (S)	**1.0**	**E+00**
microinch	meter (m)	**2.54**	**E−08**
microinch	micrometer (μm)	**2.54**	**E−02**
micron (μ)	meter (m)	**1.0**	**E−06**
micron (μ)	micrometer (μm)	**1.0**	**E+00**
mil (0.001 in)	meter (m)	**2.54**	**E−05**
mil (0.001 in)	millimeter (mm)	**2.54**	**E−02**
mil (angle)	radian (rad)	9.817 477	E−04
mil (angle)	degree (°)	**5.625**	**E−02**
mile (mi)	meter (m)	**1.609 344**	**E+03**
mile (mi)	kilometer (km)	**1.609 344**	**E+00**
mile (based on U.S. survey foot) (mi)ᵃ ...	meter (m)	1.609 347	E+03
mile (based on U.S. survey foot) (mi)ᵃ ...	kilometer (km)	1.609 347	E+00
mile, nauticalᵐ	meter (m)	**1.852**	**E+03**
mile per gallon (U.S.) (mpg) (mi/gal) ...	meter per cubic meter (m/m³)	4.251 437	E+05
mile per gallon (U.S.) (mpg) (mi/gal) ...	kilometer per liter (km/L)	4.251 437	E−01
mile per gallon (U.S.) (mpg) (mi/gal)ⁿ ...	liter per 100 kilometer (L/100 km) ...	divide 235.215 by number of miles per gallon	
mile per hour (mi/h)	meter per second (m/s)	**4.4704**	**E−01**
mile per hour (mi/h)	kilometer per hour (km/h)	**1.609 344**	**E+00**
mile per minute (mi/min)	meter per second (m/s)	**2.682 24**	**E+01**
mile per second (mi/s)	meter per second (m/s)	**1.609 344**	**E+03**
millibar (mbar)	pascal (Pa)	**1.0**	**E+02**
millibar (mbar)	kilopascal (kPa)	**1.0**	**E−01**
millimeter of mercury, conventional (mmHg)ᵉ ..	pascal (Pa)	1.333 224	E+02
millimeter of water, conventional (mmH₂O)ᵉ ...	pascal (Pa)	**9.806 65**	**E+00**
minute (angle) (')	radian (rad)	2.908 882	E−04
minute (min)	second (s)	**6.0**	**E+01**
minute (sidereal)	second (s)	5.983 617	E+01
nit	candela per meter squared (cd/m²) ...	**1.0**	**E+00**
nox	lux (lx)	**1.0**	**E−03**
oersted (Oe)	ampere per meter (A/m)	7.957 747	E+01
ohm centimeter (Ω·cm)	ohm meter (Ω·m)	**1.0**	**E−02**
ohm circular-mil per foot	ohm meter (Ω·m)	1.662 426	E−09
ohm circular-mil per foot	ohm square millimeter per meter (Ω·mm²/m) ...	1.662 426	E−03
ounce (avoirdupois) (oz)	kilogram (kg)	2.834 952	E−02
ounce (avoirdupois) (oz)	gram (g)	2.834 952	E+01
ounce (troy or apothecary) (oz)	kilogram (kg)	3.110 348	E−02
ounce (troy or apothecary) (oz)	gram (g)	3.110 348	E+01
ounce [Canadian and U.K. fluid (Imperial)] (fl oz) ...	cubic meter (m³)	2.841 306	E−05

ʲ The exact conversion factor is $10^4/\pi$.

ᵏ This conversion factor is based on 1 d = 86 400 s; and 1 Julian century = 36 525 d. (See *The Astronomical Almanac for the Year 1995*, page K6, U.S. Government Printing Office, Washington, DC, 1994.)

ˡ In 1964 the General Conference on Weights and Measures reestablished the name "liter" as a special name for the cubic decimeter. Between 1901 and 1964 the liter was slightly larger (1.000 028 dm³); when one uses high-accuracy volume data of that time, this fact must be kept in mind.

ᵐ The value of this unit, 1 nautical mile = 1852 m, was adopted by the First International Extraordinary Hydrographic Conference, Monaco, 1929, under the name "International nautical mile."

ⁿ For converting fuel economy, as used in the U.S., to fuel consumption.

To convert from	to	Multiply by	
ounce [Canadian and U.K. fluid (Imperial)] (fl oz)	milliliter (mL)	2.841 306	E+01
ounce (U.S. fluid) (fl oz)	cubic meter (m³)	2.957 353	E−05
ounce (U.S. fluid) (fl oz)	milliliter (mL)	2.957 353	E+01
ounce (avoirdupois)-force (ozf)	newton (N)	2.780 139	E−01
ounce (avoirdupois)-force inch (ozf · in)	newton meter (N · m)	7.061 552	E−03
ounce (avoirdupois)-force inch (ozf · in)	millinewton meter (mN · m)	7.061 552	E+00
ounce (avoirdupois) per cubic inch (oz/in³)	kilogram per cubic meter (kg/m³)	1.729 994	E+03
ounce (avoirdupois) per gallon [Canadian and U.K. (Imperial)] (oz/gal)	kilogram per cubic meter (kg/m³)	6.236 023	E+00
ounce (avoirdupois) per gallon [Canadian and U.K. (Imperial)] (oz/gal)	gram per liter (g/L)	6.236 023	E+00
ounce (avoirdupois) per gallon (U.S.)(oz/gal)	kilogram per cubic meter (kg/m³)	7.489 152	E+00
ounce (avoirdupois) per gallon (U.S.)(oz/gal)	gram per liter (g/L)	7.489 152	E+00
ounce (avoirdupois) per square foot (oz/ft²)	kilogram per square meter (kg/m²)	3.051 517	E−01
ounce (avoirdupois) per square inch (oz/in²)	kilogram per square meter (kg/m²)	4.394 185	E+01
ounce (avoirdupois) per square yard(oz/yd²)	kilogram per square meter (kg/m²)	3.390 575	E−02
parsec (pc)	meter (m)	3.085 678	E+16
peck (U.S.) (pk)	cubic meter (m³)	8.809 768	E−03
peck (U.S.) (pk)	liter (L)	8.809 768	E+00
pennyweight (dwt)	kilogram (kg)	1.555 174	E−03
pennyweight (dwt)	gram (g)	1.555 174	E+00
perm (0 °C)	kilogram per pascal second square meter [kg/(Pa · s · m²)]	5.721 35	E−11
perm (23 °C)	kilogram per pascal second square meter [kg/(Pa · s · m²)]	5.745 25	E−11
perm inch (0 °C)	kilogram per pascal second meter [kg/(Pa · s · m)]	1.453 22	E−12
perm inch (23 °C)	kilogram per pascal second meter [kg/(Pa · s · m)]	1.459 29	E−12
phot (ph)	lux (lx)	**1.0**	**E+04**
pica (computer) (1/6 in)	meter (m)	4.233 333	E−03
pica (computer) (1/6 in)	millimeter (mm)	4.233 333	E+00
pica (printer's)	meter (m)	4.217 518	E−03
pica (printer's)	millimeter (mm)	4.217 518	E+00
pint (U.S. dry) (dry pt)	cubic meter (m³)	5.506 105	E−04
pint (U.S. dry) (dry pt)	liter (L)	5.506 105	E−01
pint (U.S. liquid) (liq pt)	cubic meter (m³)	4.731 765	E−04
pint (U.S. liquid) (liq pt)	liter (L)	4.731 765	E−01
point (computer) (1/72 in)	meter (m)	3.527 778	E−04
point (computer) (1/72 in)	millimeter (mm)	3.527 778	E−01
point (printer's)	meter (m)	3.514 598	E−04
point (printer's)	millimeter (mm)	3.514 598	E−01
poise (P)	pascal second (Pa · s)	**1.0**	**E−01**
pound (avoirdupois) (lb)°	kilogram (kg)	4.535 924	E−01
pound (troy or apothecary) (lb)	kilogram (kg)	3.732 417	E−01
poundal	newton (N)	1.382 550	E−01
poundal per square foot	pascal (Pa)	1.488 164	E+00
poundal second per square foot	pascal second (Pa · s)	1.488 164	E+00
pound foot squared (lb · ft²)	kilogram meter squared (kg · m²)	4.214 011	E−02
pound-force (lbf)ᵖ	newton (N)	4.448 222	E+00
pound-force foot (lbf · ft)	newton meter (N · m)	1.355 818	E+00
pound-force foot per inch (lbf · ft/in)	newton meter per meter (N · m/m)	5.337 866	E+01
pound-force inch (lbf · in)	newton meter (N · m)	1.129 848	E−01
pound-force inch per inch (lbf · in/in)	newton meter per meter (N · m/m)	4.448 222	E+00
pound-force per foot (lbf/ft)	newton per meter (N/m)	1.459 390	E+01
pound-force per inch (lbf/in)	newton per meter (N/m)	1.751 268	E+02
pound-force per pound (lbf/lb) (thrust to mass ratio)	newton per kilogram (N/kg)	**9.806 65**	**E+00**

° The exact conversion factor is 4.535 923 7 E−01. All units that contain the pound refer to the avoirdupois pound unless otherwise specified.

ᵖ If the local value of the acceleration of free fall is taken as g_n=9.806 65 m/ s² (the standard value), the exact conversion factor is 4.448 221 615 260 5 E+00.

To convert from	to	Multiply by	
pound-force per square foot (lbf/ft²)	pascal (Pa)	4.788 026	E+01
pound-force per square inch (psi) (lbf/in²)	pascal (Pa)	6.894 757	E+03
pound-force per square inch (psi) (lbf/in²)	kilopascal (kPa)	6.894 757	E+00
pound-force second per square foot (lbf · s/ft²)	pascal second (Pa · s)	4.788 026	E+01
pound-force second per square inch (lbf · s/in²)	pascal second (Pa · s)	6.894 757	E+03
pound inch squared (lb · in²)	kilogram meter squared (kg · m²)	2.926 397	E−04
pound per cubic foot (lb/ft³)	kilogram per cubic meter (kg/m³)	1.601 846	E+01
pound per cubic inch (lb/in³)	kilogram per cubic meter (kg/m³)	2.767 990	E+04
pound per cubic yard (lb/yd³)	kilogram per cubic meter (kg/m³)	5.932 764	E−01
pound per foot (lb/ft)	kilogram per meter (kg/m)	1.488 164	E+00
pound per foot hour [lb/(ft · h)]	pascal second (Pa · s)	4.133 789	E−04
pound per foot second [lb/(ft · s)]	pascal second (Pa · s)	1.488 164	E+00
pound per gallon [Canadian and U.K. (Imperial)] (lb/gal)	kilogram per cubic meter (kg/m³)	9.977 637	E+01
pound per gallon [Canadian and U.K. (Imperial)] (lb/gal)	kilogram per liter (kg/L)	9.977 637	E−02
pound per gallon (U.S.) (lb/gal)	kilogram per cubic meter (kg/m³)	1.198 264	E+02
pound per gallon (U.S.) (lb/gal)	kilogram per liter (kg/L)	1.198 264	E−01
pound per horsepower hour [lb/(hp · h)]	kilogram per joule (kg/J)	1.689 659	E−07
pound per hour (lb/h)	kilogram per second (kg/s)	1.259 979	E−04
pound per inch (lb/in)	kilogram per meter (kg/m)	1.785 797	E+01
pound per minute (lb/min)	kilogram per second (kg/s)	7.559 873	E−03
pound per second (lb/s)	kilogram per second (kg/s)	4.535 924	E−01
pound per square foot (lb/ft²)	kilogram per square meter (kg/m²)	4.882 428	E+00
pound per square inch (*not* pound-force) (lb/in²)	kilogram per square meter (kg/m²)	7.030 696	E+02
pound per yard (lb/yd)	kilogram per meter (kg/m)	4.960 546	E−01
psi (pound-force per square inch) (lbf/in²)	pascal (Pa)	6.894 757	E+03
psi (pound-force per square inch) (lbf/in²)	kilopascal (kPa)	6.894 757	E+00
quad (10¹⁵ Btu$_{IT}$)^c	joule (J)	1.055 056	E+18
quart (U.S. dry) (dry qt)	cubic meter (m³)	1.101 221	E−03
quart (U.S. dry) (dry qt)	liter (L)	1.101 221	E+00
quart (U.S. liquid) (liq qt)	cubic meter (m³)	9.463 529	E−04
quart (U.S. liquid) (liq qt)	liter (L)	9.463 529	E−01
rad (absorbed dose) (rad)	gray (Gy)	**1.0**	**E−02**
rem (rem)	sievert (Sv)	**1.0**	**E−02**
revolution (r)	radian (rad)	6.283 185	E+00
revolution per minute (rpm) (r/min)	radian per second (rad/s)	1.047 198	E−01
rhe	reciprocal pascal second [(Pa · s)⁻¹]	**1.0**	**E+01**
rod (based on U.S. survey foot) (rd)^a	meter (m)	5.029 210	E+00
roentgen (R)	coulomb per kilogram (C/kg)	**2.58**	**E−04**
rpm (revolution per minute) (r/min)	radian per second (rad/s)	1.047 198	E−01
second (angle) (")	radian (rad)	4.848 137	E−06
second (sidereal)	second (s)	9.972 696	E−01
shake	second (s)	**1.0**	**E−08**
shake	nanosecond (ns)	**1.0**	**E+01**
skot	candela per meter squared (cd/m²)	3.183 098	E−04
slug (slug)	kilogram (kg)	1.459 390	E+01
slug per cubic foot (slug/ft³)	kilogram per cubic meter (kg/m³)	5.153 788	E+02
slug per foot second [slug/(ft · s)]	pascal second (Pa · s)	4.788 026	E+01
square foot (ft²)	square meter (m²)	**9.290 304**	**E−02**
square foot per hour (ft²/h)	square meter per second (m²/s)	**2.580 64**	**E−05**
square foot per second (ft²/s)	square meter per second (m²/s)	**9.290 304**	**E−02**
square inch (in²)	square meter (m²)	**6.4516**	**E−04**
square inch (in²)	square centimeter (cm²)	**6.4516**	**E+00**
square mile (mi²)	square meter (m²)	2.589 988	E+06
square mile (mi²)	square kilometer (km²)	2.589 988	E+00

To convert from	to		Multiply by	
square mile (based on U.S. survey foot) (mi²)[a]	square meter (m²)		2.589 998	E+06
square mile (based on U.S. survey foot) (mi²)[a]	square kilometer (km²)		2.589 998	E+00
square yard (yd²)	square meter (m²)		8.361 274	E−01
statampere	ampere (A)		3.335 641	E−10
statcoulomb	coulomb (C)		3.335 641	E−10
statfarad	farad (F)		1.112 650	E−12
stathenry	henry (H)		8.987 552	E+11
statmho	siemens (S)		1.112 650	E−12
statohm	ohm (Ω)		8.987 552	E+11
statvolt	volt (V)		2.997 925	E+02
stere (st)	cubic meter (m³)		**1.0**	**E+00**
stilb (sb)	candela per square meter (cd/m²)		**1.0**	**E+04**
stokes (St)	meter squared per second (m²/s)		**1.0**	**E−04**
tablespoon	cubic meter (m³)		1.478 676	E−05
tablespoon	milliliter (mL)		1.478 676	E+01
teaspoon	cubic meter (m³)		4.928 922	E−06
teaspoon	milliliter (mL)		4.928 922	E+00
tex	kilogram per meter (kg/m)		**1.0**	**E−06**
therm (EC)[q]	joule (J)		**1.055 06**	**E+08**
therm (U.S.)[q]	joule (J)		**1.054 804**	**E+08**
ton, assay (AT)	kilogram (kg)		2.916 667	E−02
ton, assay (AT)	gram (g)		2.916 667	E+01
ton-force (2000 lbf)	newton (N)		8.896 443	E+03
ton-force (2000 lbf)	kilonewton (kN)		8.896 443	E+00
ton, long (2240 lb)	kilogram (kg)		1.016 047	E+03
ton, long, per cubic yard	kilogram per cubic meter (kg/m³)		1.328 939	E+03
ton, metric (t)	kilogram (kg)		**1.0**	**E+03**
tonne (called "metric ton" in U.S.) (t)	kilogram (kg)		**1.0**	**E+03**
ton of refrigeration (12 000 Btu$_{IT}$/h)	watt (W)		3.516 853	E+03
ton of TNT (energy equivalent)[r]	joule (J)		**4.184**	**E+09**
ton, register	cubic meter (m³)		2.831 685	E+00
ton, short (2000 lb)	kilogram (kg)		9.071 847	E+02
ton, short, per cubic yard	kilogram per cubic meter (kg/m³)		1.186 553	E+03
ton, short, per hour	kilogram per second (kg/s)		2.519 958	E−01
torr (Torr)	pascal (Pa)		1.333 224	E+02
unit pole	weber (Wb)		1.256 637	E−07
watt hour (W · h)	joule (J)		**3.6**	**E+03**
watt per square centimeter (W/cm²)	watt per square meter (W/m²)		**1.0**	**E+04**
watt per square inch (W/in²)	watt per square meter (W/m²)		1.550 003	E+03
watt second (W · s)	joule (J)		**1.0**	**E+00**
yard (yd)	meter (m)		**9.144**	**E−01**
year (365 days)	second (s)		**3.1536**	**E+07**
year (sidereal)	second (s)		3.155 815	E+07
year (tropical)	second (s)		3.155 693	E+07

[q] The therm (EC) is legally defined in the Council Directive of 20 December 1979, Council of the European Communities (now the European Union, EU). The therm (U.S.) is legally defined in the Federal Register of July 27, 1968. Although the therm (EC), which is based on the International Table Btu, is frequently used by engineers in the United States, the therm (U.S.) is the legal unit used by the U.S natural gas industry.

[r] Defined (not measured) value.

CONVERSION OF TEMPERATURES

From	To	
Celsius	Fahrenheit	$t_F/°F = (9/5)\ t/°C + 32$
	Kelvin	$T/K = t/°C + 273.15$
	Rankine	$T/°R = (9/5)\ (t/°C + 273.15)$
Fahrenheit	Celsius	$t/°C = (5/9)\ [(t_F/°F) - 32]$
	Kelvin	$T/K = (5/9)\ [(t_F/°F) - 32] + 273.15$
	Rankine	$T/°R = t_F/°F + 459.67$
Kelvin	Celsius	$t/°C = T/K - 273.15$
	Rankine	$T/°R = (9/5)\ T/K$
Rankine	Fahrenheit	$t_F/°F = T/°R - 459.67$
	Kelvin	$T/K = (5/9)\ T/°R$

Definition of symbols:

T = thermodynamic (absolute) temperature

t = Celsius temperature (the symbol θ is also used for Celsius temperature)

t_F = Fahrenheit temperature

Designation of Large Numbers

	U.S.A.	Other countries
10^6	million	million
10^9	billion	milliard
10^{12}	trillion	billion
10^{15}	quadrillion	billiard
10^{18}	quintillion	trillion
100^{100}	googol	
10^{googol}	googolplex	

CONVERSION FACTORS FOR ENERGY UNITS

If greater accuracy is required, use the Energy Equivalents section of the Fundamental Physical Constants table.

	Wavenumber $\bar{\nu}$ cm^{-1}	Frequency ν MHz	Energy E aJ	Energy E eV	Energy E E_h	Molar energy E_m kJ/mol	Molar energy E_m kcal/mol	Temperature T K
$\bar{\nu}$: 1 cm^{-1}	$\doteq 1$	2.997925×10^4	1.986447×10^{-5}	1.239842×10^{-4}	4.556335×10^{-6}	11.96266×10^{-3}	2.85914×10^{-3}	1.438769
ν: 1 MHz	$\doteq 3.33564 \times 10^{-5}$	1	6.626076×10^{-10}	4.135669×10^{-9}	1.519830×10^{-10}	3.990313×10^{-7}	9.53708×10^{-8}	4.79922×10^{-5}
1 aJ	$\doteq 50341.1$	1.509189×10^9	1	6.241506	0.2293710	602.2137	143.9325	7.24292×10^4
E: 1 eV	$\doteq 8065.54$	2.417988×10^8	0.1602177	1	3.674931×10^{-2}	96.4853	23.0605	1.16045×10^4
E_h	$\doteq 219474.63$	6.579684×10^9	4.359748	27.2114	1	2625.500	627.510	3.15773×10^5
E_m: 1 kJ/mol	$\doteq 83.5935$	2.506069×10^6	1.660540×10^{-3}	1.036427×10^{-2}	3.808798×10^{-4}	1	0.239006	120.272
1 kcal/mol	$\doteq 349.755$	1.048539×10^7	6.947700×10^{-3}	4.336411×10^{-2}	1.593601×10^{-3}	4.184	1	503.217
T: 1 K	$\doteq 0.695039$	2.08367×10^4	1.380658×10^{-5}	8.61738×10^{-5}	3.16683×10^{-6}	8.31451×10^{-3}	1.98722×10^{-3}	1

Examples of the use of this table:

$$1 \text{ aJ} \doteq 50341 \text{ cm}^{-1}$$
$$1 \text{ eV} \doteq 96.4853 \text{ kJ mol}^{-1}$$

The symbol $\doteq$ should be read as meaning corresponds to or is equivalent to.

$E = h\nu = hc\bar{\nu} = kT$; $E_m = N_A E$; E_h is the Hartree energy.

CONVERSION FACTORS FOR PRESSURE UNITS

	Pa	kPa	MPa	bar	atm	Torr	μmHg	psi
Pa	1	0.001	0.000001	0.00001	9.8692×10^{-6}	0.0075006	7.5006	0.0001450377
kPa	1000	1	0.001	0.01	0.0098692	7.5006	7500.6	0.1450377
MPa	1000000	1000	1	10	9.8692	7500.6	7500600	145.0377
bar	100000	100	0.1	1	0.98692	750.06	750060	14.50377
atm	101325	101.325	0.101325	1.01325	1	760	760000	14.69594
Torr	133.322	0.133322	0.000133322	0.00133322	0.00131579	1	1000	0.01933672
μmHg	0.133322	0.000133322	1.33322×10^{-7}	1.33322×10^{-6}	1.31579×10^{-6}	0.001	1	1.933672×10^{-5}
psi	6894.757	6.894757	0.006894757	0.06894757	0.068046	51.7151	51715.1	1

To convert a pressure value from a unit in the left-hand column to a new unit, multiply the value by the factor appearing in the column for the new unit. For example:

$$1 \text{ kPa} = 9.8692 \times 10^{-3} \text{ atm}$$
$$1 \text{ Torr} = 1.33322 \times 10^{-4} \text{ MPa}$$

Notes: μmHg is often referred to as "micron"

Torr is essentially identical to mmHg

psi is an abbreviation for the unit pound–force per square inch

psia (as a term for a physical quantity) implies the true (absolute) pressure

psig implies the true pressure minus the local atmospheric pressure

CONVERSION FACTORS FOR THERMAL CONDUCTIVITY UNITS

MULTIPLY ↓ by appropriate factor to OBTAIN→	Btu_{IT} h^{-1} ft^{-1} $°F^{-1}$	Btu_{IT} in. h^{-1} ft^{-2} $°F^{-1}$	Btu_{th} h^{-1} ft^{-1} $°F^{-1}$	Btu_{th} in. h^{-1} ft^{-2} $°F^{-1}$	cal_{IT} s^{-1} cm^{-1} $°C^{-1}$	cal_{th} s^{-1} cm^{-1} $°C^{-1}$	$kcal_{th}$ h^{-1} m^{-1} $°C^{-1}$	J s^{-1} cm^{-1} K^{-1}	W cm^{-1} K^{-1}	W m^{-1} K^{-1}	mW cm^{-1} K^{-1}
Btu_{IT} h^{-1} ft^{-1} $°F^{-1}$	1	12	1.00067	12.0080	4.13379×10^{-3}	4.13656×10^{-3}	1.48916	1.73073×10^{-2}	1.73073×10^{-2}	1.73073	17.3073
Btu_{IT} in h^{-1} ft^{-2} $°F^{-1}$	8.33333×10^{-2}	1	8.33891×10^{-2}	1.00067	3.44482×10^{-4}	3.44713×10^{-4}	0.124097	1.44228×10^{-3}	1.44228×10^{-3}	0.144228	1.44228
Btu_{th} h^{-1} ft^{-1} $°F^{-1}$	0.999331	11.9920	1	12	4.13102×10^{-3}	4.13379×10^{-3}	1.48816	1.72958×10^{-2}	1.72958×10^{-2}	1.72958	17.2958
Btu_{th} in. h^{-1} ft^{-2} $°F^{-1}$	8.32776×10^{-2}	0.999331	8.33333×10^{-2}	1	3.44252×10^{-4}	3.44482×10^{-4}	0.124014	1.44131×10^{-3}	1.44131×10^{-3}	0.144131	1.44131
cal_{IT} s^{-1} cm^{-1} $°C^{-1}$	2.41909×10^{2}	2.90291×10^{3}	2.42071×10^{2}	2.90485×10^{3}	1	1.00067	3.60241×10^{2}	4.1868	4.1868	4.1868×10^{2}	4.1868×10^{3}
cal_{th} s^{-1} cm^{-1} $°C^{-1}$	2.41747×10^{2}	2.90096×10^{3}	2.41909×10^{2}	2.90291×10^{3}	0.999331	1	3.6×10^{2}	4.184	4.184	4.184×10^{2}	4.184×10^{3}
$kcal_{th}$ h^{-1} m^{-1} $°C^{-1}$	0.671520	8.05824	0.671969	8.06363	2.77592×10^{-3}	2.77778×10^{-3}	1	1.16222×10^{-2}	1.16222×10^{-2}	1.16222	11.6222
J s^{-1} cm^{-1} K^{-1}	57.7789	6.93347×10^{2}	57.8176	6.93811×10^{2}	0.238846	0.239006	86.0421	1	1	1×10^{2}	1×10^{3}
W cm^{-1} K^{-1}	57.7789	6.93347×10^{2}	57.8176	6.93811×10^{2}	0.238846	0.239006	86.0421	1	1	1×10^{2}	1×10^{3}
W m^{-1} K^{-1}	0.577789	6.93347	0.578176	6.93811	2.38846×10^{-3}	2.39006×10^{-3}	0.860421	1×10^{-2}	1×10^{-2}	1	10
mW cm^{-1} K^{-1}	5.77789×10^{-2}	0.693347	5.78176×10^{-2}	0.693811	2.38846×10^{-4}	2.39006×10^{-4}	8.60421×10^{-2}	1×10^{-3}	1×10^{-3}	0.1	1

CONVERSION FACTORS FOR ELECTRICAL RESISTIVITY UNITS

To convert FROM ↓ multiply by appropriate factor to OBTAIN →	abΩ cm	μΩ cm	Ω cm	StatΩ cm	Ω m	Ω cir. mil ft^{-1}	Ω in.	Ω ft
abohm centimeter	1	1×10^{-3}	10^{-9}	1.113×10^{-21}	10^{-11}	6.015×10^{-3}	3.937×10^{-10}	3.281×10^{-11}
microohm centimeter	10^3	1	10^{-6}	1.113×10^{-18}	10^{-8}	6.015	3.937×10^{-7}	3.281×10^{-6}
ohm centimeter	10^8	10^6	1	1.113×10^{-12}	1×10^{-2}	6.015×10^6	3.937×10^{-1}	3.281×10^{-2}
statohm centimeter (esu)	8.987×10^{20}	8.987×10^{17}	8.987×10^{11}	1	8.987×10^9	5.406×10^{18}	3.538×10^{11}	2.949×10^{10}
ohm meter	10^{11}	10^8	10^2	1.113×10^{-10}	1	6.015×10^8	3.937×10^1	3.281
ohm circular mil per foot	1.662×10^2	1.662×10^{-1}	1.662×10^{-7}	1.850×10^{-19}	1.662×10^{-9}	1	6.54×10^{-6}	5.45×10^{-9}
ohm inch	2.54×10^9	2.54×10^6	2.54	2.827×10^{-12}	2.54×10^{-2}	1.528×10^7	1	8.3×10^{-2}
ohm foot	3.048×10^{10}	3.048×10^7	3.048×10^{-1}	3.3924×10^{-11}	3.048×10^{-1}	1.833×10^8	12	1

CONVERSION FORMULAS FOR CONCENTRATION OF SOLUTIONS

A = Weight percent of solute
B = Molecular weight of solvent
E = Molecular weight of solute
F = Grams of solute per liter of solution

G = Molality
M = Molarity
N = Mole fraction
R = Density of solution in grams per milliliter

Concentration of solute—SOUGHT	Concentration of solute—GIVEN				
	A	N	G	M	F
A	—	$\dfrac{100N \times E}{N \times E + (1-N)B}$	$\dfrac{100G \times E}{1000 + G \times E}$	$\dfrac{M \times E}{10R}$	$\dfrac{F}{10R}$
N	$\dfrac{\dfrac{A}{E}}{\dfrac{A}{E} + \dfrac{100-A}{B}}$	—	$\dfrac{B \times G}{B \times G + 1000}$	$\dfrac{B \times M}{M(B-E) + 1000R}$	$\dfrac{B \times F}{F(B-E) + 1000R \times E}$
G	$\dfrac{1000A}{E(100-A)}$	$\dfrac{1000N}{B - N \times B}$	—	$\dfrac{1000M}{1000R - (M \times E)}$	$\dfrac{1000F}{E(1000R - F)}$
M	$\dfrac{10R \times A}{E}$	$\dfrac{1000R \times N}{N \times E + (1-N)B}$	$\dfrac{1000R \times G}{1000 + E \times G}$	—	$\dfrac{F}{E}$
F	$10AR$	$\dfrac{1000R \times N \times E}{N \times E + (1-N)B}$	$\dfrac{1000R \times G \times E}{1000 + G \times E}$	$M \times E$	—

CONVERSION FACTORS FOR CHEMICAL KINETICS

Equivalent Second Order Rate Constants

A \ B	$cm^3\ mol^{-1}\ s^{-1}$	$dm^3\ mol^{-1}\ s^{-1}$	$m^3\ mol^{-1}\ s^{-1}$	$cm^3\ molecule^{-1}\ s^{-1}$	$(mmHg)^{-1}\ s^{-1}$	$atm^{-1}\ s^{-1}$	$ppm^{-1}\ min^{-1}$	$m^2\ kN^{-1}\ s^{-1}$
$1\ cm^3\ mol^{-1}\ s^{-1} =$	1	10^{-3}	10^{-6}	1.66×10^{-24}	$1.604\times10^{-5}\ T^{-1}$	$1.219\times10^{-2}\ T^{-1}$	2.453×10^{-9}	$1.203\times10^{-4}\ T^{-1}$
$1\ dm^3\ mol^{-1}\ s^{-1} =$	10^3	1	10^{-3}	1.66×10^{-21}	$1.604\times10^{-2}\ T^{-1}$	$12.19\ T^{-1}$	2.453×10^{-6}	$1.203\times10^{-1}\ T^{-1}$
$1\ m^3\ mol^{-1}\ s^{-1} =$	10^6	10^3	1	1.66×10^{-18}	$16.04\ T^{-1}$	$1.219\times10^4\ T^{-1}$	2.453×10^{-3}	$120.3\ T^{-1}$
$1\ cm^3\ molecule^{-1}\ s^{-1} =$	6.023×10^{23}	6.023×10^{20}	6.023×10^{17}	1	$9.658\times10^{18}\ T^{-1}$	$7.34\times10^{21}\ T^{-1}$	1.478×10^{15}	$7.244\times10^{19}\ T^{-1}$
$1\ (mmHg)^{-1}\ s^{-1} =$	$6.236\times10^4\ T$	$62.36\ T$	$6.236\times10^{-2}\ T$	$1.035\times10^{-19}\ T$	1	760	4.56×10^{-2}	7.500
$1\ atm^{-1}\ s^{-1} =$	$82.06\ T$	$8.206\times10^{-2}\ T$	$8.206\times10^{-5}\ T$	$1.362\times10^{-22}\ T$	1.316×10^{-3}	1	6×10^{-5}	9.869×10^{-3}
$1\ ppm^{-1}\ min^{-1} =$ at 298 K, 1 atm total pressure	4.077×10^8	4.077×10^5	407.7	6.76×10^{-16}	21.93	1.667×10^4	1	164.5
$1\ m^2\ kN^{-1}\ s^{-1} =$	$8314\ T$	$8.314\ T$	$8.314\times10^{-3}\ T$	$1.38\times10^{-20}\ T$	0.1333	101.325	6.079×10^{-3}	1

To convert a rate constant from one set of units A to a new set B find the conversion factor for the row A under column B and multiply the old value by it, e.g.. to convert $cm^3\ molecule^{-1}\ s^{-1}$ to $m^3\ mol^{-1}\ s^{-1}$ multiply by 6.023×10^{17}.

Table adapted from High Temperature Reaction Rate Data No. 5, The University, Leeds (1970).

Equivalent Third Order Rate Constants

A \ B	$cm^6\ mol^{-2}\ s^{-1}$	$dm^6\ mol^{-1}\ s^{-1}$	$m^6\ mol^{-2}\ s^{-1}$	$cm^6\ molecule^{-2}\ s^{-1}$	$(mmHg)^{-2}\ s^{-1}$	$atm^{-2}\ s^{-1}$	$ppm^{-2}\ min^{-1}$	$m^4\ kN^{-2}\ s^{-1}$
$1\ cm^6\ mol^{-2}\ s^{-1} =$	1	10^{-6}	10^{-12}	2.76×10^{-48}	$2.57\times10^{-10}\ T^{-2}$	$1.48\times10^4\ T^{-2}$	1.003×10^{-19}	$1.477\times10^{-8}\ T^{-2}$
$1\ dm^6\ mol^{-1}\ s^{-1} =$	10^6	1	10^{-6}	2.76×10^{-42}	$2.57\times10^{-4}\ T^{-2}$	$148\ T^{-2}$	1.003×10^{-13}	$1.477\times10^{-2}\ T^{-2}$
$1\ m^6\ mol^{-2}\ s^{-1} =$	10^{12}	10^6	1	2.76×10^{-36}	$257\ T^{-2}$	$1.48\times10^8\ T^{-2}$	1.003×10^{-7}	$1.477\times10^4\ T^{-2}$
$1\ cm^6\ molecule^{-2}\ s^{-1} =$	3.628×10^{47}	3.628×10^{41}	3.628×10^{35}	1	$9.328\times10^{37}\ T^{-2}$	$5.388\times10^{43}\ T^{-2}$	3.64×10^{28}	$5.248\times10^{39}\ T^{-2}$
$1\ (mmHg)^{-2}\ s^{-1} =$	$3.89\times10^9\ T^2$	$3.89\times10^3\ T^2$	$3.89\times10^{-3}\ T^2$	$1.07\times10^{-38}\ T^2$	1	5.776×10^5	3.46×10^{-5}	56.25
$1\ atm^{-2}\ s^{-1} =$	$6.733\times10^3\ T^2$	$6.733\times10^{-3}\ T^2$	$6.733\times10^{-9}\ T^2$	$1.86\times10^{-44}\ T^2$	1.73×10^{-6}	1	6×10^{-11}	9.74×10^{-5}
$1\ ppm^{-2}\ min^{-1} =$ at 298K, 1 atm total pressure	9.97×10^{18}	9.97×10^{12}	9.97×10^6	2.75×10^{-29}	2.89×10^4	1.667×10^{10}	1	1.623×10^6
$1\ m^4\ kN^{-2}\ s^{-1} =$	$6.91\times10^7\ T^2$	$6.91\ T^2$	$69.1\times10^{-5}\ T^2$	$1.904\times10^{-40}\ T^2$	0.0178	1.027×10^4	6.16×10^{-7}	1

From *J. Phys. Chem. Ref. Data*, 9, 470, 1980, by permission of the authors and the copyright owner, the American Institute of Physics.

CONVERSION FACTORS FOR IONIZING RADIATION

Conversion between SI and Other Units

Quantity	Symbol for quantity	Expression in SI units	Expression in symbols for SI units	Special name for SI units	Symbols using special names	Conventional units	Symbol for conventional unit	Value of conventional unit in SI units
Activity	A	1 per second	s^{-1}	becquerel	Bq	curie	Ci	3.7×10^{10} Bq
Absorbed dose	D	joule per kilogram	$J\,kg^{-1}$	gray	Gy	rad	rad	0.01 Gy
Absorbed dose rate	$\dot{D}$	joule per kilogram second	$J\,kg^{-1}\,s^{-1}$		$Gy\,s^{-1}$	rad	$rad\,s^{-1}$	0.01 Gy s^{-1}
Average energy per ion pair	W	joule	J			electronvolt	eV	1.602×10^{-19} J
Dose equivalent	H	joule per kilogram	$J\,kg^{-1}$	sievert	Sv	rem	rem	0.01 Sv
Dose equivalent rate	$\dot{H}$	joule per kilogram second	$J\,kg^{-1}\,s^{-1}$		$Sv\,s^{-1}$	rem per second	$rem\,s^{-1}$	0.01 Sv s^{-1}
Electric current	I	ampere	A			ampere	A	1.0 A
Electric potential difference	U, V	watt per ampere	$W\,A^{-1}$	volt	V	volt	V	1.0 V
Exposure	$\dot{X}$	coulomb per kilogram	$C\,kg^{-1}$			roentgen	R	$2.58 \times 10^{-4}\,C\,kg^{-1}$
Exposure rate	X	coulomb per kilogram second	$C\,kg^{-1}\,s^{-1}$			roentgen per second	$R\,s^{-1}$	$2.58 \times 10^{-4}\,C\,kg^{-1}\,s^{-1}$
Fluence	ϕ	1 per meter squared	m^{-2}			1 per centimeter squared	cm^{-2}	$1.0 \times 10^{4}\,m^{-2}$
Fluence rate	Φ	1 per meter squared second	$m^{-2}\,s^{-1}$			1 per centimeter squared second	$cm^{-2}\,s^{-1}$	$1.0 \times 10^{4}\,m^{-2}\,s^{-1}$
Kerma	K	joule per kilogram	$J\,kg^{-1}$	gray	Gy	rad	rad	0.01 Gy
Kerma rate	$\dot{K}$	joule per kilogram second	$J\,kg^{-1}\,s^{-1}$		$Gy\,s^{-1}$	rad per second	$rad\,s^{-1}$	0.01 Gy s^{-1}
Lineal energy	y	joule per meter	$J\,m^{-1}$			kiloelectron volt per micrometer	$keV\,\mu m^{-1}$	$1.602 \times 10^{-10}\,J\,m^{-1}$
Linear energy transfer	L	joule per meter	$J\,m^{-1}$			kiloelectron volt per micrometer	$keV\,\mu m^{-1}$	$1.602 \times 10^{-10}\,J\,m^{-1}$
Mass attenuation coefficient	μ/ρ	meter squared per kilogram	$m^2\,kg^{-1}$			centimeter squared per gram	$cm^2\,g^{-1}$	$0.1\,m^2\,kg^{-1}$
Mass energy transfer coefficient	μ_{tr}/ρ	meter squared per kilogram	$m^2\,kg^{-1}$			centimeter squared per gram	$cm^2\,g^{-1}$	$0.1\,m^2\,kg^{-1}$
Mass energy absorption coefficient	μ_{en}/ρ	meter squared per kilogram	$m^2\,kg^{-1}$			centimeter squared per gram	$cm^2\,g^{-1}$	$0.1\,m^2\,kg^{-1}$
Mass stopping power	S/ρ	joule meter squared per kilogram	$J\,m^2\,kg^{-1}$			MeV centimeter squared per gram	$MeV\,cm^2\,g^{-1}$	$1.602 \times 10^{-14}\,J\,m^2\,kg^{-1}$
Power	P	joule per second	$J\,s^{-1}$	watt	W	watt	W	1.0 W
Pressure	p	newton per meter squared	$N\,m^{-2}$	pascal	Pa	torr	torr	(101325/760)Pa
Radiation chemical yield	G	mole per joule	$mol\,J^{-1}$			molecules per 100 electron volts	molecules $(100\,eV)^{-1}$	$1.04 \times 10^{-7}\,mol\,J^{-1}$
Specific energy	z	joule per kilogram	$J\,kg^{-1}$	gray	Gy	rad	rad	0.01 Gy

Conversion of Radioactivity Units from MBq to mCi and μCi

MBq	mCi	MBq	mCi	MBq	mCi	MBq	mCi	MBq	mCi
7000	189.	700	18.9	70	1.89	7	189	0.7	18.9
6000	162.	600	16.2	60	1.62	6	162	0.6	16.2
5000	135.	500	13.5	50	1.35	5	135	0.5	13.5
4000	108.	400	10.8	40	1.08	4	108	0.4	10.8
3000	81.	300	8.1	30	810	3	81	0.3	8.1
2000	54.	200	5.4	20	540	2	54	0.2	5.4
1000	27.	100	2.7	10	270	1	27	0.1	2.7
900	24.	90	2.4	9	240	0.9	24		
800	21.6	80	2.16	8	220	0.8	21.6		

Conversion of Radioactivity Units from mCi and μCi to MBq

mCi	MBq	mCi	MBq	mCi	MBq	μCi	MBq	μCi	MBq	μCi	MBq
200	7400	40	1480	5	185	1000	37.0	200	7.4	30	1.11
150	5550	30	1110	4	148	900	33.3	100	3.7	20	0.74
100	3700	20	740	3	111	800	29.6	90	3.33	10	0.37
90	3330	10	370	2	74.0	700	25.9	80	2.96	5	0.185
80	2960	9	333	1	37.0	600	22.2	70	2.59	2	0.074
70	2590	8	296			500	18.5	60	2.22	1	0.037
60	2220	7	259			400	14.8	50	1.85		
50	1850	6	222			300	11.1	40	1.48		

Conversion of Radioactivity Units

100 TBq (10^{14} Bq)	=	2.7 kCi (2.7×10^3 Ci)	100 kBq (10^5 Bq)	=	2.7 μCi (2.7×10^{-6} Ci)
10 TBq (10^{13} Bq)	=	270 Ci (2.7×10^2 Ci)	10 kBq (10^4 Bq)	=	270 nCi (2.7×10^{-7} Ci)
1 TBq (10^{12} Bq)	=	27 Ci (2.7×10^1 Ci)	1 kBq (10^3 Bq)	=	27 nCi (2.7×10^{-8} Ci)
100 GBq (10^{11} Bq)	=	2.7 Ci (2.7×10^0 Ci)	100 Bq (10^2 Bq)	=	2.7 nCi (2.7×10^{-9} Ci)
10 GBq (10^{10} Bq)	=	270 mCi (2.7×10^{-1} Ci)	10 Bq (10^1 Bq)	=	270 pCi (2.7×10^{-10} Ci)
1 GBq (10^9 Bq)	=	27 mCi (2.7×10^{-2} Ci)	1 Bq (10^0 Bq)	=	27 pCi (2.7×10^{-11} Ci)
100 MBq (10^8 Bq)	=	2.7 mCi (2.7×10^{-3} Ci)	100 mBq (10^{-1} Bq)	=	2.7 pCi (2.7×10^{-12} Ci)
10 MBq (10^7 Bq)	=	270 μCi (2.7×10^{-4} Ci)	10 mBq (10^{-2} Bq)	=	270 fCi (2.7×10^{-13} Ci)
1 MBq (10^6 Bq)	=	27 μCi (2.7×10^{-5} Ci)	1 mBq (10^{-3} Bq)	=	27 fCi (2.7×10^{-14} Ci)

Conversion of Absorbed Dose Units

SI Units		Conventional	SI Units		Conventional
100 Gy (10^2 Gy)	=	10,000 rad (10^4 rad)	100 μGy (10^{-4} Gy)	=	10 mrad (10^{-2} rad)
10 Gy (10^1 Gy)	=	1,000 rad (10^3 rad)	10 μGy (10^{-5} Gy)	=	1 mrad (10^{-3} rad)
1 Gy (10^0 Gy)	=	100 rad (10^2 rad)	1 μGy (10^{-6} Gy)	=	100 μrad (10^{-4} rad)
100 mGy (10^{-1} Gy)	=	10 rad (10^1 rad)	100 nGy (10^{-7} Gy)	=	10 μrad (10^{-5} rad)
10 mGy (10^{-2} Gy)	=	1 rad (10^0 rad)	10 nGy (10^{-8} Gy)	=	1 μrad (10^{-6} rad)
1 mGy (10^{-3} Gy)	=	100 mrad (10^{-1} rad)	1 nGy (10^{-9} Gy)	=	100 nrad (10^{-7} rad)

Conversion of Dose Equivalent Units

100 Sv (10^2 Sv)	=	10,000 rem (10^4 rem)	100 μSv (10^{-4} Sv)	=	10 mrem (10^{-2} rem)
10 Sv (10^1 Sv)	=	1,000 rem (10^3 rem)	10 μSv (10^{-5} Sv)	=	1 mrem (10^{-3} rem)
1 Sv (10^0 Sv)	=	100 rem (10^2 rem)	1 μSv (10^{-6} Sv)	=	100 μrem (10^{-4} rem)
100 mSv (10^{-1} Sv)	=	10 rem (10^1 rem)	100 nSv (10^{-7} Sv)	=	10 μrem (10^{-5} rem)
10 mSv (10^{-2} Sv)	=	1 rem (10^0 rem)	10 nSv (10^{-8} Sv)	=	1 μrem (10^{-6} rem)
1 mSv (10^{-3} Sv)	=	100 mrem (10^{-1} rem)	1 nSv (10^{-9} Sv)	=	100 nrem (10^{-7} rem)

VALUES OF THE GAS CONSTANT IN DIFFERENT UNIT SYSTEMS

In SI units the value of the gas constant, R, is:

R = 8.314472 Pa m^3 K^{-1} mol^{-1}
 = 8314.472 Pa L K^{-1} mol^{-1}
 = 0.08314472 bar L K^{-1} mol^{-1}

This table gives the appropriate value of R for use in the ideal gas equation, $PV = nRT$, when the variables are expressed in other units. The following conversion factors for pressure units were used in generating the table:

1 atm = 101325 Pa
1 psi = 6894.757 Pa

1 torr (mmHg) = 133.322 Pa [at 0 °C]
1 in Hg = 3386.38 Pa [at 0 °C]
1 in H$_2$O = 249.082 Pa [at 4 °C]
1 ft H$_2$O = 2988.98 Pa [at 4 °C]

Reference

Mohr, P. J., Taylor, B. N., and Newell, D. B., "CODATA recommended values of the fundamental physical constants: 2006", *J. Phys. Chem. Ref. Data* 37, 1187, 2008.

Units of V, T, n			Units of P						
V	T	n	kPa	atm	psi	mmHg	in Hg	in H$_2$O	ft H$_2$O
ft^3	K	mol	0.2936228	0.00289784	0.0425864	2.20236	0.0867070	1.17881	0.0982351
		lb·mol	133.1851	1.31443	19.3168	998.973	39.3296	534.704	44.5587
	°R	mol	0.1631238	0.00160990	0.0236591	1.22353	0.0481706	0.654900	0.0545751
		lb·mol	73.99170	0.730242	10.7316	554.984	21.8498	297.058	24.7548
cm^3	K	mol	8314.472	82.0574	1205.91	62363.8	2455.27	33380.4	2781.71
		lb·mol	3771381	37220.6	546993	282878000	1113690	15141100	1261760
	°R	mol	4619.151	45.5875	669.951	34646.5	1364.03	18544.7	1545.39
		lb·mol	2095211	20678.1	303885	15715400	618717	8411730	700979
L	K	mol	8.314472	0.0820574	1.20591	62.3638	2.45527	33.3804	2.78171
		lb·mol	3771.381	37.2206	546.993	28287.8	1113.69	15141.1	1261.76
	°R	mol	4.619151	0.0455875	0.669951	34.6465	1.36403	18.5447	1.54539
		lb·mol	2095.211	20.6781	303.885	15715.4	618.717	8411.73	700.979
m^3	K	mol	0.008314472	0.0000820574	0.00120591	0.0623638	0.00245527	0.0333804	0.00278171
		lb·mol	3.771381	0.0372206	0.546993	28.2878	1.11369	15.1411	1.26176
	°R	mol	0.004619151	0.0000455875	0.000669951	0.0346465	0.00136403	0.0185447	0.00154539
		lb·mol	2.095211	0.0206781	0.303885	15.7154	0.618717	8.41173	0.700979

Section 2
Symbols, Terminology, and Nomenclature

SYMBOLS AND TERMINOLOGY FOR PHYSICAL AND CHEMICAL QUANTITIES

The International Organization for Standardization (ISO), International Union of Pure and Applied Chemistry (IUPAC), and the International Union of Pure and Applied Physics (IUPAP) have jointly developed a set of recommended symbols for physical and chemical quantities. Consistent use of these recommended symbols helps assure unambiguous scientific communication. The list below is reprinted from Reference 1 with permission from IUPAC. Full details may be found in the following references:

1. Ian Mills, Ed., *Quantities, Units, and Symbols in Physical Chemistry*, Blackwell Scientific Publications, Oxford, 1988. Third Edition: RSC Publishing, Cambridge, UK, 2007.
2. E. R. Cohen and P. Giacomo, *Symbols, Units, Nomenclature, and Fundamental Constants in Physics*, Document IUPAP–25, 1987; also published in *Physica* 146A, 1–68, 1987.
3. *ISO Standards Handbook 2: Units of Measurement*, International Organization of Standardization, Geneva, 1982.

GENERAL RULES

The value of a physical quantity is expressed as the product of a numerical value and a unit, e.g.:

$T = 300$ K
$V = 26.2$ cm^3
$C_p = 45.3$ J mol^{-1} K^{-1}

The symbol for a physical quantity is always given in italic (sloping) type, while symbols for units are given in roman type. Column headings in tables and axis labels on graphs may conveniently be written as the physical quantity symbol divided by the unit symbol, e.g.:

T/K
V/cm^3
C_p/J mol^{-1} K^{-1}

The values in the table or graph axis are then pure numbers. Subscripts to symbols for physical quantities should be italic if the subscript refers to another physical quantity or to a number, e.g.:

C_p – heat capacity at constant pressure
B_n – nth virial coefficient

Subscripts that have other meanings should be in roman type:

m_p – mass of the proton
E_k – kinetic energy

The following tables give the recommended symbols for the major classes of physical and chemical quantities. The expression in the Definition column is given as an aid in identifying the quantity but is not necessarily the complete or unique definition. The SI Unit gives one (not necessarily unique) expression for the coherent SI unit for the quantity. Other equivalent unit expressions, including those that involve SI prefixes, may be used.

Name	Symbol	Definition	SI unit
Space and Time			
cartesian space coordinates	x, y, z		m
spherical polar coordinates	r, θ, ϕ		m, 1, 1
generalized coordinate	q, q_i		(varies)
position vector	r	$r = xi + yj + zk$	m
length	l		m
special symbols:			
height	h		
breadth	b		
thickness	d, δ		
distance	d		
radius	r		
diameter	d		
path length	s		
length of arc	s		
area	A, A_s, S		m^2
volume	$V, (v)$		m^3
plane angle	$\alpha, \beta, \gamma, \theta, \phi...$	$\alpha = s/r$	rad, 1
solid angle	ω, Ω	$\omega = A/r^2$	sr, 1
time	t		s
period	T	$T = t/N$	s
frequency	v, f	$v = 1/T$	Hz
circular frequency, angular frequency	ω	$\omega = 2\pi v$	rad s^{-1}, s^{-1}
characteristic time interval, relaxation time, time constant	τ, T	$\tau = |dt/d\ln x|$	s
angular velocity	ω	$\omega = d\phi/dt$	rad s^{-1}, s^{-1}
velocity	$v, u, w, c, \dot{r}$	$v = dr/dt$	m s^{-1}

Name	Symbol	Definition	SI unit
speed	v, u, w, c	$v = \|v\|$	m s^{-1}
acceleration	$a, (g)$	$a = dv/dt$	m s^{-2}

Classical Mechanics

Name	Symbol	Definition	SI unit
mass	m		kg
reduced mass	μ	$\mu = m_1 m_2/(m_1 + m_2)$	kg
density, mass density	ρ	$\rho = m/V$	kg m^{-3}
relative density	d	$d = \rho/\rho$	1
surface density	ρ_A, ρ_S	$\rho_A = m/A$	kg m^{-2}
specific volume	v	$v = V/m = 1/\rho$	m^3 kg^{-1}
momentum	p	$p = mv$	kg m s^{-1}
angular momentum, action	L	$L = r \times p$	J s
moment of inertia	I, J	$I = \Sigma m_i r_i^2$	kg m^2
force	F	$F = dp/dt = ma$	N
torque, moment of a force	$T, (M)$	$T = r \times F$	N m
energy	E		J
potential energy	E_p, V, Φ	$E_p = \int F \cdot ds$	J
kinetic energy	E_k, T, K	$E_k = 1/2 mv^2$	J
work	W, w	$W = \int F \cdot ds$	J
Hamilton function	H	$H(q, p) = T(q, p) + V(q)$	J
Lagrange function	L	$L(q, \dot{q}) = T(q, \dot{q}) - V(q)$	J
pressure	p, P	$p = F/A$	Pa, N m^{-2}
surface tension	γ, σ	$y = dW/dA$	N m^{-1}, J m^{-2}
weight	$G, (W, P)$	$G = mg$	N
gravitational constant	G	$F = Gm_1 m_2/r^2$	N m^2 kg^{-2}
normal stress	σ	$\sigma = F/A$	Pa
shear stress	τ	$\tau = F/A$	Pa
linear strain, relative elongation	ε, e	$\varepsilon = \Delta l/l$	1
modulus of elasticity, Young's modulus	E	$E = \sigma/\varepsilon$	Pa
shear strain	γ	$\gamma = \Delta x/d$	1
shear modulus	G	$G = \tau/\gamma$	Pa
volume strain, bulk strain	θ	$\theta = \Delta V/V_0$	1
bulk modulus, compression modulus	K	$K = -V_0(dp/dV)$	Pa
viscosity, dynamic viscosity	η, μ	$\tau_{x,z} = \eta(dv_x/dz)$	Pa s
fluidity	ϕ	$\phi = 1/\eta$	m kg^{-1} s
kinematic viscosity	v	$v = \eta/\rho$	m^2 s^{-1}
friction coefficient	$\mu, (f)$	$F_{frict} = \mu F_{norm}$	1
power	P	$P = dW/dt$	W
sound energy flux	P, P_a	$P = dE/dt$	W
acoustic factors			
reflection factor	ρ	$\rho = P_t/P_0$	1
acoustic absorption factor	$\alpha_a, (\alpha)$	$\alpha_a = 1 - \rho$	1
transmission factor	τ	$\tau = P_{tr}/P_0$	1
dissipation factor	δ	$\delta = \alpha_a - \tau$	1

Electricity and Magnetism

Name	Symbol	Definition	SI unit
quantity of electricity, electric charge	Q		C
charge density	ρ	$\rho = Q/V$	C m^{-3}
surface charge density	σ	$\sigma = Q/A$	C m^{-2}
electric potential	V, ϕ	$V = dW/dQ$	V, J C^{-1}
electric potential difference	$U, \Delta V, \Delta\phi$	$U = V_2 - V_1$	V
electromotive force	E	$E = \int (F/Q) \cdot ds$	V
electric field strength	E	$E = F/Q = -\text{grad } V$	V m^{-1}
electric flux	Ψ	$\Psi = \int D \cdot dA$	C
electric displacement	D	$D = \varepsilon E$	C m^{-2}
capacitance	C	$C = Q/U$	F, C V^{-1}
permittivity	ε	$D = \varepsilon E$	F m^{-1}
permittivity of vacuum	ε_0	$\varepsilon_0 = \mu_0^{-1} c_0^{-2}$	F m^{-1}
relative permittivity	ε_r	$\varepsilon_r = \varepsilon/\varepsilon_0$	1
dielectric polarization (dipole moment per volume)	P	$P = D - \varepsilon_0 E$	C m^{-2}
electric susceptibility	χ_e	$\chi_e = \varepsilon r - 1$	1
electric dipole moment	p, μ	$p = Qr$	C m

Name	Symbol	Definition	SI unit
electric current	I	$I = dQ/dt$	A
electric current density	j, J	$I = \int j \cdot dA$	A m^{-2}
magnetic flux density, magnetic induction	B	$F = Qv \times B$	T
magnetic flux	Φ	$\Phi = \int B \cdot dA$	A m^{-2}
magnetic field strength	H	$B = \mu H$	A m^{-2}
permeability	μ	$B = \mu H$	N A^{-2}, H m^{-1}
permeability of vacuum	μ_0		H m^{-1}
relative permeability	μ_r	$\mu_r = \mu/\mu_0$	1
magnetization (magnetic dipole moment per volume)	M	$M = B/\mu_0 - H$	A m^{-1}
magnetic susceptibility	$\chi, \kappa, (\chi_m)$	$\chi = \mu_r - 1$	1
molar magnetic susceptibility	χ_m	$\chi_m = V_m \chi$	m^3 mol^{-1}
magnetic dipole moment	m, μ	$E_p = -m \cdot B$	A m^2, J T^{-1}
electrical resistance	R	$R = U/I$	Ω
conductance	G	$G = 1/R$	S
loss angle	δ	$\delta = (\pi/2) + \phi_I - \phi_U$	1, rad
reactance	X	$X = (U/I)\sin \delta$	Ω
impedance (complex impedance)	Z	$Z = R + iX$	Ω
admittance (complex admittance)	Y	$Y = 1/Z$	S
susceptance	B	$Y = G + iB$	S
resistivity	ρ	$\rho = E/j$	Ω m
conductivity	κ, γ, σ	$\kappa = 1/\rho$	S m^{-1}
self-inductance	L	$E = -L(dI/dt)$	H
mutual inductance	M, L_{12}	$E_1 = L_{12}(dI_2/dt)$	H
magnetic vector potential	A	$B = \nabla \times A$	Wb m^{-1}
Poynting vector	S	$S = E \times H$	W m^{-2}

Quantum Mechanics

Name	Symbol	Definition	SI unit
momentum operator	$\hat{p}$	$\hat{p} = -ih\nabla$	m^{-1} J s
kinetic energy operator	$\hat{T}$	$\hat{T} = -(h^2/2m)\nabla^2$	J
Hamiltonian operator	$\hat{H}$	$\hat{H} = \hat{T} + V$	J
wavefunction, state function	Ψ, ψ, ϕ	$\hat{H}\psi = E\psi$	(m$^{-3/2}$)
probability density	P	$P = \psi^*\psi$	(m^{-3})
charge density of electrons	ρ	$\rho = -eP$	(C m^{-3})
probability current density	S	$S = -ih(\psi^*\nabla\psi - \psi\nabla\psi^*)/2m_e$	(m^{-2} s^{-1})
electric current density of electrons	j	$j = -eS$	(A m^{-2})
matrix element of operator $\hat{A}$	$A_{ij}, \langle i\|\hat{A}\|j\rangle$	$A_{ij} = \int \psi_i^* \hat{A} \psi_j d\tau$	(varies)
expectation value of operator $\hat{A}$	$\langle A \rangle, \bar{A}$	$\langle A \rangle = \int \psi^* \hat{A} \Psi d\tau$	(varies)
hermitian conjugate of $\hat{A}$	$\hat{A}^\dagger$	$(\hat{A}^\dagger)_{ij} = (A_{ji})^*$	(varies)
commutator of $\hat{A}$ and $\hat{B}$	$[\hat{A}, \hat{B}], [\hat{A}, \hat{B}]_-$	$[\hat{A}, \hat{B}] = \hat{A}\hat{B} - \hat{B}\hat{A}$	(varies)
anticommutator	$[\hat{A}, \hat{B}]_+$	$[\hat{A}, \hat{B}]_+ = \hat{A}\hat{B} + \hat{B}\hat{A}$	(varies)
spin wavefunction	$\alpha; \beta$		1
coulomb integral	H_{AA}	$H_{AA} = \int \psi_A^* \hat{H} \psi_A d\tau$	J
resonance integral	H_{AB}	$H_{AB} = \int \psi_A^* \hat{H} \psi_B d\tau$	J
overlap integral	S_{AB}	$S_{AB} = \int \psi_A^* \psi_B d\tau$	1

Atoms and Molecules

Name	Symbol	Definition	SI unit
nucleon number, mass number	A		1
proton number, atomic number	Z		1
neutron number	N	$N = A - Z$	1
electron rest mass	m_e		kg
mass of atom, atomic mass	m_a, m		kg
atomic mass constant	m_u	$m_u = m_a(^{12}C)/12$	kg
mass excess	Δ	$\Delta = m_a - Am_u$	kg
elementary charge, proton charge	e		C
Planck constant	h		J s
Planck constant/2π	$\hbar$	$\hbar = h/2\pi$	J s
Bohr radius	a_0	$a_0 = 4\pi\varepsilon_0 \hbar^2/m_e e^2$	m
Hartree energy	E_h	$E_h = \hbar^2/m_e a_0^2$	J
Rydberg constant	R_∞	$R_\infty = E_h/2hc$	m^{-1}
fine structure constant	α	$\alpha = e^2/4\pi\varepsilon_0 \hbar c$	1

Name	Symbol	Definition	SI unit
ionization energy	E_i		J
electron affinity	E_{ea}		J
dissociation energy	E_d, D		J
from the ground state	D_0		J
from the potential minimum	D_e		J
principal quantum number (H atom)	n	$E = -hcR/n^2$	1
angular momentum quantum numbers	see under Spectroscopy		
magnetic dipole moment of a molecule	$\boldsymbol{m}, \boldsymbol{\mu}$	$E_p = -\boldsymbol{m}\cdot\boldsymbol{B}$	J T^{-1}
magnetizability of a molecule	ξ	$\boldsymbol{m} = \xi B$	J T^{-2}
Bohr magneton	μ_B	$\mu_B = e\hbar/2m_e$	J T^{-1}
nuclear magneton	μ_N	$\mu_N = (m_e/m_p)\mu_B$	J T^{-1}
magnetogyric ratio (gyromagnetic ratio)	γ	$\gamma = \mu/L$	C kg^{-1}
g factor	g		1
Larmor circular frequency	ω_L	$\omega_L = (e/2m)B$	s^{-1}
Larmor frequency	ν_L	$\nu_L = \omega_L/2\pi$	Hz
longitudinal relaxation time	T_1		s
transverse relaxation time	T_2		s
electric dipole moment of a molecule	$\boldsymbol{p}, \boldsymbol{\mu}$	$E_p = -\boldsymbol{p}\cdot\boldsymbol{E}$	C m
quadrupole moment of a molecule	$\boldsymbol{Q}; \Theta$	$E_p = 1/2\boldsymbol{Q}: V'' = 1/3\Theta: V''$	C m^2
quadrupole moment of a nucleus	eQ	$eQ = 2\cdot\langle\Theta_{zz}\rangle$	C m^2
electric field gradient tensor	$\boldsymbol{q}$	$q_{\alpha\beta} = -\partial^2 V/\partial\alpha\partial\beta$	V m^{-2}
quadrupole interaction energy tensor	χ	$\chi_{\alpha\beta} = eQq_{\alpha\beta}$	J
electric polarizability of a molecule	α	p (induced) $= \alpha E$	C m^2 V^{-1}
activity (of a radioactive substance)	A	$A = -dN_B/dt$	Bq
decay (rate) constant, disintegration (rate) constant	λ	$A = \gamma N_B$	s^{-1}
half life	$t_{1/2}, T_{1/2}$		s
mean life	τ		s
level width	Γ	$\Gamma = \hbar/\tau$	J
disintegration energy	Q		J
cross section (of a nuclear reaction)	σ		m^2

Spectroscopy

Name	Symbol	Definition	SI unit
total term	T	$T = E_{tot}/hc$	m^{-1}
transition wavenumber	$\tilde{\nu}, (\nu)$	$\tilde{\nu} = T' - T''$	m^{-1}
transition frequency	ν	$\nu = (E' - E'')/h$	Hz
electronic term	T_e	$T_e = E_e/hc$	m^{-1}
vibrational term	G	$G = E_{vib}/hc$	m^{-1}
rotational term	F	$F = E_{rot}/hc$	m^{-1}
spin orbit coupling constant	A	$T_{s.o.} = A\langle\hat{\boldsymbol{L}}\cdot\hat{\boldsymbol{S}}\rangle$	m^{-1}
principal moments of inertia	$I_A; I_B; I_C$	$I_A \leq I_B \leq I_C$	kg m^2
rotational constants,			
in wavenumber	$\tilde{A}; \tilde{B}; \tilde{C}$	$\tilde{A} = h/8\pi^2 cI_A$	m^{-1}
in frequency	$A; B; C$	$A = h/8\pi^2 I_A$	Hz
inertial defect	Δ	$\Delta = I_C - I_A - I_B$	kg m^2
asymmetry parameter	κ	$\kappa = \dfrac{(2B - A - C)}{(A - C)}$	1
centrifugal distortion constants,			
S reduction	$D_J; D_{JK}; D_K; d_1; d_2$		m^{-1}
A reduction	$\Delta_J; \Delta_{JK}; \Delta_K; \delta_J; \delta_K$		m^{-1}
harmonic vibration wavenumber	$\omega_e; \omega_r$		m^{-1}
vibrational anharmonicity constant	$\omega_e x_e; x_{rs}; g_{u'}$		m^{-1}
vibrational quantum numbers	$v_j; l_t$		1
Coriolis zeta constant	$\zeta_{rs}^{\ a}$		1
angular momentum quantum numbers	see additional information below		
degeneracy, statistical weight	g, d, β		1
electric dipole moment of a molecule	$\boldsymbol{p}, \boldsymbol{\mu}$	$E_p = -\boldsymbol{p}\cdot\boldsymbol{E}$	C m
transition dipole moment of a molecule	$\boldsymbol{M}, \boldsymbol{R}$	$M = \int\psi' \boldsymbol{p}\psi''d\tau$	C m
molecular geometry, interatomic distances,			
equilibrium distance	r_e		m
zero–point average distance	r_z		m

Name	Symbol	Definition	SI unit
ground state distance	r_0		m
substitution structure distance	r_s		m
vibrational coordinates,			
internal coordinates	$R_i, r_i, \theta_j,$ etc.		(varies)
symmetry coordinates	S_i		(varies)
normal coordinates			
mass adjusted	Q_r		$kg^{1/2}$ m
dimensionless	q_r		1
vibrational force constants,			
diatomic	$f, (k)$	$f = \partial^2 V/\partial r^2$	$J\,m^{-2}$
polyatomic,			
internal coordinates	f_{ij}	$f_{ij} = \partial^2 V/\partial r_i \partial r_j$	(varies)
symmetry coordinates	F_{ij}	$F_{ij} = \partial^2 V/\partial S_i \partial S_j$	(varies)
dimensionless normal coordinates	$\phi_{rst...}, k_{rst...}$		m^{-1}
nuclear magnetic resonance (NMR),			
magnetogyric ratio	γ	$\gamma = \mu/I\hbar$	$C\,kg^{-1}$
shielding constant	σ_A	$B_A = (1 - \sigma_A)B$	1
chemical shift, δ scale	δ	$\delta = 10^6(v - v_0)/v_0$	1
(indirect) spin–spin coupling constant	J_{AB}	$\hat{H}/h = J_{AB}\hat{I}_A \cdot \hat{I}_B$	Hz
direct (dipolar) coupling constant	D_{AB}		Hz
longitudinal relaxation time	T_1		s
transverse relaxation time	T_2		s
electron spin resonance, electron paramagnetic resonance (ESR, EPR),			
magnetogyric ratio	γ	$\gamma = \mu/s\hbar$	$C\,kg^{-1}$
g factor	g	$hv = g\mu_B B$	1
hyperfine coupling constant,			
in liquids	a, A	$\hat{H}_{hfs}/h = a\hat{S} \cdot \hat{I}$	Hz
in solids	T	$\hat{H}_{hfs}/h = \hat{S} \cdot T \cdot \hat{I}$	Hz

Angular momentum	Operator symbol	Quantum number symbol		
		Total	Z–axis	z-axis
electron orbital	$\hat{L}$	L	M_L	Λ
one electron only	$\hat{l}$	l	m_l	λ
electron spin	$\hat{S}$	S	M_S	Σ
one electron only	$\hat{s}$	s	m_s	σ
electron orbital + spin	$\hat{L} + \hat{S}$			$\Omega = \Lambda + \Sigma$
nuclear orbital (rotational)	$\hat{R}$	R		K_R, k_R
nuclear spin	$\hat{I}$	I	M_I	
internal vibrational				
spherical top	$\hat{l}$	$l(l\zeta)$		K_l
other	$\hat{j}, \hat{\pi}$			$l(l\zeta)$
sum of $R + L(+ j)$	$\hat{N}$	N		K, k
sum of $N + S$	$\hat{J}$	J	M_J	K, k
sum of $J + I$	$\hat{F}$	F	M_F	

Electromagnetic Radiation

Name	Symbol	Definition	SI unit
wavelength	λ		m
speed of light			
in vacuum	c_0		$m\,s^{-1}$
in a medium	c	$c = c_0/n$	$m\,s^{-1}$
wavenumber in vacuum	$\tilde{v}$	$\tilde{v} = v/c_0 = 1/n\lambda$	m^{-1}
wavenumber (in a medium)	σ	$\sigma = 1/\lambda$	m^{-1}
frequency	v	$v = c/\lambda$	Hz
circular frequency, pulsatance	ω	$\omega = 2\pi v$	s^{-1}, $rad\,s^{-1}$
refractive index	n	$n = c_0/c$	1
Planck constant	h		J s

Name	Symbol	Definition	SI unit
Planck constant/2π	$\hbar$	$\hbar = h/2\pi$	J s
radiant energy	Q, W		J
radiant energy density	ρ, w	$\rho = Q/V$	J m^{-3}
spectral radiant energy density			
in terms of frequency	ρ_ν, w_ν	$\rho = d\rho/d\nu$	J m^{-3} Hz^{-1}
in terms of wavenumber	$\rho_{\tilde{\nu}}, w_{\tilde{\nu}}$	$\rho_{\tilde{\nu}} = d\rho/d\tilde{\nu}$	J m^{-2}
in terms of wavelength	ρ_λ, w_λ	$\rho_\lambda = d\rho/d\lambda$	J m^{-4}
Einstein transition probabilities			
spontaneous emission	A_{nm}	$dN_n/dt = -A_{nm}N_n$	s^{-1}
stimulated emission	B_{nm}	$dN_n/dt = -\rho_{\tilde{\nu}}(\tilde{\nu}_{nm}) \times B_{nm}N_n$	s kg^{-1}
stimulated absorption	B_{mn}	$dN_n/dt = -\rho_{\tilde{\nu}}(\tilde{\nu}_{nm}) B_{mn}N_m$	s kg^{-1}
radiant power, radiant energy per time	Φ, P	$\Phi = dQ/dt$	W
radiant intensity	I	$I = d\Phi/d\Omega$	W sr^{-1}
radiant exitance (emitted radiant flux)	M	$M = d\Phi/dA_{source}$	W m^{-2}
irradiance, (radiant flux received)	$E, (I)$	$E = d\Phi/dA$	W m^{-2}
emittance	ε	$\varepsilon = M/M_{bb}$	1
Stefan–Boltzmann constant	σ	$M_{bb} = \sigma T^4$	W m^{-2} K^{-4}
first radiation constant	c_1	$c_1 = 2\pi h c_0^2$	W m^2
second radiation constant	c_2	$c_2 = hc_0/k$	K m
transmittance, transmission factor	τ, T	$\tau = \Phi_{tr}/\Phi_0$	1
absorptance, absorption factor	α	$\alpha = \Phi_{abs}/\Phi_0$	1
reflectance, reflection factor	ρ	$\rho = \Phi_{refl}/\Phi_0$	1
(decadic) absorbance	A	$A = -\lg(1 - \alpha_i)$	1
napierian absorbance	B	$B = -\ln(1 - \alpha_i)$	1
absorption coefficient			
(linear) decadic	a, K	$a = A/l$	m^{-1}
(linear) napierian	α	$\alpha = B/l$	m^{-1}
molar (decadic)	ε	$\varepsilon = a/c = A/cl$	m^2 mol^{-1}
molar napierian	κ	$\kappa = \alpha/c = B/cl$	m^2 mol^{-1}
absorption index	k	$k = \alpha/4\pi\tilde{\nu}$	1
complex refractive index	$\hat{n}$	$\hat{n} = n + ik$	1
molar refraction	R, R_m	$R = \dfrac{(n^2-1)}{(n^2+2)}V_m$	m^3 mol^{-1}
angle of optical rotation	α		1, rad

Solid State

Name	Symbol	Definition	SI unit
lattice vector	$\boldsymbol{R}, \boldsymbol{R}_0$		m
fundamental translation vectors for the crystal lattice	$\boldsymbol{a}_1; \boldsymbol{a}_2; \boldsymbol{a}_3, \boldsymbol{a}; \boldsymbol{b}; \boldsymbol{c}$	$\boldsymbol{R} = n_1\boldsymbol{a}_1 + n_2\boldsymbol{a}_2 + n_3\boldsymbol{a}_3$	m
(circular) reciprocal lattice vector	$\boldsymbol{G}$	$\boldsymbol{G} \cdot \boldsymbol{R} = 2\pi m$	m^{-1}
(circular) fundamental translation vectors for the reciprocal lattice	$\boldsymbol{b}_1; \boldsymbol{b}_2; \boldsymbol{b}_3, \boldsymbol{a}^*; \boldsymbol{b}^*; \boldsymbol{c}^*$	$\boldsymbol{a}_i \cdot \boldsymbol{b}_k = 2\pi\delta_{ik}$	m^{-1}
lattice plane spacing	d		m
Bragg angle	θ	$n\lambda = 2d\sin\theta$	1, rad
order of reflection	n		1
order parameters			
short range	σ		1
long range	s		1
Burgers vector	$\boldsymbol{b}$		m
particle position vector	$\boldsymbol{r}, \boldsymbol{R}_j$		m
equilibrium position vector of an ion	$\boldsymbol{R}_0$		m
displacement vector of an ion	$\boldsymbol{u}$	$\boldsymbol{u} = \boldsymbol{R} - \boldsymbol{R}_0$	m
Debye–Waller factor	B, D		1
Debye circular wavenumber	q_D		m^{-1}
Debye circular frequency	ω_D		s^{-1}
Grüneisen parameter	γ, Γ	$\gamma = \alpha V/\kappa C_v$	1
Madelung constant	$\alpha, \mathcal{M}$	$E_{coul} = \dfrac{\alpha N_A z_+ z_- e^2}{4\pi\varepsilon_0 R_0}$	1
density of states	N_E	$N_E = dN(E)/dE$	J^{-1} m^{-3}
(spectral) density of vibrational modes	N_ω, g	$N_\omega = dN(\omega)/d\omega$	s m^{-3}

Name	Symbol	Definition	SI unit
resistivity tensor	ρ_{ik}	$\boldsymbol{E} = \boldsymbol{\rho} \cdot \boldsymbol{j}$	Ω m
conductivity tensor	σ_{ik}	$\sigma = \rho^{-1}$	S m^{-1}
thermal conductivity tensor	λ_{ik}	$\boldsymbol{J}_q = -\lambda \cdot \text{grad } T$	W m^{-1} K^{-1}
residual resistivity	ρ_R		Ω m
relaxation time	τ	$\tau = l/v_F$	s
Lorenz coefficient	L	$L = \lambda/\sigma T$	V^2 K^{-2}
Hall coefficient	A_H, R_H	$\boldsymbol{E} = \boldsymbol{\rho} \cdot \boldsymbol{j} + R_H(\boldsymbol{B} \times \boldsymbol{j})$	m^3 C^{-1}
thermoelectric force	E		V
Peltier coefficient	Π		V
Thomson coefficient	$\mu, (\tau)$		V K^{-1}
work function	Φ	$\Phi = E_\infty - E_F$	J
number density, number concentration	$n, (p)$		m^{-3}
gap energy	E_g		J
donor ionization energy	E_d		J
acceptor ionization energy	E_a		J
Fermi energy	E_F, ε_F		J
circular wave vector, propagation vector	$k, \boldsymbol{q}$	$k = 2\pi/\lambda$	m^{-1}
Bloch function	$u_k(\boldsymbol{r})$	$\psi(\boldsymbol{r}) = u_k(\boldsymbol{r}) \exp(\mathrm{i}\boldsymbol{k} \cdot \boldsymbol{r})$	m$^{-3/2}$
charge density of electrons	ρ	$\rho(\boldsymbol{r}) = -e\psi^*(\boldsymbol{r})\psi(\boldsymbol{r})$	C m^{-3}
effective mass	m^*		kg
mobility	μ	$\mu = v_{\text{drift}}/E$	m^2 V^{-1} s^{-1}
mobility ratio	b	$b = \mu_n/\mu_p$	1
diffusion coefficient	D	$dN/dt = -DA(dn/dx)$	m^2 s^{-1}
diffusion length	L	$L = \sqrt{D\tau}$	m
characteristic (Weiss) temperature	θ, θ_w		K
Curie temperature	T_C		K
Néel temperature	T_N		K

Statistical Thermodynamics

Name	Symbol	Definition	SI unit
number of entities	N		1
number density of entities, number concentration	n, C	$n = N/V$	m^{-3}
Avogadro constant	L, N_A		mol^{-1}
Boltzmann constant	k, k_B		J K^{-1}
gas constant (molar)	R	$R = Lk$	J K^{-1} mol^{-1}
molecular position vector	$\boldsymbol{r}\,(x, y, z)$		m
molecular velocity vector	$\boldsymbol{c}(c_x, c_y, c_z), \boldsymbol{u}(u_x, u_y, u_z)$	$\boldsymbol{c} = d\boldsymbol{r}/dt$	m s^{-1}
molecular momentum vector	$\boldsymbol{p}(p_x, p_y, p_z)$	$\boldsymbol{p} = m\boldsymbol{c}$	kg m s^{-1}
velocity distribution function (Maxwell)	$f(c_x)$	$f(c_x) = (m/2\pi kT)^{1/2} \times \exp(-mc_x^2/2kT)$	m^{-1} s
speed distribution function (Maxwell–Boltzmann)	$F(c)$	$F(c) = (m/2\pi kT)^{3/2} \times 4\pi c^2\exp(-mc^2/2kT)$	m^{-1} s
average speed	$\overline{c}, \overline{u}, \langle c\rangle, \langle u\rangle$	$\overline{c} = \int cF(c)dc$	m s^{-1}
generalized coordinate	q		(m)
generalized momentum	p	$p = \partial L/\partial \dot{q}$	(kg m s^{-1})
volume in phase space	Ω	$\Omega = (1/h)\int p dq$	1
probability	P		1
statistical weight, degeneracy	g, d, W, ω, β		1
density of states	$\rho(E)$	$\rho(E) = dN/dE$	J^{-1}
partition function, sum over states, for a single molecule	q, z	$q = \sum_i g_i\exp(-\varepsilon_i/kT)$	1
for a canonical ensemble (system, or assembly)	Q, Z		1
microcanonical ensemble	Ω		1
grand (canonical ensemble)	Ξ		1
symmetry number	σ, s		1
reciprocal temperature parameter	β	$\beta = 1/kT$	J^{-1}
characteristic temperature	Θ		K

Name	Symbol	Definition	SI unit
General Chemistry			
number of entities (e.g. molecules, atoms, ions, formula units)	N		1
amount (of substance)	n	$n_B = N_B/L$	mol
Avogadro constant	L, N_A		mol^{-1}
mass of atom, atomic mass	m_a, m		kg
mass of entity (molecule, or formula unit)	m_f, m		kg
atomic mass constant	m_u	$m_u = m_a(^{12}C)/12$	kg
molar mass	M	$M_B = m/n_B$	$kg\ mol^{-1}$
relative molecular mass (relative molar mass, molecular weight)	M_r	$M_{r,B} = m_B/m_u$	1
molar volume	V_m	$V_{m,B} = V/n_B$	$m^3\ mol^{-1}$
mass fraction	w	$w_B = m_B/\Sigma m_i$	1
volume fraction	ϕ	$\phi_B = V_B/\Sigma V_i$	1
mole fraction, amount fraction, number fraction	x, y	$x_B = n_B/\Sigma n_i$	1
(total) pressure	p, P		Pa
partial pressure	p_B	$p_B = y_B p$	Pa
mass concentration (mass density)	γ, ρ	$\gamma_B = m_B/V$	$kg\ m^{-3}$
number concentration, number density of entities	C, n	$C_B = N_B/V$	m^{-3}
amount concentration, concentration	c	$c_B = n_B/V$	$mol\ m^{-3}$
solubility	s	$s_B = c_B$ (saturated solution)	$mol\ m^{-3}$
molality (of a solute)	$m, (b)$	$m_B = n_B/m_A$	$mol\ kg^{-1}$
surface concentration	Γ	$\Gamma_B = n_B/A$	$mol\ m^{-2}$
stoichiometric number	ν		1
extent of reaction, advancement	ξ	$\Delta\xi = \Delta n_B/\nu_B$	mol
degree of dissociation	α		1
Chemical Thermodynamics			
heat	q, Q		J
work	w, W		J
internal energy	U	$\Delta U = q + w$	J
enthalpy	H	$H = U + pV$	J
thermodynamic temperature	T		K
Celsius temperature	θ, t	$\theta/°C = T/K - 273.15$	°C
entropy	S	$dS \geq dq/T$	$J\ K^{-1}$
Helmholtz energy (Helmholtz function)	A	$A = U - TS$	J
Gibbs energy (Gibbs function)	G	$G = H - TS$	J
Massieu function	J	$J = -A/T$	$J\ K^{-1}$
Planck function	Y	$Y = -G/T$	$J\ K^{-1}$
surface tension	γ, σ	$\gamma = (\partial G/\partial A_s)_{T,p}$	$J\ m^{-2}, N\ m^{-1}$
molar quantity X	X_m	$X_m = X/n$	(varies)
specific quantity X	x	$x = X/m$	(varies)
pressure coefficient	β	$\beta = (\partial p/\partial T)_V$	$Pa\ K^{-1}$
relative pressure coefficient	α_p	$\alpha_p = (1/p)(\partial p/\partial T)_V$	K^{-1}
compressibility,			
isothermal	κ_T	$\kappa_T = -(1/V)(\partial V/\partial p)_T$	Pa^{-1}
isentropic	κ_S	$\kappa_S = -(1/V)(\partial V/\partial p)_S$	Pa^{-1}
linear expansion coefficient	α_l	$\alpha_l = (1/l)(\partial l/\partial T)$	K^{-1}
cubic expansion coefficient	α, α_V, γ	$\alpha = (1/V)(\partial V/\partial T)_p$	K^{-1}
heat capacity,			
at constant pressure	C_p	$C_p = (\partial H/\partial T)_p$	$J\ K^{-1}$
at constant volume	C_V	$C_V = (\partial U/\partial T)_V$	$J\ K^{-1}$
ratio of heat capacities	$\gamma, (\kappa)$	$\gamma = C_p/C_V$	1
Joule–Thomson coefficient	μ, μ_{JT}	$\mu = (\partial T/\partial p)_H$	$K\ Pa^{-1}$
second virial coefficient	B	$pV_m = RT(1 + B/V_m + ...)$	$m^3\ mol^{-1}$
compression factor (compressibility factor)	Z	$Z = pV_m/RT$	1
partial molar quantity X	$X_B, (X_B')$	$X_B = (\partial X/\partial n_B)_{T,p,n_{j \neq B}}$	(varies)
chemical potential (partial molar Gibbs energy)	μ	$\mu_B = (\partial G/\partial n_B)_{T,p,n_{j \neq B}}$	$J\ mol^{-1}$
absolute activity	λ	$\lambda_B = \exp(\mu_B/RT)$	1

Name	Symbol	Definition	SI unit
standard chemical potential	$\mu^{\ominus}, \mu^{o}$		J mol⁻¹
standard partial molar enthalpy	$H_B^{\ominus}$	$H_B^{\ominus} = \mu_B^{\ominus} + TS_B^{\ominus}$	J mol⁻¹
standard partial molar entropy	$S_B^{\ominus}$	$S_B^{\ominus} = -(\partial\mu_B^{\ominus}/\partial T)_p$	J mol⁻¹ K⁻¹
standard reaction Gibbs energy (function)	$\Delta_r G^{\ominus}$	$\Delta_r G^{\ominus} = \sum_B \nu_B \mu_B^{\ominus}$	J mol⁻¹
affinity of reaction	$A, (\mathcal{A})$	$A = -(\partial G/\partial\xi)_{p,T} = -\sum_B \nu_B \mu_B$	J mol⁻¹
standard reaction enthalpy	$\Delta_r H^{\ominus}$	$\Delta_r H^{\ominus} = \sum_B \nu_B H_B^{\ominus}$	J mol⁻¹
standard reaction entropy	$\Delta_r S^{\ominus}$	$\Delta_r S^{\ominus} = \sum_B \nu_B S_B^{\ominus}$	J mol⁻¹ K⁻¹
equilibrium constant	$K^{\ominus}, K$	$K^{\ominus} = \exp(-\Delta_r G^{\ominus}/RT)$	1
equilibrium constant,			
pressure basis	K_p	$K_p = \prod_B p_B^{\nu_B}$	Pa$^{\Sigma\nu}$
concentration basis	K_c	$K_c = \prod_B c_B^{\nu_B}$	(mol m⁻³)$^{\Sigma\nu}$
molality basis	K_m	$K_m = \prod_B m_B^{\nu_B}$	(mol kg⁻¹)$^{\Sigma\nu}$
fugacity	$f, \tilde{p}$	$f_B = \lambda_B \lim\limits_{p\to 0}(p_B/\lambda_B)_T$	Pa
fugacity coefficient	ϕ	$\phi_B = f_B/p_B$	1
activity and activity coefficient referenced to Raoult's law, (relative) activity	a	$a_B = \exp\left[\dfrac{\mu_B - \mu_B^{*}}{RT}\right]$	1
activity coefficient	f	$f_B = a_B/x_B$	1
activities and activity coefficients referenced to Henry's law, (relative) activity,			
molality basis	a_m	$a_{m,B} = \exp\left[\dfrac{\mu_B - \mu_B^{\ominus}}{RT}\right]$	1
concentration basis	a_c	$a_{c,B} = \exp\left[\dfrac{\mu_B - \mu_B^{\ominus}}{RT}\right]$	1
mole fraction basis	a_x	$a_{x,B} = \exp\left[\dfrac{\mu_B - \mu_B^{\ominus}}{RT}\right]$	1
activity coefficient,			
molality basis	γ_m	$a_{m,B} = \gamma_{m,B} m_B/m^{\ominus}$	1
concentration basis	γ_c	$a_{c,B} = \gamma_{c,B} c_B/c^{\ominus}$	1
mole fraction basis	γ_x	$a_{x,B} = \gamma_{x,B} x_B$	1
ionic strength,			
molality basis	I_m, I	$I_m = \frac{1}{2}\Sigma m_B z_B^2$	mol kg⁻¹
concentration basis	I_c, I	$I_c = \frac{1}{2}\Sigma c_B z_B^2$	mol m⁻³
osmotic coefficient,			
molality basis	ϕ_m	$\phi_m = (\mu_A^{*} - \mu_A)/(RTM_A\Sigma m_B)$	1
mole fraction basis	ϕ_x	$\phi_x = (\mu_A - \mu_A^{*})/(RT\ln x_A)$	1
osmotic pressure	Π	$\Pi = c_B RT$ (ideal dilute solution)	Pa

(i) Symbols used as subscripts to denote a chemical process or reaction

These symbols should be printed in roman (upright) type, without a full stop (period).

vaporization, evaporation (liquid → gas)	vap
sublimation (solid → gas)	sub
melting, fusion (solid → liquid)	fus
transition (between two phases)	trs
mixing of fluids	mix
solution (of solute in solvent)	sol
dilution (of a solution)	dil
adsorption	ads
displacement	dpl
immersion	imm

reaction in general	r
atomization	at
combustion reaction	c
formation reaction	f

(ii) Recommended superscripts

standard	$\ominus$, o
pure substance	*
infinite dilution	∞
ideal	id
activated complex, transition state	$\ddagger$
excess quantity	E

Name	Symbol	Definition	SI unit
Chemical Kinetics			
rate of change of quantity X	$\dot{X}$	$\dot{X} = dX/dt$	(varies)
rate of conversion	$\dot{\xi}$	$\dot{\xi} = d\xi/dt$	mol s^{-1}
rate of concentration change (due to chemical reaction)	r_B, v_B	$r_B = dc_B/dt$	mol m^{-3} s^{-1}
rate of reaction (based on amount concentration)	v	$v = \dot{\xi}/V = v_B^{-1}dc_B/dt$	mol m^{-3} s^{-1}
partial order of reaction	n_B	$v = k\Pi c_B{}^{n_B}$	1
overall order of reaction	n	$n = \Sigma n_B$	1
rate constant, rate coefficient	k	$v = k\Pi c_B{}^{n_B}$	(mol^{-1} m^3)$^{n-1}$ s^{-1}
Boltzmann constant	k, k_B		J K^{-1}
half life	$t_{1/2}$	$c(t_{1/2}) = c_0/2$	s
relaxation time	τ	$\tau = 1/(k_1 + k_{-1})$	s
energy of activation, activation energy	E_a, E	$E_a = RT^2$ d ln k/dT	J mol^{-1}
pre-exponential factor	A	$k = A \exp(-E_a/RT)$	(mol^{-1} m^3)$^{n-1}$ s^{-1}
volume of activation	$\Delta^{\ddagger}V$	$\Delta^{\ddagger}V = -RT \times (\partial\ln k/\partial p)_T$	m^3 mol^{-1}
collision diameter	d	$d_{AB} = r_A + r_B$	m
collision cross-section	σ	$\sigma_{AB} = \pi d_{AB}{}^2$	m^2
collision frequency	Z_A		s^{-1}
collision number	Z_{AB}, Z_{AA}		m^{-3} s^{-1}
collision frequency factor	z_{AB}, z_{AA}	$z_{AB} = Z_{AB}/Lc_A c_B$	m^3 mol^{-1} s^{-1}
standard enthalpy of activation	$\Delta^{\ddagger}H^{\ominus}, \Delta H^{\ddagger}$		J mol^{-1}
standard entropy of activation	$\Delta^{\ddagger}S^{\ominus}, \Delta S^{\ddagger}$		J mol^{-1} K^{-1}
standard Gibbs energy of activation	$\Delta^{\ddagger}G^{\ominus}, \Delta G^{\ddagger}$		J mol^{-1}
quantum yield, photochemical yield	ϕ		1
Electrochemistry			
elementary charge (proton charge)	e		C
Faraday constant	F	$F = eL$	C mol^{-1}
charge number of an ion	z	$z_B = Q_B/e$	1
ionic strength	I_c, I	$I_c = \frac{1}{2}\Sigma c_i z_i^2$	mol m^{-3}
mean ionic activity	$a_{\pm}$	$a_{\pm} = m_{\pm}\gamma_{\pm}/m^{\ominus}$	1
mean ionic molality	$m_{\pm}$	$m_{\pm}{}^{(v_+ + v_-)} = m_+{}^{v_+}m_-{}^{v_-}$	mol kg^{-1}
mean ionic activity coefficient	$\gamma_{\pm}$	$\gamma_{\pm}{}^{(v_+ + v_-)} = \gamma_+{}^{v_+}\gamma_-{}^{v_-}$	1
charge number of electrochemical cell reaction	$n, (z)$		1
electric potential difference (of a galvanic cell)	$\Delta V, E, U$	$\Delta V = V_R - V_L$	V
emf, electromotive force	E	$E = \lim_{I \to 0} \Delta V$	V
standard emf, standard potential of the electrochemical cell reaction	$E^{\ominus}$	$E^{\ominus} = -\Delta_r G^{\ominus}/nF = (RT/nF)\ln K^{\ominus}$	V
standard electrode potential	$E^{\ominus}$		V
emf of the cell, potential of the electrochemical cell reaction	E	$E = E^{\ominus} - (RT/nF) \times \Sigma v_i \ln a_i$	V
pH	pH	$\text{pH} \approx -\lg\left[\dfrac{c(\mathbf{H^+})}{\text{mol dm}^{-3}}\right]$	1
inner electric potential	ϕ	$\nabla\phi = -\mathbf{E}$	V
outer electric potential	ψ	$\psi = Q/4\pi\varepsilon_0 r$	V

Name	Symbol	Definition	SI unit
surface electric potential	χ	$\chi = \phi - \psi$	V
Galvani potential difference	$\Delta\phi$	$\Delta_\alpha^{\,\beta}\phi = \phi^\beta - \phi^\alpha$	V
volta potential difference	$\Delta\psi$	$\Delta_\alpha^{\,\beta}\psi = \psi^\beta - \psi^\alpha$	V
electrochemical potential	$\tilde{\mu}$	$\tilde{\mu}_B^{\,\alpha} = (\partial G/\partial n_B^{\,\alpha})$	J mol^{-1}
electric current	I	$I = dQ/dt$	A
(electric) current density	j	$j = I/A$	A m^{-2}
(surface) charge density	σ	$\sigma = Q/A$	C m^{-2}
electrode reaction rate constant	k	$k_{ox} = I_a/(nFA\prod_i c_i^{\,n_i})$	(varies)
mass transfer coefficient, diffusion rate constant	k_d	$k_{d,B} = \lvert\nu_B\rvert I_{l,B}/nFcA$	m s^{-1}
thickness of diffusion layer	δ	$\delta_B = D_B/k_{d,B}$	m
transfer coefficient (electrochemical)	α	$\alpha_c = \dfrac{-\lvert\nu\rvert RT\partial}{nF}\dfrac{\partial \ln\lvert I_c\rvert}{\partial E}$	1
overpotential	η	$\eta = E_I - E_{I=0} - IR_u$	V
electrokinetic potential (zeta potential)	ζ		V
conductivity	$\kappa, (\sigma)$	$\kappa = j/E$	S m^{-1}
conductivity cell constant	K_{cell}	$K_{cell} = \kappa R$	m^{-1}
molar conductivity (of an electrolyte)	Λ	$\Lambda_B = \kappa/c_B$	S m^2 mol^{-1}
ionic conductivity, molar conductivity of an ion	λ	$\lambda_B = \lvert z_B\rvert Fu_B$	S m^2 mol^{-1}
electric mobility	$u, (\mu)$	$u_B = \nu_B/E$	m^2 V^{-1} s^{-1}
transport number	t	$t_B = j_B/\Sigma j_i$	1
reciprocal radius of ionic atmosphere	κ	$\kappa = (2F^2I/\varepsilon RT)^{1/2}$	m^{-1}

Colloid and Surface Chemistry

Name	Symbol	Definition	SI unit
specific surface area	a, a_s, s	$a = A/m$	m^2 kg^{-1}
surface amount of B, adsorbed amount of B	$n_B^{\,s}, n_B^{\,a}$		mol
surface excess of B	$n_B^{\,\sigma}$		mol
surface excess concentration of B	$\Gamma_B, (\Gamma_B^{\,\sigma})$	$\Gamma_B = n_B^{\,\sigma}/A$	mol m^{-2}
total surface excess concentration	$\Gamma, (\Gamma^\sigma)$	$\Gamma = \sum_i \Gamma_i$	mol m^{-2}
area per molecule	a, σ	$a_B = A/N_B^{\,\sigma}$	m^2
area per molecule in a filled monolayer	a_m, σ_m	$a_{m,B} = A/N_{m,B}$	m^2
surface coverage	θ	$\theta = N_B^{\,\sigma}/N_{m,B}$	1
contact angle	θ		1, rad
film thickness	t, h, δ		m
thickness of (surface or interfacial) layer	τ, δ, t		m
surface tension, interfacial tension	γ, σ	$\gamma = (\partial G/\partial A_s)_{T,p}$	N m^{-1}, J m^{-2}
film tension	Σ_f	$\Sigma_f = 2\gamma_f$	N m^{-1}
reciprocal thickness of the double layer	κ	$\kappa = [2F^2I_c/\varepsilon RT]^{1/2}$	m^{-1}
average molar masses			
number–average	M_n	$M_n = \Sigma n_i M_i/\Sigma n_i$	kg mol^{-1}
mass–average	M_m	$M_m = \Sigma n_i M_i^2/\Sigma n_i M_i$	kg mol^{-1}
Z–average	M_Z	$M_Z = \Sigma n_i M_i^3/\Sigma n_i M_i^2$	kg mol^{-1}
sedimentation coefficient	s	$s = \nu/a$	s
van der Waals constant	λ		J
retarded van der Waals constant	β, B		J
van der Waals–Hamaker constant	A_H		J
surface pressure	π^s, π	$\pi^s = \gamma^0 - \gamma$	N m^{-1}

Transport Properties

Name	Symbol	Definition	SI unit
flux (of a quantity X)	J_X, J	$J_X = A^{-1}\,dX/dt$	(varies)
volume flow rate	$q_V, \dot{V}$	$q_v = dV/dt$	m^3 s^{-1}
mass flow rate	$q_m, \dot{m}$	$q_m = dm/dt$	kg s^{-1}
mass transfer coefficient	k_d		m s^{-1}
heat flow rate	ϕ	$\phi = dq/dt$	W
heat flux	J_q	$J_q = \phi/A$	W m^{-2}
thermal conductance	G	$G = \phi/\Delta T$	W K^{-1}
thermal resistance	R	$R = 1/G$	K W^{-1}
thermal conductivity	λ, k	$\lambda = J_q/(dT/dl)$	W m^{-1} K^{-1}

Name	Symbol	Definition	SI unit
coefficient of heat transfer	$h, (k, K, \alpha)$	$h = J_q/\Delta T$	$\mathrm{W\ m^{-2}\ K^{-1}}$
thermal diffusivity	a	$a = \lambda/\rho c_p$	$\mathrm{m^2\ s^{-1}}$
diffusion coefficient	D	$D = J_n/(dc/dl)$	$\mathrm{m^2\ s^{-1}}$

The following symbols are used in the definitions of the dimensionless quantities: mass (m), time (t), volume (V), area (A), density (ρ), speed (v), length (l), viscosity (η), pressure (p), acceleration of free fall (g), cubic expansion coefficient (α), temperature (T), surface tension (γ), speed of sound (c), mean free path (λ), frequency (f), thermal diffusivity (a), coefficient of heat transfer (h), thermal conductivity (k), specific heat capacity at constant pressure (c_p), diffusion coefficient (D), mole fraction (x), mass transfer coefficient (k_d), permeability (μ), electric conductivity (κ), and magnetic flux density (B).

Name	Symbol	Definition	SI unit
Reynolds number	Re	$Re = \rho vl/\eta$	1
Euler number	Eu	$Eu = \Delta p/\rho v^2$	1
Froude number	Fr	$Fr = v/(lg)^{1/2}$	1
Grashof number	Gr	$Gr = l^3 g\alpha\Delta T\rho^2/\eta^2$	1
Weber number	We	$We = \rho v^2 l/\gamma$	1
Mach number	Ma	$Ma = v/c$	1
Knudsen number	Kn	$Kn = \lambda/l$	1
Strouhal number	Sr	$Sr = lf/v$	1
Fourier number	Fo	$Fo = at/l^2$	1
Péclet number	Pe	$Pe = vl/a$	1
Rayleigh number	Ra	$Ra = l^3 g\alpha\Delta T\rho/\eta a$	1
Nusselt number	Nu	$Nu = hl/k$	1
Stanton number	St	$St = h/\rho vc_p$	1
Fourier number for mass transfer	Fo^*	$Fo^* = Dt/l^2$	1
Péclet number for mass transfer	Pe^*	$Pe^* = vl/D$	1
Grashof number for mass transfer	Gr^*	$Gr^* = l^3 g\left(\dfrac{\partial p}{\partial x}\right)_{T,p}\left(\dfrac{\Delta xp}{\eta}\right)$	1
Nusselt number for mass transfer	Nu^*	$Nu^* = k_d l/D$	1
Stanton number for mass transfer	St^*	$St^* = k_d/v$	1
Prandtl number	Pr	$Pr = \eta/\rho a$	1
Schmidt number	Sc	$Sc = \eta/\rho D$	1
Lewis number	Le	$Le = a/D$	1
magnetic Reynolds number	Rm, Re_m	$Rm = v\mu\kappa l$	1
Alfvén number	Al	$Al = v(\rho\mu)^{1/2}/B$	1
Hartmann number	Ha	$Ha = Bl\,(\kappa/\eta)^{1/2}$	1
Cowling number	Co	$Co = B^2/\mu\rho v^2$	1

EXPRESSION OF UNCERTAINTY OF MEASUREMENTS

In general, the result of a measurement is only an approximation or estimate of the true value of the quantity subject to measurement, and thus the result is of limited value unless accompanied by a statement of its uncertainty. Much (but not all) of the scientific data appearing in the literature does include some indication of the uncertainty, but this may be stated in many different ways and is often explained poorly. In an effort to encourage consistency in uncertainty statements, the International Committee for Weights and Measures (CIPM) initiated a project, in collaboration with several other international organizations, to prepare a set of guidelines expressing international consensus on the recommended method of stating uncertainties. This project resulted in the publication of the *Guide to the Expression of Uncertainty in Measurement* (Reference 1), which is often referred to as *GUM*. The recommendations of *GUM* have been summarized by the National Institute of Standards and Technology in *NIST Technical Note 1297, Guidelines for Evaluating the Uncertainty of NIST Measurement Results* (Reference 2).

In the notation of *GUM*, we are concerned with the **measurand**, i.e., the quantity that is being measured. In physics and chemistry this is usually called a **physical quantity** and represents some inherent characteristic of a material, system, or process that can be expressed in numerical terms — specifically as the product of a number and a reference, commonly called a **unit.** Thus the density of water at room temperature is (approximately) 0.998 g/mL (grams per milliliter) or, alternatively 998 kg m^{-3} (kilograms per meter cubed). This statement gives the most likely value of the measurand, to this level of precision, but gives no information on how much the stated value might differ from the true value.

It is important to differentiate between the terms **error** and **uncertainty**. The error in a measurement is the difference between the measured value and the true value; the error can be stated if the true value is known (to some level of accuracy). The uncertainty is an estimate of the maximum reasonable extent to which the measured value is believed to deviate from the true value, in a situation where the true value is not known (most often the case). The result of a measurement can unknowably be very close to the true value, and thus have negligible error, even though its uncertainty is large.

The uncertainty of the result of a measurement generally consists of several components, which may be grouped in two types according to the method used to estimate their numerical values:

Type A. Those which are evaluated by statistical methods
Type B. Those which are evaluated by other means

The terms "random uncertainty" and "systematic uncertainty" are often used, but these terms do not always correspond in a simple way to the A and B categories. This is because the nature of an uncertainty component is conditioned by how the quantity appears in the mathematical model that describes the current measurement process. An uncertainty component arising from a systematic effect may in some cases be evaluated by methods of Type A while in other cases by methods of Type B.

In the *GUM* formulation, each component of uncertainty, whether in the A or B category, is represented by an estimated standard deviation, termed **standard uncertainty**, symbol u_i, and equal to the positive square root of the estimated variance u_i^2.

For an uncertainty component of Type A, $u_i = s_i$, where s_i is the statistically estimated standard deviation, as determined from a series of observations by appropriate statistical analysis. Any valid statistical method may be used. Examples are calculating the standard deviation of the mean of a series of independent observations; using the method of least squares to fit a curve to data in order to estimate parameters of the curve and their standard deviations; and carrying out an analysis of variance (ANOVA) in order to identify and quantify random effects in certain types of measurements. Details of statistical analysis are given in References 4–7 and many other places.

In a similar manner, each uncertainty component of Type B is represented by a quantity u_j, which is obtained from an assumed probability distribution based on all the available information about the measurement process. Since u_j is treated like a standard deviation, the standard uncertainty in each Type B component is simply u_j. The evaluation of u_j is usually based on scientific judgment using all the relevant information available, which may include

- previous measurement data
- experience with, or general knowledge of, the behavior and properties of relevant materials and instruments
- manufacturer's specifications
- data provided in calibrations and other reports
- uncertainties assigned to reference data taken from handbooks.

The specific approach to evaluating the standard uncertainty u_j of a Type B uncertainty will depend on the detailed model of the measurement process. The following are examples of steps that may be used:

1. Convert a quoted uncertainty (for example, in a calibration factor) that is a stated multiple of an estimated standard deviation to a standard uncertainty by dividing the quoted uncertainty by the multiplier.

2. Convert a quoted uncertainty that defines a "confidence interval" having a stated level of confidence, such as 95% or 99%, to a standard uncertainty by treating the quoted uncertainty as if a normal distribution had been used to calculate it (unless otherwise indicated) and dividing it by the appropriate factor for such a distribution. These factors are 1.960 and 2.576 for the two levels of confidence given.

3. Model knowledge of the quantity in question by a normal distribution and estimate lower and upper limits a_- and a_+ such that the best estimated value of the quantity is $(a_+ + a_-)/2$ (i.e., the midpoint of the limits) and there is 1 chance out of 2 (i.e., a 50 percent probability) that the value of the quantity lies in the interval a_- to a_+. Then $u_j \approx 1.48\,a$, where $a = (a_+ - a_-)/2$ is the half-width of the interval.

4. Model knowledge of the quantity in question by a normal distribution and estimate lower and upper limits a_- and a_+ such that the best estimated value of the quantity is $(a_+ + a_-)/2$ and there is about a 2 out of 3 chance (i.e., a 67 percent probability) that the value of the quantity lies in the interval a_- to a_+. Then $u_j \approx a$, where $a = (a_+ - a_-)/2$.

5. Estimate lower and upper limits a_- and a_+ for the value of the quantity in question such that the probability that the value lies in the interval a_- to a_+ is, for all practical purposes, 100 percent. Provided that there is no contradictory information, treat the quantity as if it is equally probable for its value to lie anywhere within the interval a_- to a_+;

that is, model it by a uniform or rectangular probability distribution. The best estimate of the value of the quantity is then $(a_+ + a_-)/2$ with $u_j = a/\sqrt{3}$ where $a = (a_+ - a_-)/2$. If the distribution used to model the quantity is triangular rather than rectangular, then $u_j = a/\sqrt{6}$. The rectangular distribution is a reasonable default model in the absence of any other information. But if it is known that values of the quantity in question near the center of the limits are more likely than values close to the limits, a triangular or a normal distribution may be a better model.

When all the standard uncertainties of Type A and Type B have been determined in this way, they should be combined to produce the **combined standard uncertainty** (suggested symbol u_c), which may be regarded as the estimated standard deviation of the measurement result. This process, often called the *law of propagation of uncertainty* or "root-sum-of-squares," involves taking the square root of the sum of the squares of all the u_i. In many practical measurement situations, the probability distribution characterized by the measurement result y and its combined standard uncertainty $u_c(y)$ is approximately normal (Gaussian). When this is the case, $u_c(y)$ defines an interval $y - u_c(y)$ to $y + u_c(y)$ about the measurement result y within which the value of the measurand Y estimated by y is believed to lie with a level of confidence of approximately 68 percent. That is, it is believed with an approximate level of confidence of 68 percent that $y - u_c(y) \leq Y \leq y + u_c(y)$, which is commonly written as $Y = y \pm u_c(y)$.

In fundamental metrological research (involving physical constants, calibration standards, and the like) the combined standard uncertainty u_c is normally used as the statement of uncertainty in a measurement. In most cases, however, it is desirable to use a measure of uncertainty that defines an interval about the measurement result y within which the value of the measurand Y is confidently believed to lie. The measure of uncertainty intended to meet this requirement is termed **expanded uncertainty,** suggested symbol U, and is obtained by multiplying $u_c(y)$ by a **coverage factor,** suggested symbol k. Thus $U = ku_c(y)$ and it is believed with high confidence that $y - U \leq Y \leq y + U$, which is commonly written as $Y = y \pm U$. The value of the coverage factor k is chosen on the basis of the desired level of confidence to be associated with the interval defined by $U = ku_c$. Typically, k is in the range 2 to 3. When the normal distribution applies, $U = 2u_c$ (i.e., $k = 2$) defines an interval having a level of confidence of approximately 95 percent, and $U = 3u_c$ defines an interval having a confidence level greater than 99 percent. In current international practice it is most common to use $k = 2$, corresponding to about 95 percent confidence, but the value of k should be stated in each case to avoid confusion. See References 1 and 2 for methods of calculating k when a value other than $k = 2$ is needed for a specific requirement.

Summary of Key Steps

- Group the uncertainty components into Type A (can be evaluated by statistical methods) and Type B (must be evaluated by other means).
- Determine the standard uncertainty for each component of Type A by statistical methods and for each component of Type B by other suitable methods, based on modeling the measurement process.
- Take the square root of the sum of the squares of all the standard uncertainties to get the combined standard uncertainty u_c.
- Specify a coverage factor k which, when multiplied by u_c, gives the expanded uncertainty U. In fundamental metrological research $k = 1$ is usually chosen; in other cases, $k = 2$ (corresponding to a confidence level of about 95%) is the most common choice.

References

1. ISO, *Guide to the Expression of Uncertainty in Measurement*, International Organization for Standardization, Geneva, Switzerland, 1993. Several supplements have been published; see Bich, W., Cox, M. C., and Harris, P. M., "Evolution of the *Guide to the Expression of Uncertainty in Measurement*," *Metrologia* 43, S161, 2006.
2. Taylor, B. N., and Kuyatt, C. E., *Guidelines for Evaluating and Expressing the Uncertainty of NIST Measurement Results*, NIST Technical Note 1297, National Institute of Standards and Technology, Gaithersburg, MD, 1994; available for free download at <physics.nist.gov/cuu/Uncertainty/bibliography.html>.
3. Bell, S., *A Beginner's Guide to Uncertainty of Measurement*, National Physical Laboratory, Teddington, Middlesex, UK, 2001; available on the Internet through <www.npl.co.uk/server.php?show=ConWebDoc.1785>.
4. Eisenhart, C., "Realistic Evaluation of the Precision and Accuracy of Instrument Calibration Systems," *J. Res. Natl. Bur. Stand.* (U.S.) 67C, 161, 1963.
5. Mandel, J., *The Statistical Analysis of Experimental Data*, Dover Publishers, New York, 1984.
6. Nantrella, M. G., *Experimental Statistics*, NBS Handbook 91, U.S. Government Printing Office, Washington, DC, 1966.
7. Box, G. E. P., Hunter, J. S., and Hunter, W. G., *Statistics for Experimenters: Design, Innovation, and Discovery, 2nd Edition*, John Wiley & Sons, Hoboken, NJ, 2005.

NOMENCLATURE FOR CHEMICAL COMPOUNDS

The International Union of Pure and Applied Chemistry (IUPAC) maintains several commissions that deal with the naming of chemical substances. In general, the approach of IUPAC is to present rules for arriving at names in a systematic manner, rather than recommending a unique name for each compound. Thus there are often several alternative "IUPAC names," depending on which nomenclature system is used, each of which may have advantages in specific applications. However, each of these names will be unambiguous.

Organizations such as the Chemical Abstacts Service and the Beilstein Institute that prepare indexes to the chemical literature must adopt a system for selecting unique names in order to avoid excessive cross referencing. Chemical Abstracts Service uses a system which groups together compounds derived from a single parent compound. Thus most index names are inverted (e.g., Benzene, bromo rather than bromobenzene; Acetic acid, sodium salt rather than sodium acetate).

Recommended names for the most common substituent groups, ligands, ions, and organic rings are given in the two following tables, "Nomenclature for Inorganic Ions and Ligands" and "Organic Substituent Groups and Ring Systems." For the basics of macromolecular nomenclature, see "Nomenclature for Organic Polymers" in Section 13.

Some of the most useful recent guides to chemical nomenclature, prepared by IUPAC and other organizations such as the International Union of Biochemistry and Molecular Biology (IUBMB) and the American Chemical Society are listed below. These books contain citations to the more detailed nomenclature documents in each area. Two very useful web sites providing links to nomenclature documents are:

www.iupac.org/publications/index.html
www.chem.qmul.ac.uk/iupac/

Inorganic Chemistry

Block, B. P., Powell, W. H., and Fernelius, W. C., *Inorganic Chemical Nomenclature, Principles and Practice*, American Chemical Society, Washington, 1990.

Nomenclature of Inorganic Chemistry - IUPAC Recommendations 2005. Connelly, N.G., Damhus, T., Hartshorn, R. M., and Hutton, A. T., The Royal Society of Chemistry, 2005.

Organic Chemistry

International Union of Pure and Applied Chemistry, *A Guide to IUPAC Nomenclature of Organic Compounds, Recommendations 1993*, Panico, R., Powell, W. H., and Richer, J.-C., Eds., Blackwell Scientific Publications, Oxford, 1993.

International Union of Pure and Applied Chemistry, *Glossary of Class Names of Organic Compounds and Reactive Intermediates Based on Structure*, Moss, G. P., Smith, P. A. S., and Tavernier, D., Eds., *Pure & Appl. Chem.* 67, 1307, 1995.

International Union of Pure and Applied Chemistry, *Basic Terminology of Stereochemistry*, Moss, G. P., Ed., *Pure & Appl. Chem.* 68, 2193, 1996.

Rhodes, P. H., *The Organic Chemist's Desk Reference*, Chapman & Hall, London, 1995.

Macromolecular Chemistry

International Union of Pure and Applied Chemistry, *Compendium of Macromolecular Nomenclature*, Metanomski, W. V., Ed., Blackwell Scientific Publications, Oxford, 1991.

International Union of Pure and Applied Chemistry, *Glossary of Basic Terms in Polymer Science*, Jenkins, A.D., Kratochvil, P., Stepto, R. F. T., and Suter, U. W., Eds., *Pure & Appl. Chem.* 68, 2287, 1996.

Biochemistry

International Union of Biochemistry and Molecular Biology, *Biochemical Nomenclature and Related Documents, 2nd Edition, 1992*, Portland Press, London, 1993; includes recommendations of the IUPAC-IUBMB Joint Commission on Biochemical Nomenclature.

International Union of Biochemistry and Molecular Biology, *Enzyme Nomenclature, 1992*, Academic Press, Orlando, FL, 1992.

IUPAC-IUBMB Joint Commission on Biochemical Nomenclature, *Nomenclature of Carbohydrates, Recommendations 1996*, McNaught, A. D., Ed., *Pure & Appl. Chem.* 68, 1919, 1996.

General

Chemical Abstracts Service, *Naming and Indexing Chemical Substances for Chemical Abstracts, Appendix IV, Chemical Abstracts 1994 Index Guide*.

Principles of Chemical Nomenclature: a Guide to IUPAC Recommendations, Leigh, G. J., Favre, H. A., and Metanomski, W. V., Blackwell Science, 1998.

NOMENCLATURE FOR INORGANIC IONS AND LIGANDS

Willem H. Koppenol

The entries below were selected from Table IX of Connelly, N. G., Damhus, T., Hartshorn, R. M. and Hutton, A. T., Eds., *Nomenclature of Inorganic Chemistry. IUPAC Recommendations 2005*, The Royal Society of Chemistry, 2005. Two changes were made: in the case of the hypohalides, the oxidohalogenate names are listed, not the new halooxygenate names. Thus, for BrO⁻ the still acceptable name "oxidobromate(1−)" is listed, not the more correct, but less palatable, "bromooxygenate(1−)." Similarly, and for reasons of consistency, ClO• is not named oxygen (mono)chloride, but chlorine mono(o) oxide. The symbol '⊃' is used for dividing names when this is made necessary by a line break. When the name is reconstructed from the name given in the table, this symbol should be omitted. Thus, all *hyphens* in the table are true parts of the names. The symbols '>' and '<' placed next to an element symbol both denote two single bonds connecting the atom in question to two other atoms. For a given compound, the various systematic names, if applicable, are given in the order: stoichiometric names, substitutive names, additive names and hydrogen names. Acceptable names that are not entirely systematic (or not formed according to any of the systems mentioned above) are given at the end after a semicolon. No order of preference is implied by the order in which formulae and names are listed. Reprinted by permission of IUPAC.

Formula for uncharged atom or group	Name			
	Uncharged atoms or molecules (including zwitterions and radicals) or substituent groups[a]	*Cations (including cation radicals) or cationic substituent groups*[a]	*Anions (including anion radicals) or anionic substituent groups*[b]	*Ligands*[c]
H	hydrogen H•, hydrogen(•), monohydrogen (natural or unspecified isotopic composition) ^{1}H•, protium(•), monoprotium ^{2}H• = D•, deuterium(•), monodeuterium ^{3}H• = T•, tritium(•), monotritium	hydrogen (general) H⁺, hydrogen(1+), hydron (natural or unspecified isotopic composition) ^{1}H⁺, protium(1+), proton ^{2}H⁺ = D⁺, deuterium(1+), deuteron ^{3}H⁺ = T⁺, tritium(1+), triton	hydride (general) H⁻, hydride (natural or unspecified isotopic composition) ^{1}H⁻, protide ^{2}H⁻ = D⁻, deuteride ^{3}H⁻ = T⁻, tritide	hydrido protido deuterido tritido
H₂	H₂, dihydrogen D₂, dideuterium T₂, ditritium	H₂•⁺, dihydrogen(•1+) ^{1}H₂•⁺, diprotium(•1+) D₂•⁺, dideuterium(•1+) T₂•⁺, ditritium(•1+)		
D, see H				
D₂, see H₂				
T, see H				
T₂, see H₂				
F	fluorine F•, fluorine(•), monofluorine −F, fluoro	fluorine (general) F⁺, fluorine(1+)	fluoride (general) F⁻, fluoride(1−); fluoride	fluorido (general) F⁻, fluorido(1−); fluorido
F₂	F₂, difluorine	F₂•⁺, difluorine(•1+)	F₂•⁻, difluoride(•1−)	F₂, difluorine
Cl	chlorine (general) Cl•, chlorine(•), monochlorine −Cl, chloro	chlorine (general) Cl•, chlorine(1+)	chloride (general) Cl⁻, chloride(1−); chloride	chlorido (general) Cl⁻, chlorido(1−); chlorido
Cl₂	Cl₂, dichlorine	Cl₂•⁺, dichlorine(•1+)	Cl₂•⁻, dichloride(•1−)	Cl₂, dichlorine Cl₂•⁻, dichlorido(•1−)
Br	bromine (general) Br•, bromine(•), monobromine −Br, bromo	bromine (general) Br⁺, bromine(1+)	bromide (general) Br⁻, bromide(1−); bromide	bromido (general) Br⁻, bromido(1−); bromido
Br₂	Br₂, dibromine	Br₂•⁺, dibromine(•1+)	Br₂•⁻, dibromide(•1−)	Br₂, dibromine
I	iodine (general) I•, iodine(•), monoiodine −I, iodo	iodine (general) I⁺, iodine(1+)	iodide (general) I⁻, iodide(1−); iodide	iodido (general) I⁻, iodido(1−); iodido
I₂	I₂, diiodine	I₂•⁺, diiodine(•1+)	I₂•⁻, diiodide(•1−)	I₂, diiodine

ClO	ClO, chlorine mon(o)oxide ClO•, oxidochlorine(•); chlorosyl –ClO, oxo-λ^3-chloranyl; chlorosyl –OCl, chlorooxy		ClO⁻, oxidochlorate(1–); hypochlorite	ClO⁻, oxidochlorato(1–); hypochlorito
ClO₂	ClO₂, chlorine dioxide ClO₂•, dioxidochlorine(•) ClOO•, chloridodioxygen ⟳ (O–O) (•), –ClO₂, dioxo-λ^5-chloranyl; chloryl –OClO, oxo-λ^3-chloranyloxy	ClO₂⁺, dioxidochlorine(1+) (*not* chloryl)	ClO₂⁻, dioxidochlorate(1–); chlorite	ClO₂⁻, dioxidochlorato(1–); chlorito
ClO₃	ClO₃, chlorine trioxide ClO₃•, trioxidochlorine(•) –ClO₃, trioxo-λ^7-chloranyl; perchloryl –OClO₂, dioxo-λ^5-chloranyloxy	ClO₃⁺, trioxidochlorine(1+) (*not* perchloryl)	ClO₃⁻, trioxidochlorate(1–); chlorate	ClO₃⁻, trioxidochlorato(1–); chlorato
ClO₄	ClO₄, chlorine tetraoxide ClO₄•, tetraoxidochlorine(•) –OClO₃, trioxo-λ^7-chloranyloxy		ClO₄⁻, tetraoxidochlorate(1–); perchlorate	ClO₄⁻, tetraoxidochlorato(1–); perchlorato
IO	IO, iodine mon(o)oxide IO•, oxidoiodine(•); iodosyl –IO, oxo-λ^3-iodanyl; iodosyl –OI, iodooxy	IO⁺, oxidoiodine(1+) (*not* iodosyl)	IO⁻, oxidoiodate(1–); hypoiodite IO•²⁻, oxidoiodate(•2–)	IO⁻, oxidoiodato(1–); hypoiodito
IO₂	IO₂, iodine dioxide IO₂•, dioxidoiodine(•) –IO₂, dioxo-λ^5-iodanyl; iodyl –OIO, oxo-λ^3-iodanyloxy	IO₂⁺, dioxidoiodine(1+) (*not* iodyl)	IO₂⁻, dioxidoiodate(1–); iodite	IO₂⁻, dioxidoiodato(1–); iodito
IO₃	IO₃, iodine trioxide IO₃•, trioxidoiodine(•) –IO₃, trioxo-λ^7-iodanyl; periodyl –OIO₂, dioxo-λ^5-iodanyloxy	IO₃⁺, trioxidoiodine(1+) (*not* periodyl)	IO₃⁻, trioxidoiodate(1–); iodate	IO₃⁻, trioxidoiodato(1–); iodato
IO₄	IO₄, iodine tetraoxide IO₄•, tetraoxidoiodine(•) –OIO₃, trioxo-λ^7-iodanyloxy		IO₄⁻, tetraoxidoiodate(1–); periodate	IO₄⁻, tetraoxidoiodato(1–); periodato
O	oxygen (general) O, monooxygen O²•, oxidanylidene, monooxygen(2•) >O, oxy, epoxy (in rings) =O, oxo	oxygen (general) O•⁺, oxygen(•1+)	oxide (general) O•⁻, oxidanidyl, oxide(•1–) O²⁻, oxide(2–); oxide –O⁻, oxido	O²⁻, oxido
O₂	O₂, dioxygen O₂²•, dioxidanediyl, dioxygen(2•) –OO–, dioxidanediyl; peroxy	O₂•⁺, dioxidanyliumyl, dioxygen(•1+) O₂²⁺, dioxidanebis(ylium), dioxygen(2+)	O₂•⁻, dioxidanidyl, dioxide(•1–); superoxide (*not* hyperoxide) O₂²⁻, dioxidanediide, dioxide(2–); peroxide	dioxido (general) O₂, dioxygen O₂•⁻, dioxido(•1–); superoxido O₂²⁻, dioxidanediido, dioxido(2–); peroxido
O₃	O₃, trioxygen; ozone –OOO–, trioxidanediyl		O₃•⁻, trioxidanidyl, trioxide(•1–); ozonide	O₃, trioxygen; ozone O₃•⁻, trioxido(•1–); ozonido
HO	HO•, oxidanyl, hydridooxygen(•); hydroxyl –OH, oxidanyl; hydroxy	HO⁺, oxidanylium, hydridooxygen(1+); hydroxylium	HO⁻, oxidanide, hydroxide	HO⁻, oxidanido; hydroxido
HO₂	HO₂•, dioxidanyl, hydridodioxygen(•) hydrogen dioxide –OOH, dioxidanyl; hydroperoxy	HO₂⁺, dioxidanylium, hydridodioxygen(1+)	HO₂⁻, dioxidanide, hydrogen(peroxide)(1–)	HO₂⁻, dioxidanido, hydrogen(peroxido)(1–)
S	sulfur (general) S, monosulfur =S, sulfanylidene; thioxo –S–, sulfanediyl	sulfur (general) S⁺, sulfur(1+)	sulfide (general) S•⁻, sulfanidyl, sulfide(•1–) S²⁻, sulfanediide, sulfide(2–); sulfide –S⁻, sulfido	sulfido (general) S•⁻, sulfanidyl, sulfido(•1–) S²⁻, sulfanediido, sulfido(2–)

S_2	S_2, disulfur $-SS-$, disulfanediyl $>S=S$, sulfanylidene-λ^4-sulfanediyl; sulfinothioyl	$S_2^{\bullet+}$, disulfur($\bullet$1+)	$S_2^{\bullet-}$, disulfanidyl, disulfide($\bullet$1−) S_2^{2-}, disulfide(2−), disulfanediide $-SS^-$, disulfanidyl	S_2^{2-}, disulfido(2−), disulfanediido
HS	HS$^{\bullet}$, sulfanyl, hydridosulfur($\bullet$) $-SH$, sulfanyl	HS$^+$, sulfanylium, hydridosulfur(1+)	HS$^-$, sulfanide, hydrogen(sulfide)(1−)	HS$^-$, sulfanido, hydrogen(sulfido)(1−)
SO	SO, sulfur mon(o)oxide [SO], oxidosulfur $>SO$, oxo-λ^4-sulfanediyl; sulfinyl	SO$^{\bullet+}$, oxidosulfur($\bullet$1+) *(not sulfinyl or thionyl)*	SO$^{\bullet-}$, oxidosulfate($\bullet$1−)	[SO], oxidosulfur
SO_2	SO_2, sulfur dioxide $[SO_2]$, dioxidosulfur $>SO_2$, dioxo-λ^6-sulfanediyl; sulfuryl, sulfonyl		$SO_2^{\bullet-}$, dioxidosulfate($\bullet$1−) SO_2^{2-}, dioxidosulfate(2−), sulfanediolate	$[SO_2]$, dioxidosulfur SO_2^{2-}, dioxidosulfato(2−), sulfanediolato
SO_3	SO_3, sulfur trioxide		$SO_3^{\bullet-}$, trioxidosulfate($\bullet$1−) SO_3^{2-}, trioxidosulfate(2−); sulfite $-S(O)_2(O^-)$, oxidodioxo-λ^6-sulfanyl; sulfonato	SO_3^{2-}, trioxidosulfato(2−); sulfito
SO_4	$-OS(O)_2O-$, sulfonylbis(oxy)		$SO_4^{\bullet-}$, tetraoxidosulfate($\bullet$1−) SO_4^{2-}, tetraoxidosulfate(2−); sulfate	SO_4^{2-}, tetraoxidosulfato(2−); sulfato
S_2O_3			$S_2O_3^{\bullet-} = SO_3S^{\bullet-}$, trioxido-1$\kappa^3O$-disulfate(*S–S*) ($\bullet$1−), trioxidosulfidosulfate($\bullet$1−) $S_2O_3^{2-} = SO_3S^{2-}$, trioxido-1κ^3O-disulfate(*S–S*) (2−), trioxidosulfidosulfate(2−); thiosulfate, sulfurothioate	$S_2O_3^{2-} = SO_3S^{2-}$, trioxido-1κ^3O-disulfato(*S–S*) (2−), trioxidosulfidosulfato(2−); thiosulfato, sulfurothioato
Se	Se (general) Se, monoselenium $>Se$, selanediyl $=Se$, selanylidene; selenoxo	selenium	selenide (general) Se$^{\bullet-}$, selanidyl, selenide($\bullet$1−) Se^{2-}, selanediide, selenide(2−); selenide	selenido (general) Se$^{\bullet-}$, selanidyl, selenido($\bullet$1−) Se^{2-}, selanediido, selenido(2−)
SeO	SeO, selenium mon(o)oxide [SeO], oxidoselenium $>SeO$, seleninyl			[SeO], oxidoselenium
SeO_2	SeO_2, selenium dioxide $[SeO_2]$, dioxidoselenium $>SeO_2$, selenonyl		SeO_2^{2-}, dioxidoselenate(2−)	$[SeO_2]$, dioxidoselenium SeO_2^{2-}, dioxidoselenato(2−)
SeO_3	SeO_3, selenium trioxide		$SeO_3^{\bullet-}$, trioxidoselenate($\bullet$1−) SeO_3^{2-}, trioxidoselenate(2−); selenite	SeO_3^{2-}, trioxidoselenato(2−); selenito
SeO_4			SeO_4^{2-}, tetraoxidoselenate(2−); selenate	SeO_4^{2-}, tetraoxidoselenato(2−); selenato
Te	tellurium $>Te$, tellanediyl $=Te$, tellanylidene; telluroxo	tellurium	telluride (general) Te$^{\bullet-}$, tellanidyl, telluride($\bullet$1−) Te^{2-}, tellanediide, telluride(2−); telluride	tellurido (general) Te$^{\bullet-}$, tellanidyl, tellurido($\bullet$1−) Te^{2-}, tellanediido, tellurido(2−)
CrO_2	CrO_2, chromium dioxide, chromium(IV) oxide			
UO_2	UO_2, uranium dioxide	UO_2^+, dioxidouranium(1+) [*not* uranyl(1+)] UO_2^{2+}, dioxidouranium(2+) [*not* uranyl(2+)]		

NpO$_2$	NpO$_2$, neptunium dioxide	NpO$_2^+$, dioxidoneptunium(1+) [*not* neptunyl(1+)] NpO$_2^{2+}$, dioxidoneptunium(2+) [*not* neptunyl(2+)]	
PuO$_2$	PuO$_2$, plutonium dioxide	PuO$_2^+$, dioxidoplutonium(1+) [not plutonyl(1+)] PuO$_2^{2+}$, dioxidoplutonium(2+) [not plutonyl (2+)]	
N	nitrogen N$^\bullet$, nitrogen($\bullet$), mononitrogen –N<, azanetriyl; nitrilo –N=, azanylylidene ≡N, azanylidyne	nitrogen (general) N$^+$, nitrogen(1+)	nitride (general) N^{3-}, nitride(3–), azanetriide; nitride =N$^-$, azanidylidene; amidylidene –N^{2-}, azanediidyl	N^{3-}, nitrido(3–), azanetriido
N$_2$	N$_2$, dinitrogen =N$^+$=N$^-$, (azanidylidene) azaniumylidene; diazo –N=N–, diazane-1,2-diylidene; hydrazinediylidene =NN=, diazene-1,2-diyl; azo	N$_2^{\bullet+}$, dinitrogen(1+) N$_2^{2+}$, dinitrogen(2+) –N$^+$≡N, diazyn-1-ium-1-yl	N$_2^{2-}$, dinitride(2–) N$_2^{4-}$, dinitride(4–), diazanetetraide; hydrazinetetraide	N$_2$, dinitrogen N$_2^{2-}$, dinitrido(2–) N$_2^{4-}$, dinitrido(4–), diazanetetraido; hydrazinetetraido
N$_3$	N$_3^\bullet$, trinitrogen($\bullet$) –N=N$^+$=N$^-$, azido		N$_3^-$, trinitride(1–); azide	N$_3^-$, trinitrido(1–); azido
NH	NH$^{2\bullet}$, azanylidene, hydridonitrogen(2$\bullet$); nitrene >NH, azanediyl =NH, azanylidene; imino	NH$^+$, azanyliumdiyl, hydridonitrogen(1+) NH^{2+}, azanebis(ylium), hydridonitrogen(2+)	NH$^-$, azanidyl, hydridonitrate(1–) NH^{2-}, azanediide, hydridonitrate(2–); imide –NH$^-$, azanidyl; amidyl	NH^{2-}, azanediido, hydridonitrato(2–); imido
NH$_2$	NH$_2^\bullet$, azanyl, dihydridonitrogen($\bullet$); aminyl –NH$_2$, azanyl; amino	NH$_2^+$, azanylium, dihydridonitrogen(1+)	NH$_2^-$, azanide, dihydridonitrate(1–); amide	NH$_2^-$, azanido, dihydridonitrato(1–), amido
NH$_3$	NH$_3$, azane (parent hydride name), amine (parent name for certain organic derivatives), trihydridonitrogen; ammonia	NH$_3^{\bullet+}$, azaniumyl, trihydridonitrogen($\bullet$1+) –NH$_3^+$, azaniumyl; ammonio	NH$_3^{\bullet-}$, azanuidyl, trihydridonitrate($\bullet$1–)	NH$_3$, ammine
NH$_4$	NH$_4^\bullet$, λ^5-azanyl, tetrahydridonitrogen($\bullet$)	NH$_4^+$, azanium; ammonium		
H$_2$NO	H$_2$NO$^\bullet$, aminooxidanyl, dihydridooxidonitrogen($\bullet$); aminoxyl HONH$^\bullet$, hydroxyazanyl, hydridohydroxidonitrogen($\bullet$) –NH(OH), hydroxyazanyl, hydroxyamino –ONH$_2$, aminooxy –NH$_2$(O), oxo-λ^5-azanyl; azinoyl		HONH$^-$, hydroxyazanide, hydridohydroxidonitrate(1–) H$_2$NO$^-$, azanolate, aminooxidanide, dihydridooxidonitrate(1–)	NHOH$^-$, hydroxyazanido, hydridohydroxidonitrato(1–) H$_2$NO$^-$, azanolato, aminooxidanido, dihydridooxidonitrato(1–)
N$_2$H$_2$	HN=NH, diazene $^-$N=NH$_2^+$, diazen-2-ium-1-ide H$_2$NN$^{2\bullet}$, diazanylidene, hydrazinylidene =NNH$_2$, diazanylidene; hydrazinylidene $^\bullet$HNNH$^\bullet$, diazane-1,2-diyl; hydrazine-1,2-diyl –HNNH–, diazane-1,2-diyl; hydrazine-1,2-diyl	HNNH^{2+}, diazynediium	HNNH^{2-}, diazane-1,2-diide, hydrazine-1,2-diide H$_2$NN^{2-}, diazane-1,1-diide, hydrazine-1,1-diide	HN=NH, diazene $^-$N=NH$_2^+$, diazen-2-ium-1-ido HNNH^{2-}, diazane-1,2-diido, hydrazine-1,2-diido H$_2$NN^{2-}, diazane-1,1-diido, hydrazine-1,1-diido

N₂H₃	H₂NNH•, diazanyl, trihydrido ⟲ dinitrogen(*N–N*)(•); hydrazinyl −NHNH₂, diazanyl; hydrazinyl ²⁻NNH₃⁺, diazan-2-ium-1,1-diide	H₂N=NH⁺, diazenium	H₂NNH⁻, diazanide, hydrazinide	²⁻NNH₃⁺, diazan-2-ium-1,1-diido H₂NNH⁻, diazanido, hydrazinido
N₂H₄	H₂NNH₂, diazane (parent hydride name), hydrazine (parent name for organic derivatives) ⁻NHNH₃⁺, diazan-2-ium-1-ide	H₂NNH₂•⁺, diazaniumyl, bis(dihydridonitrogen) ⟲ (*N–N*)(•1+); hydraziniumyl H₂N=NH₂²⁺, diazenediium		H₂NNH₂, diazane, hydrazine ⁻NHNH₃⁺, diazan-2-ium-1-ido
NO	NO, nitrogen mon(o)oxide (*not* nitric oxide) NO•, oxoazanyl, oxidonitrogen(•); nitrosyl −N=O, oxoazanyl; nitroso >N(O)⁻, oxo-λ⁵-azanyl; azoryl =N(O)⁻, oxo-λ⁵-azanylidene; azorylidene ≡N(O), oxo-λ⁵-azanylidyne; azorylidyne −O⁺=N⁻, azanidylideneoxidaniumyl	NO⁺, oxidonitrogen(1+) (*not* nitrosyl) NO•²⁺, oxidonitrogen(2+)	NO⁻, oxidonitrate(1−) NO⁽²•⁾⁻, oxidonitrate(2•1−)	NO, oxidonitrogen (general); nitrosyl = oxidonitrogen-κ*N* (general) NO⁺, oxidonitrogen(1+) NO⁻, oxidonitrato(1−)
NO₂	NO₂, nitrogen dioxide NO₂• = ONO•, nitrosooxidanyl, dioxidonitrogen(•); nitryl −NO₂, nitro −ONO, nitrosooxy	NO₂⁺, dioxidonitrogen(1+) (*not* nitryl)	NO₂⁻, dioxidonitrate(1−); nitrite NO₂•²⁻, dioxidonitrate(•2−)	NO₂⁻, dioxidonitrato(1−); nitrito NO₂•²⁻, dioxidonitrato(•2−)
NO₃	NO₃, nitrogen trioxide NO₃• = O₂NO•, nitrooxidanyl, trioxidonitrogen(•) ONOO•, nitrosodioxidanyl, (dioxido)oxidonitrogen(•) −ONO₂, nitrooxy		NO₃⁻, trioxidonitrate(1−); nitrate NO₃•²⁻, trioxidonitrate(•2−) [NO(OO)]⁻, (dioxido)oxidonitrate(1−); peroxynitrite	NO₃⁻, trioxidonitrato(1−); nitrato NO₃•²⁻, trioxidonitrato(•2−) [NO(OO)]⁻, oxidoperoxidonitrato(1−); peroxynitrito
N₂O	N₂O, dinitrogen oxide (*not* nitrous oxide) NNO, oxidodinitrogen(*N—N*) −N(O)=N−, azoxy		N₂O•⁻, oxidodinitrate(•1−)	N₂O, dinitrogen oxide (general) NNO, oxidodinitrogen(*N—N*) N₂O•⁻, oxidodinitrato(•1−)
N₂O₃	N₂O₃, dinitrogen trioxide O₂NNO, trioxido-1κ²*O*,2κ*O*-dinitrogen(*N–N*) NO⁺NO₂⁻, oxidonitrogen(1+) dioxidonitrate(1−) ONONO, dinitrosooxidane, μ-oxidobis(oxidonitrogen)		N₂O₃²⁻ = [O₂NNO]²⁻, trioxido-1κ₂*O*,2κ*O*-dinitrate(*N–N*)(2−)	
N₂O₄	N₂O₄, dinitrogen tetraoxide O₂NNO₂, bis(dioxidonitrogen) ⟲ (*N–N*) ONOONO, 1,2-dinitrosodioxidane, 2,5-diazy-1,3,4,6-tetraoxy-[6] catena NO⁺NO₃⁻, oxidonitrogen(1+) trioxidonitrate(1−)			
N₂O₅	N₂O₅, dinitrogen pentaoxide O₂NONO₂, dinitrooxidane, NO₂⁺NO₃⁻, dioxidonitrogen(1+) trioxidonitrate(1−)			
NS	NS, nitrogen monosulfide NS•, sulfidonitrogen(•) −N=S, sulfanylideneazanyl; thionitroso	NS⁺, sulfidonitrogen(1+) (*not* thionitrosyl)	NS⁻, sulfidonitrate(1−)	NS, sulfidonitrogen, sulfidonitrato, thionitrosyl (general) NS⁺, sulfidonitrogen(1+) NS⁻, sulfidonitrato(1−)

P	phosphorus (general) $P^\bullet$, phosphorus(•), monophosphorus $>P-$, phosphanetriyl	phosphorus (general) P^+, phosphorus(1+)	phosphide (general) P^-, phosphide(1−) P^{3-}, phosphide(3−), phosphanetriide; phosphide	P^{3-}, phosphido, phosphanetriido
PO	$PO^\bullet$, oxophosphanyl, oxidophosphorus(•), phosphorus mon(o)oxide; phosphoryl $>P(O)-$, oxo-λ^5-phosphanetriyl; phosphoryl $=P(O)-$, oxo-λ^5-phosphanylidene; phosphorylidene $\equiv P(O)$, oxo-λ^5-phosphanylidyne; phosphorylidyne	PO^+, oxidophosphorus(1+) (*not* phosphoryl)	PO^-, oxidophosphate(1−)	
PO_2	$-P(O)_2$, dioxo-λ^5-phosphanyl		PO_2^-, dioxidophosphate(1−)	PO_2^-, dioxidophosphato(1−)
PO_3			PO_3^-, trioxidophosphate(1−) $PO_3^{\bullet 2-}$, trioxidophosphate(•2−) PO_3^{3-}, trioxidophosphate(3−); phosphite $(PO_3^-)_n = \{P(O)_2O\}_n^{n-}$, *catena*-poly[(dioxidophosphate-μ-oxido)(1−)]; metaphosphate $-P(O)(O^-)_2$, dioxidooxo-λ^5-phosphanyl; phosphonato	PO_3^-, trioxidophosphato(1−) $PO_3^{\bullet 2-}$, trioxidophosphato(•2−) PO_3^{3-}, trioxidophosphato(3−); phosphito
PO_4			$PO_4^{\bullet 2-}$, tetraoxidophosphate(•2−) PO_4^{3-}, tetraoxidophosphate(3−); phosphate	PO_4^{3-}, tetraoxidophosphato(3−); phosphato
PS	$PS^\bullet$, sulfidophosphorus(•); $-PS$, thiophosphoryl	PS^+, sulfidophosphorus(1+) (*not* thiophosphoryl)		
AsO_3			AsO_3^{3-}, trioxidoarsenate(3−); arsenite, arsorite $-As(=O)(O^-)_2$, dioxidooxo-λ^5-arsanyl; arsonato	AsO_3^{3-}, trioxidoarsenato(3−); arsenito, arsorito
AsO_4			AsO_4^{3-}, tetraoxidoarsenate(3−); arsenate, arsorate	AsO_4^{3-}, tetraoxidoarsenato(3−); arsenato, arsorato
VO	VO, vanadium(II) oxide, vanadium mon(o)oxide	VO^{2+}, oxidovanadium(2+) (*not* vanadyl)		
CO	CO, carbon mon(o)oxide $>C=O$, carbonyl $=C=O$, carbonylidene	$CO^{\bullet+}$, oxidocarbon(•1+) CO^{2+}, oxidocarbon(2+)	$CO^{\bullet-}$, oxidocarbonate(•1−)	CO, oxidocarbon, oxidocarbonato (general); carbonyl = oxidocarbon-κC (general) $CO^{\bullet+}$, oxidocarbon(•1+) $CO^{\bullet-}$, oxidocarbonato(•1−)
CO_2	CO_2, carbon dioxide, dioxidocarbon		$CO_2^{\bullet-}$, oxidooxomethyl, dioxidocarbonate(•1−)	CO_2, dioxidocarbon $CO_2^{\bullet-}$, oxidooxomethyl, dioxidocarbonato(•1−)
CO_3			$CO_3^{\bullet-}$, trioxidocarbonate(•1−), $OCOO^{\bullet-}$, (dioxido)oxidocarbonate(•1−), oxidoperoxidocarbonate(•1−) CO_3^{2-}, trioxidocarbonate(2−); carbonate	CO_3^{2-}, trioxidocarbonato(2−); carbonato
CS	carbon monosulfide $>C=S$, carbonothioyl; thiocarbonyl $=C=S$, carbonothioylidene	$CS^{\bullet+}$, sulfidocarbon(•1+)	$CS^{\bullet-}$, sulfidocarbonate(•1−)	CS, sulfidocarbon, sulfidocarbonato, thiocarbonyl (general); $CS^{\bullet+}$, sulfidocarbon(•1+) $CS^{\bullet-}$, sulfidocarbonato(•1−)
CS_2	CS_2, disulfidocarbon, carbon disulfide		$CS_2^{\bullet-}$, sulfidothioxomethyl, disulfidocarbonate(•1−)	CS_2, disulfidocarbon $CS_2^{\bullet-}$, sulfidothioxomethyl, disulfidocarbonato(•1−)

CN	CN•, nitridocarbon(•); cyanyl −CN, cyano −NC, isocyano	CN+, azanylidynemethylium, nitridocarbon(1+)	CN−, nitridocarbonate(1−); cyanide	nitridocarbonato (general) CN−, nitridocarbonato(1−); cyanido = [nitridocarbonato(1−)-κC]
CNO	OCN•, nitridooxidocarbon(•) −OCN, cyanato −NCO, isocyanato −ONC, λ²-methylidene ◠ azanylylideneoxy −CNO, (oxo-λ⁵- azanylidynemethyl		OCN−, nitridooxidocarbonate(1−); cyanate ONC−, carbidooxidonitrate(1−); fulminate OCN•2−, nitridooxidocarbonate(•2−)	OCN−, nitridooxidocarbonato(1−); cyanato ONC−, carbidooxidonitrato(1−); fulminato
CNS	SCN•, nitridosulfidocarbon(•) −SCN, thiocyanato −NCS, isothiocyanato −SNC, λ²-methylidene ◠ azanylylidenesulfanediyl −CNS, (sulfanylidene-λ⁵- azanylidynemethyl		SCN−, nitridosulfidocarbonate(1−); thiocyanate SNC−, carbidosulfidonitrate(1−)	SCN−, nitridosulfidocarbonato(1−); thiocyanato SNC−, carbidosulfidonitrato(1−)
CNSe	SeCN•, nitridoselenidocarbon(•) −SeCN, selenocyanato −NCSe, isoselenocyanato −SeNC, λ²-methylidene ◠ azanylylideneselanediyl −CNSe, (selanylidene-λ⁵- azanylidynemethyl		SeCN−, nitridoselenidocarbonate(1−); selenocyanate SeNC−, carbidoselenidonitrate(1−)	SeCN−, nitridoselenidocarbonato(1−); selenocyanato SeNC−, carbidoselenidonitrato(1−)

^a Where an element symbol occurs in the first column, the unmodified element name is listed in the second and third columns. The unmodified name is generally used when the element appears as an electropositive constituent in the construction of a stoichiometric name (Sections IR-5.2 and IR-5.4). Names of homoatomic cations consisting of the element are also constructed using the element name, adding multiplicative prefixes and charge numbers as applicable (Sections IR-5.3.2.1 to IR-5.3.2.3). The sections mentioned refer to parts of *Nomenclature of Inorganic Chemistry. IUPAC Recommendations 2005*, see above.

^b Where an element symbol occurs in the first column, the fourth column gives the element name appropriately modified with the ending 'ide' (hydride, nitride, etc.). The 'ide' form of the element name is generally used when the element appears as an electronegative constituent in the construction of a stoichiometric name (Sections IR-5.2 and IR-5.4). Names of homoatomic anions consisting of the element in question are also constructed using this modified form, adding multiplicative prefixes and charge numbers as applicable (Sections IR-5.3.3.1 to IR-5.3.3.3). Examples are given in the Table of names of some specific anions, e.g. chloride(1−), oxide(2−), dioxide(2−). In certain cases, a particular anion has the 'ide' form itself as an accepted short name, e.g., chloride, oxide. If specific anions are named, the 'ide' form of the element name with no further modification is given as the first entry in the fourth column, with the qualifier '(general)'. The sections mentioned refer to parts of *Nomenclature of Inorganic Chemistry. IUPAC Recommendations 2005*, see above.

^c Ligand names must be placed within enclosing marks whenever necessary to avoid ambiguity, cf. Section IR-9.2.2.3. Some ligand names must always be enclosed. For example, if 'dioxido' is cited as is, it must be enclosed so as to distinguish it from two 'oxido' ligands; if combined with a multiplicative prefix it must be enclosed because it starts with a multiplicative prefix itself. A ligand name such as 'nitridocarbonato' must always be enclosed to avoid interpreting it as two separate ligand names, 'nitrido' and 'carbonato'. In this table, however, these enclosing marks are omitted for the sake of clarity. Note that the ligand names given here with a charge number can generally also be used without if it is not desired to make any implication regarding the charge of the ligand. For example, the ligand name '[dioxido(•1−)]' may be used if one wishes explicitly to consider the ligand to be the species dioxide(•1−), whereas the ligand name '(dioxido)' can be used if no such implications are desirable. The section mentioned refer to parts of *Nomenclature of Inorganic Chemistry. IUPAC Recommendations 2005*, see above.

ORGANIC SUBSTITUENT GROUPS AND RING SYSTEMS

The first part of this table lists substituent groups and their line formulas. A substituent group is defined by IUPAC as a group that replaces one or more hydrogen atoms attached to a parent structure. Such groups are sometimes called radicals, but IUPAC now reserves the term radical for a free molecular species with unpaired electrons. IUPAC does not recommend some of these names, which are marked here with asterisks (e.g., amyl*), but they are included in this list because they are often encountered in the older literature. Substituent group names that are formed by systematic rules (e.g., methyl from methane, ethyl from ethane, etc.) are included here only for the first few members of a homologous series.

In the second part of the table a number of common organic ring compounds are shown, with the conventional numbering of the ring positions indicated.

The help of Warren H. Powell in preparing this table is greatly appreciated. Pertinent references may be found in the table "Nomenclature of Chemical Compounds."

Substituent Groups

acetamido (acetylamino)	CH_3CONH-
acetoacetyl	CH_3COCH_2CO-
acetonyl	CH_3COCH_2-
acetyl	CH_3CO-
acryloyl* (1-oxo-2-propenyl)	$CH_2=CHCO-$
alanyl (from alanine)	$CH_3CH(NH_2)CO-$
β-alanyl	$H_2N(CH_2)_2CO-$
allyl (2-propenyl)	$CH_2=CHCH_2-$
allylidene (2-propenylidene)	$CH_2=CHCH=$
amidino (aminoiminomethyl)	$H_2NC(=NH)-$
amino	H_2N-
amyl* (pentyl)	$CH_3(CH_2)_4-$
anilino (phenylamino)	C_6H_5NH-
anisidino	$CH_3OC_6H_4NH-$
anthranoyl (2-aminobenzoyl)	$2-H_2NC_6H_4CO-$
arsino	AsH_2-
azelaoyl (from azelaic acid)	$-OC(CH_2)_7CO-$
azido	N_3-
azino	$=N-N=$
azo	$-N=N-$
azoxy	$-N(O)=N-$
benzal* (benzylidene)	$C_6H_5CH=$
benzamido (benzoylamino)	C_6H_5CONH-
benzhydryl (diphenylmethyl)	$(C_6H_5)_2CH-$
benzoxy* (benzoyloxy)	C_6H_5COO-
benzoyl	C_6H_5CO-
benzyl	$C_6H_5CH_2-$
benzylidene	$C_6H_5CH=$
benzylidyne	$C_6H_5C=$
biphenylyl	$C_6H_5C_6H_5-$
biphenylene	$-C_6H_4-C_6H_4-$
butoxy	C_4H_9O-
sec-butoxy (1-methylpropoxy)	$C_2H_5CH(CH_3)O-$
tert-butoxy (1,1-dimethylethoxy)	$(CH_3)_3CO-$
butyl	$CH_3(CH_2)_3-$
sec-butyl (1-methylpropyl)	$CH_3CH_2CH(CH_3)-$
tert-butyl (1,1-dimethylethyl)	$(CH_3)_3C-$
butyryl (1-oxobutyl)	$CH_3(CH_2)_2CO-$
caproyl* (hexanoyl)	$CH_3(CH_2)_4CO-$
capryl* (decanoyl)	$CH_3(CH_2)_8CO-$
capryloyl* (octanoyl)	$CH_3(CH_2)_6CO-$
carbamido (carbamoylamino)	$H_2NCONH-$
carbamoyl (aminocarbonyl)	H_2NCO-
carbamyl (aminocarbonyl)	H_2NCO-
carbazoyl (hydrazinocarbonyl)	$H_2NNHCO-$
carbethoxy (ethoxycarbonyl)	C_2H_5OCO-
carbonyl	$=C=O$
carboxy	$HOOC-$
cetyl* (hexadecyl)	$CH_3(CH_2)_{15}-$
chloroformyl (chlorcarbonyl)	$ClCO-$
cinnamoyl	$C_6H_5CH=CHCO-$
cinnamyl (3-phenyl-2-propenyl)	$C_6H_5CH=CHCH_2-$
cinnamylidene	$C_6H_5CH=CHCH=$
cresyl* (hydroxymethylphenyl)	$HO(CH_3)C_6H_4-$
crotonoyl	$CH_3CH=CHCO-$
crotyl (2-butenyl)	$CH_3CH=CHCH_2-$
cyanamido (cyanoamino)	$NCNH-$
cyanato	$NCO-$
cyano	$NC-$
decanedioyl	$-OC(CH_2)_8CO-$
decanoyl	$CH_3(CH_2)_8CO-$
diazo	$N_2=$
diazoamino	$-NHN=N-$
disilanyl	H_3SiSiH_2-
disiloxanyloxy	$H_3SiOSiH_2O-$
disulfinyl	$-S(O)S(O)-$
dithio	$-SS-$
enanthoyl* (heptanoyl)	$CH_3(CH_2)_5CO-$
epoxy	$-O-$
ethenyl (vinyl)	$CH_2=CH-$
ethynyl	$HC≡C-$
ethoxy	C_2H_5O-
ethyl	CH_3CH_2-
ethylene	$-CH_2CH_2-$
ethylidene	$CH_3CH=$
ethylthio	C_2H_5S-
formamido (formylamino)	$HCONH-$
formyl	$HCO-$
furmaroyl (from fumaric acid)	$-OCCH=CHCO-$
furfuryl (2-furanylmethyl)	$OC_4H_3CH_2-$
furfurylidene (2-furanylmethylene)	$OC_4H_3CH=$
glutamoyl (from glutamic acid)	$-OC(CH_2)_2CH(NH_2)CO-$
glutaryl (from glutaric acid)	$-OC(CH_2)_3CO-$
glycylamino	H_2NCH_2CONH-
glycoloyl; glycolyl (hydroxyacetyl)	$HOCH_2CO-$
glycyl (aminoacetyl)	H_2NCH_2CO-
glyoxyloyl; glyoxylyl (oxoacetyl)	$HCOCO-$
guanidino	$H_2NC(=NH)NH-$
guanyl (aminoiminomethyl)	$H_2NC(=NH)-$
heptadecanoyl	$CH_3(CH_2)_{15}CO-$
heptanamido	$CH_3(CH_2)_5CONH-$
heptanedioyl	$-OC(CH_2)_5CO-$
heptanoyl	$CH_3(CH_2)_5CO-$
hexadecanoyl	$CH_3(CH_2)_{14}CO-$
hexamethylene (1,6-hexanediyl)	$-(CH_2)_6-$
hexanedioyl	$-OC(CH_2)_4CO-$
hippuryl (N-benzoylglycyl)	$C_6H_5CONHCH_2CO-$
hydrazino	H_2NNH-
hydrazo	$-HNNH-$
hydrocinnamoyl	$C_6H_5(CH_2)_2CO-$

hydroperoxy	HOO-
hydroxyamino	HONH-
hydroxy	HO-
imino	HN=
iodoso* (iodosyl)	OI-
iodyl	O_2I-
isoamyl* (isopentyl; 3-methylbutyl)	$(CH_3)_2CH(CH_2)_2-$
isobutenyl (2-methyl-1-propenyl)	$(CH_3)_2C=CH-$
isobutoxy (2-methylpropoxy)	$(CH_3)_2CHCH_2O-$
isobutyl (2-methylpropyl)	$(CH_3)_2CHCH_2-$
isobutylidene (3-methylpropylidene)	$(CH_3)_2CHCH=$
isobutyryl (2-methyl-1-oxopropyl)	$(CH_3)_2CHCO-$
isocyanato	OCN-
isocyano	CN-
isohexyl (4-methylpentyl)	$(CH_3)_2CH(CH_2)_3-$
isoleucyl (from isoleucine)	$C_2H_5CH(CH_3)CH(NH_2)CO-$
isonitroso* (hydroxyamino)	HON=
isopentyl (3-methylbutyl)	$(CH_3)_2CH(CH_2)_2-$
isopentylidene (3-methylbutylidene)	$(CH_3)_2CHCH_2CH=$
isopropenyl (1-methylethenyl)	$CH_2=C(CH_3)-$
isopropoxy (1-methylethoxy)	$(CH_3)_2CHO-$
isopropyl (1-methylethyl)	$(CH_3)_2CH-$
isopropylidene (1-methylethylidene)	$(CH_3)_2C=$
isothiocyanato (isothiocyano)	SCN-
isovaleryl* (3-methyl-1-oxobutyl)	$(CH_3)_2CHCH_2CO-$
lactoyl (from lactic acid)	$CH_3CH(OH)CO-$
lauroyl (from lauric acid)	$CH_3(CH_2)_{10}CO-$
lauryl (dodecyl)	$CH_3(CH_2)_{11}-$
leucyl (from leucine)	$(CH_3)_2CHCH_2CH(NH_2)CO-$
levulinoyl (from levulinic acid)	$CH_3CO(CH_2)_2CO-$
malonyl (from malonic acid)	$-OCCH_2CO-$
mandeloyl (from mandelic acid)	$C_6H_5CH(OH)CO-$
mercapto	HS-
mesityl	$2,4,6-(CH_3)_3C_6H_2-$
methacryloyl (from methacrylic acid)	$CH_2=C(CH_3)CO-$
methallyl (2-methyl-2-propenyl)	$CH_2=C(CH_3)CH_2-$
methionyl (from methionine)	$CH_3SCH_2CH_2CH(NH_2)CO-$
methoxy	CH_3O-
methyl	H_3C-
methylene	$H_2C=$
methylthio	CH_3S-
myristoyl (from myristic acid)	$CH_3(CH_2)_{12}CO-$
myristyl (tetradecyl)	$CH_3(CH_2)_{13}-$
naphthyl	$(C_{10}H_7)-$
naphthylene	$-(C_{10}H_6)-$
neopentyl (2,2-dimethylpropyl)	$(CH_3)_3CCH_2-$
nitramino (nitroamino)	O_2NNH-
nitro	O_2N-
nitrosamino (nitrosoamino)	ONNH-
nitrosimino (nitrosoimino)	ONN=
nitroso	ON-
nonanoyl (from nonanoic acid)	$CH_3(CH_2)_7CO-$
oleoyl (from oleic acid)	$CH_3(CH_2)_7CH=CH(CH_2)_7CO-$
oxalyl (from oxalic acid)	-OCCO-
oxo	O=
palmitoyl (from palmitic acid)	$CH_3(CH_2)_{14}CO-$
pentamethylene (1,5-pentanediyl)	$-(CH_2)_5-$
pentyl	$CH_3(CH_2)_4-$
tert-pentyl	$CH_3CH_2C(CH_3)_2-$
phenacyl	$C_6H_5COCH_2-$
phenacylidene	$C_6H_5COCH=$
phenethyl (2-phenylethyl)	$C_6H_5CH_2CH_2-$
phenoxy	C_6H_5O-
phenyl	C_6H_5-

phenylene (benzenediyl)	$-C_6H_4-$
phosphino* (phosphanyl)	H_2P-
phosphinyl* (phosphinoyl)	$H_2P(O)-$
phospho	O_2P-
phosphono	$(HO)_2P(O)-$
phthaloyl (from phthalic acid)	$1,2-C_6H_4(CO-)_2$
picryl (2,4,6-trinitrophenyl)	$2,4,6-(NO_2)_3C_6H_2-$
pimeloyl (from pimelic acid)	$-OC(CH_2)_5CO-$
piperidino (1-piperidinyl)	$C_5H_{10}N-$
pivaloyl (from pivalic acid)	$(CH_3)_3CCO-$
prenyl (3-methyl-2-butenyl)	$(CH_3)_2C=CHCH_2-$
propargyl (2-propynyl)	$HC'CCH_2-$
1-propenyl	$-CH=CHCH_2$
2-propenyl (allyl)	$CH_2=CHCH_2-$
propionyl* (propanyl)	CH_3CH_2CO-
propoxy	$CH_3CH_2CH_2O-$
propyl	$CH_3CH_2CH_2-$
propylidene	$CH_3CH_2CH=$
pyrryl (pyrrolyl)	C_3H_4N-
salicyloyl (2-hydroxybenzoyl)	$2-HOC_6H_4CO-$
selenyl* (selanyl; hydroseleno)	HSe-
seryl (from serine)	$HOCH_2CH(NH_2)CO-$
siloxy	H_3SiO-
silyl	H_3Si-
silylene	$H_2Si=$
sorboyl (from sorbic acid)	$CH_3CH=CHCH=CHCO-$
stearoyl (from stearic acid)	$CH_3(CH_2)_{14}CO-$
stearyl (octadecyl)	$CH_3(CH_2)_{17}-$
styryl (2-phenylethenyl)	$C_6H_5CH=CH-$
suberoyl (from suberic acid)	$-OC(CH_2)_6CO-$
succinyl (from succinic acid)	$-OCCH_2CH_2CO-$
sulfamino (sulfoamino)	$HOSO_2NH-$
sulfamoyl (sulfamyl)	H_2NSO_2-
sulfanilyl [(4-aminophenyl)sulfonyl]	$4-H_2NC_6H_4SO_2-$
sulfeno	HOS-
sulfhydryl (mercapto)	HS-
sulfinyl	OS=
sulfo	HO_3S-
sulfonyl (sulfuryl)	$-SO_2-$
terephthaloyl	$1,4-C_6H_4(CO-)_2$
tetramethylene	$-(CH_2)_4-$
thienyl (from thiophene)	$(C_4H_3S)-$
thiocarbonyl (carbothionyl)	=CS
thiocarboxy	HOSC-
thiocyanato (thiocyano)	NCS-
thionyl* (sulfinyl)	-SO-
threonyl (from threonine)	$CH_3CH(OH)CH(NH_2)CO-$
toluidino [(methylphenyl)amino]	$CH_3C_6H_4NH-$
toluoyl (methylbenzoyl)	$CH_3C_6H_4CO-$
tolyl (methylphenyl)	$CH_3C_6H_4-$
α-tolyl (benzyl)	$C_6H_5CH_2-$
tolylene (methylphenylene)	$-(CH_3C_6H_3)-$
tosyl [(4-methylphenyl) sulfonyl)]	$4-CH_3C_6H_4SO_2-$
triazano	$H_2NNHNH-$
trimethylene (1,3-propanediyl)	$-(CH_2)_3-$
trityl (triphenylmethyl)	$(C_6H_5)_3C-$
valeryl* (pentanoyl)	$CH_3(CH_2)_3CO-$
valyl (from valine)	$(CH_3)_2CHCH(NH_2)CO-$
vinyl (ethenyl)	$CH_2=CH-$
vinylidene (ethenylidene)	$CH_2=C=$
xylidino [(dimethylphenyl)amino]	$(CH_3)_2C_6H_3NH-$
xylyl (dimethylphenyl)	$(CH_3)_2C_6H_3-$
xylylene [phenelenebis(methylene)]	$-CH_2C_6H_4CH_2-$

Organic Ring Compounds

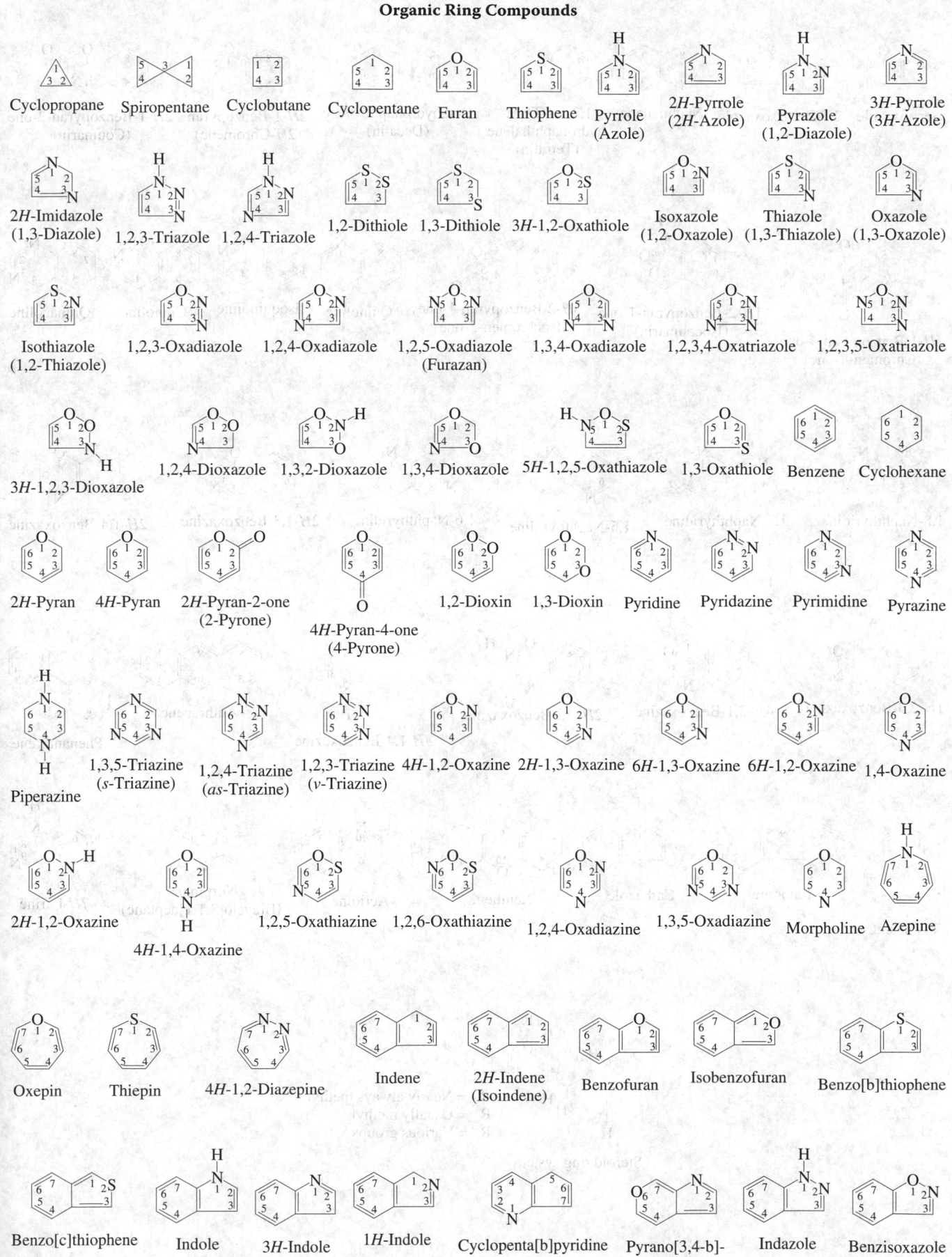

Cyclopropane Spiropentane Cyclobutane Cyclopentane Furan Thiophene Pyrrole (Azole) 2H-Pyrrole (2H-Azole) Pyrazole (1,2-Diazole) 3H-Pyrrole (3H-Azole)

2H-Imidazole (1,3-Diazole) 1,2,3-Triazole 1,2,4-Triazole 1,2-Dithiole 1,3-Dithiole 3H-1,2-Oxathiole Isoxazole (1,2-Oxazole) Thiazole (1,3-Thiazole) Oxazole (1,3-Oxazole)

Isothiazole (1,2-Thiazole) 1,2,3-Oxadiazole 1,2,4-Oxadiazole 1,2,5-Oxadiazole (Furazan) 1,3,4-Oxadiazole 1,2,3,4-Oxatriazole 1,2,3,5-Oxatriazole

3H-1,2,3-Dioxazole 1,2,4-Dioxazole 1,3,2-Dioxazole 1,3,4-Dioxazole 5H-1,2,5-Oxathiazole 1,3-Oxathiole Benzene Cyclohexane

2H-Pyran 4H-Pyran 2H-Pyran-2-one (2-Pyrone) 4H-Pyran-4-one (4-Pyrone) 1,2-Dioxin 1,3-Dioxin Pyridine Pyridazine Pyrimidine Pyrazine

Piperazine 1,3,5-Triazine (s-Triazine) 1,2,4-Triazine (as-Triazine) 1,2,3-Triazine (v-Triazine) 4H-1,2-Oxazine 2H-1,3-Oxazine 6H-1,3-Oxazine 6H-1,2-Oxazine 1,4-Oxazine

2H-1,2-Oxazine 4H-1,4-Oxazine 1,2,5-Oxathiazine 1,2,6-Oxathiazine 1,2,4-Oxadiazine 1,3,5-Oxadiazine Morpholine Azepine

Oxepin Thiepin 4H-1,2-Diazepine Indene 2H-Indene (Isoindene) Benzofuran Isobenzofuran Benzo[b]thiophene

Benzo[c]thiophene Indole 3H-Indole 1H-Indole Cyclopenta[b]pyridine Pyrano[3,4-b]-pyrrole Indazole Benzisoxazole (Indoxazene)

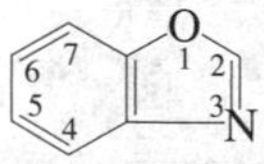

Benzoxazole 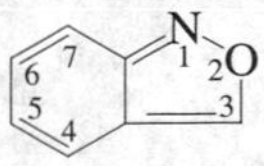   2,1-Benzisoxazole 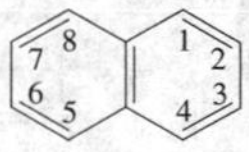   Naphthalene 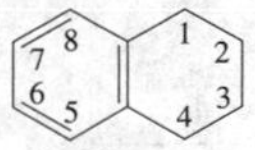   1,2,3,4-Tetra-   Octahydronaphthalene   2H-1-Benzopyran   2H-1-Benzopyran-2-one
                                                 hydronaphthalene   (Decalin)   (2H-Chromene)   (Coumarin)
                                                 (Tetralin)

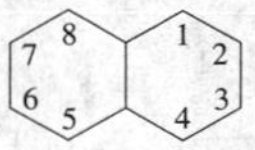

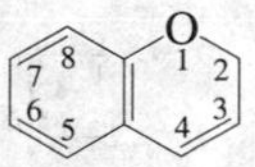

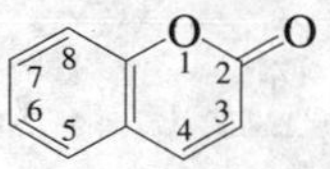

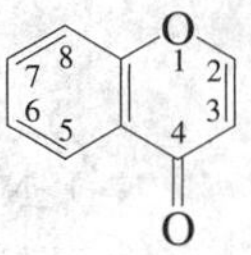

4H-1-Benzopyran-4-one   1H-2-Benzopyran-1-one   3H-2-Benzopyran-1-one 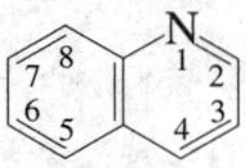   Quinoline 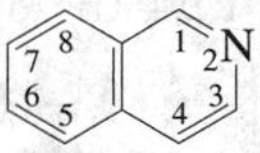   Isoquinoline 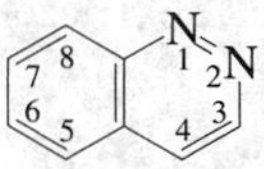   Cinnoline 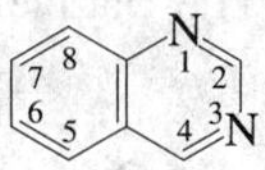   Quinazoline
(Chromen-4-one)   (Isocoumarin)   (Isochromen-3-one)

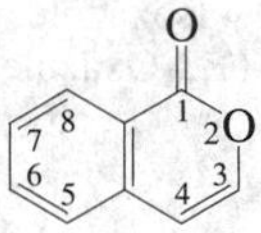

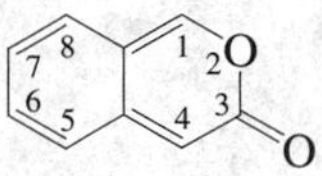

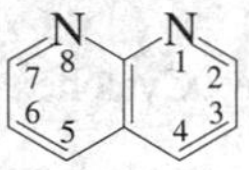

1,8-Naphthyridine 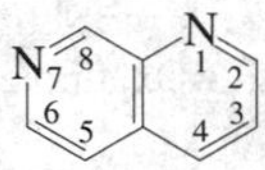   1,7-Naphthyridine 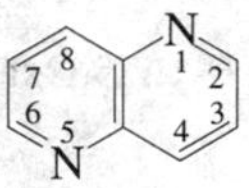   1,5-Naphthyridine 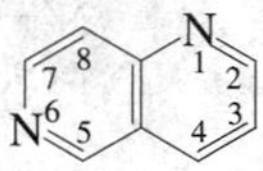   1,6-Naphthyridine   2H-1,3-Benzoxazine   2H-1,4-Benzoxazine

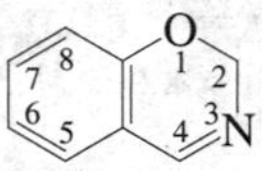

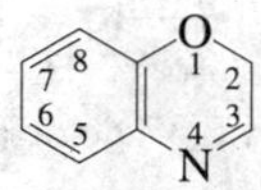

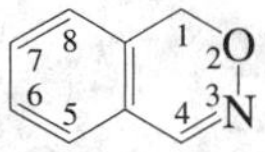

1H-2,3-Benzoxazine   4H-3,1-Benzoxazine   2H-1,2-Benzoxazine   4H-1,4-Benzoxazine 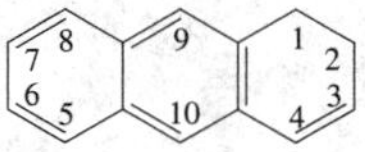   Anthracene

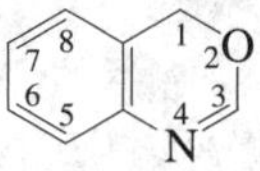

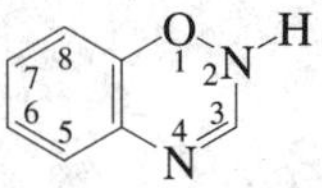

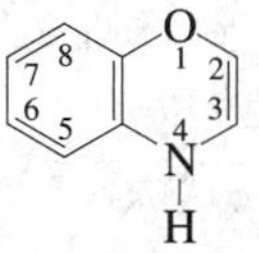

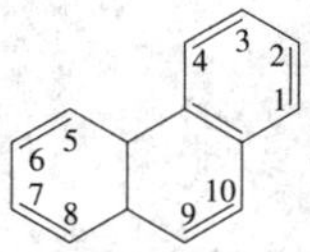

Phenanthrene

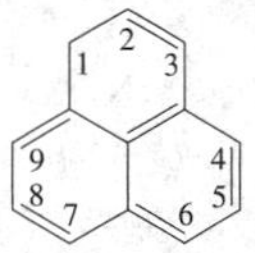

Phenalene 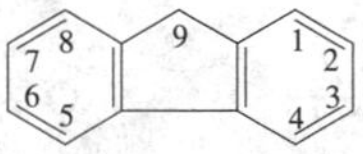   Fluorene 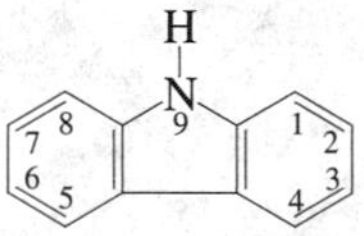   Carbazole 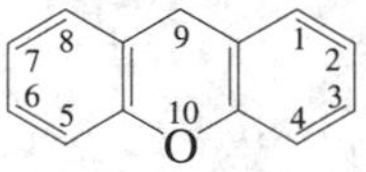   Xanthene 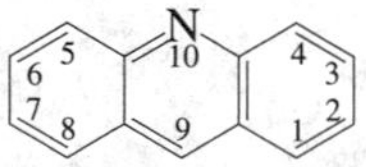   Acridine    Norpinane   7H-Purine
                                                          (Bicyclo[3.1.1]heptane)

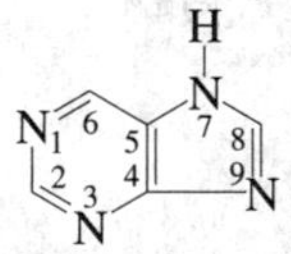

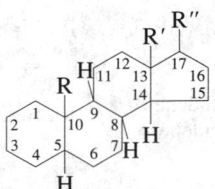

R = Nearly always methyl
R′ = Usually methyl
R″ = Various groups

Steroid ring system

REPRESENTATION OF CHEMICAL STRUCTURES WITH THE IUPAC INTERNATIONAL CHEMICAL IDENTIFIER (INCHI)

Stephen R. Heller and Alan D. McNaught

The IUPAC International Chemical Identifier (InChI) is a freely available, non-proprietary identifier for chemical substances that can be used in both printed and electronic data sources. It is generated from a computerized representation of a molecular structure diagram, which can be produced by chemical structure-drawing software. Its use enables linking of diverse data compilations and unambiguous identification of chemical substances. A full description of the Identifier and software for its generation are available from the IUPAC Web site (Ref. 1), and a helpful compilation of answers to frequently asked questions has been put together at the Unilever Centre for Molecular Science Informatics (Ref. 2). Commercial structure-drawing software that will generate the Identifier is available from several organizations, listed on the IUPAC Web site.

The conversion of structural information to the Identifier is based on a set of IUPAC structure conventions, and rules for normalization and canonicalization (conversion to a single, predictable sequence) of an input structure representation. The resulting InChI is simply a series of characters that serve to uniquely identify the structure from which it was derived. The InChI uses a layered format to represent all available structural information relevant to compound identity. InChI layers are listed below. Each layer in an InChI representation contains a specific type of structural information. These layers, automatically extracted from the input structure, are designed so that each successive layer adds additional detail to the Identifier. The specific layers generated depend on the level of structural detail available and whether or not allowance is made for tautomerism. Of course, any ambiguities or uncertainties in the original structure will remain in the InChI.

This layered structure design offers a number of advantages. If two structures for the same substance are drawn at different levels of detail, the one with the lower level of detail will, in effect, be contained within the other. Specifically, if one substance is drawn with stereo-bonds and the other without, the layers in the latter will be a subset of the former. The same will hold for compounds treated by one author as tautomers and by another as exact structures with all H-atoms fixed. This can work at a finer level. For example, if one author includes double bond and tetrahedral stereochemistry, but another omits stereochemistry, the latter InChI will be contained in the former.

The InChI layers are
1. Formula
2. Connectivity (no formal bond orders)
    a. disconnected metals
    b. connected metals
3. Isotopes
4. Stereochemistry
    a. double bond (*Z/E*)
    b. tetrahedral (sp^3)
5. Tautomers (on or off)

Charges are not part of the basic InChI, but rather are added at the end of the InChI string.

Two examples of InChI representations are given below. It is important to recognize, however, that InChI strings are intended for use by computers and end users need not understand any of their details. In fact, the open nature of InChI and its flexibility of representation, after implementation into software systems, may allow chemists to be even less concerned with the details of structure representation by computers.

guanine

InChI=1/C5H5N5O/c6-5-9-3-2(4(11)10-5)7-1-8-3/
h1H,(H4,6,7,8,9,10,11)/f/h8,10H,6H2

monosodium glutamate

InChI=1/C5H9NO4.Na/c6-3(5(9)10)1-2-4(7)8;/h3H,1-
2,6H2,(H,7,8)(H,9,10);/q;+1/p-1/t3-;/m1./s1/fC5H8NO4.Na/
h7H;/q-1;m

The layers in the InChI string are separated by the '/' character followed by a lowercase letter (except for the first layer, the chemical formula), with the layers arranged in predefined order. In the examples the following segments are included

InChI version number
/- chemical formula
/c connectivity-1.1 (excluding terminal H)
/h connectivity-1.2 (locations of terminal H, including mobile H attachment points)
/q charge
/p proton balance
/t sp^3 (tetrahedral) parity
/m parity inverted to obtain relative stereo (1 = inverted, 0 = not inverted)
/s stereo type (1 = absolute, 2 = relative, 3 = racemic)
/f chemical formula of the fixed-H structure if it is different
/h connectivity-2 (locations of fixed mobile H)
/q charge
/t sp^3 (tetrahedral) parity
/m parity inverted to obtain relative stereo (1 = inverted, 0 = not inverted, . = inversion does not affect the parity)
/s stereo type (1 = absolute, 2 = relative, 3 = racemic)

One of the most important applications of InChI is the facility to locate mention of a chemical substance using Internet-based search engines. This is made easier by using a shorter (compressed) form of InChI, known as InChIKey. The InChIKey is a 27-character representation that, because it is compressed, cannot be reconverted into the original structure, but it is not subject to the undesirable and unpredictable breaking of longer character strings by some search engines. The usefulness of the InChIKey as a search tool is enhanced by its derivation from a "standard" InChI. i.e., an InChI produced with standard option settings for features such as tautomerism and stereochemistry. An example is shown below; the "standard" InChI is denoted by the letter "S" after the version number.

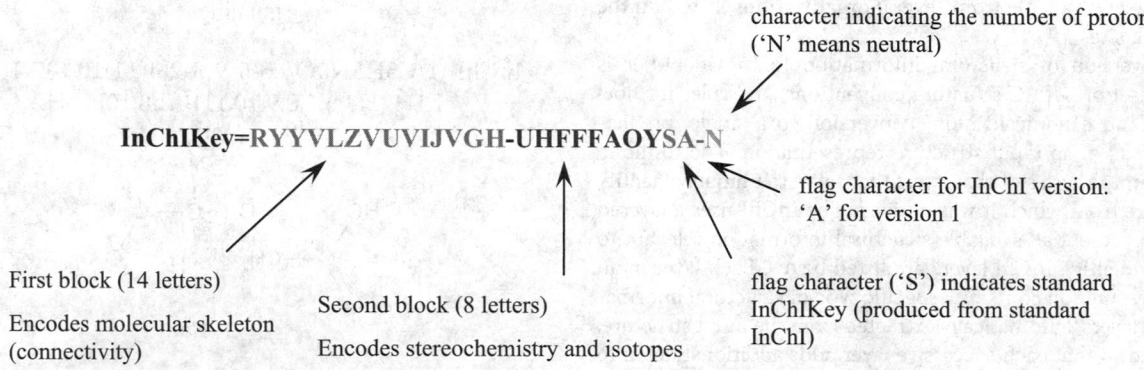

InChI=1S/C8H10N4O2/c1-10-4-9-6-5(10)7(13)12(3)8(14)11(6)2/h4H,1-3H3 (caffeine)

character indicating the number of protons ('N' means neutral)

InChIKey=RYYVLZVUVIJVGH-UHFFFAOYSA-N

First block (14 letters)

Encodes molecular skeleton (connectivity)

Second block (8 letters)

Encodes stereochemistry and isotopes

flag character for InChI version: 'A' for version 1

flag character ('S') indicates standard InChIKey (produced from standard InChI)

Use of InChIKey also allows searches based solely on atomic connectivity (first 14 characters). Software for generating InChIKey is available from the IUPAC Web site (Ref. 1).

The enormous databases compiled by organizations such as PubChem (Ref. 4), the U.S. National Cancer Institute (NCI), and ChemSpider (Ref. 5) contain millions of InChIs and InChIKeys, which allow sophisticated searching of these collections. PubChem provides InChI-based structure-search facilities for both identical and similar structures (Ref. 6), and ChemSpider offers both search facilities and Web services enabling a variety of InChI and InChIKey conversions (Ref. 7). The NCI Chemical Structure Lookup Service (Ref. 8) provides InChI-based search access to over 39 million chemical structures from over 80 different public and commercial data sources.

References

1. http://www.iupac.org/inchi
2. http://wwmm.ch.cam.ac.uk/inchifaq/
3. *Pure Appl. Chem.*, in preparation.
4. http://pubchem.ncbi.nlm.nih.gov
5. http://www.chemspider.com
6. http://pubchem.ncbi.nlm.nih.gov/search
7. http://www.chemspider.com/InChI.asmx
8. http://cholla.chemnavigator.com/cgi-bin/lookup/new/search

SCIENTIFIC ABBREVIATIONS, ACRONYMS, AND SYMBOLS

This table lists some abbreviations, acronyms, and symbols encountered in the physical sciences. Most entries in italic type are symbols for physical quantities; for more details on these, see the table "Symbols and Terminology for Physical and Chemical Quantities" in this section. Additional information on units may be found in the table "International System of Units (SI)" in Section 1. Many of the terms to which these abbreviations refer are included in the tables "Definitions of Scientific Terms" in Section 2 and "Techniques for Materials Characterization" in Section 12. Useful references for further information are given below.

Publication practices vary with regard to the use of capital or lower case letters for many abbreviations. An effort has been made to follow the most common practices in this table, but much variation is found in the literature. Likewise, policies on the use of periods in an abbreviation vary considerably. Periods are generally omitted in this table unless they are necessary for clarity. Periods should never appear in SI units. The SI prefixes (m, k, M, etc.) are included here, but they should never be used alone. Selected combinations of these prefixes with SI units (e.g., mg, kV, MW) are also included.

Abbreviations are listed in alphabetical order without regard to case. Entries beginning with Greek letters fall at the end of the table.

References

1. *Quantities, Units, and Symbols in Physical Chemistry, Third Edition*, IUPAC 2007, RSC Publishing, 2007.
2. Kotyk, A., *Quantities, Symbols, Units, and Abbreviations in the Life Sciences*, Humana Press, Totawa, NJ, 1999.
3. Rhodes, P. H., *The Organic Chemist's Desk Reference*, Chapman & Hall, London, 1995.
4. Minkin, V., Glossary of Terms used in Theoretical Organic Chemistry, *Pure Appl. Chem.* 71, 1919–1981, 1999.
5. Brown, R. D., Ed., Acronyms Used in Theoretical Chemistry, *Pure Appl. Chem.* 68, 387–456, 1996.
6. *Quantities and Units, ISO Standards Handbook, Third Edition*, International Organization for Standardization, Geneva, 1993.
7. Cohen, E. R., and Giacomo, P., Symbols, Units, Nomenclature, and Fundamental Constants in Physics, *Physica* 146A, 1–68, 1987.
8. *Chemical Acronyms Database*, Indiana University, <www.oscar.chem. indiana.edu/cfdocs/ libchem/acronyms/ acronymsearch.html>.
9. *Acronyms and Symbols*, <www3.interscience.wiley.com/stasa/>.
10. *IUPAC Compendium of Chemical Terminology* (Gold Book), <gold-book.iupac.org>.
11. IUPAC-IUB Joint Commission on Biochemical Nomenclature, *Pure & Appl. Chem.* 56, 595, 1984.

A	ampere; alanine; adenine (in genetic code)
Å	ångström
A	absorbance; area; Helmholtz energy; mass number
A_H	Hall coefficient
A_r	atomic weight (relative atomic mass)
a	atto (SI prefix for 10^{-18})
a	absorption coefficient; acceleration; activity; van der Waals constant
a_0	Bohr radius
AAA	acetoacetanilide
Aad	2-aminoadipic acid
AAF	2-(acetylamino)fluorene
AAN	aminoacetonitrile
AAO	acetaldehyde oxime
AAS	atomic absorption spectroscopy
ABA	abscisic acid; acrylonitrile-butadiene acrylate
Abe	abequose
ABL	α-acetylbutyrolactone
ABS	acrylonitrile-butadiene-styrene copolymer
abs	absolute
Abu	2-aminobutanoic acid
Ac	acetyl; acetate
ac, AC	alternating current
ACAC	acetylacetone
Aces	2-[(2-amino-2-oxoethyl)amino]ethanesulfonic acid
ACS	acrylonitrile-chlorinated polyethylene-styrene copolymer
ACT	activated complex theory
ACTH	adrenocorticotropic hormone
A/D	analog to digital
Ad	adamantyl
Ada	[(carbamoylmethyl)imino]diacetic acid
Ade	adenine
ADI	acceptable daily intake
Ado	adenosine
ADP	adenosine diphosphate; ammonium dihydrogen phosphate
ads	adsorption
AE	appearance energy
ae	eon (10^9 years)
AEP	1-(2-aminoethyl)piperazine
AEPD	2-amino-2-ethyl-1,3-propanediol
AES	atomic emission spectroscopy; Auger electron spectroscopy
AF	audio frequency
AFM	atomic force microscopy
Ahx	2-aminohexanoic acid
AI	artificial intelligence
AIBN	2,2′-azobis[isobutyronitrile]
AICA	5-amino-1*H*-imidazole-4-carboxamide
AIM	atoms in molecules (method)
AIP	aluminum isopropoxide
Al	Alfén number
Ala	alanine
alc	alcohol
ALE	atomic layer epitaxy
aliph.	aliphatic
alk.	alkaline
All	allose
Alt	altrose
AM	amplitude modulation
Am	amyl
am	amorphous solid
AMA	acrylate maleic anhydride terpolymer
AMMA	acrylate-methyl methacrylate copolymer
AMP	adenosine monophosphate
AMPD	2-amino-2-methyl-1,3-propanediol
AMS	accelerator mass spectrometry

AMTCS	amyltrichlorosilane [trichloropentylsilane]
amu	atomic mass unit (recommended symbol is u)
AN	acetonitrile; acrylonitrile
anh, anhyd	anhydrous
ANOVA	analysis of variance
antilog	antilogarithm
ANTU	1-naphthalenylthiourea
AO	atomic orbital
AOM	angular overlap model
AP	ethylene-propylene copolymer
APAD	3-acetylpyridine adenine dinucleotide
APAP	acetyl p-amino phenol (acetaminophen)
Ape	2-aminopentanoic acid
API	atmospheric pressure ionization
Api	apiose
APM	atomic probe microanalysis
Apm	2-aminopimelic acid
APO	amorphous polyolefin
APPI	atmospheric pressure photoionization
APS	appearance potential spectroscopy; adenosine phosphosulfate
APW	augmented plane wave
aq	aqueous
Ar	aryl
Ara	arabinose
Ara-ol	arabinitol
Arg	arginine
ARPES	angular resolved photoelectron spectroscopy
AS	acrylonitrile styrene copolymer
ASA	acetylsalicylic acid; acrylonitrile-styrene-acrylonitrile block copolymer
ASC	4-(acetylamino)benzenesulfonyl chloride
ASCII	American National Standard Code for Information Interchange
ASE	aromatic stabilization model
Asn	asparagine
Asp	aspartic acid
at	atomization
ATCP	4-amino-3,5,6-trichloro-2-pyridinecarboxlic acid
ATEE	N-acetyl-L-tyrosine ethyl ester
ATLC	adsorption thin layer chromatography
atm	standard atmosphere
ATP	adenosine triphosphate
ATR	attenuated total internal reflection
at.wt.	atomic weight
AU	astronomical unit (ua is also used); polyurethane
AUC	area under the time-concentration curve
av	average
avdp	avoirdupois
B	bel; asparagine or aspartic acid (unspecified)
B	magnetic flux density; second virial coefficient; susceptance
b	barn
b	van der Waals constant; molality
BA	benzyladenine
BAL	British anti-Lewisite [2,3-dimercapto-1-propanol]
BAP, BaP	benzo[a]pyrene
bar	bar (pressure unit)
bbl	barrel

BBP	benzyl butyl phthalate
BCB	bromocresol blue
bcc	body centered cubic
BCF	bioconcentration factor
BCG	bromocresol green
BCME	bis(chloromethyl) ether
BCNU	N,N'-bis(2-chloroethyl)-N-nitrosourea
BCP	bromocresol purple
BCPB	bromochlorophenol blue
BCPE	1,1-bis(4-chlorophenyl)ethanol
BCS	Bardeen-Cooper-Schrieffer (theory)
BDE	bond dissociation energy
BDEA	butyldiethanolamime
BDMA	benzyldimethylamine
Bé	Baumé
BEBO	bond energy bond order (method)
BEI	biological exposure index
BEM	biological effect monitoring
BEP	2-butyl-2-ethyl-1,3-propanediol
Bes	2-[bis(2-hydroxyethyl)amino]ethanesulfonic acid
BET	Brunauer-Emmett-Teller (isotherm)
BeV	billion electronvolt
BGE	butyl glycidyl ether
BHA	$tert$-butyl-4-hydroxyanisole
BHC	benzene hexachloride [hexachlorobenzene]
Bhn	Brinell hardness number
BHT	butylated hydroxytoluene [2,6-di-$tert$-butyl-4-methylphenol]
Bi	biot
Bicine	N,N-bis(2-hydroxyethyl)glycine
BIRD	blackbody infrared radiative dissociation
Bistris	2-[bis(2-hydroxyethyl)amino]-2-(hydroxymethyl) propane-1,3-diol
Bistris-propane	1,3-bis[tris(hydroxymethyl)methylamino]propane
BLO	γ-butyrolactone
BN	bond number; benzonitrile
BNS	nuclear backscattering spectroscopy
BO	Born-Oppenheimer (approximation); bond order
BOD	biochemical oxygen demand
BON	β-hydroxynaphthoic acid
BP	base peak (in mass spectrometry); benzo[a]pyrene
bp	boiling point; base pair
BPB	bromophenol blue
BPG	2,3-bis(phospho)-D-glycerate
BPL	β-propiolactone
BPO	benzoyl peroxide
bpy	2,2'-bipyridine
Bq	becquerel
Br	butyryl
BRE	bond resonance energy
BrUrd	5-bromouridine
BS	Birge-Sponer extrapolation
BSE	back scattered electron(s)
BSSE	basis set superposition error
BTMSA	1,2-bis(trimethylsilyl)acetylene
Btu	British thermal unit
BTX	benzene, toluene, and xylene

Bu	butyl
bu	bushel
BVE	butyl vinyl ether
Bz	benzoyl
Bzl	benzyl
C	coulomb; cysteine; cytosine (in genetic code)
°C	degree Celsius
C	capacitance; heat capacity; number concentration
c	centi (SI prefix for 10^{-2}); combustion reaction
c	amount concentration; specific heat; velocity
c_0	speed of light in vacuum
CA	collisional activation; cellulose acetate
ca.	approximately
CAB	cellulose acetate butyrate
CADD	computer-assisted drug design
cal	calorie
calc	calculated
cAMP	adenosine cyclic 3′,5′-(hydrogen phosphate)
CAN	ceric ammonium nitrate
CAR	carbon fiber
CARS	coherent anti-Stokes Raman spectroscopy
CAS	complete active space
CASRN	Chemical Abstracts Service Registry Number
CAT	computerized axial tomography; clear air turbulence
CBE	chemical beam epitaxy
CBS	complete basis set (of orbitals)
CC	coupled cluster; combustion calorimetry
cc	cubic centimeter
CCD	charge-coupled device
CD	circular dichroism
cd	candela; condensed (phase)
CDAA	2-chloro-N,N-diallylacetamide
CDNO	complete neglect of differential overlap
CDP	cytidine 5′-diphosphate
CDT	1,5,9-cyclododecatriene
CDTA	(1,2-cyclohexylenedinitrilo)tetraacetic acid monohydrate
CDW	charge density waves
CED	cohesive energy density
CEM	channel electron multiplier
CEP	counter electrophoresis
CEPA	coupled electron-pair approximation
cf.	compare
CFC	chlorofluorocarbon compound
cfm	cubic feet per minute
CFRP	carbon reinforced plastics
cgs	centimeter-gram-second system
Chaps	3-[3-(cholamidopropyl)dimethylammonio]-1-propanesulfonic acid
Ches	2-(N-cyclohexylamino)ethanesulfonic acid
CHF	coupled Hartree-Fock (method)
Chl	chlorophyll
Cho	choline
CHT	1,3,5-cycloheptatriene
Ci	curie
CI	configuration interaction; chemical ionization; color index
CID	charge-injection device; collision-induced dissociation
CIDEP	chemically induced dynamic electron polarization

CIDNP	chemically induced dynamic nuclear polarization
CIE	countercurrent immunoelectrophoresis
cir	circular
CKFF	Cotton-Kraihanzel force field
CL	cathode luminescence (spectroscopy)
CLT	central limit theorem
cm	centimeter
c.m.	center of mass
CMC	carboxymethylcellulose
c.m.c.	critical micelle concentration
CMO	canonical molecular orbital
CMP	cytidine 5′-monophosphate; chemical measurement process
CN	coordination number; cellulose nitrate
CNDO	complete neglect of differential overlap
Co	Cowling number
COC	cycloolefin copolymer
COD	chemical oxygen demand; 1,4-cyclooctadiene
conc	concentrated; concentration
const	constant
COOP	crystal orbital overlap population
cos	cosine
cosh	hyperbolic cosine
COSY	correlation spectroscopy
COT	1,3,5,7-cyclooctatetraene
cot	cotangent
coth	hyperbolic cotangent
CP	chemically pure
Cp	cyclopentadienyl
Cp*	pentamethylcyclopentadienyl
cP	centipoise
cp	candle power
CPA	coherent potential approximation
CPC	centrifugal partition chromatography
cpd	contact potential difference
CPE	chlorinated polyethylene
CPL	circular polarization of luminescence
CPR	chlorophenol red
cps	cycles per second
CPT	charge conjugation/space inversion/time inversion (theorem)
CPU	central processing unit
CPVC	chlorinated poly(vinyl chloride)
CR	chloroprene rubber (neoprene)
cr, cryst	crystalline (phase)
CRF	charge remote fragmentation
CRU	constitutional repeating unit (in polymer nomenclature)
CSA	camphorsulfonic acid
csc	cosecant
CSR	charge stripping reaction
CT	charge transfer
ct	carat
CTA	cellulose triacetate
CTEM	conventional transmission electron microscopy
CTFE	chlorotrifluoroethylene
CTP	cytidine 5′-triphosphate
CTR	controlled thermonuclear reaction
cu	cubic

CV	cyclic voltammetry		DESI	desorption electrospray ionization (in mass spectrometry)
CVD	chemical vapor deposition		det	determinant
cw	continuous wave		dev	deviation
cwt	hundredweight (112 pounds)		DFT	density functional theory
Cy	cyclohexyl		dGlc	2-deoxyglucose
Cya	cysteic acid		DHBA	2,3-dihydroxybenzoic acid
Cyd	cytidine		DHH	dehydroheliotridine
cyl	cylinder		DHR	dehydroretronecine
Cys	cysteine		DHU	dihydrouridine
Cyt	cytosine		DI	desorption ionization
D	debye unit; aspartic acid		diam	diameter
D	diffusion coefficient; dissociation energy; electric displacement		DIBA	diisobutyl adipate
d	day; deuteron; deci (SI prefix for 10^{-1})		DIBK	diisobutyl ketone
			dil	dilute; dilution
d	distance; density; dextrorotatory		DIM	diatomics in molecules (method); digital imaging microscopy
2,4-D	2,4-dichlorophenoxyacetic acid			
D/A	digital to analog		DIPA	diisopropanolamine
Da	dalton		dm	decimeter
DA	donor-acceptor (complex)		DMA	*N,N*-dimethylaniline
da	deka (SI prefix for 10^{1})		DMAB	4-(dimethylamino)azobenzene
DAA	diacetone alcohol		DMAC	*N,N*-dimethylacetamide
DAB	4-(dimethylamino)azobenzene		DMAE	*N,N*-dimethylethanolamine
Dab	2,4-diaminobutanoic acid		DMBA	7,12-dimethylbenz[a]anthracene
DACH	*trans*-1,2-diaminocyclohexane		DME	1,2-dimethoxyethane
DAIP	diallyl isophthalate plasticizer		DMF	*N,N*-dimethylformamide
DAP	diammonium phosphate		DMP	dimethyl phthalate
DART	direct analysis in real time mass spectrometry		DMS	dimethyl sulfide
dB	decibel		DMSO	dimethyl sulfoxide
DBA	dibenz[a,h]anthracene; dibenzylamine		DMT	dimethyl terephthalate; dimethyl tartrate
DBCP	1,2-dibromo-3-chloropropane		DN	donor number
DBED	dibenzyl ethylene diamine		DNA	deoxyribonucleic acid
DBM	dibutyl maleate		DNase	deoxyribonuclease
DBMC	2,4-di-*tert*-butyl-5-methylphenol		DNB	1,3-dinitrobenzene
DBMS	database management system		DNMR	dynamic NMR spectroscopy
DBP	dibutyl phthalate; 2,3-dibromo-1-propanol		DNP	dinitropyrene
DBPC	2,6-di-*tert*-butyl-*p*-cresol		Dod	dodecyl
dc, DC	direct current		DOP	dioctyl phthalate
DCB	dicyanobenzene		DOS	density of states; digital operating system; dioctyl sebacate
DCBP	4,4'-dichlorobenzophenone			
DCEE	dichloroethyl ether		doz	dozen
DCHA	dicyclohexylamine		DP, d.p.	degree of polymerization
DCM	dichloromethane		DPA	diphenylamine
DCNP	2,6-dichloro-4-nitrophenol		DPG	*N,N'*-diphenylguanidine
DCP	2,4-dichlorophenol		dpl	displacement
DCPD	dicyclopentadiene		Dpm	2,6-diaminopimelic acid
DDM	4,4'-diaminodiphenylmethane		dpm	disintegrations per minute
DDT	dichlorodiphenyltrichloroethane		dps	disintegrations per second
DE	delocalization energy; delayed extraction		DPU	*N,N'*-diphenylurea
DEA	*N,N*-diethylaniline; diethanolamine		dr	dram
Dec	decyl		DRE	Dewar resonance energy
dec	decomposes		dRib	2-deoxyribose
DEET	diethyltoluamide [*N,N*-diethyl-3-methylbenzamide]		DRIFT	diffuse reflectance infrared Fourier transform
deg	degree		DRP	dynamic reaction path
DEK	diethyl ketone		DRS	diffuse reflectance spectroscopy
den	density		DS	degree of substitution
DEP	2,2-diethyl-1,3-propanediol		DSC	differential scanning calorimetry
			DTA	differential thermal analysis
DES	diethyl sulfate		DTBP	di-*tert*-butyl peroxide

DVB	divinylbenzene
dyn	dyne
DZ	double-zeta (type of basis set)
E	exa (SI prefix for 10^{18}); glutamic acid
E	electric field strength; electromotive force; energy; Young's modulus of elasticity; entgegen (*trans* configuration)
E_h	Hartree energy
e	electron; base of natural logarithms
e	elementary charge; linear strain
EA	electron affinity
EAA	ethylene acrylic acid copolymer; ethyl acetoacetate
EAK	ethyl amyl ketone (3-octanone)
EAN	effective atomic number
EC	ethyl cellulose
ECD	electron capture dissociation
ECP	effective core potential
ECR	electron cyclotron resonance
ECTFE	ethylene-chlorotrifluoroethylene copolymer
ED	electron diffraction
EDAX	energy dispersive analysis by x-rays
EDB	ethylene dibromide [1,2-dibromoethane]
EDC	ethylene dichloride [1,2-dichloroethane]
EDI	estimated daily intake
EDS	energy-dispersive x-ray spectroscopy
EDTA	ethylenediaminetetraacetic acid
EEA	ethylene-ethyl acetate copolymer
EEDQ	ethyl 2-ethoxy-1(2*H*)-quinolinecarboxylate
EEL	environmental exposure level
EELS	electron energy loss spectroscopy
EES	excitation emission spectrum
EFF	empirical force field
EFFF	energy factored force field
EG	equilibrium in the gas phase
EGA	evolved gas analysis
EGG	Einstein-Guth-Gold equation
EHMO, EHT	extended Hückel molecular orbital (theory)
EIMS	electron impact mass spectrometry
EIS	electron impact spectroscopy; electrochemical impedance spectroscopy
ELISA	enzyme-linked immunosorbent assay
ELS	energy loss spectroscopy
EM	extended molarity; electron microscopy
EMAC	ethylene-methyl acrylate copolymer
emf	electromotive force
EMPA, EMA	electron probe microanalysis
emu	electromagnetic unit system
en	ethylenediamine
ENDOR	electron-nuclear double resonance
EOS	equation of state
EP	epoxy resin
EPDS	electron photodetachment spectroscopy
EPM	ethylene-propylene copolymer
EPR	electron paramagnetic resonance; ethylene propylene rubber
EPS	expanded polystyrene
EPT-76	provisional low temperature scale of 1976
EPTC	dipropylcarbamothioic acid, *S*-ethyl ester

EPXMA	electron probe x-ray microanalysis
eq, eqn	equation
eqQ	quadrupole coupling constant
erf	error function
erg	erg (energy unit)
ES	equilibrium in solution
ESA	electrostatic energy analyzer
ESCA	electron spectroscopy for chemical analysis
ESD	electron stimulated desorption
e.s.d.	estimated standard deviation
ESI	electrospray ionization
ESR	electron spin resonance
est	estimated
esu	electrostatic unit system
ET	ephemeris time; electron transfer
Et	ethyl
ETA	electrothermal analysis
ETFE	ethylene tetrafluoroethylene polymer
Etn	ethanolamine
ETO	ethylene oxide
ETS	electron tunneling spectroscopy
ETU	ethylene thiourea
EU	polyether polyurethane
Eu	Euler number
e.u.	entropy unit
eV	electronvolt
EVA	ethylene-vinyl acetate copolymer
EVE	ethyl vinyl ether
EXAFS	extended x-ray absorption fine structure (spectroscopy)
EXELFS	extended energy loss fine structure
exp	exponential function
expt	experimental
ext	external
F	farad; phenylalanine
°F	degree Fahrenheit
F	Faraday constant; force; angular momentum
f	formation reaction; femto (SI prefix for 10^{-15})
f	activity coefficient; aperture ratio; focal length; force constant; frequency; fugacity
FAB	fast atom bombardment
FAD	flavine adenine dinucleotide
FAIMS	high-field asymmetric waveform ion mobility spectrometry
FA-SIFT	flowing afterglow – selected ion-flow tube
fcc	face centered cubic
FD	field desorption
FEL	free electron laser
FEM	field emission microscopy
FEMO	free electron molecular orbital
FEP	fluorinated ethylene propylene
FET	field effect transistor
FI	field ionization
fid	free induction decay
FIM	field ion microscopy
FIR	far infrared
fl	fluid (phase)
FM	frequency modulation
Fo	Fourier number

fp	freezing point		Gra	glyceraldehyde
fpm	feet per minute		Gri	glyceric acid
fps	feet per second; foot-pound-second system		Grn	glycerone [dihydroxyacetone]
Fr	franklin		Gro	glycerol
Fr	Froude number		GTO	gaussian-type orbital
FRP	fibrous glass reinforced polyester; fiber reinforced plastic		GTP	guanosine 5'-triphosphate
Fru	fructose		Gua	guanine
FSGO	floating spherical gaussian orbitals		Gul	gulose
FT	Fourier transform		Guo	guanosine
ft	foot		GUT	grand unified theory
ft-lb	foot pound		GVB	generalized valence bond (method)
FTIR	Fourier transform infrared spectroscopy		GWS	Glashow-Weinberg-Salam (theory)
FTMS	Fourier transform mass spectrometry		Gy	gray; gigayear
FTNMR	Fourier transform nuclear magnetic resonance		H	henry; histidine
fus	fusion (melting)		*H*	enthalpy; Hamiltonian function; magnetic field
FVP	flash vacuum pyrolysis		H_0	Hubble constant
FWHM	full width at half maximum		h	helion; hour; hecto (SI prefix for 10^2)
G	gauss; guanine (in genetic code); giga (SI prefix for 10^9); glycine		*h*	Planck constant
G	electrical conductance; Gibbs energy; gravitational constant; sheer modulus		*Ha*	Hartmann number
			ha	hectare
g	gram; gas (phase)		HAM	hydrogenic atoms in molecules
g	acceleration due to gravity; degeneracy; Landé g-factor; statistical weight		hav	haversine
			Hb	hemoglobin
GABA	γ-aminobutyric acid		HCA	heterocyclic amine; hexachloroacetone
Gal	gal; galactose		HCB	hexachlorobenzene
gal	gallon		hcp	hexagonal closed packed
GalN	galactosamine		Hcy	homocysteine
GB	gas-phase basicity		HCZ, HCTZ	hydrochlorothiazide
GC	gas chromatography			
GC-MS	gas chromatography-mass spectroscopy		HDL	high-density lipoprotein
GDMS	glow discharge mass spectroscopy		HDPE	high-density polyethylene
GDP	guanosine 5'-diphosphate		HDS	hydrodesulfurization
gem	geminal (on the same carbon atom)		HEIS	high-energy ion scattering
GeV	gigaelectronvolt		HEP	high energy physics
GF	glass reinforced		Hepes	4-(2-hydroxyethyl)-1-piperazineethanesulfonic acid
GIAO	gauge invariant atomic orbital		Hepps	4-(2-hydroxyethyl)-1-piperazinepropanesulfonic acid
GIBMS	guided ion beam mass spectrometry		HF	high frequency; Hartree-Fock (method)
gl	glacial		HFA	hexafluoroacetone
Gla	4-carboxyglutamic acid		HFO	Hartree-Fock orbital
GLC	gas-liquid chromatography		hfs	hyperfine structure
Glc	glucose		HHPA	hexahydrophthalic anhydride
GlcA	gluconic acid		HIPS	high-impact polystyrene
GlcN	glucosamine		His	histidine
GlcNAc	*N*-acetylglucosamine		HMC	high strength molding compound
GlcU	glucuronic acid		HMDA	hexamethylenediamine
Gln	glutamine		HMO	Hückel molecular orbital
GLP	good laboratory practice		HMT	hexamethylenetetramine
Glu	glutamic acid		HMX	cyclotetramethylenetetranitramine
Glx	glutamine or glutamic acid (unspecified)		HN1	2-chloro-*N*-(2-chloroethyl)-*N*-ethylethanamine
Gly	glycine		HOAc	acetic acid
GMP	guanosine 5'-monophosphate		HOC	halogenated organic compound(s)
GMT	Greenwich mean time		HOMAS	harmonic oscillator model of aromatic stabilization
GPC	gel-permeation chromatography		HOMO	highest occupied molecular orbital
gpm	gallons per minute		HOSE	harmonic oscillator stabilization energy
gps	gallon per second		Hp	heptyl
Gr	Grashof number		hp	horsepower
gr	grain		HPLC	high-performance liquid chromatography

HPMS	high pressure mass spectrometry
HQ	*p*-hydroquinone
hr	hour
HRE	Hückel resonance energy
HREELS	high resolution electron energy loss spectroscopy
HREM	high resolution electron microscopy
HSAB	hard-soft acid-base (theory)
HSE	homodesmotic stabilization energy
Hse	homoserine
HVA	homovanillic acid
Hx	hexyl
Hyl	5-hydroxylysine
Hyp	hypoxanthine; 4-hydroxyproline
Hz	hertz
I	isoleucine; inositol; ionomer
I	electric current; ionic strength; moment of inertia; nuclear spin angular momentum; radiant intensity
i	square root of minus one
i	electric current
I/O	input/output
IAT	international atomic time
IC	integrated circuit
ICD	induced circular dichroism
ICP	inductive-coupled plasma
ICR	ion cyclotron resonance
ICVTST	improved canonical variational transition-state theory
ID	inside diameter
id	ideal (solution)
Ido	iodose
IdoA	iduronic acid
IDP	inosine 5'-diphosphate
IE	ionization energy
i.e.p.	isoelectric point
IEPA	independent electron pair approximation
IF	intermediate frequency
IGLO	individual gauge for localized orbitals
IIR	isobutylene-isoprene rubber (butyl rubber)
IKES	ion kinetic energy spectrometry
Ile	isoleucine
Im	imaginary part
IMFP	inelastic mean free path (of electrons)
imm	immersion
IMP	inosine 5'-monophosphate
IMPATT	impact ionization avalanche transit time
IMS	ion mobility spectrometry
in.	inch
InChI	IUPAC International Chemical Identifier
INDO	immediate neglect of differential overlap
Ino	inosine
INS	inelastic neutron scattering; ion neutralization spectroscopy
Ins	*myo*-inositol
int	internal
IP	ionization potential
IPA	isopropyl alcohol
IPMA	ion probe microanalysis
IPN	interpenetrating polymer network
IPR	isotope perturbation of resonance
IPTS	International Practical Temperature Scale
IQ	2-amino-3-methyl-3*H*-imidazo(4,5-f)quinoline
IR	infrared
IRAS	infrared reflection-absorption spectroscopy
IRC	intrinsic reaction coordinate
IRMPD	infrared multiphoton dissociation
IRMS	isotope ratio mass spectrometry
IRS	infrared spectroscopy
isc	intersystem crossing
ISE	ion-selective electrode; isodesmic stabilization energy
ISS	ion scattering spectroscopy
IT	ion trap; information technology
ITP	inosine 5'-triphosphate
ITS	International Temperature Scale (1990)
IU	international unit
IVE	isobutyl vinyl ether
J	joule; leucine or isoleucine (unspecified)
J	angular momentum; electric current density; flux; Massieu function
j	angular momentum; electric current density
JT	Jahn-Teller (effect)
K	kelvin; lysine
K	absorption coefficient; bulk modulus; equilibrium constant; kinetic energy
k	kilo (SI prefix for 10^3)
k	absorption index; Boltzmann constant; rate constant; thermal conductivity; wave vector
kat	katal (unit of catalytic activity)
kb	kilobar; kilobases (DNA or RNA)
KC-MS	Knudson cell mass spectrometry
kcal	kilocalorie
KDP	potassium dihydrogen phosphate
KE	kinetic energy
KERD	kinetic energy release distributions
keV	kiloelectronvolt
KG	kinetics in the gas phase
kg	kilogram
kgf	kilogram force
KIE	kinetic isotope effect
kJ	kilojoule
km	kilometer
Kn	Knudsen number
kPa	kilopascal
KS	kinetics in solution
kt	karat
KTP	potassium titanium phosphate
kV	kilovolt
kva	kilovolt ampere
kW	kilowatt
kwh	kilowatt hour
L	liter; lambert; leucine
L	Avogadro constant; inductance; Lagrange function; angular momentum
l	liter; liquid (phase)
l	angular momentum; length; mean free path; levorotatory
Lac	lactose
LAH	lithium aluminum hydride
lat.	latitude
lb	pound

lbf	pound force
LC	liquid chromatography; liquid crystal
LC-MS	liquid chromatography-mass spectrometry
lc	liquid crystal (phase)
LCAO	linear combination of atomic orbitals
LD	lethal dose; laser desorption
LDA	local density approximation; lithium diisopropylamide
LDL	low-density lipoprotein
LDPE	low-density polyethylene
LDV	laser-Doppler velocimetry
Le	Lewis function
LE	localization energy
LEC	liquid exchange chromatography
LED	light emitting diode
LEED	low-energy electron diffraction
LEIS	low-energy ion scattering
Leu	leucine
LFER	linear free energy relationships
LFL	lower flammable limit
LI	laser ionization
lim	limit
LIMS	laser ionization mass spectroscopy; laboratory information management system
liq	liquid
LIT	linear ion trap
LLCT	ligand to ligand charge transfer
lm	lumen
LMCT	ligand to metal charge transfer
LMMS	laser microprobe mass spectrometry
LMO	localized molecular orbital
LMR	laser magnetic resonance
ln	logarithm (natural)
LNDO	local neglect of differential overlap
log	logarithm (common)
LOMO	lowest occupied molecular orbital
long.	longitude
LPE	linear polyethylene
LPG	liquid petroleum gas
LPHP	laser-powered homogeneous pyrolysis
LPU	law of propagation of uncertainty
LSFE	linear field stabilization energy
LSI	liquid secondary ionization
LST	local sidereal time
LT	local time
LTE	local thermodynamic equilibrium
LUMO	lowest unoccupied molecular orbital
lx	lux
ly	langley
l.y.	light year
Lys	lysine
Lyx	lyxose
M	molar (as in 0.1 M solution); mega (SI prefix for 10^6); methionine
M	magnetization; molar mass; mutual inductance; torque; angular momentum component; median
M_r	molecular weight (relative molar mass)
m	meter; molal (as in 0.1 m solution); metastable (isotope); milli (SI prefix for 10^{-3})

m	magnetic dipole moment; mass; molality; angular momentum component; *meta* (locant on aromatic ring)
Ma	Mach number
MA	maleic anhydride
MAAc	methyl amyl acetate
Mal	maltose
Man	mannose
MASNMR	magic angle spinning nuclear magnetic resonance
max	maximum
Mb	myoglobin
MBE	molecular beam epitaxy
MBER	molecular beam electron resonance
MBK	methyl butyl ketone
MBOCA	4,4'-methylenebis[2-chloroaniline]
MBPT	many body perturbation theory
MBS	methyl methacrylate butadiene styrene terpolymer
MC	Monte Carlo (method)
MCAA	monochloroacetic acid
MCD	magnetic circular dichroism
MCP	microchannel plate
MCPA	(4-chloro-2-methylphenoxy)acetic acid
MCPF	modified coupled pair functional
MCS	Monte Carlo simulation
MCSCF	multiconfigurational self-consistent field (approximation)
MD	molecular dynamics (method)
MDI	methylene diphenylisocyanate
MDPE	medium density polyethylene
Me	methyl
MeCCNU	1-(2-chloroethyl)-3-(4-methylcyclohexyl)-1-nitrosourea
MeIQ	2-amino-3,4-dimethylimidazo[4,5-f]quinoline
MeIQx	2-amino-3,8-dimethylimidazo[4,5-f]quinoxaline
MEK	methyl ethyl ketone
MEP	molecular electrostatic potential
MERP	minimum energy reaction path
Mes	4-morpholineethanesulfonic acid
MESFET	metal-semiconductor field-effect transistor
Met	methionine
MeV	megaelectronvolt
meV	millielectronvolt
MF	molecular formula; melamine-formaldehyde resin
mg	milligram
MHD	magnetohydrodynamics
mi	mile
MIAK	methyl isoamyl ketone
MIBK	methyl isobutyl ketone
MIC	methyl isocyanate
MIK	methyl isobutyl ketone
MIKES	mass-analyzed ion kinetic energy spectrometry
min	minimum; minute
MINDO	modified INDO (method)
MIPK	methyl isopropyl ketone
MIR	mid infrared
misc	miscible
MKS	meter-kilogram-second system
MKSA	meter-kilogram-second-ampere system
mL, ml	milliliter
MM	molecular mechanics

mm	millimeter
MMDR	microwave-microwave double resonance
mmf	magnetomotive force
mmHg	millimeter of mercury
MNA	*m*-nitroaniline
MNDO	modified neglect of diatomic overlap
MNT	*m*-nitrotoluene
MNU	*N*-methyl-*N*-nitrosourea
MO	molecular orbital; methyl orange
MODR	microwave-optical double resonance
mol	mole
mol.wt.	molecular weight
mon	monomeric form
Mops	4-morpholinepropanesulfonic acid
MOS	metal-oxide semiconductor
MOSFET	metal-oxide semiconductor field-effect transistor
mp	melting point
MPa	megapascal
MPA	Mulliken population analysis
Mpc	megaparsec
MPD	2-methyl-2,4-pentanediol
MPI	multiphoton ionization
MPTP	1,2,3,6-tetrahydro-1-methyl-4-phenylpyridine
MR	methyl red
MRD	multireference double substitution (method)
MRI	magnetic resonance imaging
mRNA	messenger RNA
MS	mass spectroscopy
ms	millisecond
MSA	methanesulfonic acid
MSDS	Material Safety Data Sheet
MSF	methanesulfonyl fluoride
MS-K	mass spectroscopy – kinetic method
MSL	mean sea level
MTBE	methyl *tert*-butyl ether
MTD	maximum tolerable dose
Mur	muramic acid
mV	millivolt
MVK	methyl vinyl ketone
MW	megawatt; microwave; molecular weight
mW	milliwatt
MWD	molecular weight distribution
Mx	maxwell
N	newton; asparagine
N	angular momentum; neutron number; number density
N_A	Avogadro constant
n	neutron; nano (SI prefix for 10^{-9})
n	amount of substance; number density; principal quantum number; refractive index; normal (in chemical formulas)
NAA	nuclear activation analysis; 1-naphthaleneacetic acid
NAAD	nicotinic acid adenine dinucleotide
NAD	nicotinamide adenine dinucleotide
NADH	reduced NAD
NADP	NAD phosphate
NANA	*N*-acetylneuraminic acid
NAO	natural atomic orbital
NBO	natural bond orbital

nbp	normal boiling point
NBR	nitrile butadiene rubber [poly(butadiene-co-acrylonitrile)]
NDELA	*N*-nitrosodiethanolamine
NEDOR	nuclear electron double resonance
NEM	*N*-ethylmorpholine
Neu	neuraminic acid
NEXAFS	near-edge x-ray absorption fine structure
ng	nanogram
NHO	natural hybrid orbital
NHOMO	next-to-highest occupied molecular orbital
NICI	negative ion chemical ionization
NICS	nuclear independent chemical shift
NIR	near infrared; ribosylnicotinamide
nm	nanometer
NMN	β-nicotinamide mononucleotide
NMR	nuclear magnetic resonance
Nn	nonyl
NNDO	neglect of nonbonded differential overlap
NO	natural orbital
NOE	nuclear Overhauser effect
NOEL	no-observed-effect level
NOx	nitrogen oxides
NP	nitropyrene
NPA	natural population analysis
NQR	nuclear quadrupole resonance
NR	natural rubber
NRA	nuclear reaction analysis
ns	nanosecond
NSE	neutron spin echo
NTA	nitrilotriacetic acid
NTP	normal temperature and pressure
Nu	nucleophile
Nu	Nusselt number
o	*ortho* (locant on aromatic ring)
OAA	oxaloacetic acid
obs, obsd	observed
Oc	octyl
OD	optical density; outside diameter
ODMR	optically detected magnetic resonance
Oe	oersted
OFGF	outer valence Green's function (method)
ONA	*o*-nitroaniline
ORD	optical rotatory dispersion
Oro	orotate; orotidine
oz	ounce
P	poise; peta (SI prefix for 10^{15}); proline
P	power; pressure; probability; sound energy flux
p	proton; pico (SI prefix for 10^{-12})
p	dielectric polarization; electric dipole moment; momentum; pressure; bond order; *para* (as aromatic ring locant)
Pa	pascal
PA	proton affinity; pyrrolizidine alkaloid; polyamide (nylon)
PAA	poly(acrylic acid)
PABA	*p*-aminobenzoic acid
PABS	*p*-aminobenzenesulfonamide
PAC	photoacoustic calorimetry

PAH	polycyclic aromatic hydrocarbon(s)		PI	polyimide
PAI	polyamide-imide		pI	isoelectric point
PAL	polyaniline		PIB	polyisobutylene
PAM	polyacrylamide		PIMS	photoionization mass spectrometry
PAN	1-(2-pyridylazo)-2-naphthol; polyacrylonitrile		PIN	p-intrinsic-n (diode)
PAR	4-(2′-pyridylazo)resorcinol		Pipes	1,4-piperazinediethanesulfonic acid
PARA	polyaryl amide		PIV	particle-image velocimetry
PAS	photoacoustic spectroscopy; polyarylsulfone		PIXE	particle induced x-ray emission
PB	polybutylene		pK	negative log of ionization constant
PBA	poly(butyl acrylate)		PLM	principle of least motion
PBAN	polybutylene-acrylonitrile copolymer		PLOT	porous-layer open-tabular (column)
PBB	polybrominated biphenyl		PLS	partial least squares
PBD	poly(1,3-butadiene)		pm	picometer
PBI	polybenzimidazole		PMA	poly(methyl acrylate)
PBMA	poly(butyl methacrylate)		PMAC	phenylmercuric acetate
PBS	polybutadiene-styrene copolymer		PMMA	poly(methyl methacrylate)
PBT	poly(butylene terephthalate)		PMO	perturbation MO (theory)
PC	paper chromatography; photocalorimetry; polycarbonate		PMP	polymethylpentene
pc	parsec		PMS	polymethylstyrene; p-methylstyrene
PCB	polychlorinated biphenyl		PNA	p-nitroaniline
PCHO	paraldehyde (2,4,6-trimethyl-1,3,5-trioxane)		PNDO	partial neglect of differential overlap
PCL	polycaprolactone		PNO	pair natural orbitals
PCM	polarizable continuum model		PNRA	prompt nuclear reaction analysis
PCNB	pentachloronitrobenzene		PNT	p-nitrotoluene
PCP	pentachlorophenol		PO	polyolefin
PCR	polymerase chain reaction		POAV	π-orbital axis vector
PCT	poly(cyclohexylene terephthalate)		pol	polymeric form
PCTFE	polymonochlorotrifluoroethylene		POM	polyoxymethylene
PD	potential difference		POx	phosphorus oxides
PDB	p-dichlorobenzene		PP	polypropylene
pdl	poundal		ppb	parts per billion
PDMS	poly(dimethylsiloxane)		PPC	chlorinated polypropylene
PE	polyethylene		PPE	poly(phenylene ether)
Pe	pentyl		ppm	parts per million
Pe	Péclet number		PPO	poly(phenylene oxide)
pe	probable error		PPOX	polypropylene oxide
PEA	poly(ethyl acrylate)		PPP	Pariser-Parr-Pople (method)
PEEK	poly(ether ether ketone)		PPS	poly(phenylene sulfide)
PEG	poly(ethylene glycol)		PPSU	poly(phenylene sulfone)
PEI	polyetherimide		PPT	poly(propylene terephthalate)
PEK	polyetherketone		ppt	parts per thousand; precipitate
PEL	permissible exposure limit		Pr	propyl
PEO	poly(ethylene oxide)		*Pr*	Prandtl number
PES	photoelectron spectroscopy; potential energy surface; polyethersulfone		PRDDO	partial retention of diatomic differential overlap
			Pro	proline
PET	positron emission tomography; poly(ethylene terephthalate); pentaerythritol tetranitrate		PS	photoelectron spectroscopy; polystyrene
peth	petroleum ether		ps	picosecond
PEX	crosslinked polyethylene		PSD	photon stimulated desorption
PF	phenol-formaldehyde resin		psi	pounds per square inch
pf	power factor		psia	pounds per square inch absolute
PFOA	perfluorooctanoic acid		psig	pounds per square inch gage
pg	picogram		PT	perturbation theory
Ph	phenyl		pt	pint
pH	negative log of hydrogen ion concentration		PTFE	poly(tetrafluoroethylene)
Phe	phenylalanine		PTME	poly(tetramethylene terephthalate)
PhIP	2-amino-1-methyl-6-phenylimidazo[4,5-b]pyridine		PTMS	propyltrimethoxysilane
PHPMS	pulsed high pressure mass spectrometry		PTP	p-terphenyl

PTU	phenylthiourea
PU	polyurethane
Pu	purine
PVA	poly(vinyl alcohol)
PVAc	poly(vinyl acetate)
PVC	poly(vinyl chloride)
PVD	physical vapor deposition
PVDC	poly(vinylidene chloride)
PVDF	poly(vinylidene fluoride)
PVF	poly(vinyl fluoride)
PVK	poly(vinyl carbazole)
PVME	poly(methyl vinyl ether)
PVOH	poly(vinyl alcohol)
PVP	poly(vinyl pyrrolidone)
PVT	pressure-volume-temperature
Py	pyrimidine
PyMS	pyrolysis mass spectrometry
p.z.c.	point of zero charge
Q	electric charge; heat; partition function; quadrupole moment; radiant energy; vibrational normal coordinate; glutamine
q	electric field gradient; flow rate; heat; wave vector (phonons)
QCD	quantum chromodynamics
QCI	quadratic configuration interaction
QCT	quasi-classical trajectory (method)
QED	quantum electrodynamics
Q.E.D.	quod erat demonstrandum (which was to be proved)
QIT	quadrupole ion trap
QMRE	quantum mechanical resonance energy
QMS	quadrupole mass spectrometry
QSAR	quantitative structure-activity relations
QSO	quasi-stellar object
qt	quart
quad	quadrillion BTU ($=1.055 \cdot 10^{18}$ joules)
Qui	quinovose
q.v.	quod vide (which you should see)
R	roentgen; arginine; alkyl radical (in chemical formulas)
°R	degree Rankine
R	electrical resistance; gas constant; molar refraction; Rydberg constant; coefficient of multiple correlation reaction (as in $\Delta_r H$)
r	
r	position vector; radius
RA	right ascension
rad	radian
RAIRS	reflection-absorption infrared spectroscopy
RAM	random access memory
RBS	Rutherford back scattering
Rbu, Rul	ribulose
RCI	ring current index
RDA	rubidium dihydrogen arsenate
RDS	rate determining step
RDX	Royal Demolition Explosive (hexahydro-1,3,5-trinitro-1,3,5-triazine)
Re	real part
RE	resonance energy
RED	radial electron distribution
REELS	reflection electron energy loss spectroscopy
REM	reflection electron microscopy

rem	roentgen equivalent man
REMPI	resonance-enhanced multiphoton ionization
REPE	resonance energy per electron
RF	radiofrequency
RGA	residual gas analyzer
Rha	rhamnose
RHEED	reflection high-energy electron diffraction
RHF	restricted Hartree-Fock (theory)
RI	resonance ionization
RIA	radioimmunoassay
Rib	ribose
Ribulo	ribulose
rms	root-mean-square
RNA	ribonucleic acid
RNase	ribonuclease
ROHF	restricted open shell Hartree-Fock
ROM	read only memory
ROMP	ring opening metathesis polymerization
ROP	ring opening polymerization
RPA	random phase approximation
RPH	reaction path Hamiltonian
RPLC	reversed-phase liquid chromatography
rpm	revolutions per minute
rps	revolutions per second
RRK	Rice-Ramsperger-Kassel (theory)
RRKM	Rice-Ramsperger-Kassel-Marcus (theory)
rRNA	ribosomal RNA
RRS	resonance Raman spectroscopy
RS	Raman spectroscopy
RSC	reaction-solution calorimetry
Ry	rydberg
S	siemens; serine
S	area; entropy; probability current density; Poynting vector; symmetry coordinate; spin angular momentum
s	second; solid (phase)
s	path length; spin angular momentum; symmetry number; sedimentation coefficient; solubility; symmetrical (as stereochemical descriptor)
SAED	selected area electron diffraction
SALC	symmetry adapted linear combinations
SALI	surface analysis by laser ionization
SAM	scanning Auger microscopy
SAMS	self-assembled monolayers
SANS	small angle neutron scattering
SAR	structure-activity relationship
Sar	sarcosine
sat, satd	saturated
SAXS	small angle x-ray scattering
SB	styrene butadiene copolymer
SBS	styrene butadiene styrene block copolymer
Sc	Schmidt number
SC	spin-coupled (method)
SCD	state correlation diagram
SCE	saturated calomel electrode
SCF	self-consistent field (method); supercritical fluid
SCP	single cell protein
SCR	silicon-controlled rectifier
SCRF	self-consistent reaction field (method)

sd	standard deviation
SDA	sulfadiazine
SDW	spin density wave
SE	strain energy
SEBS	styrene ethylene butylene styrene block copolymer
SEC	size exclusion chromatography
sec	secant; second
sec	secondary (in chemical name)
SECSY	spin-echo correlated spectroscopy
Sed	sedoheptulose
SEELFS	surface extended energy loss fine structure
SEM	scanning electron microscopy; standard error of the mean
sepn	separation
Ser	serine
SERS	surface-enhanced Raman spectroscopy
SET	single electron transfer
SEXAF	surface extended x-ray absorption fine structure
SFC	supercritical fluid chromatography
Sh	Sherwood number
Shy	thiohypoxanthine
SI	International System of Units; surface ionization
SID	surface-induced dissociation
SILAR	successive ionic layer adsorption and reaction
SIM	selected ion monitoring
SIMS	secondary-ion spectroscopy
sin	sine
sinh	hyperbolic sine
SIPN	semi-interpenetrating polymer network
SIS	styrene isoprene styrene block copolymer
SLAM	scanning laser acoustic microscopy
SLUMO	second lowest unoccupied molecular orbital
SMILES	simplified molecular input line entry system
SMMA	styrene methyl methacrylate copolymer
SMO	semiempirical molecular orbital
SMOW	Standard Mean Ocean Water (Vienna)
SNMS	sputtered neutral mass spectroscopy
Sno	thiouridine
SNU	solar neutrino unit
SOJT	second-order Jahn-Teller (effect)
sol	soluble; solution
soln, sln	solution
SOMO	singly occupied molecular orbital
Sor	sorbose
sp gr	specific gravity
SPM	scanned probe microscopy
SPST	single-pulse shock tubes
sq	square
Sr	Strouhal number
sr	steradian
Srd	6-thioinosine
SSMS	source spark mass spectroscopy
St	stoke
St	Stanton number
std, stnd	standard (state)
STEL	short-term exposure limit
STEM	scanning transmission electron microscope

STM	scanning tunneling microscopy
STO	Slater-type orbital
STP	standard temperature and pressure
sub, subl	sublimes; sublimation
Suc, Sac	sucrose
Sur	thiouracil
Sv	sievert
SWIFT	stored waveform inverse Fourier transform
T	tesla; tera (SI prefix for 10^{12}); threonine
T	kinetic energy; period; term value; temperature (thermodynamic); torque; transmittance
t	metric tonne; triton
t	Celsius temperature; thickness; time; transport number
TAC	time-to-amplitude converter
TAI	International Atomic Time
Tal	talose
tan	tangent
tanh	hyperbolic tangent
Taps	3-{[2-hydroxy-1,1-bis(hydroxymethyl)ethyl]amino}-1-propanesulfonic acid
TBE	1,1,2,2-tetrabromoethane
TBP	tributyl phosphate
TC	titration calorimetry
TCA	trichloroacetic acid
TCB, TCBA	2,3,6-trichlorobenzoic acid
TCE	trichloroethylene
TCG	Geocentric Coordinated Time
TCNE	tetracyanoethylene
TCNQ	tetracyanoquinodimethane
TCP	tricresyl phosphate
TCSCF	two configuration self-consistent field
TDA	toluene-2,4-diamine
TDI	toluene diisocyanate
tDNA	transfer DNA
TE	transverse electric
TEA	triethanolamine; triethylamine
TED	transferred electron device; transmission electron diffraction
TEDA	triethylenediamine
TEELS	transmission electron energy loss spectroscopy
TEM	transverse electromagnetic; transmission electron microscope
temp	temperature
TEO	thermoplastic elastic olefin
TEPP	tetraethyl pyrophosphate
tert	tertiary (in chemical name)
Tes	2-{[2-hydroxy-1,1-bis(hydroxymethyl)ethyl]amino}-1-propanesulfonic acid
TFD	Thomas-Fermi-Dirac (method)
TFE	tetrafluoroethylene
TGA	thermogravimetric analysis
Thd	ribosylthymine
THEED	transmission high energy electron diffraction
theor	theoretical
thf, THF	tetrahydrofuran
THQ	1,2,3,4-tetrahydroquinoline
Thr	threonine
Thy	thymine
TI	thermal ionization

TIPA	triisopropanolamine
TL	thermoluminescence
TLC	thin-layer chromatography
TLV	threshold limit value
TM	transverse magnetic
TMAB	tetrabutylammonium bromide
TMAO	trimethylamine oxide
TMCP	tri-*m*-cresyl phosphate
TMEDA	*N,N,N',N'*-tetramethyl-1,2-ethanediamine
TMMV	threshold molecular weight value
TMS	tetramethylsilane
TNA	2,4,6-trinitroaniline
TNB	1,3,5-trinitrobenzene
TNM	tetranitromethane
TNT	2,4,6-trinitrotoluene
TOCP	tri-*o*-cresyl phosphate
TOF	turnover frequency
TOF-MS	time-of-flight mass spectrometer
tol	tolyl
TON	turnover number
TOPO	trioctylphosphine oxide
Torr	torr (pressure unit)
TOTP	tri-*o*-tolyl phosphate
TPE	thermoplastic elastomer
TPTA	triphenyltin acetate
TPTC	triphenyltin chloride
TRE	topological resonance energy
Tre	trehalose
Tricine	*N*-[2-hydroxy-1,1-bis(hydroxymethyl)ethyl]glycine
Tris	2-amino-2-(hydroxymethyl)-1,3-propanediol
TRMC	time-resolved microwave conductivity
tRNA	transfer RNA
Trp	tryptophan
trs	transition
TS	transition state
TSS	transition state spectroscopy
TST	generalized transition-state theory
TTF	tetrathiofulvalene
Tyr	tyrosine
U	uracil (in genetic code)
U	electric potential difference; internal energy
u	unified atomic mass unit
u	Bloch function; electric mobility; velocity
ua	astronomical unit (AU is also used)
UBFF	Urey-Bradley force field
UDMH	1,1-dimethylhydrazine
UDP	uridine 5'-diphosphate
UHF	ultrahigh frequency; unrestricted Hartree-Fock (method)
UHMWPE	ultrahigh molecular weight polyethylene
ULDPE	ultra low density polyethylene
ULPE	ultra linear polyethylene
UMP	uridine 5'-monophosphate
uns, unsym	unsymmetrical (as chemical descriptor)
UPS, UPES	ultraviolet photoelectron spectroscopy
Ura	uracil
Urd	uridine
USP	United States Pharmacopeia
UT	universal time
UTC	coordinated universal time
UTP	uridine 5'-triphosphate
UV	ultraviolet
V	volt; valine
V	electric potential; potential energy; volume
v	reaction rate; specific volume; velocity; vibrational quantum number; vicinal (as chemical descriptor)
v/v	volume per volume (volume of solute divided by volume of solution, expressed as percent)
VA	vinyl acetate, vanillic acid
Val	valine
vap	vaporization
VAT	vibration assisted tunneling
VB	valence band; valence bond (theory)
VCD	vibrational circular dichroism
VDW	van der Waals interaction
VHF	very high frequency
vic	vicinal (on adjacent carbon atom)
VIS	visible region of the spectrum
vit	vitreous (phase)
VLDPE	very low density polyethylene
VLPP	very low pressure pyrolysis
VMA	vanilmandelic acid
VOC	volatile organic compound(s)
VOFF	valence orbital force field
VPC	vapor phase chromatography
VSEPR	valence shell electron-pair repulsion (method)
VSIP	valence state ionization potential
VSLI	very large scale integrated (circuit)
VSMOW	Vienna Standard Mean Ocean Water
VTCS	vinyltrichlorosilane
VUV	vacuum ultraviolet
W	watt; tryptophan
W	radiant energy; statistical weight; work
w	energy density; mass fraction; velocity; work
w/v	weight per volume (mass of solute divided by volume of solution, usually expressed as g/100 mL)
w/w	weight per weight (mass of solute divided by mass of solution, expressed as percent)
WAXS	wide angle x-ray scattering
Wb	weber
We	Weber number
WKB	Wentzel-Kramers-Brillouin (approximation)
WLF	Williams-Landel-Ferry (equation)
WLN	Wiswesser line notation
wt	weight
X	X unit; halogen (in chemical formula)
X	reactance
x	mole fraction
X, Xaa	unspecified amino acid
XAFS	x-ray absorption fine structure
Xan	xanthine
XANES	x-ray absorption near-edge structure
Xao	xanthosine
Xle	leucine or isoleucine (unspecified)
XLPE	crosslinked polyethylene
Xlu, Xul	xylulose
XPS, XPES	x-ray photoelectron spectroscopy

XRD	x-ray diffraction
XRF	x-ray fluorescence
XRS	x-ray spectroscopy
Xyl	xylose
Y	yotta (SI prefix for 10^{24}); tyrosine
Y	admittance; Planck function; Young's modulus
y	yocto (SI prefix for 10^{-24})
y	mole fraction for gas (when x refers to liquid phase)
y, yr	year
YAG	yttrium aluminum garnet
yd	yard
YIG	yttrium iron garnet
Z	zetta (SI prefix for 10^{21}); glutamine or glutamic acid (unspecified)
Z	atomic number; compression factor; collision number; impedance; partition function; zusammen (*cis*-configuration)
z	zepto (SI prefix for 10^{-21})
z	charge number (of an ion); collision frequency factor
ZDO	zero differential overlap
ZINDO	Zerner's INDO method
ZPE, ZPVE	zero point vibrational energy
ZULU	Greenwich mean time
α	alpha particle
α	absorption coefficient; degree of dissociation; electric polarizability; expansion coefficient; fine structure constant
β	beta particle
β	reciprocal temperature parameter ($= 1/kT$)
γ	photon; gamma (obsolete mass unit $= \mu g$)
γ	activity coefficient; conductivity; magnetogyric ratio; mass concentration; ratio of heat capacities; surface tension
Γ	Grüneisen parameter; level width; surface concentration
Δ	inertial defect; mass excess
δ	chemical shift; Dirac delta function; Kronecker delta; loss angle
ε	emittance; Levi-Civita symbol; linear strain; molar absorption coefficient; permittivity

ζ	Coriolis coupling constant; electrokinetic potential
η	overpotential; viscosity
κ	compressibility; conductivity; magnetic susceptibility; molar absorption coefficient
λ	absolute activity; radioactive decay constant; thermal conductivity; wavelength
Λ	angular momentum; ionic conductivity
μ	muon; micro (SI prefix for 10^{-6})
μ	chemical potential; electric dipole moment; electric mobility; friction coefficient; Joule-Thompson coefficient; magnetic dipole moment; mobility; permeability
μF	microfarad
μg	microgram
μm	micrometer
μs	microsecond
ν	frequency; kinematic velocity; stoichiometric number
ν_e	neutrino
ν	wavenumber
Π	osmotic pressure; Peltier coefficient
π	pion
ρ	density; reflectance; resistivity
σ	electrical conductivity; cross section; normal stress; shielding constant (NMR); Stefan-Boltzmann constant; surface tension; standard deviation
τ	transmittance; chemical shift; shear stress; relaxation time
Φ	magnetic flux; potential energy; radiant power; work function
φ	electrical potential; fugacity coefficient; osmotic coefficient; quantum yield; wavefunction
χ	magnetic susceptibility
χ_e	electric susceptibility
ψ	wavefunction
Ω	ohm
Ω	axial angular momentum; solid angle
ω	circular frequency; angular velocity; harmonic vibration wavenumber; statistical weight

GREEK, RUSSIAN, AND HEBREW ALPHABETS

The following table presents the Hebrew, Greek, and Russian alphabets, their letters, the names of the letters, and the English equivalents.

Hebrew[1,3]			Greek[4]			Russian	
א	aleph	' [2]	Α α	alpha	a	А а	a
ב	beth	b, bh	Β β	beta	b	Б б	b
ג	gimel	g, gh	Γ γ	gamma	g, n	В в	v
ד	daleth	d, dh	Δ δ	delta	d	Г г	g
ה	he	h	Ε ε	epsilon	e	Д д	d
ו	waw	w	Ζ ζ	zeta	z	Е е	e
ז	zayin	z	Η η	eta	ē	Ж ж	zh
ח	heth	ḥ	Θ θ	theta	th	З з	z
ט	teth	ṭ	Ι ι	iota	i	И и Й й	i, ï
י	yodh	y	Κ κ	kappa	k	К к	k
כ ך	kaph	k, kh	Λ λ	lambda	l	Л л	l
ל	lamedh	l	Μ μ	mu	m	М м	m
מ ם	mem	m	Ν ν	nu	n	Н н	n
נ ן	nun	n	Ξ ξ	xi	x	О о	o
ס	samekh	s	Ο ο	omicron	o	П п	p
ע	ayin	ʿ	Π π	pi	p	Р р	r
פ ף	pe	p, ph	Ρ ρ	rho	r, rh	С с	s
צ ץ	sadhe	ṣ	Σ σ ς	sigma	s	Т т	t
ק	qoph	q	Τ τ	tau	t	У у	u
ר	resh	r	Υ υ	upsilon	y, u	Ф ф	f
שׂ	sin	ś	Φ φ	phi	ph	Х х	kh
שׁ	shin	sh	Χ χ	chi	ch	Ц ц	ts
ת	taw	t, th	Ψ ψ	psi	ps	Ч ч	ch
			Ω ω	omega	ō	Ш ш	sh
						Щ щ	shch
						Ъ ъ[5]	"
						Ы ы	y
						Ь ь[6]	'
						Э э	e
						Ю ю	yu
						Я я	ya

[1] Where two forms of a letter are given, the second one is the form used at the end of a word.
[2] Not represented in transliteration when initial.
[3] The Hebrew letters are primarily consonants; a few of them are also used secondarily to represent certain vowels, when provided at all, by means of a system of dots or strokes adjacent to the consonated characters.
[4] The letter gamma is transliterated "n" only before velars; the letter upsilon is transliterated "u" only as the final element in diphthongs.
[5] This sign indicates that the immediately preceding consonant is not palatized even though immediately followed by a palatized vowel.
[6] This sign indicates that the immediately preceding consonant is palatized even though not immediately followed by a palatized vowel.

DEFINITIONS OF SCIENTIFIC TERMS

Brief definitions of selected terms of importance in chemistry, physics, and related fields of science are given in this section. The selection process emphasizes the following types of terms:

- Physical quantities
- Units of measure
- Classes of chemical compounds and materials
- Important theories, laws, and basic concepts.

Individual chemical compounds are not included.

Definitions have taken wherever possible from the recommendations of international or national bodies, especially the International Union of Pure and Applied Chemistry (IUPAC) and International Organization for Standardization (ISO). For physical quantities and units, the recommended symbol is also given. The source of such definitions is indicated by the reference number in brackets following the definition. In many cases these official definitions have been edited in the interest of stylistic consistency and economy of space. The user is referred to the original source for further details.

An asterisk (*) following a term indicates that further information can be found by consulting the index of this handbook under the entry for that term.

References

1. *ISO Standards Handbook 2, Units of Measurement*, International Organization for Standardization, Geneva, 1992.
2. *Quantities, Units, and Symbols in Physical Chemistry, Second Edition*, International Union of Pure and Applied Chemistry, Blackwell Scientific Publications, Oxford, 1993.
3. *Compendium of Chemical Terminology*, International Union of Pure and Applied Chemistry, Blackwell Scientific Publications, Oxford, 1987.
4. *A Guide to IUPAC Nomenclature of Organic Compounds*, International Union of Pure and Applied Chemistry, Blackwell Scientific Publications, Oxford, 1993.
5. *Glossary of Class Names of Organic Compounds and Reactive Intermediates Based on Structure, Pure and Applied Chemistry*, 67, 1307, 1995.
6. *Compendium of Analytical Nomenclature*, International Union of Pure and Applied Chemistry, Blackwell Scientific Publications, Oxford, 1987.
7. *Nomenclature of Inorganic Chemistry*, International Union of Pure and Applied Chemistry, Blackwell Scientific Publications, Oxford, 1990.
8. *Glossary of Basic Terms in Polymer Science, Pure and Applied Chemistry*, 68, 2287, 1996.
9. *The International Temperature Scale of 1990, Metrologia*, 27, 107, 1990.
10. *Compilation of ASTM Standard Definitions*, American Society of Testing and Materials, Philadelphia, 1990.
11. *ASM Metals Reference Book*, American Society for Metals, Metals Park, OH, 1983.

Ab initio **method** - An approach to quantum-mechanical calculations on molecules which starts with the Schrödinger equation and carries out a complete integration, without introducing empirical factors derived from experimental measurement.

Absorbance (A) - Defined as $-\log(1-\alpha) = \log(1/\tau)$, where α is the absorptance and τ the transmittance of a medium through which a light beam passes. [2]

Absorbed dose (D) - For any ionizing radiation, the mean energy imparted to an element of irradiated matter divided by the mass of that element. [1]

Absorptance (α) - Ratio of the radiant or luminous flux in a given spectral interval absorbed in a medium to that of the incident radiation. Also called absorption factor. [1]

Absorption coefficient (a) - The relative decrease in the intensity of a collimated beam of electromagnetic radiation, as a result of absorption by a medium, during traversal of an infinitesimal layer of the medium, divided by the length traversed. [1]

Absorption coefficient, molar (ε) - Absorption coefficient divided by amount-of-substance concentration of the absorbing material in the sample solution ($\varepsilon = a/c$). The SI unit is m^2/mol. Also called extinction coefficient, but usually in units of $mol^{-1}dm^3cm^{-1}$. [2]

Acceleration - Rate of change of velocity with respect to time.

Acceleration due to gravity (g)* - The standard value (9.80665 m/s^2) of the acceleration experienced by a body in the earth's gravitational field. [1]

Acenes - Polycyclic aromatic hydrocarbons consisting of fused benzene rings in a rectilinear arrangement. [5]

Acid - Historically, a substance that yields an H^+ ion when it dissociates in solution, resulting in a pH<7. In the Brönsted definition, an acid is a substance that donates a proton in any type of reaction. The most general definition, due to G.N. Lewis, classifies any chemical species capable of accepting an electron pair as an acid.

Acid dissociation constant (K_a)* - The equilibrium constant for the dissociation of an acid HA through the reaction $HA + H_2O \rightleftharpoons A^- + H_3O^+$. The quantity $pK_a = -\log K_a$ is often used to express the acid dissociation constant.

Actinides - The elements of atomic number 89 through 103, e.g., Ac, Th, Pa, U, Np, Pu, Am, Cm, Bk, Cf, Es, Fm, Md, No, Lr. [7]

Activation energy* - In general, the energy that must be added to a system in order for a process to occur, even though the process may already be thermodynamically possible. In chemical kinetics, the activation energy is the height of the potential barrier separating the products and reactants. It determines the temperature dependence of the reaction rate.

Activity - For a mixture of substances, the absolute activity λ of substance B is defined as $\lambda_B = \exp(\mu_B/RT)$, where μ_B is the chemical potential of substance B, R the gas constant, and T the thermodynamic temperature. The relative activity a is defined as $a_B = \exp[(\mu_B-\mu_B°)/RT]$, where $\mu_B°$ designates the chemical potential in the standard state. [2]

Activity coefficient (γ)* - Ratio of the activity a_B of component B of a mixture to the concentration of that component. The value of γ depends on the method of stating the composition. For mole fraction x_B, the relation is $a_B = \gamma_B x_B$; for molarity c_B, it is $a_B = \gamma_B c_B/c°$, where $c°$ is the standard state composition (typically chosen as 1 mol/L); for molality m_B, it is $a_B = \gamma_B m_B/m°$, where $m°$ is the standard state molality (typically 1 mol/kg). [2]

Activity, of radioactive substance (A) - The average number of spontaneous nuclear transitions from a particular energy state occurring in an amount of a radionuclide in a small time interval divided by that interval. [1]

Acyl groups - Groups formed by removing the hydroxy groups from oxoacids that have the general structure RC(=O)(OH) and replacement analogues of such acyl groups. [5]

Adiabatic process - A thermodynamic process in which no heat enters or leaves the system.

Admittance (Y) - Reciprocal of impedance. $Y = G + iB$, where G is conductance and B is susceptance. [1]

Adsorption - A process in which molecules of gas, of dissolved substances in liquids, or of liquids adhere in an extremely thin layer to surfaces of solid bodies with which they are in contact. [10]

Albedo* - The ratio of the light reflected or scattered from a surface to the intensity of incident light. The term is often used in reference to specific types of terrain or to entire planets.

Alcohols - Compounds in which a hydroxy group, -OH, is attached to a saturated carbon atom. [5]

Aldehydes - Compounds RC(=O)H, in which a carbonyl group is bonded to one hydrogen atom and to one R group. [5]

Aldoses - Aldehydic parent sugars (polyhydroxyaldehydes H[CH(OH)]$_n$C(=O)H, $n>1$) and their intramolecular hemiacetals. [5]

Aldoximes - Oximes of aldehydes: RCH=NOH. [5]

Alfvén number (Al) - A dimensionless quantity used in plasma physics, defined by $Al = v(\rho\mu)^{1/2}/B$, where ρ is density, v is velocity, μ is permeability, and B is magnetic flux density. [2]

Alfvén waves - Very low frequency waves which can exist in a plasma in the presence of a uniform magnetic field. Also called magnetohydrodynamic waves.

Alicyclic compounds - Aliphatic compounds having a carbocyclic ring structure which may be saturated or unsaturated, but may not be a benzenoid or other aromatic system. [5]

Aliphatic compounds - Acyclic or cyclic, saturated or unsaturated carbon compounds, excluding aromatic compounds. [5]

Alkali metals - The elements lithium, sodium, potassium, rubidium, cesium, and francium.

Alkaline earth metals - The elements calcium, strontium, barium, and radium. [7]

Alkaloids - Basic nitrogen compounds (mostly heterocyclic) occurring mostly in the plant kingdom (but not excluding those of animal origin). Amino acids, peptides, proteins, nucleotides, nucleic acids, and amino sugars are not normally regarded as alkaloids. [5]

Alkanes - Acyclic branched or unbranched hydrocarbons having the general formula C_nH_{2n+2}, and therefore consisting entirely of hydrogen atoms and saturated carbon atoms. [5]

Alkenes - Acyclic branched or unbranched hydrocarbons having one carbon-carbon double bond and the general formula C_nH_{2n}. Acyclic branched or unbranched hydrocarbons having more than one double bond are alkadienes, alkatrienes, etc. [5]

Alkoxides - Compounds, ROM, derivatives of alcohols, ROH, in which R is saturated at the site of its attachment to oxygen and M is a metal or other cationic species. [5]

Alkyl groups - Univalent groups derived from alkanes by removal of a hydrogen atom from any carbon atom: C_nH_{2n+1}-. The groups derived by removal of a hydrogen atom from a terminal carbon atom of unbranched alkanes form a subclass of normal alkyl (n-alkyl) groups. The groups RCH$_2$-, R$_2$CH-, and R$_3$C- (R not equal to H) are primary, secondary, and tertiary alkyl groups, respectively. [5]

Alkynes - Acyclic branched or unbranched hydrocarbons having a carbon-carbon triple bond and the general formula C_nH_{2n-2}, RC≡CR´. Acyclic branched or unbranched hydrocarbons having more than one triple bond are known as alkadiynes, alkatriynes, etc. [5]

Allotropy - The occurrence of an element in two or more crystalline forms.

Allylic groups - The group CH$_2$=CHCH$_2$- (allyl) and derivatives formed by substitution. The term 'allylic position' or 'allylic site' refers to the saturated carbon atom. A group, such as -OH, attached at an allylic site is sometimes described as "allylic". [5]

Amagat volume unit - A non-SI unit previously used in high pressure science. It is defined as the molar volume of a real gas at one atmosphere pressure and 273.15 K. The approximate value is 22.4 L/mol.

Amides - Derivatives of oxoacids R(C=O)(OH) in which the hydroxy group has been replaced by an amino or substituted amino group. [5]

Amine oxides - Compounds derived from tertiary amines by the attachment of one oxygen atom to the nitrogen atom: R$_3$N$^+$-O$^-$. By extension the term includes the analogous derivatives of primary and secondary amines. [5]

Amines - Compounds formally derived from ammonia by replacing one, two, or three hydrogen atoms by hydrocarbyl groups, and having the general structures RNH$_2$ (primary amines), R$_2$NH (secondary amines), R$_3$N (tertiary amines). [5]

Amino acids* - Compounds containing both a carboxylic acid group (-COOH) and an amino group (-NH$_2$). The most important are the α-amino acids, in which the -NH$_2$ group in attached to the C atom adjacent to the -COOH group. In the β-amino acids, there is an intervening carbon atom. [4]

Ampere (A)* - The SI base unit of electric current. [1]

Ampere's law - The defining equation for the magnetic induction B, viz., $dF = Idl \times B$, where dF is the force produced by a current I flowing in an element of the conductor dl pointing in the direction of the current.

Ångström (Å) - A unit of length used in spectroscopy, crystallography, and molecular structure, equal to 10^{-10} m.

Angular momentum (L) - The angular momentum of a particle about a point is the vector product of the radius vector from this point to the particle and the momentum of the particle; i.e., $L = r \times p$. [1]

Angular velocity (ω) - The angle through which a body rotates per unit time.

Anilides - Compounds derived from oxoacids R(C=O)(OH) by replacing the -OH group by the -NHPh group or derivative formed by ring substitution. Also used for salts formed by replacement of a nitrogen-bound hydrogen of aniline by a metal. [5]

Anion - A negatively charged atomic or molecular particle.

Antiferroelectricity* - An effect analogous to antiferromagnetism in which electric dipoles in a crystal are ordered in two sublattices that are polarized in opposite directions, leading to zero net polarization. The effect vanishes above a critical temperature.

Antiferromagnetism* - A type of magnetism in which the magnetic moments of atoms in a solid are ordered into two antiparallel aligned sublattices. Antiferromagnets are characterized by a zero or small positive magnetic susceptibility. The susceptibility increases with temperature up to a critical value, the Néel temperature, above which the material becomes paramagnetic.

Antiparticle - A particle having the same mass as a given elementary particle and a charge equal in magnitude but opposite in sign.

Appearance potential* - The lowest energy which must be imparted to the parent molecule to cause it to produce a particular specified parent ion. This energy, usually stated in eV, may be imparted by electron impact, photon impact, or in other ways. More properly called appearance energy. [3]

Appearance potential spectroscopy (APS) - See Techniques for Materials Characterization, page 12-1.

Are (a) - A unit of area equal to 100 m². [1]

Arenes - Monocyclic and polycyclic aromatic hydrocarbons. See aromatic compounds. [5]

Aromatic compounds - Compounds whose structure includes a cyclic delocalized π-electron system. Historical use of the term implies a ring containing only carbon (e.g., benzene, naphthalene), but it is often generalized to include heterocyclic structures such as pyridine and thiophene. [5]

Arrhenius equation - A key equation in chemical kinetics which expresses the rate constant k as $k = A\exp(-E_a/RT)$, where E_a is the activation energy, R the molar gas constant, and T the temperature. A is called the preexponential factor and, for simple gas phase reactions, may be identified with the collision frequency.

Arsines - AsH_3 and compounds derived from it by substituting one, two or three hydrogen atoms by hydrocarbyl groups. $RAsH_2$, R_2AsH, R_3As (R not equal to H) are called primary, secondary and tertiary arsines, respectively. [5]

Aryl groups - Groups derived from arenes by removal of a hydrogen atom from a ring carbon atom. Groups similarly derived from heteroarenes are sometimes subsumed in this definition. [5]

Astronomical unit (AU)* - The mean distance of the earth from the sun, equal to $1.49597870 \times 10^{11}$ m.

Atomic absorption spectroscopy (AAS) - See Techniques for Materials Characterization, page 12-1.

Atomic emission spectroscopy (AES) - See Techniques for Materials Characterization, page 12-1.

Atomic force microscopy (AFM) - See Techniques for Materials Characterization, page 12-1.

Atomic mass* - The mass of a nuclide, normally expressed in unified atomic mass units (u).

Atomic mass unit (u)* - A unit of mass used in atomic, molecular, and nuclear science, defined as the mass of one atom of ^{12}C divided by 12. Its approximate value is 1.66054×10^{-27} kg. Also called the unified atomic mass unit. [1]

Atomic number (Z) - A characteristic property of an element, equal to the number of protons in the nucleus.

Atomic weight (A_r)* - The ratio of the average mass per atom of an element to 1/12 of the mass of nuclide ^{12}C. An atomic weight can be defined for a sample of any given isotopic composition. The standard atomic weight refers to a sample of normal terrestrial isotopic composition. The term relative atomic mass is synonymous with atomic weight. [2]

Attenuated total reflection (ATR) - See Techniques for Materials Characterization, page 12-1.

Auger effect - An atomic process in which an electron from a higher energy level fills a vacancy in an inner shell, transferring the released energy to another electron which is ejected.

Aurora - An atmospheric phenomenon in which streamers of light are produced when electrons from the sun are guided into the thermosphere by the earth's magnetic field. It occurs in the polar regions at altitudes of 95—300 km.

Avogadro constant (N_A)* - The number of elementary entities in one mole of a substance.

Azeotrope - A liquid mixture in a state where the variation of vapor pressure with composition at constant temperature (or, alternatively, the variation of normal boiling point with composition) shows either a maximum or a minimum. Thus when an azeotrope boils the vapor has the same composition as the liquid.

Azides - Compounds bearing the group $-N_3$, viz. $-N=N^+=N^-$; usually attached to carbon, e.g. PhN_3, phenyl azide or azidobenzene. Also used for salts of hydrazoic acid, HN_3, e.g. NaN_3, sodium azide. [5]

Azines - Condensation products, $R_2C=NN=CR_2$, of two moles of a carbonyl compound with one mole of hydrazine. [5]

Azo compounds - Derivatives of diazene (diimide), $HN=NH$, wherein both hydrogens are substituted by hydrocarbyl groups, e.g., $PhN=NPh$, azobenzene or diphenyldiazene. [5]

Balmer series - The series of lines in the spectrum of the hydrogen atom which corresponds to transitions between the state with principal quantum number $n = 2$ and successive higher states. The wavelengths are given by $1/\lambda = R_H(1/4 - 1/n^2)$, where $n = 3,4,...$ and R_H is the Rydberg constant for hydrogen. The first member of the series ($n = 2 \rightleftharpoons 3$), which is often called the H_α line, falls at a wavelength of 6563 Å.

Bar (bar) - A unit of pressure equal to 10^5 Pa.´

Bardeen-Cooper-Schrieffer (BCS) theory - A theory of superconductivity which is based upon the formation of electron pairs as a result of an electron-lattice interaction. The theory relates the superconducting transition temperature to the density of states and the Debye temperature.

Barn (b) - A unit used for expressing cross sections of nuclear processes, equal to 10^{-28} m².

Barrel - A unit of volume equal to 158.9873 L.

Baryon - Any elementary particle built up from three quarks. Examples are the proton, neutron, and various short-lived hyperons. Baryons have odd half-integer spins.

Base - Historically, a substance that yields an OH^- ion when it dissociates in solution, resulting in a pH>7. In the Brönsted definition, a base is a substance capable of accepting a proton in any type of reaction. The more general definition, due to G.N. Lewis, classifies any chemical species capable of donating an electron pair as a base.

Becquerel (Bq)* - The SI unit of radioactivity (disintegrations per unit time), equal to s^{-1}. [1]

Beer's law - An approximate expression for the change in intensity of a light beam that passes through an absorbing medium, viz., $\log(I/I_0) = -\varepsilon cl$, where I_0 is the incident intensity, I is the final intensity, ε is the molar (decadic) absorption coefficient, c is the molar concentration of the absorbing substance, and l is the path length. Also called the Beer-Lambert law

Binding energy* - A generic term for the energy required to decompose a system into two or more of its constituent parts. In nuclear physics, the binding energy is the energy difference between a nucleus and the separated nucleons of which it is composed (the energy equivalent of the mass defect). In atomic physics, it is the energy required to remove an electron from an atom.

Biot (Bi) - A name sometimes used for the unit of current in the emu system.

Birefringence - A property of certain crystals in which two refracted rays result from a single incident light ray. One, the ordinary ray, follows the normal laws of refraction, while the other, the extraordinary ray, exhibits a variable refractive index which depends on the direction in the crystal.

Black body radiation* - The radiation emitted by a perfect black body, i.e., a body which absorbs all radiation incident on it and reflects none. The wavelength dependence of the radiated energy density ρ (energy per unit volume per unit wavelength range) is given by the Planck formula

$$\rho = \frac{8\pi hc}{\lambda^5 (e^{hc/\lambda kt} - 1)}$$

where λ is the wavelength, h is Planck's constant, c is the speed of light, k is the Boltzmann constant, and T is the temperature.

Black hole - A very dense object, formed in a supernova explosion, whose gravitational field is so large that no matter or radiation can escape from the object.

Bloch wave function - A solution of the Schrödinger equation for an electron moving in a spatially periodic potential; used in the band theory of solids.

Bohr magneton (μ_B)* - The atomic unit of magnetic moment, defined as $eh/4\pi m_e$, where h is Planck's constant, m_e the electron mass, and e the elementary charge. It is the moment associated with a single electron spin.

Bohr, bohr radius (a_0)* - The radius of the lowest orbit in the Bohr model of the hydrogen atom, defined as $\varepsilon_o h^2/\pi m_e e^2$, where ε_o is the permittivity of a vacuum, h is Planck's constant, m_e the electron mass, and e the elementary charge. It is customarily taken as the unit of length when using atomic units.

Boiling point - The temperature at which the liquid and gas phases of a substance are in equilibrium at a specified pressure. The normal boiling point is the boiling point at normal atmospheric pressure (101.325 kPa).

Boltzmann constant (k)* - The molar gas constant R divided by Avogadro's constant.

Boltzmann distribution - An expression for the equilibrium distribution of molecules as a function of their energy, in which the number of molecules in a state of energy E is proportional to $\exp(-E/kT)$, where k is the Boltzmann constant and T is the temperature.

Bond strength - See Dissociation energy.

Born-Haber cycle* - A thermodynamic cycle in which a crystalline solid is converted to gaseous ions and then reconverted to the solid. The cycle permits calculation of the lattice energy of the crystal.

Bose-Einstein distribution - A modification of the Boltzmann distribution which applies to a system of particles that are bosons. The number of particles of energy E is proportional to $[e^{(E-\mu)/kT}-1]^{-1}$, where μ is a normalization constant, k is the Boltzmann constant, and T is the temperature.

Boson - A particle that obeys Bose-Einstein Statistics; specifically, any particle with spin equal to zero or an integer. This includes the photon, pion, deuteron, and all nuclei of even mass number.

Boyle's law - The empirical law, exact only for an ideal gas, which states that the volume of a gas is inversely proportional to its pressure at constant temperature.

Bragg angle (θ) - Defined by the equation $n\lambda = 2d\sin\theta$, which relates the angle θ between a crystal plane and the diffracted x-ray beam, the wavelength λ of the x-rays, the crystal plane spacing d, and the diffraction order n (any integer).

Bravais lattices* - The 14 distinct crystal lattices that can exist in three dimensions. They include three in the cubic crystal system, two in the tetragonal, four in the orthorhombic, two in the monoclinic, and one each in the triclinic, hexagonal, and trigonal systems.

Breakdown voltage - The potential difference at which an insulating substance undergoes a physical or chemical change that causes it to become a conductor, thus allowing current to flow through the sample.

Bremsstrahlung - Electromagnetic radiation generated when the velocity of a charged particle is reduced (literally, "braking radiation"). An example is the x-ray continuum resulting from collisions of electrons with the target in an x-ray tube.

Brewster angle - The angle of incidence for which the maximum degree of plane polarization occurs when a beam of unpolarized light is incident on the surface of a medium of refractive index n. At this angle, the angle between the reflected and refracted beams is 90°. The value of the Brewster angle is $\tan^{-1}n$.

Brillouin scattering - The scattering of light by acoustic phonons in a solid or liquid.

Brillouin zone - A region of allowed wave vectors and energy levels in a crystalline solid, which plays a part in the propagation of waves through the lattice.

British thermal unit (Btu) - A non-SI unit of energy, equal to approximately 1055 J. Several values of the Btu, defined in slightly different ways, have been used.

Brownian motion - The random movements of small particles suspended in a fluid, which arise from collisions with the fluid molecules.

Brunauer-Emmett-Teller method (BET) - See Techniques for Materials Characterization, page 12-1.

Buffer* - A solution designed to maintain a constant pH when small amounts of a strong acid or base are added. Buffers usually consist of a fairly weak acid and its salt with a strong base. Suitable concentrations are chosen so that the pH of the solution remains close to the pK_a of the weak acid.

Calorie (cal) - A non-SI unit of energy, originally defined as the heat required to raise the temperature of 1 g of water by 1 °C. Several calories of slightly different values have been used. The thermochemical calorie is now defined as 4.184 J.

Candela (cd)* - The SI base unit of luminous intensity. [1]

Capacitance (C) - Ratio of the charge acquired by a body to the change in potential. [1]

Carbamates - Salts or esters of carbamic acid, $H_2NC(=O)OH$, or of N-substituted carbamic acids: $R_2NC(=O)OR'$, (R' = hydrocarbyl or a cation). The esters are often called urethanes or urethans, a usage that is strictly correct only for the ethyl esters. [5]

Carbenes - The electrically neutral species H_2C: and its derivatives, in which the carbon is covalently bonded to two univalent groups of any kind or a divalent group and bears two non-bonding electrons, which may be spin-paired (singlet state) or spin-non-paired (triplet state). [5]

Carbinols - An obsolete term for substituted methanols, in which the name carbinol is synonymous with methanol. [5]

Carbohydrates - Originally, compounds such as aldoses and ketoses, having the stoichiometric formula $C_n(H_2O)_n$ (hence "hydrates of carbon"). The generic term carbohydrate now includes mono-, oligo-, and polysaccharides, as well as their reaction products and derivatives. [5]

Carboranes - A contraction of carbaboranes. Compounds in which a boron atom in a polyboron hydride is replaced by a carbon atom with maintenance of the skeletal structure. [5]

Carboxylic acids - Oxoacids having the structure RC(=O)OH. The term is used as a suffix in systematic name formation to denote the -C(=O)OH group including its carbon atom. [5]

Carnot cycle - A sequence of reversible changes in a heat engine using a perfect gas as the working substance, which is used to demonstrate that entropy is a state function. The Carnot cycle also provides a means to calculate the efficiency of a heat engine.

Catalyst - A substance that participates in a particular chemical reaction and thereby increases its rate but without a net change in the amount of that substance in the system. [3]

Catenanes, catena compounds - Hydrocarbons having two or more rings connected in the manner of links of a chain, without a covalent bond. More generally, the class catena compounds embraces functional derivatives and hetero analogues. [5]

Cation - A positively charged atomic or molecular particle.

Centipoise (cP) - A common non-SI unit of viscosity, equal to mPa s.

Centrifugal distortion - An effect in molecular spectroscopy in which rotational levels are lowered in energy, relative to the values of a rigid rotor, as the rotational angular momentum increases. The effect may be understood classically as a stretching of the bonds in the molecule as it rotates faster, thus increasing the moment of inertia.

Ceramic - A nonmetallic material of very high melting point.

Cerenkov radiation - Light emitted when a beam of charged particles travels through a medium at a speed greater than the speed of light in the medium. It is typically blue in color.

Cgs system of units - A system of units based upon the centimeter, gram, and second. The cgs system has been supplanted by the International System (SI).

Chalcogens - The Group VIA elements (oxygen, sulfur, selenium, tellurium, and polonium). Compounds of these elements are called chalcogenides. [7]

Chaotic system - A complex system whose behavior is governed by deterministic laws but whose evolution can vary drastically when small changes are made in the initial conditions.

Charge - See Electric charge.

Charles' law - The empirical law, exact only for an ideal gas, which states that the volume of a gas is directly proportional to its temperature at constant pressure.

Charm - A quantum number introduced in particle physics to account for certain properties of elementary particles and their reactions.

Chelate - A compound characterized by the presence of bonds from two or more bonding sites within the same ligand to a central metal atom. [3]

Chemical potential - For a mixture of substances, the chemical potential of constituent B is defined as the partial derivative of the Gibbs energy G with respect to the amount (number of moles) of B, with temperature, pressure, and amounts of all other constituents held constant. Also called partial molar Gibbs energy. [2]

Chemical shift* - A small change in the energy levels (and hence in the spectra associated with these levels) resulting from the effects of chemical binding in a molecule. The term is used in fields such as NMR, Mössbauer, and photoelectron spectroscopy, where the energy levels are determined primarily by nuclear or atomic effects.

Chiral molecule - A molecule which cannot be superimposed on its mirror image. A common example is an organic molecule containing a carbon atom to which four different atoms or groups are attached. Such molecules exhibit optical activity, i.e., they rotate the plane of a polarized light beam.

Chlorocarbons - Compounds consisting solely of chlorine and carbon. [5]

Chromatography* - A method for separation of the components of a sample in which the components are distributed between two phases, one of which is stationary while the other moves. In gas chromatography the gas moves over a liquid or solid stationary phase. In liquid chromatography the liquid mixture moves through another liquid, a solid, or a gel. The mechanism of separation of components may be adsorption, differential solubility, ion-exchange, permeation, or other mechanisms. [6]

Clapeyron equation - A relation between pressure and temperature of two phases of a pure substance that are in equilibrium, viz., $dp/dT = \Delta_{trs} S/\Delta_{trs} V$, where $\Delta_{trs} S$ is the difference in entropy between the phases and $\Delta_{trs} V$ the corresponding difference in volume.

Clathrates - Inclusion compounds in which the guest molecule is in a cage formed by the host molecule or by a lattice of host molecules. [5]

Clausius (Cl) - A non-SI unit of entropy or heat capacity defined as cal/K = 4.184 J/K. [2]

Clausius-Clapeyron equation - An approximation to the Clapeyron equation applicable to liquid-gas and solid-gas equilibrium, in which one assumes an ideal gas with volume much greater than the condensed phase volume. For the liquid-gas case, it takes the form $d(\ln p)/dT = \Delta_{vap} H/RT^2$, where R is the molar gas constant and $\Delta_{vap} H$ is the molar enthalpy of vaporization. For the solid-gas case, $\Delta_{vap} H$ is replaced by the molar enthalpy of sublimation, $\Delta_{sub} H$.

Clausius-Mosotti equation - A relation between the dielectric constant ε_r at optical frequencies and the polarizability α:

$$\frac{\varepsilon_r - 1}{\varepsilon_r + 2} = \frac{\rho N_A \alpha}{3 M \varepsilon_0}$$

where ρ is density, N_A is Avogadro's number, M is molar mass, and ε_0 is the permittivity of a vacuum.

Clebsch-Gordon coefficients - A set of coefficients used to describe the vector coupling of angular momenta in atomic and nuclear physics.

Codon - A set of three bases, chosen from the four primary bases found in the DNA molecule (uracil, cytosine, adenine, and guanine), which specifies the production of a particular amino acid or carries some other genetic instruction. For example, the codon UCA specifies the amino acid serine, CAG specifies glutamine, etc. There are a total of 64 codons.

Coercive force - The magnetizing force at which the magnetic flux density is equal to zero. [10]

Coercivity* - The maximum value of coercive force that can be attained when a magnetic material is symmetrically magnetized to saturation induction. [10]

Coherent anti-Stokes Raman spectroscopy (CARS) - See Techniques for Materials Characterization, page 12-1.

Colloid - Molecules or polymolecular particles dispersed in a medium that have, at least in one direction, a dimension roughly between 1 nm and 1 μm. [3]

Color center - A defect in a crystal that gives rise to optical absorption, thus changing the color of the material. A common type is the F-center, which results when an electron occupies the site of a negative ion.

Compressibility (κ)* - The fractional change of volume as pressure is increased, viz., $\kappa = -(1/V)(dV/dp)$. [1]

Compton wavelength (λ_C)* - In the scattering of electromagnetic radiation by a free particle (e.g., electron, proton), $\lambda_C = h/mc$ is the increase in wavelength, at a 90° scattering angle, corresponding to the transfer of energy from radiation to particle. Here h is Planck's constant, c the speed of light, and m the mass of the particle.

Conductance (G)* - For direct current, the reciprocal of resistance. More generally, the real part of admittance. [1]

Conductivity, electrical (σ)* - The reciprocal of the resistivity. [1]

Conductivity, thermal - See Thermal conductivity.

Congruent transformation - A phase transition (melting, vaporization, etc.) in which the substance preserves its exact chemical composition.

Constitutional repeating unit (CRU) - In polymer science, the smallest constitutional unit, the repetition of which constitutes a regular macromolecule, i.e., a macromolecule with all units connected identically with respect to directional sense. [8]

Copolymer - A polymer derived from more than one species of monomer. [8]

Coriolis effect - The deviation from simple trajectories when a mechanical system is described in a rotating coordinate system. It affects the motion of projectiles on the earth and in molecular spectroscopy leads to an important interaction between the rotational and vibrational motions. The effect may be described by an additional term in the equations of motion, called the Coriolis force.

Cosmic rays* - High energy nuclear particles, electrons, and photons, originating mostly outside the solar system, which continually bombard the earth's atmosphere.

Coulomb (C)* - The SI unit of electric charge, equal to A s. [1]

Coulomb's law - The statement that the force F between two electrical charges q_1 and q_2 separated by a distance r is $F = (4\pi\varepsilon_0)^{-1} q_1 q_2/r^2$, where ε_0 is the permittivity of a vacuum.

Covalent bond - A chemical bond between two atoms whose stability results from the sharing of two electrons, one from each atom.

Cowling number (Co) - A dimensionless quantity used in plasma physics, defined by $Co = B^2/\mu\rho v^2$, where ρ is density, v is velocity, μ is permeability, and B is magnetic flux density. [2]

CPT theorem - A theorem in particle physics which states that any local Lagrangian theory that is invariant under proper Lorentz transformations is also invariant under the combined operations of charge conjugation, C, space inversion, P, and time reversal, T, taken in any order.

Critical point* - In general, the point on the phase diagram of a two-phase system at which the two coexisting phases have identical properties and therefore represent a single phase. At the liquid-gas critical point of a pure substance, the distinction between liquid and gas vanishes, and the vapor pressure curve ends. The coordinates of this point are called the critical temperature and critical pressure. Above the critical temperature, it is not possible to liquefy the substance.

Cross section (σ)* - A measure of the probability of collision (or other interaction) between a beam of particles and a target which it encounters. In rough terms it is the effective area the target particles present to the incident ones; however, the precise definition depends on the nature of the interaction. A general definition of σ is the number of encounters per unit time divided by nv, where n is the concentration of incident particles and v their velocity.

Crosslink - In polymer science, a small region in a macromolecule from which at least four chains emanate, and formed by reactions involving sites or groups on existing macromolecules or by interactions between existing macromolecules. [8]

Crown compounds - Macrocyclic polydentate compounds, usually uncharged, in which three or more coordinating ring atoms (usually oxygen or nitrogen) are or may become suitably close for easy formation of chelate complexes with metal ions or other cationic species. [5]

Crust* - The outer layer of the solid earth, above the Mohorovicic discontinuity. Its thickness averages about 35 km on the continents and about 7 km below the ocean floor.

Cryoscopic constant (E_f)* - The constant that expresses the amount by which the freezing point T_f of a solvent is lowered by a non-dissociating solute, through the relation $\Delta T_f = E_f m$, where m is the molality of the solute.

Curie (Ci) - A non-SI unit of radioactivity (disintegrations per unit time), equal to 3.7×10^{10} s^{-1}.

Curie temperature (T_C)* - For a ferromagnetic material, the critical temperature above which the material becomes paramagnetic. Also applied to the temperature at which the spontaneous polarization disappears in a ferroelectric solid. [1]

Cyanohydrins - Alcohols substituted by a cyano group, most commonly, but not limited to, examples having a CN and an OH group attached to the same carbon atom. They are formally derived from aldehydes or ketones by the addition of hydrogen cyanide. [5]

Cycloalkanes - Saturated monocyclic hydrocarbons (with or without side chains). See alicyclic compounds. Unsaturated monocyclic hydrocarbons having one endocyclic double or one triple bond are called cycloalkenes and cycloalkynes, respectively. [5]

Cyclotron resonance - The resonant absorption of energy from a system in which electrons or ions that are orbiting in a uniform magnetic field are subjected to radiofrequency or microwave radiation. The resonance frequency is given by $\nu = eH/2\pi m^*c$, where e is the elementary charge, H is the magnetic field strength, m^* is the effective mass of the charged particle, and c is the speed of light. The effect occurs in both solids (involving electrons or holes) and in low pressure gasses (involving ions)

Dalton (Da) - A name sometimes used in biochemistry for the unified atomic mass unit (u).

De Broglie wavelength - The wavelength associated with the wave representation of a moving particle, given by h/mv, where h is Planck's constant, m the particle mass, and v the velocity.

De Haas-Van Alphen effect - An effect observed in certain metals and semiconductors at low temperatures and high magnetic fields, characterized by a periodic variation of magnetic susceptibility with field strength.

Debye equation* - The relation between the relative permittivity (dielectric constant) ε_r, polarizability α, and permanent dipole moment μ in a dielectric material whose molecules are free to rotate. It takes the form

$$\frac{\varepsilon_r - 1}{\varepsilon_r + 2} = \frac{\rho N_A}{3 M \varepsilon_0} \left(\alpha + \frac{\mu^2}{3kT} \right)$$

where ρ is density, N_A is Avogadro's number, M is molar mass, and ε_0 is the permittivity of a vacuum.

Debye length - In the Debye-Hückel theory of ionic solutions, the effective thickness of the cloud of ions of opposite charge which surrounds each given ion and shields the Coulomb potential produced by that ion.

Debye temperature (θ_D)* - In the Debye model of the heat capacity of a crystalline solid, $\theta_D = h v_D / k$, where h is Planck's constant, k is the Boltzmann constant, and v_D is the maximum vibrational frequency the crystal can support. For $T << \theta_D$, the heat capacity is proportional to T^3.

Debye unit (D) - A non-SI unit of electric dipole moment used in molecular physics, equal to 3.335641×10^{-30} C m.

Debye-Waller factor (D) - The factor by which the intensity of a diffraction line is reduced because of lattice vibrations. [1]

Defect - Any departure from the regular structure of a crystal lattice. A Frenkel defect results when an atom or ion moves to an interstitial position and leaves behind a vacancy. A Schottky defect involves either a vacancy where the atom has moved to the surface or a structure where a surface atom has moved to an interstitial position.

Degree of polymerization - The number of monomeric units in a macromolecule or an oligomer molecule. [8]

Dendrite - A tree-like crystalline pattern often observed, for example, in ice crystals and alloys in which the crystal growth branches repeatedly.

Density (ρ)* - In the most common usage, mass density or mass per unit volume. More generally, the amount of some quantity (mass, charge, energy, etc.) divided by a length, area, or volume.

Density of states (N_E, ρ) - The number of one-electron states in an infinitesimal interval of energy, divided by the range of that interval and by volume. [1]

Dew point* - The temperature at which liquid begins to condense as the temperature of a gas mixture is lowered. In meteorology, it is the temperature at which moisture begins to condense on a surface in contact with the air.

Diamagnetism - A type of magnetism characterized by a negative magnetic susceptibility, so that the material, when placed in an external magnetic field, becomes weakly magnetized in the direction opposite to the field. This magnetization is independent of temperature.

Diazo compounds - Compounds having the divalent diazo group, $=N^+=N^-$, attached to a carbon atom, e.g., $CH_2=N_2$ diazomethane. [5]

Dielectric constant (ε)* - Ratio of the electric displacement in a medium to the electric field strength. Also called permittivity. [1]

Dienes - Compounds that contain two fixed double bonds (usually assumed to be between carbon atoms). Dienes in which the two double-bond units are linked by one single bond are termed conjugated. [5]

Differential scanning calorimetry (DSC) - See Techniques for Materials Characterization, page 12-1.

Differential thermal analysis (DTA) - See Techniques for Materials Characterization, page 12-1.

Diffusion* - The migration of atoms, molecules, ions, or other particles as a result of some type of gradient (concentration, temperature, etc.).

Diopter - A unit used in optics, formally equal to m^{-1}. It is used in expressing dioptic power, which is the reciprocal of the focal length of a lens.

Dipole moment, electric (p,μ)* - For a distribution of equal positive and negative charge, the magnitude of the dipole moment vector is the positive charge multiplied by the distance between the centers of positive and negative charge distribution. The direction is given by the line from the center of negative charge to the center of positive charge.

Dipole moment, magnetic (m,μ) - Formally defined in electromagnetic theory as a vector quantity whose vector product with the magnetic flux density equals the torque. The magnetic dipole generated by a current I flowing in a small loop of area A has a magnetic moment of magnitude IA. In atomic and nuclear physics, a magnetic moment is associated with the angular momentum of a particle; e.g., an electron with orbital angular momentum l exhibits a magnetic moment of $-el/2m_e$ where e is the elementary charge and m_e the mass of the electron. [1]

Disaccharides - Compounds in which two monosaccharides are joined by a glycosidic bond. [5]

Dislocation - An extended displacement of a crystal from a regular lattice. An edge dislocation results when one portion of the crystal has partially slipped with respect to the other, resulting in an extra plane of atoms extending through part of the crystal. A screw dislocation transforms successive atomic planes into the surface of a helix.

Dispersion - Splitting of a beam of light (or other electromagnetic radiation) of mixed wavelengths into the constituent wavelengths as a result of the variation of refractive index of the medium with wavelength.

Dissociation constant* - The equilibrium constant for a chemical reaction in which a compound dissociates into its constituent parts.

Dissociation energy (D_e)* - For a diatomic molecule, the difference between the energies of the free atoms at rest and the minimum in the potential energy curve. The term bond dissociation energy (D_0), which can be applied to polyatomic molecules as well, is used for the difference between the energies of the fragments resulting when a bond is broken and the energy of the original molecule in its lowest energy state. The term bond strength implies differences in enthalpy rather than energy.

Domain - A small region of a solid in which the magnetic or electric moments of the individual units (atoms, molecules, or ions) are aligned in the same direction.

Domain wall - The transition region between adjacent ferromagnetic domains, generally a layer with a thickness of a few hundred ångström units. Also called Bloch wall.

Doppler effect - The change in the apparent frequency of a wave (sound, light, or other) when the source of the wave is moving relative to the observer.

Dose equivalent (H) - The product of the absorbed dose of radiation at a point of interest in tissue and various modifying factors which depend on the type of tissue and radiation. [1]

Drift velocity - The velocity of charge carriers (electrons, ions, etc.) moving under the influence of an electric field in a medium which subjects the carriers to some frictional force.

Dyne (dyn) - A non-SI (cgs) unit of force, equal to 10^{-5} N.

Ebullioscopic constant (E_b)* - The constant that expresses the amount by which the boiling point T_b of a solvent is raised by a non-dissociating solute, through the relation $\Delta T_b = E_b\, m$, where m is the molality of the solute.

Eddy currents - Circulating currents set up in conducting bulk materials or sheets by varying magnetic fields.

Effinghausen effect - The appearance of a temperature gradient in a current carrying conductor that is placed in a transverse magnetic field. The direction of the gradient is perpendicular to the current and the field.

Eigenvalue - An allowed value of the constant a in the equation $Au = au$, where A is an operator acting on a function u (which is called an eigenfunction). In quantum mechanics, the outcome of any observation is an eigenvalue of the corresponding operator. Also called characteristic value.

Einstein - A non-SI unit used in photochemistry, equal to one mole of photons.

Einstein temperature (θ_V) - In the Einstein theory of the heat capacity of a crystalline solid, $\theta_V = h\nu/k$, where h is Planck's constant, k is the Boltzmann constant, and ν is the vibrational frequency of the crystal.

Einstein transition probability - A constant in the Einstein relation $A_{ij} + B_{ij}\rho$ for the probability of a transition between two energy levels i and j in a radiation field of energy density ρ. The A_{ij} coefficient describes the probability of spontaneous emission, while B_{ij} and B_{ji} govern the probability of stimulated emission and absorption, respectively ($B_{ij} = B_{ji}$).

Elastic limit - The greatest stress which a material is capable of sustaining without any permanent strain remaining after complete release of the stress. [10]

Elastic modulus - See Young's modulus.

Electric charge (Q) - The quantity of electricity; i.e., the property that controls interactions between bodies through electrical forces.

Electric current (I) - The charge passing through a circuit per unit time. [1]

Electric displacement (D) - A vector quantity whose magnitude equals the electric field strength multiplied by the permittivity of the medium and whose direction is the same as that of the field strength.

Electric field strength (E) - The force exerted by an electric field on a point charge divided by the electric charge. [1]

Electric potential (V) - A scalar quantity whose gradient is equal to the negative of the electric field strength.

Electrical conductance - See Conductance.

Electrical resistance - See Resistance.

Electrical resistivity - See Resistivity.

Electrochemical series* - An arrangement of reactions which produce or consume electrons in an order based on standard electrode potentials. A common arrangement places metals in decreasing order of their tendency to give up electrons.

Electrode potential* - The electromotive force of a cell in which the electrode on the left is the standard hydrogen electrode and that on the right is the electrode in question. [2]

Electrolysis - The decomposition of a substance as a result of passing an electric current between two electrodes immersed in the sample.

Electromotive force (emf) - The energy supplied by a source divided by the charge transported through the source. [1]

Electron* - An elementary particle in the family of leptons, with negative charge and spin of 1/2.

Electron affinity* - The energy difference between the ground state of a gas-phase atom or molecule and the lowest state of the corresponding negative ion.

Electron cyclotron resonance (ECR) - See Techniques for Materials Characterization, page 12-1.

Electron energy loss spectroscopy (EELS) - See Techniques for Materials Characterization, page 12-1.

Electron nuclear double resonance (ENDOR) - See Techniques for Materials Characterization, page 12-1.

Electron paramagnetic resonance (EPR) - See Techniques for Materials Characterization, page 12-1.

Electron probe microanalysis (EPMA) - See Techniques for Materials Characterization, page 12-1.

Electron spectroscopy for chemical analysis (ESCA) - See Techniques for Materials Characterization, page 12-1.

Electron spin (s) - The quantum number, equal to 1/2, that specifies the intrinsic angular momentum of the electron.

Electron stimulated desorption (ESD) - See Techniques for Materials Characterization, page 12-1.

Electron volt (eV)* - A non-SI unit of energy used in atomic and nuclear physics, equal to approximately 1.602177×10^{-19} J. The electron volt is defined as the kinetic energy acquired by an electron upon acceleration through a potential difference of 1 V. [1]

Electronegativity* - A parameter originally introduced by Pauling which describes, on a relative basis, the power of an atom or group of atoms to attract electrons from the same molecular entity. [3]

Electrophoresis - The motion of macromolecules or colloidal particles in an electric field. [3]

Emissivity (ε)* - Ratio of the radiant flux emitted per unit area to that of an ideal black body at the same temperature. Also called emittance. [1]

Emu - The electromagnetic system of units, based upon the cm, g, and s plus the emu of current (sometimes called the abampere).

Enantiomers - A chiral molecule and its non-superposable mirror image. The two forms rotate the plane of polarized light by equal amounts in opposite directions. Also called optical isomers.

Energy (E, U)* - The characteristic of a system that enables it to do work.

Energy gap* - In the theory of solids, the region between two energy bands, in which no bound states can occur.

Enols, alkenols - The term refers specifically to vinylic alcohols, which have the structure $HOCR'=CR_2$. Enols are tautomeric with aldehydes (R´ = H) or ketones (R´ not equal to H). [5]

Enthalpy (H)* - A thermodynamic function, especially useful when dealing with constant-pressure processes, defined by $H = E + PV$, where E is energy, P pressure, and V volume. [1]

Enthalpy of combustion* - The enthalpy change in a combustion reaction. Its negative is the heat released in combustion.

Enthalpy of formation, standard* - The enthalpy change for the reaction in which a substance is formed from its constituent elements, each in its standard reference state (normally refers to 1 mol, sometimes to 1 g, of the substance).

Enthalpy of fusion* - The enthalpy change in the transition from solid to liquid state.

Enthalpy of sublimation - The enthalpy change in the transition from solid to gas state.

Enthalpy of vaporization* - The enthalpy change in the transition from liquid to gas state.

Entropy (S)* - A thermodynamic function defined such that when a small quantity of heat dQ is received by a system at temperature T, the entropy of the system is increased by dQ/T, provided that no irreversible change takes place in the system. [1]

Entropy unit (e.u.) - A non-SI unit of entropy, equal to 4.184 J/K mol.

Ephemeris time - Time measured in tropical years from January 1, 1900.

Epoxy compounds - Compounds in which an oxygen atom is directly attached to two adjacent or non-adjacent carbon atoms of a carbon chain or ring system; thus cyclic ethers. [5]

Equation of continuity - Any of a class of equations that express the fact that some quantity (mass, charge, energy, etc.) cannot be created or destroyed. Such equations typically specify that the rate of increase of the quantity in a given region of space equals the net current of the quantity flowing into the region.

Equation of state* - An equation relating the pressure, volume, and temperature of a substance or system.

Equilibrium constant (K)* - For a chemical reaction $aA + bB \rightleftharpoons cC + dD$, the equilibrium constant is defined by:

$$K = \frac{a_C^{\ c} \cdot a_D^{\ d}}{a_A^{\ a} \cdot a_B^{\ b}}$$

where a_i is the activity of component i. To a certain approximation, the activities can be replaced by concentrations. The equilibrium constant is related to $\Delta_r G°$, the standard Gibbs energy change in the reaction, by $RT \ln K = -\Delta_r G°$.

Equivalent conductance - See Conductivity, electrical

Erg (erg) - A non-SI (cgs) unit of energy, equal to 10^{-7} J.

Esters - Compounds formally derived from an oxoacid RC(=O)(OH) and an alcohol, phenol, heteroarenol, or enol by linking, with formal loss of water from an acidic hydroxy group of the former and a hydroxy group of the latter. [5]

Esu - The electrostatic system of units, based upon the cm, g, and s plus the esu of charge (sometimes called the statcoulomb or franklin).

Ethers - Compounds with formula ROR, where R is not equal to H. [5]

Euler number (Eu) - A dimensionless quantity used in fluid mechanics, defined by $Eu = \Delta p / \rho v^2$, where p is pressure, ρ is density, and v is velocity. [2]

Eutectic - The point on a two-component solid-liquid phase diagram which represents the lowest melting point of any possible mixture. A liquid having the eutectic composition will freeze at a single temperature without change of composition.

Excitance (M) - Radiant energy flux leaving an element of a surface divided by the area of that element. [1]

Exciton - A localized excited state consisting of a bound electron-hole pair in a molecular or ionic crystal. The exciton can propagate through the crystal.

Exosphere - The outermost part of the earth's atmosphere, beginning at about 500 to 1000 km above the surface. It is characterized by densities so low that air molecules can escape into outer space.

Expansion coefficient - See thermal expansion coefficient.

Extended electron energy loss fine structure (EXELFS) - See Techniques for Materials Characterization, page 12-1.

Extended x-ray absorption fine structure (EXAFS) - See Techniques for Materials Characterization, page 12-1.

Extinction coefficient - See Absorption coefficient, molar.

F-Center - See Color center.

Fahrenheit temperature (°F) - The temperature scale based on the assignment of 32°F = 0 °C and a temperature interval of °F $=(5/9)°C$; i.e., $t/°F = (9/5)t/°C + 32$.

Farad (F)* - The SI unit of electric capacitance, equal to C/V. [1]

Faraday constant (F)* - The electric charge of 1 mol of singly charged positive ions; i.e., $F = N_A e$, where N_A is Avogadro's constant and e is the elementary charge. [1]

Faraday effect* - The rotation of the plane of plane-polarized light by a medium placed in a magnetic field parallel to the direction of the light beam. The effect can be observed in solids, liquids, and gasses.

Fatty acids - Aliphatic monocarboxylic acids derived from or contained in esterified form in an animal or vegetable fat, oil, or wax. Natural fatty acids commonly have a chain of 4 to 28 carbons (usually unbranched and even-numbered), which may be saturated or unsaturated. By extension, the term is sometimes used to embrace all acyclic aliphatic carboxylic acids. [5]

Fermat's principle - The law that a ray of light traversing one or more media will follow a path which minimizes the time required to pass between two given points.

Fermi (f) - Name sometimes used in nuclear physics for the femtometer.

Fermi level - The highest energy of occupied states in a solid at zero temperature. Sometimes called Fermi energy. The Fermi surface is the surface in momentum space formed by electrons occupying the Fermi level.

Fermi resonance - An effect observed in vibrational spectroscopy when an overtone of one fundamental vibration closely coincides in energy with another fundamental of the same symmetry species. It leads to a splitting of vibrational bands.

Fermi-Dirac distribution - A modification of the Boltzmann distribution which takes into account the Pauli exclusion principle. The number of particles of energy E is proportional to $[e^{(E-\mu)/kT}+1]^{-1}$, where μ is a normalization constant, k the Boltzmann constant, and T the temperature. The distribution is applicable to a system of fermions.

Fermion - A particle that obeys Fermi-Dirac statistics. Specifically, any particle with spin equal to an odd multiple of 1/2. Examples are the electron, proton, neutron, muon, etc.

Ferrimagnetism* - A type of magnetism in which the magnetic moments of atoms in a solid are ordered into two nonequivalent sublattices with unequal magnetic moments, leading to a nonzero magnetic susceptibility.

Ferrite - A ferrimagnetic material of nominal formula MFe_2O_4, where M is a divalent metal; widely used in microwave switches and other solid state devices.

Ferroelectricity* - The retention of electric polarization by certain materials after the external field that produced the polarization has been removed.

Ferromagnetism* - A type of magnetism in which the magnetic moments of atoms in a solid are aligned within domains which can in turn be aligned with each other by a weak magnetic field. Some ferromagnetic materials can retain their magnetization when the external field is removed, as long as the temperature is below a critical value, the Curie temperature. They are characterized by a large positive magnetic susceptibility.

Fick's law - The statement that the flux J of a diffusing substance is proportional to the concentration gradient, i.e., $J = -D(dc/dx)$, where D is called the diffusion coefficient.

Field - A mathematical construct which describes the interaction between particles resulting from gravity, electromagnetism, or other physical phenomena. In classical physics a field is described by equations. Quantum field theory introduces operators to represent the physical observables.

Field emission microscopy (FEM) - See Techniques for Materials Characterization, page 12-1.

Field ion microscopy (FIM) - See Techniques for Materials Characterization, page 12-1.

Fine structure - The splitting in spectral lines that results from interactions of the electron spin with the orbital angular momentum.

Fine structure constant (α)* - Defined as $e^2/2hc\varepsilon_0$, where e is the elementary charge, h Planck's constant, c the speed of light, and ε_0 the permittivity of a vacuum. It is a measure of the strength of the electromagnetic interaction between particles.

First radiation constant (c_1)* - Constant ($= 2\pi hc^2$) in the equation for the radiant excitance M_λ of a black body:

$$M_\lambda = \frac{c_1 \lambda^{-5} \Delta\lambda}{e^{c_2/\lambda T} - 1}$$

where λ is the wavelength, T is the temperature, and $c_2 = hc/k$ is the second radiation constant.

Flash point - The lowest temperature at which vapors above a volatile combustible substance will ignite in air when exposed to a flame. [10]

Fluence (F) - Term used in photochemistry to specify the energy per unit area delivered in a given time interval, for example by a laser pulse. [2]

Fluorocarbons - Compounds consisting solely of fluorine and carbon. [5]

Fluxoid - The quantum of magnetic flux in superconductivity theory, equal to $hc/2e$, where h is Planck's constant, c the velocity of light, and e the elementary charge.

Force (F) - The rate of change of momentum with time. [1]

Force constants (f, k)* - In molecular vibrations, the coefficients in the expression of the potential energy in terms of atom displacements from their equilibrium positions. In a diatomic molecule, $f = d^2V/dr^2$, where $V(r)$ is the potential energy and r is the interatomic distance. [2]

Fourier number (Fo) - A dimensionless quantity used in fluid mechanics, defined by $Fo = at/l^2$, where a is thermal diffusivity, t is time, and l is length. [2]

Fourier transform infrared spectroscopy (FTIR) - A technique for obtaining an infrared spectrum by use of an interferometer in which the path length of one of the beams is varied. A Fourier transformation of the resulting interferogram yields the actual spectrum. The technique is also used for NMR and other types of spectroscopy.

Fractals - Geometrical objects that are self-similar under a change of scale; i.e., they appear similar at all levels of magnification. They can be considered to have fractional dimensionality. Examples occur in diverse fields such as geography (rivers and shorelines), biology (trees), and solid state physics (amorphous materials).

Franck-Condon principle - An important principle in molecular spectroscopy which states that the nuclei in a molecule remain essentially stationary while an electronic transition is taking place. The physical interpretation rests on the fact that the electrons move much more rapidly than the nuclei because of their much smaller mass.

Franklin (Fr) - Name sometimes given to the unit of charge in the esu system.

Fraunhofer diffraction - Diffraction of light in situations where the source and observation point are so far removed that the wave surfaces may be considered planar.

Fraunhofer lines - Sharp absorption lines in the spectrum of sunlight, caused by absorption of the solar blackbody radiation by atoms near the sun's surface.

Free radical - See Radicals. The term "free radical" is often used more broadly for molecules that have a paramagnetic ground state (e.g., O_2) and sometimes for any transient or highly reactive molecular species.

Freezing point - See Melting point.

Frequency (v)* - Number of cycles of a periodic phenomenon divided by time. [1]

Fresnel diffraction - Diffraction of light in a situation where the source and observation point are sufficiently close together that the curvature of the wave surfaces must be taken into account.

Froude number (Fr) - A dimensionless quantity used in fluid mechanics, defined by $Fr = v/(lg)^{1/2}$, where v is velocity, l is length, and g is acceleration due to gravity. [2]

Fugacity (f_B) - For a gas mixture, the fugacity of component B is defined as the absolute activity λ_B times the limit, as the pressure p approaches zero at constant temperature, of p_B/λ_B. [2]

Fullerenes - Compounds composed solely of an even number of carbon atoms, which form a cage-like fused-ring polycyclic system with twelve five-membered rings and the rest six-membered rings. The archetypal example is [60]fullerene, where the atoms and bonds delineate a truncated icosahedron. The term has been broadened to include any closed cage structure consisting entirely of three-coordinate carbon atoms. [5]

Fulvalenes - The hydrocarbon fulvalene and its derivatives formed by substitution (and by extension, analogues formed by replacement of one or more carbon atoms of the fulvalene skeleton by a heteroatom). [5]

Fulvenes - The hydrocarbon fulvene and its derivatives formed by substitution (and by extension, analogues formed by replacement of one or more carbon atoms of the fulvene skeleton by a heteroatom). [5]

Fundamental vibrational frequencies* - In molecular spectroscopy, the characteristic vibrational frequencies obtained when the vibrational energy is expressed in normal coordinates. They determine the primary features of the infrared and Raman spectra of the molecule.

γ - Name sometimes used for microgram.

γ-rays* - Electromagnetic radiation (photons) with energy greater than about 0.1 MeV (wavelength less than about 1 pm).

g-Factor of the electron* - The proportionality factor in the equation relating the magnetic moment μ of an electron to its total angular momentum quantum number J, i.e., $\mu = -g\mu_B J$, where μ_B is the Bohr magneton. Also called Landé factor.

Gal - A non-SI unit of acceleration, equal to 0.01 m/s. Also called galileo.

Gallon (US) - A unit of volume equal to 3.785412 L.

Gallon (UK, Imperial) - A unit of volume equal to 4.546090 L.

Gauss (G) - A non-SI unit of magnetic flux density (B) equal to 10^{-4} T.

Gaussian system of units - A hybrid system used in electromagnetic theory, which combines features of both the esu and emu systems.

Gel - A colloidal system with a finite, but usually rather small, yield stress (the sheer stress at which yielding starts abruptly). [3]

Genetic code* - The set of relations between each of the 64 codons of DNA and a specific amino acid (or other genetic instruction).

Gibbs energy (G)* - An important function in chemical thermodynamics, defined by $G = H-TS$, where H is the enthalpy, S the entropy, and T the thermodynamic temperature. Sometimes called Gibbs free energy and, in older literature, simply "free energy". [2]

Gibbs phase rule - The relation $F = C - P + 2$, where C is the number of components in a mixture, P is the number of phases, and F is the degrees of freedom, i.e., the number of intensive variables that can be changed independently without affecting the number of phases.

Glass transition temperature* - The temperature at which an amorphous polymer is transformed, in a reversible way, from a viscous or rubbery condition to a hard and relatively brittle one. [10]

Glow discharge mass spectroscopy (GDMS) - See Techniques for Materials Characterization, page 12-1.

Gluon - A hypothetical particle postulated to take part in the binding of quarks, in analogy to the role of the photon in electromagnetic interactions.

Glycerides - Esters of glycerol (propane-1,2,3-triol) with fatty acids, widely distributed in nature. They are by long-established custom subdivided into triglycerides, 1,2- or 1,3-diglycerides, and 1- or 2-monoglycerides, according to the number and positions of acyl groups. [5]

Glycols - Dihydric alcohols in which two hydroxy groups are on different carbon atoms, usually but not necessarily adjacent. Also called diols. [5]

Grain (gr) - A non-SI unit of mass, equal to 64.79891 mg.

Grain boundary - The interface between two regions of different crystal orientation.

Grashof number (Gr) - A dimensionless quantity used in fluid mechanics, defined by $Gr = l^3 g\alpha\Delta T\rho^2/\eta^2$, where T is temperature, ρ is density, l is length, η is viscosity, α is cubic expansion coefficient, and g is acceleration of gravity. [2]

Gravitational constant (G)* - The universal constant in the equation for the gravitational force between two particles, $F = Gm_1m_2/r^2$, where r is the distance between the particles and m_1 and m_2 are their masses. [1]

Gray (Gy)* - The SI unit of absorbed dose of radiation, equal to J/kg. [1]

Gregorian calendar - The modification of the Julian calendar introduced in 1582 by Pope Gregory XII which specified that a year divisible by 100 is a leap year only if divisible by 400.

Grignard reagents - Organomagnesium halides, RMgX, having a carbon–magnesium bond (or their equilibrium mixtures in solution with $R_2Mg + MgX_2$). [5]

Gruneisen parameter (γ) - Defined by $\gamma = \alpha_V/\kappa\, c_V\,\rho$, where α_V is the cubic thermal expansion coefficient, κ is the isothermal compressibility, c_V is the specific heat capacity at constant volume, and ρ is the mass density. γ is independent of temperature for most crystalline solids. [1]

Gyromagnetic ratio (γ) - Ratio of the magnetic moment of a particle to its angular momentum. Also called magnetogyric ratio.

Hadron - Any elementary particle that can take part in the strong interaction. Hadrons are subdivided into baryons, with odd half integer spins, and mesons, which have zero or integral spin.

Hall effect* - The development of a transverse potential difference V in a conducting material when subjected to a magnetic field H perpendicular to the direction of the current. The potential difference is given by $V = R_H\,BJt$, where B is the magnetic induction, J the current density, t the thickness of the specimen in the direction of the potential difference, and R_H is called the Hall coefficient.

Halocarbon - A compound containing no elements other than carbon, hydrogen, and one or more halogens. In common practice, the term is used mainly for compounds of no more than four or five carbon atoms.

Halogens - The elements F, Cl, Br, I, and At. Compounds of these elements are called halogenides or halides. [7]

Hamiltonian (H) - An expression for the total energy of a mechanical system in terms of the momenta and positions of constituent particles. In quantum mechanics, the Hamiltonian operator appears in the eigenvalue equation $H\psi = E\psi$, where E is an energy eigenvalue and ψ the corresponding eigenfunction.

Hardness* - The resistance of a material to deformation, indentation, or scratching. Hardness is measured on various scales, such as Mohs, Brinell, Knoop, Rockwell, and Vickers. [10]

Hartmann number (Ha) - A dimensionless quantity used in plasma physics, defined by $Ha = Bl(\kappa/\eta)^{1/2}$, where B is magnetic flux density, l is length, κ is electric conductivity, and η is viscosity. [2]

Hartree (E_h)* - An energy unit used in atomic and molecular science, equal to approximately $4.3597482 \times 10^{-18}$ J.

Hartree-Fock method - A iterative procedure for solving the Schrödinger equation for an atom or molecule in which the equation is solved for each electron in an initial assumed potential from all the other electrons. The new potential that results is used to repeat the calculation and the procedure continued until convergence is reached. Also called self-consistent field (SCF) method.

Heat capacity* - Defined in general as dQ/dT, where dQ is the amount of heat that must be added to a system to increase its temperature by a small amount dT. The heat capacity at constant pressure is $C_p = (\partial H/\partial T)_p$; that at constant volume is $C_V = (\partial E/\partial T)_V$, where H is enthalpy, E is internal energy, p is pressure, V is volume, and T is temperature. An upper case C normally indicates the molar heat capacity, while a lower case c is used for the specific (per unit mass) heat capacity. [1]

Heat of formation, vaporization, etc. - See corresponding terms under Enthalpy.

Hectare (ha) - A unit of area equal to 10^4 m^2. [1]

Heisenberg uncertainty principle - The statement that two observable properties of a system that are complementary, in the sense that their quantum-mechanical operators do not commute, cannot be specified simultaneously with absolute precision. An example is the position and momentum of a particle; according to this principle, the uncertainties in position Δq and momentum Δp must satisfy the relation $\Delta p \Delta q \geq h/4\pi$, where h is Planck's constant.

Heitler-London model - An early quantum-mechanical model of the hydrogen atom which introduced the concept of the exchange interaction between electrons as the primary reason for stability of the chemical bond.

Helicon - A low-frequency wave generated when a metal at low temperature is exposed to a uniform magnetic field and a circularly polarized electric field.

Helmholz energy (A) - A thermodynamic function defined by $A = E-TS$, where E is the energy, S the entropy, and T the thermodynamic temperature. [2]

Hemiacetals - Compounds having the general formula $R_2C(OH)OR'$ (R′ not equal to H). [5]

Henry (H)* - The SI unit of inductance, equal to Wb/A. [1]

Henry's law * - An expression which applies to an ideal dilute solution in which one or more gasses are dissolved, viz., $p_i = H_i x_i$, where p_i is the partial pressure of component i above the solution, x_i is its mole fraction in the solution, and H_i is the Henry's law constant (a characteristic of the given gas and solvent, as well as the temperature).

Hermitian operator - An operator A that satisfies the relation $\int u_m{}^*Au_n dx = (\int u_n{}^*Au_m\ dx)^*$, where * indicates the complex conjugate. The eigenvalues of Hermitian operators are real, and eigenfunctions belonging to different eigenvalues are orthogonal.

Hertz (Hz) - The SI unit of frequency, equal to s^{-1}. [1]

Heterocyclic compounds - Cyclic compounds having as ring members atoms of at least two different elements, e.g., quinoline, 1,2-thiazole, bicyclo[3.3.1]tetrasiloxane. [5]

Heusler alloys - Alloys of manganese, copper, aluminum, nickel, and sometimes other metals which find important uses as permanent magnets.

Holography - A technique for creating a three-dimensional image of a object by recording the interference pattern between a light beam diffracted from the object and a reference beam. The image can be reconstructed from this pattern by a suitable optical system.

Homopolymer - A polymer derived from one species of (real, implicit, or hypothetical) monomer. [8]

Hooke's law - The statement that the ratio of stress to strain is a constant in a totally elastic medium.

Horse power - A non-SI unit of energy, equal to approximately 746 W.

Hubble constant - The ratio of the recessional velocity of an extragalactic object to the distance of that object. Its value is about 2×10^{-18} s^{-1}.

Huckel theory - A simple approximation for calculating the energy of conjugated molecules in which only the resonance integrals between neighboring bonds are considered. Also called CNDO method (complete neglect of differential overlap).

Hume-Rothery rules - A set of empirical rules for predicting the occurrence of solid solutions in metallic systems. The rules involve size, crystal structure, and electronegativity.

Hund's rules - A series of rules for predicting the sequence of energy states in atoms and molecules. One of the important results is that when two electrons exist in different orbitals, the state with their spins parallel (triplet state) lies at lower energy than the state with antiparallel spins (singlet).

Hydrazines - Hydrazine (diazane), H_2NNH_2, and its hydrocarbyl derivatives. When one or more substituents are acyl groups, the compound is a hydrazide. [5]

Hydrocarbon - A compound containing only carbon and hydrogen. [5]

Hydrolysis - A reaction occurring in water in which a chemical bond is cleaved and a new bond formed with the oxygen atom of water.

Hyperfine structure - Splitting of energy levels and spectral lines into several closely spaced components as a result of interaction of nuclear spin angular momentum with other angular momenta in the atom or molecule.

Hysteresis* - An irreversible response of a system (parameter A) as a function of an external force (parameter F), usually symmetric with respect to the origin of the A vs. F graph after the initial application of the force. A common example is magnetic induction vs. magnetic field strength in a ferromagnet.

Ideal gas law - The equation of state $pV = RT$, which defines an ideal gas, where p is pressure, V molar volume, T temperature, and R the molar gas constant.

Ideal solution - A solution in which solvent-solvent and solvent-solute interactions are identical, so that properties such as volume and enthalpy are exactly additive. Ideal solutions follow Raoult's law, which states that the vapor pressure p_i of component i is $p_i = x_i p_i^*$, where x_i is the mole fraction of component i and p_i^* the vapor pressure of the pure substance i.

Ignition temperature* - The lowest temperature at which combustion of a material will occur spontaneously under specified conditions. Sometimes called autoignition temperature, kindling point. [10]

Imides - Diacyl derivatives of ammonia or primary amines, especially those cyclic compounds derived from diacids. Also used for salts having the anion RN_2^-. [5]

Impedance (Z) - The complex representation of potential difference divided by the complex representation of current. In terms of reactance X and resistance R, the impedance is given by $Z = R + iX$. [1]

Index of refraction (n)* - For a non-absorbing medium, the ratio of the velocity of electromagnetic radiation *in vacuo* to the phase velocity of radiation of a specified frequency in the medium. [1]

Inductance - The ratio of the electromagnetic force induced in a coil by a current to the rate of change of the current.

Inductive coupled plasma mass spectroscopy (ICPMS) - See Techniques for Materials Characterization, page 12-1.

Inertial defect - In molecular spectroscopy, the quantity I_c-I_a-I_b for a molecule whose equilibrium configuration is planar, where I_a, I_b, and I_c are the effective principal moments of inertia. The inertial defect for a rigid planar molecule would be zero, but vibration–rotation interactions in a real molecule lead to a positive inertial defect.

Insulator - A material in which the highest occupied energy band (valence band) is completely filled with electrons, while the next higher band (conduction band) is empty. Solids with an energy gap of 5 eV or more are generally considered as insulators at room temperature. Their conductivity is less than 10^{-6} S/m and increases with temperature.

Intercalation compounds - Compounds resulting from reversible inclusion, without covalent bonding, of one kind of molecule in a solid matrix of another compound, which has a laminar structure. The host compound, a solid, may be macromolecular, crystalline, or amorphous. [5]

International System of Units (SI)* - The unit system adopted by the General Conference on Weights and Measures in 1960. It consists of seven base units (meter, kilogram, second, ampere, kelvin, mole, candela), plus derived units and prefixes. [1]

International Temperature Scale (ITS-90)* - The official international temperature scale adopted in 1990. It consists of a set of fixed points and equations which enable the thermodynamic temperature to be determined from operational measurements. [9]

Ion - An atomic or molecular particle having a net electric charge. [3]

Ion exchange - A process involving the adsorption of one or several ionic species accompanied by the simultaneous desorption (displacement) of one or more other ionic species. [3]

Ion neutralization spectroscopy (INS) - See Techniques for Materials Characterization, page 12-1.

Ionic strength (I) - A measure of the total concentration of ions in a solution, defined by $I = 1/2\Sigma_i z_i^2 m_i$, where z_i is the charge of ionic species i and m_i is its molality. For a 1-1 electrolyte at molality m, $I = m$.

Ionization constant* - The equilibrium constant for a reaction in which a substance in solution dissociates into ions.

Ionization potential* - The minimum energy required to remove an electron from an isolated atom or molecule (in its vibrational ground state) in the gaseous phase. More properly called ionization energy. [3]

Irradiance (E) - The radiant energy flux incident on an element of a surface, divided by the area of that element. [1]

Isentropic process - A thermodynamic process in which the entropy of the system does not change.

Ising model - A model describing the coupling between two atoms in a ferromagnetic lattice, in which the interaction energy is proportional to the negative of the product of the spin components along a specified axis.

Isobar - A line connecting points of equal pressure on a graphical representation of a physical system.

Isochore - A line or surface of constant volume on a graphical representation of a physical system.

Isoelectric point* - The pH of a solution or dispersion at which the net charge on the macromolecules or colloidal particles is zero. In electrophoresis there is no motion of the particles in an electric field at the isoelectric point.

Isomers - In chemistry, compounds that have identical molecular formulas but differ in the nature or sequence of bonding of their atoms or in the arrangement of their atoms in space. In physics, nuclei of the same atomic number Z and mass number A but in different energy states. [3]

Isomorphs - Substances of different chemical nature but having the same crystal structure.

Isotactic macromolecule - A tactic macromolecule, essentially comprising only one species of repeating unit which has chiral or prochiral atoms in the main chain in a unique arrangement with respect to its adjacent constitutional units. [8]

Isotherm - A line connecting points of equal temperature on a graphical representation of a physical system.

Isothermal process - A thermodynamic process in which the temperature of the system does not change.

Isotones - Nuclides having the same neutron number N but different atomic number Z. [3]

Isotopes - Two or more nuclides with the same atomic number Z but different mass number A. The term is sometimes used synonymously with nuclide, but it is preferable to reserve the word nuclide for a species of specific Z and A. [3]

Jahn-Teller effect - An interaction of vibrational and electronic motions in a nonlinear molecule which removes the degeneracy of certain electronic energy levels. It can influence the spectrum, crystal structure, and magnetic properties of the substance.

Johnson noise - Electrical noise generated by random thermal motion of electrons in a conductor or semiconductor. Also called thermal noise.

Josephson effect - The tunneling of electron pairs through a thin insulating layer which separates two superconductors. When a potential difference is applied to the superconductors, an alternating current is generated whose frequency is precisely proportional to the potential difference. This effect has important applications in metrology and determination of fundamental physical constants.

Joule (J)* - The SI unit of energy, equal to N m. [1]

Joule-Thomson coefficient (μ) - A parameter which describes the temperature change when a gas expands adiabatically through a nozzle from a high pressure to a low pressure region. It is defined by $\mu = (\partial T/\partial p)_H$, where H is enthalpy.

Julian calendar - The calendar introduced by Julius Caesar in 46 B.C. which divided the year into 365 days with a leap year of 366 days every fourth year.

Julian date (JD) - The number of days elapsed since noon Greenwich Mean Time on January 1, 4713 B.C. Thus January 1, 2000, 0h (midnight) will be JD 2,451,543.5. This dating system was introduced by Joseph Scaliger in 1582.

Kaon - One of the elementary particles in the family of mesons. Kaons have a spin of zero and may be neutral or charged.

Kelvin (K)* - The SI base unit of thermodynamic temperature. [1]

Kepler's laws - The three laws of planetary motion, which established the elliptical shape of planetary orbits and the relation between orbital dimensions and the period of rotation.

Kerr effect* - An electrooptical effect in which birefringence is induced in a liquid or gas when a strong electric field is applied perpendicular to the direction of an incident light beam. The Kerr constant k is given by $n_1 - n_2 = k\lambda E^2$, where λ is the wavelength, E is the electric field strength, and n_1 and n_2 are the indices of refraction of the ordinary and extraordinary rays, respectively.

Ketenes - Compounds in which a carbonyl group is connected by a double bond to an alkylidene group: $R_2C=C=O$. [5]

Ketones - Compounds in which a carbonyl group is bonded to two carbon atoms: $R_1R_2C=O$ (neither R may be H). [5]

Kilogram (kg)* - The SI base unit of mass. [1]

Kinetic energy (E_k, T) - The energy associated with the motion of a system of particles in a specified reference frame. For a single particle of mass m moving at velocity v, $E_k = 1/2mv^2$.

Kirchhoff's laws - Basic rules for electric circuits, which state (a) the algebraic sum of the currents at a network node is zero and (b) the algebraic sum of the voltage drops around a closed path is zero.

Klein-Gordon equation - A relativistic extension of the Schrödinger equation.

Klein-Nishima formula - An expression for the scattering cross section of a photon by an unbound electron, based upon the Dirac electron theory.

Knight shift - The change in magnetic resonance frequency of a nucleus in a metal relative to the same nucleus in a diamagnetic solid. The effect is due to the polarization of the conduction electrons in the metal.

Knudsen number (Kn) - A dimensionless quantity used in fluid mechanics, defined by $Kn = \lambda/l$, where λ is mean free path and l is length. [2]

Kondo effect - A large increase in electrical resistance observed at low temperatures in certain dilute alloys of a magnetic metal in a nonmagnetic material.

Kramers-Kronig relation - A set of equations relating the real and imaginary parts of the index of refraction of a medium

Lactams - Cyclic amides of amino carboxylic acids, having a 1-azacycloalkan-2-one structure, or analogues having unsaturation or heteroatoms replacing one or more carbon atoms of the ring. [5]

Lactones - Cyclic esters of hydroxy carboxylic acids, containing a 1-oxacycloalkan-2-one structure, or analogues having unsaturation or heteroatoms replacing one or more carbon atoms of the ring. [5]

Lagrangian function (L) - A function used in classical mechanics, defined as the kinetic energy minus the potential energy for a system of particles.

Lamb shift - The small energy difference between the $^2S_{1/2}$ and $^2P_{1/2}$ levels in the hydrogen atom, which results from interactions between the electron and the radiation field.

Laminar flow - Smooth, uniform, non-turbulent flow of a gas or liquid in parallel layers, with little mixing between layers. It is characterized by small values of the Reynolds number.

Landé g-factor - See g-Factor of the electron

Langevin function - The mathematical function $L(x) = (e^x + e^{-x})/(e^x - e^{-x}) - 1/x$, which occurs in the expression for the average dipole moment of a group of rotating polar molecules in an electric field: $\mu_{av} = \mu L(\mu E/kT)$, where μ is the electric dipole moment of a single molecule, E is the electric field strength, k is the Boltzmann constant, and T is the temperature.

Lanthanides - The elements of atomic number 57 through 71, which share common chemical properties: La, Ce, Pr, Nd, Pm, Sm, Eu, Gd, Tb, Dy, Ho, Er, Tm, Yb, Lu. [7]

Larmor frequency (v_L) - The precession frequency of a magnetic dipole in an applied magnetic field. In particular, a nucleus in a magnetic field of strength B has a Larmor frequency of $\gamma B/2\pi$, where γ is the magnetogyric ratio of the nucleus.

Laser* - A device in which an optical cavity is filled with a medium where a population inversion can be produced by some means. When the resonant frequency of the cavity bears the proper relation to the separation of the inverted energy levels, stimulated emission occurs, producing a highly monochromatic, coherent beam of light.

Laser ionization mass spectroscopy (LIMS) - See Techniques for Materials Characterization, page 12-1.

Lattice constants* - Parameters specifying the dimensions of a unit cell in a crystal lattice, specifically the lengths of the cell edges and the angles between them.

Lattice energy* - The energy per ion pair required to separate completely the ions in a crystal lattice at a temperature of absolute zero.

Laue diagram - A diffraction pattern produced when an x-ray beam passes through a thin slice of a crystal and impinges on a detector behind the crystal.

Lenz's law - The statement that the current induced in a circuit by a change in magnetic flux is so directed as to oppose the change in flux

Leonard-Jones potential - A simple but useful function for approximating the interaction between two neutral atoms or molecules separated by a distance r by writing the potential energy as $U(r) = 4\varepsilon\{(r_0/r)^{12} - (r_0/r)^6\}$, where ε and r_0 are adjustable parameters. In this form the depth of the potential well is ε and the minimum occurs at $2^{1/6}r_0$. The $(1/r)^{12}$ term is often replaced by other powers of $1/r$.

Lepton - One of the class of elementary particles that do not take part in the strong interaction. Included are the electron, muon, and neutrino. All leptons have a spin of 1/2.

Lewis number (Le) - A dimensionless quantity used in fluid mechanics, defined by $Le = a/D$, where a is thermal diffusivity and D is diffusion coefficient. [2]

Ligand field theory - A description of the structure of crystals containing a transition metal ion surrounded by nonmetallic ions (ligands). It is based on construction of molecular orbitals involving the d-orbitals of the central metal ion and combinations of atomic orbitals of the ligands.

Light year (l.y.) - A unit of distance used in astronomy, defined as the distance light travels in one year in a vacuum. Its approximate value is 9.46073×10^{15} m.

Lignins - Macromolecular constituents of wood related to lignans, composed of phenolic propylbenzene skeletal units, linked at various sites and apparently randomly. [5]

Ligroin - The petroleum fraction consisting mostly of C_7 and C_8 hydrocarbons and boiling in the range 90-140 °C; commonly used as a laboratory solvent.

Lipids - A loosely defined term for substances of biological origin that are soluble in nonpolar solvents. They consist of saponifiable lipids, such as glycerides (fats and oils) and phospholipids, as well as nonsaponifiable lipids, principally steroids. [5]

Lipoproteins - Clathrate complexes consisting of a lipid enwrapped in a protein host without covalent binding, in such a way that the complex has a hydrophilic outer surface consisting of all the protein and the polar ends of any phospholipids. [5]

Liter (L)* - A synonym for cubic decimeter. [1]

Lithosphere* - The outer layer of the solid earth, extending from the base of the mantle to the surface of the crust.

Lorentz contraction - The reduction in length of a moving body in the direction of motion, given by the factor $(1-v^2/c^2)^{1/2}$, where v is the velocity of the body and c the velocity of light. Also known as the FitzGerald-Lorentz contraction.

Lorentz force - The force exerted on a point charge Q moving at velocity v in the presence of external fields E and B. It is given (in SI units) by $F = Q(E + v \times B)$.

Loss angle (δ) - For a dielectric material in an alternating electromagnetic field, δ is the phase difference between the current and the potential difference. The function $\tan \delta$ is a measure of the ratio of the power dissipated in the dielectric to the power stored.

Low energy electron diffraction (LEED) - See Techniques for Materials Characterization, page 12-1.

Lumen (lm)* - The SI unit of luminous flux, equal to cd sr. [1]

Luminous flux (Φ) - The intensity of light from a source multiplied by the solid angle. The SI unit is lumen. [1]

Lux (lx)* - The SI unit of illuminance, equal to cd sr m^{-2}. [1]

Lyddane-Sachs-Teller relation - A relation between the phonon frequencies and dielectric constants of an ionic crystal which states that $(\omega_T/\omega_L)^2 = \varepsilon(\infty)/\varepsilon(0)$, where ω_T is the angular frequency of transverse optical phonons, ω_L that of longitudinal optical phonons, $\varepsilon(0)$ is the static dielectric constant, and $\varepsilon(\infty)$ the dielectric constant at optical frequencies.

Lyman series - The series of lines in the spectrum of the hydrogen atom which corresponds to transitions between the ground state (principal quantum number $n = 1$) and successive excited states. The wavelengths are given by $1/\lambda = R_H(1-1/n^2)$, where $n = 2,3,4,...$ and R_H is the Rydberg constant for hydrogen. The first member of the series ($n = 1 \leftrightarrow 2$), which is often called the Lyman-α line, falls at a wavelength of 1216 Å, and the series converges at 912 Å, the ionization limit of hydrogen.

Mach number (Ma) - A dimensionless quantity used in fluid mechanics, defined by $Ma = v/c$, where v is velocity and c is the speed of sound. [2]

Macromolecule - A molecule of high relative molecular mass (molecular weight), the structure of which essentially comprises the multiple repetition of units derived, actually or conceptually, from molecules of low relative molecular mass. [8]

Madelung constant* - A constant characteristic of a particular crystalline material which gives a measure of the electrostatic energy binding the ions in the crystal.

Magnetic field strength (H) - An axial vector quantity, the curl of which is equal to the current density, including the displacement current. [1]

Magnetic induction (B) - An axial vector quantity such that the force exerted on an element of current is equal to the vector product of this element and the magnetic induction. [1]

Magnetic moment - See Dipole moment, magnetic.

Magnetic susceptibility (χ_m, κ)* - Defined by $\chi_m = (\mu-\mu_0)/\mu_0$, where μ is the permeability of the medium and μ_0 the permeability of a vacuum. [1]

Magnetization (M) - Defined by $M = (B/\mu_0)-H$, where B is magnetic induction, H magnetic field strength, and μ_0 the permeability of a vacuum. [1]

Magnetogyric ratio (γ) - Ratio of the magnetic moment of a particle to its angular momentum. Also called gyromagnetic ratio.

Magneton - See Bohr magneton, Nuclear magneton.

Magnetostriction* - The change in dimensions of a solid sample when it is placed in a magnetic field.

Magnon - A quantum of magnetic energy associated with a spin wave in a ferromagnetic or antiferromagnetic crystal.

Mantle - The layer of the earth between the crust and the liquid outer core, which begins about 2900 km below the Earth's surface.

Maser - A device in which a microwave cavity is filled with a medium where a population inversion can be produced by some means. When the resonant frequency of the cavity bears the proper relation to the separation of the inverted energy levels, the device can serve as an amplifier or oscillator at that frequency.

Mass (m)* - Quantity of matter. Mass can also be defined as "resistance to acceleration".

Mass defect (B) - Defined by $B = Zm(^1H) + Nm_n - m_a$, where Z is the atomic number, $m(^1H)$ is the mass of the hydrogen atom, N is the neutron number, m_n is the rest mass of the neutron, and m_a is the mass of the atom in question. Thus Bc^2 can be equated to the binding energy of the nucleus if the binding energy of atomic electrons is neglected. [1]

Mass excess (Δ) - Defined by $\Delta = m_a - Am_u$, where m_a is the mass of the atom, A the number of nucleons, and m_u the unified atomic mass constant ($m_u = 1$ u). [1]

Mass fraction (w_B) - The ratio of the mass of substance B to the total mass of a mixture. [1]

Mass number (A) - A characteristic property of a specific isotope of an element, equal to the sum of the number of protons and neutrons in the nucleus.

Mass spectrometry - An analytical technique in which ions are separated according to the mass/charge ratio and detected by a suitable detector. The ions may be produced by electron impact on a gas, a chemical reaction, energetic vaporization of a solid, etc. [6]

Massieu function - A thermodynamic function defined by $J = -A/T$, where A is the Helmholz energy and T the thermodynamic temperature. [2]

Matthiessen's rule - The statement that the electrical resistivity ρ of a metal can be written as $\rho = \rho_L + \rho_i$, where ρ_L is due to scattering of conduction electrons by lattice vibrations and ρ_i to scattering by impurities and imperfections. If the impurity concentration is small, ρ_i is temperature independent.

Maxwell (Mx)* - A non-SI unit of magnetic field strength (H) equal to 10^{-8} Wb. [1]

Maxwell's equations - The fundamental equations of electromagnetism. In a form appropriate to SI units, they are:

$$\text{curl } H = \partial D/\partial t + j$$
$$\text{div } B = 0$$
$$\text{curl } E = -\partial B/\partial t$$
$$\text{div } D = \rho$$

where H is the magnetic field strength, B the magnetic induction, E the electric field strength, D the electric displacement, j the current density, ρ the charge density, and t is time.

Maxwell-Boltzmann distribution - An expression for the fraction of molecules $f(v)$ in a gas that have velocity v within a specified interval. It takes the form

$$f(v) = 4\pi(M/2\pi RT)^{3/2} v^2 e^{-Mv^2/2RT}$$

where M is the molar mass, R the molar gas constant, and T the temperature.

Mean free path* - The average distance a gas molecule travels between collisions.

Meissner effect - The complete exclusion of magnetic induction from the interior of a superconductor.

Melting point* - The temperature at which the solid and liquid phases of a substance are in equilibrium at a specified pressure (normally taken to be atmospheric unless stated otherwise).

Mercaptans - A traditional term abandoned by IUPAC, synonymous with thiols. This term is still widely used. [5]

Meson - Any elementary particle that has zero or integral spin. Mesons are responsible for the forces between protons and neutrons in the nucleus.

Mesosphere - The part of the Earth's atmosphere extending from the top of the stratosphere (about 50 km above the surface) to 80–90 km. It is characterized by a decrease in temperature with increasing altitude.

Metal - A material in which the highest occupied energy band (conduction band) is only partially filled with electrons. The electrical conductivity of metals generally decreases with temperature.

Metallocenes - Organometallic coordination compounds in which one atom of a transition metal such as iron, ruthenium or osmium is bonded to and only to the face of two cyclopentadienyl ligands which lie in parallel planes. [5]

Meter (m)* - The SI base unit of length. [1]

Methine group - In organic compounds, the -C= group. [5]

Mho - An archaic name for the SI unit siemens (reciprocal ohm).

Micelle - A particle formed by the aggregation of surfactant molecules (typically, 10 to 100 molecules) in solution. For aqueous solutions, the hydrophilic end of the molecule is on the surface of the micelle, while the hydrophobic end (often a hydrocarbon chain) points toward the center. At the critical micelle concentration (cmc) the previously dissolved molecules aggregate into a micelle.

Micron (μ) - An obsolete name for micrometer.

Mie scattering - The scattering of light by spherical dielectric particles whose diameter is comparable to the wavelength of the light.

Milky way - The band of light in the night sky resulting from the stars in the galactic plane. The term is also used to denote the galaxy in which the sun is located.

Miller indices (hkl) - A set of indices used to label planes in a crystal lattice. [2]

Millimeter of mercury (mmHg) - A non-SI unit of pressure, equal to 133.322 Pa. The name is generally considered interchangeable with torr.

Mobility (μ)* - In solid state physics, the drift velocity of electrons or holes in a solid divided by the applied electric field strength. The term is used in a similar sense in other fields.

Molality (m) - A measure of concentration of a solution in which one states the amount of substance (i.e., number of moles) of solute per kilogram of solvent. Thus a 0.1 molal solution (often written as 0.1 m) has m = 0.1 mol/kg.

Molar mass - The mass of one mole of a substance. It is normally expressed in units of g/mol, in which case its numerical value is identical with the molecular weight (relative molecular mass). [1]

Molar quantity - It is often convenient to express an extensive quantity (e.g., volume, enthalpy, heat capacity, etc.) as the actual value divided by amount of substance (number of moles). The resulting quantity is called molar volume, molar enthalpy, etc.

Molar refraction (R) - A property of a dielectric defined by the equation $R = V_m[(n^2-1)/(n^2+2)]$, where n is the index of refraction of the medium (at optical wavelengths) and V_m the molar volume. It is related to the polarizability α of the molecules that make up the medium by the Lorenz-Lorentz equation, $R = N_A\alpha/3\varepsilon_0$, where N_A is Avogadro's constant and ε_0 is the permittivity of a vacuum.

Molarity (c) - A measure of concentration of a solution in which one states the amount of substance (i.e., number of moles) of solute per liter of solution. Thus a 0.1 molar solution (often referred to as 0.1 M) has a concentration c = 0.1 mol/L.

Mole (mol)* - The SI base unit of amount of substance. [1]

Mole fraction (x_B) - The ratio of the amount of substance (number of moles) of substance B to the total amount of substance in a mixture. [1]

Molecular orbital - See Orbital.

Molecular weight (M_r)* - The ratio of the average mass per molecule or specified entity of a substance to 1/12 of the mass of nuclide ^{12}C. Also called relative molar (or molecular) mass. [1]

Moment of inertia (I) - The moment of inertia of a body about an axis is the sum (or integral) of the products of its elements of mass and the squares of their distances from the axis. [1]

Momentum (p) - The product of mass and velocity. [1]

Monomer - A substance consisting of molecules which can undergo polymerization, thereby contributing constitutional units to the essential structure of a macromolecule. [8]

Monosaccharides - A term which includes aldoses, ketoses, and a wide variety of derivatives. [5]

Mössbauer effect - The recoilless emission of γ-rays from nuclei bound in a crystal under conditions where the recoil energy associated with the γ emission is taken up by the crystal as a whole. This results in a very narrow line width, which can be exploited in various types of precise measurements.

Muon* - An unstable elementary particle of spin 1/2 and mass about 200 times that of the electron.

Naphtha - The petroleum fraction consisting mostly of C_6 to C_8 hydrocarbons and boiling in the range 80–120 °C. Solvents derived from this fraction include ligroin and petroleum ether.

Nautical mile - A non-SI unit of length, equal to exactly 1852 m.

Navier-Stokes equations - A set of complex equations for the motion of a viscous fluid subject to external forces.

Néel temperature (T_N)* - The critical temperature above which an antiferromagnetic substance becomes paramagnetic. [1]

Nernst effect - The production of an electric field in a conductor subject to an applied magnetic field and containing a transverse temperature gradient. The electric field is perpendicular to the magnetic field and the temperature gradient.

Network - In polymer science, a highly ramified macromolecule in which essentially each constitutional unit is connected to each other constitutional unit and to the macroscopic phase boundary by many permanent paths through the macromolecule, the number of such paths increasing with the number of intervening bonds. The paths must on the average be coextensive with the macromolecule. [8]

Neutrino - A stable elementary particle in the lepton family. Neutrinos have zero (or at least near-zero) rest mass and spin 1/2.

Neutron* - An elementary particle on spin 1/2 and zero charge. The free neutron has a mean lifetime of 887 seconds. Neutrons and protons, which are collectively called nucleons, are the constituents of the nucleus.

Neutron activation analysis (NAA) - See Techniques for Materials Characterization, page 12-1.

Neutron number (N) - A characteristic property of a specific isotope of an element, equal to the number of neutrons in the nucleus.

Newton (N)* - The SI unit of force, equal to m kg s^{-2}. [1]

Nitriles - Compounds having the structure RC≡N; thus C-substituted derivatives of hydrocyanic acid, HC≡N. [5]

Nitrosamines - N-Nitroso amines: compounds of the structure R$_2$NNO. Compounds RNHNO are not ordinarily isolatable, but they, too, are nitrosamines. The name is a contraction of N-nitrosoamine and, as such, does not require the N locant. [5]

Nuclear magnetic resonance (NMR)* - A widely used technique in which the resonant absorption of radiofrequency radiation by magnetic nuclei in a magnetic field is measured. The results give important information on the local environment of each nucleus.

Nuclear magneton (μ_N)* - The unit of nuclear magnetic moment, defined as $eh/4\pi m_p$, where h is Planck's constant, m_p the proton mass, and e the elementary charge.

Nuclear quadrupole resonance (NQR) - See Techniques for Materials Characterization, page 12-1.

Nuclear reaction analysis (NRA) - See Techniques for Materials Characterization, page 12-1.

Nuclear spin (I) - The quantum number that specifies the intrinsic angular momentum of a particular nucleus. The magnitude of the angular momentum is given by $[I(I+1)]^{1/2} h/2\pi$, where h is Planck's constant.

Nucleic acids* - Macromolecules, the major organic matter of the nuclei of biological cells, made up of nucleotide units, and hydrolyzable into certain pyrimidine or purine bases (usually adenine, cytosine, guanine, thymine, uracil), D-ribose or 2-deoxy-D-ribose. [5]

Nucleon - A collective term for the proton and neutron.

Nucleosides - Ribosyl or deoxyribosyl derivatives (rarely, other glycosyl derivatives) of certain pyrimidine or purine bases. They are thus glycosylamines or N-glycosides related to nucleotides by the lack of phosphorylation. [5]

Nucleotides - Compounds formally obtained by esterification of the 3′ or 5′ hydroxy group of nucleosides with phosphoric acid. They are the monomers of nucleic acids and are formed from them by hydrolytic cleavage. [5]

Nuclide - A species of atoms in which each atom has identical atomic number Z and identical mass number A. [3]

Nusselt number (Nu) - A dimensionless quantity used in fluid mechanics, defined by $Nu = hl/k$, where h is coefficient of heat transfer, l is length, and k is thermal conductivity. [2]

Nyquist theorem - An expression for the mean square thermal noise voltage across a resistor, given by $4RkT\Delta f$ where R is the resistance, k the Boltzmann constant, T the temperature, and Δf the frequency band within which the voltage is measured.

Octanol-water partition coefficient (P)* - A measure of the way in which a compound will partition itself between the octanol and water phases in the two-phase octanol-water system, and thus an indicator of certain types of biological activity. Specifically, P is the ratio of the concentration (in moles per liter) of the compound in the octanol phase to that in the water phase at infinite dilution. The quantity normally reported is log P.

Oersted (Oe) - A non-SI unit of magnetic field (H), equal to 79.57747 A/m.

Ohm (Ω)* - The SI unit of electric resistance, equal to V/A. [1]

Ohm's law - A relation among electric current I, potential difference V, and resistance R, viz., $I = V/R$. At constant temperature the resistance for many materials is constant to high precision.

Olefins - Acyclic and cyclic hydrocarbons having one or more carbon-carbon double bonds, apart from the formal ones in aromatic compounds. The class olefins subsumes alkenes and cycloalkenes and the corresponding polyenes. [5]

Oligomer - A substance consisting of molecules of intermediate relative molecular mass (molecular weight), the structure of which essentially comprises the multiple repetition of units derived, actually or conceptually, from molecules of low relative molecular mass. In contrast to a polymer, the properties of an oligomer can vary significantly with the removal of one or a few of its units. [8]

Oligopeptides - Peptides containing from three to nine amino groups. [5]

Onsager relations - An important set of equations in the thermodynamics of irreversible processes. They express the symmetry between the transport coefficients describing reciprocal processes in systems with a linear dependence of flux on driving forces.

Optical rotary power - Angle by which the plane of polarization of a light beam is rotated by an optically active medium, divided by path length and by concentration of the active constituent. Depending on whether mass or molar concentration is used, the modifier "specific" or "molar" is attached. [2]

Orbital - A one-electron wavefunction. Atomic orbitals are classified as s-, p-, d-, or f-orbitals according to whether the angular momentum quantum number $l = 0$, 1, 2, or 3. Molecular orbitals, which are usually constructed as linear combinations of atomic orbitals, describe the distribution of electrons over the entire molecule.

Oscillator strength (f) - A measure of the intensity of a spectroscopic transition, defined by

$$f = \frac{8\pi^2 m_e \nu}{3he^2}\left|\mu_{ij}\right|^2$$

where ν is the frequency, μ_{ij} the transition dipole moment, m_e the mass of the electron, e the elementary charge, and h Planck's constant.

Osmosis - The flow of a solvent in a system in which two solutions of different concentration are separated by a semipermeable membrane which cannot pass solute molecules. The solvent will flow from the side of lower concentration to that of higher concentration, thus tending to equalize the concentrations. The pressure that must be applied to the more concentrated side to stop the flow is called the osmotic pressure.

Osmotic coefficient (ϕ) - Defined by $\phi = \ln a_A/(M_A \Sigma m_B)$, where M_A is the molar mass of substance A (normally the solvent), a_A is its activity, and the m_B are molalities of the solutes. [1]

Osmotic pressure (Π) - The excess pressure necessary to maintain osmotic equilibrium between a solution and the pure solvent separated by a membrane permeable only to the solvent. In an ideal dilute solution $\Pi = c_B RT$, where c_B is the amount-of-substance concentration of the solute, R is the molar gas constant, and T the temperature. [1,2]

Ostwald dilution law - A relation for the concentration dependence of the molar conductivity Λ of an electrolyte solution, viz.,

$$\frac{1}{\Lambda} = \frac{1}{\Lambda^\circ} + \frac{\Lambda c}{K(\Lambda^\circ)^2}$$

where c is the solute concentration, K is the equilibrium constant for dissociation of the solute, and Λ° is the conductivity at $c\Lambda = 0$.

Ounce (oz) - A non-SI unit of mass. The avoirdupois ounce equals 28.34952 g, while the troy ounce equals 31.10348 g.

Overpotential (η) - In an electrochemical cell, the difference between the potential of an electrode and its zero-current value.

Oximes - Compounds of structure $R_2C{=}NOH$ derived from condensation of aldehydes or ketones with hydroxylamine. Oximes from aldehydes may be called aldoximes; those from ketones may be called ketoximes. [5]

Oxo compounds - Compounds containing an oxygen atom, =O, doubly bonded to carbon or another element. The term thus embraces aldehydes, carboxylic acids, ketones, sulfonic acids, amides and esters. [5]

Ozonides - The 1,2,4-trioxolanes formed by the reaction of ozone at a carbon-carbon double bond, or the analogous compounds derived from acetylenic compounds. [5]

Pair production - A process in which a photon is converted into a particle and its antiparticle (e.g., an electron and positron) in the electromagnetic field of a nucleus.

Paraffins - Obsolescent term for saturated hydrocarbons, commonly but not necessarily acyclic. Still widely used in the petrochemical industry, where the term designates acyclic saturated hydrocarbons, and stands in contradistinction to naphthenes. [5]

Paramagnetism* - A type of magnetism characterized by a positive magnetic susceptibility, so that the material becomes weakly magnetized in the direction of an external field. The magnetization disappears when the field in removed. In the simplest approximation (Curie's law) the susceptibility is inversely proportional to temperature.

Parity - The property of a quantum-mechanical wave function that describes its behavior under the symmetry operation of coordinate inversion. A parity of +1 (or even) is assigned if the wave function does not change sign when the signs of all the coordinates are changed; the parity is −1 (or odd) if the wave function changes sign under this operation.

Parsec (pc) - A unit of distance defined as the distance at which 1 astronomical unit (AU) subtends an angle of 1 second of arc. It is equal to 206264.806 AU or 3.085678×10^{16} m.

Particle induced x-ray emission (PIXE) - See Techniques for Materials Characterization, page 12-1.

Partition function (q, z) - For a single molecule, $q = \Sigma_i g_i \exp(\varepsilon_i/kT)$, where ε_i is an energy level of degeneracy g_i, k the Boltzmann constant, and T the absolute temperature; the summation extends over all energy states. For a system of N non-interacting molecules which are indistinguishable, as in an ideal gas, the canonical partition function $Q = q^N/N!$.

Pascal (Pa)* - The SI unit of pressure, equal to N/m^2. [1]

Paschen series - The series of lines in the spectrum of the hydrogen atom which corresponds to transitions between the state with principal quantum number $n = 3$ and successive higher states. The wavelengths are given by $1/\lambda = R_H(1/9 - 1/n^2)$, where $n = 4,5,6,...$ and R_H is the Rydberg constant. The first member of the series ($n = 3{\leftrightarrow}4$), which is often called the P_α line, falls in the infrared at a wavelength of 1.875 µm.

Paschen-Back effect - In atomic spectroscopy, the decoupling of electron spin from orbital angular momentum as the strength of an external magnetic field is increased.

Pauli exclusion principle - The statement that two electrons in an atom cannot have identical quantum numbers; thus if there are two electrons in the same orbital, their spin quantum numbers must be of opposite sign.

Pearson symbol - A code for designating crystallographic information, including the crystal system, the lattice type, and the number of atoms per unit cell.

Péclet number (Pe) - A dimensionless quantity used in fluid mechanics, defined by $Pe = vl/a$, where v is velocity, l is length, and a is thermal diffussivity. [2]

Peltier effect - The absorption or generation of heat (depending on the current direction) which occurs when an electric current is passed through a junction between two materials.

Peptides - Amides derived from two or more amino carboxylic acid molecules (the same or different) by formation of a covalent bond from the carbonyl carbon of one to the nitrogen atom of another with formal loss of water. [5]

Permeability (μ) - Magnetic induction divided by magnetic field strength; i.e. $\mu = B/H$. The relative permeability $\mu_r = \mu/\mu_0$, where μ_0 is the permeability of a vacuum. [1]

Permittivity (ε) - Ratio of the electric displacement in a medium to the electric field strength. Also called dielectric constant. [1]

Peroxides - Compounds of structure ROOR in which R may be any organic group. In inorganic chemistry, salts of the anion O_2^{-2} [5]

Peroxy acids - Acids in which an acidic -OH group has been replaced by an -OOH group; e.g., $CH_3C({=}O)OOH$ peroxyacetic acid, $PhS({=}O)_2OOH$ benzeneperoxysulfonic acid. [5]

Petroleum ether - The petroleum fraction consisting of C_5 and C_6 hydrocarbons and boiling in the range 35–60 °C; commonly used as a laboratory solvent.

pH* - A convenient measure of the acid-base character of a solution, usually defined by pH = $-\log [c(H^+)/\text{mol L}^{-1})]$, where $c(H^+)$ is the concentration of hydrogen ions. The more precise definition is in terms af activity rather than concentration. [2]

Phenols - Compounds having one or more hydroxy groups attached to a benzene or other arene ring. [5]

Phonon - A quantum of energy associated with a vibrational mode of a crystal lattice.

Phosphines - PH_3 and compounds derived from it by substituting one, two or three hydrogen atoms by hydrocarbyl groups.

RPH$_2$, R$_2$PH and R$_3$P (R not equal to H) are called primary, secondary and tertiary phosphines, respectively. [5]

Phosphonium compounds - Salts (and hydroxides) [R$_4$P]$^+$X$^-$ containing tetracoordinate phosphonium ion and the associated anion. [5]

Phosphonium ylides - Compounds having the structure R$_3$P$^+$-C$^-$R$_2$ ⇌ R$_3$P=CR$_2$. Also known as Wittig reagents. [5]

Phosphorescence - The process by which a molecule is excited by light to a higher electronic state and then undergoes a radiationless transition to a state of different multiplicity from which it decays, after some delay, to the ground state. The emitted light is normally of longer wavelength than the exciting light because vibrational energy has been dissipated.

Photoelectric effect - The complete absorption of a photon by a solid with the emission of an electron.

Photon - An elementary particle of zero mass and spin 1. The photon is involved in electromagnetic interactions and is the quantum of electromagnetic radiation.

Photon stimulated desorption (PSD) - See Techniques for Materials Characterization, page 12-1.

Pinacols - Tetra(hydrocarbyl)ethane-1,2-diols, R$_2$C(OH)C(OH)R$_2$, of which the tetramethyl example is the simplest one and is itself commonly known as pinacol. [5]

Pion - An elementary particle in the family of mesons. Pions have zero spin and may be neutral or charged. They participate in the strong interaction which holds the nucleus together.

pK* - The negative logarithm (base 10) of an equilibrium constant K. For pK_a, see Acid dissociation constant.

Planck constant (h)* - The elementary quantum of action, which relates energy to frequency through the equation $E = h\nu$.

Planck distribution - See Black body radiation

Planck function - A thermodynamic function defined by $Y = -G/T$, where G is Gibbs energy and T thermodynamic temperature. [2]

Plasma - A highly ionized gas in which the charge of the electrons is balanced by the charge of the positive ions, so that the system as a whole is electrically neutral.

Plasmon - A quantum associated with a plasma oscillation in the electron gas of a solid.

Point group* - A group of symmetry operations (rotations, reflections, etc.) that leave a molecule invariant. Every molecular conformation can be assigned to a specific point group, which plays a major role in determining the spectrum of the molecule.

Poise (P) - A non-SI unit of viscosity, equal to 0.1 Pa s.

Poiseuille's equation - A formula for the rate of flow of a viscous fluid through a tube:

$$\frac{dV}{dt} = \frac{(p_1^2 - p_2^2)\pi r^4}{16l\eta p_0}$$

where V is the volume as measured at pressure p_0; p_1 and p_2 are the pressures at each end of the tube; r is the radius and l the length of the tube; and η is the viscosity.

Poisson ratio (μ) - The absolute value of the ratio of the transverse strain to the corresponding axial strain resulting from uniformly distributed axial stress below the proportional limit (i.e., where Hooke's law is valid). [10]

Polariton - A quantum associated with the coupled modes of photons and optical phonons in an ionic crystal.

Polarizability (α)* - The change in dipole moment of a molecule produced by an external electric field; specifically, $\alpha_{ab} = \partial p_a / \partial E_b$, where p_a is the dipole moment component on the a axis and E_b is the component of the electric field strength along the b axis. [2]

Polymer - A substance composed of molecules of high relative molecular mass (molecular weight), the structure of which essentially comprises the multiple repetition of units derived, actually or conceptually, from molecules of low relative molecular mass. A single molecule of a polymer is called a macromolecule. [8]

Polypeptides - Peptides containing 10 or more amino acid residues. See also Peptides. [5]

Polysaccharides - Compounds consisting of a large number of monosaccharides linked glycosidically. This term is commonly used only for those containing more than ten monosaccharide residues. Also called glycans. [5]

Porphyrins - Natural pigments containing a fundamental skeleton of four pyrrole nuclei united through the α-positions by four methine groups to form a macrocyclic structure (porphyrin is designated porphine in Chemical Abstracts indexes). [5]

Positron - The antiparticle of the electron. It has the same mass and spin as an electron, and an equal but opposite charge.

Positronium - The hydrogen-like "atom" formed from a positron nucleus and an electron. Its lifetime is very short because of annihilation of the positron and electron.

Potential - See Electric potential.

Potential energy (E_p, V, U) - The portion of the energy of a system that is associated with its position in a force field.

Pound (lb) - A non-SI unit of mass, equal to 0.4535924 kg.

Power (P) - Rate of energy transfer. For electrical circuits, this is equal to the product of current and potential difference, $P = IV$. [1]

Poynting vector (S) - For electromagnetic radiation, the vector product of the electric field strength and the magnetic field strength. [1]

Prandtl number (Pr) - A dimensionless quantity used in fluid mechanics, defined by $Pr = \eta/\rho a$, where η is viscosity, ρ is density, and a is thermal diffusivity. [2]

Pressure* - Force divided by area. [1]

Proteins - Naturally occurring and synthetic polypeptides having molecular weights greater than about 10,000 (the limit is not precise). See also Peptides. [5]

Proton* - A stable elementary particle of unit positive charge and spin 1/2. Protons and neutrons, which are collectively called nucleons, are the constituents of the nucleus.

Pulsar - A neutron star which rotates rapidly and emits electromagnetic radiation in regular pulses at a frequency related to the rotation period.

Purine bases* - Purine and its substitution derivatives, especially naturally occurring examples. [5]

Pyrimidine bases* - Pyrimidine and its substitution derivatives, especially naturally occurring examples. [5]

Q-switching - A technique for obtaining very high power from a laser by keeping the Q factor of the laser cavity low while the population inversion builds up, then suddenly increasing the Q to initiate the stimulated emission.

Quad - A unit of energy defined as 10^{15} Btu, equal to approximately 1.055056 × 10^{18} J.

Quadrupole moment - A coefficient of the third term (after monopole and dipole) in the power series expansion of the electric potential of an array of charges. A nucleus of spin greater than 1/2 has a non-vanishing nuclear quadrupole moment which can interact with the electric field gradient of the surrounding electrons. Molecular quadrupole moments have an influence on intermolecular forces.

Quality factor (Q) - The ratio of the absolute value of the reactance of an electrical system to the resistance; thus a measure of the energy stored per cycle relative to the energy dissipated.

Quantum yield - In photochemistry, the number of moles transformed in a specific process, either physically (e.g., by emission of photons) or chemically, per mole of photons absorbed by the system. [3]

Quark - An elementary entity which has not been directly observed but is considered a constituent of protons, neutrons, and other hadrons.

Quasar - An extragalactic object emitting electromagnetic radiation at a very high power level and showing a very large red shift, thus indicating that the object is receding at a speed approaching the speed of light.

Quasicrystal - A solid having conventional crystalline properties but whose lattice does not display translational periodicity.

Quaternary ammonium compounds - Derivatives of ammonium compounds, $NH_4^+ Y^-$, in which all four of the hydrogens bonded to nitrogen have been replaced with hydrocarbyl groups. Compounds having a carbon-nitrogen double bond (i.e. $R_2C=N^+R_2Y^-$) are more accurately called iminium compounds. [5]

Quinones - Compounds having a fully conjugated cyclic dione structure, such as that of benzoquinones, derived from aromatic compounds by conversion of an even number of -CH= groups into -C(=O)- groups with any necessary rearrangement of double bonds. [5]

Racemic mixture - A mixture of equal amounts of a pair of enantiomers (optical isomers); such a mixture is not optically active.

Rad - A non-SI unit of absorbed dose of radiation, equal to 0.01 Gy.

Radiance (L) - The radiant intensity in a given direction from an element of a surface, divided by the area of the orthogonal projection of this element on a plane perpendicular to the given direction. [1]

Radiant intensity (I) - The radiant energy flux leaving an element of a source within an element of solid angle, divided by that element of solid angle. [1]

Radicals - Molecular entities possessing an unpaired electron, such as $\cdot CH_3$, $\cdot SnH_3$, $\cdot Cl$. (In these formulas the dot, symbolizing the unpaired electron, should be placed so as to indicate the atom of highest spin density, if this is possible). [5]

Raman effect - The inelastic scattering of light by a molecule, in which the incident photon either gives up to, or receives energy from, one of the internal vibrational modes of the molecule. The scattered light thus has either a lower frequency (Stokes radiation) or higher frequency (anti-Stokes radiation) than the incident light. These shifts provide a measure of the normal vibrational frequencies of the molecule.

Rankine cycle - A thermodynamic cycle which can be used to calculate the ideal performance of a heat engine that uses a condensable vapor as the working fluid (e.g., a steam engine or a heat pump).

Rankine temperature - A thermodynamic temperature scale based on a temperature interval $°R = (5/9) K$; i.e., $T/°R = (9/5) T/K = t/°F + 459.67$.

Raoult's law - The expression for the vapor pressure p_i of component i in an ideal solution, viz., $p_i = x_i p_{i0}$, where x_i is the mole fraction of component i and p_{i0} the vapor pressure of the pure substance i.

Rare earth elements - The elements Sc, Y, and the lanthanides (La, Ce, Pr, Nd, Pm, Sm, Eu, Gd, Tb, Dy, Ho, Er, Tm, Yb, Lu). [7]

Rayleigh number (Ra) - A dimensionless quantity used in fluid mechanics, defined by $Ra = l^3 g\alpha\Delta T\rho/\eta a$, where l is length, g is acceleration of gravity, α is cubic expansion coefficient, T is temperature, ρ is density, η is viscosity, and a is thermal diffusivity. [2]

Rayleigh scattering - The scattering of light by particles which are much smaller than the wavelength of the light. It is characterized by a scattered intensity which varies as the inverse fourth power of the wavelength.

Rayleigh wave - A guided elastic wave along the surface of a solid; also called surface acoustic wave.

Reactance (X) - The imaginary part of impedance. For an inductive reactance L and a capacitive reactance C in series, the reactance is $X = L\omega - 1/(C\omega)$, where ω is 2π times the frequency of the current. [1]

Red shift - A displacement of a spectral line toward longer wavelengths. This can occur through the Doppler effect (e.g., in the light from receding galaxies) or, in the general theory of relativity, from the effects of a star's gravitational field.

Reflectance (ρ) - Ratio of the radiant or luminous flux at a given wavelength that is reflected to that of the incident radiation. Also called reflection factor. [1]

Reflection high energy electron diffraction (RHEED) - See Techniques for Materials Characterization, page 12-1.

Relative humidity* - The ratio of the partial pressure of water vapor in air to the saturation vapor pressure of water at the same temperature, expressed as a percentage. [10]

Relative molar mass - See Molecular weight.

Rem - A non-SI unit of dose equivalent, equal to 0.01 Sv.

Resistance (R) - Electric potential difference divided by current when there is no electromotive force in the conductor. This definition applies to direct current. More generally, resistance is defined as the real part of impedance. [1]

Resistivity (ρ) - Electric field strength divided by current density when there is no electromotive force in the conductor. Resistivity is an intrinsic property of a material. For a conductor of uniform cross section with area A and length L, and whose resistance is R, the resistivity is given by $\rho = RA/L$. [1]

Reynolds number (Re) - A dimensionless quantity used in fluid mechanics, defined by $Re = \rho vl/\eta$, where ρ is density, v is velocity, l is length, and η is viscosity. [2]

Rheology - The study of the flow of liquids and deformation of solids. Rheology addresses such phenomena as creep, stress relaxation, anelasticity, nonlinear stress deformation, and viscosity.

Ribonucleic acids (RNA) - Naturally occurring polyribonucleotides. See also nucleic acids, nucleosides, nucleotides, ribonucleotides. [5]

Ribonucleotides - Nucleotides in which the glycosyl group is a ribosyl group. See also nucleotides. [5]

Roentgen (R) - A unit used for expressing the charge (positive or negative) liberated by x-ray or γ radiation in air, divided by the mass of air. A roentgen is defined as 2.58×10^{-4} C/kg.

Rotational constants - In molecular spectroscopy, the constants appearing in the expression for the rotational energy levels as a function of the angular momentum quantum numbers. These constants are proportional to the reciprocals of the principal moments of inertia, averaged over the vibrational motion.

Rutherford back scattering (RBS) - See Techniques for Materials Characterization, page 12-1.

Rydberg constant (R_∞)* - The fundamental constant which appears in the equation for the energy levels of hydrogen-like atoms; i.e., $E_n = hcR_\infty Z^2\mu/n^2$, where h is Planck's constant, c the speed of light, Z the atomic number, μ the reduced mass of nucleus and electron, and n the principal quantum number ($n = 1, 2, ...$).

Rydberg series - A regular series of lines in the spectrum of an atom or molecule, with the spacing between successive lines becoming smaller as the frequency increases (wavelength decreases). The series eventually converges to a limit which usually corresponds to the complete removal of an electron from the atom or molecule.

Sackur-Tetrode equation* - An equation for the molar entropy S_m of an ideal monatomic gas: $S_m = R\ln(e^{5/2} V/N_A\Lambda^3)$, where R is the molar gas constant, V is the volume, and N_A is Avogadro's number. The constant Λ is given by $\Lambda = h/(2\pi mkT)^{1/2}$, where h is Planck's constant, m the atomic mass, k the Boltzmann constant, and T the temperature.

Salinity (S)* - A parameter used in oceanography to describe the concentration of dissolved salts in seawater. It is defined in terms of electrical conductivity relative to a standard solution of KCl. When expressed in units of parts per thousand, S may be roughly equated to the concentration of dissolved material in grams per kilogram of seawater.

Salt - An ionic compound formed by the reaction of an acid and a base.

Scanned probe microscopy (SPM) - See Techniques for Materials Characterization, page 12-1.

Scanning electron microscopy (SEM) - See Techniques for Materials Characterization, page 12-1.

Scanning laser acoustic microscopy (SLAM) - See Techniques for Materials Characterization, page 12-1.

Scanning transmission electron microscopy (STEM) - See Techniques for Materials Characterization, page 12-1.

Scanning tunneling microscopy (STM) - See Techniques for Materials Characterization, page 12-1.

Schiff bases - Imines bearing a hydrocarbyl group on the nitrogen atom: $R_2C=NR'$ (R' not equal to H). Considered by many to be synonymous with azomethines. [5]

Schmidt number (Sc) - A dimensionless quantity used in fluid mechanics, defined by $Sc = \eta/\rho D$, where η is viscosity, ρ is density, and D is diffusion coefficient. [2]

Schottky barrier - A potential barrier associated with a metal-semiconductor contact. It forms the basis for the rectifying device known as the Schottly diode.

Schrödinger equation - The basic equation of wave mechanics which, for systems not dependent on time, takes the form:

$$-(\hbar/2m)\nabla^2\psi + V\psi = E\psi$$

where ψ is the wavefunction, V is the potential energy expressed as a function of the spatial coordinates, E is an energy eigenvalue, ∇^2 is the Laplacian operator, $\hbar$ is Planck's constant divided by 2π, and m is the mass.

Second (s)* - The SI base unit of time. [1]

Second radiation constant (c_2)* - See First radiation constant.

Secondary ion mass spectroscopy (SIMS) - See Techniques for Materials Characterization, page 12-1.

Seebeck effect - The development of a potential difference in a circuit where two different metals or semiconductors are joined and their junctions maintained at different temperatures. It is the basis of the thermocouple.

Selenides - Compounds having the structure RSeR (R not equal to H). They are thus selenium analogues of ethers. Also used for metal salts of H_2Se. [5]

Semicarbazones - Compounds having the structure $R_2C=NNHC(=O)NH_2$, formally derived by condensation of aldehydes or ketones with semicarbazide [$NH_2NHC(=O)NH_2$]. [5]

Semiconductor - A material in which the highest occupied energy band (valence band) is completely filled with electrons at $T = 0$ K, and the energy gap to the next highest band (conduction band) ranges from 0 to 4 or 5 eV. With increasing temperature electrons are excited into the conduction band, leading to an increase in the electrical conductivity.

Semiquinones - Radical anions having the structure -O-Z-O· where Z is an ortho- or para-arylene group or analogous heteroarylene group; they are formally generated by the addition of an electron to a quinone. [5]

SI units* - The International System of Units adopted in 1960 and recommended for use in all scientific and technical fields. [1]

Siemens (S)* - The SI unit of electric conductance, equal to Ω^{-1}. [1]

Sievert (Sv)* - The SI unit of dose equivalent (of radiation), equal to J/kg. [1]

Silanes - Saturated silicon hydrides, analogues of the alkanes; i.e., compounds of the general formula Si_nH_{2n+2}. Silanes may be subdivided into silane, oligosilanes, and polysilanes. Hydrocarbyl derivatives are often referred to loosely as silanes. [5]

Silicones - Polymeric or oligomeric siloxanes, usually considered unbranched, of general formula $[-OSiR_2-]_n$ (R not equal to H). [5]

Siloxanes - Saturated silicon-oxygen hydrides with unbranched or branched chains of alternating silicon and oxygen atoms (each silicon atom is separated from its nearest silicon neighbors by single oxygen atoms). [5]

Skin effect - The concentration of high frequency alternating currents near the surface of a conductor.

Slater orbital - A particular mathematical expression for the radial part of the wave function of a single electron, which is used in quantum-mechanical calculations of the energy and other properties of atoms and molecules.

Small angle neutron scattering (SANS) - See Techniques for Materials Characterization, page 12-1.

Snell's law - The relation between the angle of incidence i and the angle of refraction r of a light beam which passes from a medium of refractive index n_0 to a medium of index n_1, viz., $\sin i/\sin r = n_1/n_0$.

Solar constant* - The mean radiant energy flux from the sun on a unit surface normal to the direction of the rays at the mean distance of the earth from the sun. The value is approximately 1373 W/m².

Solar wind - The stream of high velocity hydrogen and helium ions emitted by the sun which flows through the solar system and beyond.

Soliton - A spatially localized wave in a solid or liquid that can interact strongly with other solitons but will afterwards regain its original form.

Solubility* - A quantity expressing the maximum concentration of some material (the solute) that can exist in another liquid or solid material (the solvent) at thermodynamic equilibrium at specified temperature and pressure. Common measures of solubility include the mass of solute per unit mass of solution (mass fraction), mole fraction of solute, molality, molarity, and others.

Solubility product constant (K_{sp})* - The equilibrium constant for the dissolution of a sparsely soluble salt into its constituent ions.

Space group* - A group of symmetry operations (reflections, rotations, etc.) that leave a crystal invariant. A total of 230 space groups have been identified.

Spark source mass spectroscopy (SSMS) - See Techniques for Materials Characterization, page 12-1.

Specific gravity - Ratio of the mass density of a material to that of water. Since one must specify the temperature of both the sample and the water to have a precisely defined quantity, the use of this term is now discouraged.

Specific heat - Heat capacity divided by mass. See Heat capacity.

Specific quantity - It is often convenient to express an extensive quantity (e.g., volume, enthalpy, heat capacity, etc.) as the actual value divided by mass. The resulting quantity is called specific volume, specific enthalpy, etc.

Specific rotation $[\alpha]^\theta_\lambda$ - For an optically active substance, defined by $[\alpha]^\theta_\lambda = \alpha/\gamma l$, where α is the angle through which plane polarized light is rotated by a solution of mass concentration γ and path length l. Here θ is the Celsius temperature and λ the wavelength of the light at which the measurement is carried out. Also called specific optical rotatory power. [2]

Spin (s, I)* - A measure of the intrinsic angular momentum of a particle, which it possesses independent of its orbital motion. The symbol s is used for the spin quantum number of an electron, while I is generally used for nuclear spin.

Spiro compounds - Compounds having one atom (usually a quaternary carbon) as the only common member of two rings. [5]

Stacking fault - An error in the normal sequence of layer growth in a crystal.

Standard mean ocean water (SMOW) - A standard sample of pure water of accurately known isotopic composition which is maintained by the International Atomic Energy Agency. It is used for precise calibration of density and isotopic composition measurements.

Standard reduction potential ($E°$) - The zero-current potential of a cell in which the specified reduction reaction occurs at the right-hand electrode and the left-hand electrode is the standard hydrogen electrode. Also called Standard electrode potential.

Standard state - A defined state (specified temperature, pressure, concentration, etc.) for tabulating thermodynamic functions and carrying out thermodynamic calculations. The standard state pressure is usually taken as 100,000 Pa (1 bar), but various standard state temperatures are used. [2]

Stanton number (St) - A dimensionless quantity used in fluid mechanics, defined by $St = h/\rho v c_p$, where h is coefficient of heat transfer, ρ is density, v is velocity, and c_p is specific heat capacity at constant pressure. [2]

Stark effect - The splitting of an energy level of an atom or molecule, and hence a splitting of spectral lines arising from that level, as a result of the application of an external electric field.

Statistical weight (g) - The number of distinct states corresponding to the same energy level. Also called degeneracy.

Stefan-Boltzmann constant (σ)* - Constant in the equation for the radiant exitance M (radiant energy flux per unit area) from a black body at thermodynamic temperature T, viz. $M = \sigma T^4$. [1]

Stibines - SbH_3 and compounds derived from it by substituting one, two or three hydrogen atoms by hydrocarbyl groups: R_3Sb. $RSbH_2$, R_2SbH, and R_3Sb (R not equal to H) are called primary, secondary and tertiary stibines, respectively. [5]

Stochastic process - A process which involves random variables and whose outcome can thus be described only in terms of probabilities.

Stoichiometric number (v) - The number appearing before the symbol for each compound in the equation for a chemical reaction. By convention, it is negative for reactants and positive for products. [2]

Stokes (St) - A non-SI unit of kinematic viscosity, equal to 10^{-4} m²/s.

Stokes' law - The statement, valid under certain conditions, that the viscous force F experienced by a sphere of radius a moving at velocity v in a medium of viscosity η is given by $F = -6\pi\eta a v$.

Strain - The deformation of a body that results from an applied stress.

Stratosphere - The part of the earth's atmosphere extending from the top of the troposphere (typically 10 to 15 km above the surface) to about 50 km. It is characterized by an increase in temperature with increasing altitude.

Stress - Force per unit area (pressure) applied to a body. Tensile stress tends to stretch or compress the body in the direction of the applied force. Sheer stress results from a tangential force which tends to twist the body.

Strong interaction - The short range (order of 1 fm) attractive forces between protons, neutrons, and other hadrons which are responsible for the stability of the nucleus.

Strouhal number (Sr) - A dimensionless quantity used in fluid mechanics, defined by $Sr = lf/v$, where l is length, f is frequency, and v is velocity. [2]

Structure factor - In x-ray crystallography, the sum of the scattering factors of all the atoms in a unit cell, weighted by an appropriate phase factor. The intensity of a given reflection is proportional to the square of the structure factor.

Sublimation pressure - The pressure of a gas in equilibrium with a solid at a specified temperature.

Sulfides - Compounds having the structure RSR (R not equal to H). Such compounds were once called thioethers. In an inorganic sense, salts or other derivatives of hydrogen sulfide. [5]

Sulfones - Compounds having the structure, $RS(=O)_2R$ (R not equal to H), e.g. $C_2H_5S(=O)_2CH_3$, ethyl methyl sulfone. [5]

Sulfonic acids - $HS(=O)_2OH$, sulfonic acid, and its S-hydrocarbyl derivatives. [5]

Sulfoxides - Compounds having the structure $R_2S=O$ (R not equal to H), e.g., $Ph_2S=O$, diphenyl sulfoxide. [5]

Superconductor - A material that experiences a nearly total loss of electrical resistivity below a critical temperature T_c. The effect can occur in pure metals, alloys, semiconductors, organic compounds, and certain inorganic solids.

Superfluid - A fluid with near-zero viscosity and extremely high thermal conductivity. Liquid helium exhibits these properties below 2.186 K (the λ point).

Supernova - A star in the process of exploding because of instabilities which follow the exhaustion of its nuclear fuel.

Surface analysis by laser ionization (SALI) - See Techniques for Materials Characterization, page 12-1.

Surface tension (γ,σ)* - The force per unit length in the plane of the interface between a liquid and a gas, which resists an increase in the area of that surface. It can also be equated to the surface Gibbs energy per unit area.

Surfactant - A substance which lowers the surface tension of the medium in which it is dissolved, and/or the interfacial tension with other phases, and accordingly is positively adsorbed at the liquid-vapor or other interfaces. [3]

Susceptance (B) - Imaginary part of admittance. [1]

Svedberg - A non-SI unit of time, used to express sedimentation coefficients, equal to 10^{-13} s.

Syndiotactic macromolecule - A tactic macromolecule, essentially comprising alternating enantiomeric configurational base units which have chiral or prochiral atoms in the main chain in a unique arrangement with respect to their adjacent constitutional units. In this case the repeating unit consists of two configurational base units that are enantiomeric. [8]

Tacticity - The orderliness of the succession of configurational repeating units of a macromolecule or oligomer molecule. In a tactic macromolecule essentially all the configurational repeating units are identical with respect to directional sense. See Configurational repeating unit, Isotactic, Syndiotactic. [8]

Tautomerism - Isomerism of the general form $G-X-Y=Z \rightleftharpoons X=Y-Z-G$, where the isomers (called tautomers) are readily interconvertible; the atoms connecting the groups X, Y, Z are typically any of C, H, O, or S, and G is a group which becomes an electrofuge (i.e., a group that does not carry away the bonding electron pair when it leaves its position in the molecule) or nucleofuge (a group that does carry away the bonding electrons when leaving) during isomerization. The commonest case, when the electrofuge is H^+, is also known as prototropy. A common example, written so as to illustrate the general pattern given above, is keto-enol tautomerism, such as

$$H-O-C(CH_3)=CH-CO_2Et \text{ (enol)} \rightleftharpoons (CH_3)C(=O)-CH_2-CO_2Et \text{ (keto)}$$

In some cases the interconversion rate between tautomers is slow enough to permit isolation of the separate keto and enol forms. [5]

Tensile strength* - In tensile testing, the ratio of maximum load a body can bear before breaking to original cross-sectional area. Also called ultimate strength. [11]

Terpenes - Hydrocarbons of biological origin having carbon skeletons formally derived from isoprene $[CH_2=C(CH_3)CH=CH_2]$. [5]

Terpenoids - Natural products and related compounds formally derived from isoprene units. They contain oxygen in various functional groups. The skeleton of terpenoids may differ from strict additivity of isoprene units by the loss or shift of a methyl (or other) group. [5]

Tesla (T)* - The SI unit of magnetic flux density (B), equal to V s/m². [1]

Thermal conductivity* - Rate of heat flow divided by area and by temperature gradient. [1]

Thermal diffusivity - Thermal conductivity divided by density and by specific heat capacity at constant pressure. [1]

Thermal expansion coefficient (α)* - The linear expansion coefficient is defined by $\alpha_l = (1/l)(dl/dT)$; the volume expansion coefficient by $\alpha_V = (1/V)(dV/dT)$. [1]

Thermionic emission - The emission of electrons from a solid as a result of heat. The effect requires a high enough temperature to impart sufficient kinetic energy to the electrons to exceed the work function of the solid.

Thermodynamic laws - The foundation of the science of thermodynamics:

    First law: The internal energy of an isolated system is constant; if energy is supplied to the system in the form of heat dq and work dw, then the change in energy $dU = dq + dw$.

    Second law: No process is possible in which the only result is the transfer of heat from a reservoir and its complete conversion to work.

    Third law: The entropy of a perfect crystal approaches zero as the thermodynamic temperature approaches zero.

Thermoelectric power - For a bar of a pure material whose ends are at different temperatures, the potential difference divided by the difference in temperature of the ends. See also Seeback effect.

Thermogravimetric analysis (TGA) - See Techniques for Materials Characterization, page 12-1.

Thermosphere - The layer of the Earth's atmosphere extending from the top of the mesosphere (typically 80–90 km above the surface) to about 500 km. It is characterized by a rapid increase in temperature with increasing altitude up to about 200 km, followed by a leveling off in the 300–500 km region.

Thiols - Compounds having the structure RSH (R not equal to H). Also known by the term mercaptans (abandoned by IUPAC); e.g., CH_3CH_2SH, ethanethiol. [5]

Thomson coefficient (μ, τ) - The heat power developed in the Thomson effect (whereby heat is evolved in a conductor when a current is flowing in the presence of a temperature gradient), divided by the current and the temperature difference. [1]

Tonne (t) - An alternative name for megagram (1000 kg). [1]

Torque (T) - For a force F that produces a torsional motion, $T = r \times F$, where r is a vector from some reference point to the point of application of the force.

Torr - A non-SI unit of pressure, equal to 133.322 Pa. The name is generally considered interchangeable with millimeter of mercury.

Townsend coefficient - In a radiation counter, the number of ionizing collisions by an electron per unit path length in the direction of an applied electric field.

Transducer - Any device that converts a signal from acoustical, optical, or some other form of energy into an electrical signal (or vice versa) while preserving the information content of the original signal.

Transistor - A voltage amplifier using controlled electron currents inside a semiconductor.

Transition metals - Elements characterized by a partially filled d subshell. The First Transition Series comprises Sc, Ti, V, Cr, Mn, Fe, Co, Ni, Cu. The Second and Third Transition Series include the lanthanides and actinides, respectively. [7]

Transition probability* - See Einstein transition probability.

Transmittance (τ) - Ratio of the radiant or luminous flux at a given wavelength that is transmitted to that of the incident radiation. Also called transmission factor. [1]

Tribology - The study of frictional forces between solid surfaces.

Triple point* - The point in p,T space where the solid, liquid, and gas phases of a substance are in thermodynamic equilibrium. The corresponding temperature and pressure are called the triple point temperature and triple point pressure.

Troposphere - The lowest part of the earth's atmosphere, extending to 10–15 km above the surface. It is characterized by a decrease in temperature with increasing altitude. The exact height varies with latitude and season.

Tunnel diode - A device involving a p-n junction in which both sides are so heavily doped that the Fermi level on the p-side lies in the valence band and on the n-side in the conduction band. This leads to a current-voltage curve with a maximum, so that the device exhibits a negative resistance in some regions.

Ultraviolet photoelectron spectroscopy (UPS) - See Techniques for Materials Characterization, page 12-1.

Umklapp process - A process involving the interaction of three or more waves (lattice or electron) in a solid in which the sum of the wave vectors does not equal zero.

Unified atomic mass unit (u)* - A unit of mass used in atomic, molecular, and nuclear science, defined as the mass of one atom of ^{12}C divided by 12. Its approximate value is 1.66054×10^{-27} kg. [1]

Universal time (t_U, UT) - Mean solar time counted from midnight at the Greenwich meridian. Also called Greenwich mean time (GMT). The interval of mean solar time is based on the average, over one year, of the time between successive transits of the sun across the observer's meridian.

Vacancy - A missing atom or ion in a crystal lattice.

Van Allen belts - Two toroidal regions above the earth's atmosphere containing protons and electrons. The outer belt at about 25,000 km above the surface is probably of solar origin. The inner belt at about 3000 km contains more energetic particles from outside the solar system.

Van der Waals' equation* - An equation of state for fluids which takes the form:

$$p V_m = RT \left(\frac{1}{V_m - b} - \frac{a}{V_m^2} \right)$$

where p is pressure, V_m is molar volume, T is temperature, R is the molar gas constant, and a and b are characteristic parameters of the substance which describe the effect of attractive and repulsive intermolecular forces, respectively.

Van der Waals' force - The weak attractive force between two molecules which arises from electric dipole interactions. It can lead to the formation of stable but weakly bound dimer molecules or clusters.

Van't Hoff equation - The equation expressing the temperature dependence of the equilibrium constant K of a chemical reaction:

$$\frac{d \ln K}{dT} = \frac{\Delta_r H^\circ}{RT^2}$$

where $\Delta_r H^\circ$ is the standard enthalpy of reaction, R the molar gas constant, and T the temperature. Also called van't Hoff isochore.

Vapor pressure* - The pressure of a gas in equilibrium with a liquid (or, in some usage, a solid) at a specified temperature.

Varistor - A device that utilizes the properties of certain metal oxides with small amounts of impurities, which show abrupt nonlinearities at specific voltages where the material changes from a semiconductor to an insulator.

Velocity (v) - Rate of change of distance with time.

Verdet constants (V)* - Angle of rotation of a plane polarized light beam passing through a medium in a magnetic field, divided by the field strength and by the path length.

Virial equation of state* - An equation relating the pressure p, molar volume V_m, and temperature T of a real gas in the form of an expansion in powers of the molar volume, viz., $pV_m = RT(1 + BV_m^{-1} + CV_m^{-2} + \ldots)$, where R is the molar gas constant. B is called the second virial coefficient, C the third virial coefficient, etc. The virial coefficients are functions of temperature.

Viscosity (η)* - The proportionality factor between sheer rate and sheer stress, defined through the equation $F = \eta A(dv/dx)$, where F is the tangential force required to move a planar surface of area A at velocity v relative to a parallel surface separated from the first by a distance x. Sometimes called dynamic or absolute viscosity. The term kinematic viscosity (symbol v) is defined as η divided by the mass density.

Volt (V)* - The SI unit of electric potential, equal to W/A. [1]

Volume fraction (ϕ_j) - Defined as $V_j / \Sigma_i V_i$, where V_j is the volume of the specified component and the V_i are the volumes of all the components of a mixture prior to mixing. [2]

Watt (W)* - The SI unit of power, equal to J/s. [1]

Wave function - A function of the coordinates of all the particles in a quantum mechanical system (and, in general, of time) which fully describes the state of the system. The product of the wave function and its complex conjugate is proportional to the probability of finding a particle at a particular point in space.

Weak interaction - The weak forces (order of 10^{-12} of the strong interaction) between elementary particles which are responsible for beta decay and other nuclear effects.

Weber (Wb)* - The SI unit of magnetic flux, equal to V s. [1]

Weber number (We) - A dimensionless quantity used in fluid mechanics, defined by $We = \rho v^2 l / \gamma$, where ρ is density, v is velocity, l is length, and γ is surface tension. [2]

Weight - That force which, when applied to a body, would give it an acceleration equal to the local acceleration of gravity. [1]

Wiedeman-Franz law - The law stating that the thermal conductivity k and electrical conductivity σ of a pure metal are related by $k = L\sigma T$, where T is the temperature and L (called the Lorenz ratio) has the approximate value 2.45×10^{-8} V²/K².

Wien displacement law - The relation, which can be derived from the Planck formula for black body radiation, that

$\lambda_{max} T = 0.0028978$ m K, where λ_{max} is the wavelength of maximum radiance at temperature T.

Wigner-Seitz method - A method of calculating electron energy levels in a solid using a model in which each electron is subject to a spherically symmetric potential.

Wittig reagents - See phosphonium ylides.

Work (W) - Force multiplied by the displacement in the direction of the force. [1]

Work function (Φ)* - The energy difference between an electron at rest at infinity and an electron at the Fermi level in the interior of a substance. It is thus the minimum energy required to remove an electron from the interior of a solid to a point just outside the surface. [1]

X unit (X) - A unit of length used in x-ray crystallography, equal to approximately 1.002×10^{-13} m.

X-ray photoelectron spectroscopy (XPS) - See Techniques for Materials Characterization, page 12-1.

Yield strength - The stress at which a material exhibits a specified deviation (often chosen as 0.2% for metals) from proportionality of stress and strain. [11]

Young's modulus (E) - In tension or compression of a body below its elastic limit, the ratio of stress to corresponding strain. Since strain is normally expressed on a fractional basis, Young's modulus has dimensions of pressure. Also called elastic modulus. [11]

Zeeman effect - The splitting of an energy level of an atom or molecule, and hence a splitting of spectral lines arising from that level, as a result of the application of an external magnetic field.

Zener diode - A control device utilizing a p-n junction with a well defined reverse-bias avalanche breakdown voltage.

Zeotrope - A liquid mixture that shows no maximum or minimum when vapor pressure is plotted against composition at constant temperature. See Azeotrope.

Zero-point energy - The energy possessed by a quantum mechanical system as a result of the uncertainty principle even when it is in its lowest energy state; e.g., the difference between the lowest energy level of a harmonic oscillator and the minimum in the potential well.

Zeta potential (ζ) - The electric potential at the surface of a colloidal particle relative to the potential in the bulk medium at a long distance. Also called electrokinetic potential.

Zwitterions - Neutral compounds having formal unit electrical charges of opposite sign. Some chemists restrict the term to compounds with the charges on non-adjacent atoms. Sometimes referred to as inner salts, dipolar ions (a misnomer). [5]

THERMODYNAMIC FUNCTIONS AND RELATIONS

p = pressure $\quad\quad$ V = volume $\quad\quad$ T = temperature

n_i = amount of substance i

$x_i = n_i/\Sigma_j\, n_j$ = mole fraction of substance i

Energy	U
Entropy	S
Enthalpy	$H = U + pV$
Helmholtz energy	$A = U - TS$
Gibbs energy	$G = U + pV - TS$
Isobaric heat capacity	$C_p = (\partial H/\partial T)_p$
Isochoric heat capacity	$C_V = (\partial U/\partial T)_V$
Isobaric expansivity	$\alpha = V^{-1}(\partial V/\partial T)_p$
Isothermal compressibility	$\kappa_T = -V^{-1}(\partial V/\partial p)_T$
Isentropic compressibility	$\kappa_S = -V^{-1}(\partial V/\partial p)_S$
	$\kappa_T - \kappa_S = T\alpha^2 V/C_p$
	$C_p - C_V = T\alpha^2 V/\kappa_T$
Gibbs-Helmholtz equation	$H = G - T(\partial G/\partial T)_p$
Maxwell relations	$(\partial S/\partial p)_T = -(\partial V/\partial T)_p$
	$(\partial S/\partial V)_T = -(\partial p/\partial T)_V$
Joule-Thomson expansion	$\mu_{JT} = (\partial T/\partial p)_H = -\{V - T(\partial V/\partial T)_p\}/C_p$
	$\phi_{JT} = (\partial H/\partial p)_T = V - T(\partial V/\partial T)_p$
Partial molar quantity	$X_i = (\partial X/\partial n_i)_{T,p,nj \neq i}$
Chemical potential	$\mu_i = (\partial G/\partial n_i)_{T,p,nj \neq i}$
Perfect gas [symbol ^{pg}]	$pV = (\Sigma_i n_i)RT$
	$\mu_i^{pg} = \mu_i^\theta + RT\ln(x_i p/p^\theta)$
Fugacity	$f_i = (x_i p)\exp\{(\mu_i - \mu_i^{pg})/RT\}$
Activity coefficient	$\gamma_i = f_i/(x_i f_i^\theta)$
Gibbs-Duhem relation	$0 = SdT - Vdp + \Sigma_i n_i d\mu_i$

[Superscript θ in above equations indicates standard state]

Notation for chemical and physical changes ($X = H, S, G$, etc.):

Chemical reaction	$\Delta_r X$
Formation from elements	$\Delta_f X$
Combustion	$\Delta_c X$
Fusion (cry→liq)	$\Delta_{fus} X$
Vaporization (liq→gas)	$\Delta_{vap} X$
Sublimation (cry→gas)	$\Delta_{sub} X$
Phase transition	$\Delta_{trs} X$
Solution	$\Delta_{sol} X$
Mixing	$\Delta_{mix} X$
Dilution	$\Delta_{dil} X$

NOBEL LAUREATES IN CHEMISTRY AND PHYSICS

Full details on nationality and basis of the awards can be found at <nobelprize.org/>.

Chemistry

2011	Dan Shechtman	1956	Sir Cyril Hinshelwood, Nikolay Semenov
2010	Richard F. Heck, Ei-ichi Negishi, Akira Suzuki	1955	Vincent du Vigneaud
2009	Venkatraman Ramakrishnan, Thomas A. Steitz, Ada E. Yonath	1954	Linus Pauling
2008	Martin Chalfie, Osamu Shimomura, Roger Y. Tsien	1953	Hermann Staudinger
2007	Gerhard Ertl	1952	Archer J.P. Martin, Richard L.M. Synge
2006	Roger D. Kornberg	1951	Edwin M. McMillan, Glenn T. Seaborg
2005	Yves Chauvin, Robert H. Grubbs, Richard R. Schrock	1950	Otto Diels, Kurt Alder
2004	Aaron Ciechanover, Avram Hershko, Irwin Rose	1949	William F. Giauque
2003	Peter Agre, Roderick MacKinnon	1948	Arne Tiselius
2002	John B. Fenn, Koichi Tanaka, Kurt Wüthrich	1947	Sir Robert Robinson
2001	William S. Knowles, Ryoji Noyori, K. Barry Sharpless	1946	James B. Sumner, John H. Northrop, Wendell M. Stanley
2000	Alan Heeger, Alan G. MacDiarmid, Hideki Shirakawa	1945	Artturi Virtanen
1999	Ahmed Zewail	1944	Otto Hahn
1998	Walter Kohn, John Pople	1943	George de Hevesy
1997	Paul D. Boyer, John E. Walker, Jens C. Skou	1942	*No prize awarded*
1996	Robert F. Curl Jr., Sir Harold Kroto, Richard E. Smalley	1941	*No prize awarded*
1995	Paul J. Crutzen, Mario J. Molina, F. Sherwood Rowland	1940	*No prize awarded*
1994	George A. Olah	1939	Adolf Butenandt, Leopold Ruzicka
1993	Kary B. Mullis, Michael Smith	1938	Richard Kuhn
1992	Rudolph A. Marcus	1937	Norman Haworth, Paul Karrer
1991	Richard R. Ernst	1936	Peter Debye
1990	Elias James Corey	1935	Frédéric Joliot, Irène Joliot-Curie
1989	Sidney Altman, Thomas R. Cech	1934	Harold C. Urey
1988	Johann Deisenhofer, Robert Huber, Hartmut Michel	1933	*No prize awarded*
1987	Donald J. Cram, Jean-Marie Lehn, Charles J. Pedersen	1932	Irving Langmuir
1986	Dudley R. Herschbach, Yuan T. Lee, John C. Polanyi	1931	Carl Bosch, Friedrich Bergius
1985	Herbert A. Hauptman, Jerome Karle	1930	Hans Fischer
1984	Bruce Merrifield	1929	Arthur Harden, Hans von Euler-Chelpin
1983	Henry Taube	1928	Adolf Windaus
1982	Aaron Klug	1927	Heinrich Wieland
1981	Kenichi Fukui, Roald Hoffmann	1926	The Svedberg
1980	Paul Berg, Walter Gilbert, Frederick Sanger	1925	Richard Zsigmondy
1979	Herbert C. Brown, Georg Wittig	1924	*No prize awarded*
1978	Peter Mitchell	1923	Fritz Pregl
1977	Ilya Prigogine	1922	Francis W. Aston
1976	William Lipscomb	1921	Frederick Soddy
1975	John Cornforth, Vladimir Prelog	1920	Walther Nernst
1974	Paul J. Flory	1919	*No prize awarded*
1973	Ernst Otto Fischer, Geoffrey Wilkinson	1918	Fritz Haber
1972	Christian Anfinsen, Stanford Moore, William H. Stein	1917	*No prize awarded*
1971	Gerhard Herzberg	1916	*No prize awarded*
1970	Luis Leloir	1915	Richard Willstätter
1969	Derek Barton, Odd Hassel	1914	Theodore W. Richards
1968	Lars Onsager	1913	Alfred Werner
1967	Manfred Eigen, Ronald G.W. Norrish, George Porter	1912	Victor Grignard, Paul Sabatier
1966	Robert S. Mulliken	1911	Marie Curie
1965	Robert B. Woodward	1910	Otto Wallach
1964	Dorothy Crowfoot Hodgkin	1909	Wilhelm Ostwald
1963	Karl Ziegler, Giulio Natta	1908	Ernest Rutherford
1962	Max F. Perutz, John C. Kendrew	1907	Eduard Buchner
1961	Melvin Calvin	1906	Henri Moissan
1960	Willard F. Libby	1905	Adolf von Baeyer
1959	Jaroslav Heyrovsky	1904	Sir William Ramsay
1958	Frederick Sanger	1903	Svante Arrhenius
1957	Lord Todd	1902	Emil Fischer
		1901	Jacobus H. van't Hoff

Physics

2011	Saul Perlmutter, Brian P. Schmidt, Adam G. Riess		1959	Emilio Segrè, Owen Chamberlain
2010	Andre Geim, Konstantin Novoselov		1958	Pavel A. Cherenkov, Il'ja M. Frank, Igor Y. Tamm
2009	Charles K. Kao, Willard S. Boyle, George E. Smith		1957	Chen Ning Yang, Tsung-Dao Lee
2008	Makoto Kobayashi, Toshihide Maskawa, Yoichiro Nambu		1956	William B. Shockley, John Bardeen, Walter H. Brattain
2007	Albert Fert, Peter Grünberg		1955	Willis E. Lamb, Polykarp Kusch
2006	John C. Mather, George F. Smoot		1954	Max Born, Walther Bothe
2005	Roy J. Glauber, John L. Hall, Theodor W. Hänsch		1953	Frits Zernike
2004	David J. Gross, H. David Politzer, Frank Wilczek		1952	Felix Bloch, E. M. Purcell
2003	Alexei A. Abrikosov, Vitaly L. Ginzburg, Anthony J. Leggett		1951	John Cockcroft, Ernest T.S. Walton
			1950	Cecil Powell
2002	Raymond Davis Jr., Masatoshi Koshiba, Riccardo Giacconi		1949	Hideki Yukawa
			1948	Patrick M.S. Blackett
2001	Eric A. Cornell, Wolfgang Ketterle, Carl E. Wieman		1947	Edward V. Appleton
2000	Zhores I. Alferov, Herbert Kroemer, Jack S. Kilby		1946	Percy W. Bridgman
1999	Gerardus 't Hooft, Martinus J.G. Veltman		1945	Wolfgang Pauli
1998	Robert B. Laughlin, Horst L. Störmer, Daniel C. Tsui		1944	Isidor Isaac Rabi
1997	Steven Chu, Claude Cohen-Tannoudji, William D. Phillips		1943	Otto Stern
			1942	*No prize awarded*
1996	David M. Lee, Douglas D. Osheroff, Robert C. Richardson		1941	*No prize awarded*
			1940	*No prize awarded*
1995	Martin L. Perl, Frederick Reines		1939	Ernest Lawrence
1994	Bertram N. Brockhouse, Clifford G. Shull		1938	Enrico Fermi
1993	Russell A. Hulse, Joseph H. Taylor Jr.		1937	Clinton Davisson, George Paget Thomson
1992	Georges Charpak		1936	Victor F. Hess, Carl D. Anderson
1991	Pierre-Gilles de Gennes		1935	James Chadwick
1990	Jerome I. Friedman, Henry W. Kendall, Richard E. Taylor		1934	*No prize awarded*
1989	Norman F. Ramsey, Hans G. Dehmelt, Wolfgang Paul		1933	Erwin Schrödinger, Paul A.M. Dirac
1988	Leon M. Lederman, Melvin Schwartz, Jack Steinberger		1932	Werner Heisenberg
1987	J. Georg Bednorz, K. Alex Müller		1931	*No prize awarded*
1986	Ernst Ruska, Gerd Binnig, Heinrich Rohrer		1930	Sir Venkata Raman
1985	Klaus von Klitzing		1929	Louis de Broglie
1984	Carlo Rubbia, Simon van der Meer		1928	Owen Willans Richardson
1983	Subramanyan Chandrasekhar, William A. Fowler		1927	Arthur H. Compton, C.T.R. Wilson
1982	Kenneth G. Wilson		1926	Jean Baptiste Perrin
1981	Nicolaas Bloembergen, Arthur L. Schawlow, Kai M. Siegbahn		1925	James Franck, Gustav Hertz
			1924	Manne Siegbahn
1980	James Cronin, Val Fitch		1923	Robert A. Millikan
1979	Sheldon Glashow, Abdus Salam, Steven Weinberg		1922	Niels Bohr
1978	Pyotr Kapitsa, Arno Penzias, Robert Woodrow Wilson		1921	Albert Einstein
1977	Philip W. Anderson, Sir Nevill F. Mott, John H. van Vleck		1920	Charles Edouard Guillaume
1976	Burton Richter, Samuel C.C. Ting		1919	Johannes Stark
1975	Aage N. Bohr, Ben R. Mottelson, James Rainwater		1918	Max Planck
1974	Martin Ryle, Antony Hewish		1917	Charles Glover Barkla
1973	Leo Esaki, Ivar Giaever, Brian D. Josephson		1916	*No prize awarded*
1972	John Bardeen, Leon N. Cooper, Robert Schrieffer		1915	William Bragg, Lawrence Bragg
1971	Dennis Gabor		1914	Max von Laue
1970	Hannes Alfvén, Louis Néel		1913	Heike Kamerlingh Onnes
1969	Murray Gell-Mann		1912	Gustaf Dalén
1968	Luis Alvarez		1911	Wilhelm Wien
1967	Hans Bethe		1910	Johannes Diderik van der Waals
1966	Alfred Kastler		1909	Guglielmo Marconi, Ferdinand Braun
1965	Sin-Itiro Tomonaga, Julian Schwinger, Richard P. Feynman		1908	Gabriel Lippmann
			1907	Albert A. Michelson
1964	Charles H. Townes, Nicolay G. Basov, Aleksandr M. Prokhorov		1906	J.J. Thomson
			1905	Philipp Lenard
1963	Eugene Wigner, Maria Goeppert-Mayer, J. Hans D. Jensen		1904	Lord Rayleigh
1962	Lev Landau		1903	Henri Becquerel, Pierre Curie, Marie Curie
1961	Robert Hofstadter, Rudolf Mössbauer		1902	Hendrik A. Lorentz, Pieter Zeeman
1960	Donald A. Glaser		1901	Wilhelm Conrad Röntgen

Section 3
Physical Constants of Organic Compounds

PHYSICAL CONSTANTS OF ORGANIC COMPOUNDS

The basic physical constants and structure diagrams for about 10,900 organic compounds are presented in this table. An effort has been made to include the compounds most frequently encountered in the laboratory, the workplace, and the environment. Particular emphasis has been given to substances that are considered environmental or human health hazards. In making the selection of compounds for the table, added weight was assigned to the appearance of a compound in various lists or reference sources such as:

- Laboratory reagent lists, e.g., the ACS *Reagent Chemicals* volume (Ref. 1)
- The DIPPR list of industrially important compounds (Ref. 2) and the (much larger) TSCA Inventory of chemicals used in commerce.
- The Hazardous Substance Data Bank (Ref. 3)
- The UNEP list of Persistent Organic Pollutants (Ref. 4)
- Chemicals on Reporting Rules (CORR), a database of about 7500 regulated compounds prepared by the Environmental Protection Agency (Ref. 5)
- The EPA Integrated Risk Information System (IRIS), a database of human health effects of exposure to chemicals in the environment (Ref. 6)
- Compendia of chemicals of biochemical or medical importance, such as *The Merck Index* (Ref. 10)
- Specialized tables in this *Handbook*

It should be noted that the above lists vary widely in their choice of chemical names, and even in the use of Chemical Abstracts Registry Numbers. To the extent possible, we have attempted to systematize the names and registry numbers for this table.

Clearly, criteria of this type are somewhat subjective, and compounds considered important by some users have undoubtedly been omitted. Suggestions for additional compounds or other improvements are welcomed.

The data in the table have been derived from many sources, including both the primary literature and evaluated compilations. The *Handbook of Data on Organic Compounds, Third Edition* (Ref. 7) and the *Combined Chemical Dictionary* (Ref. 8) were important sources. Other useful sources of physical property data on organic compounds are listed in Refs. 9-19. The values in the table for the normal boiling point and the melting point that are accompanied with uncertainties (in parentheses) have been critically evaluated using the NIST ThermoData Engine (TDE, Ref. 20), designed to implement the dynamic data evaluation concept (Refs. 21-24). This concept requires large electronic databases capable of storing essentially all relevant experimental data known to date with detailed descriptions of metadata and uncertainties. The combination of these electronic databases with expert-system software, designed to automatically generate recommended property values based on available experimental and predicted data, leads to the ability to produce critically evaluated data dynamically or "to order." The uncertainties listed are combined expanded uncertainties (level of confidence, approximately 95 %) representing the most comprehensive measure of the overall data reliability (Refs. 25-28).

The table is arranged alphabetically by substance name, which generally is either an IUPAC systematic name or, in the case of pesticides, pharmaceuticals, and other complex compounds, a simple trivial name. Names in ubiquitous use, such as acetic acid and formaldehyde, are adopted rather than their systematic equivalents. Synonyms are given in the column following the primary name, and structure diagrams are given on the page facing the data listing. The explanation of the data columns follows:

No.: An identification number used in the indexes.

Name: Primary name of the substance

Synonym: A synonym in common use. When the primary name is non-systematic, a systematic name may appear here.

Mol. Form.: The molecular formula written in the Hill convention.

CAS RN.: The Chemical Abstracts Service Registry Number for the compound.

Mol. Wt: Molecular weight (relative molar mass) as calculated with the 2001 IUPAC Standard Atomic Weights.

Physical Form: A notation of the physical phase, color, crystal type, or other features of the compound at ambient temperature. Abbreviations are given below.

mp: Normal melting point in °C. A value is sometimes followed by "dec", indicating decomposition is observed at the stated temperature (so that it is probably not a true melting point). The notation "tp" indicates a triple point, where solid, liquid, and gas are in equilibrium. A number in parentheses following the melting point value is the combined expanded uncertainty (see above).

bp: Normal boiling point in °C, if it is available. This is the temperature at which the liquid phase is in equilibrium with the vapor at a pressure of 760 mmHg (101.325 kPa). A number in parentheses following the boiling point value is the combined expanded uncertainty (see above). A notation "sp" following the value indicates a sublimation point, where the vapor pressure of the solid phase reaches 760 mmHg. When a notation such as "dec" (decomposes) or "exp" (explodes) follows the value, the temperature may not be a true boiling point. A simply entry "sub" indicates the solid has a significant sublimation pressure at ambient temperatures. When the normal boiling point is not available, a boiling point at reduced pressure may be listed with a superscript indicating the pressure in mmHg.

den: Density (mass per unit volume) in g/cm³. The temperature in °C is indicated by a superscript. Values refer to the liquid or solid phase, and all values are true densities, not specific gravities. The number of decimal places gives a rough estimate of the accuracy of the value.

n_D: Refractive index, at the temperature in °C indicated by the superscript. Unless otherwise indicated, all values refer to a wavelength of 589 nm (sodium D line). Values are given only for liquids and solids.

Solubility: Qualitative indication of solubility in common solvents. Abbreviations are:
i insoluble
sl slightly soluble
s soluble
vs very soluble
msc miscible
dec decomposes

Abbreviations for solvents are given below.

In order to facilitate the location of compounds in the table, an index to synonyms follows the main table. Indexes to Molecular Formulas and CAS Registry Numbers are available in the electronic versions of the *Handbook* or as pdf files by request via e-mail (william.haynes@taylorandfrancis.com).

The assistance of members of the Thermodynamics Research Center (TRC) of the National Institute of Standards and Technology (Vladimir Diky, Rob Chirico, Andrei Kazakov) and especially Chris Muzny and Michael Frenkel in the determination of values of the normal-boiling-point and melting-point temperatures with uncertainties is greatly appreciated. The editors of the Handbook are much indebted to Chris Muzny who spent countless hours in producing these critically evaluated results. The assistance of Fiona Macdonald in checking names and formulas is gratefully acknowledged, as well as the efforts of Janice Shackleton, Trupti Desai, Nazila Kamaly, Matt Griffiths, and Lawrence Braschi in preparing the structure diagrams.

List of Abbreviations

Ac	acetyl	flr	fluorescent	pow	powder
Ac_2O	acetic anhydride	fum	fumes, fuming	Pr	propyl
AcOEt	ethyl acetate	gl	glacial	PrOH	1-propanol
ac	acid	gr	gray	pr	prisms
ace	acetone	gran	granular	purp	purple
al	alcohol (ethanol)	grn	green	py	pyridine
alk	alkali	hex	hexagonal	pym	pyramids, pyramidal
amor	amorphous	HOAc	acetic acid	reac	reacts
anh	anhydrous	hp	heptane	rhom	rhombic
aq	aqueous	hx	hexane	s	soluble
bipym	bipyramidal	hyd	hydrate	sat	saturated
bl	blue	hyg	hygroscopic	sc	scales
blk	black	i	insoluble	sl	slightly soluble
bp	boiling point	i-	iso-	soln	solution
br	brown	iso	isooctane	sp	sublimation point
bt	bright	lf	leaves	stab	stable
Bu	butyl	lig	ligroin	sub	sublimes
BuOH	1-butanol	liq	liquid	sulf	sulfuric acid
bz	benzene	lo	long	syr	syrup
chl	chloroform	mcl	monoclinic	tab	tablets
col	colorless	Me	methyl	tcl	triclinic
con, conc	concentrated	MeCN	acetonitrile	tetr	tetragonal
cry	crystals	MeOH	methanol	tfa	trifluoroacetic acid
ctc	carbon tetrachloride	misc	miscible	thf, THF	tetrahydrofuran
cy, cyhex	cyclohexane	mp	melting point	tol	toluene
dec	decomposes	n	refractive index	tp	triple point
den	density	nd	needles	trg	trigonal
dil	dilute	oct	octahedra, octahedral	unstab	unstable
diox	dioxane	oran	orange	vap	vapor
dk	dark	orth	orthorhombic	viol	violet
DMF	dimethylformamide	os	organic solvents	visc	viscous
DMSO	dimethyl sulfoxide	pa	pale	vol	volatile
efflor	efflorescent	peth	petroleum ether	vs	very soluble
Et	ethyl	Ph	phenyl	w	water
EtOH	ethanol	PhCl	chlorobenzene	wh	white
eth	diethyl ether	$PhNH_2$	aniline	xyl	xylene
exp	explodes	$PhNO_2$	nitrobenzene	ye	yellow
fl	flakes	pl	plates		

References

1. American Chemical Society, Reagent Chemicals, Tenth Edition, Oxford University Press, New York, 2005.
2. Design Institute for Physical Properties, American Institute of Chemical Engineers, <http://www.aiche.org/dippr/>.
3. National Library of Medicine, Hazardous Substances Data Bank, <http://toxnet.nlm.nih.gov/cgi-bin/sis/htmlgen?HSDB>.
4. United Nations Environmental Program, Persistent Organic Pollutants, <http://www.chem.unep.ch/pops/>.
5. Environmental Protection Agency, Chemicals on Reporting Rules, <http://www.epa.gov/opptintr/CORR>.
6. Environmental Protection Agency, Integrated Risk Information System, <http://www.epa.gov/iris/index.html>.
7. Lide, D. R., and Milne, G. W. A., Editors, Handbook of Data on Organic Compounds, Third Edition, CRC Press, Boca Raton, FL, 1993.
8. Combined Chemical Dictionary, <http://ccd.chemnetbase.com/>.
9. Linstrom, P. J., and Mallard, W. G., Editors, NIST Chemistry WebBook, NIST Standard Reference Database No. 69, February 2010, National Institute of Standards and Technology, Gaithersburg, MD 20899, <http://webbook.nist.gov>.
10. Thermodynamic Research Center, National Institute of Standards and Technology, TRC Thermodynamic Tables, <http://trc.nist.gov>.
11. O'Neil, M. J., Editor, The Merck Index, Fourteenth Edition, Merck & Co., Whitehouse Station, NJ, 2006.
12. Stevenson, R. M., and Malanowski, S., Handbook of the Thermodynamics of Organic Compounds, Elsevier, New York, 1987.
13. Riddick, J. A., Bunger, W. B., and Sakano, T. K., Organic Solvents, Fourth Edition, John Wiley & Sons, New York, 1986.
14. ChemSpider, <http://www.chemspider.com/>.
15. Crossfire Beilstein, <http://accelrys.com/products/>.
16. Springer Materials, The Landolt-Börnstein Database, <http://www.springermaterials.com>.
17. Vargaftik, N.B., Vinogradov, Y. K., and Yargin, V. S., Handbook of Physical Properties of Liquids and Gases, Third Edition, Begell House, New York, 1996.
18. Lide, D. R., and Kehiaian, H. V., Handbook of Thermophysical and Thermochemical Data, CRC Press, Boca Raton, FL, 1994.
19. Lide, D. R., Editor, Properties of Organic Compounds, <http://www.chemnetbase.com/tours/poc/intro.jsf>.
20. Frenkel, M., Chirico, R. D., Diky, V. V., Kazakov, A., and Muzny, C. D., ThermoData Engine, NIST Standard Reference Database 103b, Version 5.0 (Pure Compounds, Binary Mixtures, and Chemical Reactions, TDE-SOURCE Version 5.1), National Institute of Standards and Technology, Gaithersburg, MD – Boulder, CO, 2010, <http://www.nist.gov/srd/nist103b.cfm>.
21. Frenkel, M., Chirico, R. D., Diky, V., Yan, X., Dong, Q., and Muzny, C., J. Chem. Inf. Model. 45, 816, 2005.
22. Diky, V., Muzny, C. D., Lemmon, E. W., Chirico, R. D., and Frenkel, M., J. Chem. Inf. Model. 47, 1713, 2007.
23. Diky, V., Chirico, R. D., Kazakov, A. F., Muzny, C., and Frenkel, M., J. Chem. Inf. Model. 49, 503, 2009.
24. Diky, V., Chirico, R. D., Kazakov, A. F., Muzny, C., and Frenkel, M., J. Chem. Inf. Model. 49, 2883, 2009.
25. Chirico, R. D., Frenkel, M., Diky, V. V., March, K. N., and Wilhoit, R. C., J. Chem. Eng. Data 48, 1344, 2003.
26. Guide to the Expression of Uncertainty in Measurement, International Organization for Standardization, Geneva, Switzerland, 1993.
27. U. S. Guide to the Expression of Uncertainty in Measurement, ANSI/NCSL, Z540-2-1997, ISBN 1-58464-005-7, NCSL Int., Boulder, CO, 1997.
28. Taylor, B. N., and Kuyatt, C. E., Guidelines for the Evaluation and Expression of Uncertainty in NIST Measurement Results, NIST Tech. Note 1297, Natl. Inst. Stand. Technol., Gaithersburg, MD, 1994.

No.	Name	Synonym	Mol. Form.	CAS RN	Mol. Wt.	Physical Form	mp/°C	bp/°C	den g cm⁻³	n_D	Solubility
1	Abate	Temephos	$C_{16}H_{20}O_6P_2S_3$	3383-96-8	466.469	cry	31.6(0.5)		1.32		sl H₂O, hx; s ctc, eth, tol
2	Abietic acid		$C_{20}H_{30}O_2$	514-10-3	302.451	mcl pl (al-w)	173.5	250⁹			vs ace, bz, eth, EtOH
3	Abscisic acid		$C_{15}H_{20}O_4$	21293-29-8	264.318	cry (chl-peth)	160	120 sub			vs ace, eth, chl
4	Acacetin	5,7-Dihydroxy-2-(4-methoxyphenyl)-4H-1-benzopyran-4-one	$C_{16}H_{12}O_5$	480-44-4	284.263	ye nd (95% al)	263				vs EtOH
5	Acebutolol, (±)-		$C_{18}H_{28}N_2O_4$	37517-30-9	336.426	cry	121				
6	Acedapsone		$C_{16}H_{16}N_2O_4S$	77-46-3	332.374	pa ye nd (eth) lf (dil al)	290				sl H₂O
7	Acenaphthene	1,2-Dihydroacenaphthylene	$C_{12}H_{10}$	83-32-9	154.207		93(2)	277.5(0.8)	1.222²⁰	1.6048⁹⁵	i H₂O; sl EtOH, chl; vs bz; s HOAc
8	Acenaphthylene	Acenaphthalene	$C_{12}H_8$	208-96-8	152.192		89.4(0.3)	280	0.8987¹⁶		i H₂O; vs EtOH, eth, bz; sl chl
9	1,2-Acenaphthylenedione		$C_{12}H_6O_2$	82-86-0	182.175	ye nd (HOAc)	259(9)	sub	1.4800²⁰		i H₂O; sl EtOH, bz, HOAc; s lig
10	Acenocoumarol	Nicoumalone	$C_{19}H_{15}NO_6$	152-72-7	353.325	cry (ace aq)	198				i H₂O
11	Acephate	Phosphoramidothioic acid, acetyl-, O,S-dimethyl ester	$C_4H_{10}NO_3PS$	30560-19-1	183.166		92.0(0.5)		1.35²⁰		
12	Acepromazine		$C_{19}H_{22}N_2OS$	61-00-7	326.455	oran oil		230⁰·⁵			
13	Acesulfame		$C_4H_5NO_4S$	33665-90-6	163.153	nd (bz)	123.2				s bz, chl
14	Acetaldehyde	Ethanal	C_2H_4O	75-07-0	44.052	vol liq or gas	-123.4(0.7)	20.8(0.6)	0.7834¹⁸	1.3316²⁰	msc H₂O, EtOH, eth, bz; sl chl
15	Acetaldehyde phenylhydrazone		$C_8H_{10}N_2$	935-07-9	134.178		99.5	150⁴⁰			vs EtOH
16	Acetaldoxime	Acetaldehyde oxime	C_2H_5NO	107-29-9	59.067	nd	25(1)	115.24(0.1)	0.9656²⁰	1.4264²⁰	s H₂O, chl; msc EtOH, eth
17	Acetamide	Ethanamide	C_2H_5NO	60-35-5	59.067	trg mcl (al-eth)	80.16(0.04)	222.0	0.9986⁸⁵	1.4278	vs H₂O, EtOH
18	Acetanilide	N-Phenylacetamide	C_8H_9NO	103-84-4	135.163		114.35(0.04)	292(9)	1.2190¹⁵		sl H₂O; vs EtOH, ace; s eth, s bz, tol
19	Acetazolamide	N-[5-(Aminosulfonyl)-1,3,4-thiadiazol-2-yl]acetamide	$C_4H_6N_4O_3S_2$	59-66-5	222.246		260.5				sl H₂O
20	Acethion		$C_8H_{17}O_4PS_2$	919-54-0	272.322	liq		137¹·⁵	1.18²⁰		
21	Acetic acid	Ethanoic acid	$C_2H_4O_2$	64-19-7	60.052	col liq	17(3)	117.9(0.2)	1.0446²⁵	1.3720²⁰	msc H₂O, EtOH, eth, ace, bz; s chl, CS₂
22	Acetic acid, 2-phenylhydrazide		$C_8H_{10}N_2O$	114-83-0	150.177	hex pr (eth)	130.0				vs H₂O, EtOH; sl eth, chl, tfa; s bz
23	Acetic anhydride	Acetyl acetate	$C_4H_6O_3$	108-24-7	102.089	liq	-73.4(0.8)	139.5(0.3)	1.082²⁰	1.3901²⁰	vs H₂O; s EtOH, bz; msc eth; sl ctc
24	Acetoacetanilide		$C_{10}H_{11}NO_2$	102-01-2	177.200	pr or nd (bz or lig)	86				sl H₂O; s EtOH, eth, bz, chl, acid, lig
25	Acetoacetic acid		$C_4H_6O_3$	541-50-4	102.089	cry (eth)	36.5	100 dec			vs H₂O, eth, EtOH
26	2-Acetoacetoxyethyl methacrylate	2-(Methacryloyloxy)ethyl acetoacetate	$C_{10}H_{14}O_5$	21282-97-3	214.215	liq		100⁰·⁸	1.122	1.4560²⁰	
27	Acetochlor		$C_{14}H_{20}ClNO_2$	34256-82-1	269.768	ye liq		134⁰·⁴		1.5272²⁰	sl H₂O
28	Acetohexamide		$C_{15}H_{20}N_2O_4S$	968-81-0	324.396	cry (EtOH aq)	188				i H₂O, eth; sl EtOH, chl; s py
29	Acetohydrazide		$C_2H_6N_2O$	1068-57-1	74.081		67	137²⁵			s H₂O, EtOH; sl eth
30	Acetohydroxamic acid	N-Hydroxyacetamide	$C_2H_5NO_2$	546-88-3	75.067	hyg cry	90				
31	1-Acetonaphthone		$C_{12}H_{10}O$	941-98-0	170.206		34	297	1.1171²¹	1.6280²²	i H₂O; s EtOH, eth, ace, chl
32	2-Acetonaphthone		$C_{12}H_{10}O$	93-08-3	170.206	nd (lig, dil al)	52(1)	302			sl EtOH, ctc
33	Acetone	2-Propanone	C_3H_6O	67-64-1	58.079	liq	-94.9(0.4)	56.08(0.07)	0.7845²⁵	1.3588²⁰	msc H₂O, EtOH, eth, ace, bz, chl
34	Acetone cyanohydrin		C_4H_7NO	75-86-5	85.105	liq	-19	180(21)	0.932¹⁹	1.3992²⁰	vs H₂O, EtOH, eth; s ace, bz, chl; i peth
35	Acetone (2,4-dinitrophenyl)-hydrazone		$C_9H_{10}N_4O_4$	1567-89-1	238.200	ye nd or pl (al)	128				i H₂O; s EtOH, eth, bz, chl, AcOEt

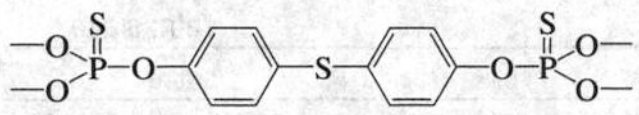

1 Abate

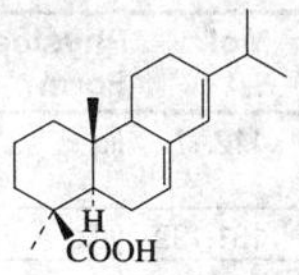

2 Abietic acid

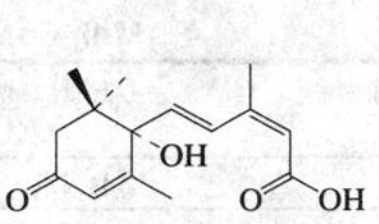

3 Abscisic acid

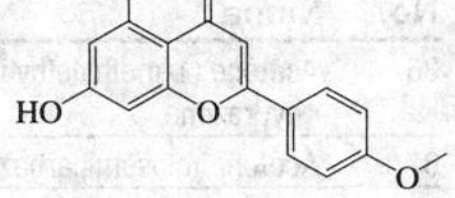

4 Acacetin

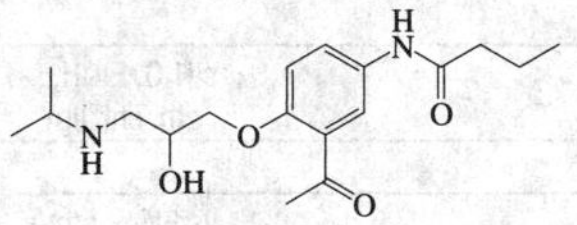

5 Acebutolol, (±)-

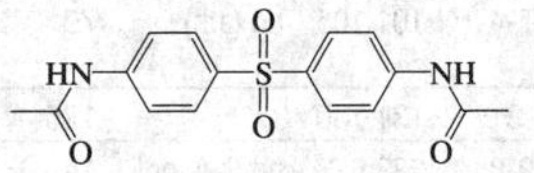

6 Acedapsone

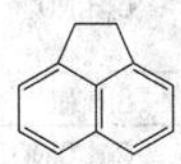

7 Acenaphthene

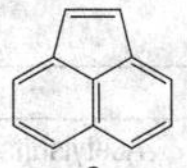

8 Acenaphthylene

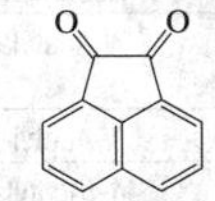

9 1,2-Acenaphthylenedione

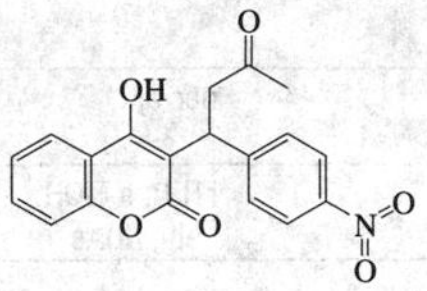

10 Acenocoumarol

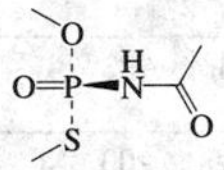

11 Acephate

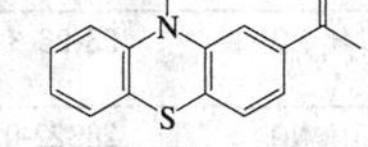

12 Acepromazine

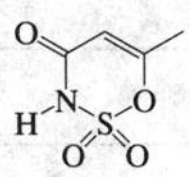

13 Acesulfame

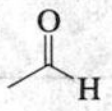

14 Acetaldehyde

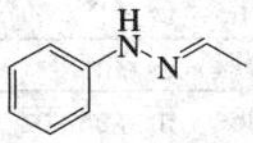

15 Acetaldehyde phenylhydrazone

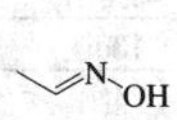

16 Acetaldoxime

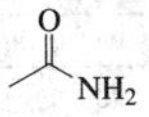

17 Acetamide

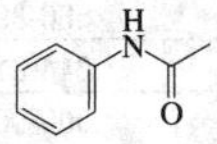

18 Acetanilide

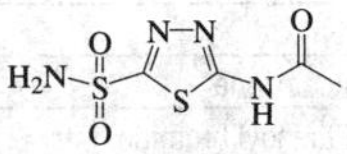

19 Acetazolamide

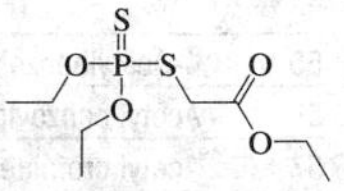

20 Acethion

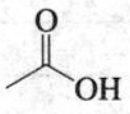

21 Acetic acid

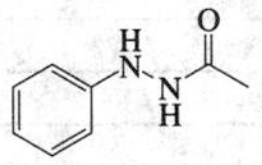

22 Acetic acid, 2-phenylhydrazide

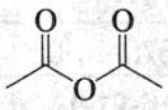

23 Acetic anhydride

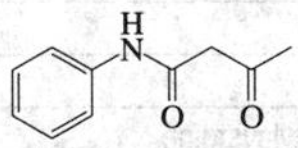

24 Acetoacetanilide

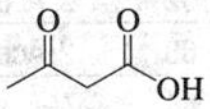

25 Acetoacetic acid

26 2-Acetoacetoxyethyl methacrylate

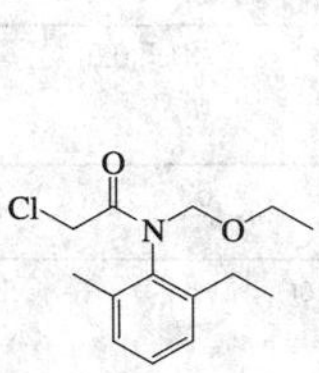

27 Acetochlor

28 Acetohexamide

29 Acetohydrazide

30 Acetohydroxamic acid

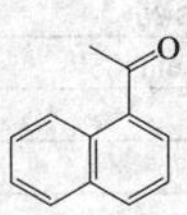

31 1-Acetonaphthone

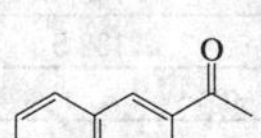

32 2-Acetonaphthone

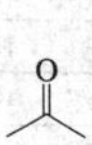

33 Acetone

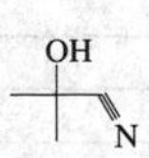

34 Acetone cyanohydrin

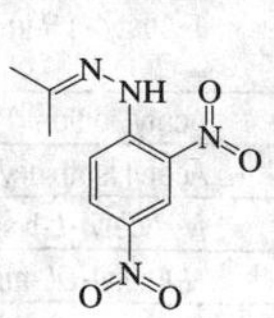

35 Acetone (2,4-dinitrophenyl)hydrazone

No.	Name	Synonym	Mol. Form.	CAS RN	Mol. Wt.	Physical Form	mp/°C	bp/°C	den g cm^{-3}	n_D	Solubility
36	Acetone (1-methylethylidene)-hydrazone	Dimethyl ketazine	C$_6$H$_{12}$N$_2$	627-70-3	112.172	liq	-12.5	133	0.8390[20]	1.4535[20]	msc H$_2$O, EtOH, eth; s ace
37	Acetone thiosemicarbazide		C$_4$H$_9$N$_3$S	1752-30-3	131.199	ye cry	176				s ace
38	Acetonitrile	Methyl cyanide	C$_2$H$_3$N	75-05-8	41.052	liq	-44(1)	81.6(0.2)	0.7857[20]	1.3442[20]	msc H$_2$O, EtOH, eth, ace, bz, ctc
39	Acetophenone	Methyl phenyl ketone	C$_8$H$_8$O	98-86-2	120.149	mcl pr or pl	19.4(0.4)	202.1(0.2)	1.0281[20]	1.5372[20]	sl H$_2$O; s EtOH, eth, ace, bz, con sulf, chl
40	Acetophenone azine	Methylphenyl ketazine	C$_{16}$H$_{16}$N$_2$	729-43-1	236.311		120				
41	Acetoxon	Acetophos	C$_6$H$_{17}$O$_5$PS	2425-25-4	256.257	liq		73[0.005]			
42	N-Acetylacetamide		C$_4$H$_7$NO$_2$	625-77-4	101.105	nd (eth)	79	223.5			s H$_2$O, EtOH, eth, chl, lig
43	N-Acetyl-L-alanine		C$_5$H$_9$NO$_3$	97-69-8	131.130		125				
44	4-(Acetylamino)benzenesulfonyl chloride	Acetylsulfanilyl chloride	C$_8$H$_8$ClNO$_3$S	121-60-8	233.673	nd (bz), pr (bz-chl)	149				vs EtOH, eth; s bz, chl
45	2-(Acetylamino)benzoic acid		C$_9$H$_9$NO$_3$	89-52-1	179.172	nd (HOAc)	187.5				sl H$_2$O; s EtOH; vs eth, ace, bz, HOAc
46	4-(Acetylamino)benzoic acid		C$_9$H$_9$NO$_3$	556-08-1	179.172	nd (HOAc)	256.5				i H$_2$O; s EtOH; sl eth, tfa
47	2-(Acetylamino)-2-deoxy-D-glucose	N-Acetyl-D-glucosamine	C$_8$H$_{15}$NO$_6$	7512-17-6	221.208		205				
48	2-(Acetylamino)-2-deoxy-D-mannose	N-Acetyl-D-mannosamine	C$_8$H$_{15}$NO$_6$	3615-17-6	221.208	cry (ace aq)	128				dec alk
49	2-(Acetylamino)fluorene		C$_{15}$H$_{13}$NO	53-96-3	223.270	cry (dil al)	193				i H$_2$O; s EtOH, eth, HOAc
50	4-(Acetylamino)fluorene		C$_{15}$H$_{13}$NO	28322-02-3	223.270	br cry (bz)	200				
51	6-(Acetylamino)hexanoic acid	ε-Acetamidocaproic acid	C$_8$H$_{15}$NO$_3$	57-08-9	173.210	cry (ace)	104.5				
52	4-Acetylanisole		C$_9$H$_{10}$O$_2$	100-06-1	150.174	pl (peth)	38.2(0.6)	254(12)	1.0818[41]	1.547[41]	sl H$_2$O; s EtOH, eth, ace, chl
53	2-Acetylbenzoic acid		C$_9$H$_8$O$_3$	577-56-0	164.158	nd (w), pr (bz)	114.5	111[2]			vs H$_2$O, eth, EtOH
54	3-Acetylbenzoic acid		C$_9$H$_8$O$_3$	586-42-5	164.158		172	111[2]			s H$_2$O; msc EtOH
55	4-Acetylbenzoic acid		C$_9$H$_8$O$_3$	586-89-0	164.158	nd (w)	208	sub			vs H$_2$O
56	Acetyl benzoylperoxide	Acetozone	C$_9$H$_8$O$_4$	644-31-5	180.158	wh nd (lig)	37	130[19]			vs eth
57	Acetyl bromide	Ethanoyl bromide	C$_2$H$_3$BrO	506-96-7	122.948	liq	-96.5(0.5)	74(1)	1.6625[16]	1.4486[20]	msc eth, bz, chl; s ace
58	Acetyl chloride	Ethanoyl chloride	C$_2$H$_3$ClO	75-36-5	78.497	liq	-112.7(0.8)	51(2)	1.1051[20]	1.3886[20]	msc eth, ace, bz, chl; s ctc
59	Acetylcholine bromide		C$_7$H$_{16}$BrNO$_2$	66-23-9	226.112	hyg cry	146				vs H$_2$O
60	Acetylcholine chloride		C$_7$H$_{16}$ClNO$_2$	60-31-1	181.661		150				s H$_2$O, EtOH; i eth
61	Acetylcholine iodide		C$_7$H$_{16}$INO$_2$	2260-50-6	273.112	hyg	163				
62	2-Acetylcyclohexanone		C$_8$H$_{12}$O$_2$	874-23-7	140.180		-11	112[18]	1.0782[25]	1.5138[20]	s ctc
63	2-Acetylcyclopentanone		C$_7$H$_{10}$O$_2$	1670-46-8	126.153			73[20]	1.0431[25]	1.4906[20]	
64	N-Acetyl-L-cysteine	Acetylcysteine	C$_5$H$_9$NO$_3$S	616-91-1	163.195	cry (w)	109.5				
65	3-Acetyldihydro-2(3H)-furanone	α-Acetylbutyrolactone	C$_6$H$_8$O$_3$	517-23-7	128.126			107[5]	1.1846[20]	1.4585[20]	vs H$_2$O
66	1-Acetyl-2,5-dihydroxybenzene	2,5-Dihydroxyacetophenone	C$_8$H$_8$O$_3$	490-78-8	152.148	ye grn nd (dil al or w)	205.3				sl H$_2$O, eth, bz; s EtOH
67	Acetylene	Ethyne	C$_2$H$_2$	74-86-2	26.037	col gas	-81.5(0.9)	-84.7 sp	0.377[25] (p>1 atm)		sl H$_2$O, EtOH, CS$_2$; s ace, bz, chl
68	N-Acetylethanolamine		C$_4$H$_9$NO$_2$	142-26-7	103.120		63.5	166[8]	1.1079[25]	1.4674[20]	msc H$_2$O; s ace; sl bz, lig
69	Acetyl fluoride	Ethanoyl fluoride	C$_2$H$_3$FO	557-99-3	62.042	vol liq or gas	-84	22(7)	1.032[25]		msc EtOH, eth; s bz, chl; sl CS$_2$
70	N-Acetylglutamic acid		C$_7$H$_{11}$NO$_5$	1188-37-0	189.166	pr (w)	199				s H$_2$O, EtOH
71	N-Acetylglycine	Aceturic acid	C$_4$H$_7$NO$_3$	543-24-8	117.104	lo nd (w, MeOH)	206				vs H$_2$O, ace, EtOH
72	trans-1-Acetyl-4-hydroxy-L-proline	Oxaceprol	C$_7$H$_{11}$NO$_4$	33996-33-7	173.167	cry (Ac)	132				vs H$_2$O, MeOH
73	1-Acetyl-1H-imidazole		C$_5$H$_6$N$_2$O	2466-76-4	110.114		104.5				sl H$_2$O; s EtOH, eth, chl, THF
74	Acetyl iodide	Ethanoyl iodide	C$_2$H$_3$IO	507-02-8	169.948			109(5)	2.0673[20]	1.5491[20]	vs eth
75	Acetyl isothiocyanate		C$_3$H$_3$NOS	13250-46-9	101.127			132.5	1.1523[13]	1.5231[18]	s eth, CS$_2$
76	N^6-Acetyl-L-lysine		C$_8$H$_{16}$N$_2$O$_3$	692-04-6	188.224		265 dec				
77	N-Acetyl-DL-methionine		C$_7$H$_{13}$NO$_3$S	1115-47-5	191.248		114.5				

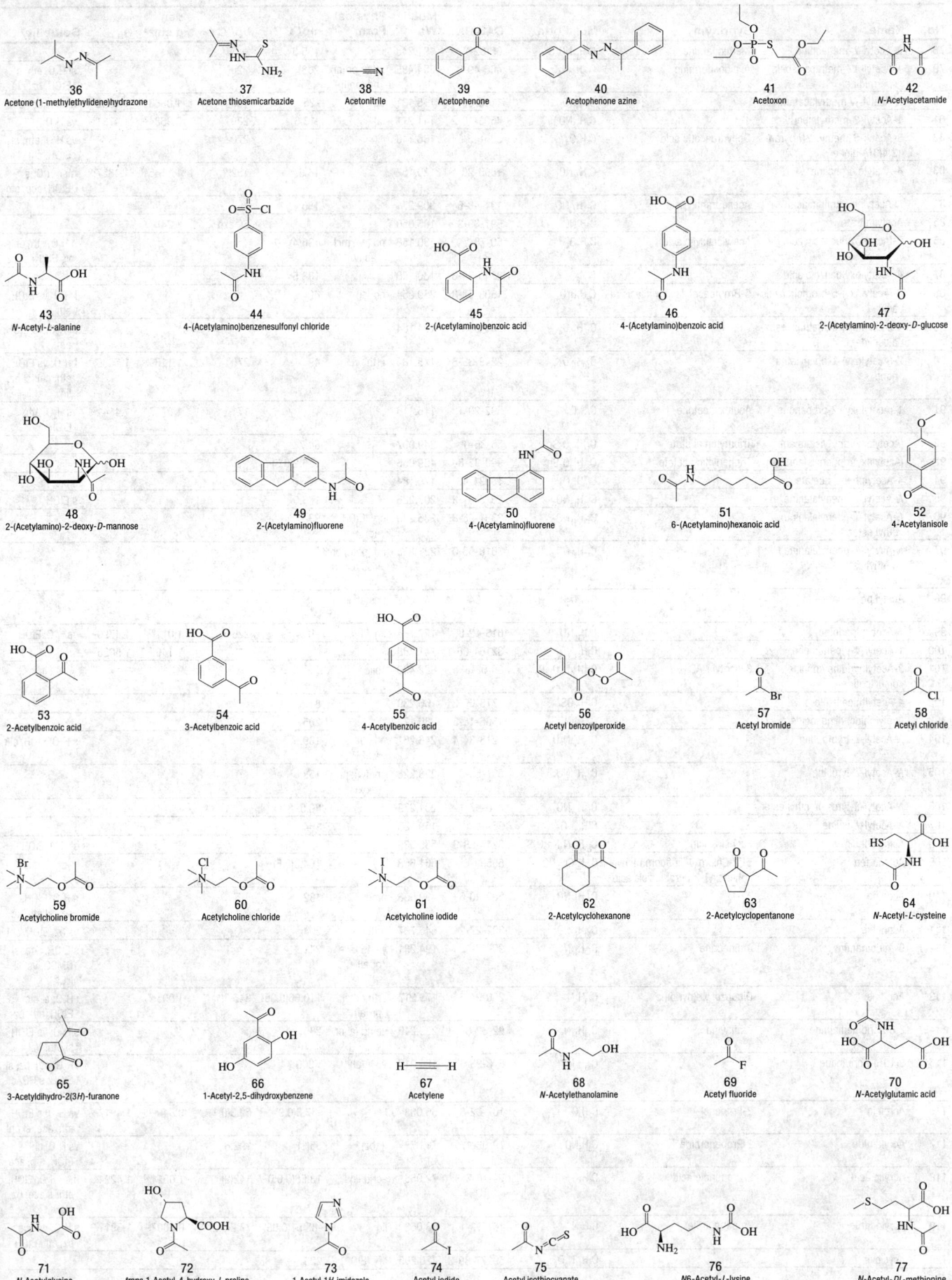

36
Acetone (1-methylethylidene)hydrazone

37
Acetone thiosemicarbazide

38
Acetonitrile

39
Acetophenone

40
Acetophenone azine

41
Acetoxon

42
N-Acetylacetamide

43
N-Acetyl-*L*-alanine

44
4-(Acetylamino)benzenesulfonyl chloride

45
2-(Acetylamino)benzoic acid

46
4-(Acetylamino)benzoic acid

47
2-(Acetylamino)-2-deoxy-*D*-glucose

48
2-(Acetylamino)-2-deoxy-*D*-mannose

49
2-(Acetylamino)fluorene

50
4-(Acetylamino)fluorene

51
6-(Acetylamino)hexanoic acid

52
4-Acetylanisole

53
2-Acetylbenzoic acid

54
3-Acetylbenzoic acid

55
4-Acetylbenzoic acid

56
Acetyl benzoylperoxide

57
Acetyl bromide

58
Acetyl chloride

59
Acetylcholine bromide

60
Acetylcholine chloride

61
Acetylcholine iodide

62
2-Acetylcyclohexanone

63
2-Acetylcyclopentanone

64
N-Acetyl-*L*-cysteine

65
3-Acetyldihydro-2(3*H*)-furanone

66
1-Acetyl-2,5-dihydroxybenzene

67
Acetylene

68
N-Acetylethanolamine

69
Acetyl fluoride

70
N-Acetylglutamic acid

71
N-Acetylglycine

72
trans-1-Acetyl-4-hydroxy-*L*-proline

73
1-Acetyl-1*H*-imidazole

74
Acetyl iodide

75
Acetyl isothiocyanate

76
*N*6-Acetyl-*L*-lysine

77
N-Acetyl-*DL*-methionine

No.	Name	Synonym	Mol. Form.	CAS RN	Mol. Wt.	Physical Form	mp/°C	bp/°C	den g cm⁻³	n_D	Solubility
78	N-Acetyl-L-methionine	Methionamine	$C_7H_{13}NO_3S$	65-82-7	191.248		105.5				
79	1-Acetyl-17-methoxyaspido-spermidine	Aspidospermine	$C_{22}H_{30}N_2O_2$	466-49-9	354.485	nd or pr (al) nd (peth)	208	220²			sl H_2O, eth; s EtOH, bz, chl
80	N-Acetyl-N-methylacetamide		$C_5H_9NO_2$	1113-68-4	115.131	liq	-25	195	1.0663²⁵	1.4502²⁵	msc H_2O; i eth
81	1-Acetyl-3-methylpiperidine		$C_8H_{15}NO$	4593-16-2	141.211	liq	-13.6	239	0.9684²⁵	1.4731²⁵	vs H_2O
82	3-Acetyl-6-methyl-2H-pyran-2,4(3H)-dione	Dehydroacetic acid	$C_8H_8O_4$	520-45-6	168.148		109	270			vs H_2O, eth; sl EtOH, chl
83	4-Acetylmorpholine		$C_6H_{11}NO_2$	1696-20-4	129.157		14.5	152⁵⁰	1.1145²⁰	1.4827²⁰	msc H_2O; s EtOH, ace, ctc
84	N-Acetylneuraminic acid	Aceneuramic acid	$C_{11}H_{19}NO_9$	131-48-6	309.271		186				
85	Acetyl nitrate		$C_2H_3NO_4$	591-09-3	105.050			60 exp	1.24¹⁵		
86	2-(Acetyloxy)benzoic acid	Acetylsalicylic acid	$C_9H_8O_4$	50-78-2	180.158	nd (w), mcl tab (w)	136(4)				s H_2O, eth, chl; vs EtOH; sl bz
87	4-(Acetyloxy)benzoic acid		$C_9H_8O_4$	2345-34-8	180.158		188.5				
88	2-(Acetyloxy)-5-bromobenzoic acid	5-Bromoacetylsalicylic acid	$C_9H_7BrO_4$	1503-53-3	259.054	nd (al)	60				i H_2O; vs EtOH, eth
89	4-(Acetyloxy)-3-methoxybenz-aldehyde		$C_{10}H_{10}O_4$	881-68-5	194.184		78				sl H_2O; vs EtOH, eth
90	2-(Acetyloxy)-1-phenyletha-none		$C_{10}H_{10}O_3$	2243-35-8	178.184	orth pl	49	270	1.1169⁶⁵	1.5036⁶⁵	i H_2O; vs EtOH, eth, chl; sl bz, lig
91	1-(Acetyloxy)-2-propanone	Acetoxyacetone	$C_5H_8O_3$	592-20-1	116.116			171	1.0757²⁰	1.4141²⁰	vs H_2O, eth, EtOH
92	(Acetyloxy)tributylstannane	Tributyltin acetate	$C_{14}H_{30}O_2Sn$	56-36-0	349.097		84.7				
93	(Acetyloxy)triphenylstannane	Triphenyltin acetate	$C_{20}H_{18}O_2Sn$	900-95-8	409.066		125.2(0.5)				
94	4-Acetylphenyl acetate		$C_{10}H_{10}O_3$	13031-43-1	178.184						s ctc, CS_2
95	N-Acetyl-L-phenylalanine		$C_{11}H_{13}NO_3$	2018-61-3	207.226		173.5				s EtOH
96	N-Acetyl-L-phenylalanine, ethyl ester		$C_{13}H_{17}NO_3$	2361-96-8	235.279	cry (EtOH aq)	93				
97	N-Acetyl-L-phenylalanine, methyl ester		$C_{12}H_{15}NO_3$	3618-96-0	221.252	nd (peth) or visc oil (chl)	91				
98	Acetyl phosphate		$C_2H_5O_5P$	590-54-5	140.032	unstab in soln					
99	1-Acetylpiperidine		$C_7H_{13}NO$	618-42-8	127.184	liq	-13.4	226.5	1.011⁹	1.4790²⁵	vs H_2O, EtOH
100	1-Acetyl-4-piperidinone		$C_7H_{11}NO_2$	32161-06-1	141.168			218	1.146²⁵	1.5026²⁰	
101	3-Acetylpyridine adenine dinucleotide	3-Acetyl NAD	$C_{22}H_{28}N_6O_{14}P_2$	86-08-8	662.436	solid					
102	4-Acetylthioanisole		$C_9H_{10}OS$	1778-09-2	166.239		81.5				
103	Acetyl thiocholine iodide		$C_7H_{16}INOS$	1866-15-5	289.177		205				
104	N-Acetyl-L-tryptophan		$C_{13}H_{14}N_2O_3$	1218-34-4	246.261	nd (dil MeOH)	189.5				s H_2O, EtOH, alk
105	N-Acetyl-L-tyrosine		$C_{11}H_{13}NO_4$	537-55-3	223.226	cry (w); pl (diox)	153				
106	N-Acetyl-L-tyrosine ethyl ester		$C_{13}H_{17}NO_4$	840-97-1	251.279		80.5				
107	N-Acetyl-L-valine		$C_7H_{13}NO_3$	96-81-1	159.183		164				
108	Acid Fuchsin	Fuchsin, acid	$C_{20}H_{17}N_3Na_2O_9S_3$	3244-88-0	585.539						sl H_2O, EtOH
109	Acifluorfen	5-[2-Chloro-4-(trifluoromethyl)-phenoxy]-2-nitrobenzoic acid	$C_{14}H_7ClF_3NO_5$	50594-66-6	361.658		164.3(0.5)				
110	Aconine		$C_{25}H_{41}NO_9$	509-20-6	499.596	amor	132				s H_2O, EtOH, chl; sl eth, lig
111	Aconitine		$C_{34}H_{47}NO_{11}$	302-27-2	645.737	orth lf	204				vs bz, EtOH, chl
112	9-Acridinamine	Aminacrine	$C_{13}H_{10}N_2$	90-45-9	194.231	ye nd (ace or al)	241				s EtOH, ace; sl DMSO; vs dil HCl
113	Acridine	Dibenzo[b,e]pyridine	$C_{13}H_9N$	260-94-6	179.217	orth nd or pr (al)	110.06(0.05)	346.9(1)	1.005²⁰		i H_2O; sl ctc; vs EtOH, eth, bz
114	3,6-Acridinediamine	Proflavine	$C_{13}H_{11}N_3$	92-62-6	209.246	ye nd (al or w)	285				s H_2O; vs EtOH; sl eth, bz
115	9(10H)-Acridinone		$C_{13}H_9NO$	578-95-0	195.216	ye lf (al)	>300				i H_2O, eth, bz; sl EtOH; s HOAc, alk
116	Acrolein	2-Propenal	C_3H_4O	107-02-8	56.063	liq	-87.8(0.9)	52.3(0.1)	0.840²⁰	1.4017²⁰	vs H_2O; s EtOH, eth, ace; sl chl
117	Acrylamide	2-Propenamide	C_3H_5NO	79-06-1	71.078	lf (bz)	85(1)	192.6			vs H_2O, chl; s EtOH, eth, ace
118	Acrylic acid	2-Propenoic acid	$C_3H_4O_2$	79-10-7	72.063	acrid liq	13.56(0.05)	142(2)	1.0511²⁰	1.4224²⁰	msc H_2O, EtOH, eth; s ace, bz, ctc
119	Acrylonitrile	Propenenitrile	C_3H_3N	107-13-1	53.063	liq	-83.51(0.05)	77.2(0.2)	0.8007²⁵	1.3911²⁰	s H_2O; vs ace, bz, eth, EtOH
120	Acyclovir		$C_8H_{11}N_5O_3$	59277-89-3	225.205	cry (EtOH)	225				

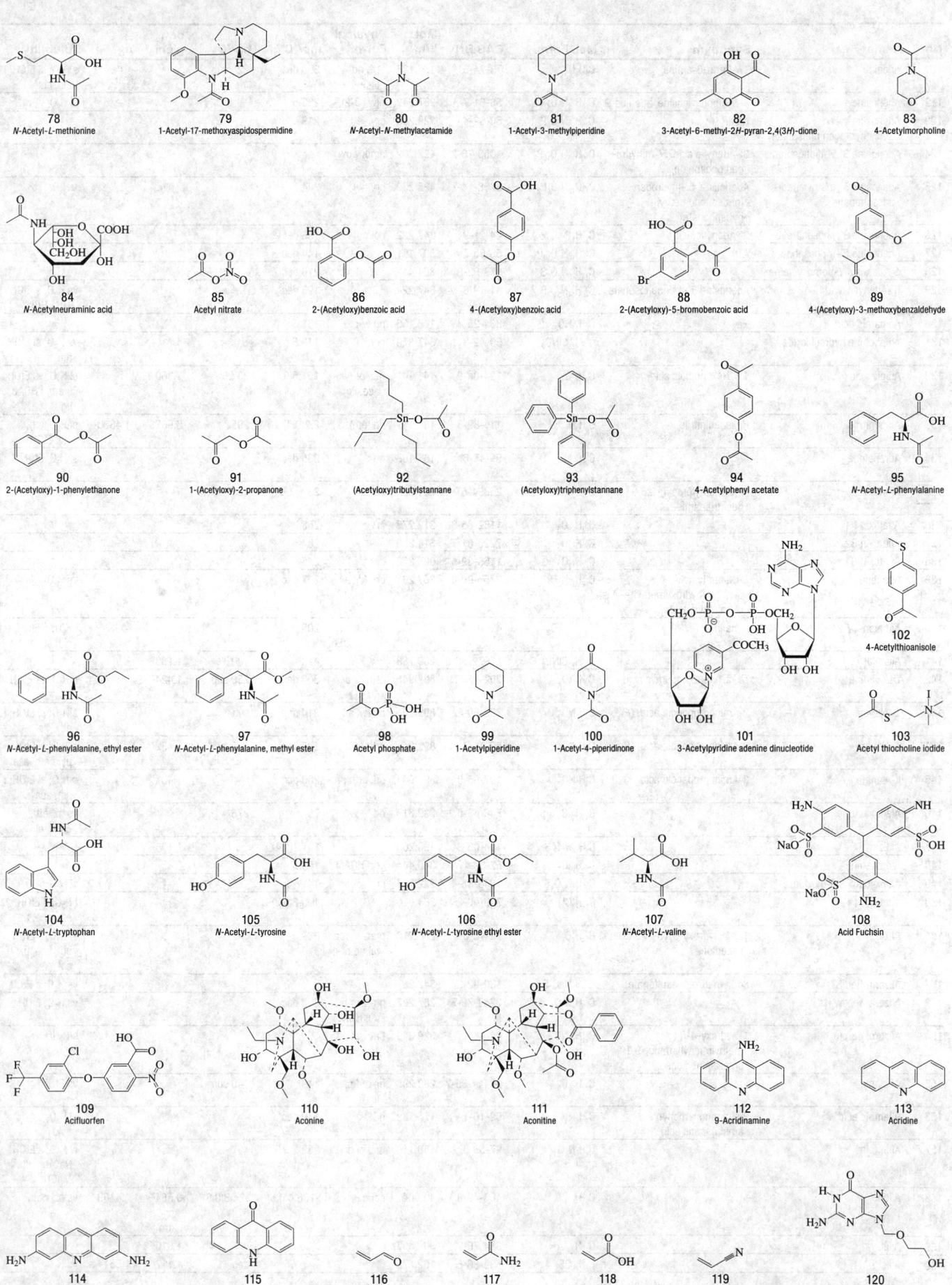

78 — N-Acetyl-L-methionine

79 — 1-Acetyl-17-methoxyaspidospermidine

80 — N-Acetyl-N-methylacetamide

81 — 1-Acetyl-3-methylpiperidine

82 — 3-Acetyl-6-methyl-2H-pyran-2,4(3H)-dione

83 — 4-Acetylmorpholine

84 — N-Acetylneuraminic acid

85 — Acetyl nitrate

86 — 2-(Acetyloxy)benzoic acid

87 — 4-(Acetyloxy)benzoic acid

88 — 2-(Acetyloxy)-5-bromobenzoic acid

89 — 4-(Acetyloxy)-3-methoxybenzaldehyde

90 — 2-(Acetyloxy)-1-phenylethanone

91 — 1-(Acetyloxy)-2-propanone

92 — (Acetyloxy)tributylstannane

93 — (Acetyloxy)triphenylstannane

94 — 4-Acetylphenyl acetate

95 — N-Acetyl-L-phenylalanine

96 — N-Acetyl-L-phenylalanine, ethyl ester

97 — N-Acetyl-L-phenylalanine, methyl ester

98 — Acetyl phosphate

99 — 1-Acetylpiperidine

100 — 1-Acetyl-4-piperidinone

101 — 3-Acetylpyridine adenine dinucleotide

102 — 4-Acetylthioanisole

103 — Acetyl thiocholine iodide

104 — N-Acetyl-L-tryptophan

105 — N-Acetyl-L-tyrosine

106 — N-Acetyl-L-tyrosine ethyl ester

107 — N-Acetyl-L-valine

108 — Acid Fuchsin

109 — Acifluorfen

110 — Aconine

111 — Aconitine

112 — 9-Acridinamine

113 — Acridine

114 — 3,6-Acridinediamine

115 — 9(10H)-Acridinone

116 — Acrolein

117 — Acrylamide

118 — Acrylic acid

119 — Acrylonitrile

120 — Acyclovir

No.	Name	Synonym	Mol. Form.	CAS RN	Mol. Wt.	Physical Form	mp/°C	bp/°C	den g cm⁻³	n_D	Solubility
121	Adenine	1H-Purin-6-amine	C₅H₅N₅	73-24-5	135.128	orth nd (+3w)	360 dec	220 sub			s H₂O; sl EtOH; i eth, chl
122	Adenosine	β-D-Ribofuranoside, adenine-9	C₁₀H₁₃N₅O₄	58-61-7	267.242	n(w+3/2)	235.5				sl H₂O; i EtOH
123	Adenosine cyclic 3',5'-(hydrogen phosphate)	cAMP	C₁₀H₁₂N₅O₆P	60-92-4	329.206	cry	219				
124	Adenosine 3',5'-diphosphate	3'-Adenylic acid, 5'-(dihydrogen phosphate)	C₁₀H₁₅N₅O₁₀P₂	1053-73-2	427.202	amor pow					
125	Adenosine 5'-methylenediphosphonate	Adenosine, 5'-[hydrogen (phosphonomethyl)phosphonate]	C₁₁H₁₇N₅O₉P₂	3768-14-7	425.229	cry (w)	204				s H₂O
126	Adenosine 3'-phosphate	3'-Adenylic acid	C₁₀H₁₄N₅O₇P	84-21-9	347.222	col nd	195 dec				
127	Adenosine 5'-triphosphate	ATP	C₁₀H₁₆N₅O₁₃P₃	56-65-5	507.181		144 dec				
128	S-Adenosyl-L-homocysteine		C₁₄H₂₀N₆O₅S	979-92-0	384.411		210 dec				
129	5'-Adenylic acid	Adenosine 5'-monophosphate	C₁₀H₁₄N₅O₇P	61-19-8	347.222		195 dec				vs H₂O; s EtOH, 10% HCl
130	Adipamic acid		C₆H₁₁NO₃	334-25-8	145.156	nd (w)	161.5				
131	Adiphenine hydrochloride		C₂₀H₂₆ClNO₂	50-42-0	347.879	cry	113.5				vs H₂O; sl EtOH, eth
132	Adipic acid	1,6-Hexanedioic acid	C₆H₁₀O₄	124-04-9	146.141	mcl pr (w, ace, lig)	151.5(0.6)	337.5	1.360²⁵		sl H₂O; vs EtOH; s eth; i HOAc, lig
133	Adiponitrile	Hexanedinitrile	C₆H₈N₂	111-69-3	108.141	nd (eth)	2.2(0.4)	295	0.9676²⁰	1.4380²⁰	sl H₂O, eth; s chl, EtOH
134	Adrenalone		C₉H₁₁NO₃	99-45-6	181.188	nd	235 dec				sl H₂O, EtOH, eth
135	Affinin	N-(2-Methylpropyl)-2,6,8-decatrienamide	C₁₄H₂₃NO	25394-57-4	221.339	ye oil	23	162⁰·⁵		1.5134²⁵	i H₂O
136	Aflatoxin B1		C₁₇H₁₂O₆	1162-65-8	312.273	cry	268				
137	Aflatoxin B2		C₁₇H₁₄O₆	7220-81-7	314.289		287.5				
138	Aflatoxin G1		C₁₇H₁₂O₇	1165-39-5	328.273	cry	245				
139	Agaritine	L-Glutamic acid, 5-[2-[4-(hydroxymethyl)phenyl]hydrazide]	C₁₂H₁₇N₃O₄	2757-90-6	267.281	cry (dil al)	207 dec				vs H₂O
140	Ajmalan-17,21-diol, (17R,21α)	Ajmaline	C₂₀H₂₆N₂O₂	4360-12-7	326.432	pl (+3.5w) (aq AcOEt)	206				i H₂O; s EtOH, chl; sl eth, bz
141	Alachlor		C₁₄H₂₀ClNO₂	15972-60-8	269.768		42(2)	100⁰·⁰²	1.133²⁵		
142	DL-Alanine	DL-2-Aminopropanoic acid	C₃H₇NO₂	302-72-7	89.094	orth pr or nd (w)	300 dec	250 sub	1.424²⁵		s H₂O; vs EtOH
143	D-Alanine	2-Aminopropanoic acid, (R)	C₃H₇NO₂	338-69-2	89.094	nd (w, al)	314 dec	sub			s H₂O; sl EtOH; i eth
144	L-Alanine	2-Aminopropanoic acid, (S)	C₃H₇NO₂	56-41-7	89.094	orth (w)	297 dec	250 sub	1.432²²		s H₂O; sl EtOH, py; i eth, ace
145	β-Alanine	3-Aminopropanoic acid	C₃H₇NO₂	107-95-9	89.094	nd, orth pr (al)	200 dec		1.437¹⁹		s H₂O; sl EtOH; i eth, ace
146	Alantolactone		C₁₅H₂₀O₂	546-43-0	232.319	nd	76	275			vs bz, eth, EtOH, chl
147	Aldicarb		C₇H₁₄N₂O₂S	116-06-3	190.263		101.1(0.4)		1.195²⁵		
148	Aldosterone		C₂₁H₂₈O₅	52-39-1	360.444	cry (HOAc)	166.5				
149	Aldoxycarb S,S-dioxide		C₇H₁₄N₂O₄S	1646-88-4	222.262	cry	141				sl H₂O
150	Aldrin		C₁₂H₈Cl₆	309-00-2	364.910		103.8(0.3)				i H₂O; s EtOH, eth, ace, bz
151	Alizarin	1,2-Dihydroxy-9,10-anthracenedione	C₁₄H₈O₄	72-48-0	240.212	oran or red tcl nd or pr (al)	289.5				sl H₂O; s EtOH, eth, ace, bz; i chl
152	Alizarin Red S	Sodium alizarinesulfonate	C₁₄H₇NaO₇S	130-22-3	342.257						vs H₂O; s EtOH
153	Alizarin Yellow R		C₁₃H₉N₃O₅	2243-76-7	287.227	oran-br nd (dil HOAc)	253 dec				vs H₂O, EtOH
154	Alizurol purple	1-Hydroxy-4-[(4-methylphenyl)amino]-9,10-anthracenedione	C₂₁H₁₅NO₃	81-48-1	329.349	flat viol nd					s H₂SO₄
155	Alkannin		C₁₆H₁₆O₅	23444-65-7	288.295	br-red pr (bz)	149	140 sub			vs EtOH
156	Allantoic acid	Bis[(aminocarbonyl)amino]acetic acid	C₄H₈N₄O₄	99-16-1	176.132	nd	170 dec				sl H₂O, os, dil acid
157	Allantoin		C₄H₆N₄O₃	97-59-6	158.116	mcl pl or	239				sl H₂O; s EtOH, NaOH; i eth, MeOH
158	Allene		C₃H₄	463-49-0	40.064	col gas	-136.4(0.5)	-34.8(0.3)	0.584²⁵ (p>1 atm)	1.4168	vs bz, peth
159	Allethrin		C₁₉H₂₆O₃	584-79-2	302.407				1.010²⁰		
160	Allicin		C₆H₁₀OS₂	539-86-6	162.272			dec	1.112²⁰	1.561²⁰	vs H₂O

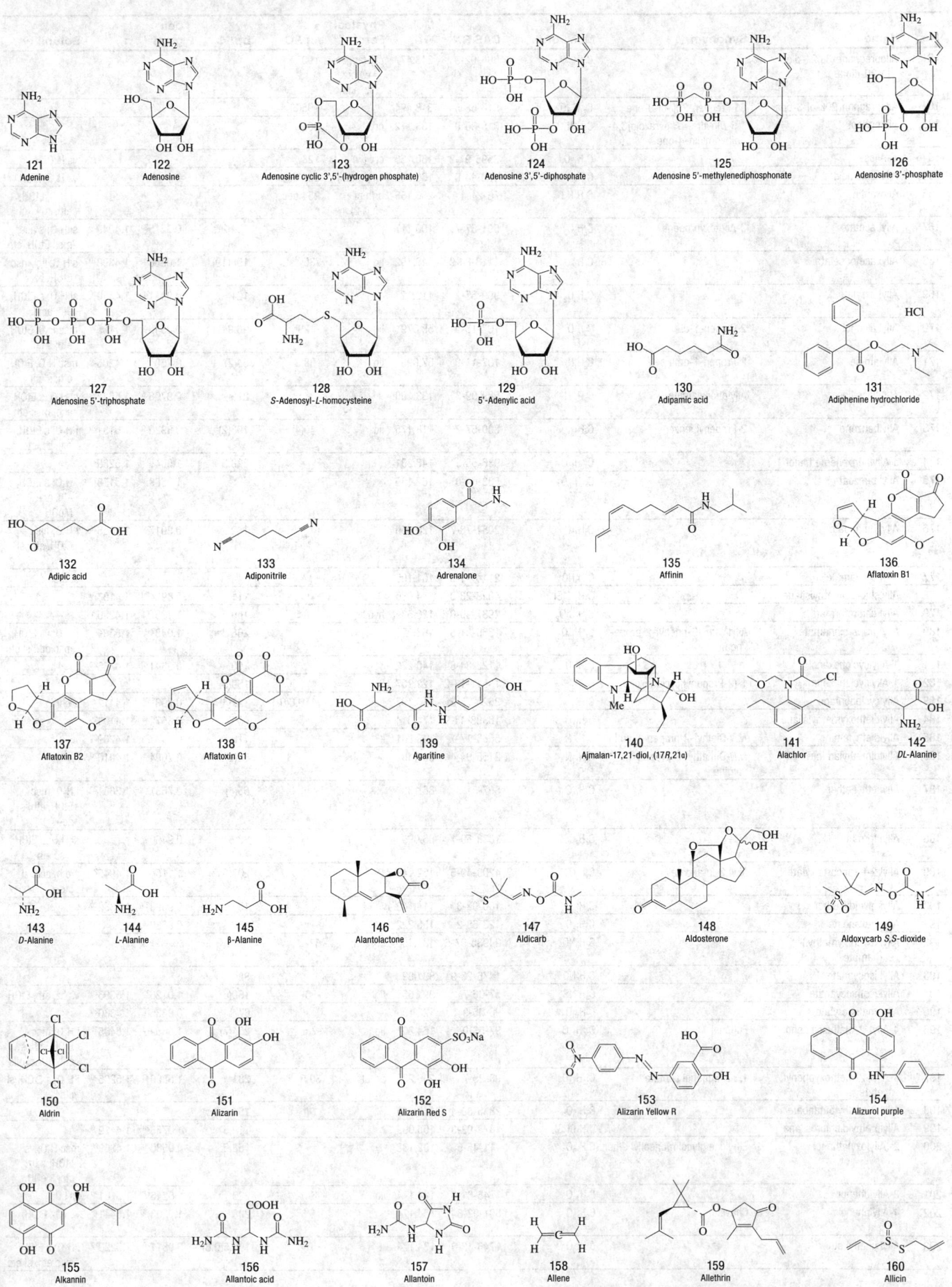

121 Adenine

122 Adenosine

123 Adenosine cyclic 3',5'-(hydrogen phosphate)

124 Adenosine 3',5'-diphosphate

125 Adenosine 5'-methylenediphosphonate

126 Adenosine 3'-phosphate

127 Adenosine 5'-triphosphate

128 S-Adenosyl-L-homocysteine

129 5'-Adenylic acid

130 Adipamic acid

131 Adiphenine hydrochloride

132 Adipic acid

133 Adiponitrile

134 Adrenalone

135 Affinin

136 Aflatoxin B1

137 Aflatoxin B2

138 Aflatoxin G1

139 Agaritine

140 Ajmalan-17,21-diol, (17R,21α)

141 Alachlor

142 DL-Alanine

143 D-Alanine

144 L-Alanine

145 β-Alanine

146 Alantolactone

147 Aldicarb

148 Aldosterone

149 Aldoxycarb S,S-dioxide

150 Aldrin

151 Alizarin

152 Alizarin Red S

153 Alizarin Yellow R

154 Alizurol purple

155 Alkannin

156 Allantoic acid

157 Allantoin

158 Allene

159 Allethrin

160 Allicin

No.	Name	Synonym	Mol. Form.	CAS RN	Mol. Wt.	Physical Form	mp/°C	bp/°C	den g cm⁻³	n_D	Solubility
161	Allopregnane-3β,21-diol-11,20-dione		$C_{21}H_{32}O_4$	566-02-9	348.477	cry (aq, ac, +w) nd (bz, ac)	190				
162	Allopregnan-20β-ol-3-one	5α-Pregnan-20β-ol-3-one	$C_{21}H_{34}O_2$	516-58-5	318.494		195(3)				
163	Allopurinol	1,5-Dihydro-4H-pyrazolo[3,4-d]pyrimidin-4-one	$C_5H_4N_4O$	315-30-0	136.112	cry	350				
164	D-Allose		$C_6H_{12}O_6$	2595-97-3	180.155	cry (w)	128				vs H_2O
165	Alloxanic acid		$C_4H_4N_2O_5$	470-44-0	160.085	tcl pr (eth)	162 dec				vs H_2O, EtOH
166	Alloxantin		$C_8H_6N_4O_8$	76-24-4	286.156	orth pr (w+2)	254 dec				sl H_2O, EtOH, eth
167	Allyl acetate	3-Acetoxypropene	$C_5H_8O_2$	591-87-7	100.117			104(2)	0.9275²⁰	1.4049²⁰	sl H_2O; s ace; msc EtOH, eth
168	Allyl acetoacetate		$C_7H_{10}O_3$	1118-84-9	142.152	liq	-85	198(19)	1.0366²⁰	1.4398²⁰	s H_2O, lig; msc EtOH, bz
169	Allyl acrylate		$C_6H_8O_2$	999-55-3	112.127			121	0.9441²⁰	1.4320²⁰	sl H_2O; s EtOH, eth, acid
170	Allyl alcohol	2-Propen-1-ol	C_3H_6O	107-18-6	58.079	liq	-129	96.9(0.5)	0.8540²⁰	1.4135²⁰	msc H_2O, EtOH, eth; s chl
171	Allylamine	2-Propen-1-amine	C_3H_7N	107-11-9	57.095	liq	-88.2	54(2)	0.758²⁰	1.4205²⁰	msc H_2O, EtOH, eth; s chl
172	N-Allylaniline	Allylphenylamine	$C_9H_{11}N$	589-09-3	133.190			219	0.9736²⁵	1.563²⁰	sl H_2O; s EtOH, ace; msc eth
173	Allylbenzene	2-Propenylbenzene	C_9H_{10}	300-57-2	118.175	liq	-40(4)	158(2)	0.8920²⁰	1.5131²⁰	i H_2O; s EtOH, eth, bz, ctc
174	α-Allylbenzenemethanol		$C_{10}H_{12}O$	936-58-3	148.201			228.5	1.004¹⁸	1.5289²¹	
175	Allyl benzoate		$C_{10}H_{10}O_2$	583-04-0	162.185				1.0569¹⁵	1.5178²⁰	i H_2O; s EtOH, eth, ace, MeOH
176	Allyl butanoate		$C_7H_{12}O_2$	2051-78-7	128.169			142	0.9017²⁰	1.4158²⁰	i H_2O; msc EtOH, eth; sl ctc
177	Allyl carbamate		$C_4H_7NO_2$	2114-11-6	101.105						sl ctc
178	Allylchlorodimethylsilane		$C_5H_{11}ClSi$	4028-23-3	134.680			111	0.8964²⁰	1.4195²⁰	
179	Allyl chloroformate		$C_4H_5ClO_2$	2937-50-0	120.535	hyg liq		109.5	1.136	1.4220²⁰	
180	Allyl trans-cinnamate	Allyl trans-3-phenyl-2-propenoate	$C_{12}H_{12}O_2$	1866-31-5	188.222			268 dec	1.048²³	1.530²⁰	i H_2O; vs EtOH; msc eth; sl ctc
181	1-Allylcyclohexanol		$C_9H_{16}O$	1123-34-8	140.222			190	0.9341²²	1.4756²²	
182	1-Allylcyclohexene	1-(2-Propenyl)cyclohexene	C_9H_{14}	13511-13-2	122.207	liq		159(9)			
183	Allylcyclopentane		C_8H_{14}	3524-75-2	110.197	liq	-110.6(0.1)	127(4)	0.793²⁵	1.4412²⁰	s chl
184	Allyldiethoxymethylsilane		$C_8H_{18}O_2Si$	18388-45-9	174.314			155	0.8572²⁵	1.4104²⁰	
185	Allyldiethylamine	N,N-Diethyl-2-propen-1-amine	$C_7H_{15}N$	5666-17-1	113.201			110	0.7477²⁵	1.4209²⁰	
186	Allyldimethylamine	N,N-Dimethyl-2-propen-1-amine	$C_5H_{11}N$	2155-94-4	85.148			62(4)	0.7094²⁵	1.4010²⁰	
187	Allyl ethyl ether		$C_5H_{10}O$	557-31-3	86.132			65(4)	0.7651²⁰	1.3881²⁰	i H_2O; msc EtOH, eth; s ace
188	Allyl formate		$C_4H_6O_2$	1838-59-1	86.090			83.6	0.9460²⁰		sl H_2O; s EtOH; msc eth
189	Allyl 2-furancarboxylate	Allyl 2-furanoate	$C_8H_8O_3$	4208-49-5	152.148			207.5	1.115²⁵	1.4945²⁰	s eth, ace; sl ctc
190	Allyl glycidyl ether		$C_6H_{10}O_2$	106-92-3	114.142			154	0.9698²⁰	1.4332²⁰	
191	Allyl hexanoate		$C_9H_{16}O_2$	123-68-2	156.222			186	0.8869²⁰		
192	Allyl (hydroxymethyl)-carbamate		$C_5H_9NO_3$	24935-97-5	131.130	cry (tol)	57				
193	Allyl isocyanate		C_4H_5NO	1476-23-9	83.089			88			
194	Allyl isothiocyanate		C_4H_5NS	57-06-7	99.155	liq	-80	160(6)	1.0126²⁰	1.5306²⁰	vs bz, eth, EtOH
195	Allyl methacrylate		$C_7H_{10}O_2$	96-05-9	126.153			67⁵⁰	0.9335²⁰	1.4360²⁰	
196	4-Allyl-2-methoxyphenol	Eugenol	$C_{10}H_{12}O_2$	97-53-0	164.201	liq	-7.5	254(7)	1.0652²⁰	1.5405²⁰	i H_2O; msc EtOH, eth; s chl, HOAc, oils
197	4-Allyl-2-methoxyphenyl acetate	1,3,4-Eugenol acetate	$C_{12}H_{14}O_3$	93-28-7	206.237	pr (al)	30.5	281	1.0806²⁰	1.5205²⁰	i H_2O; s EtOH; sl ctc
198	Allyl 3-methylbutanoate		$C_8H_{14}O_2$	2835-39-4	142.196			154			
199	Allylmethyldichlorosilane		$C_4H_8Cl_2Si$	1873-92-3	155.099			119.5	1.0758²⁰	1.4419²⁰	
200	2-(Allyloxy)ethanol	Ethylene glycol monoallyl ether	$C_5H_{10}O_2$	111-45-5	102.132			158.5	0.9580²⁰	1.4358²⁰	msc H_2O; vs EtOH; s bz, ctc, MeOH
201	2-Allylphenol		$C_9H_{10}O$	1745-81-9	134.174	liq	-6	220	1.0246¹⁵	1.5181²⁰	vs eth
202	4-Allylphenol	Chavicol	$C_9H_{10}O$	501-92-8	134.174		15.8	238	1.0203¹⁵	1.5441¹⁸	vs eth, EtOH, chl
203	Allyl phenyl ether		$C_9H_{10}O$	1746-13-0	134.174			190.5(0.8)	0.9811²⁰	1.5223²⁰	i H_2O; s EtOH; msc eth; sl ctc

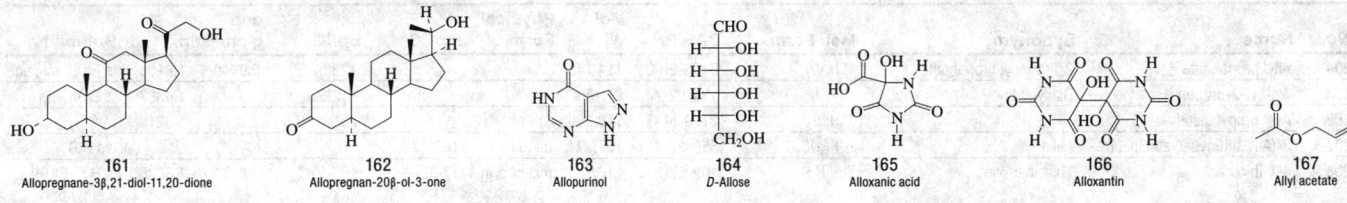

161
Allopregnane-3β,21-diol-11,20-dione

162
Allopregnan-20β-ol-3-one

163
Allopurinol

164
D-Allose

165
Alloxanic acid

166
Alloxantin

167
Allyl acetate

168
Allyl acetoacetate

169
Allyl acrylate

170
Allyl alcohol

171
Allylamine

172
N-Allylaniline

173
Allylbenzene

174
α-Allylbenzenemethanol

175
Allyl benzoate

176
Allyl butanoate

177
Allyl carbamate

178
Allylchlorodimethylsilane

179
Allyl chloroformate

180
Allyl *trans*-cinnamate

181
1-Allylcyclohexanol

182
1-Allylcyclohexene

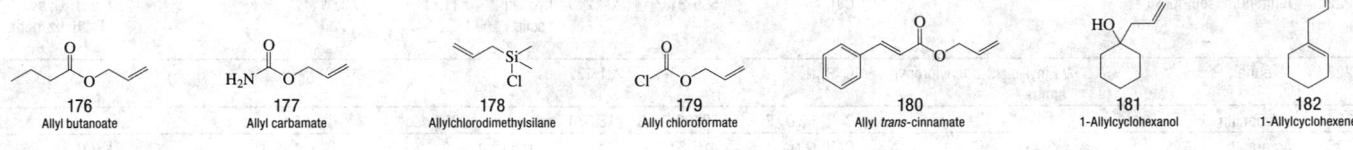

183
Allylcyclopentane

184
Allyldiethoxymethylsilane

185
Allyldiethylamine

186
Allyldimethylamine

187
Allyl ethyl ether

188
Allyl formate

189
Allyl 2-furancarboxylate

190
Allyl glycidyl ether

191
Allyl hexanoate

192
Allyl (hydroxymethyl)carbamate

193
Allyl isocyanate

194
Allyl isothiocyanate

195
Allyl methacrylate

196
4-Allyl-2-methoxyphenol

197
4-Allyl-2-methoxyphenyl acetate

198
Allyl 3-methylbutanoate

199
Allylmethyldichlorosilane

200
2-(Allyloxy)ethanol

201
2-Allylphenol

202
4-Allylphenol

203
Allyl phenyl ether

No.	Name	Synonym	Mol. Form.	CAS RN	Mol. Wt.	Physical Form	mp/°C	bp/°C	den g cm⁻³	n_D	Solubility
204	Allyl propanoate	2-Propenyl propanoate	$C_6H_{10}O_2$	2408-20-0	114.142			123(5)	0.9140[20]	1.4105[20]	s EtOH, eth, ace
205	N-Allyl-2-propen-1-amine	Diallylamine	$C_6H_{11}N$	124-02-7	97.158			112(3)		1.4387[20]	s EtOH, eth
206	Allyl propyl disulfide		$C_6H_{12}S_2$	2179-59-1	148.289			79[13]		1.5219[20]	
207	3-(Allylsulfinyl)-L-alanine, (S)-	Alliin	$C_6H_{11}NO_3S$	556-27-4	177.221	nd (dil ac)	165				vs H_2O
208	Allylthiourea	Thiosinamine	$C_4H_8N_2S$	109-57-9	116.185	mcl or orth pr (w)	77(3)		1.217[20]	1.5936[78]	s H_2O, EtOH; sl eth; i bz
209	Allyltrichlorosilane	Trichloro-2-propenylsilane	$C_3H_5Cl_3Si$	107-37-9	175.517		35	118(3)	1.2011[20]	1.4460[20]	
210	Allyltriethoxysilane		$C_9H_{20}O_3Si$	2550-04-1	204.339			100[50]	0.9030[20]	1.4072[20]	
211	Allyltrimethylsilane		$C_6H_{14}Si$	762-72-1	114.261			85	0.7158[25]	1.4074[20]	i H_2O
212	Allylurea		$C_4H_8N_2O$	557-11-9	100.119	nd (al)	85				msc H_2O, EtOH; sl eth, chl; i peth
213	Allyl vinyl ether	3-(Ethenyloxy)-1-propene	C_5H_8O	3917-15-5	84.117			66	0.7900[20]	1.4062[20]	i H_2O; s eth, ace, chl
214	Aloin A		$C_{21}H_{22}O_9$	1415-73-2	418.395		149.3				s H_2O, EtOH, ace; sl eth, bz; i chl
215	Alphaprodine		$C_{16}H_{23}NO_2$	15867-21-7	261.360	cry	103				
216	Alstonidine		$C_{22}H_{24}N_2O_4$	25394-75-6	380.437	cry (eth)	189				vs ace, EtOH
217	Alstonine		$C_{21}H_{20}N_2O_3$	642-18-2	348.395	ye nd (ace)	207 dec				
218	D-Altrose		$C_6H_{12}O_6$	1990-29-0	180.155	pr (MeOH,al)	103.5				vs H_2O
219	Aluminum 2-butoxide	2-Butanol, aluminum salt	$C_{12}H_{27}AlO_3$	2269-22-9	246.322			197[20]			
220	Aluminum distearate	Hydroxyaluminum distearate	$C_{36}H_{71}AlO_5$	300-92-5	610.928	wh pow	145				i H_2O
221	Aluminum ethanolate	Aluminum ethoxide	$C_6H_{15}AlO_3$	555-75-9	162.163	liq/wh solid	140	200[7]			dec H_2O; sl xyl
222	Aluminum isopropoxide		$C_9H_{21}AlO_3$	555-31-7	204.243	hyg wh solid	119	135[10]			reac H_2O; s EtOH, bz, peth, chl
223	Alverine	N-Ethyl-bis(3-phenylpropyl)-amine	$C_{20}H_{27}N$	150-59-4	281.435	oil		166[0.3]			
224	α-Amanitin		$C_{39}H_{54}N_{10}O_{14}S$	23109-05-9	918.970	nd	254 dec				
225	Amaranth dye		$C_{20}H_{11}N_2Na_3O_{10}S_3$	915-67-3	604.472	dk red pow					s H_2O
226	Ametryn		$C_9H_{17}N_5S$	834-12-8	227.330		83.6(0.5)				
227	Amminetrimethylboron		$C_3H_{12}BN$	1830-95-1	72.945		73.5				
228	19-Amino-8,11,13-abieta-triene		$C_{20}H_{31}N$	1446-61-3	285.467	cry	44.5				
229	2-Aminoacetamide		$C_2H_6N_2O$	598-41-4	74.081	hyg nd (chl)	67.5				vs H_2O, EtOH; sl eth, bz; s ace, chl
230	Aminoacetonitrile		$C_2H_4N_2$	540-61-4	56.066			58[15]			vs EtOH
231	Aminoacetonitrile monohydrochloride		$C_2H_5ClN_2$	6011-14-9	92.527	hyg cry (al)	165 dec				
232	α-Aminoacetophenone hydrochloride		$C_8H_{10}ClNO$	5468-37-1	171.624		194 dec				
233	1-Aminoadamantane hydrochloride	Adamantanamine hydrochloride	$C_{10}H_{18}ClN$	665-66-7	187.710	cry (al-eth)	360 dec				vs H_2O, EtOH
234	2-Aminoadipic acid		$C_6H_{11}NO_4$	626-71-1	161.156	pl (w)	207.0				sl H_2O, EtOH, eth
235	3-Aminoalanine	2,3-Diaminopropionic acid	$C_3H_8N_2O_2$	515-94-6	104.108	hyg rosettes	110				vs H_2O
236	1-Amino-9,10-anthracene-dione	1-Aminoanthraquinone	$C_{14}H_9NO_2$	82-45-1	223.227	red nd (al)	253.5	sub			vs ace, bz, EtOH, chl
237	2-Amino-9,10-anthracene-dione	2-Aminoanthraquinone	$C_{14}H_9NO_2$	117-79-3	223.227	red nd (al, HOAc)	304.5	sub			i H_2O, eth; sl EtOH; s ace, bz, chl
238	4-Aminoazobenzene		$C_{12}H_{11}N_3$	60-09-3	197.235	oran mcl nd (al)	125(1)	>360			sl H_2O, lig; s EtOH, eth, bz, chl
239	2-Aminobenzaldehyde		C_7H_7NO	529-23-7	121.137	silv lf	40.5	80[2]			sl H_2O; vs EtOH, eth; s bz, chl; i lig
240	3-Aminobenzaldehyde		C_7H_7NO	1709-44-0	121.137	nd (AcOEt)	29				s eth, acid
241	4-Aminobenzaldehyde		C_7H_7NO	556-18-3	121.137	pl (w)	71.5				s H_2O, EtOH, eth, acid
242	2-Aminobenzamide		$C_7H_8N_2O$	88-68-6	136.151		110.5 dec				s H_2O, EtOH; sl eth, bz; vs AcOEt
243	4-Aminobenzamide		$C_7H_8N_2O$	2835-68-9	136.151	ye cry (+1/4w)	183				sl H_2O; s EtOH, eth
244	α-Aminobenzeneacetic acid, (±)-	α-Phenylglycine	$C_8H_9NO_2$	2835-06-5	151.163	pl	292 dec	255 sub			s alk; sl os
245	4-Aminobenzeneacetic acid	p-Aminophenylacetic acid	$C_8H_9NO_2$	1197-55-3	151.163	pl (w)	195(1)				i H_2O; sl EtOH, DMSO

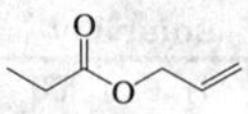

204 Allyl propanoate

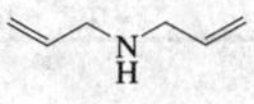

205 N-Allyl-2-propen-1-amine

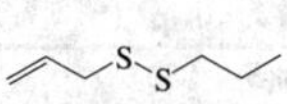

206 Allyl propyl disulfide

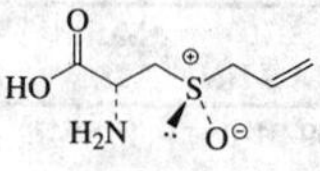

207 3-(Allylsulfinyl)-L-alanine, (S)-

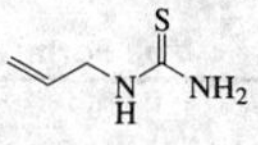

208 Allylthiourea

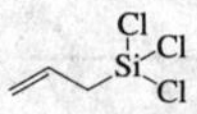

209 Allyltrichlorosilane

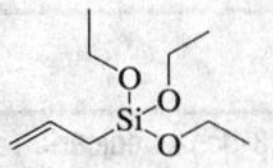

210 Allyltriethoxysilane

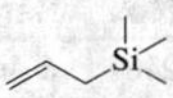

211 Allyltrimethylsilane

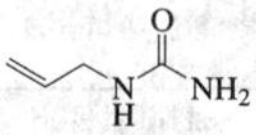

212 Allylurea

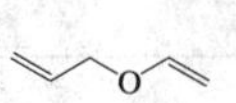

213 Allyl vinyl ether

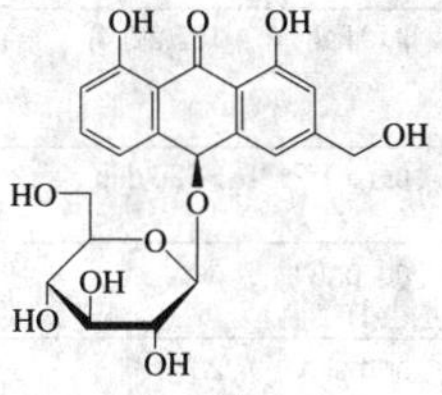

214 Aloin A

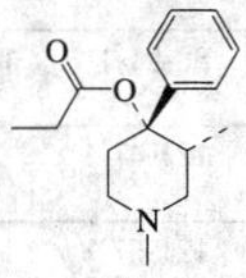

215 Alphaprodine

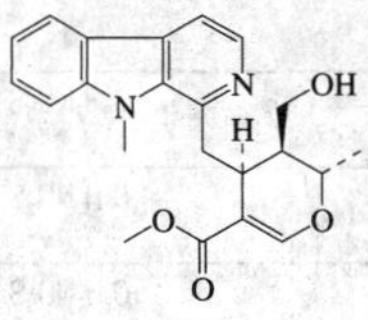

216 Alstonidine

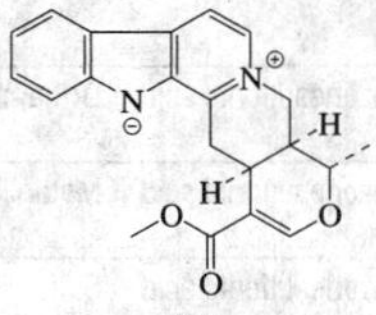

217 Alstonine

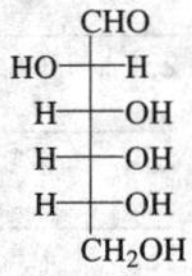

218 D-Altrose

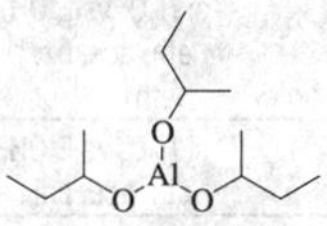

219 Aluminum 2-butoxide

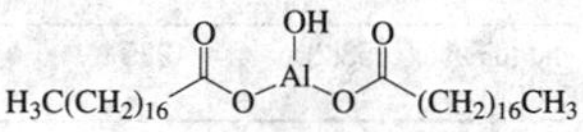

220 Aluminum distearate

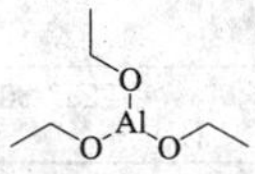

221 Aluminum ethanolate

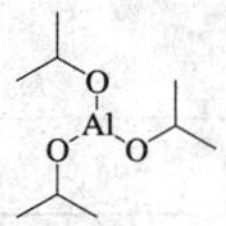

222 Aluminum isopropoxide

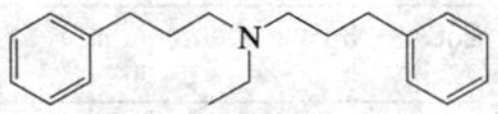

223 Alverine

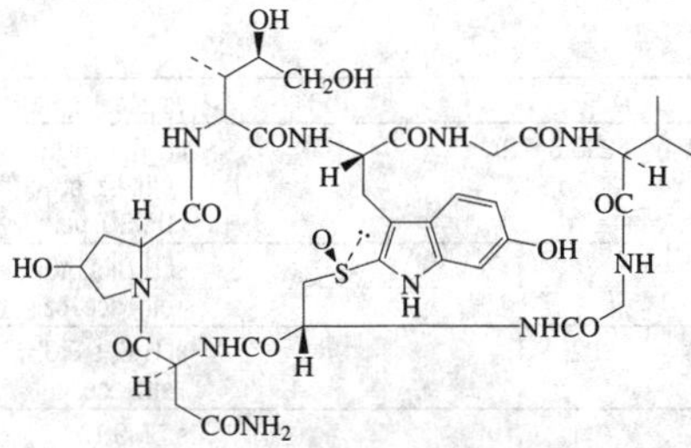

224 α-Amanitin

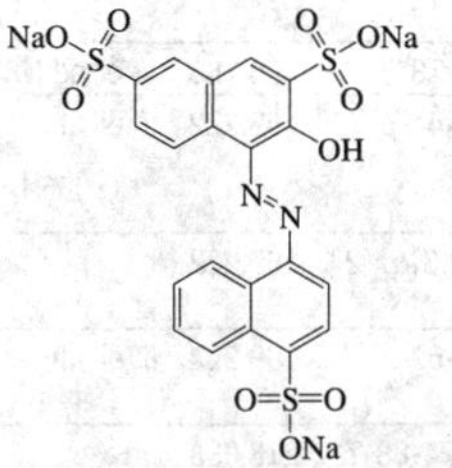

225 Amaranth dye

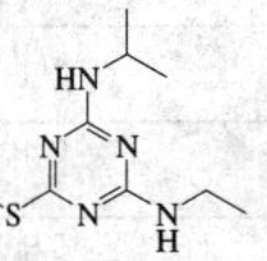

226 Ametryn

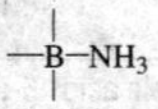

227 Amminetrimethylboron

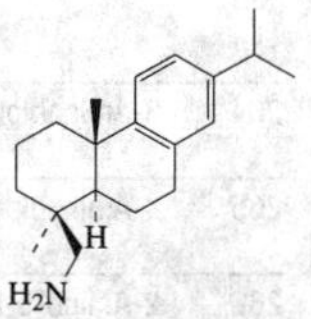

228 19-Amino-8,11,13-abietatriene

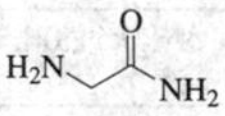

229 2-Aminoacetamide

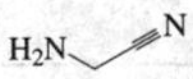

230 Aminoacetonitrile

231 Aminoacetonitrile monohydrochloride

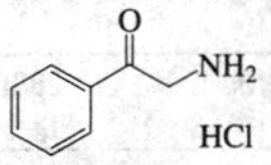

232 α-Aminoacetophenone hydrochloride

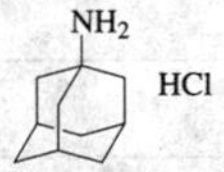

233 1-Aminoadamantane hydrochloride

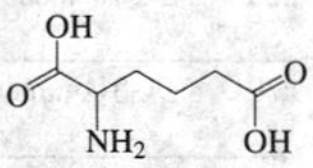

234 2-Aminoadipic acid

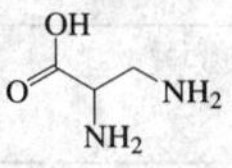

235 3-Aminoalanine

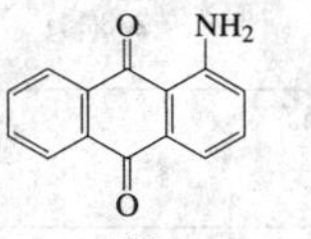

236 1-Amino-9,10-anthracenedione

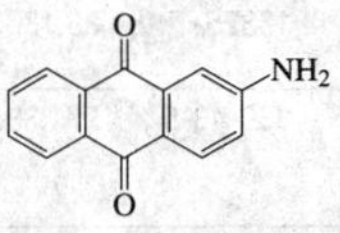

237 2-Amino-9,10-anthracenedione

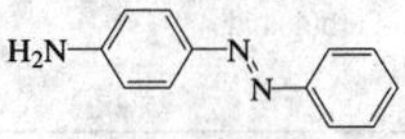

238 4-Aminoazobenzene

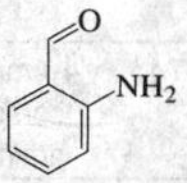

239 2-Aminobenzaldehyde

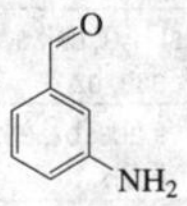

240 3-Aminobenzaldehyde

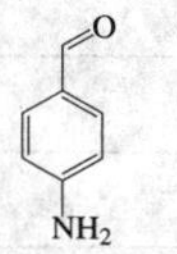

241 4-Aminobenzaldehyde

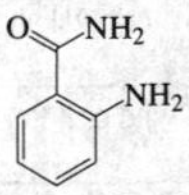

242 2-Aminobenzamide

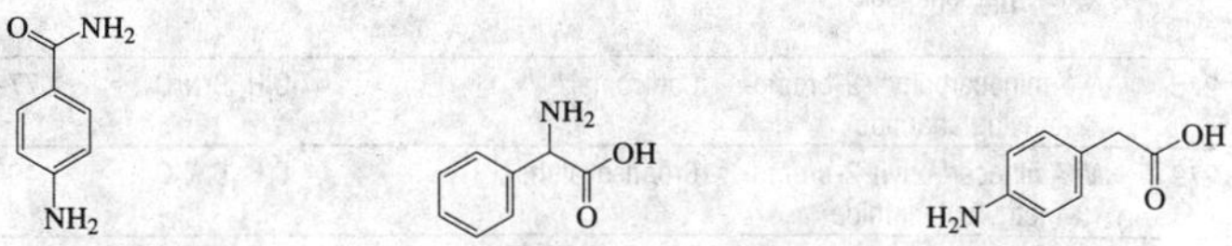

243 4-Aminobenzamide

244 α-Aminobenzeneacetic acid, (±)-

245 4-Aminobenzeneacetic acid

No.	Name	Synonym	Mol. Form.	CAS RN	Mol. Wt.	Physical Form	mp/°C	bp/°C	den g cm⁻³	n_D	Solubility
246	5-Amino-1,3-benzenedicar-boxylic acid		$C_8H_7NO_4$	99-31-0	181.147	pr(al), pl(w)	360	sub			i H_2O; sl EtOH
247	4-Aminobenzeneethanol		$C_8H_{11}NO$	104-10-9	137.179	nd (al)	108				
248	2-Aminobenzenemethanamine		$C_7H_{10}N_2$	4403-69-4	122.167		61	269			vs EtOH
249	2-Aminobenzenemethanol		C_7H_9NO	5344-90-1	123.152		83.5	273			s H_2O, EtOH, eth, HOAc; vs bz, chl
250	4-Aminobenzenesulfonamide	Sulfanilamide	$C_6H_8N_2O_2S$	63-74-1	172.205	lf (dil al)	162.2(0.4)		1.08²⁵		s H_2O, EtOH, eth, ace; sl chl, peth
251	2-Aminobenzenesulfonic acid	Orthanilic acid	$C_6H_7NO_3S$	88-21-1	173.190	pr (+ 1/2w)	>320 dec				sl H_2O; i EtOH, eth
252	3-Aminobenzenesulfonic acid	Metanilic acid	$C_6H_7NO_3S$	121-47-1	173.190	nd, pr (w +1)	dec				sl H_2O, EtOH; i eth
253	4-Aminobenzenesulfonic acid	Sulfanilic acid	$C_6H_7NO_3S$	121-57-3	173.190	orth pl or mcl (w+2)	288		1.485²⁵		sl H_2O; i EtOH, eth
254	4-Aminobenzenesulfonyl fluoride	p-Sulfanilyl fluoride	$C_6H_6FNO_2S$	98-62-4	175.181		68.5				
255	2-Aminobenzenethiol		C_6H_7NS	137-07-5	125.192		26	234		1.4606²⁰	s EtOH, eth
256	4-Aminobenzenethiol		C_6H_7NS	1193-02-8	125.192		46	143¹⁷			s H_2O, EtOH
257	2-Aminobenzonitrile		$C_7H_6N_2$	1885-29-6	118.136	ye pr (CS_2) nd (peth)	51	263			sl H_2O; vs EtOH, eth, ace, bz; i peth
258	3-Aminobenzonitrile		$C_7H_6N_2$	2237-30-1	118.136	nd (dil al or CCl_4)	53(2)	289			sl H_2O; vs EtOH, eth, ace, chl
259	4-Aminobenzonitrile		$C_7H_6N_2$	873-74-5	118.136	pr or pl (w)	86.2(0.5)				sl H_2O, ctc; vs EtOH, eth, ace, bz
260	4-Aminobenzophenone		$C_{13}H_{11}NO$	1137-41-3	197.232	lf (dil al)	123(2)	246¹³			sl H_2O, tfa; s EtOH, eth, HOAc
261	N-(4-Aminobenzoyl)-L-glutamic acid		$C_{12}H_{14}N_2O_5$	4271-30-1	266.249	cry (w)	173				
262	N-(4-Aminobenzoyl)glycine	p-Aminohippuric acid	$C_9H_{10}N_2O_3$	61-78-9	194.186	pr or nd (w)	198.5				vs ace, bz, EtOH
263	2-Aminobiphenyl		$C_{12}H_{11}N$	90-41-5	169.222	lf (dil al)	49.13(0.04)	298.3(0.2)			i H_2O; s EtOH, eth, bz; sl DMSO, peth
264	3-Aminobiphenyl		$C_{12}H_{11}N$	2243-47-2	169.222	nd	31.5				sl H_2O; s EtOH, eth, ace, bz
265	4-Aminobiphenyl	p-Biphenylamine	$C_{12}H_{11}N$	92-67-1	169.222	lf (dil al)	51.0(0.6)	302			sl H_2O; s EtOH, eth, ace, chl
266	2-Amino-5-bromobenzoic acid	5-Bromoanthranilic acid	$C_7H_6BrNO_2$	5794-88-7	216.033	nd	219.5				s DMSO
267	1-Amino-4-bromo-9,10-dihydro-9,10-dioxo-2-anthracenesulfonic acid	1-Amino-4-bromoanthraqui-none-2-sulfonic acid	$C_{14}H_8BrNO_5S$	116-81-4	382.187	red nd (w)					
268	DL-2-Aminobutanoic acid		$C_4H_9NO_2$	2835-81-6	103.120	lf (w)	304 dec	sub	1.2300²⁰		vs H_2O; sl EtOH; i eth, bz
269	L-2-Aminobutanoic acid		$C_4H_9NO_2$	1492-24-6	103.120	lf (dil al), cry (al)	292 dec				s H_2O; sl EtOH, eth; i bz
270	DL-3-Aminobutanoic acid		$C_4H_9NO_2$	2835-82-7	103.120	nd (al)	194.3				vs H_2O; i EtOH, eth, bz
271	4-Aminobutanoic acid	γ-Aminobutyric acid	$C_4H_9NO_2$	56-12-2	103.120	pr or nd (al) lf (MeOH-eth)	203 dec				vs H_2O; sl EtOH, ace; i eth, bz
272	2-Amino-1-butanol, (±)-		$C_4H_{11}NO$	13054-87-0	89.136	liq	-1.0	178(9)	0.9162²⁰	1.4489²⁵	msc H_2O, EtOH, eth; sl chl
273	4-Amino-1-butanol		$C_4H_{11}NO$	13325-10-5	89.136			203(11)	0.967¹²	1.4625²⁰	s H_2O, EtOH; i eth
274	4-Amino-N-[(butylamino)-carbonyl]benzenesulfon-amide	Carbutamide	$C_{11}H_{17}N_3O_3S$	339-43-5	271.336		144.5				
275	Aminocarb		$C_{11}H_{16}N_2O_2$	2032-59-9	208.257	cry	95.0(0.3)				sl H_2O, bz; s ace
276	N-(Aminocarbonyl)acetamide		$C_3H_6N_2O_2$	591-07-1	102.092		218	180 sub			sl H_2O, eth; s EtOH
277	[4-[(Aminocarbonyl)amino]-phenyl]arsonic acid	Carbarsone	$C_7H_9AsN_2O_4$	121-59-5	260.079	nd (w)	174				sl H_2O, DMSO, EtOH; i eth, chl; s alk
278	N-(Aminocarbonyl)-2-bromo-2-ethylbutanamide	Carbromal	$C_7H_{13}BrN_2O_2$	77-65-6	237.094	orth (dil al)	118		1.544²⁵		sl H_2O, chl; s ace, bz
279	N-(Aminocarbonyl)-2-bromo-3-methylbutanamide	Bromisovalum	$C_6H_{11}BrN_2O_2$	496-67-3	223.067	nd or lf (to)	154	sub	1.56¹⁵		vs ace, bz, eth, EtOH

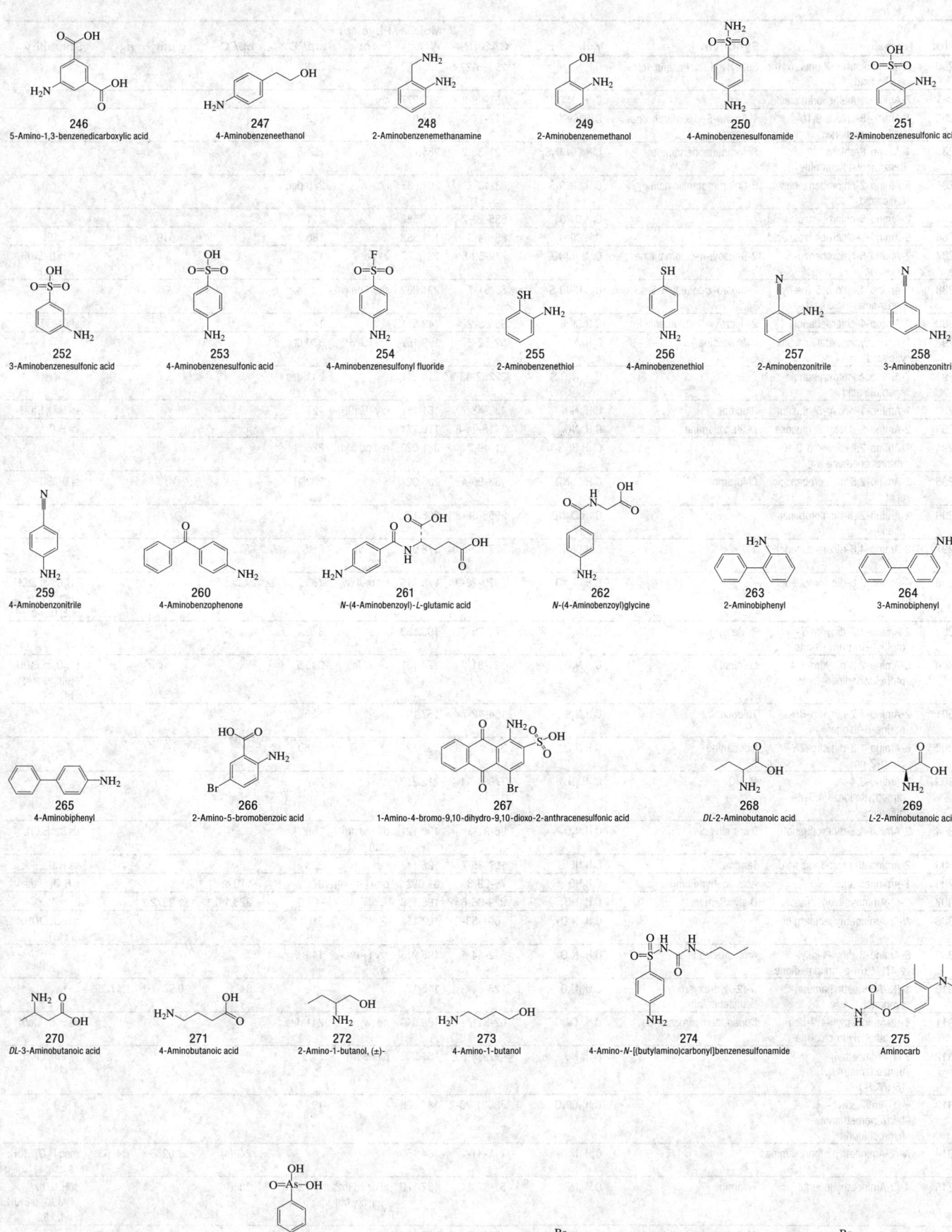

246
5-Amino-1,3-benzenedicarboxylic acid

247
4-Aminobenzeneethanol

248
2-Aminobenzenemethanamine

249
2-Aminobenzenemethanol

250
4-Aminobenzenesulfonamide

251
2-Aminobenzenesulfonic acid

252
3-Aminobenzenesulfonic acid

253
4-Aminobenzenesulfonic acid

254
4-Aminobenzenesulfonyl fluoride

255
2-Aminobenzenethiol

256
4-Aminobenzenethiol

257
2-Aminobenzonitrile

258
3-Aminobenzonitrile

259
4-Aminobenzonitrile

260
4-Aminobenzophenone

261
N-(4-Aminobenzoyl)-L-glutamic acid

262
N-(4-Aminobenzoyl)glycine

263
2-Aminobiphenyl

264
3-Aminobiphenyl

265
4-Aminobiphenyl

266
2-Amino-5-bromobenzoic acid

267
1-Amino-4-bromo-9,10-dihydro-9,10-dioxo-2-anthracenesulfonic acid

268
DL-2-Aminobutanoic acid

269
L-2-Aminobutanoic acid

270
DL-3-Aminobutanoic acid

271
4-Aminobutanoic acid

272
2-Amino-1-butanol, (±)-

273
4-Amino-1-butanol

274
4-Amino-N-[(butylamino)carbonyl]benzenesulfonamide

275
Aminocarb

276
N-(Aminocarbonyl)acetamide

277
[4-[(Aminocarbonyl)amino]phenyl]arsonic acid

278
N-(Aminocarbonyl)-2-bromo-2-ethylbutanamide

279
N-(Aminocarbonyl)-2-bromo-3-methylbutanamide

No.	Name	Synonym	Mol. Form.	CAS RN	Mol. Wt.	Physical Form	mp/°C	bp/°C	den g cm^{-3}	n_D	Solubility
280	[2-(Aminocarbonyl)phenoxy]-acetic acid	Salicylamide O-acetic acid	C$_9$H$_9$NO$_4$	25395-22-6	195.172		221				s alk
281	7-Aminocephalosporanic acid		C$_{10}$H$_{12}$N$_2$O$_5$S	957-68-6	272.277	cry					
282	1-Amino-5-chloro-9,10-anthracenedione	1-Amino-5-chloroanthraquinone	C$_{14}$H$_8$ClNO$_2$	117-11-3	257.673		212				
283	4-Amino-6-chloro-1,3-benzenedisulfonamide	Chloraminophenamide	C$_6$H$_8$ClN$_3$O$_4$S$_2$	121-30-2	285.729		254.5				
284	5-Amino-2-chlorobenzenesulfonic acid	6-Chlorometanilic acid	C$_6$H$_6$ClNO$_3$S	88-43-7	207.635	nd (w)	280 dec				
285	2-Amino-5-chlorobenzoic acid		C$_7$H$_6$ClNO$_2$	635-21-2	171.582		211				
286	5-Amino-2-chlorobenzoic acid		C$_7$H$_6$ClNO$_2$	89-54-3	171.582		188		1.519[15]		vs EtOH
287	2-Amino-5-chlorobenzophenone	2-Benzoyl-4-chloroaniline	C$_{13}$H$_{10}$ClNO	719-59-5	231.677	ye nd	100.5				vs H$_2$O, EtOH, peth, chl
288	2-Amino-4-chloro-5-methyl-benzenesulfonic acid	2-Chloro-p-toluidine-5-sulfonic acid	C$_7$H$_8$ClNO$_3$S	88-51-7	221.662	short nd (w)					
289	2-Amino-4-chlorophenol	2-Hydroxy-5-chloroaniline	C$_6$H$_6$ClNO	95-85-2	143.571		140				sl DMSO
290	1-Aminocyclopentanecarboxylic acid	Cycloleucine	C$_6$H$_{11}$NO$_2$	52-52-8	129.157	cry (al-w)	330 dec				
291	7-Aminodeacetoxycephalosporanic acid		C$_8$H$_{10}$N$_2$O$_3$S	22252-43-3	214.241		241 dec				
292	1-Amino-1-deoxy-D-glucitol	Glucamine	C$_6$H$_{15}$NO$_5$	488-43-7	181.187	cry (MeOH)	127				vs H$_2$O, EtOH
293	2-Amino-2-deoxy-D-glucose	D-Glucosamine	C$_6$H$_{13}$NO$_5$	3416-24-8	179.171						vs H$_2$O
294	1-Amino-2,4-dibromo-9,10-anthracenedione		C$_{14}$H$_7$Br$_2$NO$_2$	81-49-2	381.020	red nd (xyl)	226				
295	3-Amino-2,5-dichlorobenzoic acid	Chloramben	C$_7$H$_5$Cl$_2$NO$_2$	133-90-4	206.027		202(1)				sl DMSO
296	2-Amino-2',5-dichlorobenzophenone		C$_{13}$H$_9$Cl$_2$NO	2958-36-3	266.122		≈80				
297	2-Amino-4,6-dichlorophenol		C$_6$H$_5$Cl$_2$NO	527-62-8	178.016	long nd (CS$_2$)	95.5	70 sub			
298	4-Amino-2,6-dichlorophenol		C$_6$H$_5$Cl$_2$NO	5930-28-9	178.016	nd or lf (w, bz)	168	sub			i H$_2$O; vs EtOH, eth; s ace; sl bz, HOAc
299	2-Amino-1,7-dihydro-7-methyl-6H-purin-6-one	7-Methylguanine	C$_6$H$_7$N$_5$O	578-76-7	165.153		370				
300	5-Amino-2,3-dihydro-1,4-phthalazinedione	Luminol	C$_8$H$_7$N$_3$O$_2$	521-31-3	177.161	ye nd (al)	330.5				i H$_2$O; sl EtOH, eth; vs alk; s HOAc
301	2-Amino-1,7-dihydro-6H-purine-6-thione	Thioguanine	C$_5$H$_5$N$_5$S	154-42-7	167.193		>360				
302	6-Amino-1,3-dihydro-2H-purin-2-one	Isoguanine	C$_5$H$_5$N$_5$O	3373-53-3	151.127		>360				i H$_2$O
303	2-Amino-3,4-dimethylimidazo[4,5-f]-quinoline	Me-IQ	C$_{12}$H$_{12}$N$_4$	77094-11-2	212.250	cry	297				
304	2-Amino-4,6-dinitrophenol	Picramic acid	C$_6$H$_5$N$_3$O$_5$	96-91-3	199.121	dk red nd (al) pr (chl)	168(1)				vs bz, EtOH
305	2-Aminoethanesulfonic acid	Taurine	C$_2$H$_7$NO$_3$S	107-35-7	125.147	mcl pr (w)	328				vs H$_2$O
306	1-Aminoethanol	Acetaldehyde ammonia	C$_2$H$_7$NO	75-39-8	61.083	orth (eth-al)	97	110 dec			s H$_2$O; sl eth
307	2-(2-Aminoethoxy)ethanol	Diglycolamine	C$_4$H$_{11}$NO$_2$	929-06-6	105.136		-12.5	223.1(0.1)	1.0572[20]		
308	N-(2-Aminoethyl)acetamide		C$_4$H$_{10}$N$_2$O	1001-53-2	102.134		51				s H$_2$O, EtOH, bz; i eth
309	6-Amino-3-ethyl-1-allyl-2,4(1H,3H)-pyrimidinedione	Aminometradine	C$_9$H$_{13}$N$_3$O$_2$	642-44-4	195.218	cry (+1w, w)	143				
310	1-[(2-Aminoethyl)amino]-2-propanol	N-(2-Hydroxypropyl)-ethylenediamine	C$_5$H$_{14}$N$_2$O	123-84-2	118.177			94[3]	0.9837[25]	1.4738[20]	
311	4-(2-Aminoethyl)-1,2-benzenediol, hydrochloride	Dopamine hydrochloride	C$_8$H$_{12}$ClNO$_2$	62-31-7	189.640	nd (w)	241 dec				vs H$_2$O, MeOH
312	α-(1-Aminoethyl)-benzenemethanol, [S-(R*,R*)]-		C$_9$H$_{13}$NO	492-39-7	151.205	pl(MeOH)	77.5				vs eth, EtOH, chl
313	α-(1-Aminoethyl)-benzenemethanol, hydrochloride		C$_9$H$_{14}$ClNO	53631-70-2	187.666		198.5				s H$_2$O
314	N-(2-Aminoethyl)ethanolamine		C$_4$H$_{12}$N$_2$O	111-41-1	104.150			242(5)	1.0286[20]	1.4863[20]	msc H$_2$O, EtOH; s ace; sl bz, lig
315	4-(2-Aminoethyl)phenol	Tyramine	C$_8$H$_{11}$NO	51-67-2	137.179	pl or nd (bz, w), cry (al)	164.5	206[25]			sl H$_2$O, bz, DMSO; s EtOH, xyl; i tol
316	N-(2-Aminoethyl)-1,3-propanediamine	N-(3-Aminopropyl)-ethylenediamine	C$_5$H$_{15}$N$_3$	13531-52-7	117.193			87[3]		1.4805[25]	
317	2-Amino-2-ethyl-1,3-propanediol		C$_5$H$_{13}$NO$_2$	115-70-8	119.163		37.5	152[10]	1.099[20]	1.490[20]	msc H$_2$O

280
[2-(Aminocarbonyl)phenoxy]acetic acid

281
7-Aminocephalosporanic acid

282
1-Amino-5-chloro-9,10-anthracenedione

283
4-Amino-6-chloro-1,3-benzenedisulfonamide

284
5-Amino-2-chlorobenzenesulfonic acid

285
2-Amino-5-chlorobenzoic acid

286
5-Amino-2-chlorobenzoic acid

287
2-Amino-5-chlorobenzophenone

288
2-Amino-4-chloro-5-methylbenzenesulfonic acid

289
2-Amino-4-chlorophenol

290
1-Aminocyclopentanecarboxylic acid

291
7-Aminodeacetoxycephalosporanic acid

292
1-Amino-1-deoxy-D-glucitol

293
2-Amino-2-deoxy-D-glucose

294
1-Amino-2,4-dibromo-9,10-anthracenedione

295
3-Amino-2,5-dichlorobenzoic acid

296
2-Amino-2',5-dichlorobenzophenone

297
2-Amino-4,6-dichlorophenol

298
4-Amino-2,6-dichlorophenol

299
2-Amino-1,7-dihydro-7-methyl-6H-purin-6-one

300
5-Amino-2,3-dihydro-1,4-phthalazinedione

301
2-Amino-1,7-dihydro-6H-purine-6-thione

302
6-Amino-1,3-dihydro-2H-purin-2-one

303
2-Amino-3,4-dimethylimidazo[4,5-f]quinoline

304
2-Amino-4,6-dinitrophenol

305
2-Aminoethanesulfonic acid

306
1-Aminoethanol

307
2-(2-Aminoethoxy)ethanol

308
N-(2-Aminoethyl)acetamide

309
6-Amino-3-ethyl-1-allyl-2,4(1H,3H)-pyrimidinedione

310
1-[(2-Aminoethyl)amino]-2-propanol

311
4-(2-Aminoethyl)-1,2-benzenediol, hydrochloride

312
α-(1-Aminoethyl)benzenemethanol, [S-(R*,R*)]-

313
α-(1-Aminoethyl)benzenemethanol, hydrochloride

314
N-(2-Aminoethyl)ethanolamine

315
4-(2-Aminoethyl)phenol

316
N-(2-Aminoethyl)-1,3-propanediamine

317
2-Amino-2-ethyl-1,3-propanediol

No.	Name	Synonym	Mol. Form.	CAS RN	Mol. Wt.	Physical Form	mp/°C	bp/°C	den g cm⁻³	n_D	Solubility
318	L-2-Aminohexanedioic acid	2-Aminoadipic acid	$C_6H_{11}NO_4$	542-32-5	161.156	cry (EtOH, w)	205 dec				sl H₂O, EtOH, eth
319	6-Aminohexanenitrile	5-Cyano-1-pentylamine	$C_6H_{12}N_2$	2432-74-8	112.172	liq		118[16]			
320	6-Aminohexanoic acid	ε-Aminocaproic acid	$C_6H_{13}NO_2$	60-32-2	131.173	lf (eth)	205				vs H₂O; i EtOH; sl MeOH
321	6-Amino-1-hexanol		$C_6H_{15}NO$	4048-33-3	117.189		57	137[30]			
322	1-Amino-4-hydroxy-9,10-anthracenedione		$C_{14}H_9NO_3$	116-85-8	239.226		216.5				s EtOH, ace
323	3-Amino-4-hydroxybenzene-sulfonic acid		$C_6H_7NO_4S$	98-37-3	189.190	orth (w+1)	>300				sl H₂O; i EtOH, eth
324	4-Amino-2-hydroxybenzo-hydrazide	p-Aminosalicylic acid hydrazide	$C_7H_9N_3O_2$	6946-29-8	167.165	nd (al)	195				vs EtOH
325	2-Amino-3-hydroxybenzoic acid		$C_7H_7NO_3$	548-93-6	153.136	lf (w)	253.5				sl H₂O; s EtOH, eth, chl
326	4-Amino-2-hydroxybenzoic acid	p-Aminosalicylic acid	$C_7H_7NO_3$	65-49-6	153.136	nd, pl (al-eth)	150 dec				s H₂O, EtOH, eth, ace; i bz, peth, chl
327	5-Amino-2-hydroxybenzoic acid	Mesalamine	$C_7H_7NO_3$	89-57-6	153.136		281.0(0.5)				sl H₂O; i EtOH
328	3-Amino-4-hydroxybutanoic acid	γ-Hydroxy-β-aminobutyric acid	$C_4H_9NO_3$	589-44-6	119.119	pr	216				vs H₂O; sl EtOH, chl, eth, AcOEt
329	4-Amino-3-hydroxybutanoic acid, (±)-		$C_4H_9NO_3$	924-49-2	119.119	pr (w), cry (dil al)	218				vs H₂O
330	4-(2-Amino-1-hydroxyethyl)-1,2-benzenediol, (±)-		$C_8H_{11}NO_3$	138-65-8	169.178		189 dec				
331	1-Amino-4-hydroxy-2-methoxy-9,10-anthracene-dione		$C_{15}H_{11}NO_4$	2379-90-0	269.253						sl chl
332	4-Amino-5-(hydroxymethyl)-2(1H)-pyrimidinone	5-Hydroxymethylcytosine	$C_5H_7N_3O_2$	1123-95-1	141.129		>300 dec				
333	4-Amino-5-hydroxy-2,7-naphthalenedisulfonic acid	1-Naphthol-8-amino-3,6-disulfonic acid	$C_{10}H_9NO_7S_2$	90-20-0	319.311						sl H₂O, EtOH, eth
334	4-Amino-3-hydroxy-1-naph-thalenesulfonic acid	1-Amino-2-naphthol-4-sulfonic acid	$C_{10}H_9NO_4S$	116-63-2	239.248	gray nd					i H₂O, EtOH, bz; s alk
335	2-Amino-4-hydroxypteridine		$C_6H_5N_5O$	2236-60-4	163.137	ye cry	>360				
336	5-Amino-1H-imidazole-4-carboxamide		$C_4H_6N_4O$	360-97-4	126.117	cry (EtOH)	170				
337	O-[(Aminoiminomethyl)amino]-L-homoserine	Canavanine	$C_5H_{12}N_4O_3$	543-38-4	176.174	cry (al)	172				vs H₂O
338	(Aminoiminomethyl)urea		$C_2H_6N_4O$	141-83-3	102.095	pr	105	160 dec			s H₂O, py; sl EtOH; i eth, bz, chl, CS₂
339	2-Amino-5-iodobenzoic acid		$C_7H_6INO_2$	5326-47-6	263.033		220 dec				sl H₂O, tfa; vs EtOH, eth, ace; s bz
340	4-Amino-1H-isoindole-1,3(2H)-dione		$C_8H_6N_2O_2$	2518-24-3	162.146		269.5				
341	4-Amino-3-isoxazolidinone, (R)-	Cycloserine	$C_3H_6N_2O_2$	68-41-7	102.092		155 dec				s H₂O; sl MeOH
342	1-Amino-2-methyl-9,10-anthracenedione	1-Amino-2-methylanthraqui-none	$C_{15}H_{11}NO_2$	82-28-0	237.254		205.5				i H₂O; s EtOH, bz, chl; sl eth
343	α-(Aminomethyl)-benzenemethanol	Phenylethanolamine	$C_8H_{11}NO$	7568-93-6	137.179		56.5	160[17]			vs H₂O; s EtOH
344	β-(Aminomethyl)-benzenepropanoic acid	4-Amino-3-phenylbutyric acid	$C_{10}H_{13}NO_2$	1078-21-3	179.216		252 dec				
345	2-Amino-5-methylbenzenesul-fonic acid		$C_7H_9NO_3S$	88-44-8	187.216	lt ye nd	132 dec				vs H₂O
346	trans-4-(Aminomethyl)-cyclohexanecarboxylic acid	Tranexamic acid	$C_8H_{15}NO_2$	1197-18-8	157.211		249(4)				vs H₂O
347	4-Amino-4-methyl-2-penta-none	Diacetonamine	$C_6H_{13}NO$	625-04-7	115.173			25[0.14]			s H₂O; msc EtOH, eth
348	2-Amino-4-methylphenol		C_7H_9NO	95-84-1	123.152	cry (w), orth (bz), lf or nd	136	sub			sl H₂O, bz; s EtOH, eth, chl; i lig
349	4-Amino-2-methylphenol		C_7H_9NO	2835-96-3	123.152	nd or lf (bz)	176.5	sub			sl H₂O, bz; s EtOH, eth
350	4-Amino-3-methylphenol		C_7H_9NO	2835-99-6	123.152	pr (dil al) cry (bz)	179				sl H₂O; vs EtOH, eth; s DMSO
351	(Aminomethyl)phosphonic acid		CH_6NO_3P	1066-51-9	111.038	cry	309				
352	2-Amino-2-methyl-1,3-propanediol		$C_4H_{11}NO_2$	115-69-5	105.136		110.93(0.05)	151[10]			vs H₂O; s EtOH
353	L-3-Amino-2-methylpropanoic acid		$C_4H_9NO_2$	144-90-1	103.120	cry (w)	185				
354	2-Amino-2-methyl-1-propanol	2-Aminoisobutanol	$C_4H_{11}NO$	124-68-5	89.136		25.5	163.8(0.8)	0.934[20]	1.449[20]	msc H₂O; s ctc

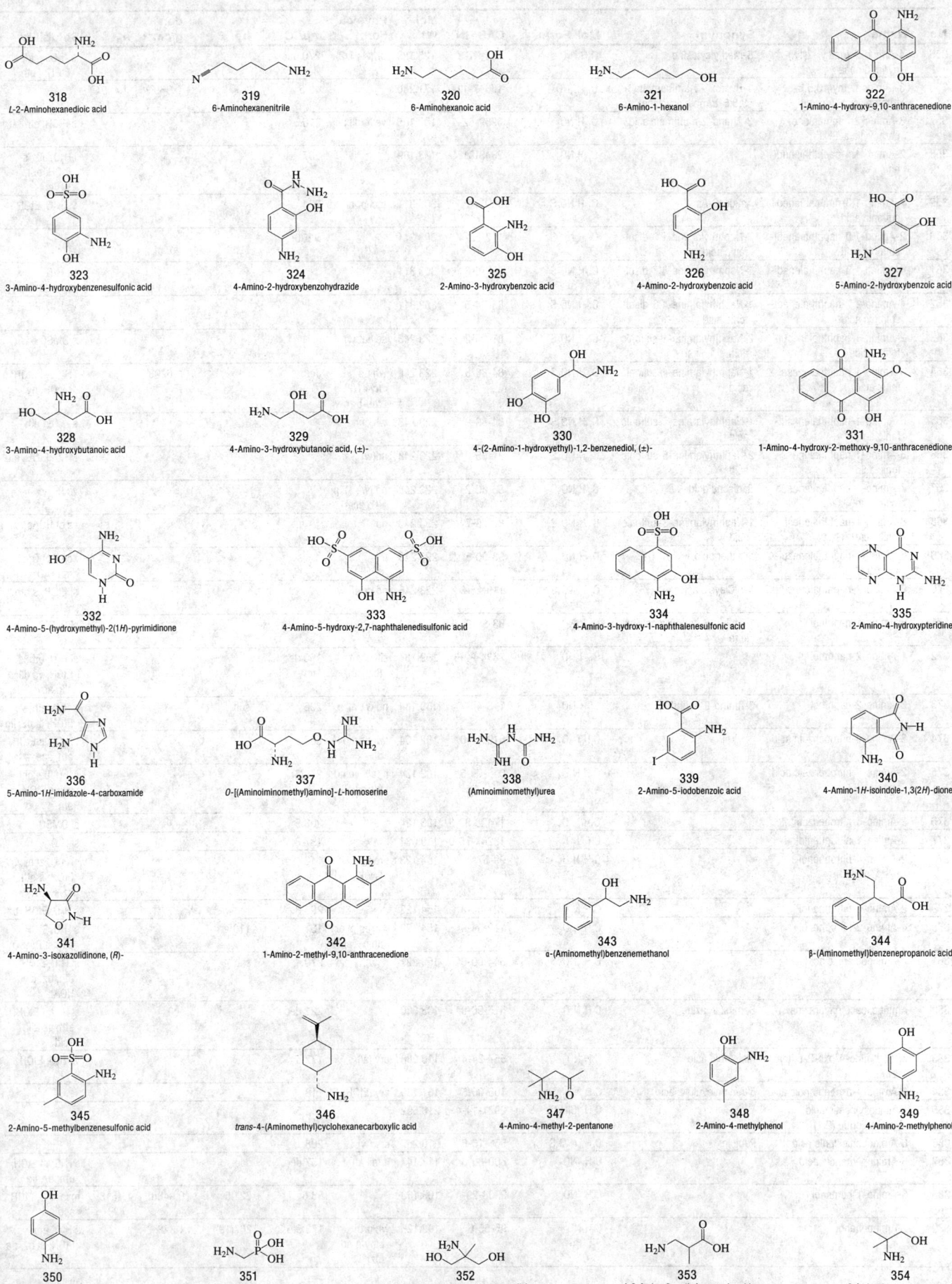

318
L-2-Aminohexanedioic acid

319
6-Aminohexanenitrile

320
6-Aminohexanoic acid

321
6-Amino-1-hexanol

322
1-Amino-4-hydroxy-9,10-anthracenedione

323
3-Amino-4-hydroxybenzenesulfonic acid

324
4-Amino-2-hydroxybenzohydrazide

325
2-Amino-3-hydroxybenzoic acid

326
4-Amino-2-hydroxybenzoic acid

327
5-Amino-2-hydroxybenzoic acid

328
3-Amino-4-hydroxybutanoic acid

329
4-Amino-3-hydroxybutanoic acid, (±)-

330
4-(2-Amino-1-hydroxyethyl)-1,2-benzenediol, (±)-

331
1-Amino-4-hydroxy-2-methoxy-9,10-anthracenedione

332
4-Amino-5-(hydroxymethyl)-2(1H)-pyrimidinone

333
4-Amino-5-hydroxy-2,7-naphthalenedisulfonic acid

334
4-Amino-3-hydroxy-1-naphthalenesulfonic acid

335
2-Amino-4-hydroxypteridine

336
5-Amino-1H-imidazole-4-carboxamide

337
O-[(Aminoiminomethyl)amino]-L-homoserine

338
(Aminoiminomethyl)urea

339
2-Amino-5-iodobenzoic acid

340
4-Amino-1H-isoindole-1,3(2H)-dione

341
4-Amino-3-isoxazolidinone, (R)-

342
1-Amino-2-methyl-9,10-anthracenedione

343
α-(Aminomethyl)benzenemethanol

344
β-(Aminomethyl)benzenepropanoic acid

345
2-Amino-5-methylbenzenesulfonic acid

346
trans-4-(Aminomethyl)cyclohexanecarboxylic acid

347
4-Amino-4-methyl-2-pentanone

348
2-Amino-4-methylphenol

349
4-Amino-2-methylphenol

350
4-Amino-3-methylphenol

351
(Aminomethyl)phosphonic acid

352
2-Amino-2-methyl-1,3-propanediol

353
L-3-Amino-2-methylpropanoic acid

354
2-Amino-2-methyl-1-propanol

No.	Name	Synonym	Mol. Form.	CAS RN	Mol. Wt.	Physical Form	mp/°C	bp/°C	den g cm⁻³	n_D	Solubility
355	4-Amino-5-methyl-2(1*H*)-pyrimidinone	5-Methylcytosine	C₅H₇N₃O	554-01-8	125.129	pr (w+1/2)	270 dec				s H₂O, acid; sl EtOH; i eth
356	3-(Aminomethyl)-3,5,5-trimethylcyclohexanol	1-Hydroxy-3-aminomethyl-3,5,5-trimethylcyclohexane	C₁₀H₂₁NO	15647-11-7	171.280		45.5	265	0.969²⁵	1.4904²⁰	
357	3-Amino-2-naphthalene-carboxylic acid	3-Amino-2-naphthoic acid	C₁₁H₉NO₂	5959-52-4	187.195	ye lf (dil al)	216.5				s EtOH, eth
358	2-Amino-1,4-naphthalenedi-one		C₁₀H₇NO₂	2348-81-4	173.169		207				i H₂O, alk; s EtOH, eth, HOAc
359	7-Amino-1,3-naphthalenedi-sulfonic acid	Amido-G-Acid	C₁₀H₉NO₆S₂	86-65-7	303.311	mcl pr or nd (w+4)	274				vs H₂O, EtOH
360	2-Amino-1,5-naphthalenedi-sulfonic acid	2-Naphthylamine-1,5-disul-fonic acid	C₁₀H₉NO₆S₂	117-62-4	303.311		>300				
361	4-Amino-1,6-naphthalenedi-sulfonic acid	1-Naphthylamine-4,7-disul-fonic acid	C₁₀H₉NO₆S₂	85-75-6	303.311						vs H₂O
362	4-Amino-1,7-naphthalenedi-sulfonic acid	1-Naphthylamine-4,6-disul-fonic acid	C₁₀H₉NO₆S₂	85-74-5	303.311						vs H₂O, EtOH
363	2-Amino-1-naphthalenesul-fonic acid	2-Naphthylamine-1-sulfonic acid	C₁₀H₉NO₃S	81-16-3	223.248	sc(hot w)					s DMSO
364	4-Amino-1-naphthalenesul-fonic acid	1-Naphthylamine-4-sulfonic acid	C₁₀H₉NO₃S	84-86-6	223.248	wh nd (w+1/2) red-br cry	dec		1.6703²⁵		i H₂O; sl EtOH; s MeOH, py
365	5-Amino-1-naphthalenesul-fonic acid	1-Naphthylamine-5-sulfonic acid	C₁₀H₉NO₃S	84-89-9	223.248	wh cry					s H₂O; i eth
366	6-Amino-1-naphthalenesul-fonic acid	2-Naphthylamine-5-sulfonic acid	C₁₀H₉NO₃S	81-05-0	223.248	nd(w)					i H₂O, EtOH, eth
367	7-Amino-1-naphthalenesul-fonic acid	Badische acid	C₁₀H₉NO₃S	86-60-2	223.248	nd (w+1), pl (aq ace)					vs HOAc
368	8-Amino-1-naphthalenesul-fonic acid	1-Naphthylamine-8-sulfonic acid	C₁₀H₉NO₃S	82-75-7	223.248	nd					vs gl HOAc
369	6-Amino-2-naphthalenesul-fonic acid	Bronner acid	C₁₀H₉NO₃S	93-00-5	223.248	lf					i cold H₂O; sl hot H₂O
370	8-Amino-2-naphthalenesul-fonic acid	1,7-Cleve's acid	C₁₀H₉NO₃S	119-28-8	223.248	nd or pr (w)					sl EtOH; s eth
371	5-Amino-1-naphthol	1-Amino-6-hydroxynaphtha-lene	C₁₀H₉NO	83-55-6	159.184		170				sl DMSO
372	1-Amino-2-naphthol		C₁₀H₉NO	2834-92-6	159.184	silvery lf (bz, eth)	150 dec				sl H₂O, eth; s EtOH; vs dil alk, acid
373	8-Amino-2-naphthol	8-Amino-β-naphthol	C₁₀H₉NO	118-46-7	159.184	nd (w, al)	206	sub			s H₂O, eth; vs EtOH; sl bz, lig
374	2-Amino-4-nitrobenzoic acid		C₇H₆N₂O₄	619-17-0	182.134	oran pr (dil al)	269				i H₂O; vs EtOH, eth, ace; s xyl
375	2-Amino-5-nitrobenzoic acid		C₇H₆N₂O₄	616-79-5	182.134	lf (al), ye nd (w, dil al)	278(5)				i H₂O, bz, chl, xyl; s EtOH, eth
376	2-Amino-5-nitrobenzonitrile		C₇H₅N₃O₂	17420-30-3	163.134		203.5				sl DMSO
377	3-Amino-1-nitroguanidine		CH₅N₅O₂	18264-75-0	119.084		187.8				sl H₂O
378	2-Amino-4-nitrophenol		C₆H₆N₂O₃	99-57-0	154.123	oran pr (+w)	146				sl H₂O, ace; vs EtOH; s eth, bz, HOAc
379	2-Amino-5-nitrophenol		C₆H₆N₂O₃	121-88-0	154.123		205.8				s H₂O, EtOH, bz
380	4-Amino-2-nitrophenol		C₆H₆N₂O₃	119-34-6	154.123	dk red pl or nd (w, al)	131	110¹²			s H₂O, EtOH, eth; sl DMSO
381	2-Aminooctanoic acid, (±)-		C₈H₁₇NO₂	644-90-6	159.227	lf (w)	270	sub			sl H₂O, EtOH, eth, bz; s HOAc
382	Aminooxoacetohydrazide	Semioxamazide	C₂H₅N₃O₂	515-96-8	103.080		221 dec				sl H₂O; i EtOH, eth; vs alk, acid
383	*cis*-4-Amino-4-oxo-2-butenoic acid	Maleamic acid	C₄H₅NO₃	557-24-4	115.088	cry (al)	172.5				vs H₂O, EtOH
384	5-Amino-4-oxopentanoic acid	5-Aminolevulinic acid	C₅H₉NO₃	106-60-5	131.130	cry (EtOH)	118				
385	(Aminooxy)acetic acid, hydrochloride (2:1)		C₄H₁₁ClN₂O₆	2921-14-4	218.592		152.5				
386	6-Aminopenicillanic acid	Penicin	C₈H₁₂N₂O₃S	551-16-6	216.257	cry (w)	208				
387	5-Aminopentanoic acid		C₅H₁₁NO₂	660-88-8	117.147	lf (dil al)	157 dec	dec			s H₂O; sl EtOH; i eth, bz, lig
388	5-Amino-1-pentanol		C₅H₁₃NO	2508-29-4	103.163		38.5	221.5	0.9488¹⁷	1.4618¹⁷	msc H₂O, EtOH, ace
389	2-Aminophenol		C₆H₇NO	95-55-6	109.126	wh orth bipym nd (bz)	173.5(0.3)	267(19)	1.328²⁵		s H₂O, eth; vs EtOH; sl bz, tfa

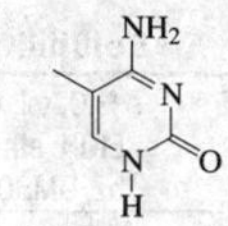

355
4-Amino-5-methyl-2(1*H*)-pyrimidinone

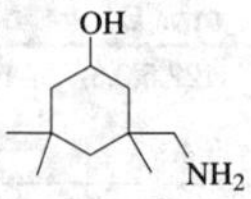

356
3-(Aminomethyl)-3,5,5-trimethylcyclohexanol

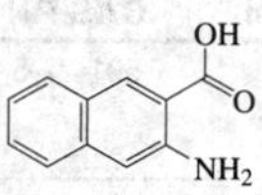

357
3-Amino-2-naphthalenecarboxylic acid

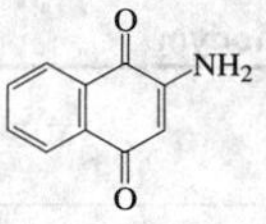

358
2-Amino-1,4-naphthalenedione

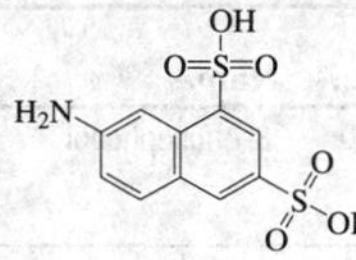

359
7-Amino-1,3-naphthalenedisulfonic acid

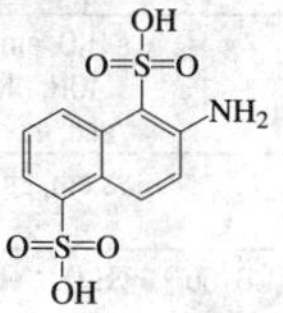

360
2-Amino-1,5-naphthalenedisulfonic acid

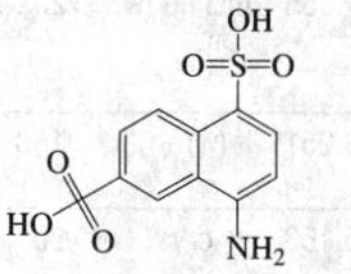

361
4-Amino-1,6-naphthalenedisulfonic acid

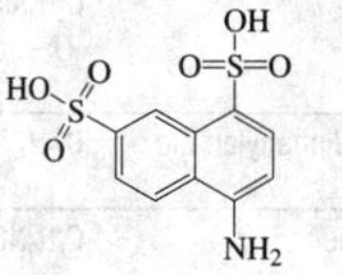

362
4-Amino-1,7-naphthalenedisulfonic acid

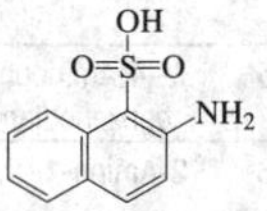

363
2-Amino-1-naphthalenesulfonic acid

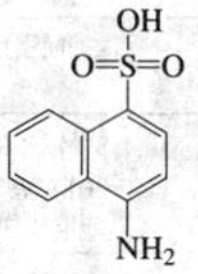

364
4-Amino-1-naphthalenesulfonic acid

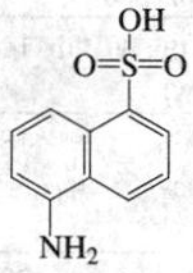

365
5-Amino-1-naphthalenesulfonic acid

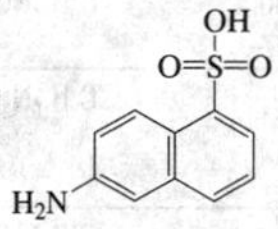

366
6-Amino-1-naphthalenesulfonic acid

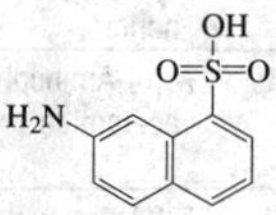

367
7-Amino-1-naphthalenesulfonic acid

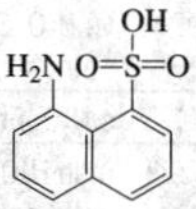

368
8-Amino-1-naphthalenesulfonic acid

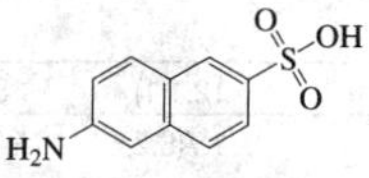

369
6-Amino-2-naphthalenesulfonic acid

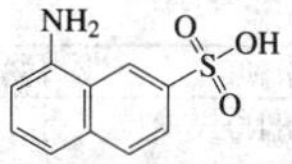

370
8-Amino-2-naphthalenesulfonic acid

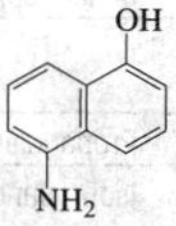

371
5-Amino-1-naphthol

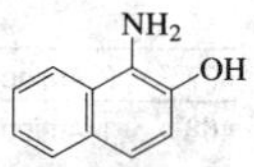

372
1-Amino-2-naphthol

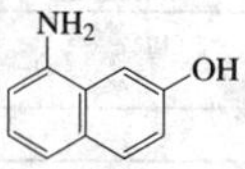

373
8-Amino-2-naphthol

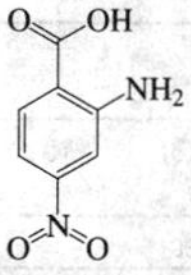

374
2-Amino-4-nitrobenzoic acid

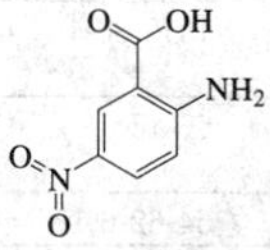

375
2-Amino-5-nitrobenzoic acid

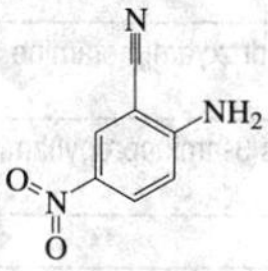

376
2-Amino-5-nitrobenzonitrile

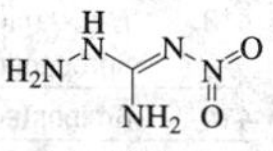

377
3-Amino-1-nitroguanidine

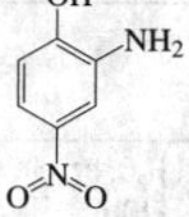

378
2-Amino-4-nitrophenol

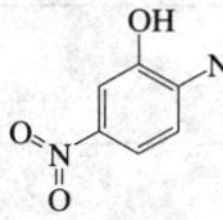

379
2-Amino-5-nitrophenol

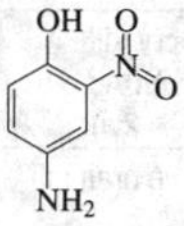

380
4-Amino-2-nitrophenol

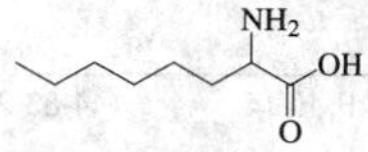

381
2-Aminooctanoic acid, (±)-

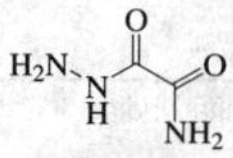

382
Aminooxoacetohydrazide

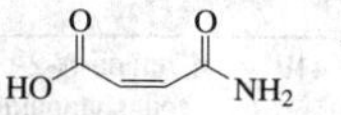

383
cis-4-Amino-4-oxo-2-butenoic acid

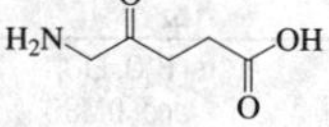

384
5-Amino-4-oxopentanoic acid

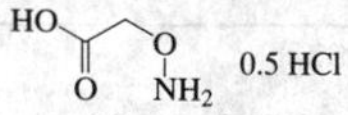

385
(Aminooxy)acetic acid, hydrochloride (2:1)

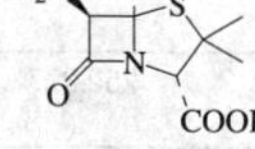

386
6-Aminopenicillanic acid

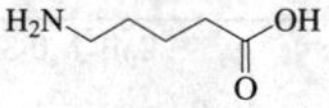

387
5-Aminopentanoic acid

388
5-Amino-1-pentanol

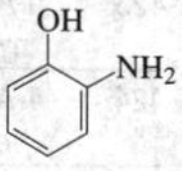

389
2-Aminophenol

No.	Name	Synonym	Mol. Form.	CAS RN	Mol. Wt.	Physical Form	mp/°C	bp/°C	den g cm⁻³	n_D	Solubility
390	3-Aminophenol		C$_6$H$_7$NO	591-27-5	109.126	pr (to)	122.5(0.3)	164[11]			s H$_2$O, tol; vs EtOH, eth; sl bz, DMSO
391	4-Aminophenol		C$_6$H$_7$NO	123-30-8	109.126	wh pl (w)	186(7)	110[0.3]			sl H$_2$O, tfa; vs EtOH; i bz, chl; s alk
392	N-(3-Aminophenyl)acetamide		C$_8$H$_{10}$N$_2$O	102-28-3	150.177	nd or pl (bz)	88				vs H$_2$O, EtOH, ace; sl eth, bz
393	N-(4-Aminophenyl)acetamide	p-Aminoacetanilide	C$_8$H$_{10}$N$_2$O	122-80-5	150.177	nd (w)	166.5	267			s H$_2$O; vs EtOH, eth
394	(4-Aminophenyl)arsonic acid	Arsanilic acid	C$_6$H$_8$AsNO$_3$	98-50-0	217.055	mcl nd (w, al)	232		1.9571[10]		s H$_2$O, eth; sl EtOH, DMSO; i ace, bz
395	N-(4-Aminophenyl)-1,4-benzenediamine	4,4'-Diaminodiphenylamine	C$_{12}$H$_{13}$N$_3$	537-65-5	199.251	lf (w)	158	dec			vs eth, EtOH
396	2-Amino-1-phenylethanone	Phenacylamine	C$_8$H$_9$NO	613-89-8	135.163	ye cry	20	251		1.6160[20]	i H$_2$O; s eth; sl ctc
397	1-(3-Aminophenyl)ethanone	m-Aminoacetophenone	C$_8$H$_9$NO	99-03-6	135.163	pa ye pl (al), lf (eth)	98.5	289.5			sl H$_2$O; s EtOH
398	1-(4-Aminophenyl)ethanone	p-Aminoacetophenone	C$_8$H$_9$NO	99-92-3	135.163	ye mcl pr (al)	105(1)	294			vs eth, EtOH
399	1-(4-Aminophenyl)-1-pentanone		C$_{11}$H$_{15}$NO	38237-74-0	177.243	cry (bz-peth)	74.5	161[3]			i H$_2$O; s EtOH, eth
400	1-(4-Aminophenyl)-1-propanone	p-Aminopropiophenone	C$_9$H$_{11}$NO	70-69-9	149.189	pl (al, w), nd (w)	140				s DMSO
401	N-[(4-Aminophenyl)sulfonyl]-acetamide	Sulfacetamide	C$_8$H$_{10}$N$_2$O$_3$S	144-80-9	214.241		182.0(0.4)				sl H$_2$O; s EtOH; i eth; vs ace, alk
402	5-[(4-Aminophenyl)sulfonyl]-2-thiazolamine	Thiazolsulfone	C$_9$H$_9$N$_3$O$_2$S$_2$	473-30-3	255.316	nd (al)	220 dec				vs ace, eth, EtOH, diox
403	4-Aminophthalimide	5-Amino-1H-isoindole-1,3(2H)-dione	C$_8$H$_6$N$_2$O$_2$	3676-85-5	162.146			224[0.5]			
404	3-Amino-1,2-propanediol, (±)-		C$_3$H$_9$NO$_2$	13552-31-3	91.109			265 dec	1.1752[20]	1.4910[25]	s H$_2$O, EtOH; i eth, bz
405	3-Aminopropanenitrile	3-Aminopropionitrile	C$_3$H$_6$N$_2$	151-18-8	70.093			185	0.9584[20]	1.4396[20]	
406	2-Amino-1-propanol, (±)-		C$_3$H$_9$NO	6168-72-5	75.109			174.5		1.4502[20]	vs H$_2$O, EtOH, eth; sl chl
407	3-Amino-1-propanol	Propanolamine	C$_3$H$_9$NO	156-87-6	75.109	liq	12.1(1)	185(4)	0.9824[26]	1.4617[20]	s H$_2$O, EtOH, eth
408	1-Amino-2-propanol	Isopropanolamine	C$_3$H$_9$NO	1674-56-2	75.109		1.7(0.2)	141(12)	0.9611[20]	1.4479[20]	msc H$_2$O, EtOH, eth, ace, bz, ctc
409	α-(1-Aminopropyl)-benzenemethanol	α-(α-Aminopropyl)benzyl alcohol	C$_{10}$H$_{15}$NO	5897-76-7	165.232	pl (bz-eth)	79.5				
410	N-(3-Aminopropyl)-N-methyl-1,3-propanediamine		C$_7$H$_{19}$N$_3$	105-83-9	145.246			232(3)	0.9023[20]	1.4705[25]	
411	Aminopropylon		C$_{16}$H$_{22}$N$_4$O$_2$	3690-04-8	302.372	pr (bz)	181				vs H$_2$O
412	4-(2-Aminopropyl)phenol, (±)-	Hydroxyamphetamine	C$_9$H$_{13}$NO	1518-86-1	151.205	cry (bz)	125.5				s H$_2$O, EtOH, bz, chl, AcOEt
413	N-(3-Aminopropyl)-1,3-propanediamine	Bis(3-aminopropyl)amine	C$_6$H$_{17}$N$_3$	56-18-8	131.219		-5.1(0.3)	151[50]	0.938[25]	1.4810[20]	s chl
414	Aminopterin		C$_{19}$H$_{20}$N$_8$O$_5$	54-62-6	440.413	ye cry	262 dec				
415	4-Amino-N-pyrazinylbenzenesulfonamide	Sulfapyrazine	C$_{10}$H$_{10}$N$_4$O$_2$S	116-44-9	250.277	nd (PhNO$_2$)	251				i H$_2$O, EtOH, eth, bz, chl; s py; sl ace
416	3-Amino-1H-pyrazole-4-carbonitrile	3-Amino-4-cyanopyrazole	C$_4$H$_4$N$_4$	16617-46-2	108.102	cry (w)	173				
417	2-Amino-3-pyridinecarboxylic acid		C$_6$H$_6$N$_2$O$_2$	5345-47-1	138.124		296 dec				sl H$_2$O
418	6-Amino-3-pyridinecarboxylic acid	6-Aminonicotinic acid	C$_6$H$_6$N$_2$O$_2$	3167-49-5	138.124	cry (dil HOAc, +2w)	312				
419	4-Amino-N-2-pyridinylbenzenesulfonamide	Sulfapyridine	C$_{11}$H$_{11}$N$_3$O$_2$S	144-83-2	249.289	ye oran (al)	190(2)				i H$_2$O, bz, ctc; s EtOH
420	5-Amino-2,4(1H,3H)-pyrimidinedione	5-Aminouracil	C$_4$H$_5$N$_3$O$_2$	932-52-5	127.102	nd (w)	dec				i H$_2$O; s alk, acid
421	6-Amino-2,4(1H,3H)-pyrimidinedione		C$_4$H$_5$N$_3$O$_2$	873-83-6	127.102	cry (w)	dec				vs H$_2$O
422	4-Amino-2(1H)-pyrimidinethione	2-Thiocytosine	C$_4$H$_5$N$_3$S	333-49-3	127.168						sl DMSO
423	5-Amino-2,4,6(1H,3H,5H)-pyrimidinetrione	Uramil	C$_4$H$_5$N$_3$O$_3$	118-78-5	143.101	nd or pl (w)	>400				s H$_2$O, chl; i eth, bz
424	4-Amino-N-2-pyrimidinyl-benzenesulfonamide	Sulfadiazine	C$_{10}$H$_{10}$N$_4$O$_2$S	68-35-9	250.277	cry (w), wh pow	261(3)				sl H$_2$O, EtOH, ace, DMSO

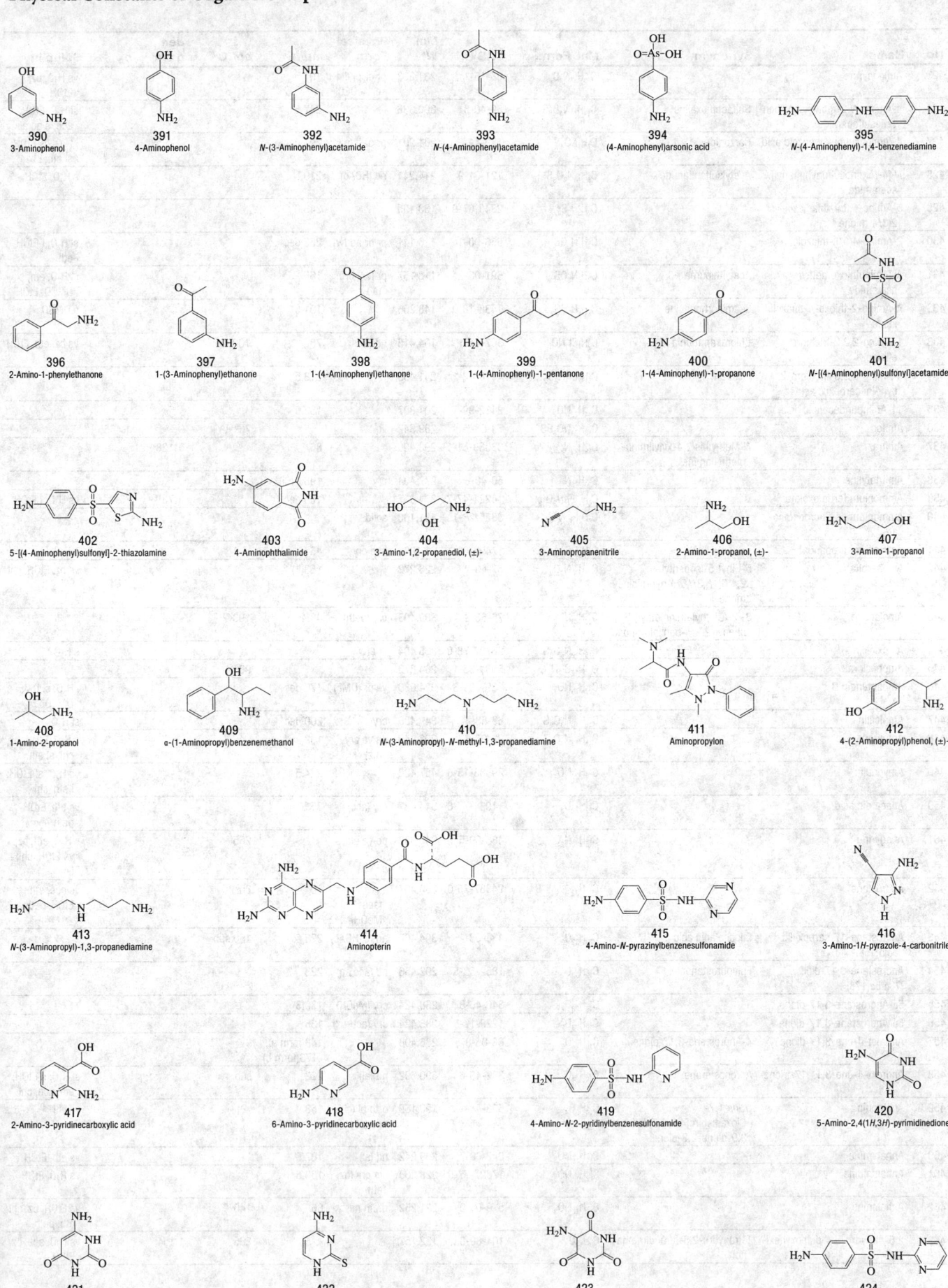

390
3-Aminophenol

391
4-Aminophenol

392
N-(3-Aminophenyl)acetamide

393
N-(4-Aminophenyl)acetamide

394
(4-Aminophenyl)arsonic acid

395
N-(4-Aminophenyl)-1,4-benzenediamine

396
2-Amino-1-phenylethanone

397
1-(3-Aminophenyl)ethanone

398
1-(4-Aminophenyl)ethanone

399
1-(4-Aminophenyl)-1-pentanone

400
1-(4-Aminophenyl)-1-propanone

401
N-[(4-Aminophenyl)sulfonyl]acetamide

402
5-[(4-Aminophenyl)sulfonyl]-2-thiazolamine

403
4-Aminophthalimide

404
3-Amino-1,2-propanediol, (±)-

405
3-Aminopropanenitrile

406
2-Amino-1-propanol, (±)-

407
3-Amino-1-propanol

408
1-Amino-2-propanol

409
α-(1-Aminopropyl)benzenemethanol

410
N-(3-Aminopropyl)-N-methyl-1,3-propanediamine

411
Aminopropylon

412
4-(2-Aminopropyl)phenol, (±)-

413
N-(3-Aminopropyl)-1,3-propanediamine

414
Aminopterin

415
4-Amino-N-pyrazinylbenzenesulfonamide

416
3-Amino-1H-pyrazole-4-carbonitrile

417
2-Amino-3-pyridinecarboxylic acid

418
6-Amino-3-pyridinecarboxylic acid

419
4-Amino-N-2-pyridinylbenzenesulfonamide

420
5-Amino-2,4(1H,3H)-pyrimidinedione

421
6-Amino-2,4(1H,3H)-pyrimidinedione

422
4-Amino-2(1H)-pyrimidinethione

423
5-Amino-2,4,6(1H,3H,5H)-pyrimidinetrione

424
4-Amino-N-2-pyrimidinylbenzenesulfonamide

No.	Name	Synonym	Mol. Form.	CAS RN	Mol. Wt.	Physical Form	mp/°C	bp/°C	den g cm⁻³	n_D	Solubility
425	Aminopyrine		$C_{13}H_{17}N_3O$	58-15-1	231.293	pr or pl (lig or AcOEt)	107.5				vs H_2O, bz, EtOH
426	4-Amino-N-2-quinoxalinylbenzenesulfonamide	Sulfaquinoxaline	$C_{14}H_{12}N_4O_2S$	59-40-5	300.336			247.5			sl H_2O, EtOH, ace; s aq alk
427	4-(Aminosulfonyl)benzoic acid	Carzenide	$C_7H_7NO_4S$	138-41-0	201.201	pr or lf (w)	291 dec				i H_2O; vs EtOH; sl eth; i bz
428	N-[4-(Aminosulfonyl)phenyl]-acetamide	Acetylsulfanilamide	$C_8H_{10}N_2O_3S$	121-61-9	214.241	nd (HOAc)	219.5				s H_2O, EtOH, ace
429	5-Amino-1,3,4-thiadiazole-2(3H)-thione		$C_2H_3N_3S_2$	2349-67-9	133.195		243.0				
430	2-Amino-4(5H)-thiazolone		$C_3H_4N_2OS$	556-90-1	116.141	pr or nd (w)	256 dec				sl H_2O; i EtOH, eth
431	N-(Aminothioxomethyl)-acetamide	Acetylthiourea	$C_3H_6N_2OS$	591-08-2	118.157	pr (w), orth (al)	165				sl H_2O, eth; s DMSO, EtOH
432	N-Amino-2-thioxo-4-thiazolidinone	3-Aminorhodanine	$C_3H_4N_2OS_2$	1438-16-0	148.206		101.5				s DMSO
433	1-Amino-2,2,2-trichloro-ethanol	Chloral ammonia	$C_2H_4Cl_3NO$	507-47-1	164.418	nd (al)	73	100 dec			vs bz, eth, EtOH
434	4-Amino-3,5,6-trichloro-2-pyridinecarboxlic acid	Picloram	$C_6H_3Cl_3N_2O_2$	1918-02-1	241.459		218.5				
435	11-Aminoundecanoic acid		$C_{11}H_{23}NO_2$	2432-99-7	201.307		189.0				
436	Amiton		$C_{10}H_{24}NO_3PS$	78-53-5	269.342	liq		76[0.01]		1.4655[27]	
437	Amitraz	N-Methylbis(2,4-xyliminomethyl)amine	$C_{19}H_{23}N_3$	33089-61-1	293.406		86		1.128[20]		
438	Amitriptyline		$C_{20}H_{23}N$	50-48-6	277.404	cry	196 (HCl)				
439	Ammonium ferric oxalate		$C_6H_{12}FeN_3O_{12}$	14221-47-7	374.017		165 dec		1.78[17.5]		vs H_2O; i EtOH
440	Ammonium perfluorooctanoate		$C_8H_4F_{15}NO_2$	3825-26-1	431.100	solid					
441	Ammonium propanoate		$C_3H_9NO_2$	17496-08-1	91.109	hyg cry	45				s H_2O
442	Amobarbital	5-Ethyl-5-isopentyl-2,4,6(1H,3H,5H)-pyrimidinetrione	$C_{11}H_{18}N_2O_3$	57-43-2	226.272		153(1)				vs bz, EtOH, chl
443	Amolanone	3-[2-(Diethylamino)ethyl]-3-phenyl-2(3H)-benzofuranone	$C_{20}H_{23}NO_2$	76-65-3	309.403	cry (peth)	43.4	193[2.0]		1.5614[25]	
444	Amoxicillin		$C_{16}H_{19}N_3O_5S$	26787-78-0	365.404	cry (w)					s H_2O
445	Amphecloral		$C_{11}H_{12}Cl_3N$	5581-35-1	264.579			96[0.5]	1.530		
446	Amphotericin B		$C_{47}H_{73}NO_{17}$	1397-89-3	924.080	ye pr (DMF)	170 dec				i H_2O; sl DMF; s DMSO
447	Ampicillin		$C_{16}H_{19}N_3O_4S$	69-53-4	349.405	cry	200 dec				sl H_2O
448	Ampyrone		$C_{11}H_{13}N_3O$	83-07-8	203.240	pa ye cry (bz)	109				s H_2O, EtOH, bz, chl; sl eth
449	Amygdalin		$C_{20}H_{27}NO_{11}$	29883-15-6	457.428		224.5				vs H_2O; sl EtOH; i eth, chl
450	Anacardic acid		$C_{22}H_{32}O_3$	11034-77-8	344.487	cry (ace)	35.5				vs eth, EtOH, peth
451	Anagyrine		$C_{15}H_{20}N_2O$	486-89-5	244.332	pe ye glass		265[12]			s H_2O, eth, bz; vs EtOH, chl; i lig
452	Androstane		$C_{19}H_{32}$	24887-75-0	260.457	lf (ace-MeOH)	50	60[0.003]			vs ace, eth, EtOH, peth
453	Androstane-17-carboxylic acid, (5β,17β)	Etiocholanic acid	$C_{20}H_{32}O_2$	438-08-4	304.467	nd (gl HOAc)	228.5	160 sub			
454	Androstane-3,17-diol, (3α,5α,17β)	Epiandrostanediol	$C_{19}H_{32}O_2$	1852-53-5	292.456	nd (ace aq)	223				
455	5α-Androstane-3,17-dione		$C_{19}H_{28}O_2$	846-46-8	288.424	cry (MeOH)	130(3)				
456	5β-Androstane-3,17-dione		$C_{19}H_{28}O_2$	1229-12-5	288.424	cry (ace-hx)	135				
457	Androst-4-ene-3,17-dione	4-Androstene-3,17-dione	$C_{19}H_{26}O_2$	63-05-8	286.408		143(form a); 173(form b)				
458	Androst-4-ene-3,11,17-trione	Adrenosterone	$C_{19}H_{24}O_3$	382-45-6	300.392	nd (al)	222	sub			sl H_2O; s EtOH, eth, ace, chl
459	Anemonin	trans-1,7-Dioxadispiro[4.0.4.2]dodeca-3,9-diene-2,8-dione	$C_{10}H_8O_4$	508-44-1	192.169	orth pl (chl) nd (al or bz)	158				vs chl
460	Anhalamine		$C_{11}H_{15}NO_3$	643-60-7	209.242	nd (al)	187.5				vs eth, EtOH
461	Anhalonidine		$C_{12}H_{17}NO_3$	17627-77-9	223.268	oct cry (bz, eth)	160.5				vs H_2O, EtOH
462	Anhalonine		$C_{12}H_{15}NO_3$	519-04-0	221.252	rhom nd	86	140[0.02]			vs EtOH, bz, chl, eth, peth
463	2,5-Anhydro-3,4-dideoxyhexitol	Tetrahydro-2,5-furandimethanol	$C_6H_{12}O_3$	104-80-3	132.157		<-50	265	1.154[20]		vs H_2O, ace, bz, EtOH

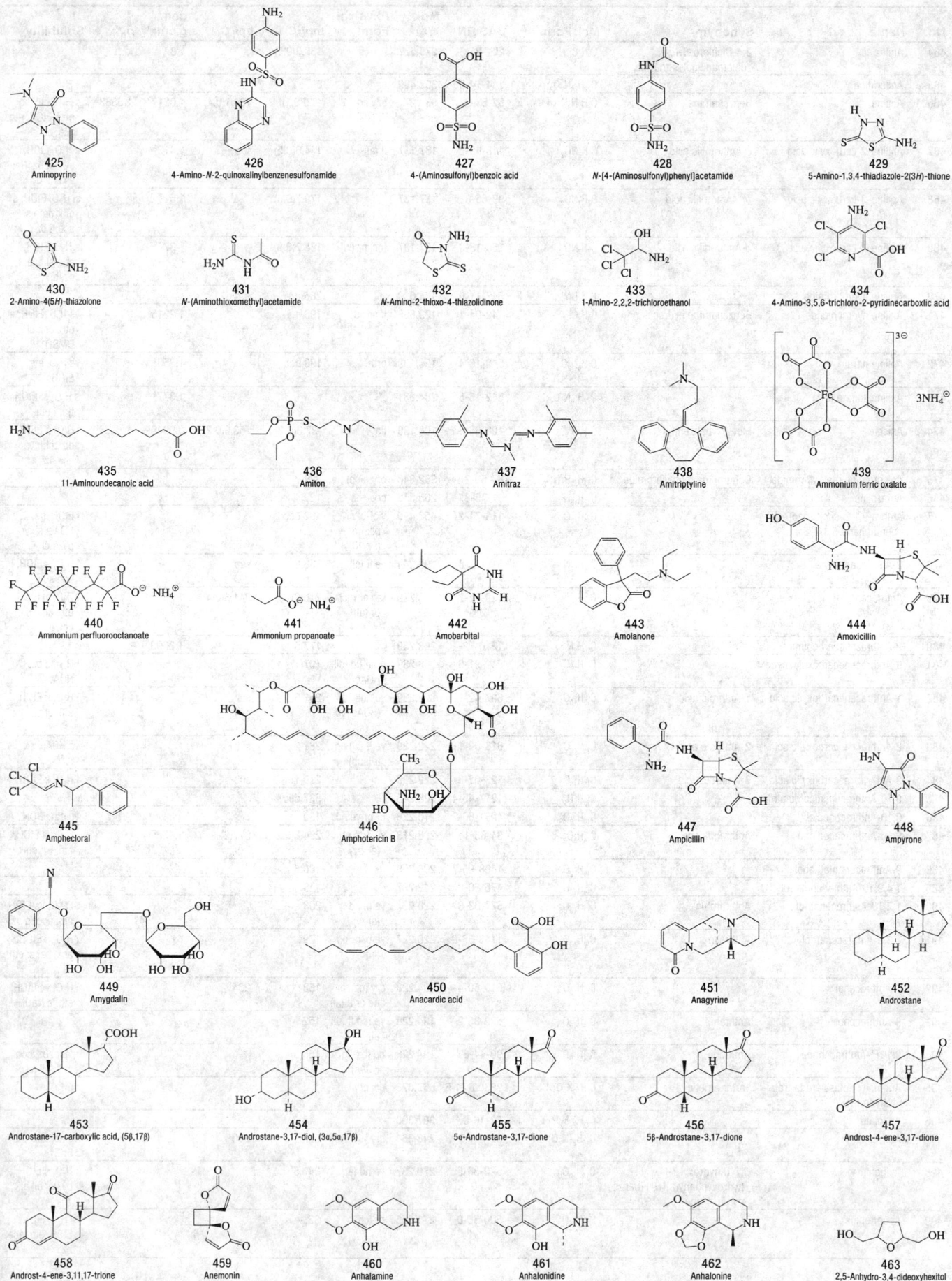

425 Aminopyrine

426 4-Amino-*N*-2-quinoxalinylbenzenesulfonamide

427 4-(Aminosulfonyl)benzoic acid

428 *N*-[4-(Aminosulfonyl)phenyl]acetamide

429 5-Amino-1,3,4-thiadiazole-2(3*H*)-thione

430 2-Amino-4(5*H*)-thiazolone

431 *N*-(Aminothioxomethyl)acetamide

432 *N*-Amino-2-thioxo-4-thiazolidinone

433 1-Amino-2,2,2-trichloroethanol

434 4-Amino-3,5,6-trichloro-2-pyridinecarboxlic acid

435 11-Aminoundecanoic acid

436 Amiton

437 Amitraz

438 Amitriptyline

439 Ammonium ferric oxalate

440 Ammonium perfluorooctanoate

441 Ammonium propanoate

442 Amobarbital

443 Amolanone

444 Amoxicillin

446 Amphotericin B

445 Amphecloral

447 Ampicillin

448 Ampyrone

449 Amygdalin

450 Anacardic acid

451 Anagyrine

452 Androstane

453 Androstane-17-carboxylic acid, (5β,17β)

454 Androstane-3,17-diol, (3α,5α,17β)

455 5α-Androstane-3,17-dione

456 5β-Androstane-3,17-dione

457 Androst-4-ene-3,17-dione

458 Androst-4-ene-3,11,17-trione

459 Anemonin

460 Anhalamine

461 Anhalonidine

462 Anhalonine

463 2,5-Anhydro-3,4-dideoxyhexitol

No.	Name	Synonym	Mol. Form.	CAS RN	Mol. Wt.	Physical Form	mp/°C	bp/°C	den g cm⁻³	n_D	Solubility
464	Anilazine	2,4-Dichloro-6-(o-chloroanilino)-s-triazine	$C_9H_5Cl_3N_4$	101-05-3	275.522		159.3(0.5)		1.8²⁰		
465	Anileridine		$C_{22}H_{28}N_2O_2$	144-14-9	352.469	cry	83				s H_2O
466	Aniline	Benzenamine	C_6H_7N	62-53-3	93.127	oily liq	-6.0(0.1)	184.1(0.4)	1.0217²⁰	1.5863²⁰	s H_2O, ctc, lig; msc EtOH, eth, ace, bz
467	Aniline-2-carboxylic acid	o-Anthranilic acid	$C_7H_7NO_2$	118-92-3	137.137	lf (al)	144.6(0.5)	sub	1.412²⁰		s H_2O, EtOH, eth; sl bz, tfa; vs chl, py
468	Aniline-3-carboxylic acid	m-Anthranilic acid	$C_7H_7NO_2$	99-05-8	137.137		179.7(0.6)		1.51²⁵		sl H_2O, EtOH; s eth, tfa; vs ace; i bz
469	Aniline-4-carboxylic acid	p-Anthranilic acid	$C_7H_7NO_2$	150-13-0	137.137	mcl pr (w)	188.2(0.6)		1.374²⁰		s H_2O, EtOH, eth; sl ace; i bz, chl
470	Aniline hydrobromide		C_6H_8BrN	542-11-0	174.039		286				
471	Aniline hydrochloride	Benzenamine hydrochloride	C_6H_8ClN	142-04-1	129.588	lf or nd	198		1.2215⁴		vs H_2O, EtOH; i eth, chl; sl DMSO
472	Aniline nitrate		$C_6H_8N_2O_3$	542-15-4	156.139	orth	190 dec		1.356⁴		vs H_2O, eth, EtOH
473	Aniline sulfate (2:1)		$C_{12}H_{16}N_2O_4S$	542-16-5	284.331				1.377⁴		s H_2O; sl EtOH, tfa; i eth
474	Anisole	Methoxybenzene	C_7H_8O	100-66-3	108.138	liq	-37.3(0.2)	153.6(0.2)	0.9940²⁰	1.5174²⁰	i H_2O; s EtOH, eth, chl; vs ace, bz
475	Anisotropine methylbromide	Octatropine methylbromide	$C_{17}H_{32}BrNO_2$	80-50-2	362.346	cry (ace)	329				
476	Antazoline		$C_{17}H_{19}N_3$	91-75-8	265.353	cry	122				
477	Anthra[9,1,2-cde]benzo[rst]pentaphene-5,10-dione		$C_{34}H_{16}O_2$	116-71-2	456.490	viol-bl or blk nd (PhNO₂)	492 dec				i EtOH, bz, HOAc; s xyl, py, sulf
478	2-Anthracenamine		$C_{14}H_{11}N$	613-13-8	193.244	ye lf (al)	238.8	sub			i H_2O; s EtOH; con sulf
479	Anthracene		$C_{14}H_{10}$	120-12-7	178.229	tab or mcl pr (al)	216(2)	341.3(0.4)	1.28²⁵		i H_2O; sl EtOH, eth, ace, bz, chl, ctc
480	9-Anthracenecarbonitrile		$C_{15}H_9N$	1210-12-4	203.239		177.5		1.3000²⁰		
481	9-Anthracenecarboxaldehyde		$C_{15}H_{10}O$	642-31-9	206.239	oran nd (dil HOAc)	107(1)				i H_2O; s bz, HOAc
482	1-Anthracenecarboxylic acid	1-Anthroic acid	$C_{15}H_{10}O_2$	607-42-1	222.239	ye nd (HOAc) ye pr (al)	251.5	sub			i H_2O; s EtOH, eth; sl bz, chl
483	2-Anthracenecarboxylic acid	2-Anthroic acid	$C_{15}H_{10}O_2$	613-08-1	222.239	ye lf (al) nd, lf (sub)	281	sub			vs HOAc
484	9-Anthracenecarboxylic acid	9-Anthroic acid	$C_{15}H_{10}O_2$	723-62-6	222.239		219(1)	sub			i H_2O; s EtOH
485	9,10-Anthracenedicarbonitrile		$C_{16}H_8N_2$	1217-45-4	228.248		337 dec				
486	9,10-Anthracenediol		$C_{14}H_{10}O_2$	4981-66-2	210.228	br or ye nd	180				vs eth, EtOH
487	9,10-Anthracenedione	Anthraquinone	$C_{14}H_8O_2$	84-65-1	208.213	ye orth nd (al, bz)	284.8(0.2)	377(2)	1.438²⁰		i H_2O; sl EtOH, eth, bz, chl
488	9-Anthracenemethanol		$C_{15}H_{12}O$	1468-95-7	208.255		160.5				
489	1,4,9,10-Anthracenetetrol		$C_{14}H_{10}O_4$	476-60-8	242.227		148				
490	1,2,10-Anthracenetriol	Anthrarobin	$C_{14}H_{10}O_3$	577-33-3	226.227	ye lf, nd (al-w)	208				sl H_2O; vs EtOH, eth, ace; s bz
491	1,8,9-Anthracenetriol	Anthralin	$C_{14}H_{10}O_3$	1143-38-0	226.227	ye pl or nd (lig)	179				i H_2O; s EtOH, ace, bz; sl eth; vs py
492	1-Anthracenol		$C_{14}H_{10}O$	610-50-4	194.228	cry (bz), br nd or lf (al)	158	234¹³			i H_2O; vs EtOH, eth; s NaOH
493	9-Anthracenol	Anthranol	$C_{14}H_{10}O$	529-86-2	194.228	ye red lf (dil al)	152				
494	9(10H)-Anthracenone	Anthrone	$C_{14}H_{10}O$	90-44-8	194.228	nd (bz-lig, HOAc)	155(3)				s ace, bz, con sulf, dil alk
495	Antimony potassium tartrate trihydrate	Tartar emetic	$C_8H_{10}K_2O_{15}Sb_2$	28300-74-5	667.873	col cry			2.6		sl H_2O
496	Apholate		$C_{12}H_{24}N_9P_3$	52-46-0	387.300		148				
497	Aphylline		$C_{15}H_{24}N_2O$	577-37-7	248.364	cry	52.5	200⁴			vs ace, bz, eth, EtOH
498	Apigenin	5,7-Dihydroxy-2-(4-hydroxyphenyl)-4H-1-benzo-pyran-4-one	$C_{15}H_{10}O_5$	520-36-5	270.237	ye nd (aq py)	347.5				i H_2O; s EtOH, py; vs dil alk
499	Apoatropine		$C_{17}H_{21}NO_2$	500-55-0	271.355	pr (chl)	62				sl H_2O, lig; vs EtOH, eth, ace, bz

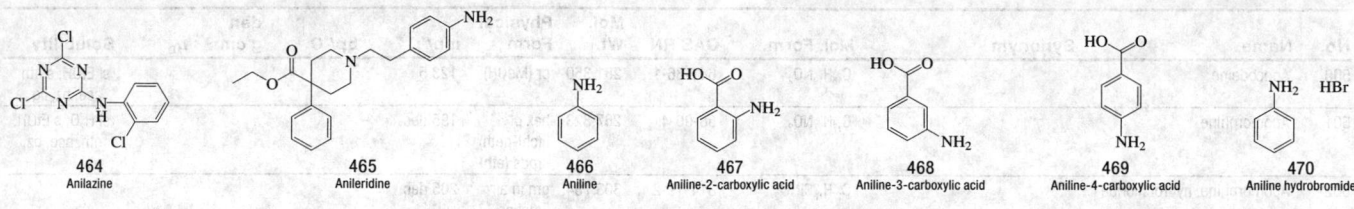

464 Anilazine

465 Anileridine

466 Aniline

467 Aniline-2-carboxylic acid

468 Aniline-3-carboxylic acid

469 Aniline-4-carboxylic acid

470 Aniline hydrobromide

471 Aniline hydrochloride

472 Aniline nitrate

473 Aniline sulfate (2:1)

474 Anisole

475 Anisotropine methylbromide

476 Antazoline

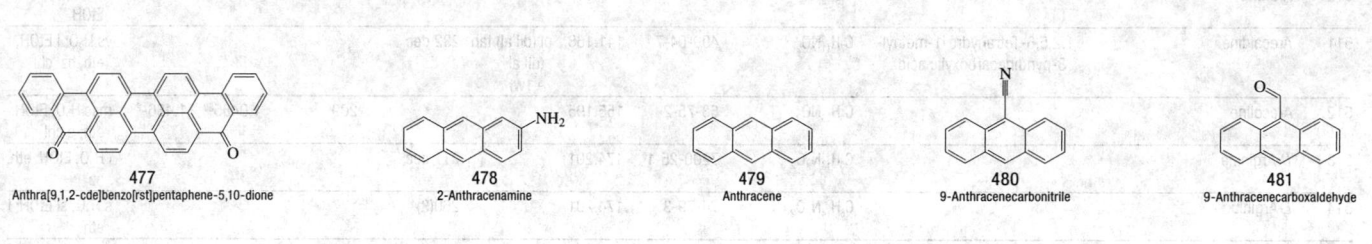

477 Anthra[9,1,2-cde]benzo[rst]pentaphene-5,10-dione

478 2-Anthracenamine

479 Anthracene

480 9-Anthracenecarbonitrile

481 9-Anthracenecarboxaldehyde

482 1-Anthracenecarboxylic acid

483 2-Anthracenecarboxylic acid

484 9-Anthracenecarboxylic acid

485 9,10-Anthracenedicarbonitrile

486 9,10-Anthracenediol

487 9,10-Anthracenedione

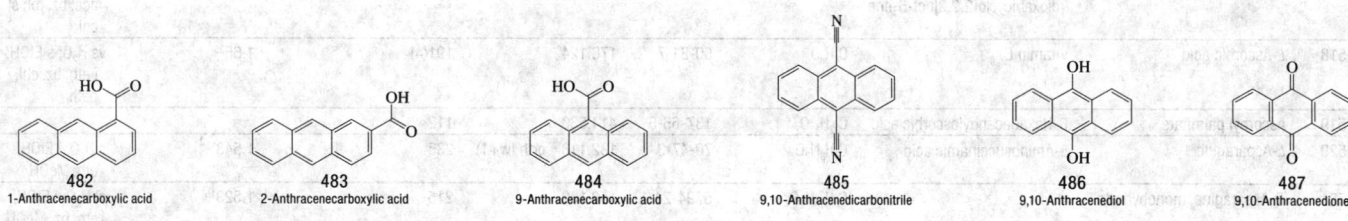

488 9-Anthracenemethanol

489 1,4,9,10-Anthracenetetrol

490 1,2,10-Anthracenetriol

491 1,8,9-Anthracenetriol

492 1-Anthracenol

493 9-Anthracenol

494 9(10H)-Anthracenone

495 Antimony potassium tartrate trihydrate

496 Apholate

497 Aphylline

498 Apigenin

499 Apoatropine

No.	Name	Synonym	Mol. Form.	CAS RN	Mol. Wt.	Physical Form	mp/°C	bp/°C	den g cm⁻³	n_D	Solubility
500	Apocodeine		$C_{18}H_{19}NO_2$	641-36-1	281.350	pr (MeOH)	123.5				sl EtOH; s eth, ace, bz, lig
501	Apomorphine		$C_{17}H_{17}NO_2$	58-00-4	267.323	hex pl (chl-peth) rods (eth)	195 dec				sl H_2O; s eth, ace, bz, alk
502	Apomorphine, hydrochloride		$C_{17}H_{18}ClNO_2$	314-19-2	303.784	grn in air mcl pr	205 dec				
503	Aprobarbital	5-Isopropyl-5-allyl-2,4,6(1H,3H,5H)-pyrimidine-trione	$C_{10}H_{14}N_2O_3$	77-02-1	210.229	cry	141				vs ace, eth, EtOH, chl
504	L-Arabinitol		$C_5H_{12}O_5$	7643-75-6	152.146		102.5				vs H_2O; sl EtOH; i eth
505	α-D-Arabinopyranose		$C_5H_{10}O_5$	608-45-7	150.130	cry (MeOH)	155.5		1.585[25]		
506	6-O-α-L-Arabinopyranosyl-D-Glucose	Vicianose	$C_{11}H_{20}O_{10}$	14116-69-9	312.271	nd (dil al)	210 dec				vs H_2O
507	DL-Arabinose		$C_5H_{10}O_5$	20235-19-2	150.130	pr, nd (al)	164.5		1.585[20]		vs H_2O; sl EtOH; i eth, bz
508	α-D-Arabinose		$C_5H_{10}O_5$	31178-68-4	150.130		156		1.585[25]		vs H_2O; sl EtOH; i eth, ace, MeOH
509	β-D-Arabinose		$C_5H_{10}O_5$	31178-69-5	150.130		156		1.625[25]		vs H_2O; sl EtOH; i eth, ace, MeOH
510	Aramite		$C_{15}H_{23}ClO_4S$	140-57-8	334.860		-37.3	195[2]	1.143[20]	1.5100[20]	vs ace, bz, eth, EtOH
511	Arecaidine	1,2,5,6-Tetrahydro-1-methyl-3-pyridinecarboxylic acid	$C_7H_{11}NO_2$	499-04-7	141.168	pl (dil al) tab (dil al +1w)	232 dec				vs H_2O; i EtOH, eth, bz, chl
512	Arecoline		$C_8H_{13}NO_2$	63-75-2	155.195			209	1.0485[20]	1.486-[20]	msc H_2O, EtOH, eth; s chl
513	D-Arginine		$C_6H_{14}N_4O_2$	7200-25-1	174.201		217 dec				i H_2O, EtOH, eth, bz
514	L-Arginine		$C_6H_{14}N_4O_2$	74-79-3	174.201		260(3)				s H_2O; sl EtOH; i eth
515	L-Arginine, monohydrochloride		$C_6H_{15}ClN_4O_2$	1119-34-2	210.662		219				
516	Artemisin	8-Hydroxysantonin	$C_{15}H_{18}O_4$	481-05-0	262.302	cry	203	260[0.1]			sl H_2O, chl; s AcOEt; i peth
517	Ascaridole	1-Methyl-4-isopropyl-2,3-dioxabicyclo[2.2.2]oct-5-ene	$C_{10}H_{16}O_2$	512-85-6	168.233	liq	3.3	exp	1.0103[20]	1.4769[20]	i H_2O; s EtOH, ace, bz, tol; sl chl
518	L-Ascorbic acid	Vitamin C	$C_6H_8O_6$	50-81-7	176.124		191(4)		1.65[25]		vs H_2O; s EtOH; i eth, bz, chl, peth
519	Ascorbyl palmitate	6-Hexadecanoylascorbic acid	$C_{22}H_{38}O_7$	137-66-6	414.533		112				
520	L-Asparagine	α-Aminosuccinamic acid	$C_4H_8N_2O_3$	70-47-3	132.118	orth (w+1)	235		1.543[15]		s H_2O; i EtOH, eth, MeOH
521	D-Asparagine, monohydrate		$C_4H_{10}N_2O_4$	5794-24-1	150.133		215		1.523[15]		sl H_2O; i EtOH, eth, bz, MeOH
522	L-Asparagine, monohydrate		$C_4H_{10}N_2O_4$	5794-13-8	150.133		234		1.543[15]		sl H_2O; i EtOH, eth, bz, MeOH
523	Aspartame	L-α-Aspartyl-L-phenylalanine, 2-methyl ester	$C_{14}H_{18}N_2O_5$	22839-47-0	294.303	nd (w)	246.5				
524	DL-Aspartic acid		$C_4H_7NO_4$	617-45-8	133.104	mcl pr (w)	277.5		1.6622[13]		sl H_2O; i EtOH, eth, bz, py
525	L-Aspartic acid	L-Aminosuccinic acid	$C_4H_7NO_4$	56-84-8	133.104	orth lf (w)	270		1.6603[13]		sl H_2O; i EtOH, eth, bz; s dil HCl, py
526	Aspergillic acid		$C_{12}H_{20}N_2O_2$	490-02-8	224.299	pa ye rods	98				vs bz, eth, EtOH
527	Astemizole		$C_{28}H_{31}FN_4O$	68844-77-9	458.570	wh cry	149.1				i H_2O; s os
528	Asulam	Methyl [(4-aminophenyl)-sulfonyl]carbamate	$C_8H_{10}N_2O_4S$	3337-71-1	230.241		144.2(0.5)				
529	Atenolol		$C_{14}H_{22}N_2O_3$	29122-68-7	266.336	cry (AcOEt)	147				sl H_2O, diox, ace; i chl; s MeOH, HOAc
530	Atisine	Anthorine	$C_{22}H_{33}NO_2$	466-43-3	343.503	orth bipym	58.5				vs eth, EtOH, chl
531	Atrazine		$C_8H_{14}ClN_5$	1912-24-9	215.684		177.0(0.5)				
532	Atropine		$C_{17}H_{23}NO_3$	51-55-8	289.370	orth nd (dil al)	118.5	95 sub			vs H_2O, EtOH; i eth; sl chl
533	Auramine hydrochloride		$C_{17}H_{22}ClN_3$	2465-27-2	303.83	ye nd (w)	267				sl H_2O
534	Aureothin		$C_{22}H_{23}NO_6$	2825-00-5	397.421	ye pr	158				vs ace, EtOH, chl

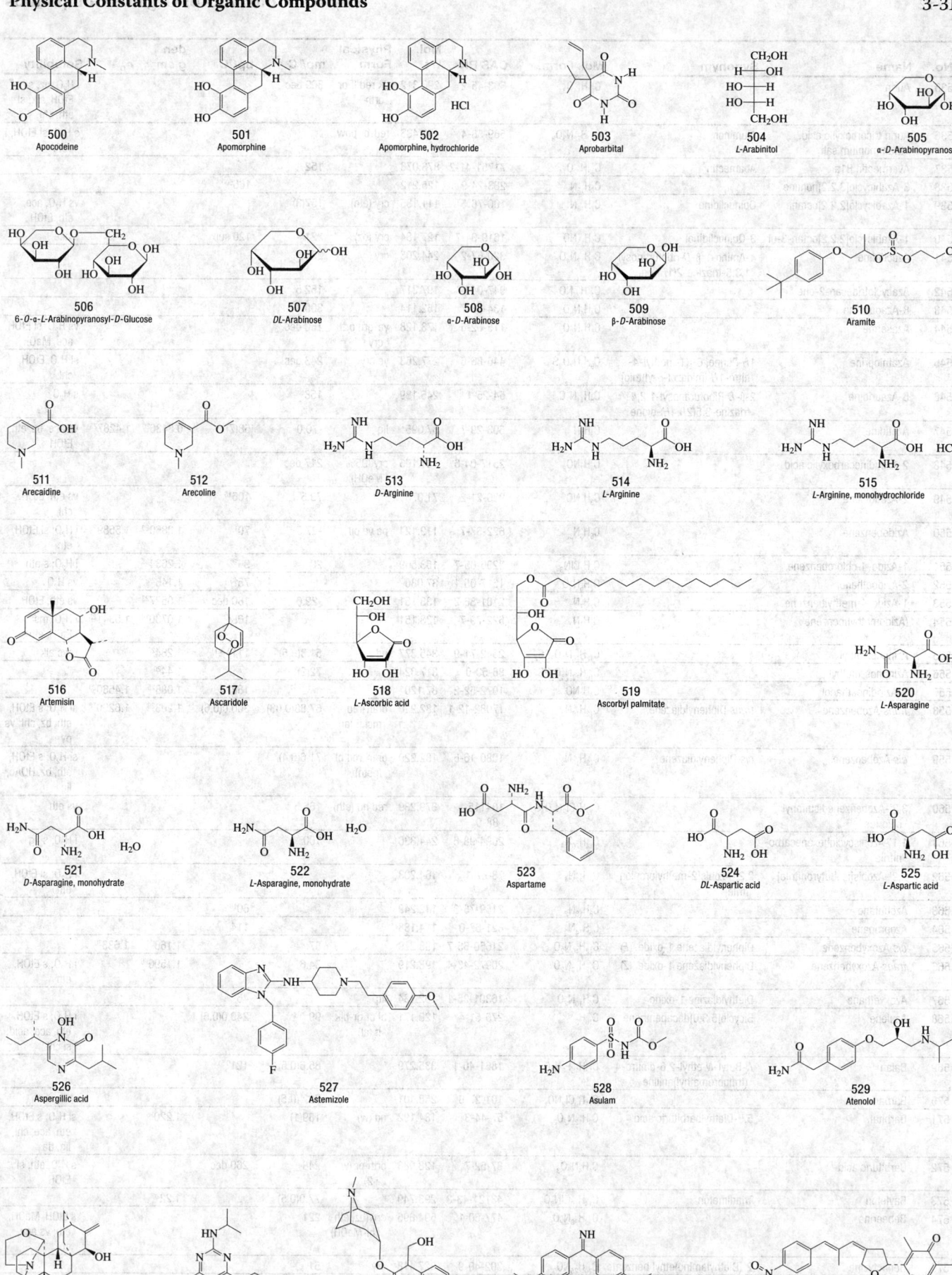

500
Apocodeine

501
Apomorphine

502
Apomorphine, hydrochloride

503
Aprobarbital

504
L-Arabinitol

505
α-D-Arabinopyranose

506
6-O-α-L-Arabinopyranosyl-D-Glucose

507
DL-Arabinose

508
α-D-Arabinose

509
β-D-Arabinose

510
Aramite

511
Arecaidine

512
Arecoline

513
D-Arginine

514
L-Arginine

515
L-Arginine, monohydrochloride

516
Artemisin

517
Ascaridole

518
L-Ascorbic acid

519
Ascorbyl palmitate

520
L-Asparagine

521
D-Asparagine, monohydrate

522
L-Asparagine, monohydrate

523
Aspartame

524
DL-Aspartic acid

525
L-Aspartic acid

526
Aspergillic acid

527
Astemizole

528
Asulam

529
Atenolol

530
Atisine

531
Atrazine

532
Atropine

533
Auramine hydrochloride

534
Aureothin

No.	Name	Synonym	Mol. Form.	CAS RN	Mol. Wt.	Physical Form	mp/°C	bp/°C	den g cm⁻³	n_D	Solubility
535	Aurin		$C_{19}H_{14}O_3$	603-45-2	290.312	dk red lf or orth	309 dec				i H_2O, bz; s EtOH, alk; sl eth, chl
536	Aurin tricarboxylic acid, triammonium salt	Aluminon	$C_{22}H_{23}N_3O_9$	569-58-4	473.433	red-br pow					s H_2O; sl EtOH; i peth
537	Avermectin B1a	Abamectin	$C_{48}H_{72}O_{14}$	71751-41-2	873.078		152				
538	3-Azabicyclo[3.2.2]nonane		$C_8H_{15}N$	283-24-9	125.212			166⁵⁰⁰			
539	1-Azabicyclo[2.2.2]octane	Quinuclidine	$C_7H_{13}N$	100-76-5	111.185	cry (eth)	157(3)				vs H_2O, ace, eth, EtOH
540	1-Azabicyclo[2.2.2]octan-3-ol	3-Quinuclidinol	$C_7H_{13}NO$	1619-34-7	127.184	cry (bz)	221	120 sub			s ace
541	Azacitidine	4-Amino-1-β-D-ribofuranosyl-1,3,5-triazine-2(1H)-one	$C_8H_{12}N_4O_5$	320-67-2	244.205	cry	229				
542	Azacyclotridecan-2-one		$C_{12}H_{23}NO$	947-04-6	197.317		152.5				
543	8-Azaguanine		$C_4H_4N_6O$	134-58-7	152.114		300				
544	Azaserine		$C_5H_7N_3O_4$	115-02-6	173.128	ye-grn orth cry	150 dec				vs H_2O; sl EtOH, ace, MeOH
545	Azathioprine	1H-Purine, 6-[(1-methyl-4-nitro-1H-imidazol-5-yl)thio]-	$C_9H_7N_7O_2S$	446-86-6	277.263	ye cry	243 dec				sl H_2O, EtOH, chl
546	6-Azauridine	2-β-D-Ribofuranosyl-1,2,4-triazine-3,5(2H,4H)-dione	$C_8H_{11}N_3O_6$	54-25-1	245.189		158				s H_2O
547	Azetidine		C_3H_7N	503-29-7	57.095	liq	-70.0	58(7)	0.8436²⁰	1.4287²⁵	vs ace, bz, eth, EtOH
548	2-Azetidinecarboxylic acid		$C_4H_7NO_2$	2517-04-6	101.105	cry (95% MeOH)	217 dec				
549	2-Azetidinone		C_3H_5NO	930-21-2	71.078		73.5	106¹⁵			vs eth, EtOH, chl
550	Azidobenzene		$C_6H_5N_3$	622-37-7	119.124	pa ye oil	-27.5	70¹¹	1.0860²⁰	1.5589²⁵	i H_2O; sl EtOH, eth
551	1-Azido-4-chlorobenzene		$C_6H_4ClN_3$	3296-05-7	153.569		20	96²⁰	1.2634²⁵		i H_2O; s eth
552	2-Azidoethanol		$C_2H_5N_3O$	1517-05-1	87.080			75⁴⁰	1.146²⁴		vs H_2O
553	1-Azido-4-methylbenzene		$C_7H_7N_3$	2101-86-2	133.151		-29.0	180 dec	1.0527²³		vs eth, EtOH
554	(Azidomethyl)benzene		$C_7H_7N_3$	622-79-7	133.151			108²³	1.0730¹⁹	1.5341²⁵	i H_2O; msc EtOH, eth
555	Azinphos ethyl		$C_{12}H_{16}N_3O_3PS_2$	2642-71-9	345.377	nd	51.3(0.5)	111⁰·⁰⁰¹	1.284²⁰		reac alk
556	Azinphos-methyl		$C_{10}H_{12}N_3O_3PS_2$	86-50-0	317.324		72(2)		1.44²⁰		
557	1-Aziridineethanol		C_4H_9NO	1072-52-2	87.120			168	1.088²⁵	1.4560²⁰	
558	trans-Azobenzene	trans-Diphenyldiazene	$C_{12}H_{10}N_2$	17082-12-1	182.220	oran-red mcl lf (al)	67.88(0.03)	300.0(0.6)	1.203²⁰	1.6266⁷⁸	sl H_2O; s EtOH, eth, bz, chl; vs py
559	cis-Azobenzene	cis-Diphenyldiazene	$C_{12}H_{10}N_2$	1080-16-6	182.220	oran-red pl (peth)	71.6(0.4)				sl H_2O; s EtOH, eth, bz, HOAc, lig
560	3,3'-Azobenzenedisulfonyl chloride		$C_{12}H_8Cl_2N_2O_4S_2$	104115-88-0	379.239	red nd (eth)	166.5				vs eth
561	1,1'-Azobiscyclohexanecarbo-nitrile		$C_{14}H_{20}N_4$	2094-98-6	244.336		100				i H_2O; s lig
562	2,2'-Azobis[isobutyronitrile]	2,2'-Azobis[2-methylpropion-itrile]	$C_8H_{12}N_4$	78-67-1	164.208						i H_2O; sl EtOH, eth
563	Azobutane		$C_8H_{18}N_2$	2159-75-3	142.242			60¹⁸			
564	Azopropane		$C_6H_{14}N_2$	821-67-0	114.188			114			
565	cis-Azoxybenzene	Diphenyldiazene 1-oxide, (E)	$C_{12}H_{10}N_2O$	21650-65-7	198.219		87		1.166²⁰	1.633²⁰	
566	trans-Azoxybenzene	Diphenyldiazene 1-oxide, (Z)	$C_{12}H_{10}N_2O$	20972-43-4	198.219		34.6		1.1590²⁶		i H_2O; s EtOH, eth
567	Azoxyethane	Diethyldiazine 1-oxide	$C_4H_{10}N_2O$	16301-26-1	102.134	liq		46			
568	Azulene	Bicyclo[5.3.0]decapentaene	$C_{10}H_8$	275-51-4	128.171	bl or gr-blk lf (al)	99	249.0(0.5)			i H_2O; s EtOH, eth, ace, acid; sl chl
569	Balan	N-Butyl-N-ethyl-2,6-dinitro-4-(trifluoromethyl)aniline	$C_{13}H_{16}F_3N_3O_4$	1861-40-1	335.279		65.6(0.5)	121⁰·⁵			
570	Barban		$C_{11}H_9Cl_2NO_2$	101-27-9	258.101		72.4(0.5)				
571	Barbital	5,5-Diethylbarbituric acid	$C_8H_{12}N_2O_3$	57-44-3	184.192	nd (w)	189(1)		1.220²⁵		sl H_2O; s EtOH, eth, ace, chl, lig, tfa
572	Barbituric acid		$C_4H_4N_2O_3$	67-52-7	128.086	orth pr (w +2)	248	260 dec			s H_2O, eth; sl EtOH
573	Bayleton	Triadimefon	$C_{14}H_{16}ClN_3O_2$	43121-43-3	293.749		77.0(0.5)		1.22²⁰		
574	Bebeerine		$C_{36}H_{38}N_2O_6$	477-60-1	594.696	cry (bz, eth, chl-MeOH)	221				s EtOH, MeOH, eth; vs ace, chl
575	Benactyzine	2-(Diethylamino)ethyl benzilate	$C_{20}H_{25}NO_3$	302-40-9	327.418	cry	51				
576	Benactyzine hydrochloride	2-Diethylaminoethyl benzilate hydrochloride	$C_{20}H_{26}ClNO_3$	57-37-4	363.878		177.5				s H_2O; i eth

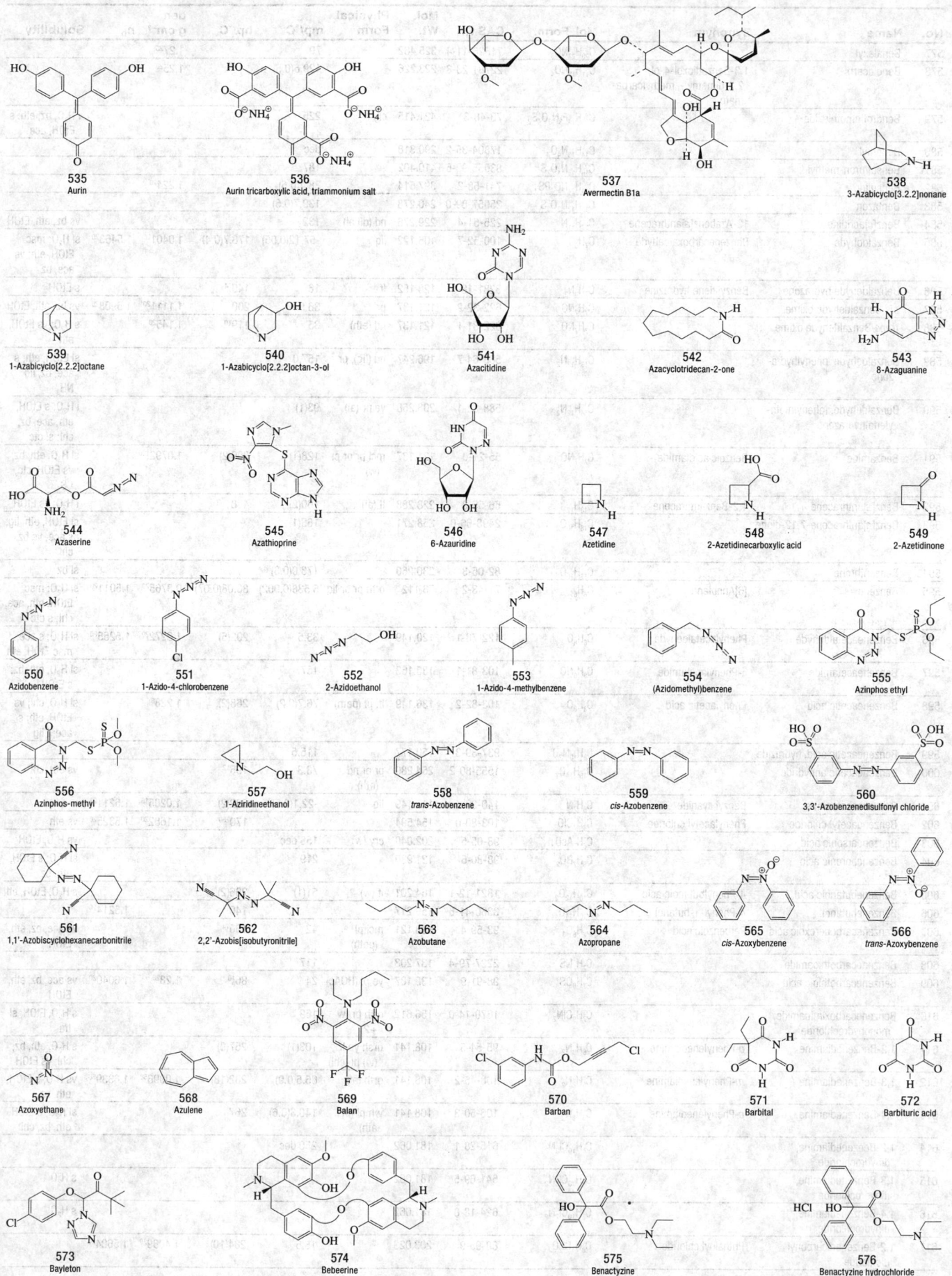

535
Aurin

536
Aurin tricarboxylic acid, triammonium salt

537
Avermectin B1a

538
3-Azabicyclo[3.2.2]nonane

539
1-Azabicyclo[2.2.2]octane

540
1-Azabicyclo[2.2.2]octan-3-ol

541
Azacitidine

542
Azacyclotridecan-2-one

543
8-Azaguanine

544
Azaserine

545
Azathioprine

546
6-Azauridine

547
Azetidine

548
2-Azetidinecarboxylic acid

549
2-Azetidinone

550
Azidobenzene

551
1-Azido-4-chlorobenzene

552
2-Azidoethanol

553
1-Azido-4-methylbenzene

554
(Azidomethyl)benzene

555
Azinphos ethyl

556
Azinphos-methyl

557
1-Aziridineethanol

558
trans-Azobenzene

559
cis-Azobenzene

560
3,3'-Azobenzenedisulfonyl chloride

561
1,1'-Azobiscyclohexanecarbonitrile

562
2,2'-Azobis[isobutyronitrile]

563
Azobutane

564
Azopropane

565
cis-Azoxybenzene

566
trans-Azoxybenzene

567
Azoxyethane

568
Azulene

569
Balan

570
Barban

571
Barbital

572
Barbituric acid

573
Bayleton

574
Bebeerine

575
Benactyzine

576
Benactyzine hydrochloride

No.	Name	Synonym	Mol. Form.	CAS RN	Mol. Wt.	Physical Form	mp/°C	bp/°C	den g cm⁻³	n_D	Solubility
577	Benalaxyl		$C_{20}H_{23}NO_3$	71626-11-4	325.402		79		1.27^{25}		
578	Bendiocarb	1,3-Benzodioxol-4-ol, 2,2-dimethyl-, methylcarbamate	$C_{11}H_{13}NO_4$	22781-23-3	223.226		129.6(0.5)		1.25^{20}		
579	Bendroflumethiazide		$C_{15}H_{14}F_3N_3O_4S_2$	73-48-3	421.415	cry	225				i H$_2$O, bz, eth; s EtOH, ace
580	Benomyl		$C_{14}H_{18}N_4O_3$	17804-35-2	290.318		dec				
581	Bensulfuron-methyl		$C_{16}H_{18}N_4O_7S$	83055-99-6	410.402		187				
582	Bensulide		$C_{14}H_{24}NO_4PS_3$	741-58-2	397.514		38.4(0.5)		1.224^{20}		
583	Bentazon		$C_{10}H_{12}N_2O_3S$	25057-89-0	240.278		139.7(0.5)				
584	Benz[c]acridine	12-Azabenz[a]anthracene	$C_{17}H_{11}N$	225-51-4	229.276	nd (dil al)	132				vs bz, eth, EtOH
585	Benzaldehyde	Benzenecarboxaldehyde	C_7H_6O	100-52-7	106.122	liq	-57.12(0.05)	178.7(0.4)	1.0401^{25}	1.5463^{20}	sl H$_2$O; msc EtOH, eth; vs ace, bz
586	Benzaldehyde hydrazone	Benzylidene hydrazine	$C_7H_8N_2$	5281-18-5	120.152	lf	16	140^{14}			s EtOH
587	cis-Benzaldehyde oxime		C_7H_7NO	622-32-2	121.137	pr	36.5	200	1.1111^{20}	1.5908^{20}	vs bz, eth, EtOH
588	trans-Benzaldehyde oxime		C_7H_7NO	622-31-1	121.137	nd (eth)	35	119^{10}	1.145^{20}		s H$_2$O; vs EtOH, eth
589	Benzaldehyde, phenylhydrazone		$C_{13}H_{12}N_2$	588-64-7	196.247	nd (lig), pr	157.0				sl EtOH, eth; s ace, bz, liq NH$_3$
590	Benzaldehyde, (phenylmethylene)hydrazone		$C_{14}H_{12}N_2$	588-68-1	208.258	ye pr (al)	93(1)				i H$_2$O; s EtOH, eth, ace, bz, chl; sl ctc
591	Benzamide	Benzoic acid amide	C_7H_7NO	55-21-0	121.137	mcl pr or pl (w)	128(1)	306(2)	1.0792^{130}		sl H$_2$O, eth, bz; vs EtOH, ctc, CS$_2$
592	Benz[a]anthracene	1,2-Benzanthracene	$C_{18}H_{12}$	56-55-3	228.288	lf (al)	160(2)	438			i H$_2$O; vs EtOH
593	Benz[a]anthracene-7,12-dione		$C_{18}H_{10}O_2$	2498-66-0	258.271		168(1)				sl EtOH, eth, lig; s ace; vs bz, chl
594	Benzanthrone		$C_{17}H_{10}O$	82-05-3	230.260		173.0(0.3)				sl bz
595	Benzene	[6]Annulene	C_6H_6	71-43-2	78.112	orth pr or liq	5.538(0.002)	80.08(0.07)	0.8765^{20}	1.5011^{20}	sl H$_2$O; msc EtOH, eth, ace, chl; s ctc
596	Benzeneacetaldehyde	Phenylacetaldehyde	C_8H_8O	122-78-1	120.149		33.5	202(5)	1.0272^{20}	1.5255^{20}	sl H$_2$O; s ace; msc EtOH, eth
597	Benzeneacetamide	α-Phenylacetamide	C_8H_9NO	103-81-1	135.163		157				sl H$_2$O, eth, bz; s EtOH
598	Benzeneacetic acid	Phenylacetic acid	$C_8H_8O_2$	103-82-2	136.149	lf, pl (peth)	76.7(0.2)	268(2)	1.228^6		sl H$_2$O, chl; vs EtOH, eth; s ace; i lig
599	Benzeneacetic acid, hydrazide		$C_8H_{10}N_2O$	937-39-3	150.177		115.5				
600	Benzeneacetic anhydride		$C_{16}H_{14}O_3$	1555-80-2	254.280	pr or nd (eth)	73.3	195^{12}			vs eth, chl
601	Benzeneacetonitrile	Benzyl cyanide	C_8H_7N	140-29-4	117.149	liq	-22.1(0.5)	232(2)	1.0205^{15}	1.5211^{25}	
602	Benzeneacetyl chloride	Phenylacetyl chloride	C_8H_7ClO	103-80-0	154.594			170^{250}	1.1682^{20}	1.5325^{20}	vs eth
603	Benzenearsonic acid		$C_6H_7AsO_3$	98-05-5	202.040	cry (w)	158 dec				vs H$_2$O, EtOH
604	Benzeneboronic acid		$C_6H_7BO_2$	98-80-6	121.930		219				sl H$_2$O; s EtOH, eth, bz
605	Benzenebutanoic acid	4-Phenylbutanoic acid	$C_{10}H_{12}O_2$	1821-12-1	164.201	lf (w)	51(1)	296(2)			s H$_2$O, EtOH, eth
606	Benzenebutanol	4-Phenyl-1-butanol	$C_{10}H_{14}O$	3360-41-6	150.217			140^{14}		1.5214^{20}	
607	Benzenecarboperoxoic acid	Perbenzoic acid	$C_7H_6O_3$	93-59-4	138.121	mcl pl (peth)	42	100^{14}			vs ace, bz, eth, EtOH
608	Benzenecarbothioamide		C_7H_7NS	2227-79-4	137.203		117				
609	Benzenecarbothioic acid		C_7H_6OS	98-91-9	138.187	ye pl (HOAc)	24	86^{10}	1.28^{20}	1.6040^{20}	vs ace, bz, eth, EtOH
610	Benzenecarboximidamide, monohydrochloride		$C_7H_9ClN_2$	1670-14-0	156.612	orth pr (w +2)	169				s H$_2$O, EtOH; sl tfa
611	1,2-Benzenediamine	o-Phenylenediamine	$C_6H_8N_2$	95-54-5	108.141	brsh ye lf (w) pl (chl)	103(1)	257(9)			s H$_2$O, eth, bz, chl; vs EtOH
612	1,3-Benzenediamine	m-Phenylenediamine	$C_6H_8N_2$	108-45-2	108.141	orth (al)	65.5(0.9)	282(18)	1.0096^{58}	1.6339^{58}	vs H$_2$O; s EtOH, eth, bz
613	1,4-Benzenediamine	p-Phenylenediamine	$C_6H_8N_2$	106-50-3	108.141	wh pl (bz, eth)	140.3(0.6)	267			sl H$_2$O; s EtOH, eth, bz, chl
614	1,2-Benzenediamine, dihydrochloride		$C_6H_{10}Cl_2N_2$	615-28-1	181.062		250 dec				
615	1,3-Benzenediamine, dihydrochloride		$C_6H_{10}Cl_2N_2$	541-69-5	181.062						s H$_2$O
616	1,4-Benzenediamine, dihydrochloride		$C_6H_{10}Cl_2N_2$	624-18-0	181.062						s H$_2$O
617	1,2-Benzenedicarbonyl dichloride	Phthaloyl chloride	$C_8H_4Cl_2O_2$	88-95-9	203.023		15.5	284(10)	1.4089^{20}	1.5684^{20}	

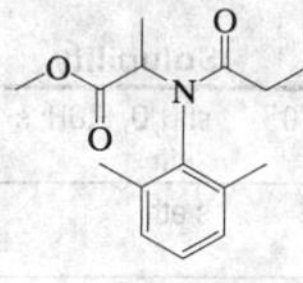

577
Benalaxyl

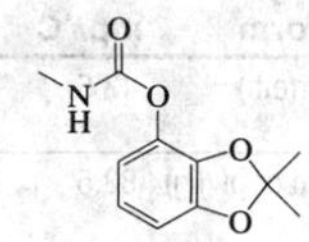

578
Bendiocarb

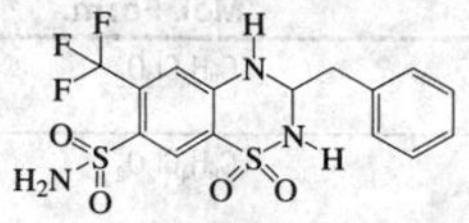

579
Bendroflumethiazide

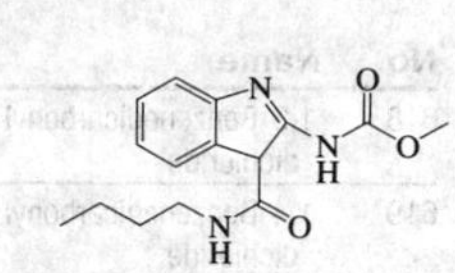

580
Benomyl

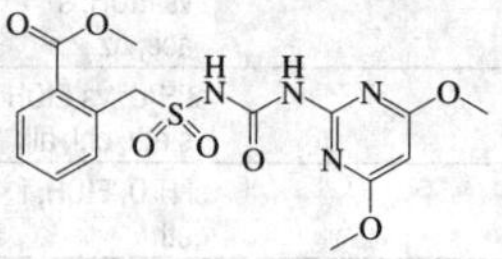

581
Bensulfuron-methyl

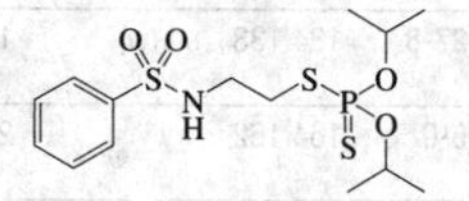

582
Bensulide

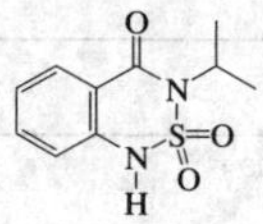

583
Bentazon

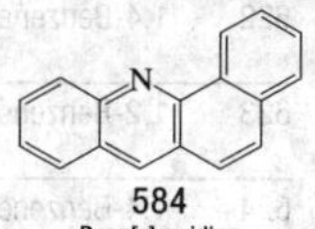

584
Benz[c]acridine

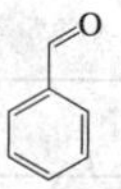

585
Benzaldehyde

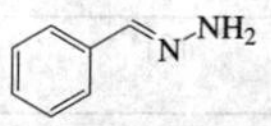

586
Benzaldehyde hydrazone

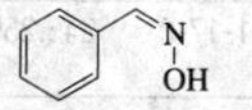

587
cis-Benzaldehyde oxime

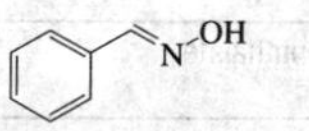

588
trans-Benzaldehyde oxime

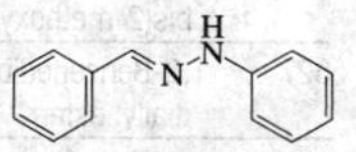

589
Benzaldehyde, phenylhydrazone

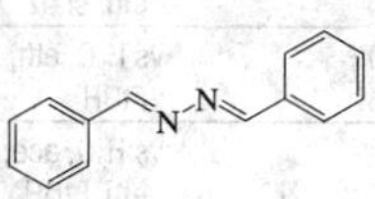

590
Benzaldehyde, (phenylmethylene)hydrazone

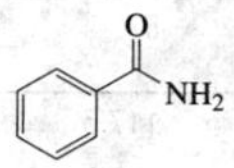

591
Benzamide

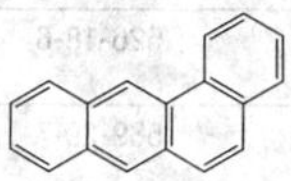

592
Benz[a]anthracene

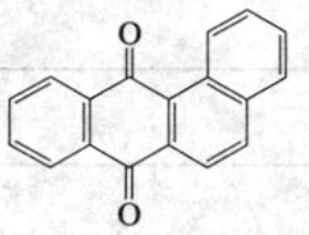

593
Benz[a]anthracene-7,12-dione

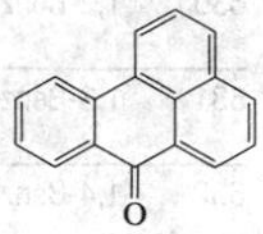

594
Benzanthrone

595
Benzene

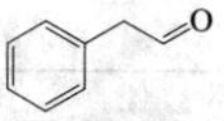

596
Benzeneacetaldehyde

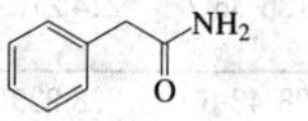

597
Benzeneacetamide

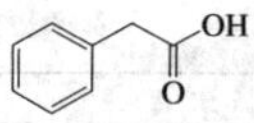

598
Benzeneacetic acid

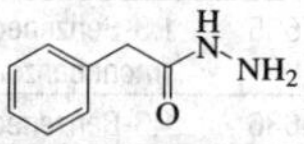

599
Benzeneacetic acid, hydrazide

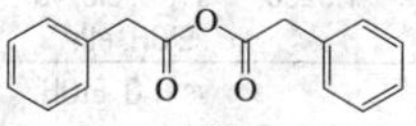

600
Benzeneacetic anhydride

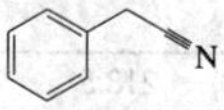

601
Benzeneacetonitrile

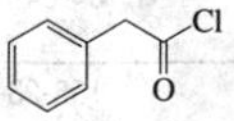

602
Benzeneacetyl chloride

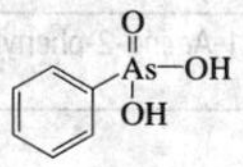

603
Benzenearsonic acid

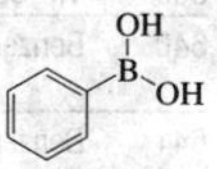

604
Benzeneboronic acid

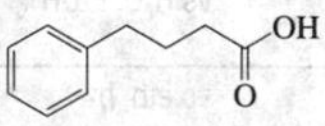

605
Benzenebutanoic acid

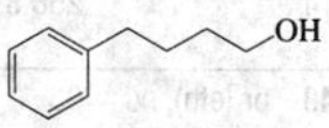

606
Benzenebutanol

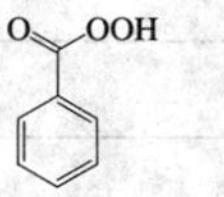

607
Benzenecarboperoxoic acid

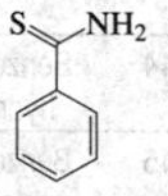

608
Benzenecarbothioamide

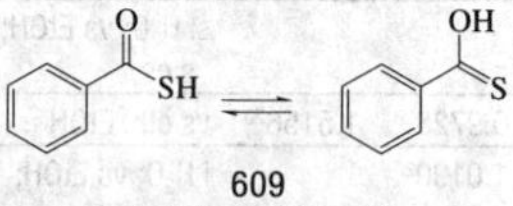

609
Benzenecarbothioic acid

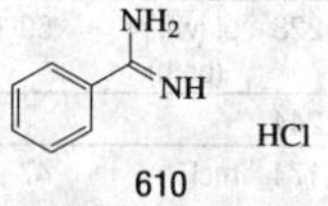

610
Benzenecarboximidamide, monohydrochloride

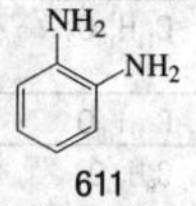

611
1,2-Benzenediamine

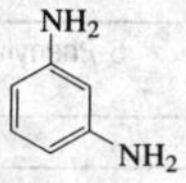

612
1,3-Benzenediamine

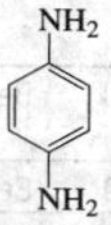

613
1,4-Benzenediamine

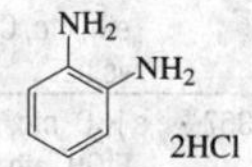

614
1,2-Benzenediamine, dihydrochloride

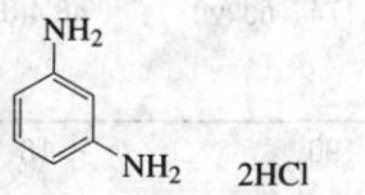

615
1,3-Benzenediamine, dihydrochloride

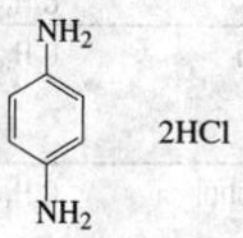

616
1,4-Benzenediamine, dihydrochloride

617
1,2-Benzenedicarbonyl dichloride

No.	Name	Synonym	Mol. Form.	CAS RN	Mol. Wt.	Physical Form	mp/°C	bp/°C	den g cm⁻³	n_D	Solubility
618	1,3-Benzenedicarbonyl dichloride		$C_8H_4Cl_2O_2$	99-63-8	203.023	pr(eth)	43.5	276	1.3880[17]	1.570[47]	sl H₂O, EtOH; s eth
619	1,4-Benzenedicarbonyl dichloride		$C_8H_4Cl_2O_2$	100-20-9	203.023	nd or pl (lig)	83.5	258			s eth
620	1,2-Benzenedicarboxaldehyde		$C_8H_6O_2$	643-79-8	134.133	ye cry or nd (lig)	55.8	83[0.8]			vs eth, EtOH
621	1,3-Benzenedicarboxaldehyde		$C_8H_6O_2$	626-19-7	134.133	nd (dil al)	89(2)	246			sl H₂O, eth, chl; vs EtOH; s ace, bz
622	1,4-Benzenedicarboxaldehyde		$C_8H_6O_2$	623-27-8	134.133	nd (w)	117	246			sl H₂O; vs EtOH; s eth, chl, alk
623	1,2-Benzenedicarboxamide	Phthalamide	$C_8H_8N_2O_2$	88-96-0	164.162	cry	222	dec			sl H₂O, EtOH; i eth
624	1,4-Benzenedicarboxamide		$C_8H_8N_2O_2$	3010-82-0	164.162	nd (w), pl (HOAc)	322.3				
625	1,2-Benzenedicarboxylic acid, bis(2-butoxyethyl) ester	Bis(2-butoxyethyl) phthalate	$C_{20}H_{30}O_6$	117-83-9	366.448			409(18)			
626	1,2-Benzenedicarboxylic acid, bis(2-methoxyethyl) ester	Bis(2-methoxyethyl) phthalate	$C_{14}H_{18}O_6$	117-82-8	282.289		-60.0	230[10]	1.1596[20]		
627	1,2-Benzenedicarboxylic acid, diallyl ester	Diallyl phthalate	$C_{14}H_{14}O_4$	131-17-9	246.259			161[4]			
628	1,2-Benzenedicarboxylic acid, dipropyl ester	Dipropyl phthalate	$C_{14}H_{18}O_4$	131-16-8	250.291	liq	-31.0	319(2)	1.0767[20]		i H₂O; s EtOH, eth
629	1,3-Benzenedimethanamine	m-Xylene diamine	$C_8H_{12}N_2$	1477-55-0	136.194			247	1.052[20]		vs H₂O, eth, EtOH
630	1,2-Benzenedimethanol		$C_8H_{10}O_2$	612-14-6	138.164	pl (eth, peth)	66(3)	145[3]			s H₂O, EtOH; vs eth; sl bz
631	1,3-Benzenedimethanol		$C_8H_{10}O_2$	626-18-6	138.164	nd (bz)	57	156[13]	1.1610[18]		vs H₂O, eth, EtOH
632	1,4-Benzenedimethanol		$C_8H_{10}O_2$	589-29-7	138.164	nd (w)	117.5	140[1]			vs H₂O, ace, eth, EtOH
633	1,2-Benzenediol, diacetate		$C_{10}H_{10}O_4$	635-67-6	194.184	nd (al)	64.5	142[9]			i H₂O; vs EtOH, eth, chl; s peth
634	1,4-Benzenediol, diacetate		$C_{10}H_{10}O_4$	1205-91-0	194.184	pl (w, al)	123.5		0.8731[25]		s H₂O; vs EtOH, eth, chl, lig
635	1,3-Benzenediol, monobenzoate		$C_{13}H_{10}O_3$	136-36-7	214.216		134.5				
636	1,3-Benzenedisulfonic acid		$C_6H_6O_6S_2$	98-48-6	238.238	hyg cry					
637	1,3-Benzenedisulfonyl dichloride		$C_6H_4Cl_2O_4S_2$	585-47-7	275.130		61.8	195[10.5]			
638	1,2-Benzenedithiol		$C_6H_6S_2$	17534-15-5	142.242		28.5	238.5			vs EtOH, eth, bz; s AcOEt
639	1,3-Benzenedithiol		$C_6H_6S_2$	626-04-0	142.242	lf	27	245			vs bz, eth, EtOH
640	Benzeneethanamine	1-Amino-2-phenylethane	$C_8H_{11}N$	64-04-0	121.180	liq	<0	204(4)	0.9640[25]	1.5290[25]	s H₂O, ctc; vs EtOH, eth
641	Benzeneethanamine, hydrochloride		$C_8H_{12}ClN$	156-28-5	157.641	pl or lf (al)	218.5				vs H₂O, EtOH
642	Benzeneethanol	Phenethyl alcohol	$C_8H_{10}O$	60-12-8	122.164	liq	-19(2)	220(3)	1.0202[20]	1.5325[20]	sl H₂O; msc EtOH, eth
643	Benzenehexacarboxylic acid	Mellitic acid	$C_{12}H_6O_{12}$	517-60-2	342.169	nd (al)	287 dec				vs H₂O; s EtOH, sulf
644	Benzenemethanamine, hydrochloride		$C_7H_{10}ClN$	3287-99-8	143.614		258.3				vs H₂O, EtOH
645	Benzenemethanesulfonyl chloride		$C_7H_7ClO_2S$	1939-99-7	190.648	pr (eth), nd (bz)	93				vs eth, bz
646	Benzenemethanesulfonyl fluoride		$C_7H_7FO_2S$	329-98-6	174.193		92.0				
647	Benzenemethanethiol	Thiobenzyl alcohol	C_7H_8S	100-53-8	124.204	liq	-30	199(1)	1.058[20]	1.5151[20]	i H₂O; vs EtOH, eth; sl ctc; s CS₂
648	Benzenepentanoic acid	5-Phenylvaleric acid	$C_{11}H_{14}O_2$	2270-20-4	178.228	pl (w), pr (peth)	59(1)	190[30]			sl H₂O; vs EtOH, s os
649	Benzenepentanol		$C_{11}H_{16}O$	10521-91-2	164.244			155[20]	0.9725[20]	1.5156[20]	vs eth, EtOH
650	Benzenepropanal	Hydrocinnamic aldehyde	$C_9H_{10}O$	104-53-0	134.174	mcl	47	224	1.0190[20]		i H₂O; vs EtOH; msc eth
651	Benzenepropanenitrile	Hydrocinnamonitrile	C_9H_9N	645-59-0	131.174	liq	-1	261	1.0016[20]	1.5266[20]	s EtOH, eth; sl chl
652	Benzenepropanethiol		$C_9H_{12}S$	24734-68-7	152.256			121[23]	1.01[25]	1.5494[20]	
653	Benzenepropanoic acid	Hydrocinnamic acid	$C_9H_{10}O_2$	501-52-0	150.174	nd (w)	48.4(0.3)	284(2)	1.0712[49]		s H₂O, EtOH, eth, ctc, CS₂; vs bz
654	Benzenepropanol	Hydrocinnamyl alcohol	$C_9H_{12}O$	122-97-4	136.190		<-18	241(4)	0.995[25]	1.5357[25]	s H₂O, ctc; msc EtOH, eth

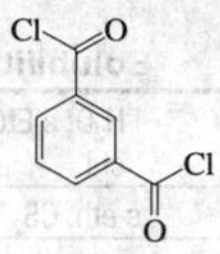

618
1,3-Benzenedicarbonyl dichloride

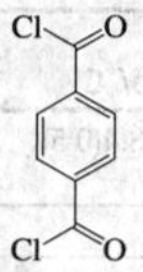

619
1,4-Benzenedicarbonyl dichloride

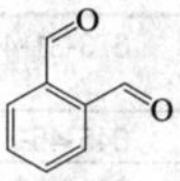

620
1,2-Benzenedicarboxaldehyde

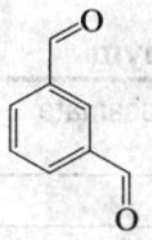

621
1,3-Benzenedicarboxaldehyde

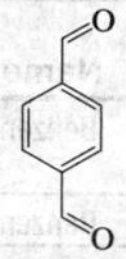

622
1,4-Benzenedicarboxaldehyde

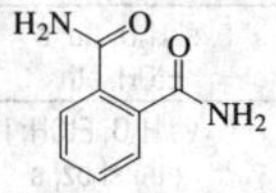

623
1,2-Benzenedicarboxamide

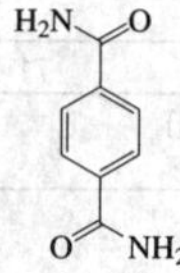

624
1,4-Benzenedicarboxamide

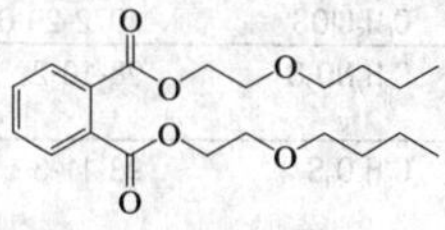

625
1,2-Benzenedicarboxylic acid, bis(2-butoxyethyl) ester

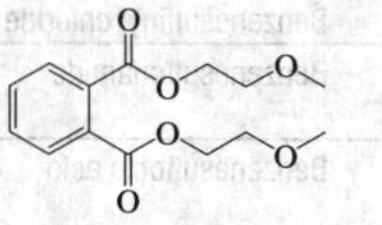

626
1,2-Benzenedicarboxylic acid, bis(2-methoxyethyl) ester

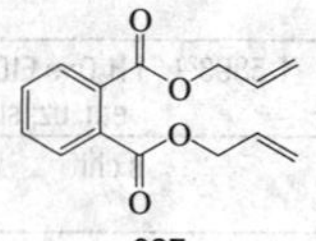

627
1,2-Benzenedicarboxylic acid, diallyl ester

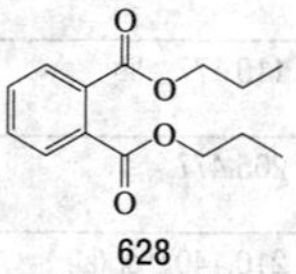

628
1,2-Benzenedicarboxylic acid, dipropyl ester

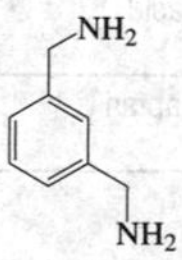

629
1,3-Benzenedimethanamine

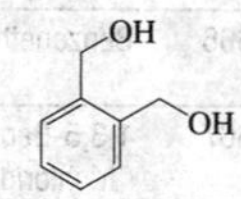

630
1,2-Benzenedimethanol

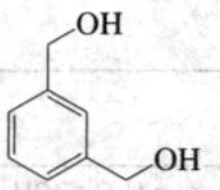

631
1,3-Benzenedimethanol

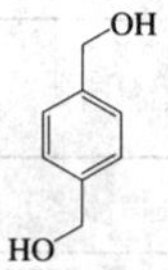

632
1,4-Benzenedimethanol

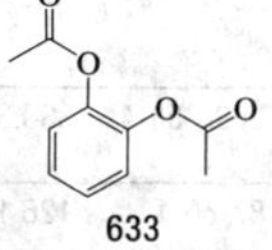

633
1,2-Benzenediol, diacetate

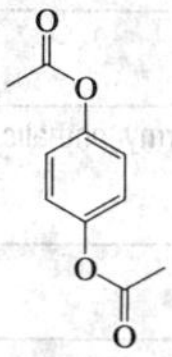

634
1,4-Benzenediol, diacetate

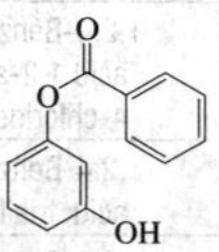

635
1,3-Benzenediol, monobenzoate

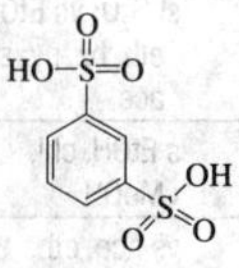

636
1,3-Benzenedisulfonic acid

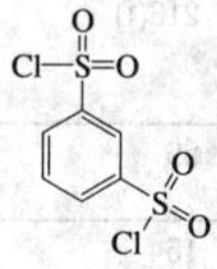

637
1,3-Benzenedisulfonyl dichloride

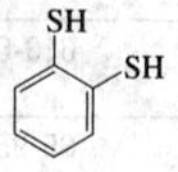

638
1,2-Benzenedithiol

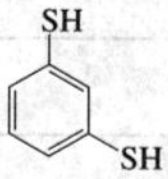

639
1,3-Benzenedithiol

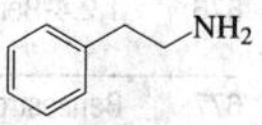

640
Benzeneethanamine

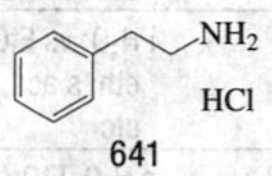

641
Benzeneethanamine, hydrochloride

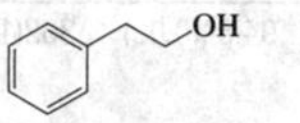

642
Benzeneethanol

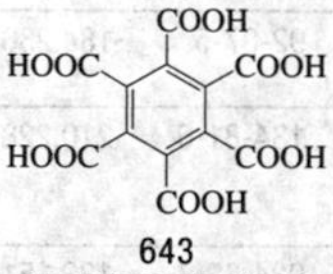

643
Benzenehexacarboxylic acid

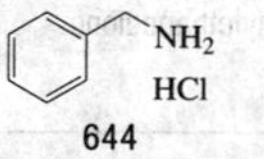

644
Benzenemethanamine, hydrochloride

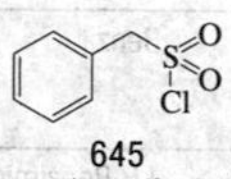

645
Benzenemethanesulfonyl chloride

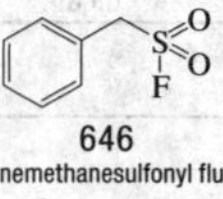

646
Benzenemethanesulfonyl fluoride

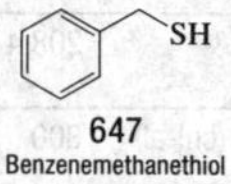

647
Benzenemethanethiol

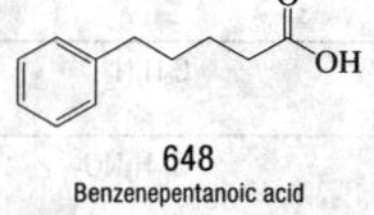

648
Benzenepentanoic acid

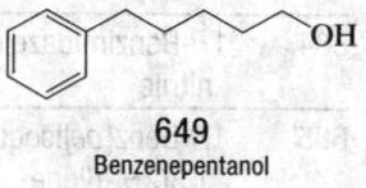

649
Benzenepentanol

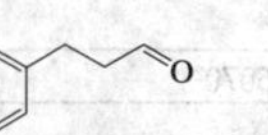

650
Benzenepropanal

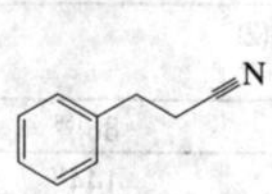

651
Benzenepropanenitrile

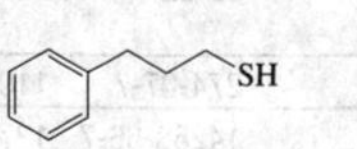

652
Benzenepropanethiol

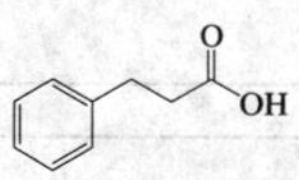

653
Benzenepropanoic acid

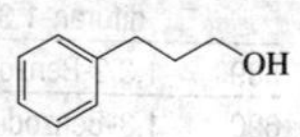

654
Benzenepropanol

No.	Name	Synonym	Mol. Form.	CAS RN	Mol. Wt.	Physical Form	mp/°C	bp/°C	den g cm⁻³	n_D	Solubility
655	Benzenepropanol carbamate	Phenprobamate	$C_{10}H_{13}NO_2$	673-31-4	179.216		103.4(0.5)				i H₂O; s EtOH, chl
656	Benzenepropanoyl chloride		C_9H_9ClO	645-45-4	168.619			225 dec	1.135[21]		s eth, CS₂
657	Benzeneseleninic acid	Phenylseleninic acid	$C_6H_6O_2Se$	6996-92-5	189.07		124.5		1.93[20]		sl H₂O; i bz; vs alk
658	Benzeneselenol		C_6H_6Se	645-96-5	157.07			183.6	1.4865[15]		i H₂O; s EtOH; vs eth, ctc
659	Benzenesulfinic acid		$C_6H_6O_2S$	618-41-7	142.176	pr (w)	84	dec			sl H₂O; s EtOH, eth, bz; i peth
660	Benzenesulfinyl chloride		C_6H_5ClOS	4972-29-6	160.621	pl (peth)	38	71[1.5]	1.3469[25]	1.3470[25]	s eth, chl
661	Benzenesulfonamide		$C_6H_7NO_2S$	98-10-2	157.191	lf, nd (w)	156(1)				sl H₂O, tfa; s EtOH, eth
662	Benzenesulfonic acid	Besylic acid	$C_6H_6O_3S$	98-11-3	158.175	nd (bz)	65				vs H₂O, EtOH; i eth; sl bz; s HOAc
663	Benzenesulfonyl chloride	Phenylsulfonyl chloride	$C_6H_5ClO_2S$	98-09-9	176.621	liq	14.5	252(7)	1.3470[15]		i H₂O; vs EtOH; s eth, ctc
664	Benzenesulfonyl fluoride	Phenylsulfonyl fluoride	$C_6H_5FO_2S$	368-43-4	160.166			203.5	1.3286[20]	1.4932[18]	s EtOH, eth
665	1,2,4,5-Benzenetetracarboxylic acid	Pyromellitic acid	$C_{10}H_6O_8$	89-05-4	254.150	tcl pr (w+2)	271(2)				sl H₂O; s EtOH
666	Benzenethiol	Phenyl mercaptan	C_6H_6S	108-98-5	110.177	liq	-14.87(0.05)	169.1(0.2)	1.0775[20]	1.5893[20]	i H₂O; s EtOH, eth, bz; sl ctc
667	1,3,5-Benzenetricarbonyl trichloride		$C_9H_3Cl_3O_3$	4422-95-1	265.477		36.3	180[16]			s chl
668	1,2,3-Benzenetricarboxylic acid	Hemimellitic acid	$C_9H_6O_6$	569-51-7	210.140	pr (al)	209(2)		1.546[20]		vs eth, EtOH
669	1,2,4-Benzenetricarboxylic acid	Trimellitic acid	$C_9H_6O_6$	528-44-9	210.140	nd (w) cry (al) cry (HOAc)	219				vs H₂O, eth, EtOH
670	1,3,5-Benzenetricarboxylic acid		$C_9H_6O_6$	554-95-0	210.140	pr or nd (w+1)	380				sl H₂O; vs EtOH, eth
671	1,2,4-Benzenetricarboxylic acid 1,2-anhydride, 4-chloride	4-(Chloroformyl)phthalic anhydride	$C_9H_3ClO_4$	1204-28-0	210.571		66				
672	1,2,4-Benzenetricarboxylic acid, triallyl ester		$C_{18}H_{18}O_6$	2694-54-4	330.332		<-30		1.164[20]		
673	1,2,3-Benzenetriol	Pyrogallol	$C_6H_6O_3$	87-66-1	126.110	lf or nd (bz)	125.5(0.5)	307(4)	1.453[4]	1.561[134]	vs H₂O, EtOH, eth, NH₃; s ace; i bz
674	1,2,4-Benzenetriol	Hydroxyhydroquinone	$C_6H_6O_3$	533-73-3	126.110	pl (eth), lf or pl (w)	140.5				vs H₂O, EtOH, eth; i bz, chl
675	1,3,5-Benzenetriol	Phloroglucinol	$C_6H_6O_3$	108-73-6	126.110	lf or pl (w+2)	216(1)	sub	1.46[25]		sl H₂O; vs EtOH, eth, bz, py; s ace
676	1,2,4-Benzenetriol triacetate		$C_{12}H_{12}O_6$	613-03-6	252.219		99	300			s EtOH, chl, MeOH
677	Benzestrol		$C_{20}H_{26}O_2$	85-95-0	298.419	cry (al)	164				vs ace, eth, EtOH, HOAc
678	Benzethonium chloride		$C_{27}H_{42}ClNO_2$	121-54-0	448.081	pl (chl/eth)	165 (hyd)				vs H₂O; s ace, chl, EtOH
679	Benzidene-3,3'-dicarboxylic acid	3,3'-Dicarboxybenzidine	$C_{14}H_{12}N_2O_4$	2130-56-5	272.256	nd	300 dec				
680	p-Benzidine	[1,1'-Biphenyl]-4,4'-diamine	$C_{12}H_{12}N_2$	92-87-5	184.236	nd (w)	127.0(0.5)	401			sl H₂O, eth, DMSO; s EtOH
681	Benzil	Diphenylethanedione	$C_{14}H_{10}O_2$	134-81-6	210.228	ye pr (al)	94.84(0.03)	332(13)	1.084[102]		i H₂O; vs EtOH, eth; s ace; sl ctc
682	1H-Benzimidazol-2-amine		$C_7H_7N_3$	934-32-7	133.151	pl (w)	231.9(0.3)				s H₂O, EtOH, ace; sl eth, bz, DMSO
683	1H-Benzimidazole	N,N'-Methenyl-o-phenylenediamine	$C_7H_6N_2$	51-17-2	118.136	orth bipym pl (w)	172.2(0.9)	>360			sl H₂O, eth; vs EtOH; i bz; s dil alk
684	1H-Benzimidazole-2-acetonitrile		$C_9H_7N_3$	4414-88-4	157.172		208.4				
685	1H-Benz[de]isoquinoline-1,3(2H)-dione		$C_{12}H_7NO_2$	81-83-4	197.190	nd (chl-al)	300				
686	Benzo[c]chrysene		$C_{22}H_{14}$	194-69-4	278.346	nd (AcOH)	125.4(0.5)				
687	Benzo[g]chrysene	Benzo[a]triphenylene	$C_{22}H_{14}$	196-78-1	278.346	nd (AcOH)	114.5				
688	1H,3H-Benzo[1,2-c:4,5-c']difuran-1,3,5,7-tetrone		$C_{10}H_2O_6$	89-32-7	218.119		286(2)				
689	1,3,2-Benzodioxaborole		$C_6H_5BO_2$	274-07-7	119.914		12	88[156]	1.2700[20]	1.5070[20]	
690	1,3-Benzodioxol-5-amine		$C_7H_7NO_2$	14268-66-7	137.137		42	144[16]			
691	1,3-Benzodioxole		$C_7H_6O_2$	274-09-9	122.122			169(1)	1.064[25]	1.5398[20]	

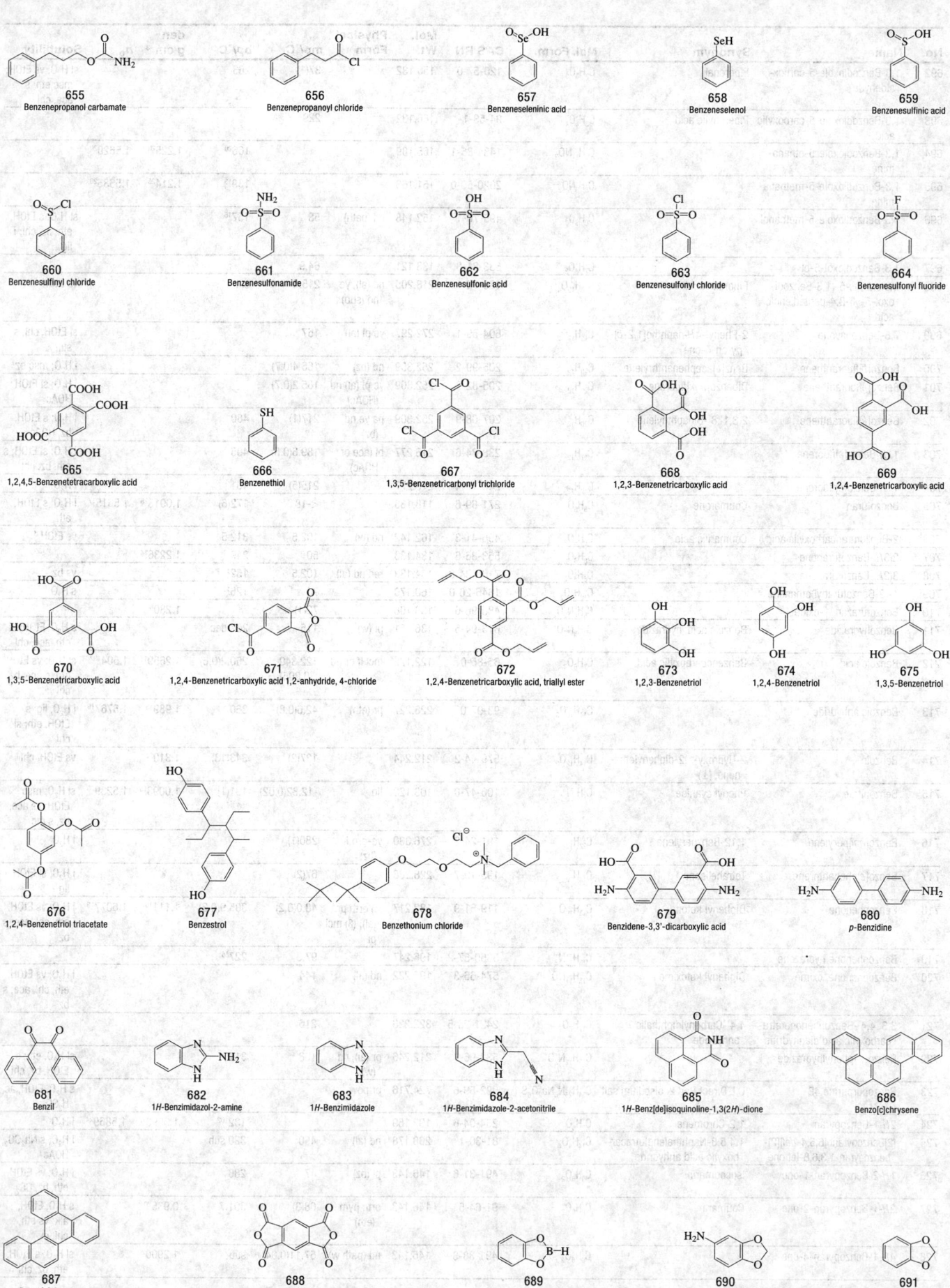

655
Benzenepropanol carbamate

656
Benzenepropanoyl chloride

657
Benzeneseleninic acid

658
Benzeneselenol

659
Benzenesulfinic acid

660
Benzenesulfinyl chloride

661
Benzenesulfonamide

662
Benzenesulfonic acid

663
Benzenesulfonyl chloride

664
Benzenesulfonyl fluoride

665
1,2,4,5-Benzenetetracarboxylic acid

666
Benzenethiol

667
1,3,5-Benzenetricarbonyl trichloride

668
1,2,3-Benzenetricarboxylic acid

669
1,2,4-Benzenetricarboxylic acid

670
1,3,5-Benzenetricarboxylic acid

671
1,2,4-Benzenetricarboxylic acid 1,2-anhydride, 4-chloride

672
1,2,4-Benzenetricarboxylic acid, triallyl ester

673
1,2,3-Benzenetriol

674
1,2,4-Benzenetriol

675
1,3,5-Benzenetriol

676
1,2,4-Benzenetriol triacetate

677
Benzestrol

678
Benzethonium chloride

679
Benzidene-3,3'-dicarboxylic acid

680
p-Benzidine

681
Benzil

682
1*H*-Benzimidazol-2-amine

683
1*H*-Benzimidazole

684
1*H*-Benzimidazole-2-acetonitrile

685
1*H*-Benz[de]isoquinoline-1,3(2*H*)-dione

686
Benzo[c]chrysene

687
Benzo[g]chrysene

688
1*H*,3*H*-Benzo[1,2-c:4,5-c']difuran-1,3,5,7-tetrone

689
1,3,2-Benzodioxaborole

690
1,3-Benzodioxol-5-amine

691
1,3-Benzodioxole

No.	Name	Synonym	Mol. Form.	CAS RN	Mol. Wt.	Physical Form	mp/°C	bp/°C	den g cm⁻³	n_D	Solubility
692	1,3-Benzodioxole-5-carbox-aldehyde	Piperonal	$C_9H_6O_3$	120-57-0	150.132		37(1)	263			sl H₂O; vs EtOH; msc eth; s ace, chl
693	1,3-Benzodioxole-5-carboxylic acid	Piperonylic acid	$C_6H_6O_4$	94-53-1	166.132		229				
694	1,3-Benzodioxole-5-ethana-mine		$C_9H_{11}NO_2$	1484-85-1	165.189			166²⁰	1.225²⁰	1.5620²⁰	
695	1,3-Benzodioxole-5-methana-mine		$C_6H_9NO_2$	2620-50-0	151.163			139¹³	1.214²⁵	1.5635²⁰	
696	1,3-Benzodioxole-5-methanol		$C_6H_8O_3$	495-76-1	152.148	nd (peth)	58	157¹⁶			sl H₂O; s EtOH, eth, bz, chl; i lig
697	1,3-Benzodioxol-5-ol		$C_7H_6O_3$	533-31-3	138.121		64.9				
698	*trans,trans*-5-(1,3-Benzodi-oxol-5-yl)-2,4-pentadienoic acid	Piperinic acid	$C_{12}H_{10}O_4$	136-72-1	218.205	nd (al), ye nd (sub)	215.8	sub			vs EtOH
699	7,8-Benzoflavone	2-Phenyl-4H-naphtho[1,2-*b*]-pyran-4-one	$C_{19}H_{12}O_2$	604-59-1	272.297	ye pl (al)	157				sl EtOH, chl; s sulf
700	Benzo[*b*]fluoranthene	Benz[*e*]acephenanthrylene	$C_{20}H_{12}$	205-99-2	252.309	nd (bz)	168.4(0.7)				i H₂O; msc bz
701	Benzo[*j*]fluoranthene	Dibenzo[*a,jk*]fluorene	$C_{20}H_{12}$	205-82-3	252.309	ye pl (al) nd (HOAc)	165.2(0.7)				i H₂O; sl EtOH, HOAc
702	Benzo[*k*]fluoranthene	2,3,1',8'-Binaphthylene	$C_{20}H_{12}$	207-08-9	252.309	pa ye nd (bz)	217(1)	480			i H₂O; s EtOH, bz, HOAc
703	11*H*-Benzo[*a*]fluorene		$C_{17}H_{12}$	238-84-6	216.277	pl (ace or HOAc)	189.6(0.5)	405			i H₂O; sl EtOH; s eth, bz, chl
704	11*H*-Benzo[*b*]fluorene		$C_{17}H_{12}$	243-17-4	216.277		215(5)	401			i H₂O
705	Benzofuran	Coumarone	C_8H_6O	271-89-6	118.133		<-18	172(6)	1.0913²⁵	1.5615¹⁷	i H₂O; s EtOH, eth
706	2-Benzofurancarboxylic acid	Coumarilic acid	$C_9H_6O_3$	496-41-3	162.142	nd (w)	192.5	312.5			vs EtOH
707	2(3*H*)-Benzofuranone		$C_8H_6O_2$	553-86-6	134.133		50	249	1.2236¹⁴		
708	3(2*H*)-Benzofuranone		$C_8H_6O_2$	7169-34-8	134.133	red nd (al)	102.5	152¹⁵			vs bz
709	1-(2-Benzofuranyl)ethanone		$C_{10}H_8O_2$	1646-26-0	160.170		76	126¹¹			s H₂O
710	Benzofurazan, 1-oxide		$C_6H_4N_2O_2$	480-96-6	136.108		72(1)		1.280⁸⁰		
711	Benzohydrazide	Benzoic acid, hydrazide	$C_7H_8N_2O$	613-94-5	136.151	pl (w)	115	267 dec			s H₂O, EtOH; sl eth, ace, chl
712	Benzoic acid	Benzenecarboxylic acid	$C_7H_6O_2$	65-85-0	122.122	mcl lf or nd	122.340 (0.005)	250.2(0.6)	1.2659¹⁵	1.504¹³²	sl H₂O; vs EtOH, eth; s ace, bz, chl
713	Benzoic anhydride		$C_{14}H_{10}O_3$	93-97-0	226.227	pr (eth)	42.6(0.8)	360	1.989¹⁵	1.5767¹⁵	i H₂O, lig; s EtOH, eth; sl chl
714	Benzoin	2-Hydroxy-1,2-diphenyletha-none, (±)	$C_{14}H_{12}O_2$	579-44-2	212.244		137(2)	343(13)	1.310²⁰		vs EtOH, chl
715	Benzonitrile	Phenyl cyanide	C_7H_5N	100-47-0	103.122	liq	-12.82(0.02)	191(1)	1.0093¹⁵	1.5289²⁰	sl H₂O; msc EtOH; vs ace, bz; s ctc
716	Benzo[*ghi*]perylene	1,12-Benzperylene	$C_{22}H_{12}$	191-24-2	276.330	ye-grn lf (bz)	280(1)				i H₂O
717	Benzo[*c*]phenanthrene	Tetrahelicene	$C_{18}H_{12}$	195-19-7	228.288		67(2)				i H₂O; sl EtOH, lig
718	Benzophenone	Diphenyl ketone	$C_{13}H_{10}O$	119-61-9	182.217	(α) orth pr (al); (β) mcl pr	48.0(0.2)	305.9(0.2)	1.111¹⁸	1.6077¹⁹	i H₂O; vs EtOH, eth, chl, ace; s bz
719	Benzophenone hydrazone		$C_{13}H_{12}N_2$	5350-57-2	196.247		97.3	227⁵⁵			
720	Benzophenone, oxime	Diphenyl ketoxime	$C_{13}H_{11}NO$	574-66-3	197.232	nd (al)	144				i H₂O; vs EtOH, eth, chl, ace; s bz
721	3,3',4,4'-Benzophenonetetra-carboxylic acid dianhydride	4,4'-Carbonyldiphthalic anhydride	$C_{17}H_6O_7$	2421-28-5	322.226		216				
722	Benzo-2-phenylhydrazide		$C_{13}H_{12}N_2O$	532-96-7	212.246	pr (al), nd (w)	168	314			sl H₂O, eth; s EtOH, bz, chl
723	Benzopurpurine 4B	C.I. Direct Red 2, disodium salt	$C_{34}H_{26}N_6Na_2O_6S_2$	992-59-6	724.716	br pow					s H₂O, EtOH, ac, H₂SO₄
724	2*H*-1-Benzopyran	1,2-Chromene	C_9H_8O	254-04-6	132.159			132¹⁰²	1.0993¹⁶	1.5869²⁴	i H₂O
725	[2]Benzopyrano[6,5,4-*def*][2]-benzopyran-1,3,6,8-tetrone	1,4,5,8-Naphthalenetetracar-boxylic acid anhydride	$C_{14}H_4O_6$	81-30-1	268.178	nd (al)	450	320 sub			i H₂O; s Na₂CO₃, HOAc
726	1*H*-2-Benzopyran-1-one	Isocoumarin	$C_9H_6O_2$	491-31-6	146.143	pl (bz)	47	286			i H₂O; vs EtOH, eth, bz, CS₂
727	2*H*-1-Benzopyran-2-one	Coumarin	$C_9H_6O_2$	91-64-5	146.143	orth pym (eth)	68(3)	301.7	0.935²⁰		s H₂O, EtOH, alk; vs eth, chl, py
728	4*H*-1-Benzopyran-4-one		$C_9H_6O_2$	491-38-3	146.143	nd (peth w)	57.11(0.04)	sub	1.2900²⁰		sl H₂O; s EtOH, eth, bz, chl

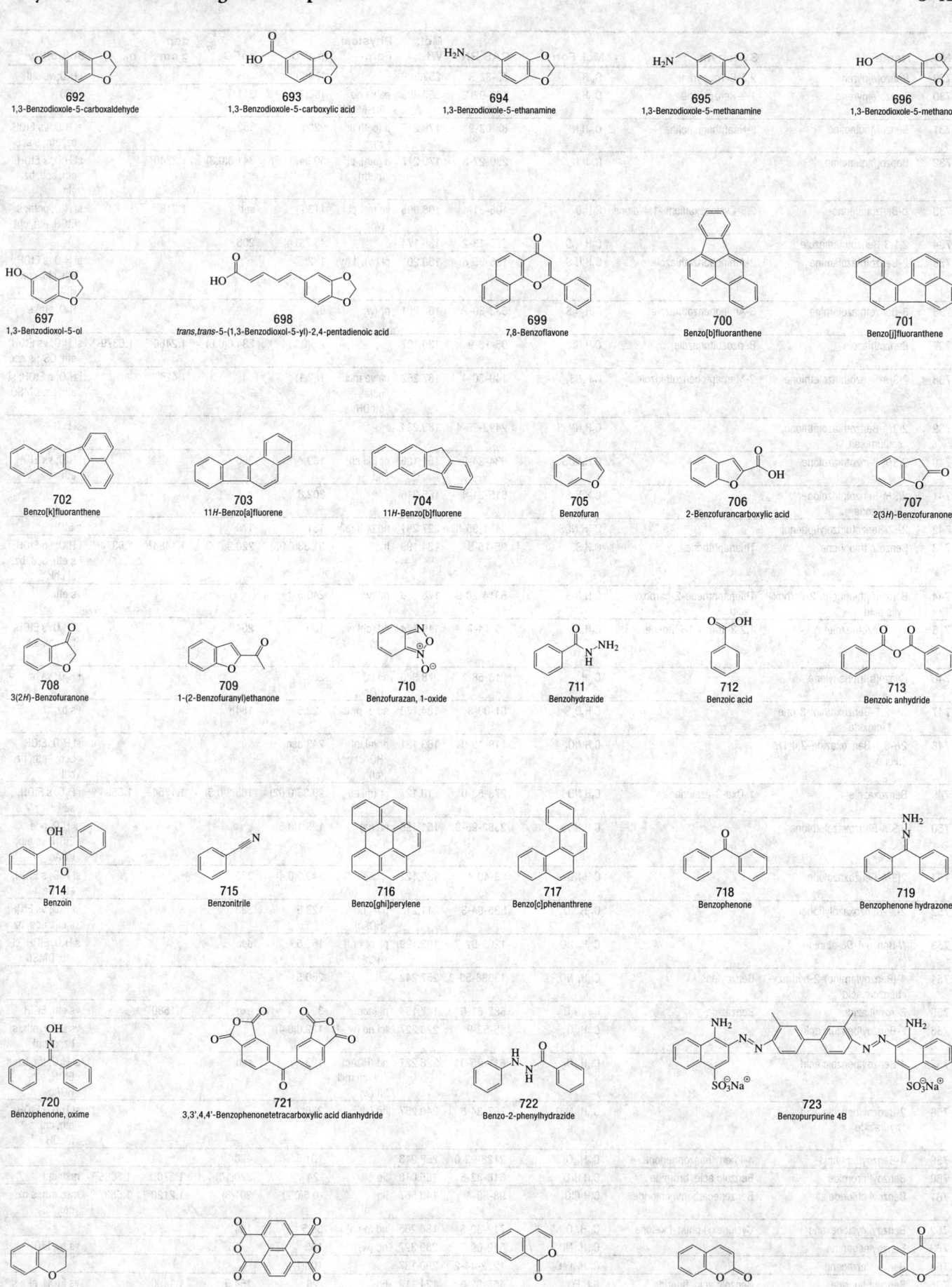

692
1,3-Benzodioxole-5-carboxaldehyde

693
1,3-Benzodioxole-5-carboxylic acid

694
1,3-Benzodioxole-5-ethanamine

695
1,3-Benzodioxole-5-methanamine

696
1,3-Benzodioxole-5-methanol

697
1,3-Benzodioxol-5-ol

698
trans,trans-5-(1,3-Benzodioxol-5-yl)-2,4-pentadienoic acid

699
7,8-Benzoflavone

700
Benzo[b]fluoranthene

701
Benzo[j]fluoranthene

702
Benzo[k]fluoranthene

703
11H-Benzo[a]fluorene

704
11H-Benzo[b]fluorene

705
Benzofuran

706
2-Benzofurancarboxylic acid

707
2(3H)-Benzofuranone

708
3(2H)-Benzofuranone

709
1-(2-Benzofuranyl)ethanone

710
Benzofurazan, 1-oxide

711
Benzohydrazide

712
Benzoic acid

713
Benzoic anhydride

714
Benzoin

715
Benzonitrile

716
Benzo[ghi]perylene

717
Benzo[c]phenanthrene

718
Benzophenone

719
Benzophenone hydrazone

720
Benzophenone, oxime

721
3,3',4,4'-Benzophenonetetracarboxylic acid dianhydride

722
Benzo-2-phenylhydrazide

723
Benzopurpurine 4B

724
2H-1-Benzopyran

725
[2]Benzopyrano[6,5,4-def][2]benzopyran-1,3,6,8-tetrone

726
1H-2-Benzopyran-1-one

727
2H-1-Benzopyran-2-one

728
4H-1-Benzopyran-4-one

No.	Name	Synonym	Mol. Form.	CAS RN	Mol. Wt.	Physical Form	mp/°C	bp/°C	den g cm^{-3}	n_D	Solubility
729	Benzo[a]pyrene	2,3-Benzopyrene	C$_{20}$H$_{12}$	50-32-8	252.309		179(2)				i H$_2$O; vs chl
730	Benzo[e]pyrene	1,2-Benzpyrene	C$_{20}$H$_{12}$	192-97-2	252.309	pa ye nd (bz-MeOH)	180(3)	311			i H$_2$O
731	Benzo[f]quinoline	β-Naphthoquinoline	C$_{13}$H$_9$N	85-02-9	179.217	lf (peth or w)	92(2)	352			sl H$_2$O; vs EtOH, bz, eth; s ace
732	Benzo[h]quinoline		C$_{13}$H$_9$N	230-27-3	179.217	lf (eth), pl (peth)	50.94(0.05)	341.3(0.3)	1.2340^{20}		sl H$_2$O; s EtOH, eth, ace, bz, ctc
733	p-Benzoquinone	2,5-Cyclohexadiene-1,4-dione	C$_6$H$_4$O$_2$	106-51-4	108.095	ye mcl pr (w)	113(2)	sub	1.318^{20}		sl H$_2$O, peth; s EtOH, eth, chl
734	2,1,3-Benzothiadiazole		C$_6$H$_4$N$_2$S	273-13-2	136.174		43.7(0.4)	206			
735	2-Benzothiazolamine	2-Aminobenzothiazole	C$_7$H$_6$N$_2$S	136-95-8	150.201	pl (w), lf (w)	132				sl H$_2$O; s EtOH, eth, chl, con HCl
736	6-Benzothiazolamine	6-Aminobenzothiazole	C$_7$H$_6$N$_2$S	533-30-2	150.201	pr (w)	87				i H$_2$O, eth; s EtOH
737	Benzothiazole	Benzosulfonazole	C$_7$H$_5$NS	95-16-9	135.187		2.5(0.1)	234.6(0.6)	1.2460^{20}	1.6379^{20}	sl H$_2$O; vs EtOH, eth, CS$_2$; s ace
738	2(3H)-Benzothiazolethione	2-Mercaptobenzothiazole	C$_7$H$_5$NS$_2$	149-30-4	167.252	pa ye mcl nd(al, MeOH)	182(1)		1.42^{20}		i H$_2$O; s EtOH; sl eth, bz, DMSO
739	2(3H)-Benzothiazolethione, sodium salt		C$_7$H$_4$NNaS$_2$	2492-26-4	189.234						sl H$_2$O
740	2(3H)-Benzothiazolone		C$_7$H$_5$NOS	934-34-9	151.186	pr (dil al), nd	139	360			i H$_2$O; vs EtOH, eth
741	2(3H)-Benzothiazolone, hydrazone		C$_7$H$_7$N$_3$S	615-21-4	165.216		202.8				
742	2-(2-Benzothiazolyl)phenol		C$_{13}$H$_9$NOS	3411-95-8	227.281	nd or lf (al)	131	179^3			s EtOH
743	Benzo[b]thiophene	Thianaphthene	C$_8$H$_6$S	95-15-8	134.199	lf	31.33(0.03)	220.9(0.4)	1.1484^{32}	1.6374^{37}	i H$_2$O; vs EtOH; s eth, ace, bz; sl chl
744	Benzo[b]thiophene-2-carboxylic acid	Thionaphthene-2-carboxylic acid	C$_9$H$_6$O$_2$S	6314-28-9	178.208	nd (w)	240.5				vs eth
745	1H-Benzotriazole	1,2,3-Triaza-1H-indene	C$_6$H$_5$N$_3$	95-14-7	119.124	nd (chl or bz)	100	204^{15}			sl H$_2$O; s EtOH, bz, chl, tol, DMF
746	Benzo[b]triphenylene		C$_{22}$H$_{14}$	215-58-7	278.346	nd (al, HOAc)	205				i H$_2$O; vs bz
747	3H-2,1-Benzoxathiol-3-one 1,1-dioxide		C$_7$H$_4$O$_4$S	81-08-3	184.170	nd or pr (bz)	129.5	184^{18}			vs bz, chl
748	2H-3,1-Benzoxazine-2,4(1H)-dione		C$_8$H$_5$NO$_3$	118-48-9	163.131	pr (al, gl HOAc) cry (al)	243 dec				sl H$_2$O, EtOH, ace; i eth, bz, chl
749	Benzoxazole	1-Oxa-3-azaindene	C$_7$H$_5$NO	273-53-0	119.121	pr (dil al)	29.36(0.02)	185.3(0.5)	1.1754^{20}	1.5594^{20}	i H$_2$O; s EtOH, sulf
750	2(3H)-Benzoxazolethione		C$_7$H$_5$NOS	2382-96-9	151.186	nd (w)	195.1(0.5)				sl H$_2$O, ace, EtOH; vs eth, HOAc
751	2(3H)-Benzoxazolone		C$_7$H$_5$NO$_2$	59-49-4	135.121		140.0(0.4)	335			sl H$_2$O; s EtOH, eth, tfa
752	2-(2-Benzoxazolyl)phenol		C$_{13}$H$_9$NO$_2$	835-64-3	211.216	pink nd (al, HOAc)	123.5	338			sl H$_2$O; vs EtOH; s eth, ace, bz
753	N-Benzoyl-DL-alanine		C$_{10}$H$_{11}$NO$_3$	1205-02-3	193.199	pl, pr or lf (eth)	165.5	dec			s H$_2$O, EtOH; sl eth, DMSO
754	4-(Benzoylamino)-2-hydroxybenzoic acid	Benzoylpas	C$_{14}$H$_{11}$NO$_4$	13898-58-3	257.242		260.5				
755	Benzoyl azide	Benzazide	C$_7$H$_5$N$_3$O	582-61-6	147.134	pl (ace)	32	exp	1.1680^{35}		vs eth, EtOH
756	2-Benzoylbenzoic acid		C$_{14}$H$_{10}$O$_3$	85-52-9	226.227	tcl nd (w+1)	129.0(0.4)				vs EtOH, eth; s bz; sl chl
757	4-Benzoylbenzoic acid		C$_{14}$H$_{10}$O$_3$	611-95-0	226.227	nd (HOAc), pl (al) mcl lf (w)	199	sub			sl H$_2$O, tfa, bz; s EtOH, eth, HOAc
758	2-Benzoylbenzoic acid, hydrazide		C$_{14}$H$_{12}$N$_2$O$_2$	787-84-8	240.257	nd (al)	238(3)				sl H$_2$O; i EtOH, eth, chl; s MeOH
759	4-Benzoylbiphenyl	4-Phenylbenzophenone	C$_{19}$H$_{14}$O	2128-93-0	258.313		101.5	420			
760	Benzoyl bromide	Benzoic acid, bromide	C$_7$H$_5$BrO	618-32-6	185.018	liq	-24	220(10)	1.570^{15}	1.5868^{25}	msc eth
761	Benzoyl chloride	Benzenecarbonyl chloride	C$_7$H$_5$ClO	98-88-4	140.567	liq	-0.5(0.2)	201(8)	1.2120^{20}	1.5537^{20}	msc eth; s bz, ctc, CS$_2$
762	Benzoyl cyclohexane	Cyclohexyl phenyl ketone	C$_{13}$H$_{16}$O	712-50-5	188.265	nd (peth)	59.5	164^{18}			
763	Benzoylecgonine		C$_{16}$H$_{19}$NO$_4$	519-09-5	289.327	nd (w)	195				vs bz, EtOH
764	Benzoylferrocene		C$_{17}$H$_{14}$FeO	1272-44-2	290.137		108(1)				
765	Benzoyl fluoride	Benzoic acid, fluoride	C$_7$H$_5$FO	455-32-3	124.112	liq	-28	154.5	1.1400^{20}		vs EtOH, eth; s ctc

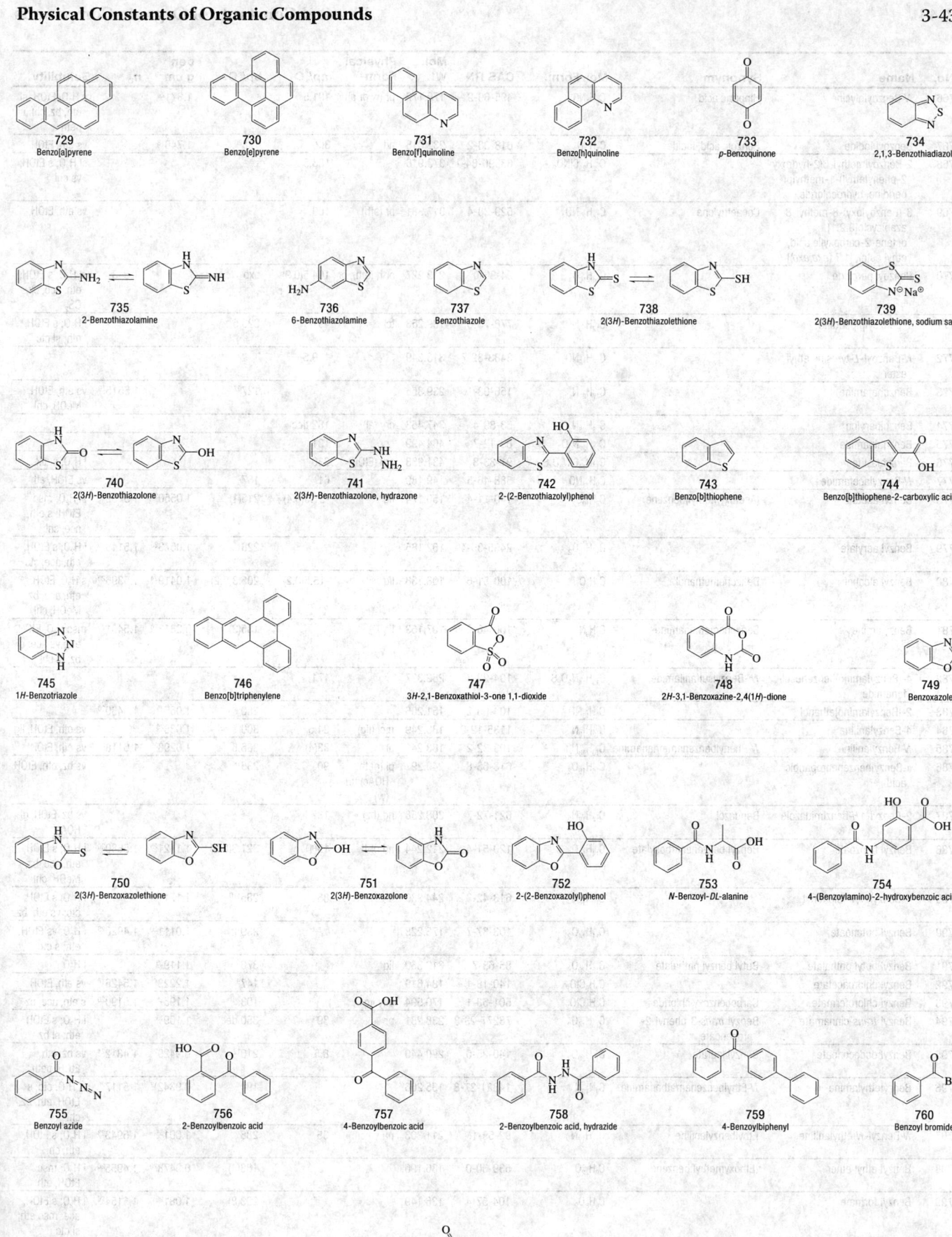

729
Benzo[a]pyrene

730
Benzo[e]pyrene

731
Benzo[f]quinoline

732
Benzo[h]quinoline

733
p-Benzoquinone

734
2,1,3-Benzothiadiazole

735
2-Benzothiazolamine

736
6-Benzothiazolamine

737
Benzothiazole

738
2(3*H*)-Benzothiazolethione

739
2(3*H*)-Benzothiazolethione, sodium salt

740
2(3*H*)-Benzothiazolone

741
2(3*H*)-Benzothiazolone, hydrazone

742
2-(2-Benzothiazolyl)phenol

743
Benzo[b]thiophene

744
Benzo[b]thiophene-2-carboxylic acid

745
1*H*-Benzotriazole

746
Benzo[b]triphenylene

747
3*H*-2,1-Benzoxathiol-3-one 1,1-dioxide

748
2*H*-3,1-Benzoxazine-2,4(1*H*)-dione

749
Benzoxazole

750
2(3*H*)-Benzoxazolethione

751
2(3*H*)-Benzoxazolone

752
2-(2-Benzoxazolyl)phenol

753
N-Benzoyl-*DL*-alanine

754
4-(Benzoylamino)-2-hydroxybenzoic acid

755
Benzoyl azide

756
2-Benzoylbenzoic acid

757
4-Benzoylbenzoic acid

758
2-Benzoylbenzoic acid, hydrazide

759
4-Benzoylbiphenyl

760
Benzoyl bromide

761
Benzoyl chloride

762
Benzoyl cyclohexane

763
Benzoylecgonine

764
Benzoylferrocene

765
Benzoyl fluoride

No.	Name	Synonym	Mol. Form.	CAS RN	Mol. Wt.	Physical Form	mp/°C	bp/°C	den g cm⁻³	n_D	Solubility
766	N-Benzoylglycine	Hippuric acid	C₉H₉NO₃	495-69-2	179.172	pr (w or al)	191.5		1.371²⁰		s H₂O, EtOH; sl eth, bz, chl; i peth
767	Benzoyl iodide	Benzoic acid, iodide	C₇H₅IO	618-38-2	232.018	nd	3(2)	128²⁰	1.746¹⁸		vs eth, EtOH
768	2-Benzoylmethyl-6(2-hydroxy-2-phenylethyl)-1-methylpiperidine, hydrochloride		C₂₂H₂₈ClNO₂	63990-84-1	373.916		183.5				sl H₂O; s EtOH; vs chl
769	3-(Benzoyloxy)-8-methyl-8-azabicyclo[3.2.1]octane-2-carboxylic acid, ethyl ester, [1R-(exo,exo)]	Cocaethylene	C₁₈H₂₃NO₄	529-38-4	317.381	pr (eth)	109				vs eth, EtOH
770	Benzoyl peroxide		C₁₄H₁₀O₄	94-36-0	242.227	orth (eth), pr	104.5(0.9)	exp		1.543	sl H₂O; s EtOH, eth, ace, bz, CS₂
771	1-Benzoylpiperidine		C₁₂H₁₅NO	776-75-0	189.253	tcl	49	320.5			i H₂O; s EtOH, eth; sl ctc
772	N-Benzoyl-L-tyrosine ethyl ester		C₁₈H₁₉NO₄	3483-82-7	313.349		119.5				
773	Benzphetamine		C₁₇H₂₁N	156-08-1	239.356			127⁰·⁰²		1.5515¹⁹	vs eth, EtOH, MeOH, chl
774	Benzpiperylon		C₂₂H₂₅N₃O	53-89-4	347.453	cry (al)	182 dec				
775	Benzquinamide		C₂₂H₃₂N₂O₅	63-12-7	404.499	cry	131				
776	Benzthiazide		C₁₅H₁₄ClN₃O₄S₃	91-33-8	431.938	cry (EtOH)	236				i H₂O; s alk
777	N-Benzylacetamide		C₉H₁₁NO	588-46-5	149.189		61	157²			vs EtOH, eth
778	Benzyl acetate	(Acetoxymethyl)benzene	C₉H₁₀O₂	140-11-4	150.174	liq	-51.5(0.4)	215(1)	1.0550²⁰	1.5232²⁰	sl H₂O; msc EtOH; s eth, ace, chl
779	Benzyl acrylate		C₁₀H₁₀O₂	2495-35-4	162.185			228	1.0573²⁰	1.5143²⁰	i H₂O; s EtOH, eth, ace, ctc
780	Benzyl alcohol	Benzenemethanol	C₇H₈O	100-51-6	108.138	liq	-15.5(0.2)	205.3(0.2)	1.0419²⁴	1.5396²⁰	s H₂O, EtOH, eth, ace, bz, MeOH, chl
781	Benzylamine	Benzenemethanamine	C₇H₉N	100-46-9	107.153	liq		185(3)	0.9813²⁰	1.5401²⁰	msc H₂O, EtOH, eth; vs ace; s bz; sl chl
782	4-(Benzylamino)benzenesulfonamide	N⁴-Benzylsulfanilamide	C₁₃H₁₄N₂O₂S	104-22-3	262.327		171				
783	2-[Benzylamino]ethanol		C₉H₁₃NO	104-63-2	151.205			280	1.065²⁵	1.5430²⁰	
784	4-Benzylaniline		C₁₃H₁₃N	1135-12-2	183.249	mcl (lig)	34.5	300	1.038²⁵		vs eth, EtOH, lig
785	N-Benzylaniline	N-Phenylbenzenemethanamine	C₁₃H₁₃N	103-32-2	183.249	pr	33(4)	306.5	1.0298⁶⁵	1.6118²⁵	vs eth, EtOH
786	α-Benzylbenzenepropanoic acid		C₁₆H₁₆O₂	618-68-8	240.297	pl (peth HOAc) nd (w)	90	235¹⁸			vs bz, eth, EtOH
787	2-Benzyl-1H-benzimidazole	Bendazol	C₁₄H₁₂N₂	621-72-7	208.258	nd (bz)	187				vs bz, EtOH, gl HOAc
788	Benzyl benzoate	Benzyl benzenecarboxylate	C₁₄H₁₂O₂	120-51-4	212.244	nd or lf	19(1)	321.3(0.9)	1.1121²⁵	1.5680²⁰	i H₂O; s EtOH, eth, ace, bz, MeOH, chl
789	4-Benzyl-1,1'-biphenyl		C₁₉H₁₆	613-42-3	244.330	lf	85	285¹¹⁰	1.171⁰		i H₂O; s EtOH, ctc; vs eth, bz
790	Benzyl butanoate		C₁₁H₁₄O₂	103-37-7	178.228			239	1.0111²⁰	1.4920²⁰	i H₂O; vs EtOH, eth; s ctc
791	Benzyl butyl phthalate	Butyl benzyl phthalate	C₁₉H₂₀O₄	85-68-7	312.360	liq		370	1.119²⁵		i H₂O
792	Benzyl chloroacetate		C₉H₉ClO₂	140-18-1	184.619			147⁹	1.2223⁴	1.5426¹⁸	vs eth, EtOH
793	Benzyl chloroformate	Carbobenzoxy chloride	C₈H₇ClO₂	501-53-1	170.594	oily liq		103²⁰	1.195²⁵	1.5190²⁰	s eth, ace, bz
794	Benzyl trans-cinnamate	Benzyl trans-3-phenyl-2-propenoate	C₁₆H₁₄O₂	78277-23-3	238.281	pr	39	350 dec	1.109¹⁵		i H₂O; s EtOH, eth; sl bz
795	Benzyl dodecanoate	Benzyl laurate	C₁₉H₃₀O₂	140-25-0	290.440		8.5	210¹²	0.9429²⁵	1.4812²⁴	vs bz, eth, EtOH, peth
796	Benzylethylamine	N-Ethylbenzenemethanamine	C₉H₁₃N	14321-27-8	135.206			194	0.9342¹⁷	1.5117²⁰	sl H₂O, ctc; s EtOH, eth, bz, chl
797	N-Benzyl-N-ethylaniline	Ethylbenzylaniline	C₁₅H₁₇N	92-59-1	211.303	pa ye oil	35	288	1.001⁵⁵	1.5943²³	i H₂O; s EtOH, eth, chl
798	Benzyl ethyl ether	(Ethoxymethyl)benzene	C₉H₁₂O	539-30-0	136.190			188(4)	0.9478²⁰	1.4955²⁰	i H₂O; msc EtOH, eth
799	Benzyl formate		C₈H₈O₂	104-57-4	136.149			203(8)	1.081²⁰	1.5154²⁰	i H₂O; s EtOH, ace; msc eth; sl ctc
800	Benzyl fumarate		C₁₈H₁₆O₄	538-64-7	296.318	cry pow	59	210⁵			vs eth, EtOH, chl
801	Benzylidene diacetate	Toluene-α,α-diol, diacetate	C₁₁H₁₂O₄	581-55-5	208.211	pl (eth)	45.1(0.5)	220	1.11²⁰		vs bz, eth, EtOH
802	Benzylimidobis(p-methoxyphenyl)methane		C₂₂H₂₁NO₂	524-96-9	331.408	pa ye cry	90				vs eth, chl

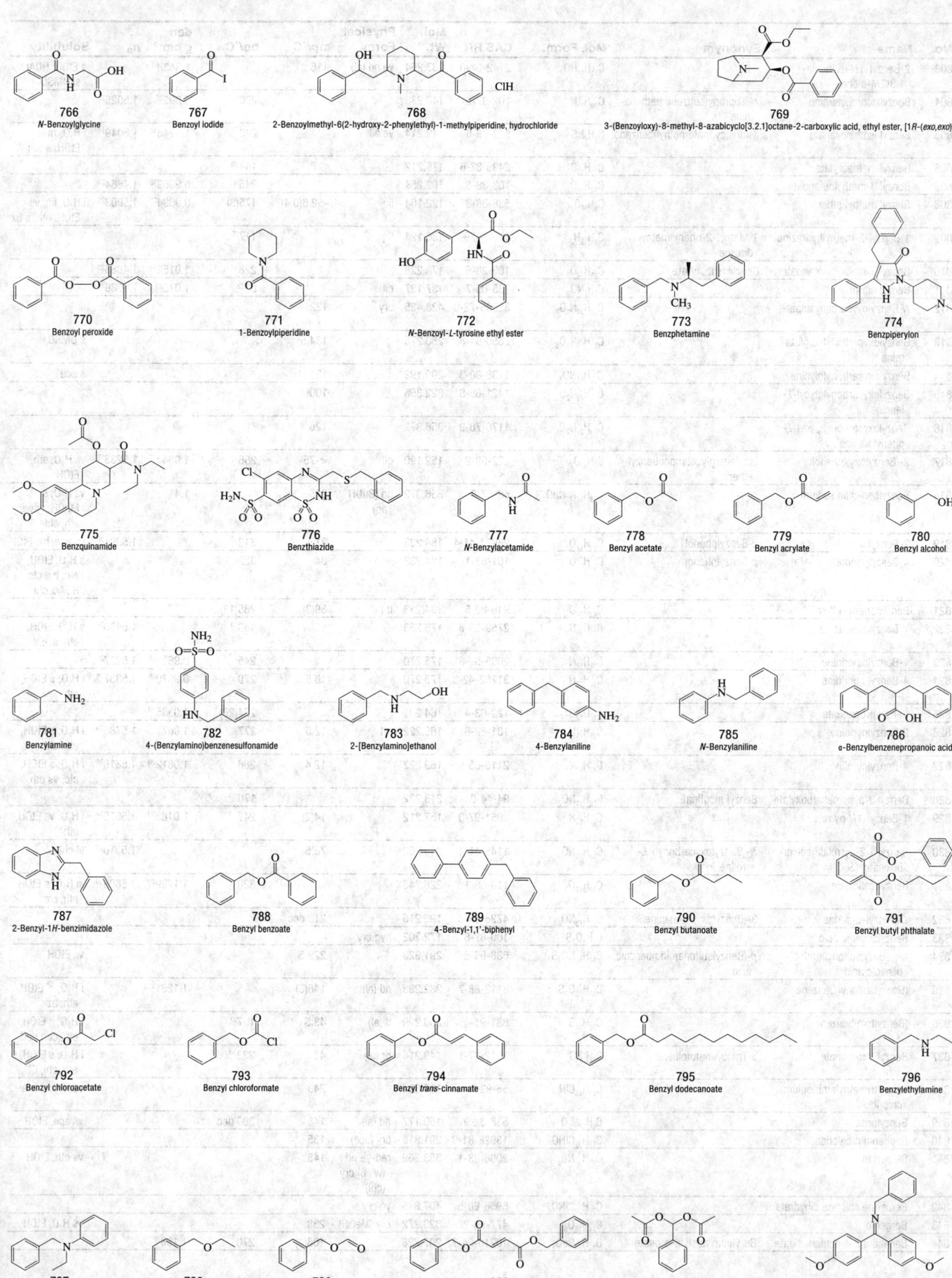

766
N-Benzoylglycine

767
Benzoyl iodide

768
2-Benzoylmethyl-6(2-hydroxy-2-phenylethyl)-1-methylpiperidine, hydrochloride

769
3-(Benzoyloxy)-8-methyl-8-azabicyclo[3.2.1]octane-2-carboxylic acid, ethyl ester, [1*R*-(*exo,exo*)]

770
Benzoyl peroxide

771
1-Benzoylpiperidine

772
N-Benzoyl-*L*-tyrosine ethyl ester

773
Benzphetamine

774
Benzpiperylon

775
Benzquinamide

776
Benzthiazide

777
N-Benzylacetamide

778
Benzyl acetate

779
Benzyl acrylate

780
Benzyl alcohol

781
Benzylamine

782
4-(Benzylamino)benzenesulfonamide

783
2-[Benzylamino]ethanol

784
4-Benzylaniline

785
N-Benzylaniline

786
α-Benzylbenzenepropanoic acid

787
2-Benzyl-1*H*-benzimidazole

788
Benzyl benzoate

789
4-Benzyl-1,1'-biphenyl

790
Benzyl butanoate

791
Benzyl butyl phthalate

792
Benzyl chloroacetate

793
Benzyl chloroformate

794
Benzyl *trans*-cinnamate

795
Benzyl dodecanoate

796
Benzylethylamine

797
N-Benzyl-*N*-ethylaniline

798
Benzyl ethyl ether

799
Benzyl formate

800
Benzyl fumarate

801
Benzylidene diacetate

802
Benzylimidobis(*p*-methoxyphenyl)methane

No.	Name	Synonym	Mol. Form.	CAS RN	Mol. Wt.	Physical Form	mp/°C	bp/°C	den g cm⁻³	n_D	Solubility
803	2-Benzyl-1H-isoindole-1,3(2H)-dione		$C_{15}H_{11}NO_2$	2142-01-0	237.254	ye nd (al)	116		1.343[18]		s EtOH, HOAc; sl DMSO
804	Benzylisopropylamine	N-Isopropylbenzenemethanamine	$C_{10}H_{15}N$	102-97-6	149.233			200	0.892[25]	1.5025[20]	
805	Benzyl isothiocyanate	(Isothiocyanatomethyl)benzene	C_8H_7NS	622-78-6	149.214	ye oil		243	1.1246[16]	1.6049[15]	i H_2O; msc EtOH; s eth
806	Benzyl methacrylate		$C_{11}H_{12}O_2$	2495-37-6	176.212			144[50]			
807	Benzyl 3-methylbutanoate		$C_{12}H_{16}O_2$	103-38-8	192.254			245	0.9983[15]	1.4884[20]	
808	Benzyl methyl ether		$C_8H_{10}O$	538-86-3	122.164	liq	-52.6(0.4)	175(3)	0.9634[20]	1.5008[20]	i H_2O, lig; vs EtOH, eth; s bz
809	1-Benzyl-2-methylhydrazine	1-Methyl-2-phenylmethylhydrazine	$C_8H_{12}N_2$	10309-79-2	136.194	liq		117[20]			
810	Benzyl 2-methylpropanoate	Benzyl isobutyrate	$C_{11}H_{14}O_2$	103-28-6	178.228			228	1.0159[18]	1.4883[20]	
811	Benzyl nitrite		$C_7H_7NO_2$	935-05-7	137.137	oil		81[35]	1.075[25]	1.4989[25]	
812	N-Benzyloxycarbonylaspartame		$C_{22}H_{24}N_2O_7$	33605-72-0	428.435	cry	122				
813	Benzyloxycarbonyl-L-glutamine		$C_{13}H_{16}N_2O_5$	2650-64-8	280.276		134.5				s DMSO
814	Benzyloxycarbonylglycine		$C_{10}H_{11}NO_4$	1138-80-3	209.199		121				s ace
815	Benzyloxycarbonylglycyl-L-leucine		$C_{16}H_{22}N_2O_5$	1421-69-8	322.356		100				
816	Benzyloxycarbonylglycyl-L-phenylalanine		$C_{19}H_{20}N_2O_5$	1170-76-9	356.372		126				
817	2-(Benzyloxy)ethanol	Ethylene glycol monobenzyl ether	$C_9H_{12}O_2$	622-08-2	152.190	oil	<-75	256	1.0640[20]	1.5233[20]	vs H_2O, eth, EtOH
818	Benzylpenicillin sodium		$C_{16}H_{17}N_2NaO_4S$	69-57-8	356.372	nd (BuOH aq)	215		1.41		vs H_2O; s MeOH; i ace, eth, chl
819	2-Benzylphenol	o-Benzylphenol	$C_{13}H_{12}O$	28994-41-4	184.233		21	312		1.5994[20]	vs ace, bz, EtOH
820	4-Benzylphenol	p-Benzylphenol	$C_{13}H_{12}O$	101-53-1	184.233		84	322			s H_2O, EtOH, eth, bz, ctc, HOAc, chl
821	Benzyl phenyl ether		$C_{13}H_{12}O$	946-80-5	184.233	lf (al)	39(3)	285(13)			
822	1-Benzylpiperazine		$C_{11}H_{16}N_2$	2759-28-6	176.258			146[12]		1.5430[28]	s H_2O, EtOH, eth; sl chl
823	1-Benzylpiperidine		$C_{12}H_{17}N$	2905-56-8	175.270			245	0.9625[16]	1.5227[20]	
824	4-Benzylpiperidine		$C_{12}H_{17}N$	31252-42-3	175.270		16.8	270	0.9970[20]	1.5337[25]	i H_2O; s EtOH, eth
825	Benzyl propanoate		$C_{10}H_{12}O_2$	122-63-4	164.201			224(20)	1.0335[20]		
826	2-Benzylpyridine		$C_{12}H_{11}N$	101-82-6	169.222	nd	12.5	277	1.067[0]	1.5785[20]	i H_2O; s EtOH, eth, chl
827	4-Benzylpyridine		$C_{12}H_{11}N$	2116-65-6	169.222		12.4	288	1.0612[20]	1.5818[20]	i H_2O; s EtOH, ctc; vs eth
828	Benzyl 3-pyridinecarboxylate	Benzyl nicotinate	$C_{13}H_{11}NO_2$	94-44-0	213.232			170[3]			
829	1-Benzyl-1H-pyrrole		$C_{11}H_{11}N$	2051-97-0	157.212		14(2)	247	1.0183[20]	1.5655[24]	i H_2O; vs EtOH, eth
830	Benzyl 1,2-pyrrolidinedicarboxylate, (S)-	N-(Benzyloxycarbonyl)-L-proline	$C_{13}H_{15}NO_4$	1148-11-4	249.263		78.5			1.5310[20]	sl chl
831	Benzyl salicylate		$C_{14}H_{12}O_3$	118-58-1	228.243			320	1.1799[20]	1.5805[20]	sl H_2O; s EtOH, eth, ctc
832	O-Benzyl-L-serine	3-(Benzyloxy)-L-alanine	$C_{10}H_{13}NO_3$	4726-96-9	195.215		218 dec				
833	Benzylsulfonic acid		$C_7H_8O_3S$	100-87-8	172.202	hyg cry					
834	4-[(Benzylsulfonyl)amino]benzoic acid	p-(Benzylsulfonamido)benzoic acid	$C_{14}H_{13}NO_4S$	536-95-8	291.323		229.5				vs EtOH
835	(Benzylsulfonyl)benzene		$C_{13}H_{12}O_2S$	3112-88-7	232.298	nd (al)	146(3)		1.1261[153]		i H_2O; sl EtOH, eth, bz
836	(Benzylthio)benzene		$C_{13}H_{12}S$	831-91-4	200.299	lf (al)	43.5	197[27]			i H_2O; s EtOH, eth, con sulf
837	Benzyl thiocyanate	α-Thiocyanatotoluene	C_8H_7NS	3012-37-1	149.214	pr (al)	43	232			i H_2O; s EtOH, eth, chl, CS_2
838	Benzyltrimethylammonium chloride		$C_{10}H_{16}ClN$	56-93-9	185.694		243				vs H_2O; s ace
839	Benzylurea		$C_8H_{10}N_2O$	538-32-9	150.177	nd (al)	148	200 dec			vs ace, EtOH
840	Bephenium chloride		$C_{17}H_{22}ClNO$	13928-81-9	291.816	cry (ace)	135				
841	Berberine		$C_{20}H_{19}NO_5$	2086-83-1	353.369	red-ye nd (w+6) cry (chl)	145				vs eth, EtOH
842	Berberine chloride dihydrate		$C_{20}H_{22}ClNO_6$	5956-60-5	407.845	ye cry					
843	Bergenin		$C_{14}H_{16}O_9$	477-90-7	328.272	cry (MeOH)	238				vs H_2O, EtOH
844	Beryllium 2,4-pentanedioate	Beryllium acetylacetonate	$C_{10}H_{14}BeO_4$	10210-64-7	207.228		108(1)	270	1.168[20]		

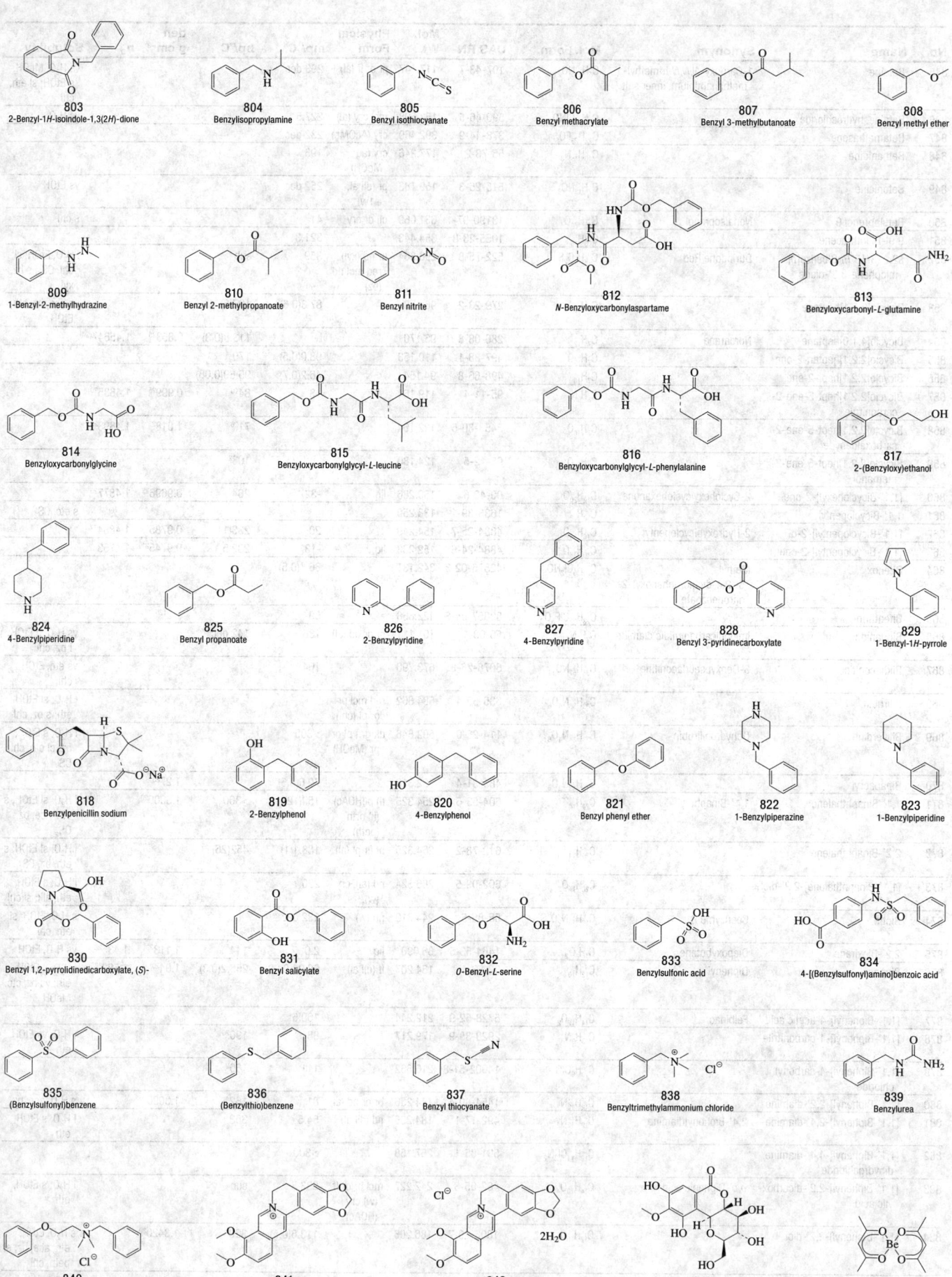

803
2-Benzyl-1*H*-isoindole-1,3(2*H*)-dione

804
Benzylisopropylamine

805
Benzyl isothiocyanate

806
Benzyl methacrylate

807
Benzyl 3-methylbutanoate

808
Benzyl methyl ether

809
1-Benzyl-2-methylhydrazine

810
Benzyl 2-methylpropanoate

811
Benzyl nitrite

812
N-Benzyloxycarbonylaspartame

813
Benzyloxycarbonyl-*L*-glutamine

814
Benzyloxycarbonylglycine

815
Benzyloxycarbonylglycyl-*L*-leucine

816
Benzyloxycarbonylglycyl-*L*-phenylalanine

817
2-(Benzyloxy)ethanol

824
4-Benzylpiperidine

825
Benzyl propanoate

826
2-Benzylpyridine

827
4-Benzylpyridine

828
Benzyl 3-pyridinecarboxylate

829
1-Benzyl-1*H*-pyrrole

818
Benzylpenicillin sodium

819
2-Benzylphenol

820
4-Benzylphenol

821
Benzyl phenyl ether

822
1-Benzylpiperazine

823
1-Benzylpiperidine

830
Benzyl 1,2-pyrrolidinedicarboxylate, (*S*)-

831
Benzyl salicylate

832
O-Benzyl-*L*-serine

833
Benzylsulfonic acid

834
4-[(Benzylsulfonyl)amino]benzoic acid

835
(Benzylsulfonyl)benzene

836
(Benzylthio)benzene

837
Benzyl thiocyanate

838
Benzyltrimethylammonium chloride

839
Benzylurea

840
Bephenium chloride

841
Berberine

842
Berberine chloride dihydrate

843
Bergenin

844
Beryllium 2,4-pentanedioate

No.	Name	Synonym	Mol. Form.	CAS RN	Mol. Wt.	Physical Form	mp/°C	bp/°C	den g cm⁻³	n_D	Solubility
845	Betaine	1-Carboxy-*N,N,N*-trimethyl-methanaminium, inner salt	$C_5H_{11}NO_2$	107-43-7	117.147	pr or lf (al)	293 dec				vs H_2O, MeOH; s EtOH; sl eth, chl
846	Betaine, hydrochloride		$C_5H_{12}ClNO_2$	590-46-5	153.608	mcl cry (al)	227.5				vs H_2O
847	Betamethasone		$C_{22}H_{29}FO_5$	378-44-9	392.460	cry (AcOMe)	232 dec				
848	Bethanidine		$C_{10}H_{15}N_3$	55-73-2	177.246	cry (aq MeOH)	196				
849	Betonicine		$C_7H_{13}NO_3$	515-25-3	159.183	pr (dil al, +1w)	252 dec				vs EtOH
850	Betulaprenol 9	Nonaisoprenol	$C_{45}H_{74}O$	13190-97-1	631.069	oil or cry	41				s chl
851	9,9'-Bianthracene		$C_{28}H_{18}$	1055-23-8	354.443		321.3				
852	Δ2,2'(3*H*,3'*H*)-Bibenzo[*b*]-thiophene-3,3'-dione	Durindone Red	$C_{16}H_8O_2S_2$	522-75-8	296.364	br nd (xyl) red mcl nd (bz)	359	sub			i H_2O, EtOH; sl chl, CS_2; s bz, xyl
853	Bicyclo[2.2.1]heptane		C_7H_{12}	279-23-2	96.170		87.3(0.6)	107(6)			vs ace, bz, eth, EtOH
854	Bicyclo[4.1.0]heptane	Norcarane	C_7H_{12}	286-08-8	96.170			113.8(0.3)	0.853²⁵	1.4564²⁰	
855	Bicyclo[2.2.1]heptan-2-one		$C_7H_{10}O$	497-38-1	110.153		98.0(0.5)	170			
856	Bicyclo[2.2.1]hept-2-ene		C_7H_{10}	498-66-8	94.154		46.2(0.7)	95.69(0.08)			
857	Bicyclo[2.2.1]hept-5-ene-2-carbonitrile		C_8H_9N	95-11-4	119.164		13	84¹⁰	0.999²⁵	1.4885²⁰	
858	Bicyclo[2.2.1]hept-5-ene-2-carboxaldehyde		$C_8H_{10}O$	5453-80-5	122.164			71²⁰	1.018²⁵	1.4893²⁰	
859	Bicyclo[2.2.1]hept-5-ene-2-methanol		$C_8H_{12}O$	95-12-5	124.180			103²⁰			
860	[1,1'-Bicyclohexyl]-2-one	2-Cyclohexylcyclohexanone	$C_{12}H_{20}O$	90-42-6	180.286	liq	-32	264	0.9696²⁵	1.4877²⁵	
861	1,1'-Bicyclopentyl		$C_{10}H_{18}$	1636-39-1	138.250						s ctc, CS_2
862	[1,1'-Bicyclopentyl]-2-ol	2-Hydroxybicyclopentyl	$C_{10}H_{18}O$	4884-25-7	154.249		20	233(11)	0.9785¹⁵	1.4884¹⁷	
863	[1,1'-Bicyclopentyl]-2-one		$C_{10}H_{16}O$	4884-24-6	152.233	liq	-13	232.5	0.9745²¹	1.4763	
864	Bifenox	Methyl 5-(2,4-dichlorophenoxy)-2-nitrobenzoate	$C_{14}H_9Cl_2NO_5$	42576-02-3	342.131		86.1(0.5)				
865	Bifenthrin		$C_{23}H_{22}ClF_3O_2$	82657-04-3	422.868		69		1.2¹²⁵		
866	Biguanide	Imidodicarbonimidic diamide	$C_2H_7N_5$	56-03-1	101.111	pr or nd (al)	136	142 dec			vs H_2O; s EtOH; i bz, chl
867	Bikhaconitine	3-Deoxypseudaconitine	$C_{36}H_{51}NO_{11}$	6078-26-8	673.790		164				vs eth, EtOH, chl
868	Bilirubin		$C_{33}H_{36}N_4O_6$	635-65-4	584.662	red mcl pr or pl (chl)					i H_2O; sl EtOH, eth; s bz, chl
869	Biliverdine	Dehydrobilirubin	$C_{33}H_{34}N_4O_6$	114-25-0	582.646	dk grn pl or pr (MeOH)	>300				i H_2O; s EtOH, bz; sl eth, chl, CS_2
870	Binapacryl		$C_{15}H_{18}N_2O_6$	485-31-4	322.313		70.6(0.5)		1.27²⁰		
871	1,1'-Binaphthalene	1,1'-Binaphthyl	$C_{20}H_{14}$	604-53-5	254.325	(i) pl(HOAc) (ii) orth (peth)	159(12)	>360	1.3000²⁰		i H_2O; sl EtOH; s eth, ace, bz, CS_2
872	2,2'-Binaphthalene		$C_{20}H_{14}$	612-78-2	254.325	bl flr pl (al)	188.1(1)	452(26)			i H_2O; sl EtOH; s eth, bz, CS_2
873	[1,1'-Binaphthalene]-2,2'-diol		$C_{20}H_{14}O_2$	602-09-5	286.324	nd (al), cry (w)	220				i H_2O; s EtOH, eth, alk; sl chl
874	Biotin	Coenzyme R	$C_{10}H_{16}N_2O_3S$	58-85-5	244.310	nd (w)	232 dec				s H_2O, EtOH; sl eth, chl
875	2,2'-Bioxirane	Diepoxybutane	$C_4H_6O_2$	1464-53-5	86.090	liq	2.0	144	1.113²⁰	1.435²⁰	vs H_2O, EtOH
876	Biphenyl	Diphenyl	$C_{12}H_{10}$	92-52-4	154.207	lf (dil al)	68.93(0.02)	255.2(0.3)	1.04²⁰	1.588⁷⁷	i H_2O; s EtOH, eth; vs bz, ctc, MeOH
877	[1,1'-Biphenyl]-4-acetic acid	Felbinac	$C_{14}H_{12}O_2$	5728-52-9	212.244		160.5				
878	[1,1'-Biphenyl]-4-carbonitrile		$C_{13}H_9N$	2920-38-9	179.217		88	190²⁰			i H_2O; vs EtOH, eth
879	[1,1'-Biphenyl]-4-carbonyl chloride		$C_{13}H_9ClO$	14002-51-8	216.662		111	160²			
880	[1,1'-Biphenyl]-2,2'-diamine		$C_{12}H_{12}N_2$	1454-80-4	184.236	pr or nd (al)	81	162⁴	1.3090²⁰		s H_2O, ace, bz
881	[1,1'-Biphenyl]-2,4'-diamine	2,4'-Biphenyldiamine	$C_{12}H_{12}N_2$	492-17-1	184.236	nd (dil al)	54.5	363			i H_2O; s EtOH, eth
882	[1,1'-Biphenyl]-4,4'-diamine, dihydrochloride		$C_{12}H_{14}Cl_2N_2$	531-85-1	257.158		>300				
883	[1,1'-Biphenyl]-2,2'-dicarbox-ylic acid	*o,o'*-Diphenic acid	$C_{14}H_{10}O_4$	482-05-3	242.227	mcl pr or lf (w) cry (HOAc)	233.5	sub			i H_2O; s EtOH, eth
884	[1,1'-Biphenyl]-2,2'-diol		$C_{12}H_{10}O_2$	1806-29-7	186.206		113.6(0.5)	320	1.3420²⁰		s H_2O, EtOH, eth, ace, bz; sl peth, chl
885	[1,1'-Biphenyl]-2,5-diol		$C_{12}H_{10}O_2$	1079-21-6	186.206	nd (dil al)	97.5				vs EtOH

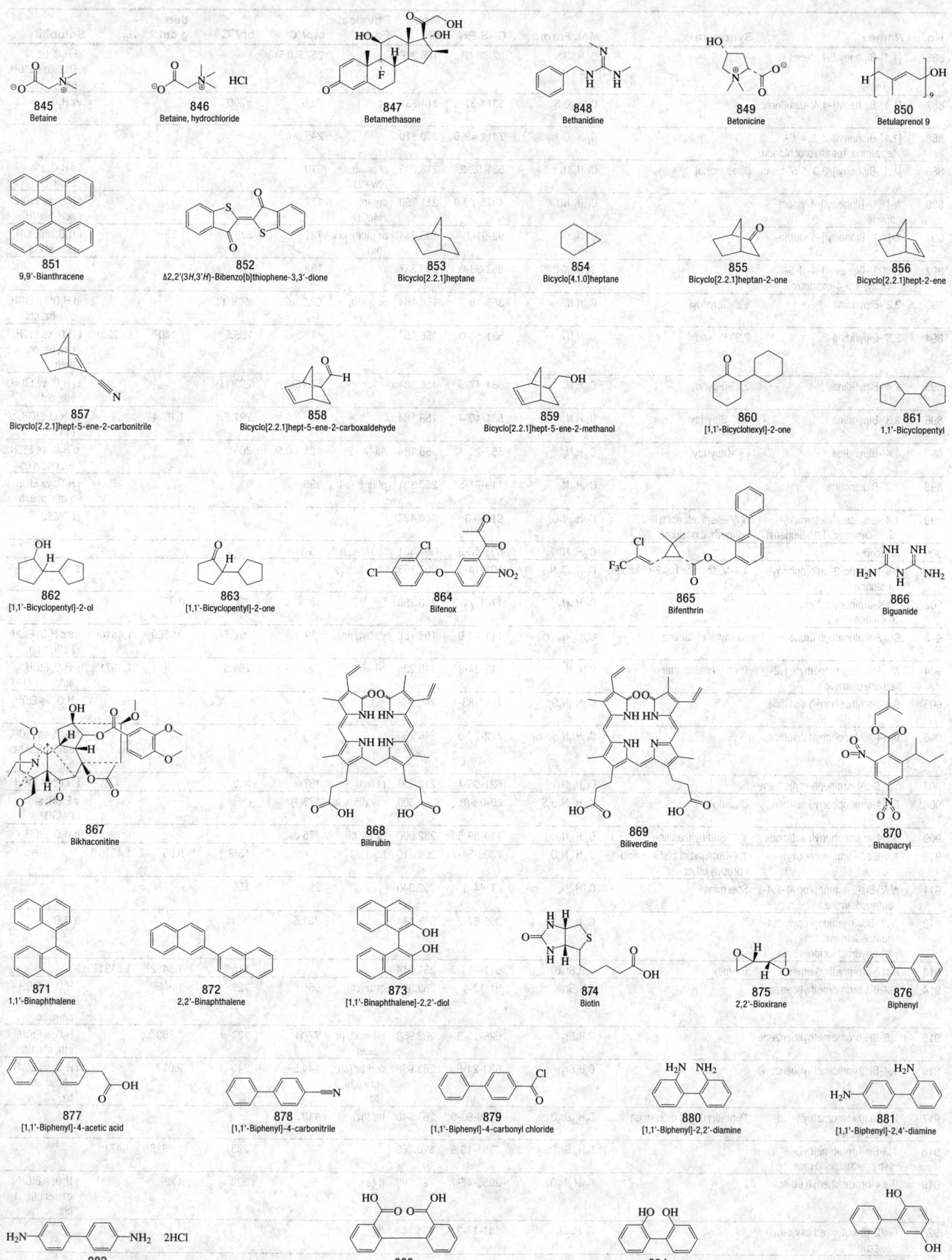

845 Betaine

846 Betaine, hydrochloride

847 Betamethasone

848 Bethanidine

849 Betonicine

850 Betulaprenol 9

851 9,9'-Bianthracene

852 Δ2,2'(3H,3'H)-Bibenzo[b]thiophene-3,3'-dione

853 Bicyclo[2.2.1]heptane

854 Bicyclo[4.1.0]heptane

855 Bicyclo[2.2.1]heptan-2-one

856 Bicyclo[2.2.1]hept-2-ene

857 Bicyclo[2.2.1]hept-5-ene-2-carbonitrile

858 Bicyclo[2.2.1]hept-5-ene-2-carboxaldehyde

859 Bicyclo[2.2.1]hept-5-ene-2-methanol

860 [1,1'-Bicyclohexyl]-2-one

861 1,1'-Bicyclopentyl

862 [1,1'-Bicyclopentyl]-2-ol

863 [1,1'-Bicyclopentyl]-2-one

864 Bifenox

865 Bifenthrin

866 Biguanide

867 Bikhaconitine

868 Bilirubin

869 Biliverdine

870 Binapacryl

871 1,1'-Binaphthalene

872 2,2'-Binaphthalene

873 [1,1'-Binaphthalene]-2,2'-diol

874 Biotin

875 2,2'-Bioxirane

876 Biphenyl

877 [1,1'-Biphenyl]-4-acetic acid

878 [1,1'-Biphenyl]-4-carbonitrile

879 [1,1'-Biphenyl]-4-carbonyl chloride

880 [1,1'-Biphenyl]-2,2'-diamine

881 [1,1'-Biphenyl]-2,4'-diamine

882 [1,1'-Biphenyl]-4,4'-diamine, dihydrochloride

883 [1,1'-Biphenyl]-2,2'-dicarboxylic acid

884 [1,1'-Biphenyl]-2,2'-diol

885 [1,1'-Biphenyl]-2,5-diol

No.	Name	Synonym	Mol. Form.	CAS RN	Mol. Wt.	Physical Form	mp/°C	bp/°C	den g cm⁻³	n_D	Solubility
886	[1,1'-Biphenyl]-4,4'-diol		$C_{12}H_{10}O_2$	92-88-6	186.206			287.6(0.5)			sl H_2O, bz, DMSO; s EtOH, eth
887	[1,1'-Biphenyl]-4,4'-disulfonic acid		$C_{12}H_{10}O_6S_2$	5314-37-4	314.333	pr	72.5	>200			vs H_2O
888	[1,1'-Biphenyl]-3,3',4,4'-tetramine, tetrahydrochloride		$C_{12}H_{18}Cl_4N_4$	7411-49-6	360.110		245 dec				
889	[1,1'-Biphenyl]-3,3',5,5'-tetrol	Diresorcinol	$C_{12}H_{10}O_4$	531-02-2	218.205	pl or nd (w+2)	310				vs H_2O, eth, EtOH
890	N-[1,1'-Biphenyl]-4-ylacetamide		$C_{14}H_{13}NO$	4075-79-0	211.259	cry (dil MeOH)	172.8				i H_2O; vs EtOH, ace, MeOH
891	1-[1,1'-Biphenyl]-4-ylethanone		$C_{14}H_{12}O$	92-91-1	196.244	pr (ace), cry (al)	121	326	1.251^0		i H_2O; vs EtOH, ace; sl chl
892	2-[1,1'-Biphenyl]-4-yl-5-phenyl-1,3,4-oxadiazole		$C_{20}H_{14}N_2O$	852-38-0	298.337		168				
893	2,2'-Bipyridine	α,α'-Dipyridyl	$C_{10}H_8N_2$	366-18-7	156.184	pr (peth)	69.9(0.2)	273(10)			sl H_2O; vs EtOH, eth, bz, chl
894	2,3'-Bipyridine	2,3'-Bipyridyl	$C_{10}H_8N_2$	581-50-0	156.184			295.5	1.140^{20}	1.6223^{20}	i H_2O; vs EtOH, eth, bz, chl; sl peth
895	2,4'-Bipyridine	2,4'-Bipyridyl	$C_{10}H_8N_2$	581-47-5	156.184		59.7(0.5)	283(11)			sl H_2O; vs EtOH, eth, chl
896	3,3'-Bipyridine	3,3'-Bipyridyl	$C_{10}H_8N_2$	581-46-4	156.184		68	291.5	1.1614^{20}		vs H_2O, EtOH; sl eth
897	4,4'-Bipyridine	γ,γ'-Dipyridyl	$C_{10}H_8N_2$	553-26-4	156.184	nd (w+2)	104.4(0.5)	305			sl H_2O; vs EtOH, bz, chl; s eth
898	2,2'-Biquinoline		$C_{18}H_{12}N_2$	119-91-5	256.301	pl or lf (al)	196				i H_2O; vs EtOH; s eth, ace, bz
899	4,4'-Bis(acetoacetamido)-3,3'-dimethyl-1,1'-biphenyl	N,N'-Bis(acetoacetyl)-3,3'-dimethylbenzidine	$C_{22}H_{24}N_2O_4$	91-96-3	380.437		212				sl DMSO
900	Bisacodyl		$C_{22}H_{19}NO_4$	603-50-9	361.391		134(1)				
901	Bis(4-amino-3-chlorophenyl)methane	4,4-Methylene bis(2-chloroaniline)	$C_{13}H_{12}Cl_2N_2$	101-14-4	267.153						s ctc
902	Bis(4-aminocyclohexyl)methane		$C_{13}H_{26}N_2$	1761-71-3	210.358		15	320	0.92^{75}		
903	Bis(2-aminoethyl)amine	Diethylenetriamine	$C_4H_{13}N_3$	111-40-0	103.166	ye hyg liq	-39	206.5(0.3)	0.9569^{20}	1.4810^{25}	msc H_2O, EtOH; i eth; s lig
904	N,N'-Bis(2-aminoethyl)-1,2-ethanediamine	Triethylenetetramine	$C_6H_{18}N_4$	112-24-3	146.234	oil	12	266.5		1.4971^{20}	s H_2O, EtOH, acid
905	Bis(2-aminophenyl) disulfide		$C_{12}H_{12}N_2S_2$	1141-88-4	248.366		93				i H_2O; vs EtOH, eth
906	Bis(4-aminophenyl) disulfide		$C_{12}H_{12}N_2S_2$	722-27-0	248.366		85				s H_2O; vs EtOH, eth, chl; sl bz, lig
907	1,2-Bis(4-aminophenyl)ethane		$C_{14}H_{16}N_2$	621-95-4	212.290	pl (w)	137	sub			i H_2O; vs EtOH
908	Bis(4-aminophenyl) sulfone	Dapsone	$C_{12}H_{12}N_2O_2S$	80-08-0	248.300	cry (95% al)	178(1)				s EtOH; sl DMSO
909	Bis(4-aminophenyl) sulfoxide	4,4'-Sulfinyldianiline	$C_{12}H_{12}N_2OS$	119-59-5	232.300	pr (w, al)	175 dec				s H_2O, EtOH
910	1,4-Bis(3-aminopropoxy)butane	1,4-Butanediol bis(3-aminopropyl) ether	$C_{10}H_{24}N_2O_2$	7300-34-7	204.310	liq		135^3	0.96^{20}	1.4619^{20}	
911	N,N'-Bis(3-aminopropyl)-1,4-butanediamine	Spermine	$C_{10}H_{26}N_4$	71-44-3	202.340		29	150^5			
912	N,N'-Bis(3-aminopropyl)-1,4-butanediamine, tetrahydrochloride		$C_{10}H_{30}Cl_4N_4$	306-67-2	348.184		301.5				s H_2O
913	Bis(2-bromoethyl) ether	Bromex	$C_4H_8Br_2O$	5414-19-7	231.914			115^{32}	1.8452^{20}	1.5131^{27}	
914	1,2-Bis(bromomethyl)benzene		$C_8H_8Br_2$	91-13-4	263.958	orth (chl)	95(1)	129$^{4.5}$	1.988^{25}		i H_2O; s EtOH, eth, ctc, chl, peth, lig
915	1,3-Bis(bromomethyl)benzene		$C_8H_8Br_2$	626-15-3	263.958	nd (chl), pr (ace)	77(1)	137^{20}	1.959^{25}		i H_2O; s EtOH, eth, chl, lig
916	1,4-Bis(bromomethyl)benzene		$C_8H_8Br_2$	623-24-5	263.958	mcl pr (al), cry (chl, bz)	144.5	245	2.012^{25}		i H_2O; vs EtOH, chl; sl eth; s bz
917	2,2-Bis(bromomethyl)-1,3-propanediol	Pentaerythritol dibromide	$C_5H_{10}Br_2O_2$	3296-90-0	261.940	nd (bz)	113				
918	1,3-Bis(bromomethyl)-tetramethyldisiloxane		$C_6H_{16}Br_2OSi_2$	2351-13-5	320.169			233	1.3918^{25}	1.4719^{25}	
919	Bis(4-bromophenyl) ether		$C_{12}H_8Br_2O$	2050-47-7	327.999	lf (al)	54(1)	339	1.8^{25}		i H_2O; s EtOH, bz; vs eth; sl chl
920	Bis(2-(2-butoxyethoxy)ethyl) adipate		$C_{22}H_{42}O_8$	141-17-3	434.563	liq			1.1^{25}		

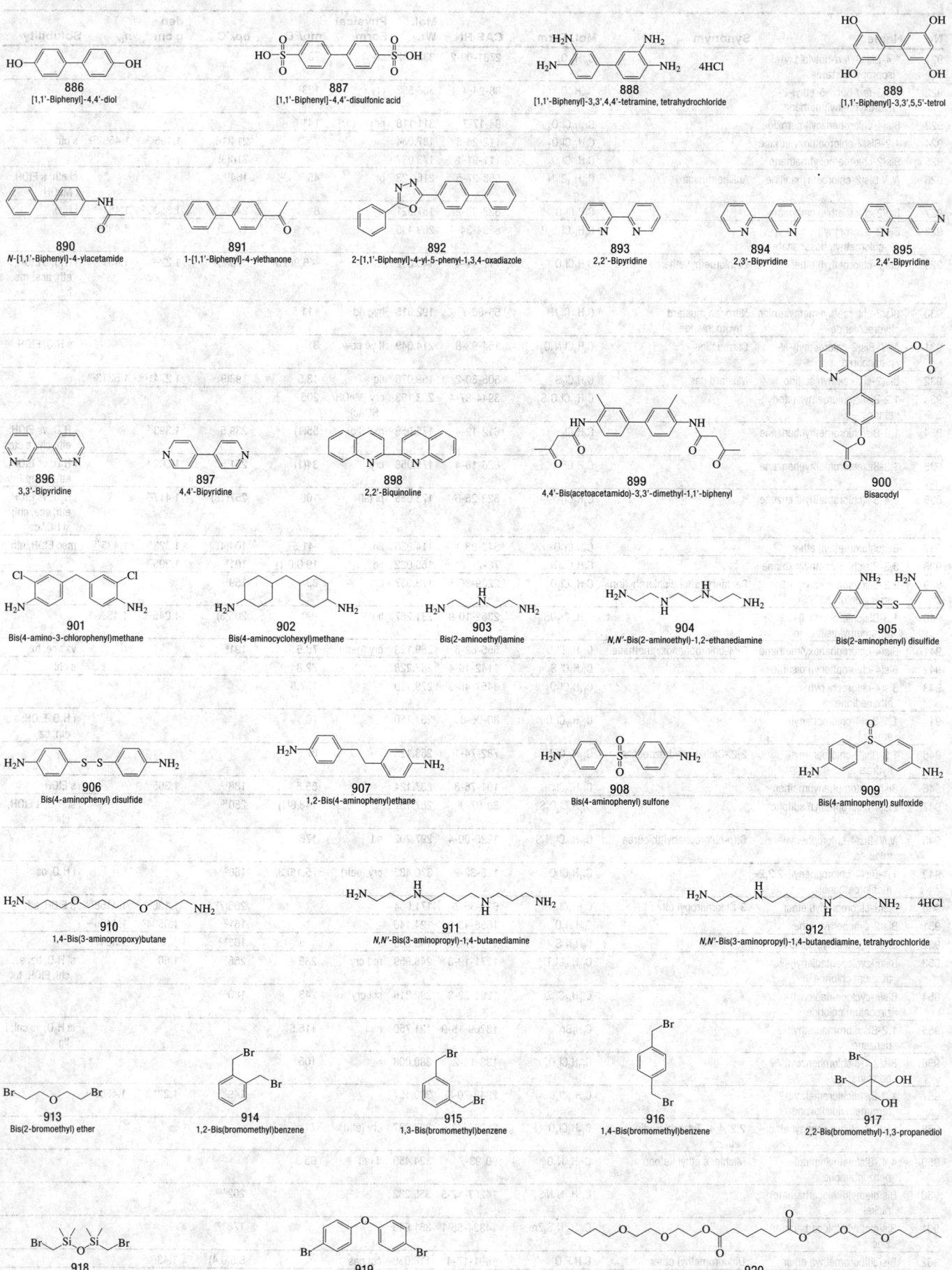

886
[1,1'-Biphenyl]-4,4'-diol

887
[1,1'-Biphenyl]-4,4'-disulfonic acid

888
[1,1'-Biphenyl]-3,3',4,4'-tetramine, tetrahydrochloride

889
[1,1'-Biphenyl]-3,3',5,5'-tetrol

890
N-[1,1'-Biphenyl]-4-ylacetamide

891
1-[1,1'-Biphenyl]-4-ylethanone

892
2-[1,1'-Biphenyl]-4-yl-5-phenyl-1,3,4-oxadiazole

893
2,2'-Bipyridine

894
2,3'-Bipyridine

895
2,4'-Bipyridine

896
3,3'-Bipyridine

897
4,4'-Bipyridine

898
2,2'-Biquinoline

899
4,4'-Bis(acetoacetamido)-3,3'-dimethyl-1,1'-biphenyl

900
Bisacodyl

901
Bis(4-amino-3-chlorophenyl)methane

902
Bis(4-aminocyclohexyl)methane

903
Bis(2-aminoethyl)amine

904
N,N'-Bis(2-aminoethyl)-1,2-ethanediamine

905
Bis(2-aminophenyl) disulfide

906
Bis(4-aminophenyl) disulfide

907
1,2-Bis(4-aminophenyl)ethane

908
Bis(4-aminophenyl) sulfone

909
Bis(4-aminophenyl) sulfoxide

910
1,4-Bis(3-aminopropoxy)butane

911
N,N'-Bis(3-aminopropyl)-1,4-butanediamine

912
N,N'-Bis(3-aminopropyl)-1,4-butanediamine, tetrahydrochloride

913
Bis(2-bromoethyl) ether

914
1,2-Bis(bromomethyl)benzene

915
1,3-Bis(bromomethyl)benzene

916
1,4-Bis(bromomethyl)benzene

917
2,2-Bis(bromomethyl)-1,3-propanediol

918
1,3-Bis(bromomethyl)tetramethyldisiloxane

919
Bis(4-bromophenyl) ether

920
Bis(2-(2-butoxyethoxy)ethyl) adipate

No.	Name	Synonym	Mol. Form.	CAS RN	Mol. Wt.	Physical Form	mp/°C	bp/°C	den g cm⁻³	n_D	Solubility
921	1,4-Bis(α-(*tert*-butyldioxy)-isopropyl)benzene		$C_{20}H_{34}O_4$	2781-00-2	338.482	cry	79				
922	Bis(3-*tert*-butyl-5-ethyl-2-hydroxyphenyl)methane		$C_{25}H_{36}O_2$	88-24-4	368.553	cry	123				
923	Bis(4-chlorobenzoyl) peroxide		$C_{14}H_8Cl_2O_4$	94-17-7	311.118	pr cry (bz)	141				
924	1,2-Bis(2-chloroethoxy)ethane		$C_6H_{12}Cl_2O_2$	112-26-5	187.064			214(18)	1.195[20]	1.4592[25]	s ctc
925	Bis(2-chloroethoxy)methane		$C_5H_{10}Cl_2O_2$	111-91-1	173.037			216(3)			
926	*N,N*-Bis(2-chloroethyl)aniline	Aniline mustard	$C_{10}H_{13}Cl_2N$	553-27-5	218.123	pr	45	164[14]			sl eth; s EtOH, MeOH
927	Bis(2-chloroethyl) carbonate		$C_5H_8Cl_2O_3$	623-97-2	187.021		8	241	1.3506[20]	1.461[20]	i H_2O
928	Bis(2-chloroethyl) 2-chloroethylphosphonate		$C_6H_{12}Cl_3O_3P$	6294-34-4	269.490			170.2[5]		1.488[25]	
929	Bis(2-chloroethyl) ether	Dichloroethyl ether	$C_4H_8Cl_2O$	111-44-4	143.012	liq	-46.9(0.5)	178(2)	1.22[20]	1.451[20]	i H_2O; s EtOH, eth, ace; msc bz
930	Bis(2-chloroethyl)methylamine hydrochloride	Nitrogen mustard hydrochloride	$C_5H_{11}Cl_3N$	55-86-7	192.515	hyg nd	111.5				
931	*N,N'*-Bis(2-chloroethyl)-*N*-nitrosourea	Carmustine	$C_5H_9Cl_2N_3O_2$	154-93-8	214.049	lt ye pow	31				vs H_2O, EtOH
932	Bis(2-chloroethyl) sulfide	Mustard gas	$C_4H_8Cl_2S$	505-60-2	159.078	liq	13.5	198(9)	1.2741[20]	1.5313[20]	
933	1,2-Bis(2-chloroethylsulfonyl)-ethane		$C_6H_{12}Cl_2O_4S_2$	3944-87-4	283.193	cry (MeOH/HOAc)	205				
934	1,2-Bis(chloromethyl)benzene		$C_8H_8Cl_2$	612-12-4	175.056	mcl (liq)	55(1)	239.5	1.393[25]		i H_2O; vs EtOH, eth, chl; s ctc
935	1,3-Bis(chloromethyl)benzene		$C_8H_8Cl_2$	626-16-4	175.056	cry	34(1)	251.5	1.302[20]		i H_2O; vs EtOH, eth; sl chl
936	1,4-Bis(chloromethyl)benzene		$C_8H_8Cl_2$	623-25-6	175.056	pl (al)	100	257(18)	1.417[25]		i H_2O; vs EtOH, eth, ace, chl; sl HOAc
937	Bis(chloromethyl) ether		$C_2H_4Cl_2O$	542-88-1	114.958	liq	-41.5	104(4)	1.323[15]	1.435[21]	msc EtOH, eth
938	3,3-Bis(chloromethyl)oxetane		$C_5H_8Cl_2O$	78-71-7	155.022	liq	19.0(0.1)	101[27]	1.295[25]		
939	2,2-Bis(chloromethyl)-1,3-propanediol	Pentaerythritol dichlorohydrin	$C_5H_{10}Cl_2O_2$	2209-86-1	173.037	cry	83	159[12]			
940	1,3-Bis(chloromethyl)-tetramethyldisiloxane		$C_6H_{16}Cl_2OSi_2$	2362-10-9	231.267	liq	-90	205(3)	1.045[20]	1.4398[20]	
941	Bis(4-chlorophenoxy)methane	Di(4-chlorophenoxy)methane	$C_{13}H_{10}Cl_2O_2$	555-89-5	269.123	cry (peth)	70.5	191[6]			vs ace, bz
942	Bis(4-chlorophenyl) disulfide		$C_{12}H_8Cl_2S_2$	1142-19-4	287.228		72.8				s chl
943	Bis(4-chlorophenyl)-ethanedione		$C_{14}H_8Cl_2O_2$	3457-46-3	279.119		197.8				
944	1,1-Bis(4-chlorophenyl)-ethanol		$C_{14}H_{12}Cl_2O$	80-06-8	267.150		70				i H_2O, EtOH; s eth, bz
945	1,2-Bis(2-chlorophenyl)-hydrazine	2,2'-Dichlorohydrazobenzene	$C_{12}H_{10}Cl_2N_2$	782-74-1	253.126		87				
946	Bis(4-chlorophenyl)methane		$C_{13}H_{10}Cl_2$	101-76-8	237.124		55.5	188[18]	1.365[17]		s EtOH
947	Bis(4-chlorophenyl) sulfone		$C_{12}H_8Cl_2O_2S$	80-07-9	287.162		148.8(1)	250[10]			sl H_2O; s EtOH, chl
948	*N,N'*-Bis(4-chlorophenyl)-thiourea	Di(p-chlorophenyl)thiourea	$C_{13}H_{10}Cl_2N_2S$	1220-00-4	297.202	nd	176				
949	1,1-Bis(4-chlorophenyl)-2,2,2-trichloroethanol		$C_{14}H_9Cl_5O$	115-32-2	370.485	cry (petr)	75.1(0.3)	180[0.1]			i H_2O, os
950	Bis(3-chloropropyl) ether	3-Chloropropyl ether	$C_6H_{12}Cl_2O$	629-36-7	171.064			205(7)	1.136[20]	1.4158[20]	s EtOH, eth
951	Bis(2-cyanoethyl) ether		$C_6H_8N_2O$	1656-48-0	124.140			161[5]	1.0504[20]	1.4405[20]	
952	Bis(2-cyanoethyl) sulfide		$C_6H_8N_2S$	111-97-7	140.206			163[1.5]		1.5047[20]	
953	Bis(η-cyclopentadienyl)-titanium chloride		$C_{10}H_{10}Cl_2Ti$	1271-19-8	248.959	red cry	289	258[10]	1.60		sl H_2O, bz; s chl, EtOH, tol
954	Bis(η-cyclopentadienyl)-zirconium chloride		$C_{10}H_{10}Cl_2Zr$	1291-32-3	292.316	col cry	248	180[0.5]			
955	1,2-Bis(dibromomethyl)-benzene		$C_8H_6Br_4$	13209-15-9	421.750	mcl	116.5				sl H_2O; vs chl; i liq
956	Bis(2,4-dichlorobenzoyl) peroxide		$C_{14}H_6Cl_4O_4$	133-14-2	380.008		106				
957	1,3-Bis(dichloromethyl)-tetramethyldisiloxane		$C_6H_{14}Cl_4OSi_2$	2943-70-6	300.157			149[50]	1.2213[20]	1.4660[20]	
958	Bis(2,4-dichlorophenyl)ether	2,2',4,4'-Tetrachlorodiphenyl ether	$C_{12}H_6Cl_4O$	28076-73-5	307.987	cry (eth)	71				
959	4,4'-Bis(diethylamino)-benzophenone	Michler's ethyl ketone	$C_{21}H_{28}N_2O$	90-93-7	324.459	lf (al)	95.3				
960	Bis(diethyldithiocarbamate)-nickel		$C_{10}H_{20}N_2NiS_4$	14267-17-5	355.232			202[0.02]			
961	Bis(diethyldithiocarbamate)-zinc		$C_{10}H_{20}N_2S_4Zn$	14324-55-1	361.948			178[0.05]			
962	Bis(difluoromethyl) ether	Difluoromethyl ether	$C_2H_2F_4O$	1691-17-4	118.030	col gas		5.5(0.4)		1.43[20]	

921
1,4-Bis(α-(tert-butyldioxy)isopropyl)benzene

922
Bis(3-tert-butyl-5-ethyl-2-hydroxyphenyl)methane

923
Bis(4-chlorobenzoyl) peroxide

924
1,2-Bis(2-chloroethoxy)ethane

925
Bis(2-chloroethoxy)methane

926
N,N-Bis(2-chloroethyl)aniline

927
Bis(2-chloroethyl) carbonate

928
Bis(2-chloroethyl) 2-chloroethylphosphonate

929
Bis(2-chloroethyl) ether

930
Bis(2-chloroethyl)methylamine hydrochloride

931
N,N'-Bis(2-chloroethyl)-N-nitrosourea

932
Bis(2-chloroethyl) sulfide

933
1,2-Bis(2-chloroethylsulfonyl)ethane

934
1,2-Bis(chloromethyl)benzene

935
1,3-Bis(chloromethyl)benzene

936
1,4-Bis(chloromethyl)benzene

937
Bis(chloromethyl) ether

938
3,3-Bis(chloromethyl)oxetane

939
2,2-Bis(chloromethyl)-1,3-propanediol

940
1,3-Bis(chloromethyl)tetramethyldisiloxane

941
Bis(4-chlorophenoxy)methane

942
Bis(4-chlorophenyl) disulfide

943
Bis(4-chlorophenyl)ethanedione

944
1,1-Bis(4-chlorophenyl)ethanol

945
1,2-Bis(2-chlorophenyl)-hydrazine

946
Bis(4-chlorophenyl)methane

947
Bis(4-chlorophenyl) sulfone

948
N,N'-Bis(4-chlorophenyl)thiourea

949
1,1-Bis(4-chlorophenyl)-2,2,2-trichloroethanol

950
Bis(3-chloropropyl) ether

951
Bis(2-cyanoethyl) ether

952
Bis(2-cyanoethyl) sulfide

953
Bis(η-cyclopentadienyl)titanium chloride

954
Bis(η-cyclopentadienyl)zirconium chloride

955
1,2-Bis(dibromomethyl)benzene

956
Bis(2,4-dichlorobenzoyl) peroxide

957
1,3-Bis(dichloromethyl)tetramethyldisiloxane

958
Bis(2,4-dichlorophenyl) ether

959
4,4'-Bis(diethylamino)benzophenone

960
Bis(diethyldithiocarbamate)nickel

961
Bis(diethyldithiocarbamate)zinc

962
Bis(difluoromethyl) ether

No.	Name	Synonym	Mol. Form.	CAS RN	Mol. Wt.	Physical Form	mp/°C	bp/°C	den g cm^{-3}	n_D	Solubility
963	Bis(2-dimethylaminoethyl) ether	2,2'-Oxybis[N,N-dimethyletha-namine]	C$_8$H$_{20}$N$_2$O	3033-62-3	160.257	liq		80^{15}			
964	Bis[4-(dimethylamino)phenyl]-methane	Michler's Base	C$_{17}$H$_{22}$N$_2$	101-61-1	254.370	pl or tab (al, lig)	91.5	390 dec			i H$_2$O; sl EtOH; vs eth, bz; s acid
965	Bis[4-(dimethylamino)phenyl]-methanethione	4,4'-Bis(dimethylamino)-thiobenzophenone	C$_{17}$H$_{20}$N$_2$S	1226-46-6	284.419	pl	204				i H$_2$O, EtOH, lig; sl eth; s bz chl, HOAc
966	Bis(4-dimethylaminophenyl)-methanol	4,4'-Bis(dimethylamino)-benzhydrol	C$_{17}$H$_{22}$N$_2$O	119-58-4	270.369		102.0				i H$_2$O; vs EtOH; s eth, bz, HOAc
967	1,3-Bis(dimethylamino)-2-propanol		C$_7$H$_{18}$N$_2$O	5966-51-8	146.230			181.5	0.8788^{20}	1.4418^{20}	vs H$_2$O
968	4,4'-Bis(dimethylamino)-triphenylmethane		C$_{23}$H$_{26}$N$_2$	129-73-7	330.465	nd or lf (al, bz)	102				vs bz, eth
969	Bis(dimethyldithiocarbamate)-copper		C$_6$H$_{12}$CuN$_2$S$_4$	137-29-1	303.978			206$^{0.01}$			
970	Bis(dimethyldithiocarbamate)-nickel		C$_6$H$_{12}$N$_2$NiS$_4$	15521-65-0	299.125			208$^{0.002}$			
971	2,5-Bis(1,1-dimethylpropyl)-1,4-benzenediol	2,5-Di-tert-pentylhydroquinone	C$_{16}$H$_{26}$O$_2$	79-74-3	250.376		180				
972	2,4-Bis(1,1-dimethylpropyl)-phenol		C$_{16}$H$_{26}$O	120-95-6	234.376		26.0	169^{22}			
973	1,2-Bis(diphenylphosphino)-ethane	Diphos	C$_{26}$H$_{24}$P$_2$	1663-45-2	398.417		143.5				
974	1,3-Bis(2,3-epoxypropoxy)-benzene	Diglycidyl resorcinol ether	C$_{12}$H$_{14}$O$_4$	101-90-6	222.237		42.5	147$^{0.4}$	1.2183^{30}	1.5408^{20}	
975	Bis(2-ethoxyethyl) phthalate		C$_{16}$H$_{22}$O$_6$	605-54-9	310.342		34	345	1.1229^{21}		
976	Bis(ethoxymethyl) ether		C$_6$H$_{14}$O$_3$	5648-29-3	134.173			136(10)			
977	N,N'-Bis(4-ethoxyphenyl)-ethanimidamide monohydrochloride	Phenacaine hydrochloride	C$_{18}$H$_{23}$ClN$_2$O$_2$	620-99-5	334.841	cry (w+1)	191				vs H$_2$O, EtOH, chl
978	Bis(ethylenediamine)copper dichloride	Cupriethylenediamine dichloride	C$_4$H$_{16}$Cl$_2$CuN$_4$	15243-01-3	254.649	dk bl cry					s EtOH
979	Bis(2-ethylhexyl) adipate	Bis(2-ethylhexyl) hexanedioate	C$_{22}$H$_{42}$O$_4$	103-23-1	370.566		-67.8	214^5	0.922^{25}	1.4474^{20}	vs ace, eth, EtOH
980	Bis(2-ethylhexyl)amine		C$_{16}$H$_{35}$N	106-20-7	241.456			161^{21}			
981	Bis(2-ethylhexyl) azelate		C$_{25}$H$_{48}$O$_4$	103-24-2	412.647		-78	237^5	0.915^{25}	1.446^{25}	i H$_2$O; s EtOH, ace, bz; sl ctc
982	Bis(2-ethylhexyl) ether	2,2'-Diethyldihexyl ether	C$_{16}$H$_{34}$O	10143-60-9	242.440			279(8)		1.4325^{20}	sl ctc
983	Bis(2-ethylhexyl) phosphate		C$_{16}$H$_{35}$O$_4$P	298-07-7	322.420	visc liq		155$^{0.015}$	0.975^{25}		sl H$_2$O; s bz, hx
984	Bis(2-ethylhexyl) phosphonate	Bis(2-ethylhexyl) phosphite	C$_{16}$H$_{35}$O$_3$P	3658-48-8	306.421	liq		150^1	0.93^{25}	1.4420^{20}	
985	Bis(2-ethylhexyl) phosphorodithioate		C$_{16}$H$_{35}$O$_2$PS$_2$	5810-88-8	354.552	cry					s bz, hp, chl
986	Bis(2-ethylhexyl) phthalate	Di-sec-octyl phthalate	C$_{24}$H$_{38}$O$_4$	117-81-7	390.557	liq	-55	384	0.981^{25}	1.4853^{20}	sl ctc
987	Bis(2-ethylhexyl) sebacate		C$_{26}$H$_{50}$O$_4$	122-62-3	426.673		-55(1)	256^5	0.912^{25}	1.451^{25}	vs ace, bz, EtOH
988	Bis(2-ethylhexyl) sodium sulfosuccinate	Docusate sodium	C$_{20}$H$_{37}$NaO$_7$S	577-11-7	444.559	waxy solid					s peth, ctc, eth, ace
989	Bis(2-ethylhexyl) terephthalate		C$_{24}$H$_{38}$O$_4$	6422-86-2	390.557			383			
990	2,2-Bis(ethylsulfonyl)butane	Sulfonethylmethane	C$_8$H$_{18}$O$_4$S$_2$	76-20-0	242.357	pl (w)	76	dec	1.199^{85}		s chl
991	Bis[4-(hexyloxy)phenyl]-diazene, 1-oxide		C$_{24}$H$_{34}$N$_2$O$_3$	2587-42-0	398.538						s chl
992	N,N'-Bis(2-hydroxybenzylidene)-1,2-ethylenediamine	Disalicylidene-1,2-ethanedi-amine	C$_{16}$H$_{16}$N$_2$O$_2$	94-93-9	268.310		125.5				sl EtOH, eth; s bz, chl
993	Bis(2-hydroxy-3-tert-butyl-5-methylphenyl)methane		C$_{23}$H$_{32}$O$_2$	119-47-1	340.499	nd (peth)	131				
994	Bis(2-hydroxy-5-chlorophenyl) sulfide	Fenticlor	C$_{12}$H$_8$Cl$_2$O$_2$S	97-24-5	287.162		174				i H$_2$O; s EtOH, eth, gl HOAc
995	2-[Bis(2-hydroxyethyl)amino]-ethanol hydrochloride	Triethanolamine hydrochloride	C$_6$H$_{16}$ClNO$_3$	637-39-8	185.649	cry (al)	179.5				vs H$_2$O
996	N,N-Bis(2-hydroxyethyl)-butylamine	Butylbis(2-hydroxyethyl)amine	C$_8$H$_{19}$NO$_2$	102-79-4	161.243			279(19)	0.9681^{20}	1.4625^{20}	s chl
997	Bis(2-hydroxyethyl) disulfide		C$_4$H$_{10}$O$_2$S$_2$	1892-29-1	154.251		26	160$^{3.5}$			
998	N,N-Bis(2-hydroxyethyl)-dodecanamide		C$_{16}$H$_{33}$NO$_3$	120-40-1	287.438	waxy solid	38.7				
999	N,N-Bis(2-hydroxyethyl)-ethylamine	N-Ethyldiethanolamine	C$_6$H$_{15}$NO$_2$	139-87-7	133.189	ye liq	-50	248(19)	1.0135^{20}	1.4663^{20}	vs H$_2$O, EtOH; sl eth
1000	N,N'-Bis(2-hydroxyethyl)-ethylenediamine		C$_6$H$_{16}$N$_2$O$_2$	4439-20-7	148.203		100.1(0.5)	136^1			s H$_2$O
1001	N,N-Bis(2-hydroxyethyl)-glycine	Bicine	C$_6$H$_{13}$NO$_4$	150-25-4	163.172	nd (al)	194 dec				vs H$_2$O; i EtOH

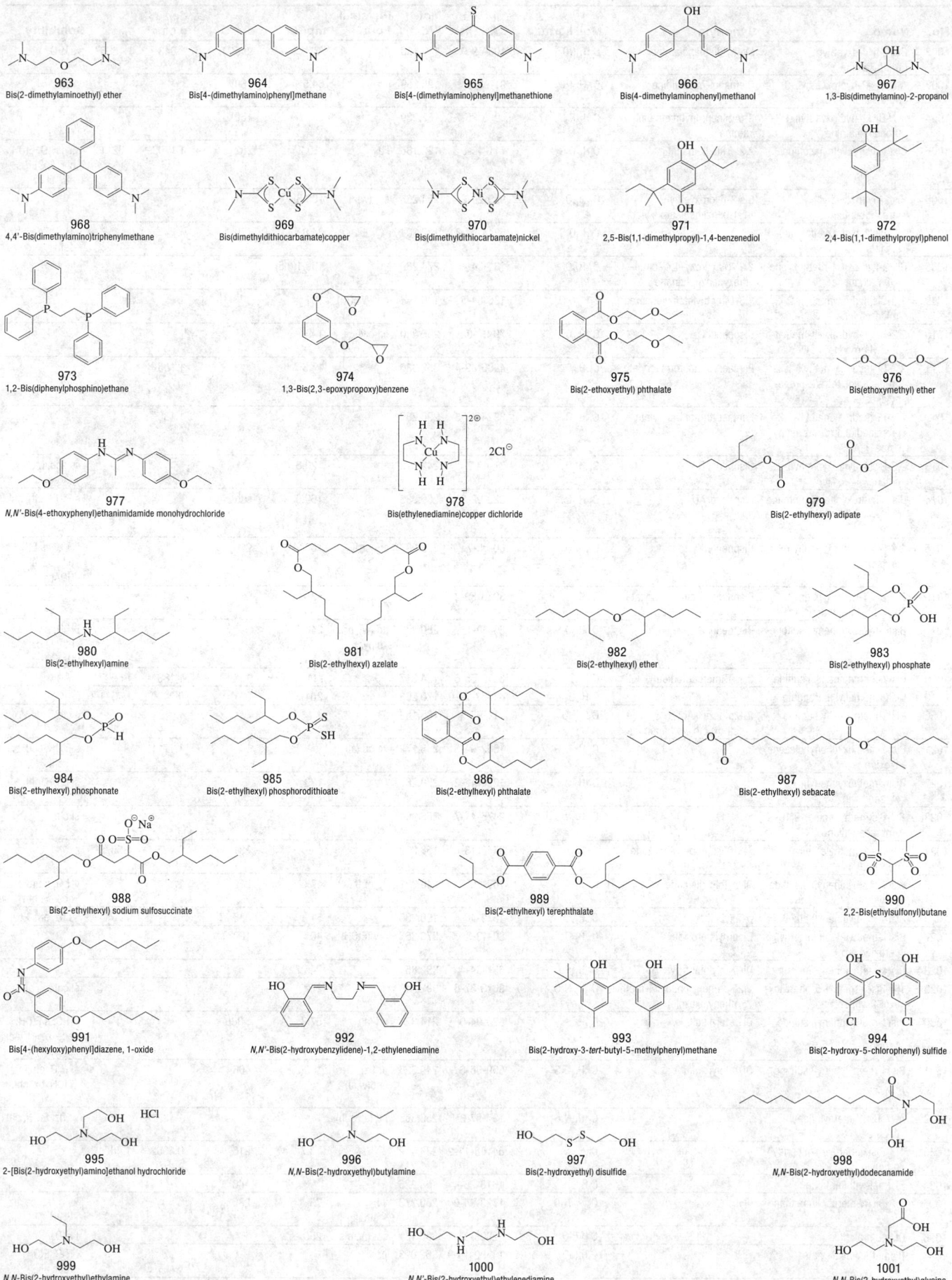

963 Bis(2-dimethylaminoethyl) ether

964 Bis[4-(dimethylamino)phenyl]methane

965 Bis[4-(dimethylamino)phenyl]methanethione

966 Bis(4-dimethylaminophenyl)methanol

967 1,3-Bis(dimethylamino)-2-propanol

968 4,4'-Bis(dimethylamino)triphenylmethane

969 Bis(dimethyldithiocarbamate)copper

970 Bis(dimethyldithiocarbamate)nickel

971 2,5-Bis(1,1-dimethylpropyl)-1,4-benzenediol

972 2,4-Bis(1,1-dimethylpropyl)phenol

973 1,2-Bis(diphenylphosphino)ethane

974 1,3-Bis(2,3-epoxypropoxy)benzene

975 Bis(2-ethoxyethyl) phthalate

976 Bis(ethoxymethyl) ether

977 *N,N'*-Bis(4-ethoxyphenyl)ethanimidamide monohydrochloride

978 Bis(ethylenediamine)copper dichloride

979 Bis(2-ethylhexyl) adipate

980 Bis(2-ethylhexyl)amine

981 Bis(2-ethylhexyl) azelate

982 Bis(2-ethylhexyl) ether

983 Bis(2-ethylhexyl) phosphate

984 Bis(2-ethylhexyl) phosphonate

985 Bis(2-ethylhexyl) phosphorodithioate

986 Bis(2-ethylhexyl) phthalate

987 Bis(2-ethylhexyl) sebacate

988 Bis(2-ethylhexyl) sodium sulfosuccinate

989 Bis(2-ethylhexyl) terephthalate

990 2,2-Bis(ethylsulfonyl)butane

991 Bis[4-(hexyloxy)phenyl]diazene, 1-oxide

992 *N,N'*-Bis(2-hydroxybenzylidene)-1,2-ethylenediamine

993 Bis(2-hydroxy-3-*tert*-butyl-5-methylphenyl)methane

994 Bis(2-hydroxy-5-chlorophenyl) sulfide

995 2-[Bis(2-hydroxyethyl)amino]ethanol hydrochloride

996 *N,N*-Bis(2-hydroxyethyl)butylamine

997 Bis(2-hydroxyethyl) disulfide

998 *N,N*-Bis(2-hydroxyethyl)dodecanamide

999 *N,N*-Bis(2-hydroxyethyl)ethylamine

1000 *N,N'*-Bis(2-hydroxyethyl)ethylenediamine

1001 *N,N*-Bis(2-hydroxyethyl)glycine

No.	Name	Synonym	Mol. Form.	CAS RN	Mol. Wt.	Physical Form	mp/°C	bp/°C	den g cm⁻³	n_D	Solubility
1002	Bis(2-hydroxyethyl)-methylamine	Methyldiethanolamine	$C_5H_{13}NO_2$	105-59-9	119.163	liq	-21	245(1)	1.043[25]	1.4685[20]	vs H_2O
1003	N,N-Bis(2-hydroxyethyl)-3-methylaniline	Diethanol-m-toluidine	$C_{11}H_{17}NO_2$	91-99-6	195.259		64.5	160[1]			sl chl
1004	N,N-Bis(2-hydroxyethyl)-1,3-propanediamine	3-(Aminopropyl)diethanol-amine	$C_7H_{18}N_2O_2$	4985-85-7	162.230			160[1]			
1005	Bis(2-hydroxyethyl) sulfide	2,2'-Thiodiethanol	$C_4H_{10}O_2S$	111-48-8	122.186	liq	-10.2	282	1.1793[25]	1.5211[20]	msc H_2O, EtOH, chl, AcOEt; s eth; sl bz
1006	Bis(2-hydroxyethyl) terephthalate	Bis(2-hydroxyethyl) 1,4-benzenedicarboxylate	$C_{12}H_{14}O_6$	959-26-2	254.235	cry (w)	109.5				
1007	1,2-Bis(2-hydroxyethylthio)-ethane		$C_6H_{14}O_2S_2$	5244-34-8	182.304		64.8	170[0.5]			s H_2O, EtOH, bz, peth
1008	Bis(2-hydroxy-4-methoxyphe-nyl)methanone	2,2'-Dihydroxy-4,4'-dime-thoxybenzophenone	$C_{15}H_{14}O_5$	131-54-4	274.269		139.1(0.5)				
1009	1,3-Bis(hydroxymethyl)-2-imidazolidone	1,3-Dimethylolethyleneurea	$C_5H_{10}N_2O_3$	136-84-5	146.144	cry (MeOH)	101				
1010	2,2-Bis(4-hydroxy-3-methyl-phenyl)propane	Bisphenol C	$C_{17}H_{20}O_2$	79-97-0	256.340	nd (xyl)	140				
1011	2,2-Bis(hydroxymethyl)-1,3-propanediol, tetra(2-prope-noyl) ester	Pentaerythritol tetraacrylate	$C_{17}H_{20}O_8$	4986-89-4	352.336		17.3		1.185[25]		
1012	2,2-Bis(hydroxymethyl)-1,3-propanediol, tri(2-propenoyl) ester	Pentaerythritol triacrylate	$C_{14}H_{18}O_7$	3524-68-3	298.289				1.180[20]		
1013	2,2-Bis(4-hydroxyphenyl)-butane	Bisphenol B	$C_{16}H_{18}O_2$	77-40-7	242.313		120.5				vs ace, MeOH
1014	Bis(4-hydroxyphenyl)methane	Bisphenol AD	$C_{13}H_{12}O_2$	620-92-8	200.233		162.5	sub			s EtOH, eth, chl, alk; sl DMSO; i CS_2
1015	2,2-Bis(4-hydroxyphenyl)-propane	Bisphenol A	$C_{15}H_{16}O_2$	80-05-7	228.287	cry or fl	160(2)	220[4]			i H_2O; vs EtOH, eth, bz, alk; s HOAc
1016	2,2-Bis(4-hydroxyphenyl)-propane dimethacrylate	Bisphenol A dimethacrylate	$C_{23}H_{24}O_4$	3253-39-2	364.435		73				
1017	Bis(4-hydroxyphenyl) sulfone	Bisphenol S	$C_{12}H_{10}O_4S$	80-09-1	250.270	nd (w), orth bipym	240.5		1.3663[15]		i H_2O; s EtOH, eth; sl bz, DMSO
1018	Bis(2-mercaptoethyl) sulfide	2,2'-Dimercaptodiethyl sulfide	$C_4H_{10}S_3$	3570-55-6	154.317		-11	135[18]	1.183[25]	1.5982[20]	
1019	Bis(2-methallyl) carbonate		$C_9H_{14}O_3$	64057-79-0	170.205		201.3	66[3]	0.943[25]	1.4371[20]	
1020	Bis(2-methoxyethyl)amine	2-Methoxy-N-(2-methoxyethyl)ethanamine	$C_6H_{15}NO_2$	111-95-5	133.189						s ctc
1021	Bis(4-methoxyphenyl)diazene, 1-oxide		$C_{14}H_{14}N_2O_3$	1562-94-3	258.272	ye nd (al)			1.1711[11]		s EtOH, ace, bz; sl chl
1022	Bis(4-methoxyphenyl)-ethanedione		$C_{16}H_{14}O_4$	1226-42-2	270.280		133				sl EtOH, chl
1023	1,4-Bis(methylamino)-9,10-anthracenedione		$C_{16}H_{14}N_2O_2$	2475-44-7	266.294						sl chl
1024	1,3-Bis(1-methylethenyl)-benzene	1,3-Diisopropenylbenzene	$C_{12}H_{14}$	3748-13-8	158.239	liq		231	0.925	1.5570[20]	
1025	Bis(4-methylphenyl) disulfide	Di-p-Tolyl disulfide	$C_{14}H_{14}S_2$	103-19-5	246.391	nd or lf (al)	47.5	212[20]	1.114[51]		i H_2O; s EtOH, ace; vs eth
1026	Bis(4-methylphenyl) ether	p-Tolyl ether	$C_{14}H_{14}O$	1579-40-4	198.260		51	285			vs bz, eth, EtOH
1027	Bis(1-methyl-1-phenylethyl) peroxide	Dicumyl peroxide	$C_{18}H_{22}O_2$	80-43-3	270.367	cry (EtOH)	40	100[0.2]			
1028	Bis(4-methylphenyl)mercury	Di-p-tolylmercury	$C_{14}H_{14}Hg$	537-64-4	382.85		245.7				
1029	1,4-Bis(4-methyl-5-phenylox-azol-2-yl)benzene	2,2'-p-Phenylenebis(4-methyl-5-phenyloxazole)	$C_{26}H_{20}N_2O_2$	3073-87-8	392.449		232				sl chl
1030	Bis(4-methylphenyl) sulfide	Di-p-tolyl sulfide	$C_{14}H_{14}S$	620-94-0	214.326	nd (al)	57.3	>300			i H_2O; s EtOH, ace, bz, HOAc; sl chl
1031	Bis(4-methylphenyl) sulfone	Di-p-tolyl sulfone	$C_{14}H_{14}O_2S$	599-66-6	246.325	pr(bz), nd(w,al)	150(8)	406			sl H_2O, eth; s EtOH, bz, chl, CS_2
1032	N,N'-Bis(2-methylphenyl)-thiourea		$C_{15}H_{16}N_2S$	137-97-3	256.366	nd (al, sub)					vs bz, EtOH, chl
1033	1,3-Bis(1-methyl-4-piperidyl)-propane		$C_{15}H_{30}N_2$	64168-11-2	238.412		13.7	215[50]	0.8962[25]	1.4804[25]	
1034	Bis(methylthio)methane		$C_3H_8S_2$	1618-26-4	108.226			148			
1035	1,2-Bis(N-morpholino)ethane		$C_{10}H_{20}N_2O_2$	1723-94-0	200.278	wh-ye (eth,lig)	75	285			vs H_2O, ace, bz, EtOH
1036	Bismuth acetate		$C_6H_9BiO_6$	22306-37-2	386.111	col tablets	250				i H_2O
1037	Bismuth subsalicylate		$C_7H_5BiO_4$	14882-18-9	362.093	pr					i H_2O, EtOH; reac alk

1002
Bis(2-hydroxyethyl)methylamine

1003
N,N-Bis(2-hydroxyethyl)-3-methylaniline

1004
N,N-Bis(2-hydroxyethyl)-1,3-propanediamine

1005
Bis(2-hydroxyethyl) sulfide

1006
Bis(2-hydroxyethyl) terephthalate

1007
1,2-Bis(2-hydroxyethylthio)ethane

1008
Bis(2-hydroxy-4-methoxyphenyl)methanone

1009
1,3-Bis(hydroxymethyl)-2-imidazolidone

1010
2,2-Bis(4-hydroxy-3-methylphenyl)propane

1011
2,2-Bis(hydroxymethyl)-1,3-propanediol, tetra(2-propenoyl) ester

1012
2,2-Bis(hydroxymethyl)-1,3-propanediol, tri(2-propenoyl) ester

1013
2,2-Bis(4-hydroxyphenyl)butane

1014
Bis(4-hydroxyphenyl)methane

1015
2,2-Bis(4-hydroxyphenyl)propane

1016
2,2-Bis(4-hydroxyphenyl)propane dimethacrylate

1017
Bis(4-hydroxyphenyl) sulfone

1018
Bis(2-mercaptoethyl) sulfide

1019
Bis(2-methallyl) carbonate

1020
Bis(2-methoxyethyl)amine

1021
Bis(4-methoxyphenyl)diazene, 1-oxide

1022
Bis(4-methoxyphenyl)ethanedione

1023
1,4-Bis(methylamino)-9,10-anthracenedione

1024
1,3-Bis(1-methylethenyl)benzene

1025
Bis(4-methylphenyl) disulfide

1026
Bis(4-methylphenyl) ether

1027
Bis(1-methyl-1-phenylethyl)peroxide

1028
Bis(4-methylphenyl)mercury

1029
1,4-Bis(4-methyl-5-phenyloxazol-2-yl)benzene

1030
Bis(4-methylphenyl) sulfide

1031
Bis(4-methylphenyl) sulfone

1032
N,N′-Bis(2-methylphenyl)thiourea

1033
1,3-Bis(1-methyl-4-piperidyl)propane

1034
Bis(methylthio)methane

1035
1,2-Bis(*N*-morpholino)ethane

1036
Bismuth acetate

1037
Bismuth subsalicylate

No.	Name	Synonym	Mol. Form.	CAS RN	Mol. Wt.	Physical Form	mp/°C	bp/°C	den g cm⁻³	n_D	Solubility
1038	Bis(2-nitrophenyl) disulfide		$C_{12}H_8N_2O_4S_2$	1155-00-6	308.333		198.5				i H_2O, eth; sl EtOH, ace, bz, HOAc
1039	Bis(3-nitrophenyl) disulfide	Nitrophenide	$C_{12}H_8N_2O_4S_2$	537-91-7	308.333		84				sl EtOH, chl; s eth
1040	Bis(4-nitrophenyl) disulfide		$C_{12}H_8N_2O_4S_2$	100-32-3	308.333		182	$255^{0.1}$			sl EtOH, HOAc
1041	1,2-Bis(4-nitrophenyl)ethane	4,4'-Dinitrobibenzyl	$C_{14}H_{12}N_2O_4$	736-30-1	272.256	ye nd (al,bz)	181.8				i EtOH; sl eth, bz, chl, HOAc
1042	N,N'-Bis(4-nitrophenyl)urea	4,4'-Dinitrocarbanilide	$C_{13}H_{10}N_4O_5$	587-90-6	302.242		312 dec				
1043	Bis(2,4-pentanedionato)cobalt	Cobalt(II) bis(acetylacetonate)	$C_{10}H_{14}CoO_4$	14024-48-7	257.149	bl-viol cry	167				
1044	Bis(1-phenylethyl)amine		$C_{16}H_{19}N$	10024-74-5	225.329			296.5	1.018^{15}	1.573	
1045	1,2-Bis(2,4,6-tribromophe-noxy)ethane		$C_{14}H_8Br_6O_2$	37853-59-1	687.637	nd (bz/ EtOH)	222				
1046	N,N'-Bis(2,2,2-trichloro-1-hydroxyethyl)urea		$C_5H_6Cl_6N_2O_3$	116-52-9	354.831		196				vs ace, EtOH
1047	1,4-Bis(trichloromethyl)-benzene		$C_8H_4Cl_6$	68-36-0	312.836	cry (bz, eth)	109				s chl
1048	Bis(trichloromethyl) carbonate	Triphosgene	$C_3Cl_6O_3$	32315-10-9	296.748	cry (eth, peth)	79	203	1.6290^{80}		
1049	Bis(tridecyl) thiodipropanoate	Ditridecyl thiodipropionate	$C_{32}H_{62}O_4S$	10595-72-9	542.897			$265^{0.25}$			vs EtOH
1050	3,5-Bis(trifluoromethyl)aniline		$C_8H_5F_6N$	328-74-5	229.123		85^{15}		1.487^{25}	1.4335^{20}	
1051	1,3-Bis(trifluoromethyl)-benzene		$C_8H_4F_6$	402-31-3	214.108			116	1.3790^{25}	1.3916^{25}	i H_2O
1052	1,4-Bis(trifluoromethyl)-benzene		$C_8H_4F_6$	433-19-2	214.108	liq		115			
1053	Bis(trifluoromethyl) disulfide		$C_2F_6S_2$	372-64-5	202.141			34.6			vs EtOH, peth
1054	1,2-Bis(trimethylsilyl)acetylene		$C_8H_{18}Si_2$	14630-40-1	170.400		26	134	0.770^{20}	1.413^{20}	
1055	Bis(2,4,6-trinitrophenyl) sulfide	Dipicryl sulfide	$C_{12}H_4N_6O_{12}S$	2217-06-3	456.258	ye cry	230				
1056	Bis[2-(vinyloxy)ethyl] ether	Diethylene glycol divinyl ether	$C_8H_{14}O_3$	764-99-8	158.195			81^{10}			
1057	Bithionol		$C_{12}H_6Cl_4O_2S$	97-18-7	356.052		188		1.73^{25}		vs ace
1058	2,2'-Bithiophene		$C_8H_6S_2$	492-97-7	166.264		31.1(0.4)	260			i H_2O; vs EtOH; s eth, ctc, HOAc
1059	Bixin		$C_{25}H_{30}O_4$	6983-79-5	394.504	viol pr (ace)	198				i H_2O; s EtOH, ace; sl eth, bz, HOAc
1060	Boldenone	Dehydrotestosterone	$C_{19}H_{26}O_2$	846-48-0	286.408		165				
1061	Boldine		$C_{19}H_{21}NO_4$	476-70-0	327.375	cry (eth)	163				vs EtOH, chl
1062	Bomyl		$C_9H_{15}O_8P$	122-10-1	282.184	ye oil		160^{17}			sl H_2O; vs ace, EtOH, xyl
1063	Borane carbonyl	Borine carbonyl	CH_3BO	13205-44-2	41.845	col gas	-137	-64			dec H_2O
1064	Borneol, (±)-		$C_{10}H_{18}O$	6627-72-1	154.249	lf (lig)	206(7)	213(7)	1.011^{20}		i H_2O; vs EtOH, eth, bz
1065	l-Bornyl acetate		$C_{12}H_{20}O_2$	5655-61-8	196.286		27	223.5	0.982^{25}	1.4626^{20}	sl H_2O; s EtOH, eth
1066	Bornylamine		$C_{10}H_{19}N$	32511-34-5	153.265		163				vs ace, bz, eth, EtOH
1067	Bornyl chloride	2-Chloro-1,7,7-trimethylbicy-clo[2.2.1]heptane, endo	$C_{10}H_{17}Cl$	464-41-5	172.695	nd	128(6)	207.5			vs bz, eth, EtOH, peth
1068	Bornyl 3-methylbutanoate, (1R)-	d-Bornyl isovalerate	$C_{15}H_{26}O_2$	53022-14-3	238.366			257.5	0.955^{25}		vs eth, EtOH
1069	Boron trifluoride - dimethyl ether complex		$C_2H_6BF_3O$	353-42-4	113.874		-14	127 dec	1.2410^{20}	1.302^{20}	
1070	Boron trifluoride etherate	Trifluoroboron etherate	$C_4H_{10}BF_3O$	109-63-7	141.927	liq	-60.4	125.5	1.125^{25}	1.348^{20}	dec H_2O; vs eth, EtOH
1071	Brilliant Green		$C_{27}H_{34}N_2O_4S$	633-03-4	482.635	small gold cry					vs H_2O, EtOH
1072	Brilliant Yellow		$C_{26}H_{20}N_4Na_2O_6S_2$	3051-11-4	626.569	ye cry (w)					s H_2O, EtOH; sl ace
1073	Brodifacoum		$C_{31}H_{23}BrO_3$	56073-10-0	523.417	off-wh pow	230				i H_2O; sl EtOH, bz; s ace, chl
1074	Bromacil	5-Bromo-3-sec-butyl-6-methyluracil	$C_9H_{13}BrN_2O_2$	314-40-9	261.115		157.8(0.5)		1.55^{25}		
1075	Bromadiolone		$C_{30}H_{23}BrO_4$	28772-56-7	527.406	ye-wh pow	205				vs DMF; sl ace, chl, EtOH, eth; i hx
1076	Bromal hydrate		$C_2H_3Br_3O_2$	507-42-6	298.756	mcl pr (w+1)	46.0(0.7)	dec	2.5661^{40}		vs eth, EtOH
1077	Bromdian	Tetrabromobisphenol A	$C_{15}H_{12}Br_4O_2$	79-94-7	543.871		179				s EtOH, eth, bz, chl
1078	N-Bromoacetamide		C_2H_4BrNO	79-15-2	137.963	nd (chl-hx)	103.5				vs eth

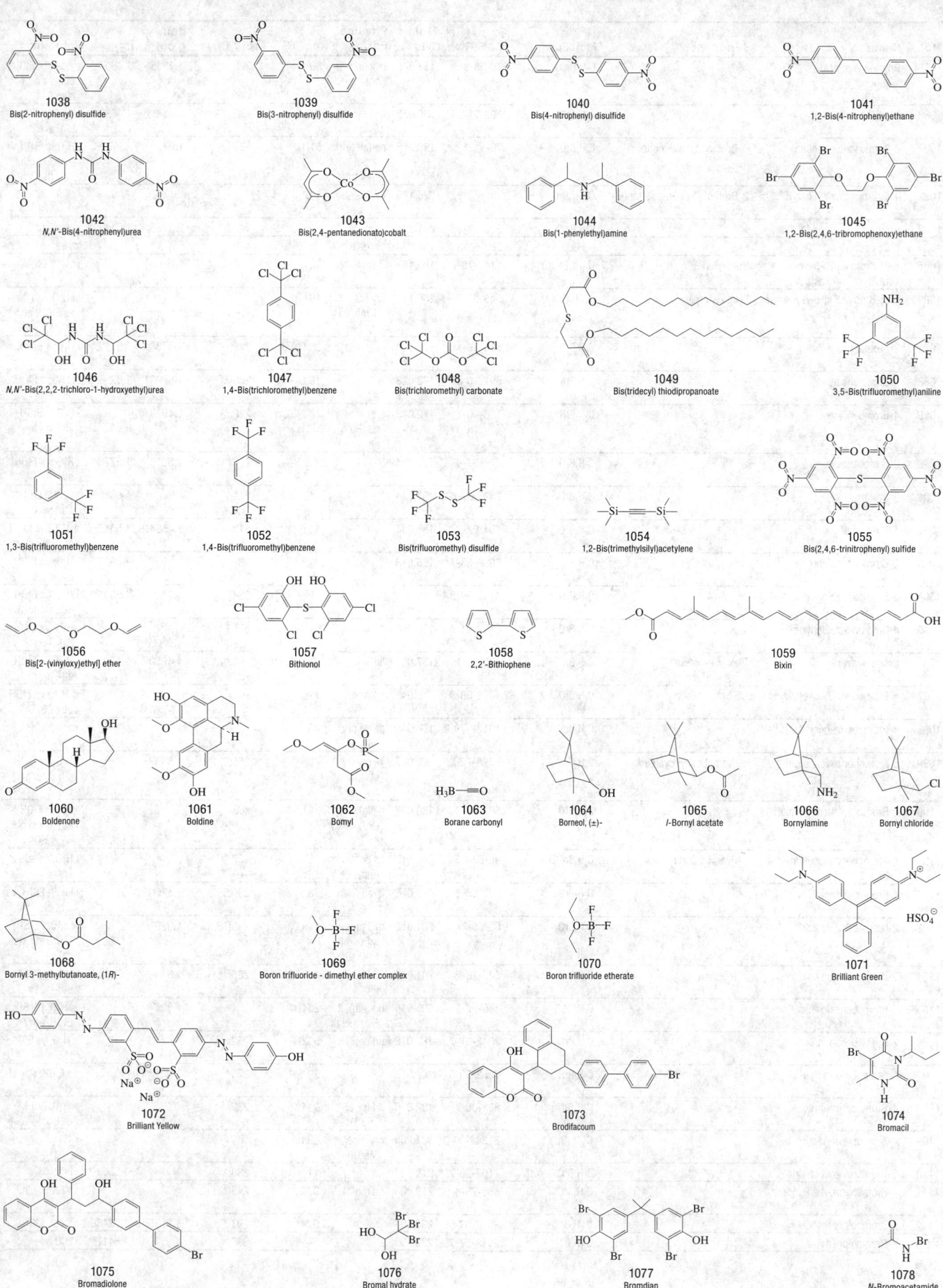

1038
Bis(2-nitrophenyl) disulfide

1039
Bis(3-nitrophenyl) disulfide

1040
Bis(4-nitrophenyl) disulfide

1041
1,2-Bis(4-nitrophenyl)ethane

1042
N,N'-Bis(4-nitrophenyl)urea

1043
Bis(2,4-pentanedionato)cobalt

1044
Bis(1-phenylethyl)amine

1045
1,2-Bis(2,4,6-tribromophenoxy)ethane

1046
N,N'-Bis(2,2,2-trichloro-1-hydroxyethyl)urea

1047
1,4-Bis(trichloromethyl)benzene

1048
Bis(trichloromethyl) carbonate

1049
Bis(tridecyl) thiodipropanoate

1050
3,5-Bis(trifluoromethyl)aniline

1051
1,3-Bis(trifluoromethyl)benzene

1052
1,4-Bis(trifluoromethyl)benzene

1053
Bis(trifluoromethyl) disulfide

1054
1,2-Bis(trimethylsilyl)acetylene

1055
Bis(2,4,6-trinitrophenyl) sulfide

1056
Bis[2-(vinyloxy)ethyl] ether

1057
Bithionol

1058
2,2'-Bithiophene

1059
Bixin

1060
Boldenone

1061
Boldine

1062
Bomyl

1063
Borane carbonyl

1064
Borneol, (±)-

1065
l-Bornyl acetate

1066
Bornylamine

1067
Bornyl chloride

1068
Bornyl 3-methylbutanoate, (1R)-

1069
Boron trifluoride - dimethyl ether complex

1070
Boron trifluoride etherate

1071
Brilliant Green

1072
Brilliant Yellow

1073
Brodifacoum

1074
Bromacil

1075
Bromadiolone

1076
Bromal hydrate

1077
Bromdian

1078
N-Bromoacetamide

No.	Name	Synonym	Mol. Form.	CAS RN	Mol. Wt.	Physical Form	mp/°C	bp/°C	den g cm^{-3}	n_D	Solubility
1079	Bromoacetic acid		$C_2H_3BrO_2$	79-08-3	138.948	hex or orth cry	50	208	1.9335[50]	1.4804[50]	msc H_2O, EtOH, eth; s ace, bz; sl chl
1080	Bromoacetone		C_3H_5BrO	598-31-2	136.975	liq	-36.5	138	1.634[23]	1.4697[15]	sl H_2O; s EtOH, eth, ace
1081	α-Bromoacetophenone	ω-Bromoacetophenone	C_8H_7BrO	70-11-1	199.045	nd (al) orth pr (al) pl(peth)	51(1)	135[18]	1.647[20]	i H_2O; s EtOH, peth; vs eth, bz, chl
1082	4-(Bromoacetyl)biphenyl	2-Bromo-4'-phenylacetophenone	$C_{14}H_{11}BrO$	135-73-9	275.140	nd (95% al)	127				
1083	Bromoacetyl bromide		$C_2H_2Br_2O$	598-21-0	201.844			148.5	2.312[22]	1.5449[20]	s ace, ctc
1084	Bromoacetylene		C_2HBr	593-61-3	104.933	col gas		4.7			vs eth
1085	5-(2-Bromoallyl)-5-sec-butylbarbituric acid	Butallylonal	$C_{11}H_{15}BrN_2O_3$	1142-70-7	303.152		131.5				vs eth, EtOH
1086	5-(2-Bromoallyl)-5-isopropyl-barbituric acid	Propallylonal	$C_{10}H_{13}BrN_2O_3$	545-93-7	289.125	cry (dil HOAc, dil al)	181				sl H_2O, eth, bz; vs EtOH, ace, HOAc
1087	2-Bromoaniline		C_6H_6BrN	615-36-1	172.023		30.9(0.5)	229	1.578[20]	1.6113[20]	i H_2O; s EtOH, eth
1088	3-Bromoaniline		C_6H_6BrN	591-19-5	172.023		18.5	250(3)	1.5793[20]	1.6260[20]	sl H_2O; s EtOH, eth
1089	4-Bromoaniline		C_6H_6BrN	106-40-1	172.023	orth bipym nd (60% al)	78.2(0.5)	220(6)	1.4970[100]		i H_2O; s EtOH, eth; sl chl
1090	2-Bromoanisole		C_7H_7BrO	578-57-4	187.034		1.3	216	1.5018[20]	1.5727[20]	i H_2O; vs EtOH, eth
1091	3-Bromoanisole		C_7H_7BrO	2398-37-0	187.034			211		1.5635[20]	i H_2O; s EtOH, eth, bz, CS_2
1092	4-Bromoanisole		C_7H_7BrO	104-92-7	187.034		13.5	215	1.4564[20]	1.5642[20]	sl H_2O; vs EtOH, eth, chl; s ctc
1093	2-Bromobenzaldehyde		C_7H_5BrO	6630-33-7	185.018		21.5	230		1.5925[20]	i H_2O; vs EtOH, bz; sl ctc
1094	3-Bromobenzaldehyde		C_7H_5BrO	3132-99-8	185.018			234		1.5935[20]	i H_2O; vs EtOH, eth; sl ctc
1095	4-Bromobenzaldehyde		C_7H_5BrO	1122-91-4	185.018	lf (dil al)	61.1(0.4)	67[2]			i H_2O; vs EtOH, bz; sl chl
1096	Bromobenzene	Phenyl bromide	C_6H_5Br	108-86-1	157.008	liq	-30.74(0.03)	155.9(0.2)	1.4950[20]	1.5597[20]	i H_2O; vs EtOH, eth, bz; s ctc
1097	4-Bromobenzeneacetic acid		$C_8H_7BrO_2$	1878-68-8	215.045	nd (w)	116	sub			sl H_2O; vs EtOH, eth, CS_2
1098	4-Bromobenzeneacetonitrile		C_8H_6BrN	16532-79-9	196.045	pa ye cry (al)	48.0				vs bz, EtOH
1099	α-Bromobenzeneacetonitrile	α-Bromobenzyl cyanide	C_8H_6BrN	5798-79-8	196.045	ye cry (dil al)	29	255(9)	1.539[29]		i H_2O; vs EtOH, eth, ace, bz, chl
1100	2-Bromo-1,4-benzenediol		$C_6H_5BrO_2$	583-69-7	189.007	lf (lig), cry (chl)	111.5	sub			vs H_2O, EtOH, eth, bz; sl chl, lig; s HOAc
1101	4-Bromobenzenesulfonyl chloride	p-Brosyl chloride	$C_6H_4BrClO_2S$	98-58-8	255.517	tcl or mcl pl (eth)	76	153[15]			i H_2O; vs eth; s chl
1102	4-Bromobenzenethiol		C_6H_5BrS	106-53-6	189.073	lf (al)	73	230.5	1.5260[83]		sl H_2O, EtOH; vs eth, ctc, chl
1103	2-Bromobenzoic acid		$C_7H_5BrO_2$	88-65-3	201.018	mcl pr (w), nd	149.0(0.9)	295(18)	1.929[25]		sl H_2O, DMSO; s EtOH, eth, ace, chl
1104	3-Bromobenzoic acid		$C_7H_5BrO_2$	585-76-2	201.018	mcl nd (dil al)	156.7(0.5)	285(15)	1.845[20]		i H_2O; s EtOH, eth
1105	4-Bromobenzoic acid		$C_7H_5BrO_2$	586-76-5	201.018	nd (eth), lf (w), mcl pr	254(1)		1.894[20]		sl H_2O, DMSO; s EtOH, eth
1106	2-Bromobenzonitrile		C_7H_4BrN	2042-37-7	182.018	nd (w)	55.5	252			s H_2O; vs EtOH; sl chl
1107	3-Bromobenzonitrile		C_7H_4BrN	6952-59-6	182.018		39.5	225			vs EtOH, eth; sl chl
1108	4-Bromobenzonitrile		C_7H_4BrN	623-00-7	182.018	nd (w, al)	114	236			s H_2O, EtOH, eth, chl
1109	6-Bromobenzo[a]pyrene		$C_{20}H_{11}Br$	21248-00-0	331.205	cry (ace/MeOH)	223				
1110	2-Bromobenzoyl chloride		C_7H_4BrClO	7154-66-7	219.463	nd	11	238(14)		1.5963[20]	sl ctc
1111	4-Bromobenzoyl chloride		C_7H_4BrClO	586-75-4	219.463	nd (peth)	37(1)	245(11)			vs EtOH, eth, bz, lig
1112	2-Bromobiphenyl		$C_{12}H_9Br$	2052-07-5	233.103		0.8	297	1.2175[26]	1.6248[25]	vs eth, EtOH
1113	3-Bromobiphenyl		$C_{12}H_9Br$	2113-57-7	233.103			300		1.6411[20]	i H_2O

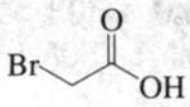

1079
Bromoacetic acid

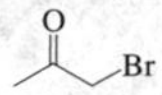

1080
Bromoacetone

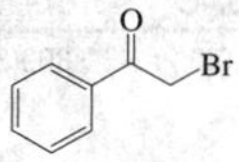

1081
α-Bromoacetophenone

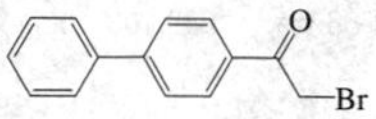

1082
4-(Bromoacetyl)biphenyl

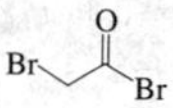

1083
Bromoacetyl bromide

1084
Bromoacetylene

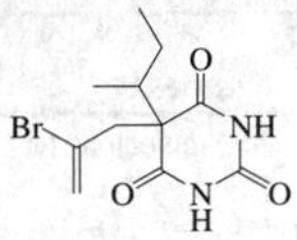

1085
5-(2-Bromoallyl)-5-*sec*-butylbarbituric acid

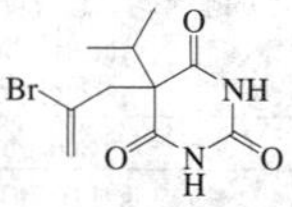

1086
5-(2-Bromoallyl)-5-isopropylbarbituric acid

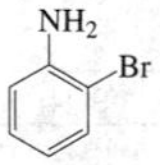

1087
2-Bromoaniline

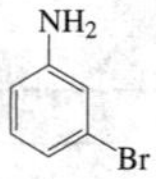

1088
3-Bromoaniline

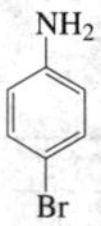

1089
4-Bromoaniline

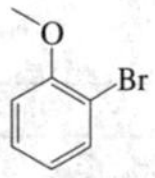

1090
2-Bromoanisole

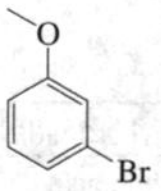

1091
3-Bromoanisole

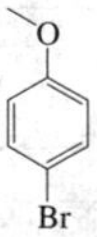

1092
4-Bromoanisole

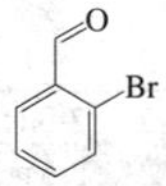

1093
2-Bromobenzaldehyde

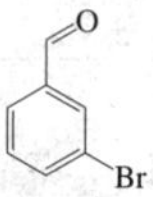

1094
3-Bromobenzaldehyde

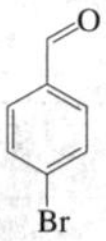

1095
4-Bromobenzaldehyde

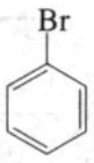

1096
Bromobenzene

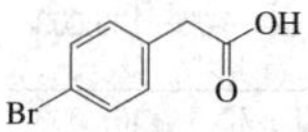

1097
4-Bromobenzeneacetic acid

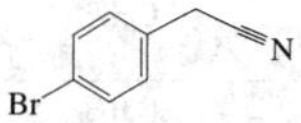

1098
4-Bromobenzeneacetonitrile

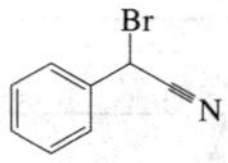

1099
α-Bromobenzeneacetonitrile

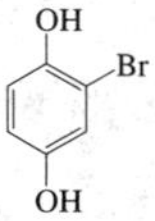

1100
2-Bromo-1,4-benzenediol

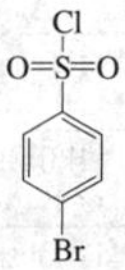

1101
4-Bromobenzenesulfonyl chloride

1102
4-Bromobenzenethiol

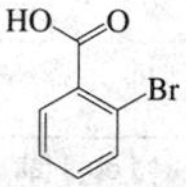

1103
2-Bromobenzoic acid

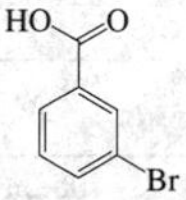

1104
3-Bromobenzoic acid

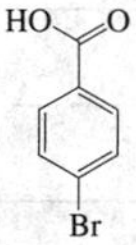

1105
4-Bromobenzoic acid

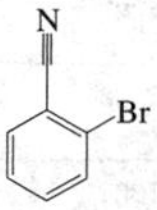

1106
2-Bromobenzonitrile

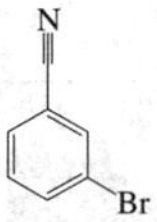

1107
3-Bromobenzonitrile

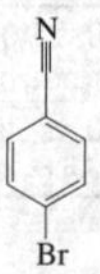

1108
4-Bromobenzonitrile

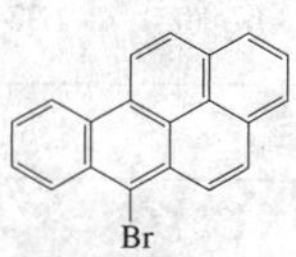

1109
6-Bromobenzo[a]pyrene

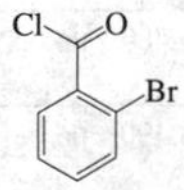

1110
2-Bromobenzoyl chloride

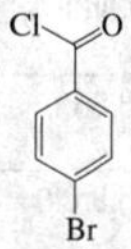

1111
4-Bromobenzoyl chloride

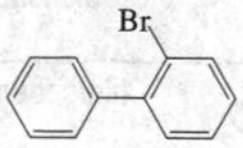

1112
2-Bromobiphenyl

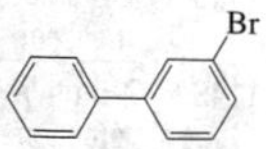

1113
3-Bromobiphenyl

No.	Name	Synonym	Mol. Form.	CAS RN	Mol. Wt.	Physical Form	mp/°C	bp/°C	den g cm⁻³	n_D	Solubility
1114	4-Bromobiphenyl		C₁₂H₉Br	92-66-0	233.103	pl (al)	87.0(0.2)	309(3)	0.9327²⁵		i H₂O; s EtOH, eth, bz, HOAc; sl chl
1115	1-Bromo-2-(bromomethyl)-benzene		C₇H₆Br₂	3433-80-5	249.931	cry (al, lig)	31	129¹⁹			vs eth, EtOH, HOAc
1116	1-Bromo-3-(bromomethyl)-benzene		C₇H₆Br₂	823-78-9	249.931	nd or lf	42	122¹²			s chl
1117	1-Bromo-4-(bromomethyl)-benzene	p-Bromobenzyl bromide	C₇H₆Br₂	589-15-1	249.931	nd (al)	63				sl H₂O; s EtOH, bz, chl; vs eth, CS₂
1118	2-Bromo-2-(bromomethyl)-pentanedinitrile	1,2-Dibromo-2,4-dicyanobutane	C₆H₆Br₂N₂	35691-65-7	265.933		52				i H₂O; vs ace, bz, DMF
1119	2-Bromo-1-(4-bromophenyl)-ethanone	p-Bromophenacyl bromide	C₈H₆Br₂O	99-73-0	277.941	nd (al)	111				i H₂O; s EtOH, eth, chl
1120	2-Bromo-1,3-butadiene		C₄H₅Br	1822-86-2	132.987			42¹⁶⁵	1.397²⁰	1.4988²⁰	vs eth, EtOH
1121	1-Bromobutane	Butyl bromide	C₄H₉Br	109-65-9	137.018	liq	-112.5(0.3)	101.4(0.7)	1.2758²⁰	1.4401²⁰	i H₂O; msc EtOH, eth, ace; sl ctc; s chl
1122	2-Bromobutane, (±)-	(±)-sec-Butyl bromide	C₄H₉Br	5787-31-5	137.018	liq	-112.6(0.2)	91(4)	1.2585²⁰	1.4366²⁰	vs ace, eth, chl
1123	Bromobutanedioic acid, (±)-	Bromosuccinic acid	C₄H₅BrO₄	584-98-5	196.985		161		2.073²⁵		s H₂O, EtOH; sl HOAc
1124	4-Bromobutanenitrile		C₄H₆BrN	5332-06-9	148.002			198(7)	1.4967²⁰	1.4818²⁰	s EtOH, eth, chl
1125	2-Bromobutanoic acid, (±)-	DL-α-Bromobutyric acid	C₄H₇BrO₂	2385-70-8	167.002		-2.0	217 dec	1.5641²⁰		s H₂O, EtOH, eth
1126	4-Bromobutanoic acid		C₄H₇BrO₂	2623-87-2	167.002		33	142²⁵			
1127	3-Bromo-2-butanone		C₄H₇BrO	814-75-5	151.002			36¹¹			
1128	cis-1-Bromo-1-butene		C₄H₇Br	31849-78-2	135.003			86(5)	1.3265¹⁵	1.4536²⁰	i H₂O; s eth, ace, bz, chl; sl ctc
1129	trans-1-Bromo-1-butene		C₄H₇Br	32620-08-9	135.003	liq	-100.3	95(3)	1.3209¹⁵	1.4527²⁰	i H₂O; s eth, ace, bz, chl; sl ctc
1130	2-Bromo-1-butene		C₄H₇Br	23074-36-4	135.003	liq	-133.4	81(3)	1.3209¹⁵	1.4527²⁰	i H₂O; s eth, ace, bz, chl; sl ctc
1131	4-Bromo-1-butene		C₄H₇Br	5162-44-7	135.003			98.5	1.3230²⁰	1.4622²⁰	sl H₂O; vs bz, eth, EtOH
1132	1-Bromo-2-butene		C₄H₇Br	4784-77-4	135.003			98(4)	1.3371²⁵	1.4822²⁰	i H₂O; s EtOH, eth, ctc; vs chl, bz
1133	cis-2-Bromo-2-butene		C₄H₇Br	3017-68-3	135.003	liq	-111.2(0.5)	89(3)	1.3416¹⁵	1.4631¹⁹	i H₂O; s EtOH, eth, ctc; vs chl, bz
1134	trans-2-Bromo-2-butene		C₄H₇Br	3017-71-8	135.003	liq	-115.4(0.5)	86(3)	1.3323¹⁵	1.4602¹⁶	i H₂O; s EtOH, eth, ctc; vs chl, bz
1135	(4-Bromobutoxy)benzene		C₁₀H₁₃BrO	1200-03-9	229.113	cry (al)	41	154¹⁸			sl EtOH, ctc
1136	1-Bromo-4-tert-butylbenzene		C₁₀H₁₃Br	3972-65-4	213.114		19	231(10)	1.2286²⁰	1.5436²⁰	i H₂O; s eth, bz, chl
1137	2-Bromo-3'-chloroacetophenone	3-Chlorophenacyl bromide	C₈H₆BrClO	41011-01-2	233.490	nd	40	397.5			vs EtOH
1138	1-Bromo-2-chlorobenzene		C₆H₄BrCl	694-80-4	191.453	liq	-12.6(0.3)	204	1.6387²⁵	1.5809²⁰	i H₂O; vs bz; sl ctc
1139	1-Bromo-3-chlorobenzene		C₆H₄BrCl	108-37-2	191.453	liq	-21.4(0.3)	196(6)	1.6302²⁰	1.5771²⁰	i H₂O; vs EtOH, eth
1140	1-Bromo-4-chlorobenzene		C₆H₄BrCl	106-39-8	191.453	nd or pl (al, eth)	64.78(0.05)	197(3)	1.576⁷¹	1.5531⁷⁰	i H₂O; sl EtOH; s eth, bz, ctc, chl
1141	1-Bromo-4-chlorobutane		C₄H₈BrCl	6940-78-9	171.464			161(11)	1.489²⁰	1.4885²⁰	i H₂O; s EtOH, eth, chl; sl ctc
1142	Bromochlorodifluoromethane	Halon 1211	CBrClF₂	353-59-3	165.365	col gas	-159.5	-3.9(0.7)			
1143	3-Bromo-1-chloro-5,5-dimethylhydantoin		C₅H₆BrClN₂O₂	126-06-7	241.471		162				
1144	1-Bromo-1-chloroethane		C₂H₄BrCl	593-96-4	143.410			101(12)	1.667¹⁰	1.4660²⁰	
1145	1-Bromo-2-chloroethane	2-Chloro-1-bromoethane	C₂H₄BrCl	107-04-0	143.410	liq	-16.7(0.3)	106(2)	1.7392²⁰	1.4908²⁰	sl H₂O; s EtOH, eth, chl
1146	Bromochlorofluoromethane		CHBrClF	593-98-6	147.374	liq	-115	39(15)	1.9771⁰	1.4144²⁵	i H₂O; s eth, ace, chl
1147	Bromochloromethane	Halon 1011	CH₂BrCl	74-97-5	129.384	liq	-87.9(0.2)	67.9(0.4)	1.9344²⁰	1.4838²⁰	i H₂O; s EtOH, eth, ace, bz
1148	1-Bromo-4-(chloromethyl)-benzene	p-Bromobenzyl chloride	C₇H₆BrCl	589-17-3	205.480	nd (al, peth)	42(4)	236			i H₂O; vs EtOH, eth; s peth
1149	2-Bromo-1-(4-chlorophenyl)-ethanone	4-Chlorophenacyl bromide	C₈H₆BrClO	536-38-9	233.490	nd	96.5				

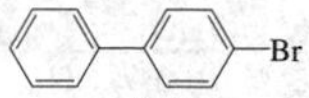

1114
4-Bromobiphenyl

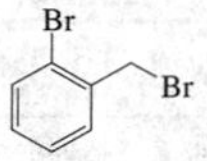

1115
1-Bromo-2-(bromomethyl)benzene

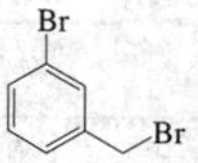

1116
1-Bromo-3-(bromomethyl)benzene

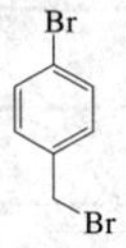

1117
1-Bromo-4-(bromomethyl)benzene

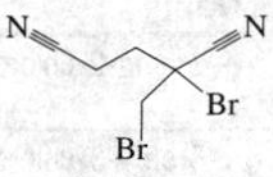

1118
2-Bromo-2-(bromomethyl)pentanedinitrile

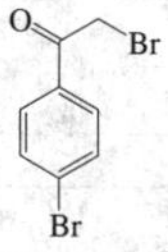

1119
2-Bromo-1-(4-bromophenyl)ethanone

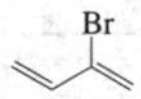

1120
2-Bromo-1,3-butadiene

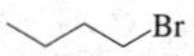

1121
1-Bromobutane

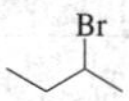

1122
2-Bromobutane, (±)-

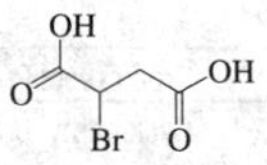

1123
Bromobutanedioic acid, (±)-

1124
4-Bromobutanenitrile

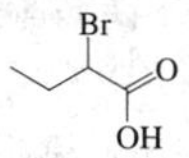

1125
2-Bromobutanoic acid, (±)-

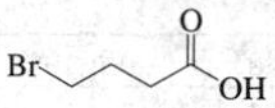

1126
4-Bromobutanoic acid

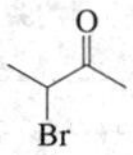

1127
3-Bromo-2-butanone

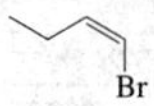

1128
cis-1-Bromo-1-butene

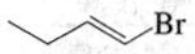

1129
trans-1-Bromo-1-butene

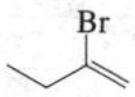

1130
2-Bromo-1-butene

1131
4-Bromo-1-butene

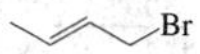

1132
1-Bromo-2-butene

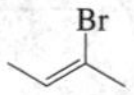

1133
cis-2-Bromo-2-butene

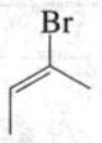

1134
trans-2-Bromo-2-butene

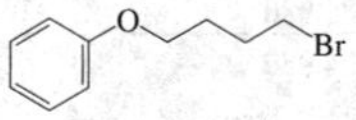

1135
(4-Bromobutoxy)benzene

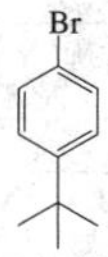

1136
1-Bromo-4-tert-butylbenzene

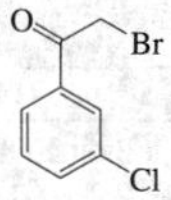

1137
2-Bromo-3'-chloroacetophenone

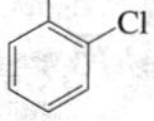

1138
1-Bromo-2-chlorobenzene

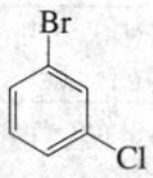

1139
1-Bromo-3-chlorobenzene

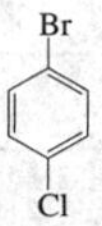

1140
1-Bromo-4-chlorobenzene

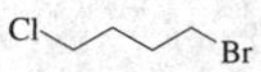

1141
1-Bromo-4-chlorobutane

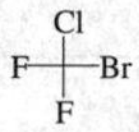

1142
Bromochlorodifluoromethane

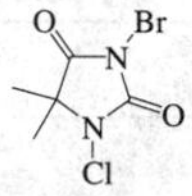

1143
3-Bromo-1-chloro-5,5-dimethylhydantoin

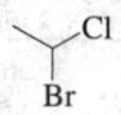

1144
1-Bromo-1-chloroethane

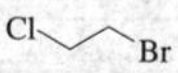

1145
1-Bromo-2-chloroethane

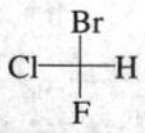

1146
Bromochlorofluoromethane

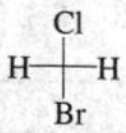

1147
Bromochloromethane

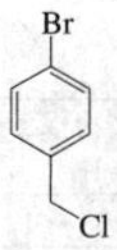

1148
1-Bromo-4-(chloromethyl)benzene

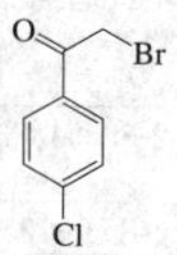

1149
2-Bromo-1-(4-chlorophenyl)ethanone

No.	Name	Synonym	Mol. Form.	CAS RN	Mol. Wt.	Physical Form	mp/°C	bp/°C	den g cm⁻³	n_D	Solubility
1150	1-Bromo-2-chloropropane		C₃H₆BrCl	3017-96-7	157.437			118	1.531²⁰	1.4745²⁰	vs ace, bz, eth, EtOH
1151	1-Bromo-3-chloropropane		C₃H₆BrCl	109-70-6	157.437	liq	-58.8(0.2)	143(6)	1.5969²⁰	1.4864²⁰	i H₂O; vs EtOH, eth, chl
1152	2-Bromo-1-chloropropane		C₃H₆BrCl	3017-95-6	157.437			117	1.537²⁰	1.4795²⁰	i H₂O; vs EtOH, eth; s ace, bz
1153	2-Bromo-2-chloropropane		C₃H₆BrCl	2310-98-7	157.437			95	1.495²⁰	1.4575²⁰	vs ace, bz, eth, EtOH
1154	1-Bromo-2-chloro-1,1,2-trifluoroethane		C₂HBrClF₃	354-06-3	197.381			52.4(0.2)	1.8574²⁵	1.3738²⁰	
1155	2-Bromo-2-chloro-1,1,1-trifluoroethane	Halothane	C₂HBrClF₃	151-67-7	197.381			50(1)	1.8563²⁵	1.3697⁰	sl H₂O; s peth
1156	Bromocresol Green	Bromcresol Green	C₂₁H₁₄Br₄O₅S	76-60-8	698.014	wh or red (+7w) ye (HOAc)	218.5				sl H₂O; vs EtOH, eth, AcOEt; s bz
1157	Bromocresol Purple	Bromcresol Purple	C₂₁H₁₆Br₂O₅S	115-40-2	540.222		241.5				
1158	Bromocycloheptane	Cycloheptyl bromide	C₇H₁₃Br	2404-35-5	177.082			101⁴⁰	1.3080²⁰	1.4996²⁰	i H₂O; vs eth, chl
1159	Bromocyclohexane	Cyclohexyl bromide	C₆H₁₁Br	108-85-0	163.055	liq	-56.28(0.07)	165.9(0.8)	1.3359²⁰	1.4957²⁰	i H₂O; msc EtOH, eth, ace, bz, lig, ctc
1160	trans-4-Bromocyclohexanol		C₆H₁₁BrO	32388-22-0	179.054	pl (hx)	81.5				
1161	2-Bromocyclohexanone		C₆H₉BrO	822-85-5	177.038			114³²	1.340²⁵	1.5085²⁵	
1162	3-Bromocyclohexene		C₆H₉Br	1521-51-3	161.039			81⁴⁰	1.3890²⁰	1.5320²⁰	i H₂O; s eth, bz, chl
1163	Bromocyclopentane	Cyclopentyl bromide	C₅H₉Br	137-43-9	149.029			137.5	1.3873²⁰	1.4886²⁰	sl ctc
1164	1-Bromodecane	Decyl bromide	C₁₀H₂₁Br	112-29-8	221.178	liq	-29.3(0.5)	240.6	1.0702²⁰	1.4557²⁰	i H₂O; vs eth, chl; s ctc
1165	2-Bromodecanoic acid		C₁₀H₁₉BrO₂	2623-95-2	251.161		2.0	140²	1.1912²⁴	1.4595²⁴	vs eth
1166	1-Bromo-3,5-dichlorobenzene		C₆H₃BrCl₂	19752-55-7	225.898	pr (al)	83	232			i H₂O; s EtOH, eth, chl; vs bz
1167	4-Bromo-1,2-dichlorobenzene		C₆H₃BrCl₂	18282-59-2	225.898	pr	25	237			i H₂O; sl EtOH; vs eth, bz, chl
1168	Bromodichlorofluoromethane	Halon 1121	CBrCl₂F	353-58-2	181.819	liq		52.8	1.95²²		
1169	Bromodichloromethane		CHBrCl₂	75-27-4	163.829	liq	-56.0(0.4)	90(2)	1.980²⁰	1.4964²⁰	i H₂O; vs EtOH, eth, ace, bz; sl ctc
1170	4-Bromo-2,5-dichlorophenol		C₆H₃BrCl₂O	1940-42-7	241.897	nd	70.8(0.5)				
1171	2-Bromo-1,1-diethoxyethane		C₆H₁₃BrO₂	2032-35-1	197.070			170	1.283²⁰	1.4387²⁰	s EtOH, eth
1172	4-Bromo-N,N-diethylaniline		C₁₀H₁₄BrN	2052-06-4	228.129	nd or pr	38	270			i H₂O; vs EtOH, eth
1173	Bromodifluoromethane		CHBrF₂	1511-62-2	130.920		-145(4)	-15.6(0.5)	1.55¹⁶		s H₂O; vs EtOH
1174	3-Bromo-4,5-dihydro-2(3H)-furanone	α-Bromo-γ-butyrolactone	C₄H₅BrO₂	5061-21-2	164.986			130²⁰	1.8²⁰	1.5059²⁰	
1175	5-Bromo-N,2-dihydroxybenzamide	5-Bromosalicylhydroxamic acid	C₇H₆BrNO₃	5798-94-7	232.032	cry (al)	232 dec				
1176	2-Bromo-1,4-dimethoxybenzene		C₈H₉BrO₂	25245-34-5	217.060	oil		262	1.445	1.5700²⁰	
1177	4-Bromo-1,2-dimethoxybenzene		C₈H₉BrO₂	2859-78-1	217.060			254.5	1.702²⁵	1.5743²⁰	
1178	2-Bromo-1,1-dimethoxyethane		C₄H₉BrO₂	7252-83-7	169.017			149	1.430²⁰	1.4450²⁰	s eth, ace, chl
1179	4-Bromo-N,N-dimethylaniline		C₈H₁₀BrN	586-77-6	200.076		55	264	1.3220¹⁰⁰		i H₂O; s EtOH; vs eth
1180	1-Bromo-2,4-dimethylbenzene		C₈H₉Br	583-70-0	185.061	liq	-17	204(8)	1.3419²⁰	1.5501²⁰	i H₂O; vs EtOH, eth, ace
1181	1-Bromo-3,5-dimethylbenzene		C₈H₉Br	556-96-7	185.061			203(8)	1.362²⁰	1.5462²²	vs eth; s ace, bz
1182	2-Bromo-1,3-dimethylbenzene		C₈H₉Br	576-22-7	185.061			204(6)		1.5552²⁰	vs eth; s ace, bz
1183	2-Bromo-1,4-dimethylbenzene		C₈H₉Br	553-94-6	185.061	lf or pl	9	207(6)	1.3582¹⁸	1.5514¹⁸	i H₂O; vs EtOH; s bz
1184	4-Bromo-1,2-dimethylbenzene		C₈H₉Br	583-71-1	185.061	liq	-0.2	215(7)	1.3708²⁰	1.5530²⁰	i H₂O; vs EtOH, eth
1185	trans-1-Bromo-3,7-dimethyl-2,6-octadiene	trans-Geranyl bromide	C₁₀H₁₇Br	6138-90-5	217.146			101¹²	1.0940²²	1.5027²⁰	
1186	1-Bromo-2,2-dimethylpropane		C₅H₁₁Br	630-17-1	151.045			108(11)	1.1997²⁰	1.4370²⁰	i H₂O; s EtOH, eth, ace, bz; vs chl
1187	2-Bromo-4,6-dinitroaniline		C₆H₄BrN₃O₄	1817-73-8	262.018	ye nd (al or HOAc)	153.5	sub			vs EtOH, ace; s HOAc
1188	1-Bromo-2,4-dinitrobenzene		C₆H₃BrN₂O₄	584-48-5	247.003	ye nd (al)	75				vs EtOH

1150
1-Bromo-2-chloropropane

1151
1-Bromo-3-chloropropane

1152
2-Bromo-1-chloropropane

1153
2-Bromo-2-chloropropane

1154
1-Bromo-2-chloro-1,1,2-trifluoroethane

1155
2-Bromo-2-chloro-1,1,1-trifluoroethane

1156
Bromocresol Green

1157
Bromocresol Purple

1158
Bromocycloheptane

1159
Bromocyclohexane

1160
trans-4-Bromocyclohexanol

1161
2-Bromocyclohexanone

1162
3-Bromocyclohexene

1163
Bromocyclopentane

1164
1-Bromodecane

1165
2-Bromodecanoic acid

1166
1-Bromo-3,5-dichlorobenzene

1167
4-Bromo-1,2-dichlorobenzene

1168
Bromodichlorofluoromethane

1169
Bromodichloromethane

1170
4-Bromo-2,5-dichlorophenol

1171
2-Bromo-1,1-diethoxyethane

1172
4-Bromo-N,N-diethylaniline

1173
Bromodifluoromethane

1174
3-Bromo-4,5-dihydro-2(3H)-furanone

1175
5-Bromo-N,2-dihydroxybenzamide

1176
2-Bromo-1,4-dimethoxybenzene

1177
4-Bromo-1,2-dimethoxybenzene

1178
2-Bromo-1,1-dimethoxyethane

1179
4-Bromo-N,N-dimethylaniline

1180
1-Bromo-2,4-dimethylbenzene

1181
1-Bromo-3,5-dimethylbenzene

1182
2-Bromo-1,3-dimethylbenzene

1183
2-Bromo-1,4-dimethylbenzene

1184
4-Bromo-1,2-dimethylbenzene

1185
trans-1-Bromo-3,7-dimethyl-2,6-octadiene

1186
1-Bromo-2,2-dimethylpropane

1187
2-Bromo-4,6-dinitroaniline

1188
1-Bromo-2,4-dinitrobenzene

No.	Name	Synonym	Mol. Form.	CAS RN	Mol. Wt.	Physical Form	mp/°C	bp/°C	den g cm⁻³	n_D	Solubility
1189	α-Bromodiphenylmethane		$C_{13}H_{11}Br$	776-74-9	247.130		45	184[20]			s EtOH, chl; vs bz
1190	1-Bromododecane	Lauryl bromide	$C_{12}H_{25}Br$	143-15-7	249.231	liq	-9.6(0.4)	275(20)	1.0399[20]	1.4583[20]	i H_2O; s EtOH, eth, ctc; msc ace
1191	2-Bromododecanoic acid		$C_{12}H_{23}BrO_2$	111-56-8	279.214	pl	32	158[2]	1.1474[74]	1.4585[24]	vs bz, eth, EtOH, lig
1192	Bromoethane	Ethyl bromide	C_2H_5Br	74-96-4	108.965	liq	-118.4(1)	38.2(0.6)	1.4604[20]	1.4239[20]	sl H_2O; msc EtOH, eth, chl
1193	2-Bromoethanol	Ethylene bromohydrin	C_2H_5BrO	540-51-2	124.964			142(4)	1.7629[20]	1.4915[20]	msc H_2O, EtOH, eth; sl lig
1194	Bromoethene	Vinyl bromide	C_2H_3Br	593-60-2	106.949	vol liq or gas	-139.5(0.2)	16(16)	1.4933[20]	1.4380[20]	i H_2O; s EtOH, eth, ace, bz, chl
1195	1-Bromo-2-ethoxybenzene		C_8H_9BrO	583-19-7	201.060			223		1.4223[20]	vs eth, EtOH
1196	1-Bromo-4-ethoxybenzene		C_8H_9BrO	588-96-5	201.060		2.0	231	1.4071[25]	1.5517[20]	i H_2O; vs EtOH, eth; s chl
1197	(2-Bromoethoxy)benzene		C_8H_9BrO	589-10-6	201.060		39	240 dec	1.3555[20]		i H_2O; vs EtOH, eth
1198	1-Bromo-2-ethoxyethane	2-Bromoethyl ethyl ether	C_4H_9BrO	592-55-2	153.017			127(4)	1.3852[20]	1.4447[20]	sl H_2O; msc EtOH, eth
1199	2-Bromoethyl acetate		$C_4H_7BrO_2$	927-68-4	167.002	liq	-13.8	159(5)	1.514[20]	1.457[23]	vs H_2O, chl; msc EtOH, eth
1200	2-Bromoethylamine hydrobromide	2-Bromoethanamine hydrobromide	$C_2H_7Br_2N$	2576-47-8	204.892		174.0				
1201	(1-Bromoethyl)benzene		C_8H_9Br	585-71-7	185.061			201(8)	1.3535[25]	1.5543[25]	
1202	(2-Bromoethyl)benzene		C_8H_9Br	103-63-9	185.061	liq	-55.9(0.2)	216(4)	1.3643[20]	1.5372[20]	i H_2O; s eth, bz; sl ctc
1203	1-Bromo-2-ethylbenzene		C_8H_9Br	1973-22-4	185.061	liq	-67.5(0.2)	202(4)	1.3548[20]	1.5472[20]	vs ace, bz, eth, EtOH
1204	1-Bromo-3-ethylbenzene		C_8H_9Br	2725-82-8	185.061			203(11)	1.3493[20]	1.5465[20]	
1205	1-Bromo-4-ethylbenzene		C_8H_9Br	1585-07-5	185.061	liq	-43.4(0.2)	204(4)	1.3423[20]	1.5445[20]	vs ace, bz, eth, EtOH
1206	(2-Bromoethyl)cyclohexane		$C_8H_{15}Br$	1647-26-3	191.109	liq	-57	212(3)	1.2357[20]	1.4899[20]	
1207	N-(2-Bromoethyl)phthalimide		$C_{10}H_8BrNO_2$	574-98-1	254.081	nd (w)	81.5(0.5)				vs eth; sl chl
1208	1-Bromo-4-ethynylbenzene		C_8H_5Br	766-96-1	181.030		64.5	89[16]			s chl
1209	1-Bromo-2-fluorobenzene		C_6H_4BrF	1072-85-1	174.998			154	1.0738[21]	1.5337[20]	
1210	1-Bromo-3-fluorobenzene		C_6H_4BrF	1073-06-9	174.998			150	1.7081[20]	1.5257[20]	s ctc
1211	1-Bromo-4-fluorobenzene		C_6H_4BrF	460-00-4	174.998	liq	-17.4	150(2)	1.593[15]	1.5310[15]	i H_2O; s EtOH, eth, chl
1212	1-Bromo-2-fluoroethane		C_2H_4BrF	762-49-2	126.955			58(12)	1.7044[25]	1.4236[20]	vs eth, EtOH
1213	Bromofluoromethane		CH_2BrF	373-52-4	112.929	vol liq or gas		23(12)			s EtOH; vs chl
1214	2-Bromofuran		C_4H_3BrO	584-12-3	146.970			103(6)	1.6500[20]	1.4980[20]	sl H_2O; s EtOH, eth, ace, bz
1215	3-Bromofuran		C_4H_3BrO	22037-28-1	146.970			106(6)	1.6606[20]	1.4958[20]	vs ace, bz, eth, EtOH
1216	5-Bromo-2-furancarboxaldehyde		$C_5H_3BrO_2$	1899-24-7	174.981	cry (50% al)	83.5	201			vs eth, EtOH
1217	1-Bromoheptadecane		$C_{17}H_{35}Br$	3508-00-7	319.364		28.4(0.4)	345(13)	0.9916[20]	1.4625[20]	i H_2O; vs chl
1218	1-Bromoheptane	Heptyl bromide	$C_7H_{15}Br$	629-04-9	179.098	liq	-56.1(0.3)	179(5)	1.1400[20]	1.4502[20]	i H_2O; vs EtOH, eth; sl ctc; s chl
1219	2-Bromoheptane	2-Heptyl bromide	$C_7H_{15}Br$	1974-04-5	179.098		47	166(6)	1.1277[20]	1.4503[20]	i H_2O; vs bz; s ctc, chl
1220	4-Bromoheptane	4-Heptyl bromide	$C_7H_{15}Br$	998-93-6	179.098			163(9)	1.1351[20]	1.4495[20]	i H_2O; s bz, ctc, chl
1221	1-Bromohexadecane		$C_{16}H_{33}Br$	112-82-3	305.337		17.5(0.4)	336	0.9991[20]	1.4618[25]	i H_2O; s eth
1222	2-Bromohexadecanoic acid		$C_{16}H_{31}BrO_2$	18263-25-7	335.320		52.8				
1223	1-Bromohexane	Hexyl bromide	$C_6H_{13}Br$	111-25-1	165.071	liq	-84.9(0.4)	156(4)	1.1744[20]	1.4478[20]	i H_2O; msc EtOH, eth; s ace; vs chl
1224	2-Bromohexane		$C_6H_{13}Br$	3377-86-4	165.071			139(9)	1.1658[20]	1.4832[25]	i H_2O; vs EtOH; s eth, ace; sl ctc
1225	3-Bromohexane		$C_6H_{13}Br$	3377-87-5	165.071			143(4)	1.1799[20]	1.4472[20]	vs ace, eth, EtOH, chl
1226	2-Bromohexanoic acid, (±)-		$C_6H_{11}BrO_2$	2681-83-6	195.054		2.0	242(11)	1.2810[33]		s EtOH, eth
1227	6-Bromohexanoic acid		$C_6H_{11}BrO_2$	4224-70-8	195.054	cry (peth)	35	167[20]			vs peth
1228	6-Bromohexanoyl chloride		$C_6H_{10}BrClO$	22809-37-6	213.499			101[6]			
1229	1-Bromo-4-(hexyloxy)benzene		$C_{12}H_{17}BrO$	30752-19-3	257.166			156[13]	1.2306[20]	1.5262[20]	
1230	5-Bromo-2-hydroxybenzaldehyde		$C_7H_5BrO_2$	1761-61-1	201.018	nd (al), lf (eth)	105.5				i H_2O; s EtOH, eth; sl chl

1189
α-Bromodiphenylmethane

1190
1-Bromododecane

1191
2-Bromododecanoic acid

1192
Bromoethane

1193
2-Bromoethanol

1194
Bromoethene

1195
1-Bromo-2-ethoxybenzene

1196
1-Bromo-4-ethoxybenzene

1197
(2-Bromoethoxy)benzene

1198
1-Bromo-2-ethoxyethane

1199
2-Bromoethyl acetate

1200
2-Bromoethylamine hydrobromide

1201
(1-Bromoethyl)benzene

1202
(2-Bromoethyl)benzene

1203
1-Bromo-2-ethylbenzene

1204
1-Bromo-3-ethylbenzene

1205
1-Bromo-4-ethylbenzene

1206
(2-Bromoethyl)cyclohexane

1207
N-(2-Bromoethyl)phthalimide

1208
1-Bromo-4-ethynylbenzene

1209
1-Bromo-2-fluorobenzene

1210
1-Bromo-3-fluorobenzene

1211
1-Bromo-4-fluorobenzene

1212
1-Bromo-2-fluoroethane

1213
Bromofluoromethane

1214
2-Bromofuran

1215
3-Bromofuran

1216
5-Bromo-2-furancarboxaldehyde

1217
1-Bromoheptadecane

1218
1-Bromoheptane

1219
2-Bromoheptane

1220
4-Bromoheptane

1221
1-Bromohexadecane

1222
2-Bromohexadecanoic acid

1223
1-Bromohexane

1224
2-Bromohexane

1225
3-Bromohexane

1226
2-Bromohexanoic acid, (±)-

1227
6-Bromohexanoic acid

1228
6-Bromohexanoyl chloride

1229
1-Bromo-4-(hexyloxy)benzene

1230
5-Bromo-2-hydroxybenzaldehyde

No.	Name	Synonym	Mol. Form.	CAS RN	Mol. Wt.	Physical Form	mp/°C	bp/°C	den g cm⁻³	n_D	Solubility
1231	4-Bromo-α-hydroxybenzeneacetic acid, (±)-	p-Bromomandelic acid	$C_8H_7BrO_3$	7021-04-7	231.044		119				vs H_2O, EtOH, eth, bz, chl
1232	5-Bromo-2-hydroxybenzene-methanol	Bromosaligenin	$C_7H_7BrO_2$	2316-64-5	203.034	lf (bz)	113				vs bz, eth, EtOH, chl
1233	5-Bromo-2-hydroxybenzoic acid		$C_7H_5BrO_3$	89-55-4	217.017	nd (w, dil al)	169.8	100 sub			sl H_2O, ace; vs EtOH, eth
1234	3-Bromo-4-hydroxy-5-methoxybenzaldehyde		$C_8H_7BrO_3$	2973-76-4	231.044	pl (HOAc), nd, pl (al)	167.0				i H_2O; s EtOH, DMSO; sl eth, bz
1235	1-Bromo-2-iodobenzene		C_6H_4BrI	583-55-1	282.904		2.1(0.5)	257	2.2570[25]	1.6618[25]	i H_2O; sl EtOH, HOAc; s ace
1236	1-Bromo-3-iodobenzene		C_6H_4BrI	591-18-4	282.904	liq	-9.3(0.6)	252			i H_2O; sl EtOH, HOAc
1237	1-Bromo-4-iodobenzene		C_6H_4BrI	589-87-7	282.904	pr or pl (eth-al)	90.37(0.05)	252			i H_2O; sl EtOH, chl; s eth
1238	Bromoiodomethane		CH_2BrI	557-68-6	220.835			136(14)	2.926[17]	1.6410[20]	vs chl
1239	1-Bromo-4-isocyanatobenzene	p-Bromophenyl isocyanate	C_7H_4BrNO	2493-02-9	198.017	nd		226			vs eth
1240	1-Bromo-4-isopropylbenzene		$C_9H_{11}Br$	586-61-8	199.087	liq	-22.4(0.2)	219(3)	1.3145[20]	1.5569[20]	i H_2O; s eth, bz, chl; sl ctc
1241	4-Bromoisoquinoline		C_9H_6BrN	1532-97-4	208.055	cry (peth)	41.5	282.5			vs eth
1242	Bromomethane	Methyl bromide	CH_3Br	74-83-9	94.939	col gas	-93.7(0.4)	3.4(0.1)	1.6755[20]	1.4218[20]	sl H_2O; msc EtOH, eth, chl, CS_2
1243	1-Bromo-2-methoxyethane		C_3H_7BrO	6482-24-2	138.991			112(4)	1.4623[20]	1.44753[20]	
1244	Bromomethoxymethane		C_2H_5BrO	13057-17-5	124.964			87	1.5976[20]	1.4562[20]	
1245	2-Bromo-4-methylaniline		C_7H_8BrN	583-68-6	186.050	lf	26	240	1.510[20]	1.5999[20]	i H_2O; s EtOH, eth
1246	4-Bromo-2-methylaniline		C_7H_8BrN	583-75-5	186.050	cry (al)	59.5	240			sl H_2O, chl; s EtOH; vs eth, HOAc
1247	(Bromomethyl)benzene	Benzyl bromide	C_7H_7Br	100-39-0	171.035	liq	-1.5	191(4)	1.4380[25]	1.5752[20]	i H_2O; msc EtOH, eth; s ctc
1248	4-(Bromomethyl)benzoic acid		$C_8H_7BrO_2$	6232-88-8	215.045		226.3				
1249	3-(Bromomethyl)benzonitrile		C_8H_6BrN	28188-41-2	196.045		96.5	130[4]			
1250	4-(Bromomethyl)benzonitrile		C_8H_6BrN	17201-43-3	196.045		114				
1251	1-Bromo-2-methylbutane, DL		$C_5H_{11}Br$	5973-11-5	151.045			119	1.2205[20]	1.4452[20]	i H_2O; s EtOH, eth; vs chl
1252	1-Bromo-3-methylbutane	Isopentyl bromide	$C_5H_{11}Br$	107-82-4	151.045	liq	-112	121(1)	1.2071[20]	1.4420[20]	i H_2O; s EtOH, eth; sl ctc; vs chl
1253	2-Bromo-2-methylbutane	tert-Pentyl bromide	$C_5H_{11}Br$	507-36-8	151.045			105(8)	1.197[18]	1.4421	
1254	3-Bromo-3-methylbutanoic acid	β-Bromoisovaleric acid	$C_5H_9BrO_2$	5798-88-9	181.028	nd (lig)	74				vs bz, eth, EtOH
1255	1-Bromo-3-methyl-2-butene		C_5H_9Br	870-63-3	149.029			121(18)	1.2930[15]	1.4930[15]	vs ace, bz, eth, EtOH
1256	1-(Bromomethyl)-2-chloro-benzene		C_7H_6BrCl	611-17-6	205.480			109[10]			
1257	(Bromomethyl)chlorodimethyl-silane		$C_3H_8BrClSi$	16532-02-8	187.539			131	1.375[25]	1.4630[25]	
1258	1-Bromo-3-methylcyclohex-ane	3-Methylcyclohexyl bromide	$C_7H_{13}Br$	13905-48-1	177.082			181	1.2676[15]	1.4979[20]	i H_2O; vs eth; s bz
1259	(Bromomethyl)cyclohexane		$C_7H_{13}Br$	2550-36-9	177.082			76[26]	1.283[20]	1.4907[30]	vs bz, eth, chl
1260	1-(Bromomethyl)-3-fluoroben-zene		C_7H_6BrF	456-41-7	189.025			88[20]		1.5474[20]	
1261	3-(Bromomethyl)heptane		$C_8H_{17}Br$	18908-66-2	193.125			67[10]			
1262	1-(Bromomethyl)-2-methyl-benzene		C_8H_9Br	89-92-9	185.061	pr	21	217	1.3811[23]	1.5730[20]	i H_2O; s EtOH, eth, ace, bz
1263	1-(Bromomethyl)-3-methyl-benzene		C_8H_9Br	620-13-3	185.061			212.5	1.3711[23]	1.5660[20]	i H_2O; vs EtOH, eth
1264	1-(Bromomethyl)-4-methyl-benzene		C_8H_9Br	104-81-4	185.061	nd (al)	34(2)	220	1.324[25]		i H_2O; s EtOH; vs eth, chl
1265	1-(Bromomethyl)naphthalene		$C_{11}H_9Br$	3163-27-7	221.093	cry (peth, al)	56(2)	183[18]			vs ace, bz, eth, EtOH
1266	2-(Bromomethyl)naphthalene		$C_{11}H_9Br$	939-26-4	221.093	lf (al)	56	213[100]			s EtOH, eth, chl, HOAc
1267	1-(Bromomethyl)-3-nitroben-zene		$C_7H_6BrNO_2$	3958-57-4	216.033	nd or pl (al)	59.3	162[13]			i H_2O; s EtOH
1268	1-(Bromomethyl)-4-nitroben-zene		$C_7H_6BrNO_2$	100-11-8	216.033	nd (al)	99(2)				sl H_2O, chl; vs EtOH, eth; s HOAc

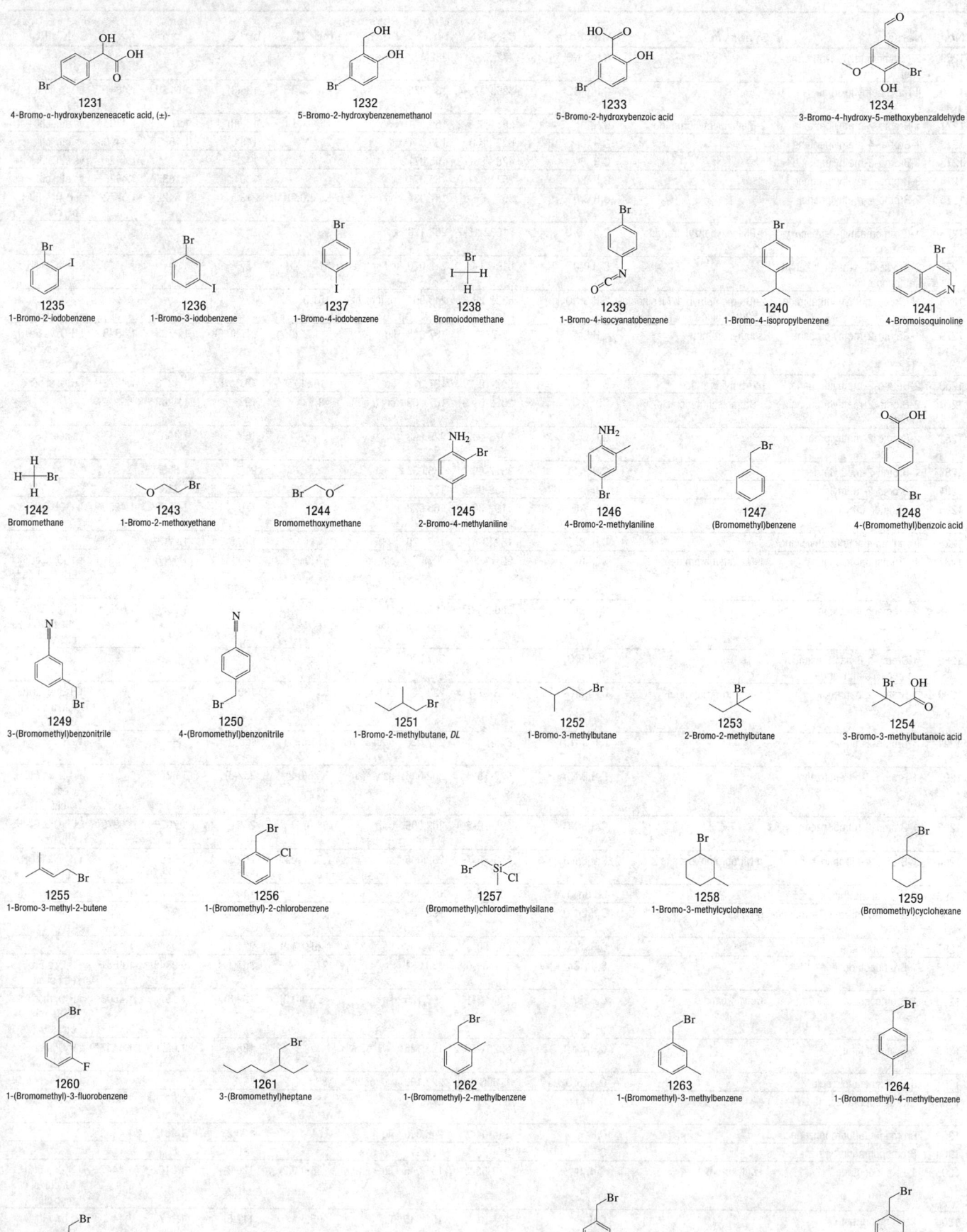

1231
4-Bromo-α-hydroxybenzeneacetic acid, (±)-

1232
5-Bromo-2-hydroxybenzenemethanol

1233
5-Bromo-2-hydroxybenzoic acid

1234
3-Bromo-4-hydroxy-5-methoxybenzaldehyde

1235
1-Bromo-2-iodobenzene

1236
1-Bromo-3-iodobenzene

1237
1-Bromo-4-iodobenzene

1238
Bromoiodomethane

1239
1-Bromo-4-isocyanatobenzene

1240
1-Bromo-4-isopropylbenzene

1241
4-Bromoisoquinoline

1242
Bromomethane

1243
1-Bromo-2-methoxyethane

1244
Bromomethoxymethane

1245
2-Bromo-4-methylaniline

1246
4-Bromo-2-methylaniline

1247
(Bromomethyl)benzene

1248
4-(Bromomethyl)benzoic acid

1249
3-(Bromomethyl)benzonitrile

1250
4-(Bromomethyl)benzonitrile

1251
1-Bromo-2-methylbutane, *DL*

1252
1-Bromo-3-methylbutane

1253
2-Bromo-2-methylbutane

1254
3-Bromo-3-methylbutanoic acid

1255
1-Bromo-3-methyl-2-butene

1256
1-(Bromomethyl)-2-chlorobenzene

1257
(Bromomethyl)chlorodimethylsilane

1258
1-Bromo-3-methylcyclohexane

1259
(Bromomethyl)cyclohexane

1260
1-(Bromomethyl)-3-fluorobenzene

1261
3-(Bromomethyl)heptane

1262
1-(Bromomethyl)-2-methylbenzene

1263
1-(Bromomethyl)-3-methylbenzene

1264
1-(Bromomethyl)-4-methylbenzene

1265
1-(Bromomethyl)naphthalene

1266
2-(Bromomethyl)naphthalene

1267
1-(Bromomethyl)-3-nitrobenzene

1268
1-(Bromomethyl)-4-nitrobenzene

No.	Name	Synonym	Mol. Form.	CAS RN	Mol. Wt.	Physical Form	mp/°C	bp/°C	den g cm⁻³	n_D	Solubility
1269	2-(Bromomethyl)-4-nitrophenol		$C_7H_6BrNO_3$	772-33-8	232.032		148				
1270	(Bromomethyl)oxirane, (±)-		C_3H_5BrO	82584-73-4	136.975	liq	-40	138(3)	1.615[14]	1.4841[20]	i H₂O; s EtOH, eth, bz, chl
1271	1-Bromo-2-methylpentane	2-Methylpentyl bromide	$C_6H_{13}Br$	25346-33-2	165.071			139(8)	1.1624[20]	1.4495[20]	vs eth, chl
1272	1-Bromo-4-methylpentane		$C_6H_{13}Br$	626-88-0	165.071			145	1.1683[20]	1.4490	vs eth, chl
1273	2-Bromo-2-methylpentane		$C_6H_{13}Br$	4283-80-1	165.071			142.5		1.442[23]	vs eth, chl
1274	3-Bromo-3-methylpentane		$C_6H_{13}Br$	25346-31-0	165.071			130	1.1835[20]	1.4525[20]	vs eth, chl
1275	2-Bromo-4-methylphenol		C_7H_7BrO	6627-55-0	187.034	nd (peth)	55.3(0.5)	213.5	1.5422[25]	1.5772[20]	sl H₂O; s EtOH, bz, chl
1276	1-(Bromomethyl)-3-phenoxy-benzene	3-Phenoxybenzyl bromide	$C_{13}H_{11}BrO$	51632-16-7	263.129	oil					
1277	2-Bromo-1-(4-methylphenyl)-ethanone		C_9H_9BrO	619-41-0	213.070	nd or lf (al)	51	157[14]			vs eth, EtOH
1278	N-(Bromomethyl)phthalimide	2-(Bromomethyl)-1H-isoindole-1,3(2H)-dione	$C_9H_6BrNO_2$	5332-26-3	240.054	pr (chl, bz)	151.5				s ace; sl bz, chl; vs AcOEt
1279	1-Bromo-2-methylpropane	Isobutyl bromide	C_4H_9Br	78-77-3	137.018	liq	-117.8(0.7)	91.3(0.5)	1.272[15]	1.4348[20]	i H₂O; vs EtOH, eth, ace, chl, bz; s ctc
1280	2-Bromo-2-methylpropane	tert-Butyl bromide	C_4H_9Br	507-19-7	137.018	liq	-16.8(0.9)	73.3	1.4278[20]	1.4278[20]	i H₂O; sl ctc
1281	2-Bromo-2-methylpropanoic acid	α-Bromoisobutyric acid	$C_4H_7BrO_2$	2052-01-9	167.002	cry (peth)	48.5	199	1.4969[60]		
1282	2-Bromo-2-methylpropanoyl bromide		$C_4H_6Br_2O$	20769-85-1	229.898			163	1.4067[14]		vs ace, CS₂
1283	1-Bromo-2-methylpropene		C_4H_7Br	3017-69-4	135.003			91(7)	1.336[20]		
1284	3-Bromo-2-methylpropene		C_4H_7Br	1458-98-6	135.003			92(18)	1.313[20]		
1285	2-(Bromomethyl)-tetrahydrofuran		C_5H_9BrO	1192-30-9	165.028			170	1.4679[20]	1.4850[20]	s EtOH, eth
1286	(Bromomethyl)trimethylsilane		$C_4H_{11}BrSi$	18243-41-9	167.120			116.5	1.170[25]	1.4460[20]	
1287	1-Bromonaphthalene	1-Naphthyl bromide	$C_{10}H_7Br$	90-11-9	207.067	oily liq	6.1(0.1)	280(2)	1.4785[20]	1.658[20]	s H₂O, ace; msc EtOH, eth, bz; sl ctc
1288	2-Bromonaphthalene		$C_{10}H_7Br$	580-13-2	207.067	pl or orth lf (al)	58(2)	281(9)	1.605[25]	1.6382[60]	i H₂O; s EtOH, eth, bz, CS₂; sl ctc
1289	4-Bromo-1,8-naphthalenedi-carboxylic anhydride		$C_{12}H_5BrO_3$	81-86-7	277.070		222				
1290	1-Bromo-2-naphthol	1-Bromo-β-naphthol	$C_{10}H_7BrO$	573-97-7	223.066	orth pr (bz-lig) nd (HOAc)	84	130			i H₂O; s EtOH, eth, bz; sl chl; vs HOAc
1291	4-Bromo-2-nitroaniline		$C_6H_5BrN_2O_2$	875-51-4	217.020	oran-ye nd (w)	111.5	sub			vs EtOH
1292	1-Bromo-2-nitrobenzene		$C_6H_4BrNO_2$	577-19-5	202.006	pa ye (al)	38.5(0.2)	253(6)	1.6245[80]		i H₂O; vs EtOH; s eth, ace, bz; sl chl
1293	1-Bromo-3-nitrobenzene		$C_6H_4BrNO_2$	585-79-5	202.006	orth	54(3)	265	1.7036[20]	1.5979[20]	sl H₂O; s EtOH, eth, bz
1294	1-Bromo-4-nitrobenzene	p-Nitrobromobenzene	$C_6H_4BrNO_2$	586-78-7	202.006	orth or mcl pr (al)	133.0(0.2)	252(5)	1.948[25]		i H₂O; s EtOH, eth, bz; sl chl
1295	Bromonitromethane		CH_2BrNO_2	563-70-2	139.937			149		1.4880[20]	vs EtOH
1296	2-Bromo-2-nitro-1,3-propane-diol	Bronopol	$C_3H_6BrNO_4$	52-51-7	199.989		131.5				
1297	1-Bromononane		$C_9H_{19}Br$	693-58-3	207.151	liq	-29.0(0.2)	221.4	1.0845[25]	1.4522[25]	
1298	1-Bromooctadecane		$C_{18}H_{37}Br$	112-89-0	333.391	cry (al)	27.6(0.4)	357(17)	0.9848[20]	1.4631[20]	i H₂O; s EtOH, eth; sl ctc
1299	1-Bromooctane	Octyl bromide	$C_8H_{17}Br$	111-83-1	193.125	liq	-55.0(0.3)	199(6)	1.1072[25]	1.4503[25]	i H₂O; msc EtOH, eth; sl ctc
1300	2-Bromooctane, (±)-		$C_8H_{17}Br$	60251-57-2	193.125			188.5	1.0878[20]	1.4442[25]	i H₂O; msc EtOH, eth
1301	8-Bromooctanoic acid		$C_8H_{15}BrO_2$	17696-11-6	223.108	nd (peth)	38.5	147[2]			vs bz, eth, EtOH
1302	1-Bromopentadecane		$C_{15}H_{31}Br$	629-72-1	291.311		18.6(0.3)	322	1.0675[20]	1.4611[20]	i H₂O; s ace; vs chl
1303	Bromopentafluorobenzene		C_6BrF_5	344-04-7	246.960	liq	-31	133(2)	1.981[25]	1.4490[20]	
1304	Bromopentafluoroethane		C_2BrF_5	354-55-2	198.917	col gas		-21	1.8098[25]		
1305	1-Bromopentane	Pentyl bromide	$C_5H_{11}Br$	110-53-2	151.045	liq	-88.0(0.2)	126(3)	1.2182[20]	1.4447[20]	i H₂O; s EtOH, bz, chl; sl ctc; msc eth
1306	2-Bromopentane		$C_5H_{11}Br$	107-81-3	151.045	liq	-95.5	117(5)	1.2075[20]	1.4413[20]	vs bz, eth, EtOH, chl
1307	3-Bromopentane		$C_5H_{11}Br$	1809-10-5	151.045	liq	-126.2	118(5)	1.214[20]	1.4441[20]	i H₂O; s EtOH, eth, bz, chl
1308	5-Bromopentanenitrile		C_5H_8BrN	5414-21-1	162.029			111[12]	1.3989[20]	1.4780[20]	

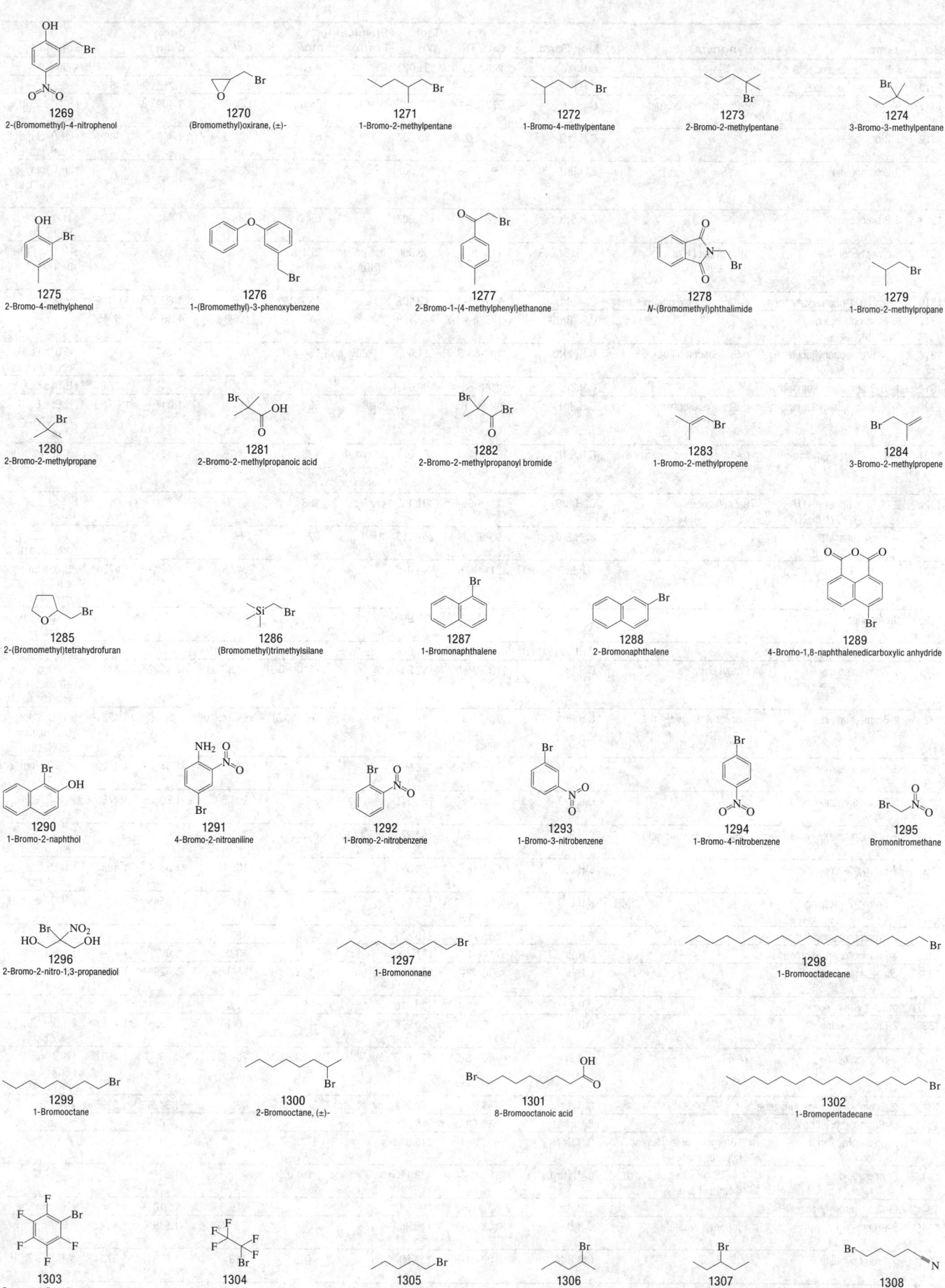

1269
2-(Bromomethyl)-4-nitrophenol

1270
(Bromomethyl)oxirane, (±)-

1271
1-Bromo-2-methylpentane

1272
1-Bromo-4-methylpentane

1273
2-Bromo-2-methylpentane

1274
3-Bromo-3-methylpentane

1275
2-Bromo-4-methylphenol

1276
1-(Bromomethyl)-3-phenoxybenzene

1277
2-Bromo-1-(4-methylphenyl)ethanone

1278
N-(Bromomethyl)phthalimide

1279
1-Bromo-2-methylpropane

1280
2-Bromo-2-methylpropane

1281
2-Bromo-2-methylpropanoic acid

1282
2-Bromo-2-methylpropanoyl bromide

1283
1-Bromo-2-methylpropene

1284
3-Bromo-2-methylpropene

1285
2-(Bromomethyl)tetrahydrofuran

1286
(Bromomethyl)trimethylsilane

1287
1-Bromonaphthalene

1288
2-Bromonaphthalene

1289
4-Bromo-1,8-naphthalenedicarboxylic anhydride

1290
1-Bromo-2-naphthol

1291
4-Bromo-2-nitroaniline

1292
1-Bromo-2-nitrobenzene

1293
1-Bromo-3-nitrobenzene

1294
1-Bromo-4-nitrobenzene

1295
Bromonitromethane

1296
2-Bromo-2-nitro-1,3-propanediol

1297
1-Bromononane

1298
1-Bromooctadecane

1299
1-Bromooctane

1300
2-Bromooctane, (±)-

1301
8-Bromooctanoic acid

1302
1-Bromopentadecane

1303
Bromopentafluorobenzene

1304
Bromopentafluoroethane

1305
1-Bromopentane

1306
2-Bromopentane

1307
3-Bromopentane

1308
5-Bromopentanenitrile

No.	Name	Synonym	Mol. Form.	CAS RN	Mol. Wt.	Physical Form	mp/°C	bp/°C	den g cm^{-3}	n_D	Solubility
1309	5-Bromopentanoic acid		$C_5H_9BrO_2$	2067-33-6	181.028		40.0	142[13]			s chl
1310	5-Bromo-1-pentene		C_5H_9Br	1119-51-3	149.029			125.5	1.2581[20]	1.4640[20]	
1311	9-Bromophenanthrene	9-Phenanthryl bromide	$C_{14}H_9Br$	573-17-1	257.125	pr (al)	64.5	>360	1.4093[10]		i H$_2$O; s EtOH, eth, CS$_2$; sl chl
1312	2-Bromophenol		C_6H_5BrO	95-56-7	173.007		5.6	194.5	1.4924[20]	1.589[20]	sl H$_2$O, chl; s EtOH, eth, alk
1313	3-Bromophenol		C_6H_5BrO	591-20-8	173.007		33	236.5			sl H$_2$O, ctc; vs EtOH, eth; s chl, alk
1314	4-Bromophenol		C_6H_5BrO	106-41-2	173.007		63(1)	238	1.840[15]		s H$_2$O, chl; vs EtOH, eth
1315	Bromophenol Blue	Bromphenol Blue	$C_{19}H_{10}Br_4O_5S$	115-39-9	669.960	hex pr (HOAc-ace)	279 dec				sl H$_2$O; s EtOH, bz, HOAc
1316	1-Bromo-4-phenoxybenzene	4-Bromophenyl phenyl ether	$C_{12}H_9BrO$	101-55-3	249.102		18.7(0.2)	126[3.5]	1.6088[20]	1.6084[20]	i H$_2$O; s eth, ctc
1317	(4-Bromophenoxy)-trimethylsilane		$C_9H_{13}BrOSi$	17878-44-3	245.188			126[25]	1.2619[20]	1.5145[20]	
1318	N-(4-Bromophenyl)acetamide	p-Bromoacetanilide	C_8H_8BrNO	103-88-8	214.060	nd (60% al)	168		1.717[25]		i H$_2$O; s EtOH, chl; sl eth, bz
1319	1-(3-Bromophenyl)ethanone		C_8H_7BrO	2142-63-4	199.045		7.5	133[19]		1.5755[20]	i H$_2$O; s ace, bz
1320	1-(4-Bromophenyl)ethanone	p-Bromoacetophenone	C_8H_7BrO	99-90-1	199.045	lf (al)	50.5	257	1.647[25]	1.647	i H$_2$O; s EtOH, eth, bz, ctc, HOAc
1321	(4-Bromophenyl)hydrazine	(p-Bromophenyl)hydrazine	$C_6H_7BrN_2$	589-21-9	187.037	nd (w), lf (lig), cry (al)	108				vs eth, EtOH, lig
1322	2-(4-Bromophenyl)-1H-indene-1,3(2H)-dione	Bromindione	$C_{15}H_9BrO_2$	1146-98-1	301.135	cry (lig)	138				
1323	(4-Bromophenyl)-phenylmethanone		$C_{13}H_9BrO$	90-90-4	261.113	lf (al)	82.5	350			i H$_2$O; sl EtOH, eth, bz, peth
1324	2-Bromo-1-phenyl-1-propa-none		C_9H_9BrO	2114-00-3	213.070			247.5	1.4298[20]	1.5720[20]	i H$_2$O; s EtOH, eth, ace, bz, ctc
1325	Bromophos		$C_8H_8BrCl_2PS$	2104-96-3	317.999	ye cry	56.3(0.3)	141[0.01]			sl H$_2$O; s eth, ctc, tol
1326	Bromophos-ethyl		$C_{10}H_{12}BrCl_2O_3PS$	4824-78-6	394.049	pale-ye liq		122[0.004]			
1327	1-Bromopropane	Propyl bromide	C_3H_7Br	106-94-5	122.992	liq	-110.1(0.3)	70.8(0.2)	1.3537[20]	1.4343[20]	sl H$_2$O; s EtOH, eth, ace, bz, chl, ctc
1328	2-Bromopropane	Isopropyl bromide	C_3H_7Br	75-26-3	122.992	liq	-88.9(0.5)	59.34(0.09)	1.3140[20]	1.4251[20]	sl H$_2$O; s ace, bz, chl; msc EtOH, eth
1329	3-Bromopropanenitrile		C_3H_4BrN	2417-90-5	133.975			92[25]	1.6152[20]	1.4800[20]	vs EtOH, eth; sl ctc
1330	2-Bromopropanoic acid, (±)-		$C_3H_5BrO_2$	10327-08-9	152.975	pr	25.7	203.5	1.7000[20]	1.4753[20]	vs H$_2$O, EtOH, eth; sl chl
1331	3-Bromopropanoic acid	β-Bromopropionic acid	$C_3H_5BrO_2$	590-92-1	152.975	pl (CCl$_4$)	62.5	141[45]	1.48[25]		s H$_2$O, EtOH, eth, bz, chl
1332	3-Bromo-1-propanol		C_3H_7BrO	627-18-9	138.991			105[185]	1.5374[20]	1.4834[25]	s H$_2$O; msc EtOH, eth
1333	1-Bromo-2-propanol		C_3H_7BrO	19686-73-8	138.991			146.5	1.5585[30]	1.4801[20]	s H$_2$O; vs EtOH, eth
1334	2-Bromopropanoyl bromide		$C_3H_4Br_2O$	563-76-8	215.871			153	2.0611[16]		
1335	2-Bromopropanoyl chloride		C_3H_4BrClO	7148-74-5	171.420			132	1.697[11]	1.4780[20]	s eth, chl; sl ctc
1336	cis-1-Bromopropene		C_3H_5Br	590-13-6	120.976	liq	-113	58(3)	1.4291[20]	1.4560[20]	i H$_2$O; s eth, ace, chl
1337	trans-1-Bromopropene		C_3H_5Br	590-15-8	120.976			61(5)			
1338	2-Bromopropene		C_3H_5Br	557-93-7	120.976	liq	-126	49(4)	1.3965[16]	1.4467[16]	i H$_2$O; s eth, ace, chl
1339	3-Bromopropene	Allyl bromide	C_3H_5Br	106-95-6	120.976	liq	-119.3(0.5)	70.1(0.5)	1.398[20]	1.4697[20]	i H$_2$O; msc EtOH, eth; s ctc, chl, CS$_2$
1340	(3-Bromo-1-propenyl)benzene		C_9H_9Br	4392-24-9	197.071	nd (al, eth)	34	130[10]	1.3428[30]	1.613[20]	vs EtOH
1341	(3-Bromopropoxy)benzene		$C_9H_{11}BrO$	588-63-6	215.086		10.7	127[18]	1.364[16]		vs eth
1342	3-Bromopropylamine hydrobromide	3-Bromo-1-propanamine hydrobromide	$C_3H_9Br_2N$	5003-71-4	218.918		171.5				
1343	Bromopropylate	4,4'-Dibromobenzilic acid isopropyl ester	$C_{17}H_{16}Br_2O_3$	18181-80-1	428.115		76.0(0.5)		1.59[20]		
1344	(3-Bromopropyl)benzene		$C_9H_{11}Br$	637-59-2	199.087			219.5	1.3106[25]	1.5440[25]	i H$_2$O; vs eth
1345	3-Bromo-1-propyne	Propargyl bromide	C_3H_3Br	106-96-7	118.960			73(14)	1.579[19]	1.4922[20]	s EtOH, eth, bz, ctc, chl
1346	2-Bromopyridine		C_5H_4BrN	109-04-6	157.997	liq	-40.1	193	1.6337[20]	1.5734[20]	sl H$_2$O; s EtOH, eth, ctc

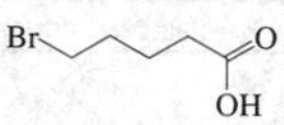

1309
5-Bromopentanoic acid

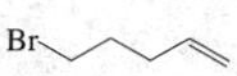

1310
5-Bromo-1-pentene

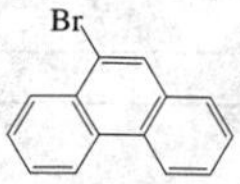

1311
9-Bromophenanthrene

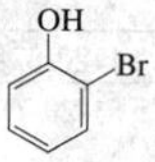

1312
2-Bromophenol

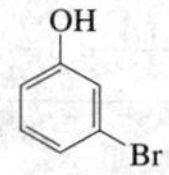

1313
3-Bromophenol

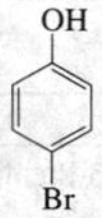

1314
4-Bromophenol

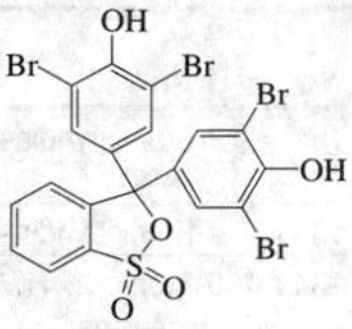

1315
Bromophenol Blue

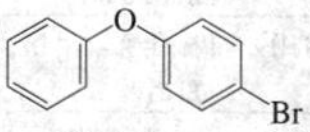

1316
1-Bromo-4-phenoxybenzene

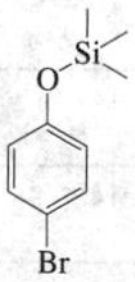

1317
(4-Bromophenoxy)trimethylsilane

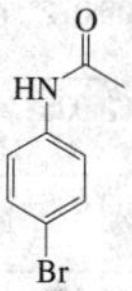

1318
N-(4-Bromophenyl)acetamide

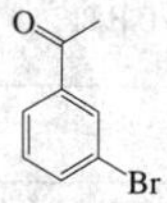

1319
1-(3-Bromophenyl)ethanone

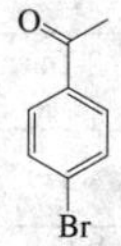

1320
1-(4-Bromophenyl)ethanone

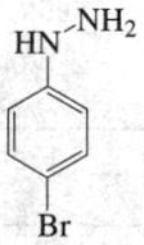

1321
(4-Bromophenyl)hydrazine

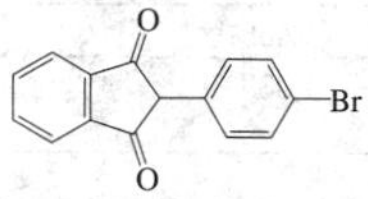

1322
2-(4-Bromophenyl)-1H-indene-1,3(2H)-dione

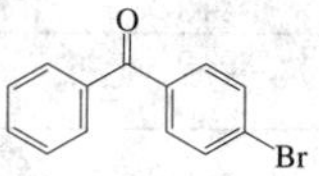

1323
(4-Bromophenyl)phenylmethanone

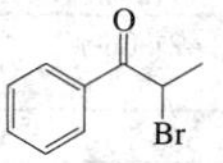

1324
2-Bromo-1-phenyl-1-propanone

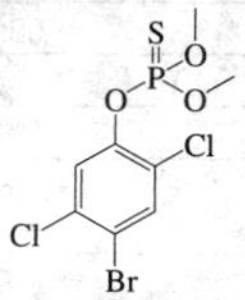

1325
Bromophos

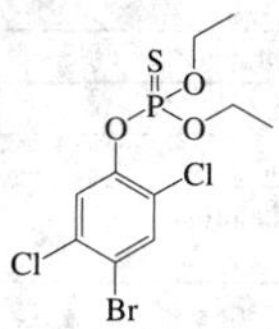

1326
Bromophos-ethyl

1327
1-Bromopropane

1328
2-Bromopropane

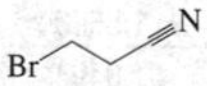

1329
3-Bromopropanenitrile

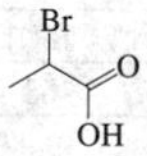

1330
2-Bromopropanoic acid, (±)-

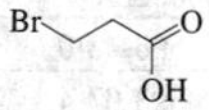

1331
3-Bromopropanoic acid

1332
3-Bromo-1-propanol

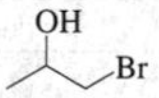

1333
1-Bromo-2-propanol

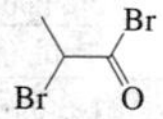

1334
2-Bromopropanoyl bromide

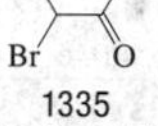

1335
2-Bromopropanoyl chloride

1336
cis-1-Bromopropene

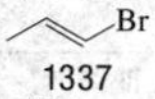

1337
trans-1-Bromopropene

1338
2-Bromopropene

1339
3-Bromopropene

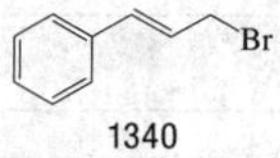

1340
(3-Bromo-1-propenyl)benzene

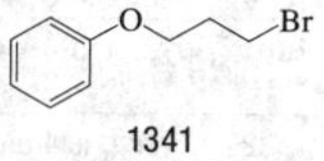

1341
(3-Bromopropoxy)benzene

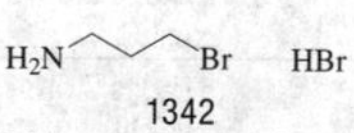

1342
3-Bromopropylamine hydrobromide

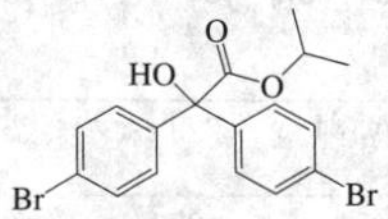

1343
Bromopropylate

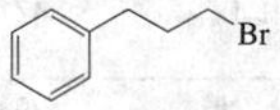

1344
(3-Bromopropyl)benzene

1345
3-Bromo-1-propyne

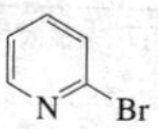

1346
2-Bromopyridine

No.	Name	Synonym	Mol. Form.	CAS RN	Mol. Wt.	Physical Form	mp/°C	bp/°C	den g cm⁻³	n_D	Solubility
1347	3-Bromopyridine		C₅H₄BrN	626-55-1	157.997	liq	-27.3	172(3)	1.645⁰	1.5694²⁰	s H₂O; vs EtOH, eth
1348	4-Bromopyridine		C₅H₄BrN	1120-87-2	157.997		0.5	29⁰·⁴	1.6450⁰	1.5694²⁰	s ace, bz
1349	5-Bromo-2,4(1H,3H)-pyrimi-dinedione	5-Bromouracil	C₄H₃BrN₂O₂	51-20-7	190.983		310				
1350	3-Bromoquinoline		C₉H₆BrN	5332-24-1	208.055	ye oil	13.3	275		1.6641²⁰	s chl; vs HOAc
1351	6-Bromoquinoline		C₉H₆BrN	5332-25-2	208.055		24	281			s EtOH, eth, acid
1352	N-Bromosuccinimide		C₄H₄BrNO₂	128-08-5	177.985	cry (bz)	174		2.098²⁵		sl H₂O, AcOEt, eth; vs ace; i hx
1353	1-Bromotetradecane		C₁₄H₂₉Br	112-71-0	277.284		5.7(0.4)	307	1.0170²⁰	1.4603²⁰	vs ace, bz, EtOH
1354	2-Bromothiazole		C₃H₂BrNS	3034-53-5	164.024			171	1.82²⁵	1.5927²⁰	
1355	1-(5-Bromo-2-thienyl)-ethanone		C₆H₅BrOS	5370-25-2	205.072	nd (al)	94.5	103⁴			sl EtOH; s ctc
1356	2-Bromothiophene	2-Thienyl bromide	C₄H₃BrS	1003-09-4	163.036			141(6)	1.684²⁰	1.5868²⁰	i H₂O; vs eth, ace; s ctc
1357	3-Bromothiophene		C₄H₃BrS	872-31-1	163.036			149(7)	1.735²⁰	1.5919²⁰	i H₂O; s ace, bz; sl chl
1358	Bromothymol Blue	Bromthymol Blue	C₂₇H₂₈Br₂O₅S	76-59-5	624.381		201				vs eth, EtOH
1359	2-Bromotoluene		C₇H₇Br	95-46-5	171.035	liq	-27.5(0.8)	182(2)	1.4232²⁰	1.5565²⁰	i H₂O; vs EtOH, eth, bz; msc ctc
1360	3-Bromotoluene		C₇H₇Br	591-17-3	171.035	liq	-38.1(0.2)	184(2)	1.4099²⁰	1.5510²⁰	i H₂O; s EtOH, ace, chl; msc eth; sl ctc
1361	4-Bromotoluene		C₇H₇Br	106-38-7	171.035	cry (al)	26.2(0.7)	184(4)	1.3959³⁵	1.5477²⁰	i H₂O; s EtOH, eth, ace, bz, chl; sl ctc
1362	Bromotrichloromethane		CBrCl₃	75-62-7	198.274	liq	-5.6(0.2)	103(5)	2.012²⁵	1.5065²⁰	vs eth, EtOH
1363	1-Bromotridecane		C₁₃H₂₇Br	765-09-3	263.257		5.9(0.3)	292	1.0234²⁵	1.4574²⁵	i H₂O; vs chl
1364	Bromotriethylsilane		C₆H₁₅BrSi	1112-48-7	195.173	liq	-49.3	163	1.143²⁰	1.4561²⁰	
1365	2-Bromo-1,1,1-trifluoroethane		C₂H₂BrF₃	421-06-7	162.936	vol liq or gas	-93.9	26	1.7881²⁰	1.3331²⁰	
1366	Bromotrifluoroethene		C₂BrF₃	598-73-2	160.920	col gas		-2.5			
1367	Bromotrifluoromethane	Halon-1301	CBrF₃	75-63-8	148.910	col gas	-174.4(0.2)	-57.8(0.4)	1.5800²⁰		i H₂O; vs chl
1368	1-Bromo-2-(trifluoromethyl)-benzene		C₇H₄BrF₃	392-83-6	225.006			167.5	1.652²⁵	1.4817²⁰	
1369	1-Bromo-3-(trifluoromethyl)-benzene		C₇H₄BrF₃	401-78-5	225.006		1	151.5	1.613²⁵	1.4716²⁰	
1370	1-Bromo-4-(trifluoromethyl)-benzene		C₇H₄BrF₃	402-43-7	225.006			160	1.607²⁵	1.4705²⁵	
1371	2-Bromo-1,3,5-trimethylben-zene		C₉H₁₁Br	576-83-0	199.087	liq	-1	225	1.3191¹⁰	1.5510²⁰	i H₂O; vs eth; s bz; sl ctc
1372	Bromotrinitromethane		CBrN₃O₆	560-95-2	229.931		18(1)	56¹⁰	2.0312²⁰	1.4808²⁰	vs EtOH, chl
1373	Bromotriphenylmethane	Triphenylmethyl bromide	C₁₉H₁₅Br	596-43-0	323.226		153	230¹⁵	1.5500²⁰		
1374	1-Bromoundecane		C₁₁H₂₃Br	693-67-4	235.205	liq	-9.9(0.4)	258.8	1.0494²⁵	1.4552²⁵	sl ctc
1375	11-Bromoundecanoic acid		C₁₁H₂₁BrO₂	2834-05-1	265.188	nd (liq)	57	188¹⁸			vs ace, bz, eth, EtOH
1376	(1-Bromovinyl)benzene		C₈H₇Br	98-81-7	183.046		-44	86¹⁴	1.4025²³	1.5881²⁰	
1377	(cis-2-Bromovinyl)benzene		C₈H₇Br	588-73-8	183.046		-7	55²	1.4322¹⁰	1.5990²²	
1378	(trans-2-Bromovinyl)benzene		C₈H₇Br	588-72-7	183.046		7	219 dec	1.4269¹⁶	1.6093²⁰	i H₂O; msc EtOH, eth; s chl
1379	1-Bromo-2-vinylbenzene		C₈H₇Br	2039-88-5	183.046	liq	-52.7(0.2)	215(2)	1.4160²⁰	1.5927²⁰	
1380	1-Bromo-3-vinylbenzene		C₈H₇Br	2039-86-3	183.046			92²⁰	1.4059²⁰	1.5933²⁰	
1381	1-Bromo-4-vinylbenzene		C₈H₇Br	2039-82-9	183.046		5(2)	213(1)	1.3984²⁰	1.5947²⁰	i H₂O; vs chl; s HOAc
1382	Brompheniramine		C₁₆H₁₉BrN₂	86-22-6	319.239	ye oily liq		150⁰·⁵			s dil acid
1383	Brucine		C₂₃H₂₆N₂O₄	357-57-3	394.463	mcl pr (w +4)	178				sl H₂O, eth, bz; vs EtOH, chl
1384	Brucine hydrochloride	2,3-Dimethoxystrychnidin-10-one, monohydrochloride	C₂₃H₂₇ClN₂O₄	5786-96-9	430.924	pr					vs H₂O, EtOH
1385	Brucine sulfate heptahydrate	2,3-Dimethoxystrychnidin-10-one, sulfate, heptahydrate	C₄₆H₆₈N₄O₁₉S	60583-39-3	1013.113	nd (w)					s H₂O; sl EtOH, chl, tfa; vs MeOH; i bz
1386	Bucolome	5-Butyl-1-cyclohexyl-2,4,6(1H,3H,5H)-pyrimidine-trione	C₁₄H₂₂N₂O₃	841-73-6	266.336	nd (MeOH)	84	186⁰·⁸			
1387	Bufotalin		C₂₆H₃₆O₆	471-95-4	444.560	cry (+1 al)	223 dec				i H₂O; s EtOH, chl
1388	Bulbocapnine		C₁₉H₁₉NO₄	298-45-3	325.359	pr (al)	199.5				i H₂O; s EtOH; vs chl

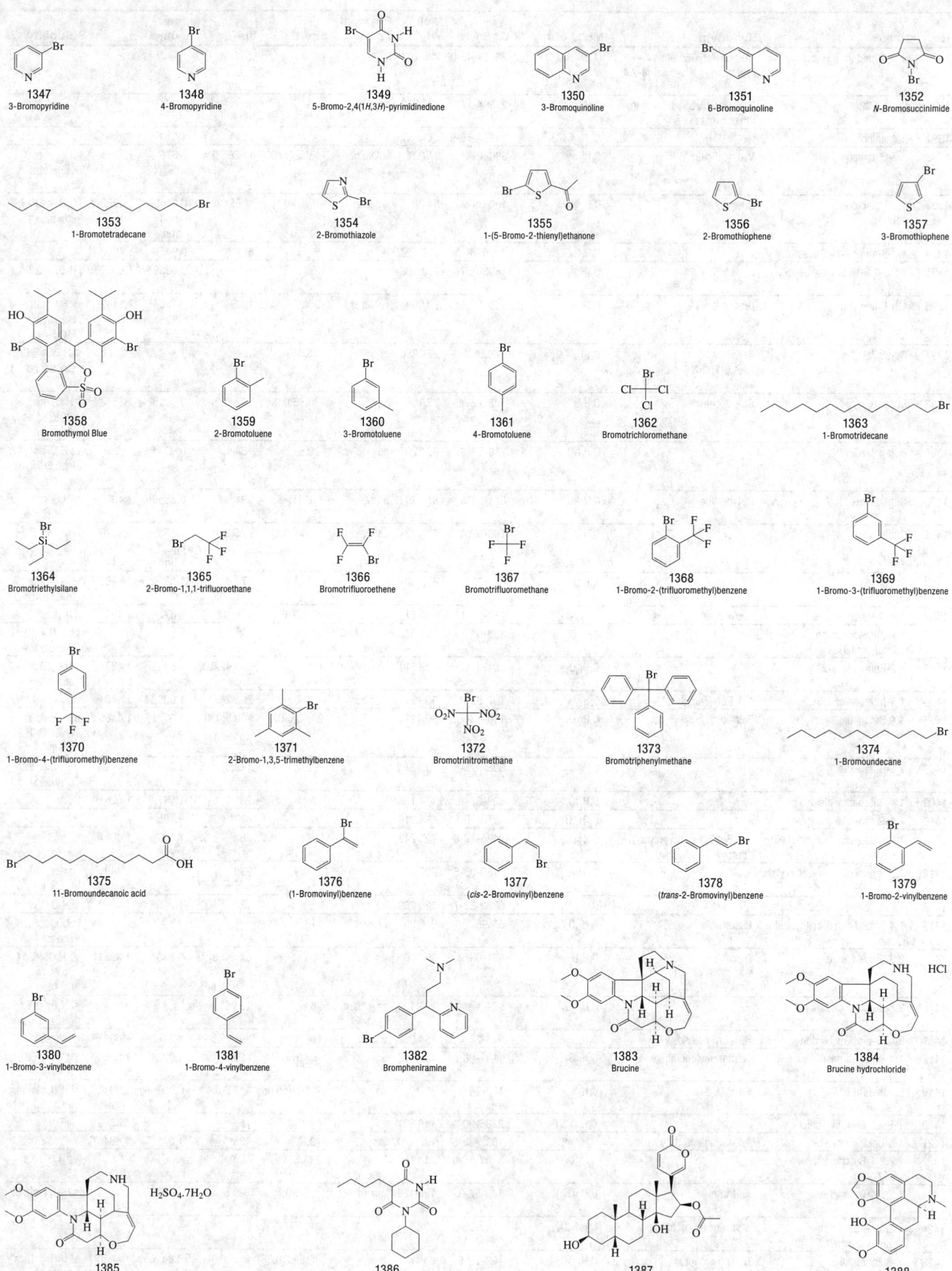

1347 3-Bromopyridine

1348 4-Bromopyridine

1349 5-Bromo-2,4(1*H*,3*H*)-pyrimidinedione

1350 3-Bromoquinoline

1351 6-Bromoquinoline

1352 *N*-Bromosuccinimide

1353 1-Bromotetradecane

1354 2-Bromothiazole

1355 1-(5-Bromo-2-thienyl)ethanone

1356 2-Bromothiophene

1357 3-Bromothiophene

1358 Bromothymol Blue

1359 2-Bromotoluene

1360 3-Bromotoluene

1361 4-Bromotoluene

1362 Bromotrichloromethane

1363 1-Bromotridecane

1364 Bromotriethylsilane

1365 2-Bromo-1,1,1-trifluoroethane

1366 Bromotrifluoroethene

1367 Bromotrifluoromethane

1368 1-Bromo-2-(trifluoromethyl)benzene

1369 1-Bromo-3-(trifluoromethyl)benzene

1370 1-Bromo-4-(trifluoromethyl)benzene

1371 2-Bromo-1,3,5-trimethylbenzene

1372 Bromotrinitromethane

1373 Bromotriphenylmethane

1374 1-Bromoundecane

1375 11-Bromoundecanoic acid

1376 (1-Bromovinyl)benzene

1377 (*cis*-2-Bromovinyl)benzene

1378 (*trans*-2-Bromovinyl)benzene

1379 1-Bromo-2-vinylbenzene

1380 1-Bromo-3-vinylbenzene

1381 1-Bromo-4-vinylbenzene

1382 Brompheniramine

1383 Brucine

1384 Brucine hydrochloride

1385 Brucine sulfate heptahydrate

1386 Bucolome

1387 Bufotalin

1388 Bulbocapnine

No.	Name	Synonym	Mol. Form.	CAS RN	Mol. Wt.	Physical Form	mp/°C	bp/°C	den g cm⁻³	n_D	Solubility
1389	sec-Bumeton	N^2-sec-Butyl-N^4-ethyl-6-methoxy-1,3,5-triazine-2,4-diamine	$C_{10}H_{19}N_5O$	26259-45-0	225.291		87				
1390	BUSAN 72A	(2-Benzothiazolylthio)methyl thiocyanate	$C_9H_6N_2S_3$	21564-17-0	238.352	liq					
1391	Butachlor		$C_{17}H_{26}ClNO_2$	23184-66-9	311.847		<-5	156.5	1.070²⁵		
1392	1,2-Butadiene	Methylallene	C_4H_6	590-19-2	54.091	vol liq or gas	-136.20(0.05)	11.0(0.2)	0.676⁰	1.4205¹	i H₂O; msc EtOH, eth; vs bz
1393	1,3-Butadiene	Divinyl	C_4H_6	106-99-0	54.091	col gas	-108.9(0.1)	-4.6(0.2)	0.6149²⁵ (p>1 atm)	1.4292⁻²⁵	i H₂O; s EtOH, eth, bz; vs ace
1394	1,3-Butadien-1-ol acetate		$C_6H_8O_2$	1515-76-0	112.127			58⁴⁰	0.945²⁵	1.4690²⁰	
1395	trans-1,3-Butadienylbenzene		$C_{10}H_{10}$	16939-57-4	130.186		2.3	76¹¹	0.9286²⁰	1.6089²⁵	i H₂O; s EtOH, eth, ace, bz
1396	1,3-Butadiyne	Diacetylene	C_4H_2	460-12-8	50.059	vol liq or gas	-35(3)	10(2)	0.7364⁰	1.4189⁵	vs H₂O, eth, ace; s chl, EtOH
1397	Butalbital	5-Isobutyl-5-allyl-2,4,6(1H,3H,5H)-pyrimidine-trione	$C_{11}H_{16}N_2O_3$	77-26-9	224.256	pr	138.5				sl H₂O; s EtOH, eth, ace, chl; i lig
1398	Butanal	Butyraldehyde	C_4H_8O	123-72-8	72.106	liq	-96.86(0.02)	74.8(0.2)	0.8016²⁰	1.3843²⁰	s H₂O; msc EtOH; vs ace, bz; sl chl
1399	Butanal oxime		C_4H_9NO	110-69-0	87.120	liq	-29.5	152(9)	0.923²⁰		vs H₂O, ace, bz; msc EtOH, eth; s chl
1400	Butanamide	Butyramide	C_4H_9NO	541-35-5	87.120	lf (bz)	116(1)	231.8(1)	0.8850¹²⁰	1.4087¹³⁰	sl H₂O, eth; i bz; s EtOH
1401	Butane		C_4H_{10}	106-97-8	58.122	col gas	-138.3(0.1)	-0.5(0.5)	0.573²⁵ (p>1 atm)	1.3326²⁰	i H₂O; vs EtOH, eth, chl
1402	Butanedial		$C_4H_6O_2$	638-37-9	86.090			170 dec	1.065²⁰	1.4262¹⁸	vs H₂O, ace, eth, EtOH
1403	1,4-Butanediamine	Putrescine	$C_4H_{12}N_2$	110-60-1	88.151	lf	21.9(0.4)	156(10)	0.877²⁵	1.4969²⁰	s H₂O
1404	1,4-Butanediamine dihydrochloride		$C_4H_{14}Cl_2N_2$	333-93-7	161.073	nd or lf (al, w)	280 dec	sub			vs H₂O, EtOH; i eth, bz, MeOH
1405	1,2-Butanediol, (±)-		$C_4H_{10}O_2$	26171-83-5	90.121			196.42(0.06)	1.0024²⁰	1.4378²⁰	s H₂O, EtOH, ace
1406	1,3-Butanediol	1,3-Butylene glycol	$C_4H_{10}O_2$	107-88-0	90.121	visc liq	-77	208.2(0.1)	1.0053²⁰	1.4401²⁰	
1407	1,4-Butanediol	Tetramethylene glycol	$C_4H_{10}O_2$	110-63-4	90.121		20.43(0.02)	229.5(0.4)	1.0171²⁰	1.4460²⁰	msc H₂O; s EtOH, DMSO; sl eth
1408	2,3-Butanediol		$C_4H_{10}O_2$	6982-25-8	90.121	cry (eth)	7(2)	178(3)	1.0033²⁰	1.4310²⁵	msc H₂O, EtOH; s eth, ace, chl
1409	1,4-Butanediol diacetate		$C_8H_{14}O_4$	628-67-1	174.195		12	233(6)	1.0479¹⁵	1.4251¹⁵	
1410	1,4-Butanediol diacrylate		$C_{10}H_{14}O_4$	1070-70-8	198.216			83⁰·³	1.105²⁵		
1411	1,4-Butanediol diglycidyl ether	1,4-Bis(2,3-epoxypropoxy)-butane	$C_{10}H_{18}O_4$	2425-79-8	202.248			266	1.1²⁵	1.4611²⁰	
1412	1,3-Butanediol dimethacrylate		$C_{12}H_{18}O_4$	1189-08-8	226.269			290		1.4495²⁵	vs ace, eth, EtOH, lig
1413	1,4-Butanediol dimethacrylate		$C_{12}H_{18}O_4$	2082-81-7	226.269	liq		133⁴	1.025²⁰	1.4560²⁰	sl H₂O
1414	1,4-Butanediol dimethylsulfonate	Busulfan	$C_6H_{14}O_6S_2$	55-98-1	246.301	cry	116				i H₂O; sl EtOH, ace
1415	2,3-Butanedione	Diacetyl	$C_4H_6O_2$	431-03-8	86.090	liq	-1.2	87.5(0.8)	0.9808¹⁸	1.3951²⁰	vs H₂O; msc EtOH, eth; s bz, ctc
1416	2,3-Butanedione monooxime		$C_4H_7NO_2$	57-71-6	101.105	pr (chl), lf (w)	76.8	185.5			sl H₂O; vs EtOH, eth, chl; s alk
1417	Butanedioyl dichloride	Succinyl chloride	$C_4H_4Cl_2O_2$	543-20-4	154.980	pl or lf	16.7(0.6)	190(5)	1.3748²⁰	1.4683²⁰	s eth, ace, bz
1418	1,4-Butanedithiol	Tetramethylenedithiol	$C_4H_{10}S_2$	1191-08-8	122.252	liq	-53.9	195.5	1.0021⁰	1.5290²⁰	i H₂O; vs EtOH, sl ctc
1419	Butanenitrile	Propyl cyanide	C_4H_7N	109-74-0	69.106	liq	-111.76(0.05)	117.6(0.4)	0.7936²⁰	1.3842²⁰	sl H₂O, ctc; msc EtOH, eth; s bz
1420	1-Butanesulfonyl chloride		$C_4H_9ClO_2S$	2386-60-9	156.631			75¹⁰		1.4559²⁰	
1421	1,4-Butane sultone	1,2-Oxathiane 2,2-dioxide	$C_4H_8O_3S$	1633-83-6	136.170	liq	13.5	135⁴	1.331²⁰	1.4640²⁰	
1422	1,2,3,4-Butanetetracarboxylic acid		$C_8H_{10}O_8$	1703-58-8	234.160	lf (w) cry (ace)	236.5				vs H₂O, EtOH
1423	1,2,3,4-Butanetetrol	Erythritol	$C_4H_{10}O_4$	149-32-6	122.120	bipym tetr pr	118.1(0.7)	330.5	1.451²⁰		s H₂O; i eth, bz
1424	1,2,3,4-Butanetetrol tetranitrate, (R*,S*)-	Erythrityl tetranitrate	$C_4H_6N_4O_{12}$	7297-25-8	302.111		61				vs EtOH
1425	1-Butanethiol	Butyl mercaptan	$C_4H_{10}S$	109-79-5	90.187	liq	-115.66(0.06)	98.4(0.5)	0.8416²⁰	1.4440²⁰	sl H₂O, chl; vs EtOH, eth

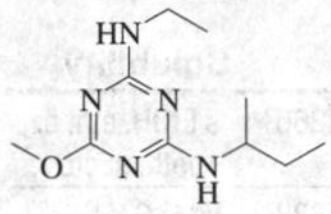

1389
sec-Bumeton

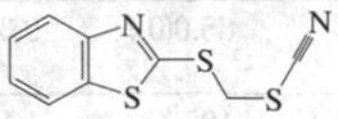

1390
BUSAN 72A

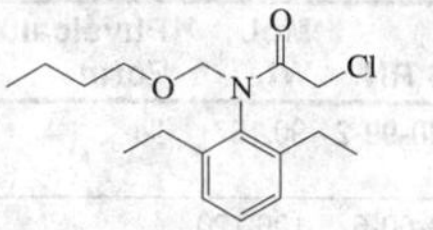

1391
Butachlor

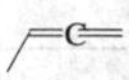

1392
1,2-Butadiene

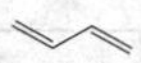

1393
1,3-Butadiene

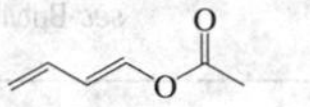

1394
1,3-Butadien-1-ol acetate

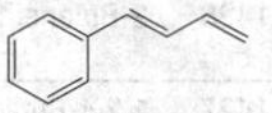

1395
trans-1,3-Butadienylbenzene

1396
1,3-Butadiyne

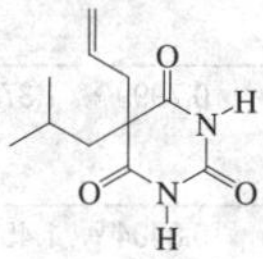

1397
Butalbital

1398
Butanal

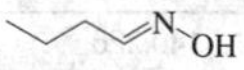

1399
Butanal oxime

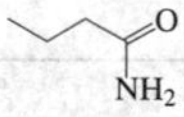

1400
Butanamide

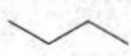

1401
Butane

1402
Butanedial

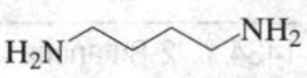

1403
1,4-Butanediamine

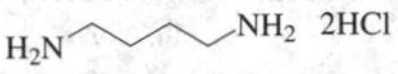

1404
1,4-Butanediamine dihydrochloride

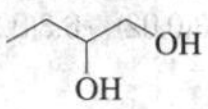

1405
1,2-Butanediol, (±)-

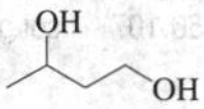

1406
1,3-Butanediol

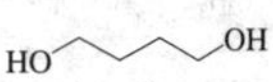

1407
1,4-Butanediol

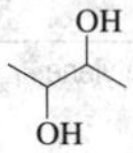

1408
2,3-Butanediol

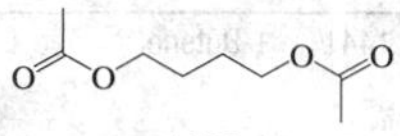

1409
1,4-Butanediol diacetate

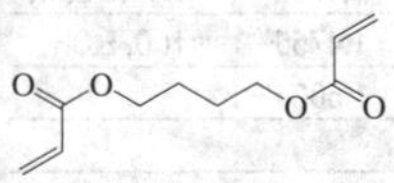

1410
1,4-Butanediol diacrylate

1411
1,4-Butanediol diglycidyl ether

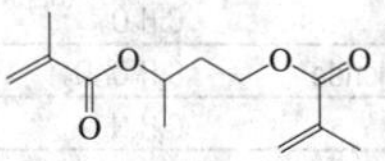

1412
1,3-Butanediol dimethacrylate

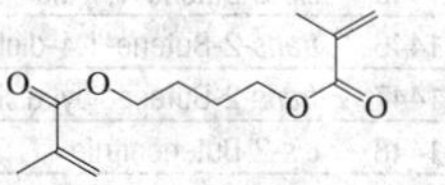

1413
1,4-Butanediol dimethacrylate

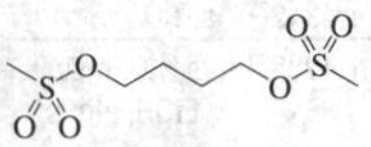

1414
1,4-Butanediol dimethylsulfonate

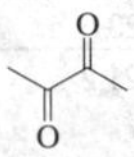

1415
2,3-Butanedione

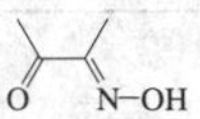

1416
2,3-Butanedione monooxime

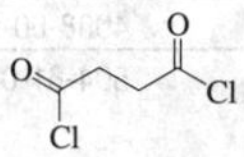

1417
Butanedioyl dichloride

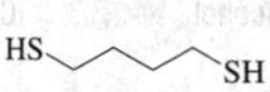

1418
1,4-Butanedithiol

1419
Butanenitrile

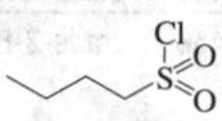

1420
1-Butanesulfonyl chloride

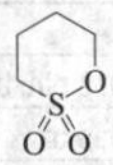

1421
1,4-Butane sultone

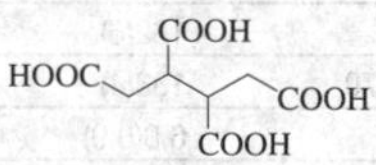

1422
1,2,3,4-Butanetetracarboxylic acid

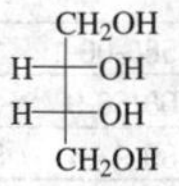

1423
1,2,3,4-Butanetetrol

1424
1,2,3,4-Butanetetrol tetranitrate, (*R**,*S**)-

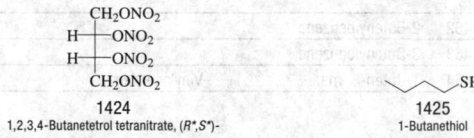

1425
1-Butanethiol

No.	Name	Synonym	Mol. Form.	CAS RN	Mol. Wt.	Physical Form	mp/°C	bp/°C	den g cm⁻³	n_D	Solubility
1426	2-Butanethiol	*sec*-Butyl mercaptan	$C_4H_{10}S$	91840-99-2	90.187	liq	-165	85.0(0.7)	0.8295²⁰	1.4366²⁰	s EtOH, eth, bz, peth; sl ctc
1427	1,2,4-Butanetriol		$C_4H_{10}O_3$	3068-00-6	106.120			190¹⁸	1.18²⁰	1.4688²⁰	vs H_2O, EtOH
1428	Butanilicaine	2-(Butylamino)-*N*-(2-chloro-6-methylphenyl)acetamide	$C_{13}H_{19}ClN_2O$	3785-21-5	254.755	cry	46	145⁰·⁰⁰¹			
1429	Butanoic acid	Butyric acid	$C_4H_8O_2$	107-92-6	88.106	liq	-5.12(0.09)	163.7(0.1)	0.9528²⁵	1.3980²⁰	msc H_2O, EtOH, eth; sl ctc
1430	Butanoic anhydride	Butyric anhydride	$C_8H_{14}O_3$	106-31-0	158.195	liq	-75.0(0.6)	195(1)	0.9668²⁰	1.4070²⁰	s eth; sl ctc
1431	1-Butanol	Butyl alcohol	$C_4H_{10}O$	71-36-3	74.121	liq	-88.60(0.02)	117.6(0.2)	0.8095²⁰	1.3988²⁰	s H_2O, bz; msc EtOH, eth; vs ace
1432	2-Butanol	*sec*-Butyl alcohol	$C_4H_{10}O$	78-92-2	74.121	liq	-88.44(0.07)	99.4(0.2)	0.8063²⁰	1.3978²⁰	vs H_2O; msc EtOH, eth; s bz, ctc
1433	2-Butanone	Methyl ethyl ketone	C_4H_8O	78-93-3	72.106	liq	-86.67(0.01)	79.6(0.2)	0.7999²⁵	1.3788²⁰	vs H_2O; msc EtOH, eth, ace, bz; s chl
1434	2-Butanone (1-methylpropylidene)hydrazone		$C_8H_{16}N_2$	5921-54-0	140.226			170(7)	0.8404²⁰	1.4511²⁰	
1435	2-Butanone oxime		C_4H_9NO	96-29-7	87.120	liq	-29.5(0.5)	151.5(0.6)	0.9232²⁰	1.4410²⁰	s H_2O, chl; msc EtOH, eth
1436	2-Butanone peroxide	Methyl ethyl ketone peroxide	$C_8H_{16}O_4$	1338-23-4	176.211	col liq		exp			sl H_2O; misc os
1437	Butanoyl chloride	*n*-Butyryl chloride	C_4H_7ClO	141-75-3	106.551	liq	-89.0(0.5)	101(3)	1.0277²⁰	1.4121²⁰	msc eth
1438	Butaperazine		$C_{24}H_{31}N_3OS$	653-03-2	409.587			275⁰·⁰⁵			
1439	Butazolamide	*N*-[5-(Aminosulfonyl)-1,3,4-thiadiazol-2-yl]butanamide	$C_6H_{10}N_4O_3S_2$	16790-49-1	250.298	cry	261 dec				
1440	*trans*-2-Butenal	*trans*-Crotonaldehyde	C_4H_6O	123-73-9	70.090	liq	-76.6(0.3)	102.2(0.3)	0.8516²⁰	1.4366²⁰	s H_2O, chl; vs EtOH, eth, ace; msc bz
1441	1-Butene	1-Butylene	C_4H_8	106-98-9	56.107	col gas	-185.33(0.02)	-6.3(0.2)	0.588²⁵ (p>1 atm)	1.3962²⁰	i H_2O; vs EtOH, eth; s bz
1442	*cis*-2-Butene		C_4H_8	590-18-1	56.107	col gas	-138.89(0.02)	3.72(0.08)	0.616²⁵ (p>1 atm)	1.3931⁻²⁵	i H_2O; vs EtOH, eth; s bz
1443	*trans*-2-Butene		C_4H_8	624-64-6	56.107	col gas	-105.52(0.02)	0.88(0.09)	0.599²⁵ (p>1 atm)	1.3848⁻²⁵	s bz
1444	*trans*-2-Butenedinitrile		$C_4H_2N_2$	764-42-1	78.072	nd (bz-peth)	96.0(0.8)	186	0.9416¹¹¹	1.4349¹¹¹	s H_2O, EtOH, eth, ace, bz, chl; sl peth
1445	*cis*-2-Butene-1,4-diol		$C_4H_8O_2$	6117-80-2	88.106		11.0(0.5)	235	1.0698²⁰	1.4782²⁰	s H_2O; vs EtOH
1446	*trans*-2-Butene-1,4-diol		$C_4H_8O_2$	821-11-4	88.106		27(1)	131¹³	1.0700²⁰	1.4755²⁰	vs H_2O, EtOH
1447	*trans*-2-Butenedioyl dichloride	Fumaric acid dichloride	$C_4H_2Cl_2O_2$	627-63-4	152.964	pa ye liq		159	1.408²⁰	1.5004¹⁸	
1448	*cis*-2-Butenenitrile	Isocrotononitrile	C_4H_5N	1190-76-7	67.090	liq		106(6)			
1449	*trans*-2-Butenenitrile	Crotononitrile	C_4H_5N	627-26-9	67.090	liq	-51.5	120	0.8239²⁰	1.4225²⁰	s eth, ace
1450	3-Butenenitrile	Allyl cyanide	C_4H_5N	109-75-1	67.090	liq	-87	117(5)	0.8341²⁰	1.4060²⁰	sl H_2O; msc EtOH, eth
1451	*cis*-2-Butenoic acid	Isocrotonic acid	$C_4H_6O_2$	503-64-0	86.090	nd or pr (peth)	15	169	1.0267²⁰	1.4450²⁰	vs H_2O; s EtOH
1452	*trans*-2-Butenoic acid	Crotonic acid	$C_4H_6O_2$	107-93-7	86.090	mcl pr or nd (w, lig)	71.3(0.2)	184.7	0.9604⁷⁷	1.4249⁷⁷	vs H_2O, EtOH; s eth, ace, lig
1453	3-Butenoic acid		$C_4H_6O_2$	625-38-7	86.090	liq	-35	169	1.0091²⁰	1.4239²⁰	s H_2O; msc EtOH, eth
1454	2-Butenoic anhydride	Crotonic acid anhydride	$C_8H_{10}O_3$	623-68-7	154.163			247	1.0397²⁰	1.4745²⁰	vs eth
1455	*cis*-2-Buten-1-ol	*cis*-Crotyl alcohol	C_4H_8O	4088-60-2	72.106			123	0.8662²⁰	1.4342²⁵	s H_2O
1456	*trans*-2-Buten-1-ol	*trans*-Crotyl alcohol	C_4H_8O	504-61-0	72.106		<-30	121.2	0.8521²⁰	1.4288²⁰	vs H_2O; msc EtOH, eth; s chl
1457	3-Buten-1-ol		C_4H_8O	627-27-0	72.106			112(5)	0.8424²⁰	1.4224²⁰	s H_2O, ace; msc EtOH, eth; sl chl
1458	3-Buten-2-ol		C_4H_8O	598-32-3	72.106			97(4)			
1459	3-Buten-2-one	Methyl vinyl ketone	C_4H_6O	78-94-4	70.090			81(4)	0.864²⁰	1.4081²⁰	s H_2O, EtOH, bz; vs eth, ace; sl ctc
1460	2-Butenoyl chloride		C_4H_5ClO	10487-71-5	104.535			121(8)	1.0905²⁰	1.460¹⁸	vs ace
1461	*trans*-1-Butenylbenzene		$C_{10}H_{12}$	1005-64-7	132.202	liq	-43.1(0.4)	201(4)	0.9019²⁰	1.5420²⁰	i H_2O; s EtOH, eth, bz, ctc
1462	2-Butenylbenzene		$C_{10}H_{12}$	1560-06-1	132.202			182(6)	0.8831²⁰	1.5101²⁰	
1463	3-Butenylbenzene		$C_{10}H_{12}$	768-56-9	132.202	liq	-70	183(4)	0.8831²⁰	1.5059²⁰	i H_2O; s eth, bz
1464	1-Buten-3-yne	Vinylacetylene	C_4H_4	689-97-4	52.075	col gas		6.0(0.9)	0.7094⁰	1.4161¹	i H_2O; s bz

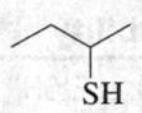

1426
2-Butanethiol

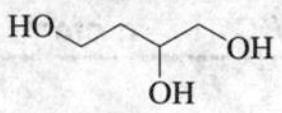

1427
1,2,4-Butanetriol

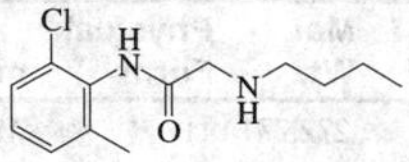

1428
Butanilicaine

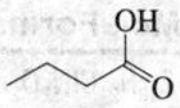

1429
Butanoic acid

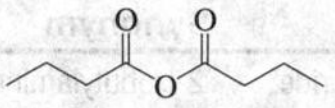

1430
Butanoic anhydride

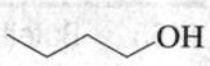

1431
1-Butanol

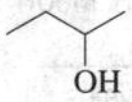

1432
2-Butanol

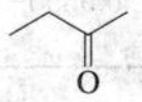

1433
2-Butanone

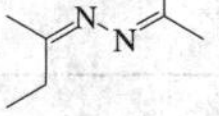

1434
2-Butanone (1-methylpropylidene)hydrazone

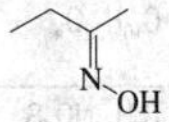

1435
2-Butanone oxime

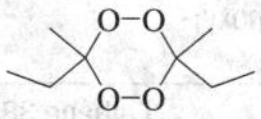

1436
2-Butanone peroxide

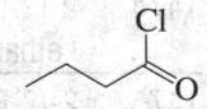

1437
Butanoyl chloride

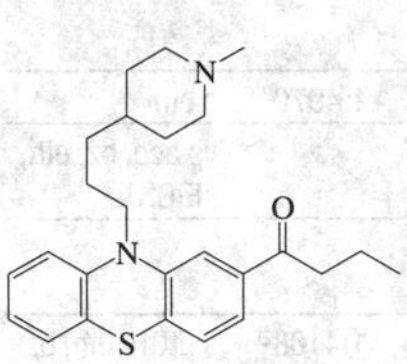

1438
Butaperazine

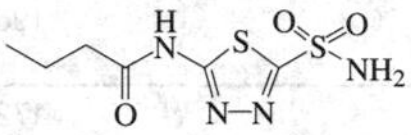

1439
Butazolamide

1440
trans-2-Butenal

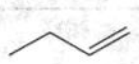

1441
1-Butene

1442
cis-2-Butene

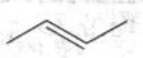

1443
trans-2-Butene

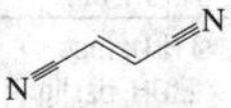

1444
trans-2-Butenedinitrile

1445
cis-2-Butene-1,4-diol

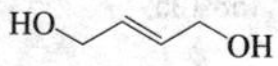

1446
trans-2-Butene-1,4-diol

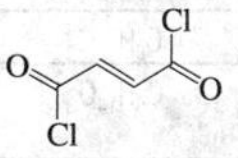

1447
trans-2-Butenedioyl dichloride

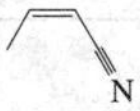

1448
cis-2-Butenenitrile

1449
trans-2-Butenenitrile

1450
3-Butenenitrile

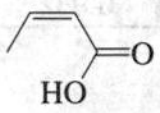

1451
cis-2-Butenoic acid

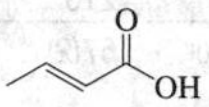

1452
trans-2-Butenoic acid

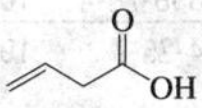

1453
3-Butenoic acid

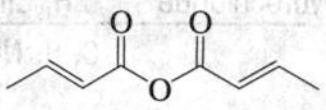

1454
2-Butenoic anhydride

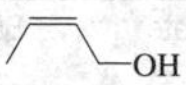

1455
cis-2-Buten-1-ol

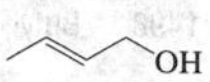

1456
trans-2-Buten-1-ol

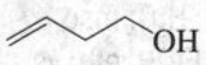

1457
3-Buten-1-ol

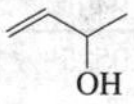

1458
3-Buten-2-ol

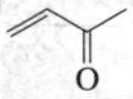

1459
3-Buten-2-one

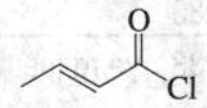

1460
2-Butenoyl chloride

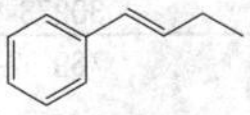

1461
trans-1-Butenylbenzene

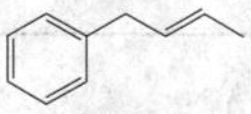

1462
2-Butenylbenzene

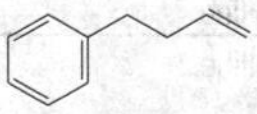

1463
3-Butenylbenzene

1464
1-Buten-3-yne

No.	Name	Synonym	Mol. Form.	CAS RN	Mol. Wt.	Physical Form	mp/°C	bp/°C	den g cm⁻³	n_D	Solubility
1465	Butethamine hydrochloride	2-Isobutylaminoethyl 4-aminobenzoate	$C_{13}H_{21}ClN_2O_2$	553-68-4	272.771	cry	194				s H_2O; sl EtOH, bz, chl; i eth
1466	Buthalital sodium		$C_{11}H_{15}N_2NaO_2S$	510-90-7	262.304						vs H_2O; sl EtOH; i eth, bz
1467	Buthiazide		$C_{11}H_{16}ClN_3O_4S_2$	2043-38-1	353.846		221.5				
1468	Buthiobate	Denmert	$C_{21}H_{28}N_2S_2$	51308-54-4	372.590	ye oil	32		1.0865²⁵	1.596²⁶	i H_2O; s os
1469	Butonate		$C_8H_{14}Cl_3O_5P$	126-22-7	327.527			129⁰·⁵			
1470	Butoxyacetylene		$C_6H_{10}O$	3329-56-4	98.142			104	0.8200²⁰	1.4067	vs eth, EtOH
1471	4-Butoxyaniline		$C_{10}H_{15}NO$	4344-55-2	165.232			132⁴			
1472	4-Butoxybenzaldehyde		$C_{11}H_{14}O_2$	5736-88-9	178.228			148¹⁰			
1473	2-Butoxyethanol	Ethylene glycol monobutyl ether	$C_6H_{14}O_2$	111-76-2	118.174	liq	-74.8	171(2)	0.9015²⁰	1.4198²⁰	msc H_2O, EtOH, eth; sl ctc
1474	2-[2-(2-Butoxyethoxy)ethoxy]-ethanol		$C_{10}H_{22}O_4$	143-22-6	206.280			278	0.9890²⁰	1.4389²⁰	vs EtOH, MeOH
1475	2-(2-Butoxyethoxy)ethyl thiocyanate	Lethane 384	$C_9H_{17}NO_2S$	112-56-1	203.302	liq		122⁰·²⁵			i H_2O; vs os
1476	1-(2-Butoxyethoxy)-2-propanol		$C_9H_{20}O_3$	124-16-3	176.253	col liq	-90	234(14)	0.931²⁰		s H_2O
1477	2-Butoxyethyl acetate	Ethylene glycol monobutyl ether acetate	$C_8H_{16}O_3$	112-07-2	160.211	liq		191.1(0.9)			
1478	2-Butoxyethyl (2,4-dichlorophenoxy)acetate	2,4-D 2-Butoxyethyl ester	$C_{14}H_{18}Cl_2O_4$	1929-73-3	321.197			159¹	1.232²⁰		
1479	2-Butoxyethyl (2,4,5-trichlorophenoxy)acetate	2,4,5-T Butoxyethyl ester	$C_{14}H_{17}Cl_3O_4$	2545-59-7	355.642			164¹	1.280²⁰		s ctc
1480	4-Butoxy-N-hydroxybenzeneacetamide	Bufexamac	$C_{12}H_{17}NO_3$	2438-72-4	223.268	nd (ace)	154				
1481	1-Butoxy-4-methylbenzene		$C_{11}H_{16}O$	10519-06-9	164.244			229.5	0.9205²⁵	1.4970²⁰	s eth
1482	4-Butoxyphenol		$C_{10}H_{14}O_2$	122-94-1	166.217		65.5	125⁴			vs ace, bz, eth, EtOH
1483	4-[3-(4-Butoxyphenoxy)propyl] morpholine	Pramoxine	$C_{17}H_{27}NO_3$	140-65-8	293.401			196⁶			
1484	1-Butoxy-2-propanol		$C_7H_{16}O_2$	5131-66-8	132.201			172(3)	0.882²⁰	1.4168²⁰	s EtOH, eth, bz, ctc, MeOH
1485	Butralin	4-tert-Butyl-N-sec-butyl-2,6-dinitroaniline	$C_{14}H_{21}N_3O_4$	33629-47-9	295.335		59.3(0.5)	135⁰·⁵			
1486	N-Butylacetamide		$C_6H_{13}NO$	1119-49-9	115.173			229	0.8960²⁵	1.4388²⁵	
1487	Butyl acetate		$C_6H_{12}O_2$	123-86-4	116.158	liq	-77.0(0.1)	126.0(0.1)	0.8825²⁰	1.3941²⁰	sl H_2O; msc EtOH, eth; s ace, chl
1488	sec-Butyl acetate	1-Methylpropyl acetate	$C_6H_{12}O_2$	105-46-4	116.158	liq	-98.9	108(4)	0.8748²⁰	1.3888²⁰	sl H_2O, ctc; s EtOH, eth
1489	tert-Butyl acetate		$C_6H_{12}O_2$	540-88-5	116.158	liq		97.9(1)	0.8665²⁰	1.3855²⁰	s EtOH, eth, chl, HOAc
1490	tert-Butylacetic acid		$C_6H_{12}O_2$	1070-83-3	116.158		6(2)	184(2)	0.9124²⁰	1.4096²⁰	s EtOH, eth
1491	Butyl acetoacetate		$C_8H_{14}O_3$	591-60-6	158.195		-35.6	127⁵⁰	0.9671²⁵	1.4137²⁰	sl H_2O; msc EtOH, bz, lig
1492	Butyl acrylate		$C_7H_{12}O_2$	141-32-2	128.169	liq	-63.6(0.5)	146.6(0.6)	0.8898²⁰	1.4185²⁰	i H_2O; s EtOH, eth, ace; sl ctc
1493	tert-Butyl acrylate		$C_7H_{12}O_2$	1663-39-4	128.169	liq		120	0.879²⁵	1.4110²⁰	
1494	Butylamine	1-Butanamine	$C_4H_{11}N$	109-73-9	73.137	liq	-49(1)	77.0(0.2)	0.7414²⁰	1.4031²⁰	msc H_2O; s EtOH, eth
1495	sec-Butylamine	2-Butanamine, (±)-	$C_4H_{11}N$	33966-50-6	73.137	liq	-104.5(0.6)	62.71(0.08)	0.7246²⁰	1.3932²⁰	s H_2O, chl; msc EtOH, eth; vs ace
1496	tert-Butylamine	2-Methyl-2-propanamine	$C_4H_{11}N$	75-64-9	73.137	liq	-66.92(0.06)	44.02(0.07)	0.6958²⁰	1.3784²⁰	msc H_2O, EtOH, eth; s chl
1497	Butylamine hydrochloride	1-Butanamine hydrochloride	$C_4H_{12}ClN$	3858-78-4	109.598		213		0.982²⁰		sl H_2O, EtOH
1498	Butyl 4-aminobenzoate	Butamben	$C_{11}H_{15}NO_2$	94-25-7	193.243	cry (al or bz)	57(2)	173⁸			i H_2O; s EtOH, eth, bz, chl
1499	2-(Butylamino)ethanol		$C_6H_{15}NO$	111-75-1	117.189			199	0.8907²⁰	1.4437²⁰	vs H_2O, EtOH, eth
1500	2-(tert-Butylamino)ethanol		$C_6H_{15}NO$	4620-70-6	117.189		44	180(16)	0.8818²⁰		
1501	N-tert-Butylaminoethyl methacrylate		$C_{10}H_{19}NO_2$	3775-90-4	185.264			102¹²			s chl
1502	2-(tert-Butylaminothio)-benzothiazole	N-tert-Butyl-2-benzothiazole-sulfenamide	$C_{11}H_{14}N_2S_2$	95-31-8	238.372		108				
1503	2-sec-Butylaniline		$C_{10}H_{15}N$	55751-54-7	149.233			120¹⁶	0.9574²⁰		s EtOH, ace, bz; sl ctc
1504	4-Butylaniline		$C_{10}H_{15}N$	104-13-2	149.233	pa ye		261	0.945²⁰		sl ctc
1505	4-sec-Butylaniline		$C_{10}H_{15}N$	30273-11-1	149.233			238	0.949¹⁵	1.5360²⁹	vs bz, eth
1506	4-tert-Butylaniline		$C_{10}H_{15}N$	769-92-6	149.233	ye rd (peth)	17	241	0.9525¹⁵	1.5380²⁰	sl H_2O; msc EtOH, eth; vs bz; s ctc

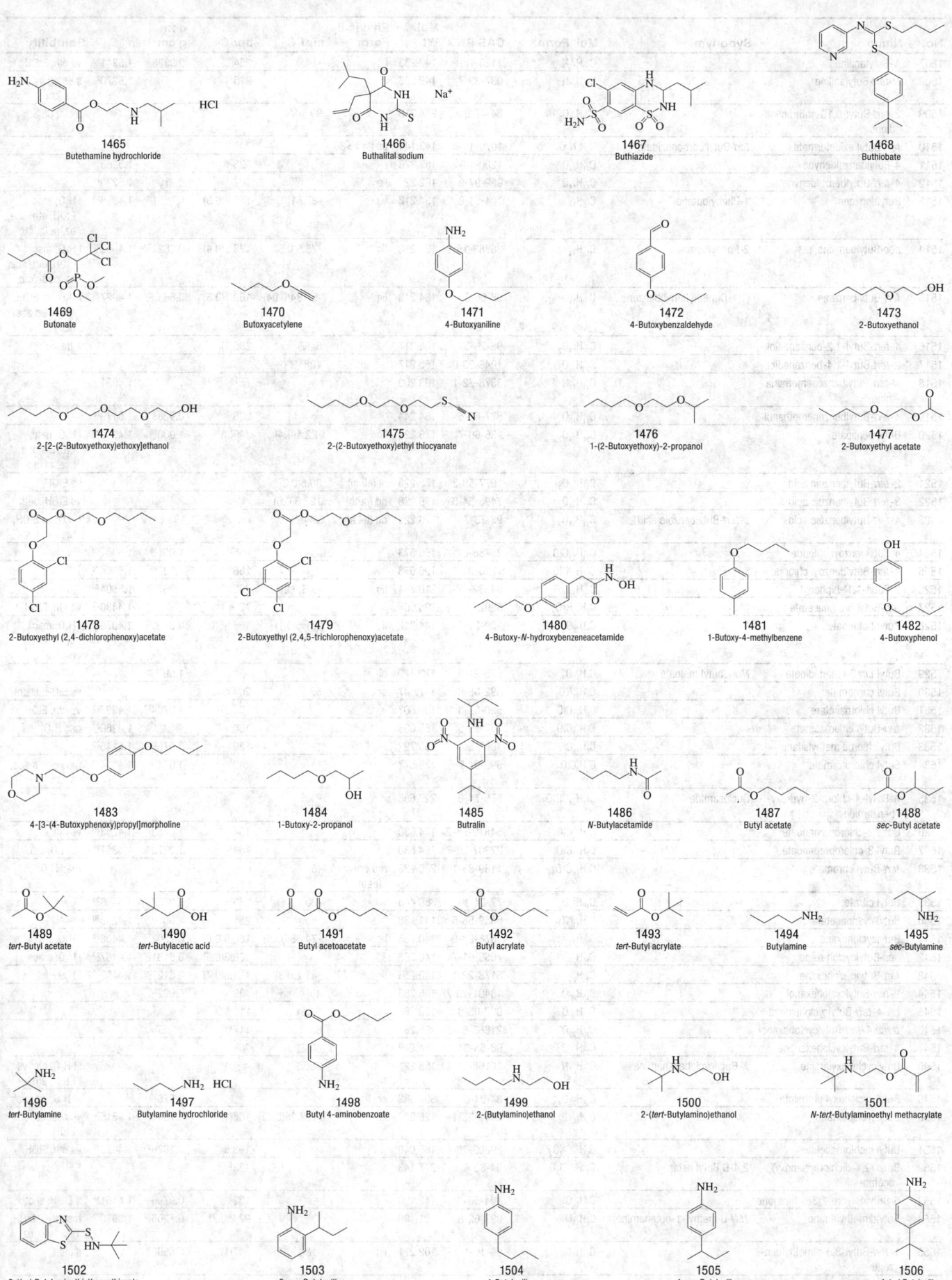

1465
Butethamine hydrochloride

1466
Buthalital sodium

1467
Buthiazide

1468
Buthiobate

1469
Butonate

1470
Butoxyacetylene

1471
4-Butoxyaniline

1472
4-Butoxybenzaldehyde

1473
2-Butoxyethanol

1474
2-[2-(2-Butoxyethoxy)ethoxy]ethanol

1475
2-(2-Butoxyethoxy)ethyl thiocyanate

1476
1-(2-Butoxyethoxy)-2-propanol

1477
2-Butoxyethyl acetate

1478
2-Butoxyethyl (2,4-dichlorophenoxy)acetate

1479
2-Butoxyethyl (2,4,5-trichlorophenoxy)acetate

1480
4-Butoxy-*N*-hydroxybenzeneacetamide

1481
1-Butoxy-4-methylbenzene

1482
4-Butoxyphenol

1483
4-[3-(4-Butoxyphenoxy)propyl]morpholine

1484
1-Butoxy-2-propanol

1485
Butralin

1486
N-Butylacetamide

1487
Butyl acetate

1488
sec-Butyl acetate

1489
tert-Butyl acetate

1490
tert-Butylacetic acid

1491
Butyl acetoacetate

1492
Butyl acrylate

1493
tert-Butyl acrylate

1494
Butylamine

1495
sec-Butylamine

1496
tert-Butylamine

1497
Butylamine hydrochloride

1498
Butyl 4-aminobenzoate

1499
2-(Butylamino)ethanol

1500
2-(*tert*-Butylamino)ethanol

1501
N-*tert*-Butylaminoethyl methacrylate

1502
2-(*tert*-Butylaminothio)benzothiazole

1503
2-*sec*-Butylaniline

1504
4-Butylaniline

1505
4-*sec*-Butylaniline

1506
4-*tert*-Butylaniline

No.	Name	Synonym	Mol. Form.	CAS RN	Mol. Wt.	Physical Form	mp/°C	bp/°C	den g cm⁻³	n_D	Solubility
1507	*N*-Butylaniline		$C_{10}H_{15}N$	1126-78-9	149.233	liq	-14.4	254(9)	0.9323[20]	1.5341[20]	vs eth, EtOH
1508	*N-tert*-Butylaniline		$C_{10}H_{15}N$	937-33-7	149.233			215		1.5270[20]	s EtOH; vs ace, bz, chl
1509	2-*tert*-Butyl-9,10-anthracene-dione		$C_{18}H_{16}O_2$	84-47-9	264.319		99				s ctc, CS_2
1510	*tert*-Butyl azidoformate	*tert*-Butyl carbonazidate	$C_5H_9N_3O_2$	1070-19-5	143.144	unstab >80		73[70]			
1511	4-Butylbenzaldehyde		$C_{11}H_{14}O$	1200-14-2	162.228			123[7]		1.5265	
1512	4-*tert*-Butylbenzaldehyde		$C_{11}H_{14}O$	939-97-9	162.228	liq		107[11]	0.970	1.5270[20]	
1513	Butylbenzene	1-Phenylbutane	$C_{10}H_{14}$	104-51-8	134.218	liq	-87.81(0.05)	183.3(0.3)	0.8601[20]	1.4898[20]	i H_2O; msc EtOH, eth, ace, bz, peth, ctc
1514	*sec*-Butylbenzene, (±)-	2-Phenylbutane	$C_{10}H_{14}$	36383-15-0	134.218	liq	-75.5(0.3)	173.3(0.4)	0.8621[20]	1.4902[20]	i H_2O; msc EtOH, eth, ace, bz, peth, ctc
1515	*tert*-Butylbenzene	(1,1-Dimethylethyl)benzene	$C_{10}H_{14}$	98-06-6	134.218	liq	-57.84(0.04)	169.1(0.3)	0.8665[20]	1.4927[20]	i H_2O; vs EtOH, eth; msc ace, bz
1516	4-*tert*-Butyl-1,2-benzenediol		$C_{10}H_{14}O_2$	98-29-3	166.217		54(2)	286(1)			s tfa
1517	2-*tert*-Butyl-1,4-benzenediol		$C_{10}H_{14}O_2$	1948-33-0	166.217		128				
1518	*N-tert*-Butylbenzenemethana-mine		$C_{11}H_{17}N$	3378-72-1	163.260			75[5]		1.4951[25]	
1519	4-*tert*-Butylbenzenemethanol		$C_{11}H_{16}O$	877-65-6	164.244			236	0.928[25]	1.5179[20]	
1520	Butyl benzoate		$C_{11}H_{14}O_2$	136-60-7	178.228	liq	-22.4(0.4)	249(3)	1.000[20]	1.4940[25]	i H_2O; msc EtOH, eth; s ace; sl ctc
1521	2-*tert*-Butylbenzoic acid		$C_{11}H_{14}O_2$	1077-58-3	178.228	pl (dil al)	80.5(0.3)				vs EtOH
1522	3-*tert*-Butylbenzoic acid		$C_{11}H_{14}O_2$	7498-54-6	178.228	nd (peth)	127.3(0.5)				vs EtOH, peth
1523	4-*tert*-Butylbenzoic acid	*p-tert*-Butylbenzoic acid	$C_{11}H_{14}O_2$	98-73-7	178.228	nd (dil al)	164(2)				i H_2O; vs EtOH, bz; s chl
1524	4-Butylbenzoyl chloride		$C_{11}H_{13}ClO$	28788-62-7	196.673			155[26]	1.051[25]	1.5351[20]	
1525	4-*tert*-Butylbenzoyl chloride		$C_{11}H_{13}ClO$	1710-98-1	196.673			266	1.007[25]	1.5364[20]	
1526	2-Butyl-1,1'-biphenyl		$C_{16}H_{18}$	54532-97-7	210.314	liq	-9.6(0.2)	292(3)	0.9676[20]	1.5604[20]	
1527	*tert*-Butyl bromoacetate		$C_6H_{11}BrO_2$	5292-43-3	195.054			73[25]		1.4430[20]	vs eth, EtOH
1528	Butyl butanoate		$C_8H_{16}O_2$	109-21-7	144.212	liq	-91.5(0.1)	164.95(0.1)	0.8700[20]	1.4075[20]	i H_2O; msc EtOH, eth; s ctc
1529	Butyl *cis*-2-butenedioate	Monobutyl maleate	$C_8H_{12}O_4$	925-21-3	172.179	oil			1.09[25]		
1530	Butyl carbamate		$C_5H_{11}NO_2$	592-35-8	117.147	pr	53	204 dec			vs EtOH; sl chl
1531	Butyl chloroacetate		$C_6H_{11}ClO_2$	590-02-3	150.603			181(3)	1.0704[20]	1.4297[20]	vs eth, EtOH
1532	*tert*-Butyl chloroacetate		$C_6H_{11}ClO_2$	107-59-5	150.603			150		1.4260[20]	dec H_2O
1533	Butylchlorodimethylsilane		$C_6H_{15}ClSi$	1000-50-6	150.722			139	0.876[20]	1.5145[20]	
1534	Butyl chloroformate		$C_5H_9ClO_2$	592-34-7	136.577			142	1.074[25]	1.4114[20]	msc eth; s ace; sl ctc
1535	*N*-Butyl-4-chloro-2-hydroxy-benzamide	Buclosamide	$C_{11}H_{14}ClNO_2$	575-74-6	227.688		91.5				
1536	Butyl 2-chloropropanoate		$C_7H_{13}ClO_2$	54819-86-2	164.630			184	1.0253[20]	1.4263[20]	vs eth
1537	Butyl 3-chloropropanoate		$C_7H_{13}ClO_2$	27387-79-7	164.630			104[22]	1.0370[20]	1.4321[20]	vs H_2O, eth
1538	*tert*-Butyl chromate		$C_8H_{18}CrO_4$	1189-85-1	230.223	red cry (peth)	-5				reac H_2O
1539	Butyl citrate		$C_{18}H_{32}O_7$	77-94-1	360.443		-20	233[22]	1.043[20]	1.4460[20]	
1540	Butyl cyanoacetate		$C_7H_{11}NO_2$	5459-58-5	141.168			231	1.0010[20]	1.4200[20]	
1541	Butylcyclohexane		$C_{10}H_{20}$	1678-93-9	140.266	liq	-74.68(0.05)	180.9(0.6)	0.7902[20]	1.4408[20]	i H_2O
1542	*sec*-Butylcyclohexane		$C_{10}H_{20}$	7058-01-7	140.266			179.3(0.5)	0.8131[20]	1.4467[20]	i H_2O; s ace
1543	*tert*-Butylcyclohexane		$C_{10}H_{20}$	3178-22-1	140.266	liq	-41.2(0.3)	171.6(0.4)	0.8127[20]	1.4469[20]	i H_2O
1544	2-*tert*-Butylcyclohexanol		$C_{10}H_{20}O$	13491-79-7	156.265		45	139[95]	0.902[25]		
1545	*cis*-4-*tert*-Butylcyclohexanol		$C_{10}H_{20}O$	937-05-3	156.265		82(3)	112[15]			
1546	*trans*-4-*tert*-Butylcyclohexanol		$C_{10}H_{20}O$	21862-63-5	156.265		83	112[15]			
1547	4-*tert*-Butylcyclohexanone		$C_{10}H_{18}O$	98-53-3	154.249		48(3)	90[9]			
1548	Butylcyclohexylamine	*N*-Butylcyclohexanamine	$C_{10}H_{21}N$	10108-56-2	155.281		208.3				sl H_2O, ctc; vs EtOH, eth
1549	Butyl cyclohexyl phthalate		$C_{18}H_{24}O_4$	84-64-0	304.382	col liq			1.076[25]		sl H_2O; misc os
1550	Butylcyclopentane		C_9H_{18}	2040-95-1	126.239	liq	-107.95(0.05)	156(1)	0.7846[20]	1.4316[20]	vs ace, bz, eth, EtOH
1551	Butyl dichloroacetate		$C_6H_{10}Cl_2O_2$	29003-73-4	185.048			193.5	1.1820[20]	1.4420[20]	vs eth, EtOH
1552	Butyl (2,4-dichlorophenoxy)acetate	2,4-D Butyl ester	$C_{12}H_{14}Cl_2O_3$	94-80-4	277.143		9	133[1]			
1553	5-Butyldihydro-2(3*H*)-furanone		$C_8H_{14}O_2$	104-50-7	142.196			132[20]	0.9796[19]	1.4451[19]	s EtOH; sl ctc
1554	Butyldimethylamine	*N,N*-Dimethyl-1-butanamine	$C_6H_{15}N$	927-62-8	101.190			92.2(0.7)	0.7206[20]	1.3970[20]	msc H_2O, EtOH, eth, ace, bz
1555	1-*tert*-Butyl-3,5-dimethylben-zene		$C_{12}H_{18}$	98-19-1	162.271	liq	-18(2)	207(1)	0.8668[20]		s ctc

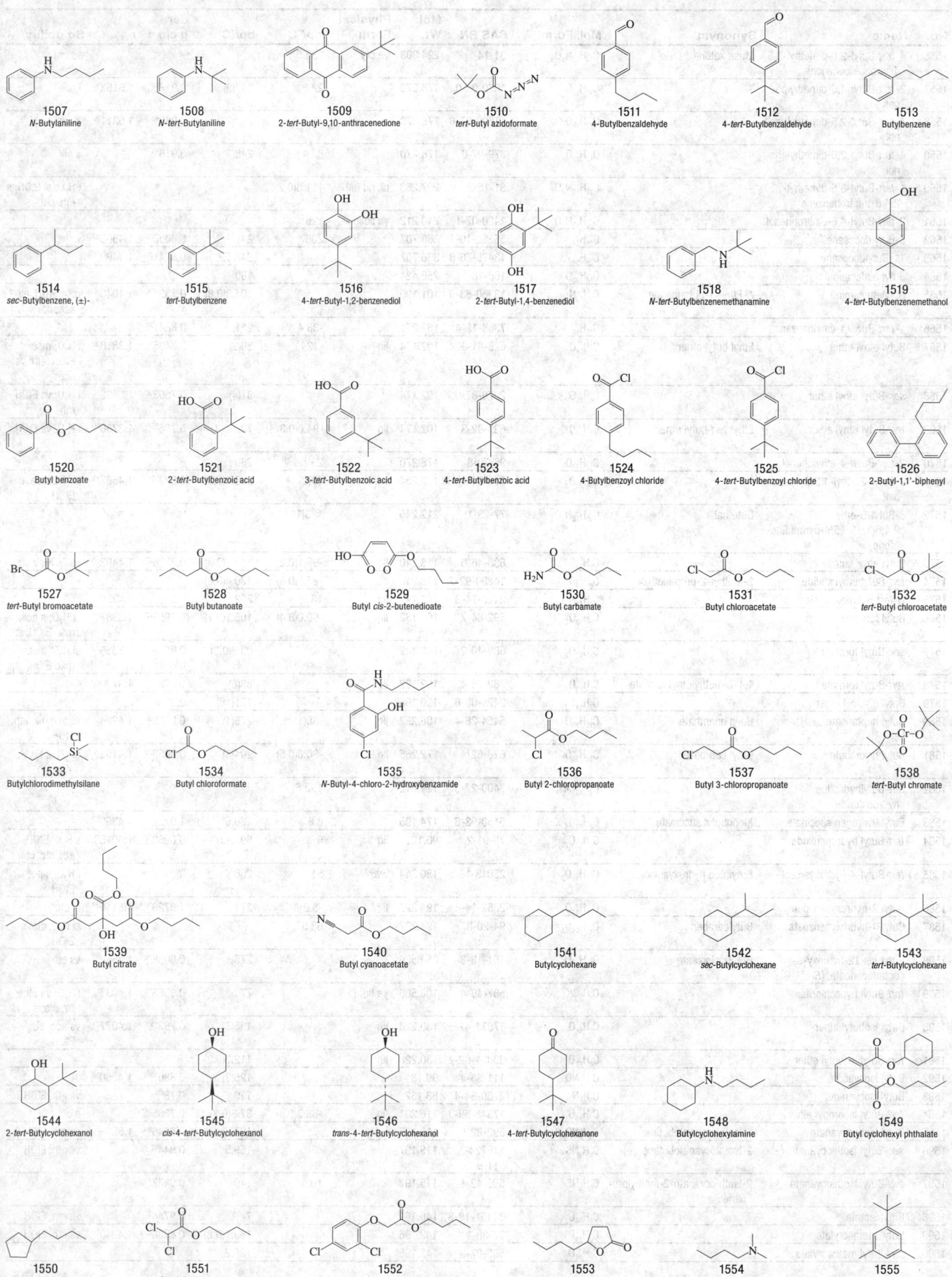

1507
N-Butylaniline

1508
N-tert-Butylaniline

1509
2-tert-Butyl-9,10-anthracenedione

1510
tert-Butyl azidoformate

1511
4-Butylbenzaldehyde

1512
4-tert-Butylbenzaldehyde

1513
Butylbenzene

1514
sec-Butylbenzene, (±)-

1515
tert-Butylbenzene

1516
4-tert-Butyl-1,2-benzenediol

1517
2-tert-Butyl-1,4-benzenediol

1518
N-tert-Butylbenzenemethanamine

1519
4-tert-Butylbenzenemethanol

1520
Butyl benzoate

1521
2-tert-Butylbenzoic acid

1522
3-tert-Butylbenzoic acid

1523
4-tert-Butylbenzoic acid

1524
4-Butylbenzoyl chloride

1525
4-tert-Butylbenzoyl chloride

1526
2-Butyl-1,1'-biphenyl

1527
tert-Butyl bromoacetate

1528
Butyl butanoate

1529
Butyl cis-2-butenedioate

1530
Butyl carbamate

1531
Butyl chloroacetate

1532
tert-Butyl chloroacetate

1533
Butylchlorodimethylsilane

1534
Butyl chloroformate

1535
N-Butyl-4-chloro-2-hydroxybenzamide

1536
Butyl 2-chloropropanoate

1537
Butyl 3-chloropropanoate

1538
tert-Butyl chromate

1539
Butyl citrate

1540
Butyl cyanoacetate

1541
Butylcyclohexane

1542
sec-Butylcyclohexane

1543
tert-Butylcyclohexane

1544
2-tert-Butylcyclohexanol

1545
cis-4-tert-Butylcyclohexanol

1546
trans-4-tert-Butylcyclohexanol

1547
4-tert-Butylcyclohexanone

1548
Butylcyclohexylamine

1549
Butyl cyclohexyl phthalate

1550
Butylcyclopentane

1551
Butyl dichloroacetate

1552
Butyl (2,4-dichlorophenoxy)acetate

1553
5-Butyldihydro-2(3H)-furanone

1554
Butyldimethylamine

1555
1-tert-Butyl-3,5-dimethylbenzene

No.	Name	Synonym	Mol. Form.	CAS RN	Mol. Wt.	Physical Form	mp/°C	bp/°C	den g cm⁻³	n_D	Solubility
1556	4-*tert*-Butyl-2,6-dimethyl-3,5-dinitroacetophenone	Musk ketone	$C_{14}H_{18}N_2O_5$	81-14-1	294.303	ye cry	135.5				vs chl
1557	2-*tert*-Butyl-4,6-dimethylphe-nol		$C_{12}H_{18}O$	1879-09-0	178.270		22.3	247(9)	0.917[80]	1.5183[20]	i alk
1558	4-*tert*-Butyl-2,5-dimethylphe-nol		$C_{12}H_{18}O$	17696-37-6	178.270		71.2	262(27)	0.939[80]	1.5311[20]	s alk
1559	4-*tert*-Butyl-2,6-dimethylphe-nol		$C_{12}H_{18}O$	879-97-0	178.270		82.4	248	0.916[80]		s alk
1560	1-*tert*-Butyl-3,5-dimethyl-2,4,6-trinitrobenzene		$C_{12}H_{15}N_3O_6$	81-15-2	297.263	pl, nd (al)	111.5(0.2)				i H$_2$O; sl EtOH; s eth, chl
1561	2-*tert*-Butyl-4,6-dinitrophenol		$C_{10}H_{12}N_2O_5$	1420-07-1	240.212	ye solid	126				
1562	5-Butyldocosane		$C_{26}H_{54}$	55282-16-1	366.707		208	244[10]	0.8058[20]	1.4503[20]	
1563	11-Butyldocosane		$C_{26}H_{54}$	13475-76-8	366.707			242.5[10]	0.8041[20]	1.4499[20]	
1564	Butyl dodecanoate		$C_{16}H_{32}O_2$	106-18-3	256.424			180[18]			
1565	Butylethylamine	N-Ethyl-1-butanamine	$C_6H_{15}N$	13360-63-9	101.190			104.8(0.8)	0.7398[20]	1.4040[20]	msc EtOH, eth, ace, bz
1566	1-*tert*-Butyl-4-ethylbenzene		$C_{12}H_{18}$	7364-19-4	162.271	liq	-38.4	211	0.8641[20]		
1567	Butyl ethyl ether	Ethyl butyl ether	$C_6H_{14}O$	628-81-9	102.174	liq	-124	89(2)	0.7495[20]	1.3818[20]	i H$_2$O; msc EtOH, eth; vs ace
1568	*sec*-Butyl ethyl ether		$C_6H_{14}O$	2679-87-0	102.174			81(4)	0.7503[20]	1.3802[20]	i H$_2$O; vs EtOH, eth
1569	*tert*-Butyl ethyl ether	Ethyl *tert*-butyl ether	$C_6H_{14}O$	637-92-3	102.174	liq	-94.0(0.3)	72.7(0.1)	0.736[25]	1.3756[20]	i H$_2$O; vs EtOH, eth
1570	2-*tert*-Butyl-4-ethylphenol		$C_{12}H_{18}O$	96-70-8	178.270		23	250			
1571	2-Butyl-2-ethyl-1,3-propane-diol		$C_9H_{20}O_2$	115-84-4	160.254	wh cry	43(2)	269.0(0.2)	0.927[50]	1.4587[25]	sl H$_2$O, ace; s EtOH
1572	5-Butyl-5-ethyl-2,4,6(1H,3H,5H)-pyrimidine-trione	Butethal	$C_{10}H_{16}N_2O_3$	77-28-1	212.245		123(1)				
1573	Butyl ethyl sulfide		$C_6H_{14}S$	638-46-0	118.240	liq	-95.1(0.2)	144.2(0.8)	0.8376[20]	1.4492[10]	vs EtOH; s chl
1574	*tert*-Butyl ethyl sulfide	2-Methyl-2-propanethiol	$C_6H_{14}S$	14290-92-7	118.240	liq	-85.9(0.3)	120.4(0.6)			
1575	N-*tert*-Butylformamide		$C_5H_{11}NO$	2425-74-3	101.147	liq	16	202	0.903	1.4330[20]	
1576	Butyl formate		$C_5H_{10}O_2$	592-84-7	102.132	liq	-90.0(0.4)	106.1(0.1)	0.8958[20]	1.3887[20]	sl H$_2$O; s ace; msc EtOH, eth
1577	*sec*-Butyl formate		$C_5H_{10}O_2$	589-40-2	102.132			93.6(0.3)	0.8846[20]	1.3865[20]	sl H$_2$O; s ace; msc EtOH, eth
1578	*tert*-Butyl formate	1,1-Dimethylethyl formate	$C_5H_{10}O_2$	762-75-4	102.132	liq		83(6)	0.872	1.3790[20]	
1579	Butyl glycidyl ether		$C_7H_{14}O_2$	2426-08-6	130.185			171(18)	0.918[20]		
1580	Butyl heptanoate	Butyl enanthate	$C_{11}H_{22}O_2$	5454-28-4	186.292	liq	-68(1)	225(4)	0.8638[20]	1.4204[20]	vs ace, bz, eth, EtOH
1581	Butyl hexanoate	Butyl caproate	$C_{10}H_{20}O_2$	626-82-4	172.265	liq	-50.0(0.5)	204(3)	0.8653[20]	1.4152[20]	i H$_2$O; s EtOH; msc eth
1582	*tert*-Butylhydrazine hydrochloride		$C_4H_{13}ClN_2$	7400-27-3	124.612		192.5				
1583	Butyl hydrogen succinate	Monobutyl succinate	$C_8H_{14}O_4$	5150-93-6	174.195		8.6	136.5[3]	1.0732[20]	1.4360[20]	
1584	*tert*-Butyl hydroperoxide		$C_4H_{10}O_2$	75-91-2	90.121	liq	6	89 dec	0.8960[20]	1.4015[20]	s H$_2$O, EtOH, eth, ctc, chl
1585	*tert*-Butyl-4-hydroxyanisole	Butylated hydroxyanisole	$C_{11}H_{16}O_2$	25013-16-5	180.244	wax	51	268			i H$_2$O; s peth, EtOH
1586	Butyl 2-hydroxybenzoate		$C_{11}H_{14}O_3$	2052-14-4	194.227	liq	-5.9	271	1.0728[20]	1.5115[20]	sl ctc
1587	Butyl 4-hydroxybenzoate	Butylparaben	$C_{11}H_{14}O_3$	94-26-8	194.227		68.5				sl H$_2$O, ctc; s EtOH
1588	Butyl *cis*-12-hydroxy-9-octadecenoate, (R)-	Butyl ricinoleate	$C_{22}H_{42}O_3$	151-13-3	354.566			275[13]	0.9058[22]	1.4566[22]	vs eth
1589	*tert*-Butyl hypochlorite		C_4H_9ClO	507-40-4	108.566	ye liq		77.5	0.9583[18]	1.403[20]	i H$_2$O; vs eth, bz; s ace
1590	Butyl isobutyl ether		$C_8H_{18}O$	17071-47-5	130.228	liq		135(8)	0.763[15]	1.4077[21]	vs ace, eth, EtOH
1591	*tert*-Butyl isobutyl ether		$C_8H_{18}O$	33021-02-2	130.228	liq		112.9(0.3)			
1592	Butyl isocyanate		C_5H_9NO	111-36-4	99.131			125(3)	0.880[20]	1.4060[20]	
1593	Butyl isocyanide		C_5H_9N	2769-64-4	83.132			120	0.78[20]		vs eth, EtOH
1594	*tert*-Butyl isopropyl ether		$C_7H_{16}O$	17348-59-3	116.201	liq	-88.3(0.4)	87.3(0.3)	0.7365[25]		s chl
1595	Butyl isothiocyanate	1-Isothiocyanatobutane	C_5H_9NS	592-82-5	115.197			167(7)	0.9546[20]	1.501[20]	vs eth, EtOH
1596	*sec*-Butyl isothiocyanate, (±)-	2-Isothiocyanatobutane, (±)	C_5H_9NS	116724-11-9	115.197			159.5	0.944[12]		vs eth, EtOH
1597	*tert*-Butyl isothiocyanate	2-Isothiocyanato-2-methylpro-pane	C_5H_9NS	590-42-1	115.197		10.5	140	0.9187[10]		
1598	Butyl lactate		$C_7H_{14}O_3$	34451-18-8	146.184			77[10]	0.9744[27]		vs eth, EtOH
1599	Butyl methacrylate		$C_8H_{14}O_2$	97-88-1	142.196			163.7(0.8)	0.8936[20]	1.4240[20]	vs eth, EtOH
1600	*tert*-Butyl methacrylate		$C_8H_{14}O_2$	585-07-9	142.196			135.2			

1556
4-*tert*-Butyl-2,6-dimethyl-3,5-dinitroacetophenone

1557
2-*tert*-Butyl-4,6-dimethylphenol

1558
4-*tert*-Butyl-2,5-dimethylphenol

1559
4-*tert*-Butyl-2,6-dimethylphenol

1560
1-*tert*-Butyl-3,5-dimethyl-2,4,6-trinitrobenzene

1561
2-*tert*-Butyl-4,6-dinitrophenol

1562
5-Butyldocosane

1563
11-Butyldocosane

1564
Butyl dodecanoate

1565
Butylethylamine

1566
1-*tert*-Butyl-4-ethylbenzene

1567
Butyl ethyl ether

1568
sec-Butyl ethyl ether

1569
tert-Butyl ethyl ether

1570
2-*tert*-Butyl-4-ethylphenol

1571
2-Butyl-2-ethyl-1,3-propanediol

1572
5-Butyl-5-ethyl-2,4,6(1*H*,3*H*,5*H*)-pyrimidinetrione

1573
Butyl ethyl sulfide

1574
tert-Butyl ethyl sulfide

1575
N-*tert*-Butylformamide

1576
Butyl formate

1577
sec-Butyl formate

1578
tert-Butyl formate

1579
Butyl glycidyl ether

1580
Butyl heptanoate

1581
Butyl hexanoate

1582
tert-Butylhydrazine hydrochloride

1583
Butyl hydrogen succinate

1584
tert-Butyl hydroperoxide

1585
tert-Butyl-4-hydroxyanisole

1586
Butyl 2-hydroxybenzoate

1587
Butyl 4-hydroxybenzoate

1588
Butyl *cis*-12-hydroxy-9-octadecenoate, (*R*)-

1589
tert-Butyl hypochlorite

1590
Butyl isobutyl ether

1591
tert-Butyl isobutyl ether

1592
Butyl isocyanate

1593
Butyl isocyanide

1594
tert-Butyl isopropyl ether

1595
Butyl isothiocyanate

1596
sec-Butyl isothiocyanate, (±)-

1597
tert-Butyl isothiocyanate

1598
Butyl lactate

1599
Butyl methacrylate

1600
tert-Butyl methacrylate

No.	Name	Synonym	Mol. Form.	CAS RN	Mol. Wt.	Physical Form	mp/°C	bp/°C	den g cm⁻³	n_D	Solubility
1601	1-tert-Butyl-4-methoxybenzene		C₁₁H₁₆O	5396-38-3	164.244			19.1(0.3)	223(5)	0.9383²⁰	1.5039²⁰	
1602	1-tert-Butyl-2-methoxy-4-methyl-3,5-dinitrobenzene		C₁₂H₁₆N₂O₅	83-66-9	268.265	pa ye lf (al)	85	185¹⁶			i H₂O; sl EtOH; s eth, chl
1603	2-tert-Butyl-4-methoxyphenol		C₁₁H₁₆O₂	121-00-6	180.244			184⁵⁰			
1604	3-tert-Butyl-4-methoxyphenol		C₁₁H₁₆O₂	88-32-4	180.244		65				
1605	Butylmethylamine	N-Methyl-1-butanamine	C₅H₁₃N	110-68-9	87.164			91(2)	0.7637¹⁵		
1606	1-tert-Butyl-2-methylbenzene	2-tert-Butyltoluene	C₁₁H₁₆	1074-92-6	148.245	liq	-50.3(0.2)	200(4)	0.8897²⁰	1.5076²⁰	vs ace, bz, eth, EtOH
1607	1-tert-Butyl-3-methylbenzene	3-tert-Butyltoluene	C₁₁H₁₆	1075-38-3	148.245	liq	-41.36(0.08)	204(5)	0.8657²⁰	1.4944²⁰	vs ace, bz, eth, EtOH
1608	1-tert-Butyl-4-methylbenzene	4-tert-Butyltoluene	C₁₁H₁₆	98-51-1	148.245	liq	-52.49(0.08)	193(3)	0.8612²⁰	1.4918²⁰	i H₂O; sl EtOH; vs eth, chl; s ace, bz
1609	Butyl 2-methylbutanoate	Butyl o-toluate	C₉H₁₈O₂	15706-73-7	158.238			179(7)	0.8620²⁰	1.4135²⁰	
1610	Butyl 3-methylbutanoate	Butyl p-toluate	C₉H₁₈O₂	109-19-3	158.238					1.4058²⁵	
1611	Butyl methyl ether		C₅H₁₂O	628-28-4	88.148	liq	-115.7(0.1)	70.1(0.3)	0.7392²⁵	1.3736²⁰	i H₂O; msc EtOH, eth; s ace
1612	sec-Butyl methyl ether		C₅H₁₂O	116783-23-4	88.148			59.1	0.7415²⁰	1.3680²⁵	vs ace, eth, EtOH
1613	2-tert-Butyl-4-methylphenol		C₁₁H₁₆O	2409-55-4	164.244		52.3(0.9)	236(7)	0.9247⁷⁵	1.4969⁷⁵	sl H₂O; s ace, bz, chl
1614	2-tert-Butyl-5-methylphenol		C₁₁H₁₆O	88-60-8	164.244		46.5	127¹¹	0.922⁸⁰	1.5250²⁰	i H₂O; s EtOH, eth, ace
1615	2-tert-Butyl-6-methylphenol		C₁₁H₁₆O	2219-82-1	164.244		29(1)	233(7)	0.9240⁸⁰	1.5195²⁰	
1616	4-tert-Butyl-2-methylphenol		C₁₁H₁₆O	98-27-1	164.244		27.5	256(4)	0.965²⁰	1.5230²⁰	i H₂O; s eth, ace, bz
1617	Butyl methyl sulfide		C₅H₁₂S	628-29-5	104.214	liq	-97.81(0.05)	123.4(0.5)	0.8426²⁰	1.4477²⁰	vs EtOH, MeOH
1618	tert-Butyl methyl sulfide		C₅H₁₂S	6163-64-0	104.214	liq		98.9(0.3)			
1619	4-Butylmorpholine		C₈H₁₇NO	1005-67-0	143.227	liq	-57.1	213.5	0.9068²⁰	1.4451²⁰	vs H₂O, ace, bz, EtOH
1620	1-Butylnaphthalene		C₁₄H₁₆	1634-09-9	184.277	liq	-19.7(0.2)	288(5)	0.9738²⁰	1.5819²⁰	i H₂O; s EtOH, eth, ace, bz
1621	2-Butylnaphthalene		C₁₄H₁₆	1134-62-9	184.277	liq	-7(4)	286(5)	0.9673²⁰	1.5777²⁰	vs ace, bz, EtOH
1622	Butyl nitrate		C₄H₉NO₃	928-45-0	119.119			133	1.0228³⁰	1.4013²³	i H₂O; s EtOH, eth; sl ctc
1623	Butyl nitrite		C₄H₉NO₂	544-16-1	103.120			78	0.9114²⁵	1.3762²⁰	msc EtOH, eth
1624	tert-Butyl nitrite		C₄H₉NO₂	540-80-7	103.120	pa ye liq		64(2)	0.8670²⁰	1.368²⁰	sl H₂O; s EtOH, eth, chl, CS₂
1625	sec-Butyl nitrite		C₄H₉NO₂	924-43-6	103.120			65(2)	0.8726²⁰	1.3710²⁰	vs eth, EtOH, chl
1626	4-(Butylnitrosoamino)-1-butanol	N-Butyl-N-(4-hydroxybutyl)-nitrosamine	C₈H₁₈N₂O₂	3817-11-6	174.241			115⁰·⁰¹			
1627	5-Butylnonane		C₁₃H₂₈	17312-63-9	184.361			219(5)	0.7635¹⁸	1.4273¹⁸	
1628	Butyl nonanoate	Butyl pelargonate	C₁₃H₂₆O₂	50623-57-9	214.344		-38.0(0.7)	123²⁰	0.8520²⁵	1.4262²⁵	
1629	Butyl octanoate		C₁₂H₂₄O₂	589-75-3	200.318	liq	-42.9(0.5)	240(4)	0.8628²⁰	1.4232²⁵	vs ace, eth, EtOH
1630	2-Butyl-1-octanol		C₁₂H₂₆O	3913-02-8	186.333			248(23)	0.891²⁰		
1631	Butyl oleate	Butyl cis-9-octadecenoate	C₂₂H₄₂O₂	142-77-8	338.567	ye cry	-26.4	227¹⁵	0.870⁴¹⁵	1.4480²⁵	vs EtOH
1632	tert-Butyl 3-oxobutanoate		C₈H₁₄O₃	1694-31-1	158.195			71.5¹¹	0.9756²⁰	1.4180²⁰	
1633	Butyl 4-oxopentanoate	Butyl levulinate	C₉H₁₆O₃	2052-15-5	172.221			237.5	0.9735²⁰	1.4290²⁰	sl chl
1634	Butyl palmitate	Butyl hexadecanoate	C₂₀H₄₀O₂	111-06-8	312.531	cry (dil al)	16(2)			1.4312⁵⁰	i H₂O; s EtOH, eth
1635	Butyl pentanoate		C₉H₁₈O₂	591-68-4	158.238	liq	-83.77(0.02)	186(5)	0.8710¹⁵	1.4128²⁰	sl H₂O; s EtOH, eth
1636	sec-Butyl pentanoate		C₉H₁₈O₂	116836-32-9	158.238			174.5	0.8605²⁰	1.4070²⁰	vs bz, eth, py, EtOH
1637	4-(1-Butylpentyl)pyridine		C₁₄H₂₃N	2961-47-9	205.340			265	0.8878²⁵	1.4846²⁵	
1638	tert-Butyl peroxybenzoate	Benzoyl tert-butyl peroxide	C₁₁H₁₄O₃	614-45-9	194.227			75⁰·²	1.021²⁵	1.4990²⁰	
1639	2-Butylphenol		C₁₀H₁₄O	3180-09-4	150.217	liq	-20(2)	234(5)	0.975²⁰	1.5180²⁵	i H₂O; s EtOH, eth, alk
1640	2-sec-Butylphenol		C₁₀H₁₄O	89-72-5	150.217		18(3)	229(3)	0.9804²⁵	1.5200²⁵	
1641	2-tert-Butylphenol		C₁₀H₁₄O	88-18-6	150.217	liq	-5.6(0.2)	224.3(0.6)	0.9783²⁰	1.5160²⁰	s EtOH, ctc, alk; vs eth
1642	3-Butylphenol		C₁₀H₁₄O	4074-43-5	150.217			249(4)	0.974²⁰		vs eth, EtOH
1643	3-tert-Butylphenol		C₁₀H₁₄O	585-34-2	150.217	nd (peth)	47(1)	240			s EtOH, alk; vs eth
1644	4-Butylphenol		C₁₀H₁₄O	1638-22-8	150.217		22	251(4)	0.976²²	1.5165²⁰	i H₂O; s EtOH, eth, alk; sl ctc
1645	4-sec-Butylphenol	4-(1-Methylpropyl)phenol	C₁₀H₁₄O	99-71-8	150.217		60(1)	243(3)	0.986²⁰	1.5182²¹	i H₂O; s EtOH, alk; vs eth

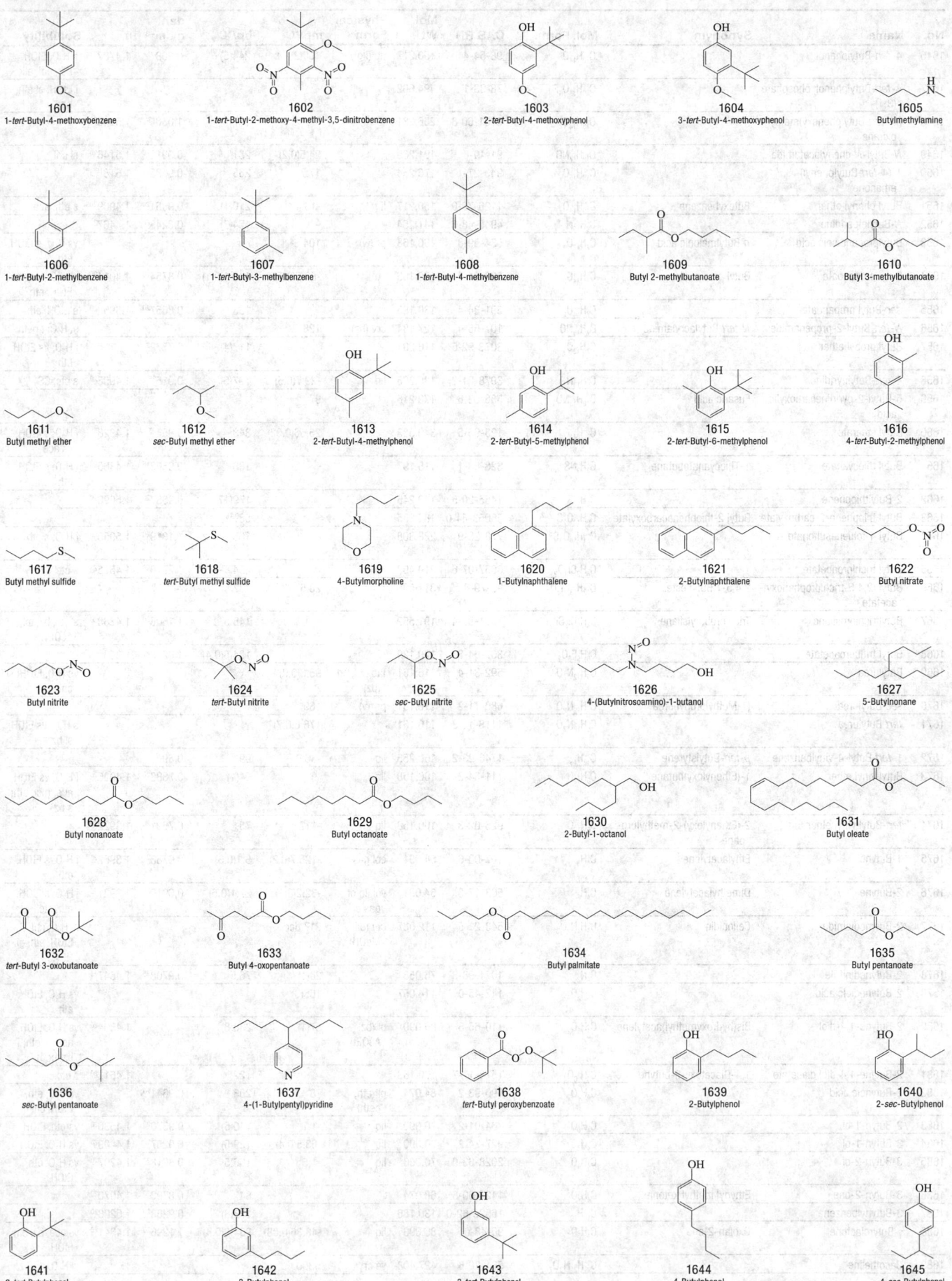

1601
1-*tert*-Butyl-4-methoxybenzene

1602
1-*tert*-Butyl-2-methoxy-4-methyl-3,5-dinitrobenzene

1603
2-*tert*-Butyl-4-methoxyphenol

1604
3-*tert*-Butyl-4-methoxyphenol

1605
Butylmethylamine

1606
1-*tert*-Butyl-2-methylbenzene

1607
1-*tert*-Butyl-3-methylbenzene

1608
1-*tert*-Butyl-4-methylbenzene

1609
Butyl 2-methylbutanoate

1610
Butyl 3-methylbutanoate

1611
Butyl methyl ether

1612
sec-Butyl methyl ether

1613
2-*tert*-Butyl-4-methylphenol

1614
2-*tert*-Butyl-5-methylphenol

1615
2-*tert*-Butyl-6-methylphenol

1616
4-*tert*-Butyl-2-methylphenol

1617
Butyl methyl sulfide

1618
tert-Butyl methyl sulfide

1619
4-Butylmorpholine

1620
1-Butylnaphthalene

1621
2-Butylnaphthalene

1622
Butyl nitrate

1623
Butyl nitrite

1624
tert-Butyl nitrite

1625
sec-Butyl nitrite

1626
4-(Butylnitrosoamino)-1-butanol

1627
5-Butylnonane

1628
Butyl nonanoate

1629
Butyl octanoate

1630
2-Butyl-1-octanol

1631
Butyl oleate

1632
tert-Butyl 3-oxobutanoate

1633
Butyl 4-oxopentanoate

1634
Butyl palmitate

1635
Butyl pentanoate

1636
sec-Butyl pentanoate

1637
4-(1-Butylpentyl)pyridine

1638
tert-Butyl peroxybenzoate

1639
2-Butylphenol

1640
2-*sec*-Butylphenol

1641
2-*tert*-Butylphenol

1642
3-Butylphenol

1643
3-*tert*-Butylphenol

1644
4-Butylphenol

1645
4-*sec*-Butylphenol

No.	Name	Synonym	Mol. Form.	CAS RN	Mol. Wt.	Physical Form	mp/°C	bp/°C	den g cm^{-3}	n_D	Solubility
1646	4-*tert*-Butylphenol		$C_{10}H_{14}O$	98-54-4	150.217	nd (lig)	100(2)	244(5)	0.908[80]	1.4787[114]	s H_2O, EtOH, eth, chl, alk
1647	4-*tert*-Butylphenol, phosphate (3:1)		$C_{30}H_{39}O_4P$	78-33-1	494.602						i EtOH; sl eth, bz
1648	[(4-*tert*-Butylphenoxy)methyl]-oxirane		$C_{13}H_{18}O_2$	3101-60-8	206.281			167[14]	1.036[25]	1.5145[20]	
1649	N-Butyl-N-phenylacetamide		$C_{12}H_{17}NO$	91-49-6	191.269		24.5(0.2)	281	0.9912[20]	1.5146[20]	sl chl
1650	1-(4-*tert*-Butylphenyl)-ethanone		$C_{12}H_{16}O$	943-27-1	176.254		17.7	263	0.9635[20]	1.518[15]	
1651	Butyl phenyl ether	Butoxybenzene	$C_{10}H_{14}O$	1126-79-0	150.217	liq	-19.4	210(1)	0.9351[20]	1.4969[20]	s eth, ace
1652	N-Butylpiperidine		$C_9H_{19}N$	4945-48-6	141.254			174(3)	0.8245[20]	1.4467[20]	
1653	Butylpropanedioic acid	*n*-Butylmalonic acid	$C_7H_{12}O_4$	534-59-8	160.168	pr (w)	104.5				vs H_2O; s EtOH, eth
1654	Butyl propanoate	Butyl propionate	$C_7H_{14}O_2$	590-01-2	130.185	liq	-89.5(0.5)	145.1(0.1)	0.8754[20]	1.4014[20]	sl H_2O, ctc; msc EtOH, eth
1655	*sec*-Butyl propanoate		$C_7H_{14}O_2$	591-34-4	130.185			133	0.8657[20]	1.3952[20]	s EtOH, eth
1656	N-*tert*-Butyl-2-propenamide	N-*tert*-Butylacrylamide	$C_7H_{13}NO$	107-58-4	127.184	cry (bz)	128				sl H_2O; i peth
1657	Butyl propyl ether		$C_7H_{16}O$	3073-92-5	116.201			117(4)	0.777[20]		i H_2O; vs EtOH, eth
1658	4-*tert*-Butylpyridine		$C_9H_{13}N$	3978-81-2	135.206	liq	-39.7(0.5)	197(5)	0.915[25]	1.4958[20]	s ctc, CS_2
1659	5-Butyl-2-pyridinecarboxylic acid	Fusaric acid	$C_{10}H_{13}NO_2$	536-69-6	179.216		97				
1660	Butyl stearate		$C_{22}H_{44}O_2$	123-95-5	340.583		26.56(0.02)	343	0.854[25]	1.4328[50]	i H_2O; s EtOH; vs ace
1661	Butyl thiocyanate	1-Thiocyanatobutane	C_5H_9NS	628-83-1	115.197			186	0.9563[15]	1.4360[20]	i H_2O; s EtOH, eth
1662	2-Butylthiophene		$C_8H_{12}S$	1455-20-5	140.246			179(17)	0.9537[20]	1.5090[20]	
1663	Butyl thiophene-2-carboxylate	Butyl 2-thiophenecarboxylate	$C_9H_{12}O_2S$	56053-84-0	184.255			58[0.15]			
1664	Butyl 4-toluenesulfonate		$C_{11}H_{16}O_3S$	778-28-9	228.308			165[6]	1.1319[20]	1.5050[20]	i H_2O; s eth; sl ctc
1665	Butyl trichloroacetate		$C_6H_9Cl_3O_2$	3657-07-6	219.493			204	1.2778[20]	1.4525[25]	s ctc
1666	Butyl (2,4,5-trichlorophenoxy)-acetate	2,4,5-T Butyl ester	$C_{12}H_{13}Cl_3O_3$	93-79-8	311.588		28.5	337			
1667	Butyltrichlorosilane	Trichlorobutylsilane	$C_4H_9Cl_3Si$	7521-80-4	191.559			148.5	1.1606[20]	1.4363[20]	s eth, bz, tol, AcOEt
1668	Butyl trifluoroacetate		$C_6H_9F_3O_2$	367-64-6	170.129			104.5(0.4)	1.0268[22]	1.353[22]	s chl
1669	Butylurea		$C_5H_{12}N_2O$	592-31-4	116.161	tab (w), nd (bz)	96.3(0.9)				vs H_2O, EtOH; sl chl
1670	*sec*-Butylurea	(1-Methylpropyl)urea	$C_5H_{12}N_2O$	689-11-2	116.161	pr (w)	169				
1671	*tert*-Butylurea		$C_5H_{12}N_2O$	1118-12-3	116.161		176.6(0.7)				s H_2O; vs EtOH; sl bz
1672	1-*tert*-Butyl-4-vinylbenzene	*p*-*tert*-Butylstyrene	$C_{12}H_{16}$	1746-23-2	160.255	liq	-36.9	99[14]	0.89[20]		
1673	Butyl vinyl ether	1-(Ethenyloxy)butane	$C_6H_{12}O$	111-34-2	100.158	liq	-92	94(1)	0.7888[20]	1.4026[20]	i H_2O; vs EtOH, ace; msc eth; s bz
1674	*tert*-Butyl vinyl ether	2-(Ethenyloxy)-2-methylpro-pane	$C_6H_{12}O$	926-02-3	100.158	liq	-112	75	0.7691[20]	1.3922[20]	
1675	1-Butyne	Ethylacetylene	C_4H_6	107-00-6	54.091	col gas	-125.7(0.2)	8.1(0.3)	0.6783[0]	1.3962[20]	i H_2O; s EtOH, eth
1676	2-Butyne	Dimethylacetylene	C_4H_6	503-17-3	54.091	vol liq or gas	-32.2(0.1)	27.1(0.5)	0.6910[20]	1.3921[20]	i H_2O; s EtOH, eth, ctc
1677	2-Butynediamide	Cellocidin	$C_4H_4N_2O_2$	543-21-5	112.087	cry (dil MeOH)	217 dec				sl H_2O, chl, EtOH, eth, gl HOAc
1678	2-Butynedinitrile		C_4N_2	1071-98-3	76.056		20(1)	76.5	0.9708[25]	1.4647[25]	
1679	2-Butynedioic acid		$C_4H_2O_4$	142-45-0	114.057		166(3)				vs H_2O, EtOH, eth
1680	2-Butyne-1,4-diol	Bis(hydroxymethyl)acetylene	$C_4H_6O_2$	110-65-6	86.090	pl (bz, AcOEt)	57(1)	238(8)		1.4804[20]	vs H_2O, EtOH, ace; sl eth; i bz, peth
1681	2-Butyne-1,4-diol diacetate	1,4-Diacetoxy-2-butyne	$C_8H_{10}O_4$	1573-17-7	170.163			122[10]		1.4611[20]	s ctc
1682	2-Butynoic acid		$C_4H_4O_2$	590-93-2	84.074	pl (eth, peth)	78	203	0.9641[20]		vs H_2O, eth, EtOH, chl
1683	2-Butyn-1-ol		C_4H_6O	764-01-2	70.090	liq	-1.1	140(5)	0.9370[20]	1.4530[20]	vs eth, EtOH
1684	3-Butyn-1-ol		C_4H_6O	927-74-2	70.090	liq	-63.5(0.4)	129(5)	0.9257[20]	1.4409[20]	vs H_2O, EtOH
1685	3-Butyn-2-ol		C_4H_6O	2028-63-9	70.090	liq	-1.5	106.5	0.8618[20]	1.4207[20]	vs H_2O, eth, EtOH
1686	3-Butyn-2-one	Ethynyl methyl ketone	C_4H_4O	1423-60-5	68.074			84	0.8793[20]	1.4070[20]	
1687	3-Butynylbenzene		$C_{10}H_{10}$	16520-62-0	130.186			177(8)	0.9258[20]	1.5208[20]	
1688	γ-Butyrolactone	Oxolan-2-one	$C_4H_6O_2$	96-48-0	86.090	liq	-43.36(0.08)	204.6(0.4)	1.1296[20]	1.4341[20]	vs ace, bz, eth, EtOH
1689	Cacotheline		$C_{21}H_{21}N_3O_7$	561-20-6	427.408	ye cry	>300				sl H_2O

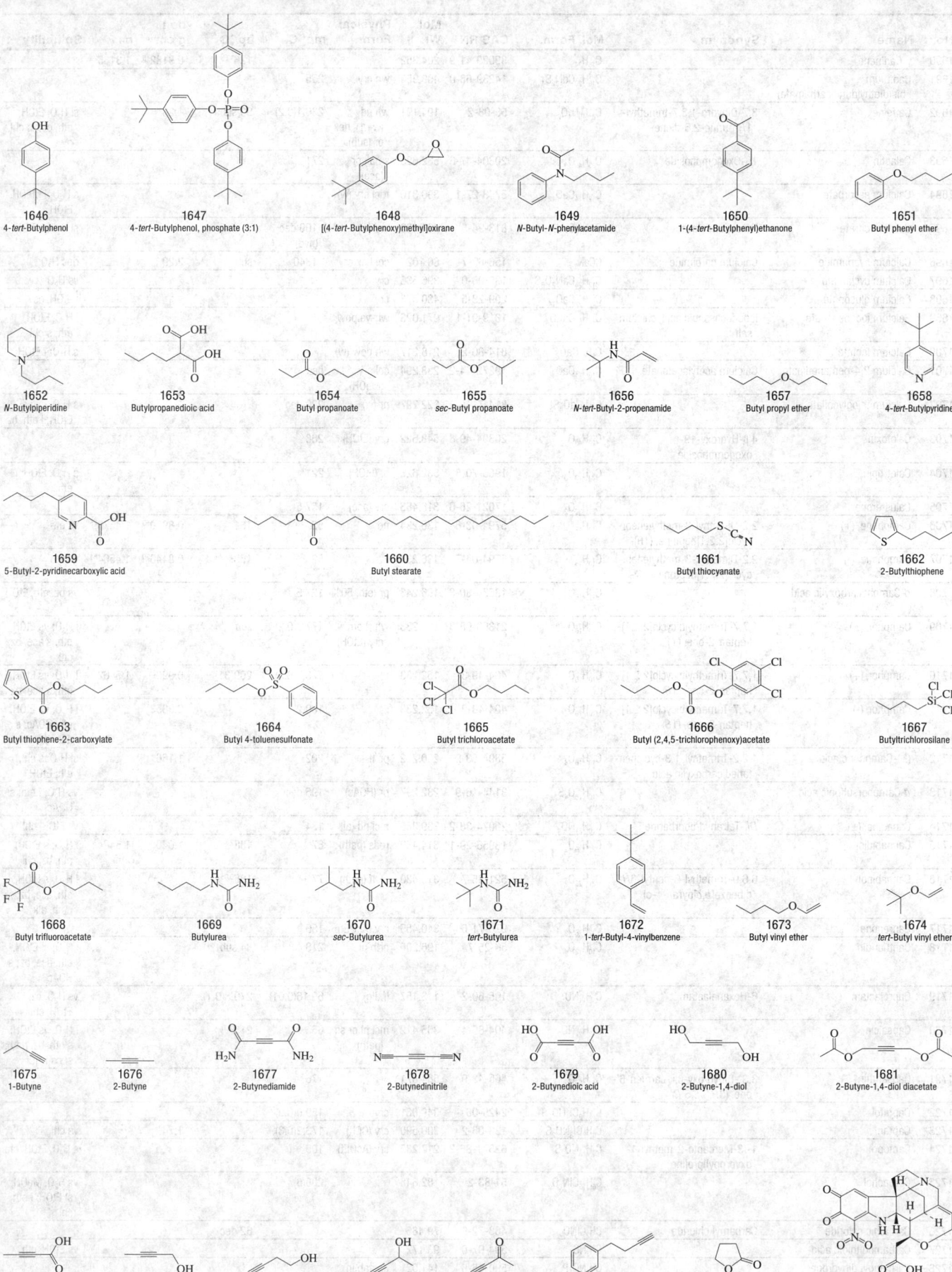

1646
4-*tert*-Butylphenol

1647
4-*tert*-Butylphenol, phosphate (3:1)

1648
[(4-*tert*-Butylphenoxy)methyl]oxirane

1649
N-Butyl-*N*-phenylacetamide

1650
1-(4-*tert*-Butylphenyl)ethanone

1651
Butyl phenyl ether

1652
N-Butylpiperidine

1653
Butylpropanedioic acid

1654
Butyl propanoate

1655
sec-Butyl propanoate

1656
N-*tert*-Butyl-2-propenamide

1657
Butyl propyl ether

1658
4-*tert*-Butylpyridine

1659
5-Butyl-2-pyridinecarboxylic acid

1660
Butyl stearate

1661
Butyl thiocyanate

1662
2-Butylthiophene

1663
Butyl thiophene-2-carboxylate

1664
Butyl 4-toluenesulfonate

1665
Butyl trichloroacetate

1666
Butyl (2,4,5-trichlorophenoxy)acetate

1667
Butyltrichlorosilane

1668
Butyl trifluoroacetate

1669
Butylurea

1670
sec-Butylurea

1671
tert-Butylurea

1672
1-*tert*-Butyl-4-vinylbenzene

1673
Butyl vinyl ether

1674
tert-Butyl vinyl ether

1675
1-Butyne

1676
2-Butyne

1677
2-Butynediamide

1678
2-Butynedinitrile

1679
2-Butynedioic acid

1680
2-Butyne-1,4-diol

1681
2-Butyne-1,4-diol diacetate

1682
2-Butynoic acid

1683
2-Butyn-1-ol

1684
3-Butyn-1-ol

1685
3-Butyn-2-ol

1686
3-Butyn-2-one

1687
3-Butynylbenzene

1688
γ-Butyrolactone

1689
Cacotheline

No.	Name	Synonym	Mol. Form.	CAS RN	Mol. Wt.	Physical Form	mp/°C	bp/°C	den g cm⁻³	n_D	Solubility
1690	γ-Cadinene		$C_{15}H_{24}$	39029-41-9	204.352			126[12]	0.9182[15]	1.3166[20]	
1691	Cadmium bis(diethyldithiocarbamate)		$C_{10}H_{20}CdN_2S_4$	14239-68-0	408.950	wh cry	255				
1692	Caffeine	3,7-Dihydro-1,3,7-trimethyl-1H-purine-2,6-dione	$C_8H_{10}N_4O_2$	58-08-2	194.191	wh nd (w+1), hex pr (sub)	236.1(0.2)	90 sub	1.23[19]		sl H₂O, EtOH; i eth, ctc; s chl, py
1693	Calactin	19-Oxogomphoside	$C_{29}H_{40}O_9$	20304-47-6	532.623	small pr (ace)	271				
1694	Calcium ascorbate		$C_{12}H_{14}CaO_{12}$	5743-27-1	390.310	tricl cry (w)					s H₂O; i MeOH, EtOH
1695	Calcium citrate	Tricalcium citrate	$C_{12}H_{10}Ca_3O_{14}$	813-94-5	498.433	cry (w)	≈100 dec (hyd)				sl H₂O; i EtOH
1696	Calcium cyanamide	Calcium carbimide	$CCaN_2$	156-62-7	80.102	col hex cry	≈1340	sub	2.29		dec H₂O
1697	Calcium cyclamate		$C_{12}H_{24}CaN_2O_6S_2$	139-06-0	396.536	cry					vs H₂O
1698	Calcium gluconate		$C_{12}H_{22}CaO_{14}$	299-28-5	430.373	cry					i EtOH, os
1699	Calcium iodobehenate	Iododocosanoic acid, calcium salt	$C_{44}H_{84}CaI_2O_4$	1319-91-1	971.023	wh-ye pow					i H₂O, EtOH, eth; s chl
1700	Calcium lactate		$C_6H_{10}CaO_6$	814-80-2	218.217	wh pow (w)					s H₂O; i EtOH
1701	Calcium 2,4-pentanedioate	Calcium acetylacetonate	$C_{10}H_{14}CaO_4$	19372-44-2	238.294	col cry (MeOH)	dec				
1702	Calcium thioglycollate		$C_4H_6CaO_4S_2$	814-71-1	222.297	pr (w)	220 dec				s H₂O, chl; sl EtOH; i eth, bz
1703	Calotoxin	4'β-Hydroxy-19-oxogomphoside	$C_{29}H_{40}O_{10}$	20304-49-8	548.622	cry (EtOH)	268				
1704	Calotropin		$C_{29}H_{40}O_9$	1986-70-5	532.623	pl (EtOH)	221				s H₂O, EtOH; i eth
1705	Calusterone		$C_{21}H_{32}O_2$	17021-26-0	316.483	cry (ace)	157.5				
1706	Camphene, (+)	2,2-Dimethyl-3-methylenebicyclo[2.2.1]heptane, (1R)-	$C_{10}H_{16}$	5794-03-6	136.234	nd	52	161	0.8950[50]	1.4570[25]	vs eth
1707	Camphene, (-)	2,2-Dimethyl-3-methylenebicyclo[2.2.1]heptane, (1S)-	$C_{10}H_{16}$	5794-04-7	136.234		52	158	0.8446[50]	1.4564[54]	vs eth
1708	d-Camphocarboxylic acid		$C_{11}H_{16}O_3$	18530-30-8	196.243	pr (eth, 50% al)	127.5				vs bz, eth, EtOH
1709	Camphor, (±)-	1,7,7-Trimethylbicyclo[2.2.1]heptan-2-one, (±)	$C_{10}H_{16}O$	21368-68-3	152.233	wh rhom cry (EtOH)	177.7(0.2)	sub			i H₂O; vs EtOH, eth; s ace, bz, ctc
1710	Camphor, (+)	1,7,7-Trimethylbicyclo[2.2.1]heptan-2-one, (1R)	$C_{10}H_{16}O$	464-49-3	152.233	pl	178.7(0.5)	209(31)	0.990[25]	1.5462	i H₂O; vs EtOH, eth; s ace, bz
1711	Camphor, (-)	1,7,7-Trimethylbicyclo[2.2.1]heptan-2-one, (1S)	$C_{10}H_{16}O$	464-48-2	152.233		180(2)		0.9853[18]		i H₂O; vs EtOH, eth, HOAc; s ace, bz
1712	(±)-Camphoric acid	1,2,2-Trimethyl-1,3-cyclopentanedicarboxylic acid	$C_{10}H_{16}O_4$	5394-83-2	200.232	pr, lf	202		1.186		sl H₂O; s chl, eth, EtOH
1713	d-Camphorsulfonic acid		$C_{10}H_{16}O_4S$	3144-16-9	232.297	pr (HOAc)	195 dec				vs H₂O; i eth; sl HOAc
1714	Canadine, (±)-	DL-Tetrahydroberberine	$C_{20}H_{21}NO_4$	29074-38-2	339.386	mcl nd (al)	134				vs EtOH, chl
1715	Cannabidiol		$C_{21}H_{30}O_2$	13956-29-1	314.462	rods (peth)	67	188[2]	1.040[40]	1.5404[20]	i H₂O; s EtOH, eth, bz, chl
1716	Cannabinol	6,6,9-Trimethyl-3-pentyl-6H-dibenzo[b,d]pyran-1-ol	$C_{21}H_{26}O_2$	521-35-7	310.430	pl, lf (peth)	77	185[0.05]			i H₂O; s EtOH, eth, ace, bz, peth, alk
1717	Canrenone		$C_{22}H_{28}O_3$	976-71-6	340.455	cry (AcOEt)	150				
1718	Cantharidin		$C_{10}H_{12}O_4$	56-25-7	196.200	orth pl	218	84 sub			i H₂O; sl EtOH, eth, ace, bz; s HOAc
1719	Caprolactam	6-Hexanelactam	$C_6H_{11}NO$	105-60-2	113.157	lf (lig)	69.16(0.01)	270.8(0.1)			vs H₂O, bz, EtOH, chl
1720	Capsaicin		$C_{18}H_{27}NO_3$	404-86-4	305.412	mcl pl or sc (peth)	65	215[0.01]			i H₂O; vs EtOH; s eth, bz, peth; sl con HCl
1721	Capsanthin	3,3'-Dihydroxy-β,κ-caroten-6'-one, (3R,3'S,5'R)	$C_{40}H_{56}O_3$	465-42-9	584.871		176				
1722	Captafol		$C_{10}H_9Cl_4NO_2S$	2425-06-1	349.061	cry	159.0(0.9)				
1723	Captan		$C_9H_8Cl_3NO_2S$	133-06-2	300.590	cry (CCl₄)	173.9(0.3)		1.74[25]		vs chl
1724	Captopril	1-(3-Mercapto-2-methyl-1-oxypropyl)proline	$C_9H_{15}NO_3S$	62571-86-2	217.285	cry (AcOEt)	105				s H₂O, EtOH, chl
1725	Carbachol		$C_6H_{15}ClN_2O_2$	51-83-2	182.648		208(6)				vs H₂O, MeOH; sl EtOH; i eth, chl
1726	Carbamic chloride	Carbamyl chloride	CH_2ClNO	463-72-9	79.486			62 dec			
1727	Carbamodithioic acid		CH_3NS_2	594-07-0	93.172						vs EtOH, eth
1728	Carbamoyl dihydrogen phosphate		CH_4NO_5P	590-55-6	141.021	unstab in soln					

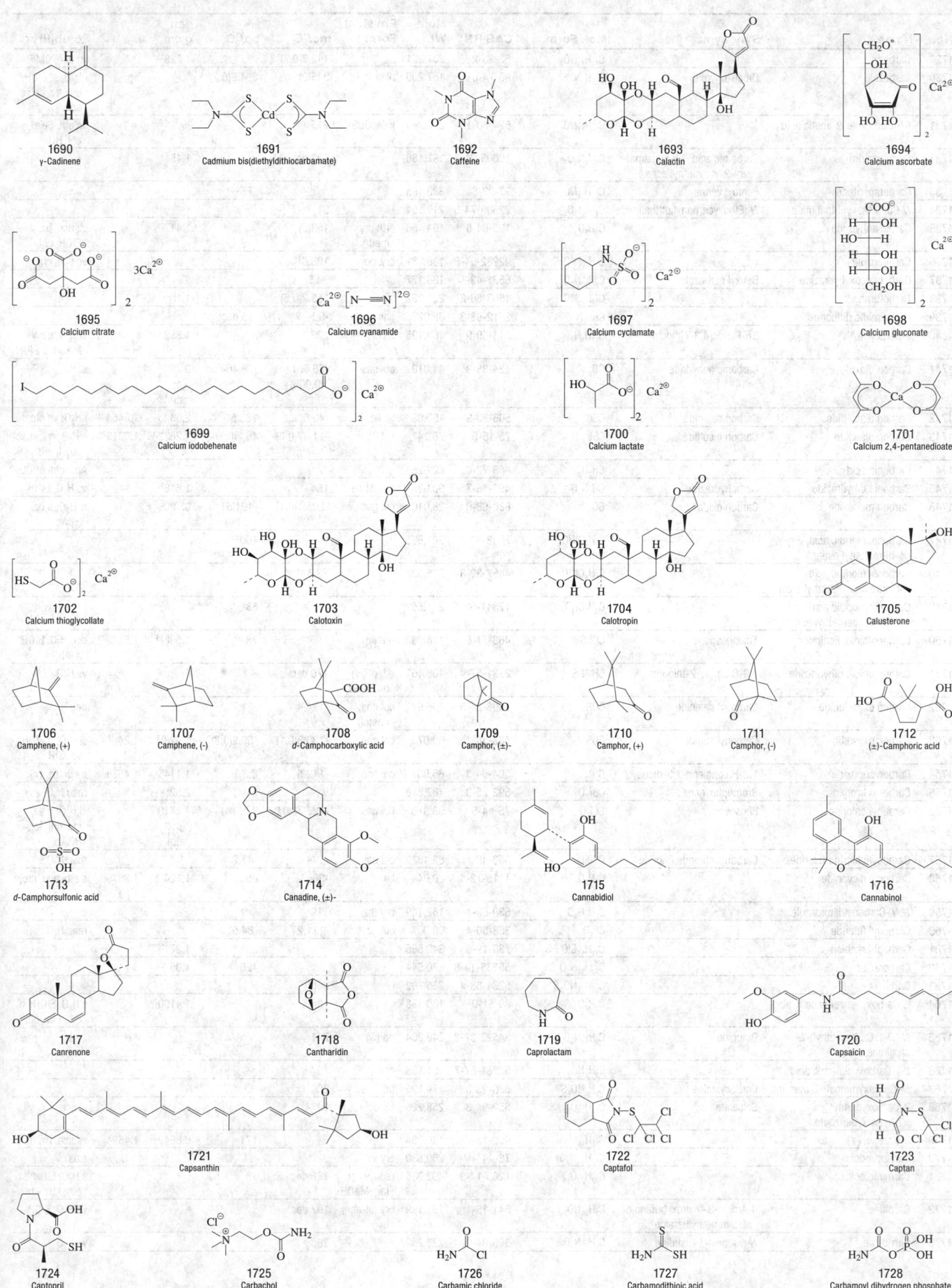

1690
γ-Cadinene

1691
Cadmium bis(diethyldithiocarbamate)

1692
Caffeine

1693
Calactin

1694
Calcium ascorbate

1695
Calcium citrate

1696
Calcium cyanamide

1697
Calcium cyclamate

1698
Calcium gluconate

1699
Calcium iodobehenate

1700
Calcium lactate

1701
Calcium 2,4-pentanedioate

1702
Calcium thioglycollate

1703
Calotoxin

1704
Calotropin

1705
Calusterone

1706
Camphene, (+)

1707
Camphene, (-)

1708
d-Camphocarboxylic acid

1709
Camphor, (±)-

1710
Camphor, (+)

1711
Camphor, (-)

1712
(±)-Camphoric acid

1713
d-Camphorsulfonic acid

1714
Canadine, (±)-

1715
Cannabidiol

1716
Cannabinol

1717
Canrenone

1718
Cantharidin

1719
Caprolactam

1720
Capsaicin

1721
Capsanthin

1722
Captafol

1723
Captan

1724
Captopril

1725
Carbachol

1726
Carbamic chloride

1727
Carbamodithioic acid

1728
Carbamoyl dihydrogen phosphate

No.	Name	Synonym	Mol. Form.	CAS RN	Mol. Wt.	Physical Form	mp/°C	bp/°C	den g cm⁻³	n_D	Solubility
1729	Carbaryl		$C_{12}H_{11}NO_2$	63-25-2	201.221		142.7(0.7)		1.228[25]		vs ace, DMF
1730	Carbazole	Dibenzopyrolle	$C_{12}H_9N$	86-74-8	167.206	pl or lf	245(2)	354.6(0.2)			i H_2O; sl EtOH, eth, bz, chl; s ace
1731	9H-Carbazole-9-acetic acid		$C_{14}H_{11}NO_2$	524-80-1	225.243	lf (AcOEt)	215				vs eth, EtOH, chl, HOAc
1732	Carbendazim	Carbamic acid, 1H-benzimid-azol-2-yl-, methyl ester	$C_9H_9N_3O_2$	10605-21-7	191.186		300 dec		1.45		
1733	Carbetapentane	Pentoxyverine	$C_{20}H_{31}NO_3$	77-23-6	333.465			165[0.01]			
1734	N-Carbethoxyphthalimide	N-(Ethoxycarbonyl)phthalimide	$C_{11}H_9NO_4$	22509-74-6	219.194		91				
1735	Carbic anhydride		$C_9H_8O_3$	129-64-6	164.158	orth cry (peth)	163(3)		1.417[25]		vs ace, bz, EtOH, chl
1736	Carbimazole		$C_7H_{10}N_2O_2S$	22232-54-8	186.231	cry, pow	123.5				vs ace, chl
1737	Carbobenzoxyhydrazine	Benzyl carbazate	$C_8H_{10}N_2O_2$	5331-43-1	166.177		69.5				
1738	Carbofuran		$C_{12}H_{15}NO_3$	1563-66-2	221.252		153.2(0.5)		1.18		
1739	Carboimidic difluoride		CHF_2N	2712-98-3	65.023	gas	-90	-13 dec			
1740	γ-Carboline	5H-Pyrido[4,3-b]indole	$C_{11}H_8N_2$	244-69-9	168.195	nd	225		1.352		sl H_2O, bz; vs MeOH; s EtOH
1741	Carbon dioxide	Carbonic anhydride	CO_2	124-38-9	44.010	col gas	-56.561 (0.008)	-78.464 sp	0.720[25] (p>1 atm)		sl H_2O
1742	Carbon diselenide	Carbon selenide	CSe_2	506-80-9	169.93	ye liq	-43.6(0.3)	125.5	2.6823[20]	1.8454[20]	i H_2O; vs ctc, tol
1743	Carbon disulfide	Carbon bisulfide	CS_2	75-15-0	76.141	col liq	-111.7(0.3)	46.2(0.1)	1.2632[20]	1.6319[20]	s H_2O, chl; msc EtOH, eth
1744	Carbonic acid		CH_2O_3	463-79-6	62.025						Aq. soln. of CO_2
1745	Carbonic dihydrazide	Carbohydrazide	CH_6N_4O	497-18-7	90.085	nd (dil al)	154		1.616[20]		vs H_2O, EtOH
1746	Carbon monoxide	Carbon oxide	CO	630-08-0	28.010	col gas	-205.1(0.1)	-191.51 (0.09)	0.7909[-19]		sl H_2O; s bz, HOAc
1747	Carbonochloridic acid, 4-nitrophenyl ester		$C_7H_4ClNO_4$	7693-46-1	201.565		80	160[19]			
1748	Carbonochloridic acid, (4-nitrophenyl)methyl ester		$C_8H_6ClNO_4$	4457-32-3	215.592		32.8				
1749	Carbonochloridic acid, 2,2,2-trichloroethyl ester		$C_3H_2Cl_4O_2$	17341-93-4	211.859			63[11]			
1750	Carbonothioic dichloride	Thiophosgene	CCl_2S	463-71-8	114.982	red liq		73	1.508[15]	1.5442[20]	dec H_2O, EtOH; s eth
1751	Carbonothioic dihydrazide	1,3-Diamino-2-thiourea	CH_6N_4S	2231-57-4	106.151	nd, pl (w) nd, pl (w)	170 dec				vs H_2O
1752	Carbon oxyselenide	Carbonyl selenide	$COSe$	1603-84-5	106.97	col gas; unstab	-124.4	-21.7			dec H_2O
1753	Carbon oxysulfide	Carbonyl sulfide	COS	463-58-1	60.075	col gas	-138.8(0.1)	-50.2(0.3)	1.028[17]	1.24[-87]	sl H_2O; s EtOH; vs KOH
1754	Carbon suboxide	1,2-Propadiene-1,3-dione	C_3O_2	504-64-3	68.031	col gas	-112.5	6.8	1.114[0]	1.4538[0]	s eth, bz, CS_2
1755	Carbonyl bromide	Bromophosgene	CBr_2O	593-95-3	187.818			64(4)	2.52[15]		reac H_2O
1756	Carbonyl chloride	Phosgene	CCl_2O	75-44-5	98.916	col gas	-127.77(0.02)	7.5(0.4)	1.3719[25] (p>1 atm)		sl H_2O; s bz, ctc, chl, tol, HOAc
1757	Carbonyl chloride fluoride	Carbonic chloride fluoride	$CClFO$	353-49-1	82.461	col gas	-148	-47.2			reac H_2O
1758	Carbonyl dicyanide		C_3N_2O	1115-12-4	80.044	liq	-36	65.5	1.124[20]	1.3919[20]	s eth, ace, ctc, chl
1759	N,N'-Carbonyldiimidazole		$C_7H_6N_4O$	530-62-1	162.149	cry (bz)	119				
1760	Carbonyl fluoride		CF_2O	353-50-4	66.007	col gas	-111.2	-84.5			reac H_2O
1761	Carbophenothion		$C_{11}H_{16}ClO_2PS_3$	786-19-6	342.866			82[0.01]	1.271[20]		
1762	Carbosulfan		$C_{20}H_{32}N_2O_3S$	55285-14-8	380.544			126	1.056[20]		
1763	Carboxin		$C_{12}H_{13}NO_2S$	5234-68-4	235.302		96(7)				
1764	2-Carboxybenzeneacetic acid		$C_9H_8O_4$	89-51-0	180.158		184.5		1.4100[20]		s H_2O, EtOH; sl eth; i bz, chl
1765	N-(D-1-Carboxyethyl)-L-arginine	Octopine	$C_9H_{18}N_4O_4$	34522-32-2	246.264	nd (w)	281				
1766	L-γ-Carboxyglutamic acid		$C_6H_9NO_6$	53861-57-7	191.138	cry	167				
1767	S-(Carboxymethyl)-L-cysteine	Carbocysteine	$C_5H_9NO_4S$	638-23-3	179.195	nd	206				
1768	2-Carboxyphenyl 2-hydroxybenzoate	Salsalate	$C_{14}H_{10}O_5$	552-94-3	258.226		147				sl ace
1769	3-Carene, (+)		$C_{10}H_{16}$	498-15-7	136.234			171	0.8549[30]	1.469[3]	vs ace, bz, eth
1770	Carisoprodol		$C_{12}H_{24}N_2O_4$	78-44-4	260.330	cry	92				s os
1771	Carminic acid		$C_{22}H_{20}O_{13}$	1260-17-9	492.386	red mcl pr (aq, MeOH)	136 dec				s H_2O, EtOH; sl eth; i bz, chl
1772	Carnitine	4-Amino-3-hydroxybutanoic acid trimethylbetaine	$C_7H_{15}NO_3$	541-15-1	161.199	cry (al-ace), hyg	197 dec				vs H_2O, EtOH
1773	Carnosine	N-β-Alanyl-L-histidine	$C_9H_{14}N_4O_3$	305-84-0	226.232		260				vs H_2O

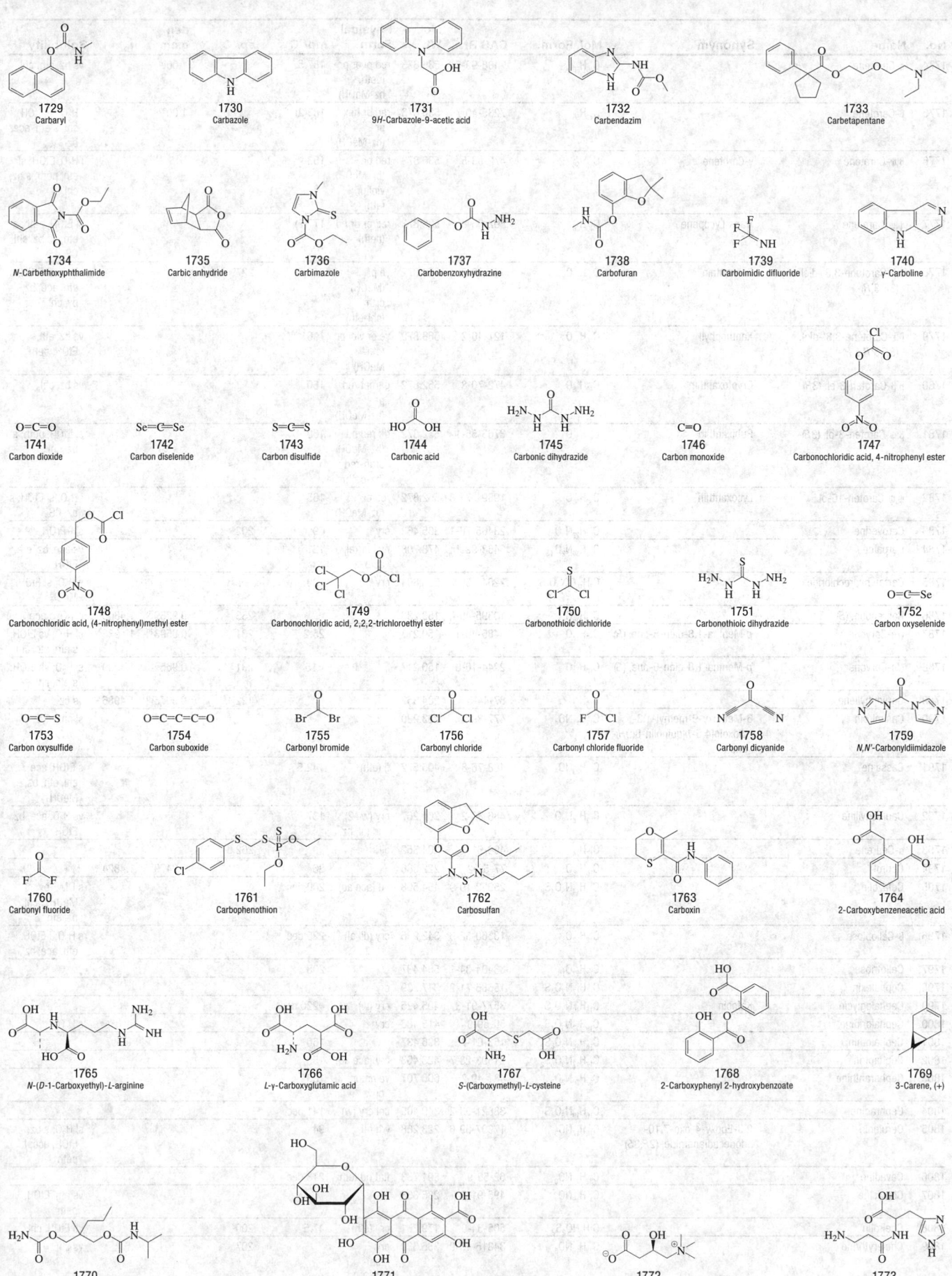

1729
Carbaryl

1730
Carbazole

1731
9*H*-Carbazole-9-acetic acid

1732
Carbendazim

1733
Carbetapentane

1734
N-Carbethoxyphthalimide

1735
Carbic anhydride

1736
Carbimazole

1737
Carbobenzoxyhydrazine

1738
Carbofuran

1739
Carboimidic difluoride

1740
γ-Carboline

1741
Carbon dioxide

1742
Carbon diselenide

1743
Carbon disulfide

1744
Carbonic acid

1745
Carbonic dihydrazide

1746
Carbon monoxide

1747
Carbonochloridic acid, 4-nitrophenyl ester

1748
Carbonochloridic acid, (4-nitrophenyl)methyl ester

1749
Carbonochloridic acid, 2,2,2-trichloroethyl ester

1750
Carbonothioic dichloride

1751
Carbonothioic dihydrazide

1752
Carbon oxyselenide

1753
Carbon oxysulfide

1754
Carbon suboxide

1755
Carbonyl bromide

1756
Carbonyl chloride

1757
Carbonyl chloride fluoride

1758
Carbonyl dicyanide

1759
N,N'-Carbonyldiimidazole

1760
Carbonyl fluoride

1761
Carbophenothion

1762
Carbosulfan

1763
Carboxin

1764
2-Carboxybenzeneacetic acid

1765
N-(*D*-1-Carboxyethyl)-*L*-arginine

1766
L-γ-Carboxyglutamic acid

1767
S-(Carboxymethyl)-*L*-cysteine

1768
2-Carboxyphenyl 2-hydroxybenzoate

1769
3-Carene, (+)

1770
Carisoprodol

1771
Carminic acid

1772
Carnitine

1773
Carnosine

No.	Name	Synonym	Mol. Form.	CAS RN	Mol. Wt.	Physical Form	mp/°C	bp/°C	den g cm⁻³	n_D	Solubility
1774	α-Carotene		$C_{40}H_{56}$	7488-99-5	536.873	red pl or pr (peth, bz-MeOH)	187.5		1.00^{20}		vs bz, eth, chl
1775	β-Carotene		$C_{40}H_{56}$	7235-40-7	536.873	red br hex pr (bz-MeOH)	183(2)		1.00^{20}		i H_2O; sl EtOH, chl; s eth, ace, bz
1776	β,ψ-Carotene	γ-Carotene	$C_{40}H_{56}$	472-93-5	536.873	red pr (bz-MeOH), viol pr (eth)	153				i H_2O, EtOH; sl eth, peth; s bz, chl
1777	ψ,ψ-Carotene	*trans*-Lycopene	$C_{40}H_{56}$	502-65-8	536.873	red pr or nd (peth)	177(2)				sl EtOH, peth; s eth; vs bz, chl, CS_2
1778	β,β-Carotene-3,3'-diol, (3*R*,3'*R*)-	Zeaxanthin	$C_{40}H_{56}O_2$	144-68-3	568.872	ye pr (MeOH) orth (chl-eth)	215.5	$227^{0.06}$			i H_2O; sl EtOH; s eth, ace, bz, py, chl
1779	β,ε-Carotene-3,3'-diol, (3*R*,3'*R*,6'*R*)-	Xanthophyll	$C_{40}H_{56}O_2$	127-40-2	568.872	ye or viol pr (eth-MeOH)	196				vs bz, eth, EtOH, peth
1780	β,β-Caroten-3-ol, (3*R*)-	Cryptoxanthin	$C_{40}H_{56}O$	472-70-8	552.872	garnet red pr (bz-MeOH)	160				vs bz, chl
1781	β,ψ-Caroten-3-ol, (3*R*)-	Rubixanthin	$C_{40}H_{56}O$	3763-55-1	552.872	dk red nd (bz-MeOH) oran-red (bz-peth)	160				sl EtOH, peth; s bz, chl
1782	ψ,ψ-Caroten-16-ol	Lycoxanthin	$C_{40}H_{56}O$	19891-74-8	552.872	red pl (bz-MeOH)	168				i H_2O; sl EtOH; s bz, CS_2
1783	Caroverine		$C_{22}H_{27}N_3O_2$	23465-76-1	365.468	cry	69	$202^{0.01}$			sl i-PrOH
1784	Carpaine		$C_{28}H_{50}N_2O_4$	3463-92-1	478.708	mcl pr (al, ace)	121				vs ace, bz, eth, EtOH
1785	Cartap hydrochloride		$C_7H_{16}ClN_3O_2S_2$	22042-59-7	273.804	cry	180				s H_2O; sl EtOH, MeOH
1786	Carvenone, (*S*)-		$C_{10}H_{16}O$	10395-45-6	152.233			233	0.9289^{20}	1.4805^{20}	i H_2O; s ace
1787	(*R*)-Carvone	*p*-Mentha-1,8-dien-6-one, (*R*)	$C_{10}H_{14}O$	6485-40-1	150.217		25.2	231	0.9593^{20}	1.4988^{20}	sl H_2O; vs EtOH; s eth, ctc, chl
1788	(*S*)-Carvone	*p*-Mentha-1,8-dien-6-one, (*S*)	$C_{10}H_{14}O$	2244-16-8	150.217		<15	231	0.965^{20}	1.4989^{20}	sl H_2O; vs EtOH; s eth, chl
1789	Caryophyllene		$C_{15}H_{24}$	87-44-5	204.352			$122^{13.5}$	0.9075^{20}	1.4986^{20}	vs bz
1790	Casimiroin	6-Methoxy-9-methyl-1,3-dioxolo[4,5-*h*]quinolin-8(9*H*)-one	$C_{12}H_{11}NO_4$	477-89-4	233.220						sl chl
1791	Cassaine		$C_{24}H_{39}NO_4$	468-76-8	405.572	fl (eth)	142.5				s EtOH, ace, chl, eth, bz, MeOH
1792	Caulophylline		$C_{12}H_{16}N_2O$	486-86-2	204.267	cry (w+2), nd (al, bz)	137				vs H_2O, ace, bz, EtOH
1793	α-Cedrene		$C_{15}H_{24}$	469-61-4	204.352	oil		262.5			
1794	Cedrol		$C_{15}H_{26}O$	77-53-2	222.366		86		0.9479^{90}	1.4824^{90}	
1795	Cefazolin		$C_{14}H_{14}N_8O_4S_3$	25953-19-9	454.508	nd (ace aq)	200 dec				s DMF, py; sl MeOH; i chl, bz, eth
1796	β-Cellobiose		$C_{12}H_{22}O_{11}$	13360-52-6	342.296	cry (dil al)	225 dec				s H_2O; i EtOH, eth, ace, bz
1797	Cellotriose		$C_{18}H_{32}O_{16}$	33404-34-1	504.437		208				
1798	Cephalexin		$C_{16}H_{17}N_3O_4S$	15686-71-2	347.389	cry					
1799	Cephaloglycin	Kafocin	$C_{18}H_{19}N_3O_6S$	3577-01-3	405.425	cry (w)	≈220 dec				
1800	Cephaloridine		$C_{19}H_{17}N_3O_4S_2$	50-59-9	415.486	cry					s H_2O
1801	Cephalothin		$C_{16}H_{16}N_2O_6S_2$	153-61-7	396.437		160				
1802	Cephapirin		$C_{17}H_{17}N_3O_6S_2$	21593-23-7	423.463	cry (ace aq)	155				
1803	Cepharanthine		$C_{37}H_{38}N_2O_6$	481-49-2	606.707	ye amor pow	150				
1804	Cephradine		$C_{16}H_{19}N_3O_4S$	38821-53-3	349.405	col cry (w)	141 dec				
1805	Cerulenin	2,3-Epoxy-4-oxo-7,10-dodecadienamide, (2*R*,3*S*)-	$C_{12}H_{17}NO_3$	17397-89-6	223.268	wh nd	94				sl H_2O; s bz, EtOH, ace; i peth
1806	Cevadine		$C_{32}H_{49}NO_9$	62-59-9	591.733	flat nd (eth)	213 dec				
1807	Chavicine		$C_{17}H_{19}NO_3$	495-91-0	285.338						vs eth, EtOH, peth
1808	Cheirolin		$C_5H_9NO_2S_2$	505-34-0	179.261	cry (eth)	47.5	200^3			vs EtOH, chl
1809	Chelerythrine		$C_{21}H_{19}NO_5$	34316-15-9	365.380	cry (chl-MeOH)		207			vs chl

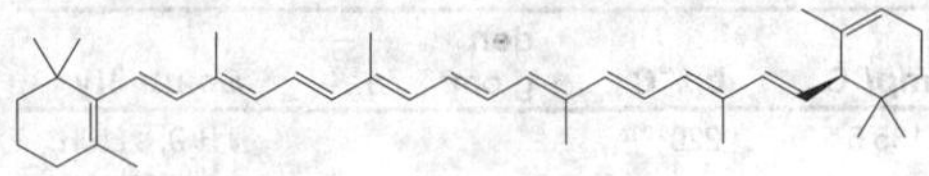

1774 α-Carotene

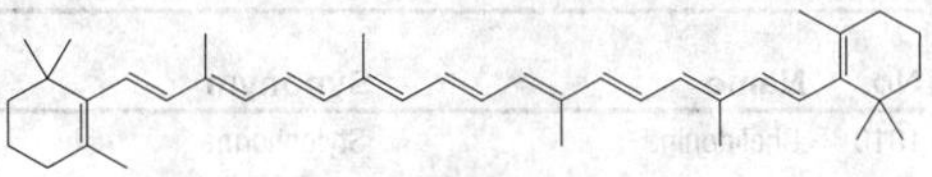

1775 β-Carotene

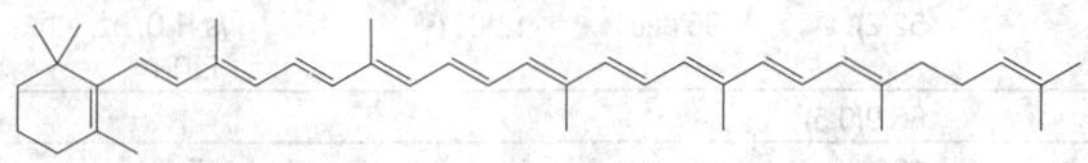

1776 β,ψ-Carotene

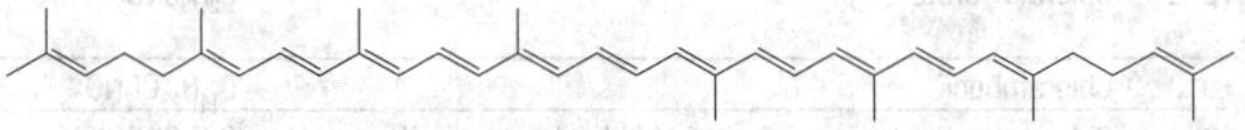

1777 ψ,ψ-Carotene

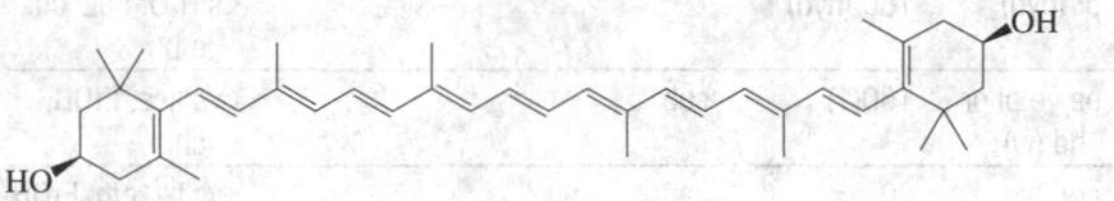

1778 β,β-Carotene-3,3'-diol, (3R,3'R)-

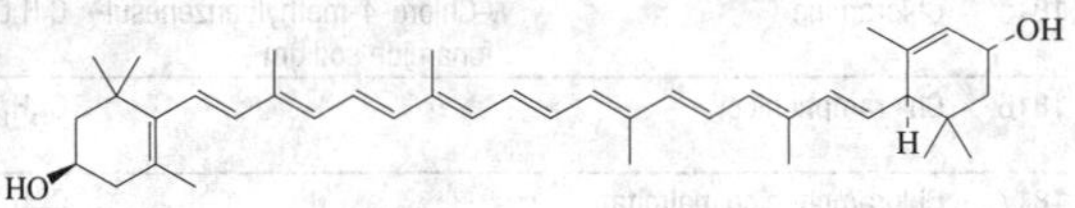

1779 β,ε-Carotene-3,3'-diol, (3R,3'R,6'R)-

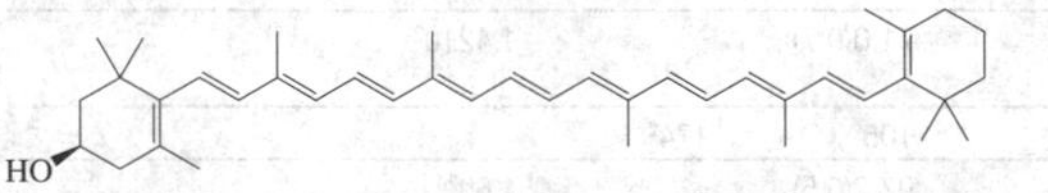

1780 β,β-Caroten-3-ol, (3R)-

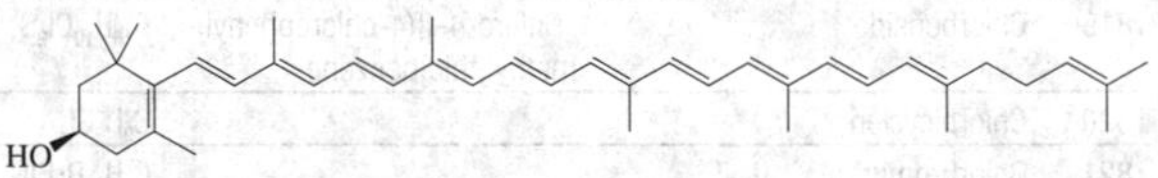

1781 β,ψ-Caroten-3-ol, (3R)-

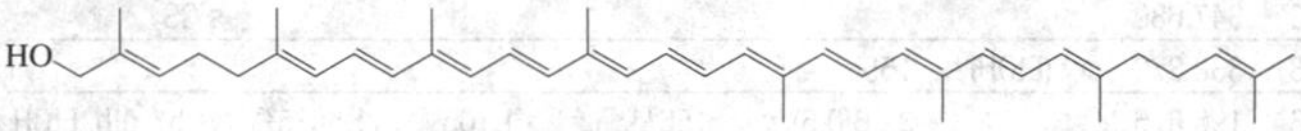

1782 ψ,ψ-Caroten-16-ol

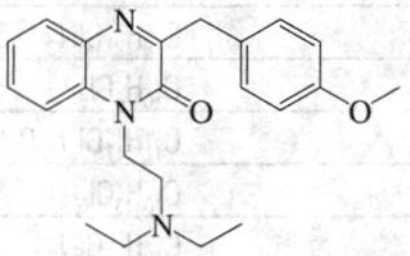

1783 Caroverine

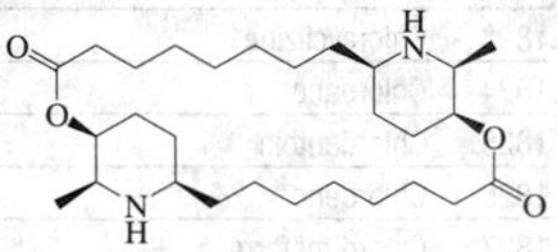

1784 Carpaine

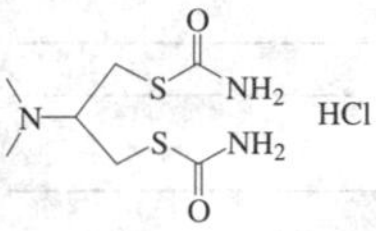

1785 Cartap hydrochloride

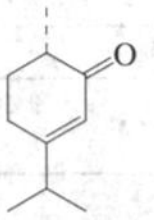

1786 Carvenone, (S)-

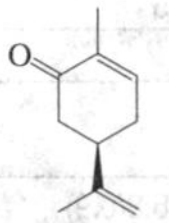

1787 (R)-Carvone

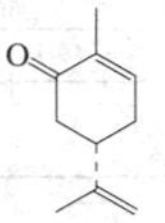

1788 (S)-Carvone

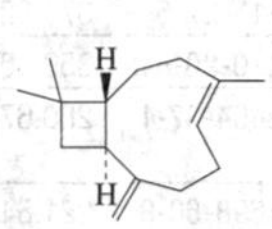

1789 Caryophyllene

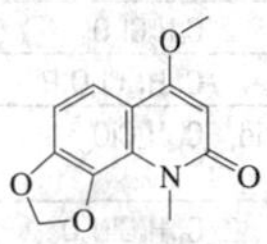

1790 Casimiroin

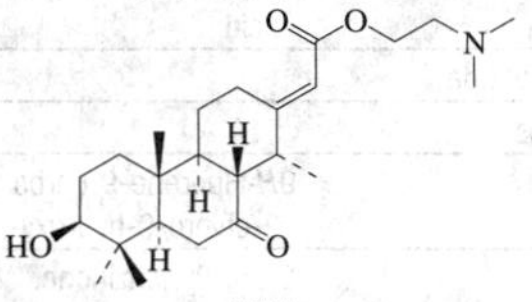

1791 Cassaine

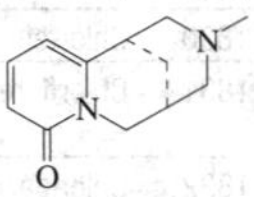

1792 Caulophylline

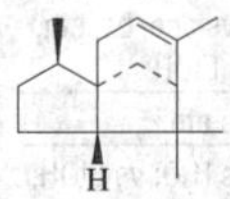

1793 α-Cedrene

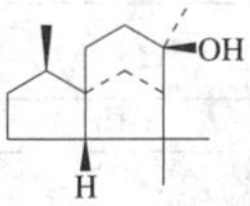

1794 Cedrol

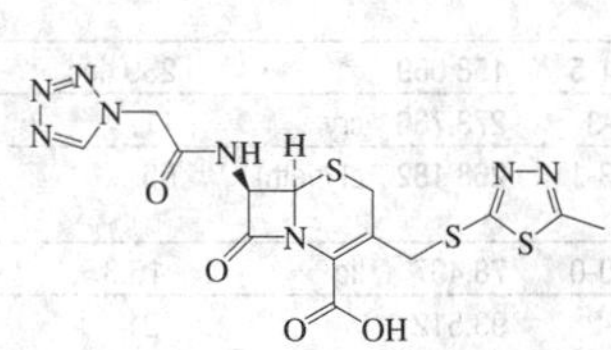

1795 Cefazolin

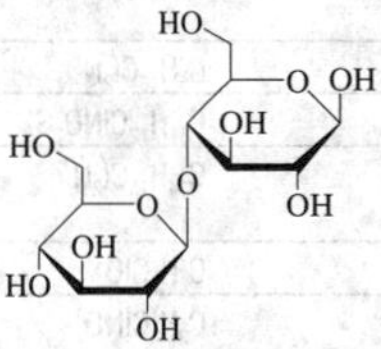

1796 β-Cellobiose

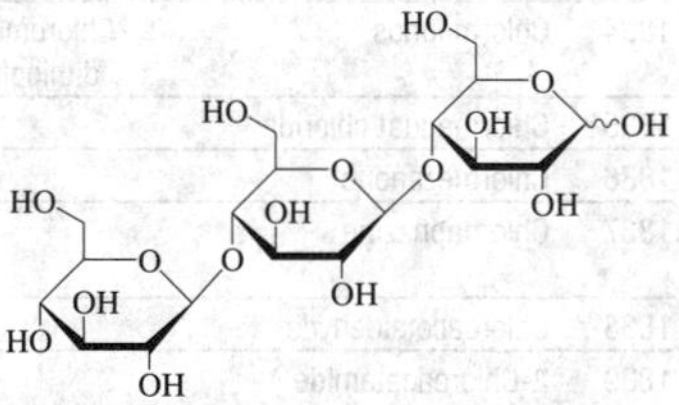

1797 Cellotriose

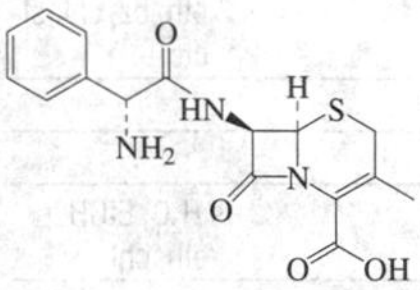

1798 Cephalexin

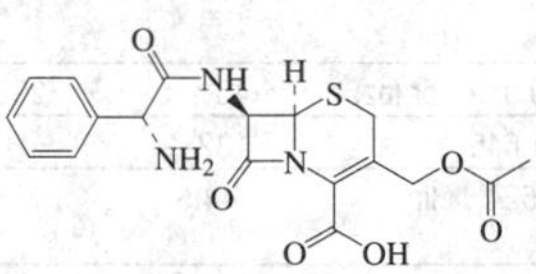

1799 Cephaloglycin

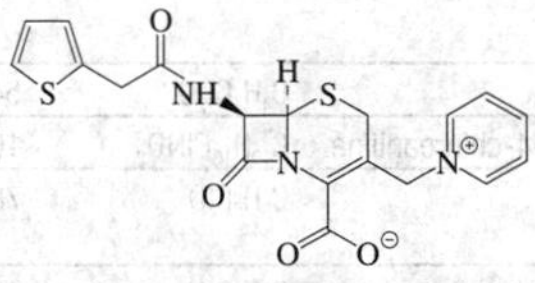

1800 Cephaloridine

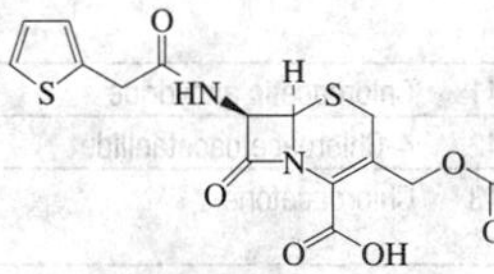

1801 Cephalothin

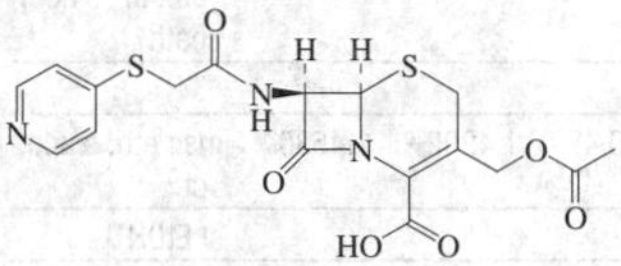

1802 Cephapirin

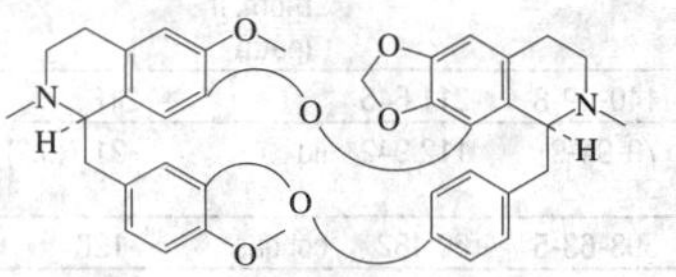

1803 Cepharanthine

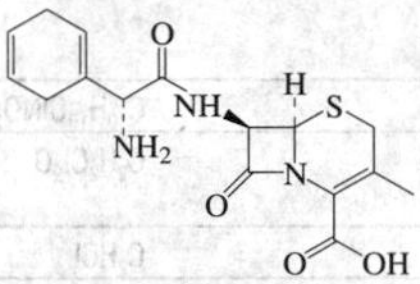

1804 Cephradine

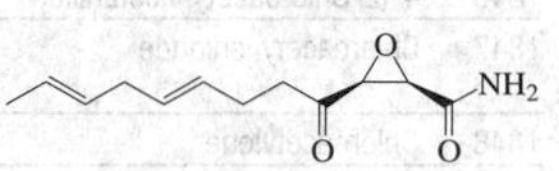

1805 Cerulenin

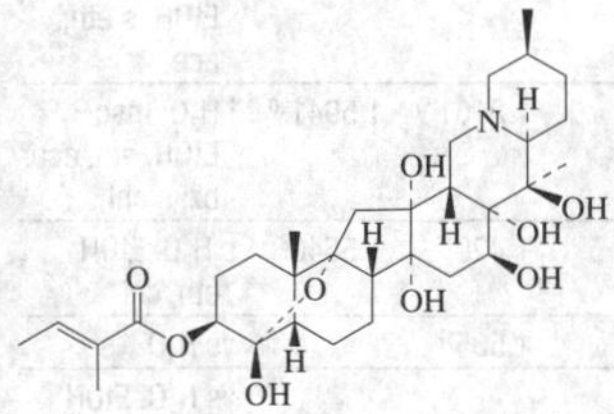

1806 Cevadine

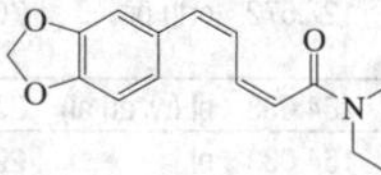

1807 Chavicine

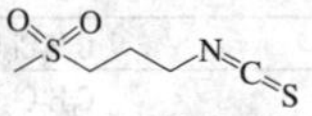

1808 Cheirolin

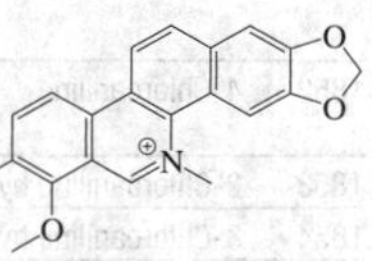

1809 Chelerythrine

No.	Name	Synonym	Mol. Form.	CAS RN	Mol. Wt.	Physical Form	mp/°C	bp/°C	den g cm⁻³	n_D	Solubility
1810	Chelidonine	Stylophorine	$C_{20}H_{19}NO_5$	476-32-4	353.369	mcl pr (al)	135.5	220[0.002]			i H_2O; s EtOH, eth, chl
1811	Chinomethionat		$C_{10}H_6N_2OS_2$	2439-01-2	234.297		170.5(0.5)				
1812	Chloral hydrate		$C_2H_3Cl_3O_2$	302-17-0	165.403		52(2)	96 dec	1.9081[20]		vs H_2O, bz, eth, EtOH
1813	Chlorambucil		$C_{14}H_{19}Cl_2NO_2$	305-03-3	304.213		66.9(0.5)				
1814	Chloramine B	N-Chlorobenzenesulfonamide sodium	$C_6H_5ClNNaO_2S$	127-52-6	213.618	pr (w)	190				sl EtOH; i chl, eth
1815	Chloramine T	N-Chloro-4-methylbenzenesulfonamide sodium	$C_7H_7ClNNaO_2S$	127-65-1	227.645	pr (hyd)	180 (hyd)				s H_2O; i bz, chl, eth
1816	Chloramphenicol		$C_{11}H_{12}Cl_2N_2O_5$	56-75-7	323.129	pa ye pl or nd (w)	150(1)	sub			vs ace, EtOH, chl
1817	Chloramphenicol palmitate		$C_{27}H_{42}Cl_2N_2O_6$	530-43-8	561.537	cry (bz)	90				vs bz, eth, EtOH
1818	Chloranilic acid	2,5-Dichloro-3,6-dihydroxy-2,5-cyclohexadiene-1,4-dione	$C_6H_2Cl_2O_4$	87-88-7	208.984	red lf (w+2)	283.5				s H_2O
1819	Chlorbenside	1-Chloro-4-[[(4-chlorophenyl)methyl]thio]benzene	$C_{13}H_{10}Cl_2S$	103-17-3	269.189		71.0(0.7)		1.4210[20]		
1820	Chlorbicyclen		$C_9H_6Cl_8$	2550-75-6	397.768	pow	105	174[2]			
1821	Chlorbromuron		$C_9H_{10}BrClN_2O_2$	13360-45-7	293.544		97.2(0.5)		1.69[20]		
1822	Chlorbufam	1-Methyl-2-propynyl(3-chlorophenyl)carbamate	$C_{11}H_{10}ClNO_2$	1967-16-4	223.656	cry	45.5				sl H_2O; s MeOH, EtOH, ace
1823	Chlorcyclizine		$C_{18}H_{21}ClN_2$	82-93-9	300.826	oil		140[0.12]			
1824	Chlordane		$C_{10}H_6Cl_8$	57-74-9	409.779		101.1(0.3)	175[1]	1.60[25]		
1825	Chlordantoin		$C_{11}H_{17}Cl_3N_2O_2S$	5588-20-5	347.689						s CS_2
1826	Chlordene		$C_{10}H_6Cl_6$	3734-48-3	338.873	cry (EtOH)	155				
1827	Chlordimeform		$C_{10}H_{13}ClN_2$	6164-98-3	196.676		32.6(0.5)	156[0.4]	1.105[25]	1.5885[25]	vs bz, eth, EtOH
1828	Chlorendic acid	1,4,5,6,7,7-Hexachloro-5-norbornene-2,3-dicarboxylic acid	$C_9H_4Cl_6O_4$	115-28-6	388.844	cry (w)	232				
1829	Chlorendic anhydride		$C_9H_2Cl_6O_3$	115-27-5	370.828		235				
1830	Chlorfenvinphos		$C_{12}H_{14}Cl_3O_4P$	470-90-6	359.569			170[0.05]			
1831	Chlorflurecol	9H-Fluorene-9-carboxylic acid, 2-chloro-9-hydroxy-	$C_{14}H_9ClO_3$	2464-37-1	260.672				1.496[20]		
1832	Chloridazon	3(2H)-Pyridazinone, 5-amino-4-chloro-2-phenyl-	$C_{10}H_8ClN_3O$	1698-60-8	221.643		206.8(0.9)				
1833	Chlorimuron-ethyl		$C_{15}H_{15}ClN_4O_6S$	90982-32-4	414.821		186				
1834	Chlormephos	Chloromethyl O,O-diethyl dithiophosphate	$C_5H_{12}ClO_2PS_2$	24934-91-6	234.705	oil		83[0.1]		1.5244	sl H_2O; misc os
1835	Chlormequat chloride		$C_5H_{13}Cl_2N$	999-81-5	158.069		239 dec				
1836	Chlormezanone		$C_{11}H_{12}ClNO_3S$	80-77-3	273.736	cry	117				sl EtOH
1837	Chlornaphazine		$C_{14}H_{15}Cl_2N$	494-03-1	268.182	pl (peth)	55	210[5]			vs ace, bz, eth, EtOH
1838	Chloroacetaldehyde		C_2H_3ClO	107-20-0	78.497	liq	-16.3	87(13)	1.19		s eth
1839	2-Chloroacetamide		C_2H_4ClNO	79-07-2	93.512		121	225			s H_2O; vs EtOH; sl eth
1840	Chloroacetic acid		$C_2H_3ClO_2$	79-11-8	94.497	mcl pl	62.0(0.7)	189.11(0.03)	1.4043[40]	1.4351[55]	vs H_2O; s EtOH, eth, bz, chl; sl ctc
1841	Chloroacetic anhydride		$C_4H_4Cl_2O_3$	541-88-8	170.979	pr (bz)	46	203	1.5497[20]		
1842	4-Chloroacetoacetanilide	N-Acetoacetyl-4-chloroaniline	$C_{10}H_{10}ClNO_2$	101-92-8	211.645		132				
1843	Chloroacetone		C_3H_5ClO	78-95-5	92.524	liq	-44.5	116(13)	1.15[20]		s H_2O, EtOH, eth, chl
1844	Chloroacetonitrile	Chloromethyl cyanide	C_2H_2ClN	107-14-2	75.497			108(5)	1.1930[20]	1.4202[25]	vs eth, EtOH
1845	α-Chloroacetophenone	ω-Chloroacetophenone	C_6H_7ClO	532-27-4	154.594	pl(dil al), rhom, lf (peth)	56.5	247	1.324[15]		i H_2O; vs EtOH, eth, bz; s ace, peth
1846	4-(2-Chloroacetyl)acetanilide		$C_{10}H_{10}ClNO_2$	140-49-8	211.645		218				
1847	Chloroacetyl chloride		$C_2H_2Cl_2O$	79-04-9	112.942	liq	-21.7(0.2)	106.0(0.4)	1.4202[20]	1.4530[20]	msc eth; s ace, ctc
1848	Chloroacetylene		C_2HCl	593-63-5	60.482	col gas	-126	-30			sl EtOH
1849	9-Chloroacridine		$C_{13}H_8ClN$	1207-69-8	213.663	nd (al)	121	sub			vs H_2O, EtOH
1850	2-Chloroaniline		C_6H_6ClN	95-51-2	127.572	liq	-2.3(0.9)	209(1)		1.5895[20]	i H_2O; msc EtOH; s eth, ace
1851	3-Chloroaniline		C_6H_6ClN	108-42-9	127.572	liq	-10.3(0.2)	230(1)	1.2161[20]	1.5941[20]	i H_2O; msc EtOH, eth, ace, bz; s chl
1852	4-Chloroaniline		C_6H_6ClN	106-47-8	127.572	orth pr	70.4(0.7)	231(4)	1.429[19]	1.5546[87]	s H_2O, EtOH, eth, chl
1853	2-Chloroaniline hydrochloride		$C_6H_7Cl_2N$	137-04-2	164.033	pl (w, aq al)	235		1.505[18]		vs H_2O
1854	3-Chloroaniline hydrochloride		$C_6H_7Cl_2N$	141-85-5	164.033	pl	222				vs H_2O, EtOH

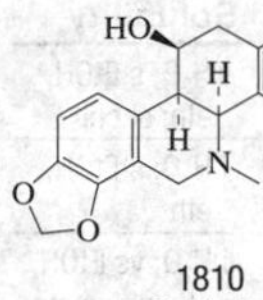

1810
Chelidonine

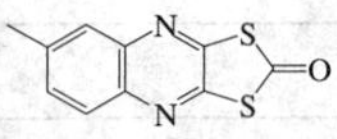

1811
Chinomethionat

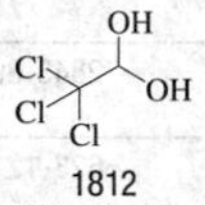

1812
Chloral hydrate

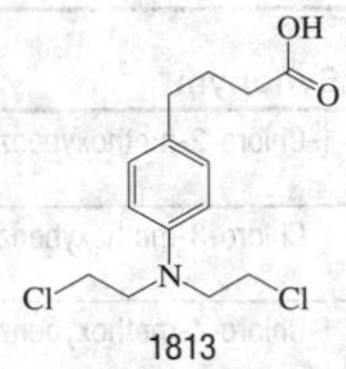

1813
Chlorambucil

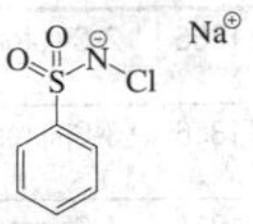

1814
Chloramine B

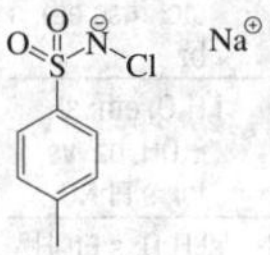

1815
Chloramine T

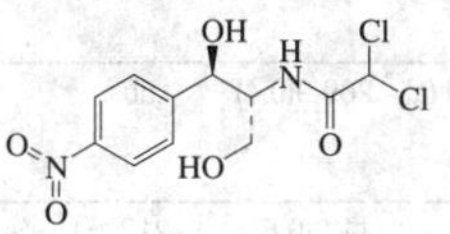

1816
Chloramphenicol

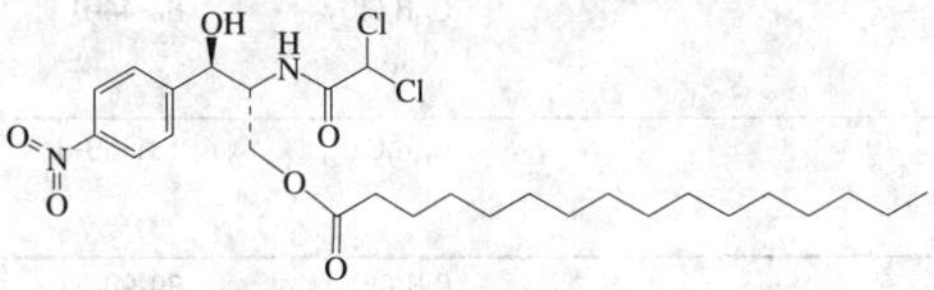

1817
Chloramphenicol palmitate

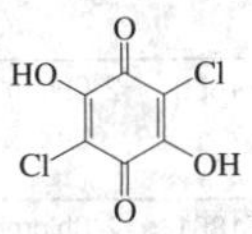

1818
Chloranilic acid

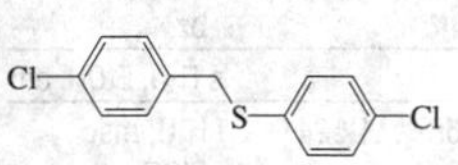

1819
Chlorbenside

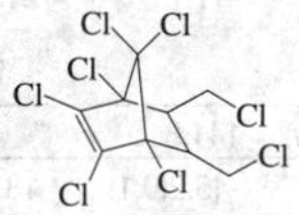

1820
Chlorbicyclen

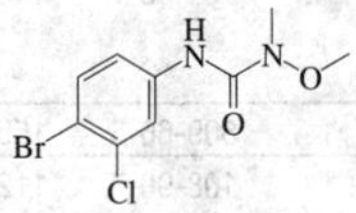

1821
Chlorbromuron

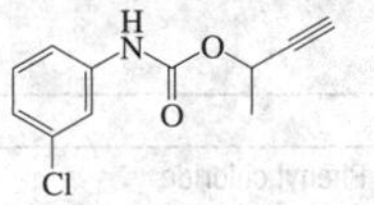

1822
Chlorbufam

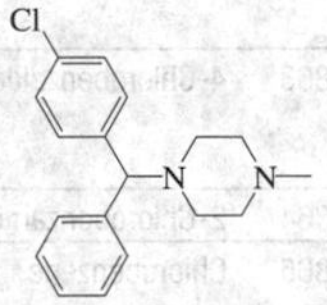

1823
Chlorcyclizine

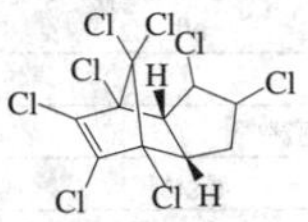

1824
Chlordane

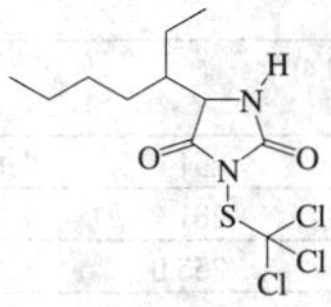

1825
Chlordantoin

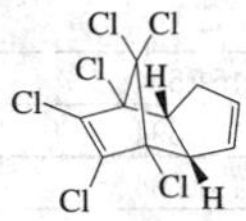

1826
Chlordene

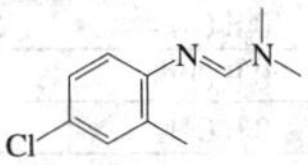

1827
Chlordimeform

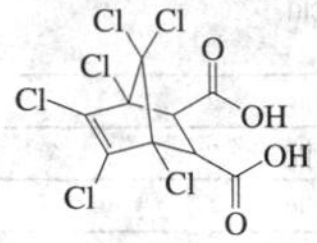

1828
Chlorendic acid

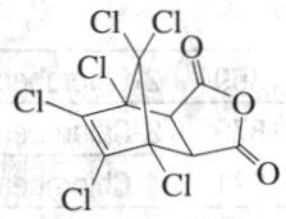

1829
Chlorendic anhydride

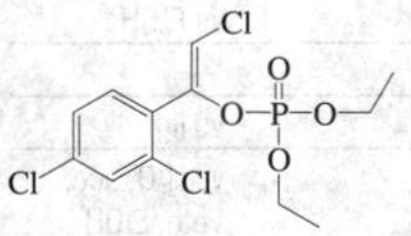

1830
Chlorfenvinphos

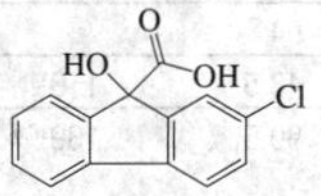

1831
Chlorflurecol

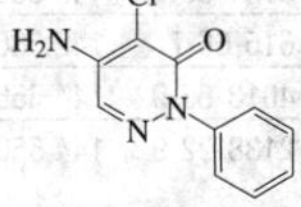

1832
Chloridazon

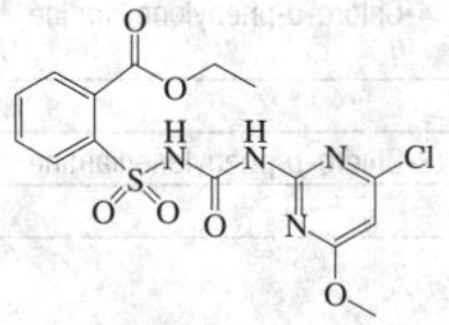

1833
Chlorimuron-ethyl

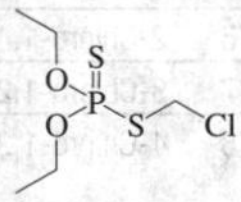

1834
Chlormephos

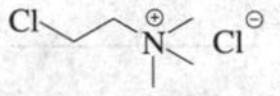

1835
Chlormequat chloride

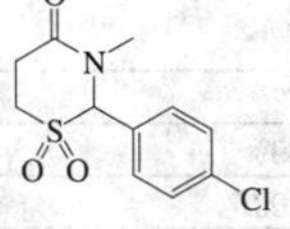

1836
Chlormezanone

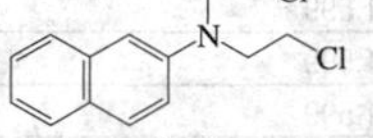

1837
Chlornaphazine

1838
Chloroacetaldehyde

1839
2-Chloroacetamide

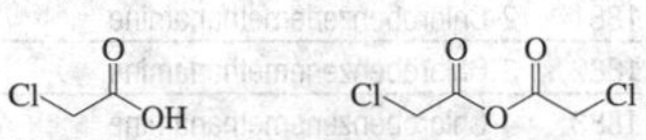

1840
Chloroacetic acid

1841
Chloroacetic anhydride

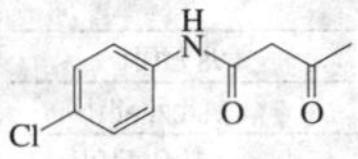

1842
4-Chloroacetoacetanilide

1843
Chloroacetone

1844
Chloroacetonitrile

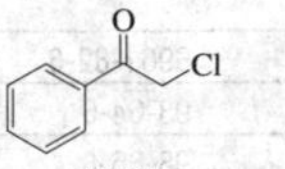

1845
α-Chloroacetophenone

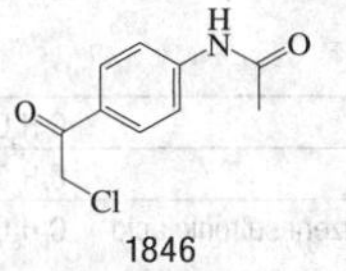

1846
4-(2-Chloroacetyl)acetanilide

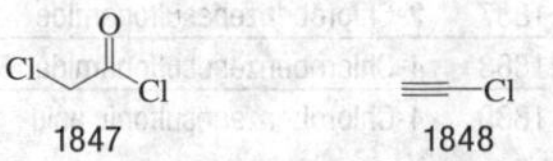

1847
Chloroacetyl chloride

1848
Chloroacetylene

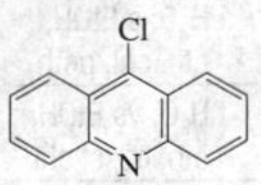

1849
9-Chloroacridine

1850
2-Chloroaniline

1851
3-Chloroaniline

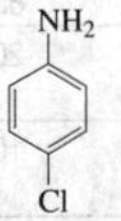

1852
4-Chloroaniline

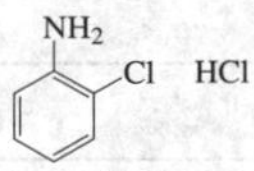

1853
2-Chloroaniline hydrochloride

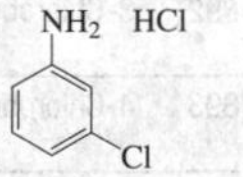

1854
3-Chloroaniline hydrochloride

No.	Name	Synonym	Mol. Form.	CAS RN	Mol. Wt.	Physical Form	mp/°C	bp/°C	den g cm⁻³	n_D	Solubility
1855	2-Chloroanisole	1-Chloro-2-methoxybenzene	C₇H₇ClO	766-51-8	142.583	liq	-26.5(0.2)	201.8(0.4)	1.1911²⁰	1.5480²⁰	i H₂O; s EtOH, eth; sl chl
1856	3-Chloroanisole	1-Chloro-3-methoxybenzene	C₇H₇ClO	2845-89-8	142.583			193.5	1.1759¹²	1.5365²⁰	i H₂O; s EtOH, eth
1857	4-Chloroanisole	1-Chloro-4-methoxybenzene	C₇H₇ClO	623-12-1	142.583		<-18	197.5	1.201²⁰	1.5390²⁰	i H₂O; vs EtOH, eth, chl; s ctc
1858	1-Chloroanthracene		C₁₄H₉Cl	4985-70-0	212.674	lf (HOAc)	83.5		1.1707¹⁰⁰	1.6959¹⁰⁰	i H₂O; s EtOH, eth, bz, ctc
1859	1-Chloro-9,10-anthracenedi-one		C₁₄H₇ClO₂	82-44-0	242.658	ye nd (to or al)	163	sub			i H₂O; sl EtOH, ctc; msc eth; s bz
1860	2-Chloro-9,10-anthracenedi-one		C₁₄H₇ClO₂	131-09-9	242.658	pa ye nd (al, HOAc)	209.9(0.4)	sub			i H₂O, eth; sl EtOH, bz; vs tol; s PhNO₂
1861	2-Chlorobenzaldehyde		C₇H₅ClO	89-98-5	140.567	nd	11.9(0.6)	212.1(1)	1.2483²⁰	1.5662²⁰	sl H₂O; s EtOH, eth, ace, bz, ctc
1862	3-Chlorobenzaldehyde		C₇H₅ClO	587-04-2	140.567	pr	17.5	213.5	1.2410²⁰	1.5650²⁰	sl H₂O, chl; s EtOH, eth, ace, bz
1863	4-Chlorobenzaldehyde		C₇H₅ClO	104-88-1	140.567	pl	47(1)	213.5	1.196⁶¹	1.555⁶¹	s H₂O, ace, chl; vs EtOH, eth, bz
1864	2-Chlorobenzamide		C₇H₆ClNO	609-66-5	155.582	orth nd (w)	141.8				s H₂O, EtOH, eth
1865	Chlorobenzene	Phenyl chloride	C₆H₅Cl	108-90-7	112.557	liq	-45.2(0.1)	131.6(0.2)	1.1058²⁰	1.5241²⁰	i H₂O; msc EtOH, eth; vs bz, ctc
1866	2-Chlorobenzeneacetic acid		C₈H₇ClO₂	2444-36-2	170.594	nd (w)	94.2(0.4)				sl H₂O; vs EtOH
1867	3-Chlorobenzeneacetic acid		C₈H₇ClO₂	1878-65-5	170.594	pl (dil al), nd (lig)	76.6(0.4)				sl H₂O, bz, ctc, EtOH; msc eth
1868	4-Chlorobenzeneacetic acid		C₈H₇ClO₂	1878-66-6	170.594	nd (w)	104.8(0.4)				s H₂O, EtOH, eth, bz
1869	2-Chlorobenzeneacetonitrile		C₈H₆ClN	2856-63-5	151.594		24	251	1.1737¹⁸		
1870	3-Chlorobenzeneacetonitrile		C₈H₆ClN	1529-41-5	151.594		11.5	261	1.1806³⁰	1.5437²⁰	
1871	4-Chlorobenzeneacetonitrile		C₈H₆ClN	140-53-4	151.594		29	265.0	1.1778³⁰		s ctc
1872	α-Chlorobenzeneacetyl chloride		C₈H₆Cl₂O	2912-62-1	189.039			120²³	1.196²⁵	1.5440²⁰	
1873	3-Chlorobenzenecarbo-peroxoic acid		C₇H₅ClO₃	937-14-4	172.566		92 dec				
1874	4-Chloro-1,2-benzenediamine	4-Chloro-o-phenylenediamine	C₆H₇ClN₂	95-83-0	142.586	pl (bz-lig) lf (w)	76				sl H₂O; vs EtOH, eth; s bz, lig
1875	4-Chloro-1,3-benzenediamine		C₆H₇ClN₂	5131-60-2	142.586	pl or nd	91				vs EtOH
1876	2-Chloro-1,4-benzenediamine	2-Chloro-p-phenylenediamine	C₆H₇ClN₂	615-66-7	142.586	nd	64				
1877	3-Chloro-1,2-benzenediol		C₆H₅ClO₂	4018-65-9	144.556	cry (lig)	48.5	110¹¹			vs lig
1878	4-Chloro-1,2-benzenediol		C₆H₅ClO₂	2138-22-9	144.556	lf (bz-peth)	90.5	139¹⁰·⁵			vs H₂O, ace, eth, EtOH
1879	4-Chloro-1,3-benzenediol		C₆H₅ClO₂	95-88-5	144.556			257			vs H₂O, EtOH, eth, ace, bz, CS₂
1880	2-Chloro-1,4-benzenediol		C₆H₅ClO₂	615-67-8	144.556	red lf (chl), nd (bz)	108	263			vs H₂O, chl; s EtOH, eth; vs bz
1881	2-Chlorobenzenemethanamine		C₇H₈ClN	89-97-4	141.599			72²		1.5594²⁵	
1882	3-Chlorobenzenemethanamine		C₇H₈ClN	4152-90-3	141.599			89²		1.5570²⁵	
1883	4-Chlorobenzenemethanamine		C₇H₈ClN	104-86-9	141.599			109¹³		1.5566²⁵	
1884	4-Chlorobenzenemethanethiol		C₇H₇ClS	6258-66-8	158.649		19.5	113¹⁷	1.202²⁵	1.5893²⁰	
1885	2-Chlorobenzenemethanol		C₇H₇ClO	17849-38-6	142.583	lf or nd (dil al)	73	230			sl H₂O; vs EtOH, eth, lig
1886	4-Chlorobenzenemethanol		C₇H₇ClO	873-76-7	142.583	nd (w), pl (bz or bz-lig)	75	235			vs bz, eth, EtOH
1887	2-Chlorobenzenesulfonamide		C₆H₆ClNO₂S	6961-82-6	191.636	lf (al)	188.3(0.5)				vs EtOH
1888	4-Chlorobenzenesulfonamide		C₆H₆ClNO₂S	98-64-6	191.636	pr or pl (eth)	146				vs bz, eth
1889	4-Chlorobenzenesulfonic acid	p-Chlorobenzenesulfonic acid	C₆H₅ClO₃S	98-66-8	192.620	nd (w+1)	67	147²⁵			s H₂O, EtOH; i eth, bz
1890	4-Chlorobenzenesulfonyl chloride		C₆H₄Cl₂O₂S	98-60-2	211.066		51	141¹⁵			vs eth, bz
1891	2-Chlorobenzenethiol		C₆H₅ClS	6320-03-2	144.622			205.5	1.2752¹⁰		sl H₂O, EtOH
1892	3-Chlorobenzenethiol		C₆H₅ClS	2037-31-2	144.622			206	1.2637¹³		i H₂O; s EtOH, eth, chl, peth
1893	4-Chlorobenzenethiol		C₆H₅ClS	106-54-7	144.622		61	206	1.1911²⁰	1.5480²⁰	i H₂O; vs EtOH, eth, bz; sl chl
1894	Chlorobenzilate		C₁₆H₁₄Cl₂O₃	510-15-6	325.186		39.0(0.5)	157⁰·⁰⁷	1.2816²⁰		

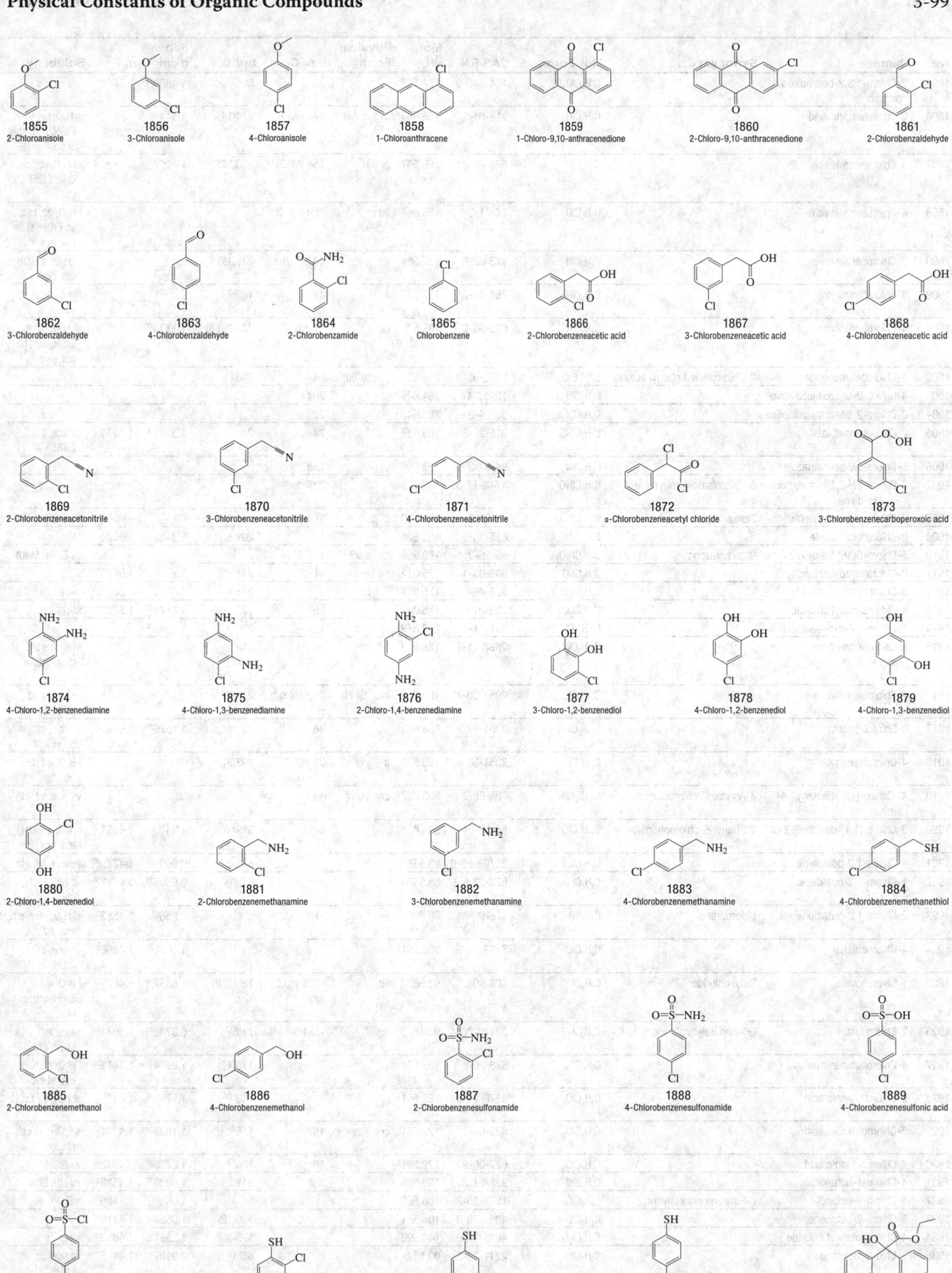

1855
2-Chloroanisole

1856
3-Chloroanisole

1857
4-Chloroanisole

1858
1-Chloroanthracene

1859
1-Chloro-9,10-anthracenedione

1860
2-Chloro-9,10-anthracenedione

1861
2-Chlorobenzaldehyde

1862
3-Chlorobenzaldehyde

1863
4-Chlorobenzaldehyde

1864
2-Chlorobenzamide

1865
Chlorobenzene

1866
2-Chlorobenzeneacetic acid

1867
3-Chlorobenzeneacetic acid

1868
4-Chlorobenzeneacetic acid

1869
2-Chlorobenzeneacetonitrile

1870
3-Chlorobenzeneacetonitrile

1871
4-Chlorobenzeneacetonitrile

1872
α-Chlorobenzeneacetyl chloride

1873
3-Chlorobenzenecarboperoxoic acid

1874
4-Chloro-1,2-benzenediamine

1875
4-Chloro-1,3-benzenediamine

1876
2-Chloro-1,4-benzenediamine

1877
3-Chloro-1,2-benzenediol

1878
4-Chloro-1,2-benzenediol

1879
4-Chloro-1,3-benzenediol

1880
2-Chloro-1,4-benzenediol

1881
2-Chlorobenzenemethanamine

1882
3-Chlorobenzenemethanamine

1883
4-Chlorobenzenemethanamine

1884
4-Chlorobenzenemethanethiol

1885
2-Chlorobenzenemethanol

1886
4-Chlorobenzenemethanol

1887
2-Chlorobenzenesulfonamide

1888
4-Chlorobenzenesulfonamide

1889
4-Chlorobenzenesulfonic acid

1890
4-Chlorobenzenesulfonyl chloride

1891
2-Chlorobenzenethiol

1892
3-Chlorobenzenethiol

1893
4-Chlorobenzenethiol

1894
Chlorobenzilate

No.	Name	Synonym	Mol. Form.	CAS RN	Mol. Wt.	Physical Form	mp/°C	bp/°C	den g cm⁻³	n_D	Solubility
1895	2-Chloro-1,3,2-benzodioxaphosphole		C₆H₄ClO₂P	1641-40-3	174.522		30	80²⁰	1.4650²⁰	1.5712²⁰	
1896	2-Chlorobenzoic acid		C₇H₅ClO₂	118-91-2	156.567	mcl pr (w)	140.4(0.7)	274(14)	1.544²⁰		s H₂O, bz; vs EtOH, eth, ace; sl CS₂
1897	3-Chlorobenzoic acid		C₇H₅ClO₂	535-80-8	156.567	pr (w)	154.2(0.2)	283(17)	1.496²⁵		sl H₂O, bz, ctc, CS₂; s EtOH, eth
1898	4-Chlorobenzoic acid		C₇H₅ClO₂	74-11-3	156.567	tcl pr (al-eth)	239.5(0.6)				i H₂O, bz, ctc; vs EtOH; sl eth, ace
1899	2-Chlorobenzonitrile		C₇H₄ClN	873-32-5	137.567	nd	43.5(0.4)	235.1(0.5)			sl H₂O; s EtOH, eth, chl
1900	3-Chlorobenzonitrile		C₇H₄ClN	766-84-7	137.567		41	100¹⁵			i H₂O; s EtOH, eth
1901	4-Chlorobenzonitrile		C₇H₄ClN	623-03-0	137.567	nd (al)	91.6(0.4)	223.0(0.4)	1.1133¹⁷		sl H₂O, lig; s EtOH, eth, bz, chl
1902	2-Chlorobenzophenone	2-Chlorophenyl phenyl ketone	C₁₃H₉ClO	5162-03-8	216.662	pl (chl-lig)	54	330			
1903	4-Chloro-2-benzothiazolamine		C₇H₅ClN₂S	19952-47-7	184.646		204				
1904	6-Chloro-2-benzothiazolamine		C₇H₅ClN₂S	95-24-9	184.646		200				
1905	2-Chlorobenzothiazole		C₇H₄ClNS	615-20-3	169.632		24	248	1.3715¹⁰	1.6338¹⁰	vs ace, eth, EtOH
1906	5-Chloro-1H-benzotriazole		C₆H₄ClN₃	94-97-3	153.569		158				
1907	6-Chloro-2H-3,1-benzoxazine-2,4(1H)-dione	5-Chloroisatoic anhydride	C₈H₄ClNO₃	4743-17-3	197.576		280 dec				
1908	5-Chloro-2-benzoxazolamine	Zoxazolamine	C₇H₅ClN₂O	61-80-3	168.580	pl (bz)	184.5				vs EtOH
1909	2-Chlorobenzoxazole		C₇H₄ClNO	615-18-9	153.566		7	201.5	1.3453¹⁸	1.5678²⁰	
1910	5-Chloro-2(3H)-benzoxazolone	Chlorzoxazone	C₇H₄ClNO₂	95-25-0	169.566	cry (ace)	191.5				vs EtOH, MeOH
1911	2-Chlorobenzoyl chloride		C₇H₄Cl₂O	609-65-4	175.012	liq	-4	241(12)		1.5726¹⁶	s ctc
1912	3-Chlorobenzoyl chloride		C₇H₄Cl₂O	618-46-2	175.012			225		1.5677²⁰	
1913	4-Chlorobenzoyl chloride		C₇H₄Cl₂O	122-01-0	175.012		16	222	1.3770²⁰	1.5756²⁰	sl chl
1914	1-Chloro-4-benzylbenzene		C₁₃H₁₁Cl	831-81-2	202.679		7.5	299	1.1247²⁰		vs ace
1915	o-Chlorobenzylidene malononitrile		C₁₀H₅ClN₂	2698-41-1	188.613	wh cry	96	312			sl H₂O; s bz, diox, EtOAc, ace
1916	2-Chlorobiphenyl		C₁₂H₉Cl	2051-60-7	188.652	mcl (dil al)	31.78(0.08)	273(7)	1.1499³²		i H₂O; vs eth, EtOH, lig
1917	3-Chlorobiphenyl		C₁₂H₉Cl	2051-61-8	188.652		16	284.5	1.1579²⁵	1.6181²⁵	vs ace, eth, EtOH
1918	4-Chlorobiphenyl		C₁₂H₉Cl	2051-62-9	188.652	lf (lig or al)	75.4(0.2)	293(3)			i H₂O; s EtOH, eth, lig
1919	4'-Chloro-[1,1'-biphenyl]-4-amine	4-Amino-4'-chlorodiphenyl	C₁₂H₁₀ClN	135-68-2	203.667	cry (peth)	134				vs ace, bz, eth
1920	3-Chloro-[1,1'-biphenyl]-2-ol	2-Phenyl-6-chlorophenol	C₁₂H₉ClO	85-97-2	204.651		6	319(4)	1.24²⁵	1.6237³⁰	i H₂O; s EtOH, eth, ace, bz
1921	4-Chloro-1,2-butadiene		C₄H₅Cl	25790-55-0	88.536			88	0.9891²⁰	1.4775²⁰	vs ace, bz, eth
1922	1-Chloro-1,3-butadiene		C₄H₅Cl	627-22-5	88.536			67(6)	0.9606²⁰	1.4712²⁰	vs eth, EtOH, chl
1923	2-Chloro-1,3-butadiene	Chloroprene	C₄H₅Cl	126-99-8	88.536	liq	-130	59(3)	0.956²⁰	1.4583²⁰	sl H₂O; msc eth, ace, bz
1924	4-Chlorobutanal		C₄H₇ClO	6139-84-0	106.551			51¹³	1.106⁸	1.4466⁸	vs ace, eth, EtOH
1925	1-Chlorobutane	Butyl chloride	C₄H₉Cl	109-69-3	92.567	liq	-123.1(0.2)	78.4(0.2)	0.8857²⁰	1.4023²⁰	i H₂O; msc EtOH, eth; sl ctc
1926	2-Chlorobutane	(±)-sec-Butyl chloride	C₄H₉Cl	53178-20-4	92.567	liq	-131.3	71(8)	0.8732²⁰	1.3971²⁰	vs bz, eth, EtOH, chl
1927	4-Chlorobutanenitrile		C₄H₆ClN	628-20-6	103.551			175(12)	1.0934¹⁵	1.4413²⁰	i H₂O; s EtOH, eth; sl ctc
1928	2-Chlorobutanoic acid		C₄H₇ClO₂	4170-24-5	122.551			189⁶²⁷	1.1796²⁰	1.441²⁰	sl H₂O; vs EtOH, eth
1929	3-Chlorobutanoic acid		C₄H₇ClO₂	625-68-3	122.551	cry (eth)	16	116²²	1.1898²⁰	1.4221²⁰	s EtOH; vs eth; sl ctc
1930	4-Chlorobutanoic acid		C₄H₇ClO₂	627-00-9	122.551		16	196²²	1.2236²⁰	1.4642²⁰	vs EtOH
1931	4-Chloro-1-butanol		C₄H₉ClO	928-51-8	108.566			84¹⁶	1.0883²⁰	1.4518²⁰	vs eth, EtOH
1932	1-Chloro-2-butanol	α-Butylene chlorohydrin	C₄H₉ClO	1873-25-2	108.566			141	1.068²⁵	1.4400²⁰	s EtOH, eth
1933	3-Chloro-2-butanone		C₄H₇ClO	4091-39-8	106.551			129(13)	1.0554²⁵	1.4219²⁰	
1934	4-Chlorobutanoyl chloride		C₄H₆Cl₂O	4635-59-0	140.996			173.5	1.2581²⁰	1.4616²⁰	s eth
1935	2-Chloro-1-butene		C₄H₇Cl	2211-70-3	90.552			58(3)	0.9107¹⁵	1.4165²¹	vs ace, bz, eth, EtOH
1936	3-Chloro-1-butene		C₄H₇Cl	563-52-0	90.552			64(3)	0.8978²⁰	1.4149²⁰	vs eth, ace; s chl

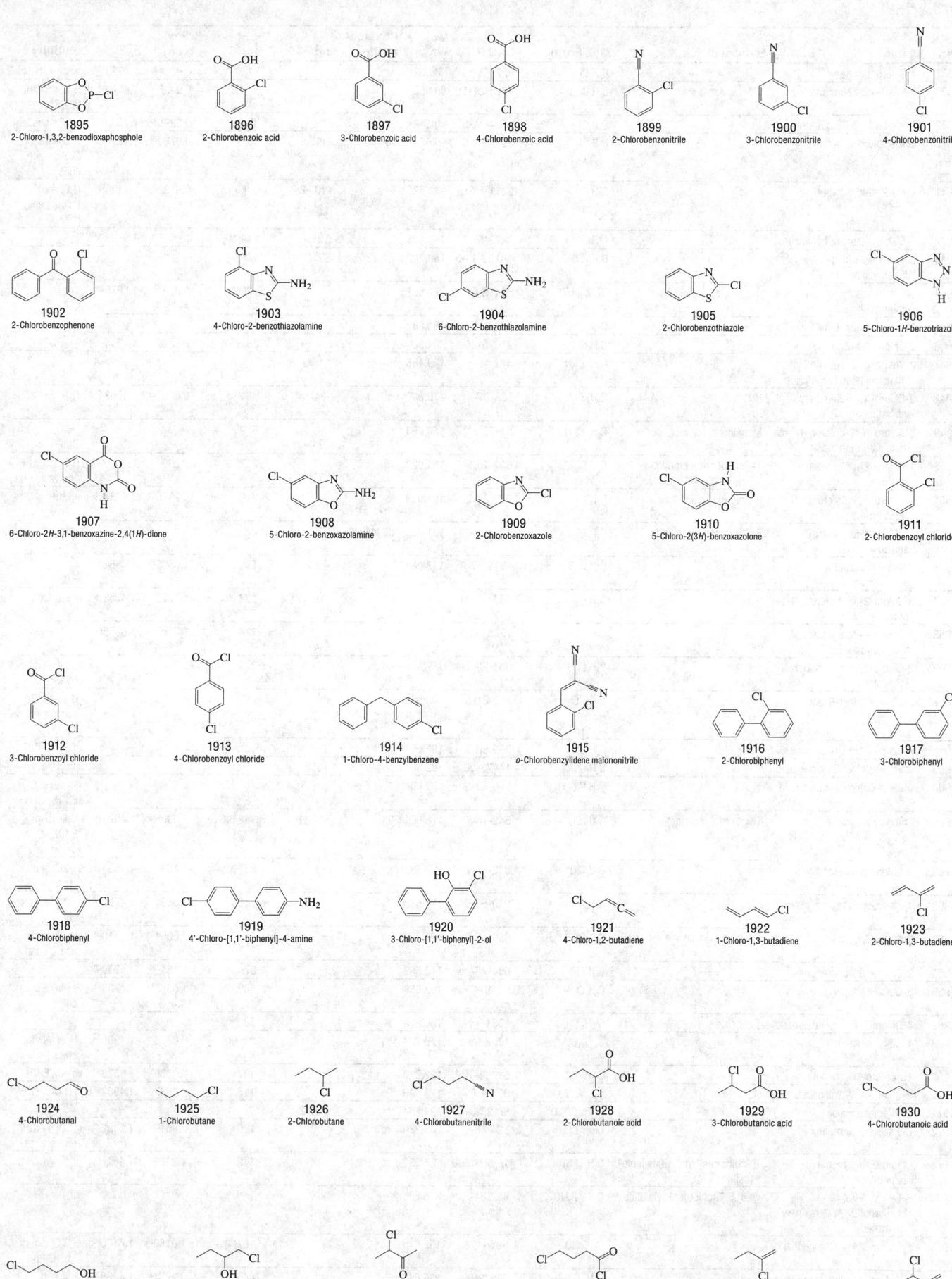

1895
2-Chloro-1,3,2-benzodioxaphosphole

1896
2-Chlorobenzoic acid

1897
3-Chlorobenzoic acid

1898
4-Chlorobenzoic acid

1899
2-Chlorobenzonitrile

1900
3-Chlorobenzonitrile

1901
4-Chlorobenzonitrile

1902
2-Chlorobenzophenone

1903
4-Chloro-2-benzothiazolamine

1904
6-Chloro-2-benzothiazolamine

1905
2-Chlorobenzothiazole

1906
5-Chloro-1*H*-benzotriazole

1907
6-Chloro-2*H*-3,1-benzoxazine-2,4(1*H*)-dione

1908
5-Chloro-2-benzoxazolamine

1909
2-Chlorobenzoxazole

1910
5-Chloro-2(3*H*)-benzoxazolone

1911
2-Chlorobenzoyl chloride

1912
3-Chlorobenzoyl chloride

1913
4-Chlorobenzoyl chloride

1914
1-Chloro-4-benzylbenzene

1915
o-Chlorobenzylidene malononitrile

1916
2-Chlorobiphenyl

1917
3-Chlorobiphenyl

1918
4-Chlorobiphenyl

1919
4'-Chloro-[1,1'-biphenyl]-4-amine

1920
3-Chloro-[1,1'-biphenyl]-2-ol

1921
4-Chloro-1,2-butadiene

1922
1-Chloro-1,3-butadiene

1923
2-Chloro-1,3-butadiene

1924
4-Chlorobutanal

1925
1-Chlorobutane

1926
2-Chlorobutane

1927
4-Chlorobutanenitrile

1928
2-Chlorobutanoic acid

1929
3-Chlorobutanoic acid

1930
4-Chlorobutanoic acid

1931
4-Chloro-1-butanol

1932
1-Chloro-2-butanol

1933
3-Chloro-2-butanone

1934
4-Chlorobutanoyl chloride

1935
2-Chloro-1-butene

1936
3-Chloro-1-butene

No.	Name	Synonym	Mol. Form.	CAS RN	Mol. Wt.	Physical Form	mp/°C	bp/°C	den g cm^{-3}	n_D	Solubility
1937	4-Chloro-1-butene		C$_4$H$_7$Cl	927-73-1	90.552			74(5)	0.9211[20]	1.4233[20]	vs ace, eth, chl
1938	*cis*-1-Chloro-2-butene		C$_4$H$_7$Cl	4628-21-1	90.552			84(4)	0.9426[20]	1.4390[20]	i H$_2$O; s EtOH, ace, chl
1939	*trans*-1-Chloro-2-butene		C$_4$H$_7$Cl	4894-61-5	90.552			85(4)	0.9295[20]	1.4350[20]	i H$_2$O; s ace, chl
1940	*cis*-2-Chloro-2-butene		C$_4$H$_7$Cl	2211-69-0	90.552	liq	-117.3	65(5)	0.9239[20]	1.4240[20]	i H$_2$O; msc EtOH; s ace, chl
1941	*trans*-2-Chloro-2-butene		C$_4$H$_7$Cl	2211-68-9	90.552	liq	-105.8	63(3)	0.9138[20]	1.4190[20]	i H$_2$O; msc EtOH; s ace, chl
1942	1-Chloro-4-*tert*-butylbenzene		C$_{10}$H$_{13}$Cl	3972-56-3	168.663			214(10)	1.0075[18]	1.5123[20]	
1943	Chloro-(*tert*-butyl)-dimethylsilane		C$_6$H$_{15}$ClSi	18162-48-6	150.722		89.5	125			
1944	Chloro(*tert*-butyl)-diphenylsilane		C$_{16}$H$_{19}$ClSi	58479-61-1	274.861			120[0.06]	1.07[20]	1.5675[20]	
1945	2-Chloro-4-*tert*-butylphenol		C$_{10}$H$_{13}$ClO	98-28-2	184.662		114[8]				
1946	3-Chloro-1-butyne		C$_4$H$_5$Cl	21020-24-6	88.536			68.5	1.4218[25]	1.4218[25]	
1947	2-Chloro-*N*-(2-chloroethyl)-ethanamine, hydrochloride		C$_4$H$_{10}$Cl$_3$N	821-48-7	178.488		215.0				
1948	2-Chloro-*N*-(2-chloroethyl)-*N*-ethylethanamine	HN1	C$_6$H$_{13}$Cl$_2$N	538-07-8	170.080	col liq	-34	66[12]	1.0861[23]	1.4653[25]	i H$_2$O
1949	2-Chloro-*N*-(2-chloroethyl)-*N*-methylethanamine	Mechlorethamine	C$_5$H$_{11}$Cl$_2$N	51-75-2	156.053		-60	87[18]			sl H$_2$O; msc ctc, DMF
1950	1-Chloro-2-(chloromethyl)-benzene	2-Chlorobenzyl chloride	C$_7$H$_6$Cl$_2$	611-19-8	161.029	liq	-17	217	1.2699[0]	1.5530[20]	i H$_2$O; sl EtOH, ctc; vs eth, bz
1951	1-Chloro-3-(chloromethyl)-benzene	3-Chlorobenzyl chloride	C$_7$H$_6$Cl$_2$	620-20-2	161.029			216	1.2695[15]	1.5554[20]	vs EtOH
1952	1-Chloro-4-(chloromethyl)-benzene	4-Chlorobenzyl chloride	C$_7$H$_6$Cl$_2$	104-83-6	161.029	nd (dil al)	31	223			sl ctc
1953	Chloro(chloromethyl)-dimethylsilane		C$_3$H$_8$Cl$_2$Si	1719-57-9	143.088			115.5	1.0865[20]	1.4360[20]	
1954	3-Chloro-2-(chloromethyl)-1-propene		C$_4$H$_6$Cl$_2$	1871-57-4	124.997	liq	-14	138	1.1782[20]	1.4753	vs EtOH, chl
1955	1-Chloro-4-[(chloromethyl)-thio]benzene		C$_7$H$_6$Cl$_2$S	7205-90-5	193.094		21.5	128[12]	1.346[25]	1.6055[20]	
1956	2-Chloro-1-(4-chlorophenyl)-ethanone		C$_8$H$_6$Cl$_2$O	937-20-2	189.039	nd (al)	101.5	270			s EtOH, bz, MeOH
1957	3-Chlorocholest-5-ene, (3β)		C$_{27}$H$_{45}$Cl	910-31-6	405.099	nd (al, ace)	96				i H$_2$O; s EtOH, ace, bz, chl; vs CS$_2$
1958	*trans-o*-Chlorocinnamic acid		C$_9$H$_7$ClO$_2$	939-58-2	182.604		212				vs eth, EtOH
1959	*trans-m*-Chlorocinnamic acid		C$_9$H$_7$ClO$_2$	14473-90-6	182.604		165				s EtOH, eth
1960	*trans-p*-Chlorocinnamic acid		C$_9$H$_7$ClO$_2$	940-62-5	182.604		249.5				vs ace, eth, EtOH
1961	Chlorocyclohexane	Cyclohexyl chloride	C$_6$H$_{11}$Cl	542-18-7	118.604	liq	-45(1)	142.6(0.5)	1.000[20]	1.4626[20]	i H$_2$O; msc EtOH, eth, ace, bz; vs chl
1962	2-Chlorocyclohexanone		C$_6$H$_9$ClO	822-87-7	132.587		23	82[15]	1.160[20]	1.4825[20]	s eth, bz, diox; sl ctc
1963	1-Chlorocyclohexene		C$_6$H$_9$Cl	930-66-5	116.588			140(4)	1.0361[19]	1.4797[20]	s eth, ace, ctc, chl
1964	Chlorocyclopentane	Cyclopentyl chloride	C$_5$H$_9$Cl	930-28-9	104.578	liq		113.1(0.6)	1.0051[20]	1.4510[20]	i H$_2$O; s eth, ace, bz, ctc
1965	2-Chlorocyclopentanone		C$_5$H$_7$ClO	694-28-0	118.562			87[19]	1.185[25]	1.4750[20]	
1966	3-Chlorocyclopentene		C$_5$H$_7$Cl	96-40-2	102.563			40[40]	1.0388[25]	1.4708[26]	vs eth, EtOH, chl
1967	4-Chloro-2-cyclopentylphenol	Dowicide 9	C$_{11}$H$_{13}$ClO	13347-42-7	196.673			183[18]			
1968	1-Chlorodecane		C$_{10}$H$_{21}$Cl	1002-69-3	176.727	liq	-31.3	225(3)	0.8696[20]	1.4380[20]	i H$_2$O; vs eth, chl; s ctc
1969	10-Chloro-1-decanol		C$_{10}$H$_{21}$ClO	51309-10-5	192.726		12.5	187[15]	0.9630[25]	1.4578[20]	vs eth, EtOH
1970	2-Chloro-*N,N*-diallylacetamide	Allidochlor	C$_8$H$_{12}$ClNO	93-71-0	173.640	liq		116[1]	1.088[25]	1.4932[25]	sl H$_2$O; s EtOH
1971	Chlorodiazepoxide		C$_{16}$H$_{14}$ClN$_3$O	58-25-3	299.754		236.2				
1972	Chlorodibromomethane		CHBr$_2$Cl	124-48-1	208.280	liq	-20	120	2.451[20]	1.5482[20]	i H$_2$O; s EtOH, eth, ace, bz
1973	Chloro(dichloromethyl)-dimethylsilane	(Dichloromethyl)dimethylchlorosilane	C$_3$H$_7$Cl$_3$Si	18171-59-0	177.533	liq	-48	149	1.2369[20]	1.461[20]	
1974	5-Chloro-*N*-(3,4-dichlorophenyl)-2-hydroxy-benzamide	3',4',5-Trichlorosalicylanilide	C$_{13}$H$_8$Cl$_3$NO$_2$	642-84-2	316.568		247				
1975	2-Chloro-1,1-diethoxyethane		C$_6$H$_{13}$ClO$_2$	621-62-5	152.619			157(4)	1.0180[20]	1.4170[20]	sl H$_2$O, ctc; msc EtOH, eth
1976	3-Chloro-1,1-diethoxypropane		C$_7$H$_{15}$ClO$_2$	35573-93-4	166.646			84[25]	0.9951[19]	1.4268[20]	vs ace, bz

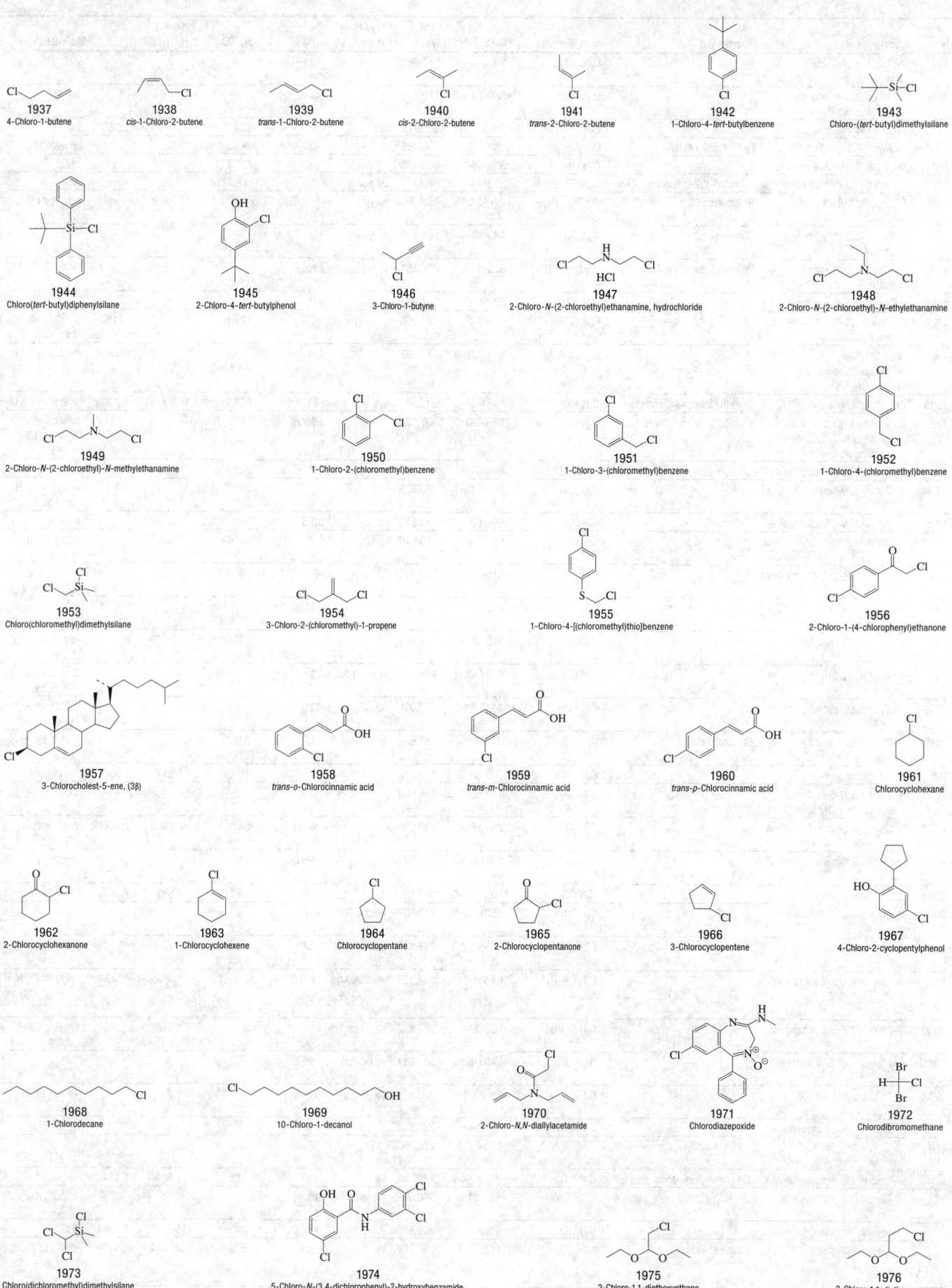

1937
4-Chloro-1-butene

1938
cis-1-Chloro-2-butene

1939
trans-1-Chloro-2-butene

1940
cis-2-Chloro-2-butene

1941
trans-2-Chloro-2-butene

1942
1-Chloro-4-*tert*-butylbenzene

1943
Chloro-(*tert*-butyl)dimethylsilane

1944
Chloro(*tert*-butyl)diphenylsilane

1945
2-Chloro-4-*tert*-butylphenol

1946
3-Chloro-1-butyne

1947
2-Chloro-*N*-(2-chloroethyl)ethanamine, hydrochloride

1948
2-Chloro-*N*-(2-chloroethyl)-*N*-ethylethanamine

1949
2-Chloro-*N*-(2-chloroethyl)-*N*-methylethanamine

1950
1-Chloro-2-(chloromethyl)benzene

1951
1-Chloro-3-(chloromethyl)benzene

1952
1-Chloro-4-(chloromethyl)benzene

1953
Chloro(chloromethyl)dimethylsilane

1954
3-Chloro-2-(chloromethyl)-1-propene

1955
1-Chloro-4-[(chloromethyl)thio]benzene

1956
2-Chloro-1-(4-chlorophenyl)ethanone

1957
3-Chlorocholest-5-ene, (3β)

1958
trans-o-Chlorocinnamic acid

1959
trans-m-Chlorocinnamic acid

1960
trans-p-Chlorocinnamic acid

1961
Chlorocyclohexane

1962
2-Chlorocyclohexanone

1963
1-Chlorocyclohexene

1964
Chlorocyclopentane

1965
2-Chlorocyclopentanone

1966
3-Chlorocyclopentene

1967
4-Chloro-2-cyclopentylphenol

1968
1-Chlorodecane

1969
10-Chloro-1-decanol

1970
2-Chloro-*N,N*-diallylacetamide

1971
Chlorodiazepoxide

1972
Chlorodibromomethane

1973
Chloro(dichloromethyl)dimethylsilane

1974
5-Chloro-*N*-(3,4-dichlorophenyl)-2-hydroxybenzamide

1975
2-Chloro-1,1-diethoxyethane

1976
3-Chloro-1,1-diethoxypropane

No.	Name	Synonym	Mol. Form.	CAS RN	Mol. Wt.	Physical Form	mp/°C	bp/°C	den g cm⁻³	n_D	Solubility
1977	2-Chloro-*N,N*-diethylacet-amide		C₆H₁₂ClNO	2315-36-8	149.618			192²⁵			
1978	2-Chloro-*N,N*-diethylethana-mine, hydrochloride		C₆H₁₅Cl₂N	869-24-9	172.096		200				sl H₂O
1979	Chlorodifluoroacetic acid		C₂HClF₂O₂	76-04-0	130.478	hyg	25	122		1.3559²⁰	s chl
1980	1-Chloro-1,1-difluoroethane	Refrigerant 142b	C₂H₃ClF₂	75-68-3	100.495	col gas	-130.43(0.02)	-9.12(0.07)	1.107²⁵		i H₂O; s bz
1981	1-Chloro-2,2-difluoroethane		C₂H₃ClF₂	338-65-8	100.495			35(7)			
1982	1-Chloro-2,2-difluoroethene	1-Chloro-2,2-difluoroethylene	C₂HClF₂	359-10-4	98.479	col gas	-138.5	-18.8(0.5)			sl H₂O
1983	Chlorodifluoromethane	Refrigerant 22	CHClF₂	75-45-6	86.469	col gas	-157.41(0.03)	-40.8(0.5)	1.4909⁻⁶⁹		sl H₂O; s eth, ace, chl
1984	7-Chloro-2,3-dihydro-1*H*-inden-4-ol	Chlorindanol	C₉H₉ClO	145-94-8	168.619	nd (peth)	92				
1985	10-Chloro-5,10-dihydrophen-arsazine	Phenarsazine chloride	C₁₂H₉AsClN	578-94-9	277.581	ye cry	195		1.65		i H₂O; sl ctc, bz, xyl
1986	5-Chloro-2,4-dimethoxyaniline		C₈H₁₀ClNO₂	97-50-7	187.624		91				
1987	2-Chloro-1,1-dimethoxy-ethane		C₄H₉ClO₂	97-97-2	124.566			127.5	1.068²⁰	1.4150²⁰	sl EtOH, eth, bz, ctc
1988	*N*-(4-Chloro-2,5-dimethoxyphenyl)-3-oxobu-tanamide		C₁₂H₁₄ClNO₄	4433-79-8	271.697		107				s chl
1989	Chlorodimethylaluminum	Dimethylaluminum chloride	C₂H₆AlCl	1184-58-3	92.504	hyg liq	-45	126	0.996		reac H₂O; s hx
1990	2-Chloro-10-(3-dimethylami-nopropyl)phenothiazine monohydrochloride	Aminazin hydrochloride	C₁₇H₂₀Cl₂N₂S	69-09-0	355.325		198(2)				s H₂O; i eth, bz; vs chl, EtOH
1991	2-Chloro-*N,N*-dimethylaniline		C₈H₁₀ClN	698-01-1	155.625			205	1.1067²⁰	1.5578²⁰	vs bz, EtOH
1992	3-Chloro-*N,N*-dimethylaniline		C₈H₁₀ClN	6848-13-1	155.625			232			sl H₂O; s EtOH, ace, bz
1993	4-Chloro-*N,N*-dimethylaniline		C₈H₁₀ClN	698-69-1	155.625	nd (al)	35.5	231	1.0480¹⁰⁰		s EtOH
1994	2-Chloro-1,4-dimethylbenzene		C₈H₉Cl	95-72-7	140.610		0.8	185(8)	1.0589¹⁵		i H₂O; s ace, ctc; vs bz
1995	4-Chloro-1,2-dimethylbenzene		C₈H₉Cl	615-60-1	140.610	liq	-6	190(13)	1.0682¹⁵		i H₂O; s ace, ctc; vs bz
1996	2-Chloro-*N,N*-dimethylethana-mine, hydrochloride		C₄H₁₁Cl₂N	4584-46-7	144.043		201.0				sl H₂O
1997	(2-Chloro-1,1-dimethylethyl)-benzene	Neophyl chloride	C₁₀H₁₃Cl	515-40-2	168.663			227(12)	1.047²⁰	1.5247²⁰	vs ace, bz, eth, EtOH
1998	4-Chloro-2,5-dimethylphenol		C₈H₉ClO	1124-06-7	156.609	silv-grn nd (lig)	74.5				sl H₂O; vs bz, EtOH, peth
1999	4-Chloro-2,6-dimethylphenol		C₈H₉ClO	1123-63-3	156.609	nd (w)	83				sl H₂O; vs bz, EtOH, HOAc
2000	4-Chloro-3,5-dimethylphenol	Chloroxylenol	C₈H₉ClO	88-04-0	156.609		115	246			sl H₂O, bz, peth; s EtOH, eth
2001	Chlorodimethylphenylsilane		C₈H₁₁ClSi	768-33-2	170.712			194(3)	1.032²⁰	1.5082²⁰	
2002	1-Chloro-*N,N*-dimethyl-2-propanamine, hydrochloride		C₅H₁₃Cl₂N	17256-39-2	158.069						s chl
2003	1-Chloro-2,2-dimethylpropane		C₅H₁₁Cl	753-89-9	106.594	liq	-20(4)	84(1)	0.8660²⁰	1.4044²⁰	vs bz, eth, EtOH, chl
2004	3-Chloro-2,2-dimethylpropa-noic acid		C₅H₉ClO₂	13511-38-1	136.577		41(4)	110¹⁰			vs ctc
2005	Chlorodimethylsilane		C₂H₇ClSi	1066-35-9	94.616	liq	-111	34.7	0.852	1.3830²⁰	
2006	2-Chloro-4,6-dinitroaniline		C₆H₄ClN₃O₄	3531-19-9	217.567	ye cry (DMF aq)	157				
2007	4-Chloro-2,6-dinitroaniline		C₆H₄ClN₃O₄	5388-62-5	217.567	oran-ye nd (al)	147				s EtOH
2008	1-Chloro-2,4-dinitrobenzene		C₆H₃ClN₂O₄	97-00-7	202.552	ye orth (eth) nd (al) ye cry	50.2(0.9)	315	1.4982⁷⁵	1.5857⁶⁰	i H₂O; sl EtOH; s eth, bz, CS₂
2009	2-Chloro-1,3-dinitrobenzene		C₆H₃ClN₂O₄	606-21-3	202.552	ye nd (al, HOAc)	88	315	1.6867¹⁶		i H₂O; s EtOH, eth, tol; sl chl
2010	1-Chloro-2,4-dinitronaphtha-lene		C₁₀H₅ClN₂O₄	2401-85-6	252.611	ye nd (bz)	146.5				
2011	4-Chloro-2,6-dinitrophenol		C₆H₃ClN₂O₅	88-87-9	218.551	pa ye cry	81		1.74²²		vs eth, EtOH, chl
2012	2-Chloro-3,5-dinitropyridine		C₅H₂ClN₃O₄	2578-45-2	203.541		66.5				
2013	2-Chloro-1,3-dinitro-5-(trifluoromethyl)benzene		C₇H₂ClF₃N₂O₄	393-75-9	270.550		57				
2014	4-Chloro-1,3-dioxolan-2-one	Chloroethylene carbonate	C₃H₃ClO₃	3967-54-2	122.507	liq	110	213	1.504	1.4540²⁰	
2015	2-Chloro-1,2-diphenyletha-none		C₁₄H₁₁ClO	447-31-4	230.689	nd (al)	68.5	dec			s EtOH; sl chl; i alk
2016	Chlorodiphenylmethane		C₁₃H₁₁Cl	90-99-3	202.679		17.0(0.4)	140³	1.140²⁵	1.5951²⁰	s chl
2017	1-Chlorododecane	Lauryl chloride	C₁₂H₂₅Cl	112-52-7	204.780	liq	-9.3	263(4)	0.8673²⁰	1.4434²⁰	i H₂O; vs EtOH; msc ace, ctc; s bz

1977
2-Chloro-N,N-diethylacetamide

1978
2-Chloro-N,N-diethylethanamine, hydrochloride

1979
Chlorodifluoroacetic acid

1980
1-Chloro-1,1-difluoroethane

1981
1-Chloro-2,2-difluoroethane

1982
1-Chloro-2,2-difluoroethene

1983
Chlorodifluoromethane

1984
7-Chloro-2,3-dihydro-1H-inden-4-ol

1985
10-Chloro-5,10-dihydrophenarsazine

1986
5-Chloro-2,4-dimethoxyaniline

1987
2-Chloro-1,1-dimethoxyethane

1988
N-(4-Chloro-2,5-dimethoxyphenyl)-3-oxobutanamide

1989
Chlorodimethylaluminum

1990
2-Chloro-10-(3-dimethylaminopropyl)phenothiazine monohydrochloride

1991
2-Chloro-N,N-dimethylaniline

1992
3-Chloro-N,N-dimethylaniline

1993
4-Chloro-N,N-dimethylaniline

1994
2-Chloro-1,4-dimethylbenzene

1995
4-Chloro-1,2-dimethylbenzene

1996
2-Chloro-N,N-dimethylethanamine, hydrochloride

1997
(2-Chloro-1,1-dimethylethyl)benzene

1998
4-Chloro-2,5-dimethylphenol

1999
4-Chloro-2,6-dimethylphenol

2000
4-Chloro-3,5-dimethylphenol

2001
Chlorodimethylphenylsilane

2002
1-Chloro-N,N-dimethyl-2-propanamine, hydrochloride

2003
1-Chloro-2,2-dimethylpropane

2004
3-Chloro-2,2-dimethylpropanoic acid

2005
Chlorodimethylsilane

2006
2-Chloro-4,6-dinitroaniline

2007
4-Chloro-2,6-dinitroaniline

2008
1-Chloro-2,4-dinitrobenzene

2009
2-Chloro-1,3-dinitrobenzene

2010
1-Chloro-2,4-dinitronaphthalene

2011
4-Chloro-2,6-dinitrophenol

2012
2-Chloro-3,5-dinitropyridine

2013
2-Chloro-1,3-dinitro-5-(trifluoromethyl)benzene

2014
4-Chloro-1,3-dioxolan-2-one

2015
2-Chloro-1,2-diphenylethanone

2016
Chlorodiphenylmethane

2017
1-Chlorododecane

No.	Name	Synonym	Mol. Form.	CAS RN	Mol. Wt.	Physical Form	mp/°C	bp/°C	den g cm⁻³	n_D	Solubility
2018	Chloroethane	Ethyl chloride	C_2H_5Cl	75-00-3	64.514	vol liq or gas	-138(2)	12.3(0.2)	0.9239[0]	1.3676[20]	sl H₂O, chl; vs EtOH; msc eth
2019	2-Chloroethanesulfonyl chloride		$C_2H_4Cl_2O_2S$	1622-32-8	163.023			201.5	1.555[20]	1.4920[20]	
2020	2-Chloroethanol	Ethylene chlorohydrin	C_2H_5ClO	107-07-3	80.513	liq	-68(2)	126(2)	1.2019[20]	1.4419[20]	msc H₂O, EtOH; sl eth; s chl
2021	2-Chloroethanol, 4-methylbenzenesulfonate		$C_9H_{11}ClO_3S$	80-41-1	234.699			210[21]			i H₂O; s ctc
2022	Chloroethene	Vinyl chloride	C_2H_3Cl	75-01-4	62.498	col gas	-153.84(0.02)	-13.8(0.3)	0.9106[20]	1.3700[20]	sl H₂O; s EtOH; vs eth
2023	1-Chloro-4-ethoxybenzene		C_8H_9ClO	622-61-7	156.609		17.1(0.2)	212(1)	1.1254[20]	1.5252[20]	s EtOH, eth, HOAc; vs bz; sl ctc
2024	(2-Chloroethoxy)benzene		C_8H_9ClO	622-86-6	156.609		28	218.5			i H₂O; vs EtOH, eth, ace, bz; sl ctc
2025	1-Chloro-1-ethoxyethane		C_4H_9ClO	7081-78-9	108.566			93.5	0.9655[20]	1.4053[20]	
2026	2-(2-Chloroethoxy)ethanol		$C_4H_9ClO_2$	628-89-7	124.566			194(3)	1.18[25]	1.4529[20]	vs H₂O; msc EtOH, eth
2027	2-Chloroethyl acetate	β-Chloroethyl acetate	$C_4H_7ClO_2$	542-58-5	122.551			129(11)	1.178[20]	1.4234[20]	i H₂O; msc EtOH, eth; s ctc
2028	2-Chloroethyl acetoacetate		$C_6H_9ClO_3$	54527-68-3	164.586			198	1.2055[21]	1.4430[20]	vs bz, eth, EtOH
2029	2-Chloroethylamine hydrochloride	2-Chloroethanamine hydrochloride	$C_2H_7Cl_2N$	870-24-6	115.990		146.3				vs H₂O, ace, EtOH
2030	(1-Chloroethyl)benzene		C_8H_9Cl	672-65-1	140.610			105[50]			
2031	(2-Chloroethyl)benzene		C_8H_9Cl	622-24-2	140.610			198(15)	1.069[25]	1.5276[20]	i H₂O; s EtOH, eth, ace, bz, CS₂
2032	1-Chloro-2-ethylbenzene		C_8H_9Cl	89-96-3	140.610	liq	-83.3(0.2)	177(3)	1.0569[20]	1.5218[20]	i H₂O; s ace, bz, ctc, chl
2033	1-Chloro-3-ethylbenzene		C_8H_9Cl	620-16-6	140.610	liq	-55.0(0.2)	180(3)	1.0529[20]	1.5195[20]	vs ace, bz, eth, EtOH
2034	1-Chloro-4-ethylbenzene		C_8H_9Cl	622-98-0	140.610	liq	-62.5(0.1)	184.4(0.9)	1.0455[20]	1.5175[20]	i H₂O; msc EtOH, eth, ace, peth; s HOAc
2035	2-Chloroethyl chloroformate		$C_3H_4Cl_2O_2$	627-11-2	142.969			155	1.3847[20]	1.4483[20]	i H₂O; s EtOH, eth, ace, bz; sl ctc
2036	1-(2-Chloroethyl)-3-cyclo-hexyl-1-nitrosourea	Lomustine	$C_9H_{16}ClN_3O_2$	13010-47-4	233.695	ye pow	90				i H₂O; s EtOH
2037	N-(2-Chloroethyl)-dibenzylamine	Dibenamine	$C_{16}H_{18}ClN$	51-50-3	259.774	oily liq		169[3]			
2038	N-(2-Chloroethyl)-dibenzylamine hydrochloride	Dibenamine hydrochloride	$C_{16}H_{19}Cl_2N$	55-43-6	296.235	cry	194				i H₂O; s EtOH, dil acid
2039	Chloroethyldimethylsilane		$C_4H_{11}ClSi$	6917-76-6	122.669			89.5	0.8675[20]	1.4105[20]	
2040	2-Chloroethyl ethyl ether		C_4H_9ClO	628-34-2	108.566			98(4)	0.9895[20]	1.4113[20]	sl H₂O; msc eth; s chl
2041	2-Chloroethyl isocyanate		C_3H_4ClNO	1943-83-5	105.523			44[17]			
2042	1-(2-Chloroethyl)-3-(4-methylcyclohexyl)-1-nitro-sourea	Semustine	$C_{10}H_{18}ClN_3O_2$	13909-09-6	247.722	cry	64 dec				
2043	5-(2-Chloroethyl)-4-methylthi-azole	Clomethiazole	C_6H_8ClNS	533-45-9	161.653	oil		92[7]	1.233[25]		
2044	N-(2-Chloroethyl)morpholine		$C_6H_{12}ClNO$	3240-94-6	149.618			42[1]			
2045	4-(2-Chloroethyl)morpholine, hydrochloride		$C_6H_{13}Cl_2NO$	3647-69-6	186.079		185				
2046	1-Chloro-2-(ethylthio)ethane		C_4H_9ClS	693-07-2	124.632			157	1.0663[25]		
2047	2-Chloroethyl vinyl ether		C_4H_7ClO	110-75-8	106.551	liq	-70	108	1.0495[20]	1.4378[20]	vs EtOH, eth; sl chl
2048	3-Chloro-4-fluoroaniline		C_6H_5ClFN	367-21-5	145.562		45.0	227.0			
2049	1-Chloro-2-fluorobenzene		C_6H_4ClF	348-51-6	130.547	liq	-43	137.6	1.2233[30]	1.4918[30]	i H₂O; s ace, bz
2050	1-Chloro-3-fluorobenzene		C_6H_4ClF	625-98-9	130.547			128(25)	1.221[25]	1.4911	
2051	1-Chloro-4-fluorobenzene		C_6H_4ClF	352-33-0	130.547	liq	-26.8	130	1.4990[15]	1.4990[15]	i H₂O; s EtOH, eth, bz
2052	1-Chloro-1-fluoroethane		C_2H_4ClF	1615-75-4	82.504	vol liq or gas		16(5)			
2053	1-Chloro-2-fluoroethane		C_2H_4ClF	762-50-5	82.504			53.1(0.2)	1.1747[20]	1.3775[20]	vs eth, EtOH
2054	Chlorofluoromethane		CH_2ClF	593-70-4	68.478	col gas	-135.1	-9.1			sl H₂O; vs chl
2055	1-Chloro-3-fluoro-2-methyl-benzene		C_7H_6ClF	443-83-4	144.574			154	1.191[25]	1.5026[20]	
2056	2-Chloro-1-fluoro-4-nitroben-zene	3-Chloro-4-fluoronitrobenzene	$C_6H_3ClFNO_2$	350-30-1	175.545		41.5	229.5			

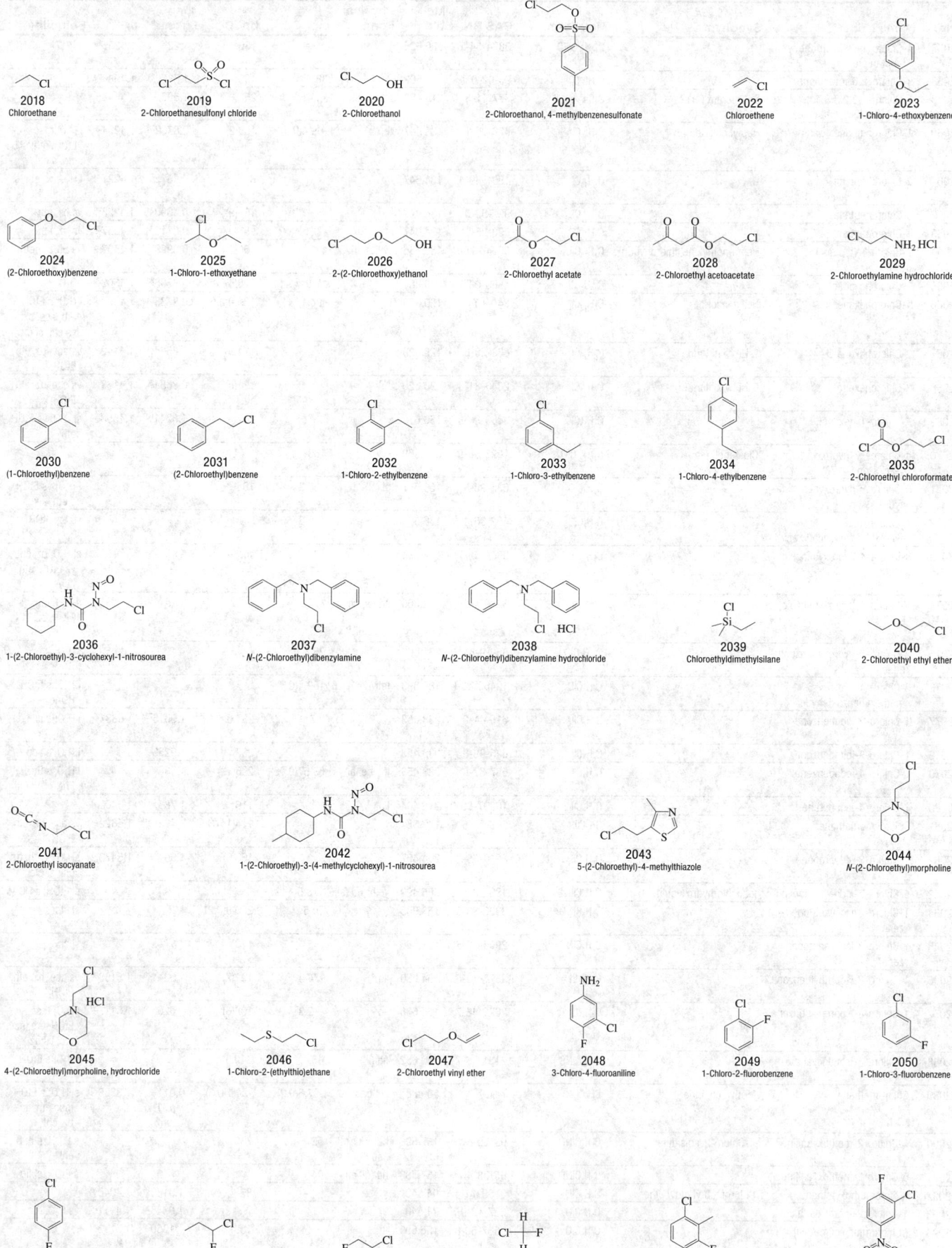

2018
Chloroethane

2019
2-Chloroethanesulfonyl chloride

2020
2-Chloroethanol

2021
2-Chloroethanol, 4-methylbenzenesulfonate

2022
Chloroethene

2023
1-Chloro-4-ethoxybenzene

2024
(2-Chloroethoxy)benzene

2025
1-Chloro-1-ethoxyethane

2026
2-(2-Chloroethoxy)ethanol

2027
2-Chloroethyl acetate

2028
2-Chloroethyl acetoacetate

2029
2-Chloroethylamine hydrochloride

2030
(1-Chloroethyl)benzene

2031
(2-Chloroethyl)benzene

2032
1-Chloro-2-ethylbenzene

2033
1-Chloro-3-ethylbenzene

2034
1-Chloro-4-ethylbenzene

2035
2-Chloroethyl chloroformate

2036
1-(2-Chloroethyl)-3-cyclohexyl-1-nitrosourea

2037
N-(2-Chloroethyl)dibenzylamine

2038
N-(2-Chloroethyl)dibenzylamine hydrochloride

2039
Chloroethyldimethylsilane

2040
2-Chloroethyl ethyl ether

2041
2-Chloroethyl isocyanate

2042
1-(2-Chloroethyl)-3-(4-methylcyclohexyl)-1-nitrosourea

2043
5-(2-Chloroethyl)-4-methylthiazole

2044
N-(2-Chloroethyl)morpholine

2045
4-(2-Chloroethyl)morpholine, hydrochloride

2046
1-Chloro-2-(ethylthio)ethane

2047
2-Chloroethyl vinyl ether

2048
3-Chloro-4-fluoroaniline

2049
1-Chloro-2-fluorobenzene

2050
1-Chloro-3-fluorobenzene

2051
1-Chloro-4-fluorobenzene

2052
1-Chloro-1-fluoroethane

2053
1-Chloro-2-fluoroethane

2054
Chlorofluoromethane

2055
1-Chloro-3-fluoro-2-methylbenzene

2056
2-Chloro-1-fluoro-4-nitrobenzene

No.	Name	Synonym	Mol. Form.	CAS RN	Mol. Wt.	Physical Form	mp/°C	bp/°C	den g cm⁻³	n_D	Solubility
2057	4-Chloro-1-(4-fluorophenyl)-1-butanone		$C_{10}H_{10}ClFO$	3874-54-2	200.636			136[6]	1.22[25]	1.5255[20]	
2058	3-Chloro-2,5-furandione		C_4HClO_3	96-02-6	132.502		33	196	1.5375[25]	1.4980[20]	
2059	1-Chloro-1,2,2,3,3,4,4-heptafluorocyclobutane	Refrigerant C317	C_4ClF_7	377-41-3	216.485	liq or gas	-39.1	25	1.602[15]		
2060	1-Chloroheptane	Heptyl chloride	$C_7H_{15}Cl$	629-06-1	134.647	liq	-69.4(0.4)	159(2)	0.8762[20]	1.4264[20]	i H₂O; msc EtOH, eth; sl ctc; s chl
2061	2-Chloroheptane		$C_7H_{15}Cl$	1001-89-4	134.647			61[32]	0.8672[20]	1.4221[20]	i H₂O; vs eth; s bz, chl, HOAc
2062	3-Chloroheptane		$C_7H_{15}Cl$	999-52-0	134.647			156(9)	0.8690[20]	1.4228[20]	vs bz, eth
2063	4-Chloroheptane		$C_7H_{15}Cl$	998-95-8	134.647			156(9)	0.8710[20]	1.4237[20]	vs bz, eth
2064	7-Chloro-1-heptanol	Heptamethylene chlorohydrin	$C_7H_{15}ClO$	55944-70-2	150.646	cry (peth, bz)	11	150[20]	0.9998[15]	1.4537[25]	vs EtOH, peth
2065	1-Chlorohexadecane		$C_{16}H_{33}Cl$	4860-03-1	260.886		17.9	326(4)	0.8635[20]	1.4503[20]	i H₂O
2066	1-Chlorohexane	Hexyl chloride	$C_6H_{13}Cl$	544-10-5	120.620	liq	-94(1)	135.0(0.5)	0.8738[25]	1.4200[20]	i H₂O; s EtOH, eth, ace, bz; vs chl; sl ctc
2067	2-Chlorohexane	2-Hexyl chloride	$C_6H_{13}Cl$	638-28-8	120.620			124(5)	0.8694[21]	1.4142[22]	vs ace, bz, eth, EtOH
2068	3-Chlorohexane	3-Hexyl chloride	$C_6H_{13}Cl$	2346-81-8	120.620			121(5)	0.8684[20]	1.4163[20]	vs ace, bz, eth, EtOH
2069	6-Chloro-1-hexanol		$C_6H_{13}ClO$	2009-83-8	136.619			107[12]	1.0241[20]	1.4550[20]	sl H₂O; vs EtOH, eth
2070	4-Chloro-17-hydroxyandrost-4-en-3-one, (17β)	Clostebol	$C_{19}H_{27}ClO_2$	1093-58-9	322.869		189				
2071	5-Chloro-2-hydroxybenzaldehyde		$C_7H_5ClO_2$	635-93-8	156.567	pl (al)	100.3	105[12]			i H₂O; vs EtOH; s eth, alk
2072	4-Chloro-α-hydroxybenzeneacetic acid		$C_8H_7ClO_3$	492-86-4	186.593		120.3				vs bz, EtOH
2073	3-Chloro-4-hydroxybenzoic acid		$C_7H_5ClO_3$	3964-58-7	172.566	nd (w)	171	sub			sl H₂O, bz, chl; vs EtOH, eth, ace
2074	5-Chloro-2-hydroxybenzoic acid		$C_7H_5ClO_3$	321-14-2	172.566	nd (w, al)	174.8				s H₂O, eth; vs EtOH, bz; sl ace
2075	2-Chloro-5-hydroxybenzophenone		$C_{13}H_9ClO_2$	85-19-8	232.662		95.3				i H₂O
2076	3-Chloro-4-hydroxy-5-methoxybenzaldehyde		$C_8H_7ClO_3$	19463-48-0	186.593	tetr	169.4(0.6)				i H₂O; s EtOH, HOAc
2077	1-Chloro-2-iodobenzene		C_6H_4ClI	615-41-8	238.453		0.7	234.5	1.9515[25]	1.6331[25]	i H₂O; s ace; sl ctc
2078	1-Chloro-3-iodobenzene		C_6H_4ClI	625-99-0	238.453			230	1.9255[20]		i H₂O; s ace
2079	1-Chloro-4-iodobenzene		C_6H_4ClI	637-87-6	238.453	lf (ace, al)	53.57(0.05)	226(19)	1.886[27]		i H₂O; s EtOH, PhNO₂; sl chl
2080	1-Chloro-4-iodobutane		C_4H_8ClI	10297-05-9	218.464	liq		116	1.785	1.5400[20]	
2081	Chloroiodomethane		CH_2ClI	593-71-5	176.384			114(14)	2.422[20]	1.5822[20]	vs ace, bz, eth, EtOH
2082	1-Chloro-3-iodopropane		C_3H_6ClI	6940-76-7	204.437			163(6)	1.904[20]	1.5472[20]	i H₂O; s eth, bz, chl; sl ctc
2083	5-Chloro-7-iodo-8-quinolinol	Iodochlorhydroxyquin	C_9H_5ClINO	130-26-7	305.499	ye br nd (al)	178.5				sl EtOH; s HOAc
2084	1-Chloro-2-isocyanatobenzene		C_7H_4ClNO	3320-83-0	153.566		30.5	200.9(0.2)			sl ctc
2085	1-Chloro-3-isocyanatobenzene		C_7H_4ClNO	2909-38-8	153.566			113[43]			sl chl
2086	1-Chloro-2-isopropylbenzene		$C_9H_{11}Cl$	2077-13-6	154.636	liq	-74.4	191(7)	1.0341[20]	1.5168[20]	vs ace, bz, eth, EtOH
2087	1-Chloro-4-isopropylbenzene		$C_9H_{11}Cl$	2621-46-7	154.636	liq	-12.3	193(6)	1.0208[20]	1.5117[20]	i H₂O; msc EtOH, eth, ace, ctc; vs bz
2088	1-Chloro-4-isothiocyanatobenzene		C_7H_4ClNS	2131-55-7	169.632	nd (al)	46	249.5			i H₂O; s EtOH
2089	Chloromethane	Methyl chloride	CH_3Cl	74-87-3	50.488	col gas	-97.6(0.2)	-24.1(0.3)	0.911[25] (p>1 atm)	1.3389[20]	sl H₂O; s EtOH; msc eth, ace, bz, chl
2090	4-Chloro-2-methoxyaniline	4-Chloro-2-anisidine	C_7H_8ClNO	93-50-5	157.598	nd or pr (dil al)	52	260			s EtOH, eth, bz, chl
2091	5-Chloro-2-methoxyaniline		C_7H_8ClNO	95-03-4	157.598	nd (dil al)	84				s EtOH; sl lig
2092	(Chloromethoxy)ethane	Chloromethyl ethyl ether	C_3H_7ClO	3188-13-4	94.540			83	1.0188[15]	1.4040[20]	
2093	1-Chloro-2-methoxyethane		C_3H_7ClO	627-42-9	94.540			79(13)	1.0345[20]	1.4111[20]	vs H₂O, eth
2094	[(Chloromethoxy)methyl]-benzene		C_8H_9ClO	3587-60-8	156.609			103[13]	1.1350[20]	1.5192[20]	
2095	1-(Chloromethoxy)propane		C_4H_9ClO	3587-57-3	108.566			109	0.9884[20]	1.4125[20]	vs eth, EtOH

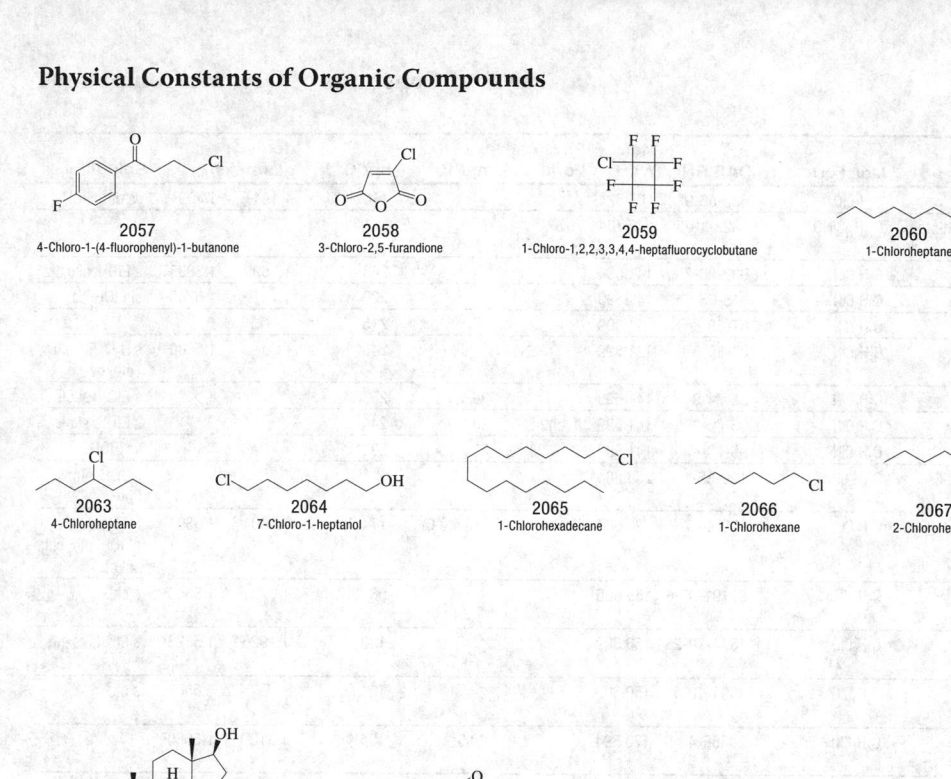

2057
4-Chloro-1-(4-fluorophenyl)-1-butanone

2058
3-Chloro-2,5-furandione

2059
1-Chloro-1,2,2,3,3,4,4-heptafluorocyclobutane

2060
1-Chloroheptane

2061
2-Chloroheptane

2062
3-Chloroheptane

2063
4-Chloroheptane

2064
7-Chloro-1-heptanol

2065
1-Chlorohexadecane

2066
1-Chlorohexane

2067
2-Chlorohexane

2068
3-Chlorohexane

2069
6-Chloro-1-hexanol

2070
4-Chloro-17-hydroxyandrost-4-en-3-one, (17β)

2071
5-Chloro-2-hydroxybenzaldehyde

2072
4-Chloro-α-hydroxybenzeneacetic acid

2073
3-Chloro-4-hydroxybenzoic acid

2074
5-Chloro-2-hydroxybenzoic acid

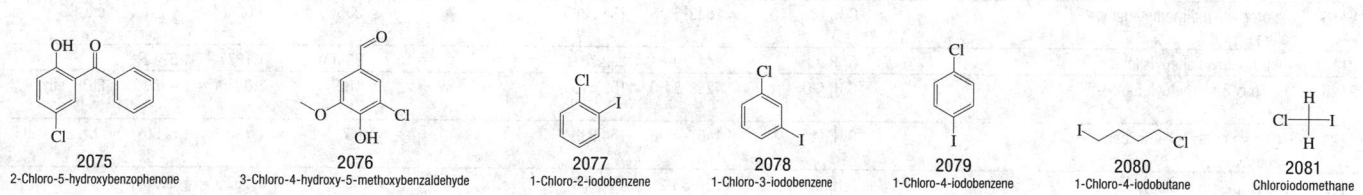

2075
2-Chloro-5-hydroxybenzophenone

2076
3-Chloro-4-hydroxy-5-methoxybenzaldehyde

2077
1-Chloro-2-iodobenzene

2078
1-Chloro-3-iodobenzene

2079
1-Chloro-4-iodobenzene

2080
1-Chloro-4-iodobutane

2081
Chloroiodomethane

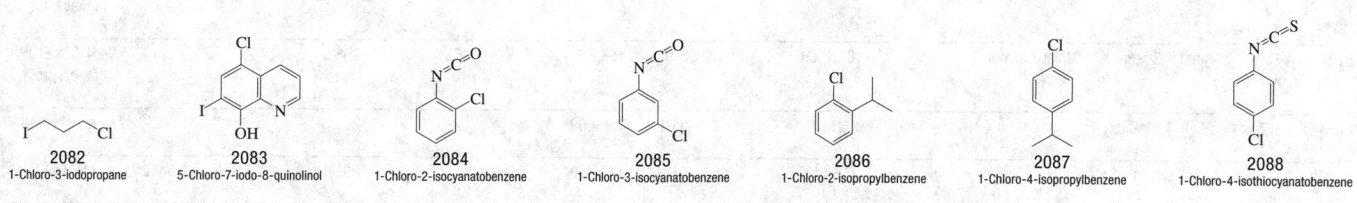

2082
1-Chloro-3-iodopropane

2083
5-Chloro-7-iodo-8-quinolinol

2084
1-Chloro-2-isocyanatobenzene

2085
1-Chloro-3-isocyanatobenzene

2086
1-Chloro-2-isopropylbenzene

2087
1-Chloro-4-isopropylbenzene

2088
1-Chloro-4-isothiocyanatobenzene

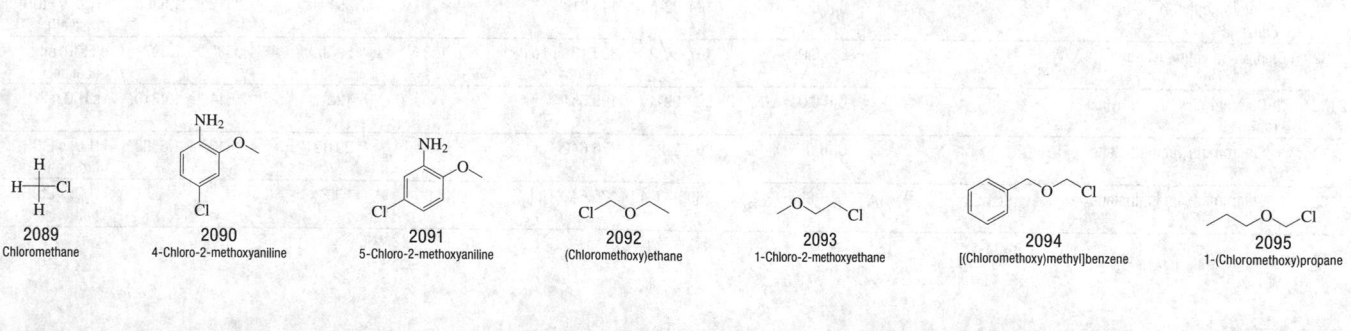

2089
Chloromethane

2090
4-Chloro-2-methoxyaniline

2091
5-Chloro-2-methoxyaniline

2092
(Chloromethoxy)ethane

2093
1-Chloro-2-methoxyethane

2094
[(Chloromethoxy)methyl]benzene

2095
1-(Chloromethoxy)propane

No.	Name	Synonym	Mol. Form.	CAS RN	Mol. Wt.	Physical Form	mp/°C	bp/°C	den g cm⁻³	n_D	Solubility
2096	Chloromethyl acetate		$C_3H_5ClO_2$	625-56-9	108.524			116	1.194[20]	1.409[20]	vs eth, EtOH
2097	5-Chloro-2-(methylamino)-benzophenone	N-Methyl-2-amino-5-chloro-benzophenone	$C_{14}H_{12}ClNO$	1022-13-5	245.704		92				
2098	4-Chloro-N-methylaniline		C_7H_8ClN	932-96-7	141.599			240	1.169[11]	1.5835[20]	s EtOH, ace, bz
2099	2-Chloro-4-methylaniline		C_7H_8ClN	615-65-6	141.599		7	220	1.151[20]	1.5748[22]	sl EtOH, bz
2100	2-Chloro-6-methylaniline		C_7H_8ClN	87-63-8	141.599			215			
2101	3-Chloro-2-methylaniline		C_7H_8ClN	87-60-5	141.599		1	245		1.5880[20]	s H_2O, EtOH; i eth, bz
2102	3-Chloro-4-methylaniline		C_7H_8ClN	95-74-9	141.599		26	233(3)			s EtOH; sl ctc
2103	4-Chloro-2-methylaniline	p-Chloro-o-toluidine	C_7H_8ClN	95-69-2	141.599	lf (al)	30.3	244			s EtOH; sl ctc
2104	5-Chloro-2-methylaniline		C_7H_8ClN	95-79-4	141.599		26	239			vs EtOH
2105	1-Chloro-2-methyl-9,10-anthracenedione		$C_{15}H_9ClO_2$	129-35-1	256.684		170.5				i EtOH, eth; sl py
2106	(Chloromethyl)benzene	Benzyl chloride	C_7H_7Cl	100-44-7	126.584	liq	-39.4(0.6)	174(7)	1.1004[20]	1.5391[20]	i H_2O; msc EtOH, eth, chl; sl ctc
2107	3-Chloro-N-methylbenzene-methanamine		$C_8H_{10}ClN$	39191-07-6	155.625			88[4]		1.5350[25]	s chl
2108	α-(Chloromethyl)-benzenemethanol		C_8H_9ClO	1674-30-2	156.609			128[17]	1.1926[20]	1.5523[20]	s EtOH; vs eth
2109	4-Chloro-α-methylbenzenemethanol		C_8H_9ClO	3391-10-4	156.609			121[15]		1.5505[20]	s ctc
2110	5-(Chloromethyl)-1,3-benzodioxole		$C_8H_7ClO_2$	20850-43-5	170.594		20.5	134[14]	1.312[25]	1.5660[20]	
2111	1-Chloro-3-methylbutane	Isopentyl chloride	$C_5H_{11}Cl$	107-84-6	106.594	liq	-104.4	99(2)	0.8750[20]	1.4084[20]	sl H_2O; msc EtOH, eth; vs chl
2112	2-Chloro-2-methylbutane		$C_5H_{11}Cl$	594-36-5	106.594	liq	-72.6(0.5)	85(1)	0.8653[20]	1.4055[20]	sl H_2O; s EtOH, eth, ctc
2113	2-Chloro-3-methylbutane		$C_5H_{11}Cl$	631-65-2	106.594			92(2)	0.878[20]		
2114	1-Chloro-3-methyl-2-butene		C_5H_9Cl	503-60-6	104.578			109	0.9273[20]	1.4485[20]	vs ace, eth, EtOH, chl
2115	3-Chloro-3-methyl-1-butyne		C_5H_7Cl	1111-97-3	102.563	liq	-61	76	0.9061[20]		
2116	(Chloromethyl)cyclopropane		C_4H_7Cl	5911-08-0	90.552	liq	-90.9	87(6)	0.98[25]	1.4350[20]	
2117	1-(Chloromethyl)-2,4-dimethylbenzene		$C_9H_{11}Cl$	824-55-5	154.636			233(18)	1.0580[19]		vs bz, eth, EtOH
2118	(Chloromethyl)dimethylphenyl-silane		$C_9H_{13}ClSi$	1833-51-8	184.738			225	1.0240[25]		s ctc, CS_2
2119	Chloromethyldiphenylsilane		$C_{13}H_{13}ClSi$	144-79-6	232.781			296(3)	1.1277[20]	1.5742[20]	
2120	1-Chloro-3-(1-methylethoxy)-2-propanol		$C_6H_{13}ClO_2$	4288-84-0	152.619			182	1.0910[20]	1.4370[25]	s EtOH, eth
2121	1-(Chloromethyl)-4-ethylbenzene		$C_9H_{11}Cl$	1467-05-6	154.636			95[15]		1.5290[25]	vs bz, EtOH, chl
2122	(1-Chloro-1-methylethyl)-benzene		$C_9H_{11}Cl$	934-53-2	154.636			98[1]	1.192[25]	1.5290[25]	
2123	1-(Chloromethyl)-2-fluorobenzene		C_7H_6ClF	345-35-7	144.574			172	1.216[25]	1.5150[20]	
2124	1-(Chloromethyl)-4-fluorobenzene		C_7H_6ClF	352-11-4	144.574			82[26]	1.2143[20]	1.5130	
2125	2-(Chloromethyl)furan		C_5H_5ClO	617-88-9	116.546			49[26]	1.1783[20]	1.4941[20]	vs bz, eth, EtOH
2126	3-(Chloromethyl)heptane	2-Ethylhexyl chloride	$C_8H_{17}Cl$	123-04-6	148.674			171(3)	0.8769[20]	1.4319[20]	i H_2O; s EtOH, eth, ace, bz; sl ctc
2127	4-Chloro-5-methyl-2-isopropylphenol	Chlorothymol	$C_{10}H_{13}ClO$	89-68-9	184.662		63	258.5			vs H_2O; s EtOH, eth, bz, ctc, peth, alk
2128	1-(Chloromethyl)-4-methoxybenzene		C_8H_9ClO	824-94-2	156.609	nd	24.5	262.5	1.261[20]	1.580[20]	vs ace, bz, eth
2129	1-(Chloromethyl)-2-methylbenzene		C_8H_9Cl	552-45-4	140.610			202(13)	1.063[25]	1.5410[25]	vs eth, EtOH
2130	1-(Chloromethyl)-3-methylbenzene		C_8H_9Cl	620-19-9	140.610			195.5	1.064[20]	1.5345[20]	i H_2O; s EtOH, eth
2131	1-(Chloromethyl)-4-methylbenzene		C_8H_9Cl	104-82-5	140.610			201	1.0512[20]	1.5380	i H_2O; s EtOH; msc eth
2132	Chloromethyl methyl ether		C_2H_5ClO	107-30-2	80.513	liq	-103.5	59(3)	1.063[10]	1.397[20]	s EtOH, eth, ace, chl
2133	2-(Chloromethyl)-2-methyloxirane		C_4H_7ClO	598-09-4	106.551			122	1.1011[20]	1.4310[20]	vs H_2O, eth
2134	1-(Chloromethyl)naphthalene		$C_{11}H_9Cl$	86-52-2	176.642	pr	32	291(21)	1.1813[20]	1.6380[20]	i H_2O; s EtOH, ctc, peth
2135	2-(Chloromethyl)naphthalene		$C_{11}H_9Cl$	2506-41-4	176.642	lf (al)	48.5	169[20]			i H_2O; s EtOH, peth

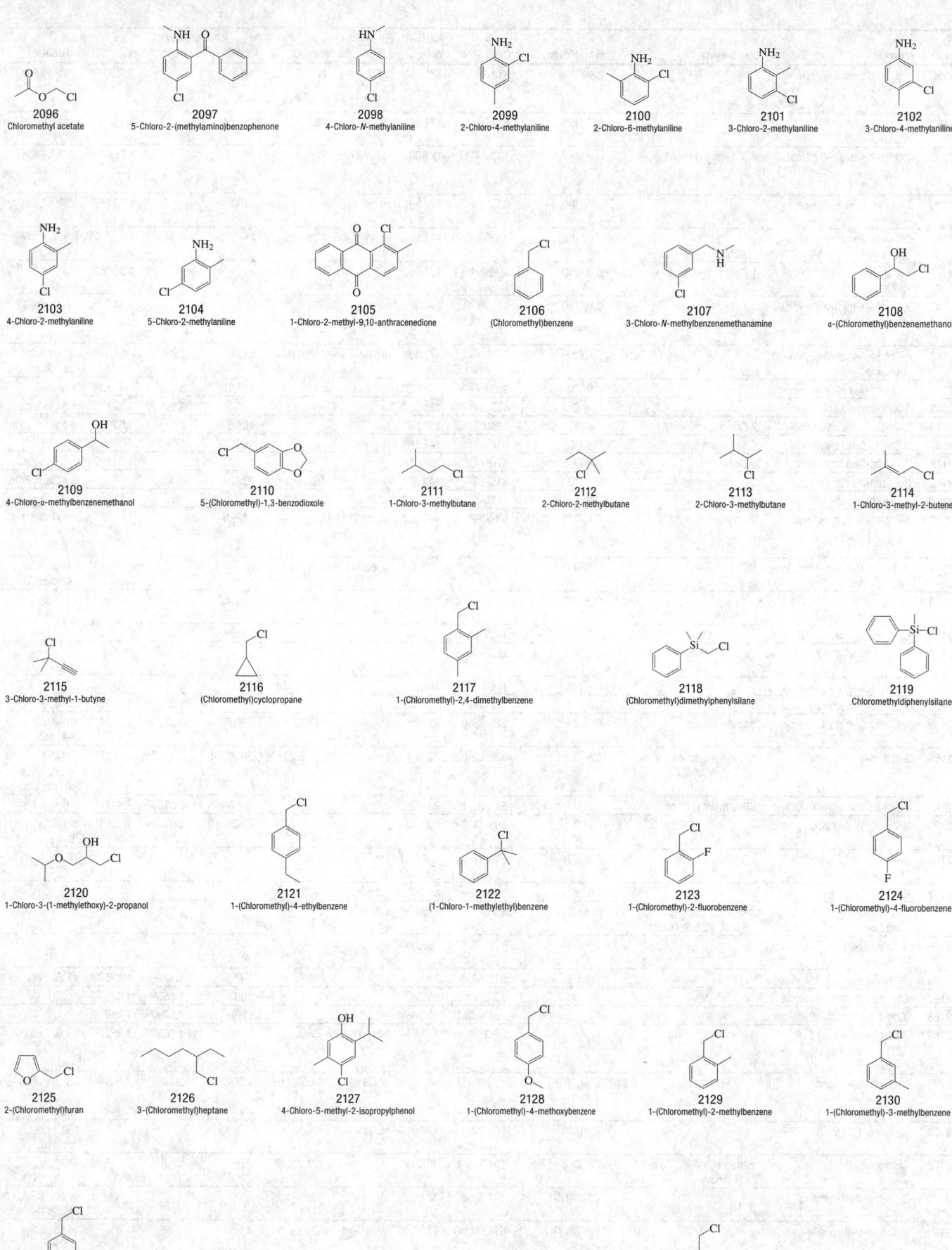

2096
Chloromethyl acetate

2097
5-Chloro-2-(methylamino)benzophenone

2098
4-Chloro-*N*-methylaniline

2099
2-Chloro-4-methylaniline

2100
2-Chloro-6-methylaniline

2101
3-Chloro-2-methylaniline

2102
3-Chloro-4-methylaniline

2103
4-Chloro-2-methylaniline

2104
5-Chloro-2-methylaniline

2105
1-Chloro-2-methyl-9,10-anthracenedione

2106
(Chloromethyl)benzene

2107
3-Chloro-*N*-methylbenzenemethanamine

2108
α-(Chloromethyl)benzenemethanol

2109
4-Chloro-α-methylbenzenemethanol

2110
5-(Chloromethyl)-1,3-benzodioxole

2111
1-Chloro-3-methylbutane

2112
2-Chloro-2-methylbutane

2113
2-Chloro-3-methylbutane

2114
1-Chloro-3-methyl-2-butene

2115
3-Chloro-3-methyl-1-butyne

2116
(Chloromethyl)cyclopropane

2117
1-(Chloromethyl)-2,4-dimethylbenzene

2118
(Chloromethyl)dimethylphenylsilane

2119
Chloromethyldiphenylsilane

2120
1-Chloro-3-(1-methylethoxy)-2-propanol

2121
1-(Chloromethyl)-4-ethylbenzene

2122
(1-Chloro-1-methylethyl)benzene

2123
1-(Chloromethyl)-2-fluorobenzene

2124
1-(Chloromethyl)-4-fluorobenzene

2125
2-(Chloromethyl)furan

2126
3-(Chloromethyl)heptane

2127
4-Chloro-5-methyl-2-isopropylphenol

2128
1-(Chloromethyl)-4-methoxybenzene

2129
1-(Chloromethyl)-2-methylbenzene

2130
1-(Chloromethyl)-3-methylbenzene

2131
1-(Chloromethyl)-4-methylbenzene

2132
Chloromethyl methyl ether

2133
2-(Chloromethyl)-2-methyloxirane

2134
1-(Chloromethyl)naphthalene

2135
2-(Chloromethyl)naphthalene

No.	Name	Synonym	Mol. Form.	CAS RN	Mol. Wt.	Physical Form	mp/°C	bp/°C	den g cm⁻³	n_D	Solubility
2136	1-(Chloromethyl)-2-nitrobenzene		$C_7H_6ClNO_2$	612-23-7	171.582	cry (liq)	47.7(0.2)	125[4]		1.5557[62]	i H$_2$O; s EtOH, eth, HOAc; vs ace, bz
2137	1-(Chloromethyl)-3-nitrobenzene		$C_7H_6ClNO_2$	619-23-8	171.582	pa ye nd (liq)	46	173[34]		1.5577[62]	vs ace, bz, eth, EtOH
2138	1-(Chloromethyl)-4-nitrobenzene	4-Nitrobenzyl chloride	$C_7H_6ClNO_2$	100-14-1	171.582	pl or nd (al)	71			1.5647[62]	i H$_2$O; s EtOH, eth; vs ace, bz, AcOEt
2139	1-Chloro-2-methyl-3-nitrobenzene		$C_7H_6ClNO_2$	83-42-1	171.582	nd (dil al)	37.8	238		1.5377[69]	i H$_2$O; s EtOH
2140	1-Chloro-2-methyl-4-nitrobenzene		$C_7H_6ClNO_2$	13290-74-9	171.582	ye cry	42.5	249			vs eth
2141	1-Chloro-4-methyl-2-nitrobenzene	4-Chloro-3-nitrotoluene	$C_7H_6ClNO_2$	89-60-1	171.582		7.2(0.2)	261		1.5572[20]	i H$_2$O; s ctc
2142	2-Chloro-1-methyl-4-nitrobenzene		$C_7H_6ClNO_2$	121-86-8	171.582	nd (al)	66.5	260		1.5470[69]	sl H$_2$O, chl; s EtOH, eth, HOAc
2143	4-Chloro-1-methyl-2-nitrobenzene		$C_7H_6ClNO_2$	89-59-8	171.582	mcl nd	36.5(1)	242	1.2559[80]		i H$_2$O; s EtOH, eth; sl chl
2144	2-Chloro-4-methylpentane		$C_6H_{13}Cl$	25346-32-1	120.620			116(10)	0.8610[20]	1.4113[20]	vs eth
2145	3-(Chloromethyl)pentane		$C_6H_{13}Cl$	4737-41-1	120.620			126(5)	0.8914[20]	1.4222[20]	vs bz, eth, chl
2146	2-Chloro-4-methylphenol	2-Chloro-p-cresol	C_7H_7ClO	6640-27-3	142.583			195.5	1.1785[27]	1.5200[27]	vs bz, eth, EtOH
2147	2-Chloro-5-methylphenol	6-Chloro-m-cresol	C_7H_7ClO	615-74-7	142.583	pr (peth)	46(3)	198(8)	1.215[15]		vs H$_2$O, EtOH
2148	2-Chloro-6-methylphenol	6-Chloro-o-cresol	C_7H_7ClO	87-64-9	142.583			189		1.5449[20]	sl H$_2$O; s eth
2149	3-Chloro-4-methylphenol	3-Chloro-p-cresol	C_7H_7ClO	615-62-3	142.583	nd (al)	55.5	228			vs bz, eth, EtOH
2150	4-Chloro-2-methylphenol	4-Chloro-o-cresol	C_7H_7ClO	1570-64-5	142.583	nd (peth)	51	223			sl H$_2$O; s peth
2151	4-Chloro-3-methylphenol	4-Chloro-m-cresol	C_7H_7ClO	59-50-7	142.583	nd (peth)	55(2)	232(9)			sl H$_2$O, chl; s EtOH, eth, peth
2152	(4-Chloro-2-methylphenoxy)acetic acid	MCPA	$C_9H_9ClO_3$	94-74-6	200.618	pl (bz, to)	119.7(0.9)				sl H$_2$O; vs EtOH, eth; s bz, ctc
2153	4-(4-Chloro-2-methylphenoxy)butanoic acid		$C_{11}H_{13}ClO_3$	94-81-5	228.672		100.4(0.5)				
2154	Chloromethylphenylsilane		C_7H_9ClSi	1631-82-9	156.685			113[100]	1.043[20]	1.5171[20]	
2155	(Chloromethyl)phosphonic acid		CH_4ClO_3P	2565-58-4	130.468	nd (bz/MeNO$_2$)	90				
2156	N-Chloromethylphthalimide		$C_9H_6ClNO_2$	17564-64-6	195.603		133.5(0.5)				
2157	2-Chloro-2-methylpropanal		C_4H_7ClO	917-93-1	106.551			90	1.053[15]	1.4160[16]	vs eth, EtOH
2158	1-Chloro-2-methylpropane	Isobutyl chloride	C_4H_9Cl	513-36-0	92.567	liq	-130.3	69(1)	0.8773[20]	1.3984[20]	sl H$_2$O, ctc; s eth, ace, chl
2159	2-Chloro-2-methylpropane	tert-Butyl chloride	C_4H_9Cl	507-20-0	92.567	liq	-25.60(0.02)	50.9(0.5)	0.8420[20]	1.3857[20]	sl H$_2$O; msc EtOH, eth; s bz, ctc, chl
2160	1-Chloro-2-methylpropene	Dimethylvinyl chloride	C_4H_7Cl	513-37-1	90.552			68(4)	0.9186[20]	1.4221[20]	sl H$_2$O; s chl
2161	3-Chloro-2-methylpropene		C_4H_7Cl	563-47-3	90.552			72(2)	0.9165[20]	1.4291[20]	msc EtOH, eth; s ace; vs chl
2162	3-(Chloromethyl)pyridine, hydrochloride		$C_6H_7Cl_2N$	6959-48-4	164.033	hyg	143.8				
2163	Chloromethylsilane		CH_5ClSi	993-00-0	80.590	col gas	-135	7			
2164	1-Chloro-4-(methylsulfonyl)benzene	4-Chlorobenzenethiol, S-methyl, S,S-dioxide	$C_7H_7ClO_2S$	98-57-7	190.648		98				
2165	1-Chloro-4-(methylthio)benzene		C_7H_7ClS	123-09-1	158.649			105[10]			
2166	1-Chloro-2-(methylthio)ethane		C_3H_7ClS	542-81-4	110.606			140	1.123[20]	1.4902[20]	s EtOH, eth, ace
2167	Chloro(methylthio)methane		C_2H_5ClS	2373-51-5	96.579			105	1.153[25]	1.4963[20]	
2168	(Chloromethyl)trimethylsilane		$C_4H_{11}ClSi$	2344-80-1	122.669			98.5	0.879[25]	1.4175[20]	
2169	1-Chloronaphthalene	1-Naphthyl chloride	$C_{10}H_7Cl$	90-13-1	162.616	oily liq	-6.0(0.2)	259(2)	1.1880[25]	1.6326[20]	i H$_2$O; s EtOH, eth, bz, CS$_2$; sl ctc
2170	2-Chloronaphthalene		$C_{10}H_7Cl$	91-58-7	162.616	pl (dil al), lf	58.02(0.05)	257(21)	1.1377[71]	1.6079[13]	i H$_2$O; s EtOH, eth, bz, chl, CS$_2$
2171	4-Chloro-1-naphthol		$C_{10}H_7ClO$	604-44-4	178.615	nd (chl, aq al)	120.5				s EtOH, eth, ace, bz, chl
2172	Chloroneb	Benzene, 1,4-dichloro-2,5-dimethoxy-	$C_8H_8Cl_2O_2$	2675-77-6	207.055		131.2(0.3)	268			
2173	2-Chloro-4-nitroaniline		$C_6H_5ClN_2O_2$	121-87-9	172.569	ye nd (w)	108.4(0.5)				vs eth, EtOH, HOAc
2174	2-Chloro-5-nitroaniline		$C_6H_5ClN_2O_2$	6283-25-6	172.569	ye nd (liq)	120.8(0.5)				vs eth, EtOH, HOAc
2175	4-Chloro-2-nitroaniline		$C_6H_5ClN_2O_2$	89-63-4	172.569	dk oran-ye pr (dil al)	116.5				vs EtOH, eth, HOAc; sl ace, liq

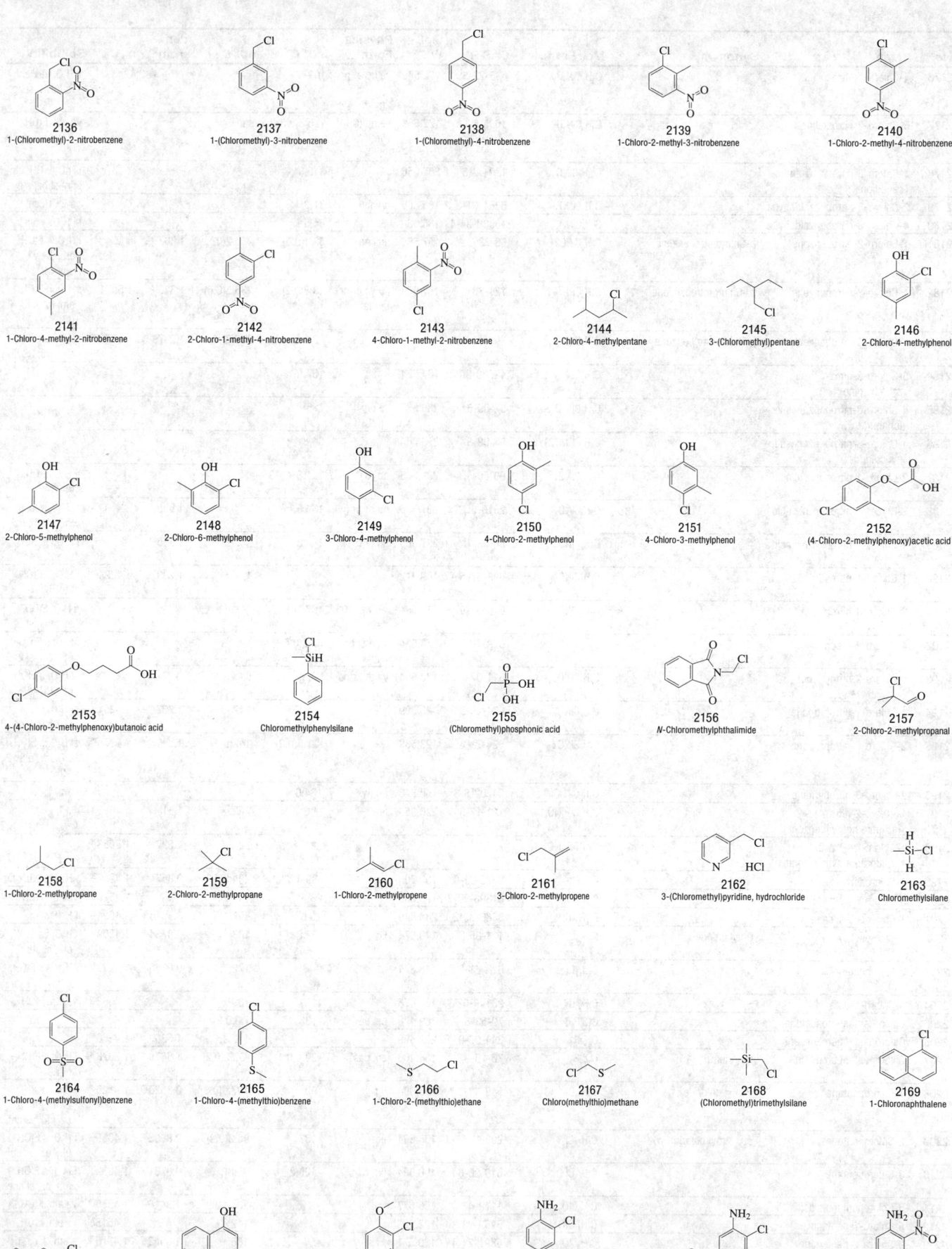

2136
1-(Chloromethyl)-2-nitrobenzene

2137
1-(Chloromethyl)-3-nitrobenzene

2138
1-(Chloromethyl)-4-nitrobenzene

2139
1-Chloro-2-methyl-3-nitrobenzene

2140
1-Chloro-2-methyl-4-nitrobenzene

2141
1-Chloro-4-methyl-2-nitrobenzene

2142
2-Chloro-1-methyl-4-nitrobenzene

2143
4-Chloro-1-methyl-2-nitrobenzene

2144
2-Chloro-4-methylpentane

2145
3-(Chloromethyl)pentane

2146
2-Chloro-4-methylphenol

2147
2-Chloro-5-methylphenol

2148
2-Chloro-6-methylphenol

2149
3-Chloro-4-methylphenol

2150
4-Chloro-2-methylphenol

2151
4-Chloro-3-methylphenol

2152
(4-Chloro-2-methylphenoxy)acetic acid

2153
4-(4-Chloro-2-methylphenoxy)butanoic acid

2154
Chloromethylphenylsilane

2155
(Chloromethyl)phosphonic acid

2156
N-Chloromethylphthalimide

2157
2-Chloro-2-methylpropanal

2158
1-Chloro-2-methylpropane

2159
2-Chloro-2-methylpropane

2160
1-Chloro-2-methylpropene

2161
3-Chloro-2-methylpropene

2162
3-(Chloromethyl)pyridine, hydrochloride

2163
Chloromethylsilane

2164
1-Chloro-4-(methylsulfonyl)benzene

2165
1-Chloro-4-(methylthio)benzene

2166
1-Chloro-2-(methylthio)ethane

2167
Chloro(methylthio)methane

2168
(Chloromethyl)trimethylsilane

2169
1-Chloronaphthalene

2170
2-Chloronaphthalene

2171
4-Chloro-1-naphthol

2172
Chloroneb

2173
2-Chloro-4-nitroaniline

2174
2-Chloro-5-nitroaniline

2175
4-Chloro-2-nitroaniline

No.	Name	Synonym	Mol. Form.	CAS RN	Mol. Wt.	Physical Form	mp/°C	bp/°C	den g cm⁻³	n_D	Solubility
2176	4-Chloro-3-nitroaniline		$C_6H_5ClN_2O_2$	635-22-3	172.569	ye nd or pr (w) nd (peth)	103				s H₂O, eth, chl; vs EtOH; sl lig
2177	5-Chloro-2-nitroaniline		$C_6H_5ClN_2O_2$	1635-61-6	172.569	ye nd (CS₂) ye lf (al, bz)	125.8(0.5)	sub			vs eth, EtOH
2178	1-Chloro-5-nitro-9,10-anthra-cenedione		$C_{14}H_6ClNO_4$	129-40-8	287.656		315.3				i H₂O, EtOH, eth, lig; sl bz; s py
2179	2-Chloro-5-nitrobenzaldehyde		$C_7H_4ClNO_3$	6361-21-3	185.565	cry (al)	81.3				vs EtOH, chl
2180	4-Chloro-3-nitrobenzaldehyde		$C_7H_4ClNO_3$	16588-34-4	185.565		64.5				sl H₂O; s chl
2181	1-Chloro-2-nitrobenzene	o-Chloronitrobenzene	$C_6H_4ClNO_2$	88-73-3	157.555	mcl nd	32.1(0.3)	246.2(0.7)	1.368²⁴²		i H₂O; s EtOH, eth, bz; vs ace, tol, py
2182	1-Chloro-3-nitrobenzene	m-Chloronitrobenzene	$C_6H_4ClNO_2$	121-73-3	157.555	pa ye orth pr (al)	43.6(0.2)	236.5(0.6)	1.343⁵⁰	1.5374⁸⁰	i H₂O; s EtOH, eth, bz, chl, CS₂
2183	1-Chloro-4-nitrobenzene	p-Chloronitrobenzene	$C_6H_4ClNO_2$	100-00-5	157.555	mcl pr	82.2(0.7)	238(3)	1.2979⁹⁰	1.5376¹⁰⁰	i H₂O; sl EtOH; s eth, chl, CS₂
2184	5-Chloro-3-nitro-1,2-ben-zenediamine		$C_6H_6ClN_3O_2$	42389-30-0	187.584		167				
2185	4-Chloro-3-nitrobenzenesul-fonamide		$C_6H_5ClN_2O_4S$	97-09-6	236.633	ye cry (EtOH)	175				
2186	4-Chloro-3-nitrobenzenesulfo-nyl chloride		$C_6H_3Cl_2NO_4S$	97-08-5	256.064		60.8				
2187	2-Chloro-4-nitrobenzoic acid		$C_7H_4ClNO_4$	99-60-5	201.565	nd (w)	141.8				s H₂O, EtOH, eth, bz
2188	2-Chloro-5-nitrobenzoic acid		$C_7H_4ClNO_4$	2516-96-3	201.565	nd or pr (w)	166.5		1.608¹⁸		sl H₂O, ace; s EtOH, eth, bz
2189	4-Chloro-3-nitrobenzoic acid		$C_7H_4ClNO_4$	96-99-1	201.565	nd or pl (w)	182.8		1.645¹⁸		i H₂O; sl EtOH, ace
2190	1-Chloro-1-nitroethane		$C_2H_4ClNO_2$	598-92-5	109.512			124.5	1.2837²⁰	1.4224²⁰	i H₂O; s EtOH, ctc, alk
2191	2-Chloro-4-nitrophenol		$C_6H_4ClNO_3$	619-08-9	173.554	wh nd (50% al)	107.5(0.4)				s H₂O, EtOH, eth, chl; sl bz
2192	4-Chloro-2-nitrophenol		$C_6H_4ClNO_3$	89-64-5	173.554	ye mcl pr (al)	87.2(0.4)				i H₂O; s EtOH, eth, chl; sl ace
2193	5-Chloro-2-nitrophenol		$C_6H_4ClNO_3$	611-07-4	173.554	ye pr or nd (w)	39(3)	sub			sl H₂O; s EtOH, eth, HOAc
2194	1-Chloro-1-nitropropane		$C_3H_6ClNO_2$	600-25-9	123.539			142	1.207²⁰	1.4251²⁰	sl H₂O, chl; s EtOH, eth, oils
2195	2-Chloro-2-nitropropane		$C_3H_6ClNO_2$	594-71-8	123.539		-21.5(0.2)	142(6)	1.20²⁰	1.4378¹⁹	sl H₂O; s EtOH, eth, ctc, oils; i KOH
2196	2-Chloro-3-nitropyridine		$C_5H_3ClN_2O_2$	5470-18-8	158.543	nd (w)	104.0				
2197	1-Chloro-2-nitro-4-(trifluoromethyl)benzene		$C_7H_3ClF_3NO_2$	121-17-5	225.553	liq	-1.3	222	1.511²⁵	1.4893²⁰	
2198	1-Chloro-4-nitro-2-(trifluoromethyl)benzene		$C_7H_3ClF_3NO_2$	777-37-7	225.553		22	232	1.527²⁵	1.5083²⁶	
2199	1-Chlorononane		$C_9H_{19}Cl$	2473-01-0	162.700	liq	-39.4	204(3)	0.8674²⁵	1.4343²⁰	i H₂O; s eth, chl
2200	9-Chloro-1-nonanol		$C_9H_{19}ClO$	51308-99-7	178.699		28	147¹⁴		1.4575²⁰	vs eth, EtOH
2201	1-Chlorooctadecane		$C_{18}H_{37}Cl$	3386-33-2	288.940		28.6	337(15)	0.8616²⁰	1.4524²⁰	i H₂O; sl ctc
2202	1-Chlorooctane	Octyl chloride	$C_8H_{17}Cl$	111-85-3	148.674	liq	-57.8	183(3)	0.8734²⁰	1.4309²⁰	i H₂O; vs EtOH, eth; sl ctc
2203	2-Chlorooctane		$C_8H_{17}Cl$	628-61-5	148.674			172(7)	0.8658¹⁷	1.4273²¹	i H₂O; vs EtOH, eth
2204	8-Chloro-1-octanol		$C_8H_{17}ClO$	23144-52-7	164.673			139¹⁹		1.4563²⁵	vs eth, EtOH
2205	Chloropentafluoroacetone		C_3ClF_5O	79-53-8	182.476	col gas	-133	7.8(0.9)			
2206	Chloropentafluorobenzene		C_6ClF_5	344-07-0	202.509			117.96	1.568²⁵	1.4256²⁰	
2207	Chloropentafluoroethane	Refrigerant 115	C_2ClF_5	76-15-3	154.466	col gas	-99.4(0.1)	-39.2(0.2)	1.5678⁻⁴²	1.2678⁻⁴²	i H₂O; s EtOH, eth
2208	1-Chloropentane	Pentyl chloride	$C_5H_{11}Cl$	543-59-9	106.594	liq	-99.0	107.9(0.3)	0.8820²⁰	1.4126²⁰	i H₂O; msc EtOH, eth; s bz, ctc; vs chl
2209	2-Chloropentane, (±)	sec-Amyl chloride	$C_5H_{11}Cl$	29882-57-3	106.594	liq	-137	96.3(0.9)	0.8698²⁰	1.4069²⁰	i H₂O; s EtOH, eth, bz; vs chl
2210	3-Chloropentane		$C_5H_{11}Cl$	616-20-6	106.594	liq	-105(2)	95(3)	0.8731²⁰	1.4082²⁰	i H₂O; s EtOH, eth, bz; sl ace
2211	5-Chloropentanoic acid		$C_5H_9ClO_2$	1119-46-6	136.577		18	234(10)	1.3416²⁵	1.4555²⁰	vs eth, EtOH
2212	5-Chloro-1-pentanol		$C_5H_{11}ClO$	5259-98-3	122.593			112¹²		1.4518²⁰	vs eth, EtOH
2213	5-Chloro-2-pentanone		C_5H_9ClO	5891-21-4	120.577			106¹¹⁰	1.0523²⁰	1.4375²⁰	s eth, ace; sl ctc
2214	1-Chloro-3-pentanone		C_5H_9ClO	32830-97-0	120.577			68²⁰		1.4361²⁰	vs eth, EtOH
2215	5-Chloropentanoyl chloride		$C_5H_8Cl_2O$	1575-61-7	155.022			83¹²	1.210¹⁸	1.4639²⁰	vs eth

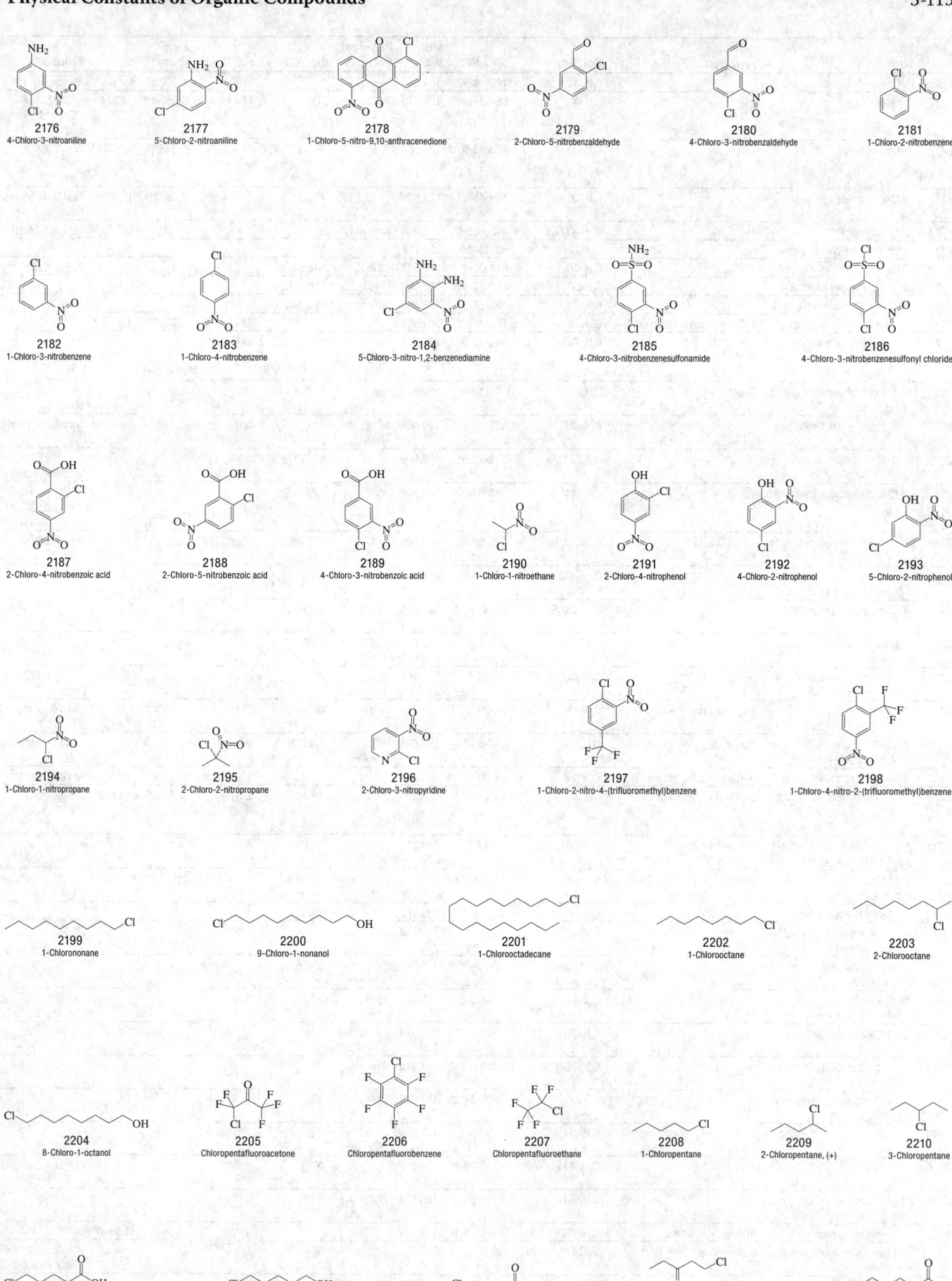

2176
4-Chloro-3-nitroaniline

2177
5-Chloro-2-nitroaniline

2178
1-Chloro-5-nitro-9,10-anthracenedione

2179
2-Chloro-5-nitrobenzaldehyde

2180
4-Chloro-3-nitrobenzaldehyde

2181
1-Chloro-2-nitrobenzene

2182
1-Chloro-3-nitrobenzene

2183
1-Chloro-4-nitrobenzene

2184
5-Chloro-3-nitro-1,2-benzenediamine

2185
4-Chloro-3-nitrobenzenesulfonamide

2186
4-Chloro-3-nitrobenzenesulfonyl chloride

2187
2-Chloro-4-nitrobenzoic acid

2188
2-Chloro-5-nitrobenzoic acid

2189
4-Chloro-3-nitrobenzoic acid

2190
1-Chloro-1-nitroethane

2191
2-Chloro-4-nitrophenol

2192
4-Chloro-2-nitrophenol

2193
5-Chloro-2-nitrophenol

2194
1-Chloro-1-nitropropane

2195
2-Chloro-2-nitropropane

2196
2-Chloro-3-nitropyridine

2197
1-Chloro-2-nitro-4-(trifluoromethyl)benzene

2198
1-Chloro-4-nitro-2-(trifluoromethyl)benzene

2199
1-Chlorononane

2200
9-Chloro-1-nonanol

2201
1-Chlorooctadecane

2202
1-Chlorooctane

2203
2-Chlorooctane

2204
8-Chloro-1-octanol

2205
Chloropentafluoroacetone

2206
Chloropentafluorobenzene

2207
Chloropentafluoroethane

2208
1-Chloropentane

2209
2-Chloropentane, (+)

2210
3-Chloropentane

2211
5-Chloropentanoic acid

2212
5-Chloro-1-pentanol

2213
5-Chloro-2-pentanone

2214
1-Chloro-3-pentanone

2215
5-Chloropentanoyl chloride

No.	Name	Synonym	Mol. Form.	CAS RN	Mol. Wt.	Physical Form	mp/°C	bp/°C	den g cm⁻³	n_D	Solubility
2216	4-Chloro-2-pentene		C_5H_9Cl	1458-99-7	104.578			97(5)	0.8988[20]	1.4322[20]	vs ace, eth, chl
2217	2-Chlorophenol		C_6H_5ClO	95-57-8	128.556	liq	8(1)	173.4(0.6)	1.2634[20]	1.5524[20]	sl H_2O, chl; s EtOH, eth; vs bz
2218	3-Chlorophenol		C_6H_5ClO	108-43-0	128.556		32.5(0.3)	210(3)	1.245[45]	1.5565[40]	sl H_2O, chl; s EtOH, eth; vs bz
2219	4-Chlorophenol		C_6H_5ClO	106-48-9	128.556		43.1(0.7)	219(4)	1.2651[40]	1.5579[40]	sl H_2O; vs EtOH, eth, bz; s alk
2220	Chlorophenol Red		$C_{19}H_{12}Cl_2O_5S$	4430-20-0	423.266	grn-br cry	261				sl H_2O; s EtOH
2221	2-Chloro-10H-phenothiazine		$C_{12}H_8ClNS$	92-39-7	233.717		198.5				
2222	2-Chlorophenoxyacetic acid		$C_8H_7ClO_3$	614-61-9	186.593	nd (w, al)	148.5				s H_2O, EtOH
2223	3-Chlorophenoxyacetic acid		$C_8H_7ClO_3$	588-32-9	186.593	cry (w)	110				i H_2O
2224	(4-Chlorophenoxy)acetic acid		$C_8H_7ClO_3$	122-88-3	186.593	pr or nd (w)	158.3(0.5)				vs H_2O; sl chl
2225	1-Chloro-4-phenoxybenzene	4-Chlorophenyl phenyl ether	$C_{12}H_9ClO$	7005-72-3	204.651			284.5	1.2026[15]	1.599	
2226	3-(4-Chlorophenoxy)-1,2-propanediol	Chlorphenesin	$C_9H_{11}ClO_3$	104-29-0	202.634	cry	78	214[19]			i H_2O; vs EtOH, eth; s bz, con sulf
2227	2-(3-Chlorophenoxy)propanoic acid	Cloprop	$C_9H_9ClO_3$	101-10-0	200.618	cry	113	100[1.5]			
2228	2-Chloro-N-phenylacetamide		C_8H_8ClNO	587-65-5	169.609	nd (dil HOAc)		sub			vs bz, eth, EtOH
2229	N-(2-Chlorophenyl)acetamide		C_8H_8ClNO	533-17-5	169.609		86.7(0.5)				i H_2O; s EtOH, bz, chl; vs eth
2230	N-(3-Chlorophenyl)acetamide		C_8H_8ClNO	588-07-8	169.609	nd	76.6(0.5)	333			sl H_2O; vs EtOH, eth, bz, CS_2; s chl
2231	N-(4-Chlorophenyl)acetamide		C_8H_8ClNO	539-03-7	169.609		178.4(0.5)	332(27)	1.385[22]		i H_2O; s EtOH; vs eth; sl ctc
2232	4-Chloro-α-phenylbenzenemethanol		$C_{13}H_{11}ClO$	119-56-2	218.678		59				sl chl
2233	4-Chlorophenyl benzenesulfonate		$C_{12}H_9ClO_3S$	80-38-6	268.715	col cry	59.3(0.5)		1.33		sl H_2O
2234	4-Chloro-1-phenyl-1-butanone		$C_{10}H_{11}ClO$	939-52-6	182.646		19.5	131[4]	1.137[25]	1.5459[20]	
2235	4-Chlorophenyl 4-chlorobenzenesulfonate	Ovex	$C_{12}H_8Cl_2O_3S$	80-33-1	303.161		88(1)				i H_2O; sl EtOH; s ace
2236	(2-Chlorophenyl)(4-chlorophenyl)methanone	2,4'-Dichlorodiphenyl ketone	$C_{13}H_8Cl_2O$	85-29-0	251.108	pr (al)	65.7(0.5)	214[22]	1.393[14]		s EtOH; sl chl
2237	N'-(4-Chlorophenyl)-N,N-dimethylurea	Monuron	$C_9H_{11}ClN_2O$	150-68-5	198.648	wh pl (MeOH)	168.3(0.3)				i H_2O; sl EtOH, ace
2238	1-(3-Chlorophenyl)ethanone	m-Chloroacetophenone	C_8H_7ClO	99-02-5	154.594			244	1.2130[40]	1.5494[20]	s EtOH, eth, ace
2239	1-(4-Chlorophenyl)ethanone	p-Chloroacetophenone	C_8H_7ClO	99-91-2	154.594		18.4(0.2)	237(1)	1.1922[20]	1.5550[20]	i H_2O; msc EtOH, eth; s chl
2240	5-(4-Chlorophenyl)-6-ethyl-2,4-pyrimidinediamine	Pyrimethamine	$C_{12}H_{13}ClN_4$	58-14-0	248.711		233.5				
2241	2-(4-Chlorophenyl)-1H-indene-1,3(2H)-dione	Clorindione	$C_{15}H_9ClO_2$	1146-99-2	256.684	dk red nd (al)	145.5				vs bz, eth, EtOH
2242	4-Chlorophenyl isocyanate		C_7H_4ClNO	104-12-1	153.566		31.3	116[45]			
2243	1-(2-Chlorophenyl)-2-methyl-2-propylamine	Clortermine	$C_{10}H_{14}ClN$	10389-73-8	183.678	liq		117[16]			
2244	N-(2-Chlorophenyl)-3-oxo-butanamide		$C_{10}H_{10}ClNO_2$	93-70-9	211.645		106.5				s EtOH; i eth, lig
2245	(4-Chlorophenyl)-phenylmethanone		$C_{13}H_9ClO$	134-85-0	216.662	nd (al)	77.5	332			s EtOH, eth, ace; sl ctc
2246	3-(2-Chlorophenyl)propanoic acid		$C_9H_9ClO_2$	1643-28-3	184.619	nd or lf (w)	102				
2247	3-(3-Chlorophenyl)propanoic acid		$C_9H_9ClO_2$	21640-48-2	184.619	lf (peth)	77				
2248	3-(4-Chlorophenyl)propanoic acid		$C_9H_9ClO_2$	2019-34-3	184.619		126				
2249	3-Chloro-1-phenyl-1-propanone	2-Chloroethyl phenyl ketone	C_9H_9ClO	936-59-4	168.619	lf (eth), cry (al, peth)	49.5	113[4]			
2250	1-(4-Chlorophenyl)-1-propanone		C_9H_9ClO	6285-05-8	168.619		37.3	135[31]			i H_2O; s EtOH, CS_2; sl chl
2251	3-(3-Chlorophenyl)-2-propynoic acid		$C_9H_5ClO_2$	7396-28-3	180.588	cry (HOAc, bz-peth)	144.5				vs HOAc
2252	Chlorophenylsilane	Phenylchlorosilane	C_6H_7ClSi	4206-75-1	142.659			162.5	1.0683[20]	1.5340[20]	
2253	1-Chloro-4-(phenylsulfonyl)benzene	Sulphenone	$C_{12}H_9ClO_2S$	80-00-2	252.716		94				i H_2O; sl EtOH; s eth; vs ace, bz
2254	5-Chloro-1-phenyltetrazole		$C_7H_5ClN_4$	14210-25-4	180.595		123				
2255	(2-Chlorophenyl)thiourea		$C_7H_7ClN_2S$	5344-82-1	186.662	nd or pl	143.3(0.5)				vs bz, EtOH

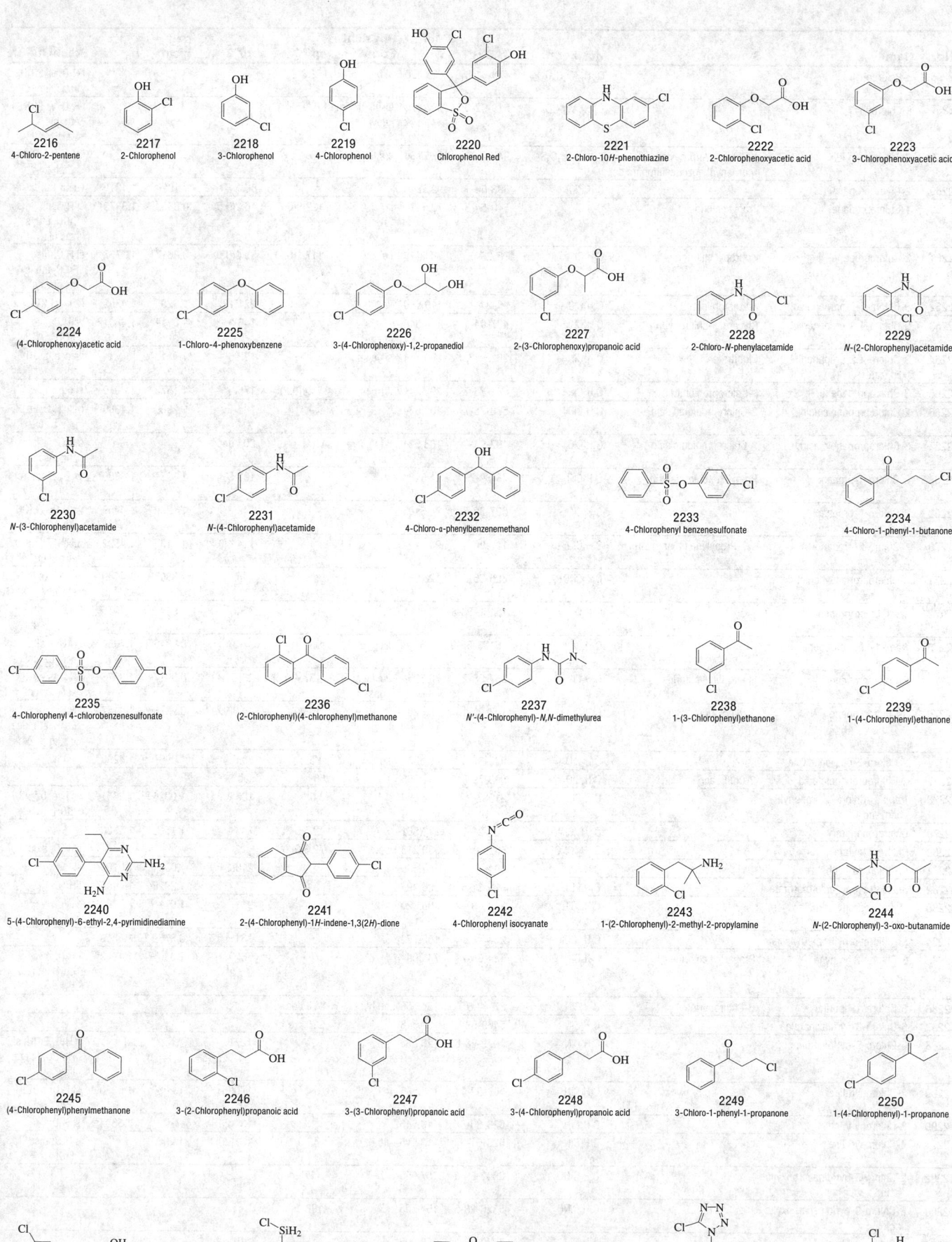

2216
4-Chloro-2-pentene

2217
2-Chlorophenol

2218
3-Chlorophenol

2219
4-Chlorophenol

2220
Chlorophenol Red

2221
2-Chloro-10H-phenothiazine

2222
2-Chlorophenoxyacetic acid

2223
3-Chlorophenoxyacetic acid

2224
(4-Chlorophenoxy)acetic acid

2225
1-Chloro-4-phenoxybenzene

2226
3-(4-Chlorophenoxy)-1,2-propanediol

2227
2-(3-Chlorophenoxy)propanoic acid

2228
2-Chloro-N-phenylacetamide

2229
N-(2-Chlorophenyl)acetamide

2230
N-(3-Chlorophenyl)acetamide

2231
N-(4-Chlorophenyl)acetamide

2232
4-Chloro-α-phenylbenzenemethanol

2233
4-Chlorophenyl benzenesulfonate

2234
4-Chloro-1-phenyl-1-butanone

2235
4-Chlorophenyl 4-chlorobenzenesulfonate

2236
(2-Chlorophenyl)(4-chlorophenyl)methanone

2237
N'-(4-Chlorophenyl)-N,N-dimethylurea

2238
1-(3-Chlorophenyl)ethanone

2239
1-(4-Chlorophenyl)ethanone

2240
5-(4-Chlorophenyl)-6-ethyl-2,4-pyrimidinediamine

2241
2-(4-Chlorophenyl)-1H-indene-1,3(2H)-dione

2242
4-Chlorophenyl isocyanate

2243
1-(2-Chlorophenyl)-2-methyl-2-propylamine

2244
N-(2-Chlorophenyl)-3-oxo-butanamide

2245
(4-Chlorophenyl)phenylmethanone

2246
3-(2-Chlorophenyl)propanoic acid

2247
3-(3-Chlorophenyl)propanoic acid

2248
3-(4-Chlorophenyl)propanoic acid

2249
3-Chloro-1-phenyl-1-propanone

2250
1-(4-Chlorophenyl)-1-propanone

2251
3-(3-Chlorophenyl)-2-propynoic acid

2252
Chlorophenylsilane

2253
1-Chloro-4-(phenylsulfonyl)benzene

2254
5-Chloro-1-phenyltetrazole

2255
(2-Chlorophenyl)thiourea

No.	Name	Synonym	Mol. Form.	CAS RN	Mol. Wt.	Physical Form	mp/°C	bp/°C	den g cm^{-3}	n_D	Solubility
2256	α-Chlorophyll		C$_{55}$H$_{72}$MgN$_4$O$_5$	479-61-8	893.490	bl blk hex pl	152.3				i H$_2$O; vs EtOH, eth; s lig
2257	β-Chlorophyll		C$_{55}$H$_{70}$MgN$_4$O$_6$	519-62-0	907.473	bl-blk or grn pow	125				i H$_2$O; vs EtOH, eth, py; s MeOH
2258	Chloropropamide	4-Chloro-N-[(propylamino)carbonyl]benzenesulfonamide	C$_{10}$H$_{13}$ClN$_2$O$_3$S	94-20-2	276.739	cry (EtOH)	128				i H$_2$O; s EtOH; sl eth, bz
2259	2-Chloropropanal		C$_3$H$_5$ClO	683-50-1	92.524			86	1.182^{15}	1.431^{17}	vs bz, eth
2260	1-Chloropropane	Propyl chloride	C$_3$H$_7$Cl	540-54-5	78.541	liq	-122.9(0.7)	46.2(0.5)	0.8899^{20}	1.3879^{20}	i H$_2$O, ctc; msc EtOH, eth; s bz, chl
2261	2-Chloropropane	Isopropyl chloride	C$_3$H$_7$Cl	75-29-6	78.541	liq	-117.1(0.2)	35.0(0.6)	0.8617^{20}	1.3777^{20}	sl H$_2$O; msc EtOH, eth; s bz, ctc, chl
2262	3-Chloro-1,2-propanediol	α-Chlorohydrin	C$_3$H$_7$ClO$_2$	96-24-2	110.540	ye liq		221(18)	1.325^{18}	1.4809^{20}	s H$_2$O, EtOH, eth
2263	2-Chloro-1,3-propanediol	Glycerol β-chlorohydrin	C$_3$H$_7$ClO$_2$	497-04-1	110.540			146^{18}	1.3219^{20}	1.4831^{20}	vs H$_2$O, ace, EtOH
2264	3-Chloro-1,2-propanediol dinitrate	Clonitrate	C$_3$H$_5$ClN$_2$O$_6$	2612-33-1	200.534	sl ye liq		192.5	1.5112^9		vs ace, EtOH, chl
2265	3-Chloropropanenitrile	β-Chloropropionitrile	C$_3$H$_4$ClN	542-76-7	89.524	liq	-51.4(0.2)	175.5	1.1573^{20}	1.4360^{20}	sl ctc
2266	2-Chloropropanoic acid	2-Chloropropionic acid	C$_3$H$_5$ClO$_2$	598-78-7	108.524			185	1.2585^{20}	1.4380^{20}	msc H$_2$O, EtOH, eth; s ace
2267	3-Chloropropanoic acid	3-Chloropropionic acid	C$_3$H$_5$ClO$_2$	107-94-8	108.524	lf (w), hyg cry (lig)	41(4)	204 dec			s H$_2$O, EtOH, chl; msc eth
2268	2-Chloro-1-propanol	Propylene chlorohydrin	C$_3$H$_7$ClO	78-89-7	94.540			133.5	1.103^{20}	1.4390^{20}	vs H$_2$O, eth, EtOH
2269	3-Chloro-1-propanol		C$_3$H$_7$ClO	627-30-5	94.540			149(8)	1.1309^{20}	1.4459^{20}	vs H$_2$O; s EtOH, eth; sl ctc
2270	1-Chloro-2-propanol	sec-Propylene chlorohydrin	C$_3$H$_7$ClO	127-00-4	94.540			124.4(0.2)	1.113^{20}	1.4392^{20}	msc H$_2$O, EtOH, eth; sl ctc
2271	3-Chloropropanoyl chloride		C$_3$H$_4$Cl$_2$O	625-36-5	126.969			144	1.3307^{13}	1.4549^{20}	sl H$_2$O; vs EtOH, eth, chl
2272	cis-1-Chloropropene		C$_3$H$_5$Cl	16136-84-8	76.525	liq	-134.8	32(2)	0.9347^{20}	1.4055^{20}	i H$_2$O; s eth, ace, bz, chl
2273	trans-1-Chloropropene		C$_3$H$_5$Cl	16136-85-9	76.525	liq	-99	37(2)	0.9349^{20}	1.4054^{20}	i H$_2$O; s eth, ace, bz, chl
2274	2-Chloropropene	Isopropenyl chloride	C$_3$H$_5$Cl	557-98-2	76.525	vol liq or gas	-137.4(0.4)	23(2)	0.9017^{20}	1.3973^{20}	i H$_2$O; s eth, ace, bz, chl
2275	3-Chloropropene	Allyl chloride	C$_3$H$_5$Cl	107-05-1	76.525	liq	-136(2)	44.8(0.4)	0.9376^{20}	1.4157^{20}	i H$_2$O; msc EtOH, eth, ace, bz, lig; sl ctc
2276	2-Chloro-2-propenenitrile		C$_3$H$_2$ClN	920-37-6	87.508	liq	-65	88.5	1.096^{25}	1.4290^{20}	
2277	2-Chloropropenoic acid	2-Chloroacrylic acid	C$_3$H$_3$ClO$_2$	598-79-8	106.508		66	sub			
2278	trans-(3-Chloro-1-propenyl)benzene		C$_9$H$_9$Cl	21087-29-6	152.620		8.5	106^{13}	1.0926^{20}	1.5851^{20}	vs ace, bz, eth, EtOH
2279	Chloropropham		C$_{10}$H$_{12}$ClNO$_2$	101-21-3	213.661		42.5(0.3)	149^2	1.18^{30}	1.5388^{20}	
2280	Chloropropylate		C$_{17}$H$_{16}$Cl$_2$O$_3$	5836-10-2	339.213	pow	72.3(0.3)				sl H$_2$O; s os
2281	(3-Chloropropyl)benzene		C$_9$H$_{11}$Cl	104-52-9	154.636			216(14)	1.056^{21}	1.5160^{25}	sl ctc
2282	3-Chloropropyl chloroformate		C$_4$H$_6$Cl$_2$O$_2$	628-11-5	156.996			177	1.2926^{25}	1.4456^{20}	i H$_2$O
2283	(3-Chloropropyl)trimethoxysilane		C$_6$H$_{15}$ClO$_3$Si	2530-87-2	198.720			91	1.077^{25}	1.4183^{25}	
2284	(3-Chloropropyl)trimethylsilane		C$_6$H$_{15}$ClSi	2344-83-4	150.722			151	0.8789^{20}	1.4319^{20}	
2285	3-Chloro-1-propyne	Propargyl chloride	C$_3$H$_3$Cl	624-65-7	74.509		-78	56(4)	1.030^{25}	1.4349^{20}	i H$_2$O; msc EtOH, eth, bz; s ctc
2286	6-Chloro-1H-purine	6-Chloropurine	C$_5$H$_3$ClN$_4$	87-42-3	154.558	nd (w)	176 dec				
2287	6-Chloro-3-pyridazinamine		C$_4$H$_4$ClN$_3$	5469-69-2	129.548		220				
2288	5-Chloro-2-pyridinamine		C$_5$H$_5$ClN$_2$	1072-98-6	128.560	pl	137	127^{11}			s H$_2$O, EtOH; sl DMSO; i peth, lig
2289	2-Chloropyridine		C$_5$H$_4$ClN	109-09-1	113.546	oil		170(2)	1.205^{15}	1.5320^{20}	sl H$_2$O; s EtOH, eth
2290	3-Chloropyridine		C$_5$H$_4$ClN	626-60-8	113.546			151(20)		1.5304^{20}	sl H$_2$O
2291	4-Chloropyridine		C$_5$H$_4$ClN	626-61-9	113.546	liq	-43.5	151(20)	1.2000^{25}		s H$_2$O; msc EtOH
2292	2-Chloro-3-pyridinecarboxylic acid		C$_6$H$_4$ClNO$_2$	2942-59-8	157.555		>175 dec				
2293	6-Chloro-3-pyridinecarboxylic acid		C$_6$H$_4$ClNO$_2$	5326-23-8	157.555		198 dec				
2294	4-Chloropyridine, hydrochloride		C$_5$H$_5$Cl$_2$N	7379-35-3	150.006			210 sub			
2295	Chloroquine		C$_{18}$H$_{26}$ClN$_3$	54-05-7	319.872		90				

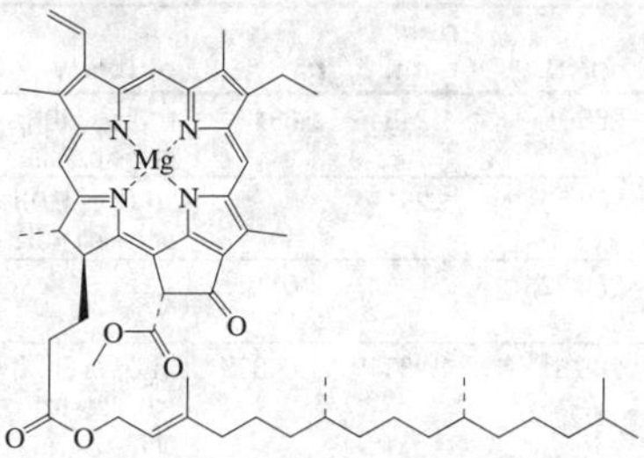

2256
α-Chlorophyll

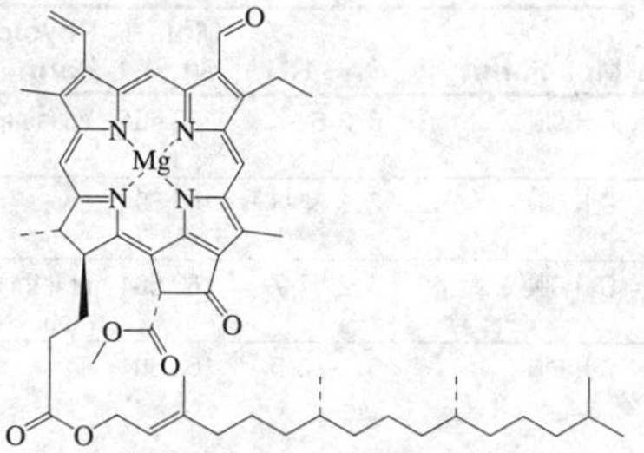

2257
β-Chlorophyll

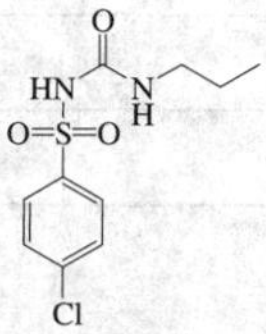

2258
Chloropropamide

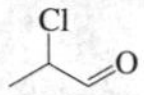

2259
2-Chloropropanal

2260
1-Chloropropane

2261
2-Chloropropane

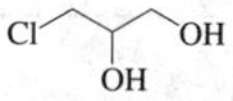

2262
3-Chloro-1,2-propanediol

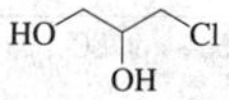

2263
2-Chloro-1,3-propanediol

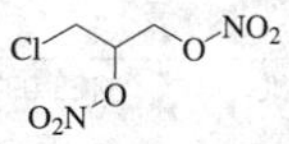

2264
3-Chloro-1,2-propanediol dinitrate

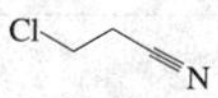

2265
3-Chloropropanenitrile

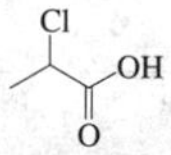

2266
2-Chloropropanoic acid

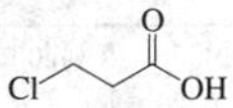

2267
3-Chloropropanoic acid

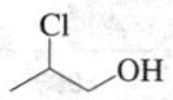

2268
2-Chloro-1-propanol

2269
3-Chloro-1-propanol

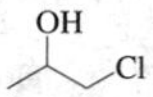

2270
1-Chloro-2-propanol

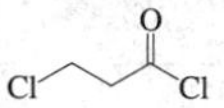

2271
3-Chloropropanoyl chloride

2272
cis-1-Chloropropene

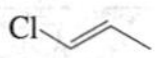

2273
trans-1-Chloropropene

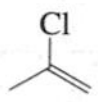

2274
2-Chloropropene

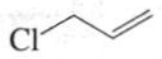

2275
3-Chloropropene

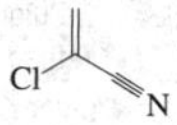

2276
2-Chloro-2-propenenitrile

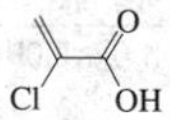

2277
2-Chloropropenoic acid

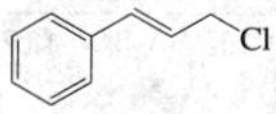

2278
trans-(3-Chloro-1-propenyl)benzene

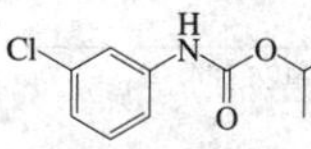

2279
Chloropropham

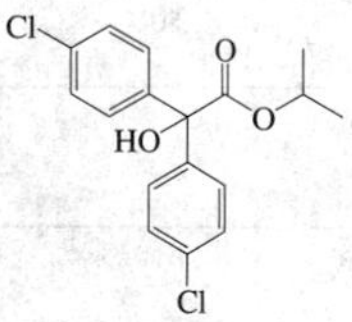

2280
Chloropropylate

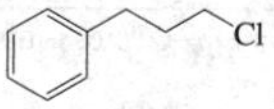

2281
(3-Chloropropyl)benzene

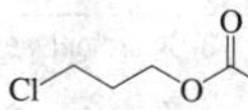

2282
3-Chloropropyl chloroformate

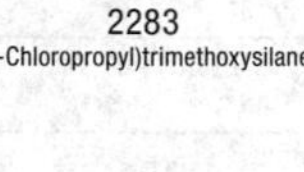

2283
(3-Chloropropyl)trimethoxysilane

2284
(3-Chloropropyl)trimethylsilane

2285
3-Chloro-1-propyne

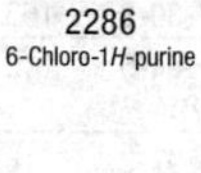

2286
6-Chloro-1H-purine

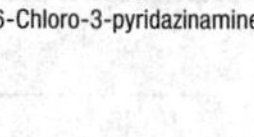

2287
6-Chloro-3-pyridazinamine

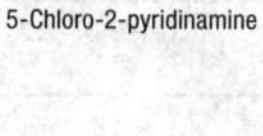

2288
5-Chloro-2-pyridinamine

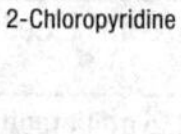

2289
2-Chloropyridine

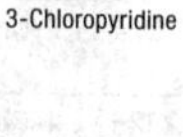

2290
3-Chloropyridine

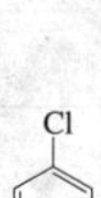

2291
4-Chloropyridine

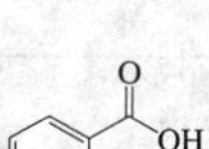

2292
2-Chloro-3-pyridinecarboxylic acid

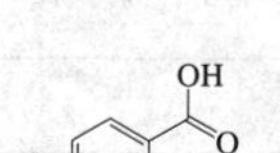

2293
6-Chloro-3-pyridinecarboxylic acid

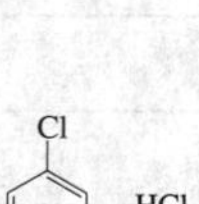

2294
4-Chloropyridine, hydrochloride

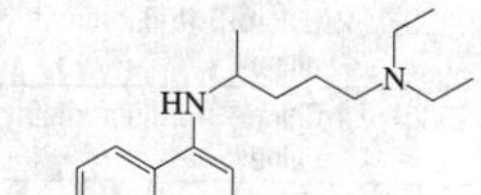

2295
Chloroquine

No.	Name	Synonym	Mol. Form.	CAS RN	Mol. Wt.	Physical Form	mp/°C	bp/°C	den g cm⁻³	n_D	Solubility
2296	2-Chloroquinoline		C_9H_6ClN	612-62-4	163.604	nd (aq al)	38.4(0.5)	266	1.2464[25]	1.6342[25]	i H$_2$O; vs EtOH, eth; s bz, chl
2297	4-Chloroquinoline		C_9H_6ClN	611-35-8	163.604	cry	34.9(0.5)	262	1.251[25]		sl H$_2$O; vs EtOH, eth; s dil HCl
2298	6-Chloroquinoline		C_9H_6ClN	612-57-7	163.604	pr (eth), nd (al)	45.9(0.5)	262(22)		1.6110[56]	
2299	8-Chloroquinoline		C_9H_6ClN	611-33-6	163.604	liq	-20	288.5	1.2834[14]	1.6408[14]	s H$_2$O; vs EtOH, eth, ace, bz, chl
2300	5-Chloro-8-quinolinol	Cloxyquin	C_9H_6ClNO	130-16-5	179.603	cry (al)	130				
2301	2-Chlorostyrene		C_8H_7Cl	2039-87-4	138.595	liq	-63.1	188.6(1)	1.1000[20]	1.5649[20]	s EtOH, eth, ace, ctc, HOAc; msc peth
2302	3-Chlorostyrene		C_8H_7Cl	2039-85-2	138.595			63[6]	1.1033[20]	1.5625[20]	i H$_2$O; s EtOH, eth
2303	4-Chlorostyrene		C_8H_7Cl	1073-67-2	138.595		15.9	191(2)	1.0868[20]	1.5660[20]	i H$_2$O; s EtOH, eth; msc ace, bz, ctc
2304	N-Chlorosuccinimide		$C_4H_4ClNO_2$	128-09-6	133.534	pl (CCl$_4$)	150		1.65[25]		sl H$_2$O, EtOH, bz, lig; s ace, HOAc
2305	1-Chlorotetradecane		$C_{14}H_{29}Cl$	2425-54-9	232.833		4.9	296(4)	0.8654[20]	1.4474[20]	i H$_2$O; s EtOH, chl; vs ace, bz; sl ctc
2306	6-Chloro-N,N,N',N'-tetraethyl-1,3,5-triazine-2,4-diamine		$C_{11}H_{20}ClN_5$	580-48-3	257.764	oily liq	27	155[9]	1.0956[20]	1.5320[20]	vs bz, chl, EtOH, lig
2307	1-Chloro-1,1,2,2-tetrafluoro-ethane		C_2HClF_4	354-25-6	136.476	col gas	-117	-13(6)			
2308	1-Chloro-1,2,2,2-tetrafluoro-ethane	HCFC-124	C_2HClF_4	2837-89-0	136.476	col gas	-199.15	-11.96(0.09)			
2309	Chlorothalonil		$C_8Cl_4N_2$	1897-45-6	265.911		253.1(0.7)	350	1.7[25]		i H$_2$O; sl ace, cyhex
2310	Chlorothen	Chloromethapyrilene	$C_{14}H_{18}ClN_3S$	148-65-2	295.831			155[10]	1.1751[25]		
2311	Chlorothiazide		$C_7H_6ClN_3O_4S_2$	58-94-6	295.724		350 dec				
2312	2-Chlorothiophene	2-Thienyl chloride	C_4H_3ClS	96-43-5	118.585	liq	-71.85(0.05)	126(2)	1.2863[20]	1.5487[20]	i H$_2$O; msc EtOH, eth; sl chl
2313	5-Chloro-2-thiophenecarbox-aldehyde		C_5H_3ClOS	7283-96-7	146.595			77.5[5]		1.6036[25]	sl chl
2314	2-Chloro-9H-thioxanthen-9-one		$C_{13}H_7ClOS$	86-39-5	246.712		153.5				
2315	2-Chlorotoluene	1-Chloro-2-methylbenzene	C_7H_7Cl	95-49-8	126.584	liq	-35.9(0.7)	158.8(0.4)	1.0825[20]	1.5268[20]	i H$_2$O; s EtOH, bz; msc eth, ace, chl
2316	3-Chlorotoluene	1-Chloro-3-methylbenzene	C_7H_7Cl	108-41-8	126.584	liq	-47.8	162.1(0.4)	1.075[20]	1.5214[19]	i H$_2$O; s EtOH, bz, ctc, chl; msc eth
2317	4-Chlorotoluene	1-Chloro-4-methylbenzene	C_7H_7Cl	106-43-4	126.584	liq	7.4(0.2)	161.8(0.2)	1.0697[20]	1.5150[20]	i H$_2$O; s EtOH, ctc, chl; msc eth
2318	6-Chloro-1,3,5-triazine-2,4-diamine		$C_3H_4ClN_5$	3397-62-4	145.551		>330				
2319	1-Chloro-2-(trichloromethyl)-benzene		$C_7H_4Cl_4$	2136-89-2	229.919		29.4(0.2)	263(3)	1.5187[20]	1.5836[20]	i H$_2$O; s eth, ace; sl ctc
2320	1-Chloro-4-(trichloromethyl)-benzene		$C_7H_4Cl_4$	5216-25-1	229.919			245	1.4463[20]		vs ace, eth
2321	Chlorotriethoxysilane		$C_6H_{15}ClO_3Si$	4667-99-6	198.720	liq	-51	156	1.030[20]	1.3999[20]	vs EtOH
2322	Chlorotriethylplumbane	Lead triethyl chloride	$C_6H_{15}ClPb$	1067-14-7	329.8		123 dec				s H$_2$O
2323	Chlorotriethylsilane		$C_6H_{15}ClSi$	994-30-9	150.722			146(3)	0.8967[20]	1.4314[20]	
2324	1-Chloro-1,1,2-trifluoroethane		$C_2H_2ClF_3$	421-04-5	118.485	vol liq or gas		16(19)			
2325	1-Chloro-1,2,2-trifluoroethane		$C_2H_2ClF_3$	431-07-2	118.485	vol liq or gas		17.3			
2326	2-Chloro-1,1,1-trifluoroethane		$C_2H_2ClF_3$	75-88-7	118.485	col gas	-105.5	6.0(0.6)	1.389[0]	1.3090[0]	
2327	Chlorotrifluoroethene	Chlorotrifluoroethylene	C_2ClF_3	79-38-9	116.469	col gas	-158.14(0.05)	-28.3(0.3)	1.54[-60]	1.38[0]	s bz, chl
2328	Chlorotrifluoromethane	Refrigerant 13	$CClF_3$	75-72-9	104.459	col gas	-181.2	-81.37			i H$_2$O
2329	2-Chloro-5-(trifluoromethyl)-aniline		$C_7H_5ClF_3N$	121-50-6	195.570			103[25]	1.428[25]	1.4975[20]	
2330	4-Chloro-3-(trifluoromethyl)-aniline		$C_7H_5ClF_3N$	320-51-4	195.570		36.5	132[27]			
2331	1-Chloro-2-(trifluoromethyl)-benzene	o-Chlorobenzotrifluoride	$C_7H_4ClF_3$	88-16-4	180.555	liq	-6(2)	153(3)	1.2540[30]	1.4513[25]	s chl

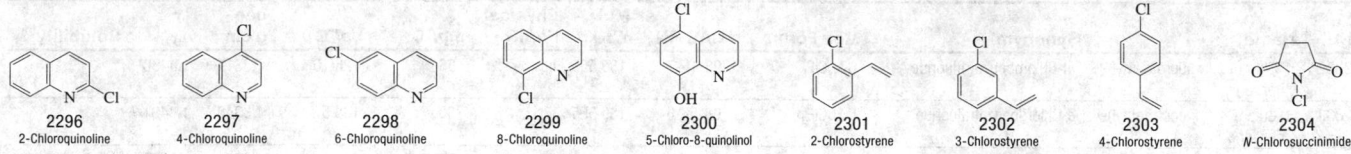

2296 2-Chloroquinoline

2297 4-Chloroquinoline

2298 6-Chloroquinoline

2299 8-Chloroquinoline

2300 5-Chloro-8-quinolinol

2301 2-Chlorostyrene

2302 3-Chlorostyrene

2303 4-Chlorostyrene

2304 *N*-Chlorosuccinimide

2305 1-Chlorotetradecane

2306 6-Chloro-*N*,*N*,*N*′,*N*′-tetraethyl-1,3,5-triazine-2,4-diamine

2307 1-Chloro-1,1,2,2-tetrafluoroethane

2308 1-Chloro-1,2,2,2-tetrafluoroethane

2309 Chlorothalonil

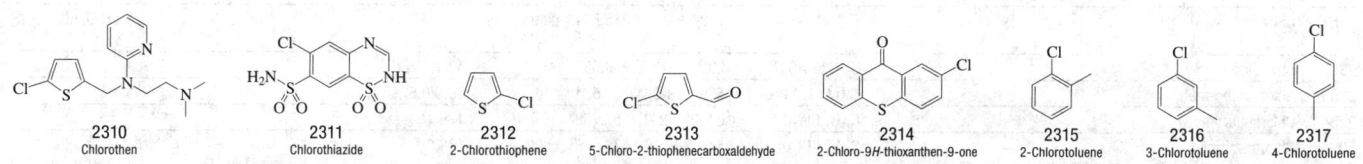

2310 Chlorothen

2311 Chlorothiazide

2312 2-Chlorothiophene

2313 5-Chloro-2-thiophenecarboxaldehyde

2314 2-Chloro-9*H*-thioxanthen-9-one

2315 2-Chlorotoluene

2316 3-Chlorotoluene

2317 4-Chlorotoluene

2318 6-Chloro-1,3,5-triazine-2,4-diamine

2319 1-Chloro-2-(trichloromethyl)benzene

2320 1-Chloro-4-(trichloromethyl)benzene

2321 Chlorotriethoxysilane

2322 Chlorotriethylplumbane

2323 Chlorotriethylsilane

2324 1-Chloro-1,1,2-trifluoroethane

2325 1-Chloro-1,2,2-trifluoroethane

2326 2-Chloro-1,1,1-trifluoroethane

2327 Chlorotrifluoroethene

2328 Chlorotrifluoromethane

2329 2-Chloro-5-(trifluoromethyl)aniline

2330 4-Chloro-3-(trifluoromethyl)aniline

2331 1-Chloro-2-(trifluoromethyl)benzene

No.	Name	Synonym	Mol. Form.	CAS RN	Mol. Wt.	Physical Form	mp/°C	bp/°C	den g cm⁻³	n_D	Solubility
2332	1-Chloro-3-(trifluoromethyl)-benzene	*m*-Chlorobenzotrifluoride	C₇H₄ClF₃	98-15-7	180.555	liq	-56	126(10)	1.3311²⁵	1.4438²⁵	
2333	1-Chloro-4-(trifluoromethyl)-benzene	*p*-Chlorobenzotrifluoride	C₇H₄ClF₃	98-56-6	180.555	liq	-33	138.5	1.3340²⁵	1.4431³⁰	
2334	3-Chloro-1,1,1-trifluoropropane		C₃H₄ClF₃	460-35-5	132.512	liq	-93.73(0.06)	45.1	1.3253²⁰	1.3350²⁰	i H₂O
2335	2-Chloro-2,4,4-trimethylpentane		C₈H₁₇Cl	6111-88-2	148.674		-26	150(8)	0.8746²⁰	1.4308²⁰	vs EtOH
2336	Chlorotrimethylstannane		C₃H₉ClSn	1066-45-1	199.266		38.5	148			s H₂O, chl, os
2337	2-Chloro-1,3,5-trinitrobenzene	Picryl chloride	C₆H₂ClN₃O₆	88-88-0	247.549	wh nd or pl (chl, al-lig)	82(1)		1.797²⁰		i H₂O; s EtOH, bz; sl eth; vs ace, tol
2338	Chlorotrinitromethane		CClN₃O₆	1943-16-4	185.480		5.8(0.2)	156(18)	1.6769²⁰	1.4500²⁰	vs eth, EtOH, chl
2339	Chlorotriphenylmethane		C₁₉H₁₅Cl	76-83-5	278.775	nd or pr (bz-peth)	109.2(0.5)	310			i H₂O; sl EtOH; vs eth, bz, chl; s ace
2340	Chlorotriphenylsilane		C₁₈H₁₅ClSi	76-86-8	294.851			241³⁵			
2341	Chlorotriphenylstannane	Triphenyltin chloride	C₁₈H₁₅ClSn	639-58-7	385.475		103.5				s chl
2342	Chlorotripropylstannane		C₉H₂₁ClSn	2279-76-7	283.426		-23.5	123¹³	1.2678²⁸	1.4910²⁸	s ctc, os
2343	Chlorovinyldimethylsilane		C₄H₉ClSi	1719-58-0	120.653			82.3(0.4)	0.8744²⁰	1.4141²⁰	
2344	Chloroxuron		C₁₅H₁₅ClN₂O₂	1982-47-4	290.745		152.0(0.9)				
2345	Chlorozotocin		C₉H₁₆ClN₃O₇	54749-90-5	313.692	cry	147 dec				s H₂O
2346	Chlorphenesin carbamate		C₁₀H₁₂ClNO₄	886-74-8	245.660	cry (bz)	90				vs ace, EtOH, diox
2347	Chlorpheniramine		C₁₆H₁₉ClN₂	132-22-9	274.788	oily liq		142¹			
2348	Chlorpheniramine maleate	Chlorprophenpyridamine	C₂₀H₂₃ClN₂O₄	113-92-8	390.861		132.5				
2349	Chlorphentermine	2-(4-Chlorobenzyl)-2-propyl-amine	C₁₀H₁₄ClN	461-78-9	183.678	liq		231			
2350	Chlorpromazine	2-Chloro-*N,N*-dimethyl-10*H*-phenothiazine-10-propan-amine	C₁₇H₁₉ClN₂S	50-53-3	318.864			202⁰·⁸			i H₂O; vs EtOH, eth, bz, chl; s dil HCl
2351	Chlorprothixene		C₁₈H₁₈ClNS	113-59-7	315.861	pale ye cry	97.1(0.3)				i H₂O, EtOH, eth, chl
2352	Chlorpyrifos		C₉H₁₁Cl₃NO₃PS	2921-88-2	350.586		43(1)				
2353	Chlorpyrifos-methyl		C₇H₇Cl₃NO₃PS	5598-13-0	322.534		46.0(0.5)				
2354	Chlorsulfuron		C₁₂H₁₂ClN₅O₄S	64902-72-3	357.773		176				
2355	Chlortetracycline		C₂₂H₂₃ClN₂O₈	57-62-5	478.879	gold-ye	168.5				i H₂O, eth; sl EtOH, ace, bz; s diox
2356	Chlorthalidone		C₁₄H₁₁ClN₂O₄S	77-36-1	338.766	wh pow or cry	225 dec				s alk, EtOH; sl eth
2357	Chlorthion		C₈H₉ClNO₅PS	500-28-7	297.653	ye cry	21	125⁰·¹	1.437²⁰	1.5661²⁰	i H₂O; vs bz, eth, EtOH
2358	Chlorthiophos		C₁₁H₁₅Cl₂O₃PS₂	21923-23-9	361.245			150⁰·⁰⁰¹			
2359	Chlortoluron	*N*-(3-Chloro-4-methylphenyl)-*N,N*-dimethylurea	C₁₀H₁₃ClN₂O	15545-48-9	212.675	cry	147				sl H₂O; s os
2360	Cholane		C₂₄H₄₂	548-98-1	330.590	pr (al)	90	190⁰·⁰⁰¹			
2361	Cholan-24-oic acid	Cholanic acid	C₂₄H₄₀O₂	25312-65-6	360.574	nd (al), cry (HOAc)	163.5				s EtOH, chl, HOAc
2362	Cholesta-3,5-diene		C₂₇H₄₄	747-90-0	368.638	wh nd (al)	80	260¹³	0.925¹⁰⁰		i H₂O; s EtOH; msc eth, bz, chl; vs lig
2363	Cholesta-5,7-dien-3-ol, (3β)	7-Dehydrocholesterol	C₂₇H₄₄O	434-16-2	384.637	pl (+1w), (eth-MeOH)	150.5				i H₂O; sl EtOH; s eth, ace
2364	Cholesta-8,24-dien-3-ol, (3β,5α)		C₂₇H₄₄O	128-33-6	384.637	pl (MeOH),nd	110	160⁰·⁰⁰¹			s ace, chl, MeOH
2365	Cholestane, (5α)	28,29,30-Trinorlanostane	C₂₇H₄₈	481-21-0	372.670	sc or pl (eth-al, ace)	78.6(0.5)	250¹	0.9090⁸⁸	1.4887⁸⁸	i H₂O; sl EtOH; vs eth, bz, chl
2366	Cholestane, (5β)	Coprostane	C₂₇H₄₈	481-20-9	372.670	orth nd (al, ace)	72		0.9119⁸⁷	1.4884⁸⁸	vs eth, chl
2367	Cholestanol	Dihydrocholesterol	C₂₇H₄₈O	80-97-7	388.669	sc (al,+1w)	141.5				vs eth, chl
2368	Cholestan-3-ol, (3α,5α)	Epicholestanol	C₂₇H₄₈O	516-95-0	388.669	nd (al)	185.5				s chl
2369	Cholest-4-en-3-ol, (3β)	Allocholesterol	C₂₇H₄₆O	517-10-2	386.653	nd (eth-MeOH)	132				i H₂O; s EtOH; vs eth, ace, bz, chl
2370	Cholest-5-en-3-ol, (3α)	Epicholesterol	C₂₇H₄₆O	474-77-1	386.653	cry (al, chl-MeOH)	141.5				sl EtOH
2371	Cholest-5-en-3-ol (3β), acetate		C₂₉H₄₈O₂	604-35-3	428.690	wh nd (ace, al)	114.6(0.5)				vs bz, eth, chl

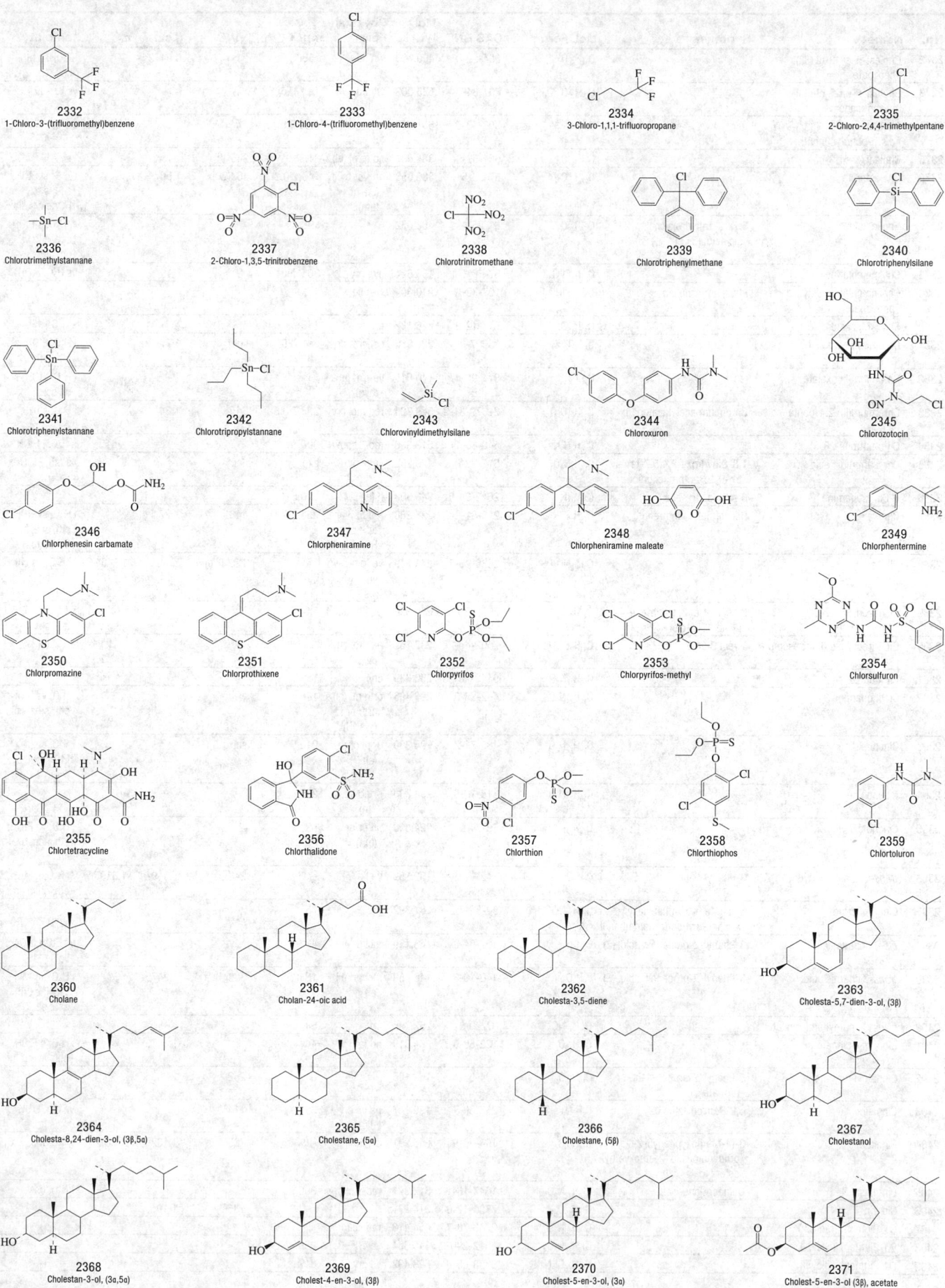

2332
1-Chloro-3-(trifluoromethyl)benzene

2333
1-Chloro-4-(trifluoromethyl)benzene

2334
3-Chloro-1,1,1-trifluoropropane

2335
2-Chloro-2,4,4-trimethylpentane

2336
Chlorotrimethylstannane

2337
2-Chloro-1,3,5-trinitrobenzene

2338
Chlorotrinitromethane

2339
Chlorotriphenylmethane

2340
Chlorotriphenylsilane

2341
Chlorotriphenylstannane

2342
Chlorotripropylstannane

2343
Chlorovinyldimethylsilane

2344
Chloroxuron

2345
Chlorozotocin

2346
Chlorphenesin carbamate

2347
Chlorpheniramine

2348
Chlorpheniramine maleate

2349
Chlorphentermine

2350
Chlorpromazine

2351
Chlorprothixene

2352
Chlorpyrifos

2353
Chlorpyrifos-methyl

2354
Chlorsulfuron

2355
Chlortetracycline

2356
Chlorthalidone

2357
Chlorthion

2358
Chlorthiophos

2359
Chlortoluron

2360
Cholane

2361
Cholan-24-oic acid

2362
Cholesta-3,5-diene

2363
Cholesta-5,7-dien-3-ol, (3β)

2364
Cholesta-8,24-dien-3-ol, (3β,5α)

2365
Cholestane, (5α)

2366
Cholestane, (5β)

2367
Cholestanol

2368
Cholestan-3-ol, (3α,5α)

2369
Cholest-4-en-3-ol, (3β)

2370
Cholest-5-en-3-ol, (3α)

2371
Cholest-5-en-3-ol (3β), acetate

No.	Name	Synonym	Mol. Form.	CAS RN	Mol. Wt.	Physical Form	mp/°C	bp/°C	den g cm⁻³	n_D	Solubility
2372	Cholest-5-en-3-ol (3β), benzoate		$C_{34}H_{50}O_2$	604-32-0	490.760	wh nd	150(1)		0.9413^{200}		i EtOH; s eth, chl
2373	Cholest-5-en-3-ol (3β)-, hexadecanoate		$C_{43}H_{76}O_2$	601-34-3	625.062	wh nd (eth al)	79.7(0.5)				vs bz, chl
2374	Cholest-5-en-3-ol (3β)-, cis-9-octadecenoate		$C_{45}H_{78}O_2$	303-43-5	651.100		47.9(0.5)				s chl
2375	Cholest-4-en-3-one		$C_{27}H_{44}O$	601-57-0	384.637	nd or pl (al)	81.5	$245^{0.03}$			
2376	Cholesterol		$C_{27}H_{46}O$	57-88-5	386.653	orth or tcl lf (al), nd (eth)	148.2(0.8)	459(20)	1.067^{20}		i H₂O; sl EtOH, ace; s bz, HOAc; vs diox
2377	Cholic acid	3,7,12-Trihydroxycholan-24-oic acid, (3α,5β,7α,12α)	$C_{24}H_{40}O_5$	81-25-4	408.572		198				sl H₂O; s EtOH, ace, alk; vs eth, chl
2378	Choline chloride		$C_5H_{14}ClNO$	67-48-1	139.624	hyg cry	305 dec				vs H₂O, EtOH
2379	Choline chloride dihydrogen phosphate	Phosphorylcholine	$C_5H_{15}ClNO_4P$	107-73-3	219.604	visc liq					
2380	Chorismic acid		$C_{10}H_{10}O_6$	617-12-9	226.182	cry	148				s H₂O
2381	Chromium carbonyl		C_6CrO_6	13007-92-6	220.056	col orth cry	dec 130	sub	1.77		i H₂O, EtOH; s eth, chl
2382	Chromium(II) oxalate		C_2CrO_4	814-90-4	140.015	ye-grn pow (hyd)					i H₂O, EtOH; s dil acid
2383	Chromium(III) 2,4-pentanedio-ate	Chromium acetylacetonate	$C_{15}H_{21}CrO_6$	21679-31-2	349.320	red mcl cry	208.7(0.5)	345	1.34		i H₂O; s bz
2384	Chromotrope 2B		$C_{16}H_9N_3Na_2O_{10}S_2$	548-80-1	513.366	red-br pow	300				s H₂O; i EtOH
2385	Chrysamminic acid	1,8-Dihydroxy-2,4,5,7-tetrani-tro-9,10-anthracenedione	$C_{14}H_4N_4O_{12}$	517-92-0	420.202	ye pl or lf	exp	dec			vs eth, EtOH
2386	6-Chrysenamine	6-Aminochrysene	$C_{18}H_{13}N$	2642-98-0	243.303	lf (al)	210.5				
2387	Chrysene	Benzo[a]phenanthrene	$C_{18}H_{12}$	218-01-9	228.288	red bl fl or orth pl (bz, HOAc)	255.0(0.1)	448	1.274^{20}		i H₂O; sl EtOH, eth, ace, bz, CS₂; s tol
2388	Ciafos		$C_9H_{10}NO_3PS$	2636-26-2	243.219	ye to red-ye liq	15	$120^{0.09}$ dec		1.5404^{32}	sl H₂O; vs chl, EtOH, ace, MeOH
2389	Cicutoxin	8,10,12-Heptadecatriene-4,6-diyne-1,14-diol	$C_{17}H_{22}O_2$	505-75-9	258.356	pr (eth/peth)	54				s hot H₂O, EtOH, eth, chl
2390	C.I. Direct Blue 6, tetrasodium salt	Direct Blue 6	$C_{32}H_{20}N_6Na_4O_{14}S_4$	2602-46-2	932.752	dk bronze pow					
2391	Cimetidine		$C_{10}H_{16}N_6S$	51481-61-9	252.339	cry	142				
2392	Cinchonamine		$C_{19}H_{24}N_2O$	482-28-0	296.406	orth nd (al) orth pr (MeOH)	186				i H₂O; vs EtOH, eth; s bz, chl
2393	Cinchonidine		$C_{19}H_{22}N_2O$	485-71-2	294.390	or pl or pr (al)	210.5	sub			i H₂O, bz; s EtOH, chl, py; sl eth
2394	Cinchonine		$C_{19}H_{22}N_2O$	118-10-5	294.390	pr nd (al, eth)	265				
2395	Cinchotoxine		$C_{19}H_{22}N_2O$	69-24-9	294.390	nd or pr (eth)	59				i H₂O; vs EtOH, eth, ace, bz, chl
2396	trans-Cinnamaldehyde	3-Phenyl-2-propenal, (E)-	C_9H_8O	14371-10-9	132.159	ye liq	-7.5	246	1.0497^{20}	1.6195^{20}	sl H₂O; s EtOH, eth, chl; i lig
2397	Cinnamedrine	α-[1-[Methyl(3-phenylallyl)-amino]ethyl]benzenemethanol	$C_{19}H_{23}NO$	90-86-8	281.392		75				
2398	cis-Cinnamic acid	3-Phenyl-2-propenoic acid, (Z)	$C_9H_8O_2$	102-94-3	148.159	mcl pr (w)	42				vs EtOH, HOAc, lig
2399	trans-Cinnamic acid	3-Phenyl-2-propenoic acid, (E)	$C_9H_8O_2$	140-10-3	148.159	mcl pr (dil al)	134(2)	300	1.2475^4		i H₂O, lig; vs EtOH; s eth, ace, bz
2400	trans-Cinnamyl anthranilate		$C_{16}H_{15}NO_2$	87-29-6	253.296	cry	64				
2401	Cinnamyl cinnamate		$C_{18}H_{16}O_2$	122-69-0	264.319	nd (al)	44		1.1565^4		i H₂O; s EtOH, chl; vs eth
2402	Cinnamyl formate	3-Phenyl-2-propen-1-ol, formate	$C_{10}H_{10}O_2$	104-65-4	162.185		0	252	1.086^{25}		
2403	Cinnoline	1,2-Benzodiazine	$C_8H_6N_2$	253-66-7	130.147	pa ye cry (lig)	38	$114^{0.3}$			vs eth, EtOH
2404	Cinoxate	3-(4-Methoxyphenyl)-2-propenoic acid, 2-ethoxyethyl ester	$C_{14}H_{18}O_4$	104-28-9	250.291	col liq	-25	185^2	1.102^{25}	1.567^{20}	i H₂O; msc EtOH
2405	Cinquasia Red	Quinacridone	$C_{20}H_{12}N_2O_2$	1047-16-1	312.321	red-viol cry	390				i H₂O, os
2406	Ciodrin		$C_{14}H_{19}O_6P$	7700-17-6	314.271			$135^{0.03}$	1.19^{25}		
2407	C.I. Pigment Red 170		$C_{26}H_{22}N_4O_4$	2786-76-7	454.478	red solid					
2408	C.I. Pigment Yellow 1		$C_{17}H_{16}N_4O_4$	2512-29-0	340.334	ye cry	256				
2409	C.I. Pigment Yellow 12		$C_{32}H_{26}Cl_2N_6O_4$	6358-85-6	629.492	ye cry	317				

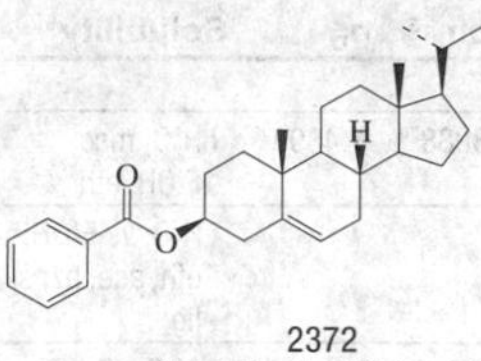

2372 Cholest-5-en-3-ol (3β), benzoate

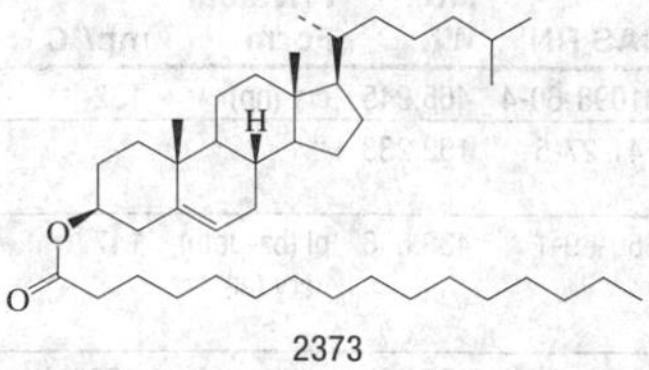

2373 Cholest-5-en-3-ol (3β)-, hexadecanoate

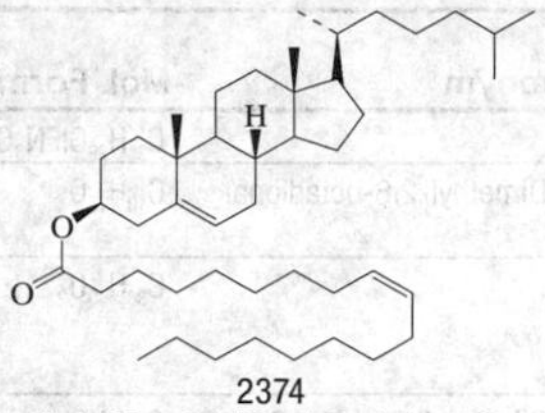

2374 Cholest-5-en-3-ol (3β)-, cis-9-octadecenoate

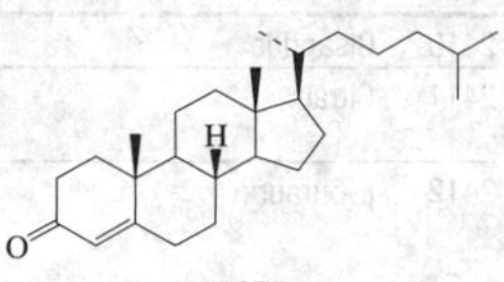

2375 Cholest-4-en-3-one

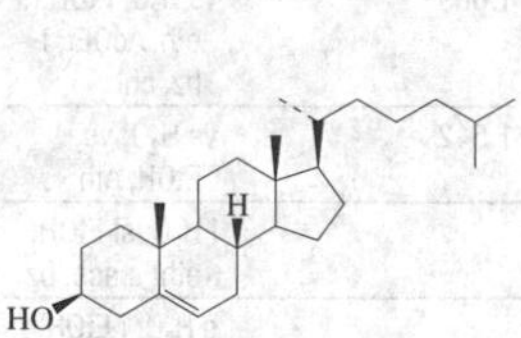

2376 Cholesterol

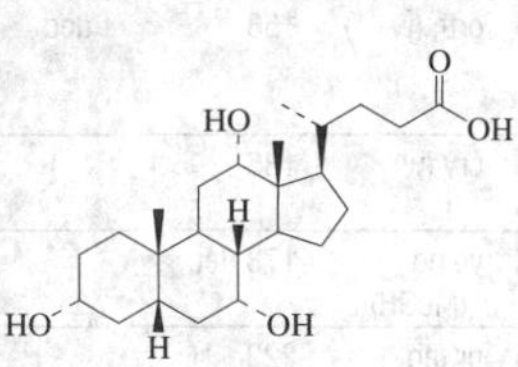

2377 Cholic acid

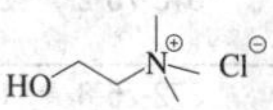

2378 Choline chloride

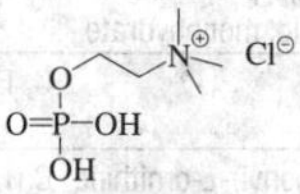

2379 Choline chloride dihydrogen phosphate

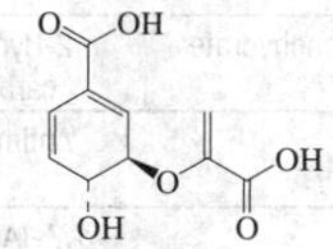

2380 Chorismic acid

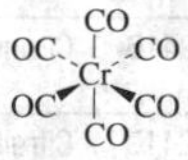

2381 Chromium carbonyl

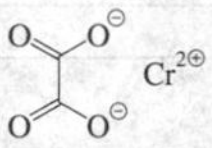

2382 Chromium(II) oxalate

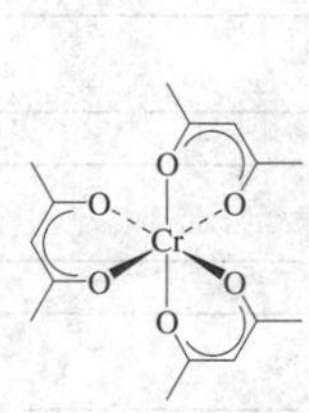

2383 Chromium(III) 2,4-pentanedioate

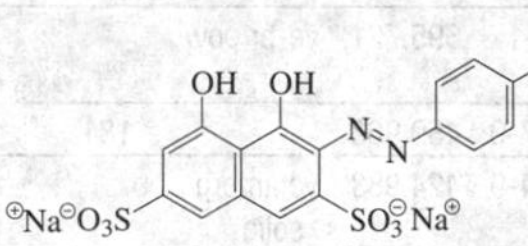

2384 Chromotrope 2B

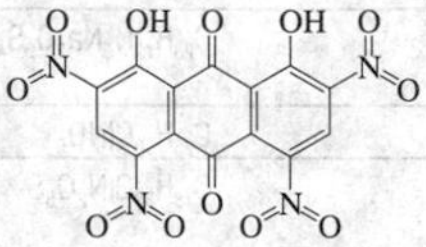

2385 Chrysamminic acid

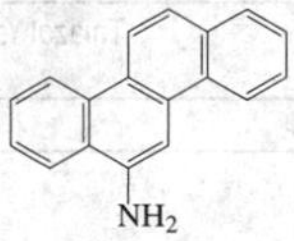

2386 6-Chrysenamine

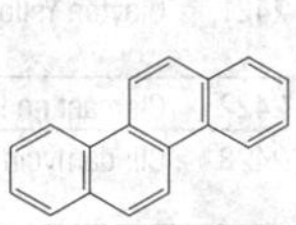

2387 Chrysene

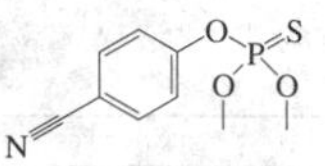

2388 Ciafos

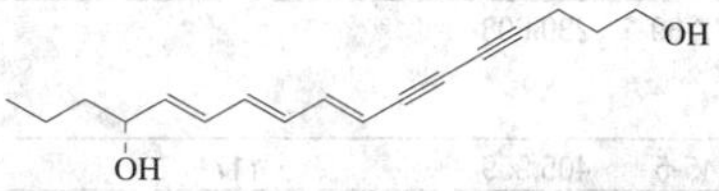

2389 Cicutoxin

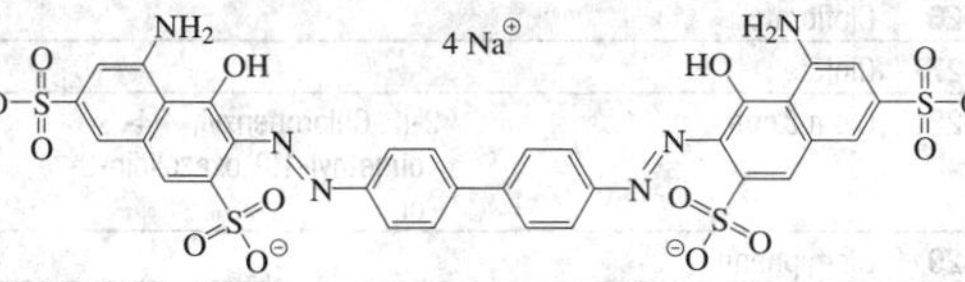

4 Na⊕

2390 C.I. Direct Blue 6, tetrasodium salt

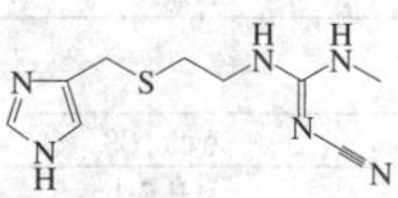

2391 Cimetidine

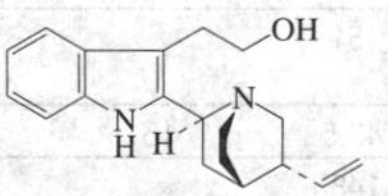

2392 Cinchonamine

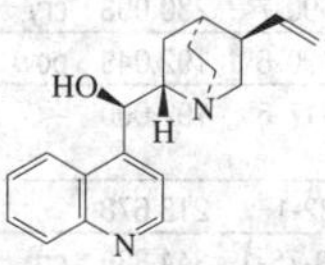

2393 Cinchonidine

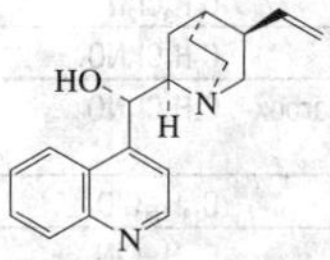

2394 Cinchonine

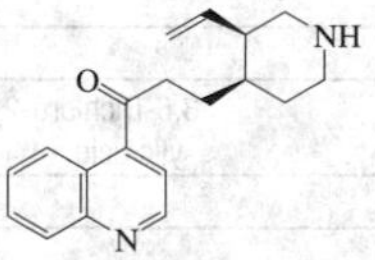

2395 Cinchotoxine

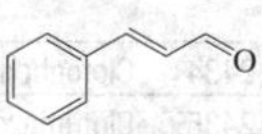

2396 trans-Cinnamaldehyde

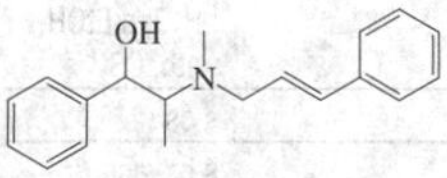

2397 Cinnamedrine

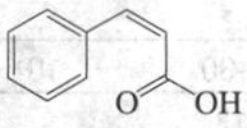

2398 cis-Cinnamic acid

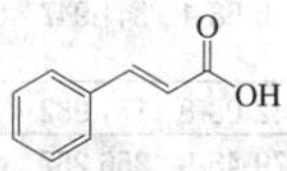

2399 trans-Cinnamic acid

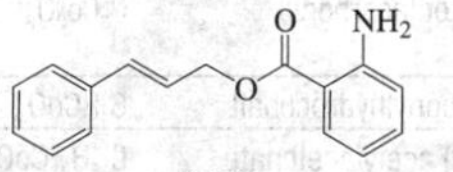

2400 trans-Cinnamyl anthranilate

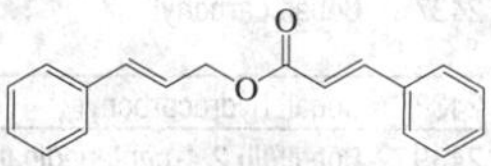

2401 Cinnamyl cinnamate

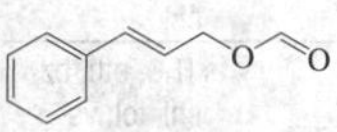

2402 Cinnamyl formate

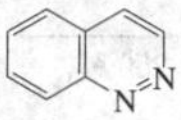

2403 Cinnoline

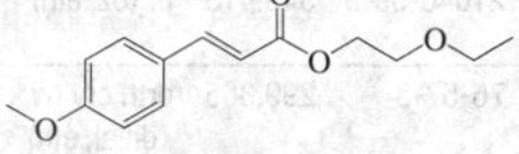

2404 Cinoxate

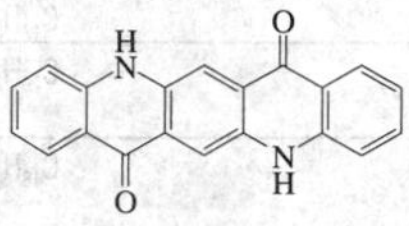

2405 Cinquasia Red

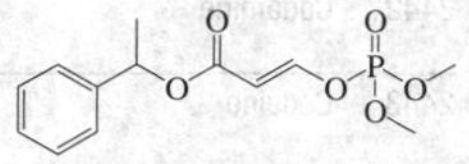

2406 Ciodrin

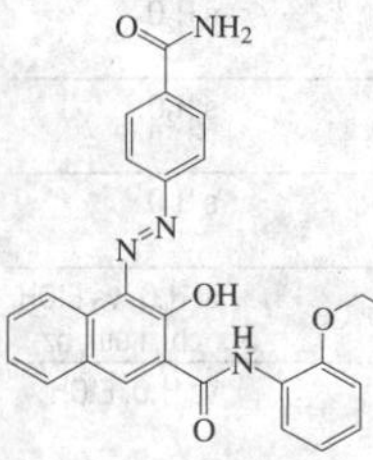

2407 C.I. Pigment Red 170

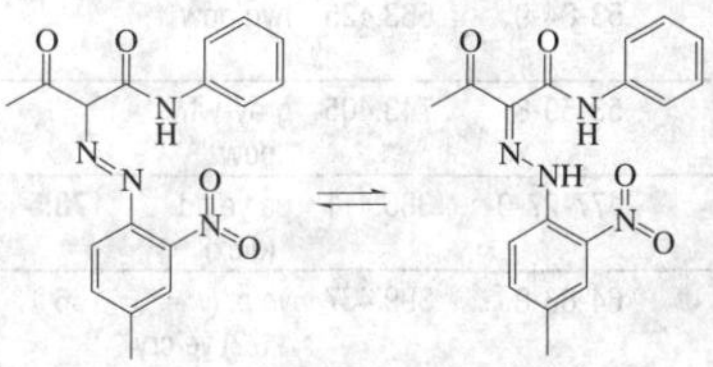

2408 C.I. Pigment Yellow 1

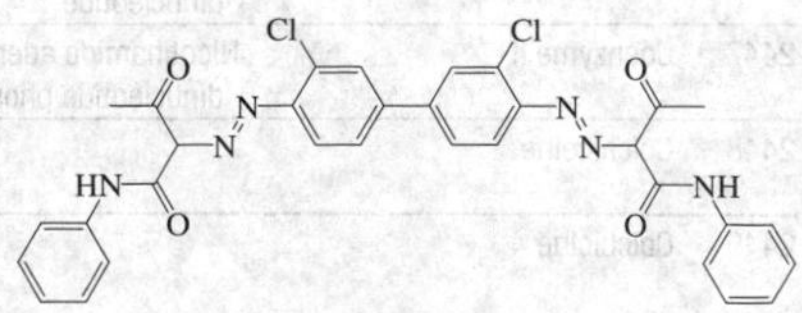

2409 C.I. Pigment Yellow 12

No.	Name	Synonym	Mol. Form.	CAS RN	Mol. Wt.	Physical Form	mp/°C	bp/°C	den g cm^{-3}	n_D	Solubility
2410	Cisapride		C$_{23}$H$_{29}$ClFN$_3$O$_4$	81098-60-4	465.945	cry (hp)	132				
2411	Citral	3,7-Dimethyl-2,6-octadienal	C$_{10}$H$_{16}$O	141-27-5	152.233			229	0.8888^{20}	1.4898^{20}	i H$_2$O; msc EtOH, eth
2412	β-Citraurin		C$_{30}$H$_{40}$O$_2$	650-69-1	432.638	pl (bz-peth), cry (al)	147				i H$_2$O; vs EtOH, eth, ace, bz; sl lig
2413	Citrazinic acid	1,2-Dihydro-6-hydroxy-2-oxo-4-pyridinecarboxylic acid	C$_6$H$_5$NO$_4$	99-11-6	155.109	ye pow	>300 dec				s H$_2$O, alk; sl HCl
2414	Citric acid	2-Hydroxy-1,2,3-propanetri-carboxylic acid	C$_6$H$_8$O$_7$	77-92-9	192.124	orth (w+1)	153	dec	1.665^{20}		vs H$_2$O, EtOH; s eth, AcOEt; i bz, chl
2415	Citric acid monohydrate	2-Hydroxy-1,2,3-propanetri-carboxylic acid, monohydrate	C$_6$H$_{10}$O$_8$	5949-29-1	210.138	cry (w)	135		1.542		vs H$_2$O; vs EtOH, eth
2416	Citrinin	Antimycin	C$_{13}$H$_{14}$O$_5$	518-75-2	250.247	ye nd (MeOH)	178 dec				i H$_2$O; sl EtOH, eth; s ace, bz
2417	Citrulline	N^5-(Aminocarbonyl)-L-ornithine	C$_6$H$_{13}$N$_3$O$_3$	372-75-8	175.185	pr (aq MeOH)	222				s H$_2$O; i EtOH, MeOH
2418	Citrus Red 2		C$_{18}$H$_{16}$N$_2$O$_3$	6358-53-8	308.331	cry	156				sl H$_2$O; s EtOH
2419	C.I. Vat Blue 6	7,16-Dichloro-6,15-dihydro-5,9,14,18-anthrazinetetrone	C$_{28}$H$_{12}$Cl$_2$N$_2$O$_4$	130-20-1	511.312	viol-bl pow					
2420	C.I. Vat Yellow 4	Anthanthrone	C$_{24}$H$_{12}$O$_2$	128-66-5	332.351	ye cry					
2421	Clayton Yellow	Thiazol Yellow G	C$_{28}$H$_{19}$N$_5$Na$_2$O$_6$S$_4$	1829-00-1	695.721	ye-br pow					s H$_2$O, EtOH, H$_2$SO$_4$
2422	Clemastine fumarate		C$_{25}$H$_{30}$ClNO$_5$	14976-57-9	459.963		181				
2423	Clindamycin		C$_{18}$H$_{33}$ClN$_2$O$_5$S	18323-44-9	424.983	ye amorp solid					
2424	Cloconazole		C$_{18}$H$_{15}$ClN$_2$O	77175-51-0	310.777		73				s EtOAc
2425	Clofentezine	3,6-Bis(2-chlorophenyl)-1,2,4,5-tetrazine	C$_{14}$H$_8$Cl$_2$N$_4$	74115-24-5	303.147		182				
2426	Clofibrate		C$_{12}$H$_{15}$ClO$_3$	637-07-0	242.698			149^{20}			
2427	Cloforex		C$_{13}$H$_{18}$ClNO$_2$	14261-75-7	255.741	cry	52.8	89$^{0.005}$			
2428	Clomazone	2-(2-Chlorobenzyl)-4,4-dimethyl-1,2-oxazolidin-3-one	C$_{12}$H$_{14}$ClNO$_2$	81777-89-1	239.698				1.192^{20}		
2429	Clomiphene		C$_{26}$H$_{28}$ClNO	911-45-5	405.959		117				
2430	Clonazepam		C$_{15}$H$_{10}$ClN$_3$O$_3$	1622-61-3	315.711	wh cry	237.5				i H$_2$O, bz; sl ace, MeOH, chl
2431	Clonidine		C$_9$H$_9$Cl$_2$N$_3$	4205-90-7	230.093	cry	137				
2432	Clopidol		C$_7$H$_7$Cl$_2$NO	2971-90-6	192.043	pow	>320				i H$_2$O
2433	Clopyralid	3,6-Dichloro-2-pyridinecarbox-ylic acid	C$_6$H$_3$Cl$_2$NO$_2$	1702-17-6	192.000		151				
2434	Clorophene		C$_{13}$H$_{11}$ClO	120-32-1	218.678		48.5	161$^{3.5}$	1.185^{58}		s ctc, CS$_2$
2435	Clotrimazole		C$_{22}$H$_{17}$ClN$_2$	23593-75-1	344.836	cry	148				sl H$_2$O, bz; s ace, chl, AcOEt, DMF
2436	Clozapine	Clozaril	C$_{18}$H$_{19}$ClN$_4$	5786-21-0	326.824	ye cry	183(1)				
2437	Cobalt carbonyl	Dicobalt octacarbonyl	C$_8$Co$_2$O$_8$	10210-68-1	341.947	oran cry	51 dec		1.78		i H$_2$O; s EtOH, eth, CS$_2$
2438	Cobalt hydrocarbonyl	Tetracarbonylhydrocobalt	C$_4$HCoO$_4$	16842-03-8	171.982	ye liq or gas	≈-30	10			s os
2439	Cobalt(III) 2,4-pentanedioate	Cobalt(III) acetylacetonate	C$_{15}$H$_{21}$CoO$_6$	21679-46-9	356.257	dark grn cry	213				s bz, ace
2440	Cocaine		C$_{17}$H$_{21}$NO$_4$	50-36-2	303.354	mcl pr (al)	98	187$^{0.1}$		1.5022^{98}	sl H$_2$O; vs EtOH, eth, bz, py; s CS$_2$
2441	Coclaurine		C$_{17}$H$_{19}$NO$_3$	486-39-5	285.338	pl (al)	220.5				
2442	Codamine		C$_{20}$H$_{25}$NO$_4$	21040-59-5	343.418	pr (bz, eth)	127				vs eth, EtOH, chl
2443	Codeine		C$_{18}$H$_{21}$NO$_3$	76-57-3	299.365	orth cry (w, dil al, eth)	157.5	250^{22}	1.32^{25}		s H$_2$O, eth, bz, chl, tol; vs EtOH; i peth
2444	Codeine phosphate		C$_{18}$H$_{24}$NO$_7$P	52-28-8	397.361	lf or pr (dil al)	227 dec				vs EtOH, chl
2445	Coenzyme A		C$_{21}$H$_{36}$N$_7$O$_{16}$P$_3$S	85-61-0	767.535	pow; unstab in air					s H$_2$O
2446	Coenzyme I	Nicotinamide adenine dinucleotide	C$_{21}$H$_{27}$N$_7$O$_{14}$P$_2$	53-84-9	663.425	hyg pow					s H$_2$O
2447	Coenzyme II	Nicotinamide adenine dinucleotide phosphate	C$_{21}$H$_{28}$N$_7$O$_{17}$P$_3$	53-59-8	743.405	gray-wh pow					s H$_2$O
2448	Colchiceine		C$_{21}$H$_{23}$NO$_6$	477-27-0	385.411	pa ye nd (diox)	178.5		1.24^{25}		sl H$_2$O; vs EtOH, chl; i eth, bz
2449	Colchicine		C$_{22}$H$_{25}$NO$_6$	64-86-8	399.437	ye pl (w + 1/2) ye cry (bz)	156				vs H$_2$O, EtOH

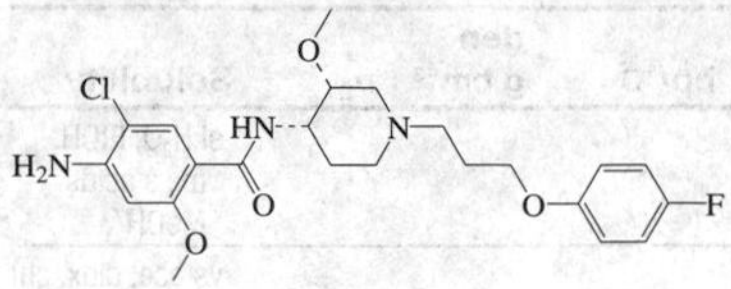

2410 Cisapride

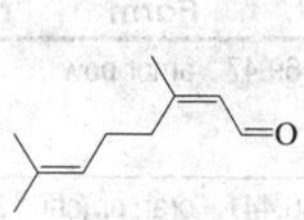

2411 Citral

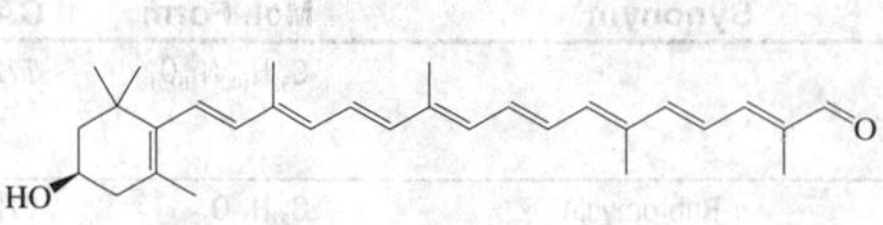

2412 β-Citraurin

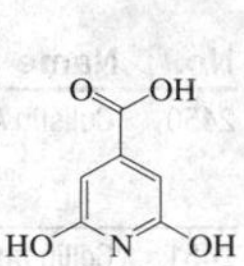

2413 Citrazinic acid

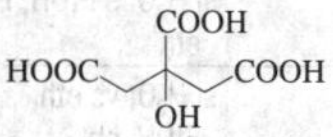

2414 Citric acid

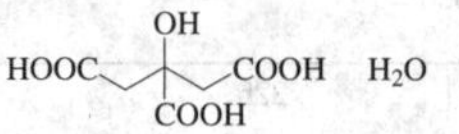

2415 Citric acid monohydrate

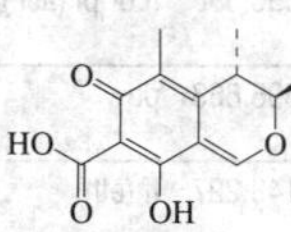

2416 Citrinin

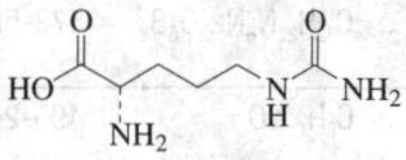

2417 Citrulline

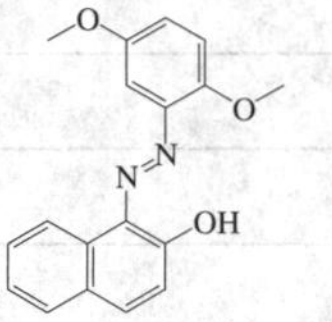

2418 Citrus Red 2

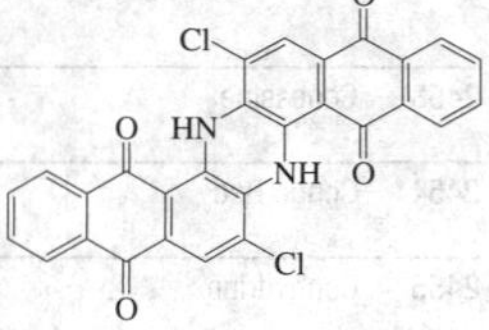

2419 C.I. Vat Blue 6

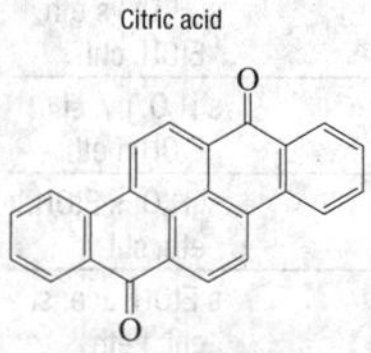

2420 C.I. Vat Yellow 4

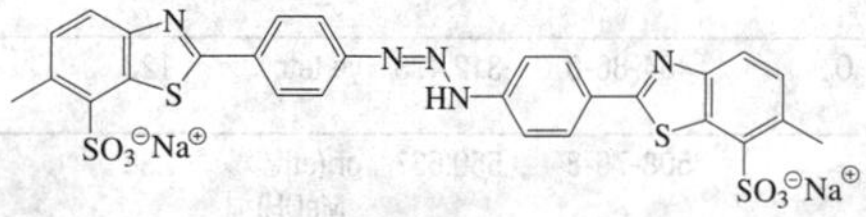

2421 Clayton Yellow

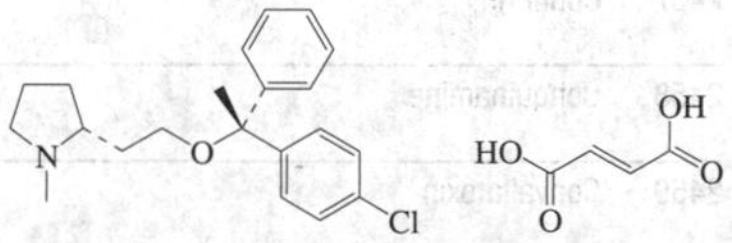

2422 Clemastine fumarate

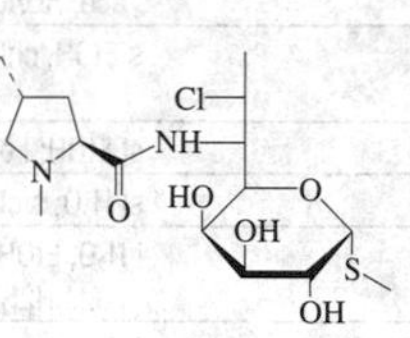

2423 Clindamycin

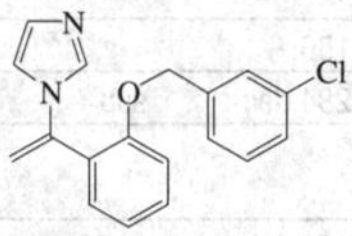

2424 Cloconazole

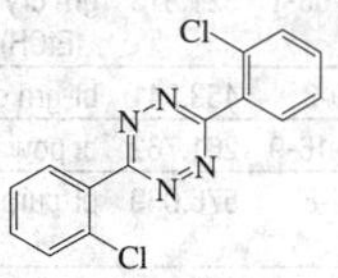

2425 Clofentezine

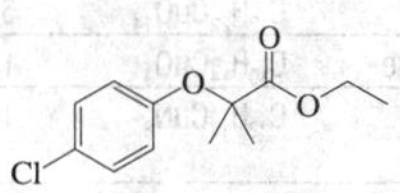

2426 Clofibrate

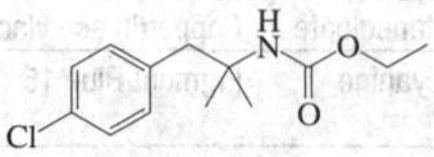

2427 Cloforex

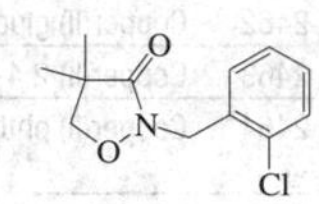

2428 Clomazone

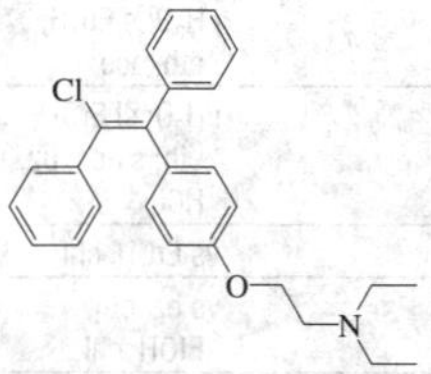

2429 Clomiphene

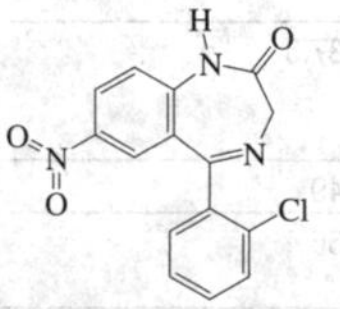

2430 Clonazepam

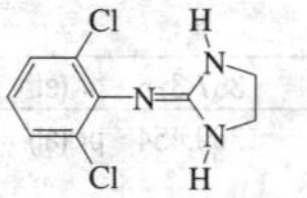

2431 Clonidine

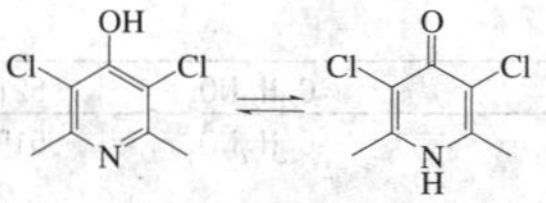

2432 Clopidol

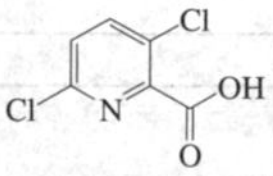

2433 Clopyralid

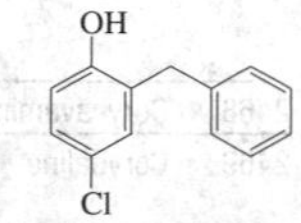

2434 Clorophene

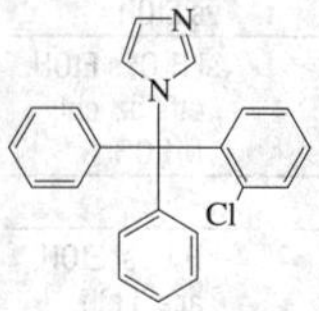

2435 Clotrimazole

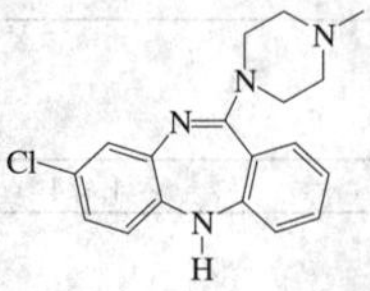

2436 Clozapine

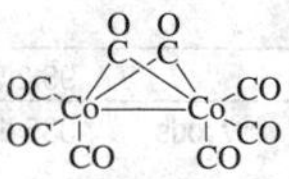

2437 Cobalt carbonyl

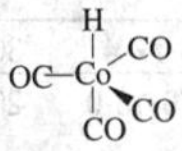

2438 Cobalt hydrocarbonyl

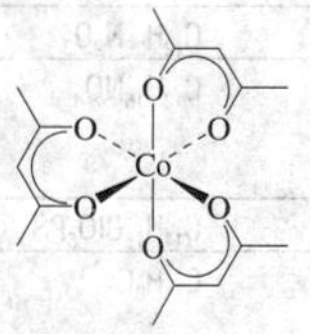

2439 Cobalt(III) 2,4-pentanedioate

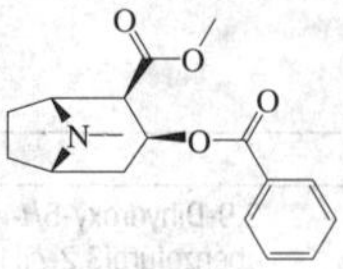

2440 Cocaine

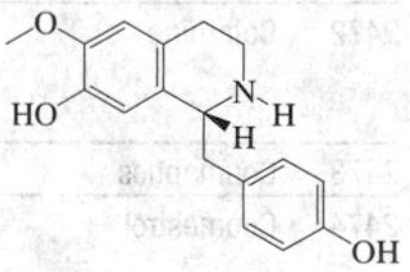

2441 Coclaurine

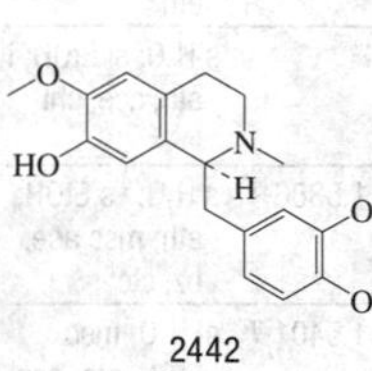

2442 Codamine

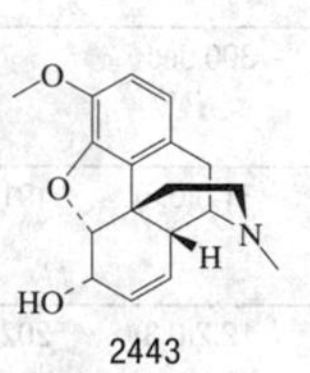

2443 Codeine

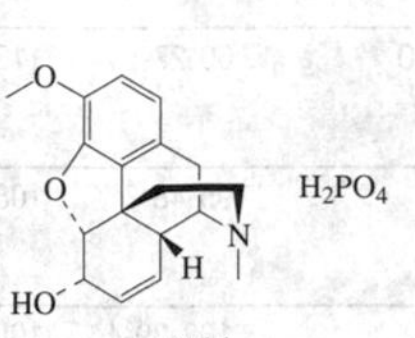

2444 Codeine phosphate

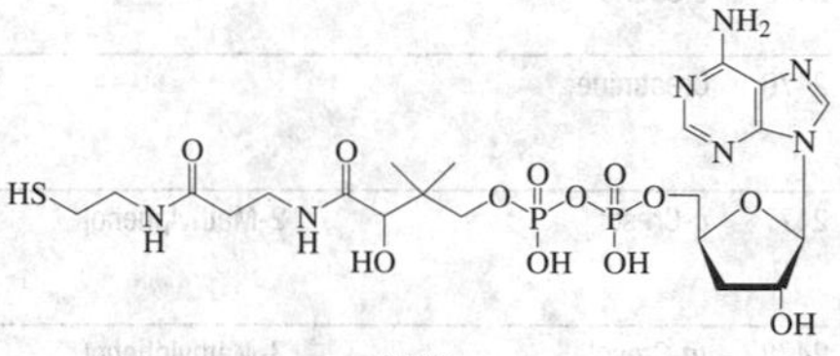

2445 Coenzyme A

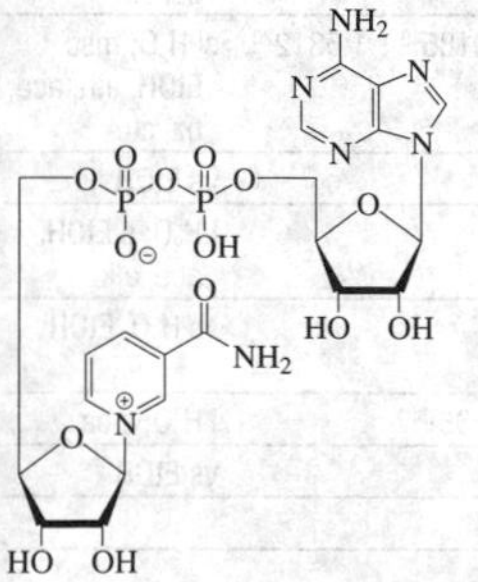

2446 Coenzyme I

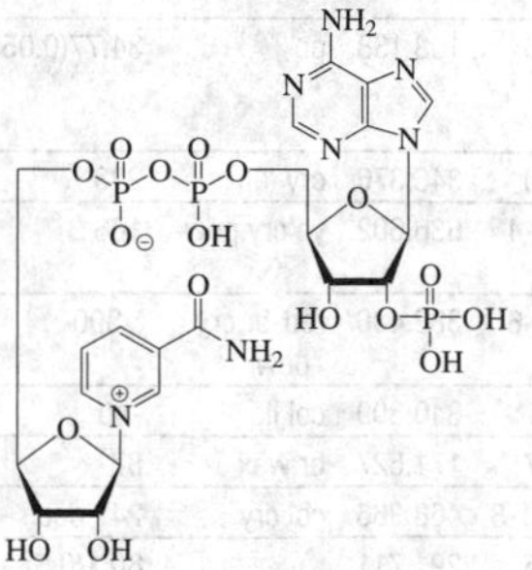

2447 Coenzyme II

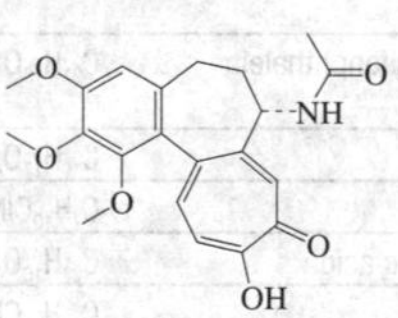

2448 Colchiceine

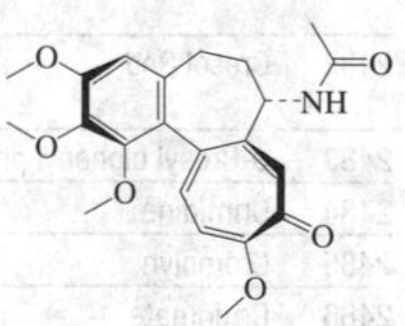

2449 Colchicine

No.	Name	Synonym	Mol. Form.	CAS RN	Mol. Wt.	Physical Form	mp/°C	bp/°C	den g cm⁻³	n_D	Solubility
2450	Colistin A		$C_{53}H_{100}N_{16}O_{13}$	7722-44-3	1169.47	amor pow					sl H_2O, EtOH, hx; s acids, MeOH
2451	Collinomycin	α-Rubromycin	$C_{27}H_{20}O_{12}$	27267-69-2	536.441	oran pr (chl-MeOH)	281				vs ace, diox, chl
2452	Columbin		$C_{20}H_{22}O_6$	546-97-4	358.385	nd (MeOH)	195.5				i H_2O; sl ace, AcOEt, MeOH; s chl
2453	Conessine		$C_{24}H_{40}N_2$	546-06-5	356.588	lf or pl (ace)	125.5	166[0.1]			sl H_2O; s chl, HOAc
2454	Congo Red		$C_{32}H_{22}N_6Na_2O_6S_2$	573-58-0	696.663	pow	>360				sl H_2O; s EtOH; i eth
2455	Conhydrine		$C_8H_{17}NO$	495-20-5	143.227	lf (eth)	121	226			sl H_2O; vs eth, EtOH, chl
2456	Conhydrine, (+)	2-(α-Hydroxypropyl)piperidine	$C_8H_{17}NO$	495-20-5	143.227	lf (eth)	121	226			sl H_2O; vs eth, EtOH, chl
2457	Coniferin		$C_{16}H_{22}O_8$	531-29-3	342.341	nd (w+2)	186				s H_2O, py; sl EtOH; i eth
2458	Conquinamine		$C_{19}H_{24}N_2O_2$	464-86-8	312.406	ye tetr	123				sl H_2O; s EtOH, eth, chl
2459	Convallatoxin		$C_{29}H_{42}O_{10}$	508-75-8	550.637	pr (eth/MeOH)	238				s EtOH, ace; sl chl; i eth
2460	Copaene		$C_{15}H_{24}$	3856-25-5	204.352			248.5	0.8996[20]	1.4894[20]	i H_2O; s eth, ace, HOAc, lig
2461	Copper(II) ethylacetoacetate	Bis(ethylacetoacetato)copper	$C_{12}H_{18}CuO_6$	14284-06-1	321.813	grn cry (EtOH)	192				s EtOH, chl
2462	Copper(II) gluconate	Cupric gluconate	$C_{12}H_{22}CuO_{14}$	527-09-3	453.841	bl-grn cry	156				sl EtOH; i os
2463	Copper(II) 2,4-pentanedioate	Copper(II) acetylacetonate	$C_{10}H_{14}CuO_4$	13395-16-9	261.762	bl pow	284 dec	sub			sl H_2O; s chl
2464	Copper(II) phthalocyanine	Pigment Blue 15	$C_{32}H_{16}CuN_8$	147-14-8	576.069	bl-purp cry					i H_2O, EtOH; s conc $H2SO4$
2465	Coronene		$C_{24}H_{12}$	191-07-1	300.352	ye nd (bz)	437.3(0.3)	525	1.371[25]		i H_2O, con sulf; sl bz
2466	Corticosterone		$C_{21}H_{30}O_4$	50-22-6	346.461	nd (al, pl) (ace)	181				i H_2O; s EtOH, eth, ace
2467	Corybulbine		$C_{21}H_{25}NO_4$	518-77-4	355.429	nd (al)	237.5				i H_2O; sl EtOH, eth; s ace, bz, HCl
2468	Corycavamine		$C_{21}H_{21}NO_5$	521-85-7	367.396	pr (eth, al)	149				vs EtOH, chl
2469	Corydaline		$C_{22}H_{27}NO_4$	518-69-4	369.454	pr (al)	136				vs bz, eth, EtOH, chl
2470	Corydine		$C_{20}H_{23}NO_4$	476-69-7	341.402	tetr pr (eth)	149				vs eth, EtOH, chl
2471	Corynantheine		$C_{22}H_{26}N_2O_3$	18904-54-6	366.452		165.5				vs EtOH
2472	Cotarnine		$C_{12}H_{15}NO_4$	82-54-2	237.252	nd (bz), cry (eth)	132 dec				sl H_2O; s EtOH, eth, bz, chl, NH_4OH
2473	Coumaphos		$C_{14}H_{16}ClO_5PS$	56-72-4	362.766		95.2(0.2)		1.474		
2474	Coumestrol	3,9-Dihydroxy-6H-benzofuro[3,2-c][1]benzopyran-6-one	$C_{15}H_8O_5$	479-13-0	268.222	cry rods	385 dec				i H_2O; sl EtOH, ace; i eth
2475	Creatine		$C_4H_9N_3O_2$	57-00-1	131.133	mcl pr (w+1)	303 dec		1.33[25]		s H_2O; sl EtOH; i eth
2476	Creatinine		$C_4H_7N_3O$	60-27-5	113.118	orth pr (w+2) lf (w)	300 dec				s H_2O; sl EtOH; i eth, ace, chl
2477	o-Cresol	2-Methylphenol	C_7H_8O	95-48-7	108.138		31.0(0.6)	191.0(0.1)	1.0327[35]	1.5386[35]	s H_2O; vs EtOH, eth; msc ace, bz, ctc
2478	m-Cresol	3-Methylphenol	C_7H_8O	108-39-4	108.138	liq	12.2(0.3)	202.2(0.1)	1.0339[20]	1.5401[20]	sl H_2O; msc EtOH, eth, ace, bz, ctc
2479	p-Cresol	4-Methylphenol	C_7H_8O	106-44-5	108.138	pr	34.77(0.05)	201.9(0.1)	1.0185[40]	1.5312[20]	sl H_2O; msc EtOH, eth, ace, bz, ctc
2480	o-Cresolphthalein		$C_{22}H_{18}O_4$	596-27-0	346.376	cry (al)	223				vs EtOH
2481	o-Cresolphthalein complexone	Metalphthalein	$C_{32}H_{32}N_2O_{12}$	2411-89-4	636.602	ye cry pow	186				i H_2O; s EtOH, ace, alk
2482	Cresol Red	o-Cresolsulfonphthalein	$C_{21}H_{18}O_5S$	1733-12-6	382.430	red-br cry pow	>300				vs H_2O, EtOH
2483	p-Cresyl diphenyl phosphate		$C_{19}H_{17}O_4P$	78-31-9	340.309	col liq	-40		1.208[25]		i H_2O; s os
2484	Crimidine		$C_7H_{10}ClN_3$	535-89-7	171.627	br wax	87	143[4]			vs EtOH
2485	Cromolyn	Cromoglicic acid	$C_{23}H_{16}O_{11}$	16110-51-3	468.366	col cry	241 dec				
2486	Crufomate		$C_{12}H_{19}ClNO_3P$	299-86-5	291.711		60.1(0.5)	118[0.01]			

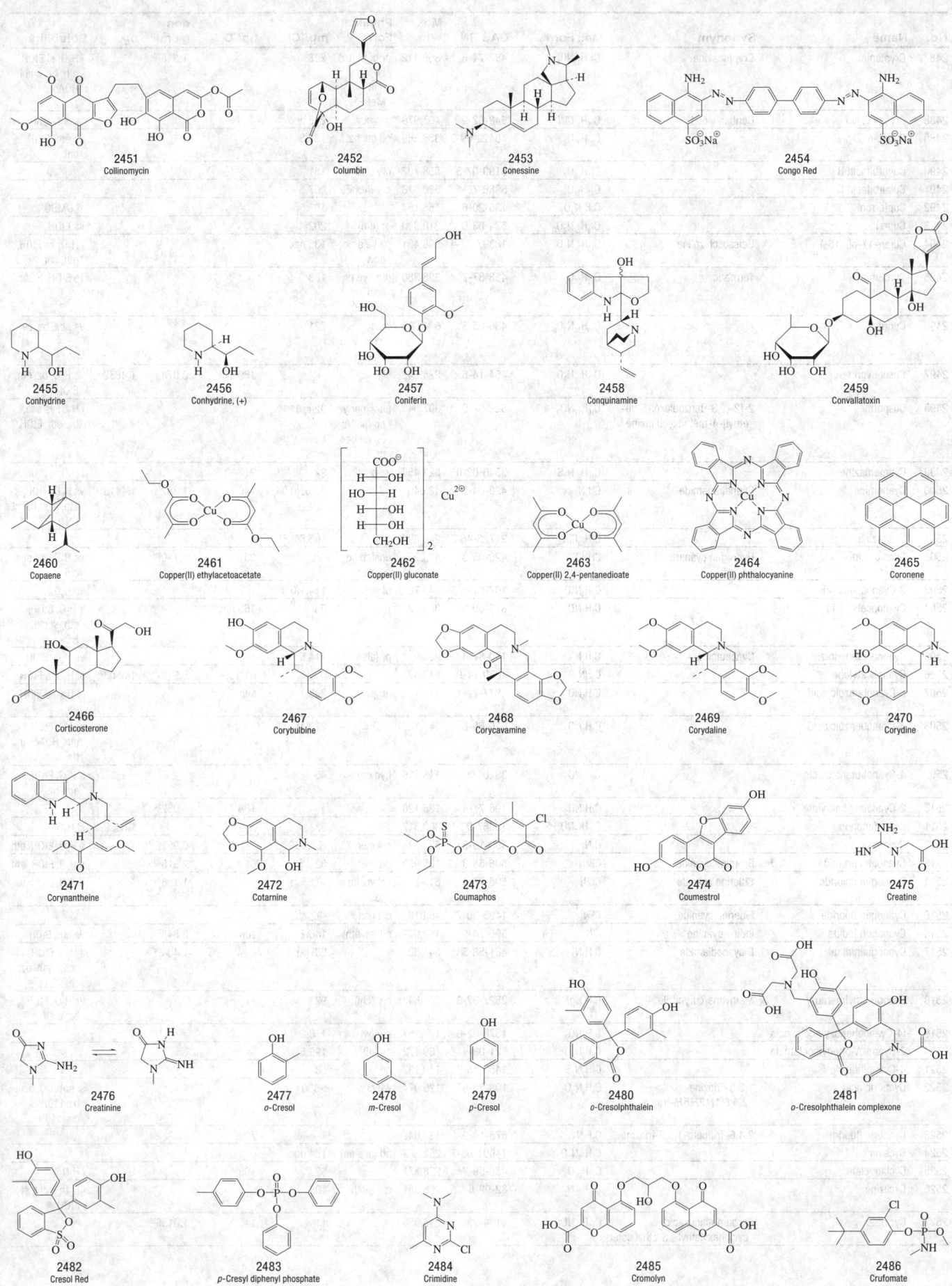

2451 Collinomycin

2452 Columbin

2453 Conessine

2454 Congo Red

2455 Conhydrine

2456 Conhydrine, (+)

2457 Coniferin

2458 Conquinamine

2459 Convallatoxin

2460 Copaene

2461 Copper(II) ethylacetoacetate

2462 Copper(II) gluconate

2463 Copper(II) 2,4-pentanedioate

2464 Copper(II) phthalocyanine

2465 Coronene

2466 Corticosterone

2467 Corybulbine

2468 Corycavamine

2469 Corydaline

2470 Corydine

2471 Corynantheine

2472 Cotarnine

2473 Coumaphos

2474 Coumestrol

2475 Creatine

2476 Creatinine

2477 o-Cresol

2478 m-Cresol

2479 p-Cresol

2480 o-Cresolphthalein

2481 o-Cresolphthalein complexone

2482 Cresol Red

2483 p-Cresyl diphenyl phosphate

2484 Crimidine

2485 Cromolyn

2486 Crufomate

No.	Name	Synonym	Mol. Form.	CAS RN	Mol. Wt.	Physical Form	mp/°C	bp/°C	den g cm⁻³	n_D	Solubility
2487	Cryptopine	Cryptocavine	$C_{21}H_{23}NO_5$	482-74-6	369.412	pr or pl (bz) nd (chl-MeOH)	223		1.315[20]		i H_2O; sl EtOH, eth, bz; s chl, HOAc
2488	Crystal Violet	Gentian violet	$C_{25}H_{30}ClN_3$	548-62-9	407.979	grn pow	215 dec				vs H_2O, chl
2489	Cubebin		$C_{20}H_{20}O_6$	18423-69-3	356.369	nd (al, bz)	131.5				vs eth, EtOH, chl
2490	Cucurbitacin B		$C_{32}H_{46}O_8$	6199-67-3	558.702	cry (EtOH)	181				
2491	Cucurbitacin C		$C_{32}H_{48}O_8$	5988-76-1	560.718	cry (AcOEt)	207.5				
2492	Cupferron		$C_6H_9N_3O_2$	135-20-6	155.154		163.5				sl DMSO
2493	Cupreine		$C_{19}H_{22}N_2O_2$	524-63-0	310.390	pr (eth)	202				vs EtOH
2494	Curan-17-ol, (16α)	Geissoschizoline	$C_{19}H_{26}N_2O$	18397-07-4	298.421	pa ye amor pow	135 dec				i H_2O; vs EtOH, eth, chl
2495	Curcumin	Turmeric	$C_{21}H_{20}O_6$	458-37-7	368.380	oran ye pr, orth pr (MeOH)	183				vs EtOH, HOAc
2496	Curine		$C_{36}H_{38}N_2O_6$	436-05-5	594.696	pr, nd (chl-MeOH)	221				vs ace, bz, py
2497	Cuscohygrine		$C_{13}H_{24}N_2O$	454-14-8	224.342	oil		169[23]	0.9733[20]	1.4832[20]	vs H_2O, bz, eth, EtOH
2498	Cusparine	2-[2-(1,3-Benzodioxol-5-yl)-ethyl]-4-methoxyquinoline	$C_{19}H_{17}NO_3$	529-92-0	307.343	(α) wh or ye nd (peth); (β) amber pr	92(α; 111(β)				i H_2O; vs ace, bz, eth, EtOH
2499	Cyamemazine		$C_{19}H_{21}N_3S$	3546-03-0	323.455	ye pow	92	212[0.25]			i H_2O; s EtOH
2500	Cyanamide	Cyanogenamide	CH_2N_2	420-04-2	42.040	nd	45.55(0.04)	140[19]	1.282[20]	1.4418[20]	vs H_2O, EtOH; s eth, ace, bz; sl CS_2
2501	Cyanazine		$C_9H_{13}ClN_6$	21725-46-2	240.692		165.6(0.5)				
2502	Cyanic acid	Hydrogen cyanate	CHNO	420-05-3	43.025	unstab liq or gas	-86	23	1.140[20]		vs H_2O, bz, eth, chl
2503	2-Cyanoacetamide		$C_3H_4N_2O$	107-91-5	84.076	pl (w)	114.1(0.3)				vs H_2O
2504	Cyanoacetic acid		$C_3H_3NO_2$	372-09-8	85.062		66	160 dec			s H_2O, EtOH, eth; sl chl, HOAc
2505	Cyanoacetohydrazide	Cyacetacide	$C_3H_5N_3O$	140-87-4	99.091	pr (al)	114.5				vs H_2O, EtOH
2506	Cyanoacetylene		C_3HN	1070-71-9	51.047		5	61(18)	0.8167[17]	1.3868[25]	sl H_2O; s EtOH
2507	3-Cyanobenzoic acid		$C_8H_5NO_2$	1877-72-1	147.132	nd (w)	223(1)	sub			sl H_2O; s EtOH, eth
2508	4-Cyanobenzoic acid		$C_8H_5NO_2$	619-65-8	147.132		220(1)				s H_2O, EtOH, eth, HOAc; sl tfa
2509	4-Cyanobutanoic acid		$C_5H_7NO_2$	39201-33-7	113.116	hyg cry	45				s H_2O, EtOH, eth, bz
2510	2-Cyanoethyl acrylate		$C_6H_7NO_2$	106-71-8	125.126			108[12]	1.062[20]		
2511	Cyanofenphos		$C_{15}H_{14}NO_2PS$	13067-93-1	303.317		83			1.5839[25]	sl H_2O
2512	Cyanogen		C_2N_2	460-19-5	52.034	col gas	-27.83	-21.1	0.9537[-21]		s H_2O, EtOH, eth
2513	Cyanogen bromide	Bromine cyanide	CBrN	506-68-3	105.922	nd	52	61.5	2.015[20]		s H_2O, EtOH, eth
2514	Cyanogen chloride	Chlorine cyanide	CClN	506-77-4	61.471	col vol liq or gas	-6.55	13	1.186[20]		s H_2O, EtOH; vs eth
2515	Cyanogen fluoride	Fluorine cyanide	CFN	1495-50-7	45.016	col gas	-82	-46			
2516	Cyanogen iodide	Iodine cyanide	CIN	506-78-5	152.922	nd (al, eth)	146.7	sub	2.84[18]		vs eth, EtOH
2517	Cyanoguanidine	Dicyanodiamide	$C_2H_4N_4$	461-58-5	84.080		207(2)		1.404[14]		s H_2O, EtOH, ace; i eth, bz, chl
2518	Cyanomethylmercury	Methylmercurynitrile	C_2H_3HgN	2597-97-9	241.64	cry (chl)	92				vs H_2O, EtOH, bz; s eth
2519	(4-Cyanophenoxy)acetic acid		$C_9H_7NO_3$	1878-82-6	177.157	cry (w)	178				
2520	2-Cyano-N-phenylacetamide		$C_9H_8N_2O$	621-03-4	160.172	nd (al)	199.5				
2521	4-Cyanothiazole		$C_4H_2N_2S$	1452-15-9	110.137	nd	58				
2522	Cyanuric acid	1,3,5-Triazine-2,4,6(1H,3H,5H)-trione	$C_3H_3N_3O_3$	108-80-5	129.074	wh cry	>330		1.75[25]		sl hot H_2O, ace, bz, EtOH; s conc HCl
2523	Cyanuric fluoride	2,4,6-Trifluoro-1,3,5-triazine	$C_3F_3N_3$	675-14-9	135.047			72.8			
2524	Cycasin		$C_8H_{16}N_2O_7$	14901-08-7	252.222	nd (ace aq)	154 dec				
2525	Cyclandelate		$C_{17}H_{24}O_3$	456-59-7	276.371		52	193[14]			i H_2O
2526	Cyclizine		$C_{18}H_{22}N_2$	82-92-8	266.381	cry (peth)	106				i H_2O; s chl; sl EtOH
2527	Cycloate	Carbamothioic acid, cyclohexylethyl-, S-ethyl ester	$C_{11}H_{21}NOS$	1134-23-2	215.356		11.5	145[10]	1.0156[30]		

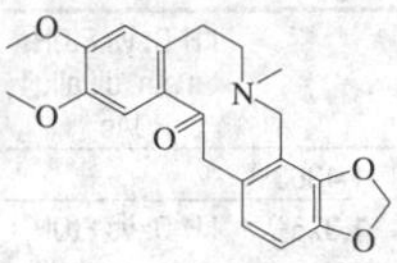

2487 Cryptopine

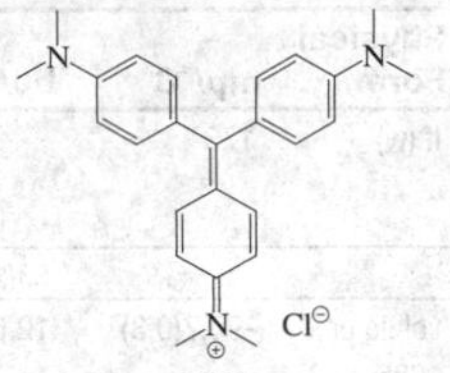

2488 Crystal Violet

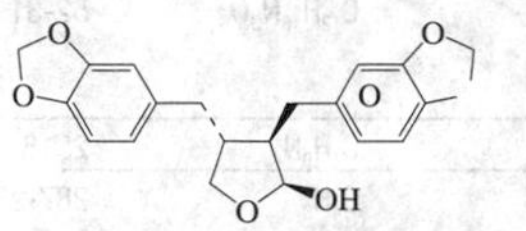

2489 Cubebin

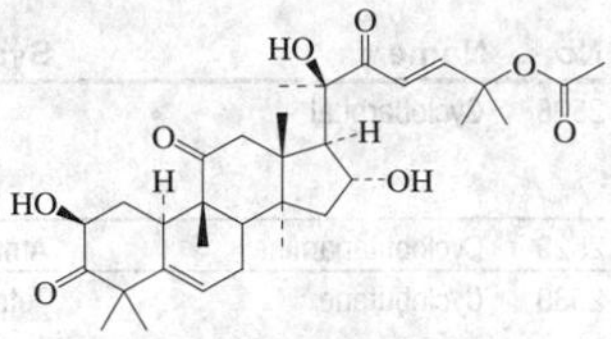

2490 Cucurbitacin B

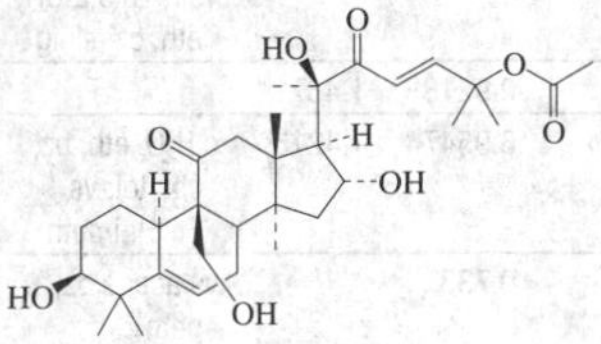

2491 Cucurbitacin C

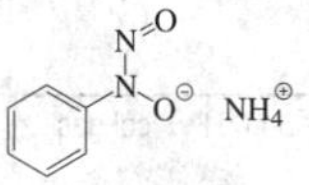

2492 Cupferron

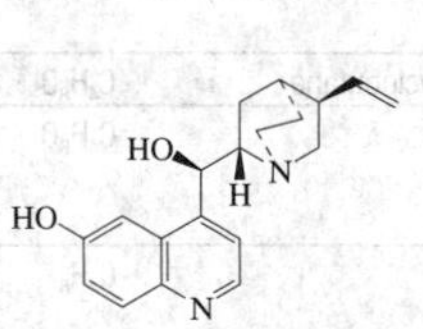

2493 Cupreine

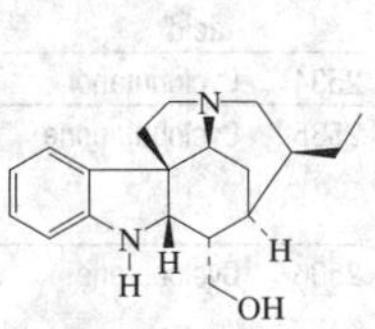

2494 Curan-17-ol, (16α)

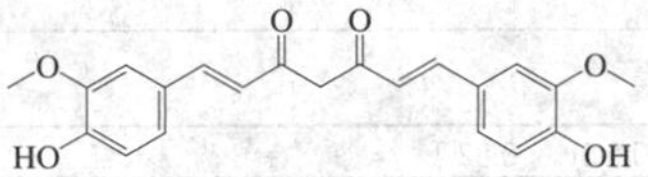

2495 Curcumin

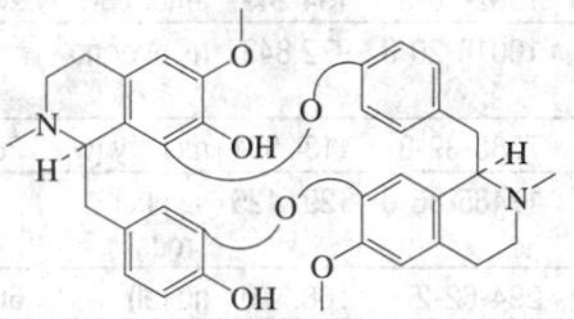

2496 Curine

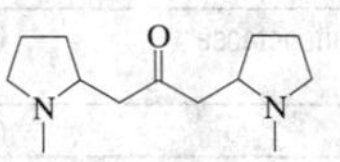

2497 Cuscohygrine

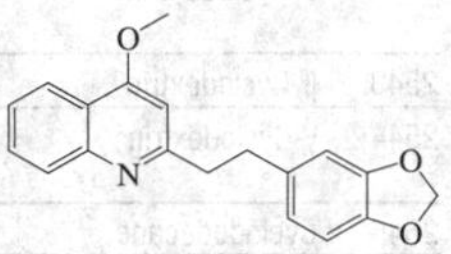

2498 Cusparine

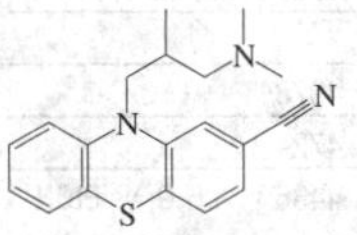

2499 Cyamemazine

2500 Cyanamide

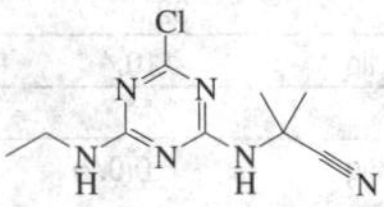

2501 Cyanazine

2502 Cyanic acid

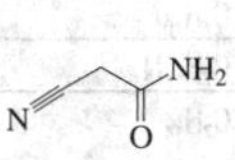

2503 2-Cyanoacetamide

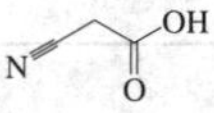

2504 Cyanoacetic acid

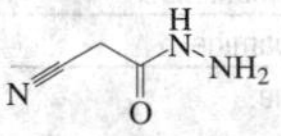

2505 Cyanoacetohydrazide

2506 Cyanoacetylene

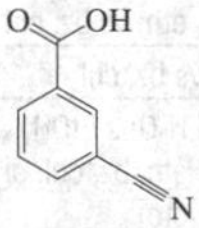

2507 3-Cyanobenzoic acid

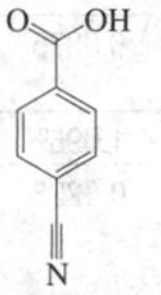

2508 4-Cyanobenzoic acid

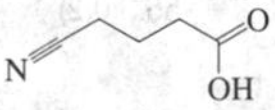

2509 4-Cyanobutanoic acid

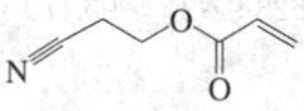

2510 2-Cyanoethyl acrylate

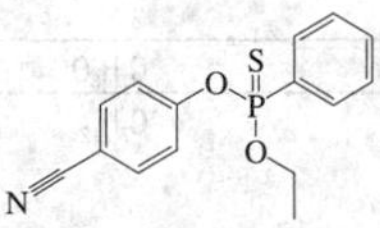

2511 Cyanofenphos

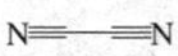

2512 Cyanogen

2513 Cyanogen bromide

2514 Cyanogen chloride

2515 Cyanogen fluoride

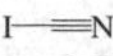

2516 Cyanogen iodide

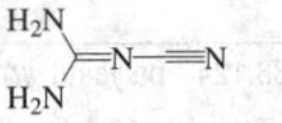

2517 Cyanoguanidine

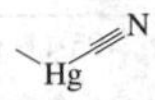

2518 Cyanomethylmercury

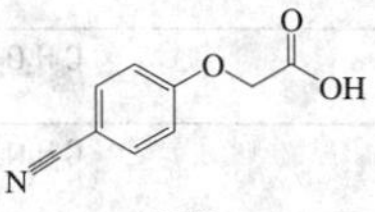

2519 (4-Cyanophenoxy)acetic acid

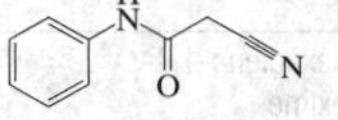

2520 2-Cyano-N-phenylacetamide

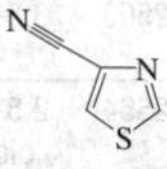

2521 4-Cyanothiazole

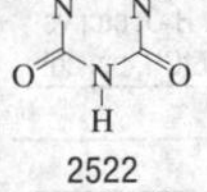

2522 Cyanuric acid

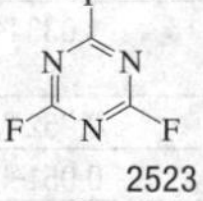

2523 Cyanuric fluoride

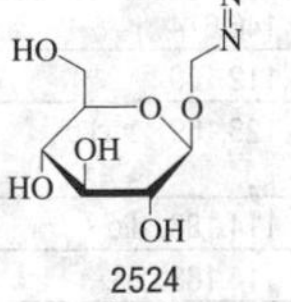

2524 Cycasin

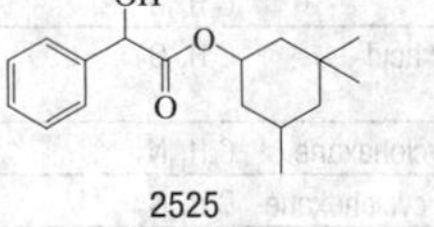

2525 Cyclandelate

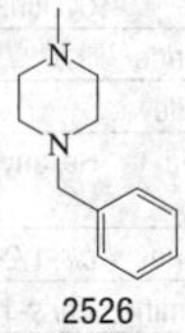

2526 Cyclizine

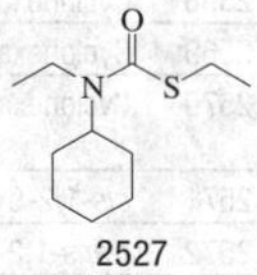

2527 Cycloate

No.	Name	Synonym	Mol. Form.	CAS RN	Mol. Wt.	Physical Form	mp/°C	bp/°C	den g cm⁻³	n_D	Solubility
2528	Cyclobarbital		$C_{12}H_{16}N_2O_3$	52-31-3	236.266	lf (w)	173				i H_2O; vs EtOH; s eth, dil alk; sl HOAc
2529	Cyclobutanamine	Aminocyclobutane	C_4H_9N	2516-34-9	71.121			83(20)	0.8328²⁰	1.4363¹⁹	
2530	Cyclobutane	Tetramethylene	C_4H_8	287-23-0	56.107	vol liq or gas	-90.7(0.3)	12.5(0.2)	0.7038⁰	1.375²⁰	i H_2O; vs EtOH, ace; msc eth; s bz
2531	Cyclobutanecarbonitrile	Cyanocyclobutane	C_5H_7N	4426-11-3	81.117			148(1)			
2532	Cyclobutanecarboxylic acid		$C_5H_8O_2$	3721-95-7	100.117	liq	-7(1)	192(4)	1.0599²⁰	1.4400²⁰	sl H_2O; msc EtOH, eth
2533	1,1-Cyclobutanedicarboxylic acid		$C_6H_8O_4$	5445-51-2	144.126	pr (w, eth)	158.0				vs H_2O; s EtOH, eth, bz; sl lig
2534	Cyclobutanol	Hydroxycyclobutane	C_4H_8O	2919-23-5	72.106			124(8)	0.9218¹⁵	1.4371²⁰	
2535	Cyclobutanone		C_4H_6O	1191-95-3	70.090	liq	-50.9	98.8(0.6)	0.9547⁰	1.4215²⁰	s H_2O, eth, bz, chl, tol; vs EtOH; i peth
2536	Cyclobutene		C_4H_6	822-35-5	54.091	col gas		2.5(0.4)	0.733⁰		vs ace; s bz, peth
2537	Cyclochlorotine		$C_{24}H_{31}Cl_2N_5O_7$	12663-46-6	572.439	nd (MeOH)	255 dec				
2538	Cyclodecane		$C_{10}H_{20}$	293-96-9	140.266		10.4(0.9)	202.3(0.3)	0.8538²⁵	1.4716²⁰	
2539	1,2-Cyclodecanedione	Sebacil	$C_{10}H_{16}O_2$	96-01-5	168.233		41(3)	104¹⁰			
2540	Cyclodecanol		$C_{10}H_{20}O$	1502-05-2	156.265		42(3)	125¹²	0.9606²⁰	1.4926²⁰	s EtOH
2541	Cyclodecanone		$C_{10}H_{18}O$	1502-06-3	154.249	amor pow	23(2)	106¹³	0.9654²⁰	1.4806²⁰	vs bz, eth, chl
2542	α-Cyclodextrin	Cyclomaltohexaose	$C_{36}H_{60}O_{30}$	10016-20-3	972.843	hx pl or nd					vs cold H_2O; i hot H_2O
2543	β-Cyclodextrin	Cyclomaltoheptaose	$C_{42}H_{70}O_{35}$	7585-39-9	1134.984	mcl cry (w)	260 dec				
2544	γ-Cyclodextrin	Cyclomaltooctaose	$C_{48}H_{80}O_{40}$	17465-86-0	1297.125	sq pl or rods					
2545	Cyclododecane		$C_{12}H_{24}$	294-62-2	168.319	nd (al)	60.8(0.4)	244.0(0.5)	0.82⁸⁰		
2546	Cyclododecanol		$C_{12}H_{24}O$	1724-39-6	184.318			286			
2547	Cyclododecanone		$C_{12}H_{22}O$	830-13-7	182.302		62.4(0.3)	127¹²	0.9059⁶⁶	1.4571⁶⁰	
2548	1,5,9-Cyclododecatriene	CDT	$C_{12}H_{18}$	4904-61-4	162.271	liq	-17	240	0.84¹⁰⁰		
2549	cis-Cyclododecene		$C_{12}H_{22}$	1129-89-1	166.303			133³⁵		1.4840²⁰	vs bz, chl
2550	trans-Cyclododecene		$C_{12}H_{22}$	1486-75-5	166.303			113¹⁷		1.4850²⁰	vs bz, chl
2551	cis-9-Cycloheptadecen-1-one	Civetone	$C_{17}H_{30}O$	542-46-1	250.419		32.5	343			
2552	1,3-Cycloheptadiene		C_7H_{10}	4054-38-0	94.154	liq	-110.4	120.5	0.868²⁵	1.4978²⁰	
2553	Cycloheptanamine		$C_7H_{15}N$	5452-35-7	113.201			54¹¹		1.4724²⁰	
2554	Cycloheptane		C_7H_{14}	291-64-5	98.186	liq	-8.0(0.2)	118.8(0.2)	0.8098²⁰	1.4436²⁰	i H_2O; vs EtOH, eth; s bz, chl
2555	1,2-Cycloheptanedione		$C_7H_{10}O_2$	3008-39-7	126.153		-40	108¹⁷	1.0583²²	1.4689²²	s EtOH
2556	Cycloheptanol		$C_7H_{14}O$	502-41-0	114.185		7.15(0.05)	185	0.9554²⁰	1.40705²⁰	sl H_2O; vs EtOH, eth
2557	Cycloheptanone	Suberone	$C_7H_{12}O$	502-42-1	112.169			180.4(0.9)	0.9508²⁰	1.4608²⁰	i H_2O; vs EtOH, eth
2558	1,3,5-Cycloheptatriene	Tropilidene	C_7H_8	544-25-2	92.139	liq; cub cry (-80°C)	-75.18(0.06)	116.3(0.7)	0.8875¹⁹	1.5343²⁰	i H_2O; s EtOH, eth; vs bz, chl
2559	2,4,6-Cycloheptatrien-1-one		C_7H_6O	539-80-0	106.122		-5(2)	113¹⁵	1.095²²	1.6172²²	vs bz, chl
2560	Cycloheptene		C_7H_{12}	628-92-2	96.170	liq	-55.3(0.2)	115(3)	0.8228²⁰	1.4552²⁰	i H_2O; s EtOH, eth, bz, chl; sl ctc
2561	1,3-Cyclohexadiene		C_6H_8	592-57-4	80.128	liq	-89	80.3(0.3)	0.8405²⁰	1.4755²⁰	i H_2O; s EtOH, bz, chl, peth; vs eth
2562	1,4-Cyclohexadiene	1,4-Dihydrobenzene	C_6H_8	628-41-1	80.128	liq	-49(1)	89.5(0.2)	0.8471²⁰	1.4725²⁰	i H_2O; msc EtOH, eth; s bz, chl, peth
2563	3,5-Cyclohexadiene-1,2-dione		$C_6H_4O_2$	583-63-1	108.095	red pl or pr	≈65 dec				s eth, ace, bz; i peth
2564	2,5-Cyclohexadiene-1,4-dione, dioxime		$C_6H_6N_2O_2$	105-11-3	138.124	pa ye nd (w)	240 dec				s H_2O
2565	Cyclohexane	Hexahydrobenzene	C_6H_{12}	110-82-7	84.159	liq	6.7(0.2)	80.7(0.7)	0.7739²⁵	1.4235²⁵	i H_2O; msc EtOH, eth, ace, bz, lig, ctc
2566	Cyclohexaneacetic acid		$C_8H_{14}O_2$	5292-21-7	142.196	nd (HCO₂H)	28.9(1)	245	1.0423¹⁸	1.4775²⁰	sl H_2O; s eth, ace
2567	Cyclohexanecarbonitrile	Cyclohexyl cyanide	$C_7H_{11}N$	766-05-2	109.169	liq	12.0(0.3)	188(3)	0.919	1.4505²⁰	
2568	Cyclohexanecarbonyl chloride		$C_7H_{11}ClO$	2719-27-9	146.614			183(6)	1.0962¹⁵	1.4711²⁹	
2569	Cyclohexanecarboxaldehyde		$C_7H_{12}O$	2043-61-0	112.169			155(12)	0.9035²⁰	1.4496²⁰	s H_2O, eth
2570	Cyclohexanecarboxylic acid	Hexahydrobenzoic acid	$C_7H_{12}O_2$	98-89-5	128.169	mcl pr	28(3)	233(6)	1.0334²²	1.4530²⁰	sl H_2O, ctc; vs EtOH, bz, chl
2571	cis-1,2-Cyclohexanediamine	cis-1,2-Diaminocyclohexane	$C_6H_{14}N_2$	1436-59-5	114.188	liq		40²	0.952²⁰	1.4951²⁰	
2572	trans-1,2-Cyclohexanediamine	trans-1,2-Diaminocyclohexane	$C_6H_{14}N_2$	1121-22-8	114.188		14.8	80¹⁵	0.951²⁰		

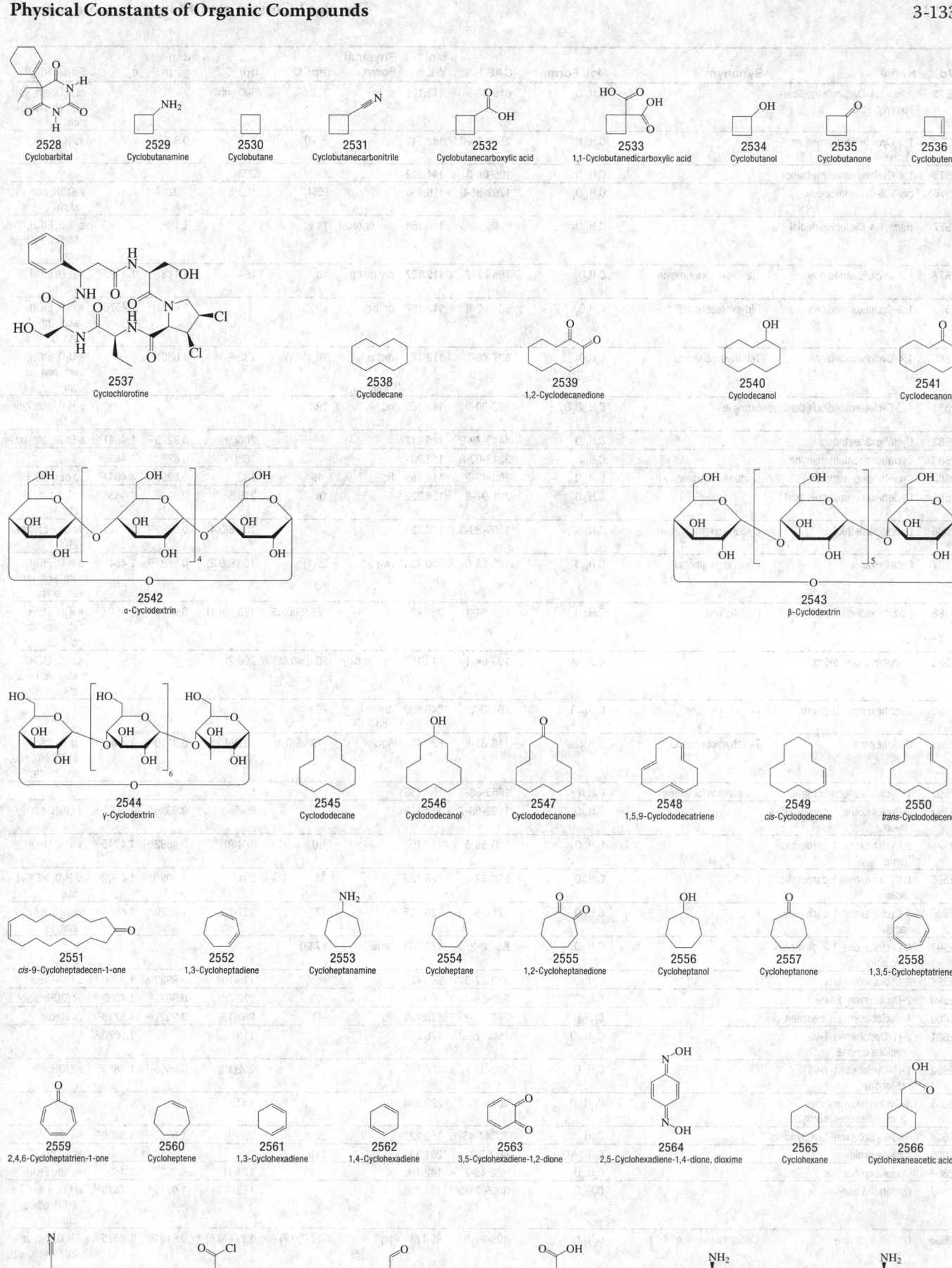

2528 Cyclobarbital

2529 Cyclobutanamine

2530 Cyclobutane

2531 Cyclobutanecarbonitrile

2532 Cyclobutanecarboxylic acid

2533 1,1-Cyclobutanedicarboxylic acid

2534 Cyclobutanol

2535 Cyclobutanone

2536 Cyclobutene

2537 Cyclochlorotine

2538 Cyclodecane

2539 1,2-Cyclodecanedione

2540 Cyclodecanol

2541 Cyclodecanone

2542 α-Cyclodextrin

2543 β-Cyclodextrin

2544 γ-Cyclodextrin

2545 Cyclododecane

2546 Cyclododecanol

2547 Cyclododecanone

2548 1,5,9-Cyclododecatriene

2549 cis-Cyclododecene

2550 trans-Cyclododecene

2551 cis-9-Cycloheptadecen-1-one

2552 1,3-Cycloheptadiene

2553 Cycloheptanamine

2554 Cycloheptane

2555 1,2-Cycloheptanedione

2556 Cycloheptanol

2557 Cycloheptanone

2558 1,3,5-Cycloheptatriene

2559 2,4,6-Cycloheptatrien-1-one

2560 Cycloheptene

2561 1,3-Cyclohexadiene

2562 1,4-Cyclohexadiene

2563 3,5-Cyclohexadiene-1,2-dione

2564 2,5-Cyclohexadiene-1,4-dione, dioxime

2565 Cyclohexane

2566 Cyclohexaneacetic acid

2567 Cyclohexanecarbonitrile

2568 Cyclohexanecarbonyl chloride

2569 Cyclohexanecarboxaldehyde

2570 Cyclohexanecarboxylic acid

2571 cis-1,2-Cyclohexanediamine

2572 trans-1,2-Cyclohexanediamine

No.	Name	Synonym	Mol. Form.	CAS RN	Mol. Wt.	Physical Form	mp/°C	bp/°C	den g cm⁻³	n_D	Solubility
2573	*trans*-1,4-Cyclohexanedicarboxylic acid		C₈H₁₂O₄	619-82-9	172.179	pr (w)	312.5	300 sub			sl H₂O, eth; vs EtOH; s ace; i chl
2574	1,3-Cyclohexanedimethanamine		C₈H₁₈N₂	2579-20-6	142.242		<-70	220	0.945²⁰		vs H₂O, eth, EtOH
2575	1,4-Cyclohexanedimethanol		C₈H₁₆O₂	105-08-8	144.212		43	283			
2576	*cis*-1,2-Cyclohexanediol		C₆H₁₂O₂	1792-81-0	116.158		95(4)	120¹⁵	1.0297¹⁰¹		s EtOH, ace, bz; sl chl
2577	*trans*-1,4-Cyclohexanediol		C₆H₁₂O₂	6995-79-5	116.158	mcl pr (ace)	143		1.18²⁰		s H₂O, EtOH, MeOH; i eth; sl ace
2578	1,2-Cyclohexanedione	1,2-Dioxocyclohexane	C₆H₈O₂	765-87-7	112.127	cry (peth)	40	194	1.1187²¹	1.4995²⁰	s H₂O, EtOH, eth, bz
2579	1,3-Cyclohexanedione	Dihydroresorcinol	C₆H₈O₂	504-02-9	112.127	pr (bz)	105(2)		1.0861⁹¹	1.4576¹⁰²	s H₂O, EtOH, ace, chl; sl eth, bz
2580	1,4-Cyclohexanedione	Tetrahydroquinone	C₆H₈O₂	637-88-7	112.127	mcl pl (w),nd (peth)	78.4(0.1)	132²⁰	1.0861⁹¹		s H₂O, EtOH, eth, ace, bz, chl
2581	1,2-Cyclohexanedione dioxime	Nioxime	C₆H₁₀N₂O₂	492-99-9	142.155	nd (w, HOAc)	192				s H₂O, ace, chl; sl tfa
2582	Cyclohexaneethanol		C₈H₁₆O	4442-79-9	128.212			208(3)	0.9229²⁰	1.4641²⁰	s EtOH, eth, bz
2583	Cyclohexanemethanamine		C₇H₁₅N	3218-02-8	113.201			162(7)	0.87²⁵	1.4630²⁰	
2584	Cyclohexanemethanol	Cyclohexylcarbinol	C₇H₁₄O	100-49-2	114.185	liq	-43	183	0.9297²⁰	1.4644²⁰	vs eth, EtOH
2585	Cyclohexanepropanoic acid		C₉H₁₆O₂	701-97-3	156.222		16	276.5	0.912²⁵	1.4638²⁰	s H₂O, eth; sl ctc
2586	Cyclohexanethiol	Cyclohexyl mercaptan	C₆H₁₂S	1569-69-3	116.224			158.8(0.4)	0.9782²⁰	1.4921²⁰	vs ace, bz, eth, EtOH
2587	Cyclohexanol	Cyclohexyl alcohol	C₆H₁₂O	108-93-0	100.158	hyg nd	26(1)	160.9(0.2)	0.9624²⁰	1.4641²⁰	s H₂O, EtOH, eth, ace; msc bz; sl chl
2588	Cyclohexanone	Pimelic ketone	C₆H₁₀O	108-94-1	98.142	liq	-27.93(0.05)	155.4(0.1)	0.9478²⁰	1.4507²⁰	s H₂O, EtOH, eth, ace, bz, chl, ctc
2589	Cyclohexanone oxime		C₆H₁₁NO	100-64-1	113.157	hex pr (lig)	89.05(0.09)	208(2)			s H₂O, EtOH, eth, MeOH; sl chl
2590	Cyclohexanone peroxide		C₁₂H₂₂O₅	78-18-2	246.300	cry or long nd	79				
2591	Cyclohexene	Tetrahydrobenzene	C₆H₁₀	110-83-8	82.143	liq	-103.5(0.4)	82.9(0.2)	0.8110²⁰	1.4465²⁰	i H₂O; msc EtOH, eth, ace, bz, lig, ctc
2592	1-Cyclohexenecarbonitrile	1-Cyanocyclohexene	C₇H₉N	1855-63-6	107.153			81¹²			
2593	1-Cyclohexene-1-carboxaldehyde		C₇H₁₀O	1192-88-7	110.153			69¹⁸	0.9694²⁰	1.5005²⁰	s EtOH, eth
2594	3-Cyclohexene-1-carboxaldehyde		C₇H₁₀O	100-50-5	110.153		1.0	164(9)	0.9692²⁰	1.4745²⁰	s ace, MeOH; sl ctc
2595	1-Cyclohexene-1-carboxylic acid		C₇H₁₀O₂	636-82-8	126.153		38	241	1.109²⁰	1.4902²⁰	sl H₂O; s EtOH, ace
2596	3-Cyclohexene-1-carboxylic acid		C₇H₁₀O₂	4771-80-6	126.153		17	237(13)	1.0820²⁰	1.4814²⁰	vs H₂O; s EtOH, ace
2597	4-Cyclohexene-1,2-dicarboxylic acid		C₈H₁₀O₄	88-98-2	170.163	pr (w)	173.0				
2598	2-Cyclohexen-1-ol		C₆H₁₀O	822-67-3	98.142			176(10)	0.9923¹⁵	1.4790²⁵	s EtOH, ace
2599	2-Cyclohexen-1-one		C₆H₈O	930-68-7	96.127	liq	-53	172.3(0.7)	0.9620²⁵	1.4883²⁰	vs EtOH; s ace
2600	1-Cyclohexen-1-ylbenzene		C₁₂H₁₄	771-98-2	158.239	liq	-11	268(15)	0.9939²⁰	1.5718²⁰	vs MeOH
2601	2-(1-Cyclohexen-1-yl)cyclohexanone		C₁₂H₁₈O	1502-22-3	178.270			116³		1.5070²⁰	
2602	1-(1-Cyclohexen-1-yl)ethanone		C₈H₁₂O	932-66-1	124.180		73	202(12)	0.9655²⁰	1.4881²⁰	s EtOH, eth
2603	3-Cyclohexenylmethyl 3-cyclohexenecarboxylate		C₁₄H₂₀O₂	2611-00-9	220.308	liq		153⁷			
2604	4-(3-Cyclohexen-1-yl)pyridine		C₁₁H₁₃N	70644-46-1	159.228		22.1	226	1.0222²⁵	1.5466²⁵	
2605	Cycloheximide		C₁₅H₂₃NO₄	66-81-9	281.349	pl (al)	119				vs EtOH
2606	Cyclohexyl acetate		C₈H₁₄O₂	622-45-7	142.196			174(4)	0.968²⁰	1.442²⁰	vs eth, EtOH
2607	Cyclohexyl acrylate		C₉H₁₄O₂	3066-71-5	154.206			183	1.0275²⁰	1.4673²⁰	i H₂O; msc EtOH, eth; s chl
2608	Cyclohexylamine	Cyclohexanamine	C₆H₁₃N	108-91-8	99.174	liq	-17.7(0.7)	133.6(0.5)	0.8191²⁰	1.4625¹⁵	s H₂O, ctc; vs EtOH; msc eth, ace, bz
2609	Cyclohexylamine hydrochloride	Cyclohexanamine hydrochloride	C₆H₁₄ClN	4998-76-9	135.635	nd (w, al-eth)	206.5				vs H₂O, EtOH

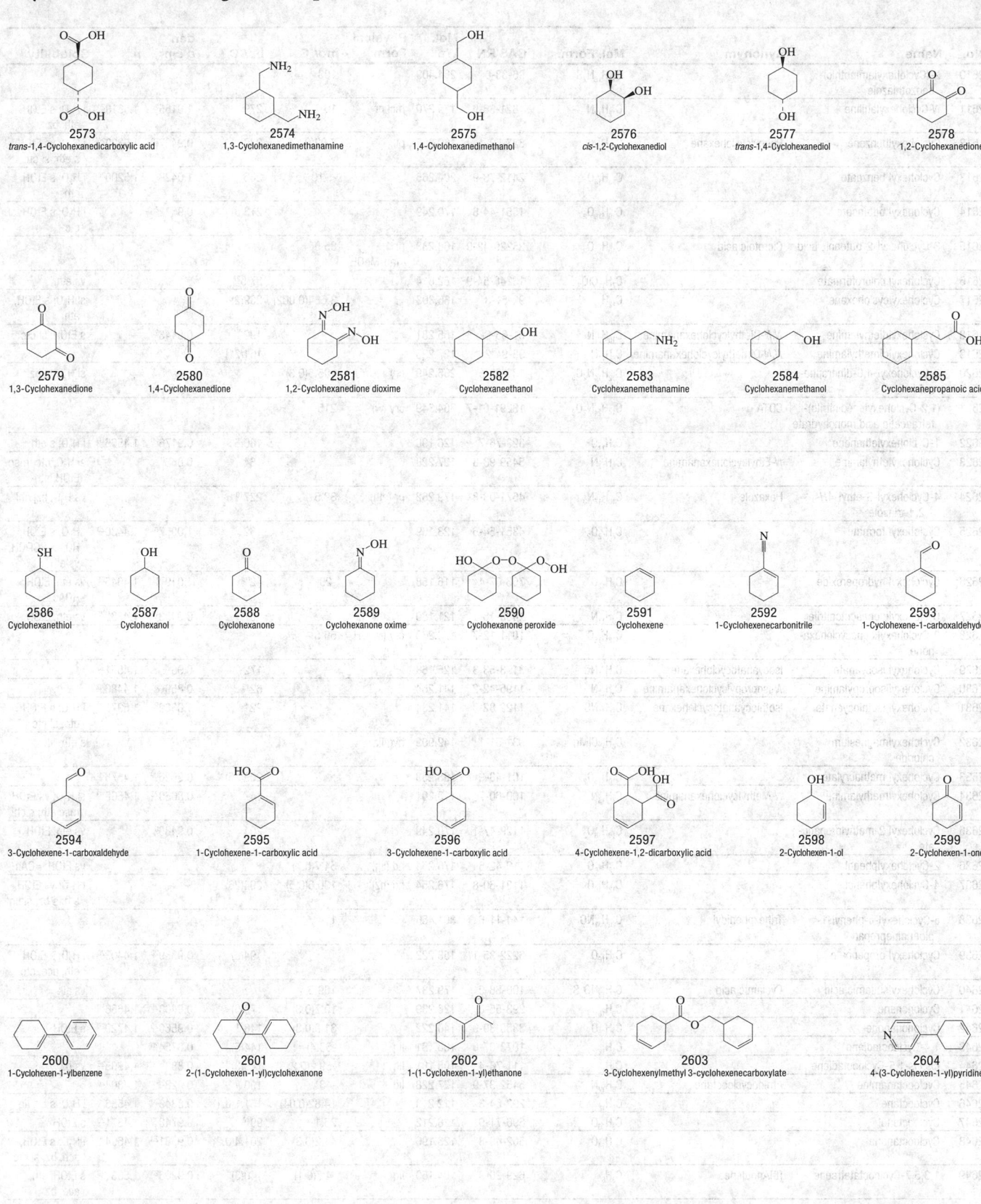

2573 *trans*-1,4-Cyclohexanedicarboxylic acid

2574 1,3-Cyclohexanedimethanamine

2575 1,4-Cyclohexanedimethanol

2576 *cis*-1,2-Cyclohexanediol

2577 *trans*-1,4-Cyclohexanediol

2578 1,2-Cyclohexanedione

2579 1,3-Cyclohexanedione

2580 1,4-Cyclohexanedione

2581 1,2-Cyclohexanedione dioxime

2582 Cyclohexaneethanol

2583 Cyclohexanemethanamine

2584 Cyclohexanemethanol

2585 Cyclohexanepropanoic acid

2586 Cyclohexanethiol

2587 Cyclohexanol

2588 Cyclohexanone

2589 Cyclohexanone oxime

2590 Cyclohexanone peroxide

2591 Cyclohexene

2592 1-Cyclohexenecarbonitrile

2593 1-Cyclohexene-1-carboxaldehyde

2594 3-Cyclohexene-1-carboxaldehyde

2595 1-Cyclohexene-1-carboxylic acid

2596 3-Cyclohexene-1-carboxylic acid

2597 4-Cyclohexene-1,2-dicarboxylic acid

2598 2-Cyclohexen-1-ol

2599 2-Cyclohexen-1-one

2600 1-Cyclohexen-1-ylbenzene

2601 2-(1-Cyclohexen-1-yl)cyclohexanone

2602 1-(1-Cyclohexen-1-yl)ethanone

2603 3-Cyclohexenylmethyl 3-cyclohexenecarboxylate

2604 4-(3-Cyclohexen-1-yl)pyridine

2605 Cycloheximide

2606 Cyclohexyl acetate

2607 Cyclohexyl acrylate

2608 Cyclohexylamine

2609 Cyclohexylamine hydrochloride

No.	Name	Synonym	Mol. Form.	CAS RN	Mol. Wt.	Physical Form	mp/°C	bp/°C	den g cm⁻³	n_D	Solubility
2610	2-(Cyclohexylaminothio)-benzothiazole		$C_{13}H_{16}N_2S_2$	95-33-0	264.409		103				
2611	N-Cyclohexylaniline		$C_{12}H_{17}N$	1821-36-9	175.270	mcl pr	16	279	1.0155[20]	1.5610[20]	i H₂O; s EtOH, eth, bz
2612	Cyclohexylbenzene	Phenylcyclohexane	$C_{12}H_{16}$	827-52-1	160.255	pl	7.02(0.1)	239(2)	0.9427[20]	1.5329[20]	i H₂O; vs EtOH; s eth; sl ctc
2613	Cyclohexyl benzoate		$C_{13}H_{16}O_2$	2412-73-9	204.265		<-10	285	1.0429[20]	1.5200[20]	i H₂O; s EtOH, eth
2614	Cyclohexyl butanoate		$C_{10}H_{18}O_2$	1551-44-6	170.249			213(8)	0.9572[20]		i H₂O; s EtOH; sl ctc
2615	3-Cyclohexyl-2-butenoic acid	Cicrotoic acid	$C_{10}H_{16}O_2$	25229-42-9	168.233	pr (aq-MeOH)	85.5				
2616	Cyclohexyl chloroformate		$C_7H_{11}ClO_2$	13248-54-9	162.614			87.5[27]			vs eth
2617	Cyclohexylcyclohexane		$C_{12}H_{22}$	92-51-3	166.303		3.684(0.002)	239(2)			sl H₂O; s EtOH, eth
2618	Cyclohexyldiethylamine	N,N-Diethylcyclohexanamine	$C_{10}H_{21}N$	91-65-6	155.281			192	0.8443[25]		s EtOH; sl ctc
2619	Cyclohexyldimethylamine	N,N-Dimethylcyclohexanamine	$C_8H_{17}N$	98-94-2	127.228			161(21)			
2620	2-Cyclohexyl-4,6-dinitrophenol		$C_{12}H_{14}N_2O_5$	131-89-5	266.249	cry	105.5(0.3)				sl H₂O; s bz, DMF
2621	(1,2-Cyclohexylenedinitrilo)-tetraacetic acid monohydrate	CDTA	$C_{14}H_{24}N_2O_9$	13291-61-7	364.349	cry (w)	215				
2622	1-Cyclohexylethanone		$C_8H_{14}O$	823-76-7	126.196			180.5	0.9176[20]	1.4565[16]	i H₂O; s eth
2623	Cyclohexylethylamine	N-Ethylcyclohexanamine	$C_8H_{17}N$	5459-93-8	127.228			164	0.868[0]		sl H₂O, ctc; msc EtOH, eth
2624	4-Cyclohexyl-3-ethyl-4H-1,2,4-triazole	Hexazole	$C_{10}H_{17}N_3$	4671-03-8	179.262	pr (eth)	89.5	227[10]			vs H₂O, bz, chl
2625	Cyclohexyl formate		$C_7H_{12}O_2$	4351-54-6	128.169			162	1.0057[0]	1.4430[20]	i H₂O; s EtOH, HOAc, HCOOH; vs eth
2626	Cyclohexyl hydroperoxide		$C_6H_{12}O_2$	766-07-4	116.158		-20	42[0.1]	1.019[20]	1.4645[25]	vs eth, EtOH, HOAc
2627	Cyclohexylideneacetonitrile		$C_8H_{11}N$	4435-18-1	121.180			107[22]	0.9483[15]	1.4382[25]	vs eth, EtOH
2628	2-Cyclohexylidenecyclohexanone		$C_{12}H_{18}O$	1011-12-7	178.270	cry (MeOH aq)	56.5				
2629	Cyclohexyl isocyanate	Isocyanatocyclohexane	$C_7H_{11}NO$	3173-53-3	125.168			172	0.98[25]	1.4551[20]	
2630	Cyclohexylisopropylamine	N-Isopropylcyclohexanamine	$C_9H_{19}N$	1195-42-2	141.254			62[12]	0.859[25]	1.4480[20]	
2631	Cyclohexyl isothiocyanate	Isothiocyanatocyclohexane	$C_7H_{11}NS$	1122-82-3	141.234			221	1.0339[20]	1.5375[20]	i H₂O; s EtOH, eth; sl ctc
2632	Cyclohexylmagnesium chloride		$C_6H_{11}ClMg$	931-51-1	142.909	hyg liq					s eth
2633	Cyclohexyl methacrylate		$C_{10}H_{16}O_2$	101-43-9	168.233			209(25)	0.9626[20]	1.4578[20]	
2634	Cyclohexylmethylamine	N-Methylcyclohexanamine	$C_7H_{15}N$	100-60-7	113.201			147	0.8660[23]	1.4560[20]	sl H₂O; vs EtOH; msc eth; s chl
2635	Cyclohexyl 2-methylpropanoate		$C_{10}H_{18}O_2$	1129-47-1	170.249			204	0.9489[0]		vs eth, EtOH
2636	2-Cyclohexylphenol		$C_{12}H_{16}O$	119-42-6	176.254	nd (lig)	54.7(0.7)				vs EtOH, HOAc
2637	4-Cyclohexylphenol		$C_{12}H_{16}O$	1131-60-8	176.254	nd (bz)	130.6(0.5)	294(26)			i H₂O; vs EtOH, eth; s bz; sl lig
2638	α-Cyclohexyl-α-phenyl-1-piperidinepropanol	Trihexphenidyl	$C_{20}H_{31}NO$	144-11-6	301.466		114				
2639	Cyclohexyl propanoate		$C_9H_{16}O_2$	6222-35-1	156.222			194(6)	0.9359[20]	1.4403[20]	i H₂O; s EtOH, eth, ace, ctc
2640	Cyclohexylsulfamic acid	Cyclamic acid	$C_6H_{13}NO_3S$	100-88-9	179.237		169.5				vs alk
2641	Cyclononane		C_9H_{18}	293-55-0	126.239		10.7(0.3)	173(5)	0.8463[25]	1.4666[20]	
2642	Cyclononanone		$C_9H_{16}O$	3350-30-9	140.222		31.9(0.3)	148[24]	0.9560[20]	1.4729[20]	s EtOH
2643	1,4-Cyclooctadiene		C_8H_{12}	1073-07-0	108.181	liq	-53(4)	144(5)	0.8754[20]		
2644	cis,cis-1,5-Cyclooctadiene		C_8H_{12}	111-78-4	108.181	liq	-69.2(0.2)	149(3)	0.883[20]	1.4905[25]	vs bz
2645	Cyclooctanamine	Aminocyclooctane	$C_8H_{17}N$	5452-37-9	127.228	liq	-48	190	0.928[25]	1.4804[20]	
2646	Cyclooctane		C_8H_{16}	292-64-8	112.213		14.82(0.04)	151.1(0.1)	0.8349[20]	1.4586[20]	i H₂O; s bz, lig
2647	Cyclooctanol		$C_8H_{16}O$	696-71-9	128.212		25.1	99[16]	0.9740[20]	1.4871[20]	s EtOH
2648	Cyclooctanone		$C_8H_{14}O$	502-49-8	126.196		44.2(0.3)	201.4(0.2)	0.9581[20]	1.4694[20]	i H₂O; s EtOH, ace, bz; sl ctc
2649	1,3,5,7-Cyclooctatetraene	[8]Annulene	C_8H_8	629-20-9	104.150	liq	-4.7(0.1)	140(3)	0.9206[20]	1.5381[20]	s EtOH, eth, ace, bz
2650	1,3,5-Cyclooctatriene		C_8H_{10}	1871-52-9	106.165	liq	-86(8)	144(11)	0.8971[25]	1.5035[25]	
2651	cis-Cyclooctene		C_8H_{14}	931-87-3	110.197	liq	-13(4)	145(3)	0.8472[20]	1.4698[20]	s EtOH, eth, ctc
2652	trans-Cyclooctene		C_8H_{14}	931-89-5	110.197	liq	-59(5)	146(5)	0.8483[20]	1.4741[25]	s EtOH, chl; sl ctc
2653	Cyclooctyne		C_8H_{12}	1781-78-8	108.181			158	0.868[20]	1.4850[20]	
2654	Cyclopamine	11-Deoxojervine	$C_{27}H_{41}NO_2$	4449-51-8	411.621	nd (EtOH)	237				
2655	Cyclopentadecane		$C_{15}H_{30}$	295-48-7	210.399	nd (MeOH)	64.0(0.6)		0.8364[61]	1.4592[61]	

2610
2-(Cyclohexylaminothio)benzothiazole

2611
N-Cyclohexylaniline

2612
Cyclohexylbenzene

2613
Cyclohexyl benzoate

2614
Cyclohexyl butanoate

2615
3-Cyclohexyl-2-butenoic acid

2616
Cyclohexyl chloroformate

2617
Cyclohexylcyclohexane

2618
Cyclohexyldiethylamine

2619
Cyclohexyldimethylamine

2620
2-Cyclohexyl-4,6-dinitrophenol

2621
(1,2-Cyclohexylenedinitrilo)tetraacetic acid monohydrate

2622
1-Cyclohexylethanone

2623
Cyclohexylethylamine

2624
4-Cyclohexyl-3-ethyl-4*H*-1,2,4-triazole

2625
Cyclohexyl formate

2626
Cyclohexyl hydroperoxide

2627
Cyclohexylideneacetonitrile

2628
2-Cyclohexylidenecyclohexanone

2629
Cyclohexyl isocyanate

2630
Cyclohexylisopropylamine

2631
Cyclohexyl isothiocyanate

2632
Cyclohexylmagnesium chloride

2633
Cyclohexyl methacrylate

2634
Cyclohexylmethylamine

2635
Cyclohexyl 2-methylpropanoate

2636
2-Cyclohexylphenol

2637
4-Cyclohexylphenol

2638
α-Cyclohexyl-α-phenyl-1-piperidinepropanol

2639
Cyclohexyl propanoate

2640
Cyclohexylsulfamic acid

2641
Cyclononane

2642
Cyclononanone

2643
1,4-Cyclooctadiene

2644
cis,cis-1,5-Cyclooctadiene

2645
Cyclooctanamine

2646
Cyclooctane

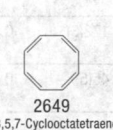

2647
Cyclooctanol

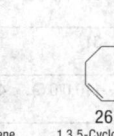

2648
Cyclooctanone

2649
1,3,5,7-Cyclooctatetraene

2650
1,3,5-Cyclooctatriene

2651
cis-Cyclooctene

2652
trans-Cyclooctene

2653
Cyclooctyne

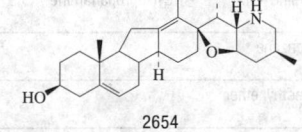

2654
Cyclopamine

2655
Cyclopentadecane

No.	Name	Synonym	Mol. Form.	CAS RN	Mol. Wt.	Physical Form	mp/°C	bp/°C	den g cm⁻³	n_D	Solubility
2656	Cyclopentadecanol	Exaltol	$C_{15}H_{30}O$	4727-17-7	226.398	cry (MeOH)	80.5	177[11]	0.930[20]	1.4555[98]	
2657	Cyclopentadecanone		$C_{15}H_{28}O$	502-72-7	224.382			65.2(0.5)	120[0.3]	0.8895[25]	1.4637[60]	sl H_2O; s EtOH, ace
2658	1,3-Cyclopentadiene	Pyropentylene	C_5H_6	542-92-7	66.102	liq	-96.54(0.05)	41(1)	0.8021[20]	1.4440[20]	i H_2O; msc EtOH, eth, bz; s ace
2659	Cyclopentane	Pentamethylene	C_5H_{10}	287-92-3	70.133	liq	-93.4(0.3)	49.2(0.1)	0.7457[20]	1.4065[20]	i H_2O; msc EtOH, eth, ace, bz, peth, ctc
2660	Cyclopentaneacetic acid		$C_7H_{12}O_2$	1123-00-8	128.169	pl	13.5	228	1.0216[18]	1.4523[18]	
2661	Cyclopentanecarbonitrile	Cyanocyclopentane	C_6H_9N	4254-02-8	95.142	liq	-76	169(2)	0.912	1.4410[20]	
2662	Cyclopentanecarboxaldehyde		$C_6H_{10}O$	872-53-7	98.142			133.5	0.9371[20]	1.4432[20]	vs H_2O, eth, EtOH
2663	Cyclopentanecarboxylic acid	Cyclopentanoic acid	$C_6H_{10}O_2$	3400-45-1	114.142	liq	-7	212	1.0527[20]	1.4532[20]	sl H_2O, ctc; s MeOH
2664	cis-1,2-Cyclopentanediol		$C_5H_{10}O_2$	5057-98-7	102.132		30	124[29]			
2665	trans-1,2-Cyclopentanediol		$C_5H_{10}O_2$	5057-99-8	102.132		46(4)	226			
2666	Cyclopentanemethanol		$C_6H_{12}O$	3637-61-4	100.158			163	0.9332[20]	1.4579[20]	
2667	Cyclopentanepropanoic acid		$C_6H_{14}O_2$	140-77-2	142.196			158[26]	1.0100[17]	1.4570[20]	
2668	Cyclopentanethiol	Cyclopentyl mercaptan	$C_5H_{10}S$	1679-07-8	102.198			132.2(0.6)	0.9550[20]		
2669	Cyclopentanol	Cyclopentyl alcohol	$C_5H_{10}O$	96-41-3	86.132	liq	-17(2)	140.4(0.2)	0.9488[20]	1.4530[20]	sl H_2O, ctc; s EtOH, eth, ace
2670	Cyclopentanone	Adipic ketone	C_5H_8O	120-92-3	84.117	liq	-51.70(0.02)	130.5(0.2)	0.9487[20]	1.4366[20]	i H_2O; s EtOH, ace, ctc, hx; msc eth
2671	Cyclopentanone oxime		C_5H_9NO	1192-28-5	99.131		57.8	196			vs H_2O, bz
2672	Cyclopentene		C_5H_8	142-29-0	68.118	liq	-135.02(0.09)	44.2(0.2)	0.7720[20]	1.4225[20]	i H_2O; s EtOH, eth, bz, ctc, peth
2673	1-Cyclopentenecarbonitrile	1-Cyanocyclopentene	C_6H_7N	3047-38-9	93.127	liq		81[30]			
2674	1-Cyclopentene-1-carboxaldehyde		C_6H_8O	6140-65-4	96.127	liq	-32	146	0.970[21]	1.4872[17]	
2675	2-Cyclopentene-1-tridecanoic acid, (S)-	Chaulmoogric acid	$C_{18}H_{32}O_2$	29106-32-9	280.446	pl or lf (al, HOAc)	68(2)	247[20]			vs eth, chl
2676	2-Cyclopentene-1-undecanoic acid, (R)-	Hydnocarpic acid	$C_{16}H_{28}O_2$	459-67-6	252.392		60.5				vs EtOH, chl, peth
2677	2-Cyclopenten-1-one		C_5H_6O	930-30-3	82.101			136	0.989[15]	1.4629[15]	vs eth, EtOH
2678	3-Cyclopenten-1-one		C_5H_6O	14320-37-7	82.101	liq		28[17]			
2679	N-(1-Cyclopenten-1-yl)-pyrrolidine	1-Pyrrolidinylcyclopentene	$C_9H_{15}N$	7148-07-4	137.222			105[15]		1.5128[20]	
2680	Cyclopenthiazide		$C_{13}H_{18}ClN_3O_4S_2$	742-20-1	379.883		238				
2681	Cyclopentobarbital		$C_{12}H_{14}N_2O_3$	76-68-6	234.250	cry (w, dil al)	139.5				sl H_2O; vs EtOH
2682	Cyclopentylamine	Cyclopentanamine	$C_5H_{11}N$	1003-03-8	85.148	liq	-82.69(0.05)	108.5(0.1)	0.8689[20]	1.4728[25]	s ace, bz, chl
2683	Cyclopentylbenzene		$C_{11}H_{14}$	700-88-9	146.229			223(4)	0.9462[20]	1.5280[20]	vs eth
2684	2-Cyclopentylidenecyclopentanone		$C_{10}H_{14}O$	825-25-2	150.217			135[25]	1.0179[18]	1.5215[18]	
2685	Cyclopentyl methyl sulfide		$C_6H_{12}S$	7133-36-0	116.224			156.2(0.8)			
2686	Cyclophosphamide	Cyclophosphane	$C_7H_{15}Cl_2N_2O_2P$	50-18-0	261.086		50.4(0.5)				vs H_2O; sl bz, chl, diox, EtOH
2687	Cycloposine		$C_{33}H_{51}NO_7$	23185-94-6	573.761		268				
2688	Cyclopropane	Trimethylene	C_3H_6	75-19-4	42.080	col gas	-127.6(0.2)	-31(2)	0.617[25] (p>1 atm)	1.3799[-42]	s H_2O, bz, peth; vs EtOH, eth
2689	Cyclopropanecarbonitrile	Cyclopropyl cyanide	C_4H_5N	5500-21-0	67.090			135(1)	0.8946[20]	1.4229[20]	s eth, hx; sl ctc
2690	Cyclopropanecarbonyl chloride		C_4H_5ClO	4023-34-1	104.535			119	1.1516[20]		
2691	Cyclopropanecarboxaldehyde	Formylcyclopropane	C_4H_6O	1489-69-6	70.090	liq		100	0.938	1.4298[20]	
2692	Cyclopropanecarboxylic acid		$C_4H_6O_2$	1759-53-1	86.090		18.5	182.20(0.02)	1.0885[20]	1.4390[20]	s H_2O, EtOH, eth; sl ctc
2693	1,1-Cyclopropanedicarboxylic acid		$C_5H_6O_4$	598-10-7	130.100	pr or nd (chl) pr (w +1)	140.5				vs H_2O, eth
2694	Cyclopropanemethanol		C_4H_8O	2516-33-8	72.106			124	0.911[25]		sl ctc
2695	Cyclopropanone		C_3H_4O	5009-27-8	56.063		stable only at low temp.				
2696	Cyclopropene		C_3H_4	2781-85-3	40.064	gas		-36(6)			
2697	Cyclopropylamine	Cyclopropanamine	C_3H_7N	765-30-0	57.095	liq	-35.38(0.05)	49.2(0.1)	0.8240[20]	1.4210[20]	msc H_2O; s EtOH, eth, chl
2698	Cyclopropylbenzene		C_9H_{10}	873-49-4	118.175	liq	-31	172(3)	0.9317[20]	1.5285[20]	i H_2O; s eth, ace, chl
2699	Cyclopropyl methyl ether		C_4H_8O	540-47-6	72.106	liq	-110.0(0.5)	45(5)	0.8100[20]	1.3802[20]	vs H_2O, bz, eth, EtOH

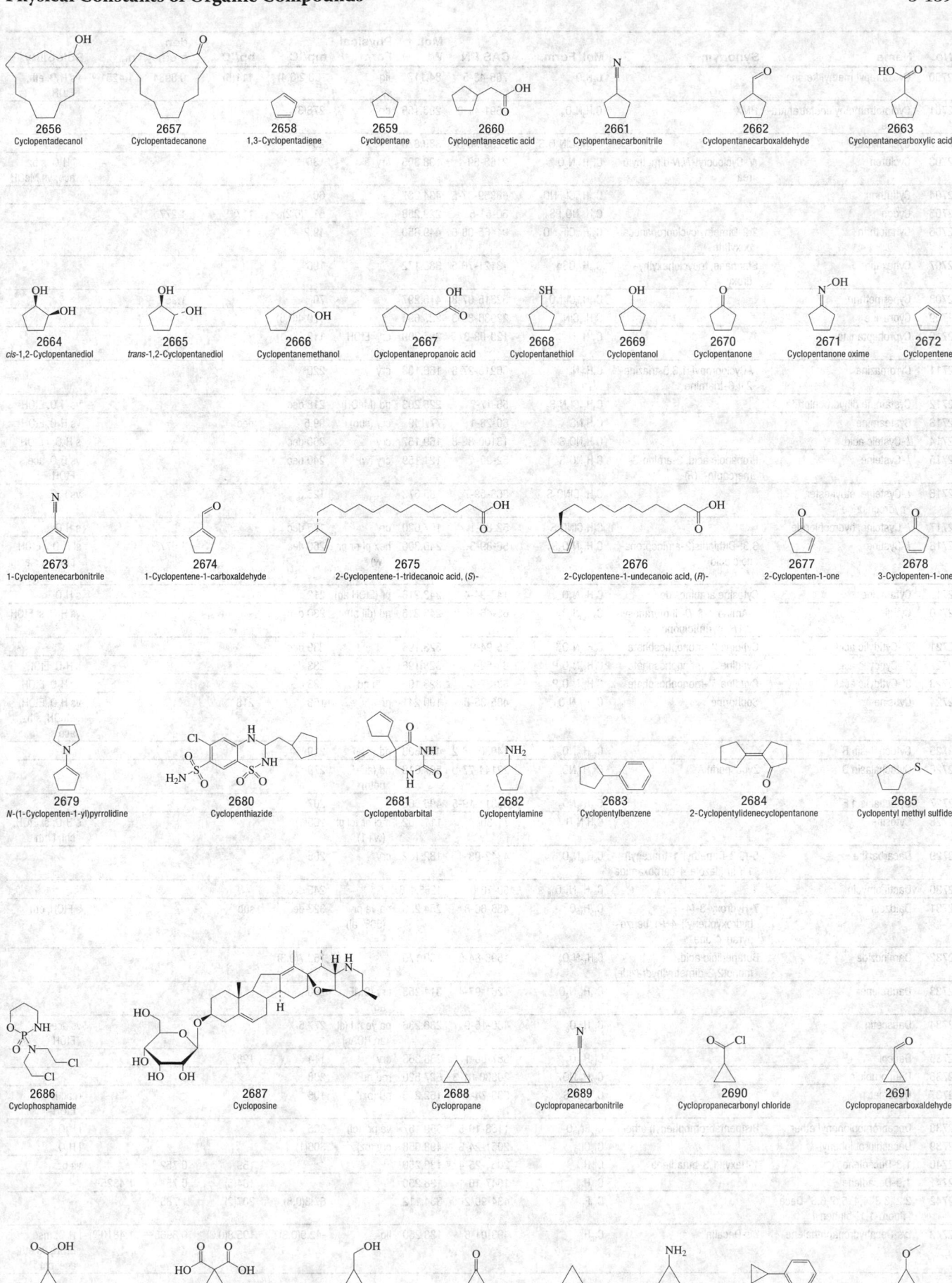

2656 Cyclopentadecanol

2657 Cyclopentadecanone

2658 1,3-Cyclopentadiene

2659 Cyclopentane

2660 Cyclopentaneacetic acid

2661 Cyclopentanecarbonitrile

2662 Cyclopentanecarboxaldehyde

2663 Cyclopentanecarboxylic acid

2664 *cis*-1,2-Cyclopentanediol

2665 *trans*-1,2-Cyclopentanediol

2666 Cyclopentanemethanol

2667 Cyclopentanepropanoic acid

2668 Cyclopentanethiol

2669 Cyclopentanol

2670 Cyclopentanone

2671 Cyclopentanone oxime

2672 Cyclopentene

2673 1-Cyclopentenecarbonitrile

2674 1-Cyclopentene-1-carboxaldehyde

2675 2-Cyclopentene-1-tridecanoic acid, (*S*)-

2676 2-Cyclopentene-1-undecanoic acid, (*R*)-

2677 2-Cyclopenten-1-one

2678 3-Cyclopenten-1-one

2679 *N*-(1-Cyclopenten-1-yl)pyrrolidine

2680 Cyclopenthiazide

2681 Cyclopentobarbital

2682 Cyclopentylamine

2683 Cyclopentylbenzene

2684 2-Cyclopentylidenecyclopentanone

2685 Cyclopentyl methyl sulfide

2686 Cyclophosphamide

2687 Cyloposine

2688 Cyclopropane

2689 Cyclopropanecarbonitrile

2690 Cyclopropanecarbonyl chloride

2691 Cyclopropanecarboxaldehyde

2692 Cyclopropanecarboxylic acid

2693 1,1-Cyclopropanedicarboxylic acid

2694 Cyclopropanemethanol

2695 Cyclopropanone

2696 Cyclopropene

2697 Cyclopropylamine

2698 Cyclopropylbenzene

2699 Cyclopropyl methyl ether

No.	Name	Synonym	Mol. Form.	CAS RN	Mol. Wt.	Physical Form	mp/°C	bp/°C	den g cm⁻³	n_D	Solubility
2700	Cyclopropyl methyl ketone		C_5H_8O	765-43-5	84.117	liq	-68.2(0.4)	111(5)	0.8984^{20}	1.4251^{20}	vs H_2O, eth, EtOH
2701	Cyclotetramethylenetetranitramine	HMX	$C_4H_8N_8O_8$	2691-41-0	296.156	cry	278(3)				
2702	Cyclothiazide		$C_{14}H_{16}ClN_3O_4S_2$	2259-96-3	389.878		234				
2703	Cycluron	N'-Cyclooctyl-N,N-dimethylurea	$C_{11}H_{22}N_2O$	2163-69-1	198.305	cry	138				sl H_2O; s bz, ace; vs MeOH
2704	Cyfluthrin		$C_{22}H_{18}Cl_2FNO_3$	68359-37-5	434.287		60				
2705	Cygon		$C_5H_{12}NO_3PS_2$	60-51-5	229.258		51.4(0.2)	117$^{0.1}$	1.277^{65}		
2706	Cyhalothrin	2,2-Dimethylcyclopropanecarboxylate	$C_{23}H_{19}ClF_3NO_3$	91465-08-6	449.850		49.2				
2707	Cyhexatin	Stannane, tricyclohexylhydroxy-	$C_{18}H_{34}OSn$	13121-70-5	385.172		196				
2708	Cypermethrin		$C_{22}H_{19}Cl_2NO_3$	52315-07-8	416.297		70		1.25^{20}		
2709	Cyprazine		$C_9H_{14}ClN_5$	22936-86-3	227.694		170.4(0.5)				
2710	Cyproheptadine		$C_{21}H_{21}N$	129-03-3	287.399	cry (EtOH aq)	113				
2711	Cyromazine	N-Cyclopropyl-1,3,5-triazine-2,4,6-triamine	$C_6H_{10}N_6$	66215-27-8	166.183	cry	220				
2712	Cystamine dihydrochloride		$C_4H_{14}Cl_2N_2S_2$	56-17-7	225.203	nd (MeOH)	218 dec				vs H_2O, EtOH
2713	Cysteamine		C_2H_7NS	60-23-1	77.149	cry (sub)	99.5	dec			vs H_2O, EtOH
2714	L-Cysteic acid		$C_3H_7NO_5S$	13100-82-8	169.157	cry	260 dec				s H_2O; i EtOH
2715	L-Cysteine	Propanoic acid, 2-amino-3-mercapto-, (R)-	$C_3H_7NO_2S$	52-90-4	121.159	cry (w)	240 dec				vs H_2O, ace, EtOH
2716	L-Cysteine, ethyl ester, hydrochloride		$C_5H_{12}ClNO_2S$	868-59-7	185.673		125.8				vs H_2O
2717	L-Cysteine, hydrochloride		$C_3H_8ClNO_2S$	52-89-1	157.620	cry	175 dec				s H_2O
2718	L-Cystine	3,3'-Dithiobis(2-aminopropanoic acid)	$C_6H_{12}N_2O_4S_2$	56-89-3	240.300	hex pl or pr (w)	260 dec		1.677^{25}		sl H_2O; i EtOH, eth, bz; s acid, alk
2719	Cytarabine	Cytosine arabinoside	$C_9H_{13}N_3O_5$	147-94-4	243.216	pr (EtOH aq)	212				s H_2O
2720	Cytidine	4-Amino-1-β-D-ribofuranosyl-2(1H)-pyrimidinone	$C_9H_{13}N_3O_5$	65-46-3	243.216	nd (dil al)	230 dec				vs H_2O; sl EtOH
2721	2'-Cytidylic acid	Cytidine 2'-monophosphate	$C_9H_{14}N_3O_8P$	85-94-9	323.196		239 dec				
2722	3'-Cytidylic acid	Cytidine 3'-monophosphate	$C_9H_{14}N_3O_8P$	84-52-6	323.196		233 dec				s H_2O, EtOH
2723	5'-Cytidylic acid	Cytidine 5'-monophosphate	$C_9H_{14}N_3O_8P$	63-37-6	323.196	orth nd	233 dec				vs H_2O, EtOH
2724	Cytisine	Sophorine	$C_{11}H_{14}N_2O$	485-35-8	190.241	pr	153	218^2			vs H_2O, EtOH, MeOH; s bz, ace
2725	Cytochalasin B		$C_{29}H_{37}NO_5$	14930-96-2	479.608	nd (ace)	219				
2726	Cytochalasin D	Zygosporin A	$C_{30}H_{37}NO_6$	22144-77-0	507.618	nd (ace/peth)	270				
2727	Cytochalasin E		$C_{28}H_{33}NO_7$	36011-19-5	495.565		207				
2728	Cytosine		$C_4H_5N_3O$	71-30-7	111.102	mcl or tcl pl (w+1)	265(1)				s H_2O; sl EtOH, chl; i eth
2729	Dacarbazine	5-(3,3-Dimethyl-1-triazenyl)-1H-imidazole-4-carboxamide	$C_6H_{10}N_6O$	4342-03-4	182.182	cry	205				
2730	Dactinomycin		$C_{62}H_{86}N_{12}O_{16}$	50-76-0	1255.416		245 dec				
2731	Daidzein	7-Hydroxy-3-(4-hydroxyphenyl)-4H-1-benzopyran-4-one	$C_{15}H_{10}O_4$	486-66-8	254.238	pa ye pr (50% al)	323 dec	sub			s EtOH, eth
2732	Daminozide	Butanedioic acid, mono(2,2-dimethylhydrazide)	$C_6H_{12}N_2O_3$	1596-84-5	160.170		152.7(0.3)				
2733	Dantrolene		$C_{14}H_{10}N_4O_5$	7261-97-4	314.253	cry (DMF aq)	280				
2734	Datiscetin		$C_{15}H_{10}O_6$	480-15-9	286.236	pa ye nd (al, aq HOAc)	277.5				vs ace, eth, EtOH
2735	Daucol		$C_{15}H_{26}O_2$	887-08-1	238.366	cry	114	128^2			
2736	Daunorubicin		$C_{27}H_{29}NO_{10}$	20830-81-3	527.520	red nd	208				
2737	Dazomet		$C_5H_{10}N_2S_2$	533-74-4	162.276	nd (bz)	106				reac H_2O; s EtOH
2738	Decabromobiphenyl ether	Bis(pentabromophenyl) ether	$C_{12}Br_{10}O$	1163-19-5	959.167	ye pr (tol)	305				i H_2O
2739	Decachlorobiphenyl		$C_{12}Cl_{10}$	2051-24-3	498.658	cry (bz)	306(1)				i H_2O
2740	1,3-Decadiene	1-Hexyl-1,3-butadiene	$C_{10}H_{18}$	2051-25-4	138.250			169	0.752^{30}		vs bz
2741	1,9-Decadiene		$C_{10}H_{18}$	1647-16-1	138.250			164(5)	0.75^{25}	1.4325^{20}	
2742	2,2',3,3',4,4',5,5',6,6'-Decafluoro-1,1'-biphenyl		$C_{12}F_{10}$	434-90-2	334.112		67.3(0.8)	207(2)	1.785^{20}		
2743	cis-Decahydronaphthalene	cis-Decalin	$C_{10}H_{18}$	493-01-6	138.250	liq	-42.9(0.3)	195.8(0.3)	0.8965^{20}	1.4810^{20}	i H_2O; msc EtOH; vs eth, ace, chl

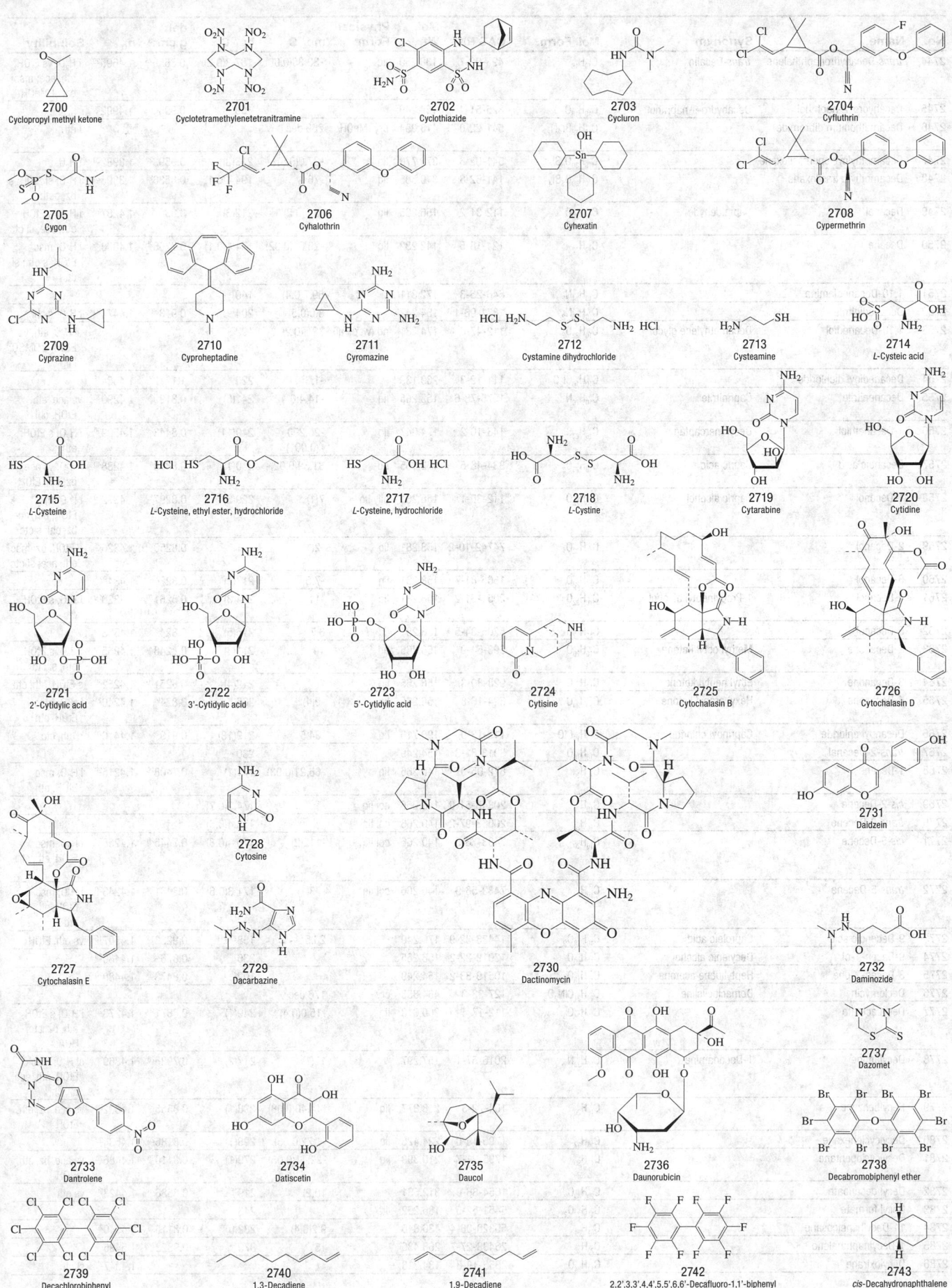

2700
Cyclopropyl methyl ketone

2701
Cyclotetramethylenetetranitramine

2702
Cyclothiazide

2703
Cycluron

2704
Cyfluthrin

2705
Cygon

2706
Cyhalothrin

2707
Cyhexatin

2708
Cypermethrin

2709
Cyprazine

2710
Cyproheptadine

2711
Cyromazine

2712
Cystamine dihydrochloride

2713
Cysteamine

2714
L-Cysteic acid

2715
L-Cysteine

2716
L-Cysteine, ethyl ester, hydrochloride

2717
L-Cysteine, hydrochloride

2718
L-Cystine

2719
Cytarabine

2720
Cytidine

2721
2'-Cytidylic acid

2722
3'-Cytidylic acid

2723
5'-Cytidylic acid

2724
Cytisine

2725
Cytochalasin B

2726
Cytochalasin D

2727
Cytochalasin E

2728
Cytosine

2729
Dacarbazine

2730
Dactinomycin

2731
Daidzein

2732
Daminozide

2733
Dantrolene

2734
Datiscetin

2735
Daucol

2736
Daunorubicin

2737
Dazomet

2738
Decabromobiphenyl ether

2739
Decachlorobiphenyl

2740
1,3-Decadiene

2741
1,9-Decadiene

2742
2,2',3,3',4,4',5,5',6,6'-Decafluoro-1,1'-biphenyl

2743
cis-Decahydronaphthalene

No.	Name	Synonym	Mol. Form.	CAS RN	Mol. Wt.	Physical Form	mp/°C	bp/°C	den g cm⁻³	n_D	Solubility
2744	*trans*-Decahydronaphthalene	*trans*-Decalin	$C_{10}H_{18}$	493-02-7	138.250	liq	-30.35(0.06)	187.3(0.2)	0.8659²⁵	1.4695²⁰	i H₂O; vs EtOH, eth, ace; msc bz; sl MeOH
2745	Decahydro-2-naphthol	Decahydro-β-naphthol	$C_{10}H_{18}O$	825-51-4	154.249			109¹⁴	0.996²⁵	1.4992²⁰	
2746	Decamethonium dibromide		$C_{16}H_{38}Br_2N_2$	541-22-0	418.294	cry (MeOH/ ace)	269 dec				i eth
2747	Decamethylcyclopentasiloxane		$C_{10}H_{30}O_5Si_5$	541-02-6	370.770	liq	-37.0(0.5)	213(3)	0.9593²⁰	1.3982²⁰	i H₂O
2748	Decamethyltetrasiloxane		$C_{10}H_{30}O_3Si_4$	141-62-8	310.685	liq	-76	194.4(0.1)	0.8536²⁵	1.3895²⁰	i H₂O; sl EtOH; s bz, peth
2749	Decanal	Capraldehyde	$C_{10}H_{20}O$	112-31-2	156.265	liq	-3.9(0.2)	212(3)	0.830¹⁵	1.4287²⁰	i H₂O; s EtOH, eth, ace; sl ctc
2750	Decane		$C_{10}H_{22}$	124-18-5	142.282	liq	-29.61(0.02)	174.1(0.1)	0.7266²⁵	1.4090²⁵	i H₂O; msc EtOH; s eth; sl ctc
2751	1,10-Decanediamine		$C_{10}H_{24}N_2$	646-25-3	172.311		59.7(0.4)	140¹²			
2752	Decanedinitrile		$C_{10}H_{16}N_2$	1871-96-1	164.247		8.0(0.5)	204¹⁶	0.913²⁰	1.4474²⁰	i H₂O; s chl
2753	1,10-Decanediol	Decamethylene glycol	$C_{10}H_{22}O_2$	112-47-0	174.281	nd (w, dil al)	72.4(0.2)	192²⁰			sl H₂O, eth; vs EtOH; s DMSO; i lig
2754	Decanedioyl dichloride		$C_{10}H_{16}Cl_2O_2$	111-19-3	239.139		-1.3	220⁷⁵	1.1212²⁰	1.4684¹⁸	
2755	Decanenitrile	Caprinitrile	$C_{10}H_{19}N$	1975-78-6	153.265	liq	-14.4(0.4)	241(5)	0.8199²⁰	1.4296²⁰	vs ace, eth, EtOH, chl
2756	1-Decanethiol	Decyl mercaptan	$C_{10}H_{22}S$	143-10-2	174.347	liq	-25.290 (0.001)	240(11)	0.8443²⁰	1.4509²⁰	i H₂O; s EtOH, eth
2757	Decanoic acid	Capric acid	$C_{10}H_{20}O_2$	334-48-5	172.265	nd	31.39(0.02)	270(1)	0.8858⁴⁰	1.4288⁴⁰	i H₂O; vs ace, bz, eth, EtOH
2758	1-Decanol	Capric alcohol	$C_{10}H_{22}O$	112-30-1	158.281	oily liq	7(1)	229(3)	0.8297²⁰	1.4372²⁰	i H₂O; msc EtOH, eth, ace, bz, chl; s ctc
2759	2-Decanol		$C_{10}H_{22}O$	74742-10-2	158.281	liq	-2(1)	211	0.8250²⁰	1.4326²⁵	s EtOH, bz; msc eth, ace; sl ctc
2760	3-Decanol		$C_{10}H_{22}O$	1565-81-7	158.281	liq	-7.5	217(7)	0.827²⁰	1.434²⁰	
2761	4-Decanol	1-Propylheptyl alcohol	$C_{10}H_{22}O$	2051-31-2	158.281	liq	-11	214(3)	0.8261²⁰	1.4320²⁰	i H₂O; s EtOH, ctc
2762	5-Decanol		$C_{10}H_{22}O$	5205-34-5	158.281	liq	8.7	216(5)	0.824²⁰	1.4333²⁰	
2763	2-Decanone	Methyl octyl ketone	$C_{10}H_{20}O$	693-54-9	156.265	nd	14	211(3)	0.8248²⁰	1.4255²⁰	i H₂O; s EtOH, eth; sl ctc
2764	3-Decanone	Ethyl heptyl ketone	$C_{10}H_{20}O$	928-80-3	156.265	liq	2(4)	212(4)	0.8251²⁰	1.4252²⁰	s EtOH, eth, ctc
2765	4-Decanone	Hexyl propyl ketone	$C_{10}H_{20}O$	624-16-8	156.265	liq	-9(4)	206.5	0.824²⁰	1.4240²¹	i H₂O; msc EtOH, eth
2766	Decanoyl chloride	Caprinoyl chloride	$C_{10}H_{19}ClO$	112-13-0	190.710	liq	-34.5	219(13)	0.919²⁵	1.4410²⁰	s eth, ctc
2767	*trans*-2-Decenal		$C_{10}H_{18}O$	3913-81-3	154.249			230			
2768	1-Decene		$C_{10}H_{20}$	872-05-9	140.266	liq	-66.21(0.03)	171(1)	0.7408²⁰	1.4215²⁰	i H₂O; msc EtOH, eth
2769	*cis*-2-Decene		$C_{10}H_{20}$	20348-51-0	140.266	col liq		174.2(0.7)			
2770	*trans*-2-Decene		$C_{10}H_{20}$	20063-97-2	140.266	col liq		173.4(0.5)			
2771	*cis*-5-Decene		$C_{10}H_{20}$	7433-78-5	140.266	col liq	-112(2)	170.4(0.8)	0.7445²⁰	1.4258²⁰	i H₂O; msc EtOH, eth; sl ctc
2772	*trans*-5-Decene		$C_{10}H_{20}$	7433-56-9	140.266	col liq	-73(1)	171.3(0.6)	0.7401²⁰	1.4243²⁰	i H₂O; msc EtOH, eth; sl ctc
2773	9-Decenoic acid	Caproleic acid	$C_{10}H_{18}O_2$	14436-32-9	170.249		26.5	158²¹	0.9238¹⁵	1.4507¹⁵	vs eth, EtOH
2774	9-Decen-1-ol	Decylenic alcohol	$C_{10}H_{20}O$	13019-22-2	156.265			236	0.876²⁵	1.4480²⁰	
2775	3-Decen-2-one	Heptylidene acetone	$C_{10}H_{18}O$	10519-33-2	154.249			102¹⁵·³	0.8473²⁰	1.4480²⁰	
2776	Declomycin	Demeclocycline	$C_{21}H_{21}ClN_2O_8$	127-33-3	464.853	cry	176 dec				i H₂O; s EtOH, eth, bz, ctc, HOAc
2777	Decyl acetate		$C_{12}H_{24}O_2$	112-17-4	200.318	liq	-15.0(0.4)	249(1)	0.8671²⁰	1.4273²⁰	i H₂O; s EtOH, eth, bz, ctc, HOAc
2778	Decylamine	1-Decanamine	$C_{10}H_{23}N$	2016-57-1	157.297		15(1)	217(2)	0.7936²⁰	1.4369²⁰	sl H₂O; msc EtOH, eth, ace, bz, chl
2779	Decylbenzene		$C_{16}H_{26}$	104-72-3	218.377	liq	-14.40(0.08)	298(1)	0.8555²⁰	1.4832²⁰	vs ace, bz, eth, EtOH
2780	Decylcyclohexane		$C_{16}H_{32}$	1795-16-0	224.425	liq	-1.72(0.05)	298(1)	0.8186²⁰	1.4534²⁰	
2781	Decylcyclopentane		$C_{15}H_{30}$	1795-21-7	210.399	liq	-22.11(0.05)	279(1)	0.8110²⁰	1.4486²⁰	vs ace, bz, eth, EtOH
2782	Decyl decanoate		$C_{20}H_{40}O_2$	1654-86-0	312.531		10(2)	219¹⁵	0.8586²⁰	1.4423²⁰	vs eth
2783	Decyl formate		$C_{11}H_{22}O_2$	5451-52-5	186.292	liq		243			
2784	11-Decylheneicosane		$C_{31}H_{64}$	55320-06-4	436.840		9.2(0.4)	282.0¹⁰	0.8116²⁰	1.4540²⁰	
2785	1-Decylnaphthalene		$C_{20}H_{28}$	26438-27-7	268.436		15	379	0.9322²⁰	1.5435²⁰	
2786	Decyloxirane		$C_{12}H_{24}O$	2855-19-8	184.318					1.4347²⁵	sl ctc

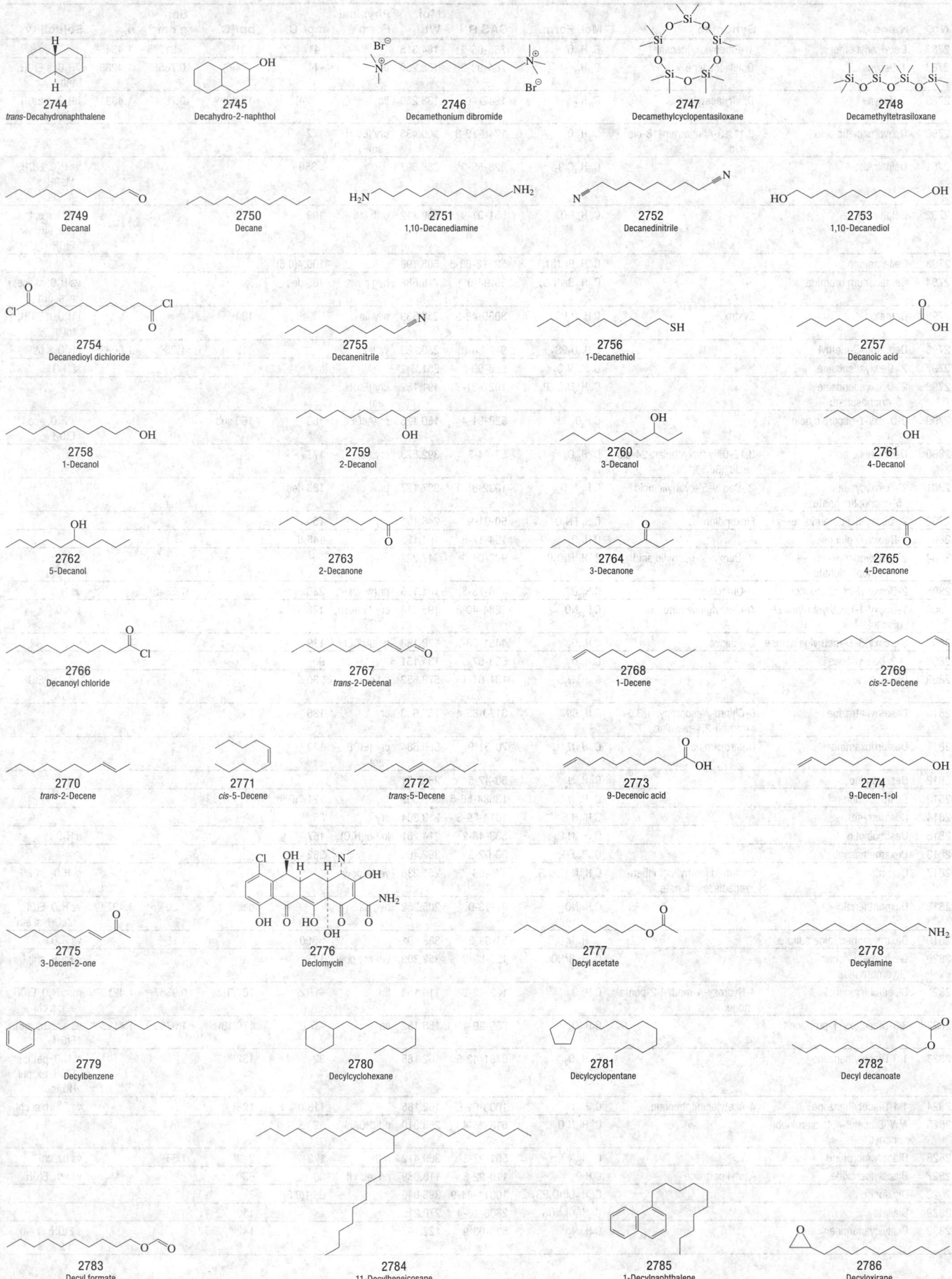

2744
trans-Decahydronaphthalene

2745
Decahydro-2-naphthol

2746
Decamethonium dibromide

2747
Decamethylcyclopentasiloxane

2748
Decamethyltetrasiloxane

2749
Decanal

2750
Decane

2751
1,10-Decanediamine

2752
Decanedinitrile

2753
1,10-Decanediol

2754
Decanedioyl dichloride

2755
Decanenitrile

2756
1-Decanethiol

2757
Decanoic acid

2758
1-Decanol

2759
2-Decanol

2760
3-Decanol

2761
4-Decanol

2762
5-Decanol

2763
2-Decanone

2764
3-Decanone

2765
4-Decanone

2766
Decanoyl chloride

2767
trans-2-Decenal

2768
1-Decene

2769
cis-2-Decene

2770
trans-2-Decene

2771
cis-5-Decene

2772
trans-5-Decene

2773
9-Decenoic acid

2774
9-Decen-1-ol

2775
3-Decen-2-one

2776
Declomycin

2777
Decyl acetate

2778
Decylamine

2779
Decylbenzene

2780
Decylcyclohexane

2781
Decylcyclopentane

2782
Decyl decanoate

2783
Decyl formate

2784
11-Decylheneicosane

2785
1-Decylnaphthalene

2786
Decyloxirane

No.	Name	Synonym	Mol. Form.	CAS RN	Mol. Wt.	Physical Form	mp/°C	bp/°C	den g cm^{-3}	n_D	Solubility
2787	Decyl vinyl ether	1-(Ethenyloxy)decane	$C_{12}H_{24}O$	765-05-9	184.318		-41	101[10]	0.812[20]	1.4346[20]	
2788	1-Decyne	Octylacetylene	$C_{10}H_{18}$	764-93-2	138.250	liq	-44	174(6)	0.7655[20]	1.4265[20]	i H_2O; s EtOH, eth
2789	5-Decyne	Dibutylacetylene	$C_{10}H_{18}$	1942-46-7	138.250	liq	-74(4)	178(3)	0.7690[20]	1.4331[20]	i H_2O; s EtOH, eth
2790	Dehydroabietic acid	8,11,13-Abietatrien-18-oic acid	$C_{20}H_{28}O_2$	1740-19-8	300.435	cry (EtOH aq)	172				
2791	Delphinidin		$C_{15}H_{11}ClO_7$	528-53-0	338.697		>350				vs H_2O, EtOH, MeOH; s AcOEt
2792	Delphinine		$C_{33}H_{45}NO_9$	561-07-9	599.712	orth (al)	199				i H_2O; s chl, ace, eth; vs EtOH
2793	Deltamethrin		$C_{22}H_{19}Br_2NO_3$	52918-63-5	505.199		100.4(0.5)				
2794	Demecarium bromide		$C_{32}H_{52}Br_2N_4O_4$	56-94-0	716.588	hyg pow	165 dec				vs H_2O; sl ace; i ace, eth
2795	Demeton	Systox	$C_8H_{19}O_3PS_2$	8065-48-3	258.339	oily liq		134[2]			i H_2O; s EtOH, tol
2796	Demeton-S-methyl		$C_6H_{15}O_3PS_2$	919-86-8	230.285	ye liq		118[1]	1.20[20]	1.5063[20]	i H_2O; s os
2797	2'-Deoxyadenosine		$C_{10}H_{13}N_5O_3$	958-09-8	251.242						sl H_2O
2798	2'-Deoxyadenosine 5'-triphosphate		$C_{10}H_{16}N_5O_{12}P_3$	1927-31-7	491.182	cry (EtOH aq)					
2799	6-Deoxy-L-ascorbic acid		$C_6H_8O_5$	528-81-4	160.125	pr (AcOEt)	168	160 sub			vs H_2O, ace, EtOH
2800	Deoxycholic acid	3,12-Dihydroxycholan-24-oic acid, (3α,5β,12α)	$C_{24}H_{40}O_4$	83-44-3	392.573	cry (al)	177				
2801	2'-Deoxycytidine 5'-monophosphate	2'-Deoxy-5'-cytidylic acid	$C_9H_{14}N_3O_7P$	1032-65-1	307.197	pow	183 dec				
2802	2'-Deoxy-5-fluorouridine	Floxuridine	$C_9H_{11}FN_2O_5$	50-91-9	246.191	cry	150				
2803	2-Deoxy-D-glucose		$C_6H_{12}O_5$	154-17-6	164.156		146.5				
2804	2'-Deoxyguanosine 5'-monophosphate	2'-Deoxy-5'-guanylic acid	$C_{10}H_{14}N_5O_7P$	902-04-5	347.222						s H_2O
2805	2-Deoxy-D-chiro-inositol	D-Quercitol	$C_6H_{12}O_5$	488-73-3	164.156	pr (w, dil al)	236			1.5845[13]	vs H_2O
2806	1-Deoxy-1-(methylamino)-D-glucitol	N-Methylglucamine	$C_7H_{17}NO_5$	6284-40-8	195.214	cry (MeOH)	128.5				s H_2O
2807	6-Deoxy-3-O-methylgalactose	Digitalose	$C_7H_{14}O_5$	4481-08-7	178.183	nd (AcOEt)	119				vs H_2O
2808	D-2-Deoxyribose		$C_5H_{10}O_4$	533-67-5	134.131		90				
2809	Deserpidine		$C_{32}H_{38}N_2O_8$	131-01-1	578.652	nd or pr	230.5				i H_2O; s EtOH, chl
2810	Desethyl atrazine	6-Chloro-N-isopropyl-1,3,5-triazine-2,4-diamine	$C_6H_{10}ClN_5$	6190-65-4	187.630	cry	136				
2811	Desferrioxamine	Deferoxamine	$C_{25}H_{48}N_6O_8$	70-51-9	560.684	cry (EtOH aq)	139				
2812	Desipramine		$C_{18}H_{22}N_2$	50-47-5	266.381			173[0.02]			
2813	Desmedipham		$C_{16}H_{16}N_2O_4$	13684-56-5	300.309		121.7(0.5)				
2814	Desmetryne		$C_8H_{15}N_5S$	1014-69-3	213.304	cry	85				
2815	Desthiobiotin		$C_{10}H_{18}N_2O_3$	533-48-2	214.261	lo nd (H_2O)	157				s H_2O
2816	Dexamethasone		$C_{22}H_{29}FO_5$	50-02-2	392.460		262				
2817	Dexon	Sodium dimethylaminoben-zenediazosulfonate	$C_8H_{10}N_3NaO_3S$	140-56-7	251.238	ye-br pow					sl H_2O; s DMF
2818	Dexpanthenol		$C_9H_{19}NO_4$	81-13-0	205.252	hyg oil		dec	1.20[20]	1.497[20]	vs H_2O, EtOH, MeOH; sl eth
2819	Dextroamphetamine sulfate		$C_{18}H_{28}N_2O_4S$	51-63-8	368.491		>300		1.15[25]		vs H_2O
2820	Dextromethorphan hydrobromide		$C_{18}H_{26}BrNO$	125-69-9	352.309	wh cry pow	123				s EtOH, chl; i eth
2821	Diacetone alcohol	4-Hydroxy-4-methyl-2-penta-none	$C_6H_{12}O_2$	123-42-2	116.158	liq	-47(2)	167.9	0.9387[20]	1.4213[20]	msc H_2O, EtOH, eth; s chl
2822	3,3-Diacetoxy-1-propene		$C_7H_{10}O_4$	869-29-4	158.152	liq	-37.6	176(18)	1.0760[20]	1.4193[20]	vs ace, bz, eth, EtOH
2823	1,3-Diacetylbenzene		$C_{10}H_{10}O_2$	6781-42-6	162.185		32	152[15]			sl H_2O, peth; s EtOH, bz, chl, HOAc
2824	1,4-Diacetylbenzene	4-Acetylacetophenone	$C_{10}H_{10}O_2$	1009-61-6	162.185		113.0	128[3]			vs EtOH; sl chl
2825	N,N'-Diacetyl-4,4'-diaminobi-phenyl		$C_{16}H_{16}N_2O_2$	613-35-4	268.310	nd (HOAc)	328.3				
2826	Diacetylmorphine		$C_{21}H_{23}NO_5$	561-27-3	369.412	orth	173	273[12]	1.56[25]		vs bz, chl
2827	Diacetylperoxide	Acetyl peroxide	$C_4H_6O_4$	110-22-5	118.089	nd (eth) lf	30	63[21]			vs eth, EtOH
2828	Dialifor		$C_{14}H_{17}ClNO_4PS_2$	10311-84-9	393.846		68.1(0.5)				
2829	Diallate		$C_{10}H_{17}Cl_2NOS$	2303-16-4	270.219			150[9]			
2830	Diallylcyanamide		$C_7H_{10}N_2$	538-08-9	122.167			142[90]			s EtOH; sl eth, ctc

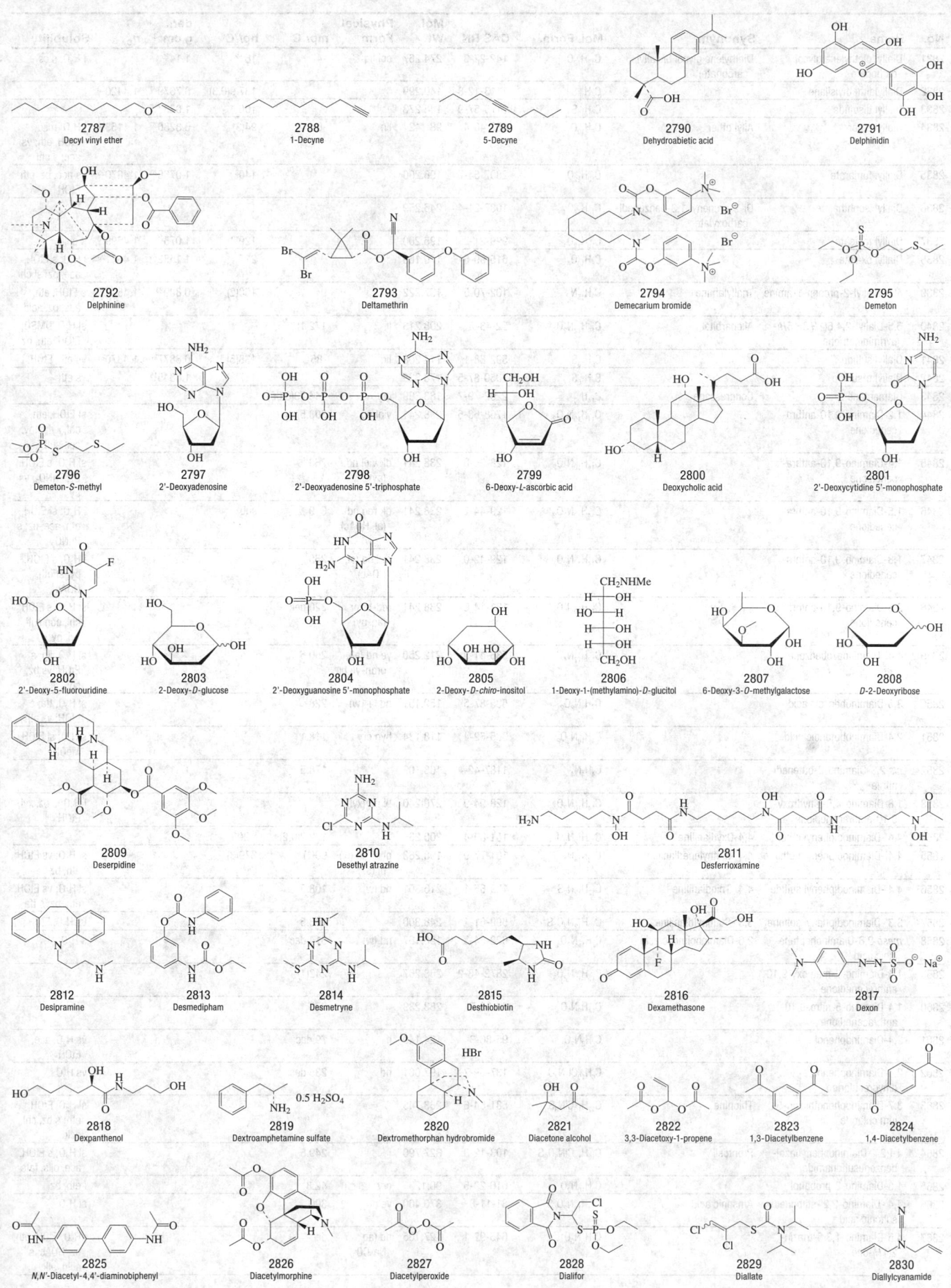

2787 Decyl vinyl ether

2788 1-Decyne

2789 5-Decyne

2790 Dehydroabietic acid

2791 Delphinidin

2792 Delphinine

2793 Deltamethrin

2794 Demecarium bromide

2795 Demeton

2796 Demeton-S-methyl

2797 2'-Deoxyadenosine

2798 2'-Deoxyadenosine 5'-triphosphate

2799 6-Deoxy-L-ascorbic acid

2800 Deoxycholic acid

2801 2'-Deoxycytidine 5'-monophosphate

2802 2'-Deoxy-5-fluorouridine

2803 2-Deoxy-D-glucose

2804 2'-Deoxyguanosine 5'-monophosphate

2805 2-Deoxy-D-chiro-inositol

2806 1-Deoxy-1-(methylamino)-D-glucitol

2807 6-Deoxy-3-O-methylgalactose

2808 D-2-Deoxyribose

2809 Deserpidine

2810 Desethyl atrazine

2811 Desferrioxamine

2812 Desipramine

2813 Desmedipham

2814 Desmetryne

2815 Desthiobiotin

2816 Dexamethasone

2817 Dexon

2818 Dexpanthenol

2819 Dextroamphetamine sulfate

2820 Dextromethorphan hydrobromide

2821 Diacetone alcohol

2822 3,3-Diacetoxy-1-propene

2823 1,3-Diacetylbenzene

2824 1,4-Diacetylbenzene

2825 N,N'-Diacetyl-4,4'-diaminobiphenyl

2826 Diacetylmorphine

2827 Diacetylperoxide

2828 Dialifor

2829 Diallate

2830 Diallylcyanamide

No.	Name	Synonym	Mol. Form.	CAS RN	Mol. Wt.	Physical Form	mp/°C	bp/°C	den g cm⁻³	n_D	Solubility
2831	Diallyl diethylene glycol carbonate	Diethylene glycol bis(allyl carbonate)	C₁₂H₁₈O₇	142-22-3	274.267	col liq	-4	161²	1.14²⁰		i H₂O; s os
2832	Diallyldimethylsilane		C₈H₁₆Si	1113-12-8	140.299			137.9(0.3)	0.7679²⁰	1.4420²⁰	
2833	Diallyl disulfide		C₆H₁₀S₂	2179-57-9	146.273			100⁴⁸	1.0237¹⁵		
2834	Diallyl ether	Allyl ether	C₆H₁₀O	557-40-4	98.142	liq	-6	94(3)	0.8260²⁰	1.4163²⁰	i H₂O; msc EtOH, eth; vs ace; s chl
2835	Diallyl fumarate		C₁₀H₁₂O₄	2807-54-7	196.200			140³	1.0768²⁰	1.4670²⁵	vs ace, bz, eth, EtOH
2836	Diallyl isophthalate	Di-2-propenyl 1,3-benzenedicarboxylate	C₁₄H₁₄O₄	1087-21-4	246.259			176⁵			
2837	Diallyl maleate		C₁₀H₁₂O₄	999-21-3	196.200			129¹⁰	1.075²⁰	1.4699²⁰	s chl
2838	Diallyl oxalate		C₈H₁₀O₄	615-99-6	170.163			217	1.1582²⁰	1.4481²⁰	i H₂O; s EtOH, ace, bz; sl chl
2839	N,N-Diallyl-2-propen-1-amine	Triallylamine	C₉H₁₅N	102-70-5	137.222		94	150(2)	0.809²⁰	1.4502²⁰	s EtOH, eth, ace, bz, acid
2840	5,5-Diallyl-2,4,6(1H,3H,5H)-pyrimidinetrione	Allobarbital	C₁₀H₁₂N₂O₃	52-43-7	208.213	lf	172(1)				sl H₂O, DMSO; s EtOH, eth, bz
2841	Diallyl sulfide		C₆H₁₀S	592-88-1	114.208	liq	-85	138(5)	0.8877²⁷	1.4870²⁵	vs eth, EtOH
2842	Diallyl trisulfide		C₆H₁₀S₃	2050-87-5	178.338			117¹⁶	1.0845¹⁵		vs eth
2843	Diamantane	Congressane	C₁₄H₂₀	2292-79-7	188.309	cry	244.73(0.05)				
2844	1,2-Diamino-9,10-anthracenedione		C₁₄H₁₀N₂O₂	1758-68-5	238.241	viol nd	303.5				sl EtOH, eth, chl, xyl; s py, con sulf
2845	1,4-Diamino-9,10-anthracenedione		C₁₄H₁₀N₂O₂	128-95-0	238.241	dk viol nd (py)	268				sl H₂O; s EtOH, bz, PhNO₂; vs py
2846	1,5-Diamino-9,10-anthracenedione		C₁₄H₁₀N₂O₂	129-44-2	238.241	dk red nd (al, HOAc)	319	sub			i H₂O; sl EtOH, eth, ace, bz; s PhNO₂
2847	1,8-Diamino-9,10-anthracenedione		C₁₄H₁₀N₂O₂	129-42-0	238.241	red nd (al, HOAc)	265				i H₂O; s EtOH, py; sl eth, HOAc
2848	2,6-Diamino-9,10-anthracenedione		C₁₄H₁₀N₂O₂	131-14-6	238.241	red-br pr (aq-py)	320 dec				sl H₂O; s EtOH, chl, con sulf, xyl, py
2849	4,4'-Diaminoazobenzene		C₁₂H₁₂N₄	538-41-0	212.250	ye nd (al), oran-ye pr (al)	250.5				sl H₂O, lig; s EtOH; vs bz, chl
2850	3,5-Diaminobenzoic acid		C₇H₈N₂O₂	535-87-5	152.151	nd (+1w)	228				sl H₂O, tfa; s EtOH; vs eth
2851	2,4-Diaminobutanoic acid		C₄H₁₀N₂O₂	305-62-4	118.134	hyg cry	118.1				s H₂O; sl EtOH, MeOH
2852	cis-2,3-Diamino-2-butenedinitrile		C₄H₄N₄	1187-42-4	108.102		178.5		1.41²⁰		
2853	1,8-Diamino-4,5-dihydroxy-9,10-anthracenedione		C₁₄H₁₀N₂O₄	128-94-9	270.240	bl nd (xyl)					i H₂O; s bz, xyl, EtOH
2854	4,4'-Diaminodiphenyl ether	4,4-Oxydianiline	C₁₂H₁₂N₂O	101-80-4	200.235		192.2(0.2)	>300			
2855	4,4'-Diaminodiphenylmethane	4,4'-Methylenedianiline	C₁₃H₁₄N₂	101-77-9	198.263	pl or nd (w) pl (bz)	90(1)	379(3)			sl H₂O; vs EtOH, eth, bz
2856	4,4'-Diaminodiphenyl sulfide	4,4'-Thiodianiline	C₁₂H₁₂N₂S	139-65-1	216.301	nd (w)	108.5				sl H₂O; vs EtOH, eth, bz; s tfa
2857	3,3'-Diaminodiphenyl sulfone	3,3'-Sulfonyldianiline	C₁₂H₁₂N₂O₂S	599-61-1	248.300		168.5				vs H₂O, EtOH
2858	meso-2,6-Diaminoheptanedioic acid	2,6-Diaminopimelic acid	C₇H₁₄N₂O₄	922-54-3	190.197	nd (w)	314 dec				s H₂O
2859	1,4-Diamino-2-methoxy-9,10-anthracenedione		C₁₅H₁₂N₂O₃	2872-48-2	268.267		242(1)				
2860	1,4-Diamino-5-nitro-9,10-anthracenedione		C₁₄H₉N₃O₄	82-33-7	283.239		278				
2861	2,4-Diaminophenol		C₆H₈N₂O	95-86-3	124.140	lf	79 dec				vs H₂O, ace, EtOH
2862	2,4-Diaminophenol, dihydrochloride		C₆H₁₀Cl₂N₂O	137-09-7	197.061	nd	235 dec				vs H₂O
2863	3,7-Diaminophenothiazin-5-ium chloride	Thionine	C₁₂H₁₀ClN₃S	581-64-6	263.745						sl H₂O, EtOH, eth; s bz, chl, acid
2864	4-[(2,4-Diaminophenyl)azo]-benzenesulfonamide	Prontosil	C₁₂H₁₄ClN₅O₂S	103-12-8	327.790		249.5				sl H₂O; s EtOH, ace, oils, fats
2865	1,3-Diamino-2-propanol		C₃H₁₀N₂O	616-29-5	90.123	cry	42.8				i eth, bz
2866	4,4'-Diamino-2,2'-stilbenedisulfonic acid	Amsonic acid	C₁₄H₁₄N₂O₆S₂	81-11-8	370.400	ye nd	300				sl H₂O
2867	4,6-Diamino-1,3,5-triazin-2(1H)-one		C₃H₅N₅O	645-92-1	127.105	nd (aq Na₂CO₃)	dec				i H₂O, EtOH, eth, bz, HOAc; s acid, alk

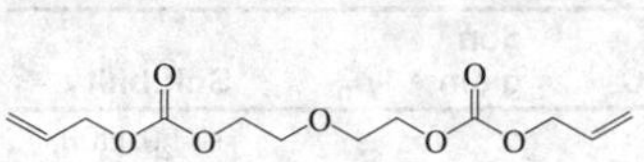

2831 Diallyl diethylene glycol carbonate

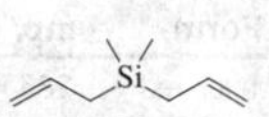

2832 Diallyldimethylsilane

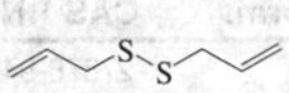

2833 Diallyl disulfide

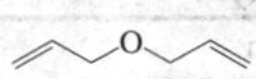

2834 Diallyl ether

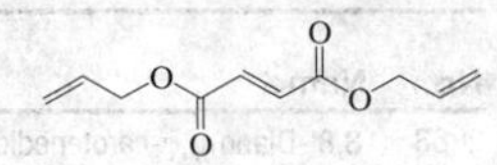

2835 Diallyl fumarate

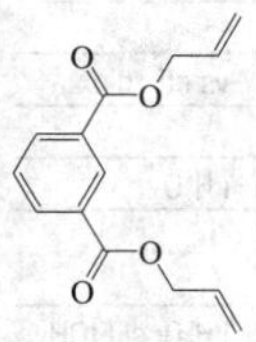

2836 Diallyl isophthalate

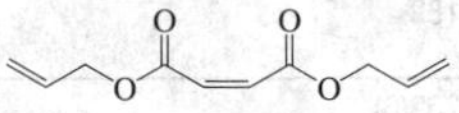

2837 Diallyl maleate

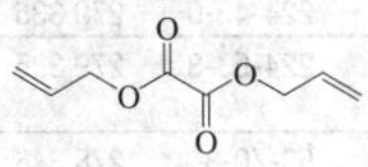

2838 Diallyl oxalate

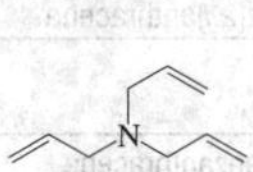

2839 N,N-Diallyl-2-propen-1-amine

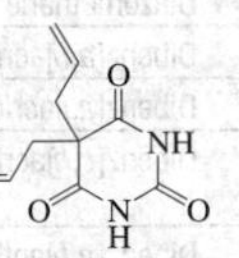

2840 5,5-Diallyl-2,4,6(1H,3H,5H)-pyrimidinetrione

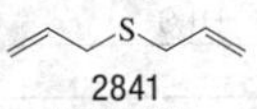

2841 Diallyl sulfide

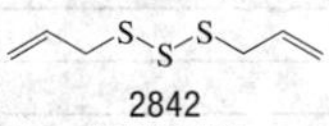

2842 Diallyl trisulfide

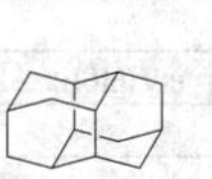

2843 Diamantane

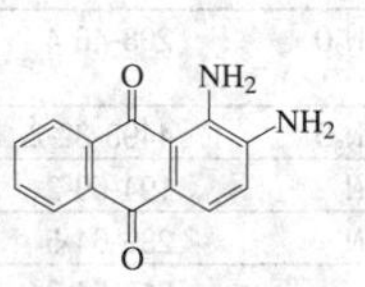

2844 1,2-Diamino-9,10-anthracenedione

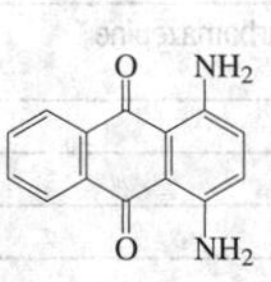

2845 1,4-Diamino-9,10-anthracenedione

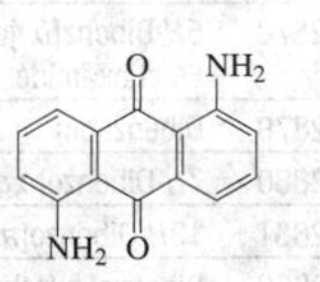

2846 1,5-Diamino-9,10-anthracenedione

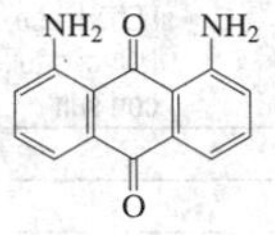

2847 1,8-Diamino-9,10-anthracenedione

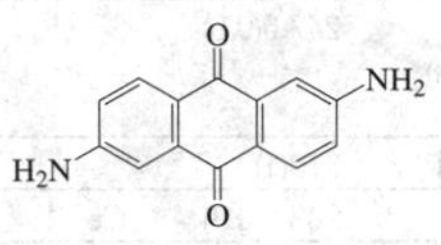

2848 2,6-Diamino-9,10-anthracenedione

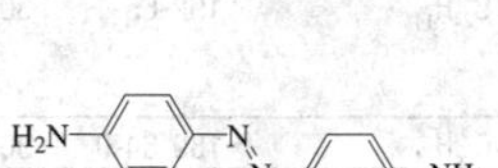

2849 4,4'-Diaminoazobenzene

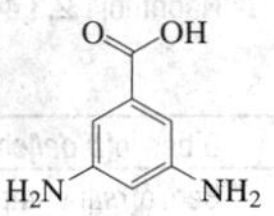

2850 3,5-Diaminobenzoic acid

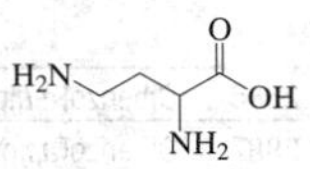

2851 2,4-Diaminobutanoic acid

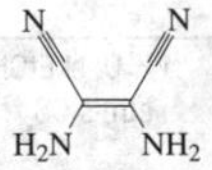

2852 cis-2,3-Diamino-2-butenedinitrile

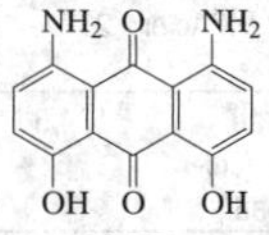

2853 1,8-Diamino-4,5-dihydroxy-9,10-anthracenedione

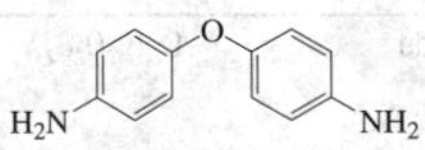

2854 4,4'-Diaminodiphenyl ether

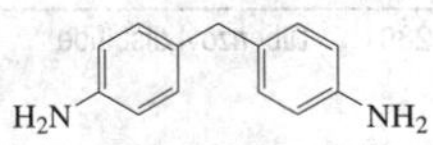

2855 4,4'-Diaminodiphenylmethane

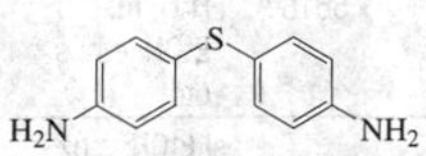

2856 4,4'-Diaminodiphenyl sulfide

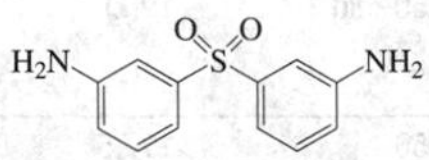

2857 3,3'-Diaminodiphenyl sulfone

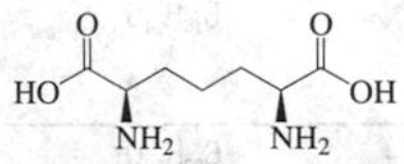

2858 meso-2,6-Diaminoheptanedioic acid

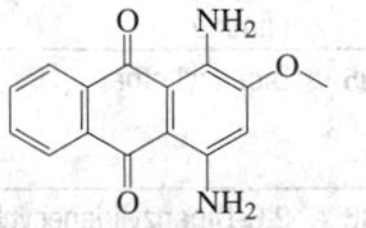

2859 1,4-Diamino-2-methoxy-9,10-anthracenedione

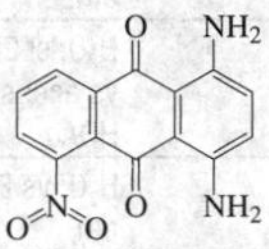

2860 1,4-Diamino-5-nitro-9,10-anthracenedione

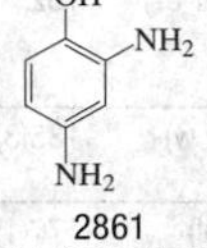

2861 2,4-Diaminophenol

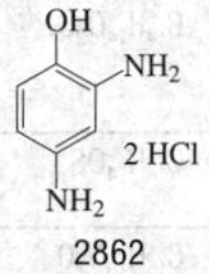

2862 2,4-Diaminophenol, dihydrochloride

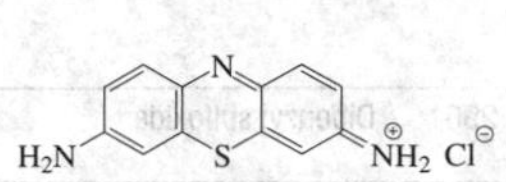

2863 3,7-Diaminophenothiazin-5-ium chloride

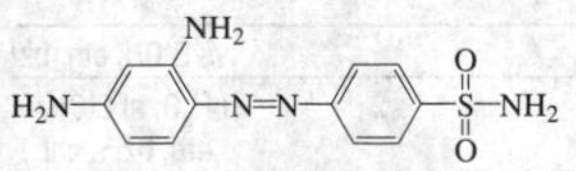

2864 4-[(2,4-Diaminophenyl)azo]benzenesulfonamide

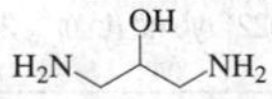

2865 1,3-Diamino-2-propanol

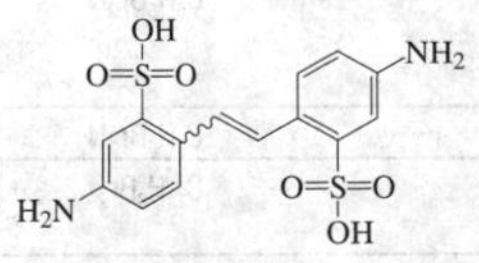

2866 4,4'-Diamino-2,2'-stilbenedisulfonic acid

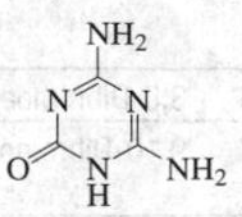

2867 4,6-Diamino-1,3,5-triazin-2(1H)-one

No.	Name	Synonym	Mol. Form.	CAS RN	Mol. Wt.	Physical Form	mp/°C	bp/°C	den g cm⁻³	n_D	Solubility
2868	8,8'-Diapo-ψ,ψ-carotenedioic acid	Crocetin	C₂₀H₂₄O₄	27876-94-4	328.403	brick red orth	286				sl H₂O, EtOH; i eth, bz; s py; vs NaOH
2869	Diatrizoic acid	N,N'-Diacetyl-3,5-diamino-2,4,6-triiodobenzoic acid	C₁₁H₉I₃N₂O₄	117-96-4	613.913	cry (EtOH aq)	300				
2870	Diazenedicarboxamide	Azodicarbonamide	C₂H₄N₄O₂	123-77-3	116.079		225(1)				
2871	Diazinon		C₁₂H₂₁N₂O₃PS	333-41-5	304.345			87[0.05]	1.1088[20]	1.4922[20]	
2872	Diazomethane		CH₂N₂	334-88-3	42.040	ye gas	-145	-23			vs eth, diox
2873	Dibenz[a,h]acridine		C₂₁H₁₃N	226-36-8	279.335	ye cry	226.6(0.7)				
2874	Dibenz[a,j]acridine	7-Azadibenz[a,j]anthracene	C₂₁H₁₃N	224-42-0	279.335		219.6(0.9)				i H₂O
2875	Dibenz[c,h]acridine		C₂₁H₁₃N	224-53-3	279.335	ye cry (EtOH)	189				
2876	Dibenz[a,h]anthracene	1,2:5,6-Dibenzanthracene	C₂₂H₁₄	53-70-3	278.346	pl (dil ace)	269(6)				i H₂O; sl EtOH; s ace, bz, CS₂
2877	Dibenz[a,j]anthracene		C₂₂H₁₄	224-41-9	278.346	oran lf or nd (bz)	198.2				i H₂O, HOAc; sl EtOH, eth, bz; s peth
2878	5H-Dibenz[b,f]azepine-5-carboxamide	Carbamazepine	C₁₅H₁₂N₂O	298-46-4	236.268		190.2				
2879	Dibenzepin		C₁₈H₂₁N₃O	4498-32-2	295.379		117	185[0.01]			
2880	7H-Dibenzo[c,g]carbazole		C₂₀H₁₃N	194-59-2	267.324	cry (EtOH)	157(1)				
2881	13H-Dibenzo[a,i]carbazole		C₂₀H₁₃N	239-64-5	267.324		221.3				i H₂O
2882	Dibenzo[b,k]chrysene		C₂₆H₁₆	217-54-9	328.405		400				
2883	Dibenzo[b,e][1,4]dioxin	Diphenylene dioxide	C₁₂H₈O₂	262-12-4	184.191	nd (MeOH)	117.5(0.2)				
2884	Dibenzofuran	2,2'-Biphenylene oxide	C₁₂H₈O	132-64-9	168.191	lf or nd (al)	82.16(0.05)	285.2(0.3)	1.0886[99]	1.6079[99]	i H₂O; s EtOH, ace, bz; vs eth, HOAc
2885	Dibenzo[a,e]pyrene	Naphtho[1,2,3,4-def]chrysene	C₂₄H₁₄	192-65-4	302.368	pa ye nd(xyl)	247.0(0.6)				sl EtOH, ace, bz, HOAc; s tol, con sulf
2886	Dibenzo[a,h]pyrene	Dibenzo[b,def]chrysene	C₂₄H₁₄	189-64-0	302.368	oran pl	318(1)				
2887	Dibenzo[a,i]pyrene	Benzo[rst]pentaphene	C₂₄H₁₄	189-55-9	302.368		283.6(0.3)	275[0.05]			
2888	Dibenzo[a,l]pyrene	Dibenzo[def,p]chrysene	C₂₄H₁₄	191-30-0	302.368	ye pl (bz/EtOH)	164.5				
2889	Dibenzothiophene		C₁₂H₈S	132-65-0	184.257	nd (dil al, lig)	98.67(0.02)	331.6(0.4)			i H₂O; s chl, MeOH; vs EtOH, bz
2890	Dibenz[c,e]oxepin-5,7-dione		C₁₄H₈O₃	6050-13-1	224.212	nd (HOAc or bz)	217	sub			i H₂O; sl eth
2891	Dibenzoyl disulfide	Benzoyl disulfide	C₁₄H₁₀O₂S₂	644-32-6	274.358	pr(al), sc(chl-peth)	134.5	dec			i H₂O; sl EtOH, eth; s CS₂
2892	Dibenzylamine	N-Benzylbenzenemethanamine	C₁₄H₁₅N	103-49-1	197.276		-26(1)	300 dec	1.0256[22]	1.5781[20]	i H₂O; vs EtOH, eth; s ctc
2893	Dibenzyl disulfide		C₁₄H₁₄S₂	150-60-7	246.391	lf (al)	68.6(0.2)				sl H₂O; s EtOH, eth, bz, MeOH
2894	N,N'-Dibenzyl-1,2-ethanedi-amine	Benzathine	C₁₆H₂₀N₂	140-28-3	240.343	oily lig	26	195[4]	1.024[20]	1.5635[20]	vs bz, eth, EtOH
2895	Dibenzyl ether	Benzyl ether	C₁₄H₁₄O	103-50-4	198.260	liq	19(2)	298	1.0428[20]	1.5618[20]	i H₂O; msc EtOH, eth; s ctc
2896	2,6-Dibenzylidenecyclohexa-none		C₂₀H₁₈O	897-78-9	274.356		117.5	190[20]			sl EtOH; s bz, HOAc
2897	Dibenzyl malonate		C₁₇H₁₆O₄	15014-25-2	284.307			187[2]	1.137[25]	1.5447[20]	
2898	Dibenzyl phosphite		C₁₄H₁₅O₃P	17176-77-1	262.241		-2.5	162[0.1]		1.5521[18]	
2899	Dibenzyl sulfide	Benzyl sulfide	C₁₄H₁₄S	538-74-9	214.326	pl (eth or chl)	48.3(0.5)	335(6)	1.0583[50]		i H₂O; s EtOH, eth, CS₂
2900	Dibenzyl sulfone		C₁₄H₁₄O₂S	620-32-6	246.325	nd (al-bz)	152	290 dec			i H₂O; sl EtOH; vs ace; s bz, HOAc
2901	Dibenzyl sulfoxide		C₁₄H₁₄OS	621-08-9	230.325	lf (al, w)	135(3)	210 dec			i H₂O; vs EtOH, eth
2902	N,N'-Dibenzylurea		C₁₅H₁₆N₂O	1466-67-7	240.300	nd (al)	170(1)				vs EtOH, HOAc
2903	Dibromoacetic acid		C₂H₂Br₂O₂	631-64-1	217.844	hyg cry	49	195[250]			vs H₂O; vs EtOH, eth
2904	Dibromoacetonitrile		C₂HBr₂N	3252-43-5	198.844			169	2.369[20]	1.5393[20]	
2905	2,4-Dibromoaniline		C₆H₅Br₂N	615-57-6	250.919	orth bipym (chl) nd or lf (al)	79.5	156[74]	2.260[20]		s EtOH, eth, chl, HOAc
2906	3,5-Dibromoaniline		C₆H₅Br₂N	626-40-4	250.919	nd (dil al)	57				vs EtOH, eth, bz
2907	9,10-Dibromoanthracene		C₁₄H₈Br₂	523-27-3	336.022	ye nd (to or xyl)	226	sub			i H₂O; sl EtOH, eth, bz; s chl

2868
8,8'-Diapo-ψ,ψ-carotenedioic acid

2869
Diatrizoic acid

2870
Diazenedicarboxamide

2871
Diazinon

2872
Diazomethane

2873
Dibenz[a,h]acridine

2874
Dibenz[a,j]acridine

2875
Dibenz[c,h]acridine

2876
Dibenz[a,h]anthracene

2877
Dibenz[a,j]anthracene

2878
5H-Dibenz[b,f]azepine-5-carboxamide

2879
Dibenzepin

2880
7H-Dibenzo[c,g]carbazole

2881
13H-Dibenzo[a,i]carbazole

2882
Dibenzo[b,k]chrysene

2883
Dibenzo[b,e][1,4]dioxin

2884
Dibenzofuran

2885
Dibenzo[a,e]pyrene

2886
Dibenzo[a,h]pyrene

2887
Dibenzo[a,i]pyrene

2888
Dibenzo[a,l]pyrene

2889
Dibenzothiophene

2890
Dibenz[c,e]oxepin-5,7-dione

2891
Dibenzoyl disulfide

2892
Dibenzylamine

2893
Dibenzyl disulfide

2894
N,N'-Dibenzyl-1,2-ethanediamine

2895
Dibenzyl ether

2896
2,6-Dibenzylidenecyclohexanone

2897
Dibenzyl malonate

2898
Dibenzyl phosphite

2899
Dibenzyl sulfide

2900
Dibenzyl sulfone

2901
Dibenzyl sulfoxide

2902
N,N'-Dibenzylurea

2903
Dibromoacetic acid

2904
Dibromoacetonitrile

2905
2,4-Dibromoaniline

2906
3,5-Dibromoaniline

2907
9,10-Dibromoanthracene

No.	Name	Synonym	Mol. Form.	CAS RN	Mol. Wt.	Physical Form	mp/°C	bp/°C	den g cm^{-3}	n_D	Solubility
2908	o-Dibromobenzene	1,2-Dibromobenzene	C$_6$H$_4$Br$_2$	583-53-9	235.904		6(2)	220.4(0.3)	1.9843[20]	1.6155[20]	i H$_2$O; s EtOH; msc eth, ace, bz, ctc
2909	m-Dibromobenzene	1,3-Dibromobenzene	C$_6$H$_4$Br$_2$	108-36-1	235.904	liq	-6.9(0.5)	214(14)	1.9523[20]	1.6083[17]	i H$_2$O; s EtOH; msc eth
2910	p-Dibromobenzene	1,4-Dibromobenzene	C$_6$H$_4$Br$_2$	106-37-6	235.904	pl	87.3(0.1)	222(3)	2.261[17]	1.5742	i H$_2$O; s EtOH; bz; vs eth, ace, CS$_2$
2911	4,4'-Dibromobenzophenone	Bis(4-bromophenyl) ketone	C$_{13}$H$_8$Br$_2$O	3988-03-2	340.010	pl (al)	177	394(24)			vs bz, HOAc, chl
2912	4,4'-Dibromo-1,1'-biphenyl		C$_{12}$H$_8$Br$_2$	92-86-4	312.000	mcl pr (MeOH)	164	357.5			i H$_2$O; sl EtOH; s bz
2913	1,3-Dibromo-2,2-bis(bromomethyl)propane	Pentaerythritol tetrabromide	C$_5$H$_8$Br$_4$	3229-00-3	387.734	cry (ace), nd (lig)	160.29(0.05)	305.5	2.596[15]		s EtOH, bz, tol; sl eth, chl
2914	3,5-Dibromo-N-(4-bromophenyl)-2-hydroxybenzamide	Tribromsalan	C$_{13}$H$_8$Br$_3$NO$_2$	87-10-5	449.921		226.4(0.5)				
2915	1,1-Dibromobutane		C$_4$H$_8$Br$_2$	62168-25-6	215.915			158	1.784[25]	1.4988[25]	
2916	1,2-Dibromobutane	α-Butylene dibromide	C$_4$H$_8$Br$_2$	533-98-2	215.915	liq	-65.4(0.4)	161(4)	1.7915[20]	1.4025[20]	i H$_2$O; s eth, chl
2917	1,3-Dibromobutane		C$_4$H$_8$Br$_2$	107-80-2	215.915			176.4(0.4)	1.800[20]	1.507[20]	i H$_2$O; s eth, chl; sl ctc
2918	1,4-Dibromobutane		C$_4$H$_8$Br$_2$	110-52-1	215.915	liq	-21.1(0.5)	197(4)	1.8199[25]	1.5167[25]	i H$_2$O; sl ctc; s chl
2919	2,3-Dibromobutane		C$_4$H$_8$Br$_2$	5408-86-6	215.915	liq	-24	158(5)	1.7893[22]	1.5133[22]	i H$_2$O; s eth
2920	trans-1,4-Dibromo-2-butene		C$_4$H$_6$Br$_2$	821-06-7	213.899	pl (peth)	53.4	203			sl H$_2$O, chl; vs EtOH, peth; s ace
2921	1,4-Dibromo-2-butyne		C$_4$H$_4$Br$_2$	2219-66-1	211.883			92[15]	2.014[18]	1.588[18]	s eth, ace; vs chl
2922	α,α'-Dibromo-d-camphor		C$_{10}$H$_{14}$Br$_2$O	514-12-5	310.025		61		1.854[21]		i H$_2$O; vs EtOH, eth, bz, chl; s AcOEt
2923	Dibromochlorofluoromethane		CBr$_2$ClF	353-55-9	226.270			80.3	2.3173[22]	1.4570[20]	
2924	1,2-Dibromo-3-chloropropane		C$_3$H$_5$Br$_2$Cl	96-12-8	236.333			200(13)	2.093[14]	1.553[14]	i H$_2$O
2925	1,2-Dibromo-1-chloro-1,2,2-trifluoroethane		C$_2$Br$_2$ClF$_3$	354-51-8	276.277		50	92.8(0.2)			
2926	2,2-Dibromo-2-cyanoacetamide		C$_3$H$_2$Br$_2$N$_2$O	10222-01-2	241.868	cry (bz)	126				
2927	trans-1,2-Dibromocyclohexane, (±)-		C$_6$H$_{10}$Br$_2$	5183-77-7	241.951		-2.0	145[100]	1.7759[20]	1.5445[19]	vs ace, bz, eth, EtOH
2928	1,10-Dibromodecane	Decamethylene dibromide	C$_{10}$H$_{20}$Br$_2$	4101-68-2	300.074	pl (al)	28	161[9]	1.335[30]	1.4927[25]	i H$_2$O; sl EtOH; s eth
2929	1,2-Dibromo-1,1-dichloroethane		C$_2$H$_2$Br$_2$Cl$_2$	75-81-0	256.751	liq	-26	195	2.135[20]	1.5662[20]	vs ace, bz, eth, EtOH
2930	1,2-Dibromo-1,2-dichloroethane		C$_2$H$_2$Br$_2$Cl$_2$	683-68-1	256.751	liq	-26	195	2.135[20]	1.5662[20]	i H$_2$O; s EtOH, eth, ace, bz
2931	Dibromodichloromethane		CBr$_2$Cl$_2$	594-18-3	242.725		38	150.2	2.42[25]		i H$_2$O; s EtOH, eth, ace, bz
2932	1,2-Dibromo-1,1-difluoroethane	Genetron 132b-B2	C$_2$H$_2$Br$_2$F$_2$	75-82-1	223.842	liq	-61.3	92.5	2.2238[20]	1.4456[20]	
2933	Dibromodifluoromethane		CBr$_2$F$_2$	75-61-6	209.816	vol liq or gas	-110.1	22.79(0.08)			s H$_2$O, eth, ace, bz
2934	1,3-Dibromo-5,5-dimethyl-2,4-imidazolidinedione	Dibromantine	C$_5$H$_6$Br$_2$N$_2$O$_2$	77-48-5	285.922		198 dec				
2935	1,3-Dibromo-2,2-dimethylpropane		C$_5$H$_{10}$Br$_2$	5434-27-5	229.941			185(11)	1.6775[20]	1.5090	
2936	1,12-Dibromododecane		C$_{12}$H$_{24}$Br$_2$	3344-70-5	328.127	nd (al,HOAc)	41	215[15]			i H$_2$O; vs EtOH, chl; s eth, HOAc
2937	1,1-Dibromoethane	Ethylidene dibromide	C$_2$H$_4$Br$_2$	557-91-5	187.861	liq	-63	109(4)	2.0555[20]	1.5128[20]	i H$_2$O; s EtOH, ace, bz; sl chl; vs eth
2938	1,2-Dibromoethane	Ethylene dibromide	C$_2$H$_4$Br$_2$	106-93-4	187.861	liq	9.8(0.1)	131.3(0.3)	2.1683[25]	1.5356[25]	vs ace, bz, eth, EtOH
2939	cis-1,2-Dibromoethene	cis-1,2-Dibromoethylene	C$_2$H$_2$Br$_2$	590-11-4	185.845	liq	-53	111(1)	2.2464[20]	1.5428[20]	i H$_2$O; vs EtOH, eth; s ace, bz, chl
2940	trans-1,2-Dibromoethene	trans-1,2-Dibromoethylene	C$_2$H$_2$Br$_2$	590-12-5	185.845	liq	-6.5	107(3)	2.2308[20]	1.5505[18]	i H$_2$O; vs EtOH, eth; s ace, bz, chl
2941	1,2-Dibromo-1-ethoxyethane		C$_4$H$_8$Br$_2$O	2983-26-8	231.914			80[20]	1.7320[20]	1.5044[20]	vs EtOH, chl
2942	1,2-Dibromoethyl acetate		C$_4$H$_6$Br$_2$O$_2$	24442-57-7	245.898	liq		89.5[16]	1.91[20]		
2943	(1,2-Dibromoethyl)benzene		C$_8$H$_8$Br$_2$	93-52-7	263.958		73(1)	133[19]			s EtOH, eth, bz, chl, HOAc, MeOH, lig

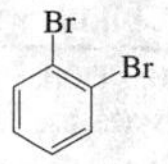

2908
o-Dibromobenzene

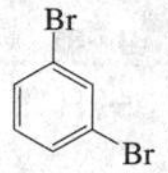

2909
m-Dibromobenzene

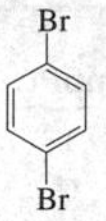

2910
p-Dibromobenzene

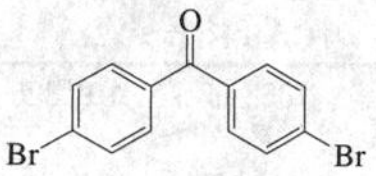

2911
4,4′-Dibromobenzophenone

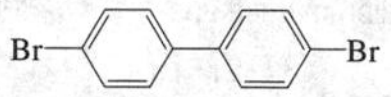

2912
4,4′-Dibromo-1,1′-biphenyl

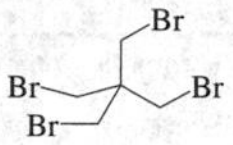

2913
1,3-Dibromo-2,2-bis(bromomethyl)propane

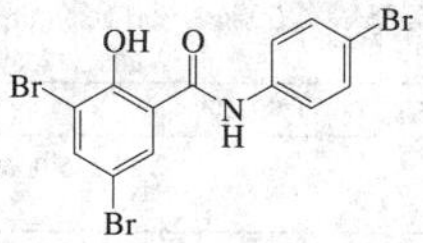

2914
3,5-Dibromo-N-(4-bromophenyl)-2-hydroxybenzamide

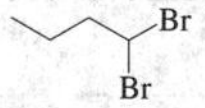

2915
1,1-Dibromobutane

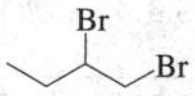

2916
1,2-Dibromobutane

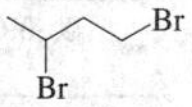

2917
1,3-Dibromobutane

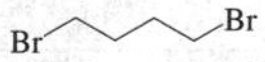

2918
1,4-Dibromobutane

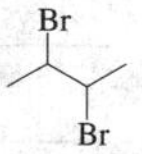

2919
2,3-Dibromobutane

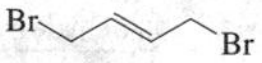

2920
trans-1,4-Dibromo-2-butene

2921
1,4-Dibromo-2-butyne

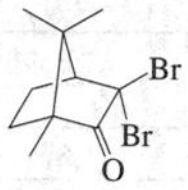

2922
α,α′-Dibromo-d-camphor

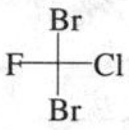

2923
Dibromochlorofluoromethane

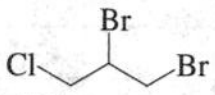

2924
1,2-Dibromo-3-chloropropane

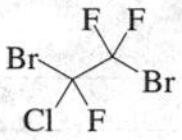

2925
1,2-Dibromo-1-chloro-1,2,2-trifluoroethane

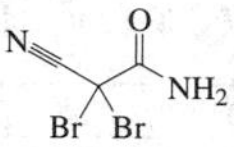

2926
2,2-Dibromo-2-cyanoacetamide

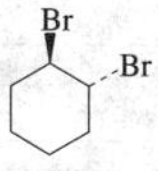

2927
trans-1,2-Dibromocyclohexane, (±)-

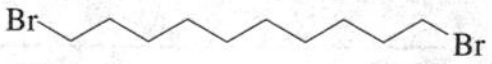

2928
1,10-Dibromodecane

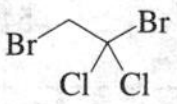

2929
1,2-Dibromo-1,1-dichloroethane

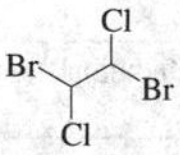

2930
1,2-Dibromo-1,2-dichloroethane

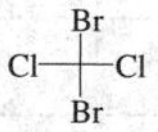

2931
Dibromodichloromethane

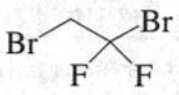

2932
1,2-Dibromo-1,1-difluoroethane

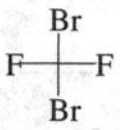

2933
Dibromodifluoromethane

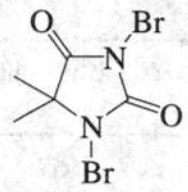

2934
1,3-Dibromo-5,5-dimethyl-2,4-imidazolidinedione

2935
1,3-Dibromo-2,2-dimethylpropane

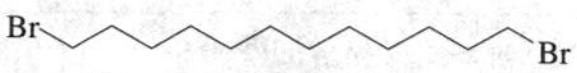

2936
1,12-Dibromododecane

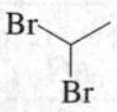

2937
1,1-Dibromoethane

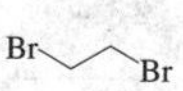

2938
1,2-Dibromoethane

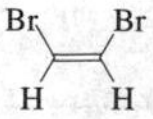

2939
cis-1,2-Dibromoethene

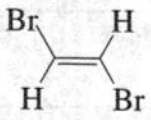

2940
trans-1,2-Dibromoethene

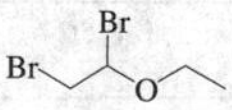

2941
1,2-Dibromo-1-ethoxyethane

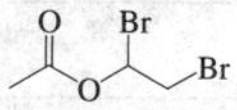

2942
1,2-Dibromoethyl acetate

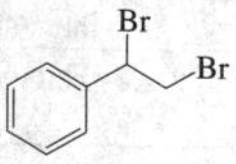

2943
(1,2-Dibromoethyl)benzene

No.	Name	Synonym	Mol. Form.	CAS RN	Mol. Wt.	Physical Form	mp/°C	bp/°C	den g cm⁻³	n_D	Solubility
2944	Dibromofluoromethane	Fluorodibromomethane	CHBr$_2$F	1868-53-7	191.825	liq	-78	64.9	2.421[20]	1.4685[20]	i H$_2$O; s EtOH, eth, ace, bz, chl
2945	1,2-Dibromoheptane		C$_7$H$_{14}$Br$_2$	42474-21-5	257.994			228	1.5086[20]	1.4986[20]	
2946	1,7-Dibromoheptane	Heptamethylene dibromide	C$_7$H$_{14}$Br$_2$	4549-31-9	257.994		41.7	247(12)	1.5306[20]	1.5034[20]	i H$_2$O; s eth, ace, bz, ctc, chl
2947	2,3-Dibromoheptane		C$_7$H$_{14}$Br$_2$	21266-88-6	257.994			101[17]	1.5139[20]	1.4992[20]	
2948	3,4-Dibromoheptane		C$_7$H$_{14}$Br$_2$	21266-90-0	257.994			107[24]	1.5182[20]	1.5010[20]	
2949	1,2-Dibromo-1,1,2,3,3,3-hexafluoropropane		C$_3$Br$_2$F$_6$	661-95-0	309.830			72.8	2.1630[20]		i H$_2$O
2950	1,2-Dibromohexane		C$_6$H$_{12}$Br$_2$	624-20-4	243.967			103[36]	1.5774[20]	1.5024[20]	vs bz, eth, chl
2951	1,6-Dibromohexane		C$_6$H$_{12}$Br$_2$	629-03-8	243.967	liq	-1.2	245.5	1.6025[25]	1.5054[25]	i H$_2$O; s eth, ace, chl; sl ctc
2952	3,4-Dibromohexane		C$_6$H$_{12}$Br$_2$	89583-12-0	243.967			80[13]	1.6027[20]	1.5043[20]	
2953	3,5-Dibromo-2-hydroxybenz-aldehyde	3,5-Dibromosalicylaldehyde	C$_7$H$_4$Br$_2$O$_2$	90-59-5	279.914	pa ye pr	86	sub			vs bz, eth, chl
2954	3,5-Dibromo-2-hydroxyben-zoic acid	3,5-Dibromosalicylic acid	C$_7$H$_4$Br$_2$O$_3$	3147-55-5	295.913	nd	228				s ace
2955	3,5-Dibromo-4-hydroxybenzo-nitrile	Bromoxynil	C$_7$H$_3$Br$_2$NO	1689-84-5	276.913		190.5(0.8)				
2956	Dibromomethane	Methylene bromide	CH$_2$Br$_2$	74-95-3	173.835	liq	-52.1(0.7)	97.0(0.6)	2.4969[20]	1.5420[20]	sl H$_2$O; msc EtOH, eth, ace; s ctc
2957	1,4-Dibromo-2-methylben-zene	2,5-Dibromotoluene	C$_7$H$_6$Br$_2$	615-59-8	249.931		6(1)	236(11)	1.8127[17]	1.5982[18]	i H$_2$O
2958	2,4-Dibromo-1-methylben-zene		C$_7$H$_6$Br$_2$	31543-75-6	249.931		-10(1)	103[11]	1.8176[25]	1.5964[25]	
2959	(Dibromomethyl)benzene		C$_7$H$_6$Br$_2$	618-31-5	249.931		1.0	156[23]	1.8365[28]	1.6147[20]	i H$_2$O; msc EtOH, eth
2960	2,3-Dibromo-2-methylbutane		C$_5$H$_{10}$Br$_2$	594-51-4	229.941		7	62[17]	1.6717[20]	1.5729[25]	
2961	2,4-Dibromo-6-methylphenol		C$_7$H$_6$Br$_2$O	609-22-3	265.930	nd (peth)	58	265 dec			s chl
2962	1,2-Dibromo-2-methylpropane		C$_4$H$_8$Br$_2$	594-34-3	215.915		10.5	139(2)	1.7827[20]	1.5119[20]	s EtOH, eth, chl
2963	1,4-Dibromonaphthalene		C$_{10}$H$_6$Br$_2$	83-53-4	285.963		83	310			i H$_2$O; s EtOH, eth; sl HOAc
2964	2,6-Dibromo-4-nitroaniline		C$_6$H$_4$Br$_2$N$_2$O$_2$	827-94-1	295.916	ye nd (al, HOAc)	207				sl H$_2$O; s HOAc
2965	2,6-Dibromo-4-nitrophenol		C$_6$H$_3$Br$_2$NO$_3$	99-28-5	296.901	pa ye pr or lf (al)	145 dec				i H$_2$O; vs EtOH, eth; sl ace, bz, HOAc
2966	1,9-Dibromononane		C$_9$H$_{18}$Br$_2$	4549-33-1	286.047	liq	-22.5	269(10)	1.4229[20]		
2967	1,4-Dibromooctafluorobutane		C$_4$Br$_2$F$_8$	335-48-8	359.838			98(25)			
2968	1,8-Dibromooctane	Octamethylene dibromide	C$_8$H$_{16}$Br$_2$	4549-32-0	272.021		15.5	271	1.4594[25]	1.4971[25]	i H$_2$O; s eth, ctc, chl
2969	1,2-Dibromopentane		C$_5$H$_{10}$Br$_2$	3234-49-9	229.941			179(13)	1.668[18]		
2970	1,4-Dibromopentane		C$_5$H$_{10}$Br$_2$	626-87-9	229.941		-34.4	146[150]	1.6222[20]	1.5086[20]	
2971	1,5-Dibromopentane		C$_5$H$_{10}$Br$_2$	111-24-0	229.941	liq	-40.0(0.4)	222.3	1.6928[25]	1.5102[25]	i H$_2$O; s bz, chl; sl ctc
2972	2,4-Dibromopentane		C$_5$H$_{10}$Br$_2$	19398-53-9	229.941			75[21]	1.6659[20]	1.4987[20]	
2973	2,4-Dibromophenol		C$_6$H$_4$Br$_2$O	615-58-7	251.903	nd (peth)	40(2)	238.5	2.0700[20]		sl H$_2$O, ctc; vs EtOH, eth, bz
2974	2,6-Dibromophenol		C$_6$H$_4$Br$_2$O	608-33-3	251.903	nd (w)	56.5	255			s H$_2$O; vs EtOH, eth
2975	1,2-Dibromopropane	Propylene dibromide	C$_3$H$_6$Br$_2$	78-75-1	201.888	liq	-55.4(0.3)	140(1)	1.9324[20]	1.5201[20]	s EtOH, eth, chl; sl ctc
2976	1,3-Dibromopropane		C$_3$H$_6$Br$_2$	109-64-8	201.888	liq	-35(1)	164(1)	1.9701[25]	1.5204[25]	i H$_2$O; s EtOH, eth, chl; sl ctc
2977	2,2-Dibromopropane		C$_3$H$_6$Br$_2$	594-16-1	201.888			113	1.880[20]		vs eth, EtOH, chl
2978	2,3-Dibromopropanoic acid		C$_3$H$_4$Br$_2$O$_2$	600-05-5	231.871		66.5	160[20]			vs bz, eth, EtOH
2979	2,3-Dibromo-1-propanol	DBP	C$_3$H$_6$Br$_2$O	96-13-9	217.887			219	2.120[20]		
2980	1,3-Dibromo-2-propanol		C$_3$H$_6$Br$_2$O	96-21-9	217.887	ye liq		219 dec	2.1364[20]	1.5495[25]	vs ace, eth, EtOH
2981	2,3-Dibromo-1-propanol, phosphate (3:1)	Tris(2,3-dibromopropyl) phosphate	C$_9$H$_{15}$Br$_6$O$_4$P	126-72-7	697.610						s chl
2982	1,3-Dibromo-2-propanone	1,3-Dibromoacetone	C$_3$H$_4$Br$_2$O	816-39-7	215.871	nd	26	97[22]	2.1670[18]		vs eth, CS$_2$
2983	1,1-Dibromo-1-propene		C$_3$H$_4$Br$_2$	13195-80-7	199.872			125	1.9767[20]	1.5260[20]	sl H$_2$O; s bz, ctc, chl
2984	1,2-Dibromo-1-propene		C$_3$H$_4$Br$_2$	26391-16-2	199.872			131.5	2.0076[20]		
2985	2,3-Dibromo-1-propene		C$_3$H$_4$Br$_2$	513-31-5	199.872			142(18)	2.0345[25]	1.5416[25]	i H$_2$O; s eth, ace, chl

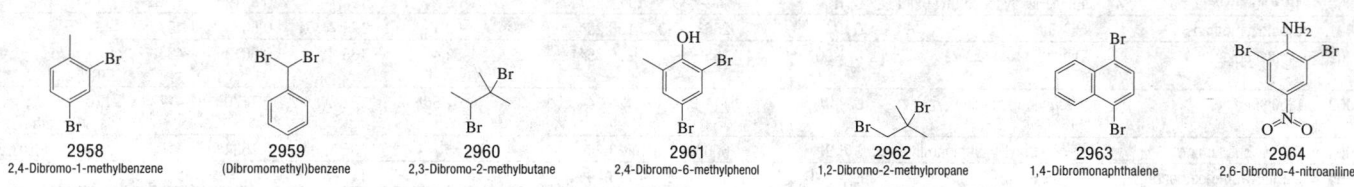

2944 Dibromofluoromethane

2945 1,2-Dibromoheptane

2946 1,7-Dibromoheptane

2947 2,3-Dibromoheptane

2948 3,4-Dibromoheptane

2949 1,2-Dibromo-1,1,2,3,3,3-hexafluoropropane

2950 1,2-Dibromohexane

2951 1,6-Dibromohexane

2952 3,4-Dibromohexane

2953 3,5-Dibromo-2-hydroxybenzaldehyde

2954 3,5-Dibromo-2-hydroxybenzoic acid

2955 3,5-Dibromo-4-hydroxybenzonitrile

2956 Dibromomethane

2957 1,4-Dibromo-2-methylbenzene

2958 2,4-Dibromo-1-methylbenzene

2959 (Dibromomethyl)benzene

2960 2,3-Dibromo-2-methylbutane

2961 2,4-Dibromo-6-methylphenol

2962 1,2-Dibromo-2-methylpropane

2963 1,4-Dibromonaphthalene

2964 2,6-Dibromo-4-nitroaniline

2965 2,6-Dibromo-4-nitrophenol

2966 1,9-Dibromononane

2967 1,4-Dibromooctafluorobutane

2968 1,8-Dibromooctane

2969 1,2-Dibromopentane

2970 1,4-Dibromopentane

2971 1,5-Dibromopentane

2972 2,4-Dibromopentane

2973 2,4-Dibromophenol

2974 2,6-Dibromophenol

2975 1,2-Dibromopropane

2976 1,3-Dibromopropane

2977 2,2-Dibromopropane

2978 2,3-Dibromopropanoic acid

2979 2,3-Dibromo-1-propanol

2980 1,3-Dibromo-2-propanol

2981 2,3-Dibromo-1-propanol, phosphate (3:1)

2982 1,3-Dibromo-2-propanone

2983 1,1-Dibromo-1-propene

2984 1,2-Dibromo-1-propene

2985 2,3-Dibromo-1-propene

No.	Name	Synonym	Mol. Form.	CAS RN	Mol. Wt.	Physical Form	mp/°C	bp/°C	den g cm⁻³	n_D	Solubility
2986	3,5-Dibromopyridine		C$_5$H$_3$Br$_2$N	625-92-3	236.893	nd (al)	112	222			sl H$_2$O; s EtOH, eth
2987	5,7-Dibromo-8-quinolinol	Broxyquinoline	C$_9$H$_5$Br$_2$NO	521-74-4	302.950	nd (al)	196	sub			i H$_2$O; s EtOH, ace, bz, chl, HOAc; sl eth
2988	2,6-Dibromoquinone-4-chlorimide	2,6-Dibromo-4-(chloroimino)-2,5-cyclohexadien-1-one	C$_6$H$_2$Br$_2$ClNO	537-45-1	299.347	ye pr (al or HOAc)	83				vs EtOH
2989	1,14-Dibromotetradecane	Tetradecamethylene dibromide	C$_{14}$H$_{28}$Br$_2$	37688-96-3	356.180	lf (al-eth) cry (al)	50.4	190^8			vs eth, EtOH, chl
2990	1,2-Dibromotetrafluoroethane	Refrigerant 114B2	C$_2$Br$_2$F$_4$	124-73-2	259.823	liq	-110(1)	47.1(0.2)	2.149^{25}	1.361^{25}	i H$_2$O
2991	2,3-Dibromothiophene		C$_4$H$_2$Br$_2$S	3140-93-0	241.932	liq	-17.5	218.5		1.6304^{22}	
2992	2,5-Dibromothiophene		C$_4$H$_2$Br$_2$S	3141-27-3	241.932	liq	-6	210.3	2.142^{23}	1.6288^{20}	i H$_2$O; vs EtOH, eth; s ctc
2993	3,4-Dibromothiophene		C$_4$H$_2$Br$_2$S	3141-26-2	241.932		4.5	217(17)			
2994	1,2-Dibromo-1,1,2-trifluoro-ethane	Halon 2302	C$_2$HBr$_2$F$_3$	354-04-1	241.832			76(4)	2.274^{27}	1.4191^{24}	
2995	2,6-Dibromo-3,4,5-trihydroxy-benzoic acid	Dibromogallic acid	C$_7$H$_4$Br$_2$O$_5$	602-92-6	327.912	nd, pr or lf (w+1)	150				vs H$_2$O, eth, EtOH
2996	3,5-Dibromo-L-tyrosine		C$_9$H$_9$Br$_2$NO$_3$	300-38-9	338.980	nd or pl	245				sl H$_2$O, EtOH; i eth; s alk, acid
2997	Dibucaine	Cinchocaine	C$_{20}$H$_{29}$N$_3$O$_2$	85-79-0	343.463	hyg cry	64				
2998	Dibucaine hydrochloride		C$_{20}$H$_{30}$ClN$_3$O$_2$	61-12-1	379.924		94 dec				s chl
2999	1,4-Dibutoxybenzene		C$_{14}$H$_{22}$O$_2$	104-36-9	222.324		45.5	158^{15}			s ctc
3000	1,2-Dibutoxyethane	Ethylene glycol dibutyl ether	C$_{10}$H$_{22}$O$_2$	112-48-1	174.281	liq	-69.1	198(10)	0.8319^{25}	1.4112^{25}	
3001	Dibutoxymethane	Butylal	C$_9$H$_{20}$O$_2$	2568-90-3	160.254	liq	-59(1)	179.7(0.7)	0.8339^{20}	1.4072^{17}	
3002	Dibutyl adipate	Dibutyl hexanedioate	C$_{14}$H$_{26}$O$_4$	105-99-7	258.354		-32.4	165^{10}	0.9613^{20}	1.4369^{20}	i H$_2$O; msc EtOH, eth
3003	Dibutylamine	N-Butylbutanamine	C$_8$H$_{19}$N	111-92-2	129.244	liq	-61.8(0.5)	162(2)	0.7670^{20}	1.4177^{20}	s H$_2$O, ace, bz; vs EtOH, eth
3004	Di-sec-butylamine	N-sec-Butyl-2-butanamine	C$_8$H$_{19}$N	626-23-3	129.244			135(7)	0.7534^{20}	1.4162^{20}	vs H$_2$O; s EtOH
3005	2-Dibutylaminoethanol		C$_{10}$H$_{23}$NO	102-81-8	173.296			114^{18}			
3006	N,N-Dibutylaniline		C$_{14}$H$_{23}$N	613-29-6	205.340	liq	-32.2(0.2)	274.8	0.9037^{20}	1.5186^{20}	i H$_2$O; msc EtOH, eth; vs ace, bz; s ctc
3007	1,4-Di-$tert$-butylbenzene		C$_{14}$H$_{22}$	1012-72-2	190.325	nd (MeOH)	77.63(0.04)	237.3(0.5)	0.9850^{20}		i H$_2$O; s EtOH, eth
3008	2,5-Di-$tert$-butyl-1,4-ben-zenediol		C$_{14}$H$_{22}$O$_2$	88-58-4	222.324	cry (aq HOAc)	213.5				
3009	Dibutylbis(dodecylthio)-stannane	Dibutyltin bis(dodecyl sulfide)	C$_{32}$H$_{68}$S$_2$Sn	1185-81-5	635.722	col liq		122$^{0.3}$	1.05^{20}		s tol, hp
3010	Dibutyl carbonate		C$_9$H$_{18}$O$_3$	542-52-9	174.237			203(4)	0.9251^{20}	1.4117^{20}	i H$_2$O; s EtOH, eth
3011	Di-$tert$-butyl carbonate		C$_9$H$_{18}$O$_3$	34619-03-9	174.237	cry (al)	40	174(14)			vs EtOH
3012	2,5-Di-$tert$-butyl-2,5-cyclo-hexadiene-1,4-dione		C$_{14}$H$_{20}$O$_2$	2460-77-7	220.308	ye cry (al)	152.5				i H$_2$O; s EtOH, eth, bz, chl, HOAc
3013	2,6-Di-$tert$-butyl-2,5-cyclo-hexadiene-1,4-dione		C$_{14}$H$_{20}$O$_2$	719-22-2	220.308		69	60$^{0.01}$			
3014	2,6-Di-$tert$-butyl-4-(dimethylaminomethyl)phenol		C$_{17}$H$_{29}$NO	88-27-7	263.418	pl (EtOH)	94	179^{40}			
3015	2,2-Dibutyl-1,3,2-dioxastan-nepin-4,7-dione		C$_{12}$H$_{20}$O$_4$Sn	78-04-6	346.995	ye solid	110				
3016	Dibutyl disulfide		C$_8$H$_{18}$S$_2$	629-45-8	178.359	oil		236(4)	0.938^{20}	1.4923^{20}	i H$_2$O; msc EtOH, eth
3017	Di-$tert$-butyl disulfide		C$_8$H$_{18}$S$_2$	110-06-5	178.359		-2.5	88^{21}	0.9226^{20}	1.4899^{20}	
3018	cis-1,2-Di-$tert$-butylethene	cis-2,2,5,5-Tetramethyl-3-hexene	C$_{10}$H$_{20}$	692-47-7	140.266	liq		144(6)	0.744^{20}	1.4270^{20}	
3019	Dibutyl ether	Butyl ether	C$_8$H$_{18}$O	142-96-1	130.228	liq	-96(3)	141.6(0.3)	0.7684^{20}	1.3992^{20}	i H$_2$O; msc EtOH, eth; vs ace; sl ctc
3020	Di-sec-butyl ether		C$_8$H$_{18}$O	6863-58-7	130.228	liq		121.9(0.3)	0.756^{25}		
3021	Di-$tert$-butyl ether		C$_8$H$_{18}$O	6163-66-2	130.228	liq		107.1(0.7)	0.7658^{20}	1.3949^{20}	
3022	N,N'-Di-$tert$-butylethylenedi-amine	N,N'-Di-$tert$-butylethanedi-amine	C$_{10}$H$_{24}$N$_2$	4062-60-6	172.311	cry	53.3	189	0.69		
3023	2,6-Di-$tert$-butyl-4-ethylphe-nol		C$_{16}$H$_{26}$O	4130-42-1	234.376		44	272			i alk
3024	N,N-Dibutylformamide		C$_9$H$_{19}$NO	761-65-9	157.253						s ctc, CS$_2$
3025	Dibutyl fumarate		C$_{12}$H$_{20}$O$_4$	105-75-9	228.285	liq	-18.0(0.4)	281(4)	0.9775^{20}	1.4469^{20}	i H$_2$O; s ace, chl
3026	N,N'-Dibutyl-1,6-hexanedi-amine		C$_{14}$H$_{32}$N$_2$	4835-11-4	228.417			138$^{3.5}$		1.4470^{25}	
3027	3,5-Di-$tert$-butyl-2-hydroxy-benzoic acid		C$_{15}$H$_{22}$O$_3$	19715-19-6	250.334		163.3				s chl

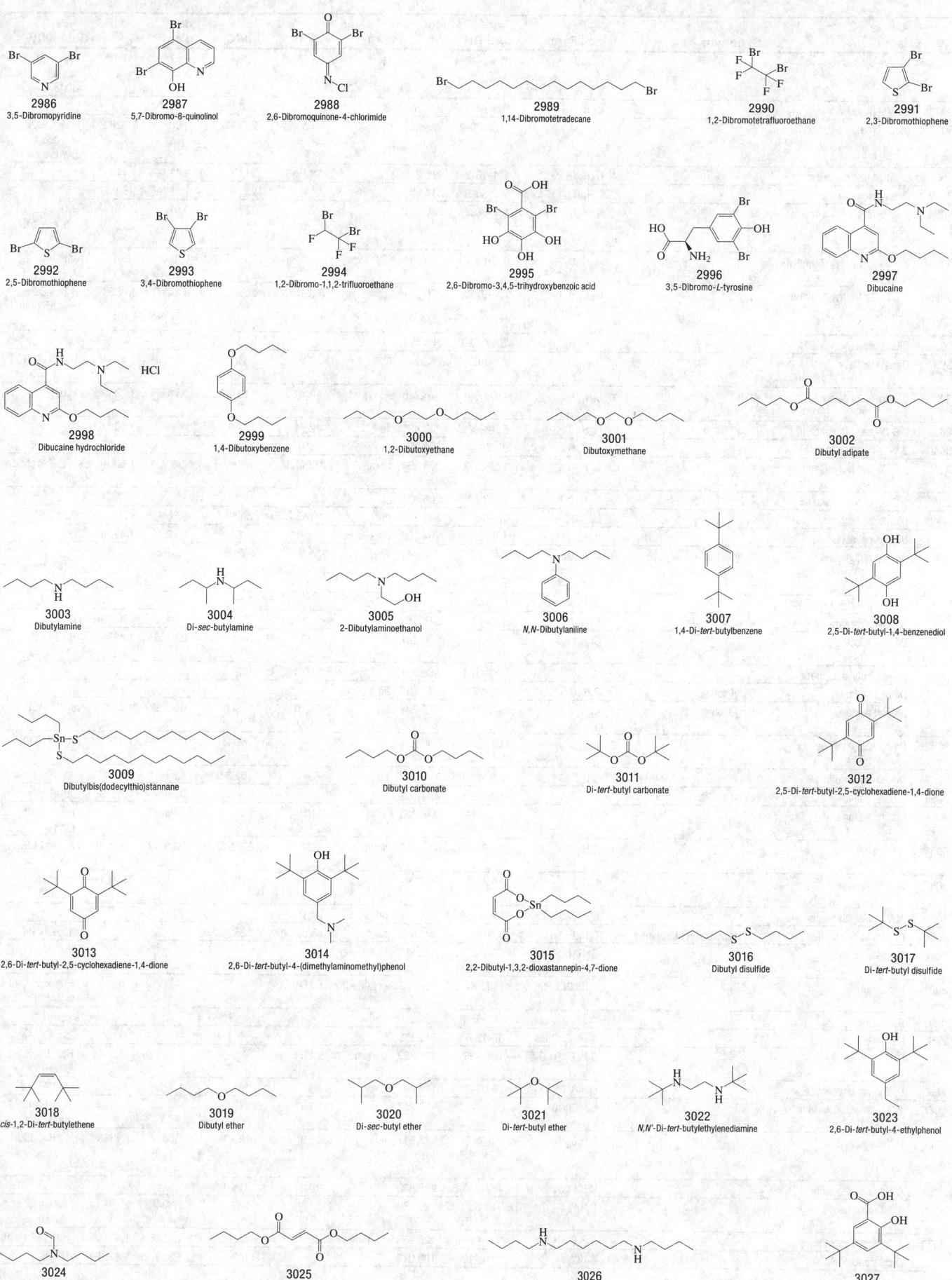

2986
3,5-Dibromopyridine

2987
5,7-Dibromo-8-quinolinol

2988
2,6-Dibromoquinone-4-chlorimide

2989
1,14-Dibromotetradecane

2990
1,2-Dibromotetrafluoroethane

2991
2,3-Dibromothiophene

2992
2,5-Dibromothiophene

2993
3,4-Dibromothiophene

2994
1,2-Dibromo-1,1,2-trifluoroethane

2995
2,6-Dibromo-3,4,5-trihydroxybenzoic acid

2996
3,5-Dibromo-L-tyrosine

2997
Dibucaine

2998
Dibucaine hydrochloride

2999
1,4-Dibutoxybenzene

3000
1,2-Dibutoxyethane

3001
Dibutoxymethane

3002
Dibutyl adipate

3003
Dibutylamine

3004
Di-sec-butylamine

3005
2-Dibutylaminoethanol

3006
N,N-Dibutylaniline

3007
1,4-Di-tert-butylbenzene

3008
2,5-Di-tert-butyl-1,4-benzenediol

3009
Dibutylbis(dodecylthio)stannane

3010
Dibutyl carbonate

3011
Di-tert-butyl carbonate

3012
2,5-Di-tert-butyl-2,5-cyclohexadiene-1,4-dione

3013
2,6-Di-tert-butyl-2,5-cyclohexadiene-1,4-dione

3014
2,6-Di-tert-butyl-4-(dimethylaminomethyl)phenol

3015
2,2-Dibutyl-1,3,2-dioxastannepin-4,7-dione

3016
Dibutyl disulfide

3017
Di-tert-butyl disulfide

3018
cis-1,2-Di-tert-butylethene

3019
Dibutyl ether

3020
Di-sec-butyl ether

3021
Di-tert-butyl ether

3022
N,N'-Di-tert-butylethylenediamine

3023
2,6-Di-tert-butyl-4-ethylphenol

3024
N,N-Dibutylformamide

3025
Dibutyl fumarate

3026
N,N'-Dibutyl-1,6-hexanediamine

3027
3,5-Di-tert-butyl-2-hydroxybenzoic acid

No.	Name	Synonym	Mol. Form.	CAS RN	Mol. Wt.	Physical Form	mp/°C	bp/°C	den g cm⁻³	n_D	Solubility
3028	Di-*tert*-butyl ketone		C₉H₁₈O	815-24-7	142.238	liq	-25.2(0.2)	152(4)	0.8240¹⁸	1.4194²⁰	i H₂O; s EtOH, eth, ace, chl, HOAc
3029	Dibutyl maleate		C₁₂H₂₀O₄	105-76-0	228.285		<-80	280			
3030	Dibutyl malonate		C₁₁H₂₀O₄	1190-39-2	216.275	liq	-95(3)	253(11)	0.9824²⁰	1.4262²⁰	i H₂O; s EtOH, eth, ace, bz, HOAc, ctc
3031	Di-*tert*-butyl malonate		C₁₁H₂₀O₄	541-16-2	216.275		-6	113³¹	1.4184²⁰	1.4184²⁹	s ace, chl
3032	Dibutylmercury		C₈H₁₈Hg	629-35-6	314.82			223	1.7779²⁰	1.5057²⁰	
3033	2,4-Di-*tert*-butyl-5-methyl-phenol	DBMC	C₁₅H₂₄O	497-39-2	220.351		62.1	282	0.912⁸⁰		i H₂O; s EtOH, eth, ace, bz, ctc
3034	2,4-Di-*tert*-butyl-6-methyl-phenol		C₁₅H₂₄O	616-55-7	220.351		51	269	0.891⁸⁰		i alk
3035	2,6-Di-*tert*-butyl-4-methyl-phenol		C₁₅H₂₄O	128-37-0	220.351		70.1(0.8)	265	0.8937⁷⁵	1.4859⁷⁵	i H₂O; s EtOH, ace, bz, peth; i alk
3036	Dibutyl nonanedioate		C₁₇H₃₂O₄	2917-73-9	300.434			170²			sl chl
3037	Dibutyl oxalate		C₁₀H₁₈O₄	2050-60-4	202.248	liq	-30.5	244(3)	0.9873²⁰	1.4234²⁰	i H₂O; s EtOH, eth
3038	Di-*tert*-butyl peroxide	DTBP	C₈H₁₈O₂	110-05-4	146.228	liq	-40	110.0(0.2)	0.704²⁰	1.3890²⁰	i H₂O; msc ace; s ctc, lig
3039	2,6-Di-*sec*-butylphenol		C₁₄H₂₂O	5510-99-6	206.324	liq	-42	257.5		1.5080²⁰	
3040	2,4-Di-*tert*-butylphenol		C₁₄H₂₂O	96-76-4	206.324		58(1)	262(1)		1.5080²⁰	sl ctc; i alk
3041	2,6-Di-*tert*-butylphenol		C₁₄H₂₂O	128-39-2	206.324	pr (al)	37.5(0.3)	161⁵⁰		1.5001²⁰	sl EtOH; s ctc; i alk
3042	3,5-Di-*tert*-butylphenol		C₁₄H₂₂O	1138-52-9	206.324		88				
3043	Dibutyl phosphate		C₈H₁₉O₄P	107-66-4	210.208	oil		136⁰·⁰⁵	1.06²⁰		s ctc, BuOH
3044	Dibutyl phosphonate		C₈H₁₉O₃P	1809-19-4	194.209	oil		230	0.985²⁵	1.4220²⁰	
3045	Dibutyl phthalate	Butyl phthalate	C₁₆H₂₂O₄	84-74-2	278.344	liq	-35	338(9)	1.0465²⁰	1.4911²⁰	i H₂O; msc EtOH, eth, bz; s ctc
3046	2,6-Di-*tert*-butylpyridine		C₁₃H₂₁N	585-48-8	191.313			120²⁰			
3047	Dibutyl sebacate	Butyl sebacate	C₁₈H₃₄O₄	109-43-3	314.461	liq	-9.2(0.5)	356(9)	0.9405¹⁵	1.4433¹⁵	i H₂O; s eth, ctc
3048	Dibutyl succinate		C₁₂H₂₂O₄	141-03-7	230.301	liq	-29.2	269(3)	0.9752²⁰	1.4299²⁰	i H₂O; s EtOH, eth, bz, ctc
3049	Di-*tert*-butyl succinate		C₁₂H₂₂O₄	926-26-1	230.301		36.5	109⁹			
3050	Dibutyl sulfate	Butyl sulfate	C₈H₁₈O₄S	625-22-9	210.292	liq		115⁶			
3051	Dibutyl sulfide	Butyl sulfide	C₈H₁₈S	544-40-1	146.294	liq	-74.97(0.05)	168(4)	0.8386²⁰	1.4530²⁰	vs eth, EtOH, chl
3052	Di-*sec*-butyl sulfide		C₈H₁₈S	626-26-6	146.294			167(1)	0.8348²⁰	1.4506²⁰	i H₂O; vs EtOH, eth
3053	Di-*tert*-butyl sulfide		C₈H₁₈S	107-47-1	146.294	liq	-9.0	152.3(0.8)	0.815²⁵	1.4506²⁰	
3054	Dibutyl sulfite	Butyl sulfite	C₈H₁₈O₃S	626-85-7	194.292			230	0.9957²⁰	1.4310²⁰	s EtOH, eth
3055	Dibutyl sulfone		C₈H₁₈O₂S	598-04-9	178.293		44(1)	291	0.9885⁴⁷		i H₂O; s EtOH, eth
3056	Dibutyl sulfoxide		C₈H₁₈OS	2168-93-6	162.293	nd (dil al)	32(3)	290(13)	0.8317²³	1.4669²⁰	i H₂O; s EtOH, eth
3057	Dibutyl tartrate		C₁₂H₂₂O₆	87-92-3	262.299	pr	22	320	1.0909²⁰	1.4451²⁰	vs H₂O, ace, EtOH
3058	*N,N'*-Dibutylthiourea		C₉H₂₀N₂S	109-46-6	188.333	nd (al)	64(1)				
3059	Dibutyltin dichloride	Dibutyldichlorostannane	C₈H₁₈Cl₂Sn	683-18-1	303.845	solid	43.0(0.2)	135¹⁰			s hx, eth, thf
3060	Dibutyltin dilaurate		C₃₂H₆₄O₄Sn	77-58-7	631.558	ye liq or cry	23				i H₂O, MeOH; s eth, bz, ctc
3061	Dicapthon		C₈H₉ClNO₅PS	2463-84-5	297.653	cry (MeOH)	50.5(0.9)				i H₂O; s ace, tol, xyl, AcOEt
3062	Dicentrine		C₂₀H₂₁NO₄	517-66-8	339.386						s chl
3063	Dichlofenthion		C₁₀H₁₃Cl₂O₃PS	97-17-6	315.153						s ctc, CS₂
3064	Dichlofluanid		C₉H₁₁Cl₂FN₂O₂S₂	1085-98-9	333.229	wh pow	105.3				i H₂O; s ace, MeOH, xyl
3065	Dichloroacetaldehyde		C₂H₂Cl₂O	79-02-7	112.942			90.5	1.436²⁵		sl EtOH
3066	2,2-Dichloroacetamide		C₂H₃Cl₂NO	683-72-7	127.957		99.4	234			s H₂O, EtOH, eth; sl ace
3067	Dichloroacetic acid		C₂H₂Cl₂O₂	79-43-6	128.942	liq	12(3)	193(3)	1.5634²⁰	1.4658²⁰	msc H₂O, EtOH, eth; s ace; sl ctc
3068	Dichloroacetic anhydride		C₄H₂Cl₄O₃	4124-30-5	239.869		18.0	215 dec	1.574²⁴		
3069	1,1-Dichloroacetone		C₃H₄Cl₂O	513-88-2	126.969			120	1.304¹⁸		sl H₂O; s EtOH; msc eth
3070	1,3-Dichloroacetone		C₃H₄Cl₂O	534-07-6	126.969	pr or nd	45	173.4	1.3826⁴⁶	1.4716⁴⁰	s H₂O, EtOH, eth
3071	Dichloroacetonitrile		C₂HCl₂N	3018-12-0	109.942			112.5	1.369²⁰	1.4391²⁵	s MeOH

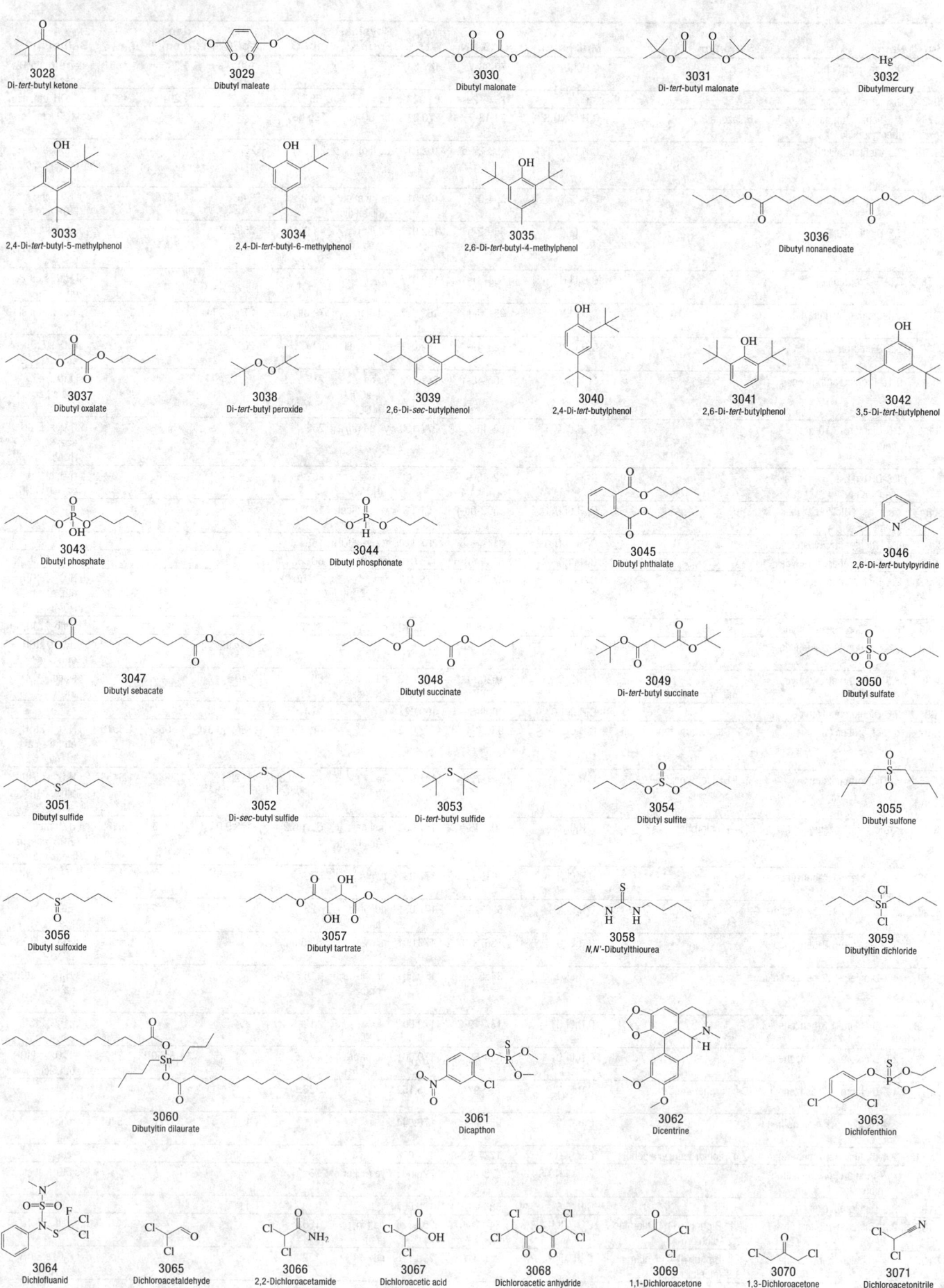

3028 Di-*tert*-butyl ketone

3029 Dibutyl maleate

3030 Dibutyl malonate

3031 Di-*tert*-butyl malonate

3032 Dibutylmercury

3033 2,4-Di-*tert*-butyl-5-methylphenol

3034 2,4-Di-*tert*-butyl-6-methylphenol

3035 2,6-Di-*tert*-butyl-4-methylphenol

3036 Dibutyl nonanedioate

3037 Dibutyl oxalate

3038 Di-*tert*-butyl peroxide

3039 2,6-Di-*sec*-butylphenol

3040 2,4-Di-*tert*-butylphenol

3041 2,6-Di-*tert*-butylphenol

3042 3,5-Di-*tert*-butylphenol

3043 Dibutyl phosphate

3044 Dibutyl phosphonate

3045 Dibutyl phthalate

3046 2,6-Di-*tert*-butylpyridine

3047 Dibutyl sebacate

3048 Dibutyl succinate

3049 Di-*tert*-butyl succinate

3050 Dibutyl sulfate

3051 Dibutyl sulfide

3052 Di-*sec*-butyl sulfide

3053 Di-*tert*-butyl sulfide

3054 Dibutyl sulfite

3055 Dibutyl sulfone

3056 Dibutyl sulfoxide

3057 Dibutyl tartrate

3058 *N,N'*-Dibutylthiourea

3059 Dibutyltin dichloride

3060 Dibutyltin dilaurate

3061 Dicapthon

3062 Dicentrine

3063 Dichlofenthion

3064 Dichlofluanid

3065 Dichloroacetaldehyde

3066 2,2-Dichloroacetamide

3067 Dichloroacetic acid

3068 Dichloroacetic anhydride

3069 1,1-Dichloroacetone

3070 1,3-Dichloroacetone

3071 Dichloroacetonitrile

No.	Name	Synonym	Mol. Form.	CAS RN	Mol. Wt.	Physical Form	mp/°C	bp/°C	den g cm^{-3}	n_D	Solubility
3072	Dichloroacetyl chloride		C_2HCl_3O	79-36-7	147.387			108	1.5315[16]	1.4591[20]	dec H_2O, EtOH; msc eth
3073	Dichloroacetylene		C_2Cl_2	7572-29-4	94.927	liq	-66	33	1.261[20]	1.42790[20]	s EtOH, eth, ace
3074	4-[(Dichloroamino)sulfonyl]-benzoic acid	Halazone	$C_7H_5Cl_2NO_4S$	80-13-7	270.091	pr (HOAc)	195 dec				sl H_2O; vs HOAc; i peth
3075	2,3-Dichloroaniline		$C_6H_5Cl_2N$	608-27-5	162.017	nd (lig)	24	252			s EtOH, ace; vs eth; sl bz, ctc, lig
3076	2,4-Dichloroaniline		$C_6H_5Cl_2N$	554-00-7	162.017	pr (ace) nd (dil al) (lig)	63(4)	245	1.567[20]		sl H_2O, chl; s EtOH, eth
3077	2,5-Dichloroaniline		$C_6H_5Cl_2N$	95-82-9	162.017	nd (lig)	44.9(0.2)	250(21)			sl H_2O; s EtOH, eth, bz, chl, CS_2
3078	2,6-Dichloroaniline		$C_6H_5Cl_2N$	608-31-1	162.017		39				sl H_2O; s EtOH, eth
3079	3,4-Dichloroaniline		$C_6H_5Cl_2N$	95-76-1	162.017	nd (lig)	72.0(0.5)	273.0(0.5)			s EtOH, eth; sl bz, chl
3080	3,5-Dichloroaniline		$C_6H_5Cl_2N$	626-43-7	162.017	nd (lig, dil al)	52	261			i H_2O; s EtOH, eth, ctc, lig
3081	9,10-Dichloroanthracene		$C_{14}H_8Cl_2$	605-48-1	247.120	ye nd (MeCOEt or CCl_4)	213.5				sl EtOH, eth, chl; s bz
3082	1,5-Dichloro-9,10-anthra-cenedione		$C_{14}H_6Cl_2O_2$	82-46-2	277.103	ye nd (to)	252				i H_2O; sl EtOH, ace; s bz, HOAc
3083	1,8-Dichloro-9,10-anthra-cenedione		$C_{14}H_6Cl_2O_2$	82-43-9	277.103	ye nd (HOAc)	202.5				i H_2O; sl EtOH; s bz, tol, $PhNO_2$
3084	trans-4,4'-Dichloroazoben-zene		$C_{12}H_8Cl_2N_2$	1602-00-2	251.111	ye nd (ace)	189				
3085	4,4'-Dichloroazoxybenzene		$C_{12}H_8Cl_2N_2O$	614-26-6	267.110	ye nd (EtOH)	158				
3086	2,3-Dichlorobenzaldehyde		$C_7H_4Cl_2O$	6334-18-5	175.012	cry (dil al)	66				vs eth, EtOH
3087	2,4-Dichlorobenzaldehyde		$C_7H_4Cl_2O$	874-42-0	175.012	pr	74.1(0.3)	105[15]			i H_2O; s EtOH, eth, bz, chl, HOAc
3088	2,6-Dichlorobenzaldehyde		$C_7H_4Cl_2O$	83-38-5	175.012	nd (lig)	69(3)				vs eth, EtOH, lig
3089	3,4-Dichlorobenzaldehyde		$C_7H_4Cl_2O$	6287-38-3	175.012		44	247.5			i H_2O; s EtOH, eth; sl ctc
3090	3,5-Dichlorobenzaldehyde		$C_7H_4Cl_2O$	10203-08-4	175.012	nd or lf (dil HOAc)	65	240			vs ace, bz, eth, EtOH
3091	2,6-Dichlorobenzamide		$C_7H_5Cl_2NO$	2008-58-4	190.027	cry	198				
3092	o-Dichlorobenzene	1,2-Dichlorobenzene	$C_6H_4Cl_2$	95-50-1	147.002	liq	-17.0(0.1)	180.2(0.3)	1.3059[20]	1.5515[20]	i H_2O; s EtOH, eth; msc ace, bz, ctc
3093	m-Dichlorobenzene	1,3-Dichlorobenzene	$C_6H_4Cl_2$	541-73-1	147.002	liq	-24.8(0.3)	172(2)	1.2884[20]	1.5459[20]	i H_2O; s EtOH, eth, bz; msc ace
3094	p-Dichlorobenzene	1,4-Dichlorobenzene	$C_6H_4Cl_2$	106-46-7	147.002	mcl pr, lf (ace)	53.1(0.2)	173.9(0.2)	1.2475[55]	1.5285[20]	i H_2O; msc EtOH, ace, bz; s eth, ctc
3095	2,5-Dichloro-1,4-benzenedi-amine		$C_6H_6Cl_2N_2$	20103-09-7	177.031	pr (w)	170				
3096	2,6-Dichloro-1,4-benzenedi-amine		$C_6H_6Cl_2N_2$	609-20-1	177.031	nd, pr (dil al)	125				s EtOH, eth, ace, bz
3097	3,5-Dichloro-1,2-benzenediol		$C_6H_4Cl_2O_2$	13673-92-2	179.001	pr	83.5(0.2)				sl H_2O; s EtOH; vs ace
3098	4,5-Dichloro-1,2-benzenediol		$C_6H_4Cl_2O_2$	3428-24-8	179.001	pr(chl-CS_2) nd(bz-peth)	116.5(0.2)				s H_2O; vs EtOH, bz
3099	4,6-Dichloro-1,3-benzenediol		$C_6H_4Cl_2O_2$	137-19-9	179.001		113	254			vs H_2O, EtOH, eth, ace; sl lig
3100	2,5-Dichloro-1,4-benzenediol		$C_6H_4Cl_2O_2$	824-69-1	179.001	nd or pr w, ace, bz)	172.5		1.8150[24]		s H_2O; vs EtOH, eth, ace
3101	4,5-Dichloro-1,3-benzenedi-sulfonamide	Dichlorphenamide	$C_6H_6Cl_2N_2O_4S_2$	120-97-8	305.159		228.7				
3102	2,4-Dichlorobenzenemethana-mine		$C_7H_7Cl_2N$	95-00-1	176.044			125[13]		1.5762[25]	s chl
3103	2,4-Dichlorobenzenemethanol	2,4-Dichlorobenzyl alcohol	$C_7H_6Cl_2O$	1777-82-8	177.028		59.5	150[25]			s chl
3104	N,N-Dichlorobenzenesulfon-amide		$C_6H_5Cl_2NO_2S$	473-29-0	226.081	ye mcl or pl	76				s EtOH; sl ctc
3105	2,5-Dichlorobenzenethiol		$C_6H_4Cl_2S$	5858-18-4	179.067			115[50]			
3106	2,2'-Dichloro-p-benzidine	[1,1'-Biphenyl]-4,4'-diamine, 2,2'-dichloro-	$C_{12}H_{10}Cl_2N_2$	84-68-4	253.126	nd (w), pr (al)	165				vs eth, EtOH
3107	3,3'-Dichloro-p-benzidine	[1,1'-Biphenyl]-4,4'-diamine, 3,3'-dichloro-	$C_{12}H_{10}Cl_2N_2$	91-94-1	253.126	nd	132.5				i H_2O; s EtOH, bz, HOAc

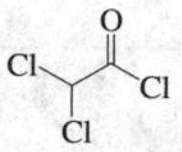

3072
Dichloroacetyl chloride

3073
Dichloroacetylene

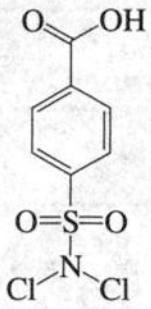

3074
4-[(Dichloroamino)sulfonyl]benzoic acid

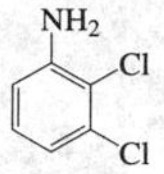

3075
2,3-Dichloroaniline

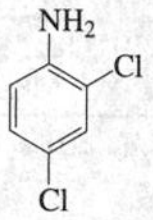

3076
2,4-Dichloroaniline

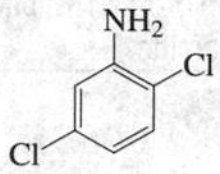

3077
2,5-Dichloroaniline

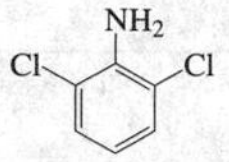

3078
2,6-Dichloroaniline

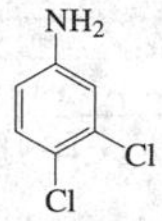

3079
3,4-Dichloroaniline

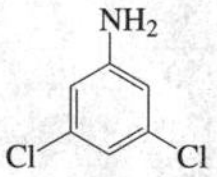

3080
3,5-Dichloroaniline

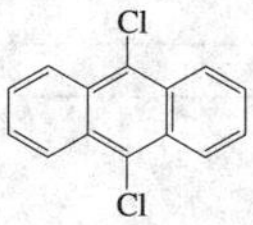

3081
9,10-Dichloroanthracene

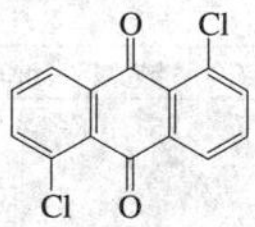

3082
1,5-Dichloro-9,10-anthracenedione

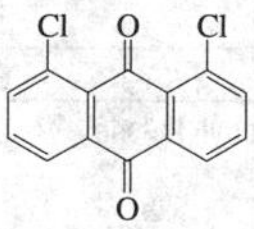

3083
1,8-Dichloro-9,10-anthracenedione

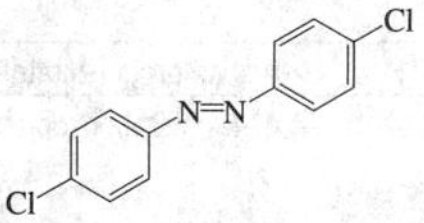

3084
trans-4,4'-Dichloroazobenzene

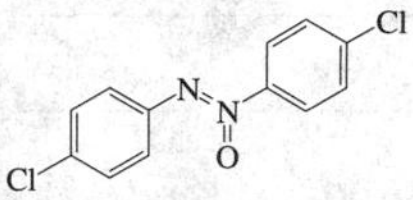

3085
4,4'-Dichloroazoxybenzene

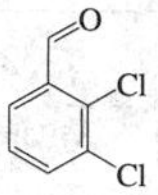

3086
2,3-Dichlorobenzaldehyde

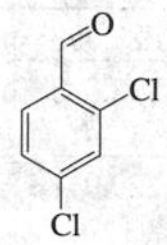

3087
2,4-Dichlorobenzaldehyde

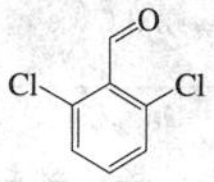

3088
2,6-Dichlorobenzaldehyde

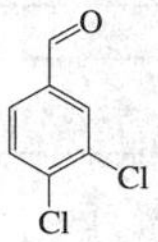

3089
3,4-Dichlorobenzaldehyde

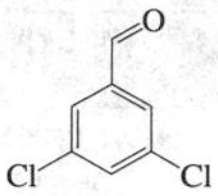

3090
3,5-Dichlorobenzaldehyde

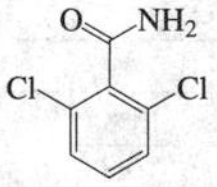

3091
2,6-Dichlorobenzamide

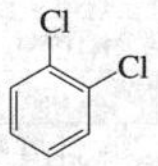

3092
o-Dichlorobenzene

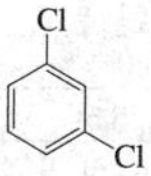

3093
m-Dichlorobenzene

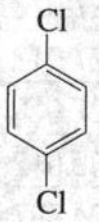

3094
p-Dichlorobenzene

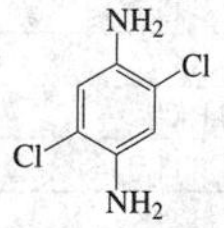

3095
2,5-Dichloro-1,4-benzenediamine

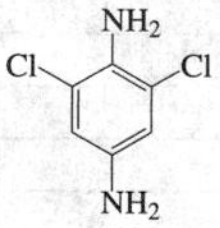

3096
2,6-Dichloro-1,4-benzenediamine

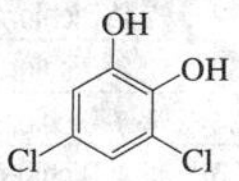

3097
3,5-Dichloro-1,2-benzenediol

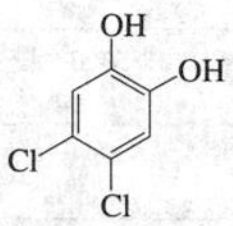

3098
4,5-Dichloro-1,2-benzenediol

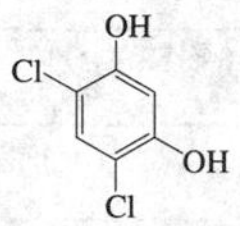

3099
4,6-Dichloro-1,3-benzenediol

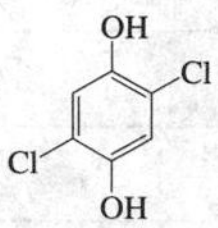

3100
2,5-Dichloro-1,4-benzenediol

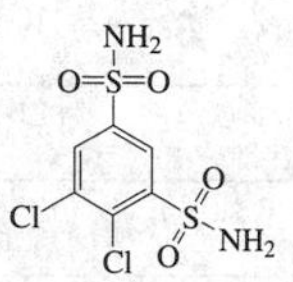

3101
4,5-Dichloro-1,3-benzenedisulfonamide

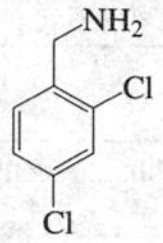

3102
2,4-Dichlorobenzenemethanamine

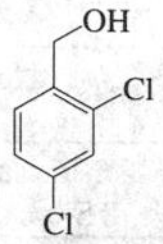

3103
2,4-Dichlorobenzenemethanol

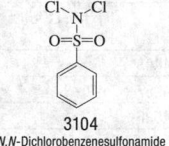

3104
N,N-Dichlorobenzenesulfonamide

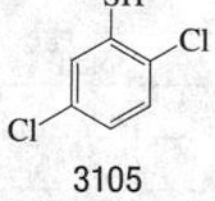

3105
2,5-Dichlorobenzenethiol

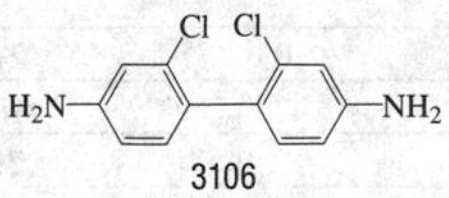

3106
2,2'-Dichloro-*p*-benzidine

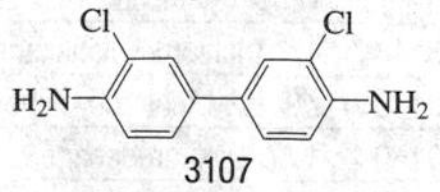

3107
3,3'-Dichloro-*p*-benzidine

No.	Name	Synonym	Mol. Form.	CAS RN	Mol. Wt.	Physical Form	mp/°C	bp/°C	den g cm⁻³	n_D	Solubility
3108	3,3'-Dichloro-p-benzidine dihydrochloride	3,3'-Dichloro-[1,1'-biphenyl]-4,4'-diamine	$C_{12}H_{12}Cl_4N_2$	612-83-9	326.048						i H_2O; vs EtOH
3109	2,4-Dichlorobenzoic acid		$C_7H_4Cl_2O_2$	50-84-0	191.012	nd (w or bz)	164.2	sub			s H_2O, EtOH, eth, bz, chl; sl ace
3110	2,5-Dichlorobenzoic acid		$C_7H_4Cl_2O_2$	50-79-3	191.012	nd (w)	154(1)	301			sl H_2O, DMSO; s EtOH, eth
3111	2,6-Dichlorobenzoic acid		$C_7H_4Cl_2O_2$	50-30-6	191.012	nd (al), pr (w)	142(2)	sub			s H_2O, EtOH, eth, bz, chl
3112	3,4-Dichlorobenzoic acid		$C_7H_4Cl_2O_2$	51-44-5	191.012	nd (w, al, bz)	204.5(0.4)				s H_2O, eth; vs EtOH; sl DMSO
3113	3,5-Dichlorobenzoic acid		$C_7H_4Cl_2O_2$	51-36-5	191.012	nd (al, w)	187.4(0.5)	sub			sl H_2O, lig, DMSO; s EtOH, eth
3114	2,6-Dichlorobenzonitrile	Dichlobenil	$C_7H_3Cl_2N$	1194-65-6	172.012	cry (peth)	143.6(0.7)	270			
3115	4,4'-Dichlorobenzophenone	Bis(4-chlorophenyl) ketone	$C_{13}H_8Cl_2O$	90-98-2	251.108	pl (al)	146.8(0.3)	353	1.4500²⁰		i H_2O; s EtOH; vs eth, chl; sl ace
3116	3,4-Dichlorobenzotrifluoride	1,2-Dichloro-4-(trifluoromethyl)benzene	$C_7H_3Cl_2F_3$	328-84-7	215.000	liq		175(3)	1.4729²⁵		
3117	2,3-Dichlorobenzoyl chloride		$C_7H_3Cl_3O$	2905-60-4	209.457	liq		140¹⁴			
3118	2,4-Dichlorobenzoyl chloride		$C_7H_3Cl_3O$	89-75-8	209.457		16.5	150³⁴		1.5895²⁰	s ctc
3119	2,5-Dichlorobenzoyl chloride		$C_7H_3Cl_3O$	2905-61-5	209.457	liq		95.4¹			
3120	3,4-Dichlorobenzoyl chloride		$C_7H_3Cl_3O$	3024-72-4	209.457		25	242			sl ctc
3121	2,5-Dichlorobiphenyl		$C_{12}H_8Cl_2$	34883-39-1	223.098			182³⁰			i H_2O
3122	2,6-Dichlorobiphenyl		$C_{12}H_8Cl_2$	33146-45-1	223.098	cry	34.7(0.5)				i H_2O
3123	3,3'-Dichlorobiphenyl		$C_{12}H_8Cl_2$	2050-67-1	223.098	nd (dil al)	29	320			vs bz, eth, EtOH
3124	4,4'-Dichlorobiphenyl		$C_{12}H_8Cl_2$	2050-68-2	223.098	pr or nd (al, to-peth)	148(2)	317	1.4420⁰		i H_2O; sl EtOH, chl; s bz
3125	1,1-Dichloro-2,2-bis(p-chloro-phenyl)ethane		$C_{14}H_{10}Cl_4$	72-54-8	320.041		109(1)	193¹			sl chl
3126	2,2-Dichloro-1,1-bis(4-chloro-phenyl)ethene		$C_{14}H_8Cl_4$	72-55-9	318.026		89.8(0.3)				
3127	2,3-Dichloro-1,3-butadiene		$C_4H_4Cl_2$	1653-19-6	122.981			101(15)	1.1829²⁰	1.4890²⁰	vs chl
3128	1,1-Dichlorobutane	Butylidene chloride	$C_4H_8Cl_2$	541-33-3	127.013			115(6)	1.0863²⁰	1.4355²⁰	i H_2O; s chl
3129	1,2-Dichlorobutane		$C_4H_8Cl_2$	616-21-7	127.013			123.9(0.8)	1.1116²⁵	1.4450²⁰	i H_2O; s eth, chl; sl ctc
3130	1,3-Dichlorobutane		$C_4H_8Cl_2$	1190-22-3	127.013			133.7(0.7)	1.1158²⁰	1.4445²⁰	i H_2O; s eth, chl; sl ctc
3131	1,4-Dichlorobutane		$C_4H_8Cl_2$	110-56-5	127.013	liq	-38.7(0.4)	155(3)	1.1331²⁵	1.4522²⁵	i H_2O; vs chl
3132	2,2-Dichlorobutane		$C_4H_8Cl_2$	4279-22-5	127.013	liq	-74	102(6)	1.1048²⁵	1.4295	i H_2O; s chl
3133	2,3-Dichlorobutane, (±)-		$C_4H_8Cl_2$	2211-67-8	127.013	liq	-80	119	1.105²⁵	1.4409²⁵	i H_2O
3134	1,4-Dichloro-2,3-butanediol		$C_4H_8Cl_2O_2$	2419-73-0	159.012		126.5	150³⁰			vs EtOH
3135	3,4-Dichloro-1-butene		$C_4H_6Cl_2$	760-23-6	124.997	liq	-61(5)	116	1.1170²⁰	1.4641²⁰	i H_2O; s EtOH, eth, ctc; vs chl, bz
3136	cis-1,3-Dichloro-2-butene		$C_4H_6Cl_2$	10075-38-4	124.997			127(16)	1.1605²⁰	1.4735²⁰	vs ace, bz, eth, EtOH
3137	trans-1,3-Dichloro-2-butene		$C_4H_6Cl_2$	7415-31-8	124.997			132	1.160²⁰	1.4719²⁰	vs ace, bz, eth, EtOH
3138	cis-1,4-Dichloro-2-butene		$C_4H_6Cl_2$	1476-11-5	124.997	liq	-42(2)	149(11)	1.188²⁵	1.4887²⁵	vs ace, bz, eth, EtOH
3139	trans-1,4-Dichloro-2-butene		$C_4H_6Cl_2$	110-57-6	124.997	col liq	3(2)	155.4	1.183²⁵	1.4871²⁵	vs ace, bz, eth, EtOH
3140	1,4-Dichloro-2-butyne		$C_4H_4Cl_2$	821-10-3	122.981			164(15)	1.258²⁰	1.5058²⁰	s eth, ace; sl ctc; vs chl
3141	2,6-Dichloro-4-(chloroimino)-2,5-cyclohexadien-1-one	Gibbs' reagent	$C_6H_2Cl_3NO$	101-38-2	210.445		66				
3142	1,2-Dichloro-4-(chloromethyl)-benzene		$C_7H_5Cl_3$	102-47-6	195.474		37.5	241			i H_2O; s EtOH, ctc
3143	2,4-Dichloro-1-(chloromethyl)-benzene		$C_7H_5Cl_3$	94-99-5	195.474			120¹³			
3144	Dichloro(chloromethyl)-methylsilane		$C_2H_5Cl_3Si$	1558-33-4	163.506			121.5	1.2858²⁰	1.4500²⁰	
3145	Dichloro(2-chlorovinyl)arsine		$C_2H_2AsCl_3$	541-25-3	207.318	liq	0.1	190	1.888²⁰		
3146	2,5-Dichloro-2,5-cyclohexadi-ene-1,4-dione		$C_6H_2Cl_2O_2$	615-93-0	176.985	pa ye mcl pr (al)	162.3				i H_2O; sl EtOH; s eth, chl
3147	2,6-Dichloro-2,5-cyclohexadi-ene-1,4-dione		$C_6H_2Cl_2O_2$	697-91-6	176.985	ye orth (lig, bz)	121.8				sl H_2O, EtOH; s chl
3148	1,1-Dichlorocyclohexane		$C_6H_{10}Cl_2$	2108-92-1	153.049	liq	-36.6(0.1)	171	1.1559²⁰	1.4803²⁰	
3149	cis-1,2-Dichlorocyclohexane		$C_6H_{10}Cl_2$	10498-35-8	153.049	liq	-5(3)	191(10)	1.2021²⁰	1.4967²⁰	vs bz
3150	1,10-Dichlorodecane		$C_{10}H_{20}Cl_2$	2162-98-3	211.172		15.6(0.5)	167²⁸	0.9945²⁵	1.4586²⁵	

3108
3,3'-Dichloro-p-benzidine dihydrochloride

3109
2,4-Dichlorobenzoic acid

3110
2,5-Dichlorobenzoic acid

3111
2,6-Dichlorobenzoic acid

3112
3,4-Dichlorobenzoic acid

3113
3,5-Dichlorobenzoic acid

3114
2,6-Dichlorobenzonitrile

3115
4,4'-Dichlorobenzophenone

3116
3,4-Dichlorobenzotrifluoride

3117
2,3-Dichlorobenzoyl chloride

3118
2,4-Dichlorobenzoyl chloride

3119
2,5-Dichlorobenzoyl chloride

3120
3,4-Dichlorobenzoyl chloride

3121
2,5-Dichlorobiphenyl

3122
2,6-Dichlorobiphenyl

3123
3,3'-Dichlorobiphenyl

3124
4,4'-Dichlorobiphenyl

3125
1,1-Dichloro-2,2-bis(p-chlorophenyl)ethane

3126
2,2-Dichloro-1,1-bis(4-chlorophenyl)ethene

3127
2,3-Dichloro-1,3-butadiene

3128
1,1-Dichlorobutane

3129
1,2-Dichlorobutane

3130
1,3-Dichlorobutane

3131
1,4-Dichlorobutane

3132
2,2-Dichlorobutane

3133
2,3-Dichlorobutane, (±)-

3134
1,4-Dichloro-2,3-butanediol

3135
3,4-Dichloro-1-butene

3136
cis-1,3-Dichloro-2-butene

3137
trans-1,3-Dichloro-2-butene

3138
cis-1,4-Dichloro-2-butene

3139
trans-1,4-Dichloro-2-butene

3140
1,4-Dichloro-2-butyne

3141
2,6-Dichloro-4-(chloroimino)-2,5-cyclohexadien-1-one

3142
1,2-Dichloro-4-(chloromethyl)benzene

3143
2,4-Dichloro-1-(chloromethyl)benzene

3144
Dichloro(chloromethyl)methylsilane

3145
Dichloro(2-chlorovinyl)arsine

3146
2,5-Dichloro-2,5-cyclohexadiene-1,4-dione

3147
2,6-Dichloro-2,5-cyclohexadiene-1,4-dione

3148
1,1-Dichlorocyclohexane

3149
cis-1,2-Dichlorocyclohexane

3150
1,10-Dichlorodecane

No.	Name	Synonym	Mol. Form.	CAS RN	Mol. Wt.	Physical Form	mp/°C	bp/°C	den g cm⁻³	n_D	Solubility
3151	2,7-Dichlorodibenzo-*p*-dioxin		$C_{12}H_6Cl_2O_2$	33857-26-0	253.081	cry	210.0(0.5)				
3152	1,2-Dichloro-4-(dichloromethyl)benzene		$C_7H_4Cl_4$	56961-84-3	229.919			257	1.515²²		vs bz, eth, EtOH
3153	Dichloro(dichloromethyl)-methylsilane		$C_2H_4Cl_4Si$	1558-31-2	197.951			149	1.4116²⁰	1.4700²⁰	
3154	2,3-Dichloro-5,6-dicyanoben-zoquinone		$C_8Cl_2N_2O_2$	84-58-2	227.004	ye-oran cry	214.5				vs bz, HOAc, diox
3155	Dichlorodiethylsilane		$C_4H_{10}Cl_2Si$	1719-53-5	157.114		-96.5	130(2)	1.0504²⁰	1.4309²⁰	
3156	1,1-Dichloro-1,2-difluoroeth-ane		$C_2H_2Cl_2F_2$	25915-78-0	134.940	col liq		48.4			
3157	1,2-Dichloro-1,1-difluoroeth-ane		$C_2H_2Cl_2F_2$	1649-08-7	134.940	liq	-101.2	47(1)	1.4163²⁰	1.36193²⁰	sl H_2O
3158	1,2-Dichloro-1,2-difluoroeth-ane		$C_2H_2Cl_2F_2$	431-06-1	134.940	liq	-101.2	59.6	1.4163²⁰	1.3619²⁰	
3159	1,1-Dichloro-2,2-difluoroeth-ene	1,1-Dichloro-2,2-difluoroethyl-ene	$C_2Cl_2F_2$	79-35-6	132.924	vol liq or gas	-116	19	1.555⁻²⁰	1.383⁻²⁰	
3160	*cis*-1,2-Dichloro-1,2-difluoro-ethene	Refrigerant 1112	$C_2Cl_2F_2$	311-81-9	132.924	vol liq	-119.6	21.1	1.495⁰		
3161	*trans*-1,2-Dichloro-1,2-difluoroethene		$C_2Cl_2F_2$	381-71-5	132.924	vol liq	-93.3	22	1.494⁰		
3162	Dichlorodifluoromethane	Refrigerant 12	CCl_2F_2	75-71-8	120.914	col gas	-157.05(0.01)	-29.8(0.1)			sl H_2O; s EtOH, eth, HOAc
3163	2,2-Dichloro-1,1-difluoro-1-methoxyethane	Methoxyflurane	$C_3H_4Cl_2F_2O$	76-38-0	164.966	col liq	-35	105	1.43²⁰	1.3861²⁰	
3164	2,2'-Dichlorodiisopropyl ether		$C_6H_{12}Cl_2O$	108-60-1	171.064			184(3)	1.103²⁰	1.4505²⁰	i H_2O; msc EtOH, eth, ace; vs bz
3165	1,4-Dichloro-2,5-dimethylben-zene		$C_8H_8Cl_2$	1124-05-6	175.056		71	222			s chl
3166	2,5-Dichloro-2,5-dimethylhex-ane		$C_6H_{16}Cl_2$	6223-78-5	183.119	lf, nd	63(3)		0.9543²⁰		vs bz, eth, EtOH, chl
3167	1,3-Dichloro-5,5-dimethyl-hydantoin		$C_5H_6Cl_2N_2O_2$	118-52-5	197.019	pr	132		1.5²⁰		sl H_2O; s chl, ctc, bz
3168	2,4-Dichloro-3,5-dimethylphe-nol	Dichloroxylenol	$C_8H_8Cl_2O$	133-53-9	191.055		83				vs eth
3169	Dichlorodimethylsilane		$C_2H_6Cl_2Si$	75-78-5	129.061	liq	-16	70.5(0.5)	1.064²⁵	1.4038²⁰	dec H_2O, EtOH
3170	2,3-Dichloro-1,4-dioxane		$C_4H_6Cl_2O_2$	95-59-0	156.996		30	81¹⁰	1.468²⁰	1.4928²⁰	i H_2O; vs eth, ace, bz, ctc, diox
3171	Dichlorodiphenylmethane		$C_{13}H_{10}Cl_2$	2051-90-3	237.124			299(17)	1.235¹⁸		s eth, bz, ctc
3172	Dichlorodiphenylsilane		$C_{12}H_{10}Cl_2Si$	80-10-4	253.199			304(3)	1.204²⁵	1.5800²⁰	s EtOH, eth, ace, bz, ctc
3173	1,1-Dichloroethane	Ethylidene dichloride	$C_2H_4Cl_2$	75-34-3	98.959	liq	-96.93(0.06)	56.3(0.7)	1.1757²⁰	1.4164²⁰	sl H_2O; vs EtOH, eth; s ace, bz
3174	1,2-Dichloroethane	Ethylene dichloride	$C_2H_4Cl_2$	107-06-2	98.959	liq	-35.6(0.2)	83.4(0.1)	1.2454²⁵	1.4422²⁵	sl H_2O; vs EtOH; msc eth; s ace, bz, chl
3175	2,2-Dichloroethanol		$C_2H_4Cl_2O$	598-38-9	114.958			147(15)	1.4040²⁵	1.4626²⁵	sl H_2O, ctc; s EtOH, eth
3176	1,1-Dichloroethene	Vinylidene chloride	$C_2H_2Cl_2$	75-35-4	96.943	liq	-122.5(0.1)	31.6(0.3)	1.213²⁰	1.4249²⁰	i H_2O; s EtOH, ace, bz; vs eth, chl
3177	*cis*-1,2-Dichloroethene	*cis*-1,2-Dichloroethylene	$C_2H_2Cl_2$	156-59-2	96.943	liq	-80.0(0.2)	60(2)	1.2837²⁰	1.4490²⁰	sl H_2O; msc EtOH, eth, ace; vs bz, chl
3178	*trans*-1,2-Dichloroethene	*trans*-1,2-Dichloroethylene	$C_2H_2Cl_2$	156-60-5	96.943	liq	-49.8(0.2)	47.64(0.08)	1.2565²⁰	1.4454²⁰	sl H_2O; msc EtOH, eth, ace; vs bz, chl
3179	1,2-Dichloro-1-ethoxyethane		$C_4H_8Cl_2O$	623-46-1	143.012			144(4)	1.1370²⁰	1.4435²⁰	sl chl
3180	1,2-Dichloroethyl acetate		$C_4H_6Cl_2O_2$	10140-87-1	156.996	liq		79³³			
3181	Dichloroethylaluminum	Ethylaluminum chloride	$C_2H_5AlCl_2$	563-43-9	126.949	hyg solid or liq	32	115⁵⁰	1.207		reac H_2O
3182	Dichloroethylmethylsilane		$C_3H_8Cl_2Si$	4525-44-4	143.088			101	1.0047²⁰	1.4197²⁰	
3183	2',7'-Dichlorofluorescein	2',7'-Dichloro-3,6-fluorandiol	$C_{20}H_{10}Cl_2O_5$	76-54-0	401.196						sl DMSO
3184	1,1-Dichloro-1-fluoroethane	HCFC-141b	$C_2H_3Cl_2F$	1717-00-6	116.949	liq	-103.5(0.5)	32.05(0.09)	1.250¹⁰	1.3600¹⁰	i H_2O
3185	1,2-Dichloro-1-fluoroethane		$C_2H_3Cl_2F$	430-57-9	116.949	liq	-60	74(4)	1.3814²⁰	1.4132²⁰	
3186	1,1-Dichloro-2-fluoroethene	1,1-Dichloro-2-fluoroethylene	C_2HCl_2F	359-02-4	114.933	liq	-108.8	38(21)	1.3732¹⁶	1.4031¹⁶	
3187	Dichlorofluoromethane	Refrigerant 21	$CHCl_2F$	75-43-4	102.923	col gas	-130.4	8.9	1.405⁹	1.3724⁹	i H_2O; s EtOH, eth, ctc, chl, HOAc
3188	(Dichlorofluoromethyl)benzene		$C_7H_5Cl_2F$	498-67-9	179.019	liq	-26.8	179	1.3138¹¹	1.5180¹¹	vs EtOH
3189	1,1-Dichloro-2-fluoropropene		$C_3H_3Cl_2F$	430-95-5	128.960			78	1.3026²⁵	1.4196²⁵	
3190	1,7-Dichloroheptane		$C_7H_{14}Cl_2$	821-76-1	169.092			124³⁵	1.0408²⁵	1.4565²⁵	

3151
2,7-Dichlorodibenzo-*p*-dioxin

3152
1,2-Dichloro-4-(dichloromethyl)benzene

3153
Dichloro(dichloromethyl)methylsilane

3154
2,3-Dichloro-5,6-dicyanobenzoquinone

3155
Dichlorodiethylsilane

3156
1,1-Dichloro-1,2-difluoroethane

3157
1,2-Dichloro-1,1-difluoroethane

3158
1,2-Dichloro-1,2-difluoroethane

3159
1,1-Dichloro-2,2-difluoroethene

3160
cis-1,2-Dichloro-1,2-difluoroethene

3161
trans-1,2-Dichloro-1,2-difluoroethene

3162
Dichlorodifluoromethane

3163
2,2-Dichloro-1,1-difluoro-1-methoxyethane

3164
2,2'-Dichlorodiisopropyl ether

3165
1,4-Dichloro-2,5-dimethylbenzene

3166
2,5-Dichloro-2,5-dimethylhexane

3167
1,3-Dichloro-5,5-dimethyl hydantoin

3168
2,4-Dichloro-3,5-dimethylphenol

3169
Dichlorodimethylsilane

3170
2,3-Dichloro-1,4-dioxane

3171
Dichlorodiphenylmethane

3172
Dichlorodiphenylsilane

3173
1,1-Dichloroethane

3174
1,2-Dichloroethane

3175
2,2-Dichloroethanol

3176
1,1-Dichloroethene

3177
cis-1,2-Dichloroethene

3178
trans-1,2-Dichloroethene

3179
1,2-Dichloro-1-ethoxyethane

3180
1,2-Dichloroethyl acetate

3181
Dichloroethylaluminum

3182
Dichloroethylmethylsilane

3183
2',7'-Dichlorofluorescein

3184
1,1-Dichloro-1-fluoroethane

3185
1,2-Dichloro-1-fluoroethane

3186
1,1-Dichloro-2-fluoroethene

3187
Dichlorofluoromethane

3188
(Dichlorofluoromethyl)benzene

3189
1,1-Dichloro-2-fluoropropene

3190
1,7-Dichloroheptane

No.	Name	Synonym	Mol. Form.	CAS RN	Mol. Wt.	Physical Form	mp/°C	bp/°C	den g cm^{-3}	n_D	Solubility
3191	1,2-Dichloro-1,2,3,3,4,4-hexafluorocyclobutane		$C_4Cl_2F_6$	356-18-3	232.939	liq	-24.2	59.5			
3192	1,2-Dichloro-3,3,4,4,5,5-hexafluorocyclopentene		$C_5Cl_2F_6$	706-79-6	244.949	liq	-105.8	90.7	1.6546[20]	1.3676[20]	
3193	1,2-Dichloro-1,1,2,3,3,3-hexafluoropropane		$C_3Cl_2F_6$	661-97-2	220.928			34(2)			i H_2O
3194	1,3-Dichloro-1,1,2,2,3,3-hexafluoropropane	Refrigerant 216	$C_3Cl_2F_6$	662-01-1	220.928	liq	-125.4	35.7	1.573[20]	1.3030[20]	
3195	1,5-Dichloro-1,1,3,3,5,5-hexamethyltrisiloxane		$C_6H_{18}Cl_2O_2Si_3$	3582-71-6	277.369	liq	-53	184	1.018[20]		dec H_2O
3196	1,2-Dichlorohexane		$C_6H_{12}Cl_2$	2162-92-7	155.065			172.2(0.9)	1.085[15]		vs eth, chl
3197	1,6-Dichlorohexane		$C_6H_{12}Cl_2$	2163-00-0	155.065			205(1)	1.0676[25]	1.4555[25]	i H_2O; s eth, ctc, chl
3198	3,5-Dichloro-2-hydroxybenz-aldehyde		$C_7H_4Cl_2O_2$	90-60-8	191.012	ye orth (HOAc)	95				i H_2O
3199	3,5-Dichloro-2-hydroxybenzoic acid		$C_7H_4Cl_2O_3$	320-72-9	207.011	nd (dil al) orth pr	220.5	sub			sl H_2O; vs EtOH, eth
3200	2,6-Dichloroindophenol, sodium salt	Tillman's reagent	$C_{12}H_6Cl_2NNaO_2$	620-45-1	290.078	dk grn cry					s H_2O, EtOH, ace
3201	5,6-Dichloro-1,3-isobenzofu-randione	4,5-Dichlorophthalic anhydride	$C_8H_2Cl_2O_3$	942-06-3	217.006	tab or pr (to)	188	313			vs eth, EtOH, tol
3202	Dichloromethane	Methylene chloride	CH_2Cl_2	75-09-2	84.933	liq	-95(2)	39.8(0.3)	1.3266[20]	1.4242[20]	sl H_2O; msc EtOH, eth; s ctc
3203	1,2-Dichloro-3-methoxyben-zene		$C_7H_6Cl_2O$	1984-59-4	177.028		31.0(0.4)				
3204	1,3-Dichloro-2-methoxyben-zene	2,6-Dichloroanisole	$C_7H_6Cl_2O$	1984-65-2	177.028	liq	10	105[20]	1.291	1.5430[20]	
3205	2,4-Dichloro-1-methoxyben-zene		$C_7H_6Cl_2O$	553-82-2	177.028	pr	28.5	232			sl chl
3206	3,6-Dichloro-2-methoxyben-zoic acid	Dicamba	$C_8H_6Cl_2O_3$	1918-00-9	221.038	cry (pent)	114.9(0.8)		1.57[25]		
3207	(Dichloromethyl)benzene	Benzal chloride	$C_7H_6Cl_2$	98-87-3	161.029	liq	-17.0(0.5)	205	1.26[25]	1.5502[20]	i H_2O; vs eth, EtOH
3208	N,N-Dichloro-4-methylben-zenesulfonamide	Dichloramine-T	$C_7H_7Cl_2NO_2S$	473-34-7	240.108	pr(chl-peth)	83				i H_2O; s EtOH, eth, bz, ctc, HOAc
3209	Dichloromethylborane	Methyldichloroborane	CH_3BCl_2	7318-78-7	96.752	col gas		11			
3210	2,3-Dichloro-2-methylbutane	Amylene dichloride	$C_5H_{10}Cl_2$	507-45-9	141.038			129	1.0696[15]	1.4450[18]	i H_2O; vs eth, EtOH
3211	1,1-Dichloromethyl methyl ether	Methoxydichloromethane	$C_2H_4Cl_2O$	4885-02-3	114.958			86(9)	1.271[25]	1.4300[20]	
3212	2,4-Dichloro-3-methylphenol		$C_7H_6Cl_2O$	17788-00-0	177.028	pr (peth)	58	236			vs eth, chl
3213	2,4-Dichloro-6-methylphenol		$C_7H_6Cl_2O$	1570-65-6	177.028	nd (w, peth)	55				sl H_2O; vs EtOH, eth, chl, CS_2
3214	2,6-Dichloro-4-methylphenol		$C_7H_6Cl_2O$	2432-12-4	177.028	nd (lig)	39	231			i H_2O; vs eth, EtOH, HOAc
3215	Dichloromethylphenylsilane		$C_7H_8Cl_2Si$	149-74-6	191.131			206(3)	1.1866[20]	1.5180[20]	
3216	Dichloromethylphosphine	Methylphosphonous dichloride	CH_3Cl_2P	676-83-5	116.915			12[50]	1.304[20]	1.4940[20]	
3217	1,2-Dichloro-2-methylpropane	1,2-Dichloroisobutane	$C_4H_8Cl_2$	594-37-6	127.013			108(2)	1.093[20]	1.4370[20]	i H_2O; msc EtOH, eth, ace, bz, ctc
3218	2,4-Dichloro-5-methylpyrimi-dine		$C_5H_4Cl_2N_2$	1780-31-0	163.004	pl (al)	26	235			sl H_2O; vs EtOH, eth, bz, chl
3219	2,4-Dichloro-6-methylpyrimi-dine		$C_5H_4Cl_2N_2$	5424-21-5	163.004	nd (lig)	46.5	219			vs bz, eth, EtOH, chl
3220	Dichloromethylsilane		CH_4Cl_2Si	75-54-7	115.035	liq	-93	40.9(0.1)	1.105[25]		
3221	1,2-Dichloronaphthalene		$C_{10}H_6Cl_2$	2050-69-3	197.061	pl (al)	36	295(26)	1.3147[49]	1.5338[49]	s EtOH, eth
3222	1,3-Dichloronaphthalene		$C_{10}H_6Cl_2$	2198-75-6	197.061	nd or pr (al)	62.3	291			s EtOH
3223	1,4-Dichloronaphthalene		$C_{10}H_6Cl_2$	1825-31-6	197.061	nd or pr (al, ace)	67.5	288	1.2997[76]	1.6228[76]	i H_2O; sl EtOH; s eth, bz, HOAc; vs ace
3224	1,5-Dichloronaphthalene		$C_{10}H_6Cl_2$	1825-30-5	197.061	nd or lf (al) pr (sub)	107	sub	1.4900[20]		i H_2O; sl EtOH; s eth
3225	1,6-Dichloronaphthalene		$C_{10}H_6Cl_2$	2050-72-8	197.061	nd or pr (al, peth)	49	sub			
3226	1,7-Dichloronaphthalene		$C_{10}H_6Cl_2$	2050-73-9	197.061	nd or pr (al, HOAc)	63.5	285.5	1.2611[100]	1.6092[100]	s EtOH, eth, bz, HOAc
3227	1,8-Dichloronaphthalene		$C_{10}H_6Cl_2$	2050-74-0	197.061	orth pl (hx) nd (al, sub)	89	sub	1.2924[100]	1.6236[100]	s EtOH, peth
3228	2,3-Dichloronaphthalene		$C_{10}H_6Cl_2$	2050-75-1	197.061	orth lf (al)	120				i H_2O; sl EtOH; vs eth

3191
1,2-Dichloro-1,2,3,3,4,4-hexafluorocyclobutane

3192
1,2-Dichloro-3,3,4,4,5,5-hexafluorocyclopentene

3193
1,2-Dichloro-1,1,2,3,3,3-hexafluoropropane

3194
1,3-Dichloro-1,1,2,2,3,3-hexafluoropropane

3195
1,5-Dichloro-1,1,3,3,5,5-hexamethyltrisiloxane

3196
1,2-Dichlorohexane

3197
1,6-Dichlorohexane

3198
3,5-Dichloro-2-hydroxybenzaldehyde

3199
3,5-Dichloro-2-hydroxybenzoic acid

3200
2,6-Dichloroindophenol, sodium salt

3201
5,6-Dichloro-1,3-isobenzofurandione

3202
Dichloromethane

3203
1,2-Dichloro-3-methoxybenzene

3204
1,3-Dichloro-2-methoxybenzene

3205
2,4-Dichloro-1-methoxybenzene

3206
3,6-Dichloro-2-methoxybenzoic acid

3207
(Dichloromethyl)benzene

3208
N,N-Dichloro-4-methylbenzenesulfonamide

3209
Dichloromethylborane

3210
2,3-Dichloro-2-methylbutane

3211
1,1-Dichloromethyl methyl ether

3212
2,4-Dichloro-3-methylphenol

3213
2,4-Dichloro-6-methylphenol

3214
2,6-Dichloro-4-methylphenol

3215
Dichloromethylphenylsilane

3216
Dichloromethylphosphine

3217
1,2-Dichloro-2-methylpropane

3218
2,4-Dichloro-5-methylpyrimidine

3219
2,4-Dichloro-6-methylpyrimidine

3220
Dichloromethylsilane

3221
1,2-Dichloronaphthalene

3222
1,3-Dichloronaphthalene

3223
1,4-Dichloronaphthalene

3224
1,5-Dichloronaphthalene

3225
1,6-Dichloronaphthalene

3226
1,7-Dichloronaphthalene

3227
1,8-Dichloronaphthalene

3228
2,3-Dichloronaphthalene

No.	Name	Synonym	Mol. Form.	CAS RN	Mol. Wt.	Physical Form	mp/°C	bp/°C	den g cm⁻³	n_D	Solubility
3229	2,6-Dichloronaphthalene		$C_{10}H_6Cl_2$	2065-70-5	197.061	nd or lf (al) pl (eth, bz)	140.5	285			sl EtOH; s eth, bz, chl, HOAc
3230	2,7-Dichloronaphthalene		$C_{10}H_6Cl_2$	2198-77-8	197.061	pl or lf (al)	115.0				vs EtOH; s hx, HOAc
3231	2,3-Dichloro-1,4-naphthal-enedione	Dichlone	$C_{10}H_4Cl_2O_2$	117-80-6	227.044	ye nd (al)	196.4(0.5)				i H_2O; sl EtOH, eth, bz; s chl
3232	2,4-Dichloro-1-naphthol	2,4-Dichloro-α-naphthol	$C_{10}H_6Cl_2O$	2050-76-2	213.060	nd (al, bz)	107.5	180			vs bz, eth, EtOH
3233	2,6-Dichloro-4-nitroaniline		$C_6H_4Cl_2N_2O_2$	99-30-9	207.014	ye nd (al, HOAc)	193.8(0.8)				s EtOH, acid; sl DMSO
3234	1,2-Dichloro-3-nitrobenzene		$C_6H_3Cl_2NO_2$	3209-22-1	192.000	mcl nd (peth, HOAc)	60.7(0.6)	257.5	1.721¹⁴		i H_2O; s EtOH, eth, ace, bz, peth; sl chl
3235	1,2-Dichloro-4-nitrobenzene		$C_6H_3Cl_2NO_2$	99-54-7	192.000	nd (al)	41.0(0.2)	255.5	1.4558⁷⁵		i H_2O; s EtOH, eth; sl ctc
3236	1,3-Dichloro-5-nitrobenzene		$C_6H_3Cl_2NO_2$	618-62-2	192.000	mcl pr or lf (HOAc, al)	65(2)		1.4000¹⁰⁰		i H_2O; s EtOH, eth
3237	1,4-Dichloro-2-nitrobenzene		$C_6H_3Cl_2NO_2$	89-61-2	192.000	pl or pr (al) pl (AcOEt)	53.5(0.2)	267(10)	1.439⁷⁵	1.4390⁷⁵	i H_2O; s EtOH, eth, bz, CS_2; sl ctc
3238	2,4-Dichloro-1-nitrobenzene		$C_6H_3Cl_2NO_2$	611-06-3	192.000	nd (al)	33(2)	258.5	1.4790⁸⁰	1.5512⁷⁰	i H_2O; s EtOH, eth; sl chl
3239	1,1-Dichloro-1-nitroethane	Ethide	$C_2H_3Cl_2NO_2$	594-72-9	143.957			123.5			s ctc
3240	2,6-Dichloro-4-nitrophenol		$C_6H_3Cl_2NO_3$	618-80-4	207.999	br nd (w)	127 exp		1.822²⁵		vs eth, chl
3241	1,1-Dichloro-1-nitropropane		$C_3H_5Cl_2NO_2$	595-44-8	157.984			145	1.312²⁰		s ctc
3242	1,9-Dichlorononane		$C_9H_{18}Cl_2$	821-99-8	197.145			241(12)	1.0173²⁵	1.4586²⁵	
3243	1,8-Dichlorooctane		$C_8H_{16}Cl_2$	2162-99-4	183.119			246(1)	1.0248²⁵	1.4572²⁵	
3244	1,3-Dichloro-1,1,2,2,3-pentafluoropropane	HCFC-225cb	$C_3HCl_2F_5$	507-55-1	202.938	liq		55(2)	1.55²⁵		
3245	3,3-Dichloro-1,1,1,2,2-pentafluoropropane	Refrigerant 225ca	$C_3HCl_2F_5$	422-56-0	202.938	liq		50(2)	1.54²⁵		
3246	1,2-Dichloropentane		$C_5H_{10}Cl_2$	1674-33-5	141.038			148.2(0.7)	1.0872²⁰	1.4485²⁰	i H_2O; s EtOH; vs chl
3247	1,5-Dichloropentane		$C_5H_{10}Cl_2$	628-76-2	141.038	liq	-72.8	182.9(0.8)	1.0956²⁵	1.4545²⁵	i H_2O; s EtOH, eth, bz, ctc
3248	2,3-Dichloropentane		$C_5H_{10}Cl_2$	600-11-3	141.038	liq	-77.3	143(4)	1.0789²⁰	1.4464²⁰	i H_2O
3249	Dichlorophene		$C_{13}H_{10}Cl_2O_2$	97-23-4	269.123	cry (bz, peth)	177.5				i H_2O; s EtOH, ace
3250	2,3-Dichlorophenol		$C_6H_4Cl_2O$	576-24-9	163.001	cry (lig, bz)	56.8(0.3)				s EtOH, eth, bz, lig
3251	2,4-Dichlorophenol		$C_6H_4Cl_2O$	120-83-2	163.001	hex nd (bz)	43(2)	210			sl H_2O; s EtOH, eth, bz, chl
3252	2,5-Dichlorophenol		$C_6H_4Cl_2O$	583-78-8	163.001	pr (bz, peth)	57.8(0.3)	222(8)			sl H_2O; vs EtOH, eth; s bz, peth
3253	2,6-Dichlorophenol		$C_6H_4Cl_2O$	87-65-0	163.001	nd (peth)	66.6(0.4)	226(4)	1.653²⁰		vs EtOH, eth; s bz, peth
3254	3,4-Dichlorophenol		$C_6H_4Cl_2O$	95-77-2	163.001	nd (bz-peth)	67.8(0.3)	253			sl H_2O; vs EtOH, eth; s bz, peth
3255	3,5-Dichlorophenol		$C_6H_4Cl_2O$	591-35-5	163.001	pr (peth)	67.8(0.3)	233			sl H_2O; vs EtOH, eth; s peth
3256	(2,4-Dichlorophenoxy)acetic acid	2,4-D	$C_8H_6Cl_2O_3$	94-75-7	221.038	cry (bz)	140(2)	160⁰·⁴			i H_2O; s EtOH; sl bz, DMSO
3257	4-(2,4-Dichlorophenoxy)-butanoic acid	Butyrac 118	$C_{10}H_{10}Cl_2O_3$	94-82-6	249.090		118(2)				
3258	2-(2,4-Dichlorophenoxy)-propanoic acid	Dichlorprop	$C_9H_8Cl_2O_3$	120-36-5	235.064		116.6(0.3)				sl H_2O, lig; s EtOH, eth
3259	Dichlorophenylarsine		$C_6H_5AsCl_2$	696-28-6	222.932	liq	-19	256(4)	1.6516²⁰	1.6386¹⁵	vs bz, eth, EtOH
3260	2,4-Dichlorophenyl benzenesulfonate	Genite	$C_{12}H_8Cl_2O_3S$	97-16-5	303.161		45.5				s ctc, CS_2
3261	2,2-Dichloro-1-phenyletha-none		$C_8H_6Cl_2O$	2648-61-5	189.039	amor	20.5	249	1.340¹⁶	1.5686²⁰	s EtOH, bz, ctc
3262	1-(2,4-Dichlorophenyl)-ethanone		$C_8H_6Cl_2O$	2234-16-4	189.039		33.5			1.5640²⁰	i H_2O
3263	1-(2,5-Dichlorophenyl)-ethanone		$C_8H_6Cl_2O$	2476-37-1	189.039		12	118¹²	1.321³⁰	1.5595³⁰	
3264	1-(3,4-Dichlorophenyl)-ethanone		$C_8H_6Cl_2O$	2642-63-9	189.039	nd (peth)	76	135¹²			i H_2O; s ctc, lig
3265	3,4-Dichlorophenyl isocyanate	1,2-Dichloro-5-isocyanatoben-zene	$C_7H_3Cl_2NO$	102-36-3	188.011	cry	42	112¹²			
3266	3,5-Dichlorophenyl isocyanate	1,3-Dichloro-5-isocyanatoben-zene	$C_7H_3Cl_2NO$	34893-92-0	188.011		33		1.380		
3267	N-(3,4-Dichlorophenyl)-2-methyl-2-propenamide	Dicryl	$C_{10}H_9Cl_2NO$	2164-09-2	230.090	cry (al-peth)	122.6(0.5)				vs ace, EtOH

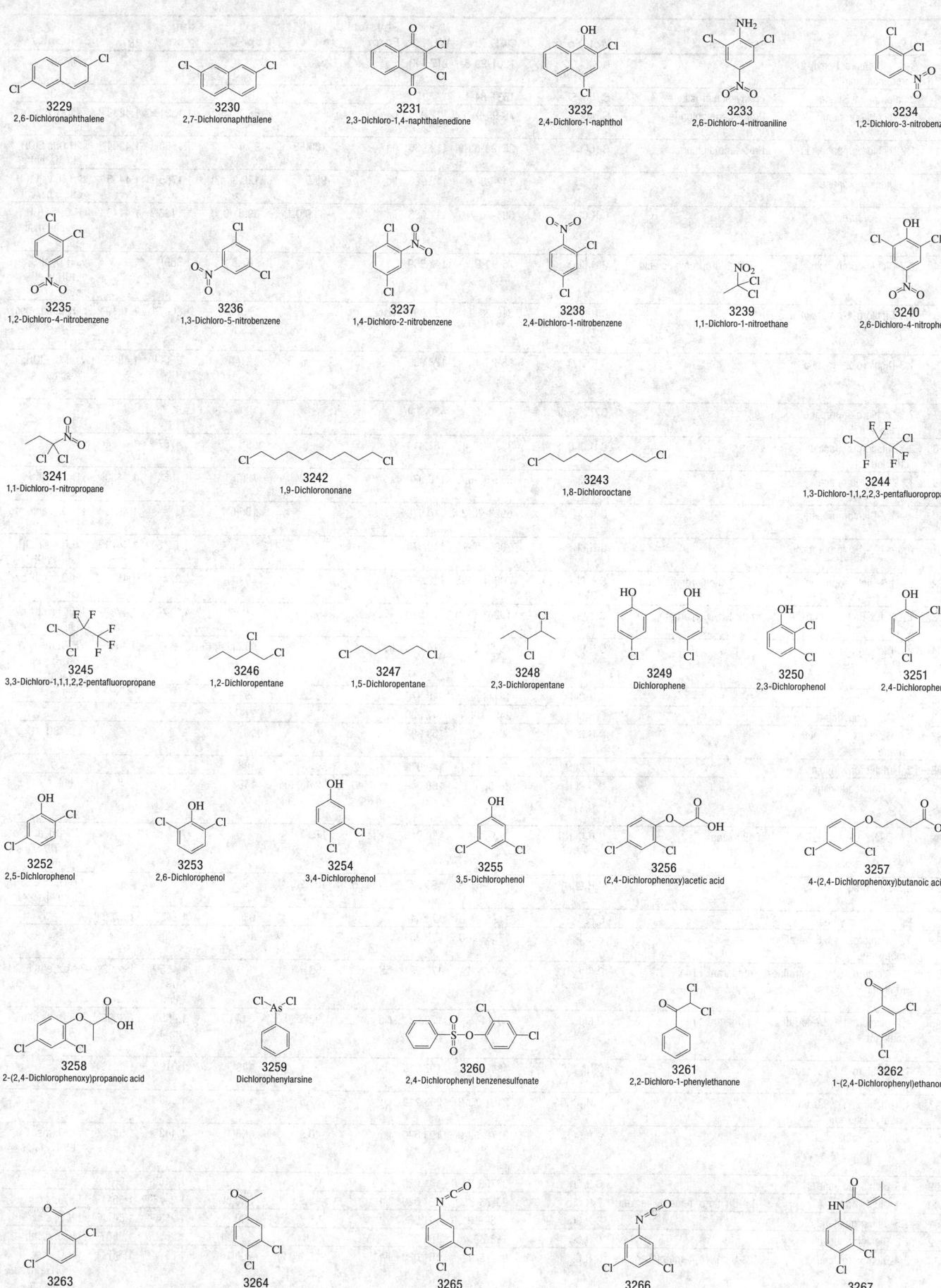

3229
2,6-Dichloronaphthalene

3230
2,7-Dichloronaphthalene

3231
2,3-Dichloro-1,4-naphthalenedione

3232
2,4-Dichloro-1-naphthol

3233
2,6-Dichloro-4-nitroaniline

3234
1,2-Dichloro-3-nitrobenzene

3235
1,2-Dichloro-4-nitrobenzene

3236
1,3-Dichloro-5-nitrobenzene

3237
1,4-Dichloro-2-nitrobenzene

3238
2,4-Dichloro-1-nitrobenzene

3239
1,1-Dichloro-1-nitroethane

3240
2,6-Dichloro-4-nitrophenol

3241
1,1-Dichloro-1-nitropropane

3242
1,9-Dichlorononane

3243
1,8-Dichlorooctane

3244
1,3-Dichloro-1,1,2,2,3-pentafluoropropane

3245
3,3-Dichloro-1,1,1,2,2-pentafluoropropane

3246
1,2-Dichloropentane

3247
1,5-Dichloropentane

3248
2,3-Dichloropentane

3249
Dichlorophene

3250
2,3-Dichlorophenol

3251
2,4-Dichlorophenol

3252
2,5-Dichlorophenol

3253
2,6-Dichlorophenol

3254
3,4-Dichlorophenol

3255
3,5-Dichlorophenol

3256
(2,4-Dichlorophenoxy)acetic acid

3257
4-(2,4-Dichlorophenoxy)butanoic acid

3258
2-(2,4-Dichlorophenoxy)propanoic acid

3259
Dichlorophenylarsine

3260
2,4-Dichlorophenyl benzenesulfonate

3261
2,2-Dichloro-1-phenylethanone

3262
1-(2,4-Dichlorophenyl)ethanone

3263
1-(2,5-Dichlorophenyl)ethanone

3264
1-(3,4-Dichlorophenyl)ethanone

3265
3,4-Dichlorophenyl isocyanate

3266
3,5-Dichlorophenyl isocyanate

3267
N-(3,4-Dichlorophenyl)-2-methyl-2-propenamide

No.	Name	Synonym	Mol. Form.	CAS RN	Mol. Wt.	Physical Form	mp/°C	bp/°C	den g cm⁻³	n_D	Solubility
3268	3-(2,4-Dichlorophenyl)-2-propenoic acid		$C_9H_6Cl_2O_2$	1201-99-6	217.049		234				s DMSO
3269	Dichlorophenylsilane	Phenyldichlorosilane	$C_6H_6Cl_2Si$	1631-84-1	177.104			181	1.221²⁵		dec H_2O
3270	1,1-Dichloropropane	Propylidene chloride	$C_3H_6Cl_2$	78-99-9	112.986			88.4(0.5)	1.1321²⁰	1.4289²⁰	s EtOH, eth, bz, chl
3271	1,2-Dichloropropane, (±)-	Propylene dichloride	$C_3H_6Cl_2$	26198-63-0	112.986	liq	-100.53	96.4	1.1560²⁰	1.4394²⁰	sl H_2O; s EtOH, eth, bz, chl
3272	1,3-Dichloropropane		$C_3H_6Cl_2$	142-28-9	112.986	liq	-99.5	120.8(0.3)	1.1785²⁰	1.4455²⁵	sl H_2O; vs EtOH, eth; s bz, chl
3273	2,2-Dichloropropane		$C_3H_6Cl_2$	594-20-7	112.986	liq	-33.9(0.1)	69.51(0.1)	1.1136²⁰	1.4148²⁰	i H_2O; s EtOH, bz, chl; msc eth
3274	2,2-Dichloropropanoic acid	2,2-Dichloropropionic acid	$C_3H_4Cl_2O_2$	75-99-0	142.969			187.5	1.389¹²		vs H_2O, alk, EtOH; s eth, ctc
3275	2,3-Dichloro-1-propanol		$C_3H_6Cl_2O$	616-23-9	128.985	visc		176(5)	1.3607²⁰	1.4819²⁰	sl H_2O, lig; msc EtOH, eth, ace, bz
3276	1,3-Dichloro-2-propanol		$C_3H_6Cl_2O$	96-23-1	128.985			171(4)	1.3506¹⁷	1.4837²⁰	vs H_2O, EtOH; msc eth; s ace, chl
3277	2,3-Dichloro-1-propanol, phosphate (3:1)		$C_9H_{15}Cl_6O_4P$	78-43-3	430.904			190⁰·¹	1.517²²		
3278	2,3-Dichloropropanoyl chloride		$C_3H_3Cl_3O$	7623-13-4	161.414			53¹⁷	1.4757²⁰	1.4764²⁰	
3279	1,1-Dichloropropene		$C_3H_4Cl_2$	563-58-6	110.970			76(5)	1.1864²⁵	1.4430²⁵	i H_2O; s eth, ace, chl
3280	cis-1,2-Dichloropropene		$C_3H_4Cl_2$	6923-20-2	110.970			105(5)		1.4549²⁰	i H_2O; s ace, bz, chl
3281	trans-1,2-Dichloropropene		$C_3H_4Cl_2$	7069-38-7	110.970			77	1.1818²⁰	1.4471²⁰	i H_2O; vs EtOH, ctc, MeOH
3282	cis-1,3-Dichloropropene	cis-1,3-Dichloropropylene	$C_3H_4Cl_2$	10061-01-5	110.970			104(1)	1.224²⁰	1.4682²⁰	i H_2O; s eth, bz, chl
3283	trans-1,3-Dichloropropene	trans-1,3-Dichloropropylene	$C_3H_4Cl_2$	10061-02-6	110.970			111(5)	1.217²⁰	1.4730²⁰	i H_2O; s eth, bz, chl
3284	2,3-Dichloropropene		$C_3H_4Cl_2$	78-88-6	110.970	liq	10	93.0(0.4)	1.211²⁰	1.4603²⁰	i H_2O; msc EtOH; s eth, bz, chl
3285	3,6-Dichloropyridazine		$C_4H_2Cl_2N_2$	141-30-0	148.978		68.8	89⁰·²			s chl
3286	2,6-Dichloropyridine		$C_5H_3Cl_2N$	2402-78-0	147.990		87	211			
3287	4,6-Dichloro-2-pyrimidin-amine		$C_4H_3Cl_2N_3$	56-05-3	163.993		215				s DMSO
3288	2,4-Dichloropyrimidine		$C_4H_2Cl_2N_2$	3934-20-1	148.978		59	198			
3289	4,7-Dichloroquinoline		$C_9H_5Cl_2N$	86-98-6	198.049	cry (MeOH), nd (80% al)	92.4(0.5)	148¹⁰			sl chl
3290	5,7-Dichloro-8-quinolinol	Chloroxine	$C_9H_5Cl_2NO$	773-76-2	214.048	cry (al)	179.5				sl EtOH, ace, chl, DMSO; s alk, bz, peth
3291	2,3-Dichloroquinoxaline		$C_8H_4Cl_2N_2$	2213-63-0	199.037	cry (al, bz)	151.2(0.4)				i H_2O; vs EtOH, bz, chl, HOAc
3292	2,5-Dichlorostyrene		$C_8H_6Cl_2$	1123-84-8	173.040		8.0	93⁵	1.246²⁰	1.5798²⁰	
3293	1,2-Dichloro-3,4,5,6-tetrafluorobenzene		$C_6Cl_2F_4$	1198-59-0	218.964			157.7			
3294	1,1-Dichloro-1,2,2,2-tetrafluoroethane	Refrigerant 114a	$C_2Cl_2F_4$	374-07-2	170.921	col gas	-56.6	3(1)	1.455²⁵ (p>1 atm)	1.3092²⁰	vs bz, eth, EtOH
3295	1,2-Dichloro-1,1,2,2-tetrafluoroethane	Refrigerant 114	$C_2Cl_2F_4$	76-14-2	170.921	col gas	-92.52(0.05)	3.6(0.5)	1.455²⁵ (p>1 atm)	1.3092²⁰	i H_2O; vs eth, EtOH
3296	1,2-Dichloro-1,1,2,2-tetramethyldisilane		$C_4H_{12}Cl_2Si_2$	4342-61-4	187.215			148	1.010²⁰	1.4548²⁰	
3297	1,3-Dichloro-1,1,3,3-tetramethyldisiloxane		$C_4H_{12}Cl_2OSi_2$	2401-73-2	203.214	liq	-37.5	138	1.038²⁰		
3298	2,5-Dichlorothiophene		$C_4H_2Cl_2S$	3172-52-9	153.030	liq	-40.5	165(4)	1.4422²⁰	1.5626²⁰	i H_2O; msc EtOH, eth; s ctc
3299	2,3-Dichlorotoluene		$C_7H_6Cl_2$	32768-54-0	161.029		6	207.5	1.2458²⁰	1.5511²⁰	vs bz
3300	2,4-Dichlorotoluene	2,4-Dichloro-1-methylbenzene	$C_7H_6Cl_2$	95-73-8	161.029	liq	-13.5	200(8)	1.2476²⁰	1.5511²⁰	i H_2O; s ctc
3301	2,5-Dichlorotoluene		$C_7H_6Cl_2$	19398-61-9	161.029		2.5	200	1.2535²⁰	1.5449²⁰	i H_2O; s bz
3302	2,6-Dichlorotoluene		$C_7H_6Cl_2$	118-69-4	161.029		25.8	194(11)	1.2686²⁰	1.5507²⁰	i H_2O; s chl
3303	3,4-Dichlorotoluene	1,2-Dichloro-4-methylbenzene	$C_7H_6Cl_2$	95-75-0	161.029	liq	-15.2(0.2)	208(18)	1.2564²⁰	1.5471²⁰	i H_2O; msc EtOH, eth, ace, bz, lig, ctc

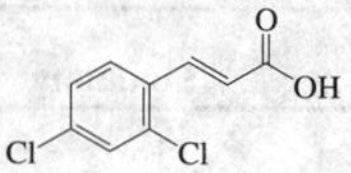

3268
3-(2,4-Dichlorophenyl)-2-propenoic acid

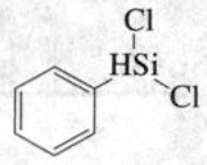

3269
Dichlorophenylsilane

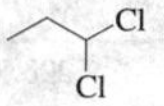

3270
1,1-Dichloropropane

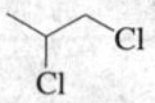

3271
1,2-Dichloropropane, (±)-

3272
1,3-Dichloropropane

3273
2,2-Dichloropropane

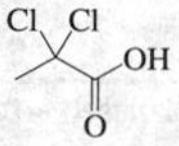

3274
2,2-Dichloropropanoic acid

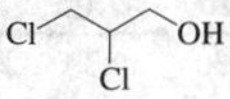

3275
2,3-Dichloro-1-propanol

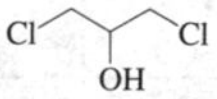

3276
1,3-Dichloro-2-propanol

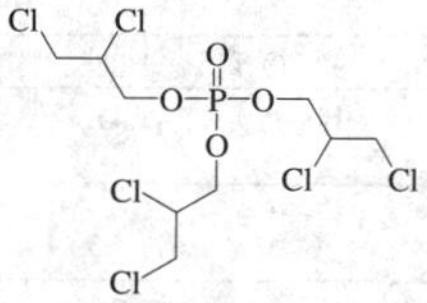

3277
2,3-Dichloro-1-propanol, phosphate (3:1)

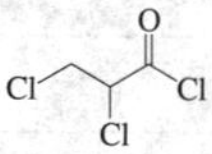

3278
2,3-Dichloropropanoyl chloride

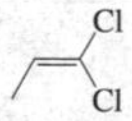

3279
1,1-Dichloropropene

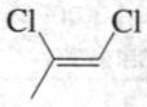

3280
cis-1,2-Dichloropropene

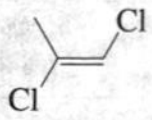

3281
trans-1,2-Dichloropropene

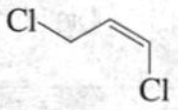

3282
cis-1,3-Dichloropropene

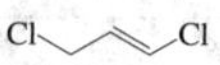

3283
trans-1,3-Dichloropropene

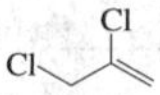

3284
2,3-Dichloropropene

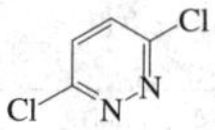

3285
3,6-Dichloropyridazine

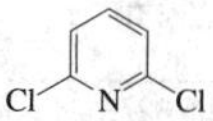

3286
2,6-Dichloropyridine

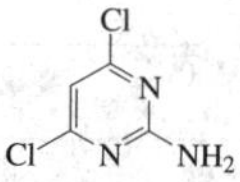

3287
4,6-Dichloro-2-pyrimidinamine

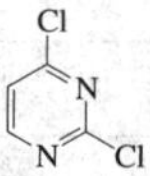

3288
2,4-Dichloropyrimidine

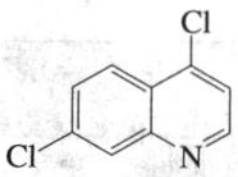

3289
4,7-Dichloroquinoline

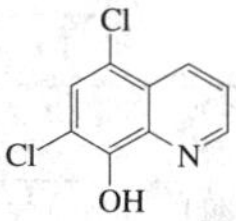

3290
5,7-Dichloro-8-quinolinol

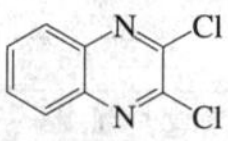

3291
2,3-Dichloroquinoxaline

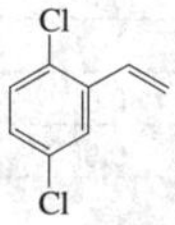

3292
2,5-Dichlorostyrene

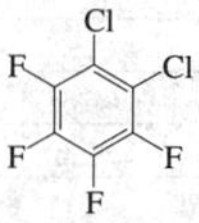

3293
1,2-Dichloro-3,4,5,6-tetrafluorobenzene

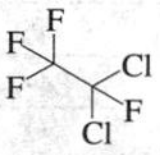

3294
1,1-Dichloro-1,2,2,2-tetrafluoroethane

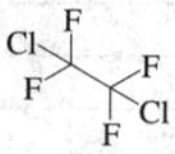

3295
1,2-Dichloro-1,1,2,2-tetrafluoroethane

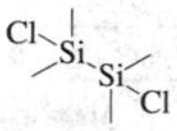

3296
1,2-Dichloro-1,1,2,2-tetramethyldisilane

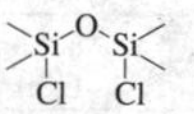

3297
1,3-Dichloro-1,1,3,3-tetramethyldisiloxane

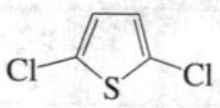

3298
2,5-Dichlorothiophene

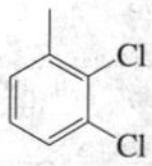

3299
2,3-Dichlorotoluene

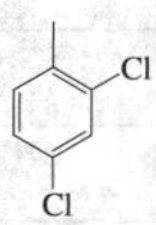

3300
2,4-Dichlorotoluene

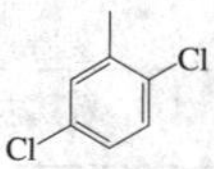

3301
2,5-Dichlorotoluene

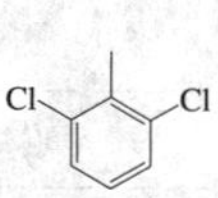

3302
2,6-Dichlorotoluene

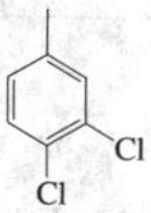

3303
3,4-Dichlorotoluene

No.	Name	Synonym	Mol. Form.	CAS RN	Mol. Wt.	Physical Form	mp/°C	bp/°C	den g cm⁻³	n_D	Solubility
3304	1,3-Dichloro-1,3,5-triazine-2,4,6(1*H*,3*H*,5*H*)-trione	Dichlorocyanuric acid	C₃HCl₂N₃O₃	2782-57-2	197.964	cry	226.6				
3305	1,2-Dichloro-4-(trichloromethyl)benzene		C₇H₃Cl₅	13014-24-9	264.364		25.8	283.1	1.5913²⁰	1.5886²⁰	
3306	1,2-Dichloro-1,1,2-trifluoro-ethane	Refrigerant 123a	C₂HCl₂F₃	354-23-4	152.930	vol liq or gas	-78	30.0(0.1)	1.50²⁵		
3307	2,2-Dichloro-1,1,1-trifluoro-ethane	HCFC-123	C₂HCl₂F₃	306-83-2	152.930	vol liq or gas	-107	27.8(0.6)	1.4638²⁵		sl H₂O
3308	1,1-Dichloro-1,2,2-trifluoro-ethane	Refrigerant 123b	C₂HCl₂F₃	812-04-4	152.930			30.2			
3309	2,4-Dichloro-1-(trifluoromethyl)benzene	2,4-Dichlorobenzotrifluoride	C₇H₃Cl₂F₃	320-60-5	215.000					1.4802²⁰	
3310	4,5-Dichloro-2-(trifluoromethyl)-1*H*-benz-imidazole	Chloroflurazole	C₈H₃Cl₂F₃N₂	3615-21-2	255.024		213.5				
3311	Dichlorovinylmethylsilane		C₃H₆Cl₂Si	124-70-9	141.072			93.7(0.5)	1.0868²⁰	1.4270²⁰	dec H₂O
3312	Dichlorvos	Phosphoric acid, 2,2-dichloro-ethenyl dimethyl ester	C₄H₇Cl₂O₄P	62-73-7	220.976			140²⁰	1.415²⁵		
3313	Diclofop-methyl	Methyl 2-[4-(2,4-dichlorophe-noxy)phenoxy]propanoate	C₁₆H₁₄Cl₂O₄	51338-27-3	341.186		42.1(0.5)	176⁰·¹			
3314	Dicrotophos		C₈H₁₆NO₅P	141-66-2	237.191			400	1.216¹⁵		
3315	Dicumarol		C₁₉H₁₂O₆	66-76-2	336.294	nd	290				
3316	Dicyanamide	Cyanocyanamide	C₂HN₃	504-66-5	67.049	aq soln only					
3317	*o*-Dicyanobenzene	*o*-Phthalodinitrile	C₈H₄N₂	91-15-6	128.131	nd (w, lig)	140.6(0.9)	150¹⁰	1.1250²⁵		sl H₂O, lig; vs EtOH, bz; s eth, ace
3318	*m*-Dicyanobenzene	*m*-Phthalodinitrile	C₈H₄N₂	626-17-5	128.131	nd(al)	162	sub	0.992⁴⁰		sl H₂O; vs EtOH; s eth, bz, chl; i peth
3319	*p*-Dicyanobenzene	*p*-Phthalodinitrile	C₈H₄N₂	623-26-7	128.131	nd (w, MeOH)	224	sub			i H₂O; sl EtOH, eth; s bz; vs HOAc
3320	Dicyclohexyl adipate	Dicyclohexyl hexanedioate	C₁₈H₃₀O₄	849-99-0	310.429		35(1)				s chl
3321	Dicyclohexylamine	*N*-Cyclohexylcyclohexanamine	C₁₂H₂₃N	101-83-7	181.318	liq	-0.1	251(4)	0.9123²⁰	1.4842²⁰	sl H₂O, ctc; s EtOH, eth, bz
3322	Dicyclohexylamine nitrite	*N*-Cyclohexylcyclohexanamine, nitrite	C₁₂H₂₄N₂O₂	3129-91-7	228.331	cry	179.5(0.6)				
3323	Dicyclohexylcarbodiimide		C₁₃H₂₂N₂	538-75-0	206.327		34.5	123⁶			
3324	Dicyclohexyl disulfide		C₁₂H₂₂S₂	2550-40-5	230.433	liq		195²⁰			
3325	Dicyclohexyl ether		C₁₂H₂₂O	4645-15-2	182.302	liq	-36	242.5	0.9227²⁰	1.4741²⁰	
3326	Dicyclohexylmethanone		C₁₃H₂₂O	119-60-8	194.313		57	159²⁰	0.986⁰	1.4860²⁰	s eth, ace, ctc
3327	Dicyclohexylphosphine		C₁₂H₂₃P	829-84-5	198.285			281	0.904²⁵	1.5163²⁰	
3328	Dicyclohexyl phthalate		C₂₀H₂₆O₄	84-61-7	330.418	pr (al)	66	225⁴	1.383²⁰	1.431²⁰	i H₂O; s EtOH, eth; sl chl
3329	*N,N*'-Dicyclohexylthiourea		C₁₃H₂₄N₂S	1212-29-9	240.408	cry (MeOH)	180				
3330	*N,N*'-Dicyclohexylurea		C₁₃H₂₄N₂O	2387-23-7	224.342		233.8				
3331	Dicyclomine hydrochloride	Dicycloverine hydrochloride	C₁₉H₃₆ClNO₂	67-92-5	345.948	cry	165				
3332	Dicyclopentadiene		C₁₀H₁₂	1755-01-7	132.202		32	170 dec	0.9302³⁵	1.5050³⁵	vs eth, EtOH
3333	Dicyclopentyl ether	Cyclopentyl ether	C₁₀H₁₈O	10137-73-2	154.249	liq		80¹³			
3334	Dicyclopropyl ketone		C₇H₁₀O	1121-37-5	110.153			161	0.977²⁵	1.4670²⁰	
3335	Didecylamine	*N*-Decyl-1-decanamine	C₂₀H₄₃N	1120-49-6	297.562			359.0			
3336	Didecyl ether		C₂₀H₄₂O	2456-28-2	298.546		16	196¹⁵·⁵	0.8187²⁰		
3337	Didecyl phthalate		C₂₈H₄₆O₄	84-77-5	446.663		2.5	240³	0.9639²⁰		
3338	3',4'-Didehydro-β,ψ-caroten-16'-oic acid	Torularhodin	C₄₀H₅₂O₂	514-92-1	564.840	purp nd (MeOH-eth)	211				vs py, chl, CS₂
3339	2',3'-Dideoxyinosine	Didanosine	C₁₀H₁₂N₄O₃	69655-05-6	236.227	wh cry (EtOH aq)	162				
3340	2,6-Dideoxy-3-*O*-methyl-*ribo*-hexose	Cymarose	C₇H₁₄O₄	579-04-4	162.184	pr (eth-peth) nd (ace)	101				vs H₂O, ace, EtOH
3341	Didodecanoyl peroxide	Lauroyl peroxide	C₂₄H₄₆O₄	105-74-8	398.620	wh pl	49				i H₂O; s chl
3342	Didodecylamine	*N*-Dodecyl-1-dodecanamine	C₂₄H₅₁N	3007-31-6	353.669		46.9(0.5)	263²⁷			vs bz, eth, EtOH, chl
3343	Didodecyl phosphate		C₂₄H₅₁O₄P	7057-92-3	434.633	cry (MeOH)	59				
3344	Didodecyl phthalate	1,2-Benzenedicarboxylic acid, didodecyl ester	C₃₂H₅₄O₄	2432-90-8	502.769		22.0	256¹	0.9389²⁰		
3345	Dieldrin		C₁₂H₈Cl₆O	60-57-1	380.909		178.8(0.3)		1.75²⁵		i H₂O; sl EtOH; s ace, bz
3346	Dienestrol		C₁₈H₁₈O₂	84-17-3	266.335	cry (dil al)	227.5	130 sub			vs ace, eth, EtOH
3347	1,2:8,9-Diepoxy-*p*-menthane	Limonene diepoxide	C₁₀H₁₆O₂	96-08-2	168.233		242				

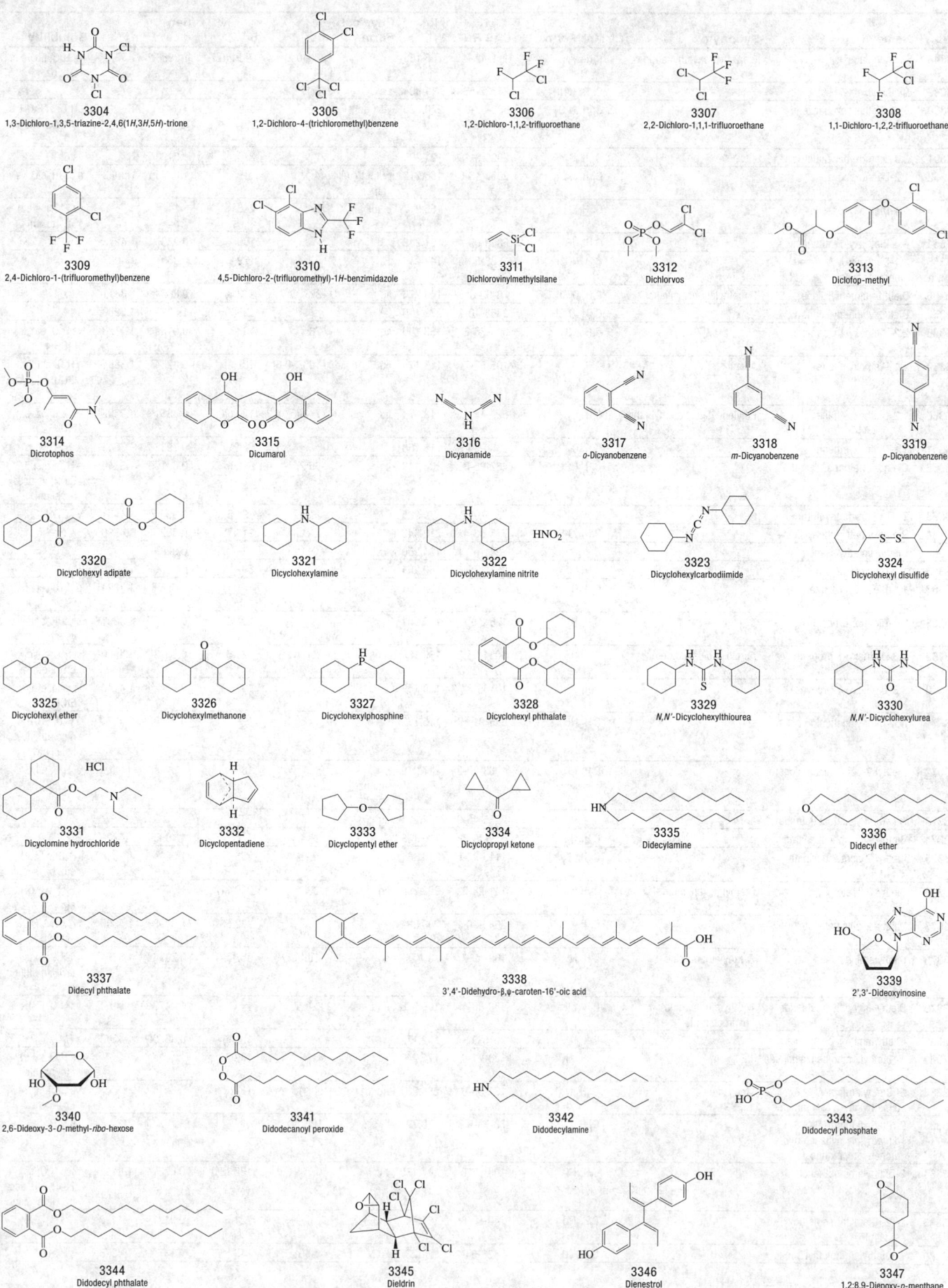

3304
1,3-Dichloro-1,3,5-triazine-2,4,6(1*H*,3*H*,5*H*)-trione

3305
1,2-Dichloro-4-(trichloromethyl)benzene

3306
1,2-Dichloro-1,1,2-trifluoroethane

3307
2,2-Dichloro-1,1,1-trifluoroethane

3308
1,1-Dichloro-1,2,2-trifluoroethane

3309
2,4-Dichloro-1-(trifluoromethyl)benzene

3310
4,5-Dichloro-2-(trifluoromethyl)-1*H*-benzimidazole

3311
Dichlorovinylmethylsilane

3312
Dichlorvos

3313
Diclofop-methyl

3314
Dicrotophos

3315
Dicumarol

3316
Dicyanamide

3317
o-Dicyanobenzene

3318
m-Dicyanobenzene

3319
p-Dicyanobenzene

3320
Dicyclohexyl adipate

3321
Dicyclohexylamine

3322
Dicyclohexylamine nitrite

3323
Dicyclohexylcarbodiimide

3324
Dicyclohexyl disulfide

3325
Dicyclohexyl ether

3326
Dicyclohexylmethanone

3327
Dicyclohexylphosphine

3328
Dicyclohexyl phthalate

3329
N,N'-Dicyclohexylthiourea

3330
N,N'-Dicyclohexylurea

3331
Dicyclomine hydrochloride

3332
Dicyclopentadiene

3333
Dicyclopentyl ether

3334
Dicyclopropyl ketone

3335
Didecylamine

3336
Didecyl ether

3337
Didecyl phthalate

3338
3',4'-Didehydro-β,ψ-caroten-16'-oic acid

3339
2',3'-Dideoxyinosine

3340
2,6-Dideoxy-3-*O*-methyl-*ribo*-hexose

3341
Didodecanoyl peroxide

3342
Didodecylamine

3343
Didodecyl phosphate

3344
Didodecyl phthalate

3345
Dieldrin

3346
Dienestrol

3347
1,2:8,9-Diepoxy-*p*-menthane

No.	Name	Synonym	Mol. Form.	CAS RN	Mol. Wt.	Physical Form	mp/°C	bp/°C	den g cm⁻³	n_D	Solubility
3348	Diethanolamine	Bis(2-hydroxyethyl)amine	$C_4H_{11}NO_2$	111-42-2	105.136		27.9(0.2)	271.2(0.7)	1.0966[20]	1.4776[20]	vs H_2O, EtOH; sl eth, bz
3349	Diethatyl, ethyl ester		$C_{16}H_{22}ClNO_3$	38727-55-8	311.804	cry	46(1)				
3350	4,4'-Diethoxyazobenzene		$C_{16}H_{18}N_2O_2$	588-52-3	270.326	ye lf (al)	162	dec			i H_2O; sl EtOH; s eth, bz, chl; vs HOAc
3351	3,4-Diethoxybenzaldehyde		$C_{11}H_{14}O_3$	2029-94-9	194.227		22	279	1.0100[22]		vs EtOH
3352	1,2-Diethoxybenzene		$C_{10}H_{14}O_2$	2050-46-6	166.217	pr (peth, dil al)	44	219	1.0075[20]	1.5083[25]	s EtOH, ctc; vs eth
3353	1,4-Diethoxybenzene		$C_{10}H_{14}O_2$	122-95-2	166.217	pl (dil al)	72	246			vs EtOH; s eth, bz, ctc, chl
3354	4,4-Diethoxy-1-butanamine		$C_8H_{19}NO_2$	6346-09-4	161.243			196	0.933[25]	1.4275[20]	
3355	1,1-Diethoxy-N,N-dimethyl-methanamine		$C_7H_{17}NO_2$	1188-33-6	147.216			129	0.859[25]	1.4007[20]	
3356	Diethoxydimethylsilane	Dimethyldiethoxysilane	$C_6H_{16}O_2Si$	78-62-6	148.276	liq	-87	113(3)	0.865[25]	1.3811[20]	s ctc
3357	Diethoxydiphenylsilane		$C_{16}H_{20}O_2Si$	2553-19-7	272.415			296(3)	1.0329[20]	1.5269[20]	
3358	2,2-Diethoxyethanamine		$C_6H_{15}NO_2$	645-36-3	133.189	liq	-78	163	0.9159[25]	1.4123[25]	vs H_2O, eth, EtOH, chl
3359	1,1-Diethoxyethane	Acetal	$C_6H_{14}O_2$	105-57-7	118.174	liq	-106.1(0.6)	102(2)	0.8254[20]	1.3834[20]	s H_2O, chl; msc EtOH, eth; vs ace
3360	1,2-Diethoxyethane	Ethylene glycol diethyl ether	$C_6H_{14}O_2$	629-14-1	118.174	liq	-74.0(0.2)	120.6(0.7)	0.8351[25]	1.3898[25]	vs ace, bz, eth, EtOH
3361	1,1-Diethoxyethene		$C_6H_{12}O_2$	2678-54-8	116.158			68[100]	0.7932[20]	1.3643[21]	
3362	Diethoxymethane		$C_5H_{12}O_2$	462-95-3	104.148	liq	-66(2)	86(2)	0.8319[20]	1.3748[18]	s H_2O; msc EtOH; vs ace, bz; sl chl
3363	2-(Diethoxymethyl)furan		$C_9H_{14}O_3$	13529-27-6	170.205			191.5	0.9976[20]	1.4451[20]	vs EtOH
3364	Diethoxymethylphenylsilane		$C_{11}H_{18}O_2Si$	775-56-4	210.346			217(3)	0.9627[20]	1.4690[20]	
3365	Diethoxymethylsilane		$C_5H_{14}O_2Si$	2031-62-1	134.250			98	0.829[25]		
3366	1,1-Diethoxypentane		$C_9H_{20}O_2$	3658-79-5	160.254			59[12]	0.829[22]	1.4029[22]	
3367	1,1-Diethoxypropane		$C_7H_{16}O_2$	4744-08-5	132.201			121(1)	0.825[20]	1.3924[19]	s H_2O, ace, bz; vs EtOH, eth
3368	2,2-Diethoxypropane		$C_7H_{16}O_2$	126-84-1	132.201			101(5)	0.8200[21]	1.3891[20]	s EtOH, ace, bz; vs eth; sl ctc
3369	3,3-Diethoxy-1-propene	Acrolein, diethyl acetal	$C_7H_{14}O_2$	3054-95-3	130.185			123.5	0.8543[15]	1.4000[20]	sl H_2O; msc EtOH, eth
3370	3,3-Diethoxy-1-propyne		$C_7H_{12}O_2$	10160-87-9	128.169			139	0.8942[22]	1.4140[20]	vs ace, eth, EtOH, chl
3371	N,N-Diethylacetamide		$C_6H_{13}NO$	685-91-6	115.173			199(10)	0.9130[17]	1.4374[17]	s H_2O, EtOH; msc eth, ace, bz; sl ctc
3372	Diethyl 2-acetamidomalonate		$C_9H_{15}NO_5$	1068-90-2	217.219	cry (al,bz-peth)	96.3	185[20]			sl H_2O, eth; s tfa, EtOH
3373	N,N-Diethylacetoacetamide		$C_8H_{15}NO_2$	2235-46-3	157.211	liq		76[13]			
3374	Diethyl acetylphosphonate		$C_6H_{13}O_4P$	919-19-7	180.138			114[20]	1.1005[20]	1.4200[26]	
3375	Diethyl 2-acetylsuccinate		$C_{10}H_{16}O_5$	1115-30-6	216.231			255	1.081[20]	1.4346[20]	i H_2O; s EtOH, eth, bz; sl chl
3376	Diethyl adipate	Diethyl hexanedioate	$C_{10}H_{18}O_4$	141-28-6	202.248	liq	-20(2)	250(11)	1.0076[20]	1.4272[20]	i H_2O; s EtOH, eth
3377	Diethyl 2-allylmalonate		$C_{10}H_{16}O_4$	2049-80-1	200.232			222.5	1.0098[20]	1.4305[20]	i H_2O; vs EtOH, eth; s ctc
3378	Diethylamine	N-Ethylethanamine	$C_4H_{11}N$	109-89-7	73.137	liq	-50(2)	55.4(0.1)	0.7056[20]	1.3864[20]	vs H_2O; msc EtOH; s eth, ctc
3379	Diethylamine hydrochloride	N-Ethylethanamine hydrochloride	$C_4H_{12}ClN$	660-68-4	109.598	lf (al-eth)	228.5		1.0477[22]		vs H_2O, EtOH
3380	(Diethylamino)acetonitrile		$C_6H_{12}N_2$	3010-02-4	112.172			169(6)	0.8660[20]	1.4260[20]	s H_2O
3381	4-(Diethylamino)benzaldehyde		$C_{11}H_{15}NO$	120-21-8	177.243	ye nd (w)	41	172[10]			vs H_2O; s EtOH, eth, bz, ctc
3382	2-(Diethylamino)-N-(2,6-dimethylphenyl)acetamide	Lidocaine	$C_{14}H_{22}N_2O$	137-58-6	234.337	nd (bz, al)	68(1)	181[4]			vs bz, eth, EtOH, chl
3383	2-(Diethylamino)-N-(2,6-dimethylphenyl)acetamide, monohydrochloride		$C_{14}H_{23}ClN_2O$	73-78-9	270.798		128				vs H_2O
3384	2-Diethylaminoethanol		$C_6H_{15}NO$	100-37-8	117.189	hyg		162.13(0.09)	0.8921[20]	1.4412[20]	msc H_2O; s EtOH, eth, ace, bz, peth; sl ctc
3385	2-[2-(Diethylamino)ethoxy]-ethanol		$C_8H_{19}NO_2$	140-82-9	161.243			221.5	0.9421[25]	1.4480[20]	
3386	2-(Diethylamino)ethyl acrylate		$C_9H_{17}NO_2$	2426-54-2	171.237		<-60	81[10]	0.937[20]	1.4376[25]	

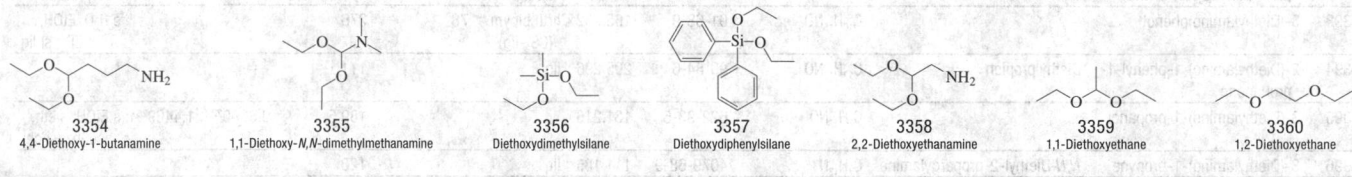

3348 Diethanolamine

3349 Diethatyl, ethyl ester

3350 4,4'-Diethoxyazobenzene

3351 3,4-Diethoxybenzaldehyde

3352 1,2-Diethoxybenzene

3353 1,4-Diethoxybenzene

3354 4,4-Diethoxy-1-butanamine

3355 1,1-Diethoxy-N,N-dimethylmethanamine

3356 Diethoxydimethylsilane

3357 Diethoxydiphenylsilane

3358 2,2-Diethoxyethanamine

3359 1,1-Diethoxyethane

3360 1,2-Diethoxyethane

3361 1,1-Diethoxyethene

3362 Diethoxymethane

3363 2-(Diethoxymethyl)furan

3364 Diethoxymethylphenylsilane

3365 Diethoxymethylsilane

3366 1,1-Diethoxypentane

3367 1,1-Diethoxypropane

3368 2,2-Diethoxypropane

3369 3,3-Diethoxy-1-propene

3370 3,3-Diethoxy-1-propyne

3371 N,N-Diethylacetamide

3372 Diethyl 2-acetamidomalonate

3373 N,N-Diethylacetoacetamide

3374 Diethyl acetylphosphonate

3375 Diethyl 2-acetylsuccinate

3376 Diethyl adipate

3377 Diethyl 2-allylmalonate

3378 Diethylamine

3379 Diethylamine hydrochloride

3380 (Diethylamino)acetonitrile

3381 4-(Diethylamino)benzaldehyde

3382 2-(Diethylamino)-N-(2,6-dimethylphenyl)acetamide

3383 2-(Diethylamino)-N-(2,6-dimethylphenyl)acetamide, monohydrochloride

3384 2-Diethylaminoethanol

3385 2-[2-(Diethylamino)ethoxy]ethanol

3386 2-(Diethylamino)ethyl acrylate

No.	Name	Synonym	Mol. Form.	CAS RN	Mol. Wt.	Physical Form	mp/°C	bp/°C	den g cm⁻³	n_D	Solubility
3387	2-Diethylaminoethyl 4-aminobenzoate	Procaine	$C_{13}H_{20}N_2O_2$	59-46-1	236.310	nd (w+2) pl (lig or eth)	61				sl H₂O; s EtOH, eth, bz, chl
3388	2-(N,N-Diethylamino)ethyl methacrylate		$C_{10}H_{19}NO_2$	105-16-8	185.264			80[10]	0.92[30]		
3389	2-(Diethylamino)ethyl 2-phenylbutanoate	Butethamate	$C_{16}H_{25}NO_2$	14007-64-8	263.376			168[11]		1.4909[20]	
3390	4-(Diethylamino)-2-hydroxy-benzaldehyde		$C_{11}H_{15}NO_2$	17754-90-4	193.243		65.0				
3391	Diethyl 2-aminomalonate		$C_7H_{13}NO_4$	6829-40-9	175.183			122[16]	1.100[16]	1.4353[16]	vs H₂O, EtOH, eth; s ace, bz; i lig
3392	7-(Diethylamino)-4-methyl-2H-1-benzopyran-2-one		$C_{14}H_{17}NO_2$	91-44-1	231.291	cry (al, bz-lig)					sl H₂O; s EtOH, eth, ace
3393	3-(Diethylamino)phenol		$C_{10}H_{15}NO$	91-68-9	165.232	orth bipym (CS₂-lig)	78	276			s H₂O, EtOH, eth, CS₂; sl lig
3394	2-(Diethylamino)-1-phenyl-1-propanone	Diethylpropion	$C_{13}H_{19}NO$	90-84-6	205.296	liq		111[14]			
3395	3-(Diethylamino)-1-propanol		$C_7H_{17}NO$	622-93-5	131.216			189.5	0.8600[20]	1.4439[20]	s EtOH; s eth, ace, bz; sl chl
3396	3-(Diethylamino)-1-propyne	N,N-Diethyl-2-propargylamine	$C_7H_{13}N$	4079-68-9	111.185	liq		120			
3397	2,6-Diethylaniline		$C_{10}H_{15}N$	579-66-8	149.233		1.5	251(8)	0.906[25]	1.5452[20]	
3398	N,N-Diethylaniline		$C_{10}H_{15}N$	91-66-7	149.233	ye oil	-21.3(0.2)	216(1)	0.9307[20]	1.5409[20]	sl H₂O; s EtOH, ace, ctc; vs eth, chl
3399	Diethylarsine		$C_4H_{11}As$	692-42-2	134.052			105	1.1338[24]	1.4709	vs ace, bz, eth, EtOH
3400	N,N-Diethylbenzamide		$C_{11}H_{15}NO$	1696-17-9	177.243			132[5]			
3401	o-Diethylbenzene	1,2-Diethylbenzene	$C_{10}H_{14}$	135-01-3	134.218	liq	-31.4(0.3)	183.4(0.4)	0.8800[20]	1.5035[20]	i H₂O; msc EtOH, eth, ace, bz, lig, ctc
3402	m-Diethylbenzene	1,3-Diethylbenzene	$C_{10}H_{14}$	141-93-5	134.218	liq	-83.9(0.2)	181.1(0.5)	0.8602[20]	1.4955[20]	i H₂O; msc EtOH, eth, ace, bz, lig, ctc
3403	p-Diethylbenzene	1,4-Diethylbenzene	$C_{10}H_{14}$	105-05-5	134.218	liq	-43.3(0.4)	184(1)	0.8620[20]	1.4967[20]	i H₂O; msc EtOH, eth, ace, bz, lig, ctc
3404	N,N-Diethyl-1,4-benzenedi-amine		$C_{10}H_{16}N_2$	93-05-0	164.247			261			vs bz
3405	Diethyl benzylidenemalonate	Diethyl benzalmalonate	$C_{14}H_{16}O_4$	5292-53-5	248.275		32	216[30]	1.1045[20]	1.5389[20]	i H₂O; s EtOH, eth, ace, bz
3406	Diethyl benzylmalonate		$C_{14}H_{18}O_4$	607-81-8	250.291			300	1.076[15]	1.4872[20]	i H₂O; sl chl
3407	Diethyl benzylphosphonate		$C_{11}H_{17}O_3P$	1080-32-6	228.225			110[2]		1.4930[20]	s ctc
3408	Diethylbromoacetamide	2-Bromo-2-ethylbutanamide	$C_6H_{12}BrNO$	511-70-6	194.069		67				sl H₂O, chl; vs EtOH, eth, bz
3409	Diethyl 2-bromomalonate	Ethyl bromomalonate	$C_7H_{11}BrO_4$	685-87-0	239.064		-54	231(12)	1.4022[25]	1.4521[20]	i H₂O; msc EtOH, eth; s ace, ctc
3410	N,N-Diethylbutanamide		$C_8H_{17}NO$	1114-76-7	143.227			206	0.8884[20]	1.4403[25]	vs H₂O, EtOH
3411	Diethyl 2-butylmalonate	Pentane-1,1-dicarboxylic acid, diethyl ester	$C_{11}H_{20}O_4$	133-08-4	216.275			237(8)	0.9764[10]	1.4250[20]	vs EtOH, eth
3412	Diethyl 2-butynedioate		$C_8H_{10}O_4$	762-21-0	170.163		0.8	184[200]	1.0075[20]	1.4425[20]	s EtOH, eth, ctc
3413	Diethylcarbamazine citrate		$C_{16}H_{29}N_3O_8$	1642-54-2	391.416	cry	138				
3414	Diethylcarbamic chloride		$C_5H_{10}ClNO$	88-10-8	135.592			186			
3415	N,N'-Diethylcarbanilide		$C_{17}H_{20}N_2O$	85-98-3	268.353	cry (al)	71(2)				i H₂O; vs EtOH; s chl
3416	Diethyl carbonate	Ethyl carbonate	$C_5H_{10}O_3$	105-58-8	118.131	liq	-43	125.9(0.9)	0.9692[25]	1.3845[20]	i H₂O; s EtOH, eth, chl
3417	O,O-Diethyl chloridothiono-phosphate	Diethyl thiophosphoryl chloride	$C_4H_{10}ClO_2PS$	2524-04-1	188.613			45[3]			s ctc
3418	Diethylchloroaluminum	Diethylaluminum chloride	$C_4H_{10}AlCl$	96-10-6	120.557	col liq	-74	127[50]	0.96		reac H2O
3419	Diethyl chloromalonate	Ethyl chloromalonate	$C_7H_{11}ClO_4$	14064-10-9	194.613			222	1.2040[20]	1.4327[20]	i H₂O; msc EtOH, eth, chl; s CS₂
3420	Diethyl chlorophosphonate	Diethoxyphosphoryl chloride	$C_4H_{10}ClO_3P$	814-49-3	172.547			93.5	1.205[19]	1.4170[20]	
3421	Diethylcyanamide		$C_5H_{10}N_2$	617-83-4	98.146	liq	-80.7(0.6)	187(15)	0.854[20]	1.4126[25]	i H₂O; s EtOH, eth
3422	Diethyl 1,1-cyclobutanedicar-boxylate		$C_{10}H_{16}O_4$	3779-29-1	200.232			223(5)	1.0456[20]	1.4330[26]	vs EtOH; sl ctc
3423	1,1-Diethylcyclohexane		$C_{10}H_{20}$	78-01-3	140.266			178(6)			
3424	Diethyl 1,1-cyclopropanedi-carboxylate		$C_9H_{14}O_4$	1559-02-0	186.205			214(7)	1.055[25]	1.4345[18]	vs EtOH, eth
3425	Diethyl dibutylmalonate		$C_{15}H_{28}O_4$	596-75-8	272.381			150[12]	0.9457[20]	1.4341[20]	i H₂O; s EtOH, eth, ctc

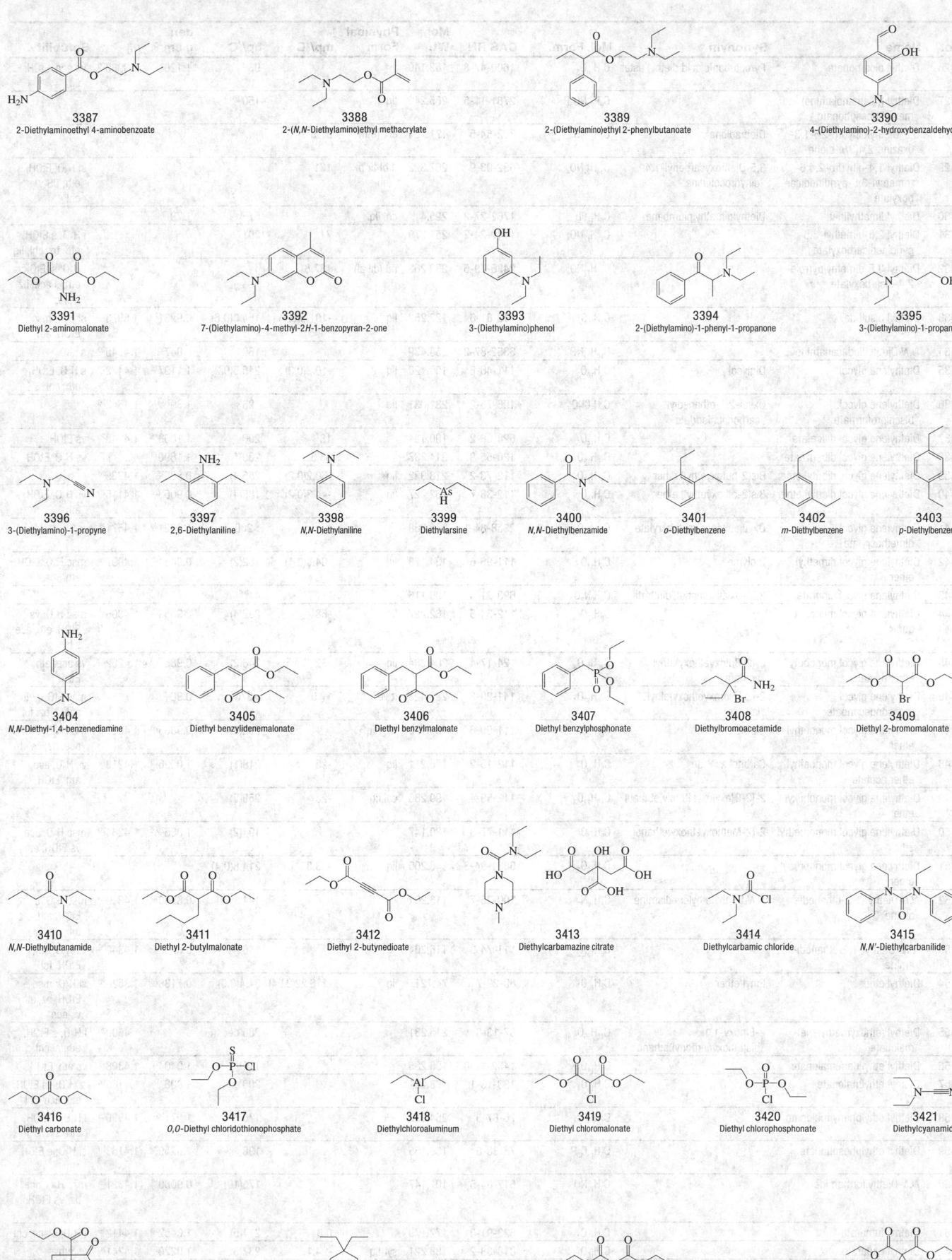

3387
2-Diethylaminoethyl 4-aminobenzoate

3388
2-(N,N-Diethylamino)ethyl methacrylate

3389
2-(Diethylamino)ethyl 2-phenylbutanoate

3390
4-(Diethylamino)-2-hydroxybenzaldehyde

3391
Diethyl 2-aminomalonate

3392
7-(Diethylamino)-4-methyl-2H-1-benzopyran-2-one

3393
3-(Diethylamino)phenol

3394
2-(Diethylamino)-1-phenyl-1-propanone

3395
3-(Diethylamino)-1-propanol

3396
3-(Diethylamino)-1-propyne

3397
2,6-Diethylaniline

3398
N,N-Diethylaniline

3399
Diethylarsine

3400
N,N-Diethylbenzamide

3401
o-Diethylbenzene

3402
m-Diethylbenzene

3403
p-Diethylbenzene

3404
N,N-Diethyl-1,4-benzenediamine

3405
Diethyl benzylidenemalonate

3406
Diethyl benzylmalonate

3407
Diethyl benzylphosphonate

3408
Diethylbromoacetamide

3409
Diethyl 2-bromomalonate

3410
N,N-Diethylbutanamide

3411
Diethyl 2-butylmalonate

3412
Diethyl 2-butynedioate

3413
Diethylcarbamazine citrate

3414
Diethylcarbamic chloride

3415
N,N'-Diethylcarbanilide

3416
Diethyl carbonate

3417
O,O-Diethyl chloridothionophosphate

3418
Diethylchloroaluminum

3419
Diethyl chloromalonate

3420
Diethyl chlorophosphonate

3421
Diethylcyanamide

3422
Diethyl 1,1-cyclobutanedicarboxylate

3423
1,1-Diethylcyclohexane

3424
Diethyl 1,1-cyclopropanedicarboxylate

3425
Diethyl dibutylmalonate

No.	Name	Synonym	Mol. Form.	CAS RN	Mol. Wt.	Physical Form	mp/°C	bp/°C	den g cm⁻³	n_D	Solubility
3426	Diethyl dicarbonate	Pyrocarbonic acid diethyl ester	C₆H₁₀O₅	1609-47-8	162.140			93[18]	1.120[20]	1.3960[20]	vs ace, EtOH, lig
3427	Diethyl [(diethanolamino)methyl]phosphonate		C₉H₂₂NO₅P	2781-11-5	255.249	liq		150[0.01]			
3428	5,5-Diethyldihydro-2H-1,3-oxazine-2,4(3H)-dione	Diethadione	C₈H₁₃NO₃	702-54-5	171.194	cry (eth)	97.5				
3429	Diethyl 1,4-dihydro-2,4,6-trimethyl-3,5-pyridinedicarboxylate	3,5-Diethoxycarbonyl-1,4-dihydrocollidine	C₁₄H₂₁NO₄	632-93-9	267.322	lt bl flr pl (al)	131				sl H₂O, EtOH, eth, CS₂; vs chl
3430	Diethyldimethyllead	Diethyldimethylplumbane	C₆H₁₆Pb	1762-27-2	295.4	col liq		51[13]	1.79[20]		
3431	Diethyl 2,6-dimethyl-3,5-pyridinedicarboxylate		C₁₃H₁₇NO₄	1149-24-2	251.279		71	301			i H₂O; s EtOH, eth, bz, chl, lig
3432	Diethyl 3,5-dimethylpyrrole-2,4-dicarboxylate		C₁₂H₁₇NO₄	2436-79-5	239.268	nd (dil al)	137.8				i H₂O; sl EtOH, eth; s ace, bz, HOAc
3433	Diethyl disulfide		C₄H₁₀S₂	110-81-6	122.252	liq	-101.5(0.1)	154.0(0.6)	0.9931[20]	1.5073[20]	sl H₂O; msc EtOH, eth
3434	N,N-Diethyldodecanamide		C₁₆H₃₃NO	3352-87-2	255.439			166[2]	0.847[25]	1.4545[20]	s chl
3435	Diethylene glycol	Diglycol	C₄H₁₀O₃	111-46-6	106.120	liq	-10.3(0.3)	245.5(0.2)	1.1197[15]	1.4472[20]	s H₂O, EtOH, eth, chl
3436	Diethylene glycol, bischloroformate	Oxydi-2,1-ethanediyl carbonochloridate	C₆H₈Cl₂O₅	106-75-2	231.031	liq		126[5]	1.39[20]	1.4542[20]	
3437	Diethylene glycol diacetate		C₈H₁₄O₅	628-68-2	190.194		18	200	1.1068[15]	1.4348[20]	vs EtOH
3438	Diethylene glycol dibenzoate		C₁₈H₁₈O₅	120-55-8	314.333		33.5	280[24]	1.1690[15]		vs H₂O, EtOH
3439	Diethylene glycol dibutyl ether	Bis(2-butoxyethyl) ether	C₁₂H₂₆O₃	112-73-2	218.332	liq	-60.2(0.2)	255(4)	0.885[25]	1.4235[20]	
3440	Diethylene glycol diethyl ether	Bis(2-ethoxyethyl) ether	C₈H₁₈O₃	112-36-7	162.227	liq	-44.3(0.2)	185(4)	0.9063[20]	1.4115[20]	vs H₂O, EtOH; s eth
3441	Diethylene glycol dimethacrylate	Oxydiethylene methacrylate	C₁₂H₁₈O₅	2358-84-1	242.268			>200	1.0821[20]	1.4571[25]	
3442	Diethylene glycol dimethyl ether	Diglyme	C₆H₁₄O₃	111-96-6	134.173	liq	-64.0(0.1)	162(2)	0.9434[20]	1.4097[20]	msc H₂O, EtOH, eth
3443	Diethylene glycol dinitrate	2,2'-Oxybisethanol, dinitrate	C₄H₈N₂O₇	693-21-0	196.116			44[0.01]			
3444	Diethylene glycol monobutyl ether		C₈H₁₈O₃	112-34-5	162.227	liq	-68	232(4)	0.9553[20]	1.4306[20]	msc H₂O; vs EtOH, eth, ace; s bz
3445	Diethylene glycol monobutyl ether acetate	2-(2-Butoxyethoxy)ethyl acetate	C₁₀H₂₀O₄	124-17-4	204.264	liq	-32	248(2)	0.985[20]	1.4262[20]	vs ace, eth, EtOH
3446	Diethylene glycol monododecanoate	2-(2-Hydroxyethoxy)ethyl laurate	C₁₆H₃₂O₄	141-20-8	288.423	lt ye	17.5	>270	0.96[25]		msc EtOH, eth, ace; s bz, tol
3447	Diethylene glycol monoethyl ether	Carbitol	C₆H₁₄O₃	111-90-0	134.173	hyg liq		202(3)	0.9885[20]	1.4300[20]	msc H₂O, EtOH, ace, bz; vs eth
3448	Diethylene glycol monoethyl ether acetate	Carbitol acetate	C₈H₁₆O₄	112-15-2	176.211	liq	-25	218(1)	1.0096[20]	1.4213[20]	vs H₂O, ace, eth, EtOH
3449	Diethylene glycol monohexyl ether	2-[2-(Hexyloxy)ethoxy]ethanol	C₁₀H₂₂O₃	112-59-4	190.280	col liq	-28	259(2)			
3450	Diethylene glycol monomethyl ether	2-(2-Methoxyethoxy)ethanol	C₅H₁₂O₃	111-77-3	120.147			194(2)	1.035[20]	1.4264[20]	msc H₂O, ace; vs EtOH, eth
3451	Diethylene glycol monopropyl ether		C₇H₁₆O₃	6881-94-3	148.200	liq	-53.3	214.8(0.4)			
3452	N,N-Diethyl-1,2-ethanediamine	N,N-Diethylethylenediamine	C₆H₁₆N₂	100-36-7	116.204			144	0.8280[20]	1.4340[20]	msc H₂O; s EtOH, eth, ctc, tol
3453	N,N'-Diethyl-1,2-ethanediamine		C₆H₁₆N₂	111-74-0	116.204			146	0.8280[20]	1.4340[20]	vs H₂O, eth, EtOH, tol
3454	Diethyl ether	Ethyl ether	C₄H₁₀O	60-29-7	74.121	liq	-116.22(0.04)	34.4(0.5)	0.7138[20]	1.3526[20]	sl H₂O; msc EtOH, bz, eth; vs ace
3455	Diethyl (ethoxymethylene)malonate	2-Ethoxy-1,1-bis(ethoxycarbonyl)ethene	C₁₀H₁₆O₅	87-13-8	216.231			280 dec		1.4600[20]	i H₂O; s EtOH, eth; sl chl
3456	Diethyl ethylidenemalonate		C₉H₁₄O₄	1462-12-0	186.205			116[17]	1.0404[20]	1.4308[17]	vs eth, EtOH
3457	Diethyl ethylmalonate		C₉H₁₆O₄	133-13-1	188.221			208	1.006[20]	1.4166[20]	sl H₂O; vs EtOH, eth, ace, chl
3458	Diethyl ethylphenylmalonate		C₁₅H₂₀O₄	76-67-5	264.318			170[19]	1.071[20]	1.4896[25]	i H₂O; s EtOH, eth; sl chl
3459	Diethyl ethylphosphonate		C₆H₁₅O₃P	78-38-6	166.155			198	1.0259[20]	1.4163[20]	sl H₂O; s EtOH, eth
3460	N,N-Diethylformamide		C₅H₁₁NO	617-84-5	101.147			175(10)	0.9080[19]	1.4321[25]	msc H₂O, ace, bz; vs EtOH, eth
3461	Diethyl fumarate		C₈H₁₂O₄	623-91-6	172.179		0.8	214(5)	1.0452[20]	1.4412[20]	i H₂O; s ace, chl
3462	Diethyl glutarate		C₉H₁₆O₄	818-38-2	188.221	syr liq	-24.1	237(3)	1.0220[20]	1.4241[20]	vs eth
3463	3,4-Diethylhexane		C₁₀H₂₂	19398-77-7	142.282			160(5)	0.7472[25]	1.4190[20]	
3464	Di-2-ethylhexyl maleate		C₂₀H₃₆O₄	142-16-5	340.498			156[7]	0.94[20]		

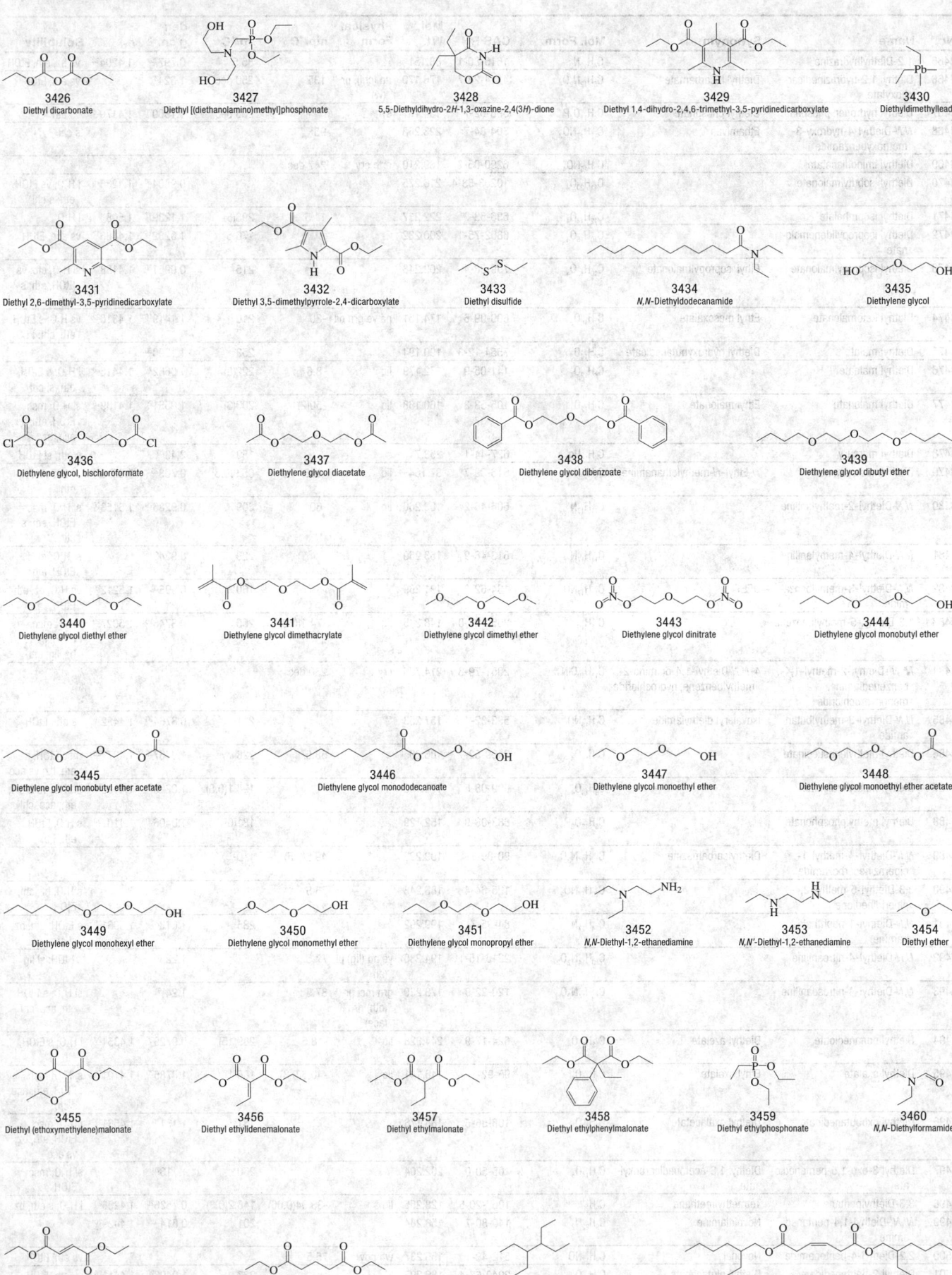

3426
Diethyl dicarbonate

3427
Diethyl [(diethanolamino)methyl]phosphonate

3428
5,5-Diethyldihydro-2H-1,3-oxazine-2,4(3H)-dione

3429
Diethyl 1,4-dihydro-2,4,6-trimethyl-3,5-pyridinedicarboxylate

3430
Diethyldimethyllead

3431
Diethyl 2,6-dimethyl-3,5-pyridinedicarboxylate

3432
Diethyl 3,5-dimethylpyrrole-2,4-dicarboxylate

3433
Diethyl disulfide

3434
N,N-Diethyldodecanamide

3435
Diethylene glycol

3436
Diethylene glycol, bischloroformate

3437
Diethylene glycol diacetate

3438
Diethylene glycol dibenzoate

3439
Diethylene glycol dibutyl ether

3440
Diethylene glycol diethyl ether

3441
Diethylene glycol dimethacrylate

3442
Diethylene glycol dimethyl ether

3443
Diethylene glycol dinitrate

3444
Diethylene glycol monobutyl ether

3445
Diethylene glycol monobutyl ether acetate

3446
Diethylene glycol monododecanoate

3447
Diethylene glycol monoethyl ether

3448
Diethylene glycol monoethyl ether acetate

3449
Diethylene glycol monohexyl ether

3450
Diethylene glycol monomethyl ether

3451
Diethylene glycol monopropyl ether

3452
N,N-Diethyl-1,2-ethanediamine

3453
N,N'-Diethyl-1,2-ethanediamine

3454
Diethyl ether

3455
Diethyl (ethoxymethylene)malonate

3456
Diethyl ethylidenemalonate

3457
Diethyl ethylmalonate

3458
Diethyl ethylphenylmalonate

3459
Diethyl ethylphosphonate

3460
N,N-Diethylformamide

3461
Diethyl fumarate

3462
Diethyl glutarate

3463
3,4-Diethylhexane

3464
Di-2-ethylhexyl maleate

No.	Name	Synonym	Mol. Form.	CAS RN	Mol. Wt.	Physical Form	mp/°C	bp/°C	den g cm⁻³	n_D	Solubility
3465	1,2-Diethylhydrazine		$C_4H_{12}N_2$	1615-80-1	88.151			85.5	0.797[26]	1.4204[20]	vs bz, eth, EtOH
3466	Diethyl 1,2-hydrazinedicar-boxylate	Diethyl bicarbamate	$C_6H_{12}N_2O_4$	4114-28-7	176.170	nd (chl), pr (w)	135	250 dec	1.324[8]		vs eth, EtOH
3467	Diethyl hydrogen phosphate	Diethyl phosphate	$C_4H_{11}O_4P$	598-02-7	154.101	syr		203 dec	1.1800[20]	1.4170[20]	vs eth
3468	N,N-Diethyl-4-hydroxy-3-methoxybenzamide	Ethamivan	$C_{12}H_{17}NO_3$	304-84-7	223.268		95				s chl
3469	Diethyl iminodiacetate		$C_8H_{15}NO_4$	6290-05-7	189.210	orth cry	247 dec				
3470	Diethyl isobutylmalonate		$C_{11}H_{20}O_4$	10203-58-4	216.275				0.9804[20]	1.4236[20]	i H_2O; vs EtOH, eth; s chl
3471	Diethyl isophthalate		$C_{12}H_{14}O_4$	636-53-3	222.237		11.5	298(6)	1.1239[17]	1.508[18]	i H_2O
3472	Diethyl isopropylidenemalo-nate		$C_{10}H_{16}O_4$	6802-75-1	200.232			176.5	1.0282[18]	1.4486[17]	vs ace, EtOH
3473	Diethyl isopropylmalonate	Ethyl isopropylmalonate	$C_{10}H_{18}O_4$	759-36-4	202.248			215	0.9961[20]	1.4188[21]	sl H_2O, ctc; vs EtOH, eth; s chl
3474	Diethyl ketomalonate	Ethyl mesoxalate	$C_7H_{10}O_5$	609-09-6	174.151	pa ye grn oil	-30	210	1.1419[16]	1.4310[22]	vs H_2O; s EtOH, eth, chl; i CS_2
3475	Diethyl malate	Diethyl hydroxybutanedioate	$C_8H_{14}O_5$	7554-12-3	190.194			253	1.1290[20]		
3476	Diethyl maleate		$C_8H_{12}O_4$	141-05-9	172.179	liq	-8.8	222(8)	1.0662[20]	1.4416[20]	i H_2O; s EtOH, eth; sl chl
3477	Diethyl malonate	Ethyl malonate	$C_7H_{12}O_4$	105-53-3	160.168	liq	-50(2)	200(3)	1.0551[20]	1.4139[20]	sl H_2O; msc EtOH, eth; vs ace, bz
3478	Diethyl mercury		$C_4H_{10}Hg$	627-44-1	258.71			159	2.43[20]		s eth; sl EtOH
3479	Diethylmethylamine	N-Ethyl-N-methylethanamine	$C_5H_{13}N$	616-39-7	87.164	liq	-196	65.9(0.3)	0.703[25]	1.3879[25]	vs H_2O, EtOH, eth
3480	N,N-Diethyl-2-methylaniline		$C_{11}H_{17}N$	606-46-2	163.260	liq	-60	209	0.9286[20]	1.5153[20]	sl H_2O; msc EtOH, eth; s ctc
3481	N,N-Diethyl-4-methylaniline		$C_{11}H_{17}N$	613-48-9	163.260			229	0.9242[16]		sl H_2O; msc EtOH, eth
3482	N,N-Diethyl-3-methylbenza-mide	DEET	$C_{12}H_{17}NO$	134-62-3	191.269			160[19]	0.996[20]	1.5212[20]	vs H_2O, bz, eth, EtOH
3483	1,3-Diethyl-5-methylbenzene		$C_{11}H_{16}$	2050-24-0	148.245	liq	-74.1(0.3)	205	0.8748[20]	1.5027[20]	i H_2O; msc EtOH, eth, ace, bz, lig, ctc
3484	N⁴,N⁴-Diethyl-2-methyl-1,4-benzenediamine, monohydrochloride	4-N,N-Diethyl-1,4-diamino-2-methylbenzene, hydrochloride	$C_{11}H_{19}ClN_2$	2051-79-8	214.735	cry	250 dec				
3485	N,N-Diethyl-3-methylbutan-amide	Isovaleryl diethylamide	$C_9H_{19}NO$	533-32-4	157.253			211	0.8764[20]	1.4422[20]	vs eth, EtOH
3486	Diethyl methylenesuccinate		$C_9H_{14}O_4$	2409-52-1	186.205		58.5	228	1.0467[20]	1.4377[20]	msc EtOH; s eth, bz; vs ace
3487	Diethyl methylmalonate		$C_8H_{14}O_4$	609-08-5	174.195			198.1(0.8)	1.0225[20]	1.4126[20]	sl H_2O; vs EtOH, eth, ace, chl
3488	Diethyl methylphosphonate		$C_5H_{13}O_3P$	683-08-9	152.129			181(6)	1.0406[30]	1.4101[30]	s H_2O, EtOH, eth; i bz
3489	N,N-Diethyl-4-methyl-1-piperazinecarboxamide	Diethylcarbamazine	$C_{10}H_{21}N_3O$	90-89-1	199.293		49.5(0.5)	110[3]			
3490	3,3-Diethyl-5-methyl-2,4-piperidinedione		$C_{10}H_{17}NO_2$	125-64-4	183.248		75.5				s H_2O, bz, chl, EtOH
3491	N,N-Diethyl-1-naphthale-namine		$C_{14}H_{17}N$	84-95-7	199.292			285	1.013[20]	1.5961[20]	s EtOH, eth, bz; sl ctc
3492	N,N-Diethyl-4-nitroaniline		$C_{10}H_{14}N_2O_2$	2216-15-1	194.230	ye nd (lig) pl (al)	77.5		1.225[25]		s EtOH; sl lig
3493	N,N-Diethyl-4-nitrosoaniline		$C_{10}H_{14}N_2O$	120-22-9	178.230	grn mcl pr (eth) grn lf (ace)	87.5		1.24[15]		sl H_2O; s EtOH, eth, ace, chl
3494	Diethyl nonanedioate	Diethyl azelate	$C_{13}H_{24}O_4$	624-17-9	244.328	liq	-18.5	289(15)	0.9729[20]	1.4351[20]	i H_2O; s EtOH, eth
3495	Diethyl oxalate	Ethyl oxalate	$C_6H_{10}O_4$	95-92-1	146.141	liq	-40.6(0.3)	186(1)	1.0785[20]	1.4101[20]	sl H_2O; msc EtOH, eth, ace; s ctc
3496	Diethyl oxobutanedioate	Diethyl oxalacetate	$C_8H_{12}O_5$	108-56-5	188.178			131[24]	1.131[20]	1.4561[17]	i H_2O; msc EtOH, eth, bz; vs ace
3497	Diethyl 3-oxo-1,5-pentanedio-ate	Diethyl 1,3-acetonedicarboxyl-ate	$C_9H_{14}O_5$	105-50-0	202.204			250	1.113[20]		sl H_2O; msc EtOH
3498	3,3-Diethylpentane	Tetraethylmethane	C_9H_{20}	1067-20-5	128.255	liq	-33.04(0.06)	146.2(0.3)	0.7536[20]	1.4206[20]	i H_2O; s eth, bz
3499	N',N'-Diethyl-1,4-pentanedi-amine	Novoldiamine	$C_9H_{22}N_2$	140-80-7	158.284			201	0.814[20]	1.4429[20]	
3500	2,2-Diethyl-4-pentenamide	Novonal	$C_9H_{17}NO$	512-48-1	155.237	wh pow	75.5				vs eth, EtOH
3501	Diethyl 2-pentenedioate	Diethyl glutaconate	$C_9H_{14}O_4$	2049-67-4	186.205			237	1.0496[20]	1.4411[20]	vs eth, EtOH

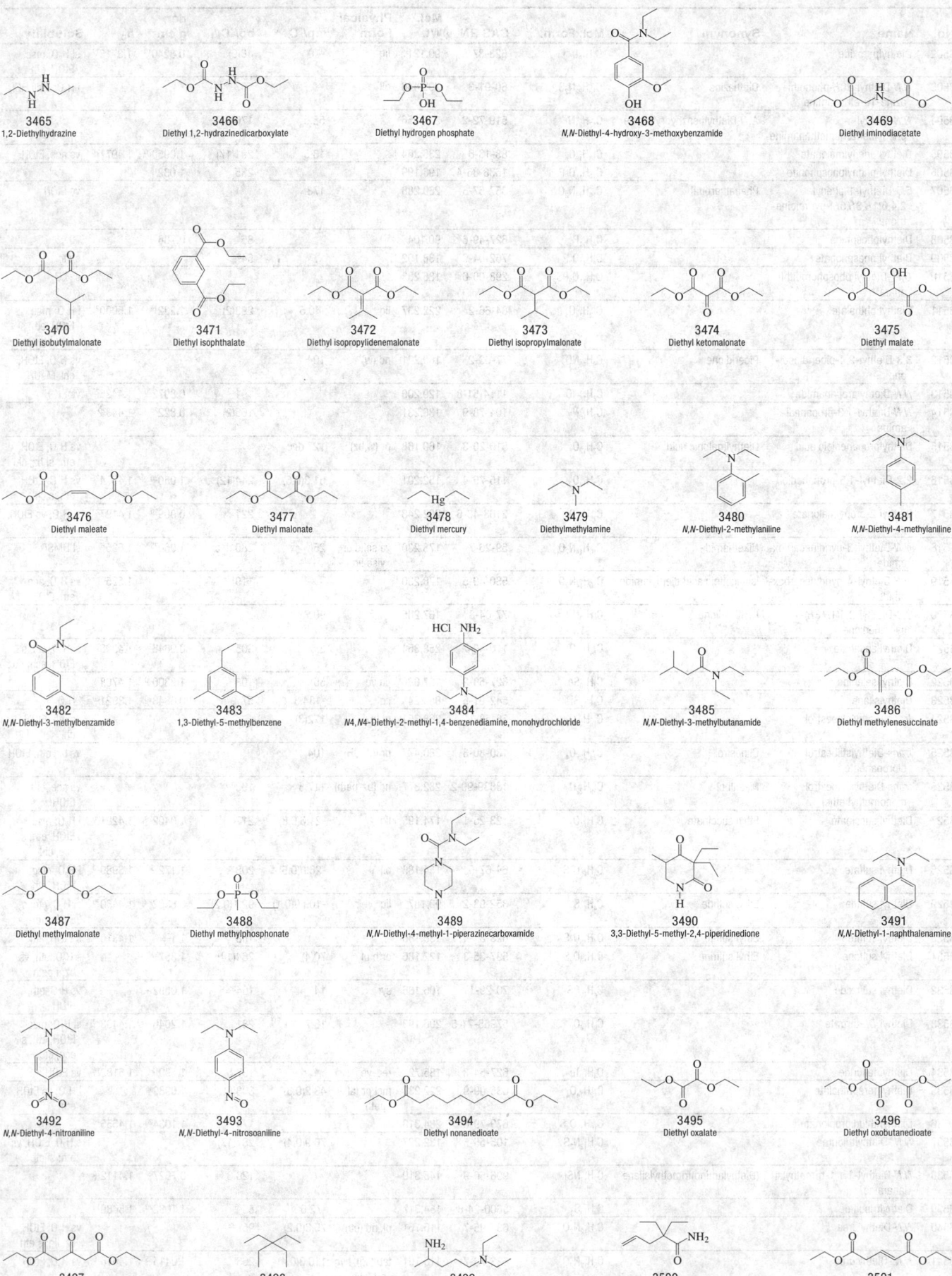

3465
1,2-Diethylhydrazine

3466
Diethyl 1,2-hydrazinedicarboxylate

3467
Diethyl hydrogen phosphate

3468
N,N-Diethyl-4-hydroxy-3-methoxybenzamide

3469
Diethyl iminodiacetate

3470
Diethyl isobutylmalonate

3471
Diethyl isophthalate

3472
Diethyl isopropylidenemalonate

3473
Diethyl isopropylmalonate

3474
Diethyl ketomalonate

3475
Diethyl malate

3476
Diethyl maleate

3477
Diethyl malonate

3478
Diethyl mercury

3479
Diethylmethylamine

3480
N,N-Diethyl-2-methylaniline

3481
N,N-Diethyl-4-methylaniline

3482
N,N-Diethyl-3-methylbenzamide

3483
1,3-Diethyl-5-methylbenzene

3484
*N*4,*N*4-Diethyl-2-methyl-1,4-benzenediamine, monohydrochloride

3485
N,N-Diethyl-3-methylbutanamide

3486
Diethyl methylenesuccinate

3487
Diethyl methylmalonate

3488
Diethyl methylphosphonate

3489
N,N-Diethyl-4-methyl-1-piperazinecarboxamide

3490
3,3-Diethyl-5-methyl-2,4-piperidinedione

3491
N,N-Diethyl-1-naphthalenamine

3492
N,N-Diethyl-4-nitroaniline

3493
N,N-Diethyl-4-nitrosoaniline

3494
Diethyl nonanedioate

3495
Diethyl oxalate

3496
Diethyl oxobutanedioate

3497
Diethyl 3-oxo-1,5-pentanedioate

3498
3,3-Diethylpentane

3499
*N*¹,*N*¹-Diethyl-1,4-pentanediamine

3500
2,2-Diethyl-4-pentenamide

3501
Diethyl 2-pentenedioate

No.	Name	Synonym	Mol. Form.	CAS RN	Mol. Wt.	Physical Form	mp/°C	bp/°C	den g cm⁻³	n_D	Solubility
3502	Diethylperoxide		$C_4H_{10}O_2$	628-37-5	90.121	liq	-70	46(9)	0.8240[19]	1.3715[17]	sl H_2O; msc EtOH, eth
3503	N,N-Diethyl-10H-phenothiazine-10-ethanamine	Diethazine	$C_{18}H_{22}N_2S$	60-91-3	298.446	oil		167[0.5]			i H_2O; s dil HCl
3504	N,N-Diethyl-α-phenylbenzenemethanamine	N,N-Diethylbenzhydrylamine	$C_{17}H_{21}N$	519-72-2	239.356		58.5	170[17]			
3505	Diethyl phenylmalonate		$C_{13}H_{16}O_4$	83-13-6	236.264		16.5	291(17)	1.0950[20]	1.4977[20]	vs ace, EtOH
3506	Diethyl phenylphosphonite		$C_{10}H_{15}O_2P$	1638-86-4	198.199			235	1.032[16]		
3507	5,5-Diethyl-1-phenyl-2,4,6(1H,3H,5H)-pyrimidinetrione	Phenetharbital	$C_{14}H_{16}N_2O_3$	357-67-5	260.288		178				vs EtOH
3508	Diethylphosphine		$C_4H_{11}P$	627-49-6	90.104			85	0.786[20]		
3509	Diethyl phosphonate		$C_4H_{11}O_3P$	762-04-9	138.102			54[6]			s ctc
3510	O,O'-Diethyl phosphorodithioate		$C_4H_{11}O_2PS_2$	298-06-6	186.233						s H_2O
3511	Diethyl phthalate		$C_{12}H_{14}O_4$	84-66-2	222.237	liq	-40.5	298(2)	1.232[14]	1.5000[21]	i H_2O; msc EtOH, eth; s ace, bz, ctc
3512	3,3-Diethyl-2,4-piperidinedione	Piperidione	$C_9H_{15}NO_2$	77-03-2	169.221	nd (w)	104				vs H_2O, EtOH, chl, MeOH
3513	N,N-Diethylpropanamide		$C_7H_{15}NO$	1114-51-8	129.200			191	0.8972[20]	1.4425[20]	vs EtOH
3514	N,N-Diethyl-1,3-propanediamine		$C_7H_{18}N_2$	104-78-9	130.231			165(2)	0.822[20]	1.443[20]	
3515	Diethylpropanedioic acid	Diethylmalonic acid	$C_7H_{12}O_4$	510-20-3	160.168	pr (w,bz)	127 dec				vs H_2O, EtOH, eth; sl bz, chl
3516	2,2-Diethyl-1,3-propanediol		$C_7H_{16}O_2$	115-76-4	132.201		61.3(0.5)	244(12)	1.050[20]	1.4574[25]	vs H_2O, EtOH, eth; s chl
3517	Diethyl 2-propylmalonate		$C_{10}H_{18}O_4$	2163-48-6	202.248			221	0.989[20]	1.4197[20]	sl H_2O; vs EtOH, eth
3518	N,N-Diethyl-3-pyridinecarboxamide	Nikethamide	$C_{10}H_{14}N_2O$	59-26-7	178.230	ye solid or visc liq	25	280 dec	1.060[25]	1.525[20]	sl DMSO
3519	N,N-Diethyl-4-pyridinecarboxamide	Isonicotinic acid diethylamide	$C_{10}H_{14}N_2O$	530-40-5	178.230			119[1]		1.525[20]	vs H_2O, ace, eth, EtOH
3520	3,3-Diethyl-2,4(1H,3H)-pyridinedione	Pyrithyldione	$C_9H_{13}NO_2$	77-04-3	167.205		90.7				
3521	Diethyl sebacate		$C_{14}H_{26}O_4$	110-40-7	258.354		2.5	305	0.9646[20]	1.4306[20]	sl H_2O, ctc; s EtOH, ace; i bz
3522	Diethyl selenide		$C_4H_{10}Se$	627-53-2	137.08	pa ye	55	108	1.2300[20]	1.4768[20]	
3523	Diethylsilane		$C_4H_{12}Si$	542-91-6	88.224	liq	-134.3	57	0.6843[20]	1.3921[20]	i H_2O
3524	trans-Diethylstilbestrol		$C_{18}H_{20}O_2$	56-53-1	268.351	pl (bz)	172(3)				vs eth, EtOH, chl
3525	trans-Diethylstilbestrol dipropanoate	Clinestrol	$C_{24}H_{28}O_4$	130-80-3	380.477	pr (MeOH)	104				vs bz, eth, EtOH
3526	trans-Diethylstilbestrol monomethyl ether	Mestilbol	$C_{19}H_{22}O_2$	18839-90-2	282.377	nd (bz-peth)	117.5	190[0.3]			vs ace, eth, EtOH
3527	Diethyl succinate	Ethyl succinate	$C_8H_{14}O_4$	123-25-1	174.195	liq	-21.6(0.8)	217(1)	1.0402[20]	1.4201[20]	i H_2O; msc EtOH, eth; s ace, chl
3528	Diethyl sulfate		$C_4H_{10}O_4S$	64-67-5	154.185	oil	-26.0(0.5)	208	1.172[25]	1.3989[20]	i H_2O; msc EtOH, eth
3529	Diethyl sulfide	Ethyl sulfide	$C_4H_{10}S$	352-93-2	90.187	liq	-103.9(0.1)	92.1(0.2)	0.8362[20]	1.4430[20]	sl H_2O, ctc; s EtOH, eth
3530	Diethyl sulfite	Ethyl sulfite	$C_4H_{10}O_3S$	623-81-4	138.185			158	1.1[20]	1.4310[20]	s EtOH, eth
3531	Diethyl sulfone	Ethyl sulfone	$C_4H_{10}O_2S$	597-35-3	122.186	orth pl	70(4)	264(14)	1.357[20]		s H_2O, eth; vs bz; i peth
3532	Diethyl sulfoxide		$C_4H_{10}OS$	70-29-1	106.186	syr	14	104[25]	1.0092[22]		vs H_2O, eth, EtOH
3533	Diethyl DL-tartrate		$C_8H_{14}O_6$	57968-71-5	206.193		18.7	281	1.2046[20]	1.4438[20]	sl H_2O; msc EtOH, eth; s ace, ctc
3534	Diethyl telluride		$C_4H_{10}Te$	627-54-3	185.72	red-ye		137.5	1.599[15]	1.5182[15]	vs EtOH
3535	Diethyl terephthalate		$C_{12}H_{14}O_4$	636-09-9	222.237	mcl pr (al, peth)	43.2(0.6)	303(8)	1.0989[45]		i H_2O; vs EtOH, eth
3536	Diethyl thiodipropionate		$C_{10}H_{18}O_4S$	673-79-0	234.313			174[15]	1.1034[20]	1.4655[20]	
3537	N,N'-Diethylthiourea		$C_5H_{12}N_2S$	105-55-5	132.227		76.9(0.4)	287(17)			s H_2O, EtOH; vs eth; sl ctc
3538	N,N-Diethyl-1,1,1-trimethylsilanamine	(Diethylamino)trimethylsilane	$C_7H_{19}NSi$	996-50-9	145.319			126.3	0.7627[20]	1.4112[20]	
3539	Diethyltrisulfide		$C_4H_{10}S_3$	3600-24-6	154.317		-72.6	85[26]	1.1082[20]	1.5689[13]	
3540	N,N-Diethylurea		$C_5H_{12}N_2O$	634-95-7	116.161	pl, nd (eth)	75.3(0.2)	95[0.02]			vs H_2O, EtOH, bz, lig; s eth
3541	N,N'-Diethylurea		$C_5H_{12}N_2O$	623-76-7	116.161	tab (lig), hyg nd (al)	110.3(0.4)	263	1.0415[25]	1.4616[40]	vs H_2O, EtOH, eth

3502
Diethylperoxide

3503
N,N-Diethyl-10H-phenothiazine-10-ethanamine

3504
N,N-Diethyl-α-phenylbenzenemethanamine

3505
Diethyl phenylmalonate

3506
Diethyl phenylphosphonite

3507
5,5-Diethyl-1-phenyl-2,4,6(1H,3H,5H)-pyrimidinetrione

3508
Diethylphosphine

3509
Diethyl phosphonate

3510
O,O'-Diethyl phosphorodithionate

3511
Diethyl phthalate

3512
3,3-Diethyl-2,4-piperidinedione

3513
N,N-Diethylpropanamide

3514
N,N-Diethyl-1,3-propanediamine

3515
Diethylpropanedioic acid

3516
2,2-Diethyl-1,3-propanediol

3517
Diethyl 2-propylmalonate

3518
N,N-Diethyl-3-pyridinecarboxamide

3519
N,N-Diethyl-4-pyridinecarboxamide

3520
3,3-Diethyl-2,4(1H,3H)-pyridinedione

3521
Diethyl sebacate

3522
Diethyl selenide

3523
Diethylsilane

3524
trans-Diethylstilbestrol

3525
trans-Diethylstilbestrol dipropanoate

3526
trans-Diethylstilbestrol monomethyl ether

3527
Diethyl succinate

3528
Diethyl sulfate

3529
Diethyl sulfide

3530
Diethyl sulfite

3531
Diethyl sulfone

3532
Diethyl sulfoxide

3533
Diethyl DL-tartrate

3534
Diethyl telluride

3535
Diethyl terephthalate

3536
Diethyl thiodipropionate

3537
N,N'-Diethylthiourea

3538
N,N-Diethyl-1,1,1-trimethylsilanamine

3539
Diethyltrisulfide

3540
N,N-Diethylurea

3541
N,N'-Diethylurea

No.	Name	Synonym	Mol. Form.	CAS RN	Mol. Wt.	Physical Form	mp/°C	bp/°C	den g cm⁻³	n_D	Solubility
3542	Diethyl vinylphosphonate		$C_6H_{13}O_3P$	682-30-4	164.139			110[2]	1.068[25]	1.4290[20]	
3543	Diethyl zinc	Zinc diethyl	$C_4H_{10}Zn$	557-20-0	123.531	col liq	-33.34(0.02)	133(6)	1.2065[20]	1.4936[20]	dec H₂O; msc eth, peth, bz
3544	Difenoconazole		$C_{19}H_{17}Cl_2N_3O_3$	119446-68-3	406.262		76	220[0.03]			
3545	Difenzoquat methyl sulfate	1H-Pyrazolium, 1,2-dimethyl-3,5-diphenyl-, methyl sulfate	$C_{18}H_{20}N_2O_4S$	43222-48-6	360.428		158.5(0.5)				
3546	Diflubenzuron	N-[[(4-Chlorophenyl)amino]-carbonyl]-2,6-difluorobenzamide	$C_{14}H_9ClF_2N_2O_2$	35367-38-5	310.683		228(1)				
3547	Difluoroacetic acid		$C_2H_2F_2O_2$	381-73-7	96.033	liq	-1	133	1.526[25]	1.3470[20]	
3548	2,4-Difluoroaniline		$C_6H_5F_2N$	367-25-9	129.108	liq	-7.5	170	1.268[25]	1.5063[20]	
3549	o-Difluorobenzene	1,2-Difluorobenzene	$C_6H_4F_2$	367-11-3	114.093	liq	-47.1(0.1)	93.9(0.5)	1.1599[18]	1.4451[18]	i H₂O; s ace, bz, chl
3550	m-Difluorobenzene	1,3-Difluorobenzene	$C_6H_4F_2$	372-18-9	114.093	liq	-69.11(0.01)	83.0(0.5)	1.1572[20]	1.4374[20]	i H₂O; s ace, bz
3551	p-Difluorobenzene	1,4-Difluorobenzene	$C_6H_4F_2$	540-36-3	114.093	liq	-23.5(0.2)	88.9(0.3)	1.1701[20]	1.4422[20]	i H₂O; s ace, bz; sl ctc
3552	4,4'-Difluoro-1,1'-biphenyl	4,4'-Difluorodiphenyl	$C_{12}H_8F_2$	398-23-2	190.189	mcl pr (al) lf (w)	90(2)	254.5			i H₂O; vs EtOH, bz, chl; s eth, ace
3553	1,1-Difluorocyclohexane		$C_6H_{10}F_2$	371-90-4	120.140	liq		99.5			
3554	3,3-Difluorocyclopropene		$C_3H_2F_2$	56830-75-2	76.045	liq		34			
3555	Difluorodimethylsilane		$C_2H_6F_2Si$	353-66-2	96.152	col gas	-87.5	2.5			
3556	1,5-Difluoro-2,4-dinitrobenzene		$C_6H_2F_2N_2O_4$	327-92-4	204.088		75.5	132[2]			sl EtOH
3557	Difluorodiphenylsilane		$C_{12}H_{10}F_2Si$	312-40-3	220.290			246	1.145[17]	1.5221[25]	
3558	1,1-Difluoroethane	Ethylidene difluoride	$C_2H_4F_2$	75-37-6	66.050	col gas	-118.6	-24.05	0.896[25] (p>1 atm)	1.3011[-72]	
3559	1,2-Difluoroethane	Ethylene difluoride	$C_2H_4F_2$	624-72-6	66.050	vol liq		26			vs bz, eth, chl
3560	1,1-Difluoroethene	Vinylidene fluoride	$C_2H_2F_2$	75-38-7	64.034	col gas	-144	-85.5(0.8)			vs eth, EtOH
3561	cis-1,2-Difluoroethene	cis-1,2-Difluoroethylene	$C_2H_2F_2$	1630-77-9	64.034	col gas		-45(19)			
3562	trans-1,2-Difluoroethene	trans-1,2-Difluoroethylene	$C_2H_2F_2$	1630-78-0	64.034	col gas		-53.1			
3563	Difluoromethane	Methylene fluoride	CH_2F_2	75-10-5	52.024	col gas	-136.8(0.2)	-51.65(0.07)	1.2139[-52]		i H₂O; s EtOH
3564	2-(Difluoromethoxy)-1,1,1-trifluoroethane	Difluoromethyl 2,2,2-trifluoroethyl ether	$C_3H_3F_5O$	1885-48-9	150.047	col liq		29.2(0.2)			
3565	Difluoromethylborane		CH_3BF_2	373-64-8	63.843	gas		-78.5[287]			reac H₂O
3566	2,4-Difluoro-1-nitrobenzene		$C_6H_3F_2NO_2$	446-35-5	159.091		9.8	207	1.4571[14]	1.5149[14]	sl chl
3567	2,2-Difluoropropane		$C_3H_6F_2$	420-45-1	80.077	col gas	-104.8	0(3)	0.9205[20] (p>1 atm)	1.2904[20]	
3568	1,3-Difluoro-2-propanol		$C_3H_6F_2O$	453-13-4	96.076			127	1.24[25]	1.3725[20]	
3569	Di-2-furanylethanedione		$C_{10}H_6O_4$	492-94-4	190.153	ye nd (al), cry (bz)	166.3				sl H₂O; s EtOH, eth, bz, chl
3570	Di-2-furanylethanedione dioxime	α-Furildioxime	$C_{10}H_8N_2O_4$	522-27-0	220.182		167				sl EtOH, eth, bz, lig
3571	1,5-Di-2-furanyl-1,4-pentadien-3-one		$C_{13}H_{10}O_3$	886-77-1	214.216	hyg pr (peth) ye pr (lig)	60.5	181[4]			vs eth, EtOH, chl
3572	Difurfuryl disulfide	Furfuryl disulfide	$C_{10}H_{10}O_2S_2$	4437-20-1	226.315		10	167[13]			vs EtOH
3573	Difurfuryl ether	Furfuryl ether	$C_{10}H_{10}O_3$	4437-22-3	178.184			101[2]	1.1405[20]	1.5088[20]	i H₂O
3574	Digitonin		$C_{56}H_{92}O_{29}$	11024-24-1	1229.312		237.5				
3575	Digitoxigenin		$C_{23}H_{34}O_4$	143-62-4	374.514		253				s EtOH; vs MeOH
3576	Digitoxin		$C_{41}H_{64}O_{13}$	71-63-6	764.939	pr (dil al)	255.5				sl H₂O; vs EtOH; s eth, chl, MeOH, py
3577	Digitoxose		$C_6H_{12}O_4$	527-52-6	148.157	cry (MeOH +eth)	112				vs H₂O, ace; s py, AcOEt
3578	Diglycidyl ether	Bis(2,3-epoxypropyl) ether	$C_6H_{10}O_3$	2238-07-5	130.141			260	1.1195[20]		
3579	Diglycolic acid	2,2'-Oxydiacetic acid	$C_4H_6O_5$	110-99-6	134.088	mcl pr (w + 1)	148	269(18)			vs H₂O, eth, EtOH
3580	Digoxigenin		$C_{23}H_{34}O_5$	1672-46-4	390.513	pr (AcOEt)	222				vs EtOH, MeOH; sl chl
3581	Digoxin		$C_{41}H_{64}O_{14}$	20830-75-5	780.939	trc pl (dil al, py)	249 dec				vs EtOH
3582	Diheptylamine	N-Heptyl-1-heptanamine	$C_{14}H_{31}N$	2470-68-0	213.403	nd	31.5	266(8)	0.7956[21]		sl H₂O; s EtOH; vs eth
3583	Diheptyl ether	Heptyl ether	$C_{14}H_{30}O$	629-64-1	214.387			258(4)	0.8008[20]	1.4275[20]	vs eth, EtOH
3584	Diheptyl phthalate		$C_{22}H_{34}O_4$	3648-21-3	362.503			360			
3585	Diheptyl sulfide	Heptyl sulfide	$C_{14}H_{30}S$	629-65-2	230.453		70	295(11)	0.8416[20]	1.4606[20]	i H₂O; s eth
3586	Dihexylamine	N-Hexyl-1-hexanamine	$C_{12}H_{27}N$	143-16-8	185.349	liq	-13.0(0.2)	236	0.7889[20]	1.4339[20]	s EtOH, eth

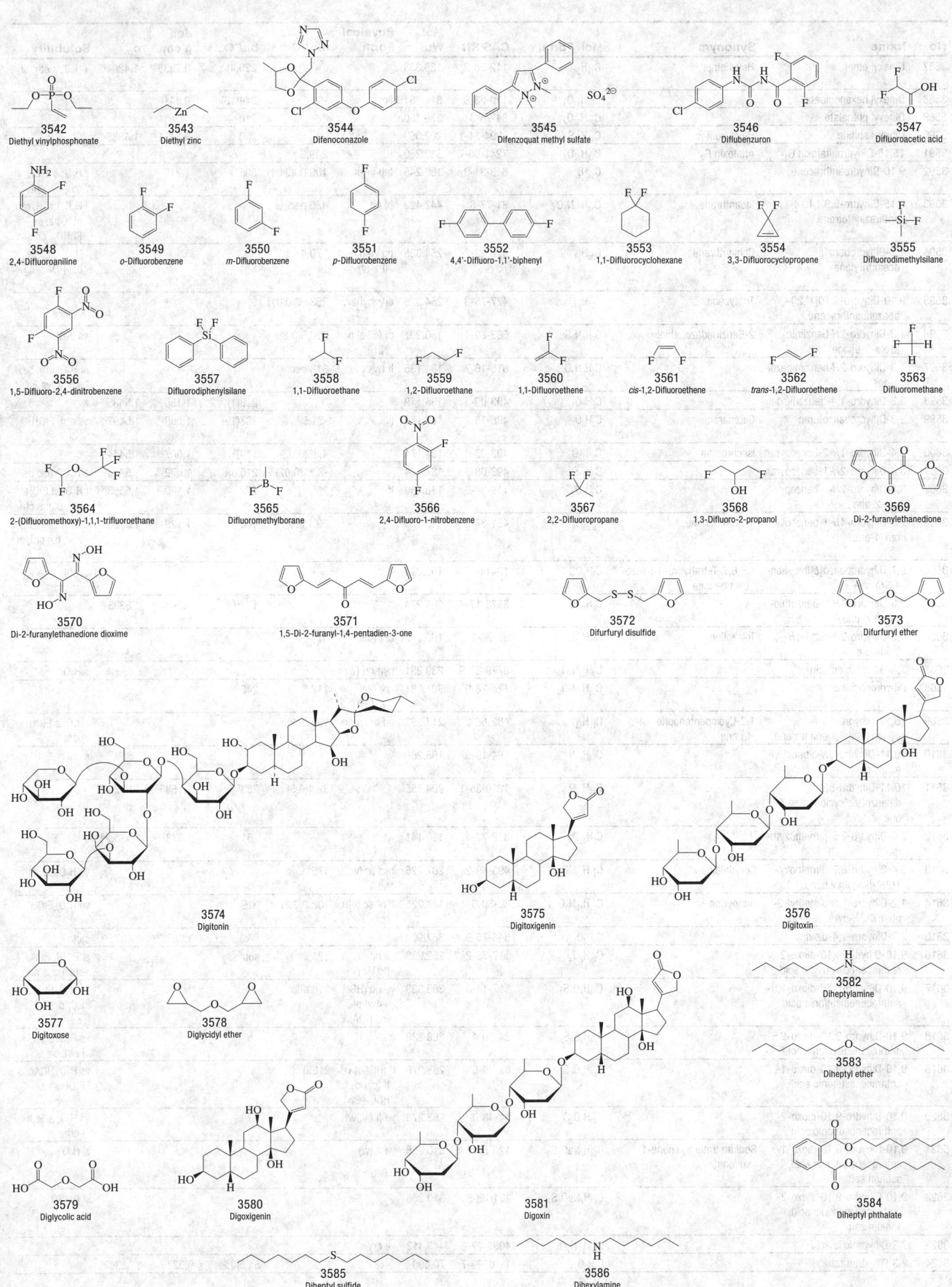

3542 Diethyl vinylphosphonate

3543 Diethyl zinc

3544 Difenoconazole

3545 Difenzoquat methyl sulfate

3546 Diflubenzuron

3547 Difluoroacetic acid

3548 2,4-Difluoroaniline

3549 *o*-Difluorobenzene

3550 *m*-Difluorobenzene

3551 *p*-Difluorobenzene

3552 4,4'-Difluoro-1,1'-biphenyl

3553 1,1-Difluorocyclohexane

3554 3,3-Difluorocyclopropene

3555 Difluorodimethylsilane

3556 1,5-Difluoro-2,4-dinitrobenzene

3557 Difluorodiphenylsilane

3558 1,1-Difluoroethane

3559 1,2-Difluoroethane

3560 1,1-Difluoroethene

3561 *cis*-1,2-Difluoroethene

3562 *trans*-1,2-Difluoroethene

3563 Difluoromethane

3564 2-(Difluoromethoxy)-1,1,1-trifluoroethane

3565 Difluoromethylborane

3566 2,4-Difluoro-1-nitrobenzene

3567 2,2-Difluoropropane

3568 1,3-Difluoro-2-propanol

3569 Di-2-furanylethanedione

3570 Di-2-furanylethanedione dioxime

3571 1,5-Di-2-furanyl-1,4-pentadien-3-one

3572 Difurfuryl disulfide

3573 Difurfuryl ether

3574 Digitonin

3575 Digitoxigenin

3576 Digitoxin

3577 Digitoxose

3578 Diglycidyl ether

3579 Diglycolic acid

3580 Digoxigenin

3581 Digoxin

3582 Diheptylamine

3583 Diheptyl ether

3584 Diheptyl phthalate

3585 Diheptyl sulfide

3586 Dihexylamine

No.	Name	Synonym	Mol. Form.	CAS RN	Mol. Wt.	Physical Form	mp/°C	bp/°C	den g cm⁻³	n_D	Solubility
3587	Dihexyl ether	Hexyl ether	C₁₂H₂₆O	112-58-3	186.333			220(4)	0.7936²⁰	1.4204²⁰	i H₂O; s eth; sl ctc
3588	Dihexyl hexanedioate		C₁₈H₃₄O₄	110-33-8	314.461	liq	-9	344(16)	0.941²⁰		
3589	Dihexyl phthalate		C₂₀H₃₀O₄	84-75-3	334.450			210⁵			
3590	Dihexyl sulfide	Hexyl sulfide	C₁₂H₂₆S	6294-31-1	202.399			230	0.8411²⁰	1.4586²⁰	
3591	15,16-Dihydroaflatoxin G₁	Aflatoxin G₂	C₁₇H₁₄O₇	7241-98-7	330.289		239.3				
3592	9,10-Dihydroanthracene		C₁₄H₁₂	613-31-0	180.245	tab or pr	109.00(0.01)	305	1.215²⁰		i H₂O; s EtOH, eth, bz, chl
3593	6,15-Dihydro-5,9,14,18-anthrazinetetrone	Indanthrene	C₂₈H₁₄N₂O₄	81-77-6	442.422	bl nd	485 dec				i H₂O, EtOH, eth, ace, bz; s PhNO₂, dil alk
3594	1,2-Dihydrobenz[j]-aceanthrylene	Cholanthrene	C₂₀H₁₄	479-23-2	254.325	pa ye lf (bz-al)	170.4				i H₂O; s EtOH, bz, HOAc, lig, tol
3595	9,10-Dihydro-9,10[1',2']-benzenoanthracene	Triptycene	C₂₀H₁₄	477-75-8	254.325	cry (cyhex)	253.99(0.01)				
3596	1,3-Dihydro-2H-benzimid-azole-2-thione	2-Benzimidazolethiol	C₇H₆N₂S	583-39-1	150.201	pl (dil al or NH₃)	316.3(0.8)				vs EtOH
3597	1,3-Dihydro-2H-benzimidazol-2-one		C₇H₆N₂O	615-16-7	134.135	lf (w or al)	318 dec				sl H₂O, eth, bz; s ace; vs EtOH
3598	2,3-Dihydro-1,4-benzodioxin		C₈H₈O₂	493-09-4	136.149			213(1)	1.180²⁰	1.5485²⁰	
3599	2,3-Dihydrobenzofuran	Coumaran	C₈H₈O	496-16-2	120.149	liq	-21.5	187(19)	1.058²⁵	1.5497²⁰	vs eth, EtOH, chl
3600	3,4-Dihydro-1H-2-benzopyran	Isochroman	C₉H₁₀O	493-05-0	134.174		4.35(0.02)	110²⁵	1.067²⁵	1.5444²⁰	
3601	3,4-Dihydro-2H-1-benzopyran		C₉H₁₀O	493-08-3	134.174		-3.31(0.02)	215.6(0.8)	1.072²⁰	1.5444²⁰	s H₂O; msc os
3602	3,4-Dihydro-2H-1-benzopyran-2-one		C₉H₈O₂	119-84-6	148.159	lf	25	272	1.169¹⁸	1.5563²⁰	i H₂O; sl EtOH, eth, ctc; s chl
3603	2,3-Dihydro-4H-1-benzopyran-4-one	4-Chromanone	C₉H₈O₂	491-37-2	148.159		39.1(0.4)	160⁵⁰	1.1291¹⁰⁰	1.5750	s EtOH; vs eth, ace, bz, chl; sl ctc
3604	6,7-Dihydrobenzo[b]thiophen-4(5H)-one	4,5,6,7-Tetrahydro-4-benzo-thiophenone	C₈H₈OS	13414-95-4	152.214						sl chl
3605	2,3-Dihydro-4H-1-benzothio-pyran-4-one		C₉H₈OS	3528-17-4	164.224		29	154¹²	1.2487¹⁴	1.6395²⁰	
3606	4,5-Dihydro-2-benzyl-1H-imidazole	Tolazoline	C₁₀H₁₂N₂	59-98-3	160.215	cry (peth)	67				
3607	7,8-Dihydrobiopterin		C₉H₁₃N₅O₃	6779-87-9	239.231	hyg nd (w)					s H₂O
3608	Dihydrocodeine		C₁₈H₂₃NO₃	125-28-0	301.381	cry (aq, MeOH)	112.5	248¹⁵			
3609	16,17-Dihydro-15H-cyclopenta[a]phenanthrene	1,2-Cyclopentenophenan-threne	C₁₇H₁₄	482-66-6	218.293	nd (al, petr)	135.5				i H₂O; s EtOH, peth
3610	10,11-Dihydro-5H-dibenz[b,f]-azepine		C₁₄H₁₃N	494-19-9	195.260						s chl
3611	10,11-Dihydro-5H-dibenzo[a,d]cyclohepten-5-one		C₁₅H₁₂O	1210-35-1	208.255		32.4(0.5)	203⁷	1.1635²⁰	1.6324²⁰	
3612	2,5-Dihydro-2,5-dimethoxyfu-ran		C₆H₁₀O₃	332-77-4	130.141			161	1.073²⁵	1.4339²⁰	
3613	3,4-Dihydro-6,7-dimethoxy-1(2H)-isoquinolinone	Corydaldine	C₁₁H₁₃NO₃	493-49-2	207.226	mcl pr (w, al)	175				vs H₂O, bz, eth, EtOH
3614	1,2-Dihydro-1,5-dimethyl-2-phenyl-3H-pyrazol-3-one	Antipyrine	C₁₁H₁₂N₂O	60-80-0	188.225	lf or sc (eth, bz)	108.0(0.2)	319			vs H₂O, EtOH
3615	2,3-Dihydro-1,4-dioxin		C₄H₆O₂	543-75-9	86.090			94.1	1.0836²⁰	1.4372²⁰	s ctc
3616	9,10-Dihydro-9,10-dioxo-2-anthracenecarboxylic acid		C₁₅H₈O₄	117-78-2	252.223	ye nd (HOAc)	291	sub			sl EtOH, HOAc; i eth, bz; s ace
3617	9,10-Dihydro-9,10-dioxo-1,5-anthracenedisulfonic acid		C₁₄H₈O₈S₂	117-14-6	368.339	ye nd (HCl +4w) pl (dil HOAc)	310 dec				vs H₂O, EtOH, HOAc
3618	9,10-Dihydro-9,10-dioxo-2,6-anthracenedisulfonic acid		C₁₄H₈O₈S₂	84-50-4	368.339						vs H₂O; s EtOH; i eth, bz
3619	9,10-Dihydro-9,10-dioxo-1-anthracenesulfonic acid		C₁₄H₈O₅S	82-49-5	288.276	lf (HOAc) ye lf (conc HCl, +3w)	216.0				vs H₂O, HOAc; s EtOH
3620	9,10-Dihydro-9,10-dioxo-2-anthracenesulfonic acid		C₁₄H₈O₅S	84-48-0	288.276	ye lf (+3w)					vs H₂O; s EtOH; i eth
3621	9,10-Dihydro-9,10-dioxo-1-anthracenesulfonic acid, sodium salt	Sodium anthraquinone-1-sulfonate	C₁₄H₇NaO₅S	128-56-3	310.258	ye lf (w)					sl H₂O
3622	9,10-Dihydro-9,10-dioxo-2-anthracenesulfonic acid, sodium salt		C₁₄H₇NaO₅S	131-08-8	310.258						sl DMSO
3623	7,8-Dihydrofolic acid		C₁₉H₂₁N₇O₆	4033-27-6	443.413	ye cry					
3624	2,3-Dihydrofuran		C₄H₆O	1191-99-7	70.090			54.5(0.2)	0.927²⁵	1.4239²⁰	

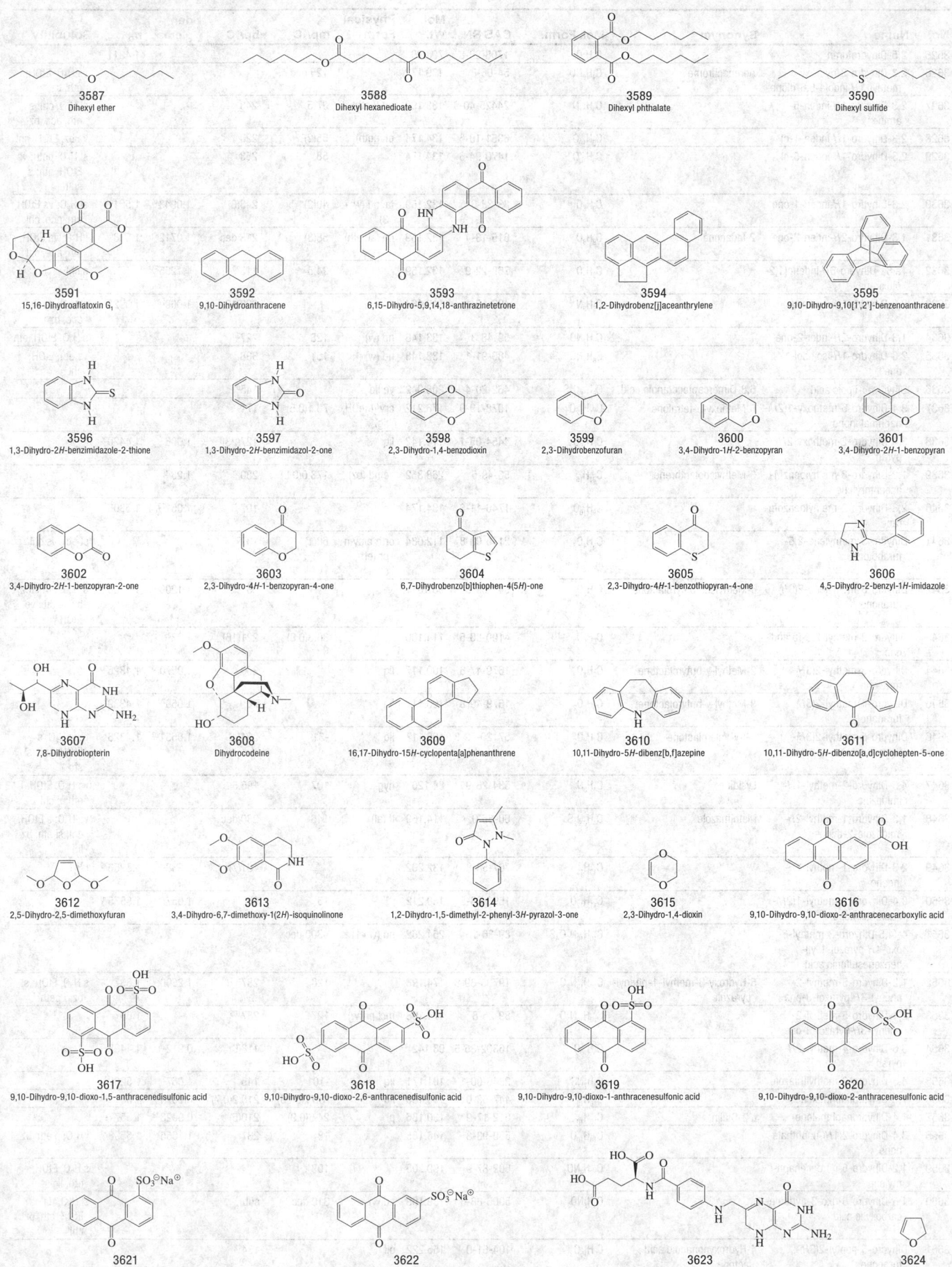

3587
Dihexyl ether

3588
Dihexyl hexanedioate

3589
Dihexyl phthalate

3590
Dihexyl sulfide

3591
15,16-Dihydroaflatoxin G₁

3592
9,10-Dihydroanthracene

3593
6,15-Dihydro-5,9,14,18-anthrazinetetrone

3594
1,2-Dihydrobenz[j]aceanthrylene

3595
9,10-Dihydro-9,10[1',2']-benzenoanthracene

3596
1,3-Dihydro-2H-benzimidazole-2-thione

3597
1,3-Dihydro-2H-benzimidazol-2-one

3598
2,3-Dihydro-1,4-benzodioxin

3599
2,3-Dihydrobenzofuran

3600
3,4-Dihydro-1H-2-benzopyran

3601
3,4-Dihydro-2H-1-benzopyran

3602
3,4-Dihydro-2H-1-benzopyran-2-one

3603
2,3-Dihydro-4H-1-benzopyran-4-one

3604
6,7-Dihydrobenzo[b]thiophen-4(5H)-one

3605
2,3-Dihydro-4H-1-benzothiopyran-4-one

3606
4,5-Dihydro-2-benzyl-1H-imidazole

3607
7,8-Dihydrobiopterin

3608
Dihydrocodeine

3609
16,17-Dihydro-15H-cyclopenta[a]phenanthrene

3610
10,11-Dihydro-5H-dibenz[b,f]azepine

3611
10,11-Dihydro-5H-dibenzo[a,d]cyclohepten-5-one

3612
2,5-Dihydro-2,5-dimethoxyfuran

3613
3,4-Dihydro-6,7-dimethoxy-1(2H)-isoquinolinone

3614
1,2-Dihydro-1,5-dimethyl-2-phenyl-3H-pyrazol-3-one

3615
2,3-Dihydro-1,4-dioxin

3616
9,10-Dihydro-9,10-dioxo-2-anthracenecarboxylic acid

3617
9,10-Dihydro-9,10-dioxo-1,5-anthracenedisulfonic acid

3618
9,10-Dihydro-9,10-dioxo-2,6-anthracenedisulfonic acid

3619
9,10-Dihydro-9,10-dioxo-1-anthracenesulfonic acid

3620
9,10-Dihydro-9,10-dioxo-2-anthracenesulfonic acid

3621
9,10-Dihydro-9,10-dioxo-1-anthracenesulfonic acid, sodium salt

3622
9,10-Dihydro-9,10-dioxo-2-anthracenesulfonic acid, sodium salt

3623
7,8-Dihydrofolic acid

3624
2,3-Dihydrofuran

No.	Name	Synonym	Mol. Form.	CAS RN	Mol. Wt.	Physical Form	mp/°C	bp/°C	den g cm⁻³	n_D	Solubility
3625	2,5-Dihydrofuran		C_4H_6O	1708-29-8	70.090					1.4311^{20}	
3626	2,3-Dihydro-3-hydroxy-1-methyl-1H-indole-5,6-dione	Adrenochrome	$C_9H_9NO_3$	54-06-8	179.172		125 dec				vs H_2O, EtOH; i eth, bz
3627	2,3-Dihydro-1H-inden-5-amine		$C_9H_{11}N$	24425-40-9	133.190	nd (peth)	37.5	248			sl H_2O, chl; s eth, ace, bz
3628	2,3-Dihydro-1H-inden-1-ol		$C_9H_{10}O$	6351-10-6	134.174	pl (peth)	55(2)	220			vs bz, EtOH, chl
3629	2,3-Dihydro-1H-inden-5-ol		$C_9H_{10}O$	1470-94-6	134.174		58	253			sl H_2O, peth; vs EtOH, eth; s sulf
3630	2,3-Dihydro-1H-inden-1-one		C_9H_8O	83-33-0	132.159	ta, nd (w + 3)	40(2)	243(8)	1.0943^{40}	1.561^{25}	sl H_2O; vs EtOH, eth, ace, chl
3631	1,3-Dihydro-2H-inden-2-one	2-Indanone	C_9H_8O	615-13-4	132.159	nd (al, eth)	58(3)	218 dec	1.0712^{69}	1.538^{67}	i H_2O; vs EtOH, eth, ace, chl
3632	1a,6a-Dihydro-6H-indeno[1,2-b]oxirene		C_9H_8O	768-22-9	132.159		24.5	113^{20}	1.1255^{24}		s chl
3633	2,3-Dihydro-1H-indole		C_8H_9N	496-15-1	119.164			230(13)	1.069^{20}	1.5923^{20}	sl H_2O; s eth, ace, bz
3634	1,3-Dihydro-2H-indol-2-one		C_8H_7NO	59-48-3	133.148	nd (w)	128	227^{23}			s H_2O, EtOH, eth
3635	2,3-Dihydro-1H-isoindol-1-one		C_8H_7NO	480-91-1	133.148	nd (w)	151	338			vs eth, EtOH, chl
3636	Dihydro-α-lipoic acid	6,8-Dimercaptooctanoic acid	$C_8H_{16}O_2S_2$	462-20-4	208.342	ye liq		$145^{0.2}$			
3637	3,4-Dihydro-6-methoxy-1(2H)-naphthalenone	6-Methoxy-α-tetralone	$C_{11}H_{12}O_2$	1078-19-9	176.212	cry (MeOH, lig)	78.1(0.5)	171^{11}			
3638	3,4-Dihydro-2-methoxy-2H-pyran		$C_6H_{10}O_2$	4454-05-1	114.142	liq		127(23)	1.006	1.4420^{20}	
3639	1,2-Dihydro-3-methylbenz[j]-aceanthrylene	3-Methylcholanthrene	$C_{21}H_{16}$	56-49-5	268.352	ye nd (bz)	178.0(0.2)	280^{80}	1.28^{20}		i H_2O
3640	2,3-Dihydro-2-methylbenzofuran		$C_9H_{10}O$	1746-11-8	134.174			197.5	1.061^{25}	1.5308	
3641	Dihydro-3-methylene-2,5-furandione		$C_5H_4O_3$	2170-03-8	112.084	orth bipym pr (eth, chl)	68(1)	139^{30}			sl eth; vs chl
3642	Dihydro-3-methylene-2(3H)-furanone	α-Methylene butyrolactone	$C_5H_6O_2$	547-65-9	98.101			85^{10}	1.1206^{20}	1.4650^{20}	s H_2O, eth, ace, bz; sl ctc; vs EtOH
3643	Dihydro-3-methyl-2,5-furandione		$C_5H_6O_3$	4100-80-5	114.100		36.8(0.8)	221(16)	1.22^{25}		
3644	Dihydro-3-methyl-2(3H)-furanone	2-Methyl-γ-butyrolactone	$C_5H_8O_2$	1679-47-6	100.117	liq		200	1.0570^{20}	1.4325^{20}	
3645	Dihydro-4-methyl-2(3H)-furanone	3-Methyl-γ-butyrolactone	$C_5H_8O_2$	1679-49-8	100.117	liq		76^{11}	1.058^{20}	1.4339^{20}	
3646	Dihydro-5-methyl-2(3H)-furanone, (±)-	(±)-γ-Valerolactone	$C_5H_8O_2$	57129-69-8	100.117	liq	-31	204(4)	1.0551^{20}	1.4328^{20}	msc H_2O; s EtOH, ace; sl ctc
3647	4,5-Dihydro-2-methyl-1H-imidazole	Lysidine	$C_4H_8N_2$	534-26-9	84.120	hyg	107	196.5			vs H_2O, EtOH; i eth; s chl
3648	1,3-Dihydro-1-methyl-2H-imidazole-2-thione	Methimazole	$C_4H_6N_2S$	60-56-0	114.169	lf (al)	146	280 dec			vs H_2O; s EtOH, chl; sl eth, bz, lig
3649	2,3-Dihydro-1-methyl-1H-indene		$C_{10}H_{12}$	767-58-8	132.202			191(7)	0.938^{25}	1.5266^{20}	i H_2O
3650	3,4-Dihydro-2-methyl-1(2H)-naphthalenone		$C_{11}H_{12}O$	1590-08-5	160.212		15	136^{16}	1.057^{25}	1.5535^{20}	
3651	4-(4,5-Dihydro-3-methyl-5-oxo-1H-pyrazol-1-yl)-benzenesulfonic acid		$C_{10}H_{10}N_2O_4S$	89-36-1	254.262	nd (w+1)	≈300 dec				
3652	1,2-Dihydro-5-methyl-2-phenyl-3H-pyrazol-3-one	5-Hydroxy-3-methyl-1-phenyl-pyrazole	$C_{10}H_{10}N_2O$	19735-89-8	174.198		128	287^{105}	1.2600^{20}	1.637	s H_2O, EtOH; sl bz; i peth
3653	2,4-Dihydro-5-methyl-2-phenyl-3H-pyrazol-3-one		$C_{10}H_{10}N_2O$	89-25-8	174.198	mcl pr (w)	127	287^{105}		1.637	
3654	3,6-Dihydro-4-methyl-2H-pyran		$C_6H_{10}O$	16302-35-5	98.142			118(6)	0.912^{25}	1.4495^{20}	
3655	4,5-Dihydro-2-methylthiazole		C_4H_7NS	2346-00-1	101.171	liq	-101	145	1.067^{25}	1.5200^{20}	
3656	1,2-Dihydronaphthalene		$C_{10}H_{10}$	447-53-0	130.186	liq	-8.77(0.05)	210.2(0.7)	0.9974^{20}	1.5814^{20}	
3657	1,4-Dihydronaphthalene	Δ 2-Dialin	$C_{10}H_{10}$	612-17-9	130.186	pl	24.6(0.9)	210(5)	0.9928^{33}	1.5577^{20}	
3658	3,4-Dihydro-2(1H)-naphthalenone		$C_{10}H_{10}O$	530-93-8	146.185		18	237	1.1055^{27}	1.5598^{20}	i H_2O; s eth, bz
3659	1,2-Dihydro-5-nitroacenaphthylene		$C_{12}H_9NO_2$	602-87-9	199.205		103				s H_2O, EtOH, eth, lig
3660	1,6-Dihydro-6-oxo-3-pyridine-carboxylic acid		$C_6H_5NO_3$	5006-66-6	139.109	nd(w)	310 dec	sub			sl H_2O, tfa; i EtOH, eth, bz, chl
3661	Dihydro-5-pentyl-2(3H)-furanone	4-Hydroxynonanoic acid lactone	$C_9H_{16}O_2$	104-61-0	156.222	oil		134^{12}			

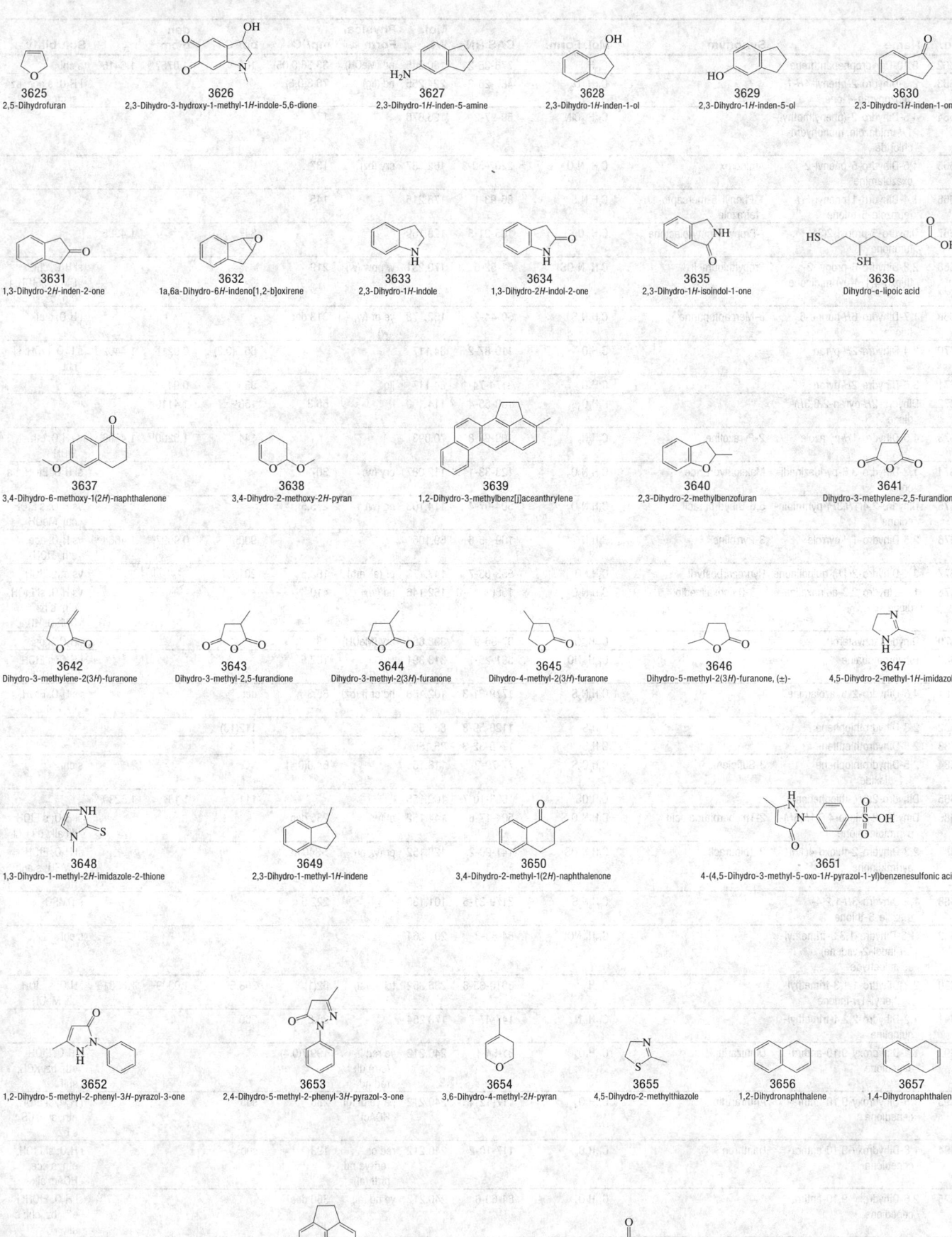

3625
2,5-Dihydrofuran

3626
2,3-Dihydro-3-hydroxy-1-methyl-1H-indole-5,6-dione

3627
2,3-Dihydro-1H-inden-5-amine

3628
2,3-Dihydro-1H-inden-1-ol

3629
2,3-Dihydro-1H-inden-5-ol

3630
2,3-Dihydro-1H-inden-1-one

3631
1,3-Dihydro-2H-inden-2-one

3632
1a,6a-Dihydro-6H-indeno[1,2-b]oxirene

3633
2,3-Dihydro-1H-indole

3634
1,3-Dihydro-2H-indol-2-one

3635
2,3-Dihydro-1H-isoindol-1-one

3636
Dihydro-α-lipoic acid

3637
3,4-Dihydro-6-methoxy-1(2H)-naphthalenone

3638
3,4-Dihydro-2-methoxy-2H-pyran

3639
1,2-Dihydro-3-methylbenz[j]aceanthrylene

3640
2,3-Dihydro-2-methylbenzofuran

3641
Dihydro-3-methylene-2,5-furandione

3642
Dihydro-3-methylene-2(3H)-furanone

3643
Dihydro-3-methyl-2,5-furandione

3644
Dihydro-3-methyl-2(3H)-furanone

3645
Dihydro-4-methyl-2(3H)-furanone

3646
Dihydro-5-methyl-2(3H)-furanone, (±)-

3647
4,5-Dihydro-2-methyl-1H-imidazole

3648
1,3-Dihydro-1-methyl-2H-imidazole-2-thione

3649
2,3-Dihydro-1-methyl-1H-indene

3650
3,4-Dihydro-2-methyl-1(2H)-naphthalenone

3651
4-(4,5-Dihydro-3-methyl-5-oxo-1H-pyrazol-1-yl)benzenesulfonic acid

3652
1,2-Dihydro-5-methyl-2-phenyl-3H-pyrazol-3-one

3653
2,4-Dihydro-5-methyl-2-phenyl-3H-pyrazol-3-one

3654
3,6-Dihydro-4-methyl-2H-pyran

3655
4,5-Dihydro-2-methylthiazole

3656
1,2-Dihydronaphthalene

3657
1,4-Dihydronaphthalene

3658
3,4-Dihydro-2(1H)-naphthalenone

3659
1,2-Dihydro-5-nitroacenaphthylene

3660
1,6-Dihydro-6-oxo-3-pyridinecarboxylic acid

3661
Dihydro-5-pentyl-2(3H)-furanone

No.	Name	Synonym	Mol. Form.	CAS RN	Mol. Wt.	Physical Form	mp/°C	bp/°C	den g cm⁻³	n_D	Solubility
3662	9,10-Dihydrophenanthrene		C₁₄H₁₂	776-35-2	180.245	nd (MeOH)	33.36(0.05)	168¹⁵	1.0757⁴⁰	1.6415²⁰	s chl
3663	2,3-Dihydro-2-phenyl-4H-1-benzopyran-4-one		C₁₅H₁₂O₂	487-26-3	224.255	nd (lig)	76.3(0.5)				i H₂O; s ace, bz; sl ctc
3664	4,5-Dihydro-2-(phenylmethyl)-1H-imidazole, monohydrochloride		C₁₀H₁₃ClN₂	59-97-2	196.676		174				
3665	4,5-Dihydro-5-phenyl-2-oxazolamine	Aminorex	C₉H₁₀N₂O	2207-50-3	162.187	cry (bz)	137				
3666	1,4-Dihydro-1-phenyl-5H-tetrazole-5-thione	1-Phenyl-5-mercapto-1H-tetrazole	C₇H₆N₄S	86-93-1	178.215		145				
3667	Dihydro-5-propyl-2(3H)-furanone	γ-Propyl-γ-butyrolactone	C₇H₁₂O₂	105-21-5	128.169			84⁵		1.4385²⁵	
3668	2,3-Dihydro-6-propyl-2-thioxo-4(1H)-pyrimidinone	Propylthiouracil	C₇H₁₀N₂OS	51-52-5	170.231	w pow (w)	219				sl H₂O, chl, DMSO, EtOH; i eth, bz
3669	1,7-Dihydro-6H-purine-6-thione	6-Mercaptopurine	C₅H₄N₄S	50-44-2	152.178	ye pr (w, + l w)	313 dec				i H₂O; s alk
3670	3,4-Dihydro-2H-pyran		C₅H₈O	110-87-2	84.117			85.5(0.2)	0.921¹⁹	1.4402¹⁹	s H₂O, EtOH; sl chl
3671	3,6-Dihydro-2H-pyran		C₅H₈O	3174-74-1	84.117	liq		95	0.94¹⁹		
3672	Dihydro-2H-pyran-2,6(3H)-dione		C₅H₆O₃	108-55-4	114.100		56.3	158¹⁵	1.4110²⁰		
3673	4,5-Dihydro-1H-pyrazole	2-Pyrazoline	C₃H₆N₂	109-98-8	70.093			144	1.0200¹⁷	1.4796¹⁷	vs H₂O, eth, EtOH
3674	1,2-Dihydro-3,6-pyridazinedione	Maleic hydrazide	C₄H₄N₂O₂	123-33-1	112.087	cry (w)	307				sl H₂O, EtOH, tfa
3675	Dihydro-2,4(1H,3H)-pyrimidinedione	5,6-Dihydrouracil	C₄H₆N₂O₂	504-07-4	114.103	nd (w)	275.5				vs H₂O; s EtOH, chl, MeOH
3676	2,5-Dihydro-1H-pyrrole	3-Pyrroline	C₄H₇N	109-96-6	69.106			90(5)	0.9097²⁰	1.4664²⁰	vs H₂O, ace, eth, EtOH
3677	3,4-Dihydro-2(1H)-quinolinone	Hydrocarbostyril	C₉H₉NO	553-03-7	147.173	pr (al, eth)	163.5	201⁴⁵			vs eth, EtOH
3678	1,4-Dihydro-2,3-quinoxalinedione	2,3-Quinoxalinediol	C₈H₆N₂O₂	15804-19-0	162.146	nd (w)	410				vs H₂O; sl EtOH, eth; s bz, DMSO, HOAc
3679	Dihydrotachysterol		C₂₈H₄₆O	67-96-9	398.664	cry (MeOH)	131				i H₂O; s os
3680	Dihydrothebaine		C₁₉H₂₃NO₃	561-25-1	313.391			162.5			i H₂O; s EtOH, bz, AcOEt
3681	4,5-Dihydro-2-thiazolamine		C₃H₆N₂S	1779-81-3	102.158	nd or lf (bz)	85.3	dec			vs H₂O, EtOH, bz, chl
3682	2,3-Dihydrothiophene		C₄H₆S	1120-59-8	86.156			112(13)			
3683	2,5-Dihydrothiophene		C₄H₆S	1708-32-3	86.156			122.4			
3684	2,5-Dihydrothiophene 1,1-dioxide	3-Sulfolene	C₄H₆O₂S	77-79-2	118.155		64.0(0.4)				s chl
3685	Dihydro-2(3H)-thiophenone		C₄H₆OS	1003-10-7	102.155			111⁵²	1.18²⁵	1.5230²⁰	
3686	Dihydro-2-thioxo-4,6(1H,5H)-pyrimidinedione	2-Thiobarbituric acid	C₄H₄N₂O₂S	504-17-6	144.152	pl (w)	235 dec				sl H₂O; s EtOH, dil alk, dil HCl
3687	2,3-Dihydro-2-thioxo-4(1H)-pyrimidinone	2-Thiouracil	C₄H₄N₂OS	141-90-2	128.152	pr (w, al)	>340 dec				sl H₂O, EtOH, DMSO; s anh HF
3688	1,2-Dihydro-3H-1,2,4-triazole-3-thione		C₂H₃N₃S	3179-31-5	101.130		222.5				s DMSO
3689	(1,3-Dihydro-1,3,3-trimethyl-2H-indol-2-ylidene)acetaldehyde		C₁₃H₁₅NO	84-83-3	201.264						s chl
3690	2,3-Dihydro-1,1,3-trimethyl-3-phenyl-1H-indene		C₁₈H₂₀	3910-35-8	236.352	tcl pr (al)	52(1)	308.5	1.0009²⁰	1.5681²⁰	i H₂O; s EtOH, bz, MeOH
3691	1,2-Dihydro-2,2,4-trimethylquinoline		C₁₂H₁₅N	147-47-7	173.254		26.5	260			
3692	1,4-Dihydroxy-9,10-anthracenedione	Quinizarin	C₁₄H₈O₄	81-64-1	240.212	ye red lf (eth) dk red nd	199.7(0.4)				s H₂O, EtOH, eth, bz, KOH, sulf
3693	1,5-Dihydroxy-9,10-anthracenedione	Anthrarufin	C₁₄H₈O₄	117-12-4	240.212	pa ye pl (gl HOAc)	280	sub			i H₂O; sl EtOH, eth, ace, CS₂; s bz
3694	1,8-Dihydroxy-9,10-anthracenedione	Danthron	C₁₄H₈O₄	117-10-2	240.212	red or red-ye nd or lf (al)	193	sub			i H₂O; sl EtOH, eth; s ace, HOAc, alk
3695	2,6-Dihydroxy-9,10-anthracenedione		C₁₄H₈O₄	84-60-6	240.212	ye nd (al)	360 dec				sl H₂O, EtOH; i eth, bz, chl; s alk
3696	2,7-Dihydroxy-9,10-anthracenedione		C₁₄H₈O₄	572-93-0	240.212	ye nd (+1w, dil al) nd (sub)	353.8	sub			i H₂O; s EtOH; sl eth, bz, chl

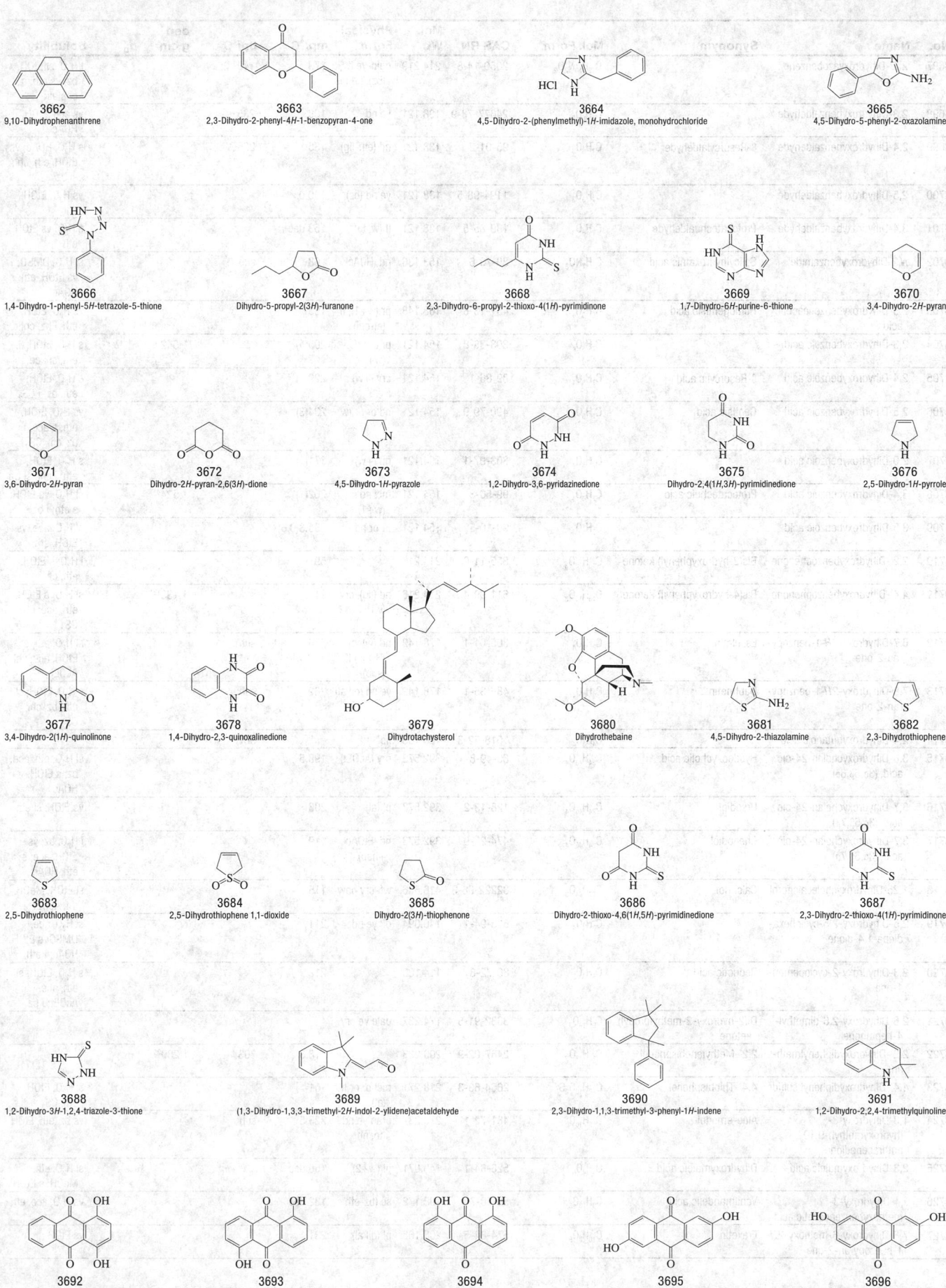

3662
9,10-Dihydrophenanthrene

3663
2,3-Dihydro-2-phenyl-4H-1-benzopyran-4-one

3664
4,5-Dihydro-2-(phenylmethyl)-1H-imidazole, monohydrochloride

3665
4,5-Dihydro-5-phenyl-2-oxazolamine

3666
1,4-Dihydro-1-phenyl-5H-tetrazole-5-thione

3667
Dihydro-5-propyl-2(3H)-furanone

3668
2,3-Dihydro-6-propyl-2-thioxo-4(1H)-pyrimidinone

3669
1,7-Dihydro-6H-purine-6-thione

3670
3,4-Dihydro-2H-pyran

3671
3,6-Dihydro-2H-pyran

3672
Dihydro-2H-pyran-2,6(3H)-dione

3673
4,5-Dihydro-1H-pyrazole

3674
1,2-Dihydro-3,6-pyridazinedione

3675
Dihydro-2,4(1H,3H)-pyrimidinedione

3676
2,5-Dihydro-1H-pyrrole

3677
3,4-Dihydro-2(1H)-quinolinone

3678
1,4-Dihydro-2,3-quinoxalinedione

3679
Dihydrotachysterol

3680
Dihydrothebaine

3681
4,5-Dihydro-2-thiazolamine

3682
2,3-Dihydrothiophene

3683
2,5-Dihydrothiophene

3684
2,5-Dihydrothiophene 1,1-dioxide

3685
Dihydro-2(3H)-thiophenone

3686
Dihydro-2-thioxo-4,6(1H,5H)-pyrimidinedione

3687
2,3-Dihydro-2-thioxo-4(1H)-pyrimidinone

3688
1,2-Dihydro-3H-1,2,4-triazole-3-thione

3689
(1,3-Dihydro-1,3,3-trimethyl-2H-indol-2-ylidene)acetaldehyde

3690
2,3-Dihydro-1,1,3-trimethyl-3-phenyl-1H-indene

3691
1,2-Dihydro-2,2,4-trimethylquinoline

3692
1,4-Dihydroxy-9,10-anthracenedione

3693
1,5-Dihydroxy-9,10-anthracenedione

3694
1,8-Dihydroxy-9,10-anthracenedione

3695
2,6-Dihydroxy-9,10-anthracenedione

3696
2,7-Dihydroxy-9,10-anthracenedione

No.	Name	Synonym	Mol. Form.	CAS RN	Mol. Wt.	Physical Form	mp/°C	bp/°C	den g cm⁻³	n_D	Solubility
3697	2,2'-Dihydroxyazobenzene		$C_{12}H_{10}N_2O_2$	2050-14-8	214.219	gold-ye lf (bz), nd (al)	173	$140^{0.001}$			i H_2O; sl EtOH, bz; vs eth; s con alk
3698	2,3-Dihydroxybenzaldehyde		$C_7H_6O_3$	24677-78-9	138.121	ye nd	108	235			vs ace, EtOH, HOAc
3699	2,4-Dihydroxybenzaldehyde	β-Resorcylaldehyde	$C_7H_6O_3$	95-01-2	138.121	nd (eth-lig)	135	226^{22}			s H_2O, HOAc; vs EtOH, eth, chl; sl bz
3700	2,5-Dihydroxybenzaldehyde		$C_7H_6O_3$	1194-98-5	138.121	ye nd (bz)	100.0				vs H_2O, EtOH, chl
3701	3,4-Dihydroxybenzaldehyde	Protocatechualdehyde	$C_7H_6O_3$	139-85-5	138.121	lf (w, to)	153 dec				s H_2O; vs EtOH, eth
3702	N,2-Dihydroxybenzamide	Salicylhydroxamic acid	$C_7H_7NO_3$	89-73-6	153.136	nd (HOAc)	168	sub			sl H_2O, DMSO; vs EtOH, eth; s HOAc
3703	2,5-Dihydroxybenzeneacetic acid	Homogentisic acid	$C_8H_8O_4$	451-13-8	168.148	pr (w+1), lf (al-chl)	153				vs H_2O, EtOH, eth; i bz, chl
3704	2,3-Dihydroxybenzoic acid		$C_7H_6O_4$	303-38-8	154.121	pr or nd (w+1)	205(4)		1.542^{20}		s H_2O, EtOH, eth; sl ace
3705	2,4-Dihydroxybenzoic acid	β-Resorcylic acid	$C_7H_6O_4$	89-86-1	154.121	cry (+w)	229(1)				s H_2O, EtOH, eth, bz; i CS_2
3706	2,5-Dihydroxybenzoic acid	Gentisic acid	$C_7H_6O_4$	490-79-9	154.121	nd or pr (w)	204(3)				vs H_2O, EtOH, eth; s ace; i bz, chl, CS_2
3707	2,6-Dihydroxybenzoic acid		$C_7H_6O_4$	303-07-1	154.121	nd (+w)	171(1)				s H_2O, EtOH, eth; i chl; sl tfa
3708	3,4-Dihydroxybenzoic acid	Protocatechuic acid	$C_7H_6O_4$	99-50-3	154.121	mcl nd (w+1)	202(1)		1.524^4		sl H_2O; vs EtOH; s eth; i bz
3709	3,5-Dihydroxybenzoic acid		$C_7H_6O_4$	99-10-5	154.121	pr or nd	235.3(0.8)				sl H_2O, ace; vs EtOH, eth
3710	2,2'-Dihydroxybenzophenone	Bis(2-hydroxyphenyl) ketone	$C_{13}H_{10}O_3$	835-11-0	214.216		59.5	333			i H_2O; s EtOH, eth, chl
3711	4,4'-Dihydroxybenzophenone	Bis(4-hydroxyphenyl) ketone	$C_{13}H_{10}O_3$	611-99-4	214.216	nd (lig), cry (w)	210		1.133^{131}		sl H_2O; s EtOH, eth, ace; i bz, CS_2
3712	6,7-Dihydroxy-2H-1-benzopyran-2-one	Esculetin	$C_9H_6O_4$	305-01-1	178.142	nd (w), pr (HOAc) lf (sub)	276	sub			sl H_2O, eth; s EtOH, ace, chl, AcOEt
3713	7,8-Dihydroxy-2H-1-benzopyran-2-one	Daphnetin	$C_9H_6O_4$	486-35-1	178.142	ye nd (dil al)	262	sub			s H_2O, EtOH; sl eth, bz, chl, CS_2
3714	2,4-Dihydroxybutanoic acid		$C_4H_8O_4$	1518-62-3	120.105	liq		96^3			
3715	3,6-Dihydroxycholan-24-oic acid, (3α,5β,6α)	Hyodeoxycholic acid	$C_{24}H_{40}O_4$	83-49-8	392.573	cry (AcOEt)	198.5				sl H_2O, eth, ace, bz; s EtOH, HOAc
3716	3,7-Dihydroxycholan-24-oic acid, (3α,5β,7β)	Ursodiol	$C_{24}H_{40}O_4$	128-13-2	392.573	pl (al)	203				vs EtOH; sl eth
3717	3,7-Dihydroxycholan-24-oic acid, (3α,5β,7α)	Chenodiol	$C_{24}H_{40}O_4$	474-25-9	392.573	nd (EtOAc +hep)	119				i H_2O, bz; vs EtOH, ace; s eth, HOAc
3718	1,25-Dihydroxycholecalciferol	Calcitriol	$C_{27}H_{44}O_3$	32222-06-3	416.636	wh cry pow	115				sl EtOH, MeOH, thf, AcOEt
3719	2,5-Dihydroxy-2,5-cyclohexadiene-1,4-dione		$C_6H_4O_4$	615-94-1	140.094	dk ye nd	211				sl H_2O, ace, DMSO; s EtOH, HOAc; i eth
3720	2,3-Dihydroxy-2-cyclopenten-1-one	Reductic acid	$C_5H_6O_3$	80-72-8	114.100		212				s H_2O, EtOH; sl eth, ace, AcOEt; i bz
3721	2,6-Dihydroxy-2,6-dimethyl-4-heptanone	Di(2-hydroxy-2-methylpropyl) ketone	$C_9H_{18}O_3$	3682-91-5	174.237	pale ye cry					
3722	2,2'-Dihydroxydiphenylmethane	2,2'-Methylenebisphenol	$C_{13}H_{12}O_2$	2467-02-9	200.233		118.3	363	1.280^{25}		
3723	4,4'-Dihydroxydiphenyl sulfide	4,4'-Thiobisphenol	$C_{12}H_{10}O_2S$	2664-63-3	218.271	mcl pr or lf (al)	151				sl H_2O, EtOH, eth, CS_2
3724	1,8-Dihydroxy-3-(hydroxymethyl)-9,10-anthracenedione	Aloe-emodol	$C_{15}H_{10}O_5$	481-72-1	270.237	oran ye nd (to, al)	223.5	sub			vs bz, eth, EtOH
3725	2,3-Dihydroxymaleic acid	Dihydroxymaleic acid	$C_4H_4O_6$	526-84-1	148.071	pl (w+2)	155 dec				sl H_2O, eth, MeOH; s EtOH
3726	α,4-Dihydroxy-3-methoxybenzeneacetic acid	Vanilmandelic acid	$C_9H_{10}O_5$	55-10-7	198.172	sc (bz-eth)	132 dec				vs H_2O, ace, eth
3727	7,8-Dihydroxy-6-methoxy-2H-1-benzopyran-2-one	Fraxetin	$C_{10}H_8O_5$	574-84-5	208.168	pl (dil al)	231				vs EtOH

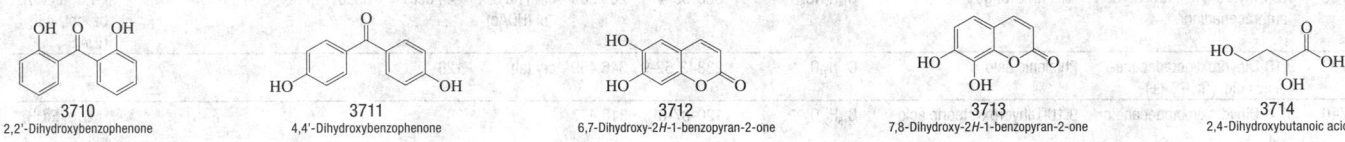

3697
2,2'-Dihydroxyazobenzene

3698
2,3-Dihydroxybenzaldehyde

3699
2,4-Dihydroxybenzaldehyde

3700
2,5-Dihydroxybenzaldehyde

3701
3,4-Dihydroxybenzaldehyde

3702
N,2-Dihydroxybenzamide

3703
2,5-Dihydroxybenzeneacetic acid

3704
2,3-Dihydroxybenzoic acid

3705
2,4-Dihydroxybenzoic acid

3706
2,5-Dihydroxybenzoic acid

3707
2,6-Dihydroxybenzoic acid

3708
3,4-Dihydroxybenzoic acid

3709
3,5-Dihydroxybenzoic acid

3710
2,2'-Dihydroxybenzophenone

3711
4,4'-Dihydroxybenzophenone

3712
6,7-Dihydroxy-2*H*-1-benzopyran-2-one

3713
7,8-Dihydroxy-2*H*-1-benzopyran-2-one

3714
2,4-Dihydroxybutanoic acid

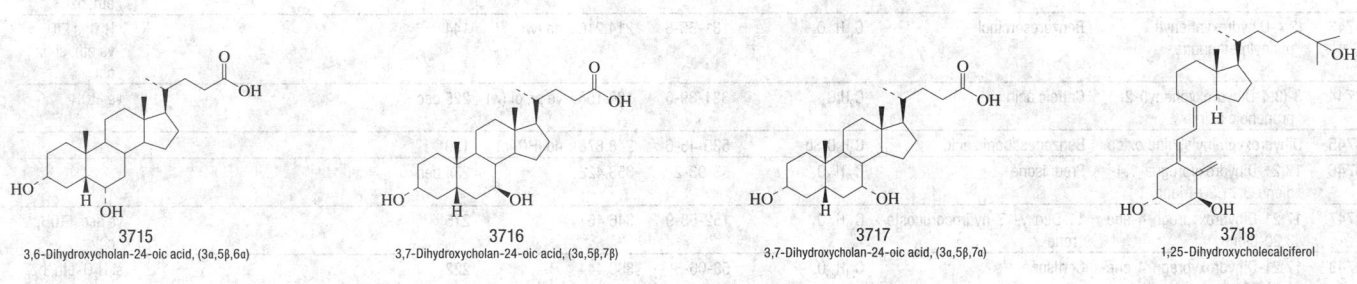

3715
3,6-Dihydroxycholan-24-oic acid, (3α,5β,6α)

3716
3,7-Dihydroxycholan-24-oic acid, (3α,5β,7β)

3717
3,7-Dihydroxycholan-24-oic acid, (3α,5β,7α)

3718
1,25-Dihydroxycholecalciferol

3719
2,5-Dihydroxy-2,5-cyclohexadiene-1,4-dione

3720
2,3-Dihydroxy-2-cyclopenten-1-one

3721
2,6-Dihydroxy-2,6-dimethyl-4-heptanone

3722
2,2'-Dihydroxydiphenylmethane

3723
4,4'-Dihydroxydiphenyl sulfide

3724
1,8-Dihydroxy-3-(hydroxymethyl)-9,10-anthracenedione

3725
2,3-Dihydroxymaleic acid

3726
α,4-Dihydroxy-3-methoxybenzeneacetic acid

3727
7,8-Dihydroxy-6-methoxy-2*H*-1-benzopyran-2-one

No.	Name	Synonym	Mol. Form.	CAS RN	Mol. Wt.	Physical Form	mp/°C	bp/°C	den g cm⁻³	n_D	Solubility
3728	5,7-Dihydroxy-3-(4-methoxyphenyl)-4H-1-benzopyran-4-one		$C_{16}H_{12}O_5$	491-80-5	284.263		214.8				
3729	(2,6-Dihydroxy-4-methoxy-phenyl)phenylmethanone	Cotoin	$C_{14}H_{12}O_4$	479-21-0	244.243	ye pr (chl) lf or nd (w)	130.5				vs ace, bz, eth, EtOH
3730	1,7-Dihydroxy-3-methoxy-9H-xanthen-9-one	Gentisin	$C_{14}H_{10}O_5$	437-50-3	258.226	ye orth	266.5				i H_2O; vs EtOH; i ace; s py
3731	1,8-Dihydroxy-3-methyl-9,10-anthracenedione	Chrysophanic acid	$C_{15}H_{10}O_4$	481-74-3	254.238	ye hex or mcl nd (sub)	196	sub	0.92^{25}		vs bz, HOAc
3732	2,4-Dihydroxy-6-methylben-zoic acid	o-Orsellinic acid	$C_8H_8O_4$	480-64-8	168.148	nd (dil HOAc, +1w)	176 dec				s EtOH, eth
3733	5,7-Dihydroxy-4-methyl-2H-1-benzopyran-2-one		$C_{10}H_8O_4$	2107-76-8	192.169	nd (al), lf (HOAc)	283				sl H_2O, eth, bz, chl; vs EtOH, alk
3734	6,7-Dihydroxy-4-methyl-2H-1-benzopyran-2-one		$C_{10}H_8O_4$	529-84-0	192.169	ye nd (dil al)	275				s H_2O, EtOH, HOAc
3735	5,8-Dihydroxy-1,4-naphthal-enedione		$C_{10}H_6O_4$	475-38-7	190.153	dk red mcl pr (bz) red-br nd (al)	243(1)	sub			sl H_2O, EtOH, eth; s HOAc
3736	4,5-Dihydroxy-2,7-naphthal-enedisulfonic acid	Chromotropic acid	$C_{10}H_8O_8S_2$	148-25-4	320.296	nd or lf (w+2)					s H_2O, alk; i EtOH, eth
3737	5,6-Dihydroxynaphtho[2,3-f]quinoline-7,12-dione	Alizarin Blue	$C_{17}H_9NO_4$	568-02-5	291.258	br-viol nd (bz)	269				vs bz, gl HOAc
3738	1,2-Dihydroxy-3-nitro-9,10-anthracenedione	Alizarin Orange	$C_{14}H_7NO_6$	568-93-4	285.209	oran nd or pl (HOAc)	244 dec	sub			sl H_2O; s EtOH, bz, chl, sulf, HOAc
3739	9,10-Dihydroxyoctadecane-dioic acid, ($R*,R*$)-(±)-	Phloionic acid	$C_{18}H_{34}O_6$	23843-52-9	346.459	cry (al)	126				
3740	9,10-Dihydroxyoctadecanoic acid	9,10-Dihydroxystearic acid	$C_{18}H_{36}O_4$	120-87-6	316.477		90				i H_2O; sl EtOH, eth
3741	5,7-Dihydroxy-2-phenyl-4H-1-benzopyran-4-one	Chrysin	$C_{15}H_{10}O_4$	480-40-0	254.238	lt ye pr (MeOH)	285(2)				i H_2O; s EtOH, ace; sl eth, bz, CS_2
3742	1-(2,4-Dihydroxyphenyl)-ethanone	Resacetophenone	$C_8H_8O_3$	89-84-9	152.148	nd or lf	146			1.18^{141}	i H_2O, chl; s EtOH, py; sl eth, bz
3743	(2,4-Dihydroxyphenyl)-phenylmethanone	Benzoresorcinol	$C_{13}H_{10}O_3$	131-56-6	214.216	nd (w)	144				i H_2O; s EtOH; vs eth; sl bz, chl
3744	3-(3,4-Dihydroxyphenyl)-2-propenoic acid	Caffeic acid	$C_9H_8O_4$	331-39-5	180.158	ye pr, pl (w)	225 dec				vs EtOH
3745	Dihydroxyphenylstibine oxide	Benzenestibonic acid	$C_6H_7O_3Sb$	535-46-6	248.878	nd (HOAc)	139				
3746	17,21-Dihydroxypregna-1,4-diene-3,11,20-trione	Prednisone	$C_{21}H_{26}O_5$	53-03-2	358.428		234 dec				
3747	17,21-Dihydroxypregn-4-ene-3,20-dione	11-Deoxy-17-hydrocorticoste-rone	$C_{21}H_{30}O_4$	152-58-9	346.461		215				vs ace, EtOH, chl
3748	17,21-Dihydroxypregn-4-ene-3,11,20-trione	Cortisone	$C_{21}H_{28}O_5$	53-06-5	360.444		222				sl H_2O, eth, bz, chl; s EtOH, ace
3749	2,3-Dihydroxypropanal, (±)-		$C_3H_6O_3$	56-82-6	90.078	nd or pr (40% MeOH)	145	$145^{0.8}$	1.453^{18}		s H_2O; sl EtOH, eth; i bz, peth, lig
3750	2,3-Dihydroxypropanoic acid, (R)-	Glyceric acid	$C_3H_6O_4$	6000-40-4	106.078	thick gum					
3751	1,3-Dihydroxy-2-propanone	Dihydroxyacetone	$C_3H_6O_3$	96-26-4	90.078		74(3)				s H_2O, EtOH, eth, ace; i lig
3752	2,3-Dihydroxypropyl decanoate	Decanoic acid glycerol monoester	$C_{13}H_{26}O_4$	2277-23-8	246.343	pr (peth)	53				
3753	2,3-Dihydroxypropyl octanoate	Octanoic acid glycerol monoester	$C_{11}H_{22}O_4$	26402-26-6	218.291	cry (peth)	40				
3754	4,8-Dihydroxy-2-quinolinecar-boxylic acid	Xanthurenic acid	$C_{10}H_7NO_4$	59-00-7	205.168	ye micry cry (w)	289				i H_2O; s EtOH, dil HCl; sl eth, bz
3755	Dihydroxytartaric acid		$C_4H_6O_8$	76-30-2	182.086		114.5				
3756	3,4-Dihydroxy-5-[(3,4,5-trihydroxybenzoyl)oxy]benzoic acid	Digallic acid	$C_{14}H_{10}O_9$	536-08-3	322.224	nf (dil al + 1w)	269 dec				vs ace, EtOH
3757	2-(3,6-Dihydroxy-9H-xanthen-9-yl)benzoic acid	Fluorescin	$C_{20}H_{14}O_5$	518-44-5	334.322	col or ye nd (eth), pl (bz)	126				i H_2O; s EtOH, eth, ace, bz, HOAc
3758	Diiodoacetylene		C_2I_2	624-74-8	277.830	orth nd (lig)	81.5	247(16)			vs ace, bz, eth, EtOH

3728
5,7-Dihydroxy-3-(4-methoxyphenyl)-4H-1-benzopyran-4-one

3729
(2,6-Dihydroxy-4-methoxyphenyl)phenylmethanone

3730
1,7-Dihydroxy-3-methoxy-9H-xanthen-9-one

3731
1,8-Dihydroxy-3-methyl-9,10-anthracenedione

3732
2,4-Dihydroxy-6-methylbenzoic acid

3733
5,7-Dihydroxy-4-methyl-2H-1-benzopyran-2-one

3734
6,7-Dihydroxy-4-methyl-2H-1-benzopyran-2-one

3735
5,8-Dihydroxy-1,4-naphthalenedione

3736
4,5-Dihydroxy-2,7-naphthalenedisulfonic acid

3737
5,6-Dihydroxynaphtho[2,3-f]quinoline-7,12-dione

3738
1,2-Dihydroxy-3-nitro-9,10-anthracenedione

3739
9,10-Dihydroxyoctadecanedioic acid, (R*,R*)-(±)-

3740
9,10-Dihydroxyoctadecanoic acid

3741
5,7-Dihydroxy-2-phenyl-4H-1-benzopyran-4-one

3742
1-(2,4-Dihydroxyphenyl)ethanone

3743
(2,4-Dihydroxyphenyl)phenylmethanone

3744
3-(3,4-Dihydroxyphenyl)-2-propenoic acid

3745
Dihydroxyphenylstibine oxide

3746
17,21-Dihydroxypregna-1,4-diene-3,11,20-trione

3747
17,21-Dihydroxypregn-4-ene-3,20-dione

3748
17,21-Dihydroxypregn-4-ene-3,11,20-trione

3749
2,3-Dihydroxypropanal, (±)-

3750
2,3-Dihydroxypropanoic acid, (R)-

3751
1,3-Dihydroxy-2-propanone

3752
2,3-Dihydroxypropyl decanoate

3753
2,3-Dihydroxypropyl octanoate

3754
4,8-Dihydroxy-2-quinolinecarboxylic acid

3755
Dihydroxytartaric acid

3756
3,4-Dihydroxy-5-[(3,4,5-trihydroxybenzoyl)oxy]benzoic acid

3757
2-(3,6-Dihydroxy-9H-xanthen-9-yl)benzoic acid

3758
Diiodoacetylene

No.	Name	Synonym	Mol. Form.	CAS RN	Mol. Wt.	Physical Form	mp/°C	bp/°C	den g cm⁻³	n_D	Solubility
3759	2,4-Diiodoaniline		$C_6H_5I_2N$	533-70-0	344.920	br nd or orth cry (al)	95.5		2.748[25]		vs ace, bz, eth, EtOH
3760	o-Diiodobenzene	1,2-Diiodobenzene	$C_6H_4I_2$	615-42-9	329.905	pl or pr (lig)	23.5(0.7)	287	2.54[20]	1.7179[20]	i H_2O; sl EtOH
3761	m-Diiodobenzene	1,3-Diiodobenzene	$C_6H_4I_2$	626-00-6	329.905	orth pl or pr (eth-al)	34.2(0.6)	285	2.47[25]		i H_2O; vs eth, EtOH, chl
3762	p-Diiodobenzene	1,4-Diiodobenzene	$C_6H_4I_2$	624-38-4	329.905	orth lf (al)	129.25(0.05)	285			i H_2O; s EtOH; vs eth; sl chl
3763	1,4-Diiodobutane		$C_4H_8I_2$	628-21-7	309.916		5.9(0.4)	125[15] dec	2.3494[25]	1.6184[25]	i H_2O; sl ctc; s os
3764	1,2-Diiodoethane		$C_2H_4I_2$	624-73-7	281.862	ye mcl pr or orth (eth)	83	200	3.325[20]	1.871[20]	sl H_2O; s EtOH, eth, ace, chl
3765	cis-1,2-Diiodoethene	cis-1,2-Diiodoethylene	$C_2H_2I_2$	590-26-1	279.846		-13(2)	72.5[16]	3.0625[20]	i H_2O; s eth, chl
3766	4,4'-Diiodofluorescein		$C_{20}H_{10}I_2O_5$	38577-97-8	584.099	oran-red pow					sl H_2O; s alk, EtOH
3767	1,6-Diiodohexane	Hexamethylene diiodide	$C_6H_{12}I_2$	629-09-4	337.968	nd	9.4(0.5)	163[17]	2.0342[25]	1.5837[25]	i H_2O; vs EtOH, eth
3768	Diiodomethane	Methylene iodide	CH_2I_2	75-11-6	267.836	ye nd or lf	6.0(0.2)	182	3.3211[20]	1.7411[20]	sl H_2O, ctc; s EtOH, eth, bz, chl
3769	2,6-Diiodo-4-nitrophenol	Disophenol	$C_6H_3I_2NO_3$	305-85-1	390.902	lt ye cry (gl HOAc)	157				vs EtOH
3770	1,5-Diiodopentane	Pentamethylene diiodide	$C_5H_{10}I_2$	628-77-3	323.942		9	149[20]	2.1692[25]	1.5987[25]	i H_2O; s eth, chl
3771	1,2-Diiodopropane		$C_3H_6I_2$	598-29-8	295.889				2.490[18]		vs eth, EtOH
3772	1,3-Diiodopropane	Trimethylene diiodide	$C_3H_6I_2$	627-31-6	295.889		-20	222(12)	2.5612[25]	1.6391[25]	i H_2O; s eth, ctc, chl
3773	5,7-Diiodo-8-quinolinol	Iodoquinol	$C_9H_5I_2NO$	83-73-8	396.951	ye nd (HOAc, xyl)	210				sl H_2O, bz, chl, eth; vs EtOH; s alk
3774	3,5-Diiodo-L-tyrosine		$C_9H_9I_2NO_3$	300-39-0	432.981	ye nd (w, 70% al)	213				sl H_2O; i EtOH, eth, bz
3775	Diisobutyl adipate	Diisobutyl hexanedioate	$C_{14}H_{26}O_4$	141-04-8	258.354			288(16)	0.9543[19]	1.4301[20]	
3776	Diisobutylaluminum chloride	Diisobutyl aluminum chloride	$C_8H_{18}AlCl$	1779-25-5	176.664	hyg col liq	-40	152[10]	0.905	1.4506[20]	s eth, hx
3777	Diisobutylaluminum hydride		$C_8H_{19}Al$	1191-15-7	142.219	liq		140[4]			s cyhex, eth, bz, tol
3778	Diisobutylamine	2-Methyl-N-(2-methylpropyl)-1-propanamine	$C_8H_{19}N$	110-96-3	129.244	liq	-73.5(0.4)	139.6		1.4090[20]	sl H_2O, ctc; s EtOH, eth, ace, bz
3779	Diisobutyl carbonate		$C_9H_{18}O_3$	539-92-4	174.237			190(6)	0.9138[20]	1.4072[20]	i H_2O; msc EtOH, eth
3780	Diisobutyl ether	1,1'-Oxybis[2-methylpropane]	$C_8H_{18}O$	628-55-7	130.228			122.7(0.7)	0.761[15]		i H_2O; msc EtOH, eth
3781	Diisobutyl phthalate		$C_{16}H_{22}O_4$	84-69-5	278.344			296.5	1.0490[15]		s ctc
3782	Diisobutyl sulfide		$C_8H_{18}S$	592-65-4	146.294	liq	-105.5(0.5)	173(4)	0.8363[10]		
3783	1,3-Diisocyanatobenzene		$C_8H_4N_2O_2$	123-61-5	160.130	cry	51	103[8]			
3784	1,4-Diisocyanatobenzene		$C_8H_4N_2O_2$	104-49-4	160.130	cry	95	117[14]			
3785	Diisodecyl phthalate	Bis(8-methylnonyl)phthalate	$C_{28}H_{46}O_4$	26761-40-0	446.663	liq	-50	253[4]	0.966[20]		i H_2O; s os
3786	Diisononyl phthalate	Bis(7-methyloctyl)phthalate	$C_{26}H_{42}O_4$	28553-12-0	418.609	col liq					i H_2O; s ace, MeOH; bz, eth
3787	Diisooctyl adipate	Diisooctyl hexanedioate	$C_{22}H_{42}O_4$	1330-86-5	370.566			210[4]			
3788	Diisooctyl phthalate		$C_{24}H_{38}O_4$	27554-26-3	390.557			370			
3789	Diisopentylamine	3-Methyl-N-isopentyl-1-butanamine	$C_{10}H_{23}N$	544-00-3	157.297	liq	-44	187(5)	0.7672[21]	1.4235[20]	i H_2O; s EtOH; msc eth
3790	Diisopentyl ether	Diisoamyl ether	$C_{10}H_{22}O$	544-01-4	158.281	col liq		172(2)	0.7777[20]	1.4085[20]	i H_2O; vs ace, EtOH, chl
3791	Diisopentyl phthalate	Diisoamyl phthalate	$C_{18}H_{26}O_4$	605-50-5	306.397			367(16)	1.0209[16]	1.4871[20]	vs EtOH
3792	Diisopentyl sulfide	Isopentyl sulfide	$C_{10}H_{22}S$	544-02-5	174.347	liq	-74.6	211	0.8323[20]	1.4520[20]	i H_2O; msc EtOH; vs eth
3793	Diisopropanolamine	1,1'-Iminobis-2-propanol	$C_6H_{15}NO_2$	110-97-4	133.189	cry	44.5	250	0.989[20]		s H_2O, EtOH; sl eth
3794	Diisopropyl adipate	Diisopropyl hexanedioate	$C_{12}H_{22}O_4$	6938-94-9	230.301		-0.6	120[6.5]	0.9569[20]	1.4247[20]	vs ace, eth, EtOH
3795	Diisopropylamine	N-Isopropyl-2-propanamine	$C_6H_{15}N$	108-18-9	101.190	liq	-61	84(3)	0.7153[20]	1.3924[20]	vs ace, bz, eth, EtOH
3796	2,6-Diisopropylaniline		$C_{12}H_{19}N$	24544-04-5	177.286	liq	-45	272(9)	0.94[25]	1.5332[20]	
3797	1,2-Diisopropylbenzene		$C_{12}H_{18}$	577-55-9	162.271	liq	-57(4)	204.7(0.6)	0.8701[20]	1.4960[20]	i H_2O; msc EtOH, eth, ace, bz, ctc
3798	1,3-Diisopropylbenzene		$C_{12}H_{18}$	99-62-7	162.271	liq	-63(2)	203(3)	0.8559[20]	1.4883[20]	i H_2O; msc EtOH, eth, ace, bz, ctc

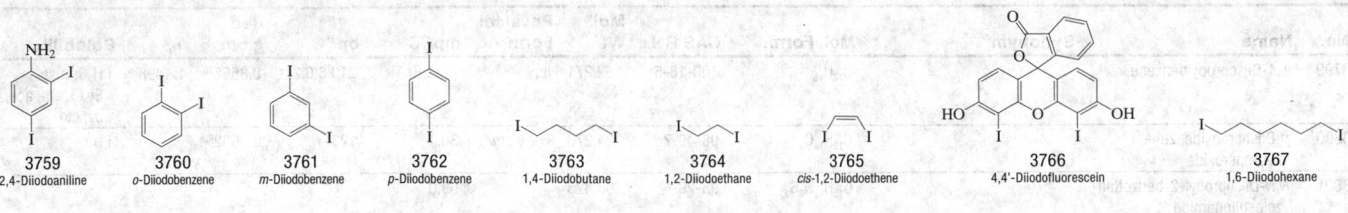

3759 2,4-Diiodoaniline

3760 o-Diiodobenzene

3761 m-Diiodobenzene

3762 p-Diiodobenzene

3763 1,4-Diiodobutane

3764 1,2-Diiodoethane

3765 cis-1,2-Diiodoethene

3766 4,4'-Diiodofluorescein

3767 1,6-Diiodohexane

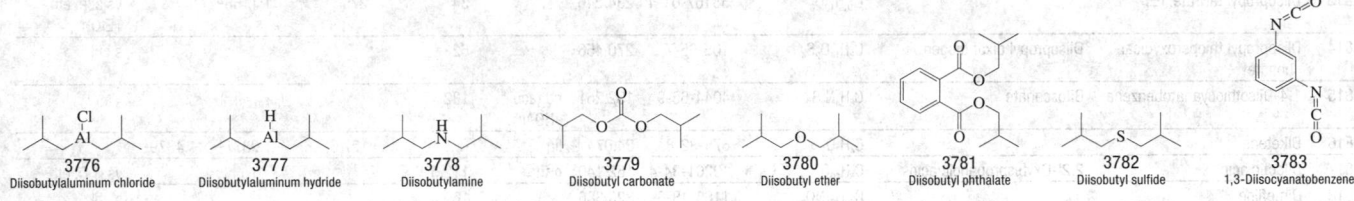

3768 Diiodomethane

3769 2,6-Diiodo-4-nitrophenol

3770 1,5-Diiodopentane

3771 1,2-Diiodopropane

3772 1,3-Diiodopropane

3773 5,7-Diiodo-8-quinolinol

3774 3,5-Diiodo-L-tyrosine

3775 Diisobutyl adipate

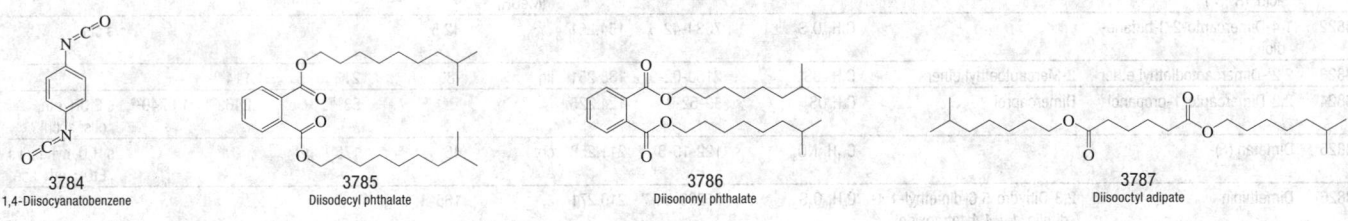

3776 Diisobutylaluminum chloride

3777 Diisobutylaluminum hydride

3778 Diisobutylamine

3779 Diisobutyl carbonate

3780 Diisobutyl ether

3781 Diisobutyl phthalate

3782 Diisobutyl sulfide

3783 1,3-Diisocyanatobenzene

3784 1,4-Diisocyanatobenzene

3785 Diisodecyl phthalate

3786 Diisononyl phthalate

3787 Diisooctyl adipate

3788 Diisooctyl phthalate

3789 Diisopentylamine

3790 Diisopentyl ether

3791 Diisopentyl phthalate

3792 Diisopentyl sulfide

3793 Diisopropanolamine

3794 Diisopropyl adipate

3795 Diisopropylamine

3796 2,6-Diisopropylaniline

3797 1,2-Diisopropylbenzene

3798 1,3-Diisopropylbenzene

No.	Name	Synonym	Mol. Form.	CAS RN	Mol. Wt.	Physical Form	mp/°C	bp/°C	den g cm^{-3}	n_D	Solubility
3799	1,4-Diisopropylbenzene		$C_{12}H_{18}$	100-18-5	162.271	liq	-17.0(0.1)	210.3(0.2)	0.8568[20]	1.4898[20]	i H$_2$O; msc EtOH, eth, ace, bz, ctc
3800	p-Diisopropylbenzene hydroperoxide		$C_{12}H_{18}O_2$	98-49-7	194.270	waxy cry	30.1	123[1]	0.9932[20]		i H$_2$O
3801	N,N-Diisopropyl-2-benzothia-zolesulfenamide		$C_{13}H_{18}N_2S_2$	95-29-4	266.425		59.0				
3802	N,N'-Diisopropylcarbodiimide		$C_7H_{14}N_2$	693-13-0	126.199			147	0.806[25]	1.4320[20]	
3803	Diisopropyl disulfide		$C_6H_{14}S_2$	4253-89-8	150.305	liq	-69.0(0.3)	177(1)	0.9435[20]	1.4916[20]	
3804	N,N-Diisopropylethanolamine	N,N-Diisopropyl-2-aminoetha-nol	$C_8H_{19}NO$	96-80-0	145.243			195(5)	0.826[25]	1.4417[20]	
3805	Diisopropyl ether	Isopropyl ether	$C_6H_{14}O$	108-20-3	102.174	liq	-85.37(0.05)	68.4(0.2)	0.7192[25]	1.3658[25]	sl H$_2$O; msc EtOH, eth; s ace, ctc
3806	Diisopropyl methylphospho-nate		$C_7H_{17}O_3P$	1445-75-6	180.182			66[3]		1.4120[16]	
3807	2,6-Diisopropylnaphthalene		$C_{16}H_{20}$	24157-81-1	212.330	cry (MeOH)	70				
3808	Diisopropyl oxalate		$C_8H_{14}O_4$	615-81-6	174.195			189(8)	1.002[20]	1.4100[20]	vs eth, EtOH
3809	Diisopropyl phosphonate		$C_6H_{15}O_3P$	1809-20-7	166.155			97[40]	0.9970[18]		
3810	O,O-Diisopropyl phosphorodi-thioate		$C_6H_{15}O_2PS_2$	107-56-2	214.286	liq		71[3]	1.09[20]		s EtOH, bz, ace, ctc, chl
3811	Diisopropyl phthalate	1,2-Benzenedicarboxylic acid, diisopropyl ester	$C_{14}H_{18}O_4$	605-45-8	250.291			130[12]	1.0615[15]	1.4900[20]	
3812	Diisopropyl sulfide		$C_6H_{14}S$	625-80-9	118.240	liq	-78.03(0.04)	120.0(0.3)	0.8142[20]	1.4438[20]	i H$_2$O; s EtOH, eth
3813	Diisopropyl tartrate, (±)-		$C_{10}H_{18}O_6$	58167-01-4	234.246		34	275	1.1166[20]		vs ace, eth, EtOH
3814	Diisopropyl thioperoxydicar-bonate	Diisopropyl dixanthogen	$C_8H_{14}O_2S_4$	105-65-7	270.456		52				s chl
3815	1,4-Diisothiocyanatobenzene	Bitoscanate	$C_8H_4N_2S_2$	4044-65-9	192.261	nd (ace, HOAc)	132				
3816	Diketene		$C_4H_4O_2$	674-82-8	84.074	liq	-6.5	127.0(0.6)	1.0877[20]	1.4379[20]	
3817	Dilactic acid	2,2'-Oxybispropanoic acid	$C_6H_{10}O_5$	19201-34-4	162.140	orth	112.5				vs H$_2$O, eth
3818	Dimefline		$C_{20}H_{21}NO_3$	1165-48-6	323.386		109.5				s chl
3819	Dimefox	Tetramethylphosphorodiamidic fluoride	$C_4H_{12}FN_2OP$	115-26-4	154.122	liq		86[15]	1.1151[20]	1.4267[20]	vs H$_2$O, bz, eth
3820	Dimemorfan	3,17-Dimethylmorphinan, (9 α,13 α,14 α)-	$C_{18}H_{25}N$	36309-01-0	255.399	ye oil	92	133[0.3]			
3821	2,3-Dimercaptobutanedioic acid, (R*,S*)	Succimer	$C_4H_6O_4S_2$	304-55-2	182.219	wh cry (MeOH)	193				
3822	1,4-Dimercapto-2,3-butane-diol		$C_4H_{10}O_2S_2$	7634-42-6	154.251		42.5				s chl
3823	2,2'-Dimercaptodiethyl ether	2-Mercaptoethyl ether	$C_4H_{10}OS_2$	2150-02-9	138.251	liq	-80	217	1.114[20]		
3824	2,3-Dimercapto-1-propanol	Dimercaprol	$C_3H_8OS_2$	59-52-9	124.225			83[0.8]	1.2463[20]	1.5749[20]	s EtOH, eth, oils; sl chl
3825	Dimetan (R)-		$C_{11}H_{17}NO_3$	122-15-6	211.258	cry	46	175[11]			s H$_2$O, cyhex; vs EtOH, eth, ace
3826	Dimethipin	2,3-Dihydro-5,6-dimethyl-1,4-dithiin, 1,1,4,4-tetraoxide	$C_6H_{10}O_4S_2$	55290-64-7	210.271		165				
3827	Dimethirimol	5-Butyl-2-(dimethylamino)-6-methylpyrimidin-4(1H)-one	$C_{11}H_{19}N_3O$	5221-53-4	209.288	nd	102				sl H$_2$O; vs chl, xyl; s EtOH, ace
3828	Dimethisoquin	2-[(3-Butyl-1-isoquinolinyl)-oxy]-N,N-dimethylethanamine	$C_{17}H_{24}N_2O$	86-80-6	272.385		146	156[3]		1.5486[20]	s H$_2$O, EtOH
3829	Dimethoxane	2,6-Dimethyl-1,3-dioxan-4-ol acetate	$C_8H_{14}O_4$	828-00-2	174.195	liq		86[10]	1.0655[20]	1.4310[20]	msc H$_2$O; s os
3830	2',5'-Dimethoxyacetophenone		$C_{10}H_{12}O_3$	1201-38-3	180.200	cry	21	156[14]	1.139	1.5441[20]	
3831	1,2-Dimethoxy-4-allylbenzene	Methyleugenol	$C_{11}H_{14}O_2$	93-15-2	178.228	liq	-2.0	254.7	1.0396[20]	1.5340[20]	i H$_2$O; s EtOH, eth
3832	4,7-Dimethoxy-5-allyl-1,3-benzodioxole	Apiole	$C_{12}H_{14}O_4$	523-80-8	222.237	nd	29.2(0.2)	294	1.015[20]	1.5360[20]	vs ace, bz, EtOH, lig
3833	2,4-Dimethoxyaniline		$C_8H_{11}NO_2$	2735-04-8	153.179	pl (lig)	33.5	262.0			sl H$_2$O, chl; s EtOH, eth, bz, lig
3834	2,5-Dimethoxyaniline		$C_8H_{11}NO_2$	102-56-7	153.179		82.5	270(21)			s H$_2$O, EtOH, chl, lig
3835	3,4-Dimethoxyaniline		$C_8H_{11}NO_2$	6315-89-5	153.179	lf (eth)	87.5	159[14]			s eth, chl
3836	2,4-Dimethoxybenzaldehyde		$C_9H_{10}O_3$	613-45-6	166.173	nd (al or lig)	72	290			i H$_2$O; s EtOH, eth, bz; sl chl
3837	2,5-Dimethoxybenzaldehyde		$C_9H_{10}O_3$	93-02-7	166.173		52	270			sl H$_2$O; s EtOH, eth
3838	3,4-Dimethoxybenzaldehyde	Veratraldehyde	$C_9H_{10}O_3$	120-14-9	166.173	nd (eth, lig, to)	43	281			sl H$_2$O, chl; vs EtOH, eth

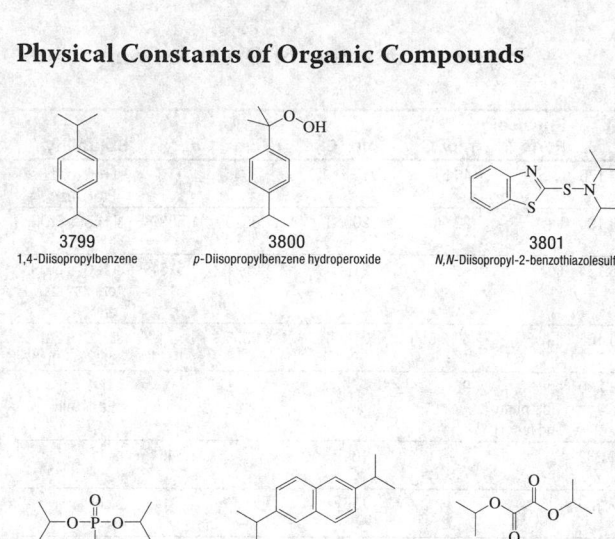

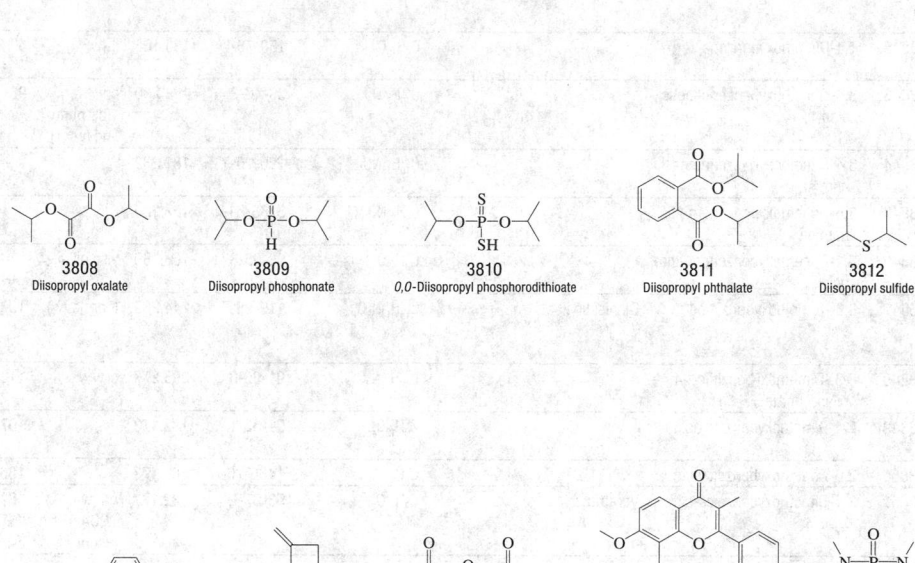

3799
1,4-Diisopropylbenzene

3800
p-Diisopropylbenzene hydroperoxide

3801
N,N-Diisopropyl-2-benzothiazolesulfenamide

3802
N,N'-Diisopropylcarbodiimide

3803
Diisopropyl disulfide

3804
N,N-Diisopropylethanolamine

3805
Diisopropyl ether

3806
Diisopropyl methylphosphonate

3807
2,6-Diisopropylnaphthalene

3808
Diisopropyl oxalate

3809
Diisopropyl phosphonate

3810
O,O-Diisopropyl phosphorodithioate

3811
Diisopropyl phthalate

3812
Diisopropyl sulfide

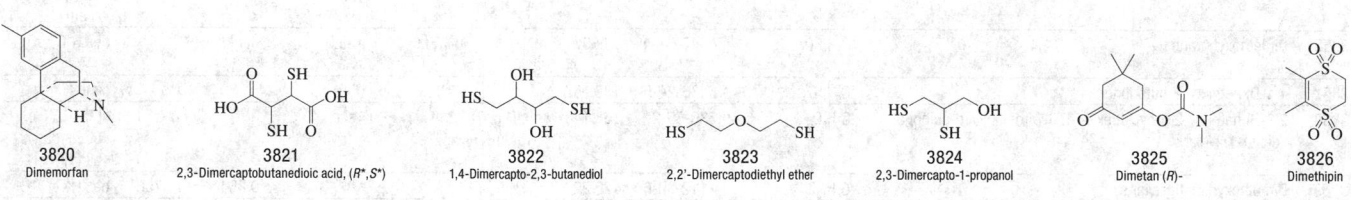

3813
Diisopropyl tartrate, (±)-

3814
Diisopropyl thioperoxydicarbonate

3815
1,4-Diisothiocyanatobenzene

3816
Diketene

3817
Dilactic acid

3818
Dimefline

3819
Dimefox

3820
Dimemorfan

3821
2,3-Dimercaptobutanedioic acid, (R*,S*)

3822
1,4-Dimercapto-2,3-butanediol

3823
2,2'-Dimercaptodiethyl ether

3824
2,3-Dimercapto-1-propanol

3825
Dimetan (R)-

3826
Dimethipin

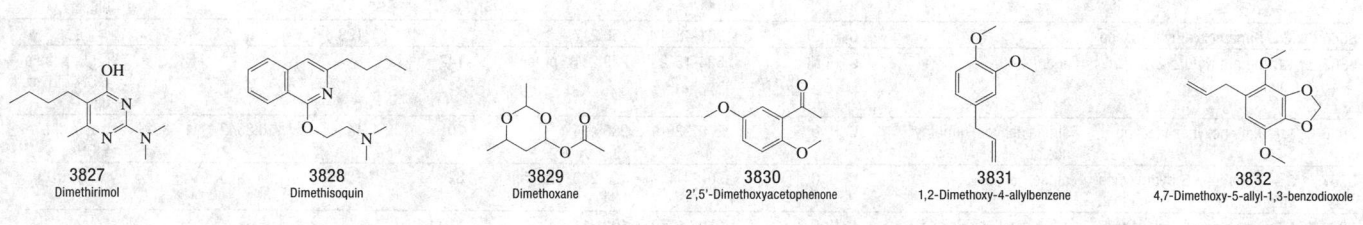

3827
Dimethirimol

3828
Dimethisoquin

3829
Dimethoxane

3830
2',5'-Dimethoxyacetophenone

3831
1,2-Dimethoxy-4-allylbenzene

3832
4,7-Dimethoxy-5-allyl-1,3-benzodioxole

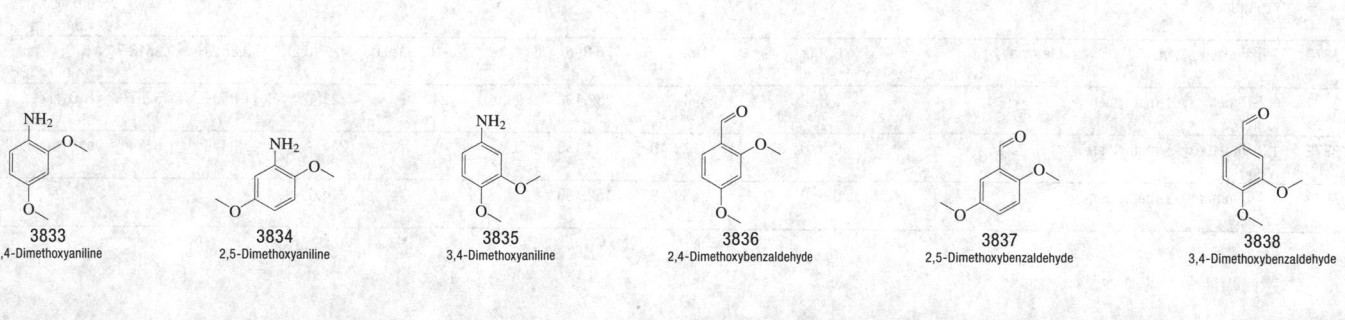

3833
2,4-Dimethoxyaniline

3834
2,5-Dimethoxyaniline

3835
3,4-Dimethoxyaniline

3836
2,4-Dimethoxybenzaldehyde

3837
2,5-Dimethoxybenzaldehyde

3838
3,4-Dimethoxybenzaldehyde

No.	Name	Synonym	Mol. Form.	CAS RN	Mol. Wt.	Physical Form	mp/°C	bp/°C	den g cm⁻³	n_D	Solubility
3839	3,5-Dimethoxybenzaldehyde		$C_9H_{10}O_3$	7311-34-4	166.173		46.3	151[16]			sl H₂O, peth; s EtOH, bz
3840	1,2-Dimethoxybenzene	Veratrole	$C_8H_{10}O_2$	91-16-7	138.164	liq	22.5(0.2)	206(1)	1.0810[25]	1.5827[21]	sl H₂O; s EtOH, eth, ctc
3841	1,3-Dimethoxybenzene		$C_8H_{10}O_2$	151-10-0	138.164	liq	-35.3(0.4)	216(5)	1.0521[25]	1.5231[20]	sl H₂O; s EtOH, eth, bz, ctc, sulf
3842	1,4-Dimethoxybenzene		$C_8H_{10}O_2$	150-78-7	138.164	lf (w)	56.2(0.7)	213(5)	1.0375[55]		sl H₂O; s EtOH, chl; vs eth, bz
3843	3,4-Dimethoxybenzeneacetic acid		$C_{10}H_{12}O_4$	93-40-3	196.200	cry (bz-peth) nd (w+1)	98				s H₂O, chl; vs EtOH, eth
3844	3,4-Dimethoxybenzeneetha-namine		$C_{10}H_{15}NO_2$	120-20-7	181.232			164[14]		1.5464[20]	s ctc
3845	3,4-Dimethoxybenzenemetha-namine		$C_9H_{13}NO_2$	5763-61-1	167.205			156[12]	1.143[25]		s chl
3846	3,4-Dimethoxybenzenemetha-nol		$C_9H_{12}O_3$	93-03-8	168.189	visc oil		298	1.178[17]	1.555[17]	s H₂O, EtOH
3847	3,3'-Dimethoxybenzidine	Dianisidine	$C_{14}H_{16}N_2O_2$	119-90-4	244.289	lf or nd (w)	137				i H₂O; s EtOH, eth, ace, bz, chl
3848	3,3'-Dimethoxybenzidine-4,4'-diisocyanate		$C_{16}H_{12}N_2O_4$	91-93-0	296.277	cry	112				
3849	2,4-Dimethoxybenzoic acid		$C_9H_{10}O_4$	91-52-1	182.173		107.2(0.4)				sl H₂O; s EtOH, eth, chl, HOAc
3850	2,6-Dimethoxybenzoic acid		$C_9H_{10}O_4$	1466-76-8	182.173		186 dec				
3851	3,4-Dimethoxybenzoic acid	Veratric acid	$C_9H_{10}O_4$	93-07-2	182.173	nd (w or HOAc) orth (sub)	181	sub			i H₂O; vs EtOH, eth; sl chl
3852	3,5-Dimethoxybenzoic acid		$C_9H_{10}O_4$	1132-21-4	182.173	nd (w), pr (al)	185.5	sub			vs eth, EtOH
3853	4,4'-Dimethoxybenzoin	p-Anisoin	$C_{16}H_{16}O_4$	119-52-8	272.296	pr (dil al)	109(3)				sl H₂O, chl, EtOH, eth; s ace
3854	5,7-Dimethoxy-2H-1-benzo-pyran-2-one	Limettin	$C_{11}H_{10}O_4$	487-06-9	206.195	pr or nd (al)	149	200 dec			sl H₂O; vs EtOH, ace, chl; i eth, lig
3855	4,4'-Dimethoxy-1,1'-biphenyl		$C_{14}H_{14}O_2$	2132-80-1	214.260	lf (bz)	175	sub			i H₂O, peth; vs EtOH, bz, chl; sl eth
3856	Dimethoxyborane		$C_2H_7BO_2$	4542-61-4	73.887	vol liq or gas	-130.6	25.9			dec H₂O
3857	4,4-Dimethoxy-2-butanone		$C_6H_{12}O_3$	5436-21-5	132.157			50[5]			s ctc
3858	2,6-Dimethoxy-2,5-cyclohexa-diene-1,4-dione	2,6-Dimethoxy-p-quinone	$C_8H_8O_4$	530-55-2	168.148	ye mcl pr (HOAc)	256	sub			sl H₂O, EtOH, eth; s tfa; vs alk, HOAc
3859	Dimethoxydimethylsilane		$C_4H_{12}O_2Si$	1112-39-6	120.223			82	0.8646[20]	1.3708[20]	dec H₂O
3860	Dimethoxydiphenylsilane		$C_{14}H_{16}O_2Si$	6843-66-9	244.362			286	1.0771[20]	1.5447[20]	
3861	1,1-Dimethoxydodecane	Lauraldehyde, dimethyl acetal	$C_{14}H_{30}O_2$	14620-52-1	230.387			133[5]		1.4310[25]	vs eth, EtOH
3862	2,2-Dimethoxyethanamine		$C_4H_{11}NO_2$	22483-09-6	105.136		-78	137[95]	0.966[25]	1.4170[20]	
3863	1,2-Dimethoxyethane	Ethylene glycol dimethyl ether	$C_4H_{10}O_2$	110-71-4	90.121	liq	-69.0(0.2)	85.0(0.1)	0.8637[25]	1.3770[25]	s H₂O, EtOH, eth, ace, bz, chl, ctc
3864	(2,2-Dimethoxyethyl)benzene		$C_{10}H_{14}O_2$	101-48-4	166.217			193.5			
3865	4,8-Dimethoxyfuro[2,3-b]quinoline	Fagarine	$C_{13}H_{11}NO_3$	524-15-2	229.231	pr (al)	142				sl H₂O, peth; s EtOH, eth, bz, chl
3866	1,1-Dimethoxyhexadecane	Palmitaldehyde, dimethyl acetal	$C_{18}H_{38}O_2$	2791-29-9	286.494		10	144[2]	0.8542[20]	1.4382[25]	vs ace, eth, EtOH
3867	2,4-Dimethoxy-6-hydroxyace-tophenone	Xanthoxylin	$C_{10}H_{12}O_4$	90-24-4	196.200	cry (al)	82	185[20]			vs eth, EtOH
3868	5,6-Dimethoxy-1-indanone		$C_{11}H_{12}O_3$	2107-69-9	192.211		119.5				sl ctc
3869	6,7-Dimethoxy-1(3H)-isoben-zofuranone	Meconin	$C_{10}H_{10}O_4$	569-31-3	194.184	wh nd (w)	102.5				sl H₂O; s EtOH, eth, ace, bz, HOAc, chl
3870	Dimethoxymethane	Methylal	$C_3H_8O_2$	109-87-5	76.095	liq	-105.11(0.03)	42.3(0.2)	0.8593[20]	1.3513[20]	s H₂O; vs ace, bz, eth, EtOH
3871	1,2-Dimethoxy-4-methylben-zene		$C_9H_{12}O_2$	494-99-5	152.190	pr (eth)	24	221(4)	1.0509[25]	1.5257[25]	i H₂O; sl ctc; vs os
3872	1,3-Dimethoxy-5-methylben-zene		$C_9H_{12}O_2$	4179-19-5	152.190			244	1.0478[15]	1.5234[20]	vs bz, eth, EtOH
3873	1,4-Dimethoxy-2-methylben-zene		$C_9H_{12}O_2$	24599-58-4	152.190		21	214.0			

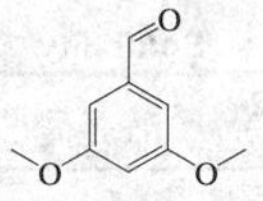

3839
3,5-Dimethoxybenzaldehyde

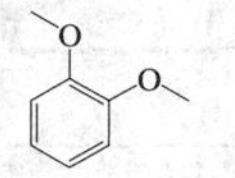

3840
1,2-Dimethoxybenzene

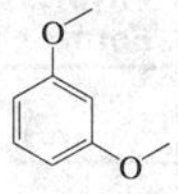

3841
1,3-Dimethoxybenzene

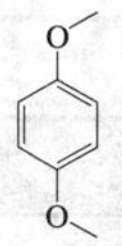

3842
1,4-Dimethoxybenzene

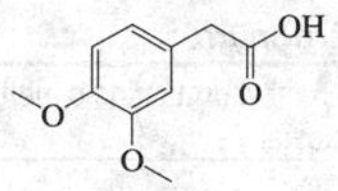

3843
3,4-Dimethoxybenzeneacetic acid

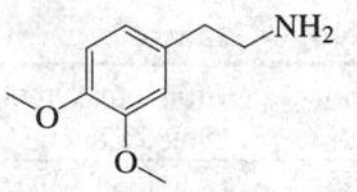

3844
3,4-Dimethoxybenzeneethanamine

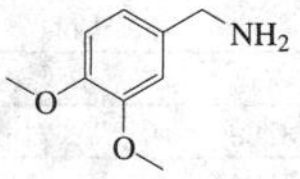

3845
3,4-Dimethoxybenzenemethanamine

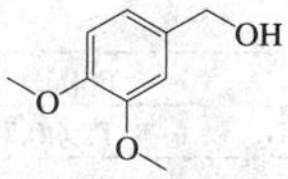

3846
3,4-Dimethoxybenzenemethanol

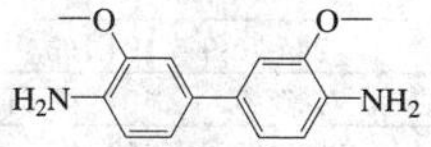

3847
3,3'-Dimethoxybenzidine

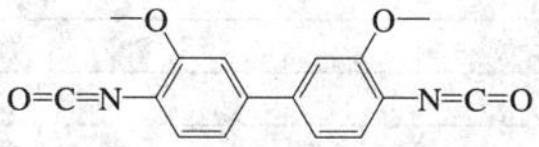

3848
3,3'-Dimethoxybenzidine-4,4'-diisocyanate

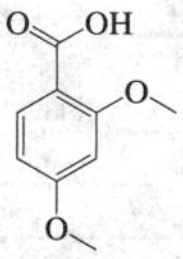

3849
2,4-Dimethoxybenzoic acid

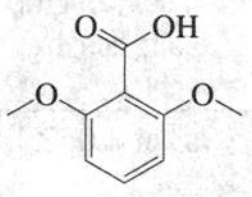

3850
2,6-Dimethoxybenzoic acid

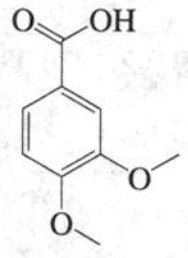

3851
3,4-Dimethoxybenzoic acid

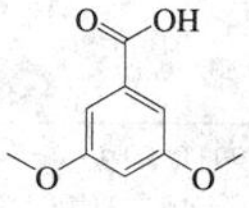

3852
3,5-Dimethoxybenzoic acid

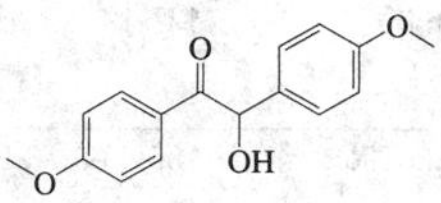

3853
4,4'-Dimethoxybenzoin

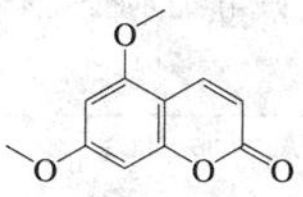

3854
5,7-Dimethoxy-2H-1-benzopyran-2-one

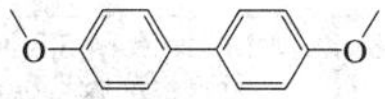

3855
4,4'-Dimethoxy-1,1'-biphenyl

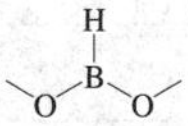

3856
Dimethoxyborane

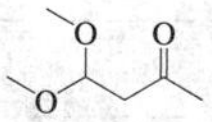

3857
4,4-Dimethoxy-2-butanone

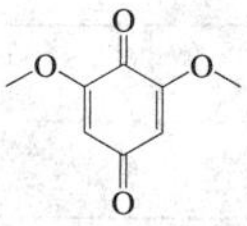

3858
2,6-Dimethoxy-2,5-cyclohexadiene-1,4-dione

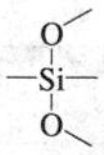

3859
Dimethoxydimethylsilane

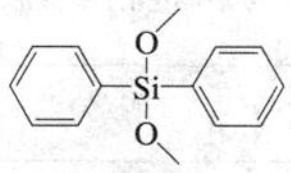

3860
Dimethoxydiphenylsilane

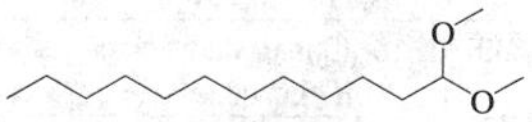

3861
1,1-Dimethoxydodecane

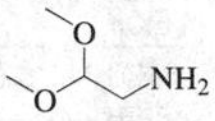

3862
2,2-Dimethoxyethanamine

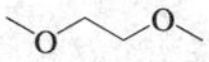

3863
1,2-Dimethoxyethane

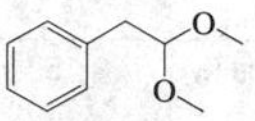

3864
(2,2-Dimethoxyethyl)benzene

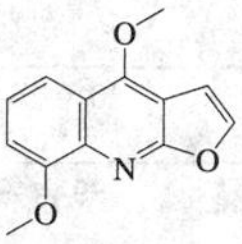

3865
4,8-Dimethoxyfuro[2,3-b]quinoline

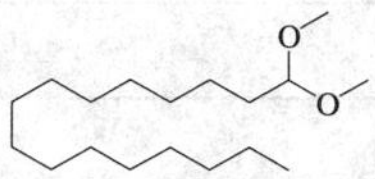

3866
1,1-Dimethoxyhexadecane

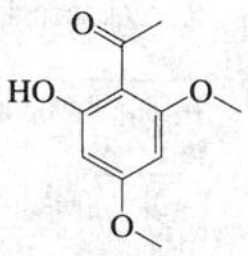

3867
2,4-Dimethoxy-6-hydroxyacetophenone

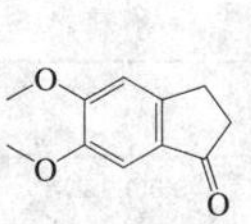

3868
5,6-Dimethoxy-1-indanone

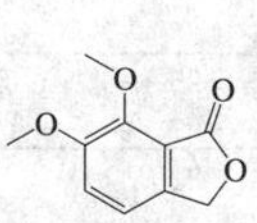

3869
6,7-Dimethoxy-1(3H)-isobenzofuranone

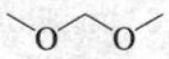

3870
Dimethoxymethane

3871
1,2-Dimethoxy-4-methylbenzene

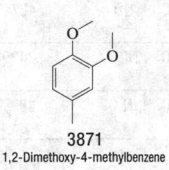

3872
1,3-Dimethoxy-5-methylbenzene

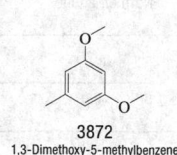

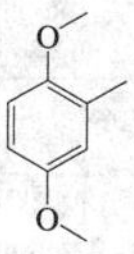

3873
1,4-Dimethoxy-2-methylbenzene

No.	Name	Synonym	Mol. Form.	CAS RN	Mol. Wt.	Physical Form	mp/°C	bp/°C	den g cm⁻³	n_D	Solubility
3874	N-(Dimethoxymethyl)-dimethylamine	Dimethylformamide dimethyl acetal	$C_5H_{13}NO_2$	4637-24-5	119.163			104	0.897²⁵	1.3972²⁰	
3875	2,2-Dimethoxy-N-methylethanamine		$C_5H_{13}NO_2$	122-07-6	119.163			140	0.928²⁵	1.4115²⁰	
3876	Dimethoxymethylphenylsilane		$C_9H_{14}O_2Si$	3027-21-2	182.292			129⁷⁹		1.4795²⁰	
3877	1,2-Dimethoxy-4-nitrobenzene		$C_8H_9NO_4$	709-09-1	183.162	ye nd (al-w)	98	230¹⁵	1.1888¹³³		i H₂O; vs EtOH, eth; s chl; sl lig
3878	1,4-Dimethoxy-2-nitrobenzene		$C_8H_9NO_4$	89-39-4	183.162	gold-ye nd (dil al)	72.5		1.1666¹³²		i H₂O; s EtOH, bz, chl, sulf
3879	2,6-Dimethoxyphenol		$C_8H_{10}O_3$	91-10-1	154.163	mcl pr (w)	56.5	261(22)			vs eth, EtOH
3880	3,5-Dimethoxyphenol		$C_8H_{10}O_3$	500-99-2	154.163		37	199³⁵			s eth, bz; sl lig
3881	1-(3,4-Dimethoxyphenyl)-ethanone		$C_{10}H_{12}O_3$	1131-62-0	180.200	pr (dil al)	51	287			vs H₂O, bz, EtOH, chl
3882	1,1-Dimethoxypropane		$C_5H_{12}O_2$	4744-10-9	104.148			88(5)	0.8648²⁰		
3883	2,2-Dimethoxypropane		$C_5H_{12}O_2$	77-76-9	104.148	liq	-47	77.4(0.7)	0.847²⁵	1.3780²⁰	
3884	3,3-Dimethoxy-1-propene		$C_5H_{10}O_2$	6044-68-4	102.132			88	0.862²⁵	1.3954²⁰	
3885	trans-1,2-Dimethoxy-4-(1-propenyl)benzene	Isoeugenyl methyl ether	$C_{11}H_{14}O_2$	6379-72-2	178.228		18	270.5	1.0521²⁰	1.5616²⁰	
3886	4,5-Dimethoxy-6-(2-propenyl)-1,3-benzodioxole	Apiole (Dill)	$C_{12}H_{14}O_4$	484-31-1	222.237	oil	29.5	285	1.1598¹⁵	1.5305¹⁷	
3887	1,2-Dimethoxy-4-vinylbenzene		$C_{10}H_{12}O_2$	6380-23-0	164.201					1.5711²⁰	s chl
3888	Dimethylacetal		$C_4H_{10}O_2$	534-15-6	90.121	liq	-113.2	63(2)	0.8501²⁰	1.3668²⁰	s H₂O, EtOH, eth, ctc, chl; vs ace
3889	N,N-Dimethylacetamide	N,N-Dimethylethanamide	C_4H_9NO	127-19-5	87.120	liq	-19(1)	165.9(0.2)	0.9372²⁵	1.4341²⁵	msc H₂O, EtOH, eth, ace, bz, chl
3890	2,7-Dimethyl-3,6-acridinediamine, monohydrochloride	Acridine Yellow	$C_{15}H_{16}ClN_3$	135-49-9	273.761	red cry pow					s hot H₂O, EtOH
3891	Dimethyl adipate	Dimethyl 1,6-hexanedioate	$C_8H_{14}O_4$	627-93-0	174.195	cry	10.3(0.5)	115¹³	1.0600²⁰	1.4283²⁰	i H₂O; s EtOH, eth, ctc, HOAc
3892	3,3-Dimethylallyl diphosphate	3-Methyl-2-butenyl pyrophosphate	$C_5H_{12}O_7P_2$	358-72-5	246.092	cry (MeOH)					
3893	Dimethylamine	N-Methylmethanamine	C_2H_7N	124-40-3	45.084	col gas	-93(2)	7.3(0.4)	0.6804⁰	1.350¹⁷	vs H₂O; s EtOH, eth
3894	Dimethylamine hydrochloride	N-Methylmethanamine hydrochloride	C_2H_8ClN	506-59-2	81.545	orth nd (al)	171				vs H₂O, EtOH, chl
3895	(Dimethylamino)acetonitrile		$C_4H_8N_2$	926-64-7	84.120			136(6)	0.8649²⁰	1.4095²⁰	vs H₂O, EtOH
3896	4'-(Dimethylamino)-acetophenone	4-Acetyl-N,N-dimethylaniline	$C_{10}H_{13}NO$	2124-31-4	163.216	nd (w, peth)	105.5				vs H₂O, eth, lig; sl chl
3897	10-[(Dimethylamino)acetyl]-10H-phenothiazine	Ahistan	$C_{16}H_{16}N_2OS$	518-61-6	284.375	cry	144.5				
3898	4-(Dimethylamino)azobenzene		$C_{14}H_{15}N_3$	60-11-7	225.289	ye lf (al)	116(1)	dec			i H₂O; vs EtOH, py; s eth; sl chl, lig
3899	2',3-Dimethyl-4-aminoazobenzene	4-o-Tolylazo-o-toluidine	$C_{14}H_{15}N_3$	97-56-3	225.289	ye lf (al)	102				vs eth, EtOH
3900	4-(Dimethylamino)-benzaldehyde	Ehrlich's reagent	$C_9H_{11}NO$	100-10-7	149.189	lf (w)	73.1(0.8)	176¹⁷	1.0254¹⁰⁰		sl H₂O, chl; s EtOH, eth, ace, bz
3901	4-(Dimethylamino)-benzalrhodanine		$C_{12}H_{12}N_2OS_2$	536-17-4	264.365	dp red nd (xyl)	270 dec				i H₂O; sl EtOH, bz; vs eth, ctc, s ace
3902	2-(Dimethylamino)benzoic acid		$C_9H_{11}NO_2$	610-16-2	165.189	pr, nd (eth)	72	sub			vs H₂O, eth, EtOH
3903	3-(Dimethylamino)benzoic acid		$C_9H_{11}NO_2$	99-64-9	165.189	nd (w)	152.5				sl H₂O, chl; s EtOH, eth
3904	4-(Dimethylamino)benzoic acid		$C_9H_{11}NO_2$	619-84-1	165.189	nd (al)	242.5				s EtOH; sl eth
3905	4,4'-Dimethylaminobenzophenonimide	Brilliant Oil Yellow	$C_{17}H_{21}N_3$	492-80-8	267.369	ye or col pl (al)	136				i H₂O; s EtOH; sl eth
3906	(Dimethylamino)dimethylborane		$C_4H_{12}BN$	1113-30-0	84.956	liq	-92	65			vs eth, ace
3907	6-(Dimethylamino)-4,4-diphenyl-3-heptanone		$C_{21}H_{27}NO$	76-99-3	309.445		99.5				vs EtOH
3908	6-(Dimethylamino)-4,4-diphenyl-3-hexanone	Normethadone	$C_{20}H_{25}NO$	467-85-6	295.419	oily liq		165³			
3909	2-(Dimethylamino)ethyl acrylate		$C_7H_{13}NO_2$	2439-35-2	143.184		<-60	95⁵⁰	0.938²⁰		
3910	3-[2-(Dimethylamino)ethyl]-1H-indol-5-ol	Bufotenine	$C_{12}H_{16}N_2O$	487-93-4	204.267	pr (EtOAc)	146.5	320⁰·¹			vs eth, EtOH

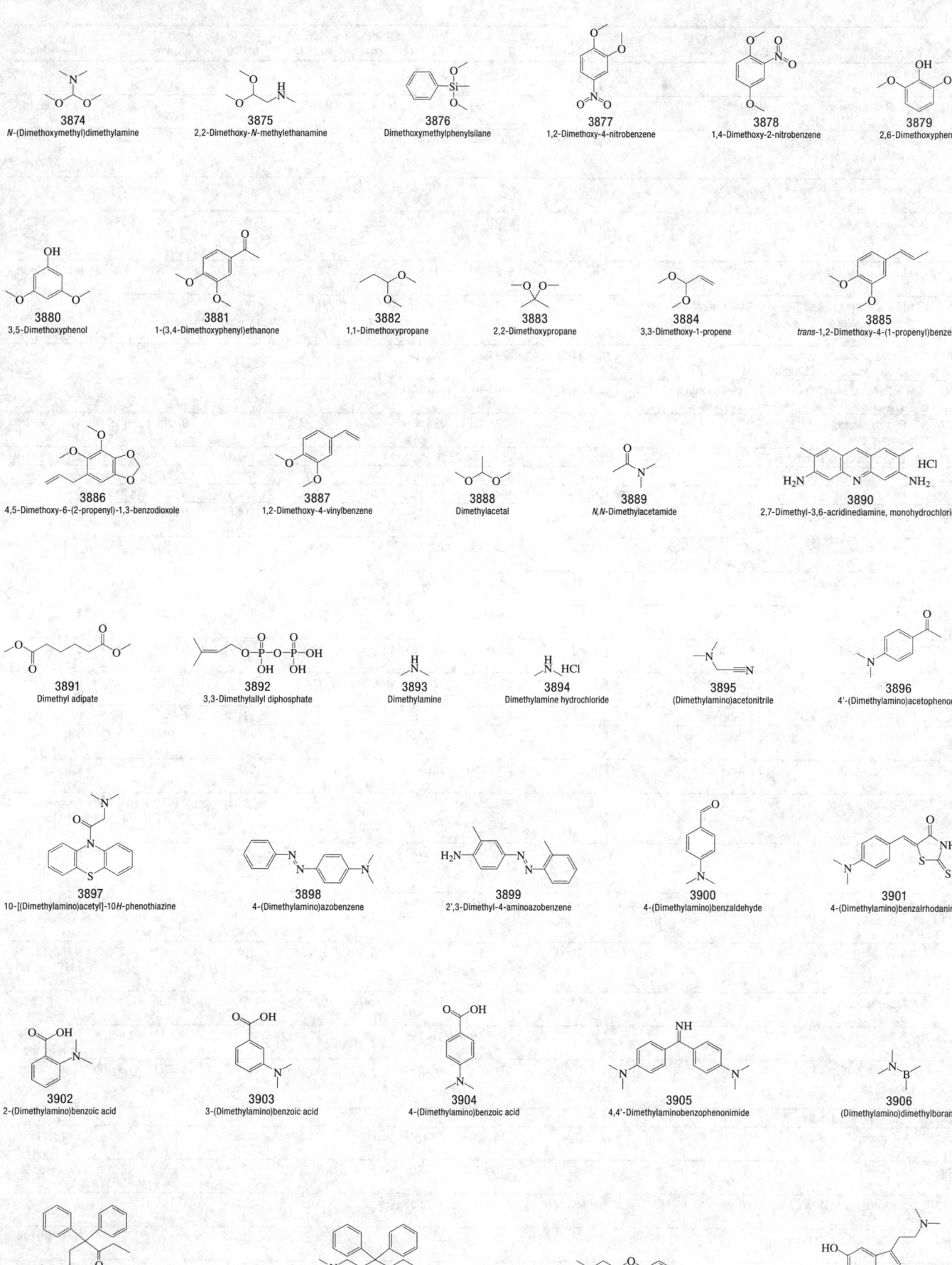

3874
N-(Dimethoxymethyl)dimethylamine

3875
2,2-Dimethoxy-N-methylethanamine

3876
Dimethoxymethylphenylsilane

3877
1,2-Dimethoxy-4-nitrobenzene

3878
1,4-Dimethoxy-2-nitrobenzene

3879
2,6-Dimethoxyphenol

3880
3,5-Dimethoxyphenol

3881
1-(3,4-Dimethoxyphenyl)ethanone

3882
1,1-Dimethoxypropane

3883
2,2-Dimethoxypropane

3884
3,3-Dimethoxy-1-propene

3885
trans-1,2-Dimethoxy-4-(1-propenyl)benzene

3886
4,5-Dimethoxy-6-(2-propenyl)-1,3-benzodioxole

3887
1,2-Dimethoxy-4-vinylbenzene

3888
Dimethylacetal

3889
N,N-Dimethylacetamide

3890
2,7-Dimethyl-3,6-acridinediamine, monohydrochloride

3891
Dimethyl adipate

3892
3,3-Dimethylallyl diphosphate

3893
Dimethylamine

3894
Dimethylamine hydrochloride

3895
(Dimethylamino)acetonitrile

3896
4'-(Dimethylamino)acetophenone

3897
10-[(Dimethylamino)acetyl]-10H-phenothiazine

3898
4-(Dimethylamino)azobenzene

3899
2',3-Dimethyl-4-aminoazobenzene

3900
4-(Dimethylamino)benzaldehyde

3901
4-(Dimethylamino)benzalrhodanine

3902
2-(Dimethylamino)benzoic acid

3903
3-(Dimethylamino)benzoic acid

3904
4-(Dimethylamino)benzoic acid

3905
4,4'-Dimethylaminobenzophenonimide

3906
(Dimethylamino)dimethylborane

3907
6-(Dimethylamino)-4,4-diphenyl-3-heptanone

3908
6-(Dimethylamino)-4,4-diphenyl-3-hexanone

3909
2-(Dimethylamino)ethyl acrylate

3910
3-[2-(Dimethylamino)ethyl]-1H-indol-5-ol

No.	Name	Synonym	Mol. Form.	CAS RN	Mol. Wt.	Physical Form	mp/°C	bp/°C	den g cm⁻³	n_D	Solubility
3911	2-(Dimethylamino)ethyl methacrylate		$C_8H_{15}NO_2$	2867-47-2	157.211			63[6]			
3912	4-[2-(Dimethylamino)ethyl]-phenol	Hordenine	$C_{10}H_{15}NO$	539-15-1	165.232	orth pr (al), nd (w)	117.5	173[11]			vs eth, EtOH, chl
3913	N-[2-(Dimethylamino)ethyl]-N,N',N'-trimethyl-1,2-ethanediamine		$C_9H_{23}N_3$	3030-47-5	173.299			84[12]		1.4413[25]	
3914	5-(Dimethylamino)-1-naphthalenesulfonyl chloride	Dansyl chloride	$C_{12}H_{12}ClNO_2S$	605-65-2	269.747		70				
3915	3-(Dimethylamino)phenol		$C_8H_{11}NO$	99-07-0	137.179	nd (lig)	86	266.5		1.5895[26]	i H_2O; s EtOH, eth, ace, bz, CS_2
3916	4-(Dimethylamino)phenol		$C_8H_{11}NO$	619-60-3	137.179		77	165[30]			sl H_2O; s EtOH, eth
3917	[4-(Dimethylamino)phenyl]-phenylmethanone	4-(Dimethylamino)-benzophenone	$C_{15}H_{15}NO$	530-44-9	225.286	ye lf (al) nd (peth)	92.5				i H_2O; sl EtOH; vs eth; s chl, peth
3918	3-(Dimethylamino)-1-phenyl-1-propanone, hydrochloride		$C_{11}H_{16}ClNO$	879-72-1	213.704		153.5				
3919	3-[4-(Dimethylamino)phenyl]-2-propenal	4-(Dimethylamino)-cinnamaldehyde	$C_{11}H_{13}NO$	6203-18-5	175.227		139.5				
3920	3-(Dimethylamino)-propanenitrile		$C_5H_{10}N_2$	1738-25-6	98.146			177(12)	0.8705[20]		
3921	2-(Dimethylamino)-1-propanol		$C_5H_{13}NO$	15521-18-3	103.163			150.3	0.8820[26]		s H_2O
3922	3-(Dimethylamino)-1-propanol		$C_5H_{13}NO$	3179-63-3	103.163			163.5	0.872[25]	1.4360[20]	s ctc
3923	1-(Dimethylamino)-2-propanol		$C_5H_{13}NO$	108-16-7	103.163			124.5	0.837[25]	1.4193[20]	s ctc
3924	3-(Dimethylamino)-1-propyne	N,N-Dimethyl-2-propargyl-amine	C_5H_9N	7223-38-3	83.132			81(3)	0.7792[20]	1.4195[20]	
3925	2-Dimethylaminopurine	N,N-Dimethyl-1H-purin-6-amine	$C_7H_9N_5$	938-55-6	163.180		263				
3926	2-(p-Dimethylaminostyryl)-benzothiazole		$C_{17}H_{16}N_2S$	1628-58-6	280.387	ye nd (MeOH)	207 dec				
3927	2,3-Dimethylaniline	2,3-Xylidine	$C_8H_{11}N$	87-59-2	121.180	liq	3(1)	223(17)	0.9931[20]	1.5684[20]	sl H_2O; vs EtOH, eth; s ctc
3928	2,4-Dimethylaniline	2,4-Xylidine	$C_8H_{11}N$	95-68-1	121.180	liq	-13(2)	215(2)	0.9723[20]	1.5569[20]	sl H_2O, ctc; s EtOH, eth, bz
3929	2,5-Dimethylaniline	2,5-Xylidine	$C_8H_{11}N$	95-78-3	121.180	ye lf (lig)	6(1)	214	0.9790[21]	1.5591[21]	sl H_2O; s eth, ctc
3930	2,6-Dimethylaniline	2,6-Xylidine	$C_8H_{11}N$	87-62-7	121.180	liq	11.0(0.4)	211(4)	0.9842[20]	1.5610[20]	vs eth, EtOH
3931	3,4-Dimethylaniline	3,4-Xylidine	$C_8H_{11}N$	95-64-7	121.180	pl or pr (lig)	49(1)	228	1.076[18]		sl H_2O, chl; s eth; vs lig
3932	3,5-Dimethylaniline	3,5-Xylidine	$C_8H_{11}N$	108-69-0	121.180		10(1)	217(5)	0.9706[20]	1.5581[20]	sl H_2O; s eth, ctc
3933	N,2-Dimethylaniline		$C_8H_{11}N$	611-21-2	121.180			207.5	0.9709[20]	1.5649[20]	i H_2O; msc EtOH, eth; s ace
3934	N,3-Dimethylaniline		$C_8H_{11}N$	696-44-6	121.180			206.5	0.9660[20]	1.5557[25]	i H_2O; msc EtOH, eth; s ace
3935	N,4-Dimethylaniline		$C_8H_{11}N$	623-08-5	121.180			215(20)	0.9348[55]	1.5568[20]	i H_2O; msc EtOH, eth; s ace
3936	N,N-Dimethylaniline		$C_8H_{11}N$	121-69-7	121.180	pa ye	2.1(0.5)	193(1)	0.9557[20]	1.5582[20]	sl H_2O; s EtOH, eth, ace, bz; vs chl
3937	N,N-Dimethylaniline hydrochloride		$C_8H_{12}ClN$	5882-44-0	157.641	hyg pl (w, bz)	90		1.1156[19]		vs H_2O, EtOH, chl
3938	2,6-Dimethylanisole		$C_9H_{12}O$	1004-66-6	136.190			183(11)	0.9619[14]	1.5053[14]	i H_2O; s EtOH, eth, bz, ctc
3939	3,5-Dimethylanisole		$C_9H_{12}O$	874-63-5	136.190			194	0.9627[15]	1.5110[20]	i H_2O; s EtOH, eth, bz, CS_2; sl ctc
3940	9,10-Dimethylanthracene		$C_{16}H_{14}$	781-43-1	206.282		186.4(0.5)	360.0			i H_2O
3941	1,4-Dimethyl-9,10-anthracenedione		$C_{16}H_{12}O_2$	1519-36-4	236.265	ye nd (al, sub)	140.5	sub			i H_2O; sl EtOH; s bz, xyl, HOAc
3942	Dimethylarsine		C_2H_7As	593-57-7	105.999	liq, ign in air	-136.1	36	1.208[29]		vs ace, bz, eth, EtOH
3943	Dimethylarsinic acid	Cacodylic acid	$C_2H_7AsO_2$	75-60-5	137.998		199.5(0.6)	>200			vs H_2O; s EtOH; i eth
3944	2,4-Dimethylbenzaldehyde		$C_9H_{10}O$	15764-16-6	134.174	liq	-9	218			s EtOH; s eth, ace, bz; sl chl
3945	2,5-Dimethylbenzaldehyde	Isoxylaldehyde	$C_9H_{10}O$	5779-94-2	134.174			220	0.9500[20]		vs EtOH; s eth, ace, bz, ctc

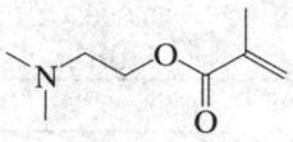

3911
2-(Dimethylamino)ethyl methacrylate

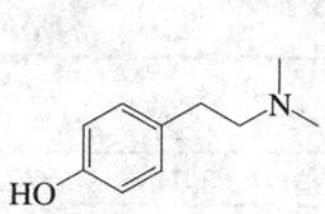

3912
4-[2-(Dimethylamino)ethyl]phenol

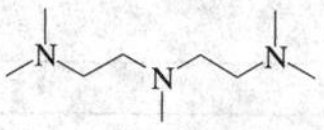

3913
N-[2-(Dimethylamino)ethyl]-N,N',N'-trimethyl-1,2-ethanediamine

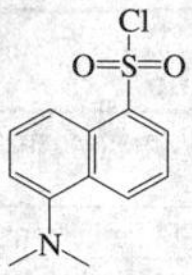

3914
5-(Dimethylamino)-1-naphthalenesulfonyl chloride

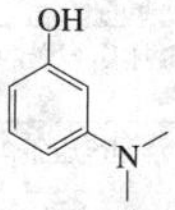

3915
3-(Dimethylamino)phenol

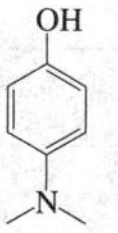

3916
4-(Dimethylamino)phenol

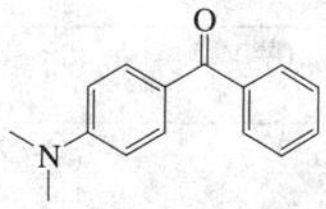

3917
[4-(Dimethylamino)phenyl]phenylmethanone

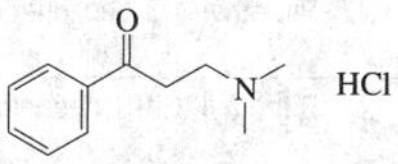

3918
3-(Dimethylamino)-1-phenyl-1-propanone, hydrochloride

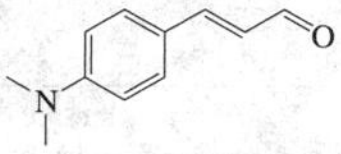

3919
3-[4-(Dimethylamino)phenyl]-2-propenal

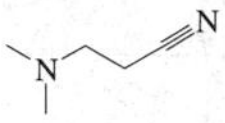

3920
3-(Dimethylamino)propanenitrile

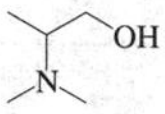

3921
2-(Dimethylamino)-1-propanol

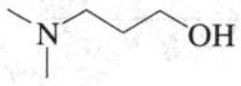

3922
3-(Dimethylamino)-1-propanol

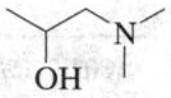

3923
1-(Dimethylamino)-2-propanol

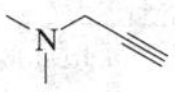

3924
3-(Dimethylamino)-1-propyne

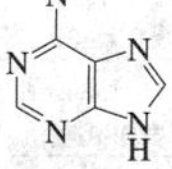

3925
2-Dimethylaminopurine

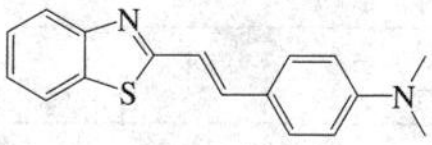

3926
2-(p-Dimethylaminostyryl)benzothiazole

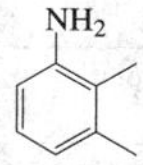

3927
2,3-Dimethylaniline

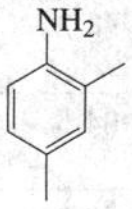

3928
2,4-Dimethylaniline

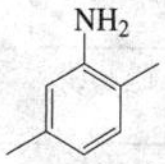

3929
2,5-Dimethylaniline

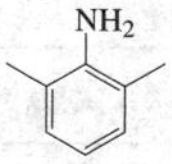

3930
2,6-Dimethylaniline

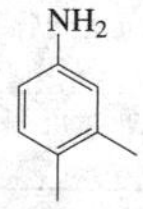

3931
3,4-Dimethylaniline

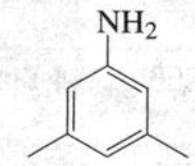

3932
3,5-Dimethylaniline

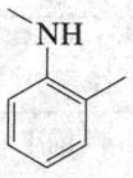

3933
N,2-Dimethylaniline

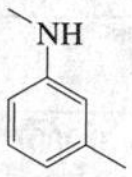

3934
N,3-Dimethylaniline

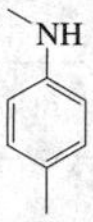

3935
N,4-Dimethylaniline

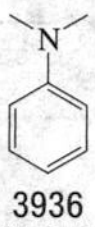

3936
N,N-Dimethylaniline

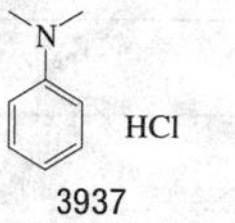

3937
N,N-Dimethylaniline hydrochloride

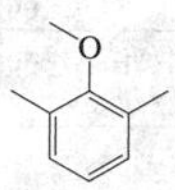

3938
2,6-Dimethylanisole

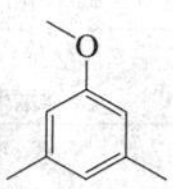

3939
3,5-Dimethylanisole

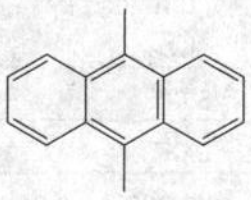

3940
9,10-Dimethylanthracene

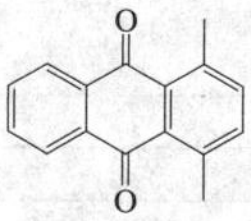

3941
1,4-Dimethyl-9,10-anthracenedione

3942
Dimethylarsine

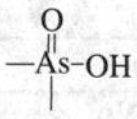

3943
Dimethylarsinic acid

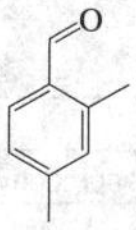

3944
2,4-Dimethylbenzaldehyde

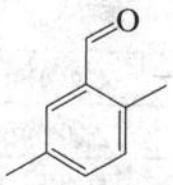

3945
2,5-Dimethylbenzaldehyde

No.	Name	Synonym	Mol. Form.	CAS RN	Mol. Wt.	Physical Form	mp/°C	bp/°C	den g cm⁻³	n_D	Solubility
3946	3,5-Dimethylbenzaldehyde		$C_9H_{10}O$	5779-95-3	134.174		9	221(12)			vs ace, bz, eth, EtOH
3947	*N,N*-Dimethylbenzamide		$C_9H_{11}NO$	611-74-5	149.189		43.8(0.5)	272.0			
3948	7,12-Dimethylbenz[*a*]anthracene	9,10-Dimethyl-1,2-benzanthracene	$C_{20}H_{16}$	57-97-6	256.341	pa ye pl (al, HOAc)	123(1)				vs ace, bz
3949	4,5-Dimethyl-1,2-benzenediamine		$C_8H_{12}N_2$	3171-45-7	136.194		128				
3950	*N,N*-Dimethyl-1,2-benzenediamine		$C_8H_{12}N_2$	2836-03-5	136.194	oil		218	0.995²²		sl H_2O; vs EtOH, eth, ace, bz
3951	*N,N*-Dimethyl-1,3-benzenediamine		$C_8H_{12}N_2$	2836-04-6	136.194		<-20	270	0.995²⁵		sl H_2O; vs EtOH, eth
3952	*N,N*-Dimethyl-1,4-benzenediamine	Dimethyl-*p*-phenylenediamine	$C_8H_{12}N_2$	99-98-9	136.194	nd (bz)	53	264(5)	1.036²⁰		s H_2O, chl; vs EtOH, eth, bz; sl lig
3953	2,5-Dimethyl-1,3-benzenediol		$C_8H_{10}O_2$	488-87-9	138.164	nd (bz), pr (w)	163	278.5			s H_2O, EtOH, eth
3954	2,6-Dimethyl-1,4-benzenediol		$C_8H_{10}O_2$	654-42-2	138.164	nd (xyl), cry (w)	152.3				vs eth, EtOH
3955	*N,β*-Dimethylbenzeneethanamine	Phenylpropylmethylamine	$C_{10}H_{15}N$	93-88-9	149.233			207.5	0.915²⁵		vs bz, eth, EtOH
3956	α,α-Dimethylbenzeneethanamine	Phentermine	$C_{10}H_{15}N$	122-09-8	149.233	oily liq		205			
3957	α,α-Dimethylbenzenemethanamine		$C_9H_{13}N$	585-32-0	135.206			196.5	0.9423²⁰	1.5181²⁵	
3958	α,4-Dimethylbenzenemethanol	1-(4-Methylphenyl)ethanol	$C_9H_{12}O$	536-50-5	136.190			215(11)	0.9668²⁵	1.5246²⁰	i H_2O; vs EtOH, eth
3959	α,α-Dimethylbenzenemethanol	α-Cumyl alcohol	$C_9H_{12}O$	617-94-7	136.190	pr	37(1)	202	0.9735²⁰	1.5325²⁰	i H_2O; s EtOH, eth, bz, HOAc
3960	α,α-Dimethylbenzenepropanol	Benzyl-*tert*-butanol	$C_{11}H_{16}O$	103-05-9	164.244	nd	24.5	121¹³	0.9626²¹	1.5077²¹	i H_2O; vs EtOH, eth, ace, bz
3961	*N,4*-Dimethylbenzenesulfonamide		$C_8H_{11}NO_2S$	640-61-9	185.244	pl (dil al)	78.5		1.340²⁵		vs eth, EtOH
3962	5,6-Dimethyl-1*H*-benzimidazole	Dimedazole	$C_9H_{10}N_2$	582-60-5	146.188	cry (eth)	205.5	sub			s H_2O, EtOH, eth, chl, DMSO
3963	2,4-Dimethylbenzoic acid		$C_9H_{10}O_2$	611-01-8	150.174	mcl or tcl nd (w)	90	268			sl H_2O; s EtOH, ace, bz, chl, HOAc, tol
3964	2,5-Dimethylbenzoic acid		$C_9H_{10}O_2$	610-72-0	150.174	nd (al)	132(3)	sub	1.069²¹		i H_2O; s EtOH, eth, ace, bz
3965	2,6-Dimethylbenzoic acid		$C_9H_{10}O_2$	632-46-2	150.174	nd (lig)	116	277(18)			sl H_2O, lig; s EtOH, eth
3966	3,4-Dimethylbenzoic acid		$C_9H_{10}O_2$	619-04-5	150.174	pr (al)	163(2)				i H_2O; s EtOH, eth, bz
3967	3,5-Dimethylbenzoic acid	Mesitylenic acid	$C_9H_{10}O_2$	499-06-9	150.174	nd (w, al)	171.1	275(19)			sl H_2O; vs EtOH, eth
3968	4,4'-Dimethylbenzophenone	Bis(4-methylphenyl) ketone	$C_{15}H_{14}O$	611-97-2	210.271	orth (al)	96.5	339(20)			vs ace, bz, eth, EtOH
3969	7,8-Dimethylbenzo[*g*]pteridine-2,4(1*H*,3*H*)-dione	Lumichrome	$C_{12}H_{10}N_4O_2$	1086-80-2	242.233	ye cry (chl)	300				sl H_2O, EtOH, chl
3970	2,5-Dimethylbenzoxazole		C_9H_9NO	5676-58-4	147.173			218.5	1.0880¹⁸	1.5412²⁰	s ctc
3971	*N,N*-Dimethylbenzylamine	Dimethylbenzylamine	$C_9H_{13}N$	103-83-3	135.206			179(15)	0.915⁰	1.5011²⁰	sl H_2O; msc EtOH, eth
3972	*N,N*-Dimethyl-*N'*-benzyl-1,2-ethanediamine	*N*-Benzyl-*N',N'*-dimethyl-1,2-ethanediamine	$C_{11}H_{18}N_2$	103-55-9	178.274			145³⁰	0.9343²⁰	1.5089²⁰	
3973	*N,N*-Dimethyl-*N'*-benzyl-*N'*-2-pyridinyl-1,2-ethanediamine	Tripelennamine	$C_{16}H_{21}N_3$	91-81-6	255.358	ye oil		140⁰·¹		1.576²⁵	misc H_2O
3974	6,6-Dimethylbicyclo[3.1.1]heptan-2-one, (1*R*)-		$C_9H_{14}O$	38651-65-9	138.206	liq	-1	209	0.9807²⁰	1.4787²⁰	vs eth, EtOH
3975	2,3-Dimethylbicyclo[2.2.1]hept-2-ene	2,3-Dimethyl-2-norbornene	C_9H_{14}	529-16-8	122.207			140.5	0.8698¹⁷	1.4688¹⁷	s eth, ace, bz
3976	6,6-Dimethylbicyclo[3.1.1]hept-2-ene-2-ethanol		$C_{11}H_{18}O$	128-50-7	166.260			235	0.973²⁵	1.4930²⁰	s chl
3977	2,2'-Dimethylbiphenyl		$C_{14}H_{14}$	605-39-0	182.261	cry (al)	19.94(0.01)	258(7)	0.9906²⁰	1.5752²⁰	i H_2O; vs EtOH, eth, bz; s ace
3978	3,3'-Dimethylbiphenyl		$C_{14}H_{14}$	612-75-9	182.261		8(4)	289.6(0.2)	0.9995²⁰	1.5946²⁰	i H_2O; vs EtOH, eth, bz; s ace
3979	4,4'-Dimethylbiphenyl		$C_{14}H_{14}$	613-33-2	182.261	mcl pr (eth)	120.9(0.4)	294(11)	0.917¹²¹		i H_2O; sl EtOH; s eth, ace, bz, CS_2
3980	4,4'-Dimethyl-2,2'-bipyridine		$C_{12}H_{12}N_2$	1134-35-6	184.236		171.5				s chl
3981	2,3-Dimethyl-1,3-butadiene	Diisopropenyl	C_6H_{10}	513-81-5	82.143	liq	-76.0(0.1)	69(1)	0.7222²⁵	1.4394²⁰	s ctc
3982	*N,N*-Dimethylbutanamide		$C_6H_{13}NO$	760-79-2	115.173	liq	-40	186	0.9064²⁵	1.4391²⁵	vs ace, bz, eth, EtOH

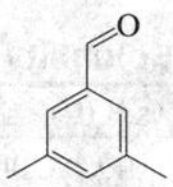

3946
3,5-Dimethylbenzaldehyde

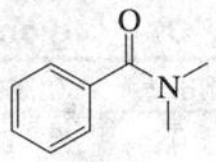

3947
N,N-Dimethylbenzamide

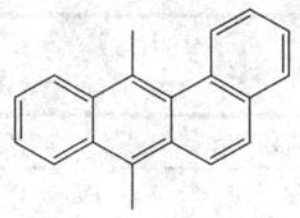

3948
7,12-Dimethylbenz[a]anthracene

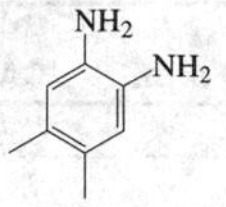

3949
4,5-Dimethyl-1,2-benzenediamine

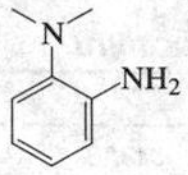

3950
N,N-Dimethyl-1,2-benzenediamine

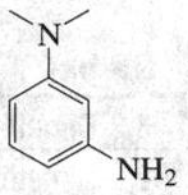

3951
N,N-Dimethyl-1,3-benzenediamine

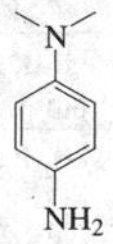

3952
N,N-Dimethyl-1,4-benzenediamine

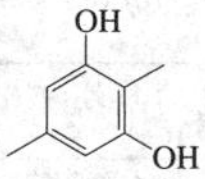

3953
2,5-Dimethyl-1,3-benzenediol

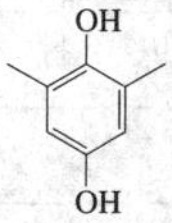

3954
2,6-Dimethyl-1,4-benzenediol

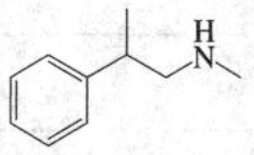

3955
N,β-Dimethylbenzeneethanamine

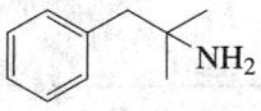

3956
α,α-Dimethylbenzeneethanamine

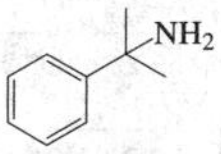

3957
α,α-Dimethylbenzenemethanamine

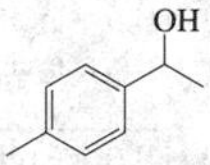

3958
α,4-Dimethylbenzenemethanol

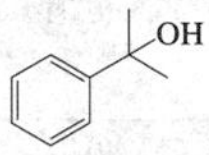

3959
α,α-Dimethylbenzenemethanol

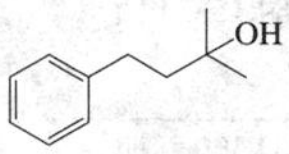

3960
α,α-Dimethylbenzenepropanol

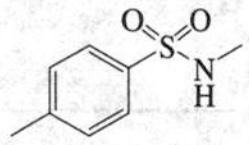

3961
N,4-Dimethylbenzenesulfonamide

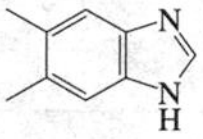

3962
5,6-Dimethyl-1*H*-benzimidazole

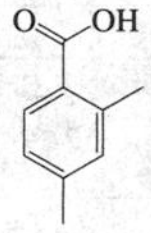

3963
2,4-Dimethylbenzoic acid

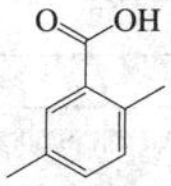

3964
2,5-Dimethylbenzoic acid

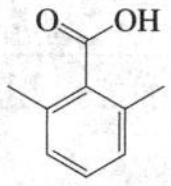

3965
2,6-Dimethylbenzoic acid

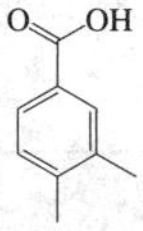

3966
3,4-Dimethylbenzoic acid

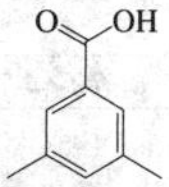

3967
3,5-Dimethylbenzoic acid

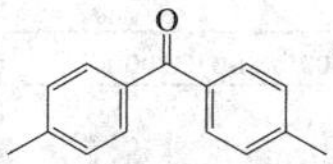

3968
4,4'-Dimethylbenzophenone

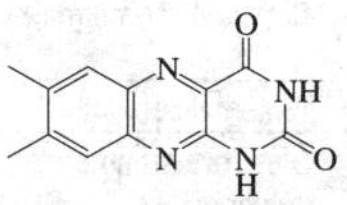

3969
7,8-Dimethylbenzo[g]pteridine-2,4(1*H*,3*H*)-dione

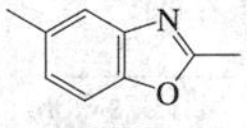

3970
2,5-Dimethylbenzoxazole

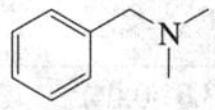

3971
N,N-Dimethylbenzylamine

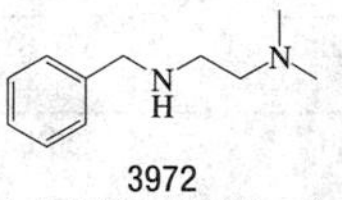

3972
N,N-Dimethyl-*N'*-benzyl-1,2-ethanediamine

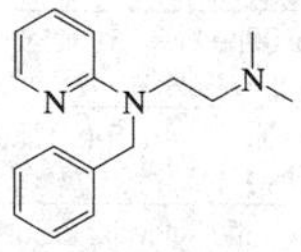

3973
N,N-Dimethyl-*N'*-benzyl-*N'*-2-pyridinyl-1,2-ethanediamine

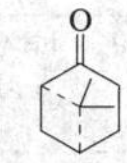

3974
6,6-Dimethylbicyclo[3.1.1]heptan-2-one, (1*R*)-

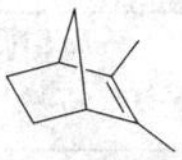

3975
2,3-Dimethylbicyclo[2.2.1]hept-2-ene

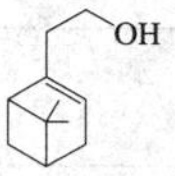

3976
6,6-Dimethylbicyclo[3.1.1]hept-2-ene-2-ethanol

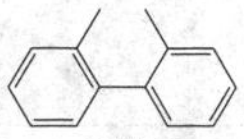

3977
2,2'-Dimethylbiphenyl

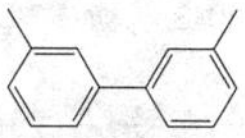

3978
3,3'-Dimethylbiphenyl

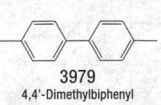

3979
4,4'-Dimethylbiphenyl

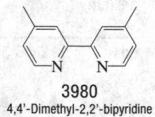

3980
4,4'-Dimethyl-2,2'-bipyridine

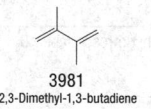

3981
2,3-Dimethyl-1,3-butadiene

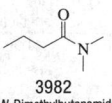

3982
N,N-Dimethylbutanamide

No.	Name	Synonym	Mol. Form.	CAS RN	Mol. Wt.	Physical Form	mp/°C	bp/°C	den g cm⁻³	n_D	Solubility
3983	3,3-Dimethyl-2-butanamine		$C_6H_{15}N$	3850-30-4	101.190	liq	-20	101(6)	0.7668[20]	1.4105[25]	vs H_2O
3984	2,2-Dimethylbutane	Neohexane	C_6H_{14}	75-83-2	86.175	liq	-99.0(0.4)	49.7(0.2)	0.6444[25]	1.3688[20]	i H_2O; s EtOH, eth; vs ace, bz, peth, ctc
3985	2,3-Dimethylbutane	Diisopropyl	C_6H_{14}	79-29-8	86.175	liq	-128.1(0.2)	58.0(0.3)	0.6616[20]	1.3750[20]	i H_2O; s EtOH, eth; vs ace, bz, peth, ctc
3986	2,3-Dimethyl-2,3-butanediol	Pinacol	$C_6H_{14}O_2$	76-09-5	118.174	nd (al,eth)	43.3(0.2)	174(3)			sl H_2O, CS_2; vs EtOH, eth
3987	2,3-Dimethyl-2-butanethiol		$C_6H_{14}S$	1639-01-6	118.240	liq		126.1(0.7)			
3988	2,2-Dimethylbutanoic acid		$C_6H_{12}O_2$	595-37-9	116.158	liq	-15.0(0.5)	187.87(0.02)	0.9276[20]	1.4145[20]	sl H_2O; s EtOH, eth
3989	2,2-Dimethyl-1-butanol		$C_6H_{14}O$	1185-33-7	102.174		<-15	137(1)	0.8283[20]	1.4208[20]	sl H_2O; s EtOH, eth
3990	3,3-Dimethyl-1-butanol	Dimbunol	$C_6H_{14}O$	624-95-3	102.174	liq	-60	143(2)	0.844[15]	1.4323[15]	sl H_2O; s EtOH, eth, ace
3991	2,3-Dimethyl-2-butanol	Isopropyldimethylcarbinol	$C_6H_{14}O$	594-60-5	102.174	liq	-10.5(0.3)	118.7(0.5)	0.8236[20]	1.4176[20]	s H_2O; msc EtOH, eth
3992	3,3-Dimethyl-2-butanol, (±)-		$C_6H_{14}O$	20281-91-8	102.174		5.6	120.4	0.8122[25]	1.4148[20]	sl H_2O; vs EtOH, eth
3993	3,3-Dimethyl-2-butanone	Pinacolone	$C_6H_{12}O$	75-97-8	100.158	liq	-51.40(0.05)	106.1(0.2)	0.7229[25]	1.3952[20]	sl H_2O; s EtOH, eth, ace, ctc
3994	3,3-Dimethylbutanoyl chloride		$C_6H_{11}ClO$	7065-46-5	134.603			129(8)	0.969[20]	1.4210[20]	vs eth
3995	2,3-Dimethyl-1-butene		C_6H_{12}	563-78-0	84.159	liq	-157.27(0.09)	55.59(0.04)	0.6803[20]	1.3995[20]	i H_2O; s EtOH, eth, ace, ctc, CS_2
3996	3,3-Dimethyl-1-butene		C_6H_{12}	558-37-2	84.159	liq	-115.2(0.3)	41.24(0.04)	0.6529[20]	1.3763[20]	i H_2O; s EtOH, eth, ctc, chl
3997	2,3-Dimethyl-2-butene		C_6H_{12}	563-79-1	84.159	liq	-74.3(0.1)	73.19(0.06)	0.7080[20]	1.4122[20]	i H_2O; s EtOH, eth, ace, chl
3998	N-(1,3-Dimethylbutyl)-N'-phenyl-1,4-benzenediamine		$C_{18}H_{24}N_2$	793-24-8	268.397		46	164[1]			
3999	3,3-Dimethyl-1-butyne	tert-Butylacetylene	C_6H_{10}	917-92-0	82.143	liq	-78.2(0.4)	38(2)	0.6623[25]	1.3736[20]	
4000	Dimethyl 2-butynedioate		$C_6H_6O_4$	762-42-5	142.110			199(9)	1.1564[20]	1.4434[20]	s EtOH, eth, ctc
4001	Dimethyl cadmium		C_2H_6Cd	506-82-1	142.480		-2.67(0.02)	105.5 (exp 150)	1.9846[18]	1.5488	s peth
4002	Dimethylcarbamic chloride	Dimethylcarbamoyl chloride	C_3H_6ClNO	79-44-7	107.539	liq	-33	154(10)	1.168[25]	1.4540[20]	
4003	Dimethylcarbamothioic chloride		C_3H_6ClNS	16420-13-6	123.605	pr	42(2)	98[10]			vs eth; s chl, peth
4004	Dimethyl carbate		$C_{11}H_{14}O_4$	39589-98-5	210.227	cry	38	137[12.5]	1.164[21]	1.4852[20]	i H_2O
4005	Dimethyl carbonate	Methyl carbonate	$C_3H_6O_3$	616-38-6	90.078		-1(10)	90.11(0.09)	1.0636[25]	1.3687[20]	i H_2O; s EtOH, eth; sl ctc
4006	Dimethylcyanamide		$C_3H_6N_2$	1467-79-4	70.093			162(5)		1.4089[19]	vs ace, eth, EtOH
4007	2,3-Dimethyl-2,5-cyclohexadiene-1,4-dione		$C_8H_8O_2$	526-86-3	136.149	ye nd	55	sub			sl H_2O; s EtOH, eth, chl
4008	2,5-Dimethyl-2,5-cyclohexadiene-1,4-dione		$C_8H_8O_2$	137-18-8	136.149	ye nd (al)	126.0				sl H_2O, EtOH; s eth, bz, chl
4009	2,6-Dimethyl-2,5-cyclohexadiene-1,4-dione		$C_8H_8O_2$	527-61-7	136.149	ye nd	72.5	sub	1.0479[28]		s chl
4010	1,1-Dimethylcyclohexane		C_8H_{16}	590-66-9	112.213	liq	-33.31(0.04)	119.5(0.3)	0.7809[20]	1.4290[20]	i H_2O; s EtOH, eth, ace, bz; msc ctc
4011	cis-1,2-Dimethylcyclohexane		C_8H_{16}	2207-01-4	112.213	liq	-49.83(0.04)	129.7(0.6)	0.7963[20]	1.4360[20]	i H_2O; s EtOH, bz, ctc; msc eth, ace
4012	trans-1,2-Dimethylcyclohexane		C_8H_{16}	6876-23-9	112.213	liq	-88.12(0.02)	123.4(0.3)	0.7760[20]	1.4270[20]	i H_2O; s EtOH, eth; msc ace, bz; vs lig
4013	cis-1,3-Dimethylcyclohexane		C_8H_{16}	638-04-0	112.213	liq	-75.51(0.03)	124.4(0.6)	0.7660[20]	1.4229[20]	i H_2O; msc EtOH, eth, ace, bz, lig, ctc
4014	trans-1,3-Dimethylcyclohexane		C_8H_{16}	2207-03-6	112.213	liq	-90.05(0.03)	120.1(0.7)	0.79[15]	1.4284[25]	
4015	cis-1,4-Dimethylcyclohexane		C_8H_{16}	624-29-3	112.213	liq	-87.4(0.3)	124.3(0.7)	0.7829[20]	1.4230[20]	i H_2O; msc EtOH, eth, ace, bz, lig, ctc
4016	trans-1,4-Dimethylcyclohexane		C_8H_{16}	2207-04-7	112.213	liq	-36.9(0.2)	119.3(0.5)	0.77[15]	1.4185[25]	i H_2O
4017	Dimethyl trans-1,4-cyclohexanedicarboxylate		$C_{10}H_{16}O_4$	3399-22-2	200.232	ndl (eth)	71				s eth
4018	5,5-Dimethyl-1,3-cyclohexanedione	5,5-Dimethyldihydroresorcinol	$C_8H_{12}O_2$	126-81-8	140.180	nd (w)	52.5(0.5)				sl H_2O, eth; s ace, ctc; vs chl, HOAc

3983
3,3-Dimethyl-2-butanamine

3984
2,2-Dimethylbutane

3985
2,3-Dimethylbutane

3986
2,3-Dimethyl-2,3-butanediol

3987
2,3-Dimethyl-2-butanethiol

3988
2,2-Dimethylbutanoic acid

3989
2,2-Dimethyl-1-butanol

3990
3,3-Dimethyl-1-butanol

3991
2,3-Dimethyl-2-butanol

3992
3,3-Dimethyl-2-butanol, (±)-

3993
3,3-Dimethyl-2-butanone

3994
3,3-Dimethylbutanoyl chloride

3995
2,3-Dimethyl-1-butene

3996
3,3-Dimethyl-1-butene

3997
2,3-Dimethyl-2-butene

3998
N-(1,3-Dimethylbutyl)-*N'*-phenyl-1,4-benzenediamine

3999
3,3-Dimethyl-1-butyne

4000
Dimethyl 2-butynedioate

4001
Dimethyl cadmium

4002
Dimethylcarbamic chloride

4003
Dimethylcarbamothioic chloride

4004
Dimethyl carbate

4005
Dimethyl carbonate

4006
Dimethylcyanamide

4007
2,3-Dimethyl-2,5-cyclohexadiene-1,4-dione

4008
2,5-Dimethyl-2,5-cyclohexadiene-1,4-dione

4009
2,6-Dimethyl-2,5-cyclohexadiene-1,4-dione

4010
1,1-Dimethylcyclohexane

4011
cis-1,2-Dimethylcyclohexane

4012
trans-1,2-Dimethylcyclohexane

4013
cis-1,3-Dimethylcyclohexane

4014
trans-1,3-Dimethylcyclohexane

4015
cis-1,4-Dimethylcyclohexane

4016
trans-1,4-Dimethylcyclohexane

4017
Dimethyl *trans*-1,4-cyclohexanedicarboxylate

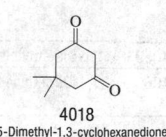

4018
5,5-Dimethyl-1,3-cyclohexanedione

No.	Name	Synonym	Mol. Form.	CAS RN	Mol. Wt.	Physical Form	mp/°C	bp/°C	den g cm^{-3}	n_D	Solubility
4019	N,α-Dimethylcyclohexaneethanamine	Propylhexedrine	C$_{10}$H$_{21}$N	101-40-6	155.281			205	0.8501[20]	1.4600[20]	vs EtOH
4020	3,3-Dimethylcyclohexanol		C$_8$H$_{16}$O	767-12-4	128.212		11(2)	186(4)	0.9128[14]	1.4606[15]	
4021	2,2-Dimethylcyclohexanone		C$_8$H$_{14}$O	1193-47-1	126.196	liq	-20.5	172	0.9145[20]	1.4486[20]	
4022	2,6-Dimethylcyclohexanone		C$_8$H$_{14}$O	2816-57-1	126.196			175(15)	0.925[25]	1.4460[20]	
4023	3,3-Dimethylcyclohexanone		C$_8$H$_{14}$O	2979-19-3	126.196			180	0.909[15]	1.4482[17]	
4024	4,4-Dimethylcyclohexanone		C$_8$H$_{14}$O	4255-62-3	126.196		39	73[14]	0.932[20]	1.4537[24]	
4025	1,2-Dimethylcyclohexene		C$_8$H$_{14}$	1674-10-8	110.197	liq	-84.1(0.2)	137(3)	0.8220[25]	1.4620[20]	
4026	1,3-Dimethylcyclohexene		C$_8$H$_{14}$	2808-76-6	110.197			126(4)	0.799[25]	1.449[20]	
4027	3,5-Dimethyl-2-cyclohexen-1-one		C$_8$H$_{12}$O	1123-09-7	124.180			208.5	0.9400[20]	1.4812[20]	s EtOH, eth
4028	1,1-Dimethylcyclopentane		C$_7$H$_{14}$	1638-26-2	98.186	liq	-69.43(0.01)	87.8(0.3)	0.7499[25]	1.4136[20]	
4029	cis-1,2-Dimethylcyclopentane		C$_7$H$_{14}$	1192-18-3	98.186	liq	-53.67(0.02)	99.5(0.3)	0.7680[25]	1.4222[20]	
4030	trans-1,2-Dimethylcyclopentane		C$_7$H$_{14}$	822-50-4	98.186	liq	-118(1)	91.9(0.4)	0.7468[25]	1.4120[20]	
4031	cis-1,3-Dimethylcyclopentane		C$_7$H$_{14}$	2532-58-3	98.186	liq	-133.67(0.03)	91.7(0.5)	0.7402[25]	1.4089[20]	
4032	trans-1,3-Dimethylcyclopentane		C$_7$H$_{14}$	1759-58-6	98.186	liq	-133.9(0.1)	90.7(0.6)	0.7443[25]	1.4107[20]	
4033	N,α-Dimethylcyclopentaneethanamine	Cyclopentamine	C$_9$H$_{19}$N	102-45-4	141.254			171		1.4500[20]	
4034	1,2-Dimethylcyclopentene		C$_7$H$_{12}$	765-47-9	96.170	liq	-90.4(0.5)	105(3)	0.7928[25]	1.4448[20]	
4035	1,5-Dimethylcyclopentene		C$_7$H$_{12}$	16491-15-9	96.170	liq	-118.5(0.5)	99	0.780[20]	1.4331[20]	
4036	1,1-Dimethylcyclopropane		C$_5$H$_{10}$	1630-94-0	70.133	vol liq or gas	-109.0(0.1)	21(2)	0.6604[20]	1.3668[20]	i H$_2$O; s EtOH; vs eth, sulf
4037	cis-1,2-Dimethylcyclopropane		C$_5$H$_{10}$	930-18-7	70.133	liq	-140.92(0.09)	37(3)	0.6889[25]	1.3829[20]	i H$_2$O; s EtOH; vs eth; sl ctc
4038	trans-1,2-Dimethylcyclopropane		C$_5$H$_{10}$	2402-06-4	70.133	vol liq or gas	-149.7(0.1)	28(2)	0.6648[25]	1.3713[20]	vs eth, EtOH
4039	Dimethyldecylamine	N,N-Dimethyl-1-decanamine	C$_{12}$H$_{27}$N	1120-24-7	185.349			234(6)			
4040	Dimethyldiacetoxysilane	Bis(acetyloxy)dimethylsilane	C$_6$H$_{12}$O$_4$Si	2182-66-3	176.243	liq	-12.5	165	1.0540[20]	1.4030[20]	
4041	trans-Dimethyldiazene	Azomethane	C$_2$H$_6$N$_2$	4143-41-3	58.082	gas	-78	1.5	0.743[0]	1.4199[19]	vs ace, EtOH, eth; s ctc, hp
4042	2,2-Dimethyl-1,3-dioxane-4,6-dione	Meldrum's acid	C$_6$H$_8$O$_4$	2033-24-1	144.126		94				
4043	cis-3,6-Dimethyl-1,4-dioxane-2,5-dione		C$_6$H$_8$O$_4$	4511-42-6	144.126	orth (eth)	96.8	150[25]			
4044	2,2-Dimethyl-1,3-dioxolane-4-methanol	Isopropylidene glycerol	C$_6$H$_{12}$O$_3$	100-79-8	132.157			82[10]	1.064[20]	1.4383[20]	
4045	Dimethyldiphenoxysilane		C$_{14}$H$_{16}$O$_2$Si	3440-02-6	244.362		-23	131[5]	1.0599[25]	1.5330[20]	
4046	2,3-Dimethyl-2,3-diphenylbutane	Dicumene	C$_{18}$H$_{22}$	1889-67-4	238.368	cry (MeOH)	119(1)				
4047	3,3'-Dimethyldiphenylmethane 4,4'-diisocyanate		C$_{17}$H$_{14}$N$_2$O$_2$	139-25-3	278.305						s chl
4048	2,9-Dimethyl-4,7-diphenyl-1,10-phenanthroline		C$_{26}$H$_{20}$N$_2$	4733-39-5	360.450		280 dec				
4049	Dimethyldiphenylsilane		C$_{14}$H$_{16}$Si	778-24-5	212.363			277	0.9867[20]	1.5644[20]	
4050	N,N'-Dimethyl-N,N'-diphenylurea		C$_{15}$H$_{16}$N$_2$O	611-92-7	240.300	pl (al)	121.6(0.7)	350			vs H$_2$O, EtOH, ace; sl eth, bz, CS$_2$
4051	Dimethyl disulfide	Methyl disulfide	C$_2$H$_6$S$_2$	624-92-0	94.199	liq	-84.67(0.1)	109.72(0.08)	1.0625[20]	1.5289[20]	i H$_2$O; msc EtOH, eth
4052	O,O-Dimethyl dithiophosphate	O,O-Dimethyl phosphorodithioonate	C$_2$H$_7$O$_2$PS$_2$	756-80-9	158.180	liq		56[4]	1.29[20]		
4053	N,N-Dimethyldodecylamine oxide		C$_{14}$H$_{31}$NO	1643-20-5	229.402	hyg nd (tol)	130.5				
4054	1,2-Dimethylenecyclohexane		C$_8$H$_{12}$	2819-48-9	108.181			125(7)	0.8361[20]	1.4718[25]	i H$_2$O; s EtOH, eth, bz, chl; vs ace
4055	N,N-Dimethyl-1,2-ethanediamine		C$_4$H$_{12}$N$_2$	108-00-9	88.151			104	0.803[25]	1.4260[20]	
4056	N,N'-Dimethyl-1,2-ethanediamine		C$_4$H$_{12}$N$_2$	110-70-3	88.151			120	0.828[15]		s EtOH, eth, dil HCl
4057	N,N-Dimethylethanolamine	Deanol	C$_4$H$_{11}$NO	108-01-0	89.136	liq	-65(1)	130.7(0.5)	0.8866[20]	1.4300[20]	msc H$_2$O, EtOH, eth; s chl
4058	Dimethyl ether	Methyl ether	C$_2$H$_6$O	115-10-6	46.068	col gas	-141.49(0.05)	-24.8(0.2)			s H$_2$O, EtOH, eth, ace, chl; sl bz
4059	(1,1-Dimethylethoxy)benzene		C$_{10}$H$_{14}$O	6669-13-2	150.217	liq	-18.3(0.4)	194(6)	0.9214[20]		
4060	[(1,1-Dimethylethoxy)methyl]-oxirane		C$_7$H$_{14}$O$_2$	7665-72-7	130.185	liq	-70	152	0.898[20]		

4019
N,α-Dimethylcyclohexaneethanamine

4020
3,3-Dimethylcyclohexanol

4021
2,2-Dimethylcyclohexanone

4022
2,6-Dimethylcyclohexanone

4023
3,3-Dimethylcyclohexanone

4024
4,4-Dimethylcyclohexanone

4025
1,2-Dimethylcyclohexene

4026
1,3-Dimethylcyclohexene

4027
3,5-Dimethyl-2-cyclohexen-1-one

4028
1,1-Dimethylcyclopentane

4029
cis-1,2-Dimethylcyclopentane

4030
trans-1,2-Dimethylcyclopentane

4031
cis-1,3-Dimethylcyclopentane

4032
trans-1,3-Dimethylcyclopentane

4033
N,α-Dimethylcyclopentaneethanamine

4034
1,2-Dimethylcyclopentene

4035
1,5-Dimethylcyclopentene

4036
1,1-Dimethylcyclopropane

4037
cis-1,2-Dimethylcyclopropane

4038
trans-1,2-Dimethylcyclopropane

4039
Dimethyldecylamine

4040
Dimethyldiacetoxysilane

4041
trans-Dimethyldiazene

4042
2,2-Dimethyl-1,3-dioxane-4,6-dione

4043
cis-3,6-Dimethyl-1,4-dioxane-2,5-dione

4044
2,2-Dimethyl-1,3-dioxolane-4-methanol

4045
Dimethyldiphenoxysilane

4046
2,3-Dimethyl-2,3-diphenylbutane

4047
3,3'-Dimethyldiphenylmethane 4,4'-diisocyanate

4048
2,9-Dimethyl-4,7-diphenyl-1,10-phenanthroline

4049
Dimethyldiphenylsilane

4050
N,*N*'-Dimethyl-*N*,*N*'-diphenylurea

4051
Dimethyl disulfide

4052
O,*O*-Dimethyl dithiophosphate

4053
N,*N*-Dimethyldodecylamine oxide

4054
1,2-Dimethylenecyclohexane

4055
N,*N*-Dimethyl-1,2-ethanediamine

4056
N,*N*'-Dimethyl-1,2-ethanediamine

4057
N,*N*-Dimethylethanolamine

4058
Dimethyl ether

4059
(1,1-Dimethylethoxy)benzene

4060
[(1,1-Dimethylethoxy)methyl]oxirane

No.	Name	Synonym	Mol. Form.	CAS RN	Mol. Wt.	Physical Form	mp/°C	bp/°C	den g cm⁻³	n_D	Solubility
4061	*N,N*-Dimethylformamide	DMF	C_3H_7NO	68-12-2	73.094	liq	-60.3(0.2)	152.8(0.5)	0.9445[25]	1.4305[20]	msc H₂O, EtOH, eth, ace, bz; sl lig
4062	Dimethyl fumarate		$C_6H_8O_4$	624-49-7	144.126		101.7(0.7)	193	1.37[20]	1.4062[111]	i H₂O; s ace, chl
4063	2,5-Dimethylfuran		C_6H_8O	625-86-5	96.127	liq	-62.8	96(3)	0.8883[20]	1.4363[20]	i H₂O; s EtOH, eth, ace, bz, HOAc, chl
4064	3,4-Dimethyl-2,5-furandione		$C_6H_6O_3$	766-39-2	126.110	pl or lf (dil al)	96	223	1.107[100]		sl H₂O; vs EtOH, eth, bz, chl
4065	Dimethyl germanium sulfide		C_2H_6GeS	16090-49-6	134.77	col cry	54.5	302			
4066	Dimethyl glutarate	Methyl glutarate	$C_7H_{12}O_4$	1119-40-0	160.168	liq	-42.5	216(4)	1.0876[20]	1.4242[20]	vs EtOH, eth; s chl
4067	*N,N*-Dimethylglycine		$C_4H_9NO_2$	1118-68-9	103.120	hyg nd (PrOH)	185.5				vs H₂O, MeOH; s EtOH, eth, ace
4068	Dimethylglyoxime		$C_4H_8N_2O_2$	95-45-4	116.119	nd (to or dil al)	245.5	234 sub			i H₂O; vs EtOH, eth; sl bz, tol
4069	2,6-Dimethyl-1,5-heptadiene		C_9H_{16}	6709-39-3	124.223	liq	-70	143(3)	0.7648[25]		
4070	2,2-Dimethylheptane		C_9H_{20}	1071-26-7	128.255	liq	-113.05(0.1)	133(1)	0.7105[20]	1.4016[20]	i H₂O; s eth, ctc; vs ace, chl; msc bz
4071	2,3-Dimethylheptane		C_9H_{20}	3074-71-3	128.255	liq	-116	140.6(0.9)	0.7260[20]	1.4088[20]	i H₂O; msc EtOH, eth, ace, bz, peth, chl
4072	2,4-Dimethylheptane		C_9H_{20}	2213-23-2	128.255			132(2)	0.7115[25]	1.4034[20]	i H₂O; msc EtOH, eth, ace, bz, chl, peth
4073	2,5-Dimethylheptane		C_9H_{20}	2216-30-0	128.255			135(2)	0.7198[20]	1.4033[20]	vs ace, bz, eth, EtOH
4074	2,6-Dimethylheptane		C_9H_{20}	1072-05-5	128.255	liq	-103.1(0.5)	135(2)	0.7089[20]	1.4011[20]	sl chl
4075	3,3-Dimethylheptane		C_9H_{20}	4032-86-4	128.255			137(3)	0.7254[20]	1.4087[20]	i H₂O; msc EtOH; s eth; vs ace, bz
4076	3,4-Dimethylheptane		C_9H_{20}	922-28-1	128.255			140(3)	0.7314[20]	1.4108[20]	i H₂O; s eth, ctc; vs ace, chl; msc bz
4077	3,5-Dimethylheptane		C_9H_{20}	926-82-9	128.255			135(3)	0.7225[20]	1.4083[20]	i H₂O; s eth, ctc; vs ace, chl; msc bz
4078	4,4-Dimethylheptane		C_9H_{20}	1068-19-5	128.255			134(3)	0.7221[20]	1.4076[20]	i H₂O; s eth, ctc; vs ace, chl; msc bz
4079	Dimethyl heptanedioate	Dimethyl pimelate	$C_9H_{16}O_4$	1732-08-7	188.221		-21	120[10]	1.0625[20]	1.4309[20]	sl H₂O; s EtOH, eth, bz
4080	2,6-Dimethyl-2-heptanol		$C_9H_{20}O$	13254-34-7	144.254			170(11)	0.8186[20]	1.4242[20]	
4081	2,6-Dimethyl-4-heptanol	Diisobutylcarbinol	$C_9H_{20}O$	108-82-7	144.254			193(6)	0.8114[20]	1.4242[20]	i H₂O; s EtOH, eth; sl ctc
4082	3,5-Dimethyl-4-heptanol		$C_9H_{20}O$	19549-79-2	144.254			179(5)	0.836[18]	1.4283[20]	sl H₂O
4083	2,6-Dimethyl-4-heptanone	Diisobutyl ketone	$C_9H_{18}O$	108-83-8	142.238	liq	-46.0(0.2)	157(3)	0.8062[20]	1.412[21]	i H₂O; msc EtOH, eth; s ctc
4084	2,6-Dimethyl-5-heptenal		$C_9H_{16}O$	106-72-9	140.222	oil		120[100]			
4085	*N*,6-Dimethyl-5-hepten-2-amine	Isometheptene	$C_9H_{19}N$	503-01-5	141.254			177			vs eth, EtOH
4086	2,5-Dimethyl-1,5-hexadiene		C_8H_{14}	627-58-7	110.197	liq	-75.6(0.2)	114(2)	0.743[20]	1.43995[21]	i H₂O; s ace, chl
4087	2,5-Dimethyl-2,4-hexadiene		C_8H_{14}	764-13-6	110.197		13.8(0.3)	135.2(0.2)	0.7577[25]	1.4785[20]	i H₂O; s EtOH, eth, bz, chl
4088	2,2-Dimethylhexane		C_8H_{18}	590-73-8	114.229	liq	-121.19(0.07)	106.8(0.4)	0.6953[20]	1.3935[20]	vs ace, bz, eth, EtOH
4089	2,3-Dimethylhexane		C_8H_{18}	584-94-1	114.229			115.6(0.5)	0.6912[25]	1.4011[20]	vs ace, bz, EtOH, lig
4090	2,4-Dimethylhexane		C_8H_{18}	589-43-5	114.229			109.4(0.4)	0.6962[25]	1.3929[25]	
4091	2,5-Dimethylhexane	Biisobutyl	C_8H_{18}	592-13-2	114.229	liq	-91.14(0.02)	109.1(0.7)	0.6901[25]	1.3925[20]	i H₂O; msc EtOH, ace, bz; s eth
4092	3,3-Dimethylhexane		C_8H_{18}	563-16-6	114.229	liq	-126.2(0.1)	111.9(0.6)	0.7100[20]	1.4001[20]	i H₂O; msc EtOH; vs eth, ace, bz
4093	3,4-Dimethylhexane		C_8H_{18}	583-48-2	114.229			117.7(0.4)	0.7151[25]	1.4041[20]	i H₂O; s eth; msc EtOH, ace, bz
4094	2,5-Dimethyl-2,5-hexanedi-amine		$C_8H_{20}N_2$	23578-35-0	144.258			184	0.8485[15]	1.4459[20]	

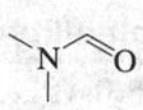

4061
N,N-Dimethylformamide

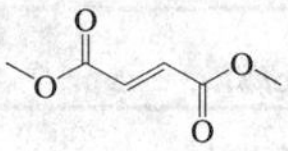

4062
Dimethyl fumarate

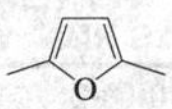

4063
2,5-Dimethylfuran

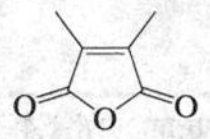

4064
3,4-Dimethyl-2,5-furandione

4065
Dimethyl germanium sulfide

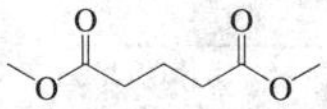

4066
Dimethyl glutarate

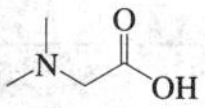

4067
N,N-Dimethylglycine

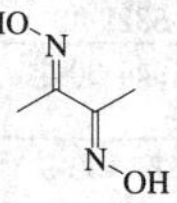

4068
Dimethylglyoxime

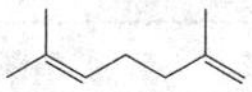

4069
2,6-Dimethyl-1,5-heptadiene

4070
2,2-Dimethylheptane

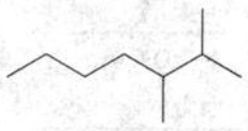

4071
2,3-Dimethylheptane

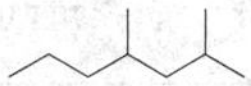

4072
2,4-Dimethylheptane

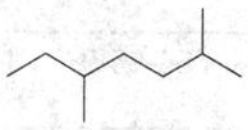

4073
2,5-Dimethylheptane

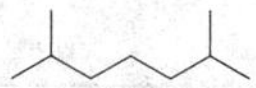

4074
2,6-Dimethylheptane

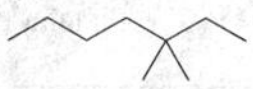

4075
3,3-Dimethylheptane

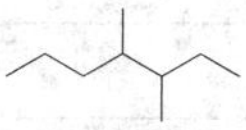

4076
3,4-Dimethylheptane

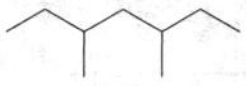

4077
3,5-Dimethylheptane

4078
4,4-Dimethylheptane

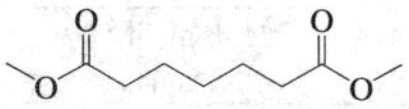

4079
Dimethyl heptanedioate

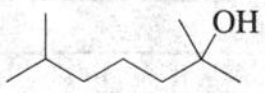

4080
2,6-Dimethyl-2-heptanol

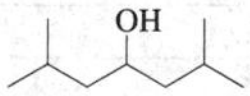

4081
2,6-Dimethyl-4-heptanol

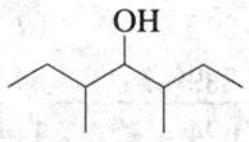

4082
3,5-Dimethyl-4-heptanol

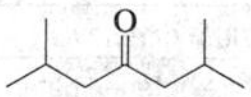

4083
2,6-Dimethyl-4-heptanone

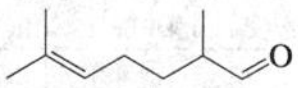

4084
2,6-Dimethyl-5-heptenal

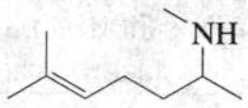

4085
N,6-Dimethyl-5-hepten-2-amine

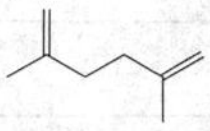

4086
2,5-Dimethyl-1,5-hexadiene

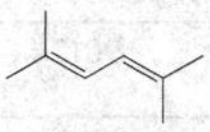

4087
2,5-Dimethyl-2,4-hexadiene

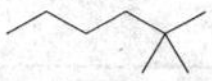

4088
2,2-Dimethylhexane

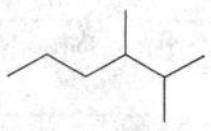

4089
2,3-Dimethylhexane

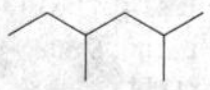

4090
2,4-Dimethylhexane

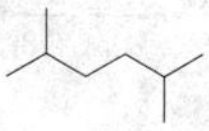

4091
2,5-Dimethylhexane

4092
3,3-Dimethylhexane

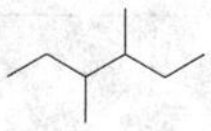

4093
3,4-Dimethylhexane

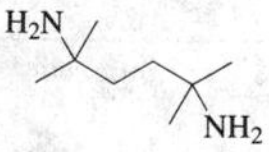

4094
2,5-Dimethyl-2,5-hexanediamine

No.	Name	Synonym	Mol. Form.	CAS RN	Mol. Wt.	Physical Form	mp/°C	bp/°C	den g cm⁻³	n_D	Solubility
4095	2,5-Dimethyl-2,5-hexanediol	1,1,4,4-Tetramethyl-1,4-butanediol	$C_8H_{18}O_2$	110-03-2	146.228	pr (AcOEt) fl (peth)	91(2)	233(13)	0.898[20]		s H_2O; vs EtOH, bz, chl
4096	2,2-Dimethyl-1-hexanol		$C_8H_{18}O$	2370-13-0	130.228			95[29]			
4097	2,3-Dimethyl-1-hexene		C_8H_{16}	16746-86-4	112.213			111(2)	0.7172[25]	1.4113[20]	
4098	5,5-Dimethyl-1-hexene		C_8H_{16}	7116-86-1	112.213			102(3)	0.705[25]	1.4049[20]	
4099	2,3-Dimethyl-2-hexene		C_8H_{16}	7145-20-2	112.213	liq	-115.0(0.1)	122(2)	0.7366[25]	1.4268[20]	
4100	2,5-Dimethyl-2-hexene		C_8H_{16}	3404-78-2	112.213			113(2)	0.7182[20]	1.4140[20]	
4101	cis-2,2-Dimethyl-3-hexene		C_8H_{16}	690-92-6	112.213	liq	-137.36(0.09)	105.4(0.8)	0.7086[25]	1.4099[20]	
4102	trans-2,2-Dimethyl-3-hexene		C_8H_{16}	690-93-7	112.213			100.9(0.2)	0.6995[25]	1.4063[20]	
4103	3,5-Dimethyl-1-hexen-3-ol		$C_8H_{16}O$	3329-48-4	128.212			146.5	0.8382[20]	1.4342[20]	
4104	1-(1,5-Dimethyl-4-hexenyl)-4-methylbenzene	α-Curcumene	$C_{15}H_{22}$	644-30-4	202.336			140[19]	0.8805[20]	1.4989[20]	i H_2O; s bz
4105	2,5-Dimethyl-3-hexyne-2,5-diol		$C_8H_{14}O_2$	142-30-3	142.196		95(2)	206(9)	0.947[20]		s H_2O, chl; vs EtOH, eth, ace, bz
4106	1,1-Dimethylhydrazine	UDMH	$C_2H_8N_2$	57-14-7	60.098	liq, fumes in air	-57.15(0.08)	62.4(0.8)	0.791[22]	1.4075[22]	vs H_2O, EtOH, eth, MeOH
4107	1,2-Dimethylhydrazine		$C_2H_8N_2$	540-73-8	60.098	fumes (air)	-8.86(0.09)	82(3)	0.8274[20]	1.4209[20]	msc H_2O, EtOH, eth
4108	1,2-Dimethylhydrazine dihydrochloride		$C_2H_{10}Cl_2N_2$	306-37-6	133.019	pr (w)	170 dec				vs H_2O, EtOH
4109	Dimethyl hydrogen phosphate	Dimethyl phosphate	$C_2H_7O_4P$	813-78-5	126.048			174 dec	1.3225[20]	1.408[25]	vs H_2O, ace, EtOH
4110	Dimethyl hydrogen phosphite		$C_2H_7O_3P$	868-85-9	110.049			170.2(0.5)	1.2002[20]	1.4036[20]	s EtOH, py; sl ctc
4111	1,2-Dimethyl-1H-imidazole		$C_5H_8N_2$	1739-84-0	96.131			206	1.0051[11]		vs H_2O, eth, EtOH
4112	2,4-Dimethyl-1H-imidazole		$C_5H_8N_2$	930-62-1	96.131		92	267			
4113	5,5-Dimethyl-2,4-imidazolidinedione		$C_5H_8N_2O_2$	77-71-4	128.130	pr (dil al)	176.1(0.5)	sub			vs H_2O, EtOH, eth, ace, bz, chl; s DMSO
4114	1,1-Dimethylindan		$C_{11}H_{14}$	4912-92-9	146.229			194.0(0.8)	0.919[20]	1.5135[25]	
4115	1,3-Dimethyl-1H-indole		$C_{10}H_{11}N$	875-30-9	145.201	nd	142	258.5			s eth
4116	2,3-Dimethyl-1H-indole		$C_{10}H_{11}N$	91-55-4	145.201		107.5	284(20)			
4117	N,N-Dimethyl-1H-indole-3-ethanamine	N,N-Dimethyltryptamine	$C_{12}H_{16}N_2$	61-50-7	188.268		46				
4118	N,N-Dimethyl-1H-indole-3-methanamine	Gramine	$C_{11}H_{14}N_2$	87-52-5	174.242	nd or pl (ace)	138.5				i H_2O; s EtOH, eth, chl; i peth
4119	Dimethyl isophthalate		$C_{10}H_{10}O_4$	1459-93-4	194.184	nd(dil al)	68(1)	285.1(0.4)	1.194[20]	1.5168[20]	sl H_2O
4120	1,4-Dimethyl-7-isopropylazulene	Guaiazulene	$C_{15}H_{18}$	489-84-9	198.304	bl-viol pl (al)	31.5	167[12]	0.973[20]		s EtOH, eth, AcOEt
4121	1,6-Dimethyl-4-isopropylnaphthalene	Cadalene	$C_{15}H_{18}$	483-78-3	198.304			294	0.9667[25]	1.5785[25]	vs oils
4122	2,4-Dimethyl-3-isopropylpentane		$C_{10}H_{22}$	13475-79-1	142.282	liq	-81.7(0.2)	157(4)	0.7545[25]	1.4246[20]	
4123	3,5-Dimethylisoxazole		C_5H_7NO	300-87-8	97.116			142(7)	0.99[25]	1.4421[20]	
4124	Dimethylmagnesium	Magnesium dimethyl	C_2H_6Mg	2999-74-8	54.374	solid	220 dec				
4125	Dimethyl maleate	Methyl cis-butenedioate	$C_6H_8O_4$	624-48-6	144.126	liq	-19(1)	202(2)	1.1606[20]	1.4416[20]	sl H_2O, lig; s eth, ctc
4126	Dimethyl malonate	Methyl malonate	$C_5H_8O_4$	108-59-8	132.116	liq	-62(1)	181.1(0.6)	1.528[20]	1.4135[20]	sl H_2O; msc EtOH; vs ace, bz; s chl
4127	Dimethylmalonic acid	Dimethylpropanedioc acid	$C_5H_8O_4$	595-46-0	132.116	pr (bz/peth)	191(2)				s hot H_2O
4128	Dimethyl mercury	Mercury dimethyl	C_2H_6Hg	593-74-8	230.66	liq		93	3.17[25]	1.5452[20]	i H_2O; vs EtOH, eth
4129	Dimethyl cis-2-methyl-2-butenedioate	Dimethyl citraconate	$C_7H_{10}O_4$	617-54-9	158.152			210.5	1.1153[20]	1.4473[20]	vs ace, eth, EtOH
4130	Dimethyl methylenesuccinate		$C_7H_{10}O_4$	617-52-7	158.152	hyg mcl (MeOH)	38	208	1.1241[18]	1.4457[20]	s EtOH, eth, MeOH; vs ace
4131	Dimethyl methylmalonate		$C_6H_{10}O_4$	609-02-9	146.141			177(15)	1.0977[20]	1.4128[20]	vs ace, eth, EtOH, chl
4132	Dimethyl methylphosphonate		$C_3H_9O_3P$	756-79-6	124.075			168.7(0.7)	1.1684[20]	1.4099[30]	s H_2O, EtOH, eth
4133	trans-2,2-Dimethyl-3-(2-methyl-1-propenyl)cyclopropanecarboxylic acid		$C_{10}H_{16}O_2$	4638-92-0	168.233	pr	20.0	245			vs eth, EtOH, chl
4134	Dimethyl 2-methylsuccinate		$C_7H_{12}O_4$	1604-11-1	160.168			196	1.076[25]	1.4200[20]	
4135	Dimethyl p-(methylthio)phenyl phosphate		$C_9H_{13}O_4PS$	3254-63-5	248.235	liq			1.273[21]		sl H_2O; s ace, EtOH, diox, ctc, xyl
4136	2,6-Dimethylmorpholine		$C_6H_{13}NO$	141-91-3	115.173	liq	-88	147(18)	0.9329[20]	1.4460[20]	msc H_2O, EtOH, bz, lig; s ace; sl chl

4095
2,5-Dimethyl-2,5-hexanediol

4096
2,2-Dimethyl-1-hexanol

4097
2,3-Dimethyl-1-hexene

4098
5,5-Dimethyl-1-hexene

4099
2,3-Dimethyl-2-hexene

4100
2,5-Dimethyl-2-hexene

4101
cis-2,2-Dimethyl-3-hexene

4102
trans-2,2-Dimethyl-3-hexene

4103
3,5-Dimethyl-1-hexen-3-ol

4104
1-(1,5-Dimethyl-4-hexenyl)-4-methylbenzene

4105
2,5-Dimethyl-3-hexyne-2,5-diol

4106
1,1-Dimethylhydrazine

4107
1,2-Dimethylhydrazine

4108
1,2-Dimethylhydrazine dihydrochloride

4109
Dimethyl hydrogen phosphate

4110
Dimethyl hydrogen phosphite

4111
1,2-Dimethyl-1H-imidazole

4112
2,4-Dimethyl-1H-imidazole

4113
5,5-Dimethyl-2,4-imidazolidinedione

4114
1,1-Dimethylindan

4115
1,3-Dimethyl-1H-indole

4116
2,3-Dimethyl-1H-indole

4117
N,N-Dimethyl-1H-indole-3-ethanamine

4118
N,N-Dimethyl-1H-indole-3-methanamine

4119
Dimethyl isophthalate

4120
1,4-Dimethyl-7-isopropylazulene

4121
1,6-Dimethyl-4-isopropylnaphthalene

4122
2,4-Dimethyl-3-isopropylpentane

4123
3,5-Dimethylisoxazole

4124
Dimethylmagnesium

4125
Dimethyl maleate

4126
Dimethyl malonate

4127
Dimethylmalonic acid

4128
Dimethyl mercury

4129
Dimethyl cis-2-methyl-2-butenedioate

4130
Dimethyl methylenesuccinate

4131
Dimethyl methylmalonate

4132
Dimethyl methylphosphonate

4133
trans-2,2-Dimethyl-3-(2-methyl-1-propenyl)cyclopropanecarboxylic acid

4134
Dimethyl 2-methylsuccinate

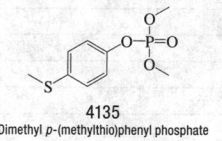

4135
Dimethyl p-(methylthio)phenyl phosphate

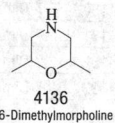

4136
2,6-Dimethylmorpholine

No.	Name	Synonym	Mol. Form.	CAS RN	Mol. Wt.	Physical Form	mp/°C	bp/°C	den g cm⁻³	n_D	Solubility
4137	Dimethyl morpho-linophosphoramidate	Dimethyl 4-morpholinylphos-phonate	C₆H₁₄NO₄P	597-25-1	195.153	liq		96[1]			
4138	1,2-Dimethylnaphthalene		C₁₂H₁₂	573-98-8	156.223		-3.0(0.7)	267(5)	1.0179[20]	1.6166[20]	i H₂O; s eth, bz
4139	1,3-Dimethylnaphthalene		C₁₂H₁₂	575-41-7	156.223	liq	-6	265(5)	1.0144[20]	1.6140[20]	i H₂O; s eth, bz
4140	1,4-Dimethylnaphthalene		C₁₂H₁₂	571-58-4	156.223		7.6(0.4)	264(5)	1.0166[20]	1.6127[20]	i H₂O; vs EtOH; msc eth, ace, bz, ctc
4141	1,5-Dimethylnaphthalene		C₁₂H₁₂	571-61-9	156.223		81.5(0.9)	267(4)			i H₂O; vs bz, eth
4142	1,6-Dimethylnaphthalene		C₁₂H₁₂	575-43-9	156.223	liq	-16.2(0.5)	263(4)	1.0021[20]	1.6166[20]	i H₂O; s eth, bz
4143	1,7-Dimethylnaphthalene		C₁₂H₁₂	575-37-1	156.223	liq	-13.9	263(6)	1.0115[20]	1.6083[20]	i H₂O; s eth, bz
4144	1,8-Dimethylnaphthalene		C₁₂H₁₂	569-41-5	156.223		63.16(0.05)	276(3)	1.003[20]		i H₂O; s eth, bz
4145	2,3-Dimethylnaphthalene	Guajen	C₁₂H₁₂	581-40-8	156.223	lf (al)	104.3(0.3)	267(2)	1.003[20]	1.5060[20]	i H₂O; vs bz, eth
4146	2,6-Dimethylnaphthalene		C₁₂H₁₂	581-42-0	156.223		110.1(0.2)	253(3)	1.003[20]		i H₂O
4147	2,7-Dimethylnaphthalene		C₁₂H₁₂	582-16-1	156.223		96(1)	262.4(0.3)	1.003[20]		
4148	N,N-Dimethyl-1-naphthyl-amine		C₁₂H₁₃N	86-56-6	171.238	viol flr cry		250	1.0423[20]	1.624[15]	i H₂O; s EtOH, eth, ctc
4149	N,N-Dimethyl-2-naphthyl-amine		C₁₂H₁₃N	2436-85-3	171.238	dk red nd	41.5(0.5)	304(15)	1.0279[60]	1.6443[53]	i H₂O; s EtOH, eth
4150	N,N-Dimethyl-2-nitroaniline		C₈H₁₀N₂O₂	610-17-3	166.177	ye-oran	-20	146[20]	1.1794[20]	1.6102[20]	s H₂O, eth; vs EtOH, chl
4151	N,N-Dimethyl-3-nitroaniline		C₈H₁₀N₂O₂	619-31-8	166.177	red mcl pr (eth)	60(2)	282.5	1.313[17]		i H₂O; s EtOH, eth
4152	N,N-Dimethyl-4-nitroaniline		C₈H₁₀N₂O₂	100-23-2	166.177	ye nd (al)	164(1)				i H₂O; s EtOH, eth, HOAc
4153	1,2-Dimethyl-3-nitrobenzene		C₈H₉NO₂	83-41-0	151.163	nd (al)	14(1)	240	1.1402[20]	1.5441[20]	i H₂O; s EtOH, ctc
4154	1,2-Dimethyl-4-nitrobenzene	4-Nitro-o-xylene	C₈H₉NO₂	99-51-4	151.163	ye pr (al)	28(1)	251	1.112[15]	1.5202[20]	i H₂O; msc EtOH
4155	1,3-Dimethyl-2-nitrobenzene		C₈H₉NO₂	81-20-9	151.163		15.2(0.7)	226	1.112[15]	1.5202[20]	i H₂O; vs EtOH; s ctc
4156	1,3-Dimethyl-5-nitrobenzene		C₈H₉NO₂	99-12-7	151.163	nd (al)	75	274			i H₂O; vs EtOH, eth
4157	1,4-Dimethyl-2-nitrobenzene		C₈H₉NO₂	89-58-7	151.163	pa ye liq	-25	240.5	1.132[15]	1.5413[20]	i H₂O; s EtOH
4158	2,4-Dimethyl-1-nitrobenzene		C₈H₉NO₂	89-87-2	151.163		9.6(0.2)	243(2)	1.135[15]	1.5473[25]	i H₂O; s eth, ace, bz, chl
4159	1,2-Dimethyl-5-nitro-1H-imidazole	Dimetridazole	C₅H₇N₃O₂	551-92-8	141.129	nd (w)	138.5				vs eth, EtOH
4160	N,N-Dimethyl-4-[2-(4-nitro-phenyl)ethenyl]aniline		C₁₆H₁₆N₂O₂	4584-57-0	268.310		258.3				
4161	N,4-Dimethyl-N-nitrosoben-zenesulfonamide	p-Tolylsulfonylmethylnitrosa-mide	C₈H₁₀N₂O₃S	80-11-5	214.241	cry	60				i H₂O; vs EtOH, eth
4162	Dimethyl nonanedioate	Methyl azelate	C₁₁H₂₀O₄	1732-10-1	216.275		-0.8	156[20]	1.0082[20]	1.4367[20]	i H₂O; s EtOH, ace, bz, ctc
4163	6,6-Dimethyl-2-norpinene-2-carboxaldehyde	Myrtenal	C₁₀H₁₄O	564-94-3	150.217	unstab oil		99[15]			
4164	cis-3,7-Dimethyl-2,6-octadi-enal		C₁₀H₁₆O	106-26-3	152.233			120[20]	0.8869[20]	1.4869[20]	i H₂O; msc EtOH, eth
4165	trans-3,7-Dimethyl-2,6-octadienal	Citral	C₁₀H₁₆O	141-27-5	152.233			229	0.8888[20]	1.4898[20]	i H₂O; msc EtOH, eth
4166	3,7-Dimethyl-1,6-octadiene	Citronellene	C₁₀H₁₈	2436-90-0	138.250				0.7601[20]	1.4362[20]	
4167	3,7-Dimethyl-2,6-octadienoic acid	Geranic acid	C₁₀H₁₆O₂	459-80-3	168.233	oil					
4168	cis-3,7-Dimethyl-2,6-octa-dien-1-ol	Nerol	C₁₀H₁₈O	106-25-2	154.249		<-15	225	0.8756[20]	1.4746[20]	vs EtOH
4169	cis-3,7-Dimethyl-2,6-octa-dien-1-ol acetate		C₁₂H₂₀O₂	141-12-8	196.286			134[25]	0.905[15]	1.452[20]	
4170	trans-3,7-Dimethyl-2,6-octadien-1-ol formate		C₁₁H₁₈O₂	105-86-2	182.260			229 dec	0.9086[25]	1.4659[20]	i H₂O; vs EtOH; s eth, ace
4171	2,2-Dimethyloctane		C₁₀H₂₂	15869-87-1	142.282			154(5)	0.7208[25]	1.4082[20]	
4172	2,3-Dimethyloctane		C₁₀H₂₂	7146-60-3	142.282			164(3)	0.7377[20]	1.4146[20]	
4173	2,4-Dimethyloctane		C₁₀H₂₂	4032-94-4	142.282			154(3)	0.7226[25]	1.4091[20]	
4174	2,5-Dimethyloctane		C₁₀H₂₂	15869-89-3	142.282			157(4)	0.7264[25]	1.4112[20]	
4175	2,6-Dimethyloctane		C₁₀H₂₂	2051-30-1	142.282			158(2)	0.7313[20]	1.4097[20]	
4176	2,7-Dimethyloctane		C₁₀H₂₂	1072-16-8	142.282	liq	-54(1)	160(1)	0.7202[25]	1.4086[20]	s eth, HOAc
4177	3,4-Dimethyloctane		C₁₀H₂₂	15869-92-8	142.282			162(7)	0.7410[25]	1.4182[20]	
4178	3,6-Dimethyloctane		C₁₀H₂₂	15869-94-0	142.282			160.8	0.7324[25]	1.4139[20]	
4179	Dimethyl octanedioate	Dimethyl suberate	C₁₀H₁₈O₄	1732-09-8	202.248	liq	6(1)	259(8)	1.0217[20]	1.4341[20]	i H₂O; s EtOH, eth, ace; sl ctc
4180	3,7-Dimethyl-1,7-octanediol		C₁₀H₂₂O₂	107-74-4	174.281			265	0.937[20]	1.4599[20]	sl bz, tol
4181	2,2-Dimethyloctanoic acid		C₁₀H₂₀O₂	29662-90-6	172.265			140[13]			
4182	2,2-Dimethyl-1-octanol		C₁₀H₂₂O	2370-14-1	158.281	liq		211(6)	0.84[20]		
4183	3,7-Dimethyl-1-octanol		C₁₀H₂₂O	106-21-8	158.281			224(11)	0.832[25]	1.438[25]	s eth

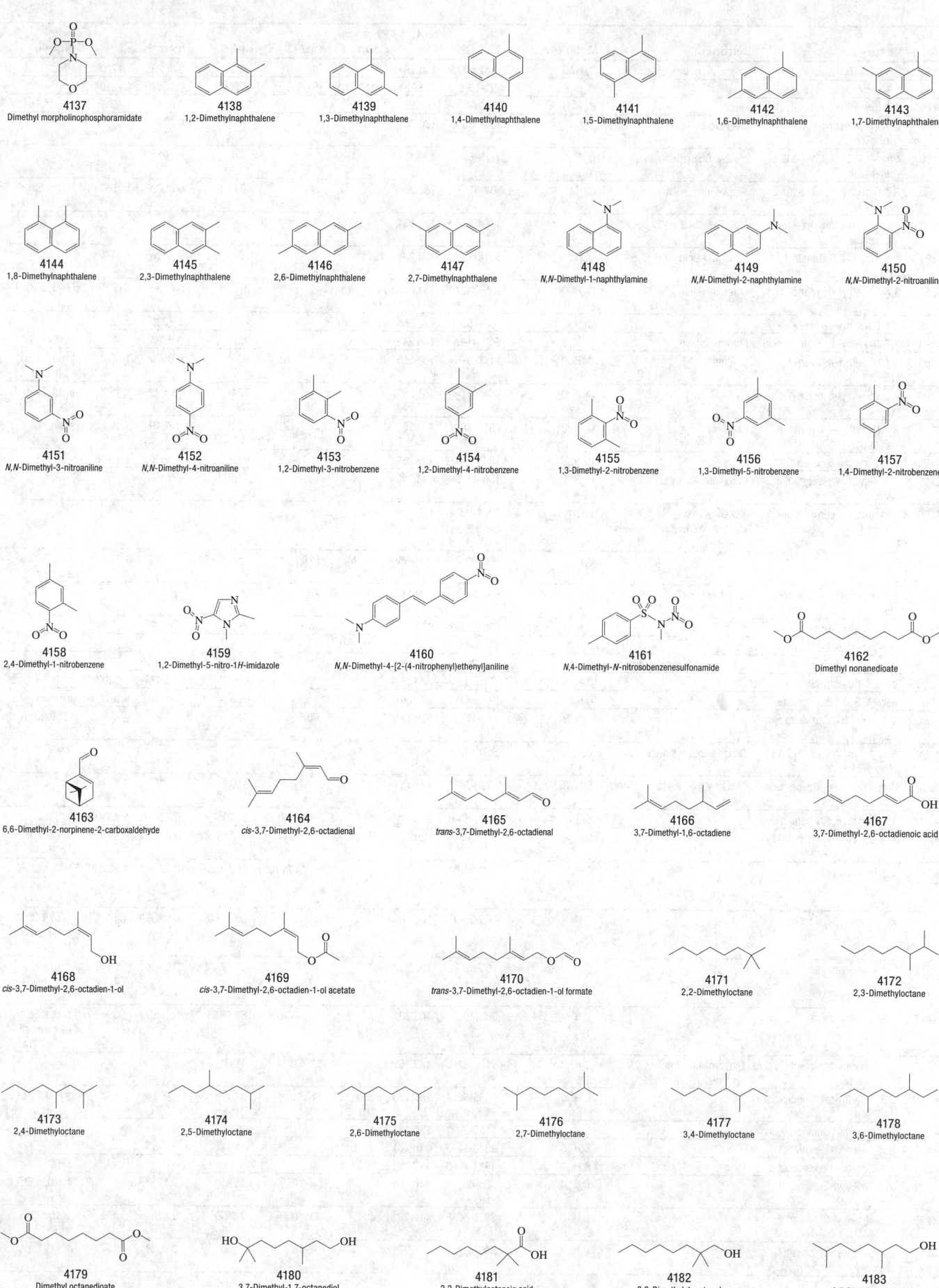

4137
Dimethyl morpholinophosphoramide

4138
1,2-Dimethylnaphthalene

4139
1,3-Dimethylnaphthalene

4140
1,4-Dimethylnaphthalene

4141
1,5-Dimethylnaphthalene

4142
1,6-Dimethylnaphthalene

4143
1,7-Dimethylnaphthalene

4144
1,8-Dimethylnaphthalene

4145
2,3-Dimethylnaphthalene

4146
2,6-Dimethylnaphthalene

4147
2,7-Dimethylnaphthalene

4148
N,N-Dimethyl-1-naphthylamine

4149
N,N-Dimethyl-2-naphthylamine

4150
N,N-Dimethyl-2-nitroaniline

4151
N,N-Dimethyl-3-nitroaniline

4152
N,N-Dimethyl-4-nitroaniline

4153
1,2-Dimethyl-3-nitrobenzene

4154
1,2-Dimethyl-4-nitrobenzene

4155
1,3-Dimethyl-2-nitrobenzene

4156
1,3-Dimethyl-5-nitrobenzene

4157
1,4-Dimethyl-2-nitrobenzene

4158
2,4-Dimethyl-1-nitrobenzene

4159
1,2-Dimethyl-5-nitro-1*H*-imidazole

4160
N,N-Dimethyl-4-[2-(4-nitrophenyl)ethenyl]aniline

4161
N,4-Dimethyl-*N*-nitrosobenzenesulfonamide

4162
Dimethyl nonanedioate

4163
6,6-Dimethyl-2-norpinene-2-carboxaldehyde

4164
cis-3,7-Dimethyl-2,6-octadienal

4165
trans-3,7-Dimethyl-2,6-octadienal

4166
3,7-Dimethyl-1,6-octadiene

4167
3,7-Dimethyl-2,6-octadienoic acid

4168
cis-3,7-Dimethyl-2,6-octadien-1-ol

4169
cis-3,7-Dimethyl-2,6-octadien-1-ol acetate

4170
trans-3,7-Dimethyl-2,6-octadien-1-ol formate

4171
2,2-Dimethyloctane

4172
2,3-Dimethyloctane

4173
2,4-Dimethyloctane

4174
2,5-Dimethyloctane

4175
2,6-Dimethyloctane

4176
2,7-Dimethyloctane

4177
3,4-Dimethyloctane

4178
3,6-Dimethyloctane

4179
Dimethyl octanedioate

4180
3,7-Dimethyl-1,7-octanediol

4181
2,2-Dimethyloctanoic acid

4182
2,2-Dimethyl-1-octanol

4183
3,7-Dimethyl-1-octanol

No.	Name	Synonym	Mol. Form.	CAS RN	Mol. Wt.	Physical Form	mp/°C	bp/°C	den g cm⁻³	n_D	Solubility
4184	2,6-Dimethyl-2-octanol	Tetrahydromyrcenol	$C_{10}H_{22}O$	18479-57-7	158.281			80.5[10]	0.8023[25]	1.4220[25]	
4185	3,6-Dimethyl-3-octanol		$C_{10}H_{22}O$	151-19-9	158.281	liq	-67.5	198(13)	0.8347[22]	1.4370[20]	
4186	3,7-Dimethyl-3-octanol		$C_{10}H_{22}O$	78-69-3	158.281			205.1	0.826[25]	1.433[25]	
4187	cis-3,7-Dimethyl-1,3,6-octatriene	cis-β-Ocimene	$C_{10}H_{16}$	3338-55-4	136.234				0.799[20]		
4188	trans-3,7-Dimethyl-1,3,6-octatriene	trans-β-Ocimene	$C_{10}H_{16}$	3779-61-1	136.234				0.799[20]		
4189	3,7-Dimethyl-1,3,7-octatriene	α-Ocimene	$C_{10}H_{16}$	502-99-8	136.234			177 dec	0.8000[20]	1.4862[20]	i H_2O; s EtOH, eth, chl, HOAc
4190	cis, cis-2,6-Dimethyl-2,4,6-octatriene	cis-allo-Ocimene	$C_{10}H_{16}$	17202-20-9	136.234	liq					
4191	trans,trans-2,6-Dimethyl-2,4,6-octatriene	trans-allo-Ocimene	$C_{10}H_{16}$	3016-19-1	136.234	liq	-35.4	188	0.8118[20]	1.5446[20]	
4192	3,7-Dimethyl-6-octenal	Citronellal	$C_{10}H_{18}O$	106-23-0	154.249	nd or orth cry		205(2)	0.853[20]	1.4473[20]	sl H_2O; s EtOH
4193	3,7-Dimethyl-1-octene		$C_{10}H_{20}$	4984-01-4	140.266	col liq		155(7)	0.7396[20]	1.4212[20]	
4194	3,7-Dimethyl-6-octenoic acid	Citronellic acid	$C_{10}H_{18}O_2$	502-47-6	170.249			257	0.9234[21]		
4195	3,7-Dimethyl-6-octen-1-ol, (R)-	Citronellol, (+)	$C_{10}H_{20}O$	1117-61-9	156.265	oil		224	0.8550[20]	1.4565[20]	sl H_2O; msc EtOH, eth
4196	3,7-Dimethyl-6-octen-1-ol, (S)-	Citronellol, (-)	$C_{10}H_{20}O$	7540-51-4	156.265	oil		224	0.859[18]	1.4576[18]	vs eth, EtOH
4197	3,7-Dimethyl-7-octen-1-ol, (S)-	Rhodinol	$C_{10}H_{20}O$	6812-78-8	156.265			114[12]	0.8549[20]	1.4556[20]	vs eth, EtOH
4198	3,7-Dimethyl-6-octen-3-ol		$C_{10}H_{20}O$	18479-51-1	156.265			94[14]	0.8695[15]	1.4569[15]	
4199	3,7-Dimethyl-6-octen-1-ol, acetate	Citronellol acetate	$C_{12}H_{22}O_2$	150-84-5	198.302			115[10]			
4200	Dimethyloldihydroxyethyleneurea	4,5-Dihydroxy-1,3-bis(hydroxymethyl)-2-imidazolidinone	$C_5H_{10}N_2O_5$	1854-26-8	178.143	hyg cry					
4201	Dimethyl oxalate		$C_4H_6O_4$	553-90-2	118.089	mcl tab	51(2)	163.4(0.5)	1.1716[60]	1.379[82]	sl H_2O; s EtOH, eth, ace, chl
4202	5,5-Dimethyl-2,4-oxazolidinedione	Dimethadione	$C_5H_7NO_3$	695-53-4	129.115		76.5				
4203	3,3-Dimethyloxetane		$C_5H_{10}O$	6921-35-3	86.132			78(6)	0.834[25]	1.3965[20]	
4204	3,3-Dimethyl-2-oxetanone		$C_5H_8O_2$	1955-45-9	100.117			58[15]			
4205	2,2-Dimethyloxirane	2-Methyl-1,2-epoxypropane	C_4H_8O	558-30-5	72.106			51(2)	0.8112[20]	1.3712[22]	s EtOH, eth
4206	cis-2,3-Dimethyloxirane		C_4H_8O	1758-33-4	72.106	liq	-83.4(0.5)	60(4)	0.8226[25]	1.3802[20]	vs eth, ace, bz
4207	trans-2,3-Dimethyloxirane		C_4H_8O	6189-41-9	72.106	liq	-85	56(10)	0.8010[25]	1.3736[20]	vs eth, ace, bz
4208	3,3-Dimethyl-2-oxobutanoic acid		$C_6H_{10}O_3$	815-17-8	130.141		90.5	189			sl H_2O; s eth, bz, chl, CS_2
4209	N-(1,1-Dimethyl-3-oxobutyl)-2-propenamide	Diacetone acrylamide	$C_9H_{15}NO_2$	2873-97-4	169.221						s chl
4210	Dimethyl 3-oxo-1,5-pentanedioate	Dimethyl 1,3-acetonedicarboxylate	$C_7H_{10}O_5$	1830-54-2	174.151			150[25]	1.185[25]	1.4434[20]	
4211	2,4-Dimethyl-1,3-pentadiene		C_7H_{12}	1000-86-8	96.170	liq	-116.0(0.3)	94(2)	0.7343[23]	1.4390[23]	
4212	N,N-Dimethylpentanamide		$C_7H_{15}NO$	6225-06-5	129.200		-51	141[100]	0.8962[25]	1.4419[25]	vs H_2O, eth, EtOH
4213	2,2-Dimethylpentane		C_7H_{16}	590-35-2	100.202	liq	-123.71(0.04)	79.2(0.3)	0.6739[20]	1.3822[20]	i H_2O; s EtOH, eth; msc ace, bz, hp, chl
4214	2,3-Dimethylpentane		C_7H_{16}	565-59-3	100.202			89.8(0.6)	0.6908[25]	1.3894[25]	i H_2O; s EtOH, eth; msc ace, bz, chl
4215	2,4-Dimethylpentane		C_7H_{16}	108-08-7	100.202	liq	-119.16(0.02)	80.4(0.5)	0.6727[20]	1.3815[20]	i H_2O; s EtOH, eth; msc ace, bz, chl, hp
4216	3,3-Dimethylpentane		C_7H_{16}	562-49-2	100.202	liq	-134.4(0.4)	86.0(0.6)	0.6936[20]	1.3909[20]	i H_2O; s EtOH, eth; msc ace, bz, hp, chl
4217	3,3-Dimethylpentanedioic acid anhydride	Dihydro-4,4-dimethyl-2H-pyran-2,6(3H)-dione	$C_7H_{10}O_3$	4160-82-1	142.152		125.8	181[25]			
4218	3,3-Dimethylpentanedioic acid		$C_7H_{12}O_4$	4839-46-7	160.168	mcl pl, nd (bz)	103.5	126[415]	1.4278[20]		vs H_2O, EtOH, eth; sl bz; i lig
4219	2,2-Dimethylpentanoic acid		$C_7H_{14}O_2$	1185-39-3	130.185	liq		98[9]	0.9189[20]		
4220	2,2-Dimethyl-1-pentanol		$C_7H_{16}O$	2370-12-9	116.201						s chl
4221	2,3-Dimethyl-2-pentanol		$C_7H_{16}O$	4911-70-0	116.201				0.804[20]		sl H_2O
4222	2,4-Dimethyl-2-pentanol		$C_7H_{16}O$	625-06-9	116.201		<-20	133(3)	0.8103[20]	1.4172[20]	sl H_2O; s EtOH, eth, ctc
4223	2,2-Dimethyl-3-pentanol		$C_7H_{16}O$	3970-62-5	116.201	liq	9.3(0.5)	136(2)	0.8253[20]	1.4223[20]	i H_2O; s EtOH, eth
4224	2,3-Dimethyl-3-pentanol		$C_7H_{16}O$	595-41-5	116.201		<-30	140(4)	0.833[20]	1.4287[20]	sl H_2O, bz; s EtOH, eth

4184 2,6-Dimethyl-2-octanol

4185 3,6-Dimethyl-3-octanol

4186 3,7-Dimethyl-3-octanol

4187 cis-3,7-Dimethyl-1,3,6-octatriene

4188 trans-3,7-Dimethyl-1,3,6-octatriene

4189 3,7-Dimethyl-1,3,7-octatriene

4190 cis, cis-2,6-Dimethyl-2,4,6-octatriene

4191 trans,trans-2,6-Dimethyl-2,4,6-octatriene

4192 3,7-Dimethyl-6-octenal

4193 3,7-Dimethyl-1-octene

4194 3,7-Dimethyl-6-octenoic acid

4195 3,7-Dimethyl-6-octen-1-ol, (R)-

4196 3,7-Dimethyl-6-octen-1-ol, (S)-

4197 3,7-Dimethyl-7-octen-1-ol, (S)-

4198 3,7-Dimethyl-6-octen-3-ol

4199 3,7-Dimethyl-6-octen-1-ol, acetate

4200 Dimethyloldihydroxyethyleneurea

4201 Dimethyl oxalate

4202 5,5-Dimethyl-2,4-oxazolidinedione

4203 3,3-Dimethyloxetane

4204 3,3-Dimethyl-2-oxetanone

4205 2,2-Dimethyloxirane

4206 cis-2,3-Dimethyloxirane

4207 trans-2,3-Dimethyloxirane

4208 3,3-Dimethyl-2-oxobutanoic acid

4209 N-(1,1-Dimethyl-3-oxobutyl)-2-propenamide

4210 Dimethyl 3-oxo-1,5-pentanedioate

4211 2,4-Dimethyl-1,3-pentadiene

4212 N,N-Dimethylpentanamide

4213 2,2-Dimethylpentane

4214 2,3-Dimethylpentane

4215 2,4-Dimethylpentane

4216 3,3-Dimethylpentane

4217 3,3-Dimethylpentanedioic acid anhydride

4218 3,3-Dimethylpentanedioic acid

4219 2,2-Dimethylpentanoic acid

4220 2,2-Dimethyl-1-pentanol

4221 2,3-Dimethyl-2-pentanol

4222 2,4-Dimethyl-2-pentanol

4223 2,2-Dimethyl-3-pentanol

4224 2,3-Dimethyl-3-pentanol

No.	Name	Synonym	Mol. Form.	CAS RN	Mol. Wt.	Physical Form	mp/°C	bp/°C	den g cm⁻³	n_D	Solubility
4225	2,4-Dimethyl-3-pentanol		$C_7H_{16}O$	600-36-2	116.201		<-70	142(2)	0.8288[20]	1.4250[20]	sl H_2O; s EtOH, eth
4226	4,4-Dimethyl-2-pentanone		$C_7H_{14}O$	590-50-1	114.185	liq	-64	124(2)	0.809[25]	1.4036[20]	
4227	2,2-Dimethyl-3-pentanone		$C_7H_{14}O$	564-04-5	114.185	liq	-45	125(3)	0.8125[20]	1.4065[20]	sl H_2O; s EtOH, eth, ace, chl
4228	2,4-Dimethyl-3-pentanone	Diisopropyl ketone	$C_7H_{14}O$	565-80-0	114.185	liq	-68.4(0.6)	125.2(0.3)	0.8108[20]	1.3999[20]	sl H_2O; msc EtOH, eth; s bz; sl ctc
4229	2,3-Dimethyl-1-pentene		C_7H_{14}	3404-72-6	98.186	liq	-137(5)	84(1)	0.7051[20]	1.4033[20]	i H_2O; msc EtOH, eth; vs dil sulf
4230	2,4-Dimethyl-1-pentene		C_7H_{14}	2213-32-3	98.186	liq	-124.1(0.1)	81.6(0.9)	0.6943[20]	1.3986[20]	i H_2O; msc EtOH, eth; s bz, ctc, chl
4231	3,3-Dimethyl-1-pentene		C_7H_{14}	3404-73-7	98.186	liq	-134.4(0.1)	77(2)	0.6974[20]	1.3984[20]	i H_2O; msc EtOH, eth; s bz, chl
4232	3,4-Dimethyl-1-pentene		C_7H_{14}	7385-78-6	98.186			82(4)	0.6934[25]	1.3992[20]	
4233	4,4-Dimethyl-1-pentene		C_7H_{14}	762-62-9	98.186	liq	-136.6(0.1)	72.5(0.2)	0.6827[20]	1.3818[20]	i H_2O; msc EtOH, eth; s bz, ctc, chl
4234	2,3-Dimethyl-2-pentene		C_7H_{14}	10574-37-5	98.186	liq	-118.3(0.1)	96(2)	0.7277[20]	1.4208[20]	i H_2O; s EtOH, eth, bz, chl
4235	2,4-Dimethyl-2-pentene		C_7H_{14}	625-65-0	98.186	liq	-127.6(0.2)	83.3(0.7)	0.6954[20]	1.4040[20]	i H_2O; s EtOH, eth, bz, ctc
4236	cis-3,4-Dimethyl-2-pentene		C_7H_{14}	4914-91-4	98.186	liq	-124.2(0.1)	92(4)	0.7092[25]	1.4104[20]	
4237	trans-3,4-Dimethyl-2-pentene		C_7H_{14}	4914-92-5	98.186	liq	-113.4(0.1)	91(3)	0.7124[25]	1.4128[20]	
4238	cis-4,4-Dimethyl-2-pentene		C_7H_{14}	762-63-0	98.186	liq	-135.5(0.1)	80.4(0.9)	0.6951[25]	1.4026[20]	
4239	trans-4,4-Dimethyl-2-pentene		C_7H_{14}	690-08-4	98.186	liq	-115.2(0.1)	76.7(0.8)	0.6889[20]	1.3982[20]	i H_2O; s EtOH, eth, bz, chl
4240	4,4-Dimethyl-1-pentyne		C_7H_{12}	13361-63-2	96.170	liq	-75.7	75(3)	0.7142[20]	1.3983[20]	vs bz, eth, chl
4241	4,4-Dimethyl-2-pentyne		C_7H_{12}	999-78-0	96.170	liq	-82(2)	83(4)	0.7176[20]	1.4071[20]	i H_2O; s eth, bz, chl; sl ctc
4242	Dimethylperoxide		$C_2H_6O_2$	690-02-8	62.068	vol liq or gas	-100	-3(12)	0.8677[0]	1.3503[0]	sl EtOH, eth; s tol, HOAc
4243	2,9-Dimethyl-1,10-phenanthroline	Neocuproine	$C_{14}H_{12}N_2$	484-11-7	208.258	cry, 1/2w (w, lig)	163(1)				
4244	3,4-Dimethylphenol phosphate (3:1)		$C_{24}H_{27}O_4P$	3862-11-1	410.442		72	261[7]			i H_2O; sl EtOH, chl, hx; s bz
4245	5-(2,5-Dimethylphenoxy)-2,2-dimethylpentanoic acid	Gemfibrozil	$C_{15}H_{22}O_3$	25812-30-0	250.334	cry	62.0(0.5)	159[0.02]			
4246	N-(2,4-Dimethylphenyl)-acetamide		$C_{10}H_{13}NO$	2050-43-3	163.216	nd (al)	129.3	170[10]			vs EtOH, chl
4247	1-[(2,4-Dimethylphenyl)azo]-2-naphthol	1-(2,4-Xylylazo)-2-naphthol	$C_{18}H_{16}N_2O$	3118-97-6	276.332	red nd (al)	166				vs eth, EtOH
4248	1-[(2,5-Dimethylphenyl)azo]-2-naphthol	1-(2,5-Xylylazo)-2-naphthol	$C_{18}H_{16}N_2O$	85-82-5	276.332	nd (al)	153				
4249	1-(2,4-Dimethylphenyl)-ethanone	2,4-Dimethylacetophenone	$C_{10}H_{12}O$	89-74-7	148.201			228	1.0121[15]	1.5340[20]	vs eth, EtOH
4250	1-(2,5-Dimethylphenyl)-ethanone	2,5-Dimethylacetophenone	$C_{10}H_{12}O$	2142-73-6	148.201	liq	-18.1(0.7)	232.5	0.9963[19]	1.5291[20]	i H_2O; vs EtOH, eth, bz, CS_2
4251	1-(3,4-Dimethylphenyl)-ethanone	3,4-Dimethylacetophenone	$C_{10}H_{12}O$	3637-01-2	148.201	liq	-5(1)	245(11)	1.0090[14]	1.5413[15]	i H_2O; vs EtOH, eth, bz; s ctc, HOAc
4252	4,4-Dimethyl-1-phenyl-1-penten-3-one		$C_{13}H_{16}O$	538-44-3	188.265		43	154[25]	0.9508[46]	1.5523[25]	
4253	2,2-Dimethyl-1-phenyl-1-propanone		$C_{11}H_{14}O$	938-16-9	162.228			220	0.963[26]	1.5086[19]	s ace
4254	3,5-Dimethyl-1-phenyl-1H-pyrazole		$C_{11}H_{12}N_2$	1131-16-4	172.226			272	1.0566[20]	1.5738[19]	vs eth, EtOH, chl
4255	4,4-Dimethyl-1-phenyl-3-pyrazolidinone	4,4-Dimethylphenidone	$C_{11}H_{14}N_2O$	2654-58-2	190.241		176				
4256	N,N-Dimethyl-γ-phenyl-2-pyridinepropanamine	Pheniramine	$C_{16}H_{20}N_2$	86-21-5	240.343			181[13]	1.0081[25]	1.5519[25]	vs bz, eth, EtOH, chl
4257	1,3-Dimethyl-3-phenyl-2,5-pyrrolidinedione	Methsuximide	$C_{12}H_{13}NO_2$	77-41-8	203.237		52.5	121[0.1]			
4258	Dimethylphenylsilane		$C_8H_{12}Si$	766-77-8	136.267			159(3)	0.8891[20]	1.4995[20]	i H_2O
4259	N,N-Dimethyl-N'-phenylurea	Fenuron	$C_9H_{12}N_2O$	101-42-8	164.203	cry (hx)	132.9(0.5)				
4260	Dimethylphosphine		C_2H_7P	676-59-5	62.051	vol liq or gas		25			i H_2O; s EtOH, eth
4261	Dimethylphosphinic acid		$C_2H_7O_2P$	3283-12-3	94.050	cry (bz)	92	377			vs H_2O, EtOH, eth; s bz

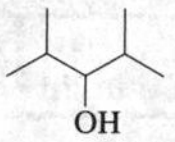

4225
2,4-Dimethyl-3-pentanol

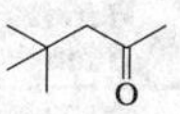

4226
4,4-Dimethyl-2-pentanone

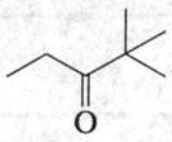

4227
2,2-Dimethyl-3-pentanone

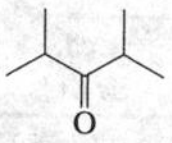

4228
2,4-Dimethyl-3-pentanone

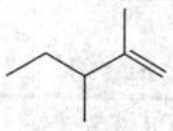

4229
2,3-Dimethyl-1-pentene

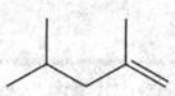

4230
2,4-Dimethyl-1-pentene

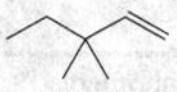

4231
3,3-Dimethyl-1-pentene

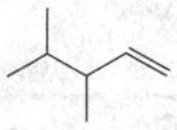

4232
3,4-Dimethyl-1-pentene

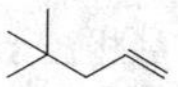

4233
4,4-Dimethyl-1-pentene

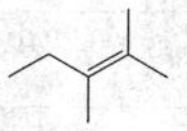

4234
2,3-Dimethyl-2-pentene

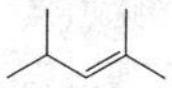

4235
2,4-Dimethyl-2-pentene

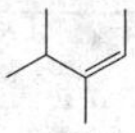

4236
cis-3,4-Dimethyl-2-pentene

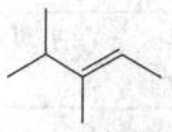

4237
trans-3,4-Dimethyl-2-pentene

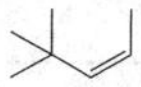

4238
cis-4,4-Dimethyl-2-pentene

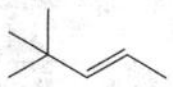

4239
trans-4,4-Dimethyl-2-pentene

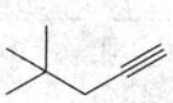

4240
4,4-Dimethyl-1-pentyne

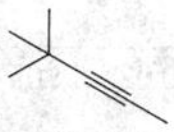

4241
4,4-Dimethyl-2-pentyne

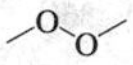

4242
Dimethylperoxide

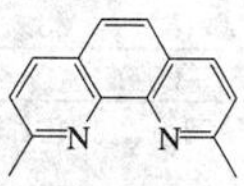

4243
2,9-Dimethyl-1,10-phenanthroline

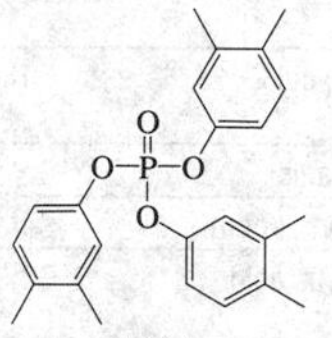

4244
3,4-Dimethylphenol phosphate (3:1)

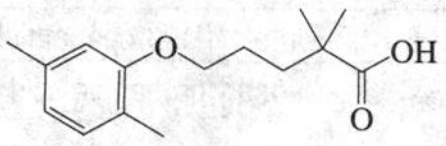

4245
5-(2,5-Dimethylphenoxy)-2,2-dimethylpentanoic acid

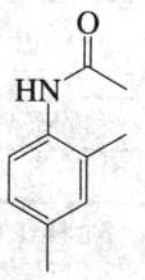

4246
N-(2,4-Dimethylphenyl)acetamide

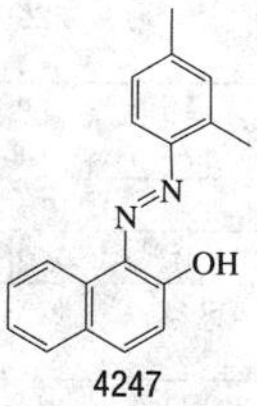

4247
1-[(2,4-Dimethylphenyl)azo]-2-naphthol

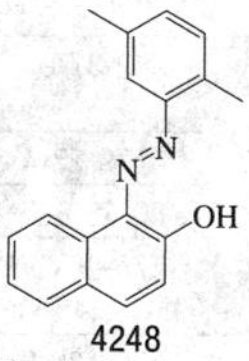

4248
1-[(2,5-Dimethylphenyl)azo]-2-naphthol

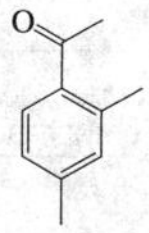

4249
1-(2,4-Dimethylphenyl)ethanone

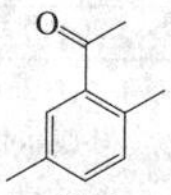

4250
1-(2,5-Dimethylphenyl)ethanone

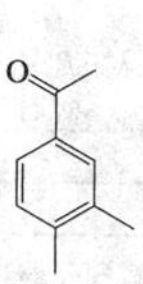

4251
1-(3,4-Dimethylphenyl)ethanone

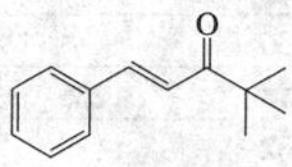

4252
4,4-Dimethyl-1-phenyl-1-penten-3-one

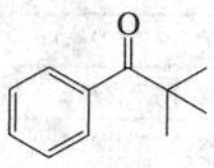

4253
2,2-Dimethyl-1-phenyl-1-propanone

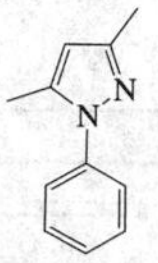

4254
3,5-Dimethyl-1-phenyl-1H-pyrazole

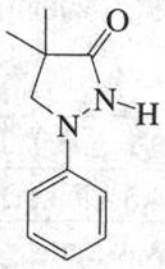

4255
4,4-Dimethyl-1-phenyl-3-pyrazolidinone

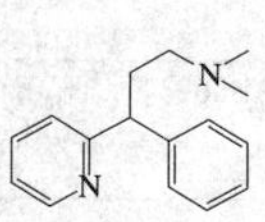

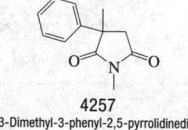

4256
N,N-Dimethyl-γ-phenyl-2-pyridinepropanamine

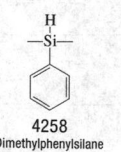

4257
1,3-Dimethyl-3-phenyl-2,5-pyrrolidinedione

4258
Dimethylphenylsilane

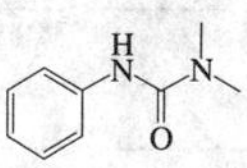

4259
N,N-Dimethyl-N'-phenylurea

4260
Dimethylphosphine

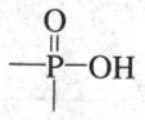

4261
Dimethylphosphinic acid

No.	Name	Synonym	Mol. Form.	CAS RN	Mol. Wt.	Physical Form	mp/°C	bp/°C	den g cm⁻³	n_D	Solubility
4262	O,O-Dimethyl phosphorochloridothioate	Dimethyl chlorothiophosphate	C₂H₆ClO₂PS	2524-03-0	160.560	hyg liq		68¹²	1.322	1.4820²⁰	
4263	Dimethyl phthalate	Methyl phthalate	C₁₀H₁₀O₄	131-11-3	194.184	pa ye	1.03(0.02)	282.7(0.2)	1.1905²⁰	1.5138²⁰	i H₂O; msc EtOH, eth; s bz; sl ctc
4264	1,4-Dimethylpiperazine		C₆H₁₄N₂	106-58-1	114.188	liq	-1.0(0.3)	129.8(0.8)	0.8600²⁰	1.4474²⁰	vs H₂O, EtOH, eth
4265	cis-2,5-Dimethylpiperazine		C₆H₁₄N₂	6284-84-0	114.188	orth bipym nd or pr (chl)	114	164(19)		1.4720²⁰	vs H₂O, EtOH, chl; sl eth, bz
4266	1,2-Dimethylpiperidine, (±)-		C₇H₁₅N	2512-81-4	113.201			127.5	0.824¹⁵	1.4395²⁰	vs H₂O, eth, EtOH
4267	2,6-Dimethylpiperidine		C₇H₁₅N	504-03-0	113.201			128(5)	0.8158²⁵	1.4377²⁰	msc H₂O, EtOH, eth; sl ctc; s acid
4268	3,5-Dimethylpiperidine	3,5-Lupetidine	C₇H₁₅N	35794-11-7	113.201			144(5)	0.853²⁵	1.4454²⁰	
4269	2,2-Dimethylpropanal	Pivaldehyde	C₅H₁₀O	630-19-3	86.132		1(2)	74(2)	0.7923¹⁷	1.3791²⁰	s EtOH, eth
4270	2,2-Dimethylpropanamide		C₅H₁₁NO	754-10-9	101.147						s tfa
4271	N,N-Dimethylpropanamide		C₅H₁₁NO	758-96-3	101.147	liq	-45	176(7)	0.9269²⁰		
4272	N,N-Dimethyl-1-propanamine	Dimethylpropylamine	C₅H₁₃N	926-63-6	87.164			65(3)	0.7152²⁰	1.3860²⁰	vs bz, eth, EtOH
4273	N,N-Dimethyl-1,3-propanediamine		C₅H₁₄N₂	109-55-7	102.178			129(13)	0.8272²⁰		
4274	2,2-Dimethyl-1,3-propanediol	Neopentyl glycol	C₅H₁₂O₂	126-30-7	104.148	nd (bz)	129.3(0.9)	207(14)			s H₂O, bz, chl; vs EtOH, eth
4275	2,2-Dimethylpropanenitrile	tert-Butyl cyanide	C₅H₉N	630-18-2	83.132		18.97(0.05)	105.2(0.2)	0.7586²⁵	1.3774²⁰	
4276	2,2-Dimethyl-1-propanethiol	Neopentyl mercaptan	C₅H₁₂S	1679-08-9	104.214	liq		103.6(0.8)			
4277	2,2-Dimethylpropanoic acid	Trimethylacetic acid	C₅H₁₀O₂	75-98-9	102.132	nd	36(1)	164.20(0.02)	0.905⁵⁰	1.3931³⁰	sl H₂O; vs EtOH, eth
4278	2,2-Dimethyl-1-propanol	Neopentyl alcohol	C₅H₁₂O	75-84-3	88.148		55(3)	112(1)	0.812²⁰		sl H₂O; vs EtOH, eth; s ctc
4279	2,2-Dimethylpropanoyl chloride	Pivalic acid chloride	C₅H₉ClO	3282-30-2	120.577			104(4)	1.003²⁰	1.4139²⁰	vs eth
4280	N,N-Dimethyl-2-propenamide	N,N-Dimethylacrylamide	C₅H₉NO	2680-03-7	99.131	liq		81²⁰	0.962²⁵	1.4730²⁰	
4281	2,2-Dimethylpropylamine	2,2-Dimethyl-1-propanamine	C₅H₁₃N	5813-64-9	87.164			82(7)	0.7455²⁰	1.4023²⁰	vs eth
4282	(1,1-Dimethylpropyl)benzene		C₁₁H₁₆	2049-95-8	148.245			191(3)	0.8748²⁰	1.4958²⁰	
4283	(2,2-Dimethylpropyl)benzene		C₁₁H₁₆	1007-26-7	148.245			186(4)	0.8581¹⁸	1.4884¹⁸	
4284	4-(1,1-Dimethylpropyl)-cyclohexanone		C₁₁H₂₀O	16587-71-6	168.276		96	125¹⁶	0.920²⁵	1.4677²⁰	
4285	1,1-Dimethylpropyl 3-methylbutanoate	tert-Pentyl isopentanoate	C₁₀H₂₀O₂	542-37-0	172.265			188(13)	0.8729⁰		vs EtOH
4286	2-(1,1-Dimethylpropyl)phenol		C₁₁H₁₆O	3279-27-4	164.244						sl ctc
4287	4-(1,1-Dimethylpropyl)phenol	p-tert-Pentylphenol	C₁₁H₁₆O	80-46-6	164.244		92.7(1)	262(9)			
4288	4,6-Dimethyl-2H-pyran-2-one		C₇H₈O₂	675-09-2	124.138	lf (eth)	51.5	245			vs H₂O, eth, EtOH
4289	2,6-Dimethyl-4H-pyran-4-one		C₇H₈O₂	1004-36-0	124.138	pl, nd (sub)	132.1(0.2)	251	0.9953¹³⁷		s H₂O, EtOH, eth, ace
4290	2,3-Dimethylpyrazine		C₆H₈N₂	5910-89-4	108.141			161(16)	1.0281⁰		s H₂O, EtOH, eth
4291	2,5-Dimethylpyrazine		C₆H₈N₂	123-32-0	108.141		15	152(4)	0.9887²⁰	1.4980²⁰	msc H₂O, EtOH, eth; s ace, chl
4292	2,6-Dimethylpyrazine		C₆H₈N₂	108-50-9	108.141	pr	47.5	155.6	0.9647⁵⁰		s H₂O, EtOH, eth; sl ctc
4293	1,3-Dimethyl-1H-pyrazole		C₅H₈N₂	694-48-4	96.131			137	0.9561¹⁷	1.4734¹⁵	vs H₂O
4294	3,5-Dimethyl-1H-pyrazole		C₅H₈N₂	67-51-6	96.131	cry (peth, al)	107.5	218	0.8839¹⁶		s H₂O, ace; vs EtOH, eth, bz, MeOH
4295	2,7-Dimethylpyrene		C₁₈H₁₄	15679-24-0	230.304		230				
4296	4,6-Dimethyl-2-pyridinamine		C₇H₁₀N₂	5407-87-4	122.167		61	235			
4297	N,N-Dimethyl-2-pyridinamine		C₇H₁₀N₂	5683-33-0	122.167		182	196	1.0149¹⁴	1.5663²⁰	s EtOH, eth, bz
4298	N,N-Dimethyl-4-pyridinamine		C₇H₁₀N₂	1122-58-3	122.167	pl (eth)	113.9(0.2)				vs H₂O, EtOH, bz, chl; s eth
4299	2,3-Dimethylpyridine	2,3-Lutidine	C₇H₉N	583-61-9	107.153			161.1(0.4)	0.9319²⁵	1.5057²⁰	s H₂O, EtOH, eth
4300	2,4-Dimethylpyridine	2,4-Lutidine	C₇H₉N	108-47-4	107.153	liq	-63.80(0.03)	158.4(0.3)	0.9309²⁰	1.5010²⁰	vs H₂O, EtOH, eth; s ace
4301	2,5-Dimethylpyridine	2,5-Lutidine	C₇H₉N	589-93-5	107.153	liq	-14.08(0.03)	157.00(0.05)	0.9297²⁰	1.5006²⁰	sl H₂O; vs EtOH; msc eth; s ace
4302	2,6-Dimethylpyridine	2,6-Lutidine	C₇H₉N	108-48-5	107.153	liq	-6.12(0.03)	144.0(0.1)	0.9226²⁰	1.4953²⁰	msc H₂O; sl EtOH; s eth, ace, chl
4303	3,4-Dimethylpyridine	3,4-Lutidine	C₇H₉N	583-58-4	107.153	liq	-10.45(0.03)	179.1(0.3)	0.9281²⁰	1.5096²⁰	sl H₂O, ctc; s EtOH, eth, ace, chl

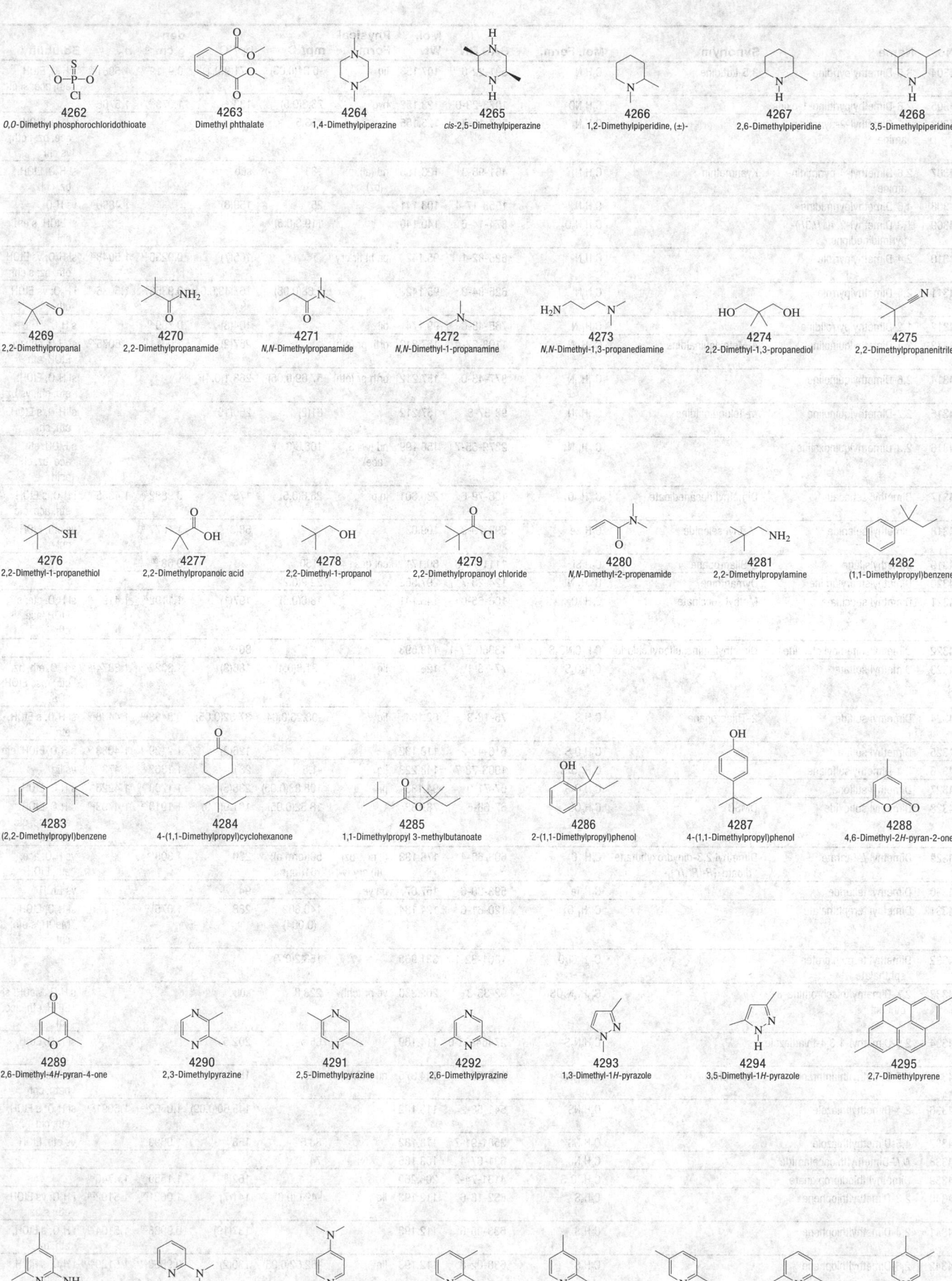

4262
O,O-Dimethyl phosphorochloridothioate

4263
Dimethyl phthalate

4264
1,4-Dimethylpiperazine

4265
cis-2,5-Dimethylpiperazine

4266
1,2-Dimethylpiperidine, (±)-

4267
2,6-Dimethylpiperidine

4268
3,5-Dimethylpiperidine

4269
2,2-Dimethylpropanal

4270
2,2-Dimethylpropanamide

4271
N,N-Dimethylpropanamide

4272
N,N-Dimethyl-1-propanamine

4273
N,N-Dimethyl-1,3-propanediamine

4274
2,2-Dimethyl-1,3-propanediol

4275
2,2-Dimethylpropanenitrile

4276
2,2-Dimethyl-1-propanethiol

4277
2,2-Dimethylpropanoic acid

4278
2,2-Dimethyl-1-propanol

4279
2,2-Dimethylpropanoyl chloride

4280
N,N-Dimethyl-2-propenamide

4281
2,2-Dimethylpropylamine

4282
(1,1-Dimethylpropyl)benzene

4283
(2,2-Dimethylpropyl)benzene

4284
4-(1,1-Dimethylpropyl)cyclohexanone

4285
1,1-Dimethylpropyl 3-methylbutanoate

4286
2-(1,1-Dimethylpropyl)phenol

4287
4-(1,1-Dimethylpropyl)phenol

4288
4,6-Dimethyl-2*H*-pyran-2-one

4289
2,6-Dimethyl-4*H*-pyran-4-one

4290
2,3-Dimethylpyrazine

4291
2,5-Dimethylpyrazine

4292
2,6-Dimethylpyrazine

4293
1,3-Dimethyl-1*H*-pyrazole

4294
3,5-Dimethyl-1*H*-pyrazole

4295
2,7-Dimethylpyrene

4296
4,6-Dimethyl-2-pyridinamine

4297
N,N-Dimethyl-2-pyridinamine

4298
N,N-Dimethyl-4-pyridinamine

4299
2,3-Dimethylpyridine

4300
2,4-Dimethylpyridine

4301
2,5-Dimethylpyridine

4302
2,6-Dimethylpyridine

4303
3,4-Dimethylpyridine

No.	Name	Synonym	Mol. Form.	CAS RN	Mol. Wt.	Physical Form	mp/°C	bp/°C	den g cm⁻³	n_D	Solubility
4304	3,5-Dimethylpyridine	3,5-Lutidine	C_7H_9N	591-22-0	107.153	liq	-6.34(0.03)	171.9(0.1)	0.9419[20]	1.5061[20]	s H_2O, EtOH, eth, ace; sl ctc
4305	2,6-Dimethylpyridine-1-oxide		C_7H_9NO	1073-23-0	123.152	hyg	23.3(0.5)	133[22]	1.073[25]	1.5706[20]	
4306	4,6-Dimethyl-2-pyrimidin-amine		$C_6H_9N_3$	767-15-7	123.155		153.5				s H_2O, EtOH, ace, bz; i eth; vs chl
4307	2,6-Dimethyl-4-pyrimidin-amine	Kyanmethin	$C_6H_9N_3$	461-98-3	123.155	nd (al), pl (bz)	183	sub			sl H_2O, EtOH, bz, chl
4308	4,6-Dimethylpyrimidine		$C_6H_8N_2$	1558-17-4	108.141		25	159(8)		1.4880[20]	vs H_2O
4309	1,3-Dimethyl-2,4(1H,3H)-pyrimidinedione		$C_6H_8N_2O_2$	874-14-6	140.140		119.3(0.5)				sl EtOH; s eth, chl
4310	2,4-Dimethylpyrrole		C_6H_9N	625-82-1	95.142	pa bl flr cry		165(7)	0.9236[20]	1.5048[20]	sl H_2O; vs EtOH, eth, bz; s chl
4311	2,5-Dimethylpyrrole		C_6H_9N	625-84-3	95.142		7.68(0.08)	167.43(0.04)	0.9353[20]	1.5036[20]	i H_2O; vs EtOH, eth
4312	1,2-Dimethylpyrrolidine		$C_6H_{13}N$	765-48-0	99.174	oil		104(8)	0.799[20]		s H_2O
4313	2,4-Dimethylquinoline	4-Methylquinaldine	$C_{11}H_{11}N$	1198-37-4	157.212	orth pr (eth)		267(2)	1.0611[15]	1.6075[20]	sl H_2O, chl; vs EtOH, eth
4314	2,6-Dimethylquinoline		$C_{11}H_{11}N$	877-43-0	157.212	orth pr (eth)	57.69(0.05)	268.1(0.4)			sl H_2O, EtOH, eth, chl; vs bz
4315	2,7-Dimethylquinoline	m-Toluquinaldine	$C_{11}H_{11}N$	93-37-8	157.212		61(2)	260(19)			sl H_2O; s EtOH, eth, chl
4316	2,3-Dimethylquinoxaline		$C_{10}H_{10}N_2$	2379-55-7	158.199	nd (w+3, ace)	106.3(0.4)				s EtOH, eth, ace, bz, chl, acid
4317	Dimethyl sebacate	Dimethyl decanedioate	$C_{12}H_{22}O_4$	106-79-6	230.301	lo pr	26.6(0.5)	175[20]	0.9882[28]	1.4355[28]	i H_2O; s EtOH, eth, ace, ctc
4318	Dimethyl selenide	Methyl selenide	C_2H_6Se	593-79-3	109.03			58(3)	1.4077[15]		vs eth, EtOH, chl
4319	Dimethylsilane	2-Silapropane	C_2H_8Si	1111-74-6	60.171	col gas	-150	-20	0.68[-80]		
4320	Dimethylstearylamine	Dymanthine	$C_{20}H_{43}N$	124-28-7	297.562		22.9(0.2)				
4321	Dimethyl succinate	Methyl succinate	$C_6H_{10}O_4$	106-65-0	146.141		18.6(0.6)	197(1)	1.1198[20]	1.4197[20]	sl H_2O, ctc; s EtOH, ace; vs eth
4322	Dimethylsulfamoyl chloride	Dimethylaminosulfonyl chloride	$C_2H_6ClNO_2S$	13360-57-1	143.593			80[16]			
4323	Dimethyl sulfate		$C_2H_6O_4S$	77-78-1	126.132	liq	-31.8(0.4)	186(3)	1.3322[20]	1.3874[20]	s H_2O, eth, bz, ctc; msc EtOH; i CS_2
4324	Dimethyl sulfide	2-Thiapropane	C_2H_6S	75-18-3	62.134	liq	-98.26(0.04)	37.32(0.05)	0.8483[20]	1.4438[20]	sl H_2O; s EtOH, eth
4325	Dimethyl sulfite		$C_2H_6O_3S$	616-42-2	110.132			126(2)	1.2129[20]	1.4083[20]	s H_2O, EtOH, eth
4326	2,4-Dimethylsulfolane		$C_6H_{12}O_2S$	1003-78-7	148.223	liq	-1.5	281	1.1362[20]	1.4732[20]	vs lig
4327	Dimethyl sulfone		$C_2H_6O_2S$	67-71-0	94.133	pr	108.83(0.05)	238(5)	1.1700[110]	1.4226	s H_2O, EtOH, bz
4328	Dimethyl sulfoxide	DMSO	C_2H_6OS	67-68-5	78.133		18.52(0.05)	191.9(0.9)	1.1010[25]	1.4793[20]	s H_2O, EtOH, eth, ace, ctc, AcOEt
4329	Dimethyl L-tartrate	Dimethyl 2,3-dihydroxybutane-dioate, [R-(R^*,R^*)]-	$C_6H_{10}O_6$	608-68-4	178.139	(i) cry (bz) (ii) cry (w)	50(form a); 61(form b)	280	1.306[45]		vs H_2O, ace, eth, EtOH
4330	Dimethyl telluride		C_2H_6Te	593-80-6	157.67	pa ye		94			vs EtOH
4331	Dimethyl terephthalate		$C_{10}H_{10}O_4$	120-61-6	194.184		140.602 (0.004)	288	1.075[141]		sl H_2O, EtOH, MeOH; s eth, chl
4332	Dimethyl tetrachloroter-ephthalate		$C_{10}H_6Cl_4O_4$	1861-32-1	331.965		158.2(0.7)				
4333	2,7-Dimethylthiachromine-8-ethanol		$C_{12}H_{14}N_4OS$	92-35-3	262.330	ye pr (chl)	228.8	sub			s H_2O, MeOH; sl EtOH, eth, ace, chl
4334	2,5-Dimethyl-1,3,4-thiadiazole		$C_4H_6N_2S$	27464-82-0	114.169		65	202.5			sl H_2O, EtOH, eth
4335	2,7-Dimethylthianthrene	Mesulphen	$C_{14}H_{12}S_2$	135-58-0	244.375	nd (HOAc,al)	123	184[3]			vs ace, eth, peth, chl
4336	2,4-Dimethylthiazole		C_5H_7NS	541-58-2	113.182			145.60(0.09)	1.0562[15]	1.5091[20]	sl H_2O; s EtOH, eth, chl
4337	4,5-Dimethylthiazole		C_5H_7NS	3581-91-7	113.182		83.5	158	1.0699[20]		vs eth, EtOH
4338	N,N-Dimethylthioacetamide		C_4H_9NS	631-67-4	103.186		74.5				
4339	Dimethyl thiodipropionate		$C_8H_{14}O_4S$	4131-74-2	206.260			162[18]	1.1559[20]	1.4740[20]	
4340	2,3-Dimethylthiophene		C_6H_8S	632-16-6	112.193	liq	-49.0(0.6)	141(7)	1.0021[20]	1.5192[20]	i H_2O; vs EtOH, eth; s bz
4341	2,4-Dimethylthiophene		C_6H_8S	638-00-6	112.193			137(19)	0.9938[20]	1.5104[20]	i H_2O; s EtOH, eth, bz
4342	2,5-Dimethylthiophene		C_6H_8S	638-02-8	112.193	liq	-62.52(0.05)	139(2)	0.9850[20]	1.5129[20]	i H_2O; s EtOH, eth, bz

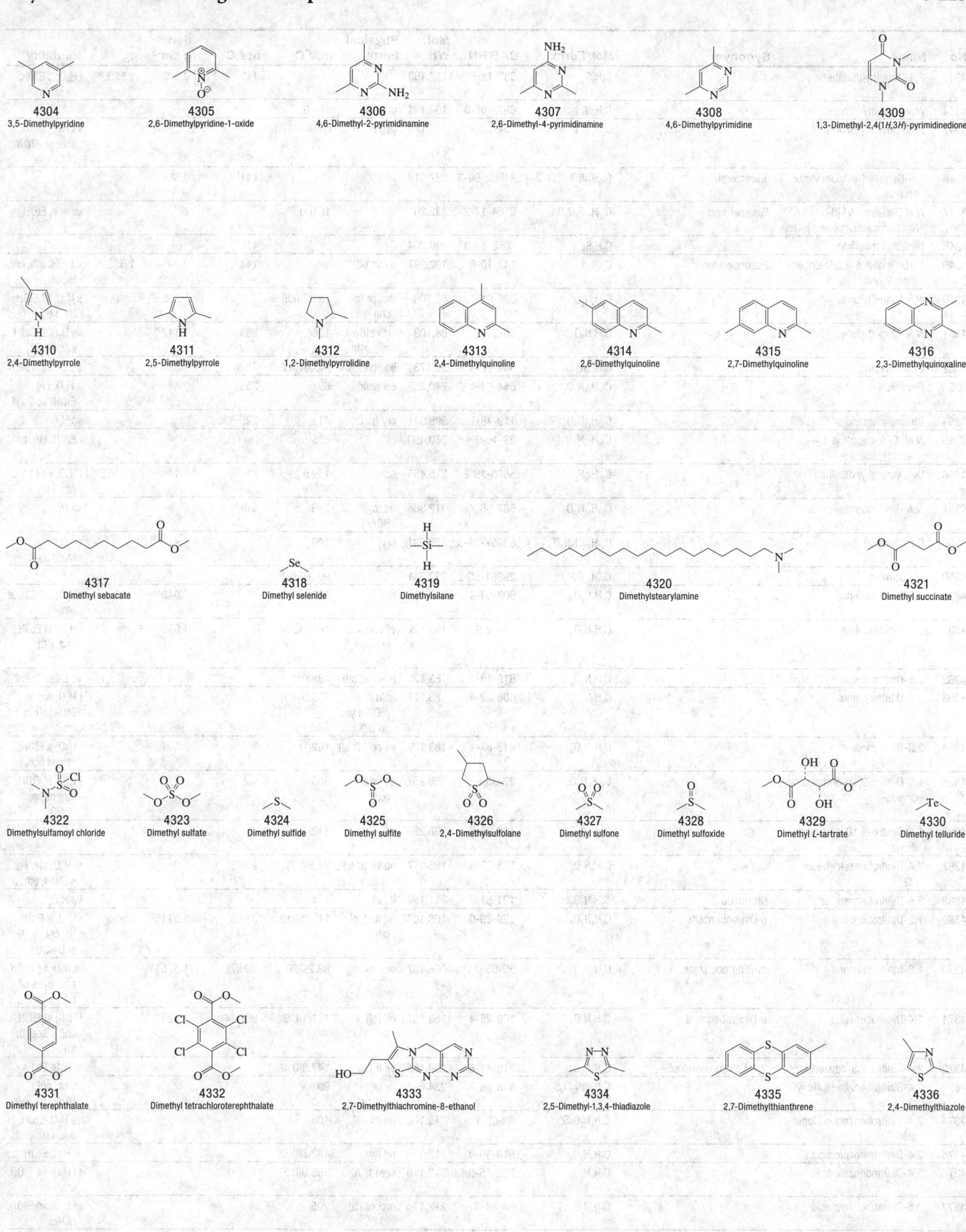

4304
3,5-Dimethylpyridine

4305
2,6-Dimethylpyridine-1-oxide

4306
4,6-Dimethyl-2-pyrimidinamine

4307
2,6-Dimethyl-4-pyrimidinamine

4308
4,6-Dimethylpyrimidine

4309
1,3-Dimethyl-2,4(1H,3H)-pyrimidinedione

4310
2,4-Dimethylpyrrole

4311
2,5-Dimethylpyrrole

4312
1,2-Dimethylpyrrolidine

4313
2,4-Dimethylquinoline

4314
2,6-Dimethylquinoline

4315
2,7-Dimethylquinoline

4316
2,3-Dimethylquinoxaline

4317
Dimethyl sebacate

4318
Dimethyl selenide

4319
Dimethylsilane

4320
Dimethylstearylamine

4321
Dimethyl succinate

4322
Dimethylsulfamoyl chloride

4323
Dimethyl sulfate

4324
Dimethyl sulfide

4325
Dimethyl sulfite

4326
2,4-Dimethylsulfolane

4327
Dimethyl sulfone

4328
Dimethyl sulfoxide

4329
Dimethyl L-tartrate

4330
Dimethyl telluride

4331
Dimethyl terephthalate

4332
Dimethyl tetrachloroterephthalate

4333
2,7-Dimethylthiachromine-8-ethanol

4334
2,5-Dimethyl-1,3,4-thiadiazole

4335
2,7-Dimethylthianthrene

4336
2,4-Dimethylthiazole

4337
4,5-Dimethylthiazole

4338
N,N-Dimethylthioacetamide

4339
Dimethyl thiodipropionate

4340
2,3-Dimethylthiophene

4341
2,4-Dimethylthiophene

4342
2,5-Dimethylthiophene

No.	Name	Synonym	Mol. Form.	CAS RN	Mol. Wt.	Physical Form	mp/°C	bp/°C	den g cm⁻³	n_D	Solubility
4343	3,4-Dimethylthiophene		C$_6$H$_8$S	632-15-5	112.193			145	0.993^{25}	1.5206^{20}	i H$_2$O; s EtOH; vs eth
4344	N,N-Dimethylthiourea		C$_3$H$_8$N$_2$S	6972-05-0	104.174	cry (w)	161.5				
4345	N,N'-Dimethylthiourea		C$_3$H$_8$N$_2$S	534-13-4	104.174	hyg pl	63.8(0.6)				vs H$_2$O, EtOH, ace; sl eth, bz; i CS$_2$
4346	2,6-Dimethyl-4-tridecylmor-pholine	Tridemorph	C$_{19}$H$_{39}$NO	24602-86-6	297.519			141$^{1.3}$	0.86		
4347	N,N-Dimethyl-N'-[3-(trifluoromethyl)phenyl]urea	Fluometuron	C$_{10}$H$_{11}$F$_3$N$_2$O	2164-17-2	232.201		161(1)				vs ace, EtOH
4348	Dimethyl trisulfide		C$_2$H$_6$S$_3$	3658-80-8	126.264			41^6			
4349	6,10-Dimethyl-3,5,9-undeca-trien-2-one	Pseudoionone	C$_{13}$H$_{20}$O	141-10-6	192.297	pa ye oil		144^{12}	0.8984^{20}	1.5335^{20}	s EtOH, eth, chl, MeOH
4350	N,N-Dimethylurea		C$_3$H$_8$N$_2$O	598-94-7	88.108	mcl pr (al, chl)	181.2(0.8)		1.2555^{25}		s H$_2$O; sl EtOH, tfa; i eth
4351	N,N'-Dimethylurea		C$_3$H$_8$N$_2$O	96-31-1	88.108	orth bipym (chl-eth)	106(2)	269	1.142^{25}		vs H$_2$O, EtOH; i eth; sl chl
4352	Dimethyl zinc		C$_2$H$_6$Zn	544-97-8	95.478	liq, ign in air	-43.01(0.02)	43(3)	1.386^{10}		s eth; msc peth
4353	Dimetilan		C$_{10}$H$_{16}$N$_4$O$_3$	644-64-4	240.259	col solid	69	205^{13}			s H$_2$O, chl, EtOH, ace, xyl
4354	Dimorpholamine		C$_{20}$H$_{38}$N$_4$O$_4$	119-48-2	398.541	cry (peth)	41.5	229$^{0.4}$			vs H$_2$O
4355	N,N'-Di-2-naphthyl-1,4-benzenediamine		C$_{26}$H$_{20}$N$_2$	93-46-9	360.450		235				i EtOH, eth, bz
4356	Di-2-naphthyl disulfide		C$_{20}$H$_{14}$S$_2$	5586-15-2	318.455	nd	139.5		1.144^{145}	1.4555^{20}	i H$_2$O; vs EtOH, eth; i lig
4357	N,N'-Di-1-naphthylurea		C$_{21}$H$_{16}$N$_2$O	607-56-7	312.364	nd (py, HOAc)	296	sub			vs py
4358	Diniconazole		C$_{15}$H$_{17}$Cl$_2$N$_3$O	83657-24-3	326.221	cry	149				s H$_2$O, ace, MeOH, xyl
4359	Dinitramine		C$_{11}$H$_{13}$F$_3$N$_4$O$_4$	29091-05-2	322.241		99.3(0.5)				
4360	2,3-Dinitroaniline		C$_6$H$_5$N$_3$O$_4$	602-03-9	183.122		128		1.646^{50}		i H$_2$O; s EtOH; sl eth
4361	2,4-Dinitroaniline		C$_6$H$_5$N$_3$O$_4$	97-02-9	183.122	ye nd (ace) grn ye tab (al)	180.1(0.5)		1.615^{14}		i H$_2$O; sl EtOH, ace, HCl
4362	2,5-Dinitroaniline		C$_6$H$_5$N$_3$O$_4$	619-18-1	183.122	oran nd (al)	138.0				vs EtOH
4363	2,6-Dinitroaniline		C$_6$H$_5$N$_3$O$_4$	606-22-4	183.122	gold lf (HOAc) ye nd (al)	138.0(0.5)				i H$_2$O, lig; sl EtOH; s eth, bz
4364	3,5-Dinitroaniline		C$_6$H$_5$N$_3$O$_4$	618-87-1	183.122	ye nd (dil al)	162(1)		1.601^{50}		i H$_2$O; s EtOH, eth; sl ace, bz
4365	1,5-Dinitro-9,10-anthracene-dione		C$_{14}$H$_6$N$_2$O$_6$	82-35-9	298.207	pa ye nd (xyl)	385	sub			i H$_2$O; sl EtOH, eth, bz; vs PhNO$_2$
4366	1,8-Dinitro-9,10-anthracene-dione		C$_{14}$H$_6$N$_2$O$_6$	129-39-5	298.207		312				
4367	2,4-Dinitrobenzaldehyde		C$_7$H$_4$N$_2$O$_5$	528-75-6	196.117	pa ye pr (al), pl (bz)	71.8(0.7)	200^{15}			sl H$_2$O, chl, lig; s EtOH, eth, bz
4368	3,5-Dinitrobenzamide	Nitromide	C$_7$H$_5$N$_3$O$_5$	121-81-3	211.132	lf (w)	184				vs H$_2$O
4369	1,2-Dinitrobenzene	o-Dinitrobenzene	C$_6$H$_4$N$_2$O$_4$	528-29-0	168.107	nd (bz), pl (al)	115.8(0.6)	319(3)	1.3119^{120}	1.565^{17}	i H$_2$O; s EtOH, bz, chl, AcOEt; sl DMSO
4370	1,3-Dinitrobenzene	m-Dinitrobenzene	C$_6$H$_4$N$_2$O$_4$	99-65-0	168.107	orth pl (al)	89.2(0.5)	296(2)	1.5751^{18}		sl H$_2$O; vs EtOH, ace, py; s eth, tol
4371	1,4-Dinitrobenzene	p-Dinitrobenzene	C$_6$H$_4$N$_2$O$_4$	100-25-4	168.107	nd (al)	171.1(0.9)	297	1.625^{18}		i H$_2$O; sl EtOH, chl; s ace, bz, tol
4372	2,4-Dinitro-1,3-benzenediol	2,4-Dinitroresorcinol	C$_6$H$_4$N$_2$O$_6$	519-44-8	200.105	ye lf (al)	147.3(0.4)				sl H$_2$O, EtOH
4373	2,4-Dinitrobenzenesulfenyl chloride		C$_6$H$_3$ClN$_2$O$_4$S	528-76-7	234.617	ye pr (bz-peth)	99				vs bz, chl, HOAc; sl peth
4374	2,4-Dinitrobenzenesulfonic acid		C$_6$H$_4$N$_2$O$_7$S	89-02-1	248.170	nd (w+3)	108				vs H$_2$O, EtOH; sl eth; i bz, peth
4375	2,4-Dinitrobenzoic acid		C$_7$H$_4$N$_2$O$_6$	610-30-0	212.116	nd (w)	182.7(0.5)		1.672^{20}		sl H$_2$O, EtOH, bz
4376	3,4-Dinitrobenzoic acid		C$_7$H$_4$N$_2$O$_6$	528-45-0	212.116	cry (dil al)	165.0(0.5)				sl H$_2$O; vs EtOH, eth
4377	3,5-Dinitrobenzoic acid		C$_7$H$_4$N$_2$O$_6$	99-34-3	212.116	mcl pr (al)	205				sl H$_2$O; vs EtOH, HOAc
4378	3,5-Dinitrobenzoyl chloride		C$_7$H$_3$ClN$_2$O$_5$	99-33-2	230.562	ye nd (bz)	74	196^{12}			s eth, chl
4379	2,2'-Dinitro-1,1'-biphenyl		C$_{12}$H$_8$N$_2$O$_4$	2436-96-6	244.203	ye mcl pr or nd (al)	127(1)	305	1.45^{25}		i H$_2$O; vs EtOH; s eth, bz; sl ace, lig

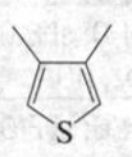

4343
3,4-Dimethylthiophene

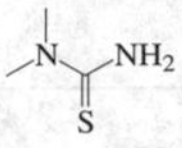

4344
N,N-Dimethylthiourea

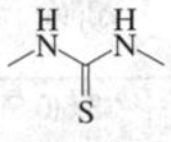

4345
N,N'-Dimethylthiourea

4346
2,6-Dimethyl-4-tridecylmorpholine

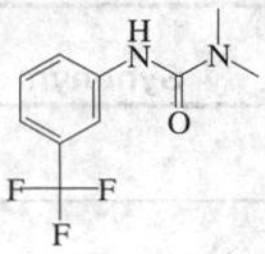

4347
N,N-Dimethyl-*N'*-[3-(trifluoromethyl)phenyl]urea

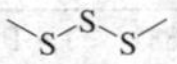

4348
Dimethyl trisulfide

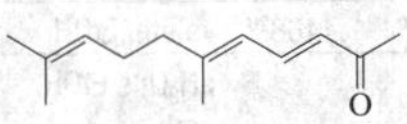

4349
6,10-Dimethyl-3,5,9-undecatrien-2-one

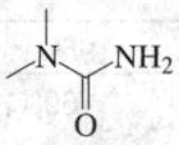

4350
N,N-Dimethylurea

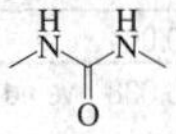

4351
N,N'-Dimethylurea

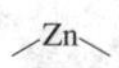

4352
Dimethyl zinc

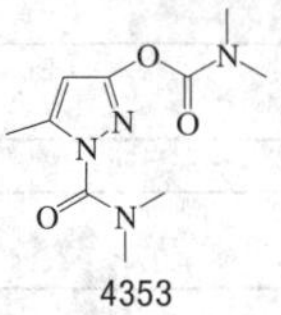

4353
Dimetilan

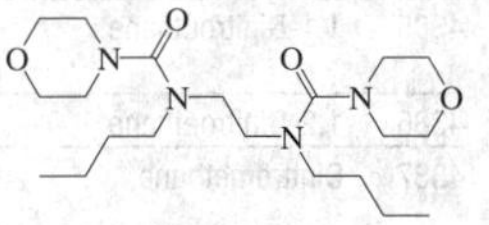

4354
Dimorpholamine

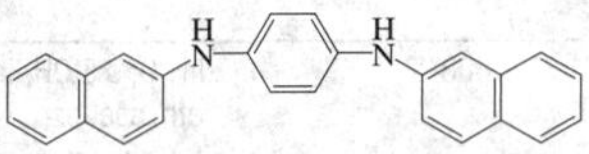

4355
N,N'-Di-2-naphthyl-1,4-benzenediamine

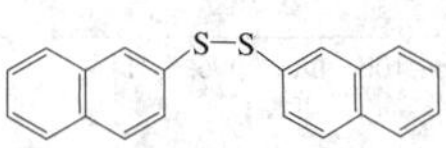

4356
Di-2-naphthyl disulfide

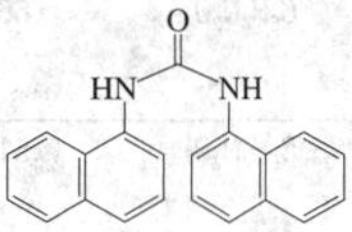

4357
N,N'-Di-1-naphthylurea

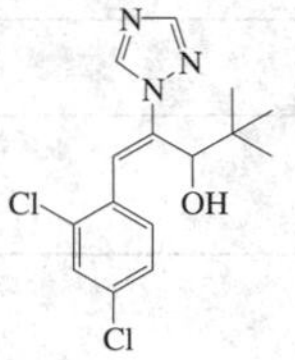

4358
Diniconazole

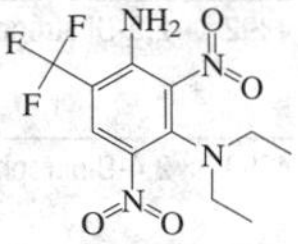

4359
Dinitramine

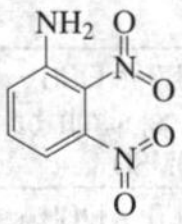

4360
2,3-Dinitroaniline

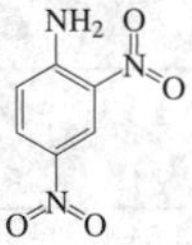

4361
2,4-Dinitroaniline

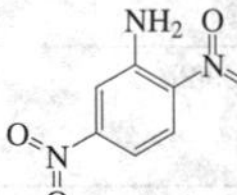

4362
2,5-Dinitroaniline

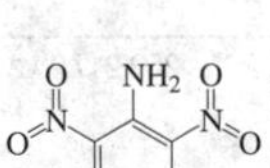

4363
2,6-Dinitroaniline

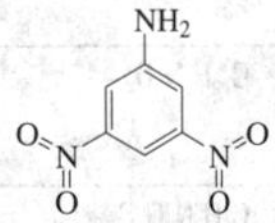

4364
3,5-Dinitroaniline

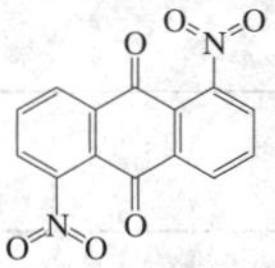

4365
1,5-Dinitro-9,10-anthracenedione

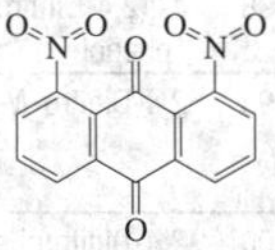

4366
1,8-Dinitro-9,10-anthracenedione

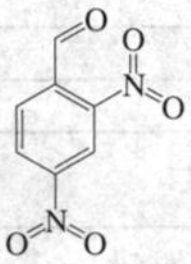

4367
2,4-Dinitrobenzaldehyde

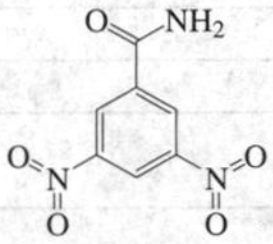

4368
3,5-Dinitrobenzamide

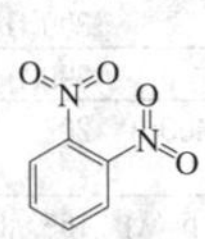

4369
1,2-Dinitrobenzene

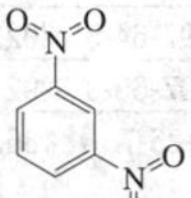

4370
1,3-Dinitrobenzene

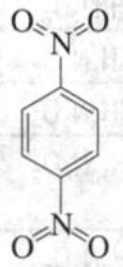

4371
1,4-Dinitrobenzene

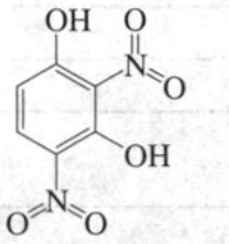

4372
2,4-Dinitro-1,3-benzenediol

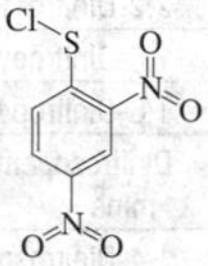

4373
2,4-Dinitrobenzenesulfenyl chloride

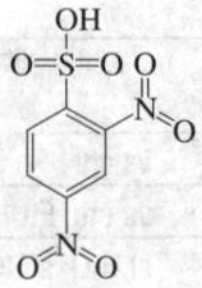

4374
2,4-Dinitrobenzenesulfonic acid

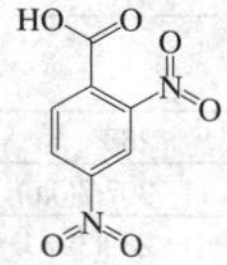

4375
2,4-Dinitrobenzoic acid

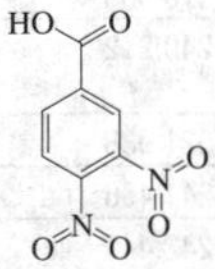

4376
3,4-Dinitrobenzoic acid

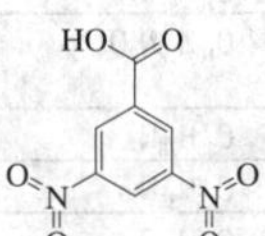

4377
3,5-Dinitrobenzoic acid

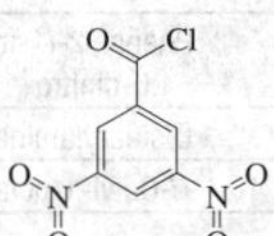

4378
3,5-Dinitrobenzoyl chloride

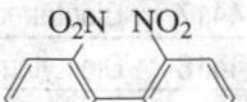

4379
2,2'-Dinitro-1,1'-biphenyl

No.	Name	Synonym	Mol. Form.	CAS RN	Mol. Wt.	Physical Form	mp/°C	bp/°C	den g cm⁻³	n_D	Solubility
4380	4,4'-Dinitro-1,1'-biphenyl		C₁₂H₈N₂O₄	1528-74-1	244.203	nd (al)	235(6)				i H₂O; sl EtOH; s bz, HOAc
4381	1,4-Dinitrobutane		C₄H₈N₂O₄	4286-49-1	148.118	pl (al)	33.5(0.5)	176[13]			i H₂O; sl EtOH; s eth, bz, MeOH
4382	4,4'-Dinitrodiphenylamine	4-Nitro-N-(4-nitrophenyl)-aniline	C₁₂H₉N₃O₄	1821-27-8	259.217	ye nd(al)	210(1)				i H₂O, tol; sl EtOH, bz; s ace, HOAc
4383	4,4'-Dinitrodiphenyl ether	Bis(4-nitrophenyl) ether	C₁₂H₈N₂O₅	101-63-3	260.202		143(1)				i H₂O; sl EtOH, eth; s bz, HOAc
4384	4,4'-Dinitrodiphenyl sulfide	Bis(4-nitrophenyl) sulfide	C₁₂H₈N₂O₄S	1223-31-0	276.268	oran pl (HOAc)	160.5				i H₂O; sl EtOH; s con sulf
4385	1,1-Dinitroethane		C₂H₄N₂O₄	600-40-8	120.064	ye mcl (bz, MeOH)		185.5	1.349[24]		sl H₂O; s EtOH, eth
4386	1,2-Dinitroethane		C₂H₄N₂O₄	7570-26-5	120.064		38.2(0.4)	95[5]	1.4597[20]	1.4468[20]	vs eth, EtOH
4387	Dinitromethane		CH₂N₂O₄	625-76-3	106.038	ye nd	<-15	169(11)			i H₂O; s EtOH, eth
4388	1,3-Dinitronaphthalene		C₁₀H₆N₂O₄	606-37-1	218.166	ye nd (bz, py-w)	148	sub			i H₂O; s EtOH, ace
4389	1,5-Dinitronaphthalene		C₁₀H₆N₂O₄	605-71-0	218.166	hex nd (ace, HOAc)	217(2)	371(3)	1.5860[20]		i H₂O; sl EtOH, ace; s bz, py; vs eth
4390	1,8-Dinitronaphthalene		C₁₀H₆N₂O₄	602-38-0	218.166	ye orth pl (chl)	172(1)	445 dec			i H₂O; sl EtOH, bz; s ace, chl, py
4391	2,4-Dinitro-1-naphthol		C₁₀H₆N₂O₅	605-69-6	234.165	ye nd (al, chl)	138(1)				
4392	2,3-Dinitrophenol		C₆H₄N₂O₅	66-56-8	184.106	ye nd (w)	145.1(0.2)		1.681[20]		sl H₂O, DMSO; vs EtOH, eth; s bz
4393	2,4-Dinitrophenol		C₆H₄N₂O₅	51-28-5	184.106	pa ye pl or lf (w)	114(2)	sub	1.683[24]		sl H₂O; s EtOH, eth, ace, bz, tol, chl, py
4394	2,5-Dinitrophenol		C₆H₄N₂O₅	329-71-5	184.106	ye mcl pr or nd (w,lig)	105.6(0.2)				vs bz, eth
4395	2,6-Dinitrophenol		C₆H₄N₂O₅	573-56-8	184.106	pa ye orth nd or lf (dil al)	62.4(0.8)				i H₂O; vs EtOH, eth; s bz, chl; sl ctc
4396	3,4-Dinitrophenol		C₆H₄N₂O₅	577-71-9	184.106	tcl nd (w)	134.4(1)		1.672[25]		vs bz, eth, EtOH
4397	2,4-Dinitrophenol, acetate		C₈H₆N₂O₆	4232-27-3	226.143	cry (MeOH)	72.5				
4398	4-[(2,4-Dinitrophenyl)amino]-phenol		C₁₂H₉N₃O₅	119-15-3	275.216	red lf	195.5				s alk
4399	2,4-Dinitro-N-phenylaniline		C₁₂H₉N₃O₄	961-68-2	259.217	ye red nd (al)	159(2)				i H₂O; s EtOH, ace; sl eth, bz, DMSO
4400	2,4-Dinitrophenyl dimethylcarbamodithioate		C₉H₉N₃O₄S₂	89-37-2	287.315		152.5		1.54[20]		i H₂O; s EtOH, ace, bz
4401	(2,4-Dinitrophenyl)hydrazine		C₆H₆N₄O₄	119-26-6	198.137	blsh-red (al)	201.6(0.4)				i H₂O; s EtOH; sl eth, bz, chl, DMSO
4402	1,1-Dinitropropane		C₃H₆N₂O₄	601-76-3	134.091	liq	-42(1)	194(4)	1.2610[25]	1.4339[20]	s alk
4403	1,3-Dinitropropane		C₃H₆N₂O₄	6125-21-9	134.091		-21.4(0.2)	103[1]	1.353[26]	1.4654[20]	i H₂O; s eth
4404	2,2-Dinitropropane		C₃H₆N₂O₄	595-49-3	134.091		52(2)	186.0(0.5)	1.30[25]		sl H₂O
4405	2,2-Dinitro-1,3-propanediol		C₃H₆N₂O₆	2736-80-3	166.089	wh pl (bz)	142				
4406	1,6-Dinitropyrene		C₁₆H₈N₂O₄	42397-64-8	292.246		309(1)				
4407	1,8-Dinitropyrene		C₁₆H₈N₂O₄	42397-65-9	292.246		299(1)				
4408	Dinitrosopentamethylenetetra-mine		C₅H₁₀N₆O₂	101-25-7	186.172	cry (MeOH)	207				
4409	1,4-Dinitrosopiperazine		C₄H₈N₄O₂	140-79-4	144.133	pa ye pl (w)	155.9(0.7)				vs EtOH
4410	4,4'-Dinitro-2,2'-stilbenedisul-fonic acid		C₁₄H₁₀N₂O₁₀S₂	128-42-7	430.366	cry (AcOH)	266				
4411	Dinobuton	Dessin	C₁₄H₁₈N₂O₇	973-21-7	326.302	ye cry (EtOH)	60				
4412	Dinocap		C₁₈H₂₄N₂O₆	6119-92-2	364.393			136[0.01]			
4413	Dinonyl adipate	Dinonyl hexanedioate	C₂₄H₄₆O₄	151-32-6	398.620			205[1]			
4414	Dinonyl ether		C₁₈H₃₈O	2456-27-1	270.494	liq		318	0.81	1.4356[20]	
4415	Dinonyl phthalate		C₂₆H₄₂O₄	84-76-4	418.609			413			
4416	Dinoseb	Phenol, 2-(1-methylpropyl)-4,6-dinitro-	C₁₀H₁₂N₂O₅	88-85-7	240.212		43.3(0.5)		1.265[45]		
4417	Dioctadecylamine	Distearylamine	C₃₆H₇₅N	112-99-2	521.988		72.9	268[2]			vs chl
4418	Dioctylamine	N-Octyl-1-octanamine	C₁₆H₃₅N	1120-48-5	241.456	nd	20(9)	307.7(0.6)	0.7963[26]	1.4415[26]	vs eth, EtOH
4419	Dioctyl ether		C₁₆H₃₄O	629-82-3	242.440	liq	-7.7(0.4)	289(3)	0.8063[20]	1.4327[20]	sl H₂O; s EtOH, eth, ctc

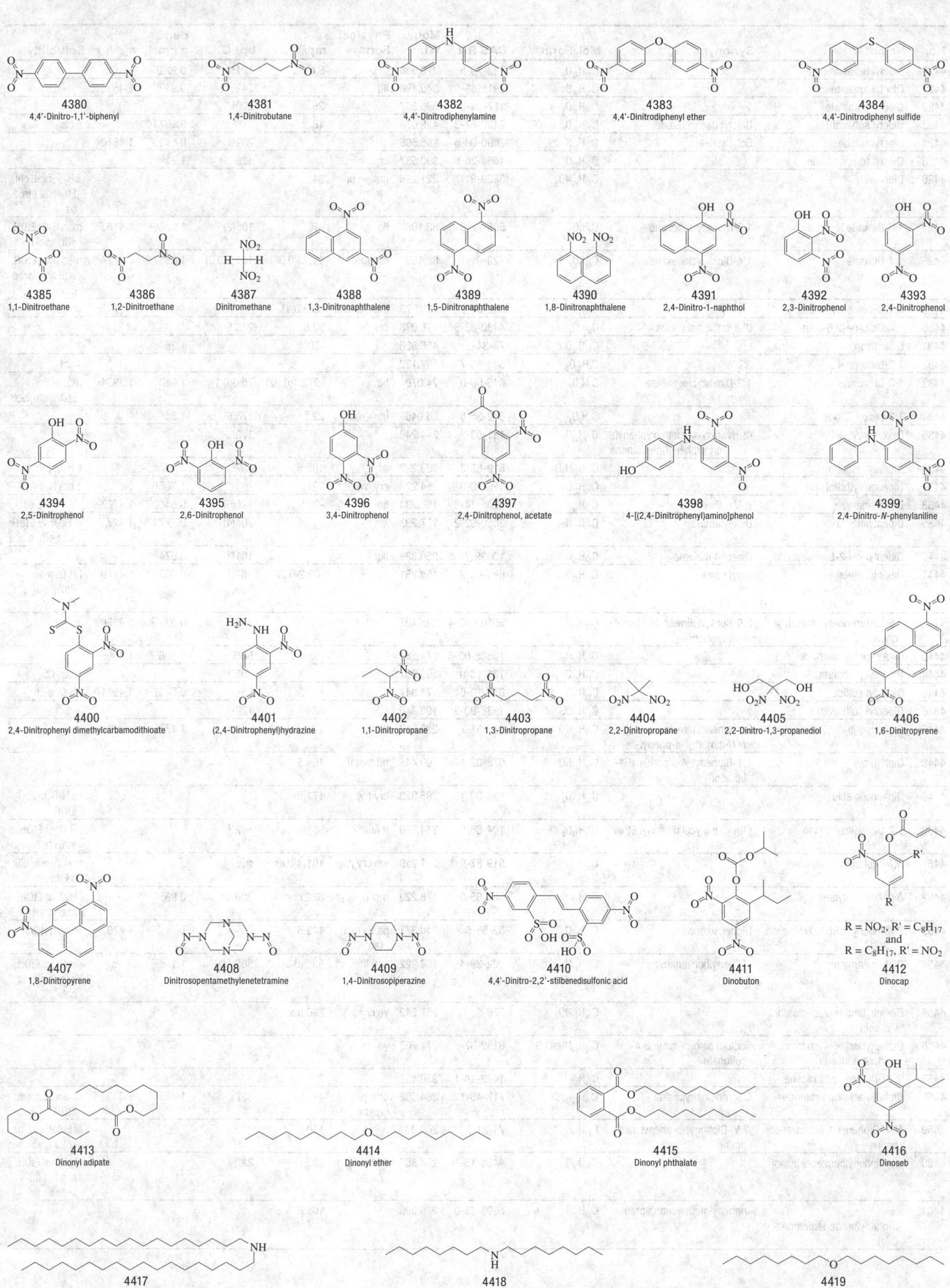

4380
4,4'-Dinitro-1,1'-biphenyl

4381
1,4-Dinitrobutane

4382
4,4'-Dinitrodiphenylamine

4383
4,4'-Dinitrodiphenyl ether

4384
4,4'-Dinitrodiphenyl sulfide

4385
1,1-Dinitroethane

4386
1,2-Dinitroethane

4387
Dinitromethane

4388
1,3-Dinitronaphthalene

4389
1,5-Dinitronaphthalene

4390
1,8-Dinitronaphthalene

4391
2,4-Dinitro-1-naphthol

4392
2,3-Dinitrophenol

4393
2,4-Dinitrophenol

4394
2,5-Dinitrophenol

4395
2,6-Dinitrophenol

4396
3,4-Dinitrophenol

4397
2,4-Dinitrophenol, acetate

4398
4-[(2,4-Dinitrophenyl)amino]phenol

4399
2,4-Dinitro-*N*-phenylaniline

4400
2,4-Dinitrophenyl dimethylcarbamodithioate

4401
(2,4-Dinitrophenyl)hydrazine

4402
1,1-Dinitropropane

4403
1,3-Dinitropropane

4404
2,2-Dinitropropane

4405
2,2-Dinitro-1,3-propanediol

4406
1,6-Dinitropyrene

4407
1,8-Dinitropyrene

4408
Dinitrosopentamethylenetetramine

4409
1,4-Dinitrosopiperazine

4410
4,4'-Dinitro-2,2'-stilbenedisulfonic acid

4411
Dinobuton

R = NO$_2$, R' = C$_8$H$_{17}$
and
R = C$_8$H$_{17}$, R' = NO$_2$

4412
Dinocap

4413
Dinonyl adipate

4414
Dinonyl ether

4415
Dinonyl phthalate

4416
Dinoseb

4417
Dioctadecylamine

4418
Dioctylamine

4419
Dioctyl ether

No.	Name	Synonym	Mol. Form.	CAS RN	Mol. Wt.	Physical Form	mp/°C	bp/°C	den g cm⁻³	n_D	Solubility
4420	Dioctyl hexanedioate		$C_{22}H_{42}O_4$	123-79-5	370.566		9.6	191[2]	0.922[25]		
4421	Dioctyl maleate		$C_{20}H_{36}O_4$	2915-53-9	340.498	liq		242[0.002]	0.94[20]	1.4539[20]	
4422	Dioctyl phthalate		$C_{24}H_{38}O_4$	117-84-0	390.557		25	220[4]			
4423	Dioctyl sebacate	Dioctyl decanedioate	$C_{26}H_{50}O_4$	2432-87-3	426.673		18	218[0.5]	0.9074[25]		s ctc
4424	Dioctyl sulfide	Octyl sulfide	$C_{16}H_{34}S$	2690-08-6	258.506			202[29]	0.842[25]	1.4610[20]	
4425	Dioctyl terephthalate		$C_{24}H_{38}O_4$	4654-26-6	390.557			425	1.21[62]		
4426	Dioscorine		$C_{13}H_{19}NO_2$	3329-91-7	221.296	grn-ye pr (eth)	34				s H_2O, ace, chl, EtOH; sl eth, bz
4427	1,3-Dioxane	1,3-Dioxacyclohexane	$C_4H_8O_2$	505-22-6	88.106	liq	-45	105(2)	1.0286[25]	1.4165[20]	msc H_2O, EtOH, eth, ace, bz
4428	1,4-Dioxane	1,4-Dioxacyclohexane	$C_4H_8O_2$	123-91-1	88.106	col liq	11.75(0.06)	101.2(0.3)	1.0337[20]	1.4224[20]	msc H_2O, EtOH, eth, ace, bz; s ctc
4429	1,4-Dioxane-2,5-dione		$C_4H_4O_4$	502-97-6	116.073	lf (al, al-chl)	83.0(0.1)				vs ace
4430	1,4-Dioxane-2,6-dione	Diglycollic anhydride	$C_4H_4O_4$	4480-83-5	116.073	cry (bz)	92.5	240.5			
4431	Dioxathion		$C_{12}H_{26}O_6P_2S_4$	78-34-2	456.538		-20		1.257[26]		
4432	1,3-Dioxepane		$C_5H_{10}O_2$	505-65-7	102.132						s chl
4433	1,3-Dioxolane	1,3-Dioxacyclopentane	$C_3H_6O_2$	646-06-0	74.079	liq	-97.21(0.02)	75.3(0.1)	1.060[20]	1.3974[20]	msc H_2O; s EtOH, eth, ace
4434	1,3-Dioxol-2-one		$C_3H_2O_3$	872-36-6	86.046	liq	22	178(6)	1.35[25]		
4435	Dioxybenzone	(2-Hydroxy-4-methoxyphenyl)-(2-hydroxyphenyl)methanone	$C_{14}H_{12}O_4$	131-53-3	244.243			172[1]			
4436	Dioxypyramidon		$C_{13}H_{17}N_3O_3$	519-65-3	263.292	pr	105.5	197[2]			s H_2O, EtOH
4437	Dipentaerythritol		$C_{10}H_{22}O_7$	126-58-9	254.278	cry (w)	221		1.366[15]		s hot H_2O
4438	Dipentene	p-Menthadiene	$C_{10}H_{16}$	7705-14-8	136.234	liq	-95.5	176(5)	0.8402[21]	1.4727[20]	
4439	Dipentylamine	Diamylamine	$C_{10}H_{23}N$	2050-92-2	157.297			204(4)	0.7771[20]	1.4272[20]	sl H_2O; vs EtOH; msc eth; s ace
4440	Dipentyl cis-2-butenedioate	Dipentyl maleate	$C_{14}H_{24}O_4$	10099-71-5	256.339	liq		161[10]	0.974[20]		
4441	Dipentyl ether	Amyl ether	$C_{10}H_{22}O$	693-65-2	158.281	liq	-69.2(0.5)	187(2)	0.7833[20]	1.4119[20]	i H_2O; msc EtOH, eth; s chl
4442	2,6-Di-tert-pentyl-4-methyl-phenol	2,6-Bis(1,1-dimethylpropyl)-4-methylphenol	$C_{17}H_{28}O$	56103-67-4	248.403			283	0.931[25]	1.4950[20]	
4443	Di-tert-pentyl peroxide		$C_{10}H_{22}O_2$	10508-09-5	174.281		-55	58[14]	0.808[20]	1.4095[20]	
4444	Dipentyl phthalate		$C_{18}H_{26}O_4$	131-18-0	306.397			205[11]			s ctc, CS_2
4445	Dipentyl sulfide		$C_{10}H_{22}S$	872-10-6	174.347		-51.3	86[3.7]	0.8407[20]	1.4561[20]	i H_2O; s eth
4446	Dipentyl sulfoxide		$C_{10}H_{22}OS$	1986-90-9	190.346		58	120[1]			
4447	Diphenamid	Benzeneacetamide, N,N-dimethyl-α-phenyl-	$C_{16}H_{17}NO$	957-51-7	239.312		133(2)		1.17[23.3]		
4448	Diphenidol	1,1-Diphenyl-4-piperidinyl-1-butanol	$C_{21}H_{27}NO$	972-02-1	309.445	nd (peth)	104.5				
4449	Diphenolic acid		$C_{17}H_{18}O_4$	126-00-1	286.323	cry (w)	171.5				vs H_2O, ace, EtOH
4450	1,2-Diphenoxyethane	Ethylene glycol diphenyl ether	$C_{14}H_{14}O_2$	104-66-5	214.260	lf (al)	98	182[12]			i H_2O; sl EtOH; s eth, chl
4451	N,N-Diphenylacetamide		$C_{14}H_{13}NO$	519-87-9	211.259	wh cry pow	101.4(0.6)	sub			sl H_2O, eth, chl; s EtOH
4452	Diphenylacetylene		$C_{14}H_{10}$	501-65-5	178.229	mcl pr or pl (al)	62(2)	300	0.9657[100]		i H_2O; sl EtOH, chl; vs eth
4453	2-(Diphenylacetyl)-1H-indene-1,3(2H)-dione	Diphenadione	$C_{23}H_{16}O_3$	82-66-6	340.371	pa ye mcl (al)	146.5			1.670	vs ace, HOAc
4454	Diphenylamine	N-Phenylbenzenamine	$C_{12}H_{11}N$	122-39-4	169.222	mcl lf(dil al)	53.2(0.3)	305.1(1)	1.158[22]		i H_2O; vs EtOH, ace; s eth; sl chl
4455	Diphenylamine-2,2'-dicarbox-ylic acid		$C_{14}H_{11}NO_4$	579-92-0	257.242	ye cry (al)	296 dec				
4456	Diphenylamine-4-sulfonic acid, sodium salt	Sodium diphenylamine-4-sulfonate	$C_{12}H_{10}NNaO_3S$	6152-67-6	271.267	ye cry					
4457	9,10-Diphenylanthracene		$C_{26}H_{18}$	1499-10-1	330.421		237.0(0.2)				
4458	Diphenylarsinous chloride	Chlorodiphenylarsine	$C_{12}H_{10}AsCl$	712-48-1	264.582	orth pl (peth)	44	337	1.4820[16]	1.6332[56]	vs ace, bz, eth, EtOH
4459	N,N'-Diphenyl-1,4-benzenedi-amine	N,N'-Diphenyl-p-phenylenedi-amine	$C_{18}H_{16}N_2$	74-31-7	260.333		150	222[0.5]			sl EtOH, eth, bz, chl; i acid
4460	α,α-Diphenylbenzeneethanol		$C_{20}H_{18}O$	4428-13-1	274.356	nd(bz-lig) pr (peth)	89.5	222[11]			i H_2O; vs EtOH; sl eth, chl, peth
4461	α,α-Diphenylbenzenemethane-thiol	Triphenylmethyl mercaptan	$C_{19}H_{16}S$	3695-77-0	276.395		105.8				

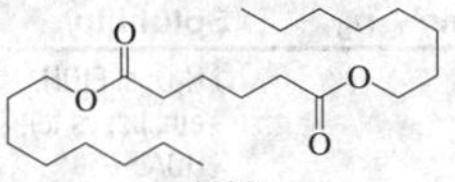

4420
Dioctyl hexanedioate

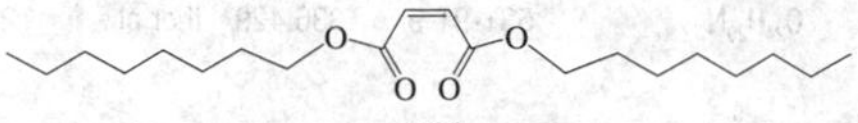

4421
Dioctyl maleate

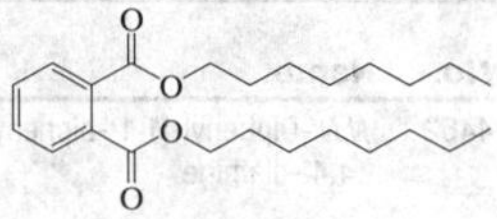

4422
Dioctyl phthalate

4423
Dioctyl sebacate

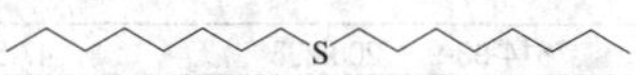

4424
Dioctyl sulfide

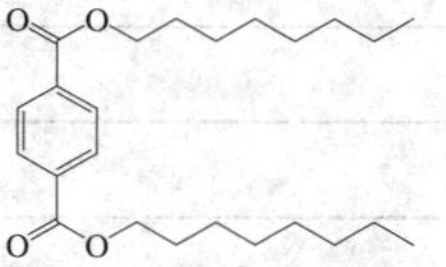

4425
Dioctyl terephthalate

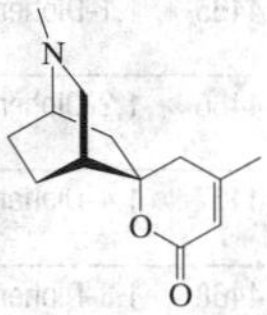

4426
Dioscorine

4427
1,3-Dioxane

4428
1,4-Dioxane

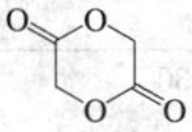

4429
1,4-Dioxane-2,5-dione

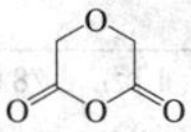

4430
1,4-Dioxane-2,6-dione

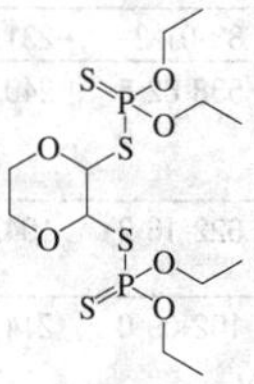

4431
Dioxathion

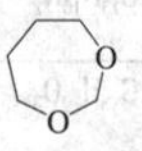

4432
1,3-Dioxepane

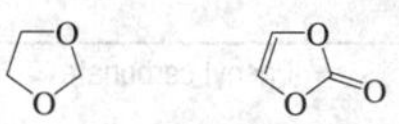

4433
1,3-Dioxolane

4434
1,3-Dioxol-2-one

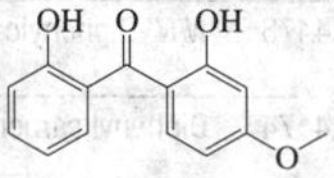

4435
Dioxybenzone

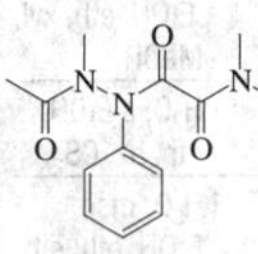

4436
Dioxypyramidon

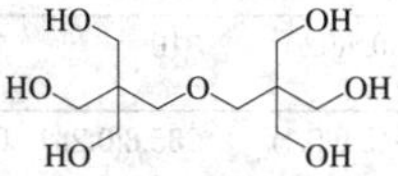

4437
Dipentaerythritol

4438
Dipentene

4439
Dipentylamine

4440
Dipentyl *cis*-2-butenedioate

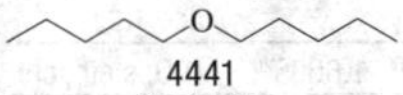

4441
Dipentyl ether

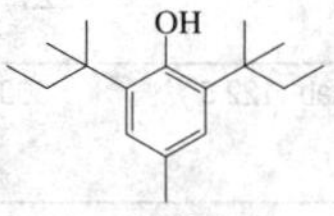

4442
2,6-Di-*tert*-pentyl-4-methylphenol

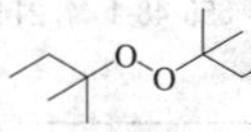

4443
Di-*tert*-pentyl peroxide

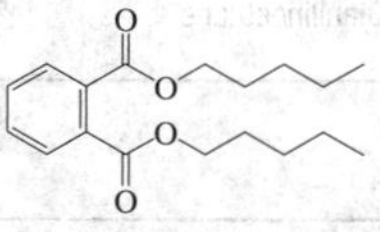

4444
Dipentyl phthalate

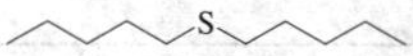

4445
Dipentyl sulfide

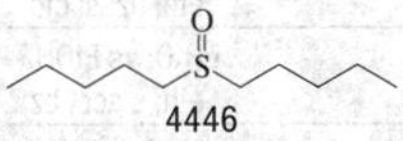

4446
Dipentyl sulfoxide

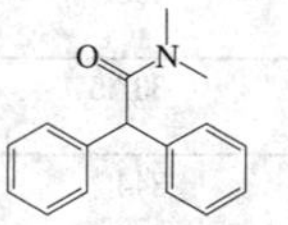

4447
Diphenamid

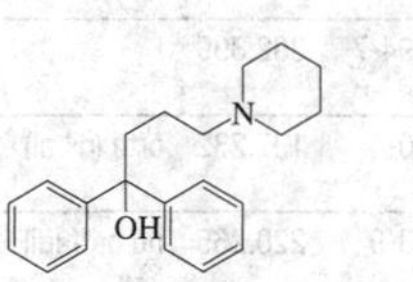

4448
Diphenidol

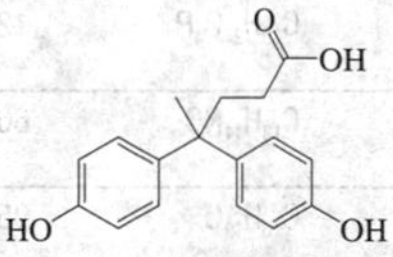

4449
Diphenolic acid

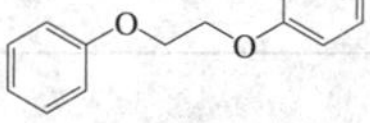

4450
1,2-Diphenoxyethane

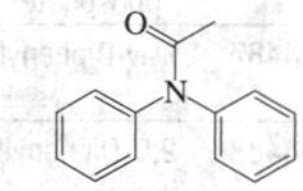

4451
N,N-Diphenylacetamide

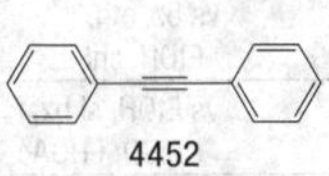

4452
Diphenylacetylene

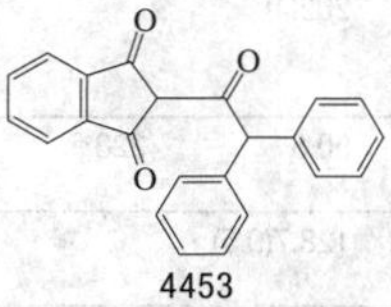

4453
2-(Diphenylacetyl)-1*H*-indene-1,3(2*H*)-dione

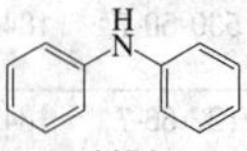

4454
Diphenylamine

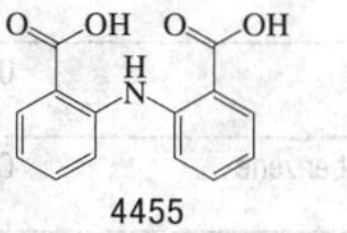

4455
Diphenylamine-2,2'-dicarboxylic acid

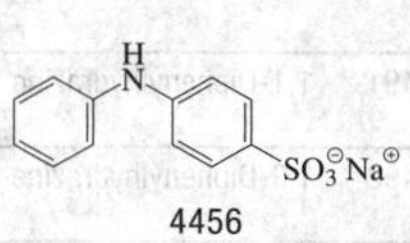

4456
Diphenylamine-4-sulfonic acid, sodium salt

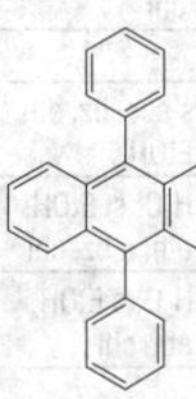

4457
9,10-Diphenylanthracene

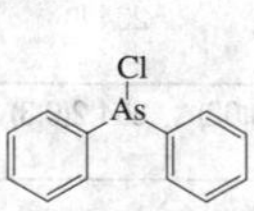

4458
Diphenylarsinous chloride

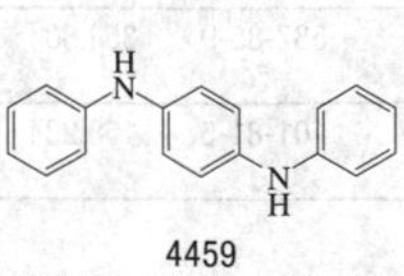

4459
N,N'-Diphenyl-1,4-benzenediamine

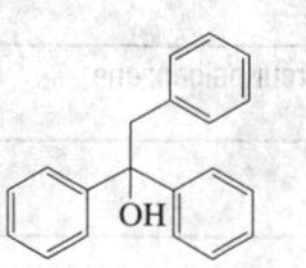

4460
α,α-Diphenylbenzeneethanol

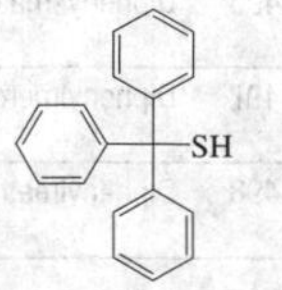

4461
α,α-Diphenylbenzenemethanethiol

No.	Name	Synonym	Mol. Form.	CAS RN	Mol. Wt.	Physical Form	mp/°C	bp/°C	den g cm^{-3}	n_D	Solubility
4462	N,N'-Diphenyl-[1,1'-biphenyl]-4,4'-diamine	N,N'-Diphenylbenzidine	C$_{24}$H$_{20}$N$_2$	531-91-9	336.429	lf or pl	247				i H$_2$O; sl EtOH, eth, bz; vs tol, HOAc
4463	trans,trans-1,4-Diphenyl-1,3-butadiene		C$_{16}$H$_{14}$	538-81-8	206.282	lf (al, HOAc)	149(4)	352			vs bz, eth, EtOH, peth
4464	1,4-Diphenyl-1,3-butadiyne	Diphenyldiacetylene	C$_{16}$H$_{10}$	886-66-8	202.250		85.9(0.5)				
4465	1,1-Diphenylbutane		C$_{16}$H$_{18}$	719-79-9	210.314		27	287	0.9928^{20}	1.5664^{20}	i H$_2$O; s EtOH, eth, bz, chl
4466	1,2-Diphenylbutane		C$_{16}$H$_{18}$	5223-59-6	210.314			291(9)	0.9673^{20}	1.5554^{20}	i H$_2$O; s EtOH, eth, bz, chl
4467	1,4-Diphenylbutane		C$_{16}$H$_{18}$	1083-56-3	210.314		52.2(0.1)	316(3)	0.9880^{20}		i H$_2$O; s EtOH, eth, chl
4468	1,3-Diphenyl-1-butene		C$_{16}$H$_{16}$	7614-93-9	208.298		47.5	311	0.9996^{20}	1.590^{15}	
4469	trans-1,4-Diphenyl-2-butene-1,4-dione		C$_{16}$H$_{12}$O$_2$	959-28-4	236.265	ye nd (al, bz)	111				sl EtOH; s bz, HOAc; vs chl; i lig
4470	1,3-Diphenyl-2-buten-1-one	Dypnone	C$_{16}$H$_{14}$O	495-45-4	222.281			342.5	1.1080^{15}	1.6343^{20}	vs eth, EtOH
4471	Diphenylcarbamic chloride		C$_{13}$H$_{10}$ClNO	83-01-2	231.677	lf (al)	84.5				
4472	Diphenylcarbazone		C$_{13}$H$_{12}$N$_4$O	538-62-5	240.260	oran oran nd (bz) pr (al)	157 dec				i H$_2$O; vs EtOH, bz, chl
4473	N,N'-Diphenylcarbodiimide		C$_{13}$H$_{10}$N$_2$	622-16-2	194.231		169	331			sl H$_2$O, EtOH, eth; s bz
4474	Diphenyl carbonate	Phenyl carbonate	C$_{13}$H$_{10}$O$_3$	102-09-0	214.216	nd (al, bz)	78.9(0.3)	306	1.1215^{87}		i H$_2$O; s EtOH, eth, ctc, HOAc
4475	2,2'-Diphenylcarbonic dihydrazide	sym-Diphenylcarbazide	C$_{13}$H$_{14}$N$_4$O	140-22-7	242.276	cry (al + 1) cry (HOAc)	170	dec			sl H$_2$O, eth; s EtOH, ace, bz
4476	Diphenyl chlorophosphonate		C$_{12}$H$_{10}$ClO$_3$P	2524-64-3	268.632			314^{272}	1.296^{25}	1.5500^{20}	s tfa
4477	Diphenyl diselenide	Phenyl diselenide	C$_{12}$H$_{10}$Se$_2$	1666-13-3	312.13	ye nd	63.5	202^{11}	1.557^{80}	1.743^{20}	s EtOH, eth, xyl, MeOH
4478	Diphenyl disulfide	Phenyl disulfide	C$_{12}$H$_{10}$S$_2$	882-33-7	218.337	nd(al) or eth	60.4(0.5)	310	1.353^{20}		i H$_2$O; s EtOH, eth, bz, CS$_2$
4479	1,1-Diphenylethane		C$_{14}$H$_{14}$	612-00-0	182.261	liq	-18.0(0.1)	285.8(0.2)	0.9997^{20}	1.5756^{20}	i H$_2$O; msc EtOH, eth; s bz
4480	1,2-Diphenylethane	Dibenzyl	C$_{14}$H$_{14}$	103-29-7	182.261	mcl pr (MeOH)	51.18(0.06)	280(3)	0.9780^{25}	1.5476^{60}	i H$_2$O; s EtOH, eth, CS$_2$
4481	N,N'-Diphenylethanediamide		C$_{14}$H$_{12}$N$_2$O$_2$	620-81-5	240.257	lf (bz)	254	>360			vs bz
4482	N,N'-Diphenyl-1,2-ethanedi-amine	1,2-Dianilinoethane	C$_{14}$H$_{16}$N$_2$	150-61-8	212.290	cry (dil al)	74	229^{12}			i H$_2$O; s EtOH, eth; sl tfa
4483	1,2-Diphenyl-1,2-ethanediol, (R*,R*)-(±)-		C$_{14}$H$_{14}$O$_2$	655-48-1	214.260	nd (w,al),tab (eth)	122.5	>300			i H$_2$O, lig; vs EtOH, eth; s ace
4484	1,1-Diphenylethene		C$_{14}$H$_{12}$	530-48-3	180.245		8.2(0.4)	277(17)	1.0232^{20}	1.6085^{20}	i H$_2$O; s eth, chl
4485	Diphenyl ether	Oxybisbenzene	C$_{12}$H$_{10}$O	101-84-8	170.206		26.865 (0.003)	258.0(0.1)	1.0661^{30}	1.5787^{25}	i H$_2$O; s EtOH, eth, bz, HOAc; sl chl
4486	Diphenyl 2-ethylhexyl phosphate		C$_{20}$H$_{27}$O$_4$P	1241-94-7	362.399			232^5	1.090^{25}	1.510^{25}	
4487	N,N-Diphenylformamide		C$_{13}$H$_{11}$NO	607-00-1	197.232	orth (dil al)	73.5	337.5			i H$_2$O; s EtOH, eth, bz; sl ctc
4488	2,5-Diphenylfuran		C$_{16}$H$_{12}$O	955-83-9	220.265	nd or lf (dil al)	91	344			i H$_2$O; vs EtOH, eth; s ace, bz
4489	N,N'-Diphenylguanidine	1,3-Diphenylguanidine	C$_{13}$H$_{13}$N$_3$	102-06-7	211.262	mcl nd (al, to)	147(1)	170 dec	1.13^{20}		sl H$_2$O; s EtOH, ctc chl, tol; vs eth
4490	1,6-Diphenyl-1,3,5-hexatriene		C$_{18}$H$_{16}$	1720-32-7	232.320	lf (ace)	202(4)				i H$_2$O, EtOH, eth, HOAc; s ace; sl bz, chl
4491	1,1-Diphenylhydrazine		C$_{12}$H$_{12}$N$_2$	530-50-7	184.236	tab (lig)	50.5	220^{40}	1.190^{16}		vs bz, eth, EtOH, chl
4492	1,2-Diphenylhydrazine	Hydrazobenzene	C$_{12}$H$_{12}$N$_2$	122-66-7	184.236	tab (al-eth)	128.7(0.5)		1.158^{16}		vs EtOH; sl bz, DMSO; i HOAc
4493	5,5-Diphenyl-4-imidazolidi-none	Doxenitoin	C$_{15}$H$_{14}$N$_2$O	3254-93-1	238.284	pl (MeOH)	183				
4494	Diphenyl isophthalate		C$_{20}$H$_{14}$O$_4$	744-45-6	318.323		138				s chl
4495	Diphenylketene	Diphenylethenone	C$_{14}$H$_{10}$O	525-06-4	194.228	red-ye liq		267.5	1.1107^{13}	1.615^{14}	
4496	Diphenyl maleate		C$_{16}$H$_{12}$O$_4$	7242-17-3	268.264	pl (lig)	73	226^{15}			vs ace, bz, eth, EtOH
4497	Diphenylmercury	Mercuriodibenzene	C$_{12}$H$_{10}$Hg	587-85-9	354.80			204^{10}	2.318^{25}		i H$_2$O; sl EtOH, eth; s bz, chl
4498	Diphenylmethane		C$_{13}$H$_{12}$	101-81-5	168.234	pr nd	25.22(0.02)	264.2(0.3)	1.001^{26}	1.5753^{20}	i H$_2$O; s EtOH, eth, chl

4462
N,N'-Diphenyl-[1,1'-biphenyl]-4,4'-diamine

4463
trans,trans-1,4-Diphenyl-1,3-butadiene

4464
1,4-Diphenyl-1,3-butadiyne

4465
1,1-Diphenylbutane

4466
1,2-Diphenylbutane

4467
1,4-Diphenylbutane

4468
1,3-Diphenyl-1-butene

4469
trans-1,4-Diphenyl-2-butene-1,4-dione

4470
1,3-Diphenyl-2-buten-1-one

4471
Diphenylcarbamic chloride

4472
Diphenylcarbazone

4473
N,N'-Diphenylcarbodiimide

4474
Diphenyl carbonate

4475
2,2'-Diphenylcarbonic dihydrazide

4476
Diphenyl chlorophosphonate

4477
Diphenyl diselenide

4478
Diphenyl disulfide

4479
1,1-Diphenylethane

4480
1,2-Diphenylethane

4481
N,N'-Diphenylethanediamide

4482
N,N'-Diphenyl-1,2-ethanediamine

4483
1,2-Diphenyl-1,2-ethanediol, (R*,R*)-(±)-

4484
1,1-Diphenylethene

4485
Diphenyl ether

4486
Diphenyl 2-ethylhexyl phosphate

4487
N,N-Diphenylformamide

4488
2,5-Diphenylfuran

4489
N,N'-Diphenylguanidine

4490
1,6-Diphenyl-1,3,5-hexatriene

4491
1,1-Diphenylhydrazine

4492
1,2-Diphenylhydrazine

4493
5,5-Diphenyl-4-imidazolidinone

4494
Diphenyl isophthalate

4495
Diphenylketene

4496
Diphenyl maleate

4497
Diphenylmercury

4498
Diphenylmethane

No.	Name	Synonym	Mol. Form.	CAS RN	Mol. Wt.	Physical Form	mp/°C	bp/°C	den g cm⁻³	n_D	Solubility
4499	4,4'-Diphenylmethane diisocyanate	Methylene diphenyl diisocyanate	$C_{15}H_{10}N_2O_2$	101-68-8	250.252		40.41(0.01)	196[5]	1.197[70]	1.5906[50]	s ace, bz, PhNO₂
4500	Diphenylmethanethione		$C_{13}H_{10}S$	1450-31-3	198.283		53.5	174[14]			sl EtOH, eth, peth; vs bz, chl
4501	N,N'-Diphenylmethanimidam-ide		$C_{13}H_{12}N_2$	622-15-1	196.247	nd (al)	142	>250			sl H₂O, peth; s EtOH, ace, bz; vs eth
4502	Diphenylmethanol	Benzohydrol	$C_{13}H_{12}O$	91-01-0	184.233	nd (lig)	65(1)	302(6)			sl H₂O; vs EtOH, eth, ctc, chl; s HOAc
4503	2-(Diphenylmethoxy)-N,N-dimethylethanamine	Diphenhydramine	$C_{17}H_{21}NO$	58-73-1	255.355	oil		165[3]			
4504	Diphenyl methylphosphonate		$C_{13}H_{13}O_3P$	7526-26-3	248.214		35	205[13]	1.2051[20]		i H₂O
4505	2-(Diphenylmethyl)-1-piperi-dineethanol	Diphemethoxidine	$C_{20}H_{25}NO$	13862-07-2	295.419		106.5	180[0.1]			
4506	2,5-Diphenyloxazole		$C_{15}H_{11}NO$	92-71-7	221.254	nd (lig)	71(3)	360	1.0940[100]	1.6231[100]	i H₂O; vs EtOH, eth; sl chl
4507	1,5-Diphenyl-1,4-pentadien-3-one	Dibenzalacetone	$C_{17}H_{14}O$	538-58-9	234.292	pl or lf (ace, AcOEt)	100.0(0.2)	dec			i H₂O; sl EtOH, eth; s ace, chl
4508	4,7-Diphenyl-1,10-phenanth-roline		$C_{24}H_{16}N_2$	1662-01-7	332.397		220 dec				
4509	Diphenylphosphinous chloride	Chlorodiphenylphosphine	$C_{12}H_{10}ClP$	1079-66-9	220.634	hyg ye liq		320	1.229	1.6360[20]	
4510	Diphenyl phosphonate		$C_{12}H_{11}O_3P$	4712-55-4	234.187		12	218[26]	1.223[25]	1.5575[20]	
4511	Diphenyl phthalate	Phenyl phthalate	$C_{20}H_{14}O_4$	84-62-8	318.323	pr (al, lig)	73	253[14]			i H₂O; sl EtOH, eth, ctc
4512	α,α-Diphenyl-2-piperidinemethanol	Pipradrol	$C_{18}H_{21}NO$	467-60-7	267.366	cry (hx)	97.5				
4513	1,3-Diphenylpropane		$C_{15}H_{16}$	1081-75-0	196.288	liq	6	300	1.007[20]	1.5760[20]	
4514	2,2-Diphenylpropane		$C_{15}H_{16}$	778-22-3	196.288		29.1(0.3)	281(6)	0.9980[20]		
4515	1,3-Diphenyl-1,3-propanedi-one	Dibenzoylmethane	$C_{15}H_{12}O_2$	120-46-7	224.255		77(3)				s EtOH, eth, chl, dil NaOH
4516	1,3-Diphenyl-1-propanone	Phenethyl phenyl ketone	$C_{15}H_{14}O$	1083-30-3	210.271	lf (EtOH)	72.5	360			
4517	1,1-Diphenyl-2-propanone	1,1-Diphenylacetone	$C_{15}H_{14}O$	781-35-1	210.271		46	307		1.5361[16]	s EtOH, eth, bz, chl, lig
4518	1,3-Diphenyl-2-propanone	Dibenzyl ketone	$C_{15}H_{14}O$	102-04-5	210.271	cry (al, peth)	34(1)	329(14)	1.195[0]		i H₂O; s EtOH, eth, peth
4519	3,3-Diphenyl-2-propenal	β-Phenylcinnamaldehyde	$C_{15}H_{12}O$	1210-39-5	208.255	pa ye pr (lig)	44.8	205[14]			
4520	1,1-Diphenyl-1-propene		$C_{15}H_{14}$	778-66-5	194.272		49(5)	289(6)	1.0250[20]	1.5880[20]	i H₂O; s EtOH, bz
4521	trans-1,3-Diphenyl-2-propen-1-one	Chalcone	$C_{15}H_{12}O$	614-47-1	208.255	pa ye lf, pr, nd (peth)	56(4)	346 dec	1.0712[62]		i H₂O; sl EtOH; s eth, bz, chl, CS₂
4522	1-(3,3-Diphenylpropyl)-piperidine	Fenpiprane	$C_{20}H_{25}N$	3540-95-2	279.420		41.5	215[8]			
4523	3,5-Diphenyl-1H-pyrazole		$C_{15}H_{12}N_2$	1145-01-3	220.269	cry (al)	200				
4524	1,4-Diphenyl-3,5-pyrazolidin-edione	Phenopyrazone	$C_{15}H_{12}N_2O_2$	3426-01-5	252.268	cry (EtOAc, Diox)	233.5				
4525	Diphenyl selenide		$C_{12}H_{10}Se$	1132-39-4	233.17	ye nd (bz)	1.3	301.5	1.351[20]	1.5500[20]	i H₂O; msc EtOH, eth; s bz, xyl
4526	Diphenylsilane		$C_{12}H_{12}Si$	775-12-2	184.309			134[16]	0.9969[20]	1.5800[20]	s ctc, CS₂
4527	Diphenylsilanediol		$C_{12}H_{12}O_2Si$	947-42-2	216.308						sl DMSO
4528	Diphenyl succinate		$C_{16}H_{14}O_4$	621-14-7	270.280	lf (al)	121	330			i H₂O; s EtOH, eth, ace, bz
4529	Diphenyl sulfide	Phenyl sulfide	$C_{12}H_{10}S$	139-66-2	186.272	liq	-15.35(0.05)	294.2(0.5)	1.1136[20]	1.6334[20]	i H₂O; s EtOH, ctc; msc eth, bz, CS₂
4530	Diphenyl sulfone		$C_{12}H_{10}O_2S$	127-63-9	218.271	mcl pr(bz), pl(al)	127(3)	379	1.252[20]		i H₂O; s EtOH, eth, bz
4531	Diphenyl sulfoxide		$C_{12}H_{10}OS$	945-51-7	202.271	pr(lig)	70.2(0.2)	340[16]			vs EtOH, eth, bz, HOAc; sl chl; i peth
4532	N,N'-Diphenylthiourea	sym-Diphenylthiourea	$C_{13}H_{12}N_2S$	102-08-9	228.312		153(1)		1.32[25]		sl H₂O; vs EtOH, eth, chl, oils
4533	1,3-Diphenyl-1-triazene	Diazoaminobenzene	$C_{12}H_{11}N_3$	136-35-6	197.235	ye lf or pr (al)	98				i H₂O; vs EtOH, eth, bz, py
4534	N,N-Diphenylurea		$C_{13}H_{12}N_2O$	603-54-3	212.246	tab (al)	189	dec	1.276[25]		sl H₂O; s EtOH, eth, chl
4535	N,N'-Diphenylurea	Carbanilide	$C_{13}H_{12}N_2O$	102-07-8	212.246	orth pr (al)	236(3)	260 dec	1.239[25]		sl H₂O, EtOH; s eth, py, HOAc; i bz

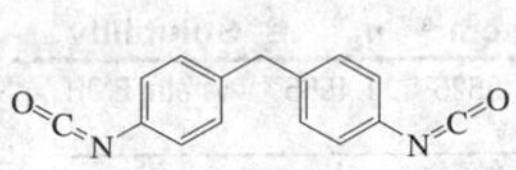

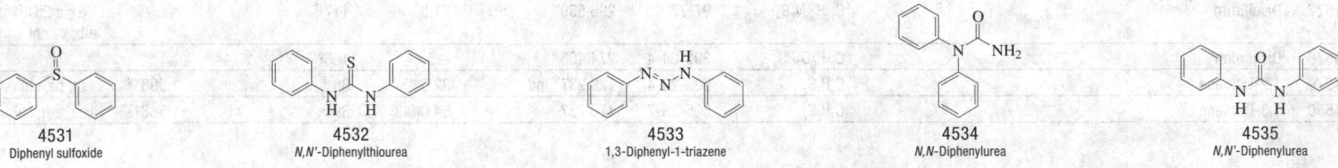

4499
4,4'-Diphenylmethane diisocyanate

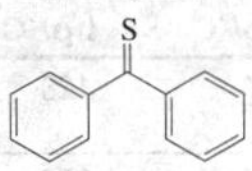

4500
Diphenylmethanethione

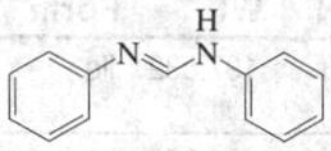

4501
N,N'-Diphenylmethanimidamide

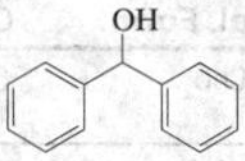

4502
Diphenylmethanol

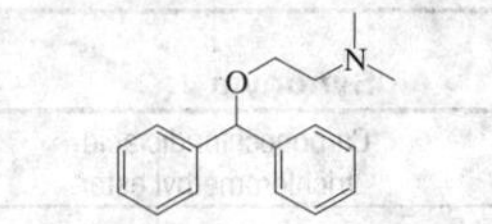

4503
2-(Diphenylmethoxy)-N,N-dimethylethanamine

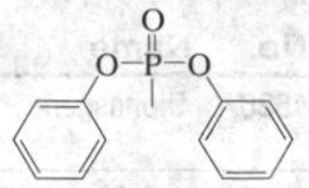

4504
Diphenyl methylphosphonate

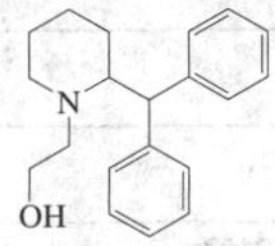

4505
2-(Diphenylmethyl)-1-piperidineethanol

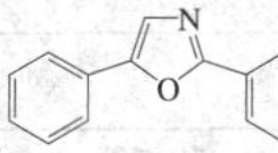

4506
2,5-Diphenyloxazole

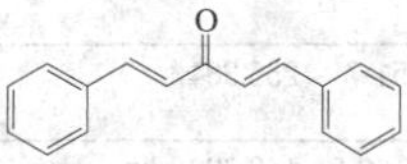

4507
1,5-Diphenyl-1,4-pentadien-3-one

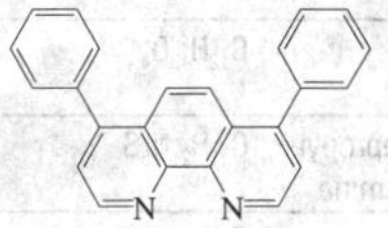

4508
4,7-Diphenyl-1,10-phenanthroline

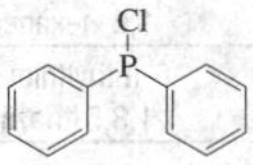

4509
Diphenylphosphinous chloride

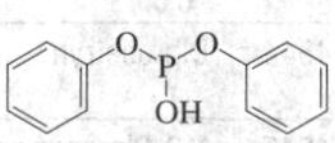

4510
Diphenyl phosphonate

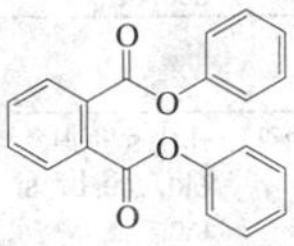

4511
Diphenyl phthalate

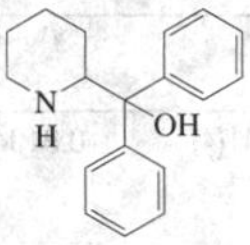

4512
α,α-Diphenyl-2-piperidinemethanol

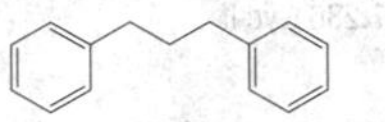

4513
1,3-Diphenylpropane

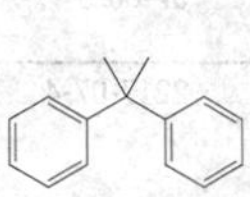

4514
2,2-Diphenylpropane

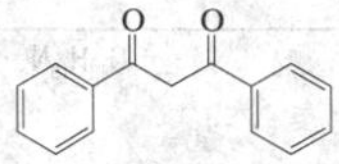

4515
1,3-Diphenyl-1,3-propanedione

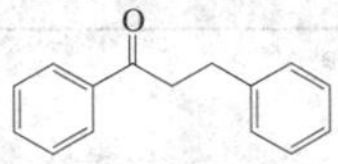

4516
1,3-Diphenyl-1-propanone

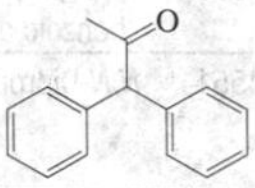

4517
1,1-Diphenyl-2-propanone

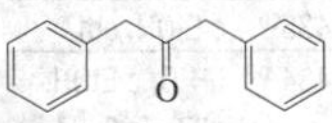

4518
1,3-Diphenyl-2-propanone

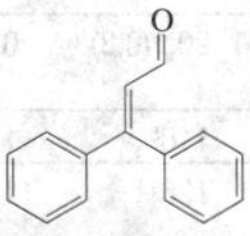

4519
3,3-Diphenyl-2-propenal

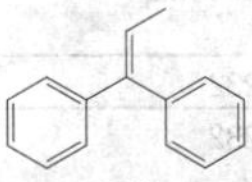

4520
1,1-Diphenyl-1-propene

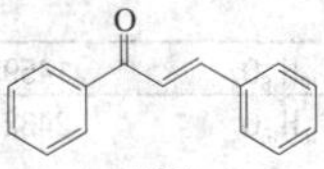

4521
trans-1,3-Diphenyl-2-propen-1-one

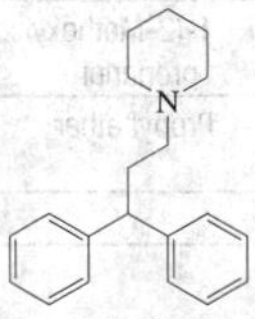

4522
1-(3,3-Diphenylpropyl)piperidine

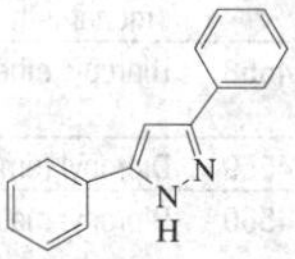

4523
3,5-Diphenyl-1H-pyrazole

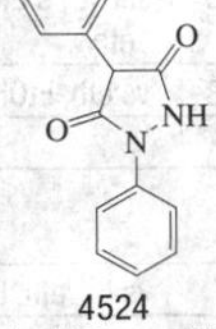

4524
1,4-Diphenyl-3,5-pyrazolidinedione

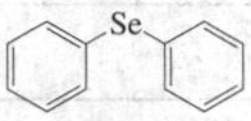

4525
Diphenyl selenide

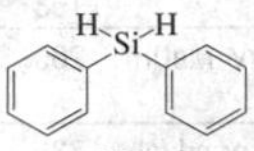

4526
Diphenylsilane

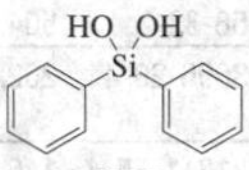

4527
Diphenylsilanediol

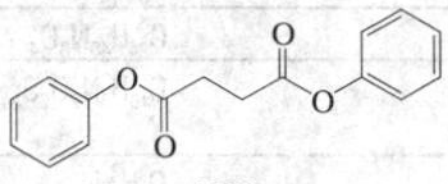

4528
Diphenyl succinate

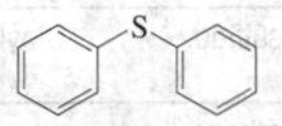

4529
Diphenyl sulfide

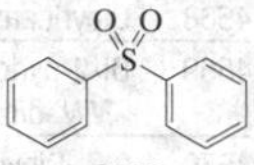

4530
Diphenyl sulfone

4531
Diphenyl sulfoxide

4532
N,N'-Diphenylthiourea

4533
1,3-Diphenyl-1-triazene

4534
N,N-Diphenylurea

4535
N,N'-Diphenylurea

No.	Name	Synonym	Mol. Form.	CAS RN	Mol. Wt.	Physical Form	mp/°C	bp/°C	den g cm⁻³	n_D	Solubility
4536	Diphosgene	Carbonochloridic acid, trichloromethyl ester	$C_2Cl_4O_2$	503-38-8	197.832	liq	-57	128	1.6525[14]	1.4566[22]	vs eth, EtOH
4537	1,2-Dipiperidinoethane		$C_{12}H_{24}N_2$	1932-04-3	196.332	liq	-0.5	265	0.9160[25]	1.4853[25]	
4538	1,1'-Dipiperidinomethane	1,1'-Methylenedipiperidine	$C_{11}H_{22}N_2$	880-09-1	182.306			230	0.9269[20]	1.4820[20]	
4539	1,3-Di-4-piperidylpropane	4,4'-Trimethylenedipiperidine	$C_{13}H_{26}N_2$	16898-52-5	210.358		67.1	329			vs H_2O
4540	Diploicin		$C_{16}H_{10}Cl_4O_5$	527-93-5	424.059		232				
4541	Di-2-propenoyldiethyleneglycol		$C_{10}H_{14}O_5$	4074-88-8	214.215			200	1.1110[25]	1.4595[25]	
4542	Di-2-propenoyl-2,2-dimethyl-1,3-propanediol	2-Propenoic acid, 2,2-dimethyl-1,3-propanediyl ester	$C_{11}H_{16}O_4$	2223-82-7	212.243					1.4542[25]	
4543	Di-2-propenoyl-1,6-hexanediol	2-Propenoic acid, 1,6-hexanediyl ester	$C_{12}H_{18}O_4$	13048-33-4	226.269				1.010[25]		
4544	Dipropetryn	6-(Ethylthio)-N,N-diisopropyl-1,3,5-triazine-2,4-diamine	$C_{11}H_{21}N_5S$	4147-51-7	255.384		105.0(0.5)				
4545	1,2-Dipropoxyethane		$C_8H_{18}O_2$	18854-56-3	146.228	liq		161(13)	0.8312[25]	1.4013[25]	
4546	Dipropoxymethane	Formaldehyde, dipropyl acetal	$C_7H_{16}O_2$	505-84-0	132.201	liq	-97.3	140.5	0.8345[20]	1.3939[19]	vs ace, bz, eth, EtOH
4547	N,N-Dipropylacetamide		$C_8H_{17}NO$	1116-24-1	143.227			227(13)	0.8992[17]	1.4419[17]	vs EtOH
4548	Dipropyl adipate	Dipropyl hexanedioate	$C_{12}H_{22}O_4$	106-19-4	230.301		-15.7	151[11]	0.9790[20]	1.4314[20]	vs eth, EtOH, chl
4549	Dipropylamine	N-Propyl-1-propanamine	$C_6H_{15}N$	142-84-7	101.190	liq	-63	107.5(0.9)	0.7400[20]	1.4050[20]	s H_2O, EtOH; msc eth; vs ace, bz
4550	4-[(Dipropylamino)sulfonyl]benzoic acid	Probenecid	$C_{13}H_{19}NO_4S$	57-66-9	285.360		195				
4551	N,N-Dipropylaniline		$C_{12}H_{19}N$	2217-07-4	177.286	ye lf		241(4)	0.9104[20]	1.5271[20]	i H_2O; s EtOH, eth, ace, bz; sl ctc
4552	Dipropylcarbamothioic acid, S-ethyl ester	EPTC	$C_9H_{19}NOS$	759-94-4	189.318			127[20]	0.9546[30]		
4553	Dipropyl carbonate		$C_7H_{14}O_3$	623-96-1	146.184			166(4)	0.9435[20]	1.4008[20]	sl H_2O; msc EtOH, eth
4554	Dipropyl disulfide		$C_6H_{14}S_2$	629-19-6	150.305	liq	-85.4(0.1)	196(1)	0.9599[20]	1.4981[20]	
4555	Dipropylene glycol		$C_6H_{14}O_3$	25265-71-8	134.173			231(2)	1.0206[20]		msc H_2O; s EtOH
4556	Dipropylene glycol dibenzoate		$C_{20}H_{22}O_5$	27138-31-4	342.386			197[1]			
4557	Dipropylene glycol monomethyl ether	1-(2-Methoxyisopropoxy)-2-propanol	$C_7H_{16}O_3$	34590-94-8	148.200	liq	-80	203(14)	0.95	1.4190[20]	
4558	Dipropyl ether	Propyl ether	$C_6H_{14}O$	111-43-3	102.174	liq	-114.8(0.4)	90.1(0.3)	0.7466[20]	1.3809[20]	sl H_2O; vs eth, EtOH
4559	Dipropyl fumarate		$C_{10}H_{16}O_4$	14595-35-8	200.232			110[5]	1.0129[20]	1.4435[20]	s EtOH, eth
4560	Dipropyl maleate		$C_{10}H_{16}O_4$	2432-63-5	200.232			126[12]	1.0245[20]	1.4434[20]	i H_2O; s EtOH, eth, ace, bz
4561	Dipropyl oxalate		$C_8H_{14}O_4$	615-98-5	174.195	liq	-44.3	212(3)	1.0188[20]	1.4158[20]	sl H_2O; msc EtOH; s eth
4562	5,5-Dipropyl-2,4-oxazolidinedione		$C_9H_{15}NO_3$	512-12-9	185.220		42.5	149[3]			
4563	Dipropyl succinate		$C_{10}H_{18}O_4$	925-15-5	202.248	liq	-5.9	248(7)	1.0020[20]	1.4250[20]	vs ace, bz, eth
4564	Dipropyl sulfate		$C_6H_{14}O_4S$	598-05-0	182.238			120[20]	1.1064[20]	1.4135[20]	vs peth
4565	Dipropyl sulfide		$C_6H_{14}S$	111-47-7	118.240	liq	-102.68(0.04)	142.8(0.8)	0.814[17]	1.4487[20]	i H_2O; s EtOH, eth
4566	Dipropyl sulfone		$C_6H_{14}O_2S$	598-03-8	150.239	cry	28(2)		1.0278[50]	1.4456[30]	sl H_2O; s EtOH, eth
4567	Dipropyl sulfoxide		$C_6H_{14}OS$	4253-91-2	134.239	nd	25(1)	80[2]	0.9654[20]	1.4663[20]	vs eth, EtOH
4568	Dipyridamole		$C_{24}H_{40}N_8O_4$	58-32-2	504.627		163				
4569	Di-2-pyridinyl disulfide, N,N'-dioxide	Dipyrithione	$C_{10}H_8N_2O_2S_2$	3696-28-4	252.313	cry (MeOH)	205				
4570	2,2'-Dipyrrolylmethane		$C_9H_{10}N_2$	21211-65-4	146.188	lf or nd (al)	73	164[12]			vs bz, eth, EtOH
4571	Diquat		$C_{12}H_{12}N_2$	2764-72-9	184.236	Cation					
4572	Diquat dibromide		$C_{12}H_{12}Br_2N_2$	85-00-7	344.044		337		1.24[20]		
4573	Disodium calcium EDTA	Edetate calcium disodium	$C_{10}H_{12}CaN_2Na_2O_8$	62-33-9	374.268	pow					s H_2O
4574	Disodium hydrogen citrate	Sodium acid citrate	$C_6H_6Na_2O_7$	144-33-2	236.088	wh pow (w)	149 dec				vs H_2O
4575	Disperse Blue No. 1	1,4,5,8-Tetraamino-9,10-anthracenedione	$C_{14}H_{12}N_4O_2$	2475-45-8	268.271	red-br nd	331				
4576	Distearyl thiodipropionate	Dioctadecyl thiobispropanoate	$C_{42}H_{82}O_4S$	693-36-7	683.163	cry	61				
4577	Disulfiram		$C_{10}H_{20}N_2S_4$	97-77-8	296.539		71.5	117[17]			i H_2O; s EtOH; sl eth; vs chl
4578	Disulfoton		$C_8H_{19}O_2PS_3$	298-04-4	274.405		-25	108[0.01]	1.144[20]		
4579	1,2-Dithiane		$C_4H_8S_2$	505-20-4	120.237	nd	32.5	80[14]		1.5981[25]	s eth, bz, chl
4580	1,3-Dithiane		$C_4H_8S_2$	505-23-7	120.237		54.0(0.3)	89[14]		1.5981[25]	vs bz, eth, chl

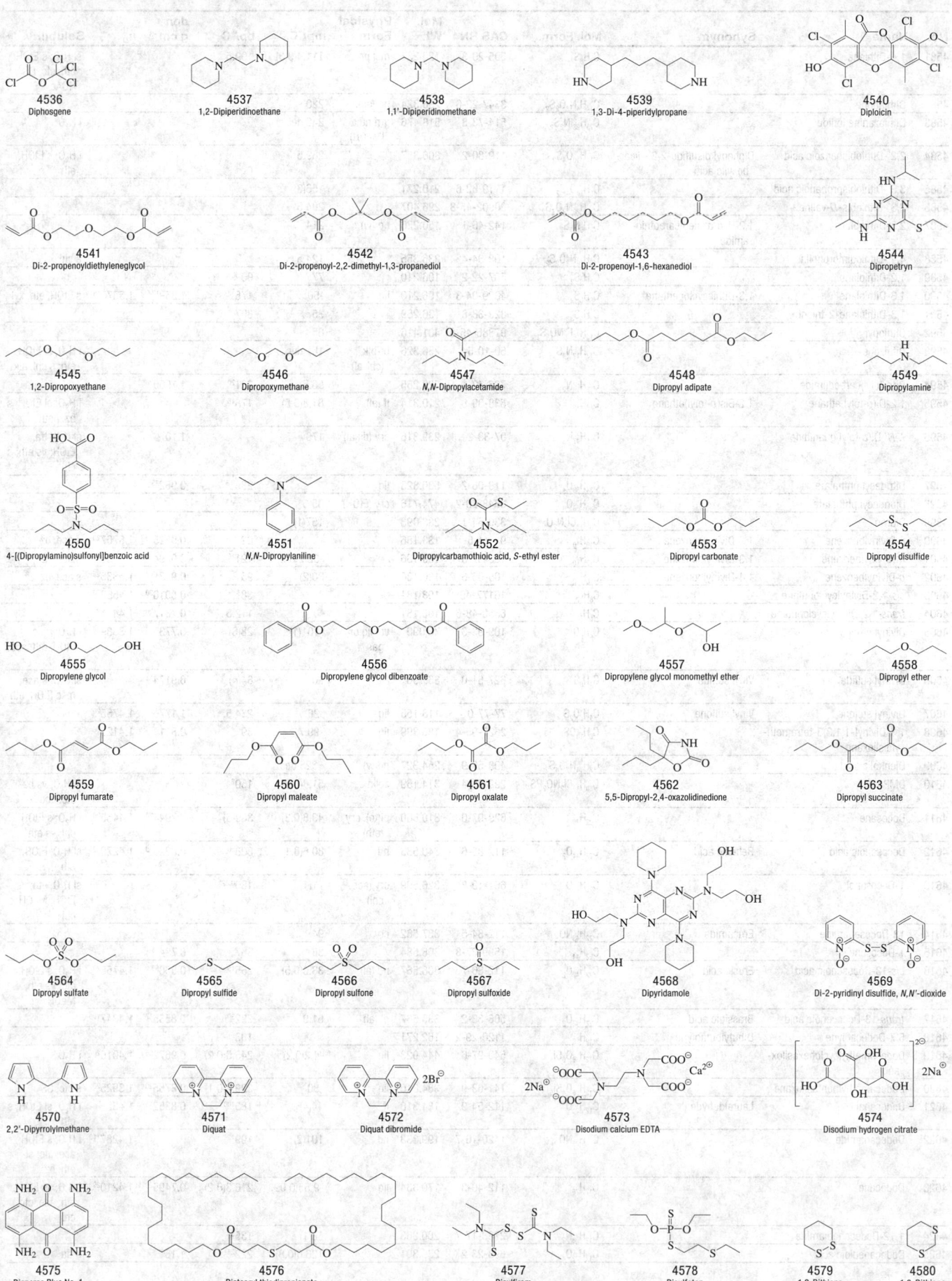

4536 Diphosgene

4537 1,2-Dipiperidinoethane

4538 1,1'-Dipiperidinomethane

4539 1,3-Di-4-piperidylpropane

4540 Diploicin

4541 Di-2-propenoyldiethyleneglycol

4542 Di-2-propenoyl-2,2-dimethyl-1,3-propanediol

4543 Di-2-propenoyl-1,6-hexanediol

4544 Dipropetryn

4545 1,2-Dipropoxyethane

4546 Dipropoxymethane

4547 N,N-Dipropylacetamide

4548 Dipropyl adipate

4549 Dipropylamine

4550 4-[(Dipropylamino)sulfonyl]benzoic acid

4551 N,N-Dipropylaniline

4552 Dipropylcarbamothioic acid, S-ethyl ester

4553 Dipropyl carbonate

4554 Dipropyl disulfide

4555 Dipropylene glycol

4556 Dipropylene glycol dibenzoate

4557 Dipropylene glycol monomethyl ether

4558 Dipropyl ether

4559 Dipropyl fumarate

4560 Dipropyl maleate

4561 Dipropyl oxalate

4562 5,5-Dipropyl-2,4-oxazolidinedione

4563 Dipropyl succinate

4564 Dipropyl sulfate

4565 Dipropyl sulfide

4566 Dipropyl sulfone

4567 Dipropyl sulfoxide

4568 Dipyridamole

4569 Di-2-pyridinyl disulfide, N,N'-dioxide

4570 2,2'-Dipyrrolylmethane

4571 Diquat

4572 Diquat dibromide

4573 Disodium calcium EDTA

4574 Disodium hydrogen citrate

4575 Disperse Blue No. 1

4576 Distearyl thiodipropionate

4577 Disulfiram

4578 Disulfoton

4579 1,2-Dithiane

4580 1,3-Dithiane

No.	Name	Synonym	Mol. Form.	CAS RN	Mol. Wt.	Physical Form	mp/°C	bp/°C	den g cm^{-3}	n_D	Solubility
4581	1,4-Dithiane		$C_4H_8S_2$	505-29-3	120.237	mcl pr	111.4(0.3)	199.5			sl H$_2$O; s EtOH, eth, ctc, CS$_2$, HOAc
4582	Dithianone		$C_{14}H_4N_2O_2S_2$	3347-22-6	296.324	nd (ace)	220				
4583	Dithiazanine iodide		$C_{23}H_{23}IN_2S_2$	514-73-8	518.476	grn nd (MeOH)	248 dec				i H$_2$O
4584	2,2'-Dithiobisbenzoic acid	Diphenyl disulfide-2,2'-dicarboxylic acid	$C_{14}H_{10}O_4S_2$	119-80-2	306.357		289.5				i H$_2$O; s EtOH, eth
4585	3,3'-Dithiobispropanoic acid		$C_6H_{10}O_4S_2$	1119-62-6	210.271		156(3)				
4586	3,3'-Dithiobis-D-valine		$C_{10}H_{20}N_2O_4S_2$	20902-45-8	296.407		204.5				
4587	2,5-Dithiobiurea	1,2-Hydrazinedicarbothioamide	$C_2H_6N_4S_2$	142-46-1	150.226	nd (w)	214				
4588	4,4'-Dithiodimorpholine		$C_8H_{16}N_2O_2S_2$	103-34-4	236.355		124.5				s chl
4589	1,2-Dithiolane		$C_3H_6S_2$	557-22-2	106.210		77	90^{27}			
4590	1,3-Dithiolane	1,3-Dithiacyclopentane	$C_3H_6S_2$	4829-04-3	106.210	liq	-50	175	1.259^{17}	1.5975^{15}	s EtOH, eth, xyl
4591	1,3-Dithiolane-2-thione		$C_3H_4S_3$	822-38-8	136.259		35	307			
4592	Dithiopyr		$C_{15}H_{16}F_5NO_2S_2$	97886-45-8	401.416		65				
4593	Dithizone		$C_{13}H_{12}N_4S$	60-10-6	256.326	bl-blk (chl-al)	167 dec				i H$_2$O; sl EtOH, eth; s chl, alk
4594	Di(p-tolyl)carbodiimide		$C_{15}H_{14}N_2$	726-42-1	222.285		58.5	221^{20}		1.1500^{20}	
4595	1,2-Di(p-tolyl)ethane	1,2-Bis(p-tolyl)ethane	$C_{16}H_{18}$	538-39-6	210.314	lf (al)	81.8(0.5)	178^{18}			i H$_2$O; sl EtOH; s bz, peth
4596	N,N'-Di(o-tolyl)guanidine		$C_{15}H_{17}N_3$	97-39-2	239.316	cry (dil al)	179			1.10^{20}	sl H$_2$O, tfa, EtOH; vs eth; s chl
4597	Ditridecyl phthalate		$C_{34}H_{58}O_4$	119-06-2	530.823	liq		285$^{3.5}$	0.952^{25}		
4598	Diundecyl phthalate		$C_{30}H_{50}O_4$	3648-20-2	474.716	cry (EtOH)	35.5				
4599	Diuron		$C_9H_{10}Cl_2N_2O$	330-54-1	233.093		157(1)				
4600	o-Divinylbenzene	1,2-Divinylbenzene	$C_{10}H_{10}$	91-14-5	130.186			82^{14}	0.9325^{22}	1.5767^{20}	s ace, bz
4601	m-Divinylbenzene	1,3-Divinylbenzene	$C_{10}H_{10}$	108-57-6	130.186		-52.2(0.2)	121^{76}	0.9294^{20}	1.5760^{20}	s ace, bz
4602	p-Divinylbenzene	1,4-Divinylbenzene	$C_{10}H_{10}$	105-06-6	130.186		30(2)	95^{18}	0.913^{40}	1.5835^{25}	s ace, bz
4603	cis-1,2-Divinylcyclobutane		C_8H_{12}	16177-46-1	108.181			38^{38}	0.8010^{20}	1.4563^{20}	
4604	trans-1,2-Divinylcyclobutane		C_8H_{12}	6553-48-6	108.181			112.5	0.7817^{20}	1.4451^{20}	
4605	Divinyl ether		C_4H_6O	109-93-3	70.090	vol liq or gas	-101(1)	28(3)	0.773^{20}	1.3989^{20}	i H$_2$O; msc EtOH, eth, ace, chl
4606	Divinyl sulfide	Vinyl sulfide	C_4H_6S	627-51-0	86.156		20	84(4)	0.9174^{15}		sl H$_2$O; s ace; msc EtOH, eth
4607	Divinyl sulfone	Vinyl sulfone	$C_4H_6O_2S$	77-77-0	118.155	liq	-26	234.5	1.177^{25}	1.4765^{20}	
4608	1,3-Divinyl-1,1,3,3-tetramethyldisiloxane		$C_8H_{18}OSi_2$	2627-95-4	186.399	liq	-99.7	39	0.811^{20}	1.4123^{20}	
4609	Djenkolic acid		$C_7H_{14}N_2O_4S_2$	498-59-9	254.327	nd(w)	≈325 dec				
4610	DMPA		$C_{10}H_{14}Cl_2NO_2PS$	299-85-4	314.169	solid	51.8(0.3)	150^2			sl H$_2$O; vs bz, ctc, ace
4611	Docosane		$C_{22}H_{46}$	629-97-0	310.600	pl(to), cry (eth)	43.8(0.3)	369(5)	0.7944^{20}	1.4455^{20}	i H$_2$O; s EtOH, chl; vs eth
4612	Docosanoic acid	Behenic acid	$C_{22}H_{44}O_2$	112-85-6	340.583	nd	80.6(0.1)	306^{60}	0.8223^{90}	1.4270^{100}	sl H$_2$O, EtOH, eth
4613	1-Docosanol		$C_{22}H_{46}O$	661-19-8	326.599	cry (ace, chl)	71(1)	180$^{0.22}$			sl H$_2$O, eth; vs EtOH, MeOH; s chl
4614	13-Docosenamide	Erucamide	$C_{22}H_{43}NO$	112-84-5	337.582	cry	94				
4615	1-Docosene		$C_{22}H_{44}$	1599-67-3	308.584		38	367	0.794^{25}		
4616	cis-13-Docosenoic acid	Erucic acid	$C_{22}H_{42}O_2$	112-86-7	338.567	nd (al)	33.0(0.5)	265^{15}	0.860^{55}	1.4758^{20}	i H$_2$O; s EtOH, ctc; vs eth, MeOH
4617	trans-13-Docosenoic acid	Brassidic acid	$C_{22}H_{42}O_2$	506-33-2	338.567	pl (al)	61.9	282^{30}	0.8585^{57}	1.4347^{100}	
4618	5,7-Dodecadiyne	Dibutylbutadiyne	$C_{12}H_{18}$	1120-29-2	162.271			103^8			
4619	Dodecamethylcyclohexasiloxane		$C_{12}H_{36}O_6Si_6$	540-97-6	444.923	liq	-4.2(0.2)	245.0(0.2)	0.9672^{25}	1.4015^{20}	i H$_2$O
4620	Dodecamethylpentasiloxane		$C_{12}H_{36}O_4Si_5$	141-63-9	384.840	liq	-80	229.9(0.1)	0.8755^{20}	1.3925^{20}	s ctc, CS$_2$
4621	Dodecanal	Lauraldehyde	$C_{12}H_{24}O$	112-54-9	184.318	lf	44(4)	185^{100}	0.8352^{15}	1.435^{22}	i H$_2$O; sl EtOH; s eth
4622	Dodecanamide		$C_{12}H_{25}NO$	1120-16-7	199.333	nd	101(2)	199^{12}		1.4287^{110}	i H$_2$O; s EtOH, ace, ctc; sl eth, bz
4623	Dodecane		$C_{12}H_{26}$	112-40-3	170.334	liq	-9.55(0.02)	216.3(0.2)	0.7495^{20}	1.4210^{20}	i H$_2$O; vs EtOH, eth, ace, ctc, chl
4624	1,12-Dodecanediamine		$C_{12}H_{28}N_2$	2783-17-7	200.363		68(1)	135^3			
4625	Dodecanedioic acid		$C_{12}H_{22}O_4$	693-23-2	230.301		126.6(0.8)	222^{25}	1.15^{25}		s tfa

4581 1,4-Dithiane

4582 Dithianone

4583 Dithiazanine iodide

4584 2,2'-Dithiobisbenzoic acid

4585 3,3'-Dithiobispropanoic acid

4586 3,3'-Dithiobis-*D*-valine

4587 2,5-Dithiobiurea

4588 4,4'-Dithiodimorpholine

4589 1,2-Dithiolane

4590 1,3-Dithiolane

4591 1,3-Dithiolane-2-thione

4592 Dithiopyr

4593 Dithizone

4594 Di(*p*-tolyl)carbodiimide

4595 1,2-Di(*p*-tolyl)ethane

4596 *N,N'*-Di(*o*-tolyl)guanidine

4597 Ditridecyl phthalate

4598 Diundecyl phthalate

4599 Diuron

4600 *o*-Divinylbenzene

4601 *m*-Divinylbenzene

4602 *p*-Divinylbenzene

4603 *cis*-1,2-Divinylcyclobutane

4604 *trans*-1,2-Divinylcyclobutane

4605 Divinyl ether

4606 Divinyl sulfide

4607 Divinyl sulfone

4608 1,3-Divinyl-1,1,3,3-tetramethyldisiloxane

4609 Djenkolic acid

4610 DMPA

4611 Docosane

4612 Docosanoic acid

4613 1-Docosanol

4614 13-Docosenamide

4615 1-Docosene

4616 *cis*-13-Docosenoic acid

4617 *trans*-13-Docosenoic acid

4618 5,7-Dodecadiyne

4619 Dodecamethylcyclohexasiloxane

4620 Dodecamethylpentasiloxane

4621 Dodecanal

4622 Dodecanamide

4623 Dodecane

4624 1,12-Dodecanediamine

4625 Dodecanedioic acid

No.	Name	Synonym	Mol. Form.	CAS RN	Mol. Wt.	Physical Form	mp/°C	bp/°C	den g cm⁻³	n_D	Solubility
4626	1,12-Dodecanediol		$C_{12}H_{26}O_2$	5675-51-4	202.333	cry (bz, dil al)	81.3	189[12]			s tfa
4627	Dodecanenitrile	Lauronitrile	$C_{12}H_{23}N$	2437-25-4	181.318		4.0(0.2)	276.1(0.5)	0.8240[20]	1.4361[20]	i H₂O; msc EtOH, eth, ace, bz, chl
4628	1-Dodecanethiol		$C_{12}H_{26}S$	112-55-0	202.399	liq	-6.7	277(3)	0.844[20]	1.4589[20]	i H₂O; s EtOH, eth, chl
4629	Dodecanoic acid	Lauric acid	$C_{12}H_{24}O_2$	143-07-7	200.318	nd (al)	43.82(0.02)	225[100]	0.8679[50]	1.4183[82]	i H₂O; vs EtOH, eth; s ace; msc bz
4630	Dodecanoic anhydride		$C_{24}H_{46}O_3$	645-66-9	382.620	lf (al, eth)	41.8		0.8533[70]	1.4292[70]	vs EtOH
4631	1-Dodecanol	Lauryl alcohol	$C_{12}H_{26}O$	112-53-8	186.333	lf (dil al)	24.2(0.3)	264.1(0.3)	0.8309[24]		i H₂O; s EtOH, eth; sl bz
4632	2-Dodecanol		$C_{12}H_{26}O$	10203-28-8	186.333		19	249(11)	0.8286[20]	1.4400[20]	
4633	2-Dodecanone	Decyl methyl ketone	$C_{12}H_{24}O$	6175-49-1	184.318		21	247(6)	0.8198[20]	1.4330[20]	i H₂O; s EtOH, eth, ace; sl ctc
4634	Dodecanoyl chloride		$C_{12}H_{23}ClO$	112-16-3	218.763		-17	145[18]	0.9169[25]	1.4458[20]	vs eth
4635	1-Dodecene		$C_{12}H_{24}$	112-41-4	168.319	liq	-35.19(0.05)	213.4(0.9)	0.7584[20]	1.4300[20]	i H₂O; s EtOH, eth, ace, ctc, peth
4636	*trans*-2-Dodecenedioic acid	Traumatic acid	$C_{12}H_{20}O_4$	6402-36-4	228.285	cry (al,ace)	165.5				vs eth, EtOH, chl
4637	2-Dodecenylsuccinic anhydride		$C_{16}H_{26}O_3$	19780-11-1	266.375	hyg cry	42	181[5]			
4638	Dodecyl acetate		$C_{14}H_{28}O_2$	112-66-3	228.371		1.2(0.2)	265	0.8652[22]	1.4439[20]	
4639	Dodecyl acrylate	Lauryl 2-propenoate	$C_{15}H_{28}O_2$	2156-97-0	240.382		4	120[0.8]	0.8727[20]		
4640	Dodecylamine	1-Dodecanamine	$C_{12}H_{27}N$	124-22-1	185.349		27.8(0.7)	255(2)	0.8015[20]	1.4421[20]	sl H₂O; msc EtOH, eth, bz, chl
4641	Dodecylamine, acetate	1-Dodecanamine, acetate	$C_{14}H_{31}NO_2$	2016-56-0	245.402		69.5				vs H₂O, EtOH
4642	Dodecylamine hydrochloride	Lauryl amine hydrochloride	$C_{12}H_{28}ClN$	929-73-7	221.810		182(2)				vs H₂O, EtOH
4643	4-Dodecylaniline		$C_{18}H_{31}N$	104-42-7	261.446		41.5	211[10]			
4644	Dodecylbenzene	Laurylbenzene	$C_{18}H_{30}$	123-01-3	246.431		-2(5)	329(4)	0.8551[20]	1.4824[20]	i H₂O
4645	4-Dodecylbenzenesulfonic acid		$C_{18}H_{30}O_3S$	121-65-3	326.494			>205			
4646	Dodecylcyclohexane		$C_{18}H_{36}$	1795-17-1	252.479		12.6(0.4)	331	0.8223[20]	1.4559[20]	
4647	Dodecyl mercaptoacetate		$C_{14}H_{28}O_2S$	3746-39-2	260.436		1.5	171[3]			
4648	Dodecyl methacrylate		$C_{16}H_{30}O_2$	142-90-5	254.408			142[4]	0.866[20]		
4649	Dodecyloxirane	1,2-Epoxytetradecane	$C_{14}H_{28}O$	3234-28-4	212.371	oil		95[0.4]	0.845	1.4408[20]	
4650	2-(Dodecyloxy)ethanol		$C_{14}H_{30}O_2$	4536-30-5	230.387			143[0.8]			
4651	4-Dodecyloxy-2-hydroxyben-zophenone		$C_{25}H_{34}O_3$	2985-59-3	382.536		43.5				
4652	4-Dodecylphenol		$C_{18}H_{30}O$	104-43-8	262.430	nd (bz)	66	175[2]			
4653	1-Dodecylpiperidine		$C_{17}H_{35}N$	5917-47-5	253.467	pa ye		161[5]	0.8378[20]	1.4588[20]	
4654	Dodecyl sulfate	Lauryl sulfate	$C_{12}H_{26}O_4S$	151-41-7	266.397	cry					s H₂O
4655	Dodecyltetraethylene glycol monoether	3,6,9,12-Tetraoxatetracosan-1-ol	$C_{20}H_{42}O_5$	5274-68-0	362.544			247[10]			
4656	Dodecyl 3,4,5-trihydroxyben-zoate		$C_{19}H_{30}O_5$	1166-52-5	338.438		96.5				s ace
4657	Dodecyltrimethylammonium chloride		$C_{15}H_{34}ClN$	112-00-5	263.891		246 dec				vs H₂O, ace, EtOH, chl
4658	1-Dodecyne	Decylacetylene	$C_{12}H_{22}$	765-03-7	166.303	liq	-19	215	0.7788[20]	1.4340[20]	
4659	6-Dodecyne		$C_{12}H_{22}$	6975-99-1	166.303			213(9)	0.785[20]	1.4442[20]	vs ace, eth, EtOH
4660	Dodine	Dodecylguanidine, monoacetate	$C_{15}H_{33}N_3O_2$	2439-10-3	287.442		137.5(0.5)				
4661	Dopamine	4(2-Aminoethyl)-1,2-benzene-diol	$C_8H_{11}NO_2$	51-61-6	153.179	pr					
4662	Dothiepin		$C_{19}H_{21}NS$	113-53-1	295.442		56	172[0.05]			
4663	Dotriacontane	Bicetyl	$C_{32}H_{66}$	544-85-4	450.866	pl (bz,chl, HOAc,eth)	69.7(0.3)	470(14)	0.8124[20]	1.4550[20]	i H₂O; sl EtOH, chl; s eth, ctc; vs bz
4664	Doxepin		$C_{19}H_{21}NO$	1668-19-5	279.376	oily liq		155[0.03]			
4665	Doxorubicin	Adriamycin	$C_{27}H_{29}NO_{11}$	23214-92-8	543.519	cry	230				
4666	Doxorubicin hydrochloride	Adriamycin hydrochloride	$C_{27}H_{30}ClNO_{11}$	25316-40-9	579.980	oran-red nd	204 dec				s H₂O, MeOH; i ace, bz, chl, eth, peth
4667	Doxylamine		$C_{17}H_{22}N_2O$	469-21-6	270.369	liq		139[0.5]			
4668	Drimenin		$C_{15}H_{22}O_2$	2326-89-8	234.335	cry	133	110[0.1]			i H₂O
4669	Dromostanolone propanoate	2-Methyl-17-(1-oxopropoxy)-androstan-3-one, (2α,5α,17β)	$C_{23}H_{36}O_3$	521-12-0	360.530		130(1)				

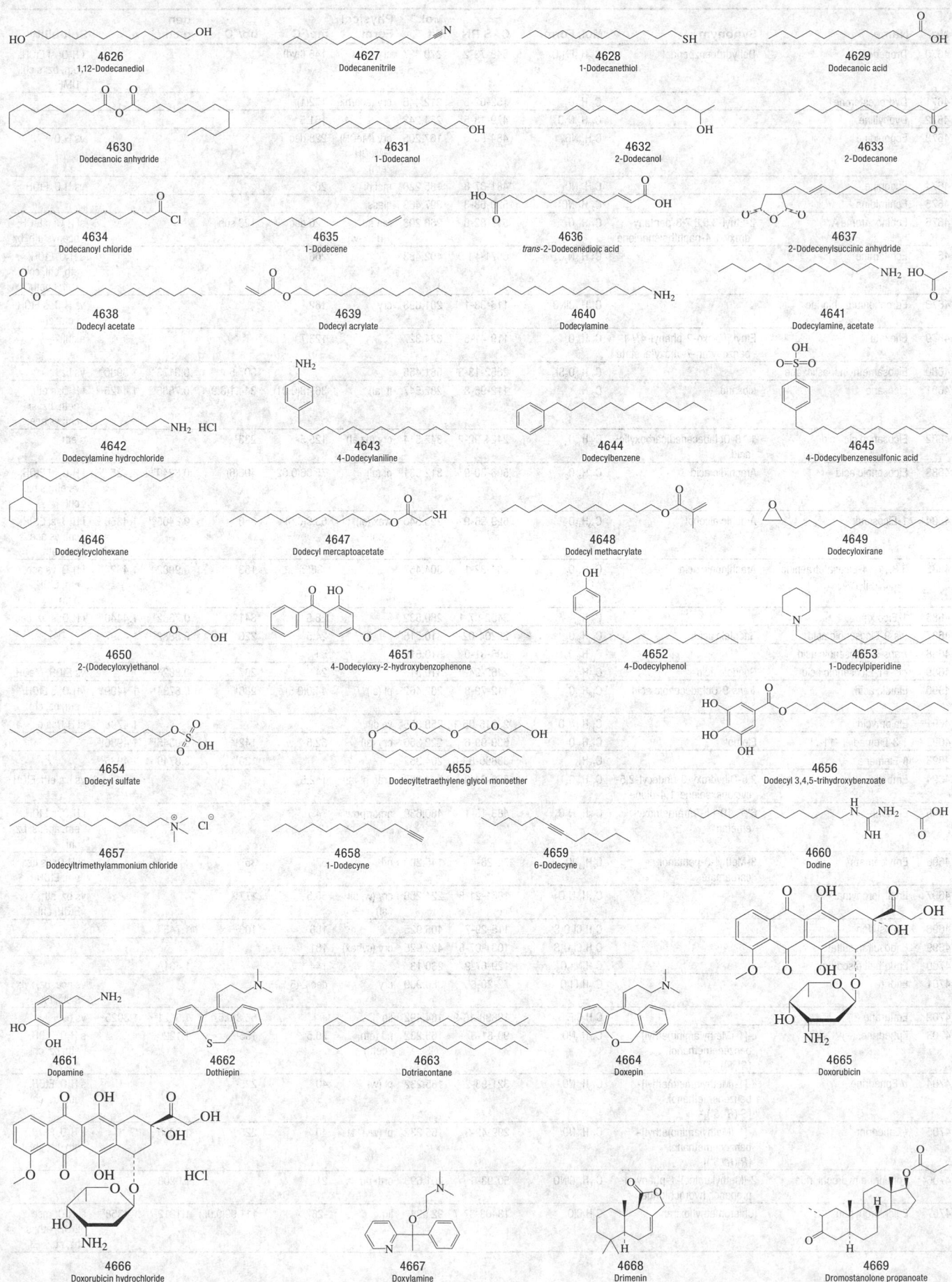

4626 1,12-Dodecanediol

4627 Dodecanenitrile

4628 1-Dodecanethiol

4629 Dodecanoic acid

4630 Dodecanoic anhydride

4631 1-Dodecanol

4632 2-Dodecanol

4633 2-Dodecanone

4634 Dodecanoyl chloride

4635 1-Dodecene

4636 *trans*-2-Dodecenedioic acid

4637 2-Dodecenylsuccinic anhydride

4638 Dodecyl acetate

4639 Dodecyl acrylate

4640 Dodecylamine

4641 Dodecylamine, acetate

4642 Dodecylamine hydrochloride

4643 4-Dodecylaniline

4644 Dodecylbenzene

4645 4-Dodecylbenzenesulfonic acid

4646 Dodecylcyclohexane

4647 Dodecyl mercaptoacetate

4648 Dodecyl methacrylate

4649 Dodecyloxirane

4650 2-(Dodecyloxy)ethanol

4651 4-Dodecyloxy-2-hydroxybenzophenone

4652 4-Dodecylphenol

4653 1-Dodecylpiperidine

4654 Dodecyl sulfate

4655 Dodecyltetraethylene glycol monoether

4656 Dodecyl 3,4,5-trihydroxybenzoate

4657 Dodecyltrimethylammonium chloride

4658 1-Dodecyne

4659 6-Dodecyne

4660 Dodine

4661 Dopamine

4662 Dothiepin

4663 Dotriacontane

4664 Doxepin

4665 Doxorubicin

4666 Doxorubicin hydrochloride

4667 Doxylamine

4668 Drimenin

4669 Dromostanolone propanoate

No.	Name	Synonym	Mol. Form.	CAS RN	Mol. Wt.	Physical Form	mp/°C	bp/°C	den g cm⁻³	n_D	Solubility
4670	Droperidol	Dehydrobenzperidol	$C_{22}H_{22}FN_3O_2$	548-73-2	379.427	cry (w)	146 (hyd)				i H₂O; sl EtOH, eth, bz; s chl, DMF
4671	Dydrogesterone		$C_{21}H_{28}O_2$	152-62-5	312.446	cry (ace/hx)	172(1)				
4672	Dyphylline		$C_{10}H_{14}N_4O_4$	479-18-5	254.243		161.5				
4673	Ecgonidine		$C_9H_{13}NO_2$	484-93-5	167.205	cry (MeOH) (MeOH-eth)	228 dec				vs H₂O
4674	Ecgonine		$C_9H_{15}NO_3$	481-37-8	185.220	mcl pr	205				vs H₂O, EtOH
4675	Echimidine		$C_{20}H_{31}NO_7$	520-68-3	397.463	glass					
4676	Echinochrome A	2-Ethyl-3,5,6,7,8-pentahy-droxy-1,4-naphthalenedione	$C_{12}H_{10}O_7$	517-82-8	266.203	red nd (Diox-w)	220 dec	120 sub			sl H₂O; s EtOH, ace; vs eth, bz
4677	Echitamine		$C_{22}H_{30}N_2O_5$	6871-44-9	402.483		206				s H₂O, EtOH, eth, chl, con sulf; i peth
4678	Edrophonium chloride		$C_{10}H_{16}ClNO$	116-38-1	201.693	cry	162				vs H₂O; s EtOH; i eth, chl
4679	Efloxate	Ethyl [(4-oxo-2-phenyl-4H-1-benzopyran-7-yl)oxy]acetate	$C_{19}H_{16}O_5$	119-41-5	324.327		123.7				s chl
4680	Eicosamethylnonasiloxane		$C_{20}H_{60}O_8Si_9$	2652-13-3	681.455			307.5	0.9173²⁰	1.3980²⁰	vs bz
4681	Eicosane	Icosane	$C_{20}H_{42}$	112-95-8	282.547	lf (al)	36.48(0.01)	344.1(0.9)	0.7886²⁰	1.4425²⁰	i H₂O; s eth, peth, bz; sl chl; vs ace
4682	Eicosanedioic acid	1,18-Octadecanedicarboxylic acid	$C_{20}H_{38}O_4$	2424-92-2	342.514	cry (bz,al)	125.5	233²			s eth
4683	Eicosanoic acid	Arachidic acid	$C_{20}H_{40}O_2$	506-30-9	312.531	pl (al)	75.06(0.02)	400(6)	0.8240¹⁰⁰	1.425¹⁰⁰	i H₂O; sl EtOH; vs eth; s bz, chl
4684	1-Eicosanol	Arachic alcohol	$C_{20}H_{42}O$	629-96-9	298.546	wax (al), cry (chl)	63.9(0.9)	356	0.8405²⁰	1.4350²⁰	i H₂O; sl EtOH, chl; vs ace; s bz, peth
4685	5,8,11,14-Eicosatetraenoic acid, (all-cis)	Arachidonic acid	$C_{20}H_{32}O_2$	506-32-1	304.467		-38(2)	163¹	0.9082²⁰	1.4824²⁰	i H₂O; vs ace, eth, EtOH, peth
4686	1-Eicosene		$C_{20}H_{40}$	3452-07-1	280.532		28.5	341	0.7882³⁰	1.4440³⁰	i H₂O; s bz, peth
4687	cis-9-Eicosenoic acid	Gadoleic acid	$C_{20}H_{38}O_2$	29204-02-2	310.515		24.5	220⁶	0.8882²⁵	1.4597²⁵	
4688	trans-9-Eicosenoic acid		$C_{20}H_{38}O_2$	506-31-0	310.515		54				
4689	cis-11-Eicosenoic acid	Gondoic acid	$C_{20}H_{38}O_2$	2462-94-4	310.515		24	267¹⁵	0.8826²⁵		vs EtOH, MeOH
4690	Elaidic acid	trans-9-Octadecenoic acid	$C_{18}H_{34}O_2$	112-79-8	282.462	pl (al)	44.0(0.5)	288¹⁰⁰	0.8734⁴⁵	1.4499⁴⁵	i H₂O; s EtOH, eth, bz, chl
4691	Elaiomycin		$C_{13}H_{26}N_2O_3$	23315-05-1	258.356	ye oil				1.4798²⁵	sl H₂O; s os
4692	1,3-Elemadien-11-ol	Elemol	$C_{15}H_{26}O$	639-99-6	222.366	cry (al)	52.5	142¹²	0.9345¹⁸	1.4980¹⁸	
4693	β-Elemene		$C_{15}H_{24}$	33880-83-0	204.352			120¹⁶	0.8749²⁰	1.4935²⁰	
4694	Embelin	2,5-Dihydroxy-3-undecyl-2,5-cyclohexadiene-1,4-dione	$C_{17}H_{26}O_4$	550-24-3	294.386	oran pl (al)	142.5				vs bz, eth, EtOH
4695	Emetine	6',7',10,11-Tetramethoxy-emetan	$C_{29}H_{40}N_2O_4$	483-18-1	480.639	amor pow	74				i H₂O; s EtOH, eth, ace; sl bz, chl
4696	Emylcamate	3-Methyl-3-pentanol, carbamate	$C_7H_{15}NO_2$	78-28-4	145.200	nd	57	35¹			sl H₂O; vs bz, eth, EtOH
4697	Enallylpropymal		$C_{11}H_{16}N_2O_3$	1861-21-8	224.256	cry (w, dil al)	56.5	177¹²			vs bz, eth, EtOH, chl
4698	Endosulfan		$C_9H_6Cl_6O_3S$	115-29-7	406.925		106	106⁰·⁷	1.745²⁰		
4699	Endosulfan sulfate		$C_9H_6Cl_6O_4S$	1031-07-8	422.925	cry (cyhex)	181				
4700	Endothall disodium		$C_8H_8Na_2O_5$	129-67-9	230.13		144		1.431²⁰		
4701	Endrin		$C_{12}H_8Cl_6O$	72-20-8	380.909	cry	dec 245				vs ace, bz, xyl; s ctc, hx
4702	Enflurane		$C_3H_2ClF_5O$	13838-16-9	184.492	liq		56.8(0.5)	1.5121²⁵	1.3025²⁰	vs os
4703	Ephedrine, (±)-	α-[1-(Methylamino)ethyl]-benzenemethanol, (R*,S*)-(±)-	$C_{10}H_{15}NO$	90-81-3	165.232	nd (eth, peth)	76.5	135¹²	1.1220²⁰		s H₂O, EtOH, eth, bz, chl
4704	d-Ephedrine	α-[1-(Methylamino)ethyl]-benzenemethanol, [S-(R*,S*)]-	$C_{10}H_{15}NO$	321-98-2	165.232	pl (w)	40	225			s H₂O, EtOH, eth, bz, chl
4705	l-Ephedrine	α-[1-(Methylamino)ethyl]-benzenemethanol, [R-(R*,S*)]-	$C_{10}H_{15}NO$	299-42-3	165.232	pl (w + 1)	40	225	1.0085²²		s H₂O, EtOH, eth, bz, chl
4706	Ephedrine hydrochloride	2-(Methylamino)-1-phenyl-1-propanol, hydrochloride	$C_{10}H_{16}ClNO$	50-98-6	201.693	orth nd	219		1.0208²⁰		
4707	Epichlorohydrin	(Chloromethyl)oxirane	C_3H_5ClO	13403-37-7	92.524	liq	-26	111.99(0.05)	1.1812²⁰	1.4358²⁵	sl H₂O; msc EtOH, eth; s bz, ctc

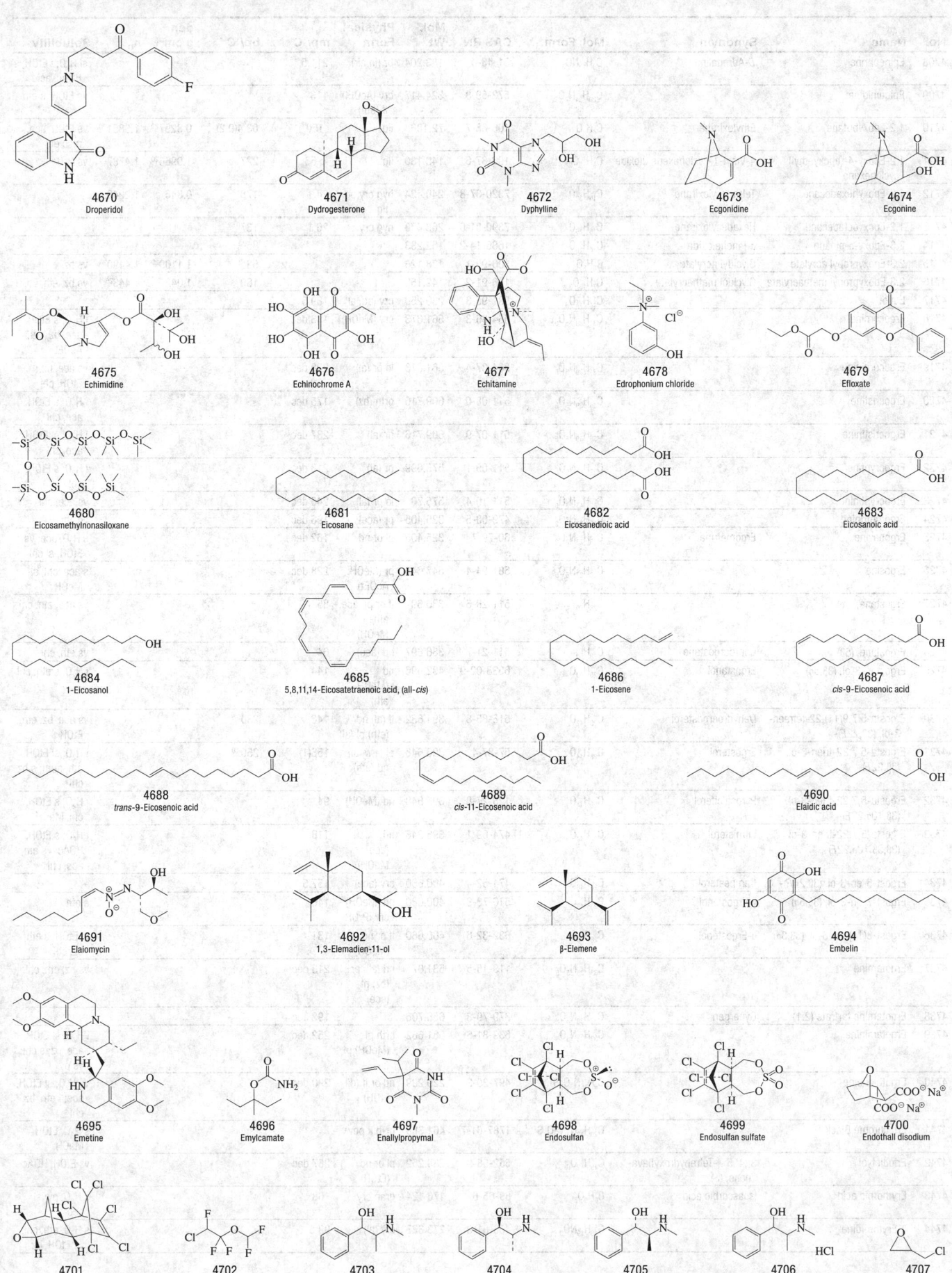

4670
Droperidol

4671
Dydrogesterone

4672
Dyphylline

4673
Ecgonidine

4674
Ecgonine

4675
Echimidine

4676
Echinochrome A

4677
Echitamine

4678
Edrophonium chloride

4679
Efloxate

4680
Eicosamethylnonasiloxane

4681
Eicosane

4682
Eicosanedioic acid

4683
Eicosanoic acid

4684
1-Eicosanol

4685
5,8,11,14-Eicosatetraenoic acid, (all-*cis*)

4686
1-Eicosene

4687
cis-9-Eicosenoic acid

4688
trans-9-Eicosenoic acid

4689
cis-11-Eicosenoic acid

4690
Elaidic acid

4691
Elaiomycin

4692
1,3-Elemadien-11-ol

4693
β-Elemene

4694
Embelin

4695
Emetine

4696
Emylcamate

4697
Enallylpropymal

4698
Endosulfan

4699
Endosulfan sulfate

4700
Endothall disodium

4701
Endrin

4702
Enflurane

4703
Ephedrine, (±)-

4704
d-Ephedrine

4705
l-Ephedrine

4706
Ephedrine hydrochloride

4707
Epichlorohydrin

No.	Name	Synonym	Mol. Form.	CAS RN	Mol. Wt.	Physical Form	mp/°C	bp/°C	den g cm^{-3}	n_D	Solubility
4708	Epinephrine	*D*-Adrenaline	C$_9$H$_{13}$NO$_3$	51-43-4	183.204	br (in air)	211.5				sl H$_2$O; i EtOH; s HOAc, acid
4709	Epiquinidine		C$_{20}$H$_{24}$N$_2$O$_2$	572-59-8	324.417	cry (AcOEt) lf (eth)	113				vs EtOH; s eth
4710	1,2-Epoxybutane	Ethyloxirane	C$_4$H$_8$O	106-88-7	72.106	liq	-150	63.4(0.2)	0.8297^{20}	1.3851^{20}	vs EtOH, ace; msc eth
4711	1,2-Epoxy-4-(epoxyethyl) cyclohexane	4-Vinyl-1-cyclohexene dioxide	C$_8$H$_{12}$O$_2$	106-87-6	140.180	liq	<-55	227	1.0966^{20}	1.4787^{20}	vs H$_2$O
4712	1,2-Epoxyhexadecane	Tetradecyloxirane	C$_{16}$H$_{32}$O	7320-37-8	240.424	hyg cry or liq	24.1	178^{12}	0.846	1.2240	
4713	1,2-Epoxyoctadecane	Hexadecyloxirane	C$_{18}$H$_{36}$O	7390-81-0	268.478	hyg cry	26.1	137$^{0.5}$			
4714	2,3-Epoxy-α-pinane	α-Pinene oxide	C$_{10}$H$_{16}$O	1686-14-2	152.233			85^{24}			
4715	2,3-Epoxypropyl acrylate	Glycidyl acrylate	C$_6$H$_8$O$_3$	106-90-1	128.126			53^{10}	1.1109^{20}	1.4490^{20}	vs bz
4716	2,3-Epoxypropyl methacrylate	Glycidol methacrylate	C$_7$H$_{10}$O$_3$	106-91-2	142.152			189	1.042^{20}	1.448^{25}	vs bz, eth, EtOH
4717	Equol		C$_{14}$H$_{14}$O$_3$	531-95-3	230.259	cry (aq, al)	189.5				
4718	Ergocornine		C$_{31}$H$_{39}$N$_5$O$_5$	564-36-3	561.673	cry (MeOH)	183 dec				i H$_2$O; s EtOH, ace, bz, chl, AcOEt
4719	Ergocorninine		C$_{31}$H$_{39}$N$_5$O$_5$	564-37-4	561.673	lo pr (al)	228 dec				vs ace, bz, EtOH, chl
4720	Ergocristine		C$_{35}$H$_{39}$N$_5$O$_5$	511-08-0	609.716	orth (bz)	175 dec				i H$_2$O; s EtOH, ace, chl
4721	Ergocristinine		C$_{35}$H$_{39}$N$_5$O$_5$	511-07-9	609.716	pr (al)	237 dec				i H$_2$O; sl EtOH, ace, chl
4722	Ergocryptine		C$_{32}$H$_{41}$N$_5$O$_5$	511-09-1	575.699	pr (al)	213 dec				i H$_2$O; s EtOH, chl
4723	Ergocryptinine		C$_{32}$H$_{41}$N$_5$O$_5$	511-10-4	575.70	lo pr (al)	245 dec				vs ace, chl
4724	Ergometrinine		C$_{19}$H$_{23}$N$_3$O$_2$	479-00-5	325.405	pr (ace)	196 dec				vs chl
4725	Ergonovine	Ergometrine	C$_{19}$H$_{23}$N$_3$O$_2$	60-79-7	325.405	pl or nd	162 dec				s H$_2$O, ace; vs EtOH; sl chl
4726	Ergosine		C$_{30}$H$_{37}$N$_5$O$_5$	561-94-4	547.646	pr (MeOH, AcOEt)	228 dec				s ace, chl; sl MeOH
4727	Ergostane, (5α)		C$_{28}$H$_{50}$	511-20-6	386.697	lf or pl (ace, eth-MeOH)	85				vs ace, eth, chl
4728	Ergostane, (5β)	Coproergostane	C$_{28}$H$_{50}$	511-21-7	386.697	nd (ace)	64				vs eth, chl
4729	Ergostan-3-ol, (3β,5α)	Ergostanol	C$_{28}$H$_{50}$O	6538-02-9	402.696	nd (MeOH-eth)	144.5				i H$_2$O; s eth, chl
4730	Ergosta-5,7,9(11),22-tetraen-3-ol, (3β,22*E*)-	Dehydroergosterol	C$_{28}$H$_{42}$O	516-85-8	394.632	lf (al) nd (eth),pl (al)	146	230$^{0.5}$			vs ace, bz, eth, EtOH
4731	Ergosta-5,7,22-trien-3-ol, (3β,22*E*)-	Ergosterol	C$_{28}$H$_{44}$O	57-87-4	396.648	pl (+w, al) nd (eth)	158(1)	250$^{0.01}$			i H$_2$O; sl EtOH, eth, peth; s bz, chl
4732	Ergosta-5,7,22-trien-3-ol, (3β,10α,22*E*)-	Pyrocalciferol	C$_{28}$H$_{44}$O	128-27-8	396.648	nd (MeOH)	94				i H$_2$O; s EtOH, chl, MeOH
4733	Ergosta-5,7,22-trien-3-ol, (3β,9β,10α,22*E*)-	Lumisterol	C$_{28}$H$_{44}$O	474-69-1	396.648	nd (ace-MeOH)	118				i H$_2$O; s EtOH, HOAc; vs eth, ace, chl
4734	Ergost-5-en-3-ol, (3β,24*R*)-	Campesterol	C$_{28}$H$_{48}$O	474-62-4	400.680	cry (ace)	157.5				
4735	Ergost-7-en-3-ol, (3β,5α)	γ-Ergostenol	C$_{28}$H$_{48}$O	516-78-9	400.680	nd (MeOH) cry (PrOH)	146				s eth
4736	Ergost-8(14)-en-3-ol, (3β,5α)	α-Ergostenol	C$_{28}$H$_{48}$O	632-32-6	400.680	lf or nd (MeOH)	131				sl EtOH; s eth, bz, chl
4737	Ergotamine		C$_{33}$H$_{35}$N$_5$O$_5$	113-15-5	581.67	nd (al), pr (bz) pl (ace)	213 dec				vs bz, eth, chl
4738	Ergotamine tartrate (2:1)	Gynergen	C$_{35}$H$_{38}$N$_5$O$_8$	379-79-3	656.706		192 dec				
4739	Ergotaminine		C$_{33}$H$_{35}$N$_5$O$_5$	639-81-6	581.662	orth pl (MeOH) pl (al)	252 dec				i H$_2$O; sl EtOH, ace, bz; s chl; vs py
4740	Ergothioneine		C$_9$H$_{15}$N$_3$O$_2$S	497-30-3	229.299	nd or lf (dil EtOH)	290 dec				vs H$_2$O; sl EtOH, ace; i eth, bz, chl
4741	Eriochrome Black T		C$_{20}$H$_{12}$N$_3$NaO$_7$S	1787-61-7	461.380	br-blk pow					s H$_2$O, EtOH, MeOH
4742	Eriodictyol	3',4',5,7-Tetrahydroxyflavanone, (*S*)	C$_{15}$H$_{12}$O$_6$	552-58-9	288.252	pl or nd (EtOH)	267 dec				vs EtOH, HOAc
4743	Erythorbic acid	Isoascorbic acid	C$_6$H$_8$O$_6$	89-65-6	176.124	gran cry	168				s H$_2$O, py; sl ace
4744	β-Erythroidine		C$_{16}$H$_{19}$NO$_3$	466-81-9	273.327	cry (al)	99.5				s H$_2$O, eth, chl; vs EtOH, bz

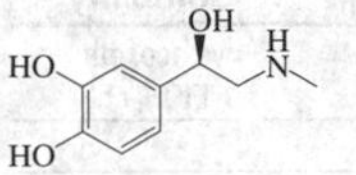

4708 Epinephrine

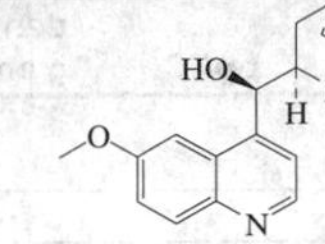

4709 Epiquinidine

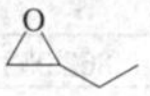

4710 1,2-Epoxybutane

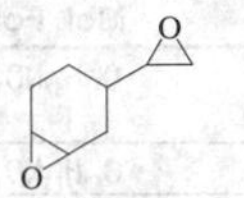

4711 1,2-Epoxy-4-(epoxyethyl)cyclohexane

4712 1,2-Epoxyhexadecane

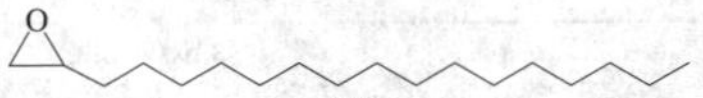

4713 1,2-Epoxyoctadecane

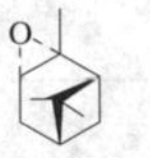

4714 2,3-Epoxy-α-pinane

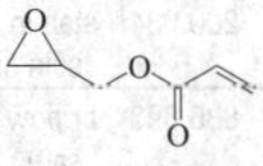

4715 2,3-Epoxypropyl acrylate

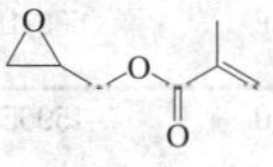

4716 2,3-Epoxypropyl methacrylate

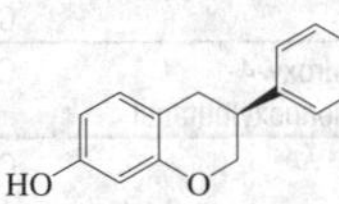

4717 Equol

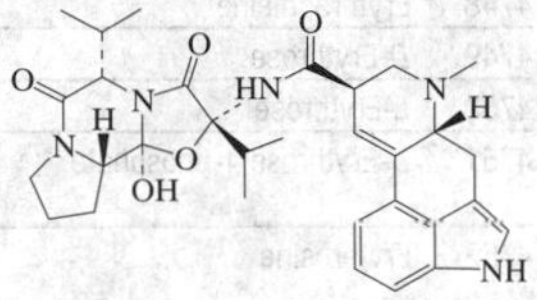

4718 Ergocornine

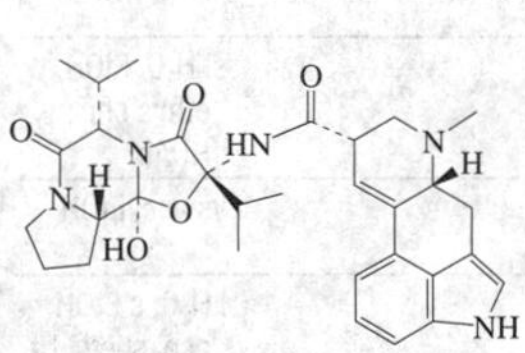

4719 Ergocorninine

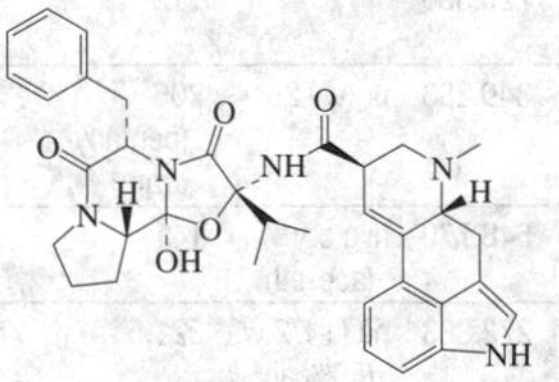

4720 Ergocristine

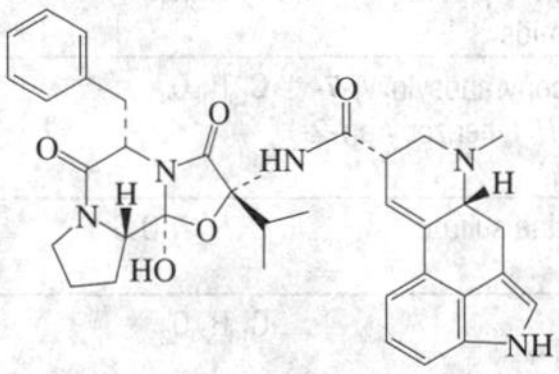

4721 Ergocristinine

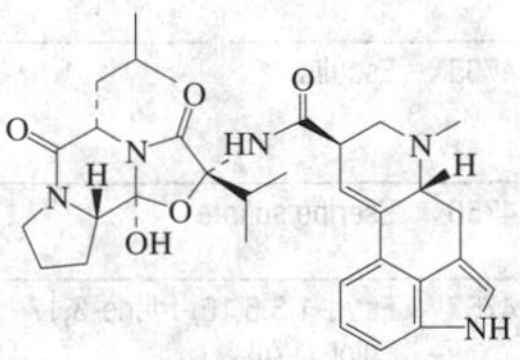

4722 Ergocryptine

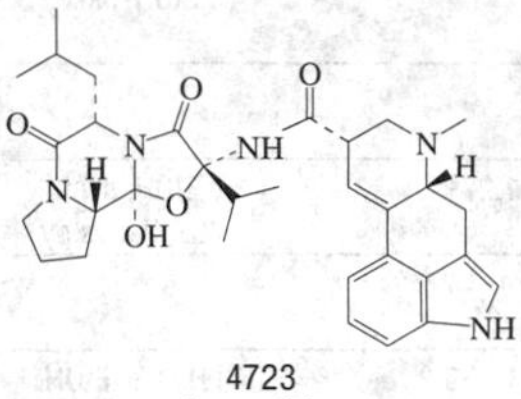

4723 Ergocryptinine

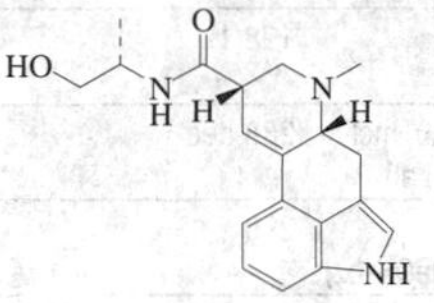

4724 Ergometrinine

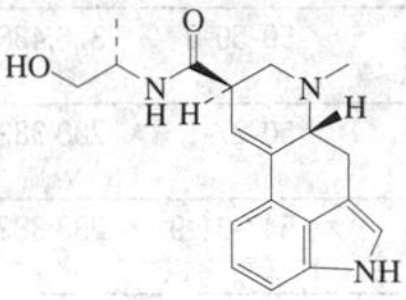

4725 Ergonovine

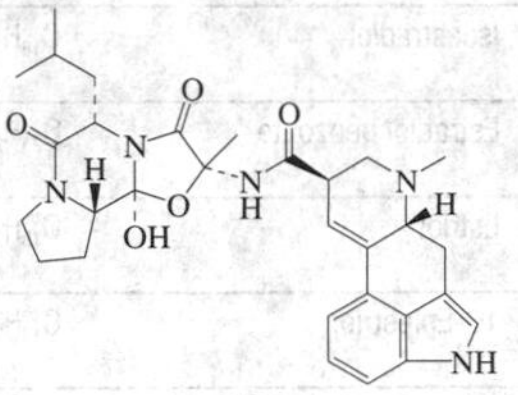

4726 Ergosine

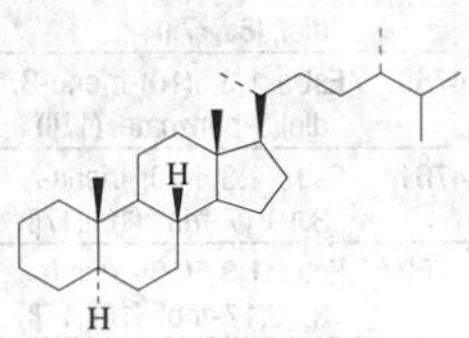

4727 Ergostane, (5α)

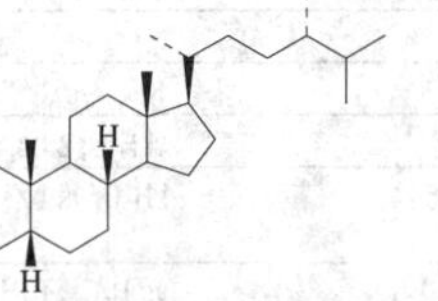

4728 Ergostane, (5β)

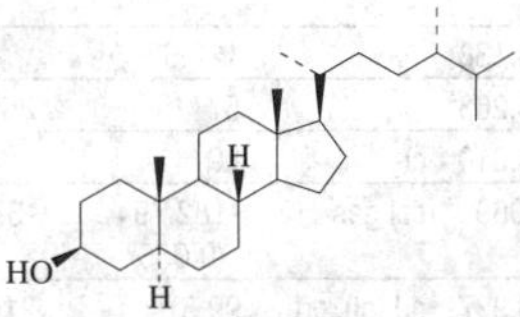

4729 Ergostan-3-ol, (3β,5α)-

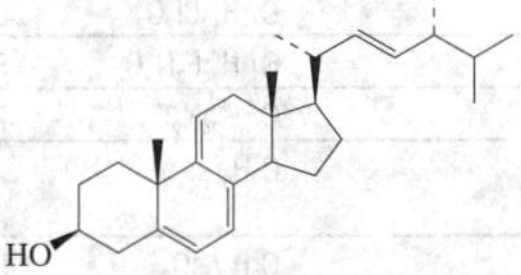

4730 Ergosta-5,7,9(11),22-tetraen-3-ol, (3β,22E)-

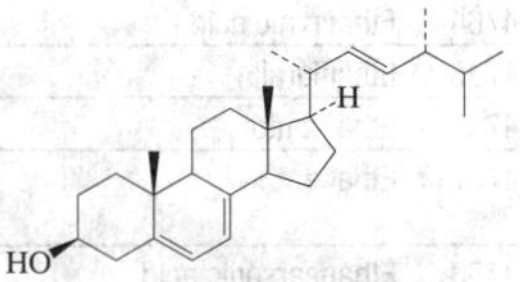

4731 Ergosta-5,7,22-trien-3-ol, (3β,22E)-

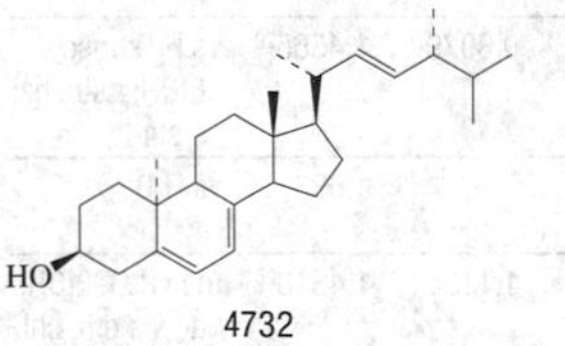

4732 Ergosta-5,7,22-trien-3-ol, (3β,10α,22E)-

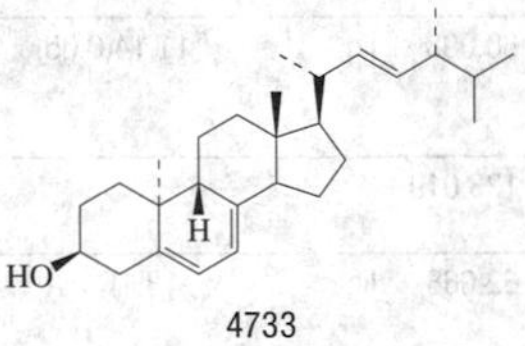

4733 Ergosta-5,7,22-trien-3-ol, (3β,9β,10α,22E)-

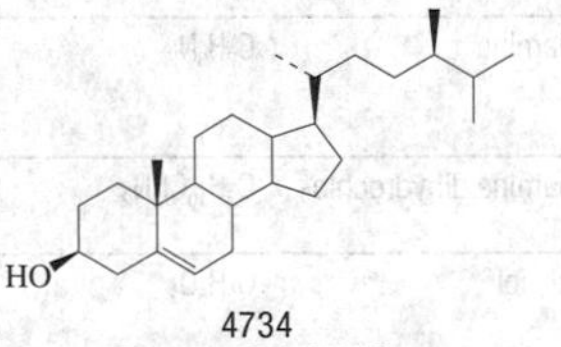

4734 Ergost-5-en-3-ol, (3β,24R)-

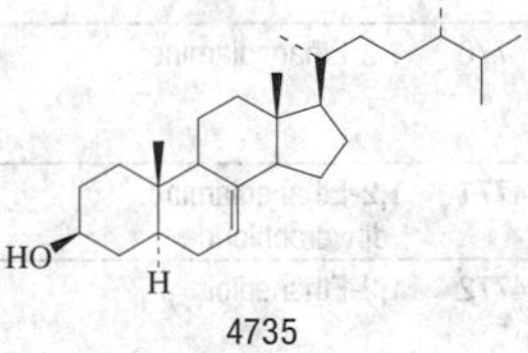

4735 Ergost-7-en-3-ol, (3β,5α)

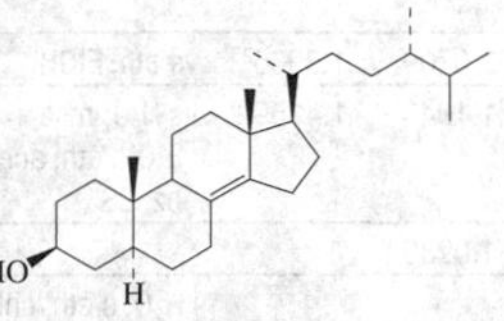

4736 Ergost-8(14)-en-3-ol, (3β,5α)

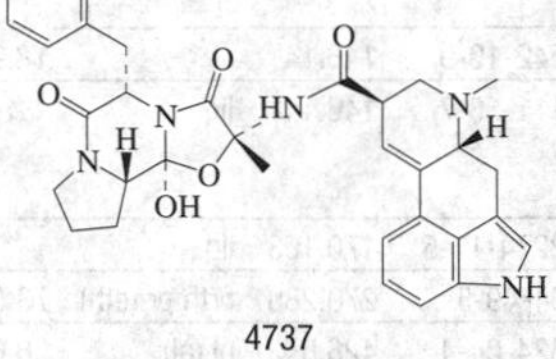

4737 Ergotamine

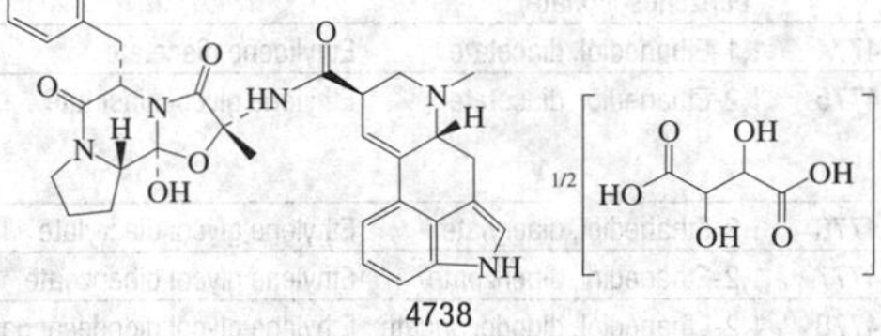

4738 Ergotamine tartrate (2:1)

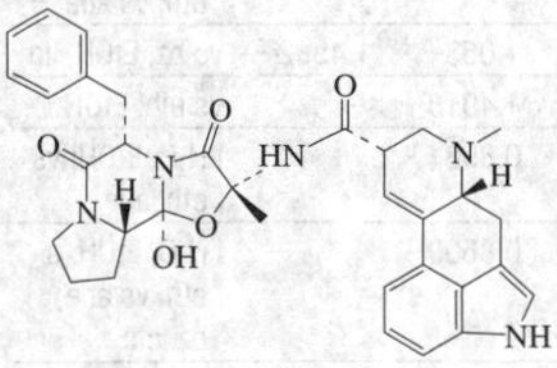

4739 Ergotaminine

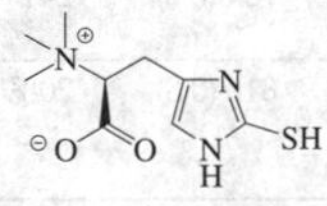

4740 Ergothioneine

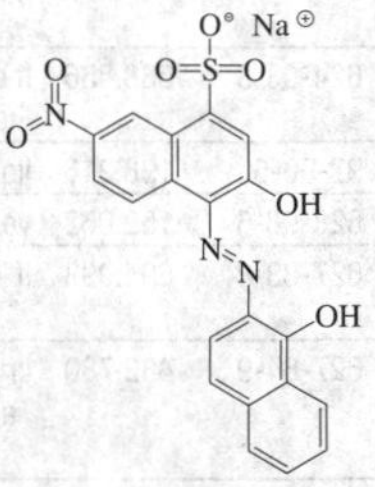

4741 Eriochrome Black T

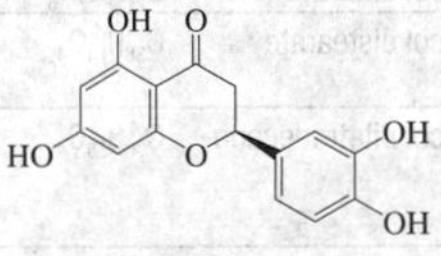

4742 Eriodictyol

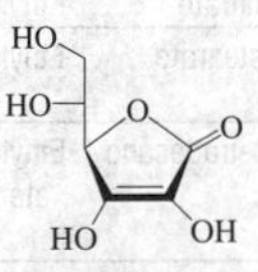

4743 Erythorbic acid

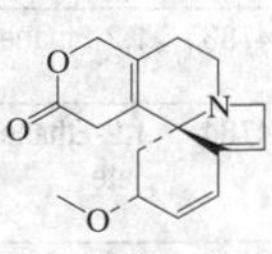

4744 β-Erythroidine

No.	Name	Synonym	Mol. Form.	CAS RN	Mol. Wt.	Physical Form	mp/°C	bp/°C	den g cm⁻³	n_D	Solubility
4745	Erythromycin	Propiocine	$C_{37}H_{67}NO_{13}$	114-07-8	733.927	cry (w)	191				vs ace, eth, EtOH, chl
4746	Erythromycin ethyl succinate		$C_{43}H_{75}NO_{16}$	1264-62-6	862.053	cry (ace aq)	222				
4747	Erythromycin stearate		$C_{55}H_{103}NO_{15}$	643-22-1	1018.405	cry	92				i H_2O; sl EtOH, eth, chl
4748	Erythrophleine	Norcassamidine	$C_{24}H_{39}NO_5$	36150-73-9	421.571	glass	115				s H_2O, EtOH
4749	D-Erythrose		$C_4H_9O_4$	583-50-6	120.105	syr					s H_2O; vs EtOH
4750	L-Erythrose		$C_4H_8O_4$	533-49-3	120.105	syr					vs H_2O, EtOH
4751	D-Erythrose 4-phosphate	2,3-Dihydroxy-4-(phosphonooxy)butanal	$C_4H_9O_7P$	585-18-2	200.084	stab in aq soln only					s H_2O
4752	Erythrosine		$C_{20}H_8I_4O_5$	15905-32-5	835.893	br pow (Na salt)					s H_2; vs eth, EtOH
4753	L-Erythrulose		$C_4H_8O_4$	533-50-6	120.105	syr	dec	dec			vs H_2O, EtOH
4754	Esaprazole	N-Cyclohexyl-1-pipera-zineacetamide	$C_{12}H_{23}N_3O$	64204-55-3	225.330		112	$190^{0.5}$			
4755	Esculin	6-(β-D-Glucopyranosyloxy)-7-hydroxy-2H-1-benzopyran-2-one	$C_{15}H_{16}O_9$	531-75-9	340.283	pr (w+2)	205 (pentahy-drate)				sl H_2O, EtOH, eth; s chl, py, HOAc
4756	Eserine sulfate	Physostigmine sulfate	$C_{30}H_{44}N_6O_8S$	64-47-1	648.770	hyg cry (ace-eth)	141				vs ace, EtOH
4757	Estra-1,3,5(10)-triene-3,17-diol, (17α)	α-Estradiol	$C_{18}H_{24}O_2$	57-91-0	272.383	nd (+1/2 w) (80% al)	221.5				i H_2O; s EtOH, ace; sl eth, bz
4758	Estra-1,3,5(10)-triene-3,17-diol (17β)	β-Estradiol	$C_{18}H_{24}O_2$	50-28-2	272.383	pr (80% al),	180(1)				vs ace, EtOH, Diox
4759	Estra-1,3,5(10)-triene-3,17-diol, (8α,17β)	Isoestradiol	$C_{18}H_{24}O_2$	517-04-4	272.383	cry (dil MeOH-chl)	181				s EtOH, diox
4760	Estra-1,3,5(10)-triene-3,17-diol 3-benzoate, (17β)	Estradiol benzoate	$C_{25}H_{28}O_3$	50-50-0	376.488		198(1)				
4761	Estra-1,3,5(10)-triene-3,16,17-triol, (16α,17β)	Estriol	$C_{18}H_{24}O_3$	50-27-1	288.382	lf (al), mcl (dil al)	288 dec		1.27^{25}		s EtOH; sl eth, bz, tfa; vs py
4762	Estra-1,3,5(10)-triene-3,16,17-triol, (16β,17β)	16-Epiestriol	$C_{18}H_{24}O_3$	547-81-9	288.382	cry (MeOH-bz)	290				
4763	Estrone		$C_{18}H_{22}O_2$	53-16-7	270.367	mcl, orth (al)	260.2		1.236^{25}		i H_2O; sl EtOH, eth, bz; s ace, diox
4764	Ethacrynic acid		$C_{13}H_{12}Cl_2O_4$	58-54-8	303.138		122.5				
4765	Ethalfluralin		$C_{13}H_{14}F_3N_3O_4$	55283-68-6	333.263		57	256 dec			
4766	Ethambutol		$C_{10}H_{24}N_2O_2$	74-55-5	204.310	cry	89				sl H_2O; s bz, chl
4767	Ethane		C_2H_6	74-84-0	30.069	col gas	-182.794 (0.005)	-88.6(0.4)	0.5446^{-89}		i H_2O; vs bz
4768	Ethanearsonic acid		$C_2H_7AsO_3$	507-32-4	153.997	nd (al), orth nd (w)	99.5	210^{12}			vs H_2O, EtOH
4769	Ethanedial dioxime		$C_2H_4N_2O_2$	557-30-2	88.065	orth pl (w)	178 dec	sub			vs H_2O, EtOH, eth
4770	1,2-Ethanediamine	Ethylenediamine	$C_2H_8N_2$	107-15-3	60.098	liq	11.14(0.05)	116.9(0.5)	0.8979^{20}	1.4565^{20}	vs H_2O; msc EtOH; i eth, bz; s ctc
4771	1,2-Ethanediamine, dihydrochloride	Ethylenediamine dihydrochloride	$C_2H_{10}Cl_2N_2$	333-18-6	133.019					1.633	vs H_2O
4772	1,2-Ethanediol	Ethylene glycol	$C_2H_6O_2$	107-21-1	62.068	liq	-13(1)	197.5(0.1)	1.1135^{20}	1.4318^{20}	msc H_2O, EtOH, ace; s eth, chl; sl bz
4773	1,2-Ethanediol, bis(4-methyl-benzenesulfonate)		$C_{16}H_{18}O_6S_2$	6315-52-2	370.440	cry (bz)	126.5(0.5)				
4774	1,1-Ethanediol, diacetate	Ethylidene diacetate	$C_6H_{10}O_4$	542-10-9	146.141		18.9	168(3)	1.070^{25}	1.3985^{25}	vs eth, EtOH
4775	1,2-Ethanediol, diacetate	Ethylene glycol diacetate	$C_6H_{10}O_4$	111-55-7	146.141	liq	-31	184(4)	1.1043^{20}	1.4159^{20}	vs H_2O; msc EtOH, eth, ace, bz, CS_2
4776	1,2-Ethanediol, diacrylate	Ethylene glycol diacrylate	$C_8H_{10}O_4$	2274-11-5	170.163	liq		$55^{0.6}$	1.0935^{26}		
4777	1,2-Ethanediol, dibenzoate	Ethylene glycol dibenzoate	$C_{16}H_{14}O_4$	94-49-5	270.280	orth pr (eth)	73.5	360 dec			i H_2O; s eth, chl
4778	1,2-Ethanediol, didodecanoate	Ethylene glycol didodecanoate	$C_{26}H_{50}O_4$	624-04-4	426.673	pl (al)	56.6	188^{20}			vs eth, EtOH
4779	1,2-Ethanediol, diformate	Ethylene glycol diformate	$C_4H_6O_4$	629-15-2	118.089			174	1.193^0	1.3580	sl H_2O; s EtOH, eth
4780	1,2-Ethanediol, dihexadecano-ate	Ethylene glycol dipalmitate	$C_{34}H_{66}O_4$	624-03-3	538.886	lf or nd (al-chl)	69.1(0.5)		0.8594^{78}		i H_2O, EtOH; s eth; vs ace
4781	1,2-Ethanediol, dimethacrylate	Ethylene glycol dimethacrylate	$C_{10}H_{14}O_4$	97-90-5	198.216	liq	-40	260	1.053^{20}	1.4532^{25}	vs bz, EtOH, lig
4782	1,2-Ethanediol, dinitrate	Ethylene glycol dinitrate	$C_2H_4N_2O_6$	628-96-6	152.062	ye liq	-22.5(0.6)	199(3)	1.4918^{20}		vs eth, EtOH
4783	1,2-Ethanediol, distearate	Ethylene glycol distearate	$C_{38}H_{74}O_4$	627-83-8	594.993	lf	75.3(0.5)	241^{20}	0.8581^{78}		i H_2O, EtOH; vs eth, ace
4784	1,2-Ethanediol, ditetradecano-ate	Ethylene glycol ditetradecano-ate	$C_{30}H_{58}O_4$	627-84-9	482.780	cry (eth, ace)	61.7(0.5)	208^{20}	0.8600^{80}		i H_2O, EtOH; s eth; vs ace, bz, ctc

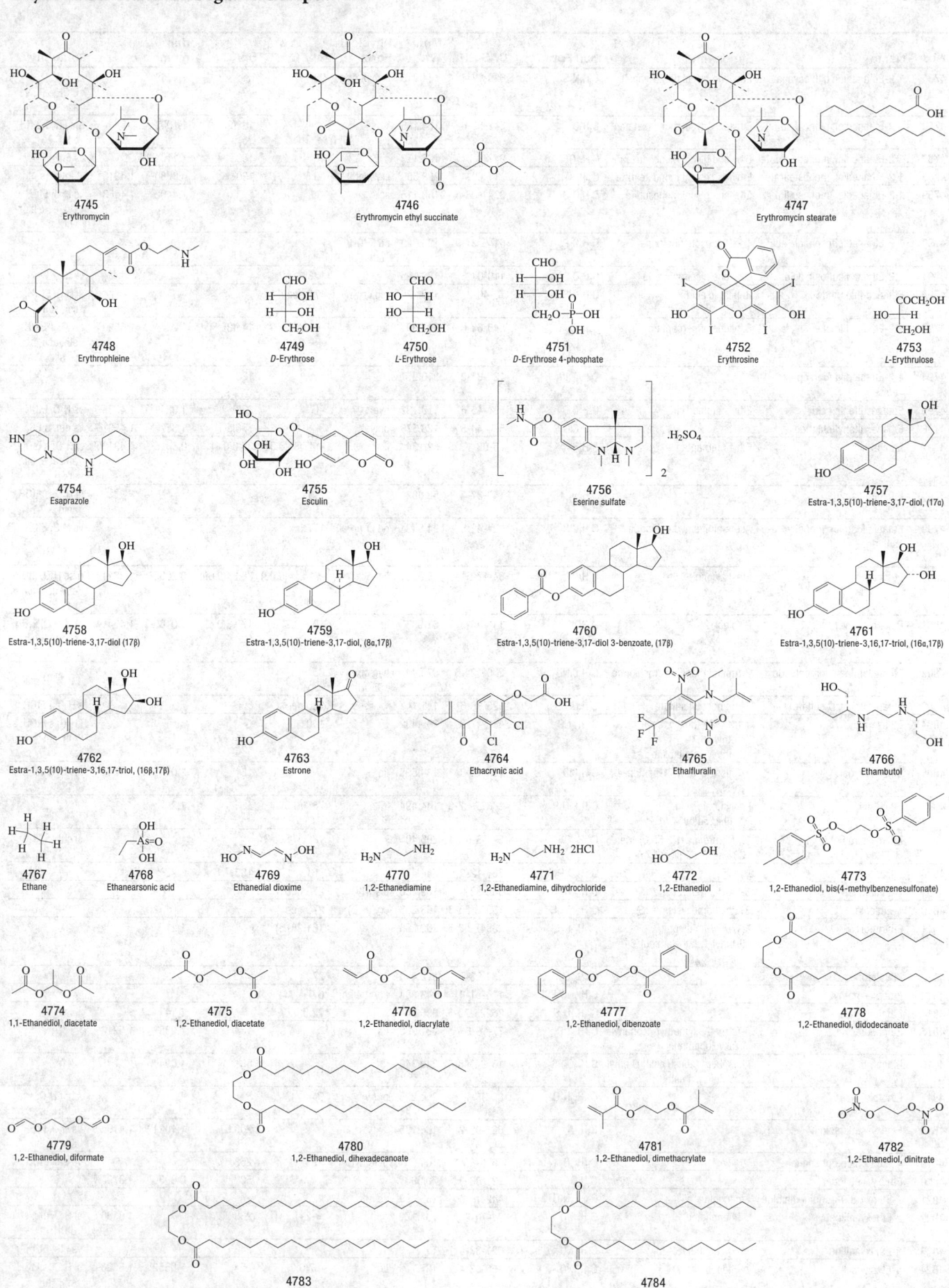

4745
Erythromycin

4746
Erythromycin ethyl succinate

4747
Erythromycin stearate

4748
Erythrophleine

4749
D-Erythrose

4750
L-Erythrose

4751
D-Erythrose 4-phosphate

4752
Erythrosine

4753
L-Erythrulose

4754
Esaprazole

4755
Esculin

4756
Eserine sulfate

4757
Estra-1,3,5(10)-triene-3,17-diol, (17α)

4758
Estra-1,3,5(10)-triene-3,17-diol (17β)

4759
Estra-1,3,5(10)-triene-3,17-diol, (8α,17β)

4760
Estra-1,3,5(10)-triene-3,17-diol 3-benzoate, (17β)

4761
Estra-1,3,5(10)-triene-3,16,17-triol, (16α,17β)

4762
Estra-1,3,5(10)-triene-3,16,17-triol, (16β,17β)

4763
Estrone

4764
Ethacrynic acid

4765
Ethalfluralin

4766
Ethambutol

4767
Ethane

4768
Ethanearsonic acid

4769
Ethanedial dioxime

4770
1,2-Ethanediamine

4771
1,2-Ethanediamine, dihydrochloride

4772
1,2-Ethanediol

4773
1,2-Ethanediol, bis(4-methylbenzenesulfonate)

4774
1,1-Ethanediol, diacetate

4775
1,2-Ethanediol, diacetate

4776
1,2-Ethanediol, diacrylate

4777
1,2-Ethanediol, dibenzoate

4778
1,2-Ethanediol, didodecanoate

4779
1,2-Ethanediol, diformate

4780
1,2-Ethanediol, dihexadecanoate

4781
1,2-Ethanediol, dimethacrylate

4782
1,2-Ethanediol, dinitrate

4783
1,2-Ethanediol, distearate

4784
1,2-Ethanediol, ditetradecanoate

No.	Name	Synonym	Mol. Form.	CAS RN	Mol. Wt.	Physical Form	mp/°C	bp/°C	den g cm⁻³	n_D	Solubility
4785	1,2-Ethanediol, dithiocyanate	Ethylene glycol dithiocyanate	$C_4H_4N_2S_2$	629-17-4	144.218	orth pl or nd (w)	90	dec	1.4200[0]		sl H_2O, bz; s EtOH, eth; vs ace
4786	1,2-Ethanediol, monoacetate	Ethylene glycol monoacetate	$C_4H_8O_3$	542-59-6	104.105			188(2)	1.108[15]		msc H_2O, EtOH, eth
4787	1,2-Ethanediol, monobenzoate	Ethylene glycol monobenzoate	$C_9H_{10}O_3$	94-33-7	166.173		45	150[10]	1.1101[30]		vs EtOH
4788	1,2-Ethanediol, monostearate	Ethylene glycol monostearate	$C_{20}H_{40}O_3$	111-60-4	328.530	cry (peth)	60.5	190[3]	0.8780[60]	1.4310[60]	sl EtOH; s eth
4789	1,2-Ethanediol, monosulfite	Ethylene glycol monosulfite	$C_2H_4O_3S$	3741-38-6	108.116	liq	-11	173	1.4402[20]	1.4463[20]	vs H_2O, EtOH, eth, ace, bz, AcOEt; sl chl
4790	1,2-Ethanediphosphonic acid	1,2-Diphosphonoethane	$C_2H_8O_6P_2$	6145-31-9	190.029	nd (EtOH/ eth)	223				
4791	1,2-Ethanedisulfonic acid	Ethylene disulfonic acid	$C_2H_6O_6S_2$	110-04-3	190.195		173				vs diox
4792	Ethanedithioamide	Rubeanic acid	$C_2H_4N_2S_2$	79-40-3	120.196	red cry	170 dec				sl H_2O, EtOH; s con sulf
4793	1,2-Ethanedithiol	Ethylene dimercaptan	$C_2H_6S_2$	540-63-6	94.199	liq	-41.2	144(3)	1.234[20]	1.5590[20]	i H_2O; s EtOH, eth, ace, bz; vs alk
4794	1,2-Ethanediyl mercaptoac-etate		$C_6H_{10}O_4S_2$	123-81-9	210.271			138[1.5]			
4795	Ethanesulfonic acid	Ethylsulfonic acid	$C_2H_6O_3S$	594-45-6	110.132	hyg	-17	123[1]	1.3341[25]	1.4335[20]	vs H_2O, EtOH
4796	Ethanesulfonyl chloride		$C_2H_5ClO_2S$	594-44-5	128.578	pa ye		176.5(1)	1.357[22]	1.4531[20]	vs eth; s CS_2
4797	Ethanethiol	Ethyl mercaptan	C_2H_6S	75-08-1	62.134	liq	-147.89(0.02)	35.0(0.1)	0.8315[25]	1.4310[20]	sl H_2O; s EtOH, eth, ace, dil alk
4798	Ethanimidamide		$C_2H_6N_2$	143-37-3	58.082		-35				sl H_2O; s EtOH, acid
4799	Ethanimidamide monohydro-chloride	Acetamidine hydrochloride	$C_2H_7ClN_2$	124-42-5	94.543	nd or pr (al) hyg lo pr (al)	177.5				vs H_2O, EtOH
4800	Ethanol	Ethyl alcohol	C_2H_6O	64-17-5	46.068	liq	-114.14(0.03)	78.24(0.09)	0.7893[20]	1.3611[20]	msc H_2O, EtOH, eth, ace, chl; s bz
4801	Ethanolamine	Glycinol	C_2H_7NO	141-43-5	61.083	liq	10.4(0.2)	170.3(0.4)	1.0180[20]	1.4541[20]	msc H_2O, EtOH; sl eth, lig, bz; s chl
4802	Ethanolamine hydrochloride	2-Aminoethanol hydrochloride	C_2H_8ClNO	2002-24-6	97.544	hyg cry (EtOH)	85				
4803	Ethanolamine O-sulfate	2-Aminoethyl sulfate	$C_2H_7NO_4S$	926-39-6	141.147		230 dec				s H_2O; i EtOH
4804	Ethaverine	1-[(3,4-Diethoxyphenyl)-methyl]-6,7-diethoxyisoquinoline	$C_{24}H_{29}NO_4$	486-47-5	395.492		100				i H_2O; s EtOH; sl eth, chl
4805	Ethchlorvynol	1-Chloro-3-ethyl-1-penten-4-yn-ol	C_7H_9ClO	113-18-8	144.598	liq		181	1.07[25]	1.474[25]	i H_2O; s os
4806	Ethephon	Phosphonic acid, (2-chloroethyl)-	$C_2H_6ClO_3P$	16672-87-0	144.494		76.7(0.5)		1.2		
4807	Ethinylestradiol	19-Norpregna-1,3,5(10)-trien-20-yne-3,17-diol, (17α)-	$C_{20}H_{24}O_2$	57-63-6	296.404		186(1)				sl chl
4808	Ethion		$C_9H_{22}O_4P_2S_4$	563-12-2	384.476		-13	165[0.3]	1.22[20]		
4809	D-Ethionine	3-Ethylhomocysteine, (R)	$C_6H_{13}NO_2S$	535-32-0	163.238	cry (H_2O)	278 dec				
4810	L-Ethionine	3-Ethylhomocysteine, (S)	$C_6H_{13}NO_2S$	13073-35-3	163.238	cry (H_2O)	273 dec				
4811	Ethirimol	4(1H)-Pyrimidinone, 5-butyl-2-(ethylamino)-6-methyl-	$C_{11}H_{19}N_3O$	23947-60-6	209.288		161.2(0.5)		1.21[25]		
4812	Ethisterone		$C_{21}H_{28}O_2$	434-03-7	312.446		272				
4813	Ethoate-methyl		$C_6H_{14}NO_3PS_2$	116-01-8	243.284	cry (tol/hp)	67				
4814	Ethofumesate		$C_{13}H_{18}O_5S$	26225-79-6	286.344		72.5(0.5)		1.14		
4815	Ethoheptazine	4-Carbethoxymethyl-4-phenyl-azacycloheptane	$C_{16}H_{23}NO_2$	77-15-6	261.360	liq		134[1]	1.038[26]	1.5210[26]	
4816	Ethoprop	Phosphorodithioic acid, O-ethyl S,S-dipropyl ester	$C_8H_{19}O_2PS_2$	13194-48-4	242.340			88[0.2]	1.094[20]		
4817	Ethotoin		$C_{11}H_{12}N_2O_2$	86-35-1	204.225	pr (w)	94				s hot H_2O; vs EtOH, bz, eth
4818	Ethoxyacetic acid		$C_4H_8O_3$	627-03-2	104.105			206.5	1.1021[20]	1.4194[20]	vs H_2O, EtOH, eth; s chl
4819	4'-Ethoxyacetophenone		$C_{10}H_{12}O_2$	1676-63-7	164.201	pl (eth)	39	268			vs eth, EtOH
4820	Ethoxyacetylene		C_4H_6O	927-80-0	70.090			50	0.8000[20]	1.3796[20]	
4821	7-Ethoxy-3,9-acridinediamine	Ethacridine	$C_{15}H_{15}N_3O$	442-16-0	253.299	ye nd	226				
4822	2-Ethoxyaniline	o-Phenetidine	$C_8H_{11}NO$	94-70-2	137.179		<-21	233(7)		1.5560[20]	sl H_2O, ctc; s EtOH, eth
4823	3-Ethoxyaniline	m-Phenetidine	$C_8H_{11}NO$	621-33-0	137.179			248			vs eth, EtOH
4824	4-Ethoxyaniline	p-Phenetidine	$C_8H_{11}NO$	156-43-4	137.179	liq	4.7(0.2)	254	1.0652[16]	1.5528[20]	sl H_2O; s EtOH, eth, chl

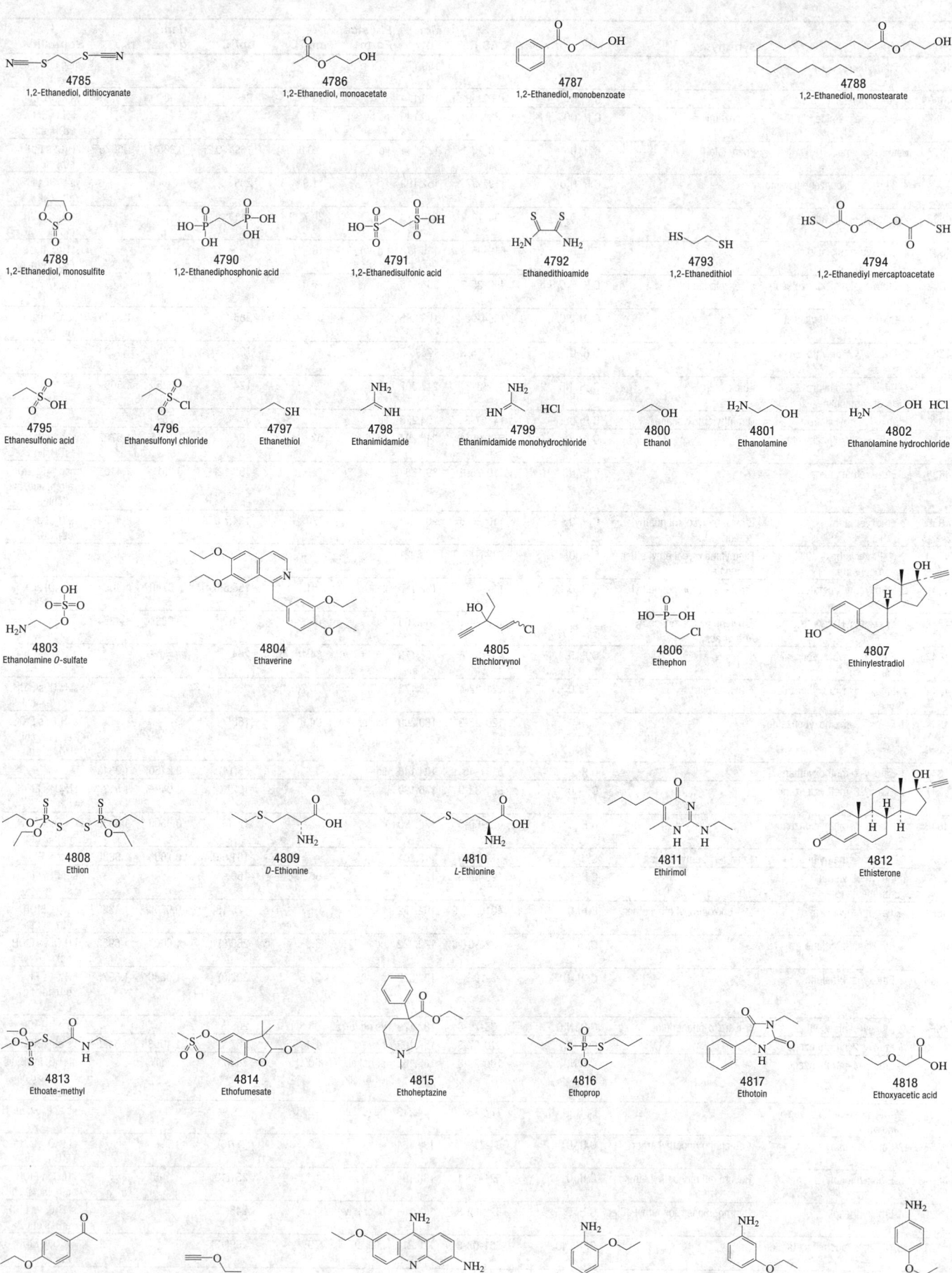

4785
1,2-Ethanediol, dithiocyanate

4786
1,2-Ethanediol, monoacetate

4787
1,2-Ethanediol, monobenzoate

4788
1,2-Ethanediol, monostearate

4789
1,2-Ethanediol, monosulfite

4790
1,2-Ethanediphosphonic acid

4791
1,2-Ethanedisulfonic acid

4792
Ethanedithioamide

4793
1,2-Ethanedithiol

4794
1,2-Ethanediyl mercaptoacetate

4795
Ethanesulfonic acid

4796
Ethanesulfonyl chloride

4797
Ethanethiol

4798
Ethanimidamide

4799
Ethanimidamide monohydrochloride

4800
Ethanol

4801
Ethanolamine

4802
Ethanolamine hydrochloride

4803
Ethanolamine *O*-sulfate

4804
Ethaverine

4805
Ethchlorvynol

4806
Ethephon

4807
Ethinylestradiol

4808
Ethion

4809
D-Ethionine

4810
L-Ethionine

4811
Ethirimol

4812
Ethisterone

4813
Ethoate-methyl

4814
Ethofumesate

4815
Ethoheptazine

4816
Ethoprop

4817
Ethotoin

4818
Ethoxyacetic acid

4819
4'-Ethoxyacetophenone

4820
Ethoxyacetylene

4821
7-Ethoxy-3,9-acridinediamine

4822
2-Ethoxyaniline

4823
3-Ethoxyaniline

4824
4-Ethoxyaniline

No.	Name	Synonym	Mol. Form.	CAS RN	Mol. Wt.	Physical Form	mp/°C	bp/°C	den g cm⁻³	n_D	Solubility
4825	2-Ethoxybenzaldehyde		C₉H₁₀O₂	613-69-4	150.174		21	248			msc EtOH, eth; sl chl
4826	4-Ethoxybenzaldehyde		C₉H₁₀O₂	10031-82-0	150.174		13.5	249	1.08²¹		vs EtOH, eth, bz
4827	2-Ethoxybenzamide	Ethenzamide	C₉H₁₁NO₂	938-73-8	165.189	nd (w, al)	133				sl H₂O, chl; vs EtOH, eth
4828	Ethoxybenzene	Phenetole	C₈H₁₀O	103-73-1	122.164	liq	-29.6(0.4)	169.8(0.2)	0.9651²⁰	1.5076²⁰	i H₂O; s EtOH, eth, ctc
4829	4-Ethoxy-1,2-benzenediamine		C₈H₁₂N₂O	1197-37-1	152.193		71.5	295			vs H₂O; s EtOH, eth, chl
4830	2-Ethoxybenzoic acid		C₉H₁₀O₃	134-11-2	166.173		20.7	211³⁹			sl H₂O, EtOH, ctc
4831	4-Ethoxybenzoic acid		C₉H₁₀O₃	619-86-3	166.173	nd (w)	199(1)				sl H₂O, tfa; s EtOH, eth, bz
4832	6-Ethoxy-2-benzothiazolesulfonamide	Ethoxzolamide	C₉H₁₀N₂O₃S₂	452-35-7	258.316		191(1)				
4833	3-Ethoxy-N,N-diethylaniline		C₁₂H₁₉NO	1864-92-2	193.285			286		1.5325²⁵	s EtOH, bz, HOAc
4834	2-Ethoxy-3,4-dihydro-2H-pyran		C₇H₁₂O₂	103-75-3	128.169			132	0.9658²⁵	1.4394²⁰	
4835	6-Ethoxy-1,2-dihydro-2,2,4-trimethylquinoline	Ethoxyquin	C₁₄H₁₉NO	91-53-2	217.307			124²	1.026²⁵	1.569²⁵	
4836	Ethoxydimethylsilane	Dimethylethoxysilane	C₄H₁₂OSi	14857-34-2	104.223	liq		54	0.76²⁰		
4837	2-Ethoxy-1,2-diphenylethanone		C₁₆H₁₆O₂	574-09-4	240.297	nd (lig)	62	194²⁰	1.1016¹⁷	1.5727¹⁷	vs bz, eth, EtOH, lig
4838	2-Ethoxyethanamine		C₄H₁₁NO	110-76-9	89.136			108(12)	0.8512²⁰	1.4101²⁰	msc H₂O, EtOH, eth; s ace, bz; sl chl
4839	2-Ethoxyethanol	Ethylene glycol monoethyl ether	C₄H₁₀O₂	110-80-5	90.121	liq	-70	134.7(0.2)	0.9253²⁵	1.4054²⁵	vs H₂O, ace, eth, EtOH
4840	2-(2-Ethoxyethoxy)ethyl 2-propenoate	Diethylene glycol ethyl ether acrylate	C₉H₁₆O₄	7328-17-8	188.221				1.13²⁵		
4841	2-Ethoxyethyl acetate	Ethylene glycol monoethyl ether acetate	C₆H₁₂O₃	111-15-9	132.157	liq	-61.7	156.6(0.4)	0.9740²⁰	1.4054²⁰	vs H₂O, ace, eth, EtOH
4842	2-Ethoxyethyl acrylate	Ethylene glycol monoethyl ether acrylate	C₇H₁₂O₃	106-74-1	144.168	liq	-47.3(0.6)	174(2)	0.983²⁰	1.4274²⁰	
4843	3-Ethoxy-2-hydroxybenzaldehyde		C₉H₁₀O₃	492-88-6	166.173		64.1(0.4)	264			
4844	3-Ethoxy-4-hydroxybenzaldehyde	Ethyl vanillin	C₉H₁₀O₃	121-32-4	166.173		76.7(0.4)	285			sl H₂O; s EtOH, eth, bz, chl
4845	4-Ethoxy-3-methoxybenzaldehyde		C₁₀H₁₂O₃	120-25-2	180.200	mcl pr	64.5	168¹³			sl H₂O; s EtOH, eth, bz, chl, HOAc
4846	1-Ethoxy-2-methoxyethane		C₅H₁₂O₂	5137-45-1	104.148	liq		88(14)	0.8460²⁵	1.3843²⁵	
4847	1-Ethoxy-3-methylbenzene		C₉H₁₂O	621-32-9	136.190			192	0.949²⁰	1.513²⁰	i H₂O; s EtOH, eth
4848	1-Ethoxy-4-methylbenzene		C₉H₁₂O	622-60-6	136.190			195(6)	0.9509¹⁸	1.5058¹⁸	i H₂O; s EtOH, eth; sl ctc
4849	2-Ethoxy-2-methylbutane	Ethyl tert-pentyl ether	C₇H₁₆O	919-94-8	116.201			101.5(0.4)	0.7606²⁵	1.3886²⁵	vs eth, EtOH
4850	(Ethoxymethylene)propanedinitrile		C₆H₆N₂O	123-06-8	122.124		66	160¹²			s EtOH, eth; sl chl
4851	(Ethoxymethyl)oxirane	2,3-Epoxypropyl ethyl ether	C₅H₁₀O₂	4016-11-9	102.132			126(18)	0.9700²⁰	1.4320²⁰	s H₂O, EtOH, eth; sl ctc
4852	1-Ethoxynaphthalene		C₁₂H₁₂O	5328-01-8	172.222	nd	5.5	280(8)	1.060²⁰	1.5953²⁵	i H₂O; vs EtOH, eth
4853	2-Ethoxynaphthalene		C₁₂H₁₂O	93-18-5	172.222	pl (al)	37.5	282(8)	1.0640²⁰	1.5975³⁶	i H₂O; s EtOH, eth, tol, lig, CS₂
4854	2-Ethoxy-5-nitroaniline	5-Nitro-o-phenetidine	C₈H₁₀N₂O₃	136-79-8	182.176	ye nd (dil al)	96.5	205¹⁴			vs eth, EtOH
4855	1-Ethoxy-2-nitrobenzene		C₈H₉NO₃	610-67-3	167.162	br ye	1.1	267	1.1903¹⁵	1.5425²⁰	vs eth, EtOH
4856	1-Ethoxy-4-nitrobenzene		C₈H₉NO₃	100-29-8	167.162	pr (dil al, eth)	60(1)	283(7)	1.1176¹⁰⁰		sl H₂O, EtOH; vs eth; msc ace, bz; s peth
4857	N-(4-Ethoxy-3-nitrophenyl)acetamide		C₁₀H₁₂N₂O₄	1777-84-0	224.213	nd (dil al)	124.0				vs ace, bz, EtOH
4858	2-Ethoxyphenol	Catechol monoethyl ether	C₈H₁₀O₂	94-71-3	138.164		29	217(6)	1.0903²⁵		sl H₂O, ctc; msc EtOH, eth
4859	3-Ethoxyphenol	Resorcinol monoethyl ether	C₈H₁₀O₂	621-34-1	138.164			251(19)	1.105¹⁵		i H₂O; s EtOH, eth, bz; sl chl
4860	4-Ethoxyphenol	Hydroquinone monoethyl ether	C₈H₁₀O₂	622-62-8	138.164	pr or lf (w)	66.5	248(19)			sl H₂O; vs EtOH, eth; s chl
4861	N-(2-Ethoxyphenyl)acetamide		C₁₀H₁₃NO₂	581-08-8	179.216	lf(dil al)	79	299(13)			i H₂O; s EtOH, eth, chl

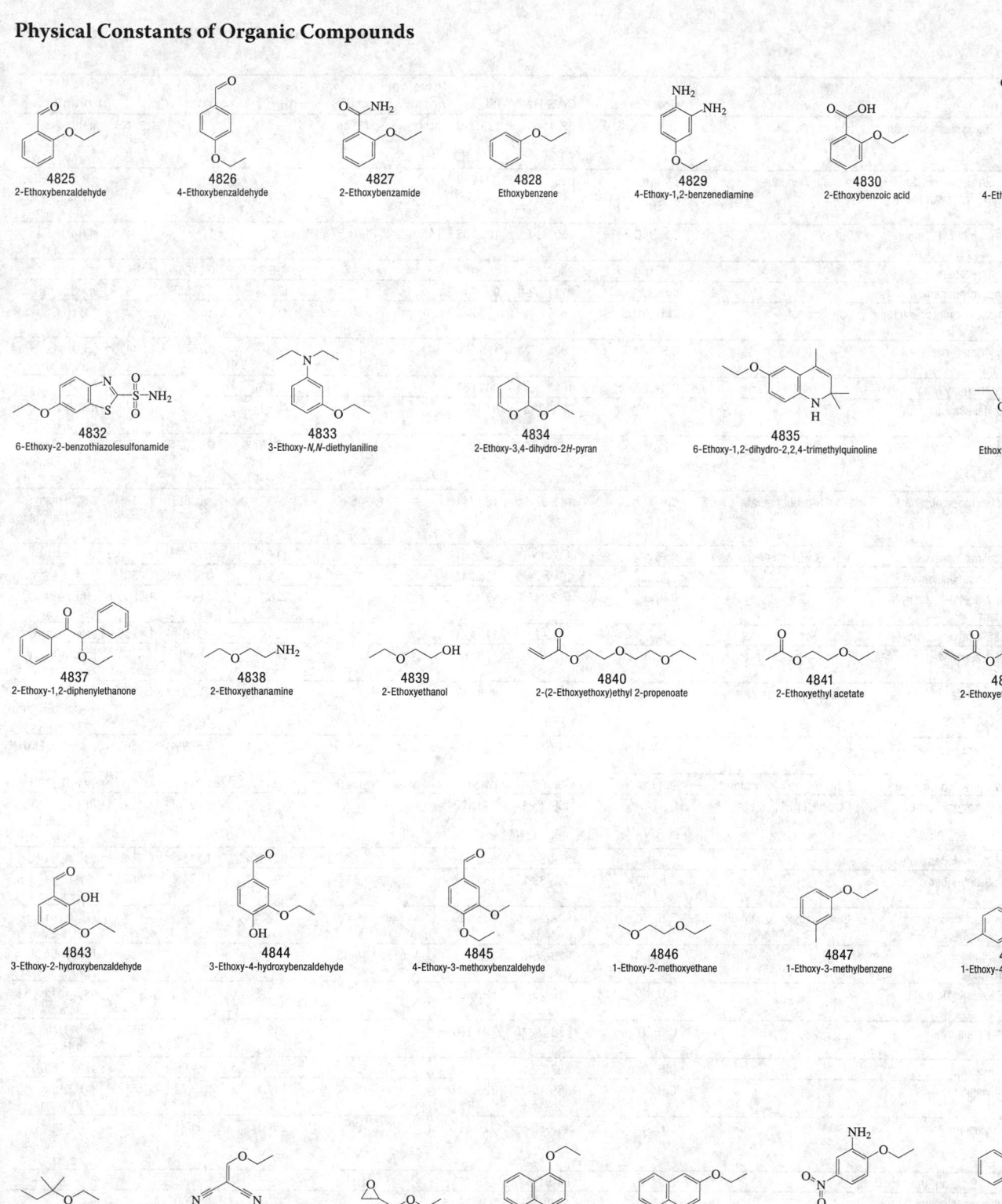

4825 2-Ethoxybenzaldehyde

4826 4-Ethoxybenzaldehyde

4827 2-Ethoxybenzamide

4828 Ethoxybenzene

4829 4-Ethoxy-1,2-benzenediamine

4830 2-Ethoxybenzoic acid

4831 4-Ethoxybenzoic acid

4832 6-Ethoxy-2-benzothiazolesulfonamide

4833 3-Ethoxy-*N,N*-diethylaniline

4834 2-Ethoxy-3,4-dihydro-2*H*-pyran

4835 6-Ethoxy-1,2-dihydro-2,2,4-trimethylquinoline

4836 Ethoxydimethylsilane

4837 2-Ethoxy-1,2-diphenylethanone

4838 2-Ethoxyethanamine

4839 2-Ethoxyethanol

4840 2-(2-Ethoxyethoxy)ethyl 2-propenoate

4841 2-Ethoxyethyl acetate

4842 2-Ethoxyethyl acrylate

4843 3-Ethoxy-2-hydroxybenzaldehyde

4844 3-Ethoxy-4-hydroxybenzaldehyde

4845 4-Ethoxy-3-methoxybenzaldehyde

4846 1-Ethoxy-2-methoxyethane

4847 1-Ethoxy-3-methylbenzene

4848 1-Ethoxy-4-methylbenzene

4849 2-Ethoxy-2-methylbutane

4850 (Ethoxymethylene)propanedinitrile

4851 (Ethoxymethyl)oxirane

4852 1-Ethoxynaphthalene

4853 2-Ethoxynaphthalene

4854 2-Ethoxy-5-nitroaniline

4855 1-Ethoxy-2-nitrobenzene

4856 1-Ethoxy-4-nitrobenzene

4857 *N*-(4-Ethoxy-3-nitrophenyl)acetamide

4858 2-Ethoxyphenol

4859 3-Ethoxyphenol

4860 4-Ethoxyphenol

4861 *N*-(2-Ethoxyphenyl)acetamide

No.	Name	Synonym	Mol. Form.	CAS RN	Mol. Wt.	Physical Form	mp/°C	bp/°C	den g cm⁻³	n_D	Solubility
4862	N-(4-Ethoxyphenyl)acetamide	Phenacetin	C₁₀H₁₃NO₂	62-44-2	179.216	mcl pr	135(3)			1.571	sl H₂O, eth, bz; s EtOH, ace; vs py
4863	N-(4-Ethoxyphenyl)-2-hydroxypropanamide	p-Lactophenetide	C₁₁H₁₅NO₃	539-08-2	209.242		118				s H₂O; vs EtOH; sl eth, bz, chl, peth
4864	(4-Ethoxyphenyl)urea	Dulcin	C₉H₁₂N₂O₂	150-69-6	180.203	lf (dil al), pl (w)	173.5	dec			sl H₂O; s EtOH; vs AcOEt
4865	3-Ethoxypropanal		C₅H₁₀O₂	2806-85-1	102.132			135.2	0.9165²⁰		
4866	3-Ethoxypropanenitrile		C₅H₉NO	2141-62-0	99.131			171(3)	0.9285¹⁵	1.4068²⁰	vs eth, EtOH
4867	8-Ethoxy-5-quinolinesulfonic acid	Actinoquinol	C₁₁H₁₁NO₄S	15301-40-3	253.275	br nd (w)	286 dec				s alk
4868	Ethoxytrimethylsilane		C₅H₁₄OSi	1825-62-3	118.250			76(3)	0.7573²⁰	1.3741²⁰	i H₂O; s EtOH, eth, ace
4869	Ethoxytriphenylsilane		C₂₀H₂₀OSi	1516-80-9	304.458		65	344(3)			s chl
4870	N-Ethylacetamide		C₄H₉NO	625-50-3	87.120			205(1)	0.942⁴	1.4338²⁰	msc H₂O, EtOH; s chl, HOAc
4871	Ethyl acetate		C₄H₈O₂	141-78-6	88.106	liq	-83.8(0.3)	77.1(0.2)	0.9003²⁰	1.3723²⁰	s H₂O; msc EtOH, eth; vs ace, bz
4872	Ethyl acetoacetate	Ethyl 3-oxobutanoate	C₆H₁₀O₃	141-97-9	130.141	liq	-45	180(2)	1.0368¹⁰	1.4171²⁰	s H₂O; msc EtOH, eth; s bz, chl
4873	4'-Ethylacetophenone	1-(4-Ethylphenyl)ethanone	C₁₀H₁₂O	937-30-4	148.201			114¹¹			
4874	Ethyl 2-acetylhexanoate		C₁₀H₁₈O₃	1540-29-0	186.248			221.5	0.9523²⁰	1.4301²⁰	vs ace, eth
4875	Ethyl 2-acetyl-3-methylbutanoate		C₉H₁₆O₃	1522-46-9	172.221			220(14)	0.9648¹⁸	1.4256¹⁸	i H₂O; msc EtOH, eth
4876	Ethyl 2-acetylpentanoate		C₉H₁₆O₃	1540-28-9	172.221			224	0.9661²⁰	1.4255²⁰	vs eth, EtOH
4877	Ethyl 2-acetyl-4-pentenoate	Ethyl 2-allylacetoacetate	C₉H₁₄O₃	610-89-9	170.205			208	0.9898²⁰	1.4388¹⁸	msc EtOH, eth, bz
4878	Ethyl acrylate	Ethyl propenoate	C₅H₈O₂	140-88-5	100.117	liq	-71(2)	98.9(0.6)	0.9234²⁰	1.4068²⁰	sl H₂O, DMSO; msc EtOH, eth; s chl
4879	Ethylamine	Ethanamine	C₂H₇N	75-04-7	45.084	vol liq or gas	-81(2)	16.6(0.2)	0.689¹⁵	1.3663²⁰	msc H₂O, EtOH, eth
4880	Ethylamine hydrochloride	Ethanamine hydrochloride	C₂H₆ClN	557-66-4	81.545	mcl pl (al)	109.5		1.2160²⁰		vs H₂O, EtOH
4881	Ethyl 2-aminoacetate	Glycine, ethyl ester	C₄H₉NO₂	459-73-4	103.120			149	1.0275¹⁰	1.4242¹⁰	msc H₂O, EtOH, eth, ace, bz; vs lig
4882	Ethyl 2-aminobenzoate		C₉H₁₁NO₂	87-25-2	165.189		14.3(0.2)	271(20)	1.1174²⁰	1.5646²⁰	vs eth, EtOH
4883	Ethyl 3-aminobenzoate		C₉H₁₁NO₂	582-33-2	165.189			294	1.171²⁰	1.5600²²	sl H₂O; vs EtOH, eth; s ctc
4884	Ethyl 4-aminobenzoate	Ethyl aminobenzoate	C₉H₁₁NO₂	94-09-7	165.189	nd (w), orth (eth)	89.6(0.3)	310			i H₂O; vs EtOH, eth; s chl, acid
4885	Ethyl (aminocarbonyl)carbamate		C₄H₈N₂O₃	626-36-8	132.118	nd (w, bz)	196.5	dec			i H₂O, eth; sl EtOH, bz, tfa
4886	2-(Ethylamino)ethanol		C₄H₁₁NO	110-73-6	89.136			178(8)	0.914²⁰	1.444²⁰	vs H₂O, EtOH, eth; s chl
4887	2-Ethylaniline		C₈H₁₁N	578-54-1	121.180	liq	-47(1)	213(4)	0.983²²	1.5584²²	sl H₂O, chl; vs EtOH, eth
4888	3-Ethylaniline		C₈H₁₁N	587-02-0	121.180	liq	-64	214(7)	0.9896²⁵		vs eth, EtOH
4889	4-Ethylaniline		C₈H₁₁N	589-16-2	121.180	liq	-5.1(0.3)	217.2(0.2)	0.9679²⁰	1.5554²⁰	sl H₂O, ctc; vs EtOH, eth
4890	N-Ethylaniline		C₈H₁₁N	103-69-5	121.180	liq	-63.4(0.4)	204(1)	0.9625²⁰	1.5559²⁰	i H₂O; msc EtOH, eth; vs ace, bz; s ctc
4891	2-Ethyl-9,10-anthracenedione		C₁₆H₁₂O₂	84-51-5	236.265		108.8				
4892	4-Ethylbenzaldehyde		C₉H₁₀O	4748-78-1	134.174			221	0.9790²⁰		
4893	N-Ethylbenzamide		C₉H₁₁NO	614-17-5	149.189	nd (w)	70.5				
4894	Ethylbenzene	Phenylethane	C₈H₁₀	100-41-4	106.165	liq	-94.95(0.02)	136.2(0.4)	0.8626²⁵	1.4930²⁵	i H₂O; msc EtOH, eth; sl chl
4895	α-Ethylbenzeneacetamide	α-Phenylbutyramide	C₁₀H₁₃NO	90-26-6	163.216	cry	86	185¹⁶			s H₂O, ctc; sl ace
4896	α-Ethylbenzeneacetic acid		C₁₀H₁₂O₂	90-27-7	164.201	pl (eth)	47.5	271			s eth, bz, ctc
4897	α-Ethylbenzeneacetonitrile		C₁₀H₁₁N	769-68-6	145.201			241(8)	0.977¹⁴		i H₂O; s EtOH, eth, bz
4898	4-Ethyl-1,3-benzenediol		C₈H₁₀O₂	2896-60-8	138.164	pr (chl, bz)	98.5	160²⁴			sl H₂O, EtOH, eth
4899	α-Ethylbenzenemethanol	α-Ethylbenzyl alcohol	C₉H₁₂O	93-54-9	136.190			219	0.9915²⁵	1.5169²³	vs bz, eth, EtOH, MeOH
4900	Ethyl benzenesulfonate		C₈H₁₀O₃S	515-46-8	186.228			156¹⁵	1.2167²⁰	1.5081²⁰	sl H₂O; s EtOH; vs eth, chl

4862
N-(4-Ethoxyphenyl)acetamide

4863
N-(4-Ethoxyphenyl)-2-hydroxypropanamide

4864
(4-Ethoxyphenyl)urea

4865
3-Ethoxypropanal

4866
3-Ethoxypropanenitrile

4867
8-Ethoxy-5-quinolinesulfonic acid

4868
Ethoxytrimethylsilane

4869
Ethoxytriphenylsilane

4870
N-Ethylacetamide

4871
Ethyl acetate

4872
Ethyl acetoacetate

4873
4'-Ethylacetophenone

4874
Ethyl 2-acetylhexanoate

4875
Ethyl 2-acetyl-3-methylbutanoate

4876
Ethyl 2-acetylpentanoate

4877
Ethyl 2-acetyl-4-pentenoate

4878
Ethyl acrylate

4879
Ethylamine

4880
Ethylamine hydrochloride

4881
Ethyl 2-aminoacetate

4882
Ethyl 2-aminobenzoate

4883
Ethyl 3-aminobenzoate

4884
Ethyl 4-aminobenzoate

4885
Ethyl (aminocarbonyl)carbamate

4886
2-(Ethylamino)ethanol

4887
2-Ethylaniline

4888
3-Ethylaniline

4889
4-Ethylaniline

4890
N-Ethylaniline

4891
2-Ethyl-9,10-anthracenedione

4892
4-Ethylbenzaldehyde

4893
N-Ethylbenzamide

4894
Ethylbenzene

4895
α-Ethylbenzeneacetamide

4896
α-Ethylbenzeneacetic acid

4897
α-Ethylbenzeneacetonitrile

4898
4-Ethyl-1,3-benzenediol

4899
α-Ethylbenzenemethanol

4900
Ethyl benzenesulfonate

No.	Name	Synonym	Mol. Form.	CAS RN	Mol. Wt.	Physical Form	mp/°C	bp/°C	den g cm⁻³	n_D	Solubility
4901	4-Ethylbenzenesulfonic acid		$C_8H_{10}O_3S$	98-69-1	186.228				1.23		
4902	2-Ethyl-1H-benzimidazole		$C_9H_{10}N_2$	1848-84-6	146.188		176.5				sl chl
4903	Ethyl benzoate	Ethyl benzenecarboxylate	$C_9H_{10}O_2$	93-89-0	150.174	liq	-34.5(0.3)	212.5(0.2)	1.0415[25]	1.5007[20]	i H₂O; s EtOH, ace, bz; msc eth; sl ctc
4904	Ethyl 1,3-benzodioxole-5-carboxylate		$C_{10}H_{10}O_4$	6951-08-2	194.184	pr	18.5	285.5			vs eth, EtOH, peth
4905	Ethyl benzoylacetate		$C_{11}H_{12}O_3$	94-02-0	192.211		<0	267 dec	1.1202[15]	1.5317[15]	sl H₂O; s EtOH, eth
4906	Ethyl N-benzylglycinate		$C_{11}H_{15}NO_2$	6436-90-4	193.243			177[50]		1.5041[20]	vs EtOH, eth, bz
4907	Ethyl 2-benzylideneacetoacetate		$C_{13}H_{14}O_3$	620-80-4	218.248	orth pl (dil al)	60.5	296			i H₂O; sl EtOH, eth, bz; vs chl
4908	Ethyl bromoacetate		$C_4H_7BrO_2$	105-36-2	167.002			159(4)	1.5032[20]	1.4489[20]	i H₂O; msc EtOH, eth; s ace; sl ctc
4909	Ethyl 4-bromoacetoacetate		$C_6H_9BrO_3$	13176-46-0	209.037			115[14]	1.5278[18]	1.5281[20]	vs eth, EtOH
4910	Ethyl 4-bromobenzoate		$C_9H_9BrO_2$	5798-75-4	229.070	liq	-18	263	1.4332[17]	1.5438[17]	sl H₂O; s EtOH, eth, ace, bz
4911	Ethyl 2-bromobutanoate		$C_6H_{11}BrO_2$	533-68-6	195.054			177	1.3273[20]	1.4475[20]	i H₂O; msc EtOH, eth; s chl
4912	Ethyl 4-bromobutanoate		$C_6H_{11}BrO_2$	2969-81-5	195.054			192	1.3540[20]	1.4559[20]	
4913	Ethyl trans-4-bromo-2-butenoate		$C_6H_9BrO_2$	37746-78-4	193.038			100[14]	1.402[16]	1.4925[20]	vs EtOH
4914	Ethyl 6-bromohexanoate	Ethyl 6-bromocaproate	$C_8H_{15}BrO_2$	25542-62-5	223.108	cry (peth)	33	126[21]	1.238[23]	1.4566[21]	
4915	Ethyl 2-bromo-3-methylbutanoate		$C_7H_{13}BrO_2$	609-12-1	209.081			186	1.2760[20]	1.4496[20]	vs eth, EtOH
4916	Ethyl 2-bromo-2-methylpropanoate		$C_6H_{11}BrO_2$	600-00-0	195.054			178(5)	1.3263[20]	1.4446[20]	i H₂O; s EtOH; msc eth
4917	Ethyl 3-bromo-2-oxopropanoate	Ethyl 3-bromopyruvate	$C_5H_7BrO_3$	70-23-5	195.012			87[9]			
4918	Ethyl 2-bromopentanoate		$C_7H_{13}BrO_2$	615-83-8	209.081			191	1.226[18]	1.4496[20]	i H₂O; s EtOH, eth
4919	Ethyl 5-bromopentanoate		$C_7H_{13}BrO_2$	14660-52-7	209.081			129[35]	1.3085[20]	1.4543[20]	sl ctc
4920	Ethyl 2-bromopropanoate	Ethyl α-bromopropionate	$C_5H_9BrO_2$	535-11-5	181.028			152(13)	1.4135[20]	1.4490[20]	i H₂O; msc EtOH, eth; s chl
4921	Ethyl 3-bromopropanoate		$C_5H_9BrO_2$	539-74-2	181.028			163(13)	1.4123[18]	1.4516[20]	s EtOH, eth, ace; sl ctc
4922	2-Ethylbutanal	Diethylacetaldehyde	$C_6H_{12}O$	97-96-1	100.158			118[160]	0.8110[20]	1.4025[20]	sl H₂O, ctc; msc EtOH, eth
4923	Ethyl butanoate	Ethyl butyrate	$C_6H_{12}O_2$	105-54-4	116.158	liq	-97(6)	121.1(0.4)	0.8735[25]	1.3898[25]	sl H₂O, ctc; s EtOH, eth
4924	2-Ethylbutanoic acid	Diethylacetic acid	$C_6H_{12}O_2$	88-09-5	116.158	liq	-31.8(0.5)	193(3)	0.9239[20]	1.4132[20]	sl H₂O, ctc; msc EtOH, eth
4925	2-Ethylbutanoic acid, triethyleneglycol diester		$C_{18}H_{34}O_6$	95-08-9	346.459			181[3.5]			
4926	2-Ethyl-1-butanol	2-Ethylbutyl alcohol	$C_6H_{14}O$	97-95-0	102.174	liq	<-15	155(2)	0.8326[20]	1.4220[20]	sl H₂O; s EtOH, eth, chl
4927	2-Ethylbutanoyl chloride		$C_6H_{11}ClO$	2736-40-5	134.603			140(8)	0.9825[20]	1.4234[20]	vs eth
4928	2-Ethyl-1-butene		C_6H_{12}	760-21-4	84.159	liq	-132.0(0.2)	64.7(0.1)	0.6894[20]	1.3969[20]	i H₂O; s eth, ace, bz, chl
4929	Ethyl cis-2-butenoate	Ethyl isocrotonate	$C_6H_{10}O_2$	6776-19-8	114.142			136	0.9182[20]	1.4242[20]	vs ace, eth, EtOH
4930	Ethyl trans-2-butenoate	Ethyl crotonate	$C_6H_{10}O_2$	623-70-1	114.142			140(5)	0.9175[20]	1.4243[20]	i H₂O; s EtOH, eth
4931	Ethyl 3-butenoate		$C_6H_{10}O_2$	1617-18-1	114.142			119	0.9122[20]	1.4105[20]	s EtOH
4932	2-Ethylbutyl acetate		$C_8H_{16}O_2$	10031-87-5	144.212		<-100	161(9)	0.8790[20]	1.4109[20]	i H₂O; s EtOH, eth, ctc
4933	2-Ethylbutyl acrylate		$C_9H_{16}O_2$	3953-10-4	156.222	liq		80[20]			
4934	2-Ethylbutylamine	2-Ethyl-1-butanamine	$C_6H_{15}N$	617-79-8	101.190	liq	125				
4935	Ethyl N-butylcarbamate		$C_7H_{15}NO_2$	591-62-8	145.200	liq	-22	221(12)	0.9434[26]	1.4278[26]	
4936	Ethyl 2-butynoate		$C_6H_8O_2$	4341-76-8	112.127			163	0.9641[20]	1.4372[20]	
4937	Ethyl carbamate	Urethane	$C_3H_7NO_2$	51-79-6	89.094	pr (bz, to)	48.25(0.04)	185	0.9862[15]	1.4144[51]	vs H₂O, EtOH, eth, bz, chl, py; sl lig
4938	9-Ethyl-9H-carbazol-3-amine		$C_{14}H_{14}N_2$	132-32-1	210.274		99				
4939	9-Ethyl-9H-carbazole		$C_{14}H_{13}N$	86-28-2	195.260	nd (al)	67(1)	190[10]	1.059[80]	1.6394[80]	i H₂O; vs EtOH, eth
4940	Ethyl chloroacetate		$C_4H_7ClO_2$	105-39-5	122.551	liq	-21	144(3)	1.1585[20]	1.4215[20]	i H₂O; msc EtOH, eth, ace; s bz
4941	Ethyl 4-chloroacetoacetate		$C_6H_9ClO_3$	638-07-3	164.586		-8	220 dec	1.218[25]	1.4520[20]	

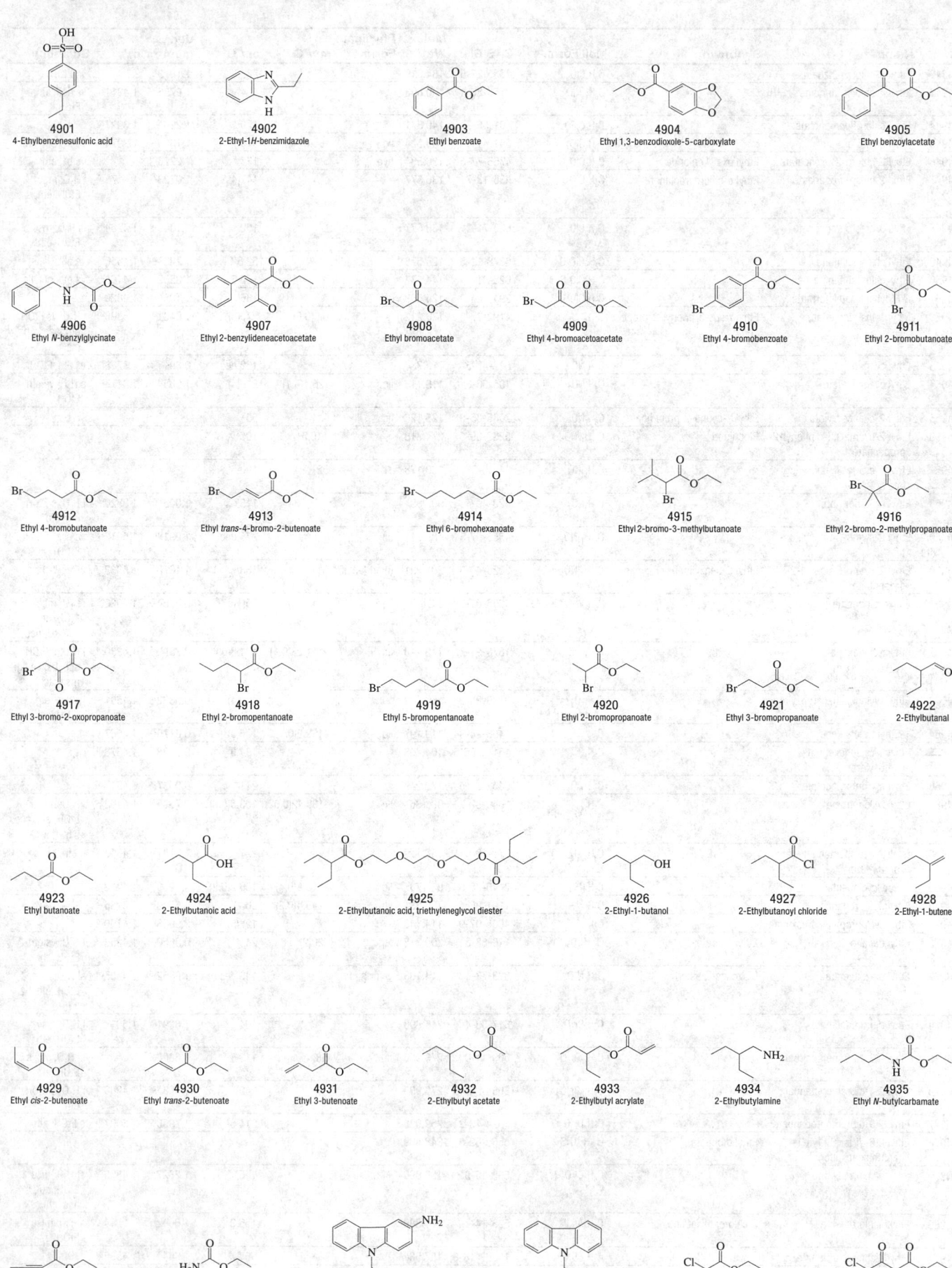

4901
4-Ethylbenzenesulfonic acid

4902
2-Ethyl-1*H*-benzimidazole

4903
Ethyl benzoate

4904
Ethyl 1,3-benzodioxole-5-carboxylate

4905
Ethyl benzoylacetate

4906
Ethyl *N*-benzylglycinate

4907
Ethyl 2-benzylideneacetoacetate

4908
Ethyl bromoacetate

4909
Ethyl 4-bromoacetoacetate

4910
Ethyl 4-bromobenzoate

4911
Ethyl 2-bromobutanoate

4912
Ethyl 4-bromobutanoate

4913
Ethyl *trans*-4-bromo-2-butenoate

4914
Ethyl 6-bromohexanoate

4915
Ethyl 2-bromo-3-methylbutanoate

4916
Ethyl 2-bromo-2-methylpropanoate

4917
Ethyl 3-bromo-2-oxopropanoate

4918
Ethyl 2-bromopentanoate

4919
Ethyl 5-bromopentanoate

4920
Ethyl 2-bromopropanoate

4921
Ethyl 3-bromopropanoate

4922
2-Ethylbutanal

4923
Ethyl butanoate

4924
2-Ethylbutanoic acid

4925
2-Ethylbutanoic acid, triethyleneglycol diester

4926
2-Ethyl-1-butanol

4927
2-Ethylbutanoyl chloride

4928
2-Ethyl-1-butene

4929
Ethyl *cis*-2-butenoate

4930
Ethyl *trans*-2-butenoate

4931
Ethyl 3-butenoate

4932
2-Ethylbutyl acetate

4933
2-Ethylbutyl acrylate

4934
2-Ethylbutylamine

4935
Ethyl *N*-butylcarbamate

4936
Ethyl 2-butynoate

4937
Ethyl carbamate

4938
9-Ethyl-9*H*-carbazol-3-amine

4939
9-Ethyl-9*H*-carbazole

4940
Ethyl chloroacetate

4941
Ethyl 4-chloroacetoacetate

No.	Name	Synonym	Mol. Form.	CAS RN	Mol. Wt.	Physical Form	mp/°C	bp/°C	den g cm⁻³	n_D	Solubility
4942	Ethyl 4-chlorobenzoate		$C_9H_9ClO_2$	7335-27-5	184.619			237.5	1.1873[14]		vs EtOH
4943	Ethyl 4-chlorobutanoate		$C_6H_{11}ClO_2$	3153-36-4	150.603			184	1.0756[20]	1.4311[20]	vs ace, eth, EtOH
4944	Ethyl chlorofluoroacetate		$C_4H_6ClFO_2$	401-56-9	140.541			129	1.225[20]	1.3927[20]	
4945	Ethyl chloroformate		$C_3H_5ClO_2$	541-41-3	108.524	liq	-80.6	91(2)	1.1352[20]	1.3974[20]	vs bz, eth, chl
4946	Ethyl 2-chloro-2-oxoacetate	Ethyl oxalyl chloride	$C_4H_5ClO_3$	4755-77-5	136.534	hyg		137	1.2226[20]		vs bz, eth
4947	Ethyl 2-chloropropanoate	Ethyl α-chloropropionate	$C_5H_9ClO_2$	535-13-7	136.577			147(4)	1.0793[20]	1.4178[20]	i H₂O; msc EtOH, eth; sl ctc
4948	Ethyl 3-chloropropanoate		$C_5H_9ClO_2$	623-71-2	136.577			162	1.1086[20]	1.4254[20]	sl H₂O; msc EtOH, eth
4949	Ethyl chlorosulfinate		$C_2H_5ClO_2S$	6378-11-6	128.578			52.5[44]	1.2837[20]	1.4550[25]	vs eth
4950	Ethyl chlorosulfonate		$C_2H_5ClO_3S$	625-01-4	144.577			152.5	1.3502[25]	1.416[20]	vs eth, chl, lig
4951	S-Ethyl chlorothioformate		C_3H_5ClOS	2941-64-2	124.589	liq		136	1.195[20]	1.4820[20]	
4952	Ethyl *trans*-cinnamate	Ethyl *trans*-3-phenyl-2-propenoate	$C_{11}H_{12}O_2$	4192-77-2	176.212		6.5(0.5)	270(3)	1.0491[20]	1.5598[20]	i H₂O; vs EtOH, eth, ace; s bz, ctc
4953	Ethyl cyanate		C_3H_5NO	627-48-5	71.078			162 dec	0.89[20]	1.3788[25]	vs eth, EtOH
4954	Ethyl cyanoacetate		$C_5H_7NO_2$	105-56-6	113.116	liq	-26.1(0.1)	216(4)	1.0654[20]	1.4175[20]	s H₂O; vs eth, EtOH
4955	Ethyl 2-cyanoacrylate	Ethyl 2-cyano-2-propenoate	$C_6H_7NO_2$	7085-85-0	125.126	liq		55[3]			
4956	Ethyl 2-cyano-3,3-diphenyl-2-propenoate	Etocrilene	$C_{18}H_{15}NO_2$	5232-99-5	277.318		110.5	195[3]			
4957	Ethyl 2-cyano-3-ethoxyacrylate		$C_8H_{11}NO_3$	94-05-3	169.178		52	190.5			
4958	Ethyl cyanoformate		$C_4H_5NO_2$	623-49-4	99.089			115.5	1.003[25]	1.3820[20]	i H₂O; s EtOH, eth, ctc
4959	Ethyl 2-cyano-2-phenylacetate		$C_{11}H_{11}NO_2$	4553-07-5	189.211	oil		275 dec	1.091[20]	1.5012[25]	vs ace, bz, eth, EtOH
4960	Ethyl 2-cyano-3-phenyl-2-propenoate	Ethyl 2-benzylidene-2-cyanoacetate	$C_{12}H_{11}NO_2$	2025-40-3	201.221	(i) nd (al) (ii) oil	51	188[15]	1.1076[25]	1.5033	vs ace, chl
4961	Ethylcyclobutane		C_6H_{12}	4806-61-5	84.159	liq	-142.9(0.2)	70(1)	0.7284[20]	1.4020[20]	i H₂O; msc EtOH, eth; s ace, bz, peth
4962	Ethylcyclohexane		C_8H_{16}	1678-91-7	112.213	liq	-111.28(0.1)	131.8(0.4)	0.7880[20]	1.4330[20]	i H₂O; s EtOH, ace, bz; vs lig; msc ctc
4963	Ethyl cyclohexanecarboxylate		$C_9H_{16}O_2$	3289-28-9	156.222			196(4)	0.9362[20]	1.4501[15]	vs ace, eth, EtOH, chl
4964	1-Ethylcyclohexene		C_8H_{14}	1453-24-3	110.197	liq	-109.9(0.1)	136(3)	0.8176[25]	1.4567[20]	
4965	Ethyl 3-cyclohexene-1-carboxylate		$C_9H_{14}O_2$	15111-56-5	154.206			194.5	0.9688[20]	1.4578[20]	
4966	Ethyl cyclohexylacetate		$C_{10}H_{18}O_2$	5452-75-5	170.249			211	0.9537[14]	1.451[14]	
4967	Ethylcyclopentane		C_7H_{14}	1640-89-7	98.186	liq	-138.42(0.02)	103.5(0.6)	0.7665[20]	1.4198[20]	i H₂O; msc EtOH, eth, ace; s bz, tol
4968	Ethyl 2-cyclopentanone-1-carboxylate		$C_8H_{12}O_3$	611-10-9	156.179			221	1.0781[21]	1.4519[20]	s eth, bz
4969	1-Ethylcyclopentene		C_7H_{12}	2146-38-5	96.170	liq	-118.8(0.4)	106.2(0.2)	0.7936[25]	1.4412[20]	
4970	Ethylcyclopropane		C_5H_{10}	1191-96-4	70.133	liq	-149.4(0.2)	36(2)	0.6790[25]	1.3786[20]	
4971	Ethyl cyclopropanecarboxylate		$C_6H_{10}O_2$	4606-07-9	114.142			128(5)	0.9608[15]	1.4190[20]	
4972	Ethyl decanoate	Ethyl caprate	$C_{12}H_{24}O_2$	110-38-3	200.318	liq	-20(2)	242(1)	0.8650[20]	1.4256[20]	i H₂O; vs eth, EtOH, chl
4973	Ethyl diazoacetate	Diazoacetic ester	$C_4H_6N_2O_2$	623-73-4	114.103	ye orth cry	-22	140 dec	1.0852[18]	1.4605[20]	sl H₂O; msc EtOH, eth, bz, lig
4974	Ethyl dibromoacetate		$C_4H_6Br_2O_2$	617-33-4	245.898			194	1.8991[20]	1.5017[13]	i H₂O; msc EtOH, eth
4975	Ethyl 2,3-dibromobutanoate		$C_6H_{10}Br_2O_2$	609-11-0	273.950	nd	58.5	113[30]	1.6800[20]		sl H₂O, ctc; s EtOH, eth
4976	Ethyl 2,4-dibromobutanoate		$C_6H_{10}Br_2O_2$	36847-51-5	273.950			149[52]	1.6987[20]	1.4960[20]	i H₂O; s EtOH, eth
4977	Ethyl 2,3-dibromopropanoate		$C_5H_8Br_2O_2$	3674-13-3	259.925			214.5	1.7966[20]	1.5007[20]	s EtOH, eth
4978	Ethyl 3,6-di(*tert*-butyl)-1-naphthalenesulfonate	Ethyl dibunate	$C_{20}H_{28}O_3S$	5560-69-0	348.499						s chl
4979	Ethyl dichloroacetate		$C_4H_6Cl_2O_2$	535-15-9	156.996			155	1.2827[20]	1.4386[20]	sl H₂O; msc EtOH, eth; s ace, chl
4980	Ethyldichloroarsine	Dichloroethylarsine	$C_2H_5AsCl_2$	598-14-1	174.889			155.3	1.66[20]		s H₂O; misc EtOH, bz
4981	Ethyl dichlorocarbamate		$C_3H_5Cl_2NO_2$	13698-16-3	157.984			66[18]	1.304[30]	1.4595[20]	
4982	Ethyl 2,3-dichloropropanoate		$C_5H_8Cl_2O_2$	6628-21-3	171.022			183(7)	1.2401[20]	1.4482[20]	vs eth, EtOH
4983	Ethyl diethoxyacetate		$C_8H_{16}O_4$	6065-82-3	176.211			199	0.985[25]	1.4100[20]	

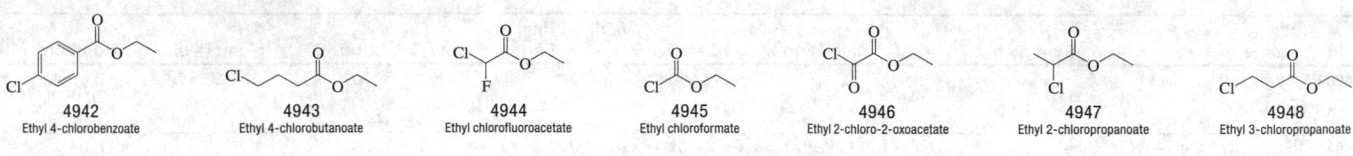

4942	4943	4944	4945	4946	4947	4948
Ethyl 4-chlorobenzoate	Ethyl 4-chlorobutanoate	Ethyl chlorofluoroacetate	Ethyl chloroformate	Ethyl 2-chloro-2-oxoacetate	Ethyl 2-chloropropanoate	Ethyl 3-chloropropanoate

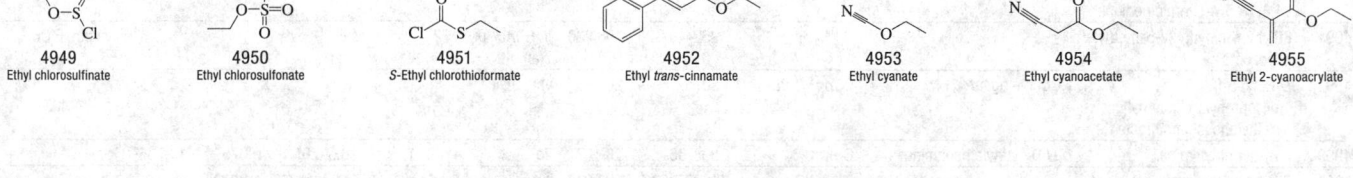

4949	4950	4951	4952	4953	4954	4955
Ethyl chlorosulfinate	Ethyl chlorosulfonate	S-Ethyl chlorothioformate	Ethyl trans-cinnamate	Ethyl cyanate	Ethyl cyanoacetate	Ethyl 2-cyanoacrylate

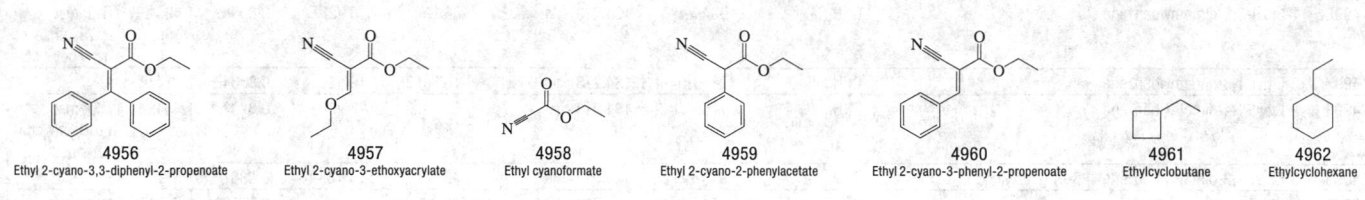

4956	4957	4958	4959	4960	4961	4962
Ethyl 2-cyano-3,3-diphenyl-2-propenoate	Ethyl 2-cyano-3-ethoxyacrylate	Ethyl cyanoformate	Ethyl 2-cyano-2-phenylacetate	Ethyl 2-cyano-3-phenyl-2-propenoate	Ethylcyclobutane	Ethylcyclohexane

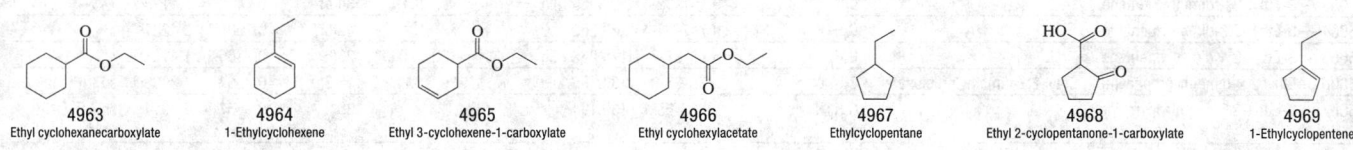

4963	4964	4965	4966	4967	4968	4969
Ethyl cyclohexanecarboxylate	1-Ethylcyclohexene	Ethyl 3-cyclohexene-1-carboxylate	Ethyl cyclohexylacetate	Ethylcyclopentane	Ethyl 2-cyclopentanone-1-carboxylate	1-Ethylcyclopentene

4970	4971	4972	4973	4974	4975	4976
Ethylcyclopropane	Ethyl cyclopropanecarboxylate	Ethyl decanoate	Ethyl diazoacetate	Ethyl dibromoacetate	Ethyl 2,3-dibromobutanoate	Ethyl 2,4-dibromobutanoate

4977	4978	4979	4980	4981	4982	4983
Ethyl 2,3-dibromopropanoate	Ethyl 3,6-di(tert-butyl)-1-naphthalenesulfonate	Ethyl dichloroacetate	Ethyldichloroarsine	Ethyl dichlorocarbamate	Ethyl 2,3-dichloropropanoate	Ethyl diethoxyacetate

No.	Name	Synonym	Mol. Form.	CAS RN	Mol. Wt.	Physical Form	mp/°C	bp/°C	den g cm⁻³	n_D	Solubility
4984	Ethyl diethylmalonate		$C_{11}H_{20}O_4$	77-25-8	216.275			230	0.9643[30]	1.4240[20]	i H$_2$O; msc EtOH, eth; s ctc
4985	Ethyl difluoroacetate		$C_4H_6F_2O_2$	454-31-9	124.087			100	1.1765[20]		i H$_2$O
4986	Ethyldifluoroarsine		$C_2H_5AsF_2$	430-40-0	141.980	liq, fumes in air	-38.7	94.3	1.708[17]		
4987	5-Ethyldihydro-5-sec-butyl-2-thioxo-4,6(1H,5H)-pyrimidin-edione	Thiobutabarbital	$C_{10}H_{16}N_2O_2S$	2095-57-0	228.311		169				
4988	5-Ethyldihydro-2(3H)-furanone		$C_6H_{10}O_2$	695-06-7	114.142	liq	-18	218(4)	1.0261[20]	1.4495[20]	vs H$_2$O, EtOH
4989	Ethyl dihydrogen phosphate		$C_2H_7O_4P$	1623-14-9	126.048	hyg cry		dec	1.430[25]	1.427	vs H$_2$O, ace, eth, EtOH
4990	5-Ethyldihydro-5-phenyl-4,6(1H,5H)-pyrimidinedione	Primidone	$C_{12}H_{14}N_2O_2$	125-33-7	218.251		281.5				
4991	Ethyl 2,4-dihydroxy-6-methyl-benzoate		$C_{10}H_{12}O_4$	2524-37-0	196.200	lf (HOAc), pr (al)	132	sub			vs eth, EtOH
4992	O-Ethyl S-[2-(diisopropylamino)ethyl] methylphosphonothioate	VX Nerve agent	$C_{11}H_{26}NO_2PS$	50782-69-9	267.369	very toxic liq					
4993	Ethyldimethylamine	N,N-Dimethylethanamine	$C_4H_{11}N$	598-56-1	73.137	liq	-140(4)	36.5(0.6)	0.675[20]	1.3705[25]	
4994	Ethyl 4-(dimethylamino)benzoate		$C_{11}H_{15}NO_2$	10287-53-3	193.243		66.5	190[14]	1.0099[100]		
4995	1-Ethyl-2,4-dimethylbenzene		$C_{10}H_{14}$	874-41-9	134.218	liq	-62.9(0.2)	188(2)	0.8763[20]	1.5038[20]	vs ace, bz, eth, EtOH
4996	1-Ethyl-3,5-dimethylbenzene		$C_{10}H_{14}$	934-74-7	134.218	liq	-84.3(0.1)	184(2)	0.8608[25]	1.4981[20]	i H$_2$O; msc EtOH, eth, ace, bz; s peth, ctc
4997	2-Ethyl-1,3-dimethylbenzene		$C_{10}H_{14}$	2870-04-4	134.218	liq	-16.3(0.3)	190(2)	0.8864[25]	1.5107[20]	
4998	2-Ethyl-1,4-dimethylbenzene		$C_{10}H_{14}$	1758-88-9	134.218	liq	-53.7(0.3)	186(2)	0.8732[25]	1.5043[20]	i H$_2$O; msc EtOH, eth, ace, bz; s peth, ctc
4999	3-Ethyl-1,2-dimethylbenzene	3-Ethyl-oxylene	$C_{10}H_{14}$	933-98-2	134.218	liq	-49.5(0.3)	194(2)	0.8881[25]	1.5117[20]	
5000	4-Ethyl-1,2-dimethylbenzene		$C_{10}H_{14}$	934-80-5	134.218	liq	-66.9(0.2)	189(2)	0.8706[25]	1.5031[20]	i H$_2$O; msc EtOH, eth, ace, bz; s peth, ctc
5001	N'-Ethyl-N,N-dimethyl-1,2-ethanediamine		$C_6H_{16}N_2$	123-83-1	116.204			134.5	0.738[25]	1.4222[20]	
5002	Ethyl 4,4-dimethyl-3-oxopen-tanoate	Ethyl pivaloylacetate	$C_9H_{16}O_3$	17094-34-7	172.221	liq		83[17]	0.97[18]		
5003	3-Ethyl-2,2-dimethylpentane		C_9H_{20}	16747-32-3	128.255	liq	-99.4(0.1)	134(4)	0.7438[20]	1.4123[20]	
5004	3-Ethyl-2,3-dimethylpentane		C_9H_{20}	16747-33-4	128.255			142(5)	0.7508[25]	1.4221[20]	
5005	3-Ethyl-2,4-dimethylpentane		C_9H_{20}	1068-87-7	128.255	liq	-122.38(0.09)	123(6)	0.7365[20]	1.4131[20]	
5006	Ethyl 2,2-dimethylpropanoate	Ethyl 2,2-dimethylpropionate	$C_7H_{14}O_2$	3938-95-2	130.185	liq	-89.5(0.2)	118.3(0.4)	0.856[20]	1.3906[20]	s EtOH, eth
5007	3-Ethyl-2,5-dimethylpyrazine		$C_8H_{12}N_2$	13360-65-1	136.194			180.5	0.9657[24]	1.5014[24]	sl H$_2$O, EtOH, eth
5008	3-Ethyl-2,4-dimethyl-1H-pyrrole		$C_8H_{13}N$	517-22-6	123.196	pr	0	197(6)	0.913[20]	1.4961[20]	sl H$_2$O; s EtOH, eth, bz, chl
5009	Ethyl 3,5-dimethylpyrrole-2-carboxylate		$C_9H_{13}NO_2$	2199-44-2	167.205	cry (al)	125	135[10.5]			s EtOH, ace
5010	Ethyl 2,4-dimethylpyrrole-3-carboxylate		$C_9H_{13}NO_2$	2199-51-1	167.205	cry (eth-lig, peth)	78.5	291			vs eth, EtOH
5011	Ethyl 2,5-dimethylpyrrole-3-carboxylate		$C_9H_{13}NO_2$	2199-52-2	167.205	orth (al)	117.5	291			vs EtOH
5012	Ethyl 4,5-dimethylpyrrole-3-carboxylate		$C_9H_{13}NO_2$	2199-53-3	167.205	cry (dil al)	111.3				vs eth, EtOH, chl
5013	Ethyl 2,4-dioxopentanoate		$C_7H_{10}O_4$	615-79-2	158.152		18	214	1.1251[20]	1.4757[17]	vs eth, EtOH
5014	O-Ethyl dithiocarbonate	Xanthogenic acid	$C_3H_6OS_2$	151-01-9	122.209	unstab liq	-53	25			
5015	Ethylene	Ethene	C_2H_4	74-85-1	28.053	col gas	-169.18(0.02)	-103.8(0.3)	0.5678[104]	1.363[-100]	i H$_2$O; sl EtOH, bz, ace; s eth
5016	Ethylenebisdithiocarbamic acid		$C_4H_8N_2S_4$	111-54-6	212.380	unstab liq					
5017	Ethylene carbonate	Vinylene carbonate	$C_3H_4O_3$	96-49-1	88.062	mcl pl (al)	36.331 (0.004)	246(1)	1.3214[39]	1.4148[50]	msc H$_2$O, EtOH, eth, bz, chl, AcOEt
5018	Ethylenediaminetetraacetic acid	EDTA	$C_{10}H_{16}N_2O_8$	60-00-4	292.242	cry (w)	245 dec				
5019	Ethylenediaminetetraacetic acid, disodium salt, dihydrate	EDTA disodium	$C_{10}H_{18}N_2Na_2O_{10}$	6381-92-6	372.237		242 dec				
5020	N,N'-Ethylene distearylamide	N,N'-Dioctadecanoylethanedi-amine	$C_{38}H_{76}N_2O_2$	110-30-5	593.022	cry (EtOH)	149				
5021	Ethyleneimine	Aziridine	C_2H_5N	151-56-4	43.068	liq	-78.0(0.5)	54(1)	0.832[25]		msc H$_2$O; s EtOH; vs eth; sl chl

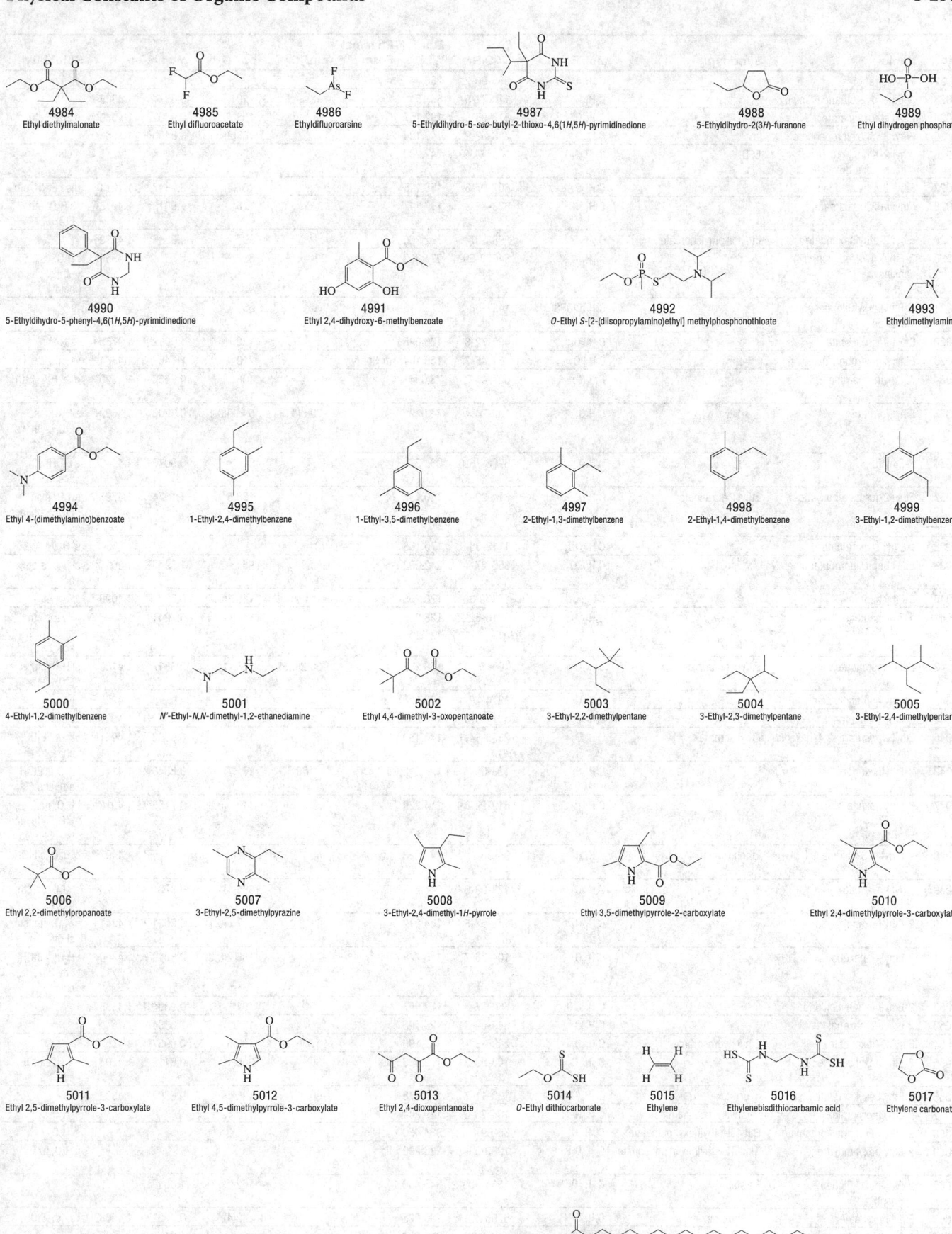

4984
Ethyl diethylmalonate

4985
Ethyl difluoroacetate

4986
Ethyldifluoroarsine

4987
5-Ethyldihydro-5-*sec*-butyl-2-thioxo-4,6(1*H*,5*H*)-pyrimidinedione

4988
5-Ethyldihydro-2(3*H*)-furanone

4989
Ethyl dihydrogen phosphate

4990
5-Ethyldihydro-5-phenyl-4,6(1*H*,5*H*)-pyrimidinedione

4991
Ethyl 2,4-dihydroxy-6-methylbenzoate

4992
O-Ethyl *S*-[2-(diisopropylamino)ethyl] methylphosphonothioate

4993
Ethyldimethylamine

4994
Ethyl 4-(dimethylamino)benzoate

4995
1-Ethyl-2,4-dimethylbenzene

4996
1-Ethyl-3,5-dimethylbenzene

4997
2-Ethyl-1,3-dimethylbenzene

4998
2-Ethyl-1,4-dimethylbenzene

4999
3-Ethyl-1,2-dimethylbenzene

5000
4-Ethyl-1,2-dimethylbenzene

5001
N′-Ethyl-*N*,*N*-dimethyl-1,2-ethanediamine

5002
Ethyl 4,4-dimethyl-3-oxopentanoate

5003
3-Ethyl-2,2-dimethylpentane

5004
3-Ethyl-2,3-dimethylpentane

5005
3-Ethyl-2,4-dimethylpentane

5006
Ethyl 2,2-dimethylpropanoate

5007
3-Ethyl-2,5-dimethylpyrazine

5008
3-Ethyl-2,4-dimethyl-1*H*-pyrrole

5009
Ethyl 3,5-dimethylpyrrole-2-carboxylate

5010
Ethyl 2,4-dimethylpyrrole-3-carboxylate

5011
Ethyl 2,5-dimethylpyrrole-3-carboxylate

5012
Ethyl 4,5-dimethylpyrrole-3-carboxylate

5013
Ethyl 2,4-dioxopentanoate

5014
O-Ethyl dithiocarbonate

5015
Ethylene

5016
Ethylenebisdithiocarbamic acid

5017
Ethylene carbonate

5018
Ethylenediaminetetraacetic acid

5019
Ethylenediaminetetraacetic acid, disodium salt, dihydrate

5020
N,*N*′-Ethylene distearylamide

5021
Ethyleneimine

No.	Name	Synonym	Mol. Form.	CAS RN	Mol. Wt.	Physical Form	mp/°C	bp/°C	den g cm⁻³	n_D	Solubility
5022	Ethylestrenol		$C_{20}H_{32}O$	965-90-2	288.467	cry	77				
5023	N-Ethyl-1,2-ethanediamine		$C_4H_{12}N_2$	110-72-5	88.151			129	0.837²⁵	1.4385²⁰	
5024	Ethyl ethoxyacetate		$C_6H_{12}O_3$	817-95-8	132.157			158	0.9702²⁰	1.4039²⁰	s EtOH, eth, ace
5025	Ethyl 3-ethoxypropanoate		$C_7H_{14}O_3$	763-69-9	146.184			168(2)	0.9490²⁰	1.4065²⁰	
5026	Ethyl 2-ethoxy-1(2H)-quinolinecarboxylate	EEDQ	$C_{14}H_{17}NO_3$	16357-59-8	247.290		56.5	126⁰·¹			s chl
5027	Ethyl 2-ethylacetoacetate		$C_8H_{14}O_3$	607-97-6	158.195			197(4)	0.9847¹⁶	1.4214²⁰	msc EtOH, eth
5028	Ethyl ethylcarbamate		$C_5H_{11}NO_2$	623-78-9	117.147			176	0.9813²⁰	1.4215²⁰	vs H₂O, eth, EtOH
5029	Ethyl 2-ethylhexanoate	Ethyl 2-ethylcaproate	$C_{10}H_{20}O_2$	2983-37-1	172.265			90²⁸	0.8586²⁵	1.4123²⁰	
5030	2-Ethyl-N-(2-ethylphenyl)-aniline		$C_{16}H_{19}N$	64653-59-4	225.329		29	336⁶⁰³		1.5550²⁵	i H₂O; vs EtOH, eth; sl chl; s acid
5031	O-Ethyl ethylthiophosphonyl chloride		$C_4H_{10}ClOPS$	1497-68-3	172.613	liq		35⁰·⁷	1.15²⁰		
5032	Ethyl fluoroacetate		$C_4H_7FO_2$	459-72-3	106.096			120	1.0912²⁰	1.3755²⁰	vs H₂O
5033	Ethyl 4-fluorobenzoate		$C_9H_9FO_2$	451-46-7	168.164	mcl pr (w)	26	210	1.146²⁵	1.4864²⁰	vs eth, EtOH
5034	N-Ethylformamide		C_3H_7NO	627-45-2	73.094			198	0.9552²⁰	1.4320²⁰	msc H₂O, EtOH, eth
5035	Ethyl formate		$C_3H_6O_2$	109-94-4	74.079	liq	-79.6(0.5)	54.09(0.1)	0.9208²⁰	1.3609²⁰	s H₂O; msc EtOH, eth; vs ace; sl ctc
5036	2-Ethylfuran		C_6H_8O	3208-16-0	96.127			92(3)	0.9018²⁰	1.4403²⁰	s EtOH, eth, ace, bz
5037	Ethyl 2-furancarboxylate	Ethyl 2-furanoate	$C_7H_8O_3$	614-99-3	140.137	lf or pr	34.5	196.8	1.1174²¹	1.4797²¹	i H₂O; msc EtOH, eth, ace; s bz
5038	γ-Ethyl L-glutamate		$C_7H_{13}NO_4$	1119-33-1	175.183		191				sl H₂O
5039	Ethyl heptafluorobutanoate		$C_6H_5F_7O_2$	356-27-4	242.092			95	1.394²⁰	1.3011²⁰	sl H₂O; s eth, ace
5040	3-Ethylheptane		C_9H_{20}	15869-80-4	128.255	liq	-114.9	143.1(0.7)	0.7225²⁵	1.4093²⁰	
5041	4-Ethylheptane		C_9H_{20}	2216-32-2	128.255			141(4)	0.7241²⁵	1.4096²⁰	i H₂O; s eth; msc EtOH, ace, bz
5042	Ethyl heptanoate	Ethyl oenanthate	$C_9H_{18}O_2$	106-30-9	158.238	liq	-66.2(0.5)	188(2)	0.8817²⁰	1.4100²⁰	sl H₂O, ctc; s EtOH, eth
5043	2-Ethylheptanoic acid		$C_9H_{18}O_2$	3274-29-1	158.238	liq		153³¹		1.4255²⁷	
5044	4-Ethyl-4-heptanol		$C_9H_{20}O$	597-90-0	144.254			179(7)	0.8350²⁰	1.4332²⁰	vs eth, EtOH
5045	Ethyl trans,trans-2,4-hexadienoate	Ethyl sorbate	$C_8H_{12}O_2$	2396-84-1	140.180			195.5	0.9506²⁰	1.4951²⁰	vs eth, EtOH, chl
5046	2-Ethylhexanal		$C_8H_{16}O$	123-05-7	128.212	liq	<-100	161(7)	0.8540²⁰	1.4142²⁰	i H₂O; s EtOH, eth; sl ctc
5047	3-Ethylhexane		C_8H_{18}	619-99-8	114.229			118.5(0.5)	0.7136²⁰	1.4018²⁰	i H₂O; msc EtOH, eth, ace, bz, chl; s ctc
5048	2-Ethyl-1,3-hexanediol	Ethohexadiol	$C_8H_{18}O_2$	94-96-2	146.228	liq	-40	243(8)	0.9325²²	1.4497²⁰	sl H₂O; s EtOH, eth
5049	Ethyl hexanoate		$C_8H_{16}O_2$	123-66-0	144.212	liq	-67.6(0.4)	165(1)	0.873²⁰	1.4073²⁰	sl H₂O; vs eth, EtOH
5050	2-Ethylhexanoic acid		$C_8H_{16}O_2$	149-57-5	144.212			227.5(0.1)	0.9031²⁵	1.4241²⁰	s H₂O, eth, ctc; sl EtOH
5051	2-Ethyl-1-hexanol		$C_8H_{18}O$	104-76-7	130.228	liq	-70	186.2(0.2)	0.8319²⁵	1.4300²⁰	i H₂O; s EtOH, eth, ace, bz, chl
5052	2-Ethylhexanoyl chloride		$C_8H_{15}ClO$	760-67-8	162.657			101⁴⁰	0.939²⁵	1.4335²⁰	
5053	2-Ethyl-2-hexenal		$C_8H_{14}O$	645-62-5	126.196			175(1)	0.8554²⁰		
5054	Ethyl 3-hexenoate	Ethyl hydrosorbate	$C_8H_{14}O_2$	2396-83-0	142.196			166.5	0.8957²⁰	1.4255²⁰	
5055	2-Ethylhexyl acetate		$C_{10}H_{20}O_2$	103-09-3	172.265	liq	-80	200(1)	0.8718²⁰	1.4204²⁰	i H₂O; s EtOH, eth
5056	2-Ethylhexyl acrylate		$C_{11}H_{20}O_2$	103-11-7	184.276		-90	125⁶⁰	0.880²⁵	1.4332²⁵	
5057	2-Ethylhexylamine	2-Ethyl-1-hexanamine	$C_8H_{19}N$	104-75-6	129.244			172(12)			sl H₂O
5058	2-Ethylhexyl butyl phthalate	Butyl 2-ethylhexyl phthalate	$C_{20}H_{30}O_4$	85-69-8	334.450	col liq					sl H₂O
5059	2-Ethylhexyl dihydrogen phosphate	Mono(2-ethylhexyl) phosphate	$C_8H_{19}O_4P$	1070-03-7	210.208	liq					s H₂O, bz
5060	2-Ethylhexyl diphenyl phosphite	Forstab	$C_{20}H_{27}O_3P$	15647-08-2	346.400			152⁰·¹⁵	1.054²⁰	1.5207²⁷	
5061	Ethyl hexyl ether	1-Ethoxyhexane	$C_8H_{18}O$	5756-43-4	130.228			142(4)	0.7722²⁰	1.4008²⁰	vs eth, EtOH
5062	2-Ethylhexyl 2-hydroxybenzoate	Octisalate	$C_{15}H_{22}O_3$	118-60-5	250.334	liq		190²¹	1.01		
5063	2-Ethylhexyl methacrylate		$C_{12}H_{22}O_2$	688-84-6	198.302			120¹⁸	0.880²⁵	1.436²⁵	
5064	2-[(2-Ethylhexyl)oxy]ethanol	Ethylene glycol mono(2-ethylhexyl) ether	$C_{10}H_{22}O_2$	1559-35-9	174.281			231(4)			

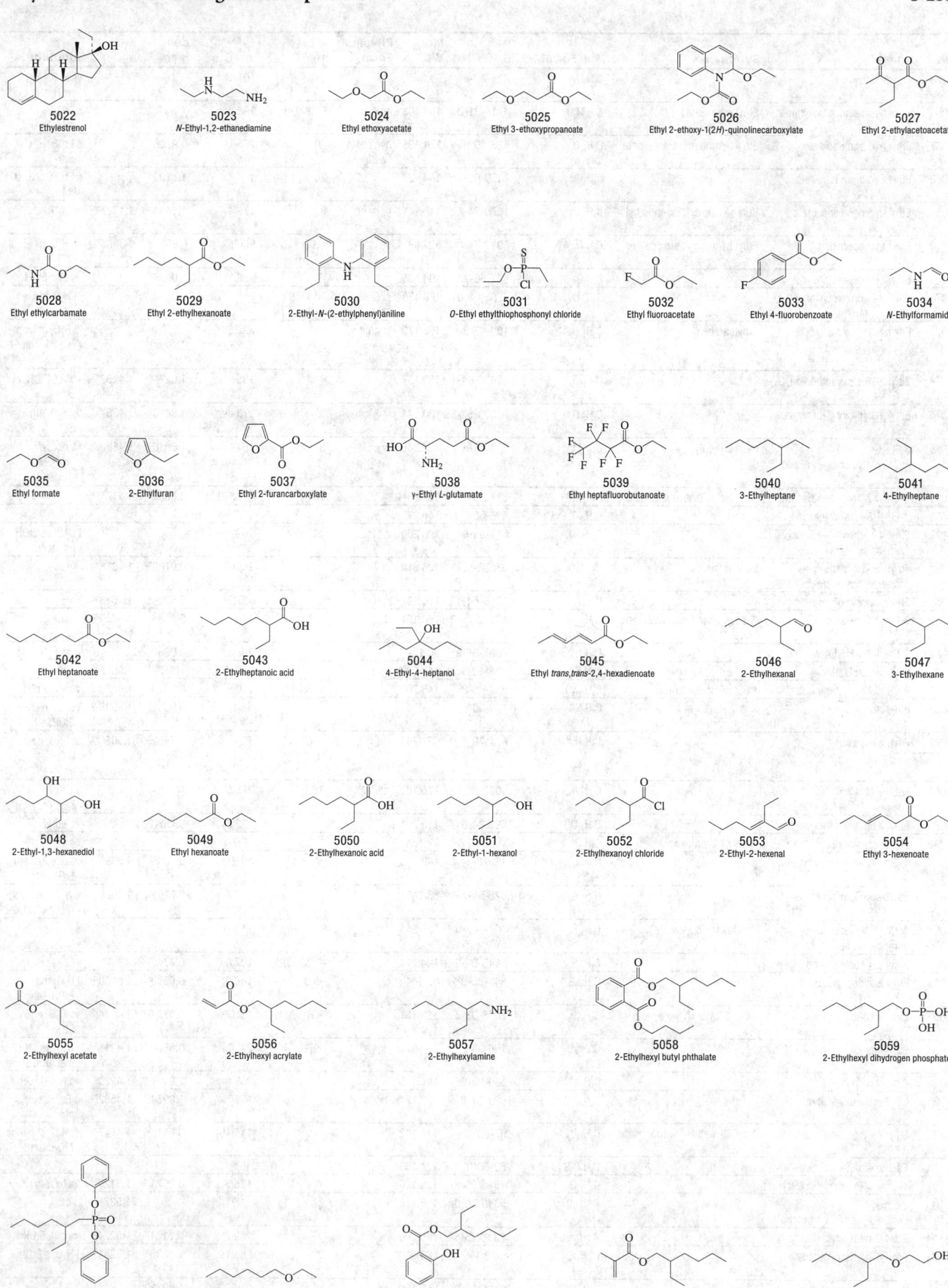

5022
Ethylestrenol

5023
N-Ethyl-1,2-ethanediamine

5024
Ethyl ethoxyacetate

5025
Ethyl 3-ethoxypropanoate

5026
Ethyl 2-ethoxy-1(2*H*)-quinolinecarboxylate

5027
Ethyl 2-ethylacetoacetate

5028
Ethyl ethylcarbamate

5029
Ethyl 2-ethylhexanoate

5030
2-Ethyl-*N*-(2-ethylphenyl)aniline

5031
O-Ethyl ethylthiophosphonyl chloride

5032
Ethyl fluoroacetate

5033
Ethyl 4-fluorobenzoate

5034
N-Ethylformamide

5035
Ethyl formate

5036
2-Ethylfuran

5037
Ethyl 2-furancarboxylate

5038
γ-Ethyl *L*-glutamate

5039
Ethyl heptafluorobutanoate

5040
3-Ethylheptane

5041
4-Ethylheptane

5042
Ethyl heptanoate

5043
2-Ethylheptanoic acid

5044
4-Ethyl-4-heptanol

5045
Ethyl *trans,trans*-2,4-hexadienoate

5046
2-Ethylhexanal

5047
3-Ethylhexane

5048
2-Ethyl-1,3-hexanediol

5049
Ethyl hexanoate

5050
2-Ethylhexanoic acid

5051
2-Ethyl-1-hexanol

5052
2-Ethylhexanoyl chloride

5053
2-Ethyl-2-hexenal

5054
Ethyl 3-hexenoate

5055
2-Ethylhexyl acetate

5056
2-Ethylhexyl acrylate

5057
2-Ethylhexylamine

5058
2-Ethylhexyl butyl phthalate

5059
2-Ethylhexyl dihydrogen phosphate

5060
2-Ethylhexyl diphenyl phosphite

5061
Ethyl hexyl ether

5062
2-Ethylhexyl 2-hydroxybenzoate

5063
2-Ethylhexyl methacrylate

5064
2-[(2-Ethylhexyl)oxy]ethanol

No.	Name	Synonym	Mol. Form.	CAS RN	Mol. Wt.	Physical Form	mp/°C	bp/°C	den g cm⁻³	n_D	Solubility
5065	Ethylhydrazine		$C_2H_8N_2$	624-80-6	60.098			101			vs H_2O, ace, eth, EtOH
5066	Ethyl hydrazinecarboxylate	Ethyl carbazate	$C_3H_8N_2O_2$	4114-31-2	104.108	cry	45.8(0.4)	198 dec			s EtOH, eth; sl chl
5067	Ethyl hydrogen adipate	Ethyl hydrogen hexanedioate	$C_8H_{14}O_4$	626-86-8	174.195	hyg cry (eth, peth)	29	285	0.9796²⁰	1.4311²⁰	s EtOH, eth, peth
5068	Ethyl hydrogen fumarate		$C_6H_8O_4$	2459-05-4	144.126		70	147¹⁶	1.1109⁸⁷		s EtOH, ace; sl chl
5069	Ethyl hydrogen succinate	Butanedioic acid, monoethyl ester	$C_6H_{10}O_4$	1070-34-4	146.141	pr or nd	8	172⁴²	1.1466²⁰	1.4327²⁰	vs H_2O, eth, EtOH
5070	Ethyl hydroperoxide	Ethyl hydrogen peroxide	$C_2H_6O_2$	3031-74-1	62.068	liq	-100	95(8)	0.9332²⁰	1.3800²⁰	vs H_2O, bz, eth, EtOH
5071	Ethyl hydroxyacetate		$C_4H_8O_3$	623-50-7	104.105			156(3)	1.0826²³	1.4180²⁰	vs eth, EtOH
5072	Ethyl 3-hydroxybenzoate		$C_9H_{10}O_3$	7781-98-8	166.173	pl (bz)	74		1.0680¹³¹		sl H_2O, chl; s EtOH, eth
5073	Ethyl 4-hydroxybenzoate	Ethylparaben	$C_9H_{10}O_3$	120-47-8	166.173	cry (dil al)	117	297.5			sl H_2O, chl, tfa; vs EtOH, eth; i CS_2
5074	Ethyl 3-hydroxybutanoate, (±)-		$C_6H_{12}O_3$	35608-64-1	132.157			185	1.017²⁰	1.4182²⁰	s H_2O, EtOH; sl ctc
5075	Ethyl 2-hydroxy-3-butenoate		$C_6H_{10}O_3$	91890-87-8	130.141			173 dec	1.0470¹⁵	1.436¹³	vs H_2O, eth, EtOH
5076	α-Ethyl-1-hydroxycyclohexaneacetic acid	Cyclobutyrol	$C_{10}H_{18}O_3$	512-16-3	186.248	cry (eth-peth)	81.5	164²⁴	1.0010¹⁸	1.4680¹⁸	vs ace, eth, EtOH, chl
5077	N-Ethyl-N-hydroxyethanamine	N,N-Diethylhydroxylamine	$C_4H_{11}NO$	3710-84-7	89.136		10	131.0(0.3)	0.8669²⁰	1.4195²⁰	
5078	2-Ethyl-3-hydroxyhexanal		$C_8H_{16}O_2$	496-03-7	144.212			138⁵⁰			
5079	Ethyl 4-hydroxy-3-methoxybenzoate		$C_{10}H_{12}O_4$	617-05-0	196.200	nd (dil al)	44	292			i H_2O; vs EtOH, eth; s chl
5080	Ethyl cis-12-hydroxy-9-octadecenoate, (R)-	Ethyl ricinoleate	$C_{20}H_{38}O_3$	55066-53-0	326.514			258¹³	0.9180²⁰	1.4618²²	
5081	Ethylidenecyclohexane		C_8H_{14}	1003-64-1	110.197			137(3)	0.822²⁵	1.4618²⁰	
5082	5-Ethylidene-2-norbornene	5-Ethylidenebicyclo[2.2.1]-hept-2-ene	C_9H_{12}	16219-75-3	120.191	liq		146	0.893	1.4900²⁰	
5083	1-Ethyl-1H-imidazole		$C_5H_8N_2$	7098-07-9	96.131			208	0.999²⁵		msc H_2O
5084	Ethyl iodoacetate		$C_4H_7IO_2$	623-48-3	214.002	oil		179(7)	1.8173¹³	1.5079¹³	s EtOH, eth
5085	Ethyl isobutylcarbamate	Isobutyl urethane	$C_7H_{15}NO_2$	539-89-9	145.200		<-65	110³⁰	0.9432²⁰	1.4288²⁰	vs eth, EtOH
5086	Ethyl isocyanate		C_3H_5NO	109-90-0	71.078			56(1)	0.9031²⁰	1.3808²⁰	i H_2O; msc EtOH, eth
5087	Ethyl isocyanide		C_3H_5N	624-79-3	55.079		<-66	79	0.7402²⁰	1.3622²⁰	vs H_2O; msc EtOH, eth; s ace
5088	N-Ethyl-1H-isoindole-1,3(2H)-dione		$C_{10}H_9NO_2$	5022-29-7	175.184	nd (al)	79	285.5			s EtOH, eth
5089	Ethyl isopentyl ether		$C_7H_{16}O$	628-04-6	116.201			112(5)	0.7688²¹		vs eth, EtOH
5090	Ethylisopropylamine	N-Ethyl-2-propanamine	$C_5H_{13}N$	19961-27-4	87.164			68(7)		1.3872²⁵	
5091	1-Ethyl-2-isopropylbenzene		$C_{11}H_{16}$	18970-44-0	148.245			193(7)	0.888²⁰	1.508²⁰	vs ace, bz, eth, EtOH
5092	Ethyl isopropyl ether		$C_5H_{12}O$	625-54-7	88.148			54(3)	0.720²⁵	1.3698²⁵	s H_2O, ace, chl; msc EtOH, eth
5093	N-Ethyl-N-isopropyl-2-propanamine	Ethyldiisopropylamine	$C_8H_{19}N$	7087-68-5	129.244			114(10)	0.742²⁵	1.4138²⁰	s ctc
5094	Ethyl isopropyl sulfide		$C_5H_{12}S$	5145-99-3	104.214	liq	-122.2(0.2)	107.3(0.6)	0.8246²⁰		
5095	Ethyl isothiocyanate		C_3H_5NS	542-85-8	87.144	liq	-5.9	140(9)	0.9990²⁰	1.5130²⁰	i H_2O; msc EtOH, eth
5096	Ethyl lactate	Ethyl 2-hydroxypropionate	$C_5H_{10}O_3$	2676-33-7	118.131	liq	-26	151(7)	1.0328²⁰	1.4124²⁰	vs H_2O, eth, EtOH
5097	Ethyl laurate		$C_{14}H_{28}O_2$	106-33-2	228.371	liq	-9(9)	275(6)	0.8618²⁰	1.4311²⁰	i H_2O; vs EtOH; msc eth; sl ctc
5098	Ethyl levulinate		$C_7H_{12}O_3$	539-88-8	144.168			205.8	1.0111²⁰	1.4229²⁰	vs H_2O, EtOH
5099	Ethyl mercaptoacetate		$C_4H_8O_2S$	623-51-8	120.171			157	1.0964¹⁵	1.4582²⁰	s EtOH, eth; sl ctc
5100	Ethyl methacrylate	Ethyl 2-methyl-2-propenoate	$C_6H_{10}O_2$	97-63-2	114.142			116(19)	0.9135²⁰	1.4147²⁰	sl H_2O, chl; msc EtOH, eth
5101	Ethyl methanesulfonate		$C_3H_8O_3S$	62-50-0	124.159			86¹⁰			
5102	1-Ethyl-4-methoxybenzene		$C_9H_{12}O$	1515-95-3	136.190			196(8)	0.9624¹⁵	1.5120²⁰	vs bz, eth
5103	α-Ethyl-4-methoxybenzenemethanol		$C_{10}H_{14}O_2$	5349-60-0	166.217			143²⁰		1.5277²⁰	s ctc
5104	Ethyl 2-methoxybenzoate		$C_{10}H_{12}O_3$	7335-26-4	180.200			262(7)	1.1124²⁰	1.5224²⁰	vs eth, EtOH
5105	Ethyl 4-methoxybenzoate		$C_{10}H_{12}O_3$	94-30-4	180.200		7.5	266(19)	1.1038²⁰	1.5254²⁰	i H_2O; s EtOH, eth
5106	4-Ethyl-2-methoxyphenol		$C_9H_{12}O_2$	2785-89-9	152.190	liq	-7	236.5	1.0931¹⁸		

5065 Ethylhydrazine

5066 Ethyl hydrazinecarboxylate

5067 Ethyl hydrogen adipate

5068 Ethyl hydrogen fumarate

5069 Ethyl hydrogen succinate

5070 Ethyl hydroperoxide

5071 Ethyl hydroxyacetate

5072 Ethyl 3-hydroxybenzoate

5073 Ethyl 4-hydroxybenzoate

5074 Ethyl 3-hydroxybutanoate, (±)-

5075 Ethyl 2-hydroxy-3-butenoate

5076 α-Ethyl-1-hydroxycyclohexaneacetic acid

5077 N-Ethyl-N-hydroxyethanamine

5078 2-Ethyl-3-hydroxyhexanal

5079 Ethyl 4-hydroxy-3-methoxybenzoate

5080 Ethyl cis-12-hydroxy-9-octadecenoate, (R)-

5081 Ethylidenecyclohexane

5082 5-Ethylidene-2-norbornene

5083 1-Ethyl-1H-imidazole

5084 Ethyl iodoacetate

5085 Ethyl isobutylcarbamate

5086 Ethyl isocyanate

5087 Ethyl isocyanide

5088 N-Ethyl-1H-isoindole-1,3(2H)-dione

5089 Ethyl isopentyl ether

5090 Ethylisopropylamine

5091 1-Ethyl-2-isopropylbenzene

5092 Ethyl isopropyl ether

5093 N-Ethyl-N-isopropyl-2-propanamine

5094 Ethyl isopropyl sulfide

5095 Ethyl isothiocyanate

5096 Ethyl lactate

5097 Ethyl laurate

5098 Ethyl levulinate

5099 Ethyl mercaptoacetate

5100 Ethyl methacrylate

5101 Ethyl methanesulfonate

5102 1-Ethyl-4-methoxybenzene

5103 α-Ethyl-4-methoxybenzenemethanol

5104 Ethyl 2-methoxybenzoate

5105 Ethyl 4-methoxybenzoate

5106 4-Ethyl-2-methoxyphenol

No.	Name	Synonym	Mol. Form.	CAS RN	Mol. Wt.	Physical Form	mp/°C	bp/°C	den g cm⁻³	n_D	Solubility
5107	Ethyl (4-methoxyphenyl)-acetate		$C_{11}H_{14}O_3$	14062-18-1	194.227			139[70]	1.097[25]	1.5075[20]	
5108	Ethyl 2-methylacetoacetate		$C_7H_{12}O_3$	609-14-3	144.168			187	0.9941[20]	1.4185[20]	sl H_2O; s EtOH, eth; vs ace
5109	N-Ethyl-2-methylallylamine	N-Ethyl-2-methyl-2-propen-1-amine	$C_6H_{13}N$	18328-90-0	99.174	liq		104.7	0.753	1.4221[20]	msc H_2O
5110	5-Ethyl-5-(2-methylallyl)-2-thiobarbituric acid	Methallatal	$C_{10}H_{14}N_2O_2S$	115-56-0	226.295		160.5				
5111	Ethylmethylamine	N-Methylethanamine	C_3H_9N	624-78-2	59.110			34.2(0.2)			vs H_2O, ace, eth, EtOH
5112	Ethylmethylamine hydrochloride	N-Methylethanamine hydrochloride	$C_3H_{10}ClN$	624-60-2	95.571	pl (al-eth)	128		1.0874[20]		vs H_2O, EtOH; i eth; s chl
5113	2-Ethyl-6-methylaniline		$C_9H_{13}N$	24549-06-2	135.206	liq	-33	231	0.968[25]	1.5525[20]	
5114	N-Ethyl-2-methylaniline		$C_9H_{13}N$	94-68-8	135.206		<-15	216	0.948[25]	1.5456[20]	s EtOH, eth
5115	N-Ethyl-3-methylaniline		$C_9H_{13}N$	102-27-2	135.206			221	0.9263[15]	1.5451[20]	s EtOH, eth
5116	N-Ethyl-4-methylaniline	N-Ethyl-4-toluidine	$C_9H_{13}N$	622-57-1	135.206			217	0.9391[16]		s EtOH, eth
5117	N-Ethyl-N-methylaniline		$C_9H_{13}N$	613-97-8	135.206			204(5)	0.92[55]		i H_2O; msc EtOH, eth; s ctc
5118	N-Ethyl-α-methylbenzeneethanamine	N-Ethylamphetamine	$C_{11}H_{17}N$	457-87-4	163.260			105[14]		1.4986[25]	
5119	N-Ethyl-4-methylbenzenesulfonamide		$C_9H_{13}NO_2S$	80-39-7	199.270		64				s EtOH
5120	1-Ethyl-2-methyl-1H-benzimidazole		$C_{10}H_{12}N_2$	5805-76-5	160.215		51	296	1.073[25]		
5121	Ethyl 2-methylbenzoate		$C_{10}H_{12}O_2$	87-24-1	164.201		<-10	227(8)	1.0325[21]	1.507[22]	i H_2O; msc EtOH, eth
5122	Ethyl 4-methylbenzoate		$C_{10}H_{12}O_2$	94-08-6	164.201			236(7)	1.0269[18]	1.5089[18]	i H_2O; msc EtOH, eth
5123	Ethyl 3-methylbutanoate	Ethyl isovalerate	$C_7H_{14}O_2$	108-64-5	130.185	liq	-99.3	135(3)	0.8656[20]	1.3962[20]	sl H_2O; vs EtOH, eth
5124	2-Ethyl-2-methylbutanoic acid		$C_7H_{14}O_2$	19889-37-3	130.185		<-20	207		1.4250[20]	vs EtOH
5125	2-Ethyl-3-methyl-1-butene		C_7H_{14}	7357-93-9	98.186			86.3(0.5)	0.7150[20]	1.410[20]	i H_2O; s eth, ace, bz, chl
5126	Ethyl trans-2-methyl-2-butenoate		$C_7H_{12}O_2$	5837-78-5	128.169			156	0.9200[20]	1.4340[20]	
5127	Ethyl 3-methyl-2-butenoate		$C_7H_{12}O_2$	638-10-8	128.169			153.5	0.9199[21]	1.4345[20]	
5128	5-Ethyl-5-(1-methylbutyl)-2,4,6(1H,3H,5H)-pyrimidine-trione		$C_{11}H_{18}N_2O_3$	76-74-4	226.272		130(1)				sl H_2O; s EtOH, eth
5129	Ethyl N-methylcarbamate		$C_4H_9NO_2$	105-40-8	103.120			169(5)	1.0115[20]	1.4183[20]	vs H_2O, EtOH
5130	Ethyl methyl carbonate		$C_4H_8O_3$	623-53-0	104.105	liq	-14	107.5	1.012[20]	1.3778[20]	vs eth, EtOH
5131	trans-1-Ethyl-4-methylcyclohexane		C_9H_{18}	6236-88-0	126.239	liq	-80.8	147(6)	0.7798[20]	1.4304[20]	
5132	1-Ethyl-1-methylcyclopentane		C_8H_{16}	16747-50-5	112.213	liq	-143.8(0.2)	121.5(0.6)	0.7767[25]	1.4272[20]	vs ace, bz, eth, EtOH
5133	cis-1-Ethyl-2-methylcyclopentane		C_8H_{16}	930-89-2	112.213	liq	-105.9(0.1)	128(1)	0.7852[20]	1.4293[20]	
5134	trans-1-Ethyl-2-methylcyclopentane		C_8H_{16}	930-90-5	112.213	liq	-109(5)	121(4)	0.7649[25]	1.4219[20]	
5135	cis-1-Ethyl-3-methylcyclopentane		C_8H_{16}	2613-66-3	112.213			121(4)	0.7724[20]	1.4203[20]	vs ace, bz, eth, EtOH
5136	trans-1-Ethyl-3-methylcyclopentane		C_8H_{16}	2613-65-2	112.213	liq	-108	121(4)	0.7619[20]	1.4186[20]	
5137	1-Ethyl-1-methylcyclopropane		C_6H_{12}	53778-43-1	84.159	liq	-130.2(0.2)	57(4)	0.6968[25]	1.3887[20]	
5138	2-Ethyl-2-methyl-1,3-dioxolane		$C_6H_{12}O_2$	126-39-6	116.158			119(3)	0.9360[20]		
5139	Ethyl methyl ether		C_3H_8O	540-67-0	60.095	col gas	-113	6(2)	0.7251[0]	1.3420[4]	s H_2O, ace, chl; msc EtOH, eth
5140	3-Ethyl-2-methylhexane		C_9H_{20}	16789-46-1	128.255			138(4)	0.7310[20]	1.4106[20]	
5141	3-Ethyl-3-methylhexane		C_9H_{20}	3074-76-8	128.255			140(3)	0.7371[25]	1.4140[20]	
5142	3-Ethyl-4-methylhexane	2,3-Diethylpentane	C_9H_{20}	3074-77-9	128.255			140(4)	0.7420[20]	1.4134[20]	
5143	4-Ethyl-2-methylhexane		C_9H_{20}	3074-75-7	128.255			134(3)	0.7195[20]	1.4063[20]	
5144	Ethyl 4-methylhexanoate	Ethyl 4-methylcaproate	$C_9H_{18}O_2$	1561-10-0	158.238			180	0.8708[20]	1.4051[20]	
5145	Ethyl 4-methyl-3-oxopentanoate		$C_8H_{14}O_3$	7152-15-0	158.195	liq	-9	173	0.98[25]	1.250[20]	
5146	3-Ethyl-2-methylpentane	2-Methyl-3-ethylpentane	C_8H_{18}	609-26-7	114.229	liq	-115.0(0.1)	115.6(0.6)	0.7193[20]	1.4040[20]	i H_2O; s eth; msc EtOH, ace, bz
5147	3-Ethyl-3-methylpentane	3-Methyl-3-ethylpentane	C_8H_{18}	1067-08-9	114.229	liq	-90.8(0.1)	118.2(0.9)	0.7274[20]	1.4078[20]	i H_2O; s eth; msc EtOH, ace, bz
5148	Ethyl 4-methylpentanoate		$C_8H_{16}O_2$	25415-67-2	144.212			163	0.8705[20]	1.4050[20]	

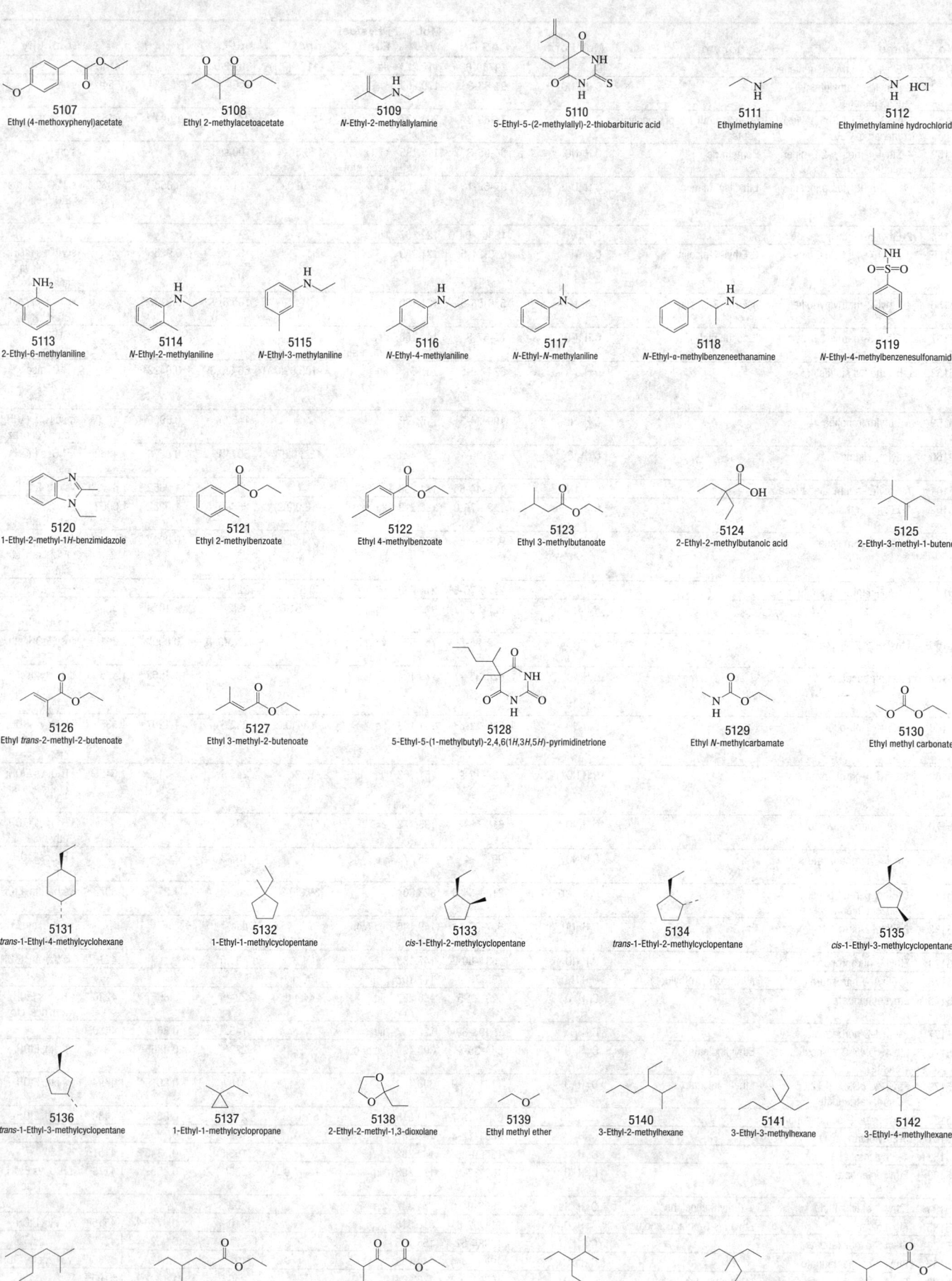

5107
Ethyl (4-methoxyphenyl)acetate

5108
Ethyl 2-methylacetoacetate

5109
N-Ethyl-2-methylallylamine

5110
5-Ethyl-5-(2-methylallyl)-2-thiobarbituric acid

5111
Ethylmethylamine

5112
Ethylmethylamine hydrochloride

5113
2-Ethyl-6-methylaniline

5114
N-Ethyl-2-methylaniline

5115
N-Ethyl-3-methylaniline

5116
N-Ethyl-4-methylaniline

5117
N-Ethyl-*N*-methylaniline

5118
N-Ethyl-α-methylbenzeneethanamine

5119
N-Ethyl-4-methylbenzenesulfonamide

5120
1-Ethyl-2-methyl-1*H*-benzimidazole

5121
Ethyl 2-methylbenzoate

5122
Ethyl 4-methylbenzoate

5123
Ethyl 3-methylbutanoate

5124
2-Ethyl-2-methylbutanoic acid

5125
2-Ethyl-3-methyl-1-butene

5126
Ethyl *trans*-2-methyl-2-butenoate

5127
Ethyl 3-methyl-2-butenoate

5128
5-Ethyl-5-(1-methylbutyl)-2,4,6(1*H*,3*H*,5*H*)-pyrimidinetrione

5129
Ethyl *N*-methylcarbamate

5130
Ethyl methyl carbonate

5131
trans-1-Ethyl-4-methylcyclohexane

5132
1-Ethyl-1-methylcyclopentane

5133
cis-1-Ethyl-2-methylcyclopentane

5134
trans-1-Ethyl-2-methylcyclopentane

5135
cis-1-Ethyl-3-methylcyclopentane

5136
trans-1-Ethyl-3-methylcyclopentane

5137
1-Ethyl-1-methylcyclopropane

5138
2-Ethyl-2-methyl-1,3-dioxolane

5139
Ethyl methyl ether

5140
3-Ethyl-2-methylhexane

5141
3-Ethyl-3-methylhexane

5142
3-Ethyl-4-methylhexane

5143
4-Ethyl-2-methylhexane

5144
Ethyl 4-methylhexanoate

5145
Ethyl 4-methyl-3-oxopentanoate

5146
3-Ethyl-2-methylpentane

5147
3-Ethyl-3-methylpentane

5148
Ethyl 4-methylpentanoate

No.	Name	Synonym	Mol. Form.	CAS RN	Mol. Wt.	Physical Form	mp/°C	bp/°C	den g cm⁻³	n_D	Solubility
5149	3-Ethyl-2-methyl-1-pentene		C_8H_{16}	19780-66-6	112.213	liq	-112.9(0.1)	109.2(0.9)	0.7262²⁰	1.4140²⁰	
5150	2-[Ethyl(3-methylphenyl)-amino]ethanol		$C_{11}H_{17}NO$	91-88-3	179.259			118¹·⁵		1.5540²⁰	s ctc
5151	Ethyl 3-methyl-3-phenyloxi-ranecarboxylate	Ethyl 3-methyl-3-phenylglyci-date	$C_{12}H_{14}O_3$	77-83-8	206.237			273.5	1.044²⁰	1.5182²⁰	
5152	4-Ethyl-4-methyl-2,6-piperi-dinedione	Bemegride	$C_8H_{13}NO_2$	64-65-3	155.195	pl (w, ace-eth)	126.5	100 sub			s chl
5153	Ethyl 2-methylpropanoate	Ethyl isobutanoate	$C_6H_{12}O_2$	97-62-1	116.158	liq	-97.8(0.4)	111(2)	0.868²⁰	1.3869¹⁸	sl H₂O, ctc; msc EtOH, eth; s ace
5154	2-Ethyl-5-methylpyrazine		$C_7H_{10}N_2$	13360-64-0	122.167	liq		79⁵⁶			
5155	3-Ethyl-4-methylpyridine	3-Ethyl-4-picoline	$C_8H_{11}N$	529-21-5	121.180			198	0.9286¹⁷		sl H₂O; s EtOH, eth, chl; vs ace
5156	4-Ethyl-2-methylpyridine	4-Ethyl-2-picoline	$C_8H_{11}N$	536-88-9	121.180			180(6)	0.9130²⁵		vs ace, bz, eth, EtOH
5157	3-Ethyl-3-methyl-2,5-pyrro-lidinedione	Ethosuximide	$C_7H_{11}NO_2$	77-67-8	141.168	cry (ace-eth)	64.5				vs H₂O
5158	Ethyl methyl sulfide		C_3H_8S	624-89-5	76.161	liq	-105.89(0.1)	66.6(0.3)	0.8422²⁰	1.4404²⁰	i H₂O; msc EtOH; s eth, chl
5159	N-Ethylmorpholine		$C_6H_{13}NO$	100-74-3	115.173			145(5)	0.8996²⁰	1.4400²⁰	msc H₂O, EtOH, eth; s ace, bz
5160	Ethyl myristate		$C_{16}H_{32}O_2$	124-06-1	256.424		12.3(0.8)	307(4)	0.8573²⁵	1.4362²⁰	i H₂O; s EtOH, ctc, lig; sl eth
5161	N-Ethyl-1-naphthalenamine		$C_{12}H_{13}N$	118-44-5	171.238			305	1.0652¹⁵	1.6477¹⁵	vs eth, EtOH
5162	1-Ethylnaphthalene		$C_{12}H_{12}$	1127-76-0	156.223	liq	-13.9(0.2)	258(3)	1.0082²⁰	1.6062²⁰	i H₂O; msc EtOH, eth
5163	2-Ethylnaphthalene		$C_{12}H_{12}$	939-27-5	156.223	liq	-7.4(0.9)	259(2)	0.9922²⁰	1.5999²⁰	i H₂O; msc EtOH, eth; sl chl
5164	Ethyl 1-naphthylacetate		$C_{14}H_{14}O_2$	2122-70-5	214.260	oil	88.5	222²⁰			s EtOH, eth
5165	Ethyl nitrate		$C_2H_5NO_3$	625-58-1	91.066	liq	-94.51(0.1)	89(2)	1.1084²⁰	1.3852²⁰	s H₂O; msc EtOH, eth
5166	Ethyl nitrite		$C_2H_5NO_2$	109-95-5	75.067	ye vol liq or gas		17.5(0.4)	0.899¹⁵	1.3418¹⁰	msc EtOH, eth
5167	Ethyl nitroacetate		$C_4H_7NO_4$	626-35-7	133.104			106²⁵	1.1953²⁰	1.4250²⁰	sl H₂O; msc EtOH; vs eth; s dil alk
5168	1-Ethyl-2-nitrobenzene		$C_8H_9NO_2$	612-22-6	151.163	liq	-12.2(0.2)	232.5	1.1207²⁰	1.5356²⁰	i H₂O; vs EtOH, eth; s ace; sl ctc
5169	1-Ethyl-4-nitrobenzene		$C_8H_9NO_2$	100-12-9	151.163	liq	-12.3	245.5	1.1192²⁰	1.5455²⁰	i H₂O; vs EtOH, eth; s ace; sl ctc
5170	Ethyl 3-nitrobenzoate		$C_9H_9NO_4$	618-98-4	195.172		47	297			i H₂O; vs EtOH, eth
5171	Ethyl 4-nitrobenzoate		$C_9H_9NO_4$	99-77-4	195.172		57	186.3			i H₂O; s EtOH, eth
5172	O-Ethyl O-p-nitrophenyl benzenethiophosphonate		$C_{14}H_{14}NO_4PS$	2104-64-5	323.304		39.2(0.3)		1.27²⁵	1.5978³⁰	vs bz, eth, EtOH
5173	2-Ethyl-2-nitro-1,3-propane-diol		$C_5H_{11}NO_4$	597-09-1	149.146	nd (w)	57.5	dec			vs H₂O, eth, EtOH
5174	Ethyl 2-nitropropanoate		$C_5H_9NO_4$	2531-80-8	147.130			190.5		1.4210²⁰	vs bz, eth, EtOH
5175	N-Ethyl-N-nitrosourea	N-Nitroso-N-ethylurea	$C_3H_7N_3O_2$	759-73-9	117.107		100 dec				s chl
5176	Ethyl nonanoate		$C_{11}H_{22}O_2$	123-29-5	186.292	liq	-44.4(0.3)	224(5)	0.8657²⁰	1.4220²⁰	i H₂O; s EtOH, eth, ace, ctc
5177	5-Ethyl-2-norbornene		C_9H_{14}	15403-89-1	122.207	liq		143.6	0.86	1.4630²⁰	
5178	Ethyl cis,cis-9,12-octadecadi-enoate	Ethyl linoleate	$C_{20}H_{36}O_2$	544-35-4	308.499	ye or col		272¹⁸⁰	0.8865²⁰		vs eth, EtOH
5179	Ethyl cis,cis,cis-9,12,15-octadecatrienoate	Ethyl linolenate	$C_{20}H_{34}O_2$	1191-41-9	306.483			218¹⁵	0.8919²⁰	1.4694²⁰	vs eth, EtOH
5180	Ethyl trans-9-octadecenoate		$C_{20}H_{38}O_2$	6114-18-7	310.515		5.1(0.3)	218¹⁵	0.8664²⁵	1.4480²⁵	vs eth, EtOH
5181	3-Ethyloctane		$C_{10}H_{22}$	5881-17-4	142.282			165(6)	0.7359²⁵	1.4156²⁰	
5182	4-Ethyloctane		$C_{10}H_{22}$	15869-86-0	142.282			163(6)	0.7343²⁵	1.4151²⁰	
5183	Ethyl octanoate		$C_{10}H_{20}O_2$	106-32-1	172.265	liq	-44.7(0.3)	206(1)	0.866¹⁸	1.4178²⁰	i H₂O; vs EtOH, eth; sl ctc
5184	Ethyl 1-octyl sulfide	1-(Ethylthio)octane	$C_{10}H_{22}S$	3698-94-0	174.347	liq		109¹⁴			
5185	Ethyl oleate	Ethyl cis-9-octadecenoate	$C_{20}H_{38}O_2$	111-62-6	310.515			216¹⁵	0.8720²⁰	1.4515²⁰	vs eth, EtOH
5186	Ethyl 5-oxohexanoate		$C_8H_{14}O_3$	13984-57-1	158.195			221.5	0.989²⁵	1.4277²⁰	
5187	Ethyl 3-oxopentanoate		$C_7H_{12}O_3$	4949-44-4	144.168			191	1.0120²⁰	1.4230²⁰	vs bz, eth, EtOH
5188	Ethyl 2-oxo-2-phenylacetate	Ethyl phenylglyoxylate	$C_{10}H_{10}O_3$	1603-79-8	178.184			256.5	1.1222²⁵	1.5190²⁵	

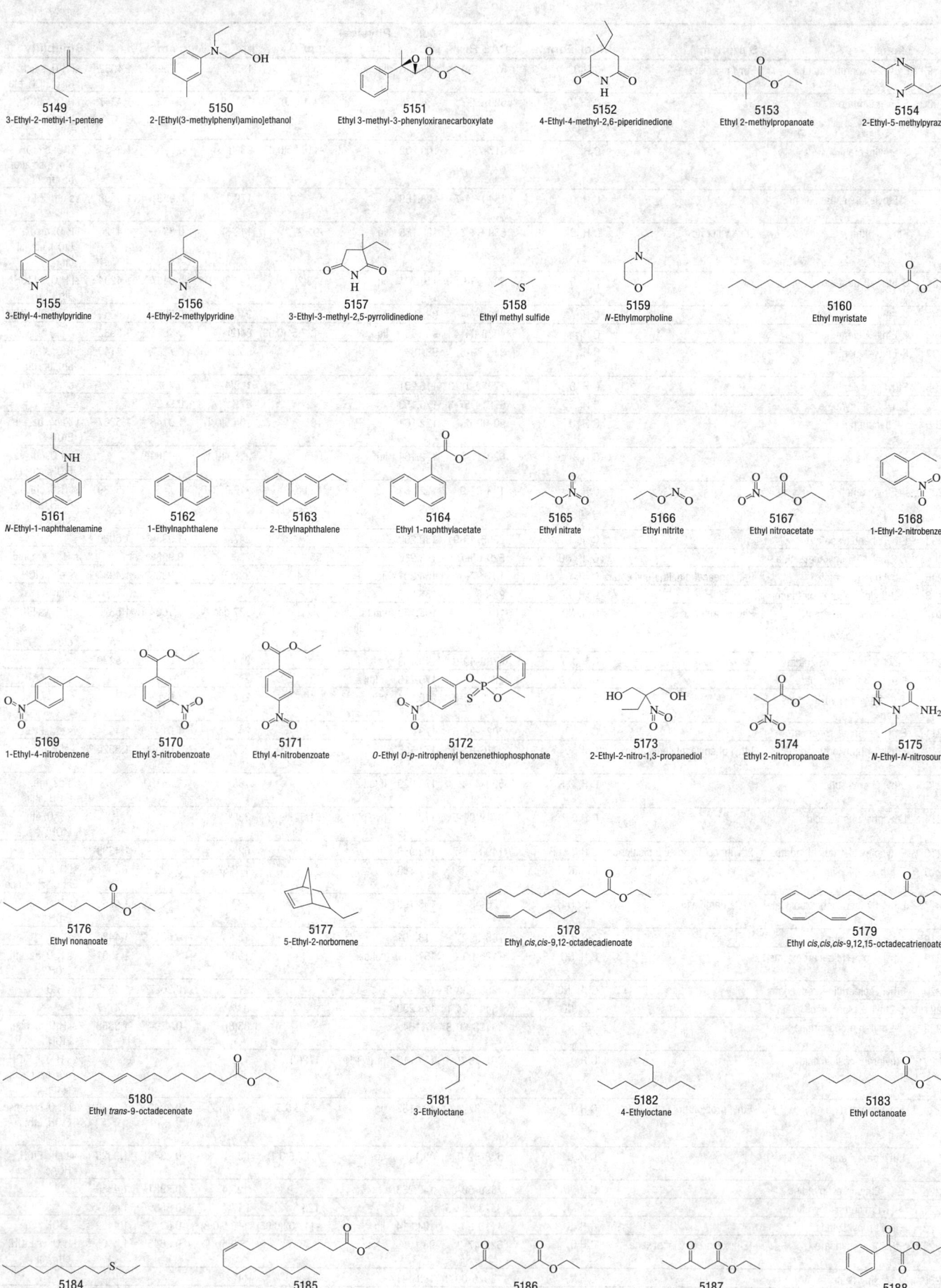

5149
3-Ethyl-2-methyl-1-pentene

5150
2-[Ethyl(3-methylphenyl)amino]ethanol

5151
Ethyl 3-methyl-3-phenyloxiranecarboxylate

5152
4-Ethyl-4-methyl-2,6-piperidinedione

5153
Ethyl 2-methylpropanoate

5154
2-Ethyl-5-methylpyrazine

5155
3-Ethyl-4-methylpyridine

5156
4-Ethyl-2-methylpyridine

5157
3-Ethyl-3-methyl-2,5-pyrrolidinedione

5158
Ethyl methyl sulfide

5159
N-Ethylmorpholine

5160
Ethyl myristate

5161
N-Ethyl-1-naphthalenamine

5162
1-Ethylnaphthalene

5163
2-Ethylnaphthalene

5164
Ethyl 1-naphthylacetate

5165
Ethyl nitrate

5166
Ethyl nitrite

5167
Ethyl nitroacetate

5168
1-Ethyl-2-nitrobenzene

5169
1-Ethyl-4-nitrobenzene

5170
Ethyl 3-nitrobenzoate

5171
Ethyl 4-nitrobenzoate

5172
O-Ethyl O-p-nitrophenyl benzenethiophosphonate

5173
2-Ethyl-2-nitro-1,3-propanediol

5174
Ethyl 2-nitropropanoate

5175
N-Ethyl-N-nitrosourea

5176
Ethyl nonanoate

5177
5-Ethyl-2-norbornene

5178
Ethyl cis,cis-9,12-octadecadienoate

5179
Ethyl cis,cis,cis-9,12,15-octadecatrienoate

5180
Ethyl trans-9-octadecenoate

5181
3-Ethyloctane

5182
4-Ethyloctane

5183
Ethyl octanoate

5184
Ethyl 1-octyl sulfide

5185
Ethyl oleate

5186
Ethyl 5-oxohexanoate

5187
Ethyl 3-oxopentanoate

5188
Ethyl 2-oxo-2-phenylacetate

No.	Name	Synonym	Mol. Form.	CAS RN	Mol. Wt.	Physical Form	mp/°C	bp/°C	den g cm⁻³	n_D	Solubility
5189	Ethyl 2-oxopropanoate	Ethyl pyruvate	$C_5H_8O_3$	617-35-6	116.116	liq	-50	155(4)	1.0596[15]	1.4052[20]	sl H₂O; s ace; msc EtOH, eth
5190	Ethyl palmitate		$C_{18}H_{36}O_2$	628-97-7	284.478	nd	24.1(0.3)	191[10]	0.8577[25]	1.4347[34]	i H₂O; s EtOH, eth, ace, bz, chl
5191	3-Ethylpentane		C_7H_{16}	617-78-7	100.202	liq	-118.55(0.01)	93.4(0.4)	0.6982[20]	1.3934[20]	i H₂O; s EtOH, eth; msc ace, bz, hp, chl
5192	3-Ethyl-2,4-pentanedione		$C_7H_{12}O_2$	1540-34-7	128.169			179(11)	0.9531[19]	1.4408[19]	vs eth, EtOH, chl
5193	Ethyl pentanoate	Ethyl valerate	$C_7H_{14}O_2$	539-82-2	130.185	liq	-91.2	142(3)	0.8770[20]	1.4120[20]	i H₂O; msc EtOH, eth; sl ctc
5194	3-Ethyl-3-pentanol		$C_7H_{16}O$	597-49-9	116.201	liq	-13(2)	142(3)	0.8407[22]	1.4294[20]	sl H₂O; s EtOH, eth
5195	2-Ethyl-1-pentene		C_7H_{14}	3404-71-5	98.186			94(2)	0.7079[20]	1.405[20]	vs bz, eth, EtOH
5196	3-Ethyl-1-pentene		C_7H_{14}	4038-04-4	98.186	liq	-127.51(0.09)	84(2)	0.6917[25]	1.3982[20]	
5197	3-Ethyl-2-pentene		C_7H_{14}	816-79-5	98.186			95(1)	0.7204[20]	1.4148[20]	i H₂O; s EtOH, eth, bz, chl
5198	Ethyl pentyl ether		$C_7H_{16}O$	17952-11-3	116.201			118(4)	0.7622[20]	1.3927[20]	vs eth, EtOH
5199	Ethyl 2-pentynoate		$C_7H_{10}O_2$	55314-57-3	126.153			67[18]	0.962[25]		
5200	2-Ethylphenol		$C_8H_{10}O$	90-00-6	122.164		18	204.5(0.1)	1.0146[25]	1.5367[20]	vs ace, bz, eth, EtOH
5201	3-Ethylphenol		$C_8H_{10}O$	620-17-7	122.164	liq	0(8)	218.4(0.1)	1.0283[20]		sl H₂O, chl; vs EtOH, eth
5202	4-Ethylphenol		$C_8H_{10}O$	123-07-9	122.164	nd	45.0(0.3)	217.97(0.06)		1.5239[25]	sl H₂O, chl; vs EtOH, eth, bz; s ace
5203	Ethyl phenoxyacetate		$C_{10}H_{12}O_3$	2555-49-9	180.200			247	1.0958[30]	1.5080[20]	
5204	N-Ethyl-N-phenylacetamide		$C_{10}H_{13}NO$	529-65-7	163.216		55	260	0.9938[60]		s H₂O, eth, ctc
5205	Ethyl phenylacetate	Benzeneacetic acid, ethyl ester	$C_{10}H_{12}O_2$	101-97-3	164.201	liq	-29.4	228(2)	1.0333[20]	1.4980[20]	vs eth, EtOH
5206	2-(Ethylphenylamino)ethanol		$C_{10}H_{15}NO$	92-50-2	165.232						s chl
5207	Ethyl phenylcarbamate	Phenylurethane	$C_9H_{11}NO_2$	101-99-5	165.189	wh nd (w) pl (dil al)	53(1)	237 dec	1.1064[30]	1.5376[30]	i H₂O; vs EtOH, eth; s bz; sl ctc
5208	Ethyl N-phenylformimidate		$C_9H_{11}NO$	6780-49-0	149.189			214	1.0051[20]	1.5279[20]	s eth, bz
5209	Ethyl N-phenylglycinate		$C_{10}H_{13}NO_2$	2216-92-4	179.216	lf (dil al)	58	273.5			vs eth, EtOH
5210	1-(4-Ethylphenyl)-2-phenyle-thane		$C_{16}H_{18}$	7439-15-8	210.314	cry		294	1.028[50]		
5211	Ethyl 3-phenylpropanoate		$C_{11}H_{14}O_2$	2021-28-5	178.228			247(3)	1.0147[20]	1.4954[20]	vs eth, EtOH
5212	Ethyl 3-phenylpropynoate	Ethyl phenylacetylenecarboxylate	$C_{11}H_{10}O_2$	2216-94-6	174.196			265	1.055[25]	1.5520[20]	s eth
5213	Ethyl phenyl sulfone		$C_8H_{10}O_2S$	599-70-2	170.229	lf (dil al)	42	160[12]	1.1410[20]		vs bz, eth, EtOH, chl
5214	Ethylphosphonic acid		$C_2H_7O_3P$	6779-09-5	110.049	hyg pl or nd	61.5	335[8]			vs H₂O, eth, EtOH
5215	Ethyl phosphorodichloridate	Ethylphosphoric acid dichloride	$C_2H_5Cl_2O_2P$	1498-51-7	162.940			62[10]		1.4338[20]	
5216	5-Ethyl-2-picoline		$C_8H_{11}N$	104-90-5	121.180			178(3)	0.9202[20]	1.4971[20]	sl H₂O; s EtOH, eth, bz; vs ace
5217	Ethyl 1-piperazinecarboxylate	1-Carbethoxypiperazine	$C_7H_{14}N_2O_2$	120-43-4	158.198			237		1.4760[25]	vs H₂O, eth, EtOH
5218	1-Ethylpiperidine		$C_7H_{15}N$	766-09-6	113.201			133(3)	0.8237[20]	1.4480[20]	
5219	Ethyl 4-piperidinecarboxylate		$C_8H_{15}NO_2$	1126-09-6	157.211	col oil		100[10]		1.4591[20]	vs H₂O, bz, eth, EtOH
5220	Ethyl 1-piperidinepropanoate		$C_{10}H_{19}NO_2$	19653-33-9	185.264			217	0.9627[25]	1.4525[25]	vs H₂O
5221	1-Ethyl-3-piperidinol		$C_7H_{15}NO$	13444-24-1	129.200			94[15]		1.4777[14]	
5222	N-Ethyl-1-propanamine		$C_5H_{13}N$	20193-20-8	87.164			85(8)	0.7204[17]	1.3858[25]	sl H₂O; vs ace, EtOH
5223	Ethylpropanedioic acid		$C_5H_8O_4$	601-75-2	132.116	pr (w+1)	112(2)	180[0.05]			vs H₂O; s EtOH, eth, bz; i ace; sl tfa
5224	Ethyl propanoate	Ethyl propionate	$C_5H_{10}O_2$	105-37-3	102.132	liq	-73.6(0.5)	98.9(0.2)	0.8843[25]	1.3839[20]	sl H₂O, ctc; msc EtOH, eth; s ace
5225	Ethyl propyl ether		$C_5H_{12}O$	628-32-0	88.148	liq	-127.5(0.1)	63(3)	0.7386[20]	1.3695[20]	vs eth, EtOH, HOAc
5226	2-(1-Ethylpropyl)pyridine		$C_{10}H_{15}N$	7399-50-0	149.233			195.4	0.8981[20]	1.4850[25]	
5227	4-(1-Ethylpropyl)pyridine		$C_{10}H_{15}N$	35182-51-5	149.233		125.5	217	0.9085[25]	1.40905[25]	
5228	Ethyl propyl sulfide		$C_5H_{12}S$	4110-50-3	104.214	liq	-117.03(0.05)	118.5(0.8)	0.8370[20]	1.4462[20]	s EtOH
5229	Ethyl 2-propynoate	(Ethoxycarbonyl)acetylene	$C_5H_6O_2$	623-47-2	98.101			119(5)	0.9645[16]	1.4105[20]	i H₂O; vs EtOH, eth, chl
5230	2-Ethylpyrazine		$C_6H_8N_2$	13925-00-3	108.141			112[200]			

5189
Ethyl 2-oxopropanoate

5190
Ethyl palmitate

5191
3-Ethylpentane

5192
3-Ethyl-2,4-pentanedione

5193
Ethyl pentanoate

5194
3-Ethyl-3-pentanol

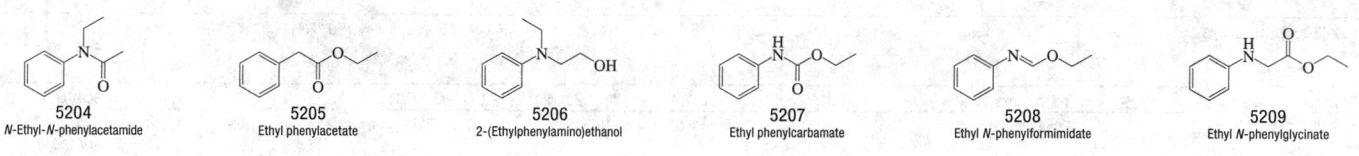

5195
2-Ethyl-1-pentene

5196
3-Ethyl-1-pentene

5197
3-Ethyl-2-pentene

5198
Ethyl pentyl ether

5199
Ethyl 2-pentynoate

5200
2-Ethylphenol

5201
3-Ethylphenol

5202
4-Ethylphenol

5203
Ethyl phenoxyacetate

5204
N-Ethyl-N-phenylacetamide

5205
Ethyl phenylacetate

5206
2-(Ethylphenylamino)ethanol

5207
Ethyl phenylcarbamate

5208
Ethyl N-phenylformimidate

5209
Ethyl N-phenylglycinate

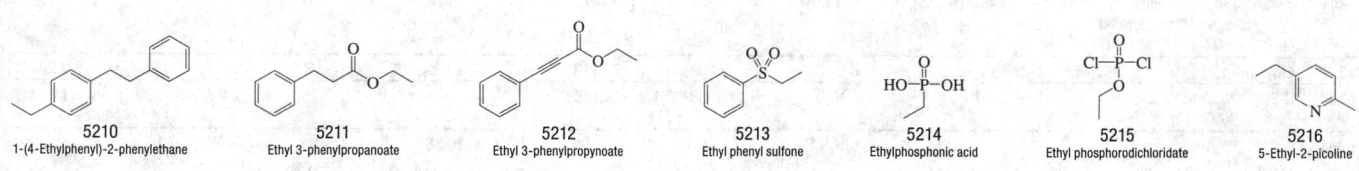

5210
1-(4-Ethylphenyl)-2-phenylethane

5211
Ethyl 3-phenylpropanoate

5212
Ethyl 3-phenylpropynoate

5213
Ethyl phenyl sulfone

5214
Ethylphosphonic acid

5215
Ethyl phosphorodichloridate

5216
5-Ethyl-2-picoline

5217
Ethyl 1-piperazinecarboxylate

5218
1-Ethylpiperidine

5219
Ethyl 4-piperidinecarboxylate

5220
Ethyl 1-piperidinepropanoate

5221
1-Ethyl-3-piperidinol

5222
N-Ethyl-1-propanamine

5223
Ethylpropanedioic acid

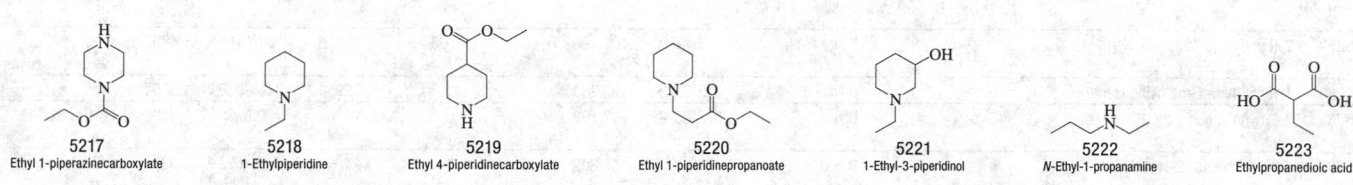

5224
Ethyl propanoate

5225
Ethyl propyl ether

5226
2-(1-Ethylpropyl)pyridine

5227
4-(1-Ethylpropyl)pyridine

5228
Ethyl propyl sulfide

5229
Ethyl 2-propynoate

5230
2-Ethylpyrazine

No.	Name	Synonym	Mol. Form.	CAS RN	Mol. Wt.	Physical Form	mp/°C	bp/°C	den g cm⁻³	n_D	Solubility
5231	2-Ethylpyridine		C₇H₉N	100-71-0	107.153	liq	-63.0(0.5)	149(1)	0.9502²⁵	1.4964²⁰	s H₂O; msc EtOH; vs eth, ace; sl ctc
5232	3-Ethylpyridine		C₇H₉N	536-78-7	107.153	liq	-76.9(0.5)	166(2)	0.9539²⁵	1.5021²⁰	s H₂O, EtOH, eth; vs ace; sl ctc
5233	4-Ethylpyridine		C₇H₉N	536-75-4	107.153	liq	-90.5(0.5)	168(2)	0.9417²⁰	1.5009²⁰	s H₂O, EtOH, eth; vs ace; sl ctc
5234	2-Ethyl-4-pyridinecarbothio-amide	Ethionamide	C₈H₁₀N₂S	536-33-4	166.243		163				
5235	Ethyl 2-pyridinecarboxylate	Ethyl 2-picolinate	C₈H₉NO₂	2524-52-9	151.163	ye cry in air	1	243	1.1194²⁰	1.5104²⁰	vs H₂O, eth, EtOH
5236	Ethyl 3-pyridinecarboxylate	Ethyl nicotinate	C₈H₉NO₂	614-18-6	151.163		8.5	221(20)	1.1070²⁰	1.5024²⁰	vs H₂O, EtOH, eth, bz; sl ctc
5237	Ethyl 4-pyridinecarboxylate		C₈H₉NO₂	1570-45-2	151.163		23	221(11)	1.0091¹⁵	1.5017²⁰	sl H₂O; s EtOH, bz; vs eth, chl
5238	N-Ethylpyridinium bromide		C₇H₁₀BrN	1906-79-2	188.065	cry (al)	111.5				s H₂O, EtOH; i eth
5239	1-Ethyl-1H-pyrrole		C₆H₉N	617-92-5	95.142			130(5)	0.9009²⁰	1.4841²⁰	vs EtOH
5240	1-Ethyl-1H-pyrrole-2,5-dione	N-Ethylmaleimide	C₆H₇NO₂	128-53-0	125.126	cry (bz)	45.5				sl H₂O; vs EtOH, eth; s chl
5241	1-Ethyl-2-pyrrolidinemethana-mine		C₇H₁₆N₂	26116-12-1	128.215			59¹⁶	0.887²⁵	1.4665²⁰	
5242	Ethyl Red	2-(4-Diethylaminophenylazo)-benzoic acid	C₁₇H₁₉N₃O₂	76058-33-8	297.352		135				
5243	Ethyl salicylate		C₉H₁₀O₃	118-61-6	166.173		45	150¹⁰	1.1326²⁰	1.5296²⁰	i H₂O; msc EtOH; vs eth; s ctc
5244	Ethyl silicate	Tetraethoxysilane	C₈H₂₀O₄Si	78-10-4	208.329	liq	-82.2(0.1)	168(1)	0.9320²⁰	1.3928²⁰	dec H₂O
5245	Ethyl stearate	Ethyl octadecanoate	C₂₀H₄₀O₂	111-61-5	312.531		32.9(0.8)	199¹⁰	1.057²⁰	1.4349⁴⁰	i H₂O; s EtOH, eth, chl; vs ace
5246	2-Ethylstyrene		C₁₀H₁₂	7564-63-8	132.202	liq	-75.6(0.5)	189(1)	0.9017²⁵	1.5380²⁰	
5247	3-Ethylstyrene		C₁₀H₁₂	7525-62-4	132.202	liq	-101.3(0.6)	190(2)	0.8945²⁰	1.5351²⁰	
5248	4-Ethylstyrene		C₁₀H₁₂	3454-07-7	132.202	liq	-49.7(0.5)	192(9)	0.8884²⁵	1.5376²⁰	
5249	Ethyl sulfate		C₂H₆O₄S	540-82-9	126.132			280 dec	1.3657²⁰	1.4105²⁰	vs H₂O
5250	2-(Ethylsulfonyl)ethanol	Ethylsulfonylethyl alcohol	C₄H₁₀O₃S	513-12-2	138.185						sl chl
5251	2-Ethyl-5-(3-sulfophenyl)-isoxazolium hydroxide, inner salt	Woodward's Reagent K	C₁₁H₁₁NO₄S	4156-16-5	253.275		dec 207				
5252	Ethyl tartrate	Ethyl tartrate, acid	C₆H₁₀O₆	608-89-9	178.139		90				vs H₂O, EtOH
5253	2-Ethyltetrahydrofuran		C₆H₁₂O	1003-30-1	100.158			107(7)	0.8570¹⁹	1.4147¹⁹	vs ace, bz, eth, EtOH
5254	5-Ethyl-1,3,4-thiadiazol-2-amine		C₄H₇N₃S	14068-53-2	129.184		200.8				
5255	S-Ethyl thioacetate		C₄H₈OS	625-60-5	104.171			114(3)	0.9792²⁰	1.4583²¹	i H₂O; vs EtOH, eth
5256	(Ethylthio)acetic acid		C₄H₈O₂S	627-04-3	120.171		-8.5	164⁸³	1.1497²⁰		vs H₂O, EtOH, eth
5257	(Ethylthio)benzene	Thiophenetole	C₈H₁₀S	622-38-8	138.230			207(1)	1.0211²⁰	1.5670²⁰	s EtOH
5258	Ethyl thiocyanate		C₃H₅NS	542-90-5	87.144	liq	-85.5(0.5)	144(7)	1.007²³	1.4684¹⁵	i H₂O; msc EtOH, eth; s chl
5259	2-(Ethylthio)ethanol		C₄H₁₀OS	110-77-0	106.186	liq	-100	175(3)	1.0166²⁰	1.4867²⁰	sl H₂O; s EtOH; vs ace
5260	1-(Ethylthio)-4-methylbenzene		C₉H₁₂S	622-63-9	152.256			220	0.9996²⁰	1.555²⁰	
5261	2-Ethylthiophene		C₆H₈S	872-55-9	112.193			136(1)	0.9930²⁰	1.5122²⁰	i H₂O; vs EtOH, eth
5262	Ethyl thiophene-2-carboxylate		C₇H₈O₂S	2810-04-0	156.203			218	1.1623¹⁶	1.5248²⁰	s EtOH, ace; sl ctc
5263	3-Ethyl-2-thioxo-4-thiazolidi-none	3-Ethylrhodanine	C₅H₇NOS₂	7648-01-3	161.246		35.5				
5264	2-Ethyltoluene		C₉H₁₂	611-14-3	120.191	liq	-80.7(0.4)	165.1(0.4)	0.8807²⁰	1.5046²⁰	i H₂O; msc EtOH, eth, ace, bz, peth, ctc
5265	3-Ethyltoluene		C₉H₁₂	620-14-4	120.191	liq	-95.7(0.2)	161.3(0.5)	0.8645²⁰	1.4966²⁰	i H₂O; vs EtOH, eth; msc ace, bz
5266	4-Ethyltoluene		C₉H₁₂	622-96-8	120.191	liq	-62.7(0.5)	162.0(0.6)	0.8614²⁰	1.4959²⁰	i H₂O; vs EtOH, eth; msc ace, bz

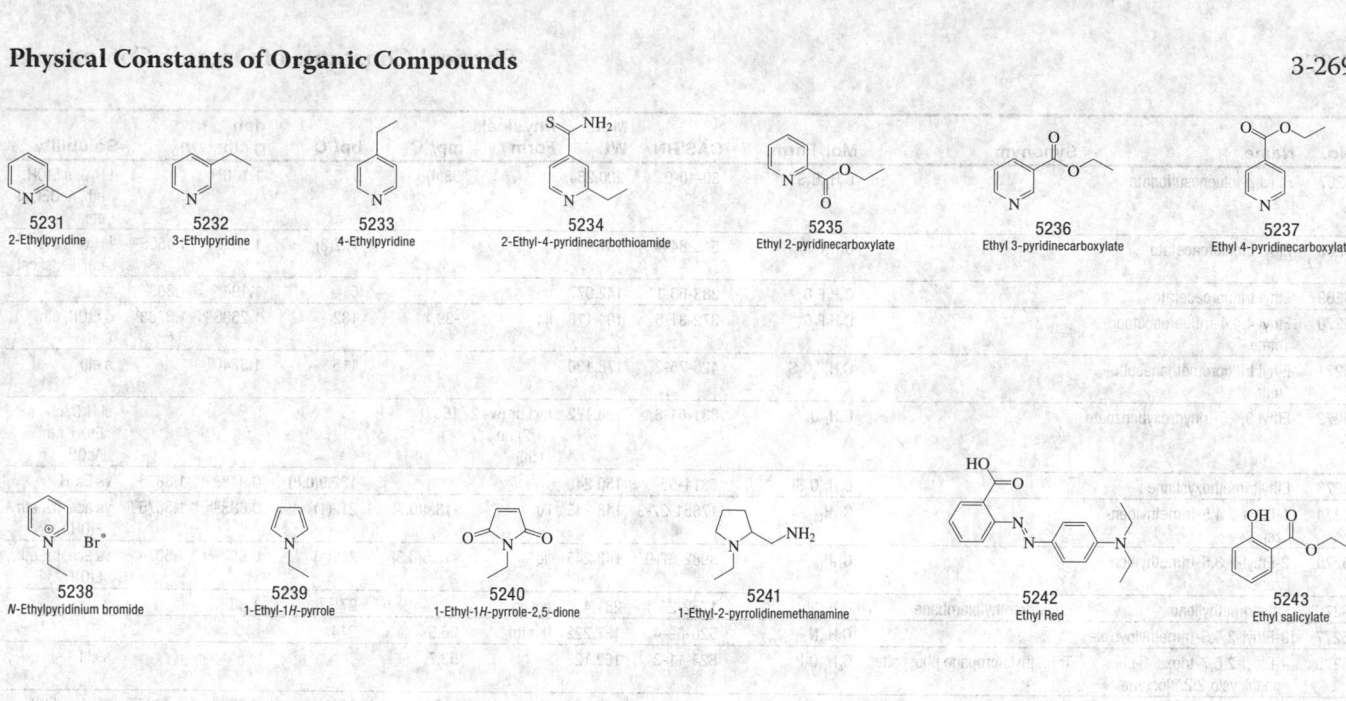

5231 2-Ethylpyridine

5232 3-Ethylpyridine

5233 4-Ethylpyridine

5234 2-Ethyl-4-pyridinecarbothioamide

5235 Ethyl 2-pyridinecarboxylate

5236 Ethyl 3-pyridinecarboxylate

5237 Ethyl 4-pyridinecarboxylate

5238 N-Ethylpyridinium bromide

5239 1-Ethyl-1H-pyrrole

5240 1-Ethyl-1H-pyrrole-2,5-dione

5241 1-Ethyl-2-pyrrolidinemethanamine

5242 Ethyl Red

5243 Ethyl salicylate

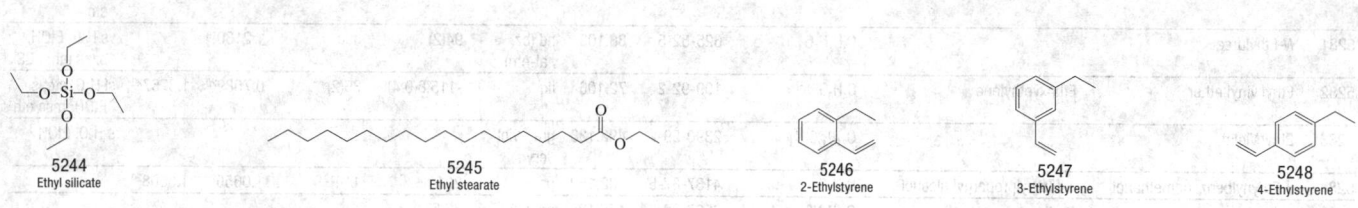

5244 Ethyl silicate

5245 Ethyl stearate

5246 2-Ethylstyrene

5247 3-Ethylstyrene

5248 4-Ethylstyrene

5249 Ethyl sulfate

5250 2-(Ethylsulfonyl)ethanol

5251 2-Ethyl-5-(3-sulfophenyl)isoxazolium hydroxide, inner salt

5252 Ethyl tartrate

5253 2-Ethyltetrahydrofuran

5254 5-Ethyl-1,3,4-thiadiazol-2-amine

5255 S-Ethyl thioacetate

5256 (Ethylthio)acetic acid

5257 (Ethylthio)benzene

5258 Ethyl thiocyanate

5259 2-(Ethylthio)ethanol

5260 1-(Ethylthio)-4-methylbenzene

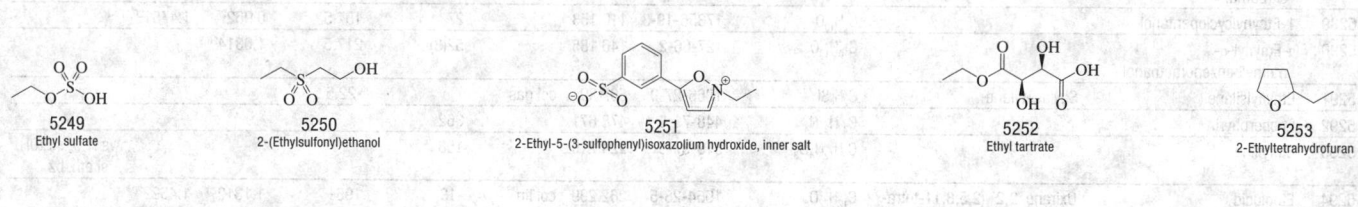

5261 2-Ethylthiophene

5262 Ethyl thiophene-2-carboxylate

5263 3-Ethyl-2-thioxo-4-thiazolidinone

5264 2-Ethyltoluene

5265 3-Ethyltoluene

5266 4-Ethyltoluene

No.	Name	Synonym	Mol. Form.	CAS RN	Mol. Wt.	Physical Form	mp/°C	bp/°C	den g cm⁻³	n_D	Solubility
5267	Ethyl p-toluenesulfonate		$C_9H_{12}O_3S$	80-40-0	200.254		33(4)	173¹⁵	1.166⁴⁸		i H₂O; s EtOH, eth, AcOEt; sl ctc
5268	Ethyl trichloroacetate		$C_4H_5Cl_3O_2$	515-84-4	191.441			167(3)	1.3836²⁰	1.4505²⁰	i H₂O; s EtOH, eth, bz; sl chl
5269	Ethyl trifluoroacetate		$C_4H_5F_3O_2$	383-63-1	142.077			61	1.194²⁰	1.308²⁰	
5270	Ethyl 4,4,4-trifluoroacetoac-etate		$C_6H_7F_3O_3$	372-31-6	184.113	liq	-39.1	132	1.2586¹⁵	1.3783¹⁵	s EtOH, eth
5271	Ethyl trifluoromethanesulfo-nate		$C_3H_5F_3O_3S$	425-75-2	178.130			115	1.3740⁰		s eth
5272	Ethyl 3,4,5-trihydroxybenzoate		$C_9H_{10}O_5$	831-61-8	198.172	mcl pr (w+2 1/2) nd (chl)	163.0				sl H₂O, chl; s EtOH, eth, AcOEt
5273	Ethyltrimethoxysilane		$C_5H_{14}O_3Si$	5314-55-6	150.249			122.9(0.8)	0.9488²⁰	1.3838²⁰	vs EtOH
5274	1-Ethyl-2,4,5-trimethylben-zene		$C_{11}H_{16}$	17851-27-3	148.245	liq	-13.4(0.7)	212(1)	0.883²⁰	1.5075²⁰	vs ace, bz, eth, EtOH
5275	2-Ethyl-1,3,5-trimethylben-zene		$C_{11}H_{16}$	3982-67-0	148.245	liq	-15.5(0.3)	211(1)	0.883²⁰	1.5074²⁰	vs ace, bz, eth, EtOH
5276	Ethyltrimethyllead	Ethyltrimethylplumbane	$C_5H_{14}Pb$	1762-26-1	281.4	col liq		27¹⁰·⁵	1.88²⁰		
5277	3-Ethyl-2,4,5-trimethylpyrrole		$C_9H_{15}N$	520-69-4	137.222	lf (eth)	66.5	214			
5278	4-Ethyl-2,6,7-trioxa-1-phos-phabicyclo[2.2.2]octane	Trimethylolpropane phosphite	$C_6H_{11}O_3P$	824-11-3	162.123		53.7				s chl
5279	Ethyl undecanoate	Ethyl undecylate	$C_{13}H_{26}O_2$	627-90-7	214.344		-16(6)	131¹⁴	0.8633²⁰	1.4285²⁰	i H₂O; s EtOH, eth, ace, bz
5280	Ethyl 10-undecenoate		$C_{13}H_{24}O_2$	692-86-4	212.329	liq	-38	264.5	0.8827¹⁵	1.4449²⁵	i H₂O; s EtOH, eth, HOAc; sl ctc
5281	N-Ethylurea		$C_3H_8N_2O$	625-52-5	88.108	nd (bz, al-eth)	94(2)	dec	1.2130¹⁸		vs H₂O, EtOH, bz; s eth; i CS₂
5282	Ethyl vinyl ether	Ethoxyethylene	C_4H_8O	109-92-2	72.106	liq	-115.8(0.4)	36(2)	0.7589²⁰	1.3767²⁰	sl H₂O, ctc; s EtOH; msc eth
5283	Ethyl Violet		$C_{31}H_{42}ClN_3$	2390-59-2	492.138	gray-viol cry					s H₂O, EtOH
5284	α-Ethynylbenzenemethanol	1-Phenylpropargyl alcohol	C_9H_8O	4187-87-5	132.159	pr	22	114¹²	1.0655²⁰	1.5508²⁰	
5285	α-Ethynylbenzenemethanol carbamate	Carfimate	$C_{10}H_9NO_2$	3567-38-2	175.184	cry (al)	86.5				
5286	1-Ethynylcyclohexanamine		$C_8H_{13}N$	30389-18-5	123.196			65²⁰	0.913²⁵	1.4817²⁰	
5287	1-Ethynylcyclohexanol		$C_8H_{12}O$	78-27-3	124.180	cry (peth)	31(3)	174	0.9873²⁰	1.4822²⁰	i H₂O; s EtOH, bz, peth; sl chl
5288	1-Ethynylcyclohexanol, carbamate	Ethinamate	$C_9H_{13}NO_2$	126-52-3	167.205	nd	97	120³		1.4441²¹	sl H₂O; vs EtOH; s hx
5289	1-Ethynylcyclopentanol		$C_7H_{10}O$	17356-19-3	110.153		27	157.5	0.962²⁵	1.4751²⁰	
5290	α-Ethynyl-α-methylbenzenemethanol		$C_{10}H_{10}O$	127-66-2	146.185		52(3)	217.5	1.0314²⁰		
5291	Ethynylsilane	Silylacetylene	C_2H_4Si	1066-27-9	56.139	col gas		-22.5			
5292	Etioporphyrin		$C_{32}H_{38}N_4$	448-71-5	478.671		362				
5293	Etofylline		$C_9H_{12}N_4O_3$	519-37-9	224.216		158				vs H₂O; s EtOH; sl eth, bz
5294	Etoglucid	Oxirane, 2,2'-(2,5,8,11-tetra-oxadodecane-1,12-diyl)bis-	$C_{12}H_{22}O_6$	1954-28-5	262.299	col liq	-13	196²	1.1312²⁰	1.4622²⁰	
5295	Etoposide		$C_{29}H_{32}O_{13}$	33419-42-0	588.556	cry (MeOH)	≈243				s MeOH
5296	Etrimfos		$C_{10}H_{17}N_2O_4PS$	38260-54-7	292.291		-1.7		1.195²⁰		
5297	Eucalyptol	Cineole	$C_{10}H_{18}O$	470-82-6	154.249	oil	1.4(0.3)	176(4)	0.9267²⁰	1.4586²⁰	i H₂O; s EtOH, eth, chl; sl ctc
5298	Euparin	1-[6-Hydroxy-2-(1-methylvinyl)-5-benzofuranyl]ethanone	$C_{13}H_{12}O_3$	532-48-9	216.232		121.5				s eth, bz, chl; sl NaOH
5299	Evan's Blue		$C_{34}H_{24}N_6Na_4O_{14}S_4$	314-13-6	960.806						s H₂O, EtOH, acid
5300	Evodiamine		$C_{19}H_{17}N_3O$	518-17-2	303.357	ye lf (al)	28				i EtOH, chl; vs DMF; s HOAc; sl MeOH
5301	Famotidine		$C_8H_{15}N_7O_2S_3$	76824-35-6	337.446	cry	170.6(0.6)				
5302	Famphur		$C_{10}H_{16}NO_5PS_2$	52-85-7	325.342		57(2)				
5303	α-Farnesene		$C_{15}H_{24}$	502-61-4	204.352			130¹²	0.8410²⁰	1.4836²⁰	i H₂O; s eth, ace; msc peth, lig
5304	β-Farnesene		$C_{15}H_{24}$	18794-84-8	204.352			121⁹	0.8363²⁰	1.4899²⁰	vs ace, eth, chl
5305	Farnesic acid		$C_{15}H_{24}O_2$	7548-13-2	236.351	oil		204¹⁶			
5306	2-cis,6-trans-Farnesol		$C_{15}H_{26}O$	3790-71-4	222.366	oil		156¹²	0.8908²⁰	1.4877²⁰	vs ace, eth, EtOH
5307	2-trans,6-trans-Farnesol		$C_{15}H_{26}O$	106-28-5	222.366	oil		160¹⁰	0.888²⁰	1.4877²⁰	i H₂O; vs EtOH; s eth, ace

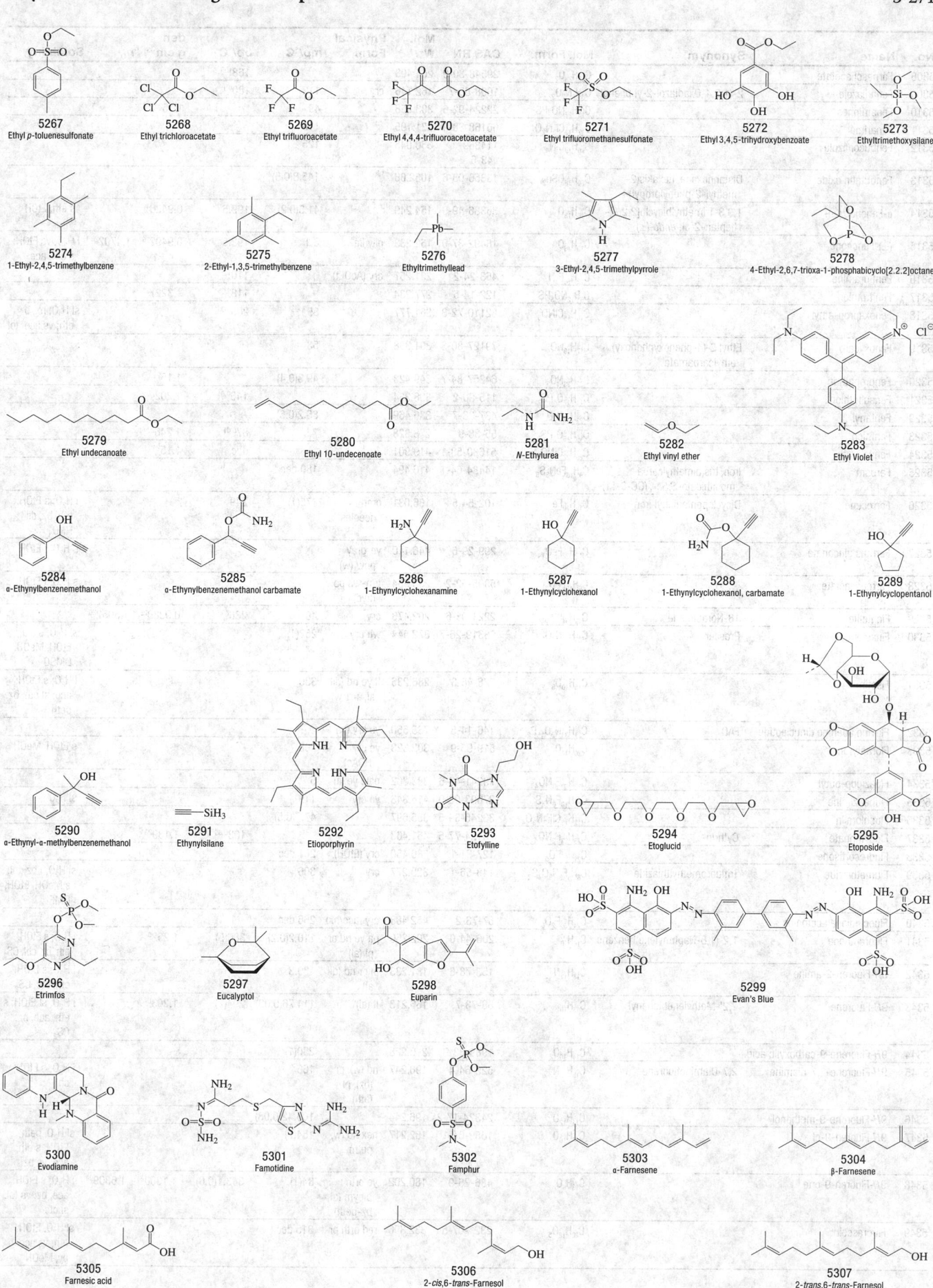

5267 Ethyl *p*-toluenesulfonate

5268 Ethyl trichloroacetate

5269 Ethyl trifluoroacetate

5270 Ethyl 4,4,4-trifluoroacetoacetate

5271 Ethyl trifluoromethanesulfonate

5272 Ethyl 3,4,5-trihydroxybenzoate

5273 Ethyltrimethoxysilane

5274 1-Ethyl-2,4,5-trimethylbenzene

5275 2-Ethyl-1,3,5-trimethylbenzene

5276 Ethyltrimethyllead

5277 3-Ethyl-2,4,5-trimethylpyrrole

5278 4-Ethyl-2,6,7-trioxa-1-phosphabicyclo[2.2.2]octane

5279 Ethyl undecanoate

5280 Ethyl 10-undecenoate

5281 *N*-Ethylurea

5282 Ethyl vinyl ether

5283 Ethyl Violet

5284 α-Ethylbenzenemethanol

5285 α-Ethynylbenzenemethanol carbamate

5286 1-Ethynylcyclohexanamine

5287 1-Ethynylcyclohexanol

5288 1-Ethynylcyclohexanol, carbamate

5289 1-Ethynylcyclopentanol

5290 α-Ethynyl-α-methylbenzenemethanol

5291 Ethynylsilane

5292 Etioporphyrin

5293 Etofylline

5294 Etoglucid

5295 Etoposide

5296 Etrimfos

5297 Eucalyptol

5298 Euparin

5299 Evan's Blue

5300 Evodiamine

5301 Famotidine

5302 Famphur

5303 α-Farnesene

5304 β-Farnesene

5305 Farnesic acid

5306 2-*cis*,6-*trans*-Farnesol

5307 2-*trans*,6-*trans*-Farnesol

No.	Name	Synonym	Mol. Form.	CAS RN	Mol. Wt.	Physical Form	mp/°C	bp/°C	den g cm⁻³	n_D	Solubility
5308	Farnesol acetate		$C_{17}H_{28}O_2$	29548-30-9	264.403			168[10]			
5309	Fenadiazole	2-(1,2,4-Oxadiazol-2-yl)phenol	$C_8H_6N_2O_2$	1008-65-7	162.146	cry	112	180[0.1]			
5310	Fenamiphos		$C_{13}H_{22}NO_3PS$	22224-92-6	303.358		49		1.15[20]		
5311	Fenarimol		$C_{17}H_{12}Cl_2N_2O$	60168-88-9	331.195		118				
5312	Fenbuconazole		$C_{20}H_{19}ClN_4$	114369-43-6	350.845		125				
5313	Fenbutatin oxide	Distannoxane, hexakis(2-methyl-2-phenylpropyl)-	$C_{60}H_{78}OSn_2$	13356-08-6	1052.68		145.8(0.5)				
5314	α-Fenchol, (±)-	1,3,3-Trimethylbicyclo[2.2.1]-heptan-2-ol, endo-(±)	$C_{10}H_{18}O$	36386-49-9	154.249		41.5(0.2)	199.5	0.9420[40]		vs eth, EtOH
5315	(±)-Fenchone		$C_{10}H_{16}O$	18492-37-0	152.233	oily liq	6.1	193.5	0.9492[15]	1.4702[20]	i H_2O; vs EtOH; s eth, ace
5316	Fenfluramine		$C_{12}H_{16}F_3N$	458-24-2	231.257	cry (AcOEt)		110[12]			
5317	Fenitrothion		$C_9H_{12}NO_5PS$	122-14-5	277.234			118[0.05]	1.3227[25]		
5318	Fenoxaprop-ethyl		$C_{18}H_{16}ClNO_5$	82110-72-3	361.777		85	200[0.001]			sl H_2O, hx; s eth; vs ace, tol
5319	Fenoxycarb	Ethyl 2-(4-phenoxyphenoxy)ethylcarbamate	$C_{17}H_{19}NO_4$	79127-80-3	301.338		53				
5320	Fenpropathrin		$C_{22}H_{23}NO_3$	64257-84-7	349.423		49.3(0.4)		1.15[25]		
5321	Fensulfothion		$C_{11}H_{17}O_4PS_2$	115-90-2	308.354			140[0.01]	1.202[20]		
5322	Fentanyl		$C_{22}H_{28}N_2O$	437-38-7	336.469		85.2(0.5)				
5323	Fenthion		$C_{10}H_{15}O_3PS_2$	55-38-9	278.328		7.5	87[0.01]	1.246[20]		
5324	Fenvalerate		$C_{25}H_{22}ClNO_3$	51630-58-1	419.901		dec		1.15[25]		
5325	Ferbam	Iron, tris(dimethylcarba-modithioato-S,S')-, (OC-6-11)-	$C_9H_{18}FeN_3S_6$	14484-64-1	416.494		180 dec				
5326	Ferrocene	Dicyclopentadienyl iron	$C_{10}H_{10}Fe$	102-54-5	186.031	oran needles	175(1)	249			i H_2O; s EtOH, eth, bz, dil HNO_3
5327	Ferrous gluconate		$C_{12}H_{22}FeO_{14}$	299-29-6	446.140	ye-gray pow (w)					s H_2O; i EtOH
5328	Ferrous lactate		$C_6H_{10}FeO_6$	5905-52-2	233.984	grn-wh pow (hyd)					s H_2O; i EtOH
5329	Fichtelite	18-Norabietane	$C_{19}H_{34}$	2221-95-6	262.473	cry	46	236[43]	0.9380[22]	1.5052[20]	
5330	Finasteride	Proscar	$C_{23}H_{36}N_2O_2$	98319-26-7	372.544	wh cry	252(1)				sl H_2O; s chl, EtOH, MeOH, DMSO
5331	Fisetin		$C_{15}H_{10}O_6$	528-48-3	286.236	lt ye nd (dil al, + 1 w)	330				i H_2O; s EtOH, ace; sl eth, bz, peth
5332	Flavine adenine dinucleotide	FAD	$C_{27}H_{33}N_9O_{15}P_2$	146-14-5	785.550	ye cry (w)					
5333	Florantyrone		$C_{20}H_{14}O_3$	519-95-9	302.323	ye cry (HOAc)	208				s EtOH, MeOH
5334	Fluazipop-butyl		$C_{19}H_{20}F_3NO_4$	79241-46-6	383.362	pale ye liq	5				
5335	Flubenzimine		$C_{17}H_{10}F_6N_4S$	37893-02-0	416.343	ye cry	119				sl H_2O
5336	Fluchloralin		$C_{12}H_{13}ClF_3N_3O_4$	33245-39-5	355.697		47.7(0.5)				
5337	Flucythrinate	Cythrin	$C_{26}H_{23}F_2NO_4$	70124-77-5	451.463			108[0.35]	1.189[22]		
5338	Fludrocortisone		$C_{21}H_{29}FO_5$	127-31-1	380.450	cry (EtOH)	261 dec				
5339	Flumethiazide	Trifluoromethylthiazide	$C_8H_6F_3N_3O_4S_2$	148-56-1	329.277	cry	306				sl H_2O; i bz, tol; s MeOH, EtOH, DMF
5340	Fluocinolone acetonide		$C_{24}H_{30}F_2O_6$	67-73-2	452.488	cry (ace/hx)	266 dec				
5341	Fluoranthene	1,2-(1,8-Naphthylene)benzene	$C_{16}H_{10}$	206-44-0	202.250	pa ye nd or pl (al)	110.2(0.2)	380(5)	1.252[0]		i H_2O; s EtOH, eth, bz, chl, CS_2
5342	9H-Fluoren-2-amine		$C_{13}H_{11}N$	153-78-6	181.233	pl or nd (dil al)	130.3				i H_2O; s EtOH, eth, ctc, CS_2
5343	9H-Fluorene	2,2'-Methylenebiphenyl	$C_{13}H_{10}$	86-73-7	166.218	lf (al)	114.76(0.03)	294(2)	1.203[0]		i H_2O; sl EtOH; s eth, ace, bz, CS_2
5344	9H-Fluorene-9-carboxylic acid		$C_{14}H_{10}O_2$	1989-33-9	210.228		230(1)				
5345	9H-Fluorene-2,7-diamine	2,7-Diaminofluorene	$C_{13}H_{12}N_2$	525-64-4	196.247	nd (w), pr (bz), pl (eth)	166				i H_2O; s EtOH, chl
5346	9H-Fluorene-9-methanol		$C_{14}H_{12}O$	24324-17-2	196.244			103.42(0.08)			
5347	9H-Fluoren-9-ol		$C_{13}H_{10}O$	1689-64-1	182.217	hex nd (w, peth)	154(2)				sl H_2O, peth, EtOH; s eth, ace; vs bz
5348	9H-Fluoren-9-one		$C_{13}H_8O$	486-25-9	180.202	ye orth bipym (al, bz-peth)	84(1)	343.1(0.6)	1.1300[99]	1.6309[99]	i H_2O; s EtOH, ace, bz; vs tol; sl ctc
5349	Fluorescein		$C_{20}H_{12}O_5$	2321-07-5	332.306	red orth pr	315 dec				sl H_2O, EtOH, eth; vs ace; s py, MeOH

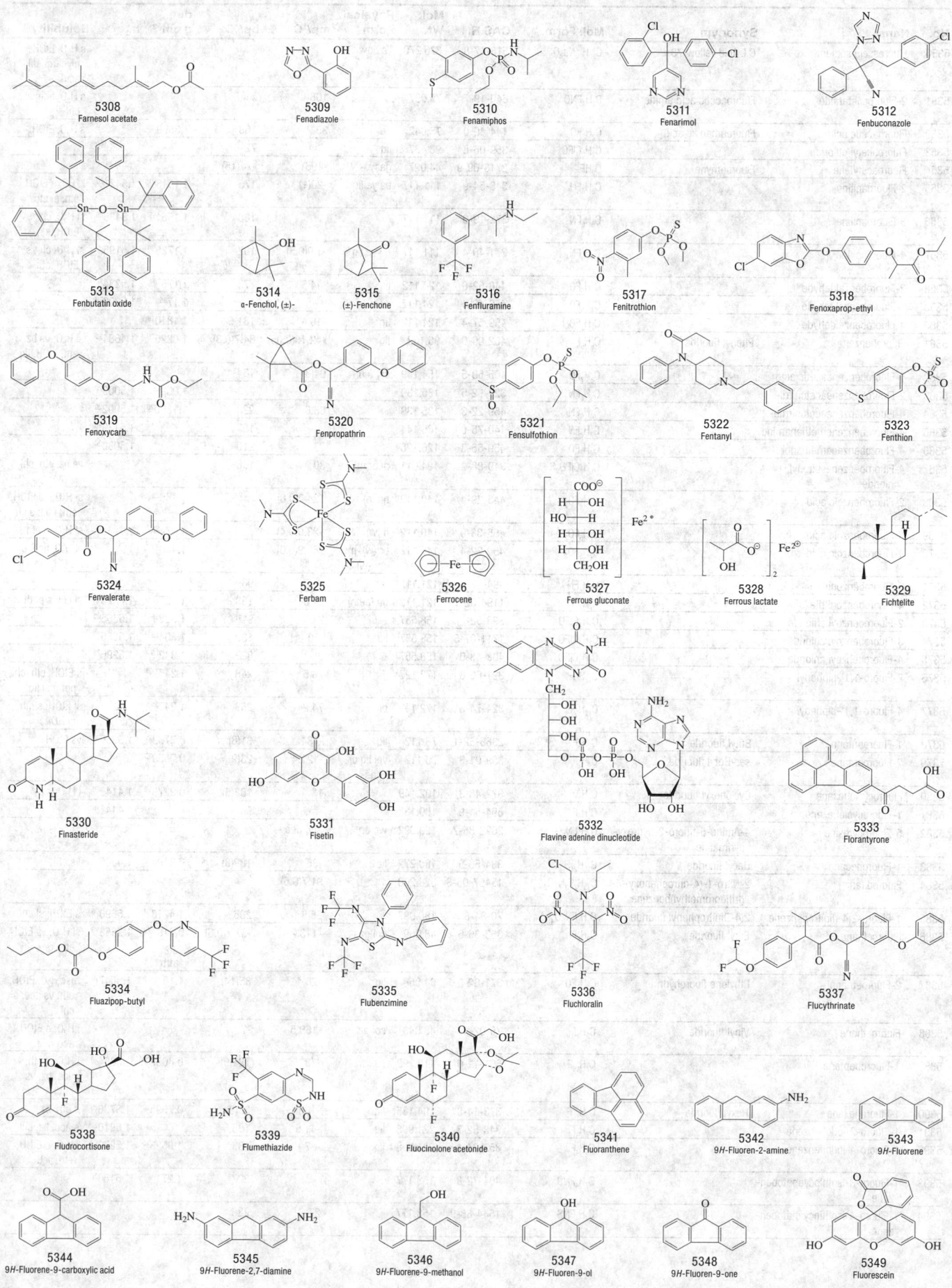

5308 Farnesol acetate

5309 Fenadiazole

5310 Fenamiphos

5311 Fenarimol

5312 Fenbuconazole

5313 Fenbutatin oxide

5314 α-Fenchol, (±)-

5315 (±)-Fenchone

5316 Fenfluramine

5317 Fenitrothion

5318 Fenoxaprop-ethyl

5319 Fenoxycarb

5320 Fenpropathrin

5321 Fensulfothion

5322 Fentanyl

5323 Fenthion

5324 Fenvalerate

5325 Ferbam

5326 Ferrocene

5327 Ferrous gluconate

5328 Ferrous lactate

5329 Fichtelite

5330 Finasteride

5331 Fisetin

5332 Flavine adenine dinucleotide

5333 Florantyrone

5334 Fluazipop-butyl

5335 Flubenzimine

5336 Fluchloralin

5337 Flucythrinate

5338 Fludrocortisone

5339 Flumethiazide

5340 Fluocinolone acetonide

5341 Fluoranthene

5342 9*H*-Fluoren-2-amine

5343 9*H*-Fluorene

5344 9*H*-Fluorene-9-carboxylic acid

5345 9*H*-Fluorene-2,7-diamine

5346 9*H*-Fluorene-9-methanol

5347 9*H*-Fluoren-9-ol

5348 9*H*-Fluoren-9-one

5349 Fluorescein

No.	Name	Synonym	Mol. Form.	CAS RN	Mol. Wt.	Physical Form	mp/°C	bp/°C	den g cm⁻³	n_D	Solubility
5350	Fluorescein sodium	C.I. Acid Yellow 73	$C_{20}H_{10}Na_2O_5$	518-47-8	376.270	ye pow					s H₂O, EtOH, glycerol, dil acid
5351	2-Fluoroacetamide	Fluoroacetic acid amide	C_2H_4FNO	640-19-7	77.057		108	sub			s H₂O, ace; sl chl
5352	Fluoroacetic acid	Fluoroethanoic acid	$C_2H_3FO_2$	144-49-0	78.042	nd	35.2	168	1.3693³⁶		s H₂O, EtOH
5353	Fluoroacetyl chloride		C_2H_2ClFO	359-06-8	96.487	liq		72			
5354	Fluoroacetylene	Fluoroethyne	C_2HF	2713-09-9	44.027	gas	-196	-74(15)			
5355	2-Fluoroaniline		C_6H_6FN	348-54-9	111.117	pa ye liq	-29(1)	175	1.1513²¹	1.5421²⁰	i H₂O; s EtOH, eth; sl ctc
5356	3-Fluoroaniline		C_6H_6FN	372-19-0	111.117			188	1.1561¹⁹	1.5436²⁰	sl H₂O, chl; s EtOH, eth
5357	4-Fluoroaniline		C_6H_6FN	371-40-4	111.117	pa ye liq	-1.9(0.5)	182	1.1725²⁰	1.5195²⁰	sl H₂O, ctc; s EtOH, eth
5358	2-Fluorobenzaldehyde		C_7H_5FO	446-52-6	124.112	liq	-44.5	175	1.178²⁵	1.5234²⁰	
5359	3-Fluorobenzaldehyde		C_7H_5FO	456-48-4	124.112			173	1.17²⁵	1.5206²⁰	
5360	4-Fluorobenzaldehyde		C_7H_5FO	459-57-4	124.112	liq	-10	181.5	1.1810¹⁹		
5361	Fluorobenzene	Phenyl fluoride	C_6H_5F	462-06-6	96.102	liq	-42.18(0.05)	84.7(0.3)	1.0225²⁰	1.4684³⁰	sl H₂O; vs bz, eth, EtOH, lig
5362	4-Fluorobenzeneacetic acid		$C_8H_7FO_2$	405-50-5	154.139	cry (chl)	94(3)	164²			
5363	2-Fluorobenzeneacetonitrile		C_8H_6FN	326-62-5	135.139			232	1.059²⁵	1.5009²⁰	
5364	4-Fluorobenzeneacetonitrile		C_8H_6FN	459-22-3	135.139		86.0	228	1.1390²⁰	1.5002²⁰	
5365	4-Fluorobenzenemethanamine		C_7H_8FN	140-75-0	125.144			183		1.5139²⁰	
5366	4-Fluorobenzenemethanol		C_7H_7FO	459-56-3	126.128		23	210		1.5080²⁰	
5367	4-Fluorobenzenesulfonyl chloride		$C_6H_4ClFO_2S$	349-88-2	194.611	pl or nd	30	106⁹			vs bz, eth, chl
5368	2-Fluorobenzoic acid		$C_7H_5FO_2$	445-29-4	140.112	nd (a)	124.2(0.8)		1.460²⁵		sl H₂O; vs EtOH, eth; i bz; s chl
5369	3-Fluorobenzoic acid		$C_7H_5FO_2$	455-38-9	140.112	lf (w)	123.6(0.4)		1.474²⁵		sl H₂O; s eth
5370	4-Fluorobenzoic acid		$C_7H_5FO_2$	456-22-4	140.112	pr (w), mcl pr (w)	183.9(0.6)		1.479²⁵		sl H₂O, ace; s EtOH, eth
5371	2-Fluorobenzonitrile		C_7H_4FN	394-47-8	121.112			93²²			
5372	4-Fluorobenzonitrile		C_7H_4FN	1194-02-1	121.112	nd (peth)	34.8	188.8	1.1070⁵⁵	1.4925⁵⁵	sl chl; s peth
5373	2-Fluorobenzoyl chloride		C_7H_4ClFO	393-52-2	158.557		2.0	91¹⁵	1.328²⁵	1.5365²⁰	
5374	3-Fluorobenzoyl chloride		C_7H_4ClFO	1711-07-5	158.557	liq	-30	189	1.304²⁵	1.5285²⁰	
5375	4-Fluorobenzoyl chloride		C_7H_4ClFO	403-43-0	158.557		9	82²⁰	1.342²⁵	1.5296²⁰	
5376	2-Fluoro-1,1'-biphenyl		$C_{12}H_9F$	321-60-8	172.197		73.5	248	1.2452²⁵		s EtOH, eth, chl, peth; sl lig
5377	4-Fluoro-1,1'-biphenyl		$C_{12}H_9F$	324-74-3	172.197	pr	74.2	253	1.247²⁵		sl EtOH; s eth, gl HOAc
5378	1-Fluorobutane	Butyl fluoride	C_4H_9F	2366-52-1	76.112	liq	-134	31(8)	0.7789²⁰	1.3396²⁰	vs EtOH
5379	2-Fluorobutane	sec-Butyl fluoride	C_4H_9F	359-01-3	76.112	vol liq or gas	-121.4	23(9)	0.7559²⁵		
5380	Fluorocyclohexane	Cyclohexyl fluoride	$C_6H_{11}F$	372-46-3	102.149		13	103(5)	0.9279²⁰	1.4146²⁰	i H₂O; s py
5381	1-Fluorocyclohexene		C_6H_9F	694-51-9	100.133			96.5		1.4441²⁵	
5382	5-Fluorocytosine	4-Amino-5-fluoro-2-hydroxy-pyrimidine	$C_4H_4FN_3O$	2022-85-7	129.092	wh cry	296 dec				
5383	1-Fluorodecane	Decyl fluoride	$C_{10}H_{21}F$	334-56-5	160.272	liq	-35	189(8)	0.8194²⁰	1.4085	vs eth
5384	Fluorodifen	2-Nitro-1-(4-nitrophenoxy)-4-(trifluoromethyl)benzene	$C_{13}H_7F_3N_2O_5$	15457-05-3	328.200		91.7(0.5)				
5385	1-Fluoro-2,4-dinitrobenzene	2,4-Dinitrophenyl fluoride	$C_6H_3FN_2O_4$	70-34-8	186.097		25.8	296	1.4718⁵⁴	1.5690²⁰	s EtOH; sl chl
5386	Fluoroethane	Ethyl fluoride	C_2H_5F	353-36-6	48.059	col gas	-143.2	-37.7(0.3) (p>1 atm)	0.7182²⁰	1.2656²⁰	sl H₂O; vs EtOH, eth
5387	2-Fluoroethanol	Ethylene fluorohydrin	C_2H_5FO	371-62-0	64.058	liq	-26.4	85(14)	1.1040²⁰	1.3647¹⁸	msc H₂O, EtOH, eth; vs ace; sl chl
5388	Fluoroethene	Vinyl fluoride	C_2H_3F	75-02-5	46.043	col gas	-160.5	-72			i H₂O; s EtOH, ace
5389	1-Fluoroheptane		$C_7H_{15}F$	661-11-0	118.192	liq	-73	122(9)	0.8062²⁰	1.3854²⁰	i H₂O; s eth, ace, bz; vs peth
5390	1-Fluorohexane	Hexyl fluoride	$C_6H_{13}F$	373-14-8	104.165	liq	-103	88(8)	0.7995²⁰	1.3738²⁰	s eth, bz
5391	1-Fluoro-2-iodobenzene		C_6H_4FI	348-52-7	221.998	liq	-41.5	188.6		1.5910²⁰	s ace, bz, chl
5392	1-Fluoro-4-iodobenzene		C_6H_4FI	352-34-1	221.998	liq	-27	183	1.9523¹⁵	1.5270²²	i H₂O; s EtOH, eth, ace
5393	1-Fluoro-3-isothiocyanatobenzene		C_7H_4FNS	404-72-8	153.177			227	1.27²⁵	1.6186²⁰	
5394	1-Fluoro-4-isothiocyanatobenzene		C_7H_4FNS	1544-68-9	153.177		27	228			

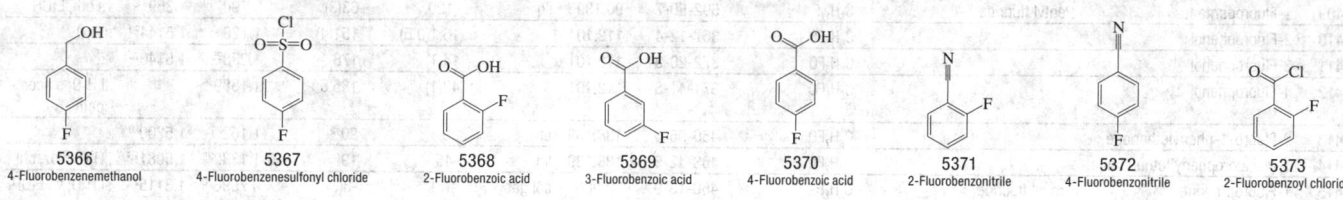

5350
Fluorescein sodium

5351
2-Fluoroacetamide

5352
Fluoroacetic acid

5353
Fluoroacetyl chloride

5354
Fluoroacetylene

5355
2-Fluoroaniline

5356
3-Fluoroaniline

5357
4-Fluoroaniline

5358
2-Fluorobenzaldehyde

5359
3-Fluorobenzaldehyde

5360
4-Fluorobenzaldehyde

5361
Fluorobenzene

5362
4-Fluorobenzeneacetic acid

5363
2-Fluorobenzeneacetonitrile

5364
4-Fluorobenzeneacetonitrile

5365
4-Fluorobenzenemethanamine

5366
4-Fluorobenzenemethanol

5367
4-Fluorobenzenesulfonyl chloride

5368
2-Fluorobenzoic acid

5369
3-Fluorobenzoic acid

5370
4-Fluorobenzoic acid

5371
2-Fluorobenzonitrile

5372
4-Fluorobenzonitrile

5373
2-Fluorobenzoyl chloride

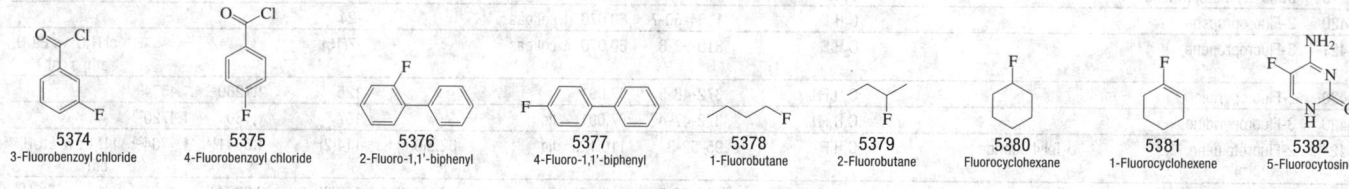

5374
3-Fluorobenzoyl chloride

5375
4-Fluorobenzoyl chloride

5376
2-Fluoro-1,1'-biphenyl

5377
4-Fluoro-1,1'-biphenyl

5378
1-Fluorobutane

5379
2-Fluorobutane

5380
Fluorocyclohexane

5381
1-Fluorocyclohexene

5382
5-Fluorocytosine

5383
1-Fluorodecane

5384
Fluorodifen

5385
1-Fluoro-2,4-dinitrobenzene

5386
Fluoroethane

5387
2-Fluoroethanol

5388
Fluoroethene

5389
1-Fluoroheptane

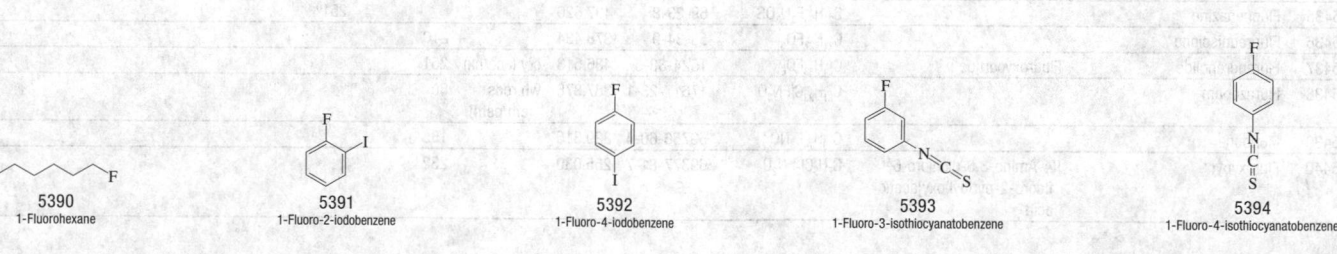

5390
1-Fluorohexane

5391
1-Fluoro-2-iodobenzene

5392
1-Fluoro-4-iodobenzene

5393
1-Fluoro-3-isothiocyanatobenzene

5394
1-Fluoro-4-isothiocyanatobenzene

No.	Name	Synonym	Mol. Form.	CAS RN	Mol. Wt.	Physical Form	mp/°C	bp/°C	den g cm⁻³	n_D	Solubility
5395	Fluoromethane	Methyl fluoride	CH₃F	593-53-3	34.033	col gas	-143.3	-78.4	0.5557[25] (p>1 atm)	1.1674[25]	sl H₂O, bz, chl; vs EtOH, eth
5396	1-Fluoro-2-methoxybenzene		C₇H₇FO	321-28-8	126.128	liq	-39	154.5	1.5489[17]	1.4969[17]	i H₂O; s eth, ctc
5397	1-Fluoro-3-methoxybenzene		C₇H₇FO	456-49-5	126.128	liq	-35	159	1.104[25]	1.4876[20]	
5398	1-Fluoro-4-methoxybenzene		C₇H₇FO	459-60-9	126.128	liq	-45	157	1.1781[18]	1.4886[18]	s eth
5399	4-Fluoro-2-methylaniline		C₇H₈FN	452-71-1	125.144		14.2	94[16]	1.1263[18]	1.5363[18]	s eth, ace, bz, ctc
5400	(Fluoromethyl)benzene		C₇H₇F	350-50-5	110.129	liq	-35	142(2)	1.0228[25]	1.4892[25]	s ctc
5401	2-Fluoro-4-methyl-1-nitrobenzene	3-Fluoro-4-nitrotoluene	C₇H₆FNO₂	446-34-4	155.127	nd (al)	53.2	97[3]	1.4380[25]		
5402	2-Fluoro-2-methylpropane	tert-Butyl fluoride	C₄H₉F	353-61-7	76.112	col gas		12.1			
5403	1-Fluoronaphthalene		C₁₀H₇F	321-38-0	146.161	liq	-14.0(0.2)	213(14)	1.1322[20]	1.5939[20]	i H₂O; s EtOH, eth, bz, chl, HOAc
5404	2-Fluoronaphthalene		C₁₀H₇F	323-09-1	146.161	nd (al)	58(1)	212			i H₂O; s EtOH, eth, bz, chl, HOAc
5405	1-Fluoro-2-nitrobenzene	o-Fluoronitrobenzene	C₆H₄FNO₂	1493-27-2	141.100	ye liq	-6	215 dec	1.3285[18]	1.5489[17]	vs eth, EtOH
5406	1-Fluoro-3-nitrobenzene	m-Fluoronitrobenzene	C₆H₄FNO₂	402-67-5	141.100	ye cry	41	197(9)	1.3254[19]	1.5262[15]	i H₂O; s EtOH, eth; sl bz
5407	1-Fluoro-4-nitrobenzene	p-Fluoronitrobenzene	C₆H₄FNO₂	350-46-9	141.100	ye nd	26.5(0.5)	205	1.3300[20]	1.5316[20]	i H₂O; s EtOH, eth; sl ctc
5408	1-Fluorooctane	Octyl fluoride	C₈H₁₇F	463-11-6	132.219	liq	-64	146(9)	0.8116[20]	1.3946[20]	
5409	1-Fluoropentane	Pentyl fluoride	C₅H₁₁F	592-50-7	90.139	liq	-120	63(3)	0.7907[20]	1.3591[2-]	vs eth, EtOH
5410	2-Fluorophenol		C₆H₅FO	367-12-4	112.101		16.1(0.5)	151(4)	1.120[25]	1.5144[20]	s H₂O
5411	3-Fluorophenol		C₆H₅FO	372-20-3	112.101		14(1)	178	1.238[25]	1.5140[20]	
5412	4-Fluorophenol		C₆H₅FO	371-41-5	112.101		48(1)	185.5	1.1889[56]		sl H₂O; s ace, peth
5413	2-Fluoro-1-phenylethanone		C₈H₇FO	450-95-3	138.139	pl	29	90[12]	1.152[20]	1.5200[20]	
5414	1-(4-Fluorophenyl)ethanone		C₈H₇FO	403-42-9	138.139	liq	-45	196	1.1382[25]	1.5081[25]	i H₂O; s bz, chl
5415	1-Fluoropropane	Propyl fluoride	C₃H₇F	460-13-9	62.086	col gas	-159	-3(2)	0.7596[20] (p>1 atm)	1.3115[20]	sl H₂O; vs EtOH, eth
5416	2-Fluoropropane	Isopropyl fluoride	C₃H₇F	420-26-8	62.086	gas		-10(3)			sl H₂O
5417	1-Fluoro-2-propanone	Fluoroacetone	C₃H₅FO	430-51-3	76.069			77	1.0288[20]	1.3700[20]	
5418	cis-1-Fluoropropene		C₃H₅F	19184-10-2	60.070	col gas		-15(16)			
5419	trans-1-Fluoropropene		C₃H₅F	20327-65-5	60.070	col gas	≈-20				
5420	2-Fluoropropene		C₃H₅F	1184-60-7	60.070	col gas		-24			
5421	3-Fluoropropene		C₃H₅F	818-92-8	60.070	col gas		-7(15)			sl H₂O; vs EtOH, eth; s chl
5422	2-Fluoropyridine		C₅H₄FN	372-48-5	97.091			125	1.1280[20]	1.4574[20]	
5423	3-Fluoropyridine		C₅H₄FN	372-47-4	97.091	liq		107	1.130	1.4720[20]	
5424	2-Fluorotoluene	o-Tolyl fluoride	C₇H₇F	95-52-3	110.129	liq	-62.5(0.5)	114(2)	1.0041[13]	1.4704[20]	i H₂O; vs EtOH, eth
5425	3-Fluorotoluene	m-Tolyl fluoride	C₇H₇F	352-70-5	110.129	liq	-89.2(0.4)	116(2)	0.9974[20]	1.4691[20]	i H₂O; vs EtOH, eth
5426	4-Fluorotoluene	p-Tolyl fluoride	C₇H₇F	352-32-9	110.129	liq	-56.6(0.1)	116.6(0.4)	0.9975[20]	1.4699[20]	i H₂O; vs EtOH, eth
5427	1-Fluoro-2-(trichloromethyl)-benzene		C₇H₄Cl₃F	488-98-2	213.464			95[12]	1.453[25]	1.5432[20]	
5428	1-Fluoro-2-(trifluoromethyl)-benzene		C₇H₄F₄	392-85-8	164.101			114.5	1.293[25]	1.4040[25]	
5429	1-Fluoro-3-(trifluoromethyl)-benzene		C₇H₄F₄	401-80-9	164.101	liq	-81.5	100.6(1)	1.3021[17]		
5430	1-Fluoro-4-(trifluoromethyl)-benzene		C₇H₄F₄	402-44-8	164.101	liq	-41.7	103.5	1.293[25]	1.4025[20]	
5431	Fluorotrimethylsilane	Trimethylsilyl fluoride	C₃H₉FSi	420-56-4	92.187	vol liq or gas		16.4			
5432	5-Fluorouracil	5-Fluoro-2,4(1H,3H)-Pyrimidinedione	C₄H₃FN₂O₂	51-21-8	130.077	cry (w, MeOH-eth)	284(4)	369(11)			
5433	Fluoxetine		C₁₇H₁₈F₃NO	54910-89-3	309.326	oil					
5434	Fluoxymesterone		C₂₀H₂₉FO₃	76-43-7	336.440		270				
5435	Fluphenazine		C₂₂H₂₆F₃N₃OS	69-23-8	437.520			251[0.3]			
5436	Fluprednisolone		C₂₁H₂₇FO₅	53-34-9	378.434		210				
5437	Flurandrenolide	Fludroxycortide	C₂₄H₃₃FO₆	1524-88-5	436.513	cry (ace/hx)	251				
5438	Flurazepam		C₂₁H₂₃ClFN₃O	17617-23-1	387.878	wh rods (eth/peth)	80				
5439	Fluridone		C₁₉H₁₄F₃NO	59756-60-4	329.315		155				
5440	Fluroxypyr	[(4-Amino-3,5-dichloro-6-fluoro-2-pyridyl)oxy]acetic acid	C₇H₅Cl₂FN₂O₃	69377-81-7	255.030		232				

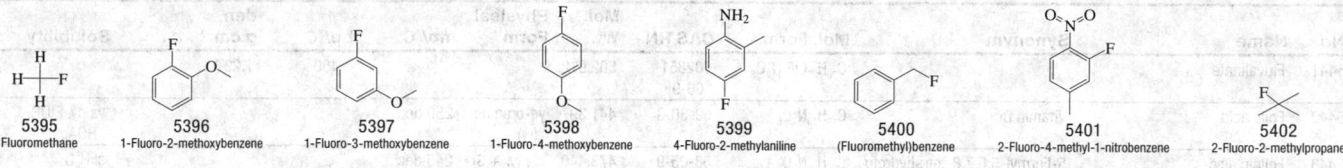

5395 Fluoromethane

5396 1-Fluoro-2-methoxybenzene

5397 1-Fluoro-3-methoxybenzene

5398 1-Fluoro-4-methoxybenzene

5399 4-Fluoro-2-methylaniline

5400 (Fluoromethyl)benzene

5401 2-Fluoro-4-methyl-1-nitrobenzene

5402 2-Fluoro-2-methylpropane

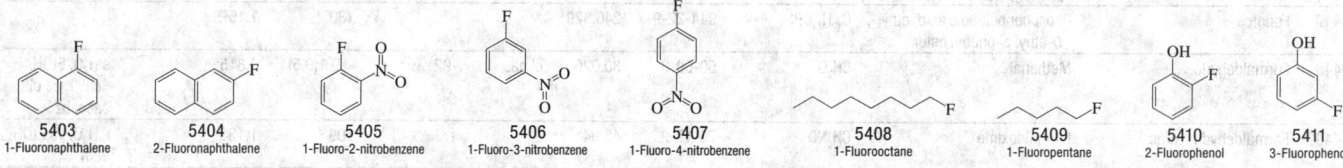

5403 1-Fluoronaphthalene

5404 2-Fluoronaphthalene

5405 1-Fluoro-2-nitrobenzene

5406 1-Fluoro-3-nitrobenzene

5407 1-Fluoro-4-nitrobenzene

5408 1-Fluorooctane

5409 1-Fluoropentane

5410 2-Fluorophenol

5411 3-Fluorophenol

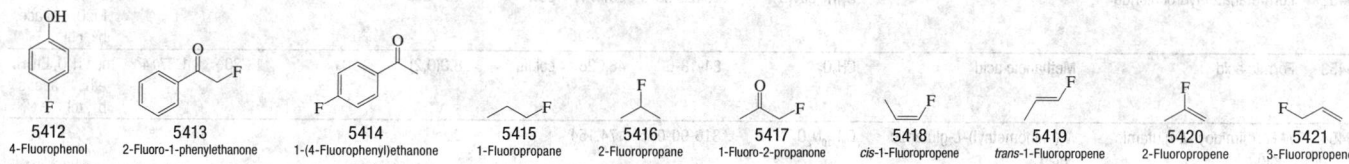

5412 4-Fluorophenol

5413 2-Fluoro-1-phenylethanone

5414 1-(4-Fluorophenyl)ethanone

5415 1-Fluoropropane

5416 2-Fluoropropane

5417 1-Fluoro-2-propanone

5418 *cis*-1-Fluoropropene

5419 *trans*-1-Fluoropropene

5420 2-Fluoropropene

5421 3-Fluoropropene

5422 2-Fluoropyridine

5423 3-Fluoropyridine

5424 2-Fluorotoluene

5425 3-Fluorotoluene

5426 4-Fluorotoluene

5427 1-Fluoro-2-(trichloromethyl)benzene

5428 1-Fluoro-2-(trifluoromethyl)benzene

5429 1-Fluoro-3-(trifluoromethyl)benzene

5430 1-Fluoro-4-(trifluoromethyl)benzene

5431 Fluorotrimethylsilane

5432 5-Fluorouracil

5433 Fluoxetine

5434 Fluoxymesterone

5435 Fluphenazine

5436 Fluprednisolone

5437 Flurandrenolide

5438 Flurazepam

5439 Fluridone

5440 Fluroxypyr

No.	Name	Synonym	Mol. Form.	CAS RN	Mol. Wt.	Physical Form	mp/°C	bp/°C	den g cm⁻³	n_D	Solubility
5441	Fluvalinate		$C_{26}H_{22}ClF_3N_2O_3$	102851-06-9	502.912			>450	1.29^{25}		
5442	Folic acid	Vitamin Bc	$C_{19}H_{19}N_7O_6$	59-30-3	441.397	ye-oran nd (w)	250 dec				vs py, EtOH, HOAc
5443	Folinic acid	5-Formyl-5,6,7,8-tetrahydrofolic acid	$C_{20}H_{23}N_7O_7$	58-05-9	473.440	cry (w + 3)	245 dec				sl H_2O
5444	Folpet	1H-Isoindole-1,3(2H)-dione, 2-[(trichloromethyl)thio]-	$C_9H_4Cl_3NO_2S$	133-07-3	296.558		181.0(0.6)				
5445	Fomesafen		$C_{15}H_{10}ClF_3N_2O_6S$	72178-02-0	438.762		220		1.28^{20}		
5446	Fomocaine	4-[3-[4-(Phenoxymethyl)phenyl]propyl]morpholine	$C_{20}H_{25}NO_2$	17692-39-6	311.419	col cry	53	$239^{1.1}$			
5447	Fonofos	Phosphonodithioic acid, ethyl-, O-ethyl S-phenyl ester	$C_{10}H_{15}OPS_2$	944-22-9	246.329			$130^{0.1}$	1.16^{25}		
5448	Formaldehyde	Methanal	CH_2O	50-00-0	30.026	col gas	-92	-19.1(0.5)	0.815^{-20}		s H_2O, EtOH, chl; msc eth, ace, bz
5449	Formaldehyde oxime	Formaldoxime	CH_3NO	75-17-2	45.041		1.3	109^{15}	1.133^{25}		s H_2O; vs EtOH, eth
5450	Formamide	Methanamide	CH_3NO	75-12-7	45.041	col liq	2.57(0.02)	217(3)	1.1334^{20}	1.4472^{20}	msc H_2O, EtOH; sl eth; s ace; i bz, chl
5451	Formamidinesulfinic acid	Aminoiminomethanesulfinic acid	$CH_4N_2O_2S$	1758-73-2	108.120	nd (al)	144 dec				vs H_2O; i eth, bz
5452	Formetanate hydrochloride		$C_{11}H_{16}ClN_3O_2$	23422-53-9	257.717	pow	201 dec				vs H_2O; s MeOH; sl ace, hx, chl
5453	Formic acid	Methanoic acid	CH_2O_2	64-18-6	46.026	col liq	8.3(0.2)	101	1.220^{20}	1.3714^{20}	msc H_2O, EtOH, eth; vs ace; s bz, tol
5454	N-Formimidoyl-L-glutamic acid	N-(Iminomethyl)-L-glutamic acid	$C_6H_{10}N_2O_4$	816-90-0	174.154		90				
5455	Formononetin	7-Hydroxy-3-(4-methoxyphenyl)-4H-1-benzopyran-4-one	$C_{16}H_{12}O_4$	485-72-3	268.264		256.5				
5456	Formothion		$C_6H_{12}NO_4PS_2$	2540-82-1	257.267	visc ye oil	25.5		1.361^{20}	1.5541^{20}	sl H_2O; misc os
5457	2-Formylbenzoic acid		$C_8H_6O_3$	119-67-5	150.132		100.5(0.2)		1.404^{25}		s H_2O; vs EtOH, eth
5458	3-Formylbenzoic acid		$C_8H_6O_3$	619-21-6	150.132	nd (w)	175.0(0.2)				vs H_2O, eth, EtOH
5459	4-Formylbenzoic acid		$C_8H_6O_3$	619-66-9	150.132		249.8(0.5)				sl H_2O; vs EtOH; s eth, chl
5460	3-Formylbenzonitrile		C_8H_5NO	24964-64-5	131.132		76.5	210			vs H_2O, EtOH, eth, chl
5461	4-Formylbenzonitrile		C_8H_5NO	105-07-7	131.132		100.5	133^{12}			s H_2O; vs EtOH, eth, chl
5462	6-Formyl-2,3-dimethoxybenzoic acid	Opianic acid	$C_{10}H_{10}O_5$	519-05-1	210.183	nd (w)	150				s EtOH, eth
5463	Formylferrocene		$C_{11}H_{10}FeO$	12093-10-6	214.041		118.5	$70^{0.1}$			
5464	Formyl fluoride	Fluoroformaldehyde	CHFO	1493-02-3	48.016	col gas	-142.2	-26.5	1.1950^{-30}		
5465	N-(4-Formylphenyl)acetamide		$C_9H_9NO_2$	122-85-0	163.173	pr (w)	158.0				vs H_2O, bz
5466	Fosetyl-Al	Aluminum tris(O-ethylphosphonate)	$C_6H_{18}AlO_9P_3$	39148-24-8	354.105		>300				
5467	Fosthietan		$C_6H_{12}NO_3PS_2$	21548-32-3	241.268	ye oil			1.3^{25}	1.5348^{25}	s ace, chl, MeOH, tol
5468	Fraxin		$C_{16}H_{18}O_{10}$	524-30-1	370.308	ye nd (al)	205				
5469	DL-Fructose	α-Acrose	$C_6H_{12}O_6$	6035-50-3	180.155	nd	130		1.665^{16}		
5470	L-Fructose		$C_6H_{12}O_6$	7776-48-9	180.155	wh cry	102				s H_2O
5471	β-D-Fructose	β-Levulose	$C_6H_{12}O_6$	53188-23-1	180.155	pr or nd (w) orth pr (al)	103 dec		1.60^{20}		vs H_2O, ace; s EtOH, MeOH, py
5472	D-Fructose 6-phosphate	Hexose monophosphate	$C_6H_{13}O_9P$	643-13-0	260.135						vs H_2O
5473	Fucoxanthin		$C_{42}H_{58}O_6$	3351-86-8	658.906	red pl (eth) hex pl (dil al)	168				vs eth, EtOH
5474	Fulminic acid	Carbyloxime	CHNO	506-85-4	43.025		unstable in pure form				s eth
5475	Fulvene		C_6H_6	497-20-1	78.112			7^{56}	0.8241^{20}	1.4920^{20}	i H_2O; s bz, chl
5476	Fumaric acid	trans-2-Butenedioic acid	$C_4H_4O_4$	110-17-8	116.073	nd, mcl pr or lf (w)	122(103)	165 sub	1.635^{20}		sl H_2O, eth, ace; s EtOH, con sulf
5477	Fumigatin	3-Hydroxy-2-methoxy-5-methyl-2,5-cyclohexadiene-1,4-dione	$C_8H_8O_4$	484-89-9	168.148	br nd or pl (peth)	116				vs ace, bz, eth, EtOH

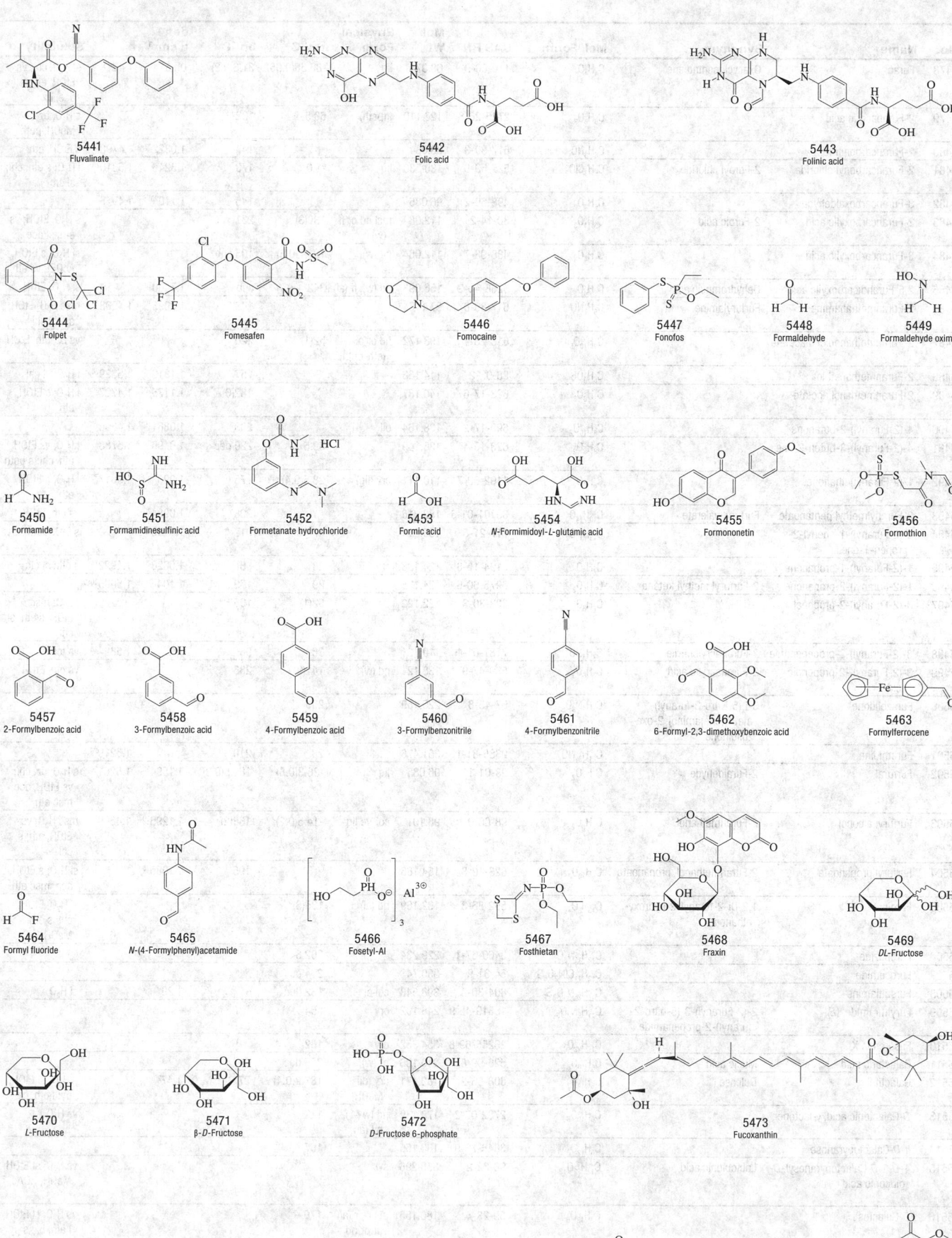

5441 Fluvalinate

5442 Folic acid

5443 Folinic acid

5444 Folpet

5445 Fomesafen

5446 Fomocaine

5447 Fonofos

5448 Formaldehyde

5449 Formaldehyde oxime

5450 Formamide

5451 Formamidinesulfinic acid

5452 Formetanate hydrochloride

5453 Formic acid

5454 *N*-Formimidoyl-*L*-glutamic acid

5455 Formononetin

5456 Formothion

5457 2-Formylbenzoic acid

5458 3-Formylbenzoic acid

5459 4-Formylbenzoic acid

5460 3-Formylbenzonitrile

5461 4-Formylbenzonitrile

5462 6-Formyl-2,3-dimethoxybenzoic acid

5463 Formylferrocene

5464 Formyl fluoride

5465 *N*-(4-Formylphenyl)acetamide

5466 Fosetyl-Al

5467 Fosthietan

5468 Fraxin

5469 *DL*-Fructose

5470 *L*-Fructose

5471 β-*D*-Fructose

5472 *D*-Fructose 6-phosphate

5473 Fucoxanthin

5474 Fulminic acid

5475 Fulvene

5476 Fumaric acid

5477 Fumigatin

No.	Name	Synonym	Mol. Form.	CAS RN	Mol. Wt.	Physical Form	mp/°C	bp/°C	den g cm⁻³	n_D	Solubility
5478	Furan	Oxacyclopentadiene	C_4H_4O	110-00-9	68.074	liq	-85.58(0.05)	31.3(0.2)	0.9514[20]	1.4214[20]	sl H_2O, chl; vs EtOH, eth; s ace, bz
5479	2-Furanacetic acid		$C_6H_6O_3$	2745-26-8	126.110	lf(peth)	68.5	102[0.4]			s H_2O, bz, MeOH, peth
5480	2-Furancarbonitrile		C_5H_3NO	617-90-3	93.084			147	1.0822[20]	1.4798[20]	s EtOH, eth
5481	2-Furancarbonyl chloride	2-Furoyl chloride	$C_5H_3ClO_2$	527-69-5	130.530	liq	-1.0	173	1.324[25]	1.5310[20]	i H_2O; s eth, chl; sl ctc
5482	3-Furancarboxaldehyde		$C_5H_4O_2$	498-60-2	96.085			145	1.110[20]	1.4945[20]	
5483	2-Furancarboxylic acid	2-Furoic acid	$C_5H_4O_3$	88-14-2	112.084	mcl nd or lf (w)	130(3)	231			s H_2O, EtOH; eth; sl ace
5484	3-Furancarboxylic acid		$C_5H_4O_3$	488-93-7	112.084	nd (w)	121.7(0.6)	105 sub			sl H_2O; s EtOH, AcOEt; vs eth
5485	2,5-Furandicarboxylic acid	Dehydromucic acid	$C_6H_4O_5$	3238-40-2	156.093	nd (w), lf (al)	342	sub	1.7400[20]		sl H_2O, EtOH
5486	2-Furanmethanamine	Furfurylamine	C_5H_7NO	617-89-0	97.116			145.5	1.0995[20]	1.4908[20]	msc H_2O, EtOH; s eth, chl
5487	2-Furanmethanediol diacetate		$C_9H_{10}O_5$	613-75-2	198.172	nd or pl (eth-peth)	52(1)	220			vs bz, eth, EtOH
5488	2-Furanmethanethiol		C_5H_6OS	98-02-2	114.166			157	1.1319[20]	1.5329[20]	i H_2O; sl chl
5489	2-Furanmethanol acetate		$C_7H_8O_3$	623-17-6	140.137			182(6)	1.1175[20]	1.4327[20]	i H_2O; s EtOH, eth
5490	4-(2-Furanyl)-2-butanone		$C_8H_{10}O_2$	699-17-2	138.164	oil		203	1.0361[19]	1.4696[17]	
5491	4-(2-Furanyl)-3-buten-2-one		$C_8H_8O_2$	623-15-4	136.149		39.5	229 dec	1.0496[57]	1.5788[45]	i H_2O; vs EtOH, eth, chl; s peth
5492	1-(2-Furanyl)ethanone		$C_6H_6O_2$	1192-62-7	110.111	cry (lig)	28.5(0.4)	175	1.098[20]	1.5017[20]	i H_2O; s EtOH, eth
5493	2-Furanylmethyl pentanoate	Furfuryl valerate	$C_{10}H_{14}O_3$	36701-01-6	182.216			228	1.0284[20]		vs eth, EtOH
5494	3-(2-Furanyl)-1-phenyl-2-propen-1-one		$C_{13}H_{10}O_2$	717-21-5	198.217		47	317	1.1140[20]		s EtOH, eth
5495	1-(2-Furanyl)-1-propanone		$C_7H_8O_2$	3194-15-8	124.138	cry	28	88[14]	1.0626[28]	1.4922[25]	s eth; sl ctc
5496	1-(2-Furanyl)-2-propanone	2-Furfuryl methyl ketone	$C_7H_8O_2$	6975-60-6	124.138		29	179.5	1.104[20]	1.5035[20]	
5497	3-(2-Furanyl)-2-propenal		$C_7H_6O_2$	623-30-3	122.122		52(1)	135[14]			i H_2O; msc EtOH; s eth; sl chl
5498	3-(2-Furanyl)-2-propenenitrile	2-Furanacrylonitrile	C_7H_5NO	7187-01-1	119.121		38	96		1.5824[25]	vs tol
5499	3-(2-Furanyl)-2-propenoic acid	2-Furanacrylic acid	$C_7H_6O_3$	539-47-9	138.121	nd (w)	141(3)	286			vs eth, EtOH
5500	Furazolidone	3-[[(5-Nitro-2-furanyl)methylene]amino]-2-oxa-zolidinone	$C_8H_7N_3O_5$	67-45-8	225.159		255				
5501	Furethidine		$C_{21}H_{31}NO_4$	2385-81-1	361.476		28	210[0.5]		1.5219[20]	
5502	Furfural	2-Furaldehyde	$C_5H_4O_2$	98-01-1	96.085	liq	-38.3(0.8)	161.5(0.3)	1.1594[20]	1.5261[20]	s H_2O, bz, chl; vs EtOH, ace; msc eth
5503	Furfuryl alcohol	2-Furanmethanol	$C_5H_6O_2$	98-00-0	98.101	col-ye liq	-14.5(0.2)	168(2)	1.1296[20]	1.4869[20]	msc H_2O; vs EtOH, eth; s chl
5504	Furfuryl propanoate	2-Furanmethanol, propanoate	$C_8H_{10}O_3$	623-19-8	154.163			195	1.1085[20]		sl H_2O; s EtOH, ace; msc eth
5505	Furoin	1,2-Di-2-furanyl-2-hydroxy-ethanone	$C_{10}H_8O_4$	552-86-3	192.169	nd (al)	135(3)				sl H_2O, EtOH, chl; s eth, MeOH
5506	Furonazide		$C_{12}H_{11}N_3O_2$	3460-67-1	229.234		202.3				
5507	Furosemide		$C_{12}H_{11}ClN_2O_5S$	54-31-9	330.743		204 dec				
5508	Fursultiamine		$C_{17}H_{26}N_4O_3S_2$	804-30-8	398.543	col pr	132 dec		1.29		sl H_2O
5509	Furylfuramide, (E)-	2-(2-Furanyl)-3-(5-nitro-2-furanyl)-2-propenamide	$C_{11}H_8N_2O_5$	18819-45-9	248.192	cry	154				
5510	Fusarenon X		$C_{17}H_{22}O_8$	23255-69-8	354.352	cry	182				
5511	Galactaric acid	Mucic acid	$C_6H_{10}O_8$	526-99-8	210.138	pr (w)	255 dec				
5512	Galactitol	Dulcose	$C_6H_{14}O_6$	608-66-2	182.171	cry (dil MeOH)	187.2(0.4)	277[1]	1.47[20]		s H_2O; sl EtOH, py; i eth, bz
5513	D-Galactonic acid, γ-lactone		$C_6H_{10}O_6$	2782-07-2	178.139	nd (w+1), nd (al)	112				vs H_2O
5514	α-D-Galactopyranose		$C_6H_{12}O_6$	3646-73-9	180.155		167				
5515	4-O-β-D-Galactopyranosyl-D-gluconic acid	Lactobionic acid	$C_{12}H_{22}O_{12}$	96-82-2	358.296	syr					vs H_2O; sl EtOH, MeOH, HOAc; i eth
5516	D-Galactose		$C_6H_{12}O_6$	59-23-4	180.155	pl or pr (al) pr or nd (w+1)	170				vs H_2O; sl EtOH; i eth, bz; s py
5517	D-Galacturonic acid		$C_6H_{10}O_7$	685-73-4	194.139	nd (w)	166 (β)				s H_2O, EtOH; i eth

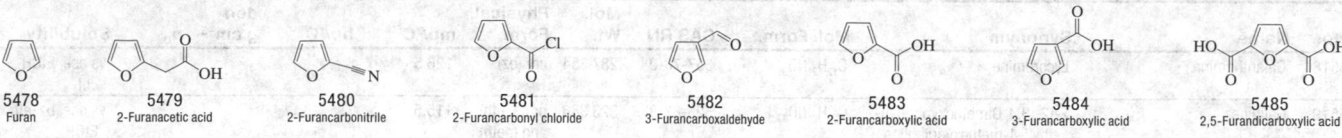

5478	5479	5480	5481	5482	5483	5484	5485
Furan	2-Furanacetic acid	2-Furancarbonitrile	2-Furancarbonyl chloride	3-Furancarboxaldehyde	2-Furancarboxylic acid	3-Furancarboxylic acid	2,5-Furandicarboxylic acid

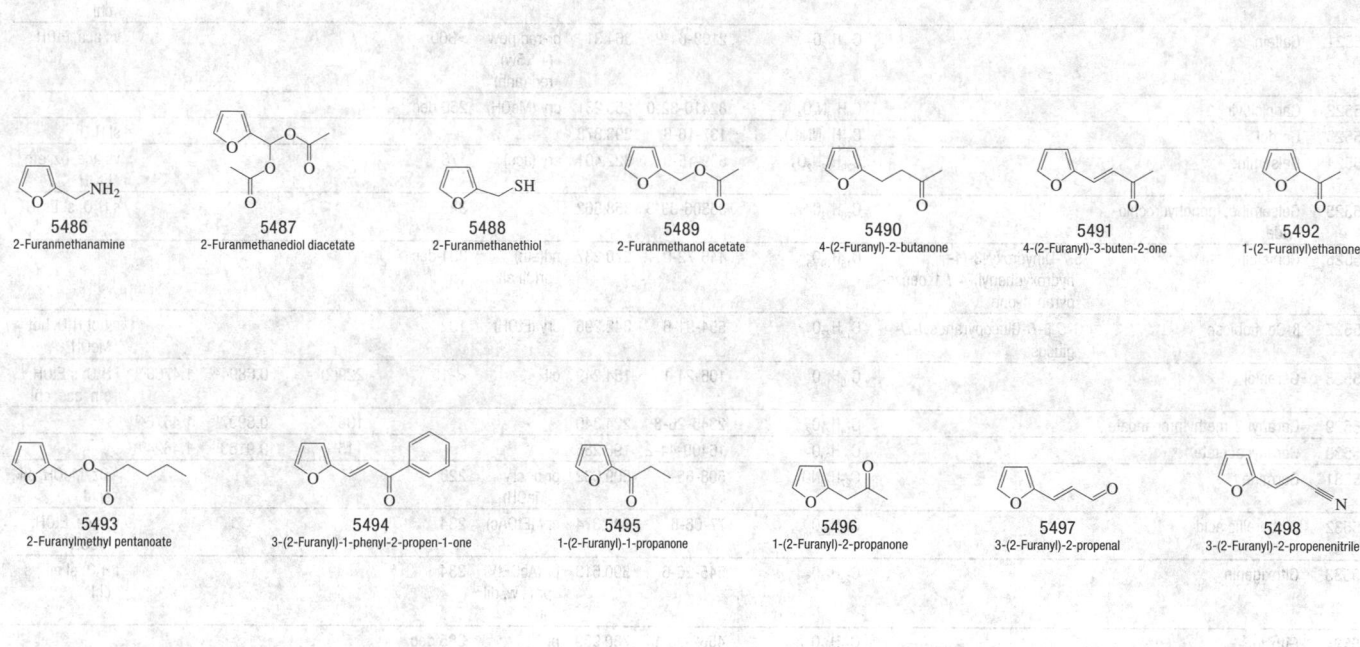

5486	5487	5488	5489	5490	5491	5492
2-Furanmethanamine	2-Furanmethanediol diacetate	2-Furanmethanethiol	2-Furanmethanol acetate	4-(2-Furanyl)-2-butanone	4-(2-Furanyl)-3-buten-2-one	1-(2-Furanyl)ethanone

5493	5494	5495	5496	5497	5498
2-Furanylmethyl pentanoate	3-(2-Furanyl)-1-phenyl-2-propen-1-one	1-(2-Furanyl)-1-propanone	1-(2-Furanyl)-2-propanone	3-(2-Furanyl)-2-propenal	3-(2-Furanyl)-2-propenenitrile

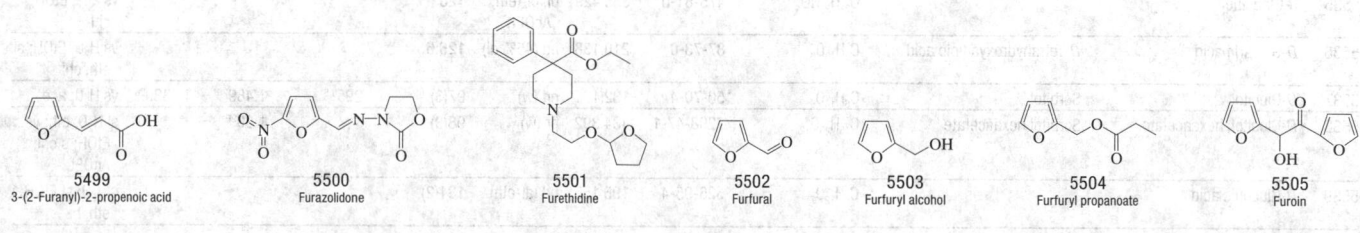

5499	5500	5501	5502	5503	5504	5505
3-(2-Furanyl)-2-propenoic acid	Furazolidone	Furethidine	Furfural	Furfuryl alcohol	Furfuryl propanoate	Furoin

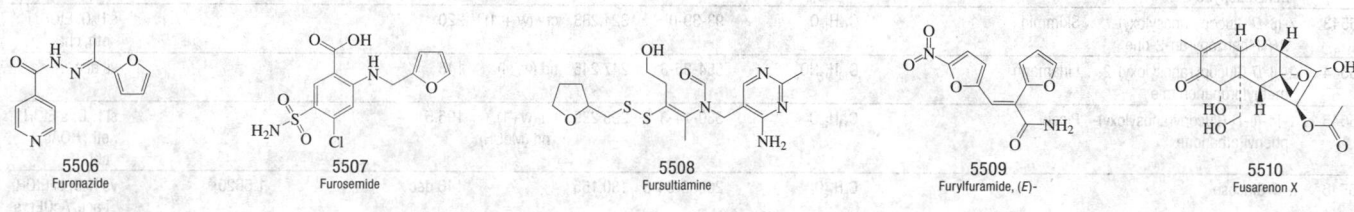

5506	5507	5508	5509	5510
Furonazide	Furosemide	Fursultiamine	Furylfuramide, (E)-	Fusarenon X

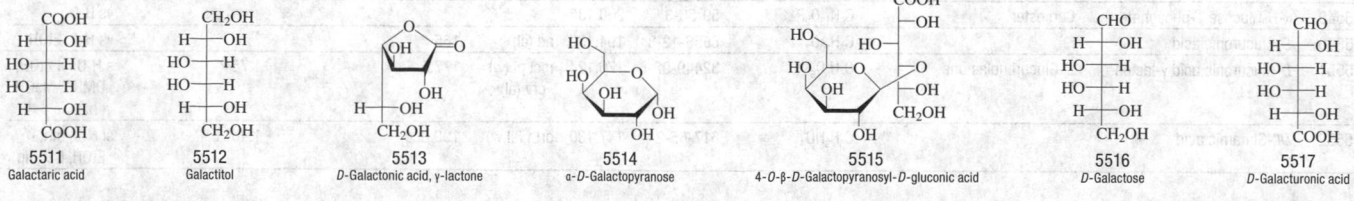

5511	5512	5513	5514	5515	5516	5517
Galactaric acid	Galactitol	D-Galactonic acid, γ-lactone	α-D-Galactopyranose	4-O-β-D-Galactopyranosyl-D-gluconic acid	D-Galactose	D-Galacturonic acid

No.	Name	Synonym	Mol. Form.	CAS RN	Mol. Wt.	Physical Form	mp/°C	bp/°C	den g cm⁻³	n_D	Solubility
5518	Galanthamine	Lycoremine	$C_{17}H_{21}NO_3$	357-70-0	287.354	cry (bz)	126.5				vs ace, EtOH, chl
5519	Galipine	2-[2-(3,4-Dimethoxyphenyl)-ethyl]-4-methoxyquinoline	$C_{20}H_{21}NO_3$	525-68-8	323.386	pr (al, eth) nd (peth)	115.5				vs ace, bz, eth, EtOH
5520	Gallamine triethiodide		$C_{30}H_{60}I_3N_3O_3$	65-29-2	891.528		147.5				vs H_2O, EtOH; sl eth, ace, bz, chl
5521	Gallein		$C_{20}H_{12}O_7$	2103-64-2	364.31	br-red pow (+1.5w) red (anh)	>300				vs ace, EtOH
5522	Ganciclovir		$C_9H_{13}N_5O_4$	82410-32-0	255.231	cry (MeOH)	250 dec				
5523	Gardol		$C_{15}H_{28}NNaO_3$	137-16-6	293.378						sl H_2O
5524	Gelsemine		$C_{20}H_{22}N_2O_2$	509-15-9	322.401	cry (ace)	178				vs ace, bz, eth, EtOH
5525	Gelsemine, monohydrochloride		$C_{20}H_{23}ClN_2O_2$	35306-33-3	358.862		326				s H_2O; sl EtOH
5526	Genistein	5,7-Dihydroxy-3-(4-hydroxyphenyl)-4H-1-benzopyran-4-one	$C_{15}H_{10}O_5$	446-72-0	270.237	nd(eth), pr(dil al)	301 dec				
5527	β-Gentiobiose	6-O-β-D-Glucopyranosyl-D-glucose	$C_{12}H_{22}O_{11}$	554-91-6	342.296	cry (EtOH)	192				s hot H_2O, hot MeOH
5528	Geraniol		$C_{10}H_{18}O$	106-24-1	154.249	oil	<-15	229(2)	0.8894²⁰	1.4766²⁰	i H_2O; s EtOH, eth, ace, chl
5529	Geranyl 2-methylpropanoate		$C_{14}H_{24}O_2$	2345-26-8	224.340			136¹³	0.8997¹⁵	1.4576²⁰	
5530	Geranyl acetate		$C_{12}H_{20}O_2$	16409-44-2	196.286			115¹²	0.9163¹⁵	1.4624²⁰	
5531	Germine		$C_{27}H_{43}NO_8$	508-65-6	509.632	pr or cry (MeOH)	220				s bz, MeOH, alk, acid
5532	Gibberellic acid		$C_{19}H_{22}O_6$	77-06-5	346.374	cry (EtOAc)	234				vs ace, EtOH, MeOH
5533	Gitoxigenin		$C_{23}H_{34}O_5$	545-26-6	390.513	pr (AcOEt) pr (+w, dil al)	234				i H_2O; sl eth; s chl
5534	Gitoxin		$C_{41}H_{64}O_{14}$	4562-36-1	780.939	pr (chl-MeOH)	285 dec				
5535	d-Glaucine		$C_{21}H_{25}NO_4$	475-81-0	355.429	pl, pr (eth, AcOEt)	120				vs ace, EtOH, chl
5536	D-Glucaric acid	D-Tetrahydroxyadipic acid	$C_6H_{10}O_8$	87-73-0	210.138	nd (45% al)	125.5				vs H_2O, EtOH; sl eth, chl
5537	D-Glucitol	Sorbitol	$C_6H_{14}O_6$	50-70-4	182.171	nd (w)	97(3)	295³·⁵	1.489²⁰	1.3330²⁰	vs H_2O, ace
5538	D-Glucitol, hexaacetate	Sorbitol hexaacetate	$C_{18}H_{26}O_{12}$	7208-47-1	434.392	pr (w)	98(3)		1.30²⁰		sl H_2O, eth; vs EtOH; s chl, AcOEt
5539	D-Gluconic acid		$C_6H_{12}O_7$	526-95-4	196.155	nd (al-eth)	131(2)				s H_2O; sl EtOH; i eth, bz
5540	β-D-Glucopyranose		$C_6H_{12}O_6$	492-61-5	180.155	cry (hot EtOH)	149				
5541	6-O-α-D-Glucopyranosyl-D-fructose	Palatinose	$C_{12}H_{22}O_{11}$	13718-94-0	342.296						s H_2O
5542	2-(β-D-Glucopyranosyloxy)-benzaldehyde	Helicin	$C_{13}H_{16}O_7$	618-65-5	284.262	nd (w)	175				vs H_2O, EtOH
5543	7-(β-D-Glucopyranosyloxy)-2H-1-benzopyran-2-one	Skimmin	$C_{15}H_{16}O_8$	93-39-0	324.283	cry (w + 1)	220				s H_2O, EtOH; i eth, chl
5544	2-(β-D-Glucopyranosyloxy)-2-methylpropanenitrile	Linamarin	$C_{10}H_{17}NO_6$	554-35-8	247.245	nd (w, al)	145				vs ace
5545	1-[4-(β-D-Glucopyranosyloxy)-phenyl]ethanone	Picein	$C_{14}H_{18}O_7$	530-14-3	298.289	nd (w+1), nd (MeOH)	195.5				sl H_2O; s EtOH, eth, HOAc; i chl
5546	α-D-Glucose		$C_6H_{12}O_6$	26655-34-5	180.155		146 dec		1.5620¹⁸		vs H_2O; sl EtOH; i ace, AcOEt; s py
5547	α-D-Glucose pentaacetate		$C_{16}H_{22}O_{11}$	604-68-2	390.339	pl or nd (al)	132.0(0.5)	sub			sl H_2O, EtOH, CS_2; s eth, chl, HOAc
5548	β-D-Glucose pentaacetate		$C_{16}H_{22}O_{11}$	604-69-3	390.339	nd (al)	134	sub	1.2740²⁰		i H_2O; sl EtOH, peth, eth; s bz; msc chl
5549	α-D-Glucose 1-phosphate	Cori ester	$C_6H_{13}O_9P$	59-56-3	260.135						vs H_2O
5550	D-Glucuronic acid		$C_6H_{10}O_7$	6556-12-3	194.139	nd (al)	165				vs H_2O, EtOH
5551	D-Glucuronic acid γ-lactone	D-Glucuronolactone	$C_6H_8O_6$	32449-92-6	176.124	mcl pl (w) cry (al)	177.5		1.76²⁰		s H_2O; sl EtOH, DMSO, MeOH; i bz
5552	DL-Glutamic acid		$C_5H_9NO_4$	617-65-2	147.130	orth (al,w)	199 dec		1.4601²⁰		sl H_2O, eth; i EtOH, CS_2, lig

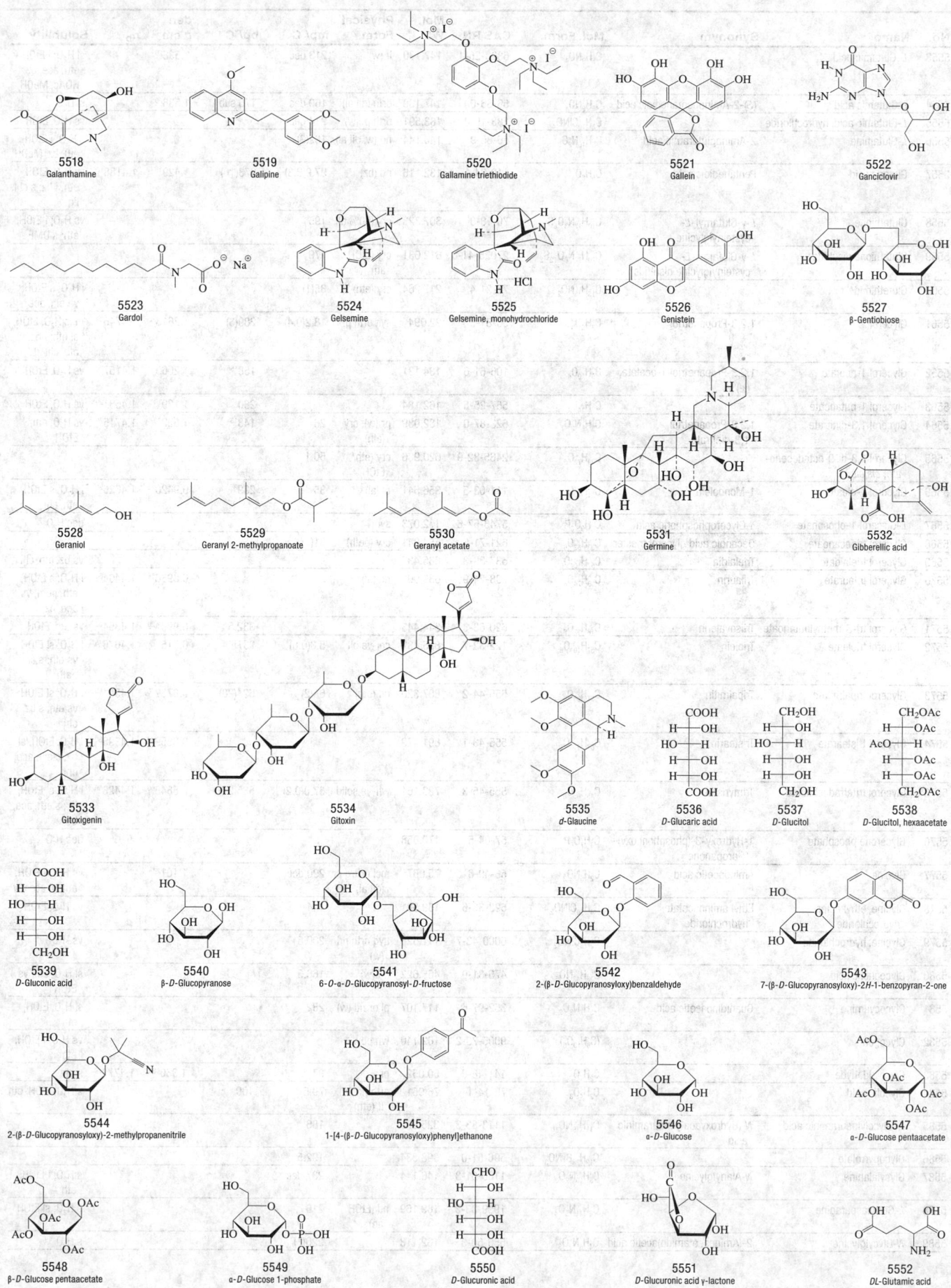

5518 Galanthamine

5519 Galipine

5520 Gallamine triethiodide

5521 Gallein

5522 Ganciclovir

5523 Gardol

5524 Gelsemine

5525 Gelsemine, monohydrochloride

5526 Genistein

5527 β-Gentiobiose

5528 Geraniol

5529 Geranyl 2-methylpropanoate

5530 Geranyl acetate

5531 Germine

5532 Gibberellic acid

5533 Gitoxigenin

5534 Gitoxin

5535 d-Glaucine

5536 D-Glucaric acid

5537 D-Glucitol

5538 D-Glucitol, hexaacetate

5539 D-Gluconic acid

5540 β-D-Glucopyranose

5541 6-O-α-D-Glucopyranosyl-D-fructose

5542 2-(β-D-Glucopyranosyloxy)benzaldehyde

5543 7-(β-D-Glucopyranosyloxy)-2H-1-benzopyran-2-one

5544 2-(β-D-Glucopyranosyloxy)-2-methylpropanenitrile

5545 1-[4-(β-D-Glucopyranosyloxy)phenyl]ethanone

5546 α-D-Glucose

5547 α-D-Glucose pentaacetate

5548 β-D-Glucose pentaacetate

5549 α-D-Glucose 1-phosphate

5550 D-Glucuronic acid

5551 D-Glucuronic acid γ-lactone

5552 DL-Glutamic acid

No.	Name	Synonym	Mol. Form.	CAS RN	Mol. Wt.	Physical Form	mp/°C	bp/°C	den g cm⁻³	n_D	Solubility
5553	*D*-Glutamic acid		$C_5H_9NO_4$	6893-26-1	147.130	lf (w)	213 dec		1.538^{20}		sl H_2O; i EtOH, eth, ace, bz, HOAc, MeOH
5554	*L*-Glutamic acid	(*S*)-2-Aminopentanedioic acid	$C_5H_9NO_4$	56-86-0	147.130	orth (dil al)	160 dec	175 sub	1.538^{20}		sl H_2O
5555	*L*-Glutamic acid, hydrochloride		$C_5H_{10}ClNO_4$	138-15-8	183.591	orth pl (w)	214 dec				vs H_2O, EtOH
5556	*L*-Glutamine	2-Aminoglutaramic acid	$C_5H_{10}N_2O_3$	56-85-9	146.144	nd (w, dil al)	182(3)				s H_2O; i EtOH, eth, bz, MeOH
5557	Glutaric acid	Pentanedioic acid	$C_5H_8O_4$	110-94-1	132.116	nd (bz)	97.9(0.3)	273(10)	1.429^{15}	1.4188^{106}	vs H_2O, EtOH, eth; i bz; s chl, lig
5558	Glutathione	*L*-γ-Glutamyl-*L*-cysteinylglycine	$C_{10}H_{17}N_3O_6S$	70-18-8	307.323	cry (50% al)	195				vs H_2O; i EtOH, eth; s DMF
5559	Glutathione disulfide	*L*-γ-Glutamyl-*L*-cysteinylglycine disulfide	$C_{20}H_{32}N_6O_{12}S_2$	27025-41-8	612.631	cry (EtOH aq)	179				
5560	Glutethimide		$C_{13}H_{15}NO_2$	77-21-4	217.264	cry (eth)	85(1)				i H_2O; s EtOH; vs eth, ace
5561	Glycerol	1,2,3-Propanetriol	$C_3H_8O_3$	56-81-5	92.094	syr, orth pl	18.2(0.4)	289(3)	1.2613^{20}	1.4746^{20}	msc H_2O, EtOH; sl eth; i bz, ctc, chl
5562	Glycerol 1-acetate	1,2,3-Propanetriol 1-acetate, (±)	$C_5H_{10}O_4$	106-61-6	134.131			158^{165}	1.2060^{20}	1.4157^{20}	vs H_2O, EtOH
5563	Glycerol 1-butanoate		$C_7H_{14}O_4$	557-25-5	162.184			280	1.129^{18}	1.4531^{20}	vs H_2O, EtOH
5564	Glycerol 1,3-dinitrate	1,2,3-Propanetriol, 1,3-dinitrate	$C_3H_6N_2O_7$	623-87-0	182.089	pr (w), cry (eth)	26	148^{15}	1.523^{20}	1.4715^{20}	vs H_2O, eth, EtOH
5565	Glycerol 1,3-di-9-octadecenoate, *cis,cis*		$C_{39}H_{72}O_5$	2465-32-9	620.986	cry (eth/EtOH)	50.1				
5566	Glycerol 1-oleate	1-Monoolein	$C_{21}H_{40}O_4$	111-03-5	356.541	pl (al)	35	239^3	0.9420^{20}	1.4626^{20}	i H_2O; s EtOH, eth, chl
5567	*L*-Glycerol 1-phosphate	α-Glycerophosphoric acid	$C_3H_9O_6P$	5746-57-6	172.073	syr					dec H_2O
5568	Glycerol tridecanoate	Decanoic acid glycerol triester	$C_{33}H_{62}O_6$	621-71-6	554.841	cry (peth)	31(1)				
5569	Glycerol trielaidate	Trielaidin	$C_{57}H_{104}O_6$	537-39-3	885.432						vs bz, eth, chl
5570	Glycerol trilaurate	Trilaurin	$C_{39}H_{74}O_6$	538-24-9	639.001	nd (al)			0.8986^{55}	1.4404^{60}	i H_2O; s EtOH, eth, peth; vs ace, bz
5571	Glycerol tri-3-methylbutanoate	Triisovalerin	$C_{18}H_{32}O_6$	620-63-3	344.443			332.5	0.9984^{20}	1.4354^{20}	vs eth, EtOH
5572	Glycerol trioleate	Triolein	$C_{57}H_{104}O_6$	122-32-7	885.432	col-ye oil	5.3(0.6)	237^{18}	0.915^{15}	1.4676^{15}	i H_2O; sl EtOH; vs eth; s chl, peth
5573	Glycerol tripalmitate	Tripalmitin	$C_{51}H_{98}O_6$	555-44-2	807.320	nd (eth)	66(2)	624(28)	0.8752^{70}	1.4381^{80}	i H_2O; sl EtOH; vs eth; s bz, chl
5574	Glycerol tristearate	Tristearin	$C_{57}H_{110}O_6$	555-43-1	891.479				0.8559^{90}	1.4395^{80}	i H_2O, EtOH; sl bz, ctc; s ace, chl
5575	Glycerol tritetradecanoate	Trimyristin	$C_{45}H_{86}O_6$	555-45-3	723.161	wh-ye solid	57.0(0.2)	585(27)	0.8848^{60}	1.4428^{60}	i H_2O; sl EtOH, lig; s eth, ace, bz
5576	Glycerone phosphate	1-Hydroxy-3-(phosphonooxy)-2-propanone	$C_3H_7O_6P$	57-04-5	170.058						dec H_2O
5577	Glycine	Aminoacetic acid	$C_2H_5NO_2$	56-40-6	75.067	mcl or trg pr (dil al)	290 dec		1.161^{20}		vs H_2O; i EtOH, eth; sl ace, py
5578	Glycine, ethyl ester, hydrochloride	Ethyl aminoacetate hydrochloride	$C_4H_{10}ClNO_2$	623-33-6	139.581		144				vs H_2O, EtOH
5579	Glycine, hydrochloride		$C_2H_6ClNO_2$	6000-43-7	111.528	hyg orth nd (w)	200.5				vs H_2O
5580	Glycocholic acid		$C_{26}H_{43}NO_6$	475-31-0	465.622	nd (w)	166.5				sl H_2O, eth; vs EtOH
5581	Glycocyamine	Guanidinoacetic acid	$C_3H_7N_3O_2$	352-97-6	117.107	pl or nd (w)	282				sl H_2O, EtOH, eth
5582	Glycogen		$(C_6H_{10}O_5)_x$	9005-79-2	162.140	wh pow					vs H_2O; i EtOH, eth
5583	Glycolaldehyde		$C_2H_4O_2$	141-46-8	60.052	pl	97		1.366^{100}	1.4772^{19}	s chl
5584	Glycolic acid		$C_2H_4O_3$	79-14-1	76.051	orth nd (w) lf (eth)	79.5	100			s H_2O, EtOH, eth
5585	*N*-Glycolylneuraminic acid	*N*-(Hydroxyacetyl)neuraminic acid	$C_{11}H_{19}NO_{10}$	1113-83-3	325.270		186				
5586	Glycopyrrolate		$C_{19}H_{28}BrNO_3$	596-51-0	398.334		192.5				
5587	Glycylalanine	*N*-Alanylglycine	$C_5H_{10}N_2O_3$	1188-01-8	146.144		237 dec				s H_2O; i EtOH, eth
5588	*L*-Glycylasparagine		$C_6H_{11}N_3O_4$	1999-33-3	189.169	nd (EtOH aq)	216				s H_2O; sl EtOH
5589	*N*-Glycylglycine	2-(Aminoacetamido)acetic acid	$C_4H_8N_2O_3$	556-50-3	132.118		220(1)				s H_2O

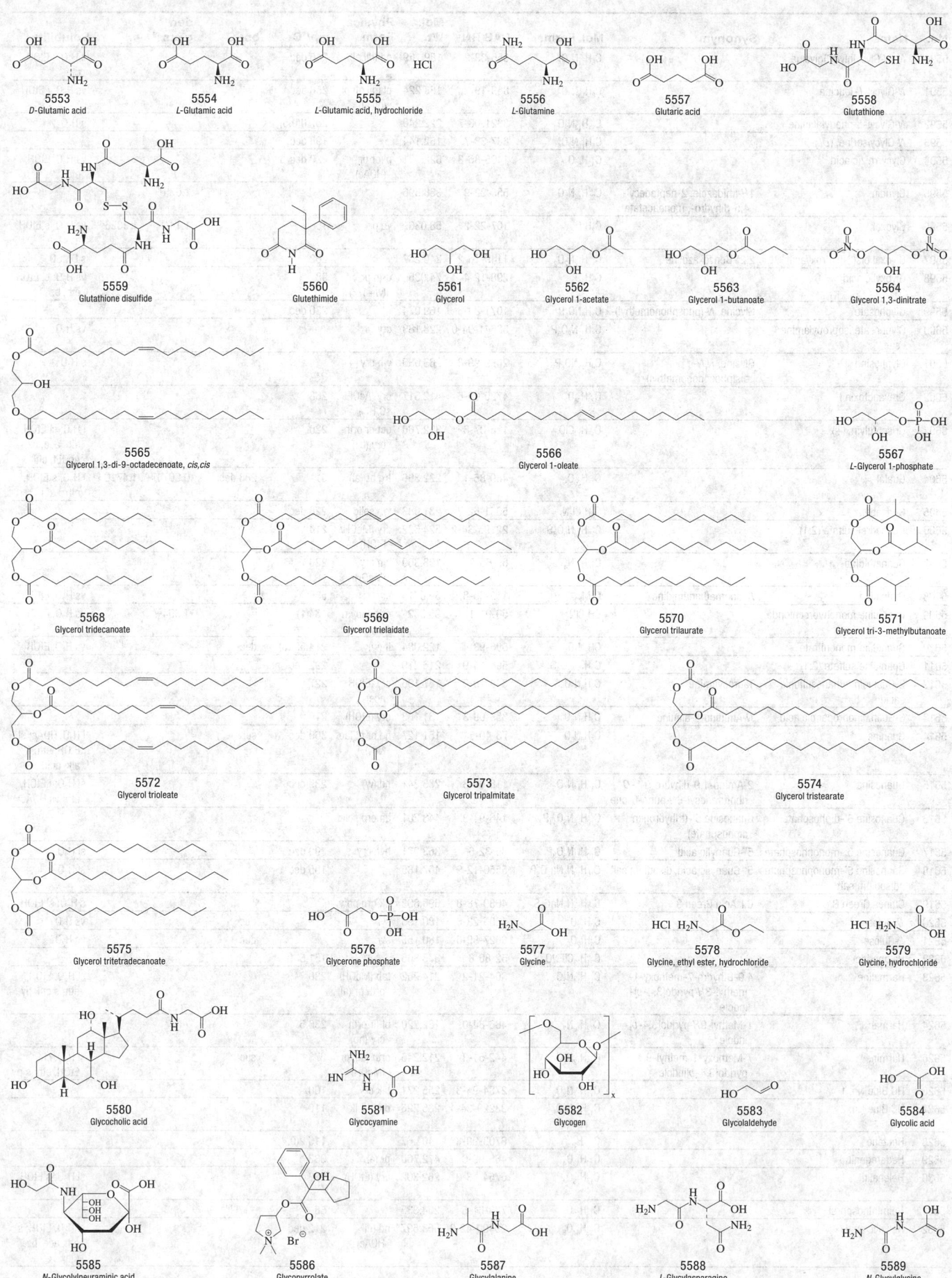

5553 D-Glutamic acid

5554 L-Glutamic acid

5555 L-Glutamic acid, hydrochloride

5556 L-Glutamine

5557 Glutaric acid

5558 Glutathione

5559 Glutathione disulfide

5560 Glutethimide

5561 Glycerol

5562 Glycerol 1-acetate

5563 Glycerol 1-butanoate

5564 Glycerol 1,3-dinitrate

5565 Glycerol 1,3-di-9-octadecenoate, cis,cis

5566 Glycerol 1-oleate

5567 L-Glycerol 1-phosphate

5568 Glycerol tridecanoate

5569 Glycerol trielaidate

5570 Glycerol trilaurate

5571 Glycerol tri-3-methylbutanoate

5572 Glycerol trioleate

5573 Glycerol tripalmitate

5574 Glycerol tristearate

5575 Glycerol tritetradecanoate

5576 Glycerone phosphate

5577 Glycine

5578 Glycine, ethyl ester, hydrochloride

5579 Glycine, hydrochloride

5580 Glycocholic acid

5581 Glycocyamine

5582 Glycogen

5583 Glycolaldehyde

5584 Glycolic acid

5585 N-Glycolylneuraminic acid

5586 Glycopyrrolate

5587 Glycylalanine

5588 L-Glycylasparagine

5589 N-Glycylglycine

No.	Name	Synonym	Mol. Form.	CAS RN	Mol. Wt.	Physical Form	mp/°C	bp/°C	den g cm⁻³	n_D	Solubility
5590	N-(N-Glycylglycyl)glycine		$C_6H_{11}N_3O_4$	556-33-2	189.169	nd (dil al)	246 dec				s H_2O; i EtOH, eth
5591	N-Glycyl-L-leucine		$C_8H_{16}N_2O_3$	869-19-2	188.224	pl (dil al) pl (dil al)	256 dec				vs H_2O; i EtOH
5592	N-Glycyl-L-phenylalanine		$C_{11}H_{14}N_2O_3$	3321-03-7	222.240		267.0(0.5)				s H_2O
5593	N-Glycylserine, (DL)-		$C_5H_{10}N_2O_4$	687-38-7	162.144		198 dec				
5594	Glycyrrhizic acid		$C_{42}H_{62}O_{16}$	1405-86-3	822.931	pl or pr (HOAc)	220 dec				vs H_2O, EtOH; i eth
5595	Glyodin	1H-Imidazole, 2-heptadecyl-4,5-dihydro-, monoacetate	$C_{22}H_{44}N_2O_2$	556-22-9	368.596				1.035[20]		
5596	Glyoxal		$C_2H_2O_2$	107-22-2	58.036	ye pr	15	50.4	1.14[20]	1.3826[20]	vs H_2O; s EtOH, eth
5597	Glyoxal bis(2-hydroxyanil)	2,2'-Benzoxazoline	$C_{14}H_{12}N_2O_2$	1149-16-2	240.257		202				s DMSO
5598	Glyoxylic acid		$C_2H_2O_3$	298-12-4	74.035	orth pr (w+1/2)	98				vs H_2O; sl EtOH, eth, bz
5599	Glyphosate	Glycine, N-(phosphonomethyl)-	$C_3H_8NO_5P$	1071-83-6	169.074		230 dec				
5600	Glyphosate isopropylamine salt		$C_6H_{17}N_2O_5P$	38641-94-0	228.183	cry					vs H_2O
5601	Glyphosine	Glycine, N,N-bis(phosphonomethyl)-	$C_4H_{11}NO_8P_2$	2439-99-8	263.080	wh cry					s H_2O
5602	Grayanotoxin I		$C_{22}H_{36}O_7$	4720-09-6	412.517	cry (AcOEt/ C_5H_{12})	268				
5603	Griseofulvin, (+)		$C_{17}H_{17}ClO_6$	126-07-8	352.766	oct or orth cry (bz)	220				i H_2O; sl EtOH, eth, ace, bz, AcOEt, chl
5604	Guaiol		$C_{15}H_{26}O$	489-86-1	222.366	trg pr (al)	91	288 dec	0.9074[100]	1.4716[100]	i H_2O; s EtOH, eth
5605	Guanabenz		$C_8H_8Cl_2N_4$	5051-62-7	231.083	wh solid	228 dec				
5606	Guanadrel sulfate (2:1)		$C_{20}H_{40}N_6O_8S$	22195-34-2	524.632	cry (MeOH/ EtOH)	214				
5607	Guanethidine		$C_{10}H_{22}N_4$	55-65-2	198.309	wh cry (MeOH)	226				
5608	Guanidine	Aminomethanamidine	CH_5N_3	113-00-8	59.071	cry	50				vs H_2O, EtOH
5609	Guanidine monohydrochloride		CH_6ClN_3	50-01-1	95.532	orth bipym (al)	184(1)		1.354[20]		vs H_2O, EtOH
5610	Guanidine mononitrate		$CH_6N_4O_3$	506-93-4	122.084	lf (w)	214.5(0.6)	dec			vs H_2O, EtOH
5611	Guanidine sulfate (2:1)		$C_2H_{12}N_6O_4S$	594-14-9	216.219		292 dec				
5612	2-Guanidinoethanesulfonic acid	Taurocyamine	$C_3H_9N_3O_3S$	543-18-0	167.186	cry (EtOH, ace)	227				
5613	3-Guanidinopropanoic acid	N-Amidino-β-alanine	$C_4H_9N_3O_2$	353-09-3	131.133	cry (EtOH)	210				
5614	Guanine		$C_5H_5N_5O$	73-40-5	151.127	nd or pl (aq NH_3)	360 dec	sub			i H_2O, HOAc; sl EtOH, eth; s alk, acid
5615	Guanosine	2-Amino-1,9-dihydro-9-β-D-ribofuranosyl-6H-purin-6-one	$C_{10}H_{13}N_5O_5$	118-00-3	283.241	nd (w)	239 dec				sl H_2O; i EtOH, eth; vs HOAc
5616	Guanosine 5'-diphosphate	Guanosine 5'-(trihydrogen diphosphate)	$C_{10}H_{15}N_5O_{11}P_2$	146-91-8	443.201	amorp solid					
5617	Guanosine 5'-monophosphate	5'-Guanylic acid	$C_{10}H_{14}N_5O_8P$	85-32-5	363.221	hyg cry	190 dec				sl H_2O
5618	Guanosine 5'-monophosphate, disodium salt	5'-Guanylic acid, disodium salt	$C_{10}H_{12}N_5Na_2O_8P$	5550-12-9	407.185		195 dec				sl H_2O
5619	Guinea Green B	C.I. Acid Green 3	$C_{37}H_{35}N_3NaO_6S_2$	4680-78-8	690.803	dk grn pow					s H_2O; sl EtOH
5620	D-Gulose		$C_6H_{12}O_6$	4205-23-6	180.155	syr		dec			vs H_2O
5621	L-Gulose		$C_6H_{12}O_6$	6027-89-0	180.155	syr		dec			vs H_2O
5622	Haloperidol		$C_{21}H_{23}ClFNO_2$	52-86-8	375.865		151.5				
5623	Harmaline	4,9-Dihydro-7-methoxy-1-methyl-3H-pyrido[3,4-b]-indole	$C_{13}H_{14}N_2O$	304-21-2	214.262	tab (MeOH) orth pr (al)	230				sl H_2O, EtOH, eth; s chl, py
5624	Harman	1-Methyl-9H-pyrido[3,4-b]-indole	$C_{12}H_{10}N_2$	486-84-0	182.220	bl flr orth cry (hp)	236.5				
5625	Harmine	7-Methoxy-1-methyl-9H-pyrido[3,4-b]indole	$C_{13}H_{12}N_2O$	442-51-3	212.246	orth (al), pr (MeOH)	273	sub			sl H_2O, chl, EtOH, eth; s py
5626	HC Blue No. 1		$C_{11}H_{17}N_3O_4$	2784-94-3	255.271	blk cry	100				
5627	HC Blue No. 2		$C_{12}H_{19}N_3O_5$	33229-34-4	285.296	dk bl-blk cry	110				
5628	Hectane		$C_{100}H_{202}$	6703-98-6	1404.67		115.2(0.4)				
5629	Hederagenin		$C_{30}H_{48}O_4$	465-99-6	472.700	pr (al)	333				
5630	Helenalin		$C_{15}H_{18}O_4$	6754-13-8	262.302	cry (EtOH)	226				sl H_2O; s EtOH, chl
5631	Helminthosporal		$C_{15}H_{22}O_2$	723-61-5	234.335		58	117[0.015]			
5632	Helvolic acid		$C_{33}H_{44}O_8$	29400-42-8	568.697	nd (dil HOAc)	212 dec				sl H_2O, EtOH; eth, ace, bz, diox

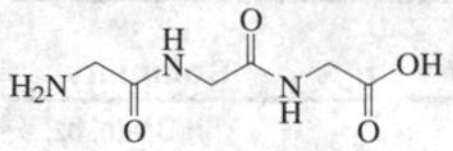

5590
N-(N-Glycylglycyl)glycine

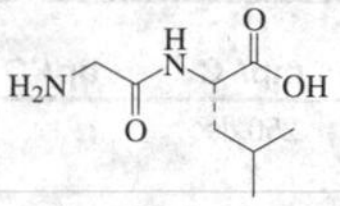

5591
N-Glycyl-L-leucine

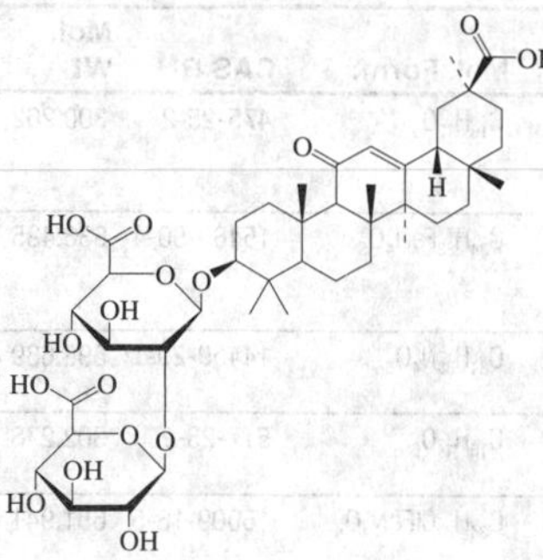

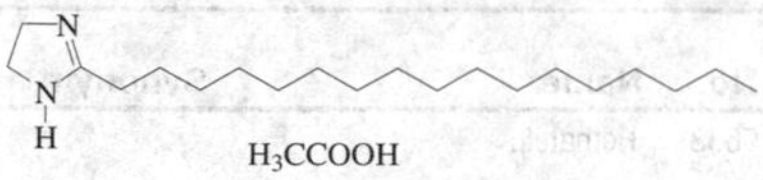

5595
Glyodin

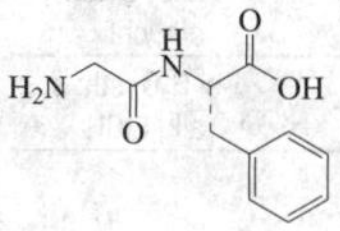

5592
N-Glycyl-L-phenylalanine

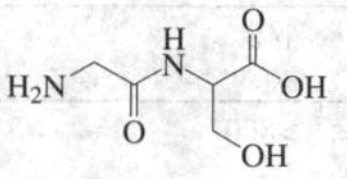

5593
N-Glycylserine, (DL)-

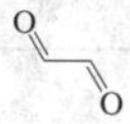

5594
Glycyrrhizic acid

5596
Glyoxal

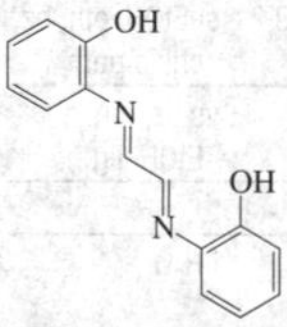

5597
Glyoxal bis(2-hydroxyanil)

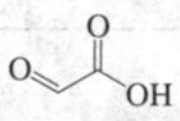

5598
Glyoxylic acid

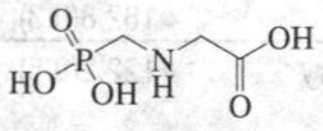

5599
Glyphosate

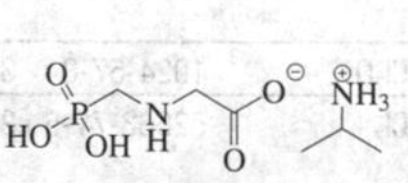

5600
Glyphosate isopropylamine salt

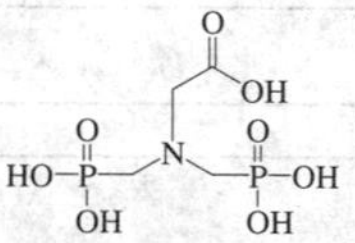

5601
Glyphosine

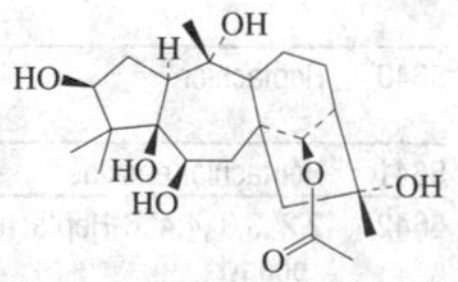

5602
Grayanotoxin I

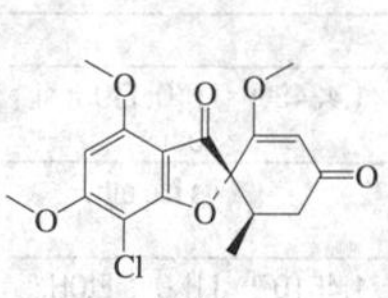

5603
Griseofulvin, (+)

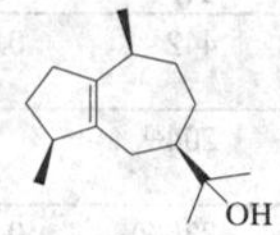

5604
Guaiol

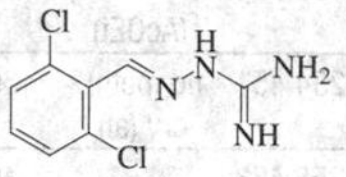

5605
Guanabenz

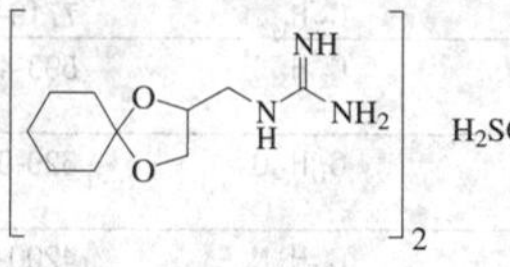

5606
Guanadrel sulfate (2:1)

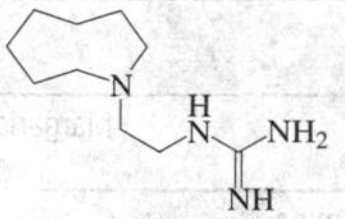

5607
Guanethidine

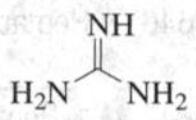

5608
Guanidine

5609
Guanidine monohydrochloride

5610
Guanidine mononitrate

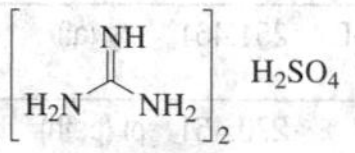

5611
Guanidine sulfate (2:1)

5612
2-Guanidinoethanesulfonic acid

5613
3-Guanidinopropanoic acid

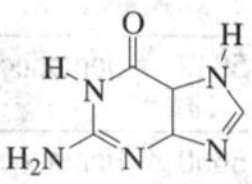

5614
Guanine

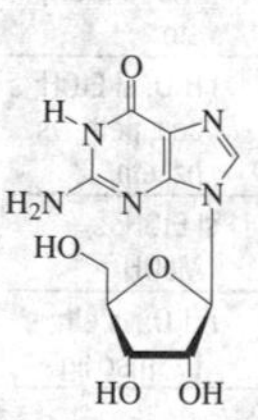

5615
Guanosine

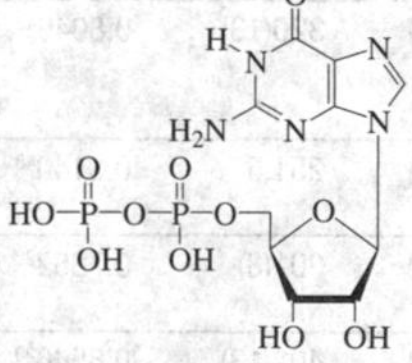

5616
Guanosine 5'-diphosphate

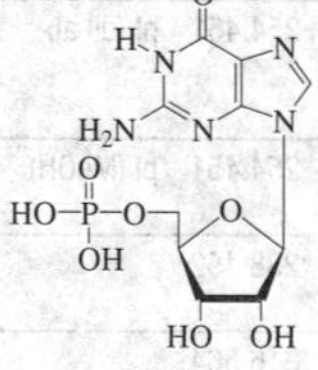

5617
Guanosine 5'-monophosphate

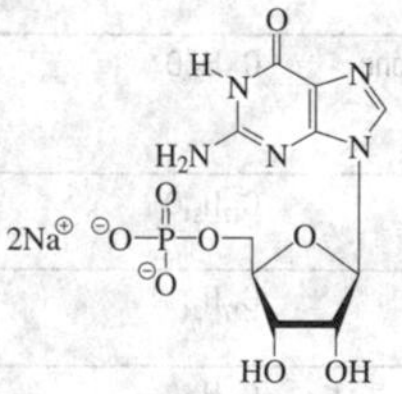

5618
Guanosine 5'-monophosphate, disodium salt

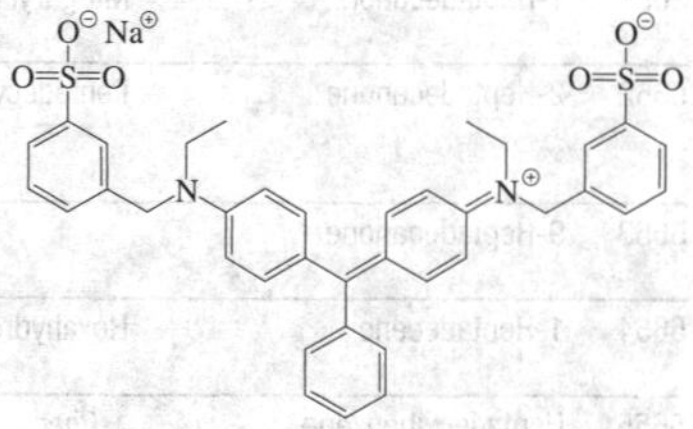

5619
Guinea Green B

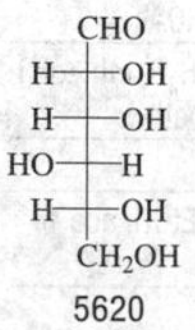

5620
D-Gulose

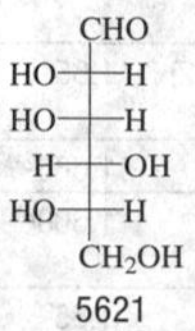

5621
L-Gulose

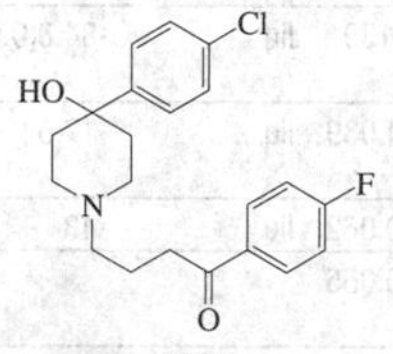

5622
Haloperidol

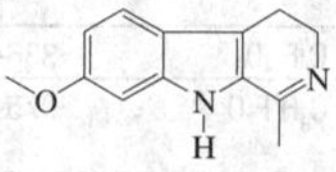

5623
Harmaline

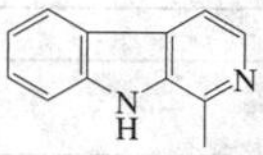

5624
Harman

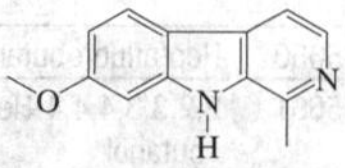

5625
Harmine

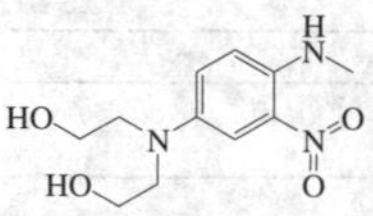

5626
HC Blue No. 1

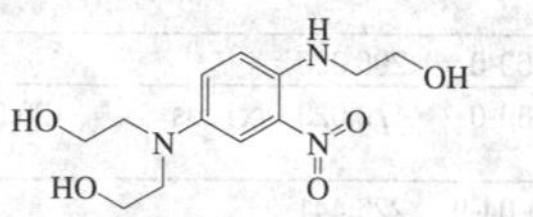

5627
HC Blue No. 2

H₃C(CH₂)₉₈CH₃
5628
Hectane

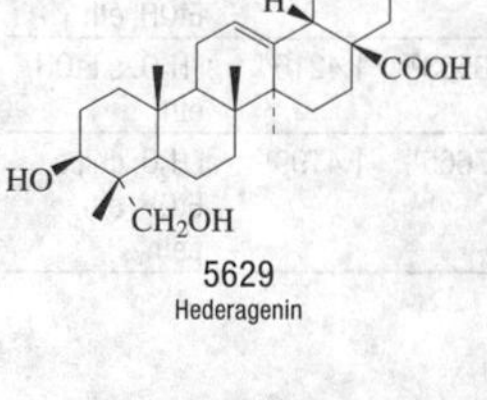

5629
Hederagenin

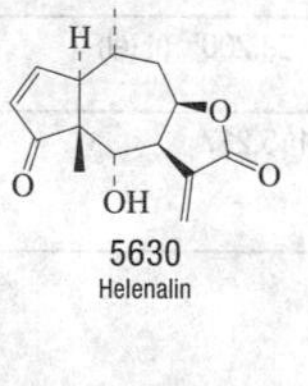

5630
Helenalin

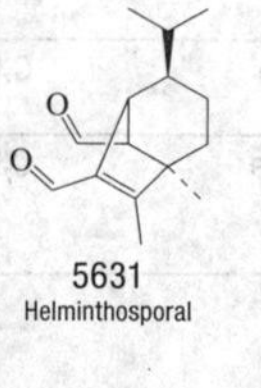

5631
Helminthosporal

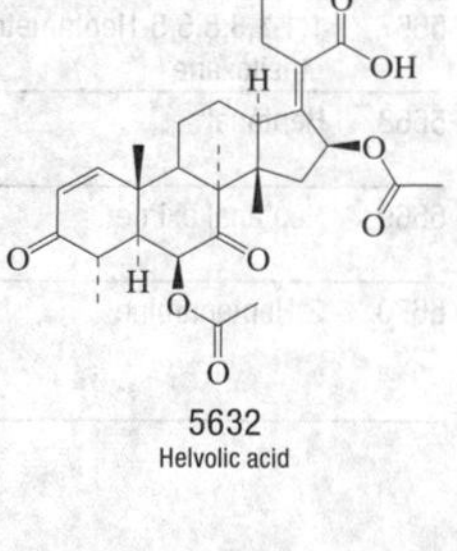

5632
Helvolic acid

No.	Name	Synonym	Mol. Form.	CAS RN	Mol. Wt.	Physical Form	mp/°C	bp/°C	den g cm⁻³	n_D	Solubility
5633	Hematein		$C_{16}H_{12}O_6$	475-25-2	300.262	red-br cry	250 dec				i H_2O, eth, bz, chl; sl EtOH, HOAc
5634	Hematin		$C_{34}H_{33}FeN_4O_5$	15489-90-4	633.495	br pow (py)	>200				i H_2O; eth; s EtOH, alk; sl py, HOAc
5635	Hematoporphyrin		$C_{34}H_{38}N_4O_6$	14459-29-1	598.689	deep red cry	172.5				i H_2O; s EtOH; sl eth, chl
5636	Hematoxylin		$C_{16}H_{14}O_6$	517-28-2	302.278	ye cry	140				sl H_2O, eth; s alk, EtOH
5637	Hemin		$C_{34}H_{32}ClFeN_4O_4$	16009-13-5	651.941	long blades (gl HOAc)	>300				
5638	Heneicosane		$C_{21}H_{44}$	629-94-7	296.574	cry (w)	40.4(0.3)	359(6)	0.7919²⁰	1.4441²⁰	i H_2O; sl EtOH; s peth
5639	Hentriacontane	Untriacontane	$C_{31}H_{64}$	630-04-6	436.840	lf (AcOEt)	68(1)	458	0.781⁶⁸	1.4278⁹⁰	sl EtOH, eth, bz, chl; s peth
5640	Heptachlor		$C_{10}H_5Cl_7$	76-44-8	373.318	wh cry	95.8(0.3)		1.57⁹		vs bz, eth, EtOH, lig
5641	Heptachlor epoxide		$C_{10}H_5Cl_7O$	1024-57-3	389.317		162.8(0.3)				
5642	2,2',3,3',4,4',6-Heptachlorobi-phenyl		$C_{12}H_3Cl_7$	52663-71-5	395.323	cry	122.2(0.5)				i H_2O
5643	1,1,1,2,3,3,3-Heptachloropro-pane		C_3HCl_7	3849-33-0	285.211		11	249	1.7921³⁴	1.5427²¹	vs chl
5644	Heptacontane		$C_{70}H_{142}$	7719-93-9	983.876		106(4)	647			
5645	Heptacosane		$C_{27}H_{56}$	593-49-7	380.734	cry (al, bz) lf (AcOEt)	58.8(0.1)	442	0.7796⁶⁰	1.4345⁶⁵	i H_2O, EtOH; sl eth
5646	Heptadecanal	Margaric aldehyde	$C_{17}H_{34}O$	629-90-3	254.451	nd (peth), cry (al)	36(3)	204²⁶			vs bz, eth
5647	1-Heptadecanamine		$C_{17}H_{37}N$	4200-95-7	255.483		49(1)	336(2)	0.8510²⁰	1.4510²⁰	i H_2O; s EtOH, eth
5648	Heptadecane		$C_{17}H_{36}$	629-78-7	240.468	hex lf	21.97(0.05)	303(2)	0.7780²⁰	1.4369²⁰	i H_2O; sl EtOH, ctc; s eth
5649	Heptadecanenitrile		$C_{17}H_{33}N$	5399-02-0	251.451	cry (al)	34.0(0.7)	349	0.8315²⁰	1.4467²⁰	i H_2O; sl EtOH, chl; vs eth
5650	Heptadecanoic acid	Margaric acid	$C_{17}H_{34}O_2$	506-12-7	270.451	pl (peth)	61.08(0.04)	227¹⁰⁰	0.8532⁶⁰	1.4342⁶⁰	i H_2O; sl EtOH; s eth, ace, bz, chl
5651	1-Heptadecanol	Margaryl alcohol	$C_{17}H_{36}O$	1454-85-9	256.467	lf (al), cry (ace)	53(1)	324	0.8475²⁰		i H_2O; s EtOH, eth
5652	2-Heptadecanone	Pentadecyl methyl ketone	$C_{17}H_{34}O$	2922-51-2	254.451	pl (dil al)	48.2(0.3)	320(13)	0.8049⁴⁸		i H_2O; sl EtOH; s ace, peth; vs bz, eth
5653	9-Heptadecanone		$C_{17}H_{34}O$	540-08-9	254.451	pl (MeOH)	50.7(0.2)	251.5	0.8140⁴⁸		sl EtOH; s MeOH
5654	1-Heptadecene	Hexahydroaplotaxene	$C_{17}H_{34}$	6765-39-5	238.452		10.7(0.6)	301(3)	0.7852²⁰	1.4432²⁰	i H_2O; vs eth; s bz; msc lig
5655	Heptadecylbenzene	1-Phenylheptadecane	$C_{23}H_{40}$	14752-75-1	316.564		19.7(0.4)	402(13)	0.8546²⁰	1.4810²⁰	
5656	*trans,trans*-2,4-Heptadienal		$C_7H_{10}O$	4313-03-5	110.153			84.5	0.881²⁵	1.5315²⁰	
5657	1,6-Heptadiene		C_7H_{12}	3070-53-9	96.170	liq		89(3)			
5658	1,6-Heptadiyne		C_7H_8	2396-63-6	92.139	liq	-84.8(0.6)	111(1)	0.8164¹⁷	1.451¹⁷	i H_2O; s bz, HOAc
5659	Heptafluorobutanoic acid		$C_4HF_7O_2$	375-22-4	214.039	liq	-17.5	121.5(0.2)	1.651²⁰	1.295²⁵	s H_2O, eth, tol; i peth
5660	Heptafluorobutanoic anhydride		$C_8F_{14}O_3$	336-59-4	410.062	liq	-43	106.5	1.665²⁰	1.285²⁰	
5661	2,2,3,3,4,4,4-Heptafluoro-1-butanol		$C_4H_3F_7O$	375-01-9	200.055			95(4)	1.600²⁰	1.294²⁰	s EtOH, ace
5662	Heptafluorobutanoyl chloride		C_4ClF_7O	375-16-6	232.484			39(10)	1.55²⁰	1.288²⁰	
5663	6,6,7,7,8,8,8-Heptafluoro-2,2-dimethyl-3,5-octanedione		$C_{10}H_{11}F_7O_2$	17587-22-3	296.182		38	46⁵	1.273²⁵	1.3766²⁰	
5664	Heptafluoro-2-iodopropane	Perfluoroisopropyl iodide	C_3F_7I	677-69-0	295.925			38	1.3298²⁰		
5665	1,1,1,2,3,3,3-Heptafluoropro-pane	Refrigerant 227ea	C_3HF_7	431-89-0	170.029	col gas	-126.8	-16.34			
5666	2,2,4,4,6,8,8-Heptamethyl-nonane		$C_{16}H_{34}$	4390-04-9	226.441			246(1)			
5667	1,1,1,3,5,5,5-Heptamethyl-trisiloxane		$C_7H_{22}O_2Si_3$	1873-88-7	222.506			143(1)	0.8194²⁰	1.3818²⁰	
5668	Heptanal	Heptaldehyde	$C_7H_{14}O$	111-71-7	114.185	liq	-43.94(0.02)	153(3)	0.8132²⁵	1.4113²⁰	sl H_2O, ctc; msc EtOH, eth
5669	Heptanal oxime	Enanthaldoxime	$C_7H_{15}NO$	629-31-2	129.200	pl (al)	57.5	195	0.8583⁵⁵	1.4210²⁰	sl H_2O; s EtOH, eth
5670	2-Heptanamine	Tuaminoheptane	$C_7H_{17}N$	123-82-0	115.217			141(4)	0.7665¹⁹	1.4199¹⁹	sl H_2O, chl; s EtOH, eth, peth

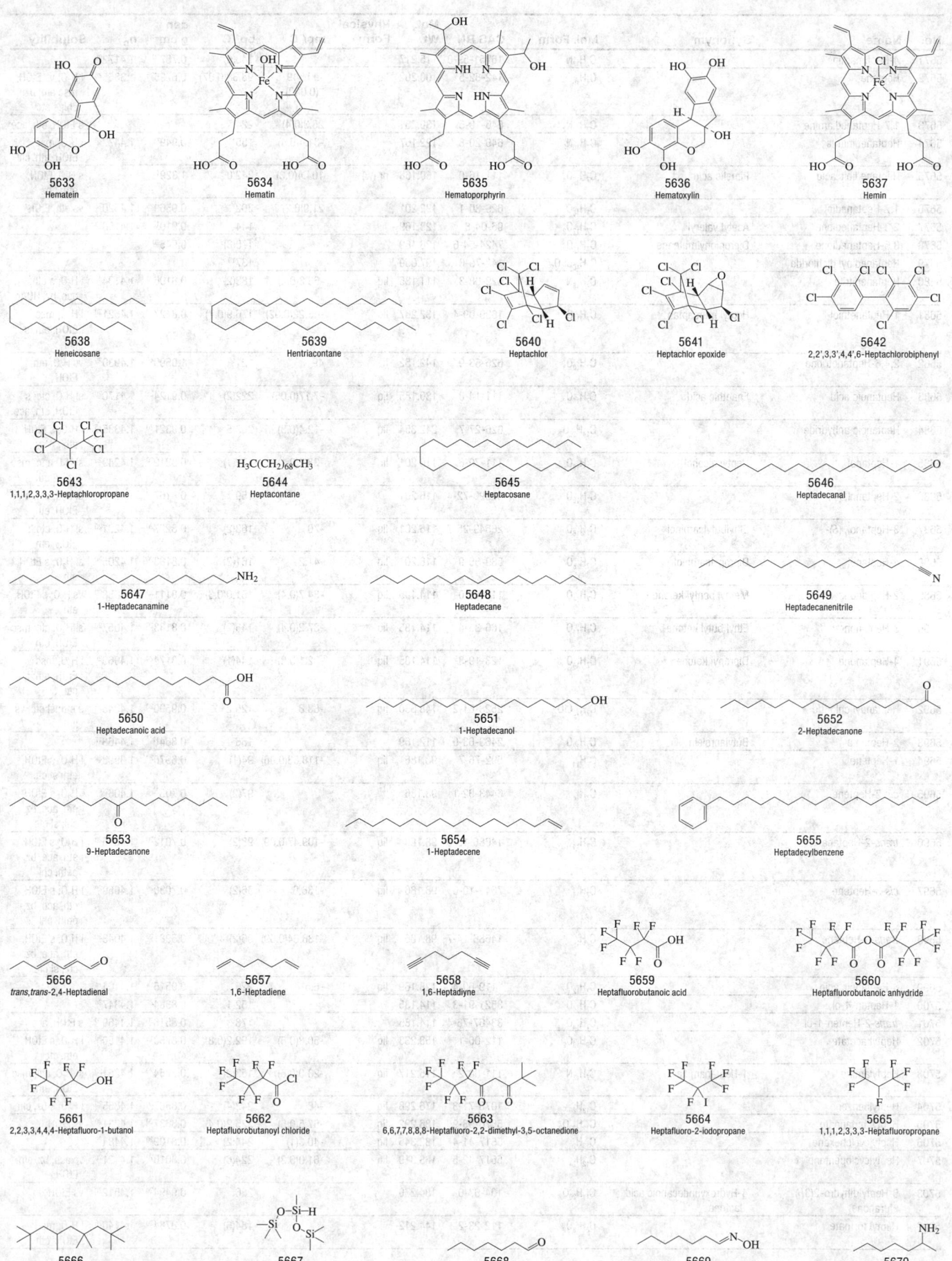

5633 Hematein

5634 Hematin

5635 Hematoporphyrin

5636 Hematoxylin

5637 Hemin

5638 Heneicosane

5639 Hentriacontane

5640 Heptachlor

5641 Heptachlor epoxide

5642 2,2',3,3',4,4',6-Heptachlorobiphenyl

5643 1,1,1,2,3,3,3-Heptachloropropane

5644 Heptacontane

5645 Heptacosane

5646 Heptadecanal

5647 1-Heptadecanamine

5648 Heptadecane

5649 Heptadecanenitrile

5650 Heptadecanoic acid

5651 1-Heptadecanol

5652 2-Heptadecanone

5653 9-Heptadecanone

5654 1-Heptadecene

5655 Heptadecylbenzene

5656 *trans,trans*-2,4-Heptadienal

5657 1,6-Heptadiene

5658 1,6-Heptadiyne

5659 Heptafluorobutanoic acid

5660 Heptafluorobutanoic anhydride

5661 2,2,3,3,4,4,4-Heptafluoro-1-butanol

5662 Heptafluorobutanoyl chloride

5663 6,6,7,7,8,8,8-Heptafluoro-2,2-dimethyl-3,5-octanedione

5664 Heptafluoro-2-iodopropane

5665 1,1,1,2,3,3,3-Heptafluoropropane

5666 2,2,4,4,6,8,8-Heptamethylnonane

5667 1,1,1,3,5,5,5-Heptamethyltrisiloxane

5668 Heptanal

5669 Heptanal oxime

5670 2-Heptanamine

No.	Name	Synonym	Mol. Form.	CAS RN	Mol. Wt.	Physical Form	mp/°C	bp/°C	den g cm⁻³	n_D	Solubility
5671	4-Heptanamine		C$_7$H$_{17}$N	16751-59-0	115.217			140(6)	0.767^{20}	1.4172^{20}	
5672	Heptane		C$_7$H$_{16}$	142-82-5	100.202	liq	-90.549 (0.002)	98.38(0.07)	0.6795^{25}	1.3855^{25}	i H$_2$O; vs EtOH; msc eth, bz, chl; s ctc
5673	1,7-Heptanediamine		C$_7$H$_{18}$N$_2$	646-19-5	130.231		25.3(0.4)	224			s EtOH, eth, ace
5674	Heptanedinitrile		C$_7$H$_{10}$N$_2$	646-20-8	122.167		-31.4(0.4)	155^{14}	0.949^{18}	1.4472^{20}	i H$_2$O; msc EtOH, eth, chl
5675	Heptanedioic acid	Pimelic acid	C$_7$H$_{12}$O$_4$	111-16-0	160.168	pr (w)	104.4(0.3)	342.0	1.329^{15}		s H$_2$O, EtOH, eth; i bz
5676	1,7-Heptanediol		C$_7$H$_{16}$O$_2$	629-30-1	132.201		21.9(0.2)	262	0.9569^{25}	1.4520^{25}	vs eth, EtOH
5677	2,3-Heptanedione	Acetyl valeryl	C$_7$H$_{12}$O$_2$	96-04-8	128.169			144	0.919^{18}	1.4150^{18}	
5678	3,5-Heptanedione	Dipropionylmethane	C$_7$H$_{12}$O$_2$	7424-54-6	128.169			170(8)	0.945^{20}		
5679	Heptanedioyl dichloride		C$_7$H$_{10}$Cl$_2$O$_2$	142-79-0	197.059			137^{15}			
5680	Heptanenitrile		C$_7$H$_{13}$N	629-08-3	111.185	liq	-63.8(0.4)	183(2)	0.8106^{20}	1.4104^{30}	i H$_2$O; s eth, ace, bz, HOAc
5681	1-Heptanethiol	Heptyl mercaptan	C$_7$H$_{16}$S	1639-09-4	132.267	liq	-43.22(0.02)	176.9(0.7)	0.8427^{20}	1.4521^{20}	i H$_2$O; msc EtOH, eth; s chl
5682	2,4,6-Heptanetrione		C$_7$H$_{10}$O$_3$	626-53-9	142.152	lf	49	121^{10}	1.0599^{40}	1.4930^{20}	vs H$_2$O, eth, EtOH
5683	Heptanoic acid	Enanthic acid	C$_7$H$_{14}$O$_2$	111-14-8	130.185	liq	-7.17(0.05)	222(2)	0.9124^{25}	1.4170^{20}	sl H$_2$O, ctc; s EtOH, eth, ace
5684	Heptanoic anhydride		C$_{14}$H$_{26}$O$_3$	626-27-7	242.354	liq	-12.4(0.5)	269.5	0.9321^{20}	1.4335^{15}	i H$_2$O; s EtOH, eth
5685	1-Heptanol	Heptyl alcohol	C$_7$H$_{16}$O	111-70-6	116.201	liq	-33.2(0.1)	178(1)	0.8219^{20}	1.4249^{20}	sl H$_2$O, ctc; msc EtOH, eth
5686	2-Heptanol, (±)-		C$_7$H$_{16}$O	52390-72-4	116.201			159	0.8167^{20}	1.4210^{20}	sl H$_2$O, ctc; s EtOH, eth
5687	3-Heptanol, (S)-	Ethylbutylcarbinol	C$_7$H$_{16}$O	26549-25-7	116.201	liq	-70	163(2)	0.8227^{20}	1.4201^{20}	sl H$_2$O, ctc; s EtOH, eth
5688	4-Heptanol	Dipropylcarbinol	C$_7$H$_{16}$O	589-55-9	116.201	liq	-41.2	161(2)	0.8183^{20}	1.4205^{20}	sl H$_2$O; s EtOH, eth
5689	2-Heptanone	Methyl pentyl ketone	C$_7$H$_{14}$O	110-43-0	114.185	liq	-34.7(0.4)	151.0(0.3)	0.8111^{20}	1.4088^{20}	vs H$_2$O; s EtOH, eth
5690	3-Heptanone	Ethyl butyl ketone	C$_7$H$_{14}$O	106-35-4	114.185	liq	-37.2(0.4)	146(2)	0.8183^{20}	1.4057^{20}	sl H$_2$O, ctc; s EtOH, eth
5691	4-Heptanone	Dipropyl ketone	C$_7$H$_{14}$O	123-19-3	114.185	liq	-32.1(0.8)	144(1)	0.8174^{20}	1.4069^{20}	i H$_2$O; msc EtOH, eth; s ctc
5692	Heptanoyl chloride		C$_7$H$_{13}$ClO	2528-61-2	148.630	liq	-83.8	125.2	0.9590^{20}	1.4345^{18}	s eth; sl ctc; vs lig
5693	2-Heptenal	Butylacrolein	C$_7$H$_{12}$O	2463-63-0	112.169			166	0.864^{17}	1.4468^{17}	
5694	1-Heptene		C$_7$H$_{14}$	592-76-7	98.186	liq	-118.83(0.06)	94(1)	0.6970^{20}	1.3998^{20}	i H$_2$O; s EtOH, eth; sl ctc
5695	cis-2-Heptene		C$_7$H$_{14}$	6443-92-1	98.186			97(2)	0.708^{20}	1.406^{20}	i H$_2$O; s EtOH, eth, ace, bz, chl; sl ctc
5696	trans-2-Heptene		C$_7$H$_{14}$	14686-13-6	98.186	liq	-109.47(0.09)	98(2)	0.7012^{20}	1.4045^{20}	i H$_2$O; s EtOH, eth, ace, bz, peth, chl
5697	cis-3-Heptene		C$_7$H$_{14}$	7642-10-6	98.186	liq	-136.6	96(2)	0.7030^{20}	1.4059^{20}	i H$_2$O; s EtOH, eth, ace, bz, peth, chl
5698	trans-3-Heptene		C$_7$H$_{14}$	14686-14-7	98.186	liq	-136.64(0.09)	96(2)	0.6981^{20}	1.4043^{20}	i H$_2$O; s EtOH, eth, ace, bz, chl; sl ctc
5699	6-Heptenoic acid		C$_7$H$_{12}$O$_2$	1119-60-4	128.169	liq	-6.5	226	0.9515^{14}	1.4404^{14}	
5700	1-Hepten-4-ol		C$_7$H$_{14}$O	3521-91-3	114.185			152.1	0.8384^{22}	1.4347^{20}	
5701	trans-2-Hepten-1-ol		C$_7$H$_{14}$O	33467-76-4	114.185			178	0.8516^{20}	1.4460^{20}	s EtOH, ace
5702	Heptyl acetate		C$_9$H$_{18}$O$_2$	112-06-1	158.238	liq	-50.3(0.6)	192.2(0.8)	0.8750^{15}	1.4150^{20}	i H$_2$O; s EtOH, eth, ctc
5703	Heptylamine	1-Heptanamine	C$_7$H$_{17}$N	111-68-2	115.217	liq	-23.0(0.6)	153(2)	0.7754^{20}	1.4251^{20}	sl H$_2$O, chl; msc EtOH, eth
5704	Heptylbenzene		C$_{13}$H$_{20}$	1078-71-3	176.298	liq	-48	242(4)	0.8567^{20}	1.4865^{20}	i H$_2$O; s bz, chl
5705	Heptyl butanoate		C$_{11}$H$_{22}$O$_2$	5870-93-9	186.292	liq	-57.5(0.5)	225.2(0.4)	0.8637^{20}	1.4231^{20}	vs EtOH
5706	Heptylcyclohexane		C$_{13}$H$_{26}$	5617-41-4	182.345	liq	-40.3(1)	244(2)	0.8109^{20}	1.4484^{20}	
5707	Heptylcyclopentane		C$_{12}$H$_{24}$	5617-42-5	168.319	liq	-61.0(0.2)	224(6)	0.8010^{20}	1.4421^{20}	vs ace, bz, eth, EtOH
5708	5-Heptyldihydro-2(3H)-furanone	4-Hydroxyundecanoic acid lactone	C$_{11}$H$_{20}$O$_2$	104-67-6	184.276			286	0.9494^{20}	1.4512^{20}	vs EtOH
5709	Heptyl formate		C$_8$H$_{16}$O$_2$	112-23-2	144.212			184(6)	0.8784^{20}	1.4140^{20}	i H$_2$O; msc EtOH, eth

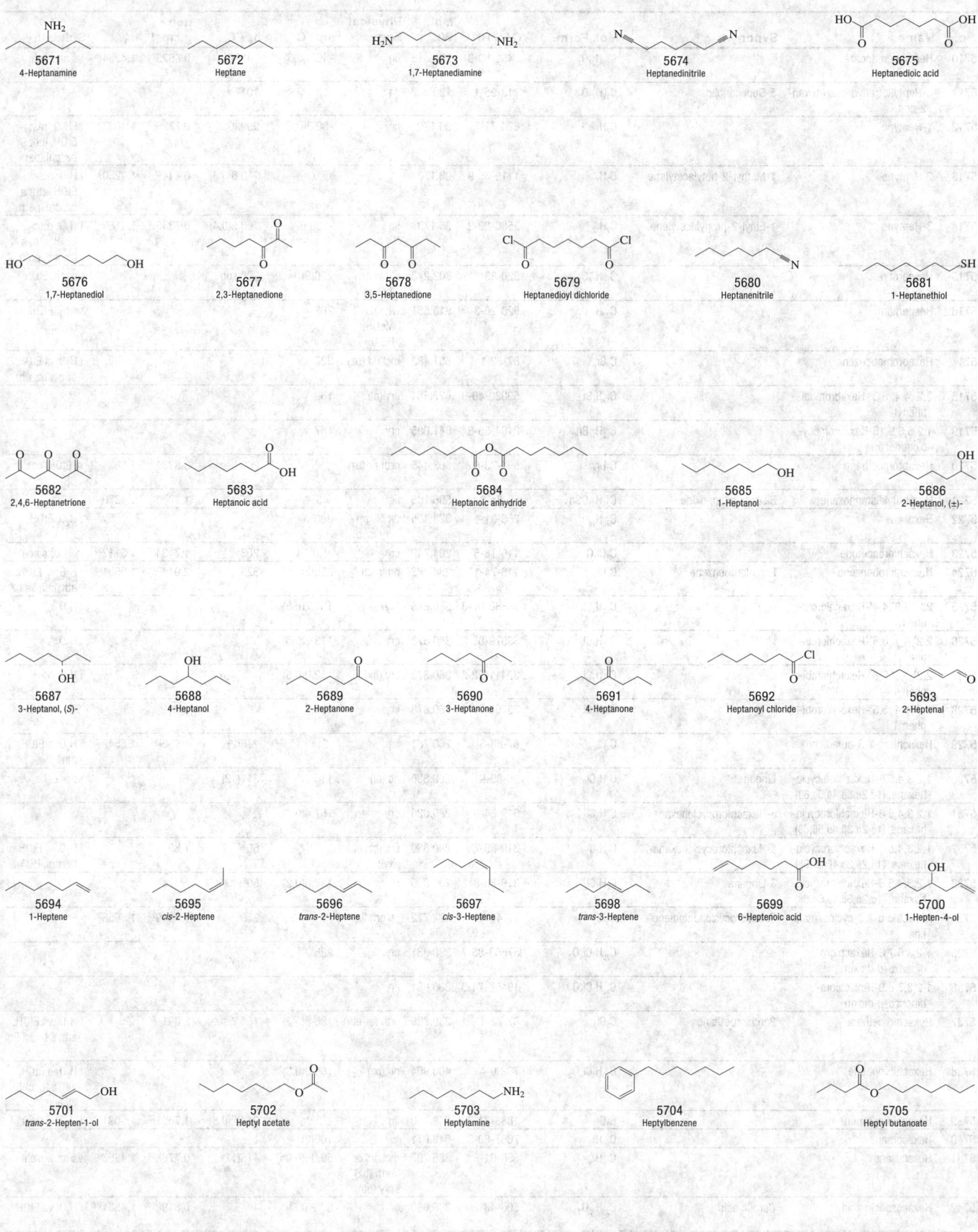

5671
4-Heptanamine

5672
Heptane

5673
1,7-Heptanediamine

5674
Heptanedinitrile

5675
Heptanedioic acid

5676
1,7-Heptanediol

5677
2,3-Heptanedione

5678
3,5-Heptanedione

5679
Heptanedioyl dichloride

5680
Heptanenitrile

5681
1-Heptanethiol

5682
2,4,6-Heptanetrione

5683
Heptanoic acid

5684
Heptanoic anhydride

5685
1-Heptanol

5686
2-Heptanol, (±)-

5687
3-Heptanol, (S)-

5688
4-Heptanol

5689
2-Heptanone

5690
3-Heptanone

5691
4-Heptanone

5692
Heptanoyl chloride

5693
2-Heptenal

5694
1-Heptene

5695
cis-2-Heptene

5696
trans-2-Heptene

5697
cis-3-Heptene

5698
trans-3-Heptene

5699
6-Heptenoic acid

5700
1-Hepten-4-ol

5701
trans-2-Hepten-1-ol

5702
Heptyl acetate

5703
Heptylamine

5704
Heptylbenzene

5705
Heptyl butanoate

5706
Heptylcyclohexane

5707
Heptylcyclopentane

5708
5-Heptyldihydro-2(3H)-furanone

5709
Heptyl formate

No.	Name	Synonym	Mol. Form.	CAS RN	Mol. Wt.	Physical Form	mp/°C	bp/°C	den g cm⁻³	n_D	Solubility
5710	Heptyl pentanoate		$C_{12}H_{24}O_2$	5451-80-9	200.318	liq	-46.4(0.5)	241(4)	0.8623[20]	1.4254[15]	vs ace, eth, EtOH
5711	6-Heptyltetrahydro-2H-pyran-2-one	5-Dodecanolide	$C_{12}H_{22}O_2$	713-95-1	198.302	liq	-12	101[0.03]			
5712	1-Heptyne		C_7H_{12}	628-71-7	96.170	liq	-80.9(0.2)	99.8(0.3)	0.7328[20]	1.4087[20]	sl H₂O; msc EtOH, eth; s bz, chl, peth
5713	2-Heptyne	1-Methyl-2-butylacetylene	C_7H_{12}	1119-65-9	96.170			113.6(0.3)	0.744[25]	1.4230[20]	i H₂O; msc EtOH, eth; s bz, chl, peth
5714	3-Heptyne	1-Ethyl-2-propylacetylene	C_7H_{12}	2586-89-2	96.170	liq	-130.5	109.2(0.4)	0.7336[25]	1.4189[20]	i H₂O; msc EtOH, eth; s bz, chl, peth
5715	Hesperetin		$C_{16}H_{14}O_6$	520-33-2	302.278	pl (dil al + 1/2 w)	226(2)	205 sub			vs eth, EtOH
5716	Hesperidin		$C_{28}H_{34}O_{15}$	520-26-3	610.561	wh nd (dil MeOH, HOAc)	262				vs py, EtOH, HOAc
5717	Hexabromobenzene		C_6Br_6	87-82-1	551.488	mcl nd (bz)	327				i H₂O; sl EtOH, eth; s bz, chl
5718	2,2',4,4',5,5'-Hexabromobiphenyl		$C_{12}H_4Br_6$	59080-40-9	627.584	cry (ctc)	160				
5719	1,2,5,6,9,10-Hexabromocyclododecane		$C_{12}H_{18}Br_6$	3194-55-6	641.695	cry	167				
5720	Hexabromoethane		C_2Br_6	594-73-0	503.445	orth pr (bz)		200 dec	3.823[20]	1.863	sl EtOH, eth, CS₂
5721	Hexabutyldistannoxane	Bis(tributyltin) oxide	$C_{24}H_{54}OSn_2$	56-35-9	596.105	liq	-45	225[10]	1.17[20]	1.4870	
5722	Hexacene		$C_{26}H_{16}$	258-31-1	328.405	dk bl-grn cry (sub)	380	sub			i H₂O, EtOH
5723	Hexachloroacetone		C_3Cl_6O	116-16-5	264.749	liq	-1.0	203	1.7434[12]	1.5112[20]	sl H₂O; s ace
5724	Hexachlorobenzene	Perchlorobenzene	C_6Cl_6	118-74-1	284.782	nd (sub)	230(3)	325	2.044[23]	1.5691[23]	i H₂O; sl EtOH; s eth, chl; vs bz
5725	2,2',3,3',4,4'-Hexachlorobiphenyl		$C_{12}H_4Cl_6$	38380-07-3	360.878	cry	151.7(0.5)				i H₂O
5726	2,2',4,4',6,6'-Hexachlorobiphenyl		$C_{12}H_4Cl_6$	33979-03-2	360.878	cry	113.7(0.5)				i H₂O
5727	2,2',3,3',6,6'-Hexachlorobiphenyl		$C_{12}H_4Cl_6$	38411-22-2	360.878	cry (hx)	112.0(0.5)				i H₂O
5728	2,2',4,4',5,5'-Hexachlorobiphenyl		$C_{12}H_4Cl_6$	35065-27-1	360.878	cry	103.5				
5729	Hexachloro-1,3-butadiene		C_4Cl_6	87-68-3	260.761	liq	-21	216(2)	1.556[25]	1.5542[20]	i H₂O; s EtOH, eth
5730	1,2,3,4,5,6-Hexachlorocyclohexane, (1α,2α,3β,4α,5α,6β)	Lindane	$C_6H_6Cl_6$	58-89-9	290.830	nd (al)	115(2)	311(12)			vs ace, bz
5731	1,2,3,4,5,6-Hexachlorocyclohexane, (1α,2α,3β,4α,5β,6β)	α-Hexachlorocyclohexane	$C_6H_6Cl_6$	319-84-6	290.830	cry	157.4(0.7)				
5732	1,2,3,4,5,6-Hexachlorocyclohexane, (1α,2β,3α,4β,5α,6β)	β-Hexachlorocyclohexane	$C_6H_6Cl_6$	319-85-7	290.830	cry (bz, al, xyl)		60[0.50]	1.89[19]		i H₂O; sl EtOH, bz, chl, HOAc
5733	1,2,3,4,5,6-Hexachlorocyclohexane, (1α,2α,3α,4β,5α,6β)	δ-Lindane	$C_6H_6Cl_6$	319-86-8	290.830	pl	137.0(0.2)	60[0.36]			
5734	Hexachloro-1,3-cyclopentadiene	Perchlorocyclopentadiene	C_5Cl_6	77-47-4	272.772	ye grn liq	-9	239	1.7019[25]	1.5658[20]	
5735	1,2,3,6,7,8-Hexachlorodibenzo-p-dioxin		$C_{12}H_2Cl_6O_2$	57653-85-7	390.861	cry	285				
5736	1,2,3,7,8,9-Hexachlorodibenzo-p-dioxin		$C_{12}H_2Cl_6O_2$	19408-74-3	390.861	cry	243				
5737	Hexachloroethane	Perchloroethane	C_2Cl_6	67-72-1	236.739	orth (al-eth)	186.8(0.2)	184.7 sp	2.091[20]		i H₂O; vs EtOH, eth; s bz; sl liq HF
5738	Hexachlorophene		$C_{13}H_6Cl_6O_2$	70-30-4	406.904	nd (bz)	165.0(0.3)				i H₂O; s EtOH, eth, ace, chl, dil alk
5739	Hexachloropropene		C_3Cl_6	1888-71-7	248.750	liq	-72.9	214.2(0.8)	1.7632[20]	1.5091[20]	i H₂O; s ctc, chl
5740	Hexacontane		$C_{60}H_{122}$	7667-80-3	843.611		100(2)				
5741	Hexacosane		$C_{26}H_{54}$	630-01-3	366.707	mcl, tcl or orth (bz) cry (eth)	56.09(0.04)	415(11)	0.7783[60]	1.4357[60]	vs bz, lig, chl
5742	Hexacosanoic acid	Cerotic acid	$C_{26}H_{52}O_2$	506-46-7	396.690		87.7(0.5)		0.8198[100]	1.4301[100]	i H₂O; vs EtOH, eth
5743	1-Hexacosanol		$C_{26}H_{54}O$	506-52-5	382.706	orth pl (dil al)	80(1)	305[20] dec			i H₂O; s EtOH, eth
5744	Hexadecamethylheptasiloxane		$C_{16}H_{48}O_6Si_7$	541-01-5	533.147	liq	-78	286.8(0.2)	0.9012[20]	1.3965[20]	vs bz, lig
5745	Hexadecanal		$C_{16}H_{32}O$	629-80-1	240.424	pl (eth), nd (peth)	36(5)	200[29]			i H₂O; s EtOH, eth, ace, bz

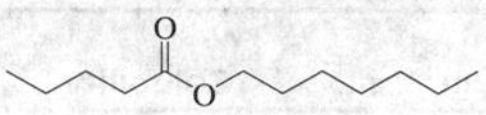

5710
Heptyl pentanoate

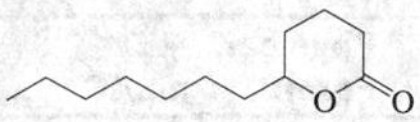

5711
6-Heptyltetrahydro-2H-pyran-2-one

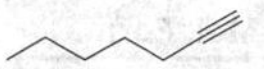

5712
1-Heptyne

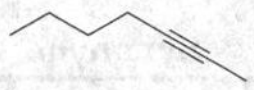

5713
2-Heptyne

5714
3-Heptyne

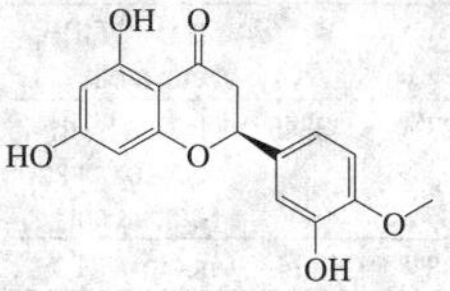

5715
Hesperetin

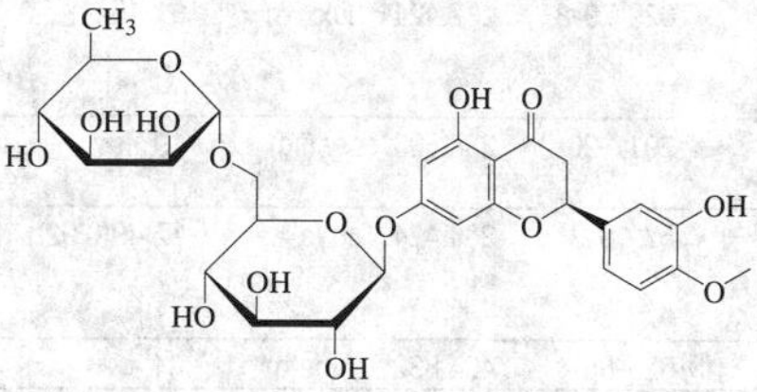

5716
Hesperidin

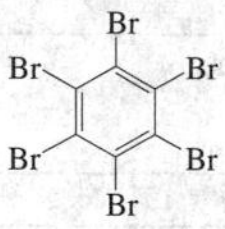

5717
Hexabromobenzene

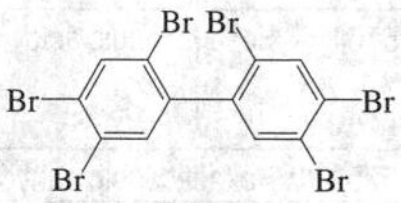

5718
2,2',4,4',5,5'-Hexabromobiphenyl

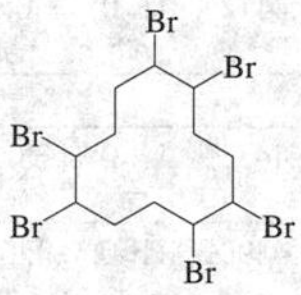

5719
1,2,5,6,9,10-Hexabromocyclododecane

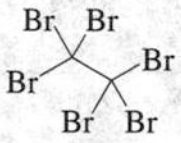

5720
Hexabromoethane

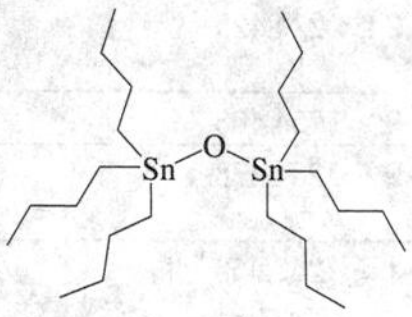

5721
Hexabutyldistannoxane

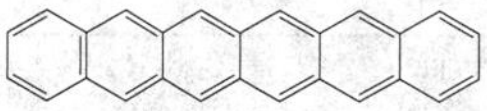

5722
Hexacene

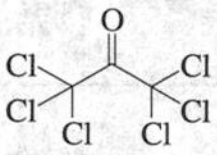

5723
Hexachloroacetone

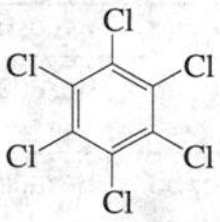

5724
Hexachlorobenzene

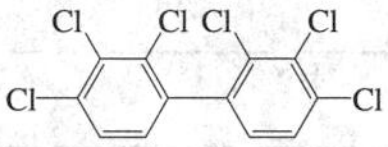

5725
2,2',3,3',4,4'-Hexachlorobiphenyl

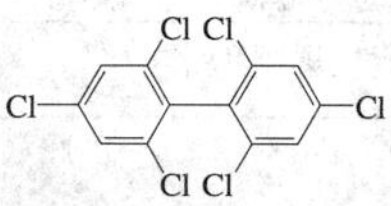

5726
2,2',4,4',6,6'-Hexachlorobiphenyl

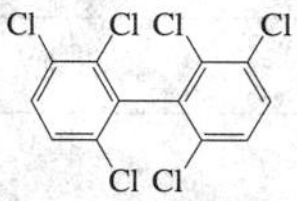

5727
2,2',3,3',6,6'-Hexachlorobiphenyl

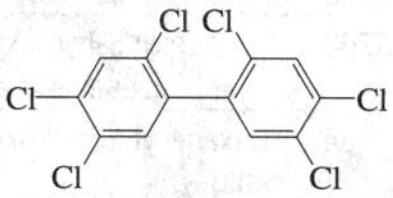

5728
2,2',4,4',5,5'-Hexachlorobiphenyl

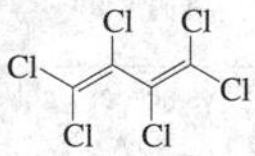

5729
Hexachloro-1,3-butadiene

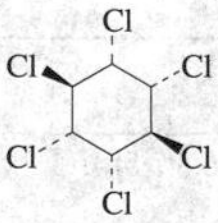

5730
1,2,3,4,5,6-Hexachlorocyclohexane, (1α,2α,3β,4α,5α,6β)

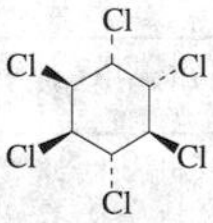

5731
1,2,3,4,5,6-Hexachlorocyclohexane, (1α,2α,3β,4α,5β,6β)

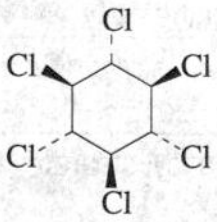

5732
1,2,3,4,5,6-Hexachlorocyclohexane, (1α,2β,3α,4β,5α,6β)

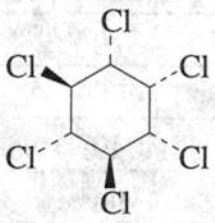

5733
1,2,3,4,5,6-Hexachlorocyclohexane, (1α,2α,3α,4β,5α,6β)

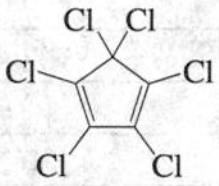

5734
Hexachloro-1,3-cyclopentadiene

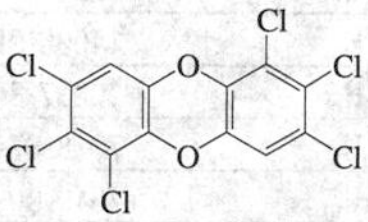

5735
1,2,3,6,7,8-Hexachlorodibenzo-p-dioxin

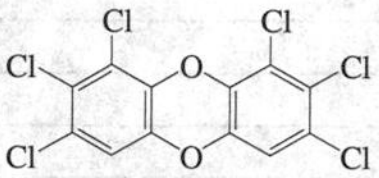

5736
1,2,3,7,8,9-Hexachlorodibenzo-p-dioxin

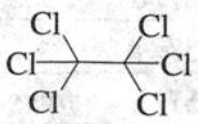

5737
Hexachloroethane

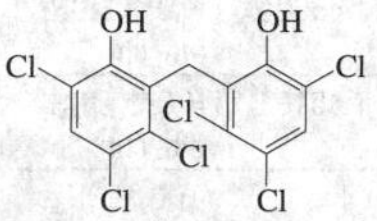

5738
Hexachlorophene

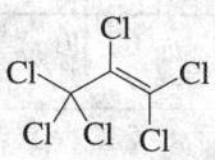

5739
Hexachloropropene

5740
Hexacontane
$H_3C(CH_2)_{58}CH_3$

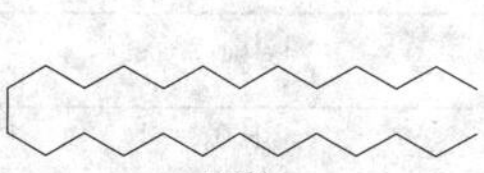

5741
Hexacosane

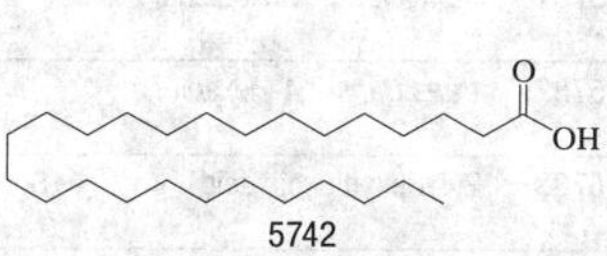

5742
Hexacosanoic acid

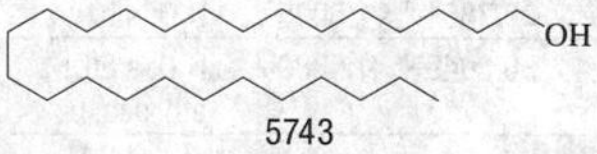

5743
1-Hexacosanol

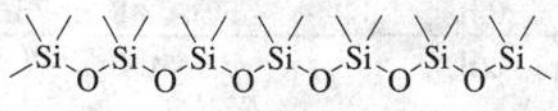

5744
Hexadecamethylheptasiloxane

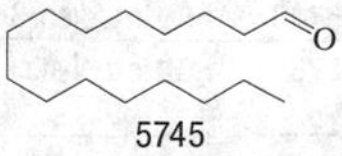

5745
Hexadecanal

No.	Name	Synonym	Mol. Form.	CAS RN	Mol. Wt.	Physical Form	mp/°C	bp/°C	den g cm⁻³	n_D	Solubility
5746	Hexadecanamide		$C_{16}H_{33}NO$	629-54-9	255.439	lf	105(3)	236[12]	1.0000[20]		i H₂O; sl EtOH, bz, ace, eth
5747	Hexadecane	Cetane	$C_{16}H_{34}$	544-76-3	226.441	lf (HOAc)	18.18(0.02)	286.9(0.7)	0.7701[25]	1.4329[25]	i H₂O; sl EtOH; msc eth; s ctc
5748	Hexadecanedioic acid		$C_{16}H_{30}O_4$	505-54-4	286.407	pl (al)	122(1)				vs ace, EtOH
5749	Hexadecanenitrile		$C_{16}H_{31}N$	629-79-8	237.424	hex	31.4(0.4)	333	0.8303[20]	1.4450[20]	i H₂O; vs EtOH, eth, ace, bz, chl
5750	1-Hexadecanethiol	Cetyl mercaptan	$C_{16}H_{34}S$	2917-26-2	258.506	cry (lig)	19	125[0.5]			i H₂O; sl EtOH, ctc; s eth
5751	Hexadecanoic acid	Palmitic acid	$C_{16}H_{32}O_2$	57-10-3	256.424	nd (al)	62.49(0.02)	351(6)	0.8527[62]	1.43345[60]	i H₂O; s EtOH, ace, bz; msc eth; vs chl
5752	Hexadecanoic anhydride		$C_{32}H_{62}O_3$	623-65-4	494.832	lf (peth)	64		0.8388[83]	1.4364[68]	vs eth
5753	1-Hexadecanol	Cetyl alcohol	$C_{16}H_{34}O$	36653-82-4	242.440	fl (AcOEt)	49.30(0.01)	325(2)	0.8187[50]	1.4283[79]	i H₂O; sl EtOH; vs eth, bz, chl; s ace
5754	3-Hexadecanone		$C_{16}H_{32}O$	18787-64-9	240.424	lf (peth)	43	184[17]			s chl
5755	Hexadecanoyl chloride		$C_{16}H_{31}ClO$	112-67-4	274.869		12	199[20]	0.9016[25]	1.4514[20]	
5756	1-Hexadecene	1-Cetene	$C_{16}H_{32}$	629-73-2	224.425	lf	4.2(0.1)	285(1)	0.7811[20]	1.4412[20]	i H₂O; s EtOH, eth, ctc, peth
5757	cis-9-Hexadecenoic acid	Palmitoleic acid	$C_{16}H_{30}O_2$	373-49-9	254.408		2(2)	182[1]			
5758	Hexadecyl acetate		$C_{18}H_{36}O_2$	629-70-9	284.478		-18.5	222[205]	0.8574[25]	1.4438[20]	i H₂O; sl EtOH; s ctc
5759	Hexadecylamine	1-Hexadecanamine	$C_{16}H_{35}N$	143-27-1	241.456	lf	46(2)	321(3)	0.8129[20]	1.4496[20]	i H₂O; vs EtOH, eth, bz; s ace
5760	Hexadecylbenzene		$C_{22}H_{38}$	1459-09-2	302.537		26.4(0.4)	385	0.8547[20]	1.4813[20]	i H₂O; sl EtOH; vs eth, bz, CS₂
5761	Hexadecyldimethylamine	N,N-Dimethyl-1-hexadecanamine	$C_{18}H_{39}N$	112-69-6	269.510			330.0			
5762	Hexadecyl hexadecanoate	Cetyl palmitate	$C_{32}H_{64}O_2$	540-10-3	480.849	mcl lf	43.9(0.2)		0.989[20]	1.4398[70]	vs eth, EtOH
5763	Hexadecyl 3-hydroxy-2-naphthalenecarboxylate	Hexadecyl 3-hydroxy-2-naphthoate	$C_{27}H_{40}O_3$	531-84-0	412.605	grn-wh fl	72.5				vs bz, HOAc
5764	Hexadecyl 2-hydroxypropanoate	Cetyl lactate	$C_{19}H_{38}O_3$	35274-05-6	314.503	wax	41(1)	219[10]		1.4410[40]	
5765	Hexadecyl 2-methyl-2-propenoate		$C_{20}H_{38}O_2$	2495-27-4	310.515		24	183[2]	0.87[20]		
5766	3-(Hexadecyloxy)-1,2-propanediol, (S)-	Chimyl alcohol	$C_{19}H_{40}O_3$	506-03-6	316.519	lf (hx)	64	120[0.005]			vs ace, peth, chl
5767	1-Hexadecylpyridinium bromide		$C_{21}H_{38}BrN$	140-72-7	384.438		61				
5768	1-Hexadecylpyridinium chloride	Cetylpyridinium chloride	$C_{21}H_{38}ClN$	123-03-5	339.987	wh pow	81(3)				vs H₂O, chl
5769	Hexadecyl stearate	Cetyl stearate	$C_{34}H_{68}O_2$	1190-63-2	508.903	lf or pl (eth, HOAc)	56.8(0.5)			1.4410[70]	vs ace, eth, chl
5770	Hexadecyltrichlorosilane		$C_{16}H_{33}Cl_3Si$	5894-60-0	359.878			269			
5771	Hexadecyl vinyl ether	1-(Ethenyloxy)hexadecane	$C_{18}H_{36}O$	822-28-6	268.478		16	160[2]	0.821[27]	1.4444[25]	
5772	1-Hexadecyne		$C_{16}H_{30}$	629-74-3	222.409		15(2)	284	0.7965[20]	1.4440[20]	vs bz
5773	trans,trans-2,4-Hexadienal	Sorbinaldehyde	C_6H_8O	142-83-6	96.127	liq	-16.5	174	0.898[20]	1.5384[20]	
5774	1,2-Hexadiene	Propylallene	C_6H_{10}	592-44-9	82.143			76(3)	0.7149[20]	1.4282[20]	vs eth, chl
5775	cis-1,3-Hexadiene		C_6H_{10}	14596-92-0	82.143			73(3)	0.7033[25]	1.4379[20]	
5776	trans-1,3-Hexadiene		C_6H_{10}	20237-34-7	82.143	liq	-102.4(0.6)	71.5(0.8)	0.6995[25]	1.4406[20]	
5777	cis-1,4-Hexadiene		C_6H_{10}	7318-67-4	82.143			70(13)	0.695[25]	1.4049[20]	vs eth
5778	trans-1,4-Hexadiene		C_6H_{10}	7319-00-8	82.143	liq	-138.7	65.0(0.5)	0.695[25]	1.4104[20]	
5779	1,5-Hexadiene	Biallyl	C_6H_{10}	592-42-7	82.143	liq	-140.7(0.1)	59.2(0.4)	0.6878[25]	1.4042[20]	i H₂O; s EtOH, eth, bz, chl; sl ctc
5780	cis,cis-2,4-Hexadiene		C_6H_{10}	6108-61-8	82.143	liq		85(3)	0.7298[25]	1.4606[20]	i H₂O; s EtOH, eth, chl
5781	trans,cis-2,4-Hexadiene		C_6H_{10}	5194-50-3	82.143	liq	-96.1(0.4)	83(3)	0.7185[25]	1.4560[20]	i H₂O; s EtOH, eth, chl
5782	trans,trans-2,4-Hexadiene		C_6H_{10}	5194-51-4	82.143	liq	-44.9(0.4)	82.4(0.7)	0.7101[25]	1.4510[20]	i H₂O; s EtOH, eth, chl
5783	2,4-Hexadienoic acid	Sorbic acid	$C_6H_8O_2$	110-44-1	112.127	nd (dil al) nd (w)	134.5	228 dec	1.204[19]		s H₂O, EtOH, chl; vs eth
5784	2,4-Hexadien-1-ol	Sorbic alcohol	$C_6H_{10}O$	111-28-4	98.142	nd	29(2)	76[12]	0.8967[23]	1.4981[20]	i H₂O; s EtOH, eth
5785	trans,trans-2,4-Hexadienoyl chloride		C_6H_7ClO	2614-88-2	130.572			82[22]	1.0666[19]	1.5545[20]	vs ace
5786	1,5-Hexadien-3-yne	Divinylacetylene	C_6H_6	821-08-9	78.112	liq	-88	85(3)	0.7851[20]	1.5035[20]	i H₂O; s bz
5787	1,5-Hexadiyne	Bipropargyl	C_6H_6	628-16-0	78.112	liq	-5.5(0.2)	88(3)	0.8049[20]	1.4380[23]	i H₂O; s EtOH, eth, ace, bz

5746
Hexadecanamide

5747
Hexadecane

5748
Hexadecanedioic acid

5749
Hexadecanenitrile

5750
1-Hexadecanethiol

5751
Hexadecanoic acid

5752
Hexadecanoic anhydride

5753
1-Hexadecanol

5754
3-Hexadecanone

5755
Hexadecanoyl chloride

5756
1-Hexadecene

5757
cis-9-Hexadecenoic acid

5758
Hexadecyl acetate

5759
Hexadecylamine

5760
Hexadecylbenzene

5761
Hexadecyldimethylamine

5762
Hexadecyl hexadecanoate

5763
Hexadecyl 3-hydroxy-2-naphthalenecarboxylate

5764
Hexadecyl 2-hydroxypropanoate

5765
Hexadecyl 2-methyl-2-propenoate

5766
3-(Hexadecyloxy)-1,2-propanediol, (S)-

5767
1-Hexadecylpyridinium bromide

5768
1-Hexadecylpyridinium chloride

5769
Hexadecyl stearate

5770
Hexadecyltrichlorosilane

5771
Hexadecyl vinyl ether

5772
1-Hexadecyne

5773
trans,trans-2,4-Hexadienal

5774
1,2-Hexadiene

5775
cis-1,3-Hexadiene

5776
trans-1,3-Hexadiene

5777
cis-1,4-Hexadiene

5778
trans-1,4-Hexadiene

5779
1,5-Hexadiene

5780
cis,cis-2,4-Hexadiene

5781
trans,cis-2,4-Hexadiene

5782
trans,trans-2,4-Hexadiene

5783
2,4-Hexadienoic acid

5784
2,4-Hexadien-1-ol

5785
trans,trans-2,4-Hexadienoyl chloride

5786
1,5-Hexadien-3-yne

5787
1,5-Hexadiyne

No.	Name	Synonym	Mol. Form.	CAS RN	Mol. Wt.	Physical Form	mp/°C	bp/°C	den g cm⁻³	n_D	Solubility
5788	2,4-Hexadiyne	Dimethyldiacetylene	C_6H_6	2809-69-0	78.112	pr (sub)	64.67(0.01)	129.5(0.2)			vs EtOH, eth
5789	Hexaethylbenzene		$C_{18}H_{30}$	604-88-6	246.431	mcl pr (al or bz)	129(1)	298(3)	0.8305[130]	1.4736[130]	i H_2O; s EtOH, sulf; vs eth, bz
5790	Hexaethyldisiloxane		$C_{12}H_{30}OSi_2$	994-49-0	246.536			252(7)	0.8457[20]	1.4340[20]	
5791	Hexaethyl tetraphosphate	Ethyl tetraphosphate	$C_{12}H_{30}O_{13}P_4$	757-58-4	506.253	hyg	-40	150 dec	1.2917[27]	1.4273[27]	vs ace, bz, EtOH
5792	Hexafluorenium bromide		$C_{36}H_{42}Br_2N_2$	317-52-2	662.539	cry (PrOH)	188				
5793	Hexafluoroacetylacetone		$C_5H_2F_6O_2$	1522-22-1	208.059			69(2)	1.485[20]	1.3333[20]	
5794	Hexafluorobenzene	Perfluorobenzene	C_6F_6	392-56-3	186.054	liq	5.10(0.01)	80.2(0.2)	1.6175[20]	1.3777[20]	
5795	1,1,2,3,4,4-Hexafluoro-1,3-butadiene		C_4F_6	685-63-2	162.033	col gas	-132	5.4(0.2)	1.473[5]	1.378[-20]	
5796	1,1,1,4,4,4-Hexafluoro-2-butyne		C_4F_6	692-50-2	162.033	col gas	-117.4	-24.6			s EtOH, eth, ace, ctc, HOAc
5797	Hexafluorocyclobutene		C_4F_6	697-11-0	162.033	col gas	-60	2(5)	1.602[20]	1.298[20]	
5798	Hexafluoroethane	Perfluoroethane	C_2F_6	76-16-4	138.011	col gas	-100.015 (0.005)	-78.1(0.1)	1.590[-78]		i H_2O; sl EtOH, eth
5799	1,1,1,2,3,3-Hexafluoropropane	Refrigerant 236ea	$C_3H_2F_6$	431-63-0	152.038	col gas		6.2	1.5026[0]		
5800	1,1,1,3,3,3-Hexafluoropropane	Refrigerant 236fa	$C_3H_2F_6$	690-39-1	152.038	col gas	-93.6	-1.4(0.2)	1.4343[0]		
5801	1,1,1,3,3,3-Hexafluoro-2-propanol		$C_3H_2F_6O$	920-66-1	168.037	liq	-2.0	59(3)	1.4600[21]		
5802	Hexahydro-1H-azepine	Hexamethylenimine	$C_6H_{13}N$	111-49-9	99.174			136(2)	0.8643[22]	1.4631[20]	s H_2O; vs EtOH, eth
5803	Hexahydro-1H-1,4-diazepine		$C_5H_{12}N_2$	505-66-8	100.162	hyg	40.5	169			
5804	1,5a,6,9,9a,9b-Hexahydro-4a(4H)-dibenzofurancarbox-aldehyde		$C_{13}H_{16}O_2$	126-15-8	204.265	liq	-80	307	1.10[20]	1.5254[20]	i H_2O
5805	cis-1,2,3,5,6,8a-Hexahydro-4,7-dimethyl-1-isopropyl-naphthalene, (1S)-		$C_{15}H_{24}$	483-76-1	204.352			125[12]	0.9160[15]	1.5089[15]	
5806	1,2,4a,5,8,8a-Hexahydro-4,7-dimethyl-1-isopropylnaph-thalene, [1S-(1α,4aβ,8aα)]		$C_{15}H_{24}$	523-47-7	204.352			274	0.9230[20]	1.5059[20]	vs eth, lig
5807	Hexahydro-1,3-isobenzofuran-dione	Hexahydrophthalic anhydride	$C_8H_{10}O_3$	85-42-7	154.163		32	145[18]			
5808	Hexahydro-1-methyl-1H-1,4-diazepine		$C_6H_{14}N_2$	4318-37-0	114.188			154	0.9111[20]	1.4769[20]	
5809	2,3,4,6,7,8-Hexahydro-pyrrolo[1,2-a]pyrimidine		$C_7H_{12}N_2$	3001-72-7	124.183			96[7]	1.005[25]	1.5190[20]	
5810	Hexahydro-1,3,5-trinitro-1,3,5-triazine	Cyclonite	$C_3H_6N_6O_6$	121-82-4	222.116	orth cry (ace)	203.4(0.9)		1.82[20]		i H_2O, EtOH, bz; sl eth, MeOH; s ace, HOAc
5811	Hexahydro-1,3,5-triphenyl-1,3,5-triazine		$C_{21}H_{21}N_3$	91-78-1	315.412		144	185			i H_2O; sl EtOH; s eth, ace, bz, tol
5812	1,2,3,5,6,7-Hexahydroxy-9,10-anthracenedione	Rufigallol	$C_{14}H_8O_8$	82-12-2	304.209	red rhom, red-ye nd (sub)		sub			i H_2O; sl EtOH, eth; s ace, alk
5813	Hexamethylbenzene	Mellitene	$C_{12}H_{18}$	87-85-4	162.271	orth pr or nd (al)	165.6(0.7)	268(3)	1.0630[25]		i H_2O; s EtOH, eth, ace, bz, HOAc, chl
5814	2,2,4,4,6,6-Hexamethylcyclo-trisilazane		$C_6H_{21}N_3Si_3$	1009-93-4	219.508	liq	-18.7(0.2)	187(1)	0.9196[20]	1.448[20]	
5815	Hexamethylcyclotrisiloxane	Dimethylsiloxane cyclic trimer	$C_6H_{18}O_3Si_3$	541-05-9	222.462		63.7(0.4)	135.1(0.4)	1.1200[20]		i H_2O
5816	Hexamethyldisilane		$C_6H_{18}Si_2$	1450-14-2	146.378		14.4(0.6)	112.65(0.07)	0.7247[22]	1.4229[20]	i H_2O; s eth, ace, bz; dec alk
5817	Hexamethyldisilathiane		$C_6H_{18}SSi_2$	3385-94-2	178.443			162.5	0.851[20]		
5818	Hexamethyldisilazane		$C_6H_{19}NSi_2$	999-97-3	161.393			125(1)	0.7741[25]	1.4090[20]	
5819	Hexamethyldisiloxane		$C_6H_{18}OSi_2$	107-46-0	162.377	liq	-68.2(0.1)	100.5(0.3)	0.7638[20]	1.3774[20]	i H_2O
5820	Hexamethylenediamine carbamate	(6-Aminohexyl)carbamic acid	$C_7H_{16}N_2O_2$	143-06-6	160.214	cry	150				
5821	Hexamethylene diisocyanate		$C_8H_{12}N_2O_2$	822-06-0	168.193			122[10]	1.0528[20]	1.4585[20]	
5822	Hexamethylenetetramine	Methenamine	$C_6H_{12}N_4$	100-97-0	140.186	orth (al)	>250	sub	1.331[-5]		vs H_2O; s EtOH, ace, chl; sl eth, bz
5823	Hexamethylolmelamine		$C_9H_{18}N_6O_6$	531-18-0	306.275		137				vs H_2O
5824	Hexamethylphosphoric triamide	Tris(dimethylamino)phosphine oxide	$C_6H_{18}N_3OP$	680-31-9	179.200	col liq	7.2	235(13)	1.03[20]	1.4579[20]	s EtOH, eth
5825	Hexamethylphosphorous triamide	Tris(dimethylamino)phosphine	$C_6H_{18}N_3P$	1608-26-0	163.201						s chl
5826	2,6,10,15,19,23-Hexamethyl-tetracosane	Squalane	$C_{30}H_{62}$	111-01-3	422.813	liq	-38	420(6)	0.8115[15]	1.4530[15]	i H_2O; sl EtOH, ace; s eth, chl; msc bz
5827	Hexanal	Caproaldehyde	$C_6H_{12}O$	66-25-1	100.158	liq	-58.2(0.2)	129.6(0.4)	0.8335[20]	1.4039[20]	sl H_2O; vs EtOH, eth; s ace, bz

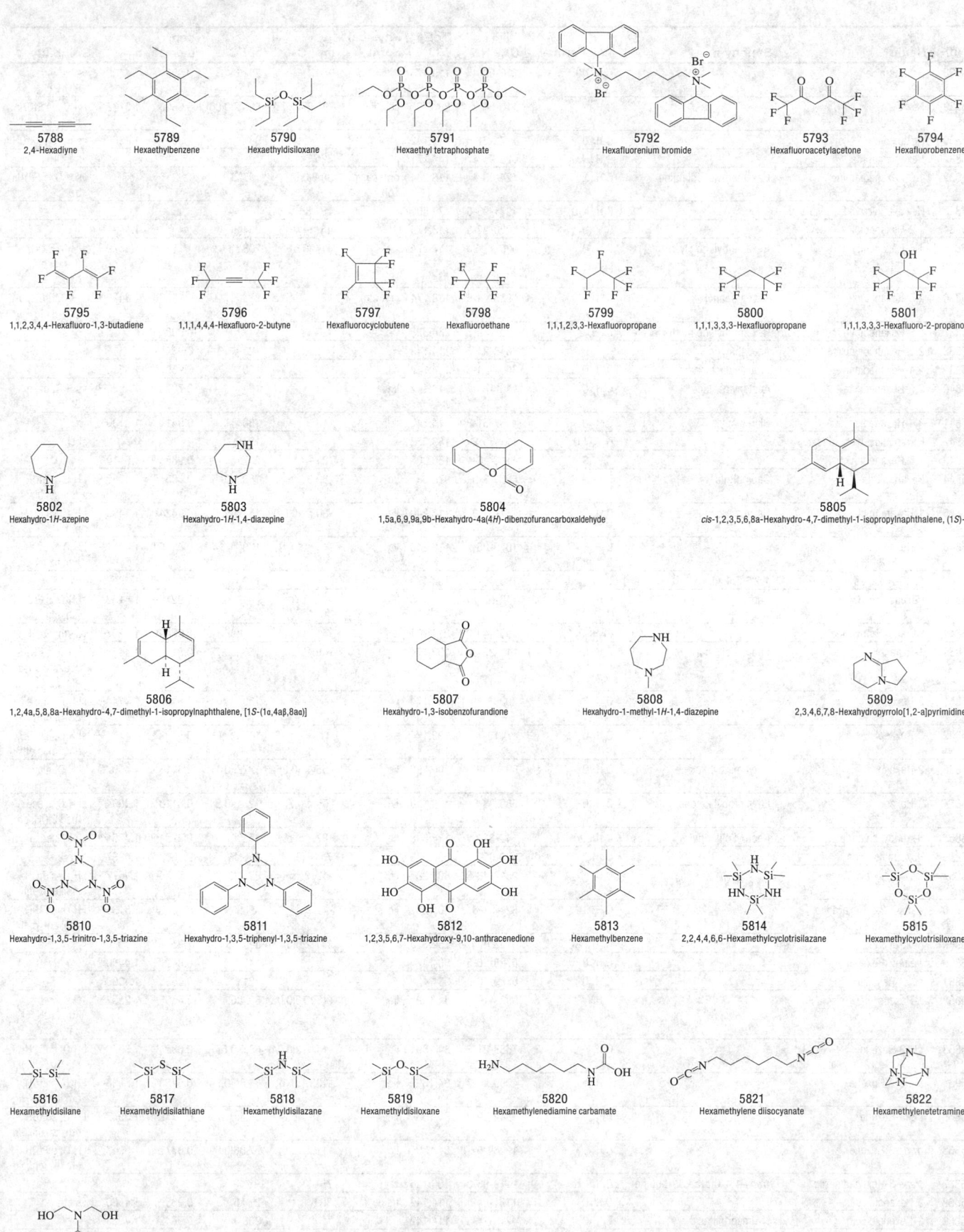

5788
2,4-Hexadiyne

5789
Hexaethylbenzene

5790
Hexaethyldisiloxane

5791
Hexaethyl tetraphosphate

5792
Hexafluorenium bromide

5793
Hexafluoroacetylacetone

5794
Hexafluorobenzene

5795
1,1,2,3,4,4-Hexafluoro-1,3-butadiene

5796
1,1,1,4,4,4-Hexafluoro-2-butyne

5797
Hexafluorocyclobutene

5798
Hexafluoroethane

5799
1,1,1,2,3,3-Hexafluoropropane

5800
1,1,1,3,3,3-Hexafluoropropane

5801
1,1,1,3,3,3-Hexafluoro-2-propanol

5802
Hexahydro-1*H*-azepine

5803
Hexahydro-1*H*-1,4-diazepine

5804
1,5a,6,9,9a,9b-Hexahydro-4a(4*H*)-dibenzofurancarboxaldehyde

5805
cis-1,2,3,5,6,8a-Hexahydro-4,7-dimethyl-1-isopropylnaphthalene, (1*S*)-

5806
1,2,4a,5,8,8a-Hexahydro-4,7-dimethyl-1-isopropylnaphthalene, [1*S*-(1α,4aβ,8aα)]

5807
Hexahydro-1,3-isobenzofurandione

5808
Hexahydro-1-methyl-1*H*-1,4-diazepine

5809
2,3,4,6,7,8-Hexahydropyrrolo[1,2-a]pyrimidine

5810
Hexahydro-1,3,5-trinitro-1,3,5-triazine

5811
Hexahydro-1,3,5-triphenyl-1,3,5-triazine

5812
1,2,3,5,6,7-Hexahydroxy-9,10-anthracenedione

5813
Hexamethylbenzene

5814
2,2,4,4,6,6-Hexamethylcyclotrisilazane

5815
Hexamethylcyclotrisiloxane

5816
Hexamethyldisilane

5817
Hexamethyldisilathiane

5818
Hexamethyldisilazane

5819
Hexamethyldisiloxane

5820
Hexamethylenediamine carbamate

5821
Hexamethylene diisocyanate

5822
Hexamethylenetetramine

5823
Hexamethylolmelamine

5824
Hexamethylphosphoric triamide

5825
Hexamethylphosphorous triamide

5826
2,6,10,15,19,23-Hexamethyltetracosane

5827
Hexanal

No.	Name	Synonym	Mol. Form.	CAS RN	Mol. Wt.	Physical Form	mp/°C	bp/°C	den g cm⁻³	n_D	Solubility
5828	Hexanamide		$C_6H_{13}NO$	628-02-4	115.173	cry (ace)	101(1)	258(6)	0.999²⁰	1.4200¹¹⁰	vs bz, eth, EtOH, chl
5829	Hexane		C_6H_{14}	110-54-3	86.175	liq	-95.27(0.02)	68.72(0.06)	0.6606²⁵	1.3727²⁵	i H₂O; vs EtOH; s eth, chl
5830	Hexanedial		$C_6H_{10}O_2$	1072-21-5	114.142		-8	93⁹	1.003¹⁹	1.4350²⁰	vs bz, eth, EtOH
5831	Hexanediamide		$C_6H_{12}N_2O_2$	628-94-4	144.171	pl	225.9(0.4)				vs EtOH
5832	1,6-Hexanediamine	Hexamethylenediamine	$C_6H_{16}N_2$	124-09-4	116.204	orth bipym pl	38.8(0.6)	197(2)			vs H₂O; s EtOH, bz
5833	Hexanedioic acid, dihydrazide		$C_6H_{14}N_4O_2$	1071-93-8	174.201		181.8				
5834	1,2-Hexanediol		$C_6H_{14}O_2$	6920-22-5	118.174		45	234(8)		1.4431²⁰	
5835	1,6-Hexanediol	Hexamethylene glycol	$C_6H_{14}O_2$	629-11-8	118.174		41.5(0.5)	208		1.4579²⁵	s H₂O, EtOH, ace; sl eth; i bz
5836	2,5-Hexanediol	Diisopropanol	$C_6H_{14}O_2$	2935-44-6	118.174	cry (eth)	43	229(7)	0.9610²⁰	1.4475²⁰	s H₂O, EtOH, eth; sl ctc
5837	1,6-Hexanediol dimethacrylate	Hexamethylene methacrylate	$C_{14}H_{22}O_4$	6606-59-3	254.323				0.998²⁵		
5838	2,3-Hexanedione	Acetylbutyryl	$C_6H_{10}O_2$	3848-24-6	114.142			128	0.934¹⁹		
5839	2,4-Hexanedione	Propionylacetone	$C_6H_{10}O_2$	3002-24-2	114.142	oil		160	0.959²⁰	1.4516²⁰	
5840	2,5-Hexanedione	Acetonylacetone	$C_6H_{10}O_2$	110-13-4	114.142	liq	-5.5	194	0.7370²⁰	1.4232²⁰	vs H₂O, bz, eth, EtOH
5841	3,4-Hexanedione	Bipropionyl	$C_6H_{10}O_2$	4437-51-8	114.142	liq	-10	130	0.941²¹	1.4130²¹	
5842	Hexanedioyl dichloride		$C_6H_8Cl_2O_2$	111-50-2	183.033			126¹²			sl chl
5843	1,6-Hexanedithiol		$C_6H_{14}S_2$	1191-43-1	150.305	liq	-21	237	0.9886²⁵	1.5110²⁰	
5844	Hexanenitrile	Capronitrile	$C_6H_{11}N$	628-73-9	97.158	liq	-80.3	163.5(0.3)	0.8051²⁰	1.4068²⁰	i H₂O; s EtOH, eth; sl chl
5845	1-Hexanethiol	Hexyl mercaptan	$C_6H_{14}S$	111-31-9	118.240	liq	-80.52(0.02)	152.7(0.6)	0.8424²⁰	1.4496²⁰	i H₂O; vs EtOH, eth
5846	2-Hexanethiol		$C_6H_{14}S$	1679-06-7	118.240	liq	-147.0(0.4)	139(1)	0.8345²⁰	1.4451²⁰	i H₂O; s EtOH, eth, bz
5847	1,2,6-Hexanetriol	1,2,6-Trihydroxyhexane	$C_6H_{14}O_3$	106-69-4	134.173			170³	1.1049²⁰	1.58²⁰	
5848	Hexanoic acid	Caproic acid	$C_6H_{12}O_2$	142-62-1	116.158	liq	-4.1(0.7)	204.9(0.6)	0.9212²⁵	1.4163²⁰	sl H₂O; s EtOH, eth, chl
5849	Hexanoic anhydride		$C_{12}H_{22}O_3$	2051-49-2	214.301		-41	267(9)	0.9240¹⁵	1.4297²⁰	vs eth, EtOH
5850	1-Hexanol	Caproyl alcohol	$C_6H_{14}O$	111-27-3	102.174	liq	-46.4(0.9)	156.9(0.7)	0.8136²⁰	1.4178²⁰	sl H₂O; s EtOH, ace, chl; msc eth, bz
5851	2-Hexanol		$C_6H_{14}O$	20281-86-1	102.174			138(6)	0.8159²⁰	1.4144²⁰	sl H₂O, ctc; s EtOH, eth
5852	3-Hexanol		$C_6H_{14}O$	17015-11-1	102.174			143(2)	0.8182²⁰	1.4167²⁰	sl H₂O; s EtOH, ace; msc eth
5853	2-Hexanone	Butyl methyl ketone	$C_6H_{12}O$	591-78-6	100.158	liq	-55.45(0.05)	127.6(0.1)	0.8113²⁰	1.4007²⁰	sl H₂O; s ace; msc EtOH, eth
5854	3-Hexanone	Ethyl propyl ketone	$C_6H_{12}O$	589-38-8	100.158	liq	-55.4(0.2)	123.5(0.3)	0.8118²⁰	1.4004²⁰	sl H₂O; s ace; msc EtOH, eth
5855	Hexanoyl chloride	Caproyl chloride	$C_6H_{11}ClO$	142-61-0	134.603	liq	-87	153	0.9784²⁰	1.4264²⁰	s eth, ace
5856	Hexatriacontane		$C_{36}H_{74}$	630-06-8	506.973		75.81(0.04)	298.4³	0.7803⁸⁰	1.4397⁸⁰	
5857	cis-1,3,5-Hexatriene		C_6H_8	2612-46-6	80.128	liq	-12	82.2(0.9)	0.7175²⁰	1.4577²⁰	i H₂O; s EtOH, ace, chl, peth
5858	trans-1,3,5-Hexatriene		C_6H_8	821-07-8	80.128	liq	-12	79(4)	0.7369¹⁵	1.5135²⁰	i H₂O; s EtOH, ace, chl, peth
5859	Hexazinone		$C_{12}H_{20}N_4O_2$	51235-04-2	252.313		117.2(0.5)	dec	1.25		
5860	trans-2-Hexenal		$C_6H_{10}O$	6728-26-3	98.142			146.5	0.8491²⁰	1.4480²⁰	
5861	cis-3-Hexenal		$C_6H_{10}O$	6789-80-6	98.142			121	0.8533²²	1.4300²¹	
5862	1-Hexene		C_6H_{12}	592-41-6	84.159	liq	-139.76(0.05)	63.4(0.1)	0.6685²⁵	1.3852²⁵	i H₂O; vs bz, eth, EtOH, peth
5863	cis-2-Hexene		C_6H_{12}	7688-21-3	84.159	liq	-141.12(0.04)	68.9(0.5)	0.6824²⁵	1.3979²⁰	i H₂O; s EtOH, eth, bz, chl, lig
5864	trans-2-Hexene		C_6H_{12}	4050-45-7	84.159	liq	-133.1(0.3)	67.85(0.09)	0.6733²⁵	1.3936²⁰	i H₂O; s EtOH, eth, bz, chl, lig
5865	cis-3-Hexene		C_6H_{12}	7642-09-3	84.159	liq	-138.7(0.7)	66.4(0.5)	0.6778²⁰	1.3947²⁰	i H₂O; s EtOH, eth, bz, chl, lig
5866	trans-3-Hexene		C_6H_{12}	13269-52-8	84.159	liq	-113.7(0.5)	67.06(0.09)	0.6772²⁰	1.3943²⁰	i H₂O; s EtOH, eth, bz, chl, lig
5867	trans-3-Hexenedinitrile	trans-1,4-Dicyano-2-butene	$C_6H_6N_2$	1119-85-3	106.125	cry	76				
5868	2-Hexenoic acid		$C_6H_{10}O_2$	1191-04-4	114.142	nd (w, al)	36.5	216.5	0.965²⁰	1.4460⁴⁰	vs eth
5869	3-Hexenoic acid	Hydrosorbic acid	$C_6H_{10}O_2$	4219-24-3	114.142		12	208	0.9640²³	1.4935²⁰	
5870	5-Hexenoic acid	5-Hexanoic acid	$C_6H_{10}O_2$	1577-22-6	114.142	liq	-37	203	0.9610²⁰	1.4343²⁰	vs eth, EtOH
5871	1-Hexen-3-ol		$C_6H_{12}O$	4798-44-1	100.158			134(7)	0.834²²	1.4297¹⁸	sl H₂O; vs ace, eth, EtOH

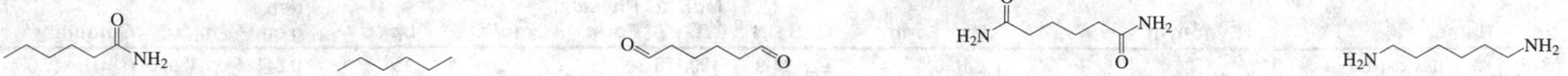

5828 Hexanamide

5829 Hexane

5830 Hexanedial

5831 Hexanediamide

5832 1,6-Hexanediamine

5833 Hexanedioic acid, dihydrazide

5834 1,2-Hexanediol

5835 1,6-Hexanediol

5836 2,5-Hexanediol

5837 1,6-Hexanediol dimethacrylate

5838 2,3-Hexanedione

5839 2,4-Hexanedione

5840 2,5-Hexanedione

5841 3,4-Hexanedione

5842 Hexanedioyl dichloride

5843 1,6-Hexanedithiol

5844 Hexanenitrile

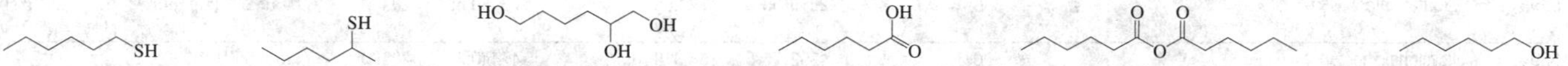

5845 1-Hexanethiol

5846 2-Hexanethiol

5847 1,2,6-Hexanetriol

5848 Hexanoic acid

5849 Hexanoic anhydride

5850 1-Hexanol

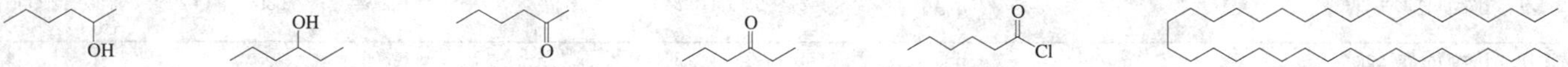

5851 2-Hexanol

5852 3-Hexanol

5853 2-Hexanone

5854 3-Hexanone

5855 Hexanoyl chloride

5856 Hexatriacontane

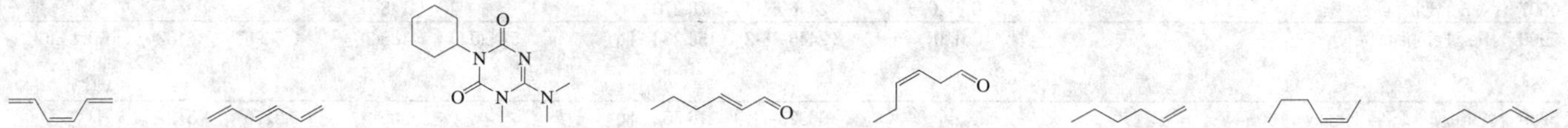

5857 *cis*-1,3,5-Hexatriene

5858 *trans*-1,3,5-Hexatriene

5859 Hexazinone

5860 *trans*-2-Hexenal

5861 *cis*-3-Hexenal

5862 1-Hexene

5863 *cis*-2-Hexene

5864 *trans*-2-Hexene

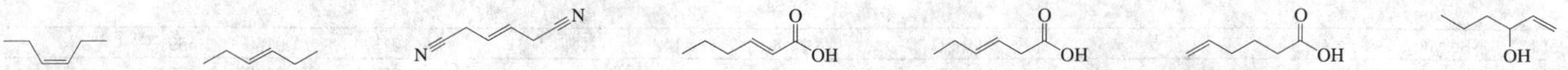

5865 *cis*-3-Hexene

5866 *trans*-3-Hexene

5867 *trans*-3-Hexenedinitrile

5868 2-Hexenoic acid

5869 3-Hexenoic acid

5870 5-Hexenoic acid

5871 1-Hexen-3-ol

No.	Name	Synonym	Mol. Form.	CAS RN	Mol. Wt.	Physical Form	mp/°C	bp/°C	den g cm⁻³	n_D	Solubility
5872	cis-2-Hexen-1-ol		C₆H₁₂O	928-94-9	100.158			157	0.8472²⁰	1.4397²⁰	s H₂O; vs EtOH; s eth, ace; sl ctc
5873	trans-2-Hexen-1-ol		C₆H₁₂O	928-95-0	100.158			172(8)	0.8490¹⁶	1.4340²⁰	
5874	cis-3-Hexen-1-ol		C₆H₁₂O	928-96-1	100.158			157(9)	0.8478²²	1.4380²⁰	s H₂O; vs EtOH, eth
5875	trans-3-Hexen-1-ol		C₆H₁₂O	928-97-2	100.158			153(7)		1.4374²⁰	
5876	trans-4-Hexen-1-ol		C₆H₁₂O	928-92-7	100.158			159	0.8513²⁰	1.4402²⁰	
5877	4-Hexen-2-ol		C₆H₁₂O	52387-50-5	100.158			137.5	0.8405¹⁸	1.4392²⁰	sl H₂O
5878	5-Hexen-2-ol		C₆H₁₂O	626-94-8	100.158			139(6)	0.842¹⁶		sl H₂O
5879	cis-3-Hexen-1-ol, acetate		C₈H₁₄O₂	3681-71-8	142.196	liq		66¹²			
5880	trans-2-Hexen-1-ol, acetate		C₈H₁₄O₂	2497-18-9	142.196	liq		165(11)	0.898	1.4270²⁰	
5881	5-Hexen-2-one		C₆H₁₀O	109-49-9	98.142			129.1(0.5)	0.833²⁷	1.4178²⁷	
5882	4-Hexen-3-one		C₆H₁₀O	2497-21-4	98.142			138(7)	0.8559²⁰	1.4388²⁰	s EtOH, eth; vs ace
5883	Hexestrol		C₁₈H₂₂O₂	84-16-2	270.367	nd (bz)	186.5				vs ace, eth, EtOH
5884	Hexobarbital		C₁₂H₁₆N₂O₃	56-29-1	236.266		145(1)				
5885	Hexocyclium methyl sulfate		C₂₁H₃₆N₂O₅S	115-63-9	428.586	cry	205				sl chl; i eth
5886	Hexyl acetate		C₈H₁₆O₂	142-92-7	144.212	liq	-61.0(0.2)	171.1(0.7)	0.8779¹⁵	1.4092²⁰	i H₂O; vs eth, EtOH
5887	sec-Hexyl acetate	4-Methyl-2-pentyl acetate	C₈H₁₆O₂	108-84-9	144.212			147.5	0.8805²⁵	1.3980²⁰	sl H₂O; vs eth, EtOH
5888	Hexyl acrylate		C₉H₁₆O₂	2499-95-8	156.222		-45	40¹	0.878²⁰		
5889	Hexylamine	1-Hexanamine	C₆H₁₅N	111-26-2	101.190	liq	-21(1)	132(1)	0.7660²⁰	1.4180²⁰	sl H₂O; msc EtOH, eth; s chl
5890	Hexylbenzene		C₁₂H₁₈	1077-16-3	162.271	liq	-63.4(0.2)	226(2)	0.8575²⁰	1.4864²⁰	i H₂O; msc eth; s bz, peth
5891	4-Hexyl-1,3-benzenediol	4-Hexylresorcinol	C₁₂H₁₈O₂	136-77-6	194.270	nd (bz)	68.3(0.2)	334			vs ace, eth, EtOH, chl
5892	Hexyl benzoate		C₁₃H₁₈O₂	6789-88-4	206.281			272	0.9793²⁰		i H₂O; s EtOH, ace
5893	Hexyl butanoate		C₁₀H₂₀O₂	2639-63-6	172.265	liq	-78.0(0.5)	207(3)	0.8652²⁰	1.4160¹⁵	i H₂O; s EtOH; sl chl
5894	Hexylcyclohexane		C₁₂H₂₄	4292-75-5	168.319	liq	-47.5(0.2)	225(1)	0.8076²⁰	1.4462²⁰	
5895	Hexylcyclopentane		C₁₁H₂₂	4457-00-5	154.293	liq	-73	206(10)	0.7965²⁰	1.4392²⁰	vs ace, bz, eth, EtOH
5896	2-Hexyldecanoic acid		C₁₆H₃₂O₂	25354-97-6	256.424	visc oil		145⁰·⁰²		1.4432²⁴	
5897	Hexyl formate		C₇H₁₄O₂	629-33-4	130.185	liq	-62.6(0.4)	154(5)	0.8813²⁰	1.4071²⁰	i H₂O; msc EtOH, eth
5898	Hexyl hexanoate	Hexyl caproate	C₁₂H₂₄O₂	6378-65-0	200.318	liq	-55.2(0.4)	241(4)	0.865¹⁸	1.4264¹⁵	vs ace, bz, eth, EtOH
5899	Hexyl isocyanate		C₇H₁₃NO	2525-62-4	127.184			44⁷			
5900	Hexyl methacrylate		C₁₀H₁₈O₂	142-09-6	170.249			162	0.880²⁵	1.429²⁵	vs ace, bz, eth, EtOH
5901	Hexyl methyl ether		C₇H₁₆O	4747-07-3	116.201			125(4)			
5902	1-Hexylnaphthalene		C₁₆H₂₀	2876-53-1	212.330	liq	-18	322(9)	0.9566²⁰	1.5647²⁰	
5903	Hexyl octanoate		C₁₄H₂₈O₂	1117-55-1	228.371	liq	-31(1)	277(5)	0.8603²⁰	1.4323²⁵	i H₂O; s EtOH, eth, ace
5904	4-(Hexyloxy)benzoic acid		C₁₃H₁₈O₃	1142-39-8	222.280	cry	106				
5905	2-(Hexyloxy)ethanol	Ethylene glycol monohexyl ether	C₈H₁₈O₂	112-25-4	146.228	liq	-45.1	208(2)	0.8878²⁰	1.4291²⁰	sl H₂O; vs EtOH, eth
5906	Hexyl pentanoate		C₁₁H₂₂O₂	1117-59-5	186.292	liq	-63.0(0.5)	225(4)	0.8635²⁰	1.4228¹⁵	vs ace, eth, EtOH
5907	4-Hexylphenol		C₁₂H₁₈O	2446-69-7	178.270			148⁹			
5908	Hexyl propanoate		C₉H₁₈O₂	2445-76-3	158.238	liq	-57.5(0.5)	188(3)	0.8698²⁰	1.4162¹⁵	i H₂O; s EtOH, eth, ace, AcOEt
5909	1-Hexyl-1,2,3,4-tetrahydro-naphthalene		C₁₆H₂₄	66325-11-9	216.362	liq		305	0.9176²⁵	1.5127²⁵	
5910	1-Hexyne	Butylacetylene	C₆H₁₀	693-02-7	82.143	liq	-132.1(0.4)	71.2(0.3)	0.7155²⁵	1.3989²⁰	i H₂O; s EtOH, eth, bz, chl; sl ctc
5911	2-Hexyne	1-Methyl-2-propylacetylene	C₆H₁₀	764-35-2	82.143	liq	-89.5(0.4)	84.3(0.5)	0.7315²⁰	1.4138²⁰	i H₂O; msc EtOH, eth; s bz, chl, peth
5912	3-Hexyne	Diethylacetylene	C₆H₁₀	928-49-4	82.143	liq	-104(3)	81.5(0.6)	0.7231²⁰	1.4115²⁰	i H₂O; s EtOH, eth, bz, chl, peth
5913	3-Hexyne-2,5-diol		C₆H₁₀O₂	3031-66-1	114.142			121¹⁵	1.0180²⁰	1.4691²⁰	
5914	3-Hexyn-1-ol	3-Hexynol	C₆H₁₀O	1002-28-4	98.142			162	0.8982²⁰	1.4530²⁰	

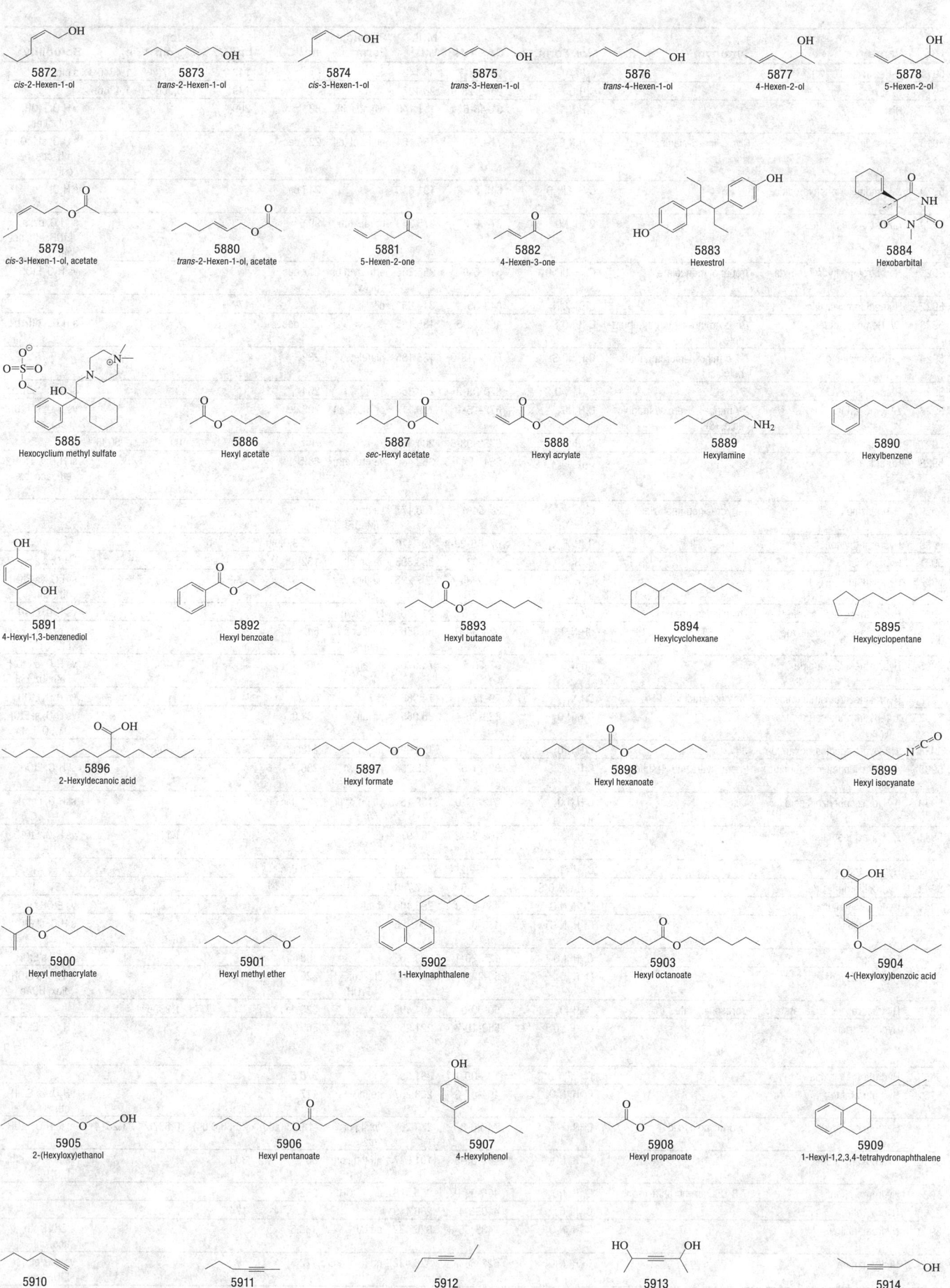

5872 cis-2-Hexen-1-ol

5873 trans-2-Hexen-1-ol

5874 cis-3-Hexen-1-ol

5875 trans-3-Hexen-1-ol

5876 trans-4-Hexen-1-ol

5877 4-Hexen-2-ol

5878 5-Hexen-2-ol

5879 cis-3-Hexen-1-ol, acetate

5880 trans-2-Hexen-1-ol, acetate

5881 5-Hexen-2-one

5882 4-Hexen-3-one

5883 Hexestrol

5884 Hexobarbital

5885 Hexocyclium methyl sulfate

5886 Hexyl acetate

5887 sec-Hexyl acetate

5888 Hexyl acrylate

5889 Hexylamine

5890 Hexylbenzene

5891 4-Hexyl-1,3-benzenediol

5892 Hexyl benzoate

5893 Hexyl butanoate

5894 Hexylcyclohexane

5895 Hexylcyclopentane

5896 2-Hexyldecanoic acid

5897 Hexyl formate

5898 Hexyl hexanoate

5899 Hexyl isocyanate

5900 Hexyl methacrylate

5901 Hexyl methyl ether

5902 1-Hexylnaphthalene

5903 Hexyl octanoate

5904 4-(Hexyloxy)benzoic acid

5905 2-(Hexyloxy)ethanol

5906 Hexyl pentanoate

5907 4-Hexylphenol

5908 Hexyl propanoate

5909 1-Hexyl-1,2,3,4-tetrahydronaphthalene

5910 1-Hexyne

5911 2-Hexyne

5912 3-Hexyne

5913 3-Hexyne-2,5-diol

5914 3-Hexyn-1-ol

No.	Name	Synonym	Mol. Form.	CAS RN	Mol. Wt.	Physical Form	mp/°C	bp/°C	den g cm⁻³	n_D	Solubility
5915	1-Hexyn-3-ol		$C_6H_{10}O$	105-31-7	98.142	liq	-80	142	0.8704[20]	1.4340[25]	s ctc
5916	5-Hexyn-2-one		C_6H_8O	2550-28-9	96.127			149	0.9065[20]	1.4366[20]	
5917	Histamine		$C_5H_9N_3$	51-45-6	111.145	wh nd (chl)	83	209[18]			s H_2O, EtOH, chl; sl eth
5918	L-Histidine	Glyoxaline-5-alanine	$C_6H_9N_3O_2$	71-00-1	155.154	nd or pl (dil al)	287 dec				s H_2O; sl EtOH; i eth, ace, bz, chl
5919	L-Histidine, monohydrochloride		$C_6H_{10}ClN_3O_2$	645-35-2	191.615		245 dec				s H_2O
5920	Homatropine		$C_{16}H_{21}NO_3$	87-00-3	275.343	pr (al, eth)	99.5				sl H_2O, bz; s EtOH, eth, ace, chl
5921	Homatropine hydrobromide	Tropanol mandelate	$C_{16}H_{22}BrNO_3$	51-56-9	356.255	orth pym or pl (w)	217 dec				vs H_2O, EtOH
5922	Homochlorocyclizine		$C_{19}H_{23}ClN_2$	848-53-3	314.852	oil		177[0.8]			
5923	DL-Homocysteine	DL-2-Amino-4-mercaptobutanoic acid	$C_4H_9NO_2S$	454-29-5	135.185		272 dec				s H_2O; i eth, bz
5924	L-Homocysteine	L-2-Amino-4-mercaptobutanoic acid	$C_4H_9NO_2S$	6027-13-0	135.185	platelets	232				
5925	Homocystine		$C_8H_{16}N_2O_4S_2$	870-93-9	268.354		264				sl H_2O; i eth, bz
5926	L-Homoserine	2-Amino-4-hydroxybutanoic acid, (S)	$C_4H_9NO_3$	672-15-1	119.119	pr (90% al)	203 dec				vs H_2O; sl EtOH; i eth, bz
5927	Humulene		$C_{15}H_{24}$	6753-98-6	204.352			123[10]	0.8905[20]	1.5038[20]	
5928	Humulon		$C_{21}H_{30}O_5$	26472-41-3	362.460	ye cry (eth)	66.5				sl H_2O; s EtOH, eth, ace, bz, alk
5929	Hydralazine	1-Hydrazinophthalazine	$C_8H_8N_4$	86-54-4	160.177	ye cry (MeOH)	172				s acid
5930	Hydramethylnon		$C_{25}H_{24}F_6N_4$	67485-29-4	494.476		193.3(0.5)				
5931	Hydrastine		$C_{21}H_{21}NO_6$	118-08-1	383.395	ye pr (al)	132				i H_2O; s ace, bz
5932	Hydrastinine		$C_{11}H_{13}NO_3$	6592-85-4	207.226	nd (lig), cry (eth)	116.5				s H_2O; vs EtOH, eth, chl
5933	Hydrazinecarbothioamide	Thiosemicarbazide	CH_5N_3S	79-19-6	91.136	lo nd (w)	183				vs H_2O, EtOH
5934	Hydrazinecarboxaldehyde		CH_4N_2O	624-84-0	60.055	ye lf or nd (al)	54				vs bz, eth, EtOH, chl
5935	Hydrazinecarboxamide		CH_5N_3O	57-56-7	75.070	pr (al)	96		1.484[8]		vs H_2O; s EtOH; i eth, bz, chl
5936	Hydrazinecarboximidamide	Aminoguanidine	CH_6N_4	79-17-4	74.086	cry	dec				vs H_2O, EtOH
5937	1,2-Hydrazinedicarboxaldehyde		$C_2H_4N_2O_2$	628-36-4	88.065	pr (al)	161.0				vs H_2O; sl EtOH, DMSO; i eth
5938	1,2-Hydrazinedicarboxamide		$C_2H_6N_4O_2$	110-21-4	118.095	pl (w)	248(1)		1.604[17]		
5939	4-Hydrazinobenzenesulfonic acid	Phenylhydrazine-4-sulfonic acid	$C_6H_8N_2O_3S$	98-71-5	188.204	nd, lf (w)	286				sl H_2O, EtOH
5940	4-Hydrazinobenzoic acid		$C_7H_8N_2O_2$	619-67-0	152.151	ye nd or pl (w)	221 dec				sl H_2O; i eth
5941	2-Hydrazinoethanol		$C_2H_8N_2O$	109-84-2	76.097	liq	-70	219	1.119[25]		vs H_2O, EtOH, MeOH
5942	Hydrindantin		$C_{18}H_{10}O_6$	5103-42-4	322.268	pr (ace)	250 dec				
5943	Hydrochlorothiazide		$C_7H_8ClN_3O_4S_2$	58-93-5	297.740		274				
5944	Hydrocinchonidine		$C_{19}H_{24}N_2O$	485-64-3	296.406	lf (al)	229				vs EtOH
5945	Hydrocinchonine		$C_{19}H_{24}N_2O$	485-65-4	296.406	pr	268.5				s H_2O; sl EtOH; i eth
5946	Hydrocodone		$C_{18}H_{21}NO_3$	125-29-1	299.365		198				i H_2O; s EtOH
5947	Hydrocortisone		$C_{21}H_{30}O_5$	50-23-7	362.460	pl (al or i-PrOH)	220				sl H_2O; s EtOH, diox, HOAc
5948	Hydrocortisone 21-acetate	Cortisol acetate	$C_{23}H_{32}O_6$	50-03-3	404.496		223 dec		1.289[20]		
5949	Hydrocotarnine		$C_{12}H_{15}NO_3$	550-10-7	221.252		56				i H_2O; s EtOH, eth, ace, bz, chl
5950	Hydroflumethiazide		$C_8H_8F_3N_3O_4S_2$	135-09-1	331.293		270.5				
5951	Hydrofuramide		$C_{15}H_{12}N_2O_3$	494-47-3	268.267	nd (al)	117				i H_2O; vs EtOH, eth
5952	Hydrogen cyanide	Hydrocyanic acid	CHN	74-90-8	27.026	vol liq or gas	-13.28(0.09)	25.63(0.04)	0.6876[20]	1.2614[20]	msc H_2O, EtOH, eth
5953	Hydrohydrastinine		$C_{11}H_{13}NO_2$	494-55-3	191.227	nd (lig), cry (peth)	66	303			vs ace, bz, eth, EtOH
5954	Hydromorphone	7,8-Dihydromorphin-6-one	$C_{17}H_{19}NO_3$	466-99-9	285.338	cry (EtOH)	266.5				
5955	Hydroprene		$C_{17}H_{30}O_2$	41096-46-2	266.419			174[19]	0.8955[20]		
5956	Hydroquinidine		$C_{20}H_{26}N_2O_2$	1435-55-8	326.432	nd (al)	168.5				s EtOH, eth, ace, chl
5957	Hydroquinine		$C_{20}H_{26}N_2O_2$	522-66-7	326.432	nd (eth, chl)	172.5				vs ace, eth, EtOH, chl

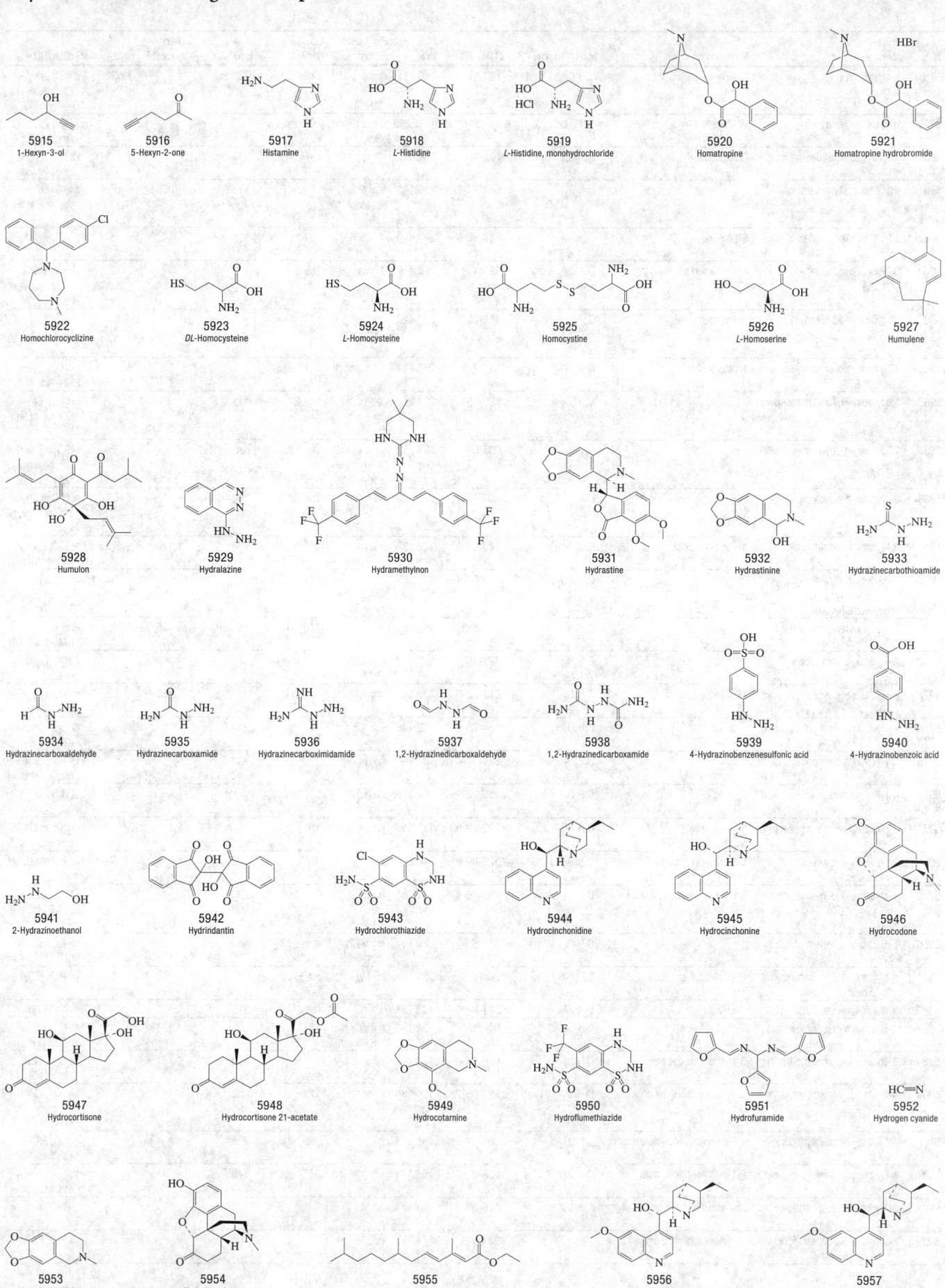

5915 1-Hexyn-3-ol

5916 5-Hexyn-2-one

5917 Histamine

5918 *L*-Histidine

5919 *L*-Histidine, monohydrochloride

5920 Homatropine

5921 Homatropine hydrobromide

5922 Homochlorocyclizine

5923 *DL*-Homocysteine

5924 *L*-Homocysteine

5925 Homocystine

5926 *L*-Homoserine

5927 Humulene

5928 Humulon

5929 Hydralazine

5930 Hydramethylnon

5931 Hydrastine

5932 Hydrastinine

5933 Hydrazinecarbothioamide

5934 Hydrazinecarboxaldehyde

5935 Hydrazinecarboxamide

5936 Hydrazinecarboximidamide

5937 1,2-Hydrazinedicarboxaldehyde

5938 1,2-Hydrazinedicarboxamide

5939 4-Hydrazinobenzenesulfonic acid

5940 4-Hydrazinobenzoic acid

5941 2-Hydrazinoethanol

5942 Hydrindantin

5943 Hydrochlorothiazide

5944 Hydrocinchonidine

5945 Hydrocinchonine

5946 Hydrocodone

5947 Hydrocortisone

5948 Hydrocortisone 21-acetate

5949 Hydrocotarnine

5950 Hydroflumethiazide

5951 Hydrofuramide

5952 Hydrogen cyanide

5953 Hydrohydrastinine

5954 Hydromorphone

5955 Hydroprene

5956 Hydroquinidine

5957 Hydroquinine

No.	Name	Synonym	Mol. Form.	CAS RN	Mol. Wt.	Physical Form	mp/°C	bp/°C	den g cm⁻³	n_D	Solubility
5958	*p*-Hydroquinone	1,4-Benzenediol	$C_6H_6O_2$	123-31-9	110.111	mcl pr (sub) nd(w) pr (MeOH)	173(2)	288(5)	1.330[20]	1.632[25]	s H_2O, eth; vs EtOH, ace; i bz
5959	Hydroxocobalamin	Vitamin B-12a	$C_{62}H_{89}CoN_{13}O_{15}P$	13422-51-0	1346.355	red cry (ace aq)	200 dec				s H_2O, EtOH; i ace, eth, bz
5960	Hydroxyacetonitrile	Glyconitrile	C_2H_3NO	107-16-4	57.051		<-72	183 dec		1.4117[19]	vs H_2O, EtOH, eth; i bz, chl
5961	(Hydroxyacetyl)benzene		$C_8H_8O_2$	582-24-1	136.149	hex pl (al), pl (w or dil al)	90	125[12]	1.0963[99]		s H_2O, EtOH, eth, chl; sl lig
5962	17-Hydroxyandrostan-3-one, (5α,17β)	Stanolone	$C_{19}H_{30}O_2$	521-18-6	290.440		181	135 sub			
5963	3-Hydroxyandrostan-17-one, (3α,5α)	Androsterone	$C_{19}H_{30}O_2$	53-41-8	290.440	lf or nd (al, ace)	185				sl H_2O, chl; s EtOH, eth, ace, bz
5964	3-Hydroxyandrostan-17-one, (3β,5α)	Epiandrosterone	$C_{19}H_{30}O_2$	481-29-8	290.440	cry (bz-peth, ace)	178				
5965	17-Hydroxyandrost-4-en-3-one, (17β)	Testosterone	$C_{19}H_{28}O_2$	58-22-0	288.424	nd (dil ace)	151.0(0.3)				i H_2O; s EtOH, eth, ace
5966	1-Hydroxy-9,10-anthracenedione		$C_{14}H_8O_3$	129-43-1	224.212	red-oran nd (al)	193.8	sub			i H_2O; s EtOH, eth, bz; sl liq NH_3
5967	2-Hydroxy-9,10-anthracenedione		$C_{14}H_8O_3$	605-32-3	224.212	ye pl or nd (al or HOAc)	306	sub			i H_2O; s EtOH, eth, aq NH_3, KOH
5968	3-Hydroxybenzaldehyde	3-Formylphenol	$C_7H_6O_2$	100-83-4	122.122	nd (w)	106.0(0.2)	240	1.1179[130]		sl H_2O; s EtOH, eth, ace, bz; i lig
5969	4-Hydroxybenzaldehyde	4-Formylphenol	$C_7H_6O_2$	123-08-0	122.122	nd (w)	116.0(0.2)		1.129[130]	1.5705[130]	sl H_2O, ace; vs EtOH, eth; s bz
5970	2-Hydroxybenzaldehyde, [(2-hydroxyphenyl)methylene]hydrazone		$C_{14}H_{12}N_2O_2$	959-36-4	240.257		214				i H_2O; s EtOH, chl; vs bz, alk
5971	2-Hydroxybenzamide	Salicylamide	$C_7H_7NO_2$	65-45-2	137.137		140(2)	181.5[14]	1.175[140]		sl H_2O, eth, DMSO; s EtOH
5972	*N*-Hydroxybenzamide		$C_7H_7NO_2$	495-18-1	137.137	orth ta, lf (eth)	131 exp				s H_2O, EtOH; sl eth, bz
5973	α-Hydroxybenzeneacetic acid, (±)-	*DL*-Mandelic acid	$C_8H_8O_3$	611-72-3	152.148	orth pl	118.6(1)		1.2890[20]		s H_2O, eth, EtOH, i-PrOH
5974	2-Hydroxybenzeneacetic acid		$C_8H_8O_3$	614-75-5	152.148		147(3)	240			sl H_2O, chl; s eth
5975	3-Hydroxybenzeneacetic acid		$C_8H_8O_3$	621-37-4	152.148	nd (bz-lig)	132	190[11]			vs H_2O, EtOH, eth; s bz; sl lig
5976	4-Hydroxybenzeneacetic acid		$C_8H_8O_3$	156-38-7	152.148	nd (w)	150.2(0.8)	sub			sl H_2O; vs EtOH, eth
5977	α-Hydroxybenzeneacetonitrile	Mandelonitrile	C_8H_7NO	532-28-5	133.148	ye oily liq	-10		1.12		i H_2O; vs chl, eth, EtOH
5978	2-Hydroxybenzenecarbodithioic acid	Dithiosalicylic acid	$C_7H_6OS_2$	527-89-9	170.252	oran-ye nd	49				vs bz, eth, EtOH
5979	4-Hydroxy-1,3-benzenedicarboxylic acid	4-Hydroxyisophthalic acid	$C_8H_6O_5$	636-46-4	182.131	nd(w), lf (dil al)	310				i H_2O, chl; vs EtOH, eth; s HOAc
5980	5-Hydroxy-1,3-benzenedicarboxylic acid		$C_8H_6O_5$	618-83-7	182.131	nd(w+2) cr(aq-al)		sub			vs bz, eth, EtOH
5981	4-Hydroxy-1,3-benzenedisulfonic acid	Phenoldisulfonic acid	$C_6H_6O_7S_2$	96-77-5	254.238	nd (w)	>100 dec				vs H_2O, EtOH
5982	4-Hydroxybenzeneethanol		$C_9H_{10}O_2$	501-94-0	138.164		90(3)	310.0			
5983	2-Hydroxybenzenemethanol	Salicyl alcohol	$C_7H_8O_2$	90-01-7	124.138	lf (bz), nd or pl (w, eth)	87	sub	1.1613[25]		s H_2O, EtOH, eth, bz; vs chl
5984	3-Hydroxybenzenemethanol	3-Hydroxybenzyl alcohol	$C_7H_8O_2$	620-24-6	124.138	nd (bz), cry (CCl_4)	73	300 dec	1.161[25]		vs H_2O, EtOH, eth; sl chl
5985	4-Hydroxybenzenemethanol	4-Hydroxybenzyl alcohol	$C_7H_8O_2$	623-05-2	124.138	pr or nd (w)	124.5	252			vs H_2O, EtOH, bz, chl; s eth; sl DMSO
5986	4-Hydroxybenzenepropanoic acid	*p*-Hydroxyhydrocinnamic acid	$C_9H_{10}O_3$	501-97-3	166.173		129.3(0.5)	209[14]			s H_2O, EtOH, eth, bz; i CS_2
5987	α-Hydroxybenzenepropanoic acid, (±)-	(±)-3-Phenyllactic acid	$C_9H_{10}O_3$	828-01-3	166.173	cry (chl, bz), pr (w)	98	149[15]			vs H_2O, ace, eth, EtOH
5988	3-Hydroxybenzenesulfonic acid	*m*-Phenolsulfonic acid	$C_6H_6O_4S$	585-38-6	174.175	nd (w+2)					
5989	4-Hydroxybenzenesulfonic acid	*p*-Phenolsulfonic acid	$C_6H_6O_4S$	98-67-9	174.175	nd					vs H_2O, EtOH

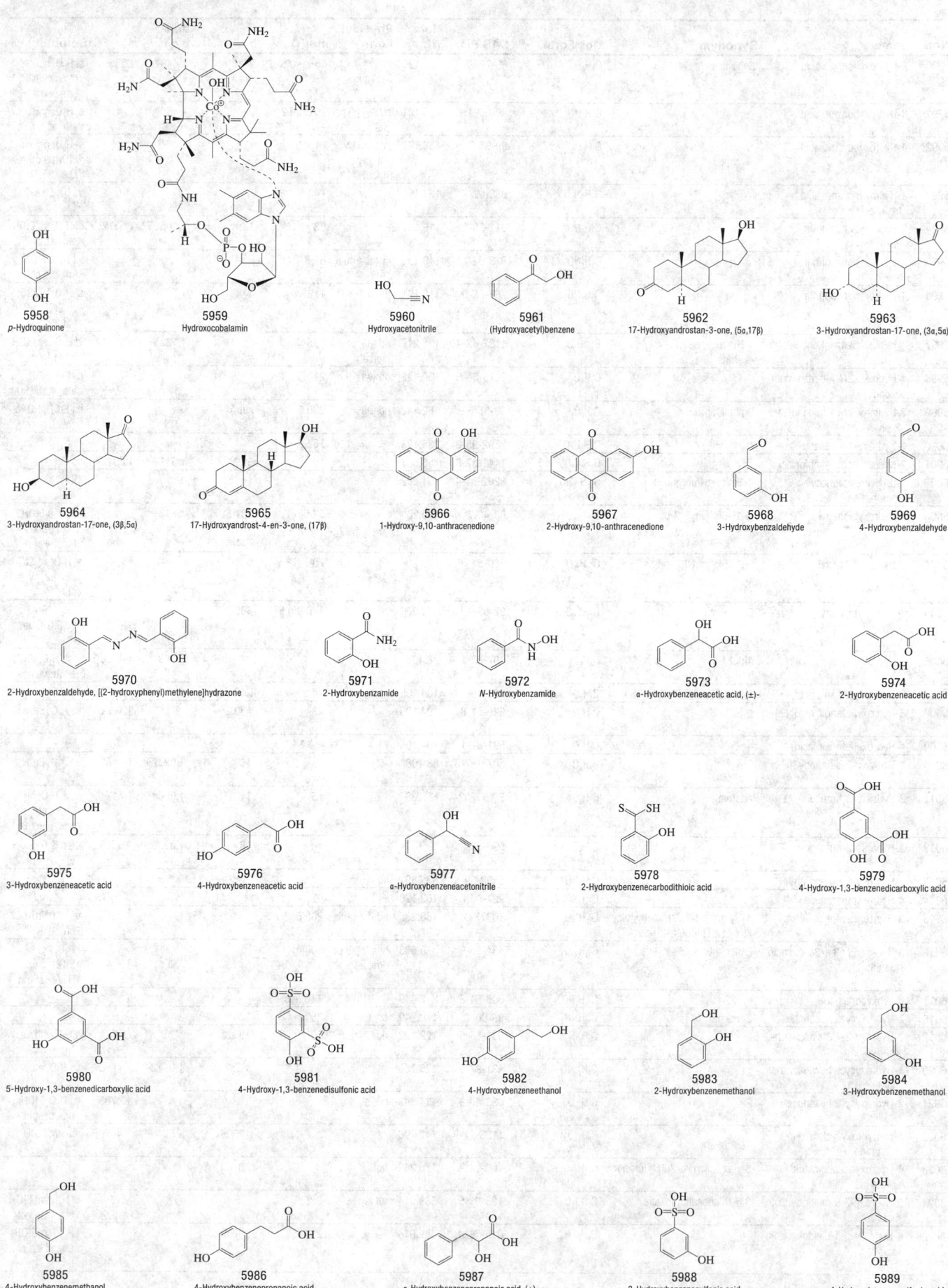

5958
p-Hydroquinone

5959
Hydroxocobalamin

5960
Hydroxyacetonitrile

5961
(Hydroxyacetyl)benzene

5962
17-Hydroxyandrostan-3-one, (5α,17β)

5963
3-Hydroxyandrostan-17-one, (3α,5α)

5964
3-Hydroxyandrostan-17-one, (3β,5α)

5965
17-Hydroxyandrost-4-en-3-one, (17β)

5966
1-Hydroxy-9,10-anthracenedione

5967
2-Hydroxy-9,10-anthracenedione

5968
3-Hydroxybenzaldehyde

5969
4-Hydroxybenzaldehyde

5970
2-Hydroxybenzaldehyde, [(2-hydroxyphenyl)methylene]hydrazone

5971
2-Hydroxybenzamide

5972
N-Hydroxybenzamide

5973
α-Hydroxybenzeneacetic acid, (±)-

5974
2-Hydroxybenzeneacetic acid

5975
3-Hydroxybenzeneacetic acid

5976
4-Hydroxybenzeneacetic acid

5977
α-Hydroxybenzeneacetonitrile

5978
2-Hydroxybenzenecarbodithioic acid

5979
4-Hydroxy-1,3-benzenedicarboxylic acid

5980
5-Hydroxy-1,3-benzenedicarboxylic acid

5981
4-Hydroxy-1,3-benzenedisulfonic acid

5982
4-Hydroxybenzeneethanol

5983
2-Hydroxybenzenemethanol

5984
3-Hydroxybenzenemethanol

5985
4-Hydroxybenzenemethanol

5986
4-Hydroxybenzenepropanoic acid

5987
α-Hydroxybenzenepropanoic acid, (±)-

5988
3-Hydroxybenzenesulfonic acid

5989
4-Hydroxybenzenesulfonic acid

No.	Name	Synonym	Mol. Form.	CAS RN	Mol. Wt.	Physical Form	mp/°C	bp/°C	den g cm⁻³	n_D	Solubility
5990	2-Hydroxybenzoic acid	Salicylic acid	C₇H₆O₃	69-72-7	138.121	nd (w), mcl pr (al)	158.6(0.5)	211²⁰	1.443²⁰	1.565	sl H₂O, bz, chl, ctc; vs EtOH, eth, ace
5991	3-Hydroxybenzoic acid		C₇H₆O₃	99-06-9	138.121	nd (w) pl, pr (al)	201.3(0.2)		1.485²⁵		sl H₂O; s EtOH, eth, ace; i bz
5992	4-Hydroxybenzoic acid		C₇H₆O₃	99-96-7	138.121	pr or pl (w, al) cry (ace)	213(2)		1.46²⁵		sl H₂O, bz; vs EtOH; s eth, ace
5993	2-Hydroxybenzoic acid, hydrazide		C₇H₈N₂O₂	936-02-7	152.151		148				vs bz, EtOH
5994	2-Hydroxybenzonitrile		C₇H₅NO	611-20-1	119.121		98	149¹⁴	1.1052¹⁰⁰	1.5372¹⁰⁰	sl H₂O; vs EtOH, eth, bz, chl
5995	3-Hydroxybenzonitrile		C₇H₅NO	873-62-1	119.121	pr (al, eth) lf (w)	82.8				vs H₂O, EtOH, eth, bz, chl
5996	4-Hydroxybenzonitrile		C₇H₅NO	767-00-0	119.121	lf (w)	113	148¹			sl H₂O, DMSO; vs EtOH, eth, chl
5997	4-Hydroxybenzophenone	4-Hydroxyphenyl phenyl ketone	C₁₃H₁₀O₂	1137-42-4	198.217	nd (al), pr (dil al)	135		1.133¹⁷²		sl H₂O; vs EtOH, eth, HOAc
5998	4-Hydroxy-2H-1-benzopyran-2-one		C₉H₆O₃	1076-38-6	162.142	nd (w)	213.5				s H₂O, EtOH, eth; sl DMSO
5999	7-Hydroxy-2H-1-benzopyran-2-one	Umbelliferone	C₉H₆O₃	93-35-6	162.142	nd (w)	230.5	sub			vs EtOH, HOAc, chl
6000	1-Hydroxy-1H-benzotriazole		C₆H₅N₃O	2592-95-2	135.123		157.8				
6001	2-Hydroxybenzoyl chloride		C₇H₅ClO₂	1441-87-8	156.567		19	92¹⁵	1.3112²⁰	1.5812²⁰	vs eth
6002	4-(2-Hydroxybenzoyl)-morpholine	4-Salicyloylmorpholine	C₁₁H₁₃NO₃	3202-84-4	204.202						s DMSO
6003	2-Hydroxybiphenyl	[1,1'-Biphenyl]-2-ol	C₁₂H₁₀O	90-43-7	170.206		57.6(0.7)	281(3)	1.213²⁵		i H₂O; s EtOH, ace, bz; vs eth, py
6004	3-Hydroxybiphenyl	[1,1'-Biphenyl]-3-ol	C₁₂H₁₀O	580-51-8	170.206		78	>300			sl H₂O; vs EtOH, eth, bz, py; s chl
6005	4-Hydroxybiphenyl	[1,1'-Biphenyl]-4-ol	C₁₂H₁₀O	92-69-3	170.206		170.0(0.5)	305			sl H₂O, DMSO; vs EtOH, eth, chl, py
6006	3-Hydroxybutanal	Aldol	C₄H₈O₂	107-89-1	88.106			83²⁰	1.103²⁰	1.4238²⁰	msc H₂O, EtOH; s eth; vs ace
6007	2-Hydroxybutanoic acid, (±)-		C₄H₈O₃	600-15-7	104.105		44.2	260 dec	1.125²⁰		s H₂O, EtOH, eth
6008	3-Hydroxybutanoic acid, (±)-		C₄H₈O₃	625-71-8	104.105		49	130¹²		1.4424²⁰	vs H₂O, EtOH, eth; i bz
6009	4-Hydroxybutanoic acid		C₄H₈O₃	591-81-1	104.105		<-17	180 dec			
6010	1-Hydroxy-2-butanone		C₄H₈O₂	5077-67-8	88.106			160	1.0272²⁰	1.4189²⁰	vs H₂O, EtOH, eth
6011	3-Hydroxy-2-butanone, (±)-	Acetoin	C₄H₈O₂	52217-02-4	88.106		15	148	1.0044²⁰	1.4171²⁰	msc H₂O; sl EtOH, eth; s ace, chl; i lig
6012	4-Hydroxy-2-butanone		C₄H₈O₂	590-90-9	88.106			182	1.0233²⁰	1.4585¹⁴	msc H₂O, EtOH, eth; vs ace
6013	2-Hydroxy-3-butenenitrile		C₄H₅NO	5809-59-6	83.089	liq		94¹⁷			
6014	4-Hydroxybutyramide		C₄H₉NO₂	927-60-6	103.120		52				
6015	3-Hydroxycamphor	3-Hydroxy-1,7,7-trimethylbicy-clo[2.2.1]heptan-2-one	C₁₀H₁₆O₂	10373-81-6	168.233	nd (bz-peth)	205.5				vs eth, EtOH, chl
6016	3-Hydroxycholan-24-oic acid, (3α,5β)	Lithocholic acid	C₂₄H₄₀O₃	434-13-9	376.573	hex lf (al) pr (dil al)	188(4)				i H₂O, lig; s EtOH, chl, HOAc; sl eth
6017	Hydroxycodeinone		C₁₈H₁₉NO₄	508-54-3	313.349		275 dec				
6018	2-Hydroxycyclodecanone	Sebacoin	C₁₀H₁₈O₂	96-00-4	170.249	cry (peth)	38.5	136¹⁴			
6019	2-Hydroxy-2,4,6-cyclohepta-trien-1-one		C₇H₆O₂	533-75-5	122.122	nd	50(1)	40 sub			s H₂O, eth, ace
6020	1-Hydroxycyclohexanecarbo-nitrile		C₇H₁₁NO	931-97-5	125.168		35	132²⁰	1.0172²⁰	1.4693²⁰	vs H₂O, eth
6021	2-Hydroxycyclohexanone		C₆H₁₀O₂	533-60-8	114.142	nd (al)				1.4785²¹	vs H₂O, EtOH; i eth, bz, peth
6022	1-(1-Hydroxycyclohexyl)-ethanone		C₈H₁₄O₂	1123-27-9	142.196			125.5	1.0248²⁵	1.4670²⁵	vs eth, EtOH
6023	4-Hydroxydecanoic acid γ-lactone	5-Hexyldihydro-2(3H)-furanone	C₁₀H₁₈O₂	706-14-9	170.249	liq		301(8)			
6024	2-Hydroxy-3,5-diiodobenzoic acid	3,5-Diiodosalicylic acid	C₇H₄I₂O₃	133-91-5	389.914	nd (al)	235.5				sl H₂O; vs EtOH, eth; i bz, chl
6025	4-Hydroxy-3,5-diiodobenzoic acid		C₇H₄I₂O₃	618-76-8	389.914		237	260 dec			i H₂O; vs EtOH, eth; sl bz, chl, lig

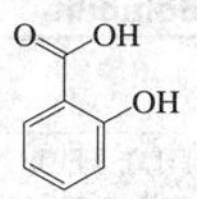

5990
2-Hydroxybenzoic acid

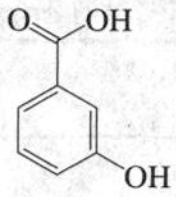

5991
3-Hydroxybenzoic acid

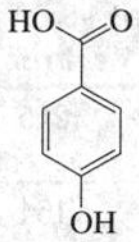

5992
4-Hydroxybenzoic acid

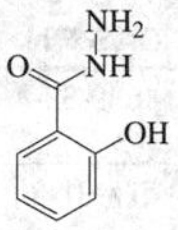

5993
2-Hydroxybenzoic acid, hydrazide

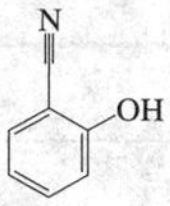

5994
2-Hydroxybenzonitrile

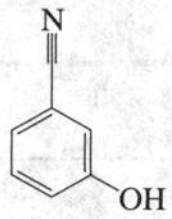

5995
3-Hydroxybenzonitrile

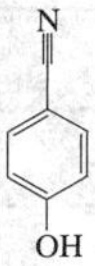

5996
4-Hydroxybenzonitrile

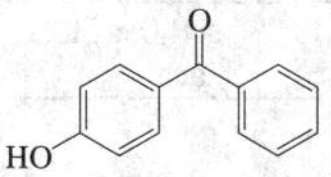

5997
4-Hydroxybenzophenone

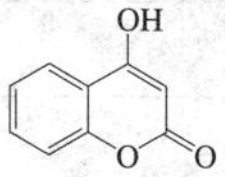

5998
4-Hydroxy-2*H*-1-benzopyran-2-one

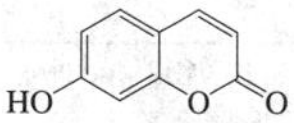

5999
7-Hydroxy-2*H*-1-benzopyran-2-one

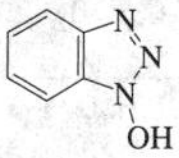

6000
1-Hydroxy-1*H*-benzotriazole

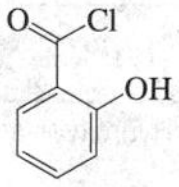

6001
2-Hydroxybenzoyl chloride

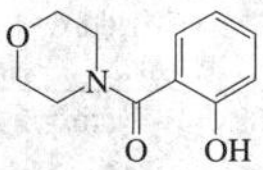

6002
4-(2-Hydroxybenzoyl)morpholine

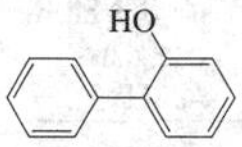

6003
2-Hydroxybiphenyl

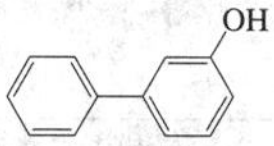

6004
3-Hydroxybiphenyl

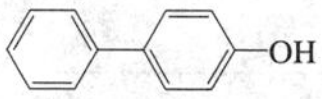

6005
4-Hydroxybiphenyl

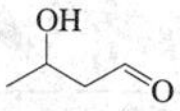

6006
3-Hydroxybutanal

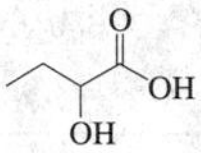

6007
2-Hydroxybutanoic acid, (±)-

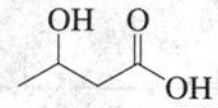

6008
3-Hydroxybutanoic acid, (±)-

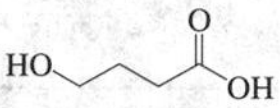

6009
4-Hydroxybutanoic acid

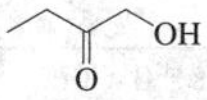

6010
1-Hydroxy-2-butanone

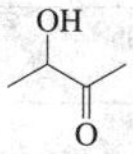

6011
3-Hydroxy-2-butanone, (±)-

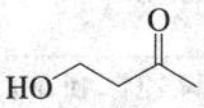

6012
4-Hydroxy-2-butanone

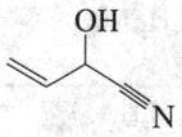

6013
2-Hydroxy-3-butenenitrile

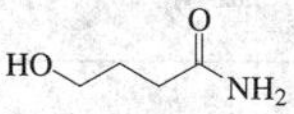

6014
4-Hydroxybutyramide

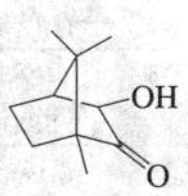

6015
3-Hydroxycamphor

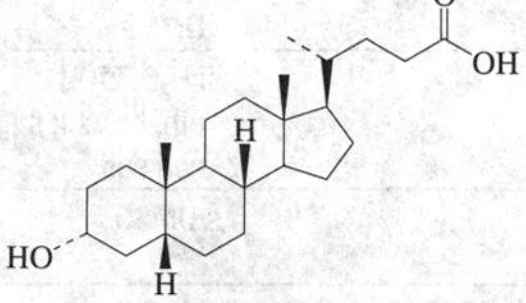

6016
3-Hydroxycholan-24-oic acid, (3α,5β)

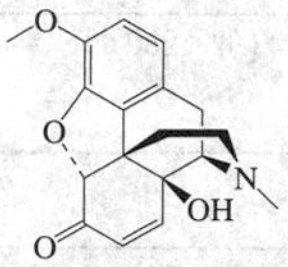

6017
Hydroxycodeinone

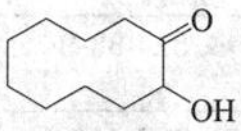

6018
2-Hydroxycyclodecanone

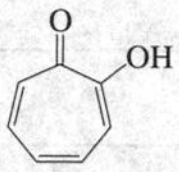

6019
2-Hydroxy-2,4,6-cycloheptatrien-1-one

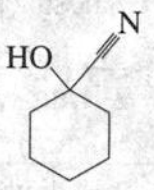

6020
1-Hydroxycyclohexanecarbonitrile

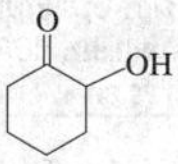

6021
2-Hydroxycyclohexanone

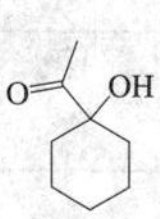

6022
1-(1-Hydroxycyclohexyl)ethanone

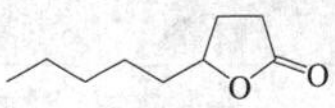

6023
4-Hydroxydecanoic acid γ-lactone

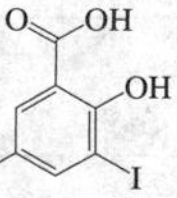

6024
2-Hydroxy-3,5-diiodobenzoic acid

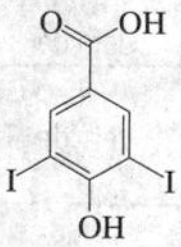

6025
4-Hydroxy-3,5-diiodobenzoic acid

No.	Name	Synonym	Mol. Form.	CAS RN	Mol. Wt.	Physical Form	mp/°C	bp/°C	den g cm⁻³	n_D	Solubility
6026	4-Hydroxy-3,5-diiodobenzonitrile		$C_7H_3I_2NO$	1689-83-4	370.914		215.8(0.5)				
6027	4-Hydroxy-3,5-diiodo-α-phenylbenzenepropanoic acid	Iodoalphionic acid	$C_{15}H_{12}I_2O_3$	577-91-3	494.063		164				i H_2O; s EtOH, eth; sl bz, chl
6028	2-Hydroxy-4,6-dimethoxybenzaldehyde		$C_9H_{10}O_4$	708-76-9	182.173		70	193[25]			i H_2O; vs EtOH, eth, bz, chl, HOAc
6029	4-Hydroxy-3,5-dimethoxybenzaldehyde	Syringaldehyde	$C_9H_{10}O_4$	134-96-3	182.173	br nd (lig)	113	192[14]			sl H_2O, lig; vs EtOH, eth, bz, chl
6030	4-Hydroxy-3,5-dimethoxybenzoic acid		$C_9H_{10}O_5$	530-57-4	198.172	nd (w)	204.5				sl H_2O; vs EtOH
6031	7-Hydroxy-3,7-dimethyloctanal		$C_{10}H_{20}O_2$	107-75-5	172.265			103[3]	0.9220[20]	1.4494[20]	sl H_2O; s EtOH, ace
6032	3-Hydroxy-2,2-dimethylpropanal	Hydroxypivaldehyde	$C_5H_{10}O_2$	597-31-9	102.132	nd (w)	89.5	173			
6033	2-Hydroxy-3,5-dinitrobenzoic acid		$C_7H_4N_2O_7$	609-99-4	228.116	ye nd or pl (+1w)	182				s H_2O, EtOH, eth, bz
6034	11-Hydroxy-9,15-dioxoprosta-5,13-dien-1-oic acid, (5Z,11α,13E)-	15-Oxo-prostaglandin E2	$C_{20}H_{30}O_5$	26441-05-4	350.449	cry					
6035	1-Hydroxy-1,1-diphosphonoethane	Etidronic acid	$C_2H_8O_7P_2$	2809-21-4	206.028	cry (w)	105				s H_2O, EtOH, MeOH
6036	3-Hydroxyestra-1,3,5,7,9-pentaen-17-one	Equilenin	$C_{18}H_{18}O_2$	517-09-9	266.335		258.5	170 sub			sl EtOH, ace, chl
6037	3-Hydroxyestra-1,3,5(10),7-tetraen-17-one	Equilin	$C_{18}H_{20}O_2$	474-86-2	268.351	pl (AcOEt)	239	170 sub			sl H_2O; s EtOH, ace, diox, AcOEt
6038	2-Hydroxyethyl acrylate	2-Hydroxyethyl 2-propenoate	$C_5H_8O_3$	818-61-1	116.116	liq		191	1.011[23]		
6039	N-(2-Hydroxyethyl)-dodecanamide		$C_{14}H_{29}NO_2$	142-78-9	243.386		88.5				
6040	N-(2-Hydroxyethyl)-ethylenediaminetriacetic acid		$C_{10}H_{18}N_2O_7$	150-39-0	278.259	cry	165 dec				
6041	2-Hydroxyethyl 2-hydroxybenzoate	Glycol salicylate	$C_9H_{10}O_4$	87-28-5	182.173		37	173[15]	1.2526[15]		sl H_2O; vs EtOH, eth, bz, chl
6042	2-Hydroxyethyl methacrylate	Ethylene glycol monomethacrylate	$C_6H_{10}O_3$	868-77-9	130.141			103[13]	1.079[20]	1.4515[20]	
6043	N-(2-Hydroxyethyl)phthalimide		$C_{10}H_9NO_3$	3891-07-4	191.183	nd (al), lf (w)	130.3				sl H_2O
6044	1-(2-Hydroxyethyl)-2-pyrrolidinone		$C_6H_{11}NO_2$	3445-11-2	129.157		20	295	1.1435[20]		
6045	4-Hydroxy-4H-furo[3,2-c]-pyran-2(6H)-one	Patulin	$C_7H_6O_4$	149-29-1	154.121	pl or pr (eth, chl)	111				s H_2O, EtOH, eth, ace, bz; i peth
6046	16-Hydroxyhexadecanoic acid	16-Hydroxypalmitic acid	$C_{16}H_{32}O_3$	506-13-8	272.423		96.5				i H_2O; s EtOH, ace; sl eth, bz
6047	2-Hydroxyhexanoic acid		$C_6H_{12}O_3$	6064-63-7	132.157	pr (eth)	60				vs H_2O
6048	6-Hydroxyhexanoic acid		$C_6H_{12}O_3$	1191-25-9	132.157	liq					
6049	3-Hydroxy-2-(hydroxymethyl)-2-methylpropanoic acid	Dimethylolpropionic acid	$C_5H_{10}O_4$	4767-03-7	134.131		195.5(0.5)				
6050	5-Hydroxy-2-(hydroxymethyl)-4H-pyran-4-one	Kojic acid	$C_6H_6O_4$	501-30-4	142.110	pr nd (ace)	153.5				sl H_2O, bz; s EtOH, eth, ace, DMSO
6051	8-Hydroxy-7-iodo-5-quinolinesulfonic acid	Ferron	$C_9H_6INO_4S$	547-91-1	351.118	ye pr, lf (al)	260 dec				sl H_2O, EtOH; i eth, bz, chl; s con sulf
6052	2-Hydroxy-1H-isoindole-1,3(2H)-dione		$C_8H_5NO_3$	524-38-9	163.131		232				s DMSO
6053	2-Hydroxy-4-isopropyl-2,4,6-cycloheptatrien-1-one		$C_{10}H_{12}O_2$	499-44-5	164.201	pa ye (peth)	51(2)	137[10]	1.0606[65]		sl H_2O, bz, lig; s ctc
6054	Hydroxylupanine		$C_{15}H_{24}N_2O_2$	15358-48-2	264.364	cry (ace)	169.5				vs H_2O, EtOH, chl
6055	N-Hydroxymethanamine	N-Methylhydroxylamine	CH_5NO	593-77-1	47.057	hyg nd	87.5	62.5[15]	1.0003[20]	1.4164[20]	vs H_2O, EtOH
6056	2-Hydroxy-3-methoxybenzaldehyde		$C_8H_8O_3$	148-53-8	152.148	lt ye lf, grn nd (w, lig)	40.5(0.4)	265.5			sl H_2O, lig; vs EtOH, eth, ctc
6057	2-Hydroxy-4-methoxybenzaldehyde		$C_8H_8O_3$	673-22-3	152.148	nd (w), cry (al)	42.0				s EtOH, eth, bz, lig
6058	2-Hydroxy-5-methoxybenzaldehyde		$C_8H_8O_3$	672-13-9	152.148	ye liq (w)	4	247.5			vs eth, EtOH
6059	3-Hydroxy-4-methoxybenzaldehyde		$C_8H_8O_3$	621-59-0	152.148		117(3)	179[15]	1.196[25]		sl H_2O; s EtOH, eth, bz, HOAc; vs chl

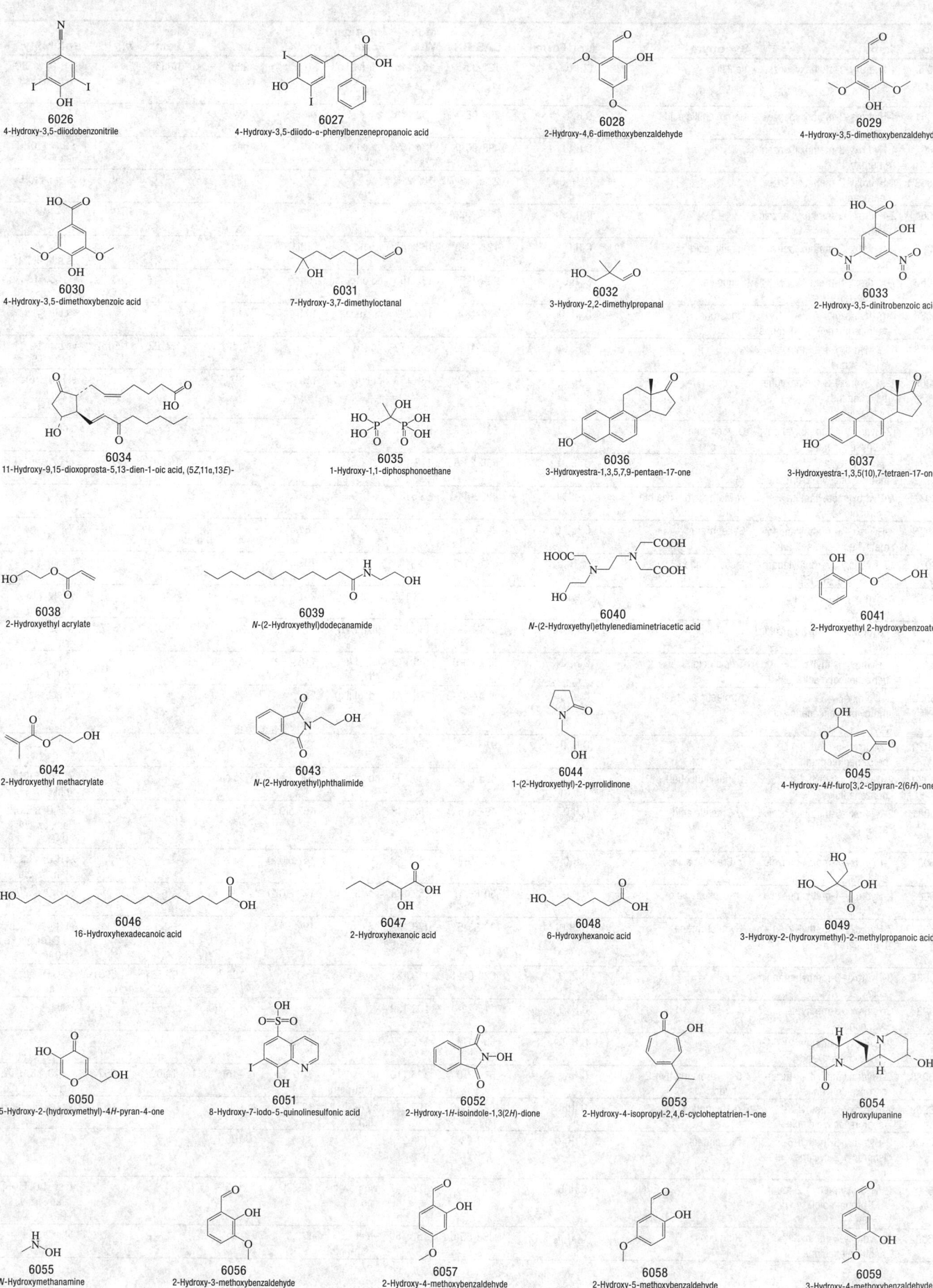

6026
4-Hydroxy-3,5-diiodobenzonitrile

6027
4-Hydroxy-3,5-diiodo-α-phenylbenzenepropanoic acid

6028
2-Hydroxy-4,6-dimethoxybenzaldehyde

6029
4-Hydroxy-3,5-dimethoxybenzaldehyde

6030
4-Hydroxy-3,5-dimethoxybenzoic acid

6031
7-Hydroxy-3,7-dimethyloctanal

6032
3-Hydroxy-2,2-dimethylpropanal

6033
2-Hydroxy-3,5-dinitrobenzoic acid

6034
11-Hydroxy-9,15-dioxoprosta-5,13-dien-1-oic acid, (5Z,11α,13E)-

6035
1-Hydroxy-1,1-diphosphonoethane

6036
3-Hydroxyestra-1,3,5,7,9-pentaen-17-one

6037
3-Hydroxyestra-1,3,5(10),7-tetraen-17-one

6038
2-Hydroxyethyl acrylate

6039
N-(2-Hydroxyethyl)dodecanamide

6040
N-(2-Hydroxyethyl)ethylenediaminetriacetic acid

6041
2-Hydroxyethyl 2-hydroxybenzoate

6042
2-Hydroxyethyl methacrylate

6043
N-(2-Hydroxyethyl)phthalimide

6044
1-(2-Hydroxyethyl)-2-pyrrolidinone

6045
4-Hydroxy-4H-furo[3,2-c]pyran-2(6H)-one

6046
16-Hydroxyhexadecanoic acid

6047
2-Hydroxyhexanoic acid

6048
6-Hydroxyhexanoic acid

6049
3-Hydroxy-2-(hydroxymethyl)-2-methylpropanoic acid

6050
5-Hydroxy-2-(hydroxymethyl)-4H-pyran-4-one

6051
8-Hydroxy-7-iodo-5-quinolinesulfonic acid

6052
2-Hydroxy-1H-isoindole-1,3(2H)-dione

6053
2-Hydroxy-4-isopropyl-2,4,6-cycloheptatrien-1-one

6054
Hydroxylupanine

6055
N-Hydroxymethanamine

6056
2-Hydroxy-3-methoxybenzaldehyde

6057
2-Hydroxy-4-methoxybenzaldehyde

6058
2-Hydroxy-5-methoxybenzaldehyde

6059
3-Hydroxy-4-methoxybenzaldehyde

No.	Name	Synonym	Mol. Form.	CAS RN	Mol. Wt.	Physical Form	mp/°C	bp/°C	den g cm⁻³	n_D	Solubility
6060	4-Hydroxy-3-methoxybenzal-dehyde	Vanillin	$C_8H_8O_3$	121-33-5	152.148	tetr (w, lig)	81(1)	285	1.056[25]		sl H₂O; vs EtOH, eth, ace; s bz, lig
6061	4-Hydroxy-3-methoxybenze-neacetic acid	Homovanillic acid	$C_9H_{10}O_4$	306-08-1	182.173		143.5				
6062	4-Hydroxy-3-methoxybenze-emethanol		$C_8H_{10}O_3$	498-00-0	154.163	pr (w), nd (bz)	115	dec			s H₂O, EtOH, eth, bz
6063	4-Hydroxy-3-methoxybenze-nepropanol		$C_{10}H_{14}O_3$	2305-13-7	182.216		65	197[15]		1.5545[25]	vs eth, EtOH
6064	2-Hydroxy-5-methoxybenzoic acid		$C_8H_8O_4$	2612-02-4	168.148		142				
6065	4-Hydroxy-3-methoxybenzoic acid	Vanillic acid	$C_8H_8O_4$	121-34-6	168.148	wh nd	212(1)	sub			sl H₂O; vs EtOH; s eth, DMSO
6066	7-Hydroxy-6-methoxy-2H-1-benzopyran-2-one	Scopoletin	$C_{10}H_8O_4$	92-61-5	192.169	nd or pr (al)	204				sl H₂O, EtOH; s chl; i bz, CS₂
6067	4-(4-Hydroxy-3-methoxyphenyl)-2-butanone	Zingerone	$C_{11}H_{14}O_3$	122-48-5	194.227	cry (ace, eth)	40.5	187[14]			vs eth
6068	1-(2-Hydroxy-4-methoxyphe-nyl)ethanone		$C_9H_{10}O_3$	552-41-0	166.173	nd (al)	52.5	158[20]	1.3102[81]	1.5452[81]	vs bz, eth, EtOH, chl
6069	1-(4-Hydroxy-3-methoxyphe-nyl)ethanone	Apocynin	$C_9H_{10}O_3$	498-02-2	166.173	pr (w)	115	297			sl H₂O; s EtOH, ace, bz; vs eth, chl
6070	(2-Hydroxy-4-methoxyphenyl)-phenylmethanone	Oxybenzone	$C_{14}H_{12}O_3$	131-57-7	228.243		65.5				s ctc
6071	3-(4-Hydroxy-3-methoxyphenyl)-2-propenal		$C_{10}H_{10}O_3$	458-36-6	178.184	cry (bz)	84		1.1562[102]		vs bz, eth, EtOH
6072	N-Hydroxymethylamine hydrochloride	N-Methylhydroxylamine hydrochloride	CH_6ClNO	4229-44-1	83.518		83.5				
6073	4-Hydroxy-α-[(methylamino)-methyl]benzenemethanol	Synephrine	$C_9H_{13}NO_2$	94-07-5	167.205		184.5				
6074	17-Hydroxy-17-methylandro-stan-3-one, (5α,17β)	Mestanolone	$C_{20}H_{32}O_2$	521-11-9	304.467		192.5				sl AcOEt
6075	N-Hydroxy-4-methylaniline		C_7H_9NO	623-10-9	123.152	lf (bz)	96	117 dec			vs eth, EtOH, chl
6076	2-Hydroxy-5-methylbenzalde-hyde		$C_8H_8O_2$	613-84-3	136.149	pl (aq, al)	55.1(0.2)	217.5	1.0913[59]	1.547[59]	vs eth, EtOH, chl
6077	α-(Hydroxymethyl)-benzeneacetic acid, (±)-	Tropic acid	$C_9H_{10}O_3$	552-63-6	166.173	nd, pl (al, bz, w)	118	dec			vs H₂O, eth, EtOH
6078	α-Hydroxy-α-methylbenzeneacetic acid, (±)-	Atrolactic acid	$C_9H_{10}O_3$	4607-38-9	166.173	nd, pl (lig)	94				vs ace, bz
6079	2-Hydroxy-5-methyl-1,3-benzenedimethanol		$C_9H_{12}O_3$	91-04-3	168.189		130.5				
6080	2-(Hydroxymethyl)-1,4-benzenediol	Gentisyl alcohol	$C_7H_8O_3$	495-08-9	140.137	nd (chl)	100	75 sub			vs H₂O, EtOH, chl
6081	2-Hydroxy-5-methylbenzoic acid	p-Cresotic acid	$C_8H_8O_3$	89-56-5	152.148		152.5(0.2)				sl H₂O; s EtOH, eth, bz, chl; i CS₂
6082	2-Hydroxy-3-methylbenzoic acid	o-Cresotic acid	$C_8H_8O_3$	83-40-9	152.148		167.0(0.2)				sl H₂O; s EtOH, eth, bz, chl
6083	2-Hydroxy-4-methylbenzoic acid	m-Cresotic acid	$C_8H_8O_3$	50-85-1	152.148	cry, lf	177.8(0.2)				sl H₂O; s EtOH, bz, chl; vs eth
6084	7-Hydroxy-4-methyl-2H-1-benzopyran-2-one	Hymecromone	$C_{10}H_8O_3$	90-33-5	176.169	nd (al)	194.5				sl H₂O, eth, chl; s EtOH, alk, HOAc
6085	3-Hydroxy-3-methylbutanoic acid		$C_5H_{10}O_3$	625-08-1	118.131		<-32	162[12]	0.9384[20]	1.5081[20]	vs H₂O, eth, EtOH
6086	3-Hydroxy-3-methyl-2-butanone		$C_5H_{10}O_2$	115-22-0	102.132			148.4(0.5)	0.9526[20]		s chl
6087	2-Hydroxy-3-methyl-2-cyclopenten-1-one		$C_6H_8O_2$	80-71-7	112.127		104.8				
6088	5-(Hydroxymethyl)-2-furan-carboxaldehyde	5-(Hydroxymethyl)-2-furalde-hyde	$C_6H_6O_3$	67-47-0	126.110	nd (eth-peth)	31.5	115[1]	1.2062[25]	1.5627[18]	s H₂O, EtOH, bz, chl; sl eth, ctc
6089	2-Hydroxy-6-methyl-3-isopropylbenzoic acid	o-Thymotic acid	$C_{11}H_{14}O_3$	548-51-6	194.227	nd (w, bz, lig)	127	sub			vs bz, eth, EtOH
6090	2-Hydroxy-3-methyl-6-isopropyl-2-cyclohexen-1-one	Diosphenol	$C_{10}H_{16}O_2$	490-03-9	168.233		83	109[10]			
6091	2-(Hydroxymethyl)-2-methyl-1,3-propanediol		$C_5H_{12}O_3$	77-85-0	120.147	wh pow or nd (al)	199(2)	136[15]			msc H₂O, EtOH; i eth, bz; vs HOAc
6092	2-Hydroxy-3-methyl-1,4-naphthalenedione	Phthiocol	$C_{11}H_8O_3$	483-55-6	188.180	ye pr (eth-peth)	173.5	sub			vs ace, eth

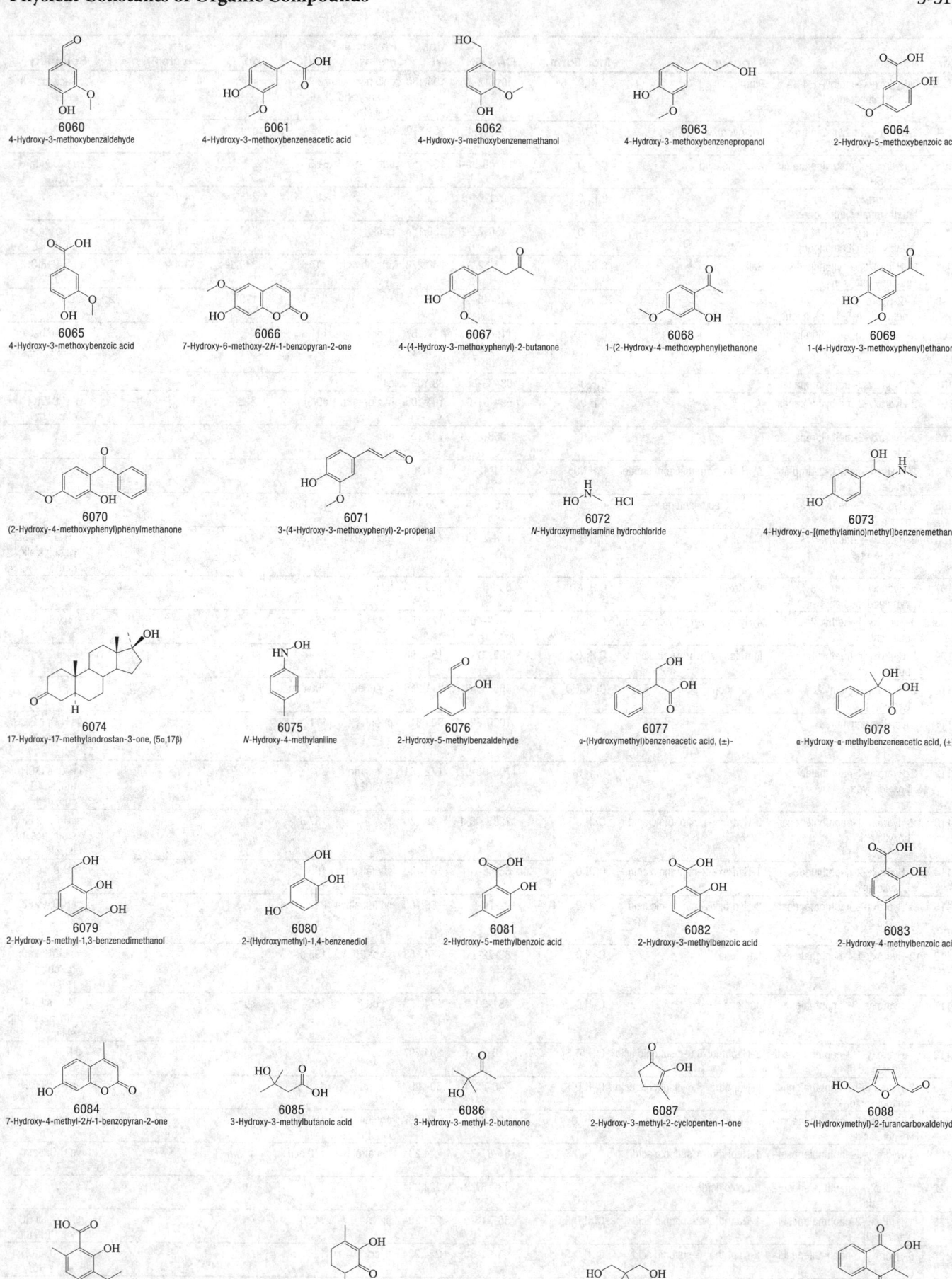

6060
4-Hydroxy-3-methoxybenzaldehyde

6061
4-Hydroxy-3-methoxybenzeneacetic acid

6062
4-Hydroxy-3-methoxybenzenemethanol

6063
4-Hydroxy-3-methoxybenzenepropanol

6064
2-Hydroxy-5-methoxybenzoic acid

6065
4-Hydroxy-3-methoxybenzoic acid

6066
7-Hydroxy-6-methoxy-2H-1-benzopyran-2-one

6067
4-(4-Hydroxy-3-methoxyphenyl)-2-butanone

6068
1-(2-Hydroxy-4-methoxyphenyl)ethanone

6069
1-(4-Hydroxy-3-methoxyphenyl)ethanone

6070
(2-Hydroxy-4-methoxyphenyl)phenylmethanone

6071
3-(4-Hydroxy-3-methoxyphenyl)-2-propenal

6072
N-Hydroxymethylamine hydrochloride

6073
4-Hydroxy-α-[(methylamino)methyl]benzenemethanol

6074
17-Hydroxy-17-methylandrostan-3-one, (5α,17β)-

6075
N-Hydroxy-4-methylaniline

6076
2-Hydroxy-5-methylbenzaldehyde

6077
α-(Hydroxymethyl)benzeneacetic acid, (±)-

6078
α-Hydroxy-α-methylbenzeneacetic acid, (±)-

6079
2-Hydroxy-5-methyl-1,3-benzenedimethanol

6080
2-(Hydroxymethyl)-1,4-benzenediol

6081
2-Hydroxy-5-methylbenzoic acid

6082
2-Hydroxy-3-methylbenzoic acid

6083
2-Hydroxy-4-methylbenzoic acid

6084
7-Hydroxy-4-methyl-2H-1-benzopyran-2-one

6085
3-Hydroxy-3-methylbutanoic acid

6086
3-Hydroxy-3-methyl-2-butanone

6087
2-Hydroxy-3-methyl-2-cyclopenten-1-one

6088
5-(Hydroxymethyl)-2-furancarboxaldehyde

6089
2-Hydroxy-6-methyl-3-isopropylbenzoic acid

6090
2-Hydroxy-3-methyl-6-isopropyl-2-cyclohexen-1-one

6091
2-(Hydroxymethyl)-2-methyl-1,3-propanediol

6092
2-Hydroxy-3-methyl-1,4-naphthalenedione

No.	Name	Synonym	Mol. Form.	CAS RN	Mol. Wt.	Physical Form	mp/°C	bp/°C	den g cm⁻³	n_D	Solubility
6093	5-Hydroxy-2-methyl-1,4-naphthalenedione	Plumbagin	$C_{11}H_8O_3$	481-42-5	188.180	gold pr or oran-ye nd (dil al)	78.5	sub			vs ace, bz, eth, EtOH
6094	2-(Hydroxymethyl)-2-nitro-1,3-propanediol	Tris(hydroxymethyl)-nitromethane	$C_4H_9NO_5$	126-11-4	151.118	nd or pr	165	dec			vs H_2O, eth, EtOH
6095	2-Hydroxy-4-methylpentanoic acid, (S)-	L-Leucic acid	$C_6H_{12}O_3$	13748-90-8	132.157	orth (eth)	81.5				vs H_2O, eth, EtOH
6096	1-(2-Hydroxy-4-methylphenyl)ethanone		$C_9H_{10}O_2$	6921-64-8	150.174		21	245	1.1012[10]	1.5527[13]	
6097	1-(2-Hydroxy-5-methylphenyl)ethanone		$C_9H_{10}O_2$	1450-72-2	150.174	pr (lig)	50	210	1.0797[53]		vs bz, eth, EtOH, chl
6098	2-(Hydroxymethyl)phenyl-β-D-glucopyranoside	Salicin	$C_{13}H_{18}O_7$	138-52-3	286.278	orth nd or lf (w)	207	240 dec	1.434[20]		vs H_2O, EtOH, HOAc
6099	1-(2-Hydroxy-5-methylphenyl)-1-propanone		$C_{10}H_{12}O_2$	938-45-4	164.201		1.0	129[16.5]	1.0841[14]	1.549[13]	s chl
6100	N-(Hydroxymethyl)phthalimide		$C_9H_7NO_3$	118-29-6	177.157	lf, pr (to)	141.5				i H_2O, eth, ctc; sl EtOH, bz; s tol
6101	3-Hydroxy-2-methylpropanal		$C_4H_8O_2$	38433-80-6	88.106	oil					
6102	2-Hydroxy-2-methylpropanoic acid		$C_4H_8O_3$	594-61-6	104.105	hyg pr (eth) nd (bz)	80(3)	212			vs H_2O, EtOH, eth; sl bz
6103	3-Hydroxy-2-methylpropanoic acid		$C_4H_8O_3$	2068-83-9	104.105	oil					
6104	N-(Hydroxymethyl)-2-propen-amide	N-(Hydroxymethyl)acrylamide	$C_4H_7NO_2$	924-42-5	101.105	cry	76				
6105	4-Hydroxy-6-methyl-2H-pyran-2-one	Triacetic acid lactone	$C_6H_6O_3$	675-10-5	126.110		189 dec				
6106	3-Hydroxy-2-methyl-4H-pyran-4-one	Maltol	$C_6H_6O_3$	118-71-8	126.110	mcl pr (chl)	161.5	93 sub			sl H_2O, eth, bz; vs chl; s alk; peth
6107	5-Hydroxy-6-methyl-3,4-pyridinedimethanol	Pyridoxin	$C_8H_{11}NO_3$	65-23-6	169.178	nd (HOAc)	160	140[0.0001]			
6108	4-Hydroxy-1-methyl-2-quinolinone	4-Hydroxy-N-methylcarbostyril	$C_{10}H_9NO_2$	1677-46-9	175.184		265				sl DMSO
6109	2-Hydroxy-4-(methylthio)-butanoic acid	Methionine hydroxy analog	$C_5H_{10}O_3S$	583-91-5	150.196	oil					
6110	3-Hydroxy-α-methyl-L-tyrosine	Methyldopa	$C_{10}H_{13}NO_4$	555-30-6	211.215	cry (MeOH)	300 dec				
6111	(Hydroxymethyl)urea		$C_2H_6N_2O_2$	1000-82-4	90.081	pr (al)	111				vs H_2O; s EtOH, MeOH, HOAc; i eth
6112	2-Hydroxy-1-naphthalenecar-boxaldehyde		$C_{11}H_8O_2$	708-06-5	172.181	pr (al), nd (AcOEt)	83	192[27]			i H_2O; s EtOH, eth, aq alk, sulf, peth
6113	2-Hydroxy-1-naphthalenecar-boxylic acid	2-Hydroxy-1-naphthoic acid	$C_{11}H_8O_3$	2283-08-1	188.180		157.3				sl H_2O; vs EtOH; s eth, ace, bz, lig, chl
6114	1-Hydroxy-2-naphthalenecar-boxylic acid	1-Hydroxy-2-naphthoic acid	$C_{11}H_8O_3$	86-48-6	188.180	cry (al) nd (al, eth, bz)	195				sl H_2O; vs EtOH, eth; s bz
6115	3-Hydroxy-2-naphthalenecar-boxylic acid	3-Hydroxy-2-naphthoic acid	$C_{11}H_8O_3$	92-70-6	188.180	nd (dil al) ye lf (dil al)	222.5				sl H_2O; vs EtOH, eth; s bz, chl, tol
6116	2-Hydroxy-1,4-naphthalenedi-one	Lawsone	$C_{10}H_6O_3$	83-72-7	174.153	ye pr (HOAc)	195 dec				vs EtOH; i eth, bz, chl; s HOAc
6117	5-Hydroxy-1,4-naphthalenedi-one	Juglone	$C_{10}H_6O_3$	481-39-0	174.153	ye nd (bz) peth)	155	sub			i H_2O; s EtOH, eth, bz; vs chl; sl lig
6118	7-Hydroxy-1,3-naphthalenedi-sulfonic acid	2-Naphthol-6,8-disulfonic acid	$C_{10}H_8O_7S_2$	118-32-1	304.297						s H_2O
6119	3-Hydroxy-2,7-naphthalenedi-sulfonic acid	2-Naphthol-3,6-disulfonic acid	$C_{10}H_8O_7S_2$	148-75-4	304.297	hyg nd	dec				vs H_2O, EtOH
6120	6-Hydroxy-2-naphthalenepro-panoic acid	Allenolic acid	$C_{13}H_{12}O_3$	553-39-9	216.232	cry (dil MeOH)	180.5				vs py, EtOH, MeOH
6121	4-Hydroxy-1-naphthalenesul-fonic acid	1-Naphthol-4-sulfonic acid	$C_{10}H_8O_4S$	84-87-7	224.234	tab or pl (w)	170 dec				vs H_2O; i eth
6122	7-Hydroxy-1-naphthalenesul-fonic acid	Croceic acid	$C_{10}H_8O_4S$	132-57-0	224.234						s H_2O
6123	1-Hydroxy-2-naphthalenesul-fonic acid	1-Naphthol-2-sulfonic acid	$C_{10}H_8O_4S$	567-18-0	224.234	pl (w)	>250				sl H_2O, dil HCl; s EtOH; i eth
6124	6-Hydroxy-2-naphthalenesul-fonic acid	2-Naphthol-6-sulfonic acid	$C_{10}H_8O_4S$	93-01-6	224.234	lf, cry (w+1)	125				vs H_2O, EtOH; i eth; s HOAc
6125	Hydroxynaphthol blue, trisodium salt		$C_{20}H_{14}N_2Na_3O_{11}S_3$	63451-35-4	620.471	dk red cry					

6093
5-Hydroxy-2-methyl-1,4-naphthalenedione

6094
2-(Hydroxymethyl)-2-nitro-1,3-propanediol

6095
2-Hydroxy-4-methylpentanoic acid, (S)-

6096
1-(2-Hydroxy-4-methylphenyl)ethanone

6097
1-(2-Hydroxy-5-methylphenyl)ethanone

6098
2-(Hydroxymethyl)phenyl-β-D-glucopyranoside

6099
1-(2-Hydroxy-5-methylphenyl)-1-propanone

6100
N-(Hydroxymethyl)phthalimide

6101
3-Hydroxy-2-methylpropanal

6102
2-Hydroxy-2-methylpropanoic acid

6103
3-Hydroxy-2-methylpropanoic acid

6104
N-(Hydroxymethyl)-2-propenamide

6105
4-Hydroxy-6-methyl-2H-pyran-2-one

6106
3-Hydroxy-2-methyl-4H-pyran-4-one

6107
5-Hydroxy-6-methyl-3,4-pyridinedimethanol

6108
4-Hydroxy-1-methyl-2-quinolinone

6109
2-Hydroxy-4-(methylthio)butanoic acid

6110
3-Hydroxy-α-methyl-L-tyrosine

6111
(Hydroxymethyl)urea

6112
2-Hydroxy-1-naphthalenecarboxaldehyde

6113
2-Hydroxy-1-naphthalenecarboxylic acid

6114
1-Hydroxy-2-naphthalenecarboxylic acid

6115
3-Hydroxy-2-naphthalenecarboxylic acid

6116
2-Hydroxy-1,4-naphthalenedione

6117
5-Hydroxy-1,4-naphthalenedione

6118
7-Hydroxy-1,3-naphthalenedisulfonic acid

6119
3-Hydroxy-2,7-naphthalenedisulfonic acid

6120
6-Hydroxy-2-naphthalenepropanoic acid

6121
4-Hydroxy-1-naphthalenesulfonic acid

6122
7-Hydroxy-1-naphthalenesulfonic acid

6123
1-Hydroxy-2-naphthalenesulfonic acid

6124
6-Hydroxy-2-naphthalenesulfonic acid

6125
Hydroxynaphthol blue, trisodium salt

No.	Name	Synonym	Mol. Form.	CAS RN	Mol. Wt.	Physical Form	mp/°C	bp/°C	den g cm⁻³	n_D	Solubility
6126	N-(2-Hydroxy-1-naphthyl)-acetamide		$C_{12}H_{11}NO_2$	117-93-1	201.221	lf (w, dil al)	235 dec	sub			vs ace, bz, eth, EtOH
6127	1-(1-Hydroxy-2-naphthyl)-ethanone		$C_{12}H_{10}O_2$	711-79-5	186.206	pr (bz, lig) grn-ye nd (al)	98.6(0.2)	325 dec			vs bz, HOAc
6128	2-Hydroxy-3-nitrobenzalde-hyde		$C_7H_5NO_4$	5274-70-4	167.120	nd (HOAc)	109.5				vs bz, EtOH
6129	2-Hydroxy-5-nitrobenzalde-hyde		$C_7H_5NO_4$	97-51-8	167.120	cry (dil HOAc)	127.0				s ace
6130	2-Hydroxy-3-nitrobenzoic acid	3-Nitrosalicylic acid	$C_7H_5NO_5$	85-38-1	183.119	ye nd (HOAc, w+1)	148				sl H₂O; vs EtOH, eth; s ace, bz, chl
6131	2-Hydroxy-5-nitrobenzoic acid	5-Nitrosalicylic acid	$C_7H_5NO_5$	96-97-9	183.119	nd (w)	229.5		1.650²⁰		sl H₂O; vs EtOH, eth, ace, bz; s chl
6132	2-Hydroxy-1,2,3-nonadecane-tricarboxylic acid	Agaricic acid	$C_{22}H_{40}O_7$	666-99-9	416.549	cry pow	142 dec				s H₂O; sl EtOH, eth; i bz, chl
6133	12-Hydroxyoctadecanoic acid	12-Hydroxysteric acid	$C_{18}H_{36}O_3$	106-14-9	300.477	cry (al)	82				i H₂O; s EtOH, eth, chl
6134	cis-12-Hydroxy-9-octadece-noic acid, (R)-	Ricinoleic acid	$C_{18}H_{34}O_3$	141-22-0	298.461	visc liq	-8.28(0.02)	227¹⁰	0.9450²¹	1.4716²¹	i H₂O; vs eth, EtOH
6135	2-Hydroxyoctanoic acid		$C_8H_{16}O_3$	617-73-2	160.211	pl	70	162¹⁰			sl H₂O, chl; vs EtOH, eth
6136	5-Hydroxy-4-octanone	Butyroin	$C_8H_{16}O_2$	496-77-5	144.212	liq	-10	185	0.9107¹⁶	1.4345¹⁶	
6137	[2-Hydroxy-4-(octyloxy)-phenyl]phenylmethanone	Octabenzone	$C_{21}H_{26}O_3$	1843-05-6	326.429		48.5				
6138	3-Hydroxy-2-oxopropanoic acid	Hydroxypyruvic acid	$C_3H_4O_4$	1113-60-6	104.062		81 dec				
6139	3-Hydroxy-4-oxo-4H-pyran-2,6-dicarboxylic acid	Meconic acid	$C_7H_4O_7$	497-59-6	200.103	orth pl (w, dil HCl) (+3w)	120 dec				sl H₂O, MeOH, ace, eth; s EtOH, bz
6140	2-Hydroxypentanoic acid		$C_5H_{10}O_3$	617-31-2	118.131	hyg pl	34	sub			s H₂O, EtOH, eth
6141	5-Hydroxy-2-pentanone		$C_5H_{10}O_2$	1071-73-4	102.132			209	1.0071²⁰	1.4390²⁰	msc H₂O; s EtOH, eth
6142	7-Hydroxy-3H-phenoxazin-3-one	Resorufine	$C_{12}H_7NO_3$	635-78-9	213.189	br nd (PhNO₂) pr (HCl)					i H₂O; sl EtOH; i eth; vs alk
6143	N-(2-Hydroxyphenyl)-acetamide		$C_8H_9NO_2$	614-80-2	151.163	pl (dil al)	209				sl H₂O; vs EtOH, eth, bz; s DMSO
6144	N-(3-Hydroxyphenyl)-acetamide		$C_8H_9NO_2$	621-42-1	151.163	nd (w)	148.5				vs H₂O, EtOH; sl eth, bz, chl, DMSO
6145	N-(4-Hydroxyphenyl)-acetamide	Acetaminophen	$C_8H_9NO_2$	103-90-2	151.163	mcl pr (w)	168.0(0.5)		1.293²¹		i H₂O; vs EtOH
6146	2-[(4-Hydroxyphenyl)azo]-benzoic acid		$C_{13}H_{10}N_2O_3$	1634-82-8	242.229		206				sl DMSO
6147	2-Hydroxy-N-phenylbenza-mide	Salicylanilide	$C_{13}H_{11}NO_2$	87-17-2	213.232	pr (w, al)	136.5				s H₂O; sl EtOH, eth, bz, chl
6148	N-Hydroxy-N-phenylbenza-mide		$C_{13}H_{11}NO_2$	304-88-1	213.232		121(1)				
6149	α-Hydroxy-α-phenylbenzeneacetic acid	Benzilic acid	$C_{14}H_{12}O_3$	76-93-7	228.243	mcl nd (w)	149(2)	180 dec			sl H₂O, ace; vs EtOH, eth; s con sulf
6150	3-Hydroxy-2-phenyl-4H-1-benzopyran-4-one		$C_{15}H_{10}O_3$	577-85-5	238.238	pa ye nd (al)	169.5				s EtOH
6151	N-(4-Hydroxyphenyl)-butanamide	4'-Hydroxybutyranilide	$C_{10}H_{13}NO_2$	101-91-7	179.216	nd (w)	139.5				vs H₂O, EtOH
6152	4-(4-Hydroxyphenyl)-2-butanone		$C_{10}H_{12}O_2$	5471-51-2	164.201		82.5				
6153	1-(2-Hydroxyphenyl)ethanone	2-Hydroxyacetophenone	$C_8H_8O_2$	118-93-4	136.149		2.5	218	1.1307²⁰	1.5584²⁰	vs eth, EtOH, HOAc
6154	1-(3-Hydroxyphenyl)ethanone	3-Hydroxyacetophenone	$C_8H_8O_2$	121-71-1	136.149	nd or lf	94(3)	296	1.0992¹⁰⁹	1.5348¹⁰⁹	sl H₂O; vs EtOH, eth, bz, chl; i lig
6155	1-(4-Hydroxyphenyl)ethanone	4-Hydroxyacetophenone	$C_8H_8O_2$	99-93-4	136.149	nd (eth, dil al)	108.2(0.5)	147³	1.1090¹⁰⁹	1.5577¹⁰⁹	sl H₂O, DMSO; vs EtOH, eth
6156	4-Hydroxyphenyl-β-D-glucopyranoside	Arbutin	$C_{12}H_{16}O_7$	497-76-7	272.251	nd (w+1)	199.5				vs H₂O; s EtOH; sl eth; i bz, chl, CS₂
6157	2-(4-Hydroxyphenyl)-D-glycine	Oxfenicine	$C_8H_9NO_3$	22818-40-2	167.162	cry	240 dec				

6126
N-(2-Hydroxy-1-naphthyl)acetamide

6127
1-(1-Hydroxy-2-naphthyl)ethanone

6128
2-Hydroxy-3-nitrobenzaldehyde

6129
2-Hydroxy-5-nitrobenzaldehyde

6130
2-Hydroxy-3-nitrobenzoic acid

6131
2-Hydroxy-5-nitrobenzoic acid

6132
2-Hydroxy-1,2,3-nonadecanetricarboxylic acid

6133
12-Hydroxyoctadecanoic acid

6134
cis-12-Hydroxy-9-octadecenoic acid, (R)-

6135
2-Hydroxyoctanoic acid

6136
5-Hydroxy-4-octanone

6137
[2-Hydroxy-4-(octyloxy)phenyl]phenylmethanone

6138
3-Hydroxy-2-oxopropanoic acid

6139
3-Hydroxy-4-oxo-4H-pyran-2,6-dicarboxylic acid

6140
2-Hydroxypentanoic acid

6141
5-Hydroxy-2-pentanone

6142
7-Hydroxy-3H-phenoxazin-3-one

6143
N-(2-Hydroxyphenyl)acetamide

6144
N-(3-Hydroxyphenyl)acetamide

6145
N-(4-Hydroxyphenyl)acetamide

6146
2-[(4-Hydroxyphenyl)azo]benzoic acid

6147
2-Hydroxy-N-phenylbenzamide

6148
N-Hydroxy-N-phenylbenzamide

6149
α-Hydroxy-α-phenylbenzeneacetic acid

6150
3-Hydroxy-2-phenyl-4H-1-benzopyran-4-one

6151
N-(4-Hydroxyphenyl)butanamide

6152
4-(4-Hydroxyphenyl)-2-butanone

6153
1-(2-Hydroxyphenyl)ethanone

6154
1-(3-Hydroxyphenyl)ethanone

6155
1-(4-Hydroxyphenyl)ethanone

6156
4-Hydroxyphenyl-β-D-glucopyranoside

6157
2-(4-Hydroxyphenyl)-D-glycine

No.	Name	Synonym	Mol. Form.	CAS RN	Mol. Wt.	Physical Form	mp/°C	bp/°C	den g cm^{-3}	n_D	Solubility
6158	N-(4-Hydroxyphenyl)glycine		$C_8H_9NO_3$	122-87-2	167.162	lf (w) pl (w)	246 dec				sl H_2O, EtOH; i eth; s AcOEt, chl
6159	2-(2-Hydroxyphenyl)-2-(4-hydroxyphenyl)propane	2,4'-Isopropylidenediphenol	$C_{15}H_{16}O_2$	837-08-1	228.287	cry (bz)	111				
6160	2-[[(2-Hydroxyphenyl)imino]-methyl]phenol	N-Salicylidene-o-aminophenol	$C_{13}H_{11}NO_2$	1761-56-4	213.232		185				
6161	N-Hydroxy-N-(phenylmethyl)-benzenemethanamine		$C_{14}H_{15}NO$	621-07-8	213.275		122.5				s chl
6162	N-(4-Hydroxyphenyl)-octadecanamide		$C_{24}H_{41}NO_2$	103-99-1	375.589		133.8	239.5[10]			i H_2O; sl eth, bz, chl; s ace
6163	3-(4-Hydroxyphenyl)-2-oxopropanoic acid	4-Hydroxy-α-oxobenzenepropanoic acid	$C_9H_8O_4$	156-39-8	180.158	cry (w)	220 dec				s H_2O; dec alk
6164	(2-Hydroxyphenyl)-phenylmethanone	2-Hydroxybenzophenone	$C_{13}H_{10}O_2$	117-99-7	198.217	pl (dil al)	35(3)	250[560]			i H_2O; vs EtOH, eth, bz; sl chl, peth
6165	1-(2-Hydroxyphenyl)-3-phenyl-2-propen-1-one	2'-Hydroxychalcone	$C_{15}H_{12}O_2$	1214-47-7	224.255		90				
6166	2-Hydroxy-1-phenyl-1-propa-none		$C_9H_{10}O_2$	5650-40-8	150.174	ye oil		251	1.1085[18]	1.536[23]	
6167	1-(2-Hydroxyphenyl)-1-propanone		$C_9H_{10}O_2$	610-99-1	150.174			150[80]		1.5501[20]	sl H_2O; s EtOH, eth, ctc, alk
6168	1-(4-Hydroxyphenyl)-1-propanone	Paroxypropione	$C_9H_{10}O_2$	70-70-2	150.174	wh nd or pl (w)	149				sl H_2O, ace; s EtOH, eth, alk
6169	3-(4-Hydroxyphenyl)-2-propenoic acid	p-Coumaric acid	$C_9H_8O_3$	7400-08-0	164.158	nd	212(11)				vs eth, EtOH
6170	3-Hydroxy-2-phenyl-4-quino-linecarboxylic acid	Oxycinchophen	$C_{16}H_{11}NO_3$	485-89-2	265.263	ye pr (al)	206 dec				vs bz, EtOH, HOAc
6171	N-Hydroxypiperidine	1-Piperidinol	$C_5H_{11}NO$	4801-58-5	101.147	hyg	39.3	110[55]			
6172	3-Hydroxypregnan-20-one, (3α,5α)	Allopregnan-3α-ol-20-one	$C_{21}H_{34}O_2$	516-54-1	318.494	cry (al)	177				
6173	3-Hydroxypregnan-20-one, (3β,5α)	Allopregnan-3β-ol-20-one	$C_{21}H_{34}O_2$	516-55-2	318.494		170(15)				
6174	17-Hydroxypregn-4-ene-3,20-dione	17α-Hydroxyprogesterone	$C_{21}H_{30}O_3$	68-96-2	330.461						sl chl
6175	21-Hydroxypregn-4-ene-3,20-dione	Deoxycorticosterone	$C_{21}H_{30}O_3$	64-85-7	330.461	pl (eth)	141.5				sl H_2O, eth; vs EtOH, ace; s chl
6176	21-Hydroxypregn-4-ene-3,11,20-trione	11-Dehydrocorticosterone	$C_{21}H_{28}O_4$	72-23-1	344.445	pr (ace-w, al, ace-eth)	183.5				i H_2O; s EtOH, ace, bz
6177	cis-4-Hydroxy-L-proline		$C_5H_9NO_3$	618-27-9	131.130	nd (w+1)	239.5				vs H_2O
6178	trans-4-Hydroxy-L-proline		$C_5H_9NO_3$	51-35-4	131.130	lf (dil al) pr (w)	274				vs H_2O; sl EtOH
6179	3-Hydroxypropanal	Hydracrolein	$C_3H_6O_2$	2134-29-4	74.079			90[18]			vs ace, eth, EtOH
6180	Hydroxypropanedioic acid	Tartronic acid	$C_3H_4O_5$	80-69-3	120.061	pr (w+1)	157	sub			s H_2O, EtOH; sl eth
6181	2-Hydroxypropanenitrile	Acetaldehyde cyanohydrin	C_3H_5NO	78-97-7	71.078	liq	-40.0(0.5)	184(3)	0.9877[20]	1.4058[18]	msc H_2O, EtOH; s eth, chl; i CS_2, peth
6182	3-Hydroxypropanenitrile	Hydracrylonitrile	C_3H_5NO	109-78-4	71.078	liq	-46	218(5)	1.0404[25]	1.4248[20]	msc H_2O, EtOH; sl eth; s chl; i CS_2
6183	3-Hydroxypropanoic acid	Hydracrylic acid	$C_3H_6O_3$	503-66-2	90.078	syr		dec		1.4489[20]	vs H_2O; s EtOH; msc eth
6184	1-Hydroxy-2-propanone	Acetone alcohol	$C_3H_6O_2$	116-09-6	74.079	hyg liq	-17	145.5	1.0805[20]	1.4295[20]	vs H_2O, EtOH, eth
6185	4-(3-Hydroxy-1-propenyl)-2-methoxyphenol	Coniferyl alcohol	$C_{10}H_{12}O_3$	458-35-5	180.200	pr (eth-lig)	74	164[3]			i H_2O; s EtOH, alk; vs eth
6186	2-Hydroxypropyl acrylate		$C_6H_{10}O_3$	999-61-1	130.141	liq		70[2]			
6187	(2-Hydroxypropyl)-trimethylammonium chloride		$C_6H_{16}ClNO$	2382-43-6	153.650	pr (Bu OH)	165	dec			vs H_2O, EtOH
6188	3-Hydroxy-1H-pyridin-2-one		$C_5H_5NO_2$	16867-04-2	111.100		245 dec				
6189	1-Hydroxy-2,5-pyrrolidinedi-one	N-Hydroxysuccinimide	$C_4H_5NO_3$	6066-82-6	115.088	hyg	96.3				sl DMSO
6190	4-Hydroxy-2-quinolinecarbox-ylic acid	Kynurenic acid	$C_{10}H_7NO_3$	492-27-3	189.168	ye nd (+w, dil al)	282.5				sl H_2O; s EtOH; i eth; vs alk
6191	8-Hydroxy-5-quinolinesulfonic acid		$C_9H_7NO_4S$	84-88-8	225.222	ye lf, nd (+1w) (dil HCl)	322.5				sl H_2O
6192	4-Hydroxy-2-quinolinone	2,4-Quinolinediol	$C_9H_7NO_2$	86-95-3	161.158		353(4)				sl EtOH, PhNO$_2$, gl HOAc

6158
N-(4-Hydroxyphenyl)glycine

6159
2(2-Hydroxyphenyl)-2(4-hydroxyphenyl)propane

6160
2-[[(2-Hydroxyphenyl)imino]methyl]phenol

6161
N-Hydroxy-N-(phenylmethyl)benzenemethanamine

6162
N-(4-Hydroxyphenyl)octadecanamide

6163
3-(4-Hydroxyphenyl)-2-oxopropanoic acid

6164
(2-Hydroxyphenyl)phenylmethanone

6165
1-(2-Hydroxyphenyl)-3-phenyl-2-propen-1-one

6166
2-Hydroxy-1-phenyl-1-propanone

6167
1-(2-Hydroxyphenyl)-1-propanone

6168
1-(4-Hydroxyphenyl)-1-propanone

6169
3-(4-Hydroxyphenyl)-2-propenoic acid

6170
3-Hydroxy-2-phenyl-4-quinolinecarboxylic acid

6171
N-Hydroxypiperidine

6172
3-Hydroxypregnan-20-one, (3α,5α)

6173
3-Hydroxypregnan-20-one, (3β,5α)

6174
17-Hydroxypregn-4-ene-3,20-dione

6175
21-Hydroxypregn-4-ene-3,20-dione

6176
21-Hydroxypregn-4-ene-3,11,20-trione

6177
cis-4-Hydroxy-L-proline

6178
trans-4-Hydroxy-L-proline

6179
3-Hydroxypropanal

6180
Hydroxypropanedioic acid

6181
2-Hydroxypropanenitrile

6182
3-Hydroxypropanenitrile

6183
3-Hydroxypropanoic acid

6184
1-Hydroxy-2-propanone

6185
4-(3-Hydroxy-1-propenyl)-2-methoxyphenol

6186
2-Hydroxypropyl acrylate

6187
(2-Hydroxypropyl)trimethylammonium chloride

6188
3-Hydroxy-1H-pyridin-2-one

6189
1-Hydroxy-2,5-pyrrolidinedione

6190
4-Hydroxy-2-quinolinecarboxylic acid

6191
8-Hydroxy-5-quinolinesulfonic acid

6192
4-Hydroxy-2-quinolinone

No.	Name	Synonym	Mol. Form.	CAS RN	Mol. Wt.	Physical Form	mp/°C	bp/°C	den g cm^{-3}	n_D	Solubility
6193	3-Hydroxyspirostan-12-one, (3β,5α,25R)-	Hecogenin	C$_{27}$H$_{42}$O$_4$	467-55-0	430.620	pl (eth)	266.5				vs ace, eth, EtOH
6194	4-Hydroxystyrene	4-Vinylphenol	C$_8$H$_8$O	2628-17-3	120.149		72(2)				
6195	2-Hydroxy-5-sulfobenzoic acid	5-Sulfosalicylic acid	C$_7$H$_6$O$_6$S	97-05-2	218.184	hyg nd	120				vs H$_2$O; vs EtOH, eth
6196	2-Hydroxy-5-sulfobenzoic acid dihydrate	5-Sulfosalicylic acid dihydrate	C$_7$H$_{10}$O$_8$S	5965-83-3	254.214	wh cry (w)					vs H$_2$O; vs EtOH, eth
6197	4-Hydroxy-2,2,6,6-tetramethylpiperidine	2,2,6,6-Tetramethyl-4-piperidinol	C$_9$H$_{19}$NO	2403-88-5	157.253		130	213.5			
6198	5-Hydroxytryptamine	3-(2-Aminoethyl)indol-5-ol	C$_{10}$H$_{12}$N$_2$O	50-67-9	176.214						s H$_2$O
6199	5-Hydroxy-DL-tryptophan		C$_{11}$H$_{12}$N$_2$O$_3$	114-03-4	220.224	rod or nd (al)	300 dec				
6200	Hydroxyurea		CH$_4$N$_2$O$_2$	127-07-1	76.055	nd (al)	141	dec			vs H$_2$O
6201	Hydroxyzine		C$_{21}$H$_{27}$ClN$_2$O$_2$	68-88-2	374.904	oil		220[0.5]			
6202	Hymecromone O,O-diethyl phosphorothioate		C$_{14}$H$_{17}$O$_5$PS	299-45-6	328.321	nd	38	210[1.0] dec	1.260[38]	1.5685[37]	vs H$_2$O; sl peth
6203	Hymenoxone		C$_{15}$H$_{22}$O$_5$	57377-32-9	282.333	cry					
6204	Hyoscyamine	Tropine tropate	C$_{17}$H$_{23}$NO$_3$	101-31-5	289.370	tetr nd (dil al)	108.5				sl H$_2$O, eth, bz; vs EtOH, chl
6205	Hypoglycin A		C$_7$H$_{11}$NO$_2$	156-56-9	141.168	ye pl (Me aq)	282				
6206	Hypoxanthine		C$_5$H$_4$N$_4$O	68-94-0	136.112	oct nd (w)	150 dec				sl H$_2$O; s alk, dil acid
6207	Ibuprofen	2-(4-Isobutylphenyl)propanoic acid	C$_{13}$H$_{18}$O$_2$	15687-27-1	206.281	col cry	75(1)				sl H$_2$O; s os
6208	Icosylamine	1-Eicosanamine	C$_{20}$H$_{43}$N	10525-37-8	297.562			389(15)			
6209	D-Idose		C$_6$H$_{12}$O$_6$	5978-95-0	180.155	syr					vs H$_2$O
6210	L-Idose		C$_6$H$_{12}$O$_6$	5934-56-5	180.155	syr					vs H$_2$O
6211	Imazalil		C$_{14}$H$_{14}$Cl$_2$N$_2$O	35554-44-0	297.179		50.2(0.5)	dec	1.243[23]		
6212	Imazapyr		C$_{13}$H$_{15}$N$_3$O$_3$	81334-34-1	261.276		171				
6213	Imazaquin		C$_{17}$H$_{17}$N$_3$O$_3$	81335-37-7	311.335		221				
6214	Imazethapyr		C$_{15}$H$_{19}$N$_3$O$_3$	81335-77-5	289.330		173				
6215	Imidazole	1,3-Diazole	C$_3$H$_4$N$_2$	288-32-4	68.077	mcl pr (bz)	89.52(0.04)	257	1.0303[101]	1.4801[101]	vs H$_2$O, EtOH; s eth, ace, py; sl bz
6216	1H-Imidazole-4,5-dicarboxylic acid		C$_5$H$_4$N$_2$O$_4$	570-22-9	156.097	pr	290 dec		1.749[25]		sl H$_2$O, py; i EtOH, eth, bz
6217	1H-Imidazole-4-ethanamine, dihydrochloride		C$_5$H$_{11}$Cl$_2$N$_3$	56-92-8	184.066	pl (eth-HOAc), pr (w)	251.3		1.43[20]		vs H$_2$O, MeOH
6218	2,4-Imidazolidinedione	Hydantoin	C$_3$H$_4$N$_2$O$_2$	461-72-3	100.076	nd (MeOH), lf (w)	220				s H$_2$O, EtOH, alk; sl eth; i peth
6219	2-Imidazolidinethione	Ethylene thiourea	C$_3$H$_6$N$_2$S	96-45-7	102.158	nd (al), pr (al)	203				vs H$_2$O; s EtOH; i eth, bz, chl; sl DMSO
6220	Imidazolidinetrione	Parabanic acid	C$_3$H$_2$N$_2$O$_3$	120-89-8	114.059	mcl nd (w)	247(3)	100 sub			s H$_2$O; vs EtOH
6221	2-Imidazolidinone	Ethylene urea	C$_3$H$_6$N$_2$O	120-93-4	86.092		131.7(0.5)				vs H$_2$O, EtOH; sl eth, chl
6222	Imidodicarbonic diamide	Biuret	C$_2$H$_5$N$_3$O$_2$	108-19-0	103.080	pl (al), nd (w+1)	190 dec				sl H$_2$O; vs EtOH; i eth
6223	3,3'-Iminobispropanenitrile	Bis(2-cyanoethyl)amine	C$_6$H$_9$N$_3$	111-94-4	123.155		-6(1)	162[5]	1.0165[20]		
6224	Iminodiacetic acid	Diglycine	C$_4$H$_7$NO$_4$	142-73-4	133.104	orth pr	247.5				sl H$_2$O; i EtOH, eth
6225	Iminodiacetic acid, dinitrile	2,2'-Iminobisacetonitrile	C$_4$H$_5$N$_3$	628-87-5	95.103		78				s H$_2$O, EtOH; sl eth, bz, chl
6226	Imipramine		C$_{19}$H$_{24}$N$_2$	50-49-7	280.407			160[0.1]			
6227	Imipramine hydrochloride	Tofranil	C$_{19}$H$_{25}$ClN$_2$	113-52-0	316.868		174.5				vs H$_2$O; s EtOH; sl ace
6228	Imperatorin		C$_{16}$H$_{14}$O$_4$	482-44-0	270.280	cry (al)	102				sl H$_2$O; s EtOH, eth, bz, peth; vs chl
6229	Indaconitine		C$_{34}$H$_{47}$NO$_{10}$	4491-19-4	629.738	cry	202 dec				vs eth, EtOH, chl
6230	Indalone	Butopyronoxyl	C$_{12}$H$_{18}$O$_4$	532-34-3	226.269	ye-red liq		263	1.057[20]	1.475[25]	i H$_2$O; vs EtOH, eth, chl
6231	Indan		C$_9$H$_{10}$	496-11-7	118.175	liq	-51.34(0.02)	177.8(0.4)	0.9639[20]	1.5378[20]	i H$_2$O; msc EtOH, eth; sl chl
6232	1-Indanamine	1-Aminoindane	C$_9$H$_{11}$N	34698-41-4	133.190			221	1.038[15]	1.5613[20]	sl H$_2$O; s eth, ace, bz
6233	1H-Indazole	1H-Benzopyrazole	C$_7$H$_6$N$_2$	271-44-3	118.136	nd (al, w)	146.1(0.6)	269			s H$_2$O, EtOH, eth

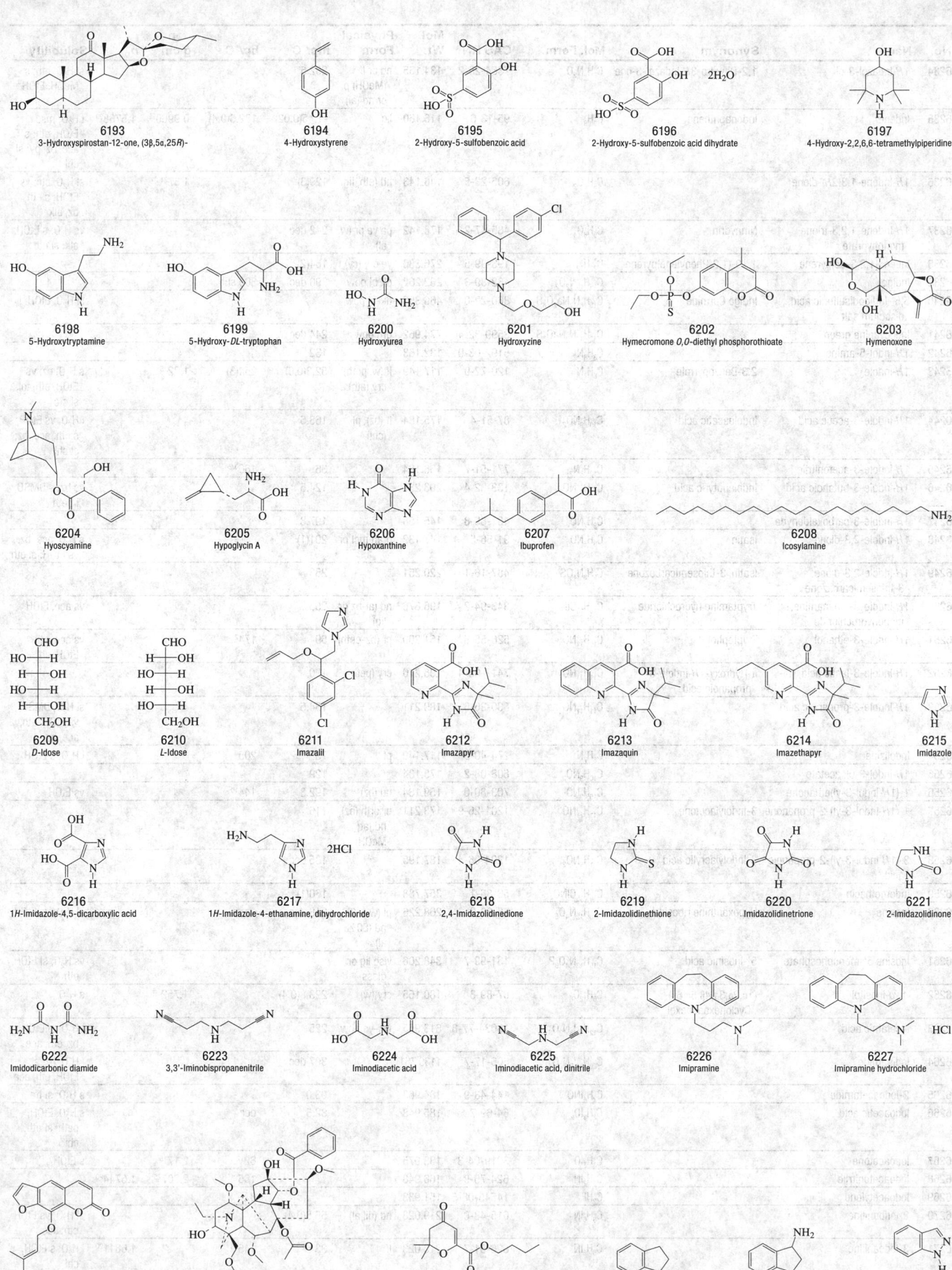

6193
3-Hydroxyspirostan-12-one, (3β,5α,25R)-

6194
4-Hydroxystyrene

6195
2-Hydroxy-5-sulfobenzoic acid

6196
2-Hydroxy-5-sulfobenzoic acid dihydrate

6197
4-Hydroxy-2,2,6,6-tetramethylpiperidine

6198
5-Hydroxytryptamine

6199
5-Hydroxy-*DL*-tryptophan

6200
Hydroxyurea

6201
Hydroxyzine

6202
Hymecromone *O,O*-diethyl phosphorothioate

6203
Hymenoxone

6204
Hyoscyamine

6205
Hypoglycin A

6206
Hypoxanthine

6207
Ibuprofen

6208
Icosylamine

6209
D-Idose

6210
L-Idose

6211
Imazalil

6212
Imazapyr

6213
Imazaquin

6214
Imazethapyr

6215
Imidazole

6216
1*H*-Imidazole-4,5-dicarboxylic acid

6217
1*H*-Imidazole-4-ethanamine, dihydrochloride

6218
2,4-Imidazolidinedione

6219
2-Imidazolidinethione

6220
Imidazolidinetrione

6221
2-Imidazolidinone

6222
Imidodicarbonic diamide

6223
3,3'-Iminobispropanenitrile

6224
Iminodiacetic acid

6225
Iminodiacetic acid, dinitrile

6226
Imipramine

6227
Imipramine hydrochloride

6228
Imperatorin

6229
Indaconitine

6230
Indalone

6231
Indan

6232
1-Indanamine

6233
1*H*-Indazole

No.	Name	Synonym	Mol. Form.	CAS RN	Mol. Wt.	Physical Form	mp/°C	bp/°C	den g cm⁻³	n_D	Solubility
6234	1H-Indazol-3-ol	1,2-Dihydro-3H-indazol-3-one	C₇H₆N₂O	7364-25-2	134.135	nd or lf (MeOH) pl or nd (al)	252.5				sl H₂O, eth; s MeOH, EtOH
6235	Indene	Indonaphthene	C₉H₈	95-13-6	116.160	liq	-1.45(0.02)	182.5(0.4)	0.9960²⁵	1.5768²⁰	i H₂O; msc EtOH, eth; s ace, bz, py; sl chl
6236	1H-Indene-1,3(2H)-dione		C₉H₆O₂	606-23-5	146.143	nd (eth, lig)	129(3)		1.37²¹		sl H₂O, ctc; vs EtOH; s eth, bz, alk
6237	1H-Indene-1,2,3-trione monohydrate	Ninhydrin	C₉H₆O₄	485-47-2	178.142	pa ye pr (w, al)	242 dec				vs H₂O; s EtOH, alk; sl eth
6238	Indeno[1,2,3-cd]pyrene	1,10-(1,2-Phenylene)pyrene	C₂₂H₁₂	193-39-5	276.330	ye cry (cy)	164(2)				
6239	Indigo		C₁₆H₁₀N₂O₂	482-89-3	262.262	dk bl pow	390 dec	300 sub			
6240	5,5'-Indigodisulfonic acid, disodium salt	Indigo Carmine	C₁₆H₈N₂Na₂O₆S₂	860-22-0	466.353	dk-bl pow					sl H₂O, EtOH; i os
6241	Indocyanine green		C₄₃H₄₇N₂NaO₆S₂	3599-32-4	774.962	grn pow	244 dec				
6242	1H-Indol-5-amine		C₈H₈N₂	5192-03-0	132.163		132				
6243	1H-Indole	2,3-Benzopyrrole	C₈H₇N	120-72-9	117.149	lf (w, peth) cry (eth)	52.3(0.6)	254(3)	1.22²⁵		s H₂O, bz; vs EtOH, eth, tol; sl ctc
6244	1H-Indole-3-acetic acid	Indoleacetic acid	C₁₀H₉NO₂	87-51-4	175.184	lf (bz), pl (chl)	168.5				i H₂O; vs EtOH; s eth, ace, bz; sl chl
6245	1H-Indole-3-acetonitrile		C₁₀H₈N₂	771-51-7	156.184		36	160⁰·²			
6246	1H-Indole-3-butanoic acid	Indolebutyric acid	C₁₂H₁₃NO₂	133-32-4	203.237		124.5				vs bz; s DMSO; i peth
6247	1H-Indole-3-carboxaldehyde		C₉H₇NO	487-89-8	145.158		197.8				
6248	1H-Indole-2,3-dione	Isatin	C₈H₅NO₂	91-56-5	147.132	oran mcl pr	201(1)				s H₂O, ace, bz; vs EtOH; sl eth
6249	1H-Indole-2,3-dione, 3-thiosemicarbazone	Isatin, 3-thiosemicarbazone	C₉H₈N₄OS	487-16-1	220.251		283				
6250	1H-Indole-3-ethanamine, monohydrochloride	Tryptamine hydrochloride	C₁₀H₁₃ClN₂	343-94-2	196.676	nd (al-bz or lig)	255				vs ace, EtOH
6251	1H-Indole-3-ethanol	Tryptophol	C₁₀H₁₁NO	526-55-6	161.200	pr (bz-peth)	59	174²			vs ace, eth, EtOH, chl
6252	1H-Indole-3-lactic acid, (S)-	α-Hydroxy-1H-indole-3-propanoic acid	C₁₁H₁₁NO₃	7417-65-4	205.210	cry (peth)	100				
6253	1H-Indole-3-propanoic acid		C₁₁H₁₁NO₂	830-96-6	189.211		134.5				sl H₂O, DMSO; vs EtOH, eth, ace, bz
6254	Indolizine		C₈H₇N	274-40-8	117.149	pl	75	205			i H₂O; s EtOH
6255	1H-Indol-3-ol, acetate		C₁₀H₉NO₂	608-08-2	175.184		129				
6256	1-(1H-Indol-3-yl)ethanone		C₁₀H₉NO	703-80-0	159.184	nd (bz)	192.3	144¹⁰			vs EtOH
6257	1-(1H-Indol-3-yl)-2-propanone	3-Indolylacetone	C₁₁H₁₁NO	1201-26-9	173.211	br orth (bz), nd (aq MeOH)	116				
6258	3-(1H-Indol-3-yl)-2-propenoic acid	3-Indolylacrylic acid	C₁₁H₉NO₂	1204-06-4	187.195		185 dec				
6259	Indomethacin		C₁₉H₁₆ClNO₄	53-86-1	357.788		160(1)				
6260	Inosine	Hypoxanthine riboside	C₁₀H₁₂N₄O₅	58-63-9	268.226	pl (w + 2), nd (80% al)	218 dec				sl H₂O; vs EtOH
6261	Inosine 5'-monophosphate	5'-Inosinic acid	C₁₀H₁₃N₄O₈P	131-99-7	348.206	visc liq or glass					vs H₂O; sl EtOH, eth
6262	myo-Inositol	(1α,2α,3α,4β,5α,6β)-Cyclohexanehexol	C₆H₁₂O₆	87-89-8	180.155	cry (w)	223.8(0.4)		1.752		s H₂O
6263	Iocetamic acid		C₁₂H₁₃I₃N₂O₃	16034-77-8	613.955	wh-ye pow	225				i H₂O; sl EtOH, bz, eth, ace
6264	Iodipamide		C₂₀H₁₄I₆N₂O₆	606-17-7	1139.761		307 dec				i H₂O; bz; sl EtOH, eth, ace
6265	2-Iodoacetamide		C₂H₄INO	144-48-9	184.963		93.0				s H₂O; sl tfa
6266	Iodoacetic acid		C₂H₃IO₂	64-69-7	185.948		82.5	dec			s H₂O, EtOH, peth; sl eth, chl
6267	Iodoacetone		C₃H₅IO	3019-04-3	183.975			62¹²	2.17¹⁵		s EtOH
6268	Iodoacetonitrile		C₂H₂IN	624-75-9	166.948			185	2.307²⁵	1.5744²⁰	
6269	Iodoacetylene		C₂HI	14545-08-5	151.933			32			
6270	2-Iodoaniline		C₆H₆IN	615-43-0	219.023	nd (dil al)	56.5(0.4)				sl H₂O; vs EtOH, eth, ace
6271	3-Iodoaniline		C₆H₆IN	626-01-7	219.023	lf	33	145¹⁵		1.6811²⁰	i H₂O; s EtOH, chl

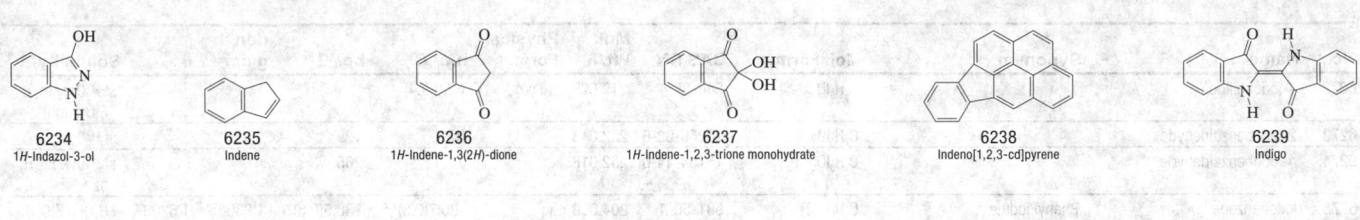

6234
1*H*-Indazol-3-ol

6235
Indene

6236
1*H*-Indene-1,3(2*H*)-dione

6237
1*H*-Indene-1,2,3-trione monohydrate

6238
Indeno[1,2,3-cd]pyrene

6239
Indigo

6240
5,5'-Indigodisulfonic acid, disodium salt

6241
Indocyanine green

6242
1*H*-Indol-5-amine

6243
1*H*-Indole

6244
1*H*-Indole-3-acetic acid

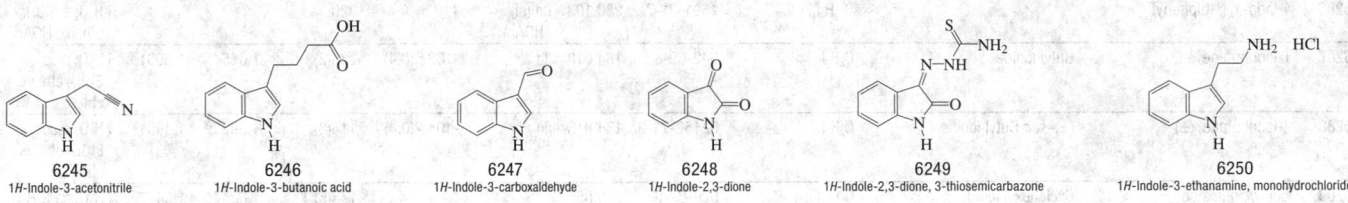

6245
1*H*-Indole-3-acetonitrile

6246
1*H*-Indole-3-butanoic acid

6247
1*H*-Indole-3-carboxaldehyde

6248
1*H*-Indole-2,3-dione

6249
1*H*-Indole-2,3-dione, 3-thiosemicarbazone

6250
1*H*-Indole-3-ethanamine, monohydrochloride

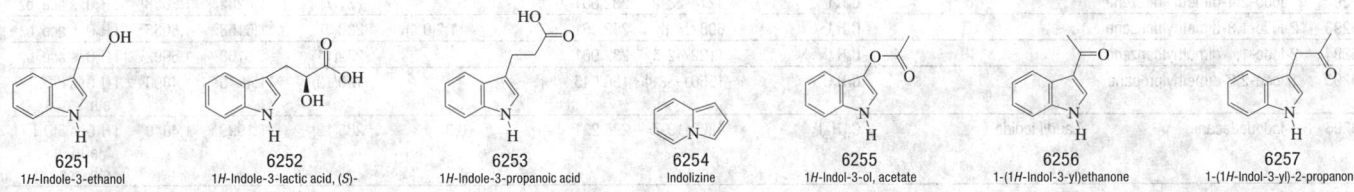

6251
1*H*-Indole-3-ethanol

6252
1*H*-Indole-3-lactic acid, (*S*)-

6253
1*H*-Indole-3-propanoic acid

6254
Indolizine

6255
1*H*-Indol-3-ol, acetate

6256
1-(1*H*-Indol-3-yl)ethanone

6257
1-(1*H*-Indol-3-yl)-2-propanone

6258
3-(1*H*-Indol-3-yl)-2-propenoic acid

6259
Indomethacin

6260
Inosine

6261
Inosine 5'-monophosphate

6262
myo-Inositol

6263
Iocetamic acid

6264
Iodipamide

6265
2-Iodoacetamide

6266
Iodoacetic acid

6267
Iodoacetone

6268
Iodoacetonitrile

6269
Iodoacetylene

6270
2-Iodoaniline

6271
3-Iodoaniline

No.	Name	Synonym	Mol. Form.	CAS RN	Mol. Wt.	Physical Form	mp/°C	bp/°C	den g cm⁻³	n_D	Solubility
6272	4-Iodoaniline		C₆H₆IN	540-37-4	219.023	nd (w)	62.9(0.4)				sl H₂O, peth; s EtOH, eth
6273	2-Iodobenzaldehyde		C₇H₅IO	26260-02-6	232.018		37	129[14]			sl H₂O; s ace
6274	4-Iodobenzaldehyde		C₇H₅IO	15164-44-0	232.018		77.5	265			sl H₂O; s EtOH, bz
6275	Iodobenzene	Phenyl iodide	C₆H₅I	591-50-4	204.008	liq	-30.7(0.5)	188.5(0.6)	1.8308[20]	1.6200[20]	i H₂O; s EtOH; msc eth, ace, bz, ctc
6276	2-Iodobenzenemethanol		C₇H₇IO	5159-41-1	234.034		92	148[32]		1.6349[20]	
6277	4-Iodobenzenesulfonyl chloride	Pipsyl chloride	C₆H₄ClIO₂S	98-61-3	302.517		85				
6278	2-Iodobenzoic acid		C₇H₅IO₂	88-67-5	248.018	nd (w)	161.6(0.5)	exp	2.25[25]		sl H₂O, ace; vs EtOH, eth
6279	3-Iodobenzoic acid		C₇H₅IO₂	618-51-9	248.018	mcl pr (ace)	186.7(0.8)	329(19)			sl H₂O, eth; vs EtOH
6280	4-Iodobenzoic acid		C₇H₅IO₂	619-58-9	248.018	mcl pr (dil al) lf (sub)	270.6(0.3)	316(17)	2.184[20]		i H₂O; sl EtOH; eth, DMSO
6281	4-Iodobenzonitrile		C₇H₄IN	3058-39-7	229.018		127.5				
6282	2-Iodobenzoyl chloride		C₇H₄ClIO	609-67-6	266.463		38.3	159[27]			
6283	4-Iodobenzoyl chloride		C₇H₄ClIO	1711-02-0	266.463		64.5(0.8)	164[32]			
6284	2-Iodo-1,1'-biphenyl		C₁₂H₉I	2113-51-1	280.103			190[36]	1.5511[25]	1.6620[20]	i H₂O; s EtOH, eth, bz, HOAc
6285	3-Iodo-1,1'-biphenyl		C₁₂H₉I	20442-79-9	280.103		26.5	188[16]	1.5967[25]		
6286	4-Iodo-1,1'-biphenyl		C₁₂H₉I	1591-31-7	280.103	nd (al, HOAc)	113.5	320			i H₂O; s EtOH, eth, bz, HOAc
6287	1-Iodobutane	Butyl iodide	C₄H₉I	542-69-8	184.018	liq	-103.5(0.4)	130(2)	1.6154[20]	1.5001[20]	i H₂O; msc EtOH, eth; vs chl
6288	2-Iodobutane, (±)-	(±)-sec-Butyl iodide	C₄H₉I	52152-71-3	184.018	liq	-104.2(0.4)	118(3)	1.5920[20]	1.4991[20]	i H₂O; msc EtOH, eth; vs chl
6289	Iodocyclohexane	Cyclohexyl iodide	C₆H₁₁I	626-62-0	210.055			178(5)	1.6244[20]	1.5477[20]	i H₂O; s EtOH, eth, ace, bz
6290	Iodocyclopentane	Cyclopentyl iodide	C₅H₉I	1556-18-9	196.029			166.5	1.7096[20]	1.5447[20]	i H₂O; s eth, bz; sl ctc
6291	1-Iododecane		C₁₀H₂₁I	2050-77-3	268.178	liq	-16.3	263.7	1.2546[20]	1.4858[20]	i H₂O; s EtOH, eth, ctc
6292	1-Iodo-2,4-dimethylbenzene		C₈H₉I	4214-28-2	232.061			231(7)	1.6282[16]	1.6008[16]	i H₂O; s ace, bz
6293	2-Iodo-1,3-dimethylbenzene		C₈H₉I	608-28-6	232.061	oil	11.2(0.4)	228(3)	1.6158[20]	1.6035[20]	i H₂O; s ace, bz
6294	2-Iodo-1,4-dimethylbenzene		C₈H₉I	1122-42-5	232.061			232(11)	1.6168[17]	1.5992[17]	i H₂O; s ace, bz
6295	1-Iodo-2,2-dimethylpropane		C₅H₁₁I	15501-33-4	198.045			130(13)	1.4940[20]	1.4890[20]	i H₂O; s EtOH, eth
6296	1-Iodododecane	Lauryl iodide	C₁₂H₂₅I	4292-19-7	296.231		0.3	285(12)	1.1999[20]	1.4840[20]	i H₂O; s EtOH, MeOH; msc eth, ace, ctc
6297	Iodoethane	Ethyl iodide	C₂H₅I	75-03-6	155.965	liq	-111.0(0.4)	72(1)	1.9357[20]	1.5133[20]	sl H₂O; msc EtOH; s eth, chl
6298	2-Iodoethanol		C₂H₅IO	624-76-0	171.964			155(15)	2.1967[20]	1.5713[20]	vs H₂O, eth, EtOH
6299	Iodoethene	Vinyl iodide	C₂H₃I	593-66-8	153.949			57(4)	2.037[20]	1.5385[20]	vs eth, EtOH
6300	(2-Iodoethyl)benzene		C₈H₉I	17376-04-4	232.061	liq		122[13]	1.603	1.6010[20]	
6301	2-(1-Iodoethyl)-1,3-dioxolane-4-methanol	Iodinated glycerol	C₆H₁₁IO₃	5634-39-9	258.053	pale ye liq			1.797	1.547	s eth, chl, thf, AcOEt
6302	Iodofenphos		C₈H₈Cl₂IO₃PS	18181-70-9	412.997	wh cry	76				i H₂O; s ace, xyl; sl EtOH
6303	1-Iodoheptane		C₇H₁₅I	4282-40-0	226.098	liq	-48.2	201(6)	1.3719[25]	1.4904[20]	i H₂O; s EtOH, eth, ace, chl; sl ctc
6304	3-Iodoheptane		C₇H₁₅I	31294-92-5	226.098			89[30]	1.3676[20]		
6305	1-Iodohexadecane		C₁₆H₃₃I	544-77-4	352.337	pa ye liq	20(3)	357	1.1213[25]	1.4797[20]	i H₂O; sl EtOH; s eth, ace; msc bz; vs chl
6306	1-Iodohexane	Hexyl iodide	C₆H₁₃I	638-45-9	212.071	liq	-74.1(0.6)	182(3)	1.4305[25]	1.4928[20]	i H₂O
6307	Iodomethane	Methyl iodide	CH₃I	74-88-4	141.939	liq	-66(2)	42.4(0.2)	2.2789[20]	1.5308[20]	sl H₂O; s ace, bz, chl; msc EtOH, eth
6308	1-Iodo-2-methoxybenzene	o-Iodoanisole	C₇H₇IO	529-28-2	234.034			241	1.8[20]		vs EtOH, eth, ace, bz, chl, lig
6309	1-Iodo-3-methoxybenzene	m-Iodoanisole	C₇H₇IO	766-85-8	234.034			244.5	1.9650[20]		vs EtOH, eth
6310	1-Iodo-4-methoxybenzene	p-Iodoanisole	C₇H₇IO	696-62-8	234.034	lf (al), nd (MeOH)	53	247(4)			s EtOH, eth, chl

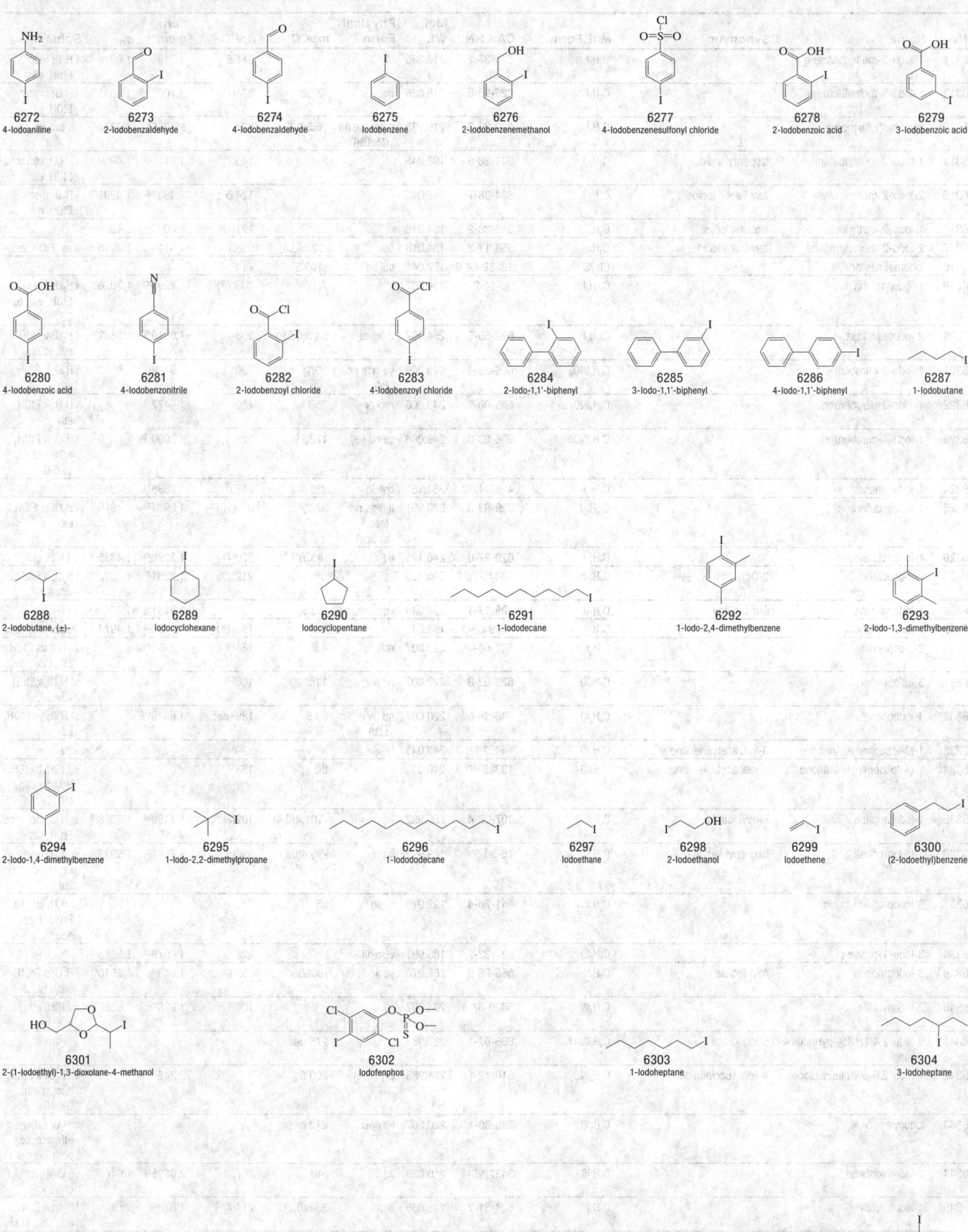

6272
4-Iodoaniline

6273
2-Iodobenzaldehyde

6274
4-Iodobenzaldehyde

6275
Iodobenzene

6276
2-Iodobenzenemethanol

6277
4-Iodobenzenesulfonyl chloride

6278
2-Iodobenzoic acid

6279
3-Iodobenzoic acid

6280
4-Iodobenzoic acid

6281
4-Iodobenzonitrile

6282
2-Iodobenzoyl chloride

6283
4-Iodobenzoyl chloride

6284
2-Iodo-1,1'-biphenyl

6285
3-Iodo-1,1'-biphenyl

6286
4-Iodo-1,1'-biphenyl

6287
1-Iodobutane

6288
2-Iodobutane, (±)-

6289
Iodocyclohexane

6290
Iodocyclopentane

6291
1-Iododecane

6292
1-Iodo-2,4-dimethylbenzene

6293
2-Iodo-1,3-dimethylbenzene

6294
2-Iodo-1,4-dimethylbenzene

6295
1-Iodo-2,2-dimethylpropane

6296
1-Iodododecane

6297
Iodoethane

6298
2-Iodoethanol

6299
Iodoethene

6300
(2-Iodoethyl)benzene

6301
2-(1-Iodoethyl)-1,3-dioxolane-4-methanol

6302
Iodofenphos

6303
1-Iodoheptane

6304
3-Iodoheptane

6305
1-Iodohexadecane

6306
1-Iodohexane

6307
Iodomethane

6308
1-Iodo-2-methoxybenzene

6309
1-Iodo-3-methoxybenzene

6310
1-Iodo-4-methoxybenzene

No.	Name	Synonym	Mol. Form.	CAS RN	Mol. Wt.	Physical Form	mp/°C	bp/°C	den g cm⁻³	n_D	Solubility
6311	1-Iodo-2-methylbenzene		C₇H₇I	615-37-2	218.035			211.5	1.713²⁰	1.6079²⁰	i H₂O; msc EtOH, eth
6312	1-Iodo-3-methylbenzene		C₇H₇I	625-95-6	218.035	liq	-27.2	218(19)	1.705²⁰	1.6053²⁰	i H₂O; msc EtOH, eth
6313	(Iodomethyl)benzene		C₇H₇I	620-05-3	218.035	col or ye nd (MeOH)	26.3(0.2)	93¹⁰	1.7335²⁵	1.6334²⁵	vs bz, eth, EtOH
6314	1-Iodo-3-methylbutane	Isopentyl iodide	C₅H₁₁I	541-28-6	198.045			145(2)	1.5118²⁰	1.4939²⁰	sl H₂O, ctc; msc EtOH, eth
6315	2-Iodo-2-methylbutane	tert-Pentyl iodide	C₅H₁₁I	594-38-7	198.045			124.5	1.4937²⁰	1.4981²⁰	i H₂O; msc EtOH, eth
6316	1-Iodo-2-methylpropane	Isobutyl iodide	C₄H₉I	513-38-2	184.018			121(1)	1.6035²⁰	1.4959²⁰	
6317	2-Iodo-2-methylpropane	tert-Butyl iodide	C₄H₉I	558-17-8	184.018	liq	-33.6(0.4)	98(6)	1.571²⁵	1.4918²⁰	msc EtOH, eth
6318	Iodomethylsilane		CH₃ISi	18089-64-0	172.041	col liq	-109.5	71.8			
6319	1-Iodonaphthalene		C₁₀H₇I	90-14-2	254.067		7(1)	303(77)	1.7399²⁰	1.7026²⁰	i H₂O; msc EtOH, eth, bz, CS₂
6320	2-Iodonaphthalene		C₁₀H₇I	612-55-5	254.067	lf (dil al)	54.3(0.5)	308	1.6319⁹⁹	1.6662⁹⁹	i H₂O; vs EtOH, eth, HOAc
6321	1-Iodo-2-nitrobenzene		C₆H₄INO₂	609-73-4	249.006	ye orth nd (al)	50(1)	290	1.9186⁷⁵		i H₂O; s EtOH, eth
6322	1-Iodo-3-nitrobenzene		C₆H₄INO₂	645-00-1	249.006	mcl pr	36(1)	280	1.9477⁵⁰		i H₂O; s EtOH, eth
6323	1-Iodo-4-nitrobenzene		C₆H₄INO₂	636-98-6	249.006	ye nd (al)	173(2)	288	1.8090¹⁵⁵		i H₂O; s EtOH, HOAc; sl DMSO
6324	1-Iodononane		C₉H₁₉I	4282-42-2	254.151	col liq	-20	245.0	1.2836²⁵	1.4848²⁵	
6325	1-Iodooctadecane		C₁₈H₃₇I	629-93-6	380.391	lf (lig), nd (ace, al-ace)	34(2)	379(20)	1.0994²⁰	1.4810²⁰	i H₂O; sl EtOH, eth
6326	1-Iodooctane		C₈H₁₇I	629-27-6	240.125	liq	-45.7(0.5)	226(4)	1.3298²⁰	1.4885²⁰	s EtOH, eth
6327	2-Iodooctane, (±)-	2-Octyl iodide, (±)	C₈H₁₇I	36049-78-2	240.125			212(12)	1.3251²⁰	1.4896²⁰	i H₂O; s EtOH, eth, lig
6328	1-Iodopentane	Amyl iodide	C₅H₁₁I	628-17-1	198.045	liq	-85.6	156(8)	1.5161²⁰	1.4959²⁰	s chl
6329	3-Iodopentane		C₅H₁₁I	1809-05-8	198.045			141(11)	1.5176²⁰	1.4974²⁰	vs ace, bz, eth
6330	2-Iodophenol		C₆H₅IO	533-58-4	220.007	nd	43	186¹⁶⁰	1.8757⁸⁰		s H₂O; vs EtOH, eth, CS₂
6331	3-Iodophenol		C₆H₅IO	626-02-8	220.007	nd (lig)	118	186¹⁰⁰			sl H₂O; s EtOH, eth
6332	4-Iodophenol		C₆H₅IO	540-38-5	220.007	nd (w or sub)	93.5	139⁵ dec	1.8573¹¹²		sl H₂O; vs EtOH, eth
6333	1-(3-Iodophenyl)ethanone	3-Iodoacetophenone	C₈H₇IO	14452-30-3	246.045			129⁸		1.622²⁰	s bz
6334	1-(4-Iodophenyl)ethanone	4-Iodoacetophenone	C₈H₇IO	13329-40-3	246.045		86	153¹⁸			s EtOH, bz, CS₂, HOAc; sl lig, eth
6335	1-Iodopropane	Propyl iodide	C₃H₇I	107-08-4	169.992	liq	-101.4(0.4)	102(2)	1.7489²⁰	1.5058²⁰	sl H₂O, ctc; msc EtOH, eth
6336	2-Iodopropane	Isopropyl iodide	C₃H₇I	75-30-9	169.992	liq	-90.4(0.9)	89(3)	1.7042²⁰	1.5028²⁰	sl H₂O; msc EtOH, eth, bz, chl
6337	3-Iodopropanoic acid		C₃H₅IO₂	141-76-4	199.975	lf (w)	85				sl H₂O, chl; vs EtOH; s eth, ace
6338	3-Iodo-1-propanol		C₃H₇IO	627-32-7	185.991	visc oil		226	1.9976²⁰	1.5585²⁰	
6339	3-Iodopropene	Allyl iodide	C₃H₅I	556-56-9	167.976	ye liq	-98(2)	102(9)	1.848¹²	1.5540²¹	i H₂O; s EtOH, eth, chl
6340	2-Iodopyridine		C₅H₄IN	5029-67-4	204.997			100¹⁵	1.928²⁵	1.6366²⁰	s EtOH, eth, ace, bz
6341	5-Iodo-2,4(1H,3H)-pyrimidine-dione	5-Iodouracil	C₄H₃IN₂O₂	696-07-1	237.983		275 dec				
6342	1-Iodo-2,5-pyrrolidinedione	N-Iodosuccinimide	C₄H₄INO₂	516-12-1	224.985	cry (ace)	200.5		2.245²⁵		vs H₂O; s EtOH, ace; sl eth, DMSO
6343	Iodosylbenzene		C₆H₅IO	536-80-1	220.007	ye pow	210 exp				s H₂O, EtOH; i eth, ace, bz, peth
6344	2-Iodothiophene		C₄H₃IS	3437-95-4	210.036	liq	-40	167(8)	2.0595²⁵	1.6465²⁵	vs EtOH, eth; sl chl
6345	4-Iodotoluene		C₇H₇I	624-31-7	218.035	lf (al)	33.6(0.2)	214(4)	1.678²⁰		i H₂O; s EtOH, eth, CS₂; sl chl
6346	L-3-Iodotyrosine		C₉H₁₀INO₃	70-78-0	307.084	cry (w)	205 dec				
6347	trans-α-Ionone, (±)-		C₁₃H₂₀O	30685-95-1	192.297			146²⁸	0.9298²¹	1.5041²⁰	vs ace, eth, EtOH

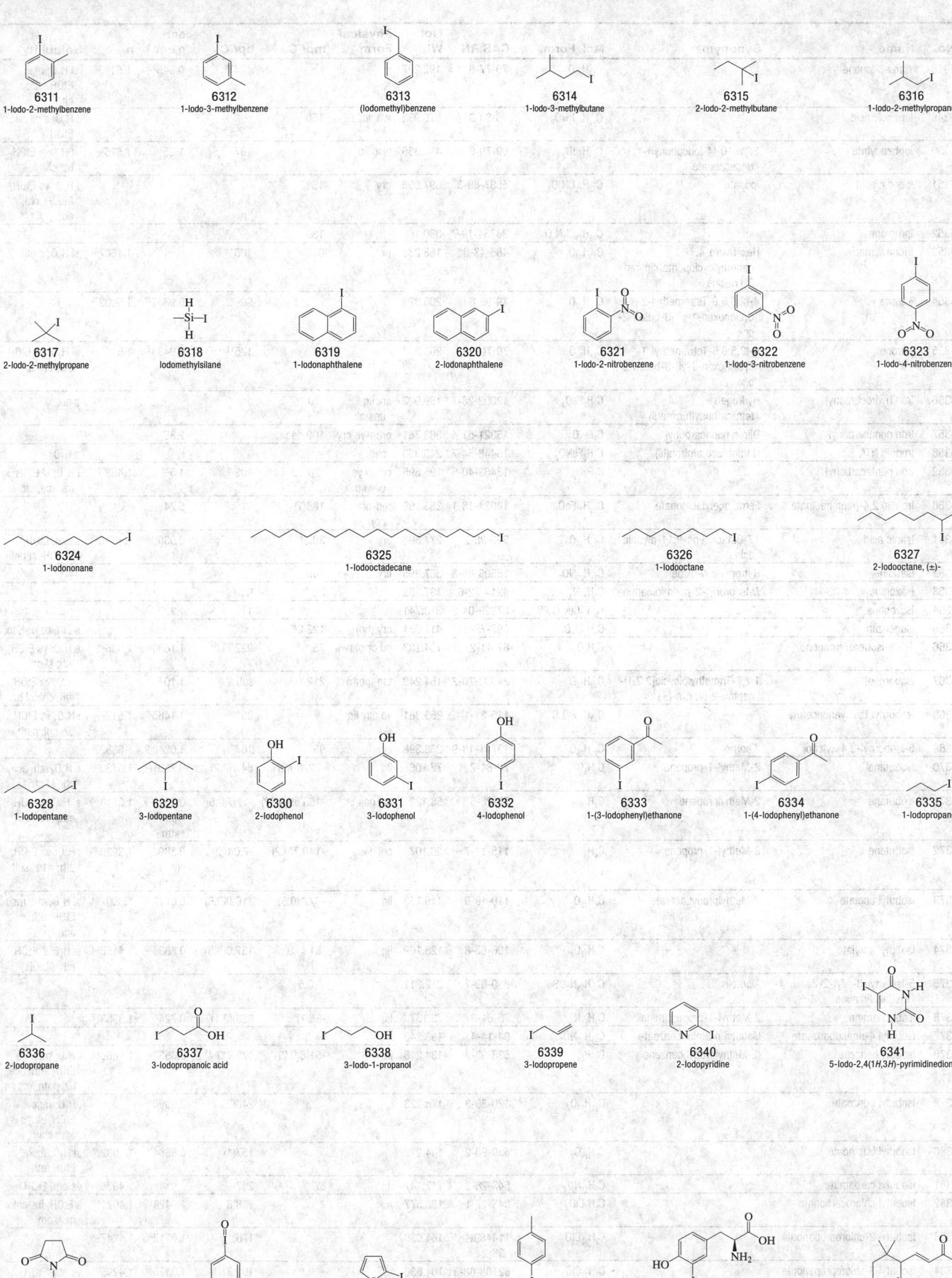

6311
1-iodo-2-methylbenzene

6312
1-iodo-3-methylbenzene

6313
(Iodomethyl)benzene

6314
1-iodo-3-methylbutane

6315
2-iodo-2-methylbutane

6316
1-iodo-2-methylpropane

6317
2-iodo-2-methylpropane

6318
Iodomethylsilane

6319
1-Iodonaphthalene

6320
2-Iodonaphthalene

6321
1-Iodo-2-nitrobenzene

6322
1-Iodo-3-nitrobenzene

6323
1-Iodo-4-nitrobenzene

6324
1-Iodononane

6325
1-Iodooctadecane

6326
1-Iodooctane

6327
2-Iodooctane, (±)-

6328
1-Iodopentane

6329
3-Iodopentane

6330
2-Iodophenol

6331
3-Iodophenol

6332
4-Iodophenol

6333
1-(3-Iodophenyl)ethanone

6334
1-(4-Iodophenyl)ethanone

6335
1-Iodopropane

6336
2-Iodopropane

6337
3-Iodopropanoic acid

6338
3-Iodo-1-propanol

6339
3-Iodopropene

6340
2-Iodopyridine

6341
5-Iodo-2,4(1H,3H)-pyrimidinedione

6342
1-Iodo-2,5-pyrrolidinedione

6343
Iodosylbenzene

6344
2-Iodothiophene

6345
4-Iodotoluene

6346
L-3-Iodotyrosine

6347
trans-α-Ionone, (±)-

No.	Name	Synonym	Mol. Form.	CAS RN	Mol. Wt.	Physical Form	mp/°C	bp/°C	den g cm⁻³	n_D	Solubility
6348	*trans*-β-Ionone		$C_{13}H_{20}O$	79-77-6	192.297			124[10]	0.945[20]	1.5198[20]	sl H_2O; msc EtOH, eth; s chl
6349	Iopanoic acid		$C_{11}H_{12}I_3NO_2$	96-83-3	570.932	wh solid	156				i H_2O; s dil alk, EtOH
6350	Iophendylate	Ethyl 10-(4-iodophenyl)-undecanoate	$C_{19}H_{29}IO_2$	99-79-6	416.336	visc liq		197[1]	1.25[20]	1.525[25]	sl H_2O; s EtOH, bz, chl
6351	Iopodic acid	Ipodate	$C_{12}H_{13}I_3N_2O_2$	5587-89-3	597.956	cry	168				i H_2O; vs EtOH, MeOH, chl, ace
6352	Iprodione		$C_{13}H_{13}Cl_2N_3O_3$	36734-19-7	330.166		136				
6353	Iridomyrmecin	Hexahydro-4,7-dimethylcyclopenta[*c*]pyran-3(1*H*)-one	$C_{10}H_{16}O_2$	485-43-8	168.233	pr	61	106[1.5]		1.4607[65]	sl H_2O; s eth
6354	α-Irone	4-(2,5,6,6-Tetramethyl-2-cyclohexen-1-yl)-3-buten-2-one	$C_{14}H_{22}O$	79-69-6	206.324			90[0.4]	0.9362[20]	1.5002[20]	
6355	β-Irone	4-(2,5,6,6-Tetramethyl-1-cyclohexen-1-yl)-3-buten-2-one	$C_{14}H_{22}O$	79-70-9	206.324			125[11]	0.9434[21]	1.5162[25]	sl H_2O; vs EtOH, eth, bz, chl
6356	Iron hydrocarbonyl	Hydrogen tetracarbonylferrate(II)	$C_4H_2FeO_4$	12002-28-7	169.902	col liq; unstab	-70	dec			s alk
6357	Iron nonacarbonyl	Diiron nonacarbonyl	$C_9Fe_2O_9$	15321-51-4	363.781	oran-ye cry	100 dec		2.85		
6358	Iron(III) NTA	Nitrilotriacetatoiron(III)	$C_6H_6FeNO_6$	16448-54-7	243.960	solid					s H_2O
6359	Iron pentacarbonyl		C_5FeO_5	13463-40-6	195.896	col to ye oily liq	-20	103	1.5[20]	1.453[22]	i H_2O; sl EtOH; s bz, ace, ctc
6360	Iron(III) 2,4-pentanedioate	Ferric acetylacetonate	$C_{15}H_{21}FeO_6$	14024-18-1	353.169	red-oran cry	188(1)		5.24		
6361	Isanic acid	17-Octadecene-9,11-diynoic acid	$C_{18}H_{26}O_2$	506-25-2	274.398	cry	39.5		0.9309[45]	1.49148[50]	s ace, EtOH, i-PrOH; sl peth
6362	Isatidine	Retrorsine *N*-oxide	$C_{18}H_{25}NO_7$	15503-86-3	367.395	cry	145				
6363	Isaxonine	*N*-Isopropyl-2-pyrimidinamine	$C_7H_{11}N_3$	4214-72-6	137.182		28	93[12]			
6364	Isazophos		$C_9H_{17}ClN_3O_3PS$	67329-04-8	313.741			170	1.22[20]		
6365	Isobenzan		$C_9H_4Cl_8O$	297-78-9	411.751	cry (hp)	122.2(0.3)				s eth, bz, xyl, tol
6366	1(3*H*)-Isobenzofuranone		$C_8H_6O_2$	87-41-2	134.133	nd or pl (w)	75	292(2)	1.1636[99]	1.536[99]	s H_2O; vs EtOH, eth; sl chl
6367	Isoborneol	1,7,7-Trimethylbicyclo[2.2.1]-heptan-2-ol, *exo*-(±)	$C_{10}H_{18}O$	24393-70-2	154.249	tab (peth)	212	sub	1.10[20]		i H_2O; vs EtOH, eth, chl; sl bz
6368	Isobornyl thiocyanoacetate		$C_{13}H_{19}NO_2S$	115-31-1	253.361	ye oily liq		95[0.06]	1.1465[25]	1.512[25]	i H_2O; vs EtOH, bz, chl, peth
6369	6-Isobornyl-3,4-xylenol	Xibornol	$C_{18}H_{26}O$	13741-18-9	258.398	cry	95	167[3]	1.0240[20]	1.5382[20]	
6370	Isobutanal	2-Methyl-1-propanal	C_4H_8O	78-84-2	72.106	liq	-72.1(0.2)	64.1(0.2)	0.7891[20]	1.3730[20]	s H_2O, eth, ace, chl; sl ctc
6371	Isobutane	2-Methylpropane	C_4H_{10}	75-28-5	58.122	col gas	-159.59(0.02)	-11.7(0.5)	0.5510[25] (p>1 atm)	1.3518[-25]	sl H_2O; s EtOH, eth, chl
6372	Isobutene	2-Methyl-1-propene	C_4H_8	115-11-7	56.107	col gas	-140.7(0.2)	-7.0(0.2)	0.589[25] (p>1 atm)	1.3926[-25]	i H_2O; vs EtOH, eth; s bz, sulf
6373	Isobutyl acetate	2-Methylpropyl acetate	$C_6H_{12}O_2$	110-19-0	116.158	liq	-97.1(0.5)	116.9(0.6)	0.8712[20]	1.3902[20]	sl H_2O, ctc; msc EtOH, eth; s ace
6374	Isobutyl acrylate		$C_7H_{12}O_2$	106-63-8	128.169	liq	-61	137.0(0.8)	0.8896[20]	1.4150[20]	sl H_2O; s EtOH, eth, MeOH
6375	5-Isobutyl-3-allyl-2-thioxo-4-imidazolidinone	Albutoin	$C_{10}H_{16}N_2OS$	830-89-7	212.311		210.5				
6376	Isobutylamine	2-Methyl-1-propanamine	$C_4H_{11}N$	78-81-9	73.137	liq	-86(1)	68.8(0.1)	0.724[25]	1.3988[19]	
6377	Isobutyl 4-aminobenzoate	Isobutyl *p*-aminobenzoate	$C_{11}H_{15}NO_2$	94-14-4	193.243		64.5				
6378	Isobutylbenzene	(2-Methylpropyl)benzene	$C_{10}H_{14}$	538-93-2	134.218	liq	-51.6(0.2)	172.7(0.4)	0.8532[20]	1.4866[20]	i H_2O; msc EtOH, eth, ace, bz, peth, ctc
6379	Isobutyl benzoate		$C_{11}H_{14}O_2$	120-50-3	178.228			240(2)	0.9990[20]		i H_2O; msc EtOH, eth; s ace, chl
6380	Isobutyl butanoate		$C_8H_{16}O_2$	539-90-2	144.212			157(1)	0.8364[18]	1.4032[20]	sl H_2O; msc EtOH, eth
6381	Isobutyl carbamate		$C_5H_{11}NO_2$	543-28-2	117.147	lf	67	207		1.4098[76]	vs eth, EtOH
6382	Isobutyl chlorocarbonate		$C_5H_9ClO_2$	543-27-1	136.577			128.8	1.0426[18]	1.4071[18]	s EtOH, bz, chl; msc eth
6383	Isobutyl 2-chloropropanoate		$C_7H_{13}ClO_2$	114489-96-2	164.630			176	1.0312[20]	1.4247[20]	
6384	Isobutyl 3-chloropropanoate		$C_7H_{13}ClO_2$	62108-68-3	164.630			191.3	1.0323[20]	1.4295[20]	vs eth, EtOH

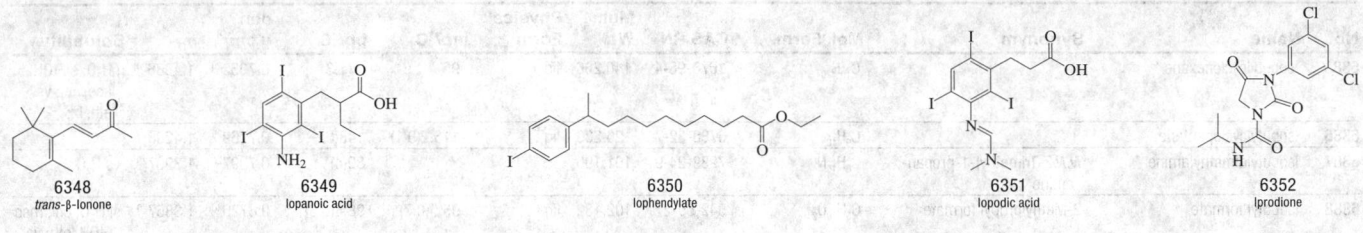

6348
trans-β-Ionone

6349
Iopanoic acid

6350
Iophendylate

6351
Iopodic acid

6352
Iprodione

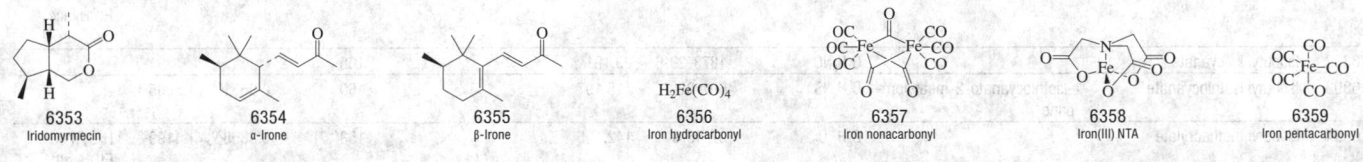

6353
Iridomyrmecin

6354
α-Irone

6355
β-Irone

6356
Iron hydrocarbonyl

6357
Iron nonacarbonyl

6358
Iron(III) NTA

6359
Iron pentacarbonyl

6360
Iron(III) 2,4-pentanedioate

6361
Isanic acid

6362
Isatidine

6363
Isaxonine

6364
Isazophos

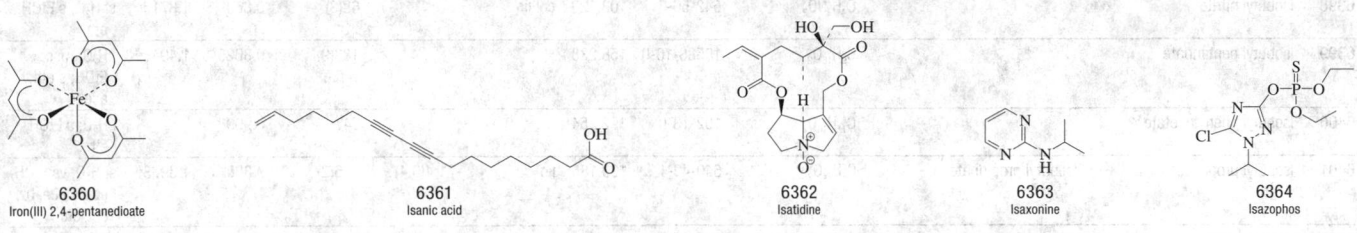

6365
Isobenzan

6366
1(3H)-Isobenzofuranone

6367
Isoborneol

6368
Isobornyl thiocyanoacetate

6369
6-Isobornyl-3,4-xylenol

6370
Isobutanal

6371
Isobutane

6372
Isobutene

6373
Isobutyl acetate

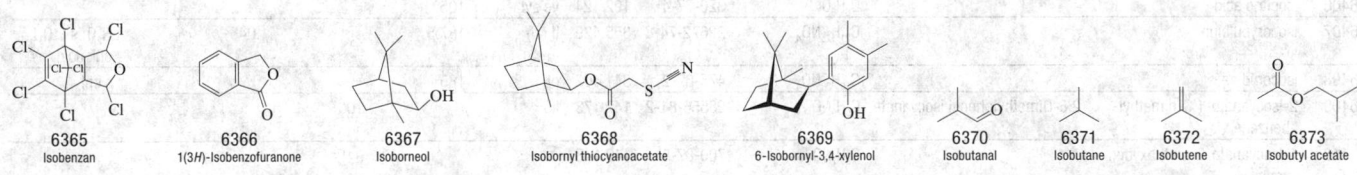

6374
Isobutyl acrylate

6375
5-Isobutyl-3-allyl-2-thioxo-4-imidazolidinone

6376
Isobutylamine

6377
Isobutyl 4-aminobenzoate

6378
Isobutylbenzene

6379
Isobutyl benzoate

6380
Isobutyl butanoate

6381
Isobutyl carbamate

6382
Isobutyl chlorocarbonate

6383
Isobutyl 2-chloropropanoate

6384
Isobutyl 3-chloropropanoate

No.	Name	Synonym	Mol. Form.	CAS RN	Mol. Wt.	Physical Form	mp/°C	bp/°C	den g cm⁻³	n_D	Solubility
6385	Isobutylcyclohexane		$C_{10}H_{20}$	1678-98-4	140.266	liq	-95	171.3	0.7952[20]	1.4386[20]	i H_2O; s EtOH, ace, chl; vs eth, bz
6386	Isobutylcyclopentane		C_9H_{18}	3788-32-7	126.239	liq	-115.2(0.1)	148(5)	0.7769[25]	1.4298[20]	
6387	Isobutyldimethylamine	N,N,2-Trimethyl-1-propan-amine	$C_6H_{15}N$	7239-24-9	101.190			80(4)	0.7097[20]	1.3907[20]	vs H_2O
6388	Isobutyl formate	2-Methylpropyl formate	$C_5H_{10}O_2$	542-55-2	102.132	liq	-95.5(0.7)	98.4(0.3)	0.8776[20]	1.3857[20]	sl H_2O, chl; msc EtOH, eth; vs ace
6389	Isobutyl heptanoate	Isobutyl enanthate	$C_{11}H_{22}O_2$	7779-80-8	186.292			208	0.8593[20]		vs ace, bz, eth, EtOH
6390	Isobutyl 2-hydroxybenzoate	Isobutyl salicylate	$C_{11}H_{14}O_3$	87-19-4	194.227		5.9	261	1.0639[20]	1.5087[20]	i H_2O; s EtOH, eth, ctc
6391	Isobutyl isobutanoate		$C_8H_{16}O_2$	97-85-8	144.212	liq	-80.6(0.2)	148(3)	0.8542[20]	1.3999[20]	sl H_2O, ctc; s EtOH, ace; msc eth
6392	Isobutyl isocyanate		C_5H_9NO	1873-29-6	99.131			106			
6393	Isobutyl isothiocyanate	1-Isothiocyanato-2-methylpro-pane	C_5H_9NS	591-82-2	115.197			160	0.9631[14]	1.5005[14]	
6394	Isobutyl methacrylate		$C_8H_{14}O_2$	97-86-9	142.196			153(17)	0.8858[20]	1.4199[20]	i H_2O; msc EtOH, eth
6395	Isobutyl 3-methylbutanoate	Isobutyl isovalerate	$C_9H_{18}O_2$	589-59-3	158.238			169(3)	0.853[20]	1.4057[20]	i H_2O; msc EtOH, eth; vs ace; s chl
6396	Isobutyl methyl ether		$C_5H_{12}O$	625-44-5	88.148			58.6	0.7311[20]		vs eth, EtOH
6397	Isobutyl nitrate		$C_4H_9NO_3$	543-29-3	119.119			123.4	1.0152[20]	1.4028[20]	
6398	Isobutyl nitrite		$C_4H_9NO_2$	542-56-3	103.120	col liq		68(3)	0.8699[22]	1.3715[22]	sl H_2O; s EtOH, eth
6399	Isobutyl pentanoate		$C_9H_{18}O_2$	10588-10-0	158.238			183(9)	0.8625[25]	1.4046[20]	i H_2O; msc EtOH; s eth, ace
6400	Isobutyl phenylacetate		$C_{12}H_{16}O_2$	102-13-6	192.254			247	0.999[18]		i H_2O; s EtOH, eth
6401	Isobutyl propanoate	Isobutyl propionate	$C_7H_{14}O_2$	540-42-1	130.185	liq	-71.4(0.4)	136(2)	0.888[0]	1.3973[20]	sl H_2O; vs EtOH, eth; s ace, bz, chl, ctc
6402	Isobutyl stearate		$C_{22}H_{44}O_2$	646-13-9	340.583	wax	28.9	223[15]	0.8498[20]		vs eth
6403	Isobutyl thiocyanate		C_5H_9NS	591-84-4	115.197	liq	-58.9(0.5)	175(7)			vs eth, EtOH
6404	Isobutyl trichloroacetate		$C_6H_9Cl_3O_2$	33560-15-5	219.493			188	1.2636[20]	1.4483[20]	vs bz, eth, EtOH
6405	Isobutyl vinyl ether		$C_6H_{12}O$	109-53-5	100.158	liq	-112	83(3)	0.7645[20]	1.3966[20]	sl H_2O; vs EtOH, ace, bz; msc eth
6406	Isocitric acid		$C_6H_8O_7$	320-77-4	192.124	ye syr	105				
6407	Isocorybulbine		$C_{21}H_{25}NO_4$	22672-74-8	355.429	lf (al)	187.5		1.045[20]		i H_2O; s EtOH, chl, acid
6408	Isocorydine		$C_{20}H_{23}NO_4$	475-67-2	341.402	pl	185				vs chl
6409	2-Isocyanato-1,3-dimethyl-benzene	2,6-Dimethylphenyl isocyanate	C_9H_9NO	28556-81-2	147.173	liq		100[13]			
6410	1-Isocyanato-2-methoxyben-zene		$C_8H_7NO_2$	700-87-8	149.148			94[17]			
6411	1-Isocyanato-3-methoxyben-zene		$C_8H_7NO_2$	18908-07-1	149.148			102[15]			
6412	1-Isocyanato-2-methylben-zene	2-Tolyl isocyanate	C_8H_7NO	614-68-6	133.148			185		1.5282[20]	i H_2O; s eth
6413	1-Isocyanato-3-methylben-zene		C_8H_7NO	621-29-4	133.148			196.5	1.0330[20]		vs bz, eth
6414	1-Isocyanato-4-methylben-zene		C_8H_7NO	622-58-2	133.148			187			vs bz, eth
6415	2-Isocyanato-2-methylpro-pane	tert-Butyl isocyanate	C_5H_9NO	1609-86-5	99.131		85.5		0.8670[7]	1.4061[20]	
6416	1-Isocyanatonaphthalene	1-Naphthyl isocyanate	$C_{11}H_7NO$	86-84-0	169.180			269	1.1774[20]		s eth, bz
6417	1-Isocyanato-2-nitrobenzene	2-Nitrophenyl isocyanate	$C_7H_4N_2O_3$	3320-86-3	164.118	wh nd (peth)	41	137[18]			vs bz, eth, chl
6418	1-Isocyanato-3-nitrobenzene	3-Nitrophenyl isocyanate	$C_7H_4N_2O_3$	3320-87-4	164.118	wh lf (lig)	51	130[11]			vs bz, eth, chl
6419	1-Isocyanato-4-nitrobenzene	4-Nitrophenyl isocyanate	$C_7H_4N_2O_3$	100-28-7	164.118	pa ye nd	57	162[20]			vs bz, eth, chl
6420	2-Isocyanatopropane	Isopropyl isocyanate	C_4H_7NO	1795-48-8	85.105			74.5	0.866[25]	1.3825[20]	
6421	1-Isocyanato-3-(trifluoromethyl)benzene	3-(Trifluoromethyl)phenyl isocyanate	$C_8H_4F_3NO$	329-01-1	187.119			54[11]	1.3455[20]	1.4690[20]	
6422	Isocyanobenzene	Phenyl isocyanide	C_7H_5N	931-54-4	103.122	unstab liq		80[40]	0.98[15]		
6423	Isocyanomethane	Methyl isocyanide	C_2H_3N	593-75-9	41.052		-45(3)	exp	0.756[4]		
6424	(Isocyanomethyl)benzene	Benzyl isocyanide	C_8H_7N	10340-91-7	117.149			199 dec	0.972[15]	1.5193[20]	
6425	2-Isocyanopropane	Isopropyl isocyanide	C_4H_7N	598-45-8	69.106			87	0.7596[25]		i H_2O; msc EtOH, eth

6385 Isobutylcyclohexane

6386 Isobutylcyclopentane

6387 Isobutyldimethylamine

6388 Isobutyl formate

6389 Isobutyl heptanoate

6390 Isobutyl 2-hydroxybenzoate

6391 Isobutyl isobutanoate

6392 Isobutyl isocyanate

6393 Isobutyl isothiocyanate

6394 Isobutyl methacrylate

6395 Isobutyl 3-methylbutanoate

6396 Isobutyl methyl ether

6397 Isobutyl nitrate

6398 Isobutyl nitrite

6399 Isobutyl pentanoate

6400 Isobutyl phenylacetate

6401 Isobutyl propanoate

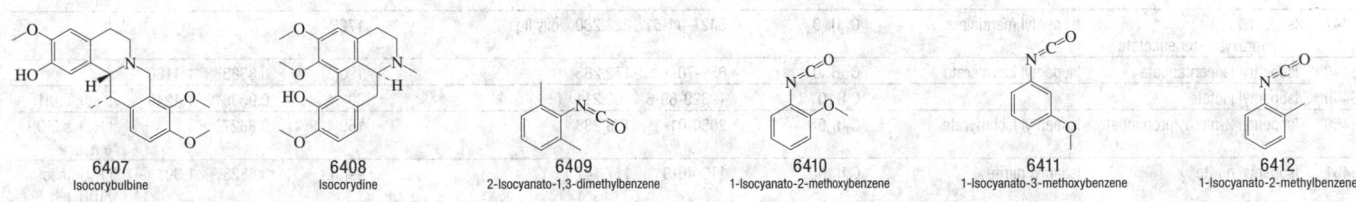

6402 Isobutyl stearate

6403 Isobutyl thiocyanate

6404 Isobutyl trichloroacetate

6405 Isobutyl vinyl ether

6406 Isocitric acid

6407 Isocorybulbine

6408 Isocorydine

6409 2-Isocyanato-1,3-dimethylbenzene

6410 1-Isocyanato-2-methoxybenzene

6411 1-Isocyanato-3-methoxybenzene

6412 1-Isocyanato-2-methylbenzene

6413 1-Isocyanato-3-methylbenzene

6414 1-Isocyanato-4-methylbenzene

6415 2-Isocyanato-2-methylpropane

6416 1-Isocyanatonaphthalene

6417 1-Isocyanato-2-nitrobenzene

6418 1-Isocyanato-3-nitrobenzene

6419 1-Isocyanato-4-nitrobenzene

6420 2-Isocyanatopropane

6421 1-Isocyanato-3-(trifluoromethyl)benzene

6422 Isocyanobenzene

6423 Isocyanomethane

6424 (Isocyanomethyl)benzene

6425 2-Isocyanopropane

No.	Name	Synonym	Mol. Form.	CAS RN	Mol. Wt.	Physical Form	mp/°C	bp/°C	den g cm^{-3}	n_D	Solubility
6426	Isodecyl acrylate		C$_{13}$H$_{24}$O$_2$	1330-61-6	212.329		-100	158^{50}	0.885^{20}	1.4416^{20}	
6427	Isodecyl diphenyl phosphate		C$_{22}$H$_{31}$O$_4$P	29761-21-5	390.452			249^{10} dec			
6428	Isodecyl methacrylate		C$_{14}$H$_{26}$O$_2$	29964-84-9	226.355			126^{10}	0.876^{20}		
6429	8-Isoestrone		C$_{18}$H$_{22}$O$_2$	517-06-6	270.367	pr (MeOH)	254				vs eth, Diox
6430	Isoeugenol		C$_{10}$H$_{12}$O$_2$	97-54-1	164.201			264(6)	1.080^{25}	1.5739^{19}	vs eth, EtOH
6431	Isofenphos		C$_{15}$H$_{24}$NO$_4$PS	25311-71-1	345.395		<-12	120$^{0.01}$	1.134^{20}		
6432	Isoflurophate		C$_6$H$_{14}$FO$_3$P	55-91-4	184.145			62^9	1.055^{25}	1.3830^{25}	sl H$_2$O, lig; s eth; vs oils
6433	1H-Isoindole-1,3(2H)-dione	Phthalimide	C$_8$H$_5$NO$_2$	85-41-6	147.132	nd (w), pr (HOAc) lf (sub)	238(1)				vs bz
6434	Isolan		C$_{10}$H$_{17}$N$_3$O$_2$	119-38-0	211.261	col liq		118$^{2.5}$	1.07^{20}		msc H$_2$O; s EtOH, xyl
6435	DL-Isoleucine		C$_6$H$_{13}$NO$_2$	443-79-8	131.173		292 dec				
6436	L-Isoleucine	2-Amino-3-methylpentanoic acid	C$_6$H$_{13}$NO$_2$	73-32-5	131.173		284 dec				s H$_2$O; i EtOH
6437	Isolongifolene		C$_{15}$H$_{24}$	1135-66-6	204.352	liq		82$^{0.4}$			
6438	Isolysergic acid		C$_{16}$H$_{16}$N$_2$O$_2$	478-95-5	268.310	cry (w+2)	218 dec				sl H$_2$O, EtOH; s py
6439	α-Isomaltose	6-O-α-D-Glucopyranosyl-D-glucose	C$_{12}$H$_{22}$O$_{11}$	499-40-1	342.296		120				
6440	Isoniazid	4-Pyridinecarboxylic acid hydrazide	C$_6$H$_7$N$_3$O	54-85-3	137.139	cry (al)	171.4				vs H$_2$O, EtOH
6441	Isopentane	2-Methylbutane	C$_5$H$_{12}$	78-78-4	72.149	vol liq or gas	-159.8(0.2)	27.83(0.06)	0.6201^{20}	1.3537^{20}	i H$_2$O; msc EtOH, eth
6442	Isopentyl acetate	Isoamyl acetate	C$_7$H$_{14}$O$_2$	123-92-2	130.185	liq	-78.5	141.6(0.7)	0.876^{15}	1.4000^{20}	sl H$_2$O; msc EtOH, eth; s ace, chl
6443	Isopentylbenzene		C$_{11}$H$_{16}$	2049-94-7	148.245			196(7)	0.856^{20}	1.4867^{10}	i H$_2$O; s EtOH, eth; vs bz
6444	Isopentyl butanoate		C$_9$H$_{18}$O$_2$	106-27-4	158.238			184.8(0.3)	0.865^{19}	1.4110^{20}	i H$_2$O; vs EtOH, eth
6445	Isopentyl formate		C$_6$H$_{12}$O$_2$	110-45-2	116.158	liq	-93.5	124.0(0.8)	0.877^{20}	1.3967^{20}	sl H$_2$O, ctc; s EtOH; msc eth
6446	Isopentyl hexanoate	Isopentyl caproate	C$_{11}$H$_{22}$O$_2$	2198-61-0	186.292			225.5	0.861^{20}		i H$_2$O; s EtOH, eth
6447	Isopentyl α-hydroxybenzeneacetate	Isopentyl mandelate	C$_{13}$H$_{18}$O$_3$	5421-04-5	222.280	oily liq		172^{11}			
6448	Isopentyl isopentanoate	Isopentyl isovalerate	C$_{10}$H$_{20}$O$_2$	659-70-1	172.265			190(3)	0.8583^{19}	1.4130^{19}	
6449	Isopentyl lactate		C$_8$H$_{16}$O$_3$	19329-89-6	160.211			202.4	0.9589^{25}	1.4240^{25}	vs eth, EtOH
6450	Isopentyl 2-methylpropanoate	Isopentyl isobutyrate	C$_9$H$_{18}$O$_2$	2050-01-3	158.238			169(4)	0.8627^{20}		sl H$_2$O; s EtOH, eth, ace
6451	Isopentyl nitrite	Isoamyl nitrite	C$_5$H$_{11}$NO$_2$	110-46-3	117.147			99(3)	0.8828^{20}	1.3918^{20}	sl H$_2$O; msc EtOH, eth
6452	Isopentyl pentanoate		C$_{10}$H$_{20}$O$_2$	2050-09-1	172.265			193(6)			
6453	Isopentyl propanoate		C$_8$H$_{16}$O$_2$	105-68-0	144.212			173(4)	0.8697^{20}	1.4069^{20}	vs eth, EtOH
6454	Isopentyl salicylate		C$_{12}$H$_{16}$O$_3$	87-20-7	208.253			278	1.0535^{20}	1.5080^{20}	i H$_2$O; vs EtOH; s eth, chl; sl ctc
6455	Isopentyl trichloroacetate		C$_7$H$_{11}$Cl$_3$O$_2$	57392-55-9	233.520			217	1.2314^{20}	1.4521^{20}	vs eth, EtOH
6456	Isophorone	3,5,5-Trimethyl-2-cyclohexen-1-one	C$_9$H$_{14}$O	78-59-1	138.206	liq	-8.1	214.8(0.7)	0.9255^{20}	1.4766^{18}	
6457	Isophorone diisocyanate		C$_{12}$H$_{18}$N$_2$O$_2$	4098-71-9	222.283	liq	60	217^{100}	1.062^{20}		
6458	Isophthalic acid	1,3-Benzenedicarboxylic acid	C$_8$H$_6$O$_4$	121-91-5	166.132	nd (w, al)	348.0(0.2)	sub			sl H$_2$O; s EtOH, HOAc; i eth, bz, lig
6459	Isopilosine		C$_{16}$H$_{18}$N$_2$O$_3$	491-88-3	286.325	pl (al), pr (w, dil al)	187				vs EtOH
6460	Isopropalin	Benzenamine, 4-(1-methylethyl)-2,6-dinitro-N,N-dipropyl-	C$_{15}$H$_{23}$N$_3$O$_4$	33820-53-0	309.362	red-oran liq					i H$_2$O; s os
6461	Isopropamide iodide		C$_{23}$H$_{33}$IN$_2$O	71-81-8	480.424	cry or pow	190				s H$_2$O, EtOH, MeOH; i chl
6462	Isopropenyl acetate		C$_5$H$_8$O$_2$	108-22-5	100.117	liq	-92.9	97.4(0.9)	0.9090^{20}	1.4033^{20}	sl H$_2$O; s EtOH, chl, ace; vs eth
6463	Isopropenylbenzene	α-Methyl styrene	C$_9$H$_{10}$	98-83-9	118.175	liq	-22.36(0.05)	165.4(0.3)	0.9106^{20}	1.5386^{20}	i H$_2$O; s EtOH, eth; msc ace, bz, ctc
6464	p-Isopropenylisopropylbenzene		C$_{12}$H$_{16}$	2388-14-9	160.255	liq	-30.6	222(1)	0.8936^{20}	1.5238^{20}	vs ace, bz, eth, EtOH
6465	p-Isopropenylstyrene		C$_{11}$H$_{12}$	16262-48-9	144.213	liq		242	0.93	1.5684^{20}	
6466	4-Isopropoxydiphenylamine	4-Isopropoxy-N-phenylaniline	C$_{15}$H$_{17}$NO	101-73-5	227.302		83				

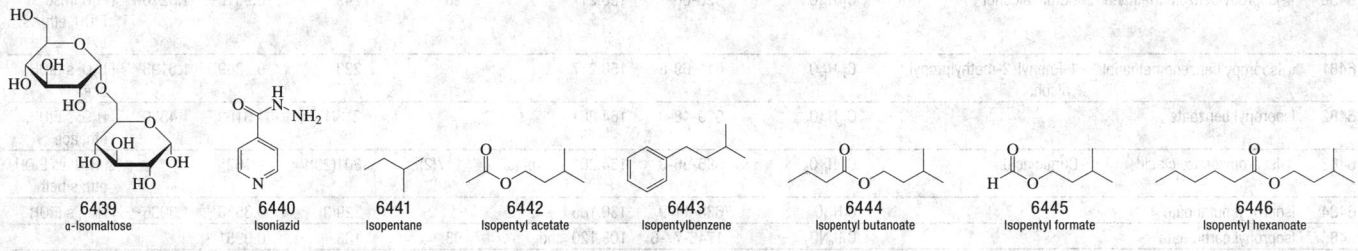

6426 Isodecyl acrylate

6427 Isodecyl diphenyl phosphate

6428 Isodecyl methacrylate

6429 8-Isoestrone

6430 Isoeugenol

6431 Isofenphos

6432 Isoflurophate

6433 1H-Isoindole-1,3(2H)-dione

6434 Isolan

6435 DL-Isoleucine

6436 L-Isoleucine

6437 Isolongifolene

6438 Isolysergic acid

6439 α-Isomaltose

6440 Isoniazid

6441 Isopentane

6442 Isopentyl acetate

6443 Isopentylbenzene

6444 Isopentyl butanoate

6445 Isopentyl formate

6446 Isopentyl hexanoate

6447 Isopentyl α-hydroxybenzeneacetate

6448 Isopentyl isopentanoate

6449 Isopentyl lactate

6450 Isopentyl 2-methylpropanoate

6451 Isopentyl nitrite

6452 Isopentyl pentanoate

6453 Isopentyl propanoate

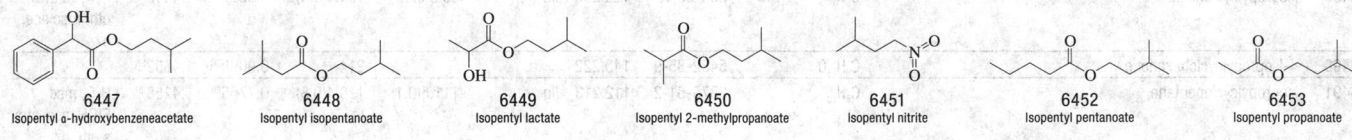

6454 Isopentyl salicylate

6455 Isopentyl trichloroacetate

6456 Isophorone

6457 Isophorone diisocyanate

6458 Isophthalic acid

6459 Isopilosine

6460 Isopropalin

6461 Isopropamide iodide

6462 Isopropenyl acetate

6463 Isopropenylbenzene

6464 p-Isopropenylisopropylbenzene

6465 p-Isopropenylstyrene

6466 4-Isopropoxydiphenylamine

No.	Name	Synonym	Mol. Form.	CAS RN	Mol. Wt.	Physical Form	mp/°C	bp/°C	den g cm⁻³	n_D	Solubility
6467	2-Isopropoxyethanol		C₅H₁₂O₂	109-59-1	104.148			141.6(0.4)	0.9030²⁰	1.4095²⁰	msc H₂O, EtOH, eth; s ace
6468	3-Isopropoxypropanenitrile	1-Cyano-2-isopropoxyethane	C₆H₁₁NO	110-47-4	113.157			65¹⁰			s chl
6469	Isopropyl acetate	1-Methylethyl acetate	C₅H₁₀O₂	108-21-4	102.132	liq	-73.4	88.6(0.2)	0.8662²⁵	1.3746²⁵	s H₂O, EtOH, ace, chl; msc eth
6470	Isopropyl acrylate	Isopropyl 2-propenoate	C₆H₁₀O₂	689-12-3	114.142	liq		51¹⁰³			
6471	Isopropylamine	2-Propanamine	C₃H₉N	75-31-0	59.110	liq	-95.119 (0.001)	31.8(0.2)	0.6891²⁰	1.3742²⁰	msc H₂O, EtOH, eth; vs ace; s bz, chl
6472	Isopropylamine hydrochloride	2-Propanamine hydrochloride	C₃H₁₀ClN	15572-56-2	95.571		164				s DMSO
6473	2-(Isopropylamino)ethanol		C₅H₁₃NO	109-56-8	103.163		128.5	169(4)	0.8970²⁰	1.4395²⁰	msc H₂O, EtOH, eth
6474	2-Isopropylaniline		C₉H₁₃N	643-28-7	135.206			221	0.9760¹²		i H₂O; s eth, bz, ctc
6475	4-Isopropylaniline	Cumidine	C₉H₁₃N	99-88-7	135.206			228(3)	0.953²⁰		
6476	N-Isopropylaniline		C₉H₁₃N	768-52-5	135.206			203	0.9526²⁵	1.5380²⁰	s EtOH, eth, ace, bz
6477	4-Isopropylbenzaldehyde	Cuminaldehyde	C₁₀H₁₂O	122-03-2	148.201			226(5)	0.9755²⁰	1.5301²⁰	i H₂O; s EtOH, eth; sl ctc
6478	Isopropylbenzene	Cumene	C₉H₁₂	98-82-8	120.191	liq	-96.01(0.05)	152.4(0.2)	0.8640²⁵	1.4915²⁰	i H₂O; msc EtOH, eth, ace, bz, peth, ctc
6479	Isopropylbenzene hydroperoxide	Cumene hydroperoxide	C₉H₁₂O₂	80-15-9	152.190	liq		153	1.03²⁰		
6480	4-Isopropylbenzenemethanol	Cumic alcohol	C₁₀H₁₄O	536-60-7	150.217		28	249	0.9818²⁰	1.5210²⁰	i H₂O; msc EtOH, eth; vs bz
6481	α-Isopropylbenzenemethanol	1-Phenyl-2-methylpropyl alcohol	C₁₀H₁₄O	611-69-8	150.217			223	0.9869¹⁴	1.5193¹⁴	i H₂O; s EtOH, ace
6482	Isopropyl benzoate		C₁₀H₁₂O₂	939-48-0	164.201			218(1)	1.0163¹⁵	1.4890²⁰	i H₂O; s EtOH, eth, ace
6483	4-Isopropylbenzoic acid	Cumic acid	C₁₀H₁₂O₂	536-66-3	164.201	tcl pl (al)	117(2)	301(20)	1.162⁴		sl H₂O; vs EtOH, eth; s peth
6484	Isopropyl butanoate		C₇H₁₄O₂	638-11-9	130.185			129(3)	0.8588²⁰	1.3936²⁰	i H₂O; s EtOH
6485	Isopropyl carbamate		C₄H₉NO₂	1746-77-6	103.120	nd	93	183	0.9951⁶⁰		
6486	Isopropyl chloroacetate		C₅H₉ClO₂	105-48-6	136.577			150(5)	1.0888²⁰	1.4382²⁰	vs eth
6487	Isopropyl chloroformate		C₄H₇ClO₂	108-23-6	122.551			105		1.4013²⁰	vs eth
6488	Isopropyl 2-chloropropanoate		C₆H₁₁ClO₂	40058-87-5	150.603			151.5	1.0315²⁰	1.4149²⁰	i H₂O; s EtOH, eth
6489	Isopropylcyclohexane		C₉H₁₈	696-29-7	126.239	liq	-89.9(0.5)	154.4(0.4)	0.8023²⁰	1.4410²⁰	i H₂O; vs EtOH, eth; msc ace, bz
6490	4-Isopropylcyclohexanone		C₉H₁₆O	5432-85-9	140.222			214	0.9099³⁰	1.4552²⁵	
6491	Isopropylcyclopentane		C₈H₁₆	3875-51-2	112.213	liq	-111.5(0.1)	126.4(0.8)	0.7765²⁰	1.4258²⁰	i H₂O; msc EtOH, ace, ctc; s eth, bz
6492	Isopropylcyclopropane		C₆H₁₂	3638-35-5	84.159	liq	-113.00(0.09)	58(2)	0.6936²⁵	1.3865²⁰	
6493	Isopropyl (2,4-dichlorophenoxy)acetate		C₁₁H₁₂Cl₂O₃	94-11-1	263.117		5	140¹	1.26²⁵	1.5209²⁵	
6494	N-Isopropyl-4,4-diphenylcyclohexanamine	Pramiverin	C₂₁H₂₇N	14334-40-8	293.446		70	165⁰·⁰⁵			
6495	Isopropyl dodecanoate	Isopropyl laurate	C₁₅H₃₀O₂	10233-13-3	242.398			196⁶⁰	0.8536²⁰	1.4280²⁵	vs eth, EtOH
6496	Isopropyl formate		C₄H₈O₂	625-55-8	88.106			68(2)	0.8728²⁰	1.3678²⁰	sl H₂O; msc EtOH, eth; vs ace; s chl
6497	Isopropyl 2-furancarboxylate	Isopropyl 2-furanoate	C₈H₁₀O₃	6270-34-4	154.163			198.5	1.0655²⁴	1.4682²⁴	i H₂O; s EtOH, eth, ace, bz
6498	Isopropyl glycidyl ether	(1-Methylethoxy)methyloxirane	C₆H₁₂O₂	4016-14-2	116.158			138(18)	0.9186²⁰		s H₂O, ace, EtOH
6499	4-Isopropylheptane		C₁₀H₂₂	52896-87-4	142.282			161(6)	0.7354²⁵	1.4153²⁰	
6500	Isopropylhydrazine		C₃H₁₀N₂	2257-52-5	74.124	liq		107			s H₂O, bz, EtOH; sl eth
6501	Isopropyl 2-hydroxybenzoate	Isopropyl salicylate	C₁₀H₁₂O₃	607-85-2	180.200			238	1.0729²⁰	1.5065²⁰	i H₂O; msc EtOH, eth
6502	Isopropyl isobutanoate	Isopropyl isobutyrate	C₇H₁₄O₂	617-50-5	130.185			123(9)	0.8471²¹		i H₂O; s EtOH, eth, ace
6503	Isopropyl lactate		C₆H₁₂O₃	617-51-6	132.157			167	0.9980²⁰	1.4082²⁵	vs H₂O, bz, eth, EtOH
6504	Isopropyl methacrylate	Isopropyl 2-methyl-2-propenoate	C₇H₁₂O₂	4655-34-9	128.169			130(17)	0.8847²⁰	1.4122²⁰	vs ace, bz, eth, EtOH
6505	Isopropyl methanesulfonate		C₄H₁₀O₃S	926-06-7	138.185			82⁶			
6506	Isopropylmethylamine	Methylisopropylamine	C₄H₁₁N	4747-21-1	73.137			53.3(0.3)			

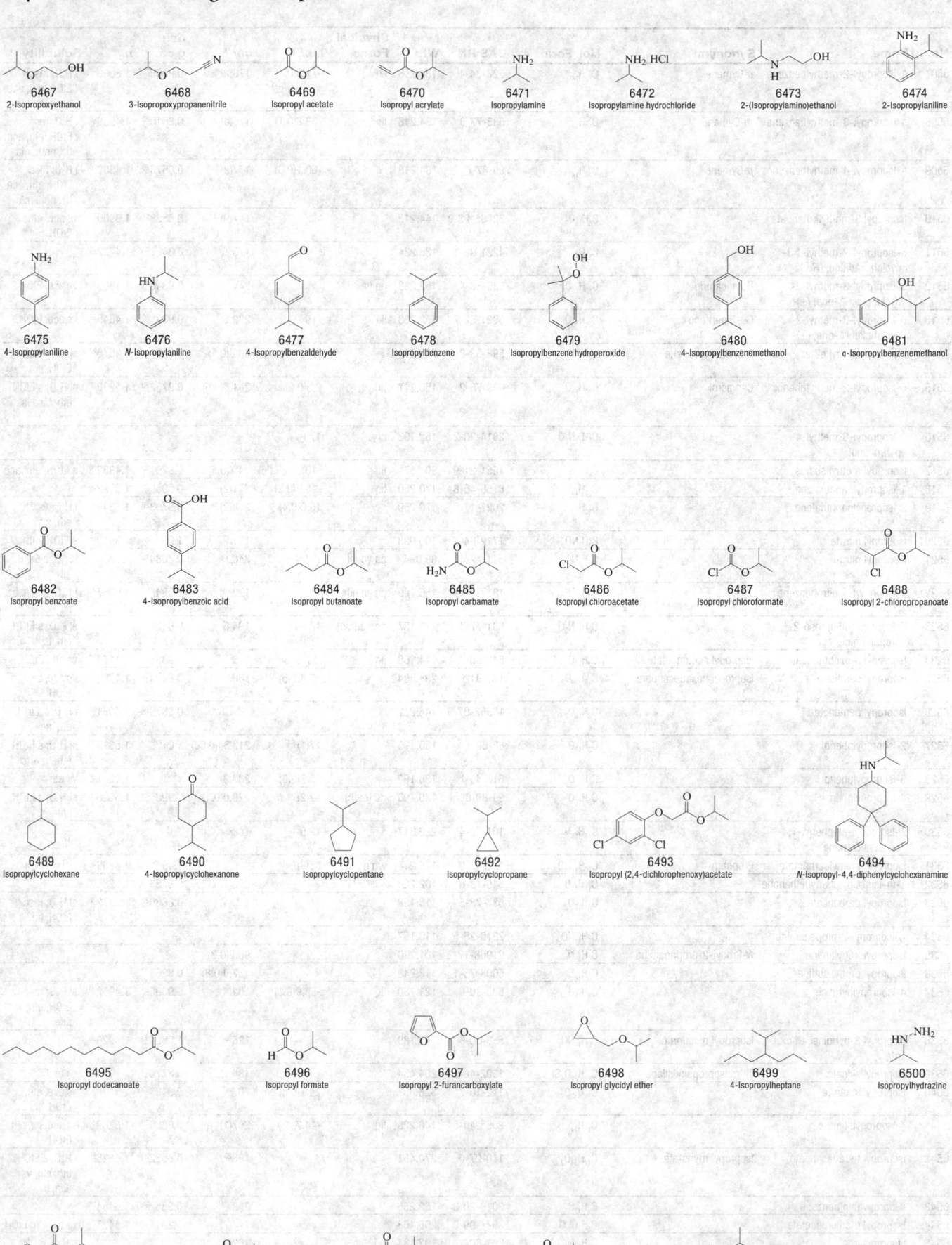

6467
2-Isopropoxyethanol

6468
3-Isopropoxypropanenitrile

6469
Isopropyl acetate

6470
Isopropyl acrylate

6471
Isopropylamine

6472
Isopropylamine hydrochloride

6473
2-(Isopropylamino)ethanol

6474
2-Isopropylaniline

6475
4-Isopropylaniline

6476
N-Isopropylaniline

6477
4-Isopropylbenzaldehyde

6478
Isopropylbenzene

6479
Isopropylbenzene hydroperoxide

6480
4-Isopropylbenzenemethanol

6481
α-Isopropylbenzenemethanol

6482
Isopropyl benzoate

6483
4-Isopropylbenzoic acid

6484
Isopropyl butanoate

6485
Isopropyl carbamate

6486
Isopropyl chloroacetate

6487
Isopropyl chloroformate

6488
Isopropyl 2-chloropropanoate

6489
Isopropylcyclohexane

6490
4-Isopropylcyclohexanone

6491
Isopropylcyclopentane

6492
Isopropylcyclopropane

6493
Isopropyl (2,4-dichlorophenoxy)acetate

6494
N-Isopropyl-4,4-diphenylcyclohexanamine

6495
Isopropyl dodecanoate

6496
Isopropyl formate

6497
Isopropyl 2-furancarboxylate

6498
Isopropyl glycidyl ether

6499
4-Isopropylheptane

6500
Isopropylhydrazine

6501
Isopropyl 2-hydroxybenzoate

6502
Isopropyl isobutanoate

6503
Isopropyl lactate

6504
Isopropyl methacrylate

6505
Isopropyl methanesulfonate

6506
Isopropylmethylamine

No.	Name	Synonym	Mol. Form.	CAS RN	Mol. Wt.	Physical Form	mp/°C	bp/°C	den g cm⁻³	n_D	Solubility
6507	1-Isopropyl-2-methylbenzene	o-Cymene	C₁₀H₁₄	527-84-4	134.218	liq	-71.5(0.1)	178(2)	0.8766²⁰	1.5006²⁰	i H₂O; msc EtOH, eth, ace, bz, peth, ctc
6508	1-Isopropyl-3-methylbenzene	m-Cymene	C₁₀H₁₄	535-77-3	134.218	liq	-63.8(0.1)	175(4)	0.8610²⁰	1.4930²⁰	i H₂O; msc EtOH, eth, ace, bz, peth, ctc
6509	1-Isopropyl-4-methylbenzene	p-Cymene	C₁₀H₁₄	99-87-6	134.218	liq	-68.1(0.3)	177(2)	0.8573²⁰	1.4909²⁰	i H₂O; msc EtOH, eth, ace, bz, peth, ctc
6510	Isopropyl 3-methylbutanoate		C₈H₁₆O₂	32665-23-9	144.212			147(9)	0.8538¹⁷	1.3960²⁰	vs ace, eth, EtOH
6511	5-Isopropyl-2-methyl-1,3-cyclohexadiene, (R)-		C₁₀H₁₆	4221-98-1	136.234			173	0.8421²⁰	1.4772¹⁹	
6512	5-Isopropyl-3-methyl-2-cyclohexen-1-one, (±)-	Homocamfin	C₁₀H₁₆O	535-86-4	152.233	pa ye		244	0.9340²¹	1.4865²¹	vs ace, EtOH
6513	6-Isopropyl-3-methyl-2-cyclohexen-1-one, (±)-	(±)-Piperitone	C₁₀H₁₆O	6091-52-7	152.233	liq	-19	232.5	0.9331²⁰	1.4845²⁰	vs ace, EtOH
6514	Isopropyl methyl ether	2-Methoxypropane	C₄H₁₀O	598-53-8	74.121			30.8(0.5)	0.7237¹⁵	1.3576²⁰	sl H₂O; msc EtOH, eth
6515	5-Isopropyl-2-methylphenol	Carvacrol	C₁₀H₁₄O	499-75-2	150.217	nd	2.5(0.2)	241.5(0.5)	0.9772²⁰	1.5230²⁰	sl H₂O; s EtOH, eth, ctc; vs ace
6516	2-Isopropyl-6-methyl-4-pyrimidinol		C₈H₁₂N₂O	2814-20-2	152.193	cry	173				
6517	Isopropyl methyl sulfide		C₄H₁₀S	1551-21-9	90.187	liq	-101.47(0.06)	84.7(0.5)	0.8291²⁰	1.4932²⁰	s EtOH, eth, ace
6518	1-Isopropylnaphthalene		C₁₃H₁₄	6158-45-8	170.250	liq	-15.6(0.2)	271(2)	0.9956²⁰	1.5952²⁰	
6519	2-Isopropylnaphthalene		C₁₃H₁₄	2027-17-0	170.250		15.0(0.4)	268(2)	0.9753²⁰	1.5848²⁰	i H₂O; vs EtOH, eth; s bz
6520	Isopropyl nitrate		C₃H₇NO₃	1712-64-7	105.093			101(2)	1.034¹⁹	1.3912¹⁶	s EtOH, eth
6521	Isopropyl nitrite		C₃H₇NO₂	541-42-4	89.094	pa ye oil		34(2)	0.8684¹⁵		i H₂O; s EtOH, eth
6522	1-Isopropyl-4-nitrobenzene		C₉H₁₁NO₂	1817-47-6	165.189	pa ye oil		122⁹	1.084²⁰	1.5367²⁰	i H₂O; s ace, bz, lig
6523	N-Isopropyl-N-nitroso-2-propanamine		C₆H₁₄N₂O	601-77-4	130.187	cry (eth,w)	48	194.5	0.9422²⁰		sl H₂O; s EtOH, eth, bz
6524	Isopropyl 3-oxobutanoate	Isopropyl acetoacetate	C₇H₁₂O₃	542-08-5	144.168	liq	-27.3	186	0.9835²⁰	1.4173²⁰	vs eth, EtOH, lig
6525	Isopropyl palmitate	Isopropyl hexadecanoate	C₁₉H₃₈O₂	142-91-6	298.504		12.8(0.5)	160²	0.8404³⁸	1.4364²⁵	vs ace, bz, eth, EtOH
6526	Isopropyl pentanoate		C₈H₁₆O₂	18362-97-5	144.212				0.8579²⁰	1.4061²⁰	i H₂O; s EtOH, eth, ace
6527	2-Isopropylphenol		C₉H₁₂O	88-69-7	136.190		17(1)	213.84(0.08)	1.012²⁰	1.5315²⁰	sl H₂O; s EtOH, eth, bz, ctc
6528	3-Isopropylphenol		C₉H₁₂O	618-45-1	136.190		25.7(0.5)	230(3)		1.5261²⁰	vs eth
6529	4-Isopropylphenol		C₉H₁₂O	99-89-8	136.190	nd (peth)	62(2)	228.0(0.1)	0.990²⁰	1.5228²⁰	sl H₂O; s EtOH, chl
6530	N-Isopropyl-N′-phenyl-1,4-benzenediamine		C₁₅H₁₈N₂	101-72-4	226.317		72.5	148²			
6531	Isopropyl phenylcarbamate	Propham	C₁₀H₁₃NO₂	122-42-9	179.216	wh nd (al)	85(2)		1.09²⁰	1.4989⁹¹	vs bz, EtOH
6532	1-(4-Isopropylphenyl)ethanone		C₁₁H₁₄O	645-13-6	162.228			250(13)	0.9753¹⁵	1.5235²⁰	
6533	Isopropyl propanoate		C₆H₁₂O₂	637-78-5	116.158			110(4)	0.8660²⁰	1.3872²⁰	sl H₂O; msc EtOH, eth
6534	N-Isopropyl-2-propenamide		C₆H₁₁NO	2210-25-5	113.157		64.5	110¹⁵			
6535	Isopropylpropylamine	N-Propyl-2-propanamine	C₆H₁₅N	21968-17-2	101.190			96.2(0.3)			
6536	Isopropyl propyl sulfide		C₆H₁₄S	5008-73-1	118.240			132.0(0.8)	0.8269²⁰		
6537	4-Isopropylpyridine		C₈H₁₁N	696-30-0	121.180	liq	-54.9(0.5)	182(4)	0.9382²⁵	1.4962²⁰	sl H₂O; msc EtOH, eth; vs ace
6538	Isopropyl 3-pyridinecarboxylate	Isopropyl nicotinate	C₉H₁₁NO₂	553-60-6	165.189			126³⁰	1.0624²⁰	1.4926²⁰	
6539	Isopropyl silicate	Tetra(isopropoxy)silane	C₁₂H₂₈O₄Si	1992-48-9	264.434			184	0.8770²⁰		s ctc, CS₂
6540	Isopropyl stearate		C₂₁H₄₂O₂	112-10-7	326.557		24.1(0.5)	207⁶	0.8403³⁸		vs ace, eth, EtOH, chl
6541	4-Isopropylstyrene		C₁₁H₁₄	2055-40-5	146.229	liq	-44.7	207(1)	0.8850²⁰	1.5289²⁰	vs ace, bz, eth, EtOH
6542	Isopropyl tetradecanoate	Isopropyl myristate	C₁₇H₃₄O₂	110-27-0	270.451			193²⁰	0.8532²⁰	1.4325²⁵	i H₂O; s EtOH, eth, chl; vs ace, bz
6543	(Isopropylthio)benzene		C₉H₁₂S	3019-20-3	152.256			208	0.9852²⁰	1.5464²⁰	
6544	Isopropyl trichloroacetate		C₅H₇Cl₃O₂	3974-99-0	205.468			174(7)	1.2911²⁵	1.4428²⁰	vs bz, eth, EtOH
6545	Isopropylurea		C₄H₁₀N₂O	691-60-1	102.134	nd		103⁰·¹			s H₂O, EtOH, chl, ace; sl eth
6546	Isopropyl vinyl ether	2-(Ethenyloxy)propane	C₅H₁₀O	926-65-8	86.132	liq	-140	56(4)	0.7534²⁰	1.3840²⁰	vs ace, bz, eth, EtOH

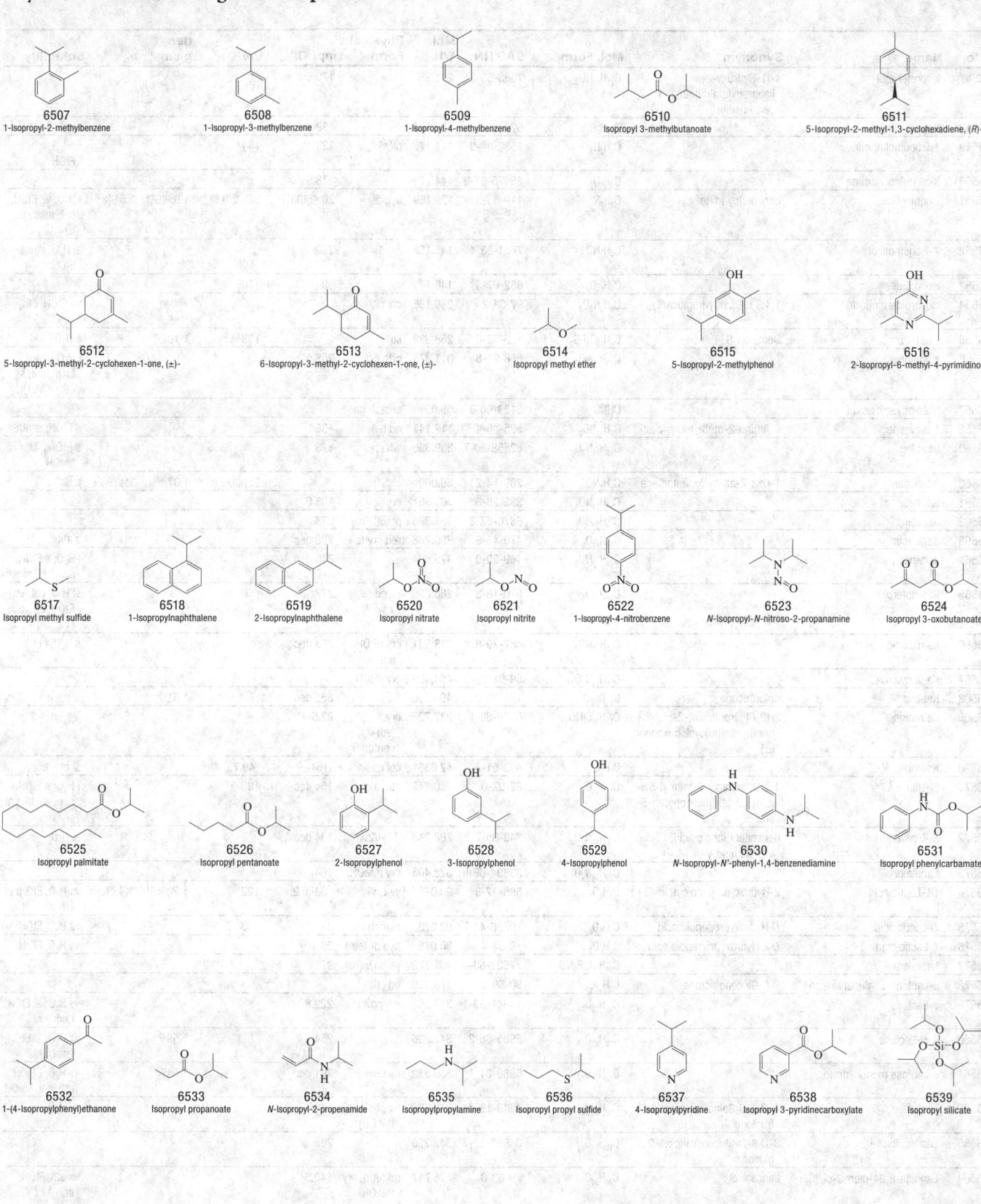

6507
1-Isopropyl-2-methylbenzene

6508
1-Isopropyl-3-methylbenzene

6509
1-Isopropyl-4-methylbenzene

6510
Isopropyl 3-methylbutanoate

6511
5-Isopropyl-2-methyl-1,3-cyclohexadiene, (R)-

6512
5-Isopropyl-3-methyl-2-cyclohexen-1-one, (±)-

6513
6-Isopropyl-3-methyl-2-cyclohexen-1-one, (±)-

6514
Isopropyl methyl ether

6515
5-Isopropyl-2-methylphenol

6516
2-Isopropyl-6-methyl-4-pyrimidinol

6517
Isopropyl methyl sulfide

6518
1-Isopropylnaphthalene

6519
2-Isopropylnaphthalene

6520
Isopropyl nitrate

6521
Isopropyl nitrite

6522
1-Isopropyl-4-nitrobenzene

6523
N-Isopropyl-N-nitroso-2-propanamine

6524
Isopropyl 3-oxobutanoate

6525
Isopropyl palmitate

6526
Isopropyl pentanoate

6527
2-Isopropylphenol

6528
3-Isopropylphenol

6529
4-Isopropylphenol

6530
N-Isopropyl-N'-phenyl-1,4-benzenediamine

6531
Isopropyl phenylcarbamate

6532
1-(4-Isopropylphenyl)ethanone

6533
Isopropyl propanoate

6534
N-Isopropyl-2-propenamide

6535
Isopropylpropylamine

6536
Isopropyl propyl sulfide

6537
4-Isopropylpyridine

6538
Isopropyl 3-pyridinecarboxylate

6539
Isopropyl silicate

6540
Isopropyl stearate

6541
4-Isopropylstyrene

6542
Isopropyl tetradecanoate

6543
(Isopropylthio)benzene

6544
Isopropyl trichloroacetate

6545
Isopropylurea

6546
Isopropyl vinyl ether

No.	Name	Synonym	Mol. Form.	CAS RN	Mol. Wt.	Physical Form	mp/°C	bp/°C	den g cm⁻³	n_D	Solubility
6547	Isoproterenol	4-[1-Hydroxy-2-[isopropylamino]ethyl]-1,2-benzenediol	$C_{11}H_{17}NO_3$	7683-59-2	211.258			170.5			
6548	Isopsoralen		$C_{11}H_6O_3$	523-50-2	186.164			139			
6549	1-Isoquinolinamine		$C_9H_8N_2$	1532-84-9	144.173	pl(w)	123	164[8]			sl H_2O, eth; vs EtOH
6550	3-Isoquinolinamine		$C_9H_8N_2$	25475-67-6	144.173			178.5			
6551	Isoquinoline	Benzo[c]pyridine	C_9H_7N	119-65-3	129.159	hyg pl	26.46(0.01)	243.2(0.6)	1.0910[30]	1.6148[20]	i H_2O; vs EtOH, chl; msc eth, bz
6552	7-Isoquinolinol		C_9H_7NO	7651-83-4	145.158			230			sl H_2O, eth; s EtOH
6553	Isosorbide		$C_6H_{10}O_4$	652-67-5	146.141		63	170[2]			
6554	Isosorbide dinitrate	1,4:3,6-Dianhydroglucitol	$C_6H_8N_2O_8$	87-33-2	236.136	col cry	52				vs EtOH, eth, ace
6555	Isosystox	Demeton-S	$C_8H_{19}O_3PS_2$	126-75-0	258.339	liq		133[2]	1.132[21]		s H_2O
6556	Isothebaine		$C_{19}H_{21}NO_3$	568-21-8	311.375	orth cry (al)	203.5				i H_2O; msc EtOH, chl; sl eth; s MeOH
6557	Isothiocyanic acid		CHNS	3129-90-6	59.091	unstab gas					
6558	L-Isovaline	2-Amino-2-methylbutyric acid	$C_5H_{11}NO_2$	595-40-4	117.147	nd (w)	≈300				s EtOH; sl eth
6559	Isoxaben		$C_{18}H_{24}N_2O_4$	82558-50-7	332.395	wh cry	173				s EtOAc, MeCN, MeOH
6560	Isoxazole	1-Oxa-2-azacyclopentadiene	C_3H_3NO	288-14-2	69.062			95.4(0.7)	1.078[20]	1.4298[17]	s H_2O
6561	Isoxsuprine		$C_{18}H_{23}NO_3$	395-28-8	301.381	cry	103.0				
6562	Jacobine		$C_{18}H_{25}NO_6$	6870-67-3	351.395	pl (EtOH)	228				
6563	Javanicin		$C_{15}H_{14}O_6$	476-45-9	290.268	red cry (al)	208 dec				s alk
6564	Jervine		$C_{27}H_{39}NO_3$	469-59-0	425.604		243 dec				i H_2O; s EtOH, ace, chl; sl eth
6565	Kaempferol		$C_{15}H_{10}O_6$	520-18-3	286.236	ye nd (al, + 1 w)	277				sl H_2O, chl; vs EtOH, eth, ace; i bz
6566	Kainic acid		$C_{10}H_{15}NO_4$	487-79-6	213.231	cry (EtOH aq)	253 dec				s H_2O; i EtOH
6567	Kanamycin A		$C_{18}H_{36}N_4O_{11}$	59-01-8	484.499	cry (EtOH)					
6568	Kepone	Chlordecone	$C_{10}Cl_{10}O$	143-50-0	490.636			350 dec	1.61[25]		
6569	Ketamine	2-(2-Chlorophenyl)-2-(methylamino)cyclohexanone, (±)	$C_{13}H_{16}ClNO$	6740-88-1	237.725	cry (eth-pentane)	92.5				
6570	Ketene		C_2H_2O	463-51-4	42.036	col gas	-151	-49.7(0.4)			sl eth, ace
6571	Khellin	4,9-Dimethoxy-7-methyl-5H-furo[3,2-g][1]benzopyran-5-one	$C_{14}H_{12}O_5$	82-02-0	260.242	eth, al	154 dec	190[0.05]			i H_2O; s EtOH, ace; sl eth, chl
6572	L-Kynurenine	Benzenebutanoic acid, α,2-diamino-γ-oxo-	$C_{10}H_{12}N_2O_3$	343-65-7	208.213	lf (+l/2w)	194 dec				sl H_2O
6573	Labetalol		$C_{19}H_{24}N_2O_3$	36894-69-6	328.405	cry (MeOH)	164				
6574	DL-Lactic acid	2-Hydroxypropanoic acid, (±)	$C_3H_6O_3$	598-82-3	90.078	ye cry	16.9(0.2)	122[15]	1.2060[21]	1.4392[20]	vs H_2O, EtOH; sl eth
6575	D-Lactic acid	D-Hydroxypropanoic acid	$C_3H_6O_3$	10326-41-7	90.078	pl (chl)	53	103[2]			vs H_2O, EtOH
6576	L-Lactic acid	L-2-Hydroxypropanoic acid	$C_3H_6O_3$	79-33-4	90.078	hyg pr (eth)	53				vs H_2O, EtOH
6577	Lactofen		$C_{19}H_{15}ClF_3NO_7$	77501-63-4	461.773	ye pow (bz)	93				
6578	δ-Lactone-D-gluconic acid	δ-D-Gluconolactone	$C_6H_{10}O_6$	90-80-2	178.139	nd (al)					
6579	α-Lactose		$C_{12}H_{22}O_{11}$	14641-93-1	342.296	wh pow	222.8				vs H_2O; sl EtOH; i eth, chl
6580	β-D-Lactose		$C_{12}H_{22}O_{11}$	5965-66-2	342.296		254		1.59[20]		vs H_2O; sl EtOH; i eth, chl
6581	α-Lactose monohydrate		$C_{12}H_{24}O_{12}$	5989-81-1	360.312	mcl (w)	201 dec		1.547[20]		vs H_2O; i EtOH, eth, chl, MeOH
6582	Lactulose	4-O-β-D-Galactopyranosyl-D-fructose	$C_{12}H_{22}O_{11}$	4618-18-2	342.296	hx pl (MeOH)	169				vs H_2O
6583	Laminaribiose	3-O-β-D-Glucopyranosyl-D-glucose	$C_{12}H_{22}O_{11}$	34980-39-7	342.296		205				
6584	Lanosta-8,24-dien-3-ol, (3β)	Lanosterol	$C_{30}H_{50}O$	79-63-0	426.717	nd (eth), cry (MeOH-ace)	140.5				vs eth, EtOH, chl
6585	Lantadene A	Rehmannic acid	$C_{35}H_{52}O_5$	467-81-2	552.785	cry (MeOH)	297				
6586	Lantadene B		$C_{35}H_{52}O_5$	467-82-3	552.785	cry (EtOH)	302				
6587	L-Lanthionine	L-Cysteine, S-(2-amino-2-carboxyethyl)-, (R)-	$C_6H_{12}N_2O_4S$	922-55-4	208.235	hex pl	294 dec				sl H_2O
6588	Lapachol	2-Hydroxy-3-(3-methyl-2-butenyl)-1,4-naphthalenedione	$C_{15}H_{14}O_3$	84-79-7	242.270	ye pr (eth, bz) pl (al)	139.5				i H_2O; s EtOH, eth, bz, chl; vs HOAc

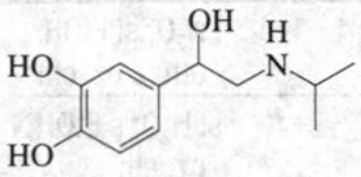

6547 Isoproterenol

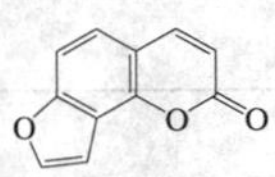

6548 Isopsoralen

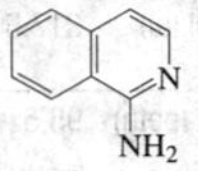

6549 1-Isoquinolinamine

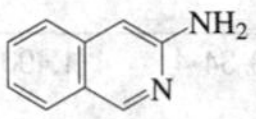

6550 3-Isoquinolinamine

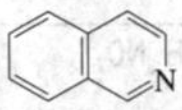

6551 Isoquinoline

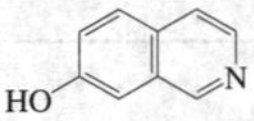

6552 7-Isoquinolinol

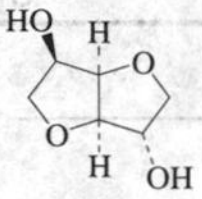

6553 Isosorbide

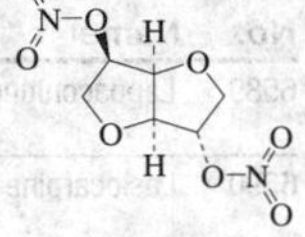

6554 Isosorbide dinitrate

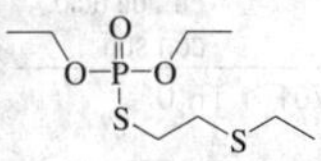

6555 Isosystox

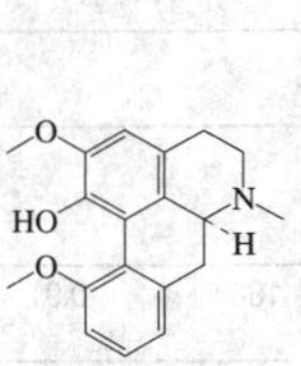

6556 Isothebaine

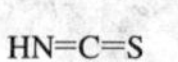

6557 Isothiocyanic acid

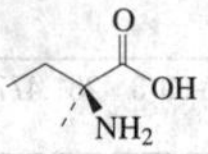

6558 L-Isovaline

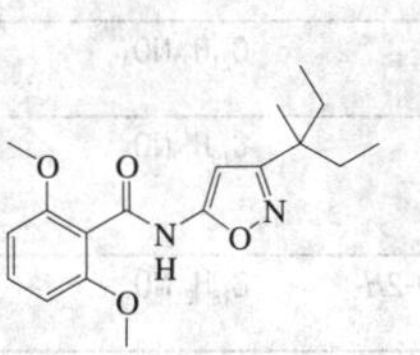

6559 Isoxaben

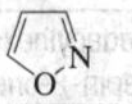

6560 Isoxazole

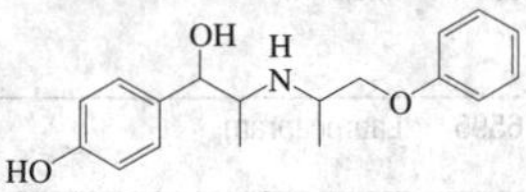

6561 Isoxsuprine

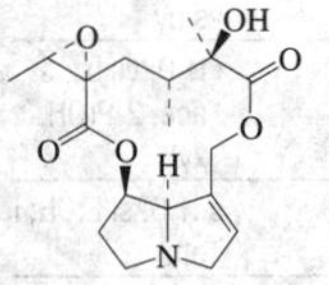

6562 Jacobine

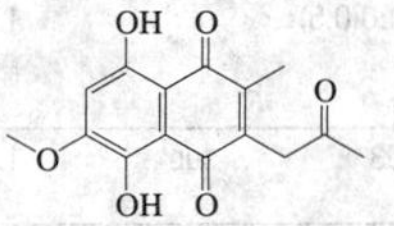

6563 Javanicin

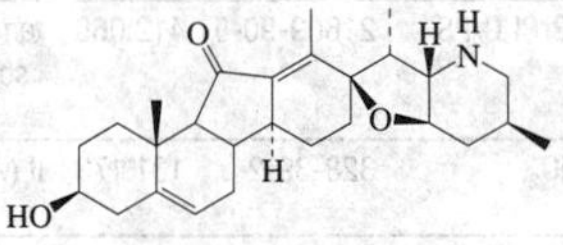

6564 Jervine

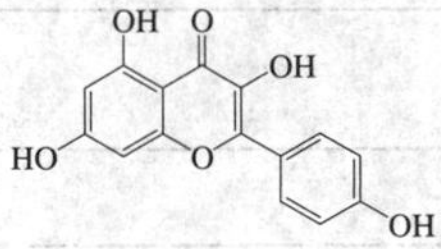

6565 Kaempferol

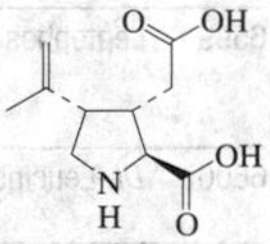

6566 Kainic acid

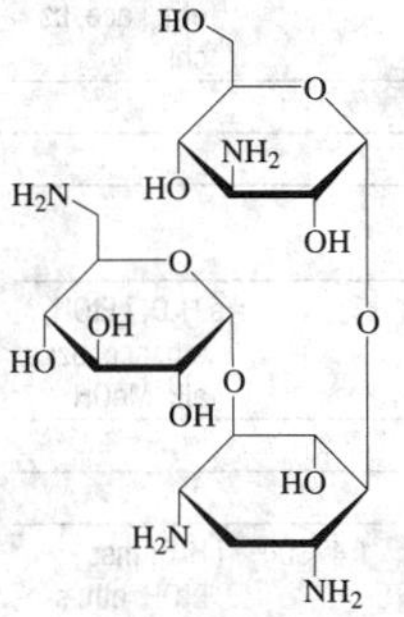

6567 Kanamycin A

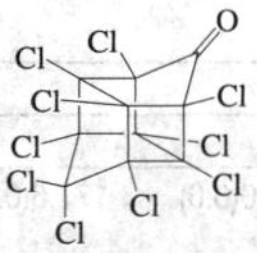

6568 Kepone

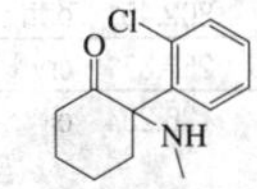

6569 Ketamine

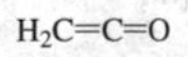

6570 Ketene

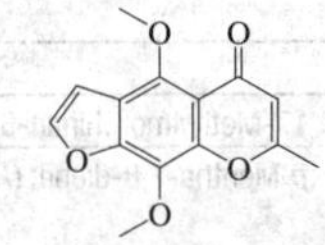

6571 Khellin

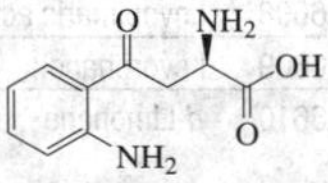

6572 L-Kynurenine

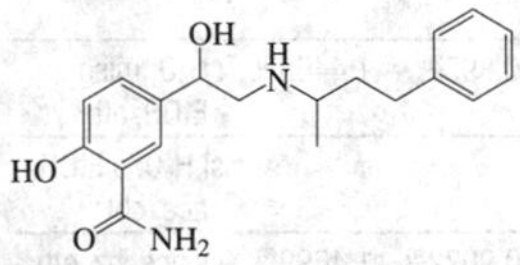

6573 Labetalol

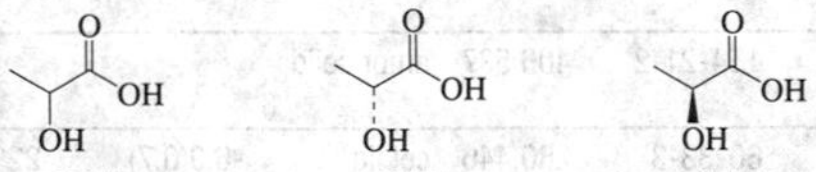

6574 DL-Lactic acid

6575 D-Lactic acid

6576 L-Lactic acid

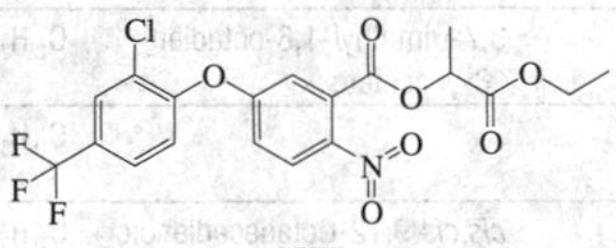

6577 Lactofen

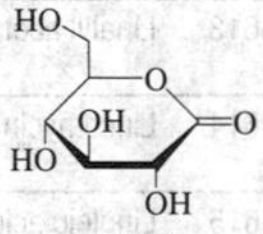

6578 δ-Lactone-D-gluconic acid

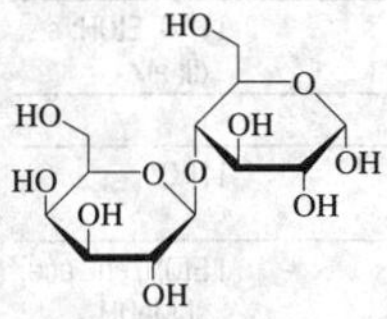

6579 α-Lactose

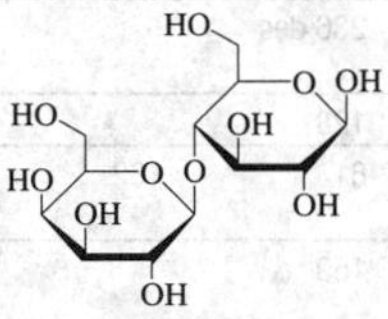

6580 β-D-Lactose

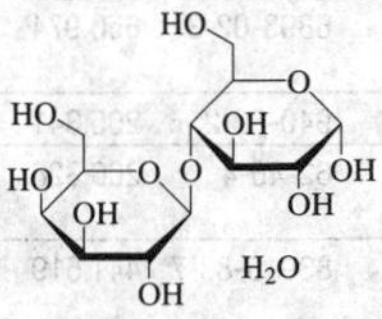

6581 α-Lactose monohydrate

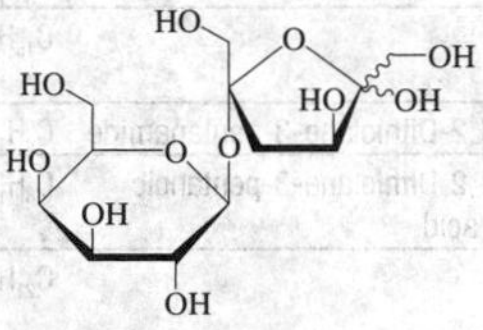

6582 Lactulose

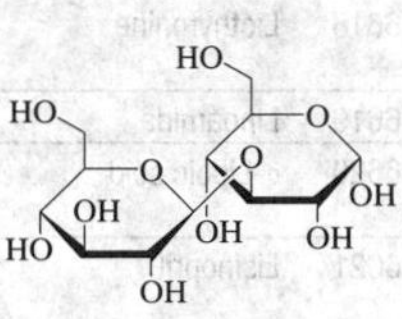

6583 Laminaribiose

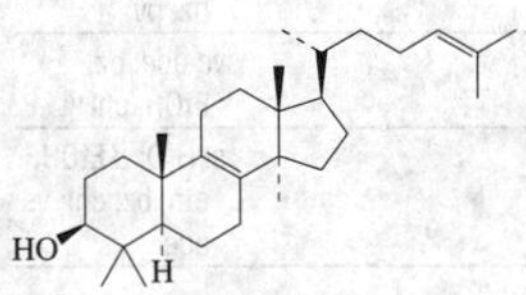

6584 Lanosta-8,24-dien-3-ol, (3β)

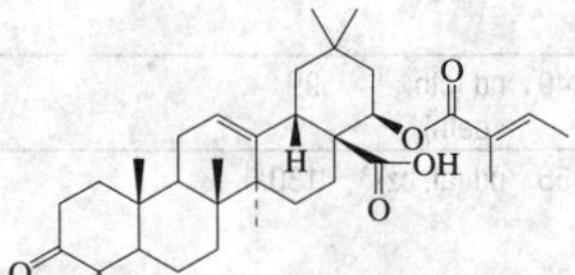

6585 Lantadene A

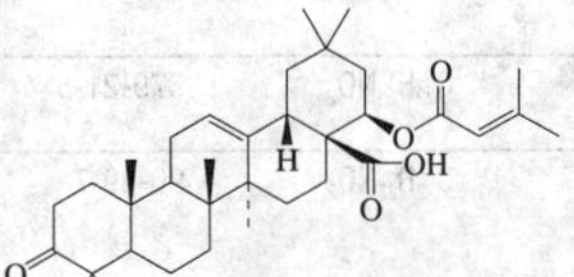

6586 Lantadene B

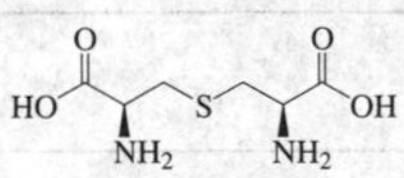

6587 L-Lanthionine

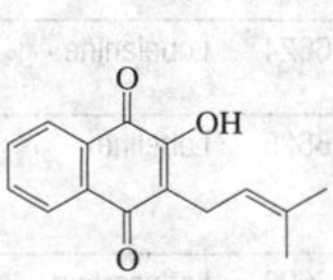

6588 Lapachol

No.	Name	Synonym	Mol. Form.	CAS RN	Mol. Wt.	Physical Form	mp/°C	bp/°C	den g cm⁻³	n_D	Solubility
6589	Lappaconitine		$C_{32}H_{44}N_2O_8$	32854-75-4	584.699	hex pl (al)	217.5				i H_2O; sl EtOH, eth; s bz, chl
6590	Lasiocarpine		$C_{21}H_{33}NO_7$	303-34-4	411.490	col pl (peth)	95.5				sl H_2O; s EtOH, bz, eth
6591	Laudanidine		$C_{20}H_{25}NO_4$	301-21-3	343.418	hex pr (al)	184.5				vs H_2O, bz
6592	Laudanine		$C_{20}H_{25}NO_4$	85-64-3	343.418	ye wh pr (dil al, al-chl)	167		1.26^{20}		sl H_2O, EtOH, eth; s bz, chl
6593	Laudanosine		$C_{21}H_{27}NO_4$	2688-77-9	357.444	nd (peth), pr (al)	89				vs ace, eth, EtOH, chl
6594	Laureline		$C_{19}H_{19}NO_3$	81-38-9	309.359	tab (al) cubes (peth)	114				i H_2O; s EtOH, eth, dil acid, con sulf
6595	Laurocapram	1-Dodecylhexahydro-2H-azepin-2-one	$C_{18}H_{35}NO$	59227-89-3	281.477	col liq	-7	160^{50}	0.91	1.4701	i H_2O
6596	Lead bis(dimethyldithiocarbamate)		$C_6H_{12}N_2PbS_4$	19010-66-3	447.6	pale ye nd	258				
6597	Ledol		$C_{15}H_{26}O$	577-27-5	222.366	nd (al)	105	292	0.9078^{100}	1.4667^{110}	vs ace, eth, EtOH
6598	Lenacil		$C_{13}H_{18}N_2O_2$	2164-08-1	234.294		290		1.32^{25}		vs py
6599	Leptophos		$C_{13}H_{10}BrCl_2O_2PS$	21609-90-5	412.066	tan waxy solid	74.0(0.5)		1.53^{25}		i H_2O; vs bz; s ace, 2-PrOH, xyl
6600	DL-Leucine		$C_6H_{13}NO_2$	328-39-2	131.173	lf (w)	293	sub	1.293^{18}		s H_2O; sl EtOH; i eth
6601	D-Leucine		$C_6H_{13}NO_2$	328-38-1	131.173	pl (al)	293	sub			sl H_2O
6602	L-Leucine	2-Amino-4-methylpentanoic acid	$C_6H_{13}NO_2$	61-90-5	131.173	hex pl (dil al)	293	sub	1.293^{18}		sl H_2O; i EtOH, eth
6603	N-Leucylglycine		$C_8H_{16}N_2O_3$	686-50-0	188.224		248 dec				s H_2O; sl EtOH, eth; i ace, bz, chl
6604	Leuprolide		$C_{59}H_{84}N_{16}O_{12}$	53714-56-0	1209.398	fluffy solid					
6605	Leurosine		$C_{46}H_{56}N_4O_9$	23360-92-1	808.959	cry	203				
6606	Levallorphan	17-Allylmorphinan-3-ol	$C_{19}H_{25}NO$	152-02-3	283.408	cry (EtOH aq)	181				
6607	Levodopa	L-3,4-Dihydroxyphenylalanine	$C_9H_{11}NO_4$	59-92-7	197.188	pl (dil al) pr or nd (w+SO_2)	277(2)				s H_2O; i EtOH, eth, ace, bz; s alk, MeOH
6608	Levopimaric acid		$C_{20}H_{30}O_2$	79-54-9	302.451	orth cry	150				
6609	Levorphanol	17-Methylmorphinan-3-ol	$C_{17}H_{23}NO$	77-07-6	257.371	cry	198				
6610	d-Limonene	p-Mentha-1,8-diene, (R)	$C_{10}H_{16}$	5989-27-5	136.234	oil	-74.0(0.6)	177.6(0.5)	0.8411^{20}	1.4730^{20}	i H_2O; msc EtOH, eth; s ctc
6611	l-Limonene	p-Mentha-1,8-diene, (S)	$C_{10}H_{16}$	5989-54-8	136.234	oil		178(1)	0.843^{20}	1.4746^{20}	i H_2O; vs eth, EtOH
6612	Linalol	3,7-Dimethyl-1,6-octadien-3-ol, (±)-	$C_{10}H_{18}O$	22564-99-4	154.249			198	0.870^{15}	1.4627	
6613	Linalyl acetate	3,7-Dimethyl-1,6-octadien-3-yl acetate	$C_{12}H_{20}O_2$	115-95-7	196.286	liq		221(6)	0.895^{20}	1.4460^{20}	i H_2O; misc EtOH, eth
6614	Lincomycin		$C_{18}H_{34}N_2O_6S$	154-21-2	406.537	amor solid					sl H_2O; s EtOH, ace, chl
6615	Linoleic acid	cis,cis-9,12-Octadecadienoic acid	$C_{18}H_{32}O_2$	60-33-3	280.446	col liq	-6.9(0.7)	229^{16}	0.9022^{20}	1.4699^{20}	vs ace, bz, eth, EtOH
6616	Linolenic acid	cis,cis,cis-9,12,15-Octadeca-trienoic acid	$C_{18}H_{30}O_2$	463-40-1	278.430		-10(2)	231^{17}	0.9164^{20}	1.4800^{20}	i H_2O; s EtOH, eth; sl bz
6617	Linuron		$C_9H_{10}Cl_2N_2O_2$	330-55-2	249.093		92.9(1)				
6618	Liothyronine		$C_{15}H_{12}I_3NO_4$	6893-02-3	650.974	cry	236 dec				i H_2O, EtOH; s dil alk
6619	Lipoamide	1,2-Dithiolane-3-pentanamide	$C_8H_{15}NOS_2$	940-69-2	205.341	cry	128				
6620	α-Lipoic acid	1,2-Dithiolane-3-pentanoic acid	$C_8H_{14}O_2S_2$	62-46-4	206.326	ye nd	61	162			i H_2O
6621	Lisinopril		$C_{21}H_{35}N_3O_7$	83915-83-7	441.519	wh cry pow	159				i EtOH, chl, ace; sl MeOH
6622	Lithium oxalate		$C_2Li_2O_4$	30903-87-8	101.901	col cry	dec		2.121^{17}		s H_2O; i EtOH, eth
6623	Lobelanidine		$C_{22}H_{29}NO_2$	552-72-7	339.471	sc (al, eth)	150				i H_2O; s EtOH; sl eth; vs ace, bz, py
6624	Lobelanine		$C_{22}H_{25}NO_2$	579-21-5	335.440	nd (eth, peth)	99				vs ace, bz, EtOH, chl
6625	Lobeline		$C_{22}H_{27}NO_2$	90-69-7	337.455	nd (al, bz)	130.5				sl H_2O; s EtOH, eth, bz, chl; vs ace
6626	Loflucarban		$C_{13}H_9Cl_2FN_2S$	790-69-2	315.192		163.5				

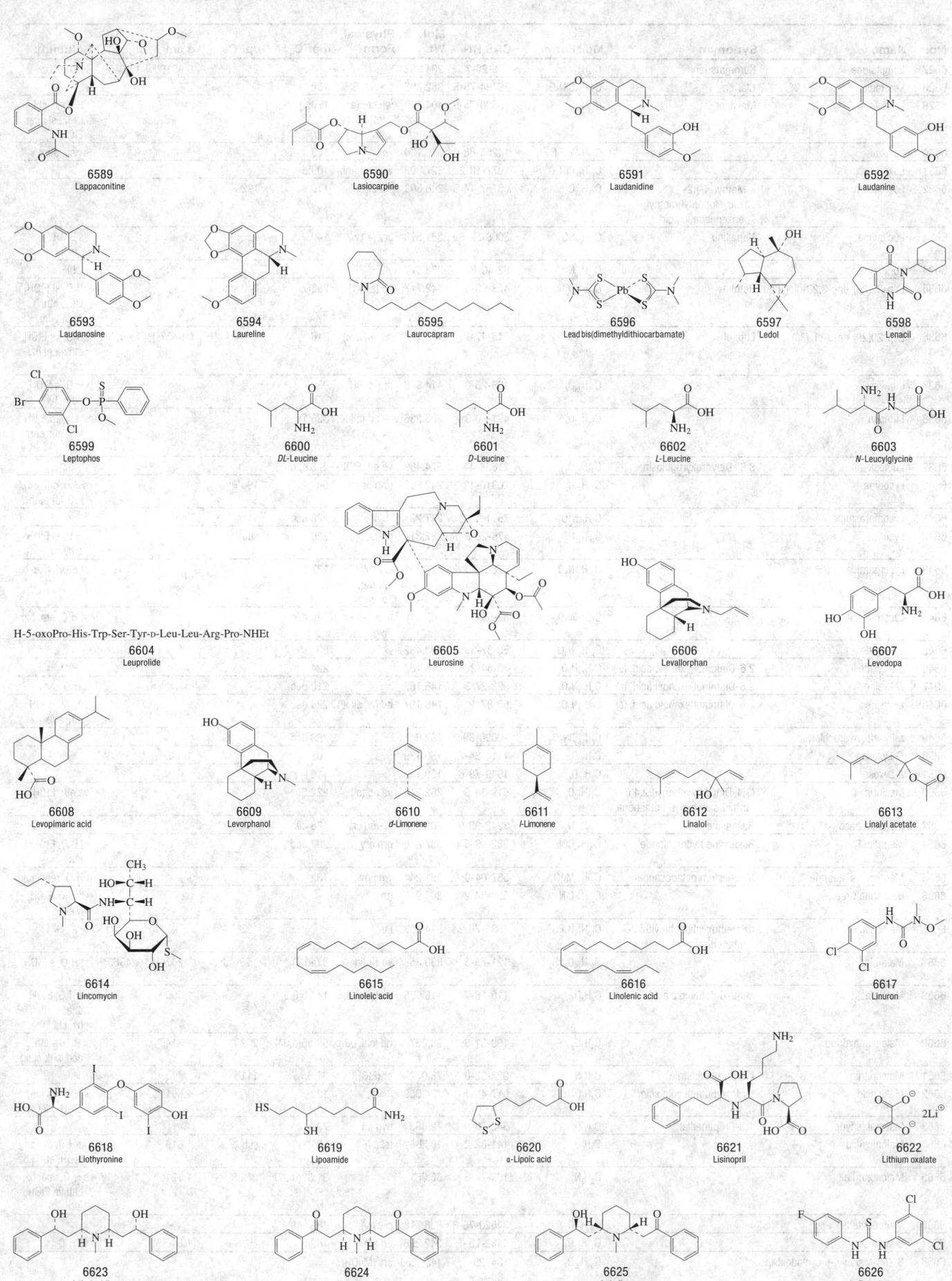

6589 Lappaconitine

6590 Lasiocarpine

6591 Laudanidine

6592 Laudanine

6593 Laudanosine

6594 Laureline

6595 Laurocapram

6596 Lead bis(dimethyldithiocarbamate)

6597 Ledol

6598 Lenacil

6599 Leptophos

6600 DL-Leucine

6601 D-Leucine

6602 L-Leucine

6603 N-Leucylglycine

H-5-oxoPro-His-Trp-Ser-Tyr-D-Leu-Leu-Arg-Pro-NHEt

6604 Leuprolide

6605 Leurosine

6606 Levallorphan

6607 Levodopa

6608 Levopimaric acid

6609 Levorphanol

6610 d-Limonene

6611 l-Limonene

6612 Linalol

6613 Linalyl acetate

6614 Lincomycin

6615 Linoleic acid

6616 Linolenic acid

6617 Linuron

6618 Liothyronine

6619 Lipoamide

6620 α-Lipoic acid

6621 Lisinopril

6622 Lithium oxalate

6623 Lobelanidine

6624 Lobelanine

6625 Lobeline

6626 Loflucarban

No.	Name	Synonym	Mol. Form.	CAS RN	Mol. Wt.	Physical Form	mp/°C	bp/°C	den g cm⁻³	n_D	Solubility
6627	Longifolene	Kuromatsuene	$C_{15}H_{24}$	475-20-7	204.352			258	0.9319^{18}	1.5040^{20}	i H₂O; s bz
6628	Loratadine	Claritin	$C_{22}H_{23}ClN_2O_2$	79794-75-5	382.883	cry (MeCN)	132				
6629	Lovastatin	Mevacor	$C_{24}H_{36}O_5$	75330-75-5	404.540	wh cry (ace aq)	179(1)				i H₂O; vs chl; s DMF; sl ace, EtOH
6630	Lovozal		$C_{15}H_7Cl_2F_3N_2O_2$	14255-88-0	375.130	ye cry	103				s ace, diox
6631	Loxapine		$C_{18}H_{18}ClN_3O$	1977-10-2	327.808	ye cry (peth)	109.5				
6632	Loxoprofen	α-Methyl-4-[(2-oxocyclopentyl)methyl]benzeneacetic acid	$C_{15}H_{18}O_3$	68767-14-6	246.302	col oil	110	$192^{0.3}$			
6633	Luciculine	Napelline	$C_{22}H_{35}NO_3$	5008-52-6	361.518	cry (+1w, ace)	149	$165^{0.02}$			vs EtOH
6634	Lunacrine		$C_{16}H_{19}NO_3$	82-40-6	273.327						s chl
6635	Lup-20(29)-ene-3,28-diol, (3β)	Betulin	$C_{30}H_{50}O_2$	473-98-3	442.717	nd (al +1)	256(2)	240 sub			i H₂O; sl EtOH, bz; s eth, AcOEt, lig
6636	Lup-20(29)-en-3-ol, (3β)	Lupeol	$C_{30}H_{50}O$	545-47-1	426.717	nd (al, ace)	216		0.9457^{218}	1.4910^{218}	i H₂O; vs EtOH, eth, ace, bz, chl
6637	Lupulon		$C_{26}H_{38}O_4$	468-28-0	414.578	pr (MeOH)	93				i H₂O; s EtOH, peth, hx
6638	Luteolin		$C_{15}H_{10}O_6$	491-70-3	286.236	ye nd (dil al + 1 w)	328(1)				sl H₂O; s EtOH, eth, alk, con sulf
6639	Luteoskyrin	8,8'-Dihydroxyrugulosin	$C_{30}H_{22}O_{12}$	21884-44-6	574.489	ye nd (EtOH)	278 dec				
6640	Lycodine		$C_{16}H_{22}N_2$	20316-18-1	242.359	orth pr	99	$190^{1.0}$			s H₂O, chl, eth, EtOH; i peth
6641	Lycomarasmine		$C_9H_{15}N_3O_7$	7611-43-0	277.231		228 dec				
6642	Lycorine		$C_{16}H_{17}NO_4$	476-28-8	287.311	pr (al, py)	280	sub			i H₂O; sl EtOH, eth, chl
6643	Lysergamide		$C_{16}H_{17}N_3O$	478-94-4	267.325	cry (MeOH), pr (aq, ace)	137.5				sl EtOH, ace, os
6644	Lysergic acid		$C_{16}H_{16}N_2O_2$	82-58-6	268.310	lf or hex sc (w)	240 dec				sl H₂O, eth, bz; s EtOH, py
6645	Lysergide		$C_{20}H_{25}N_3O$	50-37-3	323.432		82				
6646	DL-Lysine	2,6-Diaminohexanoic acid, (±)	$C_6H_{14}N_2O_2$	70-54-2	146.187		224				sl H₂O
6647	D-Lysine	2,6-Diaminohexanoic acid, (D)	$C_6H_{14}N_2O_2$	923-27-3	146.187		218 dec				s H₂O
6648	L-Lysine	2,6-Diaminohexanoic acid, (L)	$C_6H_{14}N_2O_2$	56-87-1	146.187	nd (w, dil al)	224 dec				s H₂O; i EtOH, eth, ace, bz
6649	L-Lysine, hydrochloride		$C_6H_{15}ClN_2O_2$	10098-89-2	182.648		263 dec				
6650	D-Lyxose		$C_5H_{10}O_5$	1114-34-7	150.130		108			1.545^{20}	
6651	L-Lyxose		$C_5H_{10}O_5$	1949-78-6	150.130		110				
6652	Maclurin	(3,4-Dihydroxyphenyl)(2,4,6-trihydroxyphenyl)methanone	$C_{13}H_{10}O_6$	519-34-6	262.214	ye nd (al)	222.5				vs eth, EtOH
6653	Magenta base	Rosaniline	$C_{20}H_{19}N_3$	3248-93-9	301.385	br-red cry	186 dec				
6654	Magenta I	Rosaniline hydrochloride	$C_{20}H_{20}ClN_3$	632-99-5	337.846	grn cry	200 dec				sl H₂O, EtOH; i eth
6655	Magnesium stearate	Magnesium octadecanoate	$C_{36}H_{70}MgO_4$	557-04-0	591.244	wh pow	132				i H₂O; reac acid
6656	Malachite Green		$C_{23}H_{25}ClN_2$	569-64-2	364.911	grn cry					vs H₂O, EtOH, MeOH
6657	Malaoxon	(Dimethoxyphosphinylthio)-butanedioic acid	$C_{10}H_{19}O_7PS$	1634-78-2	314.293	liq		$132^{0.1}$			
6658	Malathion		$C_{10}H_{19}O_6PS_2$	121-75-5	330.358	ye-br liq	3.0(0.4)	$156^{0.7}$ dec	1.2076^{20}	1.4960^{20}	sl H₂O; s EtOH, eth, bz
6659	Maleic acid	cis-2-Butenedioic acid	$C_4H_4O_4$	110-16-7	116.073	mcl pr (w)	143.5(0.5)		1.590^{20}		vs H₂O, EtOH, ace; s eth; i bz, chl
6660	Maleic anhydride		$C_4H_2O_3$	108-31-6	98.057	nd (chl, eth)	52.56(0.04)	202	1.314^{60}		s H₂O; s eth, ace, chl; sl lig
6661	Maleonitrile	cis-Butenedinitrile	$C_4H_2N_2$	928-53-0	78.072	pr (EtOH)	31.5	111^{20}			
6662	Malic acid	Hydroxybutanedioic acid	$C_4H_6O_5$	617-48-1	134.088		132		1.601^{20}		s H₂O; vs eth, EtOH, MeOH
6663	Malonaldehyde	1,3-Propanedial	$C_3H_4O_2$	542-78-9	72.063	hyg nd	73				
6664	Malonic acid		$C_3H_4O_4$	141-82-2	104.062	tcl (al)	135(1)	sub	1.619^{10}		vs H₂O, py; s EtOH, eth; i bz
6665	Malononitrile		$C_3H_2N_2$	109-77-3	66.061		31.83(0.02)	219(3)	1.1910^{20}	1.4146^{34}	s H₂O, ace, bz, chl; vs EtOH, eth
6666	Maltopentaose		$C_{30}H_{52}O_{26}$	34620-76-3	828.718	cry (w)	78 (hyd)				
6667	α-Maltose		$C_{12}H_{22}O_{11}$	4482-75-1	342.296	nd (al)	162.5		1.546^{20}		vs H₂O
6668	6-O-α-Maltosyl-β-cyclodextrin		$C_{54}H_{90}O_{45}$	104723-60-6	1459.266	cry (MeOH)					

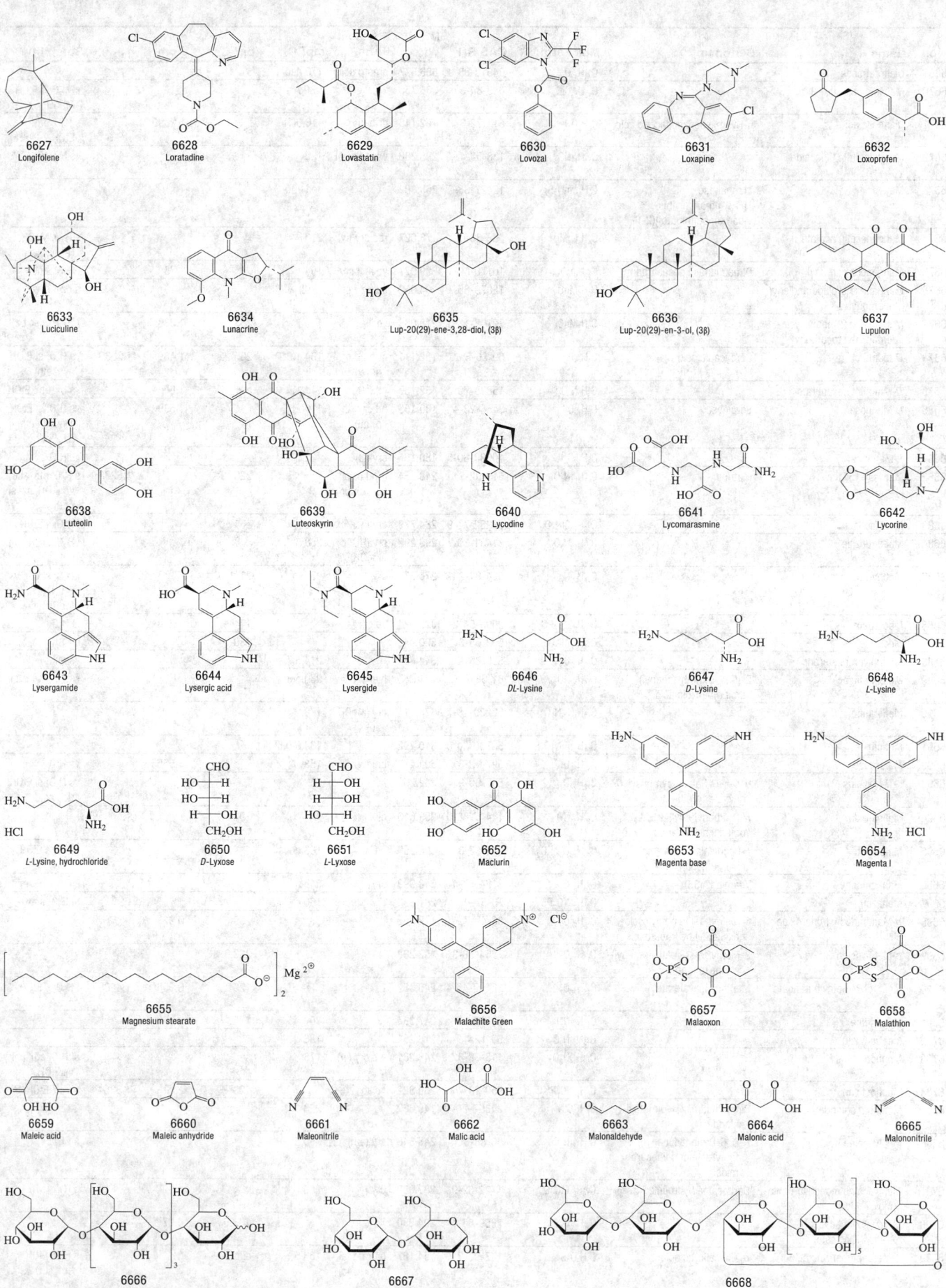

6627 Longifolene

6628 Loratadine

6629 Lovastatin

6630 Lovozal

6631 Loxapine

6632 Loxoprofen

6633 Luciculine

6634 Lunacrine

6635 Lup-20(29)-ene-3,28-diol, (3β)

6636 Lup-20(29)-en-3-ol, (3β)

6637 Lupulon

6638 Luteolin

6639 Luteoskyrin

6640 Lycodine

6641 Lycomarasmine

6642 Lycorine

6643 Lysergamide

6644 Lysergic acid

6645 Lysergide

6646 *DL*-Lysine

6647 *D*-Lysine

6648 *L*-Lysine

6649 *L*-Lysine, hydrochloride

6650 *D*-Lyxose

6651 *L*-Lyxose

6652 Maclurin

6653 Magenta base

6654 Magenta I

6655 Magnesium stearate

6656 Malachite Green

6657 Malaoxon

6658 Malathion

6659 Maleic acid

6660 Maleic anhydride

6661 Maleonitrile

6662 Malic acid

6663 Malonaldehyde

6664 Malonic acid

6665 Malononitrile

6666 Maltopentaose

6667 α-Maltose

6668 6-*O*-α-Maltosyl-β-cyclodextrin

No.	Name	Synonym	Mol. Form.	CAS RN	Mol. Wt.	Physical Form	mp/°C	bp/°C	den g cm⁻³	n_D	Solubility
6669	Maltotetraose		$C_{24}H_{42}O_{21}$	34612-38-9	666.577	amorp solid	170 dec				
6670	Malvidin chloride		$C_{17}H_{15}ClO_7$	643-84-5	366.750		>300				sl H_2O; s EtOH, MeOH
6671	Mandelic acid	α-Hydroxybenzeneacetic acid	$C_8H_8O_3$	611-72-3	152.148	orth pl	118.6(1)		1.2890²⁰		s H_2O, eth, EtOH, i-PrOH
6672	Mandelonitrile glucoside		$C_{14}H_{17}NO_6$	138-53-4	295.288	wh nd or pl (al)	122				vs H_2O, EtOH
6673	Maneb	Manganese, [[1,2-ethanediylbis-[carbamodithioato]](2-)]-	$C_4H_6MnN_2S_4$	12427-38-2	265.302		dec 200				
6674	Manganese(II) acetate		$C_4H_6O_4Mn$	638-38-0	173.027	red cry (w)	210				s H_2O, MeOH, HOAc; i ace
6675	Manganese carbonyl	Dimanganese decacarbonyl	$C_{10}Mn_2O_{10}$	10170-69-1	389.977	ye mcl cry	154		1.75		i H_2O; s os
6676	Manganese cyclopentadienyl tricarbonyl		$C_8H_5MnO_3$	12079-65-1	204.062	pale ye cry	77(2)				s os
6677	Manganese 2-methylcyclo-pentadienyl tricarbonyl		$C_9H_7MnO_3$	12108-13-3	218.088	ye liq	1.5	233	1.388²⁰		i H_2O; misc bz
6678	D-Mannitol	Cordycepic acid	$C_6H_{14}O_6$	69-65-8	182.171	orth nd or pr (w)	164.1(0.1)	295³·⁵	1.489²⁰	1.3330	vs H_2O; sl EtOH, py; i eth
6679	D-Mannitol hexanitrate		$C_6H_8N_6O_{18}$	15825-70-4	452.157	nd (al)	111(1)	exp	1.8²⁰		vs bz, eth, EtOH
6680	D-Mannose	Seminose	$C_6H_{12}O_6$	3458-28-4	180.155	nd or orth pr (al)	118(2)		1.539²⁰		vs H_2O; sl EtOH, MeOH; i eth, bz
6681	L-Mannose		$C_6H_{12}O_6$	10030-80-5	180.155	cry (al)	132				vs H_2O
6682	Matridin-15-one	Matrine	$C_{15}H_{24}N_2O$	519-02-8	248.364	α-nd or pl; β-orth pr		223⁶		1.5286²⁵	s H_2O, eth, ace; vs EtOH, bz; sl peth
6683	Mazindol		$C_{16}H_{13}ClN_2O$	22232-71-9	284.739	cry (ace/hx)	198				i H_2O; s EtOH
6684	Mebendazole		$C_{16}H_{13}N_3O_3$	31431-39-7	295.292	cry (HOAc/ MeOH)	288.5				i H_2O, EtOH, eth, chl
6685	Mebhydroline		$C_{19}H_{20}N_2$	524-81-2	276.375	cry	95	211¹			i H_2O; sl eth; vs EtOH, ace, MeOH
6686	Mecarbam		$C_{10}H_{20}NO_5PS_2$	2595-54-2	329.374	ye oil		144⁰·⁰²	1.223²⁰		sl H_2O
6687	Meclizine		$C_{25}H_{27}ClN_2$	569-65-3	390.948			230			s CS_2
6688	Medroxyprogesterone		$C_{22}H_{32}O_3$	520-85-4	344.487		214.5				vs chl
6689	Mefenamic acid	2-[(2,3-Dimethylphenyl)-amino]benzoic acid	$C_{15}H_{15}NO_2$	61-68-7	241.286	hyg cry	230 dec				s alk; sl eth, chl
6690	Mefloquine		$C_{17}H_{16}F_6N_2O$	53230-10-7	378.311	cry (MeOH aq)	178.2				
6691	Mefluidide		$C_{11}H_{13}F_3N_2O_3S$	53780-34-0	310.292		184.3(0.5)				
6692	Melezitose		$C_{18}H_{32}O_{16}$	597-12-6	504.437	cry (w+2)	153		1.5565²⁵		vs H_2O
6693	α-D-Melibiose	6-O-α-D-Galactopyranosyl-D-glucose	$C_{12}H_{22}O_{11}$	585-99-9	342.296						vs H_2O; sl EtOH; dec acid
6694	Melinamide	N-(1-Phenylethyl)-9,12-octadecadieneamide, (Z,Z)-	$C_{26}H_{41}NO$	14417-88-0	383.610	oil	<4	202⁰·⁰⁷		1.5050²³	
6695	Melphalan	L-Phenylalanine, 4-[bis(2-chloroethyl)amino]-	$C_{13}H_{18}Cl_2N_2O_2$	148-82-3	305.200	nd	183 dec				i H_2O; s EtOH
6696	Menaquinone 7	Vitamin K_2(35)	$C_{46}H_{64}O_2$	2124-57-4	648.999	cry	54				
6697	Menazon		$C_6H_{12}N_5O_2PS_2$	78-57-9	281.296	cry (MeOH)	160				sl H_2O; s thf
6698	p-Menthane hydroperoxide	1-Methyl-1-(4-methylcyclo-hexyl)ethyl hydroperoxide	$C_{10}H_{20}O_2$	80-47-7	172.265			259	0.92		
6699	p-Menth-8-en-2-one	2-Methyl-5-(1-methylethenyl)-cyclohexanone	$C_{10}H_{16}O$	7764-50-3	152.233			223.0			
6700	Menthol 3-methylbutanoate	Menthol, isovalerate	$C_{15}H_{28}O_2$	16409-46-4	240.382			129⁹	0.908¹⁵	1.4486²⁰	i H_2O; s EtOH, ace
6701	Meperidine	Pethidine	$C_{15}H_{21}NO_2$	57-42-1	247.334		30	155⁵			
6702	Mephenytoin		$C_{12}H_{14}N_2O_2$	50-12-4	218.251		136				
6703	Mephobarbital		$C_{13}H_{14}N_2O_3$	115-38-8	246.261	wh cry (w)	176				sl H_2O, eth, chl; vs EtOH
6704	Mephosfolan		$C_8H_{16}NO_3PS_2$	950-10-7	269.322	ye liq		120⁰·⁰⁰¹		1.5354²⁶	s ace, EtOH, bz
6705	Mepiquat chloride	Piperidinium, 1,1-dimethyl-, chloride	$C_7H_{16}ClN$	24307-26-4	149.662		223				
6706	Mepivacaine	N-(2,6-Dimethylphenyl)-1-methyl-2-piperidinecarbox-amide	$C_{15}H_{22}N_2O$	96-88-8	246.348	cry (eth)	150.5				s CS_2
6707	Mepivacaine monohydrochlo-ride	Carbocaine hydrochloride	$C_{15}H_{23}ClN_2O$	1722-62-9	282.809	cry	263				s H_2O
6708	Mercaptoacetic acid, 2-ethylhexyl ester		$C_{10}H_{20}O_2S$	7659-86-1	204.330			133.5	0.97²⁰		
6709	2-Mercaptobenzoic acid	o-Thiosalicylic acid	$C_7H_6O_2S$	147-93-3	154.187	lf or nd (al, w, HOAc)	168.5	sub			s H_2O, EtOH, eth; sl DMSO, lig

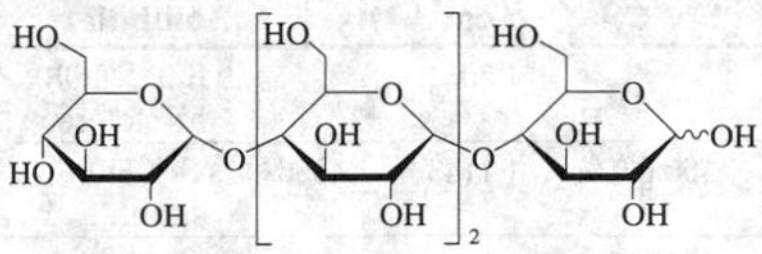

6669 Maltotetraose

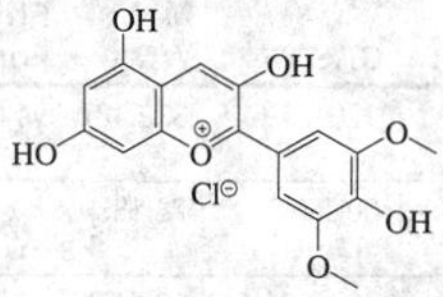

6670 Malvidin chloride

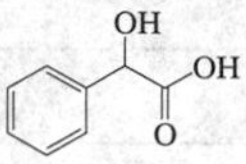

6671 Mandelic acid

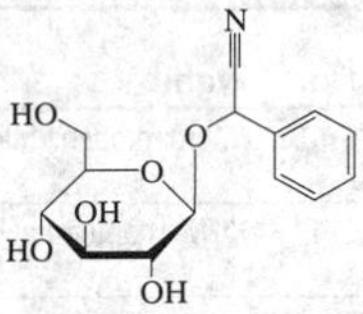

6672 Mandelonitrile glucoside

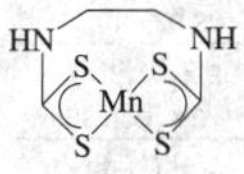

6673 Maneb

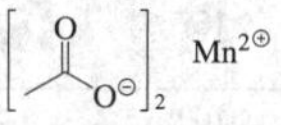

6674 Manganese(II) acetate

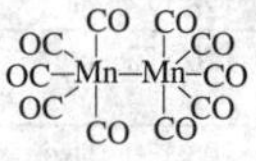

6675 Manganese carbonyl

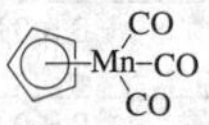

6676 Manganese cyclopentadienyl tricarbonyl

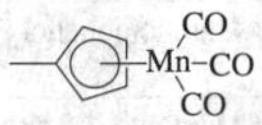

6677 Manganese 2-methylcyclopentadienyl tricarbonyl

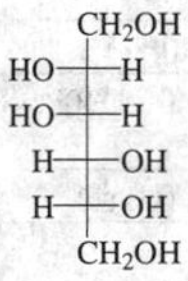

6678 *D*-Mannitol

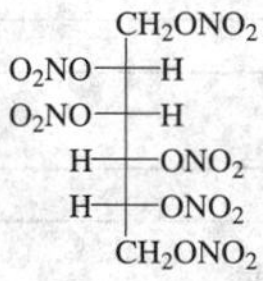

6679 *D*-Mannitol hexanitrate

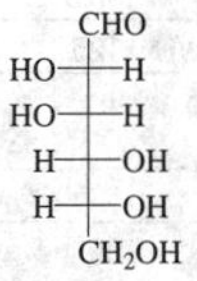

6680 *D*-Mannose

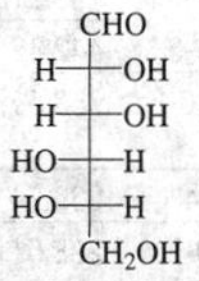

6681 *L*-Mannose

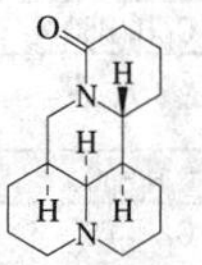

6682 Matridin-15-one

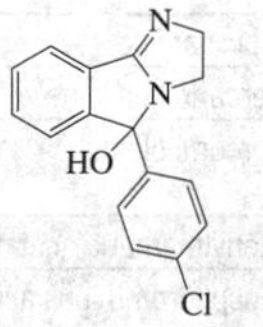

6683 Mazindol

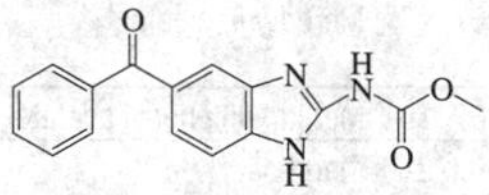

6684 Mebendazole

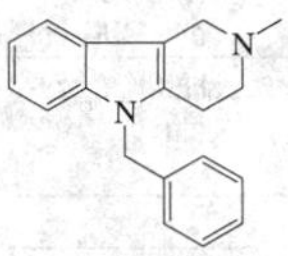

6685 Mebhydroline

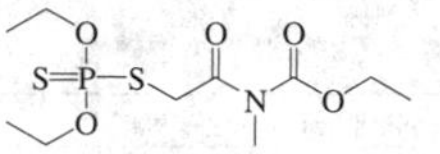

6686 Mecarbam

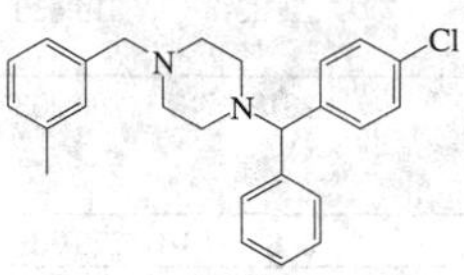

6687 Meclizine

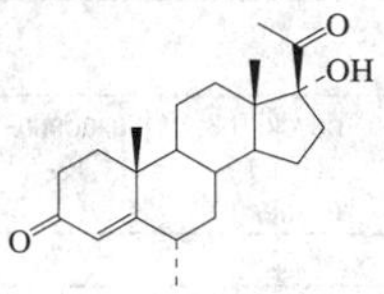

6688 Medroxyprogesterone

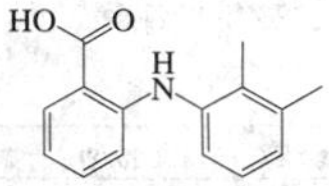

6689 Mefenamic acid

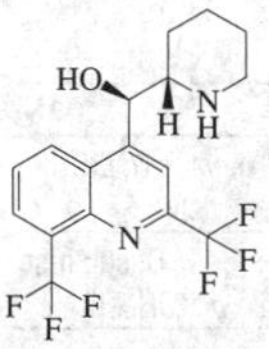

6690 Mefloquine

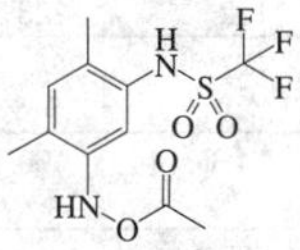

6691 Mefluidide

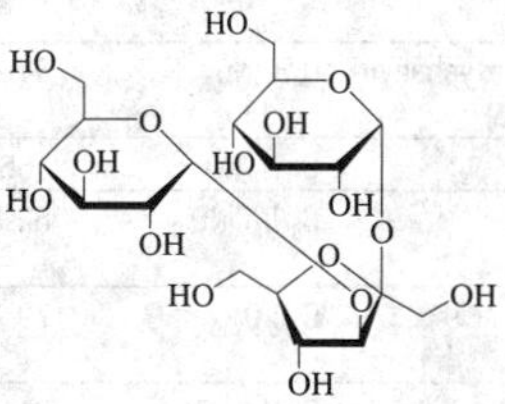

6692 Melezitose

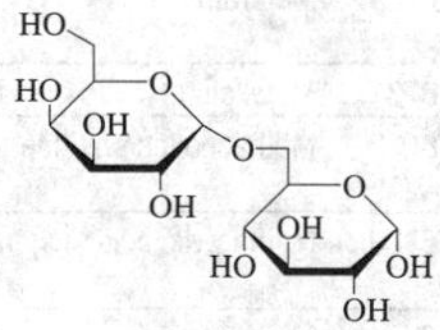

6693 α-*D*-Melibiose

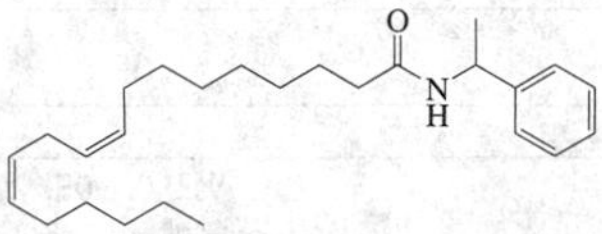

6694 Melinamide

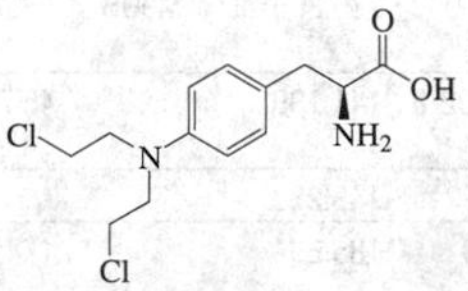

6695 Melphalan

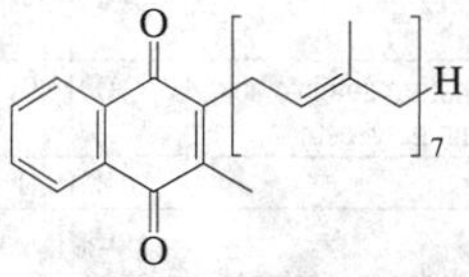

6696 Menaquinone 7

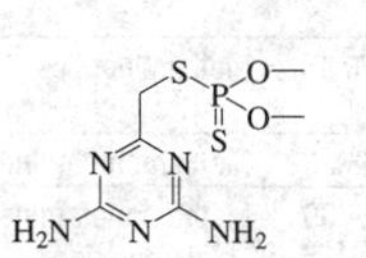

6697 Menazon

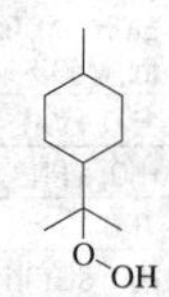

6698 *p*-Menthane hydroperoxide

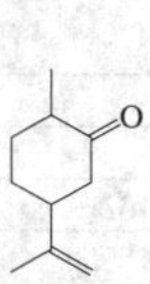

6699 *p*-Menth-8-en-2-one

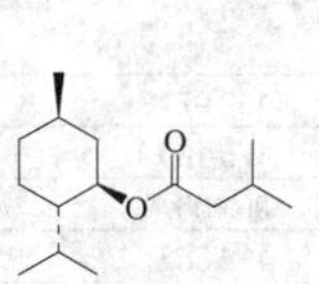

6700 Menthol 3-methylbutanoate

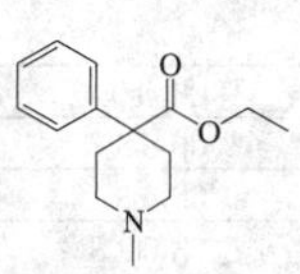

6701 Meperidine

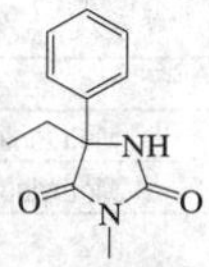

6702 Mephenytoin

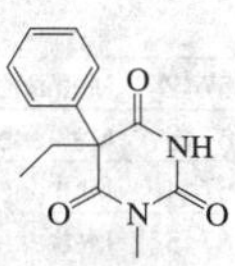

6703 Mephobarbital

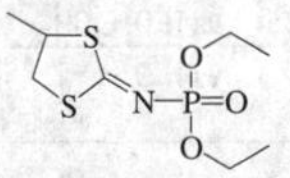

6704 Mephosfolan

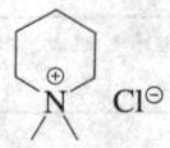

6705 Mepiquat chloride

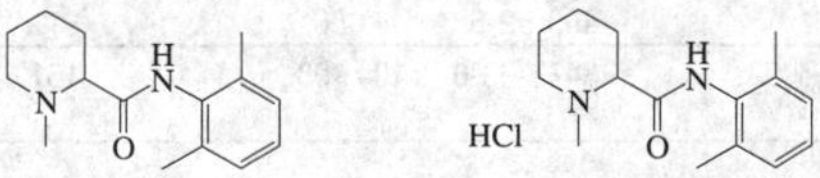

6706 Mepivacaine

6707 Mepivacaine monohydrochloride

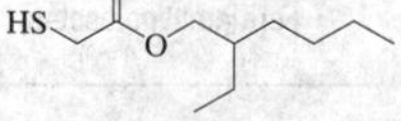

6708 Mercaptoacetic acid, 2-ethylhexyl ester

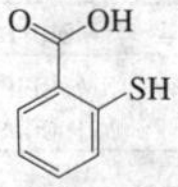

6709 2-Mercaptobenzoic acid

No.	Name	Synonym	Mol. Form.	CAS RN	Mol. Wt.	Physical Form	mp/°C	bp/°C	den g cm⁻³	n_D	Solubility
6710	Mercaptobenzthiazyl ether	2,2'-Dithiobis[benzothiazole]	$C_{14}H_8N_2S_4$	120-78-5	332.487	ye nd	180		1.50		i H_2O; sl EtOH, bz, ctc, ace
6711	2-Mercaptoethanol		C_2H_6OS	60-24-2	78.133			150.0(0.9)	1.1143²⁰	1.4996²⁰	s H_2O, EtOH, eth, bz
6712	2-Mercapto-2-methylpropanoic acid		$C_4H_8O_2S$	4695-31-2	120.171		47	101¹⁵			vs H_2O
6713	2-Mercapto-N-2-naphthylacetamide	Thionalide	$C_{12}H_{11}NOS$	93-42-5	217.286		111.5				i H_2O; vs EtOH, os
6714	2-Mercaptophenol		C_6H_6OS	1121-24-0	126.176	oil	5.5	217	1.2371⁰		vs bz, eth, EtOH
6715	4-Mercaptophenol		C_6H_6OS	637-89-8	126.176	cry	29.5	167⁴⁵	1.1285²⁵	1.5101²⁵	s H_2O, EtOH, alk, con sulf
6716	3-Mercapto-1,2-propanediol	Thioglycerol	$C_3H_8O_2S$	96-27-5	108.160	visc		100¹	1.2455²⁰	1.5268²⁰	sl H_2O, eth, bz, chl; msc EtOH; vs ace
6717	3-Mercaptopropanoic acid		$C_3H_6O_2S$	107-96-0	106.144	amor	16.5(0.5)	111¹⁵	1.218²¹	1.494²⁰	s H_2O, EtOH, eth, ctc
6718	3-Mercapto-D-valine	Penicillamine	$C_5H_{11}NO_2S$	52-67-5	149.212		198.5				
6719	Mercury(II) benzoate	Mercuric benzoate	$C_{14}H_{10}HgO_4$	583-15-3	442.82	cry pow (w)	≈125				i EtOH
6720	Mercury(II) oleate	Mercuric oleate	$C_{36}H_{66}HgO_4$	1191-80-6	763.50	ye-br solid					i H_2O; sl EtOH, eth
6721	Mercury(II) phenyl acetate	Phenylmercuric acetate	$C_8H_8HgO_2$	62-38-4	336.74		153				i H_2O; s chl
6722	Merphos	Phosphorotrithious acid, S,S,S-tributyl ester	$C_{12}H_{27}PS_3$	150-50-5	298.511		100	137⁰·⁷	1.02²⁰		
6723	Mesityl oxide	Isobutenyl methyl ketone	$C_6H_{10}O$	141-79-7	98.142	liq	-52.8(0.4)	129.7(0.4)	0.8653²⁰	1.4440²⁰	s H_2O, ace; msc EtOH, eth
6724	Mesoridazine		$C_{21}H_{26}N_2OS_2$	5588-33-0	386.573	oil					
6725	Mestranol		$C_{21}H_{26}O_2$	72-33-3	310.430	cry	154(1)				i H_2O; s diox, eth, EtOH, chl
6726	[2.2]Metacyclophane	Tricyclo[9.3.1.1]hexadeca-1(15),4,6,8(16),11,13-hexaene	$C_{16}H_{16}$	2319-97-3	208.298	orth pr	131(1)	290			sl EtOH; s bz, eth
6727	Metalaxyl		$C_{15}H_{21}NO_4$	57837-19-1	279.333		73.0(0.5)				
6728	Metaldehyde	Metacetaldehyde (polymer)	$(C_2H_4O)_x$	37273-91-9	44.052	tetr nd or pr (al)	246	115 sub			i H_2O, ace; sl EtOH, eth, bz, chl
6729	Metanil Yellow		$C_{18}H_{14}N_3NaO_3S$	587-98-4	375.377	br-ye pow					vs H_2O, EtOH; s bz, eth; sl ace
6730	Metaraminol	2-Amino-1-(3-hydroxyphenyl)-1-propanol, (1R,2S)	$C_9H_{13}NO_2$	54-49-9	167.205	hyg cry (HCl)					s H_2O
6731	Metaxalone		$C_{12}H_{15}NO_3$	1665-48-1	221.252	cry (AcOEt)	122	223¹·⁵			
6732	Methacholine chloride		$C_8H_{18}ClNO_2$	62-51-1	195.688	hyg cry	172				vs H_2O, EtOH, chl
6733	Methacrylic acid	2-Methylpropenoic acid	$C_4H_6O_2$	79-41-4	86.090	pr	14.6(0.8)	160(1)	1.0153²⁰	1.4314²⁰	s H_2O, chl; msc EtOH, eth
6734	Methacycline		$C_{22}H_{22}N_2O_8$	914-00-1	442.418	cry	205 dec				
6735	Methadone hydrochloride	6-(Dimethylamino)-4,4-diphenyl-3-heptanone hydrochloride	$C_{21}H_{28}ClNO$	1095-90-5	345.906	pl (al-eth)	235				vs H_2O, EtOH
6736	Methallenestril		$C_{18}H_{22}O_3$	517-18-0	286.366	cry (MeOH aq)	139				s eth
6737	Methamidophos	Phosphoramidothioic acid, O,S-dimethyl ester	$C_2H_8NO_2PS$	10265-92-6	141.130		46.7(0.5)		1.31²⁰		
6738	Methamphetamine		$C_{10}H_{15}N$	537-46-2	149.233			212			
6739	Methamphetamine hydrochloride	N,α-Dimethylbenzeneethanamine, hydrochloride, (S)-	$C_{10}H_{16}ClN$	51-57-0	185.694		174(1)				vs H_2O, EtOH, chl
6740	Methandrostenolone		$C_{20}H_{28}O_2$	72-63-9	300.435		166				
6741	Methane		CH_4	74-82-8	16.043	col gas	-182.4566 (0.0003)	-161.5(0.2)	0.4228·¹⁶²		sl H_2O, ace; s EtOH, eth, bz, tol, MeOH
6742	Methanearsonic acid		CH_5AsO_3	124-58-3	139.971		160.5				s H_2O, EtOH
6743	Methanedisulfonic acid	Methionic acid	$CH_4O_6S_2$	503-40-2	176.169		98				i H_2O; s HNO_3
6744	Methanesulfonic acid	Methylsulfonic acid	CH_4O_3S	75-75-2	96.106		20	167¹⁰	1.4812¹⁸	1.4317¹⁸	s H_2O
6745	Methanesulfonyl chloride		CH_3ClO_2S	124-63-0	114.552			159(3)	1.4805¹⁸	1.4573²⁰	i H_2O; s EtOH, eth
6746	Methanesulfonyl fluoride		CH_3FO_2S	558-25-8	98.097			123.5			
6747	Methanethiol	Methyl mercaptan	CH_4S	74-93-1	48.108	col gas	-122.98(0.09)	6.0(0.1)	0.8665²⁰		sl H_2O, chl; vs EtOH, eth
6748	Methanimidamide	Formamidine	CH_4N_2	463-52-5	44.056	pr	81	dec			vs H_2O, EtOH
6749	Methanimidamide, monoacetate	Formamidine acetate	$C_3H_8N_2O_2$	3473-63-0	104.108		161.5				vs H_2O

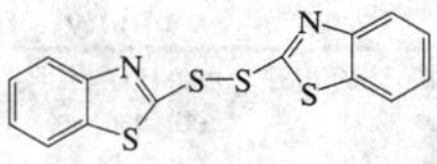

6710 Mercaptobenzthiazyl ether

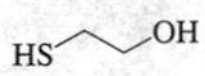

6711 2-Mercaptoethanol

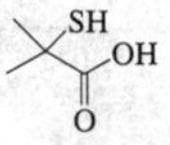

6712 2-Mercapto-2-methylpropanoic acid

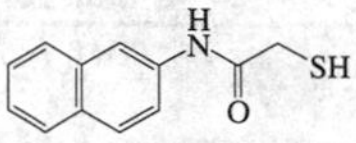

6713 2-Mercapto-N-2-naphthylacetamide

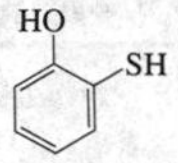

6714 2-Mercaptophenol

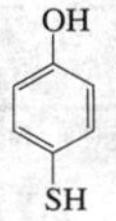

6715 4-Mercaptophenol

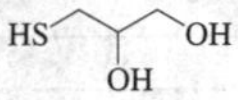

6716 3-Mercapto-1,2-propanediol

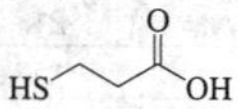

6717 3-Mercaptopropanoic acid

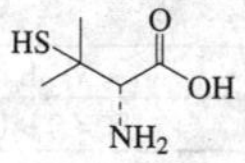

6718 3-Mercapto-D-valine

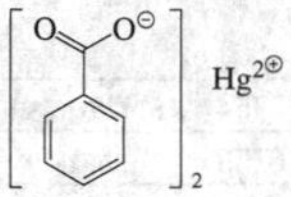

6719 Mercury(II) benzoate

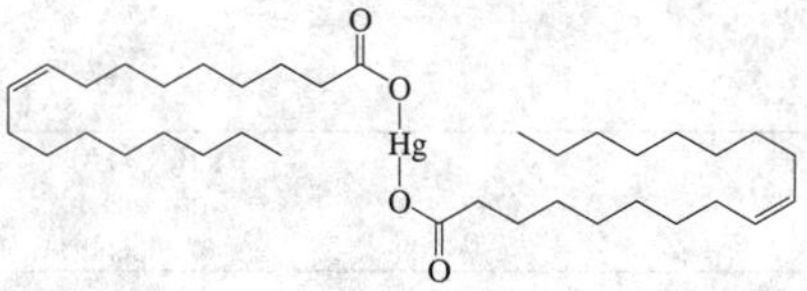

6720 Mercury(II) oleate

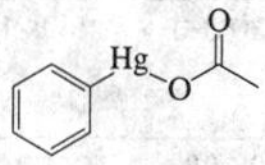

6721 Mercury(II) phenyl acetate

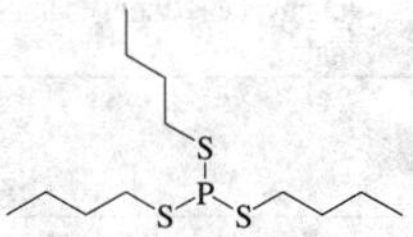

6722 Merphos

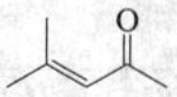

6723 Mesityl oxide

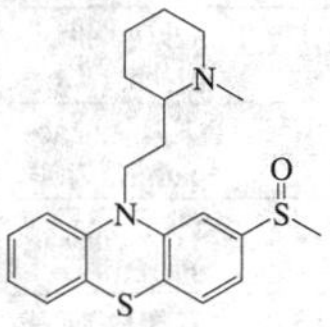

6724 Mesoridazine

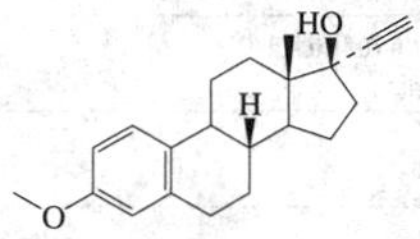

6725 Mestranol

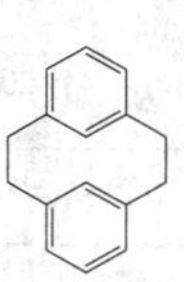

6726 [2.2]Metacyclophane

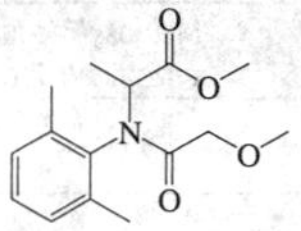

6727 Metalaxyl

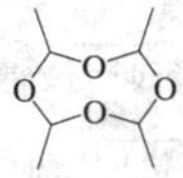

6728 Metaldehyde

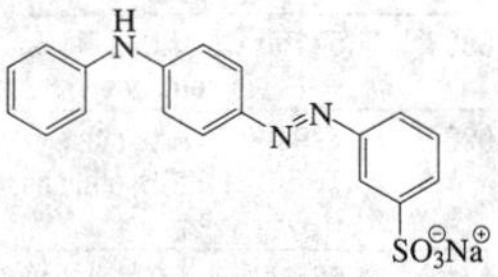

6729 Metanil Yellow

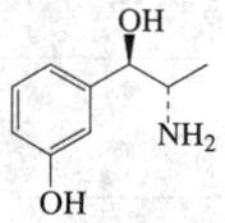

6730 Metaraminol

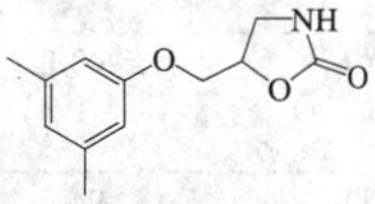

6731 Metaxalone

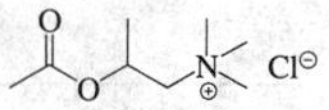

6732 Methacholine chloride

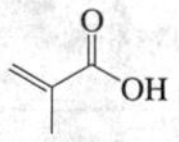

6733 Methacrylic acid

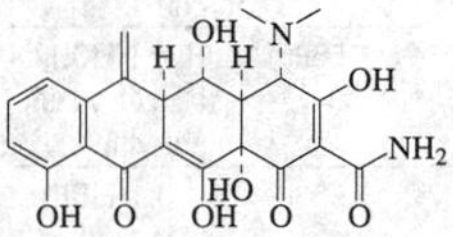

6734 Methacycline

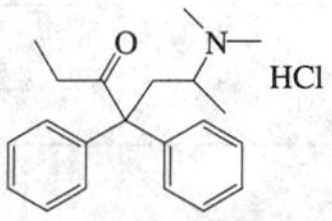

6735 Methadone hydrochloride

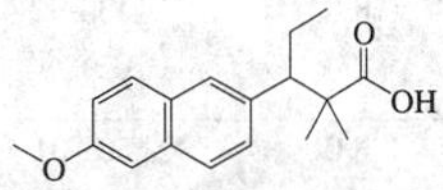

6736 Methallenestril

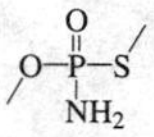

6737 Methamidophos

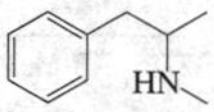

6738 Methamphetamine

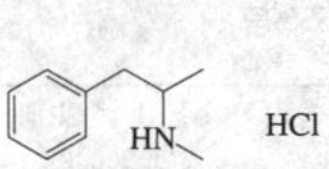

6739 Methamphetamine hydrochloride

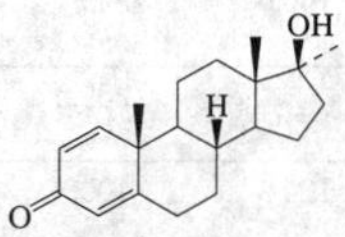

6740 Methandrostenolone

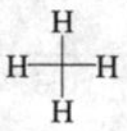

6741 Methane

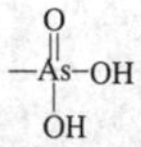

6742 Methanearsonic acid

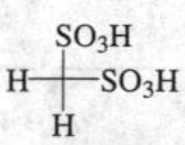

6743 Methanedisulfonic acid

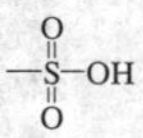

6744 Methanesulfonic acid

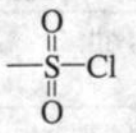

6745 Methanesulfonyl chloride

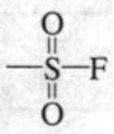

6746 Methanesulfonyl fluoride

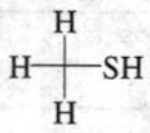

6747 Methanethiol

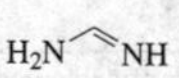

6748 Methanimidamide

6749 Methanimidamide, monoacetate

No.	Name	Synonym	Mol. Form.	CAS RN	Mol. Wt.	Physical Form	mp/°C	bp/°C	den g cm⁻³	n_D	Solubility
6750	Methanol	Methyl alcohol	CH₄O	67-56-1	32.042	liq	-97.5(0.1)	64.5(0.7)	0.7914[20]	1.3288[20]	msc H₂O, EtOH, eth, ace; vs bz; s chl
6751	Methantheline bromide		C₂₁H₂₆BrNO₃	53-46-3	420.340	cry (i-PrOH)	174.5				s H₂O, EtOH, chl; i eth
6752	Methapyrilene		C₁₄H₁₉N₃S	91-80-5	261.386			174[3]		1.5915[20]	
6753	Metharbital	5,5-Diethyl-1-methyl-2,4,6(1H,3H,5H)-pyrimidine-trione	C₉H₁₄N₂O₃	50-11-3	198.218	nd	150.5				s H₂O; sl chl
6754	Methazolamide		C₅H₈N₄O₃S₂	554-57-4	236.273	cry (w)	213 dec				
6755	Methazole		C₉H₆Cl₂N₂O₃	20354-26-1	261.061		123.9(0.5)		1.24[25]		
6756	Methenamine allyl iodide	Allylhexamethylenetetramine iodide	C₉H₁₇IN₄	36895-62-2	308.162	cry	148 dec				vs H₂O; i chl, eth
6757	Methestrol		C₂₀H₂₆O₂	130-73-4	298.419	cry (dil HOAc)	145				
6758	Methidathion		C₆H₁₁N₂O₄PS₃	950-37-8	302.330		42.5(0.5)				
6759	Methiocarb	Phenol, 3,5-dimethyl-4-(methylthio)-, methylcarbamate	C₁₁H₁₅NO₂S	2032-65-7	225.308		121.5(0.4)				
6760	L-Methionine		C₅H₁₁NO₂S	63-68-3	149.212	hex pl (dil al)	281 dec				s H₂O; i EtOH, eth, ace, bz, peth; sl HOAc
6761	Methocarbamol	Guaifenesin-1-carbamate	C₁₁H₁₅NO₅	532-03-6	241.241	cry (bz)	93				s EtOH
6762	Methomyl		C₅H₁₀N₂O₂S	16752-77-5	162.210		79(2)		1.2946[24]		
6763	Methoprene		C₁₉H₃₄O₃	40596-69-8	310.471			100[0.05]	0.926[20]		
6764	Methoprotryne		C₁₁H₂₁N₅OS	841-06-5	271.383	cry	69				sl H₂O; s os
6765	Methotrexate		C₂₀H₂₂N₈O₅	59-05-2	454.440	ye cry (w)	190 dec				
6766	Methoxamine hydrochloride		C₁₁H₁₈ClNO₃	61-16-5	247.719	cry	214				vs H₂O; i eth, bz, chl
6767	Methoxsalen	9-Methoxy-7H-furo[3,2-g][1]-benzopyran-7-one	C₁₂H₈O₄	298-81-7	216.190	pr (dil al) nd (peth)	148				sl H₂O, eth, ace, peth; vs EtOH
6768	Methoxyacetaldehyde		C₃H₆O₂	10312-83-1	74.079			92(4)	1.005[25]	1.3950[20]	vs H₂O, ace, eth, EtOH
6769	Methoxyacetic acid		C₃H₆O₃	625-45-6	90.078	hyg		204(3)	1.1768[20]	1.4168[20]	s H₂O, EtOH, eth
6770	Methoxyacetonitrile		C₃H₅NO	1738-36-9	71.078			121(4)	0.9492[20]	1.3831[20]	sl H₂O; s EtOH, eth, ace, chl, alk, acid
6771	Methoxyacetyl chloride		C₃H₅ClO₂	38870-89-2	108.524			112(7)	1.1871[20]	1.4199[20]	s eth, ace, ctc; vs chl
6772	2-Methoxyaniline	o-Anisidine	C₇H₉NO	90-04-0	123.152	ye liq	6.2	221(7)	1.0923[20]	1.5715[10]	sl H₂O; s EtOH, eth, ace, bz
6773	3-Methoxyaniline	m-Anisidine	C₇H₉NO	536-90-3	123.152	liq	-1	251	1.096[20]	1.5794[20]	sl H₂O, ctc; s EtOH, eth, ace, bz
6774	4-Methoxyaniline	p-Anisidine	C₇H₉NO	104-94-9	123.152	orth pl	57.8(0.9)	243(6)	1.071[57]	1.5559[60]	s H₂O, ace, bz; vs EtOH, eth
6775	2-Methoxyaniline hydrochloride	o-Anisidine hydrochloride	C₇H₁₀ClNO	134-29-2	159.613	nd	225				
6776	1-Methoxy-9,10-anthracene-dione		C₁₅H₁₀O₃	82-39-3	238.238		170.3				sl EtOH; vs bz, chl
6777	2-Methoxybenzaldehyde		C₈H₈O₂	135-02-4	136.149	pr	37.5(0.5)	243(4)	1.1326[20]	1.5600[20]	i H₂O; s EtOH, bz, ctc; vs eth, ace, chl
6778	3-Methoxybenzaldehyde		C₈H₈O₂	591-31-1	136.149			220(5)	1.1187[20]	1.5530[20]	i H₂O; s EtOH, bz; vs eth, ace, chl
6779	4-Methoxybenzaldehyde	p-Anisaldehyde	C₈H₈O₂	123-11-5	136.149	liq	0	255(5)	1.119[15]	1.5730[20]	i H₂O; msc EtOH, eth; vs ace, chl; s bz
6780	4-Methoxybenzamide		C₈H₉NO₂	3424-93-9	151.163	nd or tab (w)	166.5	295			vs H₂O, EtOH
6781	4-Methoxybenzeneacetalde-hyde		C₉H₁₀O₂	5703-26-4	150.174			255.5	1.096[20]	1.5359[20]	
6782	2-Methoxybenzeneacetic acid		C₉H₁₀O₃	93-25-4	166.173	nd (w)	124	100[2]			s H₂O; vs EtOH, eth, ace, bz, chl
6783	4-Methoxybenzeneacetic acid		C₉H₁₀O₃	104-01-8	166.173	pl (w)	84.9(0.5)	138[2]			i H₂O; vs EtOH; s eth, bz; sl chl, lig
6784	4-Methoxybenzeneacetonitrile		C₉H₉NO	104-47-2	147.173			286.5	1.0845[20]	1.5309[20]	s EtOH, eth, chl
6785	4-Methoxy-1,2-benzenedi-amine	4-Methoxy-o-phenylenedi-amine	C₇H₁₀N₂O	102-51-2	138.166	grn pl	51	200[21]			vs eth
6786	4-Methoxy-1,3-benzenedi-amine	4-Methoxy-m-phenylenedi-amine	C₇H₁₀N₂O	615-05-4	138.166	nd (eth)	67.5				s EtOH, eth; sl DMSO

6750
Methanol

6751
Methantheline bromide

6752
Methapyrilene

6753
Metharbital

6754
Methazolamide

6755
Methazole

6756
Methenamine allyl iodide

6757
Methestrol

6758
Methidathion

6759
Methiocarb

6760
L-Methionine

6761
Methocarbamol

6762
Methomyl

6763
Methoprene

6764
Methoprotryne

6765
Methotrexate

6766
Methoxamine hydrochloride

6767
Methoxsalen

6768
Methoxyacetaldehyde

6769
Methoxyacetic acid

6770
Methoxyacetonitrile

6771
Methoxyacetyl chloride

6772
2-Methoxyaniline

6773
3-Methoxyaniline

6774
4-Methoxyaniline

6775
2-Methoxyaniline hydrochloride

6776
1-Methoxy-9,10-anthracenedione

6777
2-Methoxybenzaldehyde

6778
3-Methoxybenzaldehyde

6779
4-Methoxybenzaldehyde

6780
4-Methoxybenzamide

6781
4-Methoxybenzeneacetaldehyde

6782
2-Methoxybenzeneacetic acid

6783
4-Methoxybenzeneacetic acid

6784
4-Methoxybenzeneacetonitrile

6785
4-Methoxy-1,2-benzenediamine

6786
4-Methoxy-1,3-benzenediamine

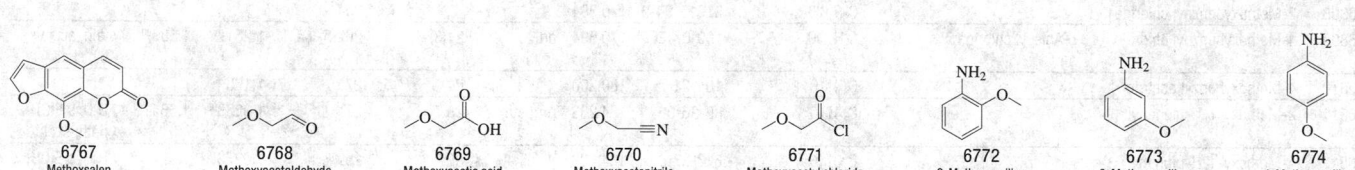

No.	Name	Synonym	Mol. Form.	CAS RN	Mol. Wt.	Physical Form	mp/°C	bp/°C	den g cm⁻³	n_D	Solubility
6787	2-Methoxy-1,4-benzenedi-amine	2,5-Diaminoanisole	$C_7H_{10}N_2O$	5307-02-8	138.166	cry	107				
6788	3-Methoxy-1,2-benzenediol		$C_7H_8O_3$	934-00-9	140.137	nd	42.8	163[48]			s chl
6789	4-Methoxybenzeneethana-mine		$C_9H_{13}NO$	55-81-2	151.205			139[20]		1.5379[20]	
6790	4-Methoxybenzeneethanol		$C_9H_{12}O_2$	702-23-8	152.190		29	335			
6791	2-Methoxybenzenemethana-mine		$C_8H_{11}NO$	6850-57-3	137.179			228	1.051[25]	1.5475[20]	
6792	4-Methoxybenzenemethana-mine		$C_8H_{11}NO$	2393-23-9	137.179			236.5	1.050[15]	1.5462[20]	sl H₂O, EtOH, eth
6793	2-Methoxybenzenemethanol		$C_8H_{10}O_2$	612-16-8	138.164			249	1.0386[25]	1.5455[20]	i H₂O; s EtOH; msc eth
6794	3-Methoxybenzenemethanol		$C_8H_{10}O_2$	6971-51-3	138.164		30	252	1.112[25]	1.5440[20]	
6795	4-Methoxybenzenemethanol	Anise alcohol	$C_8H_{10}O_2$	105-13-5	138.164	nd	17(2)	254(3)	1.109[26]	1.5420[25]	s H₂O, ctc; vs EtOH, eth
6796	4-Methoxybenzenesulfonyl chloride		$C_7H_7ClO_3S$	98-68-0	206.647	nd or pr (bz)	42.5	103[0.25]			s EtOH, eth, bz
6797	3-Methoxybenzenethiol		C_7H_8OS	15570-12-4	140.203			224.5		1.5874[20]	s chl
6798	4-Methoxybenzenethiol		C_7H_8OS	696-63-9	140.203			228	1.1313[25]	1.5801[25]	s EtOH, eth, bz; sl chl
6799	2-Methoxybenzoic acid		$C_8H_8O_3$	579-75-9	152.148	pl (w)	100.9(0.5)	200			sl H₂O; vs EtOH, eth, chl; s bz, ctc
6800	3-Methoxybenzoic acid		$C_8H_8O_3$	586-38-9	152.148	nd (w)	107	170[10]			sl H₂O, ctc; s EtOH, eth, bz; vs chl
6801	4-Methoxybenzoic acid	p-Anisic acid	$C_8H_8O_3$	100-09-4	152.148		184(2)	276.5			i H₂O; vs EtOH, MeOH, eth; s chl
6802	2-Methoxybenzonitrile		C_8H_7NO	6609-56-9	133.148		24.5	255.5	1.1063[20]		s EtOH; vs eth
6803	3-Methoxybenzonitrile		C_8H_7NO	1527-89-5	133.148			140[34]	1.089[25]	1.5402[20]	
6804	4-Methoxybenzonitrile		C_8H_7NO	874-90-8	133.148	nd (w) lf (al)	59.5(0.6)	256.5			i H₂O; vs EtOH, eth; s bz
6805	7-Methoxy-2H-1-benzopyran-2-one		$C_{10}H_8O_3$	531-59-9	176.169	lf (w, MeOH)	118.3				sl H₂O; s EtOH, eth, con sulf, alk
6806	6-Methoxy-2-benzothiazol-amine		$C_8H_8N_2OS$	1747-60-0	180.227		166				
6807	2-(4-Methoxybenzoyl)benzoic acid	o-(p-Anisoyl)benzoic acid	$C_{15}H_{12}O_4$	1151-15-1	256.254	lf (w), cry (al, to)	146				vs eth, EtOH, tol
6808	2-Methoxybenzoyl chloride		$C_8H_7ClO_2$	21615-34-9	170.594			254			
6809	4-Methoxybenzoyl chloride	p-Anisoyl chloride	$C_8H_7ClO_2$	100-07-2	170.594	nd	21(3)	262.5	1.261[20]	1.580[20]	s eth, ace; vs bz; sl ctc
6810	4-Methoxybenzyl acetate		$C_{10}H_{12}O_3$	104-21-2	180.200		84	270	1.105[25]		s ctc
6811	2-Methoxy-1,1'-biphenyl		$C_{13}H_{12}O$	86-26-0	184.233	pr (peth)	29	274	1.0233[99]	1.5641[99]	i H₂O; s EtOH, peth; sl ctc
6812	4-Methoxy-1,1'-biphenyl		$C_{13}H_{12}O$	613-37-6	184.233	pl (al)	90	157[10]	1.0278[100]	1.5744[100]	i H₂O; s EtOH, eth
6813	1-Methoxy-1,3-butadiene		C_5H_8O	3036-66-6	84.117			92(2)	0.8296[20]	1.4594[20]	s H₂O, EtOH
6814	2-Methoxy-1,3-butadiene		C_5H_8O	3588-30-5	84.117			74(6)	0.8272[20]	1.4442[20]	vs ace, bz, eth, EtOH
6815	3-Methoxy-1-butanol		$C_5H_{12}O_2$	2517-43-3	104.148			159(7)	0.923[23]	1.4148[25]	vs EtOH, ace; s eth; sl chl
6816	1-Methoxy-1-buten-3-yne		C_5H_6O	2798-73-4	82.101			123 dec	0.906[20]	1.4818[20]	i H₂O; s chl
6817	Methoxychlor		$C_{16}H_{15}Cl_3O_2$	72-43-5	345.648	cry (dil al)	89(1)		1.41[25]		i H₂O; s EtOH, ctc; vs eth, bz
6818	Methoxycyclohexane		$C_7H_{14}O$	931-56-6	114.185	liq	-74.3(0.4)	132(5)	0.8756[20]	1.4355[20]	vs eth, EtOH
6819	1-Methoxy-2,4-dinitrobenzene		$C_7H_6N_2O_5$	119-27-7	198.133	nd (al or w)	95.9(0.5)	206[12]	1.3364[131]	1.546[15]	sl H₂O; s EtOH, eth, ace, bz; vs py
6820	1-Methoxy-3,5-dinitrobenzene	3,5-Dinitroanisole	$C_7H_6N_2O_5$	5327-44-6	198.133	nd (al)	105.3		1.558[12]		vs ace, bz, MeOH
6821	2-Methoxy-1,2-diphenyletha-none		$C_{15}H_{14}O_2$	3524-62-7	226.271	nd (lig)	49.5	188[15]	1.1278[14]		vs bz, eth, EtOH
6822	2-Methoxyethanol	Ethylene glycol monomethyl ether	$C_3H_8O_2$	109-86-4	76.095	liq	-85.1	124.3(0.1)	0.9647[20]	1.4024[20]	msc H₂O, eth, bz; vs EtOH; s ace; sl chl
6823	(2-Methoxyethoxy)ethene		$C_5H_{10}O_2$	1663-35-0	102.132			107(9)			
6824	2-[2-(2-Methoxyethoxy)-ethoxy]ethanol	Triethyleneglycol monomethyl ether	$C_7H_{16}O_4$	112-35-6	164.200			243(3)			
6825	2-Methoxyethyl acetate	Ethylene glycol monomethyl ether acetate	$C_5H_{10}O_3$	110-49-6	118.131	liq	-70	142(3)	1.0074[19]	1.4002[20]	s H₂O, EtOH, eth; sl ctc
6826	2-Methoxyethyl acrylate	2-Methoxyethyl 2-propenoate	$C_6H_{10}O_3$	3121-61-7	130.141			67[16]	1.012[20]		

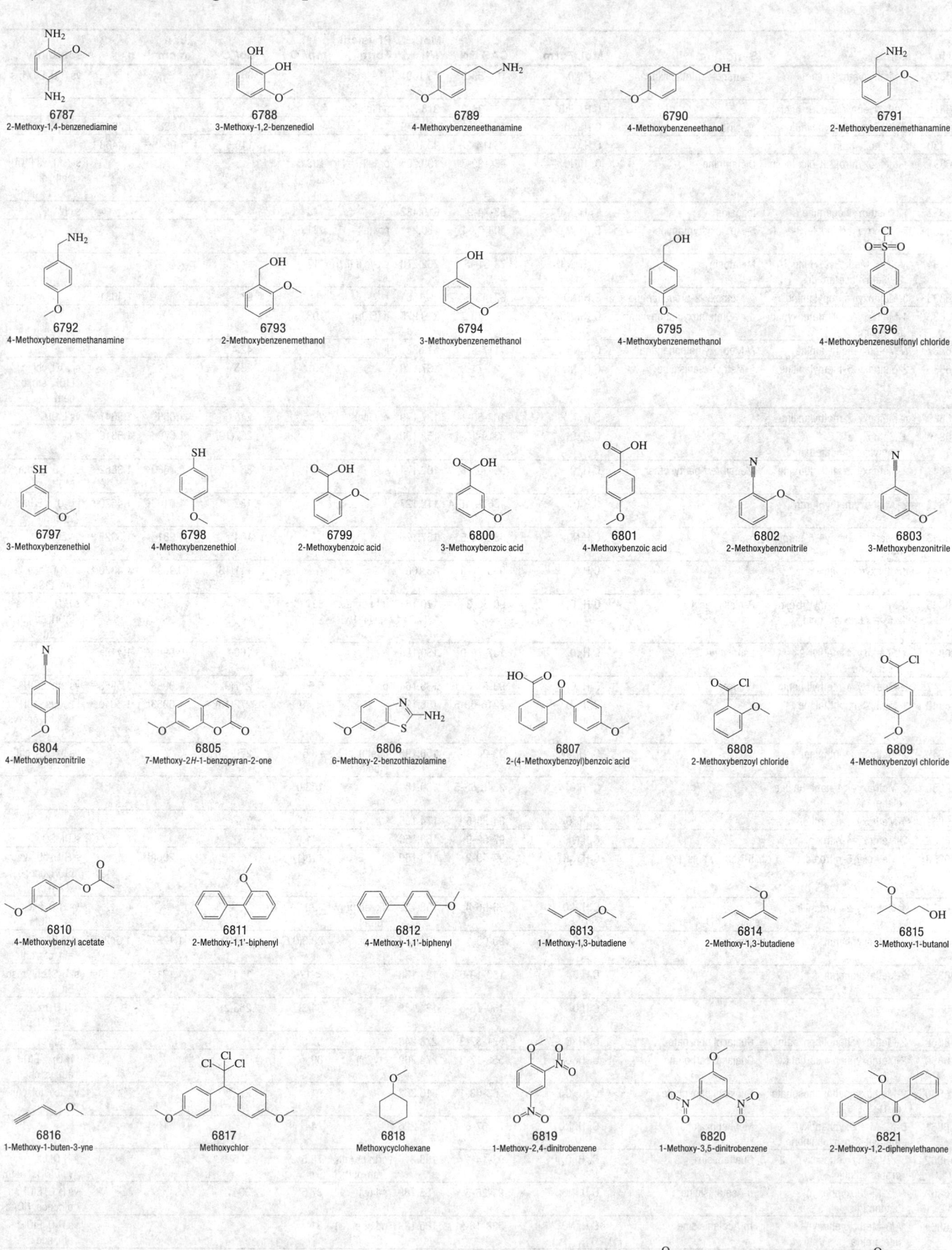

6787 2-Methoxy-1,4-benzenediamine

6788 3-Methoxy-1,2-benzenediol

6789 4-Methoxybenzeneethanamine

6790 4-Methoxybenzeneethanol

6791 2-Methoxybenzenemethanamine

6792 4-Methoxybenzenemethanamine

6793 2-Methoxybenzenemethanol

6794 3-Methoxybenzenemethanol

6795 4-Methoxybenzenemethanol

6796 4-Methoxybenzenesulfonyl chloride

6797 3-Methoxybenzenethiol

6798 4-Methoxybenzenethiol

6799 2-Methoxybenzoic acid

6800 3-Methoxybenzoic acid

6801 4-Methoxybenzoic acid

6802 2-Methoxybenzonitrile

6803 3-Methoxybenzonitrile

6804 4-Methoxybenzonitrile

6805 7-Methoxy-2H-1-benzopyran-2-one

6806 6-Methoxy-2-benzothiazolamine

6807 2-(4-Methoxybenzoyl)benzoic acid

6808 2-Methoxybenzoyl chloride

6809 4-Methoxybenzoyl chloride

6810 4-Methoxybenzyl acetate

6811 2-Methoxy-1,1'-biphenyl

6812 4-Methoxy-1,1'-biphenyl

6813 1-Methoxy-1,3-butadiene

6814 2-Methoxy-1,3-butadiene

6815 3-Methoxy-1-butanol

6816 1-Methoxy-1-buten-3-yne

6817 Methoxychlor

6818 Methoxycyclohexane

6819 1-Methoxy-2,4-dinitrobenzene

6820 1-Methoxy-3,5-dinitrobenzene

6821 2-Methoxy-1,2-diphenylethanone

6822 2-Methoxyethanol

6823 (2-Methoxyethoxy)ethene

6824 2-[2-(2-Methoxyethoxy)ethoxy]ethanol

6825 2-Methoxyethyl acetate

6826 2-Methoxyethyl acrylate

No.	Name	Synonym	Mol. Form.	CAS RN	Mol. Wt.	Physical Form	mp/°C	bp/°C	den g cm⁻³	n_D	Solubility
6827	2-Methoxyethylamine	1-Amino-2-methoxyethane	C₃H₉NO	109-85-3	75.109			95			vs H₂O, EtOH; sl chl
6828	Methoxyethylmercuric acetate		C₅H₁₀HgO₃	151-38-2	318.72	nd (peth)	42				
6829	2-(2-Methoxyethyl)pyridine	Metyridine	C₈H₁₁NO	114-91-0	137.179			203	0.988²⁰	1.4975²⁰	vs H₂O, EtOH
6830	2-Methoxyfuran		C₅H₆O₂	25414-22-6	98.101			110.5	1.0646²⁵	1.4468²⁵	
6831	4-Methoxyfuro[2,3-b]quinoline	Dictamnine	C₁₂H₉NO₂	484-29-7	199.205	pr (al)	133.5				sl H₂O; vs EtOH; s eth, chl, AcOEt
6832	12-Methoxyibogamine	Ibogaine	C₂₀H₂₆N₂O	83-74-9	310.432		148				s chl
6833	5-Methoxy-1H-indole-3-ethanamine	5-Methoxytryptamine	C₁₁H₁₄N₂O	608-07-1	190.241	cry (al)	121.5				
6834	N-[2-(5-Methoxy-1H-indol-3-yl)ethyl]acetamide	Melatonin	C₁₃H₁₆N₂O₂	73-31-4	232.278	pa ye lf (bz)	117				
6835	3-Methoxyisopropylamine	1-Methoxy-2-propanamine	C₄H₁₁NO	37143-54-7	89.136			97		1.4031²⁵	
6836	4-Methoxy-N-(4-methoxyphenyl)aniline	4,4'-Dimethoxydiphenylamine	C₁₄H₁₅NO₂	101-70-2	229.275	lf (EtOH)	103				
6837	N-Methoxymethylamine	N-Methoxymethanamine	C₂H₇NO	1117-97-1	61.083	liq		42.4			
6838	2-Methoxy-5-methylaniline	5-Methyl-o-anisidine	C₈H₁₁NO	120-71-8	137.179		53	235			sl H₂O, chl; s EtOH, eth, bz, peth
6839	4-Methoxy-2-methylaniline		C₈H₁₁NO	102-50-1	137.179	cry (lig)	29.5	248.5	1.065²⁵	1.5647²⁰	vs EtOH
6840	4-Methoxy-α-methylbenzenemethanol		C₉H₁₂O₂	3319-15-1	152.190			257(19)	1.0794²⁰	1.5310²⁵	s ctc
6841	2-Methoxy-2-methylbutane	Methyl tert-pentyl ether	C₆H₁₄O	994-05-8	102.174			86.4(0.1)	0.7660²⁵	1.3862²⁵	sl H₂O; vs eth, EtOH
6842	2-(Methoxymethyl)furan		C₆H₈O₂	13679-46-4	112.127			132	1.0163²⁰	1.4570²⁰	i H₂O; s EtOH; vs eth
6843	2-(Methoxymethyl)-5-nitrofuran		C₆H₇NO₄	586-84-5	157.125			104³	1.281²⁰	1.5325²⁰	vs EtOH
6844	(Methoxymethyl)oxirane		C₄H₈O₂	930-37-0	88.106			111(18)	0.9890²⁰	1.4320²⁰	vs H₂O, ace, eth, EtOH
6845	3-Methoxy-5-methyl-4-oxo-2,5-hexadienoic acid	Penicillic acid	C₈H₁₀O₄	90-65-3	170.163	orth or hex pl (+ 1w)	83				s H₂O, ace; vs EtOH, eth, bz; sl peth
6846	4-Methoxy-4-methyl-2-pentanone	Pentoxone	C₇H₁₄O₂	107-70-0	130.185			160	0.8980²⁵	1.418²⁰	
6847	2-Methoxy-4-methylphenol	Creosol	C₈H₁₀O₂	93-51-6	138.164	pr	5.5	217(9)	1.098²⁰	1.5353²⁵	vs eth, EtOH
6848	1-Methoxynaphthalene		C₁₁H₁₀O	2216-69-5	158.196		<-10	272(19)	1.0963¹⁴	1.6940²⁵	i H₂O; s EtOH, eth, bz, chl; vs CS₂
6849	2-Methoxynaphthalene		C₁₁H₁₀O	93-04-9	158.196	lf (eth), pl (peth)	73.5	270(18)			vs bz, eth, chl
6850	2-Methoxy-1,4-naphthalenedione		C₁₁H₈O₃	2348-82-5	188.180		183.0				
6851	4-Methoxy-1-naphthol		C₁₁H₁₀O₂	84-85-5	174.196		129.8				
6852	2-Methoxy-4-nitroaniline		C₇H₈N₂O₃	97-52-9	168.150		141.0				s DMSO
6853	2-Methoxy-5-nitroaniline	5-Nitro-o-anisidine	C₇H₈N₂O₃	99-59-2	168.150		118(1)		1.2068¹⁵		s H₂O, eth; vs EtOH, ace, bz; sl lig
6854	4-Methoxy-2-nitroaniline		C₇H₈N₂O₃	96-96-8	168.150	dk red pr (w or al)	129				vs H₂O, ace, eth, EtOH
6855	2-Methoxyphenol	Guaiacol	C₇H₈O₂	90-05-1	124.138	hex pr	28.3(0.6)	204(1)	1.1287²¹	1.5429²⁰	sl H₂O; s EtOH, eth, ctc, chl
6856	3-Methoxyphenol		C₇H₈O₂	150-19-6	124.138		<-17	114⁵	1.131²⁵	1.5510²⁰	sl H₂O, chl; msc EtOH, eth
6857	4-Methoxyphenol		C₇H₈O₂	150-76-5	124.138	pl	54(3)	253(2)			s H₂O, bz, ctc; vs EtOH, eth
6858	2-Methoxyphenol benzoate	Guaiacol benzoate	C₁₄H₁₂O₃	531-37-3	228.243		57.5				vs eth, chl
6859	2-Methoxyphenol carbonate (2:1)	Guaiacol carbonate	C₁₅H₁₄O₅	553-17-3	274.269	cry (al)	89				i H₂O; sl EtOH; s eth; vs chl
6860	2-Methoxyphenol phosphate (3:1)	Guaiacol phosphate	C₂₁H₂₁O₇P	563-03-1	416.362		91	277³			vs ace, tol, chl
6861	5-[(2-Methoxyphenoxy)methyl]-2-oxazolidinone	Mephenoxalone	C₁₁H₁₃NO₄	70-07-5	223.226		144				
6862	3-(2-Methoxyphenoxy)-1,2-propanediol	Guaifenesin	C₁₀H₁₄O₄	93-14-1	198.216	orth pr (eth, eth-peth)	78.5	215¹⁹			s H₂O, bz, chl; vs EtOH; i peth
6863	N-(2-Methoxyphenyl)acetamide	o-Acetanisidine	C₉H₁₁NO₂	93-26-5	165.189	nd (w)	87.5	304			vs H₂O, EtOH; s eth, ace, HOAc
6864	N-(3-Methoxyphenyl)acetamide	m-Acetanisidine	C₉H₁₁NO₂	588-16-9	165.189	nd or pl (w)	81				vs H₂O, EtOH; s eth, ace
6865	N-(4-Methoxyphenyl)acetamide	p-Acetanisidine	C₉H₁₁NO₂	51-66-1	165.189	pl (w)	127.2(0.5)				vs ace, EtOH, chl

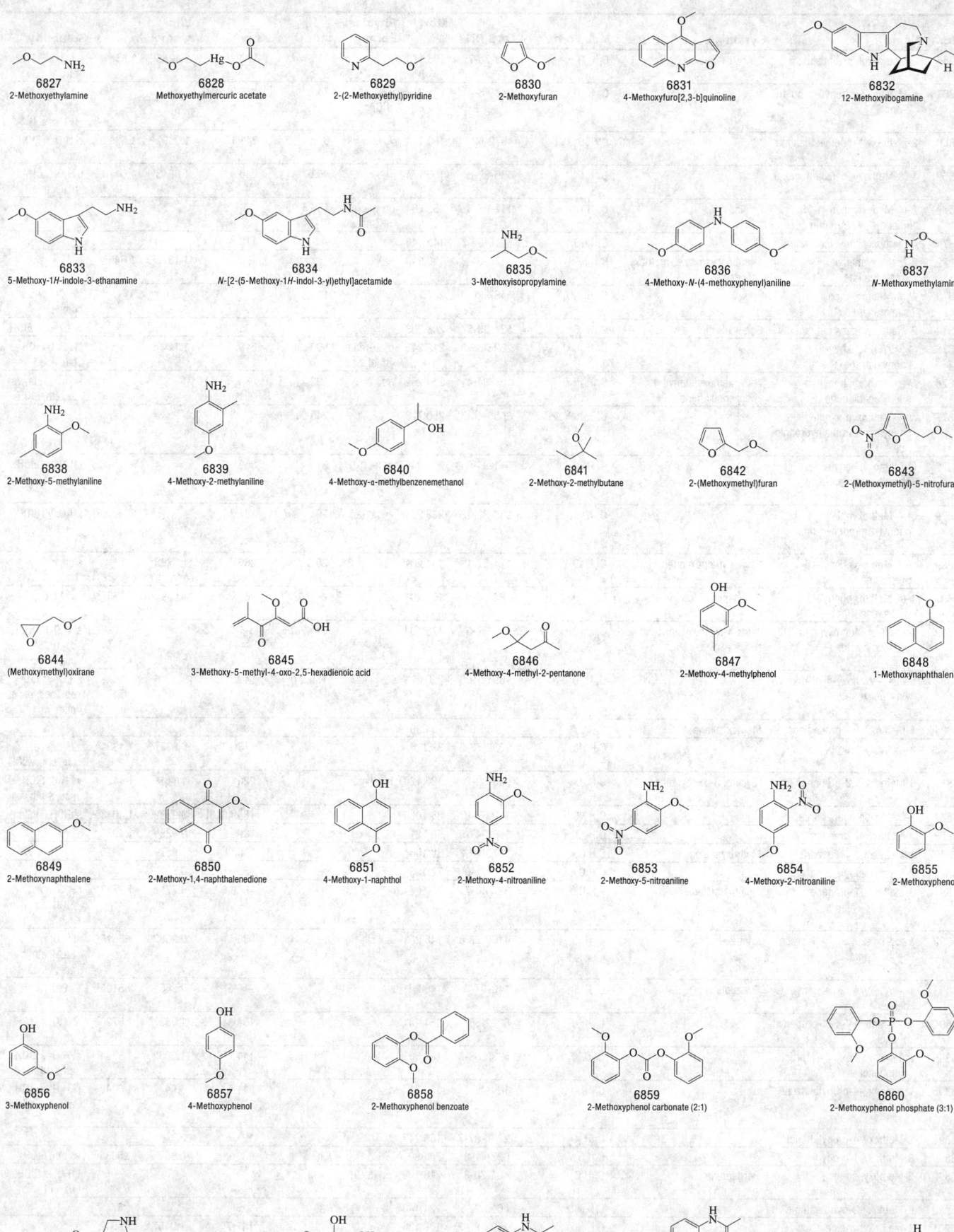

6827 2-Methoxyethylamine

6828 Methoxyethylmercuric acetate

6829 2-(2-Methoxyethyl)pyridine

6830 2-Methoxyfuran

6831 4-Methoxyfuro[2,3-b]quinoline

6832 12-Methoxyibogamine

6833 5-Methoxy-1H-indole-3-ethanamine

6834 N-[2-(5-Methoxy-1H-indol-3-yl)ethyl]acetamide

6835 3-Methoxyisopropylamine

6836 4-Methoxy-N-(4-methoxyphenyl)aniline

6837 N-Methoxymethylamine

6838 2-Methoxy-5-methylaniline

6839 4-Methoxy-2-methylaniline

6840 4-Methoxy-α-methylbenzenemethanol

6841 2-Methoxy-2-methylbutane

6842 2-(Methoxymethyl)furan

6843 2-(Methoxymethyl)-5-nitrofuran

6844 (Methoxymethyl)oxirane

6845 3-Methoxy-5-methyl-4-oxo-2,5-hexadienoic acid

6846 4-Methoxy-4-methyl-2-pentanone

6847 2-Methoxy-4-methylphenol

6848 1-Methoxynaphthalene

6849 2-Methoxynaphthalene

6850 2-Methoxy-1,4-naphthalenedione

6851 4-Methoxy-1-naphthol

6852 2-Methoxy-4-nitroaniline

6853 2-Methoxy-5-nitroaniline

6854 4-Methoxy-2-nitroaniline

6855 2-Methoxyphenol

6856 3-Methoxyphenol

6857 4-Methoxyphenol

6858 2-Methoxyphenol benzoate

6859 2-Methoxyphenol carbonate (2:1)

6860 2-Methoxyphenol phosphate (3:1)

6861 5-[(2-Methoxyphenoxy)methyl]-2-oxazolidinone

6862 3-(2-Methoxyphenoxy)-1,2-propanediol

6863 N-(2-Methoxyphenyl)acetamide

6864 N-(3-Methoxyphenyl)acetamide

6865 N-(4-Methoxyphenyl)acetamide

No.	Name	Synonym	Mol. Form.	CAS RN	Mol. Wt.	Physical Form	mp/°C	bp/°C	den g cm⁻³	n_D	Solubility
6866	2-Methoxyphenyl acetate	2-Acetoxyanisole	C₉H₁₀O₃	613-70-7	166.173		31.5	123[13]	1.1285[25]	1.5101[25]	i H₂O; s EtOH, eth
6867	4-(4-Methoxyphenyl)-3-buten-2-one		C₁₁H₁₂O₂	943-88-4	176.212	lf (al, eth, HOAc)	74.0	187.5[19]			i H₂O; vs EtOH, eth; s bz, HOAc, sulf
6868	2-Methoxy-1-phenylethanone		C₉H₁₀O₂	4079-52-1	150.174	ye liq	8	241(19)	1.0897[20]	1.5393[20]	sl H₂O; s EtOH, ace
6869	1-(3-Methoxyphenyl)ethanone		C₉H₁₀O₂	586-37-8	150.174		95.5	240	1.0343[19]	1.5410[20]	s H₂O, EtOH, ace, ctc
6870	2-(4-Methoxyphenyl)-1H-indene-1,3(2H)-dione	Anisindione	C₁₆H₁₂O₃	117-37-3	252.264	pa ye cry (HOAc, al)	156.5				
6871	4-Methoxyphenyl isocyanate		C₈H₇NO₂	5416-93-3	149.148			110[10]			
6872	2-Methoxyphenyl isothiocyanate	1-Isothiocyanato-2-methoxybenzene	C₈H₇NOS	3288-04-8	165.213			264	1.1878[20]	1.6458[20]	
6873	N-(4-Methoxyphenyl)-3-oxobutanamide		C₁₁H₁₃NO₃	5437-98-9	207.226		117.3				s EtOH, chl; sl eth
6874	2-Methoxyphenyl pentanoate	Guaiacol valerate	C₁₂H₁₆O₃	531-39-5	208.253			265	1.05[25]		vs bz, eth, EtOH
6875	(4-Methoxyphenyl)-phenyldiazene		C₁₃H₁₂N₂O	2396-60-3	212.246	oran-red pl, lf (al, peth)	56	340	1.12[75]		i H₂O; s EtOH, eth, ace
6876	N-(p-Methoxyphenyl)-p-phenylenediamine	N-(4-Methoxyphenyl)-1,4-benzenediamine	C₁₃H₁₄N₂O	101-64-4	214.262	nd	102	238[12]			sl H₂O, peth; vs bz, eth, EtOH
6877	N-(4-Methoxyphenyl)-p-phenylenediamine hydrochloride		C₁₃H₁₅ClN₂O	3566-44-7	250.723	cry	245 dec				
6878	(4-Methoxyphenyl)-phenylmethanone		C₁₄H₁₂O₂	611-94-9	212.244	pr (eth)	61.5	355			i H₂O; vs EtOH, eth; s ace, bz, HOAc
6879	3-(4-Methoxyphenyl)-1-phenyl-2-propen-1-one		C₁₆H₁₄O₂	959-33-1	238.281	ye nd (al)	79	187[19]			i H₂O; vs EtOH; s eth, ctc, chl, HOAc
6880	1-(4-Methoxyphenyl)-1-propanone	Ethyl 4-methoxyphenyl ketone	C₁₀H₁₂O₂	121-97-1	164.201		25.5	266	1.0798[16]		s ctc
6881	1-(4-Methoxyphenyl)-2-propanone	Anisyl methyl ketone	C₁₀H₁₂O₂	122-84-9	164.201		<−15	268	1.0694[17]	1.5253[20]	vs eth, EtOH
6882	trans-3-(4-Methoxyphenyl)-2-propenoic acid	trans-4-Methoxycinnamic acid	C₁₀H₁₀O₃	943-89-5	178.184		174.2(0.5)				sl H₂O, EtOH, bz, DMSO; s ctc, HOAc
6883	trans-1-Methoxy-4-(2-phenylvinyl)benzene		C₁₅H₁₄O	1694-19-5	210.271		136.5	142.5[15]			i H₂O; vs EtOH, eth, ace, bz; s peth
6884	1-Methoxy-1,2-propadiene	Methoxyallene	C₄H₆O	13169-00-1	70.090	oil		50(19)			
6885	3-Methoxy-1-propanamine		C₄H₁₁NO	5332-73-0	89.136			113(12)	0.8727[20]	1.4391[20]	s H₂O, ace, bz, ctc, chl, MeOH
6886	3-Methoxy-1,2-propanediol	Glycerol 3-methyl ether	C₄H₁₀O₃	623-39-2	106.120	hyg liq		220	1.114[20]	1.442[25]	vs H₂O, EtOH, ace; s eth
6887	3-Methoxypropanenitrile		C₄H₇NO	110-67-8	85.105			165.4(0.6)	0.9379[20]	1.4043[20]	s EtOH, eth, chl
6888	2-Methoxy-1-propanol		C₄H₁₀O₂	1589-47-5	90.121			130	0.938[20]	1.4070[20]	
6889	1-Methoxy-2-propanone	Methoxyacetone	C₄H₈O₂	5878-19-3	88.106			110(6)	0.957[25]	1.3970[20]	
6890	2-Methoxy-1-propene		C₄H₈O	116-11-0	72.106			35.7(0.3)	0.7372[20]		
6891	3-Methoxy-1-propene	Allyl methyl ether	C₄H₈O	627-40-7	72.106			46(3)	0.77[11]	1.3778[20]	i H₂O; msc EtOH, eth; s ace
6892	trans-1-Methoxy-4-(1-propenyl)benzene	Anethole	C₁₀H₁₂O	4180-23-8	148.201	col oily liq	22.5	236(5)	0.9882[20]	1.5615[20]	sl H₂O; msc EtOH, eth; s ace; vs bz
6893	1-Methoxy-4-(2-propenyl)benzene	Estragole	C₁₀H₁₂O	140-67-0	148.201			216(5)	0.965[25]	1.5195[20]	vs EtOH, chl
6894	cis-2-Methoxy-4-(1-propenyl)phenol		C₁₀H₁₂O₂	5912-86-7	164.201			134[13]	1.0837[20]	1.5726[20]	sl H₂O; s EtOH, eth
6895	trans-2-Methoxy-4-(1-propenyl)phenol		C₁₀H₁₂O₂	5932-68-3	164.201		33.5	141[13]	1.0852[20]	1.5784[20]	sl H₂O; s EtOH, eth, chl
6896	1-Methoxy-4-propylbenzene		C₁₀H₁₄O	104-45-0	150.217			211.5	0.9472[20]	1.5045[20]	sl H₂O; s EtOH, ace, bz, chl; vs eth
6897	2-Methoxy-4-propylphenol		C₁₀H₁₄O₂	2785-87-7	166.217			121[10]			
6898	3-Methoxy-1-propyne		C₄H₆O	627-41-8	70.090			63	0.83[12]	1.5035[20]	vs eth, EtOH
6899	5-Methoxypsoralen	Bergaptene	C₁₂H₈O₄	484-20-8	216.190	nd (EtOH)	188				i H₂O; sl EtOH, bz, chl
6900	6-Methoxy-3-pyridinamine		C₆H₈N₂O	6628-77-9	124.140		30	125[10]		1.5745[20]	
6901	2-Methoxypyridine		C₆H₇NO	1628-89-3	109.126			142(7)	1.0457[20]	1.5042[20]	
6902	3-Methoxypyridine		C₆H₇NO	7295-76-3	109.126	liq		178.5	1.083	1.5180[20]	
6903	4-Methoxypyridine		C₆H₇NO	620-08-6	109.126			193(14)			msc H₂O

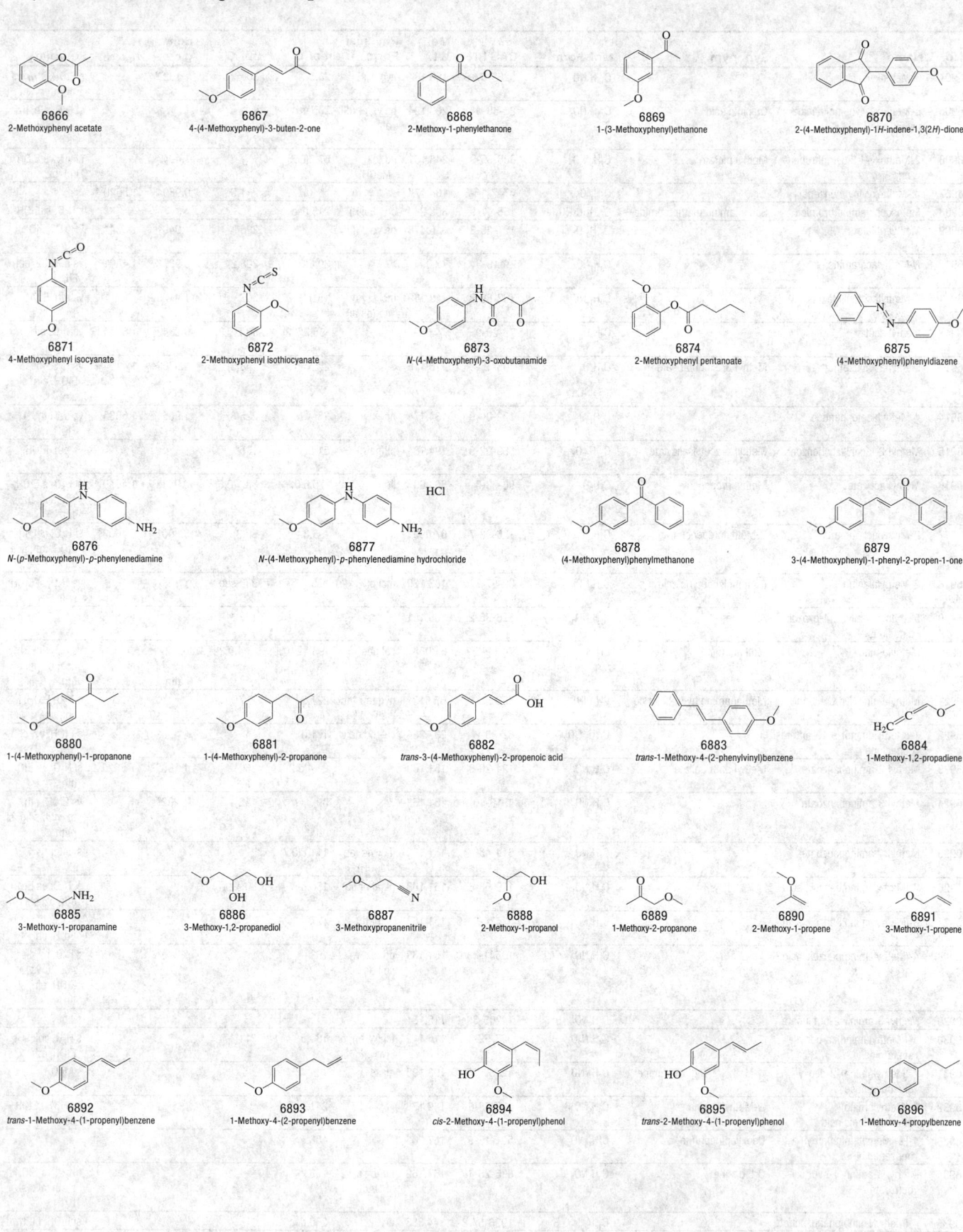

6866 2-Methoxyphenyl acetate

6867 4-(4-Methoxyphenyl)-3-buten-2-one

6868 2-Methoxy-1-phenylethanone

6869 1-(3-Methoxyphenyl)ethanone

6870 2-(4-Methoxyphenyl)-1*H*-indene-1,3(2*H*)-dione

6871 4-Methoxyphenyl isocyanate

6872 2-Methoxyphenyl isothiocyanate

6873 *N*-(4-Methoxyphenyl)-3-oxobutanamide

6874 2-Methoxyphenyl pentanoate

6875 (4-Methoxyphenyl)phenyldiazene

6876 *N*-(*p*-Methoxyphenyl)-*p*-phenylenediamine

6877 *N*-(4-Methoxyphenyl)-*p*-phenylenediamine hydrochloride

6878 (4-Methoxyphenyl)phenylmethanone

6879 3-(4-Methoxyphenyl)-1-phenyl-2-propen-1-one

6880 1-(4-Methoxyphenyl)-1-propanone

6881 1-(4-Methoxyphenyl)-2-propanone

6882 *trans*-3-(4-Methoxyphenyl)-2-propenoic acid

6883 *trans*-1-Methoxy-4-(2-phenylvinyl)benzene

6884 1-Methoxy-1,2-propadiene

6885 3-Methoxy-1-propanamine

6886 3-Methoxy-1,2-propanediol

6887 3-Methoxypropanenitrile

6888 2-Methoxy-1-propanol

6889 1-Methoxy-2-propanone

6890 2-Methoxy-1-propene

6891 3-Methoxy-1-propene

6892 *trans*-1-Methoxy-4-(1-propenyl)benzene

6893 1-Methoxy-4-(2-propenyl)benzene

6894 *cis*-2-Methoxy-4-(1-propenyl)phenol

6895 *trans*-2-Methoxy-4-(1-propenyl)phenol

6896 1-Methoxy-4-propylbenzene

6897 2-Methoxy-4-propylphenol

6898 3-Methoxy-1-propyne

6899 5-Methoxypsoralen

6900 6-Methoxy-3-pyridinamine

6901 2-Methoxypyridine

6902 3-Methoxypyridine

6903 4-Methoxypyridine

No.	Name	Synonym	Mol. Form.	CAS RN	Mol. Wt.	Physical Form	mp/°C	bp/°C	den g cm⁻³	n_D	Solubility
6904	6-Methoxyquinoline		$C_{10}H_9NO$	5263-87-6	159.184	hyg lf	26.5	306	1.152²⁰		s EtOH, eth, chl, dil HCl
6905	6-Methoxy-4-quinolinecarbox-ylic acid	Quininic acid	$C_{11}H_9NO_3$	86-68-0	203.194	pa ye pr (dil al)	285 dec	sub			sl H_2O, eth, bz, tfa; i chl; s EtOH
6906	2-Methoxy-1,3,5-trinitroben-zene	Methyl picrate	$C_7H_5N_3O_7$	606-35-9	243.131	nd (dil MeOH)	67.6(0.3)		1.4947⁸⁰		i H_2O; vs EtOH, chl, bz; s eth
6907	(2-Methoxyvinyl)benzene		$C_9H_{10}O$	4747-15-3	134.174			211.5	0.9894²³	1.5620²⁴	
6908	Methscopolamine bromide	Scopolamine methobromide	$C_{18}H_{24}BrNO_4$	155-41-9	398.293	cry (EtOH)	215 dec				s H_2O; sl EtOH
6909	Methyl abietate		$C_{21}H_{32}O_2$	127-25-3	316.478	pa ye lf (liq)		225¹⁶	1.049²⁰	1.5344	i H_2O; s EtOH, HOAc
6910	N-Methylacetamide		C_3H_7NO	79-16-3	73.094		30.6(0.1)	208(2)	0.9371²⁵	1.4301²⁰	vs ace, bz, eth, EtOH
6911	4-Methylacetanilide		$C_9H_{11}NO$	103-89-9	149.189	mcl cry or nd (dil al)	151(1)	307	1.2120¹⁵		vs eth, EtOH
6912	Methyl acetate		$C_3H_6O_2$	79-20-9	74.079	liq	-98.2(0.2)	56.7(0.2)	0.9342²⁰	1.3614²⁰	vs H_2O, eth, EtOH
6913	Methyl acetoacetate	Methyl 3-oxobutanoate	$C_5H_8O_3$	105-45-3	116.116		27.5	168(3)	1.0762²⁰	1.4184²⁰	vs H_2O; msc EtOH, eth; s ctc
6914	4-Methylacetophenone		$C_9H_{10}O$	122-00-9	134.174	nd	28	225(8)	1.0051²⁰	1.5335²⁰	vs bz, eth, EtOH, chl
6915	Methyl 2-(acetyloxy)benzoate	Methyl o-acetylsalicylate	$C_{10}H_{10}O_4$	580-02-9	194.184	pl (peth)	51.5	135⁹			vs eth, EtOH, chl
6916	Methyl acrylate	Methyl propenoate	$C_4H_6O_2$	96-33-3	86.090	liq	-75.6(0.3)	80.1(0.6)	0.9535²⁰	1.4040²⁰	sl H_2O; s EtOH, eth, ace, bz, chl
6917	2-Methylacrylonitrile	2-Methylpropenenitrile	C_4H_5N	126-98-7	67.090	liq	-35.8	90(3)	0.8001²⁰	1.4003²⁰	sl H_2O, chl; msc EtOH, eth, ace, tol
6918	2-Methylalanine	α-Aminoisobutyric acid	$C_4H_9NO_2$	62-57-7	103.120	mcl pr	335	280 sub			vs H_2O; sl EtOH; i eth
6919	5-Methyl-3-allyl-2,4-oxazoli-dinedione	Aloxidone	$C_7H_9NO_3$	526-35-2	155.151			138³⁵		1.4688²⁵	
6920	Methylamine	Methanamine	CH_5N	74-89-5	31.058	col gas	-93.42(0.09)	-6.4(0.3)	0.656²⁵ (p>1 atm)		vs H_2O; s EtOH, ace, bz; msc eth
6921	Methylamine hydrochloride	Methanamine hydrochloride	CH_6ClN	593-51-1	67.519	hyg tetr tab (al)	227.5	227¹⁵			s H_2O, EtOH; i chl, ace
6922	1-(Methylamino)-9,10-anthra-cenedione		$C_{15}H_{11}NO_2$	82-38-2	237.254	ye-red nd	171.0				s EtOH, bz, chl, HOAc
6923	Methyl 2-aminobenzoate	Methyl anthranilate	$C_8H_9NO_2$	134-20-3	151.163		24.4(0.2)	256	1.1682¹⁰	1.5810	sl H_2O; vs EtOH, eth
6924	Methyl 3-aminobenzoate		$C_8H_9NO_2$	4518-10-9	151.163		39	152¹¹	1.232²⁰		vs EtOH, eth, bz, chl; s lig; sl peth
6925	Methyl 4-aminobenzoate		$C_8H_9NO_2$	619-45-4	151.163	lf or nd (aq MeOH)	111.8(0.7)				s chl
6926	2-(Methylamino)benzoic acid		$C_8H_9NO_2$	119-68-6	151.163	pl (al or lig)	180.5	80⁰·⁰¹			sl H_2O; vs EtOH, eth, bz, chl
6927	3-(Methylamino)benzoic acid		$C_8H_9NO_2$	51524-84-6	151.163	pl (peth)	127				vs ace, bz, EtOH, chl
6928	4-(Methylamino)benzoic acid		$C_8H_9NO_2$	10541-83-0	151.163	nd (bz, w, dil al)	168				s H_2O, bz, AcOEt; vs EtOH, eth; sl tfa
6929	Methyl 3-amino-2-butenoate		$C_5H_9NO_2$	14205-39-1	115.131						s chl
6930	N-[(Methylamino)carbonyl]-acetamide		$C_4H_8N_2O_2$	623-59-6	116.119	tcl (w, al), pr (w)	180.5	311(17)			s H_2O, chl; sl EtOH, eth
6931	2-(Methylamino)-2-deoxy-α-L-glucopyranose	N-Methyl-α-L-glucosamine	$C_7H_{15}NO_5$	42852-95-9	193.198	glass					s MeOH
6932	2-(Methylamino)-ethanesulfonic acid	N-Methyltaurine	$C_3H_9NO_3S$	107-68-6	139.173		241.5				vs H_2O; i EtOH, eth
6933	4-[2-(Methylamino)ethyl]-1,2-benzenediol	Deoxyepinephrine	$C_9H_{13}NO_2$	501-15-5	167.205		188.5				
6934	Methyl 3-amino-4-hydroxy-benzoate	Orthocaine	$C_8H_9NO_3$	536-25-4	167.162	nd (bz or HOAc)	143				i H_2O; vs EtOH; s eth, alk; sl bz
6935	4-(Methylamino)phenol sulfate		$C_{14}H_{20}N_2O_6S$	1936-57-8	344.383	cry	260 dec				sl EtOH; i eth
6936	3-(Methylamino)propanenitrile		$C_4H_8N_2$	693-05-0	84.120			102⁴⁹	0.8992²⁰	1.4320²⁰	s H_2O, ace, bz, chl, MeOH
6937	4-[2-(Methylamino)propyl]-phenol	Pholedrine	$C_{10}H_{15}NO$	370-14-9	165.232	cry (MeOH)	161				vs eth, EtOH

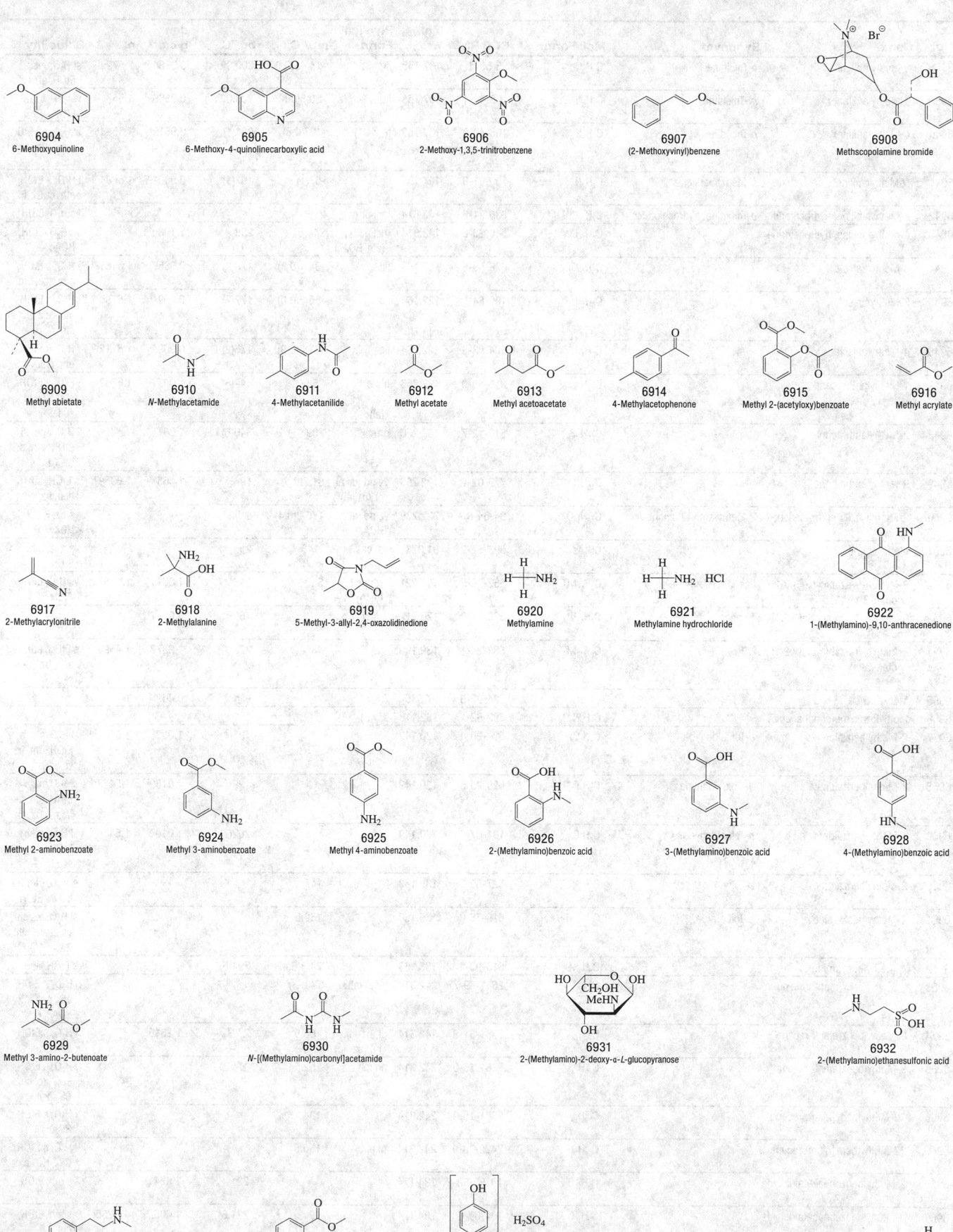

6904
6-Methoxyquinoline

6905
6-Methoxy-4-quinolinecarboxylic acid

6906
2-Methoxy-1,3,5-trinitrobenzene

6907
(2-Methoxyvinyl)benzene

6908
Methscopolamine bromide

6909
Methyl abietate

6910
N-Methylacetamide

6911
4-Methylacetanilide

6912
Methyl acetate

6913
Methyl acetoacetate

6914
4-Methylacetophenone

6915
Methyl 2-(acetyloxy)benzoate

6916
Methyl acrylate

6917
2-Methylacrylonitrile

6918
2-Methylalanine

6919
5-Methyl-3-allyl-2,4-oxazolidinedione

6920
Methylamine

6921
Methylamine hydrochloride

6922
1-(Methylamino)-9,10-anthracenedione

6923
Methyl 2-aminobenzoate

6924
Methyl 3-aminobenzoate

6925
Methyl 4-aminobenzoate

6926
2-(Methylamino)benzoic acid

6927
3-(Methylamino)benzoic acid

6928
4-(Methylamino)benzoic acid

6929
Methyl 3-amino-2-butenoate

6930
N-[(Methylamino)carbonyl]acetamide

6931
2-(Methylamino)-2-deoxy-α-L-glucopyranose

6932
2-(Methylamino)ethanesulfonic acid

6933
4-[2-(Methylamino)ethyl]-1,2-benzenediol

6934
Methyl 3-amino-4-hydroxybenzoate

6935
4-(Methylamino)phenol sulfate

6936
3-(Methylamino)propanenitrile

6937
4-[2-(Methylamino)propyl]phenol

No.	Name	Synonym	Mol. Form.	CAS RN	Mol. Wt.	Physical Form	mp/°C	bp/°C	den g cm⁻³	n_D	Solubility
6938	2-Methylaniline	o-Toluidine	C₇H₉N	95-53-4	107.153	liq	-14.41(0.02)	200.0(0.4)	0.9984²⁰	1.5725²⁰	sl H₂O; msc EtOH, eth, ctc
6939	3-Methylaniline	m-Toluidine	C₇H₉N	108-44-1	107.153	liq	-30.8(0.5)	203.3(0.5)	0.9889²⁰	1.5681²⁰	vs ace, bz, eth, EtOH
6940	4-Methylaniline	p-Toluidine	C₇H₉N	106-49-0	107.153	lf (w+1)	43.3(0.8)	201(1)	0.9619²⁰	1.5534⁴⁵	sl H₂O; vs EtOH, py; s eth, ace, ctc
6941	N-Methylaniline	N-Methylbenzenamine	C₇H₉N	100-61-8	107.153	liq	-57(1)	197(1)	0.9891²⁰	1.5684²⁰	i H₂O; s EtOH, eth, ctc, chl
6942	2-Methylaniline, hydrochloride	o-Toluidine, hydrochloride	C₇H₁₀ClN	636-21-5	143.614	mcl pr (w)	215				vs H₂O, EtOH
6943	4-Methylaniline, hydrochloride		C₇H₁₀ClN	540-23-8	143.614	mcl nd (eth-HOAc)	244.5	258	1.1930¹⁸		vs H₂O, EtOH, HOAc
6944	2-Methylanisole		C₈H₁₀O	578-58-5	122.164	liq	-34.1(0.3)	173(2)	0.985²⁵	1.5161²⁰	i H₂O; s EtOH, eth, ace, ctc
6945	3-Methylanisole		C₈H₁₀O	100-84-5	122.164	liq	-55.5(0.8)	177(2)	0.969²⁵	1.5130²⁰	i H₂O; s EtOH, eth, ace, bz; sl ctc
6946	4-Methylanisole		C₈H₁₀O	104-93-8	122.164	liq	-31.6(0.9)	175(2)	0.969²⁵	1.5112²⁰	i H₂O; s EtOH, eth, chl
6947	1-Methylanthracene		C₁₅H₁₂	610-48-0	192.256	bl nd (MeOH) lf (al)	85.5	342(6)	1.0471⁹⁹	1.6802⁹⁹	i H₂O; s EtOH, eth, bz, chl, sulf
6948	2-Methylanthracene		C₁₅H₁₂	613-12-7	192.256	grn bl flr lf (sub)	209	340(6)	1.80⁰		i H₂O, ace; sl EtOH, eth; s bz, CS₂
6949	9-Methylanthracene		C₁₅H₁₂	779-02-2	192.256	ye nd (dil al) pr (bz, al)	81.7(0.5)	196¹²	1.065⁹⁹	1.6959⁹⁹	s EtOH, eth, ace, bz, chl
6950	2-Methyl-9,10-anthracenedione	2-Methylanthraquinone	C₁₅H₁₀O₂	84-54-8	222.239	ye nd (al, HOAc)	176.2(0.4)	sub			vs bz, EtOH, HOAc
6951	Methylarsine		CH₅As	593-52-2	91.973	col gas	-143	2			vs ace, eth, EtOH
6952	9-Methyl-9-azabicyclo[3.3.1]-nonan-3-one	Pseudopelletierine	C₉H₁₅NO	552-70-5	153.221	orth pr (peth)	54	246	1.001¹⁰⁰	1.4760¹⁰⁰	vs H₂O, eth, EtOH
6953	8-Methyl-8-azabicyclo[3.2.1]-octane	Tropane	C₈H₁₅N	529-17-9	125.212			166	0.9251¹⁵		
6954	8-Methyl-8-azabicyclo[3.2.1]-octan-3-one		C₈H₁₃NO	532-24-1	139.195		43	227	1.9872¹⁰⁰	1.4598¹⁰⁰	s EtOH, eth, ace, bz, peth; sl chl
6955	Methyl azide		CH₃N₃	624-90-8	57.055			exp	0.869¹⁵		
6956	Methylazoxymethanol acetate		C₄H₈N₂O₃	592-62-1	132.118			191			
6957	2-Methylbenzaldehyde	o-Tolualdehyde	C₈H₈O	529-20-4	120.149			195(4)	1.0328²⁰	1.5462²⁰	sl H₂O, ctc; s EtOH, eth, bz; vs ace
6958	3-Methylbenzaldehyde	m-Tolualdehyde	C₈H₈O	620-23-5	120.149			199	1.0189²¹	1.5413²¹	sl H₂O; msc EtOH, eth; vs ace; s bz, chl
6959	4-Methylbenzaldehyde	p-Tolualdehyde	C₈H₈O	104-87-0	120.149			202(7)	1.0194¹⁷	1.5454²⁰	sl H₂O; msc EtOH, eth, ace; vs chl
6960	2-Methylbenzamide	o-Toluamide	C₈H₉NO	527-85-5	135.163		147				sl H₂O, eth, tfa, bz; vs EtOH
6961	4-Methylbenzamide	p-Toluamide	C₈H₉NO	619-55-6	135.163		162.5				sl H₂O, bz, chl; vs EtOH, eth; s tfa
6962	N-Methylbenzamide		C₈H₉NO	613-93-4	135.163		74.3(0.4)	294(5)			s EtOH, ace
6963	7-Methylbenz[a]anthracene		C₁₉H₁₄	2541-69-7	242.314	ye pl (al)	141				i H₂O; s EtOH, eth, ace, ctc, HOAc, CS₂
6964	8-Methylbenz[a]anthracene		C₁₉H₁₄	2381-31-9	242.314	pl (bz-al), nd (bz-lig)	156.5	272³	1.2310⁰		i H₂O; s EtOH, eth, bz, xyl
6965	9-Methylbenz[a]anthracene		C₁₉H₁₄	2381-16-0	242.314	nd (al)	152.5				i H₂O; s EtOH, eth, ctc, chl, CS₂, xyl
6966	10-Methylbenz[a]anthracene		C₁₉H₁₄	2381-15-9	242.314		184				i H₂O; s EtOH, HOAc
6967	12-Methylbenz[a]anthracene		C₁₉H₁₄	2422-79-9	242.314	pl (al)	150.5				i H₂O; s EtOH, CS₂, HOAc
6968	2-Methylbenzeneacetaldehyde		C₉H₁₀O	10166-08-2	134.174			221	1.0241¹⁰		vs eth, EtOH, chl
6969	4-Methylbenzeneacetaldehyde		C₉H₁₀O	104-09-6	134.174		40	221.5	1.0052²⁰	1.5255²⁰	vs eth, EtOH, chl
6970	α-Methylbenzeneacetaldehyde		C₉H₁₀O	93-53-8	134.174			206.4(0.8)	1.0089²⁰	1.5176²⁰	vs EtOH
6971	2-Methylbenzeneacetic acid		C₉H₁₀O₂	644-36-0	150.174	nd (w)	89				s H₂O, chl
6972	3-Methylbenzeneacetic acid		C₉H₁₀O₂	621-36-3	150.174	nd (w)	62	121²⁶			s H₂O, chl

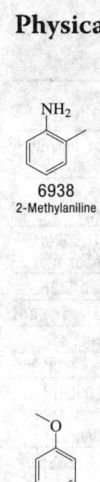

6938
2-Methylaniline

6939
3-Methylaniline

6940
4-Methylaniline

6941
N-Methylaniline

6942
2-Methylaniline, hydrochloride

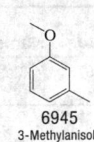

6943
4-Methylaniline, hydrochloride

6944
2-Methylanisole

6945
3-Methylanisole

6946
4-Methylanisole

6947
1-Methylanthracene

6948
2-Methylanthracene

6949
9-Methylanthracene

6950
2-Methyl-9,10-anthracenedione

6951
Methylarsine

6952
9-Methyl-9-azabicyclo[3.3.1]nonan-3-one

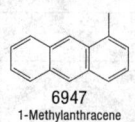

6953
8-Methyl-8-azabicyclo[3.2.1]octane

6954
8-Methyl-8-azabicyclo[3.2.1]octan-3-one

6955
Methyl azide

6956
Methylazoxymethanol acetate

6957
2-Methylbenzaldehyde

6958
3-Methylbenzaldehyde

6959
4-Methylbenzaldehyde

6960
2-Methylbenzamide

6961
4-Methylbenzamide

6962
N-Methylbenzamide

6963
7-Methylbenz[a]anthracene

6964
8-Methylbenz[a]anthracene

6965
9-Methylbenz[a]anthracene

6966
10-Methylbenz[a]anthracene

6967
12-Methylbenz[a]anthracene

6968
2-Methylbenzeneacetaldehyde

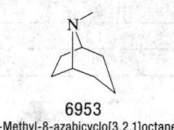

6969
4-Methylbenzeneacetaldehyde

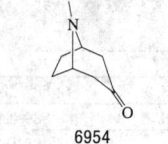

6970
α-Methylbenzeneacetaldehyde

6971
2-Methylbenzeneacetic acid

6972
3-Methylbenzeneacetic acid

No.	Name	Synonym	Mol. Form.	CAS RN	Mol. Wt.	Physical Form	mp/°C	bp/°C	den g cm⁻³	n_D	Solubility
6973	4-Methylbenzeneacetic acid		C₉H₁₀O₂	622-47-9	150.174	nd or pl (al, w)	93	265			vs bz, eth, EtOH
6974	α-Methylbenzeneacetic acid, (±)-		C₉H₁₀O₂	2328-24-7	150.174		<-20	263	1.1⁰	1.5237²⁰	
6975	4-Methylbenzeneacetonitrile		C₉H₉N	2947-61-7	131.174		18	242.5	0.992²⁵	1.5190²⁰	i H₂O; s EtOH, eth, bz, ctc
6976	α-Methylbenzeneacetonitrile		C₉H₉N	1823-91-2	131.174			231	0.9854²⁰	1.5095²⁵	vs eth, EtOH
6977	3-Methyl-1,2-benzenediamine	Toluene-2,3-diamine	C₇H₁₀N₂	2687-25-4	122.167		63.5	255			vs ace, bz, EtOH
6978	4-Methyl-1,2-benzenediamine	Toluene-3,4-diamine	C₇H₁₀N₂	496-72-0	122.167	pl (lig)	89.5	265			vs H₂O; s lig
6979	2-Methyl-1,3-benzenediamine	Toluene-2,6-diamine	C₇H₁₀N₂	823-40-5	122.167	pr (bz, w)	106				s H₂O, EtOH, bz
6980	2-Methyl-1,4-benzenediamine	Toluene-2,5-diamine	C₇H₁₀N₂	95-70-5	122.167	pl (bz)	64	273.5			s H₂O, EtOH, eth; sl bz, HOAc
6981	3-Methyl-1,2-benzenediol		C₇H₈O₂	488-17-5	124.138	lf (bz)	68	248			s H₂O, EtOH, bz, chl
6982	4-Methyl-1,2-benzenediol		C₇H₈O₂	452-86-8	124.138	lf (bz-lig), pr (bz)	65	258	1.1287⁷⁴	1.5425⁷⁴	s H₂O, EtOH, eth, ace, chl; sl lig
6983	2-Methyl-1,3-benzenediol		C₇H₈O₂	608-25-3	124.138	pr (bz)	120	265			vs H₂O, bz, eth, EtOH
6984	4-Methyl-1,3-benzenediol		C₇H₈O₂	496-73-1	124.138	cry (bz-peth)	105	270			s H₂O, EtOH, eth; sl bz, peth
6985	5-Methyl-1,3-benzenediol	Orcinol	C₇H₈O₂	504-15-4	124.138	pr(w+1), lf(chl)	107	287	1.290⁴		s H₂O, EtOH, eth, bz; sl lig, peth
6986	2-Methyl-1,4-benzenediol		C₇H₈O₂	95-71-6	124.138		125	283			vs H₂O, EtOH, eth; s ace; sl bz, lig
6987	4-Methyl-1,2-benzenedithiol	Toluene-3,4-dithiol	C₇H₈S₂	496-74-2	156.269		29				s chl
6988	β-Methylbenzeneethanamine		C₉H₁₃N	582-22-9	135.206			210	0.9433⁴	1.5255²⁰	vs bz, eth, EtOH
6989	N-Methylbenzeneethanamine		C₉H₁₃N	589-08-2	135.206			206	0.93²⁵	1.5162²⁰	
6990	2-Methylbenzeneethanol		C₉H₁₂O	19819-98-8	136.190		1.0	243.5	1.016²⁵	1.5355²⁰	
6991	4-Methylbenzeneethanol		C₉H₁₂O	699-02-5	136.190			240(11)	1.0028²⁰	1.5267²⁰	
6992	2-Methylbenzenemethanamine		C₈H₁₁N	89-93-0	121.180	liq	-30	206	0.9766¹⁹	1.5436¹⁹	
6993	3-Methylbenzenemethanamine		C₈H₁₁N	100-81-2	121.180			203.5	0.966²⁵	1.5360²⁰	
6994	4-Methylbenzenemethanamine		C₈H₁₁N	104-84-7	121.180		12.5	195	0.952²⁰	1.5340²⁰	
6995	N-Methylbenzenemethanamine		C₈H₁₁N	103-67-3	121.180			180.5	0.9442¹⁸		vs H₂O
6996	α-Methylbenzenemethanol	1-Phenylethanol	C₈H₁₀O	98-85-1	122.164		15(2)	205(4)	1.013²⁵	1.5265²⁰	i H₂O; vs EtOH, eth
6997	2-Methylbenzenemethanol	o-Tolyl alcohol	C₈H₁₀O	89-95-2	122.164	nd (peth-eth)	38	224	1.023⁴⁰		vs eth, EtOH, chl
6998	3-Methylbenzenemethanol	m-Tolyl alcohol	C₈H₁₀O	587-03-1	122.164		<-20	215.5	0.9157¹⁷		sl H₂O; vs EtOH, eth; s chl
6999	4-Methylbenzenemethanol	p-Tolyl alcohol	C₈H₁₀O	589-18-4	122.164	nd (lig)	58.72(0.04)	217	0.978²²		vs eth, EtOH
7000	α-Methylbenzenemethanol, acetate		C₁₀H₁₂O₂	93-92-5	164.201	oil		109¹⁸			
7001	4-Methylbenzenepropanal		C₁₀H₁₂O	5406-12-2	148.201			223	0.999¹⁴	1.525¹⁴	
7002	α-Methylbenzenepropanamine	1-Methyl-3-phenylpropylamine	C₁₀H₁₅N	22374-89-6	149.233		143	223	0.9289¹⁵	1.5152²⁰	vs EtOH
7003	β-Methylbenzenepropanoic acid, (±)-		C₁₀H₁₂O₂	772-17-8	164.201		46.5	168¹⁴	1.0701²⁰	1.5155²⁰	sl H₂O; s peth
7004	α-Methylbenzenepropanol		C₁₀H₁₄O	2344-70-9	150.217			239	0.9899¹⁶	1.517¹⁶	
7005	4-Methylbenzenesulfinic acid	p-Toluenesulfinic acid	C₇H₈O₂S	536-57-2	156.203	orth pl or nd (w)	86.5				s H₂O; vs EtOH, eth; sl bz
7006	4-Methylbenzenesulfinyl chloride		C₇H₇ClOS	10439-23-3	174.648	nd	57	113³·⁵			vs chl
7007	2-Methylbenzenesulfonamide		C₇H₉NO₂S	88-19-7	171.217	oct cry (al), pr (w)	156.3(0.5)	214¹⁰			sl H₂O, eth, DMSO; s EtOH
7008	4-Methylbenzenesulfonamide	p-Toluenesulfonamide	C₇H₉NO₂S	70-55-3	171.217	mcl pl (w+2)	138	214¹⁰			sl H₂O, eth; s EtOH
7009	Methyl benzenesulfonate		C₇H₈O₃S	80-18-2	172.202		4.5	150¹⁵	1.2730¹⁷	1.5151²⁰	sl H₂O; vs EtOH, eth, chl
7010	2-Methylbenzenesulfonic acid		C₇H₈O₃S	88-20-0	172.202	hyg pl (w+2)	67.5	128.8²⁵			vs H₂O; s EtOH; i eth
7011	2-Methylbenzenesulfonyl chloride	o-Toluenesulfonyl chloride	C₇H₇ClO₂S	133-59-5	190.648		11(1)	154³⁶	1.3383²⁰	1.5565²⁰	i H₂O; s EtOH, eth, bz, ctc
7012	2-Methylbenzenethiol		C₇H₈S	137-06-4	124.204		15	192(20)	1.041²⁰	1.570²⁰	i H₂O; s EtOH; vs eth
7013	3-Methylbenzenethiol		C₇H₈S	108-40-7	124.204	liq	-20	196(20)	1.044²⁰	1.572²⁰	i H₂O; s EtOH; msc eth

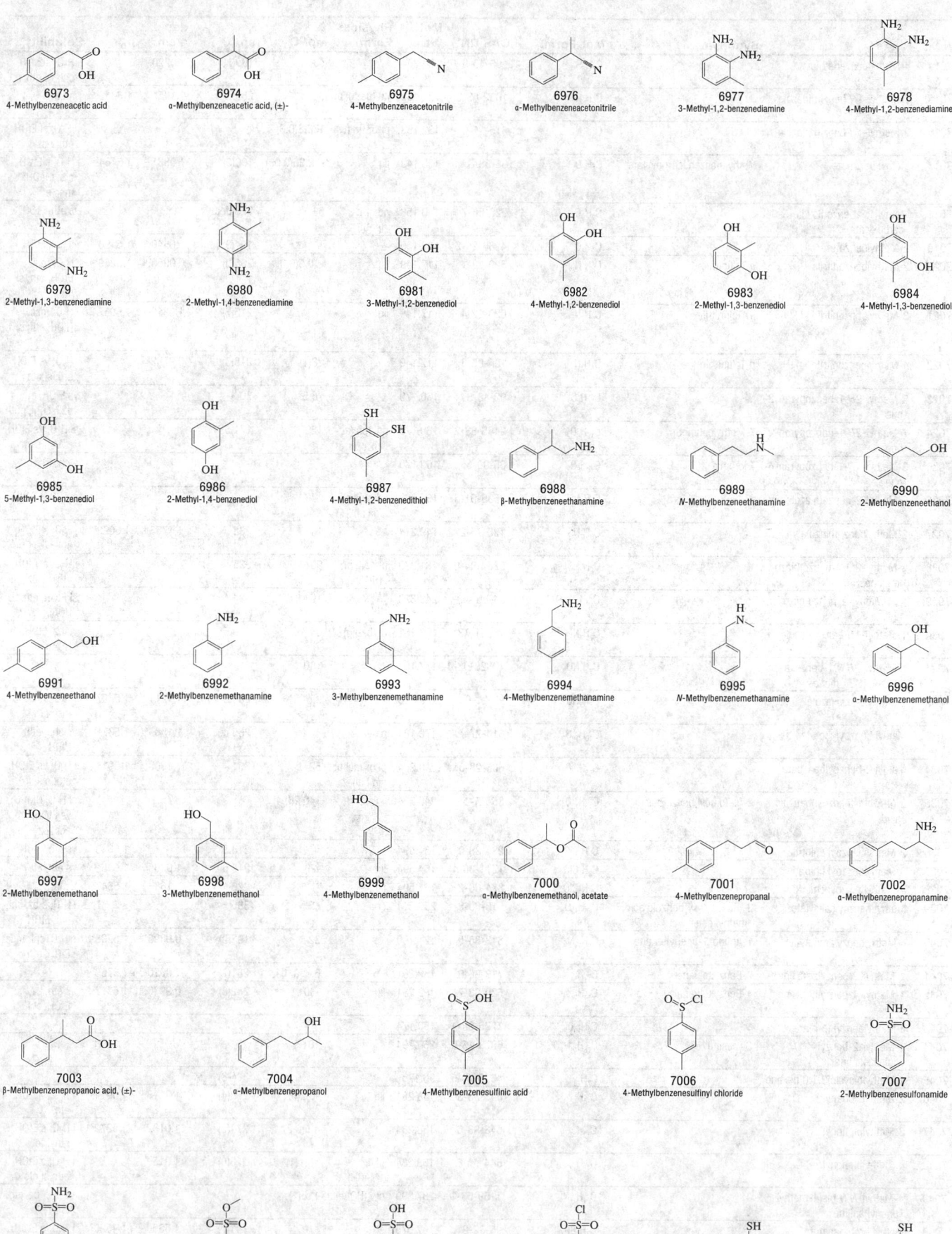

6973
4-Methylbenzeneacetic acid

6974
α-Methylbenzeneacetic acid, (±)-

6975
4-Methylbenzeneacetonitrile

6976
α-Methylbenzeneacetonitrile

6977
3-Methyl-1,2-benzenediamine

6978
4-Methyl-1,2-benzenediamine

6979
2-Methyl-1,3-benzenediamine

6980
2-Methyl-1,4-benzenediamine

6981
3-Methyl-1,2-benzenediol

6982
4-Methyl-1,2-benzenediol

6983
2-Methyl-1,3-benzenediol

6984
4-Methyl-1,3-benzenediol

6985
5-Methyl-1,3-benzenediol

6986
2-Methyl-1,4-benzenediol

6987
4-Methyl-1,2-benzenedithiol

6988
β-Methylbenzeneethanamine

6989
N-Methylbenzeneethanamine

6990
2-Methylbenzeneethanol

6991
4-Methylbenzeneethanol

6992
2-Methylbenzenemethanamine

6993
3-Methylbenzenemethanamine

6994
4-Methylbenzenemethanamine

6995
N-Methylbenzenemethanamine

6996
α-Methylbenzenemethanol

6997
2-Methylbenzenemethanol

6998
3-Methylbenzenemethanol

6999
4-Methylbenzenemethanol

7000
α-Methylbenzenemethanol, acetate

7001
4-Methylbenzenepropanal

7002
α-Methylbenzenepropanamine

7003
β-Methylbenzenepropanoic acid, (±)-

7004
α-Methylbenzenepropanol

7005
4-Methylbenzenesulfinic acid

7006
4-Methylbenzenesulfinyl chloride

7007
2-Methylbenzenesulfonamide

7008
4-Methylbenzenesulfonamide

7009
Methyl benzenesulfonate

7010
2-Methylbenzenesulfonic acid

7011
2-Methylbenzenesulfonyl chloride

7012
2-Methylbenzenethiol

7013
3-Methylbenzenethiol

No.	Name	Synonym	Mol. Form.	CAS RN	Mol. Wt.	Physical Form	mp/°C	bp/°C	den g cm^{-3}	n_D	Solubility
7014	4-Methylbenzenethiol		C$_7$H$_8$S	106-45-6	124.204		43	199(20)	1.0220[51]		i H$_2$O; s EtOH, chl; vs eth
7015	1-Methyl-1H-benzimidazole		C$_8$H$_8$N$_2$	1632-83-3	132.163	nd (peth), pl (al)	66	286	1.1254[20]	1.6013[7]	s peth
7016	2-Methyl-1H-benzimidazole		C$_8$H$_8$N$_2$	615-15-6	132.163	pr or nd (w)	178.3(0.3)				s H$_2$O; sl EtOH, eth; i bz
7017	Methyl benzoate	Methyl benzenecarboxylate	C$_8$H$_8$O$_2$	93-58-3	136.149	liq	-12.35(0.06)	199(2)	1.0837[25]	1.5164[20]	i H$_2$O; s EtOH, ctc, MeOH; msc eth
7018	Methyl 1,3-benzodioxole-5-carboxylate		C$_9$H$_8$O$_4$	326-56-7	180.158	nd or lf (peth)	53	273 dec			vs eth, EtOH
7019	2-Methylbenzofuran		C$_9$H$_8$O	4265-25-2	132.159			194(3)	1.0540[20]	1.5495[22]	vs eth, EtOH
7020	2-Methylbenzonitrile	o-Tolunitrile	C$_8$H$_7$N	529-19-1	117.149	liq	-10.5(0.3)	205(5)	0.9955[20]	1.5279[20]	i H$_2$O; msc EtOH, eth; sl ctc
7021	3-Methylbenzonitrile	m-Tolunitrile	C$_8$H$_7$N	620-22-4	117.149	liq	-23	215(7)	1.0316[20]	1.5252[20]	i H$_2$O; msc EtOH, eth; sl ctc
7022	4-Methylbenzonitrile	p-Tolunitrile	C$_8$H$_7$N	104-85-8	117.149		28(1)	218(3)	0.9762[30]		i H$_2$O; vs EtOH, eth; sl ctc
7023	6-Methyl-2H-1-benzopyran-2-one		C$_{10}$H$_8$O$_2$	92-48-8	160.170		76.5	304			vs EtOH, eth, bz; sl chl, peth
7024	7-Methyl-2H-1-benzopyran-2-one	7-Methylcoumarin	C$_{10}$H$_8$O$_2$	2445-83-2	160.170	nd, (pl) (aq al)	128	171.5[11]			sl H$_2$O; vs EtOH, HOAc; s eth
7025	3-Methyl-4H-1-benzopyran-4-one	Tricromyl	C$_{10}$H$_8$O$_2$	85-90-5	160.170						s chl
7026	6-Methyl-2-benzothiazolamine		C$_8$H$_8$N$_2$S	2536-91-6	164.228	nd (w) pr (dil al)	142				sl H$_2$O; s EtOH
7027	2-Methylbenzothiazole		C$_8$H$_7$NS	120-75-2	149.214		14	238	1.1763[19]	1.6092[19]	i H$_2$O; s EtOH, chl
7028	3-Methyl-2(3H)-benzothiazolethione		C$_8$H$_7$NS$_2$	2254-94-6	181.279	nd (al), pr (HOAc)	90	335			i H$_2$O; sl EtOH, eth; vs bz, chl
7029	4-(6-Methyl-2-benzothiazolyl)-aniline		C$_{14}$H$_{12}$N$_2$S	92-36-4	240.323		194.8	434			sl EtOH, eth, bz, HOAc
7030	1-Methyl-1H-benzotriazole		C$_7$H$_7$N$_3$	13351-73-0	133.151	pl (bz-lig)	64.5	270.5			vs bz, EtOH, HOAc
7031	1-Methyl-2H-3,1-benzoxazine-2,4(1H)-dione		C$_9$H$_7$NO$_3$	10328-92-4	177.157		180				
7032	2-Methylbenzoxazole		C$_8$H$_7$NO	95-21-6	133.148		9.5	200.5	1.1211[20]	1.5497[20]	i H$_2$O; vs EtOH; msc eth
7033	Methyl benzoylacetate		C$_{10}$H$_{10}$O$_3$	614-27-7	178.184	pa ye		265 dec	1.158[29]	1.537[20]	vs ace, eth, EtOH
7034	Methyl 2-benzoylbenzoate		C$_{15}$H$_{12}$O$_3$	606-28-0	240.254	pl or mcl pr (dil al)	52	351	1.1903[19]	1.591[20]	i H$_2$O; vs EtOH, eth; s sulf
7035	2-(4-Methylbenzoyl)benzoic acid	2-(p-Toluoyl)benzoic acid	C$_{15}$H$_{12}$O$_3$	85-55-2	240.254		140.2(0.4)				sl H$_2$O, DMSO; vs EtOH, eth, ace, bz
7036	2-Methylbenzoyl chloride		C$_8$H$_7$ClO	933-88-0	154.594			213.5		1.5549[20]	vs eth, EtOH
7037	3-Methylbenzoyl chloride		C$_8$H$_7$ClO	1711-06-4	154.594	liq	-23	219.5	1.0265[21]	1.505[22]	vs eth, EtOH
7038	4-Methylbenzoyl chloride		C$_8$H$_7$ClO	874-60-2	154.594	liq	-1.5	226	1.1686[20]	1.5547[20]	s ctc
7039	Methyl benzoylsalicylate	2-(Benzoyloxy)benzoic acid, methyl ester	C$_{15}$H$_{12}$O$_4$	610-60-6	256.254	cry	85	385			i H$_2$O; s bz, chl, eth, EtOH
7040	α-Methylbenzylamine, (±)-	1-Amino-1-phenylethane	C$_8$H$_{11}$N	618-36-0	121.180		32	193(2)	0.9395[15]	1.5238[25]	s H$_2$O, chl; msc EtOH, eth
7041	1-Methyl-2-benzylbenzene	2-Benzyltoluene	C$_{14}$H$_{14}$	713-36-0	182.261		6.61(0.01)	281(7)	1.0020[20]	1.5763[20]	
7042	1-Methyl-4-benzylbenzene	4-Benzyltoluene	C$_{14}$H$_{14}$	620-83-7	182.261	liq	-30	282(6)	0.9976[20]	1.5712[20]	vs eth, bz, EtOH, chl
7043	α-Methylbenzyl formate		C$_9$H$_{10}$O$_2$	7775-38-4	150.174	liq					
7044	1-Methyl-2-benzyl-4(1H)-quinazolinone	Glycosine	C$_{16}$H$_{14}$N$_2$O	6873-15-0	250.294		161.5				
7045	1-Methylbicyclo[3.1.0]hexane		C$_7$H$_{12}$	4625-24-5	96.170			88(18)			
7046	2-Methylbiphenyl		C$_{13}$H$_{12}$	643-58-3	168.234	liq	-0.2(0.1)	258(4)	1.0113[20]	1.5914[20]	i H$_2$O; s EtOH, eth
7047	3-Methylbiphenyl		C$_{13}$H$_{12}$	643-93-6	168.234		4.5(0.2)	274(2)	1.0182[17]	1.5972[20]	i H$_2$O; s EtOH, eth, ctc
7048	4-Methylbiphenyl		C$_{13}$H$_{12}$	644-08-6	168.234	pl (lig, MeOH)	48.1(1)	273(6)	1.015[27]		i H$_2$O; s EtOH, eth; sl ctc
7049	4-Methyl-N,N-bis(4-methylphenyl)aniline		C$_{21}$H$_{21}$N	1159-53-1	287.399	cry (HOAc)	115(1)				vs ace, bz, eth, chl
7050	Methyl bromoacetate		C$_3$H$_5$BrO$_2$	96-32-2	152.975			141(7)	1.6350[20]	1.4520[20]	i H$_2$O; s EtOH, eth, ace, bz
7051	Methyl 2-bromobenzoate		C$_8$H$_7$BrO$_2$	610-94-6	215.045			244			i H$_2$O; s EtOH

7014
4-Methylbenzenethiol

7015
1-Methyl-1*H*-benzimidazole

7016
2-Methyl-1*H*-benzimidazole

7017
Methyl benzoate

7018
Methyl 1,3-benzodioxole-5-carboxylate

7019
2-Methylbenzofuran

7020
2-Methylbenzonitrile

7021
3-Methylbenzonitrile

7022
4-Methylbenzonitrile

7023
6-Methyl-2*H*-1-benzopyran-2-one

7024
7-Methyl-2*H*-1-benzopyran-2-one

7025
3-Methyl-4*H*-1-benzopyran-4-one

7026
6-Methyl-2-benzothiazolamine

7027
2-Methylbenzothiazole

7028
3-Methyl-2(3*H*)-benzothiazolethione

7029
4-(6-Methyl-2-benzothiazolyl)aniline

7030
1-Methyl-1*H*-benzotriazole

7031
1-Methyl-2*H*-3,1-benzoxazine-2,4(1*H*)-dione

7032
2-Methylbenzoxazole

7033
Methyl benzoylacetate

7034
Methyl 2-benzoylbenzoate

7035
2-(4-Methylbenzoyl)benzoic acid

7036
2-Methylbenzoyl chloride

7037
3-Methylbenzoyl chloride

7038
4-Methylbenzoyl chloride

7039
Methyl benzoylsalicylate

7040
α-Methylbenzylamine, (±)-

7041
1-Methyl-2-benzylbenzene

7042
1-Methyl-4-benzylbenzene

7043
α-Methylbenzyl formate

7044
1-Methyl-2-benzyl-4(1*H*)-quinazolinone

7045
1-Methylbicyclo[3,1,0]hexane

7046
2-Methylbiphenyl

7047
3-Methylbiphenyl

7048
4-Methylbiphenyl

7049
4-Methyl-*N*,*N*-bis(4-methylphenyl)aniline

7050
Methyl bromoacetate

7051
Methyl 2-bromobenzoate

No.	Name	Synonym	Mol. Form.	CAS RN	Mol. Wt.	Physical Form	mp/°C	bp/°C	den g cm⁻³	n_D	Solubility
7052	Methyl 3-bromobenzoate		$C_8H_7BrO_2$	618-89-3	215.045	pl	32	125[15]			sl H_2O; s EtOH, eth
7053	Methyl 4-bromobenzoate		$C_8H_7BrO_2$	619-42-1	215.045	lf (dil al), nd (eth)	81		1.689[25]		s EtOH, eth, ace, peth; vs bz, chl
7054	Methyl 2-bromobutanoate		$C_5H_9BrO_2$	3196-15-4	181.028			168	1.4528[20]	1.4029[25]	vs EtOH
7055	Methyl 4-bromobutanoate		$C_5H_9BrO_2$	4897-84-1	181.028			186.5	1.4[25]	1.4567[25]	vs EtOH
7056	Methyl 4-bromo-2-butenoate		$C_5H_7BrO_2$	1117-71-1	179.013			84[12]	1.490[19]	1.498[19]	
7057	Methyl 5-bromopentanoate		$C_6H_{11}BrO_2$	5454-83-1	195.054	liq		101[14]	1.363	1.4630[20]	
7058	Methyl 3-bromopropanoate		$C_4H_7BrO_2$	3395-91-3	167.002			105[60]	1.4123[18]	1.4542[20]	s EtOH, eth, ace
7059	3-Methyl-1,2-butadiene		C_5H_8	598-25-4	68.118	liq	-113.61(0.08)	40.8(0.7)	0.6806[25]	1.4203[20]	vs ace, bz, eth, EtOH
7060	2-Methyl-1,3-butadiene	Isoprene	C_5H_8	78-79-5	68.118	liq	-146.1(0.5)	34.0(0.3)	0.679[20]	1.4219[20]	i H_2O; msc EtOH, eth, ace, bz
7061	3-Methylbutanal	Isovaleraldehyde	$C_5H_{10}O$	590-86-3	86.132	liq	-51	92.5(0.3)	0.7977[20]	1.3902[20]	sl H_2O; s EtOH, eth
7062	3-Methylbutanamide	Isovaleramide	$C_5H_{11}NO$	541-46-8	101.147	mcl lf (al)	137	226			s H_2O, EtOH, eth; vs peth
7063	3-Methyl-1-butanamine	Isopentylamine	$C_5H_{13}N$	107-85-7	87.164			96(4)	0.7505[20]	1.4083[20]	msc H_2O, EtOH, eth; s ace, chl
7064	2-Methyl-2-butanamine		$C_5H_{13}N$	594-39-8	87.164	liq	-105.0(0.6)	77(4)	0.731[25]	1.3954[25]	vs H_2O, ace, eth, EtOH
7065	3-Methyl-2-butanamine		$C_5H_{13}N$	598-74-3	87.164	liq	-50	81(6)	0.7574[19]	1.4096[18]	vs H_2O; s EtOH
7066	3-Methyl-1,3-butanediol		$C_5H_{12}O_2$	2568-33-4	104.148			208(5)	0.9448[20]	1.4452[20]	s H_2O, EtOH
7067	2-Methylbutanenitrile		C_5H_9N	18936-17-9	83.132			123(3)	0.7913[15]	1.3933[20]	vs eth, EtOH
7068	3-Methylbutanenitrile	Isobutyl cyanide	C_5H_9N	625-28-5	83.132	liq	-100.8(0.3)	129(4)	0.7914[20]	1.3927[20]	sl H_2O; msc EtOH, eth; vs ace
7069	2-Methyl-1-butanethiol, (+)		$C_5H_{12}S$	20089-07-0	104.214	liq		119.0(0.8)	0.8420[20]	1.4440[20]	
7070	3-Methyl-1-butanethiol	Isopentyl mercaptan	$C_5H_{12}S$	541-31-1	104.214	liq		118.3(0.8)	0.8350[20]	1.4412[20]	i H_2O; msc EtOH, eth; s ctc
7071	2-Methyl-2-butanethiol		$C_5H_{12}S$	1679-09-0	104.214	liq		99.1(0.5)	0.8120[20]	1.4385[20]	
7072	3-Methyl-2-butanethiol		$C_5H_{12}S$	2084-18-6	104.214	liq	-127.09(0.05)	109.8			
7073	Methyl butanoate		$C_5H_{10}O_2$	623-42-7	102.132	liq	-85.8	101.9(0.1)	0.8984[20]	1.3878[20]	sl H_2O, ctc; msc EtOH, eth
7074	2-Methylbutanoic acid	(±)-2-Methylbutyric acid	$C_5H_{10}O_2$	600-07-7	102.132		<-80	177	0.934[20]	1.4051[20]	sl H_2O; msc EtOH, eth; s chl
7075	3-Methylbutanoic acid	Isovaleric acid	$C_5H_{10}O_2$	503-74-2	102.132	liq	-29.6(0.7)	176.5(0.2)	0.931[20]	1.4033[20]	s H_2O; msc EtOH, eth, chl
7076	3-Methylbutanoic anhydride		$C_{10}H_{18}O_3$	1468-39-9	186.248			215	0.9327[20]	1.4043[20]	vs eth
7077	2-Methyl-1-butanol, (±)-		$C_5H_{12}O$	34713-94-5	88.148			129.0(0.4)	0.8152[25]	1.4092[20]	sl H_2O; msc EtOH, eth; vs ace
7078	3-Methyl-1-butanol	Isopentyl alcohol	$C_5H_{12}O$	123-51-3	88.148	liq	-117.2	130.8(0.3)	0.8104[20]	1.4053[20]	sl H_2O; vs ace, eth, EtOH
7079	2-Methyl-2-butanol	tert-Amyl alcohol	$C_5H_{12}O$	75-85-4	88.148	liq	-8.7(0.6)	102.4	0.8096[20]	1.4052[20]	s H_2O, bz, chl; msc EtOH, eth; vs ace
7080	3-Methyl-2-butanol, (±)-	Isopropylethanol	$C_5H_{12}O$	70116-68-6	88.148	liq		113.7(0.4)	0.8180[20]	1.4089[20]	sl H_2O; msc EtOH, eth; vs ace; s bz, ctc
7081	2-Methyl-1-butanol acetate		$C_7H_{14}O_2$	624-41-9	130.185			139(3)	0.8740[20]	1.4040[20]	vs ace, eth, EtOH
7082	3-Methyl-2-butanone	Methyl isopropyl ketone	$C_5H_{10}O$	563-80-4	86.132	liq	-93.13(0.05)	94.2(0.2)	0.8051[20]	1.3880[20]	sl H_2O; msc EtOH, eth; vs ace; s ctc
7083	2-Methylbutanoyl chloride, (±)-		C_5H_9ClO	57526-28-0	120.577			116	0.9917[20]	1.4170[20]	
7084	3-Methylbutanoyl chloride	Isovaleryl chloride	C_5H_9ClO	108-12-3	120.577			114	0.9844[20]	1.4149[20]	s eth
7085	trans-2-Methyl-2-butenal	Tiglic aldehyde	C_5H_8O	497-03-0	84.117	liq		114(3)	0.8710[20]	1.4475[20]	sl H_2O; vs EtOH
7086	3-Methyl-2-butenal	Senecialdehyde	C_5H_8O	107-86-8	84.117			134	0.8722[20]	1.4528[20]	s H_2O, EtOH, eth
7087	2-Methyl-1-butene		C_5H_{10}	563-46-2	70.133	liq	-137.53(0.02)	31.1(0.4)	0.6504[20]	1.3778[20]	i H_2O; s EtOH, eth, bz, ctc
7088	3-Methyl-1-butene		C_5H_{10}	563-45-1	70.133	vol liq or gas	-168.41(0.02)	20.1(0.2)	0.6213[25]	1.3643[20]	i H_2O; msc EtOH, eth; s bz
7089	2-Methyl-2-butene		C_5H_{10}	513-35-9	70.133	liq	-133.72(0.02)	38.5(0.4)	0.6623[20]	1.3874[20]	i H_2O; s EtOH, eth, bz, ctc; vs liq

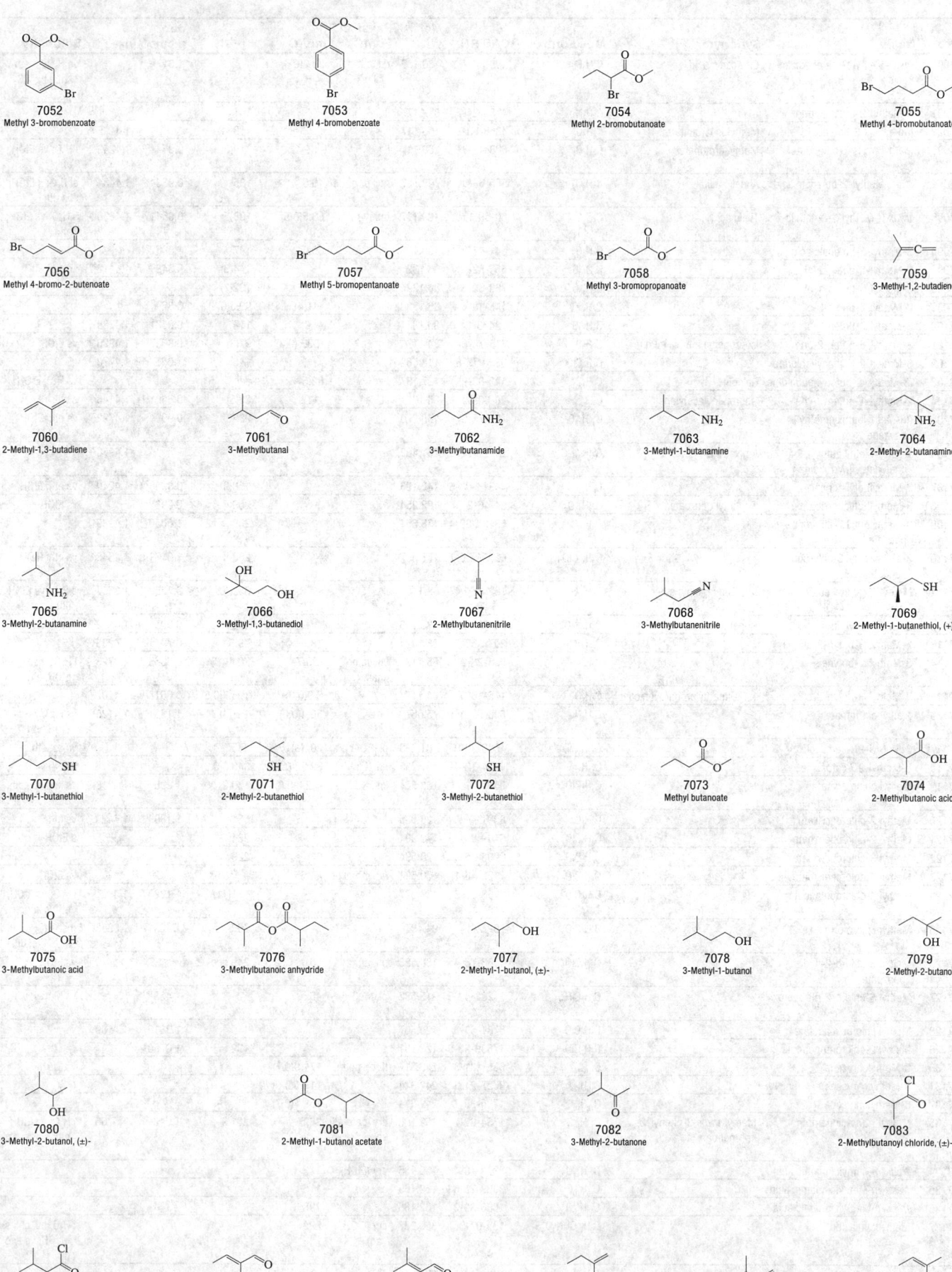

7052
Methyl 3-bromobenzoate

7053
Methyl 4-bromobenzoate

7054
Methyl 2-bromobutanoate

7055
Methyl 4-bromobutanoate

7056
Methyl 4-bromo-2-butenoate

7057
Methyl 5-bromopentanoate

7058
Methyl 3-bromopropanoate

7059
3-Methyl-1,2-butadiene

7060
2-Methyl-1,3-butadiene

7061
3-Methylbutanal

7062
3-Methylbutanamide

7063
3-Methyl-1-butanamine

7064
2-Methyl-2-butanamine

7065
3-Methyl-2-butanamine

7066
3-Methyl-1,3-butanediol

7067
2-Methylbutanenitrile

7068
3-Methylbutanenitrile

7069
2-Methyl-1-butanethiol, (+)

7070
3-Methyl-1-butanethiol

7071
2-Methyl-2-butanethiol

7072
3-Methyl-2-butanethiol

7073
Methyl butanoate

7074
2-Methylbutanoic acid

7075
3-Methylbutanoic acid

7076
3-Methylbutanoic anhydride

7077
2-Methyl-1-butanol, (±)-

7078
3-Methyl-1-butanol

7079
2-Methyl-2-butanol

7080
3-Methyl-2-butanol, (±)-

7081
2-Methyl-1-butanol acetate

7082
3-Methyl-2-butanone

7083
2-Methylbutanoyl chloride, (±)-

7084
3-Methylbutanoyl chloride

7085
trans-2-Methyl-2-butenal

7086
3-Methyl-2-butenal

7087
2-Methyl-1-butene

7088
3-Methyl-1-butene

7089
2-Methyl-2-butene

No.	Name	Synonym	Mol. Form.	CAS RN	Mol. Wt.	Physical Form	mp/°C	bp/°C	den g cm⁻³	n_D	Solubility
7090	cis-2-Methyl-2-butenedioic acid	Citraconic acid	$C_5H_6O_4$	498-23-7	130.100	nd (eth-lig) tcl pr (eth-bz)	83.2(0.6)		1.617[25]		vs H_2O; sl eth, chl; i bz, CS_2
7091	3-Methyl-2-butenenitrile		C_5H_7N	4786-24-7	81.117	liq		141			
7092	Methyl cis-2-butenoate	Methyl isocrotonate	$C_5H_8O_2$	4358-59-2	100.117			118		1.4175[20]	
7093	Methyl trans-2-butenoate	Methyl crotonate	$C_5H_8O_2$	623-43-8	100.117	liq	-42	119(6)	0.9444[20]	1.4242[20]	i H_2O; vs EtOH, eth
7094	cis-2-Methyl-2-butenoic acid	Angelic acid	$C_5H_8O_2$	565-63-9	100.117	mcl pr or nd	45.5	185	0.9834[49]	1.4434[47]	sl H_2O; s EtOH; vs eth
7095	trans-2-Methyl-2-butenoic acid	Tiglic acid	$C_5H_8O_2$	80-59-1	100.117	tab (w)	63.5(0.5)	198.5	0.9641[76]	1.4330[76]	s H_2O; vs EtOH, eth
7096	3-Methyl-2-butenoic acid		$C_5H_8O_2$	541-47-9	100.117		69.5	197	1.0062[24]		
7097	3-Methyl-2-buten-1-ol		$C_5H_{10}O$	556-82-1	86.132			144(6)	0.848[25]	1.4412[20]	
7098	3-Methyl-3-buten-1-ol		$C_5H_{10}O$	763-32-6	86.132			132(2)			
7099	2-Methyl-3-buten-2-ol		$C_5H_{10}O$	115-18-4	86.132	liq	-41(4)	99.3(0.8)	0.82[20]		
7100	3-Methyl-3-buten-2-ol		$C_5H_{10}O$	10473-14-0	86.132			114	0.8531[17]	1.4288[17]	
7101	3-Methyl-3-buten-2-one	Isopropenyl methyl ketone	C_5H_8O	814-78-8	84.117	liq	-53.6(0.3)	97(3)	0.8527[20]	1.4220[20]	vs EtOH
7102	3-Methyl-2-butenoyl chloride		C_5H_7ClO	3350-78-5	118.562			146	1.065[25]	1.4770[20]	
7103	(3-Methyl-2-butenyl)guanidine	Galegine	$C_6H_{13}N_3$	543-83-9	127.187	hyg	62.5	dec			vs H_2O, EtOH
7104	2-Methyl-1-buten-3-yne	Isopropenylacetylene	C_5H_6	78-80-8	66.102	liq	-113	33(5)	0.6801[11]	1.4140[20]	s chl
7105	[(3-Methylbutoxy)methyl]-benzene		$C_{12}H_{18}O$	122-73-6	178.270			236	0.909[20]	1.4792[20]	vs eth, EtOH
7106	1-[2-(3-Methylbutoxy)-2-phenylethyl]pyrrolidine	Amixetrine	$C_{17}H_{27}NO$	24622-72-8	261.402			121[2]		1.4978[22]	
7107	2-Methylbutyl acrylate		$C_8H_{14}O_2$	44914-03-6	142.196			158(9)	0.8936[20]	1.4240[20]	vs eth, EtOH
7108	3-Methylbutyl benzoate	Isopentyl benzoate	$C_{12}H_{16}O_2$	94-46-2	192.254			259(4)	0.993[15]		vs EtOH
7109	3-Methylbutyl 2-chloropropanoate		$C_8H_{15}ClO_2$	62108-69-4	178.657			208	1.0050[20]	1.4289[20]	
7110	3-Methylbutyl 3-chloropropanoate		$C_8H_{15}ClO_2$	62108-70-7	178.657			208	1.0171[20]	1.4343[20]	vs eth, EtOH
7111	Methyl tert-butyl ether	tert-Butyl methyl ether	$C_5H_{12}O$	1634-04-4	88.148	liq	-108.6(0.1)	55.1(0.1)	0.7353[25]	1.3664[25]	s H_2O; vs EtOH, eth
7112	3-Methylbutyl nitrate	Isopentyl nitrate	$C_5H_{11}NO_3$	543-87-3	133.146			148	0.996[22]	1.4122[21]	
7113	2-Methyl-3-butyn-2-amine		C_5H_9N	2978-58-7	83.132		18	79.5	0.79[25]	1.4235[20]	
7114	3-Methyl-1-butyne		C_5H_8	598-23-2	68.118	vol liq or gas	-89.7	28(4)	0.6660[20]	1.3723[20]	i H_2O; msc EtOH, eth
7115	2-Methyl-3-butyn-2-ol	1,1-Dimethylpropargyl alcohol	C_5H_8O	115-19-5	84.117		3.0(0.3)	105.7(0.5)	0.8618[20]	1.4207[20]	
7116	Methyl carbamate		$C_2H_5NO_2$	598-55-0	75.067	nd	55.4(0.6)	191(4)	1.1361[56]	1.4125[56]	vs H_2O; vs EtOH, eth
7117	3-Methyl-9H-carbazole		$C_{13}H_{11}N$	4630-20-0	181.233	pl (HOAc)	204(3)	365			vs bz, eth
7118	9-Methyl-9H-carbazole		$C_{13}H_{11}N$	1484-12-4	181.233	nd, lf (al)	89.32(0.03)	343.8(0.3)			vs eth
7119	Methyl chloroacetate		$C_3H_5ClO_2$	96-34-4	108.524	liq	-32.3(0.6)	130(2)	1.236[20]	1.4218[20]	vs ace, bz, eth, EtOH
7120	Methyl 2-chloroacrylate		$C_4H_5ClO_2$	80-63-7	120.535			52[51]	1.189[20]	1.4420[20]	vs eth
7121	Methyl 2-chlorobenzoate		$C_8H_7ClO_2$	610-96-8	170.594			234			s EtOH
7122	Methyl 3-chlorobenzoate		$C_8H_7ClO_2$	2905-65-9	170.594		21	229			
7123	Methyl 4-chlorobenzoate		$C_8H_7ClO_2$	1126-46-1	170.594	nd or mcl pr	43.5		1.382[20]		vs EtOH
7124	Methyl 4-chlorobutanoate		$C_5H_9ClO_2$	3153-37-5	136.577			161(11)	1.1293[20]	1.4321[20]	i H_2O; vs EtOH, eth; s ace
7125	Methyl chloroformate		$C_2H_3ClO_2$	79-22-1	94.497			70.5	1.2231[20]	1.3868[20]	msc EtOH, eth; s bz, ctc, chl
7126	Methyl 5-chloro-2-hydroxybenzoate		$C_8H_7ClO_3$	4068-78-4	186.593	nd (al)	50	249 dec			vs EtOH
7127	Methyl 5-chloro-2-nitrobenzoate		$C_8H_6ClNO_4$	51282-49-6	215.592	pl (MeOH)	48.5		1.453[18]		vs MeOH
7128	Methyl chlorooxoacetate		$C_3H_3ClO_3$	5781-53-3	122.507			119	1.3316[20]	1.4189[20]	
7129	Methyl 2-chloropropanoate		$C_4H_7ClO_2$	17639-93-9	122.551			132.5	1.0750[25]		
7130	3-Methylchrysene		$C_{19}H_{14}$	3351-31-3	242.314	lf (bz-peth)	171.9(0.7)				vs EtOH
7131	5-Methylchrysene		$C_{19}H_{14}$	3697-24-3	242.314		117.5(0.5)				i H_2O
7132	6-Methylchrysene		$C_{19}H_{14}$	1705-85-7	242.314		159.4(0.7)				
7133	Methyl trans-cinnamate	Methyl trans-3-phenyl-2-propenoate	$C_{10}H_{10}O_2$	1754-62-7	162.185	cry (peth, dil al)	36.5	261.9	1.042[36]	1.5766[22]	i H_2O; vs EtOH, eth; s ace, bz; sl chl
7134	trans-o-Methylcinnamic acid		$C_{10}H_{10}O_2$	2373-76-4	162.185	cry (EtOH)	175				
7135	trans-m-Methylcinnamic acid		$C_{10}H_{10}O_2$	3029-79-6	162.185	cry (w)	115				
7136	trans-p-Methylcinnamic acid		$C_{10}H_{10}O_2$	1866-39-3	162.185		198.5				
7137	Methyclothiazide		$C_9H_{11}Cl_2N_3O_4S_2$	135-07-9	360.237	cry (EtOH aq)	225				i H_2O, bz, chl; sl MeOH; vs ace, py
7138	Methyl cyanate		C_2H_3NO	1768-34-9	57.051	unstab gas	-30				
7139	Methyl cyanoacetate		$C_4H_5NO_2$	105-34-0	99.089	liq	-13.1(0.2)	203(4)	1.1225[25]	1.4176[20]	vs eth, EtOH

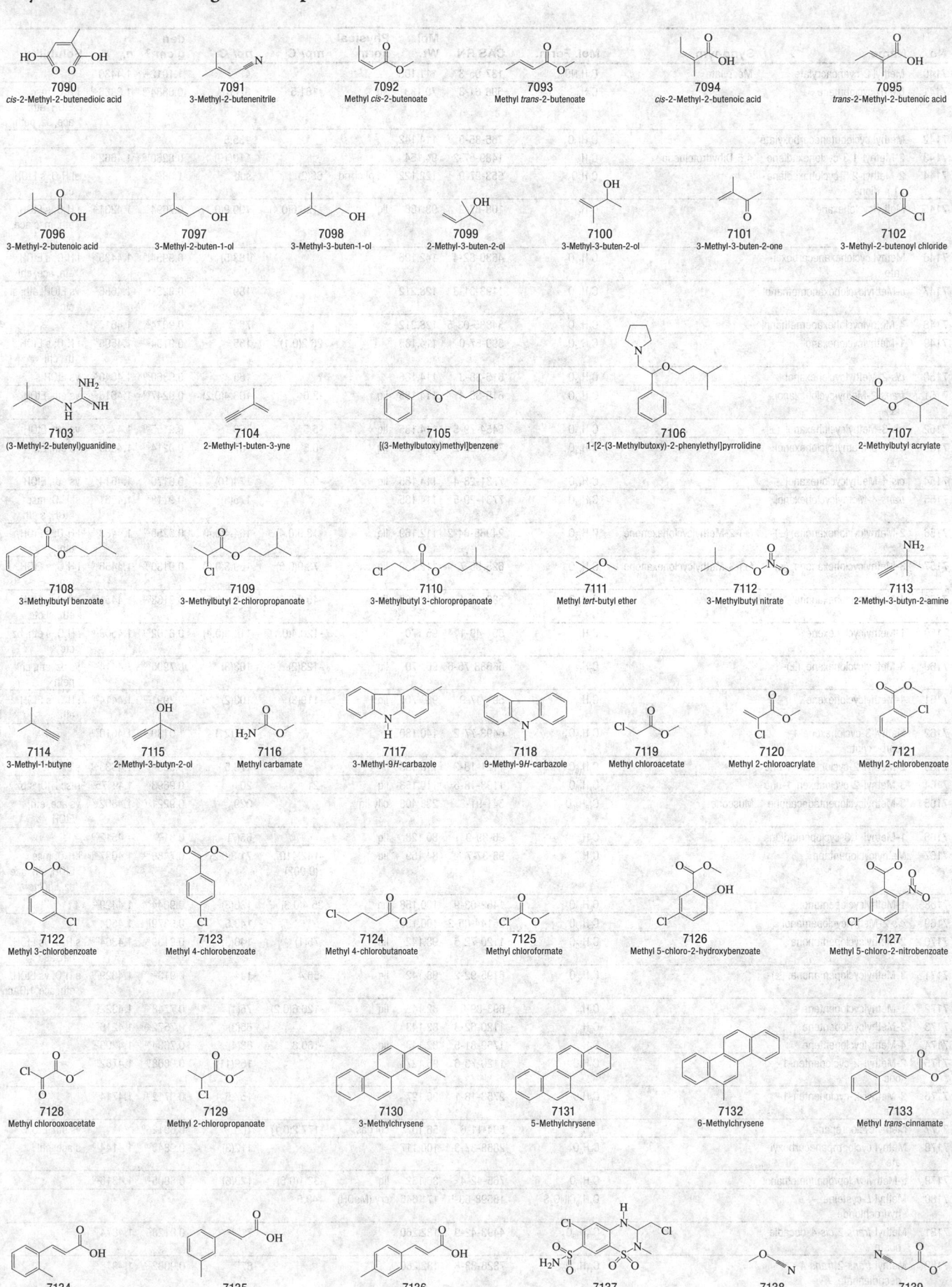

7090
cis-2-Methyl-2-butenedioic acid

7091
3-Methyl-2-butenenitrile

7092
Methyl cis-2-butenoate

7093
Methyl trans-2-butenoate

7094
cis-2-Methyl-2-butenoic acid

7095
trans-2-Methyl-2-butenoic acid

7096
3-Methyl-2-butenoic acid

7097
3-Methyl-2-buten-1-ol

7098
3-Methyl-3-buten-1-ol

7099
2-Methyl-3-buten-2-ol

7100
3-Methyl-3-buten-2-ol

7101
3-Methyl-3-buten-2-one

7102
3-Methyl-2-butenoyl chloride

7103
(3-Methyl-2-butenyl)guanidine

7104
2-Methyl-1-buten-3-yne

7105
[(3-Methylbutoxy)methyl]benzene

7106
1-[2-(3-Methylbutoxy)-2-phenylethyl]pyrrolidine

7107
2-Methylbutyl acrylate

7108
3-Methylbutyl benzoate

7109
3-Methylbutyl 2-chloropropanoate

7110
3-Methylbutyl 3-chloropropanoate

7111
Methyl tert-butyl ether

7112
3-Methylbutyl nitrate

7113
2-Methyl-3-butyn-2-amine

7114
3-Methyl-1-butyne

7115
2-Methyl-3-butyn-2-ol

7116
Methyl carbamate

7117
3-Methyl-9H-carbazole

7118
9-Methyl-9H-carbazole

7119
Methyl chloroacetate

7120
Methyl 2-chloroacrylate

7121
Methyl 2-chlorobenzoate

7122
Methyl 3-chlorobenzoate

7123
Methyl 4-chlorobenzoate

7124
Methyl 4-chlorobutanoate

7125
Methyl chloroformate

7126
Methyl 5-chloro-2-hydroxybenzoate

7127
Methyl 5-chloro-2-nitrobenzoate

7128
Methyl chlorooxoacetate

7129
Methyl 2-chloropropanoate

7130
3-Methylchrysene

7131
5-Methylchrysene

7132
6-Methylchrysene

7133
Methyl trans-cinnamate

7134
trans-o-Methylcinnamic acid

7135
trans-m-Methylcinnamic acid

7136
trans-p-Methylcinnamic acid

7137
Methyclothiazide

7138
Methyl cyanate

7139
Methyl cyanoacetate

No.	Name	Synonym	Mol. Form.	CAS RN	Mol. Wt.	Physical Form	mp/°C	bp/°C	den g cm⁻³	n_D	Solubility
7140	Methyl 2-cyanoacrylate	Mecrylate	$C_5H_5NO_2$	137-05-3	111.100			47[2]	1.1012[20]	1.4430	
7141	Methylcyclobutane		C_5H_{10}	598-61-8	70.133	liq	-161.5	37(3)	0.6884[20]	1.3866[20]	i H₂O; msc EtOH, eth; s ace, bz, peth
7142	Methyl cyclobutanecarboxylate		$C_6H_{10}O_2$	765-85-5	114.142			135.5			
7143	2-Methyl-1,3-cyclohexadiene	4,5-Dihydrotoluene	C_7H_{10}	1489-57-2	94.154			110(19)	0.8260[18]	1.4662[18]	
7144	2-Methyl-2,5-cyclohexadiene-1,4-dione		$C_7H_6O_2$	553-97-9	122.122	ye pl or nd	68(2)	sub	1.08[75]		sl H₂O; s EtOH, eth
7145	Methylcyclohexane		C_7H_{14}	108-87-2	98.186	liq	-126.6(0.4)	100.9(0.1)	0.7694[20]	1.4231[20]	i H₂O; s EtOH, eth; msc ace, bz, lig
7146	Methyl cyclohexanecarboxylate		$C_8H_{14}O_2$	4630-82-4	142.196			183(5)	0.9954[15]	1.4433[20]	i H₂O; s EtOH, eth, ace, chl
7147	α-Methylcyclohexanemethanol		$C_8H_{16}O$	1193-81-3	128.212			189	0.928[25]	1.4656[20]	vs EtOH, eth; sl ctc
7148	4-Methylcyclohexanemethanol		$C_8H_{16}O$	34885-03-5	128.212			75[2.5]	0.9074[20]	1.4617[20]	
7149	1-Methylcyclohexanol		$C_7H_{14}O$	590-67-0	114.185		26.2(0.1)	155	0.9194[20]	1.4595[20]	i H₂O; s EtOH, bz, chl
7150	cis-2-Methylcyclohexanol		$C_7H_{14}O$	615-38-3	114.185		7	165	0.9360[20]	1.4640[20]	vs EtOH
7151	trans-2-Methylcyclohexanol, (±)-		$C_7H_{14}O$	615-39-4	114.185	liq	-2.0	168.4(0.2)	0.9247[20]	1.4616[20]	vs eth, EtOH
7152	cis-3-Methylcyclohexanol, (±)-		$C_7H_{14}O$	5454-79-5	114.185	liq	-5.5	168	0.9155[20]	1.4752[20]	vs eth, EtOH
7153	trans-3-Methylcyclohexanol, (±)-		$C_7H_{14}O$	7443-55-2	114.185	liq	-0.5	167	0.9214[30]	1.4580[20]	vs eth, EtOH
7154	cis-4-Methylcyclohexanol		$C_7H_{14}O$	7731-28-4	114.185	liq	-9.2	174(10)	0.9170[20]	1.4614[20]	vs eth, EtOH
7155	trans-4-Methylcyclohexanol		$C_7H_{14}O$	7731-29-5	114.185			175(3)	0.9118[21]	1.4561[20]	sl H₂O; msc EtOH; s eth
7156	2-Methylcyclohexanone, (±)-	(±)-2-Methylcyclohexanone	$C_7H_{12}O$	24965-84-2	112.169	liq	-13.9(0.4)	164.4(0.4)	0.9250[20]	1.4483[25]	i H₂O; s EtOH, eth
7157	3-Methylcyclohexanone, (±)-	(±)-3-Methylcyclohexanone	$C_7H_{12}O$	625-96-7	112.169	liq	-73.9(0.5)	169.3(0.6)	0.9136[20]	1.4456[20]	i H₂O; s EtOH, eth
7158	4-Methylcyclohexanone		$C_7H_{12}O$	589-92-4	112.169	liq	-40.6(0.5)	170.9(0.5)	0.9138[20]	1.4451[20]	i H₂O; s EtOH, eth; sl ctc
7159	1-Methylcyclohexene		C_7H_{12}	591-49-1	96.170	liq	-120.4(0.1)	110.3(0.4)	0.8102[20]	1.4503[20]	i H₂O; s eth, bz, ctc
7160	3-Methylcyclohexene, (±)-		C_7H_{12}	56688-75-6	96.170	liq	-123(2)	103(3)	0.7990[20]	1.4414[20]	vs bz, eth, chl, peth
7161	4-Methylcyclohexene		C_7H_{12}	591-47-9	96.170	liq	-119(3)	103(2)	0.7991[20]	1.4414[20]	i H₂O; s EtOH, eth
7162	Methyl 3-cyclohexene-1-carboxylate		$C_8H_{12}O_2$	6493-77-2	140.180			181(21)	1.0130[20]	1.4610[20]	
7163	2-Methyl-2-cyclohexen-1-one		$C_7H_{10}O$	1121-18-2	110.153			178.5	0.966[20]	1.4833[20]	s bz
7164	3-Methyl-2-cyclohexen-1-one		$C_7H_{10}O$	1193-18-6	110.153	liq	-21	201	0.9693[20]	1.49475[20]	msc H₂O; s bz
7165	3-Methylcyclopentadecanone	Muscone	$C_{16}H_{30}O$	541-91-3	238.408	oily liq		329	0.9221[17]	1.4802[17]	vs ace, eth, EtOH
7166	1-Methyl-1,3-cyclopentadiene		C_6H_8	96-39-9	80.128	liq		65(7)	0.81[20]	1.4512[20]	
7167	Methylcyclopentane		C_6H_{12}	96-37-7	84.159	liq	-142.419 (0.002)	71.8(0.2)	0.7486[20]	1.4097[20]	i H₂O; msc EtOH, eth, ace, bz, lig, ctc
7168	1-Methylcyclopentanol		$C_6H_{12}O$	1462-03-9	100.158	nd	35.4(0.3)	136(8)	0.9044[23]	1.4429[23]	
7169	cis-2-Methylcyclopentanol		$C_6H_{12}O$	25144-05-2	100.158			148.5	0.9379[16]	1.4504[16]	
7170	2-Methylcyclopentanone		$C_6H_{10}O$	1120-72-5	98.142	liq	-75(1)	140(3)	0.9139[20]	1.4364[20]	s H₂O; vs EtOH, eth, ace
7171	3-Methylcyclopentanone, (±)-		$C_6H_{10}O$	6195-92-2	98.142	liq	-58.4	144	0.913[22]	1.4329[20]	s H₂O; vs EtOH, eth, ace, HOAc
7172	1-Methylcyclopentene		C_6H_{10}	693-89-0	82.143	liq	-126.6(0.2)	75(1)	0.7748[25]	1.4322[20]	
7173	3-Methylcyclopentene		C_6H_{10}	1120-62-3	82.143			65(3)	0.7572[25]	1.4216[20]	
7174	4-Methylcyclopentene		C_6H_{10}	1759-81-5	82.143	liq	-160.8	68(4)	0.7634[25]	1.4209[20]	
7175	2-Methyl-2-cyclopenten-1-one		C_6H_8O	1120-73-6	96.127			159(11)	0.9808[16]	1.4762[15]	
7176	3-Methyl-2-cyclopenten-1-one		C_6H_8O	2758-18-1	96.127			157.5	0.9712[20]	1.4714[20]	
7177	Methylcyclopropane		C_4H_8	594-11-6	56.107	col gas	-177.2(0.1)	1(2)	0.6912[-20]		vs eth, EtOH
7178	Methyl cyclopropanecarboxylate		$C_5H_8O_2$	2868-37-3	100.117			112(3)	0.9848[20]	1.4144[19]	s ace, chl
7179	α-Methylcyclopropanemethanol		$C_5H_{10}O$	765-42-4	86.132	liq	-32.1(0.6)	123(6)	0.8805[20]	1.4316[20]	
7180	Methyl L-cysteine hydrochloride		$C_4H_{10}ClNO_2S$	18598-63-5	171.646	cry (MeOH)	140.5				
7181	Methyl trans-2,cis-4-decadienoate		$C_{11}H_{18}O_2$	4493-42-9	182.260			71[0.15]	0.9128[22]	1.4874[22]	
7182	Methyl trans-2,trans-4-decadienoate		$C_{11}H_{18}O_2$	7328-33-8	182.260			87[13]	0.9082[22]	1.4918[22]	

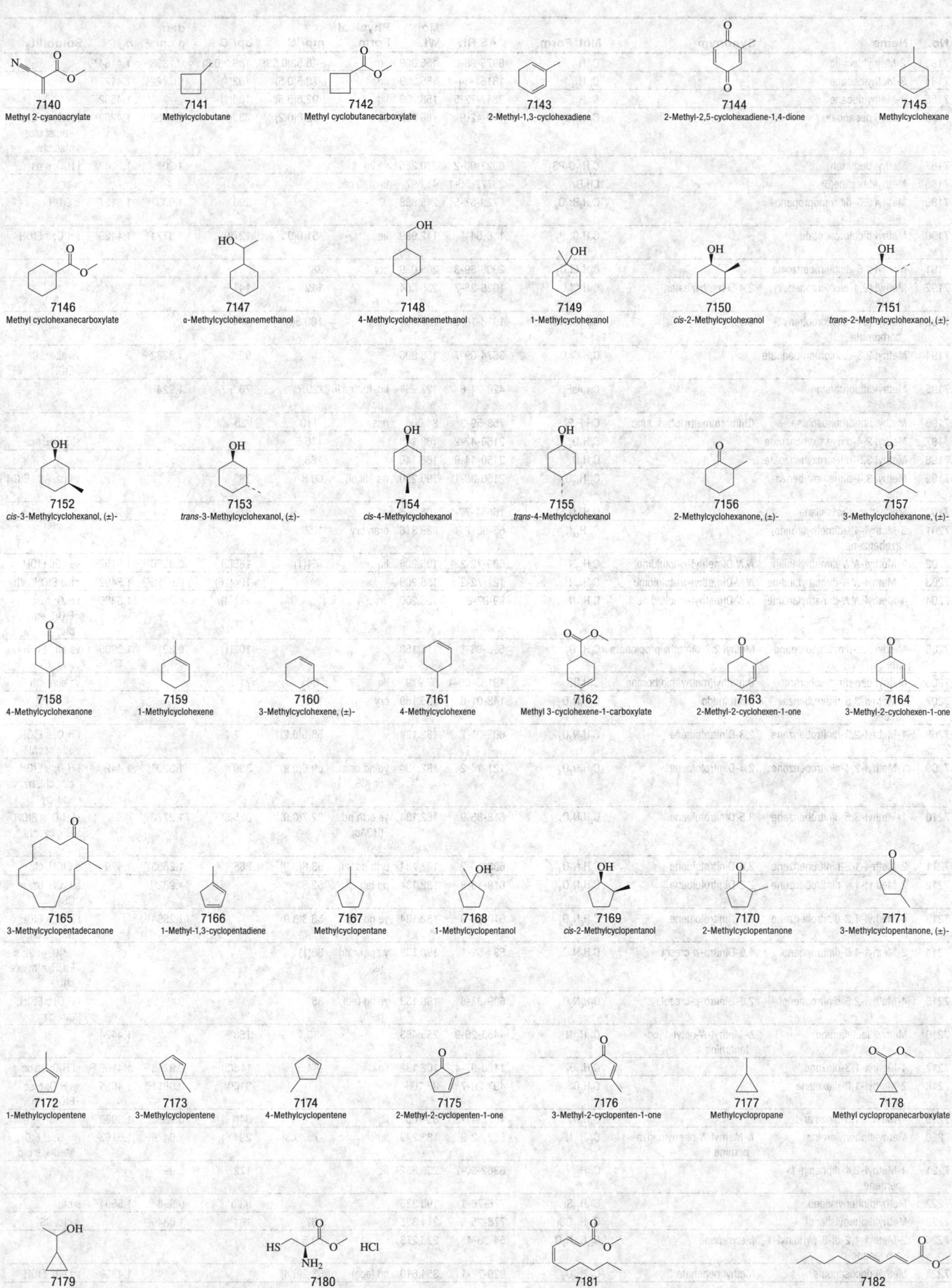

7140
Methyl 2-cyanoacrylate

7141
Methylcyclobutane

7142
Methyl cyclobutanecarboxylate

7143
2-Methyl-1,3-cyclohexadiene

7144
2-Methyl-2,5-cyclohexadiene-1,4-dione

7145
Methylcyclohexane

7146
Methyl cyclohexanecarboxylate

7147
α-Methylcyclohexanemethanol

7148
4-Methylcyclohexanemethanol

7149
1-Methylcyclohexanol

7150
cis-2-Methylcyclohexanol

7151
trans-2-Methylcyclohexanol, (±)-

7152
cis-3-Methylcyclohexanol, (±)-

7153
trans-3-Methylcyclohexanol, (±)-

7154
cis-4-Methylcyclohexanol

7155
trans-4-Methylcyclohexanol

7156
2-Methylcyclohexanone, (±)-

7157
3-Methylcyclohexanone, (±)-

7158
4-Methylcyclohexanone

7159
1-Methylcyclohexene

7160
3-Methylcyclohexene, (±)-

7161
4-Methylcyclohexene

7162
Methyl 3-cyclohexene-1-carboxylate

7163
2-Methyl-2-cyclohexen-1-one

7164
3-Methyl-2-cyclohexen-1-one

7165
3-Methylcyclopentadecanone

7166
1-Methyl-1,3-cyclopentadiene

7167
Methylcyclopentane

7168
1-Methylcyclopentanol

7169
cis-2-Methylcyclopentanol

7170
2-Methylcyclopentanone

7171
3-Methylcyclopentanone, (±)-

7172
1-Methylcyclopentene

7173
3-Methylcyclopentene

7174
4-Methylcyclopentene

7175
2-Methyl-2-cyclopenten-1-one

7176
3-Methyl-2-cyclopenten-1-one

7177
Methylcyclopropane

7178
Methyl cyclopropanecarboxylate

7179
α-Methylcyclopropanemethanol

7180
Methyl L-cysteine hydrochloride

7181
Methyl trans-2,cis-4-decadienoate

7182
Methyl trans-2,trans-4-decadienoate

No.	Name	Synonym	Mol. Form.	CAS RN	Mol. Wt.	Physical Form	mp/°C	bp/°C	den g cm⁻³	n_D	Solubility
7183	2-Methyldecane		$C_{11}H_{24}$	6975-98-0	156.309	liq	-48.83(0.03)	189.2(0.6)	0.7368[20]	1.4154[20]	
7184	3-Methyldecane		$C_{11}H_{24}$	13151-34-3	156.309	liq	-79.5(0.5)	192(1)	0.7422[20]	1.4177[20]	
7185	4-Methyldecane		$C_{11}H_{24}$	2847-72-5	156.309	liq	-92.8(0.5)	188(1)		1.4352[20]	
7186	Methyl decanoate		$C_{11}H_{22}O_2$	110-42-9	186.292	liq	-12.8(0.2)	233(1)	0.8730[20]	1.4259[20]	i H₂O; vs EtOH, eth; sl ctc; msc chl
7187	Methyl demeton		$C_6H_{15}O_3PS_2$	8022-00-2	230.285	ye liq			1.20[20]	1.5063[20]	i H₂O; s os
7188	Methyldiborane(6)		CH_8B_2	23777-55-1	41.697	unstab gas					s eth
7189	Methyl 2,3-dibromopropano-ate		$C_4H_6Br_2O_2$	1729-67-5	245.898			206	1.9333[20]	1.5127[20]	s EtOH
7190	Methyl dichloroacetate		$C_3H_4Cl_2O_2$	116-54-1	142.969	liq	-51.9(0.2)	127(9)	1.3774[20]	1.4429[20]	i H₂O; s EtOH, ctc
7191	Methyl 2,5-dichlorobenzoate		$C_8H_6Cl_2O_2$	2905-69-3	205.039	cry	38				
7192	Methyl (2,4-dichlorophenoxy)-acetate	2,4-D methyl ester	$C_9H_8Cl_2O_3$	1928-38-7	235.064		119	141[18]			
7193	Methyl (3,4-dichlorophenyl)-carbamate	Swep	$C_8H_7Cl_2NO_2$	1918-18-9	220.054	nd	109.6(0.5)				
7194	Methyl 2,3-dichloropropanoate		$C_4H_6Cl_2O_2$	3674-09-7	156.996			92[50]	1.3282[20]		vs ace, eth, EtOH
7195	Methyldifluoroarsine		CH_3AsF_2	420-24-6	127.954	liq, fumes in air	-29.7	76.5	1.924[18]		
7196	Methyldifluorophosphine	(Difluoro)methylphosphine	CH_3F_2P	753-59-3	84.006	gas	-110	-28			
7197	Methyl 2,4-dihydroxybenzoate		$C_8H_8O_4$	2150-47-2	168.148		116.5				sl EtOH, ace
7198	Methyl 3,5-dihydroxybenzoate		$C_8H_8O_4$	2150-44-9	168.148		165				
7199	Methyl 3,4-dimethoxybenzo-ate		$C_{10}H_{12}O_4$	2150-38-1	196.200	nd (dil al)	60.8	283			vs bz, eth, EtOH
7200	Methyldimethoxysilane		$C_3H_{10}O_2Si$	16881-77-9	106.196			61			
7201	3-Methyl-4'-(dimethylamino)-azobenzene		$C_{15}H_{17}N_3$	55-80-1	239.316	oran cry	122				
7202	2-Methyl-N,N-dimethylaniline	N,N-Dimethyl-o-toluidine	$C_9H_{13}N$	609-72-3	135.206	liq	-61(1)	185(2)	0.9286[20]	1.5152[20]	vs eth, EtOH
7203	3-Methyl-N,N-dimethylaniline	N,N-Dimethyl-m-toluidine	$C_9H_{13}N$	121-72-2	135.206			194(16)	0.9410[20]	1.5492[20]	msc EtOH, eth
7204	4-Methyl-N,N-dimethylaniline	N,N-Dimethyl-p-toluidine	$C_9H_{13}N$	99-97-8	135.206			211(4)	0.9366[20]	1.5366[20]	i H₂O; msc EtOH, eth; s ctc
7205	Methyl 2,2-dimethylpropano-ate	Methyl 2,2-dimethylpropionate	$C_6H_{12}O_2$	598-98-1	116.158			102(1)	0.891[0]	1.3905[20]	vs eth, EtOH
7206	Methyl dimethylthioborane	Dimethyl(methylthio)borane	C_3H_9BS	19163-05-4	87.979	liq	-84	71			vs ace, eth
7207	2-Methyl-3,5-dinitrobenza-mide	Dinitolmide	$C_8H_7N_3O_5$	148-01-6	225.159	cry	181				
7208	1-Methyl-2,3-dinitrobenzene	2,3-Dinitrotoluene	$C_7H_6N_2O_4$	602-01-7	182.134		56.6(0.2)				i H₂O; s EtOH, eth; sl chl
7209	1-Methyl-2,4-dinitrobenzene	2,4-Dinitrotoluene	$C_7H_6N_2O_4$	121-14-2	182.134	ye nd or mcl pr (CS₂)	69.6(0.9)	300 dec	1.3208[71]	1.442	i H₂O; s EtOH, eth, chl, bz; vs ace, py
7210	1-Methyl-3,5-dinitrobenzene	3,5-Dinitrotoluene	$C_7H_6N_2O_4$	618-85-9	182.134	ye orth nd (HOAc)	82.7(0.9)	325(5)	1.2772[111]		sl H₂O; s EtOH, eth, bz, chl, CS₂
7211	2-Methyl-1,3-dinitrobenzene	2,6-Dinitrotoluene	$C_7H_6N_2O_4$	606-20-2	182.134	orth nd (al)	63.8(0.9)	285	1.2833[111]	1.479	s EtOH, chl
7212	2-Methyl-1,4-dinitrobenzene	2,5-Dinitrotoluene	$C_7H_6N_2O_4$	619-15-8	182.134	nd (al)	52.5		1.282[111]		s EtOH, bz; vs CS₂
7213	4-Methyl-1,2-dinitrobenzene	3,4-Dinitrotoluene	$C_7H_6N_2O_4$	610-39-9	182.134	ye nd (CS₂)	58.3(0.9)		1.2594[111]		i H₂O; s EtOH, CS₂; sl chl
7214	2-Methyl-4,6-dinitrophenol	4,6-Dinitro-o-cresol	$C_7H_6N_2O_5$	534-52-1	198.133	ye pr or nd (al)	86(1)				sl H₂O, peth; s EtOH, eth, ace, chl
7215	4-Methyl-2,6-dinitrophenol	2,6-Dinitro-p-cresol	$C_7H_6N_2O_5$	609-93-8	198.133	ye nd (eth, peth)	85				i H₂O; s EtOH, eth, bz
7216	Methyldioctylamine	N-Methyl-N-octyl-1-oc-tanamine	$C_{17}H_{37}N$	4455-26-9	255.483		-30.1	158[10]		1.4424[20]	
7217	4-Methyl-1,3-dioxane		$C_5H_{10}O_2$	1120-97-4	102.132	liq	-44.5	115(3)	0.9758[20]	1.4159[20]	sl H₂O; vs os
7218	2-Methyl-1,3-dioxolane		$C_4H_8O_2$	497-26-7	88.106			81(3)	0.9811[20]	1.4035[17]	vs H₂O; msc EtOH, eth
7219	4-Methyl-1,3-dioxolane		$C_4H_8O_2$	1072-47-5	88.106	liq		84(19)	0.99[20]	1.3980[20]	
7220	Methyldiphenylamine	N-Methyl-N-phenylbenze-namine	$C_{13}H_{13}N$	552-82-9	183.249	liq	-7.6(0.3)	294(6)	1.0476[20]	1.6193[20]	i H₂O; sl EtOH, MeOH; s ctc
7221	4-Methyl-2,4-diphenyl-1-pentene		$C_{18}H_{20}$	6362-80-7	236.352	liq		172[8]	0.99[25]		
7222	Methyldiphenylsilane		$C_{13}H_{14}Si$	776-76-1	198.336			93.5[1]	0.996[20]	1.5694[20]	s ctc
7223	Methyldiphenylsilanol		$C_{13}H_{14}OSi$	778-25-6	214.335		167	184[24]	1.0840[25]		s ctc, CS₂
7224	2-Methyl-1,2-di-3-pyridinyl-1-propanone	Metyrapone	$C_{14}H_{14}N_2O$	54-36-4	226.273		50.5				
7225	Methyl docosanoate	Methyl behenate	$C_{23}H_{46}O_2$	929-77-1	354.610	nd (ace)	53.2(0.4)			1.4339[60]	vs eth, EtOH

7183
2-Methyldecane

7184
3-Methyldecane

7185
4-Methyldecane

7186
Methyl decanoate

7187
Methyl demeton

7188
Methyldiborane(6)

7189
Methyl 2,3-dibromopropanoate

7190
Methyl dichloroacetate

7191
Methyl 2,5-dichlorobenzoate

7192
Methyl (2,4-dichlorophenoxy)acetate

7193
Methyl (3,4-dichlorophenyl)carbamate

7194
Methyl 2,3-dichloropropanoate

7195
Methyldifluoroarsine

7196
Methyldifluorophosphine

7197
Methyl 2,4-dihydroxybenzoate

7198
Methyl 3,5-dihydroxybenzoate

7199
Methyl 3,4-dimethoxybenzoate

7200
Methyldimethoxysilane

7201
3-Methyl-4'-(dimethylamino)azobenzene

7202
2-Methyl-*N,N*-dimethylaniline

7203
3-Methyl-*N,N*-dimethylaniline

7204
4-Methyl-*N,N*-dimethylaniline

7205
Methyl 2,2-dimethylpropanoate

7206
Methyl dimethylthioborane

7207
2-Methyl-3,5-dinitrobenzamide

7208
1-Methyl-2,3-dinitrobenzene

7209
1-Methyl-2,4-dinitrobenzene

7210
1-Methyl-3,5-dinitrobenzene

7211
2-Methyl-1,3-dinitrobenzene

7212
2-Methyl-1,4-dinitrobenzene

7213
4-Methyl-1,2-dinitrobenzene

7214
2-Methyl-4,6-dinitrophenol

7215
4-Methyl-2,6-dinitrophenol

7216
Methyldioctylamine

7217
4-Methyl-1,3-dioxane

7218
2-Methyl-1,3-dioxolane

7219
4-Methyl-1,3-dioxolane

7220
Methyldiphenylamine

7221
4-Methyl-2,4-diphenyl-1-pentene

7222
Methyldiphenylsilane

7223
Methyldiphenylsilanol

7224
2-Methyl-1,2-di-3-pyridinyl-1-propanone

7225
Methyl docosanoate

No.	Name	Synonym	Mol. Form.	CAS RN	Mol. Wt.	Physical Form	mp/°C	bp/°C	den g cm⁻³	n_D	Solubility
7226	Methyl *cis*-13-docosenoate		C$_{23}$H$_{44}$O$_2$	1120-34-9	352.594		-1.2	220[5]			i H$_2$O; msc EtOH, eth, ace, bz; s chl, ctc
7227	Methyl dodecanoate	Methyl laurate	C$_{13}$H$_{26}$O$_2$	111-82-0	214.344		5.0(0.2)	267(2)	0.8702[20]	1.4319[20]	
7228	2-Methyldodecanoic acid		C$_{13}$H$_{26}$O$_2$	2874-74-0	214.344	pl	22	153[1]	0.890[18]		
7229	Methyl eicosanoate	Methyl arachidate	C$_{21}$H$_{42}$O$_2$	1120-28-1	326.557	lf (MeOH)	46.3(0.3)	215[10]		1.4317[60]	vs bz, eth, EtOH, chl
7230	(Methyleneamino)acetonitrile		C$_3$H$_4$N$_2$	109-82-0	68.077		129				
7231	α-Methylenebenzeneacetic acid	Atropic acid	C$_9$H$_8$O$_2$	492-38-6	148.159	lf (al), nd (w)	106.5	267 dec			sl H$_2$O; s EtOH, eth, bz, chl, CS$_2$
7232	Methylenebis(4-cyclohexyliso-cyanate)		C$_{15}$H$_{22}$N$_2$O$_2$	5124-30-1	262.348	liq			1.066	1.4970[20]	
7233	4,4'-Methylenebis[2,6-di-*tert*-butylphenol]	Bis(3,5-di-*tert*-butyl-4-hydroxyphenyl)methane	C$_{29}$H$_{44}$O$_2$	118-82-1	424.658		154	289[40]			
7234	4,4'-Methylenebis(*N*-methyl-aniline)	*N,N*'-Dimethyl-4,4'-diaminodi-phenylmethane	C$_{15}$H$_{18}$N$_2$	1807-55-2	226.317						s ctc, CS$_2$
7235	Methylene blue		C$_{16}$H$_{18}$ClN$_3$S	61-73-4	319.852	dk grn cry or pow (chl-eth)					s H$_2$O, EtOH, chl; i eth; sl py
7236	Methylenecyclobutane		C$_5$H$_8$	1120-56-5	68.118	liq	-134.52(0.07)	42(1)	0.7401[20]	1.4210[20]	
7237	Methylenecyclohexane		C$_7$H$_{12}$	1192-37-6	96.170	liq	-104.7(0.1)	103(3)	0.8074[20]	1.4523[20]	i H$_2$O; s eth, bz, chl
7238	2-Methylenecyclohexanol		C$_7$H$_{12}$O	4065-80-9	112.169			83[13]	0.955[20]	1.4843[20]	
7239	Methylenecyclopentane		C$_6$H$_{10}$	1528-30-9	82.143			75(3)	0.7787[20]	1.4355[20]	s bz, chl
7240	Methylenecyclopropene		C$_4$H$_4$	4095-06-1	52.075	solid stab at -196					
7241	2,4'-Methylenedianiline	2,4'-Diaminodiphenylmethane	C$_{13}$H$_{14}$N$_2$	1208-52-2	198.263	lf (bz)	88(1)	222[9]			
7242	5,5'-Methylenedisalicylic acid		C$_{15}$H$_{12}$O$_6$	122-25-8	288.252	nd (bz)	243.5				vs ace, eth, EtOH
7243	5-Methylene-2(5*H*)-furanone	Protoanemonin	C$_5$H$_4$O$_2$	108-28-1	96.085	pa ye oil		73[11]			sl H$_2$O; s chl
7244	3-Methyleneheptane		C$_8$H$_{16}$	1632-16-2	112.213			120(3)	0.7270[20]	1.4157[20]	i H$_2$O; vs eth, bz, peth
7245	4-Methylene-1-isopropylbicy-clo[3.1.0]hexan-3-ol, [1*S*-(1α,3β,5α)]	4(10)-Thujene-3-ol	C$_{10}$H$_{16}$O	471-16-9	152.233			208	0.9488[19]	1.4871[25]	s eth
7246	4-Methylene-1-isopropylcy-clohexene		C$_{10}$H$_{16}$	99-84-3	136.234			173.5	0.838[22]	1.4754[22]	
7247	2-Methylenepentanedinitrile	2,4-Dicyano-1-butene	C$_6$H$_6$N$_2$	1572-52-7	106.125			103[5]		1.4561[20]	s chl
7248	Methylene thiocyanate	Dithiocyanatomethane	C$_3$H$_2$N$_2$S$_2$	6317-18-6	130.191	solid	102				
7249	2-Methylene-1,3,3-trimethyl-indoline	Fischer's base	C$_{12}$H$_{15}$N	118-12-7	173.254			244			sl H$_2$O; s EtOH, eth, bz, chl
7250	*N*-Methylephedrine, [*R*-(*R**,*S**)]	(1*R*,2*S*)-*N*-Methylephedrine	C$_{11}$H$_{17}$NO	552-79-4	179.259	nd or pl (al, eth)	87.5				i H$_2$O; s EtOH, eth, MeOH
7251	Methylergonovine	Methylergometrine	C$_{20}$H$_{25}$N$_3$O$_2$	113-42-8	339.432	pr (MeOH, ace)	172				i H$_2$O; s EtOH, ace
7252	*N*-Methyl-1,2-ethanediamine		C$_3$H$_{10}$N$_2$	109-81-9	74.124			115.5(0.6)	0.841[25]	1.4395[20]	
7253	*N*-Methyl-2-ethanolamine		C$_3$H$_9$NO	109-83-1	75.109			159.24(0.04)	0.937[20]	1.4385[20]	msc H$_2$O, EtOH, eth
7254	1-(1-Methylethoxy)butane	Butyl isopropyl ether	C$_7$H$_{16}$O	1860-27-1	116.201			107(9)	0.7594[15]	1.3870[15]	i H$_2$O; s EtOH, eth, ace, con sulf
7255	2-[2-(1-Methylethoxy)ethyl]-pyridine		C$_{10}$H$_{15}$NO	70715-19-4	165.232			133[50]	0.9502[25]	1.4820[25]	vs H$_2$O
7256	1-(1-Methylethoxy)propane		C$_6$H$_{14}$O	627-08-7	102.174			82(8)	0.7370[20]	1.376[21]	sl H$_2$O; vs EtOH; s eth, ace
7257	1-(1-Methylethoxy)-2-propa-nol	1-Isopropoxy-2-propanol	C$_6$H$_{14}$O$_2$	3944-36-3	118.174			137.5	0.879[20]	1.4070[20]	
7258	Methyl 2-ethylacetoacetate		C$_7$H$_{12}$O$_3$	51756-08-2	144.168			182	0.995[14]		vs ace, eth, EtOH
7259	5-(1-Methylethylidene)-1,3-cyclopentadiene		C$_8$H$_{10}$	2175-91-9	106.165		1(2)	155	0.881[20]	1.5474[20]	
7260	1-Methyl-9*H*-fluorene		C$_{14}$H$_{12}$	1730-37-6	180.245		87				
7261	9-Methyl-9*H*-fluorene		C$_{14}$H$_{12}$	2523-37-7	180.245	pr	45(1)	155[15]	1.0263[66]	1.610[66]	i H$_2$O; s EtOH, eth, ace, bz, chl
7262	Methyl fluorosulfonate		CH$_3$FO$_3$S	421-20-5	114.096	col liq	-95	93	1.412	1.3326[20]	
7263	*N*-Methylformamide		C$_2$H$_5$NO	123-39-7	59.067	liq	-2.5(0.7)	186(5)	1.011[19]	1.4319[20]	vs H$_2$O, ace, EtOH
7264	Methyl formate		C$_2$H$_4$O$_2$	107-31-3	60.052	liq	-99.7(0.4)	31.6(0.3)	0.9713[20]	1.3419[20]	vs H$_2$O; msc EtOH; s eth, chl, MeOH
7265	Methyl 4-formylbenzoate		C$_9$H$_8$O$_3$	1571-08-0	164.158	nd (w)	63	265			

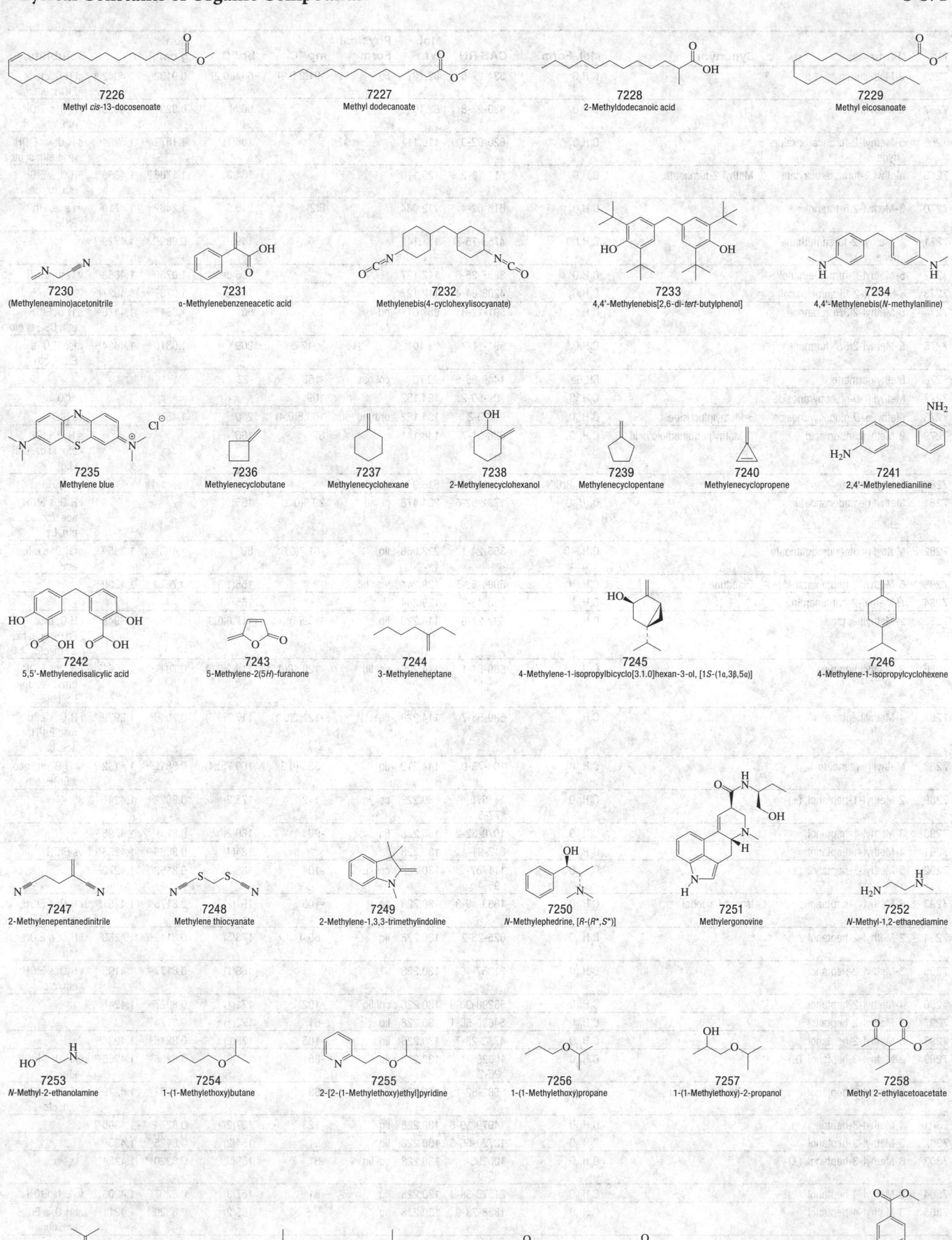

7226
Methyl *cis*-13-docosenoate

7227
Methyl dodecanoate

7228
2-Methyldodecanoic acid

7229
Methyl eicosanoate

7230
(Methyleneamino)acetonitrile

7231
α-Methylenebenzeneacetic acid

7232
Methylenebis(4-cyclohexylisocyanate)

7233
4,4'-Methylenebis[2,6-di-*tert*-butylphenol]

7234
4,4'-Methylenebis(*N*-methylaniline)

7235
Methylene blue

7236
Methylenecyclobutane

7237
Methylenecyclohexane

7238
2-Methylenecyclohexanol

7239
Methylenecyclopentane

7240
Methylenecyclopropene

7241
2,4'-Methylenedianiline

7242
5,5'-Methylenedisalicylic acid

7243
5-Methylene-2(5*H*)-furanone

7244
3-Methyleneheptane

7245
4-Methylene-1-isopropylbicyclo[3.1.0]hexan-3-ol, [1*S*-(1α,3β,5α)]

7246
4-Methylene-1-isopropylcyclohexene

7247
2-Methylenepentanedinitrile

7248
Methylene thiocyanate

7249
2-Methylene-1,3,3-trimethylindoline

7250
N-Methylephedrine, [*R*-(*R**,*S**)]

7251
Methylergonovine

7252
N-Methyl-1,2-ethanediamine

7253
N-Methyl-2-ethanolamine

7254
1-(1-Methylethoxy)butane

7255
2-[2-(1-Methylethoxy)ethyl]pyridine

7256
1-(1-Methylethoxy)propane

7257
1-(1-Methylethoxy)-2-propanol

7258
Methyl 2-ethylacetoacetate

7259
5-(1-Methylethylidene)-1,3-cyclopentadiene

7260
1-Methyl-9*H*-fluorene

7261
9-Methyl-9*H*-fluorene

7262
Methyl fluorosulfonate

7263
N-Methylformamide

7264
Methyl formate

7265
Methyl 4-formylbenzoate

No.	Name	Synonym	Mol. Form.	CAS RN	Mol. Wt.	Physical Form	mp/°C	bp/°C	den g cm⁻³	n_D	Solubility
7266	2-Methylfuran		C_5H_6O	534-22-5	82.101	liq	-91.2(0.9)	63.9(0.2)	0.9132[20]	1.4342[20]	sl H_2O, ctc; s EtOH, eth
7267	3-Methylfuran		C_5H_6O	930-27-8	82.101			66(5)	0.923[18]	1.4330[19]	i H_2O; s EtOH, eth
7268	5-Methyl-2-furancarboxalde-hyde		$C_6H_6O_2$	620-02-0	110.111			186(1)	1.1072[18]	1.5264[20]	s H_2O; vs EtOH; msc eth; sl ctc
7269	Methyl 2-furancarboxylate	Methyl 2-furanoate	$C_6H_6O_3$	611-13-2	126.110			181.3	1.1786[21]	1.4860[20]	i H_2O; s EtOH, eth, bz, chl
7270	3-Methyl-2,5-furandione		$C_5H_4O_3$	616-02-4	112.084		6(2)	213.5	1.2469[16]	1.4710[21]	vs ace, eth, EtOH
7271	N-Methyl-2-furanmethana-mine		C_6H_9NO	4753-75-7	111.141			149	0.989[25]	1.4729[20]	
7272	5-Methyl-2-furanmethanol		$C_6H_8O_2$	3857-25-8	112.127			195 dec	1.0769[20]	1.4853[20]	vs eth, EtOH
7273	α-Methyl-2-furanmethanol		$C_6H_8O_2$	4208-64-4	112.127			162.5	1.0739[25]	1.4827[15]	
7274	5-Methyl-2(3H)-furanone		$C_5H_6O_2$	591-12-8	98.101	nd	18	56[12]	1.084[20]	1.4476[20]	s H_2O, EtOH, eth, CS_2; sl ctc
7275	5-Methyl-2(5H)-furanone		$C_5H_6O_2$	591-11-7	98.101		<-17	209	1.0810[20]	1.4454[20]	msc H_2O; s EtOH, eth
7276	Methylgermane		CH_6Ge	1449-65-6	90.70	col gas	-158	-23			
7277	Methyl β-D-glucopyranoside		$C_7H_{14}O_6$	709-50-2	194.182		109				s H_2O
7278	Methyl α-D-glucopyranoside	α-Methylglucoside	$C_7H_{14}O_6$	97-30-3	194.182	orth nd (al)	163.8(0.4)	200[0.2]	1.46[30]		vs H_2O
7279	3-Methylglutaric acid	3-Methylpentanedioic acid	$C_6H_{10}O_4$	626-51-7	146.141		87	166[0.5]			s H_2O, EtOH, eth; sl bz, chl; i lig
7280	Methyl Green		$C_{27}H_{35}BrClN_3$	14855-76-6	516.944	grn pow (al)					vs H_2O
7281	Methyl heptadecanoate		$C_{18}H_{36}O_2$	1731-92-6	284.478	pl (al)	29.9(0.3)	185[9]			i H_2O; s EtOH, ace, ctc; vs eth, bz
7282	Methyl heptafluorobutanoate		$C_5H_3F_7O_2$	356-24-1	228.066	liq	-81.7(0.3)	80	1.483[20]	1.295[20]	sl H_2O; s eth, ace
7283	6-Methyl-2-heptanamine, (±)-	Octodrine	$C_8H_{19}N$	5984-58-7	129.244	visc liq		156(12)	0.767[25]	1.4209[20]	
7284	N-Methyl-2-heptanamine		$C_8H_{19}N$	540-43-2	129.244			155			
7285	2-Methylheptane		C_8H_{18}	592-27-8	114.229	liq	-109(4)	117.6(0.9)	0.6980[20]	1.3949[20]	i H_2O; msc EtOH, ace, bz; s eth, ctc
7286	3-Methylheptane		C_8H_{18}	589-81-1	114.229	col liq	-120.48(0.09)	118.9(0.6)	0.7017[25]	1.3961[25]	i H_2O; s EtOH, eth; msc ace, bz, chl
7287	4-Methylheptane		C_8H_{18}	589-53-7	114.229	liq	-121.0(0.1)	117.7(0.5)	0.7046[20]	1.3979[20]	i H_2O; s eth; msc EtOH, ace, bz
7288	Methyl heptanoate		$C_8H_{16}O_2$	106-73-0	144.212	liq	-55.7(0.3)	169.7(0.4)	0.8815[20]	1.4152[20]	sl H_2O, ctc, ace; s EtOH, eth
7289	2-Methyl-1-heptanol, (±)-		$C_8H_{18}O$	111675-77-5	130.228	col liq	-112	175(2)	0.8022[20]	1.424[20]	
7290	3-Methyl-1-heptanol		$C_8H_{18}O$	1070-32-2	130.228	liq	-90	186(2)	0.824[24]	1.4295[25]	
7291	4-Methyl-1-heptanol		$C_8H_{18}O$	817-91-4	130.228			184(4)	0.8065[25]	1.4253[25]	vs EtOH
7292	5-Methyl-1-heptanol, (±)-		$C_8H_{18}O$	111767-95-4	130.228	col liq	-104	186(2)	0.8153[25]	1.4272[25]	
7293	6-Methyl-1-heptanol	Isooctyl alcohol	$C_8H_{18}O$	1653-40-3	130.228	liq	-106	191(2)	0.8176[25]	1.4251[25]	i H_2O; s EtOH, eth
7294	2-Methyl-2-heptanol		$C_8H_{18}O$	625-25-2	130.228	liq	-50.4	173(5)	0.8142[20]	1.4250[20]	i H_2O; s EtOH, eth
7295	3-Methyl-2-heptanol		$C_8H_{18}O$	31367-46-1	130.228			166(1)	0.8177[25]	1.4199[25]	i H_2O; s EtOH, eth, ctc
7296	4-Methyl-2-heptanol		$C_8H_{18}O$	56298-90-9	130.228	col liq	-102	171(1)	0.8027[20]	1.424[20]	
7297	5-Methyl-2-heptanol		$C_8H_{18}O$	54630-50-1	130.228	liq	-61	172(1)	0.8174[21]		
7298	6-Methyl-2-heptanol		$C_8H_{18}O$	4730-22-7	130.228	liq	-105	173(2)	0.8218[20]	1.4238[10]	
7299	2-Methyl-3-heptanol, (±)-		$C_8H_{18}O$	100296-26-2	130.228	liq	-85	171(3)	0.8235[20]	1.4265[20]	sl H_2O; s EtOH, eth, ctc
7300	3-Methyl-3-heptanol	2-Ethyl-2-hexanol	$C_8H_{18}O$	5582-82-1	130.228	liq	-83	166(3)	0.8282[20]	1.4279[20]	i H_2O; s EtOH, eth, ctc
7301	4-Methyl-3-heptanol		$C_8H_{18}O$	14979-39-6	130.228	liq	-123	157(2)	0.827[25]	1.4300[20]	
7302	5-Methyl-3-heptanol		$C_8H_{18}O$	18720-65-5	130.228	liq	-91.2	154(2)	0.8425[25]	1.433[24]	
7303	6-Methyl-3-heptanol, (±)-		$C_8H_{18}O$	100295-85-0	130.228	col liq	-61	159(3)	0.8220[20]	1.4254[20]	
7304	2-Methyl-4-heptanol		$C_8H_{18}O$	21570-35-4	130.228	liq	-81	167(3)	0.8207[20]	1.4203	vs eth, EtOH
7305	3-Methyl-4-heptanol		$C_8H_{18}O$	1838-73-9	130.228	liq		165(2)	0.8329[25]	1.4211[25]	sl H_2O; s EtOH, eth, ctc
7306	4-Methyl-4-heptanol		$C_8H_{18}O$	598-01-6	130.228	liq	-82	166(3)	0.8248[20]	1.4258[20]	i H_2O; s EtOH, eth, ctc
7307	6-Methyl-2-heptanol acetate		$C_{10}H_{20}O_2$	67952-57-2	172.265			187	0.8474[20]	1.413[20]	vs EtOH

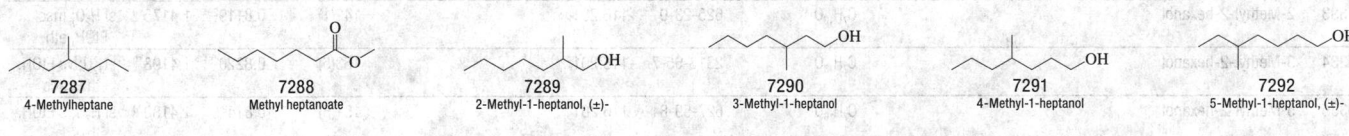

7266	7267	7268	7269	7270	7271	7272	7273
2-Methylfuran	3-Methylfuran	5-Methyl-2-furancarboxaldehyde	Methyl 2-furancarboxylate	3-Methyl-2,5-furandione	N-Methyl-2-furanmethanamine	5-Methyl-2-furanmethanol	α-Methyl-2-furanmethanol

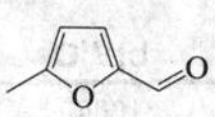

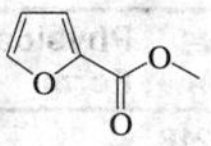

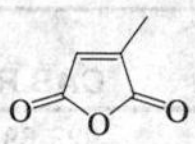

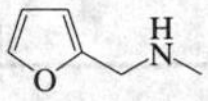

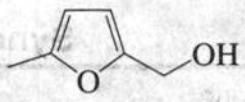

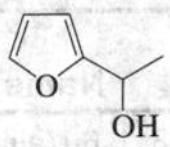

7274	7275	7276	7277	7278	7279	7280
5-Methyl-2(3H)-furanone	5-Methyl-2(5H)-furanone	Methylgermane	Methyl β-D-glucopyranoside	Methyl α-D-glucopyranoside	3-Methylglutaric acid	Methyl Green

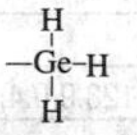

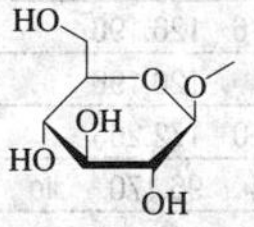

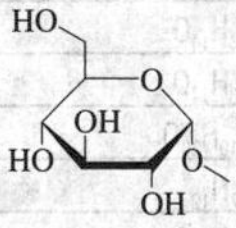

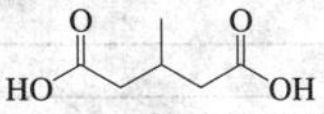

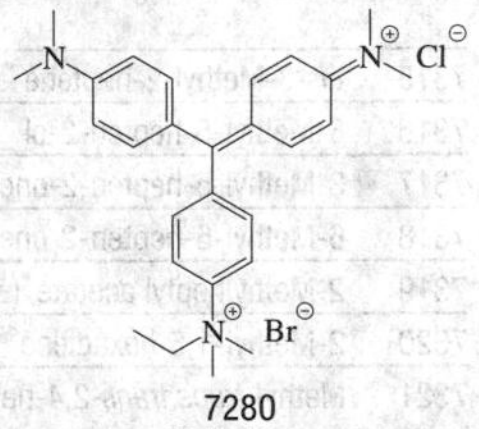

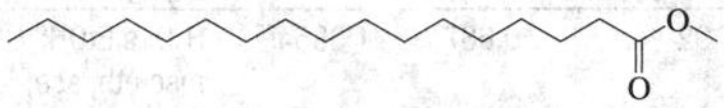

7281	7282	7283	7284	7285	7286
Methyl heptadecanoate	Methyl heptafluorobutanoate	6-Methyl-2-heptanamine, (±)-	N-Methyl-2-heptanamine	2-Methylheptane	3-Methylheptane

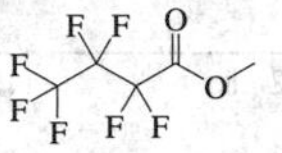

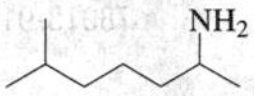

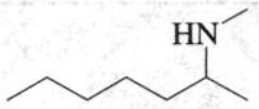

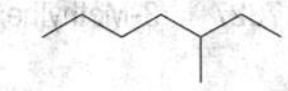

7287	7288	7289	7290	7291	7292
4-Methylheptane	Methyl heptanoate	2-Methyl-1-heptanol, (±)-	3-Methyl-1-heptanol	4-Methyl-1-heptanol	5-Methyl-1-heptanol, (±)-

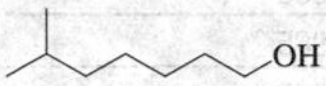

7293	7294	7295	7296	7297	7298	7299	7300
6-Methyl-1-heptanol	2-Methyl-2-heptanol	3-Methyl-2-heptanol	4-Methyl-2-heptanol	5-Methyl-2-heptanol	6-Methyl-2-heptanol	2-Methyl-3-heptanol, (±)-	3-Methyl-3-heptanol

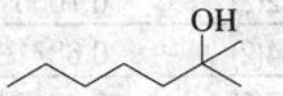

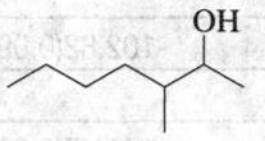

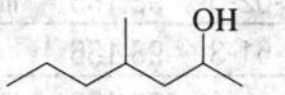

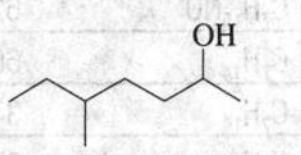

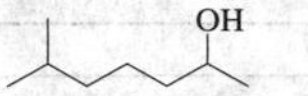

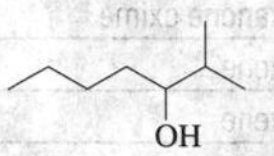

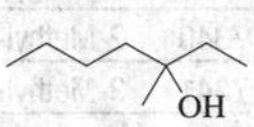

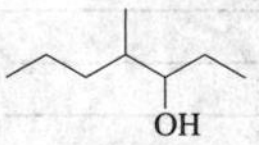

7301	7302	7303	7304	7305	7306	7307
4-Methyl-3-heptanol	5-Methyl-3-heptanol	6-Methyl-3-heptanol, (±)-	2-Methyl-4-heptanol	3-Methyl-4-heptanol	4-Methyl-4-heptanol	6-Methyl-2-heptanol acetate

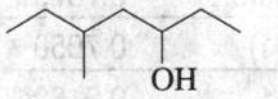

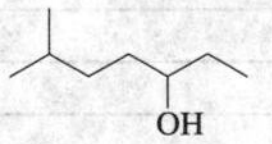

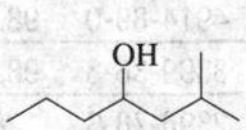

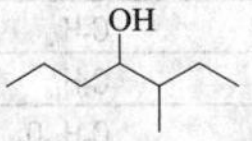

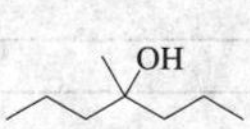

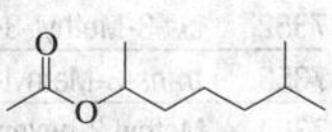

No.	Name	Synonym	Mol. Form.	CAS RN	Mol. Wt.	Physical Form	mp/°C	bp/°C	den g cm⁻³	n_D	Solubility
7308	6-Methyl-2-heptanone		$C_8H_{16}O$	928-68-7	128.212			166(6)	0.8151²⁰	1.4162²⁰	sl H₂O; vs EtOH, eth; msc ace, bz, chl
7309	5-Methyl-3-heptanone	Ethyl 2-methylbutyl ketone	$C_8H_{16}O$	541-85-5	128.212	liq		159(4)			
7310	6-Methyl-3-heptanone		$C_8H_{16}O$	624-42-0	128.212			167(10)	0.8304²⁰	1.4209²⁰	i H₂O; s EtOH, eth, bz, ctc
7311	2-Methyl-4-heptanone	Isobutyl propyl ketone	$C_8H_{16}O$	626-33-5	128.212			155(4)	0.813²²		i H₂O; s EtOH, eth
7312	2-Methyl-1-heptene		C_8H_{16}	15870-10-7	112.213	liq	-90.1(0.2)	119(2)	0.7104²⁵	1.4123²⁰	
7313	6-Methyl-1-heptene		C_8H_{16}	5026-76-6	112.213			113(1)	0.7079²⁵	1.4070²⁰	
7314	2-Methyl-2-heptene		C_8H_{16}	627-97-4	112.213			122(2)	0.7200²⁵	1.4170²⁰	i H₂O; s eth, bz, ctc, chl
7315	cis-3-Methyl-2-heptene		C_8H_{16}	22768-19-0	112.213			122	0.725²⁵	1.419²⁰	
7316	6-Methyl-5-hepten-2-ol		$C_8H_{16}O$	1569-60-4	128.212			186(5)	0.8545²⁰	1.4505²⁰	
7317	3-Methyl-5-hepten-2-one		$C_8H_{14}O$	38552-72-6	126.196			63²⁰	0.8463¹⁸	1.4345¹⁸	
7318	6-Methyl-5-hepten-2-one		$C_8H_{14}O$	110-93-0	126.196		173.5		0.8546¹⁶	1.4445²⁰	vs eth, EtOH
7319	2-Methylheptyl acetate, (±)-		$C_{10}H_{20}O_2$	74112-36-0	172.265			195	0.8626¹⁴	1.4146²⁰	vs eth, EtOH
7320	2-Methyl-1,5-hexadiene		C_7H_{12}	4049-81-4	96.170	liq	-128.9(0.4)	88(3)	0.7153²⁵	1.4183²⁰	
7321	Methyl trans,trans-2,4-hexadienoate	Methyl sorbate	$C_7H_{10}O_2$	689-89-4	126.153	lf	15	180	0.9777²⁰	1.5025²²	i H₂O; s EtOH, eth
7322	2-Methylhexanal		$C_7H_{14}O$	925-54-2	114.185	liq		141			
7323	3-Methylhexanal	3-Methylcaproaldehyde	$C_7H_{14}O$	19269-28-4	114.185			145(1)	0.8203²⁰	1.4122²⁰	i H₂O; s EtOH, eth
7324	3-Methyl-1-hexanamine		$C_7H_{17}N$	65530-93-0	115.217			149	0.772²⁶	1.4249²⁵	
7325	4-Methyl-2-hexanamine	Methylhexanamine	$C_7H_{17}N$	105-41-9	115.217			132.5	0.7655²⁰	1.4150²⁵	sl H₂O; vs EtOH, eth, chl, dil acid
7326	2-Methylhexane		C_7H_{16}	591-76-4	100.202	liq	-118.23(0.04)	90.0(0.8)	0.6787²⁰	1.3848²⁰	i H₂O; s EtOH; msc eth, ace, bz, lig, chl
7327	3-Methylhexane		C_7H_{16}	78918-91-9	100.202	liq	-119.4	92	0.687²¹	1.3854²⁵	i H₂O; s EtOH; msc eth, ace, bz, lig, chl
7328	5-Methyl-2,3-hexanedione	2-Methylhexane-4,5-dione	$C_7H_{12}O_2$	13706-86-0	128.169			138	0.908²²	1.4119²⁰	
7329	Methyl hexanoate	Methyl caproate	$C_7H_{14}O_2$	106-70-7	130.185	liq	-70(2)	151(1)	0.8846²⁰	1.4049²⁰	i H₂O; vs EtOH, eth; s ace, bz, ctc
7330	2-Methylhexanoic acid		$C_7H_{14}O_2$	4536-23-6	130.185			214(9)	0.918²⁰	1.4193²⁰	vs ace, bz, eth, EtOH
7331	2-Methyl-1-hexanol, (±)-		$C_7H_{16}O$	111768-04-8	116.201			165(4)	0.826²⁰	1.4226²⁰	vs eth, EtOH
7332	5-Methyl-1-hexanol		$C_7H_{16}O$	627-98-5	116.201			170(3)	0.8192²⁴	1.4175²⁰	vs eth, EtOH
7333	2-Methyl-2-hexanol		$C_7H_{16}O$	625-23-0	116.201			142(1)	0.8119²⁰	1.4175²⁰	sl H₂O; msc EtOH, eth
7334	3-Methyl-2-hexanol		$C_7H_{16}O$	2313-65-7	116.201			153(5)	0.8220²⁵	1.4198¹⁸	i H₂O; vs EtOH, eth; s ace
7335	5-Methyl-2-hexanol		$C_7H_{16}O$	627-59-8	116.201			151(5)	0.814²⁰	1.4180²⁰	sl H₂O; s EtOH, eth
7336	3-Methyl-3-hexanol		$C_7H_{16}O$	597-96-6	116.201			147(3)	0.8233²⁰	1.4231²⁰	sl H₂O; s EtOH, eth, ctc
7337	5-Methyl-2-hexanone	Methyl isopentyl ketone	$C_7H_{14}O$	110-12-3	114.185			139(2)	0.888²⁰	1.4062²⁰	sl H₂O; msc EtOH; vs ace, bz; s ctc
7338	2-Methyl-3-hexanone	Propyl isopropyl ketone	$C_7H_{14}O$	7379-12-6	114.185			134(3)	0.8091²⁰	1.4042²⁰	s EtOH, eth, chl; vs ace
7339	5-Methyl-2-hexanone oxime		$C_7H_{15}NO$	624-44-2	129.200			195.5	0.8881²⁰	1.4448²⁰	sl chl
7340	2-Methyl-1-hexene		C_7H_{14}	6094-02-6	98.186	liq	-102.82(0.09)	92(2)	0.7000²⁰	1.4035²⁰	
7341	3-Methyl-1-hexene		C_7H_{14}	3404-61-3	98.186			84(2)	0.6871²⁵	1.3965²⁰	
7342	4-Methyl-1-hexene		C_7H_{14}	3769-23-1	98.186	liq	-141.47(0.09)	87(2)	0.6942²⁵	1.4000²⁰	
7343	5-Methyl-1-hexene		C_7H_{14}	3524-73-0	98.186			85(1)	0.6877²⁵	1.3967²⁰	
7344	2-Methyl-2-hexene		C_7H_{14}	2738-19-4	98.186	liq	-130.36(0.09)	95(1)	0.7038²⁵	1.4106²⁰	
7345	cis-3-Methyl-2-hexene		C_7H_{14}	10574-36-4	98.186	liq	-118.5	97(3)	0.712²⁰	1.4126²⁰	
7346	cis-4-Methyl-2-hexene		C_7H_{14}	3683-19-0	98.186			86(2)	0.6952²⁵	1.4026²⁰	
7347	trans-4-Methyl-2-hexene		C_7H_{14}	3683-22-5	98.186	liq	-125.70(0.09)	88(2)	0.6925²⁵	1.4025²⁰	
7348	cis-5-Methyl-2-hexene		C_7H_{14}	13151-17-2	98.186			89(3)	0.697²⁵	1.404²⁰	
7349	trans-5-Methyl-2-hexene		C_7H_{14}	7385-82-2	98.186	liq	-124.3(0.2)	87(2)	0.6883²⁵	1.4006²⁰	
7350	cis-2-Methyl-3-hexene		C_7H_{14}	15840-60-5	98.186			86(3)	0.690²⁵	1.401²⁰	
7351	trans-2-Methyl-3-hexene		C_7H_{14}	692-24-0	98.186	liq	-141.6	88(3)	0.6853²⁵	1.4001²⁰	
7352	cis-3-Methyl-3-hexene		C_7H_{14}	4914-89-0	98.186			95.3(0.9)	0.7079²⁵	1.4126²⁰	
7353	trans-3-Methyl-3-hexene		C_7H_{14}	3899-36-3	98.186			94(1)	0.7050²⁵	1.4109²⁰	
7354	Methyl 3-hexenoate		$C_7H_{12}O_2$	2396-78-3	128.169			67³⁴	0.9132²⁵	1.4240²³	

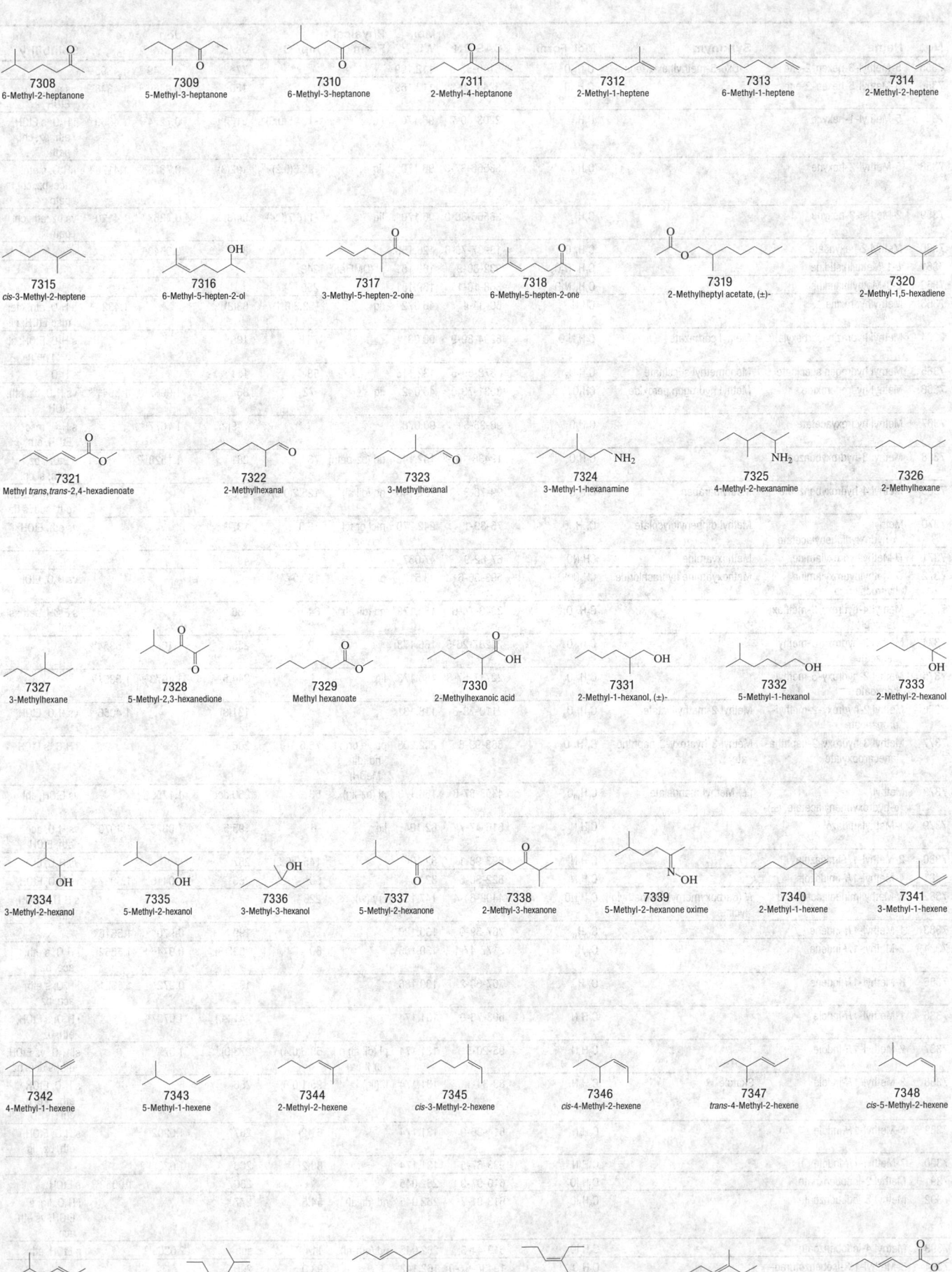

7308
6-Methyl-2-heptanone

7309
5-Methyl-3-heptanone

7310
6-Methyl-3-heptanone

7311
2-Methyl-4-heptanone

7312
2-Methyl-1-heptene

7313
6-Methyl-1-heptene

7314
2-Methyl-2-heptene

7315
cis-3-Methyl-2-heptene

7316
6-Methyl-5-hepten-2-ol

7317
3-Methyl-5-hepten-2-one

7318
6-Methyl-5-hepten-2-one

7319
2-Methylheptyl acetate, (±)-

7320
2-Methyl-1,5-hexadiene

7321
Methyl *trans,trans*-2,4-hexadienoate

7322
2-Methylhexanal

7323
3-Methylhexanal

7324
3-Methyl-1-hexanamine

7325
4-Methyl-2-hexanamine

7326
2-Methylhexane

7327
3-Methylhexane

7328
5-Methyl-2,3-hexanedione

7329
Methyl hexanoate

7330
2-Methylhexanoic acid

7331
2-Methyl-1-hexanol, (±)-

7332
5-Methyl-1-hexanol

7333
2-Methyl-2-hexanol

7334
3-Methyl-2-hexanol

7335
5-Methyl-2-hexanol

7336
3-Methyl-3-hexanol

7337
5-Methyl-2-hexanone

7338
2-Methyl-3-hexanone

7339
5-Methyl-2-hexanone oxime

7340
2-Methyl-1-hexene

7341
3-Methyl-1-hexene

7342
4-Methyl-1-hexene

7343
5-Methyl-1-hexene

7344
2-Methyl-2-hexene

7345
cis-3-Methyl-2-hexene

7346
cis-4-Methyl-2-hexene

7347
trans-4-Methyl-2-hexene

7348
cis-5-Methyl-2-hexene

7349
trans-5-Methyl-2-hexene

7350
cis-2-Methyl-3-hexene

7351
trans-2-Methyl-3-hexene

7352
cis-3-Methyl-3-hexene

7353
trans-3-Methyl-3-hexene

7354
Methyl 3-hexenoate

No.	Name	Synonym	Mol. Form.	CAS RN	Mol. Wt.	Physical Form	mp/°C	bp/°C	den g cm⁻³	n_D	Solubility
7355	5-Methyl-3-hexen-2-one	2-Oxo-5-methylhex-3-ene	$C_7H_{12}O$	5166-53-0	112.169			77[50]	0.8549[28]	1.4395[22]	
7356	5-Methyl-5-hexen-2-one		$C_7H_{12}O$	3240-09-3	112.169			150	0.8460[20]	1.4348[20]	vs ace, eth, EtOH
7357	5-Methyl-1-hexyne		C_7H_{12}	2203-80-7	96.170	liq	-124.6(0.5)	91(3)	0.7274[20]	1.4059[-20]	i H_2O; s EtOH, eth, bz, chl, peth
7358	5-Methyl-2-hexyne		C_7H_{12}	53566-37-3	96.170	liq	-92.6(0.2)	102(3)	0.7378[20]	1.4176[20]	i H_2O; s eth, ace, bz, chl, peth
7359	2-Methyl-3-hexyne		C_7H_{12}	36566-80-0	96.170	liq	-116.7(0.4)	94(3)	0.7263[20]	1.4120[20]	vs bz, eth, chl, peth
7360	Methyl 2-hexynoate		$C_7H_{10}O_2$	18937-79-6	126.153			80[23]	0.9648[25]		
7361	L-1-Methylhistidine		$C_7H_{11}N_3O_2$	332-80-9	169.181	pl (DMF aq)	249				
7362	L-3-Methylhistidine		$C_7H_{11}N_3O_2$	368-16-1	169.181		250				
7363	Methylhydrazine		CH_6N_2	60-34-4	46.072	liq	-52.3(0.1)	83(3)		1.4325[20]	s H_2O, eth, ctc; msc EtOH; i lig
7364	Methyl hydrazinecarboxylate	Methyl carbazate	$C_2H_6N_2O_2$	6294-89-9	90.081		73	108[12]			s H_2O, EtOH; sl bz; i peth
7365	Methyl hydrogen succinate	Monomethyl succinate	$C_5H_8O_4$	3878-55-5	132.116		58	151[20]			s H_2O
7366	Methyl hydroperoxide	Methyl hydrogen peroxide	CH_4O_2	3031-73-0	48.042	liq	-72	86	1.9967[15]	1.3641[15]	vs H_2O, bz, eth, EtOH
7367	Methyl hydroxyacetate		$C_3H_6O_3$	96-35-5	90.078			149(2)	1.1677[18]		s H_2O; msc EtOH, eth
7368	Methyl 3-hydroxybenzoate		$C_8H_8O_3$	19438-10-9	152.148	nd (bz-peth)	73	281	1.1528[100]		s EtOH, bz, peth; sl chl
7369	Methyl 4-hydroxybenzoate	Methylparaben	$C_8H_8O_3$	99-76-3	152.148	nd (dil al)	125.2(0.5)	275 dec			sl H_2O; vs EtOH, eth, ace; s tfa
7370	Methyl α-hydroxydiphenylacetate	Methyl diphenylglycolate	$C_{15}H_{14}O_3$	76-89-1	242.270	mcl or tcl cry (al)	75.8	187[13]			vs eth, EtOH
7371	O-Methylhydroxylamine	Methoxyamine	CH_5NO	67-62-9	47.057			49			
7372	O-Methylhydroxylamine hydrochloride	Methoxyamine hydrochloride	CH_6ClNO	593-56-6	83.518	pr	150.0				vs H_2O, EtOH
7373	Methyl 4-hydroxy-3-methoxy-benzoate		$C_9H_{10}O_4$	3943-74-6	182.173	nd (dil al)	64	286			s EtOH, peth; sl chl
7374	Methyl 2-hydroxy-3-methyl-benzoate		$C_9H_{10}O_3$	23287-26-5	166.173		29	235	1.1683[25]	1.5354[16]	
7375	Methyl 2-hydroxy-5-methyl-benzoate		$C_9H_{10}O_3$	22717-57-3	166.173	liq	-1	244.5	1.1673[25]	1.5351[15]	
7376	Methyl 2-hydroxy-2-methyl-propanoate	Methyl 2-methyllactate	$C_5H_{10}O_3$	2110-78-3	118.131			131(3)		1.4056[20]	vs H_2O, EtOH
7377	Methyl 3-hydroxy-2-naphtha-lenecarboxylate	Methyl 3-hydroxy-2-naphtho-ate	$C_{12}H_{10}O_3$	883-99-8	202.205	pa ye orth nd (dil MeOH)	75.5	206			i H_2O; s EtOH
7378	Methyl α-hydroxyphenylacetate, (±)-	(±)-Methyl mandelate	$C_9H_{10}O_3$	4358-87-6	166.173	pl (bz-lig)	58	250 dec	1.1756[20]		vs EtOH, chl
7379	1-Methylimidazol		$C_4H_6N_2$	616-47-7	82.104	liq	-6	195.5	1.0325[20]	1.4970[20]	vs H_2O, ace, eth, EtOH
7380	2-Methyl-1H-imidazole		$C_4H_6N_2$	693-98-1	82.104		145.9(0.2)	267			vs H_2O, EtOH
7381	4-Methyl-1H-imidazole		$C_4H_6N_2$	822-36-6	82.104		56	263	1.0416[14]	1.5037[14]	vs H_2O, EtOH
7382	N-Methyliminodiacetic acid	N-(Carboxymethyl)-N-methylg-lycine	$C_5H_9NO_4$	4408-64-4	147.130	cry (w)	226				s H_2O; i EtOH, eth
7383	1-Methyl-1H-indene		$C_{10}H_{10}$	767-59-9	130.186			199	0.970[25]	1.5616[20]	
7384	2-Methyl-1H-indene		$C_{10}H_{10}$	2177-47-1	130.186		80	209(18)	0.974[25]	1.5652[20]	i H_2O; s eth, ace, bz
7385	3-Methyl-1H-indene		$C_{10}H_{10}$	767-60-2	130.186			198	0.972[25]	1.5621[20]	i H_2O; s eth, ace, bz
7386	1-Methyl-1H-indole		C_9H_9N	603-76-9	131.174			241(20)	1.0707[25]		i H_2O; s EtOH, eth, bz
7387	2-Methyl-1H-indole		C_9H_9N	95-20-5	131.174	pl (dil al) nd or lf (w)	58.6(0.4)	273(9)	1.07[20]		sl H_2O; vs EtOH, eth; s ace, bz
7388	3-Methyl-1H-indole	Skatole	C_9H_9N	83-34-1	131.174	lf (lig)	95.1(0.6)	266			s H_2O, EtOH, eth, ace, bz, chl
7389	5-Methyl-1H-indole		C_9H_9N	614-96-0	131.174		61(2)	267	1.0202[78]		s H_2O, EtOH, eth, bz, lig
7390	7-Methyl-1H-indole		C_9H_9N	933-67-5	131.174		83(2)	266	1.0202[100]		
7391	Methyl 2-iodobenzoate		$C_8H_7IO_2$	610-97-9	262.045			280		1.6052[20]	s EtOH
7392	Methyl 3-iodobenzoate		$C_8H_7IO_2$	618-91-7	262.045	nd (dil al)	54.5	277			i H_2O, lig; s EtOH; vs eth, ace
7393	Methyl 4-iodobenzoate		$C_8H_7IO_2$	619-44-3	262.045	nd (eth-al)	114.8	sub	2.0200[10]		s EtOH, eth
7394	5-Methyl-1,3-isobenzofuran-dione		$C_9H_6O_3$	19438-61-0	162.142		93.0	295			

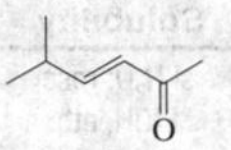

7355
5-Methyl-3-hexen-2-one

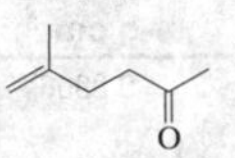

7356
5-Methyl-5-hexen-2-one

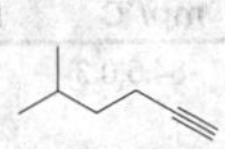

7357
5-Methyl-1-hexyne

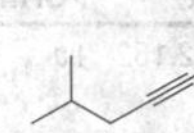

7358
5-Methyl-2-hexyne

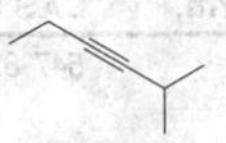

7359
2-Methyl-3-hexyne

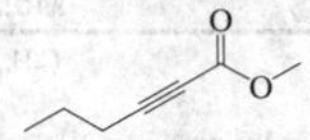

7360
Methyl 2-hexynoate

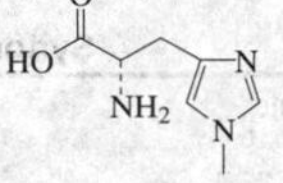

7361
L-1-Methylhistidine

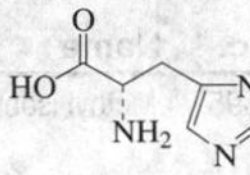

7362
L-3-Methylhistidine

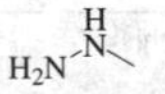

7363
Methylhydrazine

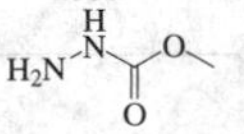

7364
Methyl hydrazinecarboxylate

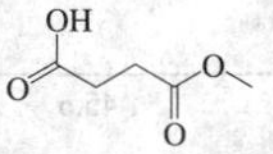

7365
Methyl hydrogen succinate

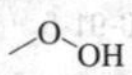

7366
Methyl hydroperoxide

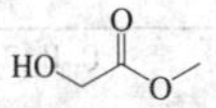

7367
Methyl hydroxyacetate

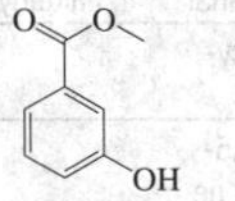

7368
Methyl 3-hydroxybenzoate

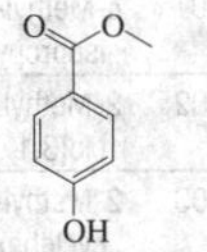

7369
Methyl 4-hydroxybenzoate

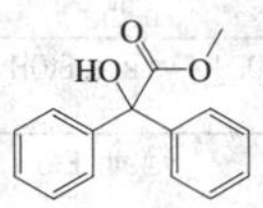

7370
Methyl α-hydroxydiphenylacetate

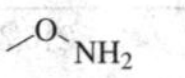

7371
O-Methylhydroxylamine

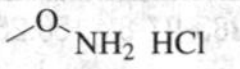

7372
O-Methylhydroxylamine hydrochloride

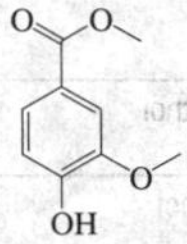

7373
Methyl 4-hydroxy-3-methoxybenzoate

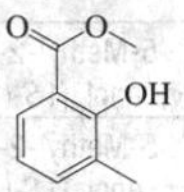

7374
Methyl 2-hydroxy-3-methylbenzoate

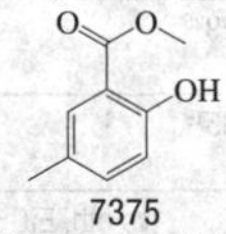

7375
Methyl 2-hydroxy-5-methylbenzoate

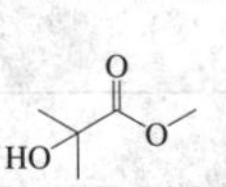

7376
Methyl 2-hydroxy-2-methylpropanoate

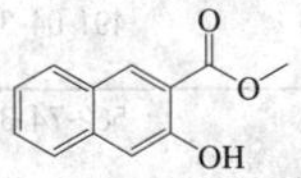

7377
Methyl 3-hydroxy-2-naphthalenecarboxylate

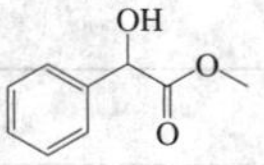

7378
Methyl α-hydroxyphenylacetate, (±)-

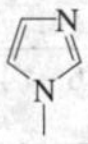

7379
1-Methylimidazol

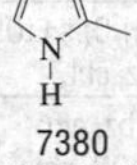

7380
2-Methyl-1*H*-imidazole

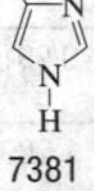

7381
4-Methyl-1*H*-imidazole

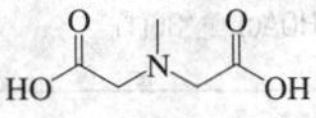

7382
N-Methyliminodiacetic acid

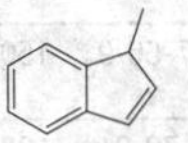

7383
1-Methyl-1*H*-indene

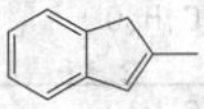

7384
2-Methyl-1*H*-indene

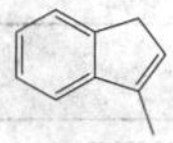

7385
3-Methyl-1*H*-indene

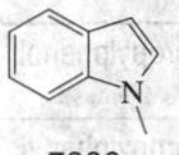

7386
1-Methyl-1*H*-indole

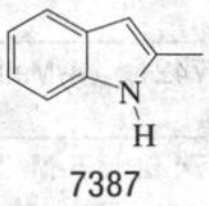

7387
2-Methyl-1*H*-indole

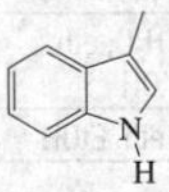

7388
3-Methyl-1*H*-indole

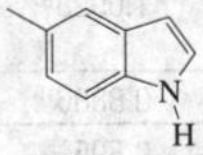

7389
5-Methyl-1*H*-indole

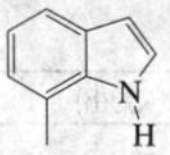

7390
7-Methyl-1*H*-indole

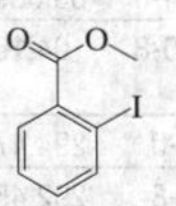

7391
Methyl 2-iodobenzoate

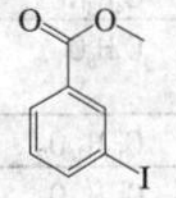

7392
Methyl 3-iodobenzoate

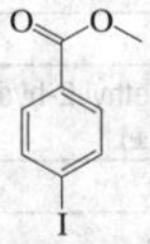

7393
Methyl 4-iodobenzoate

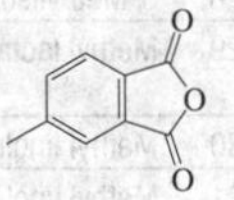

7394
5-Methyl-1,3-isobenzofurandione

No.	Name	Synonym	Mol. Form.	CAS RN	Mol. Wt.	Physical Form	mp/°C	bp/°C	den g cm⁻³	n_D	Solubility
7395	Methyl isobutanoate		$C_5H_{10}O_2$	547-63-7	102.132	liq	-84.6(0.3)	92(1)	0.8906[20]	1.3840[20]	sl H₂O; msc EtOH, eth; s ace, ctc
7396	Methyl isocyanate		C_2H_3NO	624-83-9	57.051	liq	-45	38.3(0.2)	0.9588[20]	1.3694[20]	vs H₂O
7397	2-Methyl-1H-isoindole-1,3(2H)-dione		$C_9H_7NO_2$	550-44-7	161.158	nd (al), lf (sub)	134	286			i H₂O; sl EtOH
7398	Methyl isopentanoate	Methyl isovalerate	$C_6H_{12}O_2$	556-24-1	116.158			116(2)	0.8808[20]	1.3927[20]	i H₂O; vs EtOH, eth, ace
7399	6-Methyl-N-isopentyl-2-heptanamine	Octamylamine	$C_{13}H_{29}N$	502-59-0	199.376			100[7]			
7400	2-Methyl-5-isopropylaniline		$C_{10}H_{15}N$	2051-53-8	149.233	liq	-16	241	0.9942[20]	1.5387[20]	s ctc, CS₂
7401	α-Methyl-4-isopropylbenzenepropanal	3-p-Cumenyl-2-methylpropionaldehyde	$C_{13}H_{18}O$	103-95-7	190.281			270	0.9459[20]	1.5068[20]	vs bz, eth, EtOH
7402	2-Methyl-5-isopropylbicyclo[3.1.0]hex-2-ene		$C_{10}H_{16}$	2867-05-2	136.234			156(5)	0.8301[20]	1.4515[20]	
7403	2-Methyl-5-isopropyl-2,5-cyclohexadiene-1,4-dione		$C_{10}H_{12}O_2$	490-91-5	164.201		45.5	232			s chl
7404	cis-1-Methyl-4-isopropylcyclohexane		$C_{10}H_{20}$	6069-98-3	140.266	liq	-89.8(0.1)	172(6)	0.8039[20]	1.4431[20]	i H₂O; vs EtOH, eth; s bz, peth
7405	trans-1-Methyl-4-isopropylcyclohexane	trans-p-Menthane	$C_{10}H_{20}$	1678-82-6	140.266	oil	-86.3(0.1)	171(3)	0.7928[20]	1.4366[20]	vs bz, eth, EtOH, lig
7406	1-Methyl-4-isopropylcyclohexanol		$C_{10}H_{20}O$	21129-27-1	156.265			208.5	0.90[20]	1.4619[20]	
7407	5-Methyl-2-isopropylcyclohexanol, [1S-(1α,2β,5α)]-	(+)-Menthol	$C_{10}H_{20}O$	15356-60-2	156.265		42(2)	103[9]			vs ace, bz, eth, EtOH
7408	5-Methyl-2-isopropylcyclohexanol, [1R-(1α,2β,5α)]-	(-)-Menthol	$C_{10}H_{20}O$	2216-51-5	156.265	nd (MeOH)	42.1(0.8)	214(12)	0.903[15]	1.460[22]	sl H₂O; vs EtOH, eth, ace, bz; s peth
7409	5-Methyl-2-isopropylcyclohexanol, [1S-(1α,2α,5β)]-	(+)-Neomenthol	$C_{10}H_{20}O$	2216-52-6	156.265	oil	-22	211.7	0.897[22]	1.4600[20]	vs ace, EtOH
7410	5-Methyl-2-isopropylcyclohexanol, [1S-(1α,2β,5β)]-	(+)-Isomenthol	$C_{10}H_{20}O$	23283-97-8	156.265	nd(dil al)	82.5	216(3)			vs eth, EtOH
7411	5-Methyl-2-isopropylcyclohexanol acetate, [1R-(1α,2α,5β)]		$C_{12}H_{22}O_2$	2623-23-6	198.302			223(16)	0.9244[20]	1.4469[20]	
7412	cis-5-Methyl-2-isopropylcyclohexanone	Menthone	$C_{10}H_{18}O$	491-07-6	154.249			205	0.8995[20]	1.4527[20]	
7413	trans-5-Methyl-2-isopropylcyclohexanone, (2S)-	l-Menthone	$C_{10}H_{18}O$	14073-97-3	154.249	liq	-6	207	0.8954[20]	1.4505[20]	sl H₂O; msc EtOH, eth, bz, CS₂; s ace
7414	1-Methyl-4-isopropylcyclohexene		$C_{10}H_{18}$	5502-88-5	138.250			170(9)	0.8457[15]	1.4735[20]	
7415	3-Methyl-6-isopropyl-2-cyclohexen-1-ol		$C_{10}H_{18}O$	491-04-3	154.249			97[15.5]	0.9119[25]	1.4729[25]	
7416	4-Methyl-1-isopropyl-3-cyclohexen-1-ol		$C_{10}H_{18}O$	562-74-3	154.249			209	0.926[20]	1.4785[19]	
7417	5-Methyl-2-isopropylcyclohexyl ethoxyacetate, (1α,2β,5α)		$C_{14}H_{26}O_3$	579-94-2	242.354			155[20]	0.9545[20]		vs eth, EtOH, chl
7418	1-Methyl-4-isopropyl-2-nitrobenzene		$C_{10}H_{13}NO_2$	943-15-7	179.216			126[10]	1.0744[20]	1.5301[20]	vs eth, EtOH
7419	1-Methyl-4-isopropyl-7-oxabicyclo[2.2.1]heptane		$C_{10}H_{18}O$	470-67-7	154.249		1	173.5	0.8997[20]	1.4562[20]	sl H₂O; msc EtOH, eth; s bz, lig
7420	1-Methyl-7-isopropylphenanthrene	Retene	$C_{18}H_{18}$	483-65-8	234.336		95.4(0.7)	391(24)	1.035[25]		i H₂O; s EtOH, eth, bz, CS₂, HOAc
7421	4-Methyl-2-isopropylphenol		$C_{10}H_{14}O$	4427-56-9	150.217	nd (HOAc)	36(1)	231(6)	0.9910[20]	1.5275[20]	sl H₂O; s EtOH, bz, chl
7422	5-Methyl-2-isopropylphenyl acetate	Thymol, acetate	$C_{12}H_{16}O_2$	528-79-0	192.254			245	1.009[9]		vs bz, eth, EtOH, chl
7423	1-Methylisoquinoline	Isoquinaldine	$C_{10}H_9N$	1721-93-3	143.185		10	252(9)	1.0777[20]	1.6095[20]	sl H₂O; s eth, ace, bz
7424	3-Methylisoquinoline		$C_{10}H_9N$	1125-80-0	143.185	cry (eth)	63.4(0.4)	253(1)			sl H₂O, chl; s eth, ace
7425	Methyl isothiocyanate		C_2H_3NS	556-61-6	73.117		36	119	1.0691[37]	1.5258	sl H₂O; msc EtOH; vs eth
7426	5-Methyl-3-isoxazolamine		$C_4H_6N_2O$	1072-67-9	98.103		62				
7427	4-Methylisoxazole		C_4H_5NO	6454-84-8	83.089	liq		127			
7428	5-Methylisoxazole		C_4H_5NO	5765-44-6	83.089			123(7)	1.023[20]	1.4386[20]	s DMSO
7429	Methyl lactate, (±)-	Methyl 2-hydroxypropanoate, (±)	$C_4H_8O_3$	2155-30-8	104.105	oil		144.8	1.0928[20]	1.4141[20]	vs H₂O, eth, EtOH
7430	Methyl linoleate		$C_{19}H_{34}O_2$	112-63-0	294.472		-39(8)	215[20]	0.8886[10]	1.4638[20]	vs eth, EtOH
7431	Methyl linolenate		$C_{19}H_{32}O_2$	301-00-8	292.456		-45.5	207[14]	0.895[25]	1.4709[20]	

7395 Methyl isobutanoate

7396 Methyl isocyanate

7397 2-Methyl-1H-isoindole-1,3(2H)-dione

7398 Methyl isopentanoate

7399 6-Methyl-N-isopentyl-2-heptanamine

7400 2-Methyl-5-isopropylaniline

7401 α-Methyl-4-isopropylbenzenepropanal

7402 2-Methyl-5-isopropylbicyclo[3.1.0]hex-2-ene

7403 2-Methyl-5-isopropyl-2,5-cyclohexadiene-1,4-dione

7404 cis-1-Methyl-4-isopropylcyclohexane

7405 trans-1-Methyl-4-isopropylcyclohexane

7406 1-Methyl-4-isopropylcyclohexanol

7407 5-Methyl-2-isopropylcyclohexanol, [1S-(1α,2β,5α)]-

7408 5-Methyl-2-isopropylcyclohexanol, [1R-(1α,2β,5α)]-

7409 5-Methyl-2-isopropylcyclohexanol, [1S-(1α,2α,5β)]-

7410 5-Methyl-2-isopropylcyclohexanol, [1S-(1α,2β,5β)]-

7411 5-Methyl-2-isopropylcyclohexanol acetate, [1R-(1α,2α,5β)]

7412 cis-5-Methyl-2-isopropylcyclohexanone

7413 trans-5-Methyl-2-isopropylcyclohexanone, (2S)-

7414 1-Methyl-4-isopropylcyclohexene

7415 3-Methyl-6-isopropyl-2-cyclohexen-1-ol

7416 4-Methyl-1-isopropyl-3-cyclohexen-1-ol

7417 5-Methyl-2-isopropylcyclohexyl ethoxyacetate, (1α,2β,5α)

7418 1-Methyl-4-isopropyl-2-nitrobenzene

7419 1-Methyl-4-isopropyl-7-oxabicyclo[2.2.1]heptane

7420 1-Methyl-7-isopropylphenanthrene

7421 4-Methyl-2-isopropylphenol

7422 5-Methyl-2-isopropylphenyl acetate

7423 1-Methylisoquinoline

7424 3-Methylisoquinoline

7425 Methyl isothiocyanate

7426 5-Methyl-3-isoxazolamine

7427 4-Methylisoxazole

7428 5-Methylisoxazole

7429 Methyl lactate, (±)-

7430 Methyl linoleate

7431 Methyl linolenate

No.	Name	Synonym	Mol. Form.	CAS RN	Mol. Wt.	Physical Form	mp/°C	bp/°C	den g cm^{-3}	n_D	Solubility
7432	Methyl magnesium bromide	Bromomethylmagnesium	CH$_3$BrMg	75-16-1	119.244						s eth, thf; i hx, bz
7433	Methylmagnesium chloride	Chloromethylmagnesium	CH$_3$ClMg	676-58-4	74.793	stab in thf soln					i peth, bz
7434	Methylmalonic acid	2-Methylpropanedioc acid	C$_4$H$_6$O$_4$	516-05-2	118.089	nd (bz-AcOEt) pr (eth-bz)	134(1)		1.455^{20}		vs H$_2$O, EtOH, eth; sl bz, tfa; s AcOEt
7435	Methyl mercaptoacetate		C$_3$H$_6$O$_2$S	2365-48-2	106.144			42^{10}		1.4657^{20}	vs eth, EtOH
7436	Methyl 3-mercaptopropanoate		C$_4$H$_8$O$_2$S	2935-90-2	120.171			54^{14}	1.085^{25}	1.4640^{20}	
7437	Methylmercuric dicyanamide	1-Cyano-3-(methylmercurio)-guanidine	C$_3$H$_6$HgN$_4$	502-39-6	298.70		157				
7438	Methyl methacrylate	Methyl 2-methyl-2-propenoate	C$_5$H$_8$O$_2$	80-62-6	100.117	liq	-47.55(0.02)	100.6(0.2)	0.9377^{25}	1.4142^{20}	sl H$_2$O; msc EtOH, eth, ace; s chl
7439	Methyl methanesulfonate		C$_2$H$_6$O$_3$S	66-27-3	110.132		20	185(9)	1.2943^{20}	1.4138^{20}	
7440	Methyl methoxyacetate		C$_4$H$_8$O$_3$	6290-49-9	104.105			131	1.0511^{20}	1.3962^{20}	sl H$_2$O; vs EtOH, eth, ace
7441	Methyl 2-methoxybenzoate		C$_9$H$_{10}$O$_3$	606-45-1	166.173			246.5	1.1571^{19}	1.534^{19}	i H$_2$O; s EtOH
7442	Methyl 3-methoxybenzoate		C$_9$H$_{10}$O$_3$	5368-81-0	166.173			248	1.1310^{20}	1.5224^{20}	i H$_2$O; s EtOH
7443	Methyl 4-methoxybenzoate		C$_9$H$_{10}$O$_3$	121-98-2	166.173	lf (al or eth)	49	259(2)			i H$_2$O; s EtOH, eth, chl
7444	Methyl 3-methoxy-2-(methylamino)benzoate	Damascenine	C$_{10}$H$_{13}$NO$_3$	483-64-7	195.215	pr (al)	28	271			vs bz, eth, EtOH, lig
7445	Methyl 3-methoxypropanoate		C$_5$H$_{10}$O$_3$	3852-09-3	118.131			141(2)	1.0139^{15}	1.4030^{20}	
7446	Methyl 2-methylacetoacetate		C$_6$H$_{10}$O$_3$	17094-21-2	130.141			177.4	1.0217^{25}	1.416^{24}	vs eth, EtOH
7447	Methyl 2-(methylamino)-benzoate		C$_9$H$_{11}$NO$_2$	85-91-6	165.189	cry (peth)	19	255	1.120^{15}	1.5839^{15}	i H$_2$O; s EtOH, eth
7448	Methyl 2-methylbenzoate		C$_9$H$_{10}$O$_2$	89-71-4	150.174		-44.38(0.04)	215	1.068^{20}		i H$_2$O; msc EtOH, eth
7449	Methyl 3-methylbenzoate		C$_9$H$_{10}$O$_2$	99-36-5	150.174			220(12)	1.061^{20}		i H$_2$O; s EtOH; sl ctc
7450	Methyl 4-methylbenzoate		C$_9$H$_{10}$O$_2$	99-75-2	150.174	cry (aq MeOH, peth)	33.35(0.04)	222(13)			i H$_2$O; vs EtOH, eth
7451	Methyl 2-methyl-2-butenoate, (E)-		C$_6$H$_{10}$O$_2$	6622-76-0	114.142			139	0.9349^{12}	1.4370^{20}	
7452	Methyl 3-methyl-2-butenoate		C$_6$H$_{10}$O$_2$	924-50-5	114.142		114	136.5	0.9337^{20}	1.432^{20}	
7453	3-Methyl-4-methylenehexane		C$_8$H$_{16}$	3404-67-9	112.213			113(3)	0.725^{25}	1.4142^{20}	
7454	2-Methyl-5-(1-methylethenyl)-cyclohexanone, (2R-trans)-		C$_{10}$H$_{16}$O	5524-05-0	152.233			221.5	0.928^{19}	1.4724	vs ace, eth
7455	5-Methyl-2-(1-methylethyli-dene)cyclohexanone		C$_{10}$H$_{16}$O	15932-80-6	152.233			93^{10}	0.9367^{20}	1.4869^{20}	
7456	3-Methyl-6-(1-methylethylidene)-2-cyclo-hexen-1-one	Piperitenone	C$_{10}$H$_{14}$O	491-09-8	150.217			120^{14}	0.9774^{20}	1.5294^{20}	vs EtOH, eth
7457	1-Methyl-4-(5-methyl-1-methylene-4-hexenyl)-cyclohexene, (S)-		C$_{15}$H$_{24}$	495-61-4	204.352			129^{10}	0.8673^{20}	1.4880^{20}	
7458	N-Methyl-N-(2-methylphenyl)-acetamide		C$_{10}$H$_{13}$NO	573-26-2	163.216		55.5	260			s EtOH, chl
7459	4-Methyl-N-(4-methylphenyl)-aniline		C$_{14}$H$_{15}$N	620-93-9	197.276	nd (peth)	79(1)	330.5			vs eth, peth
7460	2-Methyl-3-(2-methylphenyl)-4(3H)-quinazolinone	Methaqualone	C$_{16}$H$_{14}$N$_2$O	72-44-6	250.294		120				vs eth, EtOH, chl
7461	Methyl 3-(methylthio)-propanoate	2-Methoxycarbonylethyl methyl sulfide	C$_5$H$_{10}$O$_2$S	13532-18-8	134.197			75^{13}	1.077^{25}	1.4650^{20}	
7462	1-Methyl-4-(1-methylvinyl)-benzene		C$_{10}$H$_{12}$	1195-32-0	132.202	liq	-20	186(3)	0.8936^{23}	1.5283^{23}	
7463	1-Methyl-4-(1-methylvinyl)-cyclohexanol	β-Terpineol	C$_{10}$H$_{18}$O	138-87-4	154.249	nd	32.5	210	0.917^{20}	1.4747^{20}	
7464	5-Methyl-2-(1-methylvinyl)-cyclohexanol, [1R-(1α,2β,5α)]		C$_{10}$H$_{18}$O	89-79-2	154.249		78	93^{14}	0.911^{20}	1.4723^{20}	sl H$_2$O; s EtOH, eth
7465	5-Methyl-2-(1-methylvinyl)-cyclohexanol acetate, [1R-(1α,2β,5α)]		C$_{12}$H$_{20}$O$_2$	57576-09-7	196.286		85	113^8	0.925^{25}	1.4566^{20}	
7466	trans-5-Methyl-2-(1-methylvi-nyl)cyclohexanone		C$_{10}$H$_{16}$O	29606-79-9	152.233			100^{18}	0.9198^{20}	1.4675^{20}	
7467	2-Methyl-5-(1-methylvinyl)-2-cyclohexen-1-ol		C$_{10}$H$_{16}$O	99-48-9	152.233			228	0.9484^{25}	1.4942^{25}	
7468	4-Methylmorpholine		C$_5$H$_{11}$NO	109-02-4	101.147	liq	-64.40	121.4(1)	0.9051^{20}	1.4332^{20}	s H$_2$O, EtOH, eth
7469	α-Methyl-4-morpholineethanol		C$_7$H$_{15}$NO$_2$	2109-66-2	145.200			121^{18}	1.0174^{20}	1.4638^{20}	vs H$_2$O, ace, bz, EtOH

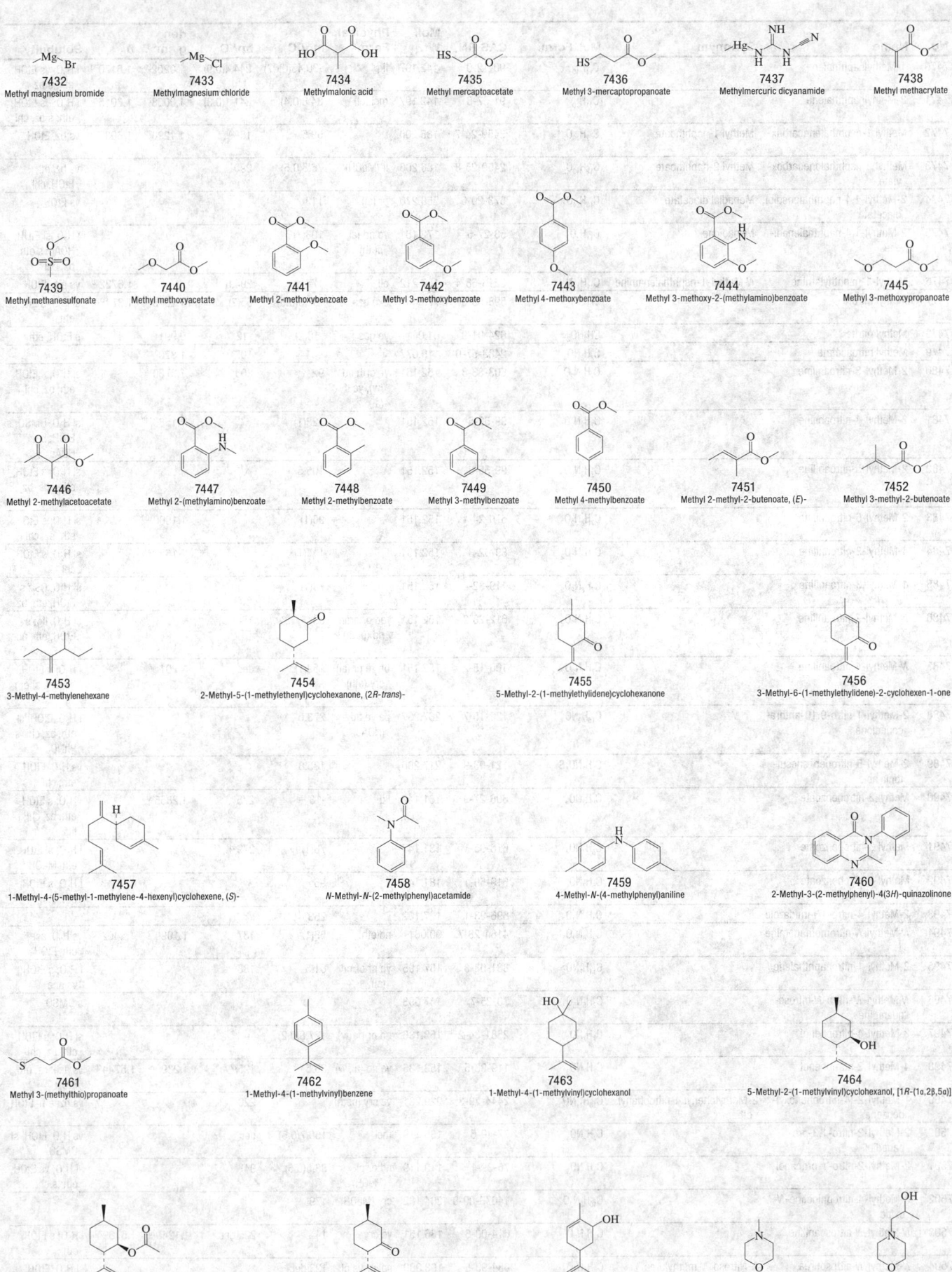

7432
Methyl magnesium bromide

7433
Methylmagnesium chloride

7434
Methylmalonic acid

7435
Methyl mercaptoacetate

7436
Methyl 3-mercaptopropanoate

7437
Methylmercuric dicyanamide

7438
Methyl methacrylate

7439
Methyl methanesulfonate

7440
Methyl methoxyacetate

7441
Methyl 2-methoxybenzoate

7442
Methyl 3-methoxybenzoate

7443
Methyl 4-methoxybenzoate

7444
Methyl 3-methoxy-2-(methylamino)benzoate

7445
Methyl 3-methoxypropanoate

7446
Methyl 2-methylacetoacetate

7447
Methyl 2-(methylamino)benzoate

7448
Methyl 2-methylbenzoate

7449
Methyl 3-methylbenzoate

7450
Methyl 4-methylbenzoate

7451
Methyl 2-methyl-2-butenoate, (E)-

7452
Methyl 3-methyl-2-butenoate

7453
3-Methyl-4-methylenehexane

7454
2-Methyl-5-(1-methylethenyl)cyclohexanone, (2R-trans)-

7455
5-Methyl-2-(1-methylethylidene)cyclohexanone

7456
3-Methyl-6-(1-methylethylidene)-2-cyclohexen-1-one

7457
1-Methyl-4-(5-methyl-1-methylene-4-hexenyl)cyclohexene, (S)-

7458
N-Methyl-N-(2-methylphenyl)acetamide

7459
4-Methyl-N-(4-methylphenyl)aniline

7460
2-Methyl-3-(2-methylphenyl)-4(3H)-quinazolinone

7461
Methyl 3-(methylthio)propanoate

7462
1-Methyl-4-(1-methylvinyl)benzene

7463
1-Methyl-4-(1-methylvinyl)cyclohexanol

7464
5-Methyl-2-(1-methylvinyl)cyclohexanol, [1R-(1α,2β,5α)]

7465
5-Methyl-2-(1-methylvinyl)cyclohexanol acetate, [1R-(1α,2β,5α)]

7466
trans-5-Methyl-2-(1-methylvinyl)cyclohexanone

7467
2-Methyl-5-(1-methylvinyl)-2-cyclohexen-1-ol

7468
4-Methylmorpholine

7469
α-Methyl-4-morpholineethanol

No.	Name	Synonym	Mol. Form.	CAS RN	Mol. Wt.	Physical Form	mp/°C	bp/°C	den g cm⁻³	n_D	Solubility
7470	1-Methylnaphthalene		$C_{11}H_{10}$	90-12-0	142.197	liq	-30.43(0.07)	244.4(0.9)	1.0202²⁰	1.6170²⁰	i H₂O; vs EtOH, eth; s bz
7471	2-Methylnaphthalene		$C_{11}H_{10}$	91-57-6	142.197	mcl (al)	34.6(0.4)	241.1(0.3)	1.0058²⁰	1.6015⁴⁰	i H₂O; vs EtOH, eth; s bz, chl
7472	Methyl 1-naphthalenecarboxylate	Methyl 1-naphthoate	$C_{12}H_{10}O_2$	2459-24-7	186.206		59.5	168²⁰	1.1290²⁰	1.6086²⁰	vs bz, EtOH
7473	Methyl 2-naphthalenecarboxylate	Methyl 2-naphthoate	$C_{12}H_{10}O_2$	2459-25-8	186.206	lf (MeOH)	76.8(0.5)	290			vs bz, eth, EtOH, chl
7474	2-Methyl-1,4-naphthalenediol diacetate	Menadiol diacetate	$C_{15}H_{14}O_4$	573-20-6	258.270	pr (al)	113				vs EtOH
7475	2-Methyl-1,4-naphthalenedione	Menadione	$C_{11}H_8O_2$	58-27-5	172.181	ye nd (al, peth)	103(1)				i H₂O; sl EtOH, HOAc; s eth, bz, chl
7476	Methyl-1-naphthylamine	N-Methyl-1-naphthalenamine	$C_{11}H_{11}N$	2216-68-4	157.212	oil	174	294.5		1.6722²⁰	vs eth, EtOH
7477	Methyl nitrate		CH_3NO_3	598-58-3	77.040	exp gas	-82.9(0.5)	65(2)	1.2075²⁰	1.3748²⁰	sl H₂O; s EtOH, eth
7478	Methyl nitrite		CH_3NO_2	624-91-9	61.041	ye gas	-16(2)	-12	0.991¹⁵		s EtOH, eth
7479	Methyl nitroacetate		$C_3H_5NO_4$	2483-57-0	119.077			107²⁸	1.320⁰		
7480	2-Methyl-3-nitroaniline		$C_7H_8N_2O_2$	603-83-8	152.151	ye orth nd (w), ye lf (al)	92	305	1.3780¹⁵		sl H₂O; s EtOH, eth, bz, chl
7481	2-Methyl-4-nitroaniline		$C_7H_8N_2O_2$	99-52-5	152.151		128(1)		1.1586¹⁴⁰		sl H₂O, DMSO; s EtOH, bz, HOAc
7482	2-Methyl-5-nitroaniline		$C_7H_8N_2O_2$	99-55-8	152.151		105.5				sl H₂O; s EtOH, eth, ace, bz, chl
7483	2-Methyl-6-nitroaniline		$C_7H_8N_2O_2$	570-24-1	152.151		96(1)		1.1900¹⁰⁰		sl H₂O; s EtOH, eth, bz, chl
7484	4-Methyl-2-nitroaniline		$C_7H_8N_2O_2$	89-62-3	152.151		117(1)		1.16¹²¹		sl H₂O; s EtOH, chl
7485	4-Methyl-3-nitroaniline		$C_7H_8N_2O_2$	119-32-4	152.151		78(1)				sl H₂O, CS₂; s EtOH, eth, bz
7486	N-Methyl-2-nitroaniline		$C_7H_8N_2O_2$	612-28-2	152.151	red or oran nd (peth)	38	158¹⁸			sl H₂O, lig; s EtOH, eth, ace, bz
7487	N-Methyl-4-nitroaniline		$C_7H_8N_2O_2$	100-15-2	152.151	br-ye pr (al) cry (eth)	152(1)	dec	1.201¹⁵⁵		i H₂O; s EtOH, bz, chl; sl eth, lig
7488	2-Methyl-1-nitro-9,10-anthracenedione		$C_{15}H_9NO_4$	129-15-7	267.237	pa ye nd (HOAc)	273.0				i H₂O, EtOH; sl eth, bz, chl; s PhNO₂
7489	2-Methyl-5-nitrobenzenesulfonic acid		$C_7H_7NO_5S$	121-03-9	217.200		135.8				vs H₂O, EtOH, eth, chl
7490	Methyl 2-nitrobenzoate		$C_8H_7NO_4$	606-27-9	181.147	liq	-13	275	1.2855²⁰		i H₂O; s EtOH, eth, bz, chl; i lig
7491	Methyl 3-nitrobenzoate		$C_8H_7NO_4$	618-95-1	181.147		78(1)	279⁶⁰			i H₂O; sl EtOH, eth, MeOH
7492	Methyl 4-nitrobenzoate		$C_8H_7NO_4$	619-50-1	181.147		96				i H₂O; s EtOH, eth, chl
7493	2-Methyl-4-nitro-1H-imidazole		$C_4H_5N_3O_2$	696-23-1	127.102		253				
7494	N-Methyl-N-nitromethanamine		$C_2H_6N_2O_2$	4164-28-7	90.081	nd(eth)	56(1)	187	1.1090⁷²	1.4462⁷²	vs H₂O, ace, eth, EtOH
7495	2-Methyl-1-nitronaphthalene		$C_{11}H_9NO_2$	881-03-8	187.195	ye pr or nd (al)	81.5	188²⁰			i H₂O; s EtOH; vs ace
7496	N-Methyl-N'-nitro-N-nitrosoguanidine		$C_2H_5N_5O_3$	70-25-7	147.093						s DMSO
7497	3-Methyl-4-nitrophenol		$C_7H_7NO_3$	2581-34-2	153.136	nd or pr (w)	127.8(0.2)				sl H₂O; s EtOH, eth, bz, chl
7498	4-Methyl-2-nitrophenol		$C_7H_7NO_3$	119-33-5	153.136	ye nd (al, w)	36.5	125²²	1.2399²⁰	1.5744⁴⁰	vs ace, bz, eth, EtOH
7499	1-Methyl-2-(4-nitrophenoxy)-benzene	2-Methylphenyl 4-nitrophenyl ether	$C_{13}H_{11}NO_3$	2444-29-3	229.231	ye cry (peth)		220²⁷			vs bz, eth, EtOH
7500	2-Methyl-2-nitro-1,3-propanediol		$C_4H_9NO_4$	77-49-6	135.119	mcl	150.7(0.5)	dec			vs H₂O, EtOH; sl DMSO
7501	2-Methyl-2-nitro-1-propanol		$C_4H_9NO_3$	76-39-1	119.119	nd or pl (MeOH)	88.5(0.5)	94¹⁰			sl H₂O; vs EtOH, eth; s chl
7502	3-Methyl-4-nitroquinoline-N-oxide		$C_{10}H_8N_2O_3$	14073-00-8	204.182	cry (MeOH)	179				
7503	N-Methyl-N-nitrosoaniline		$C_7H_8N_2O$	614-00-6	136.151	ye cry	14.7	225 dec	1.1240²⁰	1.5769²⁰	i H₂O; s EtOH, eth
7504	N-Methyl-N-nitrosourea	N-Nitroso-N-methylurea	$C_2H_5N_3O_2$	684-93-5	103.080	col or ye pl (eth)	123 dec				sl H₂O, EtOH, eth

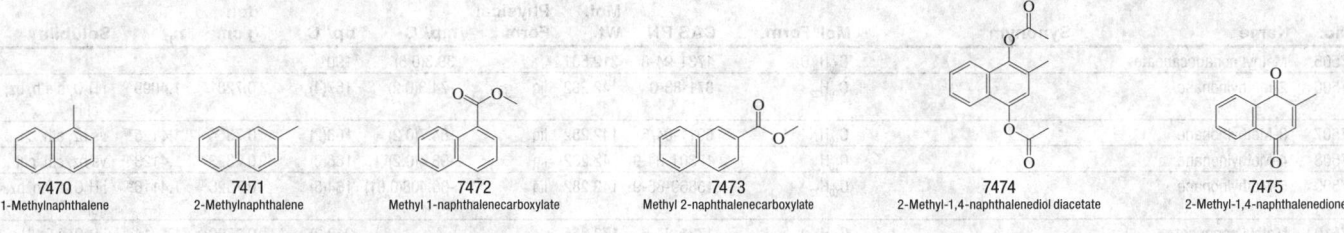

7470	7471	7472	7473	7474	7475
1-Methylnaphthalene	2-Methylnaphthalene	Methyl 1-naphthalenecarboxylate	Methyl 2-naphthalenecarboxylate	2-Methyl-1,4-naphthalenediol diacetate	2-Methyl-1,4-naphthalenedione

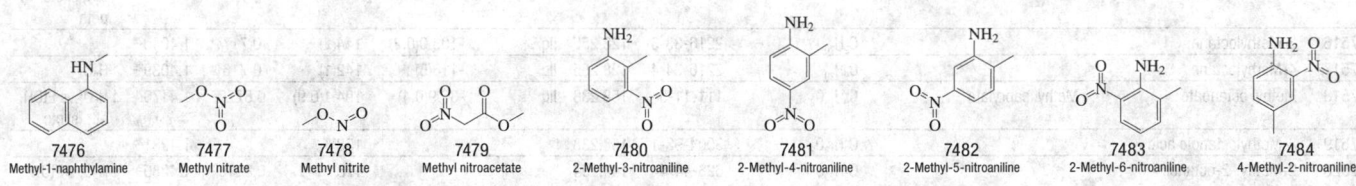

7476	7477	7478	7479	7480	7481	7482	7483	7484
Methyl-1-naphthylamine	Methyl nitrate	Methyl nitrite	Methyl nitroacetate	2-Methyl-3-nitroaniline	2-Methyl-4-nitroaniline	2-Methyl-5-nitroaniline	2-Methyl-6-nitroaniline	4-Methyl-2-nitroaniline

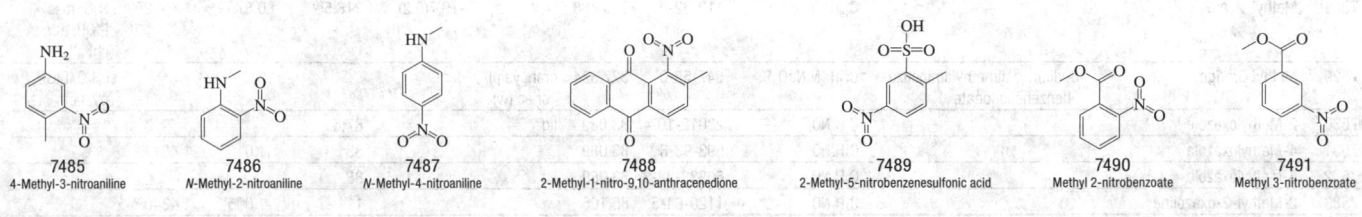

7485	7486	7487	7488	7489	7490	7491
4-Methyl-3-nitroaniline	N-Methyl-2-nitroaniline	N-Methyl-4-nitroaniline	2-Methyl-1-nitro-9,10-anthracenedione	2-Methyl-5-nitrobenzenesulfonic acid	Methyl 2-nitrobenzoate	Methyl 3-nitrobenzoate

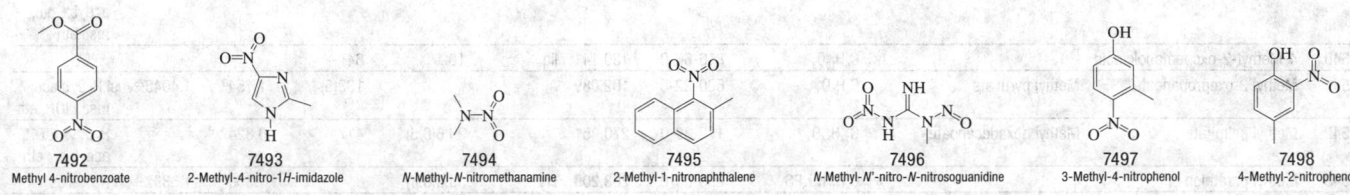

7492	7493	7494	7495	7496	7497	7498
Methyl 4-nitrobenzoate	2-Methyl-4-nitro-1H-imidazole	N-Methyl-N-nitromethanamine	2-Methyl-1-nitronaphthalene	N-Methyl-N'-nitro-N-nitrosoguanidine	3-Methyl-4-nitrophenol	4-Methyl-2-nitrophenol

7499	7500	7501	7502	7503	7504
1-Methyl-2-(4-nitrophenoxy)benzene	2-Methyl-2-nitro-1,3-propanediol	2-Methyl-2-nitro-1-propanol	3-Methyl-4-nitroquinoline-N-oxide	N-Methyl-N-nitrosoaniline	N-Methyl-N-nitrosourea

No.	Name	Synonym	Mol. Form.	CAS RN	Mol. Wt.	Physical Form	mp/°C	bp/°C	den g cm^{-3}	n_D	Solubility
7505	Methyl nonadecanoate		$C_{20}H_{40}O_2$	1731-94-8	312.531		38.3(0.6)	190[4]			
7506	2-Methylnonane		$C_{10}H_{22}$	871-83-0	142.282	liq	-74.3(0.2)	167(4)	0.7281[20]	1.4099[20]	i H_2O; s eth, bz, chl
7507	3-Methylnonane		$C_{10}H_{22}$	5911-04-6	142.282	liq	-84.7(0.2)	168(2)	0.7354[20]	1.4125[20]	vs bz, eth, chl
7508	4-Methylnonane		$C_{10}H_{22}$	17301-94-9	142.282	liq	-98.4(0.2)	166(3)	0.7323[20]	1.4123[20]	vs bz, eth, chl
7509	5-Methylnonane		$C_{10}H_{22}$	15869-85-9	142.282	liq	-86.408(0.01)	164(5)	0.7326[20]	1.4116[20]	i H_2O; s eth, bz, chl
7510	Methyl nonanoate		$C_{10}H_{20}O_2$	1731-84-6	172.265			212(3)	0.8799[15]	1.4214[20]	i H_2O; s EtOH, eth; sl ctc
7511	8-Methyl-1-nonanol		$C_{10}H_{22}O$	55505-26-5	158.281			108[10]			
7512	2-Methyl-1-nonene		$C_{10}H_{20}$	2980-71-4	140.266	liq	-64.2	167(2)	0.7412[25]	1.4241[20]	
7513	2-Methyl-2-norbornene	2-Methylbicyclo[2.2.1]hept-2-ene	C_8H_{12}	694-92-8	108.181	liq		118(7)			
7514	Methyl *trans*-9-octadecenoate		$C_{19}H_{36}O_2$	1937-62-8	296.488	liq	10.3(0.3)	218[24]	0.8730[20]	1.4513[20]	vs eth, EtOH
7515	2-Methyloctane		C_9H_{20}	3221-61-2	128.255	liq	-80.3(0.3)	143(1)	0.7095[25]	1.4031[20]	i H_2O; s EtOH, eth; sl ctc; vs peth
7516	3-Methyloctane		C_9H_{20}	2216-33-3	128.255	liq	-108.0(0.2)	144(2)	0.717[25]	1.4040[25]	
7517	4-Methyloctane		C_9H_{20}	2216-34-4	128.255	liq	-116(5)	142(1)	0.716[25]	1.4039[25]	i H_2O
7518	Methyl octanoate	Methyl caprylate	$C_9H_{18}O_2$	111-11-5	158.238	liq	-36.9(0.4)	194.1(0.9)	0.8775[20]	1.4170[20]	i H_2O; vs EtOH, eth; sl ctc
7519	2-Methyloctanoic acid		$C_9H_{18}O_2$	3004-93-1	158.238			138[14]		1.4281[25]	
7520	2-Methyl-2-octanol		$C_9H_{20}O$	628-44-4	144.254			178	0.8210[20]	1.4280[20]	i H_2O; s EtOH, eth
7521	3-Methyl-3-octanol		$C_9H_{20}O$	5340-36-3	144.254			83[18]	0.8108[25]	1.4257[25]	
7522	5-Methyl-2-octanone		$C_9H_{18}O$	58654-67-4	142.238			101[50]			
7523	2-Methyl-1-octene		C_9H_{18}	4588-18-5	126.239	liq	-77.8(0.3)	144(3)	0.7343[20]	1.4184[20]	
7524	7-Methyl-1-octene		C_9H_{18}	13151-09-9	126.239	liq		139(3)			
7525	Methyloctylamine	*N*-Methyl-1-octanamine	$C_9H_{21}N$	2439-54-5	143.270			68[8]			
7526	Methyl 2-octynoate		$C_9H_{14}O_2$	111-12-6	154.206			217	0.926[20]	1.4464[20]	
7527	3-Methyl-1-octyn-3-ol		$C_9H_{16}O$	23580-51-0	140.222			174	0.8547[20]	1.443[10]	
7528	Methyl oleate		$C_{19}H_{36}O_2$	112-62-9	296.488		-19.7(0.2)	218.5[20]	0.8739[20]	1.4522[20]	i H_2O; msc EtOH, eth; s chl
7529	Methyl Orange	Sodium *p*-dimethylaminoazo-benzenesulfonate	$C_{14}H_{14}N_3NaO_3S$	547-58-0	327.334	oran, ye pl or sc (w)	dec				sl H_2O, EtOH, py; i eth
7530	2-Methyloxazole		C_4H_5NO	23012-10-4	83.089	liq		87.5			
7531	4-Methyloxazole		C_4H_5NO	693-93-6	83.089			88	1.015[25]	1.4317[20]	
7532	5-Methyloxazole		C_4H_5NO	66333-88-8	83.089	liq		88			
7533	2-Methyl-2-oxazoline		C_4H_7NO	1120-64-5	85.105			111	1.005[25]	1.4340[20]	
7534	2-Methyloxetane		C_4H_8O	2167-39-7	72.106	hyg		59	0.841[25]	1.3885[20]	
7535	4-Methyl-2-oxetanone	3-Hydroxybutyric acid lactone	$C_4H_6O_2$	3068-88-0	86.090			86[50]	1.0555[20]		
7536	Methyloxirane	1,2-Propylene oxide	C_3H_6O	16033-71-9	58.079	liq	-111.9	35	0.859[0]	1.3660[20]	vs H_2O, EtOH, eth; s chl
7537	3-Methyl-2-oxobutanoic acid		$C_5H_8O_3$	759-05-7	116.116		31.5	170.5	0.9968[20]	1.3850[16]	s H_2O, EtOH, eth
7538	*N*-Methyl-*N*-(1-oxododecyl)-glycine	*N*-Dodecanoylsarcosine	$C_{15}H_{29}NO_3$	97-78-9	271.396		44.5				s chl
7539	Methyl 4-oxopentanoate	Methyl levulinate	$C_6H_{10}O_3$	624-45-3	130.141			196	1.0511[20]	1.4233[20]	sl H_2O; s EtOH, ace, bz, ctc; msc eth
7540	4-Methyl-2-oxopentanoic acid		$C_6H_{10}O_3$	816-66-0	130.141	liq	10	84[15]			
7541	Methyl 2-oxopropanoate	Methyl pyruvate	$C_4H_6O_3$	600-22-6	102.089			138(5)	1.154[0]	1.4046[25]	sl H_2O; s ace; msc EtOH, eth
7542	Methyl palmitate	Methyl hexadecanoate	$C_{17}H_{34}O_2$	112-39-0	270.451		29.6(0.5)	417	0.8247[75]		i H_2O; vs EtOH, ace, bz; s eth
7543	Methyl parathion		$C_8H_{10}NO_5PS$	298-00-0	263.208	cry	35.8(0.4)		1.358[20]	1.5367[25]	i H_2O; s os
7544	Methyl pentachlorophenyl sulfide	*S*-Methyl pentachlorobenzene-thiol	$C_7H_3Cl_5S$	1825-19-0	296.429	cry (EtOH)	95.5				
7545	Methyl pentadecanoate		$C_{16}H_{32}O_2$	7132-64-1	256.424	nd (dil al)	19(2)	308(7)	0.8618[25]	1.4390[25]	s EtOH, eth
7546	*cis*-2-Methyl-1,3-pentadiene		C_6H_{10}	1501-60-6	82.143	liq	-117.6	80(16)	0.714[25]	1.446[20]	
7547	3-Methyl-1,3-pentadiene		C_6H_{10}	4549-74-0	82.143			77(5)	0.730[25]	1.452[20]	
7548	4-Methyl-1,3-pentadiene	1,1-Dimethyl-1,3-butadiene	C_6H_{10}	926-56-7	82.143			76(3)	0.7181[20]	1.4532[20]	
7549	Methyl pentafluoroethyl ether	1-Methoxyperfluoroethane	$C_3H_3F_5O$	22410-44-2	150.047	col gas		5.6(0.9)			
7550	Methyl pentafluoropropanoate		$C_4H_3F_5O_2$	378-75-6	178.058			59.5	1.390[25]	1.2869[25]	
7551	2-Methylpentanal	2-Methylvaleraldehyde	$C_6H_{12}O$	123-15-9	100.158			118(4)			s H_2O; s eth, ace; sl ctc
7552	2-Methylpentane	Isohexane	C_6H_{14}	107-83-5	86.175	liq	-153.60(0.09)	60.21(0.09)	0.650[25]	1.3715[20]	i H_2O; s EtOH, eth; msc ace, bz, chl

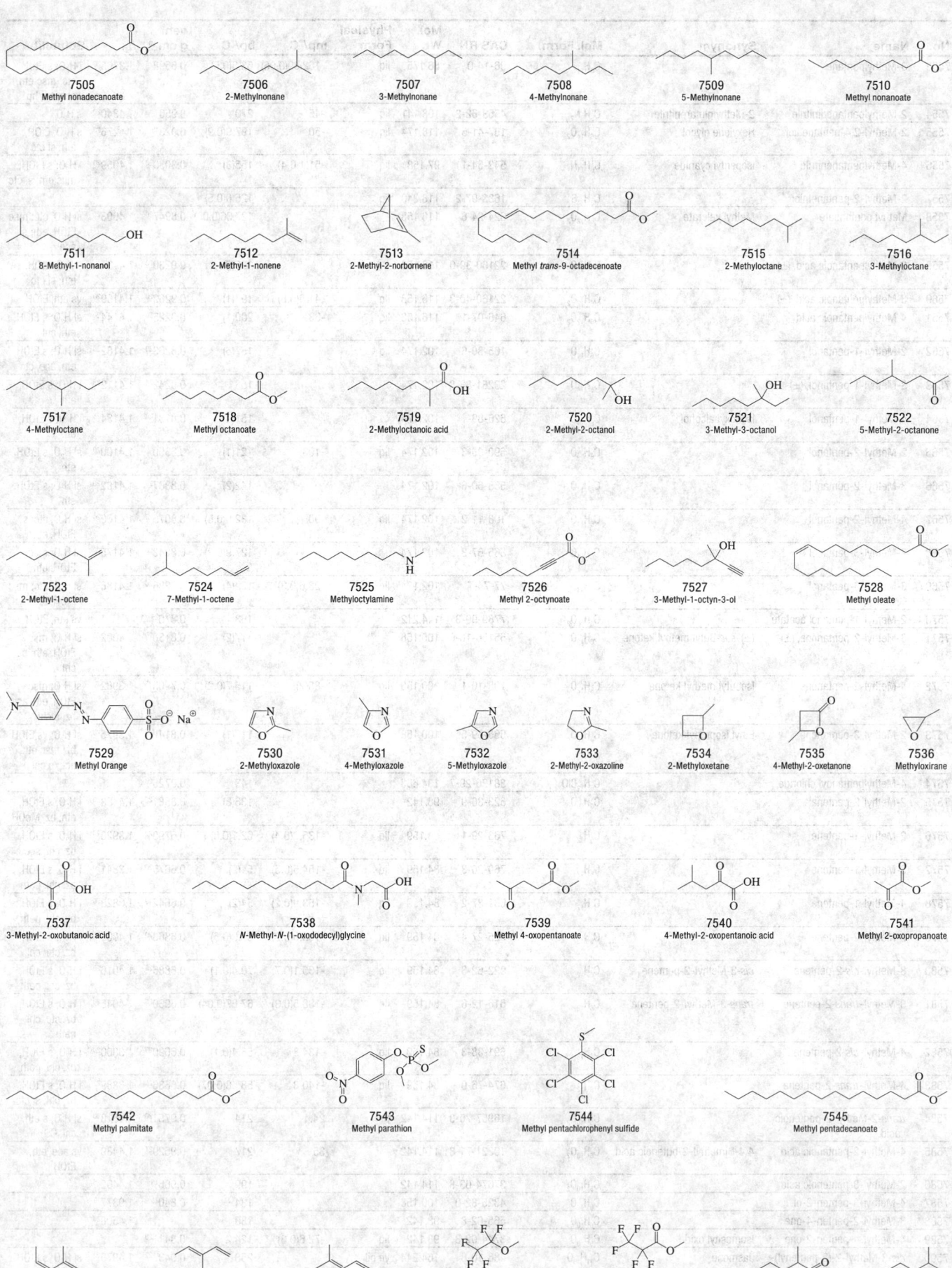

7505 Methyl nonadecanoate

7506 2-Methylnonane

7507 3-Methylnonane

7508 4-Methylnonane

7509 5-Methylnonane

7510 Methyl nonanoate

7511 8-Methyl-1-nonanol

7512 2-Methyl-1-nonene

7513 2-Methyl-2-norbornene

7514 Methyl *trans*-9-octadecenoate

7515 2-Methyloctane

7516 3-Methyloctane

7517 4-Methyloctane

7518 Methyl octanoate

7519 2-Methyloctanoic acid

7520 2-Methyl-2-octanol

7521 3-Methyl-3-octanol

7522 5-Methyl-2-octanone

7523 2-Methyl-1-octene

7524 7-Methyl-1-octene

7525 Methyloctylamine

7526 Methyl 2-octynoate

7527 3-Methyl-1-octyn-3-ol

7528 Methyl oleate

7529 Methyl Orange

7530 2-Methyloxazole

7531 4-Methyloxazole

7532 5-Methyloxazole

7533 2-Methyl-2-oxazoline

7534 2-Methyloxetane

7535 4-Methyl-2-oxetanone

7536 Methyloxirane

7537 3-Methyl-2-oxobutanoic acid

7538 *N*-Methyl-*N*-(1-oxododecyl)glycine

7539 Methyl 4-oxopentanoate

7540 4-Methyl-2-oxopentanoic acid

7541 Methyl 2-oxopropanoate

7542 Methyl palmitate

7543 Methyl parathion

7544 Methyl pentachlorophenyl sulfide

7545 Methyl pentadecanoate

7546 *cis*-2-Methyl-1,3-pentadiene

7547 3-Methyl-1,3-pentadiene

7548 4-Methyl-1,3-pentadiene

7549 Methyl pentafluoroethyl ether

7550 Methyl pentafluoropropanoate

7551 2-Methylpentanal

7552 2-Methylpentane

No.	Name	Synonym	Mol. Form.	CAS RN	Mol. Wt.	Physical Form	mp/°C	bp/°C	den g cm⁻³	n_D	Solubility
7553	3-Methylpentane		C₆H₁₄	96-14-0	86.175	liq	-162.89(0.05)	63.3(0.5)	0.6598²⁵	1.3765²⁰	i H₂O; s EtOH, ctc; msc eth, ace, bz, hp
7554	2-Methylpentanedinitrile	2-Methylglutaronitrile	C₆H₈N₂	4553-62-2	108.141	liq	-45	270	0.950	1.4340²⁰	s H₂O
7555	2-Methyl-2,4-pentanediol	Hexylene glycol	C₆H₁₄O₂	107-41-5	118.174	liq	-50	197.9(0.9)	0.923¹⁵	1.4276²⁰	s H₂O, EtOH, eth; sl ctc
7556	4-Methylpentanenitrile	Isopentyl cyanide	C₆H₁₁N	542-54-1	97.158	liq	-51.1(0.4)	155(3)	0.8030²⁰	1.4059²⁰	i H₂O; s EtOH; msc eth; sl ctc
7557	2-Methyl-2-pentanethiol		C₆H₁₄S	1633-97-2	118.240	liq		125.0(0.5)			
7558	Methyl pentanoate	Methyl valerate	C₆H₁₂O₂	624-24-8	116.158			127.36(0.06)	0.8947²⁰	1.4003²⁰	sl H₂O, ctc; msc EtOH, eth; s ace
7559	2-Methylpentanoic acid, (±)-		C₆H₁₂O₂	22160-39-0	116.158			195(2)	0.9230²⁰	1.413²⁰	s H₂O, EtOH, eth; sl ctc
7560	3-Methylpentanoic acid, (±)-		C₆H₁₂O₂	22160-40-3	116.158	liq	-41.6(0.8)	197(1)	0.9262²⁰	1.4159²⁰	vs eth, EtOH
7561	4-Methylpentanoic acid		C₆H₁₂O₂	646-07-1	116.158	liq	-33	200(1)	0.9225²⁰	1.4144²⁰	sl H₂O; s EtOH, eth, chl
7562	2-Methyl-1-pentanol		C₆H₁₄O	105-30-6	102.174	liq		157(6)	0.8263²⁰	1.4182²⁰	sl H₂O; s EtOH, eth, ace, ctc
7563	3-Methyl-1-pentanol, (±)-		C₆H₁₄O	20281-83-8	102.174			153	0.8242²⁰	1.4112²³	i H₂O; s EtOH, eth
7564	4-Methyl-1-pentanol	Isohexyl alcohol	C₆H₁₄O	626-89-1	102.174			151(2)	0.8131²⁰	1.4134²⁵	i H₂O; s EtOH, eth
7565	2-Methyl-2-pentanol		C₆H₁₄O	590-36-3	102.174	liq	-103	121(1)	0.8350¹⁶	1.4100²⁰	sl H₂O; s EtOH, eth
7566	3-Methyl-2-pentanol		C₆H₁₄O	565-60-6	102.174			142(2)	0.8307²⁰	1.4182²⁰	sl H₂O; s EtOH, eth
7567	4-Methyl-2-pentanol		C₆H₁₄O	108-11-2	102.174	liq	-90	132.0(0.5)	0.8075²⁰	1.4100²⁰	sl H₂O, ctc; s EtOH, eth
7568	2-Methyl-3-pentanol		C₆H₁₄O	565-67-3	102.174			127.9(0.2)	0.8243²⁰	1.4175²⁰	sl H₂O; msc EtOH, eth
7569	3-Methyl-3-pentanol		C₆H₁₄O	77-74-7	102.174	liq	-23.6(0.3)	129(4)	0.8286²⁰	1.4186²⁰	sl H₂O, ctc; msc EtOH, eth
7570	2-Methyl-1-pentanol acetate		C₈H₁₆O₂	7789-99-3	144.212			163	0.870²⁵		vs eth, EtOH
7571	3-Methyl-2-pentanone, (±)-	(±)-sec-Butyl methyl ketone	C₆H₁₂O	55156-16-6	100.158			117(2)	0.8130²⁰	1.4002²⁰	sl H₂O; msc EtOH, eth; s chl
7572	4-Methyl-2-pentanone	Isobutyl methyl ketone	C₆H₁₂O	108-10-1	100.158	liq	-85(2)	115.7(0.2)	0.7965²⁵	1.3962²⁰	sl H₂O; msc EtOH, eth, ace, bz; s chl
7573	2-Methyl-3-pentanone	Ethyl isopropyl ketone	C₆H₁₂O	565-69-5	100.158			118(1)	0.814¹⁸	1.3975²⁰	sl H₂O; vs EtOH, bz; msc eth, ace; s chl
7574	4-Methylpentanoyl chloride		C₆H₁₁ClO	38136-29-7	134.603			143	0.9725²⁰		
7575	2-Methyl-2-pentenal		C₆H₁₀O	623-36-9	98.142			136(8)	0.8581²⁰	1.4488²⁰	i H₂O; s EtOH, eth, bz, MeOH
7576	2-Methyl-1-pentene		C₆H₁₂	763-29-1	84.159	liq	-135.7(0.4)	62.1(0.6)	0.6799²⁰	1.3920²⁰	i H₂O; s EtOH, bz, chl; sl ctc
7577	3-Methyl-1-pentene		C₆H₁₂	760-20-3	84.159	liq	-154.5(0.5)	54(2)	0.6675²⁰	1.3841²⁰	i H₂O; s EtOH, bz, chl, peth
7578	4-Methyl-1-pentene		C₆H₁₂	691-37-2	84.159	liq	-153.9(0.2)	54(2)	0.6642²⁰	1.3828²⁰	i H₂O; s EtOH, bz, chl, peth
7579	2-Methyl-2-pentene		C₆H₁₂	625-27-4	84.159	liq	-135.0(0.2)	67.3(0.5)	0.6863²⁰	1.4004²⁰	i H₂O; s EtOH, bz, ctc, chl
7580	3-Methyl-cis-2-pentene	cis-3-Methyl-2-pentene	C₆H₁₂	922-62-3	84.159	liq	-135.1(0.7)	70.4(0.1)	0.6886²⁵	1.4016²⁰	i H₂O; s EtOH, bz, chl, peth
7581	3-Methyl-trans-2-pentene	trans-3-Methyl-2-pentene	C₆H₁₂	616-12-6	84.159	liq	-138.5(0.3)	67.67(0.04)	0.6930²⁵	1.4045²⁰	i H₂O; s EtOH, bz, ctc, chl, peth
7582	4-Methyl-cis-2-pentene		C₆H₁₂	691-38-3	84.159	liq	-134.8	56.4(0.1)	0.6690²⁰	1.3800²⁰	i H₂O; s EtOH, bz, chl, peth
7583	4-Methyl-trans-2-pentene		C₆H₁₂	674-76-0	84.159	liq	-140.8	58.58(0.07)	0.6686²⁰	1.3889²⁰	i H₂O; s EtOH, bz, chl; sl ctc
7584	trans-2-Methyl-2-pentenoic acid		C₆H₁₀O₂	16957-70-3	114.142	pr	24.4	214	0.9751²⁰	1.4513²⁰	sl H₂O; s eth, chl, CS₂
7585	4-Methyl-2-pentenoic acid	4,4-Dimethyl-2-butenoic acid	C₆H₁₀O₂	10321-71-8	114.142		35	217	0.9529²¹	1.4489²¹	vs ace, eth, EtOH
7586	2-Methyl-3-pentenoic acid		C₆H₁₀O₂	37674-63-8	114.142			199	0.966¹⁵	1.4402²⁵	
7587	4-Methyl-3-penten-2-ol		C₆H₁₂O	4325-82-0	100.158			134	0.840¹⁵	1.9377¹⁵	
7588	3-Methyl-2-penten-4-one		C₆H₁₀O	565-62-8	98.142			138		1.4508²⁰	
7589	4-Methyl-4-penten-2-one	Isomesityl oxide	C₆H₁₀O	3744-02-3	98.142	liq	-72.6(0.3)	124.2	0.8411²⁰		
7590	cis-3-Methyl-2-(2-pentenyl)-2-cyclopenten-1-one	Jasmone	C₁₁H₁₆O	488-10-8	164.244	ye oil		258	0.9437²²	1.4979²²	sl H₂O; s EtOH, eth, ctc, lig

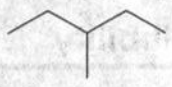

7553
3-Methylpentane

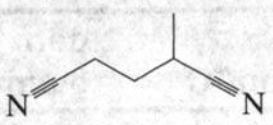

7554
2-Methylpentanedinitrile

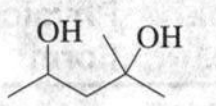

7555
2-Methyl-2,4-pentanediol

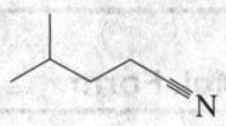

7556
4-Methylpentanenitrile

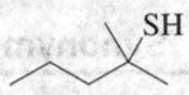

7557
2-Methyl-2-pentanethiol

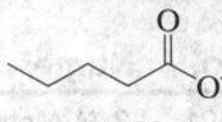

7558
Methyl pentanoate

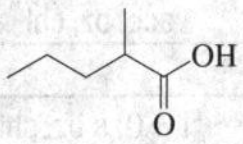

7559
2-Methylpentanoic acid, (±)-

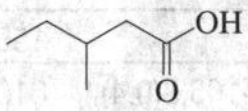

7560
3-Methylpentanoic acid, (±)-

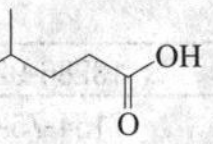

7561
4-Methylpentanoic acid

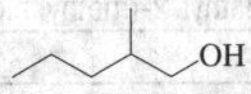

7562
2-Methyl-1-pentanol

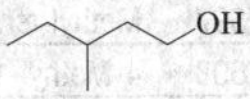

7563
3-Methyl-1-pentanol, (±)-

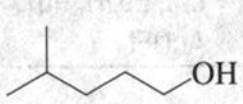

7564
4-Methyl-1-pentanol

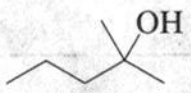

7565
2-Methyl-2-pentanol

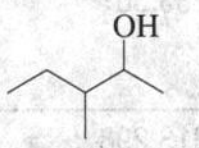

7566
3-Methyl-2-pentanol

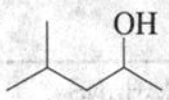

7567
4-Methyl-2-pentanol

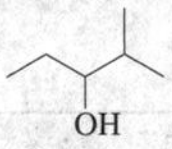

7568
2-Methyl-3-pentanol

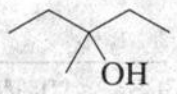

7569
3-Methyl-3-pentanol

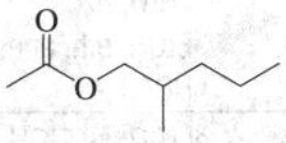

7570
2-Methyl-1-pentanol acetate

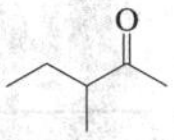

7571
3-Methyl-2-pentanone, (±)-

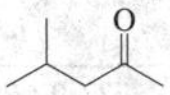

7572
4-Methyl-2-pentanone

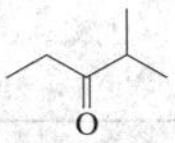

7573
2-Methyl-3-pentanone

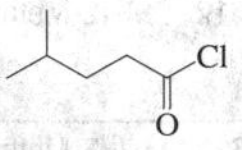

7574
4-Methylpentanoyl chloride

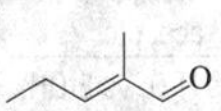

7575
2-Methyl-2-pentenal

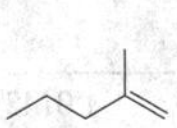

7576
2-Methyl-1-pentene

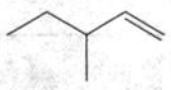

7577
3-Methyl-1-pentene

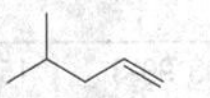

7578
4-Methyl-1-pentene

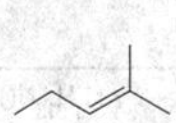

7579
2-Methyl-2-pentene

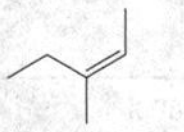

7580
3-Methyl-cis-2-pentene

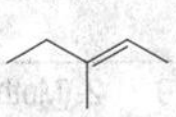

7581
3-Methyl-trans-2-pentene

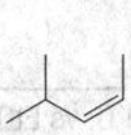

7582
4-Methyl-cis-2-pentene

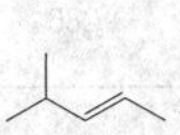

7583
4-Methyl-trans-2-pentene

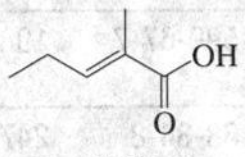

7584
trans-2-Methyl-2-pentenoic acid

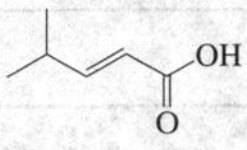

7585
4-Methyl-2-pentenoic acid

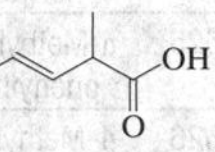

7586
2-Methyl-3-pentenoic acid

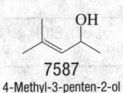

7587
4-Methyl-3-penten-2-ol

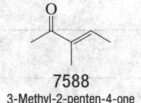

7588
3-Methyl-2-penten-4-one

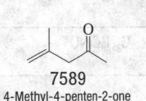

7589
4-Methyl-4-penten-2-one

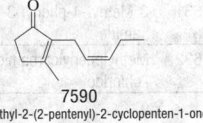

7590
cis-3-Methyl-2-(2-pentenyl)-2-cyclopenten-1-one

No.	Name	Synonym	Mol. Form.	CAS RN	Mol. Wt.	Physical Form	mp/°C	bp/°C	den g cm⁻³	n_D	Solubility
7591	3-(4-Methyl-3-pentenyl)furan		$C_{10}H_{14}O$	539-52-6	150.217			185.5	0.9017[20]	1.4705[21]	
7592	3-Methyl-3-penten-1-yne		C_6H_8	1574-33-0	80.128			69(7)	0.739[20]	1.4332[20]	s eth, bz
7593	3-Methyl-2-pentyl-2-cyclo-penten-1-one		$C_{11}H_{18}O$	1128-08-1	166.260			143[22]	0.9165[18]	1.4767[20]	
7594	Methyl pentyl ether		$C_6H_{14}O$	628-80-8	102.174			99(3)	0.759[22]	1.3862[22]	vs ace, eth, EtOH
7595	5-Methyl-2-pentylphenol	6-n-Amyl-m-cresol	$C_{12}H_{18}O$	1300-94-3	178.270		24	138[15]			vs ace, eth, EtOH
7596	Methyl pentyl sulfide		$C_6H_{14}S$	1741-83-9	118.240	liq	-94	148(4)	0.8431[20]	1.4506[20]	s EtOH, eth, ace, bz, chl
7597	Methyl tert-pentyl sulfide	2-Methyl-2-(methylthio)butane	$C_6H_{14}S$	13286-92-5	118.240	liq		150	0.84	1.4570[20]	
7598	4-Methyl-1-pentyne		C_6H_{10}	7154-75-8	82.143	liq	-105.3(0.4)	61(2)	0.7000[20]	1.3936[20]	i H_2O; s bz, chl
7599	4-Methyl-2-pentyne		C_6H_{10}	21020-27-9	82.143	liq	-110.3(0.4)	73(2)	0.7112[25]	1.4057[20]	vs bz, chl
7600	3-Methyl-1-pentyn-3-ol	Meparfynol	$C_6H_{10}O$	77-75-8	98.142		30.5	120(22)	0.8688[20]	1.4310[20]	
7601	Methyl perfluorooctanoate		$C_9H_3F_{15}O_2$	376-27-2	428.095			158	1.684[20]	1.304[27]	
7602	1-Methylphenanthrene		$C_{15}H_{12}$	832-69-9	192.256	lf, pl (dil al)	123	344(6)			i H_2O; s EtOH
7603	3-Methylphenanthrene		$C_{15}H_{12}$	832-71-3	192.256	pr or nd (al)	65	350			i H_2O; s EtOH, ace; sl chl
7604	4-Methylphenanthrene		$C_{15}H_{12}$	832-64-4	192.256	pl (90% al)	51.76(0.05)	177[10]			i H_2O; s EtOH, ctc
7605	Methylphenidate		$C_{14}H_{19}NO_2$	113-45-1	233.307			136[0.6]			i H_2O, peth; s chl, EtOH, eth, AcOEt
7606	10-Methyl-10H-phenothiazine		$C_{13}H_{11}NS$	1207-72-3	213.298		101				
7607	10-Methyl-10H-phenothi-azine-2-acetic acid	Metiazinic acid	$C_{15}H_{13}NO_2S$	13993-65-2	271.335		144				s chl
7608	Methyl phenoxyacetate		$C_9H_{10}O_3$	2065-23-8	166.173			245	1.1493[20]	1.5155[20]	vs eth, EtOH
7609	1-Methyl-3-phenoxybenzene		$C_{13}H_{12}O$	3586-14-9	184.233			272	1.051[25]	1.5727[20]	
7610	[(2-Methylphenoxy)methyl]-oxirane		$C_{10}H_{12}O_2$	2210-79-9	164.201			123[2]	1.0884[20]		
7611	3-(2-Methylphenoxy)-1,2-propanediol	Mephenesin	$C_{10}H_{14}O_3$	59-47-2	182.216		70 dec				sl H_2O, eth; s EtOH
7612	N-(2-Methylphenyl)acetamide		$C_9H_{11}NO$	120-66-1	149.189	nd (al)	110(1)	296	1.168[15]		sl H_2O, bz; s EtOH, eth, ace, HOAc
7613	N-(3-Methylphenyl)acetamide		$C_9H_{11}NO$	537-92-8	149.189	nd (w)	65.5	303	1.141[15]		sl H_2O; vs EtOH, eth; s chl
7614	N-Methyl-N-phenylacetamide	N-Methylacetanilide	$C_9H_{11}NO$	579-10-2	149.189	nd (eth), pr (al)	103	256	1.0036[105]	1.576	s H_2O, EtOH, eth, chl, lig
7615	2-Methylphenyl acetate	o-Cresyl acetate	$C_9H_{10}O_2$	533-18-6	150.174			208	1.0533[15]	1.5002[20]	vs eth, EtOH
7616	3-Methylphenyl acetate	m-Cresyl acetate	$C_9H_{10}O_2$	122-46-3	150.174		12	214(3)	1.043[20]	1.4978[20]	vs bz, eth, EtOH
7617	4-Methylphenyl acetate	p-Cresyl acetate	$C_9H_{10}O_2$	140-39-6	150.174			213(3)	1.0512[17]	1.5163[22]	sl H_2O, ctc; s EtOH, eth, chl
7618	Methyl 2-phenylacetate		$C_9H_{10}O_2$	101-41-7	150.174			215(9)	1.0622[16]	1.5075[20]	i H_2O; msc EtOH, eth; s ace, ctc
7619	2-(Methylphenylamino)ethanol		$C_9H_{13}NO$	93-90-3	151.205			218[110]	1.0143[0]		s H_2O; vs EtOH, eth, ace, bz
7620	2-[(2-Methylphenyl)amino]-ethanol		$C_9H_{13}NO$	136-80-1	151.205			285.5	1.0794[20]	1.5675[20]	vs eth, EtOH
7621	3-Methyl-N-phenylaniline		$C_{13}H_{13}N$	1205-64-7	183.249		30	316		1.6350[20]	vs bz, eth, EtOH
7622	N-(4-Methylphenyl)benzamide		$C_{14}H_{13}NO$	582-78-5	211.259	orth nd (al)	158		1.202[15]		vs eth, EtOH
7623	N-Methyl-N-phenylbenzene-methanamine		$C_{14}H_{15}N$	614-30-2	197.276						s ctc
7624	4-Methyl-α-phenylbenzenemethanol		$C_{14}H_{14}O$	1517-63-1	198.260		53(2)				
7625	α-Methyl-α-phenylbenzenemethanol		$C_{14}H_{14}O$	599-67-7	198.260			285	1.1059[15]		
7626	4-Methyl-N-phenylbenzene-sulfonamide		$C_{13}H_{13}NO_2S$	68-34-8	247.313	(α) tcl, (β) mcl pr (al, bz)	103.5				i H_2O; vs EtOH; s bz, HOAc
7627	4-Methylphenyl benzoate		$C_{14}H_{12}O_2$	614-34-6	212.244	pl (eth-al)	71.5	316			vs eth, EtOH
7628	1-Methyl-N-phenyl-N-benzyl-4-piperidinamine	Bamipine	$C_{19}H_{24}N_2$	4945-47-5	280.407	cry (MeOH)	115				
7629	Methyl 2-phenylbutanoate		$C_{11}H_{14}O_2$	2294-71-5	178.228	nd (dil al)	77.5	228			vs eth, EtOH
7630	3-Methyl-1-phenyl-1-buta-none		$C_{11}H_{14}O$	582-62-7	162.228			236.5	0.9701[16]	1.5139[15]	i H_2O; msc EtOH, eth; vs ace
7631	3-Methyl-4-phenyl-3-buten-amide	β-Benzalbutyramide	$C_{11}H_{13}NO$	7236-47-7	175.227		133				
7632	Methylphenylcarbamic chloride		C_8H_8ClNO	4285-42-1	169.609	pl (al)	88.5	280			vs eth, EtOH

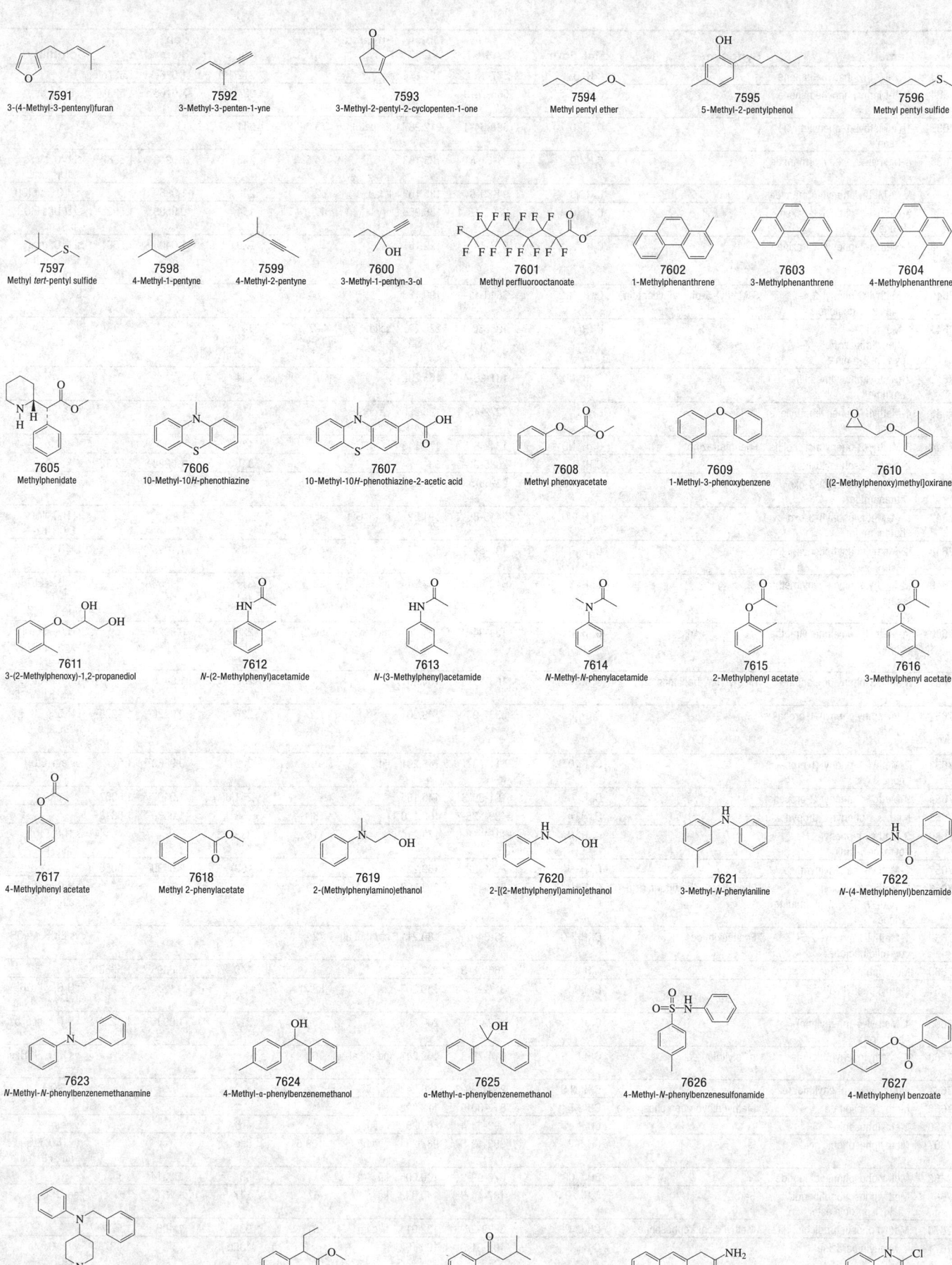

7591
3-(4-Methyl-3-pentenyl)furan

7592
3-Methyl-3-penten-1-yne

7593
3-Methyl-2-pentyl-2-cyclopenten-1-one

7594
Methyl pentyl ether

7595
5-Methyl-2-pentylphenol

7596
Methyl pentyl sulfide

7597
Methyl *tert*-pentyl sulfide

7598
4-Methyl-1-pentyne

7599
4-Methyl-2-pentyne

7600
3-Methyl-1-pentyn-3-ol

7601
Methyl perfluorooctanoate

7602
1-Methylphenanthrene

7603
3-Methylphenanthrene

7604
4-Methylphenanthrene

7605
Methylphenidate

7606
10-Methyl-10*H*-phenothiazine

7607
10-Methyl-10*H*-phenothiazine-2-acetic acid

7608
Methyl phenoxyacetate

7609
1-Methyl-3-phenoxybenzene

7610
[(2-Methylphenoxy)methyl]oxirane

7611
3-(2-Methylphenoxy)-1,2-propanediol

7612
N-(2-Methylphenyl)acetamide

7613
N-(3-Methylphenyl)acetamide

7614
N-Methyl-*N*-phenylacetamide

7615
2-Methylphenyl acetate

7616
3-Methylphenyl acetate

7617
4-Methylphenyl acetate

7618
Methyl 2-phenylacetate

7619
2-(Methylphenylamino)ethanol

7620
2-[(2-Methylphenyl)amino]ethanol

7621
3-Methyl-*N*-phenylaniline

7622
N-(4-Methylphenyl)benzamide

7623
N-Methyl-*N*-phenylbenzenemethanamine

7624
4-Methyl-α-phenylbenzenemethanol

7625
α-Methyl-α-phenylbenzenemethanol

7626
4-Methyl-*N*-phenylbenzenesulfonamide

7627
4-Methylphenyl benzoate

7628
1-Methyl-*N*-phenyl-*N*-benzyl-4-piperidinamine

7629
Methyl 2-phenylbutanoate

7630
3-Methyl-1-phenyl-1-butanone

7631
3-Methyl-4-phenyl-3-butenamide

7632
Methylphenylcarbamic chloride

No.	Name	Synonym	Mol. Form.	CAS RN	Mol. Wt.	Physical Form	mp/°C	bp/°C	den g cm⁻³	n_D	Solubility
7633	1-(2-Methylphenyl)ethanone		C₉H₁₀O	577-16-2	134.174			212(12)	1.026²⁰	1.5276²⁰	
7634	1-(3-Methylphenyl)ethanone		C₉H₁₀O	585-74-0	134.174			224(16)	1.0165⁰	1.533¹⁵	s EtOH, eth, ace; sl ctc
7635	4-(1-Methyl-1-phenylethyl)-phenol		C₁₅H₁₆O	599-64-4	212.287	pr (peth)	73.2(0.2)	341.2(0.7)			
7636	N-Methyl-N-phenylformamide		C₈H₉NO	93-61-8	135.163		14.5	243	1.0948²⁰	1.5589²⁰	sl H₂O, ctc; s EtOH, ace
7637	N-(2-Methylphenyl)formamide		C₈H₉NO	94-69-9	135.163	lf (al)	62	288	1.086⁵⁵		s H₂O; vs EtOH
7638	5-Methyl-1-phenyl-1-hexen-3-one		C₁₃H₁₆O	2892-18-4	188.265	cry	43	154²⁵	0.9509⁴⁶	1.5523²⁵	sl H₂O; s EtOH, bz, chl
7639	1-Methyl-1-phenylhydrazine		C₇H₁₀N₂	618-40-6	122.167			229.0(0.5)	1.0404²⁰	1.5691²⁰	sl H₂O; msc EtOH, eth, bz, chl
7640	3-Methyl-5-phenyl-2,4-imidazolidinedione	3-Methyl-5-phenylhydantoin	C₁₀H₁₀N₂O₂	6846-11-3	190.198		164.5				s chl
7641	1-Methyl-6-phenylimidazo[4,5-b]pyridin-2-amine	PhIP	C₁₃H₁₂N₄	105650-23-5	224.261	solid	327				
7642	2-[Methyl(phenylmethyl)-amino]ethanol		C₁₀H₁₅NO	101-98-4	165.232			134¹⁴			
7643	4-Methyl-N-(phenylmethylene)aniline		C₁₄H₁₃N	2272-45-9	195.260	ye cry	35	318			vs ace
7644	3-Methyl-2-phenylmorpholine	Phenmetrazine	C₁₁H₁₅NO	134-49-6	177.243			139¹²			
7645	2-Methyl-2-phenyloxirane		C₉H₁₀O	2085-88-3	134.174			84¹⁷	1.0228²⁰	1.5232²⁰	
7646	N-(2-Methylphenyl)-3-oxo-butanamide		C₁₁H₁₃NO₂	93-68-5	191.227	pr (AcOEt)	107.5				vs bz, EtOH
7647	N-(4-Methylphenyl)-3-oxo-butanamide		C₁₁H₁₃NO₂	2415-85-2	191.227	pr (AcOEt)	95				sl H₂O, lig; s EtOH, bz
7648	(2-Methylphenyl)phenylmetha-none		C₁₄H₁₂O	131-58-8	196.244		<-18	308	1.1098²⁰		i H₂O; vs EtOH
7649	(3-Methylphenyl)phenylmetha-none		C₁₄H₁₂O	643-65-2	196.244	oil	2	317	1.095²⁰		i H₂O; s EtOH, eth, bz, chl, HOAc
7650	(4-Methylphenyl)phenylmetha-none		C₁₄H₁₂O	134-84-9	196.244	mcl pr	54(2)	228⁷⁰	0.9926⁰		i H₂O; sl EtOH, lig; s eth, bz, chl
7651	Methyl 3-phenylpropanoate	Methyl dihydrocinnamate	C₁₀H₁₂O₂	103-25-3	164.201			238.5	1.0455²⁵		i H₂O; s EtOH, eth, bz, AcOEt
7652	1-(4-Methylphenyl)-1-propa-none		C₁₀H₁₂O	5337-93-9	148.201		7.2	236	0.9926²⁰	1.5278²⁰	i H₂O; s EtOH, eth, ace, bz, CS₂
7653	2-Methyl-1-phenyl-1-propa-none		C₁₀H₁₂O	611-70-1	148.201	liq	-0.7	220	0.9863¹¹	1.5172²⁰	vs eth, EtOH
7654	2-Methyl-3-phenyl-2-propenal		C₁₀H₁₀O	101-39-3	146.185			251.6(1)	1.0407¹⁷	1.6057¹⁷	
7655	Methyl 3-phenyl-2-propynoate		C₁₀H₈O₂	4891-38-7	160.170		18(1)	158⁴⁸	1.0830²⁵	1.5618²⁵	
7656	3-Methyl-1-phenyl-1H-pyrazol-5-amine		C₁₀H₁₁N₃	1131-18-6	173.214		116	333			s H₂O, EtOH, chl; sl bz
7657	2-Methyl-5-phenylpyridine		C₁₂H₁₁N	3256-88-0	169.222			189⁵⁰	1.0590²⁵	1.6055²⁵	
7658	5-Methyl-5-phenyl-2,4,6(1H,3H,5H)-pyrimidine-trione	Phenylmethylbarbituric acid	C₁₁H₁₀N₂O₃	76-94-8	218.208	cry	220				i H₂O; s EtOH, eth, alk
7659	1-Methyl-3-phenyl-2,5-pyrrolidinedione	Phensuximide	C₁₁H₁₁NO₂	86-34-0	189.211	cry (hot al)	72				vs EtOH, MeOH
7660	Methylphenylsilane		C₇H₁₀Si	766-08-5	122.240			140	0.8895²⁰	1.5058²⁰	
7661	Methyl phenyl sulfone		C₇H₈O₂S	3112-85-4	156.203		88(3)				i H₂O; s EtOH, bz, chl; sl ctc
7662	1-Methyl-4-(phenylthio)-benzene		C₁₃H₁₂S	3699-01-2	200.299		15.7	317	1.0986²⁵	1.6225²⁵	i H₂O; s ace, bz
7663	(2-Methylphenyl)thiourea	o-Tolylthiourea	C₈H₁₀N₂S	614-78-8	166.243	nd (dil al, w)	162				vs H₂O, EtOH; sl eth
7664	N-Methyl-N'-phenylthiourea		C₈H₁₀N₂S	2724-69-8	166.243	ta, pl	112.5				vs EtOH
7665	Methyl phosphate	Methyl dihydrogen phosphate	CH₅O₄P	812-00-0	112.022	oil					
7666	Methylphosphine		CH₅P	593-54-4	48.025	col gas		-16			vs eth
7667	Methylphosphonic acid		CH₅O₃P	993-13-5	96.023	hyg pl	108.5	dec			vs H₂O, EtOH, eth; i bz, peth
7668	Methylphosphonic difluoride		CH₃F₂OP	676-99-3	100.005	liq		98	1.3314²⁰		
7669	Methylphosphonofluoridic acid, isopropyl ester	Sarin	C₄H₁₀FO₂P	107-44-8	140.093	liq	-57	147	1.10²⁰		dec H₂O
7670	Methyl phosphorodichloridite	Methyl dichlorophosphite	CH₃Cl₂OP	3279-26-3	132.914	hyg liq	-91	93	1.406	1.4740²⁰	
7671	1-Methylpiperazine		C₅H₁₂N₂	109-01-3	100.162			135(5)		1.4378²⁰	vs H₂O, eth, EtOH
7672	2-Methylpiperazine		C₅H₁₂N₂	109-07-9	100.162	hyg lf (al)	62	153			vs H₂O; s EtOH, eth, bz, chl

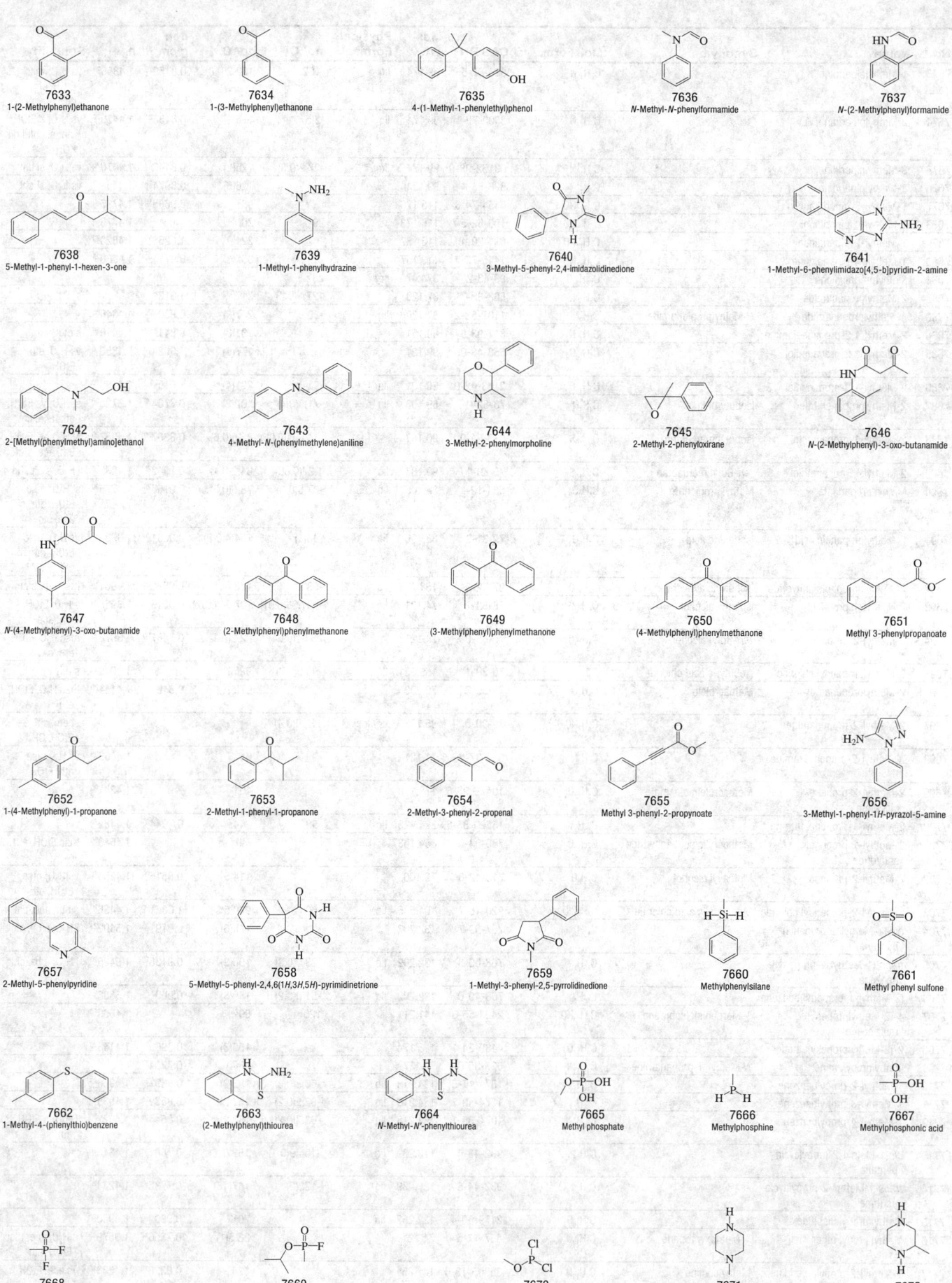

7633
1-(2-Methylphenyl)ethanone

7634
1-(3-Methylphenyl)ethanone

7635
4-(1-Methyl-1-phenylethyl)phenol

7636
N-Methyl-*N*-phenylformamide

7637
N-(2-Methylphenyl)formamide

7638
5-Methyl-1-phenyl-1-hexen-3-one

7639
1-Methyl-1-phenylhydrazine

7640
3-Methyl-5-phenyl-2,4-imidazolidinedione

7641
1-Methyl-6-phenylimidazo[4,5-b]pyridin-2-amine

7642
2-[Methyl(phenylmethyl)amino]ethanol

7643
4-Methyl-*N*-(phenylmethylene)aniline

7644
3-Methyl-2-phenylmorpholine

7645
2-Methyl-2-phenyloxirane

7646
N-(2-Methylphenyl)-3-oxo-butanamide

7647
N-(4-Methylphenyl)-3-oxo-butanamide

7648
(2-Methylphenyl)phenylmethanone

7649
(3-Methylphenyl)phenylmethanone

7650
(4-Methylphenyl)phenylmethanone

7651
Methyl 3-phenylpropanoate

7652
1-(4-Methylphenyl)-1-propanone

7653
2-Methyl-1-phenyl-1-propanone

7654
2-Methyl-3-phenyl-2-propenal

7655
Methyl 3-phenyl-2-propynoate

7656
3-Methyl-1-phenyl-1*H*-pyrazol-5-amine

7657
2-Methyl-5-phenylpyridine

7658
5-Methyl-5-phenyl-2,4,6(1*H*,3*H*,5*H*)-pyrimidinetrione

7659
1-Methyl-3-phenyl-2,5-pyrrolidinedione

7660
Methylphenylsilane

7661
Methyl phenyl sulfone

7662
1-Methyl-4-(phenylthio)benzene

7663
(2-Methylphenyl)thiourea

7664
N-Methyl-*N'*-phenylthiourea

7665
Methyl phosphate

7666
Methylphosphine

7667
Methylphosphonic acid

7668
Methylphosphonic difluoride

7669
Methylphosphonofluoridic acid, isopropyl ester

7670
Methyl phosphorodichloridite

7671
1-Methylpiperazine

7672
2-Methylpiperazine

No.	Name	Synonym	Mol. Form.	CAS RN	Mol. Wt.	Physical Form	mp/°C	bp/°C	den g cm⁻³	n_D	Solubility
7673	1-Methylpiperidine		$C_6H_{13}N$	626-67-5	99.174	liq	-102.7	102.2(0.8)	0.8159[20]	1.4355[20]	vs H_2O; msc EtOH, eth; s ctc
7674	2-Methylpiperidine, (±)-		$C_6H_{13}N$	3000-79-1	99.174	liq	-2.5	118	0.8436[24]	1.4459[20]	vs H_2O; s EtOH, eth; sl chl; i dil KOH
7675	3-Methylpiperidine, (±)-		$C_6H_{13}N$	53152-98-0	99.174	liq	-27.2(0.3)	125(2)	0.8446[26]	1.4470[20]	vs H_2O; sl chl
7676	4-Methylpiperidine		$C_6H_{13}N$	626-58-4	99.174			126(6)	0.8674[25]	1.4458[20]	vs H_2O; sl chl
7677	1-Methyl-3-piperidinol		$C_6H_{13}NO$	3554-74-3	115.173			93[26]	0.9635[16]	1.4735[20]	
7678	1-Methyl-4-piperidinol		$C_6H_{13}NO$	106-52-5	115.173		29	200		1.4775[20]	
7679	1-Methyl-2-piperidinone		$C_6H_{11}NO$	931-20-4	113.157			221	1.0263[25]	1.4820[20]	
7680	1-Methyl-4-piperidinone		$C_6H_{11}NO$	1445-73-4	113.157			85[45]	0.971[25]	1.4580[25]	
7681	Methylprednisolone		$C_{22}H_{30}O_5$	83-43-2	374.470	cry	232				
7682	2-Methylpropanamide		C_4H_9NO	563-83-7	87.120		127.1(0.4)	217	1.013[20]		s chl
7683	N-Methylpropanamide	N-Methylpropionamide	C_4H_9NO	1187-58-2	87.120	liq	-30.9	207(3)	0.9305[25]	1.4345[25]	
7684	2-Methyl-1,2-propanediamine		$C_4H_{12}N_2$	811-93-8	88.151			131(3)	0.841[25]	1.4410[20]	s ctc
7685	2-Methyl-1,2-propanediol		$C_4H_{10}O_2$	558-43-0	90.121			177(5)	1.0024[20]	1.4350[20]	vs H_2O, eth, EtOH
7686	2-Methyl-1,3-propanediol		$C_4H_{10}O_2$	2163-42-0	90.121	liq	-91	221(4)	1.015[20]	1.4450[20]	
7687	2-Methylpropanenitrile	Isobutyronitrile	C_4H_7N	78-82-0	69.106	liq	-71.5(0.4)	102(2)	0.7704[20]	1.3720[20]	sl H_2O; vs EtOH, eth, ace, chl
7688	2-Methyl-1-propanethiol	Isobutyl mercaptan	$C_4H_{10}S$	513-44-0	90.187		-144.84(0.06)	88.5(0.6)	0.8357[20]	1.4387[20]	sl H_2O; vs EtOH, eth, ace; s ctc
7689	2-Methyl-2-propanethiol	tert-Butyl mercaptan	$C_4H_{10}S$	75-66-1	90.187	liq	1.27(0.06)	64.2(0.1)	0.7943[25]	1.4232[20]	i H_2O; s ctc, hp
7690	Methyl propanoate	Methyl propionate	$C_4H_8O_2$	554-12-1	88.106	liq	-87.5(0.5)	78.6(0.2)	0.9150[20]	1.3775[20]	sl H_2O; msc EtOH, eth; s ace, ctc
7691	2-Methylpropanoic acid	Isobutyric acid	$C_4H_8O_2$	79-31-2	88.106	liq	-46(1)	154.4(0.2)	0.9681[20]	1.3930[20]	vs H_2O; msc EtOH, eth; sl ctc
7692	2-Methylpropanoic anhydride	Isobutyric anhydride	$C_8H_{14}O_3$	97-72-3	158.195	liq	-53.5	183	0.9535[20]	1.4061[19]	msc eth; s chl
7693	2-Methyl-1-propanol	Isobutyl alcohol	$C_4H_{10}O$	78-83-1	74.121	liq	-101.96(0.01)	107.84(0.07)	0.8018[20]	1.3955[20]	s H_2O, EtOH, eth, ace, ctc
7694	2-Methyl-2-propanol	tert-Butyl alcohol	$C_4H_{10}O$	75-65-0	74.121		25.81(0.04)	82.3(0.1)	0.7887[20]	1.3878[20]	msc H_2O, EtOH, cth; s chl
7695	2-Methylpropanoyl chloride	Isobutyric acid chloride	C_4H_7ClO	79-30-1	106.551	liq	-90.0(0.7)	92(5)		1.4079[20]	s eth
7696	2-Methylpropenal	Methacrolein	C_4H_6O	78-85-3	70.090			67(5)	0.840[25]	1.4144[20]	msc H_2O, EtOH, eth
7697	2-Methyl-2-propenamide		C_4H_7NO	79-39-0	85.105	cry (bz)	112.0(0.4)				sl eth, chl; s EtOH, CH_2Cl_2
7698	N-Methyl-2-propen-1-amine		C_4H_9N	627-37-2	71.121			64		1.4065[20]	vs H_2O, ace, eth, EtOH
7699	2-Methyl-2-propene-1,1-diol diacetate	Methacrolein diacetate	$C_8H_{12}O_4$	10476-95-6	172.179			191		1.4241[20]	
7700	2-Methyl-1-propene, tetramer		$C_{16}H_{32}$	15220-85-6	224.425	liq	-98	244	0.7944[20]	1.4482[20]	
7701	2-Methyl-2-propenoic anhydride	Methacrylic acid anhydride	$C_8H_{10}O_3$	760-93-0	154.163			89[5]		1.4540[20]	msc EtOH, eth
7702	2-Methyl-2-propenol	Methallyl alcohol	C_4H_8O	513-42-8	72.106			114.5	0.8515[20]	1.4255[20]	vs H_2O; msc EtOH, eth
7703	2-Methyl-2-propenoyl chloride	Methacrylic acid chloride	C_4H_5ClO	920-46-7	104.535	liq	-60	97(25)	1.0871[20]	1.4435[20]	s eth, ace, chl
7704	cis-(1-Methyl-1-propenyl)-benzene		$C_{10}H_{12}$	767-99-7	132.202			175(5)	0.9191[25]	1.5402[25]	i H_2O; s bz, chl
7705	trans-(1-Methyl-1-propenyl)-benzene		$C_{10}H_{12}$	768-00-3	132.202	liq	-23.5(0.3)	196(3)	0.9138[25]	1.5425[25]	i H_2O; s bz, chl
7706	(2-Methyl-1-propenyl)benzene		$C_{10}H_{12}$	768-49-0	132.202	liq	-51.0(0.2)	189(3)	0.900[20]	1.5388[20]	
7707	4-(2-Methylpropenyl)-morpholine	1-Morpholinoisobutene	$C_8H_{15}NO$	2403-55-6	141.211		120	89[20]		1.4663[20]	
7708	2-(2-Methylpropoxy)ethanol		$C_6H_{14}O_2$	4439-24-1	118.174			155(3)	0.8900[20]	1.4143[20]	
7709	Methylpropylamine	N-Methyl-1-propanamine	$C_4H_{11}N$	627-35-0	73.137			63(4)	0.7204[17]		
7710	1-Methyl-2-propylbenzene		$C_{10}H_{14}$	1074-17-5	134.218	liq	-60.3(0.2)	185(2)	0.8697[25]	1.4996[20]	
7711	1-Methyl-3-propylbenzene		$C_{10}H_{14}$	1074-43-7	134.218	liq	-82.5(0.2)	182(2)	0.8569[25]	1.4935[20]	
7712	1-Methyl-4-propylbenzene		$C_{10}H_{14}$	1074-55-1	134.218	liq	-63.7(0.2)	183(2)	0.8544[25]	1.4922[20]	i H_2O; s EtOH, eth
7713	cis-1-Methyl-2-propylcyclo-pentane		C_9H_{18}	932-43-4	126.239	liq	-104.9(0.7)	152.6	0.7881[25]	1.4343[20]	
7714	trans-1-Methyl-2-propylcyclo-pentane		C_9H_{18}	932-44-5	126.239	liq	-123(2)	147(16)	0.7735[25]	1.4274[20]	
7715	Methyl propyl disulfide		$C_4H_{10}S_2$	2179-60-4	122.252	liq		70[43]	0.980	1.5080[20]	
7716	Methyl propyl ether	1-Methoxypropane	$C_4H_{10}O$	557-17-5	74.121			38.5(1)	0.7356[13]	1.3579[25]	s H_2O, ace; msc EtOH, eth
7717	1-Methyl-2-propylpiperidine, (S)-	Methylconiine	$C_9H_{19}N$	35305-13-6	141.254			174	0.8326[22]	1.4538[12]	vs ace, EtOH

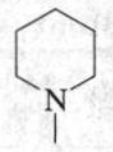

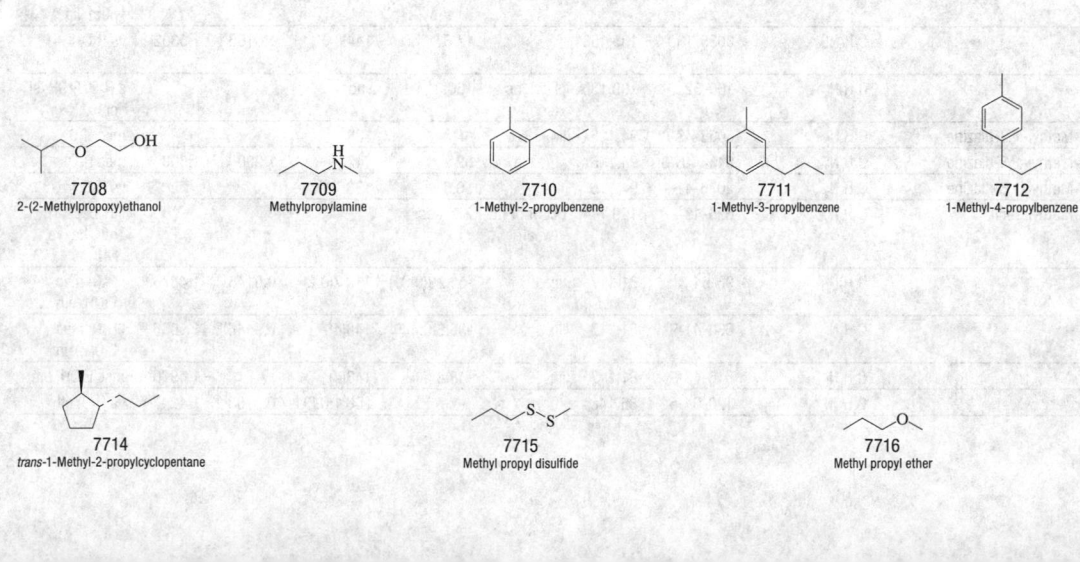

7673
1-Methylpiperidine

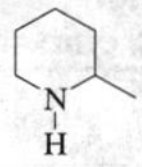

7674
2-Methylpiperidine, (±)-

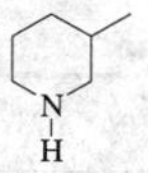

7675
3-Methylpiperidine, (±)-

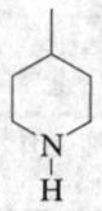

7676
4-Methylpiperidine

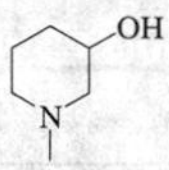

7677
1-Methyl-3-piperidinol

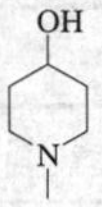

7678
1-Methyl-4-piperidinol

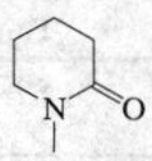

7679
1-Methyl-2-piperidinone

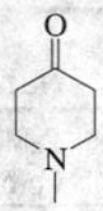

7680
1-Methyl-4-piperidinone

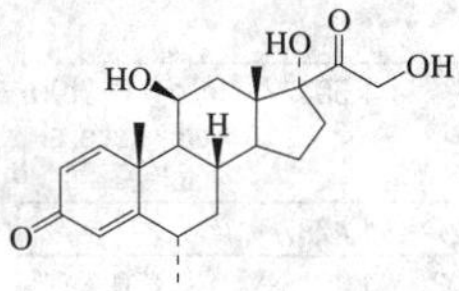

7681
Methylprednisolone

7682
2-Methylpropanamide

7683
N-Methylpropanamide

7684
2-Methyl-1,2-propanediamine

7685
2-Methyl-1,2-propanediol

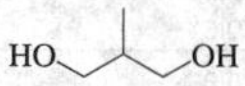

7686
2-Methyl-1,3-propanediol

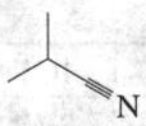

7687
2-Methylpropanenitrile

7688
2-Methyl-1-propanethiol

7689
2-Methyl-2-propanethiol

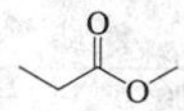

7690
Methyl propanoate

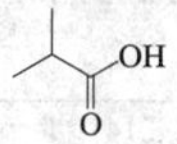

7691
2-Methylpropanoic acid

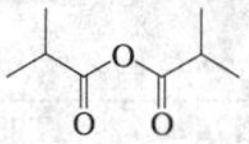

7692
2-Methylpropanoic anhydride

7693
2-Methyl-1-propanol

7694
2-Methyl-2-propanol

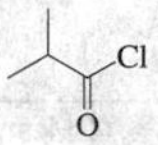

7695
2-Methylpropanoyl chloride

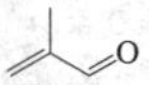

7696
2-Methylpropenal

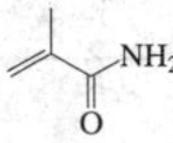

7697
2-Methyl-2-propenamide

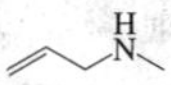

7698
N-Methyl-2-propen-1-amine

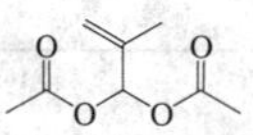

7699
2-Methyl-2-propene-1,1-diol diacetate

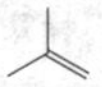

7700
2-Methyl-1-propene, tetramer

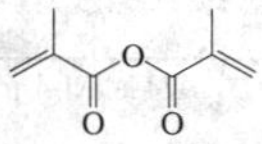

7701
2-Methyl-2-propenoic anhydride

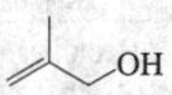

7702
2-Methyl-2-propenol

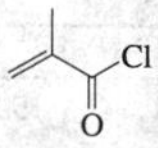

7703
2-Methyl-2-propenoyl chloride

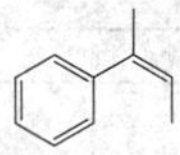

7704
cis-(1-Methyl-1-propenyl)benzene

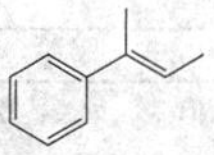

7705
trans-(1-Methyl-1-propenyl)benzene

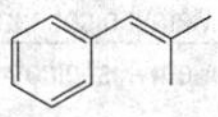

7706
(2-Methyl-1-propenyl)benzene

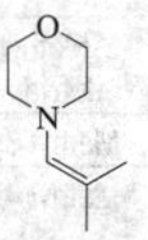

7707
4-(2-Methylpropenyl)morpholine

7708
2-(2-Methylpropoxy)ethanol

7709
Methylpropylamine

7710
1-Methyl-2-propylbenzene

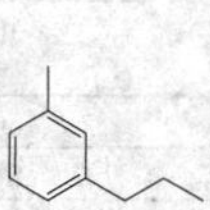

7711
1-Methyl-3-propylbenzene

7712
1-Methyl-4-propylbenzene

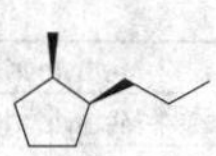

7713
cis-1-Methyl-2-propylcyclopentane

7714
trans-1-Methyl-2-propylcyclopentane

7715
Methyl propyl disulfide

7716
Methyl propyl ether

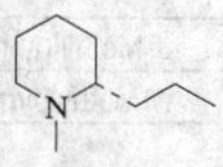

7717
1-Methyl-2-propylpiperidine, (S)-

No.	Name	Synonym	Mol. Form.	CAS RN	Mol. Wt.	Physical Form	mp/°C	bp/°C	den g cm^{-3}	n_D	Solubility
7718	2-Methyl-2-propyl-1,3-propanediol		C$_7$H$_{16}$O$_2$	78-26-2	132.201	cry (hx)	62.5	234			s H$_2$O, hx; sl chl
7719	2-Methyl-2-propyl-1,3-propanediol dicarbamate	Meprobamate	C$_9$H$_{18}$N$_2$O$_4$	57-53-4	218.250	cry (w)	105(3)				vs bz, eth, EtOH
7720	Methyl propyl sulfide		C$_4$H$_{10}$S	3877-15-4	90.187	liq	-112.96(0.06)	95.5(0.7)	0.8424[20]	1.4442[20]	s H$_2$O, EtOH, eth, ace
7721	N-Methyl-2-propyn-1-amine		C$_4$H$_7$N	35161-71-8	69.106			83	0.819[25]	1.4332[20]	
7722	N-Methyl-N-2-propynylbenzenemethanamine	Pargyline	C$_{11}$H$_{13}$N	555-57-7	159.228			96[11]	0.944[25]	1.5213[20]	
7723	3-Methylpyrazinamine	2-Amino-3-methylpyrazine	C$_5$H$_7$N$_3$	19838-08-5	109.130	nd (hx/AcOEt)	174				
7724	2-Methylpyrazine		C$_5$H$_6$N$_2$	109-08-0	94.115	liq	-29	129(4)	1.03[20]	1.5042[20]	msc H$_2$O, EtOH, eth; s ace; sl ctc
7725	1-Methyl-1H-pyrazole		C$_4$H$_6$N$_2$	930-36-9	82.104			127	0.9929[13]	1.4787[13]	
7726	3-Methyl-1H-pyrazole		C$_4$H$_6$N$_2$	1453-58-3	82.104		36.5	204	1.0203[16]	1.4915[20]	msc H$_2$O, EtOH, eth
7727	4-Methyl-1H-pyrazole	Fomepizole	C$_4$H$_6$N$_2$	7554-65-6	82.104			206	1.015[20]		
7728	3-Methyl-2-pyrazolin-5-one		C$_4$H$_6$N$_2$O	108-26-9	98.103		215				vs H$_2$O; sl EtOH
7729	1-Methylpyrene		C$_{17}$H$_{12}$	2381-21-7	216.277		74(2)	410			
7730	2-Methylpyrene		C$_{17}$H$_{12}$	3442-78-2	216.277	fl (EtOH)	143	409.8			
7731	3-Methylpyridazine	3-Methyl-1,2-diazine	C$_5$H$_6$N$_2$	1632-76-4	94.115		184	214	1.0450[26]	1.5145[20]	
7732	3-Methyl-2-pyridinamine	2-Amino-3-picoline	C$_6$H$_8$N$_2$	1603-40-3	108.141	hyg	33.5	222.8(0.5)			vs H$_2$O; s EtOH, eth, ace, bz, ctc; sl lig
7733	4-Methyl-2-pyridinamine	2-Amino-4-picoline	C$_6$H$_8$N$_2$	695-34-1	108.141	lf or pl (lig)	100	116[11]			vs H$_2$O; s EtOH, eth, ace, bz; i lig; sl chl
7734	5-Methyl-2-pyridinamine		C$_6$H$_8$N$_2$	1603-41-4	108.141		77.3(0.2)	227			
7735	6-Methyl-2-pyridinamine	2-Amino-6-picoline	C$_6$H$_8$N$_2$	1824-81-3	108.141	hyg (lig)	41	208.5			vs H$_2$O; s EtOH, eth, ace, bz, lig
7736	N-Methyl-2-pyridinamine		C$_6$H$_8$N$_2$	4597-87-9	108.141		15	202(7)	1.048[29]		s H$_2$O, bz; vs EtOH, eth, HOAc
7737	N-Methyl-4-pyridinamine		C$_6$H$_8$N$_2$	1121-58-0	108.141	pl (eth)	118.8				vs H$_2$O, ace, eth, EtOH
7738	2-Methylpyridine	2-Picoline	C$_6$H$_7$N	109-06-8	93.127	liq	-66.65(0.03)	129.4(0.2)	0.9443[20]	1.4957[20]	vs H$_2$O, ace; msc EtOH, eth; s ctc
7739	3-Methylpyridine	3-Picoline	C$_6$H$_7$N	108-99-6	93.127	liq	-18.1(0.3)	144.1(0.1)	0.9566[20]	1.5040[20]	msc H$_2$O, EtOH, eth; vs ace; s ctc
7740	4-Methylpyridine	4-Picoline	C$_6$H$_7$N	108-89-4	93.127	liq	3.68(0.02)	145.3(0.1)	0.9548[20]	1.5037[20]	msc H$_2$O, EtOH, eth; s ace, ctc
7741	6-Methyl-2-pyridinecarboxaldehyde		C$_7$H$_7$NO	1122-72-1	121.137		32	77[12]			
7742	Methyl 3-pyridinecarboxylate	Methyl nicotinate	C$_7$H$_7$NO$_2$	93-60-7	137.137	cry	39.5(0.4)	204			s H$_2$O, EtOH, bz
7743	Methyl 4-pyridinecarboxylate	Methyl isonicotinate	C$_7$H$_7$NO$_2$	2459-09-8	137.137		10.2(0.9)	209(3)	1.1599[20]	1.5135[20]	sl H$_2$O, ctc; s EtOH, eth, bz
7744	2-Methylpyridine-1-oxide		C$_6$H$_7$NO	931-19-1	109.126		49.8(0.5)	260			
7745	3-Methylpyridine-1-oxide		C$_6$H$_7$NO	1003-73-2	109.126		41.0(0.5)	148[15]			s chl
7746	4-Methylpyridine-1-oxide		C$_6$H$_7$NO	1003-67-4	109.126		185.8				
7747	1-Methyl-2(1H)-pyridinone		C$_6$H$_7$NO	694-85-9	109.126	nd	31	253(13)	1.1120[20]		msc H$_2$O; sl peth, lig
7748	1-(6-Methyl-3-pyridinyl)-ethanone		C$_8$H$_9$NO	36357-38-7	135.163		17.6	144[50]	1.0168[25]	1.5302[25]	vs H$_2$O
7749	4-Methyl-2-pyrimidinamine		C$_5$H$_7$N$_3$	108-52-1	109.130	pl (w), nd (sub)	160.3	sub			s H$_2$O, EtOH; sl chl
7750	2-Methylpyrimidine	2-Methyl-1,3-diazine	C$_5$H$_6$N$_2$	5053-43-0	94.115	liq	-4	138			msc H$_2$O
7751	4-Methylpyrimidine	4-Methyl-1,3-diazine	C$_5$H$_6$N$_2$	3438-46-8	94.115		32	142	1.030[16]	1.500[20]	msc H$_2$O
7752	5-Methylpyrimidine	5-Methyl-1,3-diazine	C$_5$H$_6$N$_2$	2036-41-1	94.115		30.5	153			vs H$_2$O
7753	6-Methyl-2,4(1H,3H)-pyrimidinedione	6-Methyluracil	C$_5$H$_6$N$_2$O$_2$	626-48-2	126.114	oct pr or nd (w, al)	275 dec				s H$_2$O, EtOH; sl eth, tfa; vs NH$_3$
7754	1-Methylpyrrole		C$_5$H$_7$N	96-54-8	81.117	liq	-56.23(0.01)	112.7(0.2)	0.9145[15]	1.4875[20]	i H$_2$O; msc EtOH, eth
7755	2-Methylpyrrole		C$_5$H$_7$N	636-41-9	81.117	liq	-35.6(0.4)	148(4)	0.9446[15]	1.5035[16]	i H$_2$O; msc EtOH, eth
7756	3-Methylpyrrole		C$_5$H$_7$N	616-43-3	81.117	liq	-48.4	143(4)		1.4970[20]	msc EtOH, eth
7757	N-Methylpyrrolidine		C$_5$H$_{11}$N	120-94-5	85.148			206.1(0.3)	0.8188[20]	1.4247[20]	vs H$_2$O, eth

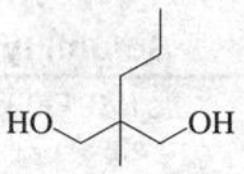

7718
2-Methyl-2-propyl-1,3-propanediol

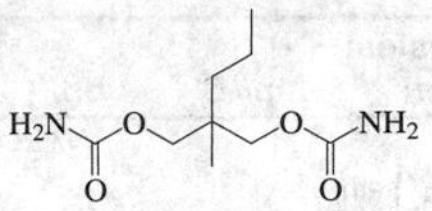

7719
2-Methyl-2-propyl-1,3-propanediol dicarbamate

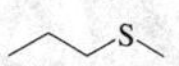

7720
Methyl propyl sulfide

7721
N-Methyl-2-propyn-1-amine

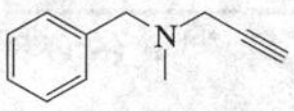

7722
N-Methyl-*N*-2-propynylbenzenemethanamine

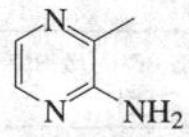

7723
3-Methylpyrazinamine

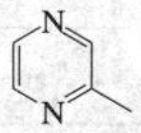

7724
2-Methylpyrazine

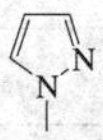

7725
1-Methyl-1*H*-pyrazole

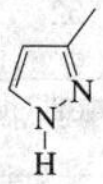

7726
3-Methyl-1*H*-pyrazole

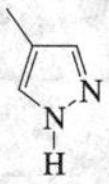

7727
4-Methyl-1*H*-pyrazole

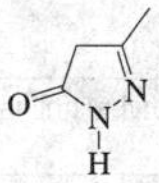

7728
3-Methyl-2-pyrazolin-5-one

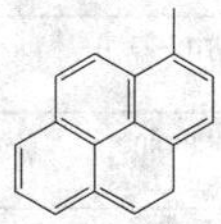

7729
1-Methylpyrene

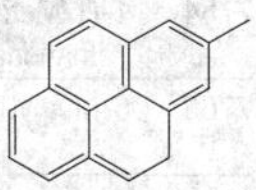

7730
2-Methylpyrene

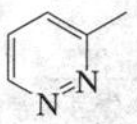

7731
3-Methylpyridazine

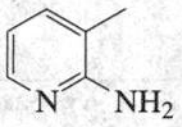

7732
3-Methyl-2-pyridinamine

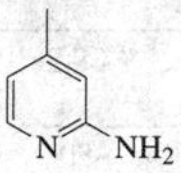

7733
4-Methyl-2-pyridinamine

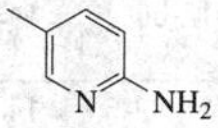

7734
5-Methyl-2-pyridinamine

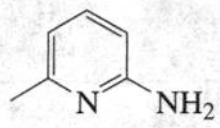

7735
6-Methyl-2-pyridinamine

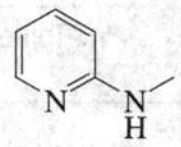

7736
N-Methyl-2-pyridinamine

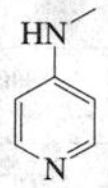

7737
N-Methyl-4-pyridinamine

7738
2-Methylpyridine

7739
3-Methylpyridine

7740
4-Methylpyridine

7741
6-Methyl-2-pyridinecarboxaldehyde

7742
Methyl 3-pyridinecarboxylate

7743
Methyl 4-pyridinecarboxylate

7744
2-Methylpyridine-1-oxide

7745
3-Methylpyridine-1-oxide

7746
4-Methylpyridine-1-oxide

7747
1-Methyl-2(1*H*)-pyridinone

7748
1-(6-Methyl-3-pyridinyl)ethanone

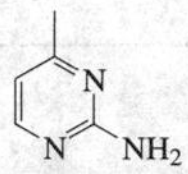

7749
4-Methyl-2-pyrimidinamine

7750
2-Methylpyrimidine

7751
4-Methylpyrimidine

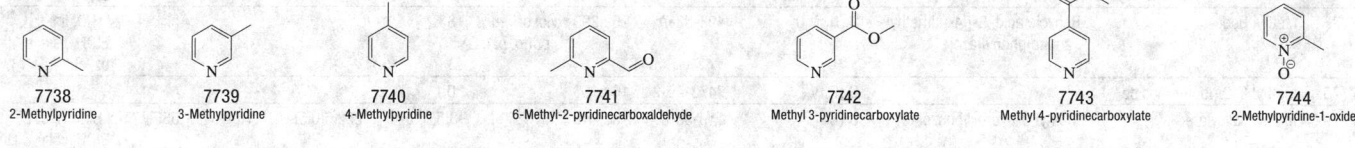

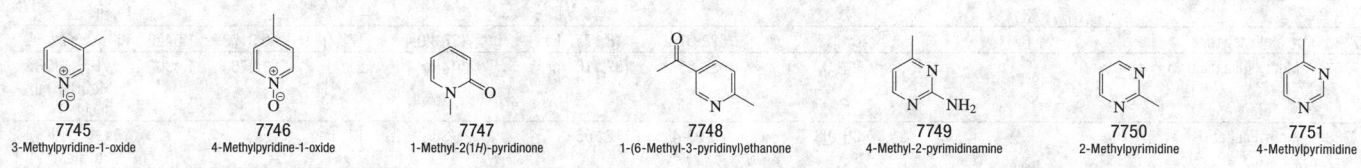

7752
5-Methylpyrimidine

7753
6-Methyl-2,4(1*H*,3*H*)-pyrimidinedione

7754
1-Methylpyrrole

7755
2-Methylpyrrole

7756
3-Methylpyrrole

7757
N-Methylpyrrolidine

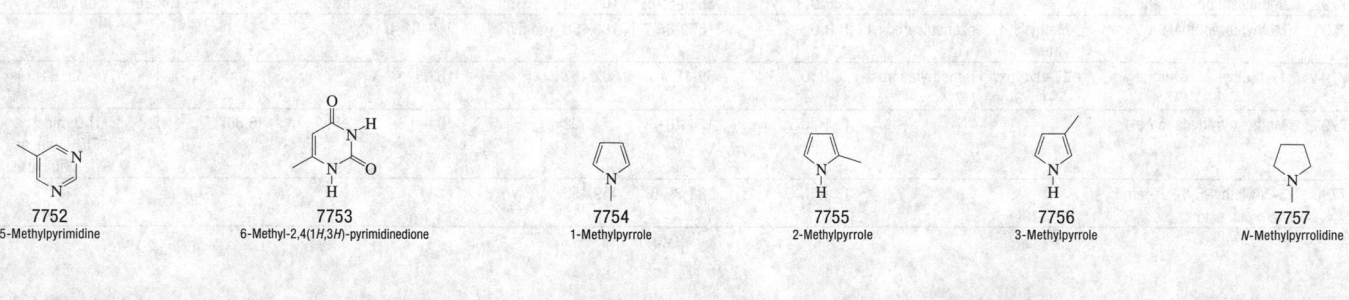

No.	Name	Synonym	Mol. Form.	CAS RN	Mol. Wt.	Physical Form	mp/°C	bp/°C	den g cm⁻³	n_D	Solubility
7758	1-Methyl-2,5-pyrrolidinedione		C₅H₇NO₂	1121-07-9	113.116	nd (eth- peth, al, ace)	71	234			s H₂O, EtOH; vs eth
7759	N-Methyl-2-pyrrolidinethione		C₅H₉NS	10441-57-3	115.197	oil		100⁰·⁰⁸			
7760	5-Methyl-2-pyrrolidinone		C₅H₉NO	108-27-0	99.131		43	248	1.0458²⁰		
7761	1-(1-Methyl-2-pyrrolidinyl)-2-propanone, (R)-	Hygrine	C₈H₁₅NO	496-49-1	141.211			76.5¹¹		1.4555²⁰	vs EtOH, chl
7762	3-(1-Methyl-2-pyrrolidinyl)-pyridine, (±)-		C₁₀H₁₄N₂	22083-74-5	162.231			244	1.0082²⁰	1.5289²⁰	msc H₂O; vs EtOH, eth, chl; s lig
7763	N-Methyl-2-pyrrolidinone	1-Methyl-2-pyrrolidinone	C₅H₉NO	872-50-4	99.131	liq	-24.0(0.6)	204.2(0.3)	1.0230²⁵	1.4684²⁰	vs H₂O; s eth, ace, chl
7764	1-(1-Methyl-1H-pyrrol-2-yl)-ethanone		C₇H₉NO	932-16-1	123.152			201²⁵²	1.0445¹⁵	1.5403¹⁵	s EtOH, bz, chl
7765	6-Methyl-8-quinolinamine	8-Amino-6-methylquinoline	C₁₀H₁₀N₂	68420-93-9	158.199	nd	73	sub			vs ace, bz, eth, EtOH
7766	2-Methylquinoline	Quinaldine	C₁₀H₉N	91-63-4	143.185	col oily liq	-2.67(0.05)	247.4(0.4)	1.06²⁵	1.6116²⁰	sl H₂O; s EtOH, eth, ace, ctc, chl
7767	3-Methylquinoline		C₁₀H₉N	612-58-8	143.185	pr	16.5	251(7)	1.0673²⁰	1.6171²⁰	vs ace, eth, EtOH
7768	4-Methylquinoline	Lepidine	C₁₀H₉N	491-35-0	143.185	col oily liq	9.5(0.3)	266(2)	1.083²⁰	1.6200²⁰	sl H₂O; s EtOH, eth, ace; i alk
7769	5-Methylquinoline		C₁₀H₉N	7661-55-4	143.185	col cry	19	257(5)	1.0832²⁰	1.6219²⁰	sl H₂O; s ace; msc EtOH, eth
7770	6-Methylquinoline		C₁₀H₉N	91-62-3	143.185	col oily liq	-22	265(1)	1.0654²⁰	1.6157²⁰	sl H₂O; s EtOH, eth, ace
7771	7-Methylquinoline	m-Toluquinoline	C₁₀H₉N	612-60-2	143.185	ye cry	39(2)	258(1)	1.0609²⁰	1.6150²⁰	sl H₂O; s EtOH, eth, ace
7772	8-Methylquinoline		C₁₀H₉N	611-32-5	143.185	col liq	-80	247.4(0.7)	1.0719²⁰	1.6164²⁰	sl H₂O; s ace; msc EtOH, eth
7773	2-Methyl-8-quinolinol		C₁₀H₉NO	826-81-3	159.184		73.8	267			i H₂O; s EtOH, eth, bz, ctc
7774	1-Methyl-2(1H)-quinolinone		C₁₀H₉NO	606-43-9	159.184	nd (lig)	74	325			sl H₂O, lig; s EtOH, eth, ace; vs bz
7775	1-Methyl-4(1H)-quinolinone	Echinopsine	C₁₀H₉NO	83-54-5	159.184	α-nd (bz); β-cry (al)	152				s H₂O; vs EtOH, bz, chl; sl eth
7776	2-Methylquinoxaline		C₉H₈N₂	7251-61-8	144.173	ye cry	180.5	244			msc H₂O, eth, ace, bz; vs EtOH; s ctc
7777	Methyl Red	Benzoic acid, 2-[[4-(dimethyl-amino)phenyl]azo]-	C₁₅H₁₅N₃O₂	493-52-7	269.299	viol or red pr (to, bz)	183				sl H₂O, lig; s EtOH; vs ace, bz, chl
7778	Methyl β-D-ribofuranoside		C₆H₁₂O₅	7473-45-2	164.156		80				
7779	Methyl salicylate	Methyl 2-hydroxybenzoate	C₈H₈O₃	119-36-8	152.148	liq	-8.5(0.5)	222.6(0.5)	1.181²⁵	1.535²⁰	sl H₂O; vs eth, EtOH, chl
7780	Methylsilane		CH₆Si	992-94-9	46.145	col gas	-156.5	-57.5			
7781	Methyl silyl ether		CH₆OSi	2171-96-2	62.144	col gas	-98.5	-21			
7782	Methylstannane		CH₆Sn	1631-78-3	136.769	col gas		1.4			dec H₂O
7783	Methyl stearate		C₁₉H₃₈O₂	112-61-8	298.504		38.7(0.6)	443	0.8498⁴⁰	1.4367⁴⁰	vs eth, chl
7784	2-Methylstyrene		C₉H₁₀	611-15-4	118.175	liq	-68.5(0.3)	170(1)	0.9077²⁵	1.5437²⁰	i H₂O; s bz, chl
7785	3-Methylstyrene		C₉H₁₀	100-80-1	118.175	liq	-86.3(0.3)	170(1)	0.9076²⁵	1.5411²⁰	i H₂O; s EtOH, eth, bz
7786	4-Methylstyrene		C₉H₁₀	622-97-9	118.175	liq	-37.8(0.4)	172(1)	0.9173²⁵	1.5420	i H₂O; s bz
7787	Methylsuccinic acid		C₅H₈O₄	636-60-2	132.116	pr	114(4)	dec	1.4200⁰	1.4303	vs H₂O, EtOH, MeOH; s eth; sl chl
7788	Methyl sulfate		CH₄O₄S	75-93-4	112.106		<-30	135 dec			vs H₂O, eth, EtOH
7789	(Methylsulfinyl)benzene		C₇H₈OS	1193-82-4	140.203		32.0	263.5		1.5885²⁰	
7790	1-(Methylsulfinyl)decane	Decyl methyl sulfoxide	C₁₁H₂₄OS	3079-28-5	204.373	cry	52.5				
7791	3-Methyl sulfolane		C₅H₁₀O₂S	872-93-5	134.197		1	277(13)	1.188²⁵	1.4772²⁰	
7792	(Methylsulfonyl)ethene		C₃H₆O₂S	3680-02-2	106.144			122²⁴	1.2117²⁰	1.4636²⁰	s eth, ace
7793	Methyl terephthalate	Methyl 1,4-benzenedicarboxylate	C₉H₈O₄	1679-64-7	180.158	nd (w)	219.4(0.4)				
7794	17-Methyltestosterone	17-Hydroxy-17-methylandrost-4-en-3-one, (17β)-	C₂₀H₃₀O₂	58-18-4	302.451		163.5				vs eth, EtOH
7795	Methyl tetradecanoate		C₁₅H₃₀O₂	124-10-7	242.398		19.0(0.5)	299(2)	0.8671²⁰	1.425⁴⁵	i H₂O; msc EtOH, eth, ace, bz, chl, ctc
7796	5-N-Methyl-5,6,7,8-tetrahydrofolic acid		C₂₀H₂₅N₇O₆	134-35-0	459.456	cry (w)					

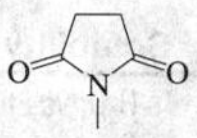

7758
1-Methyl-2,5-pyrrolidinedione

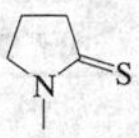

7759
N-Methyl-2-pyrrolidinethione

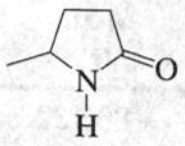

7760
5-Methyl-2-pyrrolidinone

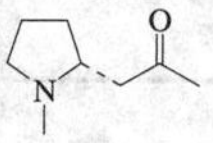

7761
1-(1-Methyl-2-pyrrolidinyl)-2-propanone, (*R*)-

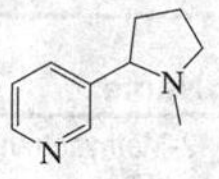

7762
3-(1-Methyl-2-pyrrolidinyl)pyridine, (±)-

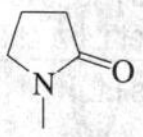

7763
N-Methyl-2-pyrrolidinone

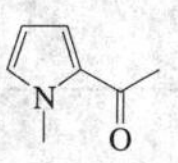

7764
1-(1-Methyl-1*H*-pyrrol-2-yl)ethanone

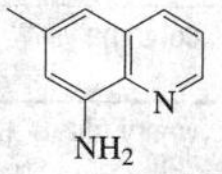

7765
6-Methyl-8-quinolinamine

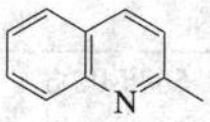

7766
2-Methylquinoline

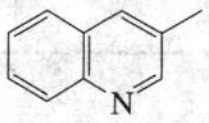

7767
3-Methylquinoline

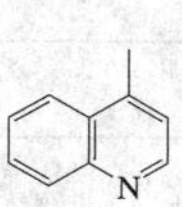

7768
4-Methylquinoline

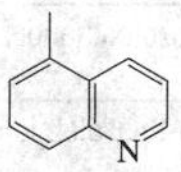

7769
5-Methylquinoline

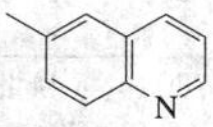

7770
6-Methylquinoline

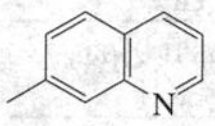

7771
7-Methylquinoline

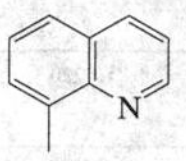

7772
8-Methylquinoline

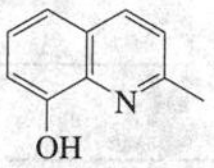

7773
2-Methyl-8-quinolinol

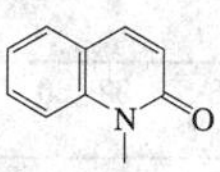

7774
1-Methyl-2(1*H*)-quinolinone

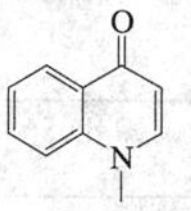

7775
1-Methyl-4(1*H*)-quinolinone

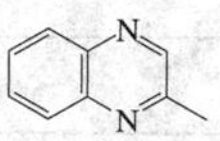

7776
2-Methylquinoxaline

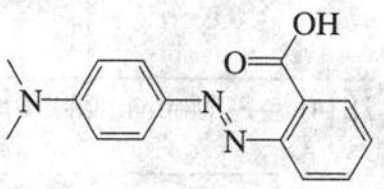

7777
Methyl Red

7778
Methyl β-*D*-ribofuranoside

7779
Methyl salicylate

7780
Methylsilane

7781
Methyl silyl ether

7782
Methylstannane

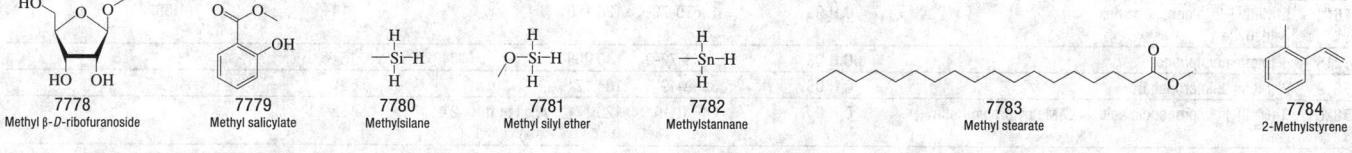

7783
Methyl stearate

7784
2-Methylstyrene

7785
3-Methylstyrene

7786
4-Methylstyrene

7787
Methylsuccinic acid

7788
Methyl sulfate

7789
(Methylsulfinyl)benzene

7790
1-(Methylsulfinyl)decane

7791
3-Methyl sulfolane

7792
(Methylsulfonyl)ethene

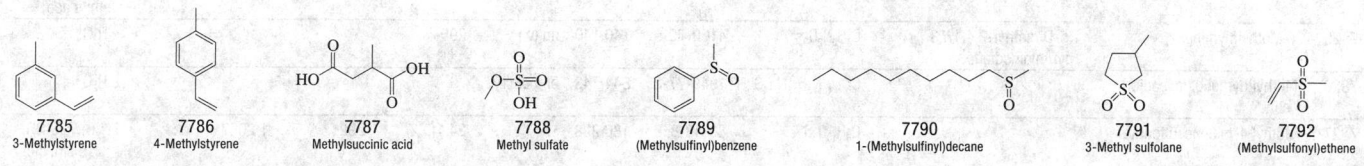

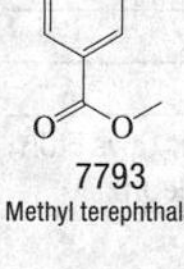

7793
Methyl terephthalate

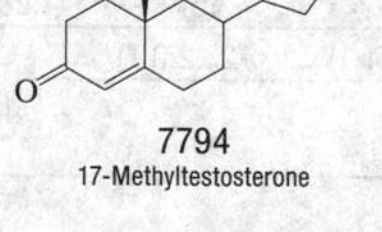

7794
17-Methyltestosterone

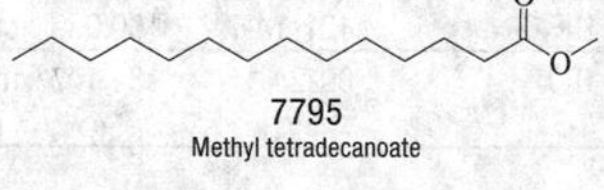

7795
Methyl tetradecanoate

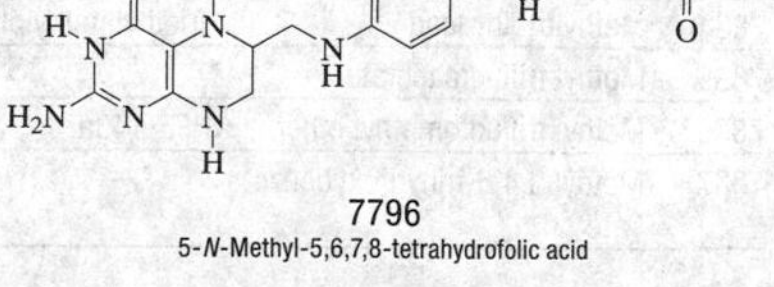

7796
5-*N*-Methyl-5,6,7,8-tetrahydrofolic acid

No.	Name	Synonym	Mol. Form.	CAS RN	Mol. Wt.	Physical Form	mp/°C	bp/°C	den g cm⁻³	n_D	Solubility
7797	2-Methyltetrahydrofuran	2-Methyloxolane	$C_5H_{10}O$	96-47-9	86.132			80(1)	0.8552[20]	1.4059[21]	s H₂O; vs EtOH, eth, ace, bz; sl ctc
7798	N-Methyl-N,2,4,6-tetra-nitroaniline	Tetryl	$C_7H_5N_5O_8$	479-45-8	287.144	ye pr (al)	128(3)	exp	1.57[10]		i H₂O; sl EtOH, eth, chl; s ace, bz, py
7799	4-Methyl-2-thiazolamine	2-Amino-4-methylthiazole	$C_4H_6N_2S$	1603-91-4	114.169		45.5	125[20]			vs H₂O, EtOH, eth
7800	2-Methylthiazole		C_4H_5NS	3581-87-1	99.155			128.77(0.05)		1.510	msc H₂O; s EtOH, ace
7801	4-Methylthiazole		C_4H_5NS	693-95-8	99.155			133.2(0.2)	1.112[25]		s H₂O, EtOH, eth
7802	4-Methyl-5-thiazoleethanol		C_6H_9NOS	137-00-8	143.206	col to pa ye		135[7]	1.196[24]		vs H₂O; s EtOH, eth, bz, chl
7803	4-Methyl-2(3H)-thiazolethione		$C_4H_5NS_2$	5685-06-3	131.220	ye cry (dil al)	89.3	188[3]			vs EtOH
7804	Methylthiirane		C_3H_6S	1072-43-1	74.145	liq	-91	72.5	0.941[20]	1.472[20]	s chl
7805	(Methylthio)acetic acid		$C_3H_6O_2S$	2444-37-3	106.144		13.0	130[27]	1.221[20]	1.495[20]	
7806	2-(Methylthio)aniline		C_7H_9NS	2987-53-3	139.218			234	1.111[25]	1.6239[20]	
7807	4-(Methylthio)aniline		C_7H_9NS	104-96-1	139.218			272.5	1.1379[20]	1.6395[20]	s EtOH, eth, ace, bz
7808	(Methylthio)benzene	Methyl phenyl sulfide	C_7H_8S	100-68-5	124.204			194.3(0.2)	1.0579[20]	1.5868[20]	i H₂O; s EtOH; vs ace
7809	2-(Methylthio)benzothiazole		$C_8H_7NS_2$	615-22-5	181.279	pr (dil al)	52	174[22]			s EtOH, chl
7810	Methyl thiocyanate		C_2H_3NS	556-64-9	73.117	col liq	-2.5	132.9	1.0678[25]	1.4669[25]	sl H₂O; msc EtOH, eth; s ctc
7811	2-(Methylthio)ethanol		C_3H_8OS	5271-38-5	92.160			70[20]	1.063[20]	1.4861[30]	vs H₂O, eth, EtOH
7812	(Methylthio)ethene	Methyl vinyl sulfide	C_3H_6S	1822-74-8	74.145			69(3)	0.9026[20]	1.4837[20]	s eth, ace, chl
7813	[(Methylthio)methyl]benzene		$C_8H_{10}S$	766-92-7	138.230	liq	-30	214(10)	1.0274[20]	1.5620[20]	
7814	4-(Methylthio)-2-oxobutanoic acid		$C_5H_8O_3S$	583-92-6	148.181	oil					
7815	2-Methylthiophene		C_5H_6S	554-14-3	98.167	liq	-63.3(0.2)	112.5(0.4)	1.0193[20]	1.5203[20]	i H₂O; msc EtOH, eth, ace, bz, hp, ctc
7816	3-Methylthiophene		C_5H_6S	616-44-4	98.167	liq	-68.91(0.06)	115.4(0.4)	1.0218[20]	1.5204[20]	i H₂O; msc EtOH, eth, ace, bz; vs chl
7817	5-Methyl-2-thiophenecarbox-aldehyde		C_6H_6OS	13679-70-4	126.176			114[25]		1.5825[20]	s chl
7818	4-(Methylthio)phenol		C_7H_8OS	1073-72-9	140.203		84	154[20]			
7819	3-(Methylthio)propanal		C_4H_8OS	3268-49-3	104.171			62[11]			
7820	3-(Methylthio)propanoic acid	S-Methylpropiothetin	$C_4H_8O_2S$	646-01-5	120.171	ye oil or fl (hx)	21	132[13]			
7821	3-(Methylthio)-1-propene		C_4H_8S	10152-76-8	88.172			94(18)	0.8767[20]	1.4714[20]	
7822	N-Methylthiosemicarbazide	N-Methylhydrazinecarbothio-amide	$C_2H_7N_3S$	6610-29-3	105.162		136.5				s H₂O, EtOH, DMSO; i eth, bz, lig
7823	Methylthiouracil		$C_5H_6N_2OS$	56-04-2	142.179		330 dec	sub			i H₂O; sl EtOH, eth, MeOH, bz
7824	Methylthiourea		$C_2H_6N_2S$	598-52-7	90.147	pr (EtOH)	119.2(0.4)				vs H₂O, EtOH; sl eth; s ace
7825	1-Methylthymine	1,5-Dimethyl-2,4(1H,3H)-pyrimidinedione	$C_6H_8N_2O_2$	4160-72-9	140.140	nd (w)	295				s H₂O
7826	Methylthymol blue, sodium salt		$C_{37}H_{40}N_2O_{13}Na_4S$	1945-77-3	844.743	bl-viol cry					s H₂O
7827	Methyl 4-toluenesulfonate		$C_8H_{10}O_3S$	80-48-8	186.228		24.1(0.2)	292	1.2087[40]		i H₂O; vs EtOH, bz; s eth, ctc; sl lig
7828	Methyltriacetoxysilane	Methylsilanetriol, triacetate	$C_7H_{12}O_6Si$	4253-34-3	220.252		40.5	111[17]	1.1750[20]	1.4083[20]	
7829	6-Methyl-1,2,4-triazine-3,5(2H,4H)-dione	6-Azathymine	$C_4H_5N_3O_2$	932-53-6	127.102	cry (w)	211				s H₂O, EtOH, ace
7830	5-Methyl-[1,2,4]triazolo[1,5-a]pyrimidin-7-ol		$C_6H_6N_4O$	2503-56-2	150.138		>245				
7831	Methyl trichloroacetate		$C_3H_3Cl_3O_2$	598-99-2	177.414	liq	-17.5(0.4)	152(8)	1.4874[20]	1.4572[20]	i H₂O; vs EtOH, eth; s ctc
7832	Methyltrichlorosilane	Trichloromethylsilane	CH_3Cl_3Si	75-79-6	149.480	liq	-75.77(0.05)	66(2)	1.273[20]	1.4106[20]	dec H₂O, EtOH
7833	Methyl tridecanoate		$C_{14}H_{28}O_2$	1731-88-0	228.371		5.9(0.7)	92[1]		1.4405[20]	msc EtOH; s ctc
7834	Methyltriethyllead	Triethylmethylplumbane	$C_7H_{18}Pb$	1762-28-3	309.4	col liq		70[16]	1.71[20]		
7835	Methyl trifluoroacetate		$C_3H_3F_3O_2$	431-47-0	128.050			43(5)	1.28[20]		
7836	Methyl trifluoromethyl ether	HFE-143a	$C_2H_3F_3O$	421-14-7	100.039	col gas	-149.1(1)	-25.2(0.7)			
7837	Methyl 3,4,5-trihydroxybenzo-ate		$C_8H_8O_5$	99-24-1	184.147	mcl pr (MeOH)	202				sl H₂O; vs EtOH, MeOH

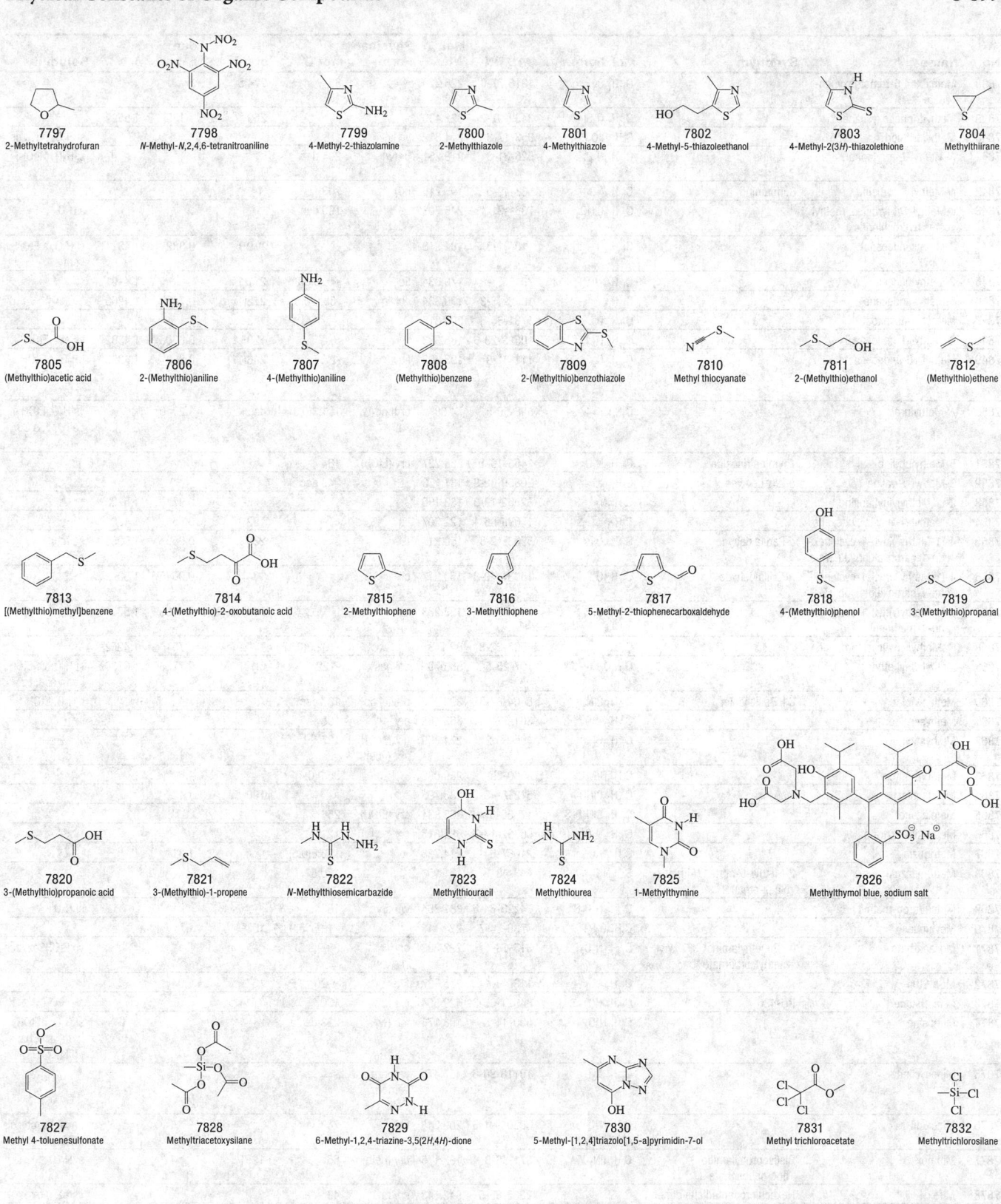

7797
2-Methyltetrahydrofuran

7798
N-Methyl-*N*,2,4,6-tetranitroaniline

7799
4-Methyl-2-thiazolamine

7800
2-Methylthiazole

7801
4-Methylthiazole

7802
4-Methyl-5-thiazoleethanol

7803
4-Methyl-2(3*H*)-thiazolethione

7804
Methylthiirane

7805
(Methylthio)acetic acid

7806
2-(Methylthio)aniline

7807
4-(Methylthio)aniline

7808
(Methylthio)benzene

7809
2-(Methylthio)benzothiazole

7810
Methyl thiocyanate

7811
2-(Methylthio)ethanol

7812
(Methylthio)ethene

7813
[(Methylthio)methyl]benzene

7814
4-(Methylthio)-2-oxobutanoic acid

7815
2-Methylthiophene

7816
3-Methylthiophene

7817
5-Methyl-2-thiophenecarboxaldehyde

7818
4-(Methylthio)phenol

7819
3-(Methylthio)propanal

7820
3-(Methylthio)propanoic acid

7821
3-(Methylthio)-1-propene

7822
N-Methylthiosemicarbazide

7823
Methylthiouracil

7824
Methylthiourea

7825
1-Methylthymine

7826
Methylthymol blue, sodium salt

7827
Methyl 4-toluenesulfonate

7828
Methyltriacetoxysilane

7829
6-Methyl-1,2,4-triazine-3,5(2*H*,4*H*)-dione

7830
5-Methyl-[1,2,4]triazolo[1,5-a]pyrimidin-7-ol

7831
Methyl trichloroacetate

7832
Methyltrichlorosilane

7833
Methyl tridecanoate

7834
Methyltriethyllead

7835
Methyl trifluoroacetate

7836
Methyl trifluoromethyl ether

7837
Methyl 3,4,5-trihydroxybenzoate

No.	Name	Synonym	Mol. Form.	CAS RN	Mol. Wt.	Physical Form	mp/°C	bp/°C	den g cm⁻³	n_D	Solubility
7838	Methyl 3,4,5-trimethoxyben-zoate		$C_{11}H_{14}O_5$	1916-07-0	226.226		83	274.5			
7839	Methyltriphenoxysilane		$C_{19}H_{18}O_3Si$	3439-97-2	322.430			269^{100}	1.135^{20}	1.5599^{20}	
7840	Methyl trithion		$C_9H_{12}ClO_2PS_3$	953-17-3	314.812	ye liq	-18				sl H_2O; misc os
7841	N-Methyl-L-tryptophan	L-Abrine	$C_{12}H_{14}N_2O_2$	526-31-8	218.251	pr (w)	295 dec				sl H_2O, MeOH; i eth; s alk
7842	N-Methyl-L-tyrosine	Surinamine	$C_{10}H_{13}NO_3$	537-49-5	195.215	nd	293				
7843	α-Methyl-DL-tyrosine, methyl ester, hydrochloride		$C_{11}H_{16}ClNO_3$	7361-31-1	245.703		190 dec				s H_2O
7844	2-Methylundecanal		$C_{12}H_{24}O$	110-41-8	184.318			119^{16}	0.832^{15}	1.4321^{20}	sl H_2O; s EtOH, eth
7845	2-Methylundecane		$C_{12}H_{26}$	7045-71-8	170.334	liq	-46.8(0.2)	211(1)		1.4191^{20}	
7846	3-Methylundecane		$C_{12}H_{26}$	1002-43-3	170.334	col liq	-58.0(0.5)	212(1)	0.7485^{25}	1.4208^{25}	
7847	Methyl undecanoate		$C_{12}H_{24}O_2$	1731-86-8	200.318			123^{10}			
7848	2-Methyl-1-undecanol		$C_{12}H_{26}O$	10522-26-6	186.333			129^{12}	0.8300^{15}	1.4382^{20}	vs eth, EtOH
7849	Methyl 10-undecenoate		$C_{12}H_{22}O_2$	111-81-9	198.302	liq	-27.5	246(11)	0.889^{15}	1.4393^{20}	i H_2O; s EtOH, eth, HOAc; sl ctc
7850	N-Methylurea		$C_2H_6N_2O$	598-50-5	74.081	orth pr (w, al)	101.3(0.8)	dec	1.2040^0		vs H_2O, EtOH; i eth, bz; s CS_2, lig
7851	5-Methyluridine	Thymine riboside	$C_{10}H_{14}N_2O_6$	1463-10-1	258.227	cry (EtOH)	184				
7852	3-Methyl-L-valine	L-tert-Leucine	$C_6H_{13}NO_2$	20859-02-3	131.173		248 dec				
7853	2-(1-Methylvinyl)aniline		$C_9H_{11}N$	52562-19-3	133.190			115^{20}	0.977^{25}	1.5722^{20}	
7854	1-Methyl-4-vinylcyclohexene		C_9H_{14}	17699-86-4	122.207	liq		152	0.85	1.4701^{20}	
7855	4-(1-Methylvinyl)-1-cyclohex-ene-1-carboxaldehyde, (R)-	d-Perillaldehyde	$C_{10}H_{14}O$	5503-12-8	150.217	oil		238	0.953^{20}	1.5058^{20}	s ctc
7856	4-(1-Methylvinyl)-1-cyclohex-ene-1-carboxaldehyde, (S)-	l-Perillaldehyde	$C_{10}H_{14}O$	18031-40-8	150.217	oil		104^{10}	0.9645^{20}	1.5072^{20}	
7857	4-(1-Methylvinyl)-1-cyclohex-ene-1-methanol		$C_{10}H_{16}O$	536-59-4	152.233			244	0.9690^{20}	1.5005^{20}	
7858	(1-Methylvinyl)cyclopropane		C_6H_{10}	4663-22-3	82.143	liq	-102.3(0.1)	70(2)	0.751^{20}	1.4252^{20}	
7859	Methyl vinyl ether		C_3H_6O	107-25-5	58.079	col gas	-122	6(14)	0.7725^0	1.3730^0	sl H_2O; vs EtOH, eth, ace, bz
7860	Methyl Violet	C.I. Basic Violet 1	$C_{24}H_{28}ClN_3$	8004-87-3	393.952	bl-viol pow	137 dec				s H_2O, EtOH
7861	Methysergide		$C_{21}H_{27}N_3O_2$	361-37-5	353.458	cry	195				
7862	Methysticin		$C_{15}H_{14}O_5$	495-85-2	274.269	nd (MeOH), pr (ace)	137				
7863	Metobromuron		$C_9H_{11}BrN_2O_2$	3060-89-7	259.099		95.9(0.4)		1.60^{20}		
7864	Metolachlor		$C_{15}H_{22}ClNO_2$	51218-45-2	283.795			$100^{0.001}$	1.12^{20}		
7865	Metolazone		$C_{16}H_{16}ClN_3O_3S$	17560-51-9	365.834	cry (EtOH)	254				
7866	Metoprolol tartrate		$C_{34}H_{56}N_2O_{12}$	56392-17-7	684.815	cry	121				
7867	Metribuzin		$C_8H_{14}N_4OS$	21087-64-9	214.288		126.8(0.5)		1.31^{20}		
7868	Metronidazole	2-Methyl-5-nitro-1H-imidaz-ole-1-ethanol	$C_6H_9N_3O_3$	443-48-1	171.153		160.5				
7869	Metsulfuron-methyl		$C_{14}H_{15}N_5O_6S$	74223-64-6	381.364	wh cry	163				sl H_2O
7870	Mevinphos		$C_7H_{13}O_6P$	7786-34-7	224.148		21 (E), 6.9 (Z)	$101^{0.3}$			
7871	Mexacarbate	4-(Dimethylamino)-3,5-xylyl methylcarbamate	$C_{12}H_{18}N_2O_2$	315-18-4	222.283	cry	88.9(0.5)				vs EtOH, bz, ace
7872	MGK 264		$C_{17}H_{25}NO_2$	113-48-4	275.387		<-20	157	1.04		
7873	Mifepristone	RU-486	$C_{29}H_{35}NO_2$	84371-65-3	429.594	cry	150				
7874	Mimosine		$C_8H_{10}N_2O_4$	500-44-7	198.176	tab (w)	228 dec				sl H_2O; i EtOH, eth, ace, bz; s dil alk
7875	Minocycline		$C_{23}H_{27}N_3O_7$	10118-90-8	457.476	ye-oran amorp solid					
7876	Minoxidil		$C_9H_{15}N_5O$	38304-91-5	209.248	cry	248				i ace, bz, chl, sl; EtOH, MeOH
7877	Mipafox	Bis(isopropylamido) fluorophosphate	$C_6H_{16}FN_2OP$	371-86-8	182.175	cry (peth)	65	125^2			sl H_2O
7878	Mirex	Hexachloropentadiene dimer	$C_{10}Cl_{12}$	2385-85-5	545.543	cry (bz)	485 dec				vs bz, diox
7879	Misoprostol		$C_{22}H_{38}O_5$	59122-46-2	382.534	ye oil					s H_2O
7880	Mithramycin	Plicamycin	$C_{52}H_{76}O_{24}$	18378-89-7	1085.145	ye cry (ace)	182				s H_2O, EtOH, AcOEt; sl bz, eth
7881	Mitomycin A		$C_{16}H_{19}N_3O_6$	4055-39-4	349.338	purp nd	160 dec				
7882	Mitomycin B		$C_{16}H_{19}N_3O_6$	4055-40-7	349.338	purp-bl nd	dec				
7883	Mitomycin C		$C_{15}H_{18}N_4O_5$	50-07-7	334.328	bl-viol cry	360				s H_2O, MeOH, ace
7884	Mitotane		$C_{14}H_{10}Cl_4$	53-19-0	320.041		76.2(0.3)				

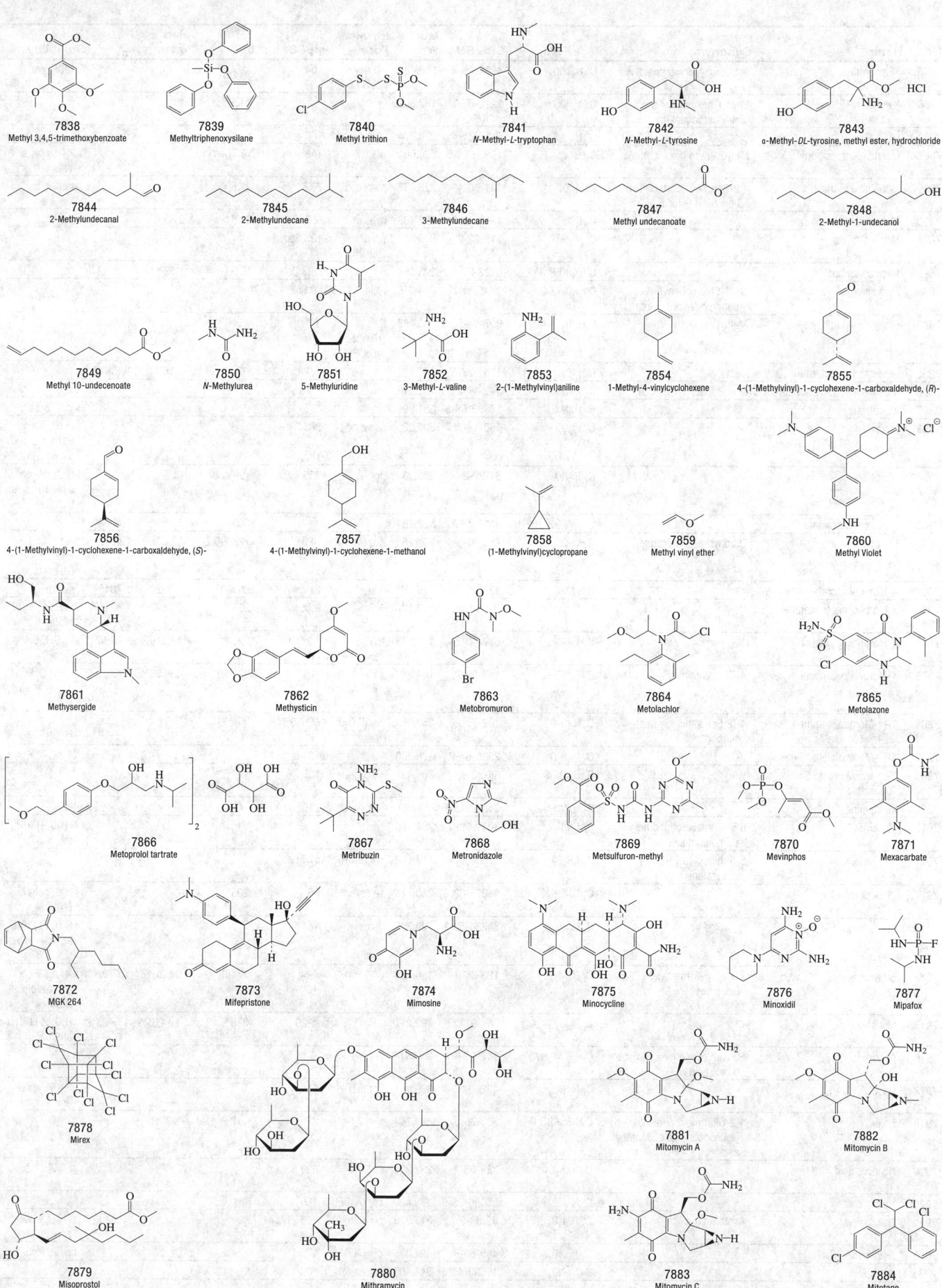

7838 Methyl 3,4,5-trimethoxybenzoate

7839 Methyltriphenoxysilane

7840 Methyl trithion

7841 N-Methyl-L-tryptophan

7842 N-Methyl-L-tyrosine

7843 α-Methyl-DL-tyrosine, methyl ester, hydrochloride

7844 2-Methylundecanal

7845 2-Methylundecane

7846 3-Methylundecane

7847 Methyl undecanoate

7848 2-Methyl-1-undecanol

7849 Methyl 10-undecenoate

7850 N-Methylurea

7851 5-Methyluridine

7852 3-Methyl-L-valine

7853 2-(1-Methylvinyl)aniline

7854 1-Methyl-4-vinylcyclohexene

7855 4-(1-Methylvinyl)-1-cyclohexene-1-carboxaldehyde, (R)-

7856 4-(1-Methylvinyl)-1-cyclohexene-1-carboxaldehyde, (S)-

7857 4-(1-Methylvinyl)-1-cyclohexene-1-methanol

7858 (1-Methylvinyl)cyclopropane

7859 Methyl vinyl ether

7860 Methyl Violet

7861 Methysergide

7862 Methysticin

7863 Metobromuron

7864 Metolachlor

7865 Metolazone

7866 Metoprolol tartrate

7867 Metribuzin

7868 Metronidazole

7869 Metsulfuron-methyl

7870 Mevinphos

7871 Mexacarbate

7872 MGK 264

7873 Mifepristone

7874 Mimosine

7875 Minocycline

7876 Minoxidil

7877 Mipafox

7878 Mirex

7879 Misoprostol

7880 Mithramycin

7881 Mitomycin A

7882 Mitomycin B

7883 Mitomycin C

7884 Mitotane

No.	Name	Synonym	Mol. Form.	CAS RN	Mol. Wt.	Physical Form	mp/°C	bp/°C	den g cm^{-3}	n_D	Solubility
7885	Mitragynine	9-Methoxycorynantheidine	C$_{23}$H$_{30}$N$_2$O$_4$	4098-40-2	398.495	wh amor pow	104	235[5]			s EtOH, chl, HOAc
7886	Molinate	Ethyl 1-hexamethylene-iminecarbothiolate	C$_9$H$_{17}$NOS	2212-67-1	187.302			202[10]	1.063[20]		
7887	Molindone		C$_{16}$H$_{24}$N$_2$O$_2$	7416-34-4	276.374	cry	180				
7888	Molybdenum carbonyl	Molybdenum hexacarbonyl	C$_6$MoO$_6$	13939-06-5	264.00	wh cry	148	155 dec	1.96		i H$_2$O; s bz; sl eth
7889	Monobutyl phthalate	1,2-Benzenedicarboxylic acid, monobutyl ester	C$_{12}$H$_{14}$O$_4$	131-70-4	222.237	pl (ace, al)	73.5				vs EtOH, chl
7890	Monobutyltin trichloride		C$_4$H$_9$Cl$_3$Sn	1118-46-3	282.183	hyg liq	-63	93[10]	0.85[20]		s bz, CH$_2$Cl$_2$
7891	Monocrotaline		C$_{16}$H$_{23}$NO$_6$	315-22-0	325.357	wh pr (EtOH)	198 dec				
7892	Monocrotophos		C$_7$H$_{14}$NO$_5$P	6923-22-4	223.164		55.3(0.4)	125[0.0005]	1.33[20]		
7893	Monolinuron	N-(4-Chlorophenyl)-N-methoxy-N-methylurea	C$_9$H$_{11}$ClN$_2$O$_2$	1746-81-2	214.648	solid	80.8(0.5)				
7894	Monomethyl adipate	Methyl adipate	C$_7$H$_{12}$O$_4$	627-91-8	160.168	lf (Me$_3$N-MeOH)	9	158[10]	1.0623[20]	1.4283[20]	s EtOH
7895	Monomethyl glutarate		C$_6$H$_{10}$O$_4$	1501-27-5	146.141			158[27]	1.169[25]	1.4381[20]	
7896	Monosodium L-glutamate	Monosodium glutamate	C$_5$H$_8$NNaO$_4$	142-47-2	169.113						s H$_2$O
7897	Moquizone		C$_{20}$H$_{21}$N$_3$O$_3$	19395-58-5	351.399		136				s chl
7898	Morin		C$_{15}$H$_{10}$O$_7$	480-16-0	302.236	pa ye nd (+ 1 w, dil al)	303.5				sl H$_2$O, eth; vs EtOH; s bz, alk; i CS$_2$
7899	Morphine		C$_{17}$H$_{19}$NO$_3$	57-27-2	285.338	pr	255	190 sub			i H$_2$O, eth, ace; s MeOH, py; sl EtOH
7900	4-Morpholinamine		C$_4$H$_{10}$N$_2$O	4319-49-7	102.134			166	1.059[25]	1.4772[20]	
7901	Morpholine	Tetrahydro-1,4-oxazine	C$_4$H$_9$NO	110-91-8	87.120	hyg liq	-4.8(0.2)	128.2(0.2)	1.0005[20]	1.4548[20]	msc H$_2$O; s EtOH, eth, ace, bz; sl chl
7902	4-Morpholinecarboxaldehyde		C$_5$H$_9$NO$_2$	4394-85-8	115.131		21	238(1)	1.1520[20]	1.4845[20]	
7903	4-Morpholineethanamine		C$_6$H$_{14}$N$_2$O	2038-03-1	130.187		25.6	205	0.9897[20]	1.4715[20]	msc H$_2$O, EtOH, bz, lig; s ace
7904	4-Morpholineethanol		C$_6$H$_{13}$NO$_2$	622-40-2	131.173	liq	1.2(0.4)	220(3)	1.0710[20]	1.4763[20]	s H$_2$O, EtOH; sl ctc
7905	4-Morpholinepropanamine	4-(3-Aminopropyl)morpholine	C$_7$H$_{16}$N$_2$O	123-00-2	144.214	liq	-15	220	0.9854[20]	1.4762[20]	msc H$_2$O, EtOH, bz, lig; s ace; sl ctc
7906	2-(4-Morpholinothio)-benzothiazole	4-(2-Benzothiazolylthio)-morpholine	C$_{11}$H$_{12}$N$_2$OS$_2$	102-77-2	252.355	cry (EtOH)	86.7(0.7)				
7907	4-(4-Morpholinyl)aniline		C$_{10}$H$_{14}$N$_2$O	2524-67-6	178.230		131.6				
7908	2-(4-Morpholinyldithio)-benzothiazole		C$_{11}$H$_{12}$N$_2$OS$_3$	95-32-9	284.420		135				
7909	Muldamine		C$_{29}$H$_{47}$NO$_3$	36069-45-1	457.688		210				
7910	Murexide	5,5'-Nitrilobarbituric acid, ammonium salt	C$_8$H$_8$N$_6$O$_6$	3051-09-0	284.186						sl H$_2$O; i EtOH, eth; s alk
7911	Muscimol	5-(Aminomethyl)-3(2H)-isoxazolone	C$_4$H$_6$N$_2$O$_2$	2763-96-4	114.103	cry (EtOH)	175 dec				
7912	Myclobutanil		C$_{15}$H$_{17}$ClN$_4$	88671-89-0	288.776	ye cry	65	205[1.0]			i H$_2$O, peth; s EtOH
7913	Mycophenolic acid		C$_{17}$H$_{20}$O$_6$	24280-93-1	320.337	nd (w)	141				i H$_2$O; vs EtOH, eth, chl; sl bz, tol
7914	β-Myrcene	7-Methyl-3-methylene-1,6-octadiene	C$_{10}$H$_{16}$	123-35-3	136.234	oil		171(3)	0.8013[15]	1.4722[20]	i H$_2$O; s EtOH, eth, bz, chl, HOAc
7915	Myristicin		C$_{11}$H$_{12}$O$_3$	607-91-0	192.211		<-20	276.5	1.1416[20]	1.5403[20]	i H$_2$O; sl EtOH; s eth, bz
7916	Nabam	Sodium ethylenebisdithiocarbamic acid	C$_4$H$_6$N$_2$Na$_2$S$_4$	142-59-6	256.344	cry (w)					s H$_2$O
7917	Nadolol		C$_{17}$H$_{27}$NO$_4$	42200-33-9	309.401	cry (bz)	≈130				s EtOH; sl chl; i ace, eth, hx
7918	Naled	1,2-Dibromo-2,2-dichloroethylphosphoric acid, dimethyl ester	C$_4$H$_7$Br$_2$Cl$_2$O$_4$P	300-76-5	380.784		27	110[0.5]	1.96[20]		
7919	Nalidixic acid		C$_{12}$H$_{12}$N$_2$O$_3$	389-08-2	232.234		229.5				sl EtOH, eth; s chl
7920	Nalmefene		C$_{21}$H$_{25}$NO$_3$	55096-26-9	339.429	cry (AcOEt)	189				
7921	Nalorphine	Acetorphin	C$_{19}$H$_{21}$NO$_3$	62-67-9	311.375	cry (eth)	208				sl H$_2$O; s alk, ace, EtOH
7922	Naloxone		C$_{19}$H$_{21}$NO$_4$	465-65-6	327.375	cry (AcOEt)	178				i peth; s chl
7923	Naltrexone		C$_{20}$H$_{23}$NO$_4$	16590-41-3	341.402	cry (ace)	169				

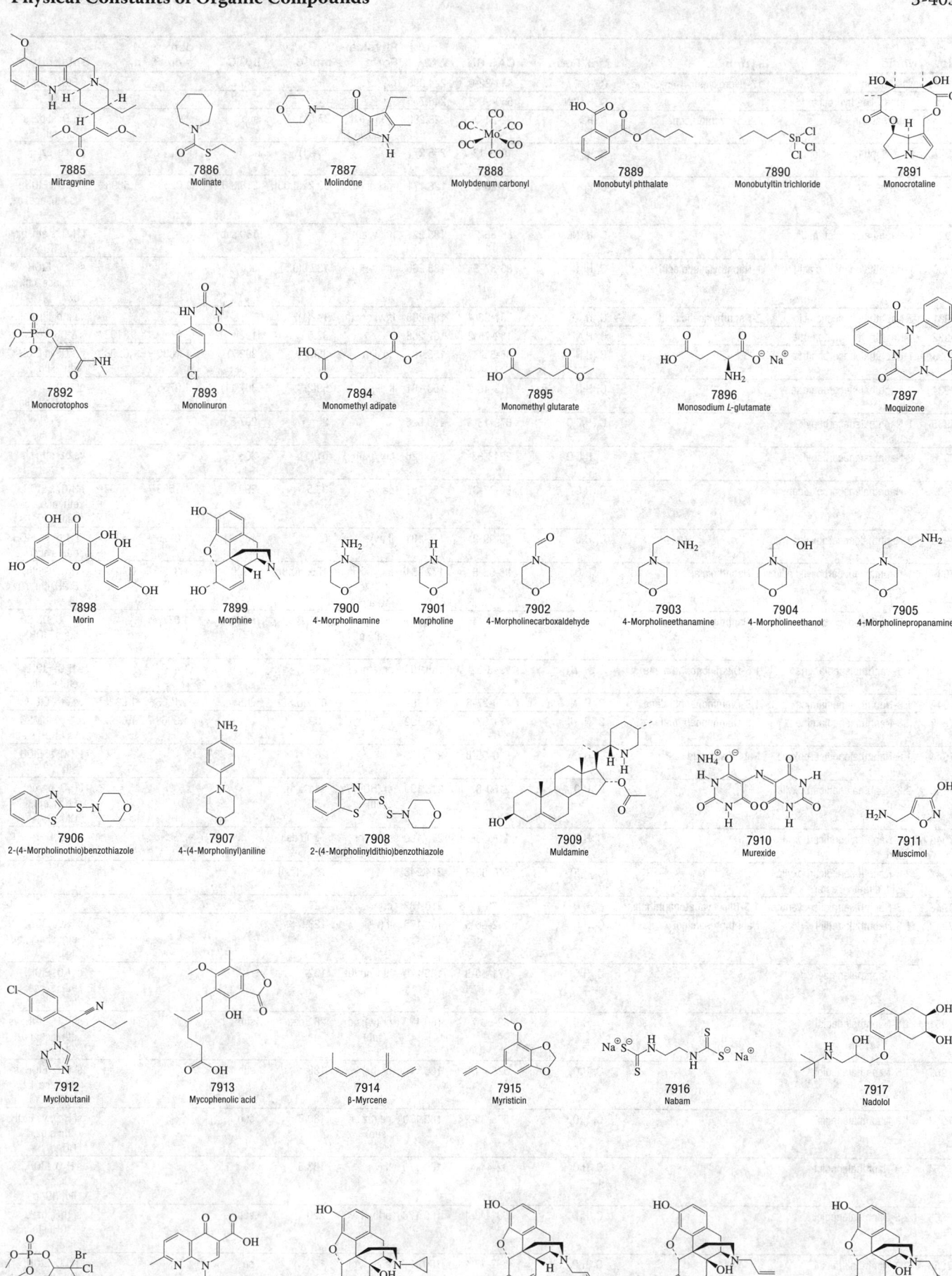

7885 Mitragynine

7886 Molinate

7887 Molindone

7888 Molybdenum carbonyl

7889 Monobutyl phthalate

7890 Monobutyltin trichloride

7891 Monocrotaline

7892 Monocrotophos

7893 Monolinuron

7894 Monomethyl adipate

7895 Monomethyl glutarate

7896 Monosodium *L*-glutamate

7897 Moquizone

7898 Morin

7899 Morphine

7900 4-Morpholinamine

7901 Morpholine

7902 4-Morpholinecarboxaldehyde

7903 4-Morpholineethanamine

7904 4-Morpholineethanol

7905 4-Morpholinepropanamine

7906 2-(4-Morpholinothio)benzothiazole

7907 4-(4-Morpholinyl)aniline

7908 2-(4-Morpholinyldithio)benzothiazole

7909 Muldamine

7910 Murexide

7911 Muscimol

7912 Myclobutanil

7913 Mycophenolic acid

7914 β-Myrcene

7915 Myristicin

7916 Nabam

7917 Nadolol

7918 Naled

7919 Nalidixic acid

7920 Nalmefene

7921 Nalorphine

7922 Naloxone

7923 Naltrexone

No.	Name	Synonym	Mol. Form.	CAS RN	Mol. Wt.	Physical Form	mp/°C	bp/°C	den g cm⁻³	n_D	Solubility
7924	Nandrolone	17-Hydroxyestr-4-en-3-one	$C_{18}H_{26}O_2$	434-22-0	274.398	cry	112				s EtOH, eth, chl
7925	Naphazoline hydrochloride		$C_{14}H_{15}ClN_2$	550-99-2	246.735						sl H_2O
7926	Naphthacene	2,3-Benzanthracene	$C_{18}H_{12}$	92-24-0	228.288	oran-ye lf (bz, xyl)	354(4)	sub			i H_2O; sl bz; s con sulf
7927	5,12-Naphthacenedione		$C_{18}H_{10}O_2$	1090-13-7	258.271		294(1)				sl ace, bz, gl HOAc
7928	Naphthalene		$C_{10}H_8$	91-20-3	128.171	mcl pl (al)	80.22(0.09)	218.0(0.1)	1.0253²⁰	1.5898²⁵	i H_2O; s EtOH; vs eth, ace, bz, CS_2
7929	1-Naphthaleneacetamide		$C_{12}H_{11}NO$	86-86-2	185.221	nd(w, al)		180 sub			i H_2O; s eth, bz, CS_2, HOAc
7930	1-Naphthaleneacetic acid	1-Naphthylacetic acid	$C_{12}H_{10}O_2$	86-87-3	186.206	nd (w)	132.8(0.5)	dec			sl H_2O, EtOH; vs eth, ace, chl; s bz
7931	2-Naphthaleneacetic acid	2-Naphthylacetic acid	$C_{12}H_{10}O_2$	581-96-4	186.206	lf(w) cry (bz)	142(1)				vs eth, lig, chl
7932	1-Naphthaleneacetonitrile		$C_{12}H_9N$	132-75-2	167.206		32.5	192¹⁸		1.6192²⁰	s EtOH
7933	1-Naphthalenecarbonitrile		$C_{11}H_7N$	86-53-3	153.181	nd (lig)	35(2)	299(7)	1.1080²⁵	1.6298¹⁸	i H_2O; vs EtOH, eth; s lig
7934	2-Naphthalenecarbonitrile		$C_{11}H_7N$	613-46-7	153.181	lf (lig)	59(2)	307(11)	1.0755⁶⁰		sl H_2O, chl; s EtOH, eth, lig
7935	1-Naphthalenecarbonyl chloride		$C_{11}H_7ClO$	879-18-5	190.626		20	297.5			
7936	2-Naphthalenecarbonyl chloride		$C_{11}H_7ClO$	2243-83-6	190.626	cry (peth)	51	305			vs bz, eth, chl
7937	1-Naphthalenecarboxaldehyde		$C_{11}H_8O$	66-77-3	156.181	pa ye	33.5	292	1.1503²⁰	1.6507²⁰	i H_2O; s EtOH, eth, ace, bz, sulf
7938	2-Naphthalenecarboxaldehyde		$C_{11}H_8O$	66-99-9	156.181	lf (w)	62	160¹⁹	1.0775⁹⁹	1.6211⁹⁹	sl H_2O; vs EtOH, eth; s ace
7939	1-Naphthalenecarboxylic acid	1-Naphthoic acid	$C_{11}H_8O_2$	86-55-5	172.181	nd (HOAc-w, w, al)	160.8(0.8)	>300	1.398²⁵	1.46	i H_2O; vs eth, EtOH, chl
7940	2-Naphthalenecarboxylic acid	2-Naphthoic acid	$C_{11}H_8O_2$	93-09-4	172.181	nd (lig, chl, sub) pl (ace)	185(1)	>300	1.077¹⁰⁰		sl H_2O, DMSO, lig; s EtOH, eth, chl
7941	1,5-Naphthalenediamine	1,5-Diaminonaphthalene	$C_{10}H_{10}N_2$	2243-62-1	158.199	pr (eth, al, w)	189(2)	sub	1.4²⁵		s H_2O, EtOH; vs chl
7942	1,8-Naphthalenediamine	1,8-Diaminonaphthalene	$C_{10}H_{10}N_2$	479-27-6	158.199		66.7(0.5)	205¹²	1.1265⁹⁰	1.6828⁹⁹	vs eth, EtOH
7943	2,3-Naphthalenediamine	2,3-Diaminonaphthalene	$C_{10}H_{10}N_2$	771-97-1	158.199	lf (eth, w)	199		1.0968²⁶	1.6392²⁶	sl H_2O, DMSO; vs EtOH; s eth
7944	1,8-Naphthalenedicarboxylic acid	Naphthalic acid	$C_{12}H_8O_4$	518-05-8	216.190		260				i H_2O; sl EtOH, eth
7945	2,3-Naphthalenedicarboxylic acid		$C_{12}H_8O_4$	2169-87-1	216.190	pr (HOAc, w, sub)	244.5				i H_2O, bz, chl; sl EtOH, eth, DMSO
7946	2,6-Naphthalenedicarboxylic acid		$C_{12}H_8O_4$	1141-38-4	216.190	nd (al or sub)	>300 dec				vs EtOH
7947	2,6-Naphthalenedicarboxylic acid, dimethyl ester		$C_{14}H_{12}O_4$	840-65-3	244.243		191.3(0.4)				
7948	1,5-Naphthalene diisocyanate	1,5-Diisocyanatonaphthalene	$C_{12}H_6N_2O_2$	3173-72-6	210.188	cry	127	183¹⁰			
7949	1,3-Naphthalenediol	Naphthoresorcinol	$C_{10}H_8O_2$	132-86-5	160.170	lf (w)	123.5				s H_2O, EtOH, eth; sl ace, bz, lig
7950	1,4-Naphthalenediol		$C_{10}H_8O_2$	571-60-8	160.170	mcl nd (bz, w)	192				s H_2O, EtOH, eth; sl ace; i bz
7951	1,5-Naphthalenediol		$C_{10}H_8O_2$	83-56-7	160.170	pr (w), nd (sub)	262 dec	sub			sl H_2O, EtOH; vs eth, ace; i bz; s HOAc
7952	1,6-Naphthalenediol		$C_{10}H_8O_2$	575-44-0	160.170	pr (bz)	138	sub			sl H_2O, EtOH; s eth, ace, bz, DMSO
7953	1,7-Naphthalenediol		$C_{10}H_8O_2$	575-38-2	160.170	nd (bz or sub)	180.5	sub			sl H_2O; vs EtOH, eth; s bz, HOAc
7954	2,3-Naphthalenediol		$C_{10}H_8O_2$	92-44-4	160.170	lf (w)	163.5				s H_2O, EtOH, eth, ace, bz, lig, HOAc
7955	2,6-Naphthalenediol		$C_{10}H_8O_2$	581-43-1	160.170	orth pl (w)	220	sub			sl H_2O, bz; s EtOH, eth, ace; i lig
7956	2,7-Naphthalenediol		$C_{10}H_8O_2$	582-17-2	160.170	nd, (w, dil al), pl (dil al)	193	sub			s H_2O, EtOH, eth, bz, chl; sl ace; i lig

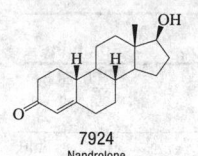

7924
Nandrolone

7925
Naphazoline hydrochloride

7926
Naphthacene

7927
5,12-Naphthacenedione

7928
Naphthalene

7929
1-Naphthaleneacetamide

7930
1-Naphthaleneacetic acid

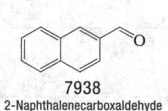

7931
2-Naphthaleneacetic acid

7932
1-Naphthaleneacetonitrile

7933
1-Naphthalenecarbonitrile

7934
2-Naphthalenecarbonitrile

7935
1-Naphthalenecarbonyl chloride

7936
2-Naphthalenecarbonyl chloride

7937
1-Naphthalenecarboxaldehyde

7938
2-Naphthalenecarboxaldehyde

7939
1-Naphthalenecarboxylic acid

7940
2-Naphthalenecarboxylic acid

7941
1,5-Naphthalenediamine

7942
1,8-Naphthalenediamine

7943
2,3-Naphthalenediamine

7944
1,8-Naphthalenedicarboxylic acid

7945
2,3-Naphthalenedicarboxylic acid

7946
2,6-Naphthalenedicarboxylic acid

7947
2,6-Naphthalenedicarboxylic acid, dimethyl ester

7948
1,5-Naphthalene diisocyanate

7949
1,3-Naphthalenediol

7950
1,4-Naphthalenediol

7951
1,5-Naphthalenediol

7952
1,6-Naphthalenediol

7953
1,7-Naphthalenediol

7954
2,3-Naphthalenediol

7955
2,6-Naphthalenediol

7956
2,7-Naphthalenediol

No.	Name	Synonym	Mol. Form.	CAS RN	Mol. Wt.	Physical Form	mp/°C	bp/°C	den g cm⁻³	n_D	Solubility
7957	1,2-Naphthalenedione	1,2-Naphthoquinone	$C_{10}H_6O_2$	524-42-5	158.154	ye-red nd (eth) oran lf (bz)	146		1.450[25]		s H_2O, EtOH, eth, sulf; sl lig
7958	1,4-Naphthalenedione	1,4-Naphthoquinone	$C_{10}H_6O_2$	130-15-4	158.154	bt ye nd (al, peth) ye (sub)	125.3(0.5)	sub			sl H_2O; vs EtOH; s eth, bz, chl, CS_2
7959	1,5-Naphthalenedisulfonic acid	Armstrong's acid	$C_{10}H_8O_6S_2$	81-04-9	288.297	pl (+4w, dil HOAc)	242 dec		1.493[25]		vs H_2O; s EtOH; i eth
7960	1,6-Naphthalenedisulfonic acid	Naphthalene-1,6-disulfonic acid	$C_{10}H_8O_6S_2$	525-37-1	288.297	oran pr (+4w, HOAc or w)	125 dec				vs H_2O; s EtOH; i eth
7961	2,7-Naphthalenedisulfonic acid	Naphthalene-2,7-disulfonic acid	$C_{10}H_8O_6S_2$	92-41-1	288.297	hyg nd (conc HCl)	199				s H_2O; sl con HCl
7962	1-Naphthalenemethanamine		$C_{11}H_{11}N$	118-31-0	157.212			292	1.0958[20]		s EtOH, eth, sulf, CS_2
7963	1-Naphthalenemethanol		$C_{11}H_{10}O$	4780-79-4	158.196	nd (w, al), cry (bz-lig)	64	304	1.1039[80]		sl H_2O; vs EtOH, eth
7964	2-Naphthalenemethanol		$C_{11}H_{10}O$	1592-38-7	158.196	lf	81.3	178[12]			sl H_2O; s EtOH, eth
7965	1-Naphthalenesulfonic acid	α-Naphthylsulfonic acid	$C_{10}H_8O_3S$	85-47-2	208.234	pr (+2 w, dil HCl)	140				s H_2O, EtOH; sl eth
7966	2-Naphthalenesulfonic acid	β-Naphthylsulfonic acid	$C_{10}H_8O_3S$	120-18-3	208.234	hyg pl (+1w), cry (+3w, HCl)	91	dec	1.441[25]		vs H_2O, EtOH; s eth; sl bz
7967	1-Naphthalenesulfonyl chloride		$C_{10}H_7ClO_2S$	85-46-1	226.680	lf (eth)	68	209[20]			vs bz, eth, EtOH
7968	2-Naphthalenesulfonyl chloride		$C_{10}H_7ClO_2S$	93-11-8	226.680	pow or lf (bz-peth)	81	201[13]			i H_2O; s EtOH, bz, chl; sl peth; vs eth
7969	1,4,5,8-Naphthalenetetracarboxylic acid		$C_{14}H_8O_8$	128-97-2	304.209	lf or nd (w, dil HCl)	320				sl H_2O, bz, chl, EtOH; vs ace
7970	1-Naphthalenethiol	1-Naphthyl mercaptan	$C_{10}H_8S$	529-36-2	160.236			285 dec	1.1607[20]	1.6802[20]	sl H_2O, dil alk; vs EtOH, eth
7971	2-Naphthalenethiol	2-Naphthyl mercaptan	$C_{10}H_8S$	91-60-1	160.236	pl (al)	81	288	1.550[25]		sl H_2O; vs EtOH, eth, lig
7972	N-(1-Naphthalenyl)-1,2-ethanediamine, dihydrochloride		$C_{12}H_{16}Cl_2N_2$	1465-25-4	259.174	hex pr	189				vs H_2O, EtOH
7973	1-Naphthalenylthiourea	ANTU	$C_{11}H_{10}N_2S$	86-88-4	202.275	pr (al)	198				i H_2O; sl EtOH, eth, ace
7974	Naphtho[2,3-c]furan-1,3-dione	2,3-Naphthalenedicarboxylic acid anhydride	$C_{12}H_6O_3$	716-39-2	198.174		246				sl EtOH, chl; s eth, bz
7975	1-Naphthol	1-Naphthalenol	$C_{10}H_8O$	90-15-3	144.170	ye nd (w)	95.1(0.6)	288(7)	1.0989[99]	1.6224[99]	i H_2O; vs EtOH, eth; s ace, bz; sl ctc
7976	2-Naphthol	2-Naphthalenol	$C_{10}H_8O$	135-19-3	144.170	mcl lf (w)	122(1)	286(1)	1.28[20]		i H_2O; vs EtOH, eth; s bz, chl; sl lig
7977	1-Naphthol, acetate	1-Naphthyl acetate	$C_{12}H_{10}O_2$	830-81-9	186.206	nd or pl (al)	46.0(0.4)	114[1]			i H_2O; s EtOH, eth
7978	2-Naphthol, acetate	2-Naphthyl acetate	$C_{12}H_{10}O_2$	1523-11-1	186.206	nd (al)	69.0(0.5)	132[2]			i H_2O; s EtOH, eth, chl
7979	p-Naphtholbenzein		$C_{27}H_{18}O_2$	145-50-6	374.431		123				
7980	1H,3H-Naphtho[1,8-cd]pyran-1,3-dione		$C_{12}H_6O_3$	81-84-5	198.174		272.6(0.5)				i H_2O, eth, bz; sl EtOH; s HOAc
7981	1-Naphthylamine	α-Naphthylamine	$C_{10}H_9N$	134-32-7	143.185		48.9(0.5)	300.7	1.0228[20]	1.6140[20]	s chl
7982	2-Naphthylamine	β-Naphthylamine	$C_{10}H_9N$	91-59-8	143.185		110(1)	306.2	1.6414[98]	1.6493[98]	s H_2O, EtOH, eth
7983	2-[(1-Naphthylamino)carbonyl]benzoic acid	Naptalam	$C_{18}H_{13}NO_3$	132-66-1	291.301		185		1.4[20]		i H_2O; sl EtOH, ace, bz, tfa
7984	2-Naphthyl benzoate	2-Naphthalenol benzoate	$C_{17}H_{12}O_2$	93-44-7	248.276	nd or pr (al)	107.8(0.6)				i H_2O; s EtOH; sl eth, HOAc
7985	N-1-Naphthalenylacetamide		$C_{12}H_{11}NO$	575-36-0	185.221		160				s H_2O, EtOH; sl eth
7986	N-1-Naphthyl-1,2-ethanediamine	N-(1-Naphthyl)ethylenediamine	$C_{12}H_{14}N_2$	551-09-7	186.252	visc liq		204[9]	1.114[25]	1.6648[25]	
7987	1-Naphthyl 2-hydroxybenzoate	1-Naphthyl salicylate	$C_{17}H_{12}O_3$	550-97-0	264.275		83				vs eth
7988	1-Naphthylhydroxylamine	N-Hydroxyl-1-naphthalenamine	$C_{10}H_9NO$	607-30-7	159.184		79				
7989	1-Naphthyl isothiocyanate	1-Isothiocyanatonaphthalene	$C_{11}H_7NS$	551-06-4	185.246	wh nd (al)	58				vs bz, eth, EtOH, chl
7990	N-2-Naphthyl-2-naphthalenamine	β,β'-Dinaphthylamine	$C_{20}H_{15}N$	532-18-3	269.340	lf(bz)	172.2	471			i H_2O; sl EtOH, bz, DMSO; s eth, HOAc

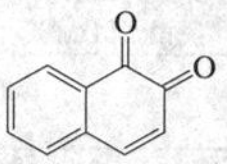

7957
1,2-Naphthalenedione

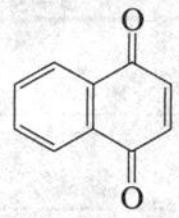

7958
1,4-Naphthalenedione

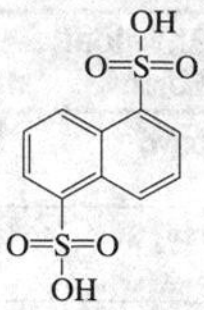

7959
1,5-Naphthalenedisulfonic acid

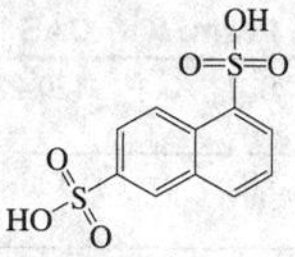

7960
1,6-Naphthalenedisulfonic acid

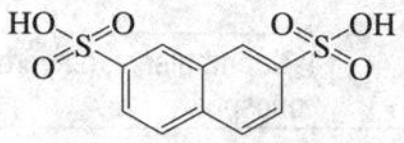

7961
2,7-Naphthalenedisulfonic acid

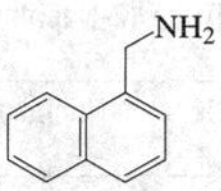

7962
1-Naphthalenemethanamine

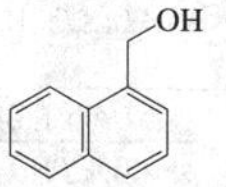

7963
1-Naphthalenemethanol

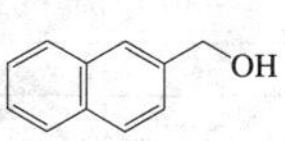

7964
2-Naphthalenemethanol

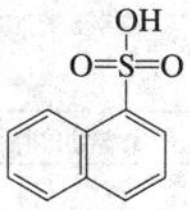

7965
1-Naphthalenesulfonic acid

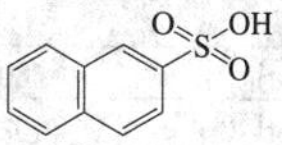

7966
2-Naphthalenesulfonic acid

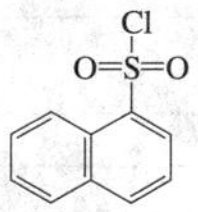

7967
1-Naphthalenesulfonyl chloride

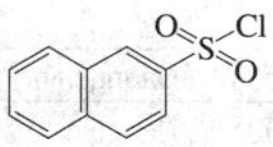

7968
2-Naphthalenesulfonyl chloride

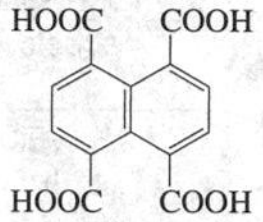

7969
1,4,5,8-Naphthalenetetracarboxylic acid

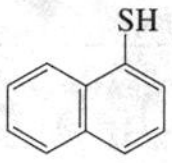

7970
1-Naphthalenethiol

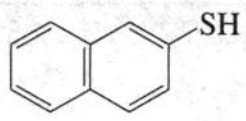

7971
2-Naphthalenethiol

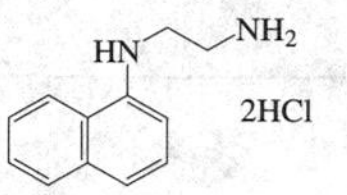

7972
N-(1-Naphthalenyl)-1,2-ethanediamine, dihydrochloride

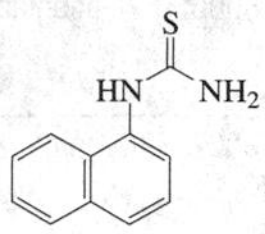

7973
1-Naphthalenylthiourea

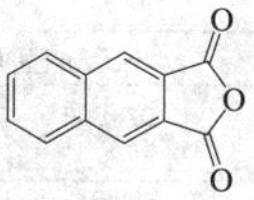

7974
Naphtho[2,3-c]furan-1,3-dione

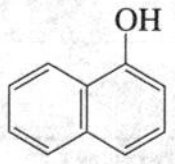

7975
1-Naphthol

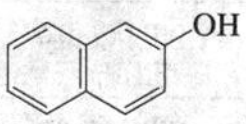

7976
2-Naphthol

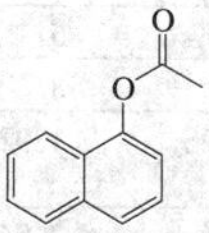

7977
1-Naphthol, acetate

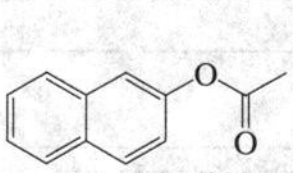

7978
2-Naphthol, acetate

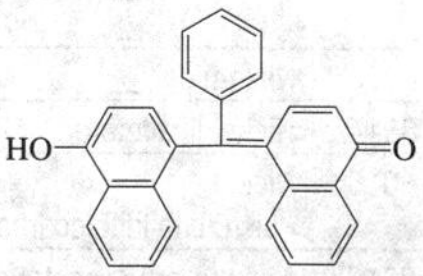

7979
p-Naphtholbenzein

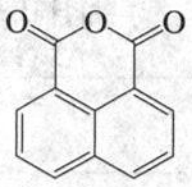

7980
1H,3H-Naphtho[1,8-cd]pyran-1,3-dione

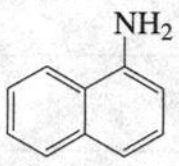

7981
1-Naphthylamine

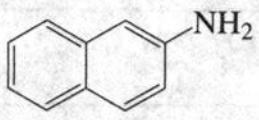

7982
2-Naphthylamine

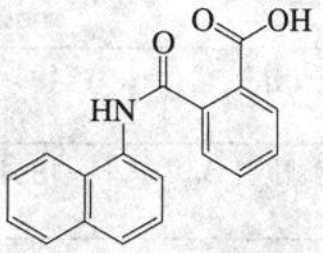

7983
2-[(1-Naphthylamino)carbonyl]benzoic acid

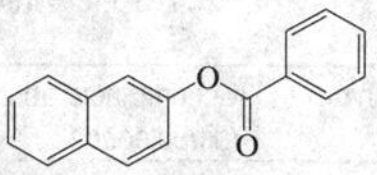

7984
2-Naphthyl benzoate

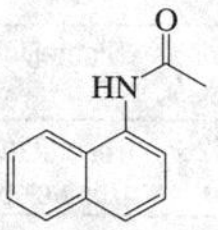

7985
N-1-Naphthalenylacetamide

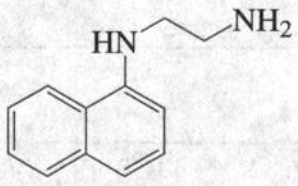

7986
N-1-Naphthyl-1,2-ethanediamine

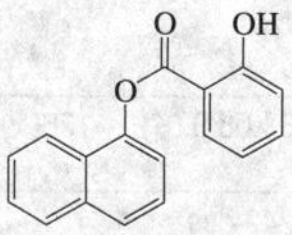

7987
1-Naphthyl 2-hydroxybenzoate

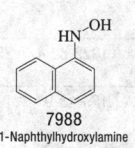

7988
1-Naphthylhydroxylamine

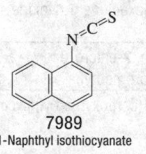

7989
1-Naphthyl isothiocyanate

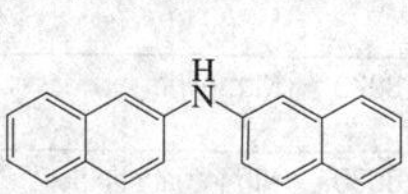

7990
N-2-Naphthyl-2-naphthalenamine

No.	Name	Synonym	Mol. Form.	CAS RN	Mol. Wt.	Physical Form	mp/°C	bp/°C	den g cm⁻³	n_D	Solubility
7991	(2-Naphthyloxy)acetic acid	2-Naphthoxyacetic acid	$C_{12}H_{10}O_3$	120-23-0	202.205	pr(w)	156.2(0.3)				s H_2O, EtOH, eth; sl DMSO
7992	1-Naphthyl phosphate	1-Naphthalenol, dihydrogen phosphate	$C_{10}H_9O_4P$	1136-89-6	224.149	cry	160				
7993	2-Naphthyl salicylate	2-Naphthyl 2-hydroxybenzoate	$C_{17}H_{12}O_3$	613-78-5	264.275	cry (al)	95.5		1.11^{116}		i H_2O; sl EtOH; s eth, bz
7994	1,5-Naphthyridine	1,5-Diazanaphthalene	$C_8H_6N_2$	254-79-5	130.147	ye nd (peth)	75	112^{12}	1.2100^{20}		
7995	1,6-Naphthyridine		$C_8H_6N_2$	253-72-5	130.147		29.5				
7996	Napropamide	Propanamide, N,N-diethyl-2-(1-naphthalenyloxy)-	$C_{17}H_{21}NO_2$	15299-99-7	271.355		73.8(0.5)				
7997	Naproxen	6-Methoxy-α-methyl-2-naphthaleneacetic acid	$C_{14}H_{14}O_3$	22204-53-1	230.259	cry (ace/hx)	155(3)				i H_2O; sl eth; s MeOH, chl
7998	Narceine		$C_{23}H_{27}NO_8$	131-28-2	445.462		138				i H_2O
7999	Narcobarbital		$C_{11}H_{15}BrN_2O_3$	125-55-3	303.152		115				sl H_2O; s EtOH, py
8000	Naringenin		$C_{15}H_{12}O_5$	480-41-1	272.253	nd (dil al)	250(1)				vs bz, eth, EtOH
8001	Naringin		$C_{27}H_{32}O_{14}$	10236-47-2	580.535	nd (w+8)					sl H_2O, EtOH; i eth, bz, chl; s HOAc
8002	Nealbarbital		$C_{12}H_{18}N_2O_3$	561-83-1	238.282		156				vs ace, eth, EtOH
8003	Nellite	Diamidafos	$C_8H_{13}N_2O_2P$	1754-58-1	200.175	cry (ctc)	103.5				sl AcOEt, bz
8004	Neoabietic acid	8(14),13(15)-Abietadien-18-oic acid	$C_{20}H_{30}O_2$	471-77-2	302.451	cry (EtOH aq)	173				
8005	Neobornylamine		$C_{10}H_{19}N$	2223-67-8	153.265	pow	184				vs ace, eth
8006	Neopentane	2,2-Dimethylpropane	C_5H_{12}	463-82-1	72.149	col gas	-16.37(0.04)	9.50(0.06)	0.5852^{25} (p>1 atm)	1.3476^6	i H_2O; s EtOH, eth, ctc
8007	Neopine		$C_{18}H_{21}NO_3$	467-14-1	299.365	nd (peth)	127.5				s H_2O, EtOH, eth, bz; vs chl; sl lig
8008	Neostigmine bromide		$C_{12}H_{19}BrN_2O_2$	114-80-7	303.195	cry (al-eth)	167 dec				vs H_2O; s EtOH
8009	Nepetalactone		$C_{10}H_{14}O_2$	490-10-8	166.217			$71^{0.05}$	1.0663^{25}	1.4859^{25}	
8010	cis-Nerolidol		$C_{15}H_{26}O$	142-50-7	222.366			276	0.8778^{20}	1.4898^{20}	vs EtOH; s eth, ace, HOAc
8011	Neurine		$C_5H_{13}NO$	463-88-7	103.163	syr					vs H_2O, eth, EtOH
8012	Neutral Red		$C_{15}H_{17}ClN_4$	553-24-2	288.776	grn pow					s H_2O, ethylene glycol, EtOH; i xyl
8013	Nialamide		$C_{16}H_{18}N_4O_2$	51-12-7	298.340		154(1)				
8014	Nickel(II) acetate		$C_4H_6NiO_4$	373-02-4	176.782						vs H_2O; s EtOH
8015	Nickel bis(dibutyldithiocarbamate)		$C_{18}H_{36}N_2NiS_4$	13927-77-0	467.445	grn cry (bz/EtOH)	91				s bz, ace
8016	Nickel bis(2,4-pentanedioate)	Nickel acetylacetonate	$C_{10}H_{14}NiO_4$	3264-82-2	256.909	grn orth cry	230	227^{11}			s H_2O, bz, chl, EtOH; i eth
8017	Nickel carbonyl	Nickel tetracarbonyl	C_4NiO_4	13463-39-3	170.734	col liq	-19.3	43 (exp 60)	1.31^{25}		i H_2O; s EtOH, bz, ace, ctc
8018	Nickelocene	Bis(η5-2,4-cyclopentadien-1-yl)nickel	$C_{10}H_{10}Ni$	1271-28-9	188.879	grn cry	173				
8019	Niclosamide		$C_{13}H_8Cl_2N_2O_4$	50-65-7	327.120		227				
8020	Nicofibrate		$C_{16}H_{16}ClNO_3$	31980-29-7	305.756		49	$180^{0.4}$			
8021	Nicosulfuron		$C_{15}H_{18}N_6O_6S$	111991-09-4	410.405		172				
8022	Nicotelline	3,2':4',3''-Terpyridine	$C_{15}H_{11}N_3$	494-04-2	233.268	prismatic nd	148	>300			sl H_2O, eth; s bz, chl, EtOH
8023	Nicotinamide hypoxanthine dinucleotide	Nicotinic acid adenine dinucleotide	$C_{21}H_{26}N_6O_{15}P_2$	1851-07-6	664.410	pow					
8024	β-Nicotinamide mononucleotide	NMN	$C_{11}H_{15}N_2O_8P$	1094-61-7	334.219	amor pow					vs H_2O; i ace
8025	L-Nicotine	3-(1-Methyl-2-pyrrolidinyl)pyridine, (S)-	$C_{10}H_{14}N_2$	54-11-5	162.231	hyg liq	-79	246(13)	1.0097^{20}	1.5282^{20}	msc H_2O; vs EtOH, eth, chl; s lig
8026	Nifurthiazole		$C_8H_6N_4O_4S$	3570-75-0	254.224	cry	215 dec				
8027	Nitralin	4-(Methylsulfonyl)-2,6-dinitro-N,N-dipropylaniline	$C_{13}H_{19}N_3O_6S$	4726-14-1	345.371		151.7(0.5)				
8028	Nitranilic acid	2,5-Dihydroxy-3,6-dinitro-2,5-cyclohexadiene-1,4-dione	$C_6H_2N_2O_8$	479-22-1	230.088	gold-ye pl (+w, dil HNO_3)	170 dec				vs H_2O, EtOH; i eth
8029	Nitrapyrin	Pyridine, 2-chloro-6-(trichloromethyl)-	$C_6H_3Cl_4N$	1929-82-4	230.907		64.08(0.01)	136^{11}			
8030	Nitrilotriacetic acid	N,N-Bis(carboxymethyl)glycine	$C_6H_9NO_6$	139-13-9	191.138	pr cry (w)	242 dec				sl H_2O, DMSO; s EtOH

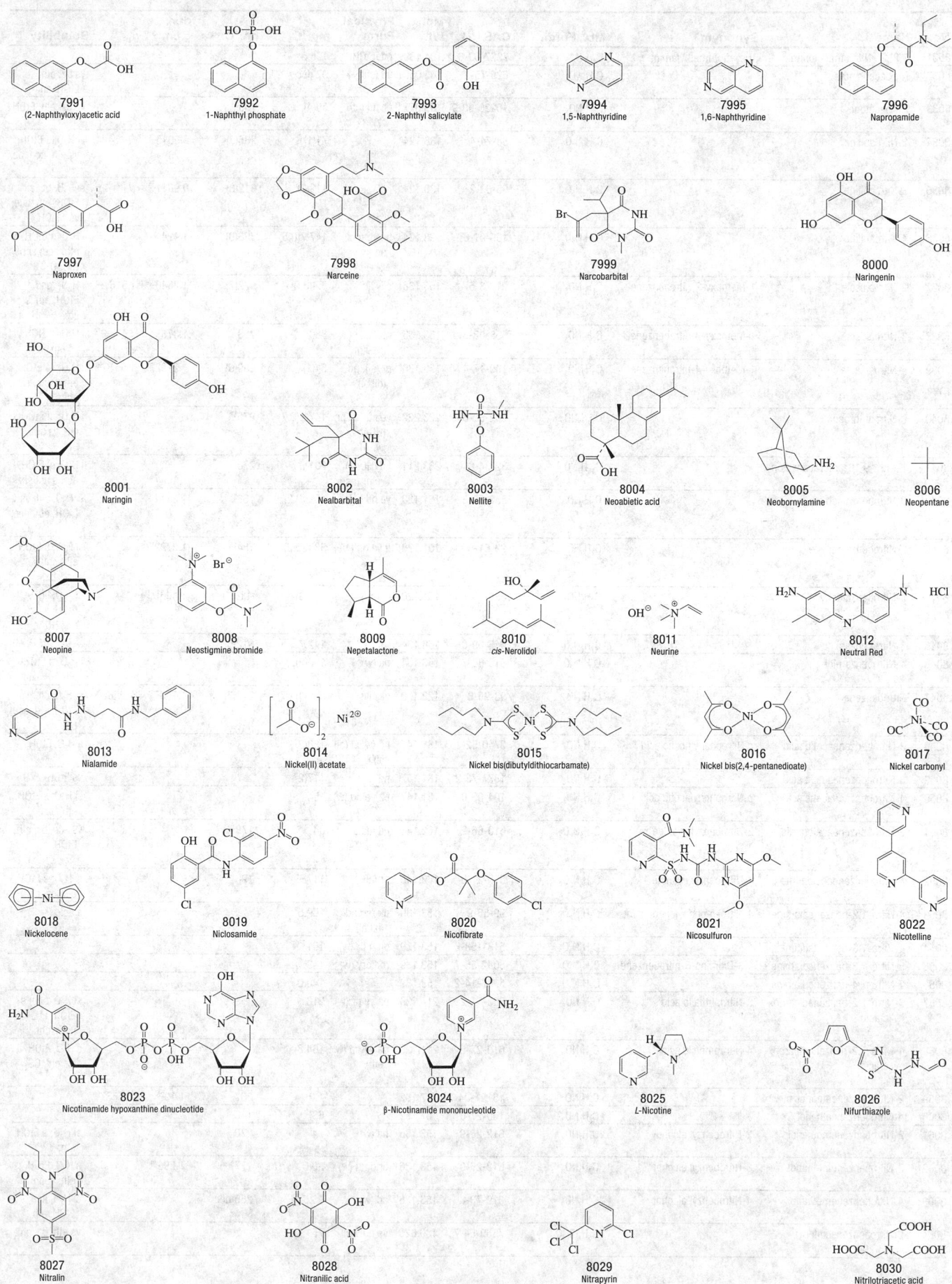

7991
(2-Naphthyloxy)acetic acid

7992
1-Naphthyl phosphate

7993
2-Naphthyl salicylate

7994
1,5-Naphthyridine

7995
1,6-Naphthyridine

7996
Napropamide

7997
Naproxen

7998
Narceine

7999
Narcobarbital

8000
Naringenin

8001
Naringin

8002
Nealbarbital

8003
Nellite

8004
Neoabietic acid

8005
Neobornylamine

8006
Neopentane

8007
Neopine

8008
Neostigmine bromide

8009
Nepetalactone

8010
cis-Nerolidol

8011
Neurine

8012
Neutral Red

8013
Nialamide

8014
Nickel(II) acetate

8015
Nickel bis(dibutyldithiocarbamate)

8016
Nickel bis(2,4-pentanedioate)

8017
Nickel carbonyl

8018
Nickelocene

8019
Niclosamide

8020
Nicofibrate

8021
Nicosulfuron

8022
Nicotelline

8023
Nicotinamide hypoxanthine dinucleotide

8024
β-Nicotinamide mononucleotide

8025
L-Nicotine

8026
Nifurthiazole

8027
Nitralin

8028
Nitranilic acid

8029
Nitrapyrin

8030
Nitrilotriacetic acid

No.	Name	Synonym	Mol. Form.	CAS RN	Mol. Wt.	Physical Form	mp/°C	bp/°C	den g cm⁻³	n_D	Solubility
8031	2,2',2"-Nitrilotriacetonitrile	Tricyanotrimethylamine	$C_6H_6N_4$	7327-60-8	134.139	nd (EtOH)	125.5				
8032	Nitroacetic acid		$C_2H_3NO_4$	625-75-2	105.050	nd (chl)	92 dec				vs bz, eth, EtOH, chl
8033	Nitroacetone		$C_3H_5NO_3$	10230-68-9	103.077	pl, nd (eth, bz)	50(4)	103[24]			vs bz, eth, EtOH
8034	2-Nitroaniline		$C_6H_6N_2O_2$	88-74-4	138.124		71(1)	285(2)	0.9015[25]		sl H_2O; s EtOH; vs eth, ace, bz, chl
8035	3-Nitroaniline		$C_6H_6N_2O_2$	99-09-2	138.124		112(2)	312(3)	0.9011[25]		sl H_2O, bz; s EtOH, eth, ace; vs MeOH
8036	4-Nitroaniline		$C_6H_6N_2O_2$	100-01-6	138.124	pa ye mcl nd (w)	147.7(0.5)	328(8)	1.424[20]		i H_2O; s EtOH, eth, ace; sl bz, DMSO
8037	2-Nitroanisole	1-Methoxy-2-nitrobenzene	$C_7H_7NO_3$	91-23-6	153.136	oily liq	9.4(0.9)	272(7)	1.2540[20]	1.5161[20]	i H_2O; msc EtOH, eth; s ctc
8038	3-Nitroanisole	1-Methoxy-3-nitrobenzene	$C_7H_7NO_3$	555-03-3	153.136	nd (al), pl (bz-lig)	38(2)	258	1.373[18]		i H_2O; s EtOH; vs eth
8039	4-Nitroanisole	1-Methoxy-4-nitrobenzene	$C_7H_7NO_3$	100-17-4	153.136	pr (al), nd (dil al)	54(1)	259(6)	1.2192[60]	1.5070[60]	i H_2O; vs EtOH, eth; s ctc; sl peth
8040	9-Nitroanthracene		$C_{14}H_9NO_2$	602-60-8	223.227	ye nd (al) pr (HOAc or xyl)	147.2(0.7)	275[17]			i H_2O; sl EtOH, chl; vs ace, CS_2
8041	1-Nitro-9,10-anthracenedione		$C_{14}H_7NO_4$	82-34-8	253.211	nd (HOAc) ye pr (ace)	231.5	270[7]			i H_2O; sl EtOH, eth; s ace, bz
8042	2-Nitrobenzaldehyde		$C_7H_5NO_3$	552-89-6	151.120	ye nd (w)	43.5(0.2)	153[23]	1.2844[20]		sl H_2O, chl; vs EtOH, eth, ace, bz
8043	3-Nitrobenzaldehyde		$C_7H_5NO_3$	99-61-6	151.120	lt ye nd (w)	58.0(0.2)	164[23]	1.2792[20]		sl H_2O; s EtOH, eth, chl; vs ace, bz
8044	4-Nitrobenzaldehyde		$C_7H_5NO_3$	555-16-8	151.120	lf, pr (w)	105(3)	sub	1.496[25]		sl H_2O, lig; vs EtOH; s bz, chl, HOAc
8045	3-Nitrobenzamide		$C_7H_6N_2O_3$	645-09-0	166.134		142.7	312.5			s H_2O, EtOH, eth
8046	4-Nitrobenzamide		$C_7H_6N_2O_3$	619-80-7	166.134	nd (w)	190(2)				i H_2O; s EtOH, eth
8047	Nitrobenzene		$C_6H_5NO_2$	98-95-3	123.110	oily liq	5.65(0.07)	210.7(0.3)	1.2037[20]	1.5562[20]	sl H_2O, ctc; vs EtOH, eth, ace, bz
8048	2-Nitrobenzeneacetic acid	o-Nitrophenylacetic acid	$C_8H_7NO_4$	3740-52-1	181.147	nd (w, pl (dil al)	138(2)				s H_2O, EtOH
8049	3-Nitrobenzeneacetic acid		$C_8H_7NO_4$	1877-73-2	181.147	nd (w)	117(2)				vs EtOH
8050	4-Nitrobenzeneacetic acid	p-Nitrophenylacetic acid	$C_8H_7NO_4$	104-03-0	181.147	pa ye nd (w)	154				sl H_2O; s EtOH, eth, bz
8051	2-Nitrobenzeneacetonitrile	2-Nitrobenzyl cyanide	$C_8H_6N_2O_2$	610-66-2	162.146	nd (dil al), pr (HOAc, al)	84	178[12]			vs ace, bz, eth, EtOH
8052	4-Nitrobenzeneacetonitrile	4-Nitrobenzyl cyanide	$C_8H_6N_2O_2$	555-21-5	162.146	pr (al)	117	196[12]			sl H_2O; s EtOH, eth, bz, chl
8053	4-Nitro-1,2-benzenediamine	4-Nitro-o-phenylenediamine	$C_6H_7N_3O_2$	99-56-9	153.139	dk red nd (dil al)	199.5				s acid
8054	4-Nitro-1,3-benzenediamine		$C_6H_7N_3O_2$	5131-58-8	153.139	oran pr (w)	161				
8055	5-Nitro-1,3-benzenediamine	1,3-Diamino-5-nitrobenzene	$C_6H_7N_3O_2$	5042-55-7	153.139	red cry (w)	143				
8056	2-Nitro-1,4-benzenediamine		$C_6H_7N_3O_2$	5307-14-2	153.139		140.0				
8057	3-Nitro-1,2-benzenedicarboxylic acid	3-Nitrophthalic acid	$C_8H_5NO_6$	603-11-2	211.129	pa ye pr (w)	218				sl H_2O, ace; s EtOH; i bz, peth, chl
8058	4-Nitro-1,2-benzenedicarboxylic acid	4-Nitrophthalic acid	$C_8H_5NO_6$	610-27-5	211.129	pa ye nd (w, eth)	164.8				s H_2O, EtOH; i bz, chl, CS_2, peth
8059	2-Nitrobenzeneethanol		$C_8H_9NO_3$	15121-84-3	167.162		1.0	267	1.19[25]	1.5637[20]	
8060	4-Nitrobenzeneethanol		$C_8H_9NO_3$	100-27-6	167.162		63	148[2]			
8061	2-Nitrobenzenemethanol	2-Nitrobenzyl alcohol	$C_7H_7NO_3$	612-25-9	153.136	nd (w)	74	270			sl H_2O; s EtOH, eth
8062	3-Nitrobenzenemethanol	3-Nitrobenzyl alcohol	$C_7H_7NO_3$	619-25-0	153.136	orth nd (w)	30.5	177[3]	1.296[19]		s H_2O, EtOH, eth; sl chl
8063	4-Nitrobenzenemethanol	4-Nitrobenzyl alcohol	$C_7H_7NO_3$	619-73-8	153.136	nd (w)	96.5	255 dec			sl H_2O, ace; s EtOH, eth
8064	2-Nitrobenzenesulfenyl chloride		$C_6H_4ClNO_2S$	7669-54-7	189.620	ye nd (bz)	75				vs eth, bz, chl

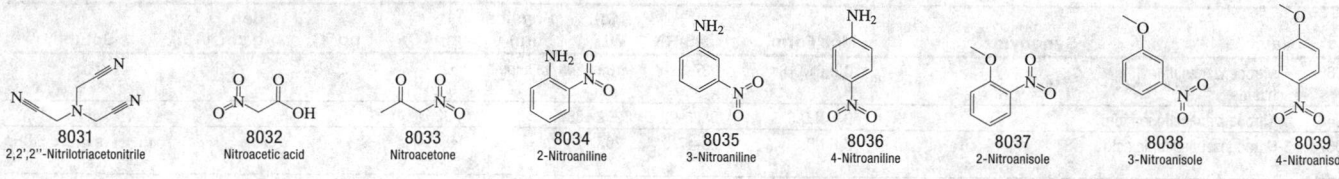

8031	8032	8033	8034	8035	8036	8037	8038	8039
2,2',2''-Nitrilotriacetonitrile	Nitroacetic acid	Nitroacetone	2-Nitroaniline	3-Nitroaniline	4-Nitroaniline	2-Nitroanisole	3-Nitroanisole	4-Nitroanisole

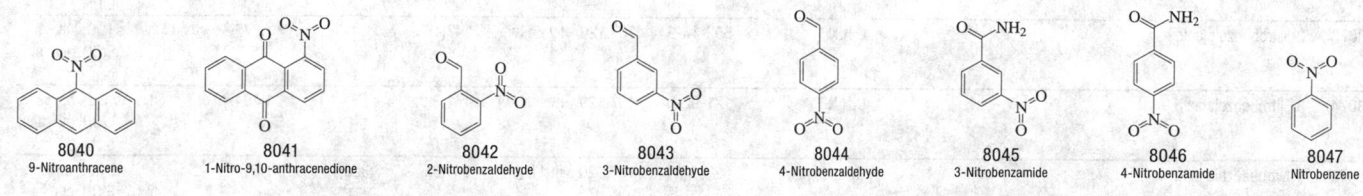

8040	8041	8042	8043	8044	8045	8046	8047
9-Nitroanthracene	1-Nitro-9,10-anthracenedione	2-Nitrobenzaldehyde	3-Nitrobenzaldehyde	4-Nitrobenzaldehyde	3-Nitrobenzamide	4-Nitrobenzamide	Nitrobenzene

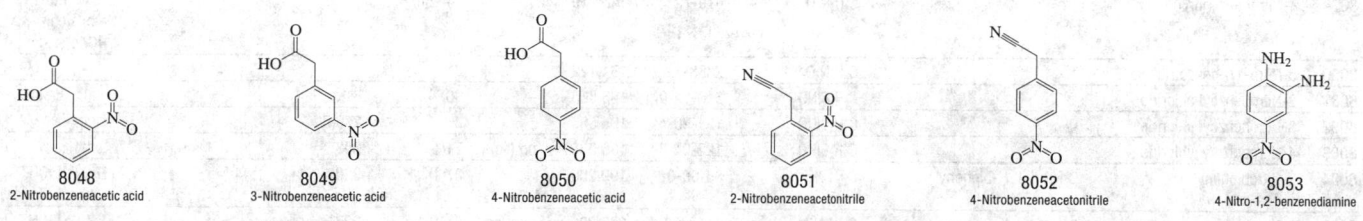

8048	8049	8050	8051	8052	8053
2-Nitrobenzeneacetic acid	3-Nitrobenzeneacetic acid	4-Nitrobenzeneacetic acid	2-Nitrobenzeneacetonitrile	4-Nitrobenzeneacetonitrile	4-Nitro-1,2-benzenediamine

8054	8055	8056	8057	8058
4-Nitro-1,3-benzenediamine	5-Nitro-1,3-benzenediamine	2-Nitro-1,4-benzenediamine	3-Nitro-1,2-benzenedicarboxylic acid	4-Nitro-1,2-benzenedicarboxylic acid

8059	8060	8061	8062	8063	8064
2-Nitrobenzeneethanol	4-Nitrobenzeneethanol	2-Nitrobenzenemethanol	3-Nitrobenzenemethanol	4-Nitrobenzenemethanol	2-Nitrobenzenesulfenyl chloride

No.	Name	Synonym	Mol. Form.	CAS RN	Mol. Wt.	Physical Form	mp/°C	bp/°C	den g cm⁻³	n_D	Solubility
8065	4-Nitrobenzenesulfenyl chloride		C$_6$H$_4$ClNO$_2$S	937-32-6	189.620	ye lf (peth)	52	125$^{0.1}$			vs bz
8066	4-Nitrobenzenesulfonamide		C$_6$H$_6$N$_2$O$_4$S	6325-93-5	202.188		-92.7(0.4)				
8067	3-Nitrobenzenesulfonic acid		C$_6$H$_5$NO$_5$S	98-47-5	203.173	pl	48				vs H$_2$O; s EtOH; i eth, bz
8068	4-Nitrobenzenesulfonic acid		C$_6$H$_5$NO$_5$S	138-42-1	203.173		95				vs H$_2$O
8069	2-Nitrobenzenesulfonyl chloride		C$_6$H$_4$ClNO$_4$S	1694-92-4	221.619	pr (lig, eth-peth)	68.5				s eth; sl peth
8070	3-Nitrobenzenesulfonyl chloride		C$_6$H$_4$ClNO$_4$S	121-51-7	221.619	mcl pr (eth) nd (lig)	64				i H$_2$O; s EtOH
8071	4-Nitrobenzenesulfonyl chloride		C$_6$H$_4$ClNO$_4$S	98-74-8	221.619	mcl pr (peth)	79.5	143$^{1.5}$			s peth
8072	5-Nitro-1H-benzimidazole		C$_7$H$_5$N$_3$O$_2$	94-52-0	163.134	nd (w)	207.8				i H$_2$O, eth, bz, chl; s acid; vs EtOH
8073	2-Nitrobenzoic acid		C$_7$H$_5$NO$_4$	552-16-9	167.120	tcl nd (w)	147(1)		1.575^{20}		s H$_2$O, eth; vs EtOH, ace; sl bz, lig
8074	3-Nitrobenzoic acid		C$_7$H$_5$NO$_4$	121-92-6	167.120	mcl pr (w)	141.3(0.4)		1.494^{20}		sl H$_2$O, bz; vs EtOH, eth, ace; s chl
8075	4-Nitrobenzoic acid		C$_7$H$_5$NO$_4$	62-23-7	167.120	mcl lf (w)	241(3)	320(14)	1.610^{20}		vs ace, eth, EtOH, chl, MeOH
8076	3-Nitrobenzoic acid, hydrazide		C$_7$H$_7$N$_3$O$_3$	618-94-0	181.149		153.5				sl H$_2$O, EtOH; i eth, bz, chl
8077	4-Nitrobenzoic acid, hydrazide		C$_7$H$_7$N$_3$O$_3$	636-97-5	181.149		215.5				sl H$_2$O, EtOH; i eth, bz, chl
8078	3-Nitrobenzonitrile		C$_7$H$_4$N$_2$O$_2$	619-24-9	148.119		116.6(0.6)	165^{16}			s H$_2$O, EtOH, bz; vs eth, ace; i peth
8079	4-Nitrobenzonitrile		C$_7$H$_4$N$_2$O$_2$	619-72-7	148.119		147.5(0.6)				sl H$_2$O, EtOH, eth; s chl, HOAc
8080	5-Nitro-1H-benzotriazole		C$_6$H$_4$N$_4$O$_2$	2338-12-7	164.122		217				
8081	2-Nitrobenzoyl chloride		C$_7$H$_4$ClNO$_3$	610-14-0	185.565		20				vs eth; sl ctc
8082	3-Nitrobenzoyl chloride		C$_7$H$_4$ClNO$_3$	121-90-4	185.565		36	276.5			vs eth
8083	4-Nitrobenzoyl chloride		C$_7$H$_4$ClNO$_3$	122-04-3	185.565	ye nd (lig)	71(1)	203^{105}			s eth
8084	2-Nitrobiphenyl	2-Nitro-1,1'-biphenyl	C$_{12}$H$_9$NO$_2$	86-00-0	199.205	pl (al, MeOH)	37(1)	320	1.44^{25}		i H$_2$O; s EtOH, eth, chl
8085	3-Nitrobiphenyl	3-Nitro-1,1'-biphenyl	C$_{12}$H$_9$NO$_2$	2113-58-8	199.205	ye pl or nd (dil al)	61(1)	227^{35}			i H$_2$O; s EtOH, eth, HOAc, lig
8086	4-Nitrobiphenyl	4-Nitro-1,1'-biphenyl	C$_{12}$H$_9$NO$_2$	92-93-3	199.205	ye nd (al)	112.9(0.7)	340			i H$_2$O; sl EtOH; s eth, bz, chl, HOAc
8087	2-Nitro-1,1-bis(p-chlorophenyl)propane		C$_{15}$H$_{13}$Cl$_2$NO$_2$	117-27-1	310.176	cry	82.0(0.5)	180$^{0.16}$			
8088	1-Nitrobutane		C$_4$H$_9$NO$_2$	627-05-4	103.120			152.8(0.4)	0.970^{25}	1.4303^{20}	sl H$_2$O; msc EtOH, eth; s alk
8089	2-Nitro-1-butanol		C$_4$H$_9$NO$_3$	609-31-4	119.119		-47	105^{10}	1.1332^{25}	1.4390^{20}	s H$_2$O, ace; msc EtOH, eth; sl ctc
8090	3-Nitro-2-butanol		C$_4$H$_9$NO$_3$	6270-16-2	119.119			91^9	1.1260^{20}	1.4414^{20}	
8091	6-Nitrochrysene		C$_{18}$H$_{11}$NO$_2$	7496-02-8	273.286	ye nd (bz)	213.5(0.9)				
8092	Nitrocyclohexane		C$_6$H$_{11}$NO$_2$	1122-60-7	129.157	liq	-34	205	1.0610^{20}	1.4612^{19}	i H$_2$O; s EtOH, lig
8093	1-Nitrodecane		C$_{10}$H$_{21}$NO$_2$	4609-87-4	187.280			86^1		1.4337^{20}	
8094	N-Nitrodiethylamine	N-Ethyl-N-nitroethanamine	C$_4$H$_{10}$N$_2$O$_2$	7119-92-8	118.134			206.5	1.057^{15}		vs eth, EtOH
8095	Nitroethane		C$_2$H$_5$NO$_2$	79-24-3	75.067	liq	-89.42(0.07)	114.1(0.2)	1.0448^{25}	1.3917^{20}	sl H$_2$O; msc EtOH, eth; s ace, chl
8096	2-Nitroethanol		C$_2$H$_5$NO$_3$	625-48-9	91.066	liq	-80	195(2)	1.270^{15}	1.4438^{19}	msc H$_2$O, EtOH, eth; i bz
8097	Nitroethene		C$_2$H$_3$NO$_2$	3638-64-0	73.051	liq	-55(2)	98.5	1.2212^{14}	1.4282^{20}	vs EtOH, eth, ace, bz, chl
8098	(2-Nitroethyl)benzene		C$_8$H$_9$NO$_2$	6125-24-2	151.163	liq	-23	250	1.126^{24}	1.5407^{19}	
8099	Nitrofen	Benzene, 2,4-dichloro-1-(4-nitrophenoxy)-	C$_{12}$H$_7$Cl$_2$NO$_3$	1836-75-5	284.095		71(2)				
8100	2-Nitro-9H-fluorene		C$_{13}$H$_9$NO$_2$	607-57-8	211.216	nd (50% HOAc ace)	155(1)				i H$_2$O; s ace, bz
8101	2-Nitro-9H-fluoren-9-one		C$_{13}$H$_7$NO$_3$	3096-52-4	225.200	ye nd or lf (HOAc)	224.3	sub			sl EtOH; s ace, sulf, HOAc

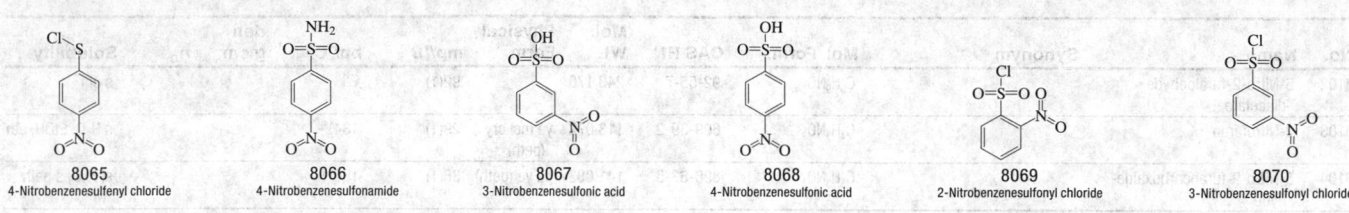

8065 4-Nitrobenzenesulfenyl chloride
8066 4-Nitrobenzenesulfonamide
8067 3-Nitrobenzenesulfonic acid
8068 4-Nitrobenzenesulfonic acid
8069 2-Nitrobenzenesulfonyl chloride
8070 3-Nitrobenzenesulfonyl chloride

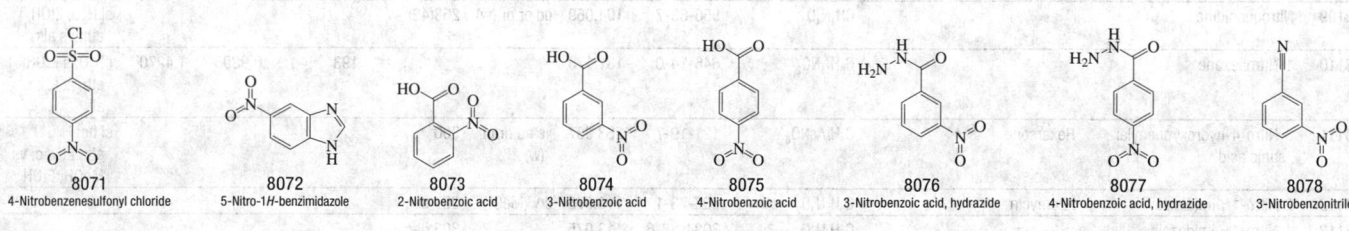

8071 4-Nitrobenzenesulfonyl chloride
8072 5-Nitro-1H-benzimidazole
8073 2-Nitrobenzoic acid
8074 3-Nitrobenzoic acid
8075 4-Nitrobenzoic acid
8076 3-Nitrobenzoic acid, hydrazide
8077 4-Nitrobenzoic acid, hydrazide
8078 3-Nitrobenzonitrile

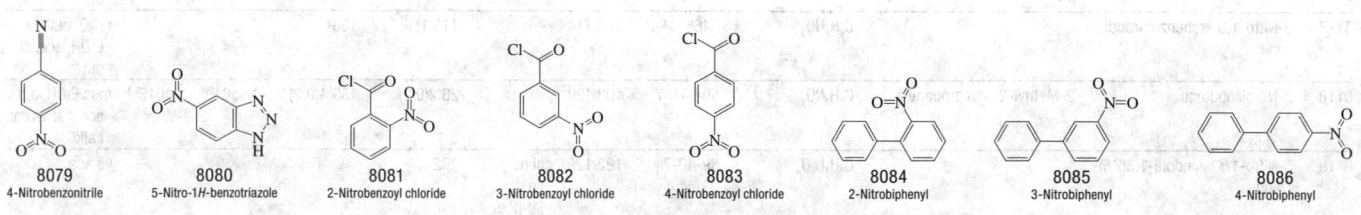

8079 4-Nitrobenzonitrile
8080 5-Nitro-1H-benzotriazole
8081 2-Nitrobenzoyl chloride
8082 3-Nitrobenzoyl chloride
8083 4-Nitrobenzoyl chloride
8084 2-Nitrobiphenyl
8085 3-Nitrobiphenyl
8086 4-Nitrobiphenyl

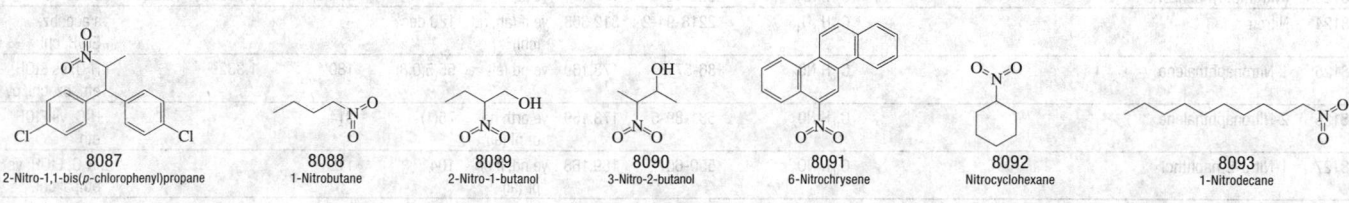

8087 2-Nitro-1,1-bis(p-chlorophenyl)propane
8088 1-Nitrobutane
8089 2-Nitro-1-butanol
8090 3-Nitro-2-butanol
8091 6-Nitrochrysene
8092 Nitrocyclohexane
8093 1-Nitrodecane

8094 N-Nitrodiethylamine
8095 Nitroethane
8096 2-Nitroethanol
8097 Nitroethene
8098 (2-Nitroethyl)benzene
8099 Nitrofen
8100 2-Nitro-9H-fluorene
8101 2-Nitro-9H-fluoren-9-one

No.	Name	Synonym	Mol. Form.	CAS RN	Mol. Wt.	Physical Form	mp/°C	bp/°C	den g cm⁻³	n_D	Solubility
8102	5-Nitro-2-furaldehyde diacetate		$C_9H_9NO_7$	92-55-7	243.170		92(1)				s chl
8103	2-Nitrofuran		$C_4H_3NO_3$	609-39-2	113.072	ye mcl cry (peth)	29(1)	134[123]			s H_2O, EtOH, eth
8104	5-Nitro-2-furancarboxalde-hyde		$C_5H_3NO_4$	698-63-5	141.083	pa ye (peth)	36(1)	130[10]			sl H_2O; s peth
8105	5-Nitro-2-furancarboxylic acid		$C_5H_3NO_5$	645-12-5	157.082	pa ye pl (w)	186(1)	sub			s H_2O, EtOH, eth; sl ace, bz; i chl
8106	Nitrofurantoin		$C_8H_6N_4O_5$	67-20-9	238.158		263				
8107	Nitrofurazone	2-[(5-Nitro-2-furanyl)-methylene]hydrazinecarbox-amide	$C_6H_6N_4O_4$	59-87-0	198.137	pa ye nd	237(1)				i H_2O, eth; sl EtOH, DMSO; s alk
8108	Nitrogen mustard *N*-oxide hydrochloride	Mechlorethamine oxide hydrochloride	$C_5H_{12}Cl_3NO$	302-70-5	208.514	pr (ace)	110				s H_2O
8109	Nitroguanidine		$CH_4N_4O_2$	556-88-7	104.069	nd or pr (w)	253(42)				sl H_2O, EtOH; i eth; vs alk
8110	1-Nitrohexane		$C_6H_{13}NO_2$	646-14-0	131.173			193	0.9396[20]	1.4270[20]	i H_2O; s EtOH, eth, ace, bz, alk
8111	3-Nitro-4-hydroxyphenylar-sonic acid	Roxarsone	$C_6H_6AsNO_6$	121-19-7	263.037	ye nd or pl (w)	300				sl hot H_2O; i eth, EtOAc; vs MeOH, EtOH
8112	2-Nitro-1*H*-imidazole	Azomycin	$C_3H_3N_3O_2$	527-73-1	113.075	cry (MeOH)	287 dec				
8113	4-Nitro-1*H*-imidazole		$C_3H_3N_3O_2$	3034-38-6	113.075		303 dec				
8114	5-Nitro-1*H*-indazole		$C_7H_5N_3O_2$	5401-94-5	163.134	ye nd or col nd (al)	208				s EtOH, eth, bz; vs ace, HOAc; i lig
8115	6-Nitro-1*H*-indazole		$C_7H_5N_3O_2$	7597-18-4	163.134	nd (w, al, ace)	181 dec				s H_2O, EtOH, eth, bz; vs ace; i lig
8116	4-Nitro-1,3-isobenzofurandi-one		$C_8H_3NO_5$	641-70-3	193.114	nd (ace, al)	163.4(0.2)				i H_2O; s EtOH, ace, HOAc; sl bz
8117	5-Nitro-1,3-isobenzofurandi-one		$C_8H_3NO_5$	5466-84-2	193.114		115(1)	196[8]			i H_2O, peth; s EtOH, ace; sl eth
8118	2-Nitroisobutane	2-Methyl-2-nitropropane	$C_4H_9NO_2$	594-70-7	103.120		26.2(0.2)	133.4(0.2)	0.9501[28]	1.4015[20]	msc EtOH, eth, ace, bz; vs chl; i alk
8119	5-Nitro-1*H*-isoindole-1,3(2*H*)-dione		$C_8H_4N_2O_4$	89-40-7	192.129	col nd (w), ye lf (al-ace)	202				vs ace
8120	Nitromersol		$C_7H_5HgNO_3$	133-58-4	351.71						i H_2O; sl ace, EtOH; s alk
8121	*N*-Nitromethanamine		$CH_4N_2O_2$	598-57-2	76.055		38	82[10]	1.2433[49]	1.4616[49]	vs H_2O, EtOH, bz, chl; s eth; sl peth
8122	Nitromethane	Nitrocarbol	CH_3NO_2	75-52-5	61.041	liq	-28.7(0.8)	101.19(0.1)	1.1371[20]	1.3817[20]	s H_2O, EtOH, eth, ace, ctc, alk
8123	(Nitromethyl)benzene		$C_7H_7NO_2$	622-42-4	137.137	ye liq		236(6)	1.1596[20]	1.5323[20]	vs ace, eth
8124	Nitron		$C_{20}H_{16}N_4$	2218-94-2	312.368	ye lf (al), nd (chl)	189 dec				vs ace, bz, EtOH, chl
8125	1-Nitronaphthalene		$C_{10}H_7NO_2$	86-57-7	173.169	ye nd (al)	55.5(0.8)	180[14]	1.332[20]		i H_2O; vs EtOH, eth, bz, chl, py
8126	2-Nitronaphthalene		$C_{10}H_7NO_2$	581-89-5	173.169	ye orth nd or pl (al)	75(1)	314			i H_2O; vs EtOH, eth
8127	1-Nitro-2-naphthol		$C_{10}H_7NO_3$	550-60-7	189.168	ye nd, lf or pr (al)	104	115[0.05]			s H_2O, EtOH; vs eth; sl chl
8128	1-Nitrooctane		$C_8H_{17}NO_2$	629-37-8	159.227		15	208.5	0.9346[20]	1.4322[20]	
8129	1-Nitropentane		$C_5H_{11}NO_2$	628-05-7	117.147			172.5	0.9525[20]	1.4175[20]	s EtOH, eth, bz
8130	3-Nitropentane		$C_5H_{11}NO_2$	551-88-2	117.147			154	0.957[0]		vs ace, eth, EtOH
8131	5-Nitro-1,10-phenanthroline		$C_{12}H_7N_3O_2$	4199-88-6	225.203		202.3				
8132	2-Nitrophenol		$C_6H_5NO_3$	88-75-5	139.109	ye nd or pr (eth, al)	44.9(0.5)	216	1.2942[40]	1.5723[50]	sl H_2O; vs EtOH, eth, ace, bz, py
8133	3-Nitrophenol		$C_6H_5NO_3$	554-84-7	139.109	ye mcl (eth, aq Hcl)	95(1)	194[70]	1.2797[100]		sl H_2O, DMSO; vs EtOH, eth, ace, bz
8134	4-Nitrophenol		$C_6H_5NO_3$	100-02-7	139.109	ye mcl pr (to)	113.8(0.2)		1.479[20]		sl H_2O; vs EtOH, eth, ace; s tol, py

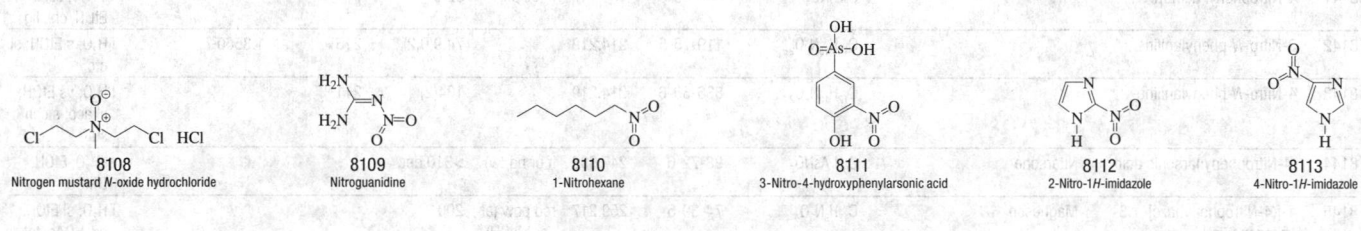

8102
5-Nitro-2-furaldehyde diacetate

8103
2-Nitrofuran

8104
5-Nitro-2-furancarboxaldehyde

8105
5-Nitro-2-furancarboxylic acid

8106
Nitrofurantoin

8107
Nitrofurazone

8108
Nitrogen mustard *N*-oxide hydrochloride

8109
Nitroguanidine

8110
1-Nitrohexane

8111
3-Nitro-4-hydroxyphenylarsonic acid

8112
2-Nitro-1*H*-imidazole

8113
4-Nitro-1*H*-imidazole

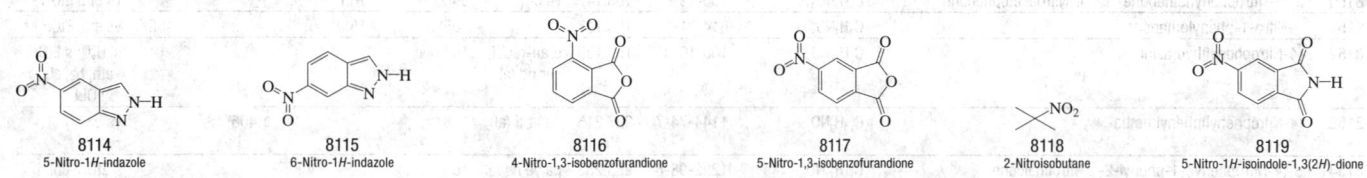

8114
5-Nitro-1*H*-indazole

8115
6-Nitro-1*H*-indazole

8116
4-Nitro-1,3-isobenzofurandione

8117
5-Nitro-1,3-isobenzofurandione

8118
2-Nitroisobutane

8119
5-Nitro-1*H*-isoindole-1,3(2*H*)-dione

8120
Nitromersol

8121
N-Nitromethanamine

8122
Nitromethane

8123
(Nitromethyl)benzene

8124
Nitron

8125
1-Nitronaphthalene

8126
2-Nitronaphthalene

8127
1-Nitro-2-naphthol

8128
1-Nitrooctane

8129
1-Nitropentane

8130
3-Nitropentane

8131
5-Nitro-1,10-phenanthroline

8132
2-Nitrophenol

8133
3-Nitrophenol

8134
4-Nitrophenol

No.	Name	Synonym	Mol. Form.	CAS RN	Mol. Wt.	Physical Form	mp/°C	bp/°C	den g cm⁻³	n_D	Solubility
8135	1-Nitro-2-phenoxybenzene		$C_{12}H_9NO_3$	2216-12-8	215.204	ye liq	<-20	235[60]	1.2539[22]	1.575[20]	vs bz, eth, EtOH, chl
8136	1-Nitro-4-phenoxybenzene		$C_{12}H_9NO_3$	620-88-2	215.204	pl (peth), MeOH)	61	320			i H_2O; sl EtOH, ctc; s eth, bz
8137	N-(2-Nitrophenyl)acetamide		$C_8H_8N_2O_3$	552-32-9	180.161		93.0(0.5)	100[0.1]	1.419[15]		s H_2O, EtOH, bz, chl, lig; vs eth
8138	N-(3-Nitrophenyl)acetamide		$C_8H_8N_2O_3$	122-28-1	180.161	wh lf (al)	154.5(0.5)	100[0.0074]			s H_2O, EtOH, chl; i eth; sl tfa
8139	N-(4-Nitrophenyl)acetamide		$C_8H_8N_2O_3$	104-04-1	180.161	ye pr (w)	216(1)	100[0.008]			sl H_2O, eth, chl; s EtOH, tfa, alk
8140	2-Nitrophenyl acetate		$C_8H_7NO_4$	610-69-5	181.147	nd or pr (lig)	40.5	253 dec			s H_2O; vs EtOH, eth, ace, bz; sl lig
8141	4-Nitrophenyl acetate		$C_8H_7NO_4$	830-03-5	181.147	lf (dil al)	82.3				vs H_2O, bz; s EtOH, chl, lig
8142	2-Nitro-N-phenylaniline		$C_{12}H_{10}N_2O_2$	119-75-5	214.219		74.9(0.2)	215[15]	1.3660[20]		i H_2O; s EtOH; sl ctc
8143	4-Nitro-N-phenylaniline		$C_{12}H_{10}N_2O_2$	836-30-6	214.219		134(2)	211[30]			i H_2O; vs EtOH; sl ace; s con sulf
8144	(4-Nitrophenyl)arsonic acid	Nitarsone	$C_6H_6AsNO_5$	98-72-6	247.038	lf or nd (w)	>310 dec				sl H_2O, EtOH, DMSO
8145	4-[(4-Nitrophenyl)azo]-1,3-benzenediol	Magneson	$C_{12}H_9N_3O_4$	74-39-5	259.217	red pow (al or MeOH)	200				i H_2O; sl EtOH, bz, HOAc, tol
8146	1-[(4-Nitrophenyl)azo]-2-naphthol		$C_{16}H_{11}N_3O_3$	6410-10-2	293.276	br-oran pl (to or bz)	257				vs bz, EtOH
8147	(3-Nitrophenyl)boronic acid		$C_6H_6BNO_4$	13331-27-6	166.928		274.5				
8148	1-(2-Nitrophenyl)ethanone	2-Nitroacetophenone	$C_8H_7NO_3$	577-59-3	165.147		28.5	178[32]	1.2370[25]	1.5468[20]	i H_2O; vs EtOH, eth, chl
8149	1-(3-Nitrophenyl)ethanone	3-Nitroacetophenone	$C_8H_7NO_3$	121-89-1	165.147	nd (al)	81(1)	202			vs H_2O, eth; sl EtOH, chl
8150	1-(4-Nitrophenyl)ethanone	4-Nitroacetophenone	$C_8H_7NO_3$	100-19-6	165.147	ye pr (al)	80(2)	165[5]			vs eth, EtOH
8151	2-Nitro-1-phenylethanone		$C_8H_7NO_3$	614-21-1	165.147		105(1)	158[16]		1.5468[30]	vs eth, EtOH
8152	(4-Nitrophenyl)hydrazine		$C_6H_7N_3O_2$	100-16-3	153.139	oran-red lf or nd (al)	158 dec				sl H_2O; s EtOH, eth, bz, chl, AcOEt
8153	(4-Nitrophenyl)phenylmethanone		$C_{13}H_9NO_3$	1144-74-7	227.215	nd or lf (al)	138		1.406[9]		vs bz
8154	3-(4-Nitrophenyl)-1-phenyl-2-propen-1-one	Nitrochalcone	$C_{15}H_{11}NO_3$	1222-98-6	253.253	pa ye nd (al) pl (bz)	164				s EtOH, chl; i eth, lig
8155	4-Nitrophenyl phosphate	4-Nitrophenyl dihydrogen phosphate	$C_6H_6NO_6P$	330-13-2	219.089	ye-wh nd	155				i cold H_2O; s EtOH, chl, bz
8156	3-(2-Nitrophenyl)propanoic acid	2-Nitrobenzenepropanoic acid	$C_9H_9NO_4$	2001-32-3	195.172	ye cry	115				
8157	3-(4-Nitrophenyl)propanoic acid	4-Nitrobenzenepropanoic acid	$C_9H_9NO_4$	16642-79-8	195.172	nd (w)	163				
8158	3-(4-Nitrophenyl)-2-propenal	4-Nitrocinnamaldehyde	$C_9H_7NO_3$	1734-79-8	177.157	nd (w, al)	141.5				s H_2O, eth, ace, bz; vs EtOH
8159	3-(2-Nitrophenyl)-2-propynoic acid	o-Nitrophenylpropiolic acid	$C_9H_5NO_4$	530-85-8	191.141		≈157 dec; may explode				sl H_2O; vs EtOH, eth; i CS_2O
8160	1-Nitro-4-(phenylthio)benzene		$C_{12}H_9NO_2S$	952-97-6	231.270	pa ye mcl pr (lig)	56	288[100]			vs eth, EtOH
8161	(4-Nitrophenyl)urea	p-Nitrophenylurea	$C_7H_7N_3O_3$	556-10-5	181.149	pr (al), nd (dil al)	238				vs H_2O, EtOH
8162	N-Nitropiperidine		$C_5H_{10}N_2O_2$	7119-94-0	130.145	liq	-5(1)	249(9)	1.1519[26]	1.4954[26]	
8163	1-Nitropropane		$C_3H_7NO_2$	108-03-2	89.094	liq	-104.3(0.6)	131.2(0.5)	0.9961[25]	1.4018[20]	sl H_2O; msc EtOH, eth; s chl
8164	2-Nitropropane	Isonitropropane	$C_3H_7NO_2$	79-46-9	89.094	liq	-91.3(0.2)	120.2(0.2)	0.9821[25]	1.3944[20]	sl H_2O; s chl
8165	3-Nitropropanoic acid		$C_3H_5NO_4$	504-88-1	119.077		62		1.59[20]		vs H_2O, EtOH, eth; s chl; i lig
8166	2-Nitro-1-propanol		$C_3H_7NO_3$	2902-96-7	105.093			120[32]	1.1841[25]	1.4379[20]	s H_2O, EtOH, eth; sl chl
8167	1-Nitro-1-propene		$C_3H_5NO_2$	3156-70-5	87.078			60[34]	1.0661[20]	1.4527[20]	s eth, ace, chl
8168	2-Nitro-1-propene		$C_3H_5NO_2$	4749-28-4	87.078	ye-grn liq		52[80]	1.0559[25]	1.4358[20]	s eth, ace, chl
8169	5-Nitro-2-propoxyaniline		$C_9H_{12}N_2O_3$	553-79-7	196.202	oran (PrOH-peth)	49				vs EtOH
8170	N-(5-Nitro-2-propoxyphenyl)-acetamide	5'-Nitro-2'-propoxyacetanilide	$C_{11}H_{14}N_2O_4$	553-20-8	238.240	cry (PrOH)	102.5				
8171	1-Nitropyrene		$C_{16}H_9NO_2$	5522-43-0	247.248	ye nd (MeCN)	152(1)				

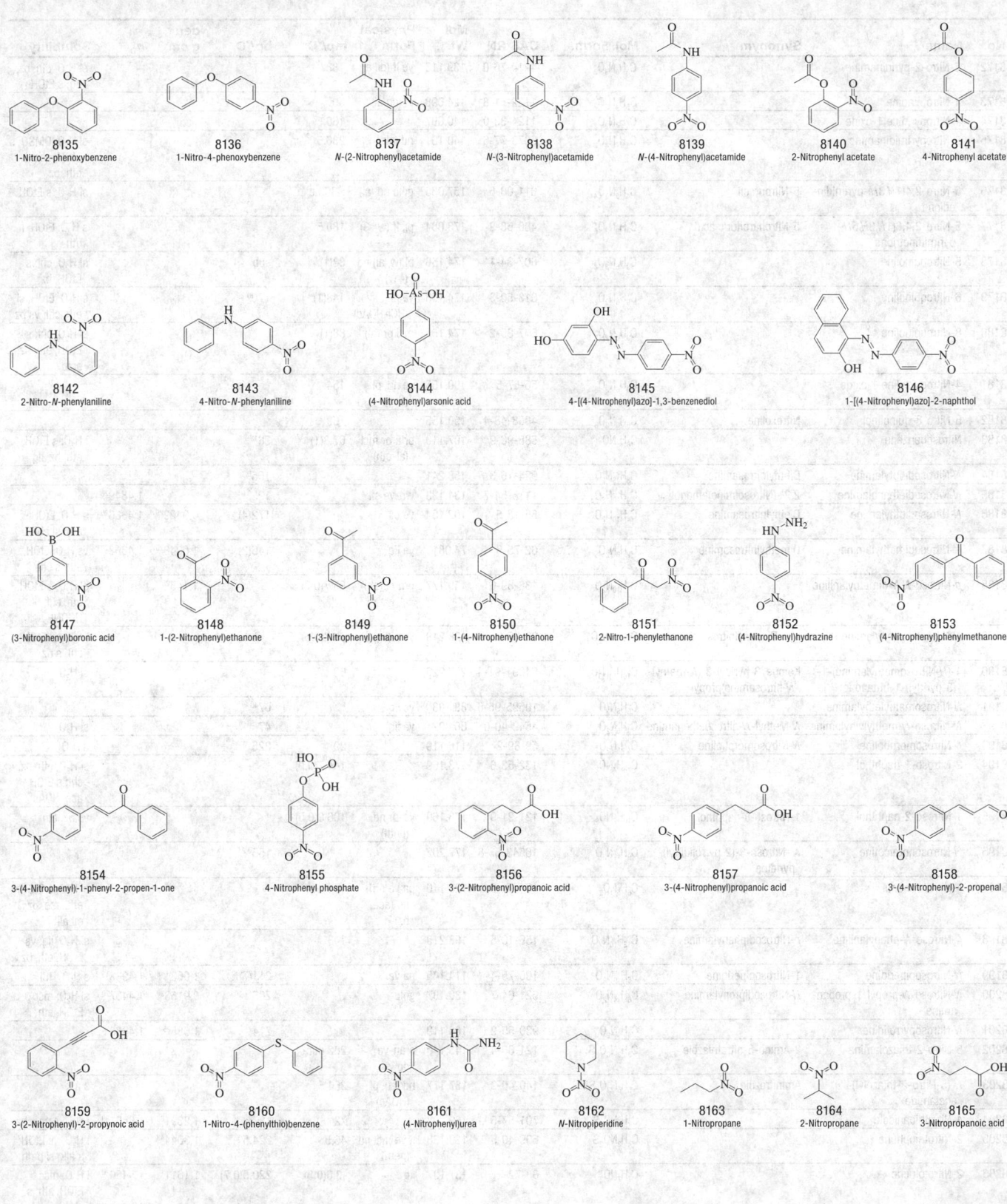

8135
1-Nitro-2-phenoxybenzene

8136
1-Nitro-4-phenoxybenzene

8137
N-(2-Nitrophenyl)acetamide

8138
N-(3-Nitrophenyl)acetamide

8139
N-(4-Nitrophenyl)acetamide

8140
2-Nitrophenyl acetate

8141
4-Nitrophenyl acetate

8142
2-Nitro-N-phenylaniline

8143
4-Nitro-N-phenylaniline

8144
(4-Nitrophenyl)arsonic acid

8145
4-[(4-Nitrophenyl)azo]-1,3-benzenediol

8146
1-[(4-Nitrophenyl)azo]-2-naphthol

8147
(3-Nitrophenyl)boronic acid

8148
1-(2-Nitrophenyl)ethanone

8149
1-(3-Nitrophenyl)ethanone

8150
1-(4-Nitrophenyl)ethanone

8151
2-Nitro-1-phenylethanone

8152
(4-Nitrophenyl)hydrazine

8153
(4-Nitrophenyl)phenylmethanone

8154
3-(4-Nitrophenyl)-1-phenyl-2-propen-1-one

8155
4-Nitrophenyl phosphate

8156
3-(2-Nitrophenyl)propanoic acid

8157
3-(4-Nitrophenyl)propanoic acid

8158
3-(4-Nitrophenyl)-2-propenal

8159
3-(2-Nitrophenyl)-2-propynoic acid

8160
1-Nitro-4-(phenylthio)benzene

8161
(4-Nitrophenyl)urea

8162
N-Nitropiperidine

8163
1-Nitropropane

8164
2-Nitropropane

8165
3-Nitropropanoic acid

8166
2-Nitro-1-propanol

8167
1-Nitro-1-propene

8168
2-Nitro-1-propene

8169
5-Nitro-2-propoxyaniline

8170
N-(5-Nitro-2-propoxyphenyl)acetamide

8171
1-Nitropyrene

No.	Name	Synonym	Mol. Form.	CAS RN	Mol. Wt.	Physical Form	mp/°C	bp/°C	den g cm^{-3}	n_D	Solubility
8172	5-Nitro-2-pyridinamine		C$_5$H$_5$N$_3$O$_2$	4214-76-0	139.113	ye lf (dil al)	188				sl H$_2$O, eth, bz, lig; s EtOH
8173	4-Nitropyridine		C$_5$H$_4$N$_2$O$_2$	1122-61-8	124.098	pl (aq al)	50				
8174	4-Nitropyridine 1-oxide		C$_5$H$_4$N$_2$O$_3$	1124-33-0	140.097		160.5				
8175	5-Nitropyrimidinamine		C$_4$H$_4$N$_4$O$_2$	3073-77-6	140.101	nd (al)	236.5				sl H$_2$O, DMSO; s EtOH, ace; i eth, bz
8176	5-Nitro-2,4(1H,3H)-pyrimidin-edione	5-Nitrouracil	C$_4$H$_3$N$_3$O$_4$	611-08-5	157.085	gold nd (al)	>300 exp				sl H$_2$O; s EtOH
8177	5-Nitro-2,4,6(1H,3H,5H)-pyrimidinetrione	5-Nitrobarbituric acid	C$_4$H$_3$N$_3$O$_5$	480-68-2	173.084	pr, lf (w+3)	180.5				s H$_2$O, EtOH; i eth
8178	5-Nitroquinoline		C$_9$H$_6$N$_2$O$_2$	607-34-1	174.156	pl (w, al) nd (+w)	88(1)	sub			sl H$_2$O, chl; s EtOH, bz
8179	6-Nitroquinoline		C$_9$H$_6$N$_2$O$_2$	613-50-3	174.156	ye pl (HCl-HOAc)	148(1)	170$^{0.2}$			s H$_2$O, EtOH; sl eth, chl; vs bz
8180	8-Nitroquinoline		C$_9$H$_6$N$_2$O$_2$	607-35-2	174.156	mcl pr (al)	88(1)				sl H$_2$O, chl; s EtOH, eth, bz, acid
8181	4-Nitroquinoline 1-oxide		C$_9$H$_6$N$_2$O$_3$	56-57-5	190.155	ye nd, pl (ace)	154				
8182	5-Nitro-8-quinolinol	Nitroxoline	C$_9$H$_6$N$_2$O$_3$	4008-48-4	190.155		180				
8183	Nitrosobenzene		C$_6$H$_5$NO	586-96-9	107.110	orth or mcl (al-eth)	67.8(1)	58^{18}			i H$_2$O; s EtOH, eth, bz, lig
8184	N-Nitrosodibutylamine	Dibutylnitrosamine	C$_8$H$_{18}$N$_2$O	924-16-3	158.241			105^8			
8185	N-Nitrosodiethanolamine	2,2'-(Nitrosoimino)ethanol	C$_4$H$_{10}$N$_2$O$_3$	1116-54-7	134.133	wh-ye oil		125$^{0.01}$		1.4849^{20}	
8186	N-Nitrosodiethylamine	Diethylnitrosamine	C$_4$H$_{10}$N$_2$O	55-18-5	102.134	ye oil		172(4)	0.9422^{20}	1.4386^{20}	s H$_2$O, EtOH; eth; sl chl
8187	N-Nitrosodimethylamine	Dimethylnitrosamine	C$_2$H$_6$N$_2$O	62-75-9	74.081	ye liq		146(2)	1.0048^{20}	1.4368^{20}	vs H$_2$O, EtOH, eth; s chl
8188	p-Nitroso-N,N-dimethylaniline		C$_8$H$_{10}$N$_2$O	138-89-6	150.177	grn pl (eth)	85.7(0.7)		1.145^{20}		sl H$_2$O; s EtOH, eth, chl, HCONH$_2$
8189	N-Nitrosodiphenylamine	N,N-Diphenylnitrosamine	C$_{12}$H$_{10}$N$_2$O	86-30-6	198.219	ye pl(lig)	66.7(0.6)				i H$_2$O; sl EtOH, chl; s bz
8190	4-(N-Nitrosomethylamino)-1-(3-pyridyl)-1-butanone	Ketone, 3-pyridyl-3-(N-methyl-N-nitrosamino)propyl	C$_{10}$H$_{13}$N$_3$O$_2$	64091-91-4	207.229		63				sl H$_2$O
8191	N-Nitrosomethylethylamine		C$_3$H$_8$N$_2$O	10595-95-6	88.108	ye liq		67^{40}			
8192	N-Nitroso-N-methylvinylamine	N-Methyl-N-nitrosoethenamine	C$_3$H$_6$N$_2$O	4549-40-0	86.092	ye liq		47			sl H$_2$O
8193	4-Nitrosomorpholine	N-Nitrosomorpholine	C$_4$H$_8$N$_2$O$_2$	59-89-2	116.119		29	225			s H$_2$O
8194	2-Nitroso-1-naphthol		C$_{10}$H$_7$NO$_2$	132-53-6	173.169		144.7(0.9)				sl H$_2$O, eth, bz, chl; s EtOH, ace, HOAc
8195	1-Nitroso-2-naphthol	1-Nitroso-β-naphthol	C$_{10}$H$_7$NO$_2$	131-91-9	173.169	ye-br nd (peth)	106.8(0.5)				vs bz, eth
8196	N-Nitrosonornicotine	N'-Nitroso-3-(2-pyrrolidinyl)-pyridine	C$_9$H$_{11}$N$_3$O	16543-55-8	177.202			155$^{0.2}$			
8197	4-Nitrosophenol		C$_6$H$_5$NO$_2$	104-91-6	123.110	pa ye orth nd (ace, bz)	144 dec				sl H$_2$O; s EtOH, eth, ace, bz, dil alk
8198	4-Nitroso-N-phenylaniline	p-Nitrosodiphenylamine	C$_{12}$H$_{10}$N$_2$O	156-10-5	198.219		143				sl H$_2$O, lig; vs EtOH, eth, bz
8199	N-Nitrosopiperidine	1-Nitrosopiperidine	C$_5$H$_{10}$N$_2$O	100-75-4	114.145	pa ye		211(7)	1.0631^{18}	1.4933^{18}	s H$_2$O, HCl
8200	N-Nitroso-N-propyl-1-propan-amine	N-Nitrosodipropylamine	C$_6$H$_{14}$N$_2$O	621-64-7	130.187	gold		206	0.9163^{20}	1.4437^{20}	sl H$_2$O; msc EtOH, eth
8201	N-Nitrosopyrrolidine		C$_4$H$_8$N$_2$O	930-55-2	100.119			214	1.085^{25}	1.4880^{25}	
8202	5-Nitro-2-thiazolamine	2-Amino-5-nitrothiazole	C$_3$H$_3$N$_3$O$_2$S	121-66-4	145.140	oran-ye pow	202 dec				
8203	N-(5-Nitro-2-thiazolyl)-acetamide	Aminitrozole	C$_5$H$_5$N$_3$O$_3$S	140-40-9	187.177	nd (al), pl (HOAc)	264.5				s alk
8204	4-Nitrothioanisole		C$_7$H$_7$NO$_2$S	701-57-5	169.202		72	137^2	1.2391^{80}	1.6401^{20}	i H$_2$O; s ace, bz
8205	2-Nitrothiophene		C$_4$H$_3$NO$_2$S	609-40-5	129.138	lt ye mcl nd (peth)	46.5	224.5	1.3644^{43}		i H$_2$O; vs EtOH; s alk; sl peth
8206	2-Nitrotoluene		C$_7$H$_7$NO$_2$	88-72-2	137.137	liq	-3.6(0.8)	220.9(0.7)	1.1611^{19}	1.5450^{20}	i H$_2$O; msc EtOH, eth; s ctc
8207	3-Nitrotoluene		C$_7$H$_7$NO$_2$	99-08-1	137.137	pa ye	15.9(0.6)	232.1(0.4)	1.1581^{20}	1.5466^{20}	i H$_2$O; s EtOH, bz, ctc; msc eth
8208	4-Nitrotoluene		C$_7$H$_7$NO$_2$	99-99-0	137.137	orth cry (al, eth)	51.7(0.3)	238.66(0.09)	1.1038^{75}		i H$_2$O; s EtOH; vs eth, ace, bz, chl

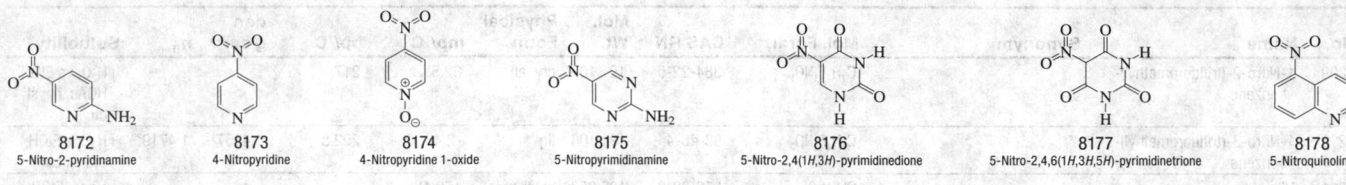

8172
5-Nitro-2-pyridinamine

8173
4-Nitropyridine

8174
4-Nitropyridine 1-oxide

8175
5-Nitropyrimidinamine

8176
5-Nitro-2,4(1H,3H)-pyrimidinedione

8177
5-Nitro-2,4,6(1H,3H,5H)-pyrimidinetrione

8178
5-Nitroquinoline

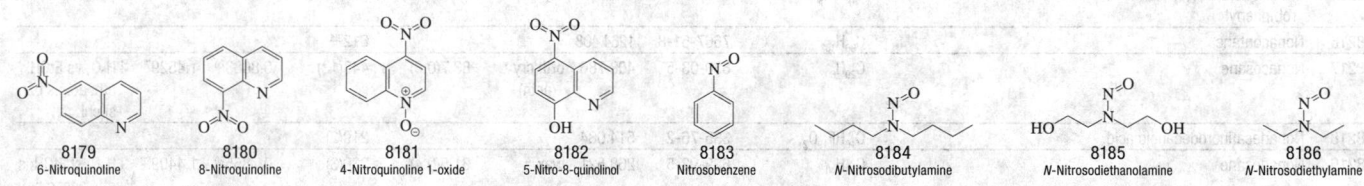

8179
6-Nitroquinoline

8180
8-Nitroquinoline

8181
4-Nitroquinoline 1-oxide

8182
5-Nitro-8-quinolinol

8183
Nitrosobenzene

8184
N-Nitrosodibutylamine

8185
N-Nitrosodiethanolamine

8186
N-Nitrosodiethylamine

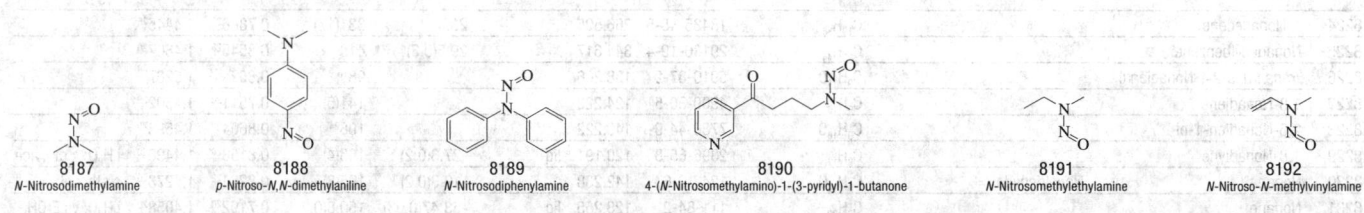

8187
N-Nitrosodimethylamine

8188
p-Nitroso-N,N-dimethylaniline

8189
N-Nitrosodiphenylamine

8190
4-(N-Nitrosomethylamino)-1-(3-pyridyl)-1-butanone

8191
N-Nitrosomethylethylamine

8192
N-Nitroso-N-methylvinylamine

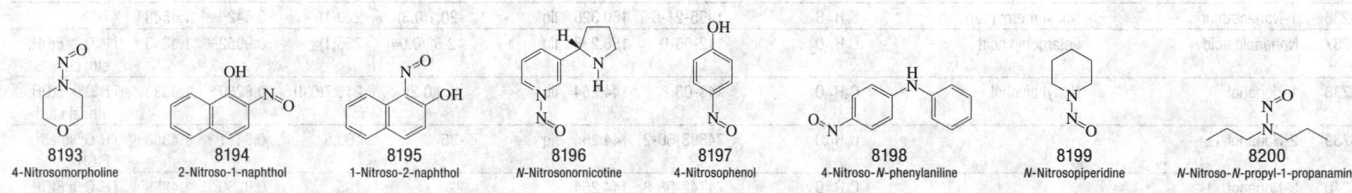

8193
4-Nitrosomorpholine

8194
2-Nitroso-1-naphthol

8195
1-Nitroso-2-naphthol

8196
N-Nitrosonornicotine

8197
4-Nitrosophenol

8198
4-Nitroso-N-phenylaniline

8199
N-Nitrosopiperidine

8200
N-Nitroso-N-propyl-1-propanamine

8201
N-Nitrosopyrrolidine

8202
5-Nitro-2-thiazolamine

8203
N-(5-Nitro-2-thiazolyl)acetamide

8204
4-Nitrothioanisole

8205
2-Nitrothiophene

8206
2-Nitrotoluene

8207
3-Nitrotoluene

8208
4-Nitrotoluene

No.	Name	Synonym	Mol. Form.	CAS RN	Mol. Wt.	Physical Form	mp/°C	bp/°C	den g cm^{-3}	n_D	Solubility
8209	1-Nitro-2-(trifluoromethyl)-benzene		$C_7H_4F_3NO_2$	384-22-5	191.108	cry (al)	32.5	217			i H_2O; vs EtOH, HOAc, bz; sl ctc
8210	1-Nitro-3-(trifluoromethyl)-benzene		$C_7H_4F_3NO_2$	98-46-4	191.108	liq	-2.4	202.8	1.4357[15]	1.4719[20]	i H_2O; s EtOH, eth; sl ctc
8211	Nitrourea		$CH_3N_3O_3$	556-89-8	105.053	pl (al-peth)	159(1)				vs ace, EtOH
8212	trans-(2-Nitrovinyl)benzene		$C_8H_7NO_2$	5153-67-3	149.148	ye pr (peth, al)	57(4)	255			i H_2O; s EtOH, ace; vs eth, chl, CS_2
8213	Nivalenol		$C_{15}H_{20}O_7$	23282-20-4	312.316	cry (MeOH)	224 dec				sl H_2O; s EtOH, MeOH
8214	Nizatidine		$C_{12}H_{21}N_5O_2S_2$	76963-41-2	331.458	cry (EtOH/AcOEt)	131				sl H_2O; s MeOH; vs chl; i bz, eth
8215	2,2',3,3',4,5,5',6,6'-Nonachlorobiphenyl		$C_{12}HCl_9$	52663-77-1	464.213	cry	182.6(0.5)				i H_2O
8216	Nonacontane		$C_{90}H_{182}$	7667-51-8	1264.408			612[200]			
8217	Nonacosane		$C_{29}H_{60}$	630-03-5	408.786	orth cry (peth)	63.7(0.6)	443(13)	0.8083[20]	1.4529[20]	i H_2O; vs EtOH, eth, ace; s bz; sl chl
8218	Nonadecafluorodecanoic acid		$C_{10}HF_{19}O_2$	335-76-2	514.084			219(2)			
8219	Nonadecane		$C_{19}H_{40}$	629-92-5	268.521	wax	31.5(0.4)	330(3)	0.7855[20]	1.4409[20]	i H_2O; sl EtOH; s eth, ace, ctc
8220	Nonadecanoic acid		$C_{19}H_{38}O_2$	646-30-0	298.504	lf (al)	68.06(0.04)	297[100]	0.8468[70]		i H_2O; vs EtOH, eth, bz, chl, lig
8221	1-Nonadecanol		$C_{19}H_{40}O$	1454-84-8	284.520	cry (ace)	61.4(0.5)	345		1.4328[75]	s eth, ace
8222	2-Nonadecanone		$C_{19}H_{38}O$	629-66-3	282.504	pr (al)	55.1(0.9)	266[110]	0.8108[56]		i H_2O; sl EtOH; s ace, bz; vs eth, ctc
8223	10-Nonadecanone		$C_{19}H_{38}O$	504-57-4	282.504	lf(al)	56.8(0.3)	334(11)			i H_2O; sl EtOH; s eth, ace, lig; vs bz
8224	1-Nonadecene		$C_{19}H_{38}$	18435-45-5	266.505		23.4	331(17)	0.7886[25]	1.4445[25]	
8225	Nonadecylbenzene		$C_{25}H_{44}$	29136-19-4	344.617		29.5(0.3)	419	0.8545[20]	1.4807[20]	
8226	trans,trans-2,4-Nonadienal		$C_9H_{14}O$	5910-87-2	138.206			98[10]	0.862[25]	1.5207[20]	
8227	1,8-Nonadiene		C_9H_{16}	4900-30-5	124.223			141(5)	0.7511[20]	1.4302[20]	
8228	2,6-Nonadien-1-ol		$C_9H_{16}O$	7786-44-9	140.222			108[24]	0.8604[25]	1.4598[25]	
8229	1,8-Nonadiyne		C_9H_{12}	2396-65-8	120.191	liq	-27.3(0.2)	165(4)	0.8158[20]	1.4490[20]	i H_2O; s eth, ace
8230	Nonanal	Nonaldehyde	$C_9H_{18}O$	124-19-6	142.238		-19.3(0.2)	195(3)	0.8264[22]	1.4273[20]	s eth, chl
8231	Nonane		C_9H_{20}	111-84-2	128.255	liq	-53.47(0.03)	150.8(0.2)	0.7192[20]	1.4058[20]	i H_2O; vs EtOH, eth; msc ace, bz, hp
8232	Nonanedioic acid	Azelaic acid	$C_9H_{16}O_4$	123-99-9	188.221	lf or nd	106.5	357.1	1.225[25]	1.4303[111]	sl H_2O, eth, bz, DMSO; s EtOH
8233	1,9-Nonanediol		$C_9H_{20}O_2$	3937-56-2	160.254	cry (bz)	46.4(0.1)	173[20]			sl H_2O; vs EtOH, eth; s bz; i lig
8234	Nonanedioyl dichloride		$C_9H_{14}Cl_2O_2$	123-98-8	225.112			166[18]	1.143	1.4680[20]	s eth; vs bz
8235	Nonanenitrile		$C_9H_{17}N$	2243-27-8	139.238	liq	-49.5(0.5)	225(5)	0.8178[20]	1.4255[20]	i H_2O; s EtOH, eth; sl ctc
8236	1-Nonanethiol	Nonyl mercaptan	$C_9H_{20}S$	1455-21-6	160.320	liq	-20.1(0.3)	220(1)	0.842[25]	1.4548[20]	
8237	Nonanoic acid	Pelargonic acid	$C_9H_{18}O_2$	112-05-0	158.238	liq	12.38(0.04)	256(1)	0.9052[20]	1.4343[19]	i H_2O; s EtOH, eth, chl
8238	1-Nonanol	Nonyl alcohol	$C_9H_{20}O$	143-08-8	144.254	liq	-5.0(0.2)	213.7(0.4)	0.8280[20]	1.4333[20]	i H_2O; s EtOH, eth; sl ctc
8239	2-Nonanol, (±)-		$C_9H_{20}O$	74683-66-2	144.254	liq	-35	193.5	0.8471[20]	1.4353[20]	i H_2O; vs eth, EtOH
8240	3-Nonanol, (±)-		$C_9H_{20}O$	74742-08-8	144.254		22	195	0.8250[20]	1.4289[20]	i H_2O; s EtOH, eth
8241	4-Nonanol		$C_9H_{20}O$	52708-03-9	144.254			192.5	0.8282[20]	1.4197[20]	i H_2O; s EtOH, eth
8242	5-Nonanol	Dibutylcarbinol	$C_9H_{20}O$	623-93-8	144.254		5.6(0.4)	195(2)	0.8220[20]	1.4289[20]	i H_2O; s EtOH
8243	2-Nonanone	Heptyl methyl ketone	$C_9H_{18}O$	821-55-6	142.238	liq	-7.4(0.2)	194(1)	0.8208[20]	1.4210[20]	i H_2O; s EtOH, eth, bz; vs ace, chl
8244	3-Nonanone	Ethyl hexyl ketone	$C_9H_{18}O$	925-78-0	142.238	liq	-8(4)	187(4)	0.8241[20]	1.4208[20]	i H_2O; s EtOH, eth, bz, chl; vs ace
8245	4-Nonanone	Pentyl propyl ketone	$C_9H_{18}O$	4485-09-0	142.238			188(4)	0.8190[25]	1.4189[20]	i H_2O; s EtOH, eth, chl; vs ace
8246	5-Nonanone	Dibutyl ketone	$C_9H_{18}O$	502-56-7	142.238	liq	-3.84(0.05)	188.4(0.3)	0.8217[20]	1.4195[20]	i H_2O; s EtOH; vs eth, chl
8247	Nonanoyl chloride		$C_9H_{17}ClO$	764-85-2	176.683	liq	-60.5	218(13)	0.9463[15]		s eth, ace

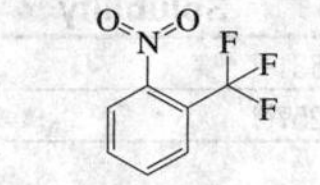

8209
1-Nitro-2-(trifluoromethyl)benzene

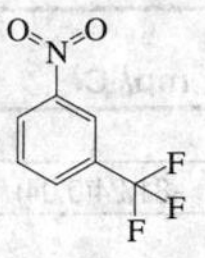

8210
1-Nitro-3-(trifluoromethyl)benzene

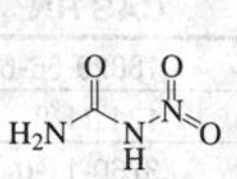

8211
Nitrourea

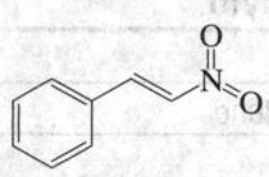

8212
trans-(2-Nitrovinyl)benzene

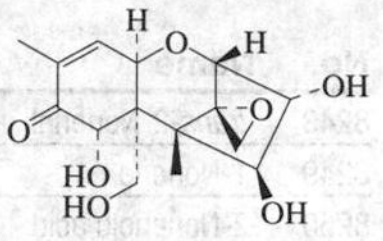

8213
Nivalenol

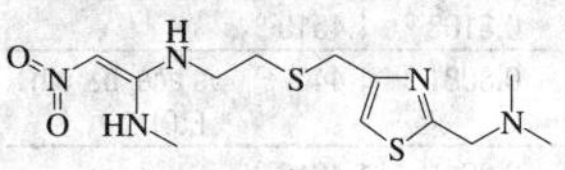

8214
Nizatidine

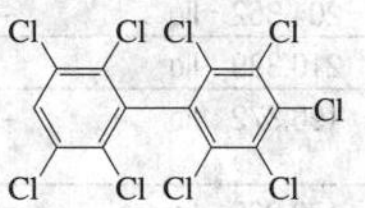

8215
2,2',3,3',4,5,5',6,6'-Nonachlorobiphenyl

8216
Nonacontane

$H_3C(CH_2)_{88}CH_3$

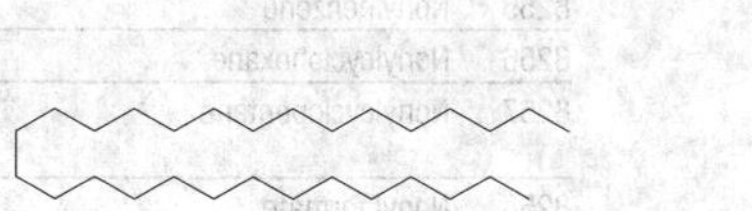

8217
Nonacosane

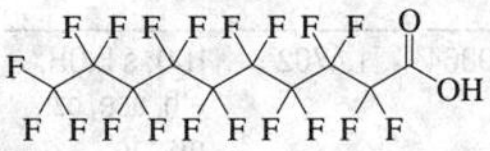

8218
Nonadecafluorodecanoic acid

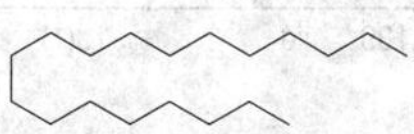

8219
Nonadecane

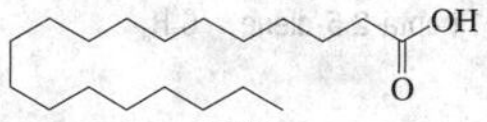

8220
Nonadecanoic acid

8221
1-Nonadecanol

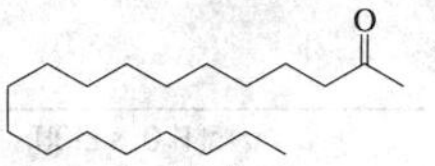

8222
2-Nonadecanone

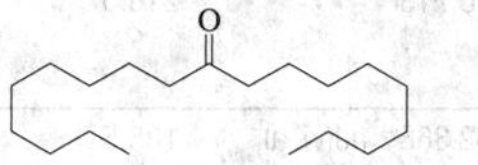

8223
10-Nonadecanone

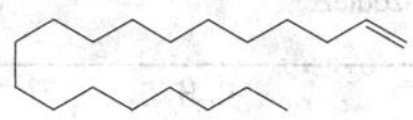

8224
1-Nonadecene

8225
Nonadecylbenzene

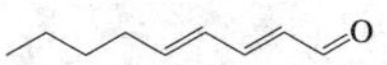

8226
trans,trans-2,4-Nonadienal

8227
1,8-Nonadiene

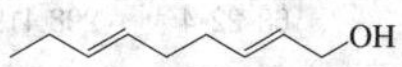

8228
2,6-Nonadien-1-ol

8229
1,8-Nonadiyne

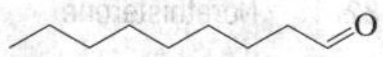

8230
Nonanal

8231
Nonane

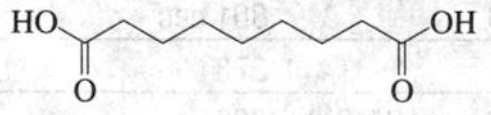

8232
Nonanedioic acid

8233
1,9-Nonanediol

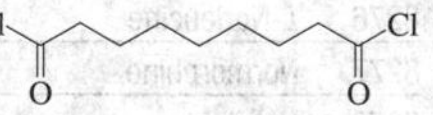

8234
Nonanedioyl dichloride

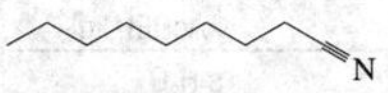

8235
Nonanenitrile

8236
1-Nonanethiol

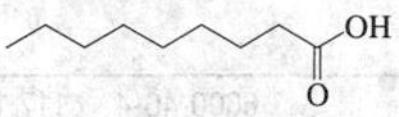

8237
Nonanoic acid

8238
1-Nonanol

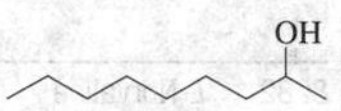

8239
2-Nonanol, (±)-

8240
3-Nonanol, (±)-

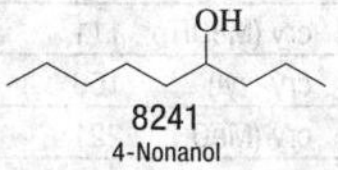

8241
4-Nonanol

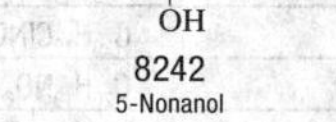

8242
5-Nonanol

8243
2-Nonanone

8244
3-Nonanone

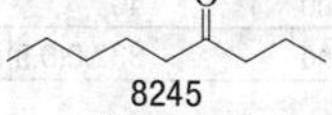

8245
4-Nonanone

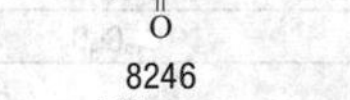

8246
5-Nonanone

8247
Nonanoyl chloride

No.	Name	Synonym	Mol. Form.	CAS RN	Mol. Wt.	Physical Form	mp/°C	bp/°C	den g cm⁻³	n_D	Solubility
8248	*trans*-2-Nonenal		C₉H₁₆O	18829-56-6	140.222	liq		101[16]	0.846	1.4531[20]	
8249	1-Nonene	1-Nonylene	C₉H₁₈	124-11-8	126.239	liq	-81.24(0.04)	146.9(0.6)	0.7253[25]	1.4257[20]	
8250	2-Nonenoic acid		C₉H₁₆O₂	3760-11-0	156.222			173[20]			
8251	3-Nonenoic acid		C₉H₁₆O₂	4124-88-3	156.222		-4.4	156[18]	0.9254[20]	1.4454[25]	
8252	1-Nonen-3-ol	1-Vinylheptanol	C₉H₁₈O	21964-44-3	142.238			193.5	0.824[21]	1.4382[15]	
8253	Nonyl acetate		C₁₁H₂₂O₂	143-13-5	186.292	liq	-26	225(8)	0.8785[15]	1.426[20]	
8254	Nonylamine	1-Nonanamine	C₉H₂₁N	112-20-9	143.270	liq	-1	198(3)	0.7886[20]	1.4336[20]	sl H₂O, chl; s EtOH, eth
8255	Nonylbenzene		C₁₅H₂₄	1081-77-2	204.352	liq	-24	280(3)	0.8584[20]	1.4816[20]	
8256	Nonylcyclohexane		C₁₅H₃₀	2883-02-5	210.399	liq	-10	281(3)	0.8163[20]	1.4519[20]	
8257	Nonylcyclopentane		C₁₄H₂₈	2882-98-6	196.372	liq	-29	262	0.8081[20]	1.4467[20]	vs ace, bz, eth, EtOH
8258	Nonyl formate		C₁₀H₂₀O₂	5451-92-3	172.265	liq	-33	216(13)	0.86	1.4216[20]	
8259	1-Nonylnaphthalene		C₁₉H₂₆	26438-26-6	254.409		8	366	0.9371[20]	1.5477[20]	
8260	4-Nonylphenol		C₁₅H₂₄O	104-40-5	220.351	visc ye liq	42	317(19)	0.950[20]	1.513[20]	i H₂O; s bz, ctc, hp
8261	1-Nonyne	Heptylacetylene	C₉H₁₆	3452-09-3	124.223	liq	-50	150.8(0.4)	0.7658[20]	1.4217[20]	i H₂O; s eth, bz, ctc
8262	Norbormide		C₃₃H₂₅N₃O₃	991-42-4	511.570	cry (eth)	194				
8263	2,5-Norbornadiene	Bicyclo[2.2.1]hepta-2,5-diene	C₇H₈	121-46-0	92.139	liq	-19(1)	90(1)	0.9064[20]	1.4702[20]	i H₂O; s EtOH, eth, ace, bz; msc tol
8264	5-Norbornene-2,3-dicarboxylic acid anhydride		C₉H₈O₃	826-62-0	164.158		166				
8265	5-Norbornene-2-methylolacrylate		C₁₁H₁₄O₂	95-39-6	178.228	col liq		104	1.029[25]		s os
8266	24-Norcholan-23-oic acid, (5β)	Norcholanic acid	C₂₃H₃₈O₂	511-18-2	346.547	nd(HOAc)	177				
8267	Nordazepam	7-Chloro-1,3-dihydro-5-phenyl-2H-1,4-benzodiazepin-2-one	C₁₅H₁₁ClN₂O	1088-11-5	270.713		216.5				
8268	Nordihydroguaiaretic acid		C₁₈H₂₂O₄	500-38-9	302.366	nd(w, al, HOAc)	185.5				sl H₂O; s EtOH, eth, ace, alk; i bz
8269	Norea		C₁₃H₂₂N₂O	18530-56-8	222.326		168.7(0.5)				
8270	Norepinephrine	Noradrenaline	C₈H₁₁NO₃	51-41-2	169.178		217 dec				sl H₂O, EtOH, eth; vs alk, dil HCl
8271	Norethisterone	19-Norpregn-4-en-20-yn-3-one, 17-hydroxy-, (17 α)-	C₂₀H₂₆O₂	68-22-4	298.419	cry	207(2)				
8272	Norethynodrel		C₂₀H₂₆O₂	68-23-5	298.419	cry (MeOH)	170				
8273	Norflurazon		C₁₂H₉ClF₃N₃O	27314-13-2	303.666		178.0(0.6)				
8274	Norhyoscyamine		C₁₆H₂₁NO₃	537-29-1	275.343	nd	140.5				vs EtOH, chl
8275	*DL*-Norleucine	2-Aminohexanoic acid, (*DL*)	C₆H₁₃NO₂	616-06-8	131.173	lf(w)	327 dec		1.172[25]		s H₂O; sl EtOH; i eth
8276	*L*-Norleucine	2-Aminohexanoic acid, (*L*)	C₆H₁₃NO₂	327-57-1	131.173		301 dec				sl H₂O
8277	Normorphine		C₁₆H₁₇NO₃	466-97-7	271.311		273				
8278	Norplant	Norgestrel, (-)	C₂₁H₂₈O₂	797-63-7	312.446	cry (MeOH)	206				
8279	19-Nortestosterone phenylpropionate	Nandrolone phenpropionate	C₂₇H₃₄O₃	62-90-8	406.557	cry	94(1)				
8280	Nortriptyline hydrochloride		C₁₉H₂₂ClN	894-71-3	299.838	cry (eth)	214				s H₂O, EtOH; i bz, eth, ace
8281	*DL*-Norvaline		C₅H₁₁NO₂	760-78-1	117.147	lf(al, w)	303	sub			s H₂O; i EtOH, eth, chl, AcOEt, lig
8282	*L*-Norvaline	2-Aminopentanoic acid, (*S*)	C₅H₁₁NO₂	6600-40-4	117.147	cry (dil al)	307				s H₂O
8283	Noscapine		C₂₂H₂₃NO₇	128-62-1	413.421	pr or nd (al)	176				i H₂O; s EtOH, bz, chl; sl eth; vs ace
8284	Novobiocin	Streptonivicin	C₃₁H₃₆N₂O₁₁	303-81-1	612.624	wh-ye orth cry	154		1.3448		i H₂O; s EtOH, EtOAc, ace, py
8285	Nuarimol		C₁₇H₁₂ClFN₂O	63284-71-9	314.740		126				
8286	Nylidrin	Buphenine	C₁₉H₂₅NO₂	447-41-6	299.408	cry (MeOH)	111				
8287	Ochratoxin A		C₂₀H₁₈ClNO₆	303-47-9	403.813	cry (xyl)	169				
8288	Ochratoxin B		C₂₀H₁₉NO₆	4825-86-9	369.368	cry (MeOH)	221				
8289	Ochratoxin C		C₂₂H₂₂ClNO₆	4865-85-4	431.866	amorp solid					
8290	Octacaine	3-(Diethylamino)-*N*-phenylbutanamide	C₁₄H₂₂N₂O	13912-77-1	234.337	cry	47	200[1]			vs EtOH, bz, eth
8291	2,2',3,3',5,5',6,6'-Octachlorobiphenyl		C₁₂H₂Cl₈	2136-99-4	429.768	cry	160.6(0.5)				i H₂O
8292	Octachlorocyclopentene	Perchlorocyclopentene	C₅Cl₈	706-78-5	343.678	nd	40	283	1.8200[50]	1.5660[50]	i H₂O; vs EtOH
8293	Octachlorodibenzo-*p*-dioxin		C₁₂Cl₈O₂	3268-87-9	459.751	nd	332.0(0.5)				

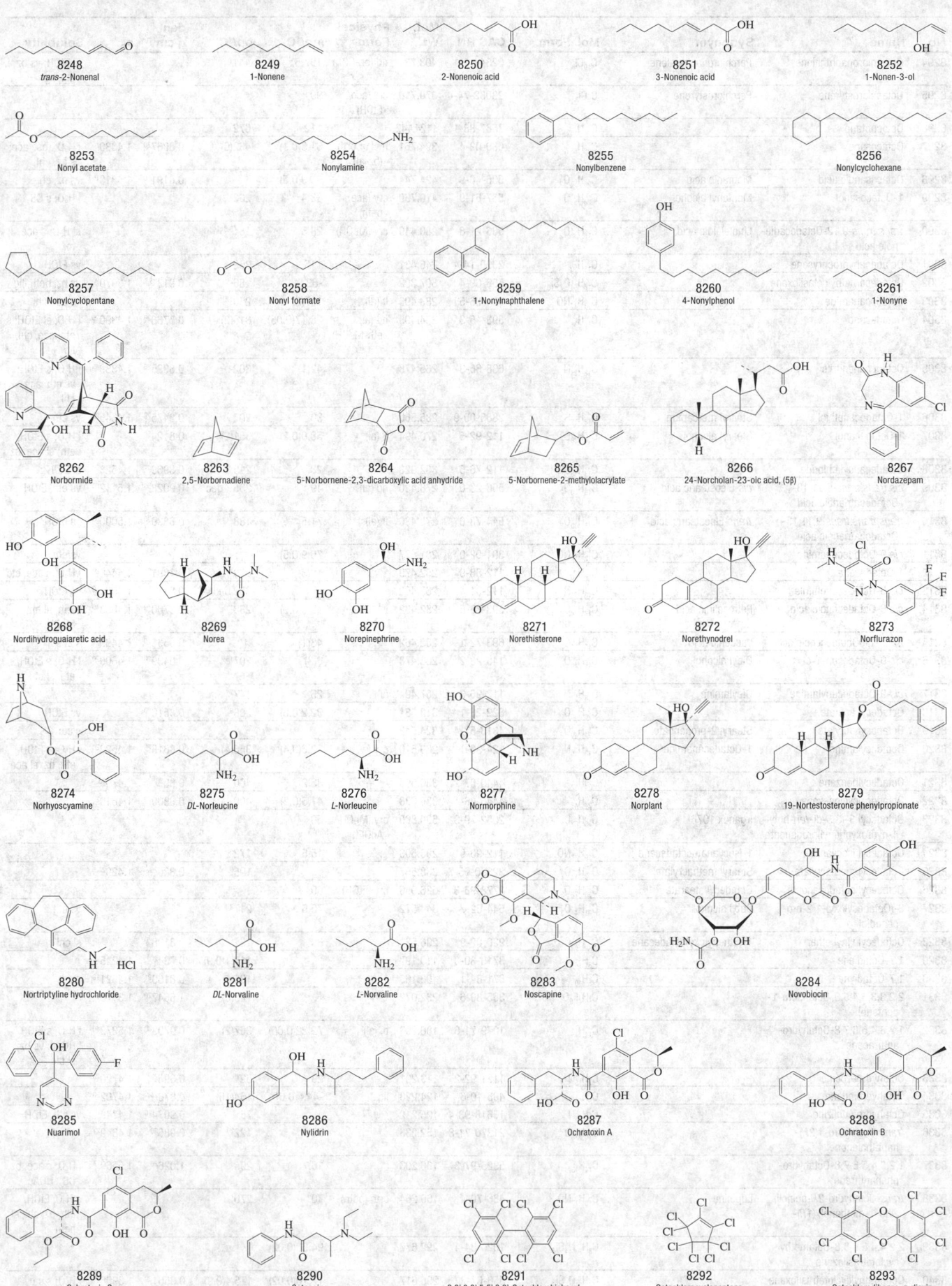

8248 *trans*-2-Nonenal

8249 1-Nonene

8250 2-Nonenoic acid

8251 3-Nonenoic acid

8252 1-Nonen-3-ol

8253 Nonyl acetate

8254 Nonylamine

8255 Nonylbenzene

8256 Nonylcyclohexane

8257 Nonylcyclopentane

8258 Nonyl formate

8259 1-Nonylnaphthalene

8260 4-Nonylphenol

8261 1-Nonyne

8262 Norbormide

8263 2,5-Norbornadiene

8264 5-Norbornene-2,3-dicarboxylic acid anhydride

8265 5-Norbornene-2-methylolacrylate

8266 24-Norcholan-23-oic acid, (5β)

8267 Nordazepam

8268 Nordihydroguaiaretic acid

8269 Norea

8270 Norepinephrine

8271 Norethisterone

8272 Norethynodrel

8273 Norflurazon

8274 Norhyoscyamine

8275 *DL*-Norleucine

8276 *L*-Norleucine

8277 Normorphine

8278 Norplant

8279 19-Nortestosterone phenylpropionate

8280 Nortriptyline hydrochloride

8281 *DL*-Norvaline

8282 *L*-Norvaline

8283 Noscapine

8284 Novobiocin

8285 Nuarimol

8286 Nylidrin

8287 Ochratoxin A

8288 Ochratoxin B

8289 Ochratoxin C

8290 Octacaine

8291 2,2',3,3',5,5',6,6'-Octachlorobiphenyl

8292 Octachlorocyclopentene

8293 Octachlorodibenzo-*p*-dioxin

No.	Name	Synonym	Mol. Form.	CAS RN	Mol. Wt.	Physical Form	mp/°C	bp/°C	den g cm⁻³	n_D	Solubility
8294	Octachloronaphthalene	Perchloronaphthalene	$C_{10}Cl_8$	2234-13-1	403.731	nd (bz-CCl₄)	197.5	441[7]			sl EtOH; vs bz, chl, lig
8295	Octachlorostyrene	Perchlorostyrene	C_8Cl_8	29082-74-4	379.710	cry (ace/ EtOH)	99				
8296	Octacontane		$C_{80}H_{162}$	7667-88-1	1124.142		112	672			
8297	Octacosane		$C_{28}H_{58}$	630-02-4	394.761	mcl or orth (bz-al)	61.3(0.1)	432(6)	0.8067[20]	1.4330[70]	i H₂O; msc ace; s bz, chl
8298	Octacosanoic acid	Montanic acid	$C_{28}H_{56}O_2$	506-48-9	424.744		90.9(0.5)		0.8191[100]	1.4313[100]	vs bz, chl
8299	1-Octacosanol	Montanyl alcohol	$C_{28}H_{58}O$	557-61-9	410.760	cry (ace, peth)	83.4	200[1]			i H₂O; s CS₂
8300	trans,trans-9,12-Octadecadie-noic acid	Linolelaidic acid	$C_{18}H_{32}O_2$	506-21-8	280.446	cry (MeOH)	28.5	181[0.8]			sl H₂O; s ace, hx
8301	Octadecahydrochrysene		$C_{18}H_{30}$	2090-14-4	246.431		115	353			vs EtOH
8302	Octadecamethyloctasiloxane		$C_{18}H_{54}O_7Si_8$	556-69-4	607.302		-63	186[20]	0.913[25]	1.3970[20]	vs bz, peth, lig
8303	Octadecanamide		$C_{18}H_{37}NO$	124-26-5	283.493	lf (al)	107(3)	250[12]			vs eth, chl
8304	Octadecane		$C_{18}H_{38}$	593-45-3	254.495	nd (al, eth-MeOH)	28.17(0.05)	316(2)	0.7768[28]	1.4390[20]	i H₂O; sl EtOH; s eth, ace, chl, lig
8305	Octadecanenitrile		$C_{18}H_{35}N$	638-65-3	265.478		42(1)	362	0.8325[20]	1.4389[45]	i H₂O; s EtOH; vs eth, ace, chl
8306	1-Octadecanethiol	Stearyl mercaptan	$C_{18}H_{38}S$	2885-00-9	286.560		30	207[11]	0.8475[20]	1.4645[20]	vs eth
8307	1-Octadecanol	Stearyl alcohol	$C_{18}H_{38}O$	112-92-5	270.494	lf (al)	58.0(0.2)	351(2)	0.8124[59]		i H₂O; s EtOH, eth; sl ace, bz
8308	Octadecanoyl chloride		$C_{18}H_{35}ClO$	112-76-5	302.923		23	215[15]	0.8969[0]	1.4523[24]	sl EtOH
8309	cis,trans,trans-9,11,13-Octadecatrienoic acid	cis-Eleostearic acid	$C_{18}H_{30}O_2$	506-23-0	278.430	nd (al)	49	235[12] dec	0.9028[50]	1.5112[50]	vs eth, EtOH
8310	trans,trans,trans-9,11,13-Octadecatrienoic acid	trans-Eleostearic acid	$C_{18}H_{30}O_2$	544-73-0	278.430	lf (al)	71.5	188[1]	0.8839[80]	1.5000[80]	vs EtOH
8311	cis-9-Octadecenamide		$C_{18}H_{35}NO$	301-02-0	281.477		75.9(0.5)				vs eth
8312	1-Octadecene		$C_{18}H_{36}$	112-88-9	252.479		18(3)	179[15]	0.7891[20]	1.4448[20]	i H₂O; s ace, ctc
8313	cis-9-Octadecenenitrile		$C_{18}H_{33}N$	112-91-4	263.462		-1	332 dec	0.847[17]	1.4566[20]	vs EtOH
8314	cis-6-Octadecenoic acid	Petroselinic acid	$C_{18}H_{34}O_2$	593-39-5	282.462	lf	30.0(0.9)	238[18]	0.8700[40]	1.4533[40]	s eth; sl hp, MeOH
8315	trans-11-Octadecenoic acid	Vaccenic acid	$C_{18}H_{34}O_2$	693-72-1	282.462		43(1)		0.8563[70]	1.4499[60]	s ace
8316	cis-9-Octadecen-1-ol	Oleyl alcohol	$C_{18}H_{36}O$	143-28-2	268.478		0(3)	207[15]	0.8489[20]	1.4606[20]	i H₂O; s EtOH, eth; sl ctc
8317	cis-9-Octadecenylamine	Oleylamine	$C_{18}H_{37}N$	112-90-3	267.494	oil	25	147[2]			
8318	Octadecyl acetate		$C_{20}H_{40}O_2$	822-23-1	312.531		32.2(0.6)	208[9]	0.8510[30]		vs EtOH
8319	Octadecyl acrylate	Stearyl 2-propenoate	$C_{21}H_{40}O_2$	4813-57-4	324.542						s ctc, CS₂
8320	Octadecylamine	1-Octadecanamine	$C_{18}H_{39}N$	124-30-1	269.510	cry (w)	52.6(0.4)	350(4)	0.8618[20]	1.4522[20]	i H₂O; s EtOH, eth, bz; sl ace
8321	Octadecylbenzene		$C_{24}H_{42}$	4445-07-2	330.590		35(2)	400	0.85[36]	1.479[36]	
8322	Octadecylcyclohexane		$C_{24}H_{48}$	4445-06-1	336.638		41.5(0.2)	409	0.8300[20]	1.4610[20]	
8323	Octadecyl 3-(3,5-di-tert-butyl-4-hydroxyphenyl)propanoate	Irganox 1076	$C_{35}H_{62}O_3$	2082-79-3	530.865	cry (MeOH/ AcOEt)	50				
8324	Octadecyl isocyanate	1-Isocyanatooctadecane	$C_{19}H_{37}NO$	112-96-9	295.503		15.5	172[5]			
8325	Octadecyl methacrylate	Stearyl methacrylate	$C_{22}H_{42}O_2$	32360-05-7	338.567			195[6]	0.880[25]	1.429[25]	
8326	Octadecyl octadecanoate	Octadecyl stearate	$C_{36}H_{72}O_2$	2778-96-3	536.956	cry (EtOH)	60				
8327	3-(Octadecyloxy)-1,2-pro-panediol	Batyl alcohol	$C_{21}H_{44}O_3$	544-62-7	344.572		70.5	217[2]			vs eth
8328	Octadecyl vinyl ether	1-(Ethenyloxy)octadecane	$C_{20}H_{40}O$	930-02-9	296.531		30	182[3]	0.8138[40]		sl chl
8329	1,7-Octadiene		C_8H_{14}	3710-30-3	110.197			117.1(0.9)	0.734[20]	1.4245[20]	
8330	1,7-Octadiyne		C_8H_{10}	871-84-1	106.165			135.5	0.8169[21]	1.4521[18]	s eth
8331	2,2,3,3,4,4,5,5-Octafluoro-1-pentanol		$C_5H_4F_8O$	355-80-6	232.072			140(6)	1.6647[20]	1.3178[20]	
8332	1,2,3,4,5,6,7,8-Octahydro-anthracene		$C_{14}H_{18}$	1079-71-6	186.293	pl (al)	72.22(0.05)	307(2)	0.9703[80]	1.5372[80]	i H₂O; s EtOH, HOAc; vs bz; sl ctc
8333	Octahydroazocine		$C_7H_{15}N$	1121-92-2	113.201		29	52[15]	0.896[25]	1.4720[20]	
8334	Octahydroindene		C_9H_{16}	496-10-6	124.223	liq	-44.4(0.6)	164(15)	0.876[25]	1.4702[20]	
8335	Octahydroindolizine		$C_8H_{15}N$	13618-93-4	125.212			75[43]	0.9074[10]	1.4748	vs eth, EtOH
8336	trans-Octahydro-1(2H)-naphthalenone		$C_{10}H_{16}O$	21370-71-8	152.233		33	122[20]	0.986[20]	1.4849[21]	
8337	1,2,3,4,5,6,7,8-Octahydro-phenanthrene		$C_{14}H_{18}$	5325-97-3	186.293		16.7	295	1.026[20]	1.5569[17]	i H₂O; s ace, bz, CS₂, HOAc
8338	trans-Octahydro-2H-quinoli-zine-1-methanol, (1R)-	Lupinine	$C_{10}H_{19}NO$	486-70-4	169.264	orth (peth)	70	270			s H₂O, EtOH, eth, bz, chl; sl peth
8339	2,2,4,4,6,6,8,8-Octamethyl-cyclotetrasilazane		$C_8H_{28}N_4Si_4$	1020-84-4	292.677		94.50(0.02)				
8340	Octamethylcyclotetrasiloxane		$C_8H_{24}O_4Si_4$	556-67-2	296.617		17.10(0.02)	175.4(0.9)	0.9561[20]	1.3968[20]	i H₂O; s ctc

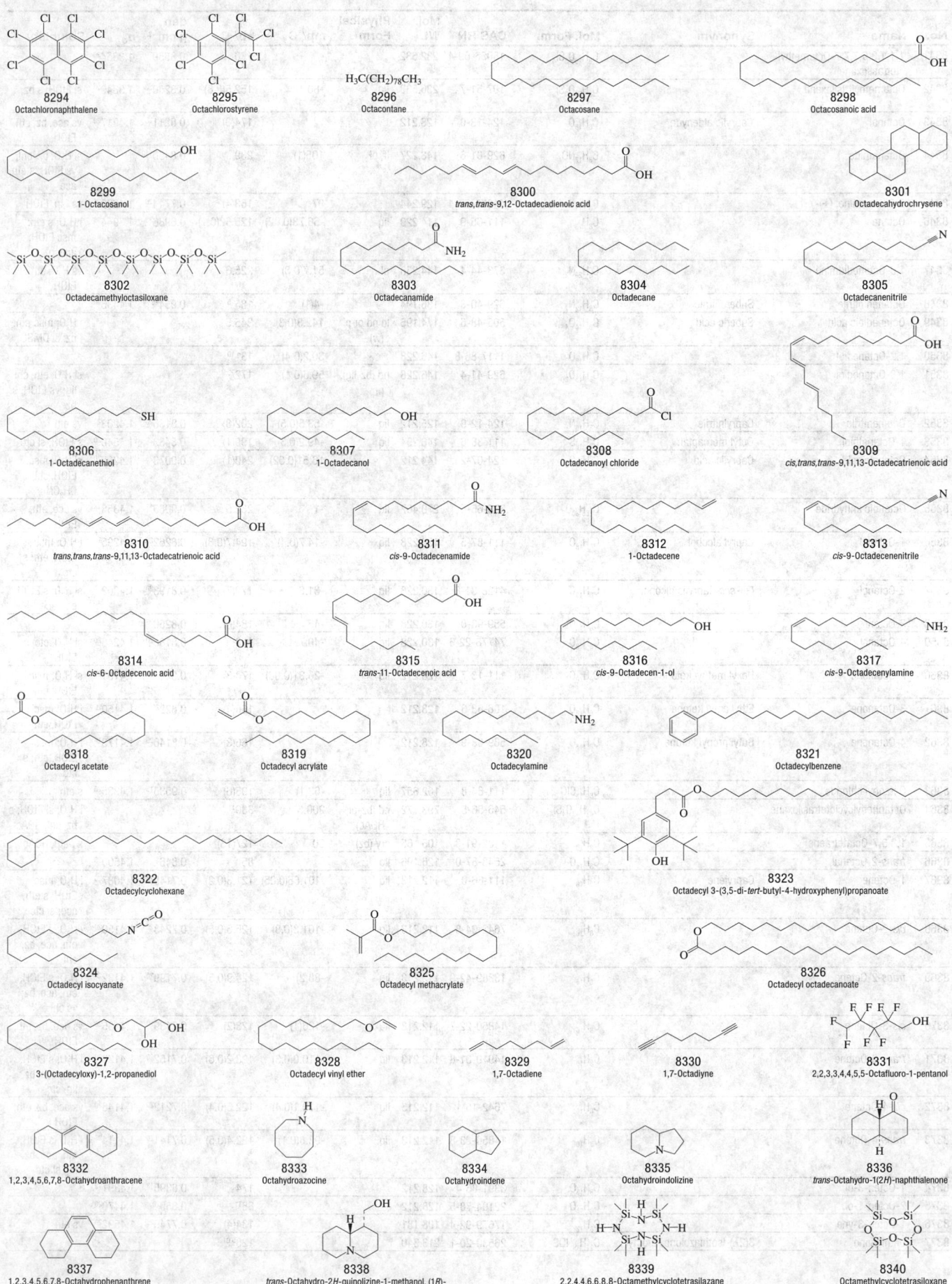

8294
Octachloronaphthalene

8295
Octachlorostyrene

8296
Octacontane
H₃C(CH₂)₇₈CH₃

8297
Octacosane

8298
Octacosanoic acid

8299
1-Octacosanol

8300
trans,trans-9,12-Octadecadienoic acid

8301
Octadecahydrochrysene

8302
Octadecamethyloctasiloxane

8303
Octadecanamide

8304
Octadecane

8305
Octadecanenitrile

8306
1-Octadecanethiol

8307
1-Octadecanol

8308
Octadecanoyl chloride

8309
cis,trans,trans-9,11,13-Octadecatrienoic acid

8310
trans,trans,trans-9,11,13-Octadecatrienoic acid

8311
cis-9-Octadecenamide

8312
1-Octadecene

8313
cis-9-Octadecenenitrile

8314
cis-6-Octadecenoic acid

8315
trans-11-Octadecenoic acid

8316
cis-9-Octadecen-1-ol

8317
cis-9-Octadecenylamine

8318
Octadecyl acetate

8319
Octadecyl acrylate

8320
Octadecylamine

8321
Octadecylbenzene

8322
Octadecylcyclohexane

8323
Octadecyl 3-(3,5-di-*tert*-butyl-4-hydroxyphenyl)propanoate

8324
Octadecyl isocyanate

8325
Octadecyl methacrylate

8326
Octadecyl octadecanoate

8327
3-(Octadecyloxy)-1,2-propanediol

8328
Octadecyl vinyl ether

8329
1,7-Octadiene

8330
1,7-Octadiyne

8331
2,2,3,3,4,4,5,5-Octafluoro-1-pentanol

8332
1,2,3,4,5,6,7,8-Octahydroanthracene

8333
Octahydroazocine

8334
Octahydroindene

8335
Octahydroindolizine

8336
trans-Octahydro-1(2H)-naphthalenone

8337
1,2,3,4,5,6,7,8-Octahydrophenanthrene

8338
trans-Octahydro-2H-quinolizine-1-methanol, (1R)-

8339
2,2,4,4,6,6,8,8-Octamethylcyclotetrasilazane

8340
Octamethylcyclotetrasiloxane

No.	Name	Synonym	Mol. Form.	CAS RN	Mol. Wt.	Physical Form	mp/°C	bp/°C	den g cm⁻³	n_D	Solubility
8341	1,1,1,3,5,7,7,7-Octamethyl-tetrasiloxane		C₈H₂₆O₃Si₄	16066-09-4	282.632			170	0.8559[20]	1.3854[20]	
8342	Octamethyltrisiloxane		C₈H₂₄O₂Si₃	107-51-7	236.533	liq	-80	152.5(0.8)	0.8200[20]	1.3840[20]	sl EtOH; s bz, peth
8343	Octanal	Caprylic aldehyde	C₈H₁₆O	124-13-0	128.212			174(3)	0.8211[20]	1.4217[20]	vs ace, bz, eth, EtOH
8344	Octanamide		C₈H₁₇NO	629-01-6	143.227	lf, pl	105(1)	239	0.8450[110]		sl H₂O, bz, chl; vs EtOH; s eth, ace
8345	2-Octanamine, (±)-		C₈H₁₉N	44855-57-4	129.244		97	163(4)	0.7744[20]	1.4232[25]	vs eth, EtOH
8346	Octane		C₈H₁₈	111-65-9	114.229	liq	-56.73(0.02)	125.62(0.1)	0.6986[25]	1.3944[25]	i H₂O; s eth; msc EtOH, ace, bz
8347	1,8-Octanediamine		C₈H₂₀N₂	373-44-4	144.258	pl	51.7(0.3)	225.6			vs H₂O, eth, EtOH
8348	Octanedinitrile	Suberonitrile	C₈H₁₂N₂	629-40-3	136.194		-4(1)	185[15]	0.954[25]	1.4436[20]	
8349	Octanedioic acid	Suberic acid	C₈H₁₄O₄	505-48-6	174.195	lo nd or pl (w)	142.3(0.3)	345.5			i H₂O; msc eth, bz; sl DMSO
8350	1,2-Octanediol		C₈H₁₈O₂	1117-86-8	146.228		30.2(0.4)	131[10]			
8351	1,8-Octanediol		C₈H₁₈O₂	629-41-4	146.228	nd (bz-lig), pr	59.6(0.1)	172[20]			sl H₂O, eth, chl, lig; vs EtOH; s bz
8352	Octanenitrile	Caprylnitrile	C₈H₁₅N	124-12-9	125.212	liq	-53.5(0.5)	202(3)	0.8136[20]	1.4203[20]	vs eth
8353	1-Octanethiol	Octyl mercaptan	C₈H₁₈S	111-88-6	146.294	liq	-49.2(0.3)	199(1)	0.8433[20]	1.4540[20]	s EtOH; sl ctc
8354	Octanoic acid	Caprylic acid	C₈H₁₆O₂	124-07-2	144.212		16.51(0.02)	240(1)	0.9073[25]	1.4285[20]	sl H₂O; msc EtOH, chl, CH₃CN
8355	Octanoic anhydride		C₁₆H₃₀O₃	623-66-5	270.407	liq	-1	282.5	0.9065[18]	1.4358[18]	vs ace, eth, EtOH
8356	1-Octanol	Capryl alcohol	C₈H₁₈O	111-87-5	130.228	liq	-14.7(0.4)	194.7(0.8)	0.8262[25]	1.4295[20]	i H₂O; msc EtOH, eth; s ctc
8357	2-Octanol	(±)-sec-Caprylic alcohol	C₈H₁₈O	4128-31-8	130.228	liq	-31.6	179(7)	0.8193[20]	1.4203[20]	sl H₂O; s EtOH, eth, ace
8358	3-Octanol		C₈H₁₈O	589-98-0	130.228	liq	-45	184(6)	0.8258[20]		
8359	4-Octanol		C₈H₁₈O	74778-22-6	130.228	liq	-40.7	176.3	0.8186[20]	1.4248[20]	sl H₂O, ctc; s EtOH
8360	2-Octanone	Hexyl methyl ketone	C₈H₁₆O	111-13-7	128.212	liq	-20.31(0.08)	173(3)	0.820[20]	1.4151[20]	sl H₂O; msc EtOH, eth
8361	3-Octanone	Ethyl amyl ketone	C₈H₁₆O	106-68-3	128.212			166(4)	0.822[25]	1.4150[20]	i H₂O; msc EtOH, eth
8362	4-Octanone	Butyl propyl ketone	C₈H₁₆O	589-63-9	128.212			166(3)	0.8146[25]	1.4173[14]	i H₂O; msc EtOH, eth; s ctc
8363	Octanoyl chloride		C₈H₁₅ClO	111-64-8	162.657	liq	-63(1)	195(6)	0.9535[15]	1.4335[20]	s eth
8364	Octaphenylcyclotetrasiloxane		C₄₈H₄₀O₄Si₄	546-56-5	793.172	nd (bz-al, HOAc)	200.5	330[1]			i H₂O; sl EtOH; s bz, chl, HOAc
8365	1,3,5,7-Octatetraene		C₈H₁₀	1482-91-3	106.165	cry (bz)	50	127(13)			s peth, HOAc
8366	trans-2-Octenal		C₈H₁₄O	2548-87-0	126.196	liq		85[19]	0.846	1.4500[20]	
8367	1-Octene	Caprylene	C₈H₁₆	111-66-0	112.213	liq	-101.66(0.05)	121.3(0.2)	0.7149[20]	1.4087[20]	i H₂O; msc EtOH; s eth, ace; sl ctc
8368	cis-2-Octene		C₈H₁₆	7642-04-8	112.213	liq	-101.3(0.8)	125.6(0.5)	0.7243[20]	1.4150[20]	i H₂O; s EtOH, eth, ace, bz, chl
8369	trans-2-Octene		C₈H₁₆	13389-42-9	112.213	liq	-88(2)	124.9(0.5)	0.7199[20]	1.4132[20]	i H₂O; s EtOH, eth, ace, bz; vs chl
8370	cis-3-Octene		C₈H₁₆	14850-22-7	112.213	liq	-126(1)	123(2)	0.7159[20]	1.4135[20]	vs ace, bz, eth, EtOH
8371	trans-3-Octene		C₈H₁₆	14919-01-8	112.213	liq	-110.0(0.3)	123.2(0.6)	0.7152[20]	1.4126[20]	i H₂O; s EtOH, eth, ace, bz, lig, ctc
8372	cis-4-Octene		C₈H₁₆	7642-15-1	112.213	liq	-119.1(0.4)	122.6(0.4)	0.7212[20]	1.4148[20]	vs ace, bz, eth, EtOH
8373	trans-4-Octene		C₈H₁₆	14850-23-8	112.213	liq	-93.8(0.1)	122.4(0.5)	0.7141[20]	1.4114[20]	i H₂O; s EtOH, eth, ace, bz, lig; sl ctc
8374	1-Octen-3-ol		C₈H₁₆O	3391-86-4	128.212			174	0.8395[13]	1.4391[12]	
8375	2-Octen-1-ol		C₈H₁₆O	22104-78-5	128.212			88[11]	0.850[20]	1.4470[20]	
8376	1-Octen-3-yne		C₈H₁₂	17679-92-4	108.181			134(4)	0.7749[20]	1.4592[20]	vs eth
8377	Octhilinone	3(2H)-Isothiazolone, 2-octyl-	C₁₁H₁₉NOS	26530-20-1	213.340			120[0.01]			

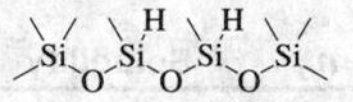

8341
1,1,1,3,5,7,7,7-Octamethyltetrasiloxane

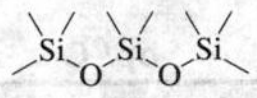

8342
Octamethyltrisiloxane

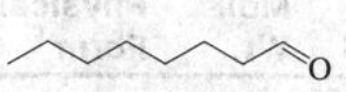

8343
Octanal

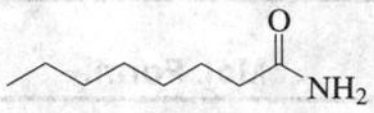

8344
Octanamide

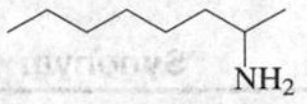

8345
2-Octanamine, (±)-

8346
Octane

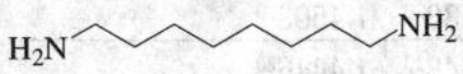

8347
1,8-Octanediamine

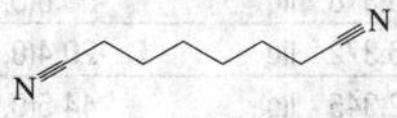

8348
Octanedinitrile

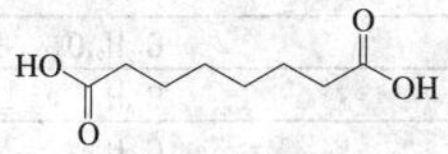

8349
Octanedioic acid

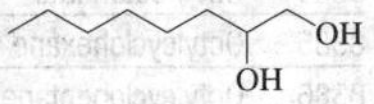

8350
1,2-Octanediol

8351
1,8-Octanediol

8352
Octanenitrile

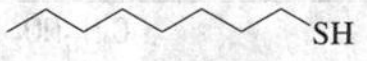

8353
1-Octanethiol

8354
Octanoic acid

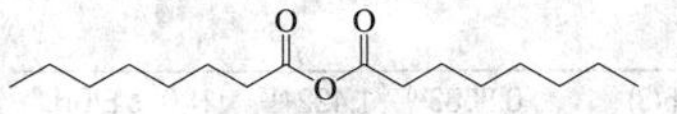

8355
Octanoic anhydride

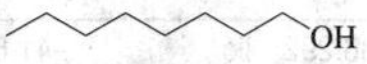

8356
1-Octanol

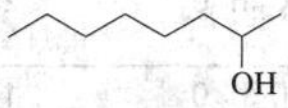

8357
2-Octanol

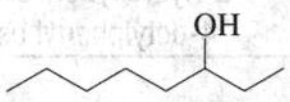

8358
3-Octanol

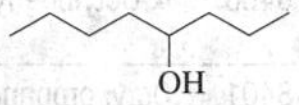

8359
4-Octanol

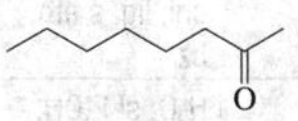

8360
2-Octanone

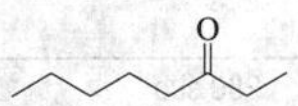

8361
3-Octanone

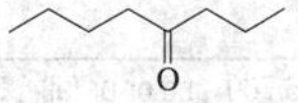

8362
4-Octanone

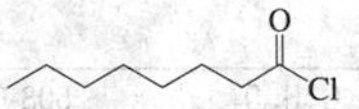

8363
Octanoyl chloride

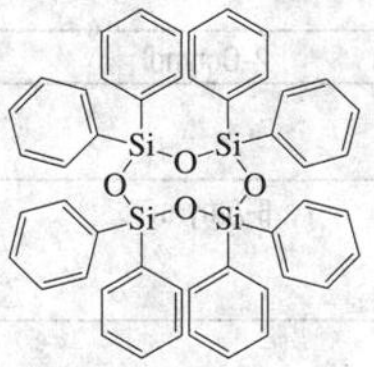

8364
Octaphenylcyclotetrasiloxane

8365
1,3,5,7-Octatetraene

8366
trans-2-Octenal

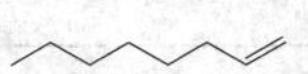

8367
1-Octene

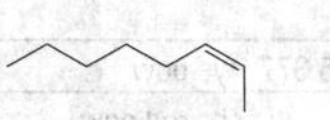

8368
cis-2-Octene

8369
trans-2-Octene

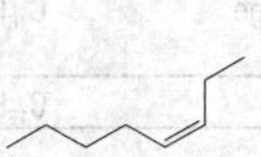

8370
cis-3-Octene

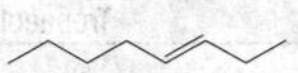

8371
trans-3-Octene

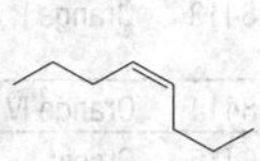

8372
cis-4-Octene

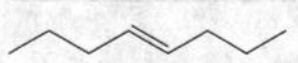

8373
trans-4-Octene

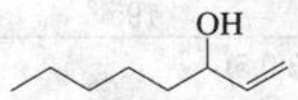

8374
1-Octen-3-ol

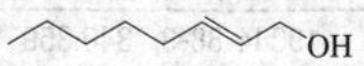

8375
2-Octen-1-ol

8376
1-Octen-3-yne

8377
Octhilinone

No.	Name	Synonym	Mol. Form.	CAS RN	Mol. Wt.	Physical Form	mp/°C	bp/°C	den g cm⁻³	n_D	Solubility
8378	Octyl acetate		$C_{10}H_{20}O_2$	112-14-1	172.265	liq	-38(2)	210(3)	0.8705[20]	1.4150[20]	i H₂O; s EtOH, eth; sl ctc
8379	Octyl acrylate	Octyl 2-propenoate	$C_{11}H_{20}O_2$	2499-59-4	184.276			227(5)	0.8810[20]		
8380	Octylamine	1-Octanamine	$C_8H_{19}N$	111-86-4	129.244		0	178.6(0.2)	0.7826[20]	1.4292[20]	sl H₂O; vs EtOH, eth; s ctc
8381	Octylamine hydrochloride	1-Octanamine hydrochloride	$C_8H_{20}ClN$	142-95-0	165.705		196.5				s H₂O
8382	4-Octylaniline		$C_{14}H_{23}N$	16245-79-7	205.340		20	310	0.9128[20]		vs eth
8383	Octylbenzene		$C_{14}H_{22}$	2189-60-8	190.325	liq	-36	263(2)	0.8562[20]	1.4845[20]	i H₂O; msc eth, bz
8384	Octyl butanoate		$C_{12}H_{24}O_2$	110-39-4	200.318	liq	-55.6(0.5)	239(4)	0.8629[20]	1.4267[15]	vs EtOH
8385	Octylcyclohexane		$C_{14}H_{28}$	1795-15-9	196.372	liq	-20.4(0.5)	263(2)	0.8138[20]	1.4503[20]	
8386	Octylcyclopentane		$C_{13}H_{26}$	1795-20-6	182.345	liq	-44.5(0.5)	242(15)	0.8048[20]	1.4446[20]	
8387	2-Octyldecanoic acid		$C_{18}H_{36}O_2$	619-39-6	284.478	nd or lf (al)	38.5	215[13]	0.8447[70]		vs eth, EtOH
8388	Octyldimethylamine	N,N-Dimethyl-1-octanamine	$C_{10}H_{23}N$	7378-99-6	157.297			192(6)			
8389	Octyl diphenyl phosphate		$C_{20}H_{27}O_4P$	115-88-8	362.399				1.09[25]		
8390	Octyl formate		$C_9H_{18}O_2$	112-32-3	158.238	liq	-39.1(0.5)	198.8	0.8744[20]	1.4208[15]	i H₂O; s EtOH; msc eth; sl ctc
8391	Octyl isocyanate		$C_9H_{17}NO$	3158-26-7	155.237			78[6]			
8392	Octyl methacrylate		$C_{12}H_{22}O_2$	2157-01-9	198.302			238(17)			
8393	Octyl nitrate		$C_8H_{17}NO_3$	629-39-0	175.226			110[20]	0.975[0]		sl H₂O; s EtOH, eth
8394	Octyl nitrite		$C_8H_{17}NO_2$	629-46-9	159.227			174.5	0.862[17]	1.4127[20]	sl H₂O; vs EtOH, eth
8395	Octyl octanoate		$C_{16}H_{32}O_2$	2306-88-9	256.424	liq	-18.1(0.3)	306.8	0.8554[20]	1.4352[20]	vs ace, eth, EtOH
8396	Octyloxirane		$C_{10}H_{20}O$	2404-44-6	156.265	liq		128[95]			
8397	4-(Octyloxy)benzaldehyde		$C_{15}H_{22}O_2$	24083-13-4	234.335			131[0.5]			
8398	4-Octylphenol		$C_{14}H_{22}O$	1806-26-4	206.324		43.0	169[10]			
8399	Octyl phenyl ether	(Octyloxy)benzene	$C_{14}H_{22}O$	1818-07-1	206.324		8	284(7)	0.9131[15]	1.4875[20]	i H₂O; s EtOH, eth
8400	4-Octylphenyl salicylate	2-Hydroxybenzoic acid, 4-octylphenyl ester	$C_{21}H_{26}O_3$	2512-56-3	326.429	wh cry	73				
8401	Octyl propanoate		$C_{11}H_{22}O_2$	142-60-9	186.292	liq	-41.6(0.5)	226(3)	0.8663[20]	1.4221[15]	i H₂O; s EtOH, eth, bz; sl ctc
8402	1-Octyne	Hexylacetylene	C_8H_{14}	629-05-0	110.197	liq	-79.4(0.2)	126.2(0.3)	0.7461[20]	1.4159[20]	i H₂O; s EtOH, eth
8403	2-Octyne	Methylpentylacetylene	C_8H_{14}	2809-67-8	110.197	liq	-61.5(0.3)	138.0(0.2)	0.7596[20]	1.4278[20]	i H₂O; s EtOH, eth
8404	3-Octyne		C_8H_{14}	15232-76-5	110.197	liq	-103.9(0.4)	135.6(0.2)	0.7529[20]	1.4250[20]	i H₂O; s EtOH, eth
8405	4-Octyne	Dipropylacetylene	C_8H_{14}	1942-45-6	110.197	liq	-102.6(0.3)	133.5(0.4)	0.7509[20]	1.4248[20]	i H₂O; s EtOH, eth
8406	2-Octyn-1-ol	2-Octynol	$C_8H_{14}O$	20739-58-6	126.196		-18	98[15]	0.8805[20]	1.4556[20]	vs eth
8407	Oleandrin		$C_{32}H_{48}O_9$	465-16-7	576.718	cry (EtOH)	250 dec				i H₂O; s EtOH, chl
8408	Olean-12-en-3-ol, (3β)	β-Amyrin	$C_{30}H_{50}O$	559-70-6	426.717	nd (lig or al)	197	260[05]			i H₂O; sl EtOH, chl, lig; s eth, bz
8409	Oleanolic acid		$C_{30}H_{48}O_3$	508-02-1	456.700	nd or pr (al)	310 dec	280 sub			i H₂O; sl EtOH, eth, ace; vs py, HOAc
8410	Oleic acid	cis-9-Octadecenoic acid	$C_{18}H_{34}O_2$	112-80-1	282.462	liq	14(1)	360	0.8935[20]	1.4582[20]	i H₂O; msc EtOH, eth, ace, bz, chl, ctc
8411	Omeprazole		$C_{17}H_{19}N_3O_3S$	73590-58-6	345.416	cry (MeCN)	156				
8412	Omethoate		$C_5H_{12}NO_4PS$	1113-02-6	213.192	oil	≈135 dec		1.32[20]	1.4987[20]	msc H₂O; i hx
8413	Orange I	1-Naphthol Orange	$C_{16}H_{11}N_2NaO_4S$	523-44-4	350.324	red-br pow					s H₂O; sl EtOH; i bz
8414	Orange IV	Tropaeolin OO	$C_{18}H_{14}N_3NaO_3S$	554-73-4	375.377	ye pow					s H₂O
8415	Orcein			1400-62-0		br-red pow					
8416	L-Ornithine	2,5-Diaminopentanoic acid, (S)	$C_5H_{12}N_2O_2$	70-26-8	132.161	micro cry (al-eth)	140				vs H₂O, EtOH
8417	L-Ornithine, monohydrochloride		$C_5H_{13}ClN_2O_2$	3184-13-2	168.622	nd	215				vs H₂O
8418	Orotic acid	1,2,3,6-Tetrahydro-2,6-dioxo-4-pyrimidinecarboxylic acid	$C_5H_4N_2O_4$	65-86-1	156.097	cry (w)	345.5				sl H₂O; i os
8419	Oroxylin A	5,7-Dihydroxy-6-methoxy-2-phenyl-4H-1-benzopyran-4-one	$C_{16}H_{12}O_5$	480-11-5	284.263	ye nd (al)	231.5				vs ace, eth, EtOH
8420	Orphenadrine		$C_{18}H_{23}NO$	83-98-7	269.382			195[12]			
8421	Oryzalin	Benzenesulfonamide, 4-(dipropylamino)-3,5-dinitro-	$C_{12}H_{18}N_4O_6S$	19044-88-3	346.359		142.3(0.5)				

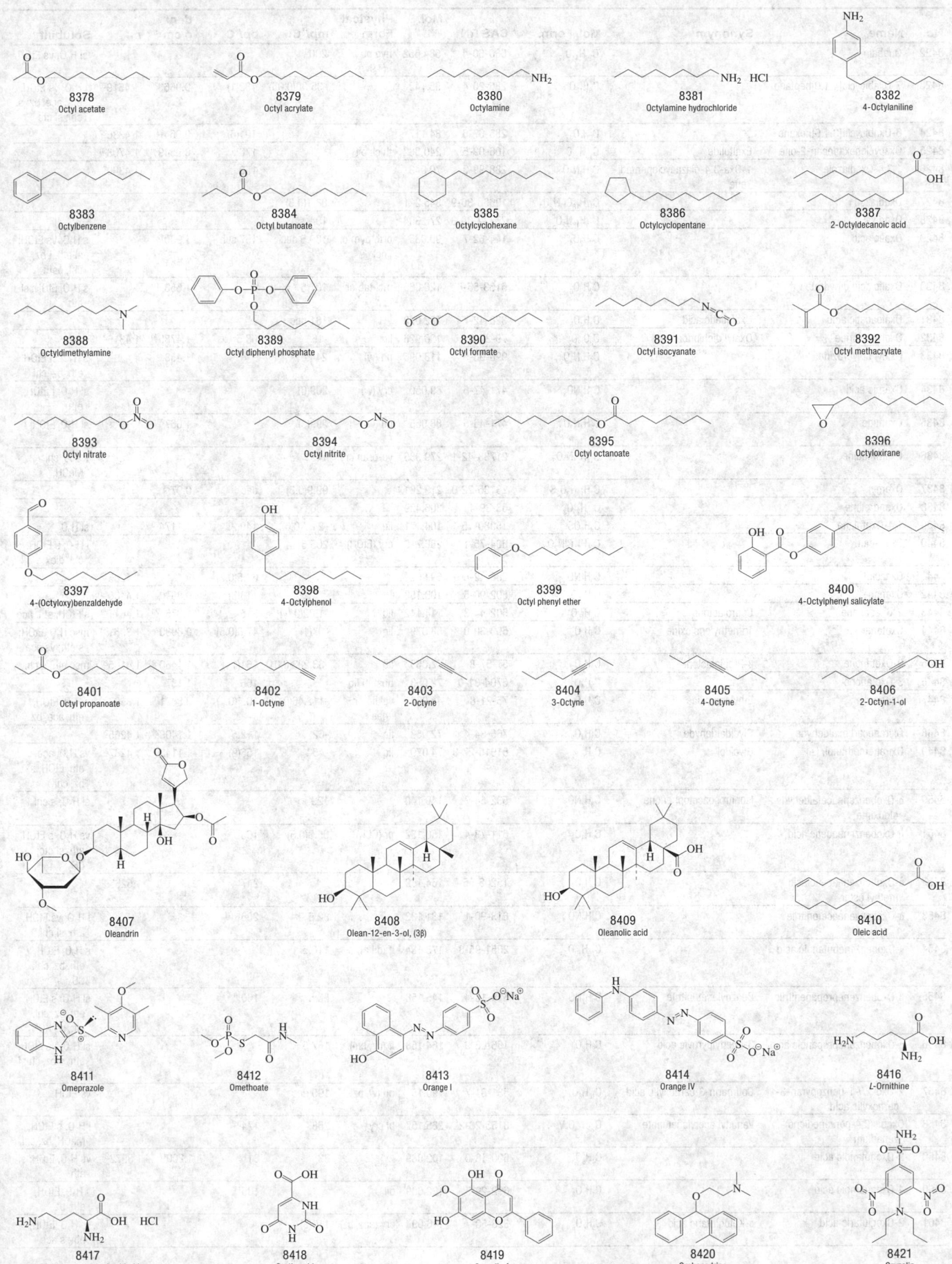

8378 Octyl acetate

8379 Octyl acrylate

8380 Octylamine

8381 Octylamine hydrochloride

8382 4-Octylaniline

8383 Octylbenzene

8384 Octyl butanoate

8385 Octylcyclohexane

8386 Octylcyclopentane

8387 2-Octyldecanoic acid

8388 Octyldimethylamine

8389 Octyl diphenyl phosphate

8390 Octyl formate

8391 Octyl isocyanate

8392 Octyl methacrylate

8393 Octyl nitrate

8394 Octyl nitrite

8395 Octyl octanoate

8396 Octyloxirane

8397 4-(Octyloxy)benzaldehyde

8398 4-Octylphenol

8399 Octyl phenyl ether

8400 4-Octylphenyl salicylate

8401 Octyl propanoate

8402 1-Octyne

8403 2-Octyne

8404 3-Octyne

8405 4-Octyne

8406 2-Octyn-1-ol

8407 Oleandrin

8408 Olean-12-en-3-ol, (3β)

8409 Oleanolic acid

8410 Oleic acid

8411 Omeprazole

8412 Omethoate

8413 Orange I

8414 Orange IV

8416 L-Ornithine

8417 L-Ornithine, monohydrochloride

8418 Orotic acid

8419 Oroxylin A

8420 Orphenadrine

8421 Oryzalin

No.	Name	Synonym	Mol. Form.	CAS RN	Mol. Wt.	Physical Form	mp/°C	bp/°C	den g cm⁻³	n_D	Solubility
8422	Ouabain		$C_{29}H_{44}O_{12}$	630-60-4	584.652	hyg pl (+9w)	200				sl H_2O; vs EtOH
8423	7-Oxabicyclo[4.1.0]heptane		$C_6H_{10}O$	286-20-4	98.142		-35.00(0.05)	132(1)	0.9663²⁰	1.4519²⁰	i H_2O; vs EtOH, eth, ace, bz; s chl; sl ctc
8424	6-Oxabicyclo[3.1.0]hexane		C_5H_8O	285-67-6	84.117			101(6)	0.964²⁵	1.4336²⁰	
8425	Oxacyclohexadecan-2-one	Exaltolide	$C_{15}H_{28}O_2$	106-02-5	240.382	thick oil		176¹⁵	0.9549²⁰	1.4708²⁰	
8426	1,3,4-Oxadiazole	1-Oxa-3,4-diazacyclopentadiene	$C_2H_2N_2O$	288-99-3	70.049			150		1.4300²⁵	
8427	Oxadiazon		$C_{15}H_{18}Cl_2N_2O_3$	19666-30-9	345.221		88.1(0.5)				
8428	Oxadixyl		$C_{14}H_{18}N_2O_4$	77732-09-3	278.304		104				
8429	Oxalic acid		$C_2H_2O_4$	144-62-7	90.035	orth pym or oct	189.5 dec	157 sub	1.900¹⁷		s H_2O; vs EtOH; sl eth; i bz, chl, peth
8430	Oxalic acid dihydrate		$C_2H_6O_6$	6153-56-6	126.065	mcl tab or pr	101.5		1.653¹⁸		s H_2O, EtOH; sl eth
8431	Oxaloacetic acid	Oxalacetic acid	$C_4H_4O_5$	328-42-7	132.072		161 dec				
8432	Oxalyl chloride	Oxalyl dichloride	$C_2Cl_2O_2$	79-37-8	126.926	liq	-16	63.5	1.4785²⁰	1.4316²⁰	s eth
8433	Oxalyl dihydrazide		$C_2H_6N_4O_2$	996-98-5	118.095	nd (w)	244.0		1.458²²		s H_2O; sl EtOH, eth, bz, chl
8434	Oxamic acid		$C_2H_3NO_3$	471-47-6	89.050	cry (w)	209(6)				sl H_2O; i EtOH, eth
8435	Oxamide		$C_2H_4N_2O_2$	471-46-5	88.065	nd (w)	299(3)		1.667²⁰		sl H_2O, EtOH; i eth
8436	Oxamniquine		$C_{14}H_{21}N_3O_3$	21738-42-1	279.335	ye-oran cry	149				s ace, chl, MeOH
8437	Oxamyl		$C_7H_{13}N_3O_3S$	23135-22-0	219.261		99.9(0.5)	dec	0.97²⁵		
8438	Oxandrolone		$C_{19}H_{30}O_3$	53-39-4	306.439		236				
8439	1,4-Oxathiane		C_4H_8OS	15980-15-1	104.171	liq	-21.7(0.6)	149(2)	1.1174²⁰		sl H_2O
8440	Oxazepam		$C_{15}H_{11}ClN_2O_2$	604-75-1	286.713	cry (EtOH)	205.5				i H_2O; s EtOH, chl, diox
8441	Oxazole		C_3H_3NO	288-42-6	69.062			69.5(0.2)		1.4285¹⁷	
8442	Oxepane		$C_6H_{12}O$	592-90-5	100.158			118(3)	0.89²⁵	1.4400²⁰	
8443	2-Oxepanone	Caprolactone	$C_6H_{10}O_2$	502-44-3	114.142	liq	-1.02(0.04)	215	1.0761²⁰	1.4611²⁰	s EtOH, eth, ace
8444	Oxetane	Trimethylene oxide	C_3H_6O	503-30-0	58.079	liq	-97(1)	47.6(0.5)	0.8930²⁵	1.3961²⁰	msc H_2O, EtOH; s eth; vs ace
8445	2-Oxetanone	β-Propiolactone	$C_3H_4O_2$	57-57-8	72.063	liq	-33.283(0.01)	161(14)	1.1460²⁰	1.4105²⁰	msc eth; s chl
8446	3-Oxetanone		$C_3H_4O_2$	6704-31-0	72.063	unstab liq		106	1.137		
8447	Oxirane	Ethylene oxide	C_2H_4O	75-21-8	44.052	vol liq or gas	-112.46(0.05)	10.4(0.1)	0.8821¹⁰	1.3597⁷	s H_2O, EtOH, eth, ace, bz
8448	Oxiranecarboxaldehyde	Glycidaldehyde	$C_3H_4O_2$	765-34-4	72.063	liq	-62	112.5	1.1403²⁰	1.4265²⁰	
8449	Oxiranemethanol, (±)-	Glycidol	$C_3H_6O_2$	61915-27-3	74.079	liq	-45	156(9)	1.1143²⁵	1.4287²⁰	vs H_2O, ace, eth, EtOH; s bz, chl
8450	α-Oxobenzeneacetaldehyde aldoxime	Isonitrosoacetophenone	$C_8H_7NO_2$	532-54-7	149.148		129				sl H_2O; s chl
8451	α-Oxobenzeneacetic acid		$C_8H_6O_3$	611-73-4	150.132	pr (CCl_4)	65.8(0.5)	163¹⁵			vs H_2O; s EtOH, eth; sl ctc; i CS_2
8452	α-Oxobenzeneacetic acid, methyl ester		$C_9H_8O_3$	15206-55-0	164.158			247		1.5268²⁰	
8453	α-Oxobenzeneacetonitrile		C_8H_5NO	613-90-1	131.132		32.5	206			i H_2O; vs EtOH, eth; sl chl
8454	γ-Oxobenzenebutanoic acid		$C_{10}H_{10}O_3$	2051-95-8	178.184	lf (dil al)	116.5				s H_2O, EtOH, eth, bz, chl, CS_2
8455	β-Oxobenzenepropanenitrile	Benzoylacetonitrile	C_9H_7NO	614-16-4	145.158		80.5	160¹⁰			sl H_2O; s EtOH, eth, bz, chl, alk, aq KCN
8456	α-Oxobenzenepropanoic acid	3-Phenylpyruvic acid	$C_9H_8O_3$	156-06-9	164.158	lf (bz, chl)	157.5				sl H_2O; vs EtOH, eth; s bz, chl; i lig
8457	2-Oxo-2H-1-benzopyran-3-carboxylic acid	Coumarin-3-carboxylic acid	$C_{10}H_6O_4$	531-81-7	190.15	nd (w, bz)	190 dec				vs EtOH
8458	Oxobis(2,4-pentanedione)-vanadium	Vanadyl acetylacetonate	$C_{10}H_{14}O_5V$	3153-26-2	265.157	bl cry	258	174⁰·²			i H_2O; s EtOH, MeOH, bz, chl
8459	2-Oxobutanoic acid		$C_4H_6O_3$	600-18-0	102.089		33	81¹⁶	1.200¹⁷	1.3972²⁰	vs H_2O, EtOH; sl eth
8460	4-Oxobutanoic acid		$C_4H_6O_3$	692-29-5	102.089	oil		135¹⁴			s H_2O, EtOH, eth, bz
8461	2-Oxoglutaric acid	α-Ketoglutaric acid	$C_5H_6O_5$	328-50-7	146.099	cry (ace-bz)	116(3)				vs H_2O, EtOH, eth; s ace

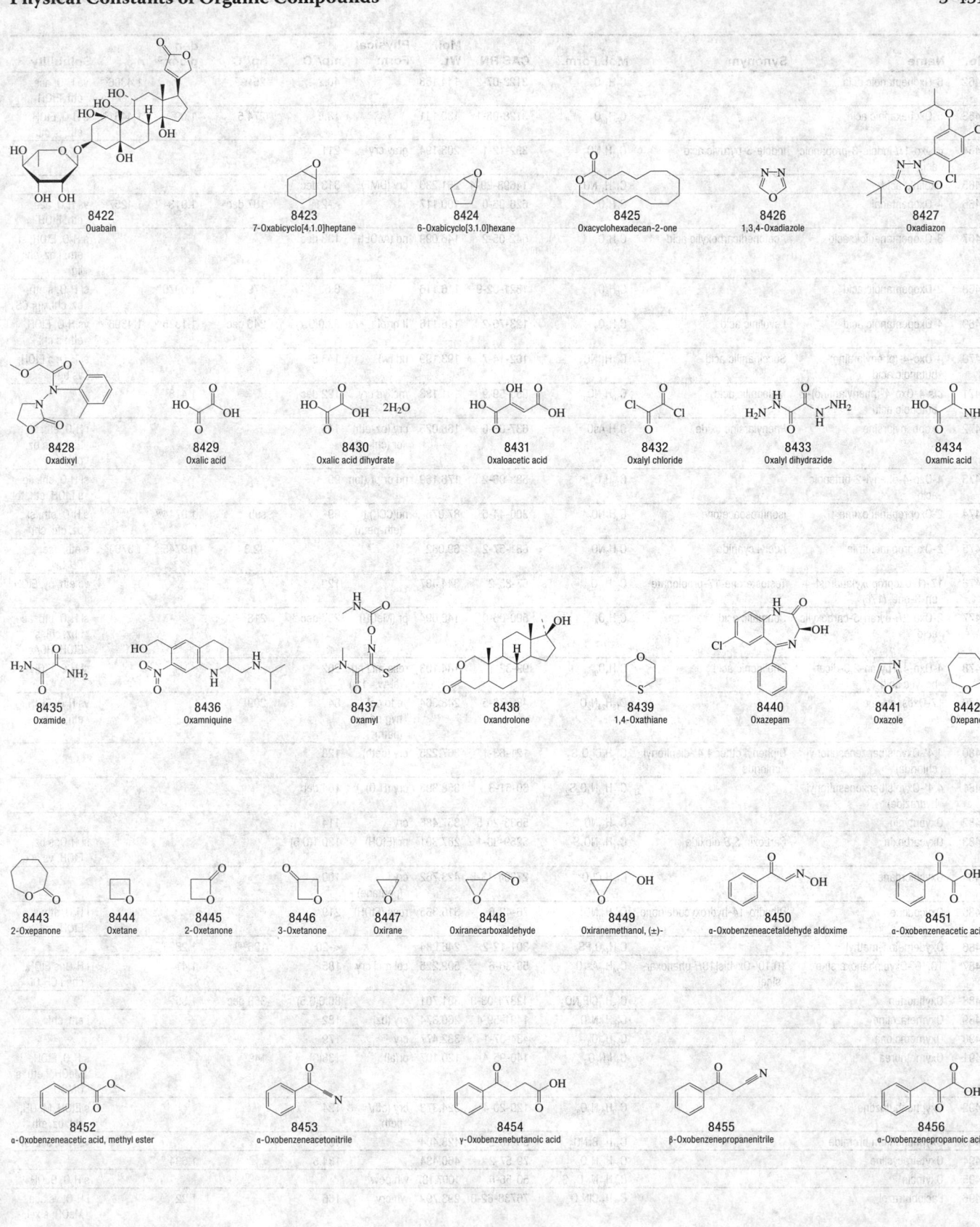

8422 Ouabain

8423 7-Oxabicyclo[4.1.0]heptane

8424 6-Oxabicyclo[3.1.0]hexane

8425 Oxacyclohexadecan-2-one

8426 1,3,4-Oxadiazole

8427 Oxadiazon

8428 Oxadixyl

8429 Oxalic acid

8430 Oxalic acid dihydrate

8431 Oxaloacetic acid

8432 Oxalyl chloride

8433 Oxalyl dihydrazide

8434 Oxamic acid

8435 Oxamide

8436 Oxamniquine

8437 Oxamyl

8438 Oxandrolone

8439 1,4-Oxathiane

8440 Oxazepam

8441 Oxazole

8442 Oxepane

8443 2-Oxepanone

8444 Oxetane

8445 2-Oxetanone

8446 3-Oxetanone

8447 Oxirane

8448 Oxiranecarboxaldehyde

8449 Oxiranemethanol, (±)-

8450 α-Oxobenzeneacetaldehyde aldoxime

8451 α-Oxobenzeneacetic acid

8452 α-Oxobenzeneacetic acid, methyl ester

8453 α-Oxobenzeneacetonitrile

8454 γ-Oxobenzenebutanoic acid

8455 β-Oxobenzenepropanenitrile

8456 α-Oxobenzenepropanoic acid

8457 2-Oxo-2H-1-benzopyran-3-carboxylic acid

8458 Oxobis(2,4-pentanedione)vanadium

8459 2-Oxobutanoic acid

8460 4-Oxobutanoic acid

8461 2-Oxoglutaric acid

No.	Name	Synonym	Mol. Form.	CAS RN	Mol. Wt.	Physical Form	mp/°C	bp/°C	den g cm⁻³	n_D	Solubility
8462	6-Oxoheptanoic acid		$C_7H_{12}O_3$	3128-07-2	144.168		40.2	251[280]		1.4306[25]	vs H_2O, ace, eth, EtOH
8463	5-Oxohexanoic acid		$C_6H_{10}O_3$	3128-06-1	130.141		13.5	274.5	1.09[25]	1.4451[20]	s H_2O, EtOH, eth; sl ctc
8464	α-Oxo-1H-indole-3-propanoic acid	Indole-3-pyruvic acid	$C_{11}H_9NO_3$	392-12-1	203.194	gray cry	211				
8465	Oxolinic acid		$C_{13}H_{11}NO_5$	14698-29-4	261.230	cry (DMF)	313 dec				
8466	4-Oxopentanal		$C_5H_8O_2$	626-96-0	100.117		<-21	187 dec	1.0134[21]	1.4257[22]	vs H_2O, ace, eth, EtOH
8467	3-Oxopentanedioic acid	Acetonedicarboxylic acid	$C_5H_6O_5$	542-05-2	146.099	nd (AcOEt)	138 dec				s H_2O, EtOH; sl eth; i bz, chl, lig
8468	2-Oxopentanoic acid		$C_5H_8O_3$	1821-02-9	116.116		6.5	179	1.0970[14]		sl H_2O; s eth, bz, chl, lig, CS_2
8469	4-Oxopentanoic acid	Levulinic acid	$C_5H_8O_3$	123-76-2	116.116	lf or pl	33.0(0.7)	245 dec	1.1335[20]	1.4396[20]	vs H_2O, EtOH, eth; s chl
8470	4-Oxo-4-(phenylamino)-butanoic acid	Succinanilic acid	$C_{10}H_{11}NO_3$	102-14-7	193.199	nd (w)	148.5				sl H_2O; s EtOH; vs eth
8471	cis-4-Oxo-4-(phenylamino)-2-butenoic acid	Maleanilic acid	$C_{10}H_9NO_3$	555-59-9	191.183	mcl ye cry	192 dec		1.418[30]		
8472	Oxophenylarsine	Phenylarsine oxide	C_6H_5AsO	637-03-6	168.025	cry (bz-eth) or (chl-eth)	145				i H_2O, eth; sl EtOH; vs bz, chl
8473	4-Oxo-4-phenyl-2-butenoic acid		$C_{10}H_8O_3$	583-06-2	176.169	nd or pr (tol)	99				sl H_2O, chl, lig; s EtOH, eth, tol
8474	2-Oxopropanal oxime	Isonitrosoacetone	$C_3H_5NO_2$	306-44-5	87.078	nd(CCl_4) lf (eth-peth)	69	sub	1.0744[67]		s H_2O, eth; sl bz, ctc, chl
8475	2-Oxopropanenitrile	Acetyl cyanide	C_3H_3NO	631-57-2	69.062			92.3	0.9745[20]	1.3764[20]	s eth, ace, CH_3CN
8476	17-(1-Oxopropoxy)androst-4-en-3-one, (17β)	Testosterone-17-propionate	$C_{22}H_{32}O_3$	57-85-2	344.487		120				vs eth, py, EtOH
8477	2-Oxo-2H-pyran-5-carboxylic acid	Coumalic acid	$C_6H_4O_4$	500-05-0	140.094	pr (MeOH)	207 dec	218[120]			sl H_2O, eth, ace; i bz, chl; s EtOH, HOAc
8478	4-Oxo-4H-pyran-2,6-dicar-boxylic acid	Chelidonic acid	$C_7H_4O_6$	99-32-1	184.103	rose mcl nd (al-w,+1w)	262				sl H_2O, EtOH
8479	17-Oxosparteine		$C_{15}H_{24}N_2O$	489-72-5	248.364	ye to col hyg nd (peth)	84	209[12]			vs H_2O, EtOH, eth; s chl
8480	4,4'-Oxybis(benzenesulfonyl chloride)	Diphenyl ether 4,4'-disulfonyl chloride	$C_{12}H_8Cl_2O_5S_2$	121-63-1	367.225	cry (peth)	128				
8481	4,4'-Oxybis(benzenesulfonyl hydrazide)		$C_{12}H_{14}N_4O_5S_2$	80-51-3	358.393	cry (H_2O)	164 dec				
8482	Oxybutynin		$C_{22}H_{31}NO_3$	5633-20-5	357.486	cry	114				
8483	Oxycarboxin	Carboxin S,S-dioxide	$C_{12}H_{13}NO_4S$	5259-88-1	267.301	pr (EtOH)	130.1(0.5)				sl H_2O; s bz, EtOH; vs ace
8484	Oxychlordane		$C_{10}H_4Cl_6O$	27304-13-8	423.762	cry (pentane)	100				
8485	Oxycodone	Dihydro-14-hydroxycodeinone	$C_{18}H_{21}NO_4$	76-42-6	315.365	rods (EtOH)	219				i H_2O, eth; s EtOH, chl
8486	Oxydemeton-methyl		$C_6H_{15}O_4PS_2$	301-12-2	246.284		<-20	106[0.01]	1.289[20]		
8487	10,10'-Oxydiphenoxarsine	10,10'-Oxybis[10H-phenoxarsine]	$C_{24}H_{16}As_2O_3$	58-36-6	502.225	col mcl cry	185		1.41		i H_2O; s EtOH, chl; i CH_2Cl_2
8488	Oxyfluorfen		$C_{15}H_{11}ClF_3NO_4$	42874-03-3	361.701		86.6(0.5)	358 dec	1.35[73]		
8489	Oxymetazoline		$C_{16}H_{24}N_2O$	1491-59-4	260.374	cry (bz)	182				i eth, chl
8490	Oxymetholone		$C_{21}H_{32}O_3$	434-07-1	332.477	cry	179				
8491	Oxymethurea		$C_3H_8N_2O_3$	140-95-4	120.107	pr(al)	138(3)	149[25]			s H_2O, EtOH, MeOH; i eth; sl DMSO
8492	Oxyphenbutazone		$C_{19}H_{20}N_2O_3$	129-20-4	324.373	cry (eth/peth)	124				s EtOH, MeOH, chl, bz, eth
8493	Oxyphenonium bromide		$C_{21}H_{34}BrNO_3$	50-10-2	428.404		191.5				vs H_2O; sl EtOH
8494	Oxytetracycline		$C_{22}H_{24}N_2O_9$	79-57-2	460.434		184.5		1.634[20]		
8495	Oxytocin		$C_{43}H_{66}N_{12}O_{12}S_2$	50-56-6	1007.187	wh pow					s H_2O, BuOH
8496	Paclobutrazol		$C_{15}H_{20}ClN_3O$	76738-62-0	293.792	wh cry	166		1.22		i H_2O; vs ace, MeOH; s xyl, hx
8497	Palustric acid		$C_{20}H_{30}O_2$	1945-53-5	302.451	cry (MeOH)	164.5				
8498	Pamoic acid		$C_{23}H_{16}O_6$	130-85-8	388.369		315				
8499	Pancuronium dibromide		$C_{35}H_{60}Br_2N_2O_4$	15500-66-0	732.670	cry	215				sl chl
8500	Panose	4-α-Isomaltosylglucose	$C_{18}H_{32}O_{16}$	33401-87-5	504.437		223 dec				
8501	Panthesin		$C_{18}H_{32}N_2O_5S$	135-44-4	388.522	pa ye pow (al)	158				vs H_2O, EtOH

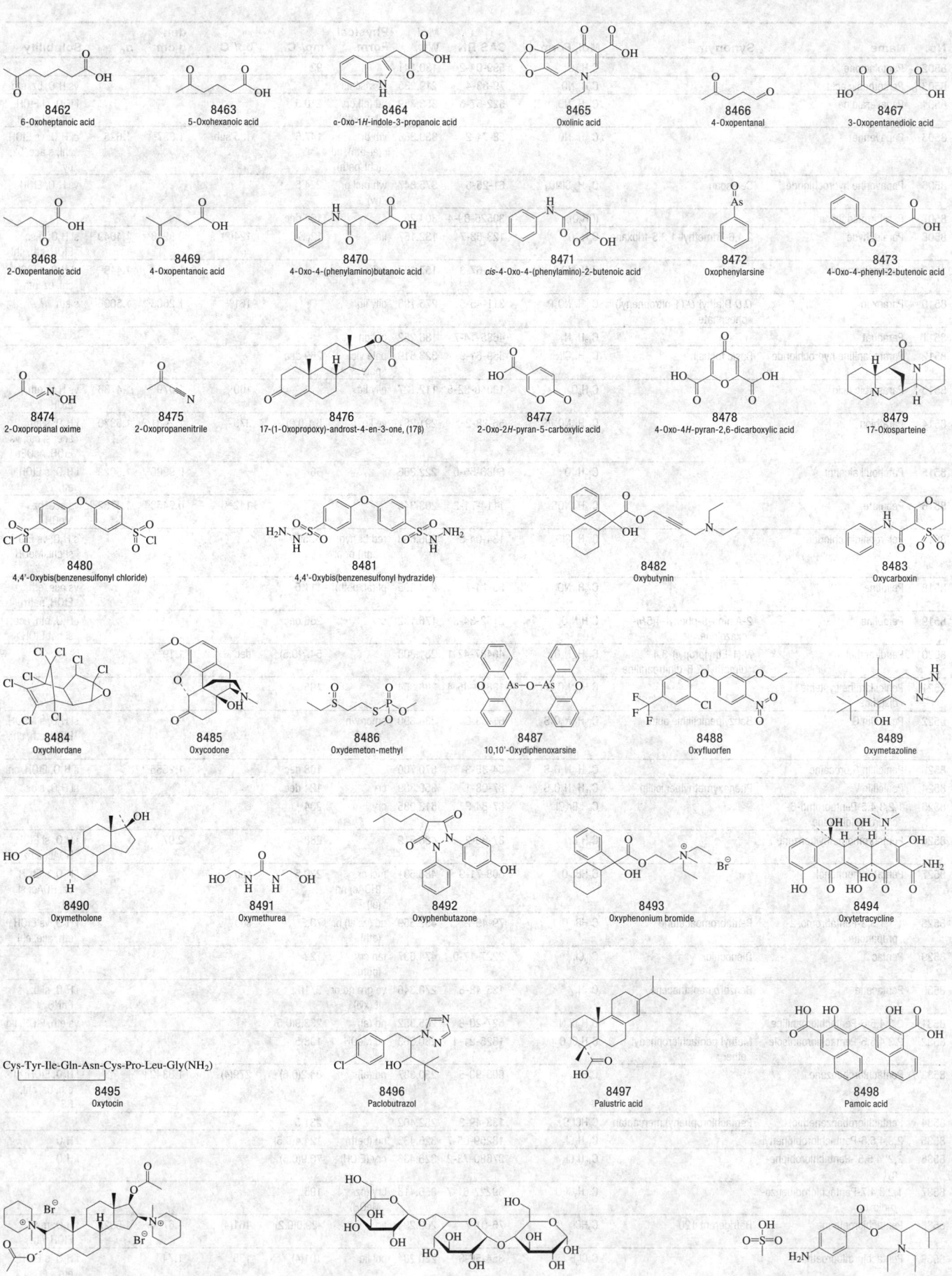

8462 6-Oxoheptanoic acid

8463 5-Oxohexanoic acid

8464 α-Oxo-1*H*-indole-3-propanoic acid

8465 Oxolinic acid

8466 4-Oxopentanal

8467 3-Oxopentanedioic acid

8468 2-Oxopentanoic acid

8469 4-Oxopentanoic acid

8470 4-Oxo-4-(phenylamino)butanoic acid

8471 *cis*-4-Oxo-4-(phenylamino)-2-butenoic acid

8472 Oxophenylarsine

8473 4-Oxo-4-phenyl-2-butenoic acid

8474 2-Oxopropanal oxime

8475 2-Oxopropanenitrile

8476 17-(1-Oxopropoxy)-androst-4-en-3-one, (17β)

8477 2-Oxo-2*H*-pyran-5-carboxylic acid

8478 4-Oxo-4*H*-pyran-2,6-dicarboxylic acid

8479 17-Oxosparteine

8480 4,4′-Oxybis(benzenesulfonyl chloride)

8481 4,4′-Oxybis(benzenesulfonyl hydrazide)

8482 Oxybutynin

8483 Oxycarboxin

8484 Oxychlordane

8485 Oxycodone

8486 Oxydemeton-methyl

8487 10,10′-Oxydiphenoxarsine

8488 Oxyfluorfen

8489 Oxymetazoline

8490 Oxymetholone

8491 Oxymethurea

8492 Oxyphenbutazone

8493 Oxyphenonium bromide

8494 Oxytetracycline

8495 Oxytocin
Cys-Tyr-Ile-Gln-Asn-Cys-Pro-Leu-Gly(NH₂)

8496 Paclobutrazol

8497 Palustric acid

8498 Pamoic acid

8499 Pancuronium dibromide

8500 Panose

8501 Panthesin

No.	Name	Synonym	Mol. Form.	CAS RN	Mol. Wt.	Physical Form	mp/°C	bp/°C	den g cm⁻³	n_D	Solubility
8502	Pantolactone		$C_6H_{10}O_3$	599-04-2	130.141		92				
8503	Pantothenic acid		$C_9H_{17}NO_5$	79-83-4	219.235	ye visc oil					vs H_2O, bz, eth
8504	Papaveraldine		$C_{20}H_{19}NO_5$	522-57-6	353.369	nd (al),cry (bz, peth)	210.5				i H_2O; sl EtOH, eth; s bz, chl
8505	Papaverine		$C_{20}H_{21}NO_4$	58-74-2	339.386	wh pr (al-eth), nd (chl-peth)	147.5	135 sub	1.337²⁰	1.625	sl H_2O; vs EtOH, chl; s ace, bz, py
8506	Papaverine hydrochloride	Cerespan	$C_{20}H_{22}ClNO_4$	61-25-6	375.847	wh mcl pr (w)	224.5				vs H_2O, EtOH
8507	Paraformaldehyde		$(CH_2O)_x$	30525-89-4	30.026		164 dec				
8508	Paraldehyde	2,4,6-Trimethyl-1,3,5-trioxane	$C_6H_{12}O_3$	123-63-7	132.157	liq	12(1)	124(2)	0.9943²⁰	1.4049²⁰	sl H_2O; msc EtOH, eth, chl
8509	Paramethadione		$C_7H_{11}NO_3$	115-67-3	157.167	liq			1.121²⁵	1.449²⁵	sl H_2O; s EtOH, chl, bz, eth
8510	Paraoxon	O,O-Diethyl O-(4-nitrophenyl) phosphate	$C_{10}H_{14}NO_6P$	311-45-5	275.195	oily liq		161⁰·⁵	1.2683²⁵	1.5096	s eth
8511	Paraquat		$C_{12}H_{14}N_2$	4685-14-7	186.252	cation					
8512	Pararosaniline hydrochloride	Basic fuchsin	$C_{19}H_{18}ClN_3$	569-61-9	323.819	pale viol pow	269 dec				
8513	Parasorbic acid		$C_6H_8O_2$	10048-32-5	112.127	oily liq		100¹⁵	1.079¹⁸	1.4736²⁰	vs H_2O, eth, EtOH
8514	Parathion		$C_{10}H_{14}NO_5PS$	56-38-2	291.261	ye liq	6.1(0.9)	375	1.2681²⁰	1.5370²⁵	i H_2O; s eth, ace; sl ctc; vs EtOH, AcOEt
8515	Patchouli alcohol		$C_{15}H_{26}O$	5986-55-0	222.366		56		0.9906⁶⁵	1.5029⁶⁵	i H_2O; s EtOH, eth
8516	Pebulate		$C_{10}H_{21}NOS$	1114-71-2	203.345			142²⁰	0.9458²⁰	1.4752²⁰	vs ace, bz, MeOH
8517	Pelargonidin chloride		$C_{15}H_{11}ClO_5$	134-04-3	306.698	red br hyg (anh) pr or pl	>350				s H_2O; vs EtOH; sl chl, MeOH
8518	Pellotine		$C_{13}H_{19}NO_3$	83-14-7	237.295	pl (al, peth)	111.5				vs ace, eth, EtOH, peth
8519	Pemoline	2-Amino-5-phenyl-4(5H)-oxazolone	$C_9H_8N_2O_2$	2152-34-3	176.172	cry	256 dec				i H_2O, eth, ace; sl hot EtOH
8520	Pendimethalin	N-(1-Ethylpropyl)-3,4-dimethyl-2,6-dinitroaniline	$C_{13}H_{19}N_3O_4$	40487-42-1	281.308		54.7(0.5)	dec	1.19²⁵		
8521	Penicillamine cysteine disulfide		$C_8H_{16}N_2O_4S_2$	18840-45-4	268.354		195				
8522	Penicillin G	Benzylpenicillinic acid	$C_{16}H_{18}N_2O_4S$	61-33-6	334.390	amor wh pow					sl H_2O; s MeOH, EtOH, eth, chl, bz, ace
8523	Penicillin G procaine		$C_{29}H_{38}N_4O_6S$	54-35-3	570.700		108 dec		1.2555²⁵		s H_2O, EtOH, chl
8524	Penicillin V	Phenoxymethylpenicillin	$C_{16}H_{18}N_2O_5S$	87-08-1	350.389	cry	124 dec				sl H_2O; s os
8525	1,2,3,4,5-Pentabromo-6-chlorocyclohexane		$C_6H_6Br_5Cl$	87-84-3	513.085	cry	204				
8526	Pentabromomethylbenzene		$C_7H_3Br_5$	87-83-2	486.619		288		2.97¹⁷		i H_2O; sl EtOH, HOAc; s bz
8527	Pentabromophenol		C_6HBr_5O	608-71-9	488.591	mcl pr (HOAc) nd (al)	229.5	sub			i H_2O; s EtOH, bz, HOAc; sl eth
8528	1,1,1,3,3-Pentabromo-2-propanone	Pentabromoacetone	C_3HBr_5O	79-49-2	452.559	nd (w, al) pr (eth)	79.5	sub			i H_2O; vs EtOH, eth, ace, chl
8529	Pentac	Dienochlor	$C_{10}Cl_{10}$	2227-17-0	474.637	tan cry (peth)	122				
8530	Pentacene	Benzo[b]naphthacene	$C_{22}H_{14}$	135-48-8	278.346	ye grn nd or lf (xyl)	271(2)				i H_2O; sl bz; s PhNO₂
8531	2,3,4,5,6-Pentachloroaniline		$C_6H_2Cl_5N$	527-20-8	265.352	nd (al)	233.9(0.5)				vs eth, EtOH, liq
8532	2,3,4,5,6-Pentachloroanisole	Methyl pentachlorophenyl ether	$C_7H_3Cl_5O$	1825-21-4	280.363	nd MeOH	108.5				
8533	Pentachlorobenzene		C_6HCl_5	608-93-5	250.337	nd (al)	84.2(0.6)	279(4)	1.8342¹⁶		i H_2O, EtOH; sl eth, bz, chl, CS_2
8534	Pentachlorobenzenethiol	Pentachlorophenyl mercaptan	C_6HCl_5S	133-49-3	282.402		231.5				
8535	2,3,4,5,6-Pentachlorobiphenyl		$C_{12}H_5Cl_5$	18259-05-7	326.433	nd (peth)	124.4(0.5)				i H_2O
8536	2,2',4,5,5'-Pentachlorobiphenyl		$C_{12}H_5Cl_5$	37680-73-2	326.433	cry (EtOH)	76.9(0.5)				i H_2O
8537	1,2,3,4,7-Pentachlorodibenzo-p-dioxin		$C_{12}H_3Cl_5O_2$	39227-61-7	356.416	cry (bz/MeOH)	195				
8538	Pentachloroethane	Refrigerant 120	C_2HCl_5	76-01-7	202.294	liq	-29.0(0.2)	161(4)	1.6796²⁰	1.5025²⁰	i H_2O; msc EtOH, eth
8539	Pentachlorofluoroethane		C_2Cl_5F	354-56-3	220.284	col liq	101(4)	138	1.74²⁵		i H_2O; s EtOH, eth

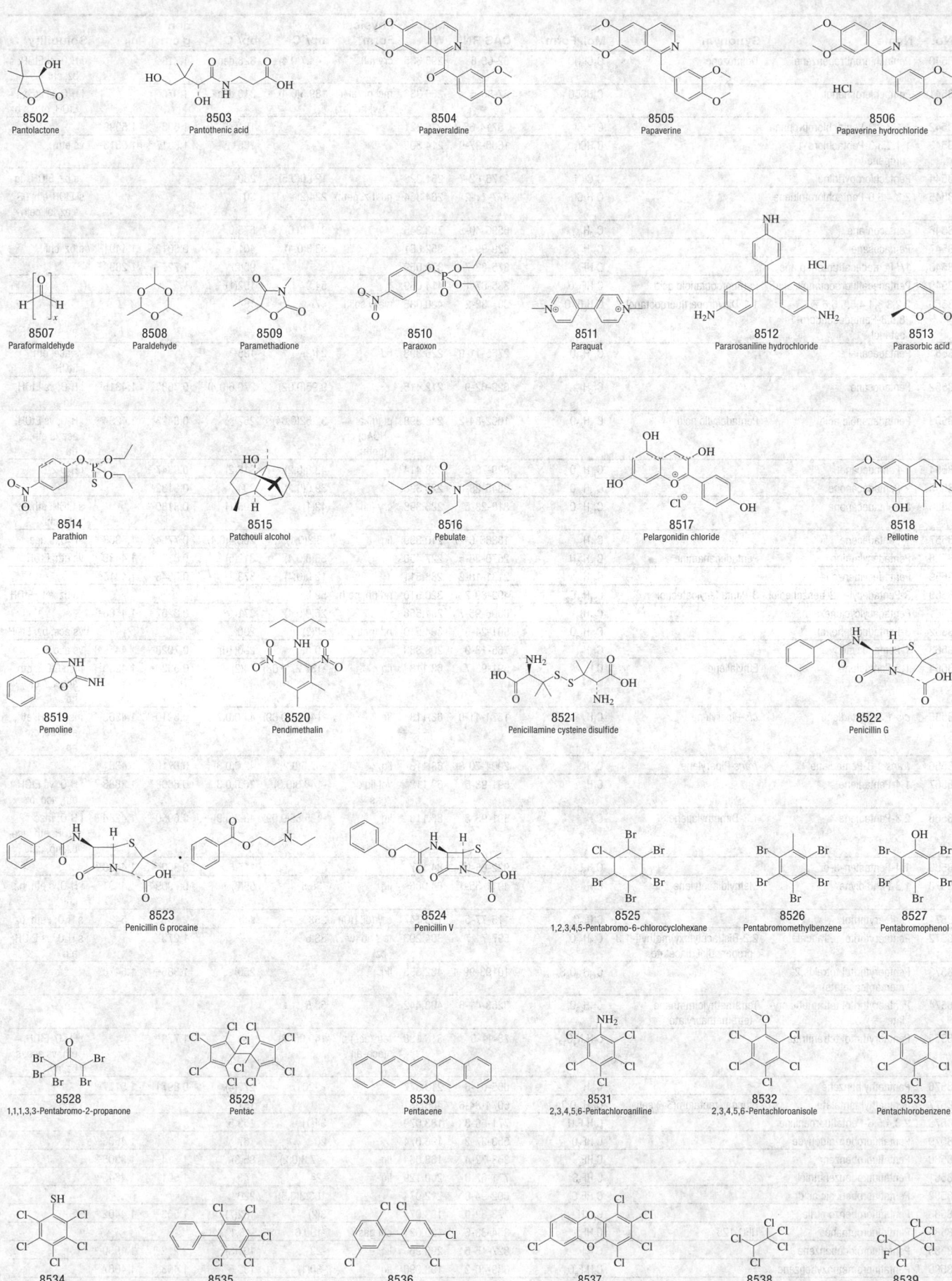

8502 Pantolactone

8503 Pantothenic acid

8504 Papaveraldine

8505 Papaverine

8506 Papaverine hydrochloride

8507 Paraformaldehyde

8508 Paraldehyde

8509 Paramethadione

8510 Paraoxon

8511 Paraquat

8512 Pararosaniline hydrochloride

8513 Parasorbic acid

8514 Parathion

8515 Patchouli alcohol

8516 Pebulate

8517 Pelargonidin chloride

8518 Pellotine

8519 Pemoline

8520 Pendimethalin

8521 Penicillamine cysteine disulfide

8522 Penicillin G

8523 Penicillin G procaine

8524 Penicillin V

8525 1,2,3,4,5-Pentabromo-6-chlorocyclohexane

8526 Pentabromomethylbenzene

8527 Pentabromophenol

8528 1,1,1,3,3-Pentabromo-2-propanone

8529 Pentac

8530 Pentacene

8531 2,3,4,5,6-Pentachloroaniline

8532 2,3,4,5,6-Pentachloroanisole

8533 Pentachlorobenzene

8534 Pentachlorobenzenethiol

8535 2,3,4,5,6-Pentachlorobiphenyl

8536 2,2',4,5,5'-Pentachlorobiphenyl

8537 1,2,3,4,7-Pentachlorodibenzo-*p*-dioxin

8538 Pentachloroethane

8539 Pentachlorofluoroethane

No.	Name	Synonym	Mol. Form.	CAS RN	Mol. Wt.	Physical Form	mp/°C	bp/°C	den g cm⁻³	n_D	Solubility
8540	Pentachloronitrobenzene	Quintozene	$C_6Cl_5NO_2$	82-68-8	295.335	cry (al)	144.7(0.4)	328 dec	1.718[25]		i H_2O; sl EtOH; s bz, chl
8541	Pentachlorophenol		C_6HCl_5O	87-86-5	266.336	mcl pr (al + 1w) nd (bz)	189.5(0.4)	310 dec	1.978[22]		i H_2O; sl lig; vs EtOH, eth; s bz
8542	1,1,2,2,3-Pentachloropropane		$C_3H_3Cl_5$	16714-68-4	216.321			181[500]	1.633[25]	1.5098[25]	
8543	1,1,2,3,3-Pentachloro-1-propene		C_3HCl_5	1600-37-9	214.305			185	1.6317[34]	1.5313[20]	vs eth
8544	Pentachloropyridine		C_5Cl_5N	2176-62-7	251.326		124.0(0.5)	280			vs bz, EtOH, lig
8545	2,3,4,5,6-Pentachlorotoluene		$C_7H_3Cl_5$	877-11-2	264.364	nd (bz, peth)	224(2)	301			sl EtOH, eth, CS_2; s bz, tol, peth
8546	Pentacontane		$C_{50}H_{102}$	6596-40-3	703.345		91.7(0.6)	575.0			
8547	Pentacosane		$C_{25}H_{52}$	629-99-2	352.681		53.3(0.4)	401.9	0.8012[20]	1.4491[20]	s bz, chl
8548	1H-Pentadecafluoroheptane		C_7HF_{15}	375-83-7	370.059			95(2)	1.725[25]	1.2690[25]	
8549	Pentadecafluorooctanoic acid	Perfluorooctanoic acid	$C_8HF_{15}O_2$	335-67-1	414.069		54.3	192(1)			
8550	2,2,3,3,4,4,5,5,6,6,7,7,8,8,8-Pentadecafluoro-1-octanol	1,1-Dihydroperfluorooctanol	$C_8H_3F_{15}O$	307-30-2	400.085	waxy solid	47	164			
8551	Pentadecanal		$C_{15}H_{30}O$	2765-11-9	226.398	nd	24(2)	185[25]			vs ace, eth, EtOH
8552	Pentadecane		$C_{15}H_{32}$	629-62-9	212.415		9.95(0.02)	270.6(0.4)	0.7685[20]	1.4315[20]	i H_2O; vs EtOH, eth
8553	Pentadecanoic acid	Pentadecylic acid	$C_{15}H_{30}O_2$	1002-84-2	242.398	pl (dil al, HOAc) cry (peth)	52.52(0.04)	257[100]	0.8423[80]	1.4254[80]	i H_2O; vs EtOH, ace; s eth; sl tfa
8554	1-Pentadecanol		$C_{15}H_{32}O$	629-76-5	228.414		43.8(0.2)	318(2)	0.8347[25]		i H_2O
8555	2-Pentadecanone		$C_{15}H_{30}O$	2345-28-0	226.398		39.0(0.2)	294	0.8182[39]		
8556	8-Pentadecanone		$C_{15}H_{30}O$	818-23-5	226.398	cry (al)	42(1)	295(1)	0.8180[39]		s EtOH, eth, bz, ctc, chl
8557	1-Pentadecene		$C_{15}H_{30}$	13360-61-7	210.399	liq	-3.8(0.1)	268.4(0.4)	0.7764[20]	1.4389[20]	i H_2O; s ace
8558	Pentadecylamine	Pentadecanamine	$C_{15}H_{33}N$	2570-26-5	227.430		36.5(0.4)	312(5)	0.8104[20]	1.4480[20]	vs eth, EtOH
8559	Pentadecylbenzene		$C_{21}H_{36}$	2131-18-2	288.511		12.0(0.4)	373	0.8548[20]	1.4815[20]	
8560	3-Pentadecyl-1,2-benzenediol	3-Pentadecylcatechol	$C_{21}H_{36}O_2$	492-89-7	320.510	nd (to, peth)	59.5				vs bz, eth, EtOH
8561	Pentadecylcyclohexane		$C_{21}H_{42}$	6006-95-7	294.558		25.0(0.4)	371(13)	0.8267[20]	1.4588[20]	
8562	3-Pentadecylphenol		$C_{21}H_{36}O$	501-24-6	304.510	nd (peth)	50(3)	230[8]			vs ace, bz, EtOH
8563	1-Pentadecyne		$C_{15}H_{28}$	765-13-9	208.383		10	280(10)	0.7928[20]	1.4419[20]	vs ace
8564	1,2-Pentadiene	Ethylallene	C_5H_8	591-95-7	68.118	liq	-137.27(0.08)	44(2)	0.6926[20]	1.4209[20]	msc EtOH, eth, ace, bz, ctc, hp
8565	cis-1,3-Pentadiene	cis-Piperylene	C_5H_8	1574-41-0	68.118	liq	-140.81(0.09)	44.0(0.7)	0.6910[20]	1.4363[20]	msc EtOH, eth, ace, bz, ctc, hp
8566	trans-1,3-Pentadiene	trans-Piperylene	C_5H_8	2004-70-8	68.118	liq	-87.5(0.2)	42.0(0.3)	0.6710[25]	1.4301[20]	
8567	1,4-Pentadiene		C_5H_8	591-93-5	68.118	vol liq or gas	-148.3(0.3)	25.9(0.3)	0.6608[20]	1.3888[20]	i H_2O; vs EtOH, eth, ace, bz
8568	2,3-Pentadiene	1,3-Dimethylallene	C_5H_8	591-96-8	68.118	liq	-125.65(0.09)	48.2(0.9)	0.6950[20]	1.4284[20]	i H_2O; msc EtOH, eth, ace, bz, hp, ctc
8569	1,4-Pentadien-3-ol		C_5H_8O	922-65-6	84.117			115.5	0.860[23]	1.4400[17]	
8570	1,3-Pentadiyne	Methyldiacetylene	C_5H_4	4911-55-1	64.086	liq	-38.5	55(7)	0.7909[20]	1.4431[21]	i H_2O; s eth, bz, chl
8571	Pentaerythritol		$C_5H_{12}O_4$	115-77-5	136.147	cry (dil HCl)	258	sub		1.548	s H_2O; i eth, bz
8572	Pentaerythritol tetraacetate	2,2-Bis[(acetyloxy)methyl]-1,3-propanediol diacetate	$C_{13}H_{20}O_8$	597-71-7	304.293	tetr nd (w, bz)	83.5		1.273[18]		s H_2O; vs EtOH, eth
8573	Pentaerythritol tetrakis(2-mercaptoacetate)		$C_{13}H_{20}O_8S_4$	10193-99-4	432.553	liq		250[1]	1.385[25]	1.5470[20]	
8574	Pentaerythritol tetramethacrylate	Tetramethylolmethane tetramethacrylate	$C_{21}H_{28}O_8$	3253-41-6	408.442		53.5				
8575	Pentaerythritol tetranitrate		$C_5H_8N_4O_{12}$	78-11-5	316.138	tetr (ace) pr (ace-al)	140.9(0.8)		1.773[20]		sl H_2O, EtOH, eth; vs ace; s bz, py
8576	Pentaethylbenzene		$C_{16}H_{26}$	605-01-6	218.377		<-20	277(4)	0.8971[19]	1.5127[20]	
8577	Pentaethyl tantalate	Ethanol, tantalum(5+) salt	$C_{10}H_{25}O_5Ta$	6074-84-6	406.251			151[1]			
8578	2,3,4,5,6-Pentafluoroaniline		$C_6H_2F_5N$	771-60-8	183.079		34(1)	153.5			
8579	Pentafluorobenzaldehyde		C_7HF_5O	653-37-2	196.074		20	167		1.4506[20]	
8580	Pentafluorobenzene		C_6HF_5	363-72-4	168.064	liq	-47.3(0.2)	85(3)	1.514[25]	1.3905[20]	
8581	Pentafluorobenzenethiol		C_6HF_5S	771-62-0	200.129	liq	-24	143	1.501[25]	1.4645[20]	
8582	Pentafluorobenzoic acid		$C_7HF_5O_2$	602-94-8	212.074		103.3(0.5)	220			
8583	Pentafluorobenzonitrile		C_7F_5N	773-82-0	193.074		3(2)	162(7)	1.563[20]	1.4402[25]	
8584	Pentafluoroethane	HFC-125	C_2HF_5	354-33-6	120.021	col gas	-100.6	-48.1			
8585	Pentafluoroiodobenzene		C_6F_5I	827-15-6	293.960	liq	-29	161(3)	2.212[20]	1.4950[25]	
8586	Pentafluoromethoxybenzene		$C_7H_3F_5O$	389-40-2	198.090	liq	-30(1)	138.5	1.493[20]	1.4087[20]	

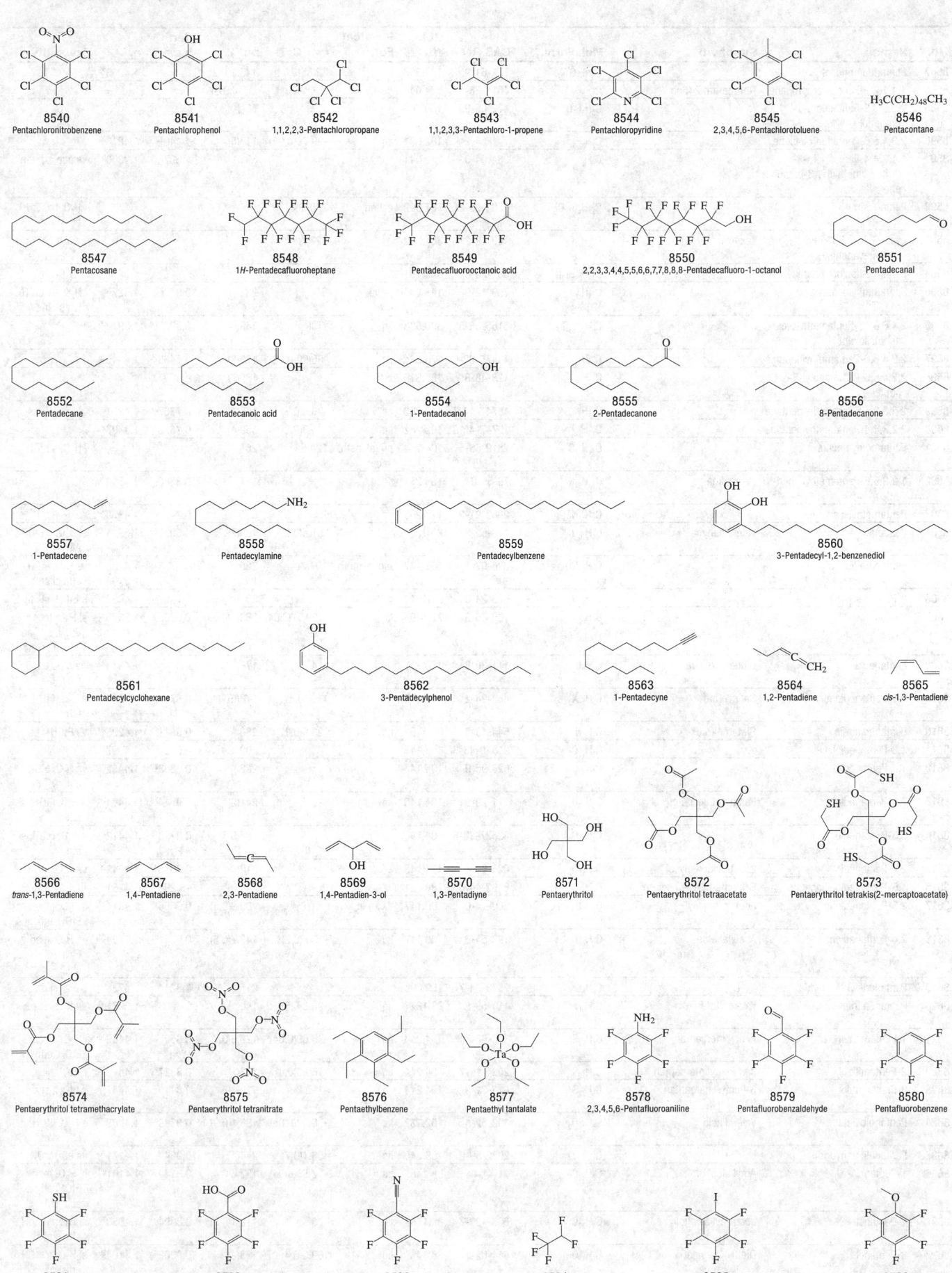

8540 Pentachloronitrobenzene

8541 Pentachlorophenol

8542 1,1,2,2,3-Pentachloropropane

8543 1,1,2,3,3-Pentachloro-1-propene

8544 Pentachloropyridine

8545 2,3,4,5,6-Pentachlorotoluene

8546 Pentacontane

8547 Pentacosane

8548 1H-Pentadecafluoroheptane

8549 Pentadecafluorooctanoic acid

8550 2,2,3,3,4,4,5,5,6,6,7,7,8,8,8-Pentadecafluoro-1-octanol

8551 Pentadecanal

8552 Pentadecane

8553 Pentadecanoic acid

8554 1-Pentadecanol

8555 2-Pentadecanone

8556 8-Pentadecanone

8557 1-Pentadecene

8558 Pentadecylamine

8559 Pentadecylbenzene

8560 3-Pentadecyl-1,2-benzenediol

8561 Pentadecylcyclohexane

8562 3-Pentadecylphenol

8563 1-Pentadecyne

8564 1,2-Pentadiene

8565 cis-1,3-Pentadiene

8566 trans-1,3-Pentadiene

8567 1,4-Pentadiene

8568 2,3-Pentadiene

8569 1,4-Pentadien-3-ol

8570 1,3-Pentadiyne

8571 Pentaerythritol

8572 Pentaerythritol tetraacetate

8573 Pentaerythritol tetrakis(2-mercaptoacetate)

8574 Pentaerythritol tetramethacrylate

8575 Pentaerythritol tetranitrate

8576 Pentaethylbenzene

8577 Pentaethyl tantalate

8578 2,3,4,5,6-Pentafluoroaniline

8579 Pentafluorobenzaldehyde

8580 Pentafluorobenzene

8581 Pentafluorobenzenethiol

8582 Pentafluorobenzoic acid

8583 Pentafluorobenzonitrile

8584 Pentafluoroethane

8585 Pentafluoroiodobenzene

8586 Pentafluoromethoxybenzene

No.	Name	Synonym	Mol. Form.	CAS RN	Mol. Wt.	Physical Form	mp/°C	bp/°C	den g cm⁻³	n_D	Solubility
8587	Pentafluorophenol		C_6HF_5O	771-61-9	184.063		32.80(0.04)	145.1(0.2)		1.4263^{20}	
8588	1,1,1,2,2-Pentafluoropropane	Refrigerant 245cb	$C_3H_3F_5$	1814-88-6	134.048	col gas		-18.0(0.3)			
8589	2,2,3,3,3-Pentafluoro-1-propanol		$C_3H_3F_5O$	422-05-9	150.047			26^{50}			
8590	2,3,4,5,6-Pentafluorotoluene		$C_7H_3F_5$	771-56-2	182.091	liq	-29.79(0.05)	117(5)	1.440^{20}	1.4016^{25}	
8591	1,1,2,4,4-Pentafluoro-3-(trifluoromethyl)-1,3-butadiene		C_5F_8	384-04-3	212.041			39	1.527^0	1.3000^0	vs ace, bz, eth
8592	Pentagastrin		$C_{37}H_{49}N_7O_9S$	5534-95-2	767.892	col nd	230 dec				i H_2O, bz, EtOH, eth
8593	trans-3,3',4',5,7-Pentahydroxyflavanone, (±)-	Taxifolin	$C_{15}H_{12}O_7$	480-18-2	304.252		227 dec				s chl
8594	Pentamethonium bromide		$C_{11}H_{28}Br_2N_2$	541-20-8	348.161		301				sl H_2O
8595	Pentamethylbenzene		$C_{11}H_{16}$	700-12-9	148.245	pr (al)	54(2)	232	0.917^{20}	1.527^{20}	i H_2O; vs EtOH, bz; s chl
8596	2,4,6,8,10-Pentamethylcyclopentasiloxane		$C_5H_{20}O_5Si_5$	6166-86-5	300.638	liq	-108	169(6)	0.9985^{20}	1.3912^{20}	
8597	2,2,4,6,6-Pentamethylheptane		$C_{12}H_{26}$	13475-82-6	170.334	liq	-66.90(0.08)	178(2)	0.7463^{20}	1.4440^{20}	
8598	2,2,4,6,6-Pentamethyl-3-heptene		$C_{12}H_{24}$	123-48-8	168.319	liq		180(7)			
8599	2,2,3,3,4-Pentamethylpentane		$C_{10}H_{22}$	16747-44-7	142.282	liq	-36.5(0.2)	166(4)	0.7767^{25}	1.4361^{20}	
8600	2,2,3,4,4-Pentamethylpentane		$C_{10}H_{22}$	16747-45-8	142.282	liq	-38.8(0.2)	159(4)	0.7636^{25}	1.4307^{20}	
8601	Pentamethylphenol		$C_{11}H_{16}O$	2819-86-5	164.244	nd (al, peth, ace)	128	267			i H_2O; s EtOH
8602	1,2,2,6,6-Pentamethylpiperidine	Pempidine	$C_{10}H_{21}N$	79-55-0	155.281			147	0.8580^0	1.4550^{21}	
8603	Pentamethylsilanamine		$C_5H_{15}NSi$	2083-91-2	117.266			86	0.7400^{20}	1.4379^{24}	
8604	Pentanal	Valeraldehyde	$C_5H_{10}O$	110-62-3	86.132	liq	-81.5(0.2)	103(2)	0.8095^{20}	1.3944^{20}	sl H_2O; s EtOH, eth
8605	Pentanamide		$C_5H_{11}NO$	626-97-1	101.147	mcl pl (peth, al)	104(2)	225	0.8735^{110}	1.4183^{110}	vs H_2O, EtOH, eth; sl chl
8606	3-Pentanamine		$C_5H_{13}N$	616-24-0	87.164			89(5)	0.7487^{20}	1.4063^{20}	s EtOH; sl chl
8607	Pentane		C_5H_{12}	109-66-0	72.149	liq	-129.67(0.04)	36.06(0.07)	0.6262^{20}	1.3575^{20}	sl H_2O; msc EtOH, eth, ace, bz, chl; s ctc
8608	Pentanedial	Glutaraldehyde	$C_5H_8O_2$	111-30-8	100.117	oil	-14	176(8)		1.4330^{25}	msc H_2O, EtOH; s bz
8609	1,5-Pentanediamine	Cadaverine	$C_5H_{14}N_2$	462-94-2	102.178		11.8(0.4)	178(6)	0.873^{25}	1.463^{20}	s H_2O, EtOH; sl eth
8610	Pentanedinitrile	Glutaronitrile	$C_5H_6N_2$	544-13-8	94.115	liq	-28.9(0.1)	286	0.9911^{15}	1.4295^{20}	vs EtOH, chl
8611	1,2-Pentanediol, (±)-		$C_5H_{12}O_2$	91049-43-3	104.148			209	0.9723^{20}	1.4397^{19}	
8612	1,4-Pentanediol		$C_5H_{12}O_2$	626-95-9	104.148			202	0.9883^{20}	1.4452^{23}	vs H_2O, EtOH, chl
8613	1,5-Pentanediol	Pentamethylene glycol	$C_5H_{12}O_2$	111-29-5	104.148	liq	-20(7)	241(2)	0.9914^{20}	1.4494^{20}	s H_2O, EtOH; sl eth, bz
8614	2,3-Pentanediol		$C_5H_{12}O_2$	42027-23-6	104.148			187.5	0.9798^{19}	1.4412^{25}	s H_2O, EtOH; sl eth
8615	2,4-Pentanediol	2,4-Amylene glycol	$C_5H_{12}O_2$	625-69-4	104.148			218(6)	0.9635^{20}	1.4349^{20}	vs H_2O, EtOH
8616	1,5-Pentanediol diacetate	Pentamethylene acetate	$C_9H_{16}O_4$	6963-44-6	188.221		2	244(9)	1.0296^{20}	1.4261^{19}	
8617	2,3-Pentanedione	Acetylpropionyl	$C_5H_8O_2$	600-14-6	100.117	dk ye liq		109.9(0.4)	0.9565^{19}	1.4014^{19}	s H_2O; msc EtOH, eth, ace
8618	2,4-Pentanedione	Acetylacetone	$C_5H_8O_2$	123-54-6	100.117	liq	-18.3(0.2)	140.7(0.5)	0.9721^{25}	1.4494^{20}	vs H_2O; msc EtOH, eth, ace, chl
8619	Pentanedioyl dichloride		$C_5H_6Cl_2O_2$	2873-74-7	169.006			217	1.324^{20}	1.4728^{20}	s eth; sl chl
8620	Pentanenitrile	Valeronitrile	C_5H_9N	110-59-8	83.132	liq	-96.2(0.5)	140(1)	0.8008^{20}	1.3971^{20}	s eth, ace, bz; sl ctc
8621	1-Pentanethiol	Pentyl mercaptan	$C_5H_{12}S$	110-66-7	104.214	liq	-75.69(0.04)	126.6(0.7)	0.850^{20}	1.4469^{20}	i H_2O; msc EtOH, eth
8622	2-Pentanethiol	sec-Pentyl mercaptan	$C_5H_{12}S$	2084-19-7	104.214	liq	-169.0(0.4)	113(1)	0.8327^{20}	1.4412^{20}	s EtOH, lig
8623	3-Pentanethiol	3-Pentyl mercaptan	$C_5H_{12}S$	616-31-9	104.214	liq	-110.8	116(10)	0.8410^{20}	1.4447^{20}	s EtOH; sl DMSO
8624	Pentanoic acid	Valeric acid	$C_5H_{10}O_2$	109-52-4	102.132	liq	-33.63(0.05)	186.1(0.3)	0.9339^{25}	1.4085^{20}	s H_2O, EtOH, eth; sl ctc
8625	Pentanoic anhydride		$C_{10}H_{18}O_3$	2082-59-9	186.248	liq	-56.1(0.7)	244(17)	0.924^{20}	1.4171^{26}	vs eth, EtOH
8626	1-Pentanol	Amyl alcohol	$C_5H_{12}O$	71-41-0	88.148	liq	-77.58(0.04)	137.6(0.4)	0.8144^{20}	1.4101^{20}	sl H_2O; msc EtOH, eth; s ace, chl
8627	2-Pentanol	sec-Amyl alcohol	$C_5H_{12}O$	6032-29-7	88.148	liq	-73	119.1(0.5)	0.8094^{20}	1.4053^{20}	sl H_2O; s EtOH, eth, ctc, chl
8628	3-Pentanol	Diethyl carbinol	$C_5H_{12}O$	584-02-1	88.148	liq	-69.9(0.4)	123(2)	0.8203^{20}	1.4104^{20}	sl H_2O; s EtOH, eth, ace, ctc

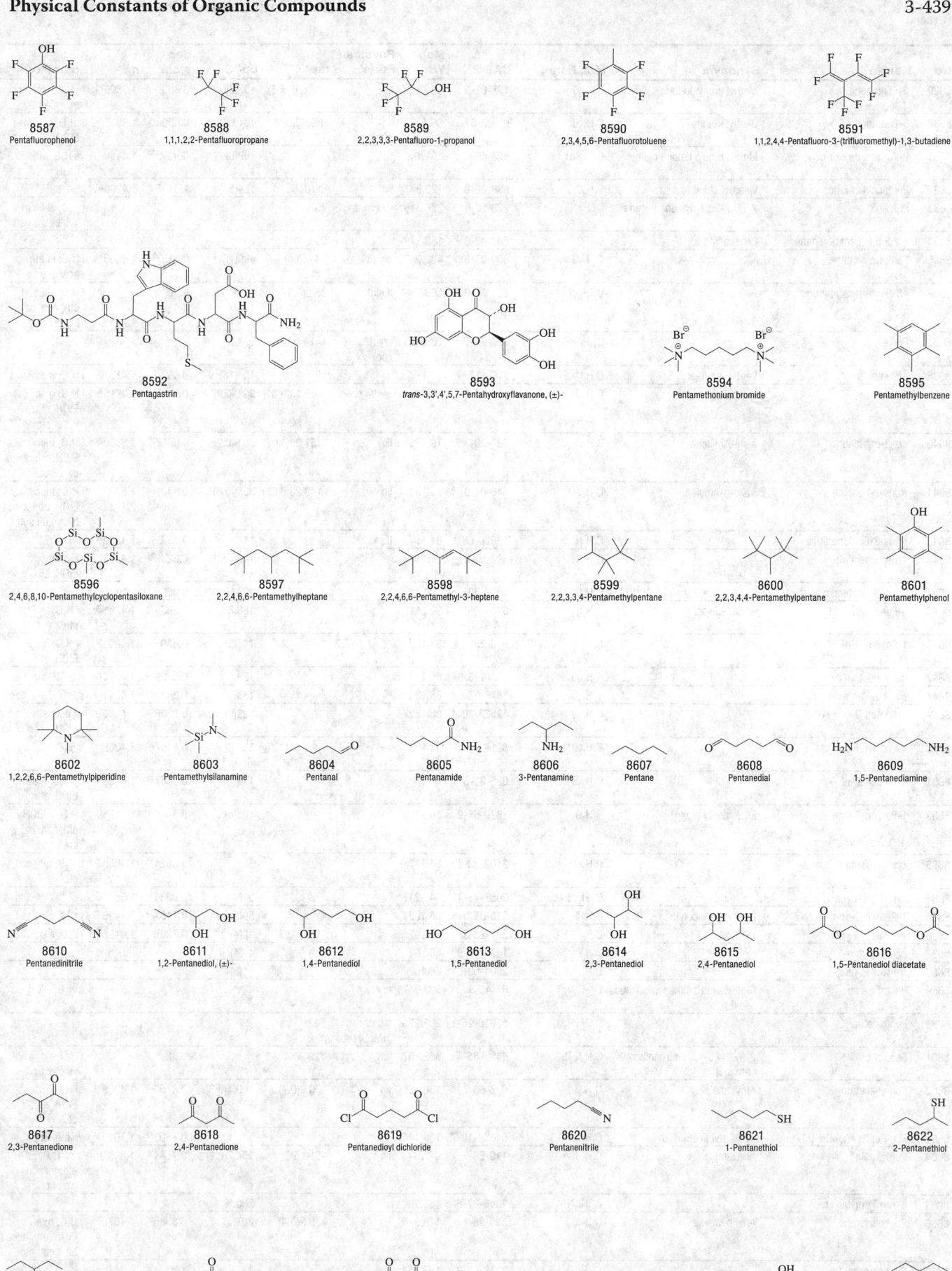

8587
Pentafluorophenol

8588
1,1,1,2,2-Pentafluoropropane

8589
2,2,3,3,3-Pentafluoro-1-propanol

8590
2,3,4,5,6-Pentafluorotoluene

8591
1,1,2,4,4-Pentafluoro-3-(trifluoromethyl)-1,3-butadiene

8592
Pentagastrin

8593
trans-3,3',4',5,7-Pentahydroxyflavanone, (±)-

8594
Pentamethonium bromide

8595
Pentamethylbenzene

8596
2,4,6,8,10-Pentamethylcyclopentasiloxane

8597
2,2,4,6,6-Pentamethylheptane

8598
2,2,4,6,6-Pentamethyl-3-heptene

8599
2,2,3,3,4-Pentamethylpentane

8600
2,2,3,4,4-Pentamethylpentane

8601
Pentamethylphenol

8602
1,2,2,6,6-Pentamethylpiperidine

8603
Pentamethylsilanamine

8604
Pentanal

8605
Pentanamide

8606
3-Pentanamine

8607
Pentane

8608
Pentanedial

8609
1,5-Pentanediamine

8610
Pentanedinitrile

8611
1,2-Pentanediol, (±)-

8612
1,4-Pentanediol

8613
1,5-Pentanediol

8614
2,3-Pentanediol

8615
2,4-Pentanediol

8616
1,5-Pentanediol diacetate

8617
2,3-Pentanedione

8618
2,4-Pentanedione

8619
Pentanedioyl dichloride

8620
Pentanenitrile

8621
1-Pentanethiol

8622
2-Pentanethiol

8623
3-Pentanethiol

8624
Pentanoic acid

8625
Pentanoic anhydride

8626
1-Pentanol

8627
2-Pentanol

8628
3-Pentanol

No.	Name	Synonym	Mol. Form.	CAS RN	Mol. Wt.	Physical Form	mp/°C	bp/°C	den g cm^{-3}	n_D	Solubility
8629	2-Pentanone	Methyl propyl ketone	$C_5H_{10}O$	107-87-9	86.132	liq	-76.83(0.05)	102.2(0.1)	0.809[20]	1.3895[20]	sl H$_2$O, ctc; msc EtOH, eth
8630	3-Pentanone	Diethyl ketone	$C_5H_{10}O$	96-22-0	86.132	liq	-38.98(0.05)	101.9(0.1)	0.8098[25]	1.3905[25]	s H$_2$O, ctc; msc EtOH, eth
8631	2-Pentanone oxime	Methyl propyl ketone oxime	$C_5H_{11}NO$	623-40-5	101.147			169(8)	0.9095[20]	1.4450[20]	vs H$_2$O, eth, EtOH
8632	Pentanoyl chloride	Valeroyl chloride	C_5H_9ClO	638-29-9	120.577	liq	-110	124(5)	1.0155[15]	1.4200[20]	
8633	Pentaphene	2,3:6,7-Dibenzphenanthrene	$C_{22}H_{14}$	222-93-5	278.346	ye grn lf(xyl)	257				i H$_2$O; sl EtOH, xyl, eth; s bz
8634	1,2,3,5,6-Pentathiepane	Lenthionine	$C_2H_4S_5$	292-46-6	188.378		60.5				
8635	Pentatriacontane		$C_{35}H_{72}$	630-07-9	492.947	cry (al)	74.4(0.4)	489(15)	0.8157[20]	1.4568[20]	i H$_2$O; sl eth; s ace
8636	18-Pentatriacontanone		$C_{35}H_{70}O$	504-53-0	506.930	lf (lig)	88.8(0.8)	270[0.1]	0.793[95]		i H$_2$O; sl EtOH, eth, ace, bz, lig, chl
8637	Pentazocine		$C_{19}H_{27}NO$	359-83-1	285.423	cry (MeOH aq)	147				
8638	4-Pentenal	Pent-4-en-1-al	C_5H_8O	2100-17-6	84.117			104(18)	0.852[20]	1.4191[20]	i H$_2$O; s eth, ace
8639	1-Pentene	α-Amylene	C_5H_{10}	109-67-1	70.133	vol liq or gas	-165.13(0.01)	30.0(0.3)	0.6405[20]	1.3715[20]	i H$_2$O; msc EtOH, eth; s bz; sl ctc
8640	cis-2-Pentene	cis-β-Amylene	C_5H_{10}	627-20-3	70.133	liq	-151.35(0.02)	36.9(0.2)	0.6556[20]	1.3830[20]	i H$_2$O; msc EtOH, eth; s bz, dil sulf
8641	trans-2-Pentene	trans-β-Amylene	C_5H_{10}	646-04-8	70.133	liq	-140.20(0.02)	36.3(0.6)	0.6431[25]	1.3793[20]	i H$_2$O; msc EtOH, eth; s bz; vs dil sulf
8642	trans-3-Pentenenitrile		C_5H_7N	16529-66-1	81.117	liq		144	0.837	1.4220[20]	
8643	4-Pentenenitrile		C_5H_7N	592-51-8	81.117			140	0.8239[24]	1.4213[14]	i H$_2$O; msc EtOH, eth
8644	trans-3-Pentenoic acid		$C_5H_8O_2$	1617-32-9	100.117			193.2	0.989[19]		
8645	4-Pentenoic acid	Allylacetic acid	$C_5H_8O_2$	591-80-0	100.117	liq	-22.5	188.5	0.9809[20]	1.4281[20]	sl H$_2$O; vs EtOH, eth
8646	1-Penten-3-ol		$C_5H_{10}O$	616-25-1	86.132			118(5)	0.839[20]	1.4239[20]	sl H$_2$O; msc EtOH, eth
8647	cis-2-Penten-1-ol		$C_5H_{10}O$	1576-95-0	86.132			138	0.8529[20]	1.4354[20]	s EtOH, eth, ace
8648	trans-2-Penten-1-ol		$C_5H_{10}O$	1576-96-1	86.132			138	0.8471[20]	1.4341[20]	s EtOH, eth, ace
8649	3-Penten-2-ol, (±)-		$C_5H_{10}O$	42569-16-4	86.132			123(9)	0.8328[25]	1.4280[20]	vs ace, eth, EtOH
8650	4-Penten-1-ol		$C_5H_{10}O$	821-09-0	86.132			136(8)	0.8457[20]	1.4309[20]	sl H$_2$O, ctc; s eth
8651	4-Penten-2-ol		$C_5H_{10}O$	625-31-0	86.132			116	0.8367[20]	1.4225[20]	vs H$_2$O; msc EtOH, eth
8652	1-Penten-3-one	Ethyl vinyl ketone	C_5H_8O	1629-58-9	84.117			103(6)	0.8468[20]	1.4195[20]	i H$_2$O; s EtOH, eth, ace, bz, chl
8653	trans-3-Penten-2-one		C_5H_8O	3102-33-8	84.117			122	0.8624[20]	1.4350[20]	s H$_2$O, eth, ace, ctc
8654	2-(3-Pentenyl)pyridine		$C_{10}H_{13}N$	2057-43-4	147.217			216	0.9234[25]	1.5076[25]	
8655	1-Penten-3-yne	Methylvinylacetylene	C_5H_6	646-05-9	66.102			59(3)	0.7401[20]	1.4496[20]	vs bz, eth
8656	1-Penten-4-yne		C_5H_6	871-28-3	66.102			42(4)	0.738[16]	1.4125[16]	i H$_2$O; s eth, bz
8657	cis-3-Penten-1-yne		C_5H_6	1574-40-9	66.102			44(4)			
8658	trans-3-Penten-1-yne		C_5H_6	2004-69-5	66.102			51(3)			
8659	Pentetic acid	Diethylenetriaminepentaacetic acid	$C_{14}H_{23}N_3O_{10}$	67-43-6	393.347	cry (w)	219				s H$_2$O, alk
8660	Pentostatin		$C_{11}H_{16}N_4O_4$	53910-25-1	268.270	wh cry (MeOH aq)	222				
8661	Pentryl	2-(N,2,4,6-Tetranitroanilino)-ethanol	$C_8H_6N_6O_{11}$	4481-55-4	362.167	wh-ye cry	129		1.82		i H$_2$O, ctc; s chl; vs eth, bz
8662	Pentyl acetate	Amyl acetate	$C_7H_{14}O_2$	628-63-7	130.185	liq	-70.9(0.7)	149.4(0.3)	0.8756[20]	1.4023[20]	sl H$_2$O; msc EtOH, eth; s ctc
8663	sec-Pentyl acetate, (R)-	sec-Amyl acetate (R)	$C_7H_{14}O_2$	54638-10-7	130.185			142	0.8803[18]	1.4012[20]	vs eth, EtOH
8664	Pentylamine	Amylamine	$C_5H_{13}N$	110-58-7	87.164	liq	-51(15)	104.7(0.2)	0.7544[20]	1.448[20]	msc H$_2$O, EtOH, eth; vs ace, bz; sl chl
8665	4-tert-Pentylaniline		$C_{11}H_{17}N$	2049-92-5	163.260			260.5			
8666	Pentylbenzene	Amylbenzene	$C_{11}H_{16}$	538-68-1	148.245	liq	-78.2(0.4)	203(3)	0.8585[20]	1.4878[20]	i H$_2$O; msc EtOH, eth, ace, bz, peth, ctc
8667	Pentyl benzoate		$C_{12}H_{16}O_2$	2049-96-9	192.254			137[15]			
8668	4-Pentylbenzoyl chloride		$C_{12}H_{15}ClO$	49763-65-7	210.699			144[10]	1.036[25]	1.5300[20]	

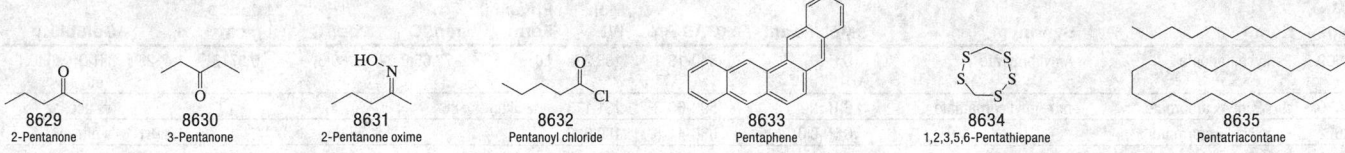

8629	8630	8631	8632	8633	8634	8635
2-Pentanone	3-Pentanone	2-Pentanone oxime	Pentanoyl chloride	Pentaphene	1,2,3,5,6-Pentathiepane	Pentatriacontane

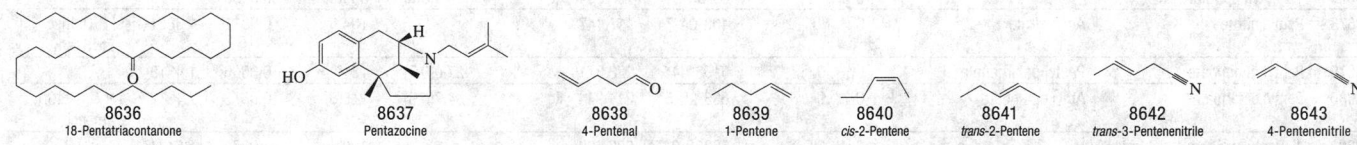

8636	8637	8638	8639	8640	8641	8642	8643
18-Pentatriacontanone	Pentazocine	4-Pentenal	1-Pentene	*cis*-2-Pentene	*trans*-2-Pentene	*trans*-3-Pentenenitrile	4-Pentenenitrile

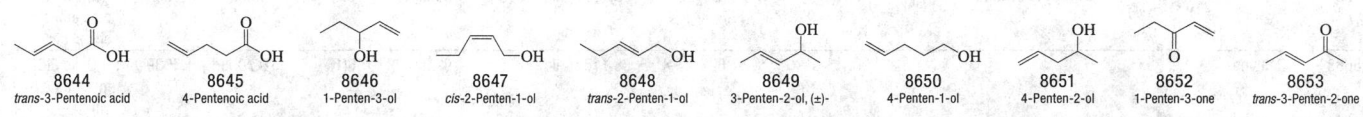

8644	8645	8646	8647	8648	8649	8650	8651	8652	8653
trans-3-Pentenoic acid	4-Pentenoic acid	1-Penten-3-ol	*cis*-2-Penten-1-ol	*trans*-2-Penten-1-ol	3-Penten-2-ol, (±)-	4-Penten-1-ol	4-Penten-2-ol	1-Penten-3-one	*trans*-3-Penten-2-one

8654	8655	8656	8657	8658	8659	8660	8661
2-(3-Pentenyl)pyridine	1-Penten-3-yne	1-Penten-4-yne	*cis*-3-Penten-1-yne	*trans*-3-Penten-1-yne	Pentetic acid	Pentostatin	Pentryl

8662	8663	8664	8665	8666	8667	8668
Pentyl acetate	*sec*-Pentyl acetate, (*R*)-	Pentylamine	4-*tert*-Pentylaniline	Pentylbenzene	Pentyl benzoate	4-Pentylbenzoyl chloride

No.	Name	Synonym	Mol. Form.	CAS RN	Mol. Wt.	Physical Form	mp/°C	bp/°C	den g cm⁻³	n_D	Solubility
8669	Pentyl butanoate	Amyl butyrate	$C_9H_{18}O_2$	540-18-1	158.238	liq	-72.66(0.02)	180(3)	0.8713[15]	1.4123[20]	i H_2O; vs EtOH, eth
8670	*tert*-Pentyl carbamate	*tert*-Amyl carbamate	$C_6H_{13}NO_2$	590-60-3	131.173	nd (dil al)	86				vs ace, bz
8671	Pentyl chloroformate		$C_6H_{11}ClO_2$	638-41-5	150.603			61[15]		1.4181[18]	s eth
8672	Pentylcyclohexane		$C_{11}H_{22}$	4292-92-6	154.293	liq	-57.5	204(1)	0.8037[20]	1.4437[20]	vs ace, bz, eth, EtOH
8673	Pentylcyclopentane		$C_{10}H_{20}$	3741-00-2	140.266	liq	-83	180(7)	0.7912[20]	1.4356[20]	i H_2O; vs ace, bz, eth, EtOH
8674	Pentyl formate	Amyl formate	$C_6H_{12}O_2$	638-49-3	116.158	liq	-73(1)	126(3)	0.8853[20]	1.3992[20]	sl H_2O; msc EtOH, eth
8675	Pentyl heptanoate	Amyl enanthate	$C_{12}H_{24}O_2$	7493-82-5	200.318	liq	-49.3(0.7)	239(5)	0.8623[20]	1.4263[15]	vs ace, bz, eth, EtOH
8676	Pentyl hexanoate	Amyl caproate	$C_{11}H_{22}O_2$	540-07-8	186.292	liq	-49(3)	225(3)	0.8612[25]	1.4202[25]	s EtOH, eth, ace; sl ctc
8677	1-Pentylnaphthalene		$C_{15}H_{18}$	86-89-5	198.304	liq	-24.5(0.3)	305(5)	0.9656[20]	1.5725[20]	
8678	Pentyl nitrite	Amyl nitrite	$C_5H_{11}NO_2$	463-04-7	117.147			104.5	0.8817[20]	1.3851[20]	sl H_2O; msc EtOH, eth
8679	Pentyl nonanoate	Pentyl pelargonate	$C_{14}H_{28}O_2$	61531-45-1	228.371		-27.0(0.7)	131[20]	0.8506[25]	1.4318[20]	
8680	Pentyl octanoate	Amyl octanoate	$C_{13}H_{26}O_2$	638-25-5	214.344	liq	-34.7(0.6)	245(8)	0.8613[20]	1.4262[25]	i H_2O; s EtOH, eth, ace
8681	4-(Pentyloxy)benzoyl chloride		$C_{12}H_{15}ClO_2$	36823-84-4	226.699			198[30]	1.087[25]	1.5434[20]	
8682	Pentyl pentanoate		$C_{10}H_{20}O_2$	2173-56-0	172.265	liq	-78.8	204(6)	0.8638[20]	1.4164[20]	sl H_2O; msc EtOH, eth
8683	4-Pentylphenol		$C_{11}H_{16}O$	14938-35-3	164.244		18(1)	250.5	0.960[20]	1.5272[25]	vs eth, EtOH
8684	Pentyl propanoate		$C_8H_{16}O_2$	624-54-4	144.212	liq	-73.1	166(4)	0.8761[25]	1.4096[15]	i H_2O; msc EtOH, eth; s bz; sl ctc
8685	Pentyl salicylate		$C_{12}H_{16}O_3$	2050-08-0	208.253			270	1.064[15]	1.506[20]	sl H_2O; msc EtOH, eth
8686	Pentyl stearate		$C_{23}H_{46}O_2$	6382-13-4	354.610	pl	29.2(0.6)			1.4342[50]	vs eth, EtOH
8687	1-Pentyne	Propylacetylene	C_5H_8	627-19-0	68.118	liq	-106.2(0.5)	39.9(1)	0.6901[20]	1.3852[20]	i H_2O; vs EtOH; msc eth; s bz, chl; sl ctc
8688	2-Pentyne		C_5H_8	627-21-4	68.118	liq	-109.3(0.2)	56(1)	0.7058[25]	1.4039[20]	i H_2O; vs EtOH; msc eth; s bz, chl
8689	4-Pentynoic acid	Propargylacetic acid	$C_5H_6O_2$	6089-09-4	98.101		55(3)	110[30]			vs eth, EtOH
8690	2-Pentyn-1-ol		C_5H_8O	6261-22-9	84.117	liq	-49.7	150(9)	0.909[20]	1.4518[17]	
8691	3-Pentyn-1-ol		C_5H_8O	10229-10-4	84.117			154	0.9002[20]	1.4454[20]	
8692	4-Pentyn-1-ol		C_5H_8O	5390-04-5	84.117			150(11)	0.913[20]	1.4414[20]	
8693	Perazine	10-[3-(4-Methyl-1-piperazinyl)propyl]-10H-phenothiazine	$C_{20}H_{25}N_3S$	84-97-9	339.498	cry	52	165[0.001]			
8694	Perfluidone		$C_{14}H_{12}F_3NO_4S_2$	37924-13-3	379.375		146.0(0.5)				
8695	Perfluoroacetone	Hexafluoroacetone	C_3F_6O	684-16-2	166.021	col gas	-125.45	-27.4(0.4)			
8696	Perfluorobutane	Decafluorobutane	C_4F_{10}	355-25-9	238.027	col gas	-129(1)	-2.1(0.8)	1.6484[25]		s bz, chl
8697	Perfluoro-2-butene		C_4F_8	360-89-4	200.030	col gas	-129	-5.9(0.9)	1.5297[25]		
8698	Perfluoro-2-butyltetrahydrofuran		$C_8F_{16}O$	335-36-4	416.059			105.2(0.6)			
8699	Perfluorocyclobutane	Octafluorocyclobutane	C_4F_8	115-25-3	200.030	col gas	-40.16(0.1)	-5.91	1.500[25] (p>1 atm)		i H_2O; s eth
8700	Perfluorocyclohexane		C_6F_{12}	355-68-0	300.045		62.9(0.3)	52.8 sp			
8701	Perfluorocyclohexene		C_6F_{10}	355-75-9	262.048			51.7(0.1)	1.6650[25]	1.293[20]	
8702	Perfluorodecalin	PFC-9-1-18	$C_{10}F_{18}$	306-94-5	462.078	liq	-8.2(0.5)	143(1)	1.9305[25]		
8703	Perfluorodecane		$C_{10}F_{22}$	307-45-9	538.072			135(3)			i H_2O
8704	Perfluorodimethoxymethane		$C_3F_8O_2$	53772-78-4	220.018	col gas	-161(2)	-10.0(0.7)			
8705	Perfluoro-2,3-dimethylbutane		C_6F_{14}	354-96-1	338.042	liq	-15.1(0.5)	59.8(0.3)			
8706	Perfluoroethyl ethyl ether		$C_4H_5F_5O$	22052-81-9	164.074	vol liq or gas		28(3)			
8707	Perfluoroethyl 2,2,2-trifluoroethyl ether		$C_4H_2F_8O$	156053-88-2	218.045	vol liq or gas		27.89			
8708	Perfluoroheptane		C_7F_{16}	335-57-9	388.049	liq	-51.3(0.1)	82.5(0.2)	1.7333[20]	1.2618[20]	i H_2O; vs ace, eth, EtOH, chl
8709	Perfluoro-1-heptene		C_7F_{14}	355-63-5	350.053			81(4)			
8710	Perfluorohexane	PFC-5-1-14	C_6F_{14}	355-42-0	338.042	liq	-86.1(0.3)	57.2(0.2)	1.6910[20]	1.2515[20]	i H_2O; s eth, bz, chl
8711	Perfluoro-1-hexene		C_6F_{12}	755-25-9	300.045			57.0			vs chl
8712	Perfluoroisobutane		C_4F_{10}	354-92-7	238.027	col gas		0			
8713	Perfluoroisobutene	Perfluoroisobutylene	C_4F_8	382-21-8	200.030	col gas	-130	7	1.592[0]		
8714	Perfluoroisopropyl methyl ether		$C_4H_3F_7O$	22052-84-2	200.055	vol liq or gas		29(1)	1.4205[20]		

8669 Pentyl butanoate

8670 *tert*-Pentyl carbamate

8671 Pentyl chloroformate

8672 Pentylcyclohexane

8673 Pentylcyclopentane

8674 Pentyl formate

8675 Pentyl heptanoate

8676 Pentyl hexanoate

8677 1-Pentylnaphthalene

8678 Pentyl nitrite

8679 Pentyl nonanoate

8680 Pentyl octanoate

8681 4-(Pentyloxy)benzoyl chloride

8682 Pentyl pentanoate

8683 4-Pentylphenol

8684 Pentyl propanoate

8685 Pentyl salicylate

8686 Pentyl stearate

8687 1-Pentyne

8688 2-Pentyne

8689 4-Pentynoic acid

8690 2-Pentyn-1-ol

8691 3-Pentyn-1-ol

8692 4-Pentyn-1-ol

8693 Perazine

8694 Perfluidone

8695 Perfluoroacetone

8696 Perfluorobutane

8697 Perfluoro-2-butene

8698 Perfluoro-2-butyltetrahydrofuran

8699 Perfluorocyclobutane

8700 Perfluorocyclohexane

8701 Perfluorocyclohexene

8702 Perfluorodecalin

8703 Perfluorodecane

8704 Perfluorodimethoxymethane

8705 Perfluoro-2,3-dimethylbutane

8706 Perfluoroethyl ethyl ether

8707 Perfluoroethyl 2,2,2-trifluoroethyl ether

8708 Perfluoroheptane

8709 Perfluoro-1-heptene

8710 Perfluorohexane

8711 Perfluoro-1-hexene

8712 Perfluoroisobutane

8713 Perfluoroisobutene

8714 Perfluoroisopropyl methyl ether

No.	Name	Synonym	Mol. Form.	CAS RN	Mol. Wt.	Physical Form	mp/°C	bp/°C	den g cm⁻³	n_D	Solubility
8715	Perfluoromethylcyclohexane		C_7F_{14}	355-02-2	350.053	liq	-39.0(0.2)	76.3(0.2)	1.7878²⁵	1.285¹⁷	s ace, bz, ctc, tol, AcOEt
8716	Perfluoro-2-methylpentane		C_6F_{14}	355-04-4	338.042			57.6(0.3)	1.7326²⁰	1.2564²²	i H₂O; s bz
8717	Perfluoro-3-methylpentane		C_6F_{14}	865-71-4	338.042	liq	-115.4(0.5)	58(9)			s bz
8718	Perfluoronaphthalene		$C_{10}F_8$	313-72-4	272.094		79(1)	200(8)			
8719	Perfluorononane		C_9F_{20}	375-96-2	488.064			117(3)	1.8001²⁰		
8720	Perfluorooctane		C_8F_{18}	307-34-6	438.057			105(2)	1.73²⁰	1.282²⁰	i H₂O
8721	Perfluorooctylsulfonyl fluoride		$C_8F_{18}O_2S$	307-35-7	502.121	liq		154			
8722	Perfluorooxetane		C_3F_6O	425-82-1	166.021		-117(2)	-28.6(0.6)			
8723	Perfluoropentane	PFC-4-1-12	C_5F_{12}	678-26-2	288.035	vol liq or gas	-10	29.2(0.2)			i H₂O
8724	Perfluoropropane	PFC-218	C_3F_8	76-19-7	188.019	col gas	-147.7(0.1)	-36.8(0.3)			i H₂O
8725	Perfluoropropene	Hexafluoropropene	C_3F_6	116-15-4	150.022	col gas	-156.5	-30.2(0.5)		1.583⁻⁴⁰	i H₂O
8726	Perfluoropropyl methyl ether	HFE-347mcc3	$C_4H_3F_7O$	375-03-1	200.055			34(1)	1.4092²⁰		
8727	Perfluoropyridine	Pentafluoropyridine	C_5F_5N	700-16-3	169.053			83(26)			
8728	Perfluorotoluene		C_7F_8	434-64-0	236.062	liq	-65.48(0.06)	104.6(0.3)	1.6616²⁵	1.3670²⁰	
8729	Perfluorotripropylamine	Tris(perfluoropropyl)amine	$C_9F_{21}N$	338-83-0	521.069			129.7(0.5)	1.822⁴	1.279²⁵	
8730	1H-Perimidine		$C_{11}H_8N_2$	204-02-4	168.195	grn cry (dil al)	223.0				i H₂O; s EtOH, eth, ace, bz; sl DMSO
8731	Permethrin		$C_{21}H_{20}Cl_2O_3$	52645-53-1	391.288	col cry or ye liq	34	200⁰·⁰¹	1.23²⁰		i H₂O; s os
8732	Peroxyacetic acid	Ethaneperoxoic acid	$C_2H_4O_3$	79-21-0	76.051	liq	-0.2	110	1.226¹⁵	1.3974²⁰	vs H₂O, eth, sulf; s EtOH
8733	Peroxypropanoic acid	Propaneperoxoic acid	$C_3H_6O_3$	4212-43-5	90.078			exp		1.4148¹⁵	
8734	Perphenazine		$C_{21}H_{26}ClN_3OS$	58-39-9	403.968		97				
8735	Perthane	Ethane, 1,1-dichloro-2,2-bis(p-ethylphenyl)-	$C_{18}H_{20}Cl_2$	72-56-0	307.258		59.7(0.9)				
8736	Perylene	Dibenz[de,kl]anthracene	$C_{20}H_{12}$	198-55-0	252.309	gold-br, ye pl (bz, HOAc)	276(13)		1.35²⁵		i H₂O; sl EtOH, eth; vs ace, chl; s bz
8737	Peucedanin	3-Methoxy-2-isopropyl-7H-furo[3,2-g][1]benzopyran-7-one	$C_{15}H_{14}O_4$	133-26-6	258.270	pr or pl (bz-peth)	85	278¹⁷			sl H₂O, bz; s EtOH, eth; vs chl, CS₂
8738	Phalloidin		$C_{35}H_{48}N_8O_{11}S$	17466-45-4	788.868	nd (w)	281 (hyd)				s EtOH, MeOH, py
8739	Phalloin		$C_{35}H_{48}N_8O_{10}S$	28227-92-1	772.869	cry (w)	250 dec				
8740	α-Phellandrene	2-Methyl-5-(1-methylethyl)-1,3-cyclohexadiene	$C_{10}H_{16}$	99-83-2	136.234		238	174.9	0.8410²⁰	1.471²⁵	i H₂O; s eth
8741	β-Phellandrene	p-Mentha-1(7),2-diene	$C_{10}H_{16}$	555-10-2	136.234			177(3)	0.8520²⁰	1.4788²⁰	i H₂O, EtOH; s eth
8742	9-Phenanthrenamine		$C_{14}H_{11}N$	947-73-9	193.244	lt ye cry (al)	138.3	sub			sl eth, bz, chl
8743	Phenanthrene		$C_{14}H_{10}$	85-01-8	178.229	mcl pl (al), lf (sub)	99(2)	338.4(0.9)	0.9800⁴	1.5943	i H₂O; s EtOH, eth, ace, bz, CS₂
8744	9,10-Phenanthrenedione	Phenanthrenequinone	$C_{14}H_8O_2$	84-11-7	208.213	oran nd (to) oran-red pl (sub)	208.5(0.5)		1.405²²		i H₂O; sl EtOH, bz; s eth
8745	Phenanthridine		$C_{13}H_9N$	229-87-8	179.217	nd (dil al)	106.56(0.02)	350.3(0.6)			sl H₂O; vs EtOH, eth, bz, CS₂; s ace
8746	1,7-Phenanthroline		$C_{12}H_8N_2$	230-46-6	180.205	pl (anh), nd (w+2)	77.2(0.5)	360			s H₂O; vs EtOH; i eth, bz, lig
8747	1,10-Phenanthroline	o-Phenanthroline	$C_{12}H_8N_2$	66-71-7	180.205	wh nd (bz) cry (w+1)	118.56(0.05)	409.2(0.8)			vs H₂O; s EtOH, ace, bz; i peth
8748	4,7-Phenanthroline		$C_{12}H_8N_2$	230-07-9	180.205	nd (w)	172.4(0.5)	100 sub			s H₂O, lig; vs EtOH; sl eth, bz, CS₂
8749	1,10-Phenanthroline monohydrate	o-Phenanthroline monohydrate	$C_{12}H_{10}N_2O$	5144-89-8	198.219	wh cry pow	93				s EtOH, ace; sl bz
8750	Phenazine	Dibenzopyrazine	$C_{12}H_8N_2$	92-82-0	180.205	ye-red nd (HOAc)	174.76(0.05)				sl H₂O, eth; s bz, EtOH
8751	2,3-Phenazinediamine	2,3-Diaminophenazine	$C_{12}H_{10}N_4$	655-86-7	210.234	ye nd	264	sub			vs bz, EtOH
8752	1-Phenazinol	Hemipyocyanine	$C_{12}H_8N_2O$	528-71-2	196.204	ye nd (bz, dil MeOH)	158	sub			sl H₂O, EtOH; s bz, py, dil alk
8753	Phenazopyridine	2,6-Diamino-3-phenylazopyridine	$C_{11}H_{11}N_5$	94-78-0	213.239	red cry	139				
8754	Phenazopyridine hydrochloride	3-(Phenylazo)-2,6-pyridinediamine, monohydrochloride	$C_{11}H_{12}ClN_5$	136-40-3	249.700	ye-red cry					sl H₂O, EtOH; i bz, ace; s HOAc
8755	Phencarbamide		$C_{19}H_{24}N_2OS$	3735-90-8	328.471		48.5	121⁰·⁰¹			vs eth, chl, MeOH, peth

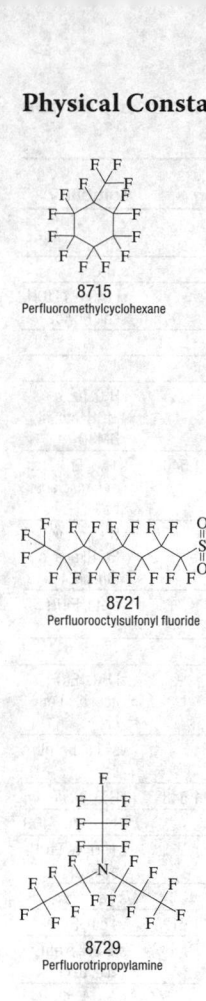

8715
Perfluoromethylcyclohexane

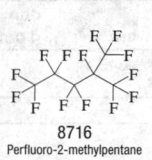

8716
Perfluoro-2-methylpentane

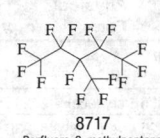

8717
Perfluoro-3-methylpentane

8718
Perfluoronaphthalene

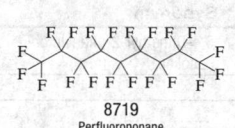

8719
Perfluorononane

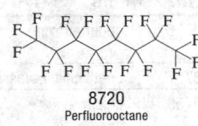

8720
Perfluorooctane

8721
Perfluorooctylsulfonyl fluoride

8722
Perfluorooxetane

8723
Perfluoropentane

8724
Perfluoropropane

8725
Perfluoropropene

8726
Perfluoropropyl methyl ether

8727
Perfluoropyridine

8728
Perfluorotoluene

8729
Perfluorotripropylamine

8730
1*H*-Perimidine

8731
Permethrin

8732
Peroxyacetic acid

8733
Peroxypropanoic acid

8734
Perphenazine

8735
Perthane

8736
Perylene

8737
Peucedanin

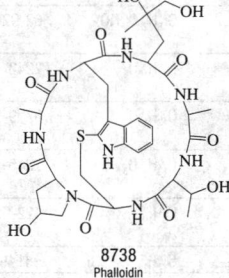

8738
Phalloidin

8739
Phalloin

8740
α-Phellandrene

8741
β-Phellandrene

8742
9-Phenanthrenamine

8743
Phenanthrene

8744
9,10-Phenanthrenedione

8745
Phenanthridine

8746
1,7-Phenanthroline

8747
1,10-Phenanthroline

8748
4,7-Phenanthroline

H₂O

8749
1,10-Phenanthroline monohydrate

8750
Phenazine

8751
2,3-Phenazinediamine

8752
1-Phenazinol

8753
Phenazopyridine

HCl

8754
Phenazopyridine hydrochloride

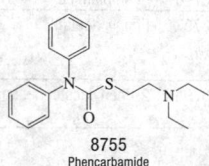

8755
Phencarbamide

No.	Name	Synonym	Mol. Form.	CAS RN	Mol. Wt.	Physical Form	mp/°C	bp/°C	den g cm^{-3}	n_D	Solubility
8756	Phendimetrazine	3,4-Dimethyl-2-phenylmorpholine	C$_{12}$H$_{17}$NO	634-03-7	191.269			134[12]			
8757	Phenethicillin potassium		C$_{17}$H$_{19}$KN$_2$O$_5$S	132-93-4	402.506	cry (ace)	235				s H$_2$O
8758	Phenicin		C$_{14}$H$_{10}$O$_6$	128-68-7	274.225	ye-br (al)	230.5				sl H$_2$O; vs EtOH, chl, HOAc
8759	Phenindamine		C$_{19}$H$_{19}$N	82-88-2	261.361	cry	91		1.17		
8760	Phenmedipham		C$_{16}$H$_{16}$N$_2$O$_4$	13684-63-4	300.309		150.8(0.8)				
8761	Phenobarbital	5-Ethyl-5-phenyl-2,4,6(1H,3H,5H)-pyrimidinetrione	C$_{12}$H$_{12}$N$_2$O$_3$	50-06-6	232.234	pl (w)	176(1)				i H$_2$O, bz; s EtOH, eth; sl DMSO
8762	Phenol	Hydroxybenzene	C$_6$H$_6$O	108-95-2	94.111		40.89(0.01)	181.8(0.1)	1.0545[45]	1.5408[41]	s H$_2$O, EtOH; vs eth; msc ace, bz
8763	Phenolphthalein	3,3-Bis(4-hydroxyphenyl)-1(3H)-isobenzofuranone	C$_{20}$H$_{14}$O$_4$	77-09-8	318.323	wh orth nd	262(2)		1.277[32]		i H$_2$O, bz; vs EtOH, ace; s eth, chl
8764	Phenolphthalin	2-[Bis(4-hydroxyphenyl)-methyl]benzoic acid	C$_{20}$H$_{16}$O$_4$	81-90-3	320.339	nd (w)	230.5				vs eth, EtOH
8765	Phenolphthalol		C$_{20}$H$_{18}$O$_3$	81-92-5	306.355	cry (dil al)	201.5				
8766	Phenol Red	Phenolsulfonphthalein	C$_{19}$H$_{14}$O$_5$S	143-74-8	354.376	dk red nd or pl	>300				sl H$_2$O, EtOH, ace, bz; i eth, chl
8767	10H-Phenothiazine	Thiodiphenylamine	C$_{12}$H$_9$NS	92-84-2	199.271	ye pr (al) ye lf or pl (tol)	184.9(0.6)	371			vs ace, bz, eth, EtOH
8768	Phenothrin		C$_{23}$H$_{26}$O$_3$	26002-80-2	350.450	col liq			1.061[25]	1.5483[25]	i H$_2$O; s ace, xyl
8769	10H-Phenoxazine		C$_{12}$H$_9$NO	135-67-1	183.205	lf (dil al, bz)	156.7(0.5)	dec			vs bz, eth, EtOH
8770	Phenoxyacetic acid		C$_8$H$_8$O$_3$	122-59-8	152.148	nd or pl (w)	98.5	285 dec			s H$_2$O; vs EtOH, eth, bz, CS$_2$
8771	Phenoxyacetyl chloride		C$_8$H$_7$ClO$_2$	701-99-5	170.594			225.5			s eth
8772	Phenoxyacetylene		C$_8$H$_6$O	4279-76-9	118.133		-36	61[25]	1.0614[20]	1.5125[20]	vs eth, EtOH
8773	2-Phenoxyaniline		C$_{12}$H$_{11}$NO	2688-84-8	185.221	cry (lig)	45.8	308			s EtOH; s eth, ace, bz
8774	3-Phenoxyaniline		C$_{12}$H$_{11}$NO	3586-12-7	185.221	pr (lig)	37	315	1.1583[25]		s EtOH, eth, ace, bz; sl lig
8775	4-Phenoxyaniline		C$_{12}$H$_{11}$NO	139-59-3	185.221	nd (w), cry (dil al)	85.5				s H$_2$O; vs EtOH, eth; sl lig
8776	3-Phenoxybenzaldehyde		C$_{13}$H$_{10}$O$_2$	39515-51-0	198.217		14.0	169[11]	1.147[25]	1.5954[20]	
8777	Phenoxybenzamine		C$_{18}$H$_{22}$ClNO	59-96-1	303.827		39				s bz
8778	Phenoxybenzamine hydrochloride		C$_{18}$H$_{23}$Cl$_2$NO	63-92-3	340.288		139				sl H$_2$O; s EtOH
8779	2-Phenoxybenzoic acid		C$_{13}$H$_{10}$O$_3$	2243-42-7	214.216	lf (dil al)	113(2)	355	1.1553[50]		i H$_2$O; vs EtOH, eth; s chl
8780	3-Phenoxybenzoic acid		C$_{13}$H$_{10}$O$_3$	3739-38-6	214.216	nd (aq al)	145.8				i H$_2$O; s EtOH, eth
8781	4-Phenoxybenzoic acid		C$_{13}$H$_{10}$O$_3$	2215-77-2	214.216	pr (chl)	161				sl H$_2$O; s EtOH, eth, chl
8782	2-Phenoxyethanol		C$_8$H$_{10}$O$_2$	122-99-6	138.164	oil	12(2)	246(3)	1.102[22]	1.534[20]	i H$_2$O; s EtOH, eth, chl, alk
8783	2-Phenoxyethyl acrylate	Phenyl Cellosolve acrylate	C$_{11}$H$_{12}$O$_3$	48145-04-6	192.211			110[2]	1.090[25]		vs ace, eth, chl
8784	2-Phenoxyethyl butanoate		C$_{12}$H$_{16}$O$_3$	23511-70-8	208.253			251	1.0388[21]		vs ace, eth, EtOH
8785	3-Phenoxyphenol		C$_{12}$H$_{10}$O$_2$	713-68-8	186.206			175[7]			
8786	4-Phenoxyphenol		C$_{12}$H$_{10}$O$_2$	831-82-3	186.206		84.0				
8787	2-(3-Phenoxyphenyl)propanoic acid, (±)-	Fenoprofen	C$_{15}$H$_{14}$O$_3$	31879-05-7	242.270	visc oil		170[0.11]		1.5742[25]	
8788	3-Phenoxy-1,2-propanediol	Phenylglyceryl ether	C$_9$H$_{12}$O$_3$	538-43-2	168.189	nd (eth, peth)	67.5	200[22]	1.225[20]		vs H$_2$O, bz, eth, EtOH
8789	2-Phenoxypropanoic acid		C$_9$H$_{10}$O$_3$	940-31-8	166.173	nd (w)	115.5	266	1.1865[20]	1.5184[20]	
8790	2-Phenoxy-1-propanol		C$_9$H$_{12}$O$_2$	4169-04-4	152.190			244	0.9801[25]	1.4760[25]	s EtOH, eth
8791	1-Phenoxy-2-propanol		C$_9$H$_{12}$O$_2$	770-35-4	152.190			233	1.0622[20]	1.5232[20]	
8792	1-Phenoxy-2-propanone	Phenoxyacetone	C$_9$H$_{10}$O$_2$	621-87-4	150.174			229.5	1.0903[20]	1.5228[20]	s eth, ace
8793	2-Phenoxypropanoyl chloride		C$_9$H$_9$ClO$_2$	122-35-0	184.619			147	1.1865[20]	1.5178[20]	s eth
8794	Phenprocoumon	3-(α-Ethylbenzyl)-4-hydroxycoumarin	C$_{18}$H$_{16}$O$_3$	435-97-2	280.318	pr (MeOH aq)	183(1)				
8795	Phenthoate		C$_{12}$H$_{17}$O$_4$PS$_2$	2597-03-7	320.364	ye oil		123[0.01]			sl H$_2$O; s hx
8796	Phentolamine		C$_{17}$H$_{19}$N$_3$O	50-60-2	281.352		175				
8797	Phenyl acetate		C$_8$H$_8$O$_2$	122-79-2	136.149			195(1)	1.0780[20]	1.5035[20]	sl H$_2$O; msc EtOH, eth, chl; s ctc
8798	2-Phenylacetophenone		C$_{14}$H$_{12}$O	451-40-1	196.244	pl (al)	56(3)	320	1.201[0]		sl H$_2$O; s EtOH, eth, ctc, chl

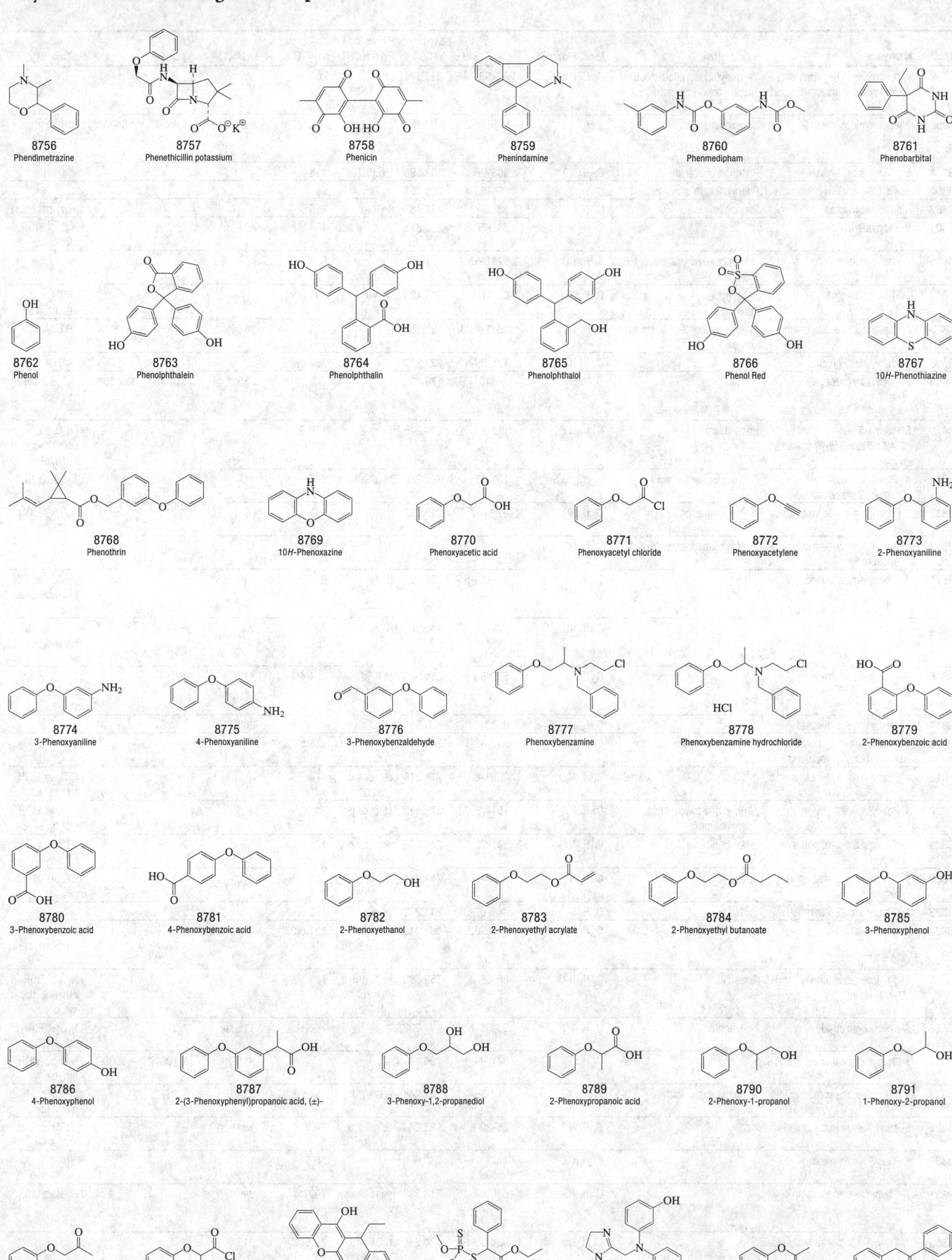

8756
Phendimetrazine

8757
Phenethicillin potassium

8758
Phenicin

8759
Phenindamine

8760
Phenmedipham

8761
Phenobarbital

8762
Phenol

8763
Phenolphthalein

8764
Phenolphthalin

8765
Phenolphthalol

8766
Phenol Red

8767
10*H*-Phenothiazine

8768
Phenothrin

8769
10*H*-Phenoxazine

8770
Phenoxyacetic acid

8771
Phenoxyacetyl chloride

8772
Phenoxyacetylene

8773
2-Phenoxyaniline

8774
3-Phenoxyaniline

8775
4-Phenoxyaniline

8776
3-Phenoxybenzaldehyde

8777
Phenoxybenzamine

8778
Phenoxybenzamine hydrochloride

8779
2-Phenoxybenzoic acid

8780
3-Phenoxybenzoic acid

8781
4-Phenoxybenzoic acid

8782
2-Phenoxyethanol

8783
2-Phenoxyethyl acrylate

8784
2-Phenoxyethyl butanoate

8785
3-Phenoxyphenol

8786
4-Phenoxyphenol

8787
2-(3-Phenoxyphenyl)propanoic acid, (±)-

8788
3-Phenoxy-1,2-propanediol

8789
2-Phenoxypropanoic acid

8790
2-Phenoxy-1-propanol

8791
1-Phenoxy-2-propanol

8792
1-Phenoxy-2-propanone

8793
2-Phenoxypropanoyl chloride

8794
Phenprocoumon

8795
Phenthoate

8796
Phentolamine

8797
Phenyl acetate

8798
2-Phenylacetophenone

No.	Name	Synonym	Mol. Form.	CAS RN	Mol. Wt.	Physical Form	mp/°C	bp/°C	den g cm^{-3}	n_D	Solubility
8799	N-(Phenylacetyl)-7-aminode-acetoxycephalosporanic acid	7-Phenylacetamidodeacetoxy-cephalosporanic acid	C$_{16}$H$_{16}$N$_2$O$_4$S	27255-72-7	332.374	cry (2-PrOH/peth)	200				
8800	Phenylacetylene	Ethynylbenzene	C$_8$H$_6$	536-74-3	102.134	liq	-45.1(0.4)	143.0(0.5)	0.9300[20]	1.5470[20]	i H$_2$O; msc EtOH, eth; s ace; sl chl
8801	(N-Phenylacetyl)glycine	Phenaceturic acid	C$_{10}$H$_{11}$NO$_3$	500-98-1	193.199	lf (EtOH)	143				
8802	Phenyl 2-(acetyloxy)benzoate	Phenyl acetylsalicylate	C$_{15}$H$_{12}$O$_4$	134-55-4	256.254		96				
8803	(Phenylacetyl)urea	Phenacemide	C$_9$H$_{10}$N$_2$O$_2$	63-98-9	178.187	cry (al)	215				vs bz, eth, EtOH
8804	9-Phenylacridine		C$_{19}$H$_{13}$N	602-56-2	255.313	ye nd, lf (al)	184	404			i H$_2$O; sl EtOH; s eth; vs bz
8805	L-Phenylalaninamide	α-Aminobenzenepropanamide, (S)-	C$_9$H$_{12}$N$_2$O	5241-58-7	164.203		82				
8806	L-Phenylalanine	α-Aminobenzenepropanoic acid, (S)	C$_9$H$_{11}$NO$_2$	63-91-2	165.189	pr (w)	283 dec				sl H$_2$O; i EtOH, eth, bz, acid
8807	L-Phenylalanine, ethyl ester	Ethyl 2-amino-3-phenylpropio-nate	C$_{11}$H$_{15}$NO$_2$	3081-24-1	193.243		136	148[13]	1.065[15]		sl H$_2$O
8808	L-Phenylalanylglycine		C$_{11}$H$_{14}$N$_2$O$_3$	721-90-4	222.240		262 dec				s H$_2$O
8809	3-Phenylallyl acetate		C$_{11}$H$_{12}$O$_2$	21040-45-9	176.212			265	1.0567[20]	1.5425[20]	i H$_2$O; s EtOH, eth, ace, bz, chl
8810	5-Phenyl-5-allyl-2,4,6(1H,3H,5H)-pyrimidine-trione	Phenallymal	C$_{13}$H$_{12}$N$_2$O$_3$	115-43-5	244.245		156.5				sl H$_2$O, bz, DMSO; vs EtOH, eth; i lig
8811	4-(Phenylamino)-benzenesulfonic acid	N-Phenylsulfanilic acid	C$_{12}$H$_{11}$NO$_3$S	101-57-5	249.285	pl (al-eth)	206				vs H$_2$O, EtOH
8812	2-(Phenylamino)benzoic acid	N-Phenylanthranilic acid	C$_{13}$H$_{11}$NO$_2$	91-40-7	213.232	lf (al)	183.5				i H$_2$O; vs EtOH; sl eth, bz
8813	Phenyl 4-amino-3-hydroxy-benzoate	Phenyl p-aminosalicylate	C$_{13}$H$_{11}$NO$_3$	133-11-9	229.231		153				
8814	3-(Phenylamino)phenol		C$_{12}$H$_{11}$NO	101-18-8	185.221	lf (w)	81.5	340			sl H$_2$O; vs EtOH, eth, ace; s bz, acid
8815	4-(Phenylamino)phenol		C$_{12}$H$_{11}$NO	122-37-2	185.221	lf (w)	73	330			sl H$_2$O; vs EtOH, eth, bz, chl; s acid
8816	9-Phenylanthracene		C$_{20}$H$_{14}$	602-55-1	254.325	bl lf (al) (HOAc)	154(4)	417			i H$_2$O; s EtOH, eth, bz, chl, CS$_2$
8817	Phenylarsonous diiodide		C$_6$H$_5$AsI$_2$	6380-34-3	405.835		15	205[14]	1.6264[15]		
8818	4-(Phenylazo)-1,3-benzenedi-amine monohydrochloride	Chrysoidine hydrochloride	C$_{12}$H$_{13}$ClN$_4$	532-82-1	248.711	red-br cry pow	118.5				vs ace
8819	4-(Phenylazo)-1,3-benzenediol		C$_{12}$H$_{10}$N$_2$O$_2$	2051-85-6	214.219	dk red nd (dil al)	170				i H$_2$O; vs EtOH, eth, bz, HOAc
8820	4-Phenylazodiphenylamine	N-Phenyl-4-(phenylazo)-benzenamine	C$_{18}$H$_{15}$N$_3$	101-75-7	273.332	ye pl or pr	84.0				i H$_2$O; vs EtOH, eth, lig
8821	4-(Phenylazo)-1-naphthale-namine	α-Naphthyl Red	C$_{16}$H$_{13}$N$_3$	131-22-6	247.294	red-viol cry (EtOH)	123				s EtOH, dil HCl, bz
8822	1-(Phenylazo)-2-naphthale-namine	Yellow AB	C$_{16}$H$_{13}$N$_3$	85-84-7	247.294	red pl (al)	103				vs EtOH, HOAc
8823	1-(Phenylazo)-2-naphthol	Sudan I	C$_{16}$H$_{12}$N$_2$O	842-07-9	248.278	ye cry	132				
8824	4-(Phenylazo)phenol		C$_{12}$H$_{10}$N$_2$O	1689-82-3	198.219	ye lf (bz) oran pr (al)	151.5(0.8)	225[20] dec			i H$_2$O; vs EtOH, eth; s bz, con sulf
8825	1-[[4-(Phenylazo)phenyl]azo]-2-naphthol	Sudan III	C$_{22}$H$_{16}$N$_4$O	85-86-9	352.388	br lf (grn lustre) (HOAc)	195				i H$_2$O; s EtOH, eth, ace, bz, xyl, chl
8826	N-Phenylbenzamide	Benzanilide	C$_{13}$H$_{11}$NO	93-98-1	197.232	lf (al)	160(1)	117 sub	1.315[25]		i H$_2$O; sl EtOH, eth, HOAc
8827	α-Phenylbenzeneacetaldehyde		C$_{14}$H$_{12}$O	947-91-1	196.244			315 dec	1.1061[21]	1.5920[21]	i H$_2$O; vs EtOH, eth, bz
8828	α-Phenylbenzeneacetic acid	Diphenylacetic acid	C$_{14}$H$_{12}$O$_2$	117-34-0	212.244	nd (w), lf (al)	147.26(0.04)	194[25]	1.257[15]		sl H$_2$O; vs EtOH; s eth, chl
8829	α-Phenylbenzeneacetonitrile		C$_{14}$H$_{11}$N	86-29-3	193.244	pr (eth), lf (dil al)	74.3	184[16]			s EtOH, chl; vs eth; sl lig
8830	α-Phenylbenzeneacetyl chloride		C$_{14}$H$_{11}$ClO	1871-76-7	230.689		56.5	170[16]			s lig
8831	N-Phenylbenzenecarbothio-amide		C$_{13}$H$_{11}$NS	636-04-4	213.298	ye pl or pr (al)	102	dec			i H$_2$O; vs EtOH; s eth, bz, chl; sl lig
8832	N-Phenyl-1,2-benzenediamine		C$_{12}$H$_{12}$N$_2$	534-85-0	184.236	nd(w)	79.5	313			sl H$_2$O, lig; s ace, bz, chl
8833	N-Phenyl-1,4-benzenediamine	p-Aminodiphenylamine	C$_{12}$H$_{12}$N$_2$	101-54-2	184.236	nd(al)	75(1)	354			sl H$_2$O, chl; vs EtOH; s eth, lig

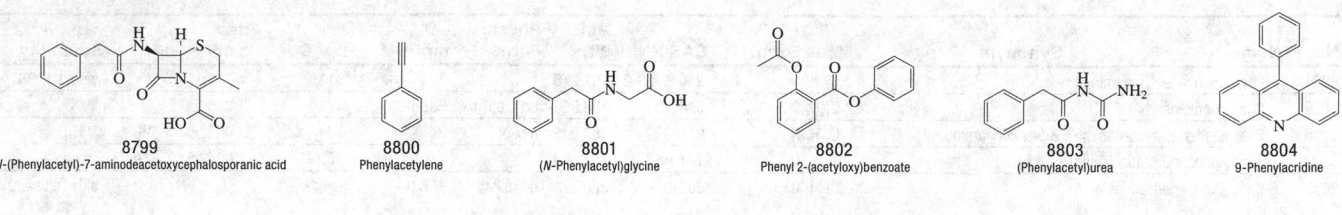

8799
N-(Phenylacetyl)-7-aminodeacetoxycephalosporanic acid

8800
Phenylacetylene

8801
(N-Phenylacetyl)glycine

8802
Phenyl 2-(acetyloxy)benzoate

8803
(Phenylacetyl)urea

8804
9-Phenylacridine

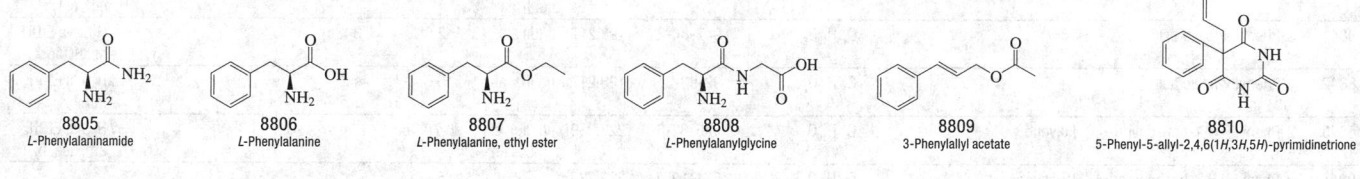

8805
L-Phenylalaninamide

8806
L-Phenylalanine

8807
L-Phenylalanine, ethyl ester

8808
L-Phenylalanylglycine

8809
3-Phenylallyl acetate

8810
5-Phenyl-5-allyl-2,4,6(1H,3H,5H)-pyrimidinetrione

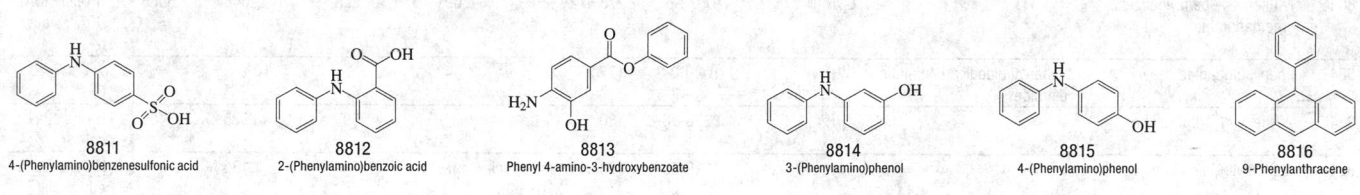

8811
4-(Phenylamino)benzenesulfonic acid

8812
2-(Phenylamino)benzoic acid

8813
Phenyl 4-amino-3-hydroxybenzoate

8814
3-(Phenylamino)phenol

8815
4-(Phenylamino)phenol

8816
9-Phenylanthracene

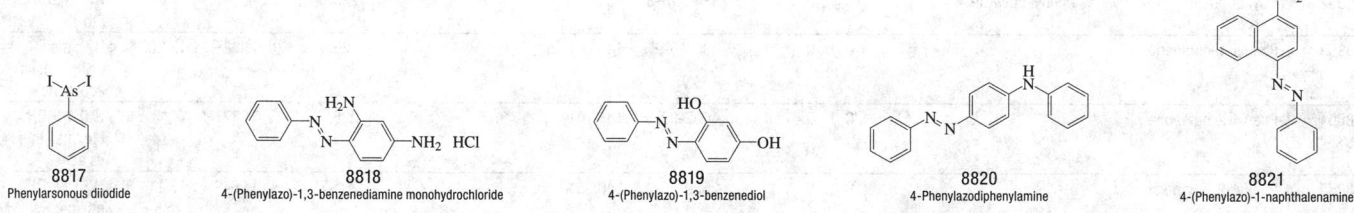

8817
Phenylarsonous diiodide

8818
4-(Phenylazo)-1,3-benzenediamine monohydrochloride

8819
4-(Phenylazo)-1,3-benzenediol

8820
4-Phenylazodiphenylamine

8821
4-(Phenylazo)-1-naphthalenamine

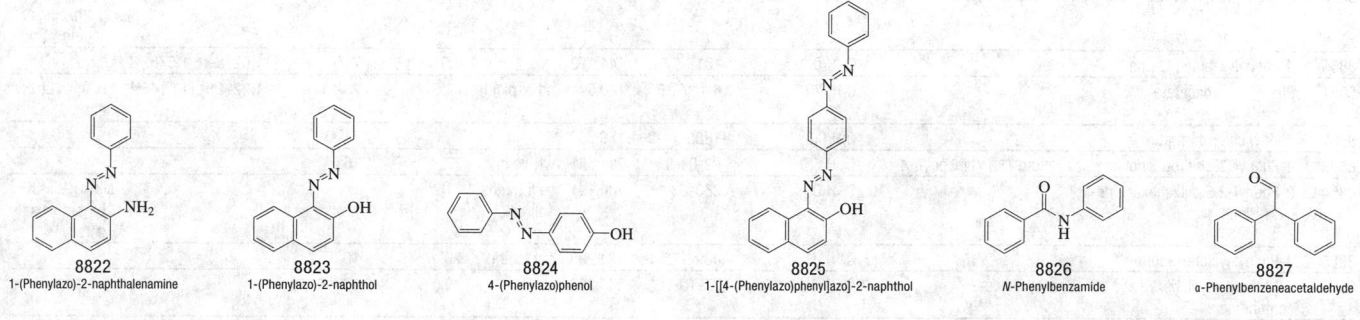

8822
1-(Phenylazo)-2-naphthalenamine

8823
1-(Phenylazo)-2-naphthol

8824
4-(Phenylazo)phenol

8825
1-[[4-(Phenylazo)phenyl]azo]-2-naphthol

8826
N-Phenylbenzamide

8827
α-Phenylbenzeneacetaldehyde

8828
α-Phenylbenzeneacetic acid

8829
α-Phenylbenzeneacetonitrile

8830
α-Phenylbenzeneacetyl chloride

8831
N-Phenylbenzenecarbothioamide

8832
N-Phenyl-1,2-benzenediamine

8833
N-Phenyl-1,4-benzenediamine

No.	Name	Synonym	Mol. Form.	CAS RN	Mol. Wt.	Physical Form	mp/°C	bp/°C	den g cm⁻³	n_D	Solubility
8834	α-Phenylbenzeneethanamine		C₁₄H₁₅N	25611-78-3	197.276			311	1.031[15]		vs eth, EtOH
8835	α-Phenylbenzeneethanol		C₁₄H₁₄O	614-29-9	198.260	nd (peth-bz)	53(2)	177[15]	1.0360[70]		
8836	α-Phenylbenzenemethanamine	Benzhydrylamine	C₁₃H₁₃N	91-00-9	183.249	hex pl	34	304	1.0633[20]	1.5963	sl H₂O; s ace
8837	α-Phenylbenzenemethanimine		C₁₃H₁₁N	1013-88-3	181.233			282	1.0847[19]	1.6191[19]	vs eth
8838	β-Phenylbenzenepropanoic acid		C₁₅H₁₄O₂	606-83-7	226.271	nd (dil al)	156.0				sl H₂O; vs EtOH; s eth, ace
8839	2-Phenylbenzimidazole	Phenzidole	C₁₃H₁₀N₂	716-79-0	194.231	pl (HOAc) (al-w) nd (bz, w)	293				sl H₂O, bz; s EtOH, chl, HOAc
8840	Phenyl benzoate		C₁₃H₁₀O₂	93-99-2	198.217	mcl pr (eth-al)	63(8)	314	1.235[20]		i H₂O; s EtOH, eth, chl
8841	2-Phenylbenzoic acid		C₁₃H₁₀O₂	947-84-2	198.217	lf (dil al)	112(3)	343.5			i H₂O; vs EtOH, bz, HOAc
8842	4-Phenylbenzoic acid		C₁₃H₁₀O₂	92-92-2	198.217	nd (bz, al)	210(5)	sub			i H₂O; s EtOH, eth, bz
8843	2-Phenyl-4H-1-benzopyran-4-one	Flavone	C₁₅H₁₀O₂	525-82-6	222.239	nd (lig), cry (30% al)	96.7(0.5)				i H₂O; s EtOH, eth, ace, bz
8844	3-Phenyl-4H-1-benzopyran-4-one	Isoflavone	C₁₅H₁₀O₂	574-12-9	222.239		148				
8845	2-Phenylbenzothiazole		C₁₃H₉NS	883-93-2	211.282	nd (dil al)	115	371			i H₂O; s EtOH, eth, CS₂
8846	N-Phenyl-N-benzylbenzene-methanamine		C₂₀H₁₉N	91-73-6	273.372		69	226[10]	1.0444[80]	1.6065[80]	i H₂O; sl EtOH, HOAc; s eth, bz
8847	Phenyl biguanide	N-Phenylimidodicarbonimidic diamide	C₈H₁₁N₅	102-02-3	177.207		143				
8848	2-Phenyl-1,3-butadiene		C₁₀H₁₀	2288-18-8	130.186			60[17]	0.925[20]	1.5489[20]	i H₂O; s eth, bz, chl
8849	N-Phenylbutanamide		C₁₀H₁₃NO	1129-50-6	163.216	mcl pr (al, bz, eth)	97	189[15]	1.134[25]		i H₂O; vs EtOH, eth; sl chl
8850	Phenylbutanedioic acid, (±)-		C₁₀H₁₀O₄	10424-29-0	194.184	lf or nd (w)	167(1)	dec			sl H₂O, chl; vs EtOH, eth, ace; i bz
8851	1-Phenyl-1,3-butanedione		C₁₀H₁₀O₂	93-91-4	162.185	pr	60(2)	257(9)	1.0599[74]	1.5678[78]	i H₂O; s eth; sl chl
8852	Phenyl butanoate	Phenyl butyrate	C₁₀H₁₂O₂	4346-18-3	164.201			228(6)	1.0382[15]		i H₂O; s EtOH, eth
8853	1-Phenyl-1-butanone		C₁₀H₁₂O	495-40-9	148.201		12	228.5	0.988[20]	1.5203[20]	i H₂O; msc EtOH, eth; vs ace; s ctc
8854	1-Phenyl-2-butanone		C₁₀H₁₂O	1007-32-5	148.201			228	0.9877[20]		i H₂O; s EtOH, ctc; msc eth; vs ace
8855	4-Phenyl-2-butanone		C₁₀H₁₂O	2550-26-7	148.201	liq	-13	235(7)	0.9849[22]	1.511[22]	i H₂O; s EtOH, eth, ctc; vs ace
8856	Phenylbutazone		C₁₉H₂₀N₂O₂	50-33-9	308.374		104(1)				
8857	2-Phenyl-1-butene	α-Ethylstyrene	C₁₀H₁₂	2039-93-2	132.202			190(6)	0.887[25]	1.5288[20]	
8858	1-Phenyl-2-buten-1-one		C₁₀H₁₀O	495-41-0	146.185		20.5	111[9]	1.025[15]	1.5626[18]	
8859	trans-4-Phenyl-3-buten-2-one	Benzilideneacetone	C₁₀H₁₀O	1896-62-4	146.185	pl	41(3)	261	1.0097[45]	1.5836[45]	i H₂O; vs EtOH; s eth, ace, bz; sl peth
8860	4-Phenyl-3-butyn-2-one		C₁₀H₈O	1817-57-8	144.170		4.5	79[2]	1.0215[20]	1.5762[20]	
8861	Phenyl chloroacetate		C₈H₇ClO₂	620-73-5	170.594	nd or pl (al)	44.5	232.5	1.2202[44]	1.5146[44]	i H₂O; vs EtOH, eth
8862	Phenyl chloroformate		C₇H₅ClO₂	1885-14-9	156.567			71[9]			
8863	4-Phenyl-2-chlorophenol	3-Chloro-(1,1'-biphenyl)-4-ol	C₁₂H₉ClO	92-04-6	204.651	wh-ye cry	77	161[7]			
8864	2-Phenyl-2,5-cyclohexadiene-1,4-dione		C₁₂H₈O₂	363-03-1	184.191	ye lf (peth, al)	114				sl H₂O; s EtOH, bz, peth; vs chl
8865	4-Phenylcyclohexanone		C₁₂H₁₄O	4894-75-1	174.238	cry (peth)	79	158[12]			
8866	1-(1-Phenylcyclohexyl)-piperidine	Phencyclidine	C₁₇H₂₅N	77-10-1	243.388		46.5	136[1.0]			
8867	3-Phenyl-2-cyclopenten-1-one		C₁₁H₁₀O	3810-26-2	158.196	liq	-23	234.2	0.9711[20]	1.5440[20]	s EtOH, ace, chl; sl eth
8868	N-Phenyl-N,N-diethanolamine		C₁₀H₁₅NO₂	120-07-0	181.232		57	200[10]	1.201[60]		vs ace, bz, eth, EtOH
8869	2-Phenyl-1,3-dioxane		C₁₀H₁₂O₂	772-01-0	164.201	nd (peth)	48(2)	253	1.6053[60]		vs EtOH, eth
8870	4-Phenyl-1,3-dioxane		C₁₀H₁₂O₂	772-00-9	164.201			247	1.1038[20]	1.5306[18]	i H₂O; s os
8871	1-Phenyl-1-dodecanone		C₁₈H₂₈O	1674-38-0	260.414		44(1)	201[9]	0.8794[18]	1.4700[18]	i H₂O; s ace; sl ctc
8872	1-Phenyl-1,2-ethanediol	Styrene glycol	C₈H₁₀O₂	93-56-1	138.164	nd (lig)	66(3)	273			vs H₂O, eth, bz, EtOH; sl lig

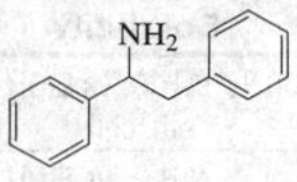

8834
α-Phenylbenzeneethanamine

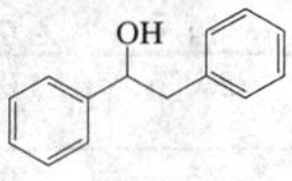

8835
α-Phenylbenzeneethanol

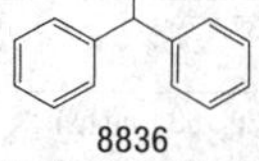

8836
α-Phenylbenzenemethanamine

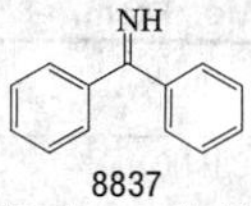

8837
α-Phenylbenzenemethanimine

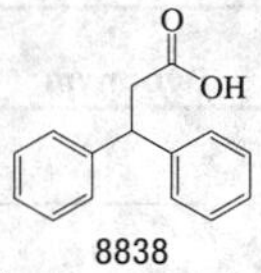

8838
β-Phenylbenzenepropanoic acid

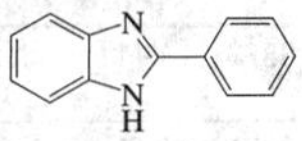

8839
2-Phenylbenzimidazole

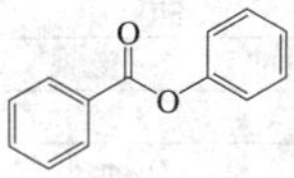

8840
Phenyl benzoate

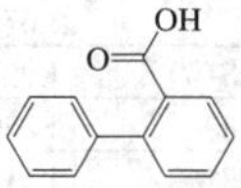

8841
2-Phenylbenzoic acid

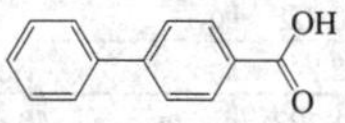

8842
4-Phenylbenzoic acid

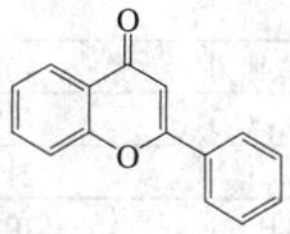

8843
2-Phenyl-4H-1-benzopyran-4-one

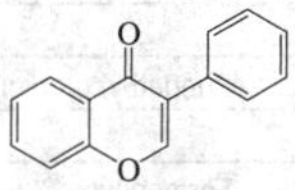

8844
3-Phenyl-4H-1-benzopyran-4-one

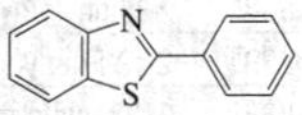

8845
2-Phenylbenzothiazole

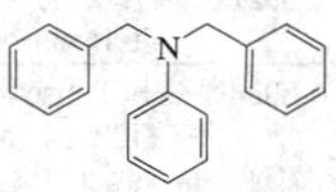

8846
N-Phenyl-N-benzylbenzenemethanamine

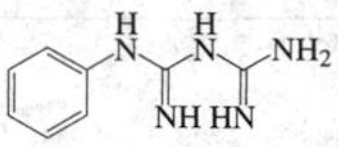

8847
Phenyl biguanide

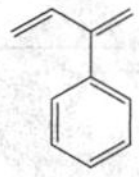

8848
2-Phenyl-1,3-butadiene

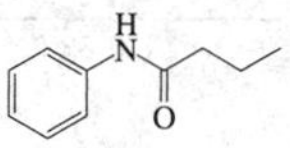

8849
N-Phenylbutanamide

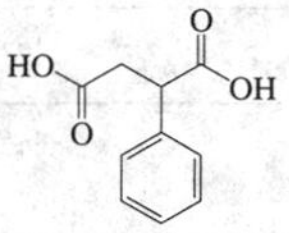

8850
Phenylbutanedioic acid, (±)-

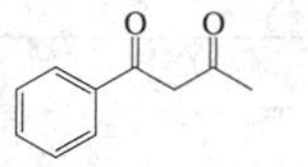

8851
1-Phenyl-1,3-butanedione

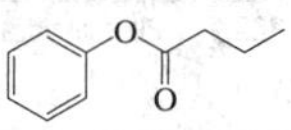

8852
Phenyl butanoate

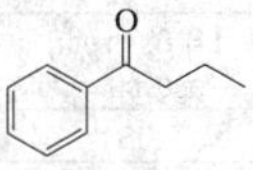

8853
1-Phenyl-1-butanone

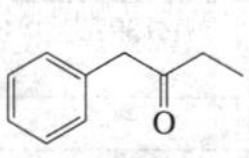

8854
1-Phenyl-2-butanone

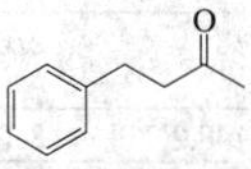

8855
4-Phenyl-2-butanone

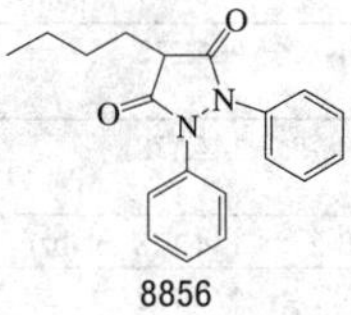

8856
Phenylbutazone

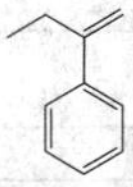

8857
2-Phenyl-1-butene

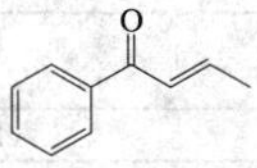

8858
1-Phenyl-2-buten-1-one

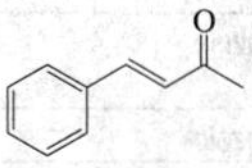

8859
trans-4-Phenyl-3-buten-2-one

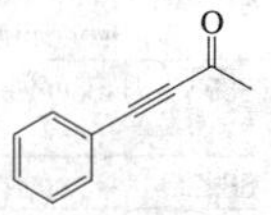

8860
4-Phenyl-3-butyn-2-one

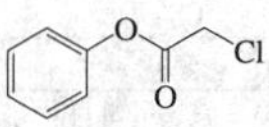

8861
Phenyl chloroacetate

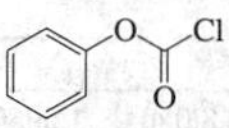

8862
Phenyl chloroformate

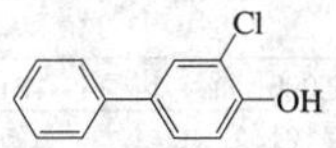

8863
4-Phenyl-2-chlorophenol

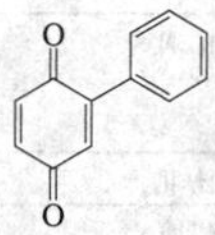

8864
2-Phenyl-2,5-cyclohexadiene-1,4-dione

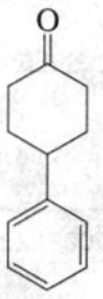

8865
4-Phenylcyclohexanone

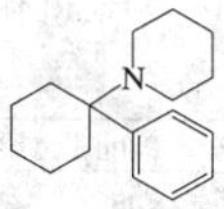

8866
1-(1-Phenylcyclohexyl)piperidine

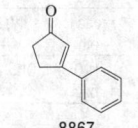

8867
3-Phenyl-2-cyclopenten-1-one

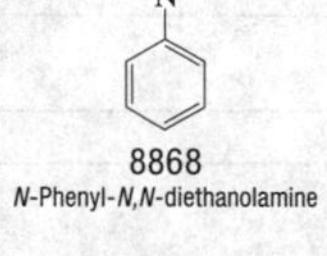

8868
N-Phenyl-N,N-diethanolamine

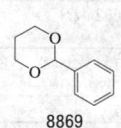

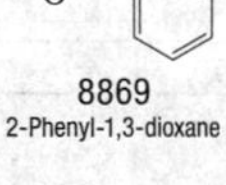

8869
2-Phenyl-1,3-dioxane

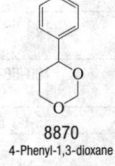

8870
4-Phenyl-1,3-dioxane

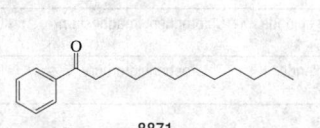

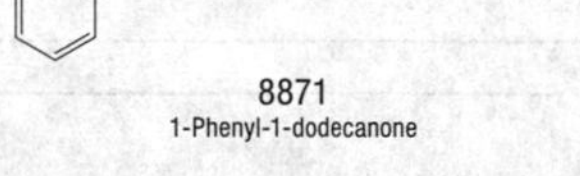

8871
1-Phenyl-1-dodecanone

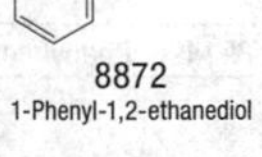

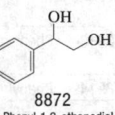

8872
1-Phenyl-1,2-ethanediol

No.	Name	Synonym	Mol. Form.	CAS RN	Mol. Wt.	Physical Form	mp/°C	bp/°C	den g cm⁻³	n_D	Solubility
8873	*N*-Phenylethanolamine		C₈H₁₁NO	122-98-5	137.179			280(3)	1.0945²⁰	1.5760²⁰	sl H₂O; vs EtOH, eth, chl
8874	1-Phenylethanone oxime		C₈H₉NO	613-91-2	135.163	nd (w)	60	245	1.0515⁷⁸		sl H₂O; vs EtOH, eth, ace, bz; s ctc
8875	2-Phenylethyl acetate		C₁₀H₁₂O₂	103-45-7	164.201	liq	-31.1	233(1)	1.0883²⁰	1.5171²⁰	vs eth, EtOH
8876	1-Phenylethyl hydroperoxide		C₈H₁₀O₂	3071-32-7	138.164	liq		50⁰·⁰¹			
8877	*N*-(2-Phenylethyl)-imidodicarbonimidic diamide, monohydrochloride	Phenformin hydrochloride	C₁₀H₁₆ClN₅	834-28-6	241.721	cry	178(1)				s H₂O
8878	2-Phenylethyl 2-methylpro-panoate	Benzylcarbinol isobutyrate	C₁₂H₁₆O₂	103-48-0	192.254			250	0.9950¹⁵	1.4871²⁰	
8879	2-Phenylethyl phenylacetate		C₁₆H₁₆O₂	102-20-5	240.297		26.5	177⁴·⁵	1.077²⁵		vs EtOH
8880	2-Phenylethyl propanoate	Phenethyl propionate	C₁₁H₁₄O₂	122-70-3	178.228	liq		231(7)	1.02²⁵	1.4950²⁰	
8881	2-(2-Phenylethyl)pyridine		C₁₃H₁₃N	2116-62-3	183.249	liq	-1.5	289	1.0465⁰		
8882	*N*-Phenylformamide	Formanilide	C₇H₇NO	103-70-8	121.137	mcl pr (lig-xyl)	46.5(0.6)	271(5)	1.1186⁵⁰		s H₂O, eth, bz; vs EtOH
8883	Phenyl formate		C₇H₆O₂	1864-94-4	122.122	liq		176(19)			
8884	2-Phenylfuran		C₁₀H₈O	17113-33-6	144.170			108¹⁸	1.083²⁰	1.5920²⁰	vs ace, bz
8885	Phenyl α-*D*-glucopyranoside		C₁₂H₁₆O₆	4630-62-0	256.251		156(1)				
8886	Phenyl glycidyl ether		C₉H₁₀O₂	122-60-1	150.174			244.2(0.9)	1.1109²¹	1.5307²¹	
8887	*N*-Phenylglycine	Phenylaminoacetic acid	C₈H₉NO₂	103-01-5	151.163		127.5				vs H₂O, EtOH
8888	1-Phenyl-1-heptanone		C₁₃H₁₈O	1671-75-6	190.281	lf	17(3)	284(13)	0.9516²⁰	1.5060²⁰	vs ace, eth, EtOH
8889	1-Phenyl-1-hexanone		C₁₂H₁₆O	942-92-7	176.254	fl	27	265	0.9576²⁰	1.5027²⁵	sl H₂O, ctc; s EtOH, eth, ace
8890	Phenylhydrazine		C₆H₈N₂	100-63-0	108.141	mcl pr or pl	20(2)	244(5)	1.0986²⁰	1.6084¹⁰	s H₂O; msc EtOH, eth, bz; vs ace
8891	2-Phenylhydrazinecarbox-amide	Phenicarbazide	C₇H₉N₃O	103-03-7	151.165		172				sl H₂O, eth, bz, lig; s EtOH, ace
8892	*N*-Phenylhydrazinecarbox-amide	4-Phenylsemicarbazide	C₇H₉N₃O	537-47-3	151.165	nd (bz), pl (w)	128				sl H₂O; vs EtOH, chl; i eth
8893	Phenylhydrazine monohydrochloride		C₆H₉ClN₂	59-88-1	144.601	lf (al)	244 dec	sub			vs H₂O, EtOH
8894	Phenylhydroxylamine	*N*-Hydroxybenzenamine	C₆H₇NO	100-65-2	109.126	nd (w, bz, peth)	79(2)				vs bz, eth, EtOH, chl
8895	Phenyl 1-hydroxy-2-naphtha-lenecarboxylate		C₁₇H₁₂O₃	132-54-7	264.275		96				vs bz, EtOH
8896	1-Phenyl-1*H*-imidazole		C₉H₈N₂	7164-98-9	144.173		13	276	1.1397¹⁵	1.6025²⁵	i H₂O; vs eth, ace, chl
8897	2-Phenyl-1*H*-imidazole		C₉H₈N₂	670-96-2	144.173	lf (bz)	148.2(0.5)	340			vs EtOH
8898	5-Phenyl-2,4-imidazolidinedi-one	5-Phenylhydantoin	C₉H₈N₂O₂	89-24-7	176.172		184.5				
8899	Phenylimidocarbonyl chloride		C₇H₅Cl₂N	622-44-6	174.028	liq		210	1.28¹⁵		
8900	2-[(Phenylimino)methyl]phenol		C₁₃H₁₁NO	779-84-0	197.232		49.5		1.087²⁵		i H₂O; s EtOH
8901	4-[(Phenylimino)methyl]phenol	*N*-(4-Hydroxybenzilidene)-aniline	C₁₃H₁₁NO	1689-73-2	197.232		196.0				i H₂O; s EtOH, eth; sl bz, chl
8902	1-Phenyl-1*H*-indene		C₁₅H₁₂	1961-96-2	192.256	oil		158⁷			
8903	2-Phenyl-1*H*-indene-1,3(2*H*)-dione	Phenindione	C₁₅H₁₀O₂	83-12-5	222.239	lf (al, bz)	150				i H₂O; s EtOH, eth, ace, bz, MeOH, chl
8904	2-Phenyl-1*H*-indole		C₁₄H₁₁N	948-65-2	193.244		188(3)	250¹⁰			sl H₂O; s eth, bz, chl, HOAc, CS₂
8905	Phenyliodine diacetate	Iodobenzene diacetate	C₁₀H₁₁IO₄	3240-34-4	322.096	cry	161				
8906	Phenyl isocyanate		C₇H₅NO	103-71-9	119.121			166.3(0.4)	1.0956²⁰	1.5368²⁰	vs eth; sl chl
8907	2-Phenyl-1*H*-isoindole-1,3(2*H*)-dione		C₁₄H₉NO₂	520-03-6	223.227	wh nd (al)	205(1)	sub			i H₂O; sl EtOH; msc chl
8908	Phenyl isopropyl ether	Isopropoxybenzene	C₉H₁₂O	2741-16-4	136.190	liq	-33.0(0.4)	180(4)	0.9408²⁵	1.4975²⁰	s H₂O, EtOH, ace, bz
8909	Phenyl isothiocyanate		C₇H₅NS	103-72-0	135.187	liq	-21	220(7)	1.1303²⁰	1.6492²³	i H₂O; s EtOH, eth, ctc
8910	3-Phenyl-2-isoxazolin-5-one		C₉H₇NO₂	1076-59-1	161.158		151				sl chl
8911	Phenyl laurate	Phenyl dodecanoate	C₁₈H₂₈O₂	4228-00-6	276.414	lf (al)	24.5	210¹⁵	0.9354³⁰		vs ace, eth, EtOH
8912	Phenylmagnesium chloride	Chlorophenylmagnesium	C₆H₅ClMg	100-59-4	136.862	cry					reac H₂O; s thf, eth
8913	Phenylmercuric chloride	Chlorophenylmercury	C₆H₅ClHg	100-56-1	313.15	pl (bz)	251				i H₂O; sl EtOH, bz
8914	Phenylmercuric nitrate		C₆H₅HgNO₃	55-68-5	339.70		≈181				

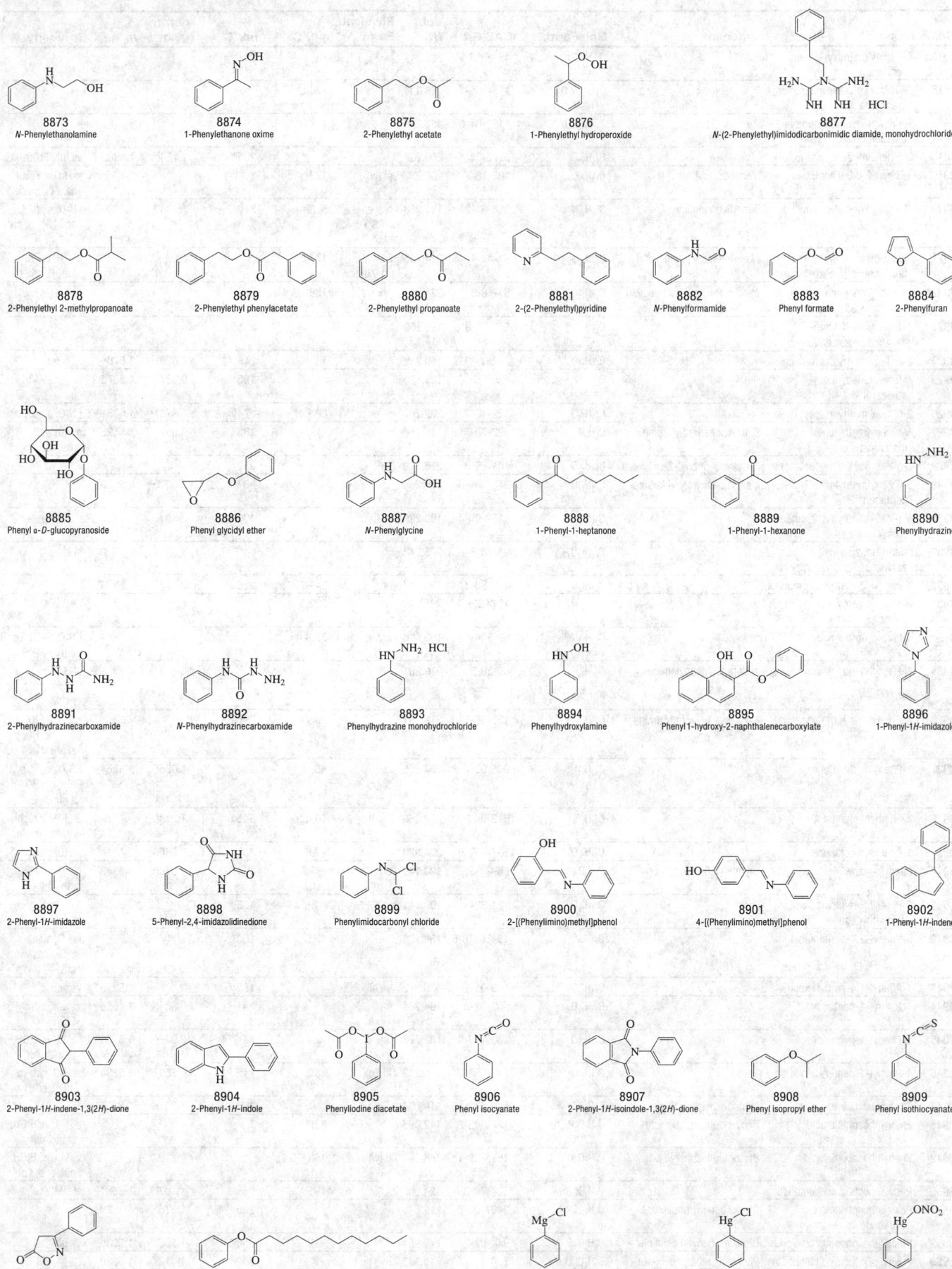

8873
N-Phenylethanolamine

8874
1-Phenylethanone oxime

8875
2-Phenylethyl acetate

8876
1-Phenylethyl hydroperoxide

8877
N-(2-Phenylethyl)imidodicarbonimidic diamide, monohydrochloride

8878
2-Phenylethyl 2-methylpropanoate

8879
2-Phenylethyl phenylacetate

8880
2-Phenylethyl propanoate

8881
2-(2-Phenylethyl)pyridine

8882
N-Phenylformamide

8883
Phenyl formate

8884
2-Phenylfuran

8885
Phenyl α-D-glucopyranoside

8886
Phenyl glycidyl ether

8887
N-Phenylglycine

8888
1-Phenyl-1-heptanone

8889
1-Phenyl-1-hexanone

8890
Phenylhydrazine

8891
2-Phenylhydrazinecarboxamide

8892
N-Phenylhydrazinecarboxamide

8893
Phenylhydrazine monohydrochloride

8894
Phenylhydroxylamine

8895
Phenyl 1-hydroxy-2-naphthalenecarboxylate

8896
1-Phenyl-1H-imidazole

8897
2-Phenyl-1H-imidazole

8898
5-Phenyl-2,4-imidazolidinedione

8899
Phenylimidocarbonyl chloride

8900
2-[(Phenylimino)methyl]phenol

8901
4-[(Phenylimino)methyl]phenol

8902
1-Phenyl-1H-indene

8903
2-Phenyl-1H-indene-1,3(2H)-dione

8904
2-Phenyl-1H-indole

8905
Phenyliodine diacetate

8906
Phenyl isocyanate

8907
2-Phenyl-1H-isoindole-1,3(2H)-dione

8908
Phenyl isopropyl ether

8909
Phenyl isothiocyanate

8910
3-Phenyl-2-isoxazolin-5-one

8911
Phenyl laurate

8912
Phenylmagnesium chloride

8913
Phenylmercuric chloride

8914
Phenylmercuric nitrate

No.	Name	Synonym	Mol. Form.	CAS RN	Mol. Wt.	Physical Form	mp/°C	bp/°C	den g cm⁻³	n_D	Solubility
8915	4-(Phenylmethoxy)-benzaldehyde		$C_{14}H_{12}O_2$	4397-53-9	212.244		73	217[13]			
8916	N^2-[(Phenylmethoxy)carbonyl]-L-arginine		$C_{14}H_{20}N_4O_4$	1234-35-1	308.334		174				
8917	N-[(Phenylmethoxy)carbonyl]-L-aspartic acid		$C_{12}H_{13}NO_6$	1152-61-0	267.234		117.0				
8918	2-(Phenylmethoxy)phenol		$C_{13}H_{12}O_2$	6272-38-4	200.233			205[20]	1.154[22]	1.5906[18]	vs eth, EtOH
8919	4-(Phenylmethoxy)phenol	Monobenzone	$C_{13}H_{12}O_2$	103-16-2	200.233	pl (w)	122				sl H_2O; vs EtOH, bz, eth; s ace
8920	N-(Phenylmethylene)aniline	Benzylideneaniline	$C_{13}H_{11}N$	538-51-2	181.233	pa ye nd (CS_2) pl (dil al)	56(5)	310	1.038[55]	1.600[100]	i H_2O; s EtOH, eth, NH_3; sl chl
8921	cis-α-(Phenylmethylene)-benzeneacetic acid	cis-α-Phenylcinnamic acid	$C_{15}H_{12}O_2$	91-47-4	224.255	silky needles	174(3)				s H_2O, EtOH, MeOH, eth, bz
8922	trans-α-(Phenylmethylene)-benzeneacetic acid	trans-α-Phenylcinnamic acid	$C_{15}H_{12}O_2$	91-48-5	224.255	prisms	138				vs H_2O; s EtOH, MeOH, eth, bz
8923	N-(Phenylmethylene)-benzenemethanamine		$C_{14}H_{13}N$	780-25-6	195.260			205[20]			
8924	2-(Phenylmethylene)butanal		$C_{11}H_{12}O$	28467-92-7	160.212		18	243	1.0201[22]	1.578[20]	
8925	N-(Phenylmethylene)-ethanamine		$C_9H_{11}N$	6852-54-6	133.190			195	0.937[20]	1.5378[15]	i H_2O; s EtOH, eth
8926	2-(Phenylmethylene)heptanal		$C_{14}H_{18}O$	122-40-7	202.292	ye oil	80	174[20]	0.9711[20]	1.5381[20]	i H_2O; s ace, ctc
8927	N-(Phenylmethylene)-methanamine	Benzylidenemethylamine	C_8H_9N	622-29-7	119.164			185	0.9671[14]	1.5526[20]	s EtOH, eth, ace, chl
8928	2-(Phenylmethylene)octanal	2-Hexyl-3-phenyl-2-propenal	$C_{15}H_{20}O$	101-86-0	216.319	liq	4	252			
8929	3-(Phenylmethylene)-2-pentanone	Methyl α-ethylstyryl ketone	$C_{12}H_{14}O$	3437-89-6	174.238			137[12]	1.0005[22]	1.5650[22]	
8930	N-(Phenylmethyl)-1,2-ethane-diamine		$C_9H_{14}N_2$	4152-09-4	150.220			130[11]			
8931	Phenylmethyl 4-hydroxyben-zoate		$C_{14}H_{12}O_3$	94-18-8	228.243						sl chl
8932	1-Phenyl-2-methyl-2-propanol		$C_{10}H_{14}O$	100-86-7	150.217	nd	24	215	0.9787[16]	1.5173[16]	
8933	N-(Phenylmethyl)-1H-purin-6-amine		$C_{12}H_{11}N_5$	1214-39-7	225.249		232.8				
8934	4-Phenylmorpholine		$C_{10}H_{13}NO$	92-53-5	163.216	cry (al-eth)	58.3				i H_2O, EtOH; vs eth
8935	N-Phenyl-1-naphthalenamine	1-Naphthylphenylamine	$C_{16}H_{13}N$	90-30-2	219.281		61				sl H_2O, ctc; s EtOH, eth, bz, HOAc
8936	N-Phenyl-2-naphthalenamine	N-Phenyl-β-naphthylamine	$C_{16}H_{13}N$	135-88-6	219.281		108(1)	395.5			i H_2O; s EtOH, eth, bz, HOAc; sl chl
8937	1-Phenylnaphthalene		$C_{16}H_{12}$	605-02-7	204.266	cry	45	334	1.096[20]	1.6664[20]	i H_2O; vs EtOH, eth, bz, HOAc; s ctc
8938	2-Phenylnaphthalene		$C_{16}H_{12}$	612-94-2	204.266	lf (al)	101(2)	345.5	1.2180[20]		s EtOH, bz, chl, HOAc; vs eth
8939	1-Phenyl-1-octanone		$C_{14}H_{20}O$	1674-37-9	204.308		22.8	300(13)	0.9360[30]		s EtOH, eth
8940	Phenyloxirane	Styrene-7,8-oxide	C_8H_8O	96-09-3	120.149	colorless liq	-35.6	194.1	1.0490[25]	1.5342[20]	i H_2O; s EtOH, eth, chl
8941	3-Phenyloxiranecarboxylic acid, ethyl ester		$C_{11}H_{12}O_3$	121-39-1	192.211			136[5]			
8942	5-Phenyl-2,4-pentadienal		$C_{11}H_{10}O$	13466-40-5	158.196		42.5	160[3]			i H_2O; msc EtOH, bz; vs eth
8943	1-Phenyl-1,4-pentanedione		$C_{11}H_{12}O_2$	583-05-1	176.212	ye oil		162[12]		1.5250[30]	vs ace
8944	1-Phenyl-1-pentanol		$C_{11}H_{16}O$	583-03-9	164.244			141[25]	0.9655[20]	1.4086[25]	vs ace, eth, EtOH
8945	1-Phenyl-1-pentanone		$C_{11}H_{14}O$	1009-14-9	162.228	liq	-9.4	247(14)	0.986[20]	1.5158[20]	i H_2O; vs EtOH, eth; sl ctc
8946	1-Phenyl-1-penten-3-one		$C_{11}H_{12}O$	3152-68-9	160.212	lf (lig)	38.5	142[12]	0.8697[20]	1.5684[20]	sl H_2O, chl; vs EtOH, eth, bz
8947	Phenylphosphine	Monophenylphosphine	C_6H_7P	638-21-1	110.094			160.5	1.001[15]	1.5796[20]	
8948	Phenylphosphinic acid	Benzenephosphinic acid	$C_6H_7O_2P$	1779-48-2	142.093		83.8				s H_2O; vs EtOH; sl eth, chl
8949	Phenylphosphonic acid	Benzenephosphonic acid	$C_6H_7O_3P$	1571-33-1	158.092	lf (w)	158.5(0.2)				vs H_2O; s EtOH, eth, ace; i bz
8950	Phenylphosphonic dichloride		$C_6H_5Cl_2OP$	824-72-6	194.983		1	258	1.197[25]	1.5581[25]	sl DMSO
8951	Phenylphosphonothioic dichloride	Dichlorophenylphosphine sulfide	$C_6H_5Cl_2PS$	3497-00-5	211.049			205[130]	1.376[13]		
8952	Phenylphosphonous dichloride	Dichlorophenylphosphine	$C_6H_5Cl_2P$	644-97-3	178.984	liq	-51	236(5)	1.356[20]	1.6030[20]	vs bz
8953	Phenyl phosphorodichloridate	Phenyl dichlorophosphate	$C_6H_5Cl_2O_2P$	770-12-7	210.983	hyg liq		242	1.412[20]	1.5230[20]	

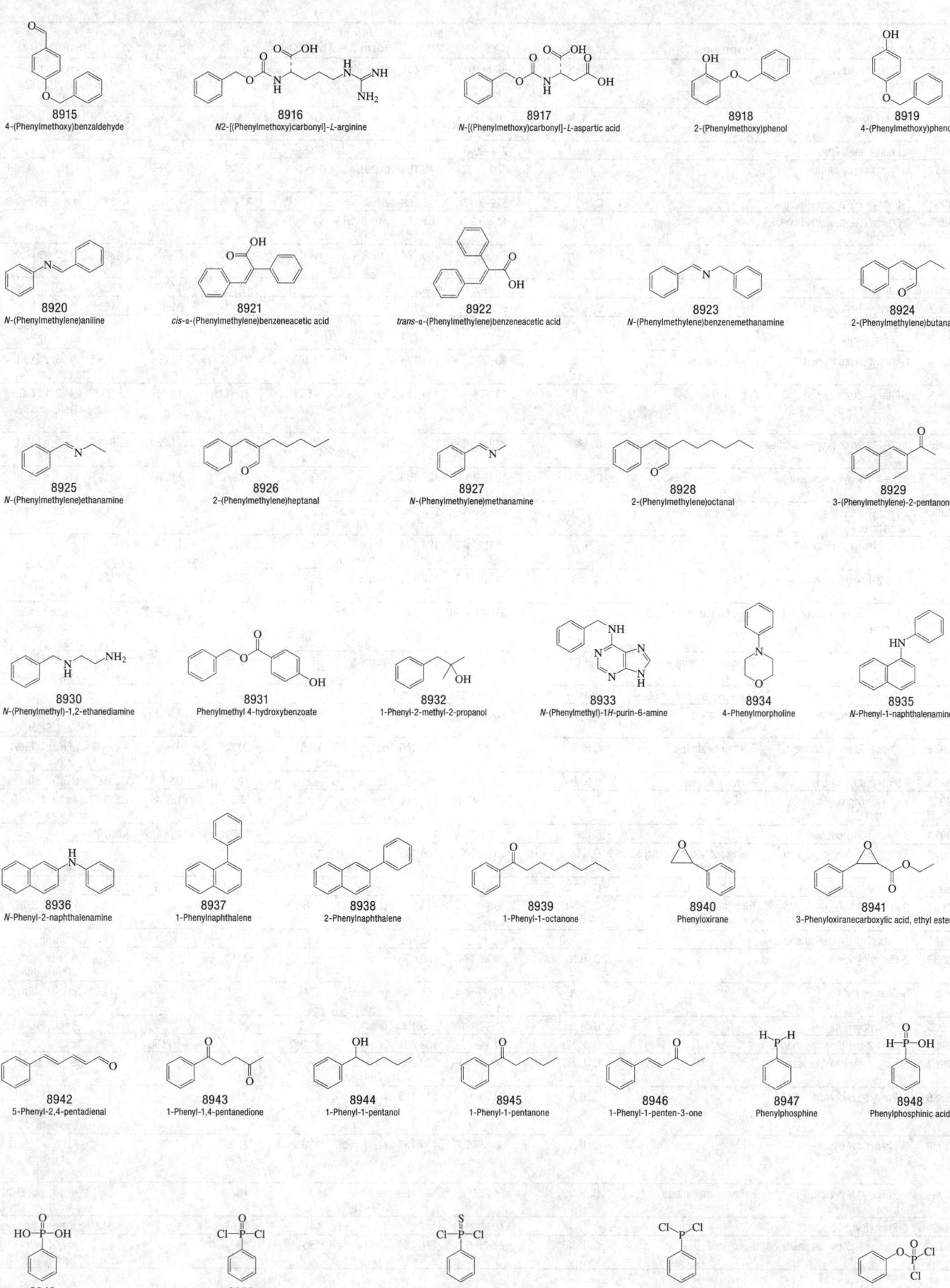

8915
4-(Phenylmethoxy)benzaldehyde

8916
N2-[(Phenylmethoxy)carbonyl]-L-arginine

8917
N-[(Phenylmethoxy)carbonyl]-L-aspartic acid

8918
2-(Phenylmethoxy)phenol

8919
4-(Phenylmethoxy)phenol

8920
N-(Phenylmethylene)aniline

8921
cis-α-(Phenylmethylene)benzeneacetic acid

8922
trans-α-(Phenylmethylene)benzeneacetic acid

8923
N-(Phenylmethylene)benzenemethanamine

8924
2-(Phenylmethylene)butanal

8925
N-(Phenylmethylene)ethanamine

8926
2-(Phenylmethylene)heptanal

8927
N-(Phenylmethylene)methanamine

8928
2-(Phenylmethylene)octanal

8929
3-(Phenylmethylene)-2-pentanone

8930
N-(Phenylmethyl)-1,2-ethanediamine

8931
Phenylmethyl 4-hydroxybenzoate

8932
1-Phenyl-2-methyl-2-propanol

8933
N-(Phenylmethyl)-1H-purin-6-amine

8934
4-Phenylmorpholine

8935
N-Phenyl-1-naphthalenamine

8936
N-Phenyl-2-naphthalenamine

8937
1-Phenylnaphthalene

8938
2-Phenylnaphthalene

8939
1-Phenyl-1-octanone

8940
Phenyloxirane

8941
3-Phenyloxiranecarboxylic acid, ethyl ester

8942
5-Phenyl-2,4-pentadienal

8943
1-Phenyl-1,4-pentanedione

8944
1-Phenyl-1-pentanol

8945
1-Phenyl-1-pentanone

8946
1-Phenyl-1-penten-3-one

8947
Phenylphosphine

8948
Phenylphosphinic acid

8949
Phenylphosphonic acid

8950
Phenylphosphonic dichloride

8951
Phenylphosphonothioic dichloride

8952
Phenylphosphonous dichloride

8953
Phenyl phosphorodichloridate

No.	Name	Synonym	Mol. Form.	CAS RN	Mol. Wt.	Physical Form	mp/°C	bp/°C	den g cm⁻³	n_D	Solubility
8954	1-Phenylpiperazine		$C_{10}H_{14}N_2$	92-54-6	162.231	pa ye oil		286.5	1.0621[20]	1.5875[20]	i H$_2$O; msc EtOH, eth; s chl
8955	1-Phenylpiperidine		$C_{11}H_{15}N$	4096-20-2	161.244		4.7	260(5)	0.9944[25]	1.5598[25]	vs EtOH, eth, bz, chl
8956	4-Phenylpiperidine		$C_{11}H_{15}N$	771-99-3	161.244		60.5	257	0.9996[16]		s chl
8957	N-Phenylpropanamide		$C_9H_{11}NO$	620-71-3	149.189	pl (eth, al, bz)	105.5	222.2	1.175[25]		sl H$_2$O; vs EtOH, eth
8958	1-Phenyl-1,2-propanedione		$C_9H_8O_2$	579-07-7	148.159	ye oil	<20	222	1.1006[20]	1.537[10]	s H$_2$O, EtOH, eth
8959	1-Phenyl-1,2-propanedione, 2-oxime		$C_9H_9NO_2$	119-51-7	163.173	wh nd (w)	115				
8960	Phenyl propanoate		$C_9H_{10}O_2$	637-27-4	150.174	pr	19(1)	207(4)	1.0436[25]	1.4980[20]	i H$_2$O; vs EtOH, eth; s bz
8961	2-Phenyl-1-propanol		$C_9H_{12}O$	1123-85-9	136.190			121[26]	0.975[25]	1.5582[2]	i H$_2$O; s EtOH
8962	1-Phenyl-2-propanol		$C_9H_{12}O$	698-87-3	136.190			125[25]	0.991[20]	1.5190[20]	
8963	Phenylpropanolamine hydrochloride		$C_9H_{14}ClNO$	154-41-6	187.666		194				vs H$_2$O; s EtOH; i eth, bz, chl
8964	1-Phenyl-1-propanone	Propiophenone	$C_9H_{10}O$	93-55-0	134.174		18.6(0.2)	217.4(0.7)	1.0096[20]	1.5269[20]	i H$_2$O; s EtOH, eth, chl
8965	1-Phenyl-2-propanone	Phenylacetone	$C_9H_{10}O$	103-79-7	134.174	liq	-15.3(0.5)	214(17)	1.0157[20]	1.5168[20]	i H$_2$O; vs EtOH, eth; msc bz, xyl; s chl
8966	cis-3-Phenyl-2-propenenitrile		C_9H_7N	24840-05-9	129.159	liq	-4.4	249	1.0289[20]	1.5843[20]	i H$_2$O; s EtOH; vs bz
8967	trans-3-Phenyl-2-propenenitrile		C_9H_7N	1885-38-7	129.159		22	263.8	1.0304[20]	1.6013[20]	i H$_2$O; s EtOH, ace, ctc
8968	3-Phenyl-2-propenoic anhydride	Cinnamic anhydride	$C_{18}H_{14}O_3$	538-56-7	278.302	nd (bz or al) pt (al)	136				vs bz
8969	cis-3-Phenyl-2-propen-1-ol		$C_9H_{10}O$	4510-34-3	134.174	wh nd (eth-peth)	34	257.5	1.0440[20]	1.5819[20]	vs eth, EtOH
8970	trans-3-Phenyl-2-propen-1-ol		$C_9H_{10}O$	4407-36-7	134.174	wh nd (eth-peth)	34.8(0.8)	259(5)	1.0440[20]	1.5819[20]	sl H$_2$O, chl; vs EtOH, eth
8971	trans-3-Phenyl-2-propen-1-ol acetate	trans-Cinnamyl acetate	$C_{11}H_{12}O_2$	21040-45-9	176.212			265	1.0567[20]	1.5425[20]	i H$_2$O; s EtOH, eth, ace, bz, chl
8972	trans-3-Phenyl-2-propenoyl chloride	Cinnamoyl chloride	C_9H_7ClO	17082 09 6	166.604	ye cry	37.5	257.5	1.1617[45]	1.614[42]	i H$_2$O; s EtOH, ctc, lig
8973	3-Phenylpropyl acetate	Benzenepropanol, acetate	$C_{11}H_{14}O_2$	122-72-5	178.228	col liq	-40	69[1]			
8974	1-Phenyl-2-propylamine, (±)-	Amphetamine	$C_9H_{13}N$	300-62-9	135.206	oil		198(12)	0.9306[25]	1.518[26]	sl H$_2$O, eth; s chl, EtOH
8975	1-Phenyl-2-propylamine, (S)-	Dexamphetamine	$C_9H_{13}N$	51-64-9	135.206	oil	27.5	203.5	0.949[15]	1.4704[20]	sl H$_2$O; s EtOH, eth
8976	Phenyl propyl ether	Propoxybenzene	$C_9H_{12}O$	622-85-5	136.190	liq	-28(1)	190.3(1)	0.9474[20]	1.5014[20]	s EtOH, eth
8977	4-(3-Phenylpropyl)pyridine		$C_{14}H_{15}N$	2057-49-0	197.276			322	1.024[25]	1.5616[25]	vs bz, eth, py, EtOH
8978	3-Phenyl-2-propynal		C_9H_6O	2579-22-8	130.143			127[28]	1.0622[20]	1.6079[12]	
8979	3-Phenyl-2-propynoic acid	Phenylacetylenecarboxylic acid	$C_9H_6O_2$	637-44-5	146.143	nd (w)	137(2)		1.28[20]		sl H$_2$O; vs EtOH, eth
8980	3-Phenyl-2-propyn-1-ol		C_9H_8O	1504-58-1	132.159			137[15]	1.078[20]	1.5873[28]	s eth, ace, bz
8981	6-Phenyl-2,4,7-pteridinetriamine	Triamterene	$C_{12}H_{11}N_7$	396-01-0	253.262	ye pl (BuOH)	316				i eth; sl EtOH, chl
8982	1-Phenyl-3-pyrazolidinone		$C_9H_{10}N_2O$	92-43-3	162.187		126				i eth, lig
8983	2-Phenylpyridine		$C_{11}H_9N$	1008-89-5	155.196			271	1.0833[25]	1.6210[20]	sl H$_2$O; msc EtOH, eth
8984	3-Phenylpyridine		$C_{11}H_9N$	1008-88-4	155.196	pa ye oil	164	274(11)		1.6123[25]	sl H$_2$O; s EtOH, eth
8985	4-Phenylpyridine		$C_{11}H_9N$	939-23-1	155.196	pl (w)	74(3)	281			s H$_2$O, EtOH, eth
8986	Phenyl-2-pyridinylmethanone		$C_{12}H_9NO$	91-02-1	183.205		43.3(0.2)	317	1.1556[20]		s chl
8987	Phenyl-4-pyridinylmethanone		$C_{12}H_9NO$	14548-46-0	183.205	nd (peth), pl (w)	75(2)	315			sl H$_2$O; s EtOH, eth, bz
8988	1-Phenyl-1H-pyrrole		$C_{10}H_9N$	635-90-5	143.185	pl (sub), red in air	61(1)	234			i H$_2$O; s EtOH, eth, ace, bz; vs peth
8989	2-Phenyl-1H-pyrrole		$C_{10}H_9N$	3042-22-6	143.185	pl (al, sub)	130(2)	270(25)			i H$_2$O; vs EtOH, eth, bz, chl; sl lig
8990	1-Phenyl-1H-pyrrole-2,5-dione	N-Phenylmaleimide	$C_{10}H_7NO_2$	941-69-5	173.169	ye nd (bz-lig)	90.5	162[12]			vs bz, eth, EtOH
8991	1-Phenylpyrrolidine		$C_{10}H_{13}N$	4096-21-3	147.217		11	119[12]	1.018[20]	1.5813[20]	s eth
8992	1-Phenyl-2,5-pyrrolidinedione	Succinanil	$C_{10}H_9NO_2$	83-25-0	175.184	mcl pr or nd (w, al)	156	400	1.356[25]		i H$_2$O; s EtOH, eth

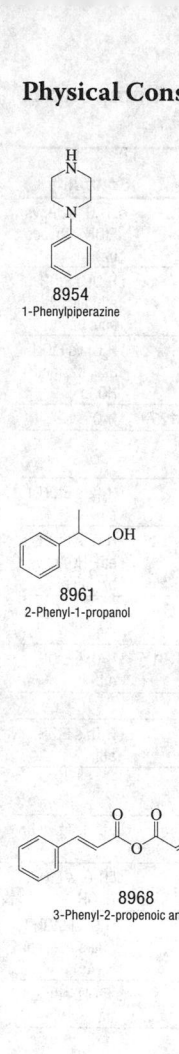

8954
1-Phenylpiperazine

8955
1-Phenylpiperidine

8956
4-Phenylpiperidine

8957
N-Phenylpropanamide

8958
1-Phenyl-1,2-propanedione

8959
1-Phenyl-1,2-propanedione, 2-oxime

8960
Phenyl propanoate

8961
2-Phenyl-1-propanol

8962
1-Phenyl-2-propanol

8963
Phenylpropanolamine hydrochloride

8964
1-Phenyl-1-propanone

8965
1-Phenyl-2-propanone

8966
cis-3-Phenyl-2-propenenitrile

8967
trans-3-Phenyl-2-propenenitrile

8968
3-Phenyl-2-propenoic anhydride

8969
cis-3-Phenyl-2-propen-1-ol

8970
trans-3-Phenyl-2-propen-1-ol

8971
trans-3-Phenyl-2-propen-1-ol acetate

8972
trans-3-Phenyl-2-propenoyl chloride

8973
3-Phenylpropyl acetate

8974
1-Phenyl-2-propylamine, (±)-

8975
1-Phenyl-2-propylamine, (*S*)-

8976
Phenyl propyl ether

8977
4-(3-Phenylpropyl)pyridine

8978
3-Phenyl-2-propynal

8979
3-Phenyl-2-propynoic acid

8980
3-Phenyl-2-propyn-1-ol

8981
6-Phenyl-2,4,7-pteridinetriamine

8982
1-Phenyl-3-pyrazolidinone

8983
2-Phenylpyridine

8984
3-Phenylpyridine

8985
4-Phenylpyridine

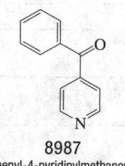

8986
Phenyl-2-pyridinylmethanone

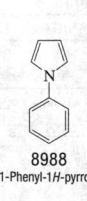

8987
Phenyl-4-pyridinylmethanone

8988
1-Phenyl-1*H*-pyrrole

8989
2-Phenyl-1*H*-pyrrole

8990
1-Phenyl-1*H*-pyrrole-2,5-dione

8991
1-Phenylpyrrolidine

8992
1-Phenyl-2,5-pyrrolidinedione

No.	Name	Synonym	Mol. Form.	CAS RN	Mol. Wt.	Physical Form	mp/°C	bp/°C	den g cm⁻³	n_D	Solubility
8993	2-Phenylquinoline		$C_{15}H_{11}N$	612-96-4	205.255	nd (dil al)	83(1)	363			sl H_2O, peth; vs EtOH, eth, ace, bz
8994	2-Phenyl-4-quinolinecarbox-ylic acid	Cinchophen	$C_{16}H_{11}NO_2$	132-60-5	249.264	nd	214.5				i H_2O; s EtOH, eth, alk; sl ace, bz
8995	Phenyl salicylate		$C_{13}H_{10}O_3$	118-55-8	214.216		41.82(0.04)	173[12]	1.2614[30]		i H_2O; vs EtOH, ace, bz; s eth, HOAc
8996	Phenylsilane		C_6H_8Si	694-53-1	108.214			119	0.8681[20]	1.5125[20]	i H_2O
8997	1-Phenylsilatrane		$C_{12}H_{17}NO_3Si$	2097-19-0	251.354	pr or nd (ace)	209				
8998	Phenyl stearate		$C_{24}H_{40}O_2$	637-55-8	360.574		52	267[15]			i H_2O; s EtOH, eth
8999	5'-Phenyl-1,1':3',1"-terphenyl		$C_{24}H_{18}$	612-71-5	306.400	orth nd (al or HOAc)	173(3)	462	1.199[30]		i H_2O; s EtOH, eth, HOAc; vs bz; sl chl
9000	5-Phenyl-2,4-thiazolediamine	Amiphenazole	$C_9H_9N_3S$	490-55-1	191.252	fl (dil al) br in air	163 dec				
9001	Phenyl-2-thienylmethanone		$C_{11}H_8OS$	135-00-2	188.246	nd (dil al)	56.5	300	1.1890[54]	1.6181[54]	i H_2O; s EtOH, eth
9002	N-Phenylthioacetamide	Thioacetanilide	C_8H_9NS	637-53-6	151.229	nd (w)	75.5	dec			i H_2O; s EtOH, eth
9003	Phenyl thiocyanate		C_7H_5NS	5285-87-0	135.187			232.5	1.153[18]		i H_2O; s EtOH, eth
9004	2-Phenylthiosemicarbazide	2-Phenylhydrazinecarbothio-amide	$C_7H_9N_3S$	645-48-7	167.231	pr (al)	200 dec				
9005	4-Phenyl-3-thiosemicarbazide	N-Phenylhydrazinecarbothio-amide	$C_7H_9N_3S$	5351-69-9	167.231	pl (al)	140 dec				i EtOH, lig; sl bz
9006	Phenylthiourea		$C_7H_8N_2S$	103-85-5	152.217	nd (w), pr	154				sl H_2O; s EtOH, NaOH
9007	3-Phenyl-2-thioxo-4-thiazolid-inone	3-Phenylrhodanine	$C_9H_7NOS_2$	1457-46-1	209.288	ye pr (HOAc) nd or pr (al)	194.5				i H_2O; sl EtOH, eth; s ace, chl, HOAc
9008	6-Phenyl-1,3,5-triazine-2,4-diamine	Benzoguanamine	$C_9H_9N_5$	91-76-9	187.201	nd, pl (al)	226.5				s EtOH, eth; sl tfa
9009	N-Phenyl-1,3,5-triazine-2,4-diamine	Amanozine	$C_9H_9N_5$	537-17-7	187.201	cry (diox, 50% al)	235.5				
9010	4-Phenyl-1,2,4-triazolidine-3,5-dione		$C_8H_7N_3O_2$	15988-11-1	177.161		205.5				
9011	Phenyltrimethylammonium iodide		$C_9H_{14}IN$	98-04-4	263.118	lf (al)	224				vs H_2O; s EtOH, HOAc; sl ace; i chl
9012	Phenyl(triphenylmethyl)-diazene		$C_{25}H_{20}N_2$	981-18-0	348.440		111 dec				
9013	Phenylurea		$C_7H_8N_2O$	64-10-8	136.151	mcl pr (w, al)	147.4(0.5)	238	1.302[25]		sl H_2O, eth, DMSO; s EtOH, AcOEt
9014	trans-5-(2-Phenylvinyl)-1,3-benzenediol	Pinosylvin	$C_{14}H_{12}O_2$	22139-77-1	212.244	nd (HOAc)	156				vs ace, bz, chl, HOAc
9015	Phenyl vinyl ether		C_8H_8O	766-94-9	120.149			155.5	0.9770[20]	1.5224[20]	i H_2O; vs eth
9016	Phenytoin	5,5-Diphenyl-2,4-imidazolidin-edione	$C_{15}H_{12}N_2O_2$	57-41-0	252.268	nd (al)	286				i H_2O; s EtOH, ace; sl eth, bz
9017	Phloretin		$C_{15}H_{14}O_5$	60-82-2	274.269	nd (dil al), cry (ace)	263 dec				sl H_2O, chl; msc EtOH, bz; i eth; s ace
9018	Phorate		$C_7H_{17}O_2PS_3$	298-02-2	260.378		<-15	119[0.8]	1.16[25]		
9019	Phorbol		$C_{20}H_{28}O_6$	17673-25-5	364.432	cry (EtOH)	250 dec				s H_2O, ace
9020	Phorone		$C_9H_{14}O$	504-20-1	138.206	ye-grn pr	25.8(0.5)	197(7)	0.8850[20]	1.4998[20]	sl H_2O; s EtOH, eth, ace, ctc
9021	Phosalone		$C_{12}H_{15}ClNO_4PS_2$	2310-17-0	367.808		48.7(0.4)				
9022	Phosfolan		$C_7H_{14}NO_3PS_2$	947-02-4	255.295		36.5	117[0.001]			vs H_2O, bz, ace; sl eth; s hx
9023	Phosmet		$C_{11}H_{12}NO_4PS_2$	732-11-6	317.321		72.0(0.2)	dec			
9024	Phosphamidon		$C_{10}H_{19}ClNO_5P$	13171-21-6	299.689	oil	-45	162[1.5]	1.2132[25]	1.4718[25]	msc H_2O; s hx
9025	N-Phospho-L-arginine		$C_6H_{15}N_4O_5P$	1189-11-3	254.181	cry (ace aq)	177				
9026	O-Phosphorylethanolamine	Ethanolamine O-phosphate	$C_2H_8NO_4P$	1071-23-4	141.063	cry (EtOH aq)	242				
9027	O-Phosphoserine		$C_3H_8NO_6P$	407-41-0	185.073	cry	166 dec				
9028	Phthalazine	2,3-Benzodiazine	$C_8H_6N_2$	253-52-1	130.147		91.3(0.2)	316			s H_2O, EtOH, bz; sl eth; i lig
9029	Phthalic acid	1,2-Benzenedicarboxylic acid	$C_8H_6O_4$	88-99-3	166.132	pl (w)	207(3)	dec	2.18[191]		sl H_2O, eth; i chl; s EtOH

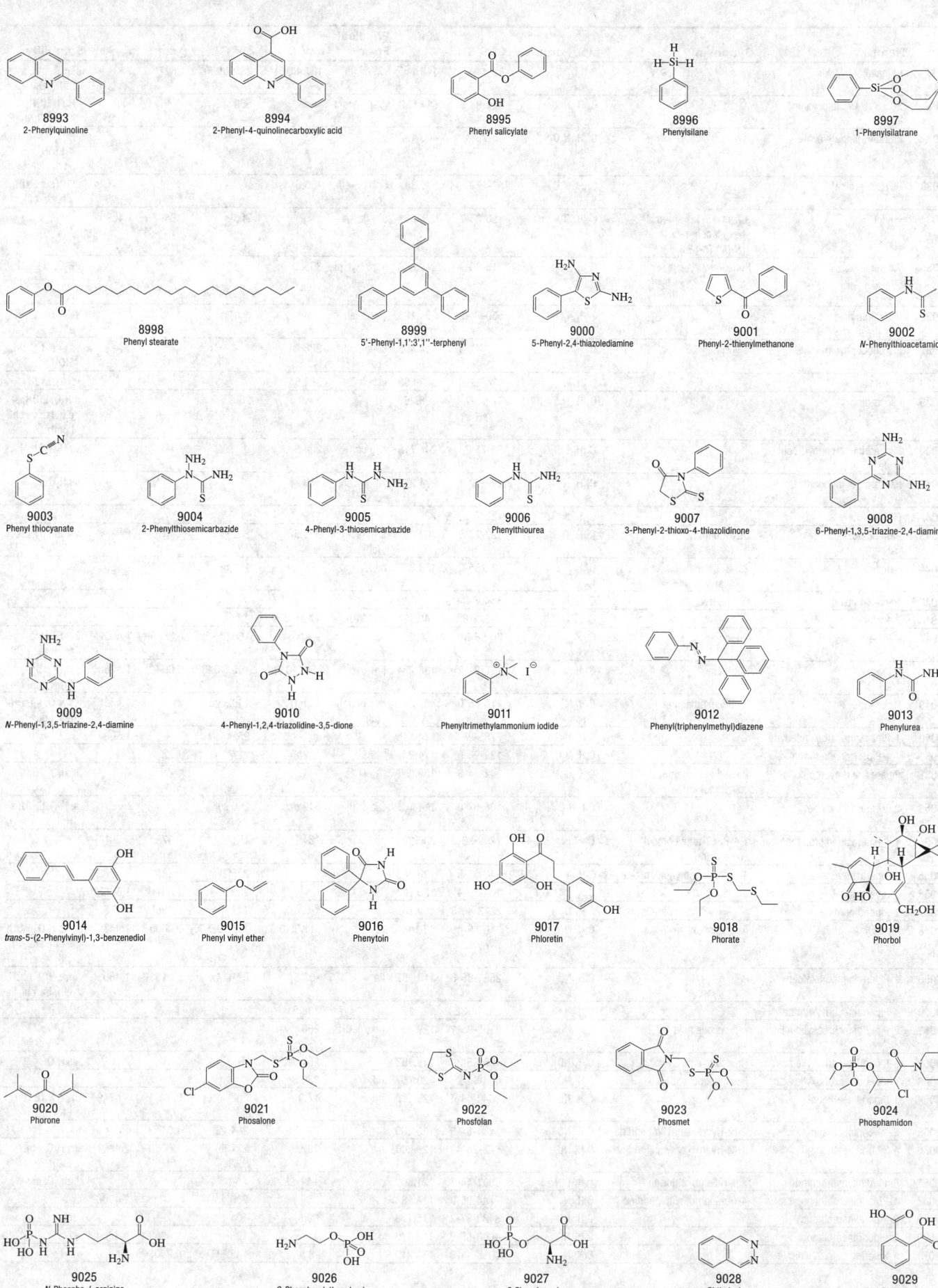

8993
2-Phenylquinoline

8994
2-Phenyl-4-quinolinecarboxylic acid

8995
Phenyl salicylate

8996
Phenylsilane

8997
1-Phenylsilatrane

8998
Phenyl stearate

8999
5'-Phenyl-1,1':3',1''-terphenyl

9000
5-Phenyl-2,4-thiazolediamine

9001
Phenyl-2-thienylmethanone

9002
N-Phenylthioacetamide

9003
Phenyl thiocyanate

9004
2-Phenylthiosemicarbazide

9005
4-Phenyl-3-thiosemicarbazide

9006
Phenylthiourea

9007
3-Phenyl-2-thioxo-4-thiazolidinone

9008
6-Phenyl-1,3,5-triazine-2,4-diamine

9009
N-Phenyl-1,3,5-triazine-2,4-diamine

9010
4-Phenyl-1,2,4-triazolidine-3,5-dione

9011
Phenyltrimethylammonium iodide

9012
Phenyl(triphenylmethyl)diazene

9013
Phenylurea

9014
trans-5-(2-Phenylvinyl)-1,3-benzenediol

9015
Phenyl vinyl ether

9016
Phenytoin

9017
Phloretin

9018
Phorate

9019
Phorbol

9020
Phorone

9021
Phosalone

9022
Phosfolan

9023
Phosmet

9024
Phosphamidon

9025
N-Phospho-L-arginine

9026
O-Phosphorylethanolamine

9027
O-Phosphoserine

9028
Phthalazine

9029
Phthalic acid

No.	Name	Synonym	Mol. Form.	CAS RN	Mol. Wt.	Physical Form	mp/°C	bp/°C	den g cm^{-3}	n_D	Solubility
9030	Phthalic anhydride		$C_8H_4O_3$	85-44-9	148.116	wh nd (al, bz)	131.4(0.3)	285.3(0.8)	1.527[4]		sl H$_2$O, eth; s EtOH, ace, bz
9031	29H,31H-Phthalocyanine		$C_{32}H_{18}N_8$	574-93-6	514.539	grsh-bl mcl (quinoline)		550 sub			i H$_2$O, EtOH, eth; s PhNH$_2$
9032	Phthalylsulphathiazole		$C_{17}H_{13}N_3O_5S_2$	85-73-4	403.432		273				i H$_2$O, eth, chl; sl EtOH; s acid, alk
9033	Physostigmine		$C_{15}H_{21}N_3O_2$	57-47-6	275.347	orth pr (eth, bz)	105.5				sl H$_2$O; s EtOH, eth, bz, chl
9034	Phytol	3,7,11,15-Tetramethyl-2-hexadecen-1-ol, [R-[R*,R*-(E)]]	$C_{20}H_{40}O$	150-86-7	296.531	oily liq		203[10]	0.8497[25]	1.4595[25]	
9035	Picene	Benzo[a]chrysene	$C_{22}H_{14}$	213-46-7	278.346	lf, pl (xyl, py, sub)	365(2)	519			i H$_2$O; sl EtOH, bz, chl; s con sulf
9036	Picrolonic acid		$C_{10}H_8N_4O_5$	550-74-3	264.195	ye nd (al)	116	dec			sl H$_2$O; s EtOH, eth, MeOH
9037	Picropodophyllin		$C_{22}H_{22}O_8$	477-47-4	414.405	col nd (al, bz)	228				vs ace, bz, eth, EtOH
9038	Picrotoxin		$C_{30}H_{34}O_{13}$	124-87-8	602.583	orth lf	203.5				vs py, EtOH
9039	Pilocarpine		$C_{11}H_{16}N_2O_2$	92-13-7	208.257	nd	34	260[5]			s H$_2$O, EtOH; sl eth, bz; vs chl; i peth
9040	Pilocarpine, monohydrochloride		$C_{11}H_{17}ClN_2O_2$	54-71-7	244.718	hyg cry	204.5				vs H$_2$O, EtOH
9041	Pilocarpine, mononitrate		$C_{11}H_{17}N_3O_5$	148-72-1	271.270	wh pow or cry (al)	178				vs H$_2$O
9042	Pilosine		$C_{16}H_{18}N_2O_3$	13640-28-3	286.325	nd (al)	179				
9043	Pimaric acid	Dextropimaric acid	$C_{20}H_{30}O_2$	127-27-5	302.451	orth (ace) pr (al)	218.5	282[18]			vs eth, py, EtOH
9044	Pinane	2,6,6-Trimethylbicyclo[3.1.1]heptane	$C_{10}H_{18}$	473-55-2	138.250	oil	-53	167(5)	0.8467[21]	1.4605[21]	
9045	trans-2-Pinanol	Pinene hydrate	$C_{10}H_{18}O$	35408-04-9	154.249		60	81[10]			
9046	Pindolol		$C_{14}H_{20}N_2O_2$	13523-86-9	248.321	cry (EtOH)	172				
9047	α-Pinene	2-Pinene	$C_{10}H_{16}$	80-56-8	136.234	liq	-74(4)	156.3(0.5)	0.8539[25]	1.4632[25]	i H$_2$O; msc EtOH, eth, chl
9048	β-Pinene	Nopinene	$C_{10}H_{16}$	127-91-3	136.234	liq	-50.0(0.2)	165.8(0.6)	0.860[25]	1.4768[25]	i H$_2$O; s bz, EtOH, chl
9049	Piperazine	Diethylenediamine	$C_4H_{10}N_2$	110-85-0	86.135	hyg pl or lf (al)	111(1)	148.63(0.05)		1.446[113]	vs H$_2$O; s EtOH, chl; i eth
9050	1-Piperazinecarboxaldehyde		$C_5H_{10}N_2O$	7755-92-2	114.145			95[0.5]		1.5094[20]	
9051	1,4-Piperazinediethanol		$C_8H_{18}N_2O_2$	122-96-3	174.241		132(2)	217[30]			
9052	Piperazine dihydrochloride	Diethylenediamine dihydrochloride	$C_4H_{12}Cl_2N_2$	142-64-3	159.057						sl H$_2$O; i EtOH
9053	2,5-Piperazinedione		$C_4H_6N_2O_2$	106-57-0	114.103	tab or pl (w)	312 dec	260 sub			sl H$_2$O, EtOH; s HCl
9054	1,4-Piperazinedipropanamine	1,4-Bis(3-aminopropyl)piperazine	$C_{10}H_{24}N_4$	7209-38-3	200.325		15	151[2]	0.973[25]	1.5015[20]	
9055	1-Piperazineethanamine	1-(2-Aminoethyl)piperazine	$C_6H_{15}N_3$	140-31-8	129.203			225(2)	0.985[25]	1.4983[20]	
9056	1-Piperazineethanol		$C_6H_{14}N_2O$	103-76-4	130.187			259(6)	1.061[25]	1.5065[20]	
9057	1-Piperidinamine		$C_5H_{12}N_2$	2213-43-6	100.162			147	0.928[25]	1.4750[20]	
9058	Piperidine	Azacyclohexane	$C_5H_{11}N$	110-89-4	85.148	liq	-11.05(0.03)	106.19(0.09)	0.8606[20]	1.4530[20]	msc H$_2$O, EtOH; s eth, ace, bz, chl
9059	1-Piperidinecarboxaldehyde		$C_6H_{11}NO$	2591-86-8	113.157	liq	-30.8	221(14)	1.0158[25]	1.4805[25]	msc H$_2$O, EtOH, eth, bz, chl, lig
9060	4-Piperidinecarboxamide		$C_6H_{12}N_2O$	39546-32-2	128.171		138.5				
9061	2-Piperidinecarboxylic acid, (S)-	L-Pipecolic acid	$C_6H_{11}NO_2$	3105-95-1	129.157	nd (MeOH/ eth)	260				
9062	3-Piperidinecarboxylic acid	Nipecotic acid	$C_6H_{11}NO_2$	498-95-3	129.157		261 dec				vs H$_2$O
9063	4-Piperidinecarboxylic acid	Isonipecotic acid	$C_6H_{11}NO_2$	498-94-2	129.157	nd	336				
9064	1-Piperidineethanol		$C_7H_{15}NO$	3040-44-6	129.200		17.9	202	0.9703[25]	1.4749[20]	msc H$_2$O; vs EtOH
9065	2-Piperidineethanol	2-(2-Hydroxyethyl)piperidine	$C_7H_{15}NO$	1484-84-0	129.200		69	204(13)	1.01[27]		vs H$_2$O
9066	4-Piperidineethanol	4-(2-Hydroxyethyl)piperidine	$C_7H_{15}NO$	622-26-4	129.200	syr	132.5	227.5	1.0059[15]	1.4907[20]	vs H$_2$O, eth, EtOH
9067	Piperidine, hydrochloride	Piperidinium chloride	$C_5H_{12}ClN$	6091-44-7	121.609		142 dec				vs H$_2$O, chl
9068	4-Piperidinemethanamine	4-(Aminomethyl)piperidine	$C_6H_{14}N_2$	7144-05-0	114.188		25	200		1.4900[20]	
9069	2-Piperidinemethanol		$C_6H_{13}NO$	3433-37-2	115.173		69(3)	104[10]			sl chl
9070	3-Piperidinemethanol		$C_6H_{13}NO$	4606-65-9	115.173		61	106[3.5]	1.0263[20]	1.4964[20]	sl chl
9071	1-Piperidinepropanenitrile		$C_8H_{14}N_2$	3088-41-3	138.210		-6.8	145[50]	0.9403[25]	1.4676[25]	

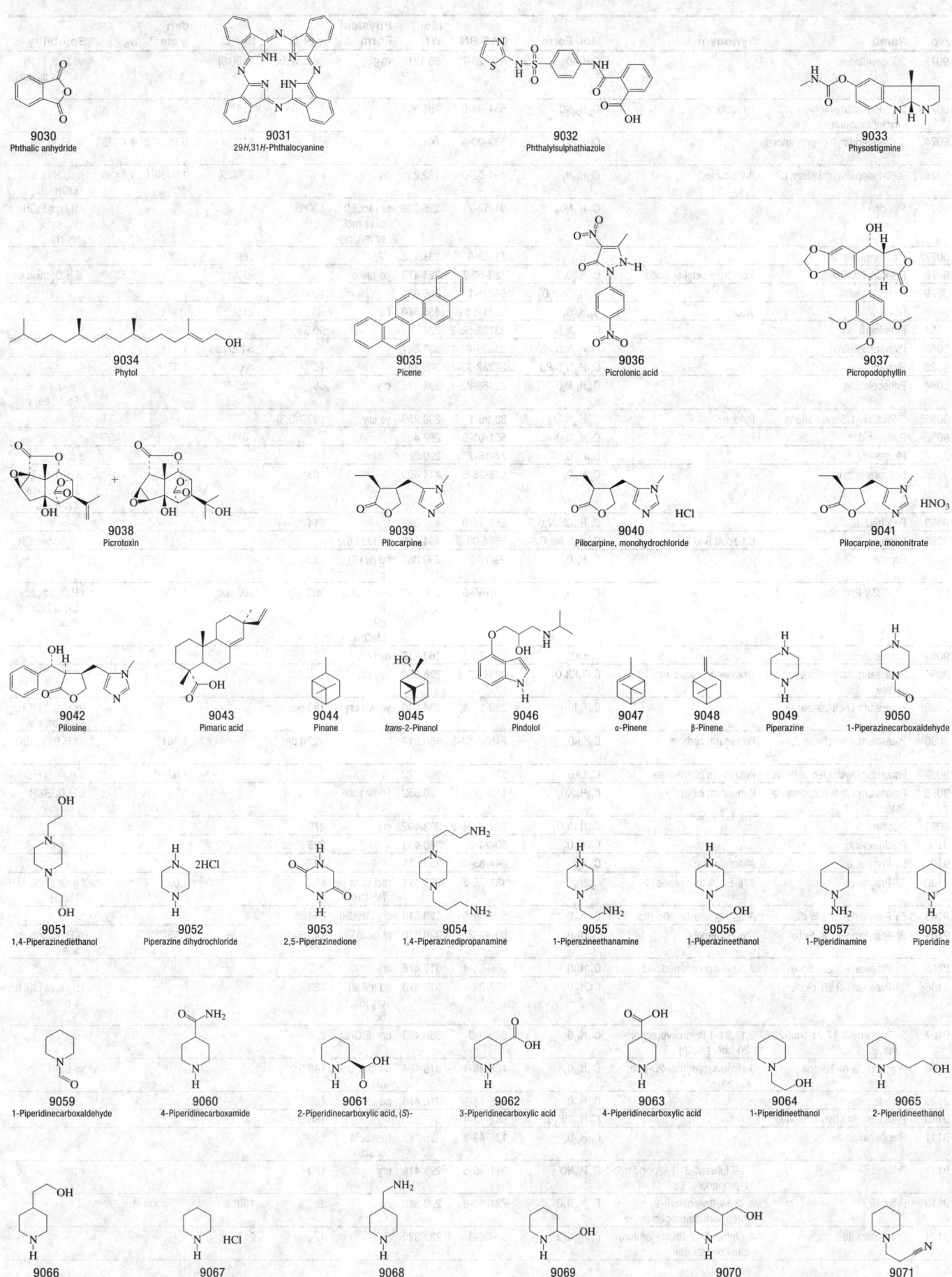

9030
Phthalic anhydride

9031
29H,31H-Phthalocyanine

9032
Phthalylsulphathiazole

9033
Physostigmine

9034
Phytol

9035
Picene

9036
Picrolonic acid

9037
Picropodophyllin

9038
Picrotoxin

9039
Pilocarpine

9040
Pilocarpine, monohydrochloride

9041
Pilocarpine, mononitrate

9042
Pilosine

9043
Pimaric acid

9044
Pinane

9045
trans-2-Pinanol

9046
Pindolol

9047
α-Pinene

9048
β-Pinene

9049
Piperazine

9050
1-Piperazinecarboxaldehyde

9051
1,4-Piperazinediethanol

9052
Piperazine dihydrochloride

9053
2,5-Piperazinedione

9054
1,4-Piperazinedipropanamine

9055
1-Piperazineethanamine

9056
1-Piperazineethanol

9057
1-Piperidinamine

9058
Piperidine

9059
1-Piperidinecarboxaldehyde

9060
4-Piperidinecarboxamide

9061
2-Piperidinecarboxylic acid, (S)-

9062
3-Piperidinecarboxylic acid

9063
4-Piperidinecarboxylic acid

9064
1-Piperidineethanol

9065
2-Piperidineethanol

9066
4-Piperidineethanol

9067
Piperidine, hydrochloride

9068
4-Piperidinemethanamine

9069
2-Piperidinemethanol

9070
3-Piperidinemethanol

9071
1-Piperidinepropanenitrile

No.	Name	Synonym	Mol. Form.	CAS RN	Mol. Wt.	Physical Form	mp/°C	bp/°C	den g cm⁻³	n_D	Solubility
9072	2-Piperidinone		C_5H_9NO	675-20-7	99.131	hyg	38.68(0.05)	260(19)			vs H₂O, EtOH, eth; s dil acid; i con alk
9073	2-(1-Piperidinylmethyl)-cyclohexanone	Pimeclone	$C_{12}H_{21}NO$	534-84-9	195.301			119[14]			
9074	1-(2-Piperidinyl)-2-propanone, (±)-		$C_8H_{15}NO$	539-00-4	141.211	oil		91[14]	0.9624[20]	1.4683[20]	vs EtOH, chl
9075	3-(2-Piperidinyl)pyridine, (S)-	Anabasine	$C_{10}H_{14}N_2$	494-52-0	162.231	liq	9	276(29)	1.0455[20]	1.5430[20]	msc H₂O; s EtOH, eth, bz
9076	Piperine		$C_{17}H_{19}NO_3$	94-62-2	285.338	pr (AcOEt) pl or mcl pr (al), cry	130(2)				i H₂O; s EtOH, bz, py; sl eth; vs chl
9077	Piperonyl butoxide		$C_{19}H_{30}O_5$	51-03-6	338.438			180[1]	1.05[25]		
9078	Piperonyl sulfoxide	Isosafrole octyl sulfoxide	$C_{18}H_{26}O_3S$	120-62-7	324.478	ye-br liq				1.530[25]	sl H₂O; misc os
9079	Pipobroman		$C_{10}H_{16}Br_2N_2O_2$	54-91-1	356.054		106				
9080	Piprotal	Tropital	$C_{24}H_{40}O_8$	5281-13-0	456.570	liq		215[0.04]			
9081	Pirimicarb		$C_{11}H_{18}N_4O_2$	23103-98-2	238.287		90.5				
9082	Pirimiphos-ethyl		$C_{13}H_{24}N_3O_3PS$	23505-41-1	333.387			>130 dec	1.14[20]		
9083	Pirimiphos-methyl		$C_{11}H_{20}N_3O_3PS$	29232-93-7	305.334		15	dec	1.17[20]		
9084	Pithecolobine		$C_{22}H_{46}N_4O_2$	22368-82-7	398.626	cry	68	230[0.007]			s H₂O, chl, eth, EtOH, peth
9085	2-Pivaloyl-1,3-indandione	Pindone	$C_{14}H_{14}O_3$	83-26-1	230.259	ye cry	109.7(0.3)				
9086	Plasmocid		$C_{17}H_{25}N_3O$	551-01-9	287.400			182[1.0]	1.0569[24]	1.5855[24]	
9087	Plumericin		$C_{15}H_{14}O_6$	77-16-7	290.268						s chl
9088	Podophyllotoxin		$C_{22}H_{22}O_8$	518-28-5	414.405		183				sl H₂O; vs EtOH; i eth; s ace, bz, HOAc
9089	Polythiazide		$C_{11}H_{13}ClF_3N_3O_4S_3$	346-18-9	439.882		214				
9090	Ponceau 3R	C.I. Food Red 6	$C_{19}H_{16}N_2Na_2O_7S_2$	3564-09-8	494.449	dk red pow					s H₂O; sl EtOH
9091	Populin		$C_{20}H_{22}O_8$	99-17-2	390.384	nd (w+2), pr (al)	180				
9092	21H,23H-Porphine		$C_{20}H_{14}N_4$	101-60-0	310.352	red or oran lf (chl-MeOH)	360	300 sub	1.336[25]		i H₂O, eth, ace, bz; sl EtOH; s diox
9093	Potassium benzoate		$C_7H_5KO_2$	582-25-2	160.212	hyg cry					
9094	Potassium dichloroisocyanurate	Troclosene potassium	$C_3Cl_2KN_3O_3$	2244-21-5	236.054	hyg cry	250 dec				
9095	Potassium D-gluconate		$C_6H_{11}KO_7$	299-27-4	234.245	ye-wh cry	183 dec				vs H₂O; i EtOH, eth, bz, chl
9096	Potassium trans,trans-2,4-hexadienoate	Potassium sorbate	$C_6H_7KO_2$	24634-61-5	150.217		>270 dec		1.361[25]		vs H₂O; s EtOH
9097	Potassium hydrogen phthalate	Potassium biphthalate	$C_8H_5KO_4$	877-24-7	204.222				1.636[25]		s H₂O; sl EtOH
9098	Potassium cis-9-octadeceno-ate	Potassium oleate	$C_{18}H_{33}KO_2$	143-18-0	320.552	ye-br solid					s H₂O, EtOH
9099	Prazosin		$C_{19}H_{21}N_5O_4$	19216-56-9	383.402	cry	279				
9100	Prednisolone		$C_{21}H_{28}O_5$	50-24-8	360.444		235				
9101	5α-Pregnane	Allopregnane	$C_{21}H_{36}$	641-85-0	288.511		84.5				
9102	5β-Pregnane	17β-Ethyletiocholane	$C_{21}H_{36}$	481-26-5	288.511	mcl sc or pl (MeOH)	83.5		1.032[15]		i H₂O; s chl, MeOH
9103	5α-Pregnane-3α,20α-diol	Allopregnane-3α,20α-diol	$C_{21}H_{36}O_2$	566-58-5	320.510	cry (MeOH)	244				
9104	5β-Pregnane-3α,20S-diol	Pregnanediol	$C_{21}H_{36}O_2$	80-92-2	320.510	pl (ace)	243.5		1.15[25]		sl EtOH, eth; s ace
9105	5α-Pregnane-3,20-dione	3,20-Allopregnanedione	$C_{21}H_{32}O_2$	566-65-4	316.478	cry	200				
9106	5β-Pregnane-3,20-dione		$C_{21}H_{32}O_2$	128-23-4	316.478	nd (dil al) cry (dil ace)	123				i H₂O; vs EtOH; s eth, ace
9107	5-Pregnane-3,17,21-triol-20-one	3,17,21-Trihydroxypregnan-20-one, (3α,5β)	$C_{21}H_{34}O_4$	68-60-0	350.493	cry (EtOAc)	226				
9108	Pregnan-3α-ol-20-one	3-Hydroxypregnan-20-one, (3α,5β)	$C_{21}H_{34}O_2$	128-20-1	318.494	nd (bz), cry (dil al)	149.5				vs EtOH
9109	Pregnenolone		$C_{21}H_{32}O_2$	145-13-1	316.478	nd (dil al)	192				
9110	Prenoxdiazine hydrochloride		$C_{23}H_{28}ClN_3O$	982-43-4	397.940		186.5				
9111	Prephenic acid		$C_{10}H_{10}O_6$	126-49-8	226.182	free acid unstab					
9112	Pridinol	1,1-Diphenyl-3-(1-piperidinyl)-1-propanol	$C_{20}H_{25}NO$	511-45-5	295.419	cry	120				s ace
9113	Prilocaine	N-(2-Methylphenyl)-2-(propylamino)propanamide	$C_{13}H_{20}N_2O$	721-50-6	220.310	nd	38	160[1]		1.5299[20]	
9114	Procainamide	4-Amino-N-[2-(diethylamino)ethyl]benzamide	$C_{13}H_{21}N_3O$	51-06-9	235.325		47	212[2]			

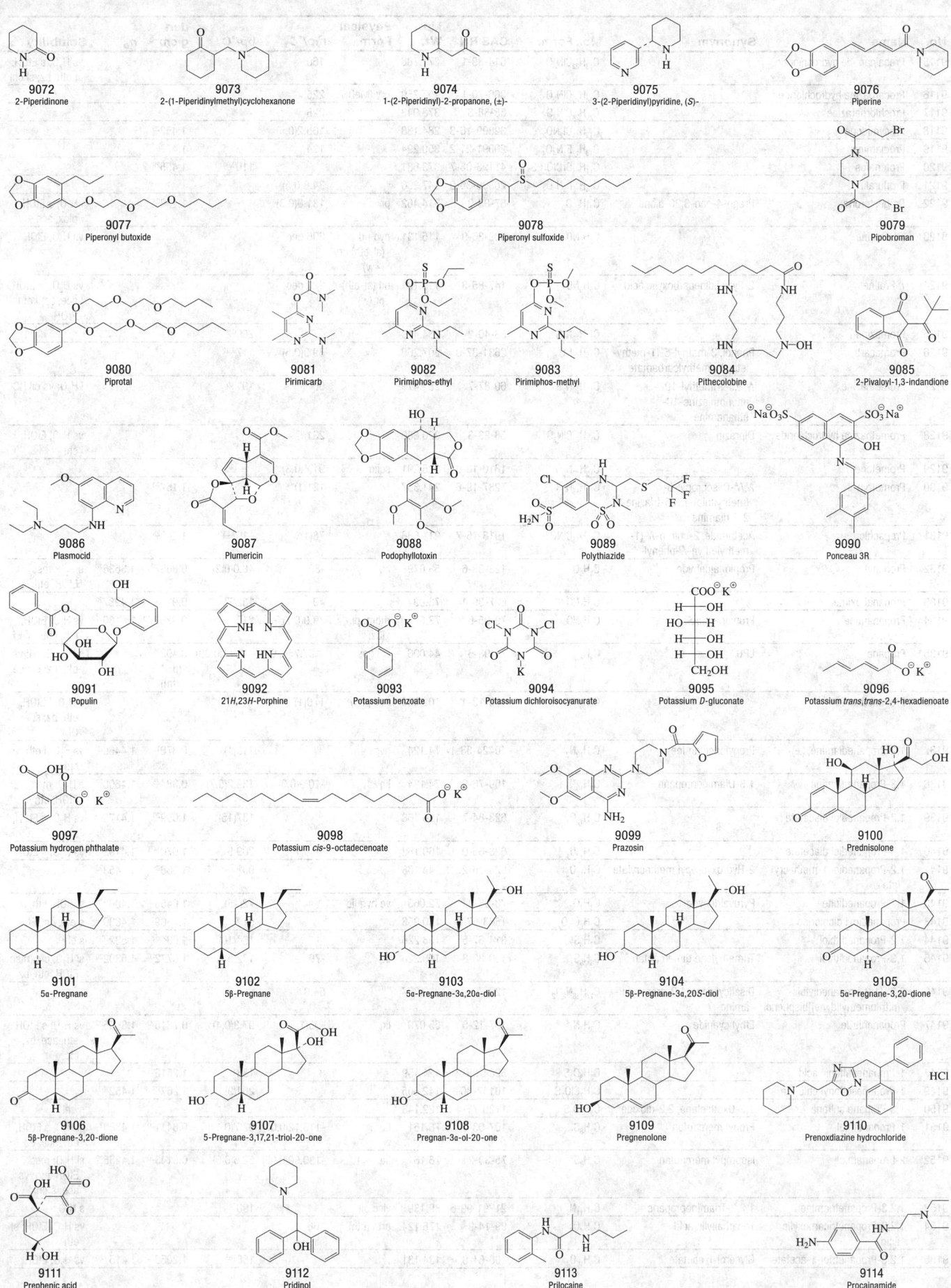

9072 2-Piperidinone

9073 2-(1-Piperidinylmethyl)cyclohexanone

9074 1-(2-Piperidinyl)-2-propanone, (±)-

9075 3-(2-Piperidinyl)pyridine, (S)-

9076 Piperine

9077 Piperonyl butoxide

9078 Piperonyl sulfoxide

9079 Pipobroman

9080 Piprotal

9081 Pirimicarb

9082 Pirimiphos-ethyl

9083 Pirimiphos-methyl

9084 Pithecolobine

9085 2-Pivaloyl-1,3-indandione

9086 Plasmocid

9087 Plumericin

9088 Podophyllotoxin

9089 Polythiazide

9090 Ponceau 3R

9091 Populin

9092 21H,23H-Porphine

9093 Potassium benzoate

9094 Potassium dichloroisocyanurate

9095 Potassium D-gluconate

9096 Potassium trans,trans-2,4-hexadienoate

9097 Potassium hydrogen phthalate

9098 Potassium cis-9-octadecenoate

9099 Prazosin

9100 Prednisolone

9101 5α-Pregnane

9102 5β-Pregnane

9103 5α-Pregnane-3α,20α-diol

9104 5β-Pregnane-3α,20S-diol

9105 5α-Pregnane-3,20-dione

9106 5β-Pregnane-3,20-dione

9107 5-Pregnane-3,17,21-triol-20-one

9108 Pregnan-3α-ol-20-one

9109 Pregnenolone

9110 Prenoxdiazine hydrochloride

9111 Prephenic acid

9112 Pridinol

9113 Prilocaine

9114 Procainamide

No.	Name	Synonym	Mol. Form.	CAS RN	Mol. Wt.	Physical Form	mp/°C	bp/°C	den g cm⁻³	n_D	Solubility
9115	Procainamide hydrochloride		$C_{13}H_{22}ClN_3O$	614-39-1	271.786		166				vs H₂O; s EtOH; i eth, bz; sl chl
9116	Procarbazine hydrochloride		$C_{12}H_{20}ClN_3O$	366-70-1	257.759	cry (MeOH)	225				
9117	Prochlorperazine		$C_{20}H_{24}ClN_3S$	58-38-8	373.943		228				
9118	Procymidone		$C_{13}H_{11}Cl_2NO_2$	32809-16-8	284.138		165.2(0.5)		1.452²⁵		
9119	Prodiamine		$C_{13}H_{17}F_3N_4O_4$	29091-21-2	350.294		124		1.47²⁵		
9120	Profenofos		$C_{11}H_{15}BrClO_3PS$	41198-08-7	373.631			110⁰·⁰⁰¹	1.455²⁰		
9121	Profluralin		$C_{14}H_{16}F_3N_3O_4$	26399-36-0	347.290		34.8(0.5)				
9122	Progesterone	Pregn-4-ene-3,20-dione	$C_{21}H_{30}O_2$	57-83-0	314.462	pr	131.0(0.3)		1.166²³		i H₂O; s EtOH, diox, ace
9123	DL-Proline		$C_5H_9NO_2$	609-36-9	115.131	hyg nd (al-eth) cry (+w)	205 dec				vs H₂O, EtOH
9124	L-Proline	2-Pyrrolidinecarboxylic acid	$C_5H_9NO_2$	147-85-3	115.131	nd (al-eth) pr (w)	221 dec				vs H₂O; sl EtOH, ace, bz; i eth, PrOH
9125	Promazine		$C_{17}H_{20}N_2S$	58-40-2	284.419			206⁰·³			
9126	Promecarb	Phenol, 3-methyl-5-(1-methylethyl)-, methylcarbamate	$C_{12}H_{17}NO_2$	2631-37-0	207.269		89.3(0.5)	117⁰·⁰¹			
9127	Promethazine	N,N,α-Trimethyl-10H-phenothiazine-10-ethanamine	$C_{17}H_{20}N_2S$	60-87-7	284.419		60	191⁰·⁵			i H₂O; vs dil HCl
9128	Promethazine hydrochloride	Diprazin	$C_{17}H_{21}ClN_2S$	58-33-3	320.880		231				vs H₂O, EtOH, chl
9129	Prometone		$C_{10}H_{19}N_5O$	1610-18-0	225.291	solid	91.4(0.3)				
9130	Prometryn	N,N'-Diisopropyl-6-(methylthio)-1,3,5-triazine-2,4-diamine	$C_{10}H_{19}N_5S$	7287-19-6	241.357		121(1)		1.157²⁰		
9131	Propachlor	Acetamide, 2-chloro-N-(1-methylethyl)-N-phenyl-	$C_{11}H_{14}ClNO$	1918-16-7	211.688		78(1)	110⁰·⁰³	1.242²⁵		
9132	Propanal	Propionaldehyde	C_3H_6O	123-38-6	58.079	liq	-80	48.0(0.2)	0.8657²⁵	1.3636²⁰	s H₂O; msc EtOH, eth
9133	Propanal oxime		C_3H_7NO	627-39-4	73.094		40	132(7)	0.9258²⁰	1.4287²⁰	
9134	Propanamide	Propionamide	C_3H_7NO	79-05-0	73.094	rhom, pl (bz)	79.9(0.7)	213	0.9262¹¹⁰	1.4180¹¹⁰	vs H₂O, EtOH, eth, chl
9135	Propane	LPG	C_3H_8	74-98-6	44.096	col gas	-187.75(0.05)	-42.11(0.09)	0.493²⁵ (p>1 atm)		s H₂O, EtOH; vs eth, bz; sl ace
9136	Propanediamide		$C_3H_6N_2O_2$	108-13-4	102.092	mcl pr(w)	170(1)				s H₂O; i EtOH, eth, bz; sl DMSO
9137	1,2-Propanediamine, (±)-	Propylenediamine	$C_3H_{10}N_2$	10424-38-1	74.124	hyg		118(1)	0.878¹⁵	1.4460²⁰	vs H₂O; i eth; vs chl
9138	1,3-Propanediamine	1,3-Diaminopropane	$C_3H_{10}N_2$	109-76-2	74.124	liq	-10.9(0.3)	139.2(0.7)	0.884²⁵	1.4600²⁰	s H₂O; msc EtOH, eth
9139	1,2-Propanediol diacetate		$C_7H_{12}O_4$	623-84-7	160.168			187(15)	1.059²⁰	1.4173²⁰	vs H₂O; s EtOH, eth
9140	1,3-Propanediol diacetate		$C_7H_{12}O_4$	628-66-0	160.168			209.5	1.070¹⁴	1.4192	vs H₂O; s EtOH
9141	1,2-Propanediol 1-methacrylate	2-Hydroxypropyl methacrylate	$C_7H_{12}O_3$	923-26-2	144.168			90⁹	1.066²⁵	1.4458²⁰	
9142	1,2-Propanedione	Pyruvaldehyde	$C_3H_4O_2$	78-98-8	72.063	ye hyg liq		72(15)	1.0455²⁰	1.4002¹⁸	s EtOH, eth, bz
9143	Propanedioyl dichloride		$C_3H_2Cl_2O_2$	1663-67-8	140.953			57²⁸	1.4509²⁰	1.4639²⁰	s eth, AcOEt
9144	1,2-Propanedithiol		$C_3H_8S_2$	814-67-5	108.226			164(10)	1.08²⁰	1.532²⁰	s chl
9145	1,3-Propanedithiol	Trimethylene dimercaptan	$C_3H_8S_2$	109-80-8	108.226	liq	-79	172.9	1.0772²⁰	1.5392²⁰	sl H₂O, ctc; msc EtOH, eth, bz
9146	2,2'-[1,3-Propanediylbis(nitrilomethylidyne)]bisphenol	Disalicylidene-1,3-propanediamine	$C_{17}H_{18}N_2O_2$	120-70-7	282.337		54.3				
9147	Propanenitrile	Ethyl cyanide	C_3H_5N	107-12-0	55.079	liq	-93(1)	97.3(0.4)	0.7818²⁰	1.3655²⁰	vs H₂O; s EtOH, eth, ace, bz, ctc
9148	1-Propanesulfonic acid		$C_3H_8O_3S$	5284-66-2	124.159		8	136¹	1.2516²⁵		
9149	1-Propanesulfonyl chloride		$C_3H_7ClO_2S$	10147-36-1	142.605			204(9)	1.267²⁰	1.452²⁰	
9150	1,3-Propane sultone	1,2-Oxathiolane, 2,2-dioxide	$C_3H_6O_3S$	1120-71-4	122.143						s chl
9151	1-Propanethiol	Propyl mercaptan	C_3H_8S	107-03-9	76.161	liq	-113.12(0.07)	67.7(0.1)	0.8411²⁰	1.4380²⁰	sl H₂O; s EtOH, eth, ace, bz
9152	2-Propanethiol	Isopropyl mercaptan	C_3H_8S	75-33-2	76.161	liq	-130.50(0.05)	52.6(0.3)	0.8143²⁰	1.4255²⁰	sl H₂O; msc EtOH, eth; vs ace; s chl
9153	1,2,3-Propanetriamine	1,2,3-Triaminopropane	$C_3H_{11}N_3$	21291-99-6	89.139	visc oil		190			s H₂O
9154	1,2,3-Propanetricarboxylic acid	Tricarballylic acid	$C_6H_8O_6$	99-14-9	176.124	orth (eth)	166				vs H₂O, EtOH; sl eth
9155	1,2,3-Propanetriol-1-acetate	Glycerol 1-acetate	$C_5H_{10}O_4$	106-61-6	134.131			158¹⁶⁵	1.2060²⁰	1.4157²⁰	vs H₂O, EtOH

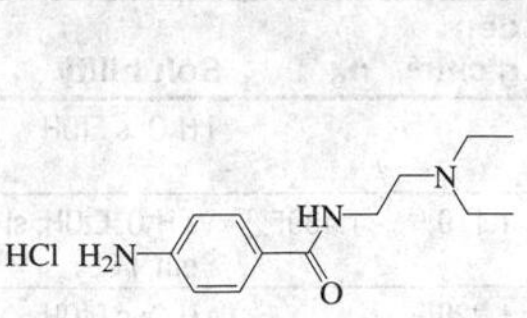

9115 Procainamide hydrochloride

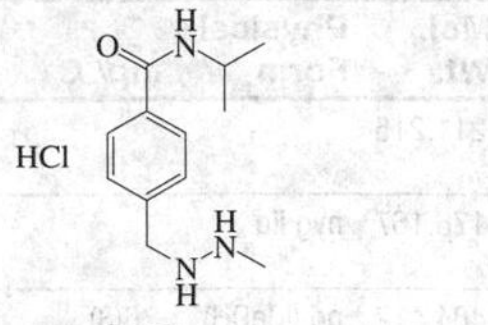

9116 Procarbazine hydrochloride

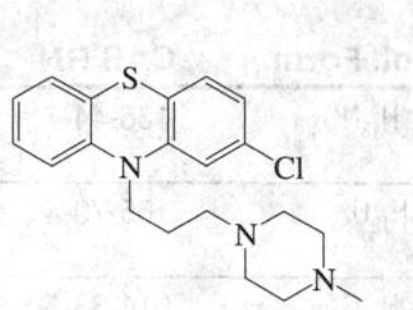

9117 Prochlorperazine

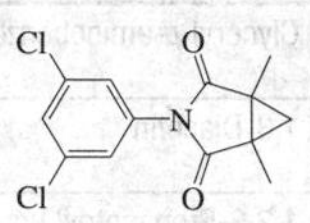

9118 Procymidone

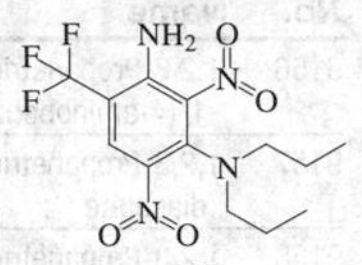

9119 Prodiamine

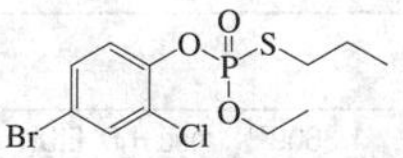

9120 Profenofos

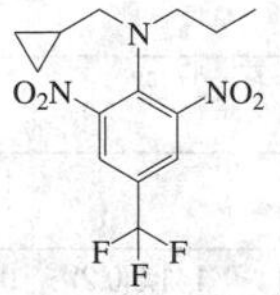

9121 Profluralin

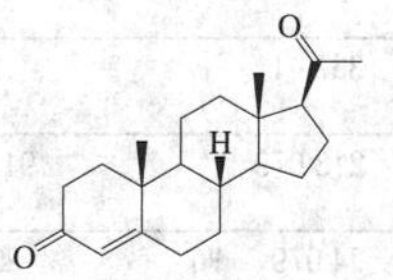

9122 Progesterone

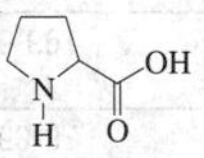

9123 DL-Proline

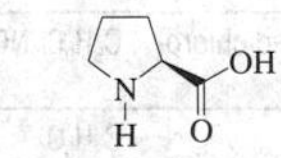

9124 L-Proline

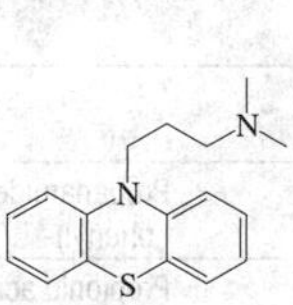

9125 Promazine

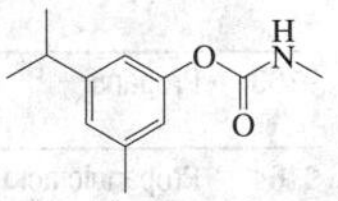

9126 Promecarb

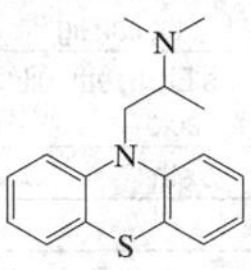

9127 Promethazine

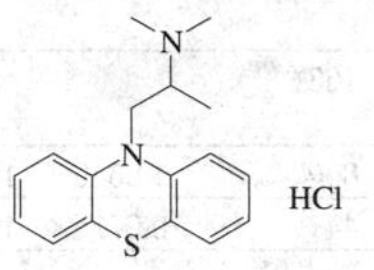

9128 Promethazine hydrochloride

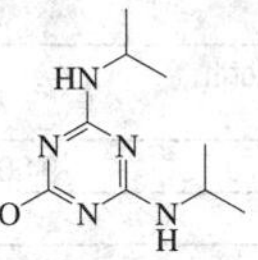

9129 Prometone

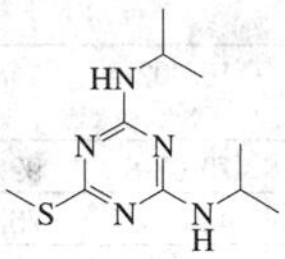

9130 Prometryn

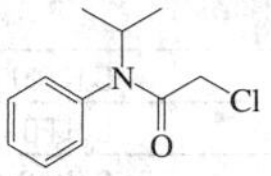

9131 Propachlor

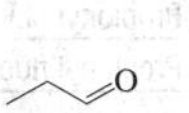

9132 Propanal

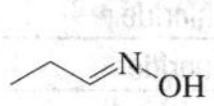

9133 Propanal oxime

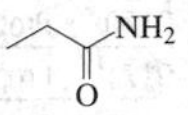

9134 Propanamide

9135 Propane

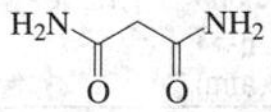

9136 Propanediamide

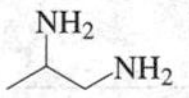

9137 1,2-Propanediamine, (±)-

9138 1,3-Propanediamine

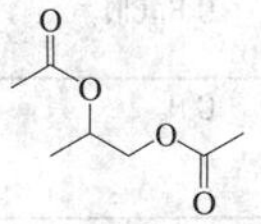

9139 1,2-Propanediol diacetate

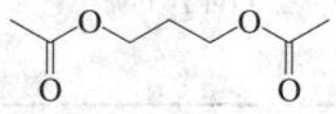

9140 1,3-Propanediol diacetate

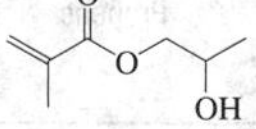

9141 1,2-Propanediol 1-methacrylate

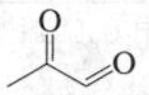

9142 1,2-Propanedione

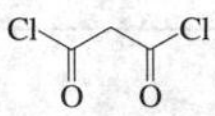

9143 Propanedioyl dichloride

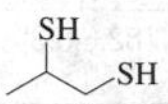

9144 1,2-Propanedithiol

9145 1,3-Propanedithiol

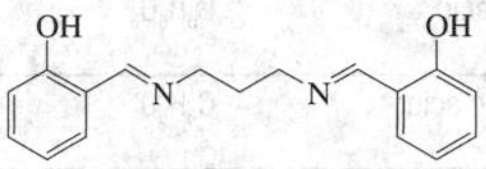

9146 2,2'-[1,3-Propanediylbis(nitrilomethylidyne)]bisphenol

9147 Propanenitrile

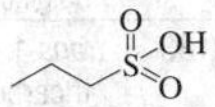

9148 1-Propanesulfonic acid

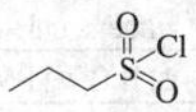

9149 1-Propanesulfonyl chloride

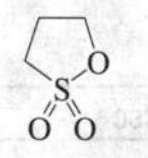

9150 1,3-Propane sultone

9151 1-Propanethiol

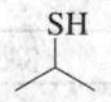

9152 2-Propanethiol

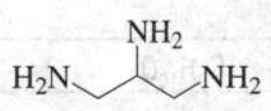

9153 1,2,3-Propanetriamine

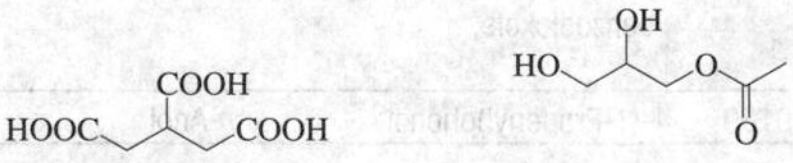

9154 1,2,3-Propanetricarboxylic acid

9155 1,2,3-Propanetriol-1-acetate

No.	Name	Synonym	Mol. Form.	CAS RN	Mol. Wt.	Physical Form	mp/°C	bp/°C	den g cm^{-3}	n_D	Solubility
9156	1,2,3-Propanetriol 1-(4-aminobenzoate)	Glyceryl p-aminobenzoate	$C_{10}H_{13}NO_4$	136-44-7	211.215						i H_2O; s EtOH
9157	1,2,3-Propanetriol-1,3-diacetate	1,3-Diacetin	$C_7H_{12}O_5$	105-70-4	176.167	hyg liq		260(7)	1.179[15]	1.4395[20]	vs H_2O, EtOH; sl eth; i CS_2
9158	1,2,3-Propanetriol tribenzoate	1,2,3-Propanetriyl benzoate	$C_{24}H_{20}O_6$	614-33-5	404.412	nd (MeOH)	76(3)		1.228[12]		i H_2O; s EtOH; vs eth, ace, bz, chl
9159	1,2,3-Propanetriol tripropanoate	1,2,3-Propanetriyl propanoate	$C_{12}H_{20}O_6$	139-45-7	260.283			175[20]	1.108[15]	1.4318[19]	i H_2O; s EtOH, chl; vs eth
9160	1,2,3-Propanetriyl hexanoate	Glycerol trihexanoate	$C_{21}H_{38}O_6$	621-70-5	386.523		-60	373.1(0.9)	0.9867[20]	1.4427[20]	i H_2O; msc EtOH, eth, bz; vs ace
9161	1,2,3-Propanetriyl octanoate	Glycerol trioctanoate	$C_{27}H_{50}O_6$	538-23-8	470.682		9.6(0.6)	447(23)	0.9540[20]	1.4482[20]	i H_2O; msc EtOH; vs eth, bz, chl, lig
9162	Propanidid		$C_{18}H_{27}NO_5$	1421-14-3	337.411			211[0.7]			i H_2O; s EtOH, chl
9163	Propanil	Propanamide, N-(3,4-dichlorophenyl)-	$C_9H_9Cl_2NO$	709-98-8	218.079		91.3(0.3)		1.25[25]		
9164	Propanoic acid	Propionic acid	$C_3H_6O_2$	79-09-4	74.079	liq	-20.5(0.5)	141.5(0.2)	0.9882[25]	1.3809[20]	msc H_2O, EtOH; s eth; sl chl
9165	Propanoic anhydride	Propionic anhydride	$C_6H_{10}O_3$	123-62-6	130.141	liq	-45.0(0.5)	168(1)	1.0110[20]	1.4038[20]	msc eth; sl ctc
9166	1-Propanol	Propyl alcohol	C_3H_8O	71-23-8	60.095	liq	-124.39(0.02)	97.04(0.09)	0.7997[25]	1.3850[20]	msc H_2O, EtOH, eth; s ace, chl; vs bz
9167	2-Propanol	Isopropyl alcohol	C_3H_8O	67-63-0	60.095	liq	-87.91(0.04)	82.21(0.09)	0.7809[25]	1.3776[20]	msc H_2O, EtOH, eth; s ace, chl; vs bz
9168	2-Propanone oxime	Acetoxime	C_3H_7NO	127-06-0	73.094	pr (al)	61	133(7)	0.9113[62]	1.4156[20]	s H_2O, EtOH, eth, chl, lig
9169	2-Propanone phenylhydrazone	Acetone, phenylhydrazone	$C_9H_{12}N_2$	103-02-6	148.204	orth	42	163[50]			s EtOH, eth, dil acid
9170	Propanoyl chloride	Propionyl chloride	C_3H_5ClO	79-03-8	92.524	liq	-94.0(0.4)	80(4)	1.0646[20]	1.4032[20]	s eth
9171	Propanoyl fluoride	Propionyl fluoride	C_3H_5FO	430-71-7	76.069			44	0.972[15]	1.329[13]	
9172	Propantheline bromide		$C_{23}H_{30}BrNO_3$	50-34-0	448.393	cry	160				vs H_2O, EtOH, chl; i eth, bz
9173	Propargite		$C_{19}H_{26}O_4S$	2312-35-8	350.472				1.10[25]		
9174	Propargyl acetate		$C_5H_6O_2$	627-09-8	98.101			124(5)	0.9982[20]	1.4187[20]	sl H_2O; s EtOH, eth
9175	Propargyl alcohol	3-Hydroxy-1-propyne	C_3H_4O	107-19-7	56.063	liq	-51.8(0.4)	113(3)	0.9478[20]	1.4322[20]	s H_2O, chl; msc EtOH, eth
9176	Propatyl nitrate	2-Ethyl-2-[(nitrooxy)methyl]-1,3-propanediol, dinitrate	$C_6H_{11}N_3O_9$	2921-92-8	269.166	wh pow	52		1.49		i H_2O; s EtOH, ace
9177	Propazine	6-Chloro-N,N-diisopropyl-1,3,5-triazine-2,4-diamine	$C_9H_{16}ClN_5$	139-40-2	229.710		217(1)		1.162[20]		
9178	Propene	Propylene	C_3H_6	115-07-1	42.080	col gas	-185.30(0.02)	-47.6(0.1)	0.505[25] (p>1 atm)	1.3567[-70]	sl H_2O; vs EtOH, HOAc
9179	trans-1-Propene-1,2-dicarboxylic acid	Mesaconic acid	$C_5H_6O_4$	498-24-8	130.100	orth nd or mcl pr (eth)	204.5	sub	1.466[20]		sl H_2O, bz, CS_2; vs EtOH; s eth, tfa
9180	1-Propene-2,3-dicarboxylic acid	Itaconic acid	$C_5H_6O_4$	97-65-4	130.100	rhom (bz)	165.7(0.9)	dec	1.632[25]		s H_2O, EtOH, ace; sl eth, bz, peth
9181	2-Propene-1-thiol		C_3H_6S	870-23-5	74.145			65	0.925[23]	1.4832[20]	i H_2O; msc EtOH, eth; s chl
9182	cis-1-Propene-1,2,3-tricarboxylic acid	cis-Aconitic acid	$C_6H_6O_6$	585-84-2	174.108	nd (w)	125				s H_2O; sl eth
9183	trans-1-Propene-1,2,3-tricarboxylic acid	trans-Aconitic acid	$C_6H_6O_6$	4023-65-8	174.108	lf (w) nd (w, eth)	182.6(0.3)				vs H_2O, EtOH
9184	1-Propen-1-one	Methylketene	C_3H_4O	6004-44-0	56.063	col gas	-80	-23			vs eth
9185	2-Propenoyl chloride	Acrylic acid chloride	C_3H_3ClO	814-68-6	90.508			75.5	1.1136[20]	1.4343[20]	vs chl
9186	cis-1-Propenylbenzene		C_9H_{10}	766-90-5	118.175	liq	-61.7(0.3)	169(4)	0.9088[20]	1.5420[20]	i H_2O; msc EtOH, eth, ace, bz, peth, ctc
9187	trans-1-Propenylbenzene		C_9H_{10}	873-66-5	118.175	liq	-29.6(0.7)	179(2)	0.9023[25]	1.5506[20]	i H_2O; msc EtOH, eth, ace, bz
9188	trans-5-(1-Propenyl)-1,3-benzodioxole		$C_{10}H_{10}O_2$	4043-71-4	162.185		6.8	253	1.1224[20]	1.5782[20]	i H_2O; msc EtOH, eth; vs ace; s chl
9189	4-(1-Propenyl)phenol	p-Anol	$C_9H_{10}O$	539-12-8	134.174	lf	94	250 dec			sl H_2O; vs DMF

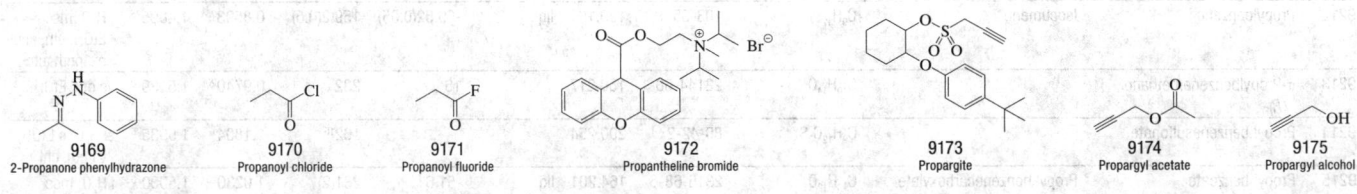

9156
1,2,3-Propanetriol 1-(4-aminobenzoate)

9157
1,2,3-Propanetriol-1,3-diacetate

9158
1,2,3-Propanetriol tribenzoate

9159
1,2,3-Propanetriol tripropanoate

9160
1,2,3-Propanetriyl hexanoate

9161
1,2,3-Propanetriyl octanoate

9162
Propanidid

9163
Propanil

9164
Propanoic acid

9165
Propanoic anhydride

9166
1-Propanol

9167
2-Propanol

9168
2-Propanone oxime

9169
2-Propanone phenylhydrazone

9170
Propanoyl chloride

9171
Propanoyl fluoride

9172
Propantheline bromide

9173
Propargite

9174
Propargyl acetate

9175
Propargyl alcohol

9176
Propatyl nitrate

9177
Propazine

9178
Propene

9179
trans-1-Propene-1,2-dicarboxylic acid

9180
1-Propene-2,3-dicarboxylic acid

9181
2-Propene-1-thiol

9182
cis-1-Propene-1,2,3-tricarboxylic acid

9183
trans-1-Propene-1,2,3-tricarboxylic acid

9184
1-Propen-1-one

9185
2-Propenoyl chloride

9186
cis-1-Propenylbenzene

9187
trans-1-Propenylbenzene

9188
trans-5-(1-Propenyl)-1,3-benzodioxole

9189
4-(1-Propenyl)phenol

No.	Name	Synonym	Mol. Form.	CAS RN	Mol. Wt.	Physical Form	mp/°C	bp/°C	den g cm⁻³	n_D	Solubility
9190	2-(1-Propenyl)piperidine	β-Coniceine	C₈H₁₅N	538-90-9	125.212		8	168	0.8716[15]		
9191	Propetamphos		C₁₀H₂₀NO₄PS	31218-83-4	281.309			88[0.005]	1.1294[20]		
9192	Propiconazole		C₁₅H₁₇Cl₂N₃O₂	60207-90-1	342.221			180[0.1]	1.27[20]		
9193	Propiomazine		C₂₀H₂₄N₂OS	362-29-8	340.482			240[0.5]			
9194	Propionyl-L-carnitine	Carnitine, O-propanoyl	C₁₀H₁₉NO₄	20064-19-1	217.263	hyg pr (2-PrOH)	147 dec				
9195	Propofol		C₁₂H₁₈O	2078-54-8	178.270		18(1)	247(8)	0.955[20]	1.5140[20]	
9196	Propoxur	Phenol, 2-(1-methylethoxy)-, methylcarbamate	C₁₁H₁₅NO₃	114-26-1	209.242		90.5(0.5)	dec	1.12[20]		
9197	2-Propoxyethanol	Ethylene glycol monopropyl ether	C₅H₁₂O₂	2807-30-9	104.148			152(3)	0.9112[20]	1.4133[20]	s H₂O; vs EtOH, eth
9198	D-Propoxyphene	Dextropropoxyphene	C₂₂H₂₉NO₂	469-62-5	339.471	cry (peth)	75.5				
9199	L-Propoxyphene	Levopropoxyphene	C₂₂H₂₉NO₂	2338-37-6	339.471	cry (peth)	75.5				
9200	1-Propoxy-2-propanol	1,2-Propylene glycol 1-propyl ether	C₆H₁₄O₂	1569-01-3	118.174			150.2(0.7)	0.8886[20]	1.4130[20]	
9201	3-Propoxy-1-propene		C₆H₁₂O	1471-03-0	100.158			91	0.7764[20]	1.3919[20]	vs ace, eth, EtOH
9202	Propranolol		C₁₆H₂₁NO₂	525-66-6	259.344	cry (cyhex)	96				
9203	Propyl acetate		C₅H₁₀O₂	109-60-4	102.132	liq	-93(2)	101.0(0.2)	0.8820[25]	1.3828[25]	sl H₂O; msc EtOH, eth; s ctc
9204	Propyl acrylate	2-Propenoic acid, propyl ester	C₆H₁₀O₂	925-60-0	114.142			121(6)			
9205	Propylamine	1-Propanamine	C₃H₉N	107-10-8	59.110	liq	-84.78(0.04)	47.21(0.08)	0.7173[20]	1.3870[20]	msc H₂O; vs EtOH, ace; s bz, chl; sl ctc
9206	Propylamine hydrochloride	1-Propanamine hydrochloride	C₃H₁₀ClN	556-53-6	95.571		163.5				s DMSO
9207	Propyl 4-aminobenzoate	Risocaine	C₁₀H₁₃NO₂	94-12-2	179.216	pr	73.1(0.7)				vs bz, eth, EtOH, chl
9208	2-(Propylamino)ethanol		C₅H₁₃NO	16369-21-4	103.163			183(13)	0.9005[20]	1.4428[20]	
9209	4-Propylaniline		C₉H₁₃N	2696-84-6	135.206			229(14)			
9210	N-Propylaniline		C₉H₁₃N	622-80-0	135.206			220(3)	0.9443[20]	1.5428[20]	vs eth, EtOH
9211	Propylarsonic acid	1-Propanearsonic acid	C₃H₉AsO₃	107-34-6	168.023	nd (al), pl (w)	134.5				vs H₂O, EtOH; i eth
9212	Propylbenzene	Isocumene	C₉H₁₂	103-65-1	120.191	liq	-99.52(0.05)	159.2(0.5)	0.8593[25]	1.4895[25]	i H₂O; msc EtOH, eth, ace, bz, peth, ctc
9213	α-Propylbenzenemethanol, (R)-		C₁₀H₁₄O	22144-60-1	150.217		16	232	0.9740[20]	1.5139[20]	vs eth, EtOH
9214	Propyl benzenesulfonate		C₉H₁₂O₃S	80-42-2	200.254			162[15]	1.1804[17]	1.5035[25]	sl H₂O; s EtOH; vs eth, chl
9215	Propyl benzoate	Propyl benzenecarboxylate	C₁₀H₁₂O₂	2315-68-6	164.201	liq	-51.6	231(2)	1.0230[20]	1.5000[20]	i H₂O; msc EtOH, eth
9216	5-Propyl-1,3-benzodioxole	Dihydrosafrole	C₁₀H₁₂O₂	94-58-6	164.201			228			s ctc
9217	Propyl butanoate		C₇H₁₄O₂	105-66-8	130.185	liq	-95.2(0.5)	144(2)	0.8730[20]	1.4001[20]	sl H₂O; msc EtOH, eth
9218	Propyl carbamate		C₄H₉NO₂	627-12-3	103.120	pr	60	196			vs ace, eth, EtOH
9219	Propyl chloroacetate		C₅H₉ClO₂	5396-24-7	136.577			160(6)	1.104[20]	1.4261[20]	vs eth
9220	Propyl 2-chlorobutanoate		C₇H₁₃ClO₂	62108-71-8	164.630			183	1.0252[20]		
9221	Propyl chlorocarbonate		C₄H₇ClO₂	109-61-5	122.551			111(2)	1.0901[20]	1.4035[20]	msc EtOH, eth
9222	Propyl 3-chloropropanoate		C₆H₁₁ClO₂	62108-66-1	150.603			180	1.0656[20]	1.4290[20]	vs eth, EtOH
9223	S-Propyl chlorothioformate	S-Propyl carbonochloridothioate	C₄H₇ClOS	13889-92-4	138.616	liq		59[26]			
9224	Propyl trans-cinnamate	Propyl trans-3-phenyl-2-propenoate	C₁₂H₁₄O₂	74513-58-9	190.238			285	1.0433[0]		i H₂O
9225	Propylcyclohexane		C₉H₁₈	1678-92-8	126.239	liq	-94.86(0.05)	156.7(0.3)	0.7936[20]	1.4370[20]	i H₂O; msc EtOH, ace, ctc; s eth, bz
9226	2-Propylcyclohexanone		C₉H₁₆O	94-65-5	140.222			197	0.927[20]	1.4538[20]	i H₂O; s EtOH, ace; vs eth, bz
9227	Propylcyclopentane		C₈H₁₆	2040-96-2	112.213	liq	-117.34(0.06)	130.9(0.8)	0.7763[20]	1.4266[20]	i H₂O; msc EtOH, eth, ace; s bz; vs ctc
9228	1-Propylcyclopentanol		C₈H₁₆O	1604-02-0	128.212	liq	-37.5	171(9)	0.9040[25]	1.4502[25]	
9229	Propylene carbonate	4-Methyl-1,3-dioxolan-2-one	C₄H₆O₃	108-32-7	102.089	liq	-48.8	241.6(0.7)	1.2047[20]	1.4189[20]	vs H₂O, EtOH, eth, ace, bz
9230	1,2-Propylene glycol	Propylene glycol	C₃H₈O₂	57-55-6	76.095	liq	-60	187.3(0.2)	1.0361[20]	1.4324[20]	msc H₂O, EtOH; s eth, bz, chl
9231	1,3-Propylene glycol	Trimethylene glycol	C₃H₈O₂	504-63-2	76.095	liq	-27.6(0.1)	214.7(0.3)	1.0538[20]	1.4398[20]	msc H₂O, EtOH; vs eth; sl bz
9232	1,2-Propylene glycol 2-tert-butyl ether	2-(1,1-Dimethylethoxy)-1-propanol	C₇H₁₆O₂	94023-15-1	132.201	liq		152	0.87		

9190
2-(1-Propenyl)piperidine

9191
Propetamphos

9192
Propiconazole

9193
Propiomazine

9194
Propionyl-*L*-carnitine

9195
Propofol

9196
Propoxur

9197
2-Propoxyethanol

9198
D-Propoxyphene

9199
L-Propoxyphene

9200
1-Propoxy-2-propanol

9201
3-Propoxy-1-propene

9202
Propranolol

9203
Propyl acetate

9204
Propyl acrylate

9205
Propylamine

9206
Propylamine hydrochloride

9207
Propyl 4-aminobenzoate

9208
2-(Propylamino)ethanol

9209
4-Propylaniline

9210
N-Propylaniline

9211
Propylarsonic acid

9212
Propylbenzene

9213
α-Propylbenzenemethanol, (*R*)-

9214
Propyl benzenesulfonate

9215
Propyl benzoate

9216
5-Propyl-1,3-benzodioxole

9217
Propyl butanoate

9218
Propyl carbamate

9219
Propyl chloroacetate

9220
Propyl 2-chlorobutanoate

9221
Propyl chlorocarbonate

9222
Propyl 3-chloropropanoate

9223
S-Propyl chlorothioformate

9224
Propyl *trans*-cinnamate

9225
Propylcyclohexane

9226
2-Propylcyclohexanone

9227
Propylcyclopentane

9228
1-Propylcyclopentanol

9229
Propylene carbonate

9230
1,2-Propylene glycol

9231
1,3-Propylene glycol

9232
1,2-Propylene glycol 2-*tert*-butyl ether

No.	Name	Synonym	Mol. Form.	CAS RN	Mol. Wt.	Physical Form	mp/°C	bp/°C	den g cm⁻³	n_D	Solubility
9233	1,2-Propylene glycol dinitrate		$C_3H_6N_2O_6$	6423-43-4	166.089	liq	exp	92[10]			
9234	1,2-Propylene glycol monomethyl ether	1-Methoxy-2-propanol	$C_4H_{10}O_2$	107-98-2	90.121			120.0(0.6)	0.9620[20]	1.4034[20]	
9235	1,2-Propylene glycol monomethyl ether acetate	2-Acetoxy-1-methoxypropane	$C_6H_{12}O_3$	108-65-6	132.157	liq		146.0(0.4)			
9236	Propyleneimine	2-Methylaziridine	C_3H_7N	75-55-8	57.095			67	0.812[16]		
9237	Propyl formate		$C_4H_8O_2$	110-74-7	88.106	liq	-92.9(0.4)	80.6(0.2)	0.9073[20]	1.377[20]	sl H₂O, ctc; msc EtOH, eth
9238	Propyl 3-(2-furyl)acrylate		$C_{10}H_{12}O_3$	623-22-3	180.200			113[16]	1.0744[20]	1.5392[24]	vs bz, eth, EtOH
9239	4-Propylheptane		$C_{10}H_{22}$	3178-29-8	142.282			167(3)	0.7321[25]	1.4135[20]	
9240	Propyl hexanoate		$C_9H_{18}O_2$	626-77-7	158.238	liq	-74.0(0.5)	185(2)	0.8672[20]	1.4170[20]	vs eth, EtOH
9241	Propyl 2-hydroxybenzoate		$C_{10}H_{12}O_3$	607-90-9	180.200		97	239	1.0979[20]	1.5161[20]	s ctc, CS₂
9242	Propyl 4-hydroxybenzoate	Propylparaben	$C_{10}H_{12}O_3$	94-13-3	180.200	pr (eth)	96.1(0.5)		1.0630[102]	1.5050[102]	i H₂O; s EtOH, eth; sl chl
9243	Propyliodone		$C_{10}H_{11}I_2NO_3$	587-61-1	447.008		186				
9244	Propyl isobutanoate		$C_7H_{14}O_2$	644-49-5	130.185			134(4)	0.884[20]	1.3955[20]	sl H₂O; s EtOH, ace; vs eth
9245	Propyl isocyanate	1-Isocyanatopropane	C_4H_7NO	110-78-1	85.105			85.4(0.4)	0.908[25]	1.3970[20]	
9246	Propyl isothiocyanate	1-Isothiocyanatopropane	C_4H_7NS	628-30-8	101.171			153	0.9781[16]	1.5085[16]	sl H₂O; msc EtOH, eth
9247	Propyl methacrylate		$C_7H_{12}O_2$	2210-28-8	128.169			140(19)	0.9022[20]	1.4190[20]	i H₂O; msc EtOH, eth
9248	Propyl 3-methylbutanoate	Propyl isopentanoate	$C_8H_{16}O_2$	557-00-6	144.212			155(3)	0.8617[20]	1.4031[20]	vs eth, EtOH
9249	1-Propylnaphthalene		$C_{13}H_{14}$	2765-18-6	170.250	liq	-8.6(0.3)	273(5)	0.9897[20]	1.5923[20]	
9250	Propyl nitrate		$C_3H_7NO_3$	627-13-4	105.093			110(1)	1.0538[20]	1.3973[20]	sl H₂O; s EtOH, eth, ctc
9251	Propyl nitrite		$C_3H_7NO_2$	543-67-9	89.094	liq		44(2)	0.886[20]	1.3604[20]	sl H₂O; s EtOH, eth
9252	Propyl octanoate		$C_{11}H_{22}O_2$	624-13-5	186.292	liq	-46(1)	222(3)	0.8659[20]	1.4191[25]	vs ace, eth, EtOH
9253	Propyl pentanoate		$C_8H_{16}O_2$	141-06-0	144.212	liq	-70.7	167(4)	0.8699[20]	1.4065[20]	i H₂O; s EtOH, eth, chl
9254	2-Propylpentanoic acid	Valproic acid	$C_8H_{16}O_2$	99-66-1	144.212	col liq		223(9)	0.904[25]	1.425[25]	sl H₂O
9255	2-Propylphenol		$C_9H_{12}O$	644-35-9	136.190		7(1)	221(4)	1.015[20]		vs eth, EtOH
9256	4-Propylphenol		$C_9H_{12}O$	645-56-7	136.190		22(2)	232(4)	1.009[20]	1.5379[25]	sl H₂O, ctc; s EtOH
9257	2-Propylpiperidine, (S)-	Coniine	$C_8H_{17}N$	458-88-8	127.228	liq	-1.0	166.5	0.8440[20]	1.4512[22]	sl H₂O, chl; msc EtOH; vs eth; s bz
9258	trans-6-Propyl-3-piperidinol, (3S)-	Pseudoconhydrine	$C_8H_{17}NO$	140-55-6	143.227	hyg nd (eth)	106	236			vs H₂O, eth, EtOH
9259	N-Propylpropanamide		$C_6H_{13}NO$	3217-86-5	115.173		154	215	0.8985[25]		sl H₂O, eth
9260	Propyl propanoate	Propyl propionate	$C_6H_{12}O_2$	106-36-5	116.158	liq	-75.9(0.4)	122.2(0.1)	0.8755[25]	1.3909[25]	sl H₂O, ctc; msc EtOH, eth; s ace
9261	2-Propylpyridine		$C_8H_{11}N$	622-39-9	121.180		1.0	170(2)	0.9119[20]	1.4925[20]	sl H₂O; msc EtOH, eth; vs ace
9262	4-Propylpyridine		$C_8H_{11}N$	1122-81-2	121.180			189(5)	0.9381[15]	1.4966[20]	vs eth, EtOH
9263	2-Propyl-4-pyridinecarbothio-amide	Protionamide	$C_9H_{12}N_2S$	14222-60-7	180.269		136.7				
9264	Propyl Red	Benzoic acid, 2-[[4-(dipropyl-amino)phenyl]azo]-	$C_{19}H_{23}N_3O_2$	2641-01-2	325.405	viol-bl or purp red cry (al)					s EtOH, KOH
9265	(Propylthio)benzene		$C_9H_{12}S$	874-79-3	152.256	liq	-45	219(7)	0.9995[20]	1.5571[20]	
9266	Propyl 4-toluenesulfonate		$C_{10}H_{14}O_3S$	599-91-7	214.281		<-20	189[9]	1.144[20]	1.4998[20]	
9267	Propyl trichloroacetate		$C_5H_7Cl_3O_2$	13313-91-2	205.468			187	1.3221[20]	1.4501[20]	vs eth, EtOH
9268	Propyl 3,4,5-trihydroxybenzo-ate	Propyl gallate	$C_{10}H_{12}O_5$	121-79-9	212.199	nd (w)	130				sl H₂O
9269	Propylurea		$C_4H_{10}N_2O$	627-06-5	102.134	pr (al)	107.8(0.4)				sl H₂O, DMSO; s EtOH
9270	Propyl vinyl ether	1-(Ethenyloxy)propane	$C_5H_{10}O$	764-47-6	86.132			64.7(0.4)	0.7674[20]	1.3908[20]	
9271	2-Propynal	Propargyl aldehyde	C_3H_2O	624-67-9	54.047			60	0.9152[20]	1.4033[25]	msc H₂O; s EtOH, eth, ace, bz, tol
9272	2-Propyn-1-amine		C_3H_5N	2450-71-7	55.079			84(3)	0.803[25]	1.4480[20]	
9273	Propyne	Methylacetylene	C_3H_4	74-99-7	40.064	col gas	-103.0(0.5)	-23.2	0.607[25] (p>1 atm)	1.3863[-40]	sl H₂O; vs EtOH; s bz, chl
9274	2-Propynoic acid	Propiolic acid	$C_3H_2O_2$	471-25-0	70.047	cry (CS₂)	9	144 dec	1.1380[20]	1.4306[20]	vs H₂O, eth, EtOH, chl
9275	1-Propynylbenzene		C_9H_8	673-32-5	116.160			186(10)	0.942[15]	1.563[15]	

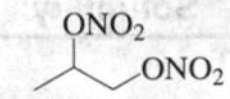

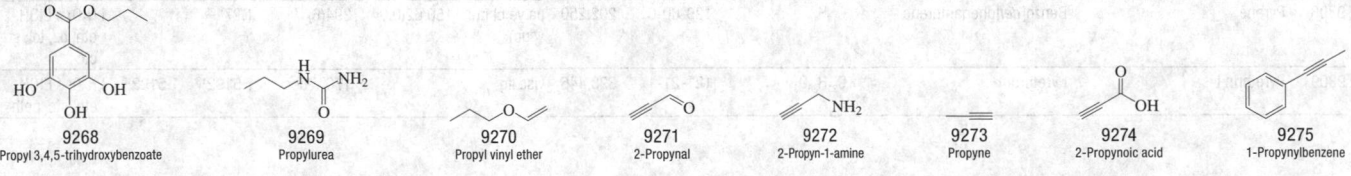

9233
1,2-Propylene glycol dinitrate

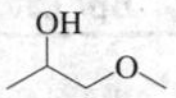

9234
1,2-Propylene glycol monomethyl ether

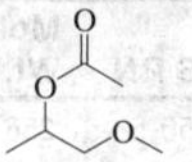

9235
1,2-Propylene glycol monomethyl ether acetate

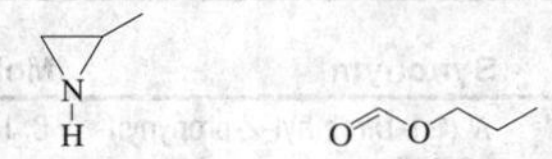

9236
Propyleneimine

9237
Propyl formate

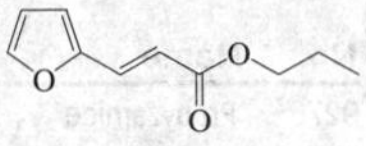

9238
Propyl 3-(2-furyl)acrylate

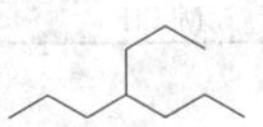

9239
4-Propylheptane

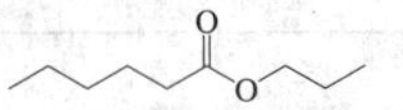

9240
Propyl hexanoate

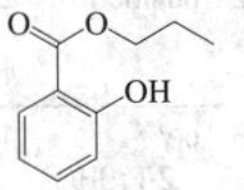

9241
Propyl 2-hydroxybenzoate

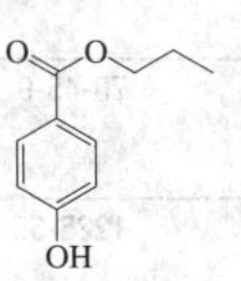

9242
Propyl 4-hydroxybenzoate

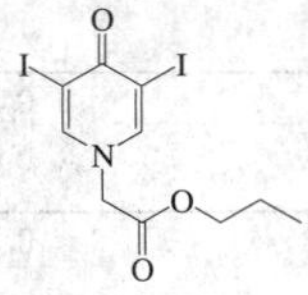

9243
Propyliodone

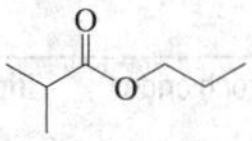

9244
Propyl isobutanoate

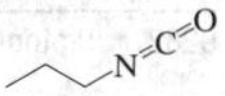

9245
Propyl isocyanate

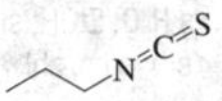

9246
Propyl isothiocyanate

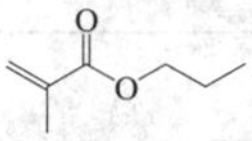

9247
Propyl methacrylate

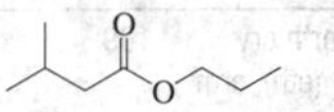

9248
Propyl 3-methylbutanoate

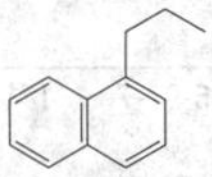

9249
1-Propylnaphthalene

9250
Propyl nitrate

9251
Propyl nitrite

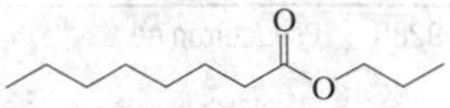

9252
Propyl octanoate

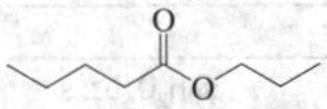

9253
Propyl pentanoate

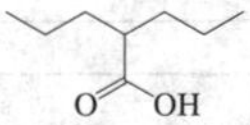

9254
2-Propylpentanoic acid

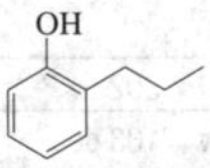

9255
2-Propylphenol

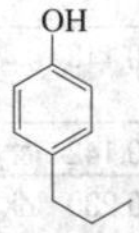

9256
4-Propylphenol

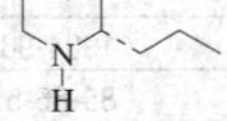

9257
2-Propylpiperidine, (S)-

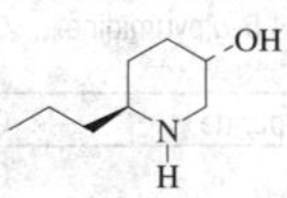

9258
trans-6-Propyl-3-piperidinol, (3S)-

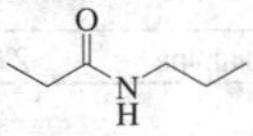

9259
N-Propylpropanamide

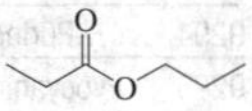

9260
Propyl propanoate

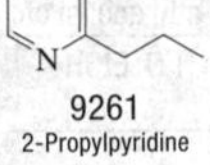

9261
2-Propylpyridine

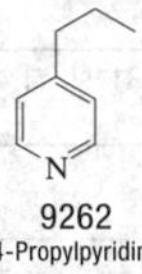

9262
4-Propylpyridine

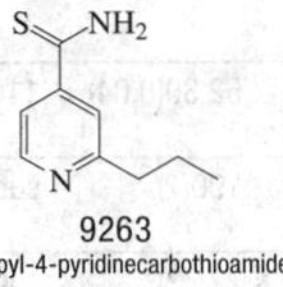

9263
2-Propyl-4-pyridinecarbothioamide

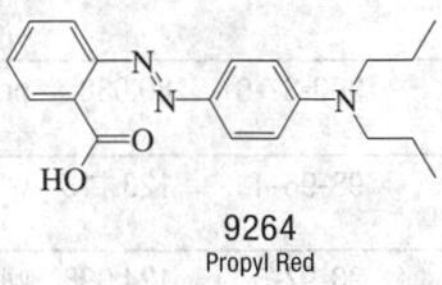

9264
Propyl Red

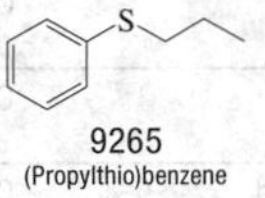

9265
(Propylthio)benzene

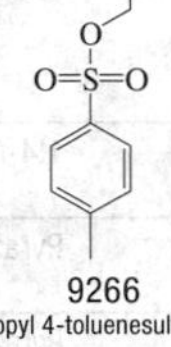

9266
Propyl 4-toluenesulfonate

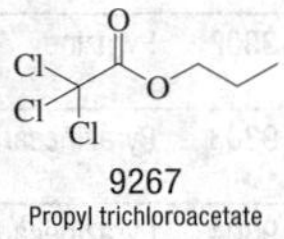

9267
Propyl trichloroacetate

9268
Propyl 3,4,5-trihydroxybenzoate

9269
Propylurea

9270
Propyl vinyl ether

9271
2-Propynal

9272
2-Propyn-1-amine

9273
Propyne

9274
2-Propynoic acid

9275
1-Propynylbenzene

No.	Name	Synonym	Mol. Form.	CAS RN	Mol. Wt.	Physical Form	mp/°C	bp/°C	den g cm^{-3}	n_D	Solubility
9276	Propyzamide	N-(1,1-Dimethyl-2-propynyl)-3,5-dichlorobenzamide	C$_{12}$H$_{11}$Cl$_2$NO	23950-58-5	256.127			156.5(0.5)			
9277	Prostaglandin E$_1$	11,15-Dihydroxy-9-oxo-13-prostenoic acid	C$_{20}$H$_{34}$O$_5$	745-65-3	354.481	cry (EtOAc)	115				s H$_2$O
9278	Prostaglandin E$_2$	11,15-Dihydroxy-9-oxo-5,13-prostadienoic acid	C$_{20}$H$_{32}$O$_5$	363-24-6	352.465	col cry	67				s H$_2$O, thf
9279	Prostaglandin F$_{2a}$	9,11,15-Trihydroxyprosta-5,13-dienoic acid	C$_{20}$H$_{34}$O$_5$	551-11-1	354.481	oil or solid	≈30				sl H$_2$O; s EtOH, MeOH, chl, AcOEt
9280	Protopine	Fumarine	C$_{20}$H$_{19}$NO$_5$	130-86-9	353.369	mcl pr (al-chl)	208				i H$_2$O; sl EtOH, eth, bz, peth; s chl
9281	Protoverine		C$_{27}$H$_{43}$NO$_9$	76-45-9	525.632	nd (MeOH)	221				i H$_2$O; s EtOH, bz, aq acid, MeOH
9282	Protriptyline hydrochloride	Triptil	C$_{19}$H$_{22}$ClN	1225-55-4	299.838	cry (2-PrOH/eth)	170				
9283	Prunetin		C$_{16}$H$_{12}$O$_5$	552-59-0	284.263		239.5				
9284	Pseudoaconitine		C$_{36}$H$_{51}$NO$_{12}$	127-29-7	689.790	tcl (MeOH)	214				vs eth, EtOH
9285	Pseudocodeine		C$_{18}$H$_{21}$NO$_3$	466-96-6	299.365	wh nd	181.5		1.290^{80}	1.574	
9286	Pseudojervine		C$_{33}$H$_{49}$NO$_8$	36069-05-3	587.744	wh nd or hex cry	304 dec				i H$_2$O, eth, bz, chl, tol, peth; s EtOH
9287	Pseudomorphine		C$_{34}$H$_{36}$N$_2$O$_6$	125-24-6	568.659	cry (aq NH$_3$, + 3 w)	282.5				i H$_2$O, EtOH, eth, chl, sulf; s py, NH$_3$
9288	Pseudotropine	8-Methyl-8-azabicyclo[3.2.1]octan-3-ol, exo	C$_8$H$_{15}$NO	135-97-7	141.211	orth cry (eth), orth bipym (peth-bz)	109	241			vs H$_2$O, EtOH; sl eth; s bz, chl
9289	Psoralen		C$_{11}$H$_6$O$_3$	66-97-7	186.164	nd (w, EtOH)	171				
9290	Pteridine	Pyrazino[2,3-d]pyrimidine	C$_6$H$_4$N$_4$	91-18-9	132.123	ye pl (bz, sub)	139.5	125 sub			vs H$_2$O; s EtOH; sl eth, bz
9291	2,4(1H,3H)-Pteridinedione	Lumazine	C$_6$H$_4$N$_4$O$_2$	487-21-8	164.122	ye-oran nd (w)	348.5				vs HOAc
9292	Pulegone		C$_{10}$H$_{16}$O	89-82-7	152.233			224(7)	0.9346^{45}	1.4894^{20}	i H$_2$O; msc EtOH, eth, chl; s ctc
9293	1H-Purine	6H-Imidazo[4,5-d]pyrimidine	C$_5$H$_4$N$_4$	120-73-0	120.113		216.5				vs H$_2$O, EtOH; sl eth, chl; s ace
9294	1H-Purine-2,6-diamine	2,6-Diaminopurine	C$_5$H$_6$N$_6$	1904-98-9	150.142	cry (dil al)	302				
9295	Pyocyanine		C$_{13}$H$_{10}$N$_2$O	85-66-5	210.230	dk bl nd (w + 1) (chl-peth)	133 dec				sl H$_2$O, bz; s EtOH, ace; i eth; vs chl
9296	4H-Pyran	1,4-Pyran	C$_5$H$_6$O	289-65-6	82.101	unstab oil		80		1.4559^{20}	s EtOH, eth, bz
9297	2H-Pyran-2-one		C$_5$H$_4$O$_2$	504-31-4	96.085		8.5	207.5	1.200^{20}	1.5270^{25}	msc H$_2$O; vs ace
9298	4H-Pyran-4-one		C$_5$H$_4$O$_2$	108-97-4	96.085		32.5	212.5	1.190^{25}	1.5238	vs H$_2$O, chl, eth; s EtOH, bz; sl CS$_2$
9299	Pyrantel		C$_{11}$H$_{14}$N$_2$S	15686-83-6	206.307	cry (MeOH)	178				
9300	4H-Pyran-4-thione		C$_5$H$_4$OS	1120-93-0	112.150		49				s H$_2$O
9301	8,16-Pyranthrenedione		C$_{30}$H$_{14}$O$_2$	128-70-1	406.431	red-ye or red-br nd (PhNO$_2$)	dec	sub			
9302	Pyrazine	1,4-Diazine	C$_4$H$_4$N$_2$	290-37-9	80.088	pr (w)	52.30(0.04)	116.3(0.1)	1.0311^{61}	1.4953^{61}	s H$_2$O, EtOH, eth, ace; sl ctc
9303	Pyrazinecarboxamide	Pyrazinamide	C$_5$H$_5$N$_3$O	98-96-4	123.113	wh nd (w, al)	190(1)	sub			s H$_2$O, EtOH
9304	Pyrazinecarboxylic acid	Pyrazinoic acid	C$_5$H$_4$N$_2$O$_2$	98-97-5	124.098	wh nd (w)	225(2)	sub			
9305	2,3-Pyrazinedicarboxylic acid	2,3-Dicarboxypyrazine	C$_6$H$_4$N$_2$O$_4$	89-01-0	168.107	pr (w+2)	193 dec				vs H$_2$O; sl EtOH, eth, bz; s ace, MeOH
9306	1H-Pyrazole	1,2-Diazole	C$_3$H$_4$N$_2$	288-13-1	68.077	nd or pr (lig)	59.9(0.3)	187(5)		1.4203	s H$_2$O, EtOH, eth, bz; sl chl
9307	1-Pyrenamine		C$_{16}$H$_{11}$N	1606-67-3	217.265	ye nd (hx) lf (dil al)	117.5				s EtOH, ace, hx, acid; sl chl
9308	Pyrene	Benzo[def]phenanthrene	C$_{16}$H$_{10}$	129-00-0	202.250	pa ye pl (to, sub)	150.62(0.04)	394(6)	1.271^{23}		i H$_2$O; s EtOH, eth, bz, tol; sl ctc
9309	Pyrethrin I	Pyrethrum	C$_{21}$H$_{28}$O$_3$	121-21-1	328.445	visc liq		170$^{0.1}$ dec	1.5192^{18}	1.5192^{18}	i H$_2$O; s EtOH, eth, ctc, peth

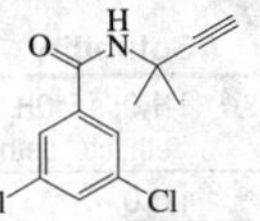

9276 Propyzamide

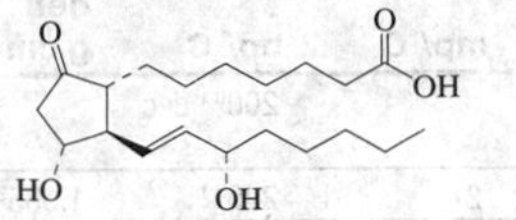

9277 Prostaglandin E₁

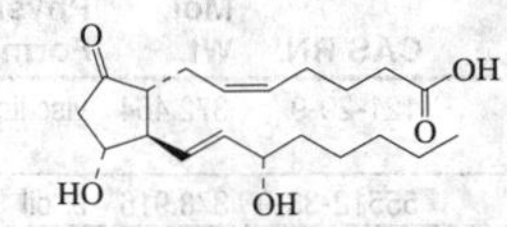

9278 Prostaglandin E₂

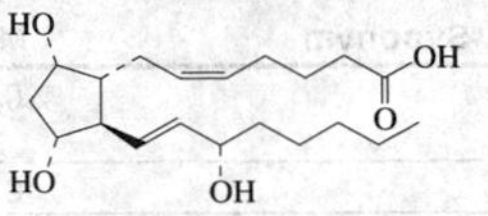

9279 Prostaglandin F₂ₐ

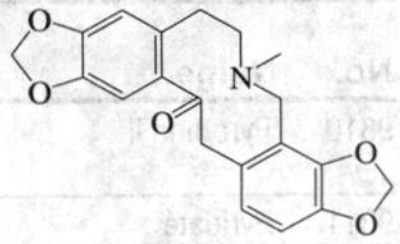

9280 Protopine

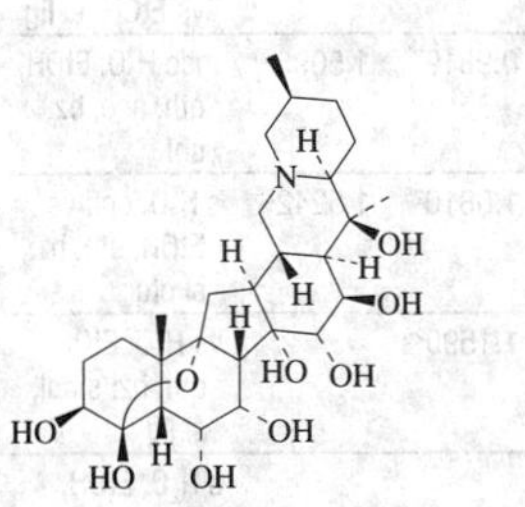

9281 Protoverine

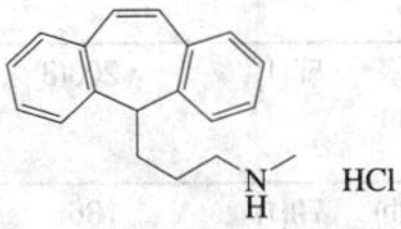

9282 Protriptyline hydrochloride

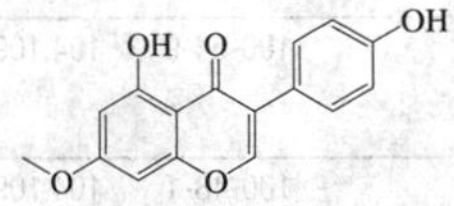

9283 Prunetin

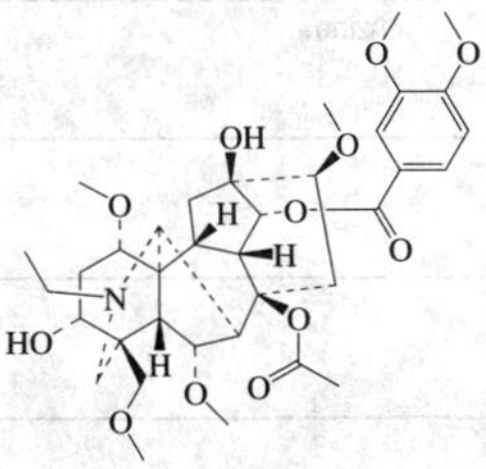

9284 Pseudoaconitine

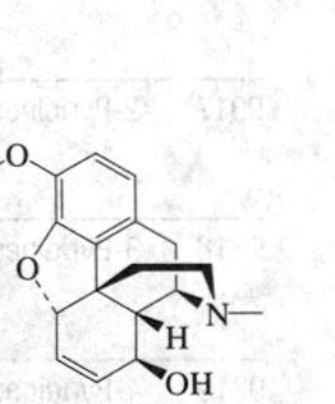

9285 Pseudocodeine

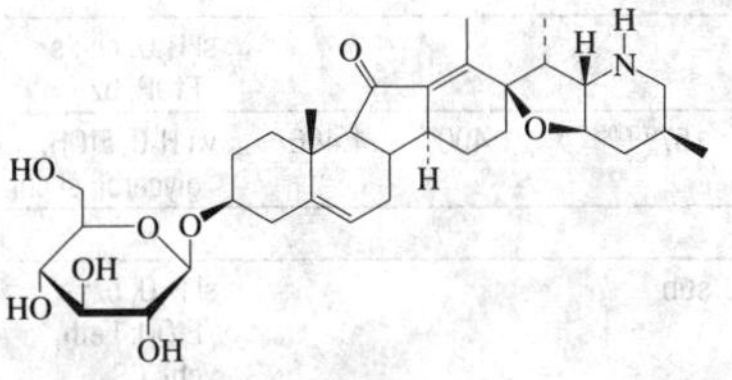

9286 Pseudojervine

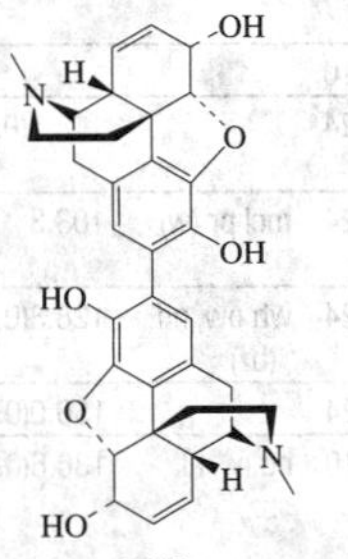

9287 Pseudomorphine

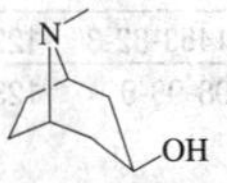

9288 Pseudotropine

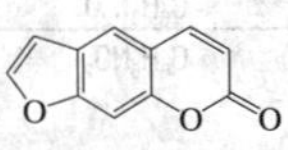

9289 Psoralen

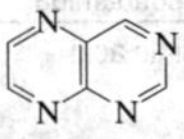

9290 Pteridine

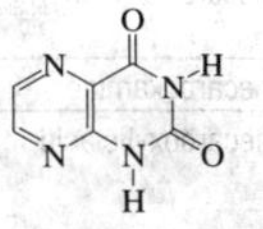

9291 2,4(1H,3H)-Pteridinedione

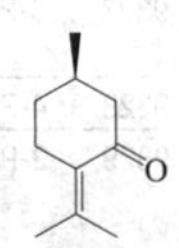

9292 Pulegone

9293 1H-Purine

9294 1H-Purine-2,6-diamine

9295 Pyocyanine

9296 4H-Pyran

9297 2H-Pyran-2-one

9298 4H-Pyran-4-one

9299 Pyrantel

9300 4H-Pyran-4-thione

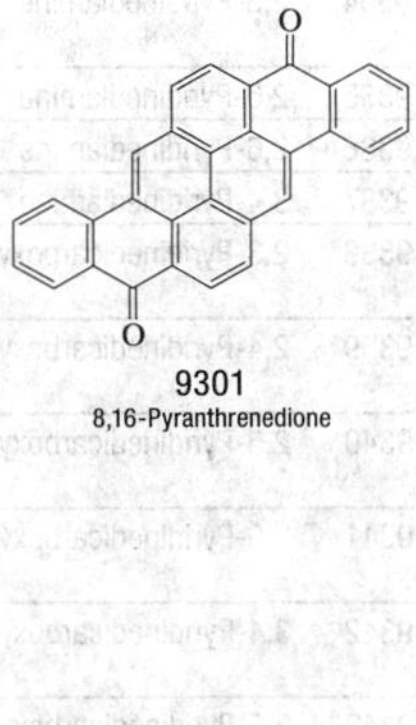

9301 8,16-Pyranthrenedione

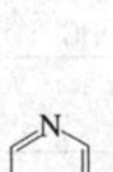

9302 Pyrazine

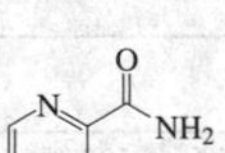

9303 Pyrazinecarboxamide

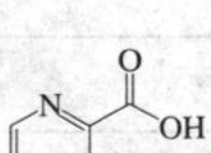

9304 Pyrazinecarboxylic acid

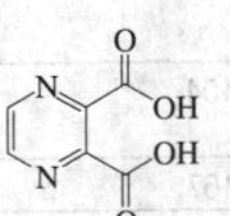

9305 2,3-Pyrazinedicarboxylic acid

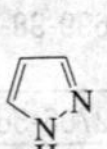

9306 1H-Pyrazole

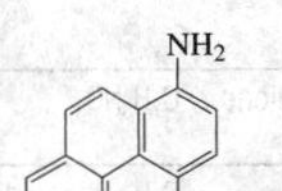

9307 1-Pyrenamine

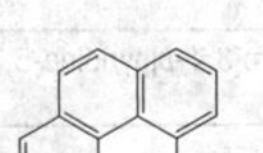

9308 Pyrene

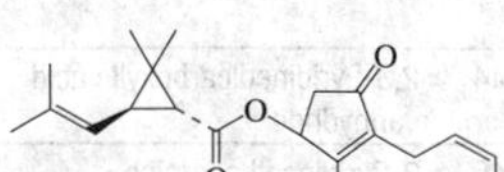

9309 Pyrethrin I

No.	Name	Synonym	Mol. Form.	CAS RN	Mol. Wt.	Physical Form	mp/°C	bp/°C	den g cm⁻³	n_D	Solubility
9310	Pyrethrin II		$C_{22}H_{28}O_5$	121-29-9	372.454	visc liq		200[0.1] dec		1.5258[20]	i H_2O; s EtOH, eth, ctc, peth
9311	Pyridate		$C_{19}H_{23}ClN_2O_2S$	55512-33-9	378.916	br oil	27	220[0.1]	1.555[20]	1.568[20]	i H_2O
9312	Pyridazine	1,2-Diazabenzene	$C_4H_4N_2$	289-80-5	80.088	liq	-8	208	1.1035[23]	1.5218[20]	msc H_2O, EtOH; vs eth, ace, bz; i peth
9313	2-Pyridinamine	2-Aminopyridine	$C_5H_6N_2$	504-29-0	94.115	lf (lig)	58.2(0.9)	105[20]			s EtOH, eth, ace, bz; sl chl
9314	3-Pyridinamine	3-Aminopyridine	$C_5H_6N_2$	462-08-8	94.115	lf (bz-lig)	64.5(0.3)	252			s H_2O, EtOH, eth; sl lig
9315	4-Pyridinamine	4-Aminopyridine	$C_5H_6N_2$	504-24-5	94.115	nd (bz)	159.0(0.5)	273			s H_2O, eth, bz; vs EtOH; sl lig
9316	Pyridine	Azine	C_5H_5N	110-86-1	79.101	liq	-41.63(0.03)	115.2(0.1)	0.9819[20]	1.5095[20]	msc H_2O, EtOH, eth, ace, bz, chl
9317	2-Pyridinecarbonitrile		$C_6H_4N_2$	100-70-9	104.109	nd or pr (eth)	29(2)	224.5	1.0810[25]	1.5242[25]	s H_2O, chl; vs EtOH, eth, bz; sl ctc
9318	3-Pyridinecarbonitrile		$C_6H_4N_2$	100-54-9	104.109	nd (lig), peth-eth	50(1)	203(3)	1.1590[25]		vs H_2O, EtOH, eth, bz; s chl; sl lig
9319	4-Pyridinecarbonitrile		$C_6H_4N_2$	100-48-1	104.109	nd(lig-eth)	79(1)	186			s H_2O, EtOH, eth, bz, chl; sl lig
9320	3-Pyridinecarbothioamide		$C_6H_6N_2S$	4621-66-3	138.190		192				
9321	4-Pyridinecarbothioamide		$C_6H_6N_2S$	2196-13-6	138.190		198 dec				
9322	2-Pyridinecarboxaldehyde		C_6H_5NO	1121-60-4	107.110			181(10)	1.1181[25]	1.5389[18]	s H_2O, EtOH, eth, AcOEt; sl ctc
9323	3-Pyridinecarboxaldehyde	Nicotinaldehyde	C_6H_5NO	500-22-1	107.110			92[23]	1.1394[25]		s H_2O, EtOH, ace, chl; sl eth, peth
9324	4-Pyridinecarboxaldehyde		C_6H_5NO	872-85-5	107.110			77[12]		1.5423[20]	s H_2O, eth, ctc
9325	2-Pyridinecarboxaldehyde oxime		$C_6H_6N_2O$	873-69-8	122.124		113.1(0.5)				
9326	2-Pyridinecarboxamide		$C_6H_6N_2O$	1452-77-3	122.124	mcl pr (w)	108.3				sl H_2O, chl; s EtOH, bz
9327	3-Pyridinecarboxamide	Niacinamide	$C_6H_6N_2O$	98-92-0	122.124	wh pw, nd (bz)	128.8(0.5)	157[0.0005]	1.400[25]	1.466	vs H_2O, EtOH, glycerol; sl chl
9328	4-Pyridinecarboxamide	Isonicotinamide	$C_6H_6N_2O$	1453-82-3	122.124		156.0(0.5)				
9329	2-Pyridinecarboxylic acid	Picolinic acid	$C_6H_5NO_2$	98-98-6	123.110	nd (w, al, bz)	136.6(0.8)	sub			sl H_2O, bz; s EtOH; i eth, chl, CS_2
9330	3-Pyridinecarboxylic acid	Nicotinic acid	$C_6H_5NO_2$	59-67-6	123.110	nd (al, w)	237(1)	sub	1.473[25]		sl H_2O, EtOH, eth
9331	4-Pyridinecarboxylic acid	Isonicotinic acid	$C_6H_5NO_2$	55-22-1	123.110	nd(w)	318(1)	260 sub			sl H_2O, EtOH, eth, bz
9332	3-Pyridinecarboxylic acid 1-oxide	Oxiniacic acid	$C_6H_5NO_3$	2398-81-4	139.109	nd	254 dec				vs H_2O, MeOH
9333	4-Pyridinecarboxylic acid 1-oxide		$C_6H_5NO_3$	13602-12-5	139.109		273 dec				
9334	2,3-Pyridinediamine		$C_5H_7N_3$	452-58-4	109.130	lf or pl (dil al)	120.8	149[5]			s H_2O, EtOH, bz
9335	2,5-Pyridinediamine	2,5-Diaminopyridine	$C_5H_7N_3$	4318-76-7	109.130	nd	110.3	182[12]			vs H_2O, EtOH
9336	2,6-Pyridinediamine		$C_5H_7N_3$	141-86-6	109.130		122(1)	285			sl H_2O, ace
9337	3,4-Pyridinediamine		$C_5H_7N_3$	54-96-6	109.130	nd or lf	219.3				
9338	2,3-Pyridinedicarboxylic acid	Quinolinic acid	$C_7H_5NO_4$	89-00-9	167.120	mcl pr (w)	228.5				sl H_2O, tfa; i EtOH, eth, bz
9339	2,4-Pyridinedicarboxylic acid	Lutidinic acid	$C_7H_5NO_4$	499-80-9	167.120	lf (w+1)	249		0.942[25]		sl H_2O; s EtOH; i eth, bz, CS_2
9340	2,5-Pyridinedicarboxylic acid	Isocinchomeronic acid	$C_7H_5NO_4$	100-26-5	167.120	lf or pr (dil HCl)	254				s H_2O, HCl; sl EtOH; i eth, bz
9341	2,6-Pyridinedicarboxylic acid	Dipicolinic acid	$C_7H_5NO_4$	499-83-2	167.120	nd (w+3/2)	252				sl H_2O, EtOH, HOAc
9342	3,4-Pyridinedicarboxylic acid	Cinchomeronic acid	$C_7H_5NO_4$	490-11-9	167.120	cry (w)	256	sub			sl H_2O, EtOH, bz; i eth, i chl
9343	3,5-Pyridinedicarboxylic acid	Dinicotinic acid	$C_7H_5NO_4$	499-81-0	167.120	cry (w)	324	sub			i H_2O; sl eth, HOAc; s DMSO, HCl
9344	2,3-Pyridinedicarboxylic acid anhydride	Furo[3,4-b]pyridine-5,7-dione	$C_7H_3NO_3$	699-98-9	149.104		138				
9345	2-Pyridineethanamine		$C_7H_{10}N_2$	2706-56-1	122.167			213	1.0220[25]	1.5335[25]	
9346	4-Pyridineethanamine		$C_7H_{10}N_2$	13258-63-4	122.167			121[10]	1.0302[25]	1.5381[25]	vs H_2O

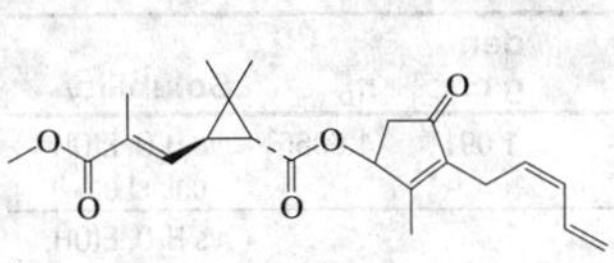

9310 Pyrethrin II

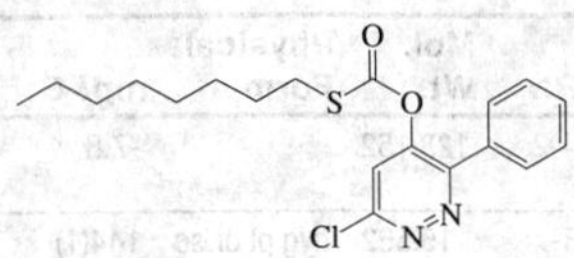

9311 Pyridate

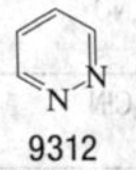

9312 Pyridazine

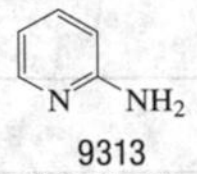

9313 2-Pyridinamine

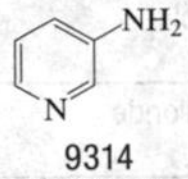

9314 3-Pyridinamine

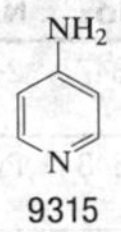

9315 4-Pyridinamine

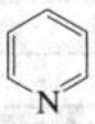

9316 Pyridine

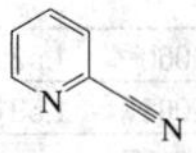

9317 2-Pyridinecarbonitrile

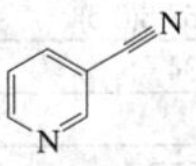

9318 3-Pyridinecarbonitrile

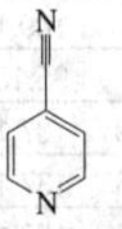

9319 4-Pyridinecarbonitrile

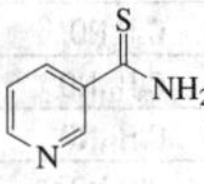

9320 3-Pyridinecarbothioamide

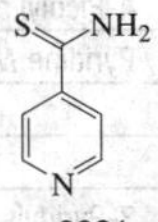

9321 4-Pyridinecarbothioamide

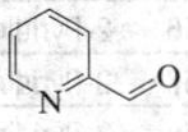

9322 2-Pyridinecarboxaldehyde

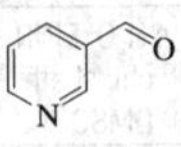

9323 3-Pyridinecarboxaldehyde

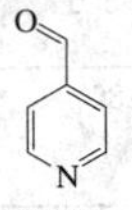

9324 4-Pyridinecarboxaldehyde

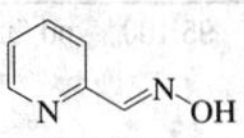

9325 2-Pyridinecarboxaldehyde oxime

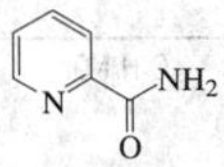

9326 2-Pyridinecarboxamide

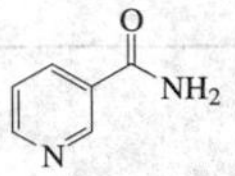

9327 3-Pyridinecarboxamide

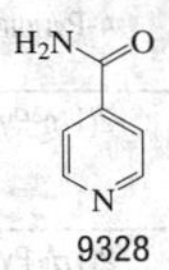

9328 4-Pyridinecarboxamide

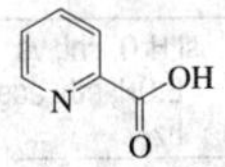

9329 2-Pyridinecarboxylic acid

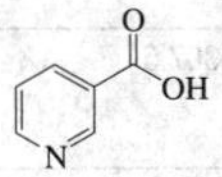

9330 3-Pyridinecarboxylic acid

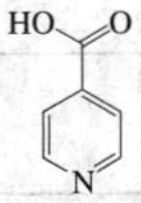

9331 4-Pyridinecarboxylic acid

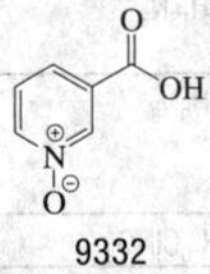

9332 3-Pyridinecarboxylic acid 1-oxide

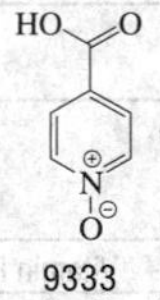

9333 4-Pyridinecarboxylic acid 1-oxide

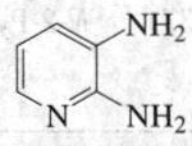

9334 2,3-Pyridinediamine

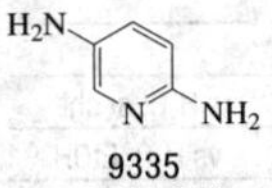

9335 2,5-Pyridinediamine

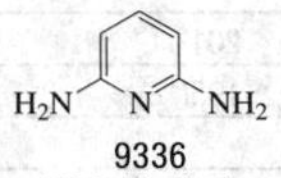

9336 2,6-Pyridinediamine

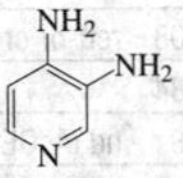

9337 3,4-Pyridinediamine

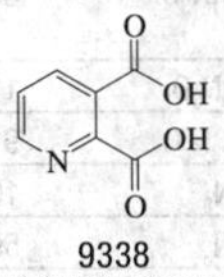

9338 2,3-Pyridinedicarboxylic acid

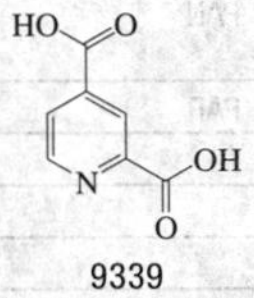

9339 2,4-Pyridinedicarboxylic acid

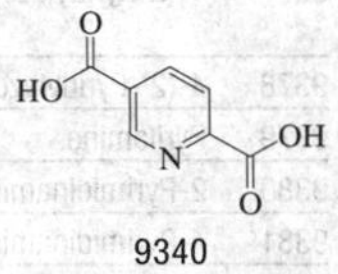

9340 2,5-Pyridinedicarboxylic acid

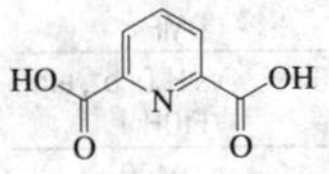

9341 2,6-Pyridinedicarboxylic acid

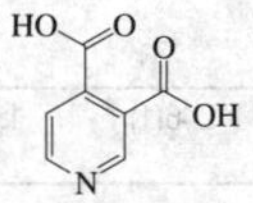

9342 3,4-Pyridinedicarboxylic acid

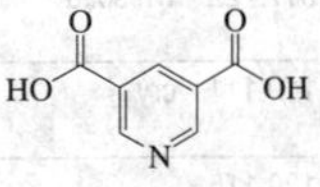

9343 3,5-Pyridinedicarboxylic acid

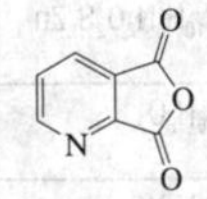

9344 2,3-Pyridinedicarboxylic acid anhydride

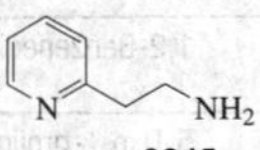

9345 2-Pyridineethanamine

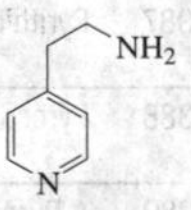

9346 4-Pyridineethanamine

No.	Name	Synonym	Mol. Form.	CAS RN	Mol. Wt.	Physical Form	mp/°C	bp/°C	den g cm⁻³	n_D	Solubility
9347	2-Pyridineethanol		C_7H_9NO	103-74-2	123.152		-7.8	190[200]	1.091[25]	1.5366[20]	vs H_2O, EtOH, chl; sl eth
9348	Pyridine hydrochloride		C_5H_6ClN	628-13-7	115.562	hyg pl or sc (al)	144(1)	222			vs H_2O, EtOH, chl
9349	2-Pyridinemethanamine		$C_6H_8N_2$	3731-51-9	108.141			203	1.0525[25]	1.5431[25]	vs H_2O
9350	3-Pyridinemethanamine		$C_6H_8N_2$	3731-52-0	108.141	liq	-21.1	226	1.064[20]	1.552[20]	vs H_2O, eth, EtOH
9351	4-Pyridinemethanamine		$C_6H_8N_2$	3731-53-1	108.141	liq	-7.6	230	1.072[20]	1.5495[25]	vs H_2O
9352	2-Pyridinemethanol		C_6H_7NO	586-98-1	109.126			112[16]	1.1317[20]	1.5444[20]	msc H_2O; vs EtOH, eth, ace, bz
9353	3-Pyridinemethanol	Nicotinyl alcohol	C_6H_7NO	100-55-0	109.126	liq	-6.5	266	1.131[20]	1.5455[20]	vs H_2O, eth
9354	4-Pyridinemethanol	4-Picolyl alcohol	C_6H_7NO	586-95-8	109.126		53	141[12]			s chl
9355	Pyridine-1-oxide	Pyridine N-oxide	C_5H_5NO	694-59-7	95.100		67.8(0.5)	146[13]			
9356	2-Pyridinepropanol		$C_8H_{11}NO$	2859-68-9	137.179		34	260.2	1.060[25]	1.5298[20]	vs H_2O
9357	3-Pyridinepropanol		$C_8H_{11}NO$	2859-67-8	137.179			284	1.063[25]	1.5313[20]	vs H_2O
9358	3-Pyridinesulfonic acid	3-Pyridylsulfonic acid	$C_5H_5NO_3S$	636-73-7	159.164	orth	357 dec		1.713[25]		vs H_2O; sl EtOH; i eth
9359	2-Pyridinethiol, 1-oxide		C_5H_5NOS	1121-31-9	127.165		70.5				
9360	2(1H)-Pyridinethione		C_5H_5NS	2637-34-5	111.166		130.0				s H_2O, EtOH, bz, chl
9361	2-Pyridinol		C_5H_5NO	72762-00-6	95.100	nd (bz)	107.8		1.3910[20]		vs H_2O, bz, EtOH
9362	3-Pyridinol		C_5H_5NO	109-00-2	95.100	nd (bz)	129				s H_2O, EtOH; sl eth, chl
9363	4-Pyridinol		C_5H_5NO	626-64-2	95.100	pr or nd (w+1)	148(1)	>350			s H_2O, EtOH; i eth, bz
9364	2(1H)-Pyridinone		C_5H_5NO	142-08-5	95.100	nd (bz)	106(3)	280	1.3910[20]		s H_2O, EtOH, bz, chl; sl eth, DMSO
9365	2(1H)-Pyridinone hydrazone	2-Pyridinylhydrazine	$C_5H_7N_3$	4930-98-7	109.130		46.6	185[140]			s chl
9366	α-[(2-Pyridinylamino)methyl]-benzenemethanol	Phenyramidol	$C_{13}H_{14}N_2O$	553-69-5	214.262	cry (dil MeOH)	83.5				
9367	1-(2-Pyridinyl)ethanone		C_7H_7NO	1122-62-9	121.137	ye in air		209(13)	1.077[25]	1.5203[20]	s EtOH, eth, HOAc; sl ctc
9368	1-(3-Pyridinyl)ethanone	Methyl pyridyl ketone	C_7H_7NO	350-03-8	121.137		13.5	225(11)		1.5341[20]	s H_2O, EtOH, eth, acid
9369	1-(4-Pyridinyl)ethanone		C_7H_7NO	1122-54-9	121.137		16	224(12)	1.097[25]	1.5282[25]	sl EtOH, eth, acid
9370	N-(2-Pyridinylmethyl)-2-pyridinemethanamine		$C_{12}H_{13}N_3$	1539-42-0	199.251			200[10]	1.1074[25]	1.5757[25]	
9371	N-2-Pyridinyl-2-pyridinamine		$C_{10}H_9N_3$	1202-34-2	171.198		90.5	307.5			sl H_2O, chl; vs EtOH, eth, ace, bz
9372	Pyridoxal hydrochloride	Vitamin B6	$C_8H_{10}ClNO_3$	65-22-5	203.623	orth	165 dec				vs H_2O; sl EtOH
9373	Pyridoxal 5-phosphate	Pyridoxal 5-(dihydrogen phosphate)	$C_8H_{10}NO_6P$	54-47-7	247.142	wh-ye pow or cry	141				
9374	Pyridoxamine	4-(Aminomethyl)-5-hydroxy-6-methyl-3-pyridinemethanol	$C_8H_{12}N_2O_2$	85-87-0	168.193	cry	198				s EtOH, acid
9375	Pyridoxamine dihydrochloride		$C_8H_{14}Cl_2N_2O_2$	524-36-7	241.115	pl (al)	226 dec				vs H_2O; sl EtOH
9376	Pyridoxine hydrochloride	5-Hydroxy-6-methyl-3,4-pyridinedimethanol hydrochloride	$C_8H_{12}ClNO_3$	58-56-0	205.639	pl (al, ace)	207	sub			vs H_2O
9377	1-(2-Pyridylazo)-2-naphthol	PAN	$C_{15}H_{11}N_3O$	85-85-8	249.267	red-br cry	130				i H_2O; s EtOH, eth, chl
9378	4-(2'-Pyridylazo)resorcinol	PAR	$C_{11}H_9N_3O_2$	1141-59-9	215.208	red-br cry	187 dec				
9379	Pyrilamine		$C_{17}H_{23}N_3O$	91-84-9	285.384			201[5]			
9380	2-Pyrimidinamine		$C_4H_5N_3$	109-12-6	95.103	nd (AcOEt)	127.5	195(18)			s H_2O; sl chl
9381	4-Pyrimidinamine		$C_4H_5N_3$	591-54-8	95.103	pl (AcOEt)	151.5				vs H_2O, EtOH
9382	Pyrimidine	1,3-Diazine	$C_4H_4N_2$	289-95-2	80.088		22	123.8		1.4998[20]	msc H_2O; s EtOH
9383	2,4,5,6(1H,3H)-Pyrimidinetetrone	Alloxan	$C_4H_2N_2O_4$	50-71-5	142.070		256 dec	sub			vs H_2O; s EtOH, ace, bz, HOAc
9384	2,4,5,6(1H,3H)-Pyrimidinetetrone 5-oxime	Violuric acid	$C_4H_3N_3O_4$	87-39-8	157.085	pa ye orth	203 dec				sl H_2O; s EtOH
9385	2,4,6-Pyrimidinetriamine		$C_4H_7N_5$	1004-38-2	125.133		248 dec				
9386	Pyriminil		$C_{13}H_{12}N_4O_3$	53558-25-1	272.259	solid	224 dec				
9387	Pyrithione zinc		$C_{10}H_8N_2O_2S_2Zn$	13463-41-7	317.722	wh solid	262				s chl, DMSO, DMF
9388	Pyrocatechol	1,2-Benzenediol	$C_6H_6O_2$	120-80-9	110.111	cry	104.6(0.3)	246(1)	1.344[20]	1.604[25]	vs H_2O, bz, eth, EtOH
9389	L-Pyroglutamic acid	5-Oxo-L-proline	$C_5H_7NO_3$	98-79-3	129.115		162				s DMSO

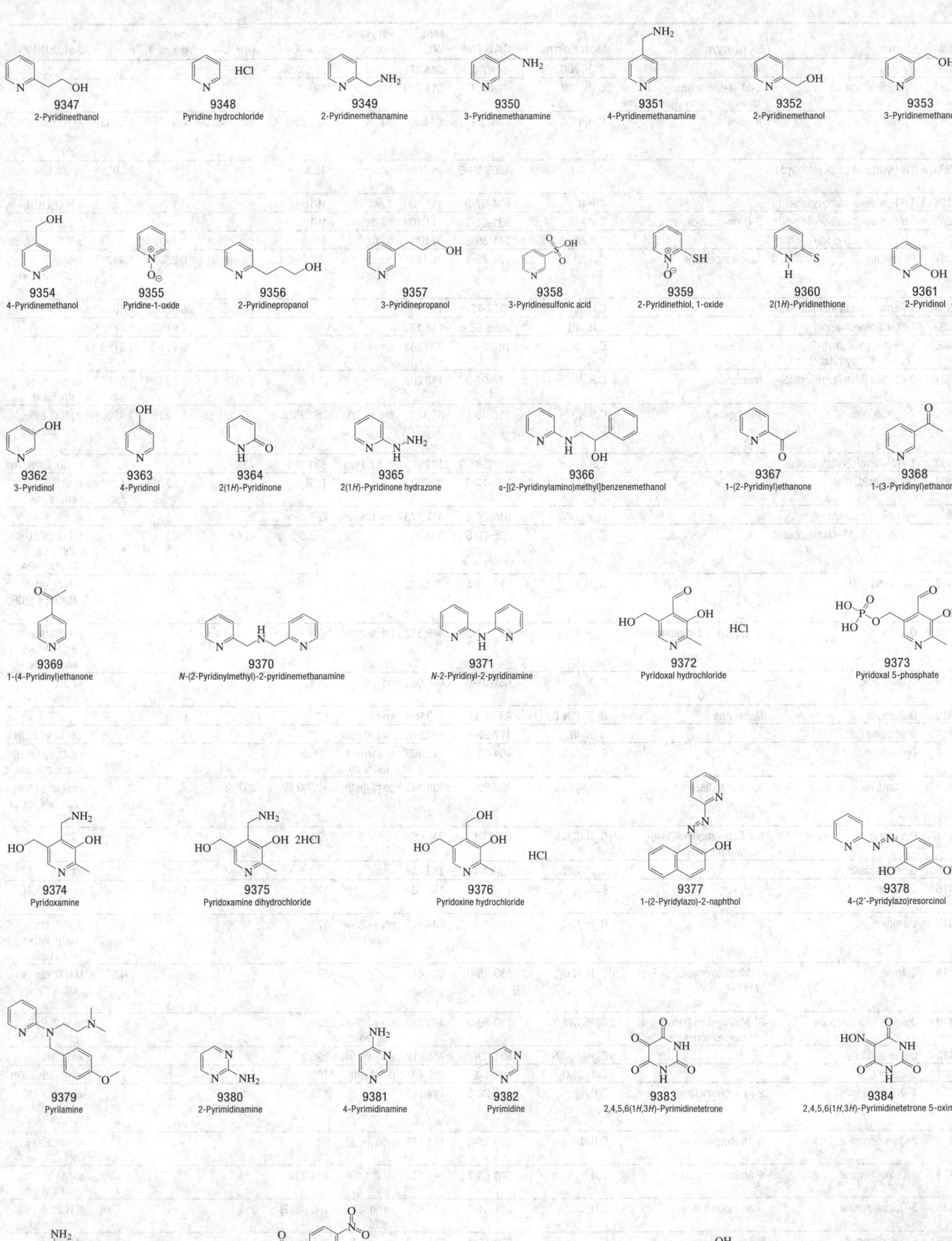

9347
2-Pyridineethanol

9348
Pyridine hydrochloride

9349
2-Pyridinemethanamine

9350
3-Pyridinemethanamine

9351
4-Pyridinemethanamine

9352
2-Pyridinemethanol

9353
3-Pyridinemethanol

9354
4-Pyridinemethanol

9355
Pyridine-1-oxide

9356
2-Pyridinepropanol

9357
3-Pyridinepropanol

9358
3-Pyridinesulfonic acid

9359
2-Pyridinethiol, 1-oxide

9360
2(1*H*)-Pyridinethione

9361
2-Pyridinol

9362
3-Pyridinol

9363
4-Pyridinol

9364
2(1*H*)-Pyridinone

9365
2(1*H*)-Pyridinone hydrazone

9366
α-[(2-Pyridinylamino)methyl]benzenemethanol

9367
1-(2-Pyridinyl)ethanone

9368
1-(3-Pyridinyl)ethanone

9369
1-(4-Pyridinyl)ethanone

9370
N-(2-Pyridinylmethyl)-2-pyridinemethanamine

9371
N-2-Pyridinyl-2-pyridinamine

9372
Pyridoxal hydrochloride

9373
Pyridoxal 5-phosphate

9374
Pyridoxamine

9375
Pyridoxamine dihydrochloride

9376
Pyridoxine hydrochloride

9377
1-(2-Pyridylazo)-2-naphthol

9378
4-(2'-Pyridylazo)resorcinol

9379
Pyrilamine

9380
2-Pyrimidinamine

9381
4-Pyrimidinamine

9382
Pyrimidine

9383
2,4,5,6(1*H*,3*H*)-Pyrimidinetetrone

9384
2,4,5,6(1*H*,3*H*)-Pyrimidinetetrone 5-oxime

9385
2,4,6-Pyrimidinetriamine

9386
Pyriminil

9387
Pyrithione zinc

9388
Pyrocatechol

9389
L-Pyroglutamic acid

No.	Name	Synonym	Mol. Form.	CAS RN	Mol. Wt.	Physical Form	mp/°C	bp/°C	den g cm⁻³	n_D	Solubility
9390	Pyrolan		$C_{13}H_{15}N_3O_2$	87-47-8	245.277		52.0(0.5)	161[0.2]			s ctc, CS_2
9391	Pyrrobutamine	1-[4-(4-Chlorophenyl)-3-phenyl-2-butenyl]pyrrolidine	$C_{20}H_{22}ClN$	91-82-7	311.849	cry	49	192[0.3]			
9392	Pyrrole	Imidole	C_4H_5N	109-97-7	67.090	liq	-23.39(0.02)	129.74(0.04)	0.9698[20]	1.5085[20]	sl H_2O; s EtOH, eth, ace, bz, chl
9393	1H-Pyrrole-2-carboxaldehyde		C_5H_5NO	1003-29-8	95.100	orth pr (peth)	46.5	218		1.5939[16]	sl chl, lig
9394	1H-Pyrrole-2-carboxylic acid		$C_5H_5NO_2$	634-97-9	111.100	lf (w)	191(5)				s H_2O, EtOH, eth
9395	1H-Pyrrole-3-carboxylic acid	3-Pyrrolecarboxylic acid	$C_5H_5NO_2$	931-03-3	111.100	nd (lig)	161.5				
9396	1H-Pyrrole-2,5-dione		$C_4H_3NO_2$	541-59-3	97.073	pl (bz)	94	sub	1.2493[106]		s H_2O, EtOH, eth
9397	Pyrrolidine	Azacyclopentane	C_4H_9N	123-75-1	71.121	col liq	-57.79(0.03)	86.6(0.1)	0.8586[20]	1.4431[20]	msc H_2O; s EtOH, eth; sl bz, chl
9398	1-Pyrrolidineethanamine		$C_6H_{14}N_2$	7154-73-6	114.188			166	0.901[25]	1.4687[20]	
9399	1-Pyrrolidineethanol		$C_6H_{13}NO$	2955-88-6	115.173			187	0.9750[20]	1.4713[20]	
9400	1-[4-(1-Pyrrolidinyl)-2-butynyl]-2-pyrrolidinone	Oxotremorine	$C_{12}H_{18}N_2O$	70-22-4	206.283	pa ye liq		124[0.1]	0.991[25]	1.5160[20]	
9401	3-(2-Pyrrolidinyl)pyridine, (S)-	Nornicotine	$C_9H_{12}N_2$	494-97-3	148.204	hyg		270	1.0737[19]	1.5378[18]	vs H_2O, ace, eth, EtOH
9402	2-Pyrrolidone	γ-Butyrolactam	C_4H_7NO	616-45-5	85.105	cry (peth)	25.92(0.01)	251.2(0.1)	1.120[20]	1.4806[30]	vs H_2O, EtOH, eth, bz, chl, CS_2
9403	1-(1H-Pyrrol-2-yl)ethanone		C_6H_7NO	1072-83-9	109.126	mcl nd (w)	89.8(0.4)	220			s H_2O, EtOH, eth
9404	Pyruvic acid		$C_3H_4O_3$	127-17-3	88.062		13.8	165 dec	1.2272[20]	1.4280[20]	msc H_2O, EtOH, eth; s ace
9405	Pyrvinium chloride		$C_{26}H_{28}ClN_3$	548-84-5	417.973	red pow (w)	250 dec				
9406	1,1':4',1":4",1'''-Quaterphenyl		$C_{24}H_{18}$	135-70-6	306.400		313.8(0.7)	428[18]			i H_2O, EtOH, eth, chl; s bz, $PhNO_2$, HOAc
9407	Quercetin		$C_{15}H_{10}O_7$	117-39-5	302.236	ye nd (dil al, + 2 w)	322(1)	sub			sl H_2O, eth, MeOH; s EtOH, ace, py
9408	Quercitrin	Quercetin-3-L-rhamnoside	$C_{21}H_{20}O_{11}$	522-12-3	448.377	pa ye nd or pl (+2w, dil al)	170				i H_2O, eth; s EtOH, HOAc, MeOH, alk
9409	Quillaic acid		$C_{30}H_{46}O_5$	631-01-6	486.683	nd (dil al)	294				vs ace, eth, py, EtOH
9410	Quinacrine	Mepacrine	$C_{23}H_{30}ClN_3O$	83-89-6	399.956	ye oil	87				
9411	Quinaldine Red		$C_{21}H_{23}IN_2$	117-92-0	430.325	dk red pow					s H_2O; vs EtOH
9412	Quinamine		$C_{19}H_{24}N_2O_2$	464-85-7	312.406	pr (bz), nd (80% al)	185.5				i H_2O; vs EtOH, bz; s eth, ace
9413	Quinazoline	1,3-Benzodiazine	$C_8H_6N_2$	253-82-7	130.147	ye pl (peth)	47.7(0.2)	237(6)			vs H_2O; s EtOH, eth, ace, bz; sl chl
9414	Quinclorac	3,7-Dichloroquinoline-8-carboxylic acid	$C_{10}H_5Cl_2NO_2$	84087-01-4	242.059		274		1.75		
9415	Quinethazone		$C_{10}H_{12}ClN_3O_3S$	73-49-4	289.738						s tfa
9416	Quinic acid		$C_7H_{12}O_6$	77-95-2	192.166		162.5		1.64[25]		vs H_2O, EtOH, HOAc
9417	Quinidine		$C_{20}H_{24}N_2O_2$	56-54-2	324.417	cry (+2.5w, dil al)	174				sl H_2O, eth; s EtOH, bz; vs chl; i peth
9418	Quinine	6'-Methoxycinchonan-9-ol, (8α,9R)	$C_{20}H_{24}N_2O_2$	130-95-0	324.417		57			1.625[15]	sl H_2O, ace; vs EtOH, py; s eth, chl
9419	Quinine hydrochloride	6'-Methoxycinchonan-9-ol monohydrochloride, (8α,9R)	$C_{20}H_{25}ClN_2O_2$	130-89-2	360.878	silky efflor nd (w)	159				vs H_2O, EtOH, chl
9420	Quinine sulfate		$C_{40}H_{50}N_4O_8S$	804-63-7	746.912	silky nd (w)	235.2				vs EtOH
9421	Quininone		$C_{20}H_{22}N_2O_2$	84-31-1	322.401	nd, lf (eth)	108				vs bz, eth, EtOH
9422	2-Quinolinamine	2-Aminoquinoline	$C_9H_8N_2$	580-22-3	144.173	lf (w)	129(2)	sub			vs H_2O; s EtOH, eth, ace, chl; sl bz
9423	3-Quinolinamine	3-Aminoquinoline	$C_9H_8N_2$	580-17-6	144.173	orth (w, dil al)	94				vs eth, EtOH, chl
9424	4-Quinolinamine	4-Aminoquinoline	$C_9H_8N_2$	578-68-7	144.173	nd (bz, dil al)	154.8	180[12]			s H_2O, bz, chl; vs EtOH, eth
9425	5-Quinolinamine	5-Aminoquinoline	$C_9H_8N_2$	611-34-7	144.173	ye nd (al) lf (eth)	106.0(0.2)	310			sl H_2O; vs EtOH, eth; s bz; i lig
9426	6-Quinolinamine	6-Aminoquinoline	$C_9H_8N_2$	580-15-4	144.173	cry (w+2), pr (eth)	116(2)	187[12]			sl H_2O, eth; s NH_3, EtOH

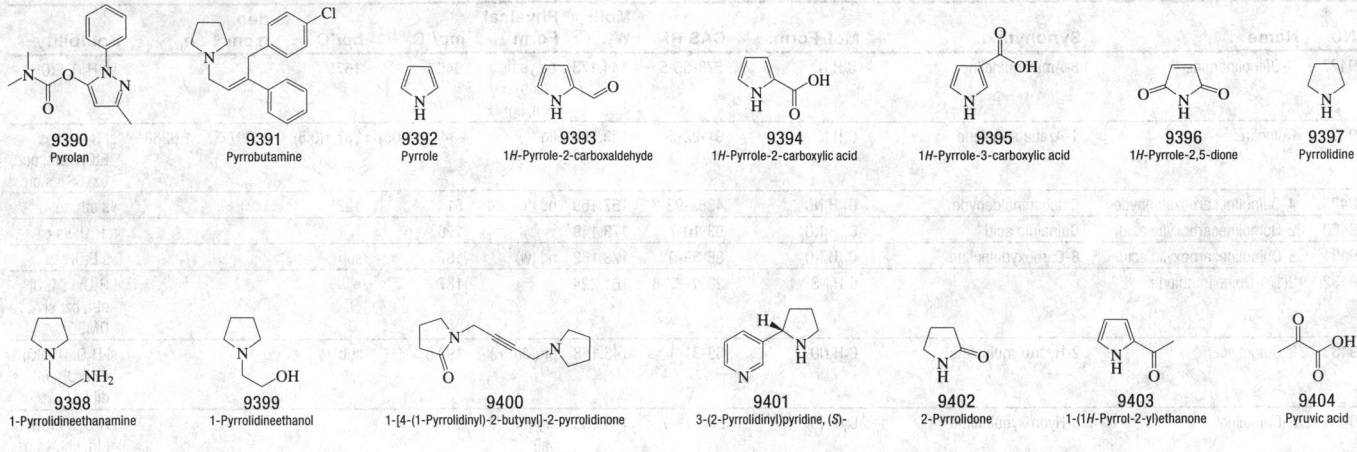

9390 Pyrolan

9391 Pyrrobutamine

9392 Pyrrole

9393 1H-Pyrrole-2-carboxaldehyde

9394 1H-Pyrrole-2-carboxylic acid

9395 1H-Pyrrole-3-carboxylic acid

9396 1H-Pyrrole-2,5-dione

9397 Pyrrolidine

9398 1-Pyrrolidineethanamine

9399 1-Pyrrolidineethanol

9400 1-[4-(1-Pyrrolidinyl)-2-butynyl]-2-pyrrolidinone

9401 3-(2-Pyrrolidinyl)pyridine, (S)-

9402 2-Pyrrolidone

9403 1-(1H-Pyrrol-2-yl)ethanone

9404 Pyruvic acid

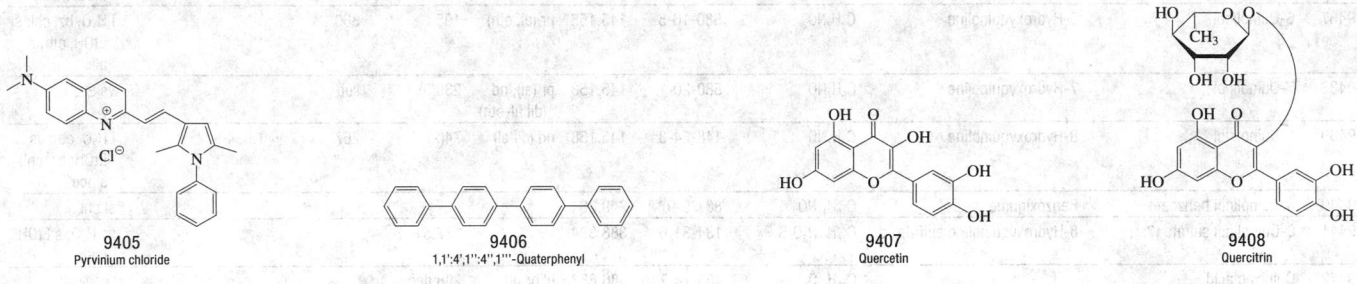

9405 Pyrvinium chloride

9406 1,1':4',1'':4'',1'''-Quaterphenyl

9407 Quercetin

9408 Quercitrin

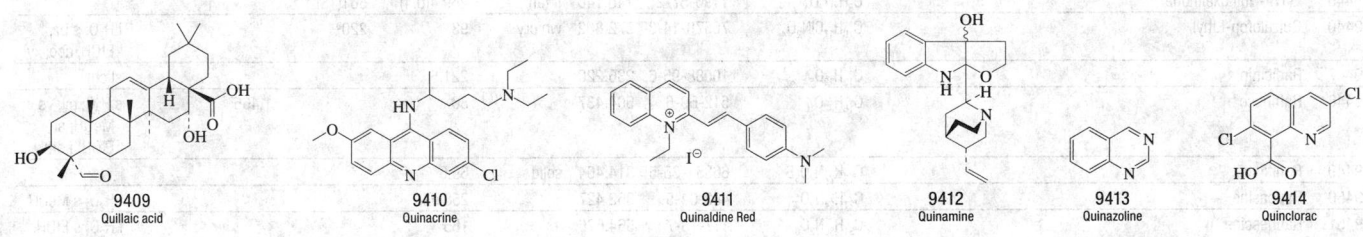

9409 Quillaic acid

9410 Quinacrine

9411 Quinaldine Red

9412 Quinamine

9413 Quinazoline

9414 Quinclorac

9415 Quinethazone

9416 Quinic acid

9417 Quinidine

9418 Quinine

9419 Quinine hydrochloride

9420 Quinine sulfate

9421 Quininone

9422 2-Quinolinamine

9423 3-Quinolinamine

9424 4-Quinolinamine

9425 5-Quinolinamine

9426 6-Quinolinamine

No.	Name	Synonym	Mol. Form.	CAS RN	Mol. Wt.	Physical Form	mp/°C	bp/°C	den g cm⁻³	n_D	Solubility
9427	8-Quinolinamine	8-Aminoquinoline	$C_9H_8N_2$	578-66-5	144.173	pa ye nd (sub) cry (al, lig)	64(2)	157[19]			vs H_2O, EtOH
9428	Quinoline	1-Azanaphthalene	C_9H_7N	91-22-5	129.159	liq	-14.78(0.05)	237.1(0.5)	1.0977[15]	1.6268[20]	sl H_2O; msc EtOH, eth, ace, bz, CS_2; s ctc
9429	4-Quinolinecarboxaldehyde	Cinchoninaldehyde	$C_{10}H_7NO$	4363-93-3	157.169	nd (to-peth)	51	122[4]			vs eth, tol
9430	2-Quinolinecarboxylic acid	Quinaldic acid	$C_{10}H_7NO_2$	93-10-7	173.169		156				s H_2O; vs bz
9431	8-Quinolinecarboxylic acid	8-Carboxyquinoline	$C_{10}H_7NO_2$	86-59-9	173.169	nd (w)	187	sub			vs EtOH
9432	2(1H)-Quinolinethione		C_9H_7NS	2637-37-8	161.224		187				i H_2O; vs EtOH, eth, bz; sl DMSO
9433	2-Quinolinol	2-Hydroxyquinoline	C_9H_7NO	59-31-4	145.158	pr (MeOH)	199.5	sub			sl H_2O, DMSO; vs EtOH, eth; s dil HCl
9434	3-Quinolinol	3-Hydroxyquinoline	C_9H_7NO	580-18-7	145.158	cry (bz, dil al)	201.3				i H_2O; s EtOH; sl eth, chl; vs bz
9435	4-Quinolinol	4-Hydroxyquinoline	C_9H_7NO	611-36-9	145.158	nd (w+3)	210				vs H_2O, EtOH; sl eth, bz, peth
9436	5-Quinolinol	5-Hydroxyquinoline	C_9H_7NO	578-67-6	145.158	nd (al), pl	224(3)	sub			s H_2O, bz, chl; sl EtOH; vs MeOH; i lig
9437	6-Quinolinol	6-Hydroxyquinoline	C_9H_7NO	580-16-5	145.158	pr (al, eth)	195	360			i H_2O, bz, chl; sl EtOH, eth; s alk
9438	7-Quinolinol	7-Hydroxyquinoline	C_9H_7NO	580-20-1	145.158	pr (al), nd (dil al-eth)	238(3)	sub			vs EtOH
9439	8-Quinolinol	8-Hydroxyquinoline	C_9H_7NO	148-24-3	145.158	nd (dil al)	74(2)	267	1.034[20]		i H_2O, eth; vs EtOH, bz, chl; s ace
9440	8-Quinolinol benzoate	Benzoxiquine	$C_{16}H_{11}NO_2$	86-75-9	249.264						sl chl
9441	8-Quinolinol sulfate (2:1)	8-Hydroxyquinoline sulfate	$C_{18}H_{16}N_2O_6S$	134-31-6	388.394		177.5				vs H_2O; s EtOH; i eth
9442	Quinovic acid		$C_{30}H_{46}O_5$	465-74-7	486.683	pl or nd	298 dec				
9443	Quinovose		$C_6H_{12}O_5$	7658-08-4	164.156	cry (AcOEt)	136(1)				vs H_2O, EtOH
9444	Quinoxaline	1,4-Benzodiazine	$C_8H_6N_2$	91-19-0	130.147	cry (peth)	32.5(0.2)	227(7)	1.1334[48]	1.6231[48]	s H_2O; msc EtOH, eth, ace, bz; sl chl
9445	2(1H)-Quinoxalinone		$C_8H_6N_2O$	1196-57-2	146.146	lf (al)	269.4(0.4)	361(15)			
9446	Quizalofop-Ethyl		$C_{19}H_{17}ClN_2O_4$	76578-14-8	372.802	wh cry	93	220[0.2]			i H_2O; s bz, EtOH, ace, xyl
9447	Radicinin		$C_{12}H_{12}O_5$	10088-95-6	236.220		221.5				sl chl
9448	Raffinose		$C_{18}H_{32}O_{16}$	512-69-6	504.437		80		1.465[25]		s H_2O, py; vs MeOH; sl EtOH; i eth
9449	Ranitidine		$C_{13}H_{22}N_4O_3S$	66357-35-5	314.404	solid	69.5				
9450	Raubasine		$C_{21}H_{24}N_2O_3$	483-04-5	352.427		258 dec				i H_2O; s MeOH
9451	Raunescine		$C_{31}H_{36}N_2O_8$	117-73-7	564.626		165				i H_2O; s EtOH, chl, HOAc
9452	Reinecke salt	Ammonium tetrathiocyanodia mminechromate(III) monohydrate	$C_4H_{12}CrN_7OS_4$	13573-16-5	354.440	red cry (w)	270 dec				s H_2O, EtOH, ace; i bz
9453	Resazurin	7-Hydroxy-3H-phenoxazin-3-one, 10-oxide	$C_{12}H_7NO_4$	550-82-3	229.189	dk red to gr pr or pl (HOAc)		sub			i H_2O, eth; sl EtOH, HOAc; s alk
9454	Rescinnamine		$C_{35}H_{42}N_2O_9$	24815-24-5	634.716	nd (bz)	238.5				i H_2O; sl EtOH; s ace, chl, AcOEt
9455	Reserpic acid		$C_{22}H_{28}N_2O_5$	83-60-3	400.467	cry (MeOH)	242				
9456	Reserpine		$C_{33}H_{40}N_2O_9$	50-55-5	608.679	lo pr (dil ace)	264.5				sl H_2O, eth, ace; s EtOH, bz, AcOEt
9457	cis-Resmethrin, (-)		$C_{22}H_{26}O_3$	10453-86-8	338.439		57.5(0.5)				
9458	Resorcinol	1,3-Benzenediol	$C_6H_6O_2$	108-46-3	110.111	nd (bz), pl (w)	109.8(0.4)	280(2)	1.278[20]	1.578[25]	vs H_2O, ctc; s EtOH, eth; sl bz, chl
9459	11-cis-Retinal	Vitamin A_1 aldehyde	$C_{20}H_{28}O$	564-87-4	284.435	cry					
9460	Retinal (all trans)		$C_{20}H_{28}O$	116-31-4	284.435	oran cry	64				i H_2O; s EtOH, chl, cy, peth
9461	13-cis-Retinoic acid	Accutane	$C_{20}H_{28}O_2$	4759-48-2	300.435	cry (EtOH)	189				
9462	13-trans-Retinoic acid		$C_{20}H_{28}O_2$	302-79-4	300.435	cry (MeOH)	181.5				
9463	Retinol	Vitamin A	$C_{20}H_{30}O$	68-26-8	286.451	ye pr (peth)	63.5	137[0.000001]			i H_2O; s EtOH, eth, ace, bz
9464	Retinyl palmitate	Retinol, hexadecanoate	$C_{36}H_{60}O_2$	79-81-2	524.860		28				

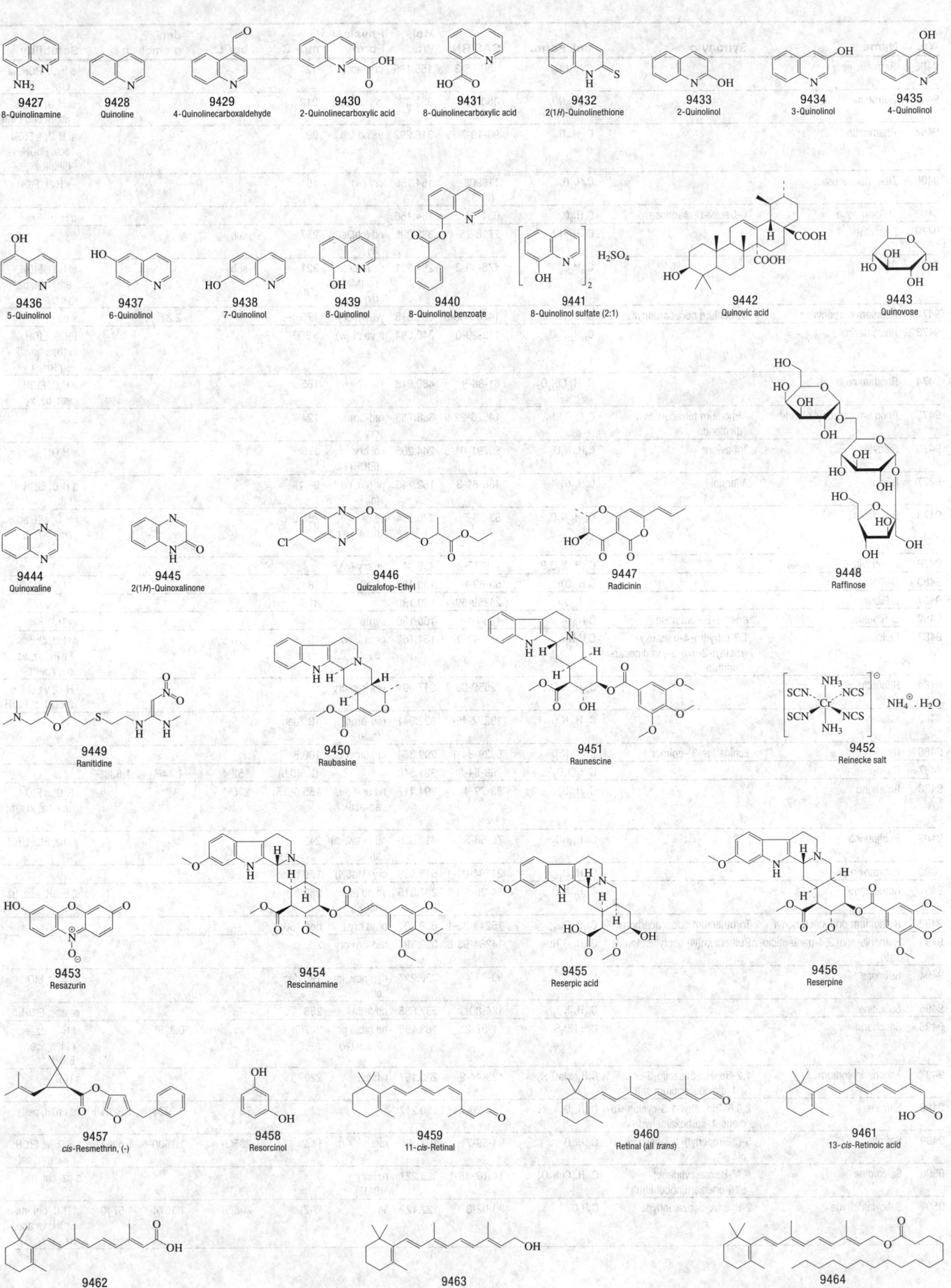

9427
8-Quinolinamine

9428
Quinoline

9429
4-Quinolinecarboxaldehyde

9430
2-Quinolinecarboxylic acid

9431
8-Quinolinecarboxylic acid

9432
2(1H)-Quinolinethione

9433
2-Quinolinol

9434
3-Quinolinol

9435
4-Quinolinol

9436
5-Quinolinol

9437
6-Quinolinol

9438
7-Quinolinol

9439
8-Quinolinol

9440
8-Quinolinol benzoate

9441
8-Quinolinol sulfate (2:1)

9442
Quinovic acid

9443
Quinovose

9444
Quinoxaline

9445
2(1H)-Quinoxalinone

9446
Quizalofop-Ethyl

9447
Radicinin

9448
Raffinose

9449
Ranitidine

9450
Raubasine

9451
Raunescine

9452
Reinecke salt

9453
Resazurin

9454
Rescinnamine

9455
Reserpic acid

9456
Reserpine

9457
cis-Resmethrin, (-)

9458
Resorcinol

9459
11-cis-Retinal

9460
Retinal (all trans)

9461
13-cis-Retinoic acid

9462
13-trans-Retinoic acid

9463
Retinol

9464
Retinyl palmitate

No.	Name	Synonym	Mol. Form.	CAS RN	Mol. Wt.	Physical Form	mp/°C	bp/°C	den g cm⁻³	n_D	Solubility
9465	Retronecine, (+)		$C_8H_{13}NO_2$	480-85-3	155.195	cry (ace)	121				s H_2O, EtOH; sl eth
9466	Retrorsine		$C_{18}H_{25}NO_6$	480-54-6	351.395	cry (AcOEt)	212				sl H_2O, ace; s EtOH, chl; i eth
9467	Rhamnetin		$C_{16}H_{12}O_7$	90-19-7	316.262	ye nd (al)	295				sl H_2O; s EtOH, ace, PhOH; vs dil alk
9468	DL-α-Rhamnose		$C_6H_{12}O_5$	116908-82-8	164.156	cry (w)	151				vs H_2O, EtOH
9469	D-Rhamnose	6-Deoxy-D-mannose	$C_6H_{12}O_5$	634-74-2	164.156						s H_2O
9470	Rheadine		$C_{21}H_{21}NO_6$	2718-25-4	383.395	nd (chl, eth, al)	257	sub			
9471	Rhein		$C_{15}H_8O_6$	478-43-3	284.221	ye or oran nd (MeOH, py)	321	sub			sl H_2O, EtOH, eth, ace, bz; vs py
9472	Rhenium carbonyl	Dirhenium decacarbonyl	$C_{10}O_{10}Re_2$	14285-68-8	652.515	ye-wh cry	170 dec		2.87		s os
9473	Rhizopterin		$C_{15}H_{12}N_6O_4$	119-20-0	340.294	lt ye pl (w)	>300				i H_2O, EtOH, eth; s aq alk, aq NH_3, py
9474	Rhodamine B		$C_{28}H_{32}ClN_2O_3$	81-88-9	480.018		165				s H_2O, EtOH, eth, bz, xyl
9475	Rhodium carbonyl chloride	Dirhodium tetracarbonyl dichloride	$C_4Cl_2O_4Rh_2$	14523-22-9	388.758	red-oran cry	124				s os
9476	Ribavirin	Tribavirin	$C_8H_{12}N_4O_5$	36791-04-5	244.205	col cry (EtOH)	175				s H_2O
9477	Ribitol	Adonitol	$C_5H_{12}O_5$	488-81-3	152.146	pr (w), nd (al)	96(1)				s H_2O, EtOH; i eth, lig
9478	Riboflavin		$C_{17}H_{20}N_4O_6$	83-88-5	376.364	ye or oran-ye nd (w)	280 dec				i H_2O, eth, ace, chl; sl EtOH
9479	Riboflavin-5'-phosphate		$C_{17}H_{21}N_4O_9P$	146-17-8	456.34	ye cry (w)					
9480	D-Ribose		$C_5H_{10}O_5$	50-69-1	150.130	pl (al)	88				s H_2O; sl EtOH
9481	L-Ribose		$C_5H_{10}O_5$	24259-59-4	150.130		81				
9482	D-Ribulose	erythro-2-Pentulose	$C_5H_{10}O_5$	488-84-6	150.130	syrup					vs H_2O
9483	Ricinine	1,2-Dihydro-4-methoxy-1-methyl-2-oxo-3-pyridinecarbonitrile	$C_8H_8N_2O_2$	524-40-3	164.162	pr or lf (w, al)	201.5	170 sub			s H_2O, chl; sl EtOH, bz; vs py; i peth
9484	Rifabutin		$C_{46}H_{62}N_4O_{11}$	72559-06-9	847.004	viol-red cry					i H_2O; vs chl; s MeOH; sl EtOH
9485	Rifampin		$C_{43}H_{58}N_4O_{12}$	13292-46-1	822.941	red-oran pl (ace)	185 dec				
9486	Rinderine	Echinatine-3'-epimer	$C_{15}H_{25}NO_5$	6029-84-1	299.364	cry (ace)	100.5				
9487	Ronnel		$C_8H_8Cl_3O_3PS$	299-84-3	321.546		42.6(0.9)	152[0.4]	1.44[32]	1.5335[35]	
9488	Rotenone		$C_{23}H_{22}O_6$	83-79-4	394.417	nd or lf (al, aq-ace)	165.0(0.5)	215[0.5]			i H_2O; s EtOH, ace, bz; sl eth; vs chl
9489	Rubijervine		$C_{27}H_{43}NO_2$	79-58-3	413.636	nd (+1w, dil al)	242				vs bz, EtOH, chl
9490	Rubratoxin B		$C_{26}H_{30}O_{11}$	21794-01-4	518.509	cry (MeCN)	169 dec				
9491	Rutecarpine		$C_{18}H_{13}N_3O$	84-26-4	287.315	ye nd (al, AcOEt)	259.5				sl EtOH, ace, bz
9492	Ruthenium dodecacarbonyl	Triruthenium dodecacarbonyl	$C_{12}O_{12}Ru_3$	15243-33-1	639.33	oran cry	dec 150				
9493	Ruthenium(III) 2,4-pentanedioate	Ruthenium(III) acetylacetonate	$C_{15}H_{21}O_6Ru$	14284-93-6	398.39	red-brn cry	230				
9494	Rutinose		$C_{12}H_{22}O_{10}$	90-74-4	326.297	hyg pow (al, eth)	190 dec				vs H_2O, EtOH
9495	Sabadine		$C_{29}H_{47}NO_8$	124-80-1	537.685	nd (eth)	258				vs ace, EtOH
9496	Saccharin		$C_7H_5NO_3S$	81-07-2	183.185	nd (ace) pr (al), lf (w)	227(2)	sub	0.828[25]		sl H_2O, bz, eth, chl; s ace, EtOH
9497	Saccharin sodium	1,2-Benzisothiazolin-3-one, 1,1-dioxide, sodium salt	$C_7H_4NNaO_3S$	128-44-9	205.167	wh cry	229				s H_2O
9498	Safranal	2,6,6-Trimethyl-1,3-cyclohexadiene-1-carboxaldehyde	$C_{10}H_{14}O$	116-26-7	150.217			70[1]	0.9734[19]	1.5281[19]	vs EtOH, peth
9499	Safrole	5-(2-Propenyl)-1,3-benzodioxole	$C_{10}H_{10}O_2$	94-59-7	162.185	mcl	11.2	235(3)	1.1000[20]	1.5381[20]	i H_2O; vs EtOH; msc eth, chl
9500	Salcomine	N,N'-Bis(salicylidene)-ethylenediaminocobalt(II)	$C_{16}H_{14}CoN_2O_2$	14167-18-1	325.227	red cry (DMF)					s bz, chl, py
9501	Salicylaldehyde	2-Hydroxybenzaldehyde	$C_7H_6O_2$	90-02-8	122.122	liq	-7(2)	208(5)	1.1674[20]	1.5740[20]	sl H_2O, chl; msc EtOH; vs ace, bz

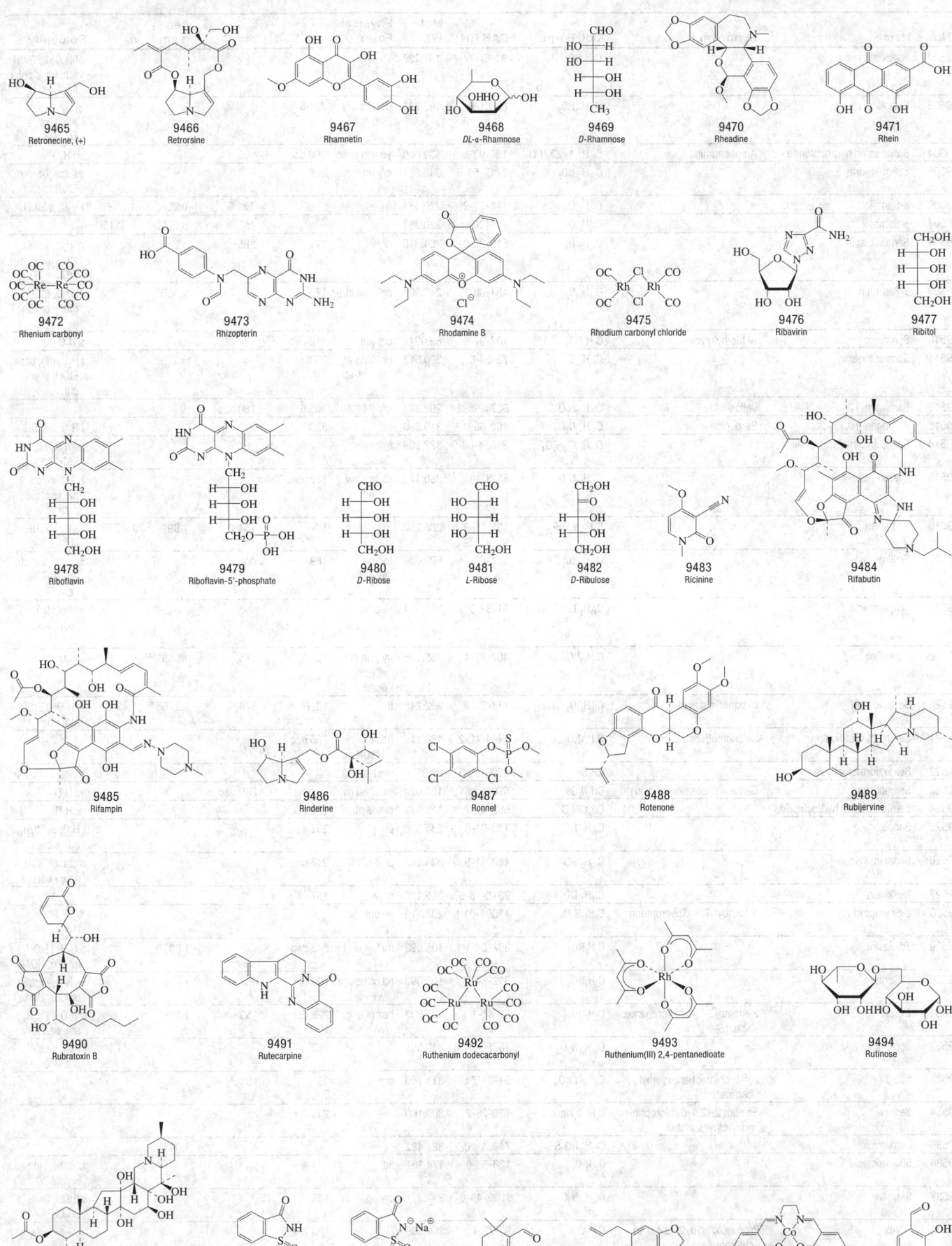

9465 Retronecine, (+)

9466 Retrorsine

9467 Rhamnetin

9468 DL-α-Rhamnose

9469 D-Rhamnose

9470 Rheadine

9471 Rhein

9472 Rhenium carbonyl

9473 Rhizopterin

9474 Rhodamine B

9475 Rhodium carbonyl chloride

9476 Ribavirin

9477 Ribitol

9478 Riboflavin

9479 Riboflavin-5'-phosphate

9480 D-Ribose

9481 L-Ribose

9482 D-Ribulose

9483 Ricinine

9484 Rifabutin

9485 Rifampin

9486 Rinderine

9487 Ronnel

9488 Rotenone

9489 Rubijervine

9490 Rubratoxin B

9491 Rutecarpine

9492 Ruthenium dodecacarbonyl

9493 Ruthenium(III) 2,4-pentanedioate

9494 Rutinose

9495 Sabadine

9496 Saccharin

9497 Saccharin sodium

9498 Safranal

9499 Safrole

9500 Salcomine

9501 Salicylaldehyde

No.	Name	Synonym	Mol. Form.	CAS RN	Mol. Wt.	Physical Form	mp/°C	bp/°C	den g cm⁻³	n_D	Solubility
9502	Salicylaldoxime		$C_7H_7NO_2$	94-67-7	137.137		57				sl H_2O; vs EtOH, eth, bz; s chl; i lig
9503	Salsoline		$C_{11}H_{15}NO_2$	89-31-6	193.243	pow or cry (al)	221.5				sl H_2O, EtOH; i eth, peth; s chl, alk
9504	Salvarsan dihydrochloride	Arsphenamine	$C_{12}H_{14}As_2Cl_2N_2O_2$	139-93-5	439.000	ye hyg pow	190 dec				vs H_2O
9505	Sanguinarine		$C_{20}H_{15}NO_5$	2447-54-3	349.337	cry (eth, al)	266				vs ace, bz, eth, EtOH
9506	α-Santalol		$C_{15}H_{24}O$	115-71-9	220.351			301.5	0.9679²⁰	1.5023²⁰	i H_2O; s EtOH
9507	β-Santalol		$C_{15}H_{24}O$	77-42-9	220.351			167¹⁰	0.9750²⁰	1.5115²⁰	
9508	Santonic acid		$C_{15}H_{20}O_4$	510-35-0	264.318	cry	171	285¹⁵			sl H_2O; s chl, eth, HOAc, EtOH
9509	α-Santonin		$C_{15}H_{18}O_3$	481-06-1	246.302	orth (w, eth)	175		1.590²⁵		sl H_2O, EtOH, eth; s bz, chl; i peth
9510	Sarcosine	N-Methylglycine	$C_3H_7NO_2$	107-97-1	89.094	cry (al)	212 dec				s H_2O
9511	Sarmentogenin		$C_{23}H_{34}O_5$	76-28-8	390.513	pr (95% al, MeOH-eth)	280				i H_2O, eth, bz; s EtOH; sl ace, chl
9512	Sarpagan-17-al	Vellosimine	$C_{19}H_{20}N_2O$	6874-98-2	292.374	cry (MeOH)	305.5	180 sub			
9513	Sarpagan-10,17-diol	Sarpagine	$C_{19}H_{22}N_2O_2$	482-68-8	310.390	nd	320				i H_2O; s EtOH
9514	Saxitoxin dihydrochloride		$C_{10}H_{19}Cl_2N_7O_4$	35554-08-6	372.209	hyg wh solid					vs H_2O, MeOH, EtOH
9515	Scarlet red		$C_{24}H_{20}N_4O$	85-83-6	380.442	dk br pow or nd	185; dec 260				i H_2O; sl ace, bz; vs chl, peth
9516	Schradan		$C_8H_{24}N_4O_3P_2$	152-16-9	286.250		17	154²·⁰	1.09²⁵	1.462²⁵	vs H_2O, EtOH, chl
9517	Scilliroside		$C_{32}H_{44}O_{12}$	507-60-8	620.684	lo pr (dil MeOH)	169	dec			sl H_2O, ace, chl; vs EtOH, diox; i eth
9518	Scopolamine		$C_{17}H_{21}NO_4$	51-34-3	303.354	visc liq					vs hot H_2O, EtOH, ace; sl bz
9519	Scopoline		$C_8H_{13}NO_2$	487-27-4	155.195	hyg nd (lig, eth, chl, peth)	108.5	248	1.0891¹³⁴		s H_2O
9520	Sebacic acid	Decanedioic acid	$C_{10}H_{18}O_4$	111-20-6	202.248	lf	131(1)	374(5)	1.2705²⁰	1.422¹³³	sl H_2O; s EtOH, eth; i bz
9521	Selenium methionine	Selenomethionine	$C_5H_{11}NO_2Se$	1464-42-2	196.11	hex pl (MeOH aq)	265 dec				
9522	Selenoformaldehyde		CH_2Se	6596-50-5	92.99	unstab gas					
9523	Selenourea	Carbamimidoselenoic acid	CH_4N_2Se	630-10-4	123.02	pr or nd (w)		200 dec			vs H_2O
9524	Semicarbazide hydrochloride		CH_6ClN_3O	563-41-7	111.531	pr (dil al)	176 dec				vs H_2O
9525	Senecionine		$C_{18}H_{25}NO_5$	130-01-8	335.396	pl	232				i H_2O; sl EtOH, eth; s chl
9526	Seneciphylline		$C_{18}H_{23}NO_5$	480-81-9	333.380	pl (AcOEt)	217 dec				s chl; sl EtOH, ace; i eth
9527	Senkirkin		$C_{19}H_{27}NO_6$	2318-18-5	365.420	pl (ace)	197				
9528	L-Sepiapterin	6-Lactoyl-7,8-dihydropterin	$C_9H_{11}N_5O_3$	17094-01-8	237.215	ye pow or cry					
9529	DL-Serine		$C_3H_7NO_3$	302-84-1	105.093	mcl pr or lf (w)	246 dec		1.603²²		s H_2O; i EtOH, eth, bz, HOAc
9530	D-Serine		$C_3H_7NO_3$	312-84-5	105.093	nd or hex pr (w)	229 dec	dec			vs H_2O; i EtOH, eth, bz, HOAc
9531	L-Serine	2-Amino-3-hydroxypropanoic acid, (S)	$C_3H_7NO_3$	56-45-1	105.093	hex pl or pr (w)	228 dec	150 sub	1.6²²		s H_2O; i EtOH, eth, bz, HOAc
9532	Serpentine alkaloid		$C_{21}H_{20}N_2O_3$	18786-24-8	348.395		175				i H_2O; s EtOH, eth, ace
9533	Sesin	2,4-Dichlorophenoxyethyl benzoate	$C_{15}H_{12}Cl_2O_3$	94-83-7	311.160	cry	66	185¹·⁵			
9534	Sesone	Sodium 2-(2,4-dichlorophenoxy)ethyl sulfate	$C_8H_7Cl_2NaO_5S$	136-78-7	309.100		245 dec				
9535	Sethoxydim		$C_{17}H_{29}NO_3S$	74051-80-2	327.482				1.043²⁵		
9536	Shikimic acid		$C_7H_{10}O_5$	138-59-0	174.151	nd	184				sl EtOH; i eth, bz, chl
9537	Siduron		$C_{14}H_{20}N_2O$	1982-49-6	232.321	cry solid	135				s EtOH, DMF, CH_2Cl_2
9538	Silvex	Propanoic acid, 2-(2,4,5-trichlorophenoxy)-	$C_9H_7Cl_3O_3$	93-72-1	269.509		180.7(0.5)				

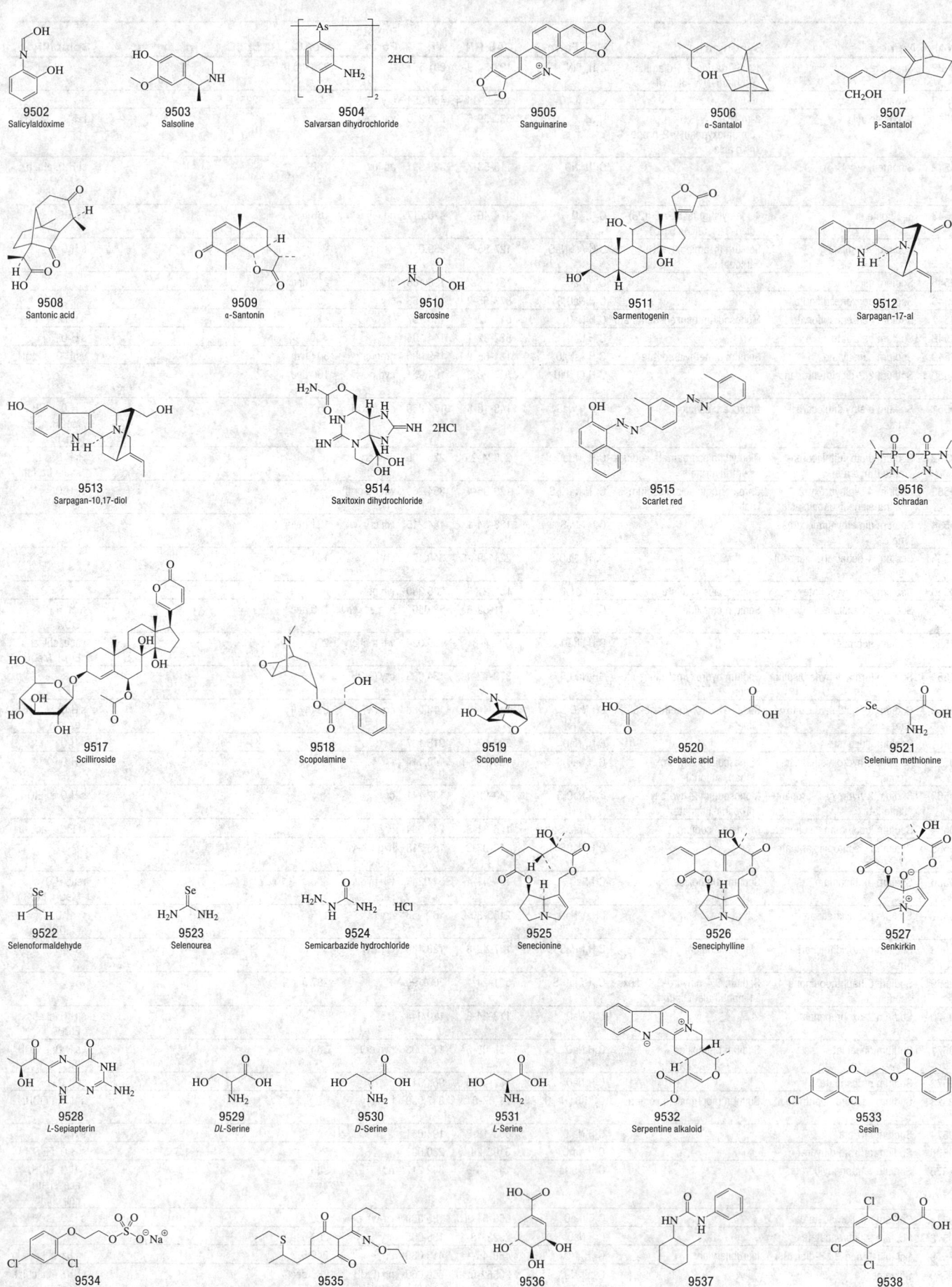

9502
Salicylaldoxime

9503
Salsoline

9504
Salvarsan dihydrochloride

9505
Sanguinarine

9506
α-Santalol

9507
β-Santalol

9508
Santonic acid

9509
α-Santonin

9510
Sarcosine

9511
Sarmentogenin

9512
Sarpagan-17-al

9513
Sarpagan-10,17-diol

9514
Saxitoxin dihydrochloride

9515
Scarlet red

9516
Schradan

9517
Scilliroside

9518
Scopolamine

9519
Scopoline

9520
Sebacic acid

9521
Selenium methionine

9522
Selenoformaldehyde

9523
Selenourea

9524
Semicarbazide hydrochloride

9525
Senecionine

9526
Seneciphylline

9527
Senkirkin

9528
L-Sepiapterin

9529
DL-Serine

9530
D-Serine

9531
L-Serine

9532
Serpentine alkaloid

9533
Sesin

9534
Sesone

9535
Sethoxydim

9536
Shikimic acid

9537
Siduron

9538
Silvex

No.	Name	Synonym	Mol. Form.	CAS RN	Mol. Wt.	Physical Form	mp/°C	bp/°C	den g cm⁻³	n_D	Solubility
9539	Simazine	1,3,5-Triazine-2,4-diamine, 6-chloro-*N,N'*-diethyl-	C₇H₁₂ClN₅	122-34-9	201.657			230.4(0.9)	1.302²⁰		
9540	Simfibrate		C₂₃H₂₆Cl₂O₆	14929-11-4	469.354	cry	52	225⁰·¹⁵			
9541	Sinapinic acid	3-(4-Hydroxy-3,5-dimethoxyphenyl)-2-propenoic acid	C₁₁H₁₂O₅	530-59-6	224.210	wh pow					i H₂O; s MeOH, ace
9542	Sinomenine		C₁₉H₂₃NO₄	115-53-7	329.391	nd (bz)	162				sl H₂O, eth, bz; s EtOH, ace, dil alk
9543	α₁-Sitosterol	4-Methylstigmasta-7,24(28)-dien-3-ol, (3β,4α,5α,24Z)	C₃₀H₅₀O	474-40-8	426.717	nd (al)	166				vs EtOH, chl
9544	Sodium arsanilate	Sodium (4-aminophenyl)-arsonate	C₆H₇AsNNaO₃	127-85-5	239.037	wh cry					s H₂O
9545	Sodium ascorbate		C₆H₇NaO₆	134-03-2	198.106	cry	218 dec				
9546	Sodium benzenesulfinate		C₆H₅NaO₂S	873-55-2	164.158	cry	300				
9547	Sodium benzenesulfonate	Monosodium benzenesulfonate	C₆H₅NaO₃S	515-42-4	180.157		>300				s H₂O; sl EtOH
9548	Sodium benzoate		C₇H₅NaO₂	532-32-1	144.104		>300				s H₂O
9549	Sodium cacodylate	Sodium dimethylarsonate	C₂H₆AsNaO₂	124-65-2	159.980	gran cry	60 (hyd)				vs H₂O; s EtOH
9550	Sodium 2,2-dichloropropanoate		C₃H₃Cl₂NaO₂	127-20-8	164.951	hyg pow	166 dec				
9551	Sodium diethyldithiocarbamate	Dithiocarb sodium	C₅H₁₀NNaS₂	148-18-5	171.260	cry (EtOH)	95				s H₂O, EtOH, MeOH, ace; i eth, bz
9552	Sodium diethyldithiocarbamate trihydrate	Diethyldithiocarbamate sodium salt trihydrate	C₅H₁₆NNaO₃S₂	20624-25-3	225.306	orth cry (ace)	95				vs H₂O; s EtOH, ace; i bz, eth
9553	Sodium 4,5-dihydroxy-2,7-naphthalenedisulfonic acid	Chromotropic acid disodium salt	C₁₀H₆Na₂O₈S₂	129-96-4	364.260	wh nd or lf (w)					vs H₂O
9554	Sodium dimethyldithiocarbamate		C₃H₇NNaS₂	128-04-1	144.215	col cry (w)	121 (hyd)				
9555	Sodium 4-dodecylbenzenesulfonate		C₁₈H₂₉NaO₃S	2211-98-5	348.476	cry	144				
9556	Sodium dodecyl sulfate	Sodium lauryl sulfate	C₁₂H₂₅NaO₄S	151-21-3	288.380	wh pow	205				
9557	Sodium ethanolate	Sodium ethoxide	C₂H₅NaO	141-52-6	68.050	hyg wh pow	260 dec				reac H₂O; s EtOH
9558	Sodium fluoroacetate		C₂H₂FNaO₂	62-74-8	100.024	wh mcl cry	200				i ace, chl; sl EtOH, MeOH
9559	Sodium formaldehyde bisulfite	Sodium hydroxymethanesulfonate	CH₃NaO₄S	870-72-4	134.088	cry (EtOH aq)					
9560	Sodium formaldehydesulfoxylate	Sodium hydroxymethanesulfinate	CH₃NaO₃S	149-44-0	118.088	cry (w)	63 (hyd)				s H₂O; i EtOH, bz, eth
9561	Sodium gluconate		C₆H₁₁NaO₇	527-07-1	218.137						s H₂O
9562	Sodium 2-hydroxyethanesulfonate	Monosodium 2-hydroxyethanesulfonate	C₂H₅NaO₄S	1562-00-1	148.114						s H₂O
9563	Sodium 2-hydroxy-2-propanesulfonate	Monosodium 2-hydroxy-2-propanesulfonate	C₃H₇NaO₄S	540-92-1	162.141	cry					s H₂O; sl EtOH
9564	Sodium iodomethanesulfonate	Methiodal sodium	CH₂INaO₃S	126-31-8	243.984	cry					sl EtOH, ace, bz
9565	Sodium *O*-isopropyl xanthate		C₄H₇NaOS₂	140-93-2	158.218	hyg wh-ye pow	150 dec				
9566	Sodium methanolate	Sodium methoxide	CH₃NaO	124-41-4	54.024	wh hyg tetr cry	300				reac H₂O; s MeOH, EtOH
9567	Sodium methylarsonate		CH₄AsNaO₃	2163-80-6	161.953	cry (w)	115				vs H₂O; s MeOH; i os
9568	Sodium methyldithiocarbamate	Metham sodium	C₂H₄NNaS₂	137-42-8	129.180	cry (w)					vs H₂O
9569	Sodium β-naphthoquinone-4-sulfonate	Sodium 3,4-dihydro-3,4-dioxo-1-naphthalenesulfonate	C₁₀H₅NaO₅S	521-24-4	260.199		287 dec				
9570	Sodium 2-oxopropanoate		C₃H₃NaO₃	113-24-6	110.044						s H₂O; sl abs EtOH
9571	Sodium phenolate	Sodium phenoxide	C₆H₅NaO	139-02-6	116.093	hyg cry	384				vs H₂O; s EtOH, thf
9572	Sodium propanoate		C₃H₅NaO₂	137-40-6	96.061						sl H₂O
9573	Sodium sulfobromophthalein	Sulfobromophthalein sodium	C₂₀H₈Br₄Na₂O₁₀S₂	71-67-0	837.998	hyg cry					s H₂O; i EtOH, ace
9574	Sodium tartrate		C₄H₄Na₂O₆	868-18-8	194.051						s H₂O
9575	Sodium tartrate dihydrate		C₄H₈Na₂O₈	6106-24-7	230.082				1.545²⁵		s H₂O; i EtOH
9576	Sodium tetraphenylborate		C₂₄H₂₀BNa	143-66-8	342.217	nd	300				s H₂O, EtOH, ace; sl eth, chl; i peth
9577	Sodium trichloroacetate		C₂Cl₃NaO₂	650-51-1	185.369	ye-wh pow	300				s H₂O, EtOH
9578	Sodium trifluoroacetate		C₂F₃NaO₂	2923-18-4	136.005	cry	207 dec				
9579	Solanid-5-ene-3,18-diol, (3β)	Isorubijervine	C₂₇H₄₃NO₂	468-45-1	413.636	pr(al)	242.5				vs bz, chl
9580	Solanine		C₄₅H₇₃NO₁₅	20562-02-1	868.060	nd (EtOH aq)	286 dec				i H₂O, eth, chl; s hot EtOH

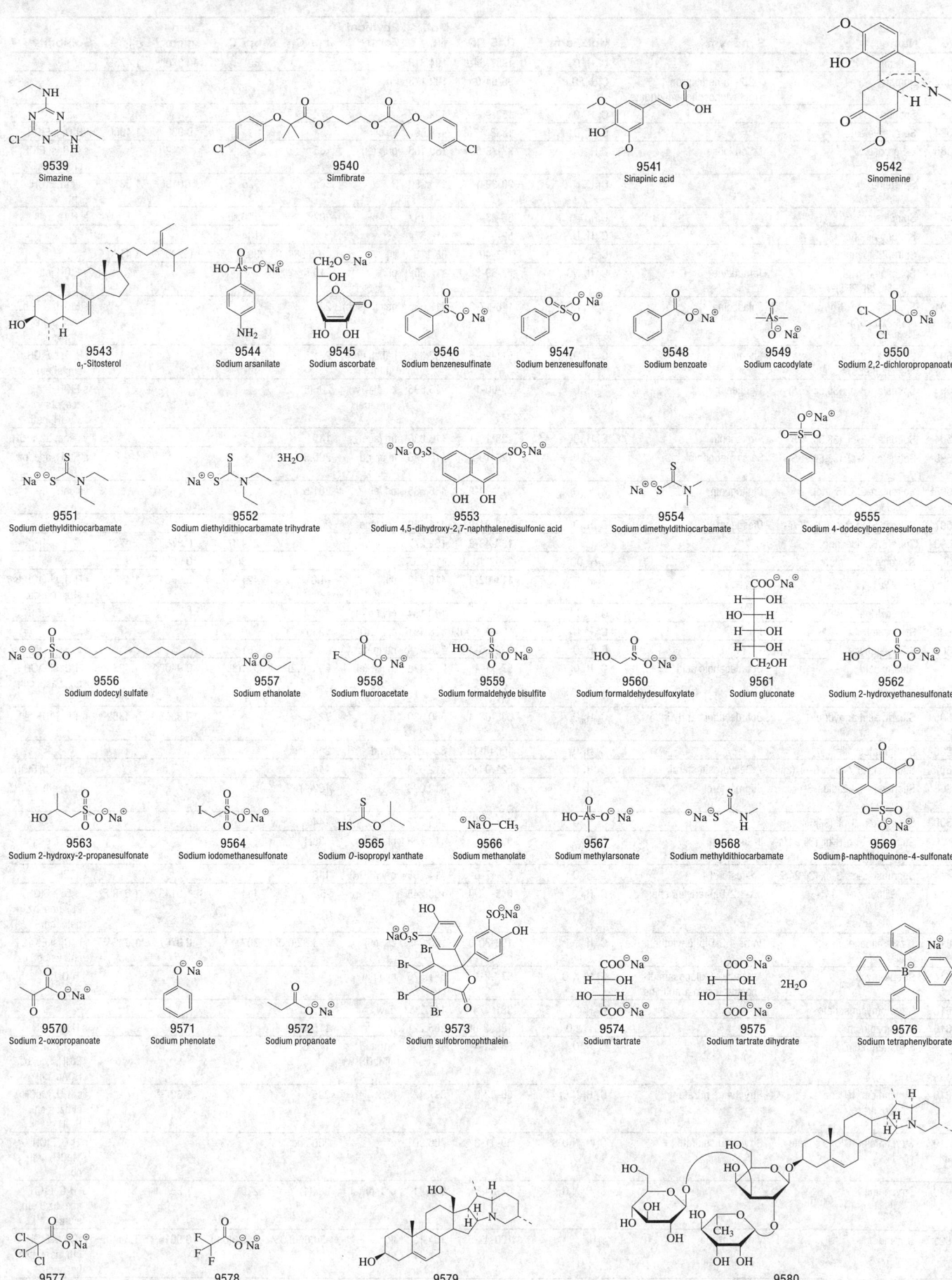

9539 Simazine

9540 Simfibrate

9541 Sinapinic acid

9542 Sinomenine

9543 α₁-Sitosterol

9544 Sodium arsanilate

9545 Sodium ascorbate

9546 Sodium benzenesulfinate

9547 Sodium benzenesulfonate

9548 Sodium benzoate

9549 Sodium cacodylate

9550 Sodium 2,2-dichloropropanoate

9551 Sodium diethyldithiocarbamate

9552 Sodium diethyldithiocarbamate trihydrate

9553 Sodium 4,5-dihydroxy-2,7-naphthalenedisulfonic acid

9554 Sodium dimethyldithiocarbamate

9555 Sodium 4-dodecylbenzenesulfonate

9556 Sodium dodecyl sulfate

9557 Sodium ethanolate

9558 Sodium fluoroacetate

9559 Sodium formaldehyde bisulfite

9560 Sodium formaldehydesulfoxylate

9561 Sodium gluconate

9562 Sodium 2-hydroxyethanesulfonate

9563 Sodium 2-hydroxy-2-propanesulfonate

9564 Sodium iodomethanesulfonate

9565 Sodium O-isopropyl xanthate

9566 Sodium methanolate

9567 Sodium methylarsonate

9568 Sodium methyldithiocarbamate

9569 Sodium β-naphthoquinone-4-sulfonate

9570 Sodium 2-oxopropanoate

9571 Sodium phenolate

9572 Sodium propanoate

9573 Sodium sulfobromophthalein

9574 Sodium tartrate

9575 Sodium tartrate dihydrate

9576 Sodium tetraphenylborate

9577 Sodium trichloroacetate

9578 Sodium trifluoroacetate

9579 Solanid-5-ene-3,18-diol, (3β)

9580 Solanine

No.	Name	Synonym	Mol. Form.	CAS RN	Mol. Wt.	Physical Form	mp/°C	bp/°C	den g cm⁻³	n_D	Solubility
9581	Solanone		$C_{13}H_{22}O$	1937-54-8	194.313			60[1]	0.870[20]	1.4755[20]	
9582	Soman	1,2,2-Trimethylpropyl methylphosphonofluoridate	$C_7H_{16}FO_2P$	96-64-0	182.173	liq					
9583	Sophoricoside		$C_{21}H_{20}O_{10}$	152-95-4	432.378		274				
9584	Sorbitan oleate		$C_{24}H_{44}O_6$	1338-43-8	428.602	ye oil			0.986	1.4800[20]	i H_2O; s EtOH
9585	L-Sorbose	L-Sorbinose	$C_6H_{12}O_6$	87-79-6	180.155	orth (al)	165		1.612[17]		s H_2O; sl EtOH, eth, MeOH
9586	Sparteine		$C_{15}H_{26}N_2$	90-39-1	234.380		30.5	325	1.0196[20]	1.5312[20]	vs eth, EtOH, chl
9587	Spinulosin		$C_8H_8O_5$	85-23-4	184.147	red-bl	202.5	120 sub			sl H_2O; s alk
9588	Spironolactone		$C_{24}H_{32}O_4S$	52-01-7	416.574			134			
9589	Spiro[2.2]pentane		C_5H_8	157-40-4	68.118	liq	-106.98(0.09)	39.0(0.3)	0.7266[20]	1.4120[20]	
9590	Spirosolan-3-ol, (3β,5α,22β,25S)-	Tomatidine	$C_{27}H_{45}NO_2$	77-59-8	415.652	pl	210.5				s EtOH, eth
9591	Spirosol-5-en-3-ol, (3β,22α,25R)-	Solasodine	$C_{27}H_{43}NO_2$	126-17-0	413.636	hex pl (sub)	202				s EtOH, ace, bz, diox, py; sl eth; vs chl
9592	Spirostan-2,3-diol, (2α,3β,5α,25R)-	Gitogenin	$C_{27}H_{44}O_4$	511-96-6	432.636	lf (bz), nd (eth)	271.5				i H_2O; s EtOH, chl; sl eth
9593	Spirostan-3-ol, (3β,5α,25R)-	Tigogenin	$C_{27}H_{44}O_3$	77-60-1	416.636	lf (al +1w) pr (ace)	205(5)				s EtOH, eth, ace, ctc, MeOH, peth
9594	Spirostan-3-ol, (3β,5β,25R)-	Smilagenin	$C_{27}H_{44}O_3$	126-18-1	416.636	nd (ace)	185				vs ace, bz, EtOH
9595	Spirostan-3-ol, (3β,5β,25S)-	Sarsasapogenin	$C_{27}H_{44}O_3$	126-19-2	416.636	lo pr, nd (ace)	200.5				s EtOH, ace, bz, chl
9596	Spirostan-2,3,15-triol, (2α,3β,5α,15β,25R)-	Digitogenin	$C_{27}H_{44}O_5$	511-34-2	448.635	nd (al)	281.5				vs chl
9597	Spirost-5-en-3-ol, (3β,25R)-	Diosgenin	$C_{27}H_{42}O_3$	512-04-9	414.620	cry (ace)	205.5				vs EtOH
9598	Spiro[5.5]undecane		$C_{11}H_{20}$	180-43-8	152.277			214(4)	0.8783[20]	1.4731	
9599	S-Propyl thioacetate		$C_5H_{10}OS$	2307-10-0	118.197			138(7)	0.9535[25]		
9600	Squalene		$C_{30}H_{50}$	111-02-4	410.718	oil	-4.8	421.3	0.8584[20]	1.4990[20]	i H_2O; sl EtOH; s eth, ace, ctc
9601	Stachydrine		$C_7H_{13}NO_2$	471-87-4	143.184	cry (w+1)	235				vs H_2O, EtOH
9602	Stanozolol		$C_{21}H_{32}N_2O$	10418-03-8	328.491	cry (EtOH)	≈236				
9603	Stearaldehyde		$C_{18}H_{36}O$	638-66-4	268.478	nd (peth)		261			
9604	Stearic acid	Octadecanoic acid	$C_{18}H_{36}O_2$	57-11-4	284.478	mcl lf (al)	69.3(0.2)	371(3)	0.9408[20]	1.4299[80]	i H_2O; sl EtOH, bz; s ace, chl, CS_2
9605	Stearic acid anhydride	Octadecanoic anhydride	$C_{36}H_{70}O_3$	638-08-4	550.939		72		0.8365[82]	1.4362[80]	i H_2O, EtOH; sl eth, bz
9606	Sterigmatocystin		$C_{18}H_{12}O_6$	10048-13-2	324.284	ye nd	246 dec				
9607	Stigmasta-5,7-dien-3-ol, (3β)	7-Dehydrositosterol	$C_{29}H_{48}O$	521-04-0	412.690		144.5				vs bz, eth, EtOH
9608	Stigmasta-5,22-dien-3-ol, (3β,22E)-	Stigmasterol	$C_{29}H_{48}O$	83-48-7	412.690		162(1)				vs bz, eth, EtOH
9609	Stigmastan-3-ol, (3β,5α)		$C_{29}H_{52}O$	83-45-4	416.722		144				
9610	Stigmast-5-en-3-ol, (3β,24R)-	β-Sitosterol	$C_{29}H_{50}O$	83-46-5	414.706	pl (al)	138(1)				s EtOH, eth, HOAc
9611	Stigmast-5-en-3-ol, (3β,24S)-	γ-Sitosterol	$C_{29}H_{50}O$	83-47-6	414.706	cry (EtOH)	148				s EtOH
9612	cis-Stilbene	cis-1,2-Diphenylethene	$C_{14}H_{12}$	645-49-8	180.245		-5	141[12]	1.0143[20]	1.6130[20]	i H_2O; s EtOH, eth, ace, bz, peth, chl
9613	trans-Stilbene	trans-1,2-Diphenylethene	$C_{14}H_{12}$	103-30-0	180.245	cry (al)	124.82(0.02)	307	0.9707[20]	1.6264[17]	i H_2O; sl EtOH, chl; vs eth, bz
9614	Streptomycin	N-Methyl-L-glucosamidi- nostreptosidostreptidine	$C_{21}H_{39}N_7O_{12}$	57-92-1	581.575	hyg pow					s H_2O
9615	Streptomycin sulfate		$C_{42}H_{84}N_{14}O_{36}S_3$	3810-74-0	1457.383	pow	≈230 dec				
9616	Streptozotocin		$C_8H_{15}N_3O_7$	18883-66-4	265.221	pl	115 dec				s H_2O, EtOH
9617	Strophanthidin		$C_{23}H_{32}O_6$	66-28-4	404.496	orth tab (MeOH-w) lf (w+2)	173 dec				i H_2O, eth; s EtOH, ace, bz, HOAc, chl
9618	Strychnidin-10-one mononitrate	Strychnine nitrate	$C_{21}H_{23}N_3O_5$	66-32-0	397.425	nd (w)	295		1.627[25]		vs H_2O, MeOH; sl bz; s chl, EtOH
9619	Strychnidin-10-one sulfate (2:1)	Strychnine sulfate	$C_{42}H_{46}N_4O_8S$	60-41-3	766.901		200 dec				s H_2O, EtOH, MeOH; i eth; sl chl
9620	Strychnine		$C_{21}H_{22}N_2O_2$	57-24-9	334.412	orth pr (al)	284(1)	270[5]	1.36[20]		sl H_2O, EtOH, ace, bz; i eth; s chl
9621	Styrene	Vinylbenzene	C_8H_8	100-42-5	104.150	liq	-30.65(0.06)	145.3(0.6)	0.9016[25]	1.5440[25]	i H_2O; s EtOH, eth, ace; msc bz; sl ctc

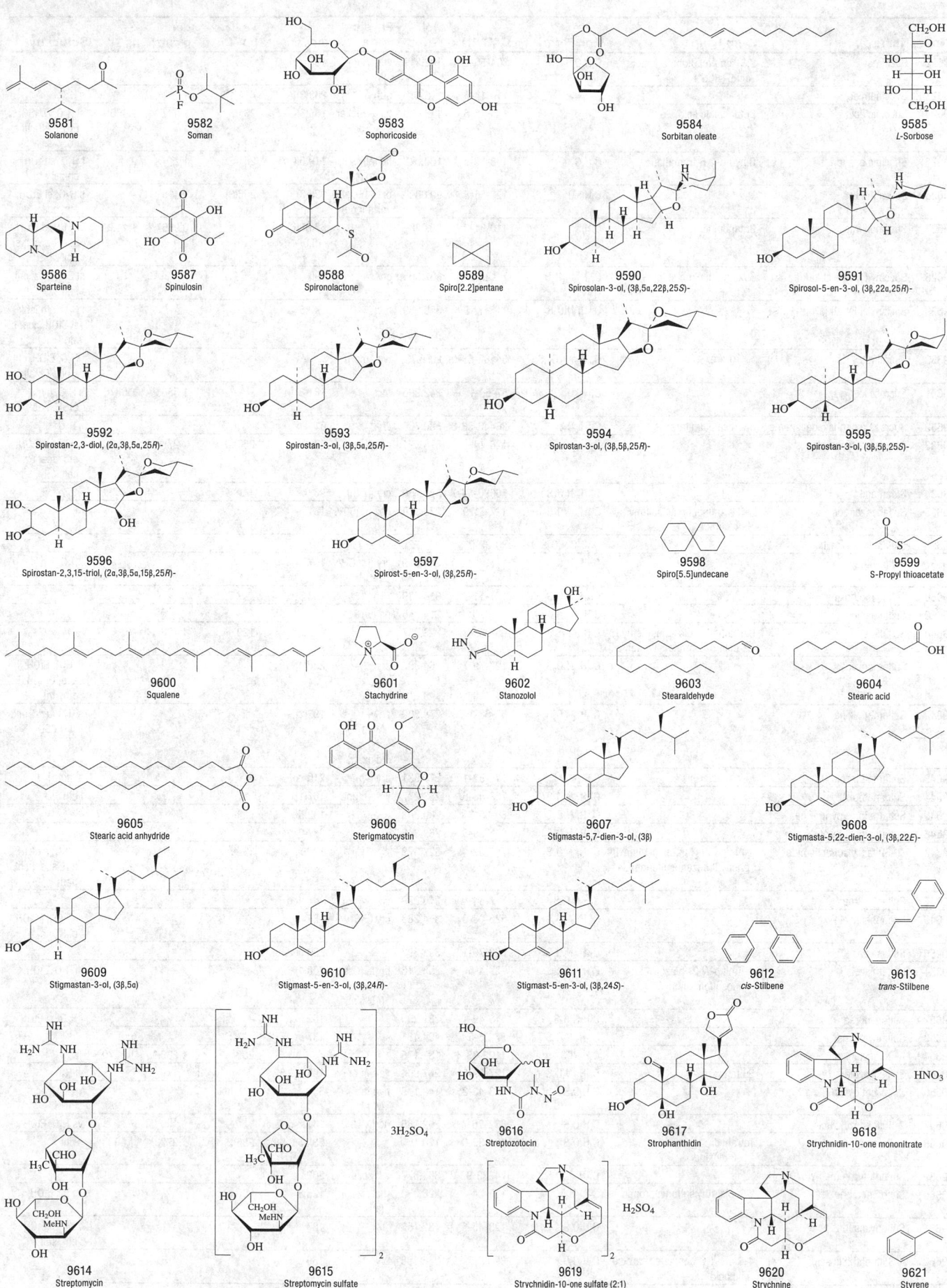

9581 Solanone

9582 Soman

9583 Sophoricoside

9584 Sorbitan oleate

9585 *L*-Sorbose

9586 Sparteine

9587 Spinulosin

9588 Spironolactone

9589 Spiro[2.2]pentane

9590 Spirosolan-3-ol, (3β,5α,22β,25S)-

9591 Spirosol-5-en-3-ol, (3β,22α,25R)-

9592 Spirostan-2,3-diol, (2α,3β,5α,25R)-

9593 Spirostan-3-ol, (3β,5α,25R)-

9594 Spirostan-3-ol, (3β,5β,25R)-

9595 Spirostan-3-ol, (3β,5β,25S)-

9596 Spirostan-2,3,15-triol, (2α,3β,5α,15β,25R)-

9597 Spirost-5-en-3-ol, (3β,25R)-

9598 Spiro[5.5]undecane

9599 S-Propyl thioacetate

9600 Squalene

9601 Stachydrine

9602 Stanozolol

9603 Stearaldehyde

9604 Stearic acid

9605 Stearic acid anhydride

9606 Sterigmatocystin

9607 Stigmasta-5,7-dien-3-ol, (3β)

9608 Stigmasta-5,22-dien-3-ol, (3β,22E)-

9609 Stigmastan-3-ol, (3β,5α)

9610 Stigmast-5-en-3-ol, (3β,24R)-

9611 Stigmast-5-en-3-ol, (3β,24S)-

9612 *cis*-Stilbene

9613 *trans*-Stilbene

9614 Streptomycin

9615 Streptomycin sulfate

9616 Streptozotocin

9617 Strophanthidin

9618 Strychnidin-10-one mononitrate

9619 Strychnidin-10-one sulfate (2:1)

9620 Strychnine

9621 Styrene

No.	Name	Synonym	Mol. Form.	CAS RN	Mol. Wt.	Physical Form	mp/°C	bp/°C	den g cm⁻³	n_D	Solubility
9622	Succimer	2,3-Dimercaptobutanedioic acid, (R*,S*)	C₄H₆O₄S₂	304-55-2	182.219	wh cry (MeOH)	193				
9623	Succinamide		C₄H₈N₂O₂	110-14-5	116.119	orth nd (w)	263(3)	432(13)			s H₂O
9624	Succinic acid	Butanedioic acid	C₄H₆O₄	110-15-6	118.089	tcl or mcl pr	185(3)	234(3)	1.572²⁵	1.450	sl H₂O, DMSO; s EtOH, eth, ace; i bz
9625	Succinic anhydride	Dihydro-2,5-furandione	C₄H₄O₃	108-30-5	100.073	nd (al), orth pym (chl)	119.5(0.7)	261	1.2²⁰		i H₂O; s EtOH, chl; sl eth
9626	Succinimide		C₄H₅NO₂	123-56-8	99.089	pl (+1w, al) orth (ace)	125(3)	296(3)	1.418²⁵		s H₂O; sl EtOH, eth, ace
9627	Succinonitrile	Butanedinitrile	C₄H₄N₂	110-61-2	80.088		57.985 (0.001)	266	0.9867⁶⁰	1.4173⁶⁰	vs H₂O; s EtOH, ace, bz, chl; sl eth
9628	Succinylcholine chloride	Suxamethonium chloride	C₁₄H₃₀Cl₂N₂O₄	71-27-2	361.305	cry (w)	190				sl EtOH, bz, chl; i eth
9629	Succinylsulphathiazole		C₁₃H₁₃N₃O₅S₂	116-43-8	355.389	cry	193.5				i H₂O, eth, chl; sl EtOH, ace; s alk
9630	Sucralfate		C₁₂H₅₄Al₁₆O₇₅S₈	54182-58-0	2086.737	wh amorp pow					i H₂O, EtOH, chl; s dil HCl, alk
9631	Sucrose		C₁₂H₂₂O₁₁	57-50-1	342.296	mcl	181(8)		1.5805¹⁷	1.5376	s H₂O, py; sl EtOH; i eth
9632	Sucrose monohexadecanoate	Sucrose palmitate	C₂₈H₅₂O₁₂	26446-38-8	580.706	cry	61				s H₂O
9633	Sucrose octaacetate		C₂₈H₃₈O₁₉	126-14-7	678.591	nd (al)	87(4)	250¹	1.27¹⁶	1.4660	sl H₂O; s EtOH, eth, ace, bz, chl
9634	Sufentanil		C₂₂H₃₀N₂O₂S	56030-54-7	386.550	cry (peth)	96.6				
9635	Sulfabenzamide	N-[(4-Aminophenyl)sulfonyl]-benzamide	C₁₃H₁₂N₂O₃S	127-71-9	276.310	hex pr (60% al)	181.5				
9636	Sulfachlorpyridazine		C₁₀H₉ClN₄O₂S	80-32-0	284.722		187				
9637	Sulfacytine		C₁₂H₁₄N₄O₃S	17784-12-2	294.329	cry (MeOH/ BuOH)	167				i H₂O; s alk
9638	Sulfadimethoxine		C₁₂H₁₄N₄O₄S	122-11-2	310.329		203.5				
9639	Sulfaguanidine		C₇H₁₀N₄O₂S	57-67-0	214.245	nd (w)	191.5				
9640	Sulfallate	Carbamodithioic acid, diethyl-, 2-chloro-2-propenyl ester	C₈H₁₄ClNS₂	95-06-7	223.787			129¹	1.088		
9641	Sulfamerazine		C₁₁H₁₂N₄O₂S	127-79-7	264.304	cry	238(4)				sl H₂O, EtOH, ace, DMSO; i eth, chl
9642	Sulfamethazine		C₁₂H₁₄N₄O₂S	57-68-1	278.330	pa ye (w+1/2) cry (diox-w)	197(2)				s H₂O, acid, alk; sl DMSO
9643	Sulfamethiazole		C₉H₁₀N₄O₂S₂	144-82-1	270.331	cry (w)	210				sl hot H₂O
9644	Sulfamethoxazole		C₁₀H₁₁N₃O₃S	723-46-6	253.277	ye-wh pow	167(1)				i eth
9645	Sulfamethoxypyridazine		C₁₁H₁₂N₄O₃S	80-35-3	280.303		182.5				
9646	Sulfamethylthiazole		C₁₀H₁₁N₃O₂S₂	515-59-3	269.343		237				vs EtOH
9647	Nᴬ-Sulfanilylsulfanilamide	4-Amino-N-[4-(aminosulfonyl)-phenyl]benzenesulfonamide	C₁₂H₁₃N₃O₄S₂	547-52-4	327.379		137				sl H₂O; s EtOH, eth, ace; i chl, peth
9648	Sulfanilylurea		C₇H₉N₃O₃S	547-44-4	215.229	cry (w)	147 dec				
9649	Sulfaphenazole		C₁₅H₁₄N₄O₂S	526-08-9	314.363	cry (EtOH)	181				sl EtOH, MeOH, gl HOAc
9650	Sulfasalazine		C₁₈H₁₄N₄O₅S	599-79-1	398.393		220 dec				
9651	Sulfathiazole	4-Amino-N-2-thiazolylben-zenesulfonamide	C₉H₉N₃O₂S₂	72-14-0	255.316	br pl, rods or pow (45% al)	200(1)				sl H₂O, EtOH, DMSO
9652	Sulfathiourea		C₇H₉N₃O₂S₂	515-49-1	231.295		182				i H₂O; sl EtOH
9653	Sulfinpyrazone		C₂₃H₂₀N₂O₃S	57-96-5	404.481		137				
9654	N-Sulfinylaniline		C₆H₅NOS	1122-83-4	139.175			200	1.236²⁵	1.6270²⁰	
9655	Sulfisoxazole	Sulfafurazole	C₁₁H₁₃N₃O₃S	127-69-5	267.304	ye-wh pr	195.0(0.5)				sl H₂O
9656	Sulfoacetic acid		C₂H₄O₅S	123-43-3	140.115	hyg tab (w+1)	85	245 dec			vs H₂O, ace, EtOH
9657	2-Sulfobenzoic acid		C₇H₆O₅S	632-25-7	202.185	nd (w+3)	141				vs H₂O, EtOH
9658	Sulfolane	Tetrahydrothiophene, 1-1-dioxide	C₄H₈O₂S	126-33-0	120.171		28.45(0.08)	286(2)	1.2723¹⁸	1.4833¹⁸	s chl
9659	Sulfometuron methyl		C₁₅H₁₆N₄O₅S	74222-97-2	364.377	wh solid	202				
9660	Sulfonmethane	2,2-Bis(ethylsulfonyl)propane	C₇H₁₆O₄S₂	115-24-2	228.330	mcl (w), pr (al)	125.8	300 dec			vs bz, EtOH, chl
9661	Sulfonyldiacetic acid		C₄H₆O₆S	123-45-5	182.152		187				vs H₂O, EtOH; s eth, sulf
9662	4-Sulfophthalic acid	4-Sulfo-1,2-benzenedicarbox-ylic acid	C₈H₆O₇S	89-08-7	246.195	cry	139				

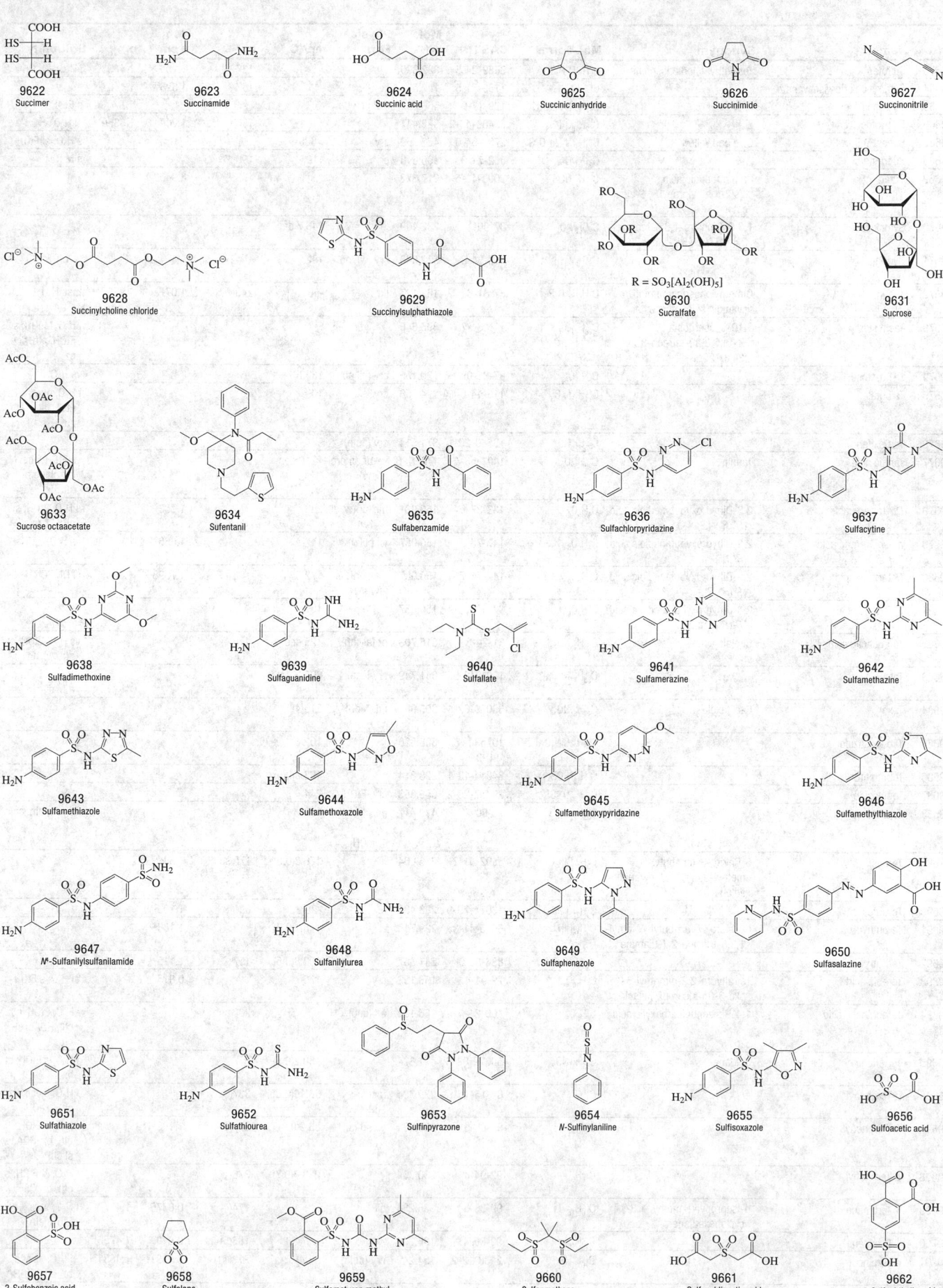

9622 Succimer

9623 Succinamide

9624 Succinic acid

9625 Succinic anhydride

9626 Succinimide

9627 Succinonitrile

9628 Succinylcholine chloride

9629 Succinylsulphathiazole

9630 Sucralfate
R = SO₃[Al₂(OH)₅]

9631 Sucrose

9633 Sucrose octaacetate

9634 Sufentanil

9635 Sulfabenzamide

9636 Sulfachlorpyridazine

9637 Sulfacytine

9638 Sulfadimethoxine

9639 Sulfaguanidine

9640 Sulfallate

9641 Sulfamerazine

9642 Sulfamethazine

9643 Sulfamethiazole

9644 Sulfamethoxazole

9645 Sulfamethoxypyridazine

9646 Sulfamethylthiazole

9647 N⁴-Sulfanilylsulfanilamide

9648 Sulfanilylurea

9649 Sulfaphenazole

9650 Sulfasalazine

9651 Sulfathiazole

9652 Sulfathiourea

9653 Sulfinpyrazone

9654 N-Sulfinylaniline

9655 Sulfisoxazole

9656 Sulfoacetic acid

9657 2-Sulfobenzoic acid

9658 Sulfolane

9659 Sulfometuron methyl

9660 Sulfonmethane

9661 Sulfonyldiacetic acid

9662 4-Sulfophthalic acid

No.	Name	Synonym	Mol. Form.	CAS RN	Mol. Wt.	Physical Form	mp/°C	bp/°C	den g cm⁻³	n_D	Solubility
9663	Sulfotep	Tetraethyl thiodiphosphate	$C_8H_{20}O_5P_2S_2$	3689-24-5	322.320			137[2]	1.196[25]	1.4753[25]	i H₂O; s EtOH
9664	Sulfuryl chloride isocyanate		CCINO₃S	1189-71-5	141.534	liq	-44	107	1.626[25]	1.4467[20]	
9665	Sulphan Blue		$C_{27}H_{31}N_2NaO_6S_2$	129-17-9	566.664	viol pow					s EtOH
9666	Sulprofos		$C_{12}H_{19}O_2PS_3$	35400-43-2	322.447			156[0.1]	1.20[20]	1.5859	sl H₂O
9667	Sunset Yellow FCF	C.I. Food Yellow 3	$C_{16}H_{10}N_2Na_2O_7S_2$	2783-94-0	452.369	cry	>300				s H₂O; sl EtOH
9668	Suprasterol II		$C_{28}H_{44}O$	562-71-0	396.648	pr	110	190[0.005]			s MeOH
9669	Sutan	Carbamothioic acid, bis(2-methylpropyl)-, S-ethyl ester	$C_{11}H_{23}NOS$	2008-41-5	217.372			138[21]	0.9402[25]		
9670	Symclosene	1,3,5-Trichloro-1,3,5-triazine-2,4,6(1H,3H,5H)-trione	$C_3Cl_3N_3O_3$	87-90-1	232.409		246.7 dec				
9671	Syringin		$C_{17}H_{24}O_9$	118-34-3	372.368	cry (w), nd (al)	192				vs EtOH
9672	Tabun	Dimethylphosphoroamido-cyanidic acid, ethyl ester	$C_5H_{11}N_2O_2P$	77-81-6	162.127	liq	-50	240	1.077	1.4250[20]	msc H₂O
9673	Tachysterol	9,10-Secoergosta-5(10),6,8,22-tetraen-3-ol, (3β,6E,22E)-	$C_{28}H_{44}O$	115-61-7	396.648						i H₂O, MeOH; s EtOH, eth, ace, bz
9674	D-Tagatose		$C_6H_{12}O_6$	87-81-0	180.155	cry (dil al)	134.5				vs H₂O
9675	Talbutal		$C_{11}H_{16}N_2O_3$	115-44-6	224.256	cry	109				i H₂O, peth; s EtOH, ace, eth, chl
9676	Tamoxifen		$C_{26}H_{29}NO$	10540-29-1	371.514	cry (peth)	97				
9677	Tannic acid	Tannin	$C_{76}H_{52}O_{46}$	1401-55-4	1701.198	ye-br amorp pow	≈210 dec				vs EtOH, ace; i bz, chl, eth, ctc
9678	DL-Tartaric acid	2,3-Dihydroxybutanedioic acid, (R*,R*)-(±)-	$C_4H_6O_6$	133-37-9	150.087	mcl pr (w, al +1w)	206		1.788[25]		s H₂O, EtOH; sl eth; i bz
9679	meso-Tartaric acid	2,3-Dihydroxybutanedioic acid, (R*,R*)-	$C_4H_6O_6$	147-73-9	150.087	tcl pl (w)	147		1.666[20]		vs H₂O, EtOH
9680	D-Tartaric acid	2,3-Dihydroxybutanedioic acid, [S-(R*,R*)]-	$C_4H_6O_6$	147-71-7	150.087	mcl, orth pr (w+1)	172.5		1.7598[20]	1.4955[20]	sl DMSO
9681	L-Tartaric acid	2,3-Dihydroxybutanedioic acid, [R-(R*,R*)]-	$C_4H_6O_6$	87-69-4	150.087		169				
9682	Taurocholic acid	Cholaic acid	$C_{26}H_{45}NO_7S$	81-24-3	515.703	pr (al-eth)	125 dec				vs H₂O, EtOH; sl eth, AcOEt
9683	Taxine A		$C_{35}H_{47}NO_{10}$	1361-49-5	641.749	cry (ace)	205				i H₂O; s EtOH, eth, chl
9684	Taxol	Paclitaxel	$C_{47}H_{51}NO_{14}$	33069-62-4	853.907	nd (MeOH aq)	215(4)				
9685	Tebuconazole		$C_{16}H_{22}ClN_3O$	107534-96-3	308.826		102.4				
9686	Tebuthiuron		$C_9H_{16}N_4OS$	34014-18-1	228.314		163.5(0.5)				
9687	Teniposide		$C_{32}H_{32}O_{13}S$	29767-20-2	656.653	cry (EtOH)	244				
9688	Tephrosin		$C_{23}H_{22}O_7$	76-80-2	410.417	pr (chl-MeOH)	198				vs ace, eth, chl
9689	Terbacil	5-Chloro-3-tert-butyl-6-methyl-2,4(1H,3H)-pyrimidin-edione	$C_9H_{13}ClN_2O_2$	5902-51-2	216.664		176.3(0.5)	175 sub	1.34[25]		
9690	Terbufos		$C_9H_{21}O_2PS_3$	13071-79-9	288.431		-29.2	69[0.01]	1.105[24]		
9691	Terbuthylazine	6-Chloro-N-tert-butyl-N'-ethyl-1,3,5-triazine-2,4-diamine	$C_9H_{16}ClN_5$	5915-41-3	229.710		176.1(0.5)		1.188[20]		
9692	Terbutryn		$C_{10}H_{19}N_5S$	886-50-0	241.357		103.8(0.5)	157[0.06]	1.115[20]		
9693	Terebic acid	Tetrahydro-2,2-dimethyl-5-oxo-3-furancarboxylic acid	$C_7H_{10}O_4$	79-91-4	158.152	cry	175		0.815		sl H₂O; s EtOH
9694	Terephthalic acid	1,4-Benzenedicarboxylic acid	$C_8H_6O_4$	100-21-0	166.132	nd (sub)		300 sub			i H₂O, EtOH, eth, chl, HOAc; sl ctc
9695	Terfenadine	Seldane	$C_{32}H_{41}NO_2$	50679-08-8	471.674		150.2(0.5)				i H₂O; s EtOH; sl hx
9696	o-Terphenyl		$C_{18}H_{14}$	84-15-1	230.304	mcl pr (MeOH)	56.19(0.05)	337(5)			i H₂O; s ace, bz, chl, MeOH
9697	m-Terphenyl		$C_{18}H_{14}$	92-06-8	230.304	ye nd (al)	86.9(0.8)	375(1)	1.199[20]		i H₂O; s EtOH, eth, bz, HOAc; sl chl
9698	p-Terphenyl		$C_{18}H_{14}$	92-94-4	230.304		213.8(0.7)	376			i H₂O; sl EtOH; s eth, bz, CS₂
9699	α-Terpinene	4-Isopropyl-1-methyl-1,3-cyclohexadiene	$C_{10}H_{16}$	99-86-5	136.234	oil		174(4)	0.8375[19]	1.477[19]	i H₂O; msc EtOH, eth
9700	γ-Terpinene		$C_{10}H_{16}$	99-85-4	136.234			183(5)	0.849[20]	1.4765[14]	
9701	α-Terpineol		$C_{10}H_{18}O$	2438-12-2	154.249	cry (peth)	35(4)	218(11)	0.9337[20]	1.4831[20]	sl H₂O; vs ace, bz, eth, EtOH

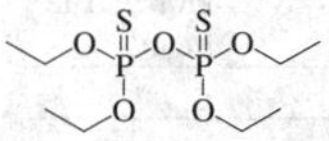

9663
Sulfotep

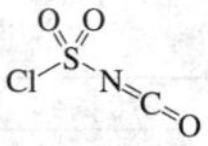

9664
Sulfuryl chloride isocyanate

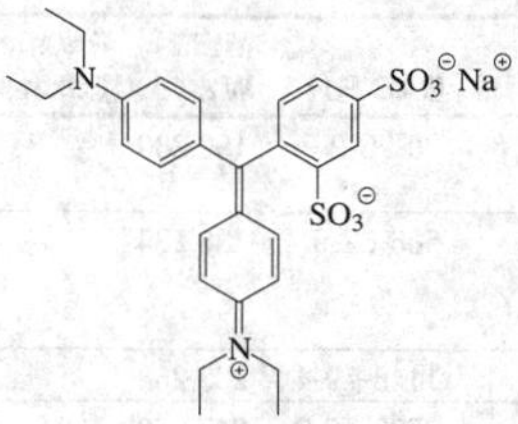

9665
Sulphan Blue

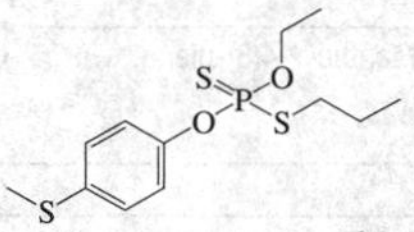

9666
Sulprofos

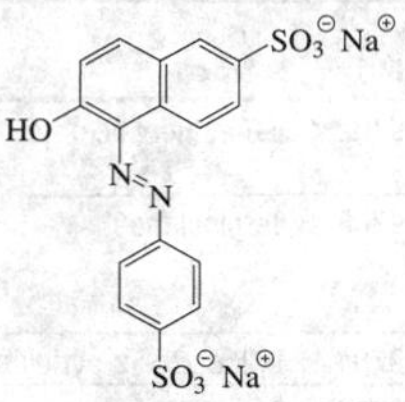

9667
Sunset Yellow FCF

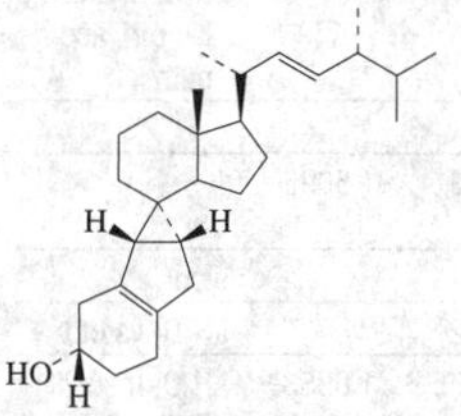

9668
Suprasterol II

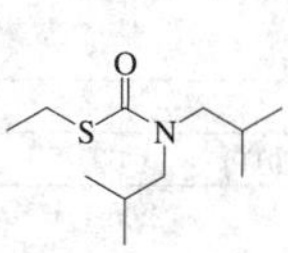

9669
Sutan

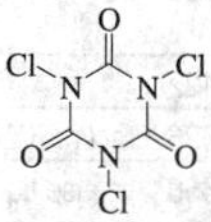

9670
Symclosene

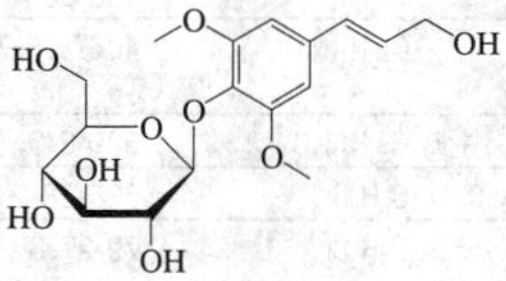

9671
Syringin

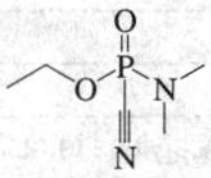

9672
Tabun

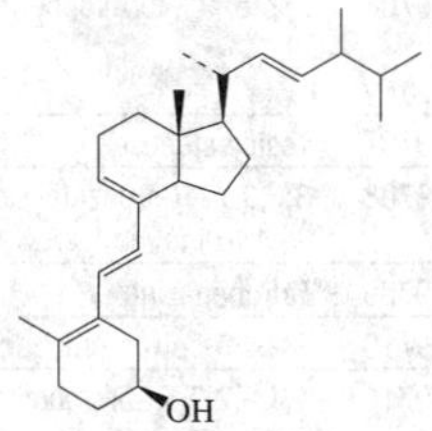

9673
Tachysterol

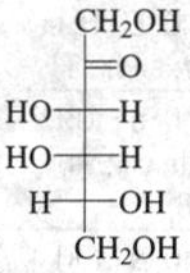

9674
D-Tagatose

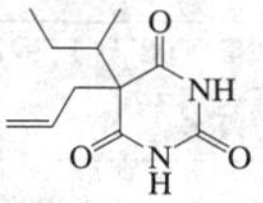

9675
Talbutal

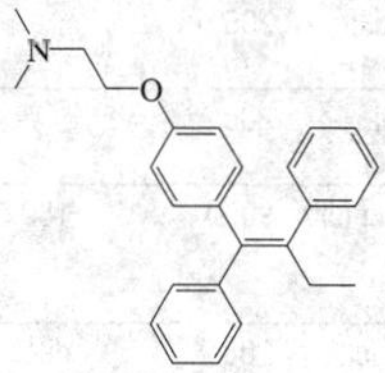

9676
Tamoxifen

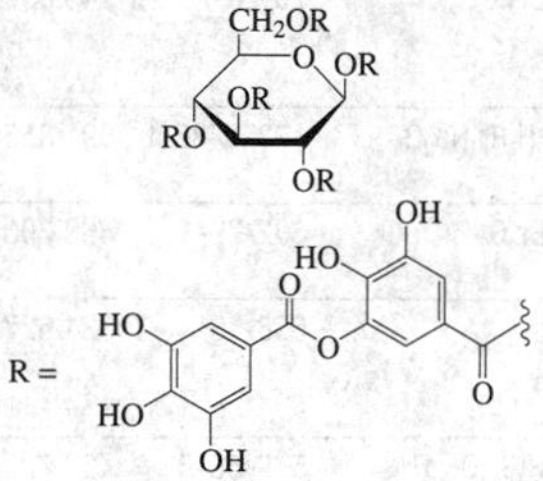

R =

9677
Tannic acid

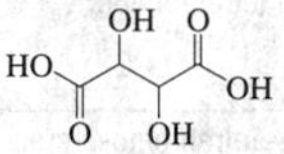

9678
DL-Tartaric acid

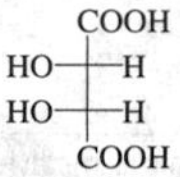

9679
meso-Tartaric acid

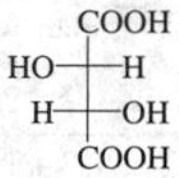

9680
D-Tartaric acid

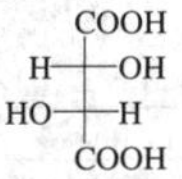

9681
L-Tartaric acid

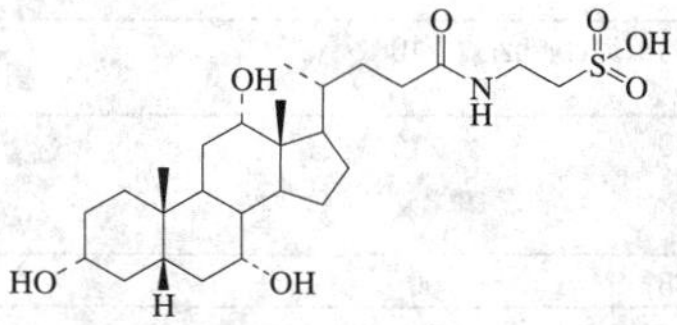

9682
Taurocholic acid

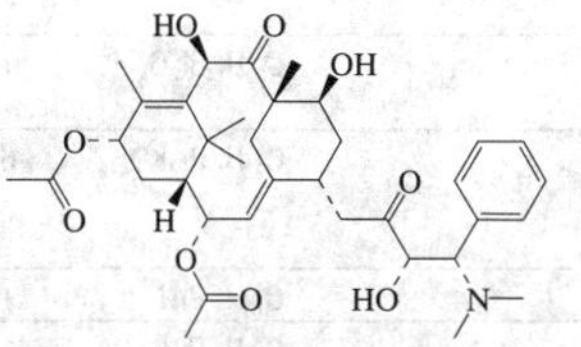

9683
Taxine A

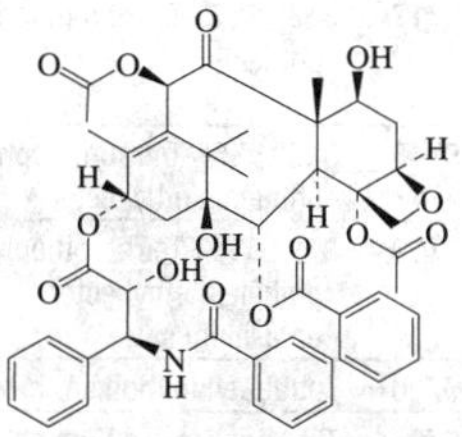

9684
Taxol

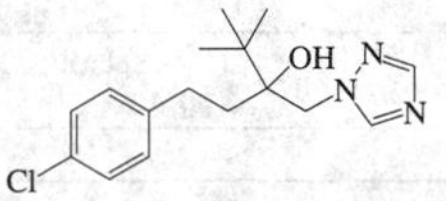

9687
Teniposide

9685
Tebuconazole

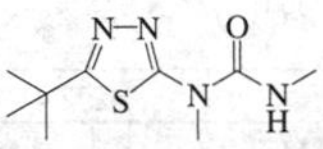

9686
Tebuthiuron

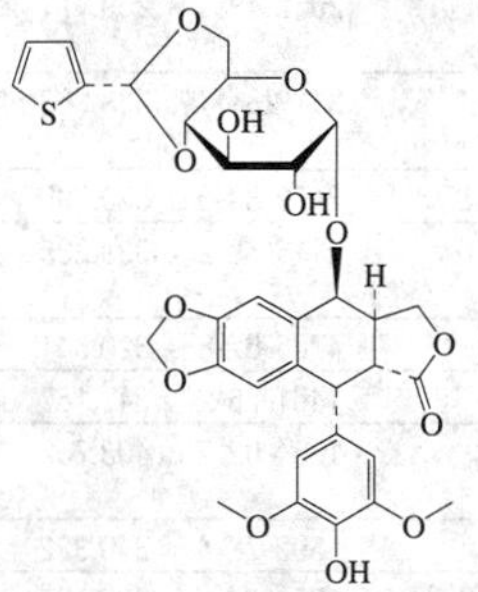

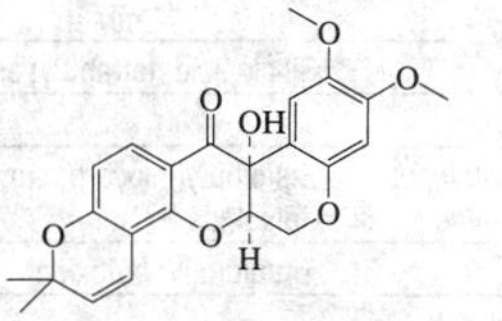

9688
Tephrosin

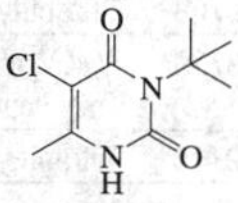

9689
Terbacil

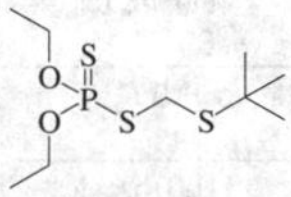

9690
Terbufos

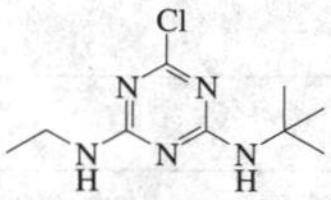

9691
Terbuthylazine

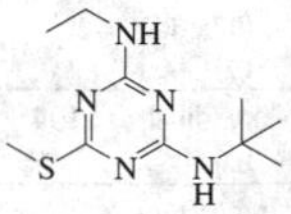

9692
Terbutryn

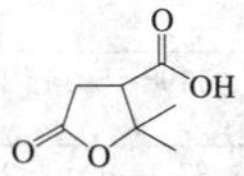

9693
Terebic acid

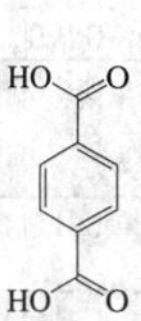

9694
Terephthalic acid

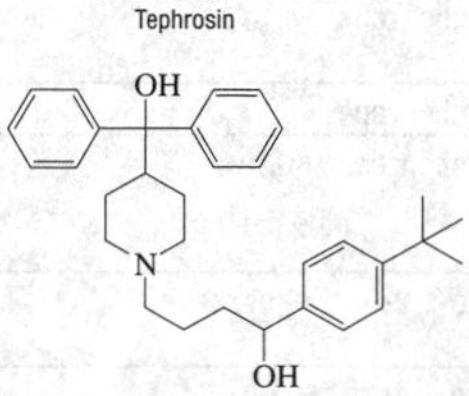

9695
Terfenadine

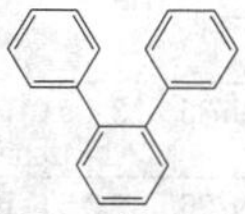

9696
o-Terphenyl

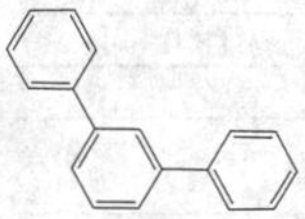

9697
m-Terphenyl

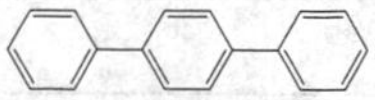

9698
p-Terphenyl

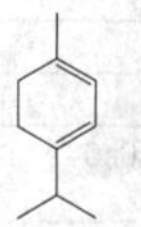

9699
α-Terpinene

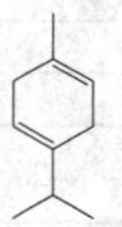

9700
γ-Terpinene

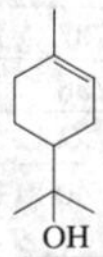

9701
α-Terpineol

No.	Name	Synonym	Mol. Form.	CAS RN	Mol. Wt.	Physical Form	mp/°C	bp/°C	den g cm⁻³	n_D	Solubility
9702	α-Terpineol acetate		$C_{12}H_{20}O_2$	80-26-2	196.286			140[40]	0.9659[21]	1.4689[21]	i H₂O; s EtOH, eth, bz
9703	Terpinolene	p-Mentha-1,4(8)-diene	$C_{10}H_{16}$	586-62-9	136.234			187(3)	0.8632[15]	1.4883[20]	i H₂O; msc EtOH, eth; s bz, ctc
9704	2,2':6',2"-Terpyridine		$C_{15}H_{11}N_3$	1148-79-4	233.268		88.0	370			
9705	Terrazole	1,2,4-Thiadiazole, 5-ethoxy-3-(trichloromethyl)-	$C_5H_5Cl_3N_2OS$	2593-15-9	247.530		19.9	95[1]	1.503[25]		
9706	2,2':5',2"-Terthiophene	α-Terthienyl	$C_{12}H_8S_3$	1081-34-1	248.387	ye-oran pl (MeOH)	93				i H₂O; sl EtOH; s bz, eth, ace, peth
9707	Testolactone		$C_{19}H_{24}O_3$	968-93-4	300.392	cry (ace)	218				
9708	3,6,9,12-Tetraazatetradecane-1,14-diamine	Pentaethylenehexamine	$C_{10}H_{28}N_6$	4067-16-7	232.369	liq			0.950	1.5096[20]	
9709	Tetrabenazine		$C_{19}H_{27}NO_3$	58-46-8	317.422		128				s chl
9710	1,2,4,5-Tetrabromobenzene		$C_6H_2Br_4$	636-28-2	393.696	mcl pr (CS₂)	180.0(0.9)	3.1[20]	3.072[20]		i H₂O; vs eth
9711	1,1,2,2-Tetrabromoethane	Acetylene tetrabromide	$C_2H_2Br_4$	79-27-6	345.653	ye visc liq	0	248(11)	2.9655[20]	1.6353[20]	i H₂O; msc EtOH, eth; s ace, bz; sl ctc
9712	Tetrabromoethene	Tetrabromoethylene	C_2Br_4	79-28-7	343.637	pl (dil al), nd (al)	56.5	226			i H₂O; s EtOH, eth, ace; vs chl
9713	2',4',5',7'-Tetrabromofluorescein, disodium salt	Eosine YS	$C_{20}H_6Br_4Na_2O_5$	17372-87-1	691.855	ye-red cry	295.5				vs EtOH
9714	4,5,6,7-Tetrabromo-1,3-isobenzofurandione		$C_8Br_4O_3$	632-79-1	463.700	nd (xyl, HOAc)	280				i H₂O, EtOH; sl bz; s PhNO₂
9715	Tetrabromomethane	Carbon tetrabromide	CBr_4	558-13-4	331.627	mcl tab (dil al)	90(2)	189.5	2.9608[100]	1.5942[100]	i H₂O; s EtOH, eth, chl; vs CS₂
9716	2,3,4,5-Tetrabromo-6-methylphenol	3,4,5,6-Tetrabromo-o-cresol	$C_7H_4Br_4O$	576-55-6	423.722	ye nd (chl, HOAc)	208	dec			i H₂O; s EtOH, eth, bz, chl; sl lig, HOAc
9717	3',3",5',5"-Tetrabromophenolphthalein		$C_{20}H_{10}Br_4O_4$	76-62-0	633.907	nd (al, eth)	296				i H₂O; sl EtOH; vs eth; s alk, HOAc
9718	3',3",5',5"-Tetrabromophenolphthalein ethyl ester		$C_{22}H_{14}Br_4O_4$	1176-74-5	661.960	ye cry (bz)	210				
9719	3',3",5',5"-Tetrabromophenolphthalein ethyl ester, potassium salt		$C_{22}H_{13}Br_4KO_4$	62637-91-6	700.050		210				
9720	Tetrabutylammonium bromide	TMAB	$C_{16}H_{36}BrN$	1643-19-2	322.368		99				s chl
9721	Tetrabutylammonium chloride		$C_{16}H_{36}ClN$	1112-67-0	277.917	cry	74				
9722	Tetrabutylammonium fluoride		$C_{16}H_{36}FN$	429-41-4	261.462	cry (w)	37				
9723	Tetrabutylammonium hydroxide	TMAH	$C_{16}H_{37}NO$	2052-49-5	259.471	stab in soln					s H₂O, MeOH
9724	Tetrabutylammonium iodide		$C_{16}H_{36}IN$	311-28-4	369.368	lf (w, bz)	145(2)				sl H₂O, chl; vs EtOH
9725	Tetrabutylammonium sulfate		$C_{32}H_{72}N_2O_4S$	2472-88-0	580.990		170				sl chl
9726	Tetrabutylphosphonium bromide		$C_{16}H_{36}BrP$	3115-68-2	339.335	cry (ace/eth)	103.5(0.5)				
9727	Tetrabutyl silicate	Silicic acid, tetrabutyl ester	$C_{16}H_{36}O_4Si$	4766-57-8	320.541			256	0.8990[20]	1.4128[20]	
9728	Tetrabutylstannane		$C_{16}H_{36}Sn$	1461-25-2	347.167	liq	-97	145[10]	1.06[20]		
9729	N,N,N',N'-Tetrabutylthioperoxydicarbonic diamide	Bis(dibutylthiocarbamoyl) disulfide	$C_{18}H_{36}N_2S_4$	1634-02-2	408.752		39.5		1.03[20]		i H₂O; sl EtOH; s eth
9730	Tetrabutyl titanate	Titanium(IV) butoxide	$C_{16}H_{36}O_4Ti$	5593-70-4	340.322			292.4			
9731	Tetracaine hydrochloride		$C_{15}H_{25}ClN_2O_2$	136-47-0	300.825		147				
9732	1,2,3,4-Tetrachlorobenzene		$C_6H_2Cl_4$	634-66-2	215.892	nd (al)	46.7(0.5)	254(3)			i H₂O; sl EtOH; vs eth, CS₂
9733	1,2,3,5-Tetrachlorobenzene		$C_6H_2Cl_4$	634-90-2	215.892	nd (al)	50.7(0.3)	245(3)			i H₂O
9734	1,2,4,5-Tetrachlorobenzene		$C_6H_2Cl_4$	95-94-3	215.892	nd, mcl pr (eth, al or bz)	139.2(0.4)	247(3)	1.858[22]		i H₂O; sl EtOH; s eth, bz, chl, CS₂
9735	3,4,5,6-Tetrachloro-1,2-benzenediol		$C_6H_2Cl_4O_2$	1198-55-6	247.891	cry (dil al, bz)	194				sl H₂O
9736	2,3,5,6-Tetrachloro-1,4-benzenediol		$C_6H_2Cl_4O_2$	87-87-6	247.891	nd (HOAc)		sub			i H₂O, bz, ctc; vs EtOH, eth; sl HOAc
9737	2,2',4',5-Tetrachlorobiphenyl		$C_{12}H_6Cl_4$	41464-40-8	291.988	cry (MeOH)	65.9(0.5)				i H₂O
9738	2,3,4,5-Tetrachlorobiphenyl		$C_{12}H_6Cl_4$	33284-53-6	291.988	cry	90.7(0.5)				i H₂O
9739	3,3',4,4'-Tetrachlorobiphenyl		$C_{12}H_6Cl_4$	32598-13-3	291.988	cry (EtOH)	180				
9740	2,2',6,6'-Tetrachlorobisphenol A		$C_{15}H_{12}Cl_4O_2$	79-95-8	366.067	cry (HOAc)	136				

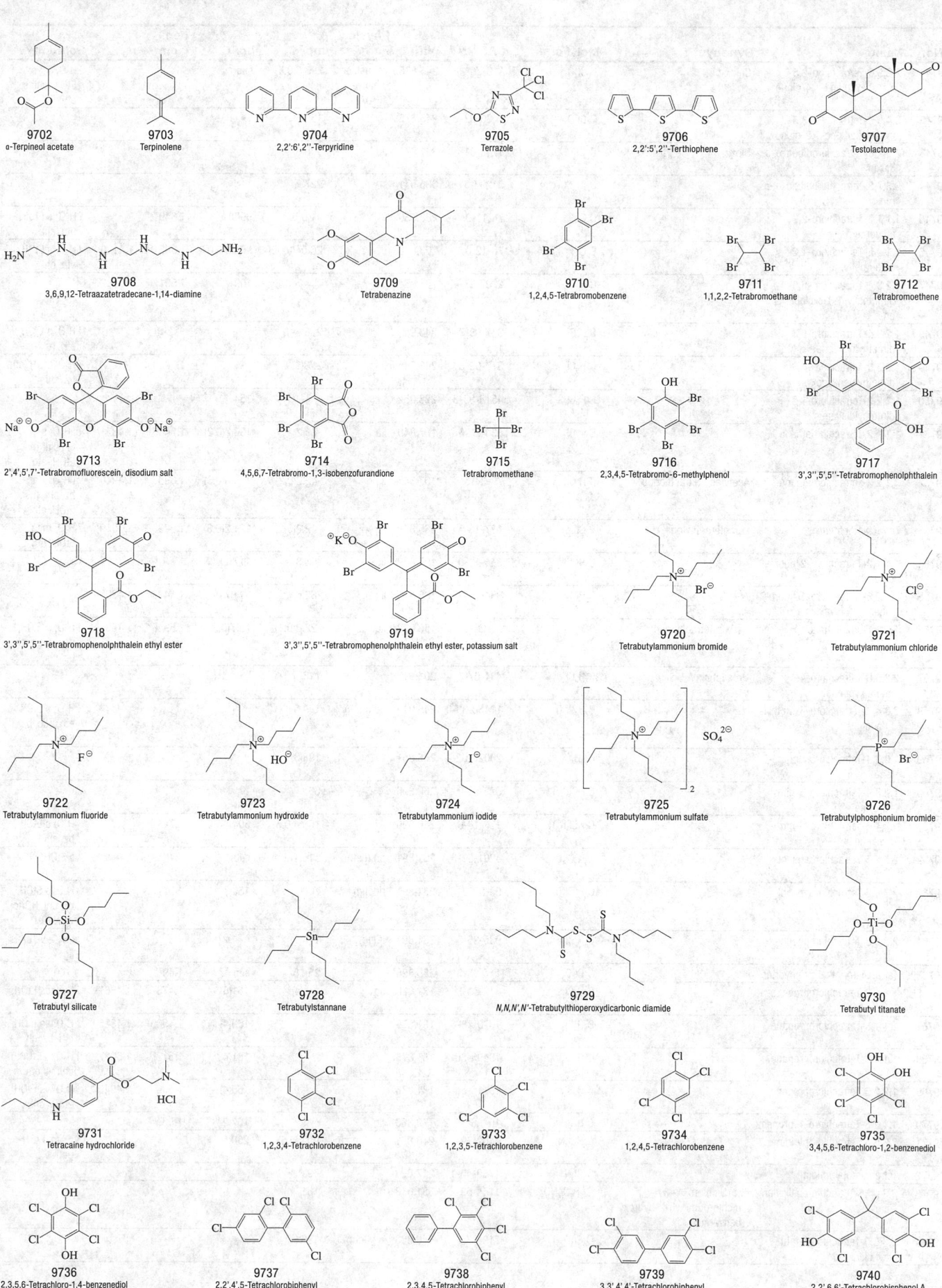

9702
α-Terpineol acetate

9703
Terpinolene

9704
2,2':6',2''-Terpyridine

9705
Terrazole

9706
2,2':5',2''-Terthiophene

9707
Testolactone

9708
3,6,9,12-Tetraazatetradecane-1,14-diamine

9709
Tetrabenazine

9710
1,2,4,5-Tetrabromobenzene

9711
1,1,2,2-Tetrabromoethane

9712
Tetrabromoethene

9713
2',4',5',7'-Tetrabromofluorescein, disodium salt

9714
4,5,6,7-Tetrabromo-1,3-isobenzofurandione

9715
Tetrabromomethane

9716
2,3,4,5-Tetrabromo-6-methylphenol

9717
3',3'',5',5''-Tetrabromophenolphthalein

9718
3',3'',5',5''-Tetrabromophenolphthalein ethyl ester

9719
3',3'',5',5''-Tetrabromophenolphthalein ethyl ester, potassium salt

9720
Tetrabutylammonium bromide

9721
Tetrabutylammonium chloride

9722
Tetrabutylammonium fluoride

9723
Tetrabutylammonium hydroxide

9724
Tetrabutylammonium iodide

9725
Tetrabutylammonium sulfate

9726
Tetrabutylphosphonium bromide

9727
Tetrabutyl silicate

9728
Tetrabutylstannane

9729
N,N,N',N'-Tetrabutylthioperoxydicarbonic diamide

9730
Tetrabutyl titanate

9731
Tetracaine hydrochloride

9732
1,2,3,4-Tetrachlorobenzene

9733
1,2,3,5-Tetrachlorobenzene

9734
1,2,4,5-Tetrachlorobenzene

9735
3,4,5,6-Tetrachloro-1,2-benzenediol

9736
2,3,5,6-Tetrachloro-1,4-benzenediol

9737
2,2',4',5-Tetrachlorobiphenyl

9738
2,3,4,5-Tetrachlorobiphenyl

9739
3,3',4,4'-Tetrachlorobiphenyl

9740
2,2',6,6'-Tetrachlorobisphenol A

No.	Name	Synonym	Mol. Form.	CAS RN	Mol. Wt.	Physical Form	mp/°C	bp/°C	den g cm⁻³	n_D	Solubility
9741	2,3,5,6-Tetrachloro-2,5-cyclohexadiene-1,4-dione	Chloranil	$C_6Cl_4O_2$	118-75-2	245.875	ye mcl, pr (bz) ye lf (HOAc)	298.3(0.5)	sub			i H_2O, liq; sl EtOH, chl; s eth
9742	3,4,5,6-Tetrachloro-3,5-cyclohexadiene-1,2-dione		$C_6Cl_4O_2$	2435-53-2	245.875		130.5				
9743	2,3,7,8-Tetrachlorodibenzo-p-dioxin	Dioxin	$C_{12}H_4Cl_4O_2$	1746-01-6	321.971	nd	305(5)				
9744	2,3,7,8-Tetrachlorodibenzofuran		$C_{12}H_4Cl_4O$	51207-31-9	305.971	cry	227				
9745	1,1,1,2-Tetrachloro-2,2-difluoroethane		$C_2Cl_4F_2$	76-11-9	203.830		41.0(0.3)	96(3)	1.649²⁵		i H_2O; s EtOH, eth, chl
9746	1,1,2,2-Tetrachloro-1,2-difluoroethane		$C_2Cl_4F_2$	76-12-0	203.830		26.54(0.02)	92.83(0.07)	1.5951⁵⁰	1.4130²⁵	i H_2O; s EtOH, eth, chl
9747	1,2,3,4-Tetrachloro-5,5-dimethoxy-1,3-cyclopentadiene		$C_7H_6Cl_4O_2$	2207-27-4	263.934			109	1.501²⁵	1.5282²⁰	
9748	1,2,3,4-Tetrachloro-5,6-dimethylbenzene		$C_8H_6Cl_4$	877-08-7	243.946		228				i H_2O; s EtOH, eth, bz
9749	1,2,3,5-Tetrachloro-4,6-dimethylbenzene		$C_8H_6Cl_4$	877-09-8	243.946		223(2)		1.703²⁵		i H_2O, EtOH, eth, bz, chl
9750	1,1,2,2-Tetrachloro-1,2-dimethyldisilane		$C_2H_6Cl_4Si_2$	4518-98-3	228.052			154			
9751	1,1,1,2-Tetrachloroethane		$C_2H_2Cl_4$	630-20-6	167.849	liq	-70.2	130.2(0.2)	1.5406²⁰	1.4821²⁰	sl H_2O; s ace, bz, chl; msc EtOH, eth
9752	1,1,2,2-Tetrachloroethane	Acetylene tetrachloride	$C_2H_2Cl_4$	79-34-5	167.849	liq	-42.3(0.2)	146.0(0.3)	1.5953²⁰	1.4940²⁰	sl H_2O; s ace, bz, chl; msc EtOH, eth
9753	Tetrachloroethene	Perchloroethylene	C_2Cl_4	127-18-4	165.833	liq	-22.2(0.1)	121.2(0.3)	1.6230²⁰	1.5059²⁰	i H_2O; msc EtOH, eth, bz
9754	1,1,1,2-Tetrachloro-2-fluoroethane		C_2HCl_4F	354-11-0	185.839	liq	-95.3	117.1			
9755	1,1,2,2-Tetrachloro-1-fluoroethane		C_2HCl_4F	354-14-3	185.839	liq	-82.6	116.7	1.5497¹⁷	1.4390²⁰	
9756	Tetrachloromethane	Carbon tetrachloride	CCl_4	56-23-5	153.823	liq	-22.8(0.1)	76.7(0.2)	1.5940²⁰	1.4601²⁰	i H_2O; s EtOH, ace; msc eth, bz, chl
9757	2,3,5,6-Tetrachloro-4-methoxyphenol	Drosophilin A	$C_7H_4Cl_4O_2$	484-67-3	261.918		116				
9758	2,3,4,6-Tetrachloro-5-methylphenol		$C_7H_4Cl_4O$	10460-33-0	245.918	nd (peth)	189.5				i H_2O; s EtOH, eth, ace, bz, KOH
9759	1,2,3,4-Tetrachloronaphthalene		$C_{10}H_4Cl_4$	20020-02-4	265.951		198(3)				
9760	1,2,3,4-Tetrachloro-5-nitrobenzene		$C_6HCl_4NO_2$	879-39-0	260.890		66				
9761	1,2,4,5-Tetrachloro-3-nitrobenzene		$C_6HCl_4NO_2$	117-18-0	260.890		100.4(0.5)	304	1.744²⁵		i H_2O; s EtOH, bz, chl
9762	2,3,4,5-Tetrachlorophenol		$C_6H_2Cl_4O$	4901-51-3	231.891	nd (peth, sub)	116(1)	sub			vs EtOH
9763	2,3,4,6-Tetrachlorophenol		$C_6H_2Cl_4O$	58-90-2	231.891	nd (lig)	70(1)	150¹⁵			i H_2O; s EtOH, bz, chl, HOAc; vs NaOH
9764	2,3,5,6-Tetrachlorophenol		$C_6H_2Cl_4O$	935-95-5	231.891	lf (lig)	115(1)				sl H_2O; vs bz; s lig
9765	Tetrachlorophthalic anhydride		$C_8Cl_4O_3$	117-08-8	285.896		254.5	sub	1.49²⁷⁵		sl eth
9766	1,1,1,2-Tetrachloropropane		$C_3H_4Cl_4$	812-03-3	181.876	liq	-64(4)	151(8)	1.473²⁰	1.4867²⁰	i H_2O; vs EtOH; s eth, chl
9767	1,1,1,3-Tetrachloropropane		$C_3H_4Cl_4$	1070-78-6	181.876			159(4)	1.4509²⁰	1.4825²⁰	i H_2O; vs EtOH, eth, bz, chl
9768	1,1,2,3-Tetrachloropropane		$C_3H_4Cl_4$	18495-30-2	181.876			181(4)	1.513¹⁷	1.5037¹⁷	i H_2O; s EtOH, chl; vs eth
9769	1,2,2,3-Tetrachloropropane		$C_3H_4Cl_4$	13116-53-5	181.876			165(4)	1.500¹⁸	1.4940¹⁸	i H_2O; vs EtOH, eth; s chl
9770	1,1,2,3-Tetrachloro-1-propene		$C_3H_2Cl_4$	10436-39-2	179.860	liq		167.2	1.55²⁰		
9771	2,3,5,6-Tetrachloropyridine		C_5HCl_4N	2402-79-1	216.881	cry (aq al)	90.0(0.5)	250.5			vs eth, EtOH, peth
9772	Tetrachloropyrimidine		$C_4Cl_4N_2$	1780-40-1	217.868		69.0				
9773	3,3',4',5-Tetrachlorosalicylanilide	3,5-Dichloro-N-(3,4-dichlorophenyl)-2-hydroxybenzamide	$C_{13}H_7Cl_4NO_2$	1154-59-2	351.013		161				
9774	2,3,5,6-Tetrachloroterphthaloyl dichloride		$C_6Cl_6O_2$	719-32-4	340.803	cry (ctc)	146.5				

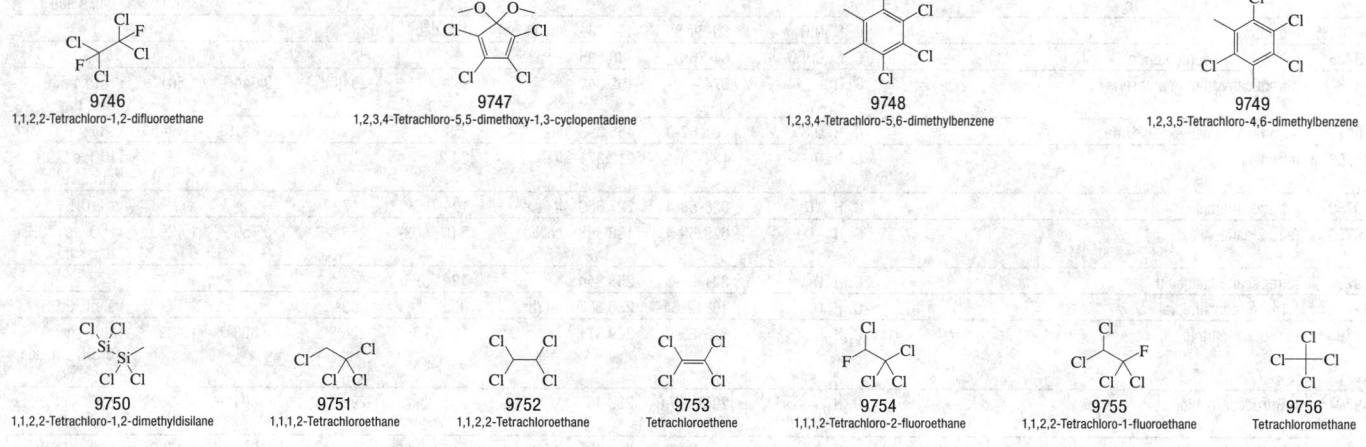

9741
2,3,5,6-Tetrachloro-2,5-cyclohexadiene-1,4-dione

9742
3,4,5,6-Tetrachloro-3,5-cyclohexadiene-1,2-dione

9743
2,3,7,8-Tetrachlorodibenzo-*p*-dioxin

9744
2,3,7,8-Tetrachlorodibenzofuran

9745
1,1,1,2-Tetrachloro-2,2-difluoroethane

9746
1,1,2,2-Tetrachloro-1,2-difluoroethane

9747
1,2,3,4-Tetrachloro-5,5-dimethoxy-1,3-cyclopentadiene

9748
1,2,3,4-Tetrachloro-5,6-dimethylbenzene

9749
1,2,3,5-Tetrachloro-4,6-dimethylbenzene

9750
1,1,2,2-Tetrachloro-1,2-dimethyldisilane

9751
1,1,1,2-Tetrachloroethane

9752
1,1,2,2-Tetrachloroethane

9753
Tetrachloroethene

9754
1,1,1,2-Tetrachloro-2-fluoroethane

9755
1,1,2,2-Tetrachloro-1-fluoroethane

9756
Tetrachloromethane

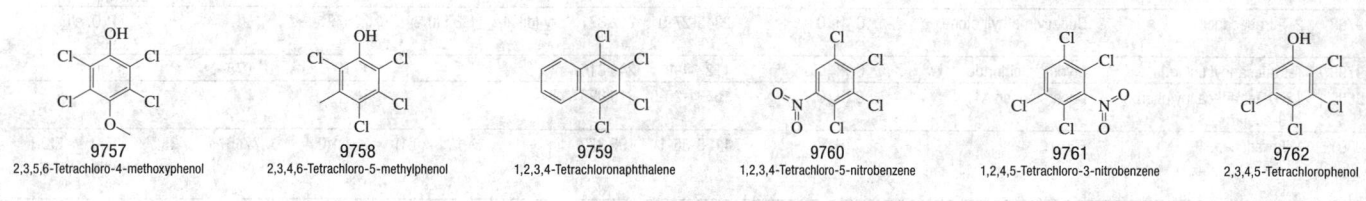

9757
2,3,5,6-Tetrachloro-4-methoxyphenol

9758
2,3,4,6-Tetrachloro-5-methylphenol

9759
1,2,3,4-Tetrachloronaphthalene

9760
1,2,3,4-Tetrachloro-5-nitrobenzene

9761
1,2,4,5-Tetrachloro-3-nitrobenzene

9762
2,3,4,5-Tetrachlorophenol

9763
2,3,4,6-Tetrachlorophenol

9764
2,3,5,6-Tetrachlorophenol

9765
Tetrachlorophthalic anhydride

9766
1,1,1,2-Tetrachloropropane

9767
1,1,1,3-Tetrachloropropane

9768
1,1,2,3-Tetrachloropropane

9769
1,2,2,3-Tetrachloropropane

9770
1,1,2,3-Tetrachloro-1-propene

9771
2,3,5,6-Tetrachloropyridine

9772
Tetrachloropyrimidine

9773
3,3',4',5-Tetrachlorosalicylanilide

9774
2,3,5,6-Tetrachloroterphthaloyl dichloride

No.	Name	Synonym	Mol. Form.	CAS RN	Mol. Wt.	Physical Form	mp/°C	bp/°C	den g cm⁻³	n_D	Solubility
9775	Tetrachlorothiophene		C_4Cl_4S	6012-97-1	221.920	nd (dil al)	30.5	233.4	1.7036[30]	1.5915[30]	i H_2O; vs EtOH; msc eth
9776	Tetrachlorovinphos		$C_{10}H_9Cl_4O_4P$	961-11-5	365.961		97				
9777	Tetracontane		$C_{40}H_{82}$	4181-95-7	563.079		81.4(0.2)	522	0.8171[25]	1.4572[25]	
9778	Tetracosamethylundecasiloxane	Tetracosamethylhendecasiloxane	$C_{24}H_{72}O_{10}Si_{11}$	107-53-9	829.764			322.8	0.9247[25]	1.3994[20]	vs bz
9779	Tetracosane		$C_{24}H_{50}$	646-31-1	338.654	cry (eth)	50.3(0.7)	391(5)	0.7991[20]	1.4283[70]	i H_2O; sl EtOH; vs eth
9780	Tetracosanoic acid	Lignoceric acid	$C_{24}H_{48}O_2$	557-59-5	368.637		82(4)	272[10]	0.8207[100]	1.4287[100]	vs bz, eth
9781	1-Tetracosanol		$C_{24}H_{50}O$	506-51-4	354.653		77	210[0.4]			
9782	cis-15-Tetracosenoic acid	Nervonic acid	$C_{24}H_{46}O_2$	506-37-6	366.621		43				
9783	Tetracyanoethene	Tetracyanoethylene	C_6N_4	670-54-2	128.091		198.8(0.3)	223	1.348[25]	1.560[25]	sl eth, bz, ctc, chl; s ace
9784	Tetracycline		$C_{22}H_{24}N_2O_8$	60-54-8	444.434	cry (+3w)	172 dec				
9785	Tetracycline hydrochloride		$C_{22}H_{25}ClN_2O_8$	64-75-5	480.895		214				
9786	Tetradecahydrophenanthrene		$C_{14}H_{24}$	5743-97-5	192.341	liq	-3	284(6)	0.944[20]	1.5011[20]	i H_2O; s eth, ace, bz
9787	Tetradecamethylhexasiloxane		$C_{14}H_{42}O_5Si_6$	107-52-8	458.993	liq	-59	259.7(0.1)	0.8910[20]	1.3948[20]	vs bz
9788	Tetradecanal		$C_{14}H_{28}O$	124-25-4	212.371	lf	23(2)				i H_2O; s EtOH, eth, ace
9789	Tetradecanamide		$C_{14}H_{29}NO$	638-58-4	227.386	lf (ace)	104(1)	217[12]			vs EtOH
9790	Tetradecane		$C_{14}H_{30}$	629-59-4	198.388	col liq	5.87(0.02)	253.5(0.4)	0.7596[20]	1.4290[20]	i H_2O; vs EtOH, eth; s ctc
9791	Tetradecanedioic acid		$C_{14}H_{26}O_4$	821-38-5	258.354		124.2(0.5)				
9792	1,14-Tetradecanediol		$C_{14}H_{30}O_2$	19812-64-7	230.387	nd (bz)	85.8	200[9]			vs eth, EtOH
9793	Tetradecanenitrile	Myristonitrile	$C_{14}H_{27}N$	629-63-0	209.371		19.1(0.3)	226[100]	0.8281[19]	1.4392[23]	i H_2O; msc EtOH, eth, ace, bz; sl ctc
9794	1-Tetradecanethiol		$C_{14}H_{30}S$	2079-95-0	230.453		7	299(10)	0.8641[20]	1.4597[20]	i H_2O; s EtOH, eth, ctc
9795	Tetradecanoic acid	Myristic acid	$C_{14}H_{28}O_2$	544-63-8	228.371	lf (eth)	54.16(0.02)	250[100]	0.8622[54]	1.4723[70]	i H_2O; s EtOH, ace, chl; sl eth; vs bz
9796	Tetradecanoic anhydride		$C_{28}H_{54}O_3$	626-29-9	438.727	lf (peth)	53.4		0.8502[70]	1.4335[70]	vs eth, EtOH
9797	1-Tetradecanol	Tetradecyl alcohol	$C_{14}H_{30}O$	112-72-1	214.387	lf	37.7(0.7)	295.8(0.4)	0.8236[38]		i H_2O; vs EtOH, eth, ace, bz, chl
9798	2-Tetradecanone	Dodecylmethylketone	$C_{14}H_{28}O$	2345-27-9	212.371	cry (dil al)	33.5(0.2)	205[100]			i H_2O; s EtOH, ace
9799	Tetradecanoyl chloride	Myristoyl chloride	$C_{14}H_{27}ClO$	112-64-1	246.816		-1	171[16]	0.9078[25]		s eth
9800	12-O-Tetradecanoylphorbol-13-acetate	Cocarcinogen A1	$C_{36}H_{56}O_8$	16561-29-8	616.825	oil					
9801	1-Tetradecene		$C_{14}H_{28}$	1120-36-1	196.372	liq	-12.88(0.1)	251.1(0.4)	0.7745[25]	1.4351[20]	i H_2O; vs EtOH, eth; s bz; sl ctc
9802	Tetradecyl acetate	1-Tetradecanol, acetate	$C_{16}H_{32}O_2$	638-59-5	256.424			173[10]			
9803	Tetradecylamine	1-Tetradecanamine	$C_{14}H_{31}N$	2016-42-4	213.403		83.1	289(3)	0.8079[20]	1.4463[20]	i H_2O; vs EtOH, eth, bz, chl; s ace
9804	Tetradecylbenzene		$C_{20}H_{34}$	1459-10-5	274.484		9.7(0.2)	347(10)	0.8549[20]	1.4818[20]	
9805	Tetradecylcyclohexane		$C_{20}H_{40}$	1795-18-2	280.532		24(1)	358(15)	0.8254[20]	1.4579[20]	
9806	Tetradifon	1,2,4-Trichloro-5-[(4-chlorophenyl)sulfonyl]benzene	$C_{12}H_6Cl_4O_2S$	116-29-0	356.052		146.9(0.3)		1.151[20]		
9807	Tetraethoxygermane	Ethanol, germanium(4+) salt	$C_8H_{20}GeO_4$	14165-55-0	252.88			139[200]			
9808	Tetraethoxymethane	Tetraethyl orthocarbonate	$C_9H_{20}O_4$	78-09-1	192.253			159.5	0.9186[20]	1.3905[25]	msc EtOH, eth; s ctc
9809	Tetraethylammonium bromide		$C_8H_{20}BrN$	71-91-0	210.156	hyg (al)	174(2)		1.3970[20]		vs H_2O, EtOH, chl, MeOH
9810	Tetraethylammonium chloride		$C_8H_{20}ClN$	56-34-8	165.705	hyg cry					vs H_2O, EtOH, ace, chl
9811	Tetraethylammonium iodide		$C_8H_{20}IN$	68-05-3	257.156	cry (w)	300 dec				s H_2O
9812	1,2,3,5-Tetraethylbenzene		$C_{14}H_{22}$	38842-05-6	190.325			250(3)			
9813	N,N,N',N'-Tetraethyl-1,2-benzenedicarboxamide	N,N,N',N'-Tetraethyl-phthalamide	$C_{16}H_{24}N_2O_2$	83-81-8	276.374		36	204[16]			
9814	Tetra(2-ethylbutyl) silicate	Silicic acid, tetrakis(2-ethylbutyl) ester	$C_{24}H_{52}O_4Si$	78-13-7	432.754	liq			0.8920[20]	1.4307[20]	i H_2O; sl EtOH, ctc; s eth, bz
9815	Tetraethylene glycol	3,6,9-Trioxaundecane-1,11-diol	$C_8H_{18}O_5$	112-60-7	194.226	liq	-9.4(0.7)	315(7)	1.1285[15]	1.4577[20]	vs H_2O; s EtOH, eth, ctc, diox
9816	Tetraethylene glycol diacrylate		$C_{14}H_{22}O_7$	17831-71-9	302.321				1.125[25]		
9817	Tetraethylene glycol dimethacrylate		$C_{16}H_{26}O_7$	109-17-1	330.373			220[1]		1.4610[25]	

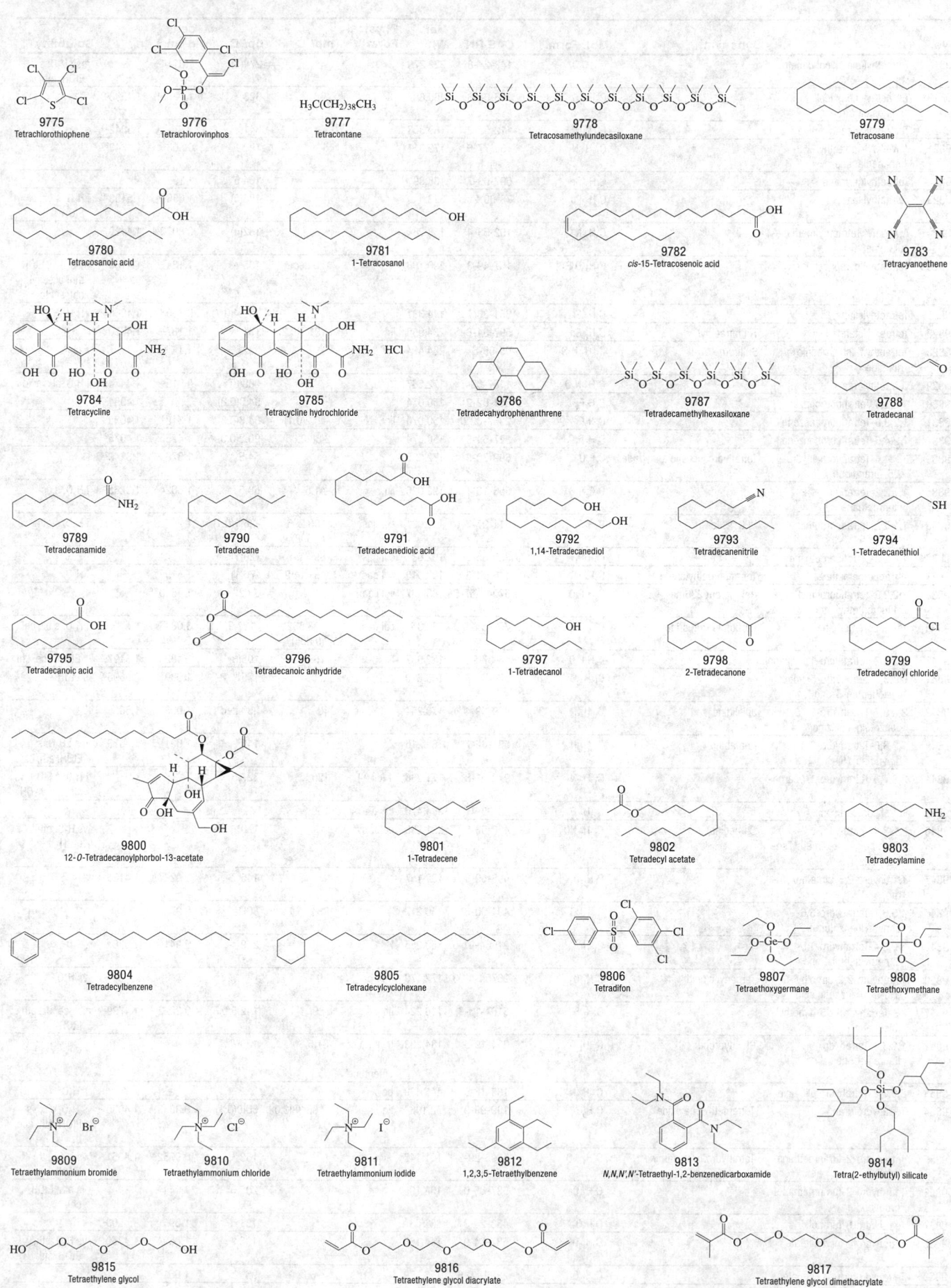

9775
Tetrachlorothiophene

9776
Tetrachlorovinphos

9777
Tetracontane
H₃C(CH₂)₃₈CH₃

9778
Tetracosamethylundecasiloxane

9779
Tetracosane

9780
Tetracosanoic acid

9781
1-Tetracosanol

9782
cis-15-Tetracosenoic acid

9783
Tetracyanoethene

9784
Tetracycline

9785
Tetracycline hydrochloride

9786
Tetradecahydrophenanthrene

9787
Tetradecamethylhexasiloxane

9788
Tetradecanal

9789
Tetradecanamide

9790
Tetradecane

9791
Tetradecanedioic acid

9792
1,14-Tetradecanediol

9793
Tetradecanenitrile

9794
1-Tetradecanethiol

9795
Tetradecanoic acid

9796
Tetradecanoic anhydride

9797
1-Tetradecanol

9798
2-Tetradecanone

9799
Tetradecanoyl chloride

9800
12-O-Tetradecanoylphorbol-13-acetate

9801
1-Tetradecene

9802
Tetradecyl acetate

9803
Tetradecylamine

9804
Tetradecylbenzene

9805
Tetradecylcyclohexane

9806
Tetradifon

9807
Tetraethoxygermane

9808
Tetraethoxymethane

9809
Tetraethylammonium bromide

9810
Tetraethylammonium chloride

9811
Tetraethylammonium iodide

9812
1,2,3,5-Tetraethylbenzene

9813
N,N,N',N'-Tetraethyl-1,2-benzenedicarboxamide

9814
Tetra(2-ethylbutyl) silicate

9815
Tetraethylene glycol

9816
Tetraethylene glycol diacrylate

9817
Tetraethylene glycol dimethacrylate

No.	Name	Synonym	Mol. Form.	CAS RN	Mol. Wt.	Physical Form	mp/°C	bp/°C	den g cm⁻³	n_D	Solubility
9818	Tetraethylene glycol dimethyl ether		$C_{10}H_{22}O_5$	143-24-8	222.279			274(1)	1.0114²⁰		msc H_2O; s EtOH, eth, ctc
9819	Tetraethylene glycol monostearate		$C_{26}H_{52}O_6$	106-07-0	460.687		40	328	1.1285¹⁵	1.4593²⁰	
9820	Tetraethylenepentamine		$C_8H_{23}N_5$	112-57-2	189.303			341.5		1.5042²⁰	s H_2O
9821	N,N,N',N'-Tetraethyl-1,2-ethanediamine		$C_{10}H_{24}N_2$	150-77-6	172.311			192	0.808²⁵	1.4343²⁰	
9822	Tetraethylgermane		$C_8H_{20}Ge$	597-63-7	188.89			164.5	1.199		
9823	Tetraethyl lead		$C_8H_{20}Pb$	78-00-2	323.4			183(5)	1.653²⁰	1.5198²⁰	i H_2O; s bz; sl EtOH
9824	N,N,N',N'-Tetraethylmethane-diamine		$C_9H_{22}N_2$	102-53-4	158.284			152(4)	0.8000²⁰	1.4420²⁵	
9825	Tetraethyl pyrophosphate		$C_8H_{20}O_7P_2$	107-49-3	290.188		170 dec	155³	1.1847²⁰	1.4180²⁰	msc H_2O, EtOH, eth, ace, xyl; chl; sl ctc
9826	Tetraethylsilane		$C_8H_{20}Si$	631-36-7	144.331			153.4(0.7)	0.7658²⁰	1.4268²⁰	i H_2O
9827	Tetraethylstannane	Tin tetraethyl	$C_8H_{20}Sn$	597-64-8	234.955	liq	-131.2(0.3)	181	1.187²⁵	1.4730²⁰	
9828	Tetraethylthiodicarbonic diamide	Sulfiram	$C_{10}H_{20}N_2S_3$	95-05-6	264.474			232³	1.12²⁰		s chl
9829	Tetraethylurea		$C_9H_{20}N_2O$	1187-03-7	172.267			208(8)	0.919²⁰	1.4474²⁰	i H_2O, alk, acid
9830	1,2,3,4-Tetrafluorobenzene		$C_6H_2F_4$	551-62-2	150.074			94.3(0.8)		1.4054²⁰	
9831	1,2,3,5-Tetrafluorobenzene		$C_6H_2F_4$	2367-82-0	150.074	liq	-46.2(0.1)	84.3(0.8)	1.319²⁵	1.4035²⁰	
9832	1,2,4,5-Tetrafluorobenzene		$C_6H_2F_4$	327-54-8	150.074		3.88(0.05)	90.2(0.3)	1.4255²⁰	1.4075²⁰	
9833	3,3,4,4-Tetrafluorodihydro-2,5-furandione	Tetrafluorosuccinic anhydride	$C_4F_4O_3$	699-30-9	172.035		54.5		1.6209²⁰	1.3240²⁰	
9834	1,1,2,2-Tetrafluoro-1,2-dinitroethane		$C_2F_4N_2O_4$	356-16-1	192.026	liq	-41.5	58.5	1.6024²⁵	1.3265²⁵	i H_2O; s ace
9835	1,1,1,2-Tetrafluoroethane	HFC-134a	$C_2H_2F_4$	811-97-2	102.031	col gas	-103.296 (0.008)	-26.1(0.1)	1.2072²⁵		i H_2O; s eth
9836	1,1,2,2-Tetrafluoroethane		$C_2H_2F_4$	359-35-3	102.031	col gas	-89	-20(1)			
9837	Tetrafluoroethene	Tetrafluoroethylene	C_2F_4	116-14-3	100.015	col gas	-131.14(0.02)	-76(1)	1.519⁻⁷⁶		i H_2O
9838	1,2,2,2-Tetrafluoroethyl difluoromethyl ether	Refrigerant 236me	$C_3H_2F_6O$	57041-67-5	168.037	vol liq or gas		23(2)	1.4540²³		
9839	Tetrafluoromethane	Carbon tetrafluoride	CF_4	75-73-0	88.005	col gas	-183.582 (0.005)	-127.9(0.1)	3.034²⁵		i H_2O; s bz, chl
9840	2,2,3,3-Tetrafluoro-1-propanol		$C_3H_4F_4O$	76-37-9	132.057	liq	-15	106(3)	1.4853²⁰	1.3197²⁰	s EtOH, ace, chl
9841	6,7,8,9-Tetrahydro-5H-benzo-cyclohepten-5-one		$C_{11}H_{12}O$	826-73-3	160.212			175⁴⁰	1.080²⁰	1.5698²⁰	s EtOH
9842	2,3,6,7-Tetrahydro-1H,5H-benzo[ij]quinolizine	Julolidine	$C_{12}H_{15}N$	479-59-4	173.254		40	280 dec	1.003²⁰	1.568²⁵	
9843	1,2,3,6-Tetrahydro-2,3'-bipyridine, (S)-	Anatabine	$C_{10}H_{12}N_2$	581-49-7	160.215			145¹⁰	1.091¹⁹	1.5676²⁰	msc H_2O; s EtOH, eth, bz
9844	2,3,4,9-Tetrahydro-1H-carba-zole		$C_{12}H_{13}N$	942-01-8	171.238	lf (dil al)	120	327.5			i H_2O; s EtOH; vs eth, bz, MeOH
9845	Tetrahydrocortisone		$C_{21}H_{32}O_5$	53-05-4	364.476	cry (EtOAc)	190				
9846	1,2,3,4-Tetrahydro-6,7-dimethoxy-1,2-dimethyliso-quinoline, (±)-	Carnegine	$C_{13}H_{19}NO_2$	490-53-9	221.296	pa br syr		170¹			vs H_2O, eth, EtOH
9847	Tetrahydro-2,5-dimethoxyfu-ran		$C_6H_{12}O_3$	696-59-3	132.157			145.7	1.02²⁵	1.4180²⁰	
9848	4,5,6,7-Tetrahydro-3,6-dimethylbenzofuran		$C_{10}H_{14}O$	494-90-6	150.217		86	80¹⁸	0.972¹⁵		
9849	1,2,3,4-Tetrahydro-1,5-dimethylnaphthalene		$C_{12}H_{16}$	21564-91-0	160.255			239	0.941²⁰	1.526²⁰	
9850	Tetrahydro-2,2-dimethyl-5-oxo-3-furanacetic acid	Terpenylic acid	$C_8H_{12}O_4$	26754-48-3	172.179	lf or pr (w+1)	90				vs H_2O
9851	cis-Tetrahydro-2,5-dimethyl-thiophene		$C_6H_{12}S$	5161-13-7	116.224	liq	-89	142.5(0.8)	0.9222²⁰	1.4799²⁰	vs ace, bz, eth, EtOH
9852	1,2,3,4-Tetrahydro-9H-fluoren-9-one	Phentydrone	$C_{13}H_{12}O$	634-19-5	184.233	lt ye nd or pr (pentane)	81.5	139⁰·⁰⁵			
9853	5,6,7,8-Tetrahydrofolic acid		$C_{19}H_{23}N_7O_6$	135-16-0	445.429	pow					s H_2O
9854	Tetrahydrofuran	Tetramethylene oxide	C_4H_8O	109-99-9	72.106	liq	-108.38(0.02)	66.0(0.1)	0.8833²⁵	1.4050²⁰	s H_2O, chl; vs EtOH, eth, ace, bz
9855	Tetrahydro-2-furanmethana-mine	Tetrahydrofurfurylamine	$C_5H_{11}NO$	4795-29-3	101.147			153	0.9752²⁰	1.4551²⁰	vs H_2O, eth, EtOH
9856	Tetrahydro-2-furanmethanol propanoate		$C_8H_{14}O_3$	637-65-0	158.195			205.5	1.044²⁰		vs eth, EtOH, chl
9857	Tetrahydro-3-furanol		$C_4H_8O_2$	453-20-3	88.106			181	1.09²⁵	1.4500²⁰	
9858	Tetrahydrofurfuryl acetate		$C_7H_{12}O_3$	637-64-9	144.168			193	1.0624²⁰	1.4350²⁵	vs H_2O, eth, EtOH, chl
9859	Tetrahydrofurfuryl acrylate		$C_8H_{12}O_3$	2399-48-6	156.179		<-60	96⁶	1.061²⁰		

9818
Tetraethylene glycol dimethyl ether

9819
Tetraethylene glycol monostearate

9820
Tetraethylenepentamine

9821
N,N,N',N'-Tetraethyl-1,2-ethanediamine

9822
Tetraethylgermane

9823
Tetraethyl lead

9824
N,N,N',N'-Tetraethylmethanediamine

9825
Tetraethyl pyrophosphate

9826
Tetraethylsilane

9827
Tetraethylstannane

9828
Tetraethylthiodicarbonic diamide

9829
Tetraethylurea

9830
1,2,3,4-Tetrafluorobenzene

9831
1,2,3,5-Tetrafluorobenzene

9832
1,2,4,5-Tetrafluorobenzene

9833
3,3,4,4-Tetrafluorodihydro-2,5-furandione

9834
1,1,2,2-Tetrafluoro-1,2-dinitroethane

9835
1,1,1,2-Tetrafluoroethane

9836
1,1,2,2-Tetrafluoroethane

9837
Tetrafluoroethene

9838
1,2,2,2-Tetrafluoroethyl difluoromethyl ether

9839
Tetrafluoromethane

9840
2,2,3,3-Tetrafluoro-1-propanol

9841
6,7,8,9-Tetrahydro-5*H*-benzocyclohepten-5-one

9842
2,3,6,7-Tetrahydro-1*H*,5*H*-benzo[ij]quinolizine

9843
1,2,3,6-Tetrahydro-2,3'-bipyridine, (*S*)-

9844
2,3,4,9-Tetrahydro-1*H*-carbazole

9845
Tetrahydrocortisone

9846
1,2,3,4-Tetrahydro-6,7-dimethoxy-1,2-dimethylisoquinoline, (±)-

9847
Tetrahydro-2,5-dimethoxyfuran

9848
4,5,6,7-Tetrahydro-3,6-dimethylbenzofuran

9849
1,2,3,4-Tetrahydro-1,5-dimethylnaphthalene

9850
Tetrahydro-2,2-dimethyl-5-oxo-3-furanacetic acid

9851
cis-Tetrahydro-2,5-dimethylthiophene

9852
1,2,3,4-Tetrahydro-9*H*-fluoren-9-one

9853
5,6,7,8-Tetrahydrofolic acid

9854
Tetrahydrofuran

9855
Tetrahydro-2-furanmethanamine

9856
Tetrahydro-2-furanmethanol propanoate

9857
Tetrahydro-3-furanol

9858
Tetrahydrofurfuryl acetate

9859
Tetrahydrofurfuryl acrylate

No.	Name	Synonym	Mol. Form.	CAS RN	Mol. Wt.	Physical Form	mp/°C	bp/°C	den g cm⁻³	n_D	Solubility
9860	Tetrahydrofurfuryl alcohol	Tetrahydro-2-furancarbinol	C₅H₁₀O₂	97-99-4	102.132	liq	<-80	176.3(0.7)	1.0524²⁰	1.4520²⁰	vs ace, eth
9861	Tetrahydrofurfuryl methacrylate		C₉H₁₄O₃	2455-24-5	170.205			265	1.040²⁵	1.4554²⁵	
9862	Tetrahydroimidazo[4,5-d] imidazole-2,5(1H,3H)-dione	Acetyleneurea	C₄H₆N₄O₂	496-46-8	142.117	nd or pr (w)	297(2)				sl H₂O; i EtOH, HOAc; s eth, HCl, alk
9863	cis-3a,4,7,7a-Tetrahydro-1,3-isobenzofurandione	4-Cyclohexene-1,2-dicarboxylic acid, anhydride	C₈H₈O₃	935-79-5	152.148	cry (peth)	98(2)				s EtOH, ace, chl, bz; sl peth
9864	4,5,6,7-Tetrahydro-1,3-isobenzofurandione	1-Cyclohexene-1,2-dicarboxylic acid, anhydride	C₈H₈O₃	2426-02-0	152.148	pl (EtOH)	70.3(0.2)		1.2¹⁰⁵		s EtOH, ace, chl; vs eth
9865	1,2,3,4-Tetrahydroisoquinoline		C₉H₁₁N	91-21-4	133.190		<-15	232.5	1.0642²⁴	1.5668²⁰	i H₂O; s EtOH, bz, acid, xyl
9866	3,4,5,6-Tetrahydro-7-methoxy-2H-azepine		C₇H₁₃NO	2525-16-8	127.184	liq		49¹⁶	0.887	1.4630²⁰	
9867	1,2,3,4-Tetrahydro-6-methoxyquinoline		C₁₀H₁₃NO	120-15-0	163.216	pr (peth, al) orth pym (w)	42.5	284		1.5718²⁰	s chl
9868	1,2,3,4-Tetrahydro-1-methylnaphthalene		C₁₁H₁₄	1559-81-5	146.229			221(6)	0.9583²⁰	1.5353²⁰	
9869	1,2,3,4-Tetrahydro-5-methylnaphthalene		C₁₁H₁₄	2809-64-5	146.229	liq	-23.0(0.3)	234(7)	0.9720²⁰	1.5439²⁰	
9870	1,2,3,4-Tetrahydro-6-methylnaphthalene		C₁₁H₁₄	1680-51-9	146.229	liq	-40	229	0.9537²⁰	1.5357²⁰	
9871	1,2,3,6-Tetrahydro-1-methyl-4-phenylpyridine	MPTP	C₁₂H₁₅N	28289-54-5	173.254	cry	41	87⁰·⁸			
9872	Tetrahydro-3-methyl-2H-thiopyran		C₆H₁₂S	5258-50-4	116.224	liq	-60	158.1(0.4)	0.9473²⁰	1.4922²⁰	
9873	5,6,7,8-Tetrahydro-1-naphthalenamine		C₁₀H₁₃N	2217-41-6	147.217		38	278(14)	1.0625¹⁶	1.5900²⁰	sl H₂O; s EtOH, eth, acid
9874	1,2,3,4-Tetrahydronaphthalene	Tetralin	C₁₀H₁₂	119-64-2	132.202	liq	-35.76(0.06)	207.2(0.3)	0.9645²⁵	1.5413²⁰	i H₂O; vs EtOH, eth; s chl, PhNH₂
9875	1,2,3,4-Tetrahydro-1-naphthol	1,2,3,4-Tetrahydro-α-naphthol	C₁₀H₁₂O	529-33-9	148.201		34.5	255	1.0996²⁰	1.5638²⁰	
9876	5,6,7,8-Tetrahydro-1-naphthol	5,6,7,8-Tetrahydro-α-naphthol	C₁₀H₁₂O	529-35-1	148.201		68(3)	266	1.0556⁷⁵		
9877	1,2,3,4-Tetrahydro-2-naphthol	Tetralol	C₁₀H₁₂O	530-91-6	148.201		15.5	140¹²			
9878	5,6,7,8-Tetrahydro-2-naphthol	5,6,7,8-Tetrahydro-β-naphthol	C₁₀H₁₂O	1125-78-6	148.201		59(3)	275.5	1.0552⁶⁵		
9879	Tetrahydro-6-pentyl-2H-pyran-2-one	5-Hydroxydecanoic acid lactone	C₁₀H₁₈O₂	705-86-2	170.249	liq	-27	121³			
9880	1,2,3,4-Tetrahydrophenanthrene		C₁₄H₁₄	1013-08-7	182.261	lf (MeOH)	29.404 (0.009)	173¹¹	1.0601⁴⁰		i H₂O; s EtOH, eth, ace, bz, HOAc, chl, lig
9881	1,2,3,6-Tetrahydrophthalimide		C₈H₉NO₂	85-40-5	151.163	cry (EtOH)	137				
9882	Tetrahydro-6-propyl-2H-pyran-2-one	5-Hydroxyoctanoic acid lactone	C₈H₁₄O₂	698-76-0	142.196	liq	-13	126¹⁵			
9883	2,3,4,5-Tetrahydro-6-propylpyridine	γ-Coniceine	C₈H₁₅N	1604-01-9	125.212			174	0.8753¹⁵	1.4661¹⁶	
9884	Tetrahydropyran	Oxane	C₅H₁₀O	142-68-7	86.132	liq	-49.1(0.2)	88.0(0.4)	0.8814²⁰	1.4200²⁰	s EtOH, eth, bz, ctc
9885	Tetrahydro-2H-pyran-2-methanol		C₆H₁₂O₂	100-72-1	116.158			189(2)	1.027²⁵	1.458²⁰	
9886	Tetrahydro-2H-pyran-2-one		C₅H₈O₂	542-28-9	100.117	liq	-10.33(0.09)	228(1)	1.1082²⁰	1.4503²⁰	s H₂O; msc EtOH, eth; sl ctc
9887	Tetrahydro-4H-pyran-4-one		C₅H₈O₂	29943-42-8	100.117			166.5	1.084²⁵	1.4520²⁰	
9888	1,2,5,6-Tetrahydropyridine	Δ 3-Piperidine	C₅H₉N	694-05-3	83.132	liq	-48	117(4)	0.911²⁵	1.4800²⁰	s chl
9889	1,2,5,6-Tetrahydro-3-pyridinecarboxylic acid	Guvacine	C₆H₉NO₂	498-96-4	127.141	pr (w), rods (+1w dil al)	295 dec				vs H₂O
9890	3,4,5,6-Tetrahydro-2(1H)-pyrimidinethione	Hexahydropyrimidine-2-thione	C₄H₈N₂S	2055-46-1	116.185		211		1.33²⁰		
9891	1,2,3,4-Tetrahydroquinoline		C₉H₁₁N	635-46-1	133.190	nd	16.76(0.05)	250.5(0.2)	1.0588²⁰	1.6062¹⁹	s H₂O, chl; msc EtOH, eth
9892	5,6,7,8-Tetrahydroquinoline	2,3-Cyclohexenopyridine	C₉H₁₁N	10500-57-9	133.190			223.5(0.9)	1.0304¹³	1.5435²⁰	sl H₂O; s EtOH, eth, ace, bz
9893	1,2,3,4-Tetrahydroquinoxaline		C₈H₁₀N₂	3476-89-9	134.178	lf (w, eth, peth)	99	289			s H₂O, chl; vs EtOH, eth, bz; sl peth
9894	6,7,8,9-Tetrahydro-5H-tetrazolo[1,5-a]azepine	Pentylenetetrazole	C₆H₁₀N₄	54-95-5	138.170	cry (bz-lig)	59.5	194¹²			vs H₂O, EtOH, ace; s eth, bz; sl chl
9895	Tetrahydrothiophene	Thiacyclopentane	C₄H₈S	110-01-0	88.172	liq	-96.13(0.05)	121.1(0.2)	0.9987²⁰	1.4871¹⁸	i H₂O; msc EtOH, eth, ace, bz; s chl

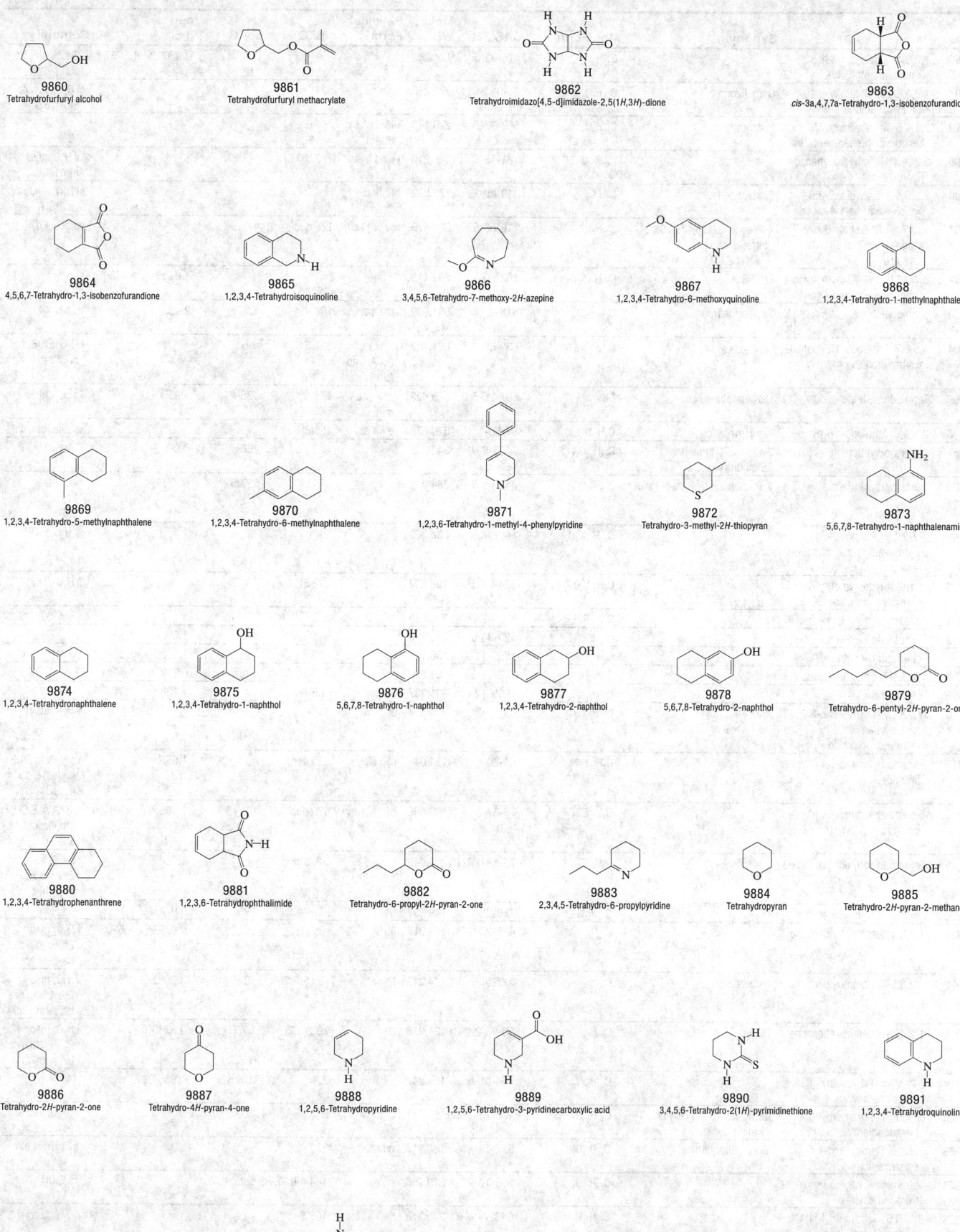

9860
Tetrahydrofurfuryl alcohol

9861
Tetrahydrofurfuryl methacrylate

9862
Tetrahydroimidazo[4,5-d]imidazole-2,5(1H,3H)-dione

9863
cis-3a,4,7,7a-Tetrahydro-1,3-isobenzofurandione

9864
4,5,6,7-Tetrahydro-1,3-isobenzofurandione

9865
1,2,3,4-Tetrahydroisoquinoline

9866
3,4,5,6-Tetrahydro-7-methoxy-2H-azepine

9867
1,2,3,4-Tetrahydro-6-methoxyquinoline

9868
1,2,3,4-Tetrahydro-1-methylnaphthalene

9869
1,2,3,4-Tetrahydro-5-methylnaphthalene

9870
1,2,3,4-Tetrahydro-6-methylnaphthalene

9871
1,2,3,6-Tetrahydro-1-methyl-4-phenylpyridine

9872
Tetrahydro-3-methyl-2H-thiopyran

9873
5,6,7,8-Tetrahydro-1-naphthalenamine

9874
1,2,3,4-Tetrahydronaphthalene

9875
1,2,3,4-Tetrahydro-1-naphthol

9876
5,6,7,8-Tetrahydro-1-naphthol

9877
1,2,3,4-Tetrahydro-2-naphthol

9878
5,6,7,8-Tetrahydro-2-naphthol

9879
Tetrahydro-6-pentyl-2H-pyran-2-one

9880
1,2,3,4-Tetrahydrophenanthrene

9881
1,2,3,6-Tetrahydrophthalimide

9882
Tetrahydro-6-propyl-2H-pyran-2-one

9883
2,3,4,5-Tetrahydro-6-propylpyridine

9884
Tetrahydropyran

9885
Tetrahydro-2H-pyran-2-methanol

9886
Tetrahydro-2H-pyran-2-one

9887
Tetrahydro-4H-pyran-4-one

9888
1,2,5,6-Tetrahydropyridine

9889
1,2,5,6-Tetrahydro-3-pyridinecarboxylic acid

9890
3,4,5,6-Tetrahydro-2(1H)-pyrimidinethione

9891
1,2,3,4-Tetrahydroquinoline

9892
5,6,7,8-Tetrahydroquinoline

9893
1,2,3,4-Tetrahydroquinoxaline

9894
6,7,8,9-Tetrahydro-5H-tetrazolo[1,5-a]azepine

9895
Tetrahydrothiophene

No.	Name	Synonym	Mol. Form.	CAS RN	Mol. Wt.	Physical Form	mp/°C	bp/°C	den g cm⁻³	n_D	Solubility
9896	1,2,3,4-Tetrahydro-1,1,6-trimethylnaphthalene		$C_{13}H_{18}$	475-03-6	174.282			240	0.9303²⁰	1.5257²⁰	s EtOH, eth, bz, chl
9897	1,2,5,8-Tetrahydroxy-9,10-anthracenedione	Quinalizarin	$C_{14}H_8O_6$	81-61-8	272.210	oran nd	>275				sl H₂O, ace, bz, EtOH, eth
9898	2,3,4,6-Tetrahydroxy-5H-benzocyclohepten-5-one	Purpurogallin	$C_{11}H_8O_5$	569-77-7	220.179	red nd (gl HOAc)	274 dec				
9899	2,2',4,4'-Tetrahydroxybenzo-phenone		$C_{13}H_{10}O_5$	131-55-5	246.215	ye nd (w+1)	198.8(0.5)				vs H₂O, ace, eth, EtOH
9900	2,3,5,6-Tetrahydroxy-2,5-cyclohexadiene-1,4-dione	Tetroquinone	$C_6H_4O_6$	319-89-1	172.092	bl-blk cry					sl H₂O, eth, ctc; vs EtOH
9901	11,17,20,21-Tetrahydroxy-pregn-4-en-3-one, (11β,20R)-	4-Pregnene-11β,17α,20β,21-tetrol-3-one	$C_{21}H_{32}O_5$	116-58-5	364.476	cry (aq ace)	125 dec				vs ace, EtOH
9902	N,N,N',N'-Tetra(2-hydroxypro-pyl)ethylenediamine	ENTPROL	$C_{14}H_{32}N_2O_4$	102-60-3	292.415				1.030²⁵	1.478²⁵	sl chl
9903	Tetraiodoethene	Tetraiodoethylene	C_2I_4	513-92-8	531.639	ye lf, pr (eth)	187	sub	2.983²⁰		vs bz, chl
9904	4,5,6,7-Tetraiodo-1,3-isoben-zofurandione		$C_8I_4O_3$	632-80-4	651.702	ye pr, nd (HOAc) nd (sub)	327.5	sub			i H₂O, EtOH, bz; sl HOAc
9905	Tetraiodomethane	Carbon tetraiodide	CI_4	507-25-5	519.629	red lf (bz, chl)	171	135¹·⁵	4.23²⁰		vs py, chl
9906	2,3,4,5-Tetraiodo-1H-pyrrole	Iodopyrrole	C_4HI_4N	87-58-1	570.676	ye nd (al)	155(6)				vs ace, eth, chl
9907	Tetraisobutyl titanate	2-Methyl-1-propanol, titanium(4+) salt	$C_{16}H_{36}O_4Ti$	7425-80-1	340.322			256⁵⁰⁰	0.960⁵⁰		dec H₂O
9908	Tetraisopropyl titanate	2-Propanol, titanium(4+) salt	$C_{12}H_{28}O_4Ti$	546-68-9	284.215			227.5	0.9711²⁰		dec H₂O; s EtOH, eth, bz, chl
9909	N,N,N',N'-Tetrakis(2-hydroxyethyl)-1,2-ethanedi-amine		$C_{10}H_{24}N_2O_4$	140-07-8	236.309						sl H₂O, EtOH
9910	Tetrakis(hydroxymethyl)-phosphonium chloride		$C_4H_{12}ClO_4P$	124-64-1	190.562		152.5				s H₂O
9911	Tetrakis(methylthio)methane		$C_5H_{12}S_4$	6156-25-8	200.409						s chl
9912	1-Tetralone		$C_{10}H_{10}O$	529-34-0	146.185		8	115⁶	1.0988¹⁶	1.5672²⁰	
9913	Tetramethoxymethane		$C_5H_{12}O_4$	1850-14-2	136.147	liq	-2.5	132(13)	1.023²⁵	1.3845²⁰	
9914	1,1,3,3-Tetramethoxypropane		$C_7H_{16}O_4$	102-52-3	164.200			183	0.997²⁵	1.4081²⁰	
9915	Tetramethrin		$C_{19}H_{25}NO_4$	7696-12-0	331.407	wh cry	≈65-80		1.108²⁰	1.5175²¹	
9916	N,N,N',N'-Tetramethyl-3,6-acridinediamine, monohydrochloride	Acridine Orange	$C_{17}H_{20}ClN_3$	65-61-2	301.814	oran-ye soln					s H₂O, EtOH
9917	Tetramethylammonium bromide		$C_4H_{12}BrN$	64-20-0	154.049	hyg bipym	230 dec		1.56²⁵		vs H₂O; sl EtOH; i eth, bz, chl; s MeOH
9918	Tetramethylammonium chloride		$C_4H_{12}ClN$	75-57-0	109.598	hyg bipym (dil al)	420 dec		1.169²⁰		s H₂O; sl EtOH; i eth, bz, chl; vs MeOH
9919	Tetramethylammonium iodide		$C_4H_{12}IN$	75-58-1	201.049		>230 dec		1.829²⁵		sl H₂O, alk, EtOH, ace; i eth, chl
9920	N,N,2,6-Tetramethylaniline		$C_{10}H_{15}N$	769-06-2	149.233	liq	-36	196	0.9147²⁰		
9921	1,2,3,4-Tetramethylbenzene		$C_{10}H_{14}$	488-23-3	134.218	liq	-6.7(1)	205(1)	0.9052²⁰	1.5203²⁰	i H₂O; msc EtOH, eth, ace, bz, peth, ctc
9922	1,2,3,5-Tetramethylbenzene	Isodurene	$C_{10}H_{14}$	527-53-7	134.218	liq	-23.8(0.2)	198(1)	0.8903²⁰	1.5130²⁰	i H₂O; msc EtOH, eth, ace, bz, peth, ctc
9923	1,2,4,5-Tetramethylbenzene	Durene	$C_{10}H_{14}$	95-93-2	134.218		79.2(0.2)	197(1)	0.8380⁸¹	1.4790⁸¹	i H₂O; msc EtOH, eth, ace, bz, peth, ctc
9924	N,N,N',N'-Tetramethyl-1,2-benzenediamine		$C_{10}H_{16}N_2$	704-01-8	164.247		8.9	215.5	0.9560²⁰		
9925	N,N,N',N'-Tetramethyl-1,4-benzenediamine	Tetramethyl-p-phenylenedi-amine	$C_{10}H_{16}N_2$	100-22-1	164.247	lf (dil al or lig)	51	260			sl H₂O; vs EtOH, eth, bz, chl
9926	2,3,5,6-Tetramethyl-1,4-benzenediol	Durohydroquinone	$C_{10}H_{14}O_2$	527-18-4	166.217	nd (al)	233				s EtOH; sl eth
9927	Tetramethyl 1,2,4,5-benzene-tetracarboxylate		$C_{14}H_{14}O_8$	635-10-9	310.256	nd (al)	144.8(0.5)	sub			vs EtOH
9928	3,3',5,5'-Tetramethyl-[1,1'-biphenyl]-4,4'-diamine		$C_{16}H_{20}N_2$	54827-17-7	240.343		168.5				
9929	N,N,N',N'-Tetramethyl-[1,1'-biphenyl]-4,4'-diamine		$C_{16}H_{20}N_2$	366-29-0	240.343		196.0				
9930	3,3',5,5'-Tetramethyl-[1,1'-biphenyl]-4,4'-diol		$C_{16}H_{18}O_2$	2417-04-1	242.313	pa ye nd or pr (HOAc)	221.8	sub			sl EtOH, bz, gl HOAc, tol; i lig

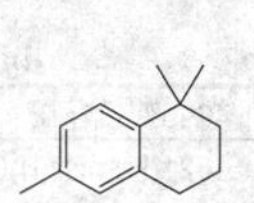

9896
1,2,3,4-Tetrahydro-1,1,6-trimethylnaphthalene

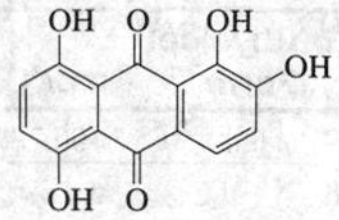

9897
1,2,5,8-Tetrahydroxy-9,10-anthracenedione

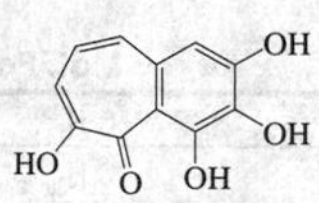

9898
2,3,4,6-Tetrahydroxy-5*H*-benzocyclohepten-5-one

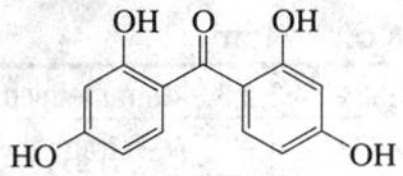

9899
2,2′,4,4′-Tetrahydroxybenzophenone

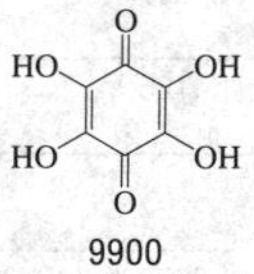

9900
2,3,5,6-Tetrahydroxy-2,5-cyclohexadiene-1,4-dione

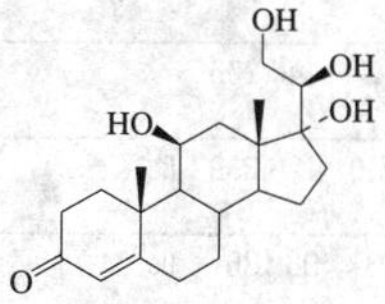

9901
11,17,20,21-Tetrahydroxypregn-4-en-3-one, (11β,20*R*)-

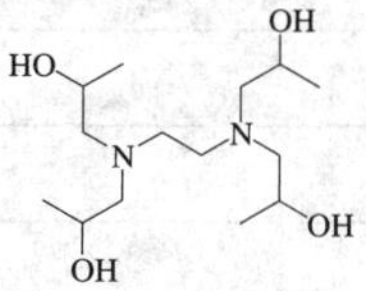

9902
N,N,N′,N′-Tetra(2-hydroxypropyl)ethylenediamine

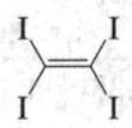

9903
Tetraiodoethene

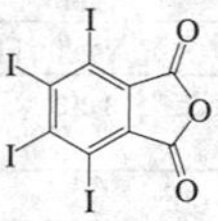

9904
4,5,6,7-Tetraiodo-1,3-isobenzofurandione

9905
Tetraiodomethane

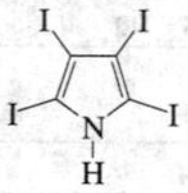

9906
2,3,4,5-Tetraiodo-1*H*-pyrrole

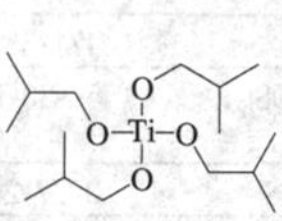

9907
Tetraisobutyl titanate

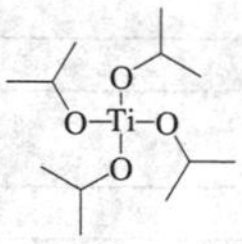

9908
Tetraisopropyl titanate

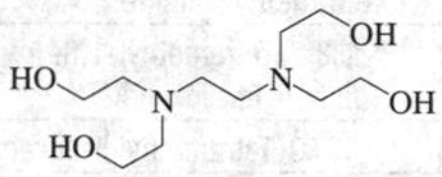

9909
N,N,N′,N′-Tetrakis(2-hydroxyethyl)-1,2-ethanediamine

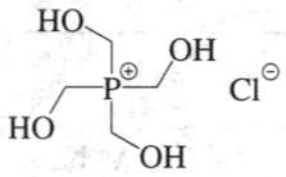

9910
Tetrakis(hydroxymethyl)phosphonium chloride

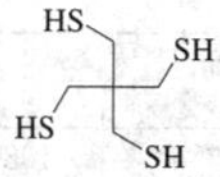

9911
Tetrakis(methylthio)methane

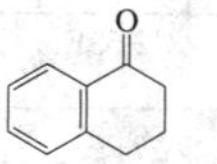

9912
1-Tetralone

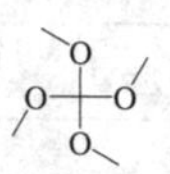

9913
Tetramethoxymethane

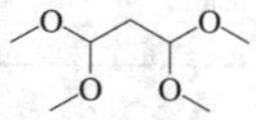

9914
1,1,3,3-Tetramethoxypropane

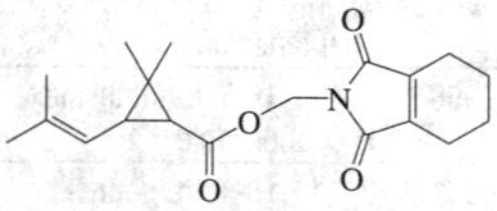

9915
Tetramethrin

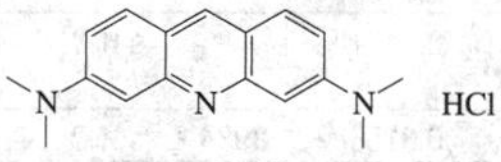

9916
N,N,N′,N′-Tetramethyl-3,6-acridinediamine, monohydrochloride

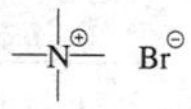

9917
Tetramethylammonium bromide

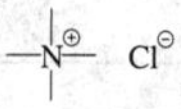

9918
Tetramethylammonium chloride

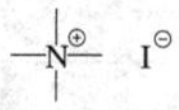

9919
Tetramethylammonium iodide

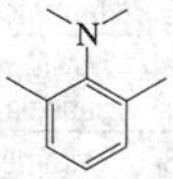

9920
N,N,2,6-Tetramethylaniline

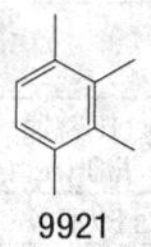

9921
1,2,3,4-Tetramethylbenzene

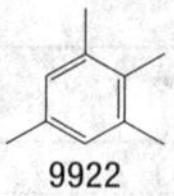

9922
1,2,3,5-Tetramethylbenzene

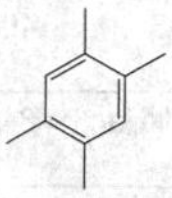

9923
1,2,4,5-Tetramethylbenzene

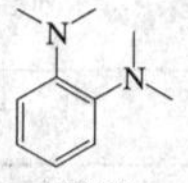

9924
N,N,N′,N′-Tetramethyl-1,2-benzenediamine

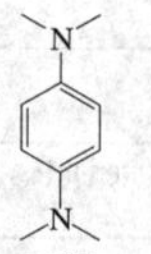

9925
N,N,N′,N′-Tetramethyl-1,4-benzenediamine

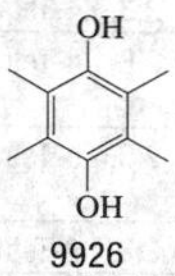

9926
2,3,5,6-Tetramethyl-1,4-benzenediol

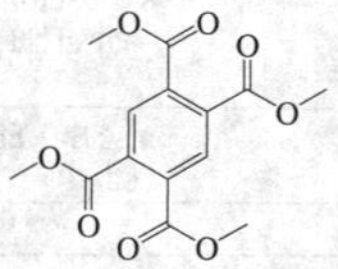

9927
Tetramethyl 1,2,4,5-benzenetetracarboxylate

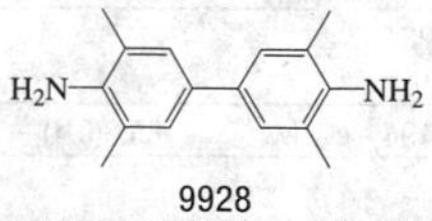

9928
3,3′,5,5′-Tetramethyl-[1,1′-biphenyl]-4,4′-diamine

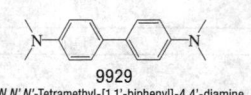

9929
N,N,N′,N′-Tetramethyl-[1,1′-biphenyl]-4,4′-diamine

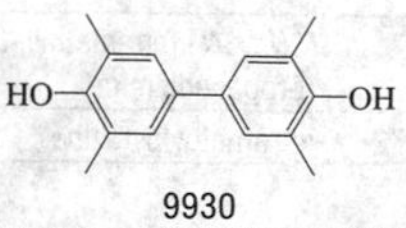

9930
3,3′,5,5′-Tetramethyl-[1,1′-biphenyl]-4,4′-diol

No.	Name	Synonym	Mol. Form.	CAS RN	Mol. Wt.	Physical Form	mp/°C	bp/°C	den g cm^{-3}	n_D	Solubility
9931	2,2,3,3-Tetramethylbutane		C$_8$H$_{18}$	594-82-1	114.229	lf (eth)	100.79(0.05)	106.32(0.04)	0.8242^{20}	1.4695^{20}	i H$_2$O; s eth, chl
9932	N,N,N',N'-Tetramethyl-1,4-butanediamine		C$_8$H$_{20}$N$_2$	111-51-3	144.258			168	0.7942^{15}	1.4621^{25}	msc H$_2$O; s EtOH, eth
9933	4-(1,1,3,3-Tetramethylbutyl)-phenol		C$_{14}$H$_{22}$O	140-66-9	206.324		85.0(0.2)	279			
9934	2,2,4,4-Tetramethyl-1,3-cyclobutanedione		C$_8$H$_{12}$O$_2$	933-52-8	140.180						s chl
9935	2,3,5,6-Tetramethyl-2,5-cyclohexadiene-1,4-dione	Duroquinone	C$_{10}$H$_{12}$O$_2$	527-17-3	164.201	ye nd (al or lig)	111.69(0.09)				i H$_2$O; s EtOH, eth, ace, bz, sulf, chl
9936	1,2,3,4-Tetramethylcyclohexane		C$_{10}$H$_{20}$	3726-45-2	140.266				0.8219^{20}	1.4531^{20}	
9937	1,1,3,3-Tetramethylcyclopentane		C$_9$H$_{18}$	50876-33-0	126.239	liq	-88.3(0.2)	118(5)	0.7469^{25}	1.4125^{20}	
9938	1,1,2,2-Tetramethylcyclopropane		C$_7$H$_{14}$	4127-47-3	98.186	liq	-80.69(0.09)	76(3)			
9939	2,4,6,8-Tetramethylcyclotetrasiloxane		C$_4$H$_{16}$O$_4$Si$_4$	2370-88-9	240.510	liq	-65	134.5	0.9912^{20}	1.3870^{20}	i H$_2$O
9940	2,4,7,9-Tetramethyl-5-decyne-4,7-diol		C$_{14}$H$_{26}$O$_2$	126-86-3	226.355		47	165^{40}			
9941	N,N,N',N'-Tetramethyl-4,4'-diaminobenzophenone	Michler's ketone	C$_{17}$H$_{20}$N$_2$O	90-94-8	268.353	lf (al), nd (bz)	179	360 dec			i H$_2$O, eth; sl EtOH; vs bz; s chl
9942	Tetramethyldiarsine	Cacodyl	C$_4$H$_{12}$As$_2$	471-35-2	209.981	liq	-6	165	1.447^{15}		vs eth, EtOH
9943	1,1,3,3-Tetramethyl-1,3-diphenyldisiloxane		C$_{16}$H$_{22}$OSi$_2$	56-33-7	286.516	liq	-80	292	0.9763^{20}	1.5176^{20}	s ctc
9944	1,1,3,3-Tetramethyldisiloxane		C$_4$H$_{14}$OSi$_2$	3277-26-7	134.324			71	0.756^{20}	1.3700^{20}	
9945	1,1,3,3-Tetramethyl-1,3-disiloxanediol		C$_4$H$_{14}$O$_3$Si$_2$	1118-15-6	166.323		66		1.095^{25}		
9946	N,N,N',N'-Tetramethyl-1,2-ethanediamine	1,2-Dimethylaminoethane	C$_6$H$_{16}$N$_2$	110-18-9	116.204	liq	-58.0(0.3)	121(1)	0.77^{25}	1.4179^{20}	
9947	Tetramethylgermane	Germanium tetramethyl	C$_4$H$_{12}$Ge	865-52-1	132.78			32^{500}	1.006		
9948	1,1,3,3-Tetramethylguanidine		C$_5$H$_{13}$N$_3$	80-70-6	115.177						s ctc
9949	2,2,6,6-Tetramethyl-3,5-heptanedione	Dipivaloylmethane	C$_{11}$H$_{20}$O$_2$	1118-71-4	184.276		93^{35}		0.883^{25}	1.4589^{20}	sl ctc
9950	3,7,11,15-Tetramethylhexadecanoic acid	Phytanic acid	C$_{20}$H$_{40}$O$_2$	14721-66-5	312.531		-65				
9951	3,7,11,15-Tetramethyl-1-hexadecen-3-ol	Isophytol	C$_{20}$H$_{40}$O	505-32-8	296.531	oil		108$^{0.01}$	0.8519^{20}	1.4571^{20}	vs bz, eth, EtOH
9952	2,2,3,3-Tetramethylhexane		C$_{10}$H$_{22}$	13475-81-5	142.282	liq	-54.0(0.1)	160(2)	0.7609^{25}	1.4282^{20}	
9953	2,2,5,5-Tetramethylhexane		C$_{10}$H$_{22}$	1071-81-4	142.282	liq	-12.6(0.1)	137(2)	0.7148^{25}	1.4055^{20}	
9954	3,3,4,4-Tetramethylhexane		C$_{10}$H$_{22}$	5171-84-6	142.282			170(4)	0.7789^{25}	1.4368^{20}	
9955	N,N,N',N'-Tetramethyl-1,6-hexanediamine		C$_{10}$H$_{24}$N$_2$	111-18-2	172.311			209.5	0.806^{25}	1.4359^{20}	
9956	Tetramethyl lead		C$_4$H$_{12}$Pb	75-74-1	267.3	liq	-30.2(0.2)	110	1.995^{20}		
9957	N,N,N',N'-Tetramethylmethanediamine		C$_5$H$_{14}$N$_2$	51-80-9	102.178			83.5(0.5)	0.7491^{18}		s H$_2$O
9958	Tetramethyloxirane		C$_6$H$_{12}$O	5076-20-0	100.158			90.4	0.8156^{16}	1.3984^{16}	s H$_2$O
9959	2,6,10,14-Tetramethylpentadecane	Pristane	C$_{19}$H$_{40}$	1921-70-6	268.521			306(6)	0.7791^{25}	1.4370^{25}	vs bz, eth, chl, peth
9960	2,2,3,3-Tetramethylpentane		C$_9$H$_{20}$	7154-79-2	128.255	liq	-9.75(0.05)	140.2(0.4)	0.7530^{25}	1.4236^{20}	
9961	2,2,3,4-Tetramethylpentane		C$_9$H$_{20}$	1186-53-4	128.255	liq	-121.3(0.7)	133.0(0.8)	0.7389^{20}	1.4147^{20}	
9962	2,2,4,4-Tetramethylpentane	Di-tert-butylmethane	C$_9$H$_{20}$	1070-87-7	128.255	liq	-66.53(0.05)	122.2(1)	0.7195^{20}	1.4069^{20}	i H$_2$O; vs EtOH, bz
9963	2,3,3,4-Tetramethylpentane		C$_9$H$_{20}$	16747-38-9	128.255	liq	-102.1(0.1)	141.5(0.7)	0.7547^{20}	1.4222^{20}	
9964	2,2,4,4-Tetramethyl-3-pentanol		C$_9$H$_{20}$O	14609-79-1	144.254		49(3)	166(5)			
9965	2,3,4,5-Tetramethylphenol	Prehnitenol	C$_{10}$H$_{14}$O	488-70-0	150.217	nd (lig, aq al)	84(3)	262(8)			sl H$_2$O, lig; vs EtOH, eth
9966	2,3,4,6-Tetramethylphenol		C$_{10}$H$_{14}$O	3238-38-8	150.217	cry (peth)	80(2)	250(7)			s EtOH
9967	2,3,5,6-Tetramethylphenol		C$_{10}$H$_{14}$O	527-35-5	150.217	nd (lig), pr (al)	118(1)	248(6)			s chl, peth, HOAc
9968	2,2,6,6-Tetramethyl-4-piperidinamine		C$_9$H$_{20}$N$_2$	36768-62-4	156.268		17	188.5	0.912^{25}	1.4706^{20}	
9969	2,2,6,6-Tetramethylpiperidine	Norpempidine	C$_9$H$_{19}$N	768-66-1	141.254		28	153(8)	0.8367^{16}	1.4455^{20}	vs eth
9970	2,2,6,6-Tetramethyl-4-piperidinone		C$_9$H$_{17}$NO	826-36-8	155.237	orth pl (eth-w) nd (eth)	35.5(0.5)	205			s H$_2$O, EtOH, eth; sl chl
9971	N,N,N',N'-Tetramethyl-1,3-propanediamine		C$_7$H$_{18}$N$_2$	110-95-2	130.231			144	0.7837^{18}		msc H$_2$O, EtOH, eth
9972	Tetramethylpyrazine		C$_8$H$_{12}$N$_2$	1124-11-4	136.194	cry (w)	85.3(0.4)	193(16)			

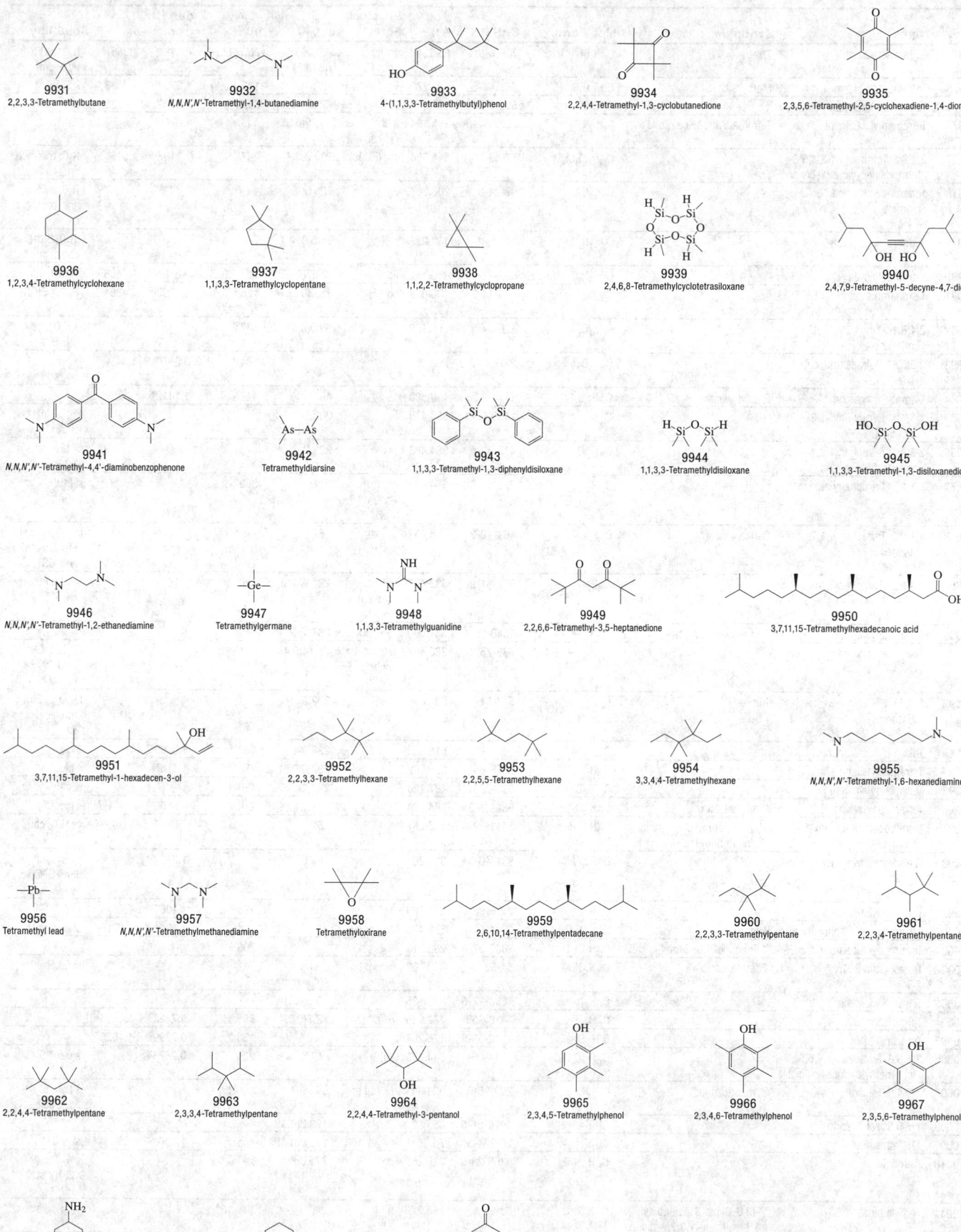

9931 2,2,3,3-Tetramethylbutane

9932 *N,N,N',N'*-Tetramethyl-1,4-butanediamine

9933 4-(1,1,3,3-Tetramethylbutyl)phenol

9934 2,2,4,4-Tetramethyl-1,3-cyclobutanedione

9935 2,3,5,6-Tetramethyl-2,5-cyclohexadiene-1,4-dione

9936 1,2,3,4-Tetramethylcyclohexane

9937 1,1,3,3-Tetramethylcyclopentane

9938 1,1,2,2-Tetramethylcyclopropane

9939 2,4,6,8-Tetramethylcyclotetrasiloxane

9940 2,4,7,9-Tetramethyl-5-decyne-4,7-diol

9941 *N,N,N',N'*-Tetramethyl-4,4'-diaminobenzophenone

9942 Tetramethyldiarsine

9943 1,1,3,3-Tetramethyl-1,3-diphenyldisiloxane

9944 1,1,3,3-Tetramethyldisiloxane

9945 1,1,3,3-Tetramethyl-1,3-disiloxanediol

9946 *N,N,N',N'*-Tetramethyl-1,2-ethanediamine

9947 Tetramethylgermane

9948 1,1,3,3-Tetramethylguanidine

9949 2,2,6,6-Tetramethyl-3,5-heptanedione

9950 3,7,11,15-Tetramethylhexadecanoic acid

9951 3,7,11,15-Tetramethyl-1-hexadecen-3-ol

9952 2,2,3,3-Tetramethylhexane

9953 2,2,5,5-Tetramethylhexane

9954 3,3,4,4-Tetramethylhexane

9955 *N,N,N',N'*-Tetramethyl-1,6-hexanediamine

9956 Tetramethyl lead

9957 *N,N,N',N'*-Tetramethylmethanediamine

9958 Tetramethyloxirane

9959 2,6,10,14-Tetramethylpentadecane

9960 2,2,3,3-Tetramethylpentane

9961 2,2,3,4-Tetramethylpentane

9962 2,2,4,4-Tetramethylpentane

9963 2,3,3,4-Tetramethylpentane

9964 2,2,4,4-Tetramethyl-3-pentanol

9965 2,3,4,5-Tetramethylphenol

9966 2,3,4,6-Tetramethylphenol

9967 2,3,5,6-Tetramethylphenol

9968 2,2,6,6-Tetramethyl-4-piperidinamine

9969 2,2,6,6-Tetramethylpiperidine

9970 2,2,6,6-Tetramethyl-4-piperidinone

9971 *N,N,N',N'*-Tetramethyl-1,3-propanediamine

9972 Tetramethylpyrazine

No.	Name	Synonym	Mol. Form.	CAS RN	Mol. Wt.	Physical Form	mp/°C	bp/°C	den g cm⁻³	n_D	Solubility
9973	Tetramethylsilane	TMS	$C_4H_{12}Si$	75-76-3	88.224	vol liq or gas	-99.063 (0.005)	26.7(0.5)	0.648[19]	1.3587[20]	i H$_2$O; vs EtOH, eth; i sulf
9974	Tetramethyl silicate	Methyl silicate	$C_4H_{12}O_4Si$	681-84-5	152.222	liq	-1.0	120.1(0.7)	1.0232[20]	1.3683[20]	vs EtOH
9975	Tetramethylstannane		$C_4H_{12}Sn$	594-27-4	178.848	liq	-55.09(0.04)	76.8(1)	1.314[25]	1.4386	i H$_2$O; s ctc, CS$_2$
9976	Tetramethylsuccinonitrile	Tetramethylbutanedinitrile	$C_8H_{12}N_2$	3333-52-6	136.194	mcl pl, lf, pr (dil al)	169.0(0.5)		1.070[25]		s EtOH
9977	2,4,6,8-Tetramethyl-2,4,6,8-tetraphenylcyclotetrasiloxane		$C_{28}H_{32}O_4Si_4$	77-63-4	544.894	cry (HOAc)	100.2(0.1)	237[115]	1.1183[20]	1.5461[20]	i H$_2$O; msc ace, hp
9978	Tetramethylthiodicarbonic diamide		$C_6H_{12}N_2S_3$	97-74-5	208.367		109.5		1.37[25]		i H$_2$O; s EtOH, ace, bz, chl; sl eth
9979	Tetramethylthiourea		$C_5H_{12}N_2S$	2782-91-4	132.227		77.3(0.4)	245			s H$_2$O, EtOH, chl; sl eth
9980	Tetramethylurea		$C_5H_{12}N_2O$	632-22-4	116.161	liq	-2.67(0.02)	177.1(0.6)	0.9687[20]	1.4496[23]	sl EtOH, eth, ctc
9981	Tetranitromethane		CN_4O_8	509-14-8	196.033	liq	13.9(0.2)	125.6(0.4)	1.6380[20]	1.4384[20]	i H$_2$O; s EtOH, eth
9982	2,4,8,10-Tetraoxaspiro[5.5]-undecane		$C_7H_{12}O_4$	126-54-5	160.168		50.0(0.5)	147[53]			vs H$_2$O, ace, eth, EtOH
9983	2,5,8,11-Tetraoxatridecan-13-ol		$C_9H_{20}O_5$	23783-42-8	208.252			164[11]	0.987[25]	1.4453[20]	
9984	Tetraphenoxysilane	Phenyl silicate	$C_{24}H_{20}O_4Si$	1174-72-7	400.500		49	417	1.1412[60]		
9985	1,1,4,4-Tetraphenyl-1,3-butadiene		$C_{28}H_{22}$	1450-63-1	358.475		203.5				s EtOH, bz, chl, HOAc
9986	2,3,4,5-Tetraphenyl-2,4-cyclopentadien-1-one		$C_{29}H_{20}O$	479-33-4	384.468	blk-viol lf (HOAc, xyl)	222.3				s EtOH, bz, xyl, HOAc
9987	1,1,2,2-Tetraphenylethane		$C_{26}H_{22}$	632-50-8	334.453	cry (bz), orth nd (chl)	211(2)	360			sl EtOH; s bz, HOAc
9988	1,1,2,2-Tetraphenyl-1,2-ethanediol	Benzopinacol	$C_{26}H_{22}O_2$	464-72-2	366.452	pr (bz), cry (ace)	182				i H$_2$O, peth; sl EtOH; s eth, ace, CS$_2$
9989	1,1,2,2-Tetraphenylethene		$C_{26}H_{20}$	632-51-9	332.437	mcl or orth (bz-eth or chl-al)	224.9(0.3)	420	1.155[0]		i H$_2$O; sl EtOH, chl, eth; vs bz
9990	Tetraphenylgermane	Germanium tetraphenyl	$C_{24}H_{20}Ge$	1048-05-1	381.06		233(1)				
9991	Tetraphenylmethane		$C_{25}H_{20}$	630-76-2	320.427	orth nd (bz, sub)	281.4(0.4)	431			i H$_2$O, EtOH, eth, lig, HOAc; s bz, tol
9992	5,6,11,12-Tetraphenylnaphthacene	Rubrene	$C_{42}H_{28}$	517-51-1	532.671	oran-red (bz-lig)	332.5				i H$_2$O; sl EtOH, eth, ace, py; s bz
9993	Tetraphenylplumbane		$C_{24}H_{20}Pb$	595-89-1	515.6		228.3	126[13]	1.5298[20]		s chl
9994	Tetraphenylsilane		$C_{24}H_{20}Si$	1048-08-4	336.502		237.6(0.6)	228[3]	1.078[20]		s ctc, CS$_2$
9995	Tetraphenylstannane		$C_{24}H_{20}Sn$	595-90-4	427.126		227(2)	420			sl chl
9996	Tetrapropoxysilane	Propyl silicate	$C_{12}H_{28}O_4Si$	682-01-9	264.434			226	0.9158[20]	1.4012[20]	s ctc, CS$_2$
9997	Tetrapropylammonium bromide	N,N,N-Tripropyl-1-propanaminium bromide	$C_{12}H_{28}BrN$	1941-30-6	266.261		252				vs H$_2$O, chl
9998	Tetrapropylammonium iodide		$C_{12}H_{28}IN$	631-40-3	313.261	orth bipym	280 dec		1.3138[25]		vs H$_2$O, chl; s EtOH, HOAc; sl eth
9999	Tetrapropylstannane		$C_{12}H_{28}Sn$	2176-98-9	291.060	liq	-109.1	228	1.1065[20]	1.4745[20]	
10000	Tetrapropyl thiodiphosphate	Aspon	$C_{12}H_{28}O_5P_2S_2$	3244-90-4	378.425	amber liq		104[0.1]	1.12[25]	1.4710[21]	sl H$_2$O, peth
10001	Tetrapropyl titanate	1-Propanol, titanium(4+) salt	$C_{12}H_{28}O_4Ti$	3087-37-4	284.215			206[100]			
10002	Tetrasodium EDTA	Edetate sodium	$C_{10}H_{12}N_2Na_4O_8$	64-02-8	380.169	amorp pow	300 (dihydrate)				sl EtOH
10003	Tetratetracontane		$C_{44}H_{90}$	7098-22-8	619.186		86.0(0.6)				
10004	Tetratriacontane		$C_{34}H_{70}$	14167-59-0	478.920	pl (eth)	72.8(0.3)	285.4[3]	0.7728[90]	1.4296[90]	
10005	Tetravinylsilane		$C_8H_{12}Si$	1112-55-6	136.267			130.2	0.7999[20]	1.4625[20]	
10006	2,4,6,8-Tetravinyl-2,4,6,8-tetramethylcyclotetrasiloxane		$C_{12}H_{24}O_4Si_4$	2554-06-5	344.659	liq	-43.5	224	0.9875[20]		s ctc, CS$_2$
10007	1,2,4,5-Tetrazine	sym-Tetrazine	$C_2H_2N_4$	290-96-0	82.064	dk red pr	99	118(17)			s H$_2$O, EtOH, eth, sulf
10008	1H-Tetrazol-5-amine		CH_3N_5	4418-61-5	85.069		198.8(0.9)				
10009	1H-Tetrazole		CH_2N_4	288-94-8	70.054	pl (al)	156.9(0.8)	sub	1.4060[20]		sl H$_2$O
10010	Tetrodotoxin		$C_{11}H_{17}N_3O_8$	4368-28-9	319.268	cry	225 dec				sl H$_2$O, eth, EtOH; s dil HOAc
10011	Thalidomide	2-(2,6-Dioxo-3-piperidinyl)-1H-isoindole-1,3(2H)-dione	$C_{13}H_{10}N_2O_4$	50-35-1	258.229	nd	270				vs py, diox
10012	Thallium(I) ethanolate	Thallous ethoxide	C_2H_5OTl	20398-06-5	249.443	cloudy liq	-3	130 dec	3.49		dec H$_2$O
10013	Thebaine		$C_{19}H_{21}NO_3$	115-37-7	311.375	pl (eth), pr (dil al)	193	91 sub	1.305[20]		i H$_2$O; vs EtOH, chl; sl eth; s bz

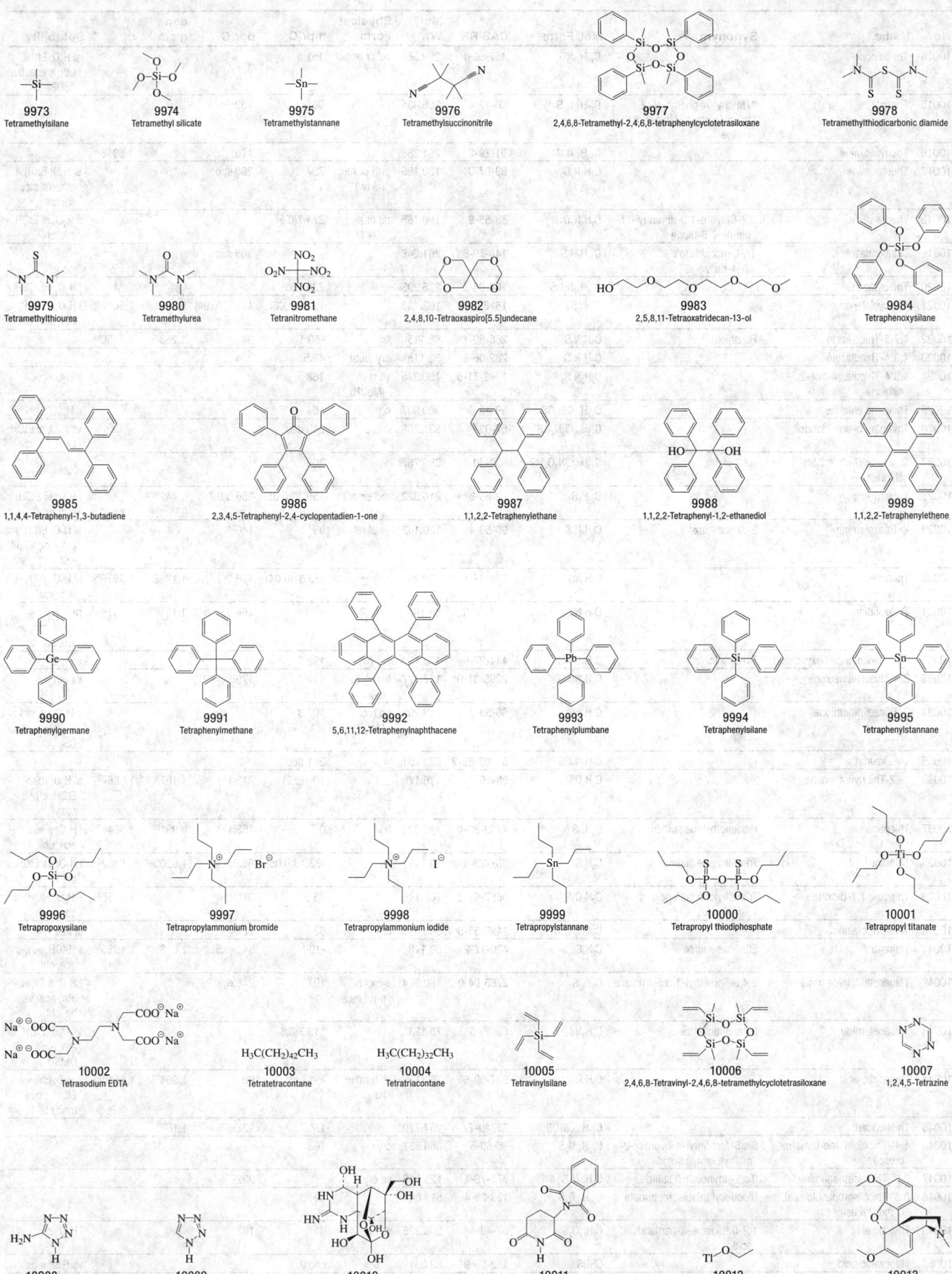

9973 Tetramethylsilane

9974 Tetramethyl silicate

9975 Tetramethylstannane

9976 Tetramethylsuccinonitrile

9977 2,4,6,8-Tetramethyl-2,4,6,8-tetraphenylcyclotetrasiloxane

9978 Tetramethylthiodicarbonic diamide

9979 Tetramethylthiourea

9980 Tetramethylurea

9981 Tetranitromethane

9982 2,4,8,10-Tetraoxaspiro[5.5]undecane

9983 2,5,8,11-Tetraoxatridecan-13-ol

9984 Tetraphenoxysilane

9985 1,1,4,4-Tetraphenyl-1,3-butadiene

9986 2,3,4,5-Tetraphenyl-2,4-cyclopentadien-1-one

9987 1,1,2,2-Tetraphenylethane

9988 1,1,2,2-Tetraphenyl-1,2-ethanediol

9989 1,1,2,2-Tetraphenylethene

9990 Tetraphenylgermane

9991 Tetraphenylmethane

9992 5,6,11,12-Tetraphenylnaphtacene

9993 Tetraphenylplumbane

9994 Tetraphenylsilane

9995 Tetraphenylstannane

9996 Tetrapropoxysilane

9997 Tetrapropylammonium bromide

9998 Tetrapropylammonium iodide

9999 Tetrapropylstannane

10000 Tetrapropyl thiodiphosphate

10001 Tetrapropyl titanate

10002 Tetrasodium EDTA

10003 Tetratetracontane

10004 Tetratriacontane

10005 Tetravinylsilane

10006 2,4,6,8-Tetravinyl-2,4,6,8-tetramethylcyclotetrasiloxane

10007 1,2,4,5-Tetrazine

10008 1H-Tetrazol-5-amine

10009 1H-Tetrazole

10010 Tetrodotoxin

10011 Thalidomide

10012 Thallium(I) ethanolate

10013 Thebaine

No.	Name	Synonym	Mol. Form.	CAS RN	Mol. Wt.	Physical Form	mp/°C	bp/°C	den g cm⁻³	n_D	Solubility
10014	Thebainone		$C_{18}H_{21}NO_3$	467-98-1	299.365	nd or pr (al)	151.5				sl H_2O, EtOH, eth; s ace, bz, AcOEt
10015	Thenaldine	1-Methyl-N-phenyl-N-(2-thienylmethyl)-4-piperidin-amine	$C_{17}H_{22}N_2S$	86-12-4	286.435		96	159$^{0.02}$			
10016	Thenyldiamine		$C_{14}H_{19}N_3S$	91-79-2	261.386			170^1		1.5915^{20}	
10017	Theobromine		$C_7H_8N_4O_2$	83-67-0	180.165	orth or mcl nd (w)	357	290 sub			sl H_2O, EtOH; i eth, bz, ctc, lig, chl
10018	Theophylline	3,7-Dihydro-1,3-dimethyl-1H-purine-2,6-dione	$C_7H_8N_4O_2$	58-55-9	180.165	nd or pl (w+1)	274.7(0.5)				s H_2O; sl EtOH, eth, chl
10019	Thiabendazole	1H-Benzimidazole, 2-(4-thiazolyl)-	$C_{10}H_7N_3S$	148-79-8	201.248			305 sub			
10020	Thiacetazone		$C_{10}H_{12}N_4OS$	104-06-3	236.293		225 dec				i H_2O, os, CS_2
10021	Thiacyclohexane		$C_5H_{10}S$	1613-51-0	102.198		19.09(0.06)	141.73(0.04)	0.9861^{20}	1.5067^{20}	i H_2O; s EtOH, eth, ace, bz
10022	1,2,5-Thiadiazole	Piazthiole	$C_2H_2N_2S$	288-39-1	86.115	liq	-50.1	94	1.268^{25}	1.5150^{25}	
10023	1,3,4-Thiadiazole		$C_2H_2N_2S$	289-06-5	86.115	cry (sub)	42.5	204			
10024	1,3,4-Thiadiazolidine-2,5-dithione		$C_2H_2N_2S_3$	1072-71-5	150.245	ye cry (MeOH)	168				s H_2O
10025	Thiamine chloride		$C_{12}H_{17}ClN_4OS$	59-43-8	300.807	cry	164				s H_2O
10026	Thiamine hydrochloride		$C_{12}H_{18}Cl_2N_4OS$	67-03-8	337.268	mcl pl	248 dec				vs H_2O; sl EtOH; i eth, bz, chl
10027	Thiamine O-phosphate, chloride		$C_{12}H_{18}ClN_4O_4PS$	532-40-1	380.787		200				
10028	Thianthrene		$C_{12}H_8S_2$	92-85-3	216.322	mcl pr or pl (al)	156.43(0.05)	366.3(0.7)	1.4420^{20}		i H_2O; sl EtOH; s eth, bz, CS_2
10029	2-Thiazolamine	2-Aminothiazole	$C_3H_4N_2S$	96-50-4	100.142	ye pl (al)	93	140^{11}			sl H_2O, EtOH, eth, chl; vs dil HCl
10030	Thiazole		C_3H_3NS	288-47-1	85.128		-33.61(0.04)	118.2(0.1)	1.1998^{17}	1.5969^{20}	sl H_2O; s EtOH, eth, ace
10031	Thiazolidine		C_3H_7NS	504-78-9	89.160			164.5	1.131^{25}	1.551^{20}	msc H_2O; s EtOH, ctc; vs eth, ace
10032	4-Thiazolidinecarboxylic acid	Timonacic	$C_4H_7NO_2S$	444-27-9	133.170	cry (w)	196.5				vs H_2O
10033	2,4-Thiazolidinedione		$C_3H_3NO_2S$	2295-31-0	117.127	pl (w), pr (al)	128	179^{19}			vs eth
10034	2-Thiazolidinethione		$C_3H_5NS_2$	96-53-7	119.209	nd (w, MeOH)	107.3				s H_2O, bz, chl; sl EtOH; i eth, CS_2
10035	Thidiazuron		$C_9H_8N_4OS$	51707-55-2	220.251		211 dec				
10036	1-(2-Thienyl)ethanone		C_6H_6OS	88-15-3	126.176		10.4(0.2)	215(16)	1.1679^{20}	1.5667^{20}	sl H_2O; msc EtOH, eth; s ctc
10037	Thiepane	Hexamethylene sulfide	$C_6H_{12}S$	4753-80-4	116.224	liq	0.5	155(17)	0.991^{20}	1.5044^{18}	i H_2O; s eth, ace, chl
10038	Thietane	Trimethylene sulfide	C_3H_6S	287-27-4	74.145	liq	-73.19(0.05)	95.0(0.4)	1.0200^{20}	1.5102^{20}	i H_2O; vs EtOH, bz; s ace
10039	Thietane 1,1-dioxide	Trimethylene sulfone	$C_3H_6O_2S$	5687-92-3	106.144		75.5	91.2^{14}		1.5156^{20}	s H_2O, EtOH; sl eth, peth
10040	Thiethylperazine		$C_{22}H_{29}N_3S_2$	1420-55-9	399.615	cry	63	227$^{0.01}$			sl ace
10041	Thiirane	Ethylene sulfide	C_2H_4S	420-12-2	60.118		-109	54.9(0.5)	1.0130^{20}	1.4935^{20}	sl EtOH, eth; s ace, chl
10042	Thioacetaldehyde trimer	2,4,6-Trimethyl-1,3,5-trithiane	$C_6H_{12}S_3$	2765-04-0	180.354	α-mcl pl; β-nd (ace)	101	246.5			i H_2O; s EtOH, eth, ace; vs bz, chl
10043	Thioacetamide	Ethanethioamide	C_2H_5NS	62-55-5	75.133		113.3(0.5)				vs H_2O, EtOH; sl eth, bz; s DMSO
10044	Thioacetic acid		C_2H_4OS	507-09-5	76.117	ye fuming liq	<-17	93	1.064^{20}	1.4648^{20}	s H_2O, chl; vs EtOH, ace; msc eth
10045	Thiobencarb		$C_{12}H_{16}ClNOS$	28249-77-6	257.779		1.7	127$^{0.008}$	1.16^{20}		
10046	4,4'-Thiobis(6-$tert$-butyl-m-cresol)	Bis(5-$tert$-butyl-4-hydroxy-2-methylphenyl) sulfide	$C_{22}H_{30}O_2S$	96-69-5	358.537	cry	163				
10047	2,2'-Thiobisethanamine	Bis(2-aminoethyl) sulfide	$C_4H_{12}N_2S$	871-76-1	120.216	ye cry		232			
10048	3,3'-Thiobispropanoic acid, didodecyl ester	Didodecyl thiobispropanoate	$C_{30}H_{58}O_4S$	123-28-4	514.845		39				
10049	Thioctic acid	1,2-Dithiolane-3-pentanoic acid	$C_9H_{14}O_2S_2$	62-46-4	206.326	ye nd	61	162			i H_2O
10050	Thiocyanic acid		CHNS	463-56-9	59.091		dec 0				vs H_2O; s os

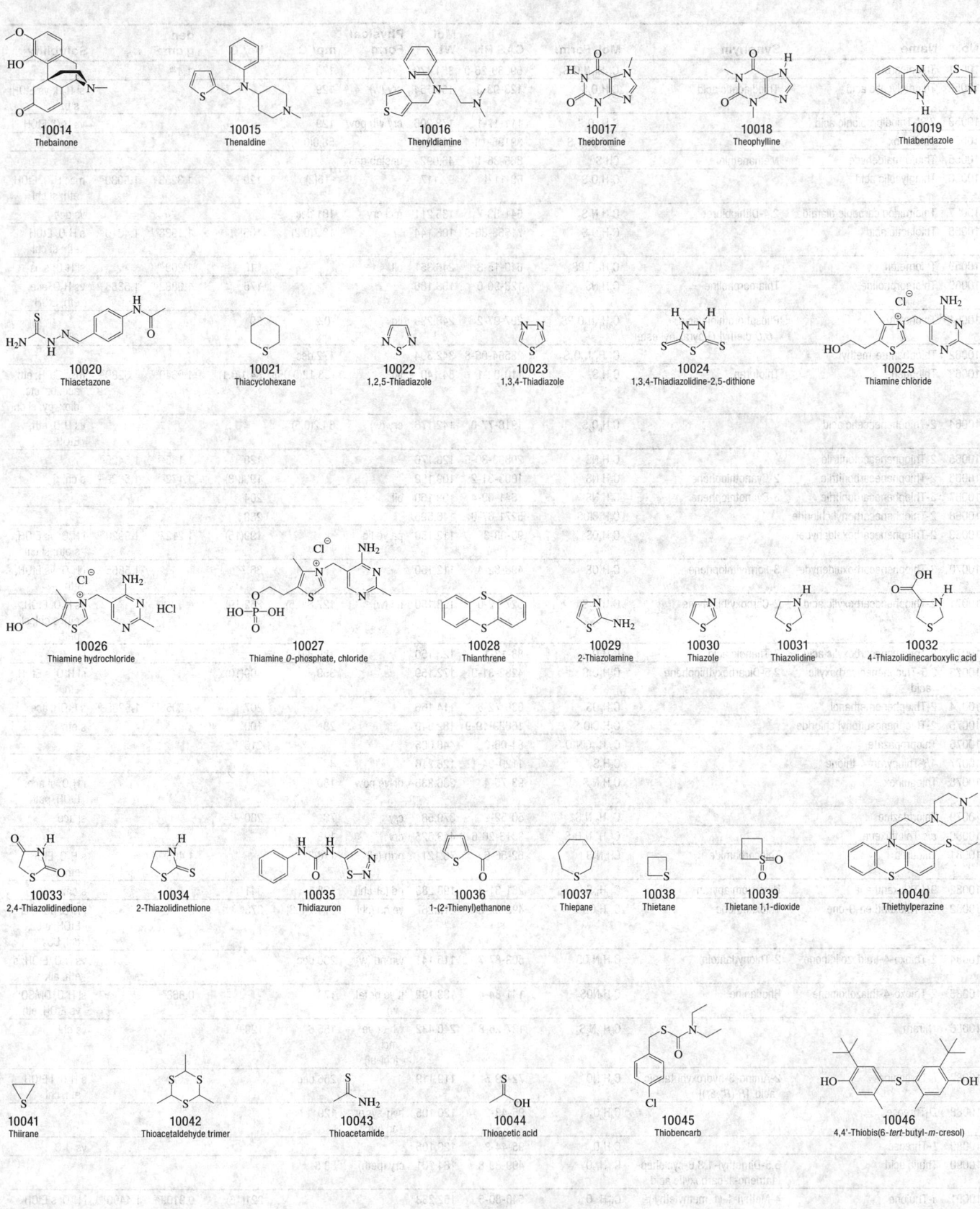

10014
Thebainone

10015
Thenaldine

10016
Thenyldiamine

10017
Theobromine

10018
Theophylline

10019
Thiabendazole

10020
Thiacetazone

10021
Thiacyclohexane

10022
1,2,5-Thiadiazole

10023
1,3,4-Thiadiazole

10024
1,3,4-Thiadiazolidine-2,5-dithione

10025
Thiamine chloride

10026
Thiamine hydrochloride

10027
Thiamine O-phosphate, chloride

10028
Thianthrene

10029
2-Thiazolamine

10030
Thiazole

10031
Thiazolidine

10032
4-Thiazolidinecarboxylic acid

10033
2,4-Thiazolidinedione

10034
2-Thiazolidinethione

10035
Thidiazuron

10036
1-(2-Thienyl)ethanone

10037
Thiepane

10038
Thietane

10039
Thietane 1,1-dioxide

10040
Thiethylperazine

10041
Thiirane

10042
Thioacetaldehyde trimer

10043
Thioacetamide

10044
Thioacetic acid

10045
Thiobencarb

10046
4,4'-Thiobis(6-tert-butyl-m-cresol)

10047
2,2'-Thiobisethanamine

10048
3,3'-Thiobispropanoic acid, didodecyl ester

10049
Thioctic acid

10050
Thiocyanic acid

No.	Name	Synonym	Mol. Form.	CAS RN	Mol. Wt.	Physical Form	mp/°C	bp/°C	den g cm⁻³	n_D	Solubility
10051	Thiodicarb		$C_{10}H_{18}N_4O_4S_3$	59669-26-0	354.470		173		1.4²⁰		
10052	Thiodiglycolic acid	Thiodiacetic acid	$C_4H_6O_4S$	123-93-3	150.154	cry (w)	129				sl H₂O; vs EtOH; s bz
10053	3,3'-Thiodipropionic acid		$C_6H_{10}O_4S$	111-17-1	178.206	cry wh pow	129				vs H₂O, EtOH
10054	Thiofanox		$C_9H_{18}N_2O_2S$	39196-18-4	218.316		58.0(0.5)				
10055	Thioformaldehyde	Methanethial	CH_2S	865-36-1	46.092	unstab gas					
10056	Thioglycolic acid		$C_2H_4O_2S$	68-11-1	92.117		-16(2)	120²⁰	1.3253²⁰	1.5080²⁰	msc H₂O, EtOH, eth; sl chl
10057	Thioimidodicarbonic diamide	2,4-Dithiobiuret	$C_2H_5N_3S_2$	541-53-7	135.211	mcl cry	181 dec				vs ace
10058	Thiolactic acid		$C_3H_6O_2S$	71563-86-5	106.144		18.7(0.2)	106¹⁵	1.1938²⁰	1.4810²⁰	s H₂O, EtOH, eth; sl chl
10059	Thiometon		$C_6H_{15}O_2PS_3$	640-15-3	246.351	oil		110⁰·¹	1.209²⁰		sl H₂O; s os
10060	Thiomorpholine	Thiamorpholine	C_4H_9NS	123-90-0	103.186		175		1.0882²⁰	1.5386²⁰	vs H₂O, ace, eth, EtOH
10061	Thionazin	Phosphorothioic acid, O,O-diethyl O-pyrazinyl ester	$C_8H_{13}N_2O_3PS$	297-97-2	248.239	liq	-0.9	80			
10062	Thiophanate-methyl		$C_{12}H_{14}N_4O_4S_2$	23564-05-8	342.394		172 dec				
10063	Thiophene	Thiofuran	C_4H_4S	110-02-1	84.140	liq	-38.12(0.05)	84.1(0.1)	1.0649²⁰	1.5289²⁰	msc EtOH, eth, ace, bz, ctc, diox, py; sl chl
10064	2-Thiopheneacetic acid		$C_6H_6O_2S$	1918-77-0	142.176	cry (w)	64.2(0.5)				vs H₂O, eth, EtOH
10065	2-Thiopheneacetonitrile		C_6H_5NS	20893-30-5	123.176			120²³	1.155²⁵	1.5425²⁰	
10066	2-Thiophenecarbonitrile	2-Cyanothiophene	C_5H_3NS	1003-31-2	109.150			193(23)	1.172²⁵	1.5629²⁰	s chl
10067	3-Thiophenecarbonitrile	3-Cyanothiophene	C_5H_3NS	1641-09-4	109.150	oil		204			
10068	2-Thiophenecarbonyl chloride		C_5H_3ClOS	5271-67-0	146.595			280			
10069	2-Thiophenecarboxaldehyde		C_5H_4OS	98-03-3	112.150	pa ye liq		199(15)	1.2127²¹	1.5920²⁰	i H₂O; vs EtOH; s eth; sl chl
10070	3-Thiophenecarboxaldehyde	3-Formylthiophene	C_5H_4OS	498-62-4	112.150		86.7²⁰			1.5855²⁰	i H₂O; vs EtOH, eth
10071	2-Thiophenecarboxylic acid	2-Carboxythiophene	$C_5H_4O_2S$	527-72-0	128.150	nd (w)	127.8(0.5)	262(19)			vs H₂O, EtOH, eth; s chl; sl peth
10072	3-Thiophenecarboxylic acid	3-Thenoic acid	$C_5H_4O_2S$	88-13-1	128.150		139.8(0.5)				s H₂O
10073	2,5-Thiophenedicarboxylic acid	2,5-Dicarboxythiophene	$C_6H_4O_4S$	4282-31-9	172.159		359	409(10)			sl H₂O; s EtOH, eth
10074	2-Thiophenemethanol		C_5H_6OS	636-72-6	114.166			207	1.2053¹⁶	1.5280²⁰	s EtOH, ace
10075	2-Thiophenesulfonyl chloride		$C_4H_3ClO_2S_2$	16629-19-9	182.649		28	100⁶			s eth
10076	Thiopropazate		$C_{23}H_{28}ClN_3O_2S$	84-06-0	446.005			216⁰·¹			
10077	4H-Thiopyran-4-thione		$C_5H_4S_2$	1120-94-1	128.216		47				
10078	Thioquinox		$C_9H_4N_2S_3$	93-75-4	236.336	br-ye pow	180				i H₂O; sl ace, EtOH, peth
10079	Thioridazine		$C_{21}H_{26}N_2S_2$	50-52-2	370.58	cry	73	230⁰·⁰²			sl ace
10080	cis-Thiothixene		$C_{23}H_{29}N_3O_2S_2$	3313-26-6	443.625	cry	148				
10081	Thiourea	Thiocarbamide	CH_4N_2S	62-56-6	76.121	orth (al)	176(3)		1.405²⁵		s H₂O, EtOH; i eth
10082	9H-Thioxanthene	Dibenzothiapyran	$C_{13}H_{10}S$	261-31-4	198.283	nd (al-chl)	128.5	341			s chl
10083	9H-Thioxanthen-9-one	Thioxanthone	$C_{13}H_8OS$	492-22-8	212.267	ye nd (chl)	214.7(0.2)	373			i H₂O, peth; sl EtOH; s bz, chl, CS₂
10084	2-Thioxo-4-imidazolidinone	2-Thiohydantoin	$C_3H_4N_2OS$	503-87-7	116.141	wh nd (w)	230 dec				vs H₂O, EtOH; s eth, alk
10085	2-Thioxo-4-thiazolidinone	Rhodanine	$C_3H_3NOS_2$	141-84-4	133.192	lt ye pr (al, w)	170		0.868²⁵		sl H₂O, DMSO; vs EtOH, eth
10086	Thiram		$C_6H_{12}N_2S_4$	137-26-8	240.432	wh or ye mcl (chl-al)	155.6	129²⁰			vs chl
10087	L-Threonine	2-Amino-3-hydroxybutanoic acid, [R-(R*,S*)]	$C_4H_9NO_3$	72-19-5	119.119		256 dec				s H₂O; i EtOH, eth, chl
10088	D-Threose		$C_4H_8O_4$	95-43-2	120.105	hyg-syr or nd (w)	129				
10089	L-Threose		$C_4H_8O_4$	95-44-3	120.105						vs H₂O
10090	Thujic acid	5,5-Dimethyl-1,3,6-cycloheptatriene-1-carboxylic acid	$C_{10}H_{12}O_2$	499-89-8	164.201	cry (peth)	88.5				
10091	α-Thujone	4-Methyl-1-(1-methylethyl)-bicyclo[3.1.0]hexan-3-one, (l)	$C_{10}H_{16}O$	546-80-5	152.233			221(16)	0.9109²⁵	1.4490¹⁵	i H₂O; s EtOH
10092	3-Thujopsene	Widdrene	$C_{15}H_{24}$	470-40-6	204.352	liq		122¹²	0.932²⁴	1.5031²⁵	
10093	Thymidine	Thymine 2-desoxyriboside	$C_{10}H_{14}N_2O_5$	50-89-5	242.228	nd (AcOEt)	187(2)				s H₂O, EtOH, ace, py, HOAc; sl chl
10094	Thymine		$C_5H_6N_2O_2$	65-71-4	126.114		325(4)				sl H₂O, EtOH, eth, DMSO

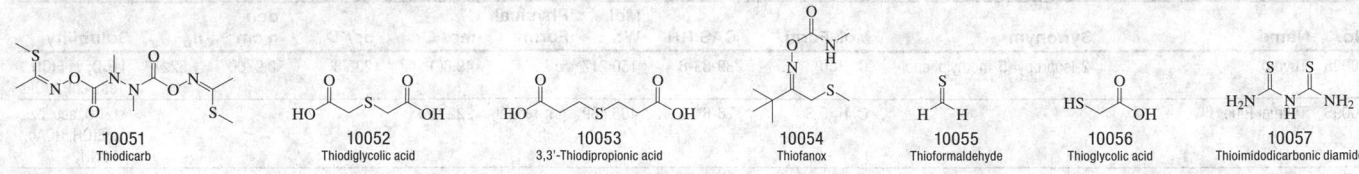

10051 Thiodicarb

10052 Thiodiglycolic acid

10053 3,3'-Thiodipropionic acid

10054 Thiofanox

10055 Thioformaldehyde

10056 Thioglycolic acid

10057 Thioimidodicarbonic diamide

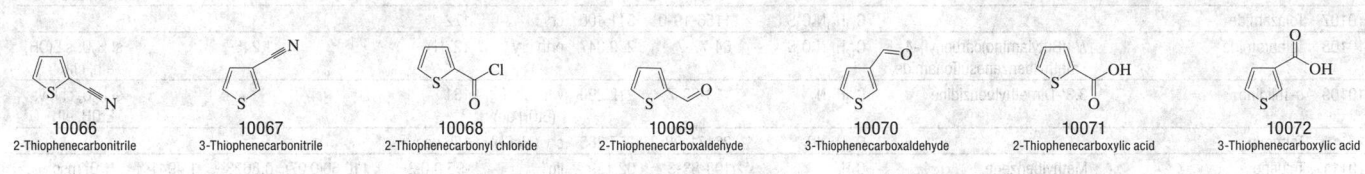

10058 Thiolactic acid

10059 Thiometon

10060 Thiomorpholine

10061 Thionazin

10062 Thiophanate-methyl

10063 Thiophene

10064 2-Thiopheneacetic acid

10065 2-Thiopheneacetonitrile

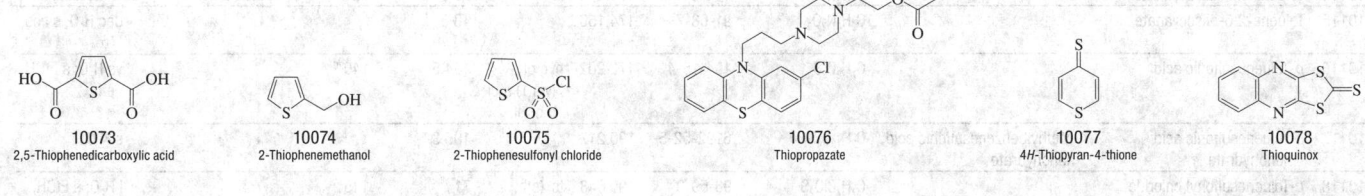

10066 2-Thiophenecarbonitrile

10067 3-Thiophenecarbonitrile

10068 2-Thiophenecarbonyl chloride

10069 2-Thiophenecarboxaldehyde

10070 3-Thiophenecarboxaldehyde

10071 2-Thiophenecarboxylic acid

10072 3-Thiophenecarboxylic acid

10073 2,5-Thiophenedicarboxylic acid

10074 2-Thiophenemethanol

10075 2-Thiophenesulfonyl chloride

10076 Thiopropazate

10077 4*H*-Thiopyran-4-thione

10078 Thioquinox

10079 Thioridazine

10080 *cis*-Thiothixene

10081 Thiourea

10082 9*H*-Thioxanthene

10083 9*H*-Thioxanthen-9-one

10084 2-Thioxo-4-imidazolidinone

10085 2-Thioxo-4-thiazolidinone

10086 Thiram

10087 *L*-Threonine

10088 *D*-Threose

10089 *L*-Threose

10090 Thujic acid

10091 α-Thujone

10092 3-Thujopsene

10093 Thymidine

10094 Thymine

No.	Name	Synonym	Mol. Form.	CAS RN	Mol. Wt.	Physical Form	mp/°C	bp/°C	den g cm^{-3}	n_D	Solubility
10095	Thymol	2-Isopropyl-5-methylphenol	$C_{10}H_{14}O$	89-83-8	150.217		49.6(0.3)	233(3)	0.970[25]	1.5227[20]	i H_2O; vs EtOH, eth, chl, AcOEt
10096	Thymol Blue		$C_{27}H_{30}O_5S$	76-61-9	466.589	grn-red (al, eth)	222 dec				sl H_2O, ace, bz; s EtOH, HOAc, PhNH$_2$
10097	Thymol iodide		$C_{20}H_{24}I_2O_2$	552-22-7	550.213	amorp					i H_2O; s eth; vs EtOH
10098	Thymolphthalein		$C_{28}H_{30}O_4$	125-20-2	430.536	pr or nd (al)	253				i H_2O; s EtOH, eth, ace; sl DMSO
10099	L-Thyroxine		$C_{15}H_{11}I_4NO_4$	51-48-9	776.871	nd	235				sl H_2O; i EtOH, bz
10100	Timolol		$C_{13}H_{24}N_4O_3S$	26839-75-8	316.420	oil					
10101	Tiocarlide		$C_{23}H_{32}N_2O_2S$	910-86-1	400.577		146				
10102	Tipepidine	3-(Di-2-thienylmethylene)-1-methylpiperidine	$C_{15}H_{17}NS_2$	5169-78-8	275.433	ye cry	65	181[4.5]			
10103	Tobramycin		$C_{18}H_{37}N_5O_9$	32986-56-4	467.516	cry					s H_2O
10104	β-Tocopherol	5,8-Dimethyltocol	$C_{28}H_{48}O_2$	148-03-8	416.680	pa ye visc oil		205[0.1]			vs ace, eth, EtOH, chl
10105	γ-Tocopherol	7,8-Dimethyltocol	$C_{28}H_{48}O_2$	7616-22-0	416.680	pa ye visc oil	-1.5	205[0.1]			i H_2O; msc EtOH, eth, ace, chl
10106	δ-Tocopherol	8-Methyltocol	$C_{27}H_{46}O_2$	119-13-1	402.653	pa ye visc oil		150[0.001]			i H_2O; vs EtOH, eth, ace, chl
10107	Tolazamide		$C_{14}H_{21}N_3O_3S$	1156-19-0	311.400	cry	172				
10108	Tolbutamide	N-[(Butylamino)carbonyl]-4-methylbenzenesulfonamide	$C_{12}H_{18}N_2O_3S$	64-77-7	270.347	orth cry	127(1)		1.245[25]		sl H_2O; s EtOH, eth, chl
10109	o-Tolidine	3,3'-Dimethylbenzidine	$C_{14}H_{16}N_2$	119-93-7	212.290	wh-red lf (EtOH aq)	131				sl H_2O, chl; vs EtOH, eth
10110	Tolmetin		$C_{15}H_{15}NO_3$	26171-23-3	257.285	cry (MeCN)	156 dec				
10111	Toluene	Methylbenzene	C_7H_8	108-88-3	92.139	liq	-95.0(0.2)	110.60(0.07)	0.8623[25]	1.4941[25]	i H_2O; msc EtOH, eth; s ace, CS$_2$
10112	Toluene-2,4-diamine	4-Methyl-1,3-benzenediamine	$C_7H_{10}N_2$	95-80-7	122.167	nd (w), cry (al)	99	292			vs H_2O, EtOH, eth, bz; s chl
10113	Toluene-3,5-diamine	5-Methyl-1,3-benzenediamine	$C_7H_{10}N_2$	108-71-4	122.167	oil		284			
10114	Toluene-2,4-diisocyanate		$C_9H_6N_2O_2$	584-84-9	174.156		20.5	251	1.2244[20]		vs ace, bz, eth
10115	Toluene-2,6-diisocyanate		$C_9H_6N_2O_2$	91-08-7	174.156		18.3				dec H_2O; s ace, bz
10116	p-Toluenesulfonic acid		$C_7H_8O_3S$	104-15-4	172.202	hyg pl (w+1) mcl lf or pl	104.5	140[20]			vs H_2O; s EtOH, eth
10117	p-Toluenesulfonic acid monohydrate	4-Methylbenzenesulfonic acid, monohydrate	$C_7H_{10}O_4S$	6192-52-5	190.217		105.3				s H_2O
10118	p-Toluenesulfonyl chloride		$C_7H_7ClO_2S$	98-59-9	190.648	tcl (eth, peth)	71	145[15]			i H_2O; s EtOH, eth, chl; vs bz
10119	o-Toluic acid		$C_8H_8O_2$	118-90-1	136.149	pr or nd (w)	103.4(0.2)	259.5(0.6)	1.062[115]	1.512[115]	i H_2O; vs EtOH, eth; s chl
10120	m-Toluic acid		$C_8H_8O_2$	99-04-7	136.149		109.3(0.7)		1.054[112]	1.509	sl H_2O, chl; vs EtOH, eth
10121	p-Toluic acid		$C_8H_8O_2$	99-94-5	136.149		180(1)				i H_2O; vs EtOH, eth, MeOH; sl tfa
10122	N-o-Tolylbiguanide	N-(2-Methylphenyl)-imidodicarbonimidic diamide	$C_9H_{13}N_5$	93-69-6	191.233	nd or pl (w+1)	145.0				sl H_2O; vs EtOH, ace; i bz, chl, eth
10123	Tomatine		$C_{50}H_{83}NO_{21}$	17406-45-0	1034.188	nd (MeOH)	270				vs EtOH, diox
10124	Tralomethrin		$C_{22}H_{19}Br_4NO_3$	66841-25-6	665.007	oran-ye solid					
10125	Tranylcypromine	2-Phenylcyclopropylamine	$C_9H_{11}N$	155-09-9	133.190	cry	44	127[32]			
10126	Trehalose		$C_{12}H_{22}O_{11}$	99-20-7	342.296	orth cry	209.4(0.5)		1.58[24]		vs H_2O; s EtOH; i eth, bz
10127	Triacetamide		$C_6H_9NO_3$	641-06-5	143.140	nd (eth)	79				vs eth
10128	Triacetin	Glycerol triacetate	$C_9H_{14}O_6$	102-76-1	218.203	col oily liq	-78	259(2)	1.1583[20]	1.4301[20]	sl H_2O; msc EtOH, eth, bz; vs ace
10129	Triacontane		$C_{30}H_{62}$	638-68-6	422.813	orth (eth, bz)	65.9(0.3)	451(7)	0.8097[20]	1.4352[70]	i H_2O; sl EtOH; s eth; vs bz
10130	Triacontanoic acid	Melissic acid	$C_{30}H_{60}O_2$	506-50-3	452.796	sc, nd (al, ace)	93.6(0.5)			1.4323[100]	vs bz, CS$_2$, chl
10131	1-Triacontanol	Myricyl alcohol	$C_{30}H_{62}O$	593-50-0	438.812	nd (eth),pl (bz)	88		0.777[95]		vs bz, eth, EtOH

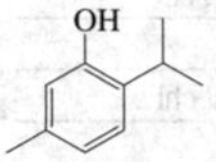

10095
Thymol

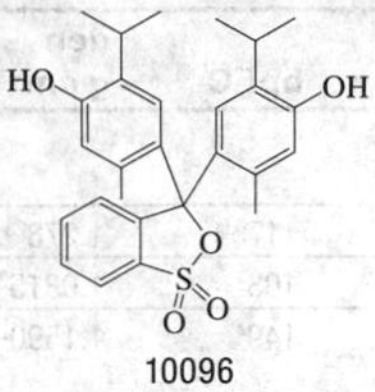

10096
Thymol Blue

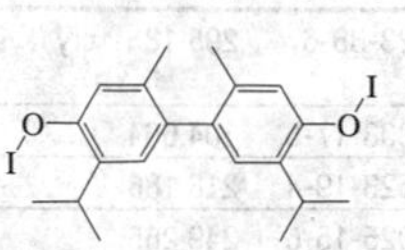

10097
Thymol iodide

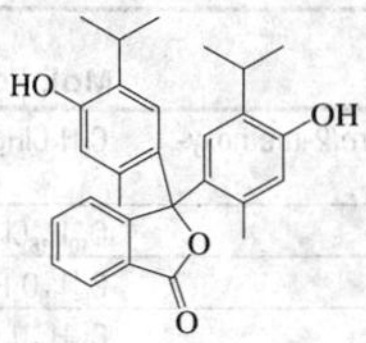

10098
Thymolphthalein

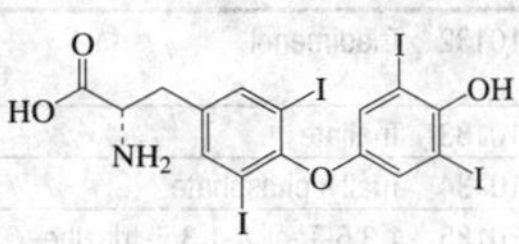

10099
L-Thyroxine

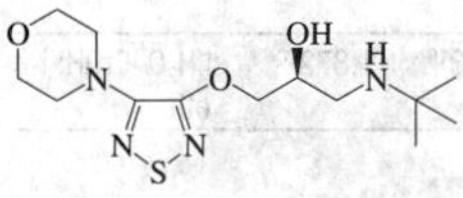

10100
Timolol

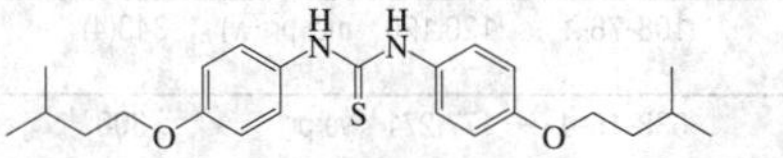

10101
Tiocarlide

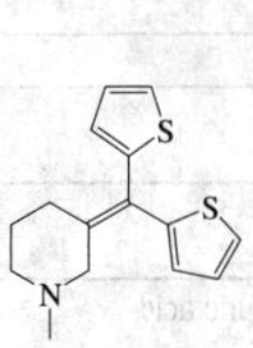

10102
Tipepidine

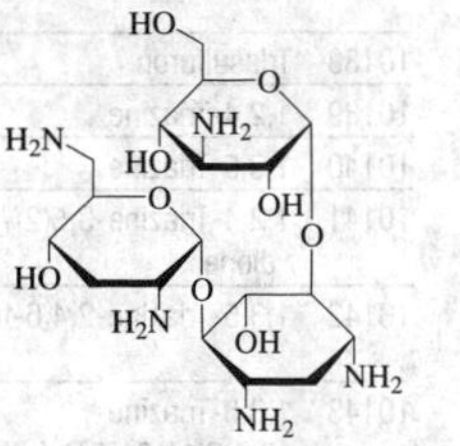

10103
Tobramycin

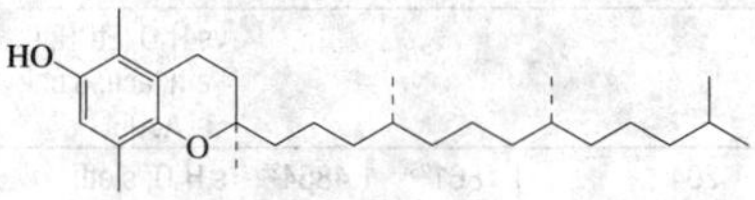

10104
β-Tocopherol

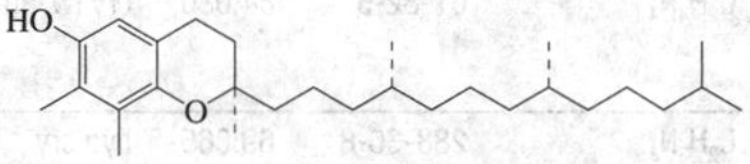

10105
γ-Tocopherol

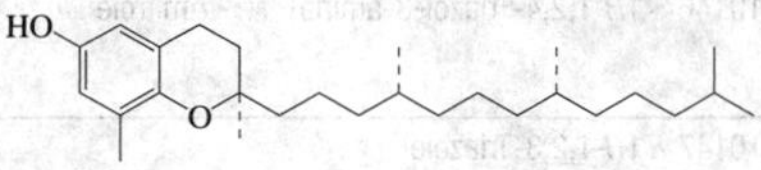

10106
δ-Tocopherol

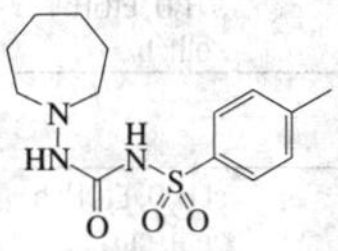

10107
Tolazamide

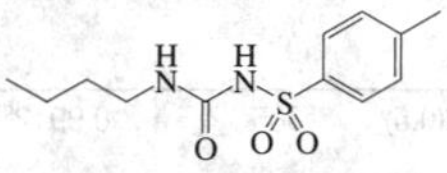

10108
Tolbutamide

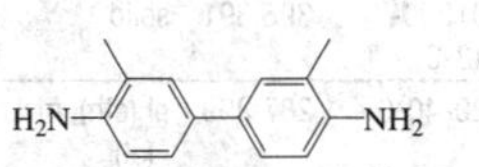

10109
o-Tolidine

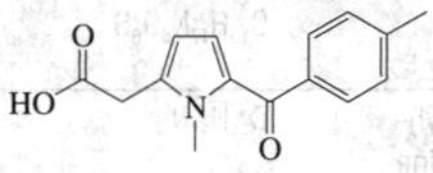

10110
Tolmetin

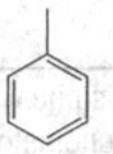

10111
Toluene

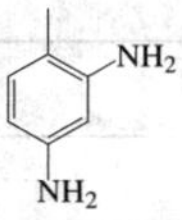

10112
Toluene-2,4-diamine

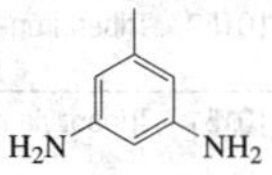

10113
Toluene-3,5-diamine

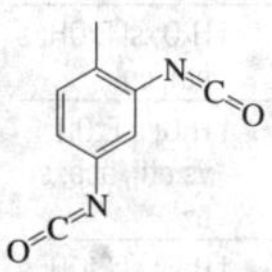

10114
Toluene-2,4-diisocyanate

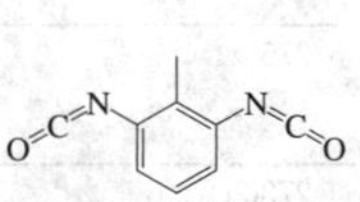

10115
Toluene-2,6-diisocyanate

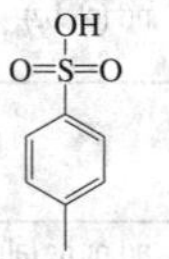

10116
p-Toluenesulfonic acid

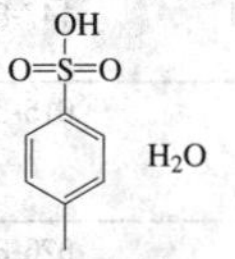

10117
p-Toluenesulfonic acid monohydrate

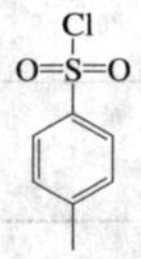

10118
p-Toluenesulfonyl chloride

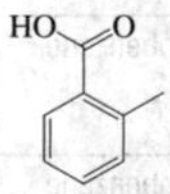

10119
o-Toluic acid

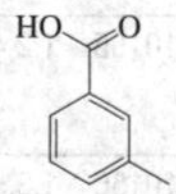

10120
m-Toluic acid

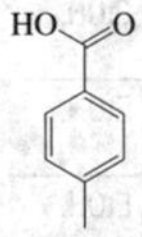

10121
p-Toluic acid

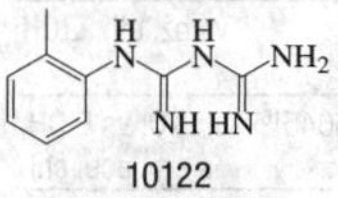

10122
N-o-Tolylbiguanide

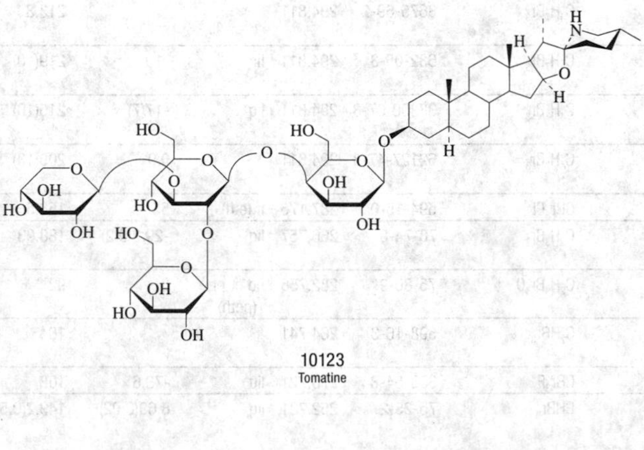

10123
Tomatine

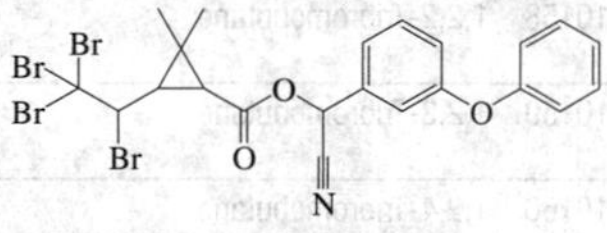

10124
Tralomethrin

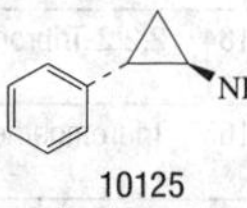

10125
Tranylcypromine

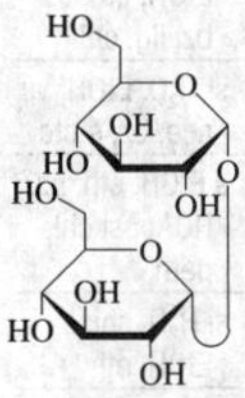

10126
Trehalose

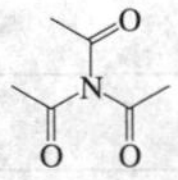

10127
Triacetamide

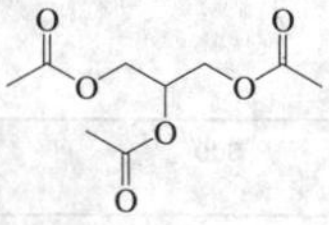

10128
Triacetin

10129
Triacontane

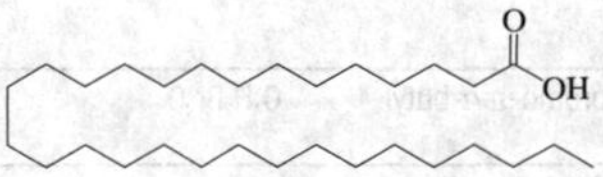

10130
Triacontanoic acid

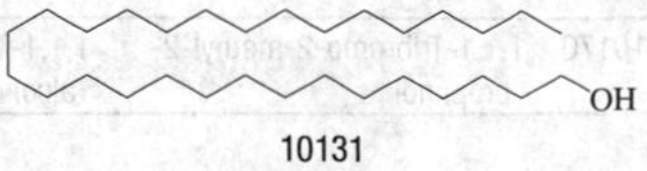

10131
1-Triacontanol

No.	Name	Synonym	Mol. Form.	CAS RN	Mol. Wt.	Physical Form	mp/°C	bp/°C	den g cm⁻³	n_D	Solubility
10132	Triadimenol	Mercury, chloro(2-methoxy-ethyl)-	C_3H_7ClHgO	123-88-6	295.13	cry	115				i H_2O; s EtOH, ace
10133	Triallate		$C_{10}H_{16}Cl_3NOS$	2303-17-5	304.664		34.0(0.5)	117[0.0003]	1.273[25]		
10134	Triallyl phosphate		$C_9H_{15}O_4P$	1623-19-4	218.186		-50	108[7]	1.0815[20]		sl chl
10135	1,3,5-Triallyl-1,3,5-triazine-2,4,6(1H,3H,5H)-trione		$C_{12}H_{15}N_3O_3$	1025-15-6	249.265		20.5	149[4]	1.1590[20]		
10136	Triamcinolone	Fluoxiprednisolone	$C_{21}H_{27}FO_6$	124-94-7	394.433	cry	270				
10137	Triamiphos		$C_{12}H_{19}N_6OP$	1031-47-6	294.292	cry (EtOH aq)	167				sl H_2O; s os
10138	Triasulfuron		$C_{14}H_{16}ClN_5O_5S$	82097-50-5	401.826		186				
10139	1,2,4-Triazine		$C_3H_3N_3$	290-38-0	81.076	pa ye oil	16.5	157		1.5149[25]	
10140	1,3,5-Triazine		$C_3H_3N_3$	290-87-9	81.076		80.3(0.5)	113(5)	1.38[25]		s EtOH, eth
10141	1,2,4-Triazine-3,5(2H,4H)-dione		$C_3H_3N_3O_2$	461-89-2	113.075		276.8				
10142	1,3,5-Triazine-2,4,6-triamine	Melamine	$C_3H_6N_6$	108-78-1	126.120	mcl pr (w)	343(4)	sub	1.573[16]	1.872[20]	sl H_2O, EtOH; i eth
10143	1,3,5-Triazine-2,4,6(1H,3H,5H)-trithione	Trithiocyanuric acid	$C_3H_3N_3S_3$	638-16-4	177.271	ye pr	>300	100[22]			
10144	Triazofos		$C_{12}H_{16}N_3O_3PS$	24017-47-8	313.312	ye-br oil	5		1.2514[20]		i H2O; s os
10145	Triazolam		$C_{17}H_{12}Cl_2N_4$	28911-01-5	343.210	tan cry (2-PrOH)	234				
10146	1H-1,2,4-Triazol-3-amine	Amitrole	$C_2H_4N_4$	61-82-5	84.080	cry (w, al)	155.8(0.6)				vs H_2O, EtOH; i eth, ace; s chl; sl AcOEt
10147	1H-1,2,3-Triazole		$C_2H_3N_3$	288-36-8	69.065	hyg cry	23	204	1.1861[25]	1.4854[25]	s H_2O; s eth, ace; i lig
10148	1H-1,2,4-Triazole	Pyrrodiazole	$C_2H_3N_3$	288-88-0	69.065	nd (bz/EtOH)	120(1)	204(6)			s H_2O, EtOH
10149	1H-1,2,4-Triazole-3,5-diamine		$C_2H_5N_5$	1455-77-2	99.095		211.5				s H_2O, EtOH; i eth, bz
10150	Tribenuron-methyl		$C_{15}H_{17}N_5O_6S$	101200-48-0	395.391	solid	141				
10151	Tribenzylamine	N,N-Bis(phenylmethyl)-benzenemethanamine	$C_{21}H_{21}N$	620-40-6	287.399	pl (eth), mcl (al)	92.2(0.5)	385	0.9912[95]		sl H_2O, EtOH; s eth, ctc
10152	Tribromoacetaldehyde	Bromal	C_2HBr_3O	115-17-3	280.740			173(5)	2.6649[25]	1.5939[20]	vs ace, eth, EtOH
10153	Tribromoacetic acid		$C_2HBr_3O_2$	75-96-7	296.740	mcl	132	245 dec			s H_2O, EtOH, eth
10154	2,4,6-Tribromoaniline		$C_6H_4Br_3N$	147-82-0	329.815	nd (al, bz)	121(2)	300	2.35[20]		i H_2O; sl EtOH; s eth, chl
10155	1,2,4-Tribromobenzene		$C_6H_3Br_3$	615-54-3	314.800		44.5	275			i H_2O; s EtOH; vs eth, ace; sl bz
10156	1,3,5-Tribromobenzene		$C_6H_3Br_3$	626-39-1	314.800	nd or pr (al)	121.8(0.2)	273(2)			i H_2O; sl EtOH; s eth, bz, chl
10157	1,1,2-Tribromobutane		$C_4H_7Br_3$	3675-68-1	294.811			219(10)	2.1835[20]	1.5626[17]	vs eth, EtOH, chl
10158	1,2,2-Tribromobutane		$C_4H_7Br_3$	3675-69-2	294.811			213.8	2.1692[20]	1.568[20]	vs eth, EtOH, chl
10159	1,2,3-Tribromobutane		$C_4H_7Br_3$	632-05-3	294.811	liq	-19	219(10)	2.1907[20]	1.5680[20]	vs eth, EtOH, chl
10160	1,2,4-Tribromobutane		$C_4H_7Br_3$	38300-67-3	294.811	liq	-17(7)	215(10)	2.170[20]	1.5608[20]	vs eth, EtOH, chl
10161	2,2,3-Tribromobutane		$C_4H_7Br_3$	62127-47-3	294.811		0.9	209(13)	2.1723[20]	1.5602[20]	i H_2O; s EtOH, eth, chl; sl ctc
10162	Tribromochloromethane		CBr_3Cl	594-15-0	287.176	lf (eth)	55	158.5	2.71[15]		vs eth
10163	1,1,2-Tribromoethane		$C_2H_3Br_3$	78-74-0	266.757	liq	-29.2(0.2)	188.93	2.6210[20]	1.5933[20]	i H_2O; s EtOH, eth, bz, ctc
10164	2,2,2-Tribromoethanol		$C_2H_3Br_3O$	75-80-9	282.756	nd or pr (peth)	81	92[10]			vs bz, eth, EtOH
10165	Tribromoethene		C_2HBr_3	598-16-3	264.741			164	2.708[20]	1.6045[16]	sl H_2O; vs EtOH; s eth, ace, chl
10166	Tribromofluoromethane		CBr_3F	353-54-8	270.721	liq	-73.6	108			i H_2O; s EtOH
10167	Tribromomethane	Bromoform	$CHBr_3$	75-25-2	252.731	liq	8.69(0.02)	149.2(0.5)	2.8788[25]	1.5948[25]	sl H_2O; msc EtOH, eth; s bz, lig, chl
10168	1,3,5-Tribromo-2-methoxy-benzene		$C_7H_5Br_3O$	607-99-8	344.826	nd (al)	88	298	2.491[25]		sl H_2O, EtOH; vs ace, bz; s ctc
10169	2,4,6-Tribromo-3-methylphenol	2,4,6-Tribromo-m-cresol	$C_7H_5Br_3O$	4619-74-3	344.826		84				s EtOH, eth, bz, HOAc; sl chl, peth
10170	1,1,1-Tribromo-2-methyl-2-propanol	1,1,1-Tribromo-tert-butyl alcohol	$C_4H_7Br_3O$	76-08-4	310.810	nd (lig) cry (dil al)	169	sub			sl H_2O, chl; s EtOH, eth

10132 Triadimenol

10133 Triallate

10134 Triallyl phosphate

10135 1,3,5-Triallyl-1,3,5-triazine-2,4,6(1*H*,3*H*,5*H*)-trione

10136 Triamcinolone

10137 Triamiphos

10138 Triasulfuron

10139 1,2,4-Triazine

10140 1,3,5-Triazine

10141 1,2,4-Triazine-3,5(2*H*,4*H*)-dione

10142 1,3,5-Triazine-2,4,6-triamine

10143 1,3,5-Triazine-2,4,6(1*H*,3*H*,5*H*)-trithione

10144 Triazofos

10145 Triazolam

10146 1*H*-1,2,4-Triazol-3-amine

10147 1*H*-1,2,3-Triazole

10148 1*H*-1,2,4-Triazole

10149 1*H*-1,2,4-Triazole-3,5-diamine

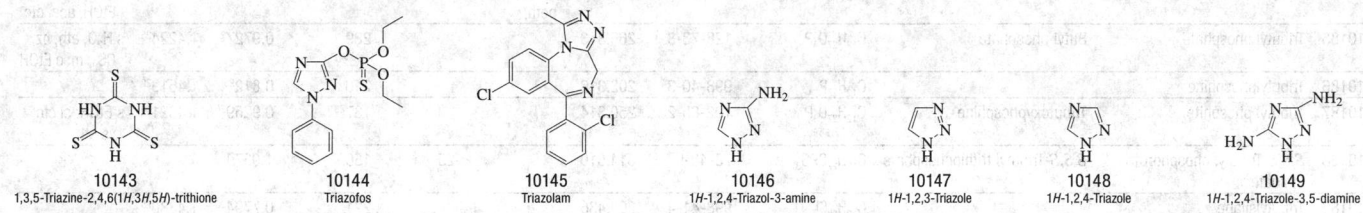

10150 Tribenuron-methyl

10151 Tribenzylamine

10152 Tribromoacetaldehyde

10153 Tribromoacetic acid

10154 2,4,6-Tribromoaniline

10155 1,2,4-Tribromobenzene

10156 1,3,5-Tribromobenzene

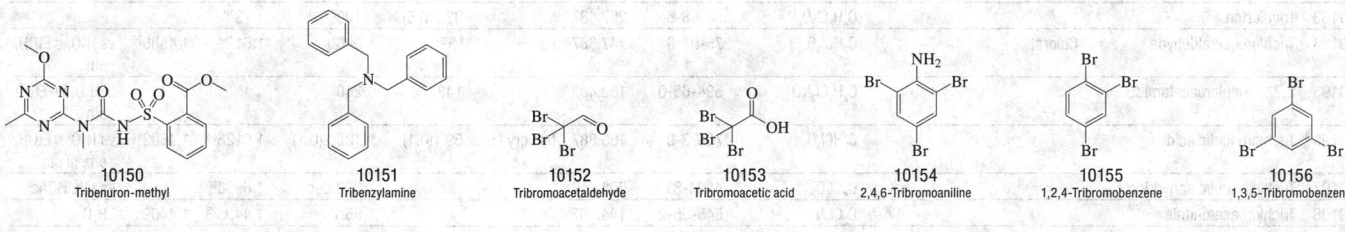

10157 1,1,2-Tribromobutane

10158 1,2,2-Tribromobutane

10159 1,2,3-Tribromobutane

10160 1,2,4-Tribromobutane

10161 2,2,3-Tribromobutane

10162 Tribromochloromethane

10163 1,1,2-Tribromoethane

10164 2,2,2-Tribromoethanol

10165 Tribromoethene

10166 Tribromofluoromethane

10167 Tribromomethane

10168 1,3,5-Tribromo-2-methoxybenzene

10169 2,4,6-Tribromo-3-methylphenol

10170 1,1,1-Tribromo-2-methyl-2-propanol

No.	Name	Synonym	Mol. Form.	CAS RN	Mol. Wt.	Physical Form	mp/°C	bp/°C	den g cm⁻³	n_D	Solubility
10171	Tribromonitromethane		CBr_3NO_2	464-10-8	297.729	pr	10	127[18]	2.811[12]	1.5790[20]	i H_2O; s EtOH, eth; vs ace, bz
10172	2,4,6-Tribromophenol		$C_6H_3Br_3O$	118-79-6	330.799	nd (al), pr (bz)	95(2)	286	2.55[20]		i H_2O; vs EtOH; s eth, bz, HOAc, chl
10173	1,1,2-Tribromopropane		$C_3H_5Br_3$	14602-62-1	280.784			199(10)	2.3547[20]	1.5790[20]	i H_2O; s EtOH, chl, HOAc; vs eth
10174	1,2,2-Tribromopropane		$C_3H_5Br_3$	14476-30-3	280.784			190.5	2.2984[20]	1.5670[20]	vs eth, EtOH, chl
10175	1,2,3-Tribromopropane		$C_3H_5Br_3$	96-11-7	280.784		14(4)	222.1(0.7)	2.4208[20]	1.5862[20]	i H_2O; vs EtOH, eth; sl ctc
10176	2,3,5-Tribromothiophene		C_4HBr_3S	3141-24-0	320.828	nd (al)	29	260			s chl
10177	Tribromotrimethyldialuminum	Methyl aluminum sesquibromide	$C_3H_9Al_2Br_3$	12263-85-3	338.778	hyg col liq		110[50]			
10178	Tributyl 2-(acetyloxy)-1,2,3-propanetricarboxylate		$C_{20}H_{34}O_8$	77-90-7	402.479			173[1]			sl chl
10179	Tributyl aluminate	1-Butanol, aluminum salt	$C_{12}H_{27}AlO_3$	3085-30-1	246.322			260[5]			
10180	Tributylaluminum		$C_{12}H_{27}Al$	1116-70-7	198.324			102[2]			
10181	Tributylamine	N,N-Dibutyl-1-butanamine	$C_{12}H_{27}N$	102-82-9	185.349	liq	-70	207(2)	0.7770[20]	1.4299[20]	sl H_2O, ctc; vs EtOH, eth; s ace, bz
10182	Tributyl borate	Butyl borate	$C_{12}H_{27}BO_3$	688-74-4	230.151	oil	<-70	233.8(1)	0.8567[20]	1.4106[18]	s EtOH, bz; vs eth, MeOH
10183	Tributylfluorostannane	Tributyltin fluoride	$C_{12}H_{27}FSn$	1983-10-4	309.050	nd	≈260				
10184	2,4,6-Tri-tert-butylphenol		$C_{18}H_{30}O$	732-26-3	262.430	cry (al, peth)	132.5(0.3)	277(6)	0.864[27]		i H_2O, alk; s EtOH, ace, ctc
10185	Tributyl phosphate	Butyl phosphate	$C_{12}H_{27}O_4P$	126-73-8	266.313			289	0.9727[25]	1.4224[25]	s H_2O, eth, bz, CS_2; msc EtOH
10186	Tributylphosphine		$C_{12}H_{27}P$	998-40-3	202.316			240	0.812[25]	1.4619[20]	
10187	Tributyl phosphite	Tributoxyphosphine	$C_{12}H_{27}O_3P$	102-85-2	250.314			137[26]	0.9259[20]	1.4321[19]	s EtOH; sl ctc; vs eth
10188	S,S,S-Tributyl phosphorotrithioate	S,S,S-Tributyl trithiophosphate	$C_{12}H_{27}OPS_3$	78-48-8	314.510		<-25	150[0.3]	1.057[20]		
10189	Tributylsilane		$C_{12}H_{28}Si$	998-41-4	200.436			221	0.7794[20]	1.4380[20]	
10190	Tributylstannane	Tributyltin hydride	$C_{12}H_{28}Sn$	688-73-3	291.060	liq		113[8]	1.103[20]		
10191	Tributyrin	Butanoic acid, 1,2,3-propanetriyl ester	$C_{15}H_{26}O_6$	60-01-5	302.363	liq	-75	307.5	1.0350[20]	1.4359[20]	i H_2O; s EtOH, ace, bz; sl ctc; vs eth
10192	Tricalcium citrate	Calcium citrate	$C_{12}H_{10}Ca_3O_{14}$	813-94-5	498.433	cry (w)	≈100 dec (hyd)				sl H_2O; i EtOH
10193	Trichlorfon		$C_4H_8Cl_3O_4P$	52-68-6	257.437		82.2(0.5)	100[0.1]	1.73[20]		
10194	Trichloroacetaldehyde	Chloral	C_2HCl_3O	75-87-6	147.387	liq	-57.5	98(2)	1.512[20]	1.4580[20]	vs H_2O; s EtOH, eth
10195	2,2,2-Trichloroacetamide		$C_2H_2Cl_3NO$	594-65-0	162.402		142	240			sl H_2O; vs EtOH, eth
10196	Trichloroacetic acid		$C_2HCl_3O_2$	76-03-9	163.387	hyg cry	59.1(0.1)	198.2(0.1)	1.6126[64]	1.4603[61]	vs H_2O; s EtOH, eth; sl ctc
10197	Trichloroacetic anhydride		$C_4Cl_6O_3$	4124-31-6	308.759			223 dec	1.6908[20]		vs eth, HOAc
10198	Trichloroacetonitrile		C_2Cl_3N	545-06-2	144.387	liq	-42	85.7	1.4403[25]	1.4409[20]	i H_2O
10199	Trichloroacetyl chloride		C_2Cl_4O	76-02-8	181.832			118.2(0.3)	1.6202[20]	1.4695[20]	msc eth
10200	2,3,4-Trichloroaniline		$C_6H_4Cl_3N$	634-67-3	196.462	nd (lig)	73	292			vs EtOH
10201	2,4,5-Trichloroaniline		$C_6H_4Cl_3N$	636-30-6	196.462	nd (lig)	96.5	270			s EtOH, eth; vs CS_2; sl lig
10202	2,4,6-Trichloroaniline		$C_6H_4Cl_3N$	634-93-5	196.462	cry (al), nd (lig or peth)	78.5	262			i H_2O; s EtOH, eth, chl; vs CS_2
10203	2,3,6-Trichlorobenzaldehyde		$C_7H_3Cl_3O$	4659-47-6	209.457	nd (lig)	87.3				vs ace, bz, eth
10204	1,2,3-Trichlorobenzene		$C_6H_3Cl_3$	87-61-6	181.447	pl (al)	53(1)	219(3)	1.4533[25]		i H_2O; sl EtOH, chl; vs eth, bz
10205	1,2,4-Trichlorobenzene		$C_6H_3Cl_3$	120-82-1	181.447	orth	17.0(0.4)	213.5(0.3)	1.459[25]	1.5717[20]	i H_2O; sl EtOH, chl; vs eth
10206	1,3,5-Trichlorobenzene		$C_6H_3Cl_3$	108-70-3	181.447	nd	62.8(0.7)	209(1)			i H_2O; sl EtOH; vs eth, bz; s chl
10207	2,3,6-Trichlorobenzeneacetic acid	Chlorfenac	$C_8H_5Cl_3O_2$	85-34-7	239.484		160.4(0.5)				
10208	3,4,5-Trichloro-1,2-benzenediol		$C_6H_3Cl_3O_2$	56961-20-7	213.446	(i) pr (HOAc) (ii) pr (bz)	130(1)				sl H_2O; vs eth, EtOH, HOAc
10209	2,3,6-Trichlorobenzoic acid		$C_7H_3Cl_3O_2$	50-31-7	225.457		129.5(0.3)				sl H_2O; s eth
10210	2,4,5-Trichlorobiphenyl		$C_{12}H_7Cl_3$	15862-07-4	257.543	cry	77(2)				i H_2O
10211	2,4,6-Trichlorobiphenyl		$C_{12}H_7Cl_3$	35693-92-6	257.543	cry (EtOH aq)	61.1(0.5)	172[15]			i H_2O

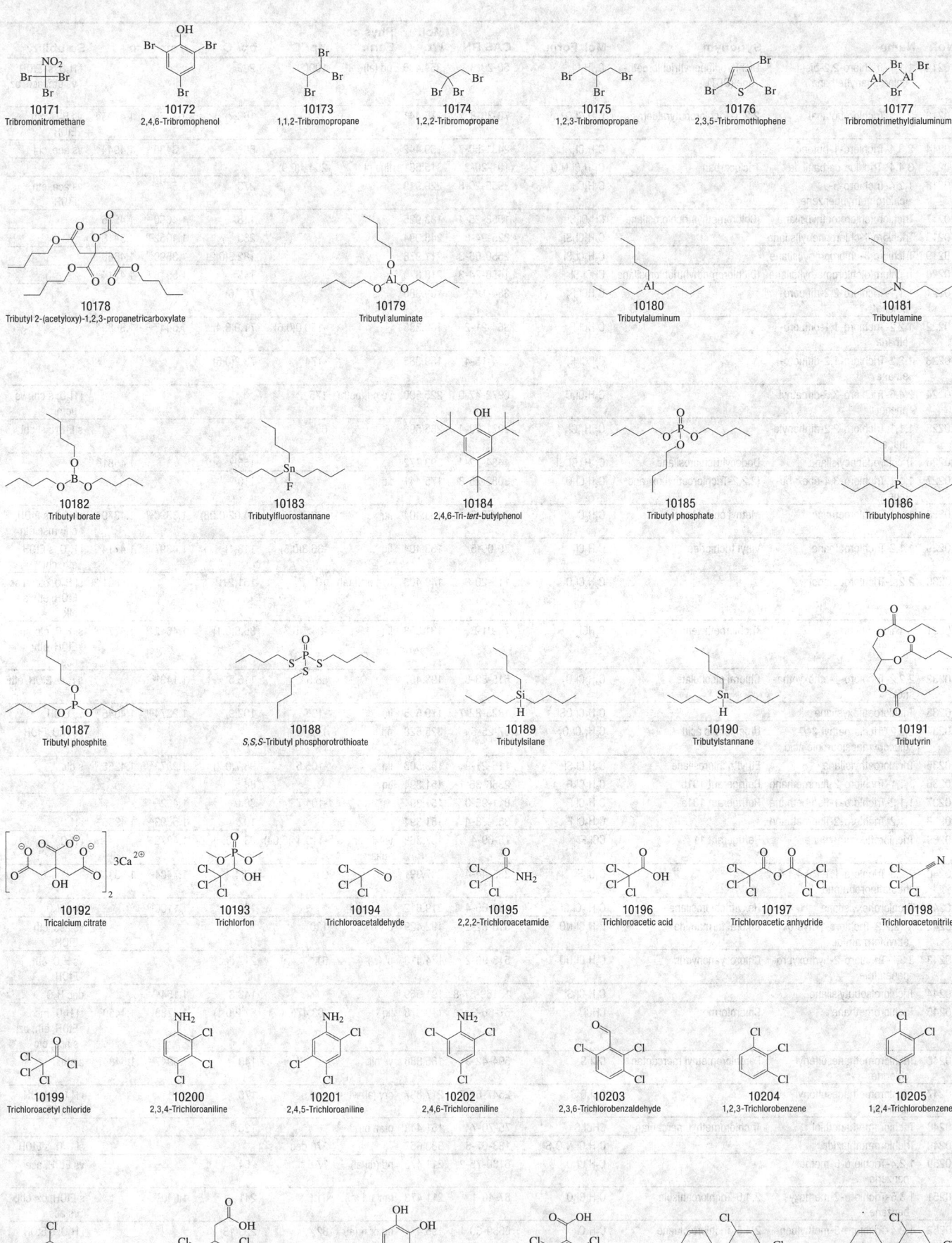

10171 Tribromonitromethane

10172 2,4,6-Tribromophenol

10173 1,1,2-Tribromopropane

10174 1,2,2-Tribromopropane

10175 1,2,3-Tribromopropane

10176 2,3,5-Tribromothiophene

10177 Tribromotrimethyldialuminum

10178 Tributyl 2-(acetyloxy)-1,2,3-propanetricarboxylate

10179 Tributyl aluminate

10180 Tributylaluminum

10181 Tributylamine

10182 Tributyl borate

10183 Tributylfluorostannane

10184 2,4,6-Tri-*tert*-butylphenol

10185 Tributyl phosphate

10186 Tributylphosphine

10187 Tributyl phosphite

10188 *S,S,S*-Tributyl phosphorotrithioate

10189 Tributylsilane

10190 Tributylstannane

10191 Tributyrin

10192 Tricalcium citrate

10193 Trichlorfon

10194 Trichloroacetaldehyde

10195 2,2,2-Trichloroacetamide

10196 Trichloroacetic acid

10197 Trichloroacetic anhydride

10198 Trichloroacetonitrile

10199 Trichloroacetyl chloride

10200 2,3,4-Trichloroaniline

10201 2,4,5-Trichloroaniline

10202 2,4,6-Trichloroaniline

10203 2,3,6-Trichlorobenzaldehyde

10204 1,2,3-Trichlorobenzene

10205 1,2,4-Trichlorobenzene

10206 1,3,5-Trichlorobenzene

10207 2,3,6-Trichlorobenzeneacetic acid

10208 3,4,5-Trichloro-1,2-benzenediol

10209 2,3,6-Trichlorobenzoic acid

10210 2,4,5-Trichlorobiphenyl

10211 2,4,6-Trichlorobiphenyl

No.	Name	Synonym	Mol. Form.	CAS RN	Mol. Wt.	Physical Form	mp/°C	bp/°C	den g cm⁻³	n_D	Solubility
10212	1,1,1-Trichloro-2,2-bis(4-chlorophenyl)ethane	Dichlorodiphenyltrichloroethane (DDT)	$C_{14}H_9Cl_5$	50-29-3	354.486	nd (al)	109(2)	260			i H_2O; sl EtOH; vs eth, ace, bz, py
10213	2,2,3-Trichlorobutanal	2,2,3-Trichlorobutyraldehyde	$C_4H_5Cl_3O$	76-36-8	175.441			166(7)	1.3956²⁰	1.4755²⁰	vs H_2O, eth, EtOH
10214	2,3,4-Trichloro-1-butene		$C_4H_5Cl_3$	2431-50-7	159.442			60²⁰	1.3430²⁰	1.4944²⁰	vs ace, chl
10215	3,4,4'-Trichlorocarbanilide	Triclocarban	$C_{13}H_9Cl_3N_2O$	101-20-2	315.581	fine pl	254.4(0.9)				
10216	1,2,4-Trichloro-5-(chloromethyl)benzene		$C_7H_4Cl_4$	3955-26-8	229.919			273	1.547²⁰		vs ace, eth, EtOH
10217	Trichloro(chloromethyl)silane	(Chloromethyl)trichlorosilane	CH_2Cl_4Si	1558-25-4	183.925			118	1.4650²⁰	1.4555²⁰	
10218	Trichloro(4-chlorophenyl)silane		$C_6H_4Cl_4Si$	825-94-5	245.994			233	1.4062²⁰	1.5418²⁰	
10219	Trichloro(3-chloropropyl)silane		$C_3H_6Cl_4Si$	2550-06-3	211.978			182.3(0.2)	1.3590²⁰	1.4668²⁰	
10220	Trichloro(dichloromethyl)silane	(Dichloromethyl)trichlorosilane	$CHCl_5Si$	1558-24-3	218.370			145	1.5518²⁰	1.4714²⁰	
10221	1,1,1-Trichloro-2,2-difluoroethane		$C_2HCl_3F_2$	354-12-1	169.385			75(16)			
10222	1,2,2-Trichloro-1,1-difluoroethane		$C_2HCl_3F_2$	354-21-2	169.385		-150.0(0.6)	71.9(0.4)	1.5447²⁰	1.3889²⁰	
10223	1,2,2-Trichloro-1,2-difluoroethane		$C_2HCl_3F_2$	354-15-4	169.385		-174	73.2(0.6)			
10224	2,4,6-Trichloro-3,5-dimethylphenol		$C_8H_7Cl_3O$	6972-47-0	225.500	ye nd (peth)	175				i H_2O; s chl; vs peth
10225	1,1,1-Trichloro-2,2-diphenylethane		$C_{14}H_{11}Cl_3$	2971-22-4	285.596		65				s EtOH; sl chl
10226	Trichlorododecylsilane	Dodecyltrichlorosilane	$C_{12}H_{25}Cl_3Si$	4484-72-4	303.772			155¹⁰		1.4581²⁰	
10227	1,1,1-Trichloro-3,4-epoxybutane	(2,2,2-Trichloroethyl)oxirane	$C_4H_5Cl_3O$	3083-25-8	175.441	liq		110¹⁰⁰			
10228	1,1,1-Trichloroethane	Methyl chloroform	$C_2H_3Cl_3$	71-55-6	133.404	liq	-30(2)	74.02(0.08)	1.3390²⁰	1.4379²⁰	sl H_2O; s EtOH, chl; msc eth
10229	1,1,2-Trichloroethane	Vinyl trichloride	$C_2H_3Cl_3$	79-00-5	133.404	liq	-36.3(0.5)	113(1)	1.4397²⁰	1.4714²⁰	i H_2O; s EtOH, eth, chl
10230	2,2,2-Trichloroethanol		$C_2H_3Cl_3O$	115-20-8	149.403	hyg orth tab or pl	19	151(21)		1.4861²⁰	sl H_2O, ctc; msc EtOH, eth; s alk
10231	Trichloroethene	Trichloroethylene	C_2HCl_3	79-01-6	131.388	liq	-84.7(0.3)	86.8(0.1)	1.4642²⁰	1.4773²⁰	sl H_2O, ctc; msc EtOH, eth; s ace
10232	2,2,2-Trichloro-1-ethoxyethanol	Chloral alcoholate	$C_4H_7Cl_3O_2$	515-83-3	193.457		56.5	115.5	1.143⁴⁰		s H_2O, EtOH, eth
10233	Trichloroethoxysilane		$C_2H_5Cl_3OSi$	1825-82-7	179.505	liq	-135	102(3)	1.2274²⁰	1.4045²⁰	vs EtOH
10234	2,2,2-Trichloroethyl-β-D-glucopyranosiduronic acid	Urochloralic acid	$C_8H_{11}Cl_3O_7$	97-25-6	325.528	nd	142				vs H_2O, EtOH
10235	Trichloroethylsilane	Ethyltrichlorosilane	$C_2H_5Cl_3Si$	115-21-9	163.506	liq	-105.6	98.7(0.7)	1.2373²⁰	1.4256²⁰	s ctc
10236	1,1,1-Trichloro-2-fluoroethane	Refrigerant 131b	$C_2H_2Cl_3F$	2366-36-1	151.394	liq		86.5			
10237	1,1,2-Trichloro-1-fluoroethane	Refrigerant 131a	$C_2H_2Cl_3F$	811-95-0	151.394		-104.7	88.0	1.492²⁰		
10238	1,1,2-Trichloro-2-fluoroethane		$C_2H_2Cl_3F$	359-28-4	151.394			102.4	1.5393²⁰	1.4390²⁰	i H_2O
10239	Trichlorofluoromethane	Refrigerant 11	CCl_3F	75-69-4	137.368	vol liq or gas	-110.44(0.04)	23.7(0.6)	1.4879²⁰		i H_2O
10240	2,2,3-Trichloro-1,1,1,3,4,4,4-heptafluorobutane		$C_4Cl_3F_7$	335-44-4	287.391		2.0	97(4)	1.7484²⁰	1.3530²⁰	
10241	Trichlorohexylsilane	Hexyltrichlorosilane	$C_6H_{13}Cl_3Si$	928-65-4	219.612			190	1.1100²⁰		dec H_2O
10242	N-(2,2,2-Trichloro-1-hydroxyethyl)formamide	Chloral formamide	$C_3H_4Cl_3NO_2$	515-82-2	192.429	cry	120				vs ace, eth, EtOH
10243	3,3,3-Trichloro-2-hydroxypropanenitrile	Chlorocyanohydrin	$C_3H_2Cl_3NO$	513-96-2	174.413	pl (w)	61	217 dec			vs H_2O, eth, EtOH
10244	Trichloroisobutylsilane		$C_4H_9Cl_3Si$	18169-57-8	191.559			143.3	1.154²⁰		dec H_2O
10245	Trichloromethane	Chloroform	$CHCl_3$	67-66-3	119.378	liq	-63.47(0.07)	61.2(0.1)	1.4788²⁵	1.4459²⁰	sl H_2O; msc EtOH, eth, bz; s ace, ctc
10246	Trichloromethanesulfenyl chloride	Perchloromethyl mercaptan	CCl_4S	594-42-3	185.888	ye oil		149	1.6947²⁰	1.5484²⁰	s eth
10247	Trichloromethanesulfonyl chloride		CCl_4O_2S	2547-61-7	217.887	cry (al-w)	140.5	170			i H_2O; s EtOH, eth, CS_2
10248	Trichloromethanethiol	Trichloromethyl mercaptan	$CHCl_3S$	75-70-7	151.443	oran oil		125¹⁵			
10249	Trichloromethiazide		$C_8H_8Cl_3N_3O_4S_2$	133-67-5	380.657		270 dec				sl H_2O; s EtOH
10250	1,2,4-Trichloro-5-methoxybenzene		$C_7H_5Cl_3O$	6130-75-2	211.473	nd (dil al)	77.5	254			vs EtOH, ace
10251	1,3,5-Trichloro-2-methoxybenzene	2,4,6-Trichloroanisole	$C_7H_5Cl_3O$	87-40-1	211.473	mcl nd (al)	61.5	241	1.640²⁵		s EtOH, bz, chl; vs ace
10252	1,2,4-Trichloro-5-methylbenzene	2,4,5-Trichlorotoluene	$C_7H_5Cl_3$	6639-30-1	195.474	nd or lf (al)	82.4	229(18)			i H_2O; s EtOH, ace
10253	(Trichloromethyl)benzene	Benzotrichloride	$C_7H_5Cl_3$	98-07-7	195.474	liq	-17.0(0.6)	221	1.3723²⁰	1.5580²⁰	i H_2O; s EtOH, eth, bz

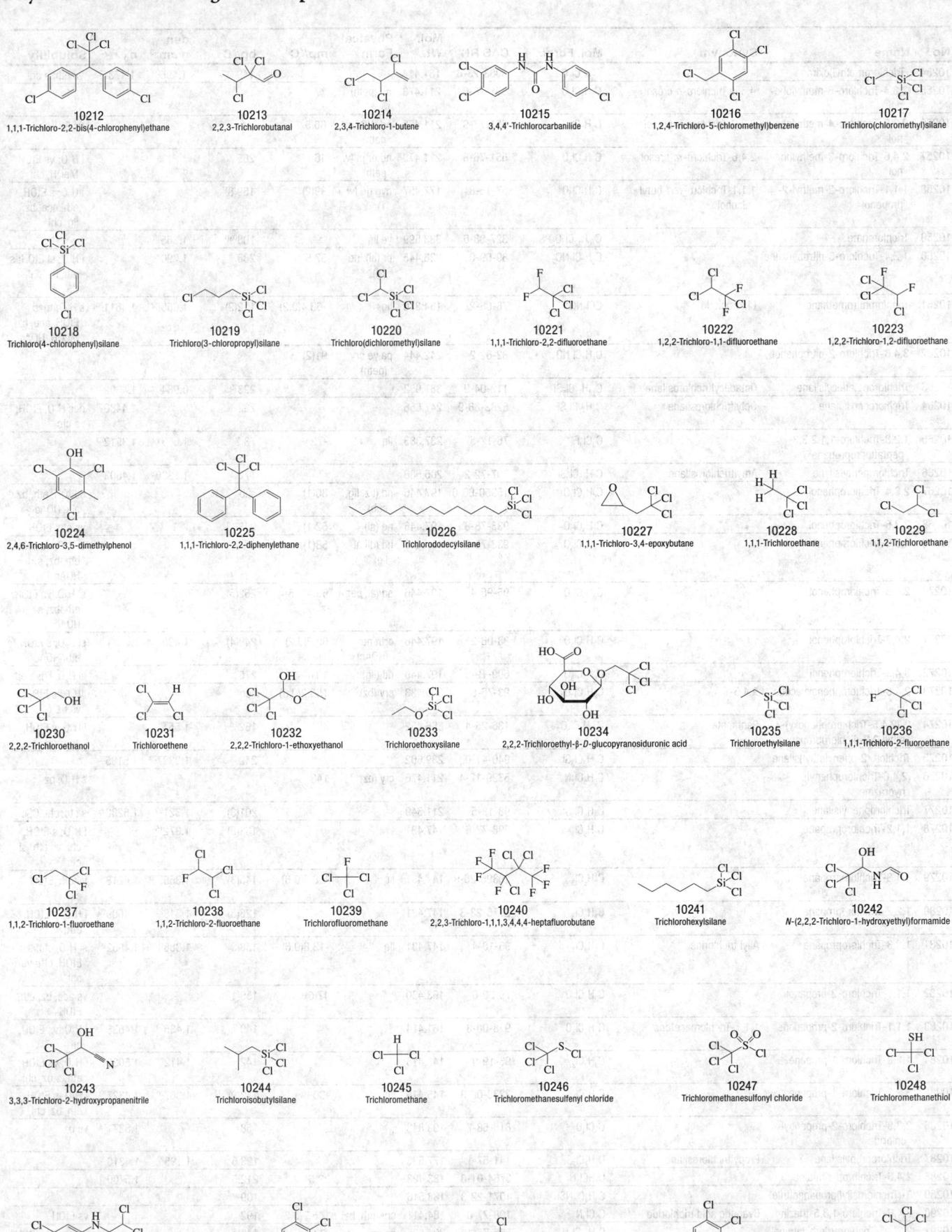

10212
1,1,1-Trichloro-2,2-bis(4-chlorophenyl)ethane

10213
2,2,3-Trichlorobutanal

10214
2,3,4-Trichloro-1-butene

10215
3,4,4'-Trichlorocarbanilide

10216
1,2,4-Trichloro-5-(chloromethyl)benzene

10217
Trichloro(chloromethyl)silane

10218
Trichloro(4-chlorophenyl)silane

10219
Trichloro(3-chloropropyl)silane

10220
Trichloro(dichloromethyl)silane

10221
1,1,1-Trichloro-2,2-difluoroethane

10222
1,2,2-Trichloro-1,1-difluoroethane

10223
1,2,2-Trichloro-1,2-difluoroethane

10224
2,4,6-Trichloro-3,5-dimethylphenol

10225
1,1,1-Trichloro-2,2-diphenylethane

10226
Trichlorododecylsilane

10227
1,1,1-Trichloro-3,4-epoxybutane

10228
1,1,1-Trichloroethane

10229
1,1,2-Trichloroethane

10230
2,2,2-Trichloroethanol

10231
Trichloroethene

10232
2,2,2-Trichloro-1-ethoxyethanol

10233
Trichloroethoxysilane

10234
2,2,2-Trichloroethyl-β-D-glucopyranosiduronic acid

10235
Trichloroethylsilane

10236
1,1,1-Trichloro-2-fluoroethane

10237
1,1,2-Trichloro-1-fluoroethane

10238
1,1,2-Trichloro-2-fluoroethane

10239
Trichlorofluoromethane

10240
2,2,3-Trichloro-1,1,1,3,4,4,4-heptafluorobutane

10241
Trichlorohexylsilane

10242
N-(2,2,2-Trichloro-1-hydroxyethyl)formamide

10243
3,3,3-Trichloro-2-hydroxypropanenitrile

10244
Trichloroisobutylsilane

10245
Trichloromethane

10246
Trichloromethanesulfenyl chloride

10247
Trichloromethanesulfonyl chloride

10248
Trichloromethanethiol

10249
Trichloromethiazide

10250
1,2,4-Trichloro-5-methoxybenzene

10251
1,3,5-Trichloro-2-methoxybenzene

10252
1,2,4-Trichloro-5-methylbenzene

10253
(Trichloromethyl)benzene

No.	Name	Synonym	Mol. Form.	CAS RN	Mol. Wt.	Physical Form	mp/°C	bp/°C	den g cm⁻³	n_D	Solubility
10254	(Trichloromethyl)oxirane		$C_3H_3Cl_3O$	3083-23-6	161.414			149	1.495²⁰	1.4737²⁵	vs eth; s chl
10255	2,3,4-Trichloro-6-methylphenol	4,5,6-Trichloro-o-cresol	$C_7H_5Cl_3O$	551-78-0	211.473	nd (peth)	77				
10256	2,3,6-Trichloro-4-methylphenol	2,3,6-Trichloro-p-cresol	$C_7H_5Cl_3O$	551-77-9	211.473	nd (HOAc, peth)	66.5				vs EtOH
10257	2,4,6-Trichloro-3-methylphenol	2,4,6-Trichloro-m-cresol	$C_7H_5Cl_3O$	551-76-8	211.473	nd or pl (w, peth)	46	265			i H₂O; vs EtOH, MeOH, chl
10258	1,1,1-Trichloro-2-methyl-2-propanol	1,1,1-Trichloro-tert-butyl alcohol	$C_4H_7Cl_3O$	57-15-8	177.457	hyg nd (w + 1)	99(2)	169(8)			i H₂O; s EtOH, eth, ace, bz, lig, chl
10259	Trichloronate		$C_{10}H_{12}Cl_3O_2PS$	327-98-0	333.599	ye liq		108⁰·⁰¹	1.365²⁰		
10260	1,2,4-Trichloro-5-nitrobenzene		$C_6H_2Cl_3NO_2$	89-69-0	226.445	pr (al), nd (al)	57.5	288	1.790²³		i H₂O; sl EtOH; s eth, bz, chl, CS₂
10261	Trichloronitromethane	Chloropicrin	CCl_3NO_2	76-06-2	164.376	liq	-69.4(0.2)	112(2)	1.6558²⁰	1.4611²⁰	s H₂O; msc EtOH, ace, bz, MeOH, HOAc
10262	3,4,6-Trichloro-2-nitrophenol		$C_6H_2Cl_3NO_3$	82-62-2	242.444	pa ye cry (peth)	91(2)				
10263	Trichlorooctadecylsilane	Octadecyltrichlorosilane	$C_{18}H_{37}Cl_3Si$	112-04-9	387.932			223¹⁰	0.984²⁵	1.4602²⁰	
10264	Trichlorooctylsilane	Octyltrichlorosilane	$C_8H_{17}Cl_3Si$	5283-66-9	247.666			232		1.4480²⁰	dec H₂O, EtOH; s ctc
10265	1,2,3-Trichloro-1,1,2,3,3-pentafluoropropane		$C_3Cl_3F_5$	76-17-5	237.383	liq	-72	73.7	1.6631²⁰	1.3512²⁰	
10266	Trichloropentylsilane	Amyltrichlorosilane	$C_5H_{11}Cl_3Si$	107-72-2	205.586			172	1.1330²⁰	1.4503²⁰	
10267	2,3,4-Trichlorophenol		$C_6H_3Cl_3O$	15950-66-0	197.446	nd (bz, lig, sub)	80(1)	sub			s EtOH, eth, bz, alk, HOAc
10268	2,3,5-Trichlorophenol		$C_6H_3Cl_3O$	933-78-8	197.446	nd (al)	62(1)	248²⁵⁰			vs eth, EtOH
10269	2,3,6-Trichlorophenol		$C_6H_3Cl_3O$	933-75-5	197.446	nd (dil al, lig)	58(1)				sl H₂O; vs EtOH, eth, bz; s HOAc
10270	2,4,5-Trichlorophenol		$C_6H_3Cl_3O$	95-95-4	197.446	nd (al, peth)	68.4(0.5)	262(5)			sl H₂O; vs EtOH, eth, bz; s HOAc
10271	2,4,6-Trichlorophenol		$C_6H_3Cl_3O$	88-06-2	197.446	orth nd (HOAc)	69.5(0.2)	249(4)	1.490¹⁷⁵		sl H₂O; s EtOH, eth, HOAc
10272	3,4,5-Trichlorophenol		$C_6H_3Cl_3O$	609-19-8	197.446	nd (lig)	101	275			sl H₂O, lig; s eth
10273	2,4,5-Trichlorophenoxyacetic acid	2,4,5-T	$C_8H_5Cl_3O_3$	93-76-5	255.483	cry (bz)	155.2(0.8)	dec			i H₂O; s EtOH; vs bz
10274	2-(2,4,5-Trichlorophenoxy)-ethyl 2,2-dichloropropanoate	Pentanate	$C_{11}H_9Cl_5O_3$	136-25-4	366.452		49	162⁰·⁵	1.55⁵⁰		i H₂O; s EtOH, ace, xyl
10275	Trichloro(2-phenylethyl)silane		$C_8H_9Cl_3Si$	940-41-0	239.602			242	1.2397²⁰	1.5185²⁰	
10276	(2,4,6-Trichlorophenyl)hydrazine		$C_6H_5Cl_3N_2$	5329-12-4	211.476	cry (bz)	143				s H₂O, bz
10277	Trichlorophenylsilane		$C_6H_5Cl_3Si$	98-13-5	211.549			201(3)	1.321²⁰	1.5230²⁰	s ctc, chl, CS₂
10278	1,1,2-Trichloropropane		$C_3H_5Cl_3$	598-77-6	147.431			133(5)	1.372¹⁵		i H₂O; s EtOH, chl; vs eth; sl ctc
10279	1,1,3-Trichloropropane		$C_3H_5Cl_3$	20395-25-9	147.431	liq	-58.9(0.6)	146(3)	1.3557²⁰	1.4718²⁰	vs eth, EtOH, chl
10280	1,2,2-Trichloropropane		$C_3H_5Cl_3$	3175-23-3	147.431			123(5)	1.318²⁵	1.4609²⁰	i H₂O; s EtOH, eth; vs chl
10281	1,2,3-Trichloropropane	Allyl trichloride	$C_3H_5Cl_3$	96-18-4	147.431	liq	-13.8(0.6)	158(2)	1.3889²⁰	1.4852²⁰	sl H₂O, ctc; s EtOH, eth; vs chl
10282	1,1,1-Trichloro-2-propanol		$C_3H_5Cl_3O$	76-00-6	163.430		47(5)	159(13)			vs ace, bz, eth, EtOH
10283	1,1,1-Trichloro-2-propanone	1,1,1-Trichloroacetone	$C_3H_3Cl_3O$	918-00-3	161.414			149	1.435²⁰	1.4635¹⁷	i H₂O; vs EtOH, eth
10284	1,2,3-Trichloro-1-propene		$C_3H_3Cl_3$	96-19-5	145.415			142	1.412²⁰	1.5030²⁰	i H₂O; vs EtOH, eth; s bz, chl
10285	3,3,3-Trichloro-1-propene		$C_3H_3Cl_3$	2233-00-3	145.415	liq	-30	112(9)	1.367²⁰	1.4827²⁰	i H₂O; s EtOH, eth, bz, chl
10286	2,3,3-Trichloro-2-propenoyl chloride		C_3Cl_4O	815-58-7	193.843			158		1.5271¹⁸	vs bz
10287	Trichloropropylsilane	Propyltrichlorosilane	$C_3H_7Cl_3Si$	141-57-1	177.533			123.5	1.195²⁰	1.4310²⁰	
10288	2,4,6-Trichloropyrimidine		$C_4HCl_3N_2$	3764-01-0	183.423		22.5	212.5		1.5700²⁰	
10289	3-(Trichlorosilyl)propanenitrile		$C_3H_4Cl_3NSi$	1071-22-3	188.516			109³⁰			
10290	2,4,6-Trichloro-1,3,5-triazine	Cyanuric acid trichloride	$C_3Cl_3N_3$	108-77-0	184.411	cry (eth, bz)	145.7(0.2)	192			vs EtOH
10291	2,2′,2″-Trichlorotriethylamine		$C_6H_{12}Cl_3N$	555-77-1	204.525	pa ye	-2.0	143¹⁵			vs bz, eth, EtOH
10292	Trichlorotriethyldialuminum	Ethylaluminum sesquichloride	$C_6H_{15}Al_2Cl_3$	12075-68-2	247.505	ye liq		115.5⁵⁰			
10293	1,3,5-Trichloro-2,4,6-trifluorobenzene		$C_6Cl_3F_3$	319-88-0	235.418			199(27)			

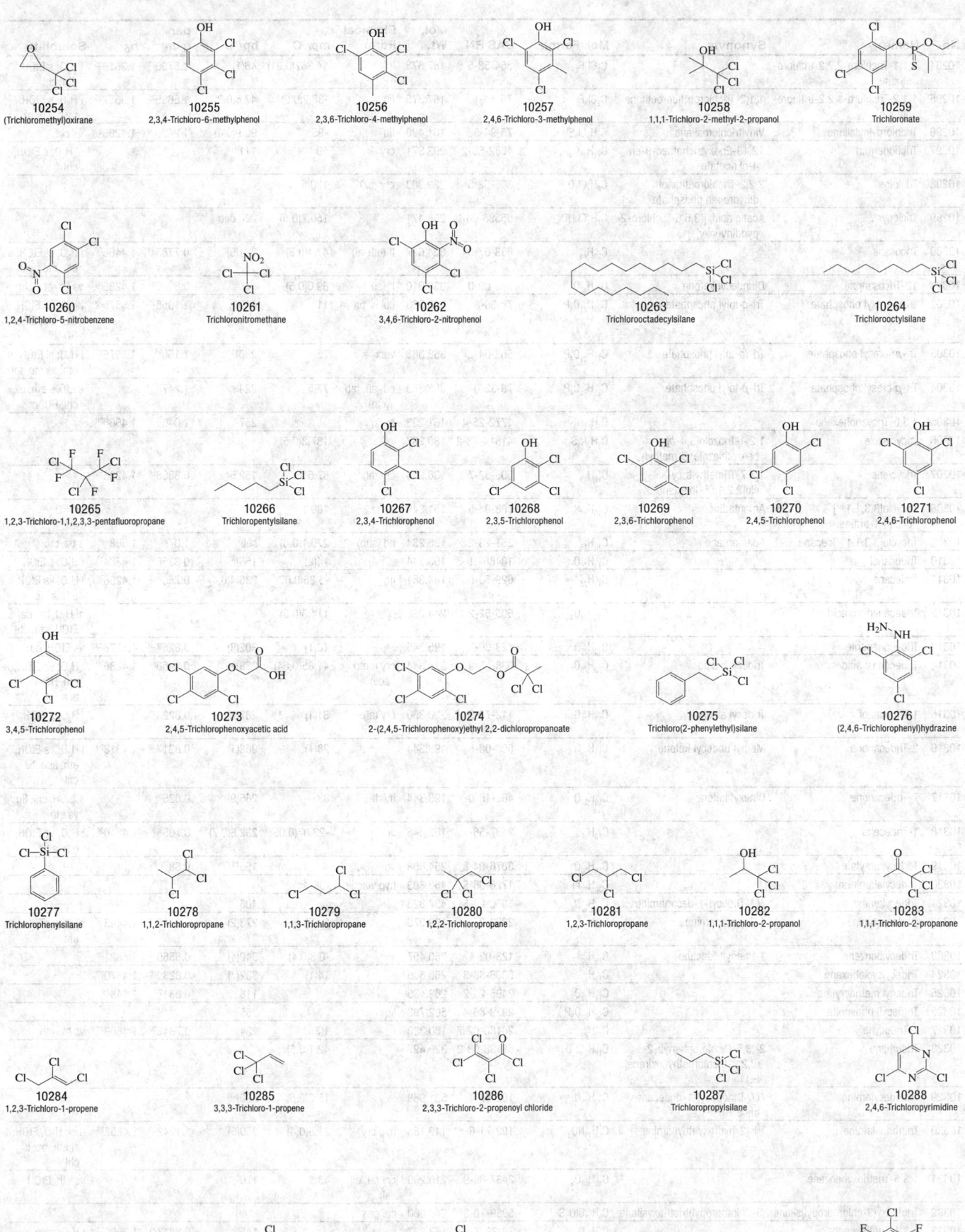

10254
(Trichloromethyl)oxirane

10255
2,3,4-Trichloro-6-methylphenol

10256
2,3,6-Trichloro-4-methylphenol

10257
2,4,6-Trichloro-3-methylphenol

10258
1,1,1-Trichloro-2-methyl-2-propanol

10259
Trichloronate

10260
1,2,4-Trichloro-5-nitrobenzene

10261
Trichloronitromethane

10262
3,4,6-Trichloro-2-nitrophenol

10263
Trichlorooctadecylsilane

10264
Trichlorooctylsilane

10265
1,2,3-Trichloro-1,1,2,3,3-pentafluoropropane

10266
Trichloropentylsilane

10267
2,3,4-Trichlorophenol

10268
2,3,5-Trichlorophenol

10269
2,3,6-Trichlorophenol

10270
2,4,5-Trichlorophenol

10271
2,4,6-Trichlorophenol

10272
3,4,5-Trichlorophenol

10273
2,4,5-Trichlorophenoxyacetic acid

10274
2-(2,4,5-Trichlorophenoxy)ethyl 2,2-dichloropropanoate

10275
Trichloro(2-phenylethyl)silane

10276
(2,4,6-Trichlorophenyl)hydrazine

10277
Trichlorophenylsilane

10278
1,1,2-Trichloropropane

10279
1,1,3-Trichloropropane

10280
1,2,2-Trichloropropane

10281
1,2,3-Trichloropropane

10282
1,1,1-Trichloro-2-propanol

10283
1,1,1-Trichloro-2-propanone

10284
1,2,3-Trichloro-1-propene

10285
3,3,3-Trichloro-1-propene

10286
2,3,3-Trichloro-2-propenoyl chloride

10287
Trichloropropylsilane

10288
2,4,6-Trichloropyrimidine

10289
3-(Trichlorosilyl)propanenitrile

10290
2,4,6-Trichloro-1,3,5-triazine

10291
2,2',2''-Trichlorotriethylamine

10292
Trichlorotriethyldialuminum

10293
1,3,5-Trichloro-2,4,6-trifluorobenzene

No.	Name	Synonym	Mol. Form.	CAS RN	Mol. Wt.	Physical Form	mp/°C	bp/°C	den g cm^{-3}	n_D	Solubility
10294	1,1,1-Trichloro-2,2,2-trifluoro-ethane		$C_2Cl_3F_3$	354-58-5	187.375		14.367(0.01)	46(1)	1.5790[20]	1.3610[35]	i H_2O; s EtOH, eth, chl
10295	1,1,2-Trichloro-1,2,2-trifluoro-ethane	1,1,2-Trichlorotrifluoroethane	$C_2Cl_3F_3$	76-13-1	187.375	liq	-36.2(0.9)	47.6(0.2)	1.5635[25]	1.3557[25]	i H_2O; s EtOH; msc eth, bz
10296	Trichlorovinylsilane	Vinyltrichlorosilane	$C_2H_3Cl_3Si$	75-94-5	161.490	liq	-95	90.9(0.4)	1.2426[20]	1.4295[20]	vs chl
10297	Trichodermin	12,13-Epoxytrichothec-9-en-4-ol acetate	$C_{17}H_{24}O_4$	4682-50-2	292.371	cry	59	111[0.05]			sl H_2O; s EtOH, chl
10298	Triclofos	2,2,2-Trichloroethanol dihydrogen phosphate	$C_2H_4Cl_3O_4P$	306-52-5	229.383	cry (bz)	120.5				
10299	Triclopyr	Acetic acid, [(3,5,6-trichloro-2-pyridinyl)oxy]-	$C_7H_4Cl_3NO_3$	55335-06-3	256.471		150.7(0.5)	290 dec			
10300	Tricosane		$C_{23}H_{48}$	638-67-5	324.627	lf (eth-al)	47.4(0.2)	381(9)	0.7785[48]	1.4468[20]	i H_2O; sl EtOH; s eth, ctc
10301	12-Tricosanone	Diundecyl ketone	$C_{23}H_{46}O$	540-09-0	338.610	lf (al)	69.0(0.5)		0.8086[69]	1.4283[80]	vs bz, eth, chl
10302	Tri-o-cresyl phosphate	Tri-o-tolyl phosphate	$C_{21}H_{21}O_4P$	78-30-8	368.363	col or pa ye	11	410	1.1955[20]	1.5575[20]	i H_2O; vs EtOH, eth, ctc, tol; s HOAc
10303	Tri-m-cresyl phosphate	Tri-m-tolyl phosphate	$C_{21}H_{21}O_4P$	563-04-2	368.363	wax	25.5	260[15]	1.150[25]	1.5575[20]	i H_2O; sl EtOH; eth; vs ctc, tol
10304	Tri-p-cresyl phosphate	Tri-p-tolyl phosphate	$C_{21}H_{21}O_4P$	78-32-0	368.363	nd (al), tab (eth)	77.5	224[35]	1.247[25]		s EtOH, eth, bz, chl, HOAc
10305	1,3,6-Tricyanohexane		$C_9H_{11}N_3$	1772-25-4	161.203	br liq		257[2]	1.040	1.4660[20]	
10306	Tricyclazole	1,2,4-Triazolo[3,4-b]-benzothiazole, 5-methyl-	$C_9H_7N_3S$	41814-78-2	189.237		187.3(0.5)				
10307	Tricyclene	1,7,7-Trimethyltricy-clo[2.2.1.0^{2,6}]heptane	$C_{10}H_{16}$	508-32-7	136.234	cry (al)	67.5	152.5	0.8668[80]	1.4296[80]	
10308	Tricyclo[3.3.1.1^{3,7}]decan-1-amine	Amantadine	$C_{10}H_{17}N$	768-94-5	151.249		180				sl H_2O
10309	Tricyclo[3.3.1.1^{3,7}]decane	Adamantane	$C_{10}H_{16}$	281-23-2	136.234	nd (sub)	270.1(0.8)	sub	1.07[25]	1.568	s bz, ctc
10310	Tridecanal		$C_{13}H_{26}O$	10486-19-8	198.344		14(2)	156[13]	0.8356[18]	1.4384[18]	i H_2O; s EtOH
10311	Tridecane		$C_{13}H_{28}$	629-50-5	184.361	liq	-5.35(0.02)	235.4(0.4)	0.7564[20]	1.4256[20]	i H_2O; vs EtOH, eth; s ctc
10312	Tridecanedioic acid		$C_{13}H_{24}O_4$	505-52-2	244.328		114.3(0.5)				sl H_2O, bz, tfa; s EtOH, eth, chl
10313	Tridecanenitrile		$C_{13}H_{25}N$	629-60-7	195.345		10(1)	302(6)	0.8257[20]	1.4378[20]	vs EtOH, eth
10314	Tridecanoic acid	Tridecylic acid	$C_{13}H_{26}O_2$	638-53-9	214.344	cry (peth ace)	41.85(0.04)	236[100]	0.8458[80]	1.4286[60]	i H_2O; vs EtOH, eth, HOAc; s ace
10315	1-Tridecanol	Tridecyl alcohol	$C_{13}H_{28}O$	112-70-9	200.360	cry (al)	31(1)	287(8)	0.8223[31]		i H_2O; s EtOH, eth
10316	2-Tridecanone	Methyl undecyl ketone	$C_{13}H_{26}O$	593-08-8	198.344		28(1)	268(1)	0.8217[30]	1.4318[20]	i H_2O; vs EtOH, eth, ace, bz, chl
10317	7-Tridecanone	Dihexyl ketone	$C_{13}H_{26}O$	462-18-0	198.344	lf (al)	33	266(9)	0.825[30]		s EtOH, chl, lig; vs eth
10318	1-Tridecene		$C_{13}H_{26}$	2437-56-1	182.345	liq	-23.07(0.09)	232.8(0.7)	0.7658[20]	1.4340[20]	i H_2O; vs EtOH, eth; s bz
10319	Tridecyl acrylate		$C_{16}H_{30}O_2$	3076-04-8	254.408	liq		150[10]	0.88[20]		
10320	Tridecylaluminum		$C_{30}H_{63}Al$	1726-66-5	450.803	hyg visc liq	-38				
10321	Tridecylamine	N,N-Didecyl-1-decanamine	$C_{30}H_{63}N$	1070-01-5	437.828			406			
10322	(Tridecyl)amine	1-Tridecanamine	$C_{13}H_{29}N$	2869-34-3	199.376		27(2)	273(2)	0.8049[20]	1.4443[20]	sl H_2O; s EtOH, eth
10323	Tridecylbenzene	1-Phenyltridecane	$C_{19}H_{32}$	123-02-4	260.457		-0.3(0.4)	340(4)	0.8550[20]	1.4821[20]	
10324	Tridecylcyclohexane		$C_{19}H_{38}$	6006-33-3	266.505		14(1)	332(7)	0.8239[20]	1.4570[20]	
10325	Tridecyl methacrylate		$C_{17}H_{32}O_2$	2495-25-2	268.435			118[1]	0.881[20]	1.448[25]	
10326	Tri(decyl) phosphite		$C_{30}H_{63}O_3P$	2929-86-4	502.793	liq		255[3]			
10327	1-Tridecyne		$C_{13}H_{24}$	26186-02-7	180.330		1(3)	234	0.7842[20]	1.4309[20]	vs bz, eth
10328	Tridiphane	2-(3,5-Dichlorophenyl)-2-(2,2,2-trichloroethyl)oxirane, (±)	$C_{10}H_7Cl_5O$	58138-08-2	320.427		42.1(0.5)				
10329	Tridodecylamine	N,N-Didodecyl-1-dodecana-mine	$C_{36}H_{75}N$	102-87-4	521.988		15.7(0.5)	220[0.03]			
10330	Triethanolamine	Tris(2-hydroxyethyl)amine	$C_6H_{15}NO_3$	102-71-6	149.188	hyg cry	21.5(0.4)	350(5)	1.1242[20]	1.4852[20]	msc H_2O, EtOH; sl eth, bz; s chl
10331	1,3,5-Triethoxybenzene		$C_{12}H_{18}O_3$	2437-88-9	210.269	cry (al, dil al)	43.5	170[24]			vs eth, EtOH
10332	Triethoxy(3-chloropropyl)silane	(3-Chloropropyl)triethoxysilane	$C_9H_{21}ClO_3Si$	5089-70-3	240.800	col gas		-149			
10333	1,1,1-Triethoxyethane		$C_8H_{18}O_3$	78-39-7	162.227			145	0.8847[25]	1.3980[20]	i H_2O; msc EtOH, eth, ctc, chl

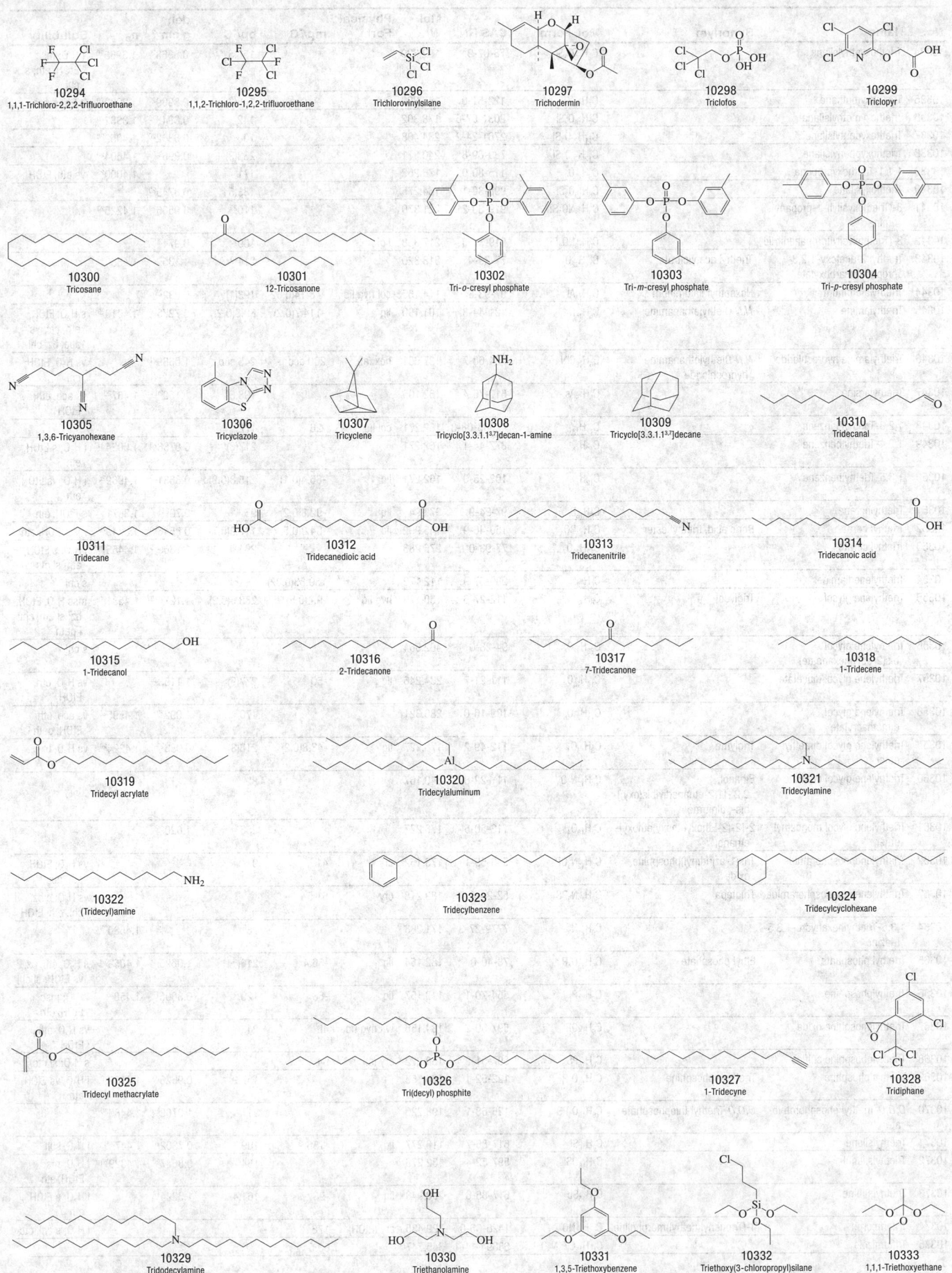

10294
1,1,1-Trichloro-2,2,2-trifluoroethane

10295
1,1,2-Trichloro-1,2,2-trifluoroethane

10296
Trichlorovinylsilane

10297
Trichodermin

10298
Triclofos

10299
Triclopyr

10300
Tricosane

10301
12-Tricosanone

10302
Tri-*o*-cresyl phosphate

10303
Tri-*m*-cresyl phosphate

10304
Tri-*p*-cresyl phosphate

10305
1,3,6-Tricyanohexane

10306
Tricyclazole

10307
Tricyclene

10308
Tricyclo[3.3.1.1³,⁷]decan-1-amine

10309
Tricyclo[3.3.1.1³,⁷]decane

10310
Tridecanal

10311
Tridecane

10312
Tridecanedioic acid

10313
Tridecanenitrile

10314
Tridecanoic acid

10315
1-Tridecanol

10316
2-Tridecanone

10317
7-Tridecanone

10318
1-Tridecene

10319
Tridecyl acrylate

10320
Tridecylaluminum

10321
Tridecylamine

10322
(Tridecyl)amine

10323
Tridecylbenzene

10324
Tridecylcyclohexane

10325
Tridecyl methacrylate

10326
Tri(decyl) phosphite

10327
1-Tridecyne

10328
Tridiphane

10329
Tridodecylamine

10330
Triethanolamine

10331
1,3,5-Triethoxybenzene

10332
Triethoxy(3-chloropropyl)silane

10333
1,1,1-Triethoxyethane

No.	Name	Synonym	Mol. Form.	CAS RN	Mol. Wt.	Physical Form	mp/°C	bp/°C	den g cm⁻³	n_D	Solubility
10334	Triethoxyethylsilane		C₈H₂₀O₃Si	78-07-9	192.329			158.5	0.8963²⁰	1.3955²⁰	i H₂O; msc EtOH, eth; s chl
10335	Triethoxymethane		C₇H₁₆O₃	122-51-0	148.200			145(3)	0.8909²⁰	1.3922²⁰	s EtOH, eth
10336	Triethoxymethylsilane		C₇H₁₈O₃Si	2031-67-6	178.302			142	0.8948²⁵	1.3832²⁰	
10337	Triethoxypentylsilane		C₁₁H₂₆O₃Si	2761-24-2	234.408			100³⁰	0.8862²⁰	1.4059²⁰	
10338	Triethoxyphenylsilane		C₁₂H₂₀O₃Si	780-69-8	240.371			233(3)	0.996²⁵	1.4604²⁰	
10339	1,1,1-Triethoxypropane		C₉H₂₀O₃	115-80-0	176.253			171		1.4000²⁵	vs eth, EtOH
10340	Triethoxysilane		C₆H₁₆O₃Si	998-30-1	164.275			133.5	0.8745²⁰		
10341	3-(Triethoxysilyl)-1-propan-amine		C₉H₂₃NO₃Si	919-30-2	221.370			119²⁹	0.9506²⁰	1.4225²⁰	
10342	3-(Triethoxysilyl)propanenitrile		C₉H₁₉NO₃Si	919-31-3	217.338	liq		109¹⁰	0.974²⁰		
10343	Triethyl 2-acetoxy-1,2,3-propanetricarboxylate	Triethyl acetylcitrate	C₁₄H₂₂O₈	77-89-4	318.320			214⁴⁰	1.135²⁵	1.4380	
10344	Triethylaluminum	Hexaethyldialuminum	C₆H₁₅Al	97-93-8	114.165	col liq liq	-48.14(0.02)	193(1)	0.832²⁵		
10345	Triethylamine	N,N-Diethylethanamine	C₆H₁₅N	121-44-8	101.190	liq	-114.7(0.2)	88.8(0.2)	0.7275²⁰	1.4010²⁰	s H₂O, EtOH, eth, ctc; vs ace, bz, chl
10346	Triethylamine hydrochloride	N,N-Diethylethanamine hydrochloride	C₆H₁₆ClN	554-68-7	137.651	hex (al)	260 dec	245 sub	1.0689²¹		vs H₂O, EtOH, chl
10347	Triethylarsine		C₆H₁₅As	617-75-4	162.105			138.5	1.150²⁰	1.467²⁰	vs ace, eth, EtOH
10348	1,2,3-Triethylbenzene		C₁₂H₁₈	42205-08-3	162.271	col liq	-26	172			
10349	1,2,4-Triethylbenzene		C₁₂H₁₈	877-44-1	162.271			217(3)	0.8738²⁰	1.5024²⁰	i H₂O; s EtOH, eth
10350	1,3,5-Triethylbenzene		C₁₂H₁₈	102-25-0	162.271	liq	-66.4(0.1)	215.8(0.9)	0.8631²⁰	1.4969²⁰	i H₂O; vs EtOH, eth
10351	Triethylborane		C₆H₁₅B	97-94-9	97.994	liq	-92.8(0.2)	95	0.70²³	1.3971	s EtOH, eth
10352	Triethyl borate	Boric acid, triethyl ester	C₆H₁₅BO₃	150-46-9	145.992	liq	-84.8(0.5)	117.9(0.7)	0.8546²⁰	1.3749²⁰	msc EtOH, eth
10353	Triethyl citrate		C₁₂H₂₀O₇	77-93-0	276.283			294	1.1369²⁰	1.4455²⁰	i H₂O; s EtOH, eth; sl ctc
10354	Triethylenediamine		C₆H₁₂N₂	280-57-9	112.172		159.83(0.02)				s chl
10355	Triethylene glycol	Triglycol	C₆H₁₄O₄	112-27-6	150.173	hyg liq	-9.4(0.5)	288.6(0.2)	1.1274¹⁵	1.4531²⁰	msc H₂O, EtOH, bz; sl eth, chl; i peth
10356	Triethylene glycol bis(2-ethylhexanoate)		C₂₂H₄₂O₆	94-28-0	402.564						s chl
10357	Triethylene glycol diacetate		C₁₀H₁₈O₆	111-21-7	234.246	liq	-50	277(9)	1.1153²⁰		vs H₂O, eth, EtOH
10358	Triethylene glycol dimethacrylate		C₁₄H₂₂O₆	109-16-0	286.321			170⁵	1.092²⁰	1.4595²⁵	vs ace, eth, EtOH, peth
10359	Triethylene glycol dimethyl ether	Triglyme	C₈H₁₈O₄	112-49-2	178.227	liq	-43.8(0.2)	218(3)	0.986²⁰	1.4224²⁰	vs H₂O, bz
10360	Triethylene glycol dinitrate	Ethanol, 2,2'-[1,2-ethanediylbis(oxy)]-bis-, dinitrate	C₆H₁₂N₂O₈	111-22-8	240.167			82⁰·⁰³			
10361	Triethylene glycol monoethyl ether	2-[2-(2-Ethoxyethoxy)ethoxy]-ethanol	C₈H₁₈O₄	112-50-5	178.227			256	1.0209²⁰		
10362	Triethylenephosphoramide	Tris(1-aziridinyl)phosphine, oxide	C₆H₁₂N₃OP	545-55-1	173.152	cry	41	91²³			vs H₂O, EtOH, eth, ace
10363	Triethylenethiophosphoramide	Thiotepa	C₆H₁₂N₃PS	52-24-4	189.218	cry	51.5				vs H₂O; s bz, chl, eth, EtOH
10364	1,3,5-Triethylhexahydro-1,3,5-triazine		C₉H₂₁N₃	7779-27-3	171.283			78⁶		1.4580²⁵	
10365	Triethyl phosphate	Ethyl phosphate	C₆H₁₅O₄P	78-40-0	182.154	liq	-56.4	216(11)	1.0695²⁰	1.4053²⁰	s H₂O, eth, bz; vs EtOH; sl chl
10366	Triethylphosphine		C₆H₁₅P	554-70-1	118.157	liq	-88	129	0.8006¹⁹	1.458¹⁵	i H₂O; msc EtOH, eth
10367	Triethylphosphine oxide		C₆H₁₅OP	597-50-2	134.156	wh hyg nd	48	243			vs H₂O, eth, EtOH
10368	Triethylphosphine sulfide		C₆H₁₅PS	597-51-3	150.222	cry (al)	94				s H₂O; sl ctc
10369	Triethyl phosphite	Triethoxyphosphine	C₆H₁₅O₃P	122-52-1	166.155			157.9	0.9629²⁰	1.4127²⁰	i H₂O; vs EtOH, eth
10370	O,O,O-Triethyl phosphorothio-ate	O,O,O-Triethyl thiophosphate	C₆H₁₅O₃PS	126-68-1	198.220			217	1.0768²⁰	1.4480²⁰	
10371	Triethylsilane		C₆H₁₆Si	617-86-7	116.277	liq	-156.9	109	0.7302²⁰	1.447²⁰	i H₂O, sulf
10372	Triethylsilanol		C₆H₁₆OSi	597-52-4	132.276			154	0.8647²⁰	1.4329²⁰	i H₂O; msc EtOH, eth
10373	Triethylstibine		C₆H₁₅Sb	617-85-6	208.943	liq	-98	161.4	1.3224¹⁵		i H₂O; s EtOH, eth
10374	Trifenmorph	4-(Triphenylmethyl)morpholine	C₂₃H₂₃NO	1420-06-0	329.435	cry (EtOH)	176				i H₂O; s chl, ctc
10375	Triflumizole		C₁₅H₁₅ClF₃N₃O	68694-11-1	345.747		63.5				

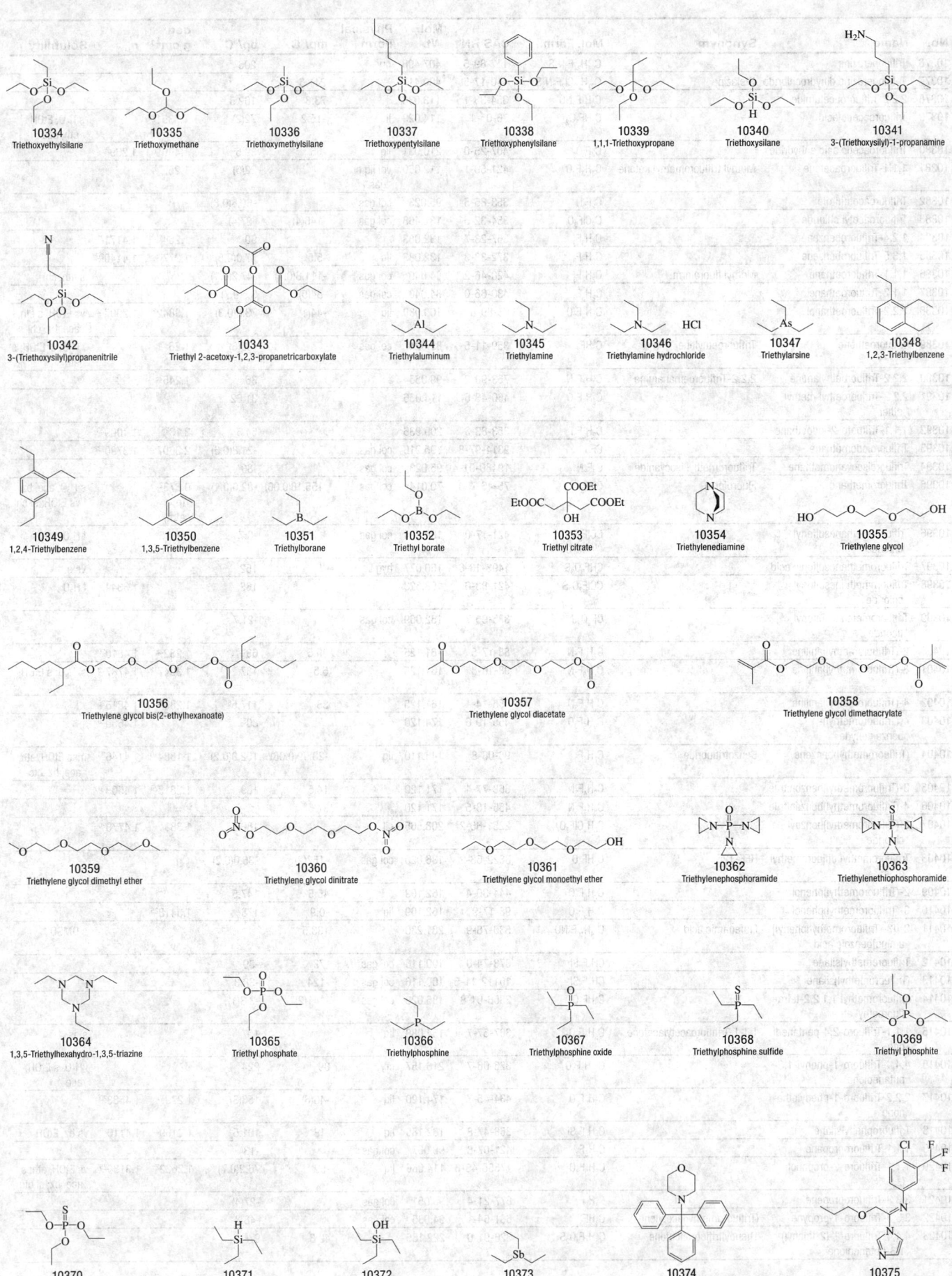

10334 Triethoxyethylsilane

10335 Triethoxymethane

10336 Triethoxymethylsilane

10337 Triethoxypentylsilane

10338 Triethoxyphenylsilane

10339 1,1,1-Triethoxypropane

10340 Triethoxysilane

10341 3-(Triethoxysilyl)-1-propanamine

10342 3-(Triethoxysilyl)propanenitrile

10343 Triethyl 2-acetoxy-1,2,3-propanetricarboxylate

10344 Triethylaluminum

10345 Triethylamine

10346 Triethylamine hydrochloride

10347 Triethylarsine

10348 1,2,3-Triethylbenzene

10349 1,2,4-Triethylbenzene

10350 1,3,5-Triethylbenzene

10351 Triethylborane

10352 Triethyl borate

10353 Triethyl citrate

10354 Triethylenediamine

10355 Triethylene glycol

10356 Triethylene glycol bis(2-ethylhexanoate)

10357 Triethylene glycol diacetate

10358 Triethylene glycol dimethacrylate

10359 Triethylene glycol dimethyl ether

10360 Triethylene glycol dinitrate

10361 Triethylene glycol monoethyl ether

10362 Triethylenephosphoramide

10363 Triethylenethiophosphoramide

10364 1,3,5-Triethylhexahydro-1,3,5-triazine

10365 Triethyl phosphate

10366 Triethylphosphine

10367 Triethylphosphine oxide

10368 Triethylphosphine sulfide

10369 Triethyl phosphite

10370 O,O,O-Triethyl phosphorothioate

10371 Triethylsilane

10372 Triethylsilanol

10373 Triethylstibine

10374 Trifenmorph

10375 Triflumizole

No.	Name	Synonym	Mol. Form.	CAS RN	Mol. Wt.	Physical Form	mp/°C	bp/°C	den g cm⁻³	n_D	Solubility
10376	Trifluoperazine		$C_{21}H_{24}F_3N_3S$	117-89-5	407.496	cry		206[0.7]			
10377	Trifluoperazine dihydrochloride	Stelazine	$C_{21}H_{26}Cl_2F_3N_3S$	440-17-5	480.417		241.5				
10378	2,2,2-Trifluoroacetamide		$C_2H_2F_3NO$	354-38-1	113.038		73.8	162.5			
10379	Trifluoroacetic acid		$C_2HF_3O_2$	76-05-1	114.023	liq	-15.2	72(2)	1.5351[25]		s H_2O, EtOH, eth, ace
10380	Trifluoroacetic acid anhydride		$C_4F_6O_3$	407-25-0	210.031	liq	-65	39.5	1.490[25]	1.269[25]	
10381	1,1,1-Trifluoroacetone	Methyl trifluoromethyl ketone	$C_3H_3F_3O$	421-50-1	112.050	vol liq or gas		22(5)	1.252[25]		
10382	Trifluoroacetonitrile		C_2F_3N	353-85-5	95.023	col gas		-68.8(0.8)			
10383	Trifluoroacetyl chloride		C_2ClF_3O	354-32-5	132.468	col gas	-146(4)	-27(4)			
10384	1,2,4-Trifluorobenzene		$C_6H_3F_3$	367-23-7	132.083			90	1.264[25]	1.4171[20]	
10385	1,3,5-Trifluorobenzene		$C_6H_3F_3$	372-38-3	132.083	liq	-5.5	77.0(0.5)	1.277[25]	1.4140[20]	
10386	1,1,1-Trifluoroethane	Methyl fluoroform	$C_2H_3F_3$	420-46-2	84.040	col gas	-111.8(0.1)	-47.2(0.1)			s eth, chl
10387	1,1,2-Trifluoroethane		$C_2H_3F_3$	430-66-0	84.040	col gas	-84(5)	3.5(0.7)			
10388	2,2,2-Trifluoroethanol		$C_2H_3F_3O$	75-89-8	100.039	liq	-44(1)	73.8(0.3)	1.3842[20]	1.2907[22]	vs EtOH; s eth, ace, bz, chl
10389	Trifluoroethene	Trifluoroethylene	C_2HF_3	359-11-5	82.024	col gas		-53(6)	1.26[-70]		i H_2O; sl EtOH; s eth
10390	2,2,2-Trifluoroethylamine	2,2,2-Trifluoroethanamine	$C_2H_4F_3N$	753-90-2	99.055			36	1.245[25]		
10391	2,2,2-Trifluoroethyl methyl ether		$C_3H_5F_3O$	460-43-5	114.066			31.62			
10392	1,1,1-Trifluoro-2-iodoethane		$C_2H_2F_3I$	353-83-3	209.936			54.5	2.13[25]	1.4009[20]	
10393	Trifluoroiodomethane		CF_3I	2314-97-8	195.910	col gas		-21.8(0.6)	2.3607[-32]	1.3790[-32]	
10394	Trifluoroisocyanomethane	Trifluoromethyl isocyanide	C_2F_3N	19480-01-4	95.023	col gas		-80			
10395	Trifluoromethane	Fluoroform	CHF_3	75-46-7	70.014	col gas	-155.18(0.06)	-82.0(0.1)	0.673[25] (p>1 atm)		s H_2O, ace, bz; vs EtOH; sl chl
10396	Trifluoromethanesulfenyl chloride		$CClF_3S$	421-17-0	136.524	col gas		-0.7			i H_2O
10397	Trifluoromethanesulfonic acid		CHF_3O_3S	1493-13-6	150.077	hyg liq	45	162			vs eth
10398	Trifluoromethanesulfonyl chloride		$CClF_3O_2S$	421-83-0	168.523			162		1.3344[20]	i H_2O
10399	Trifluoromethanesulfonyl fluoride		CF_4O_2S	335-05-7	152.069	col gas		-21.7			
10400	2-(Trifluoromethyl)aniline		$C_7H_6F_3N$	88-17-5	161.125		35.5	68[15]	1.282[25]	1.4810[20]	
10401	3-(Trifluoromethyl)aniline		$C_7H_6F_3N$	98-16-8	161.125		5.5	187	1.3047[12]	1.4787[20]	sl H_2O; s EtOH, eth
10402	4-(Trifluoromethyl)aniline		$C_7H_6F_3N$	455-14-1	161.125		38	117.5[60]	1.283[27]	1.4815[25]	
10403	4-(Trifluoromethyl)-benzaldehyde		$C_8H_5F_3O$	455-19-6	174.120			80[25]		1.4630[20]	
10404	(Trifluoromethyl)benzene	Benzotrifluoride	$C_7H_5F_3$	98-08-8	146.110	liq	-28.99(0.06)	102.0(0.2)	1.1884[20]	1.4146[20]	msc EtOH, eth, ace, bz, ctc
10405	3-(Trifluoromethyl)benzonitrile		$C_8H_4F_3N$	368-77-4	171.120		14.5	189	1.2813[20]	1.4508[20]	
10406	4-(Trifluoromethyl)benzonitrile		$C_8H_4F_3N$	455-18-5	171.120		37.5				
10407	3-(Trifluoromethyl)benzoyl chloride		$C_8H_4ClF_3O$	2251-65-2	208.565	oil		186	1.383	1.4770[20]	
10408	Trifluoromethyl difluoromethyl ether	HFE-125	C_2HF_5O	3822-68-2	136.020	col gas	-157(2)	-35.0(0.2)			
10409	2-(Trifluoromethyl)phenol		$C_7H_5F_3O$	444-30-4	162.109		45.5	147.5			
10410	3-(Trifluoromethyl)phenol		$C_7H_5F_3O$	98-17-9	162.109	liq	-0.9	178	1.3418[25]		
10411	2-[[3-(Trifluoromethyl)phenyl]-amino]benzoic acid	Flufenamic acid	$C_{14}H_{10}F_3NO_2$	530-78-9	281.230		133.5				s DMSO
10412	Trifluoromethylsilane		CH_3F_3Si	373-74-0	100.116	col gas	-73	-30			
10413	(Trifluoromethyl)silane		CH_3F_3Si	10112-11-5	100.116	col gas	-124	-38.3			
10414	Trifluoromethyl 1,1,2,2-tetra-fluoroethyl ether		C_3HF_7O	2356-61-8	186.028	col gas	-141(2)	-3.3(1)			
10415	1,1,1-Trifluoro-2,4-pentanedi-one	1,1,1-Trifluoroacetylacetone	$C_5H_5F_3O_2$	367-57-7	154.088	liq		107			s os
10416	4,4,4-Trifluoro-1-phenyl-1,3-butanedione		$C_{10}H_7F_3O_2$	326-06-7	216.157	cry	39	224			i H_2O; s EtOH, ace
10417	2,2,2-Trifluoro-1-phenyletha-none		$C_8H_5F_3O$	434-45-7	174.120	liq	-40(6)	153(3)	1.279[20]	1.4583[20]	
10418	Trifluorophenylsilane		$C_6H_5F_3Si$	368-47-8	162.185	liq	-18	101.5	1.2169[20]	1.4110[20]	vs bz, EtOH
10419	1,1,1-Trifluoropropane		$C_3H_5F_3$	421-07-8	98.067	col gas		-13			
10420	1,1,1-Trifluoro-2-propanol, (±)-		$C_3H_5F_3O$	17556-48-8	114.066	liq	-52	76.2(0.7)	1.2632[25]	1.3130[25]	vs EtOH, eth; s ace, bz; sl ctc
10421	3,3,3-Trifluoropropene		$C_3H_3F_3$	677-21-4	96.051	col gas		-27(4)			
10422	3,3,3-Trifluoro-1-propyne	(Trifluoromethyl)acetylene	C_3HF_3	661-54-1	94.035	col gas		-48.3			
10423	4,4,4-Trifluoro-1-(2-thienyl)-1,3-butanedione	Thenoyltrifluoroacetone	$C_8H_5F_3O_2S$	326-91-0	222.185		42.8	97[8]			

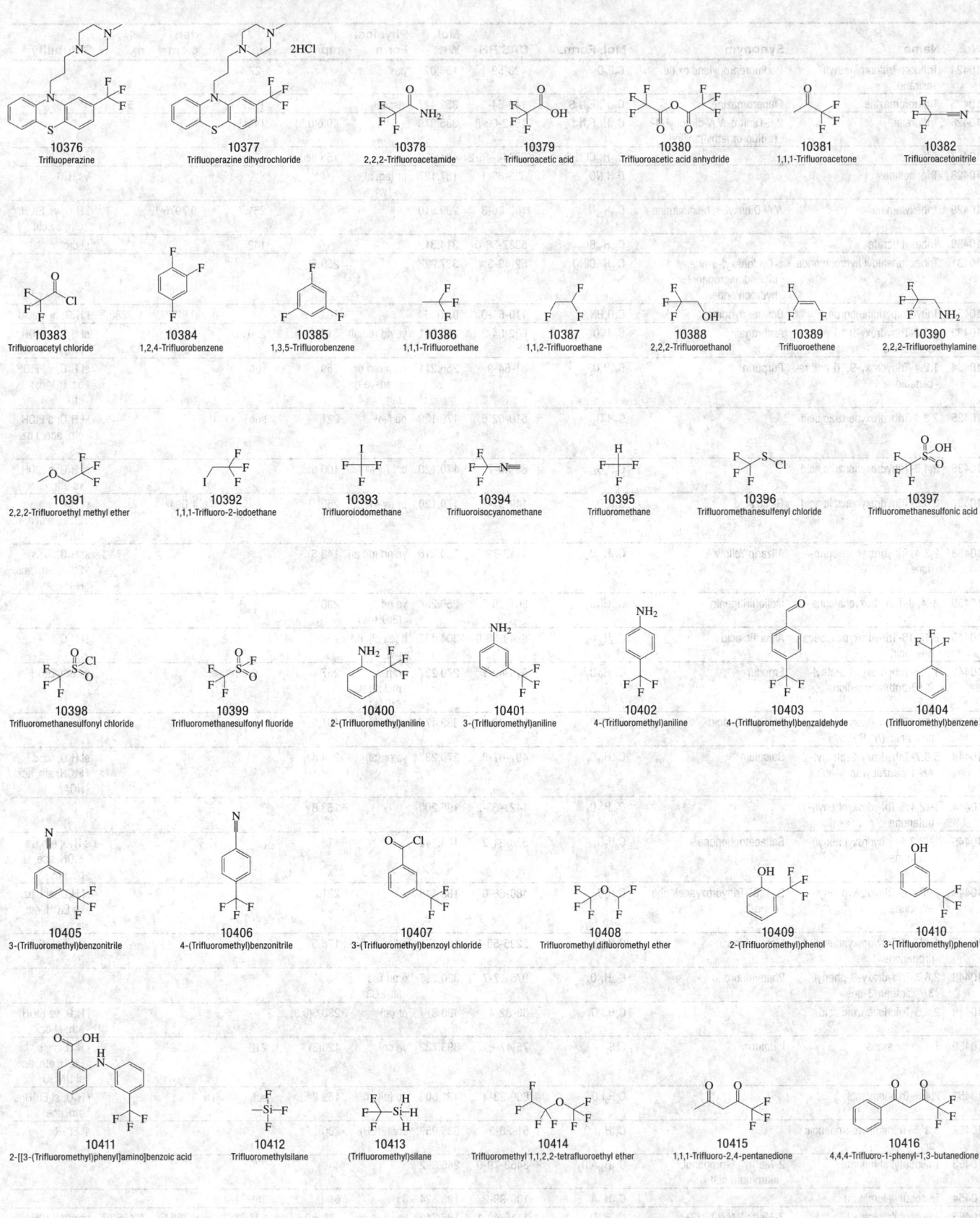

10376 Trifluoperazine

10377 Trifluoperazine dihydrochloride

10378 2,2,2-Trifluoroacetamide

10379 Trifluoroacetic acid

10380 Trifluoroacetic acid anhydride

10381 1,1,1-Trifluoroacetone

10382 Trifluoroacetonitrile

10383 Trifluoroacetyl chloride

10384 1,2,4-Trifluorobenzene

10385 1,3,5-Trifluorobenzene

10386 1,1,1-Trifluoroethane

10387 1,1,2-Trifluoroethane

10388 2,2,2-Trifluoroethanol

10389 Trifluoroethene

10390 2,2,2-Trifluoroethylamine

10391 2,2,2-Trifluoroethyl methyl ether

10392 1,1,1-Trifluoro-2-iodoethane

10393 Trifluoroiodomethane

10394 Trifluoroisocyanomethane

10395 Trifluoromethane

10396 Trifluoromethanesulfenyl chloride

10397 Trifluoromethanesulfonic acid

10398 Trifluoromethanesulfonyl chloride

10399 Trifluoromethanesulfonyl fluoride

10400 2-(Trifluoromethyl)aniline

10401 3-(Trifluoromethyl)aniline

10402 4-(Trifluoromethyl)aniline

10403 4-(Trifluoromethyl)benzaldehyde

10404 (Trifluoromethyl)benzene

10405 3-(Trifluoromethyl)benzonitrile

10406 4-(Trifluoromethyl)benzonitrile

10407 3-(Trifluoromethyl)benzoyl chloride

10408 Trifluoromethyl difluoromethyl ether

10409 2-(Trifluoromethyl)phenol

10410 3-(Trifluoromethyl)phenol

10411 2-[[3-(Trifluoromethyl)phenyl]amino]benzoic acid

10412 Trifluoromethylsilane

10413 (Trifluoromethyl)silane

10414 Trifluoromethyl 1,1,2,2-tetrafluoroethyl ether

10415 1,1,1-Trifluoro-2,4-pentanedione

10416 4,4,4-Trifluoro-1-phenyl-1,3-butanedione

10417 2,2,2-Trifluoro-1-phenylethanone

10418 Trifluorophenylsilane

10419 1,1,1-Trifluoropropane

10420 1,1,1-Trifluoro-2-propanol, (±)-

10421 3,3,3-Trifluoropropene

10422 3,3,3-Trifluoro-1-propyne

10423 4,4,4-Trifluoro-1-(2-thienyl)-1,3-butanedione

No.	Name	Synonym	Mol. Form.	CAS RN	Mol. Wt.	Physical Form	mp/°C	bp/°C	den g cm^{-3}	n_D	Solubility
10424	Trifluoro(trifluoromethyl)-oxirane	Perfluoropropylene oxide	C$_3$F$_6$O	428-59-1	166.021	gas		-27.4			
10425	Triflupromazine	Fluopromazine	C$_{18}$H$_{19}$F$_3$N$_2$S	146-54-3	352.417	visc oil		176[0.7]		1.5780[23]	
10426	Trifluralin	2,6-Dinitro-*N,N*-dipropyl-4-(trifluoromethyl)aniline	C$_{13}$H$_{16}$F$_3$N$_3$O$_4$	1582-09-8	335.279		49.6(0.8)	140[4.2]			
10427	Triforine		C$_{10}$H$_{14}$Cl$_6$N$_4$O$_2$	26644-46-2	434.962		155 dec				
10428	Trigonelline		C$_7$H$_7$NO$_2$	535-83-1	137.137	pr (aq, al, +1w)					vs H$_2$O
10429	Trihexylamine	*N,N*-Dihexyl-1-hexanamine	C$_{18}$H$_{39}$N	102-86-3	269.510			261.7	0.7976[21]		i H$_2$O; vs EtOH, eth; s acid
10430	Trihexyl borate		C$_{18}$H$_{39}$BO$_3$	5337-36-0	314.312			143[2]			sl ctc
10431	Trihexyphenidyl hydrochloride	α-Cyclohexyl-α-phenyl-1-piperidinepropanol hydrochloride	C$_{20}$H$_{32}$ClNO	52-49-3	337.927		258.5				
10432	Trihydro(pyridine)boron	Borane pyridine	C$_5$H$_8$BN	110-51-0	92.936		10.5		0.920[20]	1.5280[25]	i H$_2$O; dec acid
10433	1,2,3-Trihydroxy-9,10-anthracenedione	Anthragallol	C$_{14}$H$_8$O$_5$	602-64-2	256.211	ye nd (dil al)	313	290 sub			sl H$_2$O; s EtOH, eth, HOAc, CS$_2$
10434	1,2,4-Trihydroxy-9,10-anthracenedione	Purpurin	C$_{14}$H$_8$O$_5$	81-54-9	256.211	oran red or oran-ye nd (al)	259	sub			sl H$_2$O; vs EtOH, bz, HOAc; s eth
10435	2,3,4-Trihydroxybenzoic acid		C$_7$H$_6$O$_5$	610-02-6	170.120	nd (+w)	221	sub			sl H$_2$O; s EtOH, eth, ace; i bz, CS$_2$
10436	2,4,6-Trihydroxybenzoic acid		C$_7$H$_6$O$_5$	83-30-7	170.120	cry (w+1)	100 dec				sl H$_2$O; s EtOH; vs eth; i bz
10437	3,4,5-Trihydroxybenzoic acid	Gallic acid	C$_7$H$_6$O$_5$	149-91-7	170.120	pr (w+1)	262(1)		1.694[6]		sl H$_2$O, eth; vs EtOH; s ace; i bz, chl
10438	2,3,4-Trihydroxybenzophenone	Alizarin Yellow A	C$_{13}$H$_{10}$O$_4$	1143-72-2	230.216	ye nd (dil al)	140.5				sl H$_2$O, bz; s EtOH, eth, ace, HOAc
10439	2',4,4'-Trihydroxychalcone	Isoliquiritigenin	C$_{15}$H$_{12}$O$_4$	961-29-5	256.254	ye nd (EtOH-w)	200				
10440	9,10,16-Trihydroxyhexadecanoic acid	Aleuritic acid	C$_{16}$H$_{32}$O$_5$	6949-98-0	304.422	lf (dil al), nd (w)	102				sl H$_2$O
10441	1,3,8 Trihydroxy 6 methyl 9,10-anthracenedione	Emodin	C$_{15}$H$_{10}$O$_5$	518-82-1	270.237	oran-red mcl nd (HOAc)	257	sub			vs eth, EtOH
10442	9,10,18-Trihydroxyoctadecanoic acid, (*R**,*R**)	Phloionolic acid	C$_{18}$H$_{36}$O$_5$	583-86-8	332.476	cry (dil al)	101.5				
10443	5,6,7-Trihydroxy-2-phenyl-4*H*-1-benzopyran-4-one	Baicalein	C$_{15}$H$_{10}$O$_5$	491-67-8	270.237	ye pr (al)	264 dec				sl H$_2$O, bz; s EtOH, eth, ace, HOAc
10444	1-(2,4,5-Trihydroxyphenyl)-1-butanone		C$_{10}$H$_{12}$O$_4$	1421-63-2	196.200		153.8				
10445	1-(2,3,4-Trihydroxyphenyl)-ethanone	Gallacetophenone	C$_8$H$_8$O$_4$	528-21-2	168.148		173				s H$_2$O, eth; vs EtOH, ace; sl bz, chl
10446	1-(2,4,6-Trihydroxyphenyl)-ethanone	2',4',6'-Trihydroxyacetophenone	C$_8$H$_8$O$_4$	480-66-0	168.148		221.0				sl H$_2$O, chl, bz; vs EtOH, eth, ace
10447	1-(2,4,6-Trihydroxyphenyl)-1-propanone	Flopropione	C$_9$H$_{10}$O$_4$	2295-58-1	182.173	nd (w, +1w)	175.5				vs eth, EtOH
10448	2,6,7-Trihydroxy-9-phenyl-3*H*-xanthen-3-one	Phenylfluorone	C$_{19}$H$_{12}$O$_5$	975-17-7	320.295	oran red (al-HCl)	>300				
10449	2,3,5-Triiodobenzoic acid		C$_7$H$_3$I$_3$O$_2$	88-82-4	499.811	pr (al)	230.6(0.3)				i H$_2$O; vs EtOH, eth; sl bz
10450	Triiodomethane	Iodoform	CHI$_3$	75-47-8	393.732	ye cry	120(3)	218	4.008[25]		i H$_2$O, bz; s EtOH, eth, ace; sl DMSO
10451	2,4,6-Triiodophenol		C$_6$H$_3$I$_3$O	609-23-4	471.800	nd (dil al)	159.8	sub			i H$_2$O; sl EtOH; s eth, ace
10452	3,3',5-Triiodothyropropanoic acid		C$_{15}$H$_{11}$I$_3$O$_4$	51-26-3	635.959	cry (EtOH)	200				sl EtOH
10453	Triisobutyl aluminate	2-Methyl-1-propanol, aluminum salt	C$_{12}$H$_{27}$AlO$_3$	3453-79-0	246.322			275[50]			
10454	Triisobutylaluminum		C$_{12}$H$_{27}$Al	100-99-2	198.324	liq	6	86[10]			
10455	Triisobutylamine	2-Methyl-*N,N*-bis(2-methylpropyl)-1-propanamine	C$_{12}$H$_{27}$N	1116-40-1	185.349	liq	-21.8(0.5)	194(8)	0.7684[20]	1.4252[17]	vs eth, EtOH
10456	Triisobutylborane		C$_{12}$H$_{27}$B	1116-39-8	182.153			193.2(0.7)	0.7380[25]	1.4188[23]	vs bz, eth, EtOH
10457	Triisobutyl phosphate		C$_{12}$H$_{27}$O$_4$P	126-71-6	266.313			264	0.9681[20]	1.4193[20]	vs H$_2$O, bz, eth, EtOH

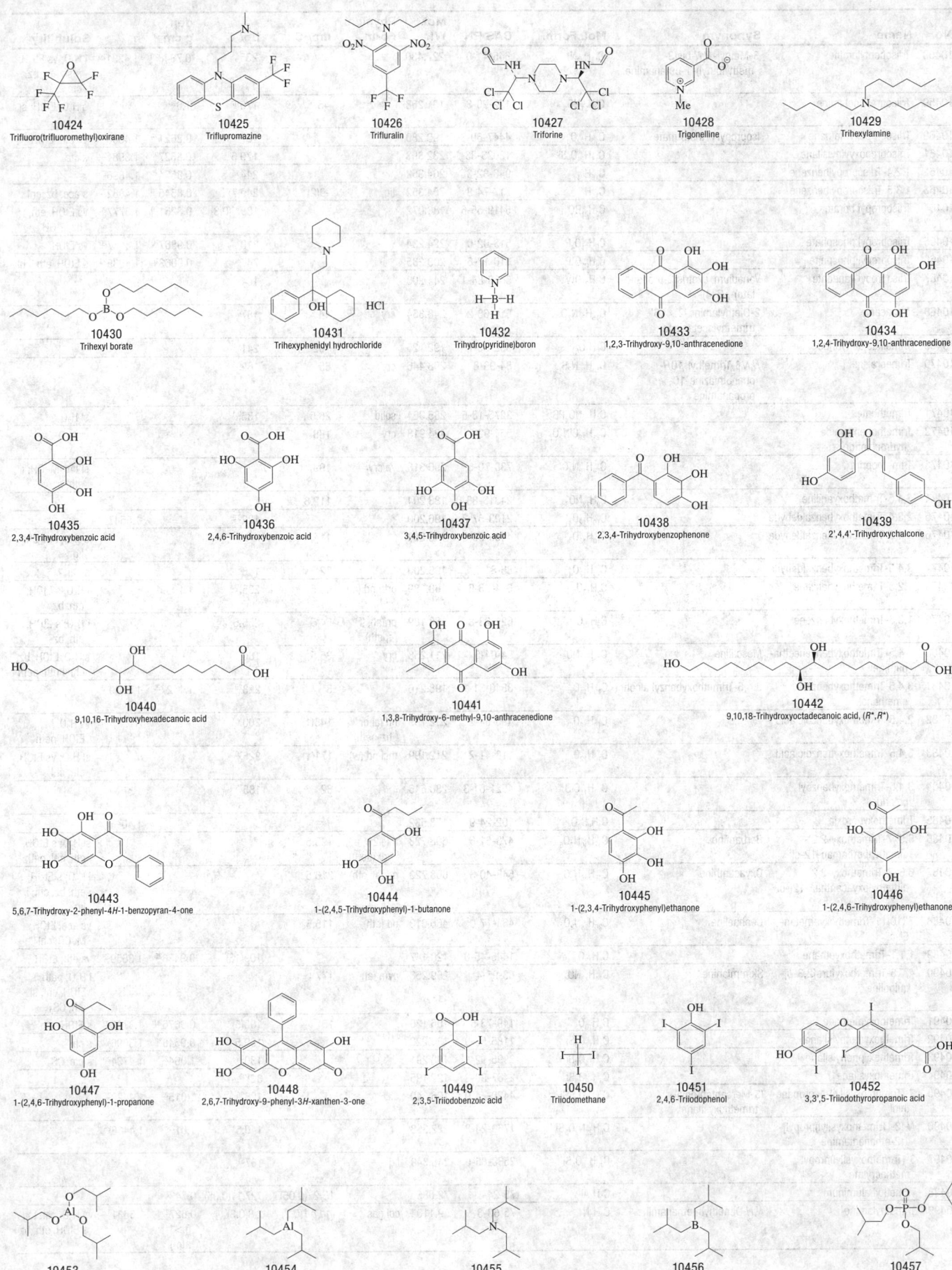

10424
Trifluoro(trifluoromethyl)oxirane

10425
Triflupromazine

10426
Trifluralin

10427
Triforine

10428
Trigonelline

10429
Trihexylamine

10430
Trihexyl borate

10431
Trihexyphenidyl hydrochloride

10432
Trihydro(pyridine)boron

10433
1,2,3-Trihydroxy-9,10-anthracenedione

10434
1,2,4-Trihydroxy-9,10-anthracenedione

10435
2,3,4-Trihydroxybenzoic acid

10436
2,4,6-Trihydroxybenzoic acid

10437
3,4,5-Trihydroxybenzoic acid

10438
2,3,4-Trihydroxybenzophenone

10439
2',4,4'-Trihydroxychalcone

10440
9,10,16-Trihydroxyhexadecanoic acid

10441
1,3,8-Trihydroxy-6-methyl-9,10-anthracenedione

10442
9,10,18-Trihydroxyoctadecanoic acid, (R*,R*)

10443
5,6,7-Trihydroxy-2-phenyl-4H-1-benzopyran-4-one

10444
1-(2,4,5-Trihydroxyphenyl)-1-butanone

10445
1-(2,3,4-Trihydroxyphenyl)ethanone

10446
1-(2,4,6-Trihydroxyphenyl)ethanone

10447
1-(2,4,6-Trihydroxyphenyl)-1-propanone

10448
2,6,7-Trihydroxy-9-phenyl-3H-xanthen-3-one

10449
2,3,5-Triiodobenzoic acid

10450
Triiodomethane

10451
2,4,6-Triiodophenol

10452
3,3',5-Triiodothyropropanoic acid

10453
Triisobutyl aluminate

10454
Triisobutylaluminum

10455
Triisobutylamine

10456
Triisobutylborane

10457
Triisobutyl phosphate

No.	Name	Synonym	Mol. Form.	CAS RN	Mol. Wt.	Physical Form	mp/°C	bp/°C	den g cm⁻³	n_D	Solubility
10458	Triisopentylamine	3-Methyl-*N,N*-bis(3-methylbutyl)-1-butanamine	C₁₅H₃₃N	645-41-0	227.430			235	0.7848²⁰	1.4331²⁰	i H₂O; vs EtOH; msc eth, bz, ctc
10459	Triisopropanolamine		C₉H₂₁NO₃	122-20-3	191.268		45	175¹⁰	1.0²⁰		s H₂O, EtOH; sl chl
10460	Triisopropoxymethane	Isopropyl orthoformate	C₁₀H₂₂O₃	4447-60-3	190.280			167	0.8621²⁰	1.4000²⁰	vs eth, EtOH
10461	Triisopropoxyvinylsilane		C₁₁H₂₄O₃Si	18023-33-1	232.393			179.5	0.8627²⁵	1.3981²⁰	s ctc
10462	1,2,4-Triisopropylbenzene		C₁₅H₂₄	948-32-3	204.352			244	0.8574²⁵	1.4896²⁵	
10463	1,3,5-Triisopropylbenzene		C₁₅H₂₄	717-74-8	204.352	liq	-9(2)	249(7)	0.8545²⁰	1.4882²⁵	s ace, bz, chl
10464	Triisopropyl borate		C₉H₂₁BO₃	5419-55-6	188.072			139.6(0.3)	0.8251²⁰	1.3777²⁰	vs EtOH, eth, bz, PrOH
10465	Triisopropyl phosphate		C₉H₂₁O₄P	513-02-0	224.234			219	0.9867²⁰	1.4057²⁰	vs EtOH
10466	Triisopropyl phosphite		C₉H₂₁O₃P	116-17-6	208.235			74²⁰	0.9063²⁰	1.4085²⁵	s EtOH, eth, chl
10467	Triisopropyl vanadate	Vanadium, oxotris(2-propanolato)-, (T-4)-	C₉H₂₁O₄V	5588-84-1	244.203			104¹⁰			
10468	Trimecaine	2-Diethylamino-2',4',6'-trimethylacetanilide	C₁₅H₂₄N₂O	616-68-2	248.364	cry	44	187⁶			
10469	Trimellitic anhydride		C₉H₄O₅	552-30-7	192.125		162	241¹⁴			
10470	Trimeprazine	*N,N,β*-Trimethyl-10*H*-phenothiazine-10-propanamine	C₁₈H₂₂N₂S	84-96-8	298.446	cry	68	162⁰·³			
10471	Trimethoate		C₉H₂₀NO₃PS₂	2275-18-5	285.364	solid	28.5	135⁰·¹			sl H₂O
10472	Trimethobenzamide hydrochloride		C₂₁H₂₉ClN₂O₅	554-92-7	424.918	cry	188				vs H₂O
10473	Trimethoprim		C₁₄H₁₈N₄O₃	738-70-5	290.318	ye cry	199				sl chl, MeOH; i eth, bz
10474	3,4,5-Trimethoxyaniline		C₉H₁₃NO₃	24313-88-0	183.204		112.8				
10475	2,3,4-Trimethoxybenzaldehyde		C₁₀H₁₂O₄	2103-57-3	196.200			122⁰·⁵		1.5547²⁰	
10476	2,4,5-Trimethoxybenzaldehyde		C₁₀H₁₂O₄	4460-86-0	196.200		114				s H₂O, eth, chl, lig
10477	3,4,5-Trimethoxybenzaldehyde		C₁₀H₁₂O₄	86-81-7	196.200		72.5	148⁵			s chl
10478	1,2,3-Trimethoxybenzene		C₉H₁₂O₃	634-36-6	168.189	orth nd (al)	48.5	235	1.1009⁴⁵		i H₂O; s EtOH, eth, bz
10479	1,3,5-Trimethoxybenzene		C₉H₁₂O₃	621-23-8	168.189	pr (al), lf (peth)	54.5	255.5			i H₂O; s EtOH, eth, bz
10480	3,4,5-Trimethoxybenzeneethanamine	Mescaline	C₁₁H₁₇NO₃	54-04-6	211.258	cry	35.5	180¹²			s H₂O, EtOH, bz, chl; i eth, peth
10481	3,4,5-Trimethoxybenzenemethanol	3,4,5-Trimethoxybenzyl alcohol	C₁₀H₁₄O₄	3840-31-1	198.216		3	228²⁵	1.1427²⁰	1.5439²⁰	
10482	2,4,5-Trimethoxybenzoic acid		C₁₀H₁₂O₅	490-64-2	212.199	nd (al or bz-peth)	145(1)	300			vs H₂O, bz, EtOH, peth
10483	3,4,5-Trimethoxybenzoic acid		C₁₀H₁₂O₅	118-41-2	212.199	mcl nd (w)	171(1)	226¹⁰			sl H₂O; vs EtOH, eth, chl
10484	3,4,5-Trimethoxybenzoyl chloride		C₁₀H₁₁ClO₄	4521-61-3	230.645		82	185¹⁸			
10485	Trimethoxyboroxin		C₃H₉B₃O₆	102-24-9	173.532					1.40²⁵	
10486	6,6',7-Trimethoxy-2,2'-dimethylberbaman-12-ol	Berbamine	C₃₇H₄₀N₂O₆	478-61-5	608.723	lf (+2w, al) cry (peth)	198.5				sl H₂O; s EtOH, eth, chl, peth
10487	6,6',7-Trimethoxy-2,2'-dimethyloxyacanthan-12'-ol	Oxyacanthine	C₃₇H₄₀N₂O₆	548-40-3	608.723	nd (al, eth)	216.5				i H₂O; s EtOH, eth, bz, chl; i lig
10488	7',10,11-Trimethoxyemetan-6'-ol	Cephaeline	C₂₈H₃₈N₂O₄	483-17-0	466.613	nd (eth)	115.5				vs ace, EtOH, MeOH, chl
10489	1,1,1-Trimethoxyethane		C₅H₁₂O₃	1445-45-0	120.147			109(13)	0.9438²⁵	1.3859²⁵	vs eth, EtOH
10490	4,7,8-Trimethoxyfuro[2,3-*b*]-quinoline	Skimmianine	C₁₄H₁₃NO₄	83-95-4	259.258	pym (al)	177				i H₂O, peth; s EtOH, chl; sl eth, CS₂
10491	Trimethoxymethane		C₄H₁₀O₃	149-73-5	106.120		15	100(5)	0.9676²⁰	1.3793²⁰	s EtOH, eth
10492	Trimethoxymethylsilane		C₄H₁₂O₃Si	1185-55-3	136.222			102.5	0.9548²⁰	1.3696²⁰	s chl
10493	Trimethoxyphenylsilane		C₉H₁₄O₃Si	2996-92-1	198.291			130⁴⁵	1.064²⁰	1.4734²⁰	s ctc, CS₂
10494	Trimethoxysilane		C₃H₁₀O₃Si	2487-90-3	122.195			32¹⁰⁰			
10495	3-(Trimethoxysilyl)-1-propanethiol	(3-Mercaptopropyl)-trimethoxysilane	C₆H₁₆O₃SSi	4420-74-0	196.340			128⁵⁰	1.015²⁵	1.4420²⁵	
10496	*N*-[3-(Trimethoxysilyl)propyl]-1,2-ethanediamine		C₈H₂₂N₂O₃Si	1760-24-3	222.358			140.5¹⁵	1.01²⁵	1.4416²⁵	
10497	3-(Trimethoxysilyl)propyl methacrylate		C₁₀H₂₀O₅Si	2530-85-0	248.349	liq		107⁵			
10498	Trimethyl aluminum		C₃H₉Al	75-24-1	72.085		15.27(0.05)	127.11(0.05)	0.743²⁰		
10499	Trimethylamine	*N,N*-Dimethylmethanamine	C₃H₉N	75-50-3	59.110	col gas	-117.1(0.2)	2.8(0.2)	0.627²⁵ (p>1 atm)	1.3631⁰	vs H₂O, chl, tol; s EtOH, eth, bz

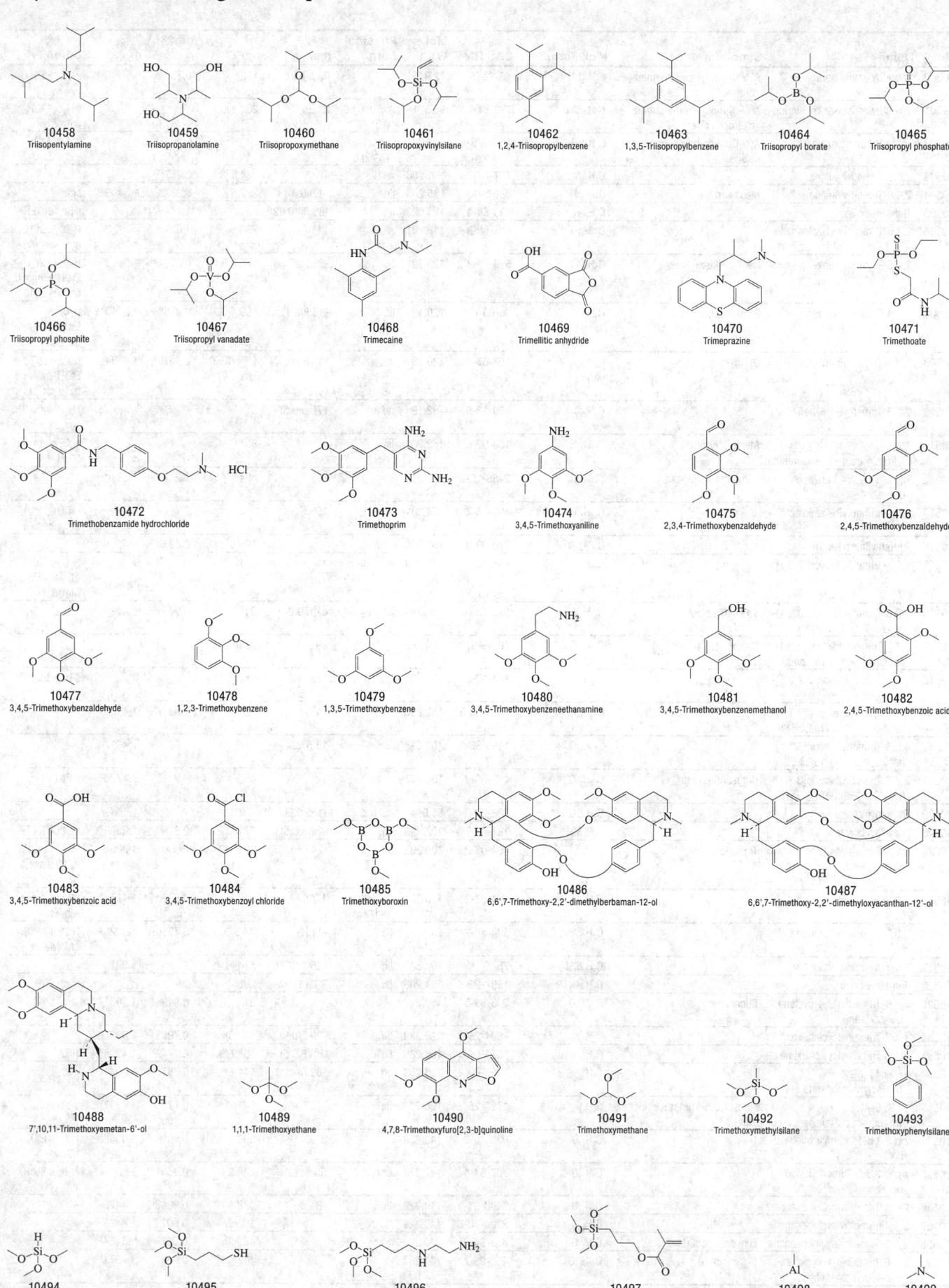

10458 Triisopentylamine

10459 Triisopropanolamine

10460 Triisopropoxymethane

10461 Triisopropoxyvinylsilane

10462 1,2,4-Triisopropylbenzene

10463 1,3,5-Triisopropylbenzene

10464 Triisopropyl borate

10465 Triisopropyl phosphate

10466 Triisopropyl phosphite

10467 Triisopropyl vanadate

10468 Trimecaine

10469 Trimellitic anhydride

10470 Trimeprazine

10471 Trimethoate

10472 Trimethobenzamide hydrochloride

10473 Trimethoprim

10474 3,4,5-Trimethoxyaniline

10475 2,3,4-Trimethoxybenzaldehyde

10476 2,4,5-Trimethoxybenzaldehyde

10477 3,4,5-Trimethoxybenzaldehyde

10478 1,2,3-Trimethoxybenzene

10479 1,3,5-Trimethoxybenzene

10480 3,4,5-Trimethoxybenzeneethanamine

10481 3,4,5-Trimethoxybenzenemethanol

10482 2,4,5-Trimethoxybenzoic acid

10483 3,4,5-Trimethoxybenzoic acid

10484 3,4,5-Trimethoxybenzoyl chloride

10485 Trimethoxyboroxin

10486 6,6',7-Trimethoxy-2,2'-dimethylberbaman-12-ol

10487 6,6',7-Trimethoxy-2,2'-dimethyloxyacanthan-12'-ol

10488 7',10,11-Trimethoxyemetan-6'-ol

10489 1,1,1-Trimethoxyethane

10490 4,7,8-Trimethoxyfuro[2,3-b]quinoline

10491 Trimethoxymethane

10492 Trimethoxymethylsilane

10493 Trimethoxyphenylsilane

10494 Trimethoxysilane

10495 3-(Trimethoxysilyl)-1-propanethiol

10496 N-[3-(Trimethoxysilyl)propyl]-1,2-ethanediamine

10497 3-(Trimethoxysilyl)propyl methacrylate

10498 Trimethyl aluminum

10499 Trimethylamine

No.	Name	Synonym	Mol. Form.	CAS RN	Mol. Wt.	Physical Form	mp/°C	bp/°C	den g cm⁻³	n_D	Solubility
10500	Trimethylamine borane	*N,N*-Dimethylmethanamine borane	$C_3H_{12}BN$	75-22-9	72.945		95.5(0.2)	172	0.792[25]		vs eth, EtOH
10501	Trimethylamine hydrochloride	*N,N*-Dimethylmethanamine hydrochloride	$C_3H_{10}ClN$	593-81-7	95.571	mcl hyg nd (al)	277.5	200 sub			vs H_2O, EtOH, chl
10502	Trimethylamine oxide	*N,N*-Dimethylmethanamine oxide	C_3H_9NO	1184-78-7	75.109	hyg nd (w+2)	256				vs H_2O, EtOH
10503	2,4,5-Trimethylaniline		$C_9H_{13}N$	137-17-7	135.206	nd (w)	68	234.5	0.957[25]		vs EtOH
10504	2,4,6-Trimethylaniline	Mesitylamine	$C_9H_{13}N$	88-05-1	135.206	liq	-4.9(0.4)	232.5	0.9633[25]	1.5495[20]	sl ctc
10505	Trimethylarsine		C_3H_9As	593-88-4	120.025	liq	-86.54(0.02)	52	1.144[15]		vs bz, eth, EtOH
10506	2,4,6-Trimethylbenzaldehyde		$C_{10}H_{12}O$	487-68-3	148.201		14	238.5	1.0154[25]		i H_2O; s EtOH, eth, ace, bz
10507	1,2,3-Trimethylbenzene	Hemimellitene	C_9H_{12}	526-73-8	120.191	liq	-25.32(0.04)	176.0(0.4)	0.8944[20]	1.5139[20]	i H_2O; msc EtOH, eth, ace, bz, peth, ctc
10508	1,2,4-Trimethylbenzene	Pseudocumene	C_9H_{12}	95-63-6	120.191	liq	-43.8(0.1)	169.4(0.3)	0.8758[20]	1.5048[20]	i H_2O; msc EtOH, eth, ace, bz, peth, ctc
10509	1,3,5-Trimethylbenzene	Mesitylene	C_9H_{12}	108-67-8	120.191	liq	-44.69(0.05)	164.7(0.3)	0.8615[25]	1.4994[20]	i H_2O; msc EtOH, eth, ace, bz, peth, ctc
10510	2,3,5-Trimethyl-1,4-benzene-diol		$C_9H_{12}O_2$	700-13-0	152.190	nd (w)	170.0(0.5)				sl H_2O; vs EtOH, eth, bz
10511	*N*,α,α-Trimethylbenzeneethanamine	Mephentermine	$C_{11}H_{17}N$	100-92-5	163.260	liq		95[9]			i H_2O; s eth; vs EtOH
10512	Trimethyl 1,2,4-benzenetricarboxylate	Trimethyl trimellitate	$C_{12}H_{12}O_6$	2459-10-1	252.219	visc oil	-13	194[12]	1.261	1.5230[20]	
10513	2,4,6-Trimethylbenzoic acid		$C_{10}H_{12}O_2$	480-63-7	164.201	pr (lig)	156.5				sl H_2O; s EtOH, eth, ace, chl
10514	Trimethylbenzylsilane		$C_{10}H_{16}Si$	770-09-2	164.320			190.5	0.8933[20]	1.4941[20]	
10515	1,7,7-Trimethylbicyclo[2.2.1]-heptane		$C_{10}H_{18}$	464-15-3	138.250	hex pl(al), pr(MeOH)		161			i H_2O; s EtOH, eth, AcOEt, MeOH
10516	1,3,3-Trimethylbicyclo[2.2.1]-heptan-2-ol, (1*S-endo*)	α-Fenchyl alcohol, (*l*)	$C_{10}H_{18}O$	512-13-0	154.249	pr	41.5(0.2)	94[20]	0.9034[84]		
10517	1,7,7-Trimethylbicyclo[2.2.1]-heptan-2-ol acetate, *endo*	Bornyl acetate	$C_{12}H_{20}O_2$	76-49-3	196.286		29	228(6)			
10518	1,7,7-Trimethylbicyclo[2.2.1]-hept-2-ene		$C_{10}H_{16}$	464-17-5	136.234	cry (al)	113	146			vs bz, eth, EtOH
10519	4,6,6-Trimethylbicyclo[3.1.1]-hept-3-en-2-ol, (1α,2α,5α)		$C_{10}H_{16}O$	1820-09-3	152.233		24	92[10]	0.9657[25]	1.4908[25]	
10520	4,6,6-Trimethylbicyclo[3.1.1]-hept-3-en-2-ol, (1α,2β,5α)		$C_{10}H_{16}O$	1845-30-3	152.233		15.5	90[10]	0.9684[25]	1.4912[25]	
10521	2,7,7-Trimethylbicyclo[3.1.1]-hept-2-en-6-one	Chrysanthenone	$C_{10}H_{14}O$	473-06-3	150.217			88[12]		1.4720[22]	vs EtOH
10522	Trimethylborane		C_3H_9B	593-90-8	55.914	col gas	-159.93(0.02)	-20.2			
10523	Trimethyl borate		$C_3H_9BO_3$	121-43-7	103.912	liq	-29.3(0.5)	67.4(0.2)	0.915[25]	1.3568[20]	vs eth, MeOH
10524	2,2,3-Trimethylbutane	Triptane	C_7H_{16}	464-06-2	100.202	liq	-24.56(0.05)	80.8(0.1)	0.6901[20]	1.3864[20]	i H_2O; s EtOH, eth; vs ace, bz, peth, ctc
10525	2,3,3-Trimethyl-2-butanol		$C_7H_{16}O$	594-83-2	116.201	cry (dil al +1/2w)	17(2)	128.3(0.8)	0.8380[25]	1.4233[22]	sl H_2O; vs ace, eth, EtOH
10526	2,3,3-Trimethyl-1-butene		C_7H_{14}	594-56-9	98.186	liq	-111(1)	77.8(0.6)	0.7050[20]	1.4025[20]	i H_2O; s eth, bz, chl, MeOH
10527	Trimethylchlorosilane		C_3H_9ClSi	75-77-4	108.642	liq	-55.17(0.02)	57.6(0.4)	0.856[25]	1.3870[20]	
10528	Trimethyl citrate		$C_9H_{14}O_7$	1587-20-8	234.203	tcl	79.3	285			vs eth, EtOH
10529	2,6,6-Trimethyl-2,4-cyclohep-tadien-1-one	Eucarvone	$C_{10}H_{14}O$	503-93-5	150.217			210	0.9490[20]	1.5087[20]	s eth, ace
10530	1,1,2-Trimethylcyclohexane		C_9H_{18}	7094-26-0	126.239	liq	-29.2(0.6)	146(3)	0.7963[25]	1.4382[20]	
10531	1,1,3-Trimethylcyclohexane		C_9H_{18}	3073-66-3	126.239	liq	-68.5(0.5)	136.6(0.4)	0.7749[25]	1.4295[20]	i H_2O
10532	1α,2β,4β-1,2,4-Trimethylcyclohexane		C_9H_{18}	7667-60-9	126.239	liq	-83.5	144(6)	0.7870[25]	1.4341[20]	
10533	1α,3α,5β-1,3,5-Trimethylcyclohexane	*trans*-1,3,5-Trimethylcyclohexane	C_9H_{18}	1795-26-2	126.239	liq	-107.4	141(2)	0.7794[20]	1.4307[20]	vs bz, eth, lig
10534	*cis*-3,3,5-Trimethylcyclohexanol		$C_9H_{18}O$	933-48-2	142.238		37.3	202	0.9006[16]	1.4550[16]	i H_2O; s EtOH, eth, chl
10535	*trans*-3,3,5-Trimethylcyclo-hexanol		$C_9H_{18}O$	767-54-4	142.238	cry (eth)	55.8	189.2	0.8631[60]		i H_2O; s EtOH, eth, chl
10536	2,2,6-Trimethylcyclohexanone		$C_9H_{16}O$	2408-37-9	140.222	liq	-31.8	178.5	0.9043[18]	1.4470[20]	
10537	2,4,4-Trimethylcyclohexanone		$C_9H_{16}O$	2230-70-8	140.222			190(9)	0.902[20]	1.4493[20]	
10538	3,3,5-Trimethylcyclohexanone	Dihydroisophorone	$C_9H_{16}O$	873-94-9	140.222	ye oil		189(9)	0.8919[19]	1.4454[15]	
10539	2,6,6-Trimethyl-1-cyclohex-ene-1-carboxaldehyde	β-Cyclocitral	$C_{10}H_{16}O$	432-25-7	152.233			112[29]	0.959[15]	1.4971[15]	

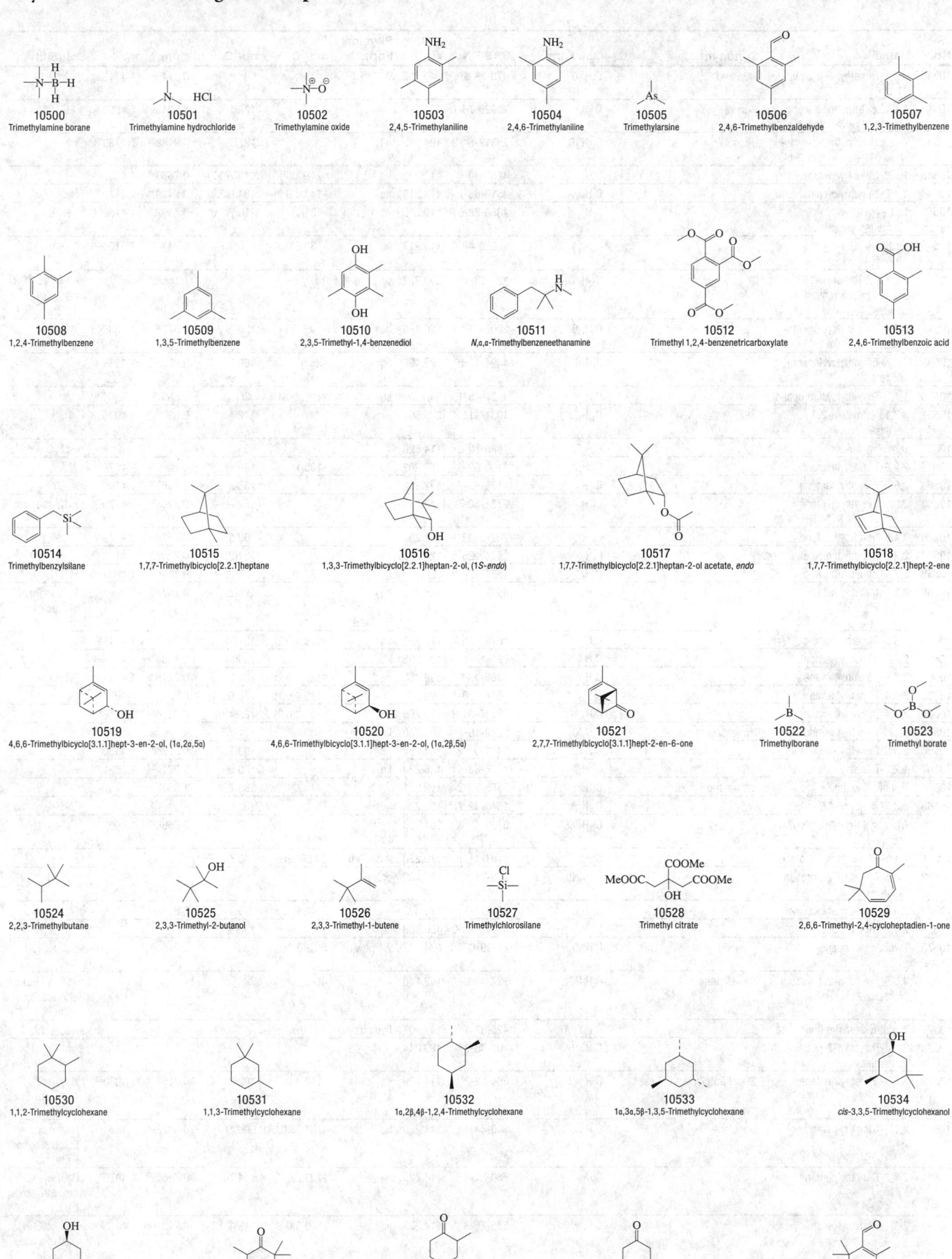

10500
Trimethylamine borane

10501
Trimethylamine hydrochloride

10502
Trimethylamine oxide

10503
2,4,5-Trimethylaniline

10504
2,4,6-Trimethylaniline

10505
Trimethylarsine

10506
2,4,6-Trimethylbenzaldehyde

10507
1,2,3-Trimethylbenzene

10508
1,2,4-Trimethylbenzene

10509
1,3,5-Trimethylbenzene

10510
2,3,5-Trimethyl-1,4-benzenediol

10511
N,α,α-Trimethylbenzeneethanamine

10512
Trimethyl 1,2,4-benzenetricarboxylate

10513
2,4,6-Trimethylbenzoic acid

10514
Trimethylbenzylsilane

10515
1,7,7-Trimethylbicyclo[2.2.1]heptane

10516
1,3,3-Trimethylbicyclo[2.2.1]heptan-2-ol, (1S-endo)

10517
1,7,7-Trimethylbicyclo[2.2.1]heptan-2-ol acetate, endo

10518
1,7,7-Trimethylbicyclo[2.2.1]hept-2-ene

10519
4,6,6-Trimethylbicyclo[3.1.1]hept-3-en-2-ol, (1α,2α,5α)

10520
4,6,6-Trimethylbicyclo[3.1.1]hept-3-en-2-ol, (1α,2β,5α)

10521
2,7,7-Trimethylbicyclo[3.1.1]hept-2-en-6-one

10522
Trimethylborane

10523
Trimethyl borate

10524
2,2,3-Trimethylbutane

10525
2,3,3-Trimethyl-2-butanol

10526
2,3,3-Trimethyl-1-butene

10527
Trimethylchlorosilane

10528
Trimethyl citrate

10529
2,6,6-Trimethyl-2,4-cycloheptadien-1-one

10530
1,1,2-Trimethylcyclohexane

10531
1,1,3-Trimethylcyclohexane

10532
1α,2β,4β-1,2,4-Trimethylcyclohexane

10533
1α,3α,5β-1,3,5-Trimethylcyclohexane

10534
cis-3,3,5-Trimethylcyclohexanol

10535
trans-3,3,5-Trimethylcyclohexanol

10536
2,2,6-Trimethylcyclohexanone

10537
2,4,4-Trimethylcyclohexanone

10538
3,3,5-Trimethylcyclohexanone

10539
2,6,6-Trimethyl-1-cyclohexene-1-carboxaldehyde

No.	Name	Synonym	Mol. Form.	CAS RN	Mol. Wt.	Physical Form	mp/°C	bp/°C	den g cm⁻³	n_D	Solubility
10540	3,5,5-Trimethyl-2-cyclohexen-1-ol	Isophorol	$C_9H_{16}O$	470-99-5	140.222			69[5]	0.914[20]	1.4717[20]	
10541	4-(2,6,6-Trimethyl-1-cyclohexen-1-yl)-3-buten-2-ol	β-Ionol	$C_{13}H_{22}O$	22029-76-1	194.313			130[14]	0.9243[20]	1.4969[20]	s EtOH, eth, ace
10542	4-(2,6,6-Trimethyl-2-cyclohexen-1-yl)-3-buten-2-ol	α-Ionol	$C_{13}H_{22}O$	25312-34-9	194.313	oil		127[14]	0.9189[20]	1.4735[20]	
10543	1,1,2-Trimethylcyclopentane		C_8H_{16}	4259-00-1	112.213	liq	-21.7(0.2)	113.7(0.2)	0.7660[20]	1.4199[20]	
10544	1,1,3-Trimethylcyclopentane		C_8H_{16}	4516-69-2	112.213	liq	-142.5(0.2)	104.9(0.6)	0.7439[25]	1.4112[20]	i H_2O
10545	1α,2α,4β-1,2,4-Trimethylcyclopentane		C_8H_{16}	4850-28-6	112.213	liq	-131(5)	116.7(0.6)	0.7592[25]	1.4186[20]	
10546	1α,2β,4α-1,2,4-Trimethylcyclopentane		C_8H_{16}	16883-48-0	112.213	liq	-130.8(0.3)	110(3)	0.7430[25]	1.4106[20]	
10547	cis-1,2,2-Trimethyl-1,3-cyclopentanedicarboxylic acid, (1R)		$C_{10}H_{16}O_4$	124-83-4	200.232	pr, lf (w)	187		1.186[20]		sl H_2O; vs EtOH, eth; s ace; i bz, chl
10548	2,2,4-Trimethylcyclopenta-none		$C_8H_{14}O$	28056-54-4	126.196	liq	-40.6	158	0.877[25]	1.4300[20]	
10549	2,4,4-Trimethylcyclopenta-none		$C_8H_{14}O$	4694-12-6	126.196	liq	-25.6	160.5	0.8785[18]	1.433[18]	
10550	1,1,2-Trimethylcyclopropane		C_6H_{12}	4127-45-1	84.159	liq	-138.2(0.1)	53(1)	0.6897[25]	1.3864[20]	
10551	3,7,11-Trimethyl-2,6,10-dodecatrienal		$C_{15}H_{24}O$	19317-11-4	220.351			172[14]	0.893[18]	1.4995	
10552	Trimethylgallium		C_3H_9Ga	1445-79-0	114.826			55.7			dec H_2O (exp)
10553	2,2,6-Trimethylheptane		$C_{10}H_{22}$	1190-83-6	142.282	liq	-105(1)	149(4)	0.7200[25]	1.4078[20]	
10554	2,5,5-Trimethylheptane		$C_{10}H_{22}$	1189-99-7	142.282			153(3)	0.7362[25]	1.4149[20]	
10555	3,3,5-Trimethylheptane		$C_{10}H_{22}$	7154-80-5	142.282			157(3)	0.7248[20]	1.4170[20]	i H_2O; s bz, ctc, chl
10556	3,4,5-Trimethylheptane		$C_{10}H_{22}$	20278-89-1	142.282			162.5	0.7519[25]	1.4229[20]	
10557	2,2,3-Trimethylhexane		C_9H_{20}	16747-25-4	128.255			134(2)	0.7257[25]	1.4106[20]	
10558	2,2,4-Trimethylhexane		C_9H_{20}	16747-26-5	128.255	liq	-122(4)	127(2)	0.711[20]	1.4033[20]	
10559	2,2,5-Trimethylhexane		C_9H_{20}	3522-94-9	128.255	liq	-105.9(0.1)	124(2)	0.7072[20]	1.3997[20]	i H_2O; vs EtOH, eth, ace, bz; s ctc
10560	2,3,3-Trimethylhexane		C_9H_{20}	16747-28-7	128.255	liq	-116.8(0.2)	137(3)	0.7345[25]	1.4141[20]	
10561	2,3,4-Trimethylhexane		C_9H_{20}	921-47-1	128.255			139(3)	0.7354[25]	1.4144[20]	
10562	2,3,5-Trimethylhexane		C_9H_{20}	1069-53-0	128.255	liq	-127.9(0.2)	131(2)	0.7218[20]	1.4051[20]	
10563	2,4,4-Trimethylhexane		C_9H_{20}	16747-30-1	128.255	liq	-113.4(0.1)	130.6(0.7)	0.7201[25]	1.4074[20]	
10564	3,3,4-Trimethylhexane		C_9H_{20}	16747-31-2	128.255	liq	-101.2(0.2)	139(4)	0.7414[25]	1.4178[20]	
10565	3,5,5-Trimethylhexanoic acid	Isononanoic acid	$C_9H_{18}O_2$	3302-10-1	158.238		121[10]				
10566	3,5,5-Trimethyl-1-hexanol		$C_9H_{20}O$	3452-97-9	144.254			193(5)	0.8236[25]	1.4300[25]	
10567	1,2,3-Trimethylindene		$C_{12}H_{14}$	4773-83-5	158.239	liq		100.5[10]	0.9714[20]	1.5521[20]	
10568	Trimethylindium	Indium trimethyl	C_3H_9In	3385-78-2	159.921			135.7	1.568[19]		
10569	2,3,3-Trimethyl-3H-indole		$C_{11}H_{13}N$	1640-39-7	159.228			107[11]			
10570	Trimethyl(4-methylphenyl)-silane		$C_{10}H_{16}Si$	3728-43-6	164.320		38	192	0.8666[20]	1.4900[20]	
10571	1,4,5-Trimethylnaphthalene		$C_{13}H_{14}$	2131-41-1	170.250	lf (MeOH)	64.0(0.6)	145[12]			i H_2O
10572	1,3,5-Trimethyl-2-nitroben-zene		$C_9H_{11}NO_2$	603-71-4	165.189	orth pr (al)	44(2)	255	1.51[25]		vs EtOH
10573	2,6,8-Trimethyl-4-nonanol		$C_{12}H_{26}O$	123-17-1	186.333			225.4	0.8178[20]		sl ctc
10574	2,4,7-Trimethyloctane		$C_{11}H_{24}$	62016-38-0	156.309			170(5)			
10575	Trimethylolpropane		$C_6H_{14}O_3$	77-99-6	134.173	wh pow or pl	60.2(0.2)	160[5]			vs H_2O, EtOH
10576	3,5,5-Trimethyl-2,4-oxazoli-dinedione	Trimethadione	$C_6H_9NO_3$	127-48-0	143.140		46	79[5]			s H_2O; vs EtOH, eth, ace, bz; i peth
10577	Trimethyloxonium fluoborate		$C_3H_9BF_4O$	420-37-1	147.907	hyg nd	148 dec				vs ace, chl
10578	2,4,4-Trimethyl-2-pen-tanamine		$C_8H_{19}N$	107-45-9	129.244						s chl
10579	2,2,3-Trimethylpentane	2-tert-Butylbutane	C_8H_{18}	564-02-3	114.229	liq	-112.4(0.3)	109.8(0.4)	0.7161[20]	1.4030[20]	i H_2O; msc EtOH, eth, ace, hp; s bz
10580	2,2,4-Trimethylpentane	Isooctane	C_8H_{18}	540-84-1	114.229	liq	-107.36(0.04)	99.2(0.2)	0.6878[25]	1.3884[25]	i H_2O; msc EtOH, ace, hp; s eth, ctc
10581	2,3,3-Trimethylpentane		C_8H_{18}	560-21-4	114.229	liq	-101.2(0.3)	114.7(0.3)	0.7262[20]	1.4075[20]	i H_2O; vs EtOH, msc eth, ace, bz, hp
10582	2,3,4-Trimethylpentane		C_8H_{18}	565-75-3	114.229	liq	-109.3(0.2)	113.4(0.3)	0.7191[20]	1.4042[20]	i H_2O; vs EtOH, msc eth, ace, bz; sl ctc
10583	2,2,4-Trimethyl-1,3-pentane-diol		$C_8H_{18}O_2$	144-19-4	146.228	pl (bz)	55.2(0.5)	230.1(0.3)	0.936[15]	1.4513[15]	sl H_2O; vs EtOH, eth; s bz, chl

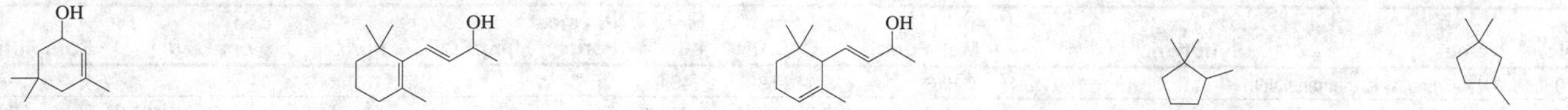

10540
3,5,5-Trimethyl-2-cyclohexen-1-ol

10541
4-(2,6,6-Trimethyl-1-cyclohexen-1-yl)-3-buten-2-ol

10542
4-(2,6,6-Trimethyl-2-cyclohexen-1-yl)-3-buten-2-ol

10543
1,1,2-Trimethylcyclopentane

10544
1,1,3-Trimethylcyclopentane

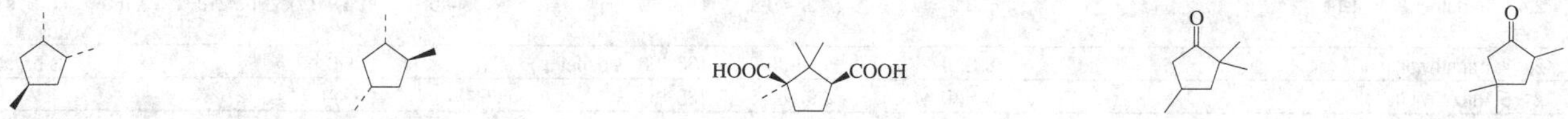

10545
1α,2α,4β-1,2,4-Trimethylcyclopentane

10546
1α,2β,4α-1,2,4-Trimethylcyclopentane

10547
cis-1,2,2-Trimethyl-1,3-cyclopentanedicarboxylic acid, (1*R*)

10548
2,2,4-Trimethylcyclopentanone

10549
2,4,4-Trimethylcyclopentanone

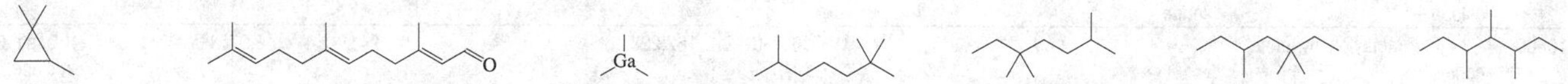

10550
1,1,2-Trimethylcyclopropane

10551
3,7,11-Trimethyl-2,6,10-dodecatrienal

10552
Trimethylgallium

10553
2,2,6-Trimethylheptane

10554
2,5,5-Trimethylheptane

10555
3,3,5-Trimethylheptane

10556
3,4,5-Trimethylheptane

10557
2,2,3-Trimethylhexane

10558
2,2,4-Trimethylhexane

10559
2,2,5-Trimethylhexane

10560
2,3,3-Trimethylhexane

10561
2,3,4-Trimethylhexane

10562
2,3,5-Trimethylhexane

10563
2,4,4-Trimethylhexane

10564
3,3,4-Trimethylhexane

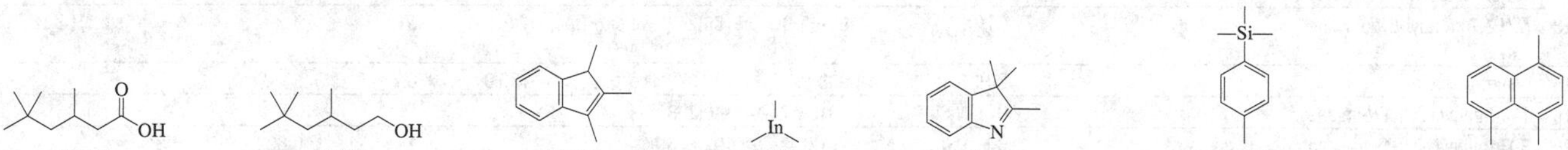

10565
3,5,5-Trimethylhexanoic acid

10566
3,5,5-Trimethyl-1-hexanol

10567
1,2,3-Trimethylindene

10568
Trimethylindium

10569
2,3,3-Trimethyl-3*H*-indole

10570
Trimethyl(4-methylphenyl)silane

10571
1,4,5-Trimethylnaphthalene

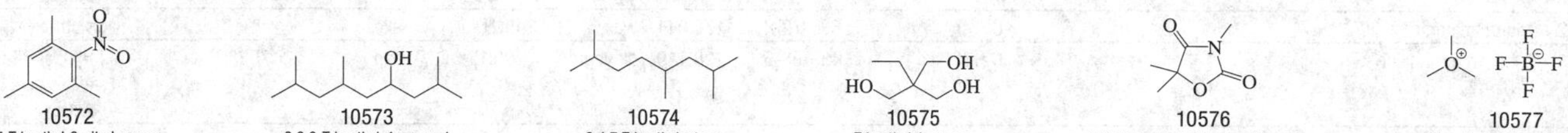

10572
1,3,5-Trimethyl-2-nitrobenzene

10573
2,6,8-Trimethyl-4-nonanol

10574
2,4,7-Trimethyloctane

10575
Trimethylolpropane

10576
3,5,5-Trimethyl-2,4-oxazolidinedione

10577
Trimethyloxonium fluoborate

10578
2,4,4-Trimethyl-2-pentanamine

10579
2,2,3-Trimethylpentane

10580
2,2,4-Trimethylpentane

10581
2,3,3-Trimethylpentane

10582
2,3,4-Trimethylpentane

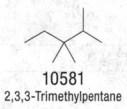

10583
2,2,4-Trimethyl-1,3-pentanediol

No.	Name	Synonym	Mol. Form.	CAS RN	Mol. Wt.	Physical Form	mp/°C	bp/°C	den g cm⁻³	n_D	Solubility
10584	2,4,4-Trimethyl-2-pentanethiol		$C_8H_{18}S$	141-59-3	146.294	liq		76[50]			
10585	2,4,4-Trimethyl-2-pentanol		$C_8H_{18}O$	690-37-9	130.228	liq	-20	146(4)	0.8225[20]	1.4284[20]	i H_2O; sl EtOH; s eth
10586	2,2,4-Trimethyl-3-pentanol		$C_8H_{18}O$	5162-48-1	130.228	liq	-13	151(3)	0.8297[20]	1.4288[20]	
10587	2,2,4-Trimethyl-3-pentanone	tert-Butyl isopropyl ketone	$C_8H_{16}O$	5857-36-3	128.212			146(3)	0.8065[20]	1.4060	i H_2O; s eth, ace
10588	2,3,3-Trimethyl-1-pentene		C_8H_{16}	560-23-6	112.213	liq	-69(3)	108(2)	0.7308[25]	1.4174[20]	
10589	2,4,4-Trimethyl-1-pentene		C_8H_{16}	107-39-1	112.213	liq	-93.7(0.3)	101.3(0.4)	0.7150[20]	1.4086[20]	i H_2O; s eth, bz, ctc, chl, lig
10590	2,3,4-Trimethyl-2-pentene		C_8H_{16}	565-77-5	112.213	liq	-113.4(0.2)	116(3)	0.7434[20]	1.4274[20]	
10591	2,4,4-Trimethyl-2-pentene		C_8H_{16}	107-40-4	112.213	liq	-106.4(0.1)	104.9(0.7)	0.7218[20]	1.4160[20]	i H_2O; s eth, bz, ctc, chl; vs lig
10592	2,3,4-Trimethylphenol		$C_9H_{12}O$	526-85-2	136.190	nd (peth)	81(1)	236(6)			vs bz, eth, EtOH
10593	2,3,5-Trimethylphenol		$C_9H_{12}O$	697-82-5	136.190		95(1)	235(4)			
10594	2,3,6-Trimethylphenol		$C_9H_{12}O$	2416-94-6	136.190		63				
10595	2,4,5-Trimethylphenol		$C_9H_{12}O$	496-78-6	136.190	nd (lig)	71.6(1)	233(4)			i H_2O; vs EtOH, eth
10596	2,4,6-Trimethylphenol		$C_9H_{12}O$	527-60-6	136.190	nd (peth, MeOH)	72.1(0.4)	221(6)			vs eth, EtOH
10597	3,4,5-Trimethylphenol		$C_9H_{12}O$	527-54-8	136.190	nd (peth)	108(2)	249(4)			
10598	Trimethylphenoxysilane		$C_9H_{14}OSi$	1529-17-5	166.292	liq	-55	119	0.8681[20]	1.5125[20]	
10599	Trimethylphenylammonium chloride	Phenyltrimethylammonium chloride	$C_9H_{14}ClN$	138-24-9	171.667						vs H_2O, EtOH
10600	1-(2,4,6-Trimethylphenyl)etha-none		$C_{11}H_{14}O$	1667-01-2	162.228			241	0.9754[20]	1.5175[20]	i H_2O; s EtOH, eth, ace, bz, ctc
10601	1,1,1-Trimethyl-N-phenylsi-lanamine	Phenyl(trimethylsilyl)amine	$C_9H_{15}NSi$	3768-55-6	165.308			206	0.940[20]		
10602	Trimethylphenylsilane		$C_9H_{14}Si$	768-32-1	150.293			169.5	0.8722[20]	1.4907[20]	s ctc, CS_2
10603	Trimethyl phosphate	Methyl phosphate	$C_3H_9O_4P$	512-56-1	140.074	liq	-46	197.2	1.2144[20]	1.3967[20]	vs H_2O; sl EtOH; s eth
10604	Trimethylphosphine		C_3H_9P	594-09-2	76.077	liq	-85	43(4)			i H_2O; s eth
10605	Trimethyl phosphite		$C_3H_9O_3P$	121-45-9	124.075			110(12)	1.0518[20]	1.4095[20]	vs EtOH, eth; sl ctc
10606	1,2,4-Trimethylpiperazine		$C_7H_{16}N_2$	120-85-4	128.215			149.5		1.4433[20]	s ctc
10607	2,2,4-Trimethylpiperidine		$C_8H_{17}N$	101257-71-0	127.228			148	0.832[15]	1.4458[20]	vs eth, EtOH
10608	Trimethylpyrazine		$C_7H_{10}N_2$	14667-55-1	122.167			87[35]			
10609	1,3,5-Trimethyl-1H-pyrazole		$C_6H_{10}N_2$	1072-91-9	110.156		37	170	0.9269[40]	1.4589[57]	
10610	2,3,6-Trimethylpyridine	2,3,6-Collidine	$C_8H_{11}N$	1462-84-6	121.180			170(15)	0.9220[25]	1.5053[20]	s H_2O, EtOH, eth, ace, bz
10611	2,4,6-Trimethylpyridine	2,4,6-Collidine	$C_8H_{11}N$	108-75-8	121.180	liq	-44.3(0.3)	170(1)	0.9166[22]	1.4959[25]	s H_2O, EtOH, eth, ace, ctc
10612	1,2,5-Trimethyl-1H-pyrrole		$C_7H_{11}N$	930-87-0	109.169			168(11)	0.807[25]	1.4969[20]	
10613	N,N,2-Trimethyl-6-quinolin-amine		$C_{12}H_{14}N_2$	92-99-9	186.252	ye pr (HOAc, AcOEt)	101	319			s ctc, CS_2
10614	Trimethylsilane		$C_3H_{10}Si$	993-07-7	74.197	col gas	-135.9	6.7			
10615	1-(Trimethylsilyl)-1H-imidazole		$C_6H_{12}N_2Si$	18156-74-6	140.258						s chl
10616	3-(Trimethylsilyl)-1-propanol		$C_6H_{16}OSi$	2917-47-7	132.276			141	0.822[25]	1.4298[20]	
10617	Trimethylstibine		C_3H_9Sb	594-10-5	166.863	liq	-62	80.6	1.523[15]	1.42[15]	i H_2O; s EtOH, eth, CS_2
10618	Trimethylsulfonium iodide		C_3H_9IS	2181-42-2	204.072	cry (eth)	211 dec				
10619	Trimethylthiourea		$C_4H_{10}N_2S$	2489-77-2	118.200	pr (bz-lig)	87.5				vs bz, EtOH, chl
10620	2,4,6-Trimethyl-2,4,6-triphen-ylcyclotrisiloxane		$C_{21}H_{24}O_3Si_3$	546-45-2	408.671		100	190[1.5]	1.1062[20]	1.5397[20]	
10621	Trimethylurea		$C_4H_{10}N_2O$	632-14-4	102.134	pr (eth)	74.0(0.2)	232.5	1.1900[20]		s H_2O, EtOH; sl eth, bz
10622	Trinitroacetonitrile		$C_2N_4O_6$	630-72-8	176.044	wax	45(1)	exp			vs eth
10623	2,4,6-Trinitroaniline		$C_6H_4N_4O_6$	489-98-5	228.119	dk ye pr (HOAc)	190(1)	exp		1.762[10]	i H_2O; sl EtOH, eth; s ace, bz, AcOEt
10624	1,3,5-Trinitrobenzene	sym-Trinitrobenzene	$C_6H_3N_3O_6$	99-35-4	213.104	orth pl (bz) lf (w)	121.3(0.4)	315	1.4775[152]		sl H_2O, EtOH, eth; vs ace; s bz, py
10625	2,4,6-Trinitro-1,3-benzenediol	Styphnic acid	$C_6H_3N_3O_8$	82-71-3	245.103	hex ye cry (dil al)	178(1)	sub			vs eth, EtOH
10626	2,4,6-Trinitrobenzoic acid		$C_7H_3N_3O_8$	129-66-8	257.114	orth (w)	229(2)				sl H_2O, bz; vs EtOH; s eth, ace
10627	2,4,7-Trinitro-9H-fluoren-9-one	2,4,7-Trinitrofluorenone	$C_{13}H_5N_3O_7$	129-79-3	315.195	pa ye nd (bz, HOAc)	175.8(0.5)				sl H_2O; vs ace, bz, chl
10628	Trinitrofluoromethane	Fluorotrinitromethane	CFN_3O_6	1840-42-2	169.025			86.3	1.59[20]		

10584
2,4,4-Trimethyl-2-pentanethiol

10585
2,4,4-Trimethyl-2-pentanol

10586
2,2,4-Trimethyl-3-pentanol

10587
2,2,4-Trimethyl-3-pentanone

10588
2,3,3-Trimethyl-1-pentene

10589
2,4,4-Trimethyl-1-pentene

10590
2,3,4-Trimethyl-2-pentene

10591
2,4,4-Trimethyl-2-pentene

10592
2,3,4-Trimethylphenol

10593
2,3,5-Trimethylphenol

10594
2,3,6-Trimethylphenol

10595
2,4,5-Trimethylphenol

10596
2,4,6-Trimethylphenol

10597
3,4,5-Trimethylphenol

10598
Trimethylphenoxysilane

10599
Trimethylphenylammonium chloride

10600
1-(2,4,6-Trimethylphenyl)ethanone

10601
1,1,1-Trimethyl-*N*-phenylsilanamine

10602
Trimethylphenylsilane

10603
Trimethyl phosphate

10604
Trimethylphosphine

10605
Trimethyl phosphite

10606
1,2,4-Trimethylpiperazine

10607
2,2,4-Trimethylpiperidine

10608
Trimethylpyrazine

10609
1,3,5-Trimethyl-1*H*-pyrazole

10610
2,3,6-Trimethylpyridine

10611
2,4,6-Trimethylpyridine

10612
1,2,5-Trimethyl-1*H*-pyrrole

10613
N,*N*,2-Trimethyl-6-quinolinamine

10614
Trimethylsilane

10615
1-(Trimethylsilyl)-1*H*-imidazole

10616
3-(Trimethylsilyl)-1-propanol

10617
Trimethylstibine

10618
Trimethylsulfonium iodide

10619
Trimethylthiourea

10620
2,4,6-Trimethyl-2,4,6-triphenylcyclotrisiloxane

10621
Trimethylurea

10622
Trinitroacetonitrile

10623
2,4,6-Trinitroaniline

10624
1,3,5-Trinitrobenzene

10625
2,4,6-Trinitro-1,3-benzenediol

10626
2,4,6-Trinitrobenzoic acid

10627
2,4,7-Trinitro-9*H*-fluoren-9-one

10628
Trinitrofluoromethane

No.	Name	Synonym	Mol. Form.	CAS RN	Mol. Wt.	Physical Form	mp/°C	bp/°C	den g cm⁻³	n_D	Solubility
10629	Trinitroglycerol	Nitroglycerin	$C_3H_5N_3O_9$	55-63-0	227.087	pa ye tcl or orth	12.8(0.2)	218 exp	1.5931[20]	1.4786[12]	sl H_2O; s EtOH, bz; msc eth; vs ace, chl
10630	Trinitromethane		CHN_3O_6	517-25-9	151.035		26(1)	exp	1.479[20]	1.4451[24]	vs ace, EtOH
10631	2,4,6-Trinitrophenol	Picric acid	$C_6H_3N_3O_7$	88-89-1	229.104	ye lf (w), pr (eth) pl (al)	121(2)	exp		1.763	sl H_2O; s EtOH, eth, bz, chl; vs ace
10632	2,4,6-Trinitrophenol, sodium salt	Sodium picrate	$C_6H_2N_3NaO_7$	3324-58-1	251.086	nd (w)	270.4				
10633	2,4,6-Trinitrotoluene	2-Methyl-1,3,5-trinitrobenzene	$C_7H_5N_3O_6$	118-96-7	227.131	orth (al)	80.9(0.8)	350(8)	1.654[25]		i H_2O; sl EtOH; s eth; vs ace, bz
10634	2,4,6-Trinitro-N-(2,4,6-trinitrophenyl)aniline	Dipicrylamine	$C_{12}H_5N_7O_{12}$	131-73-7	439.208	pa ye pr(HOAc)	251.1(0.5)				i H_2O, EtOH, bz, ctc; sl eth, ace; vs py
10635	Trioctylaluminum		$C_{24}H_{51}Al$	1070-00-4	366.644	hyg visc liq	-62		0.701		
10636	Trioctylamine	N,N-Dioctyl-1-octanamine	$C_{24}H_{51}N$	1116-76-3	353.669	liq	-34.5(0.3)	379(4)	0.8110[20]	1.4510[19]	
10637	Trioctylphosphine oxide	TOPO	$C_{24}H_{51}OP$	78-50-2	386.635		53.5(0.5)	201[2]			
10638	1,3,5-Trioxane	Formaldehyde, trimer	$C_3H_6O_3$	110-88-3	90.078	orth nd (eth)	60(2)	116(3)	1.17[65]		vs H_2O; s EtOH, eth, bz, CS_2; i peth
10639	1,3,5-Trioxane-2,4,6-triimine	Cyamelide	$C_3H_3N_3O_3$	462-02-2	129.074	amor pow	dec	dec	1.127[15]		vs eth, EtOH
10640	4,7,10-Trioxatridecane-1,13-diamine	Diethyleneglycol diaminopropyl ether	$C_{10}H_{24}N_2O_3$	4246-51-9	220.309	liq		147[4]	1.005	1.4640[20]	
10641	3,7,12-Trioxocholan-24-oic acid, (5β)	Dehydrocholic acid	$C_{24}H_{34}O_5$	81-23-2	402.524		237				i H_2O, eth; sl EtOH, bz; s ace, AcOEt
10642	Tripentylamine	N,N-Dipentyl-1-pentanamine	$C_{15}H_{33}N$	621-77-2	227.430			242.5	0.7907[20]	1.4366[20]	i H_2O; s EtOH, eth, acid
10643	Triphenylamine	N,N-Diphenylbenzenamine	$C_{18}H_{15}N$	603-34-9	245.319	mcl (MeOH, bz)	126.5(1)	362(10)	1.18[25]		i H_2O; sl EtOH; s eth, bz, MeOH
10644	Triphenylarsine		$C_{18}H_{15}As$	603-32-7	306.234		60(2)	360	1.2634[18]	1.6888[21]	i H_2O; sl EtOH; vs eth, bz; s chl
10645	Triphenylarsine oxide		$C_{18}H_{15}AsO$	1153-05-5	322.233		192	324.0			
10646	Triphenylbismuthine		$C_{18}H_{15}Bi$	603-33-8	440.292		78(2)	242[14]	1.715[75]	1.7040[75]	sl EtOH, chl; s eth, ace, bz, CS_2
10647	Triphenylborane		$C_{18}H_{15}B$	960-71-4	242.123	wh cry	142				i H_2O; sl eth; s bz, lig
10648	Triphenylene	Benzo[l]phenanthrene	$C_{18}H_{12}$	217-59-4	228.288	nd (al, chl, bz)	197.82(0.04)	425			i H_2O; s EtOH, HOAc; vs bz, chl
10649	1,1,2-Triphenylethane		$C_{20}H_{18}$	1520-42-9	258.357	mcl lf (dil al), nd (al)	54.5(0.8)				i H_2O; vs EtOH, eth, bz; sl MeOH
10650	1,1,2-Triphenylethene		$C_{20}H_{16}$	58-72-0	256.341	lf (al)	67.80(0.04)	220[14]	1.0373[78]	1.6292[78]	i H_2O; s EtOH, chl, MeOH; vs eth
10651	N,N',N''-Triphenylguanidine		$C_{19}H_{17}N_3$	101-01-9	287.358	nd or pr (al)	146.5	dec	1.163[20]		sl H_2O; s EtOH
10652	2,4,5-Triphenyl-1H-imidazole		$C_{21}H_{16}N_2$	484-47-9	296.365	nd (al)	275(2)	sub			i H_2O; s EtOH, eth
10653	Triphenylmethane		$C_{19}H_{16}$	519-73-3	244.330	orth (al)	92.0(0.7)	359	1.014[99]	1.5839[99]	i H_2O; sl EtOH; vs eth, py, chl; s bz
10654	Triphenylmethanol		$C_{19}H_{16}O$	76-84-6	260.329	pl (al), trg (bz)	162.3(0.4)	380	1.199[0]		i H_2O, peth; vs EtOH, eth; s ace, bz
10655	Triphenyl phosphate		$C_{18}H_{15}O_4P$	115-86-6	326.283	cry (lig), pr (al), nd (eth)	49.39(0.04)	245[11]	1.2055[50]		i H_2O; s EtOH; vs eth, bz, ctc, chl
10656	Triphenylphosphine		$C_{18}H_{15}P$	603-35-0	262.286		80(2)	188[1]	1.0749[80]	1.6358[80]	i H_2O; s EtOH, bz, chl; vs eth
10657	Triphenylphosphine oxide		$C_{18}H_{15}OP$	791-28-6	278.285	pr	158(3)	>360	1.2124[23]		sl H_2O, eth, chl; vs EtOH, bz
10658	Triphenyl phosphite		$C_{18}H_{15}O_3P$	101-02-0	310.284		22(2)	360	1.1842[20]	1.5900[20]	i H_2O; vs EtOH
10659	Triphenylsilane		$C_{18}H_{16}Si$	789-25-3	260.406						s ctc, CS_2
10660	Triphenylsilanol		$C_{18}H_{16}OSi$	791-31-1	276.405		154.8		1.1777[20]		s ctc, CS_2
10661	Triphenylstibine		$C_{18}H_{15}Sb$	603-36-1	353.072	pr (peth)	55(3)	>360	1.4343[25]	1.6948[42]	i H_2O; s EtOH; vs eth, ace, bz, chl
10662	Triphenyltetrazolium chloride		$C_{19}H_{15}ClN_4$	298-96-4	334.802	nd (al,chl)	243 dec				s H_2O, EtOH, ace, chl; i eth
10663	Triphenyltin hydroxide	Stannane, hydroxytriphenyl-	$C_{18}H_{16}OSn$	76-87-9	367.029		117.7(0.5)		1.54[20]		
10664	2,4,6-Triphenyl-1,3,5-triazine		$C_{21}H_{15}N_3$	493-77-6	309.364		257				

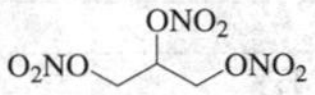

10629
Trinitroglycerol

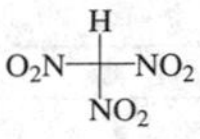

10630
Trinitromethane

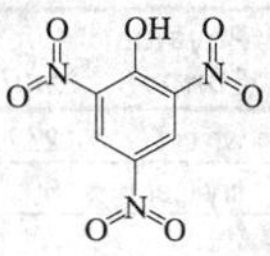

10631
2,4,6-Trinitrophenol

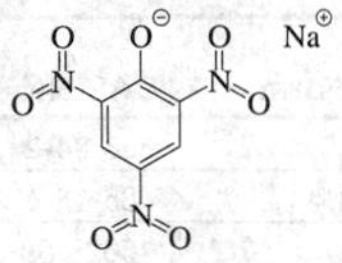

10632
2,4,6-Trinitrophenol, sodium salt

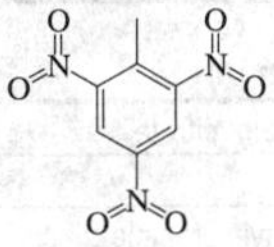

10633
2,4,6-Trinitrotoluene

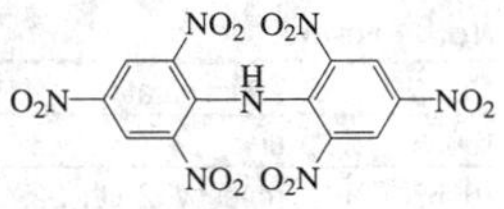

10634
2,4,6-Trinitro-*N*-(2,4,6-trinitrophenyl)aniline

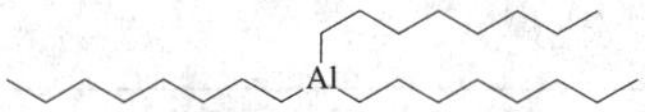

10635
Trioctylaluminum

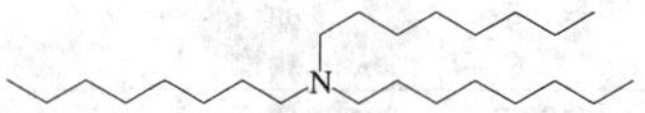

10636
Trioctylamine

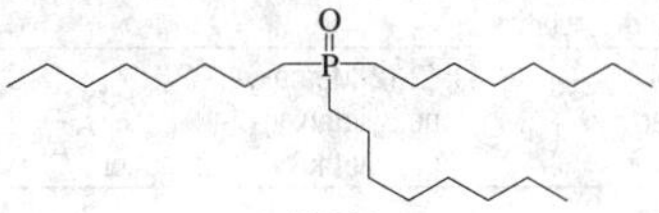

10637
Trioctylphosphine oxide

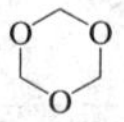

10638
1,3,5-Trioxane

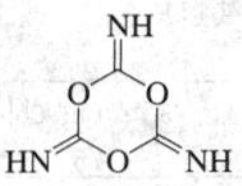

10639
1,3,5-Trioxane-2,4,6-triimine

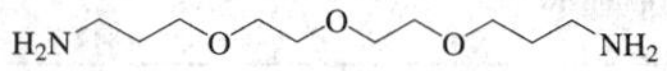

10640
4,7,10-Trioxatridecane-1,13-diamine

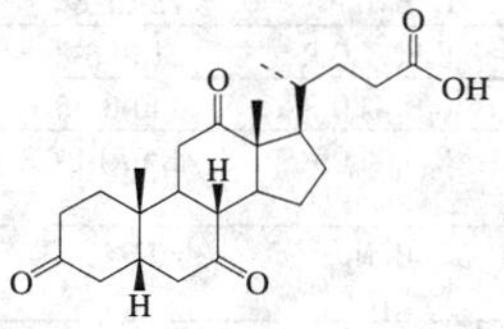

10641
3,7,12-Trioxocholan-24-oic acid, (5β)

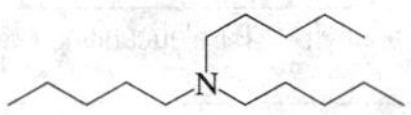

10642
Tripentylamine

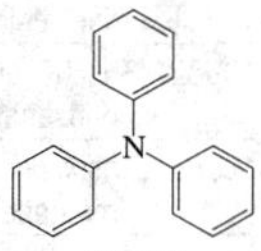

10643
Triphenylamine

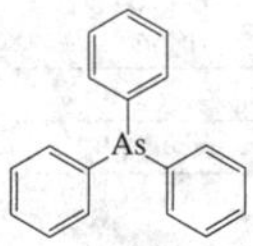

10644
Triphenylarsine

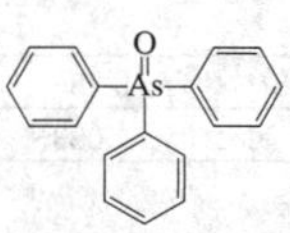

10645
Triphenylarsine oxide

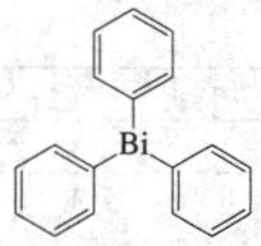

10646
Triphenylbismuthine

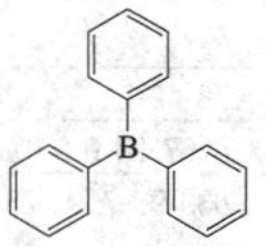

10647
Triphenylborane

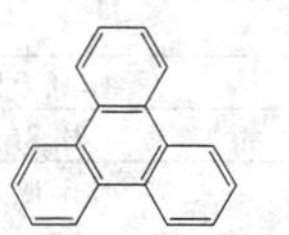

10648
Triphenylene

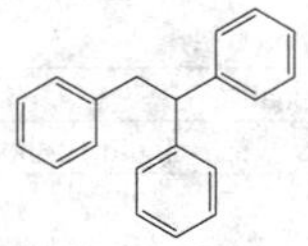

10649
1,1,2-Triphenylethane

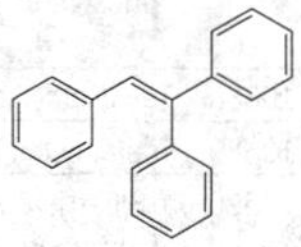

10650
1,1,2-Triphenylethene

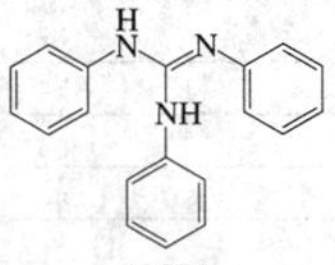

10651
N,N',N''-Triphenylguanidine

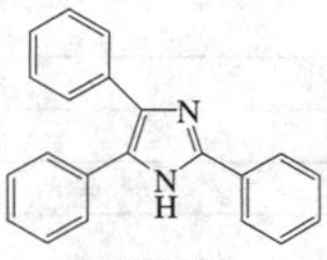

10652
2,4,5-Triphenyl-1*H*-imidazole

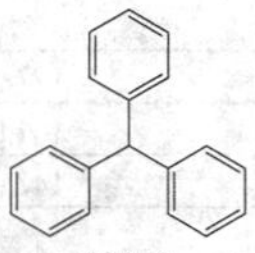

10653
Triphenylmethane

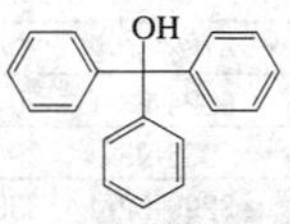

10654
Triphenylmethanol

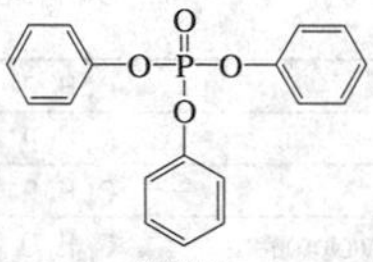

10655
Triphenyl phosphate

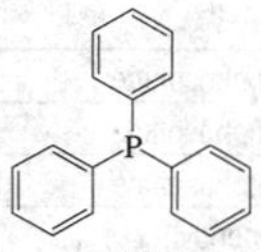

10656
Triphenylphosphine

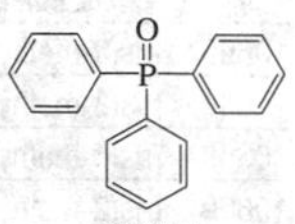

10657
Triphenylphosphine oxide

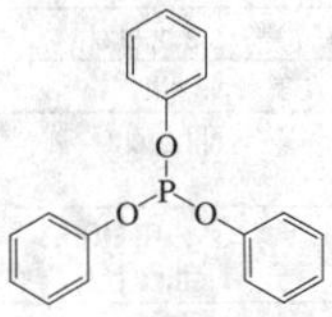

10658
Triphenyl phosphite

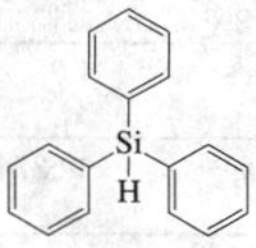

10659
Triphenylsilane

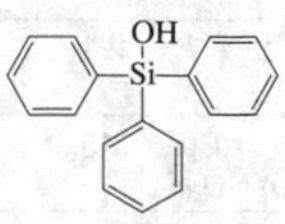

10660
Triphenylsilanol

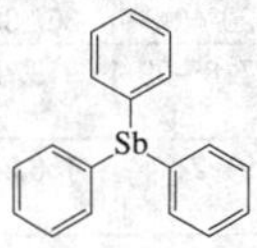

10661
Triphenylstibine

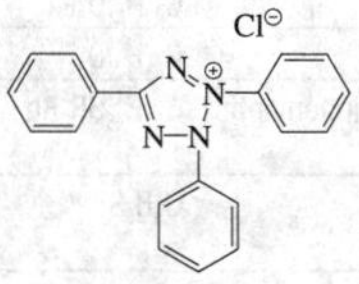

10662
Triphenyltetrazolium chloride

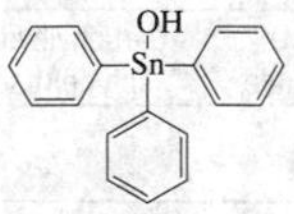

10663
Triphenyltin hydroxide

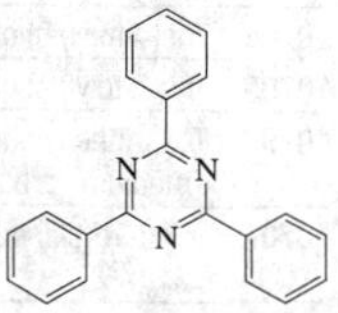

10664
2,4,6-Triphenyl-1,3,5-triazine

No.	Name	Synonym	Mol. Form.	CAS RN	Mol. Wt.	Physical Form	mp/°C	bp/°C	den g cm^{-3}	n_D	Solubility
10665	Tripotassium citrate	Potassium citrate	$C_6H_5K_3O_7$	866-84-2	306.395	wh cry (w)	275 dec				vs H$_2$O; i EtOH
10666	Triprolidine		$C_{19}H_{22}N_2$	486-12-4	278.391	cry (peth)	60				
10667	Tri-2-propenoyl-2-ethyl-2-(hydroxymethyl)-1,3-propanediol	Trimethylolpropane triacrylate	$C_{15}H_{20}O_6$	15625-89-5	296.316					1.4735^{20}	
10668	Tripropylamine	N,N-Dipropyl-1-propanamine	$C_9H_{21}N$	102-69-2	143.270	liq	-100.5(0.5)	153(2)	0.7558^{20}	1.4181^{20}	vs eth, EtOH
10669	Tripropylborane		$C_9H_{21}B$	1116-61-6	140.074	liq	-56	159	0.7204^{25}	1.4135^{22}	
10670	Tripropyl borate	Boric acid, tripropyl ester	$C_9H_{21}BO_3$	688-71-1	188.072			178.7(0.4)	0.8576^{20}	1.3948^{20}	vs EtOH; msc eth; s PrOH
10671	Tripropylene glycol	[(1-Methyl-1,2-ethanediyl)-bis(oxy)]bispropanol	$C_9H_{20}O_4$	24800-44-0	192.253	liq		262(4)	1.02^{20}	1.4440^{20}	
10672	Tripropylene glycol diacrylate		$C_{15}H_{24}O_6$	42978-66-5	300.348						
10673	Tripropylene glycol monomethyl ether	1-[2-(2-Methoxy-1-methylethoxy)-1-methylethoxy]-2-propanol	$C_{10}H_{22}O_4$	20324-33-8	206.280			241.3			
10674	Tripropyl phosphate		$C_9H_{21}O_4P$	513-08-6	224.234			252	1.0121^{20}	1.4165^{20}	sl H$_2$O, chl; s EtOH, eth, tol, CS$_2$
10675	Tripropyl phosphite	Tripropoxyphosphine	$C_9H_{21}O_3P$	923-99-9	208.235			206.5	0.9417^{20}	1.4282^{20}	vs eth, EtOH
10676	Tripropylsilane		$C_9H_{22}Si$	998-29-8	158.357			172	0.7230^{20}	1.4280^{20}	i H$_2$O
10677	Tris(4-aminophenyl)methanol	C.I. Basic Red 9	$C_{19}H_{19}N_3O$	467-62-9	305.373	purp cry	205				
10678	2,4,6-Tris(1-aziridinyl)-1,3,5-triazine	Triethylenemelamine	$C_9H_{12}N_6$	51-18-3	204.231	cry pow	139 dec				s H$_2$O
10679	Tris(2-butoxyethyl) phosphate		$C_{18}H_{39}O_7P$	78-51-3	398.473	liq		255^{10}	1.02^{25}		i H$_2$O
10680	Tris(2-chloroethyl) phosphate		$C_6H_{12}Cl_3O_4P$	115-96-8	285.489			330	1.39^{25}	1.4721^{20}	s ctc
10681	Tris(2-chloroethyl) phosphite		$C_6H_{12}Cl_3O_3P$	140-08-9	269.490			120^3	1.3443^{26}	1.4868^{20}	
10682	Tris(1,3-dichloro-2-propyl) phosphate	Fyrol FR-2	$C_9H_{15}Cl_6O_4P$	13674-87-8	430.904	visc liq		236^5		1.5022^{20}	i H$_2$O
10683	Tris(4-dimethylaminophenyl)-methane	Paraleucaniline	$C_{25}H_{31}N_3$	603-48-5	373.534	lf (al), nd (bz)	176.5				vs bz, eth, chl
10684	Tris(2,4-dimethylphenyl) phosphate	2,4-Xylenol, phosphate (3:1)	$C_{24}H_{27}O_4P$	3862-12-2	410.442			233.5	1.142^{38}	1.5550^{20}	i H$_2$O; s bz, chl, hx
10685	Tris(2,5-dimethylphenyl) phosphate	2,5-Xylenol, phosphate (3:1)	$C_{24}H_{27}O_4P$	19074-59-0	410.442		79.8	262^8	1.197^{25}		i H$_2$O; sl EtOH, hx; s eth, bz, ctc
10686	Tris(2,6-dimethylphenyl) phosphate	2,6-Xylenol, phosphate (3:1)	$C_{24}H_{27}O_4P$	121-06-2	410.442	wax	137.8	263^6			i H$_2$O; sl EtOH, hx; s bz
10687	Tris(3,5-dimethylphenyl) phosphate		$C_{24}H_{27}O_4P$	25653-16-1	410.442		46.2	290^{10}			i H$_2$O; sl EtOH, chl, hx; s HOAc
10688	Tris(2-ethylhexyl) phosphate		$C_{24}H_{51}O_4P$	78-42-2	434.633	liq		215^5	0.99^{20}		
10689	Tris(ethylthio)methane	Triethyl orthothioformate	$C_7H_{16}S_3$	6267-24-9	196.397			235 dec	1.053^{20}	1.5410^{15}	vs eth, EtOH
10690	1,3,5-Tris(2-hydroxyethyl)-isocyanuric acid		$C_9H_{15}N_3O_6$	839-90-7	261.231	cry	136				
10691	1,1,1-Tris(hydroxymethyl)-ethane trinitrate	2-Methyl-2-[(nitrooxy)methyl]-1,3-propanediol, dinitrate	$C_5H_9N_3O_9$	3032-55-1	255.140			83$^{0.05}$			
10692	N,N',N''-Tris(hydroxymethyl)-melamine	Trimethylolmelamine	$C_6H_{12}N_6O_3$	1017-56-7	216.197	cry	148				
10693	Tris(hydroxymethyl)-methylamine	2-Amino-2-(hydroxymethyl)-1,3-propanediol	$C_4H_{11}NO_3$	77-86-1	121.135		170.5(0.2)	219^{10}			vs H$_2$O; s MeOH
10694	Tris(methoxyethoxy)vinylsilane		$C_{11}H_{24}O_6Si$	1067-53-4	280.391						s ctc
10695	Tris(4-methoxyphenyl)-chloroethene	Chlorotrianisene	$C_{23}H_{21}ClO_3$	569-57-3	380.864		115				
10696	Tris(2-methylphenyl)phosphine		$C_{21}H_{21}P$	6163-58-2	304.366		127.0				
10697	Tris(3-methylphenyl)phosphine		$C_{21}H_{21}P$	6224-63-1	304.366		101.0				
10698	Tris(4-methylphenyl)phosphine		$C_{21}H_{21}P$	1038-95-5	304.366		147.0				
10699	Tris(2-methyl-2-propenoyl)-2-ethyl-2-hydroxymethyl-1,3-propanediol	1,1,1-Trimethylolpropane trimethacrylate	$C_{18}H_{26}O_6$	3290-92-4	338.395					1.470^{25}	
10700	Trisodium citrate	Sodium citrate	$C_6H_5Na_3O_7$	68-04-2	258.069	wh cry (w)	300				vs H$_2$O; i EtOH
10701	Trisodium N-hydroxyethyleth-ylenediaminetriacetate	Versen-Ol	$C_{10}H_{15}N_2Na_3O_7$	139-89-9	344.204		288 (hyd)				
10702	Tris(perfluorobutyl)amine	Trinonafluorobutylamine	$C_{12}F_{27}N$	311-89-7	671.092			178(2)	1.884^{25}	1.291^{25}	s ace
10703	2,4,6-Tris(2-pyridinyl)-1,3,5-triazine	2,4,6-Tripyridyl-s-triazine	$C_{18}H_{12}N_6$	3682-35-7	312.328		210				
10704	Tris(o-tolyl) phosphite		$C_{21}H_{21}O_3P$	2622-08-4	352.364		11	238^{11}	1.1423^{20}	1.5740^{28}	s eth; sl chl
10705	Tris(p-tolyl) phosphite		$C_{21}H_{21}O_3P$	620-42-8	352.364	pa ye	52	252^{10}	1.1280^{25}	1.5703^{28}	vs eth
10706	Tris(triphenylphosphine)-rhodium carbonyl hydride	Carbonylhydrotris(triphenylphosphine)rhodium	$C_{55}H_{46}OP_3Rh$	17185-29-4	918.781	ye cry	121		1.33		sl bz, chl
10707	1,3,5-Trithiane		$C_3H_6S_3$	291-21-4	138.275	hex (bz), pr (w) nd (al)	215(8)	sub	1.6374^{24}		sl H$_2$O, EtOH, eth; s bz

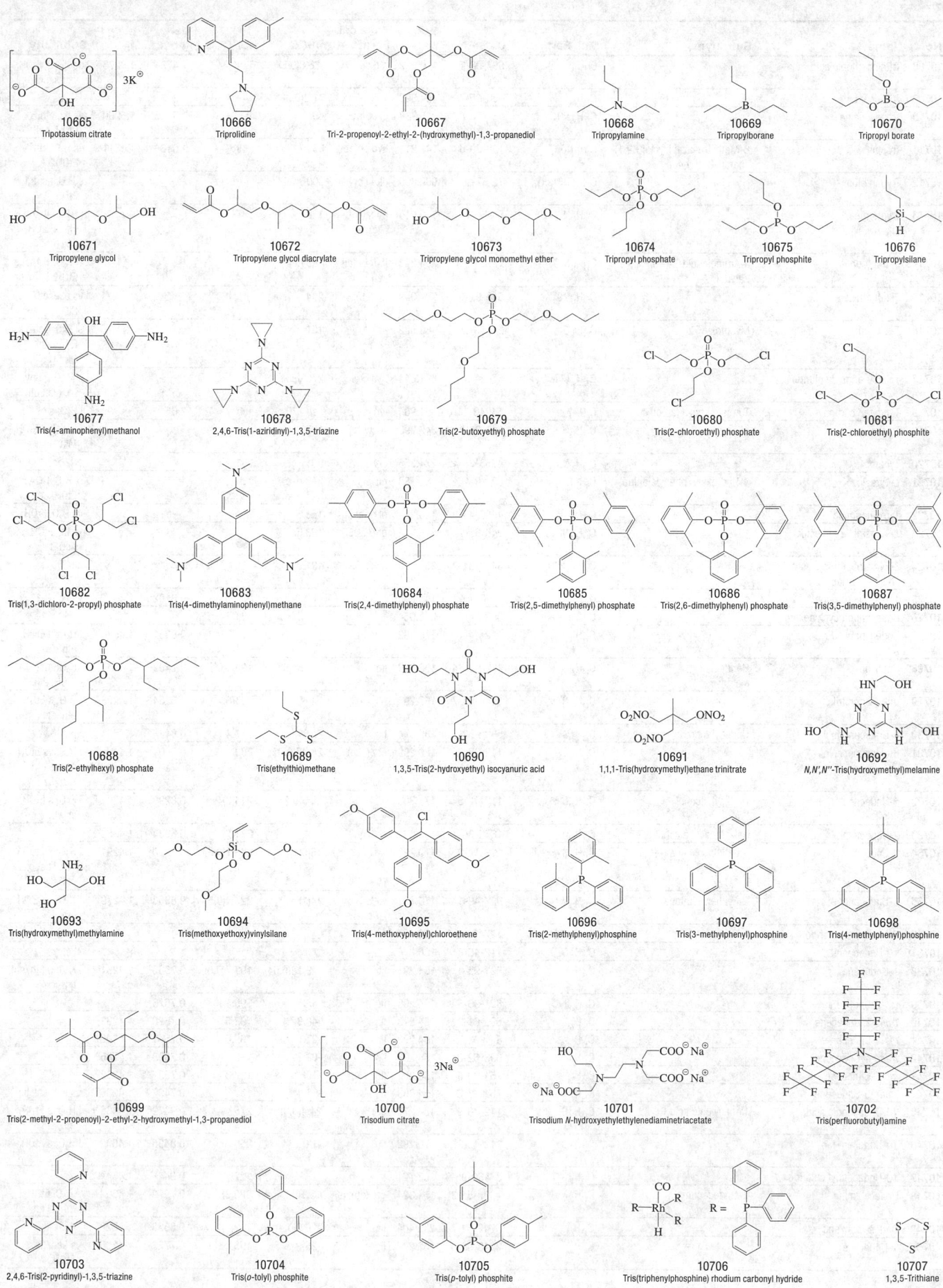

10665 Tripotassium citrate

10666 Triprolidine

10667 Tri-2-propenoyl-2-ethyl-2-(hydroxymethyl)-1,3-propanediol

10668 Tripropylamine

10669 Tripropylborane

10670 Tripropyl borate

10671 Tripropylene glycol

10672 Tripropylene glycol diacrylate

10673 Tripropylene glycol monomethyl ether

10674 Tripropyl phosphate

10675 Tripropyl phosphite

10676 Tripropylsilane

10677 Tris(4-aminophenyl)methanol

10678 2,4,6-Tris(1-aziridinyl)-1,3,5-triazine

10679 Tris(2-butoxyethyl) phosphate

10680 Tris(2-chloroethyl) phosphate

10681 Tris(2-chloroethyl) phosphite

10682 Tris(1,3-dichloro-2-propyl) phosphate

10683 Tris(4-dimethylaminophenyl)methane

10684 Tris(2,4-dimethylphenyl) phosphate

10685 Tris(2,5-dimethylphenyl) phosphate

10686 Tris(2,6-dimethylphenyl) phosphate

10687 Tris(3,5-dimethylphenyl) phosphate

10688 Tris(2-ethylhexyl) phosphate

10689 Tris(ethylthio)methane

10690 1,3,5-Tris(2-hydroxyethyl) isocyanuric acid

10691 1,1,1-Tris(hydroxymethyl)ethane trinitrate

10692 *N,N',N''*-Tris(hydroxymethyl)melamine

10693 Tris(hydroxymethyl)methylamine

10694 Tris(methoxyethoxy)vinylsilane

10695 Tris(4-methoxyphenyl)chloroethene

10696 Tris(2-methylphenyl)phosphine

10697 Tris(3-methylphenyl)phosphine

10698 Tris(4-methylphenyl)phosphine

10699 Tris(2-methyl-2-propenoyl)-2-ethyl-2-hydroxymethyl-1,3-propanediol

10700 Trisodium citrate

10701 Trisodium *N*-hydroxyethylethylenediaminetriacetate

10702 Tris(perfluorobutyl)amine

10703 2,4,6-Tris(2-pyridinyl)-1,3,5-triazine

10704 Tris(*o*-tolyl) phosphite

10705 Tris(*p*-tolyl) phosphite

10706 Tris(triphenylphosphine) rhodium carbonyl hydride

10707 1,3,5-Trithiane

No.	Name	Synonym	Mol. Form.	CAS RN	Mol. Wt.	Physical Form	mp/°C	bp/°C	den g cm⁻³	n_D	Solubility
10708	Trithiocarbonic acid		CH_2S_3	594-08-1	110.222	red oil	-26.8(0.5)	57.8	1.476²⁵	1.8225²⁰	dec H₂O, EtOH; vs tol, chl
10709	Tritriacontane		$C_{33}H_{68}$	630-05-7	464.893		71.2(0.9)				
10710	Tropacocaine		$C_{15}H_{19}NO_2$	537-26-8	245.318	pl or tab	49	dec	1.0426¹⁰⁰	1.5080¹⁰⁰	vs bz, eth, EtOH, peth
10711	Tropine	8-Methyl-8-azabicyclo[3.2.1]-octan-3-ol, *endo*	$C_8H_{15}NO$	120-29-6	141.211	hyg pl (eth)	63(2)	233	1.016¹⁰⁰	1.4811¹⁰⁰	vs H₂O, eth, EtOH
10712	Trypan blue		$C_{34}H_{24}N_6Na_4O_{14}S_4$	72-57-1	960.806	dk bl cry	300				s H₂O, acid; i EtOH
10713	Tryptamine		$C_{10}H_{12}N_2$	61-54-1	160.215	nd (al-bz, lig)	118	137⁰·¹⁵			i H₂O, eth, bz, chl; s EtOH, ace
10714	*L*-Tryptophan	α-Aminoindole-3-propionic acid, (*l*)	$C_{11}H_{12}N_2O_2$	73-22-3	204.225	lf or pl (dil al)	289 dec				sl H₂O, HOAc; s EtOH; i eth, chl
10715	Tsuduranine		$C_{18}H_{19}NO_3$	517-97-5	297.349	nd (eth)	204				vs ace, eth, EtOH
10716	T-2 Toxin	Mycotoxin T2	$C_{24}H_{34}O_9$	21259-20-1	466.522	nd	151				sl H₂O, peth; s EtOH, chl, DMSO
10717	Tubocurarine dichloride		$C_{37}H_{42}Cl_2N_2O_6$	57-94-3	681.644	hyg cry	275 dec				s MeOH; i py, bz, ace, eth
10718	Tungsten carbonyl	Tungsten hexacarbonyl	C_6O_6W	14040-11-0	351.90	wh cry	dec 170	sub	2.65		i H₂O; s os
10719	Turanose		$C_{12}H_{22}O_{11}$	547-25-1	342.296	pr (w-al, MeOH)	168				vs H₂O; s EtOH, MeOH
10720	Tybamate		$C_{13}H_{26}N_2O_4$	4268-36-4	274.356	cry	50	151⁰·⁰⁶			
10721	*L*-Tyrosine	4-Hydroxy-*L*-phenylalanine	$C_9H_{11}NO_3$	60-18-4	181.188	nd (w)	343 dec	sub			sl H₂O, HOAc; i EtOH, eth
10722	Tyrosineamide		$C_9H_{12}N_2O_2$	4985-46-0	180.203	pl or pl (al)	153.5				vs H₂O, EtOH
10723	*L*-Tyrosine, ethyl ester		$C_{11}H_{15}NO_3$	949-67-7	209.242	pr (AcOEt)	108.5				vs bz, EtOH, AcOEt
10724	*L*-Tyrosine, methyl ester, hydrochloride		$C_{10}H_{14}ClNO_3$	3417-91-2	231.676		191.0				s H₂O
10725	1,10-Undecadiyne		$C_{11}H_{16}$	4117-15-1	148.245		-17(3)	83¹²	0.8182²¹	1.453²¹	vs ace, bz
10726	Undecafluorocyclohexane		C_6HF_{11}	308-24-7	282.054			62.0			
10727	Undecanal		$C_{11}H_{22}O$	112-44-7	170.292		-4(3)	117¹⁸	0.8251²³	1.4520²⁰	i H₂O; s EtOH, eth
10728	Undecane	Hendecane	$C_{11}H_{24}$	1120-21-4	156.309	liq	-25.54(0.05)	195.9(0.3)	0.7402²⁰	1.4164²⁰	i H₂O; msc EtOH, eth
10729	Undecanenitrile	Decyl cyanide	$C_{11}H_{21}N$	2244-07-7	167.292			265(6)	0.8254³⁰	1.4293³⁰	i H₂O; s EtOH, eth, ctc
10730	1-Undecanethiol	Undecyl mercaptan	$C_{11}H_{24}S$	5332-52-5	188.374	liq	-1.5	257.4	0.8448²⁰	1.4585²⁰	
10731	Undecanoic acid		$C_{11}H_{22}O_2$	112-37-8	186.292	cry (ace)	28.47(0.04)	280	0.8907²⁰	1.4294⁴⁵	i H₂O; vs EtOH, ace; s eth; msc bz
10732	1-Undecanol	Undecyl alcohol	$C_{11}H_{24}O$	112-42-5	172.308		16.6(0.2)	246(2)	0.8298²⁰	1.4392²⁰	i H₂O; s EtOH; vs eth
10733	2-Undecanol	*sec*-Undecyl alcohol	$C_{11}H_{24}O$	1653-30-1	172.308	col liq	0	231(7)	0.8234²⁵	1.4352²⁵	
10734	2-Undecanone	Methyl nonyl ketone	$C_{11}H_{22}O$	112-12-9	170.292		16(6)	233.1(0.3)	0.8250²⁰	1.4291²⁰	i H₂O; s EtOH, eth, ace, bz, chl
10735	6-Undecanone	Butyl hexyl ketone	$C_{11}H_{22}O$	927-49-1	170.292		14(1)	227.4(0.5)	0.8308²⁰	1.4270²⁰	i H₂O; vs EtOH, eth
10736	Undecanoyl chloride		$C_{11}H_{21}ClO$	17746-05-3	204.737						sl ctc
10737	10-Undecenal		$C_{11}H_{20}O$	112-45-8	168.276						sl ctc
10738	1-Undecene		$C_{11}H_{22}$	821-95-4	154.293	liq	-49.12(0.04)	192.7(0.5)	0.7503²⁰	1.4261²⁰	i H₂O; s eth, chl, lig
10739	*cis*-2-Undecene		$C_{11}H_{22}$	821-96-5	154.293	liq	-66.5	196.1	0.7576²⁰		
10740	*trans*-2-Undecene		$C_{11}H_{22}$	693-61-8	154.293	liq	-48.3	192.5	0.7528²⁰	1.4292²⁰	
10741	*cis*-4-Undecene		$C_{11}H_{22}$	821-98-7	154.293	liq	-97	192.6	0.7541²⁰	1.4302²⁰	
10742	*trans*-4-Undecene		$C_{11}H_{22}$	693-62-9	154.293	liq	-63.7	193	0.7508²⁰	1.4285²⁰	
10743	*cis*-5-Undecene		$C_{11}H_{22}$	764-96-5	154.293	liq	-106.5	192.3	0.7537²⁰	1.4302²⁰	
10744	*trans*-5-Undecene		$C_{11}H_{22}$	764-97-6	154.293	liq	-61.1	192	0.7497²⁰	1.4285²⁰	vs eth, chl, lig
10745	10-Undecenoic acid	Undecylenic acid	$C_{11}H_{20}O_2$	112-38-9	184.276	cry	24.4(0.3)	275	0.9072²⁴	1.4486²⁴	i H₂O; s EtOH, eth; sl ctc
10746	10-Undecen-1-ol		$C_{11}H_{22}O$	112-43-6	170.292	liq	-1.0	250	0.8495¹⁵	1.4500²⁰	i H₂O; s EtOH, eth; sl ctc
10747	10-Undecenoyl chloride		$C_{11}H_{19}ClO$	38460-95-6	202.721			127¹³	0.944²⁰	1.454²⁰	
10748	Undecylamine	1-Undecanamine	$C_{11}H_{25}N$	7307-55-3	171.324	cry (eth, al)	15.1(0.7)	229(3)	0.7979²⁰	1.4398²⁰	s H₂O, EtOH; i eth; sl ctc
10749	Undecylbenzene		$C_{17}H_{28}$	6742-54-7	232.404	liq	-16.2(0.4)	312(3)	0.8553²⁰	1.4828²⁰	
10750	1-Undecyne		$C_{11}H_{20}$	2243-98-3	152.277	liq	-25	196	0.7728²⁰	1.4306²⁰	vs ace, bz, eth, EtOH

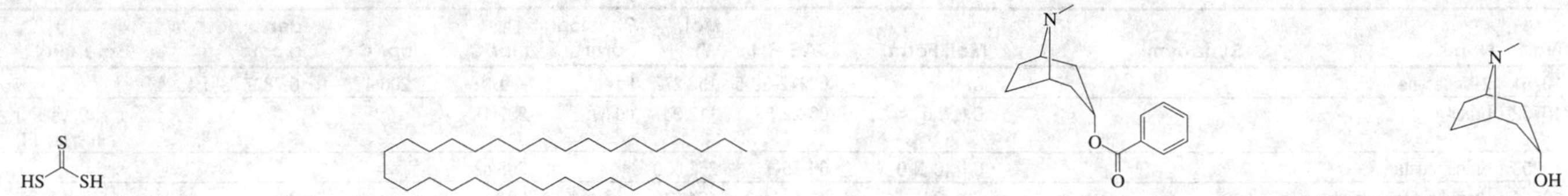

10708
Trithiocarbonic acid

10709
Tritriacontane

10710
Tropacocaine

10711
Tropine

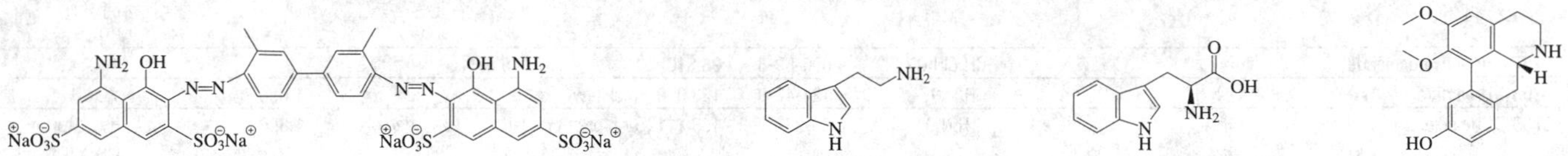

10712
Trypan blue

10713
Tryptamine

10714
L-Tryptophan

10715
Tsuduranine

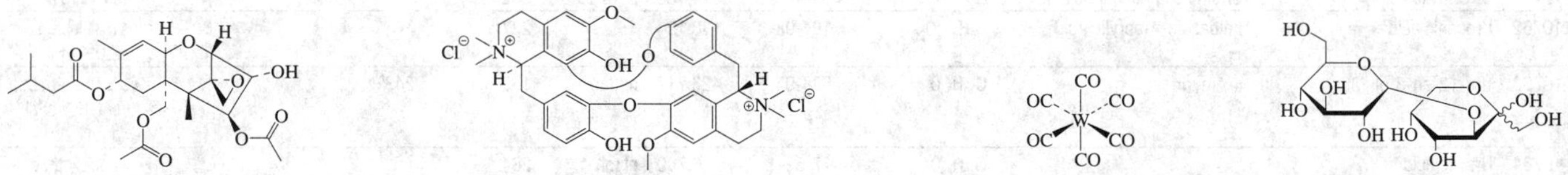

10716
T-2 Toxin

10717
Tubocurarine dichloride

10718
Tungsten carbonyl

10719
Turanose

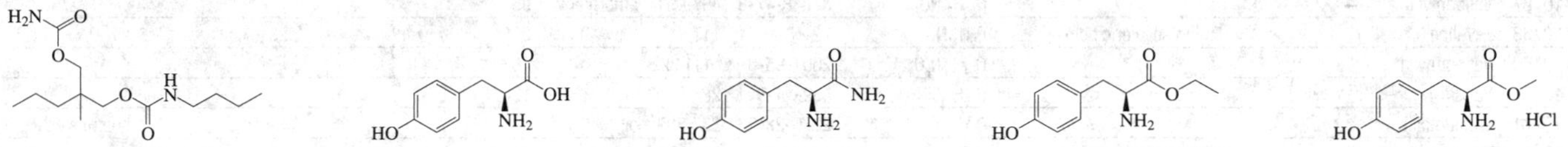

10720
Tybamate

10721
L-Tyrosine

10722
Tyrosineamide

10723
L-Tyrosine, ethyl ester

10724
L-Tyrosine, methyl ester, hydrochloride

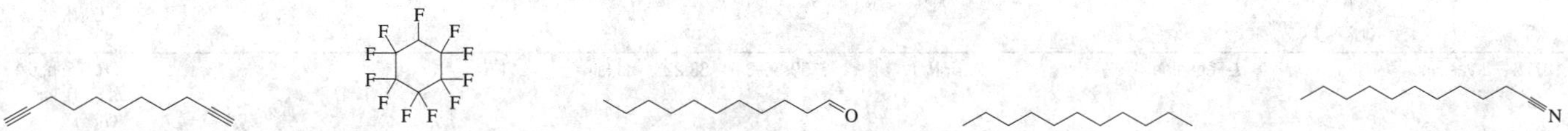

10725
1,10-Undecadiyne

10726
Undecafluorocyclohexane

10727
Undecanal

10728
Undecane

10729
Undecanenitrile

10730
1-Undecanethiol

10731
Undecanoic acid

10732
1-Undecanol

10733
2-Undecanol

10734
2-Undecanone

10735
6-Undecanone

10736
Undecanoyl chloride

10737
10-Undecenal

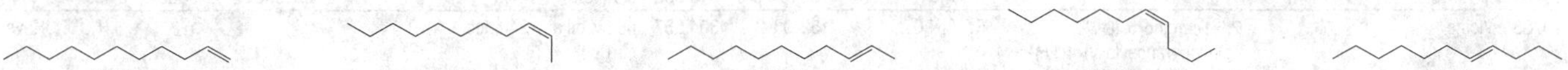

10738
1-Undecene

10739
cis-2-Undecene

10740
trans-2-Undecene

10741
cis-4-Undecene

10742
trans-4-Undecene

10743
cis-5-Undecene

10744
trans-5-Undecene

10745
10-Undecenoic acid

10746
10-Undecen-1-ol

10747
10-Undecenoyl chloride

10748
Undecylamine

10749
Undecylbenzene

10750
1-Undecyne

No.	Name	Synonym	Mol. Form.	CAS RN	Mol. Wt.	Physical Form	mp/°C	bp/°C	den g cm^{-3}	n_D	Solubility
10751	2-Undecyne		$C_{11}H_{20}$	60212-29-5	152.277	liq	-30.1	206(4)	0.7827[20]	1.4391[20]	
10752	Uracil		$C_4H_4N_2O_2$	66-22-8	112.087	nd (w)	339(1)				sl H_2O; vs EtOH, eth; s dil NH_3
10753	Uracil mustard		$C_8H_{11}Cl_2N_3O_2$	66-75-1	252.098		206 dec				sl H_2O
10754	Uranyl acetate dihydrate		$C_4H_{10}O_8U$	6159-44-0	424.146	ye cry (HOAc)	80 dec		2.89		sl EtOH
10755	Urazole		$C_2H_3N_3O_2$	3232-84-6	101.064	lf (w)	249 dec				
10756	Urea	Carbamide	CH_4N_2O	57-13-6	60.055	tetr pr (al)	132.4(0.5)	dec	1.3230[20]	1.484	vs H_2O, EtOH; i eth, bz; s HOAc, py
10757	Urea hydrochloride		CH_5ClN_2O	506-89-8	96.516		145 dec				s H_2O
10758	Urea nitrate		$CH_5N_3O_4$	124-47-0	123.069	mcl lf (w)	160(1)		1.690[20]		vs EtOH
10759	Uric acid		$C_5H_4N_4O_3$	69-93-2	168.111	orth pr or pl	dec	dec	1.89[25]		i H_2O, EtOH, eth; s alk, glycerol; sl acid
10760	Uridine	1-β-D-Ribofuranosyluracil	$C_9H_{12}N_2O_6$	58-96-8	244.200	nd (aq al)	165				s H_2O, EtOH, py
10761	5'-Uridylic acid	Uridine 5'-phosphoric acid	$C_9H_{13}N_2O_9P$	58-97-9	324.180	pr (MeOH)	202 dec				vs H_2O; s MeOH
10762	Urocanic acid	Imidazole-4-acrylic acid	$C_6H_6N_2O_2$	104-98-3	138.124		227				s H_2O, ace; i EtOH, eth
10763	Urs-12-en-3-ol, (3β)	α-Amyrin	$C_{30}H_{50}O$	638-95-9	426.717	nd (al)	186	243[0.5]			s EtOH, eth, bz, chl, HOAc; sl peth
10764	Ursolic acid		$C_{30}H_{48}O_3$	77-52-1	456.700	pl (al)	284				vs ace, eth, chl
10765	Uzarin		$C_{35}H_{54}O_{14}$	20231-81-6	698.796	pr	269				
10766	Vacciniin	D-Glucose, 6-benzoate	$C_{13}H_{16}O_7$	14200-76-1	284.262	amor (aq ace, +1w)	122				vs H_2O, ace, EtOH, eth
10767	Validamycin A		$C_{20}H_{35}NO_{13}$	37248-47-8	497.491	amorp pow	95 dec				
10768	L-Valine	2-Aminoisovaleric acid	$C_5H_{11}NO_2$	72-18-4	117.147	lf (w-al)	315	sub	1.23[25]		s H_2O
10769	Valinomycin		$C_{54}H_{90}N_6O_{18}$	2001-95-8	1111.322	cry	187				
10770	Valium		$C_{16}H_{13}ClN_2O$	439-14-5	284.739		132				
10771	Vamidothion		$C_8H_{18}NO_4PS_2$	2275-23-2	287.337	oil					i peth; s os
10772	Vanadium carbonyl	Vanadium hexacarbonyl	C_6O_6V	14024-00-1	219.002	bl-grn cry	dec 60	sub			
10773	Vanadium(III) 2,4-pentanedioate	Vanadium(III) acetylacetonate	$C_{15}H_{21}O_6V$	13476-99-8	348.266	brn cry	≈185	sub	≈1.0		s MeOH, ace, bz, chl
10774	DL-Vasicine	DL-Peganine	$C_{11}H_{12}N_2O$	6159-56-4	188.225	nd (al)	210.8				sl H_2O, eth, bz; s EtOH, ace, chl
10775	L-Vasicine	L-Peganine	$C_{11}H_{12}N_2O$	6159-55-3	188.225	nd (al)	211.5				sl H_2O, eth, bz; s EtOH, ace, chl
10776	Verapamil		$C_{27}H_{38}N_2O_4$	52-53-9	454.602	ye oil		245[0.01]		1.5448[25]	i H_2O; vs EtOH, ace; sl bz, hx
10777	Veratramine		$C_{27}H_{39}NO_2$	60-70-8	409.605	nd	206				s EtOH, bz, chl, dil acid; i dil alk
10778	Veratramine, 3-glucoside		$C_{33}H_{49}NO_7$	475-00-3	571.745	nd (aq. MeOH)	242 dec				
10779	Veratridine		$C_{36}H_{51}NO_{11}$	71-62-5	673.790	ye amorp pow	180				i H_2O; sl eth
10780	d-Verbenone		$C_{10}H_{14}O$	18309-32-5	150.217		9.8	227.5	0.9978[20]	1.4993[18]	s H_2O, EtOH, ace, bz
10781	Vernolate	Carbamothioic acid, dipropyl-, S-propyl ester	$C_{10}H_{21}NOS$	1929-77-7	203.345			150[30]	0.952[20]		
10782	Versalide		$C_{18}H_{26}O$	88-29-9	258.398	cry	46.5	130[2]			s EtOH
10783	α-Vetivone	Isonootkatone	$C_{15}H_{22}O$	15764-04-2	218.335	cry (peth)	51.5	144[2]	1.0035[20]	1.5370[20]	vs ace
10784	β-Vetivone		$C_{15}H_{22}O$	18444-79-6	218.335	cry (peth)	44.5	141[2]	1.0001[20]	1.5309[20]	s ace
10785	Vicine	2,6-Diamino-5-(β-D-glucopyranosyloxy)-4(1H)-pyrimidinone	$C_{10}H_{16}N_4O_7$	152-93-2	304.257	nd (w, dil al, +1 w)	240 dec				sl H_2O, EtOH; vs acid, alk
10786	Vidarabine	β-D-9-Arabinofuranosyladenine	$C_{10}H_{15}N_5O_5$	5536-17-4	285.257	nd (w)	257				
10787	Vinblastine		$C_{46}H_{58}N_4O_9$	865-21-4	810.975	nd (MeOH)	216				i H_2O; s EtOH, ace, chl, AcOEt
10788	Vincamine		$C_{21}H_{26}N_2O_3$	1617-90-9	354.442		231.5				
10789	Vinclozolin		$C_{12}H_9Cl_2NO_3$	50471-44-8	286.110		110.5(0.5)	131[0.05]	1.51		
10790	Vincristine		$C_{46}H_{56}N_4O_{10}$	57-22-7	824.958		219				
10791	Vinyl acetate		$C_4H_6O_2$	108-05-4	86.090	liq	-100.2(0.4)	72.6(0.3)	0.9256[25]	1.3926[25]	sl H_2O; msc EtOH; s eth, ace, bz, chl
10792	4-Vinylaniline		C_8H_9N	1520-21-4	119.164		23.5	116[9]	1.010[20]	1.6250[22]	s ace, bz
10793	α-Vinylbenzenemethanol	1-Phenylallyl alcohol	$C_9H_{10}O$	4393-06-0	134.174				1.0249[21]	1.5406[20]	sl H_2O; s EtOH, eth, bz, chl

10751
2-Undecyne

10752
Uracil

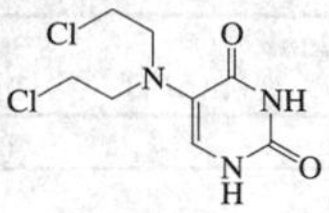

10753
Uracil mustard

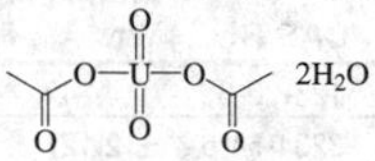

10754
Uranyl acetate dihydrate

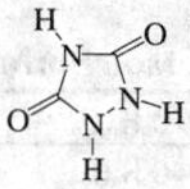

10755
Urazole

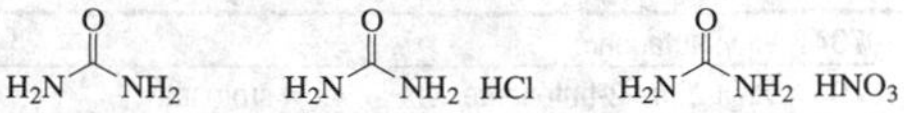

10756
Urea

10757
Urea hydrochloride

10758
Urea nitrate

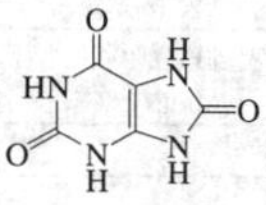

10759
Uric acid

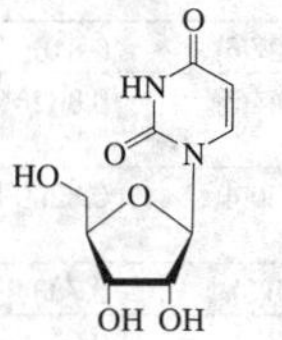

10760
Uridine

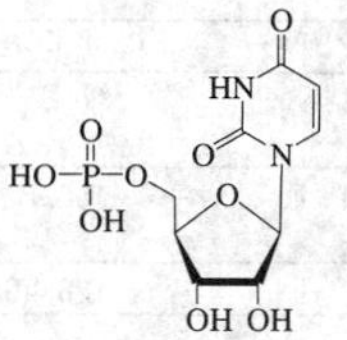

10761
5'-Uridylic acid

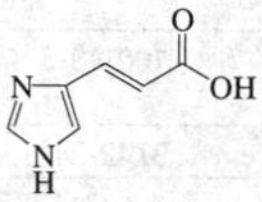

10762
Urocanic acid

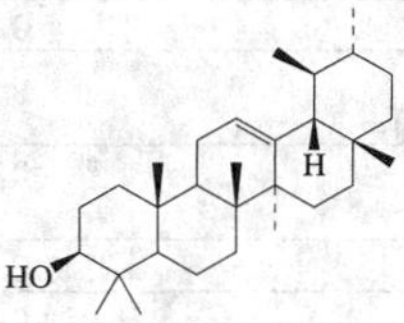

10763
Urs-12-en-3-ol, (3β)

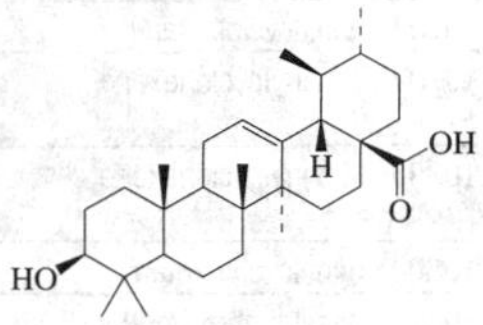

10764
Ursolic acid

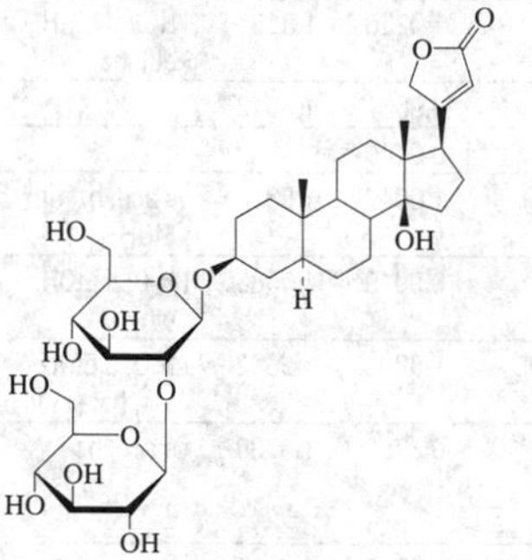

10765
Uzarin

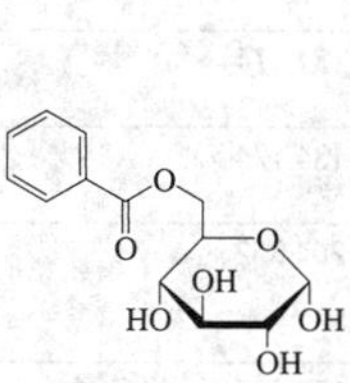

10766
Vacciniin

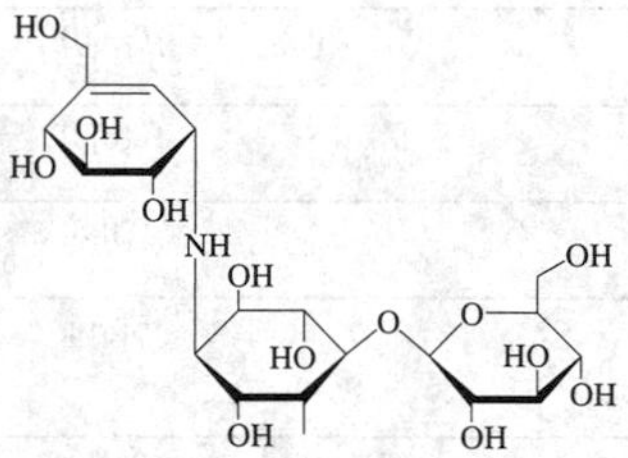

10767
Validamycin A

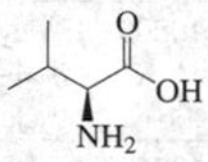

10768
L-Valine

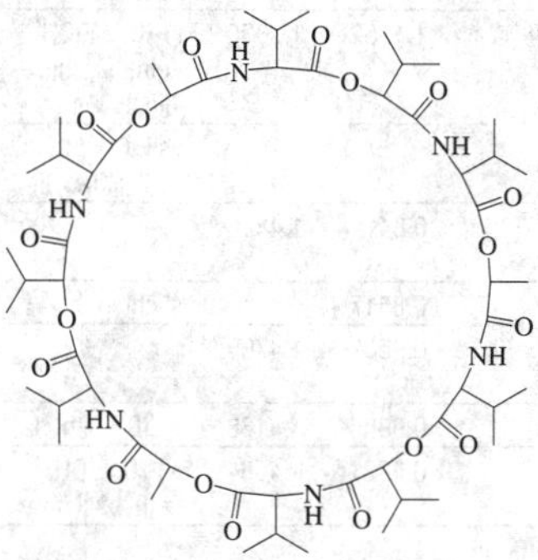

10769
Valinomycin

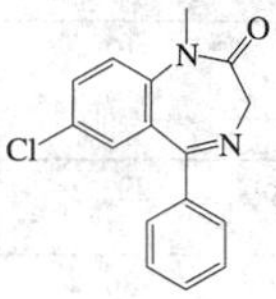

10770
Valium

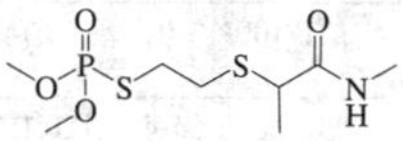

10771
Vamidothion

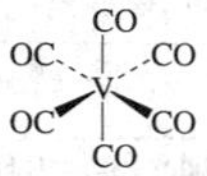

10772
Vanadium carbonyl

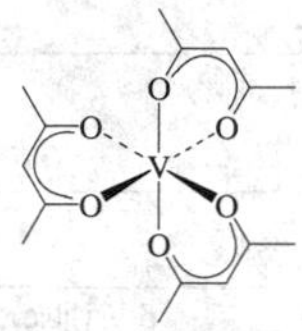

10773
Vanadium(III) 2,4-pentanedioate

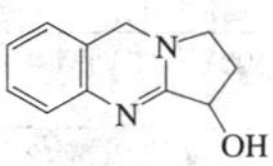

10774
DL-Vasicine

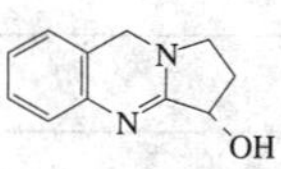

10775
L-Vasicine

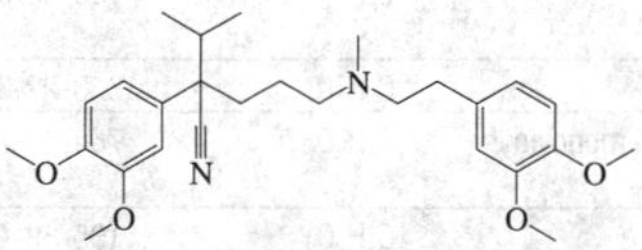

10776
Verapamil

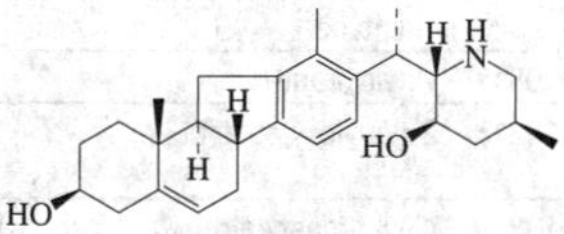

10777
Veratramine

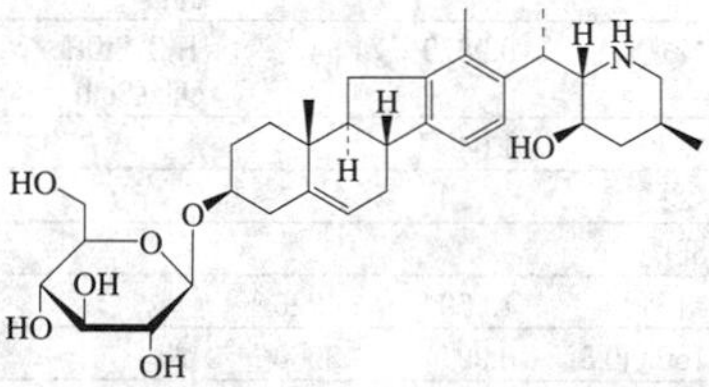

10778
Veratramine, 3-glucoside

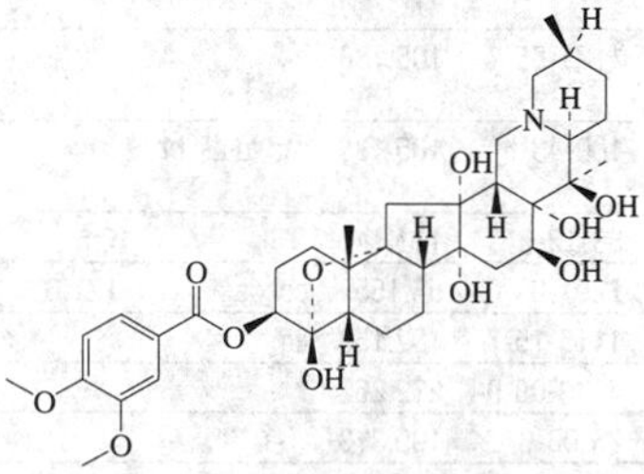

10779
Veratridine

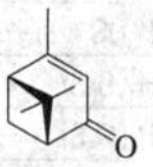

10780
d-Verbenone

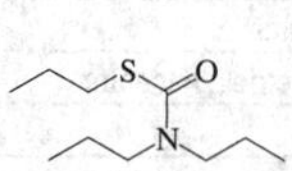

10781
Vernolate

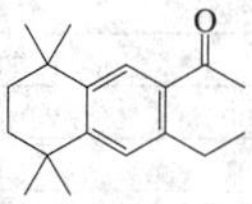

10782
Versalide

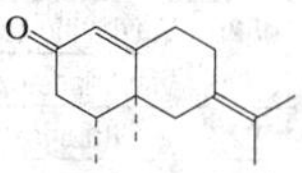

10783
α-Vetivone

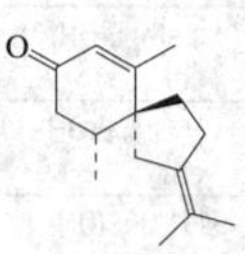

10784
β-Vetivone

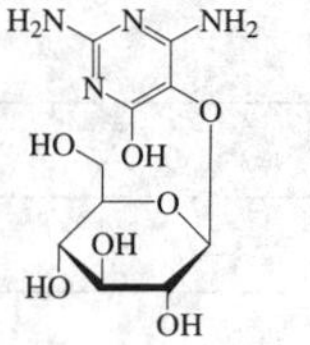

10785
Vicine

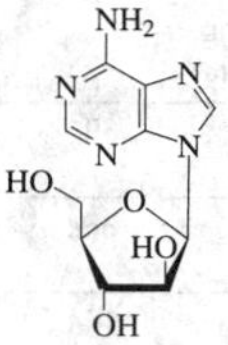

10786
Vidarabine

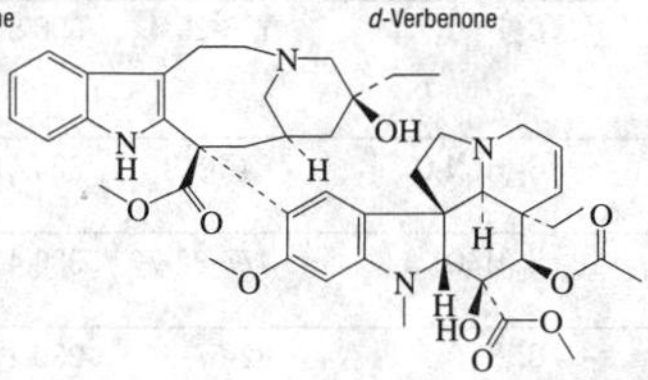

10787
Vinblastine

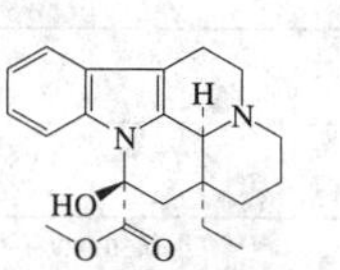

10788
Vincamine

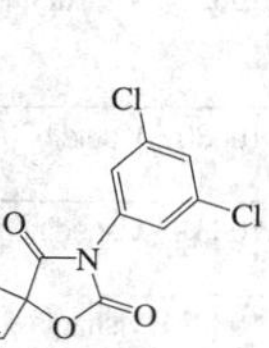

10789
Vinclozolin

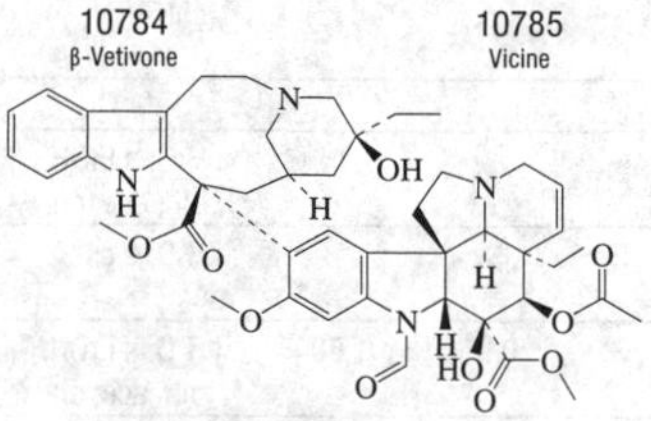

10790
Vincristine

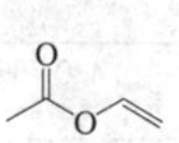

10791
Vinyl acetate

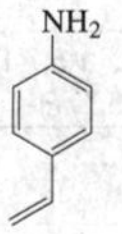

10792
4-Vinylaniline

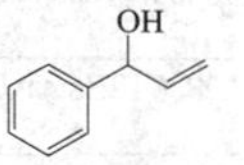

10793
α-Vinylbenzenemethanol

No.	Name	Synonym	Mol. Form.	CAS RN	Mol. Wt.	Physical Form	mp/°C	bp/°C	den g cm⁻³	n_D	Solubility
10794	Vinyl butanoate		$C_6H_{10}O_2$	123-20-6	114.142			116.7	0.9006[20]		
10795	Vinyl *trans*-2-butenoate	Vinyl crotonate	$C_6H_8O_2$	3234-54-6	112.127						s ctc
10796	9-Vinyl-9*H*-carbazole		$C_{14}H_{11}N$	1484-13-5	193.244	cry (al)	66				i H₂O; sl EtOH; vs eth
10797	Vinylcyclohexane		C_8H_{14}	695-12-5	110.197			127(6)	0.8166[19]	1.455[19]	
10798	1-Vinylcyclohexene		C_8H_{12}	2622-21-1	108.181			144(7)	0.8623[15]	1.4915[20]	i H₂O; s eth, bz; vs MeOH
10799	4-Vinylcyclohexene		C_8H_{12}	100-40-3	108.181	liq	-108.9	130(4)	0.8299[20]	1.4639[20]	i H₂O; s eth, bz, peth
10800	Vinylcyclopentane		C_7H_{12}	3742-34-5	96.170	liq	-126.4(0.2)	99(3)	0.7834[20]	1.4360[20]	
10801	Vinyldiethoxymethylsilane		$C_7H_{16}O_2Si$	5507-44-8	160.287			133	0.862[20]	1.4001[20]	
10802	Vinylethoxydimethylsilane		$C_6H_{14}OSi$	5356-83-2	130.260			99	0.790[20]	1.3983[20]	
10803	1-Vinyl-4-fluorobenzene		C_8H_7F	405-99-2	122.140		-34.5	67.4[50]	1.0220[20]	1.5150[20]	i H₂O; s EtOH, eth, bz
10804	Vinyl formate		$C_3H_4O_2$	692-45-5	72.063	visc liq	-78	42(18)	0.965[20]	1.3842[20]	
10805	2-Vinylfuran		C_6H_6O	1487-18-9	94.111	liq	-94(4)	101(3)	0.9445[19]	1.4992[19]	
10806	1-Vinyl-2-methoxybenzene		$C_9H_{10}O$	612-15-7	134.174	nd	29	215(18)	1.0049[17]	1.5388[20]	vs ace, bz, eth, EtOH
10807	1-Vinyl-3-methoxybenzene		$C_9H_{10}O$	626-20-0	134.174			91[15]	0.9919[20]	1.5586[23]	i H₂O; s EtOH, eth, bz
10808	1-Vinyl-4-methoxybenzene		$C_9H_{10}O$	637-69-4	134.174		2.0	208(19)	1.0001[13]	1.5642[13]	i H₂O; s EtOH, eth, bz; sl ctc
10809	6-Vinyl-6-methyl-1-isopropyl-3-(1-methylethylidene)-cyclohexene, (*S*)-		$C_{15}H_{24}$	5951-67-7	204.352			125[8]	0.8782[20]	1.5130[26]	vs ace, bz
10810	1-Vinylnaphthalene		$C_{12}H_{10}$	826-74-4	154.207			124[15]	1.0656[20]	1.644[20]	
10811	2-Vinylnaphthalene		$C_{12}H_{10}$	827-54-3	154.207		65(2)	135[18]			i H₂O; s EtOH, ace, bz
10812	1-Vinyl-3-nitrobenzene		$C_8H_7NO_2$	586-39-0	149.148		-10	120[11]	1.1552[32]	1.5836[20]	i H₂O; s EtOH, eth, bz, chl, lig, HOAc
10813	1-Vinyl-4-nitrobenzene		$C_8H_7NO_2$	100-13-0	149.148	pr (lig)	29	dec			vs EtOH, eth; s chl, HOAc, lig
10814	5-Vinyl-2-norbornene	5-Vinylbicyclo[2.2.1]hept-2-ene	C_9H_{12}	3048-64-4	120.191	liq	-80	140.7(0.5)	0.841	1.4810[20]	
10815	Vinyl octadecanoate	Vinyl stearate	$C_{20}H_{38}O_2$	111-63-7	310.515		29	167[2]	0.8517[20]		sl chl
10816	3-Vinyl-7-oxabicyclo[4.1.0]-heptane		$C_8H_{12}O$	106-86-5	124.180		<-100	169	0.9581[20]	1.4700[20]	
10817	Vinyloxirane		C_4H_6O	930-22-3	70.090			68(2)	0.9006[25]	1.4168[20]	s EtOH, eth, bz
10818	2-(Vinyloxy)ethanol	Ethylene glycol monovinyl ether	$C_4H_8O_2$	764-48-7	88.106			139(4)	0.9821[20]	1.4564[17]	s H₂O, EtOH, eth, bz; i lig
10819	Vinyl propanoate	Vinyl propionate	$C_5H_8O_2$	105-38-4	100.117			94.8(0.2)			
10820	2-Vinylpyridine		C_7H_7N	100-69-6	105.138			159.5	0.9983[20]	1.5495[20]	sl H₂O; vs EtOH, eth, ace, chl
10821	3-Vinylpyridine		C_7H_7N	1121-55-7	105.138			162	0.9879[20]	1.5530[20]	sl H₂O; s EtOH, eth
10822	4-Vinylpyridine		C_7H_7N	100-43-6	105.138	red to dk-br		121[150]	0.9879[20]	1.5449[20]	s H₂O, EtOH, chl; sl eth
10823	1-Vinyl-2-pyrrolidinone		C_6H_9NO	88-12-0	111.141		13.5	193[400]	1.04[20]		
10824	Vinylsilane		C_2H_6Si	7291-09-0	58.155	col gas	-171.6	-22.8			
10825	Vinyl sulfoxide	Divinyl sulfoxide	C_4H_6OS	1115-15-7	102.155	liq		86[18]			
10826	Vinyltriacetoxysilane	Vinylsilanetriol, triacetate	$C_8H_{12}O_6Si$	4130-08-9	232.263			115[10]	1.169[20]	1.4226[20]	
10827	Vinyltriethoxysilane		$C_8H_{18}O_3Si$	78-08-0	190.313			160.0(0.8)	0.901[20]	1.3960[25]	s chl
10828	Vinyltrimethylsilane		$C_5H_{12}Si$	754-05-2	100.235			55.3(0.2)	0.65[20]	1.3914[20]	i H₂O
10829	Violaxanthin		$C_{40}H_{56}O_4$	126-29-4	600.871	red pr (MeOH, al-eth)	208				s EtOH, eth, CS₂; i peth
10830	Viquidil		$C_{20}H_{24}N_2O_2$	84-55-9	324.417	red ye amor	60				vs eth, EtOH, chl
10831	Visnadine		$C_{21}H_{24}O_7$	477-32-7	388.412	nd	85.5				i H₂O; s EtOH, eth
10832	Visnagin	4-Methoxy-7-methyl-5*H*-furo[3,2-*g*][1]benzopyran-5-one	$C_{13}H_{10}O_4$	82-57-5	230.216	nd (w, MeOH)	144.5				sl H₂O, EtOH; vs chl
10833	Vitamin B12	Cyanocobalamin	$C_{63}H_{88}CoN_{14}O_{14}P$	68-19-9	1355.365		>300				
10834	Vitamin D2		$C_{28}H_{44}O$	50-14-6	396.648	pr (ace)	116.5	sub			i H₂O; s EtOH, eth, ace, chl
10835	Vitamin D3	9,10-Secocholesta-5,7,10(19)-trien-3-ol, (3β,5*Z*,7*E*)-	$C_{27}H_{44}O$	67-97-0	384.637		84.5				i H₂O; s os
10836	Vitamin E	α-Tocopherol	$C_{29}H_{50}O_2$	59-02-9	430.706	pale ye oil	3.0	210[0.1]	0.950[25]	1.5045[25]	i H₂O; s EtOH, eth, ace, chl

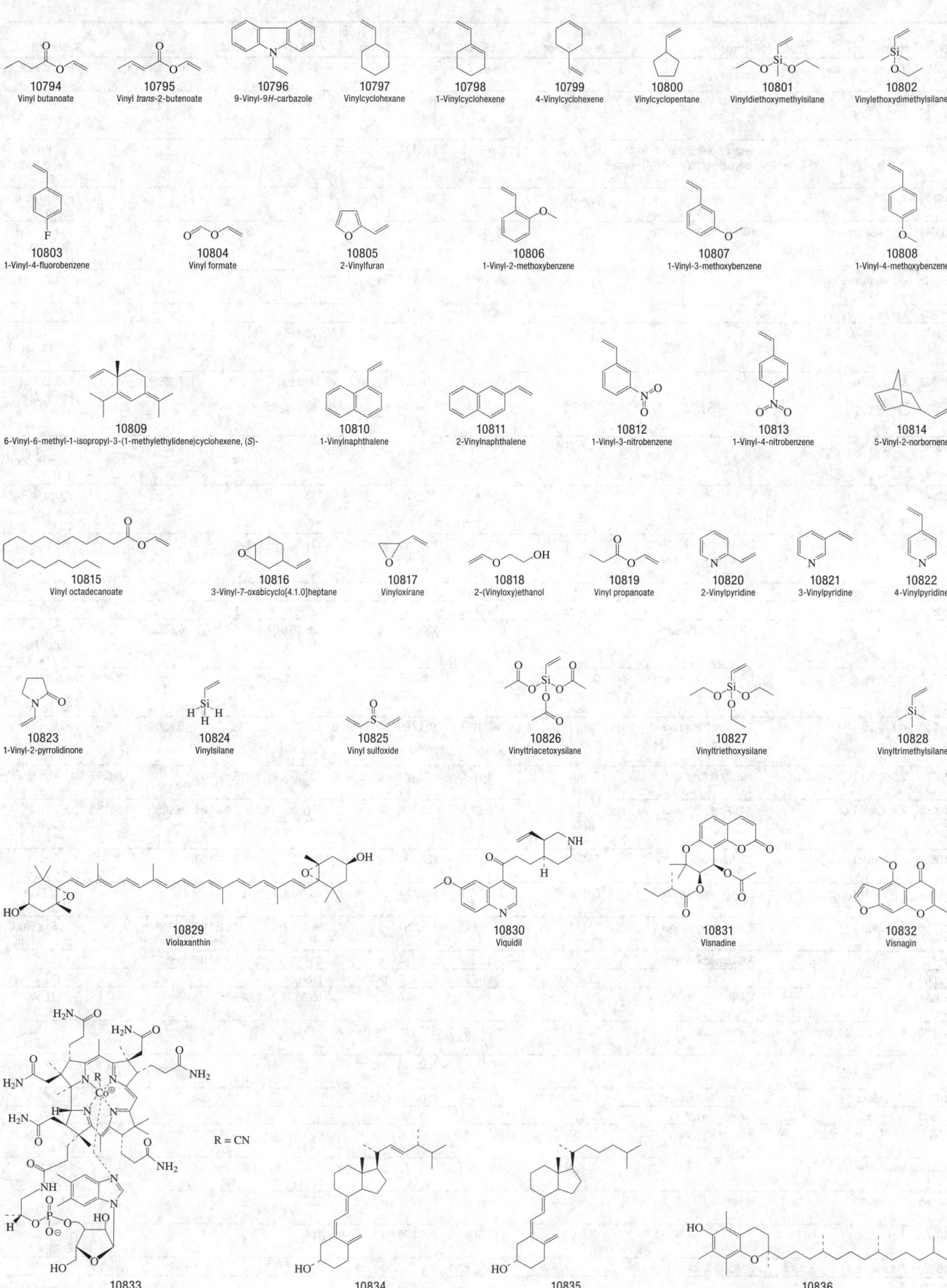

10794
Vinyl butanoate

10795
Vinyl *trans*-2-butenoate

10796
9-Vinyl-9*H*-carbazole

10797
Vinylcyclohexane

10798
1-Vinylcyclohexene

10799
4-Vinylcyclohexene

10800
Vinylcyclopentane

10801
Vinyldiethoxymethylsilane

10802
Vinylethoxydimethylsilane

10803
1-Vinyl-4-fluorobenzene

10804
Vinyl formate

10805
2-Vinylfuran

10806
1-Vinyl-2-methoxybenzene

10807
1-Vinyl-3-methoxybenzene

10808
1-Vinyl-4-methoxybenzene

10809
6-Vinyl-6-methyl-1-isopropyl-3-(1-methylethylidene)cyclohexene, (*S*)-

10810
1-Vinylnaphthalene

10811
2-Vinylnaphthalene

10812
1-Vinyl-3-nitrobenzene

10813
1-Vinyl-4-nitrobenzene

10814
5-Vinyl-2-norbornene

10815
Vinyl octadecanoate

10816
3-Vinyl-7-oxabicyclo[4.1.0]heptane

10817
Vinyloxirane

10818
2-(Vinyloxy)ethanol

10819
Vinyl propanoate

10820
2-Vinylpyridine

10821
3-Vinylpyridine

10822
4-Vinylpyridine

10823
1-Vinyl-2-pyrrolidinone

10824
Vinylsilane

10825
Vinyl sulfoxide

10826
Vinyltriacetoxysilane

10827
Vinyltriethoxysilane

10828
Vinyltrimethylsilane

10829
Violaxanthin

10830
Viquidil

10831
Visnadine

10832
Visnagin

10833
Vitamin B12

R = CN

10834
Vitamin D2

10835
Vitamin D3

10836
Vitamin E

No.	Name	Synonym	Mol. Form.	CAS RN	Mol. Wt.	Physical Form	mp/°C	bp/°C	den g cm⁻³	n_D	Solubility
10837	Vitamin E acetate		$C_{31}H_{52}O_3$	58-95-7	472.743		-27.5	184[0.01]	0.9533[21]	1.497[20]	i H_2O; sl EtOH; s eth, ace, chl
10838	Vitamin K1		$C_{31}H_{46}O_2$	84-80-0	450.696		-20	142[0.001]	0.964[25]	1.5250[25]	i H_2O; s EtOH, eth, ace, bz, peth, chl
10839	Vomicine	4-Hydroxy-19-methyl-16,19-secostrychnidine-10,16-dione	$C_{22}H_{24}N_2O_4$	125-15-5	380.437	nd (80% al) pr (ace)	282				sl EtOH, eth, ace; vs chl; s AcOEt
10840	Warfarin	Coumadin	$C_{19}H_{16}O_4$	81-81-2	308.328	cry (al)	161				i H_2O; s EtOH, ace, diox
10841	9H-Xanthene	10H-9-Oxaanthracene	$C_{13}H_{10}O$	92-83-1	182.217	ye lf (al)	101(1)	311(2)			i H_2O; sl EtOH, ctc; s eth, bz, chl
10842	9H-Xanthen-9-ol		$C_{13}H_{10}O_2$	90-46-0	198.217	nd (aq al)	125				sl H_2O; s EtOH, eth, chl
10843	Xanthine		$C_5H_4N_4O_2$	69-89-6	152.112	ye pl (w)	dec	sub			i H_2O
10844	Xanthone		$C_{13}H_8O_2$	90-47-1	196.202	nd (al)	176.4(0.4)	351			i H_2O; s EtOH, eth, bz, chl; sl peth
10845	Xanthopterin		$C_6H_5N_5O_2$	119-44-8	179.137	hyg ye amor or oran pow (HOAc)	>410 dec	99[18]	1.559[25]		i H_2O; sl EtOH, eth; vs acid, alk
10846	Xanthosine		$C_{10}H_{12}N_4O_6$	146-80-5	284.225	pr cry (w)					sl cold H_2O; vs hot H_2O; dec acid
10847	Xanthoxyletin		$C_{15}H_{14}O_4$	84-99-1	258.270	pr (MeOH, peth)	133				i H_2O; s EtOH, ace; sl eth; vs bz, alk
10848	Xanthyletin	8,8-Dimethyl-2H,8H-benzo[1,2-b:5,4-b']dipyran-2-one	$C_{14}H_{12}O_3$	553-19-5	228.243	pr (MeOH)	131.5	142[0.1]			s EtOH, peth
10849	p-Xenylcarbimide	4-Isocyanato-1,1'-biphenyl	$C_{13}H_9NO$	92-95-5	195.216	nd	56	283 dec			vs eth
10850	Xibenolol		$C_{15}H_{25}NO_2$	81584-06-7	251.366	cry	57	135[0.7]			s EtOH
10851	o-Xylene	1,2-Dimethylbenzene	C_8H_{10}	95-47-6	106.165	liq	-25.16(0.02)	144.4(0.4)	0.8755[25]	1.5018[25]	i H_2O; msc EtOH, eth, ace, bz, peth, ctc
10852	m-Xylene	1,3-Dimethylbenzene	C_8H_{10}	108-38-3	106.165	liq	-47.85(0.03)	139.1(0.4)	0.8598[25]	1.4944[25]	i H_2O; msc EtOH, eth, ace, bz; s chl
10853	p-Xylene	1,4-Dimethylbenzene	C_8H_{10}	106-42-3	106.165	mcl pr or liq	13.3(0.1)	138.3(0.5)	0.8565[25]	1.4929[25]	i H_2O; msc EtOH, eth, ace, bz; s chl
10854	2,3-Xylenol	2,3-Dimethylphenol	$C_8H_{10}O$	526-75-0	122.164	nd (w, dil al)	72.7(0.3)	216.88(0.05)		1.5420[20]	sl H_2O; s EtOH, eth
10855	2,4-Xylenol	2,4-Dimethylphenol	$C_8H_{10}O$	105-67-9	122.164	nd (w)	25(1)	210.94(0.03)	0.9650[20]	1.5420[14]	sl H_2O; msc EtOH, eth; s ctc
10856	2,5-Xylenol	2,5-Dimethylphenol	$C_8H_{10}O$	95-87-4	122.164	nd (w), pr (al-eth)	74.9(0.2)	211.14(0.08)			s H_2O, EtOH; vs eth; sl chl
10857	2,6-Xylenol	2,6-Dimethylphenol	$C_8H_{10}O$	576-26-1	122.164	lf or nd (al)	45.4(0.4)	201.03(0.05)			s H_2O, EtOH, eth, ctc
10858	3,4-Xylenol	3,4-Dimethylphenol	$C_8H_{10}O$	95-65-8	122.164		65.1(0.2)	227.31(0.05)	0.9830[20]		sl H_2O; s EtOH, ctc; msc eth
10859	3,5-Xylenol	3,5-Dimethylphenol	$C_8H_{10}O$	108-68-9	122.164	nd (w, peth)	63.4(0.3)	221.71(0.05)	0.9680[20]		s H_2O, EtOH, ctc
10860	Xylenol orange		$C_{31}H_{32}N_2O_{13}S$	1611-35-4	672.656	dk red cry	286 dec				s H_2O
10861	Xylitol	Xylite	$C_5H_{12}O_5$	87-99-0	152.146	mcl (al)	95.9(0.9)	380(15)			vs H_2O, py, EtOH
10862	6-O-β-D-Xylopyranosyl-D-glucose	Primeverose	$C_{11}H_{20}O_{10}$	26531-85-1	312.271	cry (MeOH)	210				vs H_2O, MeOH
10863	D-Xylose		$C_5H_{10}O_5$	58-86-6	150.130	mcl nd	90.5			1.525[20]	vs H_2O; s EtOH; sl eth
10864	D-Xylulose	D-threo-2-Pentulose	$C_5H_{10}O_5$	551-84-8	150.130	visc liq					s H_2O
10865	L-Xylulose	L-threo-2-Pentulose	$C_5H_{10}O_5$	527-50-4	150.130	syrup					vs H_2O
10866	3,5-Xylyl methylcarbamate	3,5-Dimethylphenyl methylcarbamate	$C_{10}H_{13}NO_2$	2655-14-3	179.216	cry	99				sl H_2O; s os
10867	Yohimbine		$C_{21}H_{26}N_2O_3$	146-48-5	354.442	nd (dil al)	241	160 sub			sl H_2O, bz; s EtOH, eth, chl
10868	Yohimbine hydrochloride	Tosanpin	$C_{21}H_{27}ClN_2O_3$	65-19-0	390.903	orth nd or pl (w, dil HCl)	302				vs H_2O
10869	Zearalenone		$C_{18}H_{22}O_5$	17924-92-4	318.365	cry	164				i H_2O; s alk, bz, EtOH, eth
10870	Zidovudine	3'-Azido-3'-deoxythymidine	$C_{10}H_{13}N_5O_4$	30516-87-1	267.242	cry (w)	121				
10871	Zinc benzoate		$C_{14}H_{10}O_4Zn$	553-72-0	307.636						sl H_2O

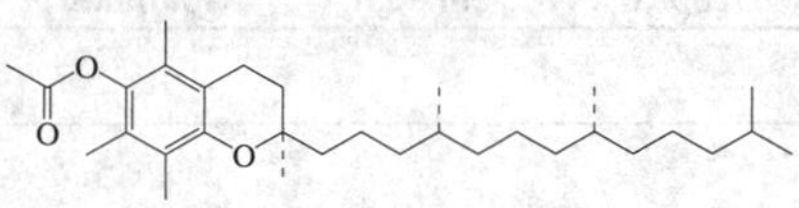

10837
Vitamin E acetate

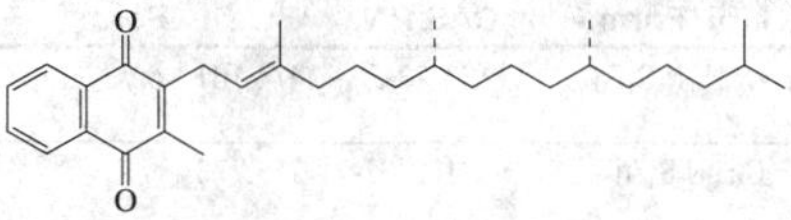

10838
Vitamin K1

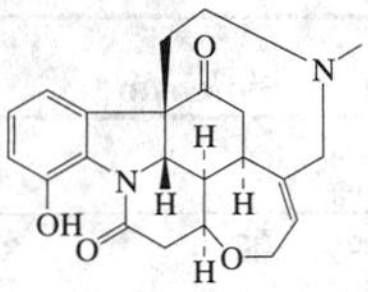

10839
Vomicine

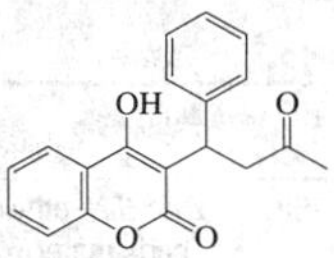

10840
Warfarin

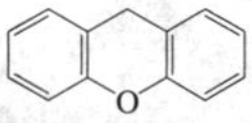

10841
9H-Xanthene

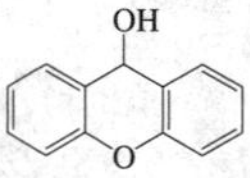

10842
9H-Xanthen-9-ol

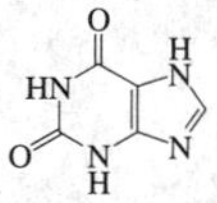

10843
Xanthine

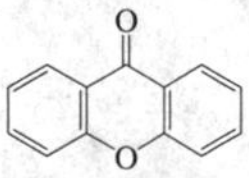

10844
Xanthone

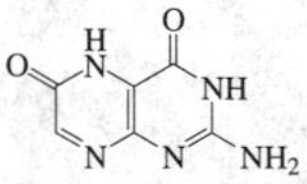

10845
Xanthopterin

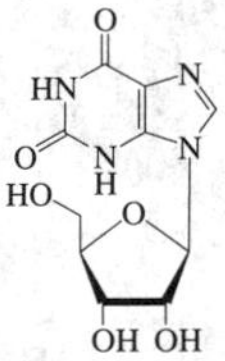

10846
Xanthosine

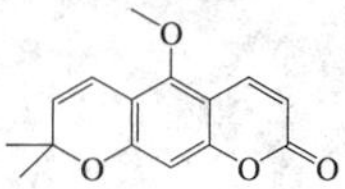

10847
Xanthoxyletin

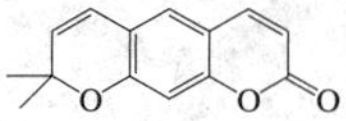

10848
Xanthyletin

10849
p-Xenylcarbimide

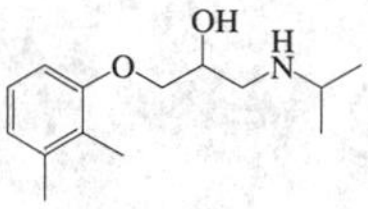

10850
Xibenolol

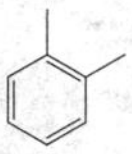

10851
o-Xylene

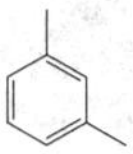

10852
m-Xylene

10853
p-Xylene

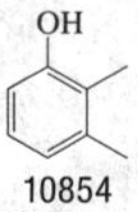

10854
2,3-Xylenol

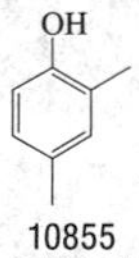

10855
2,4-Xylenol

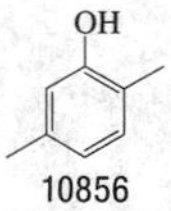

10856
2,5-Xylenol

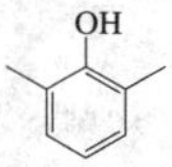

10857
2,6-Xylenol

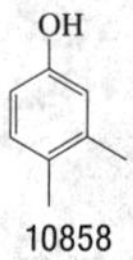

10858
3,4-Xylenol

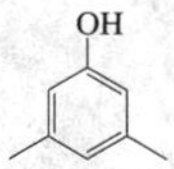

10859
3,5-Xylenol

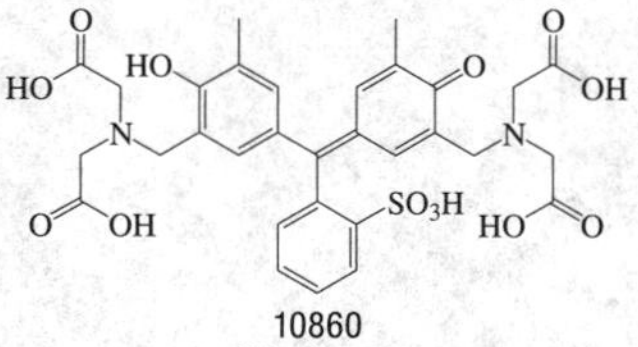

10860
Xylenol orange

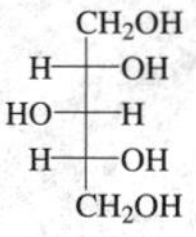

10861
Xylitol

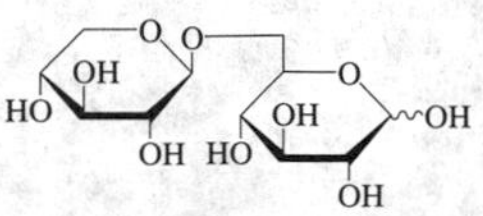

10862
6-O-β-D-Xylopyranosyl-D-glucose

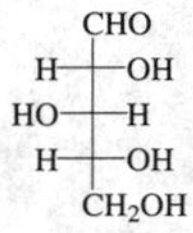

10863
D-Xylose

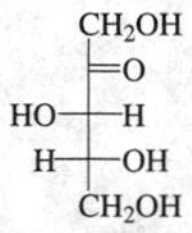

10864
D-Xylulose

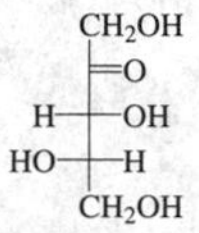

10865
L-Xylulose

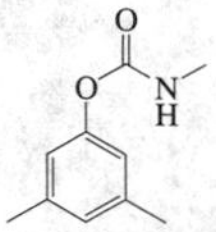

10866
3,5-Xylyl methylcarbamate

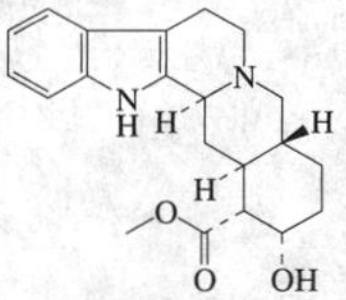

10867
Yohimbine

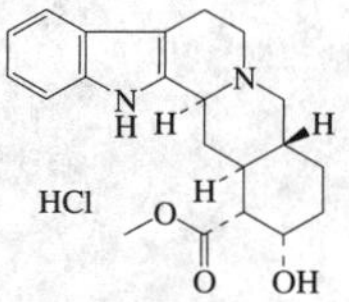

10868
Yohimbine hydrochloride

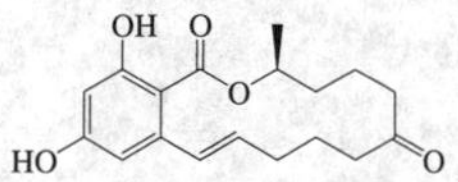

10869
Zearalenone

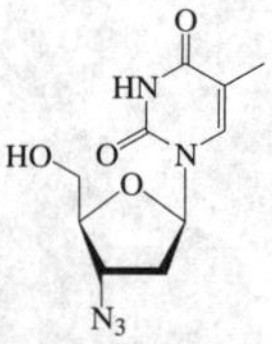

10870
Zidovudine

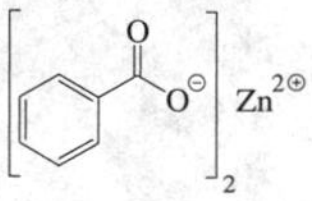

10871
Zinc benzoate

No.	Name	Synonym	Mol. Form.	CAS RN	Mol. Wt.	Physical Form	mp/°C	bp/°C	den g cm⁻³	n_D	Solubility
10872	Zinc bis(dibutyldithiocarbamate)		$C_{18}H_{36}N_2S_4Zn$	136-23-2	474.161	cry	138				
10873	Zinc *N,N*'-ethylenebisdithio-carbamate	Zineb	$C_4H_6N_2S_4Zn$	12122-67-7	275.773		157 dec				
10874	Zinc gluconate		$C_{12}H_{22}O_{14}Zn$	4468-02-4	455.704	pow					
10875	Zinc 2,4-pentanedioate	Zinc acetylacetonate	$C_{10}H_{14}O_4Zn$	14024-63-6	263.625	cry	127.3(0.2)				sl H_2O; s EtOH, DMSO
10876	Zinc propanoate		$C_6H_{10}O_4Zn$	557-28-8	211.550	hyg pl or nd					sl EtOH
10877	Ziram	Zinc, bis(dimethylcarbamodi-thioato-*S,S*')-, (T-4)-	$C_6H_{12}N_2S_4Zn$	137-30-4	305.841	cry	250		1.66²⁵		i H_2O; sl bz; s chl

10872
Zinc bis(dibutyldithiocarbamate)

10873
Zinc N,N'-ethylenebisdithiocarbamate

10874
Zinc gluconate

10875
Zinc 2,4-pentanedioate

10876
Zinc propanoate

10877
Ziram

SYNONYM INDEX OF ORGANIC COMPOUNDS

Carbonyl selenide	: 1752
Carbonyl sulfide	: 1753
Carboxin *S,S*-dioxide	: 8483
N-(Carboxymethyl)-*N*-methylglycine	: 7382
8-Carboxyquinoline	: 9431
2-Carboxythiophene	: 10071
1-Carboxy-*N,N,N*-trimethylmethana-	
minium, inner salt	: 845
Carbromal	: 278
Carbutamide	: 274
Carbyloxime	: 5474
Carfimate	: 5285
Carmustine	: 931
Carnegine	: 9846
Carnitine, *O*-propanoyl	: 9194
γ-Carotene	: 1776
Carvacrol	: 6515
Carzenide	: 427
Catechol monoethyl ether	: 4858
CDT	: 2548
CDTA	: 2621
Cellocidin	: 1677
Cephaeline	: 10488
Cerespan	: 8506
Cerotic acid	: 5742
Cetane	: 5747
1-Cetene	: 5756
Cetyl alcohol	: 5753
Cetyl lactate	: 5764
Cetyl mercaptan	: 5750
Cetyl palmitate	: 5762
Cetylpyridinium chloride	: 5768
Cetyl stearate	: 5769
Chalcone	: 4521
Chaulmoogric acid	: 2675
Chavicol	: 202
Chelidonic acid	: 8478
Chenodiol	: 3717
Chimyl alcohol	: 5766
Chloral	: 10194
Chloral alcoholate	: 10232
Chloral ammonia	: 433
Chloral formamide	: 10242
Chloramben	: 295
Chloraminophenamide	: 283
Chloranil	: 9741
Chlordecone	: 6568
Chlorfenac	: 10207
Chlorindanol	: 1984
Chlorine cyanide	: 2514
ω-Chloroacetophenone	: 1845
m-Chloroacetophenone	: 2238
p-Chloroacetophenone	: 2239
2-Chloroacrylic acid	: 2277
4-Chloro-2-anisidine	: 2090
N-Chlorobenzenesulfonamide sodium	: 1814
p-Chlorobenzenesulfonic acid	: 1889
4-Chlorobenzenethiol, *S*-methyl,	
S,S-dioxide	: 2164
o-Chlorobenzotrifluoride	: 2331
m-Chlorobenzotrifluoride	: 2332
p-Chlorobenzotrifluoride	: 2333
2-Chlorobenzyl chloride	: 1950
3-Chlorobenzyl chloride	: 1951
4-Chlorobenzyl chloride	: 1952
2-(2-Chlorobenzyl)-4,4-dimethyl-1,2-	
oxazolidin-3-one	: 2428
2-(4-Chlorobenzyl)-2-propylamine	: 2349
3-Chloro-(1,1'-biphenyl)-4-ol	: 8863

6-Chloro-*N*-*tert*-butyl-*N*'-ethyl-1,3,5-	
triazine-2,4-diamine	: 9691
5-Chloro-3-*tert*-butyl-6-methyl-	
2,4(1*H*,3*H*)-pyrimidinedione	: 9689
1-Chloro-4-[[(4-chlorophenyl)-	
methyl]thio]benzene	: 1819
4-Chloro-*o*-cresol	: 2150
6-Chloro-*o*-cresol	: 2148
4-Chloro-*m*-cresol	: 2151
6-Chloro-*m*-cresol	: 2147
2-Chloro-*p*-cresol	: 2146
3-Chloro-*p*-cresol	: 2149
Chlorocyanohydrin	: 10243
1-Chloro-2,2-difluoroethylene	: 1982
7-Chloro-1,3-dihydro-5-phenyl-2*H*-	
1,4-benzodiazepin-2-one	: 8267
6-Chloro-*N,N*'-diisopropyl-1,3,5-	
triazine-2,4-diamine	: 9177
2-Chloro-*N,N*-dimethyl-10*H*-	
phenothiazine-10-propanamine	: 2350
Chlorodiphenylarsine	: 4458
Chlorodiphenylphosphine	: 4509
2-Chloroethanamine hydrochloride	: 2029
β-Chloroethyl acetate	: 2027
Chloroethylene carbonate	: 2014
1-Chloro-3-ethyl-1-penten-4-yn-ol	: 4805
2-Chloroethyl phenyl ketone	: 2249
3-Chloro-4-fluoronitrobenzene	: 2056
Chloroflurazole	: 3310
Chloroform	: 10245
4-(Chloroformyl)phthalic anhydride	: 671
α-Chlorohydrin	: 2262
5-Chloroisatoic anhydride	: 1907
6-Chloro-*N*-isopropyl-1,3,5-triazine-	
2,4-diamine	: 2810
6-Chlorometanilic acid	: 284
Chloromethapyrilene	: 2310
1-Chloro-2-methoxybenzene	: 1855
1-Chloro-3-methoxybenzene	: 1856
1-Chloro-4-methoxybenzene	: 1857
N-Chloro-4-methylbenzenesulfon-	
amide sodium	: 1815
Chloromethyl cyanide	: 1844
Chloromethyl *O,O*-diethyl dithiophos-	
phate	: 1834
Chloromethyl ethyl ether	: 2092
Chloromethylmagnesium	: 7433
(Chloromethyl)oxirane	: 4707
N'-(3-Chloro-4-methylphenyl)-*N,N*-	
dimethylurea	: 2359
(Chloromethyl)trichlorosilane	: 10217
o-Chloronitrobenzene	: 2181
m-Chloronitrobenzene	: 2182
p-Chloronitrobenzene	: 2183
4-Chloro-3-nitrotoluene	: 2141
3-Chlorophenacyl bromide	: 1137
p-Chlorophenacyl bromide	: 1149
N'-[4-(4-Chlorophenoxy)phenyl]-*N,N*-	
dimethylurea	: 2344
N-[[(4-Chlorophenyl)amino]-	
carbonyl]-2,6-difluorobenzamide	: 3546
4-Chloro-*o*-phenylenediamine	: 1874
2-Chloro-*p*-phenylenediamine	: 1876
Chlorophenylmagnesium	: 8912
Chlorophenylmercury	: 8913
N'-(4-Chlorophenyl)-*N*-methoxy-*N*-	
methylurea	: 7893
2-(2-Chlorophenyl)-2-(methylamino)-	
cyclohexanone, (±)	: 6569

1-[4-(4-Chlorophenyl)-3-phenyl-2-	
butenyl]pyrrolidine	: 9391
4-Chlorophenyl phenyl ether	: 2225
2-Chlorophenyl phenyl ketone	: 1902
Chloropicrin	: 10261
Chloroprene	: 1923
Chloroprophenpyridamine	: 2348
2-Chloropropionic acid	: 2266
β-Chloropropionic acid	: 2267
β-Chloropropionitrile	: 2265
4-Chloro-*N*-[(propylamino)carbonyl]-	
benzenesulfonamide	: 2258
3-Chloropropyl ether	: 950
(3-Chloropropyl)triethoxysilane	: 10332
6-Chloropurine	: 2286
Chlorothymol	: 2127
p-Chloro-*o*-toluidine	: 2103
2-Chloro-*p*-toluidine-5-sulfonic acid	: 288
Chlorotrianisene	: 10695
Chlorotrifluoroethylene	: 2327
5-[2-Chloro-4-(trifluoromethyl)-	
phenoxy]-2-nitrobenzoic acid	: 109
2-Chloro-1,7,7-	
trimethylbicyclo[2.2.1]heptane, *endo*	: 1067
Chloroxine	: 3290
Chloroxylenol	: 2000
Chlorphenesin	: 2226
Chlorzoxazone	: 1910
Cholaic acid	: 9682
Cholanic acid	: 2361
Cholanthrene	: 3594
4-Chromanone	: 3603
1,2-Chromene	: 724
Chromium acetylacetonate	: 2383
Chromotropic acid	: 3736
Chromotropic acid disodium salt	: 9553
Chrysanthenone	: 10521
Chrysin	: 3741
Chrysoidine hydrochloride	: 8818
Chrysophanic acid	: 3731
C.I. Acid Green 3	: 5619
C.I. Acid Yellow 73	: 5350
C.I. Basic Red 9	: 10677
C.I. Basic Violet 1	: 7860
Cicrotoic acid	: 2615
C.I. Direct Red 2, disodium salt	: 723
C.I. Food Red 6	: 9090
C.I. Food Yellow 3	: 9667
Cinchocaine	: 2997
Cinchomeronic acid	: 9342
Cinchoninaldehyde	: 9429
Cinchophen	: 8994
Cineole	: 5297
Cinnamic anhydride	: 8968
Cinnamoyl chloride	: 8972
trans-Cinnamyl acetate	: 8971
Citraconic acid	: 7090
Citronellal	: 4192
Citronellene	: 4166
Citronellic acid	: 4194
Citronellol, (+)	: 4195
Citronellol, (-)	: 4196
Citronellol acetate	: 4199
Civetone	: 2551
Claritin	: 6628
1,7-Cleve's acid	: 370
Clinestrol	: 3525
Clomethiazole	: 2043
Clonitrate	: 2264
Cloprop	: 2227

Durindone Red	: 852	Ethanoyl iodide	: 74	Ethylene disulfonic acid	: 4791
Durohydroquinone	: 9926	Ethene	: 5015	Ethylene fluorohydrin	: 5387
Duroquinone	: 9935	1-(Ethenyloxy)butane	: 1673	Ethylene glycol	: 4772
Dymanthine	: 4320	1-(Ethenyloxy)decane	: 2787	Ethylene glycol diacetate	: 4775
Dypnone	: 4470	1-(Ethenyloxy)hexadecane	: 5771	Ethylene glycol diacrylate	: 4776
Echinatine-3'-epimer	: 9486	2-(Ethenyloxy)-2-methylpropane	: 1674	Ethylene glycol dibenzoate	: 4777
Echinopsine	: 7775	1-(Ethenyloxy)octadecane	: 8328	Ethylene glycol dibutyl ether	: 3000
Edetate calcium disodium	: 4573	1-(Ethenyloxy)propane	: 9270	Ethylene glycol didodecanoate	: 4778
Edetate sodium	: 10002	2-(Ethenyloxy)propane	: 6546	Ethylene glycol diethyl ether	: 3360
EDTA	: 5018	3-(Ethenyloxy)-1-propene	: 213	Ethylene glycol diformate	: 4779
EDTA disodium	: 5019	Ethenzamide	: 4827	Ethylene glycol dimethacrylate	: 4781
EEDQ	: 5026	Ethide	: 3239	Ethylene glycol dimethyl ether	: 3863
Ehrlich's reagent	: 3900	Ethinamate	: 5288	Ethylene glycol dinitrate	: 4782
1-Eicosanamine	: 6208	Ethionamide	: 5234	Ethylene glycol dipalmitate	: 4780
Elemol	: 4692	Ethohexadiol	: 5048	Ethylene glycol diphenyl ether	: 4450
cis-Eleostearic acid	: 8309	Ethosuximide	: 5157	Ethylene glycol distearate	: 4783
trans-Eleostearic acid	: 8310	2-Ethoxy-1,1-bis(ethoxycarbonyl)-		Ethylene glycol ditetradecanoate	: 4784
Emodin	: 10441	ethene	: 3455	Ethylene glycol dithiocyanate	: 4785
Enanthaldoxime	: 5669	(Ethoxycarbonyl)acetylene	: 5229	Ethylene glycol monoacetate	: 4786
Enanthic acid	: 5683	N-(Ethoxycarbonyl)phthalimide	: 1734	Ethylene glycol monoallyl ether	: 200
ENTPROL	: 9902	2-[2-(2-Ethoxyethoxy)ethoxy]ethanol	: 10361	Ethylene glycol monobenzoate	: 4787
Eosine YS	: 9713	1-Ethoxyhexane	: 5061	Ethylene glycol monobenzyl ether	: 817
Epiandrostanediol	: 454	Ethoxyquin	: 4835	Ethylene glycol monobutyl ether	: 1473
Epiandrosterone	: 5964	Ethoxzolamide	: 4832	Ethylene glycol monobutyl ether	
Epicholestanol	: 2368	Ethylacetylene	: 1675	acetate	: 1477
Epicholesterol	: 2370	Ethyl alcohol	: 4800	Ethylene glycol monoethyl ether	: 4839
16-Epiestriol	: 4762	Ethylallene	: 8564	Ethylene glycol monoethyl ether	
2,3-Epoxy-4-oxo-7,10-dodecadien-		Ethyl 2-allylacetoacetate	: 4877	acetate	: 4841
amide, (2R,3S)-	: 1805	Ethylaluminum chloride	: 3181	Ethylene glycol monoethyl ether	
2,3-Epoxypropyl ethyl ether	: 4851	Ethylaluminum sesquichloride	: 10292	acrylate	: 4842
1,2-Epoxytetradecane	: 4649	Ethyl aminoacetate hydrochloride	: 5578	Ethylene glycol mono(2-ethylhexyl)	
12,13-Epoxytrichothec-9-en-4-ol		Ethyl aminobenzoate	: 4884	ether	: 5064
acetate	: 10297	Ethyl 2-amino-3-phenylpropionate	: 8807	Ethylene glycol monohexyl ether	: 5905
EPTC	: 4552	N-Ethylamphetamine	: 5118	Ethylene glycol monomethacrylate	: 6042
Equilenin	: 6036	N-Ethylbenzenemethanamine	: 796	Ethylene glycol monomethyl ether	: 6822
Equilin	: 6037	α-Ethylbenzyl alcohol	: 4899	Ethylene glycol monomethyl ether	
Ergometrine	: 4725	Ethylbenzylaniline	: 797	acetate	: 6825
Ergostanol	: 4729	3-(α-Ethylbenzyl)-4-hydroxycoumarin	: 8794	Ethylene glycol monopropyl ether	: 9197
α-Ergostenol	: 4736	Ethyl 2-benzylidene-2-cyanoacetate	: 4960	Ethylene glycol monostearate	: 4788
γ-Ergostenol	: 4735	N-Ethyl-bis(3-phenylpropyl)amine	: 223	Ethylene glycol monosulfite	: 4789
Ergosterol	: 4731	Ethyl bromide	: 1192	Ethylene glycol monovinyl ether	: 10818
Erucamide	: 4614	Ethyl 6-bromocaproate	: 4914	Ethylene oxide	: 8447
Erucic acid	: 4616	Ethyl bromomalonate	: 3409	Ethylene sulfide	: 10041
Erythritol	: 1423	Ethyl α-bromopropionate	: 4920	Ethylene thiourea	: 6219
Erythrityl tetranitrate	: 1424	Ethyl 3-bromopyruvate	: 4917	Ethylene urea	: 6221
Esculetin	: 3712	N-Ethyl-1-butanamine	: 1565	N-Ethylethanamine	: 3378
α-Estradiol	: 4757	2-Ethyl-1-butanamine	: 4934	N-Ethylethanamine hydrochloride	: 3379
β-Estradiol	: 4758	Ethyl tert-butyl ether	: 1569	Ethyl ether	: 3454
Estradiol benzoate	: 4760	Ethyl butyl ketone	: 5690	Ethyl 2-ethylcaproate	: 5029
Estragole	: 6893	Ethyl caprate	: 4972	17β-Ethyletiocholane	: 9102
Estriol	: 4761	Ethyl carbazate	: 5066	Ethyl fluoride	: 5386
Ethacridine	: 4821	Ethyl carbonate	: 3416	Ethyl 2-furanoate	: 5037
Ethamivan	: 3468	Ethyl chloride	: 2018	Ethyl heptyl ketone	: 2764
Ethanal	: 14	Ethyl chloromalonate	: 3419	Ethyl 1-hexamethyleneiminecarbo-	
Ethanamide	: 17	Ethyl α-chloropropionate	: 4947	thiolate	: 7886
Ethanamine	: 4879	Ethyl crotonate	: 4930	2-Ethyl-1-hexanamine	: 5057
Ethanamine hydrochloride	: 4880	Ethyl cyanide	: 9147	2-Ethyl-2-hexanol	: 7300
Ethane, 1,1-dichloro-2,2-bis(p-		Ethyl 2-cyano-2-propenoate	: 4955	Ethyl hexyl ketone	: 8244
ethylphenyl)-	: 8735	N-Ethylcyclohexanamine	: 2623	3-Ethylhomocysteine, (R)	: 4809
Ethaneperoxoic acid	: 8732	Ethyl dibunate	: 4978	3-Ethylhomocysteine, (S)	: 4810
Ethanethioamide	: 10043	N-Ethyldiethanolamine	: 999	Ethyl hydrogen peroxide	: 5070
Ethanoic acid	: 21	Ethyl 2,2-dimethylpropionate	: 5006	Ethyl hydrosorbate	: 5054
Ethanolamine O-phosphate	: 9026	Ethylene bromohydrin	: 1193	Ethyl 2-hydroxypropionate	: 5096
Ethanol, 2,2'-[1,2-ethanediylbis(oxy)]-		Ethylene chlorohydrin	: 2020	5-Ethylidenebicyclo[2.2.1]hept-2-ene	: 5082
bis-, dinitrate	: 10360	Ethylenediamine	: 4770	Ethylidene diacetate	: 4774
Ethanol, germanium(4+) salt	: 9807	Ethylenediamine dihydrochloride	: 4771	Ethylidene dibromide	: 2937
Ethanol, tantalum(5+) salt	: 8577	Ethylene dibromide	: 2938	Ethylidene dichloride	: 3173
Ethanoyl bromide	: 57	Ethylene dichloride	: 3174	Ethylidene difluoride	: 3558
Ethanoyl chloride	: 58	Ethylene difluoride	: 3559	Ethyl iodide	: 6297
Ethanoyl fluoride	: 69	Ethylene dimercaptan	: 4793	Ethyl 10-(4-iodophenyl)undecanoate	: 6350

Kafocin	: 1799
α-Ketoglutaric acid	: 8461
Ketone, 3-pyridyl-3-(*N*-methyl-*N*-nitrosamino)propyl	: 8190
Kojic acid	: 6050
Kuromatsuene	: 6627
Kyanmethin	: 4307
Kynurenic acid	: 6190
Lactobionic acid	: 5515
p-Lactophenetide	: 4863
6-Lactoyl-7,8-dihydropterin	: 9528
Lanosterol	: 6584
Lauraldehyde	: 4621
Lauraldehyde, dimethyl acetal	: 3861
Lauric acid	: 4629
Lauronitrile	: 4627
Lauroyl peroxide	: 3341
Lauryl alcohol	: 4631
Lauryl amine hydrochloride	: 4642
Laurylbenzene	: 4644
Lauryl bromide	: 1190
Lauryl chloride	: 2017
Lauryl iodide	: 6296
Lauryl 2-propenoate	: 4639
Lauryl sulfate	: 4654
Lawsone	: 6116
Lead triethyl chloride	: 2322
Lenthionine	: 8634
Lepidine	: 7768
Lethane 384	: 1475
L-Leucic acid	: 6095
L-tert-Leucine	: 7852
Levopropoxyphene	: 9199
Levulinic acid	: 8469
β-Levulose	: 5471
Lidocaine	: 3382
Lignoceric acid	: 9780
Limettin	: 3854
Limonene diepoxide	: 3347
Linamarin	: 5544
Lindane	: 5730
δ-Lindane	: 5733
Linolelaidic acid	: 8300
Lithocholic acid	: 6016
Lomustine	: 2036
Lumazine	: 9291
Lumichrome	: 3969
Luminol	: 300
Lumisterol	: 4733
Lupeol	: 6636
3,5-Lupetidine	: 4268
Lupinine	: 8338
2,3-Lutidine	: 4299
2,4-Lutidine	: 4300
2,5-Lutidine	: 4301
2,6-Lutidine	: 4302
3,4-Lutidine	: 4303
3,5-Lutidine	: 4304
Lutidinic acid	: 9339
trans-Lycopene	: 1777
Lycoremine	: 5518
Lycoxanthin	: 1782
Lysidine	: 3647
Magnesium dimethyl	: 4124
Magnesium octadecanoate	: 6655
Magneson	: 8145
Maleamic acid	: 383
Maleanilic acid	: 8471
Maleic hydrazide	: 3674
Maltol	: 6106

DL-Mandelic acid	: 5973
Mandelonitrile	: 5977
Manganese, [[1,2-ethanediylbis[carba modithioato]](2-)]-	: 6673
Margaric acid	: 5650
Margaric aldehyde	: 5646
Margaryl alcohol	: 5651
Matrine	: 6682
MCPA	: 2152
Mechlorethamine	: 1949
Mechlorethamine oxide hydrochloride	: 8108
Meconic acid	: 6139
Meconin	: 3869
Mecrylate	: 7140
Me-IQ	: 303
Melamine	: 10142
Melatonin	: 6834
Meldrum's acid	: 4042
Mellitene	: 5813
Mellitic acid	: 643
Menadiol diacetate	: 7474
Menadione	: 7475
p-Menthadiene	: 4438
p-Mentha-1,8-diene, (*R*)	: 6610
p-Mentha-1,8-diene, (*S*)	: 6611
p-Mentha-1(7),2-diene	: 8741
p-Mentha-1,4(8)-diene	: 9703
p-Mentha-1,8-dien-6-one, (*R*)	: 1787
p-Mentha-1,8-dien-6-one, (*S*)	: 1788
trans-p-Menthane	: 7405
(+)-Menthol	: 7407
(-)-Menthol	: 7408
Menthol, isovalerate	: 6700
Menthone	: 7412
l-Menthone	: 7413
Mepacrine	: 9410
Meparfynol	: 7600
Mephenesin	: 7611
Mephenoxalone	: 6861
Mephentermine	: 10511
Meprobamate	: 7719
2-Mercaptobenzothiazole	: 738
2-Mercaptoethyl ether	: 3823
1-(3-Mercapto-2-methyl-1-oxypropyl)proline	: 1724
(3-Mercaptopropyl)trimethoxysilane	: 10495
6-Mercaptopurine	: 3669
Mercuric benzoate	: 6719
Mercuric oleate	: 6720
Mercuriodibenzene	: 4497
Mercury, chloro(2-methoxyethyl)-	: 10132
Mesaconic acid	: 9179
Mesalamine	: 327
Mescaline	: 10480
Mesitylamine	: 10504
Mesitylene	: 10509
Mesitylenic acid	: 3967
Mestanolone	: 6074
Mestilbol	: 3526
Mesulphen	: 4335
Metacetaldehyde (polymer)	: 6728
Metalphthalein	: 2481
Metanilic acid	: 252
Methacrolein	: 7696
Methacrolein diacetate	: 7699
Methacrylic acid anhydride	: 7701
Methacrylic acid chloride	: 7703
2-(Methacryloyloxy)ethyl acetoacetate	: 26
Methallatal	: 5110
Methallyl alcohol	: 7702

Metham sodium	: 9568
Methanal	: 5448
Methanamide	: 5450
Methanamine	: 6920
Methanamine hydrochloride	: 6921
Methanethial	: 10055
Methanoic acid	: 5453
Methaqualone	: 7460
Methenamine	: 5822
N,N'-Methenyl-*o*-phenylenediamine	: 683
Methimazole	: 3648
Methiodal sodium	: 9564
Methionamine	: 78
Methionic acid	: 6743
Methionine hydroxy analog	: 6109
Methoxyacetone	: 6889
Methoxyallene	: 6884
Methoxyamine	: 7371
Methoxyamine hydrochloride	: 7372
Methoxybenzene	: 474
2-Methoxycarbonylethyl methyl sulfide	: 7461
6'-Methoxycinchonan-9-ol, (8α,9*R*)	: 9418
6'-Methoxycinchonan-9-ol monohydrochloride, (8α,9*R*)	: 9419
trans-4-Methoxycinnamic acid	: 6882
9-Methoxycorynantheidine	: 7885
Methoxydichloromethane	: 3211
2-(2-Methoxyethoxy)ethanol	: 3450
2-Methoxyethyl 2-propenoate	: 6826
Methoxyflurane	: 3163
9-Methoxy-7*H*-furo[3,2-g][1]benzopyran-7-one	: 6767
1-(2-Methoxyisopropoxy)-2-propanol	: 4557
3-Methoxy-2-isopropyl-7*H*-furo[3,2-g][1]benzopyran-7-one	: 8737
N-Methoxymethanamine	: 6837
2-Methoxy-*N*-(2-methoxyethyl)-ethanamine	: 1020
6-Methoxy-9-methyl-1,3-dioxolo[4,5-h]quinolin-8(9*H*)-one	: 1790
1-[2-(2-Methoxy-1-methylethoxy)-1-methylethoxy]-2-propanol	: 10673
4-Methoxy-7-methyl-5*H*-furo[3,2-g][1]benzopyran-5-one	: 10832
6-Methoxy-α-methyl-2-naphthaleneacetic acid	: 7997
7-Methoxy-1-methyl-9*H*-pyrido[3,4-b]indole	: 5625
1-Methoxy-2-nitrobenzene	: 8037
1-Methoxy-3-nitrobenzene	: 8038
1-Methoxy-4-nitrobenzene	: 8039
1-Methoxyperfluoroethane	: 7549
N-(4-Methoxyphenyl)-1,4-benzenediamine	: 6876
4-Methoxy-*o*-phenylenediamine	: 6785
4-Methoxy-*m*-phenylenediamine	: 6786
3-(4-Methoxyphenyl)-2-propenoic acid, 2-ethoxyethyl ester	: 2404
1-Methoxy-2-propanamine	: 6835
2-Methoxypropane	: 6514
1-Methoxypropane	: 7716
1-Methoxy-2-propanol	: 9234
6-Methoxy-α-tetralone	: 3637
5-Methoxytryptamine	: 6833
Methsuximide	: 4257
N-Methylacetanilide	: 7614
Methylacetylene	: 9273
Methyl *o*-acetylsalicylate	: 6915
Methylal	: 3870

Trimethylene sulfone	: 10039	Untriacontane	: 5639	2,5-Xylidine	: 3929
Trimethylolmelamine	: 10692	Uramil	: 423	2,6-Xylidine	: 3930
Trimethylolpropane phosphite	: 5278	Urethane	: 4937	3,4-Xylidine	: 3931
Trimethylolpropane triacrylate	: 10667	Uridine 5'-phosphoric acid	: 10761	3,5-Xylidine	: 3932
1,1,1-Trimethylolpropane trimethacrylate	: 10699	Urochloralic acid	: 10234	Xylite	: 10861
		Ursodiol	: 3716	1-(2,4-Xylylazo)-2-naphthol	: 4247
6,6,9-Trimethyl-3-pentyl-6H-dibenzo[b,d]pyran-1-ol	: 1716	Vaccenic acid	: 8315	1-(2,5-Xylylazo)-2-naphthol	: 4248
		Valeraldehyde	: 8604	Yellow AB	: 8822
N,N,α-Trimethyl-10H-phenothiazine-10-ethanamine	: 9127	Valeric acid	: 8624	Zeaxanthin	: 1778
		(±)-γ-Valerolactone	: 3646	Zinc acetylacetonate	: 10875
N,N,β-Trimethyl-10H-phenothiazine-10-propanamine	: 10470	Valeronitrile	: 8620	Zinc, bis(dimethylcarbamodithioato-S,S')-, (T-4)-	: 10877
		Valeroyl chloride	: 8632		
N,N,2-Trimethyl-1-propanamine	: 6387	Valproic acid	: 9254	Zinc diethyl	: 3543
1,7,7-Trimethyltricyclo[2.2.1.0²,⁶]-heptane	: 10307	Vanadium(III) acetylacetonate	: 10773	Zineb	: 10873
		Vanadium hexacarbonyl	: 10772	Zingerone	: 6067
Trimethyl trimellitate	: 10512	Vanadium, oxotris(2-propanolato)-, (T-4)-	: 10467	Zoxazolamine	: 1908
2,4,6-Trimethyl-1,3,5-trioxane	: 8508			Zygosporin A	: 2726
2,4,6-Trimethyl-1,3,5-trithiane	: 10042	Vanadyl acetylacetonate	: 8458		
Trimyristin	: 5575	Vanillic acid	: 6065		
sym-Trinitrobenzene	: 10624	Vanillin	: 6060		
Trinonafluorobutylamine	: 10702	Vanilmandelic acid	: 3726		
28,29,30-Trinorlanostane	: 2365	Vellosimine	: 9512		
Triolein	: 5572	Veratraldehyde	: 3838		
3,6,9-Trioxaundecane-1,11-diol	: 9815	Veratric acid	: 3851		
Tripalmitin	: 5573	Veratrole	: 3840		
Tripelennamine	: 3973	Versen-Ol	: 10701		
Triphenylmethyl bromide	: 1373	Vicianose	: 506		
Triphenylmethyl mercaptan	: 4461	Vinylacetylene	: 1464		
4-(Triphenylmethyl)morpholine	: 10374	Vinylbenzene	: 9621		
Triphenyltin acetate	: 93	5-Vinylbicyclo[2.2.1]hept-2-ene	: 10814		
Triphenyltin chloride	: 2341	Vinyl bromide	: 1194		
Triphosgene	: 1048	Vinyl chloride	: 2022		
Tripropoxyphosphine	: 10675	Vinyl crotonate	: 10795		
N,N,N-Tripropyl-1-propanaminium bromide	: 9997	4-Vinyl-1-cyclohexene dioxide	: 4711		
		Vinylene carbonate	: 5017		
Triptane	: 10524	Vinyl fluoride	: 5388		
Triptil	: 9282	1-Vinylheptanol	: 8252		
Triptycene	: 3595	Vinylidene chloride	: 3176		
2,4,6-Tripyridyl-s-triazine	: 10703	Vinylidene fluoride	: 3560		
Triruthenium dodecacarbonyl	: 9492	Vinyl iodide	: 6299		
Tris(1-aziridinyl)phosphine, oxide	: 10362	4-Vinylphenol	: 6194		
Tris(2,3-dibromopropyl) phosphate	: 2981	Vinyl propionate	: 10819		
Tris(dimethylamino)phosphine	: 5825	Vinylsilanetriol, triacetate	: 10826		
Tris(dimethylamino)phosphine oxide	: 5824	Vinyl stearate	: 10815		
Tris(2-hydroxyethyl)amine	: 10330	Vinyl sulfide	: 4606		
Tris(hydroxymethyl)nitromethane	: 6094	Vinyl sulfone	: 4607		
Tristearin	: 5574	Vinyl trichloride	: 10229		
Trithiocyanuric acid	: 10143	Vinyltrichlorosilane	: 10296		
Tri-o-tolyl phosphate	: 10302	Violuric acid	: 9384		
Tri-m-tolyl phosphate	: 10303	Vitamin A	: 9463		
Tri-p-tolyl phosphate	: 10304	Vitamin A₁ aldehyde	: 9459		
Troclosene potassium	: 9094	Vitamin B6	: 9372		
Tropaeolin OO	: 8414	Vitamin B-12a	: 5959		
Tropane	: 6953	Vitamin Bc	: 5442		
Tropanol mandelate	: 5921	Vitamin C	: 518		
Tropic acid	: 6077	Vitamin K₂(35)	: 6696		
Tropilidene	: 2558	VX Nerve agent	: 4992		
Tropital	: 9080	Widdrene	: 10092		
Tryptamine hydrochloride	: 6250	Woodward's Reagent K	: 5251		
Tryptophol	: 6251	Xanthogenic acid	: 5014		
Tuaminoheptane	: 5670	Xanthophyll	: 1779		
Tungsten hexacarbonyl	: 10718	Xanthoxylin	: 3867		
Turmeric	: 2495	Xanthurenic acid	: 3754		
Tyramine	: 315	Xibornol	: 6369		
Umbelliferone	: 5999	m-Xylene diamine	: 629		
1-Undecanamine	: 10748	2,4-Xylenol, phosphate (3:1)	: 10684		
Undecyl alcohol	: 10732	2,5-Xylenol, phosphate (3:1)	: 10685		
sec-Undecyl alcohol	: 10733	2,6-Xylenol, phosphate (3:1)	: 10686		
Undecylenic acid	: 10745	2,3-Xylidine	: 3927		
Undecyl mercaptan	: 10730	2,4-Xylidine	: 3928		

DIAMAGNETIC SUSCEPTIBILITY OF SELECTED ORGANIC COMPOUNDS

When a material is placed in a magnetic field H, a magnetization M is induced in the material which is related to H by $M = \kappa H$, where κ is called the volume susceptibility. Since H and M have the same dimensions, κ is dimensionless. A more useful parameter is the molar susceptibility χ_m, defined by

$$\chi_m = \kappa V_m = \kappa M/\rho$$

where V_m is the molar volume of the substance, M the molar mass, and ρ the mass density. When the cgs system is used, the customary unit for χ_m is cm³ mol⁻¹; the corresponding SI unit is m³ mol⁻¹. Substances with no unpaired electrons are called diamagnetic; they have negative values of χ_m.

This table gives values of the diamagnetic susceptibility for about 400 common organic compounds. All values refer to room temperature and atmospheric pressure and to the physical form that is stable under these conditions. Substances are arranged by molecular formula in Hill order. A more extensive table may be found in Reference 1.

In keeping with customary practice, the molar susceptibility is given here in units appropriate to the cgs system. These values should be multiplied by 4π to obtain values for use in SI equations (where the magnetic field strength H has units of A m⁻¹).

References

1. *Landolt-Börnstein, Numerical Data and Functional Relationships in Science and Technology, New Series,* II/16, *Diamagnetic Susceptibility,* Gupta, R. R., Ed., Springer-Verlag, Heidelberg, 1986.
2. Barter, C., Meisenheimer, R. G., and Stevenson, D. P., *J. Phys. Chem.* 64, 1312, 1960.
3. Broersma, S., *J. Chem. Phys.* 17, 873, 1949.

Molecular formula	Compound	$-\chi_m/10^{-6}$ cm³ mol⁻¹
CBrCl₃	Bromotrichloromethane	73.2
CBr₄	Tetrabromomethane	93.7
CClF₃	Chlorotrifluoromethane	45.3
CClN	Cyanogen chloride	32.4
CCl₂F₂	Dichlorodifluoromethane	52.2
CCl₂O	Carbonyl chloride	47.9
CCl₃F	Trichlorofluoromethane	58.7
CCl₃NO₂	Trichloronitromethane	75.3
CCl₄	Tetrachloromethane	66.8
CHBrCl₂	Bromodichloromethane	66.3
CHBr₃	Tribromomethane	82.6
CHCl₃	Trichloromethane	58.9
CHI₃	Triiodomethane	117.1
CH₂BrCl	Bromochloromethane	55.1
CH₂Br₂	Dibromomethane	65.1
CH₂Cl₂	Dichloromethane	46.6
CH₂I₂	Diiodomethane	93.1
CH₂N₂	Cyanamide	24.8
CH₂O	Formaldehyde	18.6
CH₂O₂	Formic acid	19.9
CH₃Br	Bromomethane	42.8
CH₃Cl	Chloromethane	32.0
CH₃F	Fluoromethane	17.8
CH₃I	Iodomethane	57.2
CH₃NO	Formamide	23.0
CH₃NO₂	Nitromethane	21.0

Molecular formula	Compound	$-\chi_m/10^{-6}$ cm³ mol⁻¹
CH₄	Methane	17.4
CH₄N₂O	Urea	33.5
CH₄O	Methanol	21.4
CH₅N	Methylamine	27.0
CI₄	Tetraiodomethane	136
CN₄O₈	Tetranitromethane	43.0
C₂ClF₃	Chlorotrifluoroethylene	49.1
C₂Cl₄	Tetrachloroethylene	81.6
C₂Cl₆	Hexachloroethane	112.8
C₂HCl₃	Trichloroethylene	65.8
C₂HCl₃O	Trichloroacetaldehyde	73.0
C₂HCl₃O	Dichloroacetyl chloride	69.0
C₂HCl₃O₂	Trichloroacetic acid	73.0
C₂HCl₅	Pentachloroethane	99.1
C₂HF₃O₂	Trifluoroacetic acid	43.3
C₂H₂	Acetylene	20.8
C₂H₂Br₄	1,1,2,2-Tetrabromoethane	123.4
C₂H₂Cl₂	1,1-Dichloroethylene	49.2
C₂H₂Cl₂	*cis*-1,2-Dichloroethylene	51.0
C₂H₂Cl₂	*trans*-1,2-Dichloroethylene	48.9
C₂H₂Cl₄	1,1,2,2-Tetrachloroethane	89.8
C₂H₃Cl	Chloroethylene	35.9
C₂H₃ClO	Acetyl chloride	39.3
C₂H₃N	Acetonitrile	27.8
C₂H₄	Ethylene	18.8
C₂H₄Br₂	1,2-Dibromoethane	78.9
C₂H₄Cl₂	1,1-Dichloroethane	57.4
C₂H₄Cl₂	1,2-Dichloroethane	59.6
C₂H₄O	Acetaldehyde	22.2
C₂H₄O	Ethylene oxide	30.5
C₂H₄O₂	Acetic acid	31.8
C₂H₄O₂	Methyl formate	31.1
C₂H₅Br	Bromoethane	78.8
C₂H₅Cl	Chloroethane	69.9
C₂H₅I	Iodoethane	69.1
C₂H₅NO	Acetamide	33.9
C₂H₅NO₂	Nitroethane	35.4
C₂H₅NO₂	Glycine	39.6
C₂H₆	Ethane	26.8
C₂H₆O	Ethanol	33.7
C₂H₆O	Dimethyl ether	26.3
C₂H₆O₂	Ethylene glycol	38.9
C₂H₆S	Ethanethiol	47.0
C₂H₆S	Dimethyl sulfide	44.9
C₂H₈N₂	1,2-Ethanediamine	46.5
C₂N₂	Cyanogen	21.6
C₃H₄	Allene	25.3
C₃H₄O₂	Vinyl formate	34.7
C₃H₅Br	3-Bromopropene	58.6
C₃H₅Cl	2-Chloropropene	47.8
C₃H₅Cl	3-Chloropropene	47.8
C₃H₅N	Propanenitrile	38.6
C₃H₆	Propene	30.7
C₃H₆	Cyclopropane	39.2
C₃H₆O	Allyl alcohol	36.7
C₃H₆O	Propanal	34.2
C₃H₆O	Acetone	33.8

Molecular formula	Compound	$-\chi_m/10^{-6}$ cm^3 mol^{-1}	Molecular formula	Compound	$-\chi_m/10^{-6}$ cm^3 mol^{-1}
C_3H_6O	Methyloxirane	42.5	$C_4H_{10}O$	2-Methyl-2-propanol	56.6
$C_3H_6O_2$	Propanoic acid	43.2	$C_4H_{10}O$	Diethyl ether	55.5
$C_3H_6O_2$	Ethyl formate	42.4	$C_4H_{10}O_2$	1,3-Butanediol	61.8
C_3H_7Br	1-Bromopropane	65.6	$C_4H_{10}O_2$	1,4-Butanediol	61.8
C_3H_7Br	2-Bromopropane	65.1	$C_4H_{10}S$	1-Butanethiol	70.2
C_3H_7Cl	1-Chloropropane	56.0	$C_4H_{11}N$	Butylamine	58.9
C_3H_7I	1-Iodopropane	84.3	$C_4H_{11}N$	Isobutylamine	59.8
C_3H_7N	Allylamine	40.1	$C_4H_{11}N$	Diethylamine	56.8
$C_3H_7NO_2$	1-Nitropropane	45.0	$C_5H_4O_2$	Furfural	47.2
$C_3H_7NO_2$	2-Nitropropane	45.4	C_5H_5N	Pyridine	48.7
$C_3H_7NO_2$	Ethyl carbamate	57.0	$C_5H_6O_2$	Furfuryl alcohol	61.0
C_3H_8	Propane	38.6	$C_5H_7NO_2$	Ethyl cyanoacetate	67.3
C_3H_8O	1-Propanol	44.8	C_5H_8	2-Methyl-1,3-butadiene	46.0
C_3H_8O	2-Propanol	45.7	C_5H_8O	Cyclopentanone	51.6
$C_3H_8O_2$	1,3-Propylene glycol	50.2	$C_5H_8O_2$	Methyl methacrylate	57.3
$C_3H_8O_2$	Dimethoxymethane	47.3	$C_5H_8O_2$	2,4-Pentanedione	54.9
$C_3H_8O_3$	Glycerol	57.1	C_5H_{10}	1-Pentene	54.6
$C_4H_2O_3$	Maleic anhydride	35.8	C_5H_{10}	2-Methyl-2-butene	54.7
$C_4H_4N_2$	Pyrazine	37.8	C_5H_{10}	Cyclopentane	56.2
$C_4H_4N_2$	Pyrimidine	43.1	$C_5H_{10}O$	Cyclopentanol	64.0
C_4H_4O	Furan	43.1	$C_5H_{10}O$	Pentanal	57.5
$C_4H_4O_3$	Succinic anhydride	47.5	$C_5H_{10}O$	2-Pentanone	57.5
$C_4H_4O_4$	Maleic acid	49.6	$C_5H_{10}O$	3-Pentanone	57.7
$C_4H_4O_4$	Fumaric acid	49.1	$C_5H_{10}O_2$	Pentanoic acid	66.5
C_4H_4S	Thiophene	57.3	$C_5H_{10}O_2$	3-Methylbutanoic acid	67.7
C_4H_5N	Pyrrole	48.6	$C_5H_{10}O_2$	Butyl formate	65.8
C_4H_6	1,2-Butadiene	35.6	$C_5H_{10}O_2$	Isobutyl formate	66.8
C_4H_6	1,3-Butadiene	32.1	$C_5H_{10}O_2$	Propyl acetate	65.9
$C_4H_6O_2$	Vinyl acetate	46.4	$C_5H_{10}O_2$	Isopropyl acetate	67.0
$C_4H_6O_3$	Acetic anhydride	52.8	$C_5H_{10}O_2$	Ethyl propanoate	66.3
$C_4H_6O_4$	Succinic acid	58.0	$C_5H_{10}O_2$	Tetrahydrofurfuryl alcohol	69.4
$C_4H_6O_4$	Dimethyl oxalate	55.7	$C_5H_{10}O_3$	Diethyl carbonate	75.4
C_4H_7N	Butanenitrile	50.4	$C_5H_{11}N$	Piperidine	64.2
C_4H_8	1-Butene	41.0	C_5H_{12}	Pentane	61.5
C_4H_8	cis-2-Butene	42.6	C_5H_{12}	Isopentane	63.0
C_4H_8	trans-2-Butene	43.3	C_5H_{12}	Neopentane	63.0
C_4H_8	Isobutene	40.8	$C_5H_{12}O$	1-Pentanol	67.0
C_4H_8	Cyclobutane	40.0	$C_5H_{12}O$	2-Pentanol	69.1
C_4H_8O	Ethyl vinyl ether	47.9	$C_5H_{12}O_2$	1,5-Pentanediol	73.5
C_4H_8O	1,2-Epoxybutane	54.8	$C_5H_{13}N$	Pentylamine	69.3
C_4H_8O	Butanal	45.9	C_6Cl_6	Hexachlorobenzene	147.0
C_4H_8O	2-Butanone	45.6	$C_6H_4ClNO_2$	1-Chloro-2-nitrobenzene	75.5
$C_4H_8O_2$	Butanoic acid	55.2	$C_6H_4ClNO_2$	1-Chloro-3-nitrobenzene	77.2
$C_4H_8O_2$	2-Methylpropanoic acid	56.1	$C_6H_4ClNO_2$	1-Chloro-4-nitrobenzene	74.7
$C_4H_8O_2$	Propyl formate	55.0	$C_6H_4Cl_2$	o-Dichlorobenzene	84.4
$C_4H_8O_2$	Ethyl acetate	54.1	$C_6H_4Cl_2$	m-Dichlorobenzene	84.1
$C_4H_8O_2$	Methyl propanoate	54.5	$C_6H_4Cl_2$	p-Dichlorobenzene	81.7
$C_4H_8O_2$	1,4-Dioxane	52.2	$C_6H_4O_2$	p-Benzoquinone	36
C_4H_9Br	1-Bromobutane	77.1	C_6H_5Br	Bromobenzene	78.4
C_4H_9Br	1-Bromo-2-methylpropane	79.9	C_6H_5Cl	Chlorobenzene	69.5
C_4H_9Cl	1-Chlorobutane	67.1	C_6H_5ClO	o-Chlorophenol	77.3
C_4H_9Cl	2-Chlorobutane	67.4	C_6H_5ClO	m-Chlorophenol	77.6
C_4H_9I	1-Iodobutane	93.6	C_6H_5ClO	p-Chlorophenol	77.7
C_4H_9N	Pyrrolidine	54.8	C_6H_5F	Fluorobenzene	58.4
C_4H_9NO	Morpholine	55.0	C_6H_5I	Iodobenzene	92.0
C_4H_{10}	Butane	50.3	$C_6H_5NO_2$	Nitrobenzene	61.9
C_4H_{10}	Isobutane	50.5	$C_6H_5NO_3$	o-Nitrophenol	68.9
$C_4H_{10}O$	1-Butanol	56.4	$C_6H_5NO_3$	m-Nitrophenol	65.9
$C_4H_{10}O$	2-Butanol	57.6	$C_6H_5NO_3$	p-Nitrophenol	66.9
$C_4H_{10}O$	2-Methyl-1-propanol	57.6	C_6H_6	Benzene	54.8

Molecular formula	Compound	$-\chi_m/10^{-6}$ cm^3 mol^{-1}	Molecular formula	Compound	$-\chi_m/10^{-6}$ cm^3 mol^{-1}
C_6H_6ClN	o-Chloroaniline	79.5	$C_7H_7NO_2$	p-Nitrotoluene	73.3
C_6H_6ClN	m-Chloroaniline	76.6	C_7H_8	Toluene	65.6
C_6H_6ClN	p-Chloroaniline	76.7	C_7H_8O	o-Cresol	73.3
$C_6H_6N_2O_2$	o-Nitroaniline	67.4	C_7H_8O	m-Cresol	72.2
$C_6H_6N_2O_2$	m-Nitroaniline	69.7	C_7H_8O	p-Cresol	72.4
$C_6H_6N_2O_2$	p-Nitroaniline	68.0	C_7H_8O	Benzyl alcohol	71.8
C_6H_6O	Phenol	60.6	C_7H_8O	Anisole	72.2
$C_6H_6O_2$	p-Hydroquinone	64.7	C_7H_9N	o-Methylaniline	74.9
$C_6H_6O_2$	Pyrocatechol	68.2	C_7H_9N	m-Methylaniline	74.6
$C_6H_6O_2$	Resorcinol	67.2	C_7H_9N	p-Methylaniline	72.5
C_6H_7N	Aniline	62.4	C_7H_9N	N-Methylaniline	74.1
C_6H_7N	4-Methylpyridine	59.8	C_7H_9N	2,4-Dimethylpyridine	71.3
C_6H_8	1,4-Cyclohexadiene	48.7	C_7H_9N	2,6-Dimethylpyridine	72.5
$C_6H_8N_2$	o-Phenylenediamine	72.5	C_7H_9NO	o-Methoxyaniline [o-Anisidine]	79.1
$C_6H_8N_2$	m-Phenylenediamine	70.4			
$C_6H_8N_2$	p-Phenylenediamine	70.7	$C_7H_{12}O_4$	Diethyl malonate	92.6
C_6H_{10}	1,5-Hexadiene	55.1	C_7H_{14}	1-Heptene	77.8
C_6H_{10}	1-Hexyne	64.5	C_7H_{14}	Cycloheptane	73.9
C_6H_{10}	Cyclohexene	58.0	C_7H_{14}	Methylcyclohexane	78.9
$C_6H_{10}O$	Cyclohexanone	62.0	$C_7H_{14}O$	1-Heptanal	81.0
$C_6H_{10}O_3$	Ethyl acetoacetate	71.7	$C_7H_{14}O$	2-Heptanone	80.5
$C_6H_{10}O_4$	Diethyl oxalate	81.7	$C_7H_{14}O$	3-Heptanone	80.7
C_6H_{12}	1-Hexene	66.4	$C_7H_{14}O$	4-Heptanone	80.5
C_6H_{12}	2,3-Dimethyl-2-butene	65.9	$C_7H_{14}O$	2,4-Dimethyl-3-pentanone	81.1
C_6H_{12}	Cyclohexane	68	$C_7H_{14}O_2$	Heptanoic acid	89.0
C_6H_{12}	Methylcyclopentane	70.2	$C_7H_{14}O_2$	Pentyl acetate	88.9
$C_6H_{12}O$	Hexanal	69.4	$C_7H_{14}O_2$	Isopentyl acetate	89.4
$C_6H_{12}O$	2-Hexanone	69.2	$C_7H_{14}O_2$	Butyl propanoate	89.1
$C_6H_{12}O$	3-Hexanone	69.0	$C_7H_{14}O_2$	Ethyl 3-methylbutanoate	91.1
$C_6H_{12}O$	4-Methyl-2-pentanone	69.7	C_7H_{16}	Heptane	85.2
$C_6H_{12}O$	Cyclohexanol	73.4	C_7H_{16}	3-Ethylpentane	86.2
$C_6H_{12}O_2$	Hexanoic acid	78.1	C_7H_{16}	2,2-Dimethylpentane	87.0
$C_6H_{12}O_2$	Isopentyl formate	78.4	C_7H_{16}	2,3-Dimethylpentane	87.5
$C_6H_{12}O_2$	Isobutyl acetate	78.7	C_7H_{16}	2,4-Dimethylpentane	87.5
$C_6H_{12}O_2$	Propyl propanoate	77.7	C_7H_{16}	3,3-Dimethylpentane	89.5
$C_6H_{12}O_3$	Paraldehyde	86.1	$C_7H_{16}O$	1-Heptanol	91.7
C_6H_{14}	Hexane	74.1	$C_7H_{16}O$	4-Heptanol	92.1
C_6H_{14}	2-Methylpentane	75.3	$C_8H_4O_3$	Phthalic anhydride	66.7
C_6H_{14}	3-Methylpentane	75.5	$C_8H_6O_4$	Phthalic acid	83.6
C_6H_{14}	2,2-Dimethylbutane	76.2	$C_8H_6O_4$	Isophthalic acid	84.6
C_6H_{14}	2,3-Dimethylbutane	76.2	$C_8H_6O_4$	Terephthalic acid	83.5
$C_6H_{14}O$	1-Hexanol	79.5	C_8H_7N	Benzeneacetonitrile	76.9
$C_6H_{14}O$	4-Methyl-2-pentanol	80.4	C_8H_7N	Indole	85.0
$C_6H_{14}O$	Dipropyl ether	79.4	C_8H_8	Styrene	68.2
$C_6H_{14}O_2$	1,6-Hexanediol	84.3	C_8H_8O	Acetophenone	72.5
$C_6H_{14}O_2$	1,1-Diethoxyethane	81.4	$C_8H_8O_2$	o-Toluic acid	84.3
$C_6H_{14}O_6$	D-Glucitol	107.8	$C_8H_8O_2$	m-Toluic acid	83.0
$C_6H_{15}N$	Triethylamine	83.3	$C_8H_8O_2$	p-Toluic acid	82.4
C_7H_5N	Benzonitrile	65.2	$C_8H_8O_2$	Benzeneacetic acid	82.4
C_7H_6O	Benzaldehyde	60.7	$C_8H_8O_2$	Methyl benzoate	81.6
$C_7H_6O_2$	Salicylaldehyde	66.8	$C_8H_8O_3$	Methyl salicylate	86.6
$C_7H_6O_3$	Salicylic acid	75	C_8H_{10}	Ethylbenzene	77.3
C_7H_7Br	p-Bromotoluene	88.7	C_8H_{10}	o-Xylene	77.7
C_7H_7Cl	o-Chlorotoluene	82.4	C_8H_{10}	m-Xylene	76.4
C_7H_7Cl	m-Chlorotoluene	79.7	C_8H_{10}	p-Xylene	77.0
C_7H_7Cl	p-Chlorotoluene	80.3	$C_8H_{10}O$	Phenetole	84.5
C_7H_7Cl	(Chloromethyl)benzene	81.6	$C_8H_{11}N$	N-Ethylaniline	85.6
C_7H_7NO	Benzamide	72.0	$C_8H_{11}N$	N,N-Dimethylaniline	85.1
$C_7H_7NO_2$	o-Nitrotoluene	72.2	$C_8H_{11}N$	2,4,6-Trimethylpyridine	83.1
$C_7H_7NO_2$	m-Nitrotoluene	72.7	$C_8H_{14}O_4$	Ethyl succinate	105.0
			C_8H_{16}	1-Octene	88.8

Molecular formula	Compound	$-\chi_m/10^{-6}\ cm^3\ mol^{-1}$	Molecular formula	Compound	$-\chi_m/10^{-6}\ cm^3\ mol^{-1}$
C_8H_{16}	Cyclooctane	85.3	$C_{12}H_{24}O_2$	Dodecanoic acid	113.0
$C_8H_{16}O_2$	Octanoic acid	99.5	$C_{13}H_9N$	Acridine	118.8
$C_8H_{16}O_2$	Hexyl acetate	100.9	$C_{13}H_{10}O$	Benzophenone	109.6
$C_8H_{17}Cl$	1-Chlorooctane	114.9	$C_{13}H_{12}$	Diphenylmethane	116.0
C_8H_{18}	Octane	96.6	$C_{13}H_{28}$	Tridecane	153.7
C_8H_{18}	4-Methylheptane	97.3	$C_{14}H_8O_2$	9,10-Anthracenedione	113.0
C_8H_{18}	3-Ethylhexane	97.8	$C_{14}H_{10}$	Anthracene	129.8
C_8H_{18}	3,4-Dimethylhexane	99.1	$C_{14}H_{10}$	Phenanthrene	127.6
C_8H_{18}	2,2,4-Trimethylpentane	99.1	$C_{14}H_{10}$	Diphenylacetylene	116
C_8H_{18}	2,3,4-Trimethylpentane	99.8	$C_{14}H_{10}O_2$	Benzil	106.8
$C_8H_{18}O$	1-Octanol	101.6	$C_{14}H_{12}O_2$	Benzyl benzoate	132.2
$C_8H_{19}N$	Dibutylamine	103.7	$C_{14}H_{14}$	1,2-Diphenylethane	127.8
C_9H_7N	Quinoline	86.1	$C_{14}H_{28}O_2$	Tetradecanoic acid [Myristic acid]	176.0
C_9H_7N	Isoquinoline	83.9			
C_9H_8	Indene	83	$C_{14}H_{30}$	Tetradecane	166.2
C_9H_{10}	Isopropenylbenzene	80.0	$C_{16}H_{10}$	Pyrene	147
$C_9H_{10}O_2$	Ethyl benzoate	93.8	$C_{16}H_{32}O_2$	Hexadecanoic acid [Palmitic acid]	198.6
$C_9H_{10}O_2$	Benzyl acetate	93.2			
C_9H_{12}	Propylbenzene	89.1	$C_{16}H_{34}$	Hexadecane	187.6
C_9H_{12}	Isopropylbenzene [Cumene]	89.5	$C_{16}H_{34}O$	1-Hexadecanol	183.5
			$C_{18}H_{12}$	Chrysene	148.0
C_9H_{12}	1,3,5-Trimethylbenzene [Mesitylene]	92.3	$C_{18}H_{14}$	o-Terphenyl	150.4
			$C_{18}H_{14}$	m-Terphenyl	155.5
C_9H_{18}	1-Nonene	100.1	$C_{18}H_{14}$	p-Terphenyl	156.0
$C_9H_{18}O$	2,6-Dimethyl-4-heptanone	104.3	$C_{18}H_{34}O_2$	cis-9-Octadecenoic acid [Oleic acid]	208.5
C_9H_{20}	Nonane	108.1			
$C_{10}H_7Br$	1-Bromonaphthalene	123.6	$C_{18}H_{36}O_2$	Octadecanoic acid [Stearic acid]	220.8
$C_{10}H_7Cl$	1-Chloronaphthalene	107.6			
$C_{10}H_8$	Naphthalene	91.6	$C_{20}H_{12}$	Perylene	167.5
$C_{10}H_8$	Azulene	123.7			
$C_{10}H_8O$	1-Naphthol	96.2			
$C_{10}H_8O$	2-Naphthol	96.8			
$C_{10}H_9N$	1-Naphthalenamine	92.5			
$C_{10}H_9N$	2-Naphthalenamine	98.0			
$C_{10}H_{10}O_2$	Safrole	97.5			
$C_{10}H_{10}O_4$	Dimethyl terephthalate	101.6			
$C_{10}H_{14}$	Butylbenzene	100.7			
$C_{10}H_{14}$	tert-Butylbenzene	101.8			
$C_{10}H_{14}$	Isobutylbenzene	101.7			
$C_{10}H_{14}$	p-Cymene	102.8			
$C_{10}H_{14}$	1,2,4,5-Tetramethylbenzene	101.2			
$C_{10}H_{14}O$	p-tert-Butylphenol	108.0			
$C_{10}H_{15}N$	N,N-Diethylaniline	107.9			
$C_{10}H_{16}$	d-Limonene	98.0			
$C_{10}H_{16}$	α-Pinene	100.7			
$C_{10}H_{16}$	β-Pinene	101.9			
$C_{10}H_{16}O$	Camphor, (+)	103.0			
$C_{10}H_{18}$	cis-Decahydronaphthalene	107.0			
$C_{10}H_{18}$	trans-Decahydronaphthalene	107.6			
$C_{10}H_{22}$	Decane	119.5			
$C_{11}H_{10}$	1-Methylnaphthalene	102.9			
$C_{11}H_{10}$	2-Methylnaphthalene	102.7			
$C_{11}H_{24}$	Undecane	131.8			
$C_{12}H_8$	Acenaphthylene	111.6			
$C_{12}H_9N$	Carbazole	119.9			
$C_{12}H_{10}$	Acenaphthene	109.9			
$C_{12}H_{10}$	Biphenyl	103.3			
$C_{12}H_{10}N_2$	Azobenzene	106.8			
$C_{12}H_{11}N$	Diphenylamine	108.4			
$C_{12}H_{14}O_4$	Diethyl phthalate	127.5			
$C_{12}H_{18}$	Hexamethylbenzene	122.5			

Section 4
Properties of the Elements and Inorganic Compounds

THE ELEMENTS

C. R. Hammond

One of the most striking facts about the elements is their unequal distribution and occurrence in nature. Present knowledge of the chemical composition of the universe, obtained from the study of the spectra of stars and nebulae, indicates that hydrogen is by far the most abundant element and may account for more than 90% of the atoms or about 75% of the mass of the universe. Helium atoms make up most of the remainder. All of the other elements together contribute only slightly to the total mass.

The chemical composition of the universe is undergoing continuous change. Hydrogen is being converted into helium, and helium is being changed into heavier elements. As time goes on, the ratio of heavier elements increases relative to hydrogen. Presumably, the process is not reversible.

Burbidge, Burbidge, Fowler, and Hoyle, and more recently, Peebles, Penzias, and others have studied the synthesis of elements in stars. To explain all of the features of the nuclear abundance curve — obtained by studies of the composition of the Earth, meteorites, stars, etc. — it is necessary to postulate that the elements were originally formed by at least eight different processes: (1) hydrogen burning, (2) helium burning, (3) χ process, (4) e process, (5) s process, (6) r process, (7) p process, and (8) the X process. The X process is thought to account for the existence of light nuclei such as D, Li, Be, and B. Common metals such as Fe, Cr, Ni, Cu, Ti, Zn, etc. were likely produced early in the history of our galaxy. It is also probable that most of the heavy elements on Earth and elsewhere in the universe were originally formed in supernovae, or in the hot interior of stars.

Studies of the solar spectrum have led to the identification of 67 elements in the sun's atmosphere; however, all elements cannot be identified with the same degree of certainty. Other elements may be present in the sun, although they have not yet been detected spectroscopically. The element helium was discovered on the sun before it was found on Earth. Some elements such as scandium are relatively more plentiful in the sun and stars than here on Earth.

Minerals in lunar rocks brought back from the moon on the Apollo missions consist predominantly of *plagioclase* $\{(Ca,Na)(Al,Si)O_4O_8\}$ and *pyroxene* $\{(Ca,Mg,Fe)_2Si_2O_6\}$ — two minerals common in terrestrial volcanic rock. No new elements have been found on the moon that cannot be accounted for on Earth; however, three minerals, *armalcolite* $\{(Fe,Mg)Ti_2O_5\}$, *pyroxferroite* $\{CaFe_6(SiO_3)_7\}$, and *tranquillityite* $\{Fe_8(Zr,Y)Ti_3Si_3O_2\}$, are new. The oldest known terrestrial rocks are about 4 billion years old. One rock, known as the "Genesis Rock," brought back from the Apollo 15 Mission, is about 4.15 billion years old. This is only about one-half billion years younger than the supposed age of the moon and solar system. Lunar rocks appear to be relatively enriched in refractory elements such as chromium, titanium, zirconium, and the rare earths, and impoverished in volatile elements such as the alkali metals, in chlorine, and in noble metals such as nickel, platinum, and gold.

Even older than the "Genesis Rock" are *carbonaceous chondrites,* a type of meteorite that has fallen to Earth and has been studied. These are some of the most primitive objects of the solar system yet found. The grains making up these objects probably condensed directly out the gaseous nebula from which the sun and planets were born. Most of the condensation of the grains probably was completed within 50,000 years of the time the disk of the nebula was first formed — about 4.6 billion years ago. It is now thought that this type of meteorite may contain a small percentage of presolar dust grains. The relative abundances of the elements of these meteorites are about the same as the abundances found in the solar chromosphere.

The X-ray fluorescent spectrometer sent with the Viking I spacecraft to Mars shows that the Martian soil contains about 12 to 16% iron, 14 to 15% silicon, 3 to 8% calcium, 2 to 7% aluminum, and one-half to 2% titanium. The gas chromatograph — mass spectrometer on Viking II found no trace of organic compounds.

F. W. Clarke and others have carefully studied the composition of rocks making up the crust of the Earth. Oxygen accounts for about 47% of the crust, by weight, while silicon comprises about 28% and aluminum about 8%. These elements, plus iron, calcium, sodium, potassium, and magnesium, account for about 99% of the composition of the crust.

Many elements such as tin, copper, zinc, lead, mercury, silver, platinum, antimony, arsenic, and gold, which are so essential to our needs and civilization, are among some of the rarest elements in the Earth's crust. These are made available to us only by the processes of concentration in ore bodies. Some of the so-called *rare-earth* elements have been found to be much more plentiful than originally thought and are about as abundant as uranium, mercury, lead, or bismuth. The least abundant rare-earth or *lanthanide* element, thulium, is now believed to be more plentiful on Earth than silver, cadmium, gold, or iodine, for example. Rubidium, the 16th most abundant element, is more plentiful than chlorine while its compounds are little known in chemistry and commerce.

It is now thought that at least 24 elements are essential to living matter. The four most abundant in the human body are hydrogen, oxygen, carbon, and nitrogen. The seven next most common, in order of abundance, are calcium, phosphorus, chlorine, potassium, sulfur, sodium, and magnesium. Iron, copper, zinc, silicon, iodine, cobalt, manganese, molybdenum, fluorine, tin, chromium, selenium, and vanadium are needed and play a role in living matter. Boron is also thought essential for some plants, and it is possible that aluminum, nickel, and germanium may turn out to be necessary.

Ninety-one elements occur naturally on Earth. Minute traces of plutonium-244 have been discovered in rocks mined in southern California. This discovery supports the theory that heavy elements were produced during creation of the solar system. While technetium and promethium have not yet been found naturally on Earth, they have been found to be present in stars. Technetium has been identified in the spectra of certain "late" type stars, and promethium lines have been identified in the spectra of a faintly visible star HR465 in Andromeda. Promethium must have been made near the star's surface for no known isotope of this element has a half-life longer than 17.7 years.

It has been suggested that californium is present in certain stellar explosions known as supernovae; however, this has not been proved. At present no elements are found elsewhere in the universe that cannot be accounted for here on Earth.

All atomic mass numbers from 1 to 238 are found naturally on Earth except for masses 5 and 8. About 285 relatively stable and 67 naturally radioactive isotopes occur on Earth totaling 352. In addition, the neutron, technetium, promethium, and the transuranic elements (lying beyond uranium) have now been produced artificially. In June 1999, scientists at the Lawrence Berkeley National Laboratory reported that they had found evidence of an isotope of Element 118 and its immediate decay products of Elements 116, 114, and 112. This sequence of events tended to reinforce the theory that was predicted since the 1970s that an "island of stability" existed for nuclei with approximately 114 protons and 184 neutrons. This "island" refers to nuclei in which the decay lasts for a period of time instead of a decay that occurs instantaneously. However, on July 27, 2001, researchers at LBNL reported that their laboratory and the facilities at the GSI Laboratory in Germany and at Japanese laboratories failed to confirm the results of their earlier experiments where the fusion of a krypton atom with a lead target resulted in Element 118, with chains of decay leading to Elements 116, 114, and 112, and on down to Element 106. Therefore, the discovery was reported to be spurious. However, with the announcement it was said that different experiments at the Livermore Laboratory and Joint Institute for Nuclear Research in Dubna, Russia indicated that Element 116 had since been created directly. (See also under Elements 116 and 118.)

Laboratory processes have now extended the radioactive element mass numbers beyond 238 to about 280. Each element from atomic numbers 1 to 110 is known to have at least one radioactive isotope. As of December 2001, about 3286 isotopes and isomers were thought to be known and recognized. Many stable and radioactive isotopes are now produced and distributed by the Oak Ridge National Laboratory, Oak Ridge, TN, U.S.A., to customers licensed by the U.S. Department of Energy.

The nucleus of an atom is characterized by the number of protons it contains, denoted by Z, and by the number of neutrons, N. Isotopes of an element have the same value of Z, but different values of N. The *mass number A*, is the sum of Z and N. For example, Uranium-238 has a mass number of 238, and contains 92 protons and 146 neutrons.

There is evidence that the definition of chemical elements must be broadened to include the electron. Several compounds known as *electrides* have recently been made of alkaline metal elements and electrons. A relatively stable combination of a positron and electron, known as *positronium*, has also been studied.

The well-known proton, neutron, and electron are now thought to be members of a group that includes other fundamental particles that have been discovered or hypothesized by physicists. These very elemental particles, of which all matter is made, are now thought to belong to one of two families: namely, **quarks** or **leptons**. Each of these two families consists of six particles. Also, there are four different force carriers that lead to interactions between particles. The six members or "flavors" of the quark family are called **up**, **charm**, **top**, **down**, **strange**, and **bottom**. The force carriers for the quarks are the **gluon** and the **photon**. The six members of the lepton family are the **e neutrino**, the **mu neutrino**, the **tau neutrino**, the **electron**, the **muon particle**, and the **tau particle**. The force carriers for these are the **w boson** and the **z boson**. Furthermore, it appears that each of these particles has an anti-particle that has an opposite electrical charge from the above particles.

Quarks are not found individually, but are found with other quarks arranged to form composites known as **hadrons**. There are two basic types of hadrons: **baryons**, composed of three quarks, and **mesons**, composed of a quark and an anti-quark. Examples of baryons are the neutron and the proton. Neutrons are made of two down quarks and one up quark. Protons are made of two up quarks and one down quark. An example of the meson is the **pion**. This particle is made of an up quark and a down anti-quark. Such particles are unstable and tend to decay rapidly. The anti-particle of the proton is the anti-proton. The exception to the rule is the electron, whose anti-particle is the **positron**.

In recent years a search has been made for a hypothetical particle known as the **Higgs particle** or **Higgs boson**, suggested in 1966 by Peter Higgs of the University of Edinburgh, which could possibly explain why the carriers of the "electro-weak" field (w and z bosons) have mass. The Higgs particle is thought to be responsible possibly for the mass of objects throughout the universe.

Many physicists now hold that all matter and energy in the universe are controlled by four fundamental forces: the **electromagnetic force**, **gravity**, a **weak nuclear force**, and a **strong nuclear force**. The **gluon** binds quarks together by carrying the strong nuclear force. Each of these natural forces is passed back and forth among the basic particles of matter by the force carriers mentioned above. The electromagnetic force is carried by the photon, the weak nuclear force by the **intermediate vector boson**, and the gravity by the **graviton**.

For more complete information on these fundamental particles, please consult recent articles and books on nuclear or particle physics.

The available evidence leads to the conclusion that elements 89 (actinium) through 103 (lawrencium) are chemically similar to the rare-earth or lanthanide elements (elements 57 to 71, inclusive). These elements therefore have been named *actinides* after the first member of this series. Those elements beyond uranium that have been produced artificially have the following names and symbols: neptunium, 93 (Np); plutonium, 94 (Pu); americium, 95 (Am); curium, 96 (Cm); berkelium, 97 (Bk); californium, 98 (Cf); einsteinium, 99 (Es); fermium, 100 (Fm); mendelevium, 101 (Md); nobelium, 102 (No); lawrencium, 103 (Lr); rutherfordium, 104 (Rf); dubnium, 105 (Db); seaborgium, 106 (Sg); bohrium, 107 (Bh); hassium, 108 (Hs); meitnerium, 109 (Mt); darmstadtium, 110 (Ds); and roentgenium, 111 (Rg). As of 2005, evidence has been reported for elements 112, 113, 114, 115, 116, and 118, but these elements have not been officially recognized or named. IUPAC recommends that until the existence of a new element is proven to their satisfaction, the elements are to have names and symbols derived according to these precise and simple rules: The name is based on the digits in the element's atomic number. Each digit is replaced with these expressions, with the end using the usual –ium suffix as follows: **0 nil, 1 un, 2 bi, 3 tri, 4 quad, 5 pent, 6 hex, 7 sept, 8 oct, 9 enn**. Double letter i's are not used, as for example Ununbiium, but would be Ununbium. The symbol used would be the first letter of the three main syllables. For example, Element 126 would be Unbihexium, with the symbol Ubh. (See J. Chatt, *Pure Appl. Chem.* 51, 381, 1979; W. H. Koppenol, *Pure Appl. Chem.* 74, 787, 2002.)

There are many claims in the literature of the existence of various allotropic modifications of the elements, some of which are based on doubtful or incomplete evidence. Also, the physical properties of an element may change drastically by the presence of small amounts of impurities. With new methods of purification, which are now able to produce elements with 99.9999% purity, it has been necessary to restudy the properties of the elements. For example, the melting point of thorium changes by several hundred degrees by the presence of a small percentage of ThO_2 as an impurity. Ordinary commercial tungsten is brittle and can be worked only with difficulty. Pure tungsten, however, can be cut with a hacksaw, forged, spun, drawn, or extruded. In general, the value of a physical property given here applies to the pure element, when it is known.

Many of the chemical elements and their compounds are toxic and should be handled with due respect and care. In recent years there has been greatly increased knowledge and awareness of the health hazards associated with chemicals, radioactive materials, and other agents. Anyone working with the elements and certain of their compounds should become thoroughly familiar with the proper safeguards to be taken. Information on specific hazards and recommended exposure limits may also be found in Section 16. Reference should also be made to publications such as the following:

1. *Code of Federal Regulations, Title 29, Labor.* With additions found in issues of the *Federal Register*.
2. *Code of Federal Regulations, Title 10, Energy.* With additions found in issues of the *Federal Register*. (Published by the U.S. Government Printing Office. Supt. of Documents.)
3. *Occupational Safety and Health Reporter* (latest edition with amendments and corrections), Bureau of National Affairs, Washington, D.C.
4. *Atomic Energy Law Reporter,* Commerce Clearing House, Chicago, IL.
5. *Nuclear Regulation Reporter,* Commerce Clearing House, Chicago, IL.
6. *TLVs® Threshold Limit Values for Chemical Substances and Physical Agents* is issued annually by the American Conference of Governmental Industrial Hygienists, Cincinnati, Ohio.
7. *The Sigma Aldrich Library of Regulatory and Safety Data. Vol. 3,* Robert E. Lenga and Kristine L. Volonpal, Sigma Chemical Co. and Aldrich Chemical Co., Inc. 1993.
8. *Hazardous Chemicals Desk Reference,* 4th ed., Richard J. Lewis, Sr., John Wiley & Sons, New York, 1997.
9. *Sittig's Handbook of Toxic and Hazardous Chemicals and Carcinogens,* 3rd ed., Noyes Publications, 2001/2.
10. *Sax's Dangerous Properties of Industrial Materials,* Richard J. Lewis and N. Irving Sax, John Wiley & Sons, New York, 1999.
11. *World Wide Limits for Toxic and Hazardous Chemicals in Air, Water, and Soil,* Marshall Sittig, Noyes Publishers.

The prices of elements as indicated in this article are intended to be only a rough guide. Prices may vary, over time, widely with supplier, quantity, and purity.

The density of gases is given in grams per liter at 0 °C and a pressure of 1 atm.

Actinium — (Gr. *aktis, aktinos,* beam or ray), Ac; at. wt. (227); at. no. 89; m.p. 1050 °C, b.p. 3198 °C; sp. gr. 10.07 (calc.). Discovered by Andre Debierne in 1899 and independently by F. Giesel in 1902. Occurs naturally in association with uranium minerals. Thirty-four isotopes and isomers are now recognized. All are radioactive. Actinium-227, a decay product of uranium-235, is an alpha and beta emitter with a 21.77-year half-life. Its principal decay products are thorium-227 (18.72-day half-life), radium-223 (11.4-day half-life), and a number of short-lived products including radon, bismuth, polonium, and lead isotopes. In equilibrium with its decay products, it is a powerful source of alpha rays. Actinium metal has been prepared by the reduction of actinium fluoride with lithium vapor at about 1100 to 1300 °C. The chemical behavior of actinium is similar to that of the rare earths, particularly lanthanum. Purified actinium comes into equilibrium with its decay products at the end of 185 days, and then decays according to its 21.77-year half-life. It is about 150 times as active as radium, making it of value in the production of neutrons. Actinium-225, with a purity of 99%, is available from the Oak

Ridge National Laboratory to holders of a permit for about $500/millicurie, plus packing charges.

Aluminum — (L. *alumen, alum*), Al; at. wt. 26.9815386(8); at. no. 13; m.p. 660.32 °C; b.p. 2519 °C; sp. gr. 2.6989 (20 °C); valence 3. The ancient Greeks and Romans used *alum* in medicine as an astringent, and as a mordant in dyeing. In 1761 de Morveau proposed the name *alumine* for the base in alum, and Lavoisier, in 1787, thought this to be the oxide of a still undiscovered metal. Wohler is generally credited with having isolated the metal in 1827, although an impure form was prepared by Oersted two years earlier. In 1807, Davy proposed the name *alumium* for the metal, undiscovered at that time, and later agreed to change it to *aluminum*. Shortly thereafter, the name *aluminium* was adopted to conform with the "ium" ending of most elements, and this spelling is now in use elsewhere in the world. *Aluminium* was also the accepted spelling in the U.S. until 1925, at which time the American Chemical Society officially decided to use the name *aluminum* thereafter in their publications. The method of obtaining aluminum metal by the electrolysis of alumina dissolved in *cryolite* was discovered in 1886 by Hall in the U.S. and at about the same time by Heroult in France. Cryolite, a natural ore found in Greenland, is no longer widely used in commercial production, but has been replaced by an artificial mixture of sodium, aluminum, and calcium fluorides. *Bauxite*, an impure hydrated oxide ore, is found in large deposits in Jamaica, Australia, Suriname, Guyana, Russia, Arkansas, and elsewhere. The Bayer process is most commonly used today to refine bauxite so it can be accommodated in the Hall–Heroult refining process used to make most aluminum. Aluminum can now be produced from clay, but the process is not economically feasible at present. Aluminum is the most abundant metal to be found in the Earth's crust (8.1%), but is never found free in nature. In addition to the minerals mentioned above, it is found in feldspars, granite, and in many other common minerals. Twenty-two isotopes and isomers are known. Natural aluminum is made of one isotope, ^{27}Al. Pure aluminum, a silvery-white metal, possesses many desirable characteristics. It is light, nontoxic, has a pleasing appearance, can easily be formed, machined, or cast, has a high thermal conductivity, and has excellent corrosion resistance. It is nonmagnetic and nonsparking, stands second among metals in the scale of malleability, and sixth in ductility. It is extensively used for kitchen utensils, outside building decoration, and in thousands of industrial applications where a strong, light, easily constructed material is needed. Although its electrical conductivity is only about 60% that of copper, it is used in electrical transmission lines because of its light weight. Pure aluminum is soft and lacks strength, but it can be alloyed with small amounts of copper, magnesium, silicon, manganese, and other elements to impart a variety of useful properties. These alloys are of vital importance in the construction of modern aircraft and rockets. Aluminum, evaporated in a vacuum, forms a highly reflective coating for both visible light and radiant heat. These coatings soon form a thin layer of the protective oxide and do not deteriorate as do silver coatings. They have found application in coatings for telescope mirrors, in making decorative paper, packages, toys, and in many other uses. The compounds of greatest importance are aluminum oxide, the sulfate, and the soluble sulfate with potassium (alum). The oxide, alumina, occurs naturally as ruby, sapphire, corundum, and emery, and is used in glassmaking and refractories. Synthetic ruby and sapphire have found application in the construction of lasers for producing coherent light. In 1852, the price of aluminum was about $1200/kg, and just before Hall's discovery in 1886, about $25/kg. The price rapidly dropped to 60¢ and has been as low as 33¢/kg. The price in December 2001 was about 64¢/lb or $1.40/kg.

Americium — (the Americas), Am; at. wt. 243; at. no. 95; m.p. 1176 °C; b.p. 2011 °C; sp. gr. 12; valence 2, 3, 4, 5, or 6. Americium was the fourth transuranium element to be discovered; the isotope ^{241}Am was identified by Seaborg, James, Morgan, and Ghiorso late in 1944 at the wartime Metallurgical Laboratory of the University of Chicago as the result of successive neutron capture reactions by plutonium isotopes in a nuclear reactor:

$$^{239}Pu(n,\gamma) \rightarrow ^{240}Pu(n,\gamma) \rightarrow ^{241}Pu \xrightarrow{\beta} ^{241}Am$$

Since the isotope ^{241}Am can be prepared in relatively pure form by extraction as a decay product over a period of years from strongly neutron-bombarded plutonium, ^{241}Pu, this isotope is used for much of the chemical investigation of this element. Better suited is the isotope ^{243}Am due to its longer half-life (7.37×10^3 years as compared to 432.2 years for ^{241}Am). A mixture of the isotopes ^{241}Am, ^{242}Am, and ^{243}Am can be prepared by intense neutron irradiation of ^{241}Am according to the reactions ^{241}Am (n, γ) $\rightarrow$ ^{242}Am (n, γ) $\rightarrow$ ^{243}Am. Nearly isotopically pure, ^{243}Am can be prepared by a sequence of neutron bombardments and chemical separations as follows: neutron bombardment of ^{241}Am yields ^{242}Pu by the reactions ^{241}Am (n, γ) $\rightarrow$ ^{242}Am $\rightarrow$ ^{242}Pu, after chemical separation the ^{242}Pu can be transformed to ^{243}Am via the reactions ^{242}Pu (n, γ) $\rightarrow$ ^{243}Pu $\rightarrow$ ^{243}Am, and the ^{243}Am can be chemically separated. Fairly pure ^{242}Pu can be prepared more simply by very intense neutron irradiation of ^{239}Pu as the result of successive neutron-capture reactions. Seventeen radioactive isotopes and isomers are now recognized. Americium metal has been prepared by reducing the trifluoride with barium vapor at 1000 to 1200 °C or the dioxide by lanthanum metal. The luster of freshly prepared americium metal is white and more silvery than plutonium or neptunium prepared in the same manner. It appears to be more malleable than uranium or neptunium and tarnishes slowly in dry air at room temperature. Americium is thought to exist in two forms: an alpha form which has a double hexagonal close-packed structure and a loose-packed cubic beta form. Americium must be handled with great care to avoid personal contamination. As little as 0.03 μCi of ^{241}Am is the maximum permissible total body burden. The alpha activity from ^{241}Am is about three times that of radium. When gram quantities of ^{241}Am are handled, the intense gamma activity makes exposure a serious problem. Americium dioxide, AmO_2, is the most important oxide. AmF_3, AmF_4, $AmCl_3$, $AmBr_3$, AmI_3, and other compounds have been prepared. The isotope ^{241}Am has been used as a portable source for gamma radiography. It has also been used as a radioactive glass thickness gage for the flat glass industry, and as a source of ionization for smoke detectors. Americum-243 (99%) is available from the Oak Ridge National Laboratory at a cost of about $750/g plus packing charges.

Antimony — (Gr. *anti* plus *monos* - a metal not found alone), Sb; at. wt. 121.760(1); at. no. 51; m.p. 630.63 °C; b.p. 1587 °C; sp. gr. 6.68 (20 °C); valence 0, −3, +3, or +5. Antimony was recognized in compounds by the ancients and was known as a metal at the beginning of the 17th century and possibly much earlier.

It is not abundant, but is found in over 100 mineral species. It is sometimes found native, but more frequently as the sulfide, *stibnite* (Sb_2S_3); it is also found as antimonides of the heavy metals, and as oxides. It is extracted from the sulfide by roasting to the oxide, which is reduced by salt and scrap iron; from its oxides it is also prepared by reduction with carbon. Two allotropic forms of antimony exist: the normal stable, metallic form, and the amorphous gray form. The so-called explosive antimony is an ill-defined material always containing an appreciable amount of halogen; therefore, it no longer warrants consideration as a separate allotrope. The yellow form, obtained by oxidation of *stibine*, SbH_3, is probably impure, and is not a distinct form. Natural antimony is made of two stable isotopes, ^{121}Sb and ^{123}Sb. Forty-five other radioactive isotopes and isomers are now recognized. Metallic antimony is an extremely brittle metal of a flaky, crystalline texture. It is bluish white and has a metallic luster. It is not acted on by air at room temperature, but burns brilliantly when heated with the formation of white fumes of Sb_2O_3. It is a poor conductor of heat and electricity, and has a hardness of 3 to 3.5. Antimony, available commercially with a purity of 99.999 + %, is finding use in semiconductor technology for making infrared detectors, diodes, and Hall-effect devices. Commercial-grade antimony is widely used in alloys with percentages ranging from 1 to 20. It greatly increases the hardness and mechanical strength of lead. Batteries, antifriction alloys, type metal, small arms and tracer bullets, cable sheathing, and minor products use about half the metal produced. Compounds taking up the other half are oxides, sulfides, sodium antimonate, and antimony trichloride. These are used in manufacturing flame-proofing compounds, paints, ceramic enamels, glass, and pottery. Tartar emetic (hydrated potassium antimonyl tartrate) has been used in medicine. Antimony and many of its compounds are toxic. Antimony costs about $1.30/kg for the commercial metal or about $12/g (99.999%).

Argon — (Gr. *argos*, inactive), Ar; at. wt. 39.948(1); at. no. 18; m.p. −189.36 °C; b.p. −185.85 °C; t_c −122.28 °C; density 1.7837 g/L. Its presence in air was suspected by Cavendish in 1785, discovered by Lord Rayleigh and Sir William Ramsay in 1894. The gas is prepared by fractionation of liquid air, the atmosphere containing 0.94% argon. The atmosphere of Mars contains 1.6% of ^{40}Ar and 5 p.p.m. of ^{36}Ar. Argon is two and one half times as soluble in water as nitrogen, having about the same solubility as oxygen. It is recognized by the characteristic lines in the red end of the spectrum. It is used in electric light bulbs and in fluorescent tubes at a pressure of about 400 Pa, and in filling photo tubes, glow tubes, etc. Argon is also used as an inert gas shield for arc welding and cutting, as a blanket for the production of titanium and other reactive elements, and as a protective atmosphere for growing silicon and germanium crystals. Argon is colorless and odorless, both as a gas and liquid. It is available in high-purity form. Commercial argon is available at a cost of about 3¢ per cubic foot. Argon is considered to be a very inert gas and is not known to form true chemical compounds, as do krypton, xenon, and radon. However, it does form a hydrate having a dissociation pressure of 105 atm at 0 °C. Ion molecules such as $(ArKr)^+$, $(ArXe)^+$, $(NeAr)^+$ have been observed spectroscopically. Argon also forms a clathrate with β-hydroquinone. This clathrate is stable and can be stored for a considerable time, but a true chemical bond does not exist. Van der Waals' forces act to hold the argon. In August 2000, researchers at the University of Helsinki, Finland reported they made a new argon compound HArF by shining UV light on frozen argon that contained a small amount of HF. Naturally occurring argon is a mixture of three isotopes. Seventeen other radioactive isotopes are now known to exist. Commercial argon is priced at about $70/300 cu. ft. or 8.5 cu. meters.

Arsenic — (L. *arsenicum*, Gr. *arsenikon*, yellow orpiment, identified with *arsenikos*, male, from the belief that metals were different sexes; Arabic, *Az-zernikh*, the orpiment from Persian *zerni-zar*, gold), As; at. wt. 74.92160(2); at. no. 33; valence −3, 0, +3 or +5. Elemental arsenic occurs in two solid modifications: yellow, and gray or metallic, with specific gravities of 1.97, and 5.75, respectively. Gray arsenic, the ordinary stable form, has a triple point of 817 °C and sublimes at 616 °C and has a critical temperature of 1400 °C. Several other allotropic forms of arsenic are reported in the literature. It is believed that Albertus Magnus obtained the element in 1250 A.D. In 1649 Schroeder published two methods of preparing the element. It is found native, in the sulfides *realgar* and *orpiment*, as arsenides and sulfarsenides of heavy metals, as the oxide, and as arsenates. *Mispickel*, arsenopyrite, (FeSAs) is the most common mineral, from which on heating the arsenic sublimes leaving ferrous sulfide. The element is a steel gray, very brittle, crystalline, semimetallic solid; it tarnishes in air, and when heated is rapidly oxidized to arsenous oxide (As_2O_3) with the odor of garlic. Arsenic and its compounds are poisonous. Exposure to arsenic and its compounds should not exceed 0.01 mg/m^3 as elemental As during an 8-h work day. Arsenic is also used in bronzing, pyrotechny, and for hardening and improving the sphericity of shot. The most important compounds are white arsenic (As_2O_3), the sulfide, Paris green $3Cu(AsO_2)_2 \cdot Cu(C_2H_3O_2)_2$, calcium arsenate, and lead arsenate; the last three have been used as agricultural insecticides and poisons. Marsh's test makes use of the formation and ready decomposition of arsine (AsH_3). Arsenic is available in high-purity form. It is finding increasing uses as a doping agent in solid-state devices such as transistors. Gallium arsenide is used as a laser material to convert electricity directly into coherent light. Natural arsenic is made of one isotope ^{75}As. Thirty other radioactive isotopes and isomers are known. Arsenic (99%) costs about $75/50g. Purified arsenic (99.9995%) costs about $50/g.

Astatine — (Gr. *astatos*, unstable), At; at. wt. (210); at. no. 85; m.p. 302 °C; valence probably 1, 3, 5, or 7. Synthesized in 1940 by D. R. Corson, K. R. MacKenzie, and E. Segre at the University of California by bombarding bismuth with alpha particles. The longest-lived isotope, ^{210}At, has a half-life of only 8.1 hours. Thirty-six other isotopes and isomers are now known. Minute quantities of ^{215}At, ^{218}At, and ^{219}At exist in equilibrium in nature with naturally occurring uranium and thorium isotopes, and traces of ^{217}At are in equilibrium with ^{233}U and ^{239}Np resulting from interaction of thorium and uranium with naturally produced neutrons. The total amount of astatine present in the Earth's crust, however, is probably less than 1 oz. Astatine can be produced by bombarding bismuth with energetic alpha particles to obtain the relatively long-lived $^{209-211}At$, which can be distilled from the target by heating it in air. Only about 0.05 μg of astatine has been prepared to date. The "time of flight" mass spectrometer has been used to confirm that this highly radioactive halogen behaves chemically very much like other halogens, particularly iodine. The interhalogen compounds AtI, AtBr, and AtCl are known to form, but it is not yet known if astatine forms diatomic astatine molecules. HAt and CH$_3$At

(methyl astatide) have been detected. Astatine is said to be more metallic that iodine, and, like iodine, it probably accumulates in the thyroid gland.

Barium — (Gr. *barys,* heavy), Ba; at. wt. 137.327(7), at. no. 56; m.p. 727 °C; b.p. 1897 °C; sp. gr. 3.62 (20 °C); valence 2. Baryta was distinguished from lime by Scheele in 1774; the element was discovered by Sir Humphrey Davy in 1808. It is found only in combination with other elements, chiefly in *barite* or *heavy spar* (sulfate) and *witherite* (carbonate) and is prepared by electrolysis of the chloride. Large deposits of barite are found in China, Germany, India, Morocco, and in the U.S. Barium is a metallic element, soft, and when pure is silvery white like lead; it belongs to the alkaline earth group, resembling calcium chemically. The metal oxidizes very easily and should be kept under petroleum or other suitable oxygen-free liquids to exclude air. It is decomposed by water or alcohol. The metal is used as a "getter" in vacuum tubes. The most important compounds are the peroxide (BaO_2), chloride, sulfate, carbonate, nitrate, and chlorate. Lithopone, a pigment containing barium sulfate and zinc sulfide, has good covering power, and does not darken in the presence of sulfides. The sulfate, as permanent white or *blanc fixe*, is also used in paint, in X-ray diagnostic work, and in glassmaking. *Barite* is extensively used as a weighting agent in oilwell drilling fluids, and also in making rubber. The carbonate has been used as a rat poison, while the nitrate and chlorate give green colors in pyrotechny. The impure sulfide phosphoresces after exposure to the light. The compounds and the metal are not expensive. Barium metal (99.2 + % pure) costs about $3/g. All barium compounds that are water or acid soluble are poisonous. Naturally occurring barium is a mixture of seven stable isotopes. Thirty-six other radioactive isotopes and isomers are known to exist.

Berkelium — (*Berkeley*, home of the University of California), Bk; at. wt. (247); at. no. 97; m.p. 996 °C; valence 3 or 4; sp. gr. 14 (est.). Berkelium, the eighth member of the actinide transition series, was discovered in December 1949 by Thompson, Ghiorso, and Seaborg, and was the fifth transuranium element synthesized. It was produced by cyclotron bombardment of milligram amounts of ^{241}Am with helium ions at Berkeley, California. The first isotope produced had a mass number of 243 and decayed with a half-life of 4.5 hours. Thirteen isotopes are now known and have been synthesized. The existence of ^{249}Bk, with a half-life of 320 days, makes it feasible to isolate berkelium in weighable amounts so that its properties can be investigated with macroscopic quantities. One of the first visible amounts of a pure berkelium compound, berkelium chloride, was produced in 1962. It weighed 3 billionth of a gram. Berkelium probably has not yet been prepared in elemental form, but it is expected to be a silvery metal, easily soluble in dilute mineral acids, and readily oxidized by air or oxygen at elevated temperatures to form the oxide. X-ray diffraction methods have been used to identify the following compounds: BkO_2, BkO_3, BkF_3, BkCl, and BkOCl. As with other actinide elements, berkelium tends to accumulate in the skeletal system. The maximum permissible body burden of ^{249}Bk in the human skeleton is about 0.0004 μg. Because of its rarity, berkelium presently has no commercial or technological use. Berkelium most likely resembles terbium with respect to chemical properties. Berkelium-249 is available from O.R.N.L. at a cost of $185/μg plus packing charges.

Beryllium — (Gr. *beryllos, beryl;* also called Glucinium or Glucinum, Gr. *glykys,* sweet), Be; at. wt. 9.012182(3); at no. 4; m.p. 1287 °C; b.p. 2471 °C; sp. gr. 1.848 (20 °C); valence 2. Discovered as the oxide by Vauquelin in beryl and in emeralds in 1798. The metal was isolated in 1828 by Wohler and by Bussy independently by the action of potassium on beryllium chloride. Beryllium is found in some 30 mineral species, the most important of which are *bertrandite, beryl, chrysoberyl,* and *phenacite. Aquamarine* and *emerald* are precious forms of *beryl.* Beryllium minerals are found in the U.S., Brazil, Russia, Kazakhstan, and elsewhere. Colombia is known for its emeralds. *Beryl* ($3BeO \cdot Al_2O_3 \cdot 6SiO_2$) and *bertrandite* ($4BeO \cdot 2SiO_2 \cdot H_2O$) are the most important commercial sources of the element and its compounds. Most of the metal is now prepared by reducing beryllium fluoride with magnesium metal. Beryllium metal did not become readily available to industry until 1957. The metal, steel gray in color, has many desirable properties. It is one of the lightest of all metals, and has one of the highest melting points of the light metals. Its modulus of elasticity is about one third greater than that of steel. It resists attack by concentrated nitric acid, has excellent thermal conductivity, and is nonmagnetic. It has a high permeability to X-rays, and when bombarded by alpha particles, as from radium or polonium, neutrons are produced in the ratio of about 30 neutrons/million alpha particles. At ordinary temperatures beryllium resists oxidation in air, although its ability to scratch glass is probably due to the formation of a thin layer of the oxide. Beryllium is used as an alloying agent in producing beryllium copper, which is extensively used for springs, electrical contacts, spot-welding electrodes, and nonsparking tools. It has found application as a structural material for high-speed aircraft, missiles, spacecraft, and communication satellites. It is being used in the windshield frame, brake discs, support beams, and other structural components of the space shuttle. Because beryllium is relatively transparent to X-rays, ultra-thin Be-foil is finding use in X-ray lithography for reproduction of microminiature integrated circuits. Natural beryllium is made of ^9Be and is stable. Eight other radioactive isotopes are known.

Beryllium is used in nuclear reactors as a reflector or moderator for it has a low thermal neutron absorption cross section. It is used in gyroscopes, computer parts, and instruments where lightness, stiffness, and dimensional stability are required. The oxide has a very high melting point and is also used in nuclear work and ceramic applications. Beryllium and its salts are toxic and should be handled with the greatest of care. Beryllium and its compounds should not be tasted to verify the sweetish nature of beryllium (as did early experimenters). The metal, its alloys, and its salts can be handled safely if certain work codes are observed, but no attempt should be made to work with beryllium before becoming familiar with proper safeguards. Beryllium metal is available at a cost of about $5/g (99.5% pure).

Bismuth — (Ger. *Weisse Masse,* white mass; later *Wisuth* and *Bisemutum*), Bi; at. wt. 208.98040(1); at. no. 83; m.p. 271.4 °C; b.p. 1564 °C; sp. gr. 9.79 (20 °C); valence 3 or 5. In early times bismuth was confused with tin and lead. Claude Geoffroy the Younger showed it to be distinct from lead in 1753. It is a white crystalline, brittle metal with a pinkish tinge. It occurs native. The most important ores are *bismuthinite* or bismuth glance (Bi_2S_3) and *bismite* (Bi_2O_3). Peru, Japan, Mexico, Bolivia, and Canada are major bismuth producers. Much of the bismuth produced in the U.S. is obtained as a by-product in refining lead, copper, tin, silver, and gold ores. Bismuth is the most diamagnetic of all metals, and the thermal conductivity is lower than any metal, except mercury. It has a high electrical resistance, and has the highest Hall effect of any metal (i.e., greatest increase in electrical resistance when placed in a magnetic field). "Bismanol" is a permanent magnet of high coercive force, made of MnBi, by the U.S. Naval Surface Weapons Center. Bismuth expands 3.32% on solidification. This property makes bismuth alloys particularly suited to the making of sharp castings of objects subject to damage by high temperatures. With other metals such as tin, cadmium, etc., bismuth forms low-melting alloys that are extensively used for safety devices in fire detection and extinguishing systems. Bismuth is used in producing malleable irons and is finding use as a catalyst for making acrylic fibers. When bismuth is heated in air it burns with a blue flame, forming yellow fumes of the oxide. The metal is also used as a thermocouple material, and has found application as a carrier for U^{235} or U^{233} fuel in atomic reactors. Its soluble salts are characterized by forming insoluble basic salts on the addition of water, a property sometimes used in detection work. Bismuth oxychloride is used extensively in cosmetics. Bismuth subnitrate and subcarbonate are used in medicine. Natural bismuth contains only one isotope ^{209}Bi. Forty-four isotopes and isomers of bismuth are known. Bismuth metal (99.5%) costs about $250/kg.

Bohrium — (Named after Niels Bohr [1885–1962], Danish atomic and nuclear physicist.) Bh; at. wt. [264]. at. no. 107. Bohrium is expected to have chemical properties similar to rhenium. This element was synthesized and unambiguously identified in 1981 using the Universal Linear Accelerator (UNILAC) at the Gesellschaft für Schwerionenforschung (G.S.I.) in Darmstadt, Germany. The discovery team was led by Armbruster and Münzenberg. The reaction producing the element was proposed and applied earlier by a Dubna Group led by Oganessian in 1976. A target of ^{209}Bi was bombarded by a beam of ^{54}Cr ions. In 1983 experiments at Dubna using the 157-inch cyclotron, produced $^{262}107$ by the reaction $^{209}Bi + {}^{54}Cr$. The alpha decay of ^{246}Cf, the sixth member in the decay chain of $^{262}107$, served to establish a 1-neutron reaction channel. The IUPAC adopted the name ***Bohrium*** with the symbol Bh for Element 107 in August 1997. Five isotopes of bohrium are now recognized. One isotope of bohrium appears to have a relatively long life of 15 seconds. Work on this relatively long-lived isotope has been performed with the 88-inch cyclotron at the Lawrence-Berkeley National Laboratory.

Boron — (Ar. *Buraq,* Pers. *Burah*), B; at. wt. [10.806; 10.821]; at. no. 5; m.p. 2075 °C; b.p. 4000 °C; sp. gr. of crystals 2.34, of amorphous variety 2.37; valence 3. Boron compounds have been known for thousands of years, but the element was not discovered until 1808 by Sir Humphry Davy and by Gay-Lussac and Thenard. The element is not found free in nature, but occurs as orthoboric acid usually in certain volcanic spring waters and as borates in *borax* and *colemanite. Ulexite,* another boron mineral, is interesting as it is nature's own version of "fiber optics." Important sources of boron are the ores *rasorite (kernite)* and *tincal (borax ore).* Both of these ores are found in the Mojave Desert. *Tincal* is the most important source of boron from the Mojave. Extensive *borax* deposits are also found in Turkey. Boron exists naturally as 19.9% ^{10}B isotope and 80.1% ^{11}B isotope. Ten other isotopes of boron are known. High-purity crystalline boron may be prepared by the vapor phase reduction of boron trichloride or tribromide with hydrogen on electrically heated filaments. The impure, or amorphous, boron, a brownish-black powder, can be obtained by heating the trioxide with magnesium powder. Boron of 99.9999% purity has been produced and is available commercially. Elemental boron has an energy band gap of 1.50 to 1.56 eV, which is higher than that of either silicon or germanium. It has interesting optical characteristics, transmitting portions of the infrared, and is a poor conductor of electricity at room temperature, but a good conductor at high temperature. Amorphous boron is used in pyrotechnic flares to provide a distinctive green color, and in rockets as an igniter. By far the most commercially important boron compound in terms of dollar sales is $Na_2B_4O_7 \cdot 5H_2O$. This pentahydrate is used in very large quantities in the manufacture of insulation fiberglass and sodium perborate bleach. Boric acid is also an important boron compound with major markets in textile fiberglass and in cellulose insulation as a flame retardant. Next in order of importance is borax ($Na_2B_4O_7 \cdot 10H_2O$) which is used principally in laundry products. Use of borax as a mild antiseptic is minor in terms of dollars and tons. Boron compounds are also extensively used in the manufacture of borosilicate glasses. The isotope boron-10 is used as a control for nuclear reactors, as a shield for nuclear radiation, and in instruments used for detecting neutrons. Boron nitride has remarkable properties and can be used to make a material as hard as diamond. The nitride also behaves like an electrical insulator but conducts heat like a metal. It also has lubricating properties similar to graphite. The hydrides are easily oxidized with considerable energy liberation, and have been studied for use as rocket fuels. Demand is increasing for boron filaments, a high-strength, lightweight material chiefly employed for advanced aerospace structures. Boron is similar to carbon in that it has a capacity to form stable covalently bonded molecular networks. Carboranes, metalloboranes, phosphacarboranes, and other families comprise thousands of compounds. Crystalline boron (99.5%) costs about $6/g. Amorphous boron (94–96%) costs about $1.50/g. Elemental boron and the borates are not considered to be toxic, and they do not require special care in handling. However, some of the more exotic boron hydrogen compounds are definitely toxic and do require care.

Bromine — (Gr. *bromos*, stench), Br; at. wt. 79.904(1); at. no. 35; m.p. −7.2 °C; b.p. 58.8 °C; t_c 315 °C; density of gas 7.59 g/l, liquid 3.12 (20 °C); valence 1, 3, 5, or 7. Discovered by Balard in 1826, but not prepared in quantity until 1860. A member of the halogen group of elements, it is obtained from natural brines from wells in Michigan and Arkansas. Little bromine is extracted today from seawater, which contains only about 85 ppm. Bromine is the only liquid nonmetallic element. It is a heavy, mobile, reddish-brown liquid, volatilizing readily at room temperature to a red vapor with a strong disagreeable odor, resembling chlorine, and having a very irritating effect on the eyes and throat; it is readily soluble in water or carbon disulfide, forming a red solution, is less active than chlorine but more so than iodine; it unites readily with many elements and has a bleaching action; when spilled on the skin it produces painful sores. It presents a serious health hazard, and maximum safety precautions should be taken when handling it. Much of the bromine output in the U.S. was used in the production of ethylene dibromide, a lead scavenger used in making gasoline antiknock compounds. Lead in gasoline, however, has been drastically reduced, due to environmental considerations. This will greatly affect future production of bromine. Bromine is also used in making fumigants,

flameproofing agents, water purification compounds, dyes, medicinals, sanitizers, inorganic bromides for photography, etc. Organic bromides are also important. Natural bromine is made of two isotopes, ^{79}Br and ^{81}Br. Thirty-four isotopes and isomers are known. Bromine (99.8%) costs about $70/kg.

Cadmium — (L. *cadmia*; Gr. *kadmeia* - ancient name for calamine, zinc carbonate), Cd; at. wt. 112.411(8); at. no. 48; m.p. 321.07 °C; b.p. 767 °C; sp. gr. 8.69 (20 °C); valence 2. Discovered by Stromeyer in 1817 from an impurity in zinc carbonate. Cadmium most often occurs in small quantities associated with zinc ores, such as *sphalerite* (ZnS). *Greenockite* (CdS) is the only mineral of any consequence bearing cadmium. Almost all cadmium is obtained as a by-product in the treatment of zinc, copper, and lead ores. It is a soft, bluish-white metal which is easily cut with a knife. It is similar in many respects to zinc. It is a component of some of the lowest melting alloys; it is used in bearing alloys with low coefficients of friction and great resistance to fatigue; it is used extensively in electroplating, which accounts for about 60% of its use. It is also used in many types of solder, for standard E.M.F. cells, for Ni-Cd batteries, and as a barrier to control atomic fission. The market for Ni-Cd batteries is expected to grow significantly. Cadmium compounds are used in black and white television phosphors and in blue and green phosphors for color TV tubes. It forms a number of salts, of which the sulfate is most common; the sulfide is used as a yellow pigment. Cadmium and solutions of its compounds are toxic. Failure to appreciate the toxic properties of cadmium may cause workers to be unwittingly exposed to dangerous fumes. Some silver solders, for example, contain cadmium and should be handled with care. Serious toxicity problems have been found from long-term exposure and work with cadmium plating baths. Cadmium is present in certain phosphate rocks. This has raised concerns that the long-term use of certain phosphate fertilizers might pose a health hazard from levels of cadmium that might enter the food chain. In 1927 the International Conference on Weights and Measures redefined the meter in terms of the wavelength of the red cadmium spectral line (i.e., 1 m = 1,553,164.13 wavelengths). This definition has been changed (see under Krypton). The current price of cadmium is about 50¢/g (99.5%). It is available in high purity form for about $550/kg. Natural cadmium is made of eight isotopes. Thirty-four other isotopes and isomers are now known and recognized.

Calcium — (L. *calx*, lime), Ca; at. wt. 40.078(4); at. no. 20; m.p. 842 °C; b.p. 1484 °C; sp. gr. 1.54 (20 °C); valence 2. Though lime was prepared by the Romans in the first century under the name calx, the metal was not discovered until 1808. After learning that Berzelius and Pontin prepared calcium amalgam by electrolyzing lime in mercury, Davy was able to isolate the impure metal. Calcium is a metallic element, fifth in abundance in the Earth's crust, of which it forms more than 3%. It is an essential constituent of leaves, bones, teeth, and shells. Never found in nature uncombined, it occurs abundantly as *limestone* ($CaCO_3$), *gypsum* ($CaSO_4 \cdot 2H_2O$), and *fluorite* (CaF_2); *apatite* is the fluorophosphate or chlorophosphate of calcium. The metal has a silvery color, is rather hard, and is prepared by electrolysis of the fused chloride to which calcium fluoride is added to lower the melting point. Chemically it is one of the alkaline earth elements; it readily forms a white coating of oxide in air, reacts with water, burns with a yellow-red flame, largely forming the oxide. The metal is used as a reducing agent in preparing other metals such as thorium, uranium, zirconium, etc., and is used as a deoxidizer, desulfurizer, and inclusion modifier for various ferrous and nonferrous alloys. It is also used as an alloying agent for aluminum, beryllium, copper, lead, and magnesium alloys, and serves as a "getter" for residual gases in vacuum tubes. Its natural and prepared compounds are widely used. Quicklime (CaO), made by heating limestone and changed into slaked lime by the careful addition of water, is the great cheap base of the chemical industry with countless uses. Mixed with sand it hardens as mortar and plaster by taking up carbon dioxide from the air. Calcium from limestone is an important element in Portland cement. The solubility of the carbonate in water containing carbon dioxide causes the formation of caves with stalactites and stalagmites and is responsible for hardness in water. Other important compounds are the carbide (CaC_2), chloride ($CaCl_2$), cyanamide ($CaCN_2$), hypochlorite ($Ca(OCl)_2$), nitrate ($Ca(NO_3)_2$), and sulfide (CaS). Calcium sulfide is phosphorescent after being exposed to light. Natural calcium contains six isotopes. Sixteen other radioactive isotopes are known. Metallic calcium (99.5%) costs about $200/kg.

Californium — (State and University of California), Cf; at. wt. (251); m.p. 900 °C; sp. gr. 15.1; at. no. 98. Californium, the sixth transuranium element to be discovered, was produced by Thompson, Street, Ghioirso, and Seaborg in 1950 by bombarding microgram quantities of ^{242}Cm with 35 MeV helium ions in the Berkeley 60-inch cyclotron. Californium (III) is the only ion stable in aqueous solutions, all attempts to reduce or oxidize californium (III) having failed. The isotope ^{249}Cf results from the beta decay of ^{249}Bk while the heavier isotopes are produced by intense neutron irradiation by the reactions:

$$^{249}Bk(n,\gamma) \rightarrow\,^{250}Bk \xrightarrow{\beta}\,^{250}Cf \text{ and } ^{249}Cf(n,\gamma) \rightarrow\,^{250}Cf$$

followed by

$$^{250}Cf(n,\gamma) \rightarrow\,^{251}Cf(n,\gamma) \rightarrow\,^{252}Cf$$

The existence of the isotopes ^{249}Cf, ^{250}Cf, ^{251}Cf, and ^{252}Cf makes it feasible to isolate californium in weighable amounts so that its properties can be investigated with macroscopic quantities. Californium-252 is a very strong neutron emitter. One microgram releases 170 million neutrons per minute, which presents biological hazards. Proper safeguards should be used in handling californium. Twenty isotopes of californium are now recognized. ^{249}Cf and ^{252}Cf have half-lives of 351 years and 900 years, respectively. In 1960 a few tenths of a microgram of californium trichloride, $CfCl_3$, californium oxychloride, CfOCl, and californium oxide, Cf_2O_3, were first prepared. Reduction of californium to its metallic state has not yet been accomplished. Because californium is a very efficient source of neutrons, many new uses are expected for it. It has already found use in neutron moisture gages and in well-logging (the determination of water and oil-bearing layers). It is also being used as a portable neutron source for discovery of metals such as gold or silver by on-the-spot activation analysis. ^{252}Cf is now being offered for sale by the Oak Ridge National Laboratory (O.R.N.L.) at a cost of $60/μg and ^{249}Cf at a cost of $185/μg plus packing charges. It has been suggested that californium may be produced in certain stellar explosions, called *supernovae*, for the radioactive decay of ^{254}Cf (55-day half-life) agrees with the characteristics of the light curves of such explosions observed through telescopes. This suggestion, however, is questioned. Californium is expected to have chemical properties similar to dysprosium.

Carbon — (L. *carbo*, charcoal), C; at. wt. [12.0096; 12.0116]; at. no. 6; sublimes at 3825 °C; triple point (graphite-liquid-gas), 4489 °C; sp. gr. amorphous 1.8 to 2.1, graphite 1.9 to 2.3, diamond 3.15 to 3.53 (depending on variety); gem diamond 3.513 (25 °C); valence 2, 3, or 4. Carbon, an element of prehistoric discovery, is very widely distributed in nature. It is found in abundance in the sun, stars, comets, and atmospheres of most planets. Carbon in the form of microscopic diamonds is found in some meteorites. Natural diamonds are found in *kimberlite* or *lamporite* of ancient formations called "pipes," such as found in South Africa, Arkansas, and elsewhere. Diamonds are now also being recovered from the ocean floor off the Cape of Good Hope. About 30% of all industrial diamonds used in the U.S. are now made synthetically. The energy of the sun and stars can be attributed at least in part to the well-known carbon-nitrogen cycle. Carbon is found free in nature in three allotropic forms: amorphous, graphite, and diamond. Graphite is one of the softest known materials while diamond is one of the hardest. Graphite exists in two forms: alpha and beta. These have identical physical properties, except for their crystal structure. Naturally occurring graphites are reported to contain as much as 30% of the rhombohedral (beta) form, whereas synthetic materials contain only the alpha form. The hexagonal alpha type can be converted to the beta by mechanical treatment, and the beta form reverts to the alpha on heating it above 1000 °C. Of recent interest is the discovery of all-carbon molecules, known as "buckyballs" or fullerenes, which have a number of unusual properties. These interesting molecules, consisting of 60 or 70 carbon atoms linked together, seem capable of withstanding great pressure and trapping foreign atoms inside their network of carbon. They are said to be capable of magnetism and superconductivity and have potential as a nonlinear optical material. Buckyball films are reported to remain superconductive at temperatures as high as 45 K. In combination, carbon is found as carbon dioxide in the atmosphere of the Earth and dissolved in all natural waters. It is a component of great rock masses in the form of carbonates of calcium (limestone), magnesium, and iron. Coal, petroleum, and natural gas are chiefly hydrocarbons. Carbon is unique among the elements in the vast number and variety of compounds it can form. With hydrogen, oxygen, nitrogen, and other elements, it forms a very large number of compounds, carbon atom often being linked to carbon atom. There are close to ten million known carbon compounds, many thousands of which are vital to organic and life processes. Without carbon, the basis for life would be impossible. While it has been thought that silicon might take the place of carbon in forming a host of similar compounds, it is now not possible to form stable compounds with very long chains of silicon atoms. The atmosphere of Mars contains 96.2% CO_2. Some of the most important compounds of carbon are carbon dioxide (CO_2), carbon monoxide (CO), carbon disulfide (CS_2), chloroform ($CHCl_3$), carbon tetrachloride (CCl_4), methane (CH_4), ethylene (C_2H_4), acetylene (C_2H_2), benzene (C_6H_6), ethyl alcohol (C_2H_5OH), acetic acid (CH_3COOH), and their derivatives. Carbon has fifteen isotopes. Natural carbon consists of 98.89% ^{12}C and 1.11% ^{13}C. In 1961 the International Union of Pure and Applied Chemistry adopted the isotope carbon-12 as the basis for atomic weights. Carbon-14, an isotope with a half-life of 5715 years, has been widely used to date such materials as wood, archeological specimens, etc. A new brittle form of carbon, known as "glassy carbon," has been developed. It can be obtained with high purity. It has a high resistance to corrosion, has good thermal stability, and is structurally impermeable to both gases and liquids. It has a randomized structure, making it useful in ultra-high technology applications, such as crystal growing, crucibles for high-temperature use, etc. Glassy carbon is available at a cost of about $35/10g. Fullerene powder is available at a cost of about $55/10mg (99%$C_{10}$). Diamond powder (99.9%) costs about $40/g.

Cerium — (named for the asteroid *Ceres,* which was discovered in 1801 only 2 years before the element), Ce; at. wt. 140.116(1); at. no. 58; m.p. 799 °C; b.p. 3443 °C; sp. gr. 6.770 (25 °C); valence 3 or 4. Discovered in 1803 by Klaproth and by Berzelius and Hisinger; metal prepared by Hillebrand and Norton in 1875. Cerium is the most abundant of the metals of the so-called rare earths. It is found in a number of minerals including *allanite* (also known as *orthite*), *monazite, bastnasite, cerite,* and *samarskite.* Monazite and bastnasite are presently the two most important sources of cerium. Large deposits of monazite found on the beaches of Travancore, India, in river sands in Brazil, and deposits of *allanite* in the western United States, and *bastnasite* in Southern California will supply cerium, thorium, and the other rare-earth metals for many years to come. Metallic cerium is prepared by metallothermic reduction techniques, such as by reducing cerous fluoride with calcium, or by electrolysis of molten cerous chloride or other cerous halides. The metallothermic technique is used to produce high-purity cerium. Cerium is especially interesting because of its variable electronic structure. The energy of the inner 4f level is nearly the same as that of the outer or valence electrons, and only small amounts of energy are required to change the relative occupancy of these electronic levels. This gives rise to dual valency states. For example, a volume change of about 10% occurs when cerium is subjected to high pressures or low temperatures. It appears that the valence changes from about 3 to 4 when it is cooled or compressed. The low temperature behavior of cerium is complex. Four allotropic modifications are thought to exist: cerium at room temperature and at atmospheric pressure is known as γ cerium. Upon cooling to −16 °C, γ cerium changes to β cerium. The remaining γ cerium starts to change to α cerium when cooled to −172 °C, and the transformation is complete at −269 °C. α Cerium has a density of 8.16; δ cerium exists above 726 °C. At atmospheric pressure, liquid cerium is more dense than its solid form at the melting point. Cerium is an iron-gray lustrous metal. It is malleable, and oxidizes very readily at room temperature, especially in moist air. Except for europium, cerium is the most reactive of the "rare-earth" metals. It slowly decomposes in cold water, and rapidly in hot water. Alkali solutions and dilute and concentrated acids attack the metal rapidly. The pure metal is likely to ignite if scratched with a knife. Ceric salts are orange red or yellowish; cerous salts are usually white. Cerium is a component of misch metal, which is extensively used in the manufacture of pyrophoric alloys for cigarette lighters, etc. Natural cerium is stable and contains four isotopes. Thirty-two other radioactive isotopes and isomers are known. While cerium is not radioactive, the impure commercial grade may contain traces of thorium, which is radioactive. The oxide is an important constituent of incandescent gas mantles and it is emerging as a hydrocarbon catalyst in "self-cleaning" ovens. In this application it can be incorporated into oven walls to prevent the collection of cooking residues. As ceric sulfate it finds extensive use as a volumetric oxidizing agent in quantitative analysis. Cerium compounds are used in the manufacture of glass, both as a component and as a decolorizer. The oxide is finding increased use as a glass polishing agent instead of rouge, for it is much faster than rouge in polishing glass surfaces. Cerium compounds are finding use in automobile exhaust catalysts. Cerium is also finding use in making permanent magnets. Cerium, with other rare earths, is used in carbon-arc lighting, especially in the motion picture industry. It is also finding use as an important catalyst in petroleum refining and in metallurgical and nuclear applications. In small lots, cerium costs about $5/g (99.9%).

Cesium — (L. *caesius,* sky blue), Cs; at. wt. 132.9054519(2); at. no. 55; m.p. 28.44 °C; b.p. 671 °C; sp. gr. 1.873 (20 °C); valence 1. Cesium was discovered spectroscopically by Bunsen and Kirchhoff in 1860 in mineral water from Durkheim. Cesium, an alkali metal, occurs in *lepidolite, pollucite* (a hydrated silicate of aluminum and cesium), and in other sources. One of the world's richest sources of cesium is located at Bernic Lake, Manitoba. The deposits are estimated to contain 300,000 tons of pollucite, averaging 20% cesium. It can be isolated by electrolysis of the fused cyanide and by a number of other methods. Very pure, gas-free cesium can be prepared by thermal decomposition of cesium azide. The metal is characterized by a spectrum containing two bright lines in the blue along with several others in the red, yellow, and green. It is silvery white, soft, and ductile. It is the most electropositive and most alkaline element. Cesium, gallium, and mercury are the only three metals that are liquid at room temperature. Cesium reacts explosively with cold water, and reacts with ice at temperatures above −116 °C. Cesium hydroxide, the strongest base known, attacks glass. Because of its great affinity for oxygen the metal is used as a "getter" in electron tubes. It is also used in photoelectric cells, as well as a catalyst in the hydrogenation of certain organic compounds. The metal has recently found application in ion propulsion systems. Cesium is used in atomic clocks, which are accurate to 5 s in 300 years. A second of time is now defined as being the duration of 9,192,631,770 periods of the radiation corresponding to the transition between the two hyper-fine levels of the ground state of the cesium-133 atom. Its chief compounds are the chloride and the nitrate. Cesium has 52 isotopes and isomers with masses ranging from 112 to 148. The present price of cesium is about $50/g (99.98%) sealed in a glass ampoule.

Chlorine — (Gr. *chloros,* greenish yellow), Cl; at. wt. [35.446; 35.457]; at. no. 17; m.p. −101.5 °C; b.p. −34.04 °C; t_c 143.8 °C; density 3.214 g/L; sp. gr. 1.56 (−33.6 °C); valence 1, 3, 5, or 7. Discovered in 1774 by Scheele, who thought it contained oxygen; named in 1810 by Davy, who insisted it was an element. In nature it is found in the combined state only, chiefly with sodium as common salt (NaCl), *carnallite* ($KMgCl_3 \cdot 6H_2O$), and *sylvite* (KCl). It is a member of the halogen (salt-forming) group of elements and is obtained from chlorides by the action of oxidizing agents and more often by electrolysis; it is a greenish-yellow gas, combining directly with nearly all elements. At 10 °C one volume of water dissolves 3.10 volumes of chlorine, at 30 °C only 1.77 volumes. Chlorine is widely used in making many everyday products. It is used for producing safe drinking water the world over. Even the smallest water supplies are now usually chlorinated. It is also extensively used in the production of paper products, dyestuffs, textiles, petroleum products, medicines, antiseptics, insecticides, foodstuffs, solvents, paints, plastics, and many other consumer products. Most of the chlorine produced is used in the manufacture of chlorinated compounds for sanitation, pulp bleaching, disinfectants, and textile processing. Further use is in the manufacture of chlorates, chloroform, carbon tetrachloride, and in the extraction of bromine. Organic chemistry demands much from chlorine, both as an oxidizing agent and in substitution, since it often brings desired properties in an organic compound when substituted for hydrogen, as in one form of synthetic rubber. Chlorine is a respiratory irritant. The gas irritates the mucous membranes and the liquid burns the skin. As little as 3.5 ppm can be detected as an odor, and 1000 ppm is likely to be fatal after a few deep breaths. It was used as a war gas in 1915. Natural chlorine contains two isotopes. Twenty other isotopes and isomers are known.

Chromium — (Gr. *chroma*, color), Cr; at. wt. 51.9961(6); at. no. 24; m.p. 1907 °C; b.p. 2671 °C; sp. gr. 7.15 (20 °C); valence chiefly 2, 3, or 6. Discovered in 1797 by Vauquelin, who prepared the metal the next year, chromium is a steel-gray, lustrous, hard metal that takes a high polish. The principal ore is *chromite* ($FeCr_2O_4$), which is found in Zimbabwe, Russia, South Africa, Turkey, Iran, Albania, Finland, Democratic Republic of Madagascar, the Philippines, and elsewhere. The U.S. has no appreciable chromite ore reserves. The metal is usually produced by reducing the oxide with aluminum. Chromium is used to harden steel, to manufacture stainless steel, and to form many useful alloys. Much is used in plating to produce a hard, beautiful surface and to prevent corrosion. Chromium is used to give glass an emerald green color. It finds wide use as a catalyst. All compounds of chromium are colored; the most important are the chromates of sodium and potassium (K_2CrO_4) and the dichromates ($K_2Cr_2O_7$) and the potassium and ammonium chrome alums, as $KCr(SO_4)_2 \cdot 12H_2O$. The dichromates are used as oxidizing agents in quantitative analysis, also in tanning leather. Other compounds are of industrial value; lead chromate is chrome yellow, a valued pigment. Chromium compounds are used in the textile industry as mordants, and by the aircraft and other industries for anodizing aluminum. The refractory industry has found chromite useful for forming bricks and shapes, as it has a high melting point, moderate thermal expansion, and stability of crystalline structure. Chromium is an essential trace element for human health. Many chromium compounds, however, are acutely or chronically toxic, and some are carcinogenic. They should be handled with proper safeguards. Natural chromium contains four isotopes. Twenty other isotopes are known. Chromium metal (99.95%) costs about $1000/kg. Commercial grade chromium (99%) costs about $75/kg.

Cobalt — (*Kobald*, from the German, goblin or evil spirit, *cobalos*, Greek, mine), Co; at. wt. 58.933195(5); at. no. 27; m.p. 1495 °C; b.p. 2927 °C; sp. gr. 8.9 (20 °C); valence 2 or 3. Discovered by Brandt about 1735. Cobalt occurs in the mineral *cobaltite*, *smaltite*, and *erythrite*, and is often associated with nickel, silver, lead, copper, and iron ores, from which it is most frequently obtained as a by-product. It is also present in meteorites. Important ore deposits are found in Congo-Kinshasa, Australia, Zambia, Russia, Canada, and elsewhere. The U.S. Geological Survey has announced that the bottom of the north central Pacific Ocean may have cobalt-rich deposits at relatively shallow depths in waters close to the Hawaiian Islands and other U.S. Pacific territories. Cobalt is a brittle, hard metal, closely resembling iron and nickel in appearance. It has a magnetic permeability of about two thirds that of iron. Cobalt tends to exist as a mixture of two allotropes over a wide temperature range; the β-form predominates below 400 °C, and the α above that temperature. The transformation is sluggish and accounts in part for the wide variation in reported data on physical properties of cobalt. It is alloyed with iron, nickel and other metals to make Alnico, an alloy of unusual magnetic strength with many important uses. Stellite alloys, containing cobalt, chromium, and tungsten, are used for high-speed, heavy-duty, high-temperature cutting tools, and for dies. Cobalt is also used in other magnet steels and stainless steels, and in alloys used in jet turbines and gas turbine generators. The metal is used in electroplating because of its appearance, hardness, and resistance to oxidation. The salts have been used for centuries for the production of brilliant and permanent blue colors in porcelain, glass, pottery, tiles, and enamels. It is the principal ingredient in Sevre's and Thenard's blue. A solution of the chloride ($CoCl_2 \cdot 6H_2O$) is used as sympathetic ink. The cobalt ammines are of interest; the oxide and the nitrate are important. Cobalt carefully used in the form of the chloride, sulfate, acetate, or nitrate has been found effective in correcting a certain mineral deficiency disease in animals. Soils should contain 0.13 to 0.30 ppm of cobalt for proper animal nutrition. Cobalt is found in Vitamin B-12, which is essential for human nutrition. Cobalt of 99.9+% purity is priced at about $250/kg. Cobalt-60, an artificial isotope, is an important gamma ray source, and is extensively used as a tracer and a radiotherapeutic agent. Single compact sources of Cobalt-60 vary from about $1 to $10/curie, depending on quantity and specific activity. Thirty isotopes and isomers of cobalt are known.

Columbium — See Niobium.

Copper — (L. *cuprum*, from the island of Cyprus), Cu; at. wt. 63.546(3); at. no. 29; f.p. 1084.62 °C; b.p. 2562 °C; sp. gr. 8.96 (20 °C); valence 1 or 2. The discovery of copper dates from prehistoric times. It is said to have been mined for more than 5000 years. It is one of man's most important metals. Copper is reddish colored, takes on a bright metallic luster, and is malleable, ductile, and a good conductor of heat and electricity (second only to silver in electrical conductivity). The electrical industry is one of the greatest users of copper. Copper occasionally occurs native, and is found in many minerals such as *cuprite, malachite, azurite, chalcopyrite,* and *bornite.* Large copper ore deposits are found in the U.S., Chile, Zambia, Zaire, Peru, and Canada. The most important copper ores are the sulfides, oxides, and carbonates. From these, copper is obtained by smelting, leaching, and by electrolysis. Its alloys, brass and bronze, long used, are still very important; all American coins are now copper alloys; monel and gun metals also contain copper. The most important compounds are the oxide and the sulfate, blue vitriol; the latter has wide use as an agricultural poison and as an algicide in water purification. Copper compounds such as Fehling's solution are widely used in analytical chemistry in tests for sugar. High-purity copper (99.999 + %) is readily available commercially. The price of commercial copper has fluctuated widely. The price of copper in December 2001 was about $1.50/kg. Natural copper contains two isotopes. Twenty-six other radioactive isotopes and isomers are known.

Curium — (Pierre and Marie Curie), Cm; at. wt. (247); at. no. 96; m.p. 1345 °C; sp. gr. 13.51 (calc.); valence 3 and 4. Although curium follows americium in the periodic system, it was actually known before americium and was the third transuranium element to be discovered. It was identified by Seaborg, James, and Ghiorso in 1944 at the wartime Metallurgical Laboratory in Chicago as a result of helium-ion bombardment of ^{239}Pu in the Berkeley, California, 60-inch cyclotron. Visible amounts (30 μg) of ^{242}Cm, in the form of the hydroxide, were first isolated by Werner and Perlman of the University of California in 1947. In 1950, Crane, Wallmann, and Cunningham found that the magnetic susceptibility of microgram samples of CmF_3 was of the same magnitude as that of GdF_3. This provided direct experimental evidence for assigning an electronic configuration to Cm^{+3}. In 1951, the same workers prepared curium in its elemental form for the first time. Sixteen isotopes of curium are now known. The most stable, ^{247}Cm, with a half-life of 16 million years, is so short compared to the Earth's age that any primordial curium must have disappeared long ago from the natural scene. Minute amounts of curium probably exist in natural deposits of uranium, as a result of a sequence of neutron captures and β decays sustained by the very low flux of neutrons naturally present in uranium ores. The presence of natural curium, however, has never been detected. ^{242}Cm and ^{244}Cm are available in multigram quantities. ^{248}Cm has been produced only in milligram amounts. Curium is similar in some regards to gadolinium, its rare-earth homolog, but it has a more complex crystal structure. Curium is silver in color, is chemically reactive, and is more electropositive than aluminum. CmO_2, Cm_2O_3, CmF_3, CmF_4, $CmCl_3$, $CmBr_3$, and CmI_3 have been prepared. Most compounds of trivalent curium are faintly yellow in color. ^{242}Cm generates about three watts of thermal energy per gram. This compares to one-half watt per gram of ^{238}Pu. This suggests use for curium as a power source. ^{244}Cm is now offered for sale by the O.R.N.L. at \$185/mg plus packing charges. ^{248}Cm is available at a cost of \$160/μg, plus packing charges, from the O.R.N.L. Curium absorbed into the body accumulates in the bones, and is therefore very toxic as its radiation destroys the red-cell forming mechanism. The maximum permissible total body burden of ^{244}Cm (soluble) in a human being is 0.3 μCi (microcurie).

Darmstadtium — (Darmstadt, city in Germany), Ds. In 1987 Oganessian et al., at Dubna, claimed discovery of this element. Their experiments indicated the spontaneous fissioning nuclide ^{272}110 with a half-life of 10 ms. More recently a group led by Armbruster at G.S.I. in Darmstadt, Germany, reported evidence of ^{269}110, which was produced by bombarding lead for many days with more than 10^{18} nickel atoms. A detector searched each collision for Element 110's distinct decay sequence. On November 9, 1994, evidence of 110 was detected. In 2003 IUPAC approved the name darmstadtium, symbol Ds, for Element 110. Seven isotopes of Element 110 are now recognized.

Deuterium — an isotope of hydrogen — see Hydrogen.

Dubnium — (named after the Joint Institute of Nuclear Research in Dubna, Russia). Db; at. wt. [262]; at. no. 105. In 1967 G. N. Flerov reported that a Soviet team working at the Joint Institute for Nuclear Research at Dubna may have produced a few atoms of ^{260}105 and ^{261}105 by bombarding ^{243}Am with ^{22}Ne. Their evidence was based on time-coincidence measurements of alpha energies. More recently, it was reported that early in 1970 Dubna scientists synthesized Element 105 and that by the end of April 1970 "had investigated all the types of decay of the new element and had determined its chemical properties." In late April 1970, it was announced that Ghiorso, Nurmia, Harris, K. A. Y. Eskola, and P. L. Eskola, working at the University of California at Berkeley, had positively identi-

fied Element 105. The discovery was made by bombarding a target of ^{249}Cf with a beam of 84 MeV nitrogen nuclei in the Heavy Ion Linear Accelerator (HILAC). When a ^{15}N nucleus is absorbed by a ^{249}Cf nucleus, four neutrons are emitted and a new atom of ^{260}105 with a half-life of 1.6 s is formed. While the first atoms of Element 105 are said to have been detected conclusively on March 5, 1970, there is evidence that Element 105 had been formed in Berkeley experiments a year earlier by the method described. Ghiorso and his associates have attempted to confirm Soviet findings by more sophisticated methods without success.

In October 1971, it was announced that two new isotopes of Element 105 were synthesized with the heavy ion linear accelerator by A. Ghiorso and co-workers at Berkeley. Element ^{261}105 was produced both by bombarding ^{250}Cf with ^{15}N and by bombarding ^{249}Bk with ^{16}O. The isotope emits 8.93-MeV α particles and decays to ^{257}Lr with a half-life of about 1.8 s. Element ^{262}105 was produced by bombarding ^{249}Bk with ^{18}O. It emits 8.45 MeV α particles and decays to ^{258}Lr with a half-life of about 40 s. Nine isotopes of Dubnium are now recognized. Soon after the discovery the names *Hahnium* and *Joliotium,* named after Otto Hahn and Jean-Frederic Joliot and Mme. Joliot-Curie, were suggested as names for Element 105. The IUPAC in August 1997 finally resolved the issue, naming Element 105 Dubnium with the symbol Db. Dubnium is thought to have properties similar to tantalum.

Dysprosium — (Gr. *dysprositos,* hard to get at), Dy; at. wt. 160.500(1); at. no. 66; m.p. 1412 °C; b.p. 2567 °C; sp. gr. 8.551 (25 °C); valence 3. Dysprosium was discovered in 1886 by Lecoq de Boisbaudran, but not isolated. Neither the oxide nor the metal was available in relatively pure form until the development of ion-exchange separation and metallographic reduction techniques by Spedding and associates about 1950. Dysprosium occurs along with other so-called rare-earth or lanthanide elements in a variety of minerals such as *xenotime, fergusonite, gadolinite, euxenite, polycrase,* and *blomstrandine.* The most important sources, however, are from *monazite* and *bastnasite.* Dysprosium can be prepared by reduction of the trifluoride with calcium. The element has a metallic, bright silver luster. It is relatively stable in air at room temperature, and is readily attacked and dissolved, with the evolution of hydrogen, by dilute and concentrated mineral acids. The metal is soft enough to be cut with a knife and can be machined without sparking if overheating is avoided. Small amounts of impurities can greatly affect its physical properties. While dysprosium has not yet found many applications, its thermal neutron absorption cross-section and high melting point suggest metallurgical uses in nuclear control applications and for alloying with special stainless steels. A dysprosium oxide-nickel cermet has found use in cooling nuclear reactor rods. This cermet absorbs neutrons readily without swelling or contracting under prolonged neutron bombardment. In combination with vanadium and other rare earths, dysprosium has been used in making laser materials. Dysprosium-cadmium chalcogenides, as sources of infrared radiation, have been used for studying chemical reactions. The cost of dysprosium metal has dropped in recent years since the development of ion-exchange and solvent extraction techniques, and the discovery of large ore bodies. Thirty-two isotopes and isomers are now known. The metal costs about \$6/g (99.9% purity).

Einsteinium — (Albert Einstein [1879–1955]), Es; at. wt. (252); m.p. 860 °C (est.); at. no. 99. Einsteinium, the seventh transuranic element of the actinide series to be discovered, was identified by Ghiorso and co-workers at Berkeley in December 1952 in debris from the first large thermonuclear explosion, which took place in the Pacific in November 1952. The isotope produced was the 20-day ^{253}Es isotope. In 1961, a sufficient amount of einsteinium was produced to permit separation of a macroscopic amount of ^{253}Es. This sample weighed about 0.01 µg. A special magnetic-type balance was used in making this determination. ^{253}Es so produced was used to produce mendelevium. About 3 µg of einsteinium has been produced at Oak Ridge National Laboratories by irradiating for several years kilogram quantities of ^{239}Pu in a reactor to produce ^{242}Pu. This was then fabricated into pellets of plutonium oxide and aluminum powder, and loaded into target rods for an initial 1-year irradiation at the Savannah River Plant, followed by irradiation in a HFIR (High Flux Isotopic Reactor). After 4 months in the HFIR the targets were removed for chemical separation of the einsteinium from californium. Nineteen isotopes and isomers of einsteinium are now recognized. ^{254}Es has the longest half-life (276 days). Tracer studies using ^{253}Es show that einsteinium has chemical properties typical of a heavy trivalent, actinide element. Einsteinium is extremely radioactive. Great care must be taken when handling it.

Element 112 — In late February 1996, Siguard Hofmann and his collaborators at GSI Darmstadt announced their discovery of Element 112, having 112 protons and 165 neutrons, with an atomic mass of 277. This element was made by bombarding a lead target with high-energy zinc ions. A single nucleus of Element 112 was detected, which decayed after less than 0.001 sec by emitting an α particle, consisting of two protons and two neutrons. This created Element 110_{273}, which in turn decayed by emitting an α particle to form a new isotope of Element 108 and so on. Evidence indicates that nuclei with 162 neutrons are held together more strongly than nuclei with a smaller or larger number of neutrons. This suggests a narrow "peninsula" of relatively stable isotopes around Element 114. GSI scientists are experimenting to bombard targets with ions heavier than zinc to produce Elements 113 and 114. A name has not yet been suggested for Element 112, although the IUPAC suggested the temporary name of ununbium, with the symbol of Uub, when the element was discovered. Element 112 is expected to have properties similar to mercury.

Element 113 — (Ununtrium) See Element 115.

Element 114 — (Ununquadium) Symbol Uuq. Element 114 is the first new element to be discovered since 1996. This element was found by a Russian–American team, including Livermore researchers, by bombarding a sheet of plutonium with a rare form of calcium hoping to make the atoms stick together in a new element. Radiation showed that the new element broke into smaller pieces. Data of radiation collected at the Russian Joint Institute for Nuclear Research in November and December 1998 were analyzed in January 1999. It was found that some of the heavy atoms created when 114 decayed lived up to 30 seconds, which was longer than ever seen before for such a heavy element. This isotope decayed into a previously unknown isotope of Element 112, which itself lasted 15 minutes. That isotope, in turn, decayed to a previously undiscovered isotope of Element 108, which survived 17 minutes. Isotopes of these and those with longer life-times have been predicted for some time by theorists. It appears that these isotopes are on the edge of the "island of stability," and that some

of the isotopes in this region might last long enough for studies of their nuclear behavior and for a chemical evaluation to be made. No name has yet been suggested for Element 114; however, the temporary name of ununquadium with symbol Uuq may be used.

Element 115— (Ununpentium) On February 2, 2004, it was reported that Element 115 had been discovered at the Joint Institute for Nuclear Research (JINR) in Dubna, Russia. Four atoms of this element were produced by JINR physicists and collaborators from the Lawrence Livermore (California) Laboratory using a 248-MeV beam of calcium-48 ions striking a target of americium-243 atoms. The nuclei of these atoms are said to have a life of 90 milliseconds. The relatively long lifetime of Element 115 suggests that these experiments might be getting closer to the "island of stability" long sought to exist by some nuclear physicists. These atoms were thought to decay first to Element 113 by the emission of an alpha particle, then decay further to Element 111 by alpha emission again, and then by three more alpha decay processes to Element 105 (dubnium), which after a long delay from the time of the initial interaction, fissioned. This experiment entailed separating four atoms from trillions of other atoms. A gas-filled separator, employing chemistry, was important in this experiment. Names for Elements 115, Element 113, and Element 111 have not yet been chosen.

Element 116 — (Ununhexium) Symbol Uuh. As of January 2004 it is questionable if this element has been discovered.

Element 117 — (Ununseptium) Symbol Uus. As of January 2004, this element remains undiscovered.

Element 118 — (Ununoctium) Symbol Uuo. In June 1999 it was announced that Elements 118 and 116 had been discovered at the Lawrence Berkeley National Laboratory. A lead target was bombarded for more than 10 days with roughly 1 quintillion krypton ions. The team reported that three atoms of Element 118 were made, which quickly decayed into Elements 116, 114, and elements of lower atomic mass. It was said that the isotopes of Element 118 lasted only about 200 milliseconds, while the isotope of Element 116 lasted only 1.2 milliseconds. It was hoped that these elements might be members of "an island of stability," which had long been sought. At that time it was hoped that a target of bismuth might be bombarded with krypton ions to make Element 119, which, in turn, would decay into Elements 117, 115, and 113.

On July 27, 2001 researchers at the Lawrence Berkeley Laboratory announced that their discovery of Element 118 was being retracted because workers at the GSI Laboratory in Germany and at Japanese laboratories failed to confirm their results. However, it was reported that different experiments at the Livermore Laboratory and Joint Institute from Nuclear Research in Dubna, Russia indicated that Element 116 had since been created.

Researchers at the Australian National Laboratory suggest that super-heavy elements may be more difficult to make than previously thought. Their data suggest the best way to encourage fusion in making super-heavy elements is to combine the lightest projectiles possible with the heaviest possible targets. This would minimize a so-called "quasi-fission process" in which a projectile nucleus steals protons and neutrons from a target nucleus. In this process the two nuclei are said to fly apart without ever having actually combined.

Erbium — (*Ytterby*, a town in Sweden), Er; at. wt. 167.259(3); at. no. 68; m.p. 1529 °C; b.p. 2868 °C; sp. gr. 9.066 (25 °C); valence 3, Erbium, one of the so-called rare-earth elements of the lanthanide series, is found in the minerals mentioned under dysprosium above. In 1842 Mosander separated "yttria," found in the mineral *gadolinite*, into three fractions which he called *yttria*, *erbia*, and *terbia*. The names *erbia* and *terbia* became confused in this early period. After 1860, Mosander's *terbia* was known as *erbia*, and after 1877, the earlier known *erbia* became *terbia*. The *erbia* of this period was later shown to consist of five oxides, now known as *erbia, scandia, holmia, thulia* and *ytterbia*. By 1905 Urbain and James independently succeeded in isolating fairly pure Er_2O_3. Klemm and Bommer first produced reasonably pure erbium metal in 1934 by reducing the anhydrous chloride with potassium vapor. The pure metal is soft and malleable and has a bright, silvery, metallic luster. As with other rare-earth metals, its properties depend to a certain extent on the impurities present. The metal is fairly stable in air and does not oxidize as rapidly as some of the other rare-earth metals. Naturally occurring erbium is a mixture of six isotopes, all of which are stable. Twenty-seven radioactive isotopes of erbium are also recognized. Recent production techniques, using ion-exchange reactions, have resulted in much lower prices of the rare-earth metals and their compounds in recent years. The cost of 99.9% erbium metal is about $21/g. Erbium is finding nuclear and metallurgical uses. Added to vanadium, for example, erbium lowers the hardness and improves workability. Most of the rare-earth oxides have sharp absorption bands in the visible, ultraviolet, and near infrared. This property, associated with the electronic structure, gives beautiful pastel colors to many of the rare-earth salts. Erbium oxide gives a pink color and has been used as a colorant in glasses and porcelain enamel glazes.

Europium — (Europe), Eu; at. wt. 151.964(1); at. no. 63; m.p. 822 °C; b.p. 1596 °C; sp. gr. 5.244 (25 °C); valence 2 or 3. In 1890 Boisbaudran obtained basic fractions from samarium-gadolinium concentrates that had spark spectral lines not accounted for by samarium or gadolinium. These lines subsequently have been shown to belong to europium. The discovery of europium is generally credited to Demarcay, who separated the rare earth in reasonably pure form in 1901. The pure metal was not isolated until recent years. Europium is now prepared by mixing Eu_2O_3 with a 10% excess of lanthanum metal and heating the mixture in a tantalum crucible under high vacuum. The element is collected as a silvery-white metallic deposit on the walls of the crucible. As with other rare-earth metals, except for lanthanum, europium ignites in air at about 150 to 180 °C. Europium is about as hard as lead and is quite ductile. It is the most reactive of the rare-earth metals, quickly oxidizing in air. It resembles calcium in its reaction with water. *Bastnasite* and *monazite* are the principal ores containing europium. Europium has been identified spectroscopically in the sun and certain stars. Europium isotopes are good neutron absorbers and are being studied for use in nuclear control applications. Europium oxide is now widely used as a phosphor activator and europium-activated yttrium vanadate is in commercial use as the red phosphor in color TV tubes. Europium-doped plastic has been used as a laser material. With the development of ion-exchange techniques and special processes, the cost of the metal has been greatly reduced in recent years. Natural europium contains two stable isotopes. Thirty-five other radioactive isotopes and isomers are known. Europium is one of the rarest and most costly of the rare-earth metals. It is priced at about $60/g (99.9% pure).

Fermium — (Enrico Fermi [1901–1954], nuclear physicist), Fm; at. wt. [257]; at. no. 100; m.p. 1527 °C. Fermium, the eighth transuranium element of the actinide series to be discovered, was identified by Ghiorso and co-workers in 1952 in the debris from a thermonuclear explosion in the Pacific in work involving the University of California Radiation Laboratory, the Argonne National Laboratory, and the Los Alamos Scientific Laboratory. The isotope produced was the 20-hour ^{255}Fm. During 1953 and early 1954, while discovery of elements 99 and 100 was withheld from publication for security reasons, a group from the Nobel Institute of Physics in Stockholm bombarded ^{238}U with ^{16}O ions, and isolated a 30-min α-emitter, which they ascribed to ^{250}100, without claiming discovery of the element. This isotope has since been identified positively, and the 30-min half-life confirmed. The chemical properties of fermium have been studied solely with tracer amounts, and in normal aqueous media only the (III) oxidation state appears to exist. The isotope ^{254}Fm and heavier isotopes can be produced by intense neutron irradiation of lower elements such as plutonium by a process of successive neutron capture interspersed with beta decays until these mass numbers and atomic numbers are reached. Twenty isotopes and isomers of fermium are known to exist. ^{257}Fm, with a half-life of about 100.5 days, is the longest lived. ^{250}Fm, with a half-life of 30 min, has been shown to be a product of decay of Element ^{254}102. It was by chemical identification of ^{250}Fm that production of Element 102 (nobelium) was confirmed. Fermium would probably have chemical properties resembling erbium.

Fluorine — (L. and F. *fluere*, flow, or flux), F; at. wt. 18.9984032(5); at. no. 9; m.p. −219.67 °C (1 atm); b.p. −188.12 °C (1 atm); t_c −129.02 °C; density 1.696 g/L (0 °C, 1 atm); liq. den. at b.p. 1.50 g/cm³; valence 1. In 1529, Georgius Agricola described the use of fluorspar as a flux, and as early as 1670 Schwandhard found that glass was etched when exposed to fluorspar treated with acid. Scheele and many later investigators, including Davy, Gay-Lussac, Lavoisier, and Thenard, experimented with hydrofluoric acid, some experiments ending in tragedy. The element was finally isolated in 1886 by Moisson after nearly 74 years of continuous effort. Fluorine occurs chiefly in *fluorspar* (CaF_2) and *cryolite* (Na_3AlF_6), and is in *topaz* and other minerals. It is a member of the halogen family of elements, and is obtained by electrolyzing a solution of potassium hydrogen fluoride in anhydrous hydrogen fluoride in a vessel of metal or transparent fluorspar. Modern commercial production methods are essentially variations on the procedures first used by Moisson. Fluorine is the most electronegative and reactive of all elements. It is a pale yellow, corrosive gas, which reacts with practically all organic and inorganic substances. Finely divided metals, glass, ceramics, carbon, and even water burn in fluorine with a bright flame. Until World War II, there was no commercial production of elemental fluorine. The atom bomb project and nuclear energy applications, however, made it necessary to produce large quantities. Safe handling techniques have now been developed and it is possible at present to transport liquid fluorine by the ton. Fluorine and its compounds are used in producing uranium (from the hexafluoride) and more than 100 commercial fluorochemicals, including many well-known high-temperature plastics. Hydrofluoric acid is extensively used for etching the glass of light bulbs, etc. Fluorochlorohydrocarbons have been extensively used in air conditioning and refrigeration. However, in recent years the U.S. and other countries have been phasing out ozone-depleting substances, such as the fluorochlorohydrocarbons that have been used in these applications. It has been suggested that fluorine might be substituted for hydrogen wherever it occurs in organic compounds, which could lead to an astronomical number of new fluorine compounds. The presence of fluorine as a soluble fluoride in drinking water to the extent of 2 ppm may cause mottled enamel in teeth, when used by children acquiring permanent teeth; in smaller amounts, however, fluorides are said to be beneficial and used in water supplies to prevent dental cavities. Elemental fluorine has been studied as a rocket propellant as it has an exceptionally high specific impulse value. Compounds of fluorine with rare gases have now been confirmed. Fluorides of xenon, radon, and krypton are among those known. Elemental fluorine and the fluoride ion are highly toxic. The free element has a characteristic pungent odor, detectable in concentrations as low as 20 ppb, which is below the safe working level. The recommended maximum allowable concentration for a daily 8-hour time-weighted exposure is 1 ppm. Fluorine is known to have fourteen isotopes.

Francium — (France), Fr; at. no. 87; at. wt. [223]; m.p. 27 °C; valence 1. Discovered in 1939 by Mlle. Marguerite Perey of the Curie Institute, Paris. Francium, the heaviest known member of the alkali metal series, occurs as a result of an alpha disintegration of actinium. It can also be made artificially by bombarding thorium with protons. While it occurs naturally in uranium minerals, there is probably less than an ounce of francium at any time in the total crust of the earth. It has the highest equivalent weight of any element, and is the most unstable of the first 101 elements of the periodic system. Thirty-six isotopes and isomers of francium are recognized. The longest lived ^{223}Fr(Ac, K), a daughter of ^{227}Ac, has a half-life of 21.8 min. This is the only isotope of francium occurring in nature. Because all known isotopes of francium are highly unstable, knowledge of the chemical properties of this element comes from radiochemical techniques. No weighable quantity of the element has been prepared or isolated. The chemical properties of francium most closely resemble cesium. In 1996, researchers Orozco, Sprouse, and co-workers at the State University of New York, Stony Brook, reported that they had produced francium atoms by bombarding ^{18}O atoms at a gold target heated almost to its melting point. Collisions between gold and oxygen nuclei created atoms of francium-210 which had 87 protons and 123 neutrons. This team reported they had generated about 1 million francium-210 ions per second and held 1000 or more atoms at a time for about 20 secs in a magnetic trap they had devised before the atoms decayed or escaped. Enough francium was trapped so that a videocamera could capture the light given off by the atoms as they fluoresced. A cluster of about 10,000 francium atoms appeared as a glowing sphere about 1 mm in diameter. It is thought that the francium atoms could serve as miniature laboratories for probing interactions between electrons and quarks.

Gadolinium — (*gadolinite*, a mineral named for Gadolin, a Finnish chemist), Gd; at. wt. 157.25(3); at. no. 64; m.p. 1313 °C; b.p. 3273 °C; sp. gr. 7.901 (25 °C); valence 3. Gadolinia, the oxide of gadolinium, was separated by Marignac in 1880 and Lecoq de Boisbaudran independently isolated the element from Mosander's "yttria" in 1886. The element was named for the mineral *gadolinite* from which this rare earth was originally obtained. Gadolinium is found in several other minerals, including *monazite* and *bastnasite*, which are of commercial importance. The element has been isolated only in recent years. With the development of ion-exchange and solvent extraction techniques, the availability and price of gadolinium and the other rare-earth metals have greatly improved. Thirty-one isotopes and isomers of gadolinium are now recognized; seven are stable and occur naturally. The metal can be prepared by the reduction of the anhydrous fluoride with metallic calcium. As with other related rare-earth metals, it is silvery white, has a metallic luster, and is malleable and ductile. At room temperature, gadolinium crystallizes in the hexagonal, close-packed α form. Upon heating to 1235 °C, α gadolinium transforms into the β form, which has a body-centered cubic structure. The metal is relatively stable in dry air, but in moist air it tarnishes with the formation of a loosely adhering oxide film which splits off and exposes more surface to oxidation. The metal reacts slowly with water and is soluble in dilute acid. Gadolinium has the highest thermal neutron capture cross-section of any known element (49,000 barns). Natural gadolinium is a mixture of seven isotopes. Two of these, ^{155}Gd and ^{157}Gd, have excellent capture characteristics, but they are present naturally in low concentrations. As a result, gadolinium has a very fast burnout rate and has limited use as a nuclear control rod material. It has been used in making gadolinium yttrium garnets, which have microwave applications. Compounds of gadolinium are used in making phosphors for color TV tubes. The metal has unusual superconductive properties. As little as 1% gadolinium has been found to improve the workability and resistance of iron, chromium, and related alloys to high temperatures and oxidation. Gadolinium ethyl sulfate has extremely low noise characteristics and may find use in duplicating the performance of amplifiers, such as the maser. The metal is ferromagnetic. Gadolinium is unique for its high magnetic moment and for its special Curie temperature (above which ferromagnetism vanishes) lying just at room temperature. This suggests uses as a magnetic component that senses hot and cold. The price of the metal is about $5/g (99.9% purity).

Gallium — (L. *Gallia*, France), Ga; at. wt. 69.723(1); at. no. 31; m.p. 29.76 °C; b.p. 2204 °C; sp. gr. 5.904 (29.6 °C) solid; sp. gr. 6.095 (29.6 °C) liquid; valence 2 or 3. Predicted and described by Mendeleev as ekaaluminum, and discovered spectroscopically by Lecoq de Boisbaudran in 1875, who in the same year obtained the free metal by electrolysis of a solution of the hydroxide in KOH, Gallium is often found as a trace element in *diaspore, sphalerite, germanite, bauxite,* and *coal.* Some flue dusts from burning coal have been shown to contain as much as 1.5% gallium. It is the only metal, except for mercury, cesium, and rubidium, which can be liquid near room temperatures; this makes possible its use in high-temperature thermometers. It has one of the longest liquid ranges of any metal and has a low vapor pressure even at high temperatures. There is a strong tendency for gallium to supercool below its freezing point. Therefore, seeding may be necessary to initiate solidification. Ultra-pure gallium has a beautiful, silvery appearance, and the solid metal exhibits a conchoidal fracture similar to glass. The metal expands 3.1% on solidifying; therefore, it should not be stored in glass or metal containers, as they may break as the metal solidifies. Gallium wets glass or porcelain, and forms a brilliant mirror when it is painted on glass. It is widely used in doping semiconductors and producing solid-state devices such as transistors. High-purity gallium is attacked slowly only by mineral acids. Magnesium gallate containing divalent impurities such as Mn^{+2} is finding use in commercial ultraviolet activated powder phosphors. Gallium nitride has been used to produce blue light-emitting diodes such as those used in CD and DVD readers. Gallium has found application in the Gallex Detector Experiment located in the Gran Sasso Underground Laboratory in Italy. This underground facility has been built by the Italian Istituto Nazionale di Fisica Nucleare in the middle of a highway tunnel through the Abruzzese mountains, about 150 km east of Rome. In this experiment, 30.3 tons of gallium in the form of 110 tons of $GaCl_3$-HCl solution are being used to detect solar neutrinos. The production of ^{71}Ge from gallium is being measured. Gallium arsenide is capable of converting electricity directly into coherent light. Gallium readily alloys with most metals, and has been used as a component in low melting alloys. Its toxicity appears to be of a low order, but it should be handled with care until more data are forthcoming. Natural gallium contains two stable isotopes. Twenty-six other isotopes, one of which is an isomer, are known. The metal can be supplied in ultrapure form (99.99999+%). The cost is about $5/g (99.999%).

Germanium — (L. *Germania*, Germany), Ge; at. wt. 72.63(1); at. no. 32; m.p. 938.25 °C; b.p. 2833 °C; sp. gr. 5.323 (25 °C); valence 2 and 4. Predicted by Mendeleev in 1871 as ekasilicon, and discovered by Winkler in 1886. The metal is found in *argyrodite*, a sulfide of germanium and silver; in *germanite*, which contains 8% of the element; in zinc ores; in coal; and in other minerals. The element is frequently obtained commercially from flue dusts of smelters processing zinc ores, and has been recovered from the by-products of combustion of certain coals. Its presence in coal insures a large reserve of the element in the years to come. Germanium can be separated from other metals by fractional distillation of its volatile tetrachloride. The tetrachloride may then be hydrolyzed to give GeO_2; the dioxide can be reduced with hydrogen to give the metal. Recently developed zone-refining techniques permit the production of germanium of ultra-high purity. The element is a gray-white metalloid, and in its pure state is crystalline and brittle, retaining its luster in air at room temperature. It is a very important semiconductor material. Zone-refining techniques have led to production of crystalline germanium for semiconductor use with an impurity of only one part in 10^{10}. Doped with arsenic, gallium, or other elements, it is used as a transistor element in thousands of electronic applications. Its application in fiber optics and infrared optical systems now provides the largest use for germanium. Germanium is also finding many other applications including use as an alloying agent, as a phosphor in fluorescent lamps, and as a catalyst. Germanium and germanium oxide are transparent to the infrared and are used in infrared spectrometers and other optical equipment, including extremely sensitive infrared detectors. Germanium oxide's high index of refraction and dispersion make it useful as a component of glasses used in wide-angle camera lenses and microscope objectives. The field of organogermanium chemistry is becoming increasingly important. Certain germanium compounds have a low mammalian toxicity, but a marked activity against certain bacteria, which makes them of interest as chemotherapeutic agents. The cost of germanium is about $10/g (99.999% purity). Thirty isotopes and isomers are known, five of which occur naturally.

Gold — (Sanskrit *Jval;* Anglo-Saxon *gold*), Au (L. *aurum,* gold); at. wt. 196.966569(4); at. no. 79; m.p. 1064.18 °C; b.p. 2856 °C;

sp. gr. ~19.3 (20 °C); valence 1 or 3. Known and highly valued from earliest times, gold is found in nature as the free metal and in tellurides; it is very widely distributed and is almost always associated with quartz or pyrite. It occurs in veins and alluvial deposits, and is often separated from rocks and other minerals by sluicing and panning operations. About 25% of the world's gold output comes from South Africa, and about two thirds of the total U.S. production now comes from South Dakota and Nevada. The metal is recovered from its ores by cyaniding, amalgamating, and smelting processes. Refining is also frequently done by electrolysis. Gold occurs in sea water to the extent of 0.1 to 2 mg/ton, depending on the location where the sample is taken. As yet, no method has been found for recovering gold from sea water profitably. It is estimated that all the gold in the world, so far refined, could be placed in a single cube 60 ft on a side. Of all the elements, gold in its pure state is undoubtedly the most beautiful. It is metallic, having a yellow color when in a mass, but when finely divided it may be black, ruby, or purple. The Purple of Cassius is a delicate test for auric gold. It is the most malleable and ductile metal; 1 oz. of gold can be beaten out to 300 ft². It is a soft metal and is usually alloyed to give it more strength. It is a good conductor of heat and electricity, and is unaffected by air and most reagents. It is used in coinage and is a standard for monetary systems in many countries. It is also extensively used for jewelry, decoration, dental work, and for plating. It is used for coating certain space satellites, as it is a good reflector of infrared and is inert. Gold, like other precious metals, is measured in troy weight; when alloyed with other metals, the term *carat* is used to express the amount of gold present, 24 carats being pure gold. For many years the value of gold was set by the U.S. at $20.67/troy ounce; in 1934 this value was fixed by law at $35.00/troy ounce, 9/10th fine. On March 17, 1968, because of a gold crisis, a two-tiered pricing system was established whereby gold was still used to settle international accounts at the old $35.00/troy ounce price while the price of gold on the private market would be allowed to fluctuate. Since this time, the price of gold on the free market has fluctuated widely. The price of gold on the free market reached a price of $620/troy oz. in January 1980. More recently, the U.K. and other nations, including the I.M.F. have sold or threatened to sell a sizeable portion of their gold reserves. This has caused wide fluctuations in the price of gold. Because this has damaged the economy of some countries, a moratorium for a few years has been declared. This has tended to stabilize temporarily the price of gold. The most common gold compounds are auric chloride ($AuCl_3$) and chlorauric acid ($HAuCl_4$), the latter being used in photography for toning the silver image. Gold has forty-eight recognized isotopes and isomers; ^{198}Au, with a half-life of 2.7 days, is used for treating cancer and other diseases. Disodium aurothiomalate is administered intramuscularly as a treatment for arthritis. A mixture of one part nitric acid with three of hydrochloric acid is called *aqua regia* (because it dissolved gold, the King of Metals). Gold is available commercially with a purity of 99.999+%. For many years the temperature assigned to the freezing point of gold has been 1063.0 °C; this has served as a calibration point for the International Temperature Scales (ITS-27 and ITS-48) and the International Practical Temperature Scale (IPTS-48). In 1968, a new International Practical Temperature Scale (IPTS-68) was adopted, which demanded that the freezing point of gold be changed to 1064.43 °C. In 1990 a new International Temperature Scale (ITS-90) was adopted bringing the t.p.

(triple point) of H_2O (t_{90} (°C)) to 0.01 °C and the freezing point of gold to 1064.18 °C. The specific gravity of gold has been found to vary considerably depending on temperature, how the metal is precipitated, and cold-worked. As of December 2001, gold was priced at about $275/troy oz. ($8.50/g).

Hafnium — (*Hafnia*, Latin name for Copenhagen), Hf; at. wt. 178.49(2); at. no. 72; m.p. 2233 °C; b.p. 4603 °C; sp. gr. 13.31 (20 °C); valence 4. Hafnium was thought to be present in various minerals and concentrations many years prior to its discovery, in 1923, credited to D. Coster and G. von Hevesey. On the basis of the Bohr theory, the new element was expected to be associated with zirconium. It was finally identified in *zircon* from Norway, by means of X-ray spectroscopic analysis. It was named in honor of the city in which the discovery was made. Most zirconium minerals contain 1 to 5% hafnium. It was originally separated from zirconium by repeated recrystallization of the double ammonium or potassium fluorides by von Hevesey and Jantzen. Metallic hafnium was first prepared by van Arkel and deBoer by passing the vapor of the tetraiodide over a heated tungsten filament. Almost all hafnium metal now produced is made by reducing the tetrachloride with magnesium or with sodium (Kroll Process). Hafnium is a ductile metal with a brilliant silver luster. Its properties are considerably influenced by the impurities of zirconium present. Of all the elements, zirconium and hafnium are two of the most difficult to separate. Their chemistry is almost identical; however, the density of zirconium is about half that of hafnium. Very pure hafnium has been produced, with zirconium being the major impurity. Natural hafnium contains six isotopes, one of which is slightly radioactive. Hafnium has a total of 41 recognized isotopes and isomers. Because hafnium has a good absorption cross section for thermal neutrons (almost 600 times that of zirconium), has excellent mechanical properties, and is extremely corrosion resistant, it is used for reactor control rods. Such rods are used in nuclear submarines. Hafnium has been successfully alloyed with iron, titanium, niobium, tantalum, and other metals. Hafnium carbide is the most refractory binary composition known, and the nitride is the most refractory of all known metal nitrides (m.p. 3310 °C). Hafnium is used in gas-filled and incandescent lamps, and is an efficient "getter" for scavenging oxygen and nitrogen. Finely divided hafnium is pyrophoric and can ignite spontaneously in air. Care should be taken when machining the metal or when handling hot sponge hafnium. At 700 °C hafnium rapidly absorbs hydrogen to form the composition $HfH_{1.86}$. Hafnium is resistant to concentrated alkalis, but at elevated temperatures reacts with oxygen, nitrogen, carbon, boron, sulfur, and silicon. Halogens react directly to form tetrahalides. The price of the metal is about $2/g. The yearly demand for hafnium in the U.S. is now in excess of 50,000 kg.

Hahnium — A name previously used for Element 105, now named *dubnium*.

Hassium — (named for the German state, Hesse) Hs; at. wt. [277]; at. no. 108. This element was first synthesized and identified in 1964 by the same G.S.I. Darmstadt Group who first identified *Bohrium* and *Meitnerium*. Presumably this element has chemical properties similar to osmium. Isotope $^{265}108$ was produced using a beam of ^{58}Fe projectiles, produced by the Universal Linear Accelerator (UNILAC) to bombard a ^{208}Pb target. Discovery of *Bohrium* and *Meitnerium* was made using detection of isotopes with odd proton and neutron numbers.

Elements having even atomic numbers have been thought to be less stable against spontaneous fusion than odd elements. The production of $^{265}108$ in the same reaction as was used at G.S.I. was confirmed at Dubna with detection of the seventh member of the decay chain ^{253}Es. Isotopes of *Hassium* are believed to decay by spontaneous fission, explaining why 109 was produced before 108. Isotope $^{265}108$ and $^{266}108$ are thought to decay to $^{261}106$, which in turn decay to $^{257}104$ and $^{253}102$. The IUPAC adopted the name *Hassium* after the German state of Hesse in September 1997. In June 2001 it was announced that hassium is now the heaviest element to have its chemical properties analyzed. A research team at the UNILAC heavy-ion accelerator in Darmstadt, Germany built an instrument to detect and analyze hassium. Atoms of curium-248 were collided with atoms of magnesium-26, producing about 6 atoms of hassium with a half-life of 9 sec. This was sufficiently long to obtain data showing that hassium atoms react with oxygen to form hassium oxide molecules. These condensed at a temperature consistent with the behavior of Group 8 elements. This experiment appears to confirm hassium's location under osmium in the periodic table.

Helium — (Gr. *helios*, the sun), He; at. wt. 4.002602(2); at. no. 2; b.p. — 268.93 °C; t_c −267.96 °C; density 0.1785 g/L (0 °C, 1 atm); liquid density 0.125 g/mL at. b.p.; valence usually 0. Evidence of the existence of helium was first obtained by Janssen during the solar eclipse of 1868 when he detected a new line in the solar spectrum; Lockyer and Frankland suggested the name *helium* for the new element; in 1895, Ramsay discovered helium in the uranium mineral *cleveite*, and it was independently discovered in cleveite by the Swedish chemists Cleve and Langlet about the same time. Rutherford and Royds in 1907 demonstrated that α particles are helium nuclei. Except for hydrogen, helium is the most abundant element found throughout the universe. Helium is extracted from natural gas; all natural gas contains at least trace quantities of helium. It has been detected spectroscopically in great abundance, especially in the hotter stars, and it is an important component in both the proton–proton reaction and the carbon cycle, which account for the energy of the sun and stars. The fusion of hydrogen into helium provides the energy of the hydrogen bomb. The helium content of the atmosphere is about 1 part in 200,000. It is present in various radioactive minerals as a decay product. Much of the world's supply of helium is obtained from wells in Texas, Colorado, and Kansas. The only other known helium extraction plants, outside the United States, in 1999 were in Poland, Russia, China, Algeria, and India. The cost of helium has fallen from $2500/ft³ in 1915 to about 2.5¢/cu.ft. (.028 cu meters) in 1999. Helium has the lowest melting point of any element and has found wide use in cryogenic research, as its boiling point is close to absolute zero. Its use in the study of superconductivity is vital. Using liquid helium, Kurti and co-workers, and others, have succeeded in obtaining temperatures of a few microkelvins by the adiabatic demagnetization of copper nuclei, starting from about 0.01 K. Liquid helium (He⁴) exists in two forms: He⁴I and He⁴II, with a sharp transition point at 2.174 K (3.83 cm Hg). He⁴I (above this temperature) is a normal liquid, but He⁴II (below it) is unlike any other known substance. It expands on cooling; its conductivity for heat is enormous; and neither its heat conduction nor viscosity obeys normal rules. It has other peculiar properties. Helium is the only liquid that cannot be solidified by lowering the temperature. It remains liquid down to absolute zero at ordinary pressures, but it can readily be solidified by increasing the pressure. Solid ^{3}He and ^{4}He are unusual in that both can readily be changed in volume by more than 30% by application of pressure. The specific heat of helium gas is unusually high. The density of helium vapor at the normal boiling point is also very high, with the vapor expanding greatly when heated to room temperature. Containers filled with helium gas at 5 to 10 K should be treated as though they contained liquid helium due to the large increase in pressure resulting from warming the gas to room temperature. While helium normally has a 0 valence, it seems to have a weak tendency to combine with certain other elements. Means of preparing helium diflouride have been studied, and species such as HeNe and the molecular ions He$^+$ and He^{++} have been investigated. Helium is widely used as an inert gas shield for arc welding; as a protective gas in growing silicon and germanium crystals, and in titanium and zirconium production; as a cooling medium for nuclear reactors, and as a gas for supersonic wind tunnels. A mixture of helium and oxygen is used as an artificial atmosphere for divers and others working under pressure. Different ratios of He/O$_2$ are used for different depths at which the diver is operating. Helium is extensively used for filling balloons as it is a much safer gas than hydrogen. One of the recent largest uses for helium has been for pressurizing liquid fuel rockets. A Saturn booster such as used on the Apollo lunar missions required about 13 million ft^3 of helium for a firing, plus more for checkouts. Liquid helium's use in magnetic resonance imaging (MRI) continues to increase as the medical profession accepts and develops new uses for the equipment. This equipment is providing accurate diagnoses of problems where exploratory surgery has previously been required to determine problems. Another medical application that is being developed uses MRI to determine by blood analysis whether a patient has any form of cancer. Lifting gas applications are increasing. Various companies in addition to Goodyear, are now using "blimps" for advertising. The Navy and the Air Force are investigating the use of airships to provide early warning systems to detect low-flying cruise missiles. The Drug Enforcement Agency has used radar-equipped blimps to detect drug smugglers along the southern border of the U.S. In addition, NASA is currently using helium-filled balloons to sample the atmosphere in Antarctica to determine what is depleting the ozone layer that protects Earth from harmful U.V. radiation. Research on and development of materials which become superconductive at temperatures well above the boiling point of helium could have a major impact on the demand for helium. Less costly refrigerants having boiling points considerably higher could replace the present need to cool such superconductive materials to the boiling point of helium. Natural helium contains two stable isotopes ^{3}He and ^{4}He. ^{3}He is present in very small quantities. Six other isotopes of helium are now recognized.

Holmium — (L. *Holmia*, for Stockholm), Ho; at. wt. 164.93032(2); at. no 67; m.p. 1472 °C; b.p. 2700 °C; sp. gr. 8.795 (25 °C); valence + 3. The spectral absorption bands of holmium were noticed in 1878 by the Swiss chemists Delafontaine and Soret, who announced the existence of an "Element X." Cleve, of Sweden, later independently discovered the element while working on erbia earth. The element is named after Cleve's native city. Pure holmia, the yellow oxide, was prepared by Homberg in 1911. Holmium occurs in *gadolinite, monazite,* and in other rare-earth minerals. It is commercially obtained from monazite, occurring in that mineral to the extent of about 0.05%. It has been isolated by the reduction of its anhydrous chloride or fluoride with calcium metal. Pure holmium has a metallic to bright silver luster. It is relatively soft and malleable, and is stable in dry air at room temperature, but rapidly oxidizes in moist air and at elevated temperatures. The metal has unusual magnetic properties. Few uses have yet been found for the element. The element, as with other rare earths, seems to have a low acute toxic rating. Natural holmium consists of one isotope ^{165}Ho, which is not radioactive. Holmium has 49 other isotopes known, all of which are radioactive. The price of 99.9% holmium metal is about $20/g.

Hydrogen — (Gr. *hydro*, water, and *genes*, forming), H; at. wt. [1.00784; 100811]; at. no. 1; m.p. −259.1 °C; b.p. −252.76 °C; t_c −240.18; density 0.08988 g/L; density (liquid) 0.0708 g/mL (−253 °C); density (solid) 0.0706 g/mL (−262 °C); valence 1. Hydrogen was prepared many years before it was recognized as a distinct substance by Cavendish in 1766. It was named by Lavoisier. Hydrogen is the most abundant of all elements in the universe, and it is thought that the heavier elements were, and still are, being built from hydrogen and helium. It has been estimated that hydrogen makes up more than 90% of all the atoms or three quarters of the mass of the universe. It is found in the sun and most stars, and plays an important part in the proton–proton reaction and carbon–nitrogen cycle, which accounts for the energy of the sun and stars. It is thought that hydrogen is a major component of the planet Jupiter and that at some depth in the planet's interior the pressure is so great that solid molecular hydrogen is converted into solid metallic hydrogen. In 1973, it was reported that a group of Russian experimenters may have produced metallic hydrogen at a pressure of 2.8 Mbar. At the transition the density changed from 1.08 to 1.3 g/cm^3. Earlier, in 1972, a Livermore (California) group also reported on a similar experiment in which they observed a pressure-volume point centered at 2 Mbar. It has been predicted that metallic hydrogen may be metastable; others have predicted it would be a superconductor at room temperature. On Earth, hydrogen occurs chiefly in combination with oxygen in water, but it is also present in organic matter such as living plants, petroleum, coal, etc. It is present as the free element in the atmosphere, but only to the extent of less than 1 ppm by volume. It is the lightest of all gases, and combines with other elements, sometimes explosively, to form compounds. Great quantities of hydrogen are required commercially for the fixation of nitrogen from the air in the Haber ammonia process and for the hydrogenation of fats and oils. It is also used in large quantities in methanol production, in hydrodealkylation, hydrocracking, and hydrodesulfurization. It is also used as a rocket fuel, for welding, for production of hydrochloric acid, for the reduction of metallic ores, and for filling balloons. The lifting power of 1 ft^3 of hydrogen gas is about 0.076 lb at 0 °C, 760 mm pressure. Production of hydrogen in the U.S. alone now amounts to about 3 billion cubic feet per year. It is prepared by the action of steam on heated carbon, by decomposition of certain hydrocarbons with heat, by the electrolysis of water, or by the displacement from acids by certain metals. It is also produced by the action of sodium or potassium hydroxide on aluminum. Liquid hydrogen is important in cryogenics and in the study of superconductivity, as its melting point is only a 20 °C above absolute zero. Hydrogen consists of three isotopes, most of which is ^{1}H. The ordinary isotope of hydrogen, H, is known as *protium*. In 1932, Urey announced the discovery of a stable isotope, deuterium (^{2}H or D) with an atomic weight of 2. Deuterium is present in natural hydrogen to the extent of 0.015%. Two years later an unstable isotope, tritium (^{3}H), with an atomic weight of 3 was discovered. Tritium has a half-life of about 12.32 years. Tritium atoms are also present in natural hydrogen but in a much smaller proportion. Tritium is readily produced in nuclear reactors and is used in the production of the hydrogen bomb. It is also used as a radioactive agent in making luminous paints, and as a tracer. On August 27, 2001 Russian, French, and Japanese physicists working at the Joint Institute for Nuclear Research near Moscow reported they had made "super-heavy hydrogen," which had a nucleus with one proton and four neutrons. Using an accelerator, they used a beam of helium-6 nuclei to strike a hydrogen target, which resulted in the occasional production of a hydrogen-5 nucleus plus a helium-2 nucleus. These unstable particles quickly disintegrated. This resulted in two protons from the He-2, a triton, and two neutrons from the H-5 breakup. Deuterium gas is readily available, without permit, at about $1/l. Heavy water, deuterium oxide (D_2O), which is used as a moderator to slow down neutrons, is available without permit at a cost of 6c to $1/g, depending on quantity and purity. About 1000 tons (4,400,000 kg) of deuterium oxide (heavy water) are now in use at the Sudbury (Ontario) Neutrino Observatory. This observatory is taking data to provide new revolutionary insight into the properties of neutrinos and into the core of the sun. The heavy water is on loan from Atomic Energy of Canada, Ltd. (AECL). The observatory and detectors are located 6800 ft (2072 m) deep in the Creighton mine of the International Nickel Co., near Sudbury. The heavy water is contained in an acrylic vessel, 12 m in diameter. Neutrinos react with the heavy water to produce Cherenkov radiation. This light is then detected with 9600 photomultiplier tubes surrounding the vessel. The detector laboratory is immensely clean to reduce background radiation, which otherwise hides the very weak signals from neutrinos. Quite apart from isotopes, it has been shown that hydrogen gas under ordinary conditions is a mixture of two kinds of molecules, known as *ortho-* and *para-*hydrogen, which differ from one another by the spins of their electrons and nuclei. Normal hydrogen at room temperature contains 25% of the *para* form and 75% of the *ortho* form. The *ortho* form cannot be prepared in the pure state. Since the two forms differ in energy, the physical properties also differ. The melting and boiling points of *para*hydrogen are about 0.1 °C lower than those of normal hydrogen. Consideration is being given to an entire economy based on solar- and nuclear-generated hydrogen. Located in remote regions, power plants would electrolyze sea water; the hydrogen produced would travel to distant cities by pipelines. Pollution-free hydrogen could replace natural gas, gasoline, etc., and could serve as a reducing agent in metallurgy, chemical processing, refining, etc. It could also be used to convert trash into methane and ethylene. Public acceptance, high capital investment, and the high present cost of hydrogen with respect to current fuels are but a few of the problems facing establishment of such an economy. Hydrogen is being investigated as a substitute for deep-sea diving applications below 300 m. Hydrogen is readily available from air product suppliers.

Indium — (from the brilliant indigo line in its spectrum), In; at. wt. 114.818(3); at. no. 49; m.p. 156.60 °C; b.p. 2072 °C; sp. gr. 7.31 (20 °C); valence 1, 2, or 3. Discovered by Reich and Richter, who later isolated the metal. Indium is most frequently associated with zinc materials, and it is from these that most commercial indium is now obtained; however, it is also found in iron, lead, and copper ores. Until 1924, a gram or so constituted the world's supply of this element in isolated form. It is probably about as abundant as silver. About 4 million troy ounces of indium are now produced annually in the Free World. Canada is presently producing more than 1,000,000 troy ounces annually. The present cost of indium is about $2 to $10/g, depending on quantity and purity. It is available in ultrapure form. Indium is a very soft, silvery-white metal with a brilliant luster. The pure metal gives a high-pitched "cry" when bent. It wets glass, as does gallium. It has found application in making low-melting alloys; an alloy of 24% indium–76% gallium is liquid at room temperature. Indium is used in making bearing alloys, germanium transistors, rectifiers, thermistors, liquid crystal displays, high definition television, batteries, and photoconductors. It can be plated onto metal and evaporated onto glass, forming a mirror as good as that made with silver but with more resistance to atmospheric corrosion. There is evidence that indium has a low order of toxicity; however, care should be taken until further information is available. Seventy isotopes and isomers are now recognized (more than any other element). Natural indium contains two isotopes. One is stable. The other, ^{115}In, comprising 95.71% of natural indium is slightly radioactive with a very long half-life.

Iodine — (Gr. *iodes*, violet), I; at. wt. 126.90447(3); at. no. 53; m.p. 113.7 °C; b.p. 184.4 °C; t_c 546 °C; density of the gas 11.27 g/L; sp. gr. solid 4.93 (20 °C); valence 1, 3, 5, or 7. Discovered by Courtois in 1811. Iodine, a halogen, occurs sparingly in the form of iodides in sea water from which it is assimilated by seaweeds, in Chilean saltpeter and nitrate-bearing earth, known as *caliche* in brines from old sea deposits, and in brackish waters from oil and salt wells. Ultrapure iodine can be obtained from the reaction of potassium iodide with copper sulfate. Several other methods of isolating the element are known. Iodine is a bluish-black, lustrous solid, volatilizing at ordinary temperatures into a blue-violet gas with an irritating odor; it forms compounds with many elements, but is less active than the other halogens, which displace it from iodides. Iodine exhibits some metallic-like properties. It dissolves readily in chloroform, carbon tetrachloride, or carbon disulfide to form beautiful purple solutions. It is only slightly soluble in water. Iodine compounds are important in organic chemistry and very useful in medicine. Forty-two isotopes and isomers are recognized. Only one stable isotope, ^{127}I, is found in nature. The artificial radioisotope ^{131}I, with a half-life of 8 days, has been used in treating the thyroid gland. The most common compounds are the iodides of sodium and potassium (KI) and the iodates (KIO_3). Lack of iodine is the cause of goiter. Iodides and thyroxin, which contains iodine, are used internally in medicine, and a solution of KI and iodine in alcohol is used for external wounds. Potassium iodide finds use in photography. The deep blue color with starch solution is characteristic of the free element. Care should be taken in handling and using iodine, as contact with the skin can cause lesions; iodine vapor is intensely irritating to the eyes and mucous membranes. Elemental iodine costs about 25 to 75¢/g depending on purity and quantity.

Iridium — (L. *iris*, rainbow), Ir; at. wt. 192.217(3); at. no. 77; m.p. 2446 °C; b.p. 4428 °C; sp. gr. 22.562 (20 °C); valence 3 or 4. Discovered in 1803 by Tennant in the residue left when crude platinum is dissolved by aqua regia. The name iridium is appropriate, for its salts are highly colored. Iridium, a metal of the platinum family, is white, similar to platinum, but with a slight yellowish cast. It is very hard and brittle, making it very hard to machine, form, or work. It is the most corrosion-resistant metal known, and was used in making the standard meter bar of Paris, which is a 90% platinum–10% iridium alloy. This meter bar was replaced in 1960 as a fundamental unit of length (see under Krypton). Iridium is not attacked by any of the acids nor by aqua regia, but is attacked by molten salts, such as NaCl and NaCN. Iridium occurs uncombined in nature with platinum and other metals of this family in alluvial deposits. It is recovered as a by-product from the nickel mining industry. The largest reserves and production of the platinum group of metals, which includes iridium, is in South Africa, followed by Russia and Canada. The U.S. has only one active mine, located at Nye, MT. The presence of iridium has recently been used in examining the Cretaceous-Tertiary (K-T) boundary. Meteorites contain small amounts of iridium. Because iridium is found widely distributed at the K-T boundary, it has been suggested that a large meteorite or asteroid collided with the Earth, killing the dinosaurs, and creating a large dust cloud and crater. Searches for such a crater point to one in the Yucatan, known as Chicxulub. Iridium has found use in making crucibles and apparatus for use at high temperatures. It is also used for electrical contacts. Its principal use is as a hardening agent for platinum. With osmium, it forms an alloy that is used for tipping pens and compass bearings. The specific gravity of iridium is only very slightly lower than that of osmium, which has been generally credited as being the heaviest known element. Calculations of the densities of iridium and osmium from the space lattices give values of 22.65 and 22.61 g/cm^3, respectively. These values may be more reliable than actual physical measurements. At present, therefore, we know that either iridium or osmium is the densest known element, but the data do not yet allow selection between the two. Natural iridium contains two stable isotopes. Forty-five other isotopes, all radioactive, are now recognized. Iridium (99.9%) costs about $100/g.

Iron — (Anglo-Saxon, *iron*), Fe (L. *ferrum*); at. wt. 55.845(2); at. no. 26; m.p. 1538 °C; b.p. 2861 °C; sp. gr. 7.874 (20 °C); valence 2, 3, 4, or 6. The use of iron is prehistoric. Genesis mentions that Tubal-Cain, seven generations from Adam, was "an instructor of every artificer in brass and iron." A remarkable iron pillar, dating to about A.D. 400, remains standing today in Delhi, India. This solid shaft of wrought iron is about 7¼ m high by 40 cm in diameter. Corrosion to the pillar has been minimal although it has been exposed to the weather since its erection. Iron is a relatively abundant element in the universe. It is found in the sun and many types of stars in considerable quantity. It has been suggested that the iron we have here on Earth may have originated in a supernova. Iron is a very difficult element to produce in ordinary nuclear reactions, such as would take place in the sun. Iron is found native as a principal component of a class of iron–nickel meteorites known as *siderites*, and is a minor constituent of the other two classes of meteorites. The core of the Earth, 2150 miles in radius, is thought to be largely composed of iron with about 10% occluded hydrogen. The metal is the fourth most abundant element, by weight, making up the crust of the Earth. The most common ore is *hematite* (Fe$_2$O$_3$). Magnetite (Fe$_3$O$_4$) is frequently seen as *black sands* along beaches and banks of streams. *Lodestone* is another form of magnetite. *Taconite* is becoming increasingly important as a commercial ore. Iron is a vital constituent of plant and animal life, and appears in hemoglobin. The pure metal is not often encountered in commerce, but is usually alloyed with carbon or other metals. The pure metal is very reactive chemically, and rapidly corrodes, especially in moist air or at elevated temperatures. It has four allotropic forms, or ferrites, known as α, β, γ, and δ, with transition points at 700, 928, and 1530 °C. The α form is magnetic, but when transformed into the β form, the magnetism disappears although the lattice remains unchanged. The relations of these forms are peculiar. Pig iron is an alloy containing about 3% carbon with varying amounts of S, Si, Mn, and P. It is hard, brittle, fairly fusible, and is used to produce other alloys, including steel. Wrought iron contains only a few tenths of a percent of carbon, is tough, malleable, less fusible, and usually has a "fibrous" structure. Carbon steel is an alloy of iron with carbon, with small amounts of Mn, S, P, and Si. Alloy steels are carbon steels with other additives such as nickel, chromium, vanadium, etc. Iron is the cheapest and most abundant, useful, and important of all metals. Natural iron contains four isotopes. Twenty-six other isotopes and isomers, all radioactive, are now recognized.

Krypton — (Gr. *kryptos*, hidden), Kr; at. wt. 83.798(2); at. no. 36; m.p. –157.36 °C; b.p. –153.34 ± 0.10 °C; t_c –63.67 °C; density 3.733 g/L (0 °C); valence usually 0. Discovered in 1898 by Ramsay and Travers in the residue left after liquid air had nearly boiled away, krypton is present in the air to the extent of about 1 ppm. The atmosphere of Mars has been found to contain 0.3 ppm of krypton. It is one of the "noble" gases. It is characterized by its brilliant green and orange spectral lines. Naturally occurring krypton contains six stable isotopes. Thirty other unstable isotopes and isomers are now recognized. The spectral lines of krypton are easily produced and some are very sharp. In 1960 it was internationally agreed that the fundamental unit of length, the meter, should be defined in terms of the orange-red spectral line of ^{86}Kr. This replaced the standard meter of Paris, which was defined in terms of a bar made of a platinum-iridium alloy. In October 1983 the meter was again redefined by the International Bureau of Weights and Measures as being the length of path traveled by light in a vacuum during a time interval of 1/299,792,458 of a second. Solid krypton is a white crystalline substance with a face-centered cubic structure that is common to all the rare gases. While krypton is generally thought of as a noble gas that normally does not combine with other elements, the existence of some krypton compounds has been established. Krypton difluoride has been prepared in gram quantities and can be made by several methods. A higher fluoride of krypton and a salt of an oxyacid of krypton also have been prepared. Molecule-ions of ArKr$^+$ and KrH$^+$ have been identified and investigated, and evidence is provided for the formation of KrXe or KrXe$^+$. Krypton clathrates have been prepared with hydroquinone and phenol. ^{85}Kr has found recent application in chemical analysis. By imbedding the isotope in various solids, *kryptonates* are formed. The activity of these kryptonates is sensitive to chemical reactions at the surface. Estimates of the concentration of reactants are therefore made possible. Krypton is used in certain photographic flash lamps for high-speed photography. Uses thus far have been limited because of its high cost. Krypton gas presently costs about $690/100 L.

Kurchatovium — See Rutherfordium.

Lanthanum — (Gr. *lanthanein*, to lie hidden), La; at. wt. 138.90547(7); at. no. 57; m.p. 920 °C; b.p. 3464 °C; sp. gr. 6.145 (25 °C); valence 3. Mosander in 1839 extracted a new earth *lanthana*, from impure cerium nitrate, and recognized the new element. Lanthanum is found in rare-earth minerals such as *cerite, monazite, allanite,* and *bastnasite.* Monazite and bastnasite are principal ores in which lanthanum occurs in percentages up to 25 and 38%, respectively. Misch metal, used in making lighter flints, contains about 25% lanthanum. Lanthanum was isolated in relatively pure form in 1923. Ion-exchange and solvent extraction techniques have led to much easier isolation of the so-called "rare-earth" elements. The availability of lanthanum and other rare earths has improved greatly in recent years. The metal can be produced by reducing the anhydrous fluoride with calcium. Lanthanum is silvery white, malleable, ductile, and soft enough to be cut with a knife. It is one of the most reactive of the rare-earth metals. It oxidizes rapidly when exposed to air. Cold water attacks lanthanum slowly, and hot water attacks it much more rapidly. The metal reacts directly with elemental carbon, nitrogen, boron, selenium, silicon, phosphorus, sulfur, and with halogens. At 310 °C, lanthanum changes from a hexagonal to a face-centered cubic structure, and at 865 °C it again transforms into a body-centered cubic structure. Natural lanthanum is a mixture of two isotopes, one of which is stable and one of which is radioactive with a very long half-life. Thirty other radioactive isotopes are recognized. Rare-earth compounds containing lanthanum are extensively used in carbon lighting applications, especially by the motion picture industry for studio lighting and projection. This application consumes about 25% of the rare-earth compounds produced. La_2O_3 improves the alkali resistance of glass, and is used in making special optical glasses. Small amounts of lanthanum, as an additive, can be used to produce nodular cast iron. There is current interest in hydrogen sponge alloys containing lanthanum. These alloys take up to 400 times their own volume of hydrogen gas, and the process is reversible. Heat energy is released every time they do so; therefore these alloys have possibilities in energy conservation systems. Lanthanum and its compounds have a low to moderate acute toxicity rating; therefore, care should be taken in handling them. The metal costs about $2/g (99.9%).

Lawrencium — (Ernest O. Lawrence [1901–1958], inventor of the cyclotron), Lr; at. no. 103; at. mass no. [262]; valence + 3(?). This member of the 5f transition elements (actinide series) was discovered in March 1961 by A. Ghiorso, T. Sikkeland, A. E. Larsh, and R. M. Latimer. A 3-μg californium target, consisting of a mixture of isotopes of mass number 249, 250, 251, and 252, was bombarded with either ^{10}B or ^{11}B. The electrically charged transmutation nuclei recoiled with an atmosphere of helium and were collected on a thin copper conveyor tape which was then moved to place collected atoms in front of a series of solid-state detectors. The isotope of element 103 produced in this way decayed by emitting an 8.6-MeV alpha particle with a half-life of 8 s. In 1967, Flerov and associates of the Dubna Laboratory reported their inability to detect an alpha emitter with a half-life of 8 s which was assigned by the Berkeley group to ^{257}103. This assignment has been changed to ^{258}Lr or ^{259}Lr. In 1965, the Dubna workers found a longer-lived lawrencium isotope, ^{256}Lr, with a half-life of 35 s. In 1968, Ghiorso and associates at Berkeley were able to use a few atoms of this isotope to study the oxidation behavior of lawrencium. Using solvent extraction techniques and working very rapidly, they extracted lawrencium ions from a buffered aqueous solution into an organic solvent, completing each extraction in about 30 s. It was found that lawrencium behaves differently from dipositive nobelium and more like the tripositive elements earlier in the actinide series. Ten isotopes of lawrencium are now recognized.

Lead — (Anglo-Saxon *lead*), Pb (L. *plumbum*); at. wt. 207.2(1); at. no. 82; m.p. 327.46 °C; b.p. 1749 °C; sp. gr. 11.35 (20 °C); valence 2 or 4. Long known, mentioned in Exodus. The alchemists believed lead to be the oldest metal and associated it with the planet Saturn. Native lead occurs in nature, but it is rare. Lead is obtained chiefly from *galena* (PbS) by a roasting process. *Anglesite* ($PbSO_4$), *cerussite* ($PbCO_3$), and *minim* (Pb_3O_4) are other common lead minerals. Lead is a bluish-white metal of bright luster, is very soft, highly malleable, ductile, and a poor conductor of electricity. It is very resistant to corrosion; lead pipes bearing the insignia of Roman emperors, used as drains from the baths, are still in service. It is used in containers for corrosive liquids (such as sulfuric acid) and may be toughened by the addition of a small percentage of antimony or other metals. Natural lead is a mixture of four stable isotopes: ^{204}Pb (1.4%), ^{206}Pb (24.1%), ^{207}Pb (22.1%), and ^{208}Pb (52.4%). Lead isotopes are the end products of each of the three series of naturally occurring radioactive elements: ^{206}Pb for the uranium series, ^{207}Pb for the actinium series, and ^{208}Pb for the thorium series. Forty-three other isotopes of lead, all of which are radioactive, are recognized. Its alloys include solder, type metal, and various antifriction metals. Great quantities of lead, both as the metal and as the dioxide, are used in storage batteries. Lead is also used for cable covering, plumbing, and ammunition. The metal is very effective as a sound absorber, is used as a radiation shield around X-ray equipment and nuclear reactors, and is used to absorb vibration. Lead, alloyed with tin, is used in making organ pipes. White lead, the basic carbonate, sublimed white lead ($PbSO_4$), chrome yellow ($PbCrO_4$), red lead (Pb_3O_4), and other lead compounds are used extensively in paints, although in recent years the use of lead in paints has been drastically curtailed to eliminate or reduce health hazards. Lead oxide is used in producing fine "crystal glass" and "flint glass" of a high index of refraction for achromatic lenses. The nitrate and the acetate are soluble salts. Lead salts such as lead arsenate have been used as insecticides, but their use in recent years has been practically eliminated in favor of less harmful organic compounds. Care must be used in handling lead as it is a cumulative poison. Environmental concern with lead poisoning led to elimination of lead tetraethyl in gasoline. The U.S. Occupational Safety and Health Administration (OSHA) has recommended that industries limit airborne lead to 50 μg/cu. meter. Lead is priced at about 90¢/kg (99.9%).

Lithium — (Gr. *lithos*, stone), Li; at. wt. [6.938; 6.997]; at. no. 3; m.p. 180.5 °C; b.p. 1342 °C; sp. gr. 0.534 (20 °C); valence 1. Discovered by Arfvedson in 1817. Lithium is the lightest of all metals, with a density only about half that of water. It does not occur free in nature; combined it is found in small amounts in nearly all igneous rocks and in the waters of many mineral springs. *Lepidolite*, *spodumene*, *petalite*, and *amblygonite* are the more important minerals containing it. Lithium is presently being recovered from brines of Searles Lake, in California, and from Nevada, Chile, and Argentina. Large deposits of spodumene are found in North Carolina. The metal is produced electrolytically from the fused chloride. Lithium is silvery in appearance, much like Na and K, other members of the alkali metal series. It reacts with water, but not as vigorously as sodium. Lithium imparts a beautiful crimson color to a flame, but when the metal burns strongly the flame is a dazzling white. Since World War II, the production of lithium metal and its compounds has increased greatly. Because the metal has the highest specific heat of any solid element, it has found use in heat transfer applications; however, it is corrosive and requires special handling. The metal has been used as an alloying agent, is of interest in synthesis of organic compounds, and has nuclear applications. It ranks as a leading contender as a battery anode material because it has a high electrochemical potential. Lithium is used in special glasses and ceramics. The glass for the 200-inch telescope at Mt. Palomar contains lithium as a minor ingredient. Lithium chloride is one of the most hygroscopic materials known, and it, as well as lithium bromide, is used in air conditioning and industrial drying systems. Lithium stearate is used as an all-purpose and high-temperature lubricant. Other lithium compounds are used in dry cells and storage batteries. Seven isotopes of lithium are recognized. Natural lithium contains two isotopes. The metal is priced at about $1.50/g (99.9%).

Lutetium — (Lutetia, ancient name for Paris, sometimes called *cassiopeium* by the Germans), Lu; at. wt. 174.9668(1); at. no. 71; m.p. 1663 °C; b.p. 3402 °C; sp. gr. 9.841 (25 °C); valence 3. In 1907, Urbain described a process by which Marignac's ytterbium (1879) could be separated into the two elements, ytterbium (neoytterbium) and lutetium. These elements were identical with "aldebaranium" and "cassiopeium," independently discovered by von Welsbach about the same time. Charles James of the University of New Hampshire also independently prepared the very pure oxide, *lutecia*, at this time. The spelling of the element was changed from *lutecium* to *lutetium* in 1949. Lutetium occurs in very small amounts in nearly all minerals containing yttrium, and is present in *monazite* to the extent of about 0.003%, which is a commercial source. The pure metal has been isolated only in recent years and is one of the most difficult to prepare. It can be prepared by the reduction of anhydrous $LuCl_3$ or LuF_3 by an alkali or alkaline earth metal. The metal is silvery white and relatively stable in air. While new techniques, including ion-exchange reactions, have been developed to separate the various rare-earth elements, lutetium is still the most costly of all rare earths. It is priced at about $100/g (99.9%). ^{176}Lu occurs naturally (97.41%) with ^{175}Lu (2.59%), which is radioactive with a very long half-life of about 4×10^{10} years. Lutetium has 50 isotopes and isomers that are now recognized. Stable lutetium nuclides, which emit pure beta radiation after thermal neutron activation, can be used as catalysts in cracking, alkylation, hydrogenation, and polymerization. Virtually no other commercial uses have been found yet for lutetium. While lutetium, like other rare-earth metals, is thought to have a low toxicity rating, it should be handled with care until more information is available.

Magnesium — (*Magnesia,* district in Thessaly) Mg; at. wt. 24.3050(6); at. no. 12; m.p. 650 °C; b.p. 1090 °C; sp. gr. 1.738 (20 °C); valence 2. Compounds of magnesium have long been known. Black recognized magnesium as an element in 1755. It was isolated by Davy in 1808, and prepared in coherent form by Bussy in 1831. Magnesium is the eighth most abundant element in the Earth's crust. It does not occur uncombined, but is found in large deposits in the form of *magnesite*, *dolomite*, and other minerals. The metal is now principally obtained in the U.S. by electrolysis of fused magnesium chloride derived from brines, wells, and sea water. Magnesium is a light, silvery-white, and fairly tough metal. It tarnishes slightly in air, and finely divided magnesium readily ignites upon heating in air and burns with a dazzling white flame.

It is used in flashlight photography, flares, and pyrotechnics, including incendiary bombs. It is one third lighter than aluminum, and in alloys is essential for airplane and missile construction. The metal improves the mechanical, fabrication, and welding characteristics of aluminum when used as an alloying agent. Magnesium is used in producing nodular graphite in cast iron, and is used as an additive to conventional propellants. It is also used as a reducing agent in the production of pure uranium and other metals from their salts. The hydroxide (*milk of magnesia*), chloride, sulfate (*Epsom salts*), and citrate are used in medicine. Dead-burned magnesite is employed for refractory purposes such as brick and liners in furnaces and converters. Calcined magnesia is also used for water treatment and in the manufacture of rubber, paper, etc. Organic magnesium compounds (Grignard's reagents) are important. Magnesium is an important element in both plant and animal life. Chlorophylls are magnesium-centered porphyrins. The adult daily requirement of magnesium is about 300 mg/day, but this is affected by various factors. Great care should be taken in handling magnesium metal, especially in the finely divided state, as serious fires can occur. Water should not be used on burning magnesium or on magnesium fires. Natural magnesium contains three isotopes. Twelve other isotopes are recognized. Magnesium metal costs about $100/kg (99.8%).

Manganese — (L. *magnes*, magnet, from magnetic properties of pyrolusite; It. *manganese*, corrupt form of *magnesia*), Mn; at. wt. 54.938045(5); at. no. 25; m.p. 1246 °C; b.p. 2061 °C; sp. gr. 7.21 to 7.44, depending on allotropic form; valence 1, 2, 3, 4, 6, or 7. Recognized by Scheele, Bergman, and others as an element and isolated by Gahn in 1774 by reduction of the dioxide with carbon. Manganese minerals are widely distributed; oxides, silicates, and carbonates are the most common. The discovery of large quantities of manganese nodules on the floor of the oceans holds promise as a source of manganese. These nodules contain about 24% manganese together with many other elements in lesser abundance. Most manganese today is obtained from ores found in Ukraine, Brazil, Australia, Republic of So. Africa, Gabon, China, and India. *Pyrolusite* (MnO_2) and *rhodochrosite* ($MnCO_3$) are among the most common manganese minerals. The metal is obtained by reduction of the oxide with sodium, magnesium, aluminum, or by electrolysis. It is gray-white, resembling iron, but is harder and very brittle. The metal is reactive chemically, and decomposes in cold water slowly. Manganese is used to form many important alloys. In steel, manganese improves the rolling and forging qualities, strength, toughness, stiffness, wear resistance, hardness, and hardenability. With aluminum and antimony, especially with small amounts of copper, it forms highly ferromagnetic alloys. Manganese metal is ferromagnetic only after special treatment. The pure metal exists in four allotropic forms. The alpha form is stable at ordinary temperature; gamma manganese, which changes to alpha at ordinary temperatures, is soft, easily cut, and capable of being bent. The dioxide (pyrolusite) is used as a depolarizer in dry cells, and is used to "decolorize" glass that is colored green by impurities of iron. Manganese by itself colors glass an amethyst color, and is responsible for the color of true amethyst. The dioxide is also used in the preparation of oxygen and chlorine, and in drying black paints. The permanganate is a powerful oxidizing agent and is used in quantitative analysis and in medicine. Manganese is widely distributed throughout the animal kingdom. It is an important trace element and may be essential for utilization of vitamin B_1. Twenty-seven isotopes and isomers are known. Manganese metal (99.95%) is priced at about $800/kg. Metal of 99.6% purity is priced at about $80/kg.

Meitnerium — (Lise Meitner [1878–1968], Austrian–Swedish physicist and mathematician), Mt; at. wt [268]; at. no. 109. In 1982, Element 109 was made and identified by physicists at the Heavy Ion Research Laboratory (G.S.I.), Darmstadt, Germany, by bombarding a target of ^{209}Bi with accelerated nuclei of ^{58}Fe. The production of Element 109 has been extremely small. It took a week of target bombardment (10^{11} nuclear encounters) to produce a single atom of 109. Oganessian and his team at Dubna in 1984 repeated the Darmstadt experiment using a tenfold irradiation dose. One fission event from seven alpha decays of 109 was observed, thus indirectly confirming the existence of isotope ^{266}109. In August 1997, the IUPAC adopted the name *meitnerium* for this element, honoring L. Meitner. Four isotopes of *meitnerium* are now recognized.

Mendelevium — (Dmitri Mendeleev [1834–1907]), Md; at. wt. (258); at. no. 101; m.p. 827 °C; valence +2, +3. Mendelevium, the ninth transuranium element of the actinide series to be discovered, was first identified by Ghiorso, Harvey, Choppin, Thompson, and Seaborg early in 1955 as a result of the bombardment of the isotope ^{253}Es with helium ions in the Berkeley 60-inch cyclotron. The isotope produced was ^{256}Md, which has a half-life of 78 min. This first identification was notable in that ^{256}Md was synthesized on a one-atom-at-a-time basis. Nineteen isotopes and isomers are now recognized. ^{258}Md has a half-life of 51.5 days. This isotope has been produced by the bombardment of an isotope of einsteinium with ions of helium. It now appears possible that eventually enough ^{258}Md can be made so that some of its physical properties can be determined. ^{256}Md has been used to elucidate some of the chemical properties of mendelevium in aqueous solution. Experiments seem to show that the element possesses a moderately stable dipositive (II) oxidation state in addition to the tripositive (III) oxidation state, which is characteristic of actinide elements.

Mercury — (Planet *Mercury*), Hg (*hydrargyrum*, liquid silver); at. wt. 200.59(2); at. no. 80; t.p. –38.83 °C; b.p. 356.62 °C; t_c 1477 °C; sp. gr. 13.546 (20 °C); valence 1 or 2. Known to ancient Chinese and Hindus; found in Egyptian tombs of 1500 B.C. Mercury is the only common metal liquid at ordinary temperatures. It only rarely occurs free in nature. The chief ore is *cinnabar* (HgS). Spain and China produce about 75% of the world's supply of the metal. The commercial unit for handling mercury is the "flask," which weighs 76 lb (34.46 kg). The metal is obtained by heating cinnabar in a current of air and by condensing the vapor. It is a heavy, silvery-white metal; a rather poor conductor of heat, as compared with other metals, and a fair conductor of electricity. It easily forms alloys with many metals, such as gold, silver, and tin, which are called *amalgams*. Its ease in amalgamating with gold is made use of in the recovery of gold from its ores. The metal is widely used in laboratory work for making thermometers, barometers, diffusion pumps, and many other instruments. It is used in making mercury-vapor lamps and advertising signs, etc. and is used in mercury switches and other electrical apparatus. Other uses are in making pesticides, mercury cells for caustic soda and chlorine production, dental preparations, antifouling paint, batteries, and catalysts. The most important salts are mercuric chloride $HgCl_2$ (corrosive sublimate — a vi-

olent poison), mercurous chloride Hg_2Cl_2 (calomel, occasionally still used in medicine), mercury fulminate ($Hg(ONC)_2$), a detonator widely used in explosives, and mercuric sulfide (HgS, vermillion, a high-grade paint pigment). Organic mercury compounds are important. It has been found that an electrical discharge causes mercury vapor to combine with neon, argon, krypton, and xenon. These products, held together with van der Waals' forces, correspond to HgNe, HgAr, HgKr, and HgXe. Mercury is a virulent poison and is readily absorbed through the respiratory tract, the gastrointestinal tract, or through unbroken skin. It acts as a cumulative poison and dangerous levels are readily attained in air. Air saturated with mercury vapor at 20 °C contains a concentration that exceeds the toxic limit many times. The danger increases at higher temperatures. *It is therefore important that mercury be handled with care.* Containers of mercury should be securely covered and spillage should be avoided. If it is necessary to heat mercury or mercury compounds, it should be done in a well-ventilated hood. Methyl mercury is a dangerous pollutant and is now widely found in water and streams. The triple point of mercury, –38.8344 °C, is a fixed point on the International Temperature Scale (ITS-90). Mercury (99.98%) is priced at about $110/kg. Native mercury contains seven isotopes. Thirty-six other isotopes and isomers are known.

Molybdenum — (Gr. *molybdos*, lead), Mo; at. wt. 95.96(2); at. no. 42; m.p. 2623 °C; b.p. 4639 °C; sp. gr. 10.22 (20 °C); valence 2, 3, 4?, 5?, or 6. Before Scheele recognized molybdenite as a distinct ore of a new element in 1778, it was confused with graphite and lead ore. The metal was prepared in an impure form in 1782 by Hjelm. Molybdenum does not occur native, but is obtained principally from *molybdenite* (MoS_2). *Wulfenite* ($PbMoO_4$) and *powellite* ($Ca(MoW)O_4$) are also minor commercial ores. Molybdenum is also recovered as a by-product of copper and tungsten mining operations. The U.S., Canada, Chile, and China produce most of the world's molybdenum ores. The metal is prepared from the powder made by the hydrogen reduction of purified molybdic trioxide or ammonium molybdate. The metal is silvery white, very hard, but is softer and more ductile than tungsten. It has a high elastic modulus, and only tungsten and tantalum, of the more readily available metals, have higher melting points. It is a valuable alloying agent, as it contributes to the hardenability and toughness of quenched and tempered steels. It also improves the strength of steel at high temperatures. It is used in certain nickel-based alloys, such as the Hastelloys® which are heat-resistant and corrosion-resistant to chemical solutions. Molybdenum oxidizes at elevated temperatures. The metal has found recent application as electrodes for electrically heated glass furnaces and foreheaths. It is also used in nuclear energy applications and for missile and aircraft parts. Molybdenum is valuable as a catalyst in the refining of petroleum. It has found application as a filament material in electronic and electrical applications. Molybdenum is an essential trace element in plant nutrition. Some lands are barren for lack of this element in the soil. Molybdenum sulfide is useful as a lubricant, especially at high temperatures where oils would decompose. Almost all ultra-high strength steels with minimum yield points up to 300,000 lb/in.^{2} contain molybdenum in amounts from 0.25 to 8%. Natural molybdenum contains seven isotopes. Thirty other isotopes and isomers are known, all of which are radioactive. Molybdenum metal costs about $1/g (99.999% purity). Molybdenum metal (99.9%) costs about $160/kg.

Neodymium — (Gr. *neos*, new, and *didymos*, twin), Nd; at. wt. 144.242(3); at. no. 60; m.p. 1016 °C; b.p. 3074 °C; sp. gr. 7.008 (25 °C); valence 3. In 1841 Mosander extracted from *cerite* a new rose-colored oxide, which he believed contained a new element. He named the element *didymium*, as it was *an inseparable twin brother of lanthanum*. In 1885 von Welsbach separated didymium into two new elemental components, *neodymia* and *praseodymia*, by repeated fractionation of ammonium didymium nitrate. While the free metal is in *misch metal*, long known and used as a pyrophoric alloy for light flints, the element was not isolated in relatively pure form until 1925. Neodymium is present in misch metal to the extent of about 18%. It is present in the minerals *monazite* and *bastnasite*, which are principal sources of rare-earth metals. The element may be obtained by separating neodymium salts from other rare earths by ion-exchange or solvent extraction techniques, and by reducing anhydrous halides such as NdF_3 with calcium metal. Other separation techniques are possible. The metal has a bright silvery metallic luster. Neodymium is one of the more reactive rare-earth metals and quickly tarnishes in air, forming an oxide that splits off and exposes metal to oxidation. The metal, therefore, should be kept under light mineral oil or sealed in a plastic material. Neodymium exists in two allotropic forms, with a transformation from a double hexagonal to a body-centered cubic structure taking place at 863 °C. Natural neodymium is a mixture of seven isotopes, one of which has a very long half-life. Twenty-seven other radioactive isotopes and isomers are recognized. Didymium, of which neodymium is a component, is used for coloring glass to make welder's goggles. By itself, neodymium colors glass delicate shades ranging from pure violet through wine-red and warm gray. Light transmitted through such glass shows unusually sharp absorption bands. The glass has been used in astronomical work to produce sharp bands by which spectral lines may be calibrated. Glass containing neodymium can be used as a laser material to produce coherent light. Neodymium salts are also used as a colorant for enamels. The element is also being used with iron and boron to produce extremely strong magnets. These are the most compact magnets commercially available. The price of the metal is about $4/g. Neodymium has a low-to-moderate acute toxic rating. As with other rare earths, neodymium should be handled with care.

Neon — (Gr. *neos*, new), Ne; at. wt. 20.1797(6); at. no. 10; t.p. −248.609 °C; b.p. −246.053 °C; t_c −228.7 °C; density of gas 0.89990 g/L (1 atm, 0 °C); density of liquid at b.p. 1.204 g/cm³; valence 0. Discovered by Ramsay and Travers in 1898. Neon is a rare gaseous element present in the atmosphere to the extent of 1 part in 65,000 of air. It is obtained by liquefaction of air and separated from the other gases by fractional distillation. Natural neon is a mixture of three isotopes. Fourteen other unstable isotopes are known. It is very inert element; however, it is said to form a compound with fluorine. It is still questionable if true compounds of neon exist, but evidence is mounting in favor of their existence. The following ions are known from optical and mass spectrometric studies: Ne^+, $(NeAr)^+$, $(NeH)^+$, and $(HeNe^+)$. Neon also forms an unstable hydrate. In a vacuum discharge tube, neon glows reddish orange. Of all the rare gases, the discharge of neon is the most intense at ordinary voltages and currents. Neon is used in making the common neon advertising signs, which accounts for its largest use. It is also used to make high-voltage indicators, lightning arrestors, wave meter tubes, and TV tubes. Neon and helium are used in making gas lasers. Liquid neon is now commercially available and is finding important application as an economical cryogenic refrigerant. It has over 40 times more refrigerating capacity per unit volume than liquid helium and more than three times that of liquid hydrogen. It is compact, inert, and is less expensive than helium when it meets refrigeration requirements. Neon costs about $800/80 cu. ft. (2265 l).

Neptunium — (Planet *Neptune*), Np; at. wt. (237); at. no. 93; m.p. 644 °C; sp. gr. 20.25 (20 °C); valence 3, 4, 5, and 6. Neptunium was the first synthetic transuranium element of the actinide series discovered; the isotope ^{239}Np was produced by McMillan and Abelson in 1940 at Berkeley, California, as the result of bombarding uranium with cyclotron-produced neutrons. The isotope ^{237}Np (half-life of 2.14×10^6 years) is currently obtained in gram quantities as a by-product from nuclear reactors in the production of plutonium. Twenty-three isotopes and isomers of neptunium are now recognized. Trace quantities of the element are actually found in nature due to transmutation reactions in uranium ores produced by the neutrons which are present. Neptunium is prepared by the reduction of NpF_3 with barium or lithium vapor at about 1200 °C. Neptunium metal has a silvery appearance, is chemically reactive, and exists in at least three structural modifications: α-neptunium, orthorhombic, density 20.25 g/cm³, β-neptunium (above 280 °C), tetragonal, density (313 °C) 19.36 g/cm³; γ-neptunium (above 577 °C), cubic, density (600 °C) 18.0 g/cm³. Neptunium has four ionic oxidation states in solution: Np^{+3} (pale purple), analogous to the rare earth ion Pm^{+3}, Np^{+4} (yellow green); NpO^+ (green blue); and NpO^{++} (pale pink). These latter oxygenated species are in contrast to the rare earths that exhibit only simple ions of the (II), (III), and (IV) oxidation states in aqueous solution. The element forms tri- and tetrahalides such as NpF_3, NpF_4, $NpCl_4$, $NpBr_3$, NpI_3, and oxides of various compositions such as are found in the uranium-oxygen system, including Np_3O_8 and NpO_2.

Nickel — (Ger. *Nickel*, Satan or Old Nick's and from *kupfernickel*, Old Nick's copper), Ni; at. wt. 58.6934(4); at. no. 28; m.p. 1455 °C; b.p. 2913 °C; sp. gr. 8.902 (25 °C); valence 0, 1, 2, 3. Discovered by Cronstedt in 1751 in kupfernickel (*niccolite*). Nickel is found as a constituent in most meteorites and often serves as one of the criteria for distinguishing a meteorite from other minerals. Iron meteorites, or *siderites*, may contain iron alloyed with from 5 to nearly 20% nickel. Nickel is obtained commercially from *pentlandite* and *pyrrhotite* of the Sudbury region of Ontario, a district that produces much of the world's nickel. It is now thought that the Sudbury deposit is the result of an ancient meteorite impact. Large deposits of nickel, cobalt, and copper have recently been developed at Voisey's Bay, Labrador. Other deposits of nickel are found in Russia, New Caledonia, Australia, Cuba, Indonesia, and elsewhere. Nickel is silvery white and takes on a high polish. It is hard, malleable, ductile, somewhat ferromagnetic, and a fair conductor of heat and electricity. It belongs to the iron-cobalt group of metals and is chiefly valuable for the alloys it forms. It is extensively used for making stainless steel and other corrosion-resistant alloys such as Invar®, Monel®, Inconel®, and the Hastelloys®. Tubing made of a copper-nickel alloy is extensively used in making desalination plants for converting sea water into fresh water. Nickel is also now used extensively in coinage and in making nickel steel for armor plate and burglar-proof vaults, and is a component in Nichrome®, Permalloy®, and constantan. Nickel added to glass gives a green color. Nickel plating is often used to provide a protective coating for other metals, and finely divided nickel is a catalyst for hydrogenating vegetable oils. It is also used in ceramics, in the manufacture of Alnico magnets, and in batteries. The sulfate and the oxides are important compounds. Natural nickel is a mixture of five stable isotopes; twenty-five other unstable isotopes are known. Nickel sulfide fume and dust, as well as other nickel compounds, are carcinogens. Nickel metal (99.9%) is priced at about $2/g or less in larger quantities.

Niobium — (*Niobe*, daughter of Tantalus), Nb; or Columbium (*Columbia*, name for America); at. wt. 92.90638(2); at. no. 41; m.p. 2477 °C; b.p. 4744 °C, sp. gr. 8.57 (20 °C); valence 2, 3, 4?, 5. Discovered in 1801 by Hatchett in an ore sent to England more that a century before by John Winthrop the Younger, first governor of Connecticut. The metal was first prepared in 1864 by Blomstrand, who reduced the chloride by heating it in a hydrogen atmosphere. The name *niobium* was adopted by the International Union of Pure and Applied Chemistry in 1950 after 100 years of controversy. Most leading chemical societies and government organizations refer to it by this name. Some metallurgists and commercial producers, however, still refer to the metal as "columbium." The element is found in *niobite* (or *columbite), niobite-tantalite, pyrochlore,* and *euxenite*. Large deposits of niobium have been found associated with *carbonatites* (carbon-silicate rocks), as a constituent of *pyrochlore*. Extensive ore reserves are found in Canada, Brazil, Congo-Kinshasa, Rwanda, and Australia. The metal can be isolated from tantalum, and prepared in several ways. It is a shiny, white, soft, and ductile metal, and takes on a bluish cast when exposed to air at room temperatures for a long time. The metal starts to oxidize in air at 200 °C, and when processed at even moderate temperatures must be placed in a protective atmosphere. It is used in arc-welding rods for stabilized grades of stainless steel. Thousands of pounds of niobium have been used in advanced air frame systems such as were used in the Gemini space program. It has also found use in super-alloys for applications such as jet engine components, rocket sub-assemblies, and heat-resisting equipment. The element has superconductive properties; superconductive magnets have been made with Nb-Zr wire, which retains its superconductivity in strong magnetic fields. Natural niobium is composed of only one isotope, ^{93}Nb. Forty-seven other isotopes and isomers of niobium are now recognized. Niobium metal (99.9% pure) is priced at about 50¢/g.

Nitrogen — (L. *nitrum*, Gr. *nitron*, native soda; genes, *forming*, N; at. wt. [14.00643; 14.00728]; at. no. 7; m.p. −210.00 °C; b.p. −195.798 °C; t_c −146.94 °C; density 1.2506 g/L; sp. gr. liquid 0.808 (−195.8 °C), solid 1.026 (−252 °C); valence 3 or 5. Discovered by Daniel Rutherford in 1772, but Scheele, Cavendish, Priestley, and others about the same time studied "burnt or dephlogisticated air," as air without oxygen was then called. Nitrogen makes up 78% of the air, by volume. The atmosphere of Mars, by comparison, is 2.6% nitrogen. The estimated amount of this element in our atmosphere is more than 4000 trillion tons. From this inexhaustible source it can be obtained by liquefaction and fractional distillation. Nitrogen molecules give the orange-red, blue-green, blue-violet, and deep violet shades to the aurora. The element is so inert that Lavoisier named it *azote*, meaning without life, yet its compounds are so active as to be most important in foods, poisons, fertilizers, and explosives. Nitrogen can be also easily prepared by heating a water solution of ammonium nitrite. Nitrogen, as a gas, is colorless, odorless, and a generally inert element. As a liquid it is also colorless and odorless, and is similar in appearance to water. Two allotropic forms of solid nitrogen exist, with the transition from the α to the β form taking place at −237 °C. When nitrogen is heated, it combines directly with magnesium, lithium, or calcium; when mixed with oxygen and subjected to electric sparks, it forms first nitric oxide (NO) and then the dioxide (NO$_2$); when heated under pressure with a catalyst with hydrogen, ammonia is formed (Haber process). The ammonia thus formed is of the utmost importance as it is used in fertilizers, and it can be oxidized to nitric acid (Ostwald process). The ammonia industry is the largest consumer of nitrogen. Large amounts of gas are also used by the electronics industry, which uses the gas as a blanketing medium during production of such components as transistors, diodes, etc. Large quantities of nitrogen are used in annealing stainless steel and other steel mill products. The drug industry also uses large quantities. Nitrogen is used as a refrigerant both for the immersion freezing of food products and for transportation of foods. Liquid nitrogen is also used in missile work as a purge for components, insulators for space chambers, etc., and by the oil industry to build up great pressures in wells to force crude oil upward. Sodium and potassium nitrates are formed by the decomposition of organic matter with compounds of the metals present. In certain dry areas of the world these saltpeters are found in quantity. Ammonia, nitric acid, the nitrates, the five oxides (N$_2$O, NO, N$_2$O$_3$, NO$_2$, and N$_2$O$_5$), TNT, the cyanides, etc. are but a few of the important compounds. Nitrogen gas prices vary from 2¢ to $2.75 per 100 ft^3 (2.83 cu. meters), depending on purity, etc. Production of elemental nitrogen in the U.S. is more than 9 million short tons per year. Natural nitrogen contains two isotopes, ^{14}N and ^{15}N. Ten other isotopes are known.

Nobelium — (Alfred Nobel [1833–1896], inventor of dynamite), No; at. wt. [259]; at. no. 102; valence +2, +3. Nobelium was unambiguously discovered and identified in April 1958 at Berkeley by A. Ghiorso, T. Sikkeland, J. R. Walton, and G. T. Seaborg, who used a new double-recoil technique. A heavy-ion linear accelerator (HILAC) was used to bombard a thin target of curium (95% ^{244}Cm and 4.5% ^{246}Cm) with ^{12}C ions to produce 102^{254} according to the ^{246}Cm (^{12}C, 4n) reaction. Earlier in 1957 workers of the U.S., Britain, and Sweden announced the discovery of an isotope of Element 102 with a 10-min half-life at 8.5 MeV, as a result of bombarding ^{244}Cm with ^{13}C nuclei. On the basis of this experiment the name *nobelium* was assigned and accepted by the Commission on Atomic Weights of the International Union of Pure and Applied Chemistry. The acceptance of the name was premature, for both Russian and American efforts now completely rule out the possibility of any isotope of Element 102 having a half-life of 10 min in the vicinity of 8.5 MeV. Early work in 1957 on the search for this element, in Russia at the Kurchatov Institute, was marred by the assignment of 8.9 ± 0.4 MeV alpha radiation with a half-life of 2 to 40 sec, which was too indefinite to support claim to discovery. Confirmatory experiments at Berkeley in 1966 have shown the existence of ^{254}102 with a 55-s half-life, ^{252}102 with a 2.3-s half-life, and ^{257}102 with a 25-s half-life. Twelve isotopes are now recognized, one of which — ^{255}102 — has a half-life of 3.1 min. In view of the discoverer's traditional right to name an element, the Berkeley group, in 1967, suggested that the hastily given name *nobelium,* along with the symbol No, be retained.

Osmium — (Gr. *osme*, a smell), Os; at. wt. 190.23(3); at. no. 76; m.p. 3033 °C; b.p. 5012 °C; sp. gr. 22.587; valence 0 to +8, more usually +3, +4, +6, and +8. Discovered in 1803 by Tennant in the residue left when crude platinum is dissolved by *aqua regia*. Osmium occurs in *iridosmine* and in platinum-bearing river sands of the Urals, North America, and South America. It is also found in the nickel-bearing ores of the Sudbury, Ontario, region along with other platinum metals. While the quantity of platinum metals in these ores is very small, the large tonnages of nickel ores processed make commercial recovery possible. The metal is lustrous, bluish white, extremely hard, and brittle even at high temperatures. It has the highest melting point and the lowest vapor pressure of the platinum group. The metal is very difficult to fabricate, but the powder can be sintered in a hydrogen atmosphere at a temperature of 2000 °C. The solid metal is not affected by air at room temperature, but the powdered or spongy metal slowly gives off osmium tetroxide, which is a powerful oxidizing agent and has a strong smell. The tetroxide is highly toxic, and boils at 130 °C (760 mm). Concentrations in air as low as 10^{-7} g/m^3 can cause lung congestion, skin damage, or eye damage. The tetroxide has been used to detect fingerprints and to stain fatty tissue for microscope slides. The metal is almost entirely used to produce very hard alloys, with other metals of the platinum group, for fountain pen tips, instrument pivots, phonograph needles, and electrical contacts. The price of 99.9% pure osmium powder — the form usually supplied commercially — is about $100/g, depending on quantity and supplier. Natural osmium contains seven isotopes, one of which, ^{186}Os, is radioactive with a very long half-life. Thirty-four other isotopes and isomers are known, all of which are radioactive. The measured densities of iridium and osmium seem to indicate that osmium is slightly more dense than iridium, so osmium has generally been credited with being the heaviest known element. Calculations of the density from the space lattice, which may be more reliable for these elements than actual measurements, however, give a density of 22.65 for iridium compared to 22.61 for osmium. At present, therefore, we know either iridium or osmium is the heaviest element, but the data do not allow selection between the two.

Oxygen — (Gr. *oxys*, sharp, acid, and *genes*, forming; acid former), O; at. wt. [15.99903; 15.99977]; at. no. 8; t.p. −218.79 °C; t_c −118.56 °C; valence 2. For many centuries, workers occasionally realized air was composed of more than one component. The behavior of oxygen and nitrogen as components of air led to the advancement of the phlogiston theory of combustion, which captured the minds of chemists for a century. Oxygen was prepared by several workers, including Bayen and Borch, but they did not know how to collect it, did not study its properties, and did not recognize it as an elementary substance. Priestley is generally credited with its discovery, although Scheele also discovered it independently. Oxygen is the third most abundant element found in the sun, and it plays a part in the carbon–nitrogen cycle, one process thought to give the sun and stars their energy. Oxygen under excited conditions is responsible for the bright red and yellow-green colors of the aurora. Oxygen, as a gaseous element, forms 21% of the atmosphere by volume from which it can be obtained by liquefaction and fractional distillation. The atmosphere of Mars contains about 0.15% oxygen. The element and its compounds make up 49.2%, by weight, of the Earth's crust. About two thirds of the human body and nine tenths of water is oxygen. In the laboratory it can be prepared by the electrolysis of water or by heating potassium chlorate with manganese dioxide as a catalyst. The gas is colorless, odorless, and tasteless. The liquid and solid forms are a pale blue color and are strongly paramagnetic. Ozone (O_3), a highly active compound, is formed by the action of an electrical discharge or ultraviolet light on oxygen. Ozone's presence in the atmosphere (amounting to the equivalent of a layer 3 mm thick at ordinary pressures and temperatures) is of vital importance in preventing harmful ultraviolet rays of the sun from reaching the Earth's surface. There has been recent concern that pollutants in the atmosphere may have a detrimental effect on this ozone layer. Ozone is toxic and exposure should not exceed 0.2 mg/m^3 (8-hour time-weighted average — 40-hour work week). Undiluted ozone has a bluish color. Liquid ozone is bluish black, and solid ozone is violet-black. Oxygen is very reactive and capable of combining with most elements. It is a component of hundreds of thousands of organic compounds. It is essential for respiration of all plants and animals and for practically all combustion. In hospitals it is frequently used to aid respiration of patients. Its atomic weight was used as a standard of comparison for each of the other elements until 1961 when the International Union of Pure and Applied Chemistry adopted carbon 12 as the new basis. Oxygen has thirteen recognized isotopes. Natural oxygen is a mixture of three isotopes. Oxygen 18 occurs naturally, is stable, and is available commercially. Water (H_2O with 1.5% ^{18}O) is also available. Commercial oxygen consumption in the U.S. is estimated to be 20 million short tons per year and the demand is expected to increase substantially in the next few years. Oxygen enrichment of steel blast furnaces accounts for the greatest use of the gas. Large quantities are also used in making synthesis gas for ammonia and methanol, ethylene oxide, and for oxy-acetylene welding. Air separation plants produce about 99% of the gas, electrolysis plants about 1%. The gas costs 5¢/ft^3 ($1.75/cu. meter) in small quantities.

Palladium — (named after the asteroid *Pallas*, discovered about the same time; Gr. *Pallas*, goddess of wisdom), Pd; at. wt. 106.42(1) at. no. 46; m.p. 1554.8 °C; b.p. 2963 °C; sp. gr. 12.02 (20 °C); valence 2, 3, or 4. Discovered in 1803 by Wollaston. Palladium is found along with platinum and other metals of the platinum group in deposits of Russia, South Africa, Canada (Ontario), and elsewhere. Natural palladium contains six stable isotopes. Twenty-nine other isotopes are recognized, all of which are radioactive. It is frequently found associated with the nickel-copper deposits such as those found in Ontario. Its separation from the platinum metals depends upon the type of ore in which it is found. It is a steel-white metal, does not tarnish in air, and is the least dense and lowest melting of the platinum group of metals. When annealed, it is soft and ductile; cold working greatly increases its strength and hardness. Palladium is attacked by nitric and sulfuric acid. At room temperatures the metal has the unusual property of absorbing up to 900 times its own volume of hydrogen, possibly forming Pd_2H. It is not yet clear if this a true compound. Hydrogen readily diffuses through heated palladium and this provides a means of purifying the gas. Finely divided palladium is a good catalyst and is used for hydrogenation and dehydrogenation reactions. It is alloyed and used in jewelry trades. White gold is an alloy of gold decolorized by the addition of palladium. Like gold, palladium can be beaten into leaf as thin as 1/250,000 in. The metal is used in dentistry, watchmaking, and in making surgical instruments and electrical contacts. Palladium recently has been substituted for higher priced platinum in catalytic converters by some automobile companies. This has caused a large increase in the cost of palladium. The prices of the two metals are now, in 2002, about the same. Palladium, however, is less resistant to poisoning by sulfur and lead than platinum, but it may prove useful in controlling emissions from diesel vehicles. The metal sells for about $350/tr. oz. ($11/g).

Phosphorus — (Gr. *phosphoros*, light bearing; ancient name for the planet Venus when appearing before sunrise), P; at. wt. 30.973762(2); at. no. 15; m.p. (white) 44.15 °C; b.p. 280.5 °C; sp. gr. (white) 1.82, (red) 2.16, (black) 2.25 to 2.69; valence 3 or 5. Discovered in 1669 by Brand, who prepared it from urine. Phosphorus exists in four or more allotropic forms: white (or yellow), red, and black (or violet). White phosphorus has two modifications: α and β with a transition temperature at -3.8 °C. Never found free in nature, it is widely distributed in combination with minerals. Twenty-one isotopes of phosphorus are recognized. *Phosphate* rock, which contains the mineral *apatite*, an impure tricalcium phosphate, is an important source of the element. Large deposits are found in the Russia, China, Morocco, and in Florida, Tennessee, Utah, Idaho, and elsewhere. Phosphorus in an essential ingredient of all cell protoplasm, nervous tissue, and bones. Ordinary phosphorus is a waxy white solid; when pure it is colorless and transparent. It is insoluble in water, but soluble in carbon disulfide. It takes fire spontaneously in air, burning to the pentoxide. It is very poisonous, 50 mg constituting an approximate fatal dose. Exposure to white phosphorus should not exceed 0.1 mg/m^3 (8-hour time-weighted average — 40-hour work week). White phosphorus should be kept under water, as it is dangerously reactive in air, and it should be handled with forceps, as contact with the skin may cause severe burns. When exposed to sunlight or when heated in its own vapor to 250 °C, it is converted to the red variety, which does not phosphoresce in air as does the white variety. This form does not ignite spontaneously and it is not as dangerous as white phosphorus. It should, however, be handled with care as it does convert to the white form at some temperatures and it emits highly toxic fumes of the oxides of phosphorus when heated. The red modification is fairly stable, sublimes with a vapor pressure of 1 atm at 417 °C, and is used in the manufacture of safety matches, pyrotechnics, pesticides, incendiary shells, smoke bombs, tracer bullets, etc. White phosphorus may be made by several methods. By one process, tricalcium phosphate, the essential ingredient of phosphate rock, is heated in the presence of carbon and silica in an electric furnace or fuel-fired furnace. Elementary phosphorus is liberated as vapor and may be collected under water. If desired, the phosphorus vapor and carbon monoxide produced by the reaction can be oxidized at once in the presence of moisture to produce phosphoric acid, an important compound in making super-phosphate fertilizers. In recent years, concentrated phosphoric acids, which may contain as much as 70 to 75% P_2O_5 content, have become of great importance to agriculture and farm production. World-wide demand for fertilizers has caused record phosphate production. Phosphates are used in the production of special glasses, such as those used for sodium lamps. Bone-ash, calcium phosphate, is also used to produce fine chinaware and to produce monocalcium phosphate used in baking powder. Phosphorus is also important in the production of steels, phosphor bronze, and many other products. Trisodium phosphate is important as a cleaning agent, as a water softener, and for preventing boiler scale and corrosion of pipes and boiler tubes. Organic compounds of phosphorus are important. Amorphous (red) phosphorus costs about $70/kg (99%).

Platinum — (It. *platina*, silver), Pt; at. wt. 195.084(9); at. no. 78; m.p. 1768.2 °C; b.p. 3825 °C; sp. gr. 21.45 (20 °C); valence 1?, 2, 3, or 4. Discovered in South America by Ulloa in 1735 and by Wood in 1741. The metal was used by pre-Columbian Indians. Platinum occurs native, accompanied by small quantities of iridium, osmium, palladium, ruthenium, and rhodium, all belonging to the same group of metals. These are found in the alluvial deposits of the Ural mountains and in Columbia. *Sperrylite* ($PtAs_2$), occurring with the nickel-bearing deposits of Sudbury, Ontario, is a source of a considerable amount of metal. The large production of nickel offsets there being only one part of the platinum metals in two million parts of ore. The largest supplier of the platinum group of metals is now South Africa, followed by Russia and Canada. Platinum is a beautiful silvery-white metal, when pure, and is malleable and ductile. It has a coefficient of expansion almost equal to that of soda–lime–silica glass, and is therefore used to make sealed electrodes in glass systems. The metal does not oxidize in air at any temperature, but is corroded by halogens, cyanides, sulfur, and caustic alkalis. It is insoluble in hydrochloric and nitric acid, but dissolves when they are mixed as *aqua regia*, forming chloroplatinic acid (H_2PtCl_6), an important compound. Natural platinum contains six isotopes, one of which, ^{190}Pt, is radioactive with a long half-life. Thirty-seven other radioactive isotopes and isomers are recognized. The metal is used extensively in jewelry, wire, and vessels for laboratory use, and in many valuable instruments including thermocouple elements. It is also used for electrical contacts, corrosion-resistant apparatus, and in dentistry. Platinum–cobalt alloys have magnetic properties. One such alloy made of 76.7% Pt and 23.3% Co, by weight, is an extremely powerful magnet that offers a B-H (max) almost twice that of Alnico V. Platinum resistance wires are used for constructing high-temperature electric furnaces. The metal is used for coating missile nose cones, jet engine fuel nozzles, etc., which must perform reliably for long periods of time at high temperatures. The metal, like palladium, absorbs large volumes of hydrogen, retaining it at ordinary temperatures but giving it up at red heat. In the finely divided state platinum is an excellent catalyst, having long been used in the contact process for producing sulfuric acid. It is also used as a catalyst in cracking petroleum products. There is also much current interest in the use of platinum as a catalyst in fuel cells and in its use as antipollution devices for automobiles. Platinum anodes are extensively used in cathodic protection systems for large ships and ocean-going vessels, pipelines, steel piers, etc. Pure platinum wire will glow red hot when placed in the vapor of methyl alcohol. It acts here as a catalyst, converting the alcohol to formaldehyde. This phenomenon has been used commercially to produce cigarette lighters and hand warmers. Hydrogen and oxygen explode in the presence of platinum. The price of platinum has varied widely; more than a century ago it was used to adulterate gold. It was nearly eight times as valuable as gold in 1920. The price in January 2002 was about $430/troy oz. ($15/g), higher than the price of gold.

Plutonium — (planet *Pluto*), Pu; at. wt. (244); at. no. 94; sp. gr. (α modification) 19.84 (25 °C); m.p. 640 °C; b.p. 3228 °C; valence 3, 4, 5, or 6. Plutonium was the second transuranium element of the actinide series to be discovered. The isotope ^{238}Pu was produced in 1940 by Seaborg, McMillan, Kennedy, and Wahl by deuteron bombardment of uranium in the 60-inch cyclotron at Berkeley, California. Plutonium also exists in trace quantities in naturally occurring uranium ores. It is formed in much the same manner as neptunium, by irradiation of natural uranium with the neutrons that are present. By far of greatest importance is the isotope Pu^{239}, with a half-life of 24,100 years, produced in extensive quantities in nuclear reactors from natural uranium:

$$^{238}U(n,\gamma) \rightarrow {}^{239}U \xrightarrow{\beta} {}^{239}Np \xrightarrow{\beta} {}^{239}Pu$$

Nineteen isotopes of plutonium are now known. Plutonium has assumed the position of dominant importance among the transuranium elements because of its successful use as an explosive ingredient in nuclear weapons and the place it holds as a key material in the development of industrial use of nuclear power. One kilogram is equivalent to about 22 million kilowatt hours of heat energy. The complete detonation of a kilogram of plutonium produces an explosion equal to about 20,000 tons of chemical explosive. Its importance depends on the nuclear property of being readily fissionable with neutrons and its availability in quantity. The world's nuclear-power reactors are now producing about 20,000 kg of plutonium/yr. By 1982 it was estimated that about 300,000 kg had accumulated. The various nuclear applications of plutonium are well known. ^{238}Pu has been used in the Apollo lunar missions to power seismic and other equipment on the lunar surface. As with neptunium and uranium, plutonium metal can be prepared by reduction of the trifluoride with alkaline-earth metals. The metal has a silvery appearance and takes on a yellow tarnish when slightly oxidized. It is chemically reactive. A relatively large piece of plutonium is warm to the touch because of the energy given off in alpha decay. Larger pieces will produce enough heat to boil water. The metal readily dissolves in concentrated hydrochloric acid, hydroiodic acid, or perchloric acid with formation of the Pu^{+3} ion. The metal exhibits six allotropic modifications having various crystalline structures. The densities of these vary from 16.00 to 19.86 g/cm^3. Plutonium also exhibits four ionic valence states in aqueous solutions: Pu^{+3}(blue lavender), Pu^{+4} (yellow brown), PuO$^+$ (pink?), and PuO^{+2} (pink orange). The ion PuO$^+$ is unstable in aqueous solutions, disproportionating into Pu^{+4} and PuO^{+2}. The Pu^{+4} thus formed, however, oxidizes the PuO$^+$ into PuO^{+2}, itself being reduced to Pu^{+3}, giving finally Pu^{+3} and PuO^{+2}. Plutonium forms binary compounds with oxygen: PuO, PuO$_2$, and intermediate oxides of variable composition; with the halides: PuF$_3$, PuF$_4$, PuCl$_3$, PuBr$_3$, PuI$_3$; with carbon, nitrogen, and silicon: PuC, PuN, PuSi$_2$. Oxyhalides are also well known: PuOCl, PuOBr, PuOI. Because of the high rate of emission of alpha particles and the element being specifically absorbed by bone marrow, plutonium, as well as all of the other transuranium elements except neptunium, are radiological poisons and must be handled with very special equipment and precautions. Plutonium is a very dangerous radiological hazard. Precautions must also be taken to prevent the unintentional formation of a critical mass. Plutonium in liquid solution is more likely to become critical than solid plutonium. The shape of the mass must also be considered where criticality is concerned. Plutonium-239 is available to authorized users from the O.R.N.L. at a cost of about $4.80/mg (99.9%) plus packing costs.

Polonium — (Poland, native country of Mme. Curie [1867–1934]), Po; at. wt. (209); at. no. 84; m.p. 254 °C; b.p. 962 °C; sp. gr. 9.20; valence −2, 0, +2, +3(?), +4, and +6. Polonium was the first element discovered by Mme. Curie in 1898, while seeking the cause of radioactivity of pitchblende from Joachimsthal, Bohemia. The electroscope showed it separating with bismuth. Polonium is also called Radium F. Polonium is a very rare natural element. Uranium ores contain only about 100 µg of the element per ton. Its abundance is only about 0.2% of that of radium. In 1934, it was found that when natural bismuth (^{209}Bi) was bombarded by neutrons, ^{210}Bi, the parent of polonium, was obtained. Milligram amounts of polonium may now be prepared this way, by using the high neutron fluxes of nuclear reactors. Polonium-210 is a low-melting, fairly volatile metal, 50% of which is vaporized in air in 45 hours at 55 °C. It is an alpha emitter with a half-life of 138.39 days. A milligram emits as many alpha particles as 5 g of radium. The energy released by its decay is so large (140 W/g) that a capsule containing about half a gram reaches a temperature above 500 °C. The capsule also presents a contact gamma-ray dose rate of 0.012 Gy/h. A few curies (1 curie = 3.7×10^{10} Bq) of polonium exhibit a blue glow, caused by excitation of the surrounding gas. Because almost all alpha radiation is stopped within the solid source and its container, giving up its energy, polonium has attracted attention for uses as a lightweight heat source for thermoelectric power in space satellites. Thirty-eight isotopes and isomers of polonium are known, with atomic masses ranging from 192 to 218. All are radioactive. Polonium-210 is the most readily available. Isotopes of mass 209 (half-life 102 years) and mass 208 (half-life 2.9 years) can be prepared by alpha, proton, or deuteron bombardment of lead or bismuth in a cyclotron, but these are expensive to produce. Metallic polonium has been prepared from polonium hydroxide and some other polonium compounds in the presence of concentrated aqueous or anhydrous liquid ammonia. Two allotropic modifications are known to exist. Polonium is readily dissolved in dilute acids, but is only slightly soluble in alkalis. Polonium salts of organic acids char rapidly; halide amines are reduced to the metal. Polonium can be mixed or alloyed with beryllium to provide a source of neutrons. It has been used in devices for eliminating static charges in textile mills, etc.; however, beta sources are more commonly used and are less dangerous. It is also used on brushes for removing dust from photographic films. The polonium for these is carefully sealed and controlled, minimizing hazards to the user. Polonium-210 is very dangerous to handle in even milligram or microgram amounts, and special equipment and strict control are necessary. Damage arises from the complete absorption of the energy of the alpha particle into tissue. The maximum permissible body burden for ingested polonium is only 0.03 µCi, which represents a particle weighing only 6.8×10^{-12} g. Weight for weight it is about 2.5×10^{11} times as toxic as hydrocyanic acid. The maximum allowable concentration for soluble polonium compounds in air is about 2×10^{11} µCi/cm^3. Polonium-209 is available on special order from the Oak Ridge National Laboratory at a cost of $3600/µCi plus packing costs.

Potassium — (English, *potash* — pot ashes; L. *kalium*, Arab. *qali*, alkali), K; at. wt. 39.0983(1); at. no. 19; m.p. 63.5 °C; b.p. 759 °C; sp. gr. 0.89; valence 1. Discovered in 1807 by Davy, who obtained it from caustic potash (KOH); this was the first metal isolated by electrolysis. The metal is the seventh most abundant and makes up about 2.4% by weight of the Earth's crust. Most potassium minerals are insoluble and the metal is obtained from them only with great difficulty. Certain minerals, however, such as *sylvite, carnallite, langbeinite,* and *polyhalite* are found in ancient lake and sea beds and form rather extensive deposits from which potassium and its salts can readily be obtained. Potash is mined in Germany, New Mexico, California, Utah, and elsewhere. Large deposits of potash, found at a depth of some 1000 m in Saskatchewan, promise to be important in coming years. Potassium is also found in the ocean, but is present only in relatively small amounts compared to sodium. The greatest demand for potash has been in its use for fertilizers. Potassium is an essential constituent for plant growth and it is found in most soils. Potassium is never found free in nature, but is obtained by electrolysis of the hydroxide, much in the same manner as prepared by Davy. Thermal methods also are commonly used to produce potassium (such as by reduction of potassium compounds with CaC_2, C, Si, or Na). It is one of the most reactive and electropositive of metals. Except for lithium, it is the lightest known metal. It is soft, easily cut with a knife, and is silvery in appearance immediately after a fresh surface is exposed. It rapidly oxidizes in air and should be preserved in a mineral oil. As with other metals of the alkali group, it decomposes in water with the evolution of hydrogen. It catches fire spontaneously on water. Potassium and its salts impart a violet color to flames. Twenty-one isotopes, one of which is an isomer, of potassium are known. Ordinary potassium is composed of three isotopes, one of which is ^{40}K (0.0117%), a radioactive isotope with a half-life of 1.26×10^9 years. The radioactivity presents no appreciable hazard. An alloy of sodium and potassium (NaK) is used as a heat-transfer medium. Many potassium salts are of utmost importance, including the hydroxide, nitrate, carbonate, chloride, chlorate, bromide, iodide, cyanide, sulfate, chromate, and dichromate. Metallic potassium is available commercially for about $1200/ kg (98% purity) or $75/g (99.95% purity).

Praseodymium — (Gr. *prasios,* green, and *didymos,* twin), Pr; at. wt. 140.90765(2); at. no. 59; m.p. 931 °C; b.p. 3520 °C; sp. gr. 6.773; valence 3. In 1841 Mosander extracted the rare earth *didymia* from *lanthana;* in 1879, Lecoq de Boisbaudran isolated a new earth, *samaria,* from didymia obtained from the mineral *samarskite.* Six years later, in 1885, von Welsbach separated didymia into two others, *praseodymia* and *neodymia,* which gave salts of different colors. As with other rare earths, compounds of these elements in solution have distinctive sharp spectral absorption bands or lines, some of which are only a few Angstroms wide. The element occurs along with other rare-earth elements in a variety of minerals. *Monazite* and *bastnasite* are the two principal commercial sources of the rare-earth metals. Ion-exchange and solvent extraction techniques have led to much easier isolation of the rare earths and the cost has dropped greatly. Thirty-seven isotopes and isomers are now recognized. Praseodymium can be prepared by several methods, such as by calcium reduction of the anhydrous chloride or fluoride. Misch metal, used in making cigarette lighters, contains about 5% praseodymium metal. Praseodymium is soft, silvery, malleable, and ductile. It was prepared in relatively pure form in 1931. It is somewhat more resistant to corrosion in air than europium, lanthanum, cerium, or neodymium, but it does develop a green oxide coating that splits off when exposed to air. As with other rare-earth metals it should be kept under a light mineral oil or sealed in plastic. The rare-earth oxides, including Pr_2O_3, are among the most refractory substances known. Along with other rare earths, it is widely used as a core material for carbon arcs used

by the motion picture industry for studio lighting and projection. Salts of praseodymium are used to color glasses and enamels; when mixed with certain other materials, praseodymium produces an intense and unusually clean yellow color in glass. Didymium glass, of which praseodymium is a component, is a colorant for welder's goggles. The metal (99.9% pure) is priced at about $4/g.

Promethium — (*Prometheus,* who, according to mythology, stole fire from heaven), Pm; at. no. 61; at. wt. (145); m.p. 1042 °C; b.p. 3000 °C (est.); sp. gr. 7.264 (25 °C); valence 3. In 1902 Branner predicted the existence of an element between neodymium and samarium, and this was confirmed by Moseley in 1914. Unsuccessful searches were made for this predicted element over two decades, and various investigators proposed the names "illinium," "florentium," and "cyclonium" for this element. In 1941, workers at Ohio State University irradiated neodymium and praseodymium with neutrons, deuterons, and alpha particles, resp., and produced several new radioactivities, which most likely were those of Element 61. Wu and Segre, and Bethe, in 1942, confirmed the formation; however, chemical proof of the production of Element 61 was lacking because of the difficulty in separating the rare earths from each other at that time. In 1945, Marinsky, Glendenin, and Coryell made the first chemical identification by using ion-exchange chromatography. Their work was done by fission of uranium and by neutron bombardment of neodymium. These investigators named the newly discovered element. Searches for the element on Earth have been fruitless, and it now appears that promethium is completely missing from the Earth's crust. Promethium, however, has been reported to be in the spectrum of the star HR[465] in Andromeda. It must be formed near the star's surface, for no known isotope of promethium has a half-life longer than 17.7 years. Thirty-five isotopes and isomers of promethium, with atomic masses from 130 to 158 are now known. Promethium-145, with a half-life of 17.7 years, is the most useful. Promethium-145 has a specific activity of 940 Ci/g. It is a soft beta emitter; although no gamma rays are emitted, X-radiation can be generated when beta particles impinge on elements of a high atomic number, and great care must be taken in handling it. Promethium salts luminesce in the dark with a pale blue or greenish glow, due to their high radioactivity. Ion-exchange methods led to the preparation of about 10 g of promethium from atomic reactor fuel processing wastes in early 1963. Little is yet generally known about the properties of metallic promethium. Two allotropic modifications exist. The element has applications as a beta source for thickness gages, and it can be absorbed by a phosphor to produce light. Light produced in this manner can be used for signs or signals that require dependable operation; it can be used as a nuclear-powered battery by capturing light in photocells that convert it into electric current. Such a battery, using ^{147}Pm, would have a useful life of about 5 years. It is being used for fluorescent lighting starters and coatings for self-luminous watch dials. Promethium shows promise as a portable X-ray source, and it may become useful as a heat source to provide auxiliary power for space probes and satellites. More than 30 promethium compounds have been prepared. Most are colored.

Protactinium — (Gr. *protos,* first), Pa; at. wt. 231.03588(2); at. no. 91; m.p. 1572 °C; sp. gr. 15.37 (calc.); valence 4 or 5. The first isotope of Element 91 to be discovered was ^{234}Pa, also known as UX$_2$, a short-lived member of the naturally occurring ^{238}U decay series. It was identified by K. Fajans and O. H. Gohring in 1913 and they named the new element *brevium*. When the longer-lived isotope ^{231}Pa was identified by Hahn and Meitner in 1918, the name protoactinium was adopted as being more consistent with the characteristics of the most abundant isotope. Soddy, Cranson, and Fleck were also active in this work. The name *protoactinium* was shortened to *protactinium* in 1949. In 1927, Grosse prepared 2 mg of a white powder, which was shown to be Pa$_2$O$_5$. Later, in 1934, from 0.1 g of pure Pa$_2$O$_5$ he isolated the element by two methods, one of which was by converting the oxide to an iodide and "cracking" it in a high vacuum by an electrically heated filament by the reaction

$$2PaI_5 \rightarrow 2Pa + 5I_2$$

Protactinium has a bright metallic luster that it retains for some time in air. The element occurs in *pitchblende* to the extent of about 1 part ^{231}Pa to 10 million of ore. Ores from Congo-Kinshasa have about 3 ppm. Protactinium has twenty-eight isotopes and isomers, the most common of which is ^{231}Pr with a half-life of 32,500 years. A number of protactinium compounds are known, some of which are colored. The element is superconductive below 1.4 K. The element is a dangerous toxic material and requires precautions similar to those used when handling plutonium. In 1959 and 1961, it was announced that the Great Britain Atomic Energy Authority extracted by a 12-stage process 125 g of 99.9% protactinium, the world's only stock of the metal for many years to come. The extraction was made from 60 tons of waste material at a cost of about $500,000. Protactinium is one of the rarest and most expensive naturally occurring elements.

Radium — (L. *radius*, ray), Ra; at. wt. (226); at. no. 88; m.p. 696 °C; sp. gr. 5; valence 2. Radium was discovered in 1898 by M. and Mme. Curie in the *pitchblende* or *uraninite* of North Bohemia (Czech Republic), where it occurs. There is about 1 g of radium in 7 tons of pitchblende. The element was isolated in 1911 by Mme. Curie and Debierne by the electrolysis of a solution of pure radium chloride, employing a mercury cathode; on distillation in an atmosphere of hydrogen this amalgam yielded the pure metal. Originally, radium was obtained from the rich pitchblende ore found at Joachimsthal, Bohemia. The *carnotite* sands of Colorado furnish some radium, but richer ores are found in the Republic of Congo-Kinshasa and the Great Bear Lake region of Canada. Radium is present in all uranium minerals, and could be extracted, if desired, from the extensive wastes of uranium processing. Large uranium deposits are located in Ontario, New Mexico, Utah, Australia, and elsewhere. Radium is obtained commercially as the bromide or chloride; it is doubtful if any appreciable stock of the isolated element now exists. The pure metal is brilliant white when freshly prepared, but blackens on exposure to air, probably due to formation of the nitride. It exhibits luminescence, as do its salts; it decomposes in water and is somewhat more volatile than barium. It is a member of the alkaline-earth group of metals. Radium imparts a carmine red color to a flame. Radium emits alpha, beta, and gamma rays and when mixed with beryllium produce neutrons. One gram of ^{226}Ra undergoes 3.7×10^{10} disintegrations per s. The *curie (Ci)* is defined as that amount of radioactivity which has the same disintegration rate as 1 g of ^{226}Ra. Thirty-six isotopes are now known; radium 226, the common isotope, has a half-life of 1599 years. One gram of radium produces about 0.0001 mL (stp) of ema-nation, or radon gas, per day. This is pumped from the radium and sealed in minute tubes, which are used in the treatment of cancer and other diseases. One gram of radium yields about 4186 kJ per year. Radium is used in producing self-luminous paints, neutron sources, and in medicine for the treatment of cancer. Some of the more recently discovered radioisotopes, such as ^{60}Co, are now being used in place of radium. Some of these sources are much more powerful, and others are safer to use. Radium loses about 1% of its activity in 25 years, being transformed into elements of lower atomic weight. Lead is a final product of disintegration. Stored radium should be ventilated to prevent build-up of radon. Inhalation, injection, or body exposure to radium can cause cancer and other body disorders. The maximum permissible burden in the total body for ^{226}Ra is 7400 becquerel.

Radon — (from *radium*; called *niton* at first, L. *nitens*, shining), Rn; at. wt. (222); at. no. 86; m.p. −71 °C; b.p. −61.7 °C; t_c 104 °C; density of gas 9.73 g/L; sp. gr. liquid 4.4 at −62 °C, solid 4; valence usually 0. The element was discovered in 1900 by Dorn, who called it *radium emanation.* In 1908 Ramsay and Gray, who named it *niton*, isolated the element and determined its density, finding it to be the heaviest known gas. It is essentially inert and occupies the last place in the zero group of gases in the Periodic Table. Since 1923, it has been called radon. Thirty-seven isotopes and isomers are known. Radon-222, coming from radium, has a half-life of 3.823 days and is an alpha emitter; Radon-220, emanating naturally from thorium and called *thoron*, has a half-life of 55.6 s and is also an alpha emitter. Radon-219 emanates from actinium and is called *actinon*. It has a half-life of 3.9 s and is also an alpha emitter. It is estimated that every square mile of soil to a depth of 6 inches contains about 1 g of radium, which releases radon in tiny amounts to the atmosphere. Radon is present in some spring waters, such as those at Hot Springs, Arkansas. On the average, one part of radon is present to 1 × 10^{21} part of air. At ordinary temperatures radon is a colorless gas; when cooled below the freezing point, radon exhibits a brilliant phosphorescence which becomes yellow as the temperature is lowered and orange-red at the temperature of liquid air. It has been reported that fluorine reacts with radon, forming radon fluoride. Radon clathrates have also been reported. Radon is still produced for therapeutic use by a few hospitals by pumping it from a radium source and sealing it in minute tubes, called seeds or needles, for application to patients. This practice has now been largely discontinued as hospitals can order the seeds directly from suppliers, who make up the seeds with the desired activity for the day of use. Care must be taken in handling radon, as with other radioactive materials. The main hazard is from inhalation of the element and its solid daughters, which are collected on dust in the air. Good ventilation should be provided where radium, thorium, or actinium is stored to prevent build-up of this element. Radon build-up is a health consideration in uranium mines. Recently radon build-up in homes has been a concern. Many deaths from lung cancer are caused by radon exposure. In the U.S. it is recommended that remedial action be taken if the air from radon in homes exceeds 4 pCi/L.

Rhenium — (L. *Rhenus*, Rhine), Re; at. wt. 186.207(1); at. no. 75; m.p. 3185 °C; b.p. 5596 °C; sp. gr. 20.8 (20 °C); valence −1, +1, 2, 3, 4, 5, 6, 7. Discovery of rhenium is generally attributed to Noddack, Tacke, and Berg, who announced in 1925 they had detected the element in platinum ores and *columbite*. They also found the element in *gadolinite* and *molybdenite*. By working up 660 kg of molybdenite they were able in 1928 to extract 1 g of rhenium. The price in 1928 was $10,000/g. Rhenium does not occur free in nature or as a compound in a distinct mineral species. It is, however, widely spread throughout the Earth's crust to the extent of about 0.001 ppm. Commercial rhenium in the U.S. today is obtained from molybdenite roaster-flue dusts obtained from copper-sulfide ores mined in the vicinity of Miami, Arizona, and elsewhere in Arizona and Utah. Some molybdenites contain from 0.002 to 0.2% rhenium. It is estimated that in 1999 about 16,000 kg of rhenium was being produced. The total estimated world reserves of rhenium is 11,000,000 kg. Natural rhenium is a mixture of two isotopes, one of which has a very long half-life. Thirty-nine other unstable isotopes are recognized. Rhenium metal is prepared by reducing ammonium perrhenate with hydrogen at elevated temperatures. The element is silvery white with a metallic luster; its density is exceeded by that of only platinum, iridium, and osmium, and its melting point is exceeded by that of only tungsten and carbon. It has other useful properties. The usual commercial form of the element is a powder, but it can be consolidated by pressing and resistance-sintering in a vacuum or hydrogen atmosphere. This produces a compact shape in excess of 90% of the density of the metal. Annealed rhenium is very ductile, and can be bent, coiled, or rolled. Rhenium is used as an additive to tungsten and molybdenum-based alloys to impart useful properties. It is widely used for filaments for mass spectrographs and ion gages. Rhenium-molybdenum alloys are superconductive at 10 K. Rhenium is also used as an electrical contact material as it has good wear resistance and withstands arc corrosion. Thermocouples made of Re-W are used for measuring temperatures up to 2200 °C, and rhenium wire has been used in photoflash lamps for photography. Rhenium catalysts are exceptionally resistant to poisoning from nitrogen, sulfur, and phosphorus, and are used for hydrogenation of fine chemicals, hydrocracking, reforming, and disproportionation of olefins. Rhenium has recently become especially important as a catalyst for petroleum refining and in making super-alloys for jet engines. Rhenium costs about $16/g (99.99% pure). Little is known of its toxicity; therefore, it should be handled with care until more data are available.

Rhodium — (Gr. *rhodon*, rose), Rh; at. wt. 102.90550(2); at. no. 45; m.p. 1964 °C; b.p. 3695 °C; sp. gr. 12.41 (20 °C); valence 2, 3, 4, 5, and 6. Wollaston discovered rhodium in 1803-4 in crude platinum ore he presumably obtained from South America. Rhodium occurs native with other platinum metals in river sands of the Urals and in North and South America. It is also found with other platinum metals in the copper-nickel sulfide ores of the Sudbury, Ontario region. Although the quantity occurring here is very small, the large tonnages of nickel processed make the recovery commercially feasible. The annual world production of rhodium in 1999 was only about 9000 kg. The metal is silvery white and at red heat slowly changes in air to the sesquioxide. At higher temperatures it converts back to the element. Rhodium has a higher melting point and lower density than platinum. Its major use is as an alloying agent to harden platinum and palladium. Such alloys are used for furnace windings, thermocouple elements, bushings for glass fiber production, electrodes for aircraft spark plugs, and laboratory crucibles. It is useful as an electrical contact material as it has a low electrical resistance, a low and stable contact resistance, and is highly resistant to corrosion. Plated rhodium, produced by electroplating or evaporation, is excep-

tionally hard and is used for optical instruments. It has a high reflectance and is hard and durable. Rhodium is also used for jewelry, for decoration, and as a catalyst. Fifty-two isotopes and isomers are now known. Rhodium metal (powder) costs about $180/g (99.9%).

Roentgenium — (Wilhelm Roentgen, discoverer of X-rays), Rg. On December 20, 1994, scientists at GSI Darmstadt, Germany announced they had detected three atoms of a new element with 111 protons and 161 neutrons. This element was made by bombarding ^{83}Bi with ^{28}Ni. Signals of Element 111 appeared for less than 0.002 s, then decayed into lighter elements including Element ^{268}109 and Element ^{264}107. These isotopes had not previously been observed. In 2004 IUPAC approved the name roentgenium for Element 111. Roentgenium is expected to have properties similar to gold.

Rubidium — (L. *rubidus*, deepest red), Rb; at. wt. 85.4678(3); at. no. 37; m.p. 39.30 °C; b.p. 688 °C; sp. gr. (solid) 1.532 (20 °C), (liquid) 1.475 (39 °C); valence 1, 2, 3, 4. Discovered in 1861 by Bunsen and Kirchhoff in the mineral *lepidolite* by use of the spectroscope. The element is much more abundant than was thought several years ago. It is now considered to be the 16th most abundant element in the Earth's crust. Rubidium occurs in *pollucite, carnallite, leucite,* and *zinnwaldite,* which contains traces up to 1%, in the form of the oxide. It is found in lepidolite to the extent of about 1.5%, and is recovered commercially from this source. Potassium minerals, such as those found at Searles Lake, California, and potassium chloride recovered from brines in Michigan also contain the element and are commercial sources. It is also found along with cesium in the extensive deposits of *pollucite* at Bernic Lake, Manitoba. Rubidium can be liquid at room temperature. It is a soft, silvery-white metallic element of the alkali group and is the second most electropositive and alkaline element. It ignites spontaneously in air and reacts violently in water, setting fire to the liberated hydrogen. As with other alkali metals, it forms amalgams with mercury and it alloys with gold, cesium, sodium, and potassium. It colors a flame yellowish violet. Rubidium metal can be prepared by reducing rubidium chloride with calcium, and by a number of other methods. It must be kept under a dry mineral oil or in a vacuum or inert atmosphere. Thirty-five isotopes and isomers of rubidium are known. Naturally occurring rubidium is made of two isotopes, ^{85}Rb and ^{87}Rb. Rubidium-87 is present to the extent of 27.83% in natural rubidium and is a beta emitter with a half-life of 4.9×10^{10} years. Ordinary rubidium is sufficiently radioactive to expose a photographic film in about 30 to 60 days. Rubidium forms four oxides: Rb_2O, Rb_2O_2, Rb_2O_3, Rb_2O_4. Because rubidium can be easily ionized, it has been considered for use in "ion engines" for space vehicles; however, cesium is somewhat more efficient for this purpose. It is also proposed for use as a working fluid for vapor turbines and for use in a thermoelectric generator using the magnetohydrodynamic principle where rubidium ions are formed by heat at high temperature and passed through a magnetic field. These conduct electricity and act like an armature of a generator thereby generating an electric current. Rubidium is used as a getter in vacuum tubes and as a photocell component. It has been used in making special glasses. $RbAg_4I_5$ is important, as it has the highest room-temperature conductivity of any known ionic crystal. At 20 °C its conductivity is about the same as dilute sulfuric acid. This suggests use in thin film batteries and other applications. The present cost in small quantities is about $50/g (99.8% pure).

Ruthenium — (L. *Ruthenia*, Russia), Ru; at. wt. 101.07(2); at. no. 44, m.p. 2334 °C; b.p. 4150 °C; sp. gr. 12.1 (20 °C); valence 0, 1, 2, 3, 4, 5, 6, 7, 8. Berzelius and Osann in 1827 examined the residues left after dissolving crude platinum from the Ural mountains in *aqua regia*. While Berzelius found no unusual metals, Osann thought he found three new metals, one of which he named ruthenium. In 1844 Klaus, generally recognized as the discoverer, showed that Osann's ruthenium oxide was very impure and that it contained a new metal. Klaus obtained 6 g of ruthenium from the portion of crude platinum that is insoluble in *aqua regia*. A member of the platinum group, ruthenium occurs native with other members of the group of ores found in the Ural mountains and in North and South America. It is also found along with other platinum metals in small but commercial quantities in *pentlandite* of the Sudbury, Ontario, nickel-mining region, and in *pyroxinite* deposits of South Africa. Natural ruthenium contains seven isotopes. Twenty-eight other isotopes and isomers are known, all of which are radioactive. The metal is isolated commercially by a complex chemical process, the final stage of which is the hydrogen reduction of ammonium ruthenium chloride, which yields a powder. The powder is consolidated by powder metallurgy techniques or by argon-arc welding. Ruthenium is a hard, white metal and has four crystal modifications. It does not tarnish at room temperatures, but oxidizes in air at about 800 °C. The metal is not attacked by hot or cold acids or *aqua regia*, but when potassium chlorate is added to the solution, it oxidizes explosively. It is attacked by halogens, hydroxides, etc. Ruthenium can be plated by electrodeposition or by thermal decomposition methods. The metal is one of the most effective hardeners for platinum and palladium, and is alloyed with these metals to make electrical contacts for severe wear resistance. A ruthenium–molybdenum alloy is said to be superconductive at 10.6 K. The corrosion resistance of titanium is improved a hundredfold by addition of 0.1% ruthenium. It is a versatile catalyst. Hydrogen sulfide can be split catalytically by light using an aqueous suspension of CdS particles loaded with ruthenium dioxide. It is thought this may have application to removal of H_2S in oil refining and other industrial processes. Compounds in at least eight oxidation states have been found, but of these, the +2. +3. and +4 states are the most common. Ruthenium tetroxide, like osmium tetroxide, is highly toxic. In addition, it may explode. Ruthenium compounds show a marked resemblance to those of osmium. The metal is priced at about $25/g (99.95% pure).

Rutherfordium — (Ernest Rutherford [1871–1937], New Zealand, Canadian, and British physicist); Rf; at. wt. [261]; at. no. 104. In 1964, workers of the Joint Nuclear Research Institute at Dubna (Russia) bombarded plutonium with accelerated 113 to 115 MeV neon ions. By measuring fission tracks in a special glass with a microscope, they detected an isotope that decays by spontaneous fission. They suggested that this isotope, which has a half-life of 0.3 ± 0.1 s, might be ^{260}104, produced by the following reaction:

$$^{242}_{94}\text{Pu} + ^{22}_{10}\text{Ne} \rightarrow ^{260}104 + 4n$$

Element 104, the first *transactinide* element, is expected to have chemical properties similar to those of hafnium. It would, for example, form a relatively volatile compound with chlorine (a tetrachloride). The Soviet scientists have performed experiments aimed at chemical identification, and have attempted to show that the 0.3-s activity is more volatile than that of the relatively nonvolatile actinide trichlorides. This experiment does not fulfill the test of chemically separating the new element from all others, but it provides important evidence for evaluation. New data, reportedly issued by Soviet scientists, have reduced the half-life of the isotope they worked with from 0.3 to 0.15 s. The Dubna scientists suggest the name *kurchatovium* and symbol *Ku* for Element 104, in honor of Igor Vasilevich Kurchatov (1903–1960), late Head of Soviet Nuclear Research. The Dubna Group also has proposed the name *dubnium* for Element 104. In 1969, Ghiorso, Nurmia, Harris, K. A. Y. Eskola, and P. I. Eskola of the University of California at Berkeley reported they had positively identified two, and possibly three, isotopes of Element 104. The group also indicated that after repeated attempts so far they have been unable to produce isotope ^{260}104 reported by the Dubna groups in 1964. The discoveries at Berkeley were made by bombarding a target of ^{249}Cf with ^{12}C nuclei of 71 MeV, and ^{13}C nuclei of 69 MeV. The combination of ^{12}C with ^{249}Cf followed by instant emission of four neutrons produced Element ^{257}104. This isotope has a half-life of 4 to 5 s, decaying by emitting an alpha particle into ^{253}No, with a half-life of 105 s. The same reaction, except with the emission of three neutrons, was thought to have produced ^{258}104 with a half-life of about 1/100 s. Element ^{259}104 is formed by the merging of a ^{13}C nuclei with ^{249}Cf, followed by emission of three neutrons. This isotope has a half-life of 3 to 4 s, and decays by emitting an alpha particle into ^{255}No, which has a half-life of 185 s. Thousands of atoms of ^{257}104 and ^{259}104 have been detected. The Berkeley group believes its identification of ^{258}104 was correct. Eleven isotopes of Element 104 have now been identified. The Berkeley group proposed the name *rutherfordium* (symbol Rf) for the new element, in honor of Ernest Rutherford. This name was formally adapted by IUPAC in August 1997.

Samarium — (*Samarskite*, a mineral), Sm; at. wt. 150.36(3); at. no. 62; m.p. 1072 °C; b.p. 1794 °C; sp. gr (α) 7.520 (25 °C); valence 2 or 3. Discovered spectroscopically by its sharp absorption lines in 1879 by Lecoq de Boisbaudran in the mineral *samarskite*, named in honor of a Russian mine official, Col. Samarski. Samarium is found along with other members of the rare-earth-elements in many minerals, including *monazite* and *bastnasite*, which are commercial sources. The largest producer of rare-earth minerals is now China, followed by the U.S., India, and Russia. It occurs in monazite to the extent of 2.8%. While *misch metal* containing about 1% of samarium metal has long been used, samarium has not been isolated in relatively pure form until recently. Ion-exchange and solvent extraction techniques have recently simplified separation of the rare earths from one another; more recently, electrochemical deposition, using an electrolytic solution of lithium citrate and a mercury electrode, is said to be a simple, fast, and highly specific way to separate the rare earths. Samarium metal can be produced by reducing the oxide with barium or lanthanum. Samarium has a bright silver luster and is reasonably stable in air. Three crystal modifications of the metal exist, with transformations at 734 and 922 °C. The metal ignites in air at about 150 °C. Thirty-three isotopes and isomers of samarium are now recognized. Natural samarium is a mixture of seven isotopes, three of which are unstable but have long half-lives. Samarium, along with other rare earths, is used for carbon-arc lighting for the motion picture industry. The sulfide has excellent high-temperature stability and good thermoelectric efficiencies up to 1100 °C. $SmCo_5$ has been used in making a new permanent magnet material with the highest resistance to demagnetization of any known material. It is said to have an intrinsic coercive force as high as 2200 kA/m. Samarium oxide has been used in optical glass to absorb the infrared. Samarium is used to dope calcium fluoride crystals for use in optical masers or lasers. Compounds of the metal act as sensitizers for phosphors excited in the infrared; the oxide exhibits catalytic properties in the dehydration and dehydrogenation of ethyl alcohol. It is used in infrared absorbing glass and as a neutron absorber in nuclear reactors. The metal is priced at about $3.50/g (99.9%). Little is known of the toxicity of samarium; therefore, it should be handled carefully.

Scandium — (L. *Scandia*, Scandinavia), Sc; at. wt. 44.955912(6); at. no. 21; m.p. 1541 °C; b.p. 2836 °C; sp. gr. 2.989 (25 °C); valence 3. On the basis of the Periodic System, Mendeleev predicted the existence of *ekaboron*, which would have an atomic weight between 40 of calcium and 48 of titanium. The element was discovered by Nilson in 1878 in the minerals *euxenite* and *gadolinite*, which had not yet been found anywhere except in Scandinavia. By processing 10 kg of euxenite and other residues of rare-earth minerals, Nilson was able to prepare about 2 g of scandium oxide of high purity. Cleve later pointed out that Nilson's scandium was identical with Mendeleev's ekaboron. Scandium is apparently a much more abundant element in the sun and certain stars than here on Earth. It is about the 23rd most abundant element in the sun, compared to the 50th most abundant on Earth. It is widely distributed on Earth, occurring in very minute quantities in over 800 mineral species. The blue color of beryl (aquamarine variety) is said to be due to scandium. It occurs as a principal component in the rare mineral *thortveitite*, found in Scandinavia and Malagasy. It is also found in the residues remaining after the extraction of tungsten from Zinnwald *wolframite*, and in *wiikite* and *bazzite*. Most scandium is presently being recovered from *thortveitite* or is extracted as a by-product from uranium mill tailings. Metallic scandium was first prepared in 1937 by Fischer, Brunger, and Grieneisen, who electrolyzed a eutectic melt of potassium, lithium, and scandium chlorides at 700 to 800 °C. Tungsten wire and a pool of molten zinc served as the electrodes in a graphite crucible. Pure scandium is now produced by reducing scandium fluoride with calcium metal. The production of the first pound of 99% pure scandium metal was announced in 1960. Scandium is a silver-white metal that develops a slightly yellowish or pinkish cast upon exposure to air. It is relatively soft, and resembles yttrium and the rare-earth metals more than it resembles aluminum or titanium. It is a very light metal and has a much higher melting point than aluminum, making it of interest to designers of spacecraft. Scandium is not attacked by a 1:1 mixture of conc. HNO_3 and 48% HF. Scandium reacts rapidly with many acids. Twenty-three isotopes and isomers of scandium are recognized. The metal is expensive, costing about $200/g with a purity of about 99.9%. About 20 kg of scandium (as Sc_2O_3) are now being used yearly in the U.S. to produce high-intensity lights, and the radioactive isotope ^{46}Sc is used as a tracing agent in refinery crackers for crude oil, etc. Scandium iodide added to mercury vapor lamps produces a highly efficient light source resembling sunlight, which is important for indoor or night-time color TV. Little is yet known about the toxicity of scandium; therefore, it should be handled with care.

Seaborgium — (Glenn T. Seaborg [1912–1999], American chemist and nuclear physicist). Sg; at. wt. [266]; at no. 106. The discovery of *Seaborgium*, Element 106, took place in 1974 almost simultaneously at the Lawrence-Berkeley Laboratory and at the Joint Institute for Nuclear Research at Dubna, Russia. The Berkeley Group, under direction of Ghiorso, used the Super-Heavy Ion Linear Accelerator (Super HILAC) as a source of heavy ^{18}O ions to bombard a 259-μg target of ^{249}Cf. This resulted in the production and positive identification of $^{263}106$, which decayed with a half-life of 0.9 ± 0.2 s by the emission of alpha particles as follows:

$$^{263}106 \xrightarrow{\alpha} {}^{259}104 \xrightarrow{\alpha} {}^{255}No \xrightarrow{\alpha}.$$

The Dubna Team, directed by Flerov and Organessian, produced heavy ions of ^{54}Cr with their 310-cm heavy-ion cyclotron to bombard ^{207}Pb and ^{208}Pb and found a product that decayed with a half-life of 7 ms. They assigned $^{259}106$ to this isotope. It is now thought seven isotopes of *Seaborgium* have been identified. Two of the isotopes are believed to have half-lives of about 30 s. *Seaborgium* most likely would have properties resembling tungsten. The IUPAC adopted the name *Seaborgium* in August 1997. Normally the naming of an element is not given until after the death of the person for which the element is named; however, in this case, it was named while Dr. Seaborg was still alive.

Selenium — (Gr. *Selene*, moon), Se; at. wt. 78.96(3); at. no. 34; m.p. (gray) 221 °C; b.p. (gray) 685 °C; sp. gr. (gray) 4.79, (vitreous) 4.28; valence −2, +4, or +6. Discovered by Berzelius in 1817, who found it associated with tellurium, named for the Earth. Selenium is found in a few rare minerals, such as *crooksite* and *clausthalite*. In years past it has been obtained from flue dusts remaining from processing copper sulfide ores, but the anode muds from electrolytic copper refineries now provide the source of most of the world's selenium. Selenium is recovered by roasting the muds with soda or sulfuric acid, or by smelting them with soda and niter. Selenium exists in several allotropic forms. Three are generally recognized, but as many as six have been claimed. Selenium can be prepared with either an amorphous or crystalline structure. The color of amorphous selenium is either red, in powder form, or black, in vitreous form. Crystalline monoclinic selenium is a deep red; crystalline hexagonal selenium, the most stable variety, is a metallic gray. Natural selenium contains six stable isotopes. Twenty-nine other isotopes and isomers have been characterized. The element is a member of the sulfur family and resembles sulfur both in its various forms and in its compounds. Selenium exhibits both photovoltaic action, where light is converted directly into electricity, and photoconductive action, where the electrical resistance decreases with increased illumination. These properties make selenium useful in the production of photocells and exposure meters for photographic use, as well as solar cells. Selenium is also able to convert a.c. electricity to d.c., and is extensively used in rectifiers. Below its melting point, selenium is a p-type semiconductor and is finding many uses in electronic and solid-state applications. It is used in xerography for reproducing and copying documents, letters, etc., but recently its use in this application has been decreasing in favor of certain organic compounds. It is used by the glass industry to decolorize glass and to make ruby-colored glasses and enamels. It is also used as a photographic toner, and as an additive to stainless steel. Elemental selenium has been said to be practically nontoxic and is considered to be an essential trace element; however, hydrogen selenide and other selenium compounds are extremely toxic, and resemble arsenic in their physiological reactions. Hydrogen selenide in a concentration of 1.5 ppm is intolerable to man. Selenium occurs in some soils in amounts sufficient to produce serious effects on animals feeding on plants, such as locoweed, grown in such soils. Selenium (99.5%) is priced at about $250/kg. It is also available in high-purity form at a cost of about $350/kg (99.999%).

Silicon — (L. *silex, silicis*, flint), Si; at. wt. [28.084; 28.086]; at. no. 14; m.p. 1414 °C; b.p. 3265 °C; sp. gr. 2.33 (25 °C); valence 4. Davy in 1800 thought silica to be a compound and not an element; later in 1811, Gay Lussac and Thenard probably prepared impure amorphous silicon by heating potassium with silicon tetrafluoride. Berzelius, generally credited with the discovery, in 1824 succeeded in preparing amorphous silicon by the same general method as used earlier, but he purified the product by removing the fluosilicates by repeated washings. Deville in 1854 first prepared crystalline silicon, the second allotropic form of the element. Silicon is present in the sun and stars and is a principal component of a class of meteorites known as "aerolites." It is also a component of *tektites*, a natural glass of uncertain origin. Natural silicon contains three isotopes. Twenty-four other radioactive isotopes are recognized. Silicon makes up 25.7% of the Earth's crust, by weight, and is the second most abundant element, being exceeded only by oxygen. Silicon is not found free in nature, but occurs chiefly as the oxide and as silicates. *Sand, quartz, rock crystal, amethyst, agate, flint, jasper*, and *opal* are some of the forms in which the oxide appears. *Granite, hornblende, asbestos, feldspar, clay mica*, etc. are but a few of the numerous silicate minerals. Silicon is prepared commercially by heating silica and carbon in an electric furnace, using carbon electrodes. Several other methods can be used for preparing the element. Amorphous silicon can be prepared as a brown powder, which can be easily melted or vaporized. Crystalline silicon has a metallic luster and grayish color. The Czochralski process is commonly used to produce single crystals of silicon used for solid-state or semiconductor devices. Hyperpure silicon can be prepared by the thermal decomposition of ultra-pure trichlorosilane in a hydrogen atmosphere, and by a vacuum float zone process. This product can be doped with boron, gallium, phosphorus, or arsenic to produce silicon for use in transistors, solar cells, rectifiers, and other solid-state devices that are used extensively in the electronics and space-age industries. Hydrogenated amorphous silicon has shown promise in producing economical cells for converting solar energy into electricity. Silicon is a relatively inert element, but it is attacked by halogens and dilute alkali. Most acids, except hydrofluoric, do not affect it. Silicones are important products of silicon. They may be prepared by hydrolyzing a silicon organic chloride, such as dimethyl silicon chloride. Hydrolysis and condensation of various substituted chlorosilanes can be used to produce a very great number of polymeric products, or silicones, ranging from liquids to hard, glasslike solids with many useful properties. Elemental silicon transmits more than 95% of all wavelengths of infrared, from 1.3 to 6.7 μm. Silicon is one of man's most useful elements. In the form of sand and clay it is used to make concrete and brick; it is a useful refractory material for high-temperature work, and in the form of silicates it is used in making enamels, pottery, etc. Silica, as sand, is a principal ingredient of glass, one of the most inexpensive of materials with excellent mechanical, optical, thermal, and electrical properties. Glass can be made in a very great variety of shapes, and is used as containers, window glass, insulators, and thousands of other uses. Silicon tetrachloride can be used to iridize glass. Silicon is important

in plant and animal life. Diatoms in both fresh and salt water extract silica from the water to build up their cell walls. Silica is present in ashes of plants and in the human skeleton. Silicon is an important ingredient in steel; silicon carbide is one of the most important abrasives and has been used in lasers to produce coherent light of 4560 Å. A remarkable material, first discovered in 1930, is *Aerogel*, which is now used by NASA in their space missions to collect cometary and interplanet dust. *Aerogel* is a highly insulative material that has the lowest density of any known solid. One form of *Aerogel* is 99.9% air and 0.1% SiO_2 by volume. It is 1000 times less dense than glass. It has been called "blue smoke" or "solid smoke." A block of *Aerogel* as large as a person may weigh less than a pound and yet support the weight of 1000 lbs (455 kg). This material is expected to trap cometary particles traveling at speeds of 32 km/sec. *Aerogel* is said to be non-toxic and non-inflammable. It has high thermal insulating qualities that could be used in home insulation. Its light weight may have aircraft applications. Regular grade silicon (99.5%) costs about $160/kg. Silicon (99.9999%) pure costs about $200/kg; hyperpure silicon is available at a higher cost. Miners, stonecutters, and other engaged in work where siliceous dust is breathed in large quantities often develop a serious lung disease known as *silicosis*.

Silver — (Anglo-Saxon, *Seolfor siolfur*), Ag (L. *argentum*), at. wt. 107.8682(2); at. no. 47; m.p. 961.78 °C; b.p. 2162 °C; sp. gr. 10.50 (20 °C); valence 1, 2. Silver has been known since ancient times. It is mentioned in Genesis. Slag dumps in Asia Minor and on islands in the Aegean Sea indicate that man learned to separate silver from lead as early as 3000 B.C. Silver occurs native and in ores such as *argentite* (Ag_2S) and *horn silver* (AgCl); lead, lead-zinc, copper, gold, and copper-nickel ores are principal sources. Mexico, Canada, Peru, and the U.S. are the principal silver producers in the western hemisphere. Silver is also recovered during electrolytic refining of copper. Commercial fine silver contains at least 99.9% silver. Purities of 99.999+% are available commercially. Pure silver has a brilliant white metallic luster. It is a little harder than gold and is very ductile and malleable, being exceeded only by gold and perhaps palladium. Pure silver has the highest electrical and thermal conductivity of all metals, and possesses the lowest contact resistance. It is stable in pure air and water, but tarnishes when exposed to ozone, hydrogen sulfide, or air containing sulfur. The alloys of silver are important. Sterling silver is used for jewelry, silverware, etc. where appearance is paramount. This alloy contains 92.5% silver, the remainder being copper or some other metal. Silver is of utmost importance in photography, about 30% of the U.S. industrial consumption going into this application. It is used for dental alloys. Silver is used in making solder and brazing alloys, electrical contacts, and high capacity silver–zinc and silver–cadmium batteries. Silver paints are used for making printed circuits. It is used in mirror production and may be deposited on glass or metals by chemical deposition, electrodeposition, or by evaporation. When freshly deposited, it is the best reflector of visible light known, but is rapidly tarnishes and loses much of its reflectance. It is a poor reflector of ultraviolet. Silver fulminate ($Ag_2C_2N_2O_2$), a powerful explosive, is sometimes formed during the silvering process. Silver iodide is used in seeding clouds to produce rain. Silver chloride has interesting optical properties as it can be made transparent; it also is a cement for glass. Silver nitrate, or *lunar caustic*, the most important silver compound, is used extensively in photography. While silver itself is not considered to be toxic, most of its salts are poisonous. Natural silver contains two stable isotopes. Fifty-six other radioactive isotopes and isomers are known. Silver compounds can be absorbed in the circulatory system and reduced silver deposited in the various tissues of the body. A condition, known as *argyria*, results with a greyish pigmentation of the skin and mucous membranes. Silver has germicidal effects and kills many lower organisms effectively without harm to higher animals. Silver for centuries has been used traditionally for coinage by many countries of the world. In recent times, however, consumption of silver has at times greatly exceeded the output. In 1939, the price of silver was fixed by the U.S. Treasury at 71¢/troy oz., and at 90.5¢/troy oz. in 1946. In November 1961 the U.S. Treasury suspended sales of nonmonetized silver, and the price stabilized for a time at about $1.29, the melt-down value of silver U.S. coins. The Coinage Act of 1965 authorized a change in the metallic composition of the three U.S. subsidiary denominations to clad or composite type coins. This was the first change in U.S. coinage since the monetary system was established in 1792. Clad dimes and quarters are made of an outer layer of 75% Cu and 25% Ni bonded to a central core of pure Cu. The composition of the one- and five-cent pieces remains unchanged. One-cent coins are 95% Cu and 5% Zn. Five-cent coins are 75% Cu and 25% Ni. Old silver dollars are 90% Ag and 10% Cu. Earlier subsidiary coins of 90% Ag and 10% Cu officially were to circulate alongside the clad coins; however, in practice they have largely disappeared (Gresham's Law), as the value of the silver is now greater than their exchange value. Silver coins of other countries have largely been replaced with coins made of other metals. On June 24, 1968, the U.S. Government ceased to redeem U.S. Silver Certificates with silver. Since that time, the price of silver has fluctuated widely. As of January 2002, the price of silver was about $4.10/troy oz. (13¢/g); however the price has fluctuated considerably due to market instability. The price of silver in 2001 was only about four times the cost of the metal about 150 years ago. This has largely been caused by Central Banks disposing of some of their silver reserves and the development of more productive mines with better refining methods. Also, silver has been displaced by other metals or processes, such as digital photography.

Sodium — (English, *soda*; Medieval Latin, *sodanum*, headache remedy), Na (L. *natrium*); at. wt. 22.98976928(2); at. no. 11; m.p. 97.80 °C; b.p. 883 °C; sp. gr. 0.971 (20 °C); valence 1. Long recognized in compounds, sodium was first isolated by Davy in 1807 by electrolysis of caustic soda. Sodium is present in fair abundance in the sun and stars. The D lines of sodium are among the most prominent in the solar spectrum. Sodium is the sixth most abundant element on earth, comprising about 2.6% of the Earth's crust; it is the most abundant of the alkali group of metals of which it is a member. The most common compound is sodium chloride, but it occurs in many other minerals, such as *soda niter, cryolite, amphibole, zeolite, sodalite*, etc. It is a very reactive element and is never found free in nature. It is now obtained commercially by the electrolysis of absolutely dry fused sodium chloride. This method is much cheaper than that of electrolyzing sodium hydroxide, as was used several years ago. Sodium is a soft, bright, silvery metal that floats on water, decomposing it with the evolution of hydrogen and the formation of the hydroxide. It may or may not ignite spontaneously on water, depending on the amount of oxide and metal exposed to the water. It normally does not ignite in air at temperatures below 115 °C. Sodium

should be handled with respect, as it can be dangerous when improperly handled. Metallic sodium is vital in the manufacture of sodamide and esters, and in the preparation of organic compounds. The metal may be used to improve the structure of certain alloys, to descale metal, to purify molten metals, and as a heat transfer agent. An alloy of sodium with potassium, NaK, is also an important heat transfer agent. Sodium compounds are important to the paper, glass, soap, textile, petroleum, chemical, and metal industries. Soap is generally a sodium salt of certain fatty acids. The importance of common salt to animal nutrition has been recognized since prehistoric times. Among the many compounds that are of the greatest industrial importance are common salt ($NaCl$), soda ash (Na_2CO_3), baking soda ($NaHCO_3$), caustic soda ($NaOH$), Chile saltpeter ($NaNO_3$), di- and tri-sodium phosphates, sodium thiosulfate (hypo, $Na_2S_2O_3 \cdot 5H_2O$), and borax ($Na_2B_4O_7 \cdot 10H_2O$). Seventeen isotopes of sodium are recognized. Metallic sodium is priced at about $575/kg (99.95%). On a volume basis, it is the cheapest of all metals. Sodium metal should be handled with great care. It should be kept in an inert atmosphere and contact with water and other substances with which sodium reacts should be avoided.

Strontium — (*Strontian*, town in Scotland), Sr; at. wt. 87.62(1); at. no. 38; m.p. 777 °C; b.p. 1382 °C; sp. gr. 2.64; valence 2. Isolated by Davey by electrolysis in 1808; however, Adair Crawford in 1790 recognized a new mineral (strontianite) as differing from other barium minerals (baryta). Strontium is found chiefly as *celestite* ($SrSO_4$) and *strontianite* ($SrCO_3$). *Celestite* is found in Mexico, Turkey, Iran, Spain, Algeria, and in the U.K. The U.S. has no active *celestite* mines. The metal can be prepared by electrolysis of the fused chloride mixed with potassium chloride, or is made by reducing strontium oxide with aluminum in a vacuum at a temperature at which strontium distills off. Three allotropic forms of the metal exist, with transition points at 235 and 540 °C. Strontium is softer than calcium and decomposes water more vigorously. It does not absorb nitrogen below 380 °C. It should be kept under mineral oil to prevent oxidation. Freshly cut strontium has a silvery appearance, but rapidly turns a yellowish color with the formation of the oxide. The finely divided metal ignites spontaneously in air. Volatile strontium salts impart a beautiful crimson color to flames, and these salts are used in pyrotechnics and in the production of flares. Natural strontium is a mixture of four stable isotopes. Thirty-two other unstable isotopes and isomers are known to exist. Of greatest importance is ^{90}Sr with a half-life of 29 years. It is a product of nuclear fallout and presents a health problem. This isotope is one of the best long-lived high-energy beta emitters known, and is used in SNAP (Systems for Nuclear Auxiliary Power) devices. These devices hold promise for use in space vehicles, remote weather stations, navigational buoys, etc., where a lightweight, long-lived, nuclear-electric power source is needed. The major use for strontium at present is in producing glass for color television picture tubes. All color TV and cathode ray tubes sold in the U.S. are required by law to contain strontium in the face plate glass to block X-ray emission. Strontium also improves the brilliance of the glass and the quality of the picture. It has also found use in producing ferrite magnets and in refining zinc. Strontium titanate is an interesting optical material as it has an extremely high refractive index and an optical dispersion greater than that of diamond. It has been used as a gemstone, but it is very soft. It does not occur naturally. Strontium metal (99% pure) costs about $220/kg.

Sulfur — (Sanskrit, *sulvere*; L. *sulphurium*), S; at. wt. [32.059; 32.076]; at. no. 16; m.p. 115.21 °C; b.p. 444.61 °C; t_c 1041 °C; sp. gr. (rhombic) 2.07, (monoclinic) 2.00 (20 °C); valence 2, 4, or 6. Known to the ancients; referred to in Genesis as *brimstone*. Sulfur is found in meteorites. A dark area near the crater Aristarchus on the moon has been studied by R. W. Wood with ultraviolet light. This study suggests strongly that it is a sulfur deposit. Sulfur occurs native in the vicinity of volcanoes and hot springs. It is widely distributed in nature as *iron pyrites, galena, sphalerite, cinnabar, stibnite, gypsum, Epsom salts, celestite, barite,* etc. Sulfur is commercially recovered from wells sunk into the salt domes along the Gulf Coast of the U.S. It is obtained from these wells by the Frasch process, which forces heated water into the wells to melt the sulfur, which is then brought to the surface. Sulfur also occurs in natural gas and petroleum crudes and must be removed from these products. Formerly this was done chemically, which wasted the sulfur. New processes now permit recovery, and these sources promise to be very important. Large amounts of sulfur are being recovered from Alberta gas fields. Sulfur is a pale yellow, odorless, brittle solid that is insoluble in water but soluble in carbon disulfide. In every state, whether gas, liquid or solid, elemental sulfur occurs in more than one allotropic form or modification; these present a confusing multitude of forms whose relations are not yet fully understood. Amorphous or "plastic" sulfur is obtained by fast cooling of the crystalline form. X-ray studies indicate that amorphous sulfur may have a helical structure with eight atoms per spiral. Crystalline sulfur seems to be made of rings, each containing eight sulfur atoms that fit together to give a normal X-ray pattern. Twenty-one isotopes of sulfur are now recognized. Four occur in natural sulfur, none of which is radioactive. A finely divided form of sulfur, known as *flowers of sulfur*, is obtained by sublimation. Sulfur readily forms sulfides with many elements. Sulfur is a component of black gunpowder, and is used in the vulcanization of natural rubber and a fungicide. It is also used extensively is making phosphatic fertilizers. A tremendous tonnage is used to produce sulfuric acid, the most important manufactured chemical. It is used in making sulfite paper and other papers, as a fumigant, and in the bleaching of dried fruits. The element is a good electrical insulator. Organic compounds containing sulfur are very important. Calcium sulfate, ammonium sulfate, carbon disulfide, sulfur dioxide, and hydrogen sulfide are but a few of the many other important compounds of sulfur. Sulfur is essential to life. It is a minor constituent of fats, body fluids, and skeletal minerals. Carbon disulfide, hydrogen sulfide, and sulfur dioxide should be handled carefully. Hydrogen sulfide in small concentrations can be metabolized, but in higher concentrations it can quickly cause death by respiratory paralysis. It is insidious in that it quickly deadens the sense of smell. Sulfur dioxide is a dangerous component in atmospheric pollution. Sulfur (99.999%) costs about $575/kg.

Tantalum — (Gr. *Tantalos*, mythological character, father of *Niobe*), Ta; at. wt. 180.94788(2); at. no. 73; m.p. 3017 °C; b.p. 5458 °C; sp. gr. 16.4; valence 2?, 3, 4?, or 5. Discovered in 1802 by Ekeberg, but many chemists thought niobium and tantalum were identical elements until Rose, in 1844, and Marignac, in 1866, showed that niobic and tantalic acids were two different acids. The early investigators only isolated the impure metal. The first relatively pure ductile tantalum was produced by von Bolton in 1903. Tantalum occurs principally in the mineral *columbite-tantalite* (Fe, Mn)(Nb, Ta)$_2$O$_6$. Tantalum ores are found in Australia, Brazil, Rwanda, Zimbabwe, Congo-Kinshasa, Nigeria, and Canada. Separation of tantalum from niobium requires several complicated steps. Several methods are used to commercially produce the element, including electrolysis of molten potassium fluorotantalate, reduction of potassium fluorotantalate with sodium, or reacting tantalum carbide with tantalum oxide. Thirty-four isotopes and isomers of tantalum are known to exist. Natural tantalum contains two isotopes, one of which is radioactive with a very long half-life. Tantalum is a gray, heavy, and very hard metal. When pure, it is ductile and can be drawn into fine wire, which is used as a filament for evaporating metals such as aluminum. Tantalum is almost completely immune to chemical attack at temperatures below 150 °C, and is attacked only by hydrofluoric acid, acidic solutions containing the fluoride ion, and free sulfur trioxide. Alkalis attack it only slowly. At high temperatures, tantalum becomes much more reactive. The element has a melting point exceeded only by tungsten and rhenium. Tantalum is used to make a variety of alloys with desirable properties such as high melting point, high strength, good ductility, etc. Scientists at Los Alamos have produced a tantalum carbide graphite composite material that is said to be one of the hardest materials ever made. The compound has a melting point of 3738 °C. Tantalum has good "gettering" ability at high temperatures, and tantalum oxide films are stable and have good rectifying and dielectric properties. Tantalum is used to make electrolytic capacitors and vacuum furnace parts, which account for about 60% of its use. The metal is also widely used to fabricate chemical process equipment, nuclear reactors, and aircraft and missile parts. Tantalum is completely immune to body liquids and is a nonirritating metal. It has, therefore, found wide use in making surgical appliances. Tantalum oxide is used to make special glass with a high index of refraction for camera lenses. The metal has many other uses. The price of (99.9%) tantalum is about $2/g.

Technetium — (Gr. *technetos*, artificial), Tc; at. wt. (98); at. no. 43; m.p. 2157 °C; b.p. 4265 °C; sp. gr. 11.50 (calc.); valence 0, +2, +4, +5, +6, and +7. Element 43 was predicted on the basis of the periodic table, and was erroneously reported as having been discovered in 1925, at which time it was named *masurium*. The element was actually discovered by Perrier and Segre in Italy in 1937. It was found in a sample of molybdenum that was bombarded by deuterons in the Berkeley cyclotron, and which E. Lawrence sent to these investigators. Technetium was the first element to be produced artificially. Since its discovery, searches for the element in terrestrial materials have been made without success. If it does exist, the concentration must be very small. Technetium has been found in the spectrum of S-, M-, and N-type stars, and its presence in stellar matter is leading to new theories of the production of heavy elements in the stars. Forty-three isotopes and isomers of technetium, with mass numbers ranging from 86 to 113, are known. ^{97}Tc has a half-life of 2.6×10^6 years. ^{98}Tc has a half-life of 4.2×10^6 years. The isomeric isotope ^{95m}Tc, with a half-life of 61 days, is useful for tracer work, as it produces energetic gamma rays. Technetium metal has been produced in kilogram quantities. The metal was first prepared by passing hydrogen gas at 1100 °C over Tc$_2$S$_7$. It is now conveniently prepared by the reduction of ammonium pertechnetate with hydrogen. Technetium is a silvery-gray metal that tarnishes slowly in moist air. Until 1960, technetium was available only in small amounts and the price was as high as $2800/g, but the price is now of the order of $100/g. The chemistry of technetium is similar to that of rhenium. Technetium dissolves in nitric acid, aqua regia, and concentrated sulfuric acid, but is not soluble in hydrochloric acid of any strength. The element is a remarkable corrosion inhibitor for steel. It is reported that mild carbon steels may be effectively protected by as little as 55 ppm of $KTcO_4$ in aerated distilled water at temperatures up to 250 °C. This corrosion protection is limited to closed systems, since technetium is radioactive and must be confined. ^{99}Tc has a specific activity of 6.2×10^8 Bq/g. Activity of this level must not be allowed to spread. ^{99}Tc is a contamination hazard and should be handled in a glove box. The metal is an excellent superconductor at 11K and below.

Tellurium — (L. *tellus*, earth), Te; at. wt. 127.60(3); at. no. 52; m.p. 449.51 °C; b.p. 988 °C; sp. gr. 6.23 (20 °C); valence −2, 4, or 6. Discovered by Muller von Reichenstein in 1782; named by Klaproth, who isolated it in 1798. Tellurium is occasionally found native, but is more often found as the telluride of gold (*calaverite*), and combined with other metals. It is recovered commercially from the anode muds produced during the electrolytic refining of blister copper. The U.S., Canada, Peru, and Japan are the largest producers of the element. Crystalline tellurium has a silvery-white appearance, and when pure exhibits a metallic luster. It is brittle and easily pulverized. Amorphous tellurium is formed by precipitating tellurium from a solution of telluric or tellurous acid. Whether this form is truly amorphous, or made of minute crystals, is open to question. Tellurium is a p-type semiconductor, and shows greater conductivity in certain directions, depending on alignment of the atoms. Its conductivity increases slightly with exposure to light. It can be doped with silver, copper, gold, tin, or other elements. In air, tellurium burns with a greenish-blue flame, forming the dioxide. Molten tellurium corrodes iron, copper, and stainless steel. Tellurium and its compounds are probably toxic and should be handled with care. Workmen exposed to as little as 0.01 mg/m³ of air, or less, develop "tellurium breath," which has a garlic-like odor. Forty-two isotopes and isomers of tellurium are known, with atomic masses ranging from 106 to 138. Natural tellurium consists of eight isotopes, two of which are radioactive with very long half-lives. Tellurium improves the machinability of copper and stainless steel, and its addition to lead decreases the corrosive action of sulfuric acid on lead and improves its strength and hardness. Tellurium catalysts are used in the oxidation of organic compounds and are used in hydrogenation and halogenation reactions. Tellurium is also used in electronic and semiconductor devices. It is also used as a basic ingredient in blasting caps, and is added to cast iron for chill control. Tellurium is used in ceramics. Bismuth telluride has been used in thermoelectric devices. Tellurium costs about 50¢/g, with a purity of about 99.5%. The metal with a purity of 99.9999% costs about $5/g.

Terbium — (*Ytterby*, village in Sweden), Tb; at. wt. 158.92534(2); at. no. 65; m.p. 1356 °C; b.p. 3230 °C; sp. gr. 8.230; valence 3, 4. Discovered by Mosander in 1843. Terbium is a member of the lanthanide or "rare earth" group of elements. It is found in *cerite*, *gadolinite*, and other minerals along with other rare earths. It is recovered commercially from *monazite* in which it is present to the extent of 0.03%, from *xenotime*, and from *euxenite*, a complex oxide containing 1% or more of terbia. Terbium has been isolated only in recent years with the development of ion-exchange techniques for separating the rare-earth elements. As with other rare earths, it can be produced by reducing the anhydrous chloride or fluoride with calcium

metal in a tantalum crucible. Calcium and tantalum impurities can be removed by vacuum remelting. Other methods of isolation are possible. Terbium is reasonably stable in air. It is a silver-gray metal, and is malleable, ductile, and soft enough to be cut with a knife. Two crystal modifications exist, with a transformation temperature of 1289 °C. Forty-two isotopes and isomers are recognized. The oxide is a chocolate or dark maroon color. Sodium terbium borate is used as a laser material and emits coherent light at 0.546 μm. Terbium is used to dope calcium fluoride, calcium tungstate, and strontium molybdate, used in solid-state devices. The oxide has potential application as an activator for green phosphors used in color TV tubes. It can be used with ZrO_2 as a crystal stabilizer of fuel cells that operate at elevated temperature. Few other uses have been found. The element is priced at about $40/g (99.9%). Little is known of the toxicity of terbium. It should be handled with care as with other lanthanide elements.

Thallium — (Gr. *thallos*, a green shoot or twig), Tl; at. wt. [204.382; 204.385]; at. no. 81; m.p. 304 °C; b.p. 1473 °C; sp. gr. 11.85 (20 °C); valence 1, or 3. Thallium was discovered spectroscopically in 1861 by Crookes. The element was named after the beautiful green spectral line, which identified the element. The metal was isolated both by Crookes and Lamy in 1862 about the same time. Thallium occurs in *crooksite*, *lorandite*, and *hutchinsonite*. It is also present in *pyrites* and is recovered from the roasting of this ore in connection with the production of sulfuric acid. It is also obtained from the smelting of lead and zinc ores. Extraction is somewhat complex and depends on the source of the thallium. Manganese nodules, found on the ocean floor, contain thallium. When freshly exposed to air, thallium exhibits a metallic luster, but soon develops a bluish-gray tinge, resembling lead in appearance. A heavy oxide builds up on thallium if left in air, and in the presence of water the hydroxide is formed. The metal is very soft and malleable. It can be cut with a knife. Forty-seven isotopes of thallium, with atomic masses ranging from 179 to 210 are recognized. Natural thallium is a mixture of two isotopes. The element and its compounds are toxic and should be handled carefully. Contact of the metal with skin is dangerous, and when melting the metal adequate ventilation should be provided. Thallium is suspected of carcinogenic potential for man. Thallium sulfate has been widely employed as a rodenticide and ant killer. It is odorless and tasteless, giving no warning of its presence. Its use, however, has been prohibited in the U.S. since 1975 as a household insecticide and rodenticide. The electrical conductivity of thallium sulfide changes with exposure to infrared light, and this compound is used in photocells. Thallium bromide-iodide crystals have been used as infrared optical materials. Thallium has been used, with sulfur or selenium and arsenic, to produce low melting glasses which become fluid between 125 and 150 °C. These glasses have properties at room temperatures similar to ordinary glasses and are said to be durable and insoluble in water. Thallium oxide has been used to produce glasses with a high index of refraction. Thallium has been used in treating ringworm and other skin infections; however, its use has been limited because of the narrow margin between toxicity and therapeutic benefits. A mercury–thallium alloy, which forms a eutectic at 8.5% thallium, is reported to freeze at −60 °C, some 20° below the freezing point of mercury. Thallium metal (99.999%) costs about $2/g.

Thorium — (*Thor*, Scandinavian god of war), Th; at. wt. 232.03806(2); at. no. 90; m.p. 1750 °C; b.p. 4788 °C; sp. gr. 11.72; valence +2(?), +3(?), +4. Discovered by Berzelius in 1828. Thorium occurs in *thorite* (ThSiO$_4$) and in *thorianite* (ThO$_2$ + UO$_2$). Large deposits of thorium minerals have been reported in New England and elsewhere, but these have not yet been exploited. Thorium is now thought to be about three times as abundant as uranium and about as abundant as lead or molybdenum. The metal is a source of nuclear power. There is probably more energy available for use from thorium in the minerals of the Earth's crust than from both uranium and fossil fuels. Any sizable demand for thorium as a nuclear fuel is still several years in the future. Work has been done in developing thorium cycle converter-reactor systems. Several prototypes, including the HTGR (high-temperature gas-cooled reactor) and MSRE (molten salt converter reactor experiment), have operated. While the HTGR reactors are efficient, they are not expected to become important commercially for many years because of certain operating difficulties. Thorium is recovered commercially from the mineral *monazite*, which contains from 3 to 9% ThO$_2$ along with rare-earth minerals. Much of the internal heat the Earth produces has been attributed to thorium and uranium. Several methods are available for producing thorium metal: it can be obtained by reducing thorium oxide with calcium, by electrolysis of anhydrous thorium chloride in a fused mixture of sodium and potassium chlorides, by calcium reduction of thorium tetrachloride mixed with anhydrous zinc chloride, and by reduction of thorium tetrachloride with an alkali metal. Thorium was originally assigned a position in Group IV of the periodic table. Because of its atomic weight, valence, etc., it is now considered to be the second member of the *actinide* series of elements. When pure, thorium is a silvery-white metal which is air stable and retains its luster for several months. When contaminated with the oxide, thorium slowly tarnishes in air, becoming gray and finally black. The physical properties of thorium are greatly influenced by the degree of contamination with the oxide. The purest specimens often contain several tenths of a percent of the oxide. High-purity thorium has been made. Pure thorium is soft, very ductile, and can be cold-rolled, swaged, and drawn. Thorium is dimorphic, changing at 1400 °C from a cubic to a body-centered cubic structure. Thorium oxide has a melting point of 3300 °C, which is the highest of all oxides. Only a few elements, such as tungsten, and a few compounds, such as tantalum carbide, have higher melting points. Thorium is slowly attacked by water, but does not dissolve readily in most common acids, except hydrochloric. Powdered thorium metal is often pyrophoric and should be carefully handled. When heated in air, thorium turnings ignite and burn brilliantly with a white light. The principal use of thorium has been in the preparation of the Welsbach mantle, used for portable gas lights. These mantles, consisting of thorium oxide with about 1% cerium oxide and other ingredients, glow with a dazzling light when heated in a gas flame. Thorium is an important alloying element in magnesium, imparting high strength and creep resistance at elevated temperatures. Because thorium has a low work-function and high electron emission, it is used to coat tungsten wire used in electronic equipment. The oxide is also used to control the grain size of tungsten used for electric lamps; it is also used for high-temperature laboratory crucibles. Glasses containing thorium oxide have a high refractive index and low dispersion. Consequently, they find application in high quality lenses for cameras and scientific instruments. Thorium oxide has also found use as a catalyst in the conversion of ammonia to nitric acid, in petroleum cracking, and in producing sulfuric acid. Thorium has not found many uses due to its radioactive nature and its handling and disposal problems. Thirty isotopes of thorium are known with atomic masses ranging from 210 to 237. All are unstable. ^{232}Th occurs naturally and has a half-life of 1.4×10^{10} years. It is an alpha emitter. ^{232}Th goes through six alpha and four beta decay steps before becoming the stable isotope ^{208}Pb. ^{232}Th is sufficiently radioactive to expose a photographic plate in a few hours. Thorium disintegrates with the production of "thoron" (^{220}Rn), which is an alpha emitter and presents a radiation hazard. Good ventilation of areas where thorium is stored or handled is therefore essential. Thorium metal (99.8%) costs about $25/g.

Thulium — (*Thule*, the earliest name for Scandinavia), Tm; at. wt. 168.93421(2); at. no. 69; m.p. 1545 °C; b.p. 1950 °C; sp. gr. 9.321 (25 °C); valence 3. Discovered in 1879 by Cleve. Thulium occurs in small quantities along with other rare earths in a number of minerals. It is obtained commercially from *monazite*, which contains about 0.007% of the element. Thulium is the least abundant of the rare-earth elements, but with new sources recently discovered, it is now considered to be about as rare as silver, gold, or cadmium. Ion-exchange and solvent extraction techniques have recently permitted much easier separation of the rare earths, with much lower costs. Only a few years ago, thulium metal was not obtainable at any cost; in 1996 the oxide cost $20/g. Thulium metal powder now costs $70/g (99.9%). Thulium can be isolated by reduction of the oxide with lanthanum metal or by calcium reduction of the anhydrous fluoride. The pure metal has a bright, silvery luster. It is reasonably stable in air, but the metal should be protected from moisture in a closed container. The element is silver-gray, soft, malleable, and ductile, and can be cut with a knife. Forty-one isotopes and isomers are known, with atomic masses ranging from 146 to 176. Natural thulium, which is 100% ^{169}Tm, is stable. Because of the relatively high price of the metal, thulium has not yet found many practical applications. ^{169}Tm bombarded in a nuclear reactor can be used as a radiation source in portable X-ray equipment. ^{171}Tm is potentially useful as an energy source. Natural thulium also has possible use in *ferrites* (ceramic magnetic materials) used in microwave equipment. As with other lanthanides, thulium has a low-to-moderate acute toxicity rating. It should be handled with care.

Tin — (Anglo-Saxon, *tin*), Sn (L. *stannum*); at. wt. 118.710(7); at. no. 50; m.p. 231.93 °C; b.p. 2602 °C; sp. gr. (gray) 5.77, (white) 7.29; valence 2, 4. Known to the ancients. Tin is found chiefly in *cassiterite* (SnO$_2$). Most of the world's supply comes from China, Indonesia, Peru, Brazil, and Bolivia. The U.S. produces almost none, although occurrences have been found in Alaska and Colorado. Tin is obtained by reducing the ore with coal in a reverberatory furnace. Ordinary tin is composed of ten stable isotopes; thirty-six unstable isotopes and isomers are also known. Ordinary tin is a silver-white metal, is malleable, somewhat ductile, and has a highly crystalline structure. Due to the breaking of these crystals, a "tin cry" is heard when a bar is bent. The element has two allotropic forms at normal pressure. On warming, gray, or α tin, with a cubic structure, changes at 13.2 °C into white, or β tin, the ordinary form of the metal. White tin has a tetragonal structure. When tin is cooled below 13.2 °C, it changes slowly from white to gray. This change is affected by impurities such as aluminum and zinc, and can be prevented by small additions of antimony or bismuth. This change from the α to β form is called the tin pest. Tin–lead alloys are used to make organ pipes. There are few if any uses for gray tin. Tin takes a high polish and is used to coat other metals to prevent corrosion or other chemical action. Such tin plate over steel is used in the so-called tin can for preserving food. Alloys of tin are very important. Soft solder, type metal, fusible metal, pewter, bronze, bell metal, Babbitt metal, white metal, die casting alloy, and phosphor bronze are some of the important alloys using tin. Tin resists distilled sea and soft tap water, but is attacked by strong acids, alkalis, and acid salts. Oxygen in solution accelerates the attack. When heated in air, tin forms SnO_2, which is feebly acid, forming stannate salts with basic oxides. The most important salt is the chloride ($SnCl_2 \cdot H_2O$), which is used as a reducing agent and as a mordant in calico printing. Tin salts sprayed onto glass are used to produce electrically conductive coatings. These have been used for panel lighting and for frost-free windshields. Most window glass is now made by floating molten glass on molten tin (float glass) to produce a flat surface (Pilkington process). Of recent interest is a crystalline tin–niobium alloy that is superconductive at very low temperatures. This promises to be important in the construction of superconductive magnets that generate enormous field strengths but use practically no power. Such magnets, made of tin–niobium wire, weigh but a few pounds and produce magnetic fields that, when started with a small battery, are comparable to that of a 100 ton electromagnet operated continuously with a large power supply. The small amount of tin found in canned foods is quite harmless. The agreed limit of tin content in U.S. foods is 300 mg/kg. The trialkyl and triaryl tin compounds are used as biocides and must be handled carefully. Over the past 25 years the price of commercial tin has varied from 50¢/lb ($1.10/kg) to about $6/kg. Tin (99.99% pure) costs about $260/kg.

Titanium — (L. *Titans*, the first sons of the Earth, myth.), Ti; at. wt. 47.867(1); at. no. 22; m.p. 1668 °C; b.p. 3287 °C; sp. gr. 4.51; valence 2, 3, or 4. Discovered by Gregor in 1791; named by Klaproth in 1795. Impure titanium was prepared by Nilson and Pettersson in 1887; however, the pure metal (99.9%) was not made until 1910 by Hunter by heating $TiCl_4$ with sodium in a steel bomb. Titanium is present in meteorites and in the sun. Rocks obtained during the Apollo 17 lunar mission showed presence of 12.1% TiO_2. Analyses of rocks obtained during earlier Apollo missions show lower percentages. Titanium oxide bands are prominent in the spectra of M-type stars. The element is the ninth most abundant in the crust of the Earth. Titanium is almost always present in igneous rocks and in the sediments derived from them. It occurs in the minerals *rutile, ilmenite,* and *sphene,* and is present in titanates and in many iron ores. Deposits of ilmenite and rutile are found in Florida, California, Tennessee, and New York. Australia, Norway, Malaysia, India, and China are also large suppliers of titanium minerals. Titanium is present in the ash of coal, in plants, and in the human body. The metal was a laboratory curiosity until Kroll, in 1946, showed that titanium could be produced commercially by reducing titanium tetrachloride with magnesium. This method is largely used for producing the metal today. The metal can be purified by decomposing the iodide. Titanium, when pure, is a lustrous, white metal. It has a low density, good strength, is easily fabricated, and has excellent corrosion resistance. It is ductile only when it is free of oxygen. The metal burns in air and is the only element that

burns in nitrogen. Titanium is resistant to dilute sulfuric and hydrochloric acid, most organic acids, moist chlorine gas, and chloride solutions. Natural titanium consists of five isotopes with atomic masses from 46 to 50. All are stable. Eighteen other unstable isotopes are known. The metal is dimorphic. The hexagonal α form changes to the cubic β form very slowly at about 880 °C. The metal combines with oxygen at red heat, and with chlorine at 550 °C. Titanium is important as an alloying agent with aluminum, molybdenum, manganese, iron, and other metals. Alloys of titanium are principally used for aircraft and missiles where lightweight strength and ability to withstand extremes of temperature are important. Titanium is as strong as steel, but 45% lighter. It is 60% heavier than aluminum, but twice as strong. Titanium has potential use in desalination plants for converting sea water into fresh water. The metal has excellent resistance to sea water and is used for propeller shafts, rigging, and other parts of ships exposed to salt water. A titanium anode coated with platinum has been used to provide cathodic protection from corrosion by salt water. Titanium metal is considered to be physiologically inert; however, titanium powder may be a carcinogenic hazard. When pure, titanium dioxide is relatively clear and has an extremely high index of refraction with an optical dispersion higher than diamond. It is produced artificially for use as a gemstone, but it is relatively soft. Star sapphires and rubies exhibit their asterism as a result of the presence of TiO_2. Titanium dioxide is extensively used for both house paint and artist's paint, as it is permanent and has good covering power. Titanium oxide pigment accounts for the largest use of the element. Titanium paint is an excellent reflector of infrared, and is extensively used in solar observatories where heat causes poor seeing conditions. Titanium tetrachloride is used to iridize glass. This compound fumes strongly in air and has been used to produce smoke screens. The price of titanium metal (99.9%) is about $1100/kg.

Tungsten — (Swedish, *tung sten*, heavy stone); also known as *wolfram* (from *wolframite*, said to be named from *wolf rahm* or *spumi lupi*, because the ore interfered with the smelting of tin and was supposed to devour the tin), W; at. wt. 183.84(1); at. no. 74; m.p. 3422 °C; b.p. 5555 °C; sp. gr. 19.3 (20 °C); valence 2, 3, 4, 5, or 6. In 1779 Peter Woulfe examined the mineral now known as *wolframite* and concluded it must contain a new substance. Scheele, in 1781, found that a new acid could be made from *tung sten* (a name first applied about 1758 to a mineral now known as *scheelite*). Scheele and Berman suggested the possibility of obtaining a new metal by reducing this acid. The de Elhuyar brothers found an acid in *wolframite* in 1783 that was identical to the acid of *tungsten* (tungstic acid) of Scheele, and in that year they succeeded in obtaining the element by reduction of this acid with charcoal. Tungsten occurs in *wolframite*, (Fe, Mn)WO_4; *scheelite*, $CaWO_4$; *huebnerite*, $MnWO_4$; and *ferberite*, $FeWO_4$. Important deposits of tungsten occur in California, Colorado, Bolivia, Russia, and Portugal. China is reported to have about 75% of the world's tungsten resources. Natural tungsten contains five stable isotopes. Thirty-two other unstable isotopes and isomers are recognized. The metal is obtained commercially by reducing tungsten oxide with hydrogen or carbon. Pure tungsten is a steel-gray to tin-white metal. Very pure tungsten can be cut with a hacksaw, and can be forged, spun, drawn, and extruded. The impure metal is brittle and can be worked only with difficulty. Tungsten has the highest melting point of all metals, and at temperatures over 1650 °C has the highest tensile strength. The metal oxidizes in air and must be protected at elevated temperatures. It has excellent corrosion resistance and is attacked only slightly by most mineral acids. The thermal expansion is about the same as borosilicate glass, which makes the metal useful for glass-to-metal seals. Tungsten and its alloys are used extensively for filaments for electric lamps, electron and television tubes, and for metal evaporation work; for electrical contact points for automobile distributors; X-ray targets; windings and heating elements for electrical furnaces; and for numerous spacecraft and high-temperature applications. High-speed tool steels, Hastelloy®, Stellite®, and many other alloys contain tungsten. Tungsten carbide is of great importance to the metal-working, mining, and petroleum industries. Calcium and magnesium tungstates are widely used in fluorescent lighting; other salts of tungsten are used in the chemical and tanning industries. Tungsten disulfide is a dry, high-temperature lubricant, stable to 500 °C. Tungsten bronzes and other tungsten compounds are used in paints. Zirconium tungstate has found recent applications (see under Zirconium). Tungsten powder (99.999%) costs about $2900/kg.

Uranium — (Planet *Uranus*), U; at. wt. 238.02891(3); at. no. 92; m.p. 1135 °C; b.p. 4131 °C; sp. gr. 19.1; valence 2, 3, 4, 5, or 6. Yellow-colored glass, containing more than 1% uranium oxide and dating back to 79 A.D., has been found near Naples, Italy. Klaproth recognized an unknown element in *pitchblende* and attempted to isolate the metal in 1789. The metal apparently was first isolated in 1841 by Peligot, who reduced the anhydrous chloride with potassium. Uranium is not as rare as it was once thought. It is now considered to be more plentiful than mercury, antimony, silver, or cadmium, and is about as abundant as molybdenum or arsenic. It occurs in numerous minerals such as *pitchblende, uraninite, carnotite, autunite, uranophane, davidite,* and *tobernite*. It is also found in *phosphate rock, lignite, monazite sands*, and can be recovered commercially from these sources. Large deposits of uranium ore occur in Utah, Colorado, New Mexico, Canada, and elsewhere. Uranium can be made by reducing uranium halides with alkali or alkaline earth metals or by reducing uranium oxides by calcium, aluminum, or carbon at high temperatures. The metal can also be produced by electrolysis of KUF_5 or UF_4, dissolved in a molten mixture of $CaCl_2$ and NaCl. High-purity uranium can be prepared by the thermal decomposition of uranium halides on a hot filament. Uranium exhibits three crystallographic modifications as follows:

$$\alpha \xrightarrow{688°C} \beta \xrightarrow{776°C} \gamma$$

Uranium is a heavy, silvery-white metal that is pyrophoric when finely divided. It is a little softer than steel, and is attacked by cold water in a finely divided state. It is malleable, ductile, and slightly paramagnetic. In air, the metal becomes coated with a layer of oxide. Acids dissolve the metal, but it is unaffected by alkalis. Uranium has twenty-three isotopes, one of which is an isomer and all of which are radioactive. Naturally occurring uranium contains 99.2745% by weight ^{238}U, 0.720% ^{235}U, and 0.0055% ^{234}U. Studies show that the percentage weight of ^{235}U in natural uranium varies by as much as 0.1%, depending on the source. The U.S.D.O.E. has adopted the value of 0.711 as being their "official" percentage of ^{235}U in natural uranium. Natural uranium is sufficiently radioactive to expose a photographic plate in an hour or so. Much of the internal heat of the Earth is thought to be attributable to

the presence of uranium and thorium. ^{238}U, with a half-life of 4.46×10^9 years, has been used to estimate the age of igneous rocks. The origin of uranium, the highest member of the naturally occurring elements — except perhaps for traces of neptunium or plutonium — is not clearly understood, although it has been thought that uranium might be a decay product of elements of higher atomic weight, which may have once been present on Earth or elsewhere in the universe. These original elements may have been formed as a result of a primordial "creation," known as "the big bang," in a supernova, or in some other stellar processes. The fact that recent studies show that most trans-uranic elements are extremely rare with very short half-lives indicates that it may be necessary to find some alternative explanation for the very large quantities of radioactive uranium we find on Earth. Studies of meteorites from other parts of the solar system show a relatively low radioactive content, compared to terrestrial rocks. Uranium is of great importance as a nuclear fuel. ^{238}U can be converted into fissionable plutonium by the following reactions:

$$^{238}U(n,\gamma) \rightarrow{}^{239}U \xrightarrow{\beta}{}^{239}Np \xrightarrow{\beta}{}^{239}Pu$$

This nuclear conversion can be brought about in "breeder" reactors where it is possible to produce more new fissionable material than the fissionable material used in maintaining the chain reaction. ^{235}U is of even greater importance, for it is the key to the utilization of uranium. ^{235}U, while occurring in natural uranium to the extent of only 0.72%, is so fissionable with slow neutrons that a self-sustaining fission chain reaction can be made to occur in a reactor constructed from natural uranium and a suitable moderator, such as heavy water or graphite, alone. ^{235}U can be concentrated by gaseous diffusion and other physical processes, if desired, and used directly as a nuclear fuel, instead of natural uranium, or used as an explosive. Natural uranium, slightly enriched with ^{235}U by a small percentage, is used to fuel nuclear power reactors for the generation of electricity. Natural thorium can be irradiated with neutrons as follows to produce the important isotope ^{233}U.

$$^{232}Th(n,\gamma) \rightarrow{}^{233}Th \xrightarrow{\beta}{}^{233}Pa \xrightarrow{\beta}{}^{233}U$$

While thorium itself is not fissionable, ^{233}U is, and in this way may be used as a nuclear fuel. One pound of completely fissioned uranium has the fuel value of over 1500 tons of coal. The uses of nuclear fuels to generate electrical power, to make isotopes for peaceful purposes, and to make explosives are well known. The estimated world-wide production of the 437 nuclear power reactors in operation in 1998 amounted to about 352,000 megawatt hours. In 1998 the U.S. had about 107 commercial reactors with an output of about 100,000 megawatt-hours. Some nuclear-powered electric generating plants have recently been closed because of safety concerns. There are also serious problems with nuclear waste disposal that have not been completely resolved. Uranium in the U.S. is controlled by the U.S. Nuclear Regulatory Commission, under the Department of Energy. Uses are being found for the large quantities of "depleted" uranium now available, where uranium-235 has been lowered to about 0.2%. Depleted uranium has been used for inertial guidance devices, gyrocompasses, counterweights for aircraft control surfaces, ballast for missile reentry vehicles, and as a shielding material for tanks, etc. Concerns, however, have been raised over its low radioactive properties. Uranium metal is used for X-ray targets for production of high-energy X-rays. The nitrate has been used as photographic toner, and the acetate is used in analytical chemistry. Crystals of uranium nitrate are triboluminescent. Uranium salts have also been used for producing yellow "vaseline" glass and glazes. Uranium and its compounds are highly toxic, both from a chemical and radiological standpoint. Finely divided uranium metal, being pyrophoric, presents a fire hazard. The maximum permissible total body burden of natural uranium (based on radiotoxicity) is 0.2 μCi for soluble compounds. Recently, the natural presence of uranium and thorium in many soils has become of concern to homeowners because of the generation of radon and its daughters (see under Radon). Uranium metal is available commercially at a cost of about $6/g (99.7%) in air-tight glass under argon.

Vanadium — (Scandinavian goddess, *Vanadis*), V; at. wt. 50.9415(1); at. no. 23; m.p. 1910 °C; b.p. 3407 °C; sp. gr. 6.0 (18.7 °C); valence 2, 3, 4, or 5. Vanadium was first discovered by del Rio in 1801. Unfortunately, a French chemist incorrectly declared that del Rio's new element was only impure chromium; del Rio thought himself to be mistaken and accepted the French chemist's statement. The element was rediscovered in 1830 by Sefstrom, who named the element in honor of the Scandinavian goddess *Vanadis* because of its beautiful multicolored compounds. It was isolated in nearly pure form by Roscoe, in 1867, who reduced the chloride with hydrogen. Vanadium of 99.3 to 99.8% purity was not produced until 1927. Vanadium is found in about 65 different minerals among which *carnotite, roscoelite, vanadinite,* and *patronite* are important sources of the metal. Vanadium is also found in phosphate rock and certain iron ores, and is present in some crude oils in the form of organic complexes. It is also found in small percentages in meteorites. Commercial production from petroleum ash holds promise as an important source of the element. China, South Africa, and Russia supply much of the world's vanadium ores. High-purity ductile vanadium can be obtained by reduction of vanadium trichloride with magnesium or with magnesium–sodium mixtures. Much of the vanadium metal being produced is now made by calcium reduction of V_2O_5 in a pressure vessel, an adaptation of a process developed by McKechnie and Seybolt. Natural vanadium is a mixture of two isotopes, ^{50}V (0.25%) and ^{51}V (99.75%). ^{50}V is slightly radioactive, having a long half-life. Twenty other unstable isotopes are recognized. Pure vanadium is a bright white metal, and is soft and ductile. It has good corrosion resistance to alkalis, sulfuric and hydrochloric acid, and salt water, but the metal oxidizes readily above 660 °C. The metal has good structural strength and a low-fission neutron cross section, making it useful in nuclear applications. Vanadium is used in producing rust-resistant, spring, and high-speed tool steels. It is an important carbide stabilizer in making steels. About 80% of the vanadium now produced is used as ferrovanadium or as a steel additive. Vanadium foil is used as a bonding agent in cladding titanium to steel. Vanadium pentoxide is used in ceramics and as a catalyst. It is also used in producing a superconductive magnet with a field of 175,000 gauss. Vanadium and its compounds are toxic and should be handled with care. Ductile vanadium is commercially available. Vanadium metal (99.7%) costs about $3/g.

Wolfram — see Tungsten.

Xenon — (Gr. *xenon*, stranger), Xe; at. wt. 131.293(6); at. no. 54; m.p. −111.74 °C; b.p. −108.09 °C; t_c 16.58 °C; density (gas) 5.887 ± 0.009 g/L, sp. gr (liquid) 2.95 (−109 °C); valence usually 0. Discovered by Ramsay and Travers in 1898 in the residue left after evaporating liquid air components. Xenon is a member of the so-called noble or "inert" gases. It is present in the atmosphere to the extent of about one part in twenty million. Xenon is present in the Martian atmosphere to the extent of 0.08 ppm. The element is found in the gases evolved from certain mineral springs, and is commercially obtained by extraction from liquid air. Natural xenon is composed of nine stable isotopes. In addition to these, thirty-five unstable isotopes and isomers have been characterized. Before 1962, it had generally been assumed that xenon and other noble gases were unable to form compounds. However, it is now known that xenon, as well as other members of the zero valence elements, do form compounds. Among the compounds of xenon now reported are xenon hydrate, sodium perxenate, xenon deuterate, difluoride, tetrafluoride, hexafluoride, and $XePtF_6$ and $XeRhF_6$. Xenon trioxide, which is highly explosive, has been prepared. More than 80 xenon compounds have been made with xenon chemically bonded to fluorine and oxygen. Some xenon compounds are colored. Metallic xenon has been produced, using several hundred kilobars of pressure. Xenon in a vacuum tube produces a beautiful blue glow when excited by an electrical discharge. The gas is used in making electron tubes, stroboscopic lamps, bactericidal lamps, and lamps used to excite ruby lasers for generating coherent light. Xenon is used in the atomic energy field in bubble chambers, probes, and other applications where its high molecular weight is of value. The perxenates are used in analytical chemistry as oxidizing agents. ^{133}Xe and ^{135}Xe are produced by neutron irradiation in air-cooled nuclear reactors. ^{133}Xe has useful applications as a radioisotope. The element is available in sealed glass containers for about $20/L of gas at standard pressure. Xenon is not toxic, but its compounds are highly toxic because of their strong oxidizing characteristics.

Ytterbium — (*Ytterby*, village in Sweden), Yb; at. wt. 173.054(5); at. no. 70; m.p. 824 °C; b.p. 1196 °C; sp. gr (α) 6.903 (β) 6.966; valence 2, 3. Marignac in 1878 discovered a new component, which he called *ytterbia*, in the Earth then known as *erbia*. In 1907, Urbain separated ytterbia into two components, which he called *neoytterbia* and *lutecia*. The elements in these earths are now known as *ytterbium* and *lutetium*, respectively. These elements are identical with *aldebaranium* and *cassiopeium*, discovered independently and at about the same time by von Welsbach. Ytterbium occurs along with other rare earths in a number of rare minerals. It is commercially recovered principally from *monazite sand*, which contains about 0.03%. Ion-exchange and solvent extraction techniques developed in recent years have greatly simplified the separation of the rare earths from one another. The element was first prepared by Klemm and Bonner in 1937 by reducing ytterbium trichloride with potassium. Their metal was mixed, however, with KCl. Daane, Dennison, and Spedding prepared a much purer form in 1953 from which the chemical and physical properties of the element could be determined. Ytterbium has a bright silvery luster, is soft, malleable, and quite ductile. While the element is fairly stable, it should be kept in closed containers to protect it from air and moisture. Ytterbium is readily attacked and dissolved by dilute and concentrated mineral acids and reacts slowly with water. Ytterbium has three allotropic forms with transformation points at −13° and 795 °C. The beta form is a room-temperature, face-centered, cubic modification, while the high-temperature gamma form is a body-centered cubic form. Another body-centered cubic phase has recently been found to be stable at high pressures at room temperatures. The beta form ordinarily has metallic-type conductivity, but becomes a semiconductor when the pressure is increased above 16,000 atm. The electrical resistance increases tenfold as the pressure is increased to 39,000 atm and drops to about 80% of its standard temperature-pressure resistivity at a pressure of 40,000 atm. Natural ytterbium is a mixture of seven stable isotopes. Twenty-six other unstable isotopes and isomers are known. Ytterbium metal has possible use in improving the grain refinement, strength, and other mechanical properties of stainless steel. One isotope is reported to have been used as a radiation source as a substitute for a portable X-ray machine where electricity is unavailable. Few other uses have been found. Ytterbium metal is available with a purity of about 99.9% for about $10/g. Ytterbium has a low acute toxicity rating.

Yttrium — (*Ytterby*, village in Sweden near Vauxholm), Y; at. wt. 88.90585(2); at. no. 39; m.p. 1522 °C; b.p. 3345 °C; sp. gr. 4.469 (25 °C); valence 3. *Yttria*, which is an earth containing yttrium, was discovered by Gadolin in 1794. *Ytterby* is the site of a quarry which yielded many unusually minerals containing rare earths and other elements. This small town, near Stockholm, bears the honor of giving names to *erbium, terbium,* and *ytterbium* as well as *yttrium*. In 1843 Mosander showed that yttria could be resolved into the oxides (or earths) of three elements. The name yttria was reserved for the most basic one; the others were named *erbia* and *terbia*. Yttrium occurs in nearly all of the rare-earth minerals. Analysis of lunar rock samples obtained during the Apollo missions show a relatively high yttrium content. It is recovered commercially from *monazite sand*, which contains about 3%, and from *bastnasite*, which contains about 0.2%. Wohler obtained the impure element in 1828 by reduction of the anhydrous chloride with potassium. The metal is now produced commercially by reduction of the fluoride with calcium metal. It can also be prepared by other techniques. Yttrium has a silver-metallic luster and is relatively stable in air. Turnings of the metal, however, ignite in air if their temperature exceeds 400 °C, and finely divided yttrium is very unstable in air. Yttrium oxide is one of the most important compounds of yttrium and accounts for the largest use. It is widely used in making YVO_4 europium, and Y_2O_3 europium phosphors to give the red color in color television tubes. Many hundreds of thousands of pounds are now used in this application. Yttrium oxide also is used to produce yttrium iron garnets, which are very effective microwave filters. Yttrium iron, aluminum, and gadolinium garnets, with formulas such as $Y_3Fe_5O_{12}$ and $Y_3Al_5O_{12}$, have interesting magnetic properties. Yttrium iron garnet is also exceptionally efficient as both a transmitter and transducer of acoustic energy. Yttrium aluminum garnet, with a hardness of 8.5, is also finding use as a gemstone (simulated diamond). Small amounts of yttrium (0.1 to 0.2%) can be used to reduce the grain size in chromium, molybdenum, zirconium, and titanium, and to increase strength of aluminum and magnesium alloys. Alloys with other useful properties can be obtained by using yttrium as an additive. The metal can be used as a deoxidizer for vanadium and other nonferrous metals. The metal has a low cross section for nuclear capture. ^{90}Y, one of the isotopes of yttrium, exists in equilibrium with its parent ^{90}Sr, a product of atomic explosions. Yttrium has been considered for use as a nodulizer for producing nodular cast iron, in which the graphite forms compact nodules instead of the usual flakes. Such iron has increased ductility. Yttrium is also finding application in laser systems and as a catalyst for ethylene polymerization. It also has potential use in ceramic and glass formulas, as the oxide has a high melting point and imparts shock resistance and low expansion characteristics to glass. Natural yttrium contains but one isotope, ^{89}Y. Forty-three other unstable isotopes and isomers have been characterized. Yttrium metal of 99.9% purity is commercially available at a cost of about $5/g.

Zinc — (Ger. *Zink*, of obscure origin), Zn; at. wt. 65.38(2); at. no. 30; m.p. 419.53 °C; b.p. 907 °C; sp. gr. 7.134 (25 °C); valence 2. Centuries before zinc was recognized as a distinct element, zinc ores were used for making brass. Tubal-Cain, seven generations from Adam, is mentioned as being an "instructor in every artificer in brass and iron." An alloy containing 87% zinc has been found in prehistoric ruins in Transylvania. Metallic zinc was produced in the 13th century A.D. in India by reducing calamine with organic substances such as wool. The metal was rediscovered in Europe by Marggraf in 1746, who showed that it could be obtained by reducing *calamine* with charcoal. The principal ores of zinc are *sphalerite* or *blende* (sulfide), *smithsonite* (carbonate), *calamine* (silicate), and *franklinite* (zinc, manganese, iron oxide). Canada, Japan, Belgium, Germany, and the Netherlands are suppliers of zinc ores. Zinc is also mined in Alaska, Tennessee, Missouri, and elsewhere in the U.S. Zinc can be obtained by roasting its ores to form the oxide and by reduction of the oxide with coal or carbon, with subsequent distillation of the metal. Other methods of extraction are possible. Naturally occurring zinc contains five stable isotopes. Twenty-five other unstable isotopes and isomers are recognized. Zinc is a bluish-white, lustrous metal. It is brittle at ordinary temperatures but malleable at 100 to 150 °C. It is a fair conductor of electricity, and burns in air at high red heat with evolution of white clouds of the oxide. The metal is employed to form numerous alloys with other metals. Brass, nickel silver, typewriter metal, commercial bronze, spring brass, German silver, soft solder, and aluminum solder are some of the more important alloys. Large quantities of zinc are used to produce die castings, used extensively by the automotive, electrical, and hardware industries. An alloy called *Prestal*®, consisting of 78% zinc and 22% aluminum, is reported to be almost as strong as steel but as easy to mold as plastic. It is said to be so plastic that it can be molded into form by relatively inexpensive die casts made of ceramics and cement. It exhibits superplasticity. Zinc is also extensively used to galvanize other metals such as iron to prevent corrosion. Neither zinc nor zirconium is ferromagnetic; but $ZrZn_2$ exhibits ferromagnetism at temperatures below 35 K. Zinc oxide is a unique and very useful material to modern civilization. It is widely used in the manufacture of paints, rubber products, cosmetics, pharmaceuticals, floor coverings, plastics, printing inks, soap, storage batteries, textiles, electrical equipment, and other products. It has unusual electrical, thermal, optical, and solid-state properties that have not yet been fully investigated. Lithopone, a mixture of zinc sulfide and barium sulfate, is an important pigment. Zinc sulfide is used in making luminous dials, X-ray and TV screens, and fluorescent lights. The chloride and chromate are also important compounds. Zinc is an essential element in the growth of human beings and animals. Tests show that zinc-deficient animals require 50% more food to gain the same weight as an animal supplied with sufficient zinc. Zinc is not considered to be toxic, but when freshly formed ZnO is inhaled a disorder known as the *oxide shakes* or *zinc chills* sometimes occurs. It is recommended that where zinc oxide is encountered good ventilation be provided. The commercial price of zinc in January 2002 was roughly 40¢/lb ($90 kg). Zinc metal with a purity of 99.9999% is priced at about $5/g.

Zirconium — (Syriac, *zargun*, color of gold), Zr; at. wt. 91.224(2); at. no. 40; m.p. 1855 °C; b.p. 4409 °C; sp. gr. 6.52 (20 °C); valence +2, +3, and +4. The name *zircon* may have originated from the Syriac word *zargono*, which describes the color of certain gemstones now known as *zircon, jargon, hyacinth, jacinth,* or *ligure.* This mineral, or its variations, is mentioned in biblical writings. These minerals were not known to contain this element until Klaproth, in 1789, analyzed a *jargon* from Sri Lanka and found a new earth, which Werner named zircon (*silex circonius*), and Klaproth called *Zirkonerde (zirconia).* The impure metal was first isolated by Berzelius in 1824 by heating a mixture of potassium and potassium zirconium fluoride in a small iron tube. Pure zirconium was first prepared in 1914. Very pure zirconium was first produced in 1925 by van Arkel and de Boer by an iodide decomposition process they developed. Zirconium is found in abundance in S-type stars, and has been identified in the sun and meteorites. Analyses of lunar rock samples obtained during the various Apollo missions to the moon show a surprisingly high zirconium oxide content, compared with terrestrial rocks. Naturally occurring zirconium contains five isotopes. Thirty-one other radioactive isotopes and isomers are known to exist. *Zircon,* $ZrSiO_4$, the principal ore, is found in deposits in Florida, South Carolina, Australia, South Africa, and elsewhere. *Baddeleyite,* found in Brazil, is an important zirconium mineral. It is principally pure ZrO_2 in crystalline form having a hafnium content of about 1%. Zirconium also occurs in some 30 other recognized mineral species. Zirconium is produced commercially by reduction of the chloride with magnesium (the Kroll Process), and by other methods. It is a grayish-white lustrous metal. When finely divided, the metal may ignite spontaneously in air, especially at elevated temperatures. The solid metal is much more difficult to ig-

nite. The inherent toxicity of zirconium compounds is low. Hafnium is invariably found in zirconium ores, and the separation is difficult. Commercial-grade zirconium contains from 1 to 3% hafnium. Zirconium has a low absorption cross section for neutrons, and is therefore used for nuclear energy applications, such as for cladding fuel elements. Commercial nuclear power generation now takes more than 90% of zirconium metal production. Reactors of the size now being made may use as much as a half-million lineal feet of zirconium alloy tubing. Reactor-grade zirconium is essentially free of hafnium. *Zircaloy®* is an important alloy developed specifically for nuclear applications. Zirconium is exceptionally resistant to corrosion by many common acids and alkalis, by sea water, and by other agents. It is used extensively by the chemical industry where corrosive agents are employed. Zirconium is used as a getter in vacuum tubes, as an alloying agent in steel, in surgical appliances, photoflash bulbs, explosive primers, rayon spinnerets, lamp filaments, etc. It is used in poison ivy lotions in the form of the carbonate as it combines with *urushiol.* With niobium, zirconium is superconductive at low temperatures and is used to make superconductive magnets. Alloyed with zinc, zirconium becomes magnetic at temperatures below 35 K. Zirconium oxide (zircon) has a high index of refraction and is used as a gem material. The impure oxide, zirconia, is used for laboratory crucibles that will withstand heat shock, for linings of metallurgical furnaces, and by the glass and ceramic industries as a refractory material. Its use as a refractory material accounts for a large share of all zirconium consumed. Zirconium tungstate is an unusual material that shrinks, rather than expands, when heated. A few other compounds are known to possess this property, but they tend to shrink in one direction, while they stretch out in others in order to maintain an overall volume. Zirconium tungstate shrinks in all directions over a wide temperature range of from near absolute zero to +777 °C. It is being considered for use in composite materials where thermal expansion may be a problem. Zirconium of about 99.5% purity is available at a cost of about $2000/kg or about $4/g.

PHYSICAL CONSTANTS OF INORGANIC COMPOUNDS

The compounds in this table were selected on the basis of their laboratory and industrial importance, as well as their value in illustrating trends in the variation of physical properties with position in the periodic table. An effort has been made to include the most frequently encountered inorganic substances; a limited number of organometallics are also covered. Many, if not most, of the compounds that are solids at ambient temperature can exist in more than one crystalline modification. In the absence of other information, the data given here can be assumed to apply to the most stable or common crystalline form. In many cases, however, two or more forms are of practical importance, and separate entries will be found in the table.

Compounds are arranged primarily in alphabetical order by the most commonly used name. However, adjustments are made in many instances so as to bring closely related compounds together. For example, hydrides of elements such as boron, silicon, and germanium are grouped together immediately following the entry for the parent element, since they would otherwise be scattered throughout the table. Likewise, the oxoacids of an element are given in one group whenever a strict alphabetical order would separate them (e.g., sulfuric acid and fluorosulfuric acid). The Formula Index following the table provides another means of locating a compound. An index to CAS Registry Numbers is also available in the electronic versions of the *Handbook* or by request via e-mail (william.haynes@taylorandfrancis.com).

The following data fields appear in the table:

- **Name**: Systematic name for the substance. The valence state of a metallic element is indicated by a Roman numeral, e.g., copper in the +1 state is written as copper(I) rather than cuprous, iron in the +3 state is iron(III) rather than ferric.
- **Formula**: The simplest descriptive formula is given, but this does not necessarily specify the actual structure of the compound. For example, aluminum chloride is designated as $AlCl_3$, even though a more accurate representation of the structure in the solid phase (and, under some conditions, in the gas phase) is Al_2Cl_6. A few exceptions are made, such as the use of Hg_2^{+2} for the mercury(I) ion.
- **CAS Registry Number**: Chemical Abstracts Service Registry Number. An asterisk* following the CAS RN for a hydrate indicates that the number refers to the anhydrous compound. In most cases the generic CAS RN for the compound is given rather than the number for a specific crystalline form or mineral.
- **Mol. Weight**: Molecular weight (relative molar mass) as calculated with the 2005 IUPAC Recommended Atomic Weights. The number of decimal places corresponds to the number of places in the atomic weight of the least accurately known element (e.g., one place for lead compounds, two

places for compounds of selenium, germanium, etc.); a maximum of three places is given. For compounds of radioactive elements for which IUPAC makes no recommendation, the mass number of the isotope with longest half-life is used.

- **Physical Form**: The crystal system is given, when available, for compounds that are solid at room temperature, together with color and other descriptive features. Abbreviations are listed below.
- **mp**: Normal melting point in °C. The notation tp indicates the temperature where solid, liquid, and gas are in equilibrium at a pressure greater than one atmosphere (i.e., the normal melting point does not exist). When available, the triple point pressure is listed.
- **bp**: Normal boiling point in °C (referred to 101.325 kPa or 760 mmHg pressure). The notation sp following the number indicates the temperature where the pressure of the vapor in equilibrium with the solid reaches 101.325 kPa. See Reference 8, p. 23, for further discussion of sublimation points and triple points. A notation "sublimes" without a temperature being given indicates that there is a perceptible sublimation pressure above the solid at ambient temperatures.
- **Density**: Density values for solids and liquids are always in units of grams per cubic centimeter and can be assumed to refer to temperatures near room temperature unless otherwise stated. Values for gases are the calculated ideal gas densities in grams per liter at 25 °C and 101.325 kPa; the unit is always specified for a gas value.
- **Aqueous Solubility**: Solubility is expressed as the number of grams of the compound (excluding any water of hydration) that will dissolve in 100 grams of water. The temperature in °C is given as a superscript. Solubility at other temperatures can be found for many compounds in the table "Aqueous Solubility of Inorganic Compounds at Various Temperatures" in Section 8.
- **Qualitative Solubility**: Qualitative information on the solubility in other solvents (and in water, if quantitative data are unavailable) is given here. The abbreviations are:

i	insoluble
sl	slightly soluble
s	soluble
vs	very soluble
reac	reacts with the solvent

Data were taken from a wide variety of reliable sources, including monographs, treatises, review articles, evaluated compilations and databases, and in many cases the primary literature. Some of the most useful references for the properties covered here are listed below.

List of Abbreviations

Ac - acetyl	**blk** - black	**cub** - cubic	**exp** - explodes, explosive
ace - acetone	**brn** - brown	**cyhex** - cyclohexane	**extrap** - extrapolated
acid - acid solutions	**bz** - benzene	**dec** - decomposes	**flam** - flammable
alk - alkaline solutions	**chl** - chloroform	**dil** - dilute	**gl** - glass, glassy
amorp - amorphous	**col** - colorless	**diox** - dioxane	**grn** - green
anh - anhydrous	**conc** - concentrated	**eth** - ethyl ether	**hc** - hydrocarbon solvents
aq - aqueous	**cry** - crystals, crystalline	**EtOH** - ethanol	**hex** - hexagonal, hexane

hp - heptane
HT - high temperature
hyd - hydrate
hyg - hygroscopic
i - insoluble in
liq - liquid
LT - low temperature
MeOH - methanol
monocl - monoclinic
octahed - octahedral
oran - orange

orth - orthorhombic
os - organic solvents
peth - petroleum ether
pow - powder
prec - precipitate
pur - purple
py - pyridine
reac - reacts with
refrac - refractory
rhom - rhombohedral
r.t. - room temperature

s - soluble in
silv - silvery
sl - slightly soluble in
soln - solution
sp - sublimation point
stab - stable
subl - sublimes
temp - temperature
tetr - tetragonal
thf - tetrahydrofuran
tol - toluene

tp - triple point
trans - transition, transformation
tricl - triclinic
trig - trigonal
unstab - unstable
viol - violet
visc - viscous
vs - very soluble in
wh - white
xyl - xylene
yel - yellow

References

1. Phillips, S. L., and Perry, D.L., *Handbook of Inorganic Compounds*, CRC Press, Boca Raton, FL, 1995.
2. Trotman-Dickenson, A. F., Executive Editor, *Comprehensive Inorganic Chemistry*, Vol. 1-5, Pergamon Press, Oxford, 1973.
3. Greenwood, N. N., and Earnshaw, A., *Chemistry of the Elements, Second Edition*, Butterworth-Heinemann, Oxford, 1997.
4. Wiberg, N., Wiberg, E., and Holleman, H. F., *Inorganic Chemistry, 34th Edition*, Academic Press, San Diego, 2001.
5. *GMELIN Handbook of Inorganic and Organometallic Chemistry*, Springer-Verlag, Heidelberg.
6. Chase, M.W., Davies, C.A., Downey, J.R., Furrip, D. J., McDonald, R.A., and Syverud, A.N.; *JANAF Thermochemical Tables, Third Edition, J. Phys. Chem. Ref. Data*, Vol. 14, Suppl. 1, 1985; Chase, M. W., *NIST-JANAF Thermochemical Tables, Fourth Edition, J. Phys. Chem. Ref. Data*, Monograph No. 9, 1998.
7. *Landolt-Börnstein, Numerical Data and Functional Relationships in Science and Technology, New Series*, IV/19A, "Thermodynamic Properties of Inorganic Materials compiled by SGTE", Springer-Verlag, Heidelberg; Part 1, 1999; Part 2; 1999; Part 3, 2000; Part 4, 2001.
8. Lide, D. R., and Kehiaian, H.V., *CRC Handbook of Thermophysical and Thermochemical Data*, CRC Press, Boca Raton, FL, 1994.
9. *Kirk-Othmer Concise Encyclopedia of Chemical Technology*, Wiley-Interscience, New York, 1985.
10. *Dictionary of Inorganic Compounds*, Chapman & Hall, New York, 1992.
11. Massalski, T. B., Ed., *Binary Alloy Phase Diagrams, 2nd Edition*, ASM International, Metals Park, Ohio, 1990.
12. Dinsdale, A.T., "SGTE Data for Pure Elements", *CALPHAD*, 15, 317–425, 1991.
13. Madelung, O., *Semiconductors: Group IV Elements and III-IV Compounds*, Springer-Verlag, Heidelberg, 1991.

14. Lidin, R. A., Andreeva, L. L., and Molochko, V. A., *Constants of Inorganic Substances*, Begell House, New York, 1995.
15. Gurvich, L. V., Veyts, I. V., and Alcock, C. B., *Thermodynamic Properties of Individual Substances, Fourth Edition*, Hemisphere Publishing Corp., New York, 1989.
16. *The Combined Chemical Dictionary on CDROM*, Version 9:1, Chapman & Hall/CRC, Boca Raton, FL, 2005.
17. Macdonald, F., Editor, *Chapman & Hall/CRC Combined Chemical Dictionary*, <http://www.chemnetbase.com/scripts/ccdweb.exe>.
18. Sangeeta, G., and LaGraff, J. R., *Inorganic Materials Chemistry, Second Edition*, CRC Press, Boca Raton, FL, 2005.
19. Stern, K. H., *High Temperature Properties and Thermal Decomposition of Inorganic Salts with Oxyanions*, CRC Press, Boca Raton, FL, 2001.
20. Donnay, J.D.H., and Ondik, H.M., *Crystal Data Determinative Tables, Third Edition, Volumes 2 and 4, Inorganic Compounds*, Joint Committee on Powder Diffraction Standards, Swarthmore, PA, 1973.
21. Robie, R., Bethke, P. M., and Beardsley, K. M., *Selected X-ray Crystallographic Data, Molar Volumes, and Densities of Minerals and Related Substances*, U.S. Geological Survey Bulletin 1248, 1967.
22. Carmichael, R. S., *Practical Handbook of Physical Properties of Rocks and Minerals*, CRC Press, Boca Raton, FL, 1989.
23. Deer, W. A., Howie, R.A., and Zussman, J., An *Introduction to the Rock-Forming Minerals, 2nd Edition*, Longman Scientific & Technical, Harlow, Essex, 1992.
24. Linstrom, P. J., and Mallard, W. G., Editors, NIST Chemistry WebBook, NIST Standard Reference Database No. 69, June 2005, National Institute of Standards and Technology, Gaithersburg, MD 20899, <http://webbook.nist.gov>.
25. *Phase Diagrams for Ceramists, Volumes 1–8; ACerS-NIST Phase Equilibrium Diagrams, Volumes 9–13*, American Ceramic Society, Westerville, Ohio, 1964–2001.

No.	Name	Formula	CAS Reg No.	Mol. weight	Physical form	mp/°C	bp/°C	Density g cm^{-3}	Solubility g/100 g H$_2$O	Qualitative solubility
1	Actinium	Ac	7440-34-8	227	silv metal; cub	1050	3200	10		
2	Actinium bromide	AcBr$_3$	33689-81-5	467	wh hex cry		800 subl	5.85		s H$_2$O
3	Actinium chloride	AcCl$_3$	22986-54-5	333	wh hex cry		960 subl	4.81		
4	Actinium fluoride	AcF$_3$	33689-80-4	284	wh hex cry			7.88		i H$_2$O
5	Actinium iodide	AcI$_3$	33689-82-6	608	wh cry					s H$_2$O
6	Actinium oxide	Ac$_2$O$_3$	12002-61-8	502	wh hex cry	1977		9.19		i H$_2$O
7	Aluminum	Al	7429-90-5	26.982	silv-wh metal; cub cry	660.323	2519	2.70		i H$_2$O; s acid, alk
8	Aluminum acetate	Al(C$_2$H$_3$O$_2$)$_3$	139-12-8	204.113	wh hyg solid	dec				s H$_2$O; sl ace
9	Aluminum diacetate	Al(OH)(C$_2$H$_3$O$_2$)$_2$	142-03-0	162.078	wh amorp powder					i H$_2$O
10	Aluminum ammonium sulfate	AlNH$_4$(SO$_4$)$_2$	7784-25-0	237.146	wh powder					sl H$_2$O; i EtOH
11	Aluminum ammonium sulfate dodecahydrate	AlNH$_4$(SO$_4$)$_2$ · 12H$_2$O	7784-26-1	453.329	col cry or powder	94.5	>280 dec	1.65		s H$_2$O; i EtOH
12	Aluminum antimonide	AlSb	25152-52-7	148.742	brn cub cry	1065		4.26		
13	Aluminum arsenide	AlAs	22831-42-1	101.903	oran cub cry; hyg	1740		3.76		
14	Aluminum borate	2Al$_2$O$_3$ · B$_2$O$_3$	11121-16-7	273.543	needles	≈1050				i H$_2$O
15	Aluminum borohydride	Al(BH$_4$)$_3$	16962-07-5	71.510	flam liq	-64.5	44.5			reac H$_2$O
16	Aluminum bromate nonahydrate	Al(BrO$_3$)$_3$ · 9H$_2$O	11126-81-1*	572.826	wh hyg cry	62	>100 dec			s H$_2$O
17	Aluminum bromide	AlBr$_3$	7727-15-3	266.694	wh-yel monocl cry; hyg	97.5	255	3.2		reac H$_2$O; s bz, tol

No.	Name	Formula	CAS Reg No.	Mol. weight	Physical form	mp/°C	bp/°C	Density g cm^{-3}	Solubility g/100 g H$_2$O	Qualitative solubility
18	Aluminum bromide hexahydrate	AlBr$_3$ · 6H$_2$O	7784-11-4	374.785	col-yel hyg cry	93		2.54		s H$_2$O, EtOH, CS$_2$
19	Aluminum carbide	Al$_4$C$_3$	1299-86-1	143.958	yel hex cry	2100	>2200 dec	2.36		reac H$_2$O
20	Aluminum chlorate nonahydrate	Al(ClO$_3$)$_3$ · 9H$_2$O	15477-33-5	439.473	hyg cry					vs H$_2$O; s EtOH
21	Aluminum chloride	AlCl$_3$	7446-70-0	133.341	wh hex cry or powder; hyg	192.6	180 sp	2.48	45.1^{25}	s bz, ctc, chl
22	Aluminum chloride hexahydrate	AlCl$_3$ · 6H$_2$O	7784-13-6	241.432	col hyg cry	100 dec		2.398	45.1^{25}	s EtOH, eth
23	Dichloromethylaluminum	AlCl$_2$CH$_3$	917-65-7	112.923	cry	72.7	95^{10}			s bz, eth, hc
24	Chlorodiethylaluminum	AlCl(C$_2$H$_5$)$_2$	96-10-6	120.557	col liq	-74		0.96		reac H$_2$O
25	Chlorodiisobutylaluminum	AlCl(C$_4$H$_9$)$_2$	1779-25-5	176.664	hyg col liq	-40		0.95		s eth, hx
26	Aluminum diboride	AlB$_2$	12041-50-8	48.604	powder	>920 dec		3.19		s dil HCl
27	Aluminum dodecaboride	AlB$_{12}$	12041-54-2	156.714	yel-brn prisms	2070		2.55		s hot HNO$_3$; i acid, alk
28	Aluminum ethanolate	Al(C$_2$H$_5$O)$_3$	555-75-9	162.163	liq, condenses to wh solid	140				reac H$_2$O; sl xyl
29	Aluminum fluoride	AlF$_3$	7784-18-1	83.977	wh hex cry	2250 tp (220 MPa)	1276 sp	3.10	0.50^{25}	
30	Aluminum fluoride monohydrate	AlF$_3$ · H$_2$O	32287-65-3	101.992	orth cry			2.17	0.50^{25}	
31	Aluminum fluoride trihydrate	AlF$_3$ · 3H$_2$O	15098-87-0	138.023	wh hyg cry			1.914	0.50^{25}	
32	Aluminum hexafluorosilicate nonahydrate	Al$_2$(SiF$_6$)$_3$ · 9H$_2$O	17099-70-6	642.329	hex prisms	>500 dec				s H$_2$O
33	Aluminum hydride	AlH$_3$	7784-21-6	30.006	col hex cry	>150 dec				reac H$_2$O
34	Aluminum hydroxide	Al(OH)$_3$	21645-51-2	78.004	wh amorp powder			2.42		i H$_2$O; s alk, acid
35	Aluminum hydroxychloride	Al$_2$(OH)$_5$Cl · 2H$_2$O	1327-41-9	210.483	gl solid					s H$_2$O
36	Aluminum iodide	AlI$_3$	7784-23-8	407.695	wh leaflets	188.28	382	3.98		reac H$_2$O
37	Aluminum iodide hexahydrate	AlI$_3$ · 6H$_2$O	10090-53-6	515.786	yel hyg cry powder					vs H$_2$O; s EtOH, eth
38	Aluminum lactate	Al(C$_3$H$_5$O$_3$)$_3$	18917-91-4	294.192	powder					vs H$_2$O
39	Aluminum molybdate	Al$_2$(MoO$_4$)$_3$	15123-80-5	533.78	wh pow	≈950				
40	Aluminum nitrate	Al(NO$_3$)$_3$	13473-90-0	212.997	wh hyg solid	dec			68.9^{25}	vs EtOH; sl ace
41	Aluminum nitrate nonahydrate	Al(NO$_3$)$_3$ · 9H$_2$O	7784-27-2	375.134	wh hyg monocl cry	73	135 dec	1.72	68.9^{25}	vs EtOH; i pyr
42	Aluminum nitride	AlN	24304-00-5	40.989	blue-wh hex cry	3000		3.255		reac H$_2$O
43	Aluminum oleate	Al(C$_{18}$H$_{33}$O$_2$)$_3$	688-37-9	871.342	yel solid					i H$_2$O; s EtOH, bz
44	Aluminum oxalate monohydrate	Al$_2$(C$_2$O$_4$)$_3$ · H$_2$O	814-87-9	336.035	wh pow					i H$_2$O, EtOH; s acid
45	Aluminum oxide (α)	Al$_2$O$_3$	1344-28-1	101.961	wh powder; hex	2054	2977	3.99		i H$_2$O, os; sl alk
46	Aluminum oxide (γ)	Al$_2$O$_3$	1344-28-1	101.961	soft wh pow	trans to corundum 1200		3.97		i H$_2$O; s acid; sl alk
47	Aluminum oxyhydroxide (boehmite)	AlO(OH)	1318-23-6	59.989	wh orth cry	trans to diasphore 227		3.07		i H$_2$O; s hot acid, alk
48	Aluminum oxyhydroxide (diaspore)	AlO(OH)	14457-84-2	59.989	orth cry	dec 450		3.38		i H$_2$O; s acid, alk
49	Aluminum palmitate	Al(C$_{15}$H$_{31}$COO)$_3$	555-35-1	793.230	wh-yel powder					i H$_2$O, EtOH; s peth
50	Aluminum 2,4-pentanedioate	Al(CH$_3$COCHCOCH$_3$)$_3$	13963-57-0	324.306	pale yel prisms	194.6	315	1.27		i H20; s bz, EtOH; sl hex
51	Aluminum perchlorate	Al(ClO$_4$)$_3$	14452-39-2	325.334	wh hyg cry				55^0	s H$_2$O, eth; i ctc
52	Aluminum perchlorate nonahydrate	Al(ClO$_4$)$_3$ · 9H$_2$O	14452-39-2	487.471	wh hyg cry	82 dec		2.0	182.4^0	
53	Aluminum phosphate	AlPO$_4$	7784-30-7	121.953	wh rhomb plates	>1460		2.56		i H$_2$O; sl acid
54	Aluminum phosphate dihydrate	AlPO$_4$ · 2H$_2$O	13477-75-3	157.984	wh rhom cry	dec 1500		2.54		i H$_2$O
55	Aluminum phosphate trihydroxide	Al$_2$(OH)$_3$PO$_4$	12004-29-4	199.957	wh or yel monocl cry			2.7		
56	Aluminum metaphosphate	Al(PO$_3$)$_3$	32823-06-6	263.898	col powder; tetr	≈1525		2.78		i H$_2$O
57	Aluminum hypophosphite	Al(H$_2$PO$_2$)$_3$	7784-22-7	221.948	cry powder	220 dec				i H$_2$O; s alk, acid
58	Aluminum phosphide	AlP	20859-73-8	57.956	grn or yel cub cry	2550		2.40		reac H$_2$O
59	Aluminum selenide	Al$_2$Se$_3$	1302-82-5	290.84	yel-brown powder	960		3.437		reac H$_2$O
60	Aluminum silicate (andalusite)	Al$_2$SiO$_5$	12183-80-1	162.046	gray-grn cry			3.145		
61	Aluminum silicate (kyanite)	Al$_2$SiO$_5$	1302-76-7	162.046	blue or gray tricl cry	dec 1000		3.68		
62	Aluminum silicate (mullite)	3Al$_2$O$_3$ · 2SiO$_2$	1302-93-8	426.052	col orth cry	1750		3.17		i H$_2$O, acid, HF
63	Aluminum silicate (sillimanite)	Al$_2$SiO$_5$	12141-45-6	162.046	wh orth cry	1816		3.25		
64	Aluminum silicate dihydrate	Al$_2$O$_3$ · 2SiO$_2$ · 2H$_2$O	1332-58-7	258.161	wh-yel powder; tricl			2.59		i H$_2$O, acid, alk
65	Aluminum stearate	Al(C$_{18}$H$_{35}$O$_2$)$_3$	637-12-7	877.390	wh powder	115		1.070		i H$_2$O, EtOH, eth; s alk
66	Aluminum monostearate	Al(OH)$_2$(C$_{18}$H$_{35}$O$_2$)	7047-84-9	344.467	yel-wh pow	155		1.02		i H$_2$O
67	Aluminum distearate	Al(OH)(C$_{18}$H$_{35}$O$_2$)$_2$	300-92-5	610.928	wh pow	145				i H$_2$O
68	Aluminum sulfate	Al$_2$(SO$_4$)$_3$	10043-01-3	342.151	wh cry	1040 dec			38.5^{25}	i EtOH
69	Aluminum sulfate octadecahydrate	Al$_2$(SO$_4$)$_3$ · 18H$_2$O	7784-31-8	666.426	col monocl cry	86 dec		1.69	38.5^{25}	
70	Aluminum sulfide	Al$_2$S$_3$	1302-81-4	150.158	yel-gray powder	1100		2.02		
71	Aluminum telluride	Al$_2$Te$_3$	12043-29-7	436.76	gray-blk hex cry	≈895		4.5		

No.	Name	Formula	CAS Reg No.	Mol. weight	Physical form	mp/°C	bp/°C	Density g cm^{-3}	Solubility g/100 g H$_2$O	Qualitative solubility
72	Aluminum thiocyanate	Al(SCN)$_3$	538-17-0	201.229	yel powder					s H$_2$O; i EtOH, eth
73	Aluminum titanate	Al$_2$TiO$_5$	12004-39-6	181.827	refrac solid	1860				
74	Aluminum zirconium	Al$_3$Zr	12004-50-1	145.187	metallic solid	1645				
75	Americium	Am	7440-35-9	243	silv metal; hex or cub	1176	2011	12		s acid
76	Americium(III) oxide	Am$_2$O$_3$	12254-64-7	534	tan hex cry			11.77		s acid
77	Americium(III) bromide	AmBr$_3$	14933-38-1	483	wh orth cry			6.85		s H$_2$O
78	Americium(III) chloride	AmCl$_3$	13464-46-5	349	pink hex cry	500		5.87		
79	Americium(III) fluoride	AmF$_3$	13708-80-0	300	pink hex cry	1393		9.53		
80	Americium(III) iodide	AmI$_3$	13813-47-3	624	yel ortho cry	≈950		6.9		
81	Americium(IV) fluoride	AmF$_4$	15947-41-8	319	tan monocl cry			7.23		
82	Americium(IV) oxide	AmO$_2$	12005-67-3	275	blk cub cry	>1000 dec		11.68		s acid
83	Ammonia	NH$_3$	7664-41-7	17.031	col gas	-77.73	-33.33	0.696 g/L		vs H$_2$O; s EtOH, eth
84	Ammonium acetate	NH$_4$C$_2$H$_3$O$_2$	631-61-8	77.083	wh hyg cry	114		1.073	148[4]	s EtOH; sl ace
85	Ammonium azide	NH$_4$N$_3$	12164-94-2	60.059	orth cry; flam	160	exp	1.346	20.2[30]	
86	Ammonium benzoate	NH$_4$C$_7$H$_5$O$_2$	1863-63-4	139.152	wh cry or powder	198		1.26		s H$_2$O; sl EtOH
87	Ammonium bromate	NH$_4$BrO$_3$	13843-59-9	145.941	col hex cry	exp				vs H$_2$O
88	Ammonium bromide	NH$_4$Br	12124-97-9	97.943	wh hyg tetr cry	542 dec	396 sp	2.429	78.3[25]	s EtOH, ace; sl eth
89	Ammonium caprylate	NH$_4$C$_8$H$_{15}$O$_2$	5972-76-9	161.243	hyg monocl cry	≈75				reac H$_2$O; s EtOH; i chl, bz
90	Ammonium carbamate	NH$_2$COONH$_4$	1111-78-0	78.071	cry powder					vs H$_2$O; s EtOH
91	Ammonium carbonate	(NH$_4$)$_2$CO$_3$	506-87-6	96.086	col cry powder	58 dec			100[15]	
92	Ammonium chlorate	NH$_4$ClO$_3$	10192-29-7	101.490	wh cry	102 exp		1.80	28.7[20]	
93	Ammonium chloride	NH$_4$Cl	12125-02-9	53.492	col cub cry	520.1 tp (dec)	338 sp	1.519	39.5[25]	
94	Ammonium chromate	(NH$_4$)$_2$CrO$_4$	7788-98-9	152.071	yel cry	185 dec		1.90	37[25]	sl ace, MeOH; i EtOH
95	Ammonium chromic sulfate dodecahydrate	NH$_4$Cr(SO$_4$)$_2$ · 12H$_2$O	10022-47-6	478.343	blue-viol cry	94 dec		1.72		s H$_2$O; sl EtOH
96	Ammonium cobalt(II) phosphate	CoNH$_4$PO$_4$	14590-13-7	171.943	red-viol powder (hyd)					i H$_2$O; s acid
97	Ammonium cobalt(II) phosphate monohydrate	CoNH$_4$PO$_4$ · H$_2$O	16827-96-6	189.959	red-purp orth plates	dec 450				s acid
98	Ammonium cobalt(II) sulfate hexahydrate	(NH$_4$)$_2$Co(SO$_4$)$_2$ · 6H$_2$O	13586-38-4	395.227	red monocl prisms			1.90		s H$_2$O; i EtOH
99	Ammonium copper(II) chloride	CuCl$_2$ · 2NH$_4$Cl	10060-13-6*	241.435	yel hyg orth cry					s H$_2$O
100	Ammonium copper(II) chloride dihydrate	CuCl$_2$ · 2NH$_4$Cl · 2H$_2$O	10060-13-6	277.465	blue-grn tetr cry	110 dec		1.993		s H$_2$O, EtOH
101	Ammonium cyanide	NH$_4$CN	12211-52-8	44.056	col tetr cry	dec		1.10		vs H$_2$O
102	Ammonium dichromate	(NH$_4$)$_2$Cr$_2$O$_7$	7789-09-5	252.065	oran-red monocl cry; hyg	180 dec		2.155	35.6[20]	
103	Ammonium dihydrogen arsenate	NH$_4$H$_2$AsO$_4$	13462-93-6	158.975	tetr cry	300 dec		2.311	52.7[25]	
104	Ammonium dihydrogen phosphate	NH$_4$H$_2$PO$_4$	7722-76-1	115.026	wh tetr cry	190		1.80	40.4[25]	sl EtOH; i ace
105	Ammonium O,O-diethyldithiophosphate	(C$_2$H$_5$O)$_2$P(S)SNH$_4$	1068-22-0	203.264	cry	165				
106	Ammonium dithiocarbamate	NH$_4$NH$_2$CSS	513-74-6	110.202	yel ortho cry	99 dec		1.45		s H$_2$O
107	Ammonium ferricyanide trihydrate	(NH$_4$)$_3$Fe(CN)$_6$ · 3H$_2$O	14221-48-8*	320.110	red cry					s H$_2$O; i EtOH
108	Ammonium ferrocyanide trihydrate	(NH$_4$)$_4$Fe(CN)$_6$ · 3H$_2$O	14481-29-9*	338.149	yel cry	dec				s H$_2$O; i EtOH
109	Ammonium fluoride	NH$_4$F	12125-01-8	37.037	wh hex cry; hyg	238		1.015	83.5[25]	sl EtOH
110	Ammonium fluorosulfonate	NH$_4$SO$_3$F	13446-08-7	117.100	col needles	245				s H$_2$O, EtOH, MeOH
111	Ammonium formate	NH$_4$CHO$_2$	540-69-2	63.057	hyg cry	116		1.27	143[20]	s EtOH
112	Ammonium heptafluorotantalate	(NH$_4$)$_2$TaF$_7$	12022-02-5	350.014	hyg cry					
113	Ammonium hexabromoosmate(IV)	(NH$_4$)$_2$OsBr$_6$	24598-62-7	705.73	small blk cubes					sl H$_2$O; s glycerol; i EtOH
114	Ammonium hexabromoplatinate(IV)	(NH$_4$)$_2$PtBr$_6$	17363-02-9	710.585	powder	145 dec			0.59[20]	
115	Ammonium hexachloroiridate(III)	(NH$_4$)$_3$IrCl$_6$	15752-05-3	459.050	grn pow					
116	Ammonium hexachloroiridate(IV)	(NH$_4$)$_2$IrCl$_6$	16940-92-4	441.012	blk cry powder	dec		2.856	1.09[25]	
117	Ammonium hexachloroosmate(IV)	(NH$_4$)$_2$OsCl$_6$	12125-08-5	439.03	red cry or powder		subl	2.93		s H$_2$O, EtOH
118	Ammonium hexachloropalladate(IV)	(NH$_4$)$_2$PdCl$_6$	19168-23-1	355.22	red-brn hyg cry	dec		2.418		
119	Ammonium hexachloroplatinate(IV)	(NH$_4$)$_2$PtCl$_6$	16919-58-7	443.879	red-oran cub cry	380 dec		3.065	0.5[20]	i EtOH
120	Ammonium hexachlororuthenate(IV)	(NH$_4$)$_2$RuCl$_6$	18746-63-9	349.87	red cry					
121	Ammonium hexafluoroaluminate	(NH$_4$)$_3$AlF$_6$	7784-19-2	195.087	cub cry			1.78		s H$_2$O
122	Ammonium hexafluorogallate	(NH$_4$)$_3$GaF$_6$	14639-94-2	237.828	col cub cry	>200 dec		2.10		
123	Ammonium hexafluorogermanate	(NH$_4$)$_2$GeF$_6$	16962-47-3	222.71	wh cry	380	subl	2.564		s H$_2$O; i EtOH
124	Ammonium hexafluorophosphate	NH$_4$PF$_6$	16941-11-0	163.003	wh cub cry	58 dec		2.180		vs H$_2$O; s ace, EtOH, MeOH
125	Ammonium hexafluorosilicate	(NH$_4$)$_2$SiF$_6$	16919-19-0	178.153	wh cub or trig cry	dec		2.011	22.7[25]	i EtOH, ace
126	Ammonium hexafluorotitanate	(NH$_4$)$_2$TiF$_6$	16962-40-6	197.934	wh solid					s H$_2$O

No.	Name	Formula	CAS Reg No.	Mol. weight	Physical form	mp/°C	bp/°C	Density g cm⁻³	Solubility g/100 g H₂O	Qualitative solubility
127	Ammonium hexafluorozirconate(IV)	$(NH_4)_2ZrF_6$	16919-31-6	241.291	wh hex cry			1.154		s H_2O
128	Ammonium hydrogen arsenate	$(NH_4)_2HAsO_4$	7784-44-3	176.004	wh powder			1.99		s H_2O
129	Ammonium hydrogen carbonate	NH_4HCO_3	1066-33-7	79.056	col or wh prisms	107 dec		1.586	24.8²⁵	i EtOH, bz
130	Ammonium hydrogen citrate	$(NH_4)_2HC_6H_5O_7$	3012-65-5	226.184	col cry			1.48		vs H_2O; sl EtOH
131	Ammonium hydrogen fluoride	NH_4HF_2	1341-49-7	57.044	wh orth cry	125	240 dec	1.50	60.2²⁰	
132	Ammonium hydrogen malate	$NH_4C_4H_5O_5$	5972-71-4	151.118	orth cry	160		1.15		s H_2O; sl EtOH
133	Ammonium hydrogen oxalate monohydrate	$NH_4HC_2O_4 \cdot H_2O$	5972-72-5*	125.081	col rhomb cry	dec		1.56		sl H_2O, EtOH
134	Ammonium hydrogen phosphate	$(NH_4)_2HPO_4$	7783-28-0	132.055	wh cry	155 dec		1.619	69.5²⁵	i EtOH, ace
135	Ammonium hydrogen phosphite monohydrate	$(NH_4)_2HPO_3 \cdot H_2O$	51503-61-8	134.071	hyg cry					s H_2O
136	Ammonium hydrogen selenate	NH_4HSeO_4	10294-60-7	162.01	rhom cry	dec		2.162		
137	Ammonium hydrogen sulfate	NH_4HSO_4	7803-63-6	115.110	wh hyg cry	147		1.78	100²⁰	i EtOH, ace, py
138	Ammonium hydrogen sulfide	NH_4HS	12124-99-1	51.112	wh tetr or orth cry	dec		1.17	128⁰	sl ace; i bz, eth
139	Ammonium hydrogen sulfite	NH_4HSO_3	10192-30-0	99.110	col cry	dec		2.03	71.8⁰	
140	Ammonium hydrogen tartrate	$NH_4HC_4H_4O_6$	3095-65-6	167.117	wh cry			1.68		sl H_2O; s alk; i EtOH
141	Ammonium hydroxide	NH_4OH	1336-21-6	35.046	exists only in soln					
142	Ammonium hypophosphite	$NH_4H_2PO_2$	7803-65-8	83.028	wh hyg cry	dec				vs H_2O; sl EtOH; i ace
143	Ammonium iodate	NH_4IO_3	13446-09-8	192.941	wh powder	150		3.3	3.84²⁵	
144	Ammonium iodide	NH_4I	12027-06-4	144.943	wh tetr cry; hyg	551 dec	405 sp	2.514	178²⁵	sl EtOH, MeOH
145	Ammonium iron(II) sulfate hexahydrate	$(NH_4)_2Fe(SO_4)_2 \cdot 6H_2O$	7783-85-9	392.139	blue-grn monocl cry	≈100 dec		1.86		s H_2O; i EtOH
146	Ammonium iron(III) chromate	$NH_4Fe(CrO_4)_2$	7789-08-4	305.871	red powder					i H_2O
147	Ammonium iron(III) oxalate trihydrate	$(NH_4)_3Fe(C_2O_4)_3 \cdot 3H_2O$	13268-42-3	428.063	grn monocl cry; hyg	≈160 dec		1.780		vs H_2O; i EtOH
148	Ammonium iron(III) sulfate dodecahydrate	$NH_4Fe(SO_4)_2 \cdot 12H_2O$	7783-83-7	482.192	col to viol cry	≈37		1.71		vs H_2O; i EtOH
149	Ammonium lactate	$NH_4C_3H_5O_3$	52003-58-4	107.108	col cry	92				s H_2O, EtOH; sl MeOH; i ace, eth
150	Ammonium magnesium chloride hexahydrate	$NH_4MgCl_3 \cdot 6H_2O$	39733-35-2	256.794	hyg cry	dec 100		1.46	17²⁰	s H_2O
151	Ammonium mercuric chloride dihydrate	$(NH_4)_2HgCl_4 \cdot 2H_2O$	33445-15-7*	414.51	powder					s H_2O; sl EtOH
152	Ammonium metatungstate hexahydrate	$(NH_4)_6W_7O_{24} \cdot 6H_2O$	12028-48-7	1887.19	wh cry					s H_2O; i EtOH
153	Ammonium metavanadate	NH_4VO_3	7803-55-6	116.979	wh-yel cry	200 dec		2.326	4.8²⁰	
154	Ammonium molybdate(VI) tetrahydrate	$(NH_4)_6Mo_7O_{24} \cdot 4H_2O$	12054-85-2	1235.86	col or grn-yel cry	90 dec		2.498	43	i EtOH
155	Ammonium dimolybdate	$(NH_4)_2Mo_2O_7$	27546-07-2	339.95	cry					s H_2O
156	Ammonium molybdophosphate	$(NH_4)_3PO_4 \cdot 12MoO_3$	12026-66-3	1876.35	grn or yel cry	dec			0.02²⁰	sl H_2O; s alk
157	Ammonium nitrate	NH_4NO_3	6484-52-2	80.043	wh hyg cry; orth	169.7	dec 200-260	1.72	213²⁵	sl MeOH
158	Ammonium nitrite	NH_4NO_2	13446-48-5	64.044	wh-yel cry	60 exp		1.69	221²⁵	i eth
159	Ammonium nitroferricyanide	$(NH_4)_2Fe(CN)_5NO$	14402-70-1	252.016	red-brn cry					s H_2O, EtOH
160	Ammonium oleate	$NH_4C_{18}H_{33}O_2$	544-60-5	299.493	yel-brn paste	21				s H_2O; sl ace
161	Ammonium oxalate	$(NH_4)_2C_2O_4$	1113-38-8	124.096	col sol			1.5	5.20²⁵	
162	Ammonium oxalate monohydrate	$(NH_4)_2C_2O_4 \cdot H_2O$	6009-70-7	142.110	wh orth cry	dec		1.50	5.20²⁵	sl EtOH
163	Ammonium palmitate	$NH_4C_{15}H_{31}CO_2$	593-26-0	273.455	yel-wh powder	22				s H_2O; sl bz, xyl; i ace, EtOH, ctc
164	Ammonium pentaborate tetrahydrate	$NH_4B_5O_8 \cdot 4H_2O$	12007-89-5	272.150	wh cry				7.03¹⁸	
165	Ammonium pentachlororhodate(III) monohydrate	$(NH_4)_2RhCl_5 \cdot H_2O$	63771-33-5	334.262	red cry	dec 210				
166	Ammonium pentachlorozincate	$(NH_4)_3ZnCl_5$	14639-98-6	296.789	hyg orth cry			1.81		vs H_2O
167	Ammonium perchlorate	NH_4ClO_4	7790-98-9	117.490	wh orth cry	dec, exp		1.95	24.5²⁵	s MeOH; sl EtOH, ace; i eth
168	Ammonium permanganate	NH_4MnO_4	13446-10-1	136.975	purp rhomb cry	70 dec		2.22	7.9¹⁵	
169	Ammonium peroxydisulfate	$(NH_4)_2S_2O_8$	7727-54-0	228.202	monocl cry or wh powder	dec		1.982	83.5²⁵	
170	Ammonium perrhenate	NH_4ReO_4	13598-65-7	268.244	col powder			3.97	6.23²⁰	
171	Ammonium phosphate trihydrate	$(NH_4)_3PO_4 \cdot 3H_2O$	10361-65-6*	203.133	wh prisms				25.0²⁵	i ace
172	Ammonium phosphomolybdate monohydrate	$(NH_4)_3PO_4 \cdot 12MoO_3 \cdot H_2O$	54723-94-3	1894.36	yel cry or powder	dec			0.02	
173	Ammonium phosphotungstate dihydrate	$(NH_4)_3PO_4 \cdot 12WO_3 \cdot 2H_2O$	1311-90-6	2967.18	cry powder					sl H_2O
174	Ammonium picrate	$NH_4C_6H_2N_3O_7$	131-74-8	246.135	yel orth cry	exp		1.72		sl H_2O
175	Ammonium polysulfide	$(NH_4)_2S_x$	9080-17-5		yel unstab soln					reac acids
176	Ammonium salicylate	$NH_4C_7H_5O_3$	528-94-9	155.151	wh cry powder					vs H_2O; s EtOH

No.	Name	Formula	CAS Reg No.	Mol. weight	Physical form	mp/°C	bp/°C	Density g cm⁻³	Solubility g/100 g H₂O	Qualitative solubility
177	Ammonium selenate	$(NH_4)_2SeO_4$	7783-21-3	179.04	wh monocl cry	dec		2.194	117^{25}	i EtOH, ace
178	Ammonium selenite	$(NH_4)_2SeO_3$	7783-19-9	163.04	wh or red hyg cry	dec			121^{25}	
179	Ammonium stearate	$NH_4C_{18}H_{35}O_2$	1002-89-7	301.509	yel-wh powder	22		0.89		sl H₂O, bz; s EtOH, MeOH; i ace
180	Ammonium sulfamate	$NH_4NH_2SO_3$	7773-06-0	114.124	wh hyg cry	131	160 dec			vs H₂O; sl EtOH
181	Ammonium sulfate	$(NH_4)_2SO_4$	7783-20-2	132.140	wh or brn orth cry	280 dec		1.77	76.4^{25}	i EtOH, ace
182	Ammonium sulfide	$(NH_4)_2S$	12135-76-1	68.142	yel-oran cry	≈0 dec				s H₂O, EtOH, alk
183	Ammonium sulfite	$(NH_4)_2SO_3$	17026-44-7	116.140	wh hyg cry				64.2^{25}	
184	Ammonium sulfite monohydrate	$(NH_4)_2SO_3 \cdot H_2O$	7783-11-1	134.155	col cry	dec		1.41	64.2^{25}	i EtOH, ace
185	Ammonium tartrate	$(NH_4)_2C_4H_4O_6$	3164-29-2	184.147	wh cry	dec		1.601		s H₂O
186	Ammonium tellurate	$(NH_4)_2TeO_4$	13453-06-0	227.68	wh powder	dec		3.024		
187	Ammonium tetraborate tetrahydrate	$(NH_4)_2B_4O_7 \cdot 4H_2O$	12228-87-4	263.377	wh tetr cry	dec 87				vs H₂O; s HNO₃
188	Ammonium tetrachloroaluminate	NH_4AlCl_4	7784-14-7	186.833	wh hyg solid	304				s H₂O, eth
189	Ammonium tetrachloropalladate(II)	$(NH_4)_2PdCl_4$	13820-40-1	284.31	grn cry or red-brn pow					s H₂O
190	Ammonium tetrachloroplatinate(II)	$(NH_4)_2PtCl_4$	13820-41-2	372.973	red cry	dec		2.936		s H₂O; i EtOH
191	Ammonium tetrachlorozincate	$(NH_4)_2ZnCl_4$	14639-97-5	243.298	wh orth plates; hyg	150 dec		1.879		vs H₂O
192	Ammonium tetrafluoroantimonate	NH_4SbF_4	14972-90-8	215.793	col cry					s H₂O
193	Ammonium tetrafluoroborate	NH_4BF_4	13826-83-0	104.844	wh powder; orth	487 dec		1.871	25^{20}	
194	Ammonium tetrathiocyanodiammonochromate(III) monohydrate	$NH_4[Cr(NH_3)_2(SCN)_4] \cdot H_2O$	13573-16-5	354.440	red cry	270 dec				s H₂O, EtOH, ace; i bz
195	Ammonium tetrathiomolybdate	$(NH_4)_2MoS_4$	15060-55-6	260.28	red cry	100 dec				vs H₂O
196	Ammonium tetrathiotungstate	$(NH_4)_2WS_4$	13862-78-7	348.18	oran cry	dec		2.71		s H₂O
197	Ammonium tetrathiovanadate	$(NH_4)_3VS_4$	14693-56-2	233.317	dark viol cry					
198	Ammonium thiocyanate	NH_4SCN	1762-95-4	76.121	col hyg cry	≈149	dec	1.30	181^{25}	vs EtOH; s ace; i chl
199	Ammonium thiosulfate	$(NH_4)_2S_2O_3$	7783-18-8	148.205	wh cry	150 dec		1.678		vs H₂O; i EtOH, eth
200	Ammonium titanium oxalate monohydrate	$(NH_4)_2TiO(C_2O_4)_2 \cdot H_2O$	10580-03-7	293.996	hyg cry					vs H₂O
201	Ammonium tungstate(VI)	$(NH_4)_{10}W_{12}O_{41}$	11120-25-5	3042.44	cry powder			2.3		s H₂O; i EtOH
202	Ammonium tungstate(VI) pentahydrate	$(NH_4)_{10}W_{12}O_{41} \cdot 5H_2O$	1311-93-9	3132.52	cry pow or plates			2.3		vs H₂O; i EtOH
203	Ammonium uranate(VI)	$(NH_4)_2U_2O_7$	7783-22-4	624.131	red-yel amorp powder					i H₂O, alk; s acid
204	Ammonium uranium fluoride	$UO_2(NH_4)_3F_5$	18433-40-4	419.135	grn-yel monocl cry					s H₂O; i EtOH
205	Ammonium valerate	$NH_4C_4H_9CO_2$	42739-38-8	119.163	hyg cry	108				vs H₂O, EtOH; s eth
206	Antimony (gray)	Sb	7440-36-0	121.760	silv metal; hex	630.628	1587	6.68		i dil acid
207	Antimony (black)	Sb	7440-36-0	121.760	blk amorp solid	trans gray 0				
208	Stibine	SbH_3	7803-52-3	124.784	col gas; flam	-88	-17	5.100 g/L		sl H₂O; s EtOH
209	Trimethylstibine	$Sb(CH_3)_3$	594-10-5	166.863	col flam liq	-62	81	1.52		
210	Pentamethylstibine	$Sb(CH_3)_5$	15120-50-0	196.933	col hyg liq	-19	127			reac H₂O
211	Tetramethyldistibine	$[Sb(CH_3)_2]_2$	41422-43-9	303.658	yel flam liq or red solid	17				
212	Antimony arsenide	SbAs	12322-34-8	196.682	hex cry	≈680		6.0		
213	Antimony potassium tartrate trihydrate	$K_2(SbC_4H_2O_6)_2 \cdot 3H_2O$	28300-74-5	667.873	col cry			2.6		sl H₂O
214	Antimony(III) acetate	$Sb(C_2H_3O_2)_3$	3643-76-3	298.891	wh pow					
215	Antimony(III) bromide	$SbBr_3$	7789-61-9	361.472	yel orth cry; hyg	97	288	4.35		reac H₂O; s ace, bz, chl
216	Antimony(III) chloride	$SbCl_3$	10025-91-9	228.119	col orth cry; hyg	73.4	220.3	3.14	987^{25}	s acid, EtOH, bz, ace
217	Antimony(III) fluoride	SbF_3	7783-56-4	178.755	wh orth cry; hyg	287	376	4.38	492^{25}	
218	Antimony(III) iodide	SbI_3	7790-44-5	502.473	red rhomb cry	171	400	4.92		reac H₂O; s EtOH, ace; i ctc
219	Antimony(III) iodide sulfide	SbIS	13816-38-1	280.729	dark red prisms or needles	400				
220	Antimony(III) oxide (senarmontite)	Sb_2O_3	1309-64-4	291.518	col cub cry	570 trans	1425	5.58		sl H₂O; i os
221	Antimony(III) oxide (valentinite)	Sb_2O_3	1309-64-4	291.518	wh orth cry	655	1425	5.7		sl H₂O; i os
222	Antimony(III) oxychloride	SbOCl	7791-08-4	173.212	wh momo cry	170 dec				reac H₂O; i EtOH, eth
223	Antimony(III) phosphate	$SbPO_4$	12036-46-3	216.731	cry pow					reac H₂O
224	Antimony(III) potassium oxalate trihydrate	$K_3Sb(C_2O_4)_3 \cdot 3H_2O$	5965-33-3*	557.158	cry pow					s H₂O
225	Antimony(III) selenide	Sb_2Se_3	1315-05-5	480.40	grn orth cry	611		5.81		sl H₂O
226	Antimony(III) sulfate	$Sb_2(SO_4)_3$	7446-32-4	531.708	wh cry powder; hyg	dec		3.62		sl H₂O
227	Antimony(III) sulfide	Sb_2S_3	1345-04-6	339.715	gray-blk orth cry	550		4.562		i H₂O; s conc HCl
228	Antimony(III) telluride	Sb_2Te_3	1327-50-0	626.32	gray cry	620		6.5		

No.	Name	Formula	CAS Reg No.	Mol. weight	Physical form	mp/°C	bp/°C	Density g cm^{-3}	Solubility g/100 g H$_2$O	Qualitative solubility
229	Antimony(III,V) oxide	Sb$_2$O$_4$	1332-81-6	307.518	yel orth cry			6.64		
230	Antimony(V) chloride	SbCl$_5$	7647-18-9	299.025	col or yel liq	4	140 dec	2.34		reac H$_2$O; s chl, ctc
231	Antimony(V) fluoride	SbF$_5$	7783-70-2	216.752	hyg visc liq	8.3	141	3.10		reac H$_2$O
232	Antimony(V) dichlorotrifluoride	SbCl$_2$F$_3$	7791-16-4	249.661	visc liq					reac H$_2$O
233	Antimony(V) oxide	Sb$_2$O$_5$	1314-60-9	323.517	yel powder; cub	dec		3.78	0.3^{20}	
234	Antimony(V) sulfide	Sb$_2$S$_5$	1315-04-4	403.845	oran-yel powder	75 dec		4.120		i H$_2$O; s acid, alk
235	Argon	Ar	7440-37-1	39.948	col gas	-189.34	-185.847	1.633 g/L		sl H$_2$O
236	Arsenic (gray)	As	7440-38-2	74.922	gray metal; rhomb	817	616 sp	5.75		i H$_2$O
237	Arsenic (black)	As	7440-38-2	74.922	blk amorp solid	trans gray As 270		4.9		
238	Arsenic (yellow)	As	7440-38-2	74.922	soft yel cub cry	trans gray As 358		1.97		s CS$_2$
239	Arsine	AsH$_3$	7784-42-1	77.946	col gas	-116	-62.5	3.186 g/L		sl H$_2$O
240	Diarsine	As$_2$H$_4$	15942-63-9	153.875	unstab liq		≈100			
241	Arsenic acid	H$_3$AsO$_4$	7778-39-4	141.944	exists only in soln					
242	Arsenic acid hemihydrate	H$_3$AsO$_4$ · 0.5H$_2$O	7778-39-4*	150.951	wh hyg cry	36.1		2.5		vs H$_2$O, EtOH
243	Arsenious acid	H$_3$AsO$_3$	13464-58-9	125.944	exists only in soln					
244	Arsenic diiodide	As$_2$I$_4$	13770-56-4	657.461	red cry	137				reac H$_2$O; s os
245	Arsenic hemiselenide	As$_2$Se	1303-35-1	228.80	blk cry					i H$_2$O, os; dec acid, alk
246	Arsenic sulfide	As$_4$S$_4$	12279-90-2	427.946	red monocl cry	320	565	3.5		i H$_2$O; sl bz; s alk
247	Arsenic(III) bromide	AsBr$_3$	7784-33-0	314.634	col or yel orth cry; hyg	31.1	221	3.40		reac H$_2$O; s hc, ctc; vs eth, bz
248	Arsenic(III) chloride	AsCl$_3$	7784-34-1	181.281	col liq	-16	130	2.150		reac H$_2$O; vs chl, ctc, eth
249	Arsenic(III) ethoxide	As(C$_2$H$_5$O)$_3$	3141-12-6	210.103	liq		166	1.21		
250	Arsenic(III) fluoride	AsF$_3$	7784-35-2	131.917	col liq	-5.9	57.13	2.7		reac H$_2$O; s EtOH, eth, bz
251	Arsenic(III) iodide	AsI$_3$	7784-45-4	455.635	red hex cry	141	424	4.73		sl H$_2$O, EtOH, eth; s bz. tol
252	Arsenic(III) oxide (arsenolite)	As$_2$O$_3$	1327-53-3	197.841	wh cub cry	274	460	3.86	2.05^{25}	
253	Arsenic(III) oxide (claudetite)	As$_2$O$_3$	1327-53-3	197.841	wh monocl cry	314	460	3.74	2.05^{25}	s dil acid, alk; i EtOH
254	Arsenic(III) selenide	As$_2$Se$_3$	1303-36-2	386.72	brn-blk solid	260		4.75		i H$_2$O; s alk
255	Arsenic(III) sulfide	As$_2$S$_3$	1303-33-9	246.038	yel-oran monocl cry	312	707	3.46		i H$_2$O; s alk
256	Arsenic(III) telluride	As$_2$Te$_3$	12044-54-1	532.64	blk monocl cry	621		6.50		
257	Arsenic(V) chloride	AsCl$_5$	22441-45-8	252.187	stab at low temp	≈-50 dec				
258	Arsenic(V) fluoride	AsF$_5$	7784-36-3	169.914	col gas	-79.8	-52.8	6.945 g/L		reac H$_2$O; s EtOH, bz, eth
259	Arsenic(V) oxide	As$_2$O$_5$	1303-28-2	229.840	wh amorp powder	315		4.32	65.8^{20}	vs EtOH
260	Arsenic(V) selenide	As$_2$Se$_5$	1303-37-3	544.64	blk solid	dec				i H$_2$O, EtOH, eth; s alk
261	Arsenic(V) sulfide	As$_2$S$_5$	1303-34-0	310.168	brn-yel amorp solid	dec				i H$_2$O; s alk
262	Astatine	At	7440-68-8	210	cry	302				s HNO$_3$, os
263	Barium	Ba	7440-39-3	137.327	silv-yel metal; cub	727	1845	3.62		reac H$_2$O; sl EtOH
264	Barium acetate	Ba(C$_2$H$_3$O$_2$)$_2$	543-80-6	255.416	wh powder			2.47	79.2^{25}	
265	Barium acetate monohydrate	Ba(C$_2$H$_3$O$_2$)$_2$ · H$_2$O	5908-64-5	273.431	wh cry	110 dec		2.19	79.2^{25}	sl EtOH
266	Barium aluminate	BaAl$_2$O$_4$	12004-04-5	255.288	hex cry	1827				
267	Barium aluminide	BaAl$_4$	12672-79-6	245.253	metallic solid	1097				
268	Barium azide	Ba(N$_3$)$_2$	18810-58-7	221.367	monocl cry; exp	≈120 dec		2.936	17.3^{20}	sl EtOH; i eth
269	Barium bismuthate	BaBiO$_3$	12785-50-1	394.305	bronze cry	1040 dec				
270	Barium bromate	Ba(BrO$_3$)$_2$	13967-90-3	393.131	col monocl cry				0.79^{25}	s ace
271	Barium bromate monohydrate	Ba(BrO$_3$)$_2$ · H$_2$O	10326-26-8	411.147	wh monocl cry	260 dec		3.99	0.831^{25}	i EtOH
272	Barium bromide	BaBr$_2$	10553-31-8	297.135	wh orth cry	857	1835	4.781	100^{25}	
273	Barium bromide dihydrate	BaBr$_2$ · 2H$_2$O	7791-28-8	333.166	wh cry	75 dec		3.7	100^{25}	s MeOH; i EtOH, ace, diox
274	Barium calcium tungstate	Ba$_2$CaWO$_6$	15552-14-4	594.57	cub cry	1420				
275	Barium carbide	BaC$_2$	50813-65-5	161.348	gray tetr cry	dec		3.74		reac H$_2$O
276	Barium carbonate	BaCO$_3$	513-77-9	197.336	wh orth cry	1380 dec; 1555 (high pres.)		4.308	0.0014^{20}	s acid
277	Barium chlorate	Ba(ClO$_3$)$_2$	13477-00-4	304.229	wh cry	414			37.9^{25}	sl EtOH, ace
278	Barium chlorate monohydrate	Ba(ClO$_3$)$_2$ · H$_2$O	10294-38-9	322.245	wh monocl cry	120 dec		3.179	37.9^{25}	s acid; sl EtOH, ace
279	Barium chloride	BaCl$_2$	10361-37-2	208.233	wh orth cry; hyg	961	1560	3.9	37.0^{25}	
280	Barium chloride dihydrate	BaCl$_2$ · 2H$_2$O	10326-27-9	244.264	wh monocl cry	≈120 dec		3.097	37.0^{25}	i EtOH
281	Barium chloride fluoride	BaClF	13718-55-3	191.778	wh cry					
282	Barium chromate(V)	Ba$_3$(CrO$_4$)$_2$	12345-14-1	643.968	grn-blk hex cry			5.25		s H$_2$O
283	Barium chromate(VI)	BaCrO$_4$	10294-40-3	253.321	yel orth cry	1380		4.50	0.00026^{20}	reac acid

No.	Name	Formula	CAS Reg No.	Mol. weight	Physical form	mp/°C	bp/°C	Density g cm^{-3}	Solubility g/100 g H$_2$O	Qualitative solubility
284	Barium citrate monohydrate	Ba$_3$(C$_6$H$_5$O$_7$)$_2$ · H$_2$O	512-25-4*	808.195	gray-wh cry					s H$_2$O, acid
285	Barium copper yttrium oxide	BaCuY$_2$O$_5$	82642-06-6	458.682	grn cry; not superconductor					
286	Barium copper yttrium oxide	Ba$_2$Cu$_3$YO$_7$	109064-29-1	666.194	blk solid; HT superconductor					
287	Barium copper yttrium oxide	Ba$_2$Cu$_4$YO$_8$	114104-80-2	745.739	HT superconductor					
288	Barium copper yttrium oxide	Ba$_4$Cu$_7$Y$_2$O$_{15}$	124365-83-9	1411.933	HT superconductor					
289	Barium cyanide	Ba(CN)$_2$	542-62-1	189.361	wh cry powder					vs H$_2$O; s EtOH
290	Barium dichromate dihydrate	BaCr$_2$O$_7$ · 2H$_2$O	10031-16-0	389.346	brn-red needles	dec				reac H$_2$O
291	Barium disilicate	BaSi$_2$O$_5$	12650-28-1	273.495	wh orth cry	1420		3.70		
292	Barium dithionate dihydrate	BaS$_2$O$_6$ · 2H$_2$O	13845-17-5	333.484	wh cry	140 dec		4.54	22.1^{20}	sl EtOH
293	Barium ferrite	BaFe$_{12}$O$_{19}$	11138-11-7	1111.456	magnetic solid					
294	Barium ferrocyanide hexahydrate	Ba$_2$Fe(CN)$_6$ · 6H$_2$O	13821-06-2*	594.694	yel monocl cry	80 dec				i H$_2$O, EtOH
295	Barium fluoride	BaF$_2$	7787-32-8	175.324	wh cub cry	1368	2260	4.893	0.161^{25}	
296	Barium formate	Ba(CHO$_2$)$_2$	541-43-5	227.362	cry			3.21		s H$_2$O; i EtOH
297	Barium hexaboride	BaB$_6$	12046-08-1	202.193	blk cub cry	2070		4.36		i H$_2$O; s acid; i EtOH
298	Barium hexafluorogermanate	BaGeF$_6$		323.96	wh cry	≈665		4.56		
299	Barium hexafluorosilicate	BaSiF$_6$	17125-80-3	279.403	wh orth needles	300 dec		4.29		i H$_2$O, EtOH; sl acid
300	Barium hydride	BaH$_2$	13477-09-3	139.343	gray orth cry	1200		4.16		reac H$_2$O
301	Barium hydrogen phosphate	BaHPO$_4$	10048-98-3	233.306	wh cry powder	400 dec		4.16	0.015^{20}	s dil acid
302	Barium hydrosulfide	Ba(HS)$_2$	25417-81-6	203.473	yel hyg cry					s H$_2$O
303	Barium hydrosulfide tetrahydrate	Ba(HS)$_2$ · 4H$_2$O	12230-74-9	275.534	yel rhomb cry	50 dec				s H$_2$O
304	Barium hydroxide	Ba(OH)$_2$	17194-00-2	171.342	wh powder	408			4.91^{25}	
305	Barium hydroxide monohydrate	Ba(OH)$_2$ · H$_2$O	22326-55-2	189.357	wh powder			3.743	4.91^{25}	s acid
306	Barium hydroxide octahydrate	Ba(OH)$_2$ · 8H$_2$O	12230-71-6	315.464	wh monocl cry	78 dec		2.18	4.91^{25}	
307	Barium hypophosphite monohydrate	Ba(H$_2$PO$_2$)$_2$ · H$_2$O	14871-79-5*	285.320	monocl plates			2.90		s H$_2$O; i EtOH
308	Barium iodate	Ba(IO$_3$)$_2$	10567-69-8	487.132	wh cry powder	476 dec		5.23	0.0396^{25}	
309	Barium iodate monohydrate	Ba(IO$_3$)$_2$ · H$_2$O	7787-34-0	505.148	cry	130 dec		5.00	0.0396^{25}	s acid; i EtOH
310	Barium iodide	BaI$_2$	13718-50-8	391.136	wh orth cry	711		5.15	221^{25}	
311	Barium iodide dihydrate	BaI$_2$ · 2H$_2$O	7787-33-9	427.167	col cry	740 dec		5.0	221^{25}	s EtOH, ace
312	Barium manganate(VI)	BaMnO$_4$	7787-35-1	256.263	grn-gray hyg cry			4.85	0.00041^{20}	
313	Barium metaborate monohydrate	Ba(BO$_2$)$_2$ · H$_2$O	26124-86-7	240.962	wh powder	>900		3.3		sl H$_2$O
314	Barium metaborate dihydrate	Ba(BO$_2$)$_2$ · 2H$_2$O	23436-05-7	258.977	wh prec	dec			1.3^{25}	sl H$_2$O
315	Barium metaphosphate	Ba(PO$_3$)$_2$	13466-20-1	295.271	wh powder	1560				i H$_2$O; sl acid
316	Barium metasilicate	BaSiO$_3$	13255-26-0	213.411	col rhomb powder	1605		4.40		i H$_2$O; s acid
317	Barium molybdate	BaMoO$_4$	7787-37-3	297.27	wh powder	1450		4.975	0.0021^{20}	
318	Barium niobate	Ba(NbO$_3$)$_2$	12009-14-2	419.136	yel orth cry	1455		5.44		i H$_2$O
319	Barium nitrate	Ba(NO$_3$)$_2$	10022-31-8	261.336	wh cub cry	590		3.24	10.3^{25}	sl EtOH, ace
320	Barium nitride	Ba$_3$N$_2$	12047-79-9	439.994	yel-brn cry	>500 dec		4.78		reac H$_2$O
321	Barium nitrite	Ba(NO$_2$)$_2$	13465-94-6	229.338	col hex cry	267		3.234	79.5^{25}	
322	Barium nitrite monohydrate	Ba(NO$_2$)$_2$ · H$_2$O	7787-38-4	247.353	yel-wh hex cry	217 dec		3.18	79.5^{25}	i EtOH
323	Barium orthovanadate	Ba$_3$(VO$_4$)$_2$	39416-30-3	641.859	hex cry	707		5.14		
324	Barium oxalate	BaC$_2$O$_4$	516-02-9	225.346	wh powder	400 dec		2.658	0.0075	
325	Barium oxalate monohydrate	BaC$_2$O$_4$ · H$_2$O	13463-22-4	243.361	wh cry powder			2.66	0.0075^{20}	s acid
326	Barium oxide	BaO	1304-28-5	153.326	wh-yel powder; cub and hex	1973		5.72(cub)	1.5^{20}	s dil acid, EtOH; i ace
327	Barium 2,4-pentanedioate octahydrate	Ba(CH$_3$COCHCOCH$_3$)$_2$ · 8H$_2$O	12084-29-6*	479.665	col hyg cry	320 (anh)				
328	Barium perchlorate	Ba(ClO$_4$)$_2$	13465-95-7	336.228	col hex cry	505		3.20	312^{25}	vs EtOH
329	Barium perchlorate trihydrate	Ba(ClO$_4$)$_2$ · 3H$_2$O	10294-39-0	390.274	col cry			2.74	312^{25}	s MeOH; sl EtOH, ace; i eth
330	Barium permanganate	Ba(MnO$_4$)$_2$	7787-36-2	375.198	brn-viol cry	200 dec		3.77	62.5^{20}	reac EtOH
331	Barium peroxide	BaO$_2$	1304-29-6	169.326	gray-wh tetr cry	450 dec		4.96	0.091^{20}	reac dil acid
332	Barium plumbate	BaPbO$_3$	12047-25-5	392.5	orth cry					
333	Barium potassium chromate	BaK$_2$(CrO$_4$)$_2$	27133-66-0	447.511	yel hex cry			3.63		vs H$_2$O
334	Barium pyrophosphate	Ba$_2$P$_2$O$_7$	13466-21-2	448.597	wh powder	1430		3.9	0.0088^{20}	s acid
335	Barium selenate	BaSeO$_4$	7787-41-9	280.29	wh rhomb cry	dec		4.75	0.015^{20}	
336	Barium selenide	BaSe	1304-39-8	216.29	cub cry powder	1780		5.02		reac H$_2$O
337	Barium selenite	BaSeO$_3$	13718-59-7	264.29	solid					i H$_2$O
338	Barium silicide	BaSi$_2$	1304-40-1	193.498	gray lumps	1180				reac H$_2$O
339	Barium sodium niobate	Ba$_2$Na(NbO$_3$)$_5$	12323-03-4	1002.167	wh orth cry	1437		5.40		i H$_2$O
340	Barium stannate	BaSnO$_3$	12009-18-6	304.035	cub cry			7.24		sl H$_2$O
341	Barium stannate trihydrate	BaSnO$_3$ · 3H$_2$O	12009-18-6*	358.081	wh cry powder					sl H$_2$O; s acid
342	Barium stearate	Ba(C$_{18}$H$_{35}$O$_2$)$_2$	6865-35-6	704.266	wh powder	160		1.145		i H$_2$O, EtOH

No.	Name	Formula	CAS Reg No.	Mol. weight	Physical form	mp/°C	bp/°C	Density g cm^{-3}	Solubility g/100 g H$_2$O	Qualitative solubility
343	Barium strontium niobate	BaSr(NbO$_3$)$_4$	37185-09-4	788.57	pale yel solid					
344	Barium strontium tungstate	Ba$_2$SrWO$_6$	14871-56-8	642.11	hyg pow	1400				
345	Barium sulfate	BaSO$_4$	7727-43-7	233.390	wh orth cry	1580		4.49	0.00031[20]	i EtOH
346	Barium sulfide	BaS	21109-95-5	169.392	col cub cry or gray powder	2227		4.3	8.94[25]	
347	Barium sulfite	BaSO$_3$	7787-39-5	217.390	wh monocl cry	dec		4.44	0.0011[25]	i EtOH
348	Barium tartrate	BaC$_4$H$_4$O$_6$	5908-81-6	285.398	wh cry			2.98		s H$_2$O; i EtOH
349	Barium tetracyanoplatinate(II) tetrahydrate	BaPt(CN)$_4$ · 4H$_2$O	13755-32-3	508.543	yel powder or cry			2.076		sl H$_2$O; i EtOH
350	Barium tetraiodomercurate(II)	BaHgI$_4$	10048-99-4	845.54	yel-red hyg cry					vs H$_2$O, EtOH
351	Barium thiocyanate	Ba(SCN)$_2$	2092-17-3	253.491	hyg cry				167[25]	s ace, MeOH, EtOH
352	Barium thiocyanate dihydrate	Ba(SCN)$_2$ · 2H$_2$O	2092-17-3*	289.522	hyg wh cry				167[25]	s EtOH
353	Barium thiocyanate trihydrate	Ba(SCN)$_2$ · 3H$_2$O	68016-36-4	307.537	wh needles; hyg			2.286	167[25]	s EtOH
354	Barium thiosulfate	BaS$_2$O$_3$	35112-53-9	249.455	wh cry powder	220 dec			0.2[20]	i EtOH
355	Barium thiosulfate monohydrate	BaS$_2$O$_3$ · H$_2$O	7787-40-8	267.471	wh cry powder	dec		3.5	0.2	i EtOH
356	Barium titanate (BaTiO$_3$)	BaTiO$_3$	12047-27-7	233.192	wh tetr cry	1625		6.02		i H$_2$O
357	Barium titanate (BaTi$_2$O$_5$)	BaTi$_2$O$_5$	12009-27-7	313.058	wh solid					
358	Barium titanate (BaTi$_4$O$_9$)	BaTi$_4$O$_9$	12009-31-3	472.790	wh solid					
359	Barium titanium silicate	BaTi(SiO$_3$)$_3$	15491-35-7	413.446	rhom blue-pur cry					
360	Barium tungstate	BaWO$_4$	7787-42-0	385.17	wh tetr cry	1475	1730	5.04	0.0016[20]	
361	Barium uranium oxide	BaU$_2$O$_7$	10380-31-1	725.381	oran-yel powder					i H$_2$O; s acid
362	Barium yttrium tungsten oxide	Ba$_3$Y$_2$WO$_9$	37265-86-4	1006.53	cub cry	1470				
363	Barium zirconate	BaZrO$_3$	12009-21-1	276.549	gray-wh cub cry	2500		5.52		i H$_2$O, alk; sl acid
364	Barium zirconium silicate	BaO · ZrO$_2$ · SiO$_2$		336.634	wh pow					i H2O, alk; sl acid; s HF
365	Berkelium (α form)	Bk	7440-40-6	247	hex cry	trans to · 930		14.78		
366	Berkelium (β form)	Bk	7440-40-6	247	cub cry	986		13.25		
367	Beryllium	Be	7440-41-7	9.012	hex cry	1287	2468	1.85		s acid, alk
368	Beryllium acetate	Be(C$_2$H$_3$O$_2$)$_2$	543-81-7	127.101	wh cry	60 dec				i H$_2$O, EtOH
369	Beryllium basic acetate	Be$_4$O(C$_2$H$_3$O$_2$)$_6$	1332-52-1	406.312	wh cry	285	330	1.25		i H$_2$O; s eth, os
370	Beryllium aluminate	BeAl$_2$O$_4$	12004-06-7	126.973	orth cry			3.65		
371	Beryllium aluminum metasilicate	Be$_3$Al$_2$(SiO$_3$)$_6$	1302-52-9	537.502	col or grn-yel cry; hex			2.64		
372	Beryllium boride (BeB$_2$)	BeB$_2$	12228-40-9	30.634	refrac solid	>1970				
373	Beryllium boride (BeB$_6$)	BeB$_6$	12429-94-6	73.878	red solid	2070				
374	Beryllium boride (Be$_2$B)	Be$_2$B	12536-51-5	28.835	pink cry	1520				
375	Beryllium boride (Be$_4$B)	Be$_4$B	12536-52-6	46.860	refrac solid	1160				
376	Beryllium borohydride	Be(BH$_4$)$_2$	17440-85-6	36.682	solid	125 dec	subl			reac H$_2$O
377	Beryllium bromide	BeBr$_2$	7787-46-4	168.820	orth cry; hyg	508	473 sp	3.465		vs H$_2$O; s EtOH, pyr
378	Beryllium carbide	Be$_2$C	506-66-1	30.035	red cub cry	2127		1.90		reac H$_2$O
379	Beryllium carbonate tetrahydrate	BeCO$_3$ · 4H$_2$O	60883-64-9	93.085	wh solid	100 dec			0.36[0]	
380	Beryllium basic carbonate	Be$_3$(OH)$_2$(CO$_3$)$_2$	66104-24-3	181.069	wh powder					i H$_2$O; s acid, alk
381	Beryllium chloride	BeCl$_2$	7787-47-5	79.918	wh-yel orth cry; hyg	415	482	1.90	71.5[25]	s EtOH, eth, py; i bz, tol
382	Beryllium fluoride	BeF$_2$	7787-49-7	47.009	tetr cry or gl; hyg	552	1283	2.1		vs H$_2$O; sl EtOH
383	Beryllium formate	Be(CHO$_2$)$_2$	1111-71-3	99.047	powder	>250 dec				reac H$_2$O; i os
384	Beryllium hydride	BeH$_2$	7787-52-2	11.028	wh amorp solid	250 dec		0.65		reac H$_2$O; i eth, tol
385	Beryllium hydrogen phosphate	BeHPO$_4$	13598-15-7	104.991	cry					i H$_2$O
386	Beryllium hydroxide (α)	Be(OH)$_2$	13327-32-7	43.027	wh powder or cry	≈200 dec		1.92		sl H$_2$O, alk; s acid
387	Beryllium hydroxide (β)	Be(OH)$_2$	13327-32-7	43.027	col tetr cry	dec 138				i H$_2$O; s, acid, alk
388	Beryllium iodide	BeI$_2$	7787-53-3	262.821	hyg needles	480	590	4.32		reac H$_2$O; s EtOH
389	Beryllium nitrate trihydrate	Be(NO$_3$)$_2$ · 3H$_2$O	13597-99-4	187.068	yel-wh hyg cry	≈30	dec		107[20]	s EtOH
390	Beryllium nitride	Be$_3$N$_2$	1304-54-7	55.050	gray refrac cry; cub	2200		2.71		reac acid, alk
391	Beryllium oxalate trihydrate	BeC$_2$O$_4$ · 3H$_2$O	15771-43-4	151.077	rhom cry	dec 320				vs H$_2$O
392	Beryllium oxide	BeO	1304-56-9	25.011	wh hex cry	2578		3.01		i H$_2$O; sl acid, alk
393	Beryllium 2,4-pentanedioate	Be(CH$_3$COCHCOCH$_3$)$_2$	10210-64-7	207.228	monocl cry powder	108	270	1.168		i H$_2$O; vs EtOH, eth
394	Beryllium perchlorate tetrahydrate	Be(ClO$_4$)$_2$ · 4H$_2$O	7787-48-6	279.975	hyg cry	250 dec			198[25]	
395	Beryllium selenate tetrahydrate	BeSeO$_4$ · 4H$_2$O	10039-31-3	224.03	orth cry	100 dec		2.03		vs H$_2$O
396	Beryllium sulfate	BeSO$_4$	13510-49-1	105.075	col tetr cry; hyg	1127		2.5	41.3[25]	
397	Beryllium sulfate dihydrate	BeSO$_4$ · 2H$_2$O	14215-00-0	141.105	col cry	dec 92				
398	Beryllium sulfate tetrahydrate	BeSO$_4$ · 4H$_2$O	7787-56-6	177.136	col tetr cry	≈100 dec		1.71	41.3[25]	i EtOH
399	Beryllium sulfide	BeS	13598-22-6	41.077	col cub cry	dec		2.36		reac hot H$_2$O
400	Bismuth	Bi	7440-69-9	208.980	gray-wh soft metal	271.406	1564	9.79		s acid
401	Bismuth acetate	Bi(C$_2$H$_3$O$_2$)$_3$	22306-37-2	386.111	col tablets	250				i H$_2$O
402	Bismuth subacetate	BiOC$_2$H$_3$O$_2$	5142-76-7	284.023	thin cry plates					i H$_2$O; s dil acid
403	Bismuth antimonide	BiSb	12323-19-2	330.740	cry	475				

No.	Name	Formula	CAS Reg No.	Mol. weight	Physical form	mp/°C	bp/°C	Density g cm⁻³	Solubility g/100 g H₂O	Qualitative solubility
404	Bismuth arsenate	$BiAsO_4$	13702-38-0	347.900	wh monocl cry			7.14		i H_2O; sl conc HNO_3
405	Bismuth basic carbonate	$(BiO)_2CO_3$	5892-10-4	509.969	wh powder			6.86		i H_2O; s acid
406	Bismuth basic dichromate	$Bi_2O_3 \cdot 2CrO_3$		665.948	red-oran amorp pow					i H_2O; s acid, alk
407	Bismuth citrate	$BiC_6H_5O_7$	813-93-4	398.080	wh powder			3.458		i H_2O; sl EtOH
408	Bismuth hydride	BiH_3	18288-22-7	212.004	col gas; unstab	-67	≈17	8.665 g/L		
409	Bismuth hydroxide	$Bi(OH)_3$	10361-43-0	260.002	wh-yel amorp powder			4.962		i H_2O; s acid
410	Bismuth germanium oxide	$2Bi_2O_3 \cdot 3GeO_2$	12233-56-6	1245.84	wh pow	1044				
411	Bismuth hexafluoro-2,4-pentanedioate	$Bi(CF_3COCHCOCF_3)_3$	142617-56-9	830.132	powder	96				
412	Bismuth molybdate	Bi_2MoO_6	13565-96-3	609.90	yel solid			9.32		
413	Bismuth molybdate	$Bi_2(MoO_4)_3$	51898-99-8	897.77	monocl cry			5.95		
414	Bismuth nitrate pentahydrate	$Bi(NO_3)_3 \cdot 5H_2O$	10035-06-0	485.071	col tricl cry; hyg	≈75 dec		2.83		reac H_2O; s ace; i EtOH
415	Bismuth subnitrate	$Bi_5O(OH)_9(NO_3)_4$	1304-85-4	1461.987	hyg cry powder	260 dec		4.928		i H_2O, EtOH; s dil acid
416	Bismuth oleate	$Bi(C_{18}H_{33}O_2)_3$	52951-38-9	1053.340	soft yel-brn solid					i H_2O; s eth; sl bz
417	Bismuth oxalate	$Bi_2(C_2O_4)_3$	6591-55-5	682.018	wh powder					i H_2O, EtOH; s dil acid
418	Bismuth oxide	Bi_2O_3	1304-76-3	465.959	yel monocl cry or powder	825	1890	8.9		i H_2O; s acid
419	Bismuth tetroxide	Bi_2O_4	12048-50-9	481.959	red-oran powder	305		5.6		reac H_2O
420	Bismuth oxybromide	$BiOBr$	7787-57-7	304.883	col tetr cry	560 dec		8.08		i H_2O, EtOH; s acid
421	Bismuth oxychloride	$BiOCl$	7787-59-9	260.432	wh tetr cry	575 dec		7.72		i H_2O
422	Bismuth oxyiodide	$BiOI$	7787-63-5	351.883	red tetr cry	300 dec		7.92		i H_2O, EtOH, chl; s HCl
423	Bismuth oxynitrate	$BiONO_3$	10361-46-3	286.985	wh powder	260 dec		4.93		i H_2O, EtOH; s acid
424	Bismuth phosphate	$BiPO_4$	10049-01-1	303.951	monocl cry			6.32		sl H_2O, dil acid; i EtOH
425	Bismuth potassium iodide	BiK_4I_7	41944-01-8	1253.704	red cry					reac H_2O; s alk iodide soln
426	Bismuth selenide	Bi_2Se_3	12068-69-8	654.84	blk hex cry	710 dec		7.5		i H_2O
427	Bismuth stannate pentahydrate	$Bi_2(SnO_3)_3 \cdot 5H_2O$	12777-45-6	1008.162	wh cry					i H_2O
428	Bismuth sulfate	$Bi_2(SO_4)_3$	7787-68-0	706.149	wh needles or powder	405 dec		5.08		reac H_2O, EtOH
429	Bismuth sulfide	Bi_2S_3	1345-07-9	514.156	blk-brn orth cry	850		6.78		i H_2O; s acid
430	Bismuth telluride	Bi_2Te_3	1304-82-1	800.76	gray hex plates	580		7.74		i H_2O; s EtOH
431	Bismuth tribromide	$BiBr_3$	7787-58-8	448.692	yel cub cry	219	462	5.72		reac H_2O; s dil acid, ace; i EtOH
432	Bismuth trichloride	$BiCl_3$	7787-60-2	315.339	col or yel cub cry; hyg	234	441	4.75		reac H_2O; s acid, EtOH, ace
433	Bismuth trifluoride	BiF_3	7787-61-3	265.975	wh-gray cub cry	727	900	8.3		i H_2O
434	Bismuth pentafluoride	BiF_5	7787-62-4	303.972	wh tetr needles; hyg	151.4	230	5.55		reac H_2O
435	Bismuth triiodide	BiI_3	7787-64-6	589.693	blk-brn hex cry	408.6	542	5.778	0.00078[20]	s EtOH
436	Bismuth trimethyl	$Bi(CH_3)_3$	593-91-9	254.083	col flam liq	-86	110	2.3		
437	Bismuth titanate	$Bi_4(TiO_4)_3$	12048-51-0	1171.516	wh orth cry			7.85		
438	Bismuth tungstate	$Bi_2(WO_4)_3$	13595-87-4	1161.47	wh powder					
439	Bismuth vanadate	$BiVO_4$	14059-33-7	323.920	orth cry	trans 500		6.25		i H_2O; s acid
440	Bismuth zirconate	$2Bi_2O_3 \cdot 3ZrO_2$	37306-42-6	1301.587	wh pow					
441	Boron	B	7440-42-8	10.81	blk rhomb cry	2077	4000	2.34		i H_2O
442	Diborane	B_2H_6	19287-45-7	27.670	col gas; flam	-164.85	-92.49	1.131 g/L		reac H_2O
443	Tetraborane(10)	B_4H_{10}	18283-93-7	53.323	unstab col gas	-120	18	2.180 g/L		reac H_2O
444	Pentaborane(9)	B_5H_9	19624-22-7	63.126	flam col liq	-46.74	60.10	0.60		reac hot H_2O
445	Pentaborane(11)	B_5H_{11}	18433-84-6	65.142	col liq; unstab	-122	65			reac H_2O
446	Hexaborane(10)	B_6H_{10}	23777-80-2	74.945	col liq	-62.3	108 dec	0.67		reac hot H_2O
447	Hexaborane(12)	B_6H_{12}	12008-19-4	76.961	col liq	-82.3	≈85			reac H_2O
448	Nonaborane(15)	B_9H_{15}	19465-30-6	112.418	col liq			2.7		
449	Decaborane(14)	$B_{10}H_{14}$	17702-41-9	122.221	wh orth cry	98.78	213	0.94		sl H_2O; s EtOH, bz, CS_2, ctc
450	Decaborane(16)	$B_{10}H_{16}$	71595-75-0	124.237	col cry	≈81	dec 170	subl		
451	Dodecaborane(16)	$B_{12}H_{16}$	89711-39-7	145.859	col cry	65				s bz, hx
452	Tridecaborane(19)	$B_{13}H_{19}$	43093-20-5	159.694	yel cry	44				s hx, CH2Cl2
453	Tetradecaborane(18)	$B_{14}H_{18}$	55606-55-8	169.497	visc yel oil		dec 100			s cyhex, CS_2
454	Hexadecaborane(20)	$B_{16}H_{20}$	28265-11-4	193.135	col cry	≈110				s ctc, cyhex, thf
455	Octadecaborane(22)	$B_{18}H_{22}$	11071-61-7	216.773	yel cry	180				s os
456	Tetrabromodiborane	B_2Br_4	14355-29-4	341.238	col liq	≈1	dec 20			
457	Tetrachlorodiborane	B_2Cl_4	13701-67-2	163.434	col liq; flam	-92.6	66.5			reac H_2O

No.	Name	Formula	CAS Reg No.	Mol. weight	Physical form	mp/°C	bp/°C	Density g cm^{-3}	Solubility g/100 g H$_2$O	Qualitative solubility
458	Tetrafluorodiborane	B$_2$F$_4$	13965-73-6	97.616	col gas; flam	-56	-34.0	3.990 g/L		reac H$_2$O
459	Borane carbonyl	BH$_3$CO	13205-44-2	41.845	col gas	-137	-64	1.710 g/L		reac H$_2$O
460	Borazine	B$_3$N$_3$H$_6$	6569-51-3	80.501	col liq	-58	53	0.824		reac H$_2$O
461	Boric acid	H$_3$BO$_3$	10043-35-3	61.833	col tricl cry	170.9		1.5	5.80^{25}	sl EtOH
462	Metaboric acid (α form)	HBO$_2$	13460-50-9	43.818	col orth cry; hyg	176		1.784		s H$_2$O
463	Metaboric acid (β form)	HBO$_2$	13460-50-9	43.818	col monocl cry; hyg	201		2.045		s H$_2$O
464	Metaboric acid (γ form)	HBO$_2$	13460-50-9	43.818	col cub cry	236		2.487		s H$_2$O
465	Tetrafluoroboric acid	HBF$_4$	16872-11-0	87.813	col liq		130 dec	≈1.8		vs H$_2$O, EtOH
466	Boron arsenide	BAs	12005-69-5	85.733	brn cub cry	1100 dec		5.22		
467	Boron carbide	B$_4$C	12069-32-8	55.255	hard blk cry	2350	>3500	2.50		i H$_2$O, acid
468	Boron nitride	BN	10043-11-5	24.818	wh powder; hex or cub cry	2967		2.18		i H$_2$O, acid
469	Boron oxide	B$_2$O$_3$	1303-86-2	69.620	col gl or hex cry; hyg	450		2.55	2.2^{20}	s EtOH
470	Boron phosphide	BP	20205-91-8	41.785	red cub cry or powder	1125 dec				reac H$_2$O, acid
471	Boron silicide	B$_6$Si	12008-29-6	92.952	blk cry	1980				
472	Boron sulfide	B$_2$S$_3$	12007-33-9	117.817	yel amorp solid	563		≈1.7		
473	Boron tribromide	BBr$_3$	10294-33-4	250.523	col liq; hyg	-46	91.3	2.6		reac H$_2$O, EtOH
474	Boron trichloride	BCl$_3$	10294-34-5	117.170	col liq or gas	-107.3	12.5	4.789 g/L		reac H$_2$O, EtOH
475	Boron trifluoride	BF$_3$	7637-07-2	67.806	col gas	-126.8	-99.9	2.772 g/L		s H$_2$O
476	Boron trifluoride etherate	BF$_3$(C$_2$H$_5$)$_2$O	109-63-7	141.927	liq	-60.4	125.5	1.125^{25}		reac H$_2$O; vs eth, EtOH
477	Boron triiodide	BI$_3$	13517-10-7	391.524	wh needles	49.7	209.5	3.35		i H$_2$O
478	Bromine	Br$_2$	7726-95-6	159.808	red liq	-7.2	58.8	3.1028		sl H$_2$O
479	Bromic acid	HBrO$_3$	7789-31-3	128.910	stab only in aq soln					s H$_2$O
480	Hypobromous acid	HOBr	13517-11-8	96.911	exists aq soln					s H$_2$O
481	Bromine dioxide	BrO$_2$	21255-83-4	111.903	unstab yel cry	≈0 dec				
482	Bromine monoxide	Br$_2$O	21308-80-5	175.807	unstab brn solid	-17.5 dec				
483	Dibromine trioxide	Br$_2$O$_3$	53809-75-9	207.806	oran needles (LT)	dec -40				
484	Dibromine pentoxide	Br$_2$O$_5$	58572-43-3	239.805	col cry (low temp)	-20 dec				
485	Bromine azide	BrN$_3$	13973-87-0	121.924	red cry; exp	≈45	exp			
486	Bromine chloride	BrCl	13863-41-7	115.357	dark red liq (<5°C)	-66	5 dec			reac H$_2$O; s eth, CS$_2$
487	Bromine fluoride	BrF	13863-59-7	98.902	unstab red-brn gas	≈-33	≈20 dec	4.043 g/L		
488	Bromine trifluoride	BrF$_3$	7787-71-5	136.899	col hyg liq	8.77	125.8	2.803		reac H$_2$O
489	Bromine pentafluoride	BrF$_5$	7789-30-2	174.896	col liq	-60.5	41.3	2.460		reac H$_2$O (exp)
490	Bromosyl trifluoride	BrOF$_3$	61519-37-7	152.898	col liq	-5	dec >20			reac H$_2$O
491	Bromyl fluoride	BrO$_2$F	22585-64-4	130.901	col liq	-9	dec 55			reac H$_2$O
492	Perbromyl fluoride	BrO$_3$F	37265-91-1	146.900	col gas	-110	dec 20			reac H$_2$O
493	Cadmium	Cd	7440-43-9	112.411	silv-wh metal	321.069	767	8.69		i H$_2$O; reac acid
494	Cadmium acetate	Cd(C$_2$H$_3$O$_2$)$_2$	543-90-8	230.500	col cry	255		2.34		s H$_2$O, EtOH
495	Cadmium acetate dihydrate	Cd(C$_2$H$_3$O$_2$)$_2$ · 2H$_2$O	5743-04-4	266.529	wh cry	130 dec		2.01		vs H$_2$O; s EtOH
496	Cadmium antimonide	CdSb	12014-29-8	234.171	orth cry	456		6.92		
497	Cadmium arsenide	Cd$_3$As$_2$	12006-15-4	487.076	gray tetr cry	721		6.25		
498	Cadmium azide	Cd(N$_3$)$_2$	14215-29-3	196.451	yel-wh orth cry; exp	exp		3.24		
499	Cadmium borotungstate octadecahydrate	Cd$_5$(BW$_{12}$O$_{40}$) · 18H$_2$O	1306-26-9	3743.20	yel cry					vs H$_2$O
500	Cadmium bromide	CdBr$_2$	7789-42-6	272.219	wh-yel hex cry; hyg	568	863	5.19	115^{25}	sl ace, eth
501	Cadmium bromide tetrahydrate	CdBr$_2$ · 4H$_2$O	13464-92-1	344.281	wh-yel cry				115^{25}	s ace, EtOH
502	Cadmium carbonate	CdCO$_3$	513-78-0	172.420	wh hex cry	500 dec		5.026		i H$_2$O; s acid
503	Cadmium chlorate dihydrate	Cd(ClO$_3$)$_2$ · 2H$_2$O	22750-54-5*	315.344	col hyg cry	80 dec		2.28	2.64^0	
504	Cadmium chloride	CdCl$_2$	10108-64-2	183.317	rhom cry; hyg	568	964	4.08	120^{25}	s ace; sl EtOH; i eth
505	Cadmium chloride monohydrate	CdCl$_2$ · H$_2$O	34330-64-8	201.332	wh cry				120^{25}	
506	Cadmium chloride hemipentahydrate	CdCl$_2$ · 2.5H$_2$O	7790-78-5	228.354	wh rhomb leaflets			3.327	120^{25}	s ace
507	Cadmium chromate	CdCrO$_4$	14312-00-6	228.405	yel orth cry			4.5		i H$_2$O
508	Cadmium cyanide	Cd(CN)$_2$	542-83-6	164.445	wh cub cry			2.23	1.7^{15}	
509	Cadmium dichromate monohydrate	CdCr$_2$O$_7$ · H$_2$O	69239-51-6	346.414	oran solid					s H$_2$O
510	Cadmium 2-ethylhexanoate	Cd(C$_8$H$_{15}$O$_2$)$_2$	2420-98-6	398.818	powder					
511	Cadmium fluoride	CdF$_2$	7790-79-6	150.408	cub cry	1075	1750	6.33	4.36^{25}	s acid; i EtOH
512	Cadmium hydroxide	Cd(OH)$_2$	21041-95-2	146.426	wh trig or hex cry	130 dec		4.79	0.00015^{20}	s dil acid
513	Cadmium iodate	Cd(IO$_3$)$_2$	7790-81-0	462.216	wh powder			6.48	0.091^{25}	s HNO$_3$

No.	Name	Formula	CAS Reg No.	Mol. weight	Physical form	mp/°C	bp/°C	Density g cm⁻³	Solubility g/100 g H₂O	Qualitative solubility
514	Cadmium iodide	CdI_2	7790-80-9	366.220	col hex flakes	388	744	5.64	86.2[25]	vs H₂O; s EtOH, eth, ace
515	Cadmium metasilicate	$CdSiO_3$	13477-19-5	188.495	grn monocl cry	1252		5.10		
516	Cadmium molybdate	$CdMoO_4$	13972-68-4	272.35	col tetr cry	≈900 dec		5.4		i H₂O; s acid
517	Cadmium niobate	$Cd_2Nb_2O_7$	12187-14-3	522.631	cub cry	≈1410		6.28		i H₂O
518	Cadmium nitrate	$Cd(NO_3)_2$	10325-94-7	236.420	wh cub cry; hyg	360		3.6	156[25]	s EtOH
519	Cadmium nitrate tetrahydrate	$Cd(NO_3)_2 \cdot 4H_2O$	10022-68-1	308.482	col orth cry; hyg	59.5		2.45	156[25]	s EtOH, ace
520	Cadmium oxalate	CdC_2O_4	814-88-0	200.430	wh solid			3.32	0.0060[25]	
521	Cadmium oxalate trihydrate	$CdC_2O_4 \cdot 3H_2O$	20712-42-9	254.476	wh amorp powder	340 dec			0.0060[25]	i EtOH; s dil acid
522	Cadmium oxide	CdO	1306-19-0	128.410	brn cub cry		1559 sp	8.15		i H₂O; s dil acid
523	Cadmium 2,4-pentanedioate	$Cd(CH_3COCHCOCH_3)_2$	14689-45-3	310.627	wh solid or red cry	235				
524	Cadmium perchlorate hexahydrate	$Cd(ClO_4)_2 \cdot 6H_2O$	10326-28-0	419.404	wh hex cry			2.37	191.5[25]	
525	Cadmium phosphate	$Cd_3(PO_4)_2$	13477-17-3	527.176	powder	≈1500				i H₂O
526	Cadmium phosphide	Cd_3P_2	12014-28-7	399.181	grn tetr needles	700		5.96		s dil HCl
527	Cadmium selenate dihydrate	$CdSeO_4 \cdot 2H_2O$	10060-09-0	291.40	orth cry	100 dec		3.62	70.5[25]	
528	Cadmium selenide	$CdSe$	1306-24-7	191.37	wh cub cry	1240		5.81		i H₂O
529	Cadmium selenite	$CdSeO_3$	13814-59-0	239.37	col prisms					
530	Cadmium stearate	$Cd(C_{18}H_{35}O_2)_2$	2223-93-0	679.350	wh cry pow	134		1.21		
531	Cadmium succinate	$CdC_4H_4O_4$	141-00-4	228.484	wh pow or needles				0.37[40]	sl H₂O; i EtOH
532	Cadmium sulfate	$CdSO_4$	10124-36-4	208.474	col orth cry	1000		4.69	76.7[25]	i EtOH
533	Cadmium sulfate monohydrate	$CdSO_4 \cdot H_2O$	7790-84-3	226.489	monocl cry	105		3.79	76.7[25]	
534	Cadmium sulfate octahydrate	$CdSO_4 \cdot 8H_2O$	15244-35-6	352.596	col monocl cry	40 dec		3.08	76.7[25]	
535	Cadmium sulfide	CdS	1306-23-6	144.476	yel-oran hex cry	≈1480		4.826		i H₂O; s acid
536	Cadmium sulfite	$CdSO_3$	13477-23-1	192.474	col prisms	dec ≈400			0.05[20]	sl H₂O
537	Cadmium telluride	$CdTe$	1306-25-8	240.01	brn-blk cub cry	1042		6.2		i H₂O, dil acid
538	Cadmium tellurite	$CdTeO_3$	15851-44-2	288.01	col monocl cry	695	dec 1050			
539	Cadmium tetrafluoroborate	$Cd(BF_4)_2$	14486-19-2	286.020	col hyg liq			1.6		vs H₂O, EtOH
540	Cadmium titanate	$CdTiO_3$	12014-14-1	208.276	orth cry			6.5		
541	Cadmium tungstate	$CdWO_4$	7790-85-4	360.25	wh monocl cry			8.0		i H₂O, acid; s NH₄OH
542	Calcium	Ca	7440-70-2	40.078	silv-wh metal	842	1484	1.54		reac H₂O; i bz
543	Calcium acetate	$Ca(C_2H_3O_2)_2$	62-54-4	158.167	wh hyg cry	160 dec		1.50		s H₂O; sl EtOH
544	Calcium acetate monohydrate	$Ca(C_2H_3O_2)_2 \cdot H_2O$	5743-26-0	176.182	wh needles or powder	≈150 dec				s H₂O; sl EtOH
545	Calcium acetate dihydrate	$Ca(C_2H_3O_2)_2 \cdot 2H_2O$	14977-17-4	194.196	long col needles					s H₂O
546	Calcium aluminate	$CaAl_2O_4$	12042-68-1	158.039	wh monocl cry	1605		2.98		reac H₂O
547	Calcium aluminate (β form)	$Ca_3Al_2O_6$	12042-78-3	270.193	wh cub cry; refr	1535		3.04		i H₂O
548	Calcium arsenate	$Ca_3(AsO_4)_2$	7778-44-1	398.072	wh powder	dec		3.6	0.0036[20]	s dil acid
549	Calcium arsenite (1:1)	$CaAsO_3$	52740-16-6	162.998	wh powder					sl H₂O; s acid
550	Calcium borate hexahydrate	$CaB_4O_7 \cdot 6H_2O$	13701-64-9*	303.409	wh cry pow	1162 (anh)				
551	Calcium boride	CaB_6	12007-99-7	104.944	refrac solid	2235		2.49		
552	Calcium bromate	$Ca(BrO_3)_2$	10102-75-7	295.882	wh pow	180				
553	Calcium bromate monohydrate	$Ca(BrO_3)_2 \cdot H_2O$	10102-75-7*	313.898	wh monocl cry	dec 180		3.33		vs H₂O
554	Calcium bromide	$CaBr_2$	7789-41-5	199.886	rhom cry; hyg	742	1815	3.38	156[25]	s EtOH, ace
555	Calcium bromide dihydrate	$CaBr_2 \cdot 2H_2O$	22208-73-7	235.917	wh cry pow					vs H₂O
556	Calcium bromide hexahydrate	$CaBr_2 \cdot 6H_2O$	13477-28-6	307.977	wh hyg powder	38 dec		2.29	156[25]	
557	Calcium carbide	CaC_2	75-20-7	64.099	gray-blk orth cry	2300		2.22		reac H₂O
558	Calcium carbonate (aragonite)	$CaCO_3$	471-34-1	100.087	wh orth cry or powder	trans calcite 450		2.930	0.00066[20]	s dil acid
559	Calcium carbonate (calcite)	$CaCO_3$	471-34-1	100.087	wh hex cry or powder	dec 700-900		2.710	0.00066[20]	s dil acid
560	Calcium carbonate (vaterite)	$CaCO_3$	471-34-1	100.087	col hex cry			2.653	0.0011[25]	s dil acid
561	Calcium chlorate	$Ca(ClO_3)_2$	10137-74-3	206.980	wh cry	340			197[25]	
562	Calcium chlorate dihydrate	$Ca(ClO_3)_2 \cdot 2H_2O$	10035-05-9	243.011	wh monocl cry; hyg	100 dec		2.711	197[25]	s EtOH
563	Calcium chloride	$CaCl_2$	10043-52-4	110.984	wh cub cry or powder; hyg	775	1935	2.15	81.3[25]	vs EtOH
564	Calcium chloride monohydrate	$CaCl_2 \cdot H_2O$	13477-29-7	128.999	wh hyg cry	260 dec		2.24	81.3[25]	s EtOH
565	Calcium chloride dihydrate	$CaCl_2 \cdot 2H_2O$	10035-04-8	147.015	hyg flakes or powder	175 dec		1.85	81.3[25]	vs EtOH
566	Calcium chloride tetrahydrate	$CaCl_2 \cdot 4H_2O$	25094-02-4	183.046	col tricl cry			1.83		
567	Calcium chloride hexahydrate	$CaCl_2 \cdot 6H_2O$	7774-34-7	219.075	wh hex cry; hyg	30 dec		1.71	81.3[25]	
568	Calcium chlorite	$Ca(ClO_2)_2$	14674-72-7	174.982	wh cub cry			2.71		reac H₂O
569	Calcium chromate	$CaCrO_4$	13765-19-0	156.072	yel cry	1000 dec				sl H₂O; i EtOH, ace
570	Calcium chromate dihydrate	$CaCrO_4 \cdot 2H_2O$	10060-08-9	192.102	yel orth cry	dec 200		2.50	13.2[20]	s dil acids
571	Calcium citrate tetrahydrate	$Ca_3(C_6H_5O_7)_2 \cdot 4H_2O$	5785-44-4	570.494	wh needles or pow	dec 100			0.096[23]	i eth
572	Calcium cyanamide	$CaCN_2$	156-62-7	80.102	col hex cry	≈1340	subl	2.29		reac H₂O
573	Calcium cyanide	$Ca(CN)_2$	592-01-8	92.112	wh rhomb cry; hyg					s H₂O, EtOH

No.	Name	Formula	CAS Reg No.	Mol. weight	Physical form	mp/°C	bp/°C	Density g cm^{-3}	Solubility g/100 g H$_2$O	Qualitative solubility
574	Calcium dichromate trihydrate	CaCr$_2$O$_7$ · 3H$_2$O	14307-33-6*	310.112	red-oran cry	100 dec		2.37		vs H$_2$O; reac EtOH; i eth, ctc
575	Calcium dihydrogen phosphate monohydrate	Ca(H$_2$PO$_4$)$_2$ · H$_2$O	10031-30-8	252.068	col tricl plates	100 dec		2.220		sl H$_2$O; s dil acid
576	Calcium 2-ethylhexanoate	Ca(C$_8$H$_{15}$O$_2$)$_2$	136-51-6	326.485	powder					
577	Calcium ferrocyanide dodecahydrate	Ca$_2$Fe(CN)$_6$ · 12H$_2$O		508.289	yel tricl cry	dec		1.68	87^{25}	vs H$_2$O; i EtOH
578	Calcium fluoride	CaF$_2$	7789-75-5	78.075	wh cub cry or powder	1418	2500	3.18	0.0016^{25}	sl acid
579	Calcium fluorophosphate	Ca$_5$(PO$_4$)$_3$F	12015-73-5	504.302	col hex cry	1650		3.201		i H$_2$O
580	Calcium fluorophosphate dihydrate	CaPO$_3$F · 2H$_2$O	37809-19-1	174.079	col monocl cry				0.42^{27}	i os
581	Calcium formate	Ca(CHO$_2$)$_2$	544-17-2	130.113	orth cry	300 dec		2.02	16.6^{20}	i EtOH
582	Calcium hexaborate pentahydrate	2CaO · 3B$_2$O$_3$ · 5H$_2$O	12291-65-5	411.091	col monocl cry	dec 375 (exp)		2.42	1^{25}	sl acid
583	Calcium hexafluoro-2,4-pentanedioate	Ca(CF$_3$COCHCOCF$_3$)$_2$	121012-90-6	454.180	powder	135				
584	Calcium hexafluorosilicate dihydrate	CaSiF$_6$ · 2H$_2$O	16925-39-6	218.185	col tetr cry			2.25	0.52^{20}	i ace; reac hot H$_2$O
585	Calcium hydride	CaH$_2$	7789-78-8	42.094	gray orth cry or powder	1000		1.7		reac H$_2$O, EtOH
586	Calcium hydrogen phosphate	CaHPO$_4$	7757-93-9	136.057	wh tricl cry	dec		2.92	0.02^{25}	i EtOH
587	Calcium hydrogen phosphate dihydrate	CaHPO$_4$ · 2H$_2$O	7789-77-7	172.088	monocl cry	≈100 dec		2.31	0.02^{25}	i EtOH; s dil acid
588	Calcium hydrogen sulfite	CaH$_2$(SO$_3$)$_2$	13780-03-5	202.220				1.06		s H$_2$O
589	Calcium hydrosulfide hexahydrate	Ca(HS)$_2$ · 6H$_2$O		214.315	col cry	dec				s H$_2$O, EtOH
590	Calcium hydroxide	Ca(OH)$_2$	1305-62-0	74.093	soft hex cry			≈2.2	0.160^{20}	s acid
591	Calcium hydroxide phosphate	Ca$_5$(OH)(PO$_4$)$_3$	12167-74-7	502.311	col hex cry	dec >900		3.155		i H$_2$O
592	Calcium hypochlorite	Ca(OCl)$_2$	7778-54-3	142.983	powder	100		2.350		
593	Calcium hypophosphite	Ca(H$_2$PO$_2$)$_2$	7789-79-9	170.055	wh monocl cry	300 dec				s H$_2$O; i EtOH
594	Calcium iodate	Ca(IO$_3$)$_2$	7789-80-2	389.883	wh monocl cry			4.52	0.306^{25}	s HNO$_3$; i EtOH
595	Calcium iodide	CaI$_2$	10102-68-8	293.887	hyg hex cry	783	1100	3.96	215^{25}	s MeOH, EtOH, ace; i eth
596	Calcium iodide hexahydrate	CaI$_2$ · 6H$_2$O	71626-98-7	401.978	wh hex needles or powder	42 dec		2.55	215^{25}	vs EtOH
597	Calcium metaborate	Ca(BO$_2$)$_2$	13701-64-9	125.698	powder				0.13^{20}	
598	Calcium metasilicate	CaSiO$_3$	1344-95-2	116.162	wh monocl cry	1540		2.92		i H$_2$O
599	Calcium molybdate	CaMoO$_4$	7789-82-4	200.02	wh tetr cry	1520		4.35	0.0011^{20}	i EtOH; s conc acid
600	Calcium nitrate	Ca(NO$_3$)$_2$	10124-37-5	164.087	wh cub cry; hyg	561		2.5	144^{25}	s EtOH, MeOH, ace
601	Calcium nitrate tetrahydrate	Ca(NO$_3$)$_2$ · 4H$_2$O	13477-34-4	236.149	wh cry	≈40 dec		1.82	144^{25}	s EtOH, ace
602	Calcium nitride	Ca$_3$N$_2$	12013-82-0	148.247	red-brn cub cry	1195		2.67		s H$_2$O, acid; i EtOH
603	Calcium nitrite	Ca(NO$_2$)$_2$	13780-06-8	132.089	wh-yel hex cry; hyg	392		2.23	94.6^{25}	sl EtOH
604	Calcium nitrite monohydrate	Ca(NO$_2$)$_2$ · H$_2$O	10031-34-2	150.104	col or yel cry	dec 100				vs H$_2$O; sl EtOH
605	Calcium oleate	Ca(C$_{18}$H$_{33}$O$_2$)$_2$	142-17-6	602.985	pale yel solid	dec 140			0.04^{25}	sl H$_2$O; s bz; i EtOH, ace, eth
606	Calcium oxalate	CaC$_2$O$_4$	563-72-4	128.097	wh cry powder			2.2	0.00061^{20}	
607	Calcium oxalate monohydrate	CaC$_2$O$_4$ · H$_2$O	5794-28-5	146.112	cub cry	200 dec		2.2	0.00061^{20}	s dil acid
608	Calcium oxide	CaO	1305-78-8	56.077	gray-wh cub cry	2613		3.34		reac H$_2$O; s acid
609	Calcium oxide silicate	Ca$_3$OSiO$_4$	12168-85-3	228.317	refrac solid	2150				
610	Calcium palmitate	Ca(C$_{16}$H$_{31}$O$_2$)$_2$	542-42-7	550.910	wh-yel pow	dec 155				i H$_2$O, EtOH, eth, ace; sl bz
611	Calcium perborate heptahydrate	Ca(BO$_3$)$_2$ · 7H$_2$O		283.803	gray-wh pow					s H$_2$O, acid
612	Calcium 2,4-pentanedioate	Ca(CH$_3$COCHCOCH$_3$)$_2$	19372-44-2	238.294	cry	dec 175				
613	Calcium perchlorate	Ca(ClO$_4$)$_2$	13477-36-6	238.979	wh cry	270 dec		2.65	188^{25}	s EtOH, MeOH
614	Calcium perchlorate tetrahydrate	Ca(ClO$_4$)$_2$ · 4H$_2$O	15627-86-8	311.041	wh cry					vs H$_2$O
615	Calcium permanganate	Ca(MnO$_4$)$_2$	10118-76-0	277.949	purp hyg cry			2.4	331^{20}	reac EtOH
616	Calcium peroxide	CaO$_2$	1305-79-9	72.077	wh-yel tetr cry; hyg	≈200 dec		2.9		sl H$_2$O; s acid
617	Calcium phosphate	Ca$_3$(PO$_4$)$_2$	7758-87-4	310.177	wh amorp powder	1670		3.14	0.00012^{20}	i EtOH; s dil acid
618	Calcium phosphide	Ca$_3$P$_2$	1305-99-3	182.182	red-brn hyg cry	≈1600		2.51		reac H$_2$O; i EtOH, eth
619	Calcium phosphonate monohydrate	CaHPO$_3$ · H$_2$O	25232-60-4	138.073	col monocl cry	dec 150				sl H$_2$O; i EtOH
620	Calcium plumbate	Ca$_2$PbO$_4$	12013-69-3	351.4	oran-brn orth cry	dec		5.71		i H$_2$O; s acid
621	Calcium propanoate	Ca(C$_3$H$_5$O$_2$)$_2$	4075-81-4	186.219	monocl cry, powder					s H$_2$O; sl MeOH, EtOH; i ace, bz
622	Calcium pyrophosphate	Ca$_2$P$_2$O$_7$	7790-76-3	254.099	wh powder	1353		3.09		i H$_2$O; s dil acid
623	Calcium selenate dihydrate	CaSeO$_4$ · 2H$_2$O	7790-74-1	219.07	wh monocl cry			2.75	8.3^{18}	
624	Calcium selenide	CaSe	1305-84-6	119.04	wh-brn cub cry	1400 dec		3.8		reac H$_2$O
625	Calcium silicide (CaSi)	CaSi	12013-55-7	68.164	orth cry	1324		2.39		

No.	Name	Formula	CAS Reg No.	Mol. weight	Physical form	mp/°C	bp/°C	Density g cm^{-3}	Solubility g/100 g H$_2$O	Qualitative solubility
626	Calcium silicide (CaSi$_2$)	CaSi$_2$	12013-56-8	96.249	gray hex cry	1040		2.50		i cold H$_2$O; reac hot H$_2$O; s acid
627	Calcium stannate trihydrate	CaSnO$_3$ · 3H$_2$O	12013-46-6*	260.832	wh cry pow	dec ≈350				i H$_2$O
628	Calcium stearate	Ca(C$_{18}$H$_{35}$O$_2$)$_2$	1592-23-0	607.017	granular powder	180				i H$_2$O, EtOH
629	Calcium succinate trihydrate	CaC$_4$H$_4$O$_4$ · 3H$_2$O	140-99-8	210.196	needles					sl H$_2$O; s dil acid; i EtOH
630	Calcium sulfate	CaSO$_4$	7778-18-9	136.141	orth cry	1460		2.96	0.205^{25}	
631	Calcium sulfate hemihydrate	CaSO$_4$ · 0.5H$_2$O	10034-76-1	145.149	wh powder				0.205^{25}	
632	Calcium sulfate dihydrate	CaSO$_4$ · 2H$_2$O	10101-41-4	172.171	monocl cry or powder	150 dec		2.32	0.205^{20}	i os
633	Calcium sulfide	CaS	20548-54-3	72.143	wh-yel cub cry; hyg	2524		2.59		sl H$_2$O; i EtOH
634	Calcium sulfite dihydrate	CaSO$_3$ · 2H$_2$O	10257-55-3	156.172	wh powder				0.0070^{25}	sl EtOH; s acid
635	Calcium tartrate tetrahydrate	CaC$_4$H$_4$O$_6$ · 4H$_2$O	3164-34-9*	260.210	wh pow				0.04^{10}	s dil acid; sl EtOH
636	Calcium telluride	CaTe	12013-57-9	167.68	wh cub cry	1600 dec		4.87		
637	Calcium tetrahydroaluminate	Ca(AlH$_4$)$_2$	16941-10-9	102.105	gray powder; flam					reac H$_2$O; s thf; i eth, bz
638	Calcium thiocyanate tetrahydrate	Ca(SCN)$_2$ · 4H$_2$O	2092-16-2	228.304	hyg cry	160 dec				vs H$_2$O; s EtOH, ace
639	Calcium thiosulfate hexahydrate	CaS$_2$O$_3$ · 6H$_2$O	10124-41-1	260.298	tricl cry	45 dec		1.87		s H$_2$O; i EtOH
640	Calcium titanate	CaTiO$_3$	12049-50-2	135.943	cub cry	1980		3.98		
641	Calcium tungstate	CaWO$_4$	7790-75-2	287.92	wh tetr cry	1620		6.06	0.2^{18}	s hot acid
642	Calcium zirconate	CaZrO$_3$	12013-47-7	179.300	powder	2550				
643	Californium	Cf	7440-71-3	251	hex or cub metal	900		15.1		
644	Carbon (diamond)	C	7782-40-3	12.011	col cub cry	4440 (12.4 GPa)		3.513		i H$_2$O
645	Carbon (graphite)	C	7782-42-5	12.011	soft blk hex cry	4489 tp (10.3 MPa)	3825 sp	2.2		i H$_2$O
646	Carbon black	C	1333-86-4	12.011	fine blk pow					i H$_2$O
647	Carbon (fullerene-C$_{60}$)	C$_{60}$	99685-96-8	720.642	yel needles or plates	>280				s os
648	Carbon (fullerene-C$_{70}$)	C$_{70}$	115383-22-7	840.749	red-brn solid	>280				s bz, tol
649	Fullerene fluoride	C$_{60}$F$_{60}$	134929-59-2	1860.546	col plates	287				vs ace; s thf; i chl
650	Carbon monoxide	CO	630-08-0	28.010	col gas	-205.02	-191.5	1.145 g/L		sl H$_2$O; s chl, EtOH
651	Carbon dioxide	CO$_2$	124-38-9	44.010	col gas	-56.558 tp	-78.464 sp	1.799 g/L		s H$_2$O
652	Carbon suboxide	C$_3$O$_2$	504-64-3	68.031	col gas	-112.5	6.8	2.781 g/L		reac H$_2$O
653	Carbon disulfide	CS$_2$	75-15-0	76.141	col or yel liq	-112.1	46	1.2632^{20}		i H$_2$O; vs EtOH, bz, os
654	Carbon subsulfide	C$_3$S$_2$	627-34-9	100.162	red liq	-1	90 dec	1.27		reac H$_2$O
655	Carbon diselenide	CSe$_2$	506-80-9	169.93	yel liq	-43.7	125.5	2.6823^{20}		i H$_2$O; vs ctc, tol
656	Carbon oxysulfide	COS	463-58-1	60.075	col gas	-138.8	-50	2.456 g/L		s H$_2$O, EtOH
657	Carbon oxyselenide	COSe	1603-84-5	106.97	col gas; unstab	-124.4	-21.7	4.372 g/L		reac H$_2$O
658	Carbon sulfide selenide	CSSe	5951-19-9	123.04	yel liq	-85	84.5	1.99		i H$_2$O
659	Carbon sulfide telluride	CSTe	10340-06-4	171.68	red-yel liq; unstab	-54	20 dec			reac H$_2$O
660	Carbonyl bromide	COBr$_2$	593-95-3	187.818	col liq		64.5	2.5		reac H$_2$O
661	Carbonyl chloride	COCl$_2$	75-44-5	98.916	col gas	-127.78	8	4.043 g/L		sl H$_2$O; s bz, tol
662	Carbonyl fluoride	COF$_2$	353-50-4	66.007	col gas	-111.26	-84.57	2.698 g/L		reac H$_2$O
663	Cyanogen	C$_2$N$_2$	460-19-5	52.034	col gas	-27.83	-21.1	2.127 g/L		sl H$_2$O, eth; s EtOH
664	Cyanogen azide	N$_3$CN	764-05-6	68.038	col oily liq	exp				
665	Cyanogen bromide	BrCN	506-68-3	105.922	wh hyg needles	52	61.5	2.015		s H$_2$O, EtOH, eth
666	Cyanogen chloride	ClCN	506-77-4	61.471	col vol liq or gas	-6.55	13	2.513 g/L		s H$_2$O, EtOH, eth
667	Cyanogen fluoride	FCN	1495-50-7	45.016	col gas	-82	-46	1.840 g/L		
668	Cyanogen iodide	ICN	506-78-5	152.922	col needles	146.7		2.84		s H$_2$O, EtOH, eth
669	Cerium	Ce	7440-45-1	140.116	silv metal; cub or hex	799	3443	6.770		s dil acid
670	Cerium boride	CeB$_6$	12008-02-5	204.982	blue refrac solid; hex	2550		4.87		i H$_2$O, HCl
671	Cerium carbide	CeC$_2$	12012-32-7	164.137	red hex cry	2250		5.47		reac H$_2$O
672	Cerium carbide	Ce$_2$C$_3$	12115-63-8	316.264	yel-brn cub cry	1505		6.9		
673	Cerium nitride	CeN	25764-08-3	154.123	refrac cub cry	2557		7.89		
674	Cerium silicide	CeSi$_2$	12014-85-6	196.287	tetr cry	1420		5.31		i H$_2$O
675	Cerium(II) hydride	CeH$_2$	13569-50-1	142.132	cub cry			5.45		reac H$_2$O
676	Cerium(II) iodide	CeI$_2$	19139-47-0	393.925	bronze cry	808				
677	Cerium(II) sulfide	CeS	12014-82-3	172.181	yel cub cry	2445		5.9		
678	Cerium(III) acetate sesqihydrate	Ce(C$_2$H$_3$O$_2$)$_3$ · 1.5H$_2$O	17829-82-2		col cry	dec 115			26^{15}	s H$_2$O
679	Cerium(III) ammonium nitrate tetrahydrate	(NH$_4$)$_2$Ce(NO$_3$)$_5$ · 4H$_2$O	13083-04-0	558.279	col monocl cry	74				vs H$_2$O
680	Cerium(III) ammonium sulfate tetrahydrate	NH$_4$Ce(SO$_4$)$_2$ · 4H$_2$O	21995-38-0*	422.341	monocl cry					s H$_2$O
681	Cerium(III) bromide	CeBr$_3$	14457-87-5	379.828	wh hex cry; hyg	732	1457			s H$_2$O

No.	Name	Formula	CAS Reg No.	Mol. weight	Physical form	mp/°C	bp/°C	Density g cm⁻³	Solubility g/100 g H₂O	Qualitative solubility
682	Cerium(III) bromide heptahydrate	$CeBr_3 \cdot 7H_2O$	7789-56-2	505.935	col hyg needles	732				s H₂O, EtOH
683	Cerium(III) carbonate	$Ce_2(CO_3)_3$	537-01-9	460.259	wh pow	dec 500				i H₂O; s acid
684	Cerium(III) carbonate pentahydrate	$Ce_2(CO_3)_3 \cdot 5H_2O$	72520-94-6	550.335	wh powder					i H₂O; s dil acid
685	Cerium(III) chloride	$CeCl_3$	7790-86-5	246.475	wh hex cry	807		3.97		s H₂O, EtOH
686	Cerium(III) chloride heptahydrate	$CeCl_3 \cdot 7H_2O$	18618-55-8	372.582	yel orth cry; hyg	90 dec				vs H₂O, EtOH
687	Cerium(III) fluoride	CeF_3	7758-88-5	197.111	wh hex cry; hyg	1430	2180	6.157		i H₂O
688	Cerium(III) hydride	CeH_3	13864-02-3	143.140	blk pow or blue-blk cry	dec (flam)				reac H₂O
689	Cerium(III) hydroxide	$Ce(OH)_3$	15785-09-8	191.138	wh solid					i H₂O; s acid
690	Cerium(III) iodide	CeI_3	7790-87-6	520.829	yel orth cry; hyg	760				s H₂O
691	Cerium(III) iodide nonahydrate	$CeI_3 \cdot 9H_2O$	7790-87-6*	682.967	wh-red cry					vs H₂O; s EtOH
692	Cerium(III) nitrate hexahydrate	$Ce(NO_3)_3 \cdot 6H_2O$	10108-73-3*	434.222	col-red cry	150 dec			176²⁵	s ace
693	Cerium(III) oxalate nonahydrate	$Ce(C_2O_4)_3 \cdot 9H_2O$	13266-83-6	706.426	wh pow	dec				i H₂O, EtOH; s acid
694	Cerium(III) oxide	Ce_2O_3	1345-13-7	328.230	yel-grn cub cry	2210	3730	6.2		i H₂O; s acid
695	Cerium(III) 2,4-pentanedioate trihydrate	$Ce(CH_3COCHCOCH_3)_3 \cdot 3H_2O$	15653-01-7	491.486	yel hyg cry	≈150				vs EtOH
696	Cerium(III) perchlorate hexahydrate	$Ce(ClO_4)_3 \cdot 6H_2O$	36907-38-7	546.559	hyg col cry	dec 200				s H₂O, EtOH
697	Cerium(III) selenate	$Ce_2(SeO_4)_3$		709.11	rhom cry			4.46		s H₂O
698	Cerium(III) sulfate	$Ce_2(SO_4)_3$	13454-94-9	568.420	col hyg cry	920 dec				s H₂O
699	Cerium(III) sulfate octahydrate	$Ce_2(SO_4)_3 \cdot 8H_2O$	13454-94-9	712.542	wh orth cry	≈250 dec		2.87		s H₂O
700	Cerium(III) sulfide	Ce_2S_3	12014-93-6	376.427	red cub cry	2450		5.02		i H₂O
701	Cerium(III) tungstate	$Ce_2(WO_4)_3$	13454-74-5	1023.75	yel tetr cry	1089		6.77		i H₂O
702	Cerium(IV) ammonium nitrate	$(NH_4)_2Ce(NO_3)_6$	16774-21-3	548.223	red-oran cry					vs H₂O
703	Cerium(IV) ammonium sulfate dihydrate	$(NH_4)_4Ce(SO_4)_4 \cdot 2H_2O$	10378-47-9	632.551	cry pow	dec 450				
704	Cerium(IV) fluoride	CeF_4	10060-10-3	216.110	wh hyg powder	≈600 dec		4.77		i H₂O
705	Cerium(IV) hydroxide	$Ce(OH)_4$	12014-56-1	208.146	yel-wh pow					i H₂O; s conc acid
706	Cerium(IV) oxide	CeO_2	1306-38-3	172.115	wh-yel powder; cub	2480		7.216		i H₂O, dil acid
707	Cerium(IV) sulfate tetrahydrate	$Ce(SO_4)_2 \cdot 4H_2O$	10294-42-5	404.303	yel-oran orth cry	180 dec		3.91	9.66²⁰	
708	Cesium	Cs	7440-46-2	132.905	silv-wh metal	28.5	671	1.873		reac H₂O
709	Cesium acetate	$CsC_2H_3O_2$	3396-11-0	191.949	hyg lumps	194			10¹¹	
710	Cesium aluminum sulfate dodecahydrate	$CsAl(SO_4)_2 \cdot 12H_2O$	7784-17-0	568.196	col cub cry	117 dec		1.97		s H₂O; i EtOH
711	Cesium amide	$CsNH_2$	22205-57-8	148.928	wh tetr cry			3.70		
712	Cesium azide	CsN_3	22750-57-8	174.925	hyg tetr cry; exp	326		≈3.5	22⁴⁰	
713	Cesium bromate	$CsBrO_3$	13454-75-6	260.807	col hex cry			4.11	3.83²⁵	
714	Cesium bromide	$CsBr$	7787-69-1	212.809	wh cub cry; hyg	636	≈1300	4.43	123²⁵	s EtOH; i ace
715	Cesium carbonate	Cs_2CO_3	534-17-8	325.820	wh monocl cry; hyg	793		4.24	261¹⁵	s EtOH, eth
716	Cesium chlorate	$CsClO_3$	13763-67-2	216.356	col hex cry	342		3.57	7.78²⁵	sl H₂O
717	Cesium chloride	$CsCl$	7647-17-8	168.358	wh cub cry; hyg	646	1297	3.988	191²⁵	s EtOH
718	Cesium chromate(IV)	Cs_2CrO_4	56320-90-2	647.616	yel hex cry	982		4.24		vs H₂O
719	Cesium cyanide	$CsCN$	21159-32-0	158.923	wh cub cry; hyg	350		3.34		vs H₂O
720	Cesium dibromoiodate	$CsIBr_2$	18278-82-5	419.617	dark oran cry	dec				s H₂O
721	Cesium fluoride	CsF	13400-13-0	151.903	wh cub cry; hyg	703		4.64	573²⁵	s MeOH; i diox, py
722	Cesium fluoroborate	$CsBF_4$	18909-69-8	219.710	wh orth cry	555 dec		3.2	1.6¹⁷	sl H₂O
723	Cesium formate	$CsCHO_2$	3495-36-1	177.923	wh cry			1.017		vs H₂O
724	Cesium hexafluorogermanate	Cs_2GeF_6		452.44	wh cry	≈675		4.10		sl cold H₂O; s hot H₂O
725	Cesium hydride	CsH	58724-12-2	133.913	wh cub cry; flam	528		3.42		reac H₂O
726	Cesium hydrogen carbonate	$CsHCO_3$	15519-28-5	193.922	rhom cry	175 dec			209¹⁵	s EtOH
727	Cesium hydrogen fluoride	$CsHF_2$	12280-52-3	171.910	tetr cry	170		3.86		
728	Cesium hydrogen sulfate	$CsHSO_4$	7789-16-4	229.976	col rhom prisms	dec		3.352		s H₂O
729	Cesium hydroxide	$CsOH$	21351-79-1	149.912	wh-yel hyg cry	342.3		3.68	300³⁰	s EtOH
730	Cesium iodate	$CsIO_3$	13454-81-4	307.807	wh monocl cry			4.85	2.6²⁵	
731	Cesium iodide	CsI	7789-17-5	259.809	col cub cry; hyg	632	≈1280	4.51	84.8²⁵	s EtOH, MeOH, ace
732	Cesium metaborate	$CsBO_2$	92141-86-1	175.715	cub cry	732		≈3.7		
733	Cesium molybdate	Cs_2MoO_4	13597-64-3	425.75	wh cry	956.3			67¹⁸	s H₂O
734	Cesium nitrate	$CsNO_3$	7789-18-6	194.910	wh hex or cub cry	409		3.66	27.9²⁵	s ace; sl EtOH
735	Cesium nitrite	$CsNO_2$	13454-83-6	178.911	yel cry	406				s H₂O
736	Cesium oxide	Cs_2O	20281-00-9	281.810	yel-oran hex cry	495		4.65		vs H₂O
737	Cesium superoxide	CsO_2	12018-61-0	164.904	yel tetr cry	432		3.77		reac H₂O
738	Cesium trioxide	Cs_2O_3	12134-22-4	313.809	brn cry	≈400		4.25		reac H₂O
739	Cesium perchlorate	$CsClO_4$	13454-84-7	232.356	wh orth cry; hyg	≈600 dec		3.327	2.00²⁵	
740	Cesium periodate	$CsIO_4$	13478-04-1	323.807	wh rhom prisms			4.26	2.2¹⁵	
741	Cesium sulfate	Cs_2SO_4	10294-54-9	361.874	wh orth cry or hex prisms; hyg	1005		4.24	182²⁵	i EtOH, ace, py

No.	Name	Formula	CAS Reg No.	Mol. weight	Physical form	mp/°C	bp/°C	Density g cm⁻³	Solubility g/100 g H₂O	Qualitative solubility
742	Cesium sulfide	Cs_2S	12214-16-3	297.876	yel orth hyg cry	520				vs H_2O
743	Cesium trifluoroacetate	$Cs(C_2F_3O_2)$	21907-50-6	245.920	hyg solid	115				vs H_2O
744	Chlorine	Cl_2	7782-50-5	70.90	grn-yel gas	-101.5	-34.04	2.898 g/L		sl H_2O
745	Hypochlorous acid	$HOCl$	7790-92-3	52.460	grn-yel; stable only in aq soln					s H_2O
746	Chloric acid	$HClO_3$	7790-93-4	84.459	exists only in aq soln					vs H_2O
747	Perchloric acid	$HClO_4$	7601-90-3	100.459	col hyg liq	-112	≈90 dec	1.77		s H_2O
748	Chlorine monoxide	Cl_2O	7791-21-1	86.905	yel-brn gas	-120.6	2.2	3.552 g/L		vs H_2O
749	Chlorine dioxide	ClO_2	10049-04-4	67.452	oran-grn gas	-59	11	2.757 g/L		sl H_2O
750	Dichlorine trioxide	Cl_2O_3	17496-59-2	118.904	dark brn solid	exp <25				
751	Dichlorine hexoxide	Cl_2O_6	12442-63-6	166.902	red liq	3.5	≈200			reac H_2O
752	Dichlorine heptoxide	Cl_2O_7	10294-48-1	182.902	col oily liq; exp	-91.5	82	1.9		reac H_2O
753	Chlorine fluoride	ClF	7790-89-8	54.451	col gas	-155.6	-101.1	2.226 g/L		reac H_2O
754	Chlorine trifluoride	ClF_3	7790-91-2	92.448	gas	-76.34	11.75	3.779 g/L		reac H_2O
755	Chlorine pentafluoride	ClF_5	13637-63-3	130.445	col gas	-103	-13.1	5.332 g/L		
756	Chlorosyl trifluoride	$ClOF_3$	30708-80-6	108.447	col liq	-42	27			reac H_2O
757	Chloryl fluoride	ClO_2F	13637-83-7	86.450	col gas	-115	-6	3.534 g/L		reac H_2O
758	Chloryl trifluoride	ClO_2F_3	38680-84-1	124.447	col gas	-81.2	-21.6	5.087 g/L		reac H_2O
759	Perchloryl fluoride	ClO_3F	7616-94-6	102.449	col gas	-147	-46.75	4.187 g/L		
760	Chlorine perchlorate	$ClOClO_3$	27218-16-2	134.904	unstab yel liq	-117	≈45 dec	1.81⁰		
761	Chromium	Cr	7440-47-3	51.996	blue-wh metal; cub	1907	2671	7.15		reac dil acid
762	Chromic acid	H_2CrO_4	7738-94-5	118.010	aq soln only					s H_2O
763	Chromium antimonide	$CrSb$	12053-12-2	173.756	hex cry	1110		7.11		
764	Chromium arsenide	Cr_2As	12254-85-2	178.914	tetr cry			7.04		
765	Chromium boride (CrB)	CrB	12006-79-0	62.807	refrac orth cry	2100		6.1		
766	Chromium boride (CrB₂)	CrB_2	12007-16-8	73.618	refrac solid; hex	2200		5.22		
767	Chromium boride (Cr₂B)	Cr_2B	12006-80-3	114.803	refrac solid	1875				
768	Chromium boride (Cr₅B₃)	Cr_5B_3	12007-38-4	292.414	tetr cry	1900		6.10		
769	Chromium carbide	Cr_3C_2	12012-35-0	180.009	gray orth cry	1895		6.68		
770	Chromium carbonyl	$Cr(CO)_6$	13007-92-6	220.056	col orth cry	130 dec	subl	1.77		i H_2O, EtOH; s eth, chl
771	Chromium nitride (CrN)	CrN	24094-93-7	66.003	gray cub cry	1080 dec		5.9		
772	Chromium nitride (Cr₂N)	Cr_2N	12053-27-9	117.999	hex cry	1650		6.8		
773	Chromium phosphide	CrP	26342-61-0	82.970	orth cry			5.25		
774	Chromium selenide	$CrSe$	12053-13-3	130.96	hex cry	≈1500		6.1		
775	Chromium silicide (CrSi₂)	$CrSi_2$	12018-09-6	108.167	gray hex cry	1490		4.91		
776	Chromium silicide (Cr₃Si)	Cr_3Si	12018-36-9	184.074	cub cry	1770		6.4		
777	Chromium(II) acetate monohydrate	$Cr(C_2H_3O_2)_2 \cdot H_2O$	628-52-4*	188.100	red monocl cry			1.79		sl H_2O
778	Chromium(II) bromide	$CrBr_2$	10049-25-9	211.804	wh monocl cry; aq soln blue	842		4.236		s H_2O, EtOH
779	Chromium(II) chloride	$CrCl_2$	10049-05-5	122.902	wh hyg needles; aq soln blue	824	1120	2.88		s H_2O
780	Chromium(II) chloride tetrahydrate	$Cr(H_2O)_4Cl_2 \cdot 4H_2O$	13931-94-7	267.024	blue hyg cry	51 dec				s H_2O
781	Chromium(II) fluoride	CrF_2	10049-10-2	89.993	blue-grn monocl cry	894		3.79		sl H_2O; i EtOH
782	Chromium(II) formate monohydrate	$Cr(CHOO)_2 \cdot H_2O$	4493-37-2	160.046	red needles					s H_2O
783	Chromium(II) iodide	CrI_2	13478-28-9	305.805	red-brn cry; hyg	867		5.1		s H_2O
784	Chromium(II) oxalate monohydrate	$CrC_2O_4 \cdot H_2O$	814-90-4*	158.030	yel-grn powder			2.468		sl H_2O
785	Chromium(II) sulfate pentahydrate	$CrSO_4 \cdot 5H_2O$	13825-86-0	238.135	blue cry				21⁰	s dil acid; sl EtOH; i ace
786	Chromium(II,III) oxide	Cr_3O_4	12018-34-7	219.986	cub cry			6.1		
787	Chromium(III) acetate	$Cr(C_2H_3O_2)_3$	1066-30-4	229.127	blue-grn pwd					sl H_2O
788	Chromium(III) acetate monohydrate	$Cr(C_2H_3O_2)_3 \cdot H_2O$	25013-82-5	247.143	gray-grn pow					sl H_2O; i EtOH
789	Chromium(III) acetate hexahydrate	$Cr(C_2H_3O_2)_3 \cdot 6H_2O$	1066-30-4*	337.220	blue needles					s H_2O
790	Chromium(III) acetate hydroxide	$Cr(C_2H_3O_2)_2(OH)$	39430-51-8	187.092	viol cry pow					vs H_2O
791	Chromium(III) bromide	$CrBr_3$	10031-25-1	291.708	dark grn hex cry	812		4.68		s hot H_2O, bz
792	Chromium(III) bromide hexahydrate (α)	$CrBr_3(H_2O)_4 \cdot 2H_2O$	18721-05-6	399.799	grn hyg cry					s H_2O, EtOH
793	Chromium(III) bromide hexahydrate (β)	$Cr(H_2O)_6Br_3$	10031-25-1*	399.799	viol hyg cry					s H_2O; i EtOH, eth
794	Chromium(III) chloride	$CrCl_3$	10025-73-7	158.355	red-viol cry	1152	1300 dec	2.76		sl H_2O
795	Chromium(III) chloride hexahydrate	$[CrCl_2(H_2O)_4]Cl \cdot 2H_2O$	10060-12-5	266.446	grn monocl cry; hyg					s H_2O, EtOH; sl ace; i eth
796	Chromium(III) fluoride	CrF_3	7788-97-8	108.991	grn needles	1425		3.8		i H_2O, EtOH
797	Chromium(III) fluoride trihydrate	$CrF_3 \cdot 3H_2O$	16671-27-5	163.037	grn hex cry			2.2		sl H_2O
798	Chromium(III) hydroxide sulfate	$Cr(OH)SO_4$	12336-95-7	165.066	grn cry					
799	Chromium(III) fluoride nonahydrate	$Cr(H_2O)_6F_3 \cdot 3H_2O$	102430-09-1	271.129	rhom viol cry					sl H_2O

No.	Name	Formula	CAS Reg No.	Mol. weight	Physical form	mp/°C	bp/°C	Density g cm⁻³	Solubility g/100 g H₂O	Qualitative solubility
800	Chromium(III) hydroxide trihydrate	$Cr(OH)_3 \cdot 3H_2O$	1308-14-1	157.063	blue-grn powder					i H₂O; s acid
801	Chromium(III) iodide	CrI_3	13569-75-0	432.709	dark grn hex cry	500 dec		5.32		sl H₂O
802	Chromium(III) nitrate	$Cr(NO_3)_3$	13548-38-4	238.011	grn hyg powder	>60 dec				vs H₂O
803	Chromium(III) nitrate nonahydrate	$Cr(NO_3)_3 \cdot 9H_2O$	7789-02-8	400.148	grn-blk monocl cry	66.3	>100 dec	1.80		vs H₂O
804	Chromium(III) oxide	Cr_2O_3	1308-38-9	151.990	grn hex cry	2320	≈3000	5.22		i H₂O, EtOH; sl acid, alk
805	Chromium(III) 2,4-pentanedioate	$Cr(CH_3COCHCOCH_3)_3$	21679-31-2	349.320	red monocl cry	208	345	1.34		i H₂O; s bz
806	Chromium(III) perchlorate	$Cr(ClO_4)_3$	27535-70-2	350.348	grn-blue cry				58²⁵	vs H₂O
807	Chromium(III) phosphate	$CrPO_4$	7789-04-0	146.967	blue orth cry	>1800		4.6		i H₂O, acid, aqua regia
808	Chromium(III) phosphate hemiheptahydrate	$CrPO_4 \cdot 3.5H_2O$	84359-31-9	210.021	blue-grn powder			2.15		i H₂O; s acid
809	Chromium(III) phosphate hexahydrate	$CrPO_4 \cdot 6H_2O$	84359-31-9	255.059	viol cry	>500 dec		2.121		i H₂O; s acid, alk
810	Chromium(III) potassium oxalate trihydrate	$K_3Cr(C_2O_4)_3 \cdot 3H_2O$	15275-09-9	487.394	blue-grn monocl cry					s H₂O
811	Chromium(III) potassium sulfate dodecahydrate	$CrK(SO_4)_2 \cdot 12H_2O$	7788-99-0	499.403	viol-blk cub cry	89 dec		1.83		s H₂O; i EtOH
812	Chromium(III) sulfate	$Cr_2(SO_4)_3$	10101-53-8	392.180	red pow	dec >700		3.1	64²⁵	s H₂O; vs acid
813	Chromium(III) sulfate octadecahydrate	$Cr_2(SO_4)_3 \cdot 18H_2O$	10101-53-8*	716.455	viol cry	dec 115		1.7		reac H₂O
814	Chromium(III) sulfide	Cr_2S_3	12018-22-3	200.187	brn-blk hex cry			3.8		
815	Chromium(III) telluride	Cr_2Te_3	12053-39-3	486.79	hex cry	≈1300		7.0		
816	Chromium(IV) chloride	$CrCl_4$	15597-88-3	193.808	gas, stable at HT		>600 dec	7.922 g/L		
817	Chromium(IV) fluoride	CrF_4	10049-11-3	127.990	grn cry	277		2.89		reac H₂O
818	Chromium(IV) oxide	CrO_2	12018-01-8	83.995	brn-blk tetr powder	≈400 dec		4.89		i H₂O; s acid
819	Chromium(V) fluoride	CrF_5	14884-42-5	146.988	red orth cry	34	117			reac H₂O
820	Chromium(V) oxide	Cr_2O_5	12218-36-9	183.989	blk needles	dec 200				
821	Chromium(VI) fluoride	CrF_6	13843-28-2	165.986	yel solid; stable at low temp	-100 dec				
822	Chromium(VI) oxide	CrO_3	1333-82-0	99.994	red orth cry	197	≈250 dec	2.7	169²⁵	
823	Chromium(VI) tetrafluoride oxide	$CrOF_4$	23276-90-6	143.989	dark red solid	55				reac H₂O, ace, dmso
824	Chromium(VI) dichloride dioxide	CrO_2Cl_2	14977-61-8	154.901	red liq	-96.5	117	1.91		reac H₂O; s ctc, chl, bz
825	Chromium(VI) difluoride dioxide	CrO_2F_2	7788-96-7	121.992	red-viol cry	30	subl			reac H₂O
826	Cobalt	Co	7440-48-4	58.933	gray metal; hex or cub	1495	2927	8.86		s dil acid
827	Cobaltocene	$Co(C_5H_5)_2$	1277-43-6	189.119	blk-purp cry	173				
828	Cobalt antimonide	$CoSb$	12052-42-5	180.693	hex cry	1202		8.8		
829	Cobalt arsenic sulfide	$CoAsS$	12254-82-9	165.920	silv-wh solid			≈6.1		
830	Cobalt arsenide (CoAs)	$CoAs$	27016-73-5	133.855	orth cry	1180		8.22		
831	Cobalt arsenide (CoAs₂)	$CoAs_2$	12044-42-7	208.776	monocl cry			7.2		
832	Cobalt arsenide (CoAs₃)	$CoAs_3$	12256-04-1	283.698	cub cry	942		6.84		
833	Cobalt boride (CoB)	CoB	12006-77-8	69.744	refrac solid	1460		7.25		reac H₂O, HNO₃
834	Cobalt boride (Co₂B)	Co_2B	12045-01-1	128.677	refrac solid	1280		8.1		
835	Cobalt carbonyl	$Co_2(CO)_8$	10210-68-1	341.947	oran cry	51 dec		1.78		i H₂O; s EtOH, eth, CS₂
836	Cobalt disulfide	CoS_2	12013-10-4	123.063	cub cry			4.3		
837	Cobalt dodecacarbonyl	$Co_4(CO)_{12}$	17786-31-1	571.854	blk cry	60 dec		2.09		
838	Cobalt phosphide	Co_2P	12134-02-0	148.840	gray needles	1386		6.4		i H₂O; s HNO₃
839	Cobalt silicide	$CoSi_2$	12017-12-8	115.104	gray cub cry	1326		4.9		s hot HCl
840	Cobalt(II) acetate	$Co(C_2H_3O_2)_2$	71-48-7	177.022	pink cry					vs H₂O; s EtOH
841	Cobalt(II) acetate tetrahydrate	$Co(C_2H_3O_2)_2 \cdot 4H_2O$	6147-53-1	249.082	red monocl cry			1.705		s H₂O, EtOH, dil acid
842	Cobalt(II) aluminate	$CoAl_2O_4$	13820-62-7	176.894	blue cub cry			4.37		i H₂O
843	Cobalt(II) arsenate octahydrate	$Co_3(AsO_4)_2 \cdot 8H_2O$	24719-19-5	598.760	red monocl needles	400 dec	1000 dec	3.0		i H₂O; s dil acid
844	Cobalt(II) bromate hexahydrate	$Co(BrO_3)_2 \cdot 6H_2O$	13476-01-2	422.829	viol cry			≈2.5		vs H₂O
845	Cobalt(II) bromide	$CoBr_2$	7789-43-7	218.741	grn hex cry; hyg	678		4.91	113.2²⁰	s MeOH, EtOH, ace
846	Cobalt(II) bromide hexahydrate	$CoBr_2 \cdot 6H_2O$	13762-12-4	326.832	red hyg cry	47 dec	100 dec	2.46	113.2	
847	Cobalt(II) carbonate	$CoCO_3$	513-79-1	118.942	pink rhomb cry	dec 280		4.2	0.00014²⁰	i EtOH
848	Cobalt(II) basic carbonate	$2CoCO_3 \cdot 3Co(OH)_2 \cdot H_2O$	7542-09-8	534.743	red-viol cry	dec				i H₂O; s acid
849	Cobalt(II) chlorate hexahydrate	$Co(ClO_3)_2 \cdot 6H_2O$		333.927	dark red hyg cry	dec 61				s H₂O
850	Cobalt(II) chloride	$CoCl_2$	7646-79-9	129.839	blue hyg leaflets	737	1049	3.36	56.2²⁵	s EtOH, eth, ace, py
851	Cobalt(II) chloride dihydrate	$CoCl_2 \cdot 2H_2O$	16544-92-6	165.870	viol-blue cry			2.477	56.2²⁵	

No.	Name	Formula	CAS Reg No.	Mol. weight	Physical form	mp/°C	bp/°C	Density g cm⁻³	Solubility g/100 g H₂O	Qualitative solubility
852	Cobalt(II) chloride hexahydrate	$CoCl_2 \cdot 6H_2O$	7791-13-1	237.930	pink-red monocl cry	87 dec		1.924	56.2²⁵	s EtOH, ace, eth
853	Cobalt(II) chromate	$CoCrO_4$	24613-38-5	174.927	yel-brn orth cry			≈4.0		i H₂O; s acid
854	Cobalt(II) chromite	$CoCr_2O_4$	13455-25-9	226.923	blue-grn cub cry			5.14		i H₂O, conc acid
855	Cobalt(II) citrate dihydrate	$Co_3(C_6H_5O_7)_2 \cdot 2H_2O$	18727-04-3	265.170	rose red cry	dec 150			0.8¹⁵	
856	Cobalt(II) cyanide	$Co(CN)_2$	542-84-7	110.967	blue hyg cry			1.872		i H₂O
857	Cobalt(II) cyanide dihydrate	$Co(CN)_2 \cdot 2H_2O$	20427-11-6	146.998	pink-brn needles					i H₂O, acid
858	Cobalt(II) diiron tetroxide	$CoFe_2O_4$	12052-28-7	234.621	blk solid					s hot HCl
859	Cobalt(II) ferricyanide	$Co_3[Fe(CN)_6]_2$	14049-81-1	600.698	red needles					i H₂O, HCl; s NH₄OH
860	Cobalt(II) fluoride	CoF_2	10026-17-2	96.930	red tetr cry	1127	≈1400	4.46	1.4²⁵	s acid
861	Cobalt(II) fluoride tetrahydrate	$CoF_2 \cdot 4H_2O$	13817-37-3	168.992	red orth cry	dec		2.22	1.4²⁵	
862	Cobalt(II) formate dihydrate	$Co(CHO_2)_2 \cdot 2H_2O$	6424-20-0	184.998	red cry powder	140 dec		2.13	5.03²⁰	i EtOH
863	Cobalt(II) hexafluoro-2,4-pentanedioate	$Co(CF_3COCHCOCF_3)_2$	19648-83-0	473.035	powder	197				
864	Cobalt(II) hexafluorosilicate hexahydrate	$CoSiF_6 \cdot 6H_2O$	12021-68-0	309.100	pale red cry			2.087	76.8²²	
865	Cobalt(II) hydroxide	$Co(OH)_2$	21041-93-0	92.948	blue-grn cry	≈160 dec		3.60		sl H₂O; s acid
866	Cobalt(II) hydroxide monohydrate	$Co(OH)_2 \cdot H_2O$	35340-84-2	110.963	blue solid	136 dec				
867	Cobalt(II) iodate	$Co(IO_3)_2$	13455-28-2	408.738	blk-viol needles	200 dec		5.09	0.46²⁰	
868	Cobalt(II) iodide	CoI_2	15238-00-3	312.742	blk hex cry; hyg	520		5.60	203²⁵	
869	Cobalt(II) iodide dihydrate	$CoI_2 \cdot 2H_2O$	13455-29-3	348.773	hyg grn cry	dec 100				
870	Cobalt(II) iodide hexahydrate	$CoI_2 \cdot 6H_2O$	15238-00-3*	420.833	red hex prisms	130 dec		2.90	203²⁵	s EtOH, eth, ace
871	Cobalt(II) molybdate	$CoMoO_4$	13762-14-6	218.87	blk monocl cry	1040		4.7		
872	Cobalt(II) molybdate monohydrate	$CoMoO_4 \cdot H_2O$	18601-87-1	236.89	blk pow					
873	Cobalt(II) nitrate	$Co(NO_3)_2$	10141-05-6	182.942	pale red powder	100 dec		2.49	103²⁵	
874	Cobalt(II) nitrate hexahydrate	$Co(NO_3)_2 \cdot 6H_2O$	10026-22-9	291.034	red monocl cry; hyg	≈55		1.88	103²⁵	s EtOH
875	Cobalt(II) nitrite	$Co(NO_2)_2$	18488-96-5	150.944					0.49²⁵	
876	Cobalt(II) oleate	$Co(C_{18}H_{33}O_2)_2$	14666-94-5	621.840	brn amorp pow					i H₂O; s EtOH, eth
877	Cobalt(II) orthosilicate	Co_2SiO_4	12017-08-2	209.950	red-viol orth cry	1345		4.63		i H₂O; s dil HCl
878	Cobalt(II) oxalate	CoC_2O_4	814-89-1	146.952	pink powder	250 dec		3.02	0.0037²⁰	s acid, NH₄OH
879	Cobalt(II) oxalate dihydrate	$CoC_2O_4 \cdot 2H_2O$	5965-38-8	182.982	pink needles	dec			0.0037	sl acid; s NH₄OH
880	Cobalt(II) oxide	CoO	1307-96-6	74.932	gray cub cry	1830		6.44		i H₂O; s acid
881	Cobalt(II) 2,4-pentanedioate	$Co(CH_3COCHCOCH_3)_2$	14024-48-7	257.149	bl-viol cry	167				
882	Cobalt(II) perchlorate	$Co(ClO_4)_2$	13455-31-7	257.834	red needles			3.33	113²⁵	i EtOH, ace
883	Cobalt(II) perchlorate hexahydrate	$Co(ClO_4)_2 \cdot 6H_2O$	13478-33-6	365.926	dark red cry	dec 170		3.33		vs H₂O
884	Cobalt(II) phosphate octahydrate	$Co_3(PO_4)_2 \cdot 8H_2O$	10294-50-5	510.865	pink amorp powder			2.77		i H₂O; s acid
885	Cobalt(II) potassium sulfate hexahydrate	$CoK_2(SO_4)_2 \cdot 6H_2O$	10026-20-7	437.347	red monocl cry	75 dec		2.22		vs H₂O
886	Cobalt(II) selenate pentahydrate	$CoSeO_4 \cdot 5H_2O$	14590-19-3	291.97	red tricl cry	dec		2.51	55¹⁵	
887	Cobalt(II) selenide	$CoSe$	1307-99-9	137.89	yel hex cry	1055		7.65		i H₂O, alk; s aqua regia
888	Cobalt(II) selenite dihydrate	$CoSeO_3 \cdot 2H_2O$	19034-13-0	221.92	blue-red powder					i H₂O
889	Cobalt(II) stannate	Co_2SnO_4	12139-93-4	300.574	grn-blue cub cry			6.30		i H₂O; s alk
890	Cobalt(II) stearate	$Co(C_{18}H_{35}O_2)_2$	1002-88-6	625.872	purp solid	74		1.13		
891	Cobalt(II) sulfate	$CoSO_4$	10124-43-3	154.996	red orth cry	>700		3.71	38.3²⁵	
892	Cobalt(II) sulfate monohydrate	$CoSO_4 \cdot H_2O$	13455-34-0	173.011	red monocl cry			3.08	38.3²⁵	
893	Cobalt(II) sulfate heptahydrate	$CoSO_4 \cdot 7H_2O$	10026-24-1	281.102	pink monocl cry	41 dec		2.03	38.3²⁵	sl EtOH, MeOH
894	Cobalt(II) sulfide	CoS	1317-42-6	90.998	blk amorp powder	1117		5.45		i H₂O; s acid
895	Cobalt(II) telluride	$CoTe$	12017-13-9	186.53	hex cry			≈8.8		
896	Cobalt(II) thiocyanate	$Co(SCN)_2$	3017-60-5	175.097	yel-brn powder				103²⁵	s EtOH, MeOH, ace, eth
897	Cobalt(II) thiocyanate trihydrate	$Co(SCN)_2 \cdot 3H_2O$	97126-35-7	229.143	viol rhomb cry				103²⁵	s EtOH, eth, ace
898	Cobalt(II) titanate	$CoTiO_3$	12017-01-5	154.798	grn rhomb cry			5.0		
899	Cobalt(II) tungstate	$CoWO_4$	12640-47-0	306.77	blue monocl cry			≈7.8		i H₂O; s hot conc acid
900	Cobalt(II,III) oxide	Co_3O_4	1308-06-1	240.798	blk cub cry	900 dec		6.11		i H₂O; s acid, alk
901	Cobalt(III) acetate	$Co(C_2H_3O_2)_3$	917-69-1	236.064	grn hyg cry	100 dec				s H₂O, EtOH
902	Cobalt(III) ammonium tetranitrodiammine	$NH_4[Co(NH_3)_2(NO_2)_4]$	13600-89-0	295.054	red-brn orth cry			1.97		s H₂O
903	Cobalt(III) fluoride	CoF_3	10026-18-3	115.928	brn hex cry	927		3.88		reac H₂O; s EtOH, eth, bz
904	Cobalt(III) fluoride dihydrate	$Co_2F_6 \cdot 2H_2O$	54496-71-8	267.887	red rhomb cry			2.19		s H₂O; i EtOH
905	Cobalt(III) hexammine chloride	$Co(NH_3)_6Cl_3$	10534-89-1	267.475	red monocl cry			1.71		s H₂O; i EtOH
906	Cobalt(III) hydroxide	$Co(OH)_3$	1307-86-4	109.955	brn powder	dec		≈4		i H₂O; s acid
907	Cobalt(III) nitrate	$Co(NO_3)_3$	15520-84-0	244.948	grn cub cry; hyg			≈3.0		s H₂O; reac os
908	Cobalt(III) oxide	Co_2O_3	1308-04-9	165.864	gray-blk powder	895 dec		5.18		i H₂O; s conc acid

No.	Name	Formula	CAS Reg No.	Mol. weight	Physical form	mp/°C	bp/°C	Density g cm^{-3}	Solubility g/100 g H$_2$O	Qualitative solubility
909	Cobalt(III) oxide monohydrate	Co$_2$O$_3 \cdot$ H$_2$O	12016-80-7	183.880	brn-blk hex cry	150 dec				i H$_2$O; s acid
910	Cobalt(III) 2,4-pentanedioate	Co(CH$_3$COCHCOCH$_3$)$_3$	21679-46-9	356.257	dark grn cry	213				s bz, ace
911	Cobalt(III) potassium nitrite sesquihydrate	CoK$_3$(NO$_2$)$_6 \cdot$ 1.5H$_2$O	13782-01-9*	479.284	yel cub cry			2.6		sl H$_2$O; reac acid; i EtOH
912	Cobalt(III) sulfide	Co$_2$S$_3$	1332-71-4	214.061	blk cub cry			4.8		reac acid
913	Cobalt(III) titanate	Co$_2$TiO$_4$	12017-38-8	229.731	grn-blk cub cry			5.1		s conc HCl
914	Copper	Cu	7440-50-8	63.546	red metal; cub	1084.62	2560	8.96		sl dil acid
915	Copper arsenide	Cu$_3$As	12005-75-3	265.560	dark gray solid	827				
916	Copper nitride	Cu$_3$N	1308-80-1	204.645	cub cry	300 dec		5.84		
917	Copper phosphide	CuP$_2$	12019-11-3	125.494	monocl cry	≈900		4.20		
918	Copper silicide	Cu$_5$Si	12159-07-8	345.816	solid	825				
919	Copper(I) acetate	CuC$_2$H$_3$O$_2$	598-54-9	122.590	col cry	dec	subl			reac H$_2$O
920	Copper(I) acetylide	Cu$_2$C$_2$	1117-94-8	151.113	red amorp powder; exp					
921	Copper(I) azide	CuN$_3$	14336-80-2	105.566	tetr cry; exp					
922	Copper(I) bromide	CuBr	7787-70-4	143.450	wh cub cry; hyg	483	1345	4.98	0.0012^{20}	i ace
923	Copper(I) chloride	CuCl	7758-89-6	98.999	wh cub cry	423	1490	4.14	0.0047^{20}	i EtOH, ace
924	Copper(I) cyanide	CuCN	544-92-3	89.564	wh powder or grn orth cry	474	dec	2.9		i H$_2$O, EtOH; s KCN soln
925	Copper(I) fluoride	CuF	13478-41-6	82.544	cub cry			7.1		
926	Copper(I) hydride	CuH	13517-00-5	64.554	red-brn solid	60 dec				
927	Copper(I) iodide	CuI	7681-65-4	190.450	wh cub cry	591	≈1290	5.67	0.000020^{20}	i dil acid
928	Copper(I) mercury iodide	Cu$_2$HgI$_4$	13876-85-2	835.30	red cry powder	trans ≈60 (brn)				i H$_2$O, EtOH
929	Copper(I) oxide	Cu$_2$O	1317-39-1	143.091	red-brn cub cry	1244	1800 dec	6.0		i H$_2$O
930	Copper(I) selenide	Cu$_2$Se	20405-64-5	206.05	blue-blk tetr cry	1113		6.84		i H$_2$O; s acid
931	Copper(I) sulfide	Cu$_2$S	22205-45-4	159.157	blue-blk orth cry	1129		5.6		i H$_2$O; sl acid
932	Copper(I) sulfite hemihydrate	Cu$_2$SO$_3 \cdot$ 0.5H$_2$O	13982-53-1*	216.164	wh-yel hex cry					sl H$_2$O; s acid, alk; i EtOH, eth
933	Copper(I) sulfite monohydrate	Cu$_2$SO$_3 \cdot$ H$_2$O	35788-00-2	225.171	cry			3.83		sl H$_2$O; s HCl
934	Copper(I) telluride	Cu$_2$Te	12019-52-2	254.69	blue hex cry	1127		4.6		
935	Copper(I) thiocyanate	CuSCN	1111-67-7	121.629	wh-yel amorp powder	1084		2.85		i H$_2$O, dil acid, EtOH, ace; s eth
936	Copper(I,II) sulfite dihydrate	Cu$_2$SO$_3 \cdot$ CuSO$_3 \cdot$ 2H$_2$O	13814-81-8	386.795	red prisms or powder					i H$_2$O, EtOH; s HCl
937	Copper(II) acetate	Cu(C$_2$H$_3$O$_2$)$_2$	142-71-2	181.635	blue-grn hyg powder					
938	Copper(II) acetate monohydrate	Cu(C$_2$H$_3$O$_2$)$_2 \cdot$ H$_2$O	6046-93-1	199.650	grn monocl cry	115	240 dec	1.88		s H$_2$O, EtOH; sl eth
939	Copper(II) acetate metaarsenite	Cu(C$_2$H$_3$O$_2$)$_2 \cdot$ 3Cu(AsO$_2$)$_2$	12002-03-8	1013.795	grn cry powder					i H$_2$O; reac acid
940	Copper(II) basic acetate	Cu(C$_2$H$_3$O$_2$)$_2 \cdot$ CuO $\cdot$ 6H$_2$O	52503-64-7	369.271	blue-grn cry or powder					sl H$_2$O, EtOH; s dil acid, NH$_4$OH
941	Copper(II) acetylide	CuC$_2$	12540-13-5	87.567	brn-blk solid; exp	exp 100				
942	Copper(II) arsenate	Cu$_3$(AsO$_4$)$_2$	7778-41-8	468.476	blue-grn cry					i H$_2$O, EtOH; s dil acid
943	Copper(II) arsenite	CuHAsO$_3$	10290-12-7	187.474	yel-grn powder					i H$_2$O, EtOH; s acid
944	Copper(II) azide	Cu(N$_3$)$_2$	14215-30-6	147.586	brn orth cry; exp			≈2.6		
945	Copper(II) borate	Cu(BO$_2$)$_2$	39290-85-2	149.166	blue-grn powder			3.859		i H$_2$O; s acid
946	Copper(II) bromide	CuBr$_2$	7789-45-9	223.354	blk monocl cry; hyg	498	900	4.710	126^{25}	vs H$_2$O; s EtOH, ace; i bz, eth
947	Copper(II) butanoate monohydrate	Cu(C$_4$H$_7$O$_2$)$_2 \cdot$ H$_2$O	540-16-9	255.756	grn monocl plates					s H$_2$O, diox, bz; sl EtOH
948	Copper(II) carbonate	CuCO$_3$	1184-64-1	123.555	cry					i H$_2$O
949	Copper(II) carbonate hydroxide	CuCO$_3 \cdot$ Cu(OH)$_2$	12069-69-1	221.116	grn monocl cry	200 dec		4.0		i H$_2$O, EtOH; s dil acid
950	Copper(II) chlorate hexahydrate	Cu(ClO$_3$)$_2 \cdot$ 6H$_2$O	14721-21-2	338.540	blue-grn hyg cry	65	100 dec		164^{18}	vs EtOH
951	Copper(II) chloride	CuCl$_2$	7447-39-4	134.452	yel-brn monocl cry; hyg	598	993	3.4	75.7^{25}	s EtOH, ace
952	Copper(II) chloride dihydrate	CuCl$_2 \cdot$ 2H$_2$O	10125-13-0	170.483	grn-blue orth cry; hyg	100 dec		2.51	75.7^{20}	vs EtOH, MeOH; s ace; i eth
953	Copper(II) chloride hydroxide	Cu$_2$(OH)$_3$Cl	1332-65-6	213.567	pale grn cry					i H$_2$O; s acid
954	Copper(II) chromate	CuCrO$_4$	13548-42-0	179.540	red-brn cry					i H$_2$O; s EtOH
955	Copper(II) basic chromate	CuCrO$_4 \cdot$ 2Cu(OH)$_2$	12433-14-6	374.661	brn pow	dec 260				i H$_2$O; s HNO$_3$
956	Copper(II) chromite	CuCr$_2$O$_4$	12018-10-9	231.536	gray-blk tetr cry			5.4		i H$_2$O, dil acid
957	Copper(II) citrate hemipentahydrate	Cu$_2$C$_6$H$_4$O$_7 \cdot$ 2.5H$_2$O	10402-15-0	360.221	blue-grn cry	100 dec				sl H$_2$O; s dil acid
958	Copper(II) cyanide	Cu(CN)$_2$	14763-77-0	115.580	grn powder					i H$_2$O; s acid, alk
959	Copper(II) cyclohexanebutanoate	Cu(C$_{10}$H$_{17}$O$_2$)$_2$	2218-80-6	402.028	powder	126 dec				
960	Copper(II) dichromate dihydrate	CuCr$_2$O$_7 \cdot$ 2H$_2$O	13675-47-3	315.565	red-brn tricl cry			2.286		vs H$_2$O
961	Copper(II) ethanolate	Cu(C$_2$H$_5$O)$_2$	2850-65-9	153.667	blue hyg solid	120 dec				i os
962	Copper(II) ethylacetoacetate	Cu(C$_2$H$_5$CO$_2$CHCOCH$_3$)$_2$	14284-06-1	321.813	grn cry	192				s EtOH, chl

No.	Name	Formula	CAS Reg No.	Mol. weight	Physical form	mp/°C	bp/°C	Density g cm^{-3}	Solubility g/100 g H$_2$O	Qualitative solubility
963	Copper(II) 2-ethylhexanoate	Cu(C$_8$H$_{15}$O$_2$)$_2$	149-11-1	349.953	powder	252 dec				
964	Copper(II) ferrate	CuFe$_2$O$_4$	12018-79-0	239.234	blk cry					
965	Copper(II) ferrocyanide	Cu$_2$Fe(CN)$_6$	13601-13-3	339.041	red-br cub cry or powder			2.2		i H$_2$O, acid, os
966	Copper(II) ferrous sulfide	CuFeS$_2$	1308-56-1	183.521	yel tetr cry	950		4.2		i H$_2$O, HCl; s HNO$_3$
967	Copper(II) fluoride	CuF$_2$	7789-19-7	101.543	wh monocl cry	836	1676	4.23	0.075[25]	
968	Copper(II) fluoride dihydrate	CuF$_2$ · 2H$_2$O	13454-88-1	137.574	blue monocl cry	130 dec		2.934	0.075[25]	
969	Copper(II) formate	Cu(CHO$_2$)$_2$	544-19-4	153.581	blue cry				12.5[20]	i os
970	Copper(II) formate tetrahydrate	Cu(CHO$_2$)$_2$ · 4H$_2$O	5893-61-8	225.641	blue monocl cry				12.5	sl EtOH; i os
971	Copper(II) gluconate	CuC$_{12}$H$_{22}$O$_{14}$	527-09-3	453.841	bl-grn cry	156				sl EtOH; i os
972	Copper(II) hexafluoro-2,4-pentanedioate	Cu(CF$_3$COCHCOCF$_3$)$_2$	14781-45-4	477.648	cry	98	220 dec			s MeOH, ace, tol
973	Copper(II) hexafluorosilicate tetrahydrate	CuSiF$_6$ · 4H$_2$O	12062-24-7	277.684	blue monocl cry	dec		2.56	99.7[17]	sl EtOH
974	Copper(II) hydroxide	Cu(OH)$_2$	20427-59-2	97.561	blue-grn powder			3.37		i H$_2$O; s acid, conc alk
975	Copper(II) iodate	Cu(IO$_3$)$_2$	13454-89-2	413.351	grn monocl cry	dec		5.241	0.15[20]	s dil acid
976	Copper(II) iodate monohydrate	Cu(IO$_3$)$_2$ · H$_2$O	13454-90-5	431.367	blue tricl cry	248 dec		4.872	0.15[20]	s dil H$_2$SO$_4$
977	Copper(II) molybdate	CuMoO$_4$	13767-34-5	223.48	grn cry	≈500		3.4	0.038	
978	Copper(II) nitrate	Cu(NO$_3$)$_2$	3251-23-8	187.555	blue-grn orth cry; hyg	255	subl		145[25]	s diox; reac eth
979	Copper(II) nitrate trihydrate	Cu(NO$_3$)$_2$ · 3H$_2$O	10031-43-3	241.602	blue rhomb cry	114	170 dec	2.32	145[25]	vs EtOH
980	Copper(II) nitrate hexahydrate	Cu(NO$_3$)$_2$ · 6H$_2$O	13478-38-1	295.647	blue rhomb cry; hyg			2.07	145[25]	s EtOH
981	Copper(II) oleate	Cu(C$_{18}$H$_{33}$O$_2$)$_2$	1120-44-1	626.453	blue-grn solid					i H$_2$O; sl EtOH; s eth
982	Copper(II) oxalate	CuC$_2$O$_4$	814-91-5	151.565	blue-wh powder	310 dec			0.0026[20]	i EtOH, eth; s NH$_4$OH
983	Copper(II) oxalate hemihydrate	CuC$_2$O$_4$ · 0.5H$_2$O	814-91-5*	144.573	blue-wh cry	200 dec			0.0026[20]	s NH$_4$OH
984	Copper(II) oxide	CuO	1317-38-0	79.545	blk powder or monocl cry	1227		6.31		i H$_2$O, EtOH; s dil acid
985	Copper(II) 2,4-pentanedioate	Cu(CH$_3$COCHCOCH$_3$)$_2$	13395-16-9	261.762	blue powder	284 dec	subl			sl H$_2$O; s chl
986	Copper(II) oxychloride hemiheptahydrate	CuCl$_2$ · 3CuO · 3.5H$_2$O	1332-40-7		blue-grn pow	dec 140				i H$_2$O; s acid, NH$_4$OH
987	Copper(II) perchlorate	Cu(ClO$_4$)$_2$	13770-18-8	262.447	grn hyg cry	130 dec			146[30]	s eth, diox; i bz, ctc
988	Copper(II) perchlorate hexahydrate	Cu(ClO$_4$)$_2$ · 6H$_2$O	10294-46-9	370.539	blue monocl cry; hyg	82	120 dec	2.22	146[30]	vs EtOH, HOAc, ace; sl eth
989	Copper(II) phosphate	Cu$_3$(PO$_4$)$_2$	7798-23-4	380.581	blue-grn tricl cry					i H$_2$O; s acid, NH$_4$OH
990	Copper(II) phosphate trihydrate	Cu$_3$(PO$_4$)$_2$ · 3H$_2$O	10031-48-8	434.627	blue-grn orth cry					i H$_2$O; s acid, NH$_4$OH
991	Copper(II) phthalocyanine	CuC$_{32}$H$_{16}$N$_8$	147-14-8	576.069	bl-purp cry					i H$_2$O, EtOH; s conc H$_2$SO$_4$
992	Copper(II) selenate pentahydrate	CuSeO$_4$ · 5H$_2$O	10031-45-5	296.58	blue tricl cry	80 dec		2.56	27.4[25]	s acid, NH$_4$OH; sl ace; i EtOH
993	Copper(II) selenide	CuSe	1317-41-5	142.51	blue-blk needles or plates	550 dec		5.99		reac acid
994	Copper(II) selenite dihydrate	CuSeO$_3$ · 2H$_2$O	15168-20-4	226.54	blue orth cry			3.31		i H$_2$O; s acid, NH$_4$OH
995	Copper(II) silicate dihydrate	CuSiO$_3$ · 2H$_2$O	26318-99-0	175.661	grn-blue orth cry					
996	Copper(II) stannate	CuSnO$_3$	12019-07-7	230.254	blue pow					
997	Copper(II) stearate	Cu(C$_{18}$H$_{35}$O$_2$)$_2$	660-60-6	630.485	blue-grn amorp powder	≈250				i H$_2$O, EtOH, eth; s py
998	Copper(II) sulfate	CuSO$_4$	7758-98-7	159.609	wh-grn amorp powder or rhomb cry	560 dec		3.60	22.0[25]	i EtOH
999	Copper(II) sulfate pentahydrate	CuSO$_4$ · 5H$_2$O	7758-99-8	249.685	blue tricl cry	110 dec		2.286	22.0[25]	s MeOH; sl EtOH
1000	Copper(II) sulfate, basic	Cu$_3$(OH)$_4$SO$_4$	1332-14-5	354.730	grn rhomb cry			3.88		i H$_2$O
1001	Copper(II) sulfide	CuS	1317-40-4	95.611	blk hex cry	trans 507		4.76		i H$_2$O, EtOH, dil acid, alk
1002	Copper(II) tartrate trihydrate	CuC$_4$H$_4$O$_6$ · 3H$_2$O	815-82-7	265.663	blue-grn powder					sl H$_2$O; s acid, alk
1003	Copper(II) telluride	CuTe	12019-23-7	191.15	yel orth cry	trans ≈400		7.09		
1004	Copper(II) tellurite	CuTeO$_3$	13812-58-3	239.14	blk glassy solid					i H$_2$O
1005	Copper(II) tetrafluoroborate	Cu(BF$_4$)$_2$	14735-84-3	237.155	solid					s H$_2$O
1006	Copper(II) titanate	CuTiO$_3$	12019-08-8	159.411	gray pow					
1007	Copper(II) 1,1,1-trifluoro-2,4-pentanedioate	Cu(CF$_3$COCHCOCH$_3$)$_2$	14324-82-4	369.705	blue-purp cry	197	dec 260			s EtOH, tol
1008	Copper(II) tungstate	CuWO$_4$	13587-35-4	311.38	yel-brn powder			7.5		
1009	Copper(II) tungstate dihydrate	CuWO$_4$ · 2H$_2$O	13587-35-4*	347.41	grn powder					i H$_2$O; sl HOAc; reac conc acid
1010	Copper(II) vanadate	Cu(VO$_3$)$_2$	12789-09-2	261.425	powder					

No.	Name	Formula	CAS Reg No.	Mol. weight	Physical form	mp/°C	bp/°C	Density g cm⁻³	Solubility g/100 g H₂O	Qualitative solubility
1011	Curium	Cm	7440-51-9	247	silv metal; hex or cub	1345		13.51	
1012	Dysprosium	Dy	7429-91-6	162.500	silv metal; hex	1412	2567	8.55		s dil acid
1013	Dysprosium boride	DyB_4	12310-43-9	205.744	tetr cry	2500		6.98	
1014	Dysprosium nitride	DyN	12019-88-4	176.507	cub cry			9.93	
1015	Dysprosium silicide	$DySi_2$	12133-07-2	218.671	orth cry	1550		5.2	
1016	Dysprosium(II) bromide	$DyBr_2$	83229-05-4	322.308	blk solid				
1017	Dysprosium(II) chloride	$DyCl_2$	13767-31-2	233.406	blk cry	721 dec				reac H_2O
1018	Dysprosium(II) iodide	DyI_2	36377-94-3	416.309	purp cry	659				reac H_2O
1019	Dysprosium(III) acetate tetrahydrate	$Dy(C_2H_3O_2)_3 \cdot 4H_2O$	15280-55-4	411.693	yel needles	dec 120				s H_2O; sl EtOH
1020	Dysprosium(III) bromide	$DyBr_3$	14456-48-5	402.212	wh hyg cry	879				s H_2O
1021	Dysprosium(III) carbonate tetrahydrate	$Dy_2(CO_3)_3 \cdot 4H_2O$	38245-35-1	577.088	wh cry pow					i H_2O
1022	Dysprosium(III) chloride	$DyCl_3$	10025-74-8	268.859	wh or yel cry	718	1530	3.67		s H_2O, MeOH
1023	Dysprosium(III) chloride hexahydrate	$DyCl_3 \cdot 6H_2O$	15059-52-6	376.950	bright yel cry	dec 162			
1024	Dysprosium(III) fluoride	DyF_3	13569-80-7	219.495	grn cry	1157			
1025	Dysprosium(III) hydride	DyH_3	13537-09-2	165.524	hex cry			7.1	
1026	Dysprosium(III) hydroxide	$Dy(OH)_3$	1308-85-6	213.522	yel or wh needles	205 dec				i H_2O
1027	Dysprosium(III) iodide	DyI_3	15474-63-2	543.213	grn cry	978			
1028	Dysprosium(III) nitrate pentahydrate	$Dy(NO_3)_3 \cdot 5H_2O$	10143-38-1*	438.591	yel cry	88.6			208.4^{25}
1029	Dysprosium(III) oxide	Dy_2O_3	1308-87-8	372.998	wh cub cry	2228	3900	7.81		s acid
1030	Dysprosium(III) sulfate octahydrate	$Dy_2(SO_4)_3 \cdot 8H_2O$	10031-50-2	757.310	pale yel cry	110				sl H_2O
1031	Dysprosium(III) sulfide	Dy_2S_3	12133-10-7	421.195	red-brn monocl cry			6.08	
1032	Dysprosium(III) telluride	Dy_2Te_3	12159-43-2	707.80	solid	≈1550			
1033	Einsteinium	Es	7429-92-7	252	metal; cub	860			
1034	Erbium	Er	7440-52-0	167.259	silv metal; hex	1529	2868	9.07		i H_2O; s acid
1035	Erbium boride	ErB_4	12310-44-0	210.503	tetr cry	2450		7.0	
1036	Erbium acetate tetrahydrate	$Er(C_2H_3O_2)_3 \cdot 4H_2O$	15280-57-6	416.452	pink or wh cry			2.11		s H_2O
1037	Erbium bromide	$ErBr_3$	13536-73-7	406.971	viol hyg cry	950	≈1460			s H_2O, thf
1038	Erbium bromide hexahydrate	$ErBr_3 \cdot 6H_2O$	14890-44-9	515.062	pink cry					s H_2O
1039	Erbium chloride	$ErCl_3$	10138-41-7	273.618	viol monocl cry; hyg	776		4.1		s H_2O
1040	Erbium chloride hexahydrate	$ErCl_3 \cdot 6H_2O$	10025-75-9	381.709	pink hyg cry	dec				s H_2O; sl EtOH
1041	Erbium fluoride	ErF_3	13760-83-3	224.254	pink orth cry	1146		7.8		i H_2O
1042	Erbium hydride	ErH_3	13550-53-3	170.283	hex cry			≈7.6	
1043	Erbium hydroxide	$Er(OH)_3$	14646-16-3	218.281	pink solid					i H_2O
1044	Erbium iodide	ErI_3	13813-42-8	547.972	viol hex cry; hyg	1014		≈5.5		s H_2O
1045	Erbium nitrate pentahydrate	$Er(NO_3)_3 \cdot 5H_2O$	10168-80-6*	443.350	red cry	130 dec			240.8^{25}	s EtOH, ace
1046	Erbium nitride	ErN	12020-21-2	181.266	cub cry			10.6	
1047	Erbium oxide	Er_2O_3	12061-16-4	382.516	pink powder	2344	3920	8.64		i H_2O; s acid
1048	Erbium silicide	$ErSi_2$	12020-28-9	223.430	orth cry			7.26	
1049	Erbium sulfate	$Er_2(SO_4)_3$	13478-49-4	622.706	hyg powder	dec		3.68	13^{20}
1050	Erbium sulfate octahydrate	$Er_2(SO_4)_3 \cdot 8H_2O$	10031-52-4	766.828	pink monocl cry	dec		3.20	13^{20}
1051	Erbium sulfide	Er_2S_3	12159-66-9	430.713	red-brn monocl cry	1730		6.07	
1052	Erbium telluride	Er_2Te_3	12020-39-2	717.32	orth cry	1213		7.11	
1053	Europium	Eu	7440-53-1	151.964	soft silv metal; cub	822	1529	5.24		reac H_2O
1054	Europium boride	EuB_6	12008-05-8	216.830	cub cry	≈2600		4.91	
1055	Europium nitride	EuN	12020-58-5	165.971	cub cry			8.7	
1056	Europium silicide	$EuSi_2$	12434-24-1	208.135	tetr cry	1500		5.46	
1057	Europium(II) bromide	$EuBr_2$	13780-48-8	311.772	wh cry	683				s H_2O
1058	Europium(II) chloride	$EuCl_2$	13769-20-5	222.870	wh orth cry	731		4.9		s H_2O
1059	Europium(II) fluoride	EuF_2	14077-39-5	189.961	grn-yel cub cry	≈1380		6.5	
1060	Europium(II) iodide	EuI_2	22015-35-6	405.773	grn cry	580				s H_2O
1061	Europium(II) selenide	EuSe	12020-66-5	230.92	brn cub cry			6.45	
1062	Europium(II) sulfate	$EuSO_4$	10031-54-6	248.027	col orth cry			4.99		i H_2O
1063	Europium(II) sulfide	EuS	12020-65-4	184.029	cub cry			5.7	
1064	Europium(II) telluride	EuTe	12020-69-8	279.56	blk cub cry	1526		6.48	
1065	Europium(III) bromide	$EuBr_3$	13759-88-1	391.676	gray cry	dec				s H_2O
1066	Europium(III) chloride	$EuCl_3$	10025-76-0	258.323	grn-yel needles	623		4.89	
1067	Europium(III) chloride hexahydrate	$EuCl_3 \cdot 6H_2O$	13759-92-7	366.414	wh-yel hyg cry	850		4.89		s H_2O
1068	Europium(III) fluoride	EuF_3	13765-25-8	208.959	wh hyg cry	1276				i H_2O
1069	Europium(III) iodide	EuI_3	13759-90-5	532.677	col cry; unstab	≈875			
1070	Europium(III) nitrate hexahydrate	$Eu(NO_3)_3 \cdot 6H_2O$	10031-53-5	446.070	wh-pink hyg cry	85 dec			193^{25}
1071	Europium(III) oxalate	$Eu_2(C_2O_4)_3$	3269-12-3	567.985	wh solid					i H_2O; s acid

No.	Name	Formula	CAS Reg No.	Mol. weight	Physical form	mp/°C	bp/°C	Density g cm⁻³	Solubility g/100 g H₂O	Qualitative solubility
1072	Europium(III) oxide	Eu_2O_3	1308-96-9	351.926	pink powder	2291	3790	7.42		i H_2O; s acid
1073	Europium(III) perchlorate hexahydrate	$Eu(ClO_4)_3 \cdot 6H_2O$	36907-40-1	558.407	wh or pink cry					s H_2O, EtOH
1074	Europium(III) sulfate	$Eu_2(SO_4)_3$	13537-15-0	592.116	pale pink cry			4.99	2.1[20]	
1075	Europium(III) sulfate octahydrate	$Eu_2(SO_4)_3 \cdot 8H_2O$	10031-52-4	736.238	pink cry	375 dec			2.1[20]	
1076	Fermium	Fm	7440-72-4	257	metal	1527				
1077	Fluorine	F_2	7782-41-4	37.997	pale yel gas	-219.67	-188.11	1.553 g/L		reac H_2O
1078	Fluorine monoxide	F_2O	7783-41-7	53.996	col gas	-223.8	-144.3	2.207 g/L		sl H_2O
1079	Difluorine dioxide	F_2O_2	7783-44-0	69.996	red-oran solid, unstab gas	-163.5	-57 extrap	2.861 g/L		
1080	Fluorine tetroxide	F_2O_4	107782-11-6	101.995	red-brn solid	-191	dec -185			
1081	Fluorine nitrate	FNO_3	7789-26-6	81.003	col gas	-175	-46	3.311 g/L		reac H_2O, EtOH, eth; s ace
1082	Fluorine perchlorate	$FOClO_3$	10049-03-3	118.449	col gas; exp	-167.3	-16	4.841 g/L		reac H_2O
1083	Francium	Fr	7440-73-5	223.000	short-lived alkali metal	21				
1084	Gadolinium	Gd	7440-54-2	157.25	silv metal; hex	1313	3273	7.90		s dil acid
1085	Gadolinium boride	GdB_6	12008-06-9	222.12	blk-brn cub cry	2510		5.31		
1086	Gadolinium nitride	GdN	25764-15-2	171.26	cub cry			9.10		
1087	Gadolinium silicide	$GdSi_2$	12134-75-7	213.42	orth cry	1540		5.9		
1088	Gadolinium(II) iodide	GdI_2	13814-72-7	411.06	bronze cry	831				
1089	Gadolinium(II) selenide	GdSe	12024-81-6	236.21	cub cry	2170		8.1		
1090	Gadolinium(III) acetate tetrahydrate	$Gd(C_2H_3O_2)_3 \cdot 4H_2O$	15280-53-2	406.44	wh tricl cry	dec		1.61		s H_2O
1091	Gadolinium(III) bromide	$GdBr_3$	13818-75-2	396.96	wh monocl cry; hyg	770		4.56		
1092	Gadolinium(III) chloride	$GdCl_3$	10138-52-0	263.61	wh monocl cry; hyg	602		4.52		s H_2O
1093	Gadolinium(III) chloride hexahydrate	$GdCl_3 \cdot 6H_2O$	19423-81-5	371.70	col hyg cry			2.424		s H_2O
1094	Gadolinium(III) fluoride	GdF_3	13765-26-9	214.25	wh cry	1232				
1095	Gadolinium(III) iodide	GdI_3	13572-98-0	537.96	yel cry	930				
1096	Gadolinium(III) nitrate pentahydrate	$Gd(NO_3)_3 \cdot 5H_2O$	52788-53-1	433.34	wh cry	92 dec		2.41	190[25]	
1097	Gadolinium(III) nitrate hexahydrate	$Gd(NO_3)_3 \cdot 6H_2O$	19598-90-4	451.36	hyg tricl cry	91 dec		2.33	190[25]	s EtOH
1098	Gadolinium(III) oxalate decahydrate	$Gd_2(C_2O_4)_3 \cdot 10H_2O$	22992-15-0	758.71	wh monocl pow	dec 110				i H_2O; sl acid
1099	Gadolinium(III) oxide	Gd_2O_3	12064-62-9	362.50	wh hyg powder	2339	3900	7.41		i H_2O; s acid
1100	Gadolinium(III) sulfate	$Gd_2(SO_4)_3$	13628-54-1	602.69	col cry	500 dec		4.1	2.60[20]	sl H_2O
1101	Gadolinium(III) sulfate octahydrate	$Gd_2(SO_4)_3 \cdot 8H_2O$	13450-87-8	746.81	col monocl cry	400 dec		4.14	2.3[20]	sl H_2O
1102	Gadolinium(III) sulfide	Gd_2S_3	12134-77-9	410.70	yel cub cry			6.1		
1103	Gadolinium(III) telluride	Gd_2Te_3	12160-99-5	697.30	orth cry	1255		7.7		
1104	Gallium	Ga	7440-55-3	69.723	silv liq or gray orth cry	29.7646	2229	5.91		reac alk
1105	Gallium antimonide	GaSb	12064-03-8	191.483	brn cub cry	712		5.6137		
1106	Gallium arsenide	GaAs	1303-00-0	144.645	gray cub cry	1238		5.3176		
1107	Gallium nitride	GaN	25617-97-4	83.730	gray hex cry	>2500		6.1		
1108	Gallium phosphide	GaP	12063-98-8	100.697	yel cub cry	1457		4.138		
1109	Gallium suboxide	Ga_2O	12024-20-3	155.445	brn powder	>660	>800 dec	4.77		
1110	Gallium(II) chloride	$GaCl_2$	24597-12-4	140.629	wh orth cry	172.4	535	2.74		
1111	Gallium(II) selenide	GaSe	12024-11-2	148.68	hex cry	960		5.03		
1112	Gallium(II) sulfide	GaS	12024-10-1	101.788	hex cry	965		3.86		
1113	Gallium(II) telluride	GaTe	12024-14-5	197.32	monocl cry	824		5.44		
1114	Gallium(III) bromide	$GaBr_3$	13450-88-9	309.435	wh orth cry	123	279	3.69		
1115	Gallium(III) chloride	$GaCl_3$	13450-90-3	176.082	col needles or gl solid	77.9	201	2.47		
1116	Gallium(III) fluoride	GaF_3	7783-51-9	126.718	wh powder or col needles	>1000		4.47		i H_2O
1117	Gallium(III) fluoride trihydrate	$GaF_3 \cdot 3H_2O$	22886-66-4	180.764	wh cry	>140 dec				sl H_2O
1118	Gallium(III) hydride	GaH_3	13572-93-5	72.747	visc liq	-15	≈0 dec			
1119	Gallium(III) hydroxide	$Ga(OH)_3$	12023-99-3	120.745	unstab prec					
1120	Gallium(III) iodide	GaI_3	13450-91-4	450.436	monocl cry	212	340	4.5		
1121	Gallium(III) nitrate	$Ga(NO_3)_3$	13494-90-1	255.738	wh cry powder					s H_2O, EtOH, eth
1122	Gallium(III) oxide	Ga_2O_3	12024-21-4	187.444	wh cry	1807		≈6.0		s hot acid
1123	Gallium(III) oxide hydroxide	GaOOH	20665-52-5	102.730	orth cry			5.23		
1124	Gallium(III) 2,4-pentanedioate	$Ga(CH_3COCHCOCH_3)_3$	14405-43-7	367.047	wh powder	193	subl	1.42		
1125	Gallium(III) perchlorate hexahydrate	$Ga(ClO_4)_3 \cdot 6H_2O$	17835-81-3	476.166	cry	dec 175				
1126	Gallium(III) selenide	Ga_2Se_3	12024-24-7	376.33	cub cry	937		4.92		
1127	Gallium(III) sulfate	$Ga_2(SO_4)_3$	13494-91-2	427.634	hex cry					
1128	Gallium(III) sulfate octadecahydrate	$Ga_2(SO_4)_3 \cdot 18H_2O$	13780-42-2	751.909	octahed cry					s H_2O, EtOH
1129	Gallium(III) sulfide	Ga_2S_3	12024-22-5	235.641	monocl cry	1090		3.7		

No.	Name	Formula	CAS Reg No.	Mol. weight	Physical form	mp/°C	bp/°C	Density g cm⁻³	Solubility g/100 g H₂O	Qualitative solubility
1130	Gallium(III) telluride	Ga_2Te_3	12024-27-0	522.25	cub cry	790		5.57		
1131	Germanium	Ge	7440-56-4	72.63	gray-wh cub cry	938.25	2833	5.3234		i H_2O, dil acid, alk
1132	Germane	GeH_4	7782-65-2	76.67	col gas; flam	-165	-88.1	3.133 g/L		i H_2O
1133	Digermane	Ge_2H_6	13818-89-8	151.33	col liq; flam	-109	29	1.98⁻¹⁰⁹		
1134	Trigermane	Ge_3H_8	14691-44-2	225.98	col liq	-105.6	110.5	2.20⁻¹⁰⁵		i H_2O
1135	Tetragermane	Ge_4H_{10}	14691-47-5	300.64	col liq		176.9			i H_2O
1136	Pentagermane	Ge_5H_{12}	15587-39-0	375.30	col liq		234			i H_2O
1137	Bromogermane	GeH_3Br	13569-43-2	155.57	col liq	-32	52	2.34		reac H_2O
1138	Chlorogermane	GeH_3Cl	13637-65-5	111.12	col liq	-52	28	1.75		reac H_2O
1139	Chlorotrifluorogermane	GeF_3Cl	14188-40-0	165.09	gas	-66.2	-20.3	6.747 g/L		
1140	Dibromogermane	GeH_2Br_2	13769-36-3	234.46	col liq	-15	89	2.80		reac H_2O
1141	Dichlorogermane	GeH_2Cl_2	15230-48-5	145.56	col liq	-68	69.5	1.90		reac H_2O
1142	Dichlorodifluorogermane	GeF_2Cl_2	24422-21-7	181.54	col gas	-51.8	-2.8	7.419 g/L		
1143	Dichlorodimethylgermane	$Ge(CH_3)_2Cl_2$	1529-48-2	173.62	liq	-22	124	1.49		
1144	Fluorogermane	GeH_3F	13537-30-9	94.66	col gas			3.868 g/L		reac H_2O
1145	Iodogermane	GeH_3I	13573-02-9	202.57	liq	-15	≈90			reac H_2O
1146	Methylgermane	GeH_3CH_3	1449-65-6	90.70	col gas	-158	-23	3.706 g/L		
1147	Tribromogermane	$GeHBr_3$	14779-70-5	313.36	col liq	-25	dec			reac H_2O
1148	Trichlorogermane	$GeHCl_3$	1184-65-2	180.01	liq	-71	75.3	1.93		reac H_2O
1149	Trichlorofluorogermane	$GeCl_3F$	24422-20-6	198.00	liq	-49.8	37.5			
1150	Germanium(II) bromide	$GeBr_2$	24415-00-7	232.45	yel monocl cry	122	150 dec			reac H_2O
1151	Germanium(II) chloride	$GeCl_2$	10060-11-4	143.55	wh-yel hyg powder	dec				reac H_2O; s eth, bz
1152	Germanium(II) fluoride	GeF_2	13940-63-1	110.64	wh orth cry; hyg	110	130 dec	3.64		reac H_2O
1153	Germanium(II) iodide	GeI_2	13573-08-5	326.45	oran-yel hex cry	428	550 dec	5.4		reac H_2O
1154	Germanium(II) oxide	GeO	20619-16-3	88.64	blk solid	700 dec				
1155	Germanium(II) selenide	GeSe	12065-10-0	151.60	gray orth cry or brn powder	675		5.6		
1156	Germanium(II) sulfide	GeS	12025-32-0	104.71	gray orth cry	658		4.1		
1157	Germanium(II) telluride	GeTe	12025-39-7	200.24	cub cry	724		6.16		i H_2O; s conc HNO_3
1158	Germanium(IV) bromide	$GeBr_4$	13450-92-5	392.26	wh cry	26.1	186.35	3.132		reac H_2O
1159	Germanium(IV) chloride	$GeCl_4$	10038-98-9	214.45	col liq	-51.50	86.55	1.88		reac H_2O; s bz, eth, EtOH, ctc
1160	Germanium(IV) fluoride	GeF_4	7783-58-6	148.63	col gas	-15 tp	-36.5 sp	6.074 g/L		reac H_2O
1161	Germanium(IV) iodide	GeI_4	13450-95-8	580.26	red-oran cub cry	146	348	4.322		reac H_2O
1162	Germanium(IV) nitride	Ge_3N_4	12065-36-0	273.95	orth cry	900 dec				i H_2O, acid, aqua regia
1163	Germanium(IV) oxide	GeO_2	1310-53-8	104.64	wh hex cry	1116		4.25		i H_2O
1164	Germanium(IV) selenide	$GeSe_2$	12065-11-1	230.56	yel-oran orth cry	707 dec		4.56		
1165	Germanium(IV) sulfide	GeS_2	12025-34-2	136.77	blk orth cry	530		3.01		
1166	Gold	Au	7440-57-5	196.967	soft yel metal	1064.18	2836	19.3		s aqua regia
1167	Bromoauric(III) acid pentahydrate	$HAuBr_4 \cdot 5H_2O$	17083-68-0	607.667	red-brn hyg cry	27				s H_2O, EtOH
1168	Chloroauric(III) acid tetrahydrate	$HAuCl_4 \cdot 4H_2O$	16903-35-8	411.848	yel monocl cry; hyg			≈3.9		vs H_2O, EtOH; s eth
1169	Gold(I) bromide	AuBr	10294-27-6	276.871	yel-gray tetr cry	165 dec		8.20		i H_2O
1170	Gold(I) chloride	AuCl	10294-29-8	232.420	yel orth cry	289 dec		7.6	0.000031²⁰	
1171	Gold(I) cyanide	AuCN	506-65-0	222.985	yel hex cry	dec		7.2		i H_2O, EtOH, eth, dil acid
1172	Gold(I) iodide	AuI	10294-31-2	323.871	yel-grn powder; tetr	120 dec		8.25		i H_2O; s CN soln
1173	Gold(I) sulfide	Au_2S	1303-60-2	425.998	brn-blk cub cry; unstab	240 dec		≈11		i H_2O, acid; s aqua regia
1174	Gold(III) bromide	$AuBr_3$	10294-28-7	436.679	red-br monocl cry	≈160 dec				s H_2O, EtOH
1175	Gold(III) chloride	$AuCl_3$	13453-07-1	303.326	red monocl cry	>160 dec		4.7	68²⁰	
1176	Gold(III) cyanide trihydrate	$Au(CN)_3 \cdot 3H_2O$	535-37-5*	329.065	wh hyg cry	50 dec				vs H_2O; sl EtOH
1177	Gold(III) fluoride	AuF_3	14720-21-9	253.962	oran-yel hex cry	>300	subl	6.75		
1178	Gold(III) hydroxide	$Au(OH)_3$	1303-52-2	247.989	brn powder	≈100 dec				i H_2O; s acid
1179	Gold(III) iodide	AuI_3	31032-13-0	577.680	unstab grn powder	20 dec				
1180	Gold(III) oxide	Au_2O_3	1303-58-8	441.931	brn powder	≈150 dec				i H_2O; s acid
1181	Gold(III) selenate	$Au_2(SeO_4)_3$	10294-32-3	822.81	yel cry					i H_2O; s acid
1182	Gold(III) selenide	Au_2Se_3	1303-62-4	630.81	blk amorp solid	dec		4.65		s aqua regia
1183	Gold(III) sulfide	Au_2S_3	1303-61-3	490.128	unstab blk powder	200 dec				
1184	Hafnium	Hf	7440-58-6	178.49	gray metal; hex	2233	4600	13.3		s HF
1185	Hafnium boride	HfB_2	12007-23-7	200.11	gray hex cry	3100		10.5		
1186	Hafnium carbide	HfC	12069-85-1	190.50	refrac cub cry	≈3000		12.2		
1187	Hafnium hydride	HfH_2	12770-26-2	180.51	refrac tetr cry			11.4		
1188	Hafnium nitride	HfN	25817-87-2	192.50	yel-brn cub cry	3310		13.8		
1189	Hafnium phosphide	HfP	12325-59-6	209.46	hex cry			9.78		

No.	Name	Formula	CAS Reg No.	Mol. weight	Physical form	mp/°C	bp/°C	Density g cm^{-3}	Solubility g/100 g H$_2$O	Qualitative solubility
1190	Hafnium silicide	HfSi$_2$	12401-56-8	234.66	gray orth cry	≈1700		7.6		
1191	Hafnocene dichloride	Hf(C$_5$H$_5$)$_2$Cl$_2$	12116-66-4	379.58	col hyg cry	235				s bz, chl; sl thf, eth; i hex
1192	Hafnium(II) bromide	HfBr$_2$	13782-95-1	338.30	blue-blk cry	dec 400				
1193	Hafnium(II) chloride	HfCl$_2$	13782-92-8	249.40	blk solid	dec 400				
1194	Hafnium(III) bromide	HfBr$_3$	13782-96-2	418.20	blue-blk cry	dec 350				
1195	Hafnium(III) chloride	HfCl$_3$	13782-93-9	284.85	blk solid	dec				
1196	Hafnium(III) iodide	HfI$_3$	13779-73-2	559.20	blk cry	dec				
1197	Hafnium(IV) bromide	HfBr$_4$	13777-22-5	498.11	wh cub cry	424 tp	323 sp	4.90		reac H$_2$O
1198	Hafnium(IV) chloride	HfCl$_4$	13499-05-3	320.30	wh monocl cry	432 tp	317 sp			reac H$_2$O
1199	Hafnium(IV) fluoride	HfF$_4$	13709-52-9	254.48	wh monocl cry	1025	970 sp	7.1		reac H$_2$O
1200	Hafnium(IV) iodide	HfI$_4$	13777-23-6	686.11	yel-oran cub cry	449 tp	394 sp	5.6		reac H$_2$O
1201	Hafnium(IV) oxide	HfO$_2$	12055-23-1	210.49	wh cub cry	2800	≈5400	9.68		i H$_2$O
1202	Hafnium(IV) oxychloride octahydrate	HfOCl$_2$ · 8H$_2$O	14456-34-9	409.52	wh tetr cry	dec				s H$_2$O
1203	Hafnium(IV) selenide	HfSe$_2$	12162-21-9	336.41	brn hex cry			7.46		
1204	Hafnium(IV) silicate	HfSiO$_4$	13870-13-8	270.57	tetr cry	2758		7.0		
1205	Hafnium(IV) sulfate	Hf(SO$_4$)$_2$	15823-43-5	370.62	wh cry	>500 dec				
1206	Hafnium(IV) sulfide	HfS$_2$	18855-94-2	242.62	purp-brn hex cry			6.03		
1207	Hafnium(IV) titanate	HfTiO$_4$	12055-24-2	290.36	wh pow	1980 dec				
1208	Helium	He	7440-59-7	4.003	col gas		-268.928	0.164 g/L		sl H$_2$O; i EtOH
1209	Holmium	Ho	7440-60-0	164.930	silv metal; hex	1472	2700	8.80		s dil acid
1210	Holmium acetate	Ho(C$_2$H$_3$O$_2$)$_3$	25519-09-9	342.062	yel cry	dec 327				s H$_2$O
1211	Holmium bromide	HoBr$_3$	13825-76-8	404.642	yel hyg cry	919	1470			
1212	Holmium chloride	HoCl$_3$	10138-62-2	271.289	yel monocl cry; hyg	720	1500	3.7		s H$_2$O
1213	Holmium chloride hexahydrate	HoCl$_3$ · 6H$_2$O	14914-84-2	379.381	hyg yel cry	160 dec				s H$_2$O
1214	Holmium fluoride	HoF$_3$	13760-78-6	221.925	pink-yel orth cry; hyg	1143	>2200	7.664		s H$_2$O
1215	Holmium iodide	HoI$_3$	13813-41-7	545.643	yel hex cry	994		5.4		
1216	Holmium nitrate pentahydrate	Ho(NO$_3$)$_3$ · 5H$_2$O	14483-18-2	441.022	hyg oran cry					s H$_2$O, EtOH, ace
1217	Holmium nitride	HoN	12029-81-1	178.937	cub cry			10.6		
1218	Holmium oxalate decahydrate	Ho$_2$(C$_2$O$_4$)$_3$ · 10H$_2$O	28965-57-3	774.070	yel solid	dec 40				
1219	Holmium oxide	Ho$_2$O$_3$	12055-62-8	377.859	yel cub cry	2330	3900	8.41		s acid
1220	Holmium silicide	HoSi$_2$	12136-24-2	221.101	hex cry			7.1		
1221	Holmium sulfide	Ho$_2$S$_3$	12162-59-3	426.056	yel-oran monocl cry			5.92		
1222	Hydrazine	N$_2$H$_4$	302-01-2	32.045	col oily liq	1.54	113.55	1.0036		vs H$_2$O, EtOH, MeOH
1223	Hydrazine acetate	N$_2$H$_4$ · CH$_3$COOH	13255-48-6	92.097	cry	100				
1224	Hydrazine azide	N$_2$H$_4$ · HN$_3$	14662-04-5	75.074	hyg wh prism	75 exp				vs H$_2$O
1225	Hydrazine monohydrate	N$_2$H$_4$ · H$_2$O	7803-57-8	50.060	fuming liq	-51.7	119	1.030		vs H$_2$O, EtOH; i chl, eth
1226	Hydrazine hydrobromide	N$_2$H$_4$ · HBr	13775-80-9	112.957	wh monocl cry flakes	84	≈190 dec	2.3		s H$_2$O, EtOH
1227	Hydrazine hydrochloride	N$_2$H$_4$ · HCl	2644-70-4	68.506	wh orth cry	89	240 dec	1.5		s H$_2$O; i os
1228	Hydrazine dihydrochloride	N$_2$H$_4$ · 2HCl	5341-61-7	104.967	wh orth cry	198 dec		1.42		s H$_2$O; sl EtOH
1229	Hydrazine hydroiodide	N$_2$H$_4$ · HI	10039-55-1	159.957	hyg cry	125				s H$_2$O
1230	Hydrazine nitrate	N$_2$H$_4$ · HNO$_3$	13464-97-6	95.058	monocl cry; exp	70				vs H$_2$O
1231	Hydrazine dinitrate	N$_2$H$_4$ · 2HNO$_3$	13464-98-7	158.071	needles	104 dec			20^{35}	s H$_2$O
1232	Hydrazine perchlorate hemihydrate	N$_2$H$_4$ · HClO$_4$ · 0.5H$_2$O	13762-65-7		solid	137	exp	1.94		reac H$_2$O, s EtOH; i eth, bz
1233	Hydrazine sulfate	N$_2$H$_4$ · H$_2$SO$_4$	10034-93-2	130.124	col orth cry	254		1.378		sl H$_2$O; i EtOH
1234	Dihydrazine sulfate	(N$_2$H$_4$)$_2$ · H$_2$SO$_4$	13464-80-7	162.169	hyg wh cry flakes	104	dec >180		200^{25}	vs H$_2$O; i os
1235	Hydrazoic acid	HN$_3$	7782-79-8	43.028	col liq; exp	-80	35.7			s H$_2$O
1236	Hydroxylamine	H$_2$NOH	7803-49-8	33.030	wh orth flakes or needles	33.1	58	1.21		vs H$_2$O, MeOH
1237	Hydroxylamine hydrobromide	H$_2$NOH · HBr	41591-55-3	113.942	monocl cry			2.35		s H$_2$O
1238	Hydroxylamine hydrochloride	H$_2$NOH · HCl	5470-11-1	69.491	col monocl cry	159 dec	exp	1.68	94^{25}	vs H$_2$O
1239	Hydroxylamine perchlorate	H$_2$NOH · HClO$_4$		133.489	orth cry	88	dec 120			
1240	Hydroxylamine sulfate	2(H$_2$NOH) · H$_2$SO$_4$	10039-54-0	164.138	cry	170				vs H$_2$O
1241	Hydrogen	H$_2$	1333-74-0	2.016	col gas; flam	-259.16	-252.762	0.082 g/L		sl H$_2$O
1242	Hydrogen-d_2	D$_2$	7782-39-0	4.028	col gas	-254.42	-249.48	0.164 g/L		
1243	Hydrogen-t_2	T$_2$	10028-17-8	6.032	col gas	-252.53	-248.11	0.246 g/L		
1244	Hydrogen-d_1	HD	13983-20-5	3.022	col gas	-256.55	-251.02	0.123 g/L		
1245	Hydrogen-t_1	HT	14885-60-0	4.024	col gas	-254.7	-249.6			
1246	Hydrogen-d_1,t_1	DT	14885-61-1	5.030	col gas	-253.5	-238.9			
1247	Hydrogen bromide	HBr	10035-10-6	80.912	col gas	-86.80	-66.38	3.307 g/L		vs H$_2$O; s EtOH
1248	Hydrogen bromide-d	DBr	13536-59-9	81.918	col gas	-87.54	-66.9			s H$_2$O

No.	Name	Formula	CAS Reg No.	Mol. weight	Physical form	mp/°C	bp/°C	Density g cm⁻³	Solubility g/100 g H₂O	Qualitative solubility
1249	Hydrogen chloride	HCl	7647-01-0	36.461	col gas	-114.17	-85	1.490 g/L		vs H_2O
1250	Hydrogen chloride dihydrate	HCl · 2H₂O	13465-05-9	72.492	col liq	-17.7		1.46		
1251	Hydrogen chloride-d	DCl	7698-05-7	37.467	col gas	-114.72	-84.4			s H_2O
1252	Hydrogen cyanide	HCN	74-90-8	27.026	col liq or gas	-13.29	26	0.6876²⁰		vs H_2O, EtOH; sl eth
1253	Hydrogen fluoride	HF	7664-39-3	20.006	col gas	-83.36	20	0.818 g/L		vs H_2O, EtOH; sl eth
1254	Hydrogen iodide	HI	10034-85-2	127.912	col or yel gas	-50.76	-35.55	5.228 g/L		vs H_2O; s os
1255	Hydrogen iodide-d	DI	14104-45-1	128.918	col gas	-51.93	-36.2			s H_2O
1256	Hydrogen peroxide	H₂O₂	7722-84-1	34.015	col liq	-0.43	150.2	1.44		vs H_2O
1257	Hydrogen selenide	H₂Se	7783-07-5	80.98	col gas; flam	-65.73	-41.25	3.310 g/L		s H_2O
1258	Hydrogen sulfide	H₂S	7783-06-4	34.081	col gas; flam	-85.5	-59.55	1.393 g/L		s H_2O
1259	Hydrogen disulfide	H₂S₂	13465-07-1	66.146	col liq		70.7	1.334		
1260	Hydrogen telluride	H₂Te	7783-09-7	129.62	col gas	-49	-2	5.298 g/L		s H_2O, EtOH, alk
1261	Indium	In	7440-74-6	114.818	soft wh metal	156.6	2027	7.31		s acid
1262	Indium antimonide	InSb	1312-41-0	236.578	blk cub cry	524		5.7747		
1263	Indium arsenide	InAs	1303-11-3	189.740	gray cub cry	942		5.67		i acid
1264	Indium nitride	InN	25617-98-5	128.825	brn hex cry	1100		6.88		
1265	Indium phosphide	InP	22398-80-7	145.792	blk cub cry	1062		4.81		sl acid
1266	Indium(I) bromide	InBr	14280-53-6	194.722	oran-red orth cry	285	656	4.96		reac H_2O
1267	Indium(I) chloride	InCl	13465-10-6	150.271	yel cub cry	225	608	4.19		reac H_2O
1268	Indium(I) iodide	InI	13966-94-4	241.722	orth cry	364.4	712	5.32		
1269	Indium(II) bromide	InBr₂	21264-43-7	274.626	orth cry			4.22		reac H_2O
1270	Indium(II) chloride	InCl₂	13465-11-7	185.724	col orth cry	235		3.64		reac H_2O
1271	Indium(II) sulfide	InS	12030-14-7	146.883	red-brn orth cry	692		5.2		
1272	Indium(III) bromide	InBr₃	13465-09-3	354.530	hyg yel-wh monocl cry	420		4.74	414²⁰	
1273	Indium(III) chloride	InCl₃	10025-82-8	221.177	yel monocl cry; hyg	583		4.0	195.1²²	s EtOH
1274	Indium(III) chloride tetrahydrate	InCl₃ · 4H₂O	22519-64-8	293.239	wh cry					s H_2O
1275	Indium(III) fluoride	InF₃	7783-52-0	171.813	wh hex cry; hyg	1172	>1200	4.39		sl H_2O; s dil acid
1276	Indium(III) fluoride trihydrate	InF₃ · 3H₂O	14166-78-0	225.859	wh cry	100 dec				s H_2O
1277	Indium(III) hydroxide	In(OH)₃	20661-21-6	165.840	cub cry			4.4		
1278	Indium(III) iodide	InI₃	13510-35-5	495.531	yel-red monocl cry; hyg	207		4.69	1308²²	
1279	Indium(III) nitrate trihydrate	In(NO₃)₃ · 3H₂O	13770-61-1	354.879	col cry	dec 100				
1280	Indium(III) oxide	In₂O₃	1312-43-2	277.634	yel cub cry	1912		7.18		i H_2O; s hot acid
1281	Indium(III) perchlorate octahydrate	In(ClO₄)₃ · 8H₂O	13465-15-1	557.292	wh cry	≈80	200 dec			
1282	Indium(III) phosphate	InPO₄	14693-82-4	209.789	wh orth cry			4.9		i H_2O
1283	Indium(III) selenide	In₂Se₃	1312-42-1	466.52	blk hex cry	660		5.8		
1284	Indium(III) sulfate	In₂(SO₄)₃	13464-82-9	517.824	hyg wh powder			3.44	117²⁰	
1285	Indium(III) sulfide	In₂S₃	12030-24-9	325.831	oran cub cry	1050		4.45		
1286	Indium(III) telluride	In₂Te₃	1312-45-4	612.44	blk cub cry	667		5.75		
1287	Iodine	I₂	7553-56-2	253.809	blue-blk plates	113.7	184.4	4.933	0.03²⁰	s bz, EtOH, eth, ctc, chl
1288	Iodic acid	HIO₃	7782-68-5	175.910	col orth cry	110 dec		4.63	308²⁵	i EtOH, eth
1289	Periodic acid dihydrate	HIO₄ · 2H₂O	10450-60-9	227.940	monocl hyg cry	122 dec				s H_2O, EtOH; sl eth
1290	Iodine tetroxide	I₂O₄	12399-08-5	317.807	yel cry	130	dec >85	4.2		sl H_2O
1291	Iodine pentoxide	I₂O₅	12029-98-0	333.806	hyg wh cry	≈300 dec		4.98	253.4²⁰	i EtOH, eth, CS_2
1292	Iodine hexoxide	I₂O₆	65355-99-9	349.805	yel solid	dec 150				reac H_2O
1293	Iodine nonaoxide	I₄O₉	73560-00-6	651.613	hyg yel powder	75 dec				
1294	Iodine bromide	IBr	7789-33-5	206.808	blk orth cry	40	116 dec	4.3		s H_2O, EtOH, eth
1295	Iodine chloride	ICl	7790-99-0	162.357	red cry or oily liq	27.38	94.4 dec	3.24		reac H_2O; s EtOH
1296	Iodine trichloride	ICl₃	865-44-1	233.263	yel tricl cry; hyg	101 tp (16 atm)	64 sp dec	3.2		reac H_2O; s EtOH, bz
1297	Iodine fluoride	IF	13873-84-2	145.902	wh pow (-78°C)	-14 dec				
1298	Iodine trifluoride	IF₃	22520-96-3	183.899	yel solid, stable at low temp	-28 dec				
1299	Iodine pentafluoride	IF₅	7783-66-6	221.896	yel liq	9.43	100.5	3.19		reac H_2O
1300	Iodine heptafluoride	IF₇	16921-96-3	259.893	col gas	6.5 tp	4.8 sp	10.62 g/L		s H_2O
1301	Iodosyl trifluoride	IOF₃	19058-78-7	199.898	hyg col needles	dec >110				reac H_2O
1302	Iodosyl pentafluoride	IOF₅	16056-61-4	237.895	col liq	4.5				
1303	Iodyl trifluoride	IO₂F₃	25402-50-0	215.898	yel solid	41	subl			
1304	Periodyl fluoride	IO₃F	30708-86-2	193.900	col cry	dec >100				
1305	Iridium	Ir	7439-88-5	192.217	silv-wh metal; cub	2446	4428	22.562²⁰		s aqua regia
1306	Iridium carbonyl	Ir₄(CO)₁₂	11065-24-0	1104.989	yel cry	210 dec				
1307	Iridium(III) bromide	IrBr₃	10049-24-8	431.929	red-brn monocl cry			6.82		i H_2O, acid, alk

No.	Name	Formula	CAS Reg No.	Mol. weight	Physical form	mp/°C	bp/°C	Density g cm^{-3}	Solubility g/100 g H$_2$O	Qualitative solubility
1308	Iridium(III) bromide tetrahydrate	IrBr$_3$ · 4H$_2$O	10049-24-8*	503.991	grn-brn cry					s H$_2$O; i EtOH
1309	Iridium(III) chloride	IrCl$_3$	10025-83-9	298.576	brn monocl cry	763 dec		5.30		i H$_2$O, acid, alk
1310	Iridium(III) fluoride	IrF$_3$	23370-59-4	249.212	blk hex cry	250 dec		≈8.0		i H$_2$O, dil acid
1311	Iridium(III) iodide	IrI$_3$	7790-41-2	572.930	dark brn monocl cry			≈7.4		i H$_2$O, acid, bz, chl; s alk
1312	Iridium(III) oxide	Ir$_2$O$_3$	1312-46-5	432.432	blue-blk cry	1000 dec				i H$_2$O; sl hot HCl
1313	Iridium(III) 2,4-pentanedioate	Ir(CH$_3$COCHCOCH$_3$)$_3$	15635-87-7	489.541	oran-yel cry	270	subl			sl H$_2$O; s tol, chl, ace, MeOH
1314	Iridium(III) sulfide	Ir$_2$S$_3$	12136-42-4	480.629	orth cry			10.2		
1315	Iridium(IV) chloride	IrCl$_4$	10025-97-5	334.029	brn hyg solid	≈700 dec				s H$_2$O, EtOH
1316	Iridium(IV) oxide	IrO$_2$	12030-49-8	224.216	brn tetr cry	1100 dec		11.7		
1317	Iridium(IV) sulfide	IrS$_2$	12030-51-2	256.347	orth cry			9.3		
1318	Iridium(VI) fluoride	IrF$_6$	7783-75-7	306.207	yel cub cry; hyg	44	53.6	4.8		reac H$_2$O
1319	Iron	Fe	7439-89-6	55.845	silv-wh or gray met	1538	2861	7.87		s dil acid
1320	Ferrocene	Fe(C$_5$H$_5$)$_2$	102-54-5	186.031	oran needles	172.5	249			i H$_2$O; s EtOH, eth, bz, dil HNO$_3$
1321	Tetracarbonyldihydroiron	Fe(CO)$_4$H$_2$	12002-28-7	169.902	col liq, stab low temp	-70	dec -20			s alk
1322	Iron pentacarbonyl	Fe(CO)$_5$	13463-40-6	195.896	yel oily liq; flam	-20.5	103	1.46		i H$_2$O; s eth, bz, ace
1323	Iron nonacarbonyl	Fe$_2$(CO)$_9$	15321-51-4	363.781	oran-yel cry	100 dec		2.85		
1324	Iron dodecacarbonyl	Fe$_3$(CO)$_{12}$	12088-65-2	503.656	dark grn cry	140		2.00		
1325	Iron arsenide	FeAs	12044-16-5	130.767	gray orth cry	1030		7.85		
1326	Iron boride (FeB)	FeB	12006-84-7	66.656	refrac solid; orth	1658		≈7		
1327	Iron boride (Fe$_2$B)	Fe$_2$B	12006-86-9	122.501	refrac solid; tetr	1389		7.3		
1328	Iron carbide	Fe$_3$C	12011-67-5	179.546	gray cub cry	1227		7.694		
1329	Iron phosphide (FeP)	FeP	26508-33-8	86.819	rhom cry			6.07		
1330	Iron phosphide (Fe$_2$P)	Fe$_2$P	1310-43-6	142.664	gray hex needles	1370		6.8		i H$_2$O, dil acid, alk
1331	Iron phosphide (Fe$_3$P)	Fe$_3$P	12023-53-9	198.509	gray solid	1100		6.74		i H$_2$O
1332	Iron disulfide	FeS$_2$	1317-66-4	119.975	blk cub cry	>600 dec		5.02		i H$_2$O
1333	Iron silicide	FeSi	12022-95-6	83.931	gray cub cry	1410		6.1		
1334	Iron disilicide	FeSi$_2$	12022-99-0	112.016	gray tetr cry	1220		4.74		
1335	Iron(II) acetate	Fe(C$_2$H$_3$O$_2$)$_2$	3094-87-9	173.934	wh cry	190 dec				s H$_2$O
1336	Iron(II) acetate tetrahydrate	Fe(C$_2$H$_3$O$_2$)$_2$ · 4H$_2$O	3094-87-9*	245.994	grn cry	dec				s H$_2$O, EtOH
1337	Iron(II) aluminate	Fe(AlO$_2$)$_2$	12068-49-4	173.806	blk cub cry			4.3		
1338	Iron(II) arsenate	Fe$_3$(AsO$_4$)$_2$	10102-50-8	445.373	grn powder					i H$_2$O
1339	Iron(II) arsenate hexahydrate	Fe$_3$(AsO$_4$)$_2$ · 6H$_2$O	10102-50-8*	553.465	grn amorp powder	dec				i H$_2$O; s acid
1340	Iron(II) bromide	FeBr$_2$	7789-46-0	215.653	yel-brn hex cry; hyg	691	dec	4.636	120^{25}	vs EtOH
1341	Iron(II) bromide hexahydrate	FeBr$_2$ · 6H$_2$O	13463-12-2	323.744	grn hyg cry	27 dec		4.64	120^{25}	s EtOH
1342	Iron(II) carbonate	FeCO$_3$	563-71-3	115.854	gray-brn hex cry			3.944	0.000062^{20}	
1343	Iron(II) chloride	FeCl$_2$	7758-94-3	126.751	wh hex cry; hyg	677	1023	3.16	65.0^{25}	vs EtOH, ace; sl bz
1344	Iron(II) chloride dihydrate	FeCl$_2$ · 2H$_2$O	16399-77-2	162.782	wh-grn monocl cry	120 dec		2.39	65.0^{25}	
1345	Iron(II) chloride tetrahydrate	FeCl$_2$ · 4H$_2$O	13478-10-9	198.813	grn monocl cry	105 dec		1.93	65.0^{25}	s EtOH
1346	Iron(II) chromite	FeCr$_2$O$_4$	1308-31-2	223.835	blk cub cry			5.0		
1347	Iron(II) fluoride	FeF$_2$	7789-28-8	93.842	wh tetr cry	1100		4.09		sl H$_2$O; s dil HF; i EtOH, eth
1348	Iron(II) fluoride tetrahydrate	FeF$_2$ · 4H$_2$O	13940-89-1	165.904	col hex cry			2.20		
1349	Iron(II) hydroxide	Fe(OH)$_2$	18624-44-7	89.860	wh-grn hex cry			3.4	0.000052^{20}	
1350	Iron(II) iodide	FeI$_2$	7783-86-0	309.654	red-viol hex cry; hyg	594		5.3		s H$_2$O, EtOH, eth
1351	Iron(II) iodide tetrahydrate	FeI$_2$ · 4H$_2$O	7783-86-0*	381.716	blk hyg leaflets	90 dec		2.87		s H$_2$O, EtOH
1352	Iron(II) molybdate	FeMoO$_4$	13718-70-2	215.78	brn-yel monocl cry	1115		5.6		i H$_2$O
1353	Iron(II) nitrate	Fe(NO$_3$)$_2$	14013-86-6	179.854	grn solid				87.5^{25}	
1354	Iron(II) nitrate hexahydrate	Fe(NO$_3$)$_2$ · 6H$_2$O	14013-86-6*	287.946	grn solid	60 dec			87.5^{25}	
1355	Iron(II) orthosilicate	Fe$_2$SiO$_4$	10179-73-4	203.774	brn orth cry			4.30		
1356	Iron(II) oxalate dihydrate	FeC$_2$O$_4$ · 2H$_2$O	6047-25-2	179.894	yel cry	150 dec		2.28	0.078^{25}	s acid
1357	Iron(II) oxide	FeO	1345-25-1	71.844	blk cub cry	1377		6.0		i H$_2$O, alk; s acid
1358	Iron(II) 2,4-pentanedioate	Fe(CH$_3$COCHCOCH$_3$)$_2$	14024-17-0	254.061	oran-brn cry	170	subl			sl bz, tol
1359	Iron(II) perchlorate	Fe(ClO$_4$)$_2$	13933-23-8	254.746	grn-wh hyg needles	>100 dec			210^{25}	
1360	Iron(II) phosphate octahydrate	Fe$_3$(PO$_4$)$_2$ · 8H$_2$O	14940-41-1	501.600	gray-blue monocl cry; hyg			2.58		i H$_2$O; s acid
1361	Iron(II) selenide	FeSe	1310-32-3	134.81	blk hex cry			6.7		i H$_2$O
1362	Iron(II) sulfate	FeSO$_4$	7720-78-7	151.908	wh orth cry; hyg			3.65	29.5^{25}	
1363	Iron(II) sulfate monohydrate	FeSO$_4$ · H$_2$O	17375-41-6	169.923	wh-yel monocl cry	300 dec		3.0	29.5^{25}	

No.	Name	Formula	CAS Reg No.	Mol. weight	Physical form	mp/°C	bp/°C	Density g cm⁻³	Solubility g/100 g H₂O	Qualitative solubility
1364	Iron(II) sulfate heptahydrate	$FeSO_4 \cdot 7H_2O$	7782-63-0	278.014	blue-grn monocl cry	≈60 dec		1.895	29.5[25]	i EtOH
1365	Iron(II) sulfide	FeS	1317-37-9	87.910	col hex or tetr cry; hyg	1188	dec	4.7		i H₂O; reac acid
1366	Iron(II) tantalate	$Fe(TaO_3)_2$	12140-41-9	513.737	brn tetr cry			7.33		
1367	Iron(II) tartrate	$FeC_4H_4O_6$		203.916	wh cry				0.88	vs acid; s NH₄OH
1368	Iron(II) telluride	FeTe	12125-63-2	183.45	tetr cry	914		6.8		
1369	Iron(II) thiocyanate trihydrate	$Fe(SCN)_2 \cdot 3H_2O$	6010-09-9	226.055	grn monocl cry					s H₂O, EtOH, eth
1370	Iron(II) titanate	$FeTiO_3$	12168-52-4	151.710	blk rhomb cry	≈1470		4.72		
1371	Iron(II) tungstate	$FeWO_4$	13870-24-1	303.68	monocl cry			7.51		
1372	Iron(II,III) oxide	Fe_3O_4	1317-61-9	231.533	blk cub cry or amorp powder	1597		5.17		i H₂O; s acid
1373	Iron(III) acetate, basic	$FeOH(C_2H_3O_2)_2$	10450-55-2	190.941	brn-red amorp powder					i H₂O; s EtOH, acid
1374	Iron(III) ammonium citrate	$Fe(NH_4)_3(C_6H_5O_7)_2$	1185-57-5	488.160	red or brn pow; hyg					s H₂O; i EtOH
1375	Iron(III) arsenate dihydrate	$FeAsO_4 \cdot 2H_2O$	10102-49-5	230.795	grn-brn powder	dec		3.18		i H₂O; s dil acid
1376	Iron(III) bromide	$FeBr_3$	10031-26-2	295.557	dark red hex cry; hyg	dec		4.5	455[25]	s EtOH, eth
1377	Iron(III) chloride	$FeCl_3$	7705-08-0	162.204	grn hex cry; hyg	307.6	≈316	2.90	91.2[25]	s EtOH, eth, ace
1378	Iron(III) chloride hexahydrate	$FeCl_3 \cdot 6H_2O$	10025-77-1	270.295	yel-oran monocl cry; hyg	37 dec		1.82	91.2[25]	s EtOH, eth, ace
1379	Iron(III) chromate	$Fe_2(CrO_4)_3$	10294-52-7	459.671	yel powder					i H₂O, EtOH; s acid
1380	Iron(III) citrate pentahydrate	$FeC_6H_5O_7 \cdot 5H_2O$	3522-50-7	335.021	red-brn cry					s H₂O; i EtOH
1381	Iron(III) dichromate	$Fe_2(Cr_2O_7)_3$	10294-53-8	759.654	red-brn solid					s H₂O, acid
1382	Iron(III) ferrocyanide	$Fe_4[Fe(CN)_6]_3$	14038-43-8	859.229	dark blue powder			1.80		i H₂O, dil acid, os
1383	Iron(III) fluoride	FeF_3	7783-50-8	112.840	grn hex cry	>1000		3.87	5.92[25]	i EtOH, eth, bz
1384	Iron(III) fluoride trihydrate	$FeF_3 \cdot 3H_2O$	15469-38-2	166.886	yel-brn tetr cry			2.3	5.92[25]	
1385	Iron(III) formate	$Fe(CHO_2)_3$	555-76-0	190.897	red-yel cry powder					s H₂O; sl EtOH
1386	Iron(III) hydroxide	$Fe(OH)_3$	1309-33-7	106.867	yel monocl cry			3.12		
1387	Iron(III) hydroxide oxide	FeO(OH)	20344-49-4	88.852	red-brn orth cry			4.26		i H₂O; s acid
1388	Iron(III) metavanadate	$Fe(VO_3)_3$	65842-03-7	352.665	gray-brn powder					i H₂O, EtOH; s acid
1389	Iron(III) nitrate	$Fe(NO_3)_3$	10421-48-4	241.860	cry				82.5[20]	
1390	Iron(III) nitrate hexahydrate	$Fe(NO_3)_3 \cdot 6H_2O$	13476-08-9	349.951	viol cub cry	35 dec			82.5[20]	
1391	Iron(III) nitrate nonahydrate	$Fe(NO_3)_3 \cdot 9H_2O$	7782-61-8	403.997	viol-gray hyg cry	47 dec		1.68	82.5[20]	vs EtOH, ace
1392	Iron(III) oxalate	$Fe_2(C_2O_4)_3$	19469-07-9	375.747	yel amorp powder	100 dec				s H₂O, acid; i alk
1393	Iron(III) oxide	Fe_2O_3	1309-37-1	159.688	red-brn hex cry	1539		5.25		i H₂O; s acid
1394	Iron(III) 2,4-pentanedioate	$Fe(CH_3COCHCOCH_3)_3$	14024-18-1	353.169	red-oran cry	179		5.24		sl H₂O; s os
1395	Iron(III) perchlorate hexahydrate	$Fe(ClO_4)_3 \cdot 6H_2O$	32963-81-8	462.288	viol cry					
1396	Iron(III) phosphate dihydrate	$FePO_4 \cdot 2H_2O$	10045-86-0	186.847	gray-wh orth cry			2.87		i H₂O; s HCl
1397	Iron(III) pyrophosphate nonahydrate	$Fe_4(P_2O_7)_3 \cdot 9H_2O$	10058-44-3	907.348	yel powder					i H₂O; s acid
1398	Iron(III) hypophosphite	$Fe(H_2PO_2)_3$	7783-84-8	250.811	wh-gray powder					i H₂O
1399	Iron(III) sodium pyrophosphate	$FeNaP_2O_7$	10045-87-1	252.778	wh powder			1.5		i H₂O; s HCl
1400	Iron(III) sulfate	$Fe_2(SO_4)_3$	10028-22-5	399.878	gray-wh rhomb cry; hyg			3.10	440[20]	sl EtOH; i ace
1401	Iron(III) sulfate nonahydrate	$Fe_2(SO_4)_3 \cdot 9H_2O$	13520-56-4	562.015	yel hex cry	400 dec		2.1	440[20]	
1402	Iron(III) thiocyanate	$Fe(SCN)_3$	4119-52-2	230.092	red-viol hyg cry	dec				s H₂O, EtOH, ace; i tol, chl
1403	Krypton	Kr	7439-90-9	83.798	col gas	-157.37	-153.415	3.425 g/L		sl H₂O
1404	Krypton difluoride	KrF_2	13773-81-4	121.795	col tetr cry	≈25 dec		3.24		reac H₂O
1405	Krypton fluoride hexafluoroantimonate	$KrFSb_2F_{11}$	39578-36-4	555.299	wh solid	dec 45				
1406	Lanthanum	La	7439-91-0	138.905	silv metal; hex	920	3464	6.15		s dil acid
1407	Lanthanum aluminum oxide	$LaAlO_3$	12003-65-5	213.885	wh rhomb cry	trans cub 500				
1408	Lanthanum boride	LaB_6	12008-21-8	203.771	blk cub cry; refrac	2715		4.76		
1409	Lanthanum bromate nonahydrate	$LaBrO_3 \cdot 9H_2O$		684.749	hex cry	dec 100		5.06		vs H₂O
1410	Lanthanum bromide	$LaBr_3$	13536-79-3	378.617	wh hex cry; hyg	788		5.1		s H₂O
1411	Lanthanum carbide	LaC_2	12071-15-7	162.926	tetr cry	2360		5.29		
1412	Lanthanum carbonate octahydrate	$La_2(CO_3)_3 \cdot 8H_2O$	6487-39-4	601.960	wh cry powder			2.6		i H₂O; s dil acid
1413	Lanthanum chloride	$LaCl_3$	10099-58-8	245.264	wh hex cry; hyg	858		3.84	95.7[25]	
1414	Lanthanum chloride heptahydrate	$LaCl_3 \cdot 7H_2O$	20211-76-1	371.371	wh tricl cry; hyg	91 dec			95.7[25]	s EtOH
1415	Lanthanum fluoride	LaF_3	13709-38-1	195.900	wh hex cry; hyg	1493		5.9		i H₂O, acid
1416	Lanthanum hydride	LaH_3	13864-01-2	141.929	blk cub cry			5.36		
1417	Lanthanum hydroxide	$La(OH)_3$	14507-19-8	189.927	wh amorp solid	dec			0.000020[20]	
1418	Lanthanum iodate	$La(IO_3)_3$	13870-19-4	663.614	col cry				1.7	
1419	Lanthanum iodide	LaI_3	13813-22-4	519.619	wh orth cry; hyg	778		5.6		s H₂O
1420	Lanthanum nitrate hexahydrate	$La(NO_3)_3 \cdot 6H_2O$	10277-43-7	433.011	wh hyg tricl cry	≈40 dec			200[25]	vs EtOH; s ace
1421	Lanthanum nitride	LaN	25764-10-7	152.912	cub cry			6.73		

No.	Name	Formula	CAS Reg No.	Mol. weight	Physical form	mp/°C	bp/°C	Density g cm⁻³	Solubility g/100 g H₂O	Qualitative solubility
1422	Lanthanum oxide	La_2O_3	1312-81-8	325.809	wh amorp powder	2304	3620	6.51		i H₂O; s dil acid
1423	Lanthanum perchlorate hexahydrate	$La(ClO_4)_3 \cdot 6H_2O$	36907-37-6	475.021	hyg col cry	dec 100				vs H₂O; s EtOH
1424	Lanthanum silicide	$LaSi_2$	12056-90-5	195.076	gray tetr cry	1520		5.0		
1425	Lanthanum sulfate	$La_2(SO_4)_3$	10099-60-2	565.999	hyg wh pow	1150				sl H₂O
1426	Lanthanum sulfate octahydrate	$La_2(SO_4)_3 \cdot 8H_2O$	57804-25-8	702.058	col cry	dec		2.82		sl H₂O
1427	Lanthanum sulfate nonahydrate	$La_2(SO_4)_3 \cdot 9H_2O$	10294-62-9	728.136	hex cry			2.82	2.7²⁰	i EtOH
1428	Lanthanum monosulfide	LaS	12031-30-0	170.970	yel cub cry	2300		5.61		
1429	Lanthanum sulfide	La_2S_3	12031-49-1	374.006	red cub cry	2110		4.9		
1430	Lawrencium	Lr	22537-19-5	262	metal	1627				
1431	Lead	Pb	7439-92-1	207.2	soft silv-gray metal; cub	327.462	1749	11.3		s conc acid
1432	Plumbane	PbH_4	15875-18-0	211.2	unstab col gas		-13			
1433	Lead(II) acetate	$Pb(C_2H_3O_2)_2$	301-04-2	325.3	wh cry	280	dec	3.25	44.3²⁰	
1434	Lead(II) acetate trihydrate	$Pb(C_2H_3O_2)_2 \cdot 3H_2O$	6080-56-4	379.3	col cry	75 dec		2.55		vs H₂O; sl EtOH
1435	Lead(II) acetate, basic	$Pb(C_2H_3O_2)_2 \cdot 2Pb(OH)_2$	1335-32-6	807.7	wh powder	dec			6.3⁰	
1436	Lead(II) antimonate	$Pb_3(SbO_4)_2$	13510-89-9	993.1	oran-yel powder			6.58		i H₂O, dil acid
1437	Lead(II) arsenate	$Pb_3(AsO_4)_2$	3687-31-8	899.4	wh cry	1042 dec		5.8		i H₂O; s HNO₃
1438	Lead(II) arsenite	$Pb(AsO_2)_2$	10031-13-7	421.0	wh powder			5.85		i H₂O; s dil HNO₃
1439	Lead(II) azide	$Pb(N_3)_2$	13424-46-9	291.2	col orth needles; exp	exp ≈350		4.7	0.023¹⁸	vs HOAc
1440	Lead(II) borate monohydrate	$Pb(BO_2)_2 \cdot H_2O$	10214-39-8	310.8	wh powder	500 dec		5.6		i H₂O; s dil HNO₃
1441	Lead(II) bromate monohydrate	$Pb(BrO_3)_2 \cdot H_2O$	10031-21-7	481.0	col cry	≈180 dec		5.53	1.33²⁰	
1442	Lead(II) bromide	$PbBr_2$	10031-22-8	367.0	wh orth cry	371	892	6.69	0.975²⁵	i EtOH
1443	Lead(II) butanoate	$Pb(C_4H_7O_2)_2$	819-73-8	381.4	col solid	≈90				i H₂O; s dil HNO₃
1444	Lead(II) carbonate	$PbCO_3$	598-63-0	267.2	col orth cry	≈315 dec		6.582		i H₂O
1445	Lead(II) carbonate, basic	$Pb(OH)_2 \cdot 2PbCO_3$	1319-46-6	775.6	wh hex cry	400 dec		≈6.5		i H₂O, EtOH; s acid
1446	Lead(II) chlorate	$Pb(ClO_3)_2$	10294-47-0	374.1	col hyg cry	230 dec		3.9	144¹⁸	vs EtOH
1447	Lead(II) chloride	$PbCl_2$	7758-95-4	278.1	wh orth needles or powder	501	951	5.98	1.08²⁵	s alk
1448	Lead(II) chloride fluoride	PbClF	13847-57-9	261.7	tetr cry			7.05	0.035²⁰	
1449	Lead(II) chlorite	$Pb(ClO_2)_2$	13453-57-1	342.1	yel monocl cry	dec 126			0.2²⁵	sl H₂O; s alk
1450	Lead(II) chromate	$PbCrO_4$	7758-97-6	323.2	yel-oran monocl cry	844		6.12	0.000017²⁰	s alk, dil acid
1451	Lead(II) chromate(VI) oxide	$PbCrO_4 \cdot PbO$	18454-12-1	546.4	red powder					i H₂O
1452	Lead(II) citrate trihydrate	$Pb_3(C_6H_5O_7)_2 \cdot 3H_2O$	512-26-5	1053.8	wh cry powder					s H₂O; sl EtOH
1453	Lead(II) cyanide	$Pb(CN)_2$	592-05-2	259.2	wh-yel powder					sl H₂O; reac acid
1454	Lead(II) 2-ethylhexanoate	$Pb(C_7H_{15}CO_2)_2$	301-08-6	493.6	visc liq			1.56		
1455	Lead(II) fluoride	PbF_2	7783-46-2	245.2	wh orth cry	830	1293	8.44	0.0670²⁵	
1456	Lead(II) fluoroborate	$Pb(BF_4)_2$	13814-96-5	380.8	stab only in aq soln					s H₂O
1457	Lead(II) formate	$Pb(CHO_2)_2$	811-54-1	297.2	wh prisms or needles	190 dec		4.63	1.6¹⁶	i EtOH
1458	Lead(II) hexafluoro-2,4-pentanedioate	$Pb(CF_3COCHCOCF_3)_2$	19648-88-5	621.3	cry	155	210			
1459	Lead(II) hydrogen arsenate	$PbHAsO_4$	7784-40-9	347.1	wh monocl cry	280 dec		5.943		i H₂O; s HNO₃, alk
1460	Lead(II) hydrogen phosphate	$PbHPO_4$	15845-52-0	303.2	wh monocl cry	dec		5.66		
1461	Lead(II) hydroxide	$Pb(OH)_2$	19783-14-3	241.2	wh powder	145 dec		5.69	0.00012²⁰	s acid
1462	Lead(II) iodate	$Pb(IO_3)_2$	25659-31-8	557.0	wh orth cry			6.50	0.0025²⁵	
1463	Lead(II) iodide	PbI_2	10101-63-0	461.0	yel hex cry or powder	410	872 dec	6.16	0.076²⁵	i EtOH
1464	Lead(II) lactate	$Pb(C_3H_5O_3)_2$	18917-82-3	385.3	wh cry powder					s H₂O, hot EtOH
1465	Lead(II) molybdate	$PbMoO_4$	10190-55-3	367.1	yel tetr cry	≈1060		6.7		i H₂O; s HNO₃, NaOH
1466	Lead(II) niobate	$Pb(NbO_3)_2$	12034-88-7	489.0	rhom or tetr cry	1343		6.6		i H₂O
1467	Lead(II) nitrate	$Pb(NO_3)_2$	10099-74-8	331.2	col cub cry	470		4.53	59.7²⁵	sl EtOH
1468	Lead(II) oleate	$Pb(C_{18}H_{33}O_2)_2$	1120-46-3	770.1	wax-like solid					i H₂O; s EtOH, bz, eth
1469	Lead(II) oxalate	PbC_2O_4	814-93-7	295.2	wh powder	300 dec		5.28	0.00025²⁰	s dil HNO₃
1470	Lead(II) oxide (litharge)	PbO	1317-36-8	223.2	red tetr cry	trans to massicot 489		9.35		i H₂O, EtOH; s dil HNO₃
1471	Lead(II) oxide (massicot)	PbO	1317-36-8	223.2	yel orth cry	887		9.64		i H₂O, EtOH; s dil HNO₃
1472	Lead(II) oxide hydrate	$3PbO \cdot H_2O$	1311-11-1	687.6	wh powder			7.41		i H₂O; s dil acid
1473	Lead(II) 2,4-pentanedioate	$Pb(CH_3COCHCOCH_3)_2$	15282-88-9	405.4	cry	143				
1474	Lead(II) perchlorate	$Pb(ClO_4)_2$	13453-62-8	406.1	wh cry				441²⁵	
1475	Lead(II) perchlorate trihydrate	$Pb(ClO_4)_2 \cdot 3H_2O$	13637-76-8	460.1	wh cry	100 dec		2.6	441²⁵	s EtOH
1476	Lead(II) phosphate	$Pb_3(PO_4)_2$	7446-27-7	811.5	wh hex cry	1014		7.01		i H₂O, EtOH

No.	Name	Formula	CAS Reg No.	Mol. weight	Physical form	mp/°C	bp/°C	Density g cm⁻³	Solubility g/100 g H₂O	Qualitative solubility
1477	Lead(II) hypophosphite	$Pb(H_2PO_2)_2$	10294-58-3	337.2	hyg cry powder	dec				sl H_2O; i EtOH
1478	Lead(II) metasilicate	$PbSiO_3$	10099-76-0	283.3	wh monocl cry powder	764		6.49		i H_2O, os
1479	Lead(II) orthosilicate	Pb_2SiO_4	13566-17-1	506.5	monocl cry	743		7.60		
1480	Lead(II) hexafluorosilicate dihydrate	$PbSiF_6 \cdot 2H_2O$	1310-03-8	385.3	col cry	dec				vs H_2O
1481	Lead(II) selenate	$PbSeO_4$	7446-15-3	350.2	orth cry			6.37	0.013²⁵	s conc acid
1482	Lead(II) selenide	PbSe	12069-00-0	286.2	gray cub cry	1078		8.1		i H_2O; s HNO_3
1483	Lead(II) selenite	$PbSeO_3$	7488-51-9	334.2	wh monocl cry	≈500		7.0		i H_2O
1484	Lead(II) sodium thiosulfate	$Na_4Pb(S_2O_3)_3$	10101-94-7	635.5	wh cry					sl H_2O
1485	Lead(II) stearate	$Pb(C_{18}H_{35}O_2)_2$	1072-35-1	774.1	wh powder	≈100		1.4		i H_2O; s hot EtOH
1486	Lead(II) sulfate	$PbSO_4$	7446-14-2	303.3	orth cry	1087		6.29	0.0044²⁵	i acid; sl alk
1487	Lead(II) sulfide	PbS	1314-87-0	239.3	blk powder or silv cub cry	1113		7.60		i H_2O; s acid
1488	Lead(II) sulfite	$PbSO_3$	7446-10-8	287.3	wh powder	dec				i H_2O; s HNO_3
1489	Lead(II) tantalate	$Pb(TaO_3)_2$	12065-68-8	665.1	orth cry			7.9		i H_2O
1490	Lead(II) telluride	PbTe	1314-91-6	334.8	gray cub cry	924		8.164		i H_2O, acid
1491	Lead(II) thiocyanate	$Pb(SCN)_2$	592-87-0	323.4	wh-yel powder			3.82	0.05²⁰	
1492	Lead(II) thiosulfate	PbS_2O_3	13478-50-7	319.3	wh cry	dec		5.18		i H_2O; s acid
1493	Lead(II) titanate	$PbTiO_3$	12060-00-3	303.1	yel tetr cry			7.9		i H_2O; reac HCl
1494	Lead(II) tungstate (stolzite)	$PbWO_4$	7759-01-5	455.0	yel tetr cry	1130		8.24	0.03²⁰	s alk
1495	Lead(II) tungstate (raspite)	$PbWO_4$	7759-01-5	455.0	monocl cry	trans 400		8.46	0.03²⁰	s alk
1496	Lead(II) metavanadate	$Pb(VO_3)_2$	10099-79-3	405.1	yel powder					i H_2O; reac HNO_3
1497	Lead(II) zirconate	$PbZrO_3$	12060-01-4	346.4	col orth cry			≈8		i H_2O, alk; s acid
1498	Lead(II,IV) oxide	Pb_2O_3	1314-27-8	462.4	blk monocl cry or red amorp powder	530 dec		10.05		i H_2O; s alk; reac conc HCl
1499	Lead(II,II,IV) oxide	Pb_3O_4	1314-41-6	685.6	red tetr cry	830		8.92		i H_2O, EtOH; s hot HCl
1500	Lead(IV) acetate	$Pb(C_2H_3O_2)_4$	546-67-8	443.4	col monocl cry	≈175		2.23		reac H_2O, EtOH; s bz, chl
1501	Lead(IV) bromide	$PbBr_4$	13701-91-2	526.8	unstab liq					
1502	Lead(IV) chloride	$PbCl_4$	13463-30-4	349.0	yel oily liq	-15	≈50 dec			
1503	Lead(IV) fluoride	PbF_4	7783-59-7	283.2	wh tetr cry; hyg	≈600		6.7		
1504	Lead(IV) oxide	PbO_2	1309-60-0	239.2	red tetr cry or brn powder	290 dec		9.64		
1505	Lithium	Li	7439-93-2	6.94	soft silv-wh metal	180.50	1342	0.534		reac H_2O
1506	Lithium acetate	$LiC_2H_3O_2$	546-89-4	65.985	cry	286			45.0²⁵	vs EtOH
1507	Lithium acetate dihydrate	$LiC_2H_3O_2 \cdot 2H_2O$	6108-17-4	102.016	wh rhomb cry	58 dec		1.3	45.0²⁵	s EtOH
1508	Lithium aluminum hydride	$LiAlH_4$	16853-85-3	37.955	gray-wh monocl cry	>125 dec		0.917		reac H_2O, EtOH; s eth, thf
1509	Lithium aluminum silicate	$LiAlSi_2O_6$	12068-40-5	186.090	wh monocl cry	1430		3.188		
1510	Lithium amide	$LiNH_2$	7782-89-0	22.964	tetr cry	380		1.18		reac H_2O
1511	Lithium arsenate	Li_3AsO_4	13478-14-3	159.743	col orth cry			3.07		sl H_2O; s HOAc
1512	Lithium azide	LiN_3	19597-69-4	48.961	hyg monocl cry; exp			1.83		vs H_2O
1513	Lithium borohydride	$LiBH_4$	16949-15-8	21.784	wh-gray orth cry or powder	268	380 dec	0.66		s alk, eth, thf
1514	Lithium bromate	$LiBrO_3$	13550-28-2	134.843	hyg col orth cry	260			65.4²⁵	vs H_2O
1515	Lithium bromide	LiBr	7550-35-8	86.845	wh cub cry; hyg	550	≈1300	3.464	181²⁵	s EtOH, eth
1516	Lithium bromide monohydrate	$LiBr \cdot H_2O$	23303-71-1	104.860	wh orth cry	trans cub 33		3.46	145⁴	vs H_2O
1517	Lithium carbide	Li_2C_2	1070-75-3	37.903	wh hyg cry			1.65		reac H_2O; i os
1518	Lithium carbonate	Li_2CO_3	554-13-2	73.891	wh monocl cry	732	1300 dec	2.11	1.30²⁵	s acid; i EtOH
1519	Lithium chlorate	$LiClO_3$	13453-71-9	90.392	col hyg rhom needles	127.6	300 dec	1.119	459²⁵	vs EtOH; sl ace
1520	Lithium chloride	LiCl	7447-41-8	42.394	wh cub cry or powder; hyg	610	1383	2.07	84.5²⁵	s EtOH, ace, py
1521	Lithium chloride monohydrate	$LiCl \cdot H_2O$	16712-20-2	60.409	hyg wh tetr cry	dec 98		1.78	45.9²⁵	vs H_2O
1522	Lithium chromate dihydrate	$Li_2CrO_4 \cdot 2H_2O$	7789-01-7	165.906	yel orth cry; hyg	75 dec		2.15		vs H_2O; s EtOH
1523	Lithium citrate tetrahydrate	$Li_3C_6H_5O_7 \cdot 4H_2O$	6680-58-6	281.983	wh cry	210 (anh)			75²⁵	vs H_2O; sl EtOH
1524	Lithium cobaltite	$LiCoO_2$	12190-79-3	97.873	dark gray pow					i H_2O
1525	Lithium cyanide	LiCN	2408-36-8	32.959	wh orth cry	160				
1526	Lithium hydride-d	LiD	13587-16-1	8.955	hyg wh cry	680		0.82		reac H_2O
1527	Lithium dichromate dihydrate	$Li_2Cr_2O_7 \cdot 2H_2O$	10022-48-7	265.901	yel-red hyg cry	130 dec		2.34		vs H_2O
1528	Lithium dihydrogen phosphate	LiH_2PO_4	13453-80-0	103.928	col hyg cry	>100		2.461	126⁰	
1529	Lithium diisopropylamide	$LiN(C_3H_7)_2$	4111-54-0	107.123	hyg col cry	dec				s eth; i hc
1530	Lithium ferrosilicon	LiFeSi	64082-35-5	90.872	dark brittle cry					reac H_2O
1531	Lithium fluoride	LiF	7789-24-4	25.939	wh cub cry or powder	848.2	1673	2.640	0.134²⁵	s acid

No.	Name	Formula	CAS Reg No.	Mol. weight	Physical form	mp/°C	bp/°C	Density g cm⁻³	Solubility g/100 g H₂O	Qualitative solubility
1532	Lithium formate monohydrate	$Li(CHO_2) \cdot H_2O$	6108-23-2	69.974	col-wh cry			1.46		s H_2O
1533	Lithium hexafluoroantimonate	$LiSbF_6$	18424-17-4	242.691	hyg pow	dec				
1534	Lithium hexafluoroarsenate	$LiAsF_6$	29935-35-1	310.672	rhom wh cry; hyg					
1535	Lithium hexafluorophosphate	$LiPF_6$	21324-40-3	151.905	wh pow					
1536	Lithium hexafluorosilicate	Li_2SiF_6	17347-95-4	155.958	col hex cry	dec 350				sl ace
1537	Lithium hexafluorostannate	Li_2SnF_6	17029-16-2	246.582	wh pow					
1538	Lithium hydride	LiH	7580-67-8	7.949	gray cub cry or powder; hyg	692		0.78		reac H_2O, EtOH
1539	Lithium hydrogen carbonate	$LiHCO_3$	5006-97-3	67.958	wh pow					sl H_2O
1540	Lithium hydroxide	$LiOH$	1310-65-2	23.948	col tetr cry	473	1626	1.45	12.5²⁵	sl EtOH
1541	Lithium hydroxide monohydrate	$LiOH \cdot H_2O$	1310-66-3	41.964	wh monocl cry or powder			1.51	12.5²⁵	sl EtOH
1542	Lithium hypochlorite	$LiOCl$	13840-33-0	58.393	wh pow					vs H_2O
1543	Lithium iodate	$LiIO_3$	13765-03-2	181.843	wh hyg hex cry	450		4.502	77.9²⁵	i EtOH
1544	Lithium iodide	LiI	10377-51-2	133.845	wh cub cry; hyg	469	1171	4.06	165²⁵	
1545	Lithium iodide trihydrate	$LiI \cdot 3H_2O$	7790-22-9	187.891	wh hyg cry	73		2.38	165²⁵	vs EtOH, ace
1546	Lithium manganate	Li_2MnO_3	12163-00-7	116.818	red-brn monocl cry			3.90		i H_2O
1547	Lithium metaborate	$LiBO_2$	13453-69-5	49.751	wh monocl cry; hyg	844		2.18	2.6²⁰	sl H_2O; s EtOH
1548	Lithium metaborate dihydrate	$LiBO_2 \cdot 2H_2O$	15293-74-0	85.782	wh cry pow			1.8		s H_2O
1549	Lithium metaphosphate	$LiPO_3$	13762-75-9	85.913	wh cry or gl solid			1.8		i H_2O
1550	Lithium metasilicate	Li_2SiO_3	10102-24-6	89.966	wh orth needles	1201		2.52		i cold H_2O; reac dil acid
1551	Lithium molybdate	Li_2MoO_4	13568-40-6	173.82	hyg wh cry	702		2.66	44.8²⁵	s H_2O
1552	Lithium niobate	$LiNbO_3$	12031-63-9	147.845	wh hex cry	1240		4.30		
1553	Lithium nitrate	$LiNO_3$	7790-69-4	68.946	col hex cry; hyg	253		2.38	102²⁵	s EtOH
1554	Lithium nitride	Li_3N	26134-62-3	34.830	red hex cry	813		1.27		reac H_2O
1555	Lithium nitrite	$LiNO_2$	13568-33-7	52.947	wh hyg cry	222				vs H_2O
1556	Lithium nitrite monohydrate	$LiNO_2 \cdot H_2O$	13568-33-7*	70.962	col needles	>100		1.615	139.5²⁵	vs H_2O, EtOH
1557	Lithium orthosilicate	$LiSiO_4$	13453-84-4	99.025	wh rhom cry	1256		2.39		
1558	Lithium oxalate	$Li_2C_2O_4$	30903-87-8	101.901	col cry	dec		2.121¹⁷		s H_2O; i EtOH, eth
1559	Lithium phosphate	Li_3PO_4	10377-52-3	115.794	wh orth cry	1205		2.46	0.027²⁵	
1560	Lithium oxide	Li_2O	12057-24-8	29.881	wh cub cry	1438		2.013		
1561	Lithium perchlorate	$LiClO_4$	7791-03-9	106.392	wh orth cry or powder	236	430 dec	2.428	58.7²⁵	s EtOH, ace, eth
1562	Lithium perchlorate trihydrate	$LiClO_4 \cdot 3H_2O$	13453-78-6	160.438	wh hex cry	95 dec		1.84		vs H_2O, EtOH, ace; i eth
1563	Lithium peroxide	Li_2O_2	12031-80-0	45.881	wh hex cry			2.31		s H_2O; i EtOH
1564	Lithium selenate monohydrate	$Li_2SeO_4 \cdot H_2O$	7790-71-8	174.86	monocl cry			2.56		vs H_2O
1565	Lithium selenite monohydrate	$Li_2SeO_3 \cdot H_2O$	15593-51-8	158.86	hyg cry					
1566	Lithium stearate	$LiC_{18}H_{35}O_2$	4485-12-5	290.411	cry	≈220				
1567	Lithium sulfate	Li_2SO_4	10377-48-7	109.945	wh monocl cry; hyg	860		2.21	34.2²⁵	
1568	Lithium sulfate monohydrate	$Li_2SO_4 \cdot H_2O$	10102-25-7	127.960	col cry	130 dec		2.06	34.2²⁵	sl EtOH
1569	Lithium sulfide	Li_2S	12136-58-2	45.947	wh cub cry; hyg	1372		1.64		
1570	Lithium tantalate	$LiTaO_3$	12031-66-2	235.887	wh pow	1650				
1571	Lithium tetraborate	$Li_2B_4O_7$	12007-60-2	169.122	wh tetr cry	917			2.9²⁰	sl H_2O
1572	Lithium tetraborate pentahydrate	$Li_2B_4O_7 \cdot 5H_2O$	1303-94-2	259.198	wh cry pow	dec 200				vs H_2O; i EtOH
1573	Lithium tetracyanoplatinate pentahydrate	$Li_2Pt(CN)_4 \cdot 5H_2O$	14402-73-4	403.112	grn-yel cry					sl H_2O
1574	Lithium tetrafluoroborate	$LiBF_4$	14283-07-9	93.746	hyg wh pow	dec				vs H_2O
1575	Lithium thiocyanate	$LiSCN$	556-65-0	65.024	wh hyg cry				120²⁵	
1576	Lithium titanate	Li_2TiO_3	12031-82-2	109.747	wh pow	1325				i H_2O
1577	Lithium tungstate	Li_2WO_4	13568-45-1	261.72	wh trig pow	740		3.71		s H_2O
1578	Lithium vanadate	$LiVO_3$	15060-59-0	105.881	yel pow	subl 1400				
1579	Lithium zirconate	Li_2ZrO_3	12031-83-3	153.104	wh solid					
1580	Lutetium	Lu	7439-94-3	174.967	silv metal; hex	1663	3402	9.84		s dil acid
1581	Lutetium boride	LuB_4	12688-52-7	218.211	tetr cry	2600		≈7.0		
1582	Lutetium bromide	$LuBr_3$	14456-53-2	414.679	wh hyg cry	1025				vs H_2O
1583	Lutetium chloride	$LuCl_3$	10099-66-8	281.326	wh monocl cry; hyg	925		3.98		s H_2O
1584	Lutetium chloride hexahydrate	$LuCl_3 \cdot 6H_2O$	15230-79-2	389.417	col cry	dec 150				s H_2O, EtOH
1585	Lutetium fluoride	LuF_3	13760-81-1	231.962	orth cry	1182	2200	8.3		i H_2O
1586	Lutetium iodide	LuI_3	13813-45-1	555.680	brn hex cry; hyg	1050		≈5.6		vs H_2O
1587	Lutetium iron oxide	$Lu_3Fe_5O_{12}$	12023-71-1	996.119	cry					
1588	Lutetium nitrate	$Lu(NO_3)_3$	10099-67-9	360.982	hyg col solid					s H_2O, EtOH
1589	Lutetium nitride	LuN	12125-25-6	188.974	cub cry			11.6		

No.	Name	Formula	CAS Reg No.	Mol. weight	Physical form	mp/°C	bp/°C	Density g cm^{-3}	Solubility g/100 g H$_2$O	Qualitative solubility
1590	Lutetium oxide	Lu$_2$O$_3$	12032-20-1	397.932	wh cub cry or powder	2427	3980	9.41		
1591	Lutetium perchlorate hexahydrate	Lu(ClO$_4$)$_3$ · 6H$_2$O	14646-29-8	581.410	col cry	dec 350 (anh)				s H$_2$O, MeOH
1592	Lutetium sulfate	Lu$_2$(SO$_4$)$_3$	14986-89-1	638.122	wh pow	dec >850				vs H$_2$O
1593	Lutetium sulfate octahydrate	Lu$_2$(SO$_4$)$_3$ · 8H$_2$O	13473-77-3	782.244	col cry				42.3^{20}	s H$_2$O
1594	Lutetium sulfide	Lu$_2$S$_3$	12163-20-1	446.129	gray rhomb cry	1750 dec		6.26		
1595	Lutetium telluride	Lu$_2$Te$_3$	12163-22-3	732.73	orth cry			7.8		
1596	Magnesium	Mg	7439-95-4	24.305	silv-wh metal	650	1090	1.74		s dil acid
1597	Magnesium acetate	Mg(C$_2$H$_3$O$_2$)$_2$	142-72-3	142.394	wh orth/monocl cry	323 dec		1.50	65.6^{25}	
1598	Magnesium acetate monohydrate	Mg(C$_2$H$_3$O$_2$)$_2$ · H$_2$O	60582-92-5	160.409	orth cry			1.55		
1599	Magnesium acetate tetrahydrate	Mg(C$_2$H$_3$O$_2$)$_2$ · 4H$_2$O	16674-78-5	214.454	col monocl cry; hyg	80 dec		1.45	65.6^{25}	vs EtOH
1600	Magnesium aluminate	Mg(AlO$_2$)$_2$	12068-51-8	142.266	col cub cry	2105		3.55		i H$_2$O
1601	Magnesium aluminum silicate	Mg$_2$Al$_4$(AlSi$_5$O$_{18}$)	1302-88-1	584.953	blue cry			2.6		
1602	Magnesium amide	Mg(NH$_2$)$_2$	7803-54-5	56.350	wh powder; flam	dec		1.39		reac H$_2$O
1603	Magnesium ammonium phosphate hexahydrate	MgNH$_4$PO$_4$ · 6H$_2$O	13478-16-5	245.407	wh pow	dec		1.71		i H$_2$O, EtOH; s acid
1604	Magnesium antimonide	Mg$_3$Sb$_2$	12057-75-9	316.435	hex cry	1245		3.99		
1605	Magnesium arsenide	Mg$_3$As$_2$	12044-49-4	222.758	solid	≈1200		3.15		i H$_2$O
1606	Magnesium diboride	MgB$_2$	12007-25-9	45.927	hex cry	800 dec		2.57		
1607	Magnesium hexaboride	MgB$_6$	12008-22-9	89.171	refrac solid	1100 dec				i H$_2$O
1608	Magnesium dodecaboride	MgB$_{12}$	12230-32-9	154.037	refrac solid	1300 dec				
1609	Magnesium bromate hexahydrate	Mg(BrO$_3$)$_2$ · 6H$_2$O	7789-36-8	388.201	col cub cry	200 dec		2.29	98^{25}	
1610	Magnesium bromide	MgBr$_2$	7789-48-2	184.113	wh hex cry; hyg	711		3.72	102^{25}	
1611	Magnesium bromide hexahydrate	MgBr$_2$ · 6H$_2$O	13446-53-2	292.204	col monocl cry	165 dec		2.0	102^{25}	s EtOH
1612	Magnesium carbonate	MgCO$_3$	546-93-0	84.314	wh hex cry	990		3.010	0.18^{20}	i EtOH; s acid
1613	Magnesium carbonate dihydrate	MgCO$_3$ · 2H$_2$O	5145-48-2	120.345	col tricl cry			2.8		i H$_2$O, ace, NH$_4$OH
1614	Magnesium carbonate trihydrate	MgCO$_3$ · 3H$_2$O	14457-83-1	138.360	col monocl cry	165		1.8	0.18^{16}	
1615	Magnesium carbonate pentahydrate	MgCO$_3$ · 5H$_2$O	61042-72-6	174.390	wh monocl cry	dec >400		3.04	0.38^{16}	
1616	Magnesium carbonate hydroxide tetrahydrate	4MgCO$_3$ · Mg(OH)$_2$ · 4H$_2$O	39409-82-0	467.636	wh monocl cry			2.3		
1617	Magnesium carbonate hydroxide pentahydrate	4MgCO$_3$ · Mg(OH)$_2$ · 5H$_2$O	56378-72-4	485.652	wh pow	dec 700				i H$_2$O; s dil acid; i EtOH
1618	Magnesium carbonate dihydroxide trihydrate	MgCO$_3$ · Mg(OH)$_2$ · 3H$_2$O	12143-96-3	196.680	wh monocl cry	dec		2.04		
1619	Magnesium chlorate hexahydrate	Mg(ClO$_3$)$_2$ · 6H$_2$O	13446-19-0	299.299	wh hyg cry	≈35 dec		1.80	142^{25}	sl EtOH
1620	Magnesium chloride	MgCl$_2$	7786-30-3	95.211	wh hex leaflets; hyg	714	1412	2.325	56.0^{25}	
1621	Magnesium chloride hexahydrate	MgCl$_2$ · 6H$_2$O	7791-18-6	203.302	wh hyg cry	≈100 dec		1.56	56.0^{25}	s EtOH
1622	Magnesium chromate heptahydrate	MgCrO$_4$ · 7H$_2$O	13423-61-5*	266.405	yel rhom cry			1.695	54.8^{25}	
1623	Magnesium chromite	MgCr$_2$O$_4$	12053-26-8	192.295	deep grn cry	2390		4.4		
1624	Magnesium citrate	Mg$_3$(C$_6$H$_5$O$_7$)$_2$	3344-18-1	451.114	wh cry					sl H$_2$O
1625	Magnesium citrate tetradecahydrate	Mg$_3$(C$_6$H$_5$O$_7$)$_2$ · 14H$_2$O	3344-18-1*	703.328	wh cry pow					sl H$_2$O; s acid
1626	Magnesium fluoride	MgF$_2$	7783-40-6	62.302	wh tetr cry	1263	2227	3.148	0.013^{25}	
1627	Magnesium formate dihydrate	Mg(CHO$_2$)$_2$ · 2H$_2$O	6150-82-9	150.370	wh cry	dec				s H$_2$O; i EtOH
1628	Magnesium germanate	Mg$_2$GeO$_4$	12025-13-7	185.25	wh prec					i H$_2$O
1629	Magnesium germanide	Mg$_2$Ge	1310-52-7	121.25	cub cry	1117		3.09		
1630	Magnesium hydride	MgH$_2$	7693-27-8	26.321	wh tetr cry	327		1.45		reac H$_2$O
1631	Magnesium hydrogen phosphate trihydrate	MgHPO$_4$ · 3H$_2$O	7757-86-0	174.331	wh powder	550 dec		2.13		sl H$_2$O; s dil acid
1632	Magnesium hydroxide	Mg(OH)$_2$	1309-42-8	58.320	wh hex cry	350		2.37	0.00069^{20}	s dil acid
1633	Magnesium iodate tetrahydrate	Mg(IO$_3$)$_2$ · 4H$_2$O	7790-32-1*	446.172	col monocl cry	210 dec		3.3	11.1^{25}	
1634	Magnesium iodide	MgI$_2$	10377-58-9	278.114	wh hex cry; hyg	634		4.43	146^{25}	
1635	Magnesium iodide hexahydrate	MgI$_2$ · 6H$_2$O	66778-21-0	386.205	wh monocl cry			2.35		
1636	Magnesium iodide octahydrate	MgI$_2$ · 8H$_2$O	7790-31-0	422.236	wh orth cry; hyg	41 dec		2.10	146^{25}	s EtOH
1637	Magnesium metaborate octahydrate	Mg(BO$_2$)$_2$ · 8H$_2$O	13703-82-7*	254.047	wh pow	988 (anh)				sl H$_2$O
1638	Magnesium metasilicate	MgSiO$_3$	13776-74-4	100.389	wh monocl cry	≈1550 dec		3.19		i H$_2$O; sl HF
1639	Magnesium metatitanate	MgTiO$_3$	12032-30-3	120.170	col hex cry	1565		3.85		
1640	Magnesium molybdate	MgMoO$_4$	12013-21-7	184.24	wh pow	≈1060		2.2	15.9^{25}	s H$_2$O
1641	Magnesium nitrate	Mg(NO$_3$)$_2$	10377-60-3	148.314	wh cub cry			≈2.3	71.2^{25}	
1642	Magnesium nitrate dihydrate	Mg(NO$_3$)$_2$ · 2H$_2$O	15750-45-5	184.345	wh cry	≈100 dec		1.45	71.2^{25}	s EtOH
1643	Magnesium nitrate hexahydrate	Mg(NO$_3$)$_2$ · 6H$_2$O	13446-18-9	256.406	col monocl cry; hyg	≈95 dec		1.46	71.2^{25}	s EtOH
1644	Magnesium nitride	Mg$_3$N$_2$	12057-71-5	100.928	yel cub cry	≈1500 dec		2.71		
1645	Magnesium nitrite trihydrate	Mg(NO$_2$)$_2$ · 3H$_2$O	15070-34-5	170.362	wh hyg prisms	100 dec			129.9^{25}	s EtOH
1646	Magnesium orthosilicate	Mg$_2$SiO$_4$	26686-77-1	140.694	wh orth cry	1897		3.21		i H$_2$O

No.	Name	Formula	CAS Reg No.	Mol. weight	Physical form	mp/°C	bp/°C	Density g cm⁻³	Solubility g/100 g H₂O	Qualitative solubility
1647	Magnesium orthotitanate	Mg_2TiO_4	12032-52-9	160.475	wh cub cry	1840		3.53		
1648	Magnesium oxalate	MgC_2O_4	547-66-0	112.324	wh powder				0.038[25]	
1649	Magnesium oxalate dihydrate	$MgC_2O_4 \cdot 2H_2O$	6150-88-5	148.354	wh powder				0.038[25]	i EtOH; s dil acid
1650	Magnesium oxide	MgO	1309-48-4	40.304	wh cub cry	2825	3600	3.6		sl H₂O; i EtOH
1651	Magnesium perborate heptahydrate	$Mg(BO_3)_2 \cdot 7H_2O$	14635-87-1	268.030	wh pow					sl H₂O
1652	Magnesium perchlorate	$Mg(ClO_4)_2$	10034-81-8	223.206	wh hyg powder	250 dec		2.2	100[25]	
1653	Magnesium perchlorate hexahydrate	$Mg(ClO_4)_2 \cdot 6H_2O$	13446-19-0	331.298	wh hyg cry	190 dec		1.98	100[25]	s EtOH
1654	Magnesium permanganate hexahydrate	$Mg(MnO_4)_2 \cdot 6H_2O$	10377-62-5	370.268	blue-blk cry	dec		2.18		s H₂O
1655	Magnesium peroxide	MgO_2	1335-26-8	56.304	wh cub cry	100 dec		≈3.0		i H₂O; s dil acid
1656	Magnesium phosphate pentahydrate	$Mg_3(PO_4)_2 \cdot 5H_2O$	7757-87-1*	352.934	wh cry	400 dec			0.00009[20]	s dil acid
1657	Magnesium phosphate octahydrate	$Mg_3(PO_4)_2 \cdot 8H_2O$	13446-23-6	406.980	wh monocl cry			2.17	0.00009[20]	s acid
1658	Magnesium pyrophosphate	$Mg_2P_2O_7$	13446-24-7	222.553	col monocl plates	1395		2.56		
1659	Magnesium pyrophosphate trihydrate	$Mg_2P_2O_7 \cdot 3H_2O$	10102-34-8	276.600	wh powder	100 dec		2.56		i H₂O; s acid
1660	Magnesium phosphide	Mg_3P_2	12057-74-8	134.863	yel cub cry			2.06		reac H₂O
1661	Magnesium selenate hexahydrate	$MgSeO_4 \cdot 6H_2O$	13446-28-1	275.35	wh monocl cry			1.928	55.5[25]	
1662	Magnesium selenide	$MgSe$	1313-04-8	103.27	brn cub cry			4.2		reac H₂O
1663	Magnesium selenite hexahydrate	$MgSeO_3 \cdot 6H_2O$	15593-61-0	259.36	col hex cry			2.09		i H₂O; s dil acid
1664	Magnesium hexafluorosilicate hexahydrate	$MgSiF_6 \cdot 6H_2O$	60950-56-3	274.472	wh cry	120 dec		1.79	39.3[18]	i EtOH
1665	Magnesium silicide	Mg_2Si	22831-39-6	76.696	gray cub cry	1102		1.99		reac H₂O
1666	Magnesium stannide	Mg_2Sn	1313-08-2	167.320	blue cub cry	771		3.60		s H₂O, dil HCl
1667	Magnesium sulfate	$MgSO_4$	7487-88-9	120.368	col orth cry	1137		2.66	35.7[25]	
1668	Magnesium sulfate monohydrate	$MgSO_4 \cdot H_2O$	14168-73-1	138.383	col monocl cry	150 dec		2.57	35.7[25]	
1669	Magnesium sulfate heptahydrate	$MgSO_4 \cdot 7H_2O$	10034-99-8	246.474	col orth cry	150 dec		1.67	35.7[25]	sl EtOH
1670	Magnesium sulfide	MgS	12032-36-9	56.370	red-brn cub cry	2226		2.68		reac H₂O
1671	Magnesium sulfite trihydrate	$MgSO_3 \cdot 3H_2O$	19086-20-5	158.414	col orth cry			2.12	0.79[25]	
1672	Magnesium sulfite hexahydrate	$MgSO_3 \cdot 6H_2O$	13446-29-2	212.460	wh hex cry	200 dec		1.72	0.79[25]	i EtOH
1673	Magnesium tetrahydrogen phosphate dihydrate	$Mg(H_2PO_4)_2 \cdot 2H_2O$	15609-80-0	254.311	wh hyg cry	dec 90				s H₂O; i EtOH
1674	Magnesium thiocyanate tetrahydrate	$Mg(SCN)_2 \cdot 4H_2O$	306-61-6	212.531	wh hyg cry					vs H₂O, EtOH
1675	Magnesium thiosulfate hexahydrate	$MgS_2O_3 \cdot 6H_2O$	13446-30-5	244.525	col cry	170 dec		1.82	93[25]	i EtOH
1676	Magnesium trisilicate	$Mg_2Si_3O_8$	14987-04-3	260.862	wh powder					i H₂O, EtOH
1677	Magnesium tungstate	$MgWO_4$	13573-11-0	272.14	wh monocl cry			6.89	0.016[20]	i EtOH
1678	Magnesium vanadate	$Mg_2V_2O_7$	13568-63-3	262.489	tricl cry			3.1		
1679	Magnesium zirconate	$MgZrO_3$	12032-31-4	163.527	col cry	2060		4.23		
1680	Magnesium zirconium silicate	$MgO \cdot ZrO_2 \cdot SiO_2$	52110-05-1	223.612	wh solid					i H₂O, alk; sl acid
1681	Manganese	Mn	7439-96-5	54.938	hard gray metal	1246	2061	7.3		s dil acids
1682	Manganocene	$Mn(C_5H_5)_2$	1271-27-8	185.124	yel-brn cry	173				s py, thf; sl bz
1683	Manganese antimonide (MnSb)	$MnSb$	12032-82-5	176.698	hex cry	840		6.9		
1684	Manganese antimonide (Mn₂Sb)	Mn_2Sb	12032-97-2	231.636	tetr cry	948		7.0		
1685	Manganese boride (MnB)	MnB	12045-15-7	65.749	orth cry	1890		6.45		
1686	Manganese boride (MnB₂)	MnB_2	12228-50-1	76.560	hex cry	1827		5.3		
1687	Manganese boride (Mn₄B)	Mn_4B	12045-16-8	120.687	red-brn tetr cry	1580		7.20		
1688	Manganese carbide	Mn_3C	12266-65-8	176.825	refrac solid	1520		6.89		
1689	Manganese carbonyl	$Mn_2(CO)_{10}$	10170-69-1	389.977	yel monocl cry	154		1.75		i H₂O; s os
1690	Manganese pentacarbonyl bromide	$Mn(CO)_5Br$	14516-54-2	274.893	oran-yel cry					s os
1691	Manganese phosphide (MnP)	MnP	12032-78-9	85.912	orth cry	1147		5.49		
1692	Manganese phosphide (Mn₂P)	Mn_2P	12333-54-9	140.850	hex cry	1327		6.0		
1693	Manganese silicide	$MnSi_2$	12032-86-9	111.109	gray solid	1152 dec				
1694	Manganese(II) acetate tetrahydrate	$Mn(C_2H_3O_2)_2 \cdot 4H_2O$	6156-78-1	245.087	red monocl cry	80		1.59		s H₂O, EtOH
1695	Manganese(II) bromide	$MnBr_2$	13446-03-2	214.746	pink hex cry	698		4.385	151[25]	
1696	Manganese(II) bromide tetrahydrate	$MnBr_2 \cdot 4H_2O$	10031-20-6	286.808	red hyg cry	64 dec			151[25]	
1697	Manganese(II) carbonate	$MnCO_3$	598-62-9	114.947	pink hex cry	>200 dec		3.70	0.00008[20]	s dil acid
1698	Manganese(II) chloride	$MnCl_2$	7773-01-5	125.844	pink trig cry; hyg	650	1190	2.977	77.3[25]	s py, EtOH; i eth
1699	Manganese(II) chloride tetrahydrate	$MnCl_2 \cdot 4H_2O$	13446-34-9	197.906	red monocl cry; hyg	87.5		1.913	77.3[25]	s EtOH; i eth
1700	Manganese(II) dihydrogen phosphate dihydrate	$Mn(H_2PO_4)_2 \cdot 2H_2O$	18718-07-5	284.944	col hyg cry					s H₂O; i EtOH
1701	Manganese(II) fluoride	MnF_2	7782-64-1	92.935	red tetr cry	900		3.98	1.02[25]	i EtOH
1702	Manganese(II) hydroxide	$Mn(OH)_2$	18933-05-6	88.953	pink hex cry	dec		3.26	0.00034[20]	
1703	Manganese(II) hypophosphite monohydrate	$Mn(H_2PO_2)_2 \cdot H_2O$	10043-84-2	202.931	pink cry	>250			15[20]	s H₂O
1704	Manganese(II) iodide	MnI_2	7790-33-2	308.747	wh hex cry; hyg	638		5.04		s H₂O, EtOH

No.	Name	Formula	CAS Reg No.	Mol. weight	Physical form	mp/°C	bp/°C	Density g cm^{-3}	Solubility g/100 g H$_2$O	Qualitative solubility
1705	Manganese(II) iodide tetrahydrate	MnI$_2$ · 4H$_2$O	7790-33-2*	380.809	red cry					vs H$_2$O; s EtOH
1706	Manganese(II) metasilicate	MnSiO$_3$	7759-00-4	131.022	red orth cry	1291		3.48		i H$_2$O
1707	Manganese(II) molybdate	MnMoO$_4$	14013-15-1	214.88	yel monocl cry			4.05		
1708	Manganese(II) nitrate	Mn(NO$_3$)$_2$	10377-93-2	178.947	col orth cry; hyg			2.2	161^{25}	s diox, thf
1709	Manganese(II) nitrate tetrahydrate	Mn(NO$_3$)$_2$ · 4H$_2$O	20694-39-7	251.009	pink hyg cry	37.1 dec		2.13	161^{25}	s EtOH
1710	Manganese(II) nitrate hexahydrate	Mn(NO$_3$)$_2$ · 6H$_2$O	10377-66-9	287.039	rose monocl cry	28 dec		1.8	161^{25}	vs EtOH
1711	Manganese(II) orthosilicate	Mn$_2$SiO$_4$	13568-32-6	201.960	orth cry			4.11		i H$_2$O
1712	Manganese(II) oxalate dihydrate	MnC$_2$O$_4$ · 2H$_2$O	6556-16-7	178.987	wh cry powder	150 dec		2.45	0.032^{20}	s acid
1713	Manganese(II) oxide	MnO	1344-43-0	70.937	grn cub cry or powder	1842		5.37		i H$_2$O; s acid
1714	Manganese(II) perchlorate hexahydrate	Mn(ClO$_4$)$_2$ · 6H$_2$O	15364-94-0	361.931	pink hex cry			2.10		
1715	Manganese(II) pyrophosphate	Mn$_2$P$_2$O$_7$	53731-35-4	283.819	wh monocl cry	1196		3.71		i H$_2$O
1716	Manganese(II) selenide	MnSe	1313-22-0	133.90	gray cub cry	1460		5.45		i H$_2$O
1717	Manganese(II) sulfate	MnSO$_4$	7785-87-7	151.001	wh orth cry	700	850 dec	3.25	63.7^{25}	
1718	Manganese(II) sulfate monohydrate	MnSO$_4$ · H$_2$O	10034-96-5	169.016	red monocl cry			2.95	63.7^{25}	i EtOH
1719	Manganese(II) sulfate tetrahydrate	MnSO$_4$ · 4H$_2$O	10101-68-5	223.062	red monocl cry	38 dec		2.26	63.7^{25}	i EtOH
1720	Manganese(II) sulfide (α form)	MnS	18820-29-6	87.003	grn cub cry	1610		4.0		i H$_2$O; s dil acid
1721	Manganese(II) sulfide (β form)	MnS	18820-29-6	87.003	red cub cry			3.3		i H$_2$O; s dil acid
1722	Manganese(II) sulfide (γ form)	MnS	18820-29-6	87.003	red hex cry			≈3.3		i H$_2$O; s dil acid
1723	Manganese(II) telluride	MnTe	12032-88-1	182.54	hex cry	≈1150		6.0		
1724	Manganese(II) tetraborate octahydrate	MnB$_4$O$_7$ · 8H$_2$O	12228-91-0	354.300	red solid					i H$_2$O, EtOH; s dil acid
1725	Manganese(II) titanate	MnTiO$_3$	12032-74-5	150.803	red hex cry	1360		4.55		
1726	Manganese(II) tungstate	MnWO$_4$	13918-22-4	302.78	wh monocl cry			7.2	0.0054^{20}	
1727	Manganese(II,III) oxide	Mn$_3$O$_4$	1317-35-7	228.812	brn tetr cry	1567		4.84		i H$_2$O; s HCl
1728	Manganese(III) acetate dihydrate	Mn(C$_2$H$_3$O$_2$)$_3$ · 2H$_2$O	19513-05-4	268.100	brn cry					s eth, HOAc
1729	Manganese(III) fluoride	MnF$_3$	7783-53-1	111.933	red monocl cry; hyg	>600 dec		3.54		reac H$_2$O
1730	Manganese(III) hydroxide	MnO(OH)	1332-63-4	87.945	blk monocl cry	250 dec		≈4.3		i H$_2$O
1731	Manganese(III) oxide	Mn$_2$O$_3$	1317-34-6	157.874	blk cub cry	1080 dec		≈5.0		i H$_2$O
1732	Manganese(IV) oxide	MnO$_2$	1313-13-9	86.937	blk tetr cry	535 dec		5.08		i H$_2$O, HNO$_3$
1733	Manganese(VII) oxide	Mn$_2$O$_7$	12057-92-0	221.872	grn oil; exp	5.9	95 exp	2.40		vs H$_2$O
1734	Mendelevium	Md	7440-11-1	258	metal	827				
1735	Mercury	Hg	7439-97-6	200.59	heavy silv liq	-38.829	356.619	13.5336		i H$_2$O
1736	Dimethyl mercury	Hg(CH$_3$)$_2$	593-74-8	230.66	liq		93	3.17		i H$_2$O; vs EtOH, eth
1737	Mercury(I) acetate	Hg$_2$(C$_2$H$_3$O$_2$)$_2$	631-60-7	519.27	col scales	dec				sl H$_2$O; i EtOH, eth
1738	Mercury(I) bromate	Hg$_2$(BrO$_3$)$_2$	13465-33-3	656.98	col cry	dec				i H$_2$O; sl acid
1739	Mercury(I) bromide	Hg$_2$Br$_2$	15385-58-7	560.99	wh tetr cry or powder	345 dec		7.307		i H$_2$O, EtOH, eth
1740	Mercury(I) carbonate	Hg$_2$CO$_3$	6824-78-8	461.19	yel-brn cry	130 dec			0.0000045	i EtOH
1741	Mercury(I) chlorate	Hg$_2$(ClO$_3$)$_2$	10294-44-7	568.08	wh rhom cry	≈250 dec		6.409		sl H$_2$O; s EtOH
1742	Mercury(I) chloride	Hg$_2$Cl$_2$	10112-91-1	472.09	wh tetr cry	525 tp	383 sp	7.16	0.0004^{25}	i EtOH, eth
1743	Mercury(I) chromate	Hg$_2$CrO$_4$	13465-34-4	517.17	brn-red solid					i H$_2$O EtOH; s conc HNO$_3$
1744	Mercury(I) fluoride	Hg$_2$F$_2$	13967-25-4	439.18	yel cub cry	570 dec	subl	8.73		reac H$_2$O
1745	Mercury(I) iodate	Hg$_2$(IO$_3$)$_2$	13465-35-5	750.99	yel-wh pow	dec 175			0.0032^{20}	
1746	Mercury(I) iodide	Hg$_2$I$_2$	15385-57-6	654.99	yel amorp powder	290		7.70		i H$_2$O, EtOH, eth
1747	Mercury(I) nitrate	Hg$_2$(NO$_3$)$_2$	10415-75-5	525.19	cry					sl H$_2$O
1748	Mercury(I) nitrate dihydrate	Hg$_2$(NO$_3$)$_2$ · 2H$_2$O	14836-60-3	561.22	col cry	70 dec		4.8		sl H$_2$O
1749	Mercury(I) nitrite	Hg$_2$(NO$_2$)$_2$	13492-25-6	493.19	yel cry	100 dec		7.3		reac H$_2$O
1750	Mercury(I) oxalate	Hg$_2$C$_2$O$_4$	2949-11-3	489.20	cry					i H$_2$O; sl HNO$_3$
1751	Mercury(I) oxide	Hg$_2$O	15829-53-5	417.18	prob mixture of HgO+Hg	100 dec		9.8		i H$_2$O; s HNO$_3$
1752	Mercury(I) perchlorate tetrahydrate	Hg$_2$(ClO$_4$)$_2$ · 4H$_2$O	65202-12-2	672.14	cry	64			442^{25}	
1753	Mercury(I) sulfate	Hg$_2$SO$_4$	7783-36-0	497.24	wh-yel cry powder			7.56	0.051^{25}	s dil HNO$_3$
1754	Mercury(I) sulfide	Hg$_2$S	51595-71-2	433.25	unstab blk pow	dec				i H$_2$O
1755	Mercury(I) thiocyanate	Hg$_2$(SCN)$_2$	13465-37-7	517.34	col powder	dec			0.03^{25}	s HCl, KCNS
1756	Mercury(I) tungstate	Hg$_2$WO$_4$	38705-19-0	649.02	yel amorp solid	dec				i H$_2$O, EtOH
1757	Mercury(II) acetate	Hg(C$_2$H$_3$O$_2$)$_2$	1600-27-7	318.68	wh-yel cry or powder	179 dec		3.28	25^{10}	s EtOH
1758	Mercury(II) amide chloride	Hg(NH$_2$)Cl	10124-48-8	252.07	wh solid		subl	5.38		i H$_2$O, EtOH; s warm acid
1759	Mercury(II) benzoate monohydrate	Hg(C$_7$H$_5$O$_2$)$_2$ · H$_2$O	32839-04-6	460.83	wh cry	165			1.2^{15}	sl EtOH
1760	Mercury(II) bromate	Hg(BrO$_3$)$_2$	26522-91-8	456.39	cry	130 dec			0.15	s acid
1761	Mercury(II) bromide	HgBr$_2$	7789-47-1	360.40	wh rhomb cry or powder	241	318	6.05	0.61^{25}	sl chl; s EtOH, MeOH
1762	Mercury(II) chlorate	Hg(ClO$_3$)$_2$	13465-30-0	367.49	wh needles	dec		4.998	25	

No.	Name	Formula	CAS Reg No.	Mol. weight	Physical form	mp/°C	bp/°C	Density g cm⁻³	Solubility g/100 g H₂O	Qualitative solubility
1763	Mercury(II) chloride	HgCl₂	7487-94-7	271.50	wh orth cry	277	304	5.6	7.31²⁵	sl bz; s EtOH, MeOH, ace, eth
1764	Mercury(II) chromate	HgCrO₄	13444-75-2	316.58	red monocl cry			6.06		sl H₂O
1765	Mercury(II) cyanide	Hg(CN)₂	592-04-1	252.62	col tetr cry	320 dec		4.00	11.4²⁵	s EtOH; sl eth
1766	Mercury(II) dichromate	HgCr₂O₇	7789-10-8	416.58	red cry powder					i H₂O; s acid
1767	Mercury(II) fluoride	HgF₂	7783-39-3	238.59	wh cub cry; hyg	645 dec		8.95		reac H₂O
1768	Mercury(II) fulminate	Hg(CNO)₂	628-86-4	284.62	gray cry	exp		4.42		sl H₂O; s EtOH, NH₄OH
1769	Mercury(II) hydrogen arsenate	HgHAsO₄	7784-37-4	340.52	yel powder					i H₂O; s acid
1770	Mercury(II) iodate	Hg(IO₃)₂	7783-32-6	550.40	wh powder	175 dec				i H₂O
1771	Mercury(II) iodide (yellow)	HgI₂	7774-29-0	454.40	yel tetr cry or powder	256	351	6.28	0.0055²⁵	sl EtOH, ace, eth
1772	Mercury(II) iodide (red)	HgI₂	7774-29-0	454.40	red pow	trans to yel 127			0.006²⁵	sl EtOH, ace, eth, chl
1773	Mercury(II) nitrate	Hg(NO₃)₂	10045-94-0	324.60	col hyg cry	79		4.3		s H₂O; i EtOH
1774	Mercury(II) nitrate monohydrate	Hg(NO₃)₂ · H₂O	7783-34-8	342.62	wh-yel hyg cry			4.3		s H₂O, dil acid
1775	Mercury(II) nitrate dihydrate	Hg(NO₃)₂ · 2H₂O	22852-67-1	360.63	monocl cry			4.78		s H₂O
1776	Mercury(II) oxalate	HgC₂O₄	3444-13-1	288.61	powder	165 dec				i H₂O
1777	Mercury(II) oxide	HgO	21908-53-2	216.59	red or yel orth cry	500 dec		11.14		i H₂O, EtOH; s dil acid
1778	Mercury(II) oxide sulfate	(Hg₃O₂)SO₄	1312-03-4	729.83	yel powder					i H₂O; s acid
1779	Mercury(II) oxycyanide	Hg(CN)₂ · HgO	1335-31-5	469.21	wh orth cry	exp		4.44	11.4²⁵	
1780	Mercury(II) perchlorate trihydrate	Hg(ClO₄)₂ · 3H₂O	7616-83-3	453.54	cry					
1781	Mercury(II) phosphate	Hg₃(PO₄)₂	7782-66-3	791.71	wh-yel powder					i H₂O, EtOH; s acid
1782	Mercury(II) selenide	HgSe	20601-83-6	279.55	gray cub cry		subl	8.21		i H₂O
1783	Mercury(II) sulfate	HgSO₄	7783-35-9	296.65	wh monocl cry			6.47		reac H₂O
1784	Mercury(II) sulfide (black)	HgS	1344-48-5	232.66	blk cub cry or powder	850		7.70		i H₂O; s acid, EtOH
1785	Mercury(II) sulfide (red)	HgS	1344-48-5	232.66	red hex cry	trans to blk HgS 344		8.17		i H₂O, acid; s aqua regia
1786	Mercury(II) telluride	HgTe	12068-90-5	328.19	gray cub cry	673		8.63		
1787	Mercury(II) thiocyanate	Hg(SCN)₂	592-85-8	316.75	monocl cry	≈165 dec		3.71	0.070²⁵	s dil HCl
1788	Mercury(II) tungstate	HgWO₄	37913-38-5	448.43	yel cry	dec				i H₂O, EtOH
1789	Molybdenum	Mo	7439-98-7	95.96	gray-blk metal; cub	2622	4639	10.2		i H₂O, dil acid, alk
1790	Molybdophosphoric acid	H₃P(Mo₃O₁₀)₄	51429-74-4	1825.25	bright yel cry					
1791	Molybdenum boride (Mo₂B)	Mo₂B	12006-99-4	202.69	refrac tetr cry	2000		9.2		
1792	Molybdenum boride (Mo₂B₅)	Mo₂B₅	12007-97-5	245.94	refrac hex cry	1600		≈7.2		
1793	Molybdenum carbide (MoC)	MoC	12011-97-1	107.95	refrac solid; cub	2577				
1794	Molybdenum carbide (Mo₂C)	Mo₂C	12069-89-5	203.89	gray orth cry	2687		9.18		
1795	Molybdenum carbonyl	Mo(CO)₆	13939-06-5	264.00	wh orth cry	148	155 dec	1.96		i H₂O; s bz; sl eth
1796	Molybdenum nitride (MoN)	MoN	12033-19-1	109.95	hex cry	1750		9.20		
1797	Molybdenum nitride (Mo₂N)	Mo₂N	12033-31-7	205.89	gray cub cry	790 dec		9.46		
1798	Molybdenum phosphide	MoP	12163-69-8	126.91	blk hex cry			7.34		
1799	Molybdenum silicide (MoSi₂)	MoSi₂	12136-78-6	152.11	gray tetr cry	≈1900		6.2		i H₂O; s HF
1800	Molybdenum(II) bromide	MoBr₂	13446-56-5	255.75	yel-red cry	dec 700		4.88		i H₂O, EtOH
1801	Molybdenum(II) chloride	MoCl₂	13478-17-6	166.85	yel cry	dec 500		3.71		i H₂O
1802	Molybdenum(II) iodide	MoI₂	14055-74-4	349.75	blk hyg cry	700		5.28		i H₂O
1803	Molybdenum(III) bromide	MoBr₃	13446-57-6	335.65	grn hex cry	dec 500		4.89		i H₂O, EtOH
1804	Molybdenum(III) chloride	MoCl₃	13478-18-7	202.30	dark red monocl cry	dec 400		3.74		i H₂O, os
1805	Molybdenum(III) fluoride	MoF₃	20193-58-2	152.94	yel-brn hex cry	>600		4.64		i H₂O
1806	Molybdenum(III) iodide	MoI₃	14055-75-5	476.65	blk solid	927				i H₂O
1807	Molybdenum(III) oxide	Mo₂O₃	1313-29-7	239.88	gray-blk powder					i H₂O; sl acid
1808	Molybdenum(IV) bromide	MoBr₄	13520-59-7	415.56	blk cry	dec 110				reac H₂O
1809	Molybdenum(IV) chloride	MoCl₄	13320-71-3	237.75	blk cry	317				reac H₂O, sl chl; i eth, bz
1810	Molybdenum(IV) fluoride	MoF₄	23412-45-5	171.93	grn cry	dec				reac H₂O
1811	Molybdenum(IV) iodide	MoI₄	14055-76-6	603.56	blk cry	dec 100				i H₂O
1812	Molybdenum(IV) oxide	MoO₂	18868-43-4	127.94	brn-viol tetr cry	≈1800 dec		6.47		i H₂O, acid, alk
1813	Molybdenum(IV) selenide	MoSe₂	12058-18-3	253.86	gray hex cry	>1200		6.90		
1814	Molybdenum(IV) sulfide	MoS₂	1317-33-5	160.07	blk powder or hex cry	1750		5.06		i H₂O; s conc acid
1815	Molybdenum(IV) telluride	MoTe₂	12058-20-7	351.14	gray hex cry			7.7		
1816	Molybdenum(V) chloride	MoCl₅	10241-05-1	273.21	grn-blk monocl cry; hyg	194	268	2.93		reac H₂O; s EtOH, eth
1817	Molybdenum(V) fluoride	MoF₅	13819-84-6	190.93	yel monocl cry	67	213.6	3.5		reac H₂O
1818	Molybdenum(V) oxytrichloride	MoOCl₃	13814-74-9	218.30	blk monocl cry	310	subl			reac H₂O
1819	Molybdenum(VI) acid monohydrate	H₂MoO₄ · H₂O	7782-91-4	179.97	wh powder			3.1		sl H₂O; s alk

No.	Name	Formula	CAS Reg No.	Mol. weight	Physical form	mp/°C	bp/°C	Density g cm⁻³	Solubility g/100 g H₂O	Qualitative solubility
1820	Molybdenum(VI) dioxydichloride	MoO_2Cl_2	13637-68-8	198.85	yel-oran solid	176	250	3.31		reac H_2O
1821	Molybdenum(VI) dioxydifluoride	MoO_2F_2	13824-57-2	165.94	pale lilac cry	subl 270		3.5		i MeCN, chl; sl HF
1822	Molybdenum(VI) fluoride	MoF_6	7783-77-9	209.93	wh cub cry or col liq; hyg	17.5	34.0	2.54		reac H_2O; vs hex, ctc
1823	Molybdenum(VI) metaphosphate	$Mo(PO_3)_6$	133863-98-6	569.77	yel powder			3.28		i H_2O, acid
1824	Molybdenum(VI) oxide	MoO_3	1313-27-5	143.94	wh-yel rhomb cry	802	1155	4.70	0.14²⁰	sl H_2O; s alk, acid
1825	Molybdenum(VI) oxytetrachloride	$MoOCl_4$	13814-75-0	253.75	grn hyg powder	105	159			
1826	Molybdenum(VI) oxytetrafluoride	$MoOF_4$	14459-59-7	187.93	volatile solid	97.2	186.0			
1827	Molybdenum(VI) sulfide	MoS_3	12033-29-3	192.14	blk solid	350 dec				i H_2O, os
1828	Neodymium	Nd	7440-00-8	144.242	silv metal; hex	1016	3074	7.01		
1829	Neodymium boride	NdB_6	12008-23-0	209.108	blk cub cry	2610		4.93		
1830	Neodymium nitride	NdN	25764-11-8	158.249	blk cub cry			7.69		
1831	Neodymium(II) acetate	$Nd(C_2H_3O_2)_2$	6192-13-8	321.373	red-purp cry					s H_2O
1832	Neodymium(II) chloride	$NdCl_2$	25469-93-6	215.148	grn hyg solid	841				
1833	Neodymium(III) bromate nonahydrate	$Nd(BrO_3)_3 \cdot 9H_2O$	15162-92-2	690.086	red hex cry	66 dec				
1834	Neodymium(III) bromide	$NdBr_3$	13536-80-6	383.954	viol orth cry; hyg	682	1540	5.3		s H_2O
1835	Neodymium(III) chloride	$NdCl_3$	10024-93-8	250.601	viol hex cry	759	1600	4.13	100²⁵	vs EtOH; i eth, chl
1836	Neodymium(III) chloride hexahydrate	$NdCl_3 \cdot 6H_2O$	13477-89-9	358.692	purp cry	124 dec		2.3	100²⁵	s EtOH
1837	Neodymium(III) fluoride	NdF_3	13709-42-7	201.237	viol hex cry; hyg	1377	2300	6.51		i H_2O
1838	Neodymium(III) hydroxide	$Nd(OH)_3$	16469-17-3	195.264	blue solid	dec 210				i H_2O
1839	Neodymium(III) iodide	NdI_3	13813-24-6	524.955	grn orth cry; hyg	787		5.85		s H_2O
1840	Neodymium(III) nitrate	$Nd(NO_3)_3$	10045-95-1	330.257	viol hyg. cry				152²⁵	s EtOH
1841	Neodymium(III) nitrate hexahydrate	$Nd(NO_3)_3 \cdot 6H_2O$	14517-29-4	438.348	purp hyg cry				152²⁵	s EtOH, ace
1842	Neodymium(III) oxide	Nd_2O_3	1313-97-9	336.482	blue hex cry; hyg	2233	3760	7.24		i H_2O; s dil acid
1843	Neodymium(III) sulfate	$Nd_2(SO_4)_3$	13477-91-3	576.672	pink needles	≈700 dec			7.1²⁰	
1844	Neodymium(III) sulfate octahydrate	$Nd_2(SO_4)_3 \cdot 8H_2O$	13477-91-3	720.794	red cry	350 dec		2.85		sl H_2O
1845	Neodymium(III) sulfide	Nd_2S_3	12035-32-4	384.679	orth cry	2207		5.46		
1846	Neodymium(III) telluride	Nd_2Te_3	12035-35-7	671.28	gray orth cry	1377		7.0		
1847	Neodymium(III) tris(cyclopentadienyl)	$Nd(C_5H_5)_3$	1273-98-9	339.522	red-blue cry	380				s thf
1848	Neon	Ne	7440-01-9	20.180	col gas	-248.59	-246.053	0.825 g/L		sl H_2O
1849	Neptunium	Np	7439-99-8	237	silv metal	644	3902	20.2		s HCl
1850	Neptunium(IV) oxide	NpO_2	12035-79-9	269	grn cub cry	2547		11.1		
1851	Nickel	Ni	7440-02-0	58.693	wh metal; cub	1455	2913	8.90		i H_2O; sl dil acid
1852	Nickelocene	$Ni(C_5H_5)_2$	1271-28-9	188.879	grn cry	173				
1853	Nickel aluminide (NiAl)	NiAl	12003-78-0	85.675	metallic solid	1638				
1854	Nickel antimonide	NiSb	12035-52-8	180.453	hex cry	1147		8.74		
1855	Nickel arsenide	NiAs	27016-75-7	133.615	hex cry	967		7.77		
1856	Nickel boride (NiB)	NiB	12007-00-0	69.504	grn refrac solid	1035		7.13		
1857	Nickel boride (Ni₂B)	Ni_2B	12007-01-1	128.198	refrac solid	1125		7.90		
1858	Nickel boride (Ni₃B)	Ni_3B	12007-02-2	186.891	refrac solid	1166		8.17		
1859	Nickel carbonyl	$Ni(CO)_4$	13463-39-3	170.734	col liq	-19.3	42.1 (exp ≈60)	1.31		i H_2O; s EtOH, bz, ace, ctc
1860	Nickel phosphide	Ni_2P	12035-64-2	148.361	hex cry	1100		7.33		
1861	Nickel silicide (NiSi₂)	$NiSi_2$	12201-89-7	114.864	cub cry	993		4.83		
1862	Nickel silicide (Ni₂Si)	Ni_2Si	12059-14-2	145.473	orth cry	1255		7.40		
1863	Nickel subsulfide	Ni_3S_2	12035-72-2	240.210	yel hex cry	789		5.87		
1864	Nickel(II) acetate tetrahydrate	$Ni(C_2H_3O_2)_2 \cdot 4H_2O$	6018-89-9	248.842	grn monocl cry	250 dec		1.74	16²⁰	s H_2O, EtOH
1865	Nickel(II) ammonium chloride hexahydrate	$NH_4NiCl_3 \cdot 6H_2O$	16122-03-5*	291.182	grn hyg cry			1.65		s H_2O
1866	Nickel(II) ammonium sulfate	$Ni(NH_4)_2(SO_4)_2$	15699-18-0	286.895	blue-grn cry	dec 250				sl H_2O
1867	Nickel(II) ammonium sulfate hexahydrate	$Ni(NH_4)_2(SO_4)_2 \cdot 6H_2O$	7785-20-8	394.987	blue-grn cry	dec 130		1.92	6.5²⁰	s H_2O; i EtOH
1868	Nickel(II) arsenate octahydrate	$Ni_3(AsO_4)_2 \cdot 8H_2O$	7784-48-7	598.040	yel-grn powder	dec		4.98		i H_2O; s acid
1869	Nickel(II) bromide	$NiBr_2$	13462-88-9	218.501	yel hex cry; hyg	963	subl	5.10	131²⁰	
1870	Nickel(II) bromide trihydrate	$NiBr_2 \cdot 3H_2O$	13462-88-9*	272.547	yel-grn hyg cry	200 dec				vs H_2O; s EtOH, eth
1871	Nickel(II) carbonate	$NiCO_3$	3333-67-3	118.702	grn rhomb cry			4.389	0.0043²⁰	s dil acid
1872	Nickel(II) chlorate hexahydrate	$Ni(ClO_3)_2 \cdot 6H_2O$	13477-94-6	333.687	grn cub cry	dec 80		2.07		vs H_2O
1873	Nickel(II) chloride	$NiCl_2$	7718-54-9	129.599	yel hex cry; hyg	1031	985 sp	3.51	67.5²⁵	s EtOH
1874	Nickel(II) chloride hexahydrate	$NiCl_2 \cdot 6H_2O$	7791-20-0	237.690	grn monocl cry				67.5²⁵	s EtOH
1875	Nickel(II) chromate	$NiCrO_4$	14721-18-7	174.687	red solid					sl H_2O
1876	Nickel(II) cyanide tetrahydrate	$Ni(CN)_2 \cdot 4H_2O$	13477-95-7	182.789	grn plates	200 dec				i H_2O; sl dil acid; s NH_4OH
1877	Nickel(II) fluoride	NiF_2	10028-18-9	96.690	yel tetr cry	1380		4.7	2.56²⁵	i H_2O, eth
1878	Nickel(II) fluoride tetrahydrate	$NiF_2 \cdot 4H_2O$	13940-83-5	168.752	grn pow					sl H_2O
1879	Nickel(II) hydroxide	$Ni(OH)_2$	12054-48-7	92.708	grn hex cry	230 dec		4.1	0.00015²⁰	

No.	Name	Formula	CAS Reg No.	Mol. weight	Physical form	mp/°C	bp/°C	Density g cm^{-3}	Solubility g/100 g H$_2$O	Qualitative solubility
1880	Nickel(II) hydroxide monohydrate	Ni(OH)$_2$ · H$_2$O	36897-37-7	110.723	grn powder				0.00015^{20}	s dil acid
1881	Nickel(II) iodate	Ni(IO$_3$)$_2$	13477-98-0	408.498	yel needles			5.07	1.1^{30}	sl H$_2$O
1882	Nickel(II) iodate tetrahydrate	Ni(IO$_3$)$_2$ · 4H$_2$O	13477-99-1	480.560	yel hex cry	dec 100		5.07		sl H$_2$O
1883	Nickel(II) iodide	NiI$_2$	13462-90-3	312.502	blk hex cry; hyg	800	subl	5.22	154^{25}	
1884	Nickel(II) iodide hexahydrate	NiI$_2$ · 6H$_2$O	7790-34-3	420.593	grn monocl cry; hyg				154^{25}	vs EtOH
1885	Nickel(II) nitrate	Ni(NO$_3$)$_2$	13138-45-9	182.702	grn cry				99.2^{25}	s EtOH
1886	Nickel(II) nitrate hexahydrate	Ni(NO$_3$)$_2$ · 6H$_2$O	13478-00-7	290.794	grn monocl cry; hyg	56 dec		2.05	99.2^{25}	s EtOH
1887	Nickel(II) oxalate dihydrate	NiC$_2$O$_4$ · 2H$_2$O	6018-94-6	182.742	grn-wh solid	dec 150			0.0012^{25}	i H$_2$O; s acid, NH$_4$OH
1888	Nickel(II) oxide	NiO	1313-99-1	74.692	grn cub cry	1957		6.72		i H$_2$O; s acid
1889	Nickel(II) perchlorate hexahydrate	Ni(ClO$_4$)$_2$ · 6H$_2$O	13637-71-3*	365.686	grn hex needles	140			158.8^{25}	s EtOH, ace
1890	Nickel(II) phosphate octahydrate	Ni$_3$(PO$_4$)$_2$ · 8H$_2$O	10381-36-9*	510.145	grn plates					s acid
1891	Nickel(II) selenate hexahydrate	NiSeO$_4$ · 6H$_2$O	15060-62-5*	309.74	grn tetr cry			2.314	35.5^{20}	
1892	Nickel(II) selenide	NiSe	1314-05-2	137.65	yel-grn hex cry	980		7.2		
1893	Nickel(II) stannate dihydrate	NiSnO$_3$ · 2H$_2$O	12035-38-0	261.432	grn pow	dec 120				
1894	Nickel(II) sulfate	NiSO$_4$	7786-81-4	154.756	grn-yel orth cry	840 dec		4.01	40.4^{25}	
1895	Nickel(II) sulfate hexahydrate	NiSO$_4$ · 6H$_2$O	10101-97-0	262.847	blue-grn tetr cry	≈100 dec		2.07	40.4^{25}	sl EtOH
1896	Nickel(II) sulfate heptahydrate	NiSO$_4$ · 7H$_2$O	10101-98-1	280.862	grn orth cry			1.98	40.4^{25}	s EtOH
1897	Nickel(II) sulfide	NiS	16812-54-7	90.758	yel hex cry	976		5.5		i H$_2$O
1898	Nickel(II) thiocyanate	Ni(SCN)$_2$	13689-92-4	174.857	grn pwd				55.0^{25}	
1899	Nickel(II) titanate	NiTiO$_3$	12035-39-1	154.558	brn hex cry			5.0		
1900	Nickel(II,III) sulfide	Ni$_3$S$_4$	12137-12-1	304.340	cub cry	995		4.77		
1901	Nickel(III) oxide	Ni$_2$O$_3$	1314-06-3	165.385	gray-blk cub cry	≈600 dec				i H$_2$O; s hot acid
1902	Niobium	Nb	7440-03-1	92.906	gray metal; cub	2477	4741	8.57		i acid
1903	Niobocene dichloride	Nb(C$_5$H$_5$)$_2$Cl$_2$	12793-14-5	293.998	hyg blk cry					sl tol
1904	Niobium boride (NbB)	NbB	12045-19-1	103.717	gray orth cry	2270		7.5		
1905	Niobium boride (NbB$_2$)	NbB$_2$	12007-29-3	114.528	gray hex cry	3050		6.97		
1906	Niobium carbide (NbC)	NbC	12069-94-2	104.917	gray cub cry	3608	4300	7.82		i H$_2$O, acid
1907	Niobium carbide (Nb$_2$C)	Nb$_2$C	12011-99-3	197.824	refrac hex cry	3080		7.8		i H$_2$O
1908	Niobium nitride	NbN	24621-21-4	106.913	gray cry; cub	2300		8.47		i HCl, acid
1909	Niobium phosphide	NbP	12034-66-1	123.880	tetr cry			6.5		
1910	Niobium silicide	NbSi$_2$	12034-80-9	149.077	gray hex cry	1950		5.7		
1911	Niobium(II) oxide	NbO	12034-57-0	108.905	gray cub cry	1937		7.30		
1912	Niobium(III) bromide	NbBr$_3$	15752-41-7	332.618	dark brn solid		subl 400			
1913	Niobium(III) chloride	NbCl$_3$	13569-59-0	199.265	blk solid					
1914	Niobium(III) fluoride	NbF$_3$	15195-53-6	149.901	blue cub cry			4.2		
1915	Niobium(III) iodide	NbI$_3$	13870-20-7	473.619	blk solid	dec 510				
1916	Niobium(IV) bromide	NbBr$_4$	13842-75-6	412.522	dark brn cry		subl 300	4.72		reac H$_2$O
1917	Niobium(IV) chloride	NbCl$_4$	13569-70-5	234.718	viol-blk monocl cry	dec 800	275 subl	3.2		reac H$_2$O
1918	Niobium(IV) fluoride	NbF$_4$	13842-88-1	168.900	blk tetr cry; hyg	>350 dec		4.01		
1919	Niobium(IV) iodide	NbI$_4$	13870-21-8	600.524	gray orth cry	503		5.6		reac H$_2$O
1920	Niobium(IV) oxide	NbO$_2$	12034-59-2	124.905	wh tetr cry or powder	1901		5.9		
1921	Niobium(IV) selenide	NbSe$_2$	12034-77-4	250.83	gray hex cry	>1300		6.3		
1922	Niobium(IV) sulfide	NbS$_2$	12136-97-9	157.036	blk rhomb cry			4.4		
1923	Niobium(IV) telluride	NbTe$_2$	12034-83-2	348.11	hex cry			7.6		
1924	Niobium(V) bromide	NbBr$_5$	13478-45-0	492.426	oran orth cry	265.2	361.6	4.36		s H$_2$O, EtOH
1925	Niobium(V) chloride	NbCl$_5$	10026-12-7	270.171	yel monocl cry; hyg	205.8	247.4	2.78		reac H$_2$O; s HCl, ctc
1926	Niobium(V) dioxyfluoride	NbO$_2$F	15195-33-2	143.903	wh cub cry			4.0		
1927	Niobium(V) ethoxide	Nb(OC$_2$H$_5$)$_5$	3236-82-6	318.209	col hyg liq	5	203	1.258		reac H$_2$O; s peth
1928	Niobium(V) fluoride	NbF$_5$	7783-68-8	187.898	col monocl cry; hyg	80	234	2.70		reac H$_2$O; sl CS$_2$, chl
1929	Niobium(V) iodide	NbI$_5$	13779-92-5	727.428	yel-blk monocl cry	327		5.32		reac H$_2$O
1930	Niobium(V) oxide	Nb$_2$O$_5$	1313-96-8	265.810	wh orth cry	1500		4.47		i H$_2$O; s HF
1931	Niobium(V) oxybromide	NbOBr$_3$	14459-75-7	348.617	yel-brn cry	≈320 dec	subl			
1932	Niobium(V) oxychloride	NbOCl$_3$	13597-20-1	215.264	wh tetr cry		subl	3.72		
1933	Nitrogen	N$_2$	7727-37-9	28.014	col gas	-210.0	-195.798	1.145 g/L		sl H$_2$O; i EtOH
1934	Nitramide	NO$_2$NH$_2$	7782-94-7	62.028	unstab wh cry	72 dec				s H$_2$O, EtOH, ace, eth; i chl
1935	Nitric acid	HNO$_3$	7697-37-2	63.013	col liq; hyg	-41.6	83	1.5129^{20}		vs H$_2$O
1936	Nitrous acid	HNO$_2$	7782-77-6	47.014	stab only in soln					
1937	Nitrous oxide	N$_2$O	10024-97-2	44.012	col gas	-90.8	-88.48	1.799 g/L		sl H$_2$O; s EtOH, eth
1938	Nitric oxide	NO	10102-43-9	30.006	col gas	-163.6	-151.74	1.226 g/L		sl H$_2$O

No.	Name	Formula	CAS Reg No.	Mol. weight	Physical form	mp/°C	bp/°C	Density g cm⁻³	Solubility g/100 g H₂O	Qualitative solubility
1939	Nitrogen dioxide	NO_2	10102-44-0	46.006	brn gas; equil with N_2O_4		see N_2O_4	1.880 g/L		reac H_2O
1940	Nitrogen trioxide	N_2O_3	10544-73-7	76.011	blue solid or liq (low temp)	-101.1	≈3 dec	1.4²		reac H_2O
1941	Nitrogen tetroxide	N_2O_4	10544-72-6	92.011	col liq; equil with NO_2	-9.3	21.15	1.45²⁰		reac H_2O
1942	Nitrogen pentoxide	N_2O_5	10102-03-1	108.010	col hex cry		33 sp	2.0		s chl; sl ctc
1943	Nitrogen tribromide	NBr_3	15162-90-0	253.719	unstab solid	exp -100				
1944	Nitrogen trichloride	NCl_3	10025-85-1	120.366	yel oily liq; exp	-40	71	1.653		i H_2O; s CS_2, bz, ctc
1945	Nitrogen trifluoride	NF_3	7783-54-2	71.002	col gas	-206.79	-128.75	2.902 g/L		i H_2O
1946	Nitrogen triiodide	NI_3	13444-85-4	394.720	unstab blk cry; exp					
1947	Nitrogen chloride difluoride	$NClF_2$	13637-87-1	87.457	col gas	-195	-67	3.575 g/L		
1948	Chloramine	NH_2Cl	10599-90-3	51.476	yel liq	-66				s H_2O, EtOH, eth; sl bz, ctc
1949	Fluoramine	NH_2F	15861-05-9	35.021	unstab gas	≈-110		1.431 g/L		
1950	Difluoramine	NHF_2	10405-27-3	53.012	col gas	-116	-23	2.167 g/L		
1951	cis-Difluorodiazine	N_2F_2	13812-43-6	66.010	col gas	<-195	-105.75	2.698 g/L		
1952	trans-Difluorodiazine	N_2F_2	13776-62-0	66.010	col gas	-172	-111.45	2.698 g/L		
1953	Tetrafluorohydrazine	N_2F_4	10036-47-2	104.007	col gas	-164.5	-74	4.251 g/L		
1954	Nitrosyl bromide	$NOBr$	13444-87-6	109.910	red gas	-56	≈0	4.492 g/L		reac H_2O
1955	Nitrosyl chloride	$NOCl$	2696-92-6	65.459	yel gas	-59.6	-5.5	2.676 g/L		reac H_2O
1956	Nitrosyl fluoride	NOF	7789-25-5	49.004	col gas	-132.5	-59.9	2.003 g/L		
1957	Trifluoramine oxide	NOF_3	13847-65-9	87.001	col gas	-161	-87.5	3.556 g/L		
1958	Nitryl chloride	NO_2Cl	13444-90-1	81.459	col gas	-145	-15	3.330 g/L		
1959	Nitryl fluoride	NO_2F	10022-50-1	65.004	col gas	-166	-72.4	2.657 g/L		reac H_2O
1960	Nitrogen selenide	N_4Se_4	12033-88-4	371.87	red monocl cry; hyg	exp		4.2		i H_2O, eth, EtOH; sl bz, CS_2
1961	Nobelium	No	10028-14-5	259.000	metal	827				
1962	Osmium	Os	7440-04-2	190.23	blue-wh metal; hex	3033	5008	22.587²⁰		s aqua regia
1963	Osmocene	$Os(C_5H_5)_2$	1273-81-0	320.42	col cry	229				
1964	Osmium carbonyl	$Os_3(CO)_{12}$	15696-40-9	906.81	yel cry	224		3.48		
1965	Osmium pentacarbonyl	$Os(CO)_5$	16406-49-8	330.28	col liq	-15	dec 100			s os
1966	Osmium nonacarbonyl	$Os_2(CO)_9$	28411-13-4	632.55	oran-yel cry	65 dec				s hc
1967	Osmium(II) chloride	$OsCl_2$	13444-92-3	261.14	hyg brn solid	dec >450				s EtOH, eth
1968	Osmium(III) bromide	$OsBr_3$	59201-51-3	429.94	dark gray cry	340 dec				i H_2O, os, acid
1969	Osmium(III) chloride	$OsCl_3$	13444-93-4	296.59	gray cub cry	450 dec				i H_2O, os; s conc acid
1970	Osmium(IV) chloride	$OsCl_4$	10026-01-4	332.04	red-blk orth cry	323 dec		4.38		reac H_2O; i os
1971	Osmium(IV) fluoride	OsF_4	54120-05-7	266.22	yel cry	230				reac H_2O
1972	Osmium(IV) oxide	OsO_2	12036-02-1	222.23	yel-brn tetr cry	dec 500		11.4		i H_2O, acid
1973	Osmium(V) fluoride	OsF_5	31576-40-6	285.22	hyg blue-grn cry	70	233			reac H_2O
1974	Osmium(VI) fluoride	OsF_6	13768-38-2	304.22	yel cub cry	33.4	47.5	4.1		reac H_2O
1975	Osmium(VI) tetrachloride oxide	$OsOCl_4$	36509-15-6	348.04	dark brn hyg cry	32	200			reac H_2O; s hc
1976	Osmium(VIII) oxide	OsO_4	20816-12-0	254.23	yel monocl cry	40.6	131.2	5.1	6.44²⁰	sl H_2O; s ctc, bz, EtOH, eth
1977	Oxygen	O_2	7782-44-7	31.998	col gas	-218.79	-182.953	1.308 g/L		sl H_2O, EtOH, os
1978	Ozone	O_3	10028-15-6	47.998	blue gas	-193	-111.35	1.962 g/L		sl H_2O
1979	Palladium	Pd	7440-05-3	106.42	silv-wh metal; cub	1554.8	2963	12.0		s aqua regia
1980	Palladium(II) acetate	$Pd(C_2H_3O_2)_2$	3375-31-3	224.51	oran-brn cry	205 dec				i H_2O; s MeCN, eth, ace
1981	Palladium(II) bromide	$PdBr_2$	13444-94-5	266.23	red-blk monocl cry; hyg	250 dec		≈5.2		i H_2O
1982	Palladium(II) chloride	$PdCl_2$	7647-10-1	177.33	red rhomb cry; hyg	679		4.0		s H_2O, EtOH, ace
1983	Palladium(II) chloride dihydrate	$PdCl_2 \cdot 2H_2O$	7647-10-1*	213.36	brn cry					s H_2O, EtOH, ace
1984	Palladium(II) cyanide	$Pd(CN)_2$	2035-66-7	158.45	yel solid	dec				
1985	Palladium(II) fluoride	PdF_2	13444-96-7	144.42	viol tetr cry; hyg	952		5.76		reac H_2O
1986	Palladium(II) iodide	PdI_2	7790-38-7	360.23	blk cry	360 dec		6.0		i H_2O, EtOH, eth
1987	Palladium(II) nitrate	$Pd(NO_3)_2$	10102-05-3	230.43	brn hyg cry	dec				sl H_2O; s dil HNO_3
1988	Palladium(II) oxide	PdO	1314-08-5	122.42	grn-blk tetr cry	750 dec		8.3		i H_2O, acid; sl aqua regia
1989	Palladium(II) 2,4-pentanedioate	$Pd(CH_3COCHCOCH_3)_2$	14024-61-4	304.64	oran-yel cry	205 dec				s bz, chl
1990	Palladium(II) sulfate dihydrate	$PdSO_4 \cdot 2H_2O$	13566-03-5	238.51	grn-brn cry	dec				
1991	Palladium(II) sulfide	PdS	12125-22-3	138.49	gray tetr cry			6.7		
1992	cis-Dichlorodiamminepalladium(II)	$Pd(NH_3)_2Cl_2$	15684-18-1	211.39	yel pow				0.025²⁵	
1993	trans-Dichlorodiamminepalladium(II)	$Pd(NH_3)_2Cl_2$	13782-33-7	211.39	yel solid			2.50		

No.	Name	Formula	CAS Reg No.	Mol. weight	Physical form	mp/°C	bp/°C	Density g cm⁻³	Solubility g/100 g H₂O	Qualitative solubility
1994	Phosphorus (white)	P	7723-14-0	30.974	col waxlike cub cry	44.15	280.5	1.823		i H_2O; sl bz, EtOH, chl; s CS_2
1995	Phosphorus (red)	P	7723-14-0	30.974	red-viol amorp powder	579.2	431 sp	2.16		i H_2O, os
1996	Phosphorus (black)	P	7723-14-0	30.974	blk orth cry or amorp solid	610		2.69		i os
1997	Phosphine	PH_3	7803-51-2	33.998	col gas; flam	-133.8	-87.75	1.390 g/L		i H_2O; sl EtOH, eth
1998	Diphosphine	P_2H_4	13445-50-6	65.980	col liq	-99	63.5 dec			reac H_2O
1999	Diphosphorus tetrachloride	P_2Cl_4	13497-91-1	203.760	col oily liq	-28	≈180 dec			
2000	Diphosphorus tetrafluoride	P_2F_4	13824-74-3	137.942	col gas	-86.5	-6.2	5.638 g/L		
2001	Diphosphorus tetraiodide	P_2I_4	13455-00-0	569.566	red tricl needles	125.5	dec	3.89		
2002	Phosphonium chloride	PH_4Cl	24567-53-1	70.459	gas		-27 sp	2.880 g/L		reac H_2O
2003	Phosphonium iodide	PH_4I	12125-09-6	161.910	col tetr cry	18.5	62.5	2.86		reac H_2O, EtOH
2004	Phosphoric acid	H_3PO_4	7664-38-2	97.995	col visc liq	42.4	407		548[20]	s EtOH
2005	Phosphotungstic acid	$H_3PW_{12}O_{40}$	12067-99-1	2880.05	wh-yel cry	89				vs H_2O; s EtOH, eth
2006	Phosphonic acid	H_3PO_3	13598-36-2	81.996	wh hyg cry	74.4	200	1.65	309[0]	vs EtOH
2007	Phosphinic acid	HPH_2O_2	6303-21-5	65.997	hyg cry or col oily liq	26.5	130	1.49		vs H_2O, EtOH, eth
2008	Metaphosphoric acid	HPO_3	37267-86-0	79.980	gl solid; hyg					sl H_2O; s EtOH
2009	Hypophosphoric acid	$H_4P_2O_6$	7803-60-3	161.976	col orth cry	73 dec				vs H_2O
2010	Diphosphoric acid	$H_4P_2O_7$	2466-09-3	177.975	wh cry	71.5			709[23]	
2011	Difluorophosphoric acid	HPO_2F_2	13779-41-4	101.978	col liq	≈-94	110 dec	1.583		reac H_2O
2012	Hexafluorophosphoric acid	HPF_6	16940-81-1	145.972	col oily liq	25 dec				reac H_2O
2013	Fluorophosphonic acid	H_2PFO_3	13537-32-1	99.986	col visc liq	<-70		1.82		vs H_2O
2014	Phosphorus nitride (P_3N_5)	P_3N_5	12136-91-3	162.955	yel-brn solid	800 dec				i H_2O; s os
2015	Phosphorus sesquisulfide	P_4S_3	1314-85-8	220.090	yel-grn orth cry	173	407	2.03		i H_2O; s bz; vs CS_2
2016	Phosphorus heptasulfide	P_4S_7	12037-82-0	348.350	pale yel monocl cry	308	523	2.19		sl CS_2
2017	Phosphonitrilic chloride trimer	$(PNCl_2)_3$	940-71-6	347.659	wh hyg cry	128.8		1.98		reac H_2O
2018	Phosphorus(III) bromide	PBr_3	7789-60-8	270.686	col liq	-41.5	173.2	2.8		reac H_2O, EtOH; s ace, CS_2
2019	Phosphorus(III) dibromide fluoride	PBr_2F	15597-39-4	209.780	col liq	-115	78.5			
2020	Phosphorus(III) bromide difluoride	$PBrF_2$	15597-40-7	148.875	col gas	-133.8	-16.1	6.085 g/L		
2021	Phosphorus(III) chloride	PCl_3	7719-12-2	137.333	col liq	-93	76	1.574		reac H_2O, EtOH; s bz, chl, eth
2022	Phosphorus(III) dichloride fluoride	PCl_2F	15597-63-4	120.878	col gas	-144	13.85	4.941 g/L		
2023	Phosphorus(III) chloride difluoride	$PClF_2$	14335-40-1	104.424	col gas	-164.8	-47.3	4.268 g/L		
2024	Phosphorus(III) fluoride	PF_3	7783-55-3	87.969	col gas	-151.5	-101.8	3.596 g/L		reac H_2O
2025	Phosphorus(III) iodide	PI_3	13455-01-1	411.687	red-oran hex cry; hyg	61.2	227 dec	4.18		reac H_2O; s EtOH
2026	Phosphorus(III) oxide	P_2O_3	1314-24-5	109.946	col monocl cry or liq	23.8	173	2.13		reac H_2O
2027	Tetraphosphorus(III) hexoxide	P_4O_6	12440-00-5	219.891	soft wh cry	23.8	175.4			
2028	Phosphorus(III) selenide	P_2Se_3	1314-86-9	298.83	oran-red cry	245	≈380	1.31		reac H_2O; s bz, ctc, CS_2, ace
2029	Phosphorus(III) sulfide	P_2S_3	12165-69-4	158.143	yel solid	290	490			reac H_2O; s EtOH, eth, CS_2
2030	Phosphorus(V) bromide	PBr_5	7789-69-7	430.494	yel orth cry, hyg	≈100 dec		3.61		reac H_2O, EtOH; s CS_2, ctc
2031	Phosphorus(V) tetrabromide fluoride	PBr_4F		369.588	pale yel cry	87 dec				
2032	Phosphorus(V) dibromide trifluoride	PBr_2F_3	13445-58-4	247.777	yel-red liq	-20	15 dec			
2033	Phosphorus(V) chloride	PCl_5	10026-13-8	208.239	wh-yel tetr cry; hyg	167 tp	160 sp	2.1		reac H_2O; s CS_2, ctc
2034	Phosphorus(V) tetrachloride fluoride	PCl_4F	13498-11-8	191.784	col liq	-59	30 dec			
2035	Phosphorus(V) trichloride difluoride	PCl_3F_2	13537-23-0	175.330	col liq	-63				
2036	Phosphorus(V) dichloride trifluoride	PCl_2F_3	13454-99-4	158.875	col gas	-125	7.1	6.494 g/L		
2037	Phosphorus(V) chloride tetrafluoride	$PClF_4$	13498-11-8	142.421	col gas	-132	-43.4	5.821 g/L		
2038	Phosphorus(V) fluoride	PF_5	7647-19-0	125.966	col gas	-93.8	-84.6	5.149 g/L		reac H_2O
2039	Phosphorus(V) oxide	P_2O_5	1314-56-3	141.945	wh orth cry; hyg	562	605	2.30		reac H_2O, EtOH
2040	Phosphorus(V) selenide	P_2Se_5	1314-82-5	456.75	blk-purp amorp solid					reac hot H_2O, ctc; i CS_2
2041	Phosphorus(V) sulfide	P_2S_5	1314-80-3	222.273	grn-yel hyg cry	285	515	2.03		reac H_2O; s CS_2
2042	Phosphonic difluoride	POF_2H	14939-34-5	85.978	volatile liq	>-120	≈60 (gas unstab)			

No.	Name	Formula	CAS Reg No.	Mol. weight	Physical form	mp/°C	bp/°C	Density g cm⁻³	Solubility g/100 g H₂O	Qualitative solubility
2043	Phosphoryl bromide	POBr₃	7789-59-5	286.685	faint oran plates	55	191.7	2.822		reac H₂O; s bz, eth, chl
2044	Phosphoryl dibromide chloride	POBr₂Cl	13550-31-7	242.234	yel solid	31	165			
2045	Phosphoryl dibromide fluoride	POBr₂F	14014-19-8	225.779	col liq	-117.2	110.1			
2046	Phosphoryl bromide dichloride	POBrCl₂	13455-03-3	197.783	col liq	11	136.5	2.104¹⁴		
2047	Phosphoryl bromide difluoride	POBrF₂	14014-18-7	164.874	col liq	-84.8	31.6			
2048	Phosphoryl bromide chloride fluoride	POBrClF	14518-81-1	181.328	col liq		79			
2049	Phosphoryl chloride	POCl₃	10025-87-3	153.332	col liq	1.18	105.5	1.645		reac H₂O, EtOH
2050	Phosphoryl dichloride fluoride	POCl₂F	13769-76-1	136.877	col liq	-80.1	52.9			
2051	Phosphoryl chloride difluoride	POClF₂	13769-75-0	120.423	col gas	-96.4	3.1	4.922 g/L		
2052	Phosphoryl fluoride	POF₃	13478-20-1	103.968	col gas	-39.1 tp	-39.7 sp	4.250 g/L		reac H₂O
2053	Phosphoryl iodide	POI₃	13455-04-4	427.686	viol cry	53				
2054	Phosphorothioc tribromide	PSBr₃	3931-89-3	302.751	yel cry	37.8	212 dec	2.85		
2055	Phosphorothioc dibromide fluoride	PSBr₂F	13706-10-0	241.845	yel liq	-75.2	125.3			
2056	Phosphorothioc bromide difluoride	PSBrF₂	13706-09-7	180.940	yel liq	-136.9	35.5			
2057	Phosphorothioc trichloride	PSCl₃	3982-91-0	169.398	fuming liq	-36.2	125	1.635		reac H₂O; s bz, ctc, chl, CS₂
2058	Phosphorothioc dichloride fluoride	PSCl₂F	155698-29-6	152.943	col liq	-96.0	64.7			
2059	Phosphorothioc chloride difluoride	PSClF₂	2524-02-9	136.489	col gas	-155.2	6.3	5.579 g/L		
2060	Phosphorothioc trifluoride	PSF₃	2404-52-6	120.034	col gas	-148.8	-52.25	4.906 g/L		
2061	Phosphorothioc triiodide	PSI₃	63972-04-3	443.752	yel cry	48	dec			
2062	Platinum	Pt	7440-06-4	195.084	silv-gray metal; cub	1768.2	3825	21.5		i acid; s aqua regia
2063	Hexachloroplatinic acid	H₂PtCl₆	16941-12-1	409.818	hyg yel-brn cry	60				s H₂O, EtOH
2064	Hydrogen hexahydroxyplatinate(IV)	H₂Pt(OH)₆	51850-20-5	299.144	yel needles	dec 100				s H₂O, acid, dil alk
2065	Platinum(II) bromide	PtBr₂	13455-12-4	354.892	red-brn powder	250 dec		6.65		i H₂O
2066	Platinum(II) chloride	PtCl₂	10025-65-7	265.990	grn hex cry	581 dec		6.0		i H₂O, EtOH, eth; s HCl
2067	Platinum(II) cyanide	Pt(CN)₂	592-06-3	247.118	pale yel cry					i H₂O, acid, alk
2068	Platinum(II) iodide	PtI₂	7790-39-8	448.893	blk powder	325 dec		6.4		i H₂O
2069	Platinum(II) oxide	PtO	12035-82-4	211.083	blk tetr cry	325 dec		14.1		i H₂O, EtOH; s aqua regia
2070	Platinum(II) sulfide	PtS	12038-20-9	227.149	tetr cry			10.25		
2071	Platinum(III) bromide	PtBr₃	25985-07-3	434.796	grn-blk cry	200 dec				
2072	Platinum(III) chloride	PtCl₃	25909-39-1	301.443	grn-blk cry	435 dec		5.26		
2073	Platinum(IV) bromide	PtBr₄	68938-92-1	514.700	brn-blk cry	180 dec			0.41²⁰	sl EtOH, eth
2074	Platinum(IV) chloride	PtCl₄	37773-49-2	336.896	red-brn cub cry	327 dec		4.30	142²⁵	
2075	Platinum(IV) chloride pentahydrate	PtCl₄ · 5H₂O	13454-96-1	426.972	red cry			2.43		s H₂O, EtOH
2076	Platinum(IV) fluoride	PtF₄	13455-15-7	271.078	red cry	600				
2077	Platinum(IV) iodide	PtI₄	7790-46-7	702.702	brn-blk powder	130 dec				s H₂O
2078	Platinum(IV) oxide	PtO₂	1314-15-4	227.083	blk hex cry	450		11.8		i H₂O; s conc acid, dil alk
2079	Platinum(IV) sulfide	PtS₂	12038-21-0	259.214	hex cry			7.85		
2080	Platinum(VI) fluoride	PtF₆	13693-05-5	309.074	red cub cry	61.3	69.1	≈4.0		
2081	cis-Diamminedichloroplatinum	Pt(NH₃)₂Cl₂	15663-27-1	300.051	yel solid	270 dec			0.253²⁵	
2082	trans-Diamminedichloroplatinum	Pt(NH₃)₂Cl₂	14913-33-8	300.051	pale yel solid	270 dec			0.036²⁵	s DMF, DMSO
2083	Hexachloroplatinic acid hexahydrate	H₂PtCl₆ · 6H₂O	16941-12-1	517.909	brn-yel hyg cry	60		2.43	140¹⁸	vs EtOH
2084	Platinum silicide	PtSi	12137-83-6	223.170	orth cry	1229		12.4		
2085	Plutonium	Pu	7440-07-5	244	silv-wh metal; monocl	640	3228	19.7		
2086	Plutonium nitride	PuN	12033-54-4	258	gray cub cry	2550		14.4		
2087	Plutonium(II) oxide	PuO	12035-83-5	260	cub cry			14.0		
2088	Plutonium(III) bromide	PuBr₃	15752-46-2	484	grn orth cry	681		6.75		s H₂O
2089	Plutonium(III) chloride	PuCl₃	13569-62-5	350	grn hex cry	760		5.71		s H₂O
2090	Plutonium(III) fluoride	PuF₃	13842-83-6	301	purp hex cry	1396		9.33		i H₂O; sl acid
2091	Plutonium(III) iodide	PuI₃	13813-46-2	625	grn orth cry; hyg	777		6.92		s H₂O
2092	Plutonium(III) oxide	Pu₂O₃	12036-34-9	536	blk cub cry	2085		10.5		
2093	Plutonium(IV) fluoride	PuF₄	13709-56-3	320	red-brn monocl cry	1037		7.1		
2094	Plutonium(IV) oxide	PuO₂	12059-95-9	276	yel-brn cub cry	2390		11.5		
2095	Plutonium(VI) fluoride	PuF₆	13693-06-6	358	red-brn orth cry	51.6		5.08		
2096	Polonium	Po	7440-08-6	209	silv metal; cub	254	962	9.20		
2097	Polonium(IV) chloride	PoCl₄	10026-02-5	351	yel hyg cry	≈300	390			s H₂O, EtOH, ace
2098	Polonium(IV) oxide	PoO₂	7446-06-2	241	yel cub cry	500 dec		8.9		
2099	Potassium	K	7440-09-7	39.098	soft silv-wh metal; cub	63.5	759	0.89		reac H₂O
2100	Potassium acetate	KC₂H₃O₂	127-08-2	98.142	wh hyg cry	309		1.57	269²⁵	s EtOH; i eth

No.	Name	Formula	CAS Reg No.	Mol. weight	Physical form	mp/°C	bp/°C	Density g cm⁻³	Solubility g/100 g H₂O	Qualitative solubility
2101	Potassium aluminate trihydrate	$K_2Al_2O_4 \cdot 3H_2O$	12003-63-3*	250.204	wh orth cry			2.13		vs H_2O; i EtOH
2102	Potassium aluminum silicate	$KAlSi_3O_8$	1327-44-2	278.332	col monocl cry			2.56		i H_2O
2103	Potassium aluminum sulfate	$KAl(SO_4)_2$	10043-67-1	258.205	wh hyg powder				5.9²⁰	
2104	Potassium aluminum sulfate dodecahydrate	$KAl(SO_4)_2 \cdot 12H_2O$	7784-24-9	474.389	col cry	≈100 dec		1.72	5.9²⁰	
2105	Potassium amide	KNH_2	17242-52-3	55.121	wh/yel-grn hyg cry	335				reac H_2O, EtOH
2106	Potassium arsenate	K_3AsO_4	13464-36-3	256.215	col cry			2.8	125²⁵	
2107	Potassium arsenite	$KAsO_2$	13464-35-2	146.019	wh hyg powder					s H_2O; sl EtOH
2108	Potassium azide	KN_3	20762-60-1	81.118	tetr cry; exp			2.04	49.7¹⁷	
2109	Potassium borohydride	KBH_4	13762-51-1	53.941	wh cub cry	≈500 dec		1.11		s H_2O
2110	Potassium bromate	$KBrO_3$	7758-01-2	167.000	wh hex cry	434 dec		3.27	8.17²⁵	i EtOH
2111	Potassium bromide	KBr	7758-02-3	119.002	col cub cry; hyg	734	1435	2.74	67.8²⁵	sl EtOH
2112	Potassium carbonate	K_2CO_3	584-08-7	138.206	wh monocl cry; hyg	899	dec	2.29	111²⁵	i EtOH
2113	Potassium carbonate sesquihydrate	$K_2CO_3 \cdot 1.5H_2O$	6381-79-9	165.229	granular cry				111²⁰	
2114	Potassium chlorate	$KClO_3$	3811-04-9	122.549	wh monocl cry	357	dec	2.34	8.61²⁵	
2115	Potassium chloride	KCl	7447-40-7	74.551	wh cub cry	771		1.988	35.5²⁵	i eth, ace
2116	Potassium chlorochromate	$KCrO_3Cl$	16037-50-6	174.545	oran cry			2.5		reac H_2O; s ace, acid
2117	Potassium chromate	K_2CrO_4	7789-00-6	194.191	yel orth cry	974		2.73	65.0²⁵	
2118	Potassium citrate monohydrate	$K_3C_6H_5O_7 \cdot H_2O$	6100-05-6	324.410	col hyg cry	180 dec		1.98	172²⁰	vs H_2O; sl EtOH
2119	Potassium cobalt(II) selenate hexahydrate	$K_2Co(SeO_4)_2 \cdot 6H_2O$	28041-86-3	531.14	red monocl cry			2.51		
2120	Potassium cyanate	$KCNO$	590-28-3	81.115	wh tetr cry	≈700 dec		2.05	75²⁵	sl EtOH
2121	Potassium cyanide	KCN	151-50-8	65.116	wh cub cry; hyg	622		1.55	69.9²⁰	sl EtOH
2122	Potassium cyanoaurite	$KAu(CN)_2$	13967-50-5	288.099	col cry			3.45	14²⁰	s H_2O; sl EtOH; i eth, ace
2123	Potassium dichromate	$K_2Cr_2O_7$	7778-50-9	294.185	oran-red tricl cry	398	≈500 dec	2.68	15.1²⁵	
2124	Potassium dihydrogen arsenate	KH_2AsO_4	7784-41-0	180.034	col cry	288		2.87	19⁶	i EtOH
2125	Potassium dihydrogen phosphate	KH_2PO_4	7778-77-0	136.085	wh tetr cry	253 dec		2.34	25.0²⁵	sl EtOH
2126	Potassium dihydrogen phosphonate	KH_2PO_3	13977-65-6	120.086	col monocl hyg cry					
2127	Potassium dithionate	$K_2S_2O_6$	13455-20-4	238.323	col hex cry	dec		2.27		sl H_2O; i EtOH
2128	Potassium ferricyanide	$K_3Fe(CN)_6$	13746-66-2	329.244	red cry	dec		1.89	48.8²⁵	
2129	Potassium ferrocyanide trihydrate	$K_4Fe(CN)_6 \cdot 3H_2O$	14459-95-1	422.388	yel monocl cry	60 dec		1.85	36.0²⁵	i EtOH, eth
2130	Potassium fluoride	KF	7789-23-3	58.096	wh cub cry	858	1502	2.48	102²⁵	
2131	Potassium fluoride dihydrate	$KF \cdot 2H_2O$	13455-21-5	94.127	monocl cry	41 dec		2.5	102²⁵	
2132	Potassium fluoroborate	KBF_4	14075-53-7	125.903	col orth cry	530		2.505	0.55²⁵	sl EtOH
2133	Potassium fluorotantalate	K_2TaF_7	16924-00-8	392.134	col cry	730		5.24	0.5⁰	
2134	Potassium formate	$KCHO_2$	590-29-4	84.116	col hyg cry	167		1.91	331¹⁸	
2135	Potassium hexachloroosmate(IV)	K_2OsCl_6	16871-60-6	481.15	red cub cry					vs H_2O; sl EtOH
2136	Potassium hexachloroplatinate	K_2PtCl_6	16921-30-5	485.999	yel-oran cub cry	250 dec		3.50	0.77²⁰	i EtOH
2137	Potassium hexacyanocobaltate	$K_3Co(CN)_6$	13963-58-1	332.332	yel monocl cry	dec		1.91		vs H_2O; i EtOH
2138	Potassium hexafluoromanganate(IV)	K_2MnF_6	16962-31-5	247.125	yel hex cry					reac H_2O
2139	Potassium hexafluorosilicate	K_2SiF_6	16871-90-2	220.273	wh cry	dec		2.27	0.084²⁰	i EtOH
2140	Potassium hexafluorozirconate(IV)	K_2ZrF_6	16923-95-8	283.411	col monocl cry			3.48	0.78²	
2141	Potassium hydride	KH	7693-26-7	40.106	cub cry	619		1.43		reac H_2O
2142	Potassium hydrogen arsenate	K_2HAsO_4	21093-83-4	218.125	col monocl prisms	300 dec			18.7⁶	i EtOH
2143	Potassium hydrogen carbonate	$KHCO_3$	298-14-6	100.115	col monocl cry	≈100 dec		2.17	36.2²⁵	i EtOH
2144	Potassium hydrogen fluoride	KHF_2	7789-29-9	78.103	col tetr cry	238.8		2.37	39.2²⁰	i EtOH
2145	Potassium hydrogen iodate	$KH(IO_3)_2$	13455-24-8	389.911	col cry	dec			1.3¹⁵	sl H_2O; i EtOH
2146	Potassium hydrogen oxalate hemihydrate	$KHC_2O_4 \cdot 0.5H_2O$	127-95-7		wh cry	dec		2.09	2.5²⁰	sl EtOH
2147	Potassium hydrogen phosphate	K_2HPO_4	7758-11-4	174.176	wh hyg cry	dec			168²⁵	s EtOH
2148	Potassium hydrogen phosphite	K_2HPO_3	13492-26-7	158.177	wh hyg powder	dec			170²⁰	i EtOH
2149	Potassium hydrogen selenite	$KHSeO_3$	7782-70-9	167.06	hyg orth cry	>100 dec				s H_2O; sl EtOH
2150	Potassium hydrogen sulfate	$KHSO_4$	7646-93-7	136.169	wh monocl cry; hyg	≈200		2.32	50.6²⁵	
2151	Potassium hydrogen sulfide	KHS	1310-61-8	72.171	wh hex cry; hyg	≈450		1.69		s H_2O, EtOH
2152	Potassium hydrogen sulfide hemihydrate	$KHS \cdot 0.5H_2O$	1310-61-8*	81.179	wh-yel hyg cry	≈175		1.7		vs H_2O, EtOH
2153	Potassium hydrogen sulfite	$KHSO_3$	7773-03-7	120.169	wh cry powder	190 dec			49²⁰	i EtOH
2154	Potassium hydrogen tartrate	$KHC_4H_4O_6$	868-14-4	188.177	wh cry			1.98	0.57²⁰	s acid, alk; i EtOH
2155	Potassium hydroxide	KOH	1310-58-3	56.105	wh rhomb cry; hyg	406	1327	2.044	121²⁵	s EtOH; s MeOH
2156	Potassium hypochlorite	$KOCl$	7778-66-7	90.550	exists only in aq soln					
2157	Potassium phosphinate	KH_2PO_2	7782-87-8	104.087	wh hyg cry	dec				vs H_2O; s EtOH

No.	Name	Formula	CAS Reg No.	Mol. weight	Physical form	mp/°C	bp/°C	Density g cm⁻³	Solubility g/100 g H₂O	Qualitative solubility
2158	Potassium iodate	KIO_3	7758-05-6	214.001	wh monocl cry	560 dec		3.89	9.22[25]	
2159	Potassium iodide	KI	7681-11-0	166.003	col cub cry	681	1323	3.12	148[25]	sl EtOH
2160	Potassium iron(III) oxalate trihydrate	$K_3Fe(C_2O_4)_3 \cdot 3H_2O$		491.243	grn monocl cry	100	230 dec	2.133	4.7[0]	i EtOH
2161	Potassium manganate	K_2MnO_4	10294-64-1	197.133	grn cry	190 dec				s H₂O; reac HCl
2162	Potassium metaarsenate	$KAsO_3$	19197-73-0	162.018	wh solid	660				
2163	Potassium metabisulfite	$K_2S_2O_5$	16731-55-8	222.324	wh powder	≈150 dec		2.3	49.5[25]	reac acid; i EtOH
2164	Potassium metaborate	KBO_2	13709-94-9	81.908	wh hex cry	947		≈2.3		
2165	Potassium molybdate	K_2MoO_4	13446-49-6	238.14	wh hyg cry	919		2.3	183[25]	i EtOH
2166	Potassium niobate	$KNbO_3$	12030-85-2	180.002	wh rhomb cry	≈1100		4.64		i H₂O
2167	Potassium nitrate	KNO_3	7757-79-1	101.103	col orth cry or powder	334	400 dec	2.105	38.3[25]	i EtOH
2168	Potassium nitrite	KNO_2	7758-09-0	85.104	wh hyg cry	438	537 exp	1.915	312[25]	sl EtOH
2169	Potassium oxalate	$K_2C_2O_4$	583-52-8	166.216	wh pwd					sl H₂O
2170	Potassium oxalate monohydrate	$K_2C_2O_4 \cdot H_2O$	6487-48-5	184.231	col cry	160 dec		2.13	36.4[20]	
2171	Potassium oxide	K_2O	12136-45-7	94.196	gray cub cry	740		2.35		s H₂O, EtOH, eth
2172	Potassium perbromate	$KBrO_4$	22207-96-1	183.000	wh cry	275 dec			4.21[25]	
2173	Potassium percarbonate monohydrate	$K_2C_2O_6 \cdot H_2O$	589-97-9	216.230	oran or blue pow				6.5[20]	
2174	Potassium perchlorate	$KClO_4$	7778-74-7	138.549	col orth cry; hyg	525		2.52	2.08[25]	
2175	Potassium periodate	KIO_4	7790-21-8	230.001	col tetr cry	582	exp	3.618	0.51[25]	
2176	Potassium permanganate	$KMnO_4$	7722-64-7	158.034	purp orth cry	dec		2.7	7.60[25]	reac EtOH
2177	Potassium peroxide	K_2O_2	17014-71-0	110.196	yel amorp solid	490				reac H₂O
2178	Potassium persulfate	$K_2S_2O_8$	7727-21-1	270.322	col cry	≈100 dec		2.48	4.7[20]	
2179	Potassium phosphate	K_3PO_4	7778-53-2	212.266	wh orth cry; hyg	1340		2.564	106[25]	i EtOH
2180	Potassium pyrophosphate	$K_4P_2O_7$	7320-34-5		wh cry	dec 1300				s H₂
2181	Potassium pyrophosphate trihydrate	$K_4P_2O_7 \cdot 3H_2O$	7790-67-2	384.383	col hyg cry	dec 300		2.33		vs H₂O; i EtOH
2182	Potassium pyrosulfate	$K_2S_2O_7$	7790-62-7	254.323	col needles	≈325		2.28		s H₂O
2183	Potassium selenate	K_2SeO_4	7790-59-2	221.16	wh powder			3.07	114[25]	
2184	Potassium selenide	K_2Se	1312-74-9	157.16	red cub cry; hyg	800		2.29		s H₂O
2185	Potassium selenite	K_2SeO_3	10431-47-7	205.16	wh hyg cry	875 dec			217[25]	sl EtOH
2186	Potassium silver cyanide	$KAg(CN)_2$	506-61-6	199.000	wh cry					s H₂O
2187	Potassium sodium tartrate tetrahydrate	$KNaC_4H_4O_6 \cdot 4H_2O$	6381-59-5	282.220	wh cry	≈70 dec	anh at 130	1.79		vs H₂O; i EtOH
2188	Potassium stannate trihydrate	$K_2SnO_3 \cdot 3H_2O$	12142-33-5*	298.951	col cry			3.20		vs H₂O; i EtOH
2189	Potassium stearate	$KC_{18}H_{35}O_2$	593-29-3	322.568	wh pow					sl cold H₂O; s hot H₂O, EtOH
2190	Potassium sulfate	K_2SO_4	7778-80-5	174.260	wh orth cry	1069		2.66	12.0[25]	i EtOH
2191	Potassium sulfide	K_2S	1312-73-8	110.262	red-yel cub cry; hyg	948		1.74		s H₂O, EtOH; i eth
2192	Potassium sulfide pentahydrate	$K_2S \cdot 5H_2O$	37248-34-3	200.338	col rhomb cry	60				vs H₂O, EtOH; i eth
2193	Potassium sulfite	K_2SO_3	10117-38-1	158.260	col hex cry				106[25]	sl EtOH
2194	Potassium sulfite dihydrate	$K_2SO_3 \cdot 2H_2O$	7790-56-9	194.291	wh monocl cry	dec			107[20]	sl EtOH; dec dil acid
2195	Potassium superoxide	KO_2	12030-88-5	71.097	yel tetr cry; hyg	380		2.16		reac H₂O
2196	Potassium tellurate(VI) trihydrate	$K_2TeO_4 \cdot 3H_2O$	15571-91-2*	323.84	wh cry powder					s H₂O
2197	Potassium tellurite	K_2TeO_3	7790-58-1	253.80	wh hyg cry	≈460 dec				vs H₂O
2198	Potassium tetraborate pentahydrate	$K_2B_4O_7 \cdot 5H_2O$	1332-77-0	323.513	wh cry powder				16.5[30]	sl EtOH
2199	Potassium tetrachloroaurate dihydrate	$KAuCl_4 \cdot 2H_2O$	13682-61-6	413.908	yel monocl cry					s H₂O, EtOH, eth
2200	Potassium tetrachloroplatinate	K_2PtCl_4	10025-99-7	415.093	pink-red tetr cry	500 dec		3.38		s H₂O; i EtOH
2201	Potassium tetracyanocadmate	$K_2Cd(CN)_4$	14402-75-6	294.678	cub cry	≈450		1.85	25[20]	sl EtOH
2202	Potassium tetracyanonickelate monohydrate	$K_2[Ni(CN)_4] \cdot H_2O$	14220-17-8*	258.975	red-oran cry	dec 100				
2203	Potassium tetracyanoplatinate(II) trihydrate	$K_2Pt(CN)_4 \cdot 3H_2O$	562-76-5*	431.397	col rhomb prisms					s H₂O
2204	Potassium tetracyanozincate	$K_2Zn(CN)_4$	14244-62-3	247.676	cry pow					vs H₂O
2205	Potassium tetraiodomercurate(II)	K_2HgI_4	7783-33-7	786.40	yel hyg cry			4.29		vs H₂O; s EtOH, eth, ace
2206	Potassium thiocyanate	$KSCN$	333-20-0	97.181	col tetr cry; hyg	173	500 dec	1.88	238[25]	s EtOH
2207	Potassium thiosulfate	$K_2S_2O_3$	10294-66-3	190.325	col hyg cry				165[25]	i EtOH
2208	Potassium titanate	K_2TiO_3	12030-97-6	174.062	wh orth cry	1515		3.1		reac H₂O
2209	Potassium triiodide monohydrate	$KI_3 \cdot H_2O$	7790-42-3	437.827	brn monocl cry; hyg	225 dec		3.5		s H₂O; reac EtOH, eth
2210	Potassium triiodozincate	$KZnI_3$	7790-43-4	485.221	hyg cry					vs H₂O
2211	Potassium thiocarbonate	K_2CS_3	26750-66-3	186.403	yel-red hyg cry					vs H₂O
2212	Potassium tungstate	K_2WO_4	7790-60-5	326.04	hyg cry	921		3.12		vs H₂O; i EtOH
2213	Potassium uranate	$K_2U_2O_7$	7790-63-8	666.251	oran cub cry			6.12		i H₂O; s acid

No.	Name	Formula	CAS Reg No.	Mol. weight	Physical form	mp/°C	bp/°C	Density g cm^{-3}	Solubility g/100 g H$_2$O	Qualitative solubility
2214	Potassium uranyl nitrate	K(UO$_2$)(NO$_3$)$_3$	18078-40-5	495.140	grn-yel cry pow					vs H$_2$O
2215	Potassium uranyl sulfate dihydrate	K$_2$(UO$_2$)(SO$_4$)$_2$ · 2H$_2$O	27709-53-1	576.381	grn-yel cry pow	dec 120		3.36		vs H$_2$O
2216	Potassium zinc sulfate hexahydrate	K$_2$Zn(SO$_4$)$_2$ · 6H$_2$O	13932-17-7	443.823	cry					s H$_2$O
2217	Potassium zirconium sulfate trihydrate	K$_4$Zr(SO$_4$)$_4$ · 3H$_2$O	53608-79-0	685.914	wh cry pow					sl H$_2$O
2218	Praseodymium	Pr	7440-10-0	140.908	silv metal; hex	931	3520	6.77		
2219	Praseodymium boride	PrB$_6$	12008-27-4	205.774	blk cub cry	2610		4.84		
2220	Praseodymium nitride	PrN	25764-09-4	154.915	cub cry			7.46		
2221	Praseodymium silicide	PrSi$_2$	12066-83-0	197.079	tetr cry	1712		5.46		
2222	Praseodymium(II) iodide	PrI$_2$	65530-47-4	394.717	bronze solid	758				
2223	Praseodymium(III) bromate	Pr(BrO$_3$)$_3$	15162-93-3	524.615	grn cry					vs H$_2$O
2224	Praseodymium(III) bromide	PrBr$_3$	13536-53-3	380.620	grn hex cry; hyg	693		5.28		s H$_2$O
2225	Praseodymium(III) carbonate octahydrate	Pr$_2$(CO$_3$)$_3$ · 8H$_2$O	14948-62-0	605.964	grn silky plates	dec 420 (anh)				i H$_2$O; s acid
2226	Praseodymium(III) chloride	PrCl$_3$	10361-79-2	247.267	grn hex needles; hyg	786		4.0	96.1^{25}	s EtOH
2227	Praseodymium(III) chloride heptahydrate	PrCl$_3$ · 7H$_2$O	10025-90-8	373.374	grn cry	110 dec			96.1^{25}	s EtOH
2228	Praseodymium(III) fluoride	PrF$_3$	13709-46-1	197.903	grn hex cry	1399		6.3		
2229	Praseodymium(III) hydroxide	Pr(OH)$_3$	16469-16-2	191.930	grn solid	dec 220		3.7		i H$_2$O
2230	Praseodymium(III) iodide	PrI$_3$	13813-23-5	521.621	orth hyg cry	738		≈5.8		s H$_2$O
2231	Praseodymium(III) nitrate	Pr(NO$_3$)$_3$	10361-80-5	326.923	pale grn hyg cry				165^{25}	s EtOH
2232	Praseodymium(III) nitrate hexahydrate	Pr(NO$_3$)$_3$ · 6H$_2$O	15878-77-0	435.014	grn needles				165^{25}	s EtOH, ace
2233	Praseodymium(III) oxide	Pr$_2$O$_3$	12036-32-7	329.813	wh hex cry	2183	3760	6.9		
2234	Praseodymium(III) perchlorate hexahydrate	Pr(ClO$_4$)$_3$ · 6H$_2$O	13498-07-2*	547.351	hyg grn cry	dec 200				s H$_2$O, EtOH
2235	Praseodymium(III) sulfate octahydrate	Pr$_2$(SO$_4$)$_3$ · 8H$_2$O	13510-41-3	714.125	grn monocl cry			2.83	17^{20}	s H$_2$O
2236	Praseodymium(III) sulfide	Pr$_2$S$_3$	12038-13-0	378.010	cub cry	1765		5.1		
2237	Praseodymium(III) telluride	Pr$_2$Te$_3$	12038-12-9	664.62	cub cry	1500		≈7.0		
2238	Praseodymium(IV) fluoride	PrF$_4$	15192-24-2	216.902	yel-wh solid	dec 90				
2239	Promethium	Pm	7440-12-2	145	silv metal; hex	1042	3000	7.26		
2240	Promethium(III) bromide	PmBr$_3$	14325-78-1	385	red cry	625				s H$_2$O
2241	Promethium(III) chloride	PmCl$_3$	13779-10-7	251	pale blue hyg cry	655				s H$_2$O
2242	Promethium(III) fluoride	PmF$_3$	13709-45-0	202	pink solid	1338				s H$_2$O
2243	Promethium(III) iodide	PmI$_3$	13818-73-0	526	red solid	695				
2244	Protactinium	Pa	7440-13-3	231.036	shiny metal; tetr or cub	1572		15.4		
2245	Protactinium(V) chloride	PaCl$_5$	13760-41-3	408.301	yel monocl cry	306		3.74		
2246	Radium	Ra	7440-14-4	226	wh metal; cub	696		5		
2247	Radium bromide	RaBr$_2$	10031-23-9	386	wh orth cry	728		5.79	70.6^{20}	s EtOH
2248	Radium carbonate	RaCO$_3$	7116-98-5	286	wh orth cry					i H$_2$O
2249	Radium chloride	RaCl$_2$	10025-66-8	297	wh orth cry	1000		4.9	24.5^{20}	s EtOH
2250	Radium fluoride	RaF$_2$	20610-49-5	264	wh cub cry			6.7		
2251	Radium nitrate	Ra(NO$_3$)$_2$	10213-12-4	350	cry				13.9	
2252	Radium sulfate	RaSO$_4$	7446-16-4	322	wh cry					i H$_2$O, acid
2253	Radon	Rn	10043-92-2	222	col gas	-71	-61.7	9.074 g/L		sl H$_2$O
2254	Rhenium	Re	7440-15-5	186.207	silv-gray metal	3185	5590	20.8		i HCl
2255	Perrhenic acid	HReO$_4$	13768-11-1	251.213	exists only in soln					vs H$_2$O, os
2256	Rhenium carbonyl	Re$_2$(CO)$_{10}$	14285-68-8	652.515	yel-wh cry	170 dec		2.87		s os
2257	Rhenium pentacarbonyl bromide	Re(CO)$_5$Br	14220-21-4	406.162	wh cry	90				
2258	Rhenium pentacarbonyl chloride	Re(CO)$_5$Cl	14099-01-5	361.711	wh cry		subl 140			
2259	Rhenium(III) bromide	ReBr$_3$	13569-49-8	425.919	red-brn monocl cry		500 subl	6.10		s ace, MeOH, EtOH
2260	Rhenium(III) chloride	ReCl$_3$	13569-63-6	292.566	red-blk hyg cry	500 dec		4.81		s H$_2$O
2261	Rhenium(III) iodide	ReI$_3$	15622-42-1	566.920	blk solid	dec				
2262	Rhenium(IV) chloride	ReCl$_4$	13569-71-6	328.019	purp-blk cry; hyg	300 dec		4.9		
2263	Rhenium(IV) fluoride	ReF$_4$	15192-42-4	262.201	blue tetr cry		>300 subl	7.49		
2264	Rhenium(IV) oxide	ReO$_2$	12036-09-8	218.206	gray orth cry	900 dec		11.4		
2265	Rhenium(IV) selenide	ReSe$_2$	12038-64-1	344.13	tricl cry					
2266	Rhenium(IV) silicide	ReSi$_2$	12038-66-3	242.378	refrac solid	2000				
2267	Rhenium(IV) sulfide	ReS$_2$	12038-63-0	250.337	tricl cry			7.6		
2268	Rhenium(IV) telluride	ReTe$_2$	12067-00-4	441.41	orth cry			8.50		
2269	Rhenium(V) bromide	ReBr$_5$	30937-53-2	585.727	brn solid	110 dec				
2270	Rhenium(V) chloride	ReCl$_5$	39368-69-9	363.472	brn-blk solid	220		4.9		reac H$_2$O
2271	Rhenium(V) fluoride	ReF$_5$	30937-52-1	281.199	yel-grn solid	48	221.3			
2272	Rhenium(V) oxide	Re$_2$O$_5$	12165-05-8	452.411	blue-blk tetr cry			≈7		

No.	Name	Formula	CAS Reg No.	Mol. weight	Physical form	mp/°C	bp/°C	Density g cm⁻³	Solubility g/100 g H₂O	Qualitative solubility
2273	Rhenium(VI) chloride	$ReCl_6$	31234-26-1	398.925	red-grn solid	29				
2274	Rhenium(VI) dioxydifluoride	ReO_2F_2	81155-18-2	256.203	col cry	156				
2275	Rhenium(VI) fluoride	ReF_6	10049-17-9	300.197	yel liq or cub cry	18.5	33.8	4.06(cry)		s HNO_3
2276	Rhenium(VI) oxide	ReO_3	1314-28-9	234.205	red cub cry	400 dec		6.9		i H_2O, acid, alk
2277	Rhenium(VI) oxytetrachloride	$ReOCl_4$	13814-76-1	344.018	brn cry	29.3	223			reac H_2O
2278	Rhenium(VI) oxytetrafluoride	$ReOF_4$	17026-29-8	278.200	blue solid	108	171.7			
2279	Rhenium(VII) fluoride	ReF_7	17029-21-9	319.196	yel cub cry	48.3	73.7	4.32		
2280	Rhenium(VII) oxide	Re_2O_7	1314-68-7	484.410	yel hyg cry	327	360	6.10		s H_2O, EtOH, eth, diox, py
2281	Rhenium(VII) trioxychloride	ReO_3Cl	7791-09-5	269.658	col liq	4.5	128	3.87		reac H_2O
2282	Rhenium(VII) trioxyfluoride	ReO_3F	42246-24-2	253.203	yel solid	147	164			
2283	Rhenium(VII) dioxytrifluoride	ReO_2F_3	57246-89-6	275.201	yel solid	90	185.4			reac H_2O
2284	Rhenium(VII) oxypentafluoride	$ReOF_5$	23377-53-9	297.198	cream solid	43.8	73.0			
2285	Rhenium(VII) sulfide	Re_2S_7	12038-67-4	596.869	brn-blk tetr cry			4.87		i H_2O
2286	Rhodium	Rh	7440-16-6	102.906	silv-wh metal; cub	1963	3695	12.4		i acid, sl aqua regia
2287	Rhodium carbonyl	$Rh_6(CO)_{16}$	28407-51-4	1065.594	red-brn cry	220 dec				
2288	Rhodium carbonyl chloride	$[Rh(CO)_2Cl]_2$	14523-22-9	388.758	red-oran cry	124				s os
2289	Rhodium dodecacarbonyl	$Rh_4(CO)_{12}$	19584-30-6	747.743	red hyg cry	150 dec		2.52		reac H_2O
2290	Rhodium(III) bromide	$RhBr_3$	15608-29-4	342.618	dark brn plates	800 dec		5.56		s H_2O; i acid, os
2291	Rhodium(III) chloride	$RhCl_3$	10049-07-7	209.265	red monocl cry		717	5.38		i H_2O; s alk
2292	Rhodium(III) fluoride	RhF_3	60804-25-3	159.901	red hex cry			5.4		
2293	Rhodium(III) iodide	RhI_3	15492-38-3	483.619	blk monocl cry; hyg			6.4		
2294	Rhodium(III) nitrate	$Rh(NO_3)_3$	10139-58-9	288.921	hyg brn solid	600 dec				i H_2O
2295	Rhodium(III) nitrate dihydrate	$Rh(NO_3)_3 \cdot 2H_2O$	13465-43-5	324.951	blk solid	dec				i H_2O; s aqua regia
2296	Rhodium(III) oxide	Rh_2O_3	12036-35-0	253.809	gray hex cry	1100 dec		8.2		
2297	Rhodium(III) oxide pentahydrate	$Rh_2O_3 \cdot 5H_2O$	39373-27-8	309.010	yel pow	dec				sl H_2O; s acid
2298	Rhodium(III) sulfate	$Rh_2(SO_4)_3$	10489-46-0	493.999	red-yel solid	>500 dec				
2299	Rhodium(IV) oxide	RhO_2	12137-27-8	134.905	blk tetr cry			7.2		
2300	Rhodium(IV) oxide dihydrate	$RhO_2 \cdot 2H_2O$	12137-27-8	170.936	grn solid	dec		8.20		i H_2O, sol HCl, alk
2301	Rhodium(VI) fluoride	RhF_6	13693-07-7	216.896	blk cub cry	≈70		3.1		
2302	Rubidium	Rb	7440-17-7	85.468	soft silv metal; cub	39.30	688	1.53		reac H_2O
2303	Rubidium acetate	$RbC_2H_3O_2$	563-67-7	144.512	wh hyg cry	246				vs H_2O
2304	Rubidium aluminum sulfate	$RbAl(SO_4)_2$	13530-57-9	304.575	hex cry			≈3.1	1.60²⁰	i EtOH
2305	Rubidium aluminum sulfate dodecahydrate	$RbAl(SO_4)_2 \cdot 12H_2O$	7784-29-4	520.759	col cub cry	≈100 dec		≈1.9		s H_2O; i EtOH
2306	Rubidium azide	RbN_3	22756-36-1	127.488	tetr cry; exp	317		2.79	107¹⁶	
2307	Rubidium bromate	$RbBrO_3$	13446-70-3	213.370	cub cry	430		3.68	2.95²⁵	
2308	Rubidium bromide	RbBr	7789-39-1	165.372	wh cub cry; hyg	692	1340	3.35	116²⁵	
2309	Rubidium carbonate	Rb_2CO_3	584-09-8	230.945	col monocl cry; hyg	837			223²⁰	
2310	Rubidium chlorate	$RbClO_3$	13446-71-4	168.919	col cry	324	dec 480	3.19	6.63²⁵	sl H_2O
2311	Rubidium chloride	RbCl	7791-11-9	120.921	wh cub cry; hyg	724	1390	2.76	93.9²⁵	sl EtOH
2312	Rubidium chromate	Rb_2CrO_4	13446-72-5	286.930	yel rhom cry			3.518	76.2²⁵	
2313	Rubidium dichromate	$Rb_2Cr_2O_7$	13446-73-6	386.924	red tricl or yel monocl cry			3.1		s H_2O
2314	Rubidium cyanide	RbCN	19073-56-4	111.486	wh cub cry			2.3		s H_2O; i EtOH, eth
2315	Rubidium fluoride	RbF	13446-74-7	104.466	wh cub cry; hyg	795	1410	3.2	300²⁰	i EtOH
2316	Rubidium fluoroborate	$RbBF_4$	18909-68-7	172.273	orth cry	612 dec		2.82		sl H_2O
2317	Rubidium formate	$RbCHO_2$	3495-35-0	130.486	wh hyg cry	dec				
2318	Rubidium hexafluorogermanate	Rb_2GeF_6	16962-48-4	357.57	wh cry	696				s H_2O
2319	Rubidium hydride	RbH	13446-75-8	86.476	wh cub cry; flam	≈170 dec		2.60		reac H_2O
2320	Rubidium hydrogen carbonate	$RbHCO_3$	19088-74-5	146.485	wh rhomb cry	175 dec			116²⁰	
2321	Rubidium hydrogen fluoride	$RbHF_2$	12280-64-7	124.473	tetr cry	188		3.3		
2322	Rubidium hydrogen sulfate	$RbHSO_4$	15587-72-1	182.539	col monocl cry	208		2.9		s H_2O
2323	Rubidium hydroxide	RbOH	1310-82-3	102.475	gray-wh orth cry; hyg	385		3.2	173³⁰	s EtOH
2324	Rubidium iodate	$RbIO_3$	13446-76-9	260.370	monocl or cub cry	dec		4.33	2.44²⁵	vs HCl
2325	Rubidium iodide	RbI	7790-29-6	212.372	wh cub cry	656	1300	3.55	165²⁵	s EtOH
2326	Rubidium molybdate	Rb_2MoO_4	13718-22-4	330.87	wh cry	958				s H_2O
2327	Rubidium nitrate	$RbNO_3$	13126-12-0	147.473	wh hex cry; hyg	310		3.11	65.0²⁵	vs H_2O
2328	Rubidium nitrite	$RbNO_2$	13825-25-7	131.474	wh cry	422				vs H_2O
2329	Rubidium oxide	Rb_2O	18088-11-4	186.935	yel-brn cub cry; hyg	400 dec		4.0		reac H_2O
2330	Rubidium perchlorate	$RbClO_4$	13510-42-4	184.919	wh hyg cry	550	dec >550	2.9	1.5²⁵	
2331	Rubidium permanganate	$RbMnO_4$	13465-49-1	204.404	dark purp cry	300 dec		3.24		sl H_2O
2332	Rubidium peroxide	Rb_2O_2	23611-30-5	202.935	wh orth cry	570		3.8		reac H_2O
2333	Rubidium selenide	Rb_2Se	31052-43-4	249.90	wh cub cry	733		3.22		reac H_2O

No.	Name	Formula	CAS Reg No.	Mol. weight	Physical form	mp/°C	bp/°C	Density g cm⁻³	Solubility g/100 g H₂O	Qualitative solubility
2334	Rubidium sulfate	Rb_2SO_4	7488-54-2	266.999	wh orth cry	1066		3.61	50.8[25]	
2335	Rubidium sulfide	Rb_2S	31083-74-6	203.001	wh cub cry	425		2.91		s H₂O
2336	Rubidium superoxide	RbO_2	12137-25-6	117.467	tetr cry	412		≈3.0		
2337	Ruthenium	Ru	7440-18-8	101.07	silv-wh metal; hex	2333	4147	12.1		i acid, aqua regia
2338	Ruthenium dodecacarbonyl	$Ru_3(CO)_{12}$	15243-33-1	639.33	oran cry	150 dec				
2339	Ruthenium pentacarbonyl	$Ru(CO)_5$	16406-48-7	241.12	col liq	-22	dec 50			i H₂O; s EtOH, bz, chl, hc
2340	Ruthenium nonacarbonyl	$Ru_2(CO)_9$	63128-11-0	454.23	stab below -40					s hex
2341	Ruthenium nitrosyl chloride monohydrate	$Ru(NO)Cl_3 \cdot H_2O$	18902-42-6	255.45	hyg red cry					
2342	Hexaammineruthenium(III) chloride	$Ru(NH_3)_6Cl_3$	14282-91-8	309.61	col monocl cry					s H₂O
2343	Ruthenium(III) bromide	$RuBr_3$	14014-88-1	340.78	brn hex cry	dec 500		5.3		i H₂O, acid, EtOH
2344	Ruthenium(III) chloride	$RuCl_3$	10049-08-8	207.43	blk-brn hex cry	≈500 dec		3.1		i H₂O; sl EtOH
2345	Ruthenium(III) fluoride	RuF_3	51621-05-7	158.07	brn rhomb cry	≈600 dec		5.36		i H₂O, dil acid
2346	Ruthenium(III) iodide	RuI_3	13896-65-6	481.78	blk hex cry	dec 300		6.0		sl H₂O
2347	Ruthenium(III) 2,4-pentanedioate	$Ru(CH_3COCHCOCH_3)_3$	14284-93-6	398.39	red-brn cry	230				
2348	Ruthenium(IV) fluoride	RuF_4	71500-16-8	177.06	yel-red cry					reac H₂O
2349	Ruthenium(IV) oxide	RuO_2	12036-10-1	133.07	gray-blk tetr cry	dec 1300		7.05		i H₂O, acid
2350	Ruthenium(V) fluoride	RuF_5	14521-18-7	196.06	grn monocl cry	86.5	227	3.90		
2351	Ruthenium(VI) fluoride	RuF_6	13693-08-8	215.06	dark brn orth cry	54	200 dec	3.54		reac H₂O
2352	Ruthenium(VIII) oxide	RuO_4	20427-56-9	165.07	yel monocl prisms	25.4	40	3.29	2.03[20]	sl H₂O; vs ctc; reac EtOH
2353	Samarium	Sm	7440-19-9	150.36	silv metal; rhomb	1072	1794	7.52		
2354	Samarium boride	SmB_6	12008-30-9	215.23	refrac solid	2580		5.07		
2355	Samarium silicide	$SmSi_2$	12300-22-0	206.53	orth cry			5.14		
2356	Samarium(II) bromide	$SmBr_2$	50801-97-3	310.17	brn cry	669				reac H₂O
2357	Samarium(II) chloride	$SmCl_2$	13874-75-4	221.27	brn cry	855		3.69		reac H₂O
2358	Samarium(II) fluoride	SmF_2	15192-17-3	188.36	purp cry					reac H₂O
2359	Samarium(II) iodide	SmI_2	32248-43-4	404.17	grn cry	520				reac H₂O
2360	Samarium(III) acetate trihydrate	$Sm(C_2H_3O_2)_3 \cdot 3H_2O$	17829-86-6	381.54	hyg yel-wh solid				1.94	s H₂O
2361	Samarium(III) bromate nonahydrate	$Sm(BrO_3)_3 \cdot 9H_2O$	63427-22-5	696.20	pink hex cry	75 dec				vs H₂O; sl EtOH
2362	Samarium(III) bromide	$SmBr_3$	13759-87-0	390.07	yel cry	640				reac H₂O
2363	Samarium(III) carbonate	$Sm_2(CO_3)_3$	5895-47-6	480.75	wh-yel pow	dec >500				
2364	Samarium(III) chloride	$SmCl_3$	10361-82-7	256.72	yel cry	682		4.46	93.8[25]	
2365	Samarium(III) chloride hexahydrate	$SmCl_3 \cdot 6H_2O$	13465-55-9	364.81	yel cry	dec		2.38	93.8[25]	
2366	Samarium(III) fluoride	SmF_3	13765-24-7	207.36	wh cry	1306				reac H₂O
2367	Samarium(III) iodide	SmI_3	13813-25-7	531.07	oran cry	850				reac H₂O
2368	Samarium(III) nitrate	$Sm(NO_3)_3$	10361-83-8	336.38	yel-wh hyg solid				144[25]	s EtOH
2369	Samarium(III) nitrate hexahydrate	$Sm(NO_3)_3 \cdot 6H_2O$	13759-83-6	444.47	pale yel cry	78				s H₂O, MeOH, ace
2370	Samarium(III) oxide	Sm_2O_3	12060-58-1	348.72	yel-wh cub cry	2269	3780	7.6		
2371	Samarium(III) sulfate octahydrate	$Sm_2(SO_4)_3 \cdot 8H_2O$	13465-58-2	733.03	yel cry			2.93	2.67[20]	
2372	Samarium(III) sulfide	Sm_2S_3	12067-22-0	396.92	gray-brn cub cry	1720		5.87		
2373	Samarium(III) telluride	Sm_2Te_3	12040-00-5	683.52	orth cry			7.31		
2374	Scandium	Sc	7440-20-2	44.956	silv metal; hex	1541	2836	2.99		
2375	Scandium boride	ScB_2	12007-34-0	66.578	refrac solid	2250		3.17		
2376	Scandium bromide	$ScBr_3$	13465-59-3	284.668	wh hyg cry	969		9.33		s H₂O
2377	Scandium chloride	$ScCl_3$	10361-84-9	151.315	wh hyg cry	967		2.4		s H₂O; i EtOH
2378	Scandium fluoride	ScF_3	13709-47-2	101.951	wh powder	1552				sl H₂O
2379	Scandium hydroxide	$Sc(OH)_3$	17674-34-9	95.978	col amorp solid					i H₂O; s dil acid
2380	Scandium iodide	ScI_3	14474-33-0	425.669	hyg yel cry	953	subl			s H₂O, EtOH, CCD
2381	Scandium nitrate	$Sc(NO_3)_3$	13465-60-6	230.971	wh cry				169[25]	s EtOH
2382	Scandium oxide	Sc_2O_3	12060-08-1	137.910	wh cub cry	2489		3.864		s conc acid
2383	Scandium sulfate pentahydrate	$Sc_2(SO_4)_3 \cdot 5H_2O$	15292-44-1	468.176	col cry	dec 110				vs H₂O
2384	Scandium sulfide	Sc_2S_3	12166-29-9	186.107	yel orth cry	1775		2.91		
2385	Scandium telluride	Sc_2Te_3	12166-44-8	472.71	blk hex cry			5.29		
2386	Selenium (gray)	Se	7782-49-2	78.96	gray metallic cry; hex	220.8	685	4.809		i H₂O, CS₂
2387	Selenium (α form)	Se	7782-49-2	78.96	red monocl cry	trans gray Se >120	685	4.39		i H₂O, EtOH; sl eth
2388	Selenium (vitreous)	Se	7782-49-2	78.96	blk amorp solid	trans gray Se 180	685	4.28		i H₂O; sl CS₂
2389	Selenic acid	H_2SeO_4	7783-08-6	144.97	wh hyg solid	58	260 dec	2.95		vs H₂O; reac EtOH
2390	Pentafluoroorthoselenic acid	$HOSeF_5$	38989-47-8	190.96	col solid	38	47			
2391	Selenous acid	H_2SeO_3	7783-00-8	128.97	wh hyg cry	70 dec		3.0		vs H₂O; s EtOH
2392	Selenium dioxide	SeO_2	7446-08-4	110.96	wh tetr needles or powder	340 tp	315 sp	3.95	264[22]	s EtOH, MeOH; sl ace
2393	Selenium trioxide	SeO_3	13768-86-0	126.96	wh tetr cry; hyg	118	subl	3.44		s H₂O, os

No.	Name	Formula	CAS Reg No.	Mol. weight	Physical form	mp/°C	bp/°C	Density g cm⁻³	Solubility g/100 g H₂O	Qualitative solubility
2394	Selenium bromide	Se_2Br_2	7789-52-8	317.73	red liq	5	225 dec	3.60		reac H_2O; s CS_2, chl
2395	Selenium chloride	Se_2Cl_2	10025-68-0	228.83	yel-brn oily liq	-85	127 dec	2.774		reac H_2O; s CS_2, bz, ctc, chl
2396	Selenium tetrabromide	$SeBr_4$	7789-65-3	398.58	oran-red cry	123				reac H_2O; s CS_2, chl
2397	Selenium tetrachloride	$SeCl_4$	10026-03-6	220.77	wh-yel cry	305 tp	191.4 sp	2.6		reac H_2O
2398	Selenium tetrafluoride	SeF_4	13465-66-2	154.95	col liq	-9.5	101.6	2.75		reac H_2O; vs EtOH, eth
2399	Selenium hexafluoride	SeF_6	7783-79-1	192.95	col gas	-34.6 tp	-46.6 sp	7.887 g/L		i H_2O
2400	Selenium chloride pentafluoride	SeF_5Cl	34979-62-9	209.41	col gas	-19	4.5			
2401	Selenium oxybromide	$SeOBr_2$	7789-51-7	254.77	red-yel solid	41.6	220 dec	3.38		reac H_2O; s CS_2, bz, ctc
2402	Selenium oxychloride	$SeOCl_2$	7791-23-3	165.87	col or yel liq	8.5	177	2.44		reac H_2O; s ctc, chl, bz, tol
2403	Selenium oxyfluoride	$SeOF_2$	7783-43-9	132.96	col liq	15	125	2.8		reac H_2O
2404	Selenium oxytetrafluoride	$SeOF_4$	53319-44-1	170.95	unstab col liq	12	65			
2405	Selenium dioxydifluoride	SeO_2F_2	14984-81-7	148.96	col gas	-99.5	-8.4	6.089 g/L		reac H_2O
2406	Selenium monosulfide	SeS	7446-34-6	111.03						
2407	Selenium disulfide	SeS_2	7488-56-4	143.09	red-yel cry	100				i H_2O; s acid
2408	Selenium sulfide (Se_2S_6)	Se_2S_6	75926-26-0	350.31	oran needles	121.5		2.44		s CS_2; sl bz
2409	Selenium sulfide (Se_4S_4)	Se_4S_4	75926-28-2	444.10	red cry	113 dec		3.29		s bz; sl CS_2
2410	Selenium sulfide (Se_6S_2)	Se_6S_2	75926-30-6	537.89	oran cry	121.5				s CS_2
2411	Silicon	Si	7440-21-3	28.085	gray cry or brn amorp solid	1414	3265	2.3296		i H_2O, acid; s alk
2412	Silane	SiH_4	7803-62-5	32.118	col gas; flam	-185	-111.9	1.313 g/L		reac H_2O; i EtOH, bz
2413	Disilane	Si_2H_6	1590-87-0	62.219	col gas; flam	-129.4	-14.8	2.543 g/L		reac H_2O, ctc, chl; s EtOH, bz
2414	Trisilane	Si_3H_8	7783-26-8	92.321	flam col liq	-117.4	52.9	0.739		reac H_2O
2415	Tetrasilane	Si_4H_{10}	7783-29-1	122.421	col liq; flam	-89.9	108.1	0.792		reac H_2O
2416	2-Silyltrisilane	Si_4H_{10}	13597-87-0	122.421	col liq	-99.4	101.7	0.792		reac H_2O
2417	Cyclopentasilane	Si_5H_{10}	289-22-5	150.507	col liq	-10.5	194.3	0.963		reac H_2O
2418	Pentasilane	Si_5H_{12}	14868-53-2	152.523	col liq	-72.8	153.2	0.827		reac H_2O
2419	2-Silyltetrasilane	Si_5H_{12}	14868-54-3	152.523	col liq	-109.9	146.2	0.820		reac H_2O
2420	2,2-Disilyltrisilane	Si_5H_{12}	15947-57-6	152.523	col liq	-57.8	134.3	0.815		reac H_2O
2421	Cyclohexasilane	Si_6H_{12}	291-59-8	180.608	col liq	16.5	226			reac H_2O
2422	Hexasilane	Si_6H_{14}	14693-61-9	182.624	col liq	-44.7	193.6	0.847		reac H_2O
2423	2-Silylpentasilane	Si_6H_{14}	14868-55-4	182.624	col liq	-78.4	185.2	0.840		reac H_2O
2424	3-Silylpentasilane	Si_6H_{14}	52988-75-7	182.624	col liq	-69	179.5	0.843		reac H_2O
2425	Heptasilane	Si_7H_{16}	14693-65-3	212.726	col liq	-30.1	226.8	0.859		reac H_2O
2426	Bromosilane	SiH_3Br	13465-73-1	111.014	col gas	-94	1.9	4.538 g/L		
2427	Dibromosilane	SiH_2Br_2	13768-94-0	189.910	liq	-70.1	66			
2428	Tribromosilane	$SiHBr_3$	7789-57-3	268.806	flam liq	-73	109	2.7		reac H_2O
2429	Tetrabromosilane	$SiBr_4$	7789-66-4	347.702	col fuming liq	5.39	154	2.8		reac H_2O
2430	Bromotrichlorosilane	$SiBrCl_3$	13465-74-2	214.349	col liq	-62	80.3	1.826		reac H_2O
2431	Dibromodichlorosilane	$SiBr_2Cl_2$	13465-75-3	258.800	col liq	-45.5	104	2.172		reac H_2O
2432	Tribromochlorosilane	$SiBr_3Cl$	13465-76-4	303.251	col liq	-20.8	127	2.497		reac H_2O
2433	Hexabromosilane	Si_2Br_6	13517-13-0	535.595	col cry	95	265			
2434	Octabromotrisilane	Si_3Br_8	54804-32-9	723.489	col liq	46				
2435	Chlorosilane	SiH_3Cl	13465-78-6	66.563	col gas	-118	-30.4	2.721 g/L		
2436	Dichlorosilane	SiH_2Cl_2	4109-96-0	101.008	col gas; flam	-122	8.3	4.129 g/L		reac H_2O
2437	Trichlorosilane	$SiHCl_3$	10025-78-2	135.453	fuming liq	-128.2	33	1.331		reac H_2O
2438	Tetrachlorosilane	$SiCl_4$	10026-04-7	169.898	col fuming liq	-68.74	57.65	1.5		reac H_2O
2439	Chlorotrifluorosilane	$SiClF_3$	14049-36-6	120.534	col gas	-138	-70.0	4.927 g/L		reac H_2O
2440	Dichlorodifluorosilane	$SiCl_2F_2$	18356-71-3	136.989	col gas	-44	-32	5.599 g/L		reac H_2O
2441	Trichlorofluorosilane	$SiCl_3F$	14965-52-7	153.443	col gas		12.25	6.272 g/L		
2442	Trichloroiodosilane	$SiCl_3I$	13465-85-5	261.349	col liq	-60	113.5			reac H_2O
2443	Hexachlorodisilane	Si_2Cl_6	13465-77-5	268.889	col liq	2.5	146			reac H_2O
2444	Octachlorotrisilane	Si_3Cl_8	13596-23-1	367.881	col liq	-67	216			
2445	Fluorosilane	SiH_3F	13537-33-2	50.108	col gas		-98.6	2.048 g/L		
2446	Difluorosilane	SiH_2F_2	13824-36-7	68.099	col gas	-122	-77.8	2.783 g/L		
2447	Trifluorosilane	$SiHF_3$	13465-71-9	86.089	col gas	-131	-95	3.519 g/L		
2448	Tetrafluorosilane	SiF_4	7783-61-1	104.080	col gas	-90.2	-86	4.254 g/L		reac H_2O
2449	Hexafluorodisilane	Si_2F_6	13830-68-7	170.161	col gas	-18.7 tp @ 780 mmHg	-19.1 sp			reac H_2O
2450	Octafluorotrisilane	Si_3F_8	14521-14-3	236.244	col liq	-1.2	42			
2451	Decafluorotetrasilane	Si_4F_{10}	14521-15-4	302.326	col cry	68	85.1			
2452	Iodosilane	SiH_3I	13598-42-0	158.014	col liq	-57	45.6			

No.	Name	Formula	CAS Reg No.	Mol. weight	Physical form	mp/°C	bp/°C	Density g cm⁻³	Solubility g/100 g H₂O	Qualitative solubility
2453	Diiodosilane	SiH_2I_2	13760-02-6	283.911	col liq	-1	150			
2454	Triiodosilane	$SiHI_3$	13465-72-0	409.807	liq	8	220 dec			
2455	Tetraiodosilane	SiI_4	13465-84-4	535.704	wh powder	120.5	287.35	4.1		
2456	Hexaiododisilane	Si_2I_6	13510-43-5	817.598	pale yel cry	250				
2457	Disiloxane	$(SiH_3)_2O$	13597-73-4	78.218	gas	-144	-15.2	3.197 g/L		
2458	Hexachlorodisiloxane	$(SiCl_3)_2O$	14986-21-1	284.888	liq	-28	137			
2459	Methylsilane	SiH_3CH_3	992-94-9	46.145	col gas	-156.5	-57.5			
2460	Metasilicic acid	H_2SiO_3	7699-41-4	78.100	wh amorp powder					i H_2O; s HF
2461	Orthosilicic acid	H_4SiO_4	10193-36-9	96.116	exists only in soln					
2462	Fluorosilicic acid	H_2SiF_6	16961-83-4	144.092	stab only in aq soln					s H_2O
2463	Silicon carbide (hexagonal)	SiC	409-21-2	40.097	hard grn-black hex cry	2830		3.16		i H_2O, EtOH
2464	Silicon nitride (Si_3N_4)	Si_3N_4	12033-89-5	140.284	gray refrac solid; hex	1900		3.17		
2465	Silicon monoxide	SiO	10097-28-6	44.085	blk cub cry, stable >1200			2.18		
2466	Silicon dioxide (α-quartz)	SiO_2	14808-60-7	60.085	col hex cry	trans to beta quartz 573	2950	2.648		i H_2O, acid; s HF
2467	Silicon dioxide (β-quartz)	SiO_2	14808-60-7	60.085	col hex cry	trans to tridymite 867	2950	2.533[600]		i H_2O, acid; s HF
2468	Silicon dioxide (tridymite)	SiO_2	15468-32-3	60.085	col hex cry	trans cristobalite 1470	2950	2.265		i H_2O, acid; s HF
2469	Silicon dioxide (cristobalite)	SiO_2	14464-46-1	60.085	col hex cry	1722	2950	2.334		i H_2O, acid; s HF
2470	Silicon dioxide (vitreous)	SiO_2	60676-86-0	60.085	col amorp solid	1713	2950	2.196		i H_2O, acid; s HF
2471	Silicon monosulfide	SiS	12504-41-5	60.151	yel-red hyg powder	1090	940	1.85		reac H_2O
2472	Silicon disulfide	SiS_2	13759-10-9	92.216	wh rhomb cry	1090	subl	2.04		reac H_2O, EtOH; i bz
2473	Silicon tetraacetate	$Si(C_2H_3O_2)_4$	562-90-3	264.262	wh hyg cry	110				reac H_2O; s ace, bz
2474	Silicon tetraboride	SiB_4	12007-81-7	71.330	gray refrac solid	1870 dec		2.4		
2475	Silicotungstic acid	$H_4SiO_4 \cdot (W_3O_9)_4$	12520-88-6	2878.17	hyg yel cry					vs H_2O, EtOH
2476	Silver	Ag	7440-22-4	107.868	silv metal; cub	961.78	2162	10.5		
2477	Silver azide	AgN_3	13863-88-2	149.888	orth cry; exp	exp ≈250		4.9	0.00081[20]	
2478	Silver subfluoride	Ag_2F	1302-01-8	234.734	yel hex cry	100 dec		8.6		reac H_2O
2479	Silver(I) acetate	$AgC_2H_3O_2$	563-63-3	166.912	wh needles or powder	dec		3.26	1.04[20]	
2480	Silver(I) acetylide	Ag_2C_2	7659-31-6	239.757	wh powder; exp					
2481	Silver(I) acetylide (AgC_2H)	AgC_2H	13092-75-6	132.897	wh powder; exp					
2482	Silver(I) arsenate	Ag_3AsO_4	13510-44-6	462.524	red cub cry	dec		6.657	0.00085	s NH_4OH
2483	Silver(I) benzoate	$Ag(C_6H_5CO_2)$	532-31-0	228.982	powder				30[20]	
2484	Silver(I) bromate	$AgBrO_3$	7783-89-3	235.770	wh tetr cry	360 dec		5.21	0.193[25]	
2485	Silver(I) bromide	AgBr	7785-23-1	187.772	yel cub cry	430	1502	6.47	0.000014[25]	i H_2O, acid, EtOH
2486	Silver(I) carbonate	Ag_2CO_3	534-16-7	275.745	yel monocl cry	218		6.077	0.0036[20]	s acid
2487	Silver(I) chlorate	$AgClO_3$	7783-92-8	191.319	wh tetr cry	230	270 dec	4.430	17.6[25]	sl EtOH
2488	Silver(I) chloride	AgCl	7783-90-6	143.321	wh cub cry	455	1547	5.56	0.00019[25]	
2489	Silver(I) chlorite	$AgClO_2$	7783-91-7	175.320	yel cry	105 exp			0.55[25]	
2490	Silver(I) chromate	Ag_2CrO_4	7784-01-2	331.730	brn-red monocl cry			5.625	0.000014[0]	
2491	Silver(I) citrate	$Ag_3C_6H_5O_7$	126-45-4	512.705	wh cry powder					i H_2O; s HNO_3
2492	Silver(I) cyanide	AgCN	506-64-9	133.886	wh-gray hex cry	320 dec		3.95	0.0000011	i EtOH, dil acid
2493	Silver(I) dichromate	$Ag_2Cr_2O_7$	7784-02-3	431.724	red cry			4.770		sl H_2O
2494	Silver(I) diethyldithiocarbamate	$Ag(C_2H_5)_2NCS_2$	1470-61-7	256.138	powder	173				s py
2495	Silver(I) fluoride	AgF	7775-41-9	126.866	yel-brn cub cry; hyg	435	1159	5.852	172[20]	
2496	Silver(I) hexafluoroantimonate	$AgSbF_6$	26042-64-8	343.618	powder					
2497	Silver(I) hexafluoroarsenate	$AgAsF_6$	12005-82-2	296.780	powder					
2498	Silver(I) hexafluorophosphate	$AgPF_6$	26042-63-7	252.832	powder	102 dec				
2499	Silver(I) hydrogen fluoride	$AgHF_2$	12249-52-4	146.873	hyg cry	dec				
2500	Silver(I) iodate	$AgIO_3$	7783-97-3	282.770	wh orth cry	>200		5.53	0.053[25]	
2501	Silver(I) iodide	AgI	7783-96-2	234.772	yel powder; hex	558	1506	5.68	0.000003	i acid
2502	Silver(I) lactate monohydrate	$AgC_3H_5O_3 \cdot H_2O$	128-00-7	214.954	gray cry powder					sl H_2O, EtOH
2503	Silver(I) metaphosphate	$AgPO_3$	13465-96-8	186.840	grn glass	490		6.37		i H_2O; s HNO_3, NH_4OH
2504	Silver(I) molybdate	Ag_2MoO_4	13765-74-7	375.67	yel cub cry	483		6.18		sl H_2O
2505	Silver(I) nitrate	$AgNO_3$	7761-88-8	169.873	col rhomb cry	210	440 dec	4.35	234[25]	sl EtOH, ace
2506	Silver(I) nitrite	$AgNO_2$	7783-99-5	153.874	yel needles	140 dec		4.453	0.415[25]	i EtOH; reac acid
2507	Silver(I) oxalate	$Ag_2C_2O_4$	533-51-7	303.755	wh cry powder	exp 140		5.03	0.0043[20]	

No.	Name	Formula	CAS Reg No.	Mol. weight	Physical form	mp/°C	bp/°C	Density g cm⁻³	Solubility g/100 g H₂O	Qualitative solubility
2508	Silver(I) oxide	Ag_2O	20667-12-3	231.735	brn-blk cub cry	≈200 dec		7.2	0.0025	i EtOH; s acid, alk
2509	Silver(I) perchlorate	$AgClO_4$	7783-93-9	207.319	col cub cry; hyg	486 dec		2.806	558[25]	s bz, py, os
2510	Silver(I) perchlorate monohydrate	$AgClO_4 \cdot H_2O$	14242-05-8	225.334	hyg wh cry	43 dec			558[25]	
2511	Silver(I) permanganate	$AgMnO_4$	7783-98-4	226.804	viol monocl cry	dec		4.49	0.91[18]	reac EtOH
2512	Silver(I) phosphate	Ag_3PO_4	7784-09-0	418.576	yel powder	849		6.37	0.0064	sl dil acid
2513	Silver(I) picrate monohydrate	$AgC_6H_2N_3O_7 \cdot H_2O$	146-84-9	353.979	yel cry					sl H₂O, EtOH; i chl, eth
2514	Silver(I) selenate	Ag_2SeO_4	7784-07-8	358.69	orth cry			5.72	0.118[20]	
2515	Silver(I) selenide	Ag_2Se	1302-09-6	294.70	gray hex needles	880		8.216		i H₂O
2516	Silver(I) selenite	Ag_2SeO_3	7784-05-6	342.69	needles	530	>550 dec	5.930		sl H₂O; s acid
2517	Silver(I) sulfate	Ag_2SO_4	10294-26-5	311.799	col cry or powder	660		5.45	0.84[25]	
2518	Silver(I) sulfide	Ag_2S	21548-73-2	247.801	gray-blk orth powder	825 (high press.)		7.23		i H₂O; s acid
2519	Silver(I) sulfite	Ag_2SO_3	13465-98-0	295.799	wh cry	100 dec			0.00046[20]	s acid, NH₄OH
2520	Silver(I) telluride	Ag_2Te	12002-99-2	343.34	blk orth cry	955		8.4		
2521	Silver(I) tetraiodomercurate(II)	Ag_2HgI_4	7784-03-4	923.94	yel tetr cry	trans to red cub ≈40		6.1		i H₂O, dil acid
2522	Silver(I) thiocyanate	$AgSCN$	1701-93-5	165.951	wh powder	dec				i H₂O
2523	Silver(I) thiosulfate	$Ag_2S_2O_3$	23149-52-2	327.864	wh cry	dec				sl H₂O; s NH₄OH
2524	Silver(II) oxide	AgO	1301-96-8	123.867	gray powder; monocl or cub	>100 dec		7.5	0.0027[25]	s alk; reac acid
2525	Silver(I) tungstate	Ag_2WO_4	13465-93-5	463.57	yel cry	620			0.015	s HNO₃, NH₄OH
2526	Silver(II) fluoride	AgF_2	7783-95-1	145.865	wh or gray hyg cry	690		4.58		reac H₂O
2527	Silver(II) oxide (Ag₂O₂)	Ag_2O_2	25455-73-6	247.735	gray-blk cub cry	>100		7.44		i H₂O; s acid, NH₄OH
2528	Sodium	Na	7440-23-5	22.990	soft silv met; cub	97.794	882.940	0.97		reac H₂O
2529	Sodium acetate	$NaC_2H_3O_2$	127-09-3	82.034	col cry	328.2		1.528	50.4[25]	
2530	Sodium acetate trihydrate	$NaC_2H_3O_2 \cdot 3H_2O$	6131-90-4	136.079	col cry	58 dec		1.45	50.4[25]	sl EtOH
2531	Sodium aluminate	$NaAlO_2$	1302-42-7	81.971	wh orth cry; hyg	1650		4.63		vs H₂O; i EtOH
2532	Sodium aluminum hydride	$NaAlH_4$	13770-96-2	54.004	wh hyg solid	174 dec		1.24		i eth; s thf
2533	Sodium aluminum sulfate dodecahydrate	$NaAl(SO_4)_2 \cdot 12H_2O$	10102-71-3	458.281	col cry	≈60		1.61	39.7[20]	i EtOH
2534	Sodium amide	$NaNH_2$	7782-92-5	39.013	wh-grn orth cry	210	500 dec	1.39		reac H₂O
2535	Sodium ammonium phosphate tetrahydrate	$NaNH_4HPO_4 \cdot 4H_2O$	13011-54-6	209.069	monocl cry	≈80 dec		1.54		s H₂O; i EtOH
2536	Sodium arsenate dodecahydrate	$Na_3AsO_4 \cdot 12H_2O$	7778-43-0	424.072	col monocl prism	86 dec				s H₂O; sl EtOH; i eth
2537	Sodium arsenite	$NaAsO_2$	7784-46-5	129.911	wh-gray hyg powder			1.87		vs H₂O; i EtOH
2538	Sodium azide	NaN_3	26628-22-8	65.010	col hex cry	300 dec		1.846	40.8[20]	sl EtOH; i eth
2539	Sodium borohydride	$NaBH_4$	16940-66-2	37.833	wh cub cry; hyg	≈400 dec		1.07	55[20]	reac EtOH
2540	Sodium bromate	$NaBrO_3$	7789-38-0	150.892	col cub cry	381		3.34	39.4[25]	i EtOH
2541	Sodium bromide	$NaBr$	7647-15-6	102.894	wh cub cry	747	1390	3.200	94.6[25]	s EtOH
2542	Sodium bromide dihydrate	$NaBr \cdot 2H_2O$	13466-08-5	138.925	wh cry	36 dec		2.18	94.6[25]	sl EtOH
2543	Sodium carbonate	Na_2CO_3	497-19-8	105.989	wh hyg powder	856		2.54	30.7[25]	i EtOH
2544	Sodium carbonate monohydrate	$Na_2CO_3 \cdot H_2O$	5968-11-6	124.005	col orth cry	100 dec		2.25	30.7[25]	i EtOH
2545	Sodium carbonate decahydrate	$Na_2CO_3 \cdot 10H_2O$	6132-02-1	286.142	col cry	34 dec		1.46	30.7[25]	i EtOH
2546	Sodium chlorate	$NaClO_3$	7775-09-9	106.441	col cub cry	248	dec 630	2.5	100[25]	sl EtOH
2547	Sodium chloride	$NaCl$	7647-14-5	58.443	col cub cry	800.7	1465	2.17	36.0[25]	sl EtOH
2548	Sodium chlorite	$NaClO_2$	7758-19-2	90.442	wh hyg cry	≈180 dec			64[17]	
2549	Sodium chromate	Na_2CrO_4	7775-11-3	161.974	yel orth cry	794		2.72	87.6[25]	sl EtOH
2550	Sodium chromate tetrahydrate	$Na_2CrO_4 \cdot 4H_2O$	10034-82-9	234.035	yel hyg cry	dec			87.6[25]	sl EtOH
2551	Sodium citrate dihydrate	$Na_3C_6H_5O_7 \cdot 2H_2O$	6132-04-3	294.099	wh cry	150 dec				vs H₂O; i EtOH, eth
2552	Sodium citrate pentahydrate	$Na_3C_6H_5O_7 \cdot 5H_2O$	6858-44-2	348.145	hyg col cry	dec 150		1.86	92[25]	vs H₂O; sl EtOH; i eth
2553	Sodium cyanate	$NaCNO$	917-61-3	65.007	col needles	550		1.89		s H₂O; sl EtOH; i eth
2554	Sodium cyanide	$NaCN$	143-33-9	49.008	wh cub cry; hyg	562		1.6	58.2[20]	sl EtOH
2555	Sodium cyanoborohydride	$NaBH_3(CN)$	25895-60-7	62.843	wh hyg powder	240 dec		1.12		vs H₂O; s thf; sl EtOH; i bz, eth
2556	Sodium dichromate	$Na_2Cr_2O_7$	10588-01-9	261.968	red hyg cry	357	400 dec		187[25]	
2557	Sodium dichromate dihydrate	$Na_2Cr_2O_7 \cdot 2H_2O$	7789-12-0	297.999	oran-red monocl cry	85 dec		2.35		vs H₂O; S HOAc
2558	Sodium dihydrogen phosphate	NaH_2PO_4	7558-80-7	119.977	col monocl cry	200 dec			94.9[25]	
2559	Sodium dihydrogen phosphate monohydrate	$NaH_2PO_4 \cdot H_2O$	10049-21-5	137.993	wh hyg cry	100 dec			94.9[25]	i EtOH
2560	Sodium dihydrogen phosphate dihydrate	$NaH_2PO_4 \cdot 2H_2O$	13472-35-0	156.008	col orth cry	60 dec		1.91	94.9[25]	i EtOH
2561	Sodium dihydrogen hypophosphate hexahydrate	$Na_2H_2P_2O_6 \cdot 6H_2O$	7782-95-8	314.031	monocl plates	110 dec		1.849	2.0[25]	i EtOH

No.	Name	Formula	CAS Reg No.	Mol. weight	Physical form	mp/°C	bp/°C	Density g cm^{-3}	Solubility g/100 g H$_2$O	Qualitative solubility
2562	Sodium dihydrogen pyrophosphate	Na$_2$H$_2$P$_2$O$_7$	7758-16-9	221.939	wh powder	220 dec		≈1.9		s H$_2$O
2563	Sodium dithionate	Na$_2$S$_2$O$_4$	7775-14-6	174.108	gray-wh powder	52 dec			24.1[20]	sl EtOH
2564	Sodium dithionate dihydrate	Na$_2$S$_2$O$_6$ · 2H$_2$O	7631-94-9*	242.137	col orth cry	110 dec		2.19	15.1[20]	i EtOH
2565	Sodium ethanolate	NaC$_2$H$_5$O	141-52-6	68.050	wh-yel hyg powder	260 dec				reac H$_2$O; s EtOH
2566	Sodium ferricyanide monohydrate	Na$_3$Fe(CN)$_6$ · H$_2$O	14217-21-1*	298.933	red hyg cry					s H$_2$O; i EtOH
2567	Sodium ferrocyanide decahydrate	Na$_4$Fe(CN)$_6$ · 10H$_2$O	13601-19-9	484.061	yel monocl cry	≈50 dec		1.46	20[20]	i os
2568	Sodium fluoride	NaF	7681-49-4	41.988	col cub or tetr cry	996	1704	2.78	4.13[25]	i EtOH
2569	Sodium fluorophosphate	Na$_2$PO$_3$F	10163-15-2	143.950	powder					
2570	Sodium formate	NaCHO$_2$	141-53-7	68.008	wh hyg cry	257.3	dec	1.92	94.9[25]	sl EtOH
2571	Sodium germanate	Na$_2$GeO$_3$	12025-19-3	166.62	wh monocl hyg cry	1083		3.31		
2572	Sodium gold cyanide	NaAu(CN)$_2$	15280-09-8	271.991	wh-yel cry pow					s H$_2$O, NH$_4$OH
2573	Sodium gold thiosulfate dihydrate	Na$_3$Au(S$_2$O$_3$)$_2$ · 5H$_2$O	10233-88-2	526.223	wh needles or prisms	dec 150		3.09		vs H$_2$O; i EtOH
2574	Sodium hexabromoplatinate(IV) hexahydrate	Na$_2$PtBr$_6$ · 6H$_2$O	39277-13-9	828.579	cry					
2575	Sodium hexachloroiridate(IV) hexahydrate	Na$_2$IrCl$_6$ · 6H$_2$O	19567-78-3	559.006	cry	600 dec				
2576	Sodium hexachloroplatinate(IV)	Na$_2$PtCl$_6$	16923-58-3	453.782	yel hyg cry				53[16]	s EtOH
2577	Sodium hexachloroplatinate(IV) hexahydrate	Na$_2$PtCl$_6$ · 6H$_2$O	16923-58-3	561.873	yel cry	110 dec		2.50	53[16]	s EtOH; i eth
2578	Sodium hexafluoroaluminate	Na$_3$AlF$_6$	13775-53-6	209.941	col monocl cry; trans cub 560	1013		2.97		i H$_2$O
2579	Sodium hexafluoroantimonate	NaSbF$_6$	16925-25-0	258.740	wh cub cry			3.375	129[20]	s EtOH, ace
2580	Sodium hexafluorophosphate monohydrate	NaPF$_6$ · H$_2$O	20644-15-9	185.969	col orth cry			2.369	103[0]	s EtOH, MeOH, ace
2581	Sodium hexafluorosilicate	Na$_2$SiF$_6$	16893-85-9	188.056	wh hex cry	847		2.7	0.67[20]	i EtOH
2582	Sodium hexanitrocobaltate(III)	Na$_3$Co(NO$_2$)$_6$	14649-73-1	403.935	yel-brn cry powder					vs H$_2$O; sl EtOH
2583	Sodium hydride	NaH	7646-69-7	23.998	silv cub cry; flam	425 dec		1.39		reac H$_2$O, EtOH
2584	Sodium hydrogen arsenate	Na$_2$HAsO$_4$	7778-43-0	185.908	wh powder	≈195 dec			51[20]	sl EtOH
2585	Sodium hydrogen arsenate heptahydrate	Na$_2$HAsO$_4$ · 7H$_2$O	10048-95-0	312.014	wh monocl cry	≈50 dec		1.87	51[20]	sl EtOH
2586	Sodium hydrogen carbonate	NaHCO$_3$	144-55-8	84.007	wh monocl cry	≈50 dec		2.20	10.3[25]	i EtOH
2587	Sodium hydrogen fluoride	NaHF$_2$	1333-83-1	61.995	wh hex cry	>160 dec		2.08	3.25[20]	
2588	Sodium hydrogen phosphate	Na$_2$HPO$_4$	7558-79-4	141.959	wh hyg powder			1.7	11.8[25]	
2589	Sodium hydrogen phosphate heptahydrate	Na$_2$HPO$_4$ · 7H$_2$O	7782-85-6	268.066	col cry			≈1.7	11.8[25]	i EtOH
2590	Sodium hydrogen phosphate dodecahydrate	Na$_2$HPO$_4$ · 12H$_2$O	10039-32-4	358.143	col cry	≈35 dec		≈1.5	11.8[25]	i EtOH
2591	Sodium hydrogen sulfate	NaHSO$_4$	7681-38-1	120.061	wh hyg cry	≈315		2.43	28.5[25]	
2592	Sodium hydrogen sulfate monohydrate	NaHSO$_4$ · H$_2$O	10034-88-5	138.076	wh monocl cry			2.10	28.5[25]	reac EtOH
2593	Sodium hydrogen sulfide	NaHS	16721-80-5	56.063	col rhomb cry	350		1.79		s H$_2$O, EtOh, eth
2594	Sodium hydrogen sulfide dihydrate	NaHS · 2H$_2$O	16721-80-5	92.094	yel hyg needles	55 dec				vs H$_2$O, EtOH, eth
2595	Sodium hydrogen sulfite	NaHSO$_3$	7631-90-5	104.061	wh cry			1.48		s H$_2$O; sl EtOH
2596	Sodium hydroxide	NaOH	1310-73-2	39.997	wh orth cry; hyg	323	1388	2.13	100[25]	s EtOH, MeOH
2597	Sodium hypochlorite	NaClO	7681-52-9	74.442	stab in aq soln	anh form exp			79.9[25]	
2598	Sodium hypochlorite pentahydrate	NaOCl · 5H$_2$O	10022-70-5	164.518	pale grn orth cry	18		1.6		s H$_2$O
2599	Sodium iodate	NaIO$_3$	7681-55-2	197.892	wh orth cry	422		4.28	9.47[25]	i EtOH
2600	Sodium iodide	NaI	7681-82-5	149.894	wh cub cry; hyg	661	1304	3.67	184[25]	s EtOH, ace
2601	Sodium iodide dihydrate	NaI · 2H$_2$O	13517-06-1	185.925	hyg col monocl cry	69 dec		2.45	318[0]	vs H$_2$O
2602	Sodium bismuthate	NaBiO$_3$	12232-99-4	279.968	yel-brn hyg cry					i cold H$_2$O, reac acid
2603	Sodium metabisulfite	Na$_2$S$_2$O$_5$	7681-57-4	190.107	wh cry				66.7[25]	sl EtOH
2604	Sodium metaborate	NaBO$_2$	7775-19-1	65.800	wh hex cry	966	1434	2.46		s H$_2$O
2605	Sodium metasilicate	Na$_2$SiO$_3$	6834-92-0	122.064	wh amorp solid; hyg	1089		2.61		s cold H$_2$O; reac hot H$_2$O; i EtOH
2606	Sodium metasilicate pentahydrate	Na$_2$SiO$_3$ · 5H$_2$O	13517-24-3	212.140	wh pow	72 dec				s H$_2$O
2607	Sodium molybdate	Na$_2$MoO$_4$	7631-95-0	205.92	col cub cry	687		≈3.5	65.0[25]	
2608	Sodium molybdate dihydrate	Na$_2$MoO$_4$ · 2H$_2$O	10102-40-6	241.95	cry powder	100 dec		≈3.5	65.0[25]	
2609	Sodium molybdophosphate	Na$_3$PO$_4$ · 12MoO$_3$	1313-30-0	1891.20	hyg solid			2.83		vs H$_2$O, EtOH
2610	Sodium niobate	NaNbO$_3$	12034-09-2	163.894	rhom cry	1422		4.55		i H$_2$O
2611	Sodium nitrate	NaNO$_3$	7631-99-4	84.995	col hex cry; hyg	306.5		2.261	91.2[25]	sl EtOH, MeOH
2612	Sodium nitrite	NaNO$_2$	7632-00-0	68.996	wh orth cry; hyg	284	>320 dec	2.17	84.8[25]	sl EtOH; reac acid
2613	Sodium nitroferricyanide dihydrate	Na$_2$[Fe(CN)$_5$NO] · 2H$_2$O	13755-38-9	297.949	red cry			1.72	40[16]	sl EtOH
2614	Sodium orthovanadate	Na$_3$VO$_4$	13721-39-6	183.909	col hex prisms	860				s H$_2$O; i EtOH
2615	Sodium oxalate	Na$_2$C$_2$O$_4$	62-76-0	133.999	wh powder	≈250 dec		2.34	3.61[25]	i EtOH

No.	Name	Formula	CAS Reg No.	Mol. weight	Physical form	mp/°C	bp/°C	Density g cm⁻³	Solubility g/100 g H₂O	Qualitative solubility
2616	Sodium oxide	Na_2O	1313-59-3	61.979	wh amorp powder	1134		2.27		reac H_2O
2617	Sodium perborate tetrahydrate	$NaBO_3 \cdot 4H_2O$	7632-04-4	153.861	wh cry	60 dec				reac H_2O
2618	Sodium perchlorate	$NaClO_4$	7601-89-0	122.441	wh orth cry; hyg	482 dec		2.52	205²⁵	
2619	Sodium perchlorate monohydrate	$NaClO_4 \cdot H_2O$	7791-07-3	140.456	wh hyg cry	≈130 dec		2.02	205²⁵	
2620	Sodium periodate	$NaIO_4$	7790-28-5	213.892	wh tetr cry	≈300 dec		3.86	14.4²⁵	s acid
2621	Sodium periodate trihydrate	$NaIO_4 \cdot 3H_2O$	13472-31-6	267.938	wh hex cry	175 dec		3.22	14.4²⁵	
2622	Sodium permanganate trihydrate	$NaMnO_4 \cdot 3H_2O$	10101-50-5*	195.972	red-blk hyg cry	170 dec		2.47	144²⁰	reac EtOH
2623	Sodium peroxide	Na_2O_2	1313-60-6	77.979	yel hyg powder	675		2.805		reac H_2O
2624	Sodium perrhenate	$NaReO_4$	13472-33-8	273.195	cry	300		5.39		
2625	Sodium persulfate	$Na_2S_2O_8$	7775-27-1	238.105	wh hyg cry					vs H_2O; reac EtOH
2626	Sodium phosphate	Na_3PO_4	7601-54-9	163.940	col cry	1583		2.54	14.5²⁵	s H_2O
2627	Sodium phosphate dodecahydrate	$Na_3PO_4 \cdot 12H_2O$	10101-89-0	380.124	col hex cry	≈75		1.62	14.4²⁵	i EtOH
2628	Sodium phosphate, chlorinated	$Na_3PO_4 \cdot NaOCl$	56802-99-4	238.383	wh cry				25²⁵	
2629	Sodium phosphide	Na_3P	12058-85-4	99.943	red solid	>650				reac H_2O
2630	Sodium phosphinate	NaH_2PO_2	7681-53-0	87.979	wh cry				100²⁵	
2631	Sodium phosphinate monohydrate	$NaH_2PO_2 \cdot H_2O$	10039-56-2	105.994	col hyg cry	310 dec			100²⁵	s EtOH
2632	Sodium phosphonate pentahydrate	$Na_2HPO_3 \cdot 5H_2O$	13517-23-2	216.036	wh hex plates	dec 200			429²⁰	vs H_2O; i EtOH
2633	Sodium pyrophosphate	$Na_4P_2O_7$	7722-88-5	265.902	col cry	988		2.53	7.09²⁵	
2634	Sodium selenate	Na_2SeO_4	13410-01-0	188.94	col orth cry				58.5²⁵	
2635	Sodium selenate decahydrate	$Na_2SeO_4 \cdot 10H_2O$	10102-23-5	369.09	wh cry			1.61	58.5²⁵	
2636	Sodium selenide	Na_2Se	1313-85-5	124.94	amorp solid	>875		2.62		reac H_2O
2637	Sodium selenite	Na_2SeO_3	10102-18-8	172.94	wh tetr cry				89.8²⁵	i EtOH
2638	Sodium selenite pentahydrate	$Na_2SeO_3 \cdot 5H_2O$	26970-82-1	184.054	wh tetr cry	dec				s H_2O; i EtOH
2639	Sodium stannate trihydrate	$Na_2SnO_3 \cdot 3H_2O$	12209-98-2	266.734	col hex cry	dec 140			61¹⁵	vs H_2O; i EtOH, ace
2640	Sodium stearate	$NaC_{18}H_{35}O_2$	822-16-2	306.460	wh powder					sl H_2O, EtOH; vs hot H_2O
2641	Sodium succinate hexahydrate	$Na_2C_4H_4O_4 \cdot 6H_2O$	150-90-3	270.144	cry powder	120 dec			20	i EtOH
2642	Sodium sulfate	Na_2SO_4	7757-82-6	142.043	wh orth cry or powder	884		2.7	28.1²⁵	i EtOH
2643	Sodium sulfate heptahydrate	$Na_2SO_4 \cdot 7H_2O$	13472-39-4	204.152	wh cry	dec				vs H_2O
2644	Sodium sulfate decahydrate	$Na_2SO_4 \cdot 10H_2O$	7727-73-3	322.196	col monocl cry	32 dec		1.46	28.1²⁵	i EtOH
2645	Sodium sulfide	Na_2S	1313-82-2	78.045	wh cub cry; hyg	1172		1.856	20.6²⁵	sl EtOH; i eth
2646	Sodium sulfide pentahydrate	$Na_2S \cdot 5H_2O$	1313-83-3	168.121	col orth cry	120 dec		1.58	20.6²⁵	s EtOH; i eth
2647	Sodium sulfide nonahydrate	$Na_2S \cdot 9H_2O$	1313-84-4	240.183	wh-yel hyg cry	≈50 dec		1.43	20.6²⁵	sl EtOH; i eth
2648	Sodium sulfite	Na_2SO_3	7757-83-7	126.043	wh hex cry	911		2.63	30.7²⁵	i EtOH
2649	Sodium sulfite heptahydrate	$Na_2SO_3 \cdot 7H_2O$	10102-15-5	252.150	wh monocl cry; unstab			1.56	30.7²⁵	s EtOH
2650	Sodium superoxide	NaO_2	12034-12-7	54.989	yel cub cry	552		2.2		reac H_2O
2651	Sodium tartrate dihydrate	$Na_2C_4H_4O_6 \cdot 2H_2O$	6106-24-7	230.082				1.545		s H_2O; i EtOH
2652	Sodium tellurate	Na_2TeO_4	10101-83-4	237.58	wh powder				0.8	
2653	Sodium tellurite	Na_2TeO_3	10102-20-2	221.58	wh rhomb prisms					sl H_2O
2654	Sodium tetraborate	$Na_2B_4O_7$	1330-43-4	201.220	col gl solid; hyg	743	1575	2.4	3.17²⁵	sl MeOH
2655	Sodium tetraborate tetrahydrate	$Na_2B_4O_7 \cdot 4H_2O$	12045-87-3	273.281	wh monocl cry			1.95	3.17²⁵	
2656	Sodium tetraborate pentahydrate	$Na_2B_4O_7 \cdot 5H_2O$	12045-88-4	291.296	hex cry	dec		1.88	3.17²⁵	
2657	Sodium tetraborate decahydrate	$Na_2B_4O_7 \cdot 10H_2O$	1303-96-4	381.373	wh monocl cry	75 dec		1.73	3.17²⁵	i EtOH
2658	Sodium tetrachloroaluminate	$NaAlCl_4$	7784-16-9	191.784	orth cry			2.01		s H_2O
2659	Sodium tetrachloroaurate(III) dihydrate	$NaAuCl_4 \cdot 2H_2O$	13874-02-7	397.800	oran-yel rhom cry	100 dec			150¹⁰	s EtOH, eth
2660	Sodium tetrachloropalladate(II) trihydrate	$Na_2PdCl_4 \cdot 3H_2O$	13820-53-6	348.26	brn-red hyg cry					vs H_2O; s EtOH
2661	Sodium tetrachloroplatinate(II) tetrahydrate	$Na_2PtCl_4 \cdot 4H_2O$	10026-00-3	454.938	red prisms	100				s H_2O, EtOH
2662	Sodium tetrafluoroberyllate	Na_2BeF_4	13871-27-7	130.986	orth cry	575		2.47		sl H_2O
2663	Sodium tetrafluoroborate	$NaBF_4$	13755-29-8	109.795	wh orth prisms	384		2.47	108²⁰	sl EtOH
2664	Sodium thioantimonate nonahydrate	$Na_3SbS_4 \cdot 9H_2O$	10101-91-4	481.127	yel cry	dec 108		1.8	28²⁰	i EtOH
2665	Sodium thiocyanate	$NaSCN$	540-72-7	81.073	col hyg cry	287			151²⁵	
2666	Sodium thiophosphate dodecahydrate	$Na_3PO_3S \cdot 12H_2O$	10101-88-9	396.190	hex hyg leaflets	60				vs hot H_2O
2667	Sodium thiosulfate	$Na_2S_2O_3$	7772-98-7	158.108	col monocl cry	100 dec		1.69	76.4²⁵	i EtOH
2668	Sodium thiosulfate pentahydrate	$Na_2S_2O_3 \cdot 5H_2O$	10102-17-7	248.184	col cry	≈50 dec		1.69	76.4²⁵	i EtOH
2669	Sodium trimetaphosphate	$Na_3(PO_3)_3$	7785-84-4	305.885	wh cry			2.49	22	
2670	Sodium trimetaphosphate hexahydrate	$Na_3(PO_3)_3 \cdot 6H_2O$	7785-84-4	413.976	tricl-rhom hyg prisms	53		1.786	22	i EtOH
2671	Sodium tripolyphosphate	$Na_5P_3O_{10}$	7758-29-4	367.864	wh hyg powder	622			20²⁵	
2672	Sodium tungstate	Na_2WO_4	13472-45-2	293.82	wh rhom cry	695		4.18	74.2²⁵	
2673	Sodium tungstate dihydrate	$Na_2WO_4 \cdot 2H_2O$	10213-10-2	329.85	wh orth cry	100 dec		3.25	74.2²⁵	i EtOH
2674	Sodium uranate(VI) monohydrate	$Na_2U_2O_7 \cdot H_2O$	13721-34-1	652.049	yel powder					i H_2O; s acid

No.	Name	Formula	CAS Reg No.	Mol. weight	Physical form	mp/°C	bp/°C	Density g cm⁻³	Solubility g/100 g H₂O	Qualitative solubility
2675	Sodium vanadate(V)	$NaVO_3$	13718-26-8	121.930	col monocl prisms	630			21^{25}	
2676	Sodium vanadate(V) tetrahydrate	$NaVO_3 \cdot 4H_2O$	13718-26-8	193.992	yel-wh cry powder				21^{25}	
2677	Strontium	Sr	7440-24-6	87.62	silv-wh metal; cub	777	1377	2.64		reac H_2O; s EtOH
2678	Strontium acetate	$Sr(C_2H_3O_2)_2$	543-94-2	205.71	col hyg cry	dec		2.1	40^{25}	vs H_2O
2679	Strontium arsenite tetrahydrate	$Sr(AsO_2)_2 \cdot 4H_2O$	10378-48-0	373.52	wh powder					sl H_2O, EtOH; sol dil acid
2680	Strontium bromate monohydrate	$Sr(BrO_3)_2 \cdot H_2O$	14519-18-7	361.44	yel hyg monocl cry	120 dec		3.773	39.0^{25}	
2681	Strontium bromide	$SrBr_2$	10476-81-0	247.43	wh tetr cry	657		4.216	107^{25}	
2682	Strontium bromide hexahydrate	$SrBr_2 \cdot 6H_2O$	7789-53-9	355.52	col hyg cry	88 dec			107^{25}	s EtOH; i eth
2683	Strontium carbide	SrC_2	12071-29-3	111.64	blk tetr cry	>1700		3.19		i H_2O
2684	Strontium carbonate	$SrCO_3$	1633-05-2	147.63	wh orth cry; hyg	1494		3.785	0.00034^{20}	s dil acid
2685	Strontium chlorate	$Sr(ClO_3)_2$	7791-10-8	254.52	col cry	120 dec		3.15	176^{25}	sl EtOH
2686	Strontium chloride	$SrCl_2$	10476-85-4	158.53	wh cub cry; hyg	874	1250	3.052	54.7^{25}	
2687	Strontium chloride hexahydrate	$SrCl_2 \cdot 6H_2O$	10025-70-4	266.62	col hyg cry	100 dec		1.96	54.7^{25}	s EtOH
2688	Strontium chromate	$SrCrO_4$	7789-06-2	203.61	yel monocl cry	dec		3.9	0.106^{20}	s dil acid
2689	Strontium cyanide dihydrate	$Sr(CN)_2 \cdot 4H_2O$	52870-08-3	211.72	wh hyg cry	dec				vs H_2O
2690	Strontium ferrocyanide pentadecahydrate	$SrFe(CN)_6 \cdot 15H_2O$	14654-44-5	569.80	yel monocl cry				50	
2691	Strontium fluoride	SrF_2	7783-48-4	125.62	wh cub cry or powder	1477	2460	4.24	0.021^{25}	s dil acid
2692	Strontium formate	$Sr(CHO_2)_2$	592-89-2	177.66	wh cry	71.9		2.693	9.1^0	
2693	Strontium formate dihydrate	$Sr(CHO_2)_2 \cdot 2H_2O$	6160-34-5	213.69	col rhom cry	100 dec		2.25	9.1^{37}	i EtOH, eth
2694	Strontium hexaboride	SrB_6	12046-54-7	152.49	blk cub cry	2235		3.39		i H_2O; s HNO_3
2695	Strontium hydride	SrH_2	13598-33-9	89.64	orth cry	1050		3.26		reac H_2O
2696	Strontium hydroxide	$Sr(OH)_2$	18480-07-4	121.64	col orth cry; hyg	535	710 dec	3.625	2.25^{25}	
2697	Strontium iodate	$Sr(IO_3)_2$	13470-01-4	437.43	tricl cry			5.045	0.165^{25}	
2698	Strontium iodide	SrI_2	10476-86-5	341.43	wh hyg cry	538	1773 dec	4.55	177^{25}	
2699	Strontium iodide hexahydrate	$SrI_2 \cdot 6H_2O$	73796-25-5	449.52	wh-yel hex cry; hyg	120 dec		4.4	177^{25}	s EtOH
2700	Strontium molybdate	$SrMoO_4$	13470-04-7	247.56	wh cry pow	1040		4.54		i H_2O
2701	Strontium niobate	$SrNb_2O_6$	12034-89-8	369.43	monocl cry	1225		5.11		i H_2O
2702	Strontium nitrate	$Sr(NO_3)_2$	10042-76-9	211.63	wh cub cry	570		2.99	80.2^{25}	sl EtOH, ace
2703	Strontium nitride	Sr_3N_2	12033-82-8	290.87	refrac solid	1200				reac H_2O; s HCl
2704	Strontium nitrite	$Sr(NO_2)_2$	13470-06-9	179.63	wh-yel hyg needles	240 dec		2.8	72.1^{30}	s H_2O
2705	Strontium orthosilicate	Sr_2SiO_4	13597-55-2	267.32	orth cry			4.5		
2706	Strontium oxalate monohydrate	$SrC_2O_4 \cdot H_2O$	814-95-9	193.65	cry pow	dec 150			0.00005^{20}	sl dil acid
2707	Strontium oxide	SrO	1314-11-0	103.62	col cub cry	2531		5.1		reac H_2O
2708	Strontium perchlorate	$Sr(ClO_4)_2$	13450-97-0	286.52	col hyg cry				306^{25}	s EtOH, MeOH
2709	Strontium permanganate trihydrate	$Sr(MnO_4)_2 \cdot 3H_2O$	14446-13-0	379.54	purp cub cry	175 dec		2.75	250^{18}	
2710	Strontium peroxide	SrO_2	1314-18-7	119.62	wh tetr cry; unstab	215 dec		4.78		reac H_2O
2711	Strontium phosphate	$Sr_3(PO_4)_2$	7446-28-8	452.80	wh powder				0.000011^{20}	s acid
2712	Strontium selenate	$SrSeO_4$	7446-21-1	230.58	orth cry			4.25	0.115^{20}	s hot HCl
2713	Strontium selenide	SrSe	1315-07-7	166.58	wh cub cry	1600		4.54		
2714	Strontium silicide	$SrSi_2$	12138-28-2	143.79	silv-gray cub cry	1100		3.35		
2715	Strontium sulfate	$SrSO_4$	7759-02-6	183.68	wh orth cry	1606		3.96	0.0135^{25}	i EtOH; sl acid
2716	Strontium sulfide	SrS	1314-96-1	119.69	gray cub cry	2226		3.70		sl H_2O; s acid
2717	Strontium sulfite	$SrSO_3$	13451-02-0	167.68	col cry	dec			0.0015^{25}	s H_2SO_4, HCl
2718	Strontium telluride	SrTe	12040-08-3	215.22	wh cub cry			4.83		
2719	Strontium thiosulfate pentahydrate	$SrS_2O_3 \cdot 5H_2O$	15123-90-7	289.82	monocl needles	100 dec		2.17	36.3^{25}	i EtOH
2720	Strontium titanate	$SrTiO_3$	12060-59-2	183.49	wh cub cry	2080		5.1		i H_2O
2721	Strontium tungstate	$SrWO_4$	13451-05-3	335.46	col tetr cry	dec		6.187	0.14^{15}	i EtOH
2722	Strontium zirconate	$SrZrO_3$	12036-39-4	226.84	col cry	2600				
2723	Sulfur (rhombic)	S	7704-34-9	32.06	yel orth cry	95.2 (trans to monocl)	444.61	2.07		i H_2O; sl EtOH, bz, eth; s CS_2
2724	Sulfur (monoclinic)	S	7704-34-9	32.06	yel monocl needles, stable 95.3-120	115.21	444.61	2.00		i H_2O; sl EtOH, bz, eth; s CS_2
2725	Sulfuric acid	H_2SO_4	7664-93-9	98.079	col oily liq	10.31	337	1.8302		vs H_2O
2726	Peroxysulfuric acid	H_2SO_5	7722-86-3	114.078	wh cry; unstab	45 dec				vs H_2O
2727	Nitrosylsulfuric acid	$HNOSO_4$	7782-78-7	127.077	prisms	73 dec				reac H_2O; s H_2SO_4
2728	Chlorosulfonic acid	$SO_2(OH)Cl$	7790-94-5	116.524	col-yel liq	-80	152	1.75		reac H_2O; s py
2729	Fluorosulfonic acid	$SO_2(OH)F$	7789-21-1	100.069	col liq	-89	163	1.726		reac H_2O
2730	Sulfurous acid	H_2SO_3	7782-99-2	82.079	exists only in aq soln					
2731	Sulfamic acid	H_2NSO_3H	5329-14-6	97.094	orth cry	≈205 dec		2.15	14.7^0	sl ace; i eth
2732	Sulfur dioxide	SO_2	7446-09-5	64.064	col gas	-75.5	-10.05	2.619 g/L		s H_2O, EtOH, eth, chl

No.	Name	Formula	CAS Reg No.	Mol. weight	Physical form	mp/°C	bp/°C	Density g cm⁻³	Solubility g/100 g H₂O	Qualitative solubility
2733	Sulfur trioxide	SO_3	7446-11-9	80.063	wh needles	62.2	subl			reac H_2O
2734	Sulfur trioxide (γ-form)	SO_3	7446-11-9	80.063	col solid or liq	16.8	44.5	1.90		reac H_2O
2735	Sulfur trioxide (β-form)	SO_3	7446-11-9	80.063	wh needles	30.5	44.5			reac H_2O
2736	Sulfur bromide ($SSBr_2$)	$SSBr_2$	13172-31-1	223.938	red oily liq	-46	>25 dec	2.63		reac H_2O
2737	Sulfur chloride ($SSCl_2$)	$SSCl_2$	10025-67-9	135.036	yel-red oily liq	-77	137	1.69		reac H_2O; s EtOH, bz, eth, ctc
2738	Sulfur fluoride (SSF_2)	SSF_2	16860-99-4	102.127	col gas	-164.6	-10.6	4.174 g/L		reac H_2O
2739	Sulfur fluoride (FSSF)	FSSF	13709-35-8	102.127	col gas	-133	15	4.174 g/L		reac H_2O
2740	Sulfur dichloride	SCl_2	10545-99-0	102.971	red visc liq	-122	59.6	1.62		reac H_2O
2741	Sulfur tetrafluoride	SF_4	7783-60-0	108.059	col gas	-125	-40.45	4.417 g/L		reac H_2O
2742	Sulfur hexafluoride	SF_6	2551-62-4	146.055	col gas	-49.596 tp	-63.8 sp	5.970 g/L		sl H_2O; s EtOH
2743	Sulfur bromide pentafluoride	SF_5Br	15607-89-3	206.961	col gas	-79	3.1	8.459 g/L		
2744	Sulfur chloride pentafluoride	SF_5Cl	13780-57-9	162.510	col gas	-64	-19.05	6.642 g/L		
2745	Sulfur decafluoride	S_2F_{10}	5714-22-7	254.114	liq	-52.7	30; dec 150	2.08		i H_2O
2746	Sulfuryl amide	$(NH_2)_2SO_2$	7803-58-9	96.109	orth plates	93	250 dec			vs H_2O; sl EtOH
2747	Sulfuryl chloride	SO_2Cl_2	7791-25-5	134.970	col liq	-51	69.4	1.680		reac H_2O; s bz, tol, eth
2748	Sulfuryl fluoride	SO_2F_2	2699-79-8	102.061	col gas	-135.8	-55.4	4.172 g/L		sl H_2O, EtOH; s tol, ctc
2749	Sulfuryl bromide fluoride	SO_2BrF	13536-61-3	162.966	col liq	-86	41			reac H_2O
2750	Sulfuryl chloride fluoride	SO_2ClF	13637-84-8	118.515	col gas	-124.7	7.1	1.62⁰		reac H_2O
2751	Pyrosulfuryl chloride	$S_2O_5Cl_2$	7791-27-7	215.033	col fuming liq	-37	151	1.837		reac H_2O
2752	Thionyl bromide	$SOBr_2$	507-16-4	207.872	yel liq	-50	140			reac H_2O
2753	Thionyl chloride	$SOCl_2$	7719-09-7	118.970	yel fuming liq	-101	75.6	1.631		reac H_2O; s bz, ctc, chl
2754	Thionyl fluoride	SOF_2	7783-42-8	86.061	col gas	-129.5	-43.8	3.518 g/L		reac H_2O; s bz, eth
2755	Sulfur fluoride oxide (SOF_4)	SOF_4	13709-54-1	124.058	col gas	-99.6	-48.5	1.95⁻⁸²		reac H_2O
2756	Sulfur fluoride hypofluorite	F_5SOF	15179-32-5	162.054	col gas	-86	-35.1	6.624 g/L		
2757	Tetrasulfur tetranitride	S_4N_4	28950-34-7	184.287	yel-oran cry	178.2	subl			i H_2O; reac alk, acid
2758	Tantalum	Ta	7440-25-7	180.948	gray metal; cub	3017	5455	16.4		reac HF
2759	Tantalum aluminide	$TaAl_3$	12004-76-1	261.893	gray refrac powder	≈1400		7.02		i H_2O, acid, alk
2760	Tantalum boride (TaB)	TaB	12007-07-7	191.759	refrac orth cry	2040		14.2		
2761	Tantalum boride (TaB_2)	TaB_2	12007-35-1	202.570	blk hex cry	3100		11.2		i H_2O, acid, alk
2762	Tantalum carbide (TaC)	TaC	12070-06-3	192.959	gold-brown powder; cub	3880	4780	14.3		s HF-HNO_3 mixture
2763	Tantalum carbide (Ta_2C)	Ta_2C	12070-07-4	373.907	refrac hex cry	3327		15.1		
2764	Tantalum hydride	TaH	13981-95-8	181.956	gray metallic solid			15.1		i acid
2765	Tantalum nitride	TaN	12033-62-4	194.955	blk hex cry	3090		13.7		i H_2O; sl aqua regia; reac alk
2766	Tantalum silicide	$TaSi_2$	12039-79-1	237.119	gray powder	2200		9.14		
2767	Tantalum(III) bromide	$TaBr_3$	13842-73-4	420.660	gray-grn solid	dec 220				
2768	Tantalum(III) chloride	$TaCl_3$	13569-67-0	287.307	blk-grn solid	dec 440				s H_2O
2769	Tantalum(IV) bromide	$TaBr_4$	13842-76-7	500.564	dark blue solid	392		5.77		reac H_2O
2770	Tantalum(IV) chloride	$TaCl_4$	13569-72-7	322.760	dark grn solid	dec 300	subl	4.35		reac H_2O
2771	Tantalum(IV) iodide	TaI_4	14693-80-2	688.566	gray-blk solid	400 dec				reac H_2O
2772	Tantalum(IV) oxide	TaO_2	12036-14-5	212.947	tetr cry			10.0		
2773	Tantalum(IV) selenide	$TaSe_2$	12039-55-3	338.87	hex cry			6.7		
2774	Tantalum(IV) sulfide	TaS_2	12143-72-5	245.078	blk hex cry	>3000		6.86		i H_2O
2775	Tantalum(IV) telluride	$TaTe_2$	12067-66-2	436.15	monocl cry			9.4		
2776	Tantalum(V) bromide	$TaBr_5$	13451-11-1	580.468	yel cry powder	265.8	348.8	4.67		
2777	Tantalum(V) chloride	$TaCl_5$	7721-01-9	358.213	yel-wh monocl cry; hyg	216.6	239	3.68		reac H_2O; s EtOH
2778	Tantalum(V) fluoride	TaF_5	7783-71-3	275.940	wh monocl cry; hyg	96.9	229.5	4.74		s H_2O, eth; sl CS_2, ctc
2779	Tantalum(V) iodide	TaI_5	14693-81-3	815.470	blk hex cry; hyg	496	543	5.80		reac H_2O
2780	Tantalum(V) oxide	Ta_2O_5	1314-61-0	441.893	wh rhomb cry or powder	1875		8.24		i H_2O, EtOH, acid; s HF
2781	Technetium	Tc	7440-26-8	98	hex cry	2157	4262	11		
2782	Technetium(V) fluoride	TcF_5	31052-14-9	193	yel solid	50	dec			
2783	Technetium(VI) fluoride	TcF_6	13842-93-8	212	yel cub cry	37.4	55.3	3.0		
2784	Tellurium	Te	13494-80-9	127.60	gray-wh rhomb cry	449.51	988	6.232		i H_2O, bz, CS_2
2785	Telluric(VI) acid	H_6TeO_6	7803-68-1	229.64	wh monocl cry	136		3.07	50.1³⁰	
2786	Tellurous acid	H_2TeO_3	10049-23-7	177.61	wh cry	40 dec		3.0		sl H_2O; s dil acid, alk
2787	Tellurium dioxide	TeO_2	7446-07-3	159.60	wh orth cry	733	1245	5.9		i H_2O; s alk, acid
2788	Tellurium trioxide	TeO_3	13451-18-8	175.60	yel-oran cry	430		5.07		i H_2O

No.	Name	Formula	CAS Reg No.	Mol. weight	Physical form	mp/°C	bp/°C	Density g cm⁻³	Solubility g/100 g H₂O	Qualitative solubility
2789	Tellurium dibromide	$TeBr_2$	7789-54-0	287.41	grn-brn hyg cry	210	339			reac H₂O; s eth; sl chl
2790	Tellurium dichloride	$TeCl_2$	10025-71-5	198.51	blk amorp solid; hyg	208	328	6.9		reac H₂O; i ctc
2791	Tellurium tetrabromide	$TeBr_4$	10031-27-3	447.22	yel-oran monocl cry	380	≈420 dec	4.3		reac H₂O; s eth
2792	Tellurium tetrachloride	$TeCl_4$	10026-07-0	269.41	wh monocl cry; hyg	224	387	3.0		reac H₂O; s EtOH, tol
2793	Tellurium tetrafluoride	TeF_4	15192-26-4	203.59	col cry	129	195 dec			reac H₂O
2794	Tellurium decafluoride	Te_2F_{10}	53214-07-6	445.18	col liq	-33.7	59			
2795	Tellurium tetraiodide	TeI_4	7790-48-9	635.22	blk orth cry	280		5.05		reac H₂O; sl ace
2796	Tellurium hexafluoride	TeF_6	7783-80-4	241.59	col gas	-37.6 tp	-38.9 sp	9.875 g/L		reac H₂O
2797	Terbium	Tb	7440-27-9	158.925	silv metal; hex	1359	3230	8.23		
2798	Terbium nitride	TbN	12033-64-6	172.932	cub cry			9.55		
2799	Terbium silicide	$TbSi_2$	12039-80-4	215.096	orth cry			6.66		
2800	Terbium(III) bromide	$TbBr_3$	14456-47-4	398.637	wh hex cry	830	1490			s H₂O
2801	Terbium(III) chloride	$TbCl_3$	10042-88-3	265.284	wh orth cry; hyg	582		4.35		s H₂O
2802	Terbium(III) chloride hexahydrate	$TbCl_3 \cdot 6H_2O$	13798-24-8	373.375	hyg cry			4.35		vs H₂O
2803	Terbium(III) iodide	TbI_3	13813-40-6	539.638	hex cry; hyg	955		≈5.2		s H₂O
2804	Terbium(III) nitrate	$Tb(NO_3)_3$	10043-27-3	344.940	pink hyg solid				157[25]	s EtOH
2805	Terbium(III) nitrate hexahydrate	$Tb(NO_3)_3 \cdot 6H_2O$	13451-19-9	453.031	col needles	89				s H₂O, EtOH, ace
2806	Terbium(III) oxide	Tb_2O_3	12036-41-8	365.849	wh cub cry	2303		7.91		
2807	Terbium(III) sulfate octahydrate	$Tb_2(SO_4)_3 \cdot 8H_2O$	13842-67-6	750.161	wh cry	dec 360				sl H₂O
2808	Terbium(III) sulfide	Tb_2S_3	12138-11-3	414.046	cub cry			6.35		
2809	Terbium(III) fluoride	TbF_3	13708-63-9	215.920	wh solid	1175	2280			i H₂O
2810	Terbium(IV) fluoride	TbF_4	36781-15-4	234.919	wh monocl cry	dec 300				i H₂O
2811	Thallium	Tl	7440-28-0	204.38	soft blue-wh metal	304	1473	11.8		i H₂O; reac acid
2812	Thallium(I) acetate	$TlC_2H_3O_2$	563-68-8	263.427	hyg wh cry	131		3.68		s H₂O, EtOH
2813	Thallium(I) azide	TlN_3	13847-66-0	246.403	yel cry	334	exp			s H₂O
2814	Thallium(I) bromate	$TlBrO_3$	14550-84-6	332.285	col needles	120 dec			0.49[30]	s EtOH
2815	Thallium(I) bromide	TlBr	7789-40-4	284.287	yel cub cry	460	819	7.5	0.059[20]	
2816	Thallium(I) carbonate	Tl_2CO_3	6533-73-9	468.776	wh monocl cry	273		7.11	4.69[20]	i EtOH
2817	Thallium(I) chlorate	$TlClO_3$	13453-30-0	287.834	col hex cry	dec 500		5.5	3.92[20]	
2818	Thallium(I) chloride	TlCl	7791-12-0	239.836	wh cub cry	431	720	7.0	0.33[20]	i EtOH
2819	Thallium(I) chromate	Tl_2CrO_4	13473-75-1	524.761	yel cry				0.003[20]	sl acid, alk
2820	Thallium(I) cyanide	TlCN	13453-34-4	230.401	wh hex plates			6.523		s H₂O, acid, EtOH
2821	Thallium(I) ethanolate	TlC_2H_5O	20398-06-5	249.443	cloudy liq	-3	130 dec	3.49		reac H₂O
2822	Thallium(I) fluoride	TlF	7789-27-7	223.381	wh orth cry	326	826	8.36	245[25]	
2823	Thallium(I) formate	$TlCHO_2$	992-98-3	249.401	hyg col needles	101		4.97		vs H₂O; s MeOH
2824	Thallium(I) hexafluorophosphate	$TlPF_6$	60969-19-9	349.347	wh cub cry			4.6		
2825	Thallium(I) hydroxide	TlOH	12026-06-1	221.390	yel needles	139 dec		7.44	34.3[18]	
2826	Thallium(I) iodate	$TlIO_3$	14767-09-0	379.285	wh needles				0.058	sl HNO₃
2827	Thallium(I) iodide	TlI	7790-30-9	331.287	yel cry powder	441.7	824	7.1	0.0085[20]	i EtOH
2828	Thallium(I) molybdate	Tl_2MoO_4	34128-09-1	568.71	yel-wh cub cry					i H₂O
2829	Thallium(I) nitrate	$TlNO_3$	10102-45-1	266.388	wh cry	206	450 dec	5.55	9.55[20]	s H₂O; i EtOH
2830	Thallium(I) nitrite	$TlNO_2$	13826-63-6	250.389	yel cub cry	186		5.7	32.1[25]	s H₂O
2831	Thallium(I) oxalate	$Tl_2C_2O_4$	30737-24-7	496.786	wh powder			6.31	1.83[20]	
2832	Thallium(I) oxide	Tl_2O	1314-12-1	424.766	blk rhomb cry; hyg	579	≈1080	9.52		s H₂O, EtOH
2833	Thallium(I) perchlorate	$TlClO_4$	13453-40-2	303.834	col orth cry	501		4.89	19.7[30]	
2834	Thallium(I) selenate	Tl_2SeO_4	7446-22-2	551.73	orth cry	>400		6.875	2.8[20]	i EtOH, eth
2835	Thallium(I) selenide	Tl_2Se	15572-25-5	487.73	gray plates	340				i H₂O, acid
2836	Thallium(I) sulfate	Tl_2SO_4	7446-18-6	504.830	wh rhomb prisms	632		6.77	5.47[25]	
2837	Thallium(I) sulfide	Tl_2S	1314-97-2	440.832	blue-blk cry	457	1367	8.39	0.02[20]	sl alk; s acid
2838	Thallium(III) acetate	$Tl(C_2H_3O_2)_3$	2570-63-0	381.514	hyg wh platelets	182 dec				
2839	Thallium(III) bromide tetrahydrate	$TlBr_3 \cdot 4H_2O$	13701-90-1	516.157	yel orth cry			3.65		s H₂O, EtOH
2840	Thallium(III) chloride	$TlCl_3$	13453-32-2	310.742	monocl cry	155		4.7		vs H₂O, EtOH, eth
2841	Thallium(III) chloride tetrahydrate	$TlCl_3 \cdot 4H_2O$	13453-32-2*	382.804	orth cry			3.00		s H₂O
2842	Thallium(III) fluoride	TlF_3	7783-57-5	261.378	wh orth cry; hyg	550 dec		8.65		reac H₂O
2843	Thallium(III) nitrate	$Tl(NO_3)_3$	13746-98-0	390.398	col cry					reac H₂O
2844	Thallium(III) oxide	Tl_2O_3	1314-32-5	456.765	brn cub cry	834		10.2		i H₂O; reac acid
2845	Thallium(III) sulfate	$Tl_2(SO_4)_3$	16222-66-5	696.955	col leaflets					reac H₂O
2846	Thallium selenide	TlSe	12039-52-0	283.34	blk solid	330				i H₂O, acid
2847	Thorium	Th	7440-29-1	232.038	soft gray-wh metal; cub	1750	4785	11.7		s acid
2848	Thorium hydride	ThH_2	16689-88-6	234.054	tetr cry			9.5		
2849	Thorium boride	ThB_6	12229-63-9	296.904	refrac solid	2450		6.99		
2850	Thorium(IV) bromide	$ThBr_4$	13453-49-1	551.654	wh hyg cry	679			65[20]	

No.	Name	Formula	CAS Reg No.	Mol. weight	Physical form	mp/°C	bp/°C	Density g cm⁻³	Solubility g/100 g H₂O	Qualitative solubility
2851	Thorium carbide	ThC	12012-16-7	244.049	cub cry	2500		10.6		reac H_2O
2852	Thorium dicarbide	ThC_2	12071-31-7	256.059	yel monocl cry	≈2650		9.0		reac H_2O
2853	Thorium(IV) chloride	$ThCl_4$	10026-08-1	373.850	gray-wh tetr needles; hyg	770	921	4.59		s H_2O, EtOH
2854	Thorium(IV) fluoride	ThF_4	13709-59-6	308.032	wh monocl cry; hyg	1110	1680	6.1		
2855	Thorium(IV) iodide	ThI_4	7790-49-0	739.656	wh-yel monocl cry	566	837			
2856	Thorium(IV) nitrate	$Th(NO_3)_4$	13823-29-5	480.058	hyg wh plates	55 dec				vs H_2O, EtOH
2857	Thorium(IV) nitrate tetrahydrate	$Th(NO_3)_4 \cdot 4H_2O$	13470-07-0	552.119	wh hyg cry	500 dec			191²⁰	s EtOH
2858	Thorium nitride	ThN	12033-65-7	246.045	refrac cub cry	2820		11.6		reac H_2O
2859	Thorium(IV) oxide	ThO_2	1314-20-1	264.037	wh cub cry	3350	4400	10.0		i H_2O, alk; sl acid
2860	Thorium(IV) selenide	$ThSe_2$	60763-24-8	389.96	orth cry			8.5		
2861	Thorium orthosilicate	$ThSiO_4$	14553-44-7	324.122	brn tetr cry			6.7		
2862	Thorium silicide	$ThSi_2$	12067-54-8	288.209	tetr cry	1850		7.9		
2863	Thorium(IV) sulfate nonahydrate	$Th(SO_4)_2 \cdot 9H_2O$	10381-37-0	586.301	wh monocl cry	dec		2.8	4.2²⁰	
2864	Thorium(IV) sulfide	ThS_2	12138-07-7	296.168	dark brn cry	1905		7.30		i H_2O; s acid
2865	Thulium	Tm	7440-30-4	168.934	silv metal; hex	1545	1950	9.32		s dil acid
2866	Thulium(II) bromide	$TmBr_2$	64171-97-7	328.742	dark grn solid	619				
2867	Thulium(II) chloride	$TmCl_2$	22852-11-5	239.840	red or grn cry	718				reac H_2O
2868	Thulium(II) iodide	TmI_2	60864-26-8	422.743	blk hyg solid	756				reac H_2O
2869	Thulium(III) bromide	$TmBr_3$	14456-51-0	408.646	wh hyg cry	954				s H_2O
2870	Thulium(III) chloride	$TmCl_3$	13537-18-3	275.293	yel hyg cry	845				s H_2O
2871	Thulium(III) chloride heptahydrate	$TmCl_3 \cdot 7H_2O$	13778-39-7	401.400	hyg cry					s H_2O, EtOH
2872	Thulium(III) fluoride	TmF_3	13760-79-7	225.929	wh cry	1158				s H_2O
2873	Thulium(III) hydroxide	$Tm(OH)_3$	1311-33-7	219.956	wh or grn prec					i H_2O
2874	Thulium(III) iodide	TmI_3	13813-43-9	549.647	yel hyg cry	1021				
2875	Thulium(III) nitrate	$Tm(NO_3)_3$	14985-19-4	354.949	grn hyg solid				212²⁵	s EtOH
2876	Thulium(III) nitrate pentahydrate	$Tm(NO_3)_3 \cdot 5H_2O$	36548-87-5	445.025	grn hyg cry					s H_2O, EtOH, ace
2877	Thulium(III) oxalate hexahydrate	$Tm_2(C_2O_4)_3 \cdot 6H_2O$	26677-68-9	710.016	grn solid	dec 50				s alk oxalates
2878	Thulium(III) oxide	Tm_2O_3	12036-44-1	385.866	grn-wh cub cry	2341	3945	8.6		sl acid
2879	Tin (gray)	Sn	7440-31-5	118.710	cub cry	trans to wh Sn 13.2	2586	5.769		
2880	Tin (white)	Sn	7440-31-5	118.710	silv tetr cry	231.928	2586	7.287		
2881	Stannane	SnH_4	2406-52-2	122.742	unstab col gas	-146	-51.8	5.017 g/L		
2882	Methylstannane	SnH_3CH_3	1631-78-3	136.769	col gas		1.4	5.590 g/L		reac H_2O
2883	(Dimethylamino)trimethystannane	$Sn(CH_3)_3N(CH_3)_2$	993-50-0	207.890	liq	1	126	1.22		reac H_2O
2884	Tin monophosphide	SnP	25324-56-5	149.684	dull metallic solid	540				
2885	Tin triphosphide	Sn_4P_3	12286-33-8	567.761	wh cry	≈550		5.2		
2886	Tin(II) acetate	$Sn(C_2H_3O_2)_2$	638-39-1	236.799	wh orth cry	183	subl	2.31		s dil HCl
2887	Tin(II) bromide	$SnBr_2$	10031-24-0	278.518	yel powder	215	639	5.12	85⁰	s EtOH, eth, ace
2888	Tin(II) chloride	$SnCl_2$	7772-99-8	189.616	wh orth cry	247.0	623	3.90	178¹⁰	s EtOH, ace, eth; i xyl
2889	Tin(II) chloride dihydrate	$SnCl_2 \cdot 2H_2O$	10025-69-1	225.647	wh monocl cry	37 dec		2.71	178¹⁰	s EtOH, NaOH; vs HCl
2890	Tin(II) fluoride	SnF_2	7783-47-3	156.707	wh monocl cry; hyg	215	850	4.57		s H_2O; i EtOH, eth, chl
2891	Tin(II) hexafluorozirconate	$SnZrF_6$	12419-43-1	323.924	cry			4.21		s H_2O
2892	Tin(II) hydroxide	$Sn(OH)_2$	12026-24-3	152.725	wh amorp solid					
2893	Tin(II) iodide	SnI_2	10294-70-9	372.519	red-oran powder	320	714	5.28	0.98²⁰	s bz, chl, CS_2
2894	Tin(II) oxalate	SnC_2O_4	814-94-8	206.729	wh powder	280 dec		3.56		i H_2O; s dil HCl
2895	Tin(II) oxide	SnO	21651-19-4	134.709	blue-blk tetr cry	1080 dec		6.45		i H_2O, EtOH; s acid
2896	Tin(II) pyrophosphate	$Sn_2P_2O_7$	15578-26-4	411.363	wh amorp powder	400 dec		4.009		i H_2O; s conc acid
2897	Tin(II) selenide	SnSe	1315-06-6	197.67	gray orth cry	861		6.18		i H_2O; s aqua regia
2898	Tin(II) sulfate	$SnSO_4$	7488-55-3	214.773	wh orth cry	378 dec		4.15	18.8¹⁹	
2899	Tin(II) sulfide	SnS	1314-95-0	150.775	gray orth cry	881	1210	5.08		i H_2O; s conc acid
2900	Tin(II) tartrate	$SnC_4H_4O_6$	815-85-0	266.781	wh cry powder					s H_2O, dil HCl
2901	Tin(II) telluride	SnTe	12040-02-7	246.31	gray cub cry	806		6.5		
2902	Tin(IV) bromide	$SnBr_4$	7789-67-5	438.326	wh cry	29.1	205	3.34		vs H_2O; s EtOH
2903	Tin(IV) chloride	$SnCl_4$	7646-78-8	260.522	col fuming liq	-34.07	114.15	2.234		reac H_2O; s EtOH, ctc, bz, ace
2904	Tin(IV) chloride pentahydrate	$SnCl_4 \cdot 5H_2O$	10026-06-9	350.598	wh-yel cry	56 dec		2.04		vs H_2O; s EtOH
2905	Tin(IV) chromate	$Sn(CrO_4)_2$	38455-77-5	350.697	brn-yel cry powder	dec				s H_2O
2906	Tin(IV) fluoride	SnF_4	7783-62-2	194.704	wh tetr cry	442	705 subl	4.78		reac H_2O
2907	Tin(IV) iodide	SnI_4	7790-47-8	626.328	yel-brn cub cry	143	364.35	4.46		reac H_2O; s EtOH, bz, chl, eth
2908	Tin(IV) oxide	SnO_2	18282-10-5	150.709	gray tetr cry	1630		6.85		i H_2O, EtOH; s hot conc alk
2909	Tin(IV) selenide	$SnSe_2$	20770-09-6	276.63	red-brn cry	650		≈5.0		i H_2O; s alk, conc acid

No.	Name	Formula	CAS Reg No.	Mol. weight	Physical form	mp/°C	bp/°C	Density g cm⁻³	Solubility g/100 g H₂O	Qualitative solubility
2910	Tin(IV) selenite	$Sn(SeO_3)_2$	7446-25-5	372.63	cry powder					i H₂O; s hot HCl
2911	Tin(IV) sulfide	SnS_2	1315-01-1	182.840	gold-yel hex cry	600 dec		4.5		i H₂O; s alk, aqua regia
2912	Titanium	Ti	7440-32-6	47.867	gray metal; hex	1670	3287	4.506		
2913	Titanocene dichloride	$Ti(C_5H_5)_2Cl_2$	1271-19-8	248.959	red cry	289		1.60		sl H₂O, bz; s chl, EtOH, tol
2914	Titanium hydride	TiH_2	7704-98-5	49.883	gray-blk powder	≈450 dec		3.75		i H₂O
2915	Titanium boride	TiB_2	12045-63-5	69.489	gray refrac solid; hex	3225		4.38		
2916	Titanium carbide	TiC	12070-08-5	59.878	cub cry	3067		4.93		i H₂O; s HNO₃
2917	Titanium nitride	TiN	25583-20-4	61.874	yel-brn cub cry	2947		5.21		i H₂O; s aqua regia
2918	Titanium phosphide	TiP	12037-65-9	78.841	gray hex cry	1990		4.08		
2919	Titanium silicide	$TiSi_2$	12039-83-7	104.038	blk orth cry	1500		4.0		i H₂O, acid, alk; s HF
2920	Titanium(II) bromide	$TiBr_2$	13783-04-5	207.675	blk powder	dec 400		4.0		reac H₂O
2921	Titanium(II) chloride	$TiCl_2$	10049-06-6	118.773	blk hex cryc	1035	1500	3.13		reac H₂O; s EtOH; i chl, eth
2922	Titanium(II) iodide	TiI_2	13783-07-8	301.676	blk hex cry	dec 400		5.02		reac H₂O
2923	Titanium(II) oxide	TiO	12137-20-1	63.866	yel cub cry	1770	3227	4.95		
2924	Titanium(II) sulfide	TiS	12039-07-5	79.932	brn hex cry	1927		3.85		i H₂O; s conc acid
2925	Titanium(III) bromide	$TiBr_3$	13135-31-4	287.579	viol hex cry	dec 400				s H₂O
2926	Titanium(III) chloride	$TiCl_3$	7705-07-9	154.226	red-viol hex cry; hyg	425 dec	960	2.64		reac H₂O
2927	Titanium(III) fluoride	TiF_3	13470-08-1	104.862	viol hex cry	950 dec		2.98		i H₂O, dil acid, alk
2928	Titanium(III) iodide	TiI_3	13783-08-9	428.580	viol cry	dec 350				
2929	Titanium(III) oxide	Ti_2O_3	1344-54-3	143.732	blk hex cry	1842		4.486		s hot HF
2930	Titanium(III) sulfate	$Ti_2(SO_4)_3$	10343-61-0	383.922	grn cry					i H₂O, EtOH; s dil HCl
2931	Titanium(III) sulfide	Ti_2S_3	12039-16-6	191.929	blk hex cry			3.56		
2932	Titanium(III,IV) oxide	Ti_3O_5	12065-65-5	223.598	blk monocl cry	1777		4.24		
2933	Titanium(IV) bromide	$TiBr_4$	7789-68-6	367.483	yel-oran cub cry; hyg	38.3	233.5	3.37		reac H₂O
2934	Titanium(IV) chloride	$TiCl_4$	7550-45-0	189.679	col or yel liq	-24.12	136.45	1.73		reac H₂O; s EtOH
2935	Titanium(IV) fluoride	TiF_4	7783-63-3	123.861	wh hyg powder	377	subl 284	2.798		reac H₂O; s EtOH, py
2936	Titanium(IV) iodide	TiI_4	7720-83-4	555.485	red hyg powder	155	377	4.3		reac H₂O
2937	Titanium(IV) oxide (anatase)	TiO_2	1317-70-0	79.866	brn tetr cry	1560		3.9		
2938	Titanium(IV) oxide (brookite)	TiO_2	12188-41-9	79.866	wh orth cry			4.17		
2939	Titanium(IV) oxide (rutile)	TiO_2	1317-80-2	79.866	wh tetr cry	1843	≈3000	4.17		i H₂O, dil acid; s conc acid
2940	Titanium(IV) oxysulfate monohydrate	$TiOSO_4 \cdot H_2O$	13825-74-6*	177.944	col orth cry			2.71		reac H₂O
2941	Titanium(IV) sulfate	$Ti(SO_4)_2$	13693-11-3	239.992	wh-yel hyg cry	150 dec				s H₂O
2942	Titanium(IV) sulfide	TiS_2	12039-13-3	111.997	yel-brn hex cry; hyg			3.37		s H₂SO₄
2943	Tungsten	W	7440-33-7	183.84	gray-wh metal; cub	3414	5555	19.3		
2944	Tungstic acid	H_2WO_4	7783-03-1	249.85	yel amorp powder	100 dec		5.5		i H₂O, acid; s alk
2945	Tungsten boride (W_2B)	W_2B	12007-10-2	378.49	refrac blk powder	2670		16.0		i H₂O
2946	Tungsten boride (WB)	WB	12007-09-9	194.65	blk refrac powder	2665		15.2		i H₂O
2947	Tungsten boride (W_2B_5)	W_2B_5	12007-98-6	421.74	refrac solid	2370		11.0		i H₂O
2948	Tungsten carbide (W_2C)	W_2C	12070-13-2	379.69	refrac hex cry	≈2800		14.8		i H₂O
2949	Tungsten carbide (WC)	WC	12070-12-1	195.85	gray hex cry	2785		15.6		i H₂O; s HNO₃/HF
2950	Tungsten carbonyl	$W(CO)_6$	14040-11-0	351.90	wh cry	170 dec	subl	2.65		i H₂O; s os
2951	Tungsten nitride (WN_2)	WN_2	60922-26-1	211.85	hex cry	600 dec		7.7		
2952	Tungsten nitride (W_2N)	W_2N	12033-72-6	381.69	gray cub cry	dec		17.8		
2953	Tungsten silicide (WSi_2)	WSi_2	12039-88-2	240.01	blue-gray tetr cry	2160		9.3		i H₂O
2954	Tungsten silicide (W_5Si_3)	W_5Si_3	12039-95-1	1003.46	blue-gray refrac solid	2320		14.4		
2955	Tungsten(II) bromide	WBr_2	13470-10-5	343.65	yel powder	dec 400				i H₂O
2956	Tungsten(II) chloride	WCl_2	13470-12-7	254.75	gray solid	dec 500		5.44		sl H₂O
2957	Tungsten(II) iodide	WI_2	13470-17-2	437.65	oran-brn cry	dec 800		6.79		i H₂O
2958	Tungsten(III) bromide	WBr_3	15163-24-3	423.55	blk hex cry	dec 180				i H₂O
2959	Tungsten(III) chloride	WCl_3	20193-56-0	290.20	red solid	550 dec	subl			reac H₂O
2960	Tungsten(III) iodide	WI_3	15513-69-6	564.55	blk solid	dec r.t.				i H₂O; s ace; sl EtOH, chl
2961	Tungsten(IV) bromide	WBr_4	14055-81-3	503.46	blk orth cry		240 subl			reac H₂O
2962	Tungsten(IV) chloride	WCl_4	13470-13-8	325.65	blk hyg powder	450 dec		4.62		reac H₂O
2963	Tungsten(IV) fluoride	WF_4	13766-47-7	259.83	red-brn cry	dec 800				reac H₂O; s MeCN; i bz, tol, ctc

No.	Name	Formula	CAS Reg No.	Mol. weight	Physical form	mp/°C	bp/°C	Density g cm⁻³	Solubility g/100 g H₂O	Qualitative solubility
2964	Tungsten(IV) iodide	WI_4	14055-84-6	691.46	blk cry	dec				reac H_2O; s EtOH; i eth chl
2965	Tungsten(IV) oxide	WO_2	12036-22-5	215.84	brn monocl cry	≈1500 dec	1730	10.8		i H_2O, os
2966	Tungsten(IV) selenide	WSe_2	12067-46-8	341.76	gray hex cry			9.2		
2967	Tungsten(IV) sulfide	WS_2	12138-09-9	247.97	gray hex cry	1250 dec		7.6		i H_2O, HCl, alk
2968	Tungsten(IV) telluride	WTe_2	12067-76-4	439.04	gray orth cry	1020		9.43		
2969	Tungsten(V) bromide	WBr_5	13470-11-6	583.36	brn-blk hyg solid	286	333			reac H_2O
2970	Tungsten(V) chloride	WCl_5	13470-14-9	361.11	blk-grn hyg cry	253	286	3.88		reac H_2O
2971	Tungsten(V) ethanolate	$W(C_2H_5O)_5$	62571-53-3	409.14	powder		105(0.05 mmHg)			s EtAc
2972	Tungsten(V) fluoride	WF_5	19357-83-6	278.83	yel solid	dec 20				reac H_2O
2973	Tungsten(V) oxytribromide	$WOBr_3$	20213-56-3	439.55	dark brn tetr cry			≈5.9		
2974	Tungsten(V) oxytrichloride	$WOCl_3$	14249-98-0	306.20	grn tetr cry			≈4.6		
2975	Tungsten(VI) bromide	WBr_6	13701-86-5	663.26	blue-blk cry	309		6.9		reac H_2O
2976	Tungsten(VI) chloride	WCl_6	13283-01-7	396.56	purp hex cry; hyg	282	337	3.52		reac H_2O; s EtOH, os
2977	Tungsten(VI) dioxydibromide	WO_2Br_2	13520-75-7	375.65	red cry		440 subl			
2978	Tungsten(VI) dioxydichloride	WO_2Cl_2	13520-76-8	286.75	yel orth cry	265		4.67		i H_2O
2979	Tungsten(VI) dioxydiiodide	WO_2I_2	14447-89-3	469.65	grn monocl cry	400 dec		6.39		
2980	Tungsten(VI) fluoride	WF_6	7783-82-6	297.83	yel liq or col gas	1.9	17.1	3.44		reac H_2O; vs ctc, cyhex
2981	Tungsten(VI) oxide	WO_3	1314-35-8	231.84	yel powder	1473	≈1700	7.2		i H_2O, os; sl acid; s alk
2982	Tungsten(VI) oxytetrabromide	$WOBr_4$	13520-77-9	519.46	red tetr cry	277	327	≈5.5		reac H_2O
2983	Tungsten(VI) oxytetrachloride	$WOCl_4$	13520-78-0	341.65	red hyg cry	210	230	11.92		reac H_2O; s bz, CS_2
2984	Tungsten(VI) oxytetrafluoride	WOF_4	13520-79-1	275.83	wh monocl cry	105	185.9	5.07		reac H_2O
2985	Tungsten(VI) sulfide	WS_3	12125-19-8	280.04	brn powder					sl H_2O; s alk
2986	Uranium	U	7440-61-1	238.029	silv-wh orth cry	1135	4131	19.1		
2987	Uranium boride (UB₂)	UB_2	12007-36-2	259.651	refrac solid	2430		12.7		
2988	Uranium boride (UB₄)	UB_4	12007-84-0	281.273	refrac solid	2530		9.32		i H_2O
2989	Uranium carbide (UC)	UC	12070-09-6	250.040	gray cub cry	2790				
2990	Uranium carbide (UC₂)	UC_2	12071-33-9	262.050	gray tetr cry	2350	4370	11.3		reac H_2O; sl EtOH
2991	Uranium carbide (U₂C₃)	U_2C_3	12076-62-9	512.090	gray cub cry	≈1700 dec		12.7		
2992	Uranium nitride (UN)	UN	25658-43-9	252.036	gray cub cry	2805		14.3		i H_2O
2993	Uranium nitride (U₂N₃)	U_2N_3	12033-83-9	518.078	cub cry	dec		11.3		
2994	Uranium(III) bromide	UBr_3	13470-19-4	477.741	red hyg cry	727				s H_2O
2995	Uranium(III) chloride	UCl_3	10025-93-1	344.388	grn hyg cry	837		5.51		vs H_2O; i bz, ctc
2996	Uranium(III) fluoride	UF_3	13775-06-9	295.024	blk hex cry	1495		8.9		i H_2O; s acid
2997	Uranium(III) hydride	UH_3	13598-56-6	241.053	gray-blk cub cry			11.1		
2998	Uranium(III) iodide	UI_3	13775-18-3	618.742	blk hyg cry	766				s H_2O
2999	Uranium(IV) bromide	UBr_4	13470-20-7	557.645	brn hyg cry	519				s H_2O, EtOH
3000	Uranium(IV) chloride	UCl_4	10026-10-5	379.841	grn octahed cry	590	791	4.72		reac H_2O; s EtOH
3001	Uranium(IV) fluoride	UF_4	10049-14-6	314.023	grn monocl cry	1036	1417	6.7	0.01²⁵	s conc acid, alk
3002	Uranium(IV) iodide	UI_4	13470-22-9	745.647	blk hyg cry	506				s H_2O, EtOH
3003	Uranium(IV) oxide	UO_2	1344-57-6	270.028	brn cub cry	2847		10.97		i H_2O, dil acid; s conc acid
3004	Uranium(IV,V) oxide	U_4O_9	12037-15-9	1096.111	cub cry			11.2		
3005	Uranium(V) bromide	UBr_5	13775-16-1	637.549	brn hyg cry					reac H_2O
3006	Uranium(V) chloride	UCl_5	13470-21-8	415.294	brn hyg cry	287				reac H_2O
3007	Uranium(V) fluoride	UF_5	13775-07-0	333.021	pale blue tetr cry; hyg	348		5.81		s H_2O
3008	Uranium(V,VI) oxide	U_3O_8	1344-59-8	842.082	grn-blk orth cry	1300 dec		8.38		
3009	Uranium(VI) chloride	UCl_6	13763-23-0	450.747	grn hex cry	177		3.6		
3010	Uranium(VI) fluoride	UF_6	7783-81-5	352.019	wh monocl solid	64.06 tp	56.5 sp	5.09		reac H_2O; s ctc, chl
3011	Uranium(VI) oxide	UO_3	1344-58-7	286.027	oran-yel cry			≈7.3		i H_2O; s acid
3012	Uranium(VI) oxide monohydrate	$UO_3 \cdot H_2O$	12326-21-5	304.043	yel orth cry	570 dec		7.05		
3013	Uranium peroxide dihydrate	$UO_4 \cdot 2H_2O$	19525-15-6	338.057	yel hyg cry	115 dec				i H_2O
3014	Uranyl acetate dihydrate	$UO_2(C_2H_3O_2)_2 \cdot 2H_2O$	6159-44-0	424.146	ye cry (HOAc)	80 dec		2.89		sl EtOH
3015	Uranyl chloride	UO_2Cl_2	7791-26-6	340.934	yel orth cry; hyg	577				vs H_2O; s EtOH, ace; i bz
3016	Uranyl fluoride	UO_2F_2	13536-84-0	308.025	yel hyg solid				64.4²⁰	i bz
3017	Uranyl hydrogen phosphate tetrahydrate	$UO_2HPO_4 \cdot 4H_2O$	18433-48-2	438.068	yel cry pow					i H_2O; s acid
3018	Uranyl nitrate	$UO_2(NO_3)_2$	10102-06-4	394.037	yel cry				127²⁵	s eth
3019	Uranyl nitrate hexahydrate	$UO_2(NO_3)_2 \cdot 6H_2O$	13520-83-7	502.129	yel orth cry; hyg	60	118 dec	2.81	127²⁵	s EtOH, eth
3020	Uranyl sulfate	UO_2SO_4	1314-64-3	366.090	yel cry					
3021	Uranyl sulfate trihydrate	$UO_2SO_4 \cdot 3H_2O$	20910-28-5	420.137	yel cry			3.28	152¹⁶	sl EtOH
3022	Vanadium	V	7440-62-2	50.942	gray-wh metal; cub	1910	3407	6.0		i H_2O; s acid

No.	Name	Formula	CAS Reg No.	Mol. weight	Physical form	mp/°C	bp/°C	Density g cm⁻³	Solubility g/100 g H₂O	Qualitative solubility
3023	Vanadocene	V(C₅H₅)₂	1277-47-0	181.128	viol cry; hyg	167				s bz, thf
3024	Vanadocene dichloride	V(C₅H₅)₂Cl₂	12083-48-6	252.034	dark grn cry	205 dec				s H₂O, chl, EtOH
3025	Vanadium boride (VB)	VB	12045-27-1	61.753	refrac solid	2250				i H₂O
3026	Vanadium boride (VB₂)	VB₂	12007-37-3	72.564	refrac solid	2450				
3027	Vanadium carbide (VC)	VC	12070-10-9	62.953	refrac blk cry; cub	2810		5.77		i H₂O
3028	Vanadium carbide (V₂C)	V₂C	12012-17-8	113.894	hex cry	2167				
3029	Vanadium carbonyl	V(CO)₆	14024-00-1	219.002	blue-grn cry; flam	60 dec	subl			
3030	Vanadium nitride	VN	24646-85-3	64.949	blk powder; cub	2050		6.13		i H₂O; s aqua regia
3031	Vanadium silicide (VSi₂)	VSi₂	12039-87-1	107.113	metallic prisms			4.42		s HF
3032	Vanadium silicide (V₃Si)	V₃Si	12039-76-8	180.911	cub cry	1935		5.70		
3033	Vanadium(II) bromide	VBr₂	14890-41-6	210.750	oran-brn hex cry		800 subl	4.58		reac H₂O
3034	Vanadium(II) chloride	VCl₂	10580-52-6	121.848	grn hex plates	1350	910 subl	3.23		reac H₂O; s EtOH, eth
3035	Vanadium(II) fluoride	VF₂	13842-80-3	88.939	blue hyg cry	1490				reac H₂O
3036	Vanadium(II) iodide	VI₂	15513-84-5	304.751	red-viol hex cry		subl 800	5.44		reac H₂O
3037	Vanadium(II) oxide	VO	12035-98-2	66.941	gray-blk cry	1790		5.758		s acid
3038	Vanadium(II) sulfate heptahydrate	VSO₄ · 7H₂O	36907-42-3	273.111	viol cry					
3039	Vanadium(III) bromide	VBr₃	13470-26-3	290.654	blk-grn hyg cry	dec 500	subl	4.00		reac H₂O
3040	Vanadium(III) chloride	VCl₃	7718-98-1	157.301	red-viol hex cry; hyg	500 dec		3.00		reac H₂O; s EtOH, eth
3041	Vanadium(III) fluoride	VF₃	10049-12-4	107.937	yel-grn hex cry	1395	subl	3.363		i H₂O, EtOH
3042	Vanadium(III) fluoride trihydrate	VF₃ · 3H₂O	10049-12-4*	161.983	grn rhomb cry	≈100 dec				sl H₂O
3043	Vanadium(III) iodide	VI₃	15513-94-7	431.655	brn-blk rhomb cry; hyg	dec 300		5.21		reac H₂O
3044	Vanadium(III) oxide	V₂O₃	1314-34-7	149.881	blk powder	1957	≈3000	4.87		i H₂O
3045	Vanadium(III) 2,4-pentanedioate	V(CH₃COCHCOCH₃)₃	13476-99-8	348.266	brn cry	≈185	subl	≈1.0		s MeOH, ace, bz, chl
3046	Vanadium(III) sulfate	V₂(SO₄)₃	13701-70-7	390.071	yel powder	≈400 dec				sl H₂O
3047	Vanadium(III) sulfide	V₂S₃	1315-03-3	198.078	grn-blk powder	dec		4.7		i H₂O; s hot HCl
3048	Vanadium(IV) bromide	VBr₄	13595-30-7	370.558	unstab purp cry	-23 dec				
3049	Vanadium(IV) chloride	VCl₄	7632-51-1	192.754	red-brn liq	-28	151	1.816		reac H₂O; s EtOH, eth
3050	Vanadium(IV) fluoride	VF₄	10049-16-8	126.936	grn hyg powder	325 dec	subl	3.15		vs H₂O
3051	Vanadium(IV) oxide	VO₂	12036-21-4	82.941	blue-blk powder	1967		4.339		i H₂O; s acid, alk
3052	Vanadium(V) fluoride	VF₅	7783-72-4	145.934	col liq	19.5	48.3	2.50		reac H₂O
3053	Vanadium(V) dioxide fluoride	VO₂F	14259-82-6	101.939	brn hyg cry	350 dec				reac H₂O
3054	Vanadium(V) dioxide chloride	VO₂Cl	13759-30-3	118.394	oran hyg cry	dec 180				s thf
3055	Vanadium(V) oxide	V₂O₅	1314-62-1	181.880	yel-brn orth cry	681	1750	3.35	0.07²⁵	s conc acid, alk; i EtOH
3056	Vanadium(V) sulfide	V₂S₅	12138-17-9	262.208	grn-blk pow	dec		3.0		i H₂O; s acid, alk
3057	Vanadyl bromide	VOBr	13520-88-2	146.845	viol cry	480 dec				
3058	Vanadyl chloride	VOCl	13520-87-1	102.394	brn orth cry		127	1.72		
3059	Vanadyl dibromide	VOBr₂	13520-89-3	226.749	yel-brn cry	180 dec				
3060	Vanadyl dichloride	VOCl₂	10213-09-9	137.847	grn hyg cry	380 dec		2.88		reac H₂O; s EtOH
3061	Vanadyl difluoride	VOF₂	13814-83-0	104.938	yel cry					
3062	Vanadyl selenite hydrate	VOSeO₃ · H₂O	133578-89-9	211.92	grn tricl plates			3.506		
3063	Vanadyl sulfate dihydrate	VOSO₄ · 2H₂O	27774-13-6	199.035	blue cry powder					s H₂O
3064	Vanadyl tribromide	VOBr₃	13520-90-6	306.653	deep red liq	-59	170			reac H₂O
3065	Vanadyl trichloride	VOCl₃	7727-18-6	173.300	fuming red-yel liq	-79	127	1.829		reac H₂O; s MeOH, eth, ace
3066	Vanadyl trifluoride	VOF₃	13709-31-4	123.936	yel hyg powder	300	480	2.459		reac H₂O
3067	Water	H₂O	7732-18-5	18.015	col liq	0.00	99.974	0.9970²⁵		vs EtOH, MeOH, ace
3068	Water-d₂	D₂O	7789-20-0	20.027	col liq	3.82	101.42	1.1044²⁵		
3069	Water-t₂	T₂O	14940-65-9	22.032	col liq	4.48	101.51	1.2138²⁵		
3070	Xenon	Xe	7440-63-3	131.293	col gas	-111.75	-108.09	5.366 g/L		sl H₂O
3071	Xenon trioxide	XeO₃	13776-58-4	179.291	col orth cry	exp ≈25		4.55		s H₂O
3072	Xenon tetroxide	XeO₄	12340-14-6	195.291	yel solid or col gas; exp	-35.9	≈0 dec			
3073	Xenon difluoride	XeF₂	13709-36-9	169.290	col tetr cry	129.03 tp	114.35 sp	4.32		sl H₂O
3074	Xenon tetrafluoride	XeF₄	13709-61-0	207.287	col monocl cry	117.10 tp	115.75 sp	4.04		reac H₂O
3075	Xenon hexafluoride	XeF₆	13693-09-9	245.283	col monocl cry	49.48	75.6	3.56		reac H₂O
3076	Xenon fluoride oxide	XeOF₂	13780-64-8	185.289	yel solid, stab <-25	exp ≈0				reac H₂O
3077	Xenon oxytetrafluoride	XeOF₄	13774-85-1	223.286	col liq	-46.2		3.17⁰		reac H₂O
3078	Xenon dioxydifluoride	XeO₂F₂	13875-06-4	201.289	col orth cry	30.8 exp		4.10		
3079	Xenon difluoride trioxide	XeO₃F₂	15192-14-0	217.288	unstab at r.t.	-54.1	exp			
3080	Xenon pentafluoride hexafluoroarsenate	XeF₅AsF₆	20328-94-3	415.197	wh monocl cry	130.5		3.51		

No.	Name	Formula	CAS Reg No.	Mol. weight	Physical form	mp/°C	bp/°C	Density g cm⁻³	Solubility g/100 g H₂O	Qualitative solubility
3081	Xenon pentafluoride hexafluororuthenate	XeF_5RuF_6	39796-98-0	441.35	grn orth cry	152		3.79		
3082	Xenon fluoride hexafluoroantimonate	XeF_3SbF_6	39797-63-2	424.039	yel-grn monocl cry	≈110		3.92		
3083	Xenon fluoride hexafluoroarsenate	$Xe_2F_3AsF_6$	50432-32-1	508.494	yel-grn monocl cry	99		3.62		reac H₂O
3084	Xenon fluoride hexafluororuthenate	$XeFRuF_6$	22527-13-5	365.35	yel-grn monocl cry	110		3.78		
3085	Xenon fluoride undecafluoroantimonate	$XeFSb_2F_{11}$	15364-10-0	602.794	yel monocl cry	63		3.69		
3086	Xenon trifluoride undecafluoroantimonate	$XeF_3Sb_2F_{11}$	35718-37-7	640.791	yel-grn tricl cry	82		3.98		
3087	Ytterbium	Yb	7440-64-4	173.054	silv metal; cub	824	1196	6.90		s dil acid
3088	Ytterbium silicide	$YbSi_2$	12039-89-3	229.21	hex cry			7.54		
3089	Ytterbium(II) bromide	$YbBr_2$	25502-05-0	332.85	yel cry	673				reac H₂O
3090	Ytterbium(II) chloride	$YbCl_2$	13874-77-6	243.95	grn cry	721		5.27		reac H₂O
3091	Ytterbium(II) fluoride	YbF_2	15192-18-4	211.04	gray solid	1407				i H₂O
3092	Ytterbium(II) iodide	YbI_2	19357-86-9	426.85	blk cry	772				reac H₂O
3093	Ytterbium(III) acetate tetrahydrate	$Yb(C_2H_3O_2)_3 \cdot 4H_2O$	15280-58-7	422.23	hyg col cry	dec 70		2.09		vs H₂O
3094	Ytterbium(III) bromide	$YbBr_3$	13759-89-2	412.75	col cry	956 dec				s H₂O
3095	Ytterbium(III) chloride	$YbCl_3$	10361-91-8	279.40	wh hyg powder	854				s H₂O
3096	Ytterbium(III) chloride hexahydrate	$YbCl_3 \cdot 6H_2O$	19423-87-1	387.49	grn hyg cry	150 dec		2.57		vs H₂O
3097	Ytterbium(III) fluoride	YbF_3	13760-80-0	230.04	wh cry	1157		8.2		i H₂O
3098	Ytterbium(III) iodide	YbI_3	13813-44-0	553.75	yel cry	dec 700				s H₂O
3099	Ytterbium(III) nitrate	$Yb(NO_3)_3$	13768-67-7	359.06	col hyg solid				239[25]	s EtOH
3100	Ytterbium(III) oxide	Yb_2O_3	1314-37-0	394.08	col cub cry	2355	4070	9.2		s dil acid
3101	Ytterbium(III) sulfate octahydrate	$Yb_2(SO_4)_3 \cdot 8H_2O$	10034-98-7	778.39	col cry			3.3	38.4[20]	
3102	Yttrium	Y	7440-65-5	88.906	silv metal; hex	1522	3345	4.47		reac H₂O; s dil acid
3103	Yttrium aluminum oxide	$Y_3Al_5O_{12}$	12005-21-9	593.619	grn cub cry			≈4.5		
3104	Yttrium antimonide	YSb	12186-97-9	210.666	cub cry	2310		5.97		
3105	Yttrium arsenide	YAs	12255-48-0	163.828	cub cry			5.59		
3106	Yttrium boride	YB_6	12008-32-1	153.772	refrac solid	2600		3.72		
3107	Yttrium bromide	YBr_3	13469-98-2	328.618	col hyg cry	904			83.3[30]	
3108	Yttrium carbide	YC_2	12071-35-1	112.927	refrac solid	≈2400		4.13		
3109	Yttrium carbonate trihydrate	$Y_2(CO_3)_3 \cdot 3H_2O$	5970-44-5	411.885	red-brn powder					i H₂O; s dil acid
3110	Yttrium chloride	YCl_3	10361-92-9	195.265	wh monocl cry; hyg	721	1482	2.61	75.1[20]	vs H₂O
3111	Yttrium chloride hexahydrate	$YCl_3 \cdot 6H_2O$	10025-94-2	303.356	hyg col cry	dec 100				vs H₂O; s EtOH
3112	Yttrium fluoride	YF_3	13709-49-4	145.901	wh hyg powder	1155		4.0		i H₂O
3113	Yttrium hydroxide	$Y(OH)_3$	16469-22-0	139.928	wh prec or pow	dec 190				i H₂O
3114	Yttrium iodide	YI_3	13470-38-7	469.619	hyg wh-yel cry	997				s H₂O, ace, EtOH
3115	Yttrium iron oxide	$Y_3Fe_5O_{12}$	12063-56-8	737.936	cub cry	1555				
3116	Yttrium nitrate	$Y(NO_3)_3$	10361-93-0	274.921	wh hyg solid				149[25]	s EtOH
3117	Yttrium nitrate tetrahydrate	$Y(NO_3)_3 \cdot 4H_2O$	13773-69-8	346.982	red-wh prisms			2.68	149[25]	
3118	Yttrium nitrate hexahydrate	$Y(NO_3)_3 \cdot 6H_2O$	13494-98-9	383.012	hyg cry				149[25]	
3119	Yttrium oxide	Y_2O_3	1314-36-9	225.810	wh cry; cub	2439		5.03		s dil acid
3120	Yttrium phosphide	YP	12294-01-8	119.880	cub cry			≈4.4		
3121	Yttrium sulfate octahydrate	$Y_2(SO_4)_3 \cdot 8H_2O$	7446-33-5	610.122	red monocl cry			2.6	7.47[16]	
3122	Yttrium sulfide	Y_2S_3	12039-19-9	274.007	yel cub cry	1925		3.87		
3123	Zinc	Zn	7440-66-6	65.38	blue-wh metal; hex	419.527	907	7.134		s acid, alk
3124	Zinc acetate dihydrate	$Zn(C_2H_3O_2)_2 \cdot 2H_2O$	5970-45-6	219.527	wh powder	237 dec		1.735	30.0[20]	s EtOH
3125	Zinc ammonium sulfate	$Zn(NH_4)_2(SO_4)_2$	7783-24-6	293.611	wh cry				9.2[20]	
3126	Zinc antimonide	$ZnSb$	12039-35-9	187.169	silv-wh orth cry	565		6.33		reac H₂O
3127	Zinc arsenate	$Zn_3(AsO_4)_2$	13464-44-3	474.065	wh powder				0.000078[20]	s acid, alk
3128	Zinc arsenate octahydrate	$Zn_3(AsO_4)_2 \cdot 8H_2O$	13464-45-4	618.187	wh monocl cry			3.33	0.000078[20]	s acid, alk
3129	Zinc arsenide	Zn_3As_2	12006-40-5	346.070	powder	1015		5.528		
3130	Zinc arsenite	$Zn(AsO_2)_2$	10326-24-6	279.250	col powder					i H₂O; s acid
3131	Zinc borate	$3ZnO \cdot 2B_2O_3$	27043-84-1	383.466	wh amorp powder			3.64		sl H₂O; s dil acid
3132	Zinc borate hemiheptahydrate	$2ZnO \cdot 3B_2O_3 \cdot 3.5H_2O$	12513-27-8	434.69	wh cry	980		4.22		i H₂O
3133	Zinc borate pentahydrate	$2ZnO \cdot 3B_2O_3 \cdot 5H_2O$	12536-65-1	461.753	wh powder			3.64	0.007[25]	sl HCl
3134	Zinc bromate hexahydrate	$Zn(BrO_3)_2 \cdot 6H_2O$	13517-27-6	429.305	wh hyg solid	100		2.57		vs H₂O
3135	Zinc bromide	$ZnBr_2$	7699-45-8	225.217	wh hex cry; hyg	402	≈670	4.5	488[25]	vs EtOH; s eth
3136	Zinc caprylate	$Zn(C_8H_{15}O_2)_2$	557-09-5	351.816	wh hyg cry	136				sl H₂O
3137	Zinc carbonate	$ZnCO_3$	3486-35-9	125.418	wh hex cry	140 dec		4.434	0.000091[20]	s dil acid, alk
3138	Zinc carbonate hydroxide	$3Zn(OH)_2 \cdot 2ZnCO_3$	12070-69-8	549.107	wh powder					
3139	Zinc chlorate	$Zn(ClO_3)_2$	10361-95-2	232.311	yel hyg cry	60 dec		2.15	200[20]	
3140	Zinc chloride	$ZnCl_2$	7646-85-7	136.315	wh hyg cry	290	732	2.907	408[25]	vs H₂O; s EtOH, ace

No.	Name	Formula	CAS Reg. No.	Mol. weight	Physical form	mp/°C	bp/°C	Density g cm⁻³	Solubility g/100 g H₂O	Qualitative solubility
3141	Zinc chromate	$ZnCrO_4$	13530-65-9	181.403	yel prisms	316		3.40	3.08	s acid; i ace
3142	Zinc chromite	$ZnCr_2O_4$	12018-19-8	233.399	grn cub cry			5.29		
3143	Zinc citrate dihydrate	$Zn_3(C_6H_5O_7)_2 \cdot 2H_2O$	546-46-3	610.456	col powder					sl H₂O; s dil acid, alk
3144	Zinc cyanide	$Zn(CN)_2$	557-21-1	117.443	wh powder			1.852	0.00047[20]	reac acid
3145	Zinc diethyl	$Zn(C_2H_5)_2$	557-20-0	123.531	col liq	-28	118	1.2065		reac H₂O; msc eth, peth, bz
3146	Zinc dithionate	ZnS_2O_4	7779-86-4	193.537	wh amorp solid	200 dec			40[20]	
3147	Zinc fluoride	ZnF_2	7783-49-5	103.406	wh tetr needles; hyg	872	1500	4.9	1.55[25]	sl H₂O
3148	Zinc fluoride tetrahydrate	$ZnF_2 \cdot 4H_2O$	13986-18-0	175.468	wh orth cry			2.30	1.55[25]	
3149	Zinc fluoroborate hexahydrate	$Zn(BF_4)_2 \cdot 6H_2O$	27860-83-9	347.109	hex cry			2.12		vs H₂O; s EtOH
3150	Zinc formate dihydrate	$Zn(CHO_2)_2 \cdot 2H_2O$	5970-62-7	191.474	wh cry			2.207	5.2[20]	i EtOH
3151	Zinc hexafluorosilicate hexahydrate	$ZnSiF_6 \cdot 6H_2O$	16871-71-9	315.576	wh cry					s H₂O
3152	Zinc hydroxide	$Zn(OH)_2$	20427-58-1	99.424	col orth cry	125 dec		3.05	0.000042[20]	
3153	Zinc iodate	$Zn(IO_3)_2$	7790-37-6	415.214	wh cry powder				0.64[25]	
3154	Zinc iodide	ZnI_2	10139-47-6	319.218	wh-yel hyg cry	450	625	4.74	438[25]	vs H₂O; s EtOH, eth
3155	Zinc laurate	$Zn(C_{12}H_{23}O_2)_2$	2452-01-9	464.029	wh powder	128				sl H₂O
3156	Zinc molybdate	$ZnMoO_4$	13767-32-3	225.35	wh tetr cry	>700		4.3		i H₂O
3157	Zinc nitrate	$Zn(NO_3)_2$	7779-88-6	189.418	wh powder				120[25]	
3158	Zinc nitrate hexahydrate	$Zn(NO_3)_2 \cdot 6H_2O$	10196-18-6	297.510	col orth cry	36 dec		2.067	120[25]	vs EtOH
3159	Zinc nitride	Zn_3N_2	1313-49-1	224.240	blue-gray cub cry	700 dec		6.22		i H₂O
3160	Zinc nitrite	$Zn(NO_2)_2$	10102-02-0	157.420	hyg solid					reac H₂O
3161	Zinc oleate	$Zn(C_{18}H_{33}O_2)_2$	557-07-3	628.316	wh powder	70 dec				i H₂O; s EtOH, eth, bz
3162	Zinc oxalate	ZnC_2O_4	547-68-2	153.428	wh pwd				0.0026[25]	
3163	Zinc oxalate dihydrate	$ZnC_2O_4 \cdot 2H_2O$	4255-07-6	189.458	wh powder	100 dec		2.56	0.0026[25]	s dil acid
3164	Zinc tartrate dihydrate	$ZnC_4H_4O_6 \cdot 2H_2O$	22570-08-7	249.511	wh cry pow	dec 150			0.022[20]	
3165	Zinc oxide	ZnO	1314-13-2	81.408	wh powder; hex	1974		5.6		i H₂O; s dil acid
3166	Zinc 2,4-pentanedioate	$Zn(CH_3COCHCOCH_3)_2$	14024-63-6	263.625	cry	137 dec				sl H₂O; s EtOH, DMSO
3167	Zinc pentanoate dihydrate	$Zn(C_5H_9O_2)_2 \cdot 2H_2O$	556-38-7	303.687	scales or powder					sl H₂O; reac acid; s EtOH
3168	Zinc perchlorate hexahydrate	$Zn(ClO_4)_2 \cdot 6H_2O$	10025-64-6	372.402	wh cub cry; hyg	106 dec		2.2	121.3[25]	s EtOH
3169	Zinc permanganate hexahydrate	$Zn(MnO_4)_2 \cdot 6H_2O$	23414-72-4	411.372	blk orth cry; hyg			2.45		s H₂O; reac EtOH
3170	Zinc peroxide	ZnO_2	1314-22-3	97.408	yel-wh powder	>150 dec	212 exp	1.57		i H₂O; reac acid, EtOH, ace
3171	Zinc phosphate	$Zn_3(PO_4)_2$	7779-90-0	386.170	wh monocl cry	900		4.0		i H₂O
3172	Zinc phosphate tetrahydrate	$Zn_3(PO_4)_2 \cdot 4H_2O$	7543-51-3	458.231	col orth cry			3.04		i H₂O, EtOH; s dil acid, alk
3173	Zinc phosphide	Zn_3P_2	1314-84-7	258.175	gray tetr cry	1160		4.55		i H₂O, EtOH; reac acid; s bz
3174	Zinc pyrophosphate	$Zn_2P_2O_7$	7446-26-6	304.761	wh cry powder			3.75		i H₂O; s dil acid
3175	Zinc selenate pentahydrate	$ZnSeO_4 \cdot 5H_2O$	13597-54-1	298.44	tricl cry	50 dec		2.59	63.4[25]	
3176	Zinc selenide	$ZnSe$	1315-09-9	144.37	yel-red cub cry	>1100	subl	5.65		i H₂O; s dil acid
3177	Zinc orthosilicate	Zn_2SiO_4	13597-65-4	222.902	wh hex cry	1509		4.1		i H₂O, dil acid
3178	Zinc selenite	$ZnSeO_3$	13597-46-1	192.37	wh powder	621				
3179	Zinc stearate	$Zn(C_{18}H_{35}O_2)_2$	557-05-1	632.348	wh powder	130		1.095		i H₂O, EtOH, eth; s bz
3180	Zinc sulfate	$ZnSO_4$	7733-02-0	161.472	col orth cry	680 dec		3.8	57.7[25]	
3181	Zinc sulfate monohydrate	$ZnSO_4 \cdot H_2O$	7446-19-7	179.487	wh monocl cry	238 dec		3.20	57.7[25]	i EtOH
3182	Zinc sulfate heptahydrate	$ZnSO_4 \cdot 7H_2O$	7446-20-0	287.578	col orth cry	100 dec		1.97	57.7[25]	i EtOH
3183	Zinc sulfide (sphalerite)	ZnS	1314-98-3	97.474	gray-wh cub cry	trans wurtzite 1020		4.04		i H₂O, EtOH; s dil acid
3184	Zinc sulfide (wurtzite)	ZnS	1314-98-3	97.474	wh hex cry	1700	subl	4.09		i H₂O; s dil acid
3185	Zinc sulfite dihydrate	$ZnSO_3 \cdot 2H_2O$	7488-52-0	181.503	wh powder	200 dec			0.224[25]	i EtOH
3186	Zinc telluride	$ZnTe$	1315-11-3	193.01	red cub cry	1239		5.9		i H₂O
3187	Zinc thiocyanate	$Zn(SCN)_2$	557-42-6	181.573	wh hyg cry					sl H₂O; s EtOH
3188	Zirconium	Zr	7440-67-7	91.224	gray-wh metal; hex	1854	4406	6.52		s hot conc acid
3189	Zirconocene dichloride	$Zr(C_5H_5)_2Cl_2$	1291-32-3	292.316	col cry	248	subl 150			
3190	Zirconium boride	ZrB_2	12045-64-6	112.846	gray refrac solid; hex	3050		6.17		
3191	Zirconium carbide	ZrC	12020-14-3	103.235	gray refrac solid; cub	3532		6.73		s HF
3192	Zirconium nitride	ZrN	25658-42-8	105.231	yel cub cry	2952		7.09		s conc HF; sl dil acid
3193	Zirconium phosphide	ZrP_2	12037-80-8	153.172	orth cry			≈5.1		
3194	Zirconium silicide	$ZrSi_2$	12039-90-6	147.395	gray powder	1620		4.88		i H₂O, aqua regia; s HF

No.	Name	Formula	CAS Reg No.	Mol. weight	Physical form	mp/°C	bp/°C	Density g cm⁻³	Solubility g/100 g H₂O	Qualitative solubility
3195	Zirconium(II) bromide	$ZrBr_2$	24621-17-8	251.032	blue-blk cry	dec 400				
3196	Zirconium(II) chloride	$ZrCl_2$	13762-26-0	162.130	blk cry	772		3.16		reac H_2O
3197	Zirconium(II) fluoride	ZrF_2	13842-94-9	129.221	blk cry	902				
3198	Zirconium(II) hydride	ZrH_2	7704-99-6	93.240	gray tetr cry	800 dec		5.6		i H_2O
3199	Zirconium(II) iodide	ZrI_2	15513-85-6	345.033	blk cry	827				
3200	Zirconium(III) bromide	$ZrBr_3$	24621-18-9	330.936	dark blue cry	dec 300				
3201	Zirconium(III) chloride	$ZrCl_3$	10241-03-9	197.583	dark blue cry	627		3.05		reac H_2O
3202	Zirconium(III) fluoride	ZrF_3	13814-22-7	148.219	blue-grn cry	927		4.26		i H_2O; s acid
3203	Zirconium(III) iodide	ZrI_3	13779-87-8	471.937	dark blue cry	727				
3204	Zirconium(IV) acetate hydroxide	$Zr(C_2H_3O_2)_2(OH)_2$	14311-93-4	243.327	wh amorp solid					s H_2O
3205	Zirconium(IV) ammonium carbonate dihydrate	$Zr(NH_4)_3OH(CO_3)_3 \cdot 2H_2O$	12616-24-9*	362.404	prisms; unstab					s H_2O
3206	Zirconium(IV) bromide	$ZrBr_4$	13777-25-8	410.840	wh cub cry	450 tp	360 sp	3.98		reac H_2O
3207	Zirconium(IV) chloride	$ZrCl_4$	10026-11-6	233.036	wh monocl cry; hyg	437 tp	331 sp	2.80		reac H_2O; s EtOH, eth
3208	Zirconium(IV) fluoride	ZrF_4	7783-64-4	167.218	wh monocl cry	910	912 sp	4.43	1.5²⁵	
3209	Zirconium(IV) hydroxide	$Zr(OH)_4$	14475-63-9	159.254	wh amorp powder	dec		3.25		i H_2O; s acid
3210	Zirconium(IV) iodide	ZrI_4	13986-26-0	598.842	yel-oran cub cry	500	431 sp	4.85		vs H_2O
3211	Zirconium(IV) nitrate pentahydrate	$Zr(NO_3)_4 \cdot 5H_2O$	13746-89-9	429.320	wh hyg cry	100 dec				vs H_2O; s EtOH
3212	Zirconium(IV) orthosilicate	$ZrSiO_4$	10101-52-7	183.308	wh tetr cry	1540 dec		4.6		i H_2O, acid
3213	Zirconium(IV) oxide	ZrO_2	1314-23-4	123.223	wh amorp powder	2710	4300	5.68		i H_2O; sl acid
3214	Zirconium(IV) pyrophosphate	ZrP_2O_7	13565-97-4	265.167	wh refrac solid	dec 1550				i H_2O, dil acid; s HF
3215	Zirconium(IV) sulfate	$Zr(SO_4)_2$	14644-61-2	283.349	wh hyg cry	410 dec		3.22		s H_2O; sl EtOH
3216	Zirconium(IV) sulfate tetrahydrate	$Zr(SO_4)_2 \cdot 4H_2O$	7446-31-3	355.411	wh tetr cry	100 dec		2.80		vs H_2O
3217	Zirconium(IV) sulfide	ZrS_2	12039-15-5	155.354	red-brn hex cry	1550		3.87		i H_2O
3218	Zirconium(IV) tungstate	$Zr(WO_4)_2$	16853-74-0	586.90	grn pow					
3219	Zirconyl chloride	$ZrOCl_2$	7699-43-6	178.129	wh solid	250 dec				s H_2O, EtOH
3220	Zirconyl chloride octahydrate	$ZrOCl_2 \cdot 8H_2O$	13520-92-8	322.252	tetr cry	400 dec		1.91		vs H_2O, EtOH

FORMULA INDEX OF INORGANIC COMPOUNDS

$Fe(CO)_5$	1322	Fe_3O_4	1372	GeF_4	1160	H_2NSO_3H	2731
$FeC_6H_5O_7 \cdot 5H_2O$	1380	Fe_3P	1331	$GeHBr_3$	1147	H_2O	3067
$Fe(C_5H_5)_2$	1320	$Fe_3(PO_4)_2 \cdot 8H_2O$	1360	$GeHCl_3$	1148	H_2O_2	1256
$Fe(CH_3COCHCOCH_3)_2$	1358	$Fe_4[Fe(CN)_6]_3$	1382	GeH_2Br_2	1140	H_2PFO_3	2013
$Fe(CH_3COCHCOCH_3)_3$	1394	$Fe_4(P_2O_7)_3 \cdot 9H_2O$	1397	GeH_2Cl_2	1141	H_2PtCl_6	2063
$FeCl_2$	1343	Fm	1076	GeH_3Br	1137	$H_2PtCl_6 \cdot 6H_2O$	2083
$FeCl_2 \cdot 2H_2O$	1344	Fr	1083	GeH_3CH_3	1146	$H_2Pt(OH)_6$	2064
$FeCl_2 \cdot 4H_2O$	1345	Ga	1104	GeH_3Cl	1138	H_2S	1258
$Fe(ClO_4)_2$	1359	$GaAs$	1106	GeH_3F	1144	H_2SO_3	2730
$FeCl_3$	1377	$GaBr_3$	1114	GeH_3I	1145	H_2SO_4	2725
$FeCl_3 \cdot 6H_2O$	1378	$Ga(CH_3COCHCOCH_3)_3$	1124	GeH_4	1132	H_2SO_5	2726
$Fe(ClO_4)_3 \cdot 6H_2O$	1395	$GaCl_2$	1110	GeI_2	1153	H_2S_2	1259
$FeCr_2O_4$	1346	$GaCl_3$	1115	GeI_4	1161	H_2Se	1257
FeF_2	1347	$Ga(ClO_4)_3 \cdot 6H_2O$	1125	GeO	1154	H_2SeO_3	2391
$FeF_2 \cdot 4H_2O$	1348	GaF_3	1116	GeO_2	1163	H_2SeO_4	2389
FeF_3	1383	$GaF_3 \cdot 3H_2O$	1117	GeS	1156	H_2SiF_6	2462
$FeF_3 \cdot 3H_2O$	1384	GaH_3	1118	GeS_2	1165	H_2SiO_3	2460
$Fe(H_2PO_2)_3$	1398	GaI_3	1120	$GeSe$	1155	H_2Te	1260
FeI_2	1350	GaN	1107	$GeSe_2$	1164	H_2TeO_3	2786
$FeI_2 \cdot 4H_2O$	1351	$Ga(NO_3)_3$	1121	$GeTe$	1157	H_2WO_4	2944
$FeMoO_4$	1352	$GaOOH$	1123	Ge_2H_6	1133	H_3AsO_3	243
$Fe(NO_3)_2$	1353	$Ga(OH)_3$	1119	Ge_3H_8	1134	H_3AsO_4	241
$Fe(NO_3)_2 \cdot 6H_2O$	1354	GaP	1108	Ge_3N_4	1162	$H_3AsO_4 \cdot 0.5H_2O$	242
$Fe(NO_3)_3$	1389	GaS	1112	Ge_4H_{10}	1135	H_3BO_3	461
$Fe(NO_3)_3 \cdot 6H_2O$	1390	$GaSb$	1105	Ge_5H_{12}	1136	H_3PO_3	2006
$Fe(NO_3)_3 \cdot 9H_2O$	1391	$GaSe$	1111	$HAuBr_4 \cdot 5H_2O$	1167	H_3PO_4	2004
$Fe(NH_4)_3(C_6H_5O_7)_2$	1374	$GaTe$	1113	$HAuCl_4 \cdot 4H_2O$	1168	$H_3PW_{12}O_{40}$	2005
$FeNaP_2O_7$	1399	Ga_2O	1109	HBF_4	465	$H_3P(Mo_3O_{10})_4$	1790
FeO	1357	Ga_2O_3	1122	HBO_2	462	$H_4P_2O_6$	2009
$FeOH(C_2H_3O_2)_2$	1373	Ga_2S_3	1129	HBO_2	463	$H_4P_2O_7$	2010
$FeO(OH)$	1387	$Ga_2(SO_4)_3$	1127	HBO_2	464	H_4SiO_4	2461
$Fe(OH)_2$	1349	$Ga_2(SO_4)_3 \cdot 18H_2O$	1128	HBr	1247	$H_4SiO_4 \cdot (W_3O_9)_4$	2475
$Fe(OH)_3$	1386	Ga_2Se_3	1126	$HBrO_3$	479	H_6TeO_6	2785
FeP	1329	Ga_2Te_3	1130	HCN	1252	He	1208
$FePO_4 \cdot 2H_2O$	1396	Gd	1084	HCl	1249	Hf	1184
FeS	1365	GdB_6	1085	$HCl \cdot 2H_2O$	1250	HfB_2	1185
$FeSO_4$	1362	$GdBr_3$	1091	$HClO_3$	746	$HfBr_2$	1192
$FeSO_4 \cdot H_2O$	1363	$Gd(C_2H_3O_2)_2 \cdot 4H_2O$	1090	$HClO_4$	747	$HfBr_3$	1194
$FeSO_4 \cdot 7H_2O$	1364	$GdCl_3$	1092	HD	1244	$HfBr_4$	1197
FeS_2	1332	$GdCl_3 \cdot 6H_2O$	1093	HF	1253	HfC	1186
$Fe(SCN)_2 \cdot 3H_2O$	1369	GdF_3	1094	HI	1254	$Hf(C_5H_5)_2Cl_2$	1191
$Fe(SCN)_3$	1402	GdI_2	1088	HIO_3	1288	$HfCl_2$	1193
$FeSe$	1361	GdI_3	1095	$HIO_4 \cdot 2H_2O$	1289	$HfCl_3$	1195
$FeSi$	1333	GdN	1086	$HNOSO_4$	2727	$HfCl_4$	1198
$FeSi_2$	1334	$Gd(NO_3)_3 \cdot 5H_2O$	1096	HNO_2	1936	HfF_4	1199
$Fe(TaO_3)_2$	1366	$Gd(NO_3)_3 \cdot 6H_2O$	1097	HNO_3	1935	HfH_2	1187
$FeTe$	1368	$GdSe$	1089	HN_3	1235	HfI_3	1196
$FeTiO_3$	1370	$GdSi_2$	1087	$HOBr$	480	HfI_4	1200
$Fe(VO_3)_3$	1388	$Gd_2(C_2O_4)_3 \cdot 10H_2O$	1098	$HOCl$	745	HfN	1188
$FeWO_4$	1371	Gd_2O_3	1099	$HOSeF_5$	2390	$HfOCl_2 \cdot 8H_2O$	1202
Fe_2B	1327	Gd_2S_3	1102	HPF_6	2012	HfO_2	1201
$Fe_2(C_2O_4)_3$	1392	$Gd_2(SO_4)_3$	1100	HPH_2O_2	2007	HfP	1189
$Fe_2(CO)_9$	1323	$Gd_2(SO_4)_3 \cdot 8H_2O$	1101	HPO_2F_2	2011	HfS_2	1206
$Fe_2(CrO_4)_3$	1379	Gd_2Te_3	1103	HPO_3	2008	$Hf(SO_4)_2$	1205
$Fe_2(Cr_2O_7)_3$	1381	Ge	1131	$HReO_4$	2255	$HfSe_2$	1203
Fe_2O_3	1393	$GeBr_2$	1150	HT	1245	$HfSiO_4$	1204
Fe_2P	1330	$GeBr_4$	1158	H_2	1241	$HfSi_2$	1190
$Fe_2(SO_4)_3$	1400	$Ge(CH_3)_2Cl_2$	1143	H_2CrO_4	762	$HfTiO_4$	1207
$Fe_2(SO_4)_3 \cdot 9H_2O$	1401	$GeCl_2$	1151	$H_2MoO_4 \cdot H_2O$	1819	Hg	1735
Fe_2SiO_4	1355	$GeCl_3F$	1149	H_2NOH	1236	$HgBr_2$	1761
$Fe_3(AsO_4)_2$	1338	$GeCl_4$	1159	$H_2NOH \cdot HBr$	1237	$Hg(BrO_3)_2$	1760
$Fe_3(AsO_4)_2 \cdot 6H_2O$	1339	GeF_2	1152	$H_2NOH \cdot HCl$	1238	$Hg(CH_3)_2$	1736
Fe_3C	1328	GeF_2Cl_2	1142	$H_2NOH \cdot HClO_4$	1239	$Hg(CN)_2$	1765
$Fe_3(CO)_{12}$	1324	GeF_3Cl	1139	$2(H_2NOH) \cdot H_2SO_4$	1240	$Hg(CN)_2 \cdot HgO$	1779

Formula	No.	Formula	No.	Formula	No.	Formula	No.
RuF_3	2345	Sb_2Te_3	228	SiO	2465	$SnCl_4 \cdot 5H_2O$	2904
RuF_4	2348	Sc	2374	SiO_2	2466	$Sn(CrO_4)_2$	2905
RuF_5	2350	ScB_2	2375	SiO_2	2467	SnF_2	2890
RuF_6	2351	$ScBr_3$	2376	SiO_2	2468	SnF_4	2906
RuI_3	2346	$ScCl_3$	2377	SiO_2	2469	SnH_3CH_3	2882
$Ru(NO)Cl_3 \cdot H_2O$	2341	ScF_3	2378	SiO_2	2470	SnH_4	2881
$Ru(NH_3)_6Cl_3$	2342	ScI_3	2380	SiS	2471	SnI_2	2893
RuO_2	2349	$Sc(NO_3)_3$	2381	SiS_2	2472	SnI_4	2907
RuO_4	2352	$Sc(OH)_3$	2379	Si_2Br_6	2433	SnO	2895
$Ru_2(CO)_9$	2340	Sc_2O_3	2382	Si_2Cl_6	2443	SnO_2	2908
$Ru_3(CO)_{12}$	2338	Sc_2S_3	2384	$(SiCl_3)_2O$	2458	$Sn(OH)_2$	2892
S	2723	$Sc_2(SO_4)_3 \cdot 5H_2O$	2383	Si_2F_6	2449	SnP	2884
S	2724	Sc_2Te_3	2385	Si_2H_6	2413	SnS	2899
SCl_2	2740	Se	2386	$(SiH_3)_2O$	2457	$SnSO_4$	2898
SF_4	2741	Se	2387	Si_2I_6	2456	SnS_2	2911
SF_5Br	2743	Se	2388	Si_3Br_8	2434	$SnSe$	2897
SF_5Cl	2744	$SeBr_4$	2396	Si_3Cl_8	2444	$SnSe_2$	2909
SF_6	2742	$SeCl_4$	2397	Si_3F_8	2450	$Sn(SeO_3)_2$	2910
$SOBr_2$	2752	SeF_4	2398	Si_3H_8	2414	$SnTe$	2901
$SOCl_2$	2753	SeF_5Cl	2400	Si_3N_4	2464	$SnZrF_6$	2891
SOF_2	2754	SeF_6	2399	Si_4F_{10}	2451	$Sn_2P_2O_7$	2896
SOF_4	2755	$SeOBr_2$	2401	Si_4H_{10}	2415	Sn_4P_3	2885
SO_2	2732	$SeOCl_2$	2402	Si_5H_{10}	2416	Sr	2677
SO_2BrF	2749	$SeOF_2$	2403	Si_5H_{10}	2417	$Sr(AsO_2)_2 \cdot 4H_2O$	2679
SO_2Cl_2	2747	$SeOF_4$	2404	Si_5H_{12}	2418	SrB_6	2694
SO_2ClF	2750	SeO_2	2392	Si_5H_{12}	2419	$SrBr_2$	2681
SO_2F_2	2748	SeO_2F_2	2405	Si_5H_{12}	2420	$SrBr_2 \cdot 6H_2O$	2682
$SO_2(OH)Cl$	2728	SeO_3	2393	Si_6H_{12}	2421	$Sr(BrO_3)_2 \cdot H_2O$	2680
$SO_2(OH)F$	2729	SeS	2406	Si_6H_{14}	2422	$SrCO_3$	2684
SO_3	2733	SeS_2	2407	Si_6H_{14}	2423	SrC_2	2683
SO_3	2734	Se_2Br_2	2394	Si_6H_{14}	2424	$Sr(CHO_2)_2$	2692
SO_3	2735	Se_2Cl_2	2395	Si_7H_{16}	2425	$Sr(CHO_2)_2 \cdot 2H_2O$	2693
$SSBr_2$	2736	Se_2S_6	2408	Sm	2353	$Sr(CN)_2 \cdot 4H_2O$	2689
$SSCl_2$	2737	Se_4S_4	2409	SmB_6	2354	$SrC_2O_4 \cdot H_2O$	2706
SSF_2	2738	Se_6S_2	2410	$SmBr_2$	2356	$Sr(C_2H_3O_2)_2$	2678
S_2F_{10}	2745	Si	2411	$SmBr_3$	2362	$SrCl_2$	2686
$S_2O_5Cl_2$	2751	SiB_4	2474	$Sm(BrO_3)_3 \cdot 9H_2O$	2361	$SrCl_2 \cdot 6H_2O$	2687
S_4N_4	2757	$SiBrCl_3$	2430	$Sm(C_2H_3O_2)_3 \cdot 3H_2O$	2360	$Sr(ClO_3)_2$	2685
Sb	206	$SiBr_2Cl_2$	2431	$SmCl_2$	2357	$Sr(ClO_4)_2$	2708
Sb	207	$SiBr_3Cl$	2432	$SmCl_3$	2364	$SrCrO_4$	2688
$SbAs$	212	$SiBr_4$	2429	$SmCl_3 \cdot 6H_2O$	2365	SrF_2	2691
$SbBr_3$	215	SiC	2463	SmF_2	2358	$SrFe(CN)_6 \cdot 15H_2O$	2690
$Sb(CH_3)_3$	209	$Si(C_2H_3O_2)_4$	2473	SmF_3	2366	SrH_2	2695
$Sb(CH_3)_5$	210	$SiClF_3$	2439	SmI_2	2359	SrI_2	2698
$Sb(C_2H_3O_2)_3$	214	$SiCl_2F_2$	2440	SmI_3	2367	$SrI_2 \cdot 6H_2O$	2699
$SbCl_2F_3$	232	$SiCl_3F$	2441	$Sm(NO_3)_3$	2368	$Sr(IO_3)_2$	2697
$SbCl_3$	216	$SiCl_3I$	2442	$Sm(NO_3)_3 \cdot 6H_2O$	2369	$Sr(MnO_4)_2 \cdot 3H_2O$	2709
$SbCl_5$	230	$SiCl_4$	2438	$SmSi_2$	2355	$SrMoO_4$	2700
SbF_3	217	SiF_4	2448	$Sm_2(CO_3)_3$	2363	$Sr(NO_2)_2$	2704
SbF_5	231	$SiHBr_3$	2428	Sm_2O_3	2370	$Sr(NO_3)_2$	2702
SbH_3	208	$SiHCl_3$	2437	Sm_2S_3	2372	$SrNb_2O_6$	2701
$SbIS$	219	$SiHF_3$	2447	$Sm_2(SO_4)_3 \cdot 8H_2O$	2371	SrO	2707
SbI_3	218	$SiHI_3$	2454	Sm_2Te_3	2373	SrO_2	2710
$SbOCl$	222	SiH_2Br_2	2427	Sn	2879	$Sr(OH)_2$	2696
$SbPO_4$	223	SiH_2Cl_2	2436	Sn	2880	SrS	2716
$[Sb(CH_3)_2]_2$	211	SiH_2F_2	2446	$SnBr_2$	2887	$SrSO_3$	2717
Sb_2O_3	220	SiH_2I_2	2453	$SnBr_4$	2902	$SrSO_4$	2715
Sb_2O_3	221	SiH_3Br	2426	SnC_2O_4	2894	$SrS_2O_3 \cdot 5H_2O$	2719
Sb_2O_4	229	SiH_3Cl	2435	$Sn(CH_3)_3N(CH_3)_2$	2883	$SrSe$	2713
Sb_2O_5	233	SiH_3F	2445	$SnC_4H_4O_6$	2900	$SrSeO_4$	2712
Sb_2S_3	227	SiH_3I	2452	$Sn(C_2H_3O_2)_2$	2886	$SrSi_2$	2714
$Sb_2(SO_4)_3$	226	SiH_3CH_3	2459	$SnCl_2$	2888	$SrTe$	2718
Sb_2S_5	234	SiH_4	2412	$SnCl_2 \cdot 2H_2O$	2889	$SrTiO_3$	2720
Sb_2Se_3	225	SiI_4	2455	$SnCl_4$	2903	$SrWO_4$	2721

$SrZrO_3$	2722	ThH_2	2848	$TlOH$	2825	U_4O_9	3004
Sr_2SiO_4	2705	ThI_4	2855	$TlPF_6$	2824	V	3022
Sr_3N_2	2703	ThN	2858	$TlSe$	2846	VB	3025
$Sr_3(PO_4)_2$	2711	$Th(NO_3)_4$	2856	Tl_2CO_3	2816	VB_2	3026
T_2	1243	$Th(NO_3)_4 \cdot 4H_2O$	2857	$Tl_2C_2O_4$	2831	VBr_2	3033
T_2O	3069	ThO_2	2859	Tl_2CrO_4	2819	VBr_3	3039
Ta	2758	ThS_2	2864	Tl_2MoO_4	2828	VBr_4	3048
$TaAl_3$	2759	$Th(SO_4)_2 \cdot 9H_2O$	2863	Tl_2O	2832	VC	3027
TaB	2760	$ThSe_2$	2860	Tl_2O_3	2844	$V(CO)_6$	3029
TaB_2	2761	$ThSiO_4$	2861	Tl_2S	2837	$V(C_5H_5)_2$	3023
$TaBr_3$	2767	$ThSi_2$	2862	Tl_2SO_4	2836	$V(C_5H_5)_2Cl_2$	3024
$TaBr_4$	2769	Ti	2912	$Tl_2(SO_4)_3$	2845	$V(CH_3COCHCOCH_3)_3$	3045
$TaBr_5$	2776	TiI_3	2928	Tl_2Se	2835	VCl_2	3034
TaC	2762	TiB_2	2915	Tl_2SeO_4	2834	VCl_3	3040
$TaCl_3$	2768	$TiBr_2$	2920	Tm	2865	VCl_4	3049
$TaCl_4$	2770	$TiBr_3$	2925	$TmBr_2$	2866	VF_2	3035
$TaCl_5$	2777	$TiBr_4$	2933	$TmBr_3$	2869	VF_3	3041
TaF_5	2778	TiC	2916	$TmCl_2$	2867	$VF_3 \cdot 3H_2O$	3042
TaH	2764	$Ti(C_5H_5)_2Cl_2$	2913	$TmCl_3$	2870	VF_4	3050
TaI_4	2771	$TiCl_2$	2921	$TmCl_3 \cdot 7H_2O$	2871	VF_5	3052
TaI_5	2779	$TiCl_3$	2926	TmF_3	2872	VI_2	3036
TaN	2765	$TiCl_4$	2934	TmI_2	2868	VI_3	3043
TaO_2	2772	TiF_3	2927	TmI_3	2874	VN	3030
TaS_2	2774	TiF_4	2935	$Tm(NO_3)_3$	2875	VO	3037
$TaSe_2$	2773	TiH_2	2914	$Tm(NO_3)_3 \cdot 5H_2O$	2876	$VOBr$	3057
$TaSi_2$	2766	TiI_2	2922	$Tm(OH)_3$	2873	$VOBr_2$	3059
$TaTe_2$	2775	TiI_4	2936	$Tm_2(C_2O_4)_3 \cdot 6H_2O$	2877	$VOBr_3$	3064
Ta_2C	2763	TiN	2917	Tm_2O_3	2878	$VOCl$	3058
Ta_2O_5	2780	TiO	2923	U	2986	$VOCl_2$	3060
Tb	2797	$TiOSO_4 \cdot H_2O$	2940	UB_2	2987	$VOCl_3$	3065
$TbBr_3$	2800	TiO_2	2937	UB_4	2988	VOF_2	3061
$TbCl_3$	2801	TiO_2	2938	UBr_3	2994	VOF_3	3066
$TbCl_3 \cdot 6H_2O$	2802	TiO_2	2939	UBr_4	2999	$VOSO_4 \cdot 2H_2O$	3063
TbF_3	2809	TiP	2918	UBr_5	3005	$VOSeO_3 \cdot H_2O$	3062
TbF_4	2810	TiS	2924	UC	2989	VO_2	3051
TbI_3	2803	TiS_2	2942	UC_2	2990	VO_2Cl	3054
TbN	2798	$Ti(SO_4)_2$	2941	UCl_3	2995	VO_2F	3053
$Tb(NO_3)_3$	2804	$TiSi_2$	2919	UCl_4	3000	$VSO_4 \cdot 7H_2O$	3038
$Tb(NO_3)_3 \cdot 6H_2O$	2805	Ti_2O_3	2929	UCl_5	3006	VSi_2	3031
$TbSi_2$	2799	Ti_2S_3	2931	UCl_6	3009	V_2C	3028
Tb_2O_3	2806	$Ti_2(SO_4)_3$	2930	UF_3	2996	V_2O_3	3044
Tb_2S_3	2808	Ti_3O_5	2932	UF_4	3001	V_2O_5	3055
$Tb_2(SO_4)_3 \cdot 8H_2O$	2807	Tl	2811	UF_5	3007	V_2S_3	3047
Tc	2781	$TlBr$	2815	UF_6	3010	$V_2(SO_4)_3$	3046
TcF_5	2782	$TlBrO_3$	2814	UH_3	2997	V_2S_5	3056
TcF_6	2783	$TlBr_3 \cdot 4H_2O$	2839	UI_3	2998	V_3Si	3032
Te	2784	$TlCHO_2$	2823	UI_4	3002	W	2943
$TeBr_2$	2789	$TlCN$	2820	UN	2992	WB	2946
$TeBr_4$	2791	$TlC_2H_3O_2$	2812	UO_2	3003	WBr_2	2955
$TeCl_2$	2790	TlC_2H_5O	2821	$UO_2(C_2H_3O_2)_2 \cdot 2H_2O$	3014	WBr_3	2958
$TeCl_4$	2792	$Tl(C_2H_3O_2)_3$	2838	UO_2Cl_2	3015	WBr_4	2961
TeF_4	2793	$TlCl$	2818	UO_2F_2	3016	WBr_5	2969
TeF_6	2796	$TlClO_3$	2817	$UO_2HPO_4 \cdot 4H_2O$	3017	WBr_6	2975
TeI_4	2795	$TlClO_4$	2833	$UO_2(NO_3)_2$	3018	WC	2949
TeO_2	2787	$TlCl_3$	2840	$UO_2(NO_3)_2 \cdot 6H_2O$	3019	$W(CO)_6$	2950
TeO_3	2788	$TlCl_3 \cdot 4H_2O$	2841	$UO_2(NH_4)_3F_5$	204	$W(C_2H_5O)_5$	2971
Te_2F_{10}	2794	TlF	2822	UO_2SO_4	3020	WCl_2	2956
Th	2847	TlF_3	2842	$UO_2SO_4 \cdot 3H_2O$	3021	WCl_3	2959
ThB_6	2849	TlI	2827	UO_3	3011	WCl_4	2962
$ThBr_4$	2850	$TlIO_3$	2826	$UO_3 \cdot H_2O$	3012	WCl_5	2970
ThC	2851	$TlNO_2$	2830	$UO_4 \cdot 2H_2O$	3013	WCl_6	2976
ThC_2	2852	$TlNO_3$	2829	U_2C_3	2991	WF_4	2963
$ThCl_4$	2853	TlN_3	2813	U_2N_3	2993	WF_5	2972
ThF_4	2854	$Tl(NO_3)_3$	2843	U_3O_8	3008	WF_6	2980

PHYSICAL PROPERTIES OF THE RARE EARTH METALS

K.A. Gschneidner, Jr.

TABLE 1. Data for the Trivalent Ions of the Rare Earth Elements

Rare earth	Symbol	Atomic no.	Atomic wt.[a]	No. 4f electrons	S	L	J	Spectroscopic ground state symbol
Scandium	Sc	21	44.955910	0	—	—	—	—
Yttrium	Y	39	88.90585	0	—	—	—	—
Lanthanum	La	57	138.9055	0	—	—	—	—
Cerium	Ce	58	140.115	1	1/2	3	5/2	$^2F_{5/2}$
Praseodymium	Pr	59	140.90765	2	1	5	4	$^3H_4$
Neodymium	Nd	60	144.24	3	3/2	6	9/2	$^4I_{9/2}$
Promethium	Pm	61	(145)	4	2	6	4	$^5I_4$
Samarium	Sm	62	150.36	5	5/2	5	5/2	$^6H_{5/2}$
Europium	Eu	63	151.965	6	3	3	0	$^7F_0$
Gadolinium	Gd	64	157.25	7	7/2	0	7/2	$^8S_{7/2}$
Terbium	Tb	65	158.92534	8	3	3	6	$^7F_6$
Dysprosium	Dy	66	162.50	9	5/2	5	15/2	$^6H_{15/2}$
Holmium	Ho	67	164.93032	10	2	6	8	$^5I_8$
Erbium	Er	68	167.26	11	3/2	6	15/2	$^4I_{15/2}$
Thulium	Tm	69	168.93421	12	1	5	6	$^3H_6$
Ytterbium	Yb	70	173.04	13	1/2	3	7/2	$^2F_{7/2}$
Lutetium	Lu	71	174.967	14	—	—	—	—

Note: For additional information, see Goldschmidt, Z.B., in *Handbook on the Physics and Chemistry of Rare Earths,* Vol. 1, Gschneidner, K.A., Jr. and Eyring, L., Eds., North-Holland Physics, Amsterdam, 1978; DeLaeter, J.R., and Heumann, K.G., *J. Phys. Chem. Ref. Data,* 20, 1313 , 1991; *Pure Appl. Chem.,* 66, 2423, 1994.

[a] 1993 standard atomic weights.

TABLE 2. Crystallographic Data for the Rare Earth Metals at 24 °C (297 K) or Below

Rare earth metal	Crystal structure[a]	Lattice constants (Å) a_o	b_o	c_o	Metallic radius CN = 12 (Å)	Atomic volume (cm³/mol)	Density (g/cm³)
αSc	hcp	3.3088	—	5.2680	1.6406	15.039	2.989
αY	hcp	3.6482	—	5.7318	1.8012	19.893	4.469
αLa	dhcp	3.7740	—	12.171	1.8791	22.602	6.146
αCe[b]	fcc	4.85[b]	—	—	1.72[b]	17.2[b]	8.16[b]
βCe	dhcp	3.6810	—	11.857	1.8321	20.947	6.689
γCe[c]	fcc	5.1610	—	—	1.8247	20.696	6.770
αPr	dhcp	3.6721	—	11.8326	1.8279	20.803	6.773
αNd	dhcp	3.6582	—	11.7966	1.8214	20.583	7.008
αPm	dhcp	3.65	—	11.65	1.811	20.24	7.264
αSm	rhomb[d]	3.6290[d]	—	26.207	1.8041	20.000	7.520
Eu	bcc	4.5827	—	—	2.0418	28.979	5.244
αGd	hcp	3.6336	—	5.7810	1.8013	19.903	7.901
α′Tb[e]	ortho	3.605[e]	6.244[e]	5.706[e]	1.784[e]	19.34[e]	8.219[e]
αTb	hcp	3.6055	—	5.6966	1.7833	19.310	8.230
α′Dy[f]	ortho	3.595[f]	6.184[f]	5.678[f]	1.774[f]	19.00[f]	8.551[f]
αDy	hcp	3.5915	—	5.6501	1.7740	19.004	8.551
Ho	hcp	3.5778	—	5.6178	1.7661	18.752	8.795
Er	hcp	3.5592	—	5.5850	1.7566	18.449	9.066
Tm	hcp	3.5375	—	5.5540	1.7462	18.124	9.321
αYb[g]	hcp	3.8799[g]	—	6.3859[g]	1.9451[g]	25.067[g]	6.903[g]
βYb	fcc	5.4848	—	—	1.9392	24.841	6.966
Lu	hcp	3.5052	—	5.5494	1.7349	17.779	9.841

Note: For additional information, see Gschneidner, K.A., Jr. and Calderwood, F.W., in *Handbook on the Physics and Chemistry of Rare Earths,* Vol. 8, Gschneidner, K.A., Jr. and Eyring, L., Eds., North-Holland Physics, Amsterdam, 1986; Gschneidner, K.A., Jr., Pecharsky, V.K., Cho, Jaephil and Martin, S.W., *Scripta Mater.,* 34, 1717, 1996.

[a] hcp = hexagonal close-packed; P6₃/mmc, hP2, A3, Mg-type; dhcp = double-c hexagonal close-packed; P6₃/mmc, hP4, A3′, αLa-type; fcc = face-centered cubic; Fm$\overline{3}$m, cF4, A1, Cu-type; rhomb = rhombohedral; R$\overline{3}$m, hR3, αSm-type; bcc = body-centered cubic; Im$\overline{3}$m, cI2, A2, W-type; ortho = orthorhombic; Cmcm, oC4, α′ Dy-type.
[b] At 77 K (−196 °C).
[c] Equilibrium room temperature (standard state) phase.
[d] Rhombohedral is the primitive cell. Lattice parameters given are for the nonprimitive hexagonal cell.
[e] At 220 K (−53 °C).
[f] At 86 K (−187 °C).
[g] At 23 °C.

TABLE 3. Crystallographic Data for Rare Earth Metals at High Temperature

Rare earth metal	Structure	Lattice parameter (Å)	Temp. (°C)	Metallic radius CN = 8 (Å)	Metallic radius CN = 12 (Å)	Atomic volume (cm³/mol)	Density (g/cm³)
βSc	bcc	3.73 (est.)	1337	1.62	1.66	15.6	2.88
βY	bcc	4.10[a]	1478	1.78	1.83	20.8	4.28
βLa	fcc	5.303	325	—	1.875	22.45	6.187
γLa	bcc	4.26	887	1.84	1.90	23.3	5.97
δCe	bcc	4.12	757	1.78	1.84	21.1	6.65
βPr	bcc	4.13	821	1.79	1.84	21.2	6.64
βNd	bcc	4.13	883	1.79	1.84	21.2	6.80
βPm	bcc	4.10 (est.)	890	1.78	1.83	20.8	6.99
βSm	hcp	a = 3.6630 c = 5.8448	450[b]	—	1.8176	20.450	7.353
γSm	bcc	4.10 (est.)	922	1.77	1.82	20.8	7.25
βGd	bcc	4.06	1265	1.76	1.81	20.2	7.80
βTb	bcc	4.07[a]	1289	1.76	1.81	20.3	7.82
βDy	bcc	4.03[a]	1381	1.75	1.80	19.7	8.23
γYb	bcc	4.44	763[c]	1.92	1.98	26.4	6.57

Note: The rare earths Eu, Ho, Er, Tm, and Lu are monomorphic. For additional information, see Gschneidner, K.A., Jr. and Calderwood, F.W., in *Handbook on the Physics and Chemistry of Rare Earths*, Vol. 8, Gschneidner, K.A., Jr. and Eyring, L., Eds., North-Holland Physics, Amsterdam, 1986, 1.

[a] Determined by extrapolation to 0% solute of a vs. composition data for R-Mg alloys at 24 °C and corrected for thermal expansion to temperature given.
[b] The hcp phase was stabilized by impurities and the temperature of measurement was below the equilibrium transition temperature (see Table 4).
[c] The bcc phase was stabilized by impurities and the temperature of measurement was below the equilibrium transition temperature (see Table 4).

TABLE 4. High Temperature Transition Temperatures and Melting Point of Rare Earth Metals

Rare earth metal	Transition I (α – β)[a] Temp. (°C)	Transition I (α – β)[a] Phases	Transition II (β – γ)[a] Temp. (C°)	Transition II (β – γ)[a] Phases	Melting point (C°)
Sc	1337	hcp ⇌ bcc	—	—	1541
Y	1478	hcp ⇌ bcc	—	—	1522
La[b]	310	dhcp ⇌ fcc	865	fcc ⇌ bcc	918
Ce[c,d]	139	dhcp ⇌ fcc (β - γ)	726	fcc ⇌ bcc (γ - δ)	798
Pr	795	dhcp ⇌ bcc	—	—	931
Nd	863	dhcp ⇌ bcc	—	—	1021
Pm	890	dhcp ⇌ bcc	—	—	1042
Sm[e]	734	rhom ⇌ hcp	922	hcp ⇌ bcc	1074
Eu	—	—	—	—	822
Gd	1235	hcp ⇌ bcc	—	—	1313
Tb	1289	hcp ⇌ bcc	—	—	1356
Dy	1381	hcp ⇌ bcc	—	—	1412
Ho	—	—	—	—	1474
Er	—	—	—	—	1529
Tm	—	—	—	—	1545
Yb	795	fcc ⇌ bcc (β - γ)	—	—	819
Lu	—	—	—	—	1663

Note: For additional information, see Gschneidner, K.A., Jr. and Calderwood, F.W., in *Handbook on the Physics and Chemistry of Rare Earths*, Vol. 8, Gschneidner, K.A., Jr. and Eyring, L., Eds., North-Holland Physics, Amsterdam, 1986; Gschneidner, K.A., Jr., Pecharsky, V.K., Cho, Jaephil and Martin, S.W., *Scripta Mater.*, 34, 1717, 1996.

[a] For all the transformations listed, unless otherwise noted.
[b] On cooling, fcc → dhcp (β → α), 260 °C.
[c] The β ⇌ γ equilibrium transition temperature is 10 ± 5 °C.
[d] On cooling, fcc → dhcp (γ → β), –16 °C.
[e] On cooling, hcp → rhomb (β → α), 727 °C.

TABLE 5. Low Temperature Transition Temperatures of the Rare Earth Metals

Rare earth metal	Cooling Transformation	°C	K	Rare earth metal	Heating Transformation	°C	K
Ce	γ → β[a]	–16	257	Ce	α → β	–148	125
	γ → α	–172	101		α → β + γ	–104	169
	β → α	–228	45		β → γ[a]	139	412
Tb	α → α′	–53	220	Yb	α → β	7	280
Dy	α → α′	–187	86				
Yb	β → α	–13	260				

Note: For additional information, see Beaudry, B.J. and Gschneidner, K.A., Jr., in *Handbook on the Physics and Chemistry of Rare Earths*, Vol. 1, Gschneidner, K.A., Jr. and Eyring, L., Eds., North-Holland Physics, Amsterdam, 1978, 173; Koskenmaki, D.C. and Gschneidner, K.A., Jr., 1978, in *Handbook on the Physics and Chemistry of Rare Earths*, Vol. 1, Gschneidner, K.A., Jr. and Eyring, L., Eds., North-Holland Physics, Amsterdam, 1978, 337; Gschneidner, K.A., Jr., Pecharsky, V.K., Cho, Jaephil and Martin, S.W., *Scripta Mater.*, 34, 1717, 1996.

[a] The β ⇌ γ equilibrium transition temperature is 10 ± 5 °C (283 ± 5K).

TABLE 6. Heat Capacity, Standard Entropy, Heats of Transformation, and Fusion of the Rare Earth Metals

Rare earth metal	Heat capacity at 298 K (J/mol K)	Standard entropy $S°_{298}$ (J/mol K)	Heat of transformation (kJ/mol)				Heat of fusion (kJ/mol)
			trans. 1	ΔH_{tr}^{1}	trans. 2	ΔH_{tr}^{2}	
Sc	25.5	34.6	$\alpha \rightleftharpoons \beta$	4.00	—	—	14.1
Y	26.5	44.4	$\alpha \rightleftharpoons \beta$	4.99	—	—	11.4
La	27.1	56.9	$\alpha \rightleftharpoons \beta$	0.36	$\beta \rightleftharpoons \gamma$	3.12	6.20
Ce	26.9	72.0	$\beta \rightleftharpoons \gamma$	0.05	$\gamma \rightleftharpoons \delta$	2.99	5.46
Pr	27.2	73.2	$\alpha \rightleftharpoons \beta$	3.17	—	—	6.89
Nd	27.5	71.5	$\alpha \rightleftharpoons \beta$	3.03	—	—	7.14
Pm	27.3[a]	71.6[a]	$\alpha \rightleftharpoons \beta$	3.0[a]	—	—	7.7[a]
Sm	29.5	69.6	$\alpha \rightleftharpoons \beta$	0.2[a]	$\beta \rightleftharpoons \gamma$	3.11	8.62
Eu	27.7	77.8	—	—	—	—	9.21
Gd	37.0	68.1	$\alpha \rightleftharpoons \beta$	3.91	—	—	10.0
Tb	28.9	73.2	$\alpha \rightleftharpoons \beta$	5.02	—	—	10.79
Dy	27.7	75.6	$\alpha \rightleftharpoons \beta$	4.16	—	—	11.06
Ho	27.2	75.3	—	—	—	—	17.0[a]
Er	28.1	73.2	—	—	—	—	19.9
Tm	27.0	74.0	—	—	—	—	16.8
Yb	26.7	59.9	$\beta \rightleftharpoons \gamma$	1.75	—	—	7.66
Lu	26.9	51.0	—	—	—	—	22[a]

Note: For additional information, see Hultgren, R., Desai, P.D., Hawkins, D.T., Gleiser, M., Kelley, K.K., and Wagman, D.D., *Selected Values of the Thermodynamic Properties of the Elements,* ASM International, Metals Park, Ohio, 1973; Wagman, D.D., Evans, W.H., Parker, V.B., Schumm, R.H., Halow, I., Bailey, S.M., Churney, K.L., and Nuttall, R.L., *The NBS Tables of Chemical Thermodynamic Properties, J. Phys. Chem. Ref. Data,* Vol. 11, Suppl. 2, 1982; Amitin, E.B., Bessergenev, W.G., Kovalevskaya, Yu. A., and Paukov, I.E., *J. Chem. Thermodyn.,* 15, 181, 1983; Amitin, E.B., Bessergenev, W.G., Kovalevskaya, Yu. A., and Paukov, I.E., *J. Chem. Thermodyn.,* 15, 181, 1983.

[a] Estimated.

TABLE 7. Vapor Pressures, Boiling Points, and Heats of Sublimation of Rare Earth Metals

Rare earth metal	Temperature in °C[a] for a vapor pressure of				Boiling point[a] (°C)	Heat of sublimation at 25 °C (kJ/mol)
	10^{-8} atm (0.001 Pa)	10^{-6} atm (0.101 Pa)	10^{-4} atm (10.1Pa)	10^{-2} atm (1013 Pa)		
Sc	1036	1243	1533	1999	2836	377.8
Y	1222	1460	1812	2360	3345	424.7
La	1301	1566	1938	2506	3464	431.0
Ce	1290	1554	1926	2487	3443	422.6
Pr	1083	1333	1701	2305	3520	355.6
Nd	955	1175	1500	2029	3074	327.6
Pm	—	—	—	—	3000[b]	348[b]
Sm	508	642	835	1150	1794	206.7
Eu	399	515	685	964	1529	175.3
Gd	1167	1408	1760	2306	3273	397.5
Tb	1124	1354	1698	2237	3230	388.7
Dy	804	988	1252	1685	2567	290.4
Ho	845	1036	1313	1771	2700	300.8
Er	908	1113	1405	1896	2868	317.1
Tm	599	748	964	1300	1950	232.2
Yb	301	400	541	776	1196	152.1
Lu	1241	1483	1832	2387	3402	427.6

Note: For additional information, see Hultgren, R., Desai, P.D., Hawkins, D.T., Gleiser, M., Kelley, K.K., and Wagman, D.D., *Selected Values of the Thermodynamic Properties of the Elements,* ASM International, Metals Park, Ohio, 1973; Beaudry, B.J. and Gschneidner, K.A., Jr., in *Handbook on the Physics and Chemistry of Rare Earths,* Vol. 1, Gschneidner, K.A., Jr. and Eyring, L., Eds., North-Holland Physics, Amsterdam, 1978, 173.

[a] International Temperature Scale of 1990 (ITS-90) values.

[b] Estimated.

TABLE 8. Magnetic Properties of the Rare Earth Metals

| Rare earth metal | $\chi_A \times 10^6$ at 298 K (emu/mol) | Effective magnetic moment | | | | Easy axis | Néel temp. T_N (K) | | Curie temp. T_C (K) | θ_p (K) | | |
| | | Paramagnetic at ~298 K | | Ferromagnetic at ~0 K | | | | | | | | |
		Theory[a]	Obs.	Theory[b]	Obs.		Hex sites	Cubic sites		‖c	⊥c	Polycryst. or avg.
αSc	295.2	—	—	—	—	—	—	—	—	—	—	—
αY	187.7	—	—	—	—	—	—	—	—	—	—	—
αLa	95.9	—	—	—	—	—	—	—	—	—	—	—
βLa	105		—	—	—	—	—	—	—	—	—	—
γCe	2,270	2.54	2.52	2.14	—	—	—	14.4	—	—	—	−50
βCe	2,500	2.54	2.61	2.14	—	—	13.7	12.5	—	—	—	−41
αPr	5,530	3.58	3.56	3.20	2.7[c]	a	0.03	—	—	—	—	0
αNd	5,930	3.62	3.45	3.27	2.2[c]	b	19.9	7.5	—	0	5	3.3
αPm	—	2.68	—	2.40	—	—	—	—	—	—	—	—
αSm	1,278[d]	0.85	1.74	0.71	0.5[c]	a	109	14.0	—	—	—	—
Eu	30,900	7.94	8.48	7.0	5.9	<110>	—	90.4	—	—	—	100
αGd	185,000[e]	7.94	7.98	7.0	7.63	30° to c	—	—	293.4	317	317	317
αTb	170,000	9.72	9.77	—	—		230.0	—	—	195	239	224
a′Tb		—	—	9.0	9.34	b	—	—	219.5	—	—	—
αDy	98,000	10.64	10.83	—	—		180.2	—	—	121	169	153
a′Dy	—	—	—	10.0	10.33	a	—	—	90.5[g]	—	—	—
Ho	72,900	10.60	11.2	10.0	10.34	b	132	—	19.5	73.0	88.0	83.0
Er	48,000	9.58	9.9	9.0	9.1	30° to c	85	—	18.7	61.7	32.5	42.2
Tm	24,700	7.56	7.61	7.0	7.14	c	58	—	32.0	41.0	−17.0	2.3
βYb	67[d]	—	—	—	—	—	—	—	—	—	—	—
Lu	182.9	—	—	—	—	—	—	—	—	—	—	—

Note: For additional information, see McEwen, K.A., in *Handbook on the Physics and Chemistry of Rare Earths*, Vol. 1, Gschneidner, K.A., Jr. and Eyring, L., Eds., North-Holland Physics, Amsterdam, 1978, 411; Legvold, S., in *Ferromagnetic Materials*, Vol. 1, Wohlfarth, E.P., Ed., North-Holland Physics, Amsterdam, 1980, 183; Pecharsky, V.K., Gschneidner, K.A., Jr. and Fort, D., *Phys. Rev. B*, 47, 5063, 1993; Pecharsky, V.K., Gschneidner, K.A., Jr. and Fort, D., 1996, to be published; Steward, A.M. and Collocott, S.J., *J. Phys.: Condens. Matter*, 1, 677, 1988.

[a] $g[J(J + 1)]^{1/2}$.
[b] gJ.
[c] At 38 T and 4.2 K.
[d] At 290 K.
[e] At 350 K.
[g] On cooling T_C = 89.6 K and on warming T_C = 91.5 K.

TABLE 9. Room Temperature Coefficient of Thermal Expansion, Thermal Conductivity, Electrical Resistance, and Hall Coefficient

| Rare earth metal | Expansion ($\alpha_i \times 10^6$) (°C^{-1}) | | | Thermal conductivity (W/cm K) | Electrical resistance (μΩ·cm) | | | Hall coefficient ($R_i \times 10^{12}$) (V·cm/A·Oe) | | |
	α_a	α_c	α_{poly}		ρ_a	ρ_c	ρ_{poly}	R_a	R_c	R_{poly}
αSc	7.6	15.3	10.2	0.158	70.9	26.9	56.2[a]	—	—	−0.13
αY	6.0	19.7	10.6	0.172	72.5	35.5	59.6	−0.27	−1.6	—
aLa	4.5	27.2	12.1	0.134	—	—	61.5	—	—	−0.35
bCe	—	—	—	—	—	—	82.8	—	—	—
γCe	6.3	—	6.3	0.113	—	—	74.4	—	—	+1.81
αPr	4.5	11.2	6.7	0.125	—	—	70.0	—	—	+0.709
αNd	7.6	13.5	9.6	0.165	—	—	64.3	—	—	+0.971
αPm	9[b]	16[b]	11[b]	0.15[b]	—	—	75[b]	—	—	—
αSm	9.6	19.0	12.7	0.133	—	—	94.0	—	—	−0.21
Eu	35.0	—	35.0	0.139[b]	—	—	90.0	—	—	+24.4
αGd	9.1[c]	10.0[c]	9.4[c]	0.105	135.1	121.7	131.0	−10	−54	−4.48[d]
αTb	9.3	12.4	10.3	0.111	123.5	101.5	115.0	−1.0	−3.7	—
αDy	7.1	15.6	9.9	0.107	111.0	76.6	92.6	−0.3	−3.7	—
Ho	7.0	19.5	11.2	0.162	101.5	60.5	81.4	+0.2	−3.2	—
Er	7.9	20.9	12.2	0.145	94.5	60.3	86.0	+0.3	−3.6	—
Tm	8.8	22.2	13.3	0.169	88.0	47.2	67.6	—	—	−1.8
βYb	26.3	—	26.3	0.385	—	—	25.0	—	—	+3.77
Lu	4.8	20.0	9.9	0.164	76.6	34.7	58.2	+0.45	−2.6	−0.535

Note: For additional information, see Beaudry, B. J. and Gschneidner, K.A., Jr., in *Handbook on the Physics and Chemistry of Rare Earths*, Vol. 1, Gschneidner, K.A., Jr. and Eyring, L., Eds., North-Holland Physics, Amsterdam, 1978, 173; McEwen, K.A., in *Handbook on the Physics and Chemistry of Rare Earths*, Vol. 1, Gschneidner, K.A., Jr. and Eyring, L., Eds., North-Holland Physics, Amsterdam, 1978, 411.

[a] Calculated from single crystal values.
[b] Estimated.
[c] At 100 °C.
[d] At 77 °C.

TABLE 10. Electronic Specific Heat Constant (γ), Electron–Electron (Coulomb) Coupling Constant (μ^*), Electron–Phonon Coupling Constant (λ), Debye Temperature at 0 K(θ_D), and Superconducting Transition Temperature

Rare earth metal	γ (mJ/mol·K^2)	μ^*	λ	θ_D (K) from Heat capacity	θ_D (K) from Elastic constants	Superconducting temperature (K)
αSc	10.334	0.16	0.30	345.3	—	0.050[a]
αY	7.878	0.15	0.30	244.4	258	1.3[b]
αLa	9.45	0.08	0.76	150	154	5.10
βLa	11.5	—	—	140	—	6.00
αCe	12.8	—	—	179	—	0.022[c]
αPr	20	—	1.07[d]	155[e]	153	—
αNd	f	—	0.86[d]	157[e]	163	—
αPm	—	—	—	159[e]	—	—
αSm	8.1 ± 1.5[g]	—	0.81[d]	162[e,f]	169	—
Eu	f	—	—	f	118	—
αGd	4.48	—	0.30	169	182	—
α'Tb	3.71	—	0.34[d]	169.6	177	—
α'Dy	4.9	—	0.32[d]	192	183	—
Ho	2.1	—	0.30[d]	175[e]	190	—
Er	8.7	—	0.33[d]	176.9	188	—
Tm	f	—	0.36[d]	179[e]	200	—
αYb	3.30	—	—	117.6	118	—
βYb	8.36	—	—	109	—	—
Lu	8.194	0.14	0.31	183.2	185	0.022[h]

Note: For additional information, see Sundström, L.J., in *Handbook on the Physics and Chemistry of Rare Earths*, Vol. 1, Gschneidner, K.A., Jr., and Eyring, L., Eds., North-Holland Physics, Amsterdam, 1978, 379; Scott, T., in *Handbook on the Physics and Chemistry of Rare Earths*, Vol. 1, Gschneidner, K.A., Jr. and Eyring, L., Eds., North-Holland Physics, Amsterdam, 1978, 591; Probst, C. and Wittig, J., in *Handbook on the Physics and Chemistry of Rare Earths*, Vol. 1, Gschneidner, K.A., Jr. and Eyring, L., Eds., North-Holland Physics, Amsterdam, 1978, 749; Tsang, T.-W.E., Gschneidner, K.A., Jr., Schmidt, F.A., and Thome, D.K., *Phys. Rev.*, B, 31, 235, 1985; Collocott, S.J., Hill, R.W. and Stewart, A.M., *J. Phys. F*, 18, L223, 1988; Hill, R.W. and Gschneidner, K.A., Jr., *J. Phys. F*, 18, 2545, 1988; Skriver, H.L. and Mertig, I., *Phys. Rev. B*, 41, 6553, 1990; Collocott, S.J. and Stewart, A.M., *J. Phys.: Condens. Matter*, 4, 6743, 1992; Pecharsky, V.K., Gschneidner, K.A., Jr. and Fort, D., *Phys. Rev. B*, 47, 5063, 1993.

[a] At 18.6 GPa.
[b] At 11 GPa.
[c] At 2.2 GPa.
[d] Calculated value.
[e] Estimated.
[f] Heat capacity results have been reported, but the resultant γ and θ_D values are unreliable because of the presence of impurities and/or there was no reliable procedure or model to correct for the magnetic contribution to the heat capacity.
[g] Based on the values reported for the purer Sm sample (IV).
[h] At 4.5 GPa.

TABLE 11. Room Temperature Elastic Moduli and Mechanical Properties

Rare earth metal	Young's (elastic) modulus	Shear modulus	Bulk modulus	Poisson's ratio	Yield strength 0.2% offset	Ultimate tensile strength	Uniform elongation (%)	Reduction in area (%)	Recryst. temp. (°C)
Sc	74.4	29.1	56.6	0.279	173[a]	255[a]	5.0[a]	8.0[a]	550
Y	63.5	25.6	41.2	0.243	42	129	34.0	—	550
αLa	36.6	14.3	27.9	0.280	126[a]	130	7.9[a]	—	300
βCe	—	—	—	—	86	138	—	24.0	—
γCe	33.6	13.5	21.5	0.24	28	117	22.0	30.0	325
αPr	37.3	14.8	28.8	0.281	73	147	15.4	67.0	400
αNd	41.4	16.3	31.8	0.281	71	164	25.0	72.0	400
αPm	46[b]	18[b]	33[b]	0.28[b]	—	—	—	—	400[b]
αSm	49.7	19.5	37.8	0.274	68	156	17.0	29.5	440
Eu	18.2	7.9	8.3	0.152	—	—	—	—	300
αGd	54.8	21.8	37.9	0.259	15	118	37.0	56.0	500
αTb	55.7	22.1	38.7	0.261	—	—	—	—	500
αDy	61.4	24.7	40.5	0.247	43	139	30.0	30.0	550
Ho	64.8	26.3	40.2	0.231	—	—	—	—	520
Er	69.9	28.3	44.4	0.237	60	136	11.5	11.9	520
Tm	74.0	30.5	44.5	0.213	—	—	—	—	600
βYb	23.9	9.9	30.5	0.207	7	58	43.0	92.0	300
Lu	68.6	27.2	47.6	0.261	—	—	—	—	600

Note: For additional information, see Scott, T., in *Handbook on the Physics and Chemistry of Rare Earths*, Vol. 1, Gschneidner, K.A., Jr. and Eyring, L., Eds., North-Holland Physics, Amsterdam, 1978, 591.

[a] Value is questionable.
[b] Estimated.

TABLE 12. Liquid Metal Properties near the Melting Point

Rare earth metal	Density (g/cm³)	Surface tension (N/m)	Viscosity (centipoise)	Heat capacity (J/mol K)	Thermal conductivity (W/cm K)	Magnetic susceptibility χ × 10⁴ (emu/mol)	Electrical resistivity (μΩ·cm)	ΔV (l→s)[a] (%)	Spectral emittance at λ = 645 nm ε (%)	Temp. range (°C)
Sc	2.80	0.954	—	44.2[b]	—	—	—	—	—	—
Y	4.24	0.871	—	43.1	—	—	—	—	36.8	1522–1647
La	5.96	0.718	2.65	34.3	0.238	1.20	133	−0.6	25.4	920–1287
Ce	6.68	0.706	3.20	37.7	0.210	9.37	130	+1.1	32.2	877–1547
Pr	6.59	0.707	2.85	43.0	0.251	17.3	139	−0.02	28.4	931–1537
Nd	6.72	0.687	—	48.8	0.195	18.7	151	−0.9	39.4	1021–1567
Pm	6.9[b]	0.680[b]	—	50[b]	—	—	160[b]	—	—	—
Sm	7.16	0.431	—	50.2[b]	—	18.3	182	−3.6	43.7	1075
Eu	4.87	0.264	—	38.1	—	97	242	−4.8	—	—
Gd	7.4	0.664	—	37.2	0.149	67	195	−2.0	34.2	1313–1600
Tb	7.65	0.669	—	46.5	—	82	193	−3.1	—	—
Dy	8.2	0.648	—	49.9	0.187	95	210	−4.5	29.7	1412–1437
Ho	8.34	0.650	—	43.9	—	88	221	−7.4	—	—
Er	8.6	0.637	—	38.7	—	69	226	−9.0	37.2	1529–1587
Tm	9.0[b]	—	—	41.4	—	41	235[b]	−6.9	—	—
Yb	6.21	0.320	2.67	36.8	—	—	113	−5.1	—	—
Lu	9.3	0.940	—	47.9[b]	—	—	224	−3.6	—	—

Note: For additional information, see Van Zytveld, J., in *Handbook on the Physics and Chemistry of Rare Earths,* Vol. 12, Gschneidner, K.A., Jr. and Eyring, L., Eds., North-Holland Physics, Amsterdam, 1989, 357; Stretz, L.A. and Bautista, R.G., in *Temperature, Its Measurement and Control in Science and Industry,* Vol. 4, part I, H.H. Plumb, Ed., Instrument Society of America, Pittsburgh, 1972, 489; King, T.S., Baria, D.N., and Bautista, R.G., *Met. Trans. B,* 7, 411, 1976; Baria, D.N., King, T.S., and Bautista, R.G., *Met. Trans. B,* 7, 577, 1976.

[a] Volume change on freezing.
[b] Estimated.

TABLE 13. Ionization Potentials (Electronvolts)

Rare earth	I Neutral atom	II Singly ionized	III Doubly ionized	IV Triply ionized	V Quadruply ionized
Sc	6.56144	12.79967	24.75666	73.4894	91.65
Y	6.217	12.24	20.52	60.597	77.0
La	5.5770	11.060	19.1773	49.95	61.6
Ce	5.5387	10.85	20.198	36.758	65.55
Pr	5.464	10.55	21.624	38.98	57.53
Nd	5.5250	10.73	22.1	40.41	—
Pm	5.554	10.90	22.3	41.1	—
Sm	5.6437	11.07	23.4	41.4	—
Eu	5.6704	11.241	24.92	42.7	—
Gd	6.1500	12.09	20.63	44.0	—
Tb	5.8639	11.52	21.91	39.79	—
Dy	5.9389	11.67	22.8	41.47	—
Ho	6.0216	11.80	22.84	42.5	—
Er	6.1078	11.93	22.74	42.7	—
Tm	6.18431	12.05	23.68	42.7	—
Yb	6.25416	12.1761	25.05	43.56	—
Lu	5.42585	13.9	20.9594	45.25	66.8

Note: For references, see the table "Ionization Potentials of Atoms and Atomic Ions" in Section 10.

TABLE 14. Effective Ionic Radii (Å)[A]

Rare earth ion	R^{2+} CN = 6	CN = 8	R^{3+} CN = 6	CN = 8	CN = 12	R^{4+} CN = 6	CN = 8
Sc	—	—	0.745	0.87	1.116	—	—
Y	—	—	0.900	1.015	1.220	—	—
La	—	—	1.045	1.18	1.320	—	—
Ce	—	—	1.010	1.14	1.290	0.80	0.97
Pr	—	—	0.997	1.14	1.286	0.78	0.96
Nd	—	—	0.983	1.12	1.276	—	—
Pm	—	—	0.97	1.10	1.267	—	—
Sm	1.19	1.27	0.958	1.09	1.260	—	—
Eu	1.17	1.25	0.947	1.07	1.252	—	—
Gd	—	—	0.938	1.06	1.246	—	—
Tb	—	—	0.923	1.04	1.236	0.76	0.88
Dy	—	—	0.912	1.03	1.228	—	—
Ho	—	—	0.901	1.02	1.221	—	—
Er	—	—	0.890	1.00	1.214	—	—
Tm	—	—	0.880	0.99	1.207	—	—
Yb	1.00	1.07	0.868	0.98	1.199	—	—
Lu	—	—	0.861	0.97	1.194	—	—

Note: For additional information, see Shannon, R.D. and Prewitt, C.T., *Acta Cryst.,* 25, 925, 1969 and Shannon, R.D. and Prewitt, C.T., *Acta Cryst.,* 26, 1046, 1970.

[a] Radius of O^{2-} is 1.40 Å for a coordination number (CN) of 6.

MELTING, BOILING, TRIPLE, AND CRITICAL POINTS OF THE ELEMENTS

This table summarizes the significant points on the phase diagrams for the elements. When reliable date are available, values are given for the solid-liquid-gas triple-point temperature t_{tp} and pressure p_{tp}; normal melting point at 101.325 kPa pressure t_m; normal boiling point t_b; and critical temperature t_c and pressure p_c. All temperatures are on the ITS-90 scale.

An "sp" notation in the boiling point column indicates a sublimation point, where the vapor pressure of the solid phase reaches 101.325 kPa (1 atm). Transition temperatures between allotropic forms are included for several elements. The notation "tr" in the melting point column indicates the temperature of the transition to the crystalline form immediately below that entry. An asterisk* indicates an extrapolated or estimated value.

The major data sources are listed below.

References

1. "The International Temperature Scale of 1990", *Metrologia* 27, 3, 1990; errata in *Metrologia* 27, 107, 1990.
2. Bedford, R. E., Bonnier, G., Maas, H., and Pavese, F., *Metrologia* 33, 133, 1996.
3. Lemmon, E. W., Huber, M. L., and McLinden, M. O., NIST Standard Reference Database 23: Reference Fluid Thermodynamic and Transport Properties-REFPROP, Version 9.0, National Institute of Standards and Technology, Standard Reference Data Program, Gaithersburg, MD, 2007 (www.nist.gov/nist23.cfm). [Ar, F, H, He, Kr, N, Ne, O, Xe]
4. Dinsdale, A.T., "SGTE Data for Pure Elements", *CALPHAD* 15, 317-425, 1991.
5. *Landolt-Börnstein, Numerical Data and Functional Relationships in Science and Technology*, New Series, IV/19A, "Thermodynamic Properties of Inorganic Materials compiled by SGTE," Springer-Verlag, Heidelberg; Part 1, 1999.
6. Chase, M. W., *NIST-JANAF Thermochemical Tables, Fourth Edition, J. Phys. Chem. Ref. Data*, Monograph No. 9, 1998.
7. Gurvich, L. V., Veyts, I. V., and Alcock, C. B., *Thermodynamic Properties of Individual Substances, Fourth Edition*, Hemisphere Publishing Corp., New York, 1989.
8. Greenwood, N. N., and Earnshaw, A., *Chemistry of the Elements, Second Edition*, Butterworth-Heinemann, Oxford, 1997.
9. Hultgren, R. R., *Selected Values of Thermodynamic Properties of the Elements*, American Society of Metals, 1973.
10. Geiger, F., Busse, C. A., and Loehrke, R. I., *Int. J. Thermophys.* 8, 425, 1987. [Boiling points of In, Ag, Ga, Cu, Sn, and Au]
11. Velasco, S., Roman, F. L., White, J. A., and Mulero, A., *Fluid Phase Equilib.* 244, 11, 2006. [Critical temperatures of Ag, Be, Bi, Ge, Fe, Mn, Te]
12. Gustafson, P., *Carbon* 24, 169, 1986. [Graphite and diamond]
13. Michels, A., and Prins, C., *Physica* 28, 101, 1962. [Triple point of Ar, Kr, Xe]

Element	Symbol	t_{tp}/°C	p_{tp}/MPa	t_m/°C	t_b/°C	t_c/°C	p_c/MPa
Actinium	Ac			1050	3200*		
Aluminum	Al			660.323	2519	6427*	
Americium	Am			1176	2011*		
Antimony (gray)	Sb			630.628	1587		
Argon	Ar	−189.3442	0.06889	−189.34	−185.848	−122.463	4.863
Arsenic (gray)	As	817	3.70	817	616 sp	1400	22.3
Astatine	At			302*			
Barium	Ba			727	1845*		
Berkelium (β form)	Bk			986			
Beryllium	Be			1287	2468	4932*	
Bismuth	Bi			271.406	1564	4347*	
Boron	B			2077	4000		
Bromine	Br₂	−7.25	0.0058	−7.2	58.8	315	10.34
Cadmium	Cd			321.069	767		
Calcium	Ca			842	1484		
Californium	Cf			900			
Carbon (graphite)	C	4489	10.3		3825 sp		
Carbon (diamond)	C			4440 (12.4 GPa)ᵃ			
Cerium	Ce			799	3443		
Cesium	Cs			28.5	671	1665	9.4
Chlorine	Cl₂			−101.5	−34.04	143.8	7.991
Chromium	Cr			1907	2671		
Cobalt	Co			1495	2927		
Copper	Cu			1084.62	2560		
Curium	Cm			1345			
Dysprosium	Dy			1412	2567		
Einsteinium	Es			860			
Erbium	Er			1529	2868		
Europium	Eu			822	1529		
Fermium	Fm			1527			
Fluorine	F₂	−219.67	0.09	−219.67	−188.11	−128.74	5.1724
Francium	Fr			21			
Gadolinium	Gd			1313	3273		

Element	Symbol	t_{tp}/°C	p_{tp}/MPa	t_m/°C	t_b/°C	t_c/°C	p_c/MPa
Gallium	Ga	29.7666		29.7646	2229		
Germanium	Ge			938.25	2833	9529*	
Gold	Au			1064.18	2836		
Hafnium	Hf			2233	4600		
Helium	He	−270.973	0.005043		−268.928	−267.9547	0.22746
Holmium	Ho			1472	2700		
Hydrogen	H$_2$	−259.3467	0.007041	−259.16	−252.879	−240.212	1.2858
Indium	In	156.5936		156.60	2027		
Iodine	I$_2$	113.6	0.01211	113.7	184.4	546	
Iridium	Ir			2446	4428		
Iron	Fe			1538	2861	9067*	
Krypton	Kr	−157.375	0.07353	−157.37	−153.415	−63.67	5.525
Lanthanum	La			920	3464		
Lawrencium	Lr			1627			
Lead	Pb			327.462	1749		
Lithium	Li			180.50	1342	2950*	67*
Lutetium	Lu			1663	3402		
Magnesium	Mg			650	1090		
Manganese	Mn			1246	2061	4052*	
Mendelevium	Md			827			
Mercury	Hg	−38.8344		−38.8290	356.619	1491	167
Molybdenum	Mo			2622	4639		
Neodymium	Nd			1016	3074		
Neon	Ne	−248.594	0.04337	−248.59	−246.046	−228.6582	2.6786
Neptunium	Np			644	3902*		
Nickel	Ni			1455	2913		
Niobium	Nb			2477	4741		
Nitrogen	N$_2$	−209.999	0.01252	−210.0	−195.795	−146.958	3.3958
Nobelium	No			827			
Osmium	Os			3033	5008		
Oxygen	O$_2$	−218.789	0.0001463	−218.79	−182.962	−118.569	5.043
Palladium	Pd			1554.8	2963		
Phosphorus (white)	P			44.15	280.5	721	
Phosphorus (red)	P			579.2	431 sp	721	
Platinum	Pt			1768.2	3825		
Plutonium	Pu			640	3228		
Polonium	Po			254	962		
Potassium	K			63.5	759	1950*	16*
Praseodymium	Pr			931	3520		
Promethium	Pm			1042	3000*		
Protactinium	Pa			1572			
Radium	Ra			696			
Radon	Rn			−71	−61.7	104	6.28
Rhenium	Re			3185	5590		
Rhodium	Rh			1963	3695		
Rubidium	Rb	39.26		39.30	688	1820*	16*
Ruthenium	Ru			2333	4147		
Samarium	Sm			1072	1794		
Scandium	Sc			1541	2836		
Selenium (vitreous)	Se			180 tr	685	1493	
Selenium (gray)	Se			220.8	685	1493	27.2
Silicon	Si			1414	3265		
Silver	Ag			961.78	2162	6137*	
Sodium	Na			97.794	882.940	2300*	35*
Strontium	Sr			777	1377		
Sulfur (rhombic)	S			95.2 tr	444.61	1041	20.7
Sulfur (monoclinic)	S			115.21	444.61	1041	
Tantalum	Ta			3017	5455		
Technetium	Tc			2157	4262		
Tellurium	Te			449.51	988	2056*	
Terbium	Tb			1359	3230		

Element	Symbol	t_{tp}/°C	p_{tp}/MPa	t_m/°C	t_b/°C	t_c/°C	p_c/MPa
Thallium	Tl			304	1473		
Thorium	Th			1750	4785		
Thulium	Tm			1545	1950		
Tin (gray)	Sn			13.2 tr	2586		
Tin (white)	Sn			231.928	2586		
Titanium	Ti			1670	3287		
Tungsten	W			3414	5555		
Uranium	U			1135	4131		
Vanadium	V			1910	3407		
Xenon	Xe	−111.745	0.08177	−111.75	−108.099	16.583	5.842
Ytterbium	Yb			824	1196		
Yttrium	Y			1522	3345		
Zinc	Zn			419.527	907		
Zirconium	Zr			1854	4406		

* Extrapolated or estimated value
a This is the estimated diamond-graphite-liquid carbon triple point (Ref. 12).

HEAT CAPACITY OF THE ELEMENTS AT 25 °C

This table gives the specific heat capacity (c_p) in J/g K and the molar heat capacity (C_p) in J/mol K at a temperature of 25 °C and a pressure of 100 kPa (1 bar or 0.987 standard atmospheres) for all the elements for which reliable data are available.

Name	c_p J/g K	C_p J/mol K
Actinium	0.120	27.2
Aluminum	0.897	24.20
Antimony	0.207	25.23
Argon	0.520	20.786
Arsenic	0.329	24.64
Barium	0.204	28.07
Beryllium	1.825	16.443
Bismuth	0.122	25.52
Boron	1.026	11.087
Bromine (Br_2)	0.474	75.69
Cadmium	0.232	26.020
Calcium	0.647	25.929
Carbon (graphite)	0.709	8.517
Cerium	0.192	26.94
Cesium	0.242	32.210
Chlorine (Cl_2)	0.479	33.949
Chromium	0.449	23.35
Cobalt	0.421	24.81
Copper	0.385	24.440
Dysprosium	0.173	28.16
Erbium	0.168	28.12
Europium	0.182	27.66
Fluorine (F_2)	0.824	31.304
Gadolinium	0.236	37.03
Gallium	0.373	26.03
Germanium	0.320	23.222
Gold	0.129	25.418
Hafnium	0.144	25.73
Helium	5.193	20.786
Holmium	0.165	27.15
Hydrogen (H_2)	14.304	28.836
Indium	0.233	26.74
Iodine (I_2)	0.214	54.43
Iridium	0.131	25.10
Iron	0.449	25.10
Krypton	0.248	20.786
Lanthanum	0.195	27.11
Lead	0.130	26.84
Lithium	3.582	24.860
Lutetium	0.154	26.86
Magnesium	1.023	24.869
Manganese	0.479	26.32
Mercury	0.140	27.983

Name	c_p J/g K	C_p J/mol K
Molybdenum	0.251	24.06
Neodymium	0.190	27.45
Neon	1.030	20.786
Nickel	0.444	26.07
Niobium	0.265	24.60
Nitrogen (N_2)	1.040	29.124
Osmium	0.130	24.7
Oxygen (O_2)	0.918	29.378
Palladium	0.246	25.98
Phosphorus (white)	0.769	23.824
Platinum	0.133	25.86
Potassium	0.757	29.600
Praseodymium	0.193	27.20
Radon	0.094	20.786
Rhenium	0.137	25.48
Rhodium	0.243	24.98
Rubidium	0.363	31.060
Ruthenium	0.238	24.06
Samarium	0.197	29.54
Scandium	0.568	25.52
Selenium	0.321	25.363
Silicon	0.712	19.99
Silver	0.235	25.350
Sodium	1.228	28.230
Strontium	0.306	26.79
Sulfur (rhombic)	0.708	22.70
Tantalum	0.140	25.36
Tellurium	0.202	25.73
Terbium	0.182	28.91
Thallium	0.129	26.32
Thorium	0.118	27.32
Thulium	0.160	27.03
Tin (white)	0.227	26.99
Titanium	0.523	25.060
Tungsten	0.132	24.27
Uranium	0.116	27.665
Vanadium	0.489	24.89
Xenon	0.158	20.786
Ytterbium	0.155	26.74
Yttrium	0.298	26.53
Zinc	0.388	25.390
Zirconium	0.278	25.36

VAPOR PRESSURE OF THE METALLIC ELEMENTS — EQUATIONS

C. B. Alcock

This table gives coefficients in an equation for the vapor pressure of 65 metallic elements in both the solid and liquid states. Vapor pressures in the range 10^{-10} to 10^2 Pa (10^{-15} to 10^{-3} atm) are covered. The equation is:

for p in atmospheres: $\log(p/\text{atm}) = A + BT^{-1} + C\log T + DT^{-3}$

for p in pascals: $\log(p/\text{Pa}) = 5.006 + A + BT^{-1} + C\log T + DT^{-3}$

for p in torr (mmHg): $\log(p/\text{torr}) = 2.881 + A + BT^{-1} + C\log T + DT^{-3}$

where T is the temperature in K.

This equation reproduces the observed vapor pressures to an accuracy of 5% or better. The metals are listed alphabetically by name, and the melting point is included.

The table following this one gives values of the vapor pressure at several temperatures in the 400 K to 2400 K range, as calculated from these equations.

Reprinted with permission of the publisher, Pergamon Press.

Reference

Alcock, C. B., Itkin, V. P., and Horrigan, M. K., *Canadian Metallurgical Quarterly*, 23, 309, 1984.

Element	Phase	A	B	C	D	Range/K	mp/K
Aluminum	Solid	9.459	-17342	-0.7927		298-mp	933
Aluminum	Liquid	5.911	-16211			mp-1800	
Americium	Solid	11.311	-15059	-1.3449		298-mp	1449
Barium	Solid	12.405	-9690	-2.2890		298-mp	1000
Barium	Liquid	4.007	-8163			mp-1200	
Beryllium	Solid	8.042	-17020	-0.4440		298-mp	1560
Beryllium	Liquid	5.786	-15731			mp-1800	
Cadmium	Solid	5.939	-5799			298-mp	594
Cadmium	Liquid	5.242	-5392			mp-650	
Calcium	Solid	10.127	-9517	-1.4030		298-mp	1115
Cerium	Solid	6.139	-21752			298-mp	1071
Cerium	Liquid	5.611	-21200			mp-2450	
Cesium	Solid	4.711	-3999			298-mp	302
Cesium	Liquid	4.165	-3830			mp-550	
Chromium	Solid	6.800	-20733	0.4391	-0.4094	298-2000	2180
Cobalt	Solid	10.976	-22576	-1.0280		298-mp	1768
Cobalt	Liquid	6.488	-20578			mp-2150	
Copper	Solid	9.123	-17748	-0.7317		298-mp	1358
Copper	Liquid	5.849	-16415			mp-1850	
Curium	Solid	8.369	-20364	-0.5770		298-mp	1618
Curium	Liquid	5.223	-18292			mp-2200	
Dysprosium	Solid	9.579	-15336	-1.1114		298-mp	1685
Erbium	Solid	9.916	-16642	-1.2154		298-mp	1802
Erbium	Liquid	4.668	-14380			mp-1900	
Europium	Solid	9.240	-9459	-1.1661		298-mp	1095
Gadolinium	Solid	8.344	-20861	-0.5775		298-mp	1586
Gadolinium	Liquid	5.557	-19389			mp-2250	
Gallium	Solid	6.657	-14208			298-mp	303
Gallium	Liquid	6.754	-13984	-0.3413		mp-1600	
Gold	Solid	9.152	-19343	-0.7479		298-mp	1337
Gold	Liquid	5.832	-18024			mp-2050	
Hafnium	Solid	9.445	-32482	-0.6735		298-mp	2506
Holmium	Solid	9.785	-15899	-1.1753		298-mp	1747
Indium	Solid	5.991	-12548			298-mp	430
Indium	Liquid	5.374	-12276			mp-1500	
Iridium	Solid	10.506	-35099	-0.7500		298-2500	2719
Iron	Solid	7.100	-21723	0.4536	-0.5846	298-mp	1811
Iron	Liquid	6.347	-19574			mp-2100	
Lanthanum	Solid	7.463	-22551	-0.3142		298-mp	1191
Lanthanum	Liquid	5.911	-21855			mp-2450	
Lead	Solid	5.643	-10143			298-mp	600
Lead	Liquid	4.911	-9701			mp-1200	
Lithium	Solid	5.667	-8310			298-mp	454
Lithium	Liquid	5.055	-8023			mp-1000	
Lutetium	Solid	8.793	-22423	-0.6200		298-mp	1936
Lutetium	Liquid	5.648	-20302			mp-2350	

Element	Phase	A	B	C	D	Range/K	mp/K
Magnesium	Solid	8.489	-7813	-0.8253		298-mp	923
Manganese	Solid	12.805	-15097	-1.7896		298-mp	1519
Mercury	Liquid	5.116	-3190			298-400	234
Molybdenum	Solid	11.529	-34626	-1.1331		298-2500	2895
Neodymium	Solid	8.996	-17264	-0.9519		298-mp	1294
Neodymium	Liquid	4.912	-15824			mp-2000	
Neptunium	Solid	19.643	-24886	-3.9991		298-mp	917
Neptunium	Liquid	10.076	-23378	-1.3250		mp-2500	
Nickel	Solid	10.557	-22606	-0.8717		298-mp	1728
Nickel	Liquid	6.666	-20765			mp-2150	
Niobium	Solid	8.882	-37818	-0.2575		298-2500	2750
Osmium	Solid	9.419	-41198	-0.3896		298-2500	3306
Palladium	Solid	9.502	-19813	-0.9258		298-mp	1828
Palladium	Liquid	5.426	-17899			mp-2100	
Platinum	Solid	4.882	-29387	1.1039	-0.4527	298-mp	2041
Platinum	Liquid	6.386	-26856			mp-2500	
Plutonium	Solid	26.160	-19162	-6.6675		298-600	913
Plutonium	Solid	18.858	-18460	-4.4720		500-mp	
Plutonium	Liquid	3.666	-16658			mp-2450	
Potassium	Solid	4.961	-4646			298-mp	337
Potassium	Liquid	4.402	-4453			mp-600	
Praseodymium	Solid	8.859	-18720	-0.9512		298-mp	1204
Praseodymium	Liquid	4.772	-17315			mp-2200	
Protactinium	Solid	10.552	-34869	-1.0075		298-mp	1845
Protactinium	Liquid	6.177	-32874			mp-2500	
Rhenium	Solid	11.543	-40726	-1.1629		298-2500	3459
Rhodium	Solid	10.168	-29010	-0.7068		298-mp	2236
Rhodium	Liquid	6.802	-26792			mp-2500	
Rubidium	Solid	4.857	-4215			298-mp	312
Rubidium	Liquid	4.312	-4040			mp-550	
Ruthenium	Solid	9.755	-34154	-0.4723		298-mp	2606
Samarium	Solid	9.988	-11034	-1.3287		298-mp	1347
Scandium	Solid	6.650	-19721	0.2885	-0.3663	298-mp	1814
Scandium	Liquid	5.795	-17681			mp-2000	
Silver	Solid	9.127	-14999	-0.7845		298-mp	1235
Silver	Liquid	5.752	-13827			mp-1600	
Sodium	Solid	5.298	-5603			298-mp	371
Sodium	Liquid	4.704	-5377			mp-700	
Strontium	Solid	9.226	-8572	-1.1926		298-mp	1050
Tantalum	Solid	16.807	-41346	-3.2152	0.7437	298-2500	3280
Terbium	Solid	9.510	-20457	-0.9247		298-mp	1629
Terbium	Liquid	5.411	-18639			mp-2200	
Thallium	Solid	5.971	-9447			298-mp	577
Thallium	Liquid	5.259	-9037			mp-1100	
Thorium	Solid	8.668	-31483	-0.5288		298-mp	2023
Thorium	Liquid	-18.453	-24569	6.6473		mp-2500	
Thulium	Solid	8.882	-12270	-0.9564		298-1400	1818
Tin	Solid	6.036	-15710			298-mp	505
Tin	Liquid	5.262	-15332			mp-1850	
Titanium	Solid	11.925	-24991	-1.3376		298-mp	1943
Titanium	Liquid	6.358	-22747			mp-2400	
Tungsten	Solid	2.945	-44094	1.3677		298-2350	3687
Tungsten	Solid	54.527	-57687	-12.2231		2200-2500	
Uranium	Solid	0.770	-27729	2.6982	-1.5471	298-mp	1408
Uranium	Liquid	20.735	-28776	-4.0962		mp-2500	
Vanadium	Solid	9.744	-27132	-0.5501		298-mp	2183
Vanadium	Liquid	6.929	-25011			mp-2500	
Ytterbium	Solid	9.111	-8111	-1.0849		298-900	1092
Yttrium	Solid	9.735	-22306	-0.8705		298-mp	1795
Yttrium	Liquid	5.795	-20341			mp-2300	
Zinc	Solid	6.102	-6776			298-mp	693
Zinc	Liquid	5.378	-6286			mp-750	
Zirconium	Solid	10.008	-31512	-0.7890		298-m.p	2127
Zirconium	Liquid	6.806	-30295			mp-2500	

VAPOR PRESSURE OF THE METALLIC ELEMENTS — DATA

The following values of the vapor pressure of metallic elements are calculated from the equations in the preceding table. All values are given in pascals. For conversion, note that 1 Pa = 7.50 μmHg = 9.87·10⁻⁶ atm.

Metal	mp/K	400 K	600 K	800 K	1000 K	1200 K	1400 K	1600 K	1800 K	2000 K	2200 K	2400 K
Aluminum	933			3.06×10^{-10}	5.08×10^{-6}	0.00256	0.218	6.10	81.4			
Americium	1449			3.88×10^{-7}	0.00167	0.423	21.35					
Barium	1000		7.97×10^{-6}	0.0450	7.11	162						
Beryllium	1560			3.04×10^{-10}	4.96×10^{-6}	0.00314	0.312	9.12	113			
Cadmium	594	0.000280	18.2									
Calcium	1115		2.36×10^{-5}	0.146	25.5							
Cerium	1071				2.47×10^{-11}	8.91×10^{-8}	2.97×10^{-5}	0.00233	0.0691	1.04	9.56	60.8
Cesium	302	0.394										
Chromium	2180					2.45×10^{-8}	7.59×10^{-5}	0.0239	1.80	52.1	774	
Cobalt	1768					2.09×10^{-10}	1.00×10^{-5}	0.000419	0.0379	1.15	16.0	
Copper	1358			6.60×10^{-11}	1.53×10^{-6}	0.00122	0.135	3.94	54.4			
Curium	1618				1.90×10^{-9}	4.24×10^{-6}	0.00103	0.0629	1.17	12.1	82.1	
Dysprosium	1685			1.54×10^{-8}	8.21×10^{-5}	0.0241	1.362	27.5				
Erbium	1802			3.90×10^{-10}	4.30×10^{-6}	0.00205	0.163	4.23	52.5			
Europium	1095		1.74×10^{-5}	0.109	19.4							
Gadolinium	1586				5.70×10^{-10}	1.54×10^{-6}	0.000429	0.0279	0.618	7.39	56.2	
Gallium	303			1.94×10^{-7}	0.000565	0.114	4.98	84.4				
Gold	1337				3.72×10^{-8}	5.44×10^{-5}	0.00920	0.374	6.68	67.0		
Hafnium	2506						1.35×10^{-11}	9.81×10^{-9}	1.63×10^{-6}	9.69×10^{-5}	0.00272	0.0437
Holmium	1747			3.20×10^{-9}	2.32×10^{-5}	0.00837	0.546	12.3				
Indium	430		8.31×10^{-11}	1.08×10^{-5}	0.0127	1.413	40.9					
Iridium	2719							1.48×10^{-9}	3.72×10^{-7}	3.06×10^{-5}	0.00112	0.0225
Iron	1811				5.54×10^{-9}	2.51×10^{-5}	0.0104	0.961	32.7	36.8		
Lanthanum	1191				5.09×10^{-8}	2.02×10^{-5}	0.00181	0.0596	0.976	9.61	64.7	
Lead	601		5.54×10^{-7}	0.00618	1.64	68.1						
Lithium	454	7.90×10^{-11}	0.000489	1.08	109							
Lutetium	1936				3.28×10^{-11}	1.59×10^{-7}	6.79×10^{-5}	0.00628	0.211	3.18	26.7	
Magnesium	923	6.53×10^{-9}	0.0152	21.5								
Manganese	1519			5.55×10^{-7}	0.00221	0.524	24.9					
Mercury	234	140										
Molybdenum	2895							1.83×10^{-9}	4.07×10^{-7}	3.03×10^{-5}	0.00102	0.0189
Neodymium	1294			4.55×10^{-11}	7.62×10^{-7}	0.000483	0.0412	1.07	13.4	101		
Neptunium	917					3.31×10^{-9}	1.63×10^{-6}	0.000168	0.00604	0.105	1.06	7.28
Nickel	1728				2.19×10^{-10}	1.09×10^{-6}	0.000471	0.0438	1.37	19.5		
Niobium	2750							2.32×10^{-11}	9.54×10^{-9}	1.17×10^{-6}	5.98×10^{-5}	0.00158
Osmium	3306								1.85×10^{-10}	3.46×10^{-8}	2.49×10^{-6}	8.75×10^{-5}
Palladium	1828				8.27×10^{-9}	1.40×10^{-5}	0.00277	0.144	3.07	30.4		
Platinum	2041						2.34×10^{-8}	1.14×10^{-5}	0.00143	0.0689	0.153	1.59
Plutonium	913				1.03×10^{-8}	6.17×10^{-6}	0.000594	0.0182	0.262	2.20	12.6	53.8
Potassium	337	0.0188	96.9									
Praseodymium	1204				1.95×10^{-8}	2.16×10^{-5}	0.00257	0.0904	1.44	13.2	80.8	
Protactinium	1845							3.44×10^{-10}	8.06×10^{-8}	5.57×10^{-6}	0.000174	0.00306
Rhenium	3459								1.37×10^{-10}	2.22×10^{-8}	1.41×10^{-6}	4.45×10^{-5}
Rhodium	2236						1.69×10^{-8}	5.99×10^{-6}	0.000571	0.0217	0.422	4.41
Rubidium	312	0.165										
Ruthenium	2606							7.96×10^{-9}	1.77×10^{-6}	0.000133	0.00455	0.0858
Samarium	1347		8.17×10^{-8}	0.00221	0.942	51.0						
Scandium	1814				6.31×10^{-8}	0.000129	0.0300	1.80	43.6	91.3		
Silver	1235			1.27×10^{-7}	0.000603	0.165	7.61	131				
Sodium	371	0.000185	5.60									
Strontium	1050	4.99×10^{-11}	0.000429	1.134	121							
Tantalum	3280									3.36×10^{-10}	1.87×10^{-8}	5.21×10^{-7}
Terbium	1629				1.92×10^{-9}	4.18×10^{-6}	0.000988	0.0585	1.15	12.5	88.0	
Thallium	577		1.59×10^{-5}	0.0931	16.9							
Thorium	2023						3.33×10^{-11}	2.00×10^{-8}	2.89×10^{-6}	0.000154	0.00401	0.0610
Thulium	1818		6.03×10^{-10}	5.94×10^{-5}	0.0561	5.22	130					
Tin	505			1.26×10^{-9}	8.62×10^{-6}	0.00310	0.207	4.85	56.3			
Titanium	1943					9.69×10^{-9}	7.44×10^{-6}	0.00106	0.0493	0.978	10.6	76.9
Tungsten	3687									2.62×10^{-10}	3.01×10^{-8}	1.59×10^{-6}
Uranium	1408					9.47×10^{-10}	2.87×10^{-6}	0.000263	0.00678	0.0933	0.803	
Vanadium	2183					2.79×10^{-10}	4.35×10^{-7}	0.000107	0.00769	0.233	3.68	32.6
Ytterbium	1092	1.03×10^{-9}	0.00384	6.74								
Yttrium	1795				6.66×10^{-11}	2.96×10^{-7}	0.000117	0.0102	0.316	4.27	35.9	
Zinc	693	1.47×10^{-6}	0.653									
Zirconium	2127						1.05×10^{-10}	6.17×10^{-8}	8.68×10^{-6}	0.000450	0.0110	0.155

DENSITY OF MOLTEN ELEMENTS AND REPRESENTATIVE SALTS

This table lists the liquid density at the melting point, ρ_m, for elements that are solid at room temperature, as well as for some representative salts of these elements. Densities at higher temperatures (up to the t_{max} given in the last column) may be estimated from the equation

$$\rho(t) = \rho_m - k(t - t_m)$$

where t_m is the melting point and k is given in the fifth column of the table. If a value of t_{max} is not given, the equation should not be used to extrapolate more than about 20 °C beyond the melting point.

Data for the elements were selected from the primary literature; the assistance of Gernot Lang in compiling these data is gratefully acknowledged. The molten salt data were derived from Reference 1.

References

1. Janz, G. J., Thermodynamic and Transport Properties of Molten Salts: Correlation Equations for Critically Evaluated Density, Surface Tension, Electrical Conductance, and Viscosity Data, *J. Phys. Chem. Ref. Data*, 17, Suppl. 2, 1988.
2. Nasch, P. M., and Steinemann, S. G., *Phys. Chem. Liq.*, 29, 43, 1995.
3. Assael, M. J., Kakosimos, K., Banish, R. M., Brillo, J., Egry, I., Brooks, R., Quested, P. N., Mills, K. C., Nagashima, A., Sato, Y., and Wakeham, W. A., *J. Phys. Chem. Ref. Data* 35, 285, 2006. [Al, Fe]
4. Assael, M. J., Kalyva, A. E., Antoniadis, K. D., Banish, R. M., Egry, I., Wu, J., Kaschnitz, E., and Wakeham, W. A., *J. Phys. Chem. Ref. Data* 39, 033105-1, 2010. [Cu, Sn]

Formula	Name	t_m/°C	ρ_m/g cm^{-3}	k/g cm^{-3} °C^{-1}	t_{max}
Ag	Silver	961.78	9.320	0.0009	1500
AgBr	Silver(I) bromide	430	5.577	0.001035	667
AgCl	Silver(I) chloride	455	4.83	0.00094	627
AgI	Silver(I) iodide	558	5.58	0.00101	802
AgNO$_3$	Silver(I) nitrate	210	3.970	0.001098	360
Ag$_2$SO$_4$	Silver(I) sulfate	660	4.84	0.001089	770
Al	Aluminum	660.32	2.377	0.000311	917
AlBr$_3$	Aluminum bromide	97.5	2.647	0.002435	267
AlCl$_3$	Aluminum chloride	192.6	1.302	0.002711	296
AlI$_3$	Aluminum iodide	188.32	3.223	0.0025	240
As	Arsenic	817	5.22	0.000544	
Au	Gold	1064.18	17.31	0.001343	1200
B	Boron	2075	2.08		
Ba	Barium	727	3.338	0.000299	1550
BaBr$_2$	Barium bromide	857	3.991	0.000924	900
BaCl$_2$	Barium chloride	961	3.174	0.000681	1081
BaF$_2$	Barium fluoride	1368	4.14	0.000999	1727
BaI$_2$	Barium iodide	711	4.26	0.000977	975
Be	Beryllium	1287	1.690	0.00011	
BeCl$_2$	Beryllium chloride	415	1.54	0.0011	473
BeF$_2$	Beryllium fluoride	552	1.96	0.000015	850
Bi	Bismuth	271.406	10.05	0.00135	800
BiBr$_3$	Bismuth bromide	219	4.76	0.002637	927
BiCl$_3$	Bismuth chloride	234	3.916	0.0023	350
Ca	Calcium	842	1.378	0.000230	1484
CaBr$_2$	Calcium bromide	742	3.111	0.0005	791
CaCl$_2$	Calcium chloride	775	2.085	0.000422	950
CaF$_2$	Calcium fluoride	1418	2.52	0.000391	2027
CaI$_2$	Calcium iodide	783	3.443	0.000751	1028
Cd	Cadmium	321.069	7.996	0.001218	500
CdBr$_2$	Cadmium bromide	568	4.075	0.00108	720
CdCl$_2$	Cadmium chloride	568	3.392	0.00082	807
CdI$_2$	Cadmium iodide	388	4.396	0.001117	700
Ce	Cerium	799	6.55	0.000710	1460
CeCl$_3$	Cerium(III) chloride	807	3.25	0.00092	950
CeF$_3$	Cerium(III) fluoride	1430	4.659	0.000936	1927
Co	Cobalt	1495	7.75	0.00165	1580
Cr	Chromium	1907	6.3	0.0011	2100
Cs	Cesium	28.44	1.843	0.000556	510
CsBr	Cesium bromide	636	3.133	0.001223	860
CsCl	Cesium chloride	646	2.79	0.001065	906
CsF	Cesium fluoride	703	3.649	0.001282	912
CsI	Cesium iodide	632	3.197	0.001183	907
CsNO$_3$	Cesium nitrate	409	2.820	0.001166	491
Cs$_2$SO$_4$	Cesium sulfate	1005	3.1	0.00095	1530
Cu	Copper	1084.62	7.997	0.000819	2227
CuCl	Copper(I) chloride	423	3.692	0.00076	585

Formula	Name	t_m/°C	ρ_m/g cm^{-3}	k/g cm^{-3} °C^{-1}	t_{max}
Dy	Dysprosium	1411	8.37	0.00143	1540
DyCl$_3$	Dysprosium(III) chloride	718	3.62	0.00068	987
Er	Erbium	1529	8.86	0.00157	1700
Eu	Europium	822	5.13	0.0028	980
Fe	Iron	1537.9	7.035	0.000926	2207
FeCl$_2$	Iron(II) chloride	677	2.348	0.000555	877
Ga	Gallium	29.7666	6.08	0.00062	400
GaBr$_3$	Gallium(III) bromide	123	3.116	0.00246	135
GaCl$_3$	Gallium(III) chloride	77.9	2.053	0.002083	141
GaI$_3$	Gallium(III) iodide	212	3.630	0.002377	252
Gd	Gadolinium	1314	7.4		
GdCl$_3$	Gadolinium(III) chloride	602	3.56	0.000671	1007
GdI$_3$	Gadolinium(III) iodide	930	4.12	0.000908	1032
Ge	Germanium	938.25	5.60	0.00055	1600
Hf	Hafnium	2233	12		
HgBr$_2$	Mercury(II) bromide	241	5.126	0.003233	319
HgCl$_2$	Mercury(II) chloride	277	4.368	0.002862	304
HgI$_2$	Mercury(II) iodide	256	5.222	0.003235	354
Ho	Holmium	1472	8.34		
In	Indium	156.60	7.02	0.000836	500
InBr$_3$	Indium(III) bromide	420	3.121	0.0015	528
InCl$_3$	Indium(III) chloride	583	2.140	0.0021	666
InI$_3$	Indium(III) iodide	207	3.820	0.0015	360
Ir	Iridium	2446	19		
K	Potassium	63.38	0.828	0.000232	500
KBr	Potassium bromide	734	2.127	0.000825	930
KCl	Potassium chloride	771	1.527	0.000583	939
KF	Potassium fluoride	858	1.910	0.000651	1037
KI	Potassium iodide	681	2.448	0.000956	904
KNO$_3$	Potassium nitrate	334	1.865	0.000723	457
La	Lanthanum	920	5.94	0.00061	1600
LaBr$_3$	Lanthanum bromide	788	4.933	0.000096	912
LaCl$_3$	Lanthanum chloride	858	3.209	0.000777	973
LaF$_3$	Lanthanum fluoride	1493	4.589	0.000682	2177
LaI$_3$	Lanthanum iodide	778	4.29	0.001110	907
Li	Lithium	180.5	0.512	0.00052	285
LiBr	Lithium bromide	550	2.528	0.000652	739
LiCl	Lithium chloride	610	1.502	0.000432	781
LiF	Lithium fluoride	848.2	1.81	0.000490	1047
LiI	Lithium iodide	469	3.109	0.000917	667
LiNO$_3$	Lithium nitrate	253	1.781	0.000546	441
Li$_2$SO$_4$	Lithium sulfate	860	2.003	0.000407	1214
Lu	Lutetium	1663	9.3		
Mg	Magnesium	650	1.584	0.000234	900
MgBr$_2$	Magnesium bromide	711	2.62	0.000478	935
MgCl$_2$	Magnesium chloride	714	1.68	0.000271	826
MgI$_2$	Magnesium iodide	634	3.05	0.000651	888
Mn	Manganese	1246	5.95	0.00105	1590
MnCl$_2$	Manganese(II) chloride	650	2.353	0.000437	850
Mo	Molybdenum	2623	9.33		
Na	Sodium	97.794	0.927	0.00023	600
NaBr	Sodium bromide	747	2.342	0.000816	945
Na$_2$CO$_3$	Sodium carbonate	856	1.972	0.000448	1004
NaCl	Sodium chloride	800.7	1.556	0.000543	1027
NaF	Sodium fluoride	996	1.948	0.000636	1097
NaI	Sodium iodide	661	2.742	0.000949	912
NaNO$_3$	Sodium nitrate	306.5	1.90	0.000715	370
Na$_2$SO$_4$	Sodium sulfate	884	2.069	0.000483	1077
Nd	Neodymium	1016	6.89	0.00076	1350
Ni	Nickel	1455	7.81	0.000726	1700
NiCl$_2$	Nickel(II) chloride	1031	2.653	0.00066	1057
Os	Osmium	3033	20		
Pb	Lead	327.462	10.66	0.00122	700
PbBr$_2$	Lead(II) bromide	371	5.73	0.00165	600
PbCl$_2$	Lead(II) chloride	501	4.951	0.0015	710
PbI$_2$	Lead(II) iodide	410	5.691	0.001594	697
Pd	Palladium	1554.8	10.38	0.001169	1700

Formula	Name	$t_m/°C$	$\rho_m/\text{g cm}^{-3}$	$k/\text{g cm}^{-3}\,°C^{-1}$	t_{max}
Pr	Praseodymium	931	6.50	0.00093	1460
PrCl$_3$	Praseodymium chloride	786	3.23	0.00074	977
Pt	Platinum	1768.2	19.77	0.0024	2200
Pu	Plutonium	640	16.63	0.001419	950
Rb	Rubidium	39.31	1.46	0.000451	800
RbBr	Rubidium bromide	692	2.715	0.001072	907
Rb$_2$CO$_3$	Rubidium carbonate	837	2.84	0.000640	1007
RbCl	Rubidium chloride	724	2.248	0.000883	923
RbF	Rubidium fluoride	795	2.87	0.00102	1067
RbI	Rubidium iodide	656	2.904	0.001143	902
RbNO$_3$	Rubidium nitrate	310	2.519	0.001068	417
Rb$_2$SO$_4$	Rubidium sulfate	1066	2.56	0.000665	1545
Re	Rhenium	3185	18.9		
Rh	Rhodium	1964	10.7	0.000895	2200
Ru	Ruthenium	2334	10.65		
S	Sulfur	115.21	1.819	0.00080	160
Sb	Antimony	630.628	6.53	0.00067	745
SbCl$_3$	Antimony(III) chloride	73.4	2.681	0.002293	77
SbCl$_5$	Antimony(V) chloride	4	2.37	0.001869	77
SbI$_3$	Antimony(III) iodide	171	4.171	0.002483	322
Sc	Scandium	1541	2.80		
Se	Selenium	220.8	3.99		
Si	Silicon	1414	2.57	0.00036	1500
Sm	Samarium	1072	7.16		
Sn	Tin	231.93	6.979	0.000652	1650
SnCl$_2$	Tin(II) chloride	247	3.36	0.001253	480
SnCl$_4$	Tin(IV) chloride	−33	2.37	0.002687	138
Sr	Strontium	777	6.980		
SrBr$_2$	Strontium bromide	657	3.70	0.000745	1004
SrCl$_2$	Strontium chloride	874	2.727	0.000578	1037
SrF$_2$	Strontium fluoride	1477	3.470	0.000751	1927
SrI$_2$	Strontium iodide	538	4.085	0.000885	1026
Ta	Tantalum	3017	15		
TaCl$_5$	Tantalum(V) chloride	216.6	2.700	0.004316	457
Tb	Terbium	1359	7.65		
Te	Tellurium	449.51	5.70	0.00035	600
ThCl$_4$	Thorium chloride	770	3.363	0.0014	847
ThF$_4$	Thorium fluoride	1110	6.058	0.000759	1378
Ti	Titanium	1668	4.11		
TiCl$_4$	Titanium(IV) chloride	−25	1.807	0.001735	137
Tl	Thallium	304	11.22	0.00144	600
TlBr	Thallium(I) bromide	460	5.98	0.001755	647
TlCl	Thallium(I) chloride	431	5.628	0.0018	642
TlI	Thallium(I) iodide	441.8	6.15	0.001761	737
TlNO$_3$	Thallium(I) nitrate	206	4.91	0.001873	279
Tl$_2$SO$_4$	Thallium(I) sulfate	632	5.62	0.00130	927
Tm	Thulium	1545	8.56	0.00050	1675
U	Uranium	1135	17.3		
UCl$_3$	Uranium(III) chloride	837	4.84	0.007943	1057
UCl$_4$	Uranium(IV) chloride	590	3.572	0.001945	667
UF$_4$	Uranium(IV) fluoride	1036	6.485	0.000992	1341
V	Vanadium	1910	5.5		
W	Tungsten	3422	17.6		
Y	Yttrium	1526	4.24		
YCl$_3$	Yttrium chloride	721	2.510	0.0005	845
Yb	Ytterbium	824	6.21		
Zn	Zinc	419.53	6.57	0.0011	700
ZnBr$_2$	Zinc bromide	402	3.47	0.000959	602
ZnCl$_2$	Zinc chloride	290	2.54	0.00053	557
ZnI$_2$	Zinc iodide	450	3.878	0.00136	588
ZnSO$_4$	Zinc sulfate	680	3.14	0.00047	987
Zr	Zirconium	1854.7	5.8		
ZrCl$_4$	Zirconium chloride	437	1.643	0.007464	492

MAGNETIC SUSCEPTIBILITY OF THE ELEMENTS AND INORGANIC COMPOUNDS

When a material is placed in a magnetic field H, a magnetization (magnetic moment per unit volume) M is induced in the material which is related to H by $M = \kappa H$, where κ is called the volume susceptibility. Since H and M have the same dimensions, κ is dimensionless. A more useful parameter is the molar susceptibility χ_m, defined by

$$\chi_m = \kappa V_m = \kappa\, M / \rho$$

where V_m is the molar volume of the substance, M the molar mass, and ρ the mass density. When the cgs system is used, the customary units for χ_m are $cm^3\ mol^{-1}$; the corresponding SI units are $m^3\ mol^{-1}$.

Substances that have no unpaired electron orbital or spin angular momentum generally have negative values of χ_m and are called diamagnetic. Their molar susceptibility varies only slightly with temperature. Substances with unpaired electrons, which are termed paramagnetic, have positive χ_m and show a much stronger temperature dependence, varying roughly as $1/T$. The net susceptibility of a paramagnetic substance is the sum of the paramagnetic and diamagnetic contributions, but the former almost always dominates.

This table gives values of χ_m for the elements and selected inorganic compounds. All values refer to nominal room temperature (285 to 300 K) unless otherwise indicated. When the physical state (s = solid, l = liquid, g = gas, aq = aqueous solution) is not given, the most common crystalline form is understood. An entry of Ferro. indicates a ferromagnetic substance.

Substances are arranged in alphabetical order by the most common name, except that compounds such as hydrides, oxides, and acids are grouped with the parent element (the same ordering used in the table "Physical Constants of Inorganic Compounds").

In keeping with customary practice, the molar susceptibility is given here in units appropriate to the cgs system. These values should be multiplied by 4π to obtain values for use in SI equations (where the magnetic field strength H has units of $A\ m^{-1}$).

References

1. *Landolt-Börnstein, Numerical Data and Functional Relationships in Science and Technology, New Series*, II/16, *Diamagnetic Susceptibility*, Springer-Verlag, Heidelberg, 1986.
2. *Landolt-Börnstein, Numerical Data and Functional Relationships in Science and Technology, New Series*, III/19, Subvolumes a to i2, *Magnetic Properties of Metals*, Springer-Verlag, Heidelberg, 1986-1992.
3. *Landolt-Börnstein, Numerical Data and Functional Relationships in Science and Technology, New Series*, II/2, II/8, II/10, II/11, and II/12a, *Coordination and Organometallic Transition Metal Compounds*, Springer-Verlag, Heidelberg, 1966-1984.
4. *Tables de Constantes et Données Numérique, Volume 7, Relaxation Paramagnetique*, Masson, Paris, 1957.

Name	Formula	$\chi_m/10^{-6}\ cm^3\ mol^{-1}$
Aluminum	Al	+16.5
Aluminum trifluoride	AlF_3	−13.9
Aluminum oxide	Al_2O_3	−37
Aluminum sulfate	$Al_2(SO_4)_3$	−93
Ammonia (g)	NH_3	−16.3
Ammonia (aq)	NH_3	−18.3
Ammonium acetate	$NH_4C_2H_3O_2$	−41.1
Ammonium bromide	NH_4Br	−47
Ammonium carbonate	$(NH_4)_2CO_3$	−42.5
Ammonium chlorate	NH_4ClO_3	−42.1
Ammonium chloride	NH_4Cl	−36.7
Ammonium fluoride	NH_4F	−23
Ammonium iodate	NH_4IO_3	−62.3
Ammonium iodide	NH_4I	−66
Ammonium nitrate	NH_4NO_3	−33
Ammonium sulfate	$(NH_4)_2SO_4$	−67
Ammonium thiocyanate	NH_4SCN	−48.1
Antimony	Sb	−99
Stibine (g)	SbH_3	−34.6
Antimony(III) bromide	$SbBr_3$	−111.4
Antimony(III) chloride	$SbCl_3$	−86.7
Antimony(III) fluoride	SbF_3	−46
Antimony(III) iodide	SbI_3	−147.2
Antimony(III) oxide	Sb_2O_3	−69.4
Antimony(III) sulfide	Sb_2S_3	−86
Antimony(V) chloride	$SbCl_5$	−120.5
Argon (g)	Ar	−19.32
Arsenic (gray)	As	−5.6
Arsenic (yellow)	As	−23.2
Arsine (g)	AsH_3	−35.2
Arsenic(III) bromide	$AsBr_3$	−106
Arsenic(III) chloride	$AsCl_3$	−72.5
Arsenic(III) iodide	AsI_3	−142.2
Arsenic(III) oxide	As_2O_3	−30.34
Arsenic(III) sulfide	As_2S_3	−70
Barium	Ba	+20.6
Barium bromide	$BaBr_2$	−92
Barium bromide dihydrate	$BaBr_2 \cdot 2H_2O$	−119.3
Barium carbonate	$BaCO_3$	−58.9
Barium chloride	$BaCl_2$	−72.6
Barium chloride dihydrate	$BaCl_2 \cdot 2H_2O$	−100
Barium fluoride	BaF_2	−51
Barium hydroxide	$Ba(OH)_2$	−53.2
Barium iodate	$Ba(IO_3)_2$	−122.5
Barium iodide	BaI_2	−124.4
Barium iodide dihydrate	$BaI_2 \cdot 2H_2O$	−163
Barium nitrate	$Ba(NO_3)_2$	−66.5
Barium oxide	BaO	−29.1
Barium peroxide	BaO_2	−40.6
Barium sulfate	$BaSO_4$	−65.8
Beryllium	Be	−9.0
Beryllium chloride	$BeCl_2$	−26.5
Beryllium hydroxide	$Be(OH)_2$	−23.1
Beryllium oxide	BeO	−11.9
Beryllium sulfate	$BeSO_4$	−37
Bismuth	Bi	−280.1
Bismuth tribromide	$BiBr_3$	−147
Bismuth trichloride	$BiCl_3$	−26.5
Bismuth fluoride	BiF_3	−61.2
Bismuth hydroxide	$Bi(OH)_3$	−65.8

Name	Formula	$\chi_m/10^{-6}$ cm^3 mol^{-1}	Name	Formula	$\chi_m/10^{-6}$ cm^3 mol^{-1}
Bismuth triiodide	BiI_3	−200.5	Cesium iodide	CsI	−82.6
Bismuth nitrate pentahydrate	$Bi(NO_3)_3 \cdot 5H_2O$	−159	Cesium superoxide	CsO_2	+1534
Bismuth oxide	Bi_2O_3	−83	Cesium sulfate	Cs_2SO_4	−116
Bismuth phosphate	$BiPO_4$	−77	Chlorine (l)	Cl_2	−40.4
Bismuth sulfate	$Bi_2(SO_4)_3$	−199	Chlorine trifluoride (g)	ClF_3	−26.5
Bismuth sulfide	Bi_2S_3	−123	Chromium	Cr	+167
Boron	B	−6.7	Chromium(II) chloride	$CrCl_2$	+7230
Diborane (g)	B_2H_6	−21.0	Chromium(III) chloride	$CrCl_3$	+6350
Boric acid (orthoboric acid)	H_3BO_3	−34.1	Chromium(III) fluoride	CrF_3	+4370
Boron trichloride	BCl_3	−59.9	Chromium(III) oxide	Cr_2O_3	+1960
Boron oxide	B_2O_3	−38.7	Chromium(III) sulfate	$Cr_2(SO_4)_3$	+11800
Bromine (l)	Br_2	−56.4	Chromium(VI) oxide	CrO_3	+40
Bromine (g)	Br_2	−73.5	Cobalt	Co	Ferro.
Bromine trifluoride	BrF_3	−33.9	Cobalt(II) bromide	$CoBr_2$	+13000
Bromine pentafluoride	BrF_5	−45.1	Cobalt(II) chloride	$CoCl_2$	+12660
Cadmium	Cd	−19.7	Cobalt(II) chloride hexahydrate	$CoCl_2 \cdot 6H_2O$	+9710
Cadmium bromide	$CdBr_2$	−87.3	Cobalt(II) cyanide	$Co(CN)_2$	+3825
Cadmium bromide tetrahydrate	$CdBr_2 \cdot 4H_2O$	−131.5	Cobalt(II) fluoride	CoF_2	+9490
Cadmium carbonate	$CdCO_3$	−46.7	Cobalt(II) iodide	CoI_2	+10760
Cadmium chloride	$CdCl_2$	−68.7	Cobalt(II) sulfate	$CoSO_4$	+10000
Cadmium chromate	$CdCrO_4$	−16.8	Cobalt(II) sulfide	CoS	+225
Cadmium cyanide	$Cd(CN)_2$	−54	Cobalt(II,III) oxide	Co_3O_4	+7380
Cadmium fluoride	CdF_2	−40.6	Cobalt(III) fluoride	CoF_3	+1900
Cadmium hydroxide	$Cd(OH)_2$	−41	Cobalt(III) oxide	Co_2O_3	+4560
Cadmium iodate	$Cd(IO_3)_2$	−108.4	Copper	Cu	−5.46
Cadmium iodide	CdI_2	−117.2	Copper(I) bromide	$CuBr$	−49
Cadmium nitrate	$Cd(NO_3)_2$	−55.1	Copper(I) chloride	$CuCl$	−40
Cadmium nitrate tetrahydrate	$Cd(NO_3)_2 \cdot 4H_2O$	−140	Copper(I) cyanide	$CuCN$	−24
Cadmium oxide	CdO	−30	Copper(I) iodide	CuI	−63
Cadmium sulfate	$CdSO_4$	−59.2	Copper(I) oxide	Cu_2O	−20
Cadmium sulfide	CdS	−50	Copper(II) bromide	$CuBr_2$	+685
Calcium	Ca	+40	Copper(II) chloride	$CuCl_2$	+1080
Calcium bromide	$CaBr_2$	−73.8	Copper(II) chloride dihydrate	$CuCl_2 \cdot 2H_2O$	+1420
Calcium carbonate	$CaCO_3$	−38.2	Copper(II) fluoride	CuF_2	+1050
Calcium chloride	$CaCl_2$	−54.7	Copper(II) fluoride dihydrate	$CuF_2 \cdot 2H_2O$	+1600
Calcium fluoride	CaF_2	−28	Copper(II) hydroxide	$Cu(OH)_2$	+1170
Calcium hydroxide	$Ca(OH)_2$	−22	Copper(II) nitrate trihydrate	$Cu(NO_3)_2 \cdot 3H_2O$	+1570
Calcium iodate	$Ca(IO_3)_2$	−101.4	Copper(II) nitrate hexahydrate	$Cu(NO_3)_2 \cdot 6H_2O$	+1625
Calcium iodide	CaI_2	−109	Copper(II) oxide	CuO	+238
Calcium oxide	CaO	−15.0	Copper(II) sulfate	$CuSO_4$	+1330
Calcium sulfate	$CaSO_4$	−49.7	Copper(II) sulfate pentahydrate	$CuSO_4 \cdot 5H_2O$	+1460
Calcium sulfate dihydrate	$CaSO_4 \cdot 2H_2O$	−74	Copper(II) sulfide	CuS	−2.0
Carbon (diamond)	C	−5.9	Dysprosium (α)	Dy	+98000
Carbon (graphite)	C	−6.0	Dysprosium(III) oxide	Dy_2O_3	+89600
Carbon monoxide (g)	CO	−11.8	Dysprosium(III) sulfide	Dy_2S_3	+95200
Carbon dioxide (g)	CO_2	−21.0	Erbium	Er	+48000
Cerium (β)	Ce	+2500	Erbium oxide	Er_2O_3	+73920
Cerium(II) sulfide	CeS	+2110	Erbium sulfate octahydrate	$Er_2(SO_4)_3 \cdot 8H_2O$	+74600
Cerium(III) chloride	$CeCl_3$	+2490	Erbium sulfide	Er_2S_3	+77200
Cerium(III) fluoride	CeF_3	+2190	Europium	Eu	+30900
Cerium(III) sulfide	Ce_2S_3	+5080	Europium(II) bromide	$EuBr_2$	+26800
Cerium(IV) oxide	CeO_2	+26	Europium(II) chloride	$EuCl_2$	+26500
Cerium(IV) sulfate tetrahydrate	$Ce(SO_4)_2 \cdot 4H_2O$	−97	Europium(II) fluoride	EuF_2	+23750
Cesium	Cs	+29	Europium(II) iodide	EuI_2	+26000
Cesium bromate	$CsBrO_3$	−75.1	Europium(II) sulfide	EuS	+23800
Cesium bromide	$CsBr$	−67.2	Europium(III) oxide	Eu_2O_3	+10100
Cesium carbonate	Cs_2CO_3	−103.6	Europium(III) sulfate	$Eu_2(SO_4)_3$	+10400
Cesium chlorate	$CsClO_3$	−65	Fluorine	F_2	−9.63
Cesium chloride	$CsCl$	−56.7	Gadolinium (350 K)	Gd	+185000
Cesium fluoride	CsF	−44.5	Gadolinium(III) chloride	$GdCl_3$	+27930

Name	Formula	$\chi_m/10^{-6}$ cm^3 mol^{-1}	Name	Formula	$\chi_m/10^{-6}$ cm^3 mol^{-1}
Gadolinium(III) oxide	Gd_2O_3	+53200	Iron(II) chloride tetrahydrate	$FeCl_2 \cdot 4H_2O$	+12900
Gadolinium(III) sulfate octahydrate	$Gd_2(SO_4)_3 \cdot 8H_2O$	+53280	Iron(II) fluoride	FeF_2	+9500
Gadolinium(III) sulfide	Gd_2S_3	+55500	Iron(II) iodide	FeI_2	+13600
Gallium	Ga	−21.6	Iron(II) oxide	FeO	+7200
Gallium suboxide	Ga_2O	−34	Iron(II) sulfate	$FeSO_4$	+12400
Gallium(II) sulfide	GaS	−23	Iron(II) sulfate monohydrate	$FeSO_4 \cdot H_2O$	+10500
Gallium(III) chloride	$GaCl_3$	−63	Iron(II) sulfate heptahydrate	$FeSO_4 \cdot 7H_2O$	+11200
Gallium(III) sulfide	Ga_2S_3	−80	Iron(II) sulfide	FeS	+1074
Germanium	Ge	−11.6	Iron(III) chloride	$FeCl_3$	+13450
Germane (g)	GeH_4	−29.7	Iron(III) chloride hexahydrate	$FeCl_3 \cdot 6H_2O$	+15250
Germanium(II) oxide	GeO	−28.8	Iron(III) fluoride	FeF_3	+13760
Germanium(II) sulfide	GeS	−40.9	Iron(III) fluoride trihydrate	$FeF_3 \cdot 3H_2O$	+7870
Germanium(IV) chloride	$GeCl_4$	−72	Iron(III) nitrate nonahydrate	$Fe(NO_3)_3 \cdot 9H_2O$	+15200
Germanium(IV) fluoride	GeF_4	−50	Krypton (g)	Kr	−29.0
Germanium(IV) iodide	GeI_4	−171	Lanthanum (α)	La	+95.9
Germanium(IV) oxide	GeO_2	−34.3	Lanthanum oxide	La_2O_3	−78
Germanium(IV) sulfide	GeS_2	−53.9	Lanthanum sulfate nonahydrate	$La_2(SO_4)_3 \cdot 9H_2O$	−262
Gold	Au	−28	Lanthanum sulfide	La_2S_3	−37
Gold(I) bromide	AuBr	−61	Lead	Pb	−23
Gold(I) chloride	AuCl	−67	Lead(II) acetate	$Pb(C_2H_3O_2)_2$	−89.1
Gold(I) iodide	AuI	−91	Lead(II) bromide	$PbBr_2$	−90.6
Gold(III) chloride	$AuCl_3$	−112	Lead(II) carbonate	$PbCO_3$	−61.2
Hafnium	Hf	+71	Lead(II) chloride	$PbCl_2$	−73.8
Hafnium oxide	HfO_2	−23	Lead(II) chromate	$PbCrO_4$	−18
Helium (g)	He	−2.02	Lead(II) fluoride	PbF_2	−58.1
Holmium	Ho	+72900	Lead(II) iodate	$Pb(IO_3)_2$	−131
Holmium oxide	Ho_2O_3	+88100	Lead(II) iodide	PbI_2	−126.5
Hydrazine (l)	N_2H_4	−201	Lead(II) nitrate	$Pb(NO_3)_2$	−74
Hydrogen (l, 20.3 K)	H_2	−5.44	Lead(II) oxide	PbO	−42
Hydrogen (g)	H_2	−3.99	Lead(II) phosphate	$Pb_3(PO_4)_2$	−182
Hydrogen chloride (l)	HCl	−22.6	Lead(II) sulfate	$PbSO_4$	−69.7
Hydrogen chloride (aq)	HCl	−22	Lead(II) sulfide	PbS	−83.6
Hydrogen fluoride (l)	HF	−8.6	Lithium	Li	+14.2
Hydrogen fluoride (aq)	HF	−9.3	Lithium bromide	LiBr	−34.3
Hydrogen iodide (s, 195 K)	HI	−47.3	Lithium carbonate	Li_2CO_3	−27
Hydrogen iodide (l, 233 K)	HI	−48.3	Lithium chloride	LiCl	−24.3
Hydrogen iodide (aq)	HI	−50.2	Lithium fluoride	LiF	−10.1
Hydrogen peroxide (l)	H_2O_2	−17.3	Lithium hydride	LiH	−4.6
Hydrogen sulfide (g)	H_2S	−25.5	Lithium hydroxide (aq)	LiOH	−12.3
Indium	In	−10.2	Lithium iodide	LiI	−50
Indium(I) chloride	InCl	−30	Lithium sulfate	Li_2SO_4	−41.6
Indium(II) chloride	$InCl_2$	−56	Lutetium	Lu	+182.9
Indium(II) sulfide	InS	−28	Magnesium	Mg	+13.1
Indium(III) bromide	$InBr_3$	−107	Magnesium bromide	$MgBr_2$	−72
Indium(III) chloride	$InCl_3$	−86	Magnesium carbonate	$MgCO_3$	−32.4
Indium(III) oxide	In_2O_3	−56	Magnesium chloride	$MgCl_2$	−47.4
Indium(III) sulfide	In_2S_3	−98	Magnesium fluoride	MgF_2	−22.7
Iodine	I_2	−90	Magnesium hydroxide	$Mg(OH)_2$	−22.1
Iodic acid	HIO_3	−48	Magnesium iodide	MgI_2	−111
Iodine pentoxide	I_2O_5	−79.4	Magnesium oxide	MgO	−10.2
Iodine chloride	ICl	−54.6	Magnesium sulfate	$MgSO_4$	−42
Iodine trichloride	ICl_3	−90.2	Magnesium sulfate monohydrate	$MgSO_4 \cdot H_2O$	−61
Iodine pentafluoride	IF_5	−58.1	Magnesium sulfate heptahydrate	$MgSO_4 \cdot 7H_2O$	−135.7
Iridium	Ir	+25	Manganese	Mn	+511
Iridium(III) chloride	$IrCl_3$	−14.4	Manganese(II) bromide	$MnBr_2$	+13900
Iridium(IV) oxide	IrO_2	+224	Manganese(II) carbonate	$MnCO_3$	+11400
Iron	Fe	Ferro.	Manganese(II) chloride	$MnCl_2$	+14350
Iron(II) bromide	$FeBr_2$	+13600	Manganese(II) chloride tetrahydrate	$MnCl_2 \cdot 4H_2O$	+14600
Iron(II) carbonate	$FeCO_3$	+11300	Manganese(II) fluoride	MnF_2	+10700
Iron(II) chloride	$FeCl_2$	+14750	Manganese(II) hydroxide	$Mn(OH)_2$	+13500

Name	Formula	$\chi_m/10^{-6}$ cm³ mol⁻¹	Name	Formula	$\chi_m/10^{-6}$ cm³ mol⁻¹
Manganese(II) iodide	MnI_2	+14400	Niobium(V) oxide	Nb_2O_5	−10
Manganese(II) oxide	MnO	+4850	Nitrogen (g)	N_2	−12.0
Manganese(II) sulfate	$MnSO_4$	+13660	Nitric acid (l)	HNO_3	−19.9
Manganese(II) sulfate monohydrate	$MnSO_4 \cdot H_2O$	+14200	Nitrous oxide (g)	N_2O	−18.9
Manganese(II) sulfate tetrahydrate	$MnSO_4 \cdot 4H_2O$	+14600	Nitric oxide (s, 90 K)	NO	+19.8
Manganese(II) sulfide (α form)	MnS	+5630	Nitric oxide (l, 118 K)	NO	+114.2
Manganese(II) sulfide (β form)	MnS	+3850	Nitric oxide (g)	NO	+1461
Manganese(II,III) oxide	Mn_3O_4	+12400	Nitrogen dioxide (g, 408 K)	NO_2	+150
Manganese(III) fluoride	MnF_3	+10500	Nitrogen trioxide (g)	N_2O_3	−16
Manganese(III) oxide	Mn_2O_3	+14100	Nitrogen tetroxide (g)	N_2O_4	−23.0
Manganese(IV) oxide	MnO_2	+2280	Osmium	Os	+11
Mercury (s, 234 K)	Hg	−24.1	Oxygen (s, 54 K)	O_2	+10200
Mercury (l)	Hg	−33.5	Oxygen (l, 90 K)	O_2	+7699
Mercury(I) bromide	Hg_2Br_2	−105	Oxygen (g)	O_2	+3415
Mercury(I) chloride	Hg_2Cl_2	−120	Ozone (l)	O_3	+6.7
Mercury(I) fluoride	Hg_2F_2	−106	Palladium	Pd	+540
Mercury(I) iodide	Hg_2I_2	−166	Palladium(II) chloride	$PdCl_2$	−38
Mercury(I) nitrate	$Hg_2(NO_3)_2$	−121	Phosphorus (white)	P	−26.66
Mercury(I) oxide	Hg_2O	−76.3	Phosphorus (red)	P	−20.77
Mercury(I) sulfate	Hg_2SO_4	−123	Phosphine (g)	PH_3	−26.2
Mercury(II) bromide	$HgBr_2$	−94.2	Phosphoric acid (aq)	H_3PO_4	−43.8
Mercury(II) chloride	$HgCl_2$	−82	Phosphorous acid (aq)	H_3PO_3	−42.5
Mercury(II) cyanide	$Hg(CN)_2$	−67	Phosphorus(III) chloride (l)	PCl_3	−63.4
Mercury(II) fluoride	HgF_2	−57.3	Platinum	Pt	+193
Mercury(II) iodide	HgI_2	−165	Platinum(II) chloride	$PtCl_2$	−54
Mercury(II) nitrate	$Hg(NO_3)_2$	−74	Platinum(III) chloride	$PtCl_3$	−66.7
Mercury(II) oxide	HgO	−46	Platinum(IV) chloride	$PtCl_4$	−93
Mercury(II) sulfate	$HgSO_4$	−78.1	Platinum(IV) fluoride	PtF_4	+445
Mercury(II) sulfide	HgS	−55.4	Plutonium	Pu	+525
Mercury(II) thiocyanate	$Hg(SCN)_2$	−96.5	Plutonium(IV) fluoride	PuF_4	+1760
Molybdenum	Mo	+72	Plutonium(IV) oxide	PuO_2	+730
Molybdenum(III) bromide	$MoBr_3$	+525	Plutonium(VI) fluoride	PuF_6	+173
Molybdenum(III) chloride	$MoCl_3$	+43	Potassium	K	+20.8
Molybdenum(III) oxide	Mo_2O_3	-42.0	Potassium bromate	$KBrO_3$	−52.6
Molybdenum(IV) bromide	$MoBr_4$	+520	Potassium bromide	KBr	−49.1
Molybdenum(IV) chloride	$MoCl_4$	+1750	Potassium carbonate	K_2CO_3	−59
Molybdenum(IV) oxide	MoO_2	+41	Potassium chlorate	$KClO_3$	−42.8
Molybdenum(V) chloride	$MoCl_5$	+990	Potassium chloride	KCl	−38.8
Molybdenum(VI) fluoride	MoF_6	−26.0	Potassium chromate	K_2CrO_4	−3.9
Molybdenum(VI) oxide	MoO_3	+3	Potassium cyanide	KCN	−37
Neodymium (α)	Nd	+5930	Potassium ferricyanide	$K_3Fe(CN)_6$	+2290
Neodymium fluoride	NdF_3	+4980	Potassium ferrocyanide trihydrate	$K_4Fe(CN)_6 \cdot 3H_2O$	−172.3
Neodymium oxide	Nd_2O_3	+10200	Potassium fluoride	KF	−23.6
Neodymium sulfate	$Nd_2(SO_4)_3$	+9990	Potassium hydrogen sulfate	$KHSO_4$	−49.8
Neodymium sulfide	Nd_2S_3	+5550	Potassium hydroxide (aq)	KOH	−22
Neon (g)	Ne	−6.96	Potassium iodate	KIO_3	−63.1
Neptunium	Np	+575	Potassium iodide	KI	−63.8
Nickel	Ni	Ferro.	Potassium nitrate	KNO_3	−33.7
Nickel(II) bromide	$NiBr_2$	+5600	Potassium nitrite	KNO_2	−23.3
Nickel(II) chloride	$NiCl_2$	+6145	Potassium permanganate	$KMnO_4$	+20
Nickel(II) chloride hexahydrate	$NiCl_2 \cdot 6H_2O$	+4240	Potassium sulfate	K_2SO_4	−67
Nickel(II) fluoride	NiF_2	+2410	Potassium sulfide	K_2S	−60
Nickel(II) hydroxide	$Ni(OH)_2$	+4500	Potassium superoxide	KO_2	+3230
Nickel(II) iodide	NiI_2	+3875	Potassium thiocyanate	KSCN	−48
Nickel(II) nitrate hexahydrate	$Ni(NO_3)_2 \cdot 6H_2O$	+4300	Praseodymium (α)	Pr	+5530
Nickel(II) oxide	NiO	+660	Praseodymium chloride	$PrCl_3$	+44.5
Nickel(II) sulfate	$NiSO_4$	+4005	Praseodymium oxide	Pr_2O_3	+8994
Nickel(II) sulfide	NiS	+190	Praseodymium sulfide	Pr_2S_3	+10770
Nickel(III) sulfide	Ni_3S_2	+1030	Protactinium	Pa	+277
Niobium	Nb	+208	Rhenium	Re	+67

Name	Formula	$\chi_m/10^{-6}$ cm³ mol⁻¹	Name	Formula	$\chi_m/10^{-6}$ cm³ mol⁻¹
Rhenium(IV) oxide	ReO_2	+44	Sodium carbonate	Na_2CO_3	−41
Rhenium(IV) sulfide	ReS_2	+38	Sodium chlorate	$NaClO_3$	−34.7
Rhenium(V) chloride	$ReCl_5$	+1225	Sodium chloride	$NaCl$	−30.2
Rhenium(VI) oxide	ReO_3	+16	Sodium dichromate	$Na_2Cr_2O_7$	+55
Rhenium(VII) oxide	Re_2O_7	−16	Sodium fluoride	NaF	−15.6
Rhodium	Rh	+102	Sodium hydrogen phosphate	Na_2HPO_4	−56.6
Rhodium(III) chloride	$RhCl_3$	−7.5	Sodium hydroxide (aq)	$NaOH$	−15.8
Rhodium(III) oxide	Rh_2O_3	+104	Sodium iodate	$NaIO_3$	−53
Rubidium	Rb	+17	Sodium iodide	NaI	−57
Rubidium bromide	$RbBr$	−56.4	Sodium nitrate	$NaNO_3$	−25.6
Rubidium carbonate	Rb_2CO_3	−75.4	Sodium nitrite	$NaNO_2$	−14.5
Rubidium chloride	$RbCl$	−46	Sodium oxide	Na_2O	−19.8
Rubidium fluoride	RbF	−31.9	Sodium peroxide	Na_2O_2	−28.10
Rubidium iodide	RbI	−72.2	Sodium sulfate	Na_2SO_4	−52
Rubidium nitrate	$RbNO_3$	−41	Sodium sulfate decahydrate	$Na_2SO_4 \cdot 10H_2O$	−184
Rubidium sulfate	Rb_2SO_4	−88.4	Sodium sulfide	Na_2S	−39
Rubidium superoxide	RbO_2	+1527	Sodium tetraborate	$Na_2B_4O_7$	−85
Ruthenium	Ru	+39	Strontium	Sr	+92
Ruthenium(III) chloride	$RuCl_3$	+1998	Strontium bromide	$SrBr_2$	−86.6
Ruthenium(IV) oxide	RuO_2	+162	Strontium bromide hexahydrate	$SrBr_2 \cdot 6H_2O$	−160
Samarium (α)	Sm	+1278	Strontium carbonate	$SrCO_3$	−47
Samarium(II) bromide	$SmBr_2$	+5337	Strontium chlorate	$Sr(ClO_3)_2$	−73
Samarium(III) bromide	$SmBr_3$	+972	Strontium chloride	$SrCl_2$	−61.5
Samarium(III) oxide	Sm_2O_3	+1988	Strontium chloride hexahydrate	$SrCl_2 \cdot 6H_2O$	−145
Samarium(III) sulfate octahydrate	$Sm_2(SO_4)_3 \cdot 8H_2O$	+1710	Strontium chromate	$SrCrO_4$	−5.1
Samarium(III) sulfide	Sm_2S_3	+3300	Strontium fluoride	SrF_2	−37.2
Scandium (α)	Sc	+295.2	Strontium hydroxide	$Sr(OH)_2$	−40
Selenium	Se	−25	Strontium iodate	$Sr(IO_3)_2$	−108
Selenium dioxide	SeO_2	−27.2	Strontium iodide	SrI_2	−112
Selenium bromide	Se_2Br_2	−113	Strontium nitrate	$Sr(NO_3)_2$	−57.2
Selenium chloride (l)	Se_2Cl_2	−94.8	Strontium oxide	SrO	−35
Selenium hexafluoride (g)	SeF_6	−51	Strontium peroxide	SrO_2	−32.3
Silicon	Si	−3.12	Strontium sulfate	$SrSO_4$	−57.9
Silane (g)	SiH_4	−20.4	Sulfur (rhombic)	S	−15.5
Disilane (g)	Si_2H_6	−37.3	Sulfur (monoclinic)	S	−14.9
Tetramethylsilane (l)	$(CH_3)_4Si$	−74.80	Sulfuric acid (l)	H_2SO_4	−39
Tetraethylsilane (l)	$(C_2H_5)_4Si$	−120.2	Sulfur dioxide (g)	SO_2	−18.2
Tetrabromosilane (l)	$SiBr_4$	−126	Sulfur trioxide (l)	SO_3	−28.54
Tetrachlorosilane (l)	$SiCl_4$	−87.5	Sulfur chloride (l)	$SSCl_2$	−62.2
Silicon carbide	SiC	−12.8	Sulfur dichloride (l)	SCl_2	−49.4
Silicon dioxide	SiO_2	−29.6	Sulfur hexafluoride (g)	SF_6	−44
Silver	Ag	−19.5	Thionyl chloride (l)	$SOCl_2$	−44.3
Silver(I) bromide	$AgBr$	−61	Tantalum	Ta	+154
Silver(I) carbonate	Ag_2CO_3	−80.90	Tantalum(V) chloride	$TaCl_5$	+140
Silver(I) chloride	$AgCl$	−49	Tantalum(V) oxide	Ta_2O_5	−32
Silver(I) chromate	Ag_2CrO_4	−40	Technetium	Tc	+115
Silver(I) cyanide	$AgCN$	−43.2	Tellurium	Te	−38
Silver(I) fluoride	AgF	−36.5	Tellurium dibromide	$TeBr_2$	−106
Silver(I) iodide	AgI	−80	Tellurium dichloride	$TeCl_2$	−94
Silver(I) nitrate	$AgNO_3$	−45.7	Tellurium hexafluoride (g)	TeF_6	−66
Silver(I) nitrite	$AgNO_2$	−42	Terbium (α)	Tb	+170000
Silver(I) oxide	Ag_2O	−134	Terbium oxide	Tb_2O_3	+78340
Silver(I) phosphate	Ag_3PO_4	−120	Thallium	Tl	−50
Silver(I) sulfate	Ag_2SO_4	−92.90	Thallium(I) bromate	$TlBrO_3$	−75.9
Silver(I) thiocyanate	$AgSCN$	−61.8	Thallium(I) bromide	$TlBr$	−63.9
Silver(II) oxide	AgO	−19.6	Thallium(I) carbonate	Tl_2CO_3	−101.6
Sodium	Na	+16	Thallium(I) chlorate	$TlClO_3$	−65.5
Sodium acetate	$NaC_2H_3O_2$	−37.6	Thallium(I) chloride	$TlCl$	−57.8
Sodium bromate	$NaBrO_3$	−44.2	Thallium(I) chromate	Tl_2CrO_4	−39.3
Sodium bromide	$NaBr$	−41	Thallium(I) cyanide	$TlCN$	−49

Name	Formula	$\chi_m/10^{-6}$ cm^3 mol^{-1}	Name	Formula	$\chi_m/10^{-6}$ cm^3 mol^{-1}
Thallium(I) fluoride	TlF	−44.4	Uranium(IV) bromide	UBr_4	+3530
Thallium(I) iodate	$TlIO_3$	−86.8	Uranium(IV) chloride	UCl_4	+3680
Thallium(I) iodide	TlI	−82.2	Uranium(IV) fluoride	UF_4	+3530
Thallium(I) nitrate	$TlNO_3$	−56.5	Uranium(IV) oxide	UO_2	+2360
Thallium(I) nitrite	$TlNO_2$	−50.8	Uranium(VI) fluoride	UF_6	+43
Thallium(I) sulfate	Tl_2SO_4	−112.6	Uranium(VI) oxide	UO_3	+128
Thallium(I) sulfide	Tl_2S	−88.8	Vanadium	V	+285
Thorium	Th	+97	Vanadium(II) bromide	VBr_2	+3230
Thorium(IV) oxide	ThO_2	−16	Vanadium(II) chloride	VCl_2	+2410
Thulium	Tm	+24700	Vanadium(III) bromide	VBr_3	+2910
Thulium oxide	Tm_2O_3	+51444	Vanadium(III) chloride	VCl_3	+3030
Tin (gray)	Sn	−37.4	Vanadium(III) fluoride	VF_3	+2757
Tin(II) chloride	$SnCl_2$	−69	Vanadium(III) oxide	V_2O_3	+1976
Tin(II) chloride dihydrate	$SnCl_2 \cdot 2H_2O$	−91.4	Vanadium(III) sulfide	V_2S_3	+1560
Tin(II) oxide	SnO	−19	Vanadium(IV) chloride	VCl_4	+1215
Tin(IV) bromide	$SnBr_4$	−149	Vanadium(IV) oxide	VO_2	+99
Tin(IV) chloride (l)	$SnCl_4$	−115	Vanadium(V) oxide	V_2O_5	+128
Tin(IV) oxide	SnO_2	−41	Water (s, 273 K)	H_2O	−12.63
Titanium	Ti	+151	Water (l, 293 K)	H_2O	−12.96
Titanium(II) bromide	$TiBr_2$	+720	Water (l, 373 K)	H_2O	−13.09
Titanium(II) chloride	$TiCl_2$	+484	Water (g, 373 K))	H_2O	−13.1
Titanium(II) iodide	TiI_2	+1790	Xenon (g)	Xe	−45.5
Titanium(II) sulfide	TiS	+432	Ytterbium (β)	Yb	+67
Titanium(III) bromide	$TiBr_3$	+660	Yttrium (α)	Y	+187.7
Titanium(III) chloride	$TiCl_3$	+1110	Yttrium oxide	Y_2O_3	+44.4
Titanium(III) fluoride	TiF_3	+1300	Yttrium sulfide	Y_2S_3	+100
Titanium(III) oxide	Ti_2O_3	+132	Zinc	Zn	−9.15
Titanium(IV) chloride	$TiCl_4$	−54	Zinc carbonate	$ZnCO_3$	−34
Titanium(IV) oxide	TiO_2	+5.9	Zinc chloride	$ZnCl_2$	−55.33
Tungsten	W	+53	Zinc cyanide	$Zn(CN)_2$	−46
Tungsten carbide	WC	+10	Zinc fluoride	ZnF_2	−34.3
Tungsten(II) chloride	WCl_2	−25	Zinc hydroxide	$Zn(OH)_2$	−67
Tungsten(IV) oxide	WO_2	+57	Zinc iodide	ZnI_2	−108
Tungsten(IV) sulfide	WS_2	+5850	Zinc oxide	ZnO	−27.2
Tungsten(V) bromide	WBr_5	+270	Zinc phosphate	$Zn_3(PO_4)_2$	−141
Tungsten(V) chloride	WCl_5	+387	Zinc sulfate	$ZnSO_4$	−47.8
Tungsten(VI) chloride	WCl_6	−71	Zinc sulfate monohydrate	$ZnSO_4 \cdot H_2O$	−63
Tungsten(VI) fluoride (g)	WF_6	−53	Zinc sulfate heptahydrate	$ZnSO_4 \cdot 7H_2O$	−138
Tungsten(VI) oxide	WO_3	−15.8	Zinc sulfide	ZnS	−25
Uranium	U	+409	Zirconium	Zr	+120
Uranium(III) bromide	UBr_3	+4740	Zirconium carbide	ZrC	−26
Uranium(III) chloride	UCl_3	+3460	Zirconium nitrate pentahydrate	$Zr(NO_3)_4 \cdot 5H_2O$	−77
Uranium(III) hydride	UH_3	+6244	Zirconium(IV) oxide	ZrO_2	−13.8
Uranium(III) iodide	UI_3	+4460			

INDEX OF REFRACTION OF INORGANIC LIQUIDS

This table gives the index of refraction n of several inorganic substances in the liquid state at specified temperatures. The measurements refer to ambient atmospheric pressure except for substances whose normal boiling points are greater than the indicated temperature; in this case the pressure is the saturated vapor pressure of the substance. All values refer to a wavelength of 589 nm unless otherwise indicated. Entries are arranged in alphabetical order by chemical formula as normally written.

Data on the index of refraction at other temperatures and wavelengths may be found in Reference 1.

References

1. Wohlfarth, C., and Wohlfarth, B., *Landolt-Börnstein, Numerical Data and Functional Relationships in Science and Technology, New Series*, III/38A, Martienssen, W., Editor, Springer-Verlag, Heidelberg, 1996.
2. Francis, A.W., *J. Chem. Eng. Data*, 5, 534, 1960.

Formula	Name	$t/°C$	n
Ar	Argon	−188	1.2312
AsCl$_3$	Arsenic(III) chloride	16	1.604
BBr$_3$	Boron tribromide	16	1.312
BrF$_3$	Bromine trifluoride	25	1.4536
BrF$_5$	Bromine pentafluoride	25	1.3529
Br$_2$	Bromine	15	1.659
COS	Carbon oxysulfide	25	1.3506
CO$_2$	Carbon dioxide	24	1.6630
CS$_2$	Carbon disulfide	20	1.62774
C$_3$O$_2$	Carbon suboxide	0	1.453
Cl$_2$	Chlorine	20	1.3834
CrO$_2$Cl$_2$	Chromyl chloride	23	1.524
Fe(CO)$_5$	Iron pentacarbonyl	14	1.523
GeBr$_4$	Germanium(IV) bromide	26	1.6269
GeCl$_4$	Germanium(IV) chloride	25	1.4614
HBr	Hydrogen bromide	10	1.325
HCN	Hydrogen cyanide	20	1.26136
HCl	Hydrogen chloride	18	1.3287 [a]
HClO$_4$	Perchloric acid	50	1.3819
HF	Hydrogen fluoride	25	1.1574
HI	Hydrogen iodide	16	1.466
HNO$_3$	Nitric acid	25	1.393
H$_2$	Hydrogen	−253	1.1096
H$_2$O	Water	20	1.33336
H$_2$O$_2$	Hydrogen peroxide	28	1.4061
H$_2$S	Hydrogen sulfide	−80	1.460
		20	1.3682
H$_2$SO$_4$	Sulfuric acid	20	1.4183
H$_2$S$_2$	Hydrogen disulfide	20	1.630

Formula	Name	$t/°C$	n
He	Helium	−269	1.02451 [c]
Kr	Krypton	−157	1.3032 [c]
NH$_3$	Ammonia	−77	1.3944 [b]
		20	1.3327
NO	Nitric oxide	−90	1.330
N$_2$	Nitrogen	−196	1.19876 [b]
N$_2$H$_4$	Hydrazine	22	1.470
N$_2$O	Nitrous oxide	25	1.238
O$_2$	Oxygen	−183	1.2243 [c]
PBr$_3$	Phosphorus(III) bromide	25	1.687
PCl$_3$	Phosphorus(III) chloride	21	1.5122
PH$_3$	Phosphine	17	1.317
P$_2$O$_3$	Phosphorus(III) oxide	27	1.540
S	Sulfur	125	1.9170
SCl$_2$	Sulfur dichloride	14	1.557
SF$_6$	Sulfur hexafluoride	25	1.167
SOCl$_2$	Thionyl chloride	10	1.527
SO$_2$	Sulfur dioxide	25	1.3396
SO$_2$Cl$_2$	Sulfuryl chloride	12	1.444
SO$_3$	Sulfur trioxide	20	1.40965
SSCl$_2$	Sulfur chloride	20	1.671
SbCl$_5$	Antimony(V) chloride	22	1.5925
SiBr$_4$	Tetrabromosilane	31	1.5685
SiCl$_4$	Tetrachlorosilane	25	1.41156
SnBr$_4$	Tin(IV) bromide	31	1.6628
SnCl$_4$	Tin(IV) chloride	25	1.5086
TiCl$_4$	Titanium(IV) chloride	18	1.6076
Xe	Xenon	−112	1.3918 [c]

[a] At 581 nm
[b] At 578 nm
[c] At 546 nm

PHYSICAL AND OPTICAL PROPERTIES OF MINERALS

The chemical formula, crystal system, density, hardness, and index of refraction of some common minerals are given in this table. Entries are arranged alphabetically by mineral name. The columns are:

- **Formula:** Chemical formula for a typical sample of the mineral. Composition often varies considerably with the origin of the sample.
- **Crystal system:** tricl = triclinic; monocl = monoclinic; orth = orthorhombic; tetr = tetragonal; hex = hexagonal; rhomb = rhombohedral; cub = cubic.
- **Density:** Typical density in g/cm³. Individual samples may vary by a few percent.
- **Hardness:** On the Mohs' scale (range of 1 to 10, with talc = 1 and diamond = 10).
- **Index of refraction:** Values are given for the three coordinate axes in the order of least, intermediate, and greatest

index. For cubic crystals there is only a single value. See Reference 1 for details on the axis systems. Variations of several percent, depending on the origin and exact composition of the sample, are common.

References

1. Deer, W. A., Howie, R. A., and Zussman, J., *An Introduction to the Rock-Forming Minerals*, 2nd Edition, Longman Scientific & Technical, Harlow, Essex, 1992.
2. Carmichael, R. S., *Practical Handbook of Physical Properties of Rocks and Minerals*, CRC Press, Boca Raton, FL, 1989.
3. Donnay, J. D. H., and Ondik, H. M., *Crystal Data Determinative Tables, Third Edition, Volume 2, Inorganic Compounds*, Joint Committee on Powder Diffraction Standards, Swarthmore, PA, 1973.

Name	Formula	Crystal system	Density g/cm³	Hardness	n_α	n_β	n_γ
Acanthite	Ag_2S	orth	7.2	2.3			
Actinolite	$Ca_2(Mg,Fe)_5Si_8O_{22}(OH,F)_2$	monocl	3.23	5.5	1.624	1.655	1.664
Aegirine	$NaFe(SiO_3)_2$	monocl	3.58	6	1.763	1.800	1.815
Akermanite	$Ca_2MgSi_2O_7$	tetr	2.94	5.5	1.632	1.640	
Alabandite	MnS	cub	4.0	3.8			
Albite	$NaAlSi_3O_8$	tricl	2.63	6.3	1.527	1.531	1.538
Allanite	$(Ca,Mn,Ce,La,Y,Th)_2(Fe,Ti)(Al,Fe)O\cdot OH$ $(Si_2O_7)(SiO_4)$	monocl	3.8	5.8	1.75	1.78	1.80
Allemontite	$SbAs$	hex	6.0	3.5			
Almandine	$Fe_3Al_2Si_3O_{12}$	cub	4.32	6.8	1.830		
Altaite	$PbTe$	cub	8.16	3			
Aluminite	$Al_2(SO_4)(OH)_4\cdot 7H_2O$	monocl	1.74	1.5	1.459	1.464	1.470
Alunite	$(K,Na)Al_3(SO_4)_2(OH)_6$	rhomb	2.8	3.8	1.572	1.592	
Alunogen	$Al_2(SO_4)_3\cdot 18H_2O$	monocl	1.69	1.8	1.467	1.47	1.478
Amblygonite	$(Li,Na)Al(PO_4)(F,OH)$	tricl	3.1	5.8	1.591	1.604	1.613
Analcite	$NaAlSi_2O_6\cdot H_2O$	cub	2.27	5.5	1.486		
Anatase	TiO_2	tetr	4.23	5.8	2.488	2.561	
Andalusite	Al_2OSiO_4	orth	3.15	7.5	1.635	1.639	1.644
Andesine	$NaAlSi_3O_8\cdot CaAl_2Si_2O_8$	tricl	2.67	6.3	1.550	1.553	1.557
Andorite	$PbAgSb_3S_6$	rhomb	5.35	3.3			
Andradite	$Ca_3(Fe,Ti)_2Si_3O_{12}$	cub	3.86	6.8	1.887		
Anglesite	$PbSO_4$	orth	6.29	2.8	1.877	1.883	1.894
Anhydrite	$CaSO_4$	orth	2.96	3.5	1.570	1.575	1.614
Ankerite	$Ca(Fe,Mg,Mn)(CO_3)_2$	rhomb	3.0	3.8	1.529	1.720	
Anorthite	$CaAl_2Si_2O_8$	tricl	2.76	6.3	1.577	1.585	1.590
Anorthoclase	$(Na,K)AlSi_3O_8$	tricl	2.58	6	1.523	1.528	1.529
Anthophyllite	$(Mg,Fe)_7Si_8O_{22}(OH,F)_2$	rhomb	3.21	5.8	1.645	1.658	1.668
Apatite	$Ca_5(PO_4)_3(OH,F,Cl)$	hex	3.2	5	1.645	1.648	
Apophyllite	$KFCa_4Si_8O_{20}\cdot 8H_2O$	tetr	2.35	4.8	1.535	1.536	
Aragonite	$CaCO_3$	orth	2.83	3.5	1.531	1.680	1.686
Arcanite	K_2SO_4	orth	2.66		1.494	1.494	1.497
Argentite	Ag_2S	orth	7.2	2.3			
Arsenolite	As_2O_3	cub	3.86	1.5	1.755		
Arsenopyrite	$FeAsS$	monocl	6.1	5.8			
Atacamite	$Cu_2(OH)_3Cl$	rhomb	3.76	3.3	1.831	1.861	1.880
Augelite	$Al_2(PO_4)(OH)_3$	monocl	2.70	4.8	1.574	1.576	1.588
Augite	$(Ca,Mg,Fe,Ti,Al)_2(Si,Al)_2O_6$	monocl	3.38	6	1.703	1.707	1.738
Autunite	$Ca(UO_{22})(PO_4)_2\cdot 10H20$	tetr	3.2	2.3	1.553	1.577	
Axinite	$(Ca,Mn,Fe)_3Al_2BO_3Si_4O_{12}(OH)$	tricl	3.31	6.8	1.684	1.691	1.694

Name	Formula	Crystal system	Density g/cm³	Hardness	Index of refraction n_α	n_β	n_γ
Azurite	$Cu_3(OH)_2(CO_3)_2$	monocl	3.77	3.8	1.730	1.758	1.838
Baddeleyite	ZrO_2	monocl	5.7	6.5	2.13	2.19	2.20
Barite	$BaSO_4$	orth	4.49	3.3	1.636	1.637	1.648
Benitoite	$BaTi(SiO_3)_3$	rhomb	3.65	6.3	1.757	1.804	
Bertrandite	$Be_4Si_2O_7(OH)_2$	rhomb	2.6	6	1.589	1.602	1.613
Beryl	$Be_3Al_2(SiO_3)_6$	hex	2.64	7.8	1.582	1.589	
Beryllonite	$NaBe(PO)_4$	monocl	2.81	5.8	1.552	1.558	1.561
Biotite	$K(Mg,Fe)_3AlSi_3O_{10}(OH,F)_2$	monocl	3.0	2.8	1.595	1.651	1.651
Bismuthinite	Bi_2S_3	orth	6.78	2			
Bixbyite	$(Mn,Fe)_2O_3$	cub	4.95	6.3			
Bloedite	$Na_2Mg(SO_4)_2\cdot4H_2O$	monocl	2.25	2.8	1.483	1.486	1.487
Boehmite	$AlO(OH)$	orth	3.44	3.8	1.64	1.65	1.66
Boracite	$Mg_3B_7O_{13}Cl$	rhomb	2.94	7.3	1.66	1.66	1.67
Borax	$Na_2B_4O_7\cdot10H_2O$	monocl	1.73	2.3	1.447	1.469	1.472
Bornite	Cu_5FeS_4	cub	5.07	3			
Boulangerite	$Pb_5Sb_4S_{11}$	monocl	6.1	2.8			
Bournonite	$PbCuSbS_3$	rhomb	5.83	2.8			
Braggite	PtS	tetr	10.2				
Braunite	$(Mn,Si)_2O_3$	tetr	4.78	6.3			
Bravoite	$(Ni,Fe)S_2$	cub	4.62	5.8			
Breithauptite	$NiSb$	hex	≈8.7	5.5			
Brochantite	$Cu_4(SO_4)(OH)_6$	monocl	3.79	3.8	1.728	1.771	1.800
Bromyrite	$AgBr$	cub	6.47	2.5	2.253		
Brookite	TiO_2	orth	4.23	5.8	2.583	2.584	2.700
Brucite	$Mg(OH)_2$	hex	2.37	2.5	1.575	1.59	
Bunsenite	NiO	cub	6.72	5.5			
Cacoxenite	$Fe_4(PO_4)_3(OH)_3\cdot12H_2O$	hex	2.3	3.5	1.580	1.646	
Calcite	$CaCO_3$	hex	2.71	3	1.486	1.658	
Caledonite	$Cu_2Pb_5(SO_4)_3(CO_3)(OH)_6$	rhomb	5.76	2.8	1.818	1.866	1.909
Calomel	Hg_2Cl_2	tetr	7.16	1.5	1.973	2.656	
Cancrinite	$(Na,Ca,K)_7[Al_6Si_6O_{24}](CO_3,SO_4,Cl,OH)_2\cdot H_2O$	hex	2.42	5.5	1.495	1.509	
Carnalite	$KMgCl_3\cdot6H_2O$	rhomb	1.60	2.5	1.466	1.475	1.494
Carnotite	$K_2(UO_2)_2(VO_4)_2\cdot3H_2O$	rhomb		1.5	1.75	1.92	1.95
Cassiterite	SnO_2	tetr	6.85	6.5	2.006	2.097	
Celestite	$SrSO_4$	orth	3.96	3.3	1.622	1.624	1.631
Celsian	$BaAl_2Si_2O_8$	monocl	3.25	6.3	1.583	1.588	1.594
Cerargyrite	$AgCl$	cub	5.56	2.5	2.071		
Cerussite	$PbCO_3$	orth	6.6	3.3	1.804	2.076	2.079
Cervantite	Sb_2O_4	orth	6.64	4.5			
Chabazite	$Ca[Al_2Si_4O_{12}]\cdot6H_2O$	trig	2.08	4.5	1.482		
Chalcanthite	$CuSO_4\cdot5H_2O$	tricl	2.29	2.5	1.514	1.537	1.543
Chalcocite	Cu_2S	orth	5.6	2.8			
Chalcopyrite	$CuFeS_2$	tetr	4.2	3.8			
Chiolite	$Na_5Al_3F_{14}$	tetr	3.00	3.8	1.342	1.349	
Chlorite	$(Mg,Al,Fe)_{12}(Si,Al)_8O_{20}(OH)_{16}$	monocl	3.0	2.5	1.61	1.62	1.62
Chloritoid	$FeAl_2O_2(SiO_4)_2(OH)_4$	monocl	3.66	6.5	1.717	1.721	1.726
Chondrodite	$Mg(OH,F)_2\cdot2Mg_2SiO_4$	monocl	3.21	6.5	1.604	1.615	1.634
Chromite	$FeCr_2O_4$	cub	5.0	5.5	2.16		
Chrysoberyl	$BeAl_2O_4$	orth	3.65	8.5	1.746	1.748	1.756
Chrysocolla	$CuSiO_3\cdot2H_2O$	rhomb	2.4	2	1.575	1.597	1.598
Cinnabar	HgS	hex	8.17	2.3	2.814	3.143	
Claudetite	As_2O_3	monocl	3.74	2.5	1.87	1.92	2.01
Clinohumite	$Mg(OH,F)_2\cdot4Mg_2SiO_4$	monocl	3.21	6	1.633	1.647	1.668
Clinozoisite	$Ca_2Al_3Si_3O_{12}(OH)$	monocl	3.30	6.5	1.693	1.700	1.712
Cobaltite	$CoAsS$	cub	≈6.1	5.5			
Colemanite	$Ca_2B_6O_{11}\cdot5H_2O$	monocl	2.42	4.5	1.586	1.592	1.614
Columbite	$(Fe,Mn)(Nb,Ta)_2O_6$	rhomb	5.20	6			
Connellite	$Cu_{19}(SO_4)Cl_4(OH)_{32}\cdot3H_2O$	hex	3.36	3	1.731	1.752	
Copiapite	$(Fe,Mg)Fe_4(SO_4)_6(OH)_2\cdot20H_2O$	tricl	2.13	2.8	1.52	1.54	1.59
Coquimbite	$Fe_2(SO_4)_3\cdot9H_2O$	hex	2.1	2.5	1.54	1.56	

Name	Formula	Crystal system	Density g/cm³	Hardness	n_α	n_β	n_γ
Cordierite	$Al_3(Mg,Fe)_2Si_5AlO_{18}$	rhomb	2.66	7	1.540	1.549	1.553
Corundum	Al_2O_3	hex	3.97	9	1.761	1.769	
Cotunnite	$PbCl_2$	orth	5.98	2.5	2.199	2.217	2.260
Covellite	CuS	hex	4.8	1.8			
Cristobalite	SiO_2	hex	2.33	6.5	1.484	1.487	
Crocoite	$PbCrO_4$	monocl	6.12	2.8	2.29	2.36	2.66
Cryolite	Na_3AlF_6	monocl	2.97	2.5	1.338	1.338	1.339
Cryolithionite	$Na_3Li_3Al_2F_{12}$	cub	2.77	2.8	1.340		
Cubanite	$CuFe_2S_3$	rhomb	4.11	3.5			
Cummingtonite	$(Mg,Fe)_7Si_8O_{22}(OH)_2$	monocl	3.4	5.5	1.650	1.660	1.676
Cuprite	Cu_2O	cub	6.0	3.8			
Danburite	$CaSi_2B_2O_8$	rhomb	3.0	7	1.63	1.63	1.63
Datolite	$CaBSiO_4(OH)$	monocl	2.98	5.3	1.624	1.652	1.668
Daubreelite	Cr_2FeS_4	cub	3.81				
Derbylite	$Fe_6Ti_6Sb_2O_{23}$	rhomb	4.53	5	2.45	2.45	2.51
Diamond	C	cub	3.51	10	2.418		
Diaspore	$AlO(OH)$	orth	3.4	6.8	1.694	1.715	1.741
Digenite	$Cu_{2-x}S$	cub	5.55	2.8			
Diopside	$CaMgSi_2O_6$	monocl	3.30	6	1.680	1.687	1.708
Dioptase	$CuSiO_2(OH)_2$	rhomb	3.5	5	1.65	1.70	
Dolomite	$CaMg(CO_3)_2$	rhomb	2.86	3.5	1.500	1.679	
Douglasite	$K_2FeCl_4·2H_2O$	orth	2.16		1.488	1.500	
Dyscrasite	Ag_3Sb	rhomb	9.74	3.8			
Eddingtonite	$BaAl_2Si_3O_{10}·4H_2O$	rhomb	2.8		1.541	1.553	1.557
Eglestonite	Hg_4OCl_2	cub	8.4	2.5	2.49		
Emplectite	$CuBiS_2$	rhomb	6.38	2			
Enargite	Cu_3AsS_4	rhomb	4.5	3			
Enstatite	$MgSiO_3$	monocl	3.19	5.5	1.656	1.662	1.669
Epidote	$Ca_2Al_2(Al,Fe)OH(SiO_4)_3$	monocl	3.44	6	1.733	1.755	1.765
Epsomite	$MgSO_4·7H_2O$	orth	1.67	2.3	1.433	1.455	1.461
Erythrite	$(Co,Ni)_3(AsO_4)_2·8H_2O$	monocl	3.06	2	1.626	1.661	1.699
Eucairite	$CuAgSe$	orth	7.7	2.5			
Euclasite	$BeAlSiO_4(OH)$	monocl	3.1	7.5	1.651	1.655	1.671
Eudialite	$(Na,Ca,Ce)_5(Fe,Mn)(Zr,Ti)(Si_3O_9)_2(OH,Cl)$	hex	3.0	5.5	1.623	1.600	1.615
Eulytite	$Bi_4Si_3O_{12}$	cub	6.6	4.5	2.05		
Euxenite	$(Y,Ca,Ce,U,Th)(Nb,Ta,Ti)_2O_6$	rhomb	5.5	6	2.2		
Fayalite	Fe_2SiO_4	orth	4.30	6.5	1.827	1.869	1.879
Ferberite	$FeWO_4$	monocl	7.51	4.3			
Fergussonite	$(Y,Er,Ce,Fe)(Nb,Ta,Ti)O_4$	tetr	5.7	6	2.1		
Fluorite	CaF_2	cub	3.18	4	1.434		
Forsterite	Mg_2SiO_4	orth	3.21	7	1.635	1.651	1.670
Franklinite	$ZnFe_2O_4$	cub	5.21	6	2.36		
Gahnite	$ZnAl_2O_4$	cub	4.62	7.8	1.805		
Galaxite	$MnAl_2O_4$	cub	4.04	7.8	1.92		
Galena	PbS	cub	7.60	2.5	3.91		
Galenabismuthite	$PbBi_2S_4$	rhomb	7.04	3			
Ganomalite	$(Ca,Pb)_{10}(OH,Cl)_2(Si_2O_7)_3$	hex	5.6	3.5	1.910	1.945	
Gaylussite	$Na_2Ca(CO_3)_2·5H_2O$	monocl	1.99	2.8	1.444	1.516	1.523
Gehlenite	$Ca_2Al_2SiO_7$	tetr	3.04	5.5	1.658	1.669	
Geikielite	$MgTiO_3$	hex	3.85	5.5	1.95	2.31	
Gibbsite	$Al(OH)_3$	monocl	2.42	3	1.57	1.57	1.59
Glauberite	$Na_2Ca(SO_4)_2$	monocl	2.80	2.8	1.515	1.535	1.536
Glauconite	$(K,Na,Ca)_{1.6}(Fe,Al,Mg)_{4.0}Si_{7.3}Al_{0.7}O_{20}(OH)_4$	monocl	2.7	2	1.60	1.63	1.63
Glaucophane	$Na_2Mg_3Al_2Si_8O_{22}(OH)_2$	monocl	3.19	6	1.634	1.645	1.648
Gmelinite	$(Ca,Na_2)[Al_2Si_4O_{12}]·6H_2O$	hex	2.10	4.5	1.477	1.485	
Goethite	$FeO(OH)$	orth	4.3	5.3	2.268	2.401	2.457
Goslarite	$ZnSO_4·7H_2O$	orth	1.97	2.3	1.457	1.480	1.484
Greenockite	CdS	hex	4.8	3.3	2.506	2.529	
Grossularite	$Ca_3Al_2Si_3O_{12}$	cub	3.59	6.8	1.734		
Gummite	$UO_3·H_2O$	orth	7.05	3.8			

Name	Formula	Crystal system	Density g/cm³	Hardness	Index of refraction		
					n_α	n_β	n_γ
Gypsum	$CaSO_4 \cdot 2H_2O$	monocl	2.32	2	1.520	1.525	1.530
Halite	$NaCl$	cub	2.17	2	1.544		
Hambergite	$Be_2(OH)(BO_3)$	rhomb	2.36	7.5	1.56	1.59	1.63
Hanksite	$Na_{22}K(SO_4)_9(CO_3)_2Cl$	hex	2.56	3.3	1.461	1.481	
Harmotome	$Ba[Al_2Si_6O_{16}] \cdot 6H_2O$	monocl	2.44	4.5	1.506	1.507	1.511
Hausmannite	Mn_3O_4	tetr	4.84	5.5	2.15	2.46	
Haüyne	$(Na,Ca)_{4-8}Al_6Si_6O_{24}(SO_4,S)_{1-2}$	cub	2.47	5.8	1.502		
Hedenbergite	$CaFeSi_2O_6$	monocl	3.53	6	1.721	1.727	1.746
Helvite	$Mn_4Be_3Si_3O_{12}S$	cub	3.32	6	1.739		
Hematite	Fe_2O_3	hex	5.25	6	2.91	3.19	
Hemimorphite	$Zn_4Si_2O_7(OH)_2 \cdot H_2O$	rhomb	3.45	5	1.614	1.617	1.636
Hercynite	$Fe(AlO_2)_2$	cub	4.3	7.8	1.835		
Herderite	$CaBe(PO_4)(Fe,OH)$	monocl	2.98	5.3	1.592	1.612	1.621
Hessite	Ag_2Te	orth	8.4	2.5			
Heulandite	$(Ca,Na_2,K_2)[Al_2Si_7O_{18}] \cdot 6H_2O$	monocl	2.2	3.8	1.498	1.498	1.506
Hopeite	$Zn_3(PO_4)_2 \cdot 4H_2O$	orth	3.0	3.2	1.58	1.59	1.59
Hornblende	$Ca_2(Mg,Fe)_4Al(Si_7AlO_{22})(OH)_2$	monocl	3.24	5.5	1.67	1.67	1.69
Huebnerite	$MnWO_4$	monocl	7.2	4.3	2.17	2.22	2.32
Humite	$Mg(OH,F)_2 \cdot 3Mg_2SiO_4$	orth	3.3	6	1.625	1.636	1.657
Huntite	$Mg_3Ca(CO_3)_4$	trig	2.70				
Hydrogrossularite	$Ca_3Al_2Si_2O_8(SiO_4)_{1-m}(OH)_{4m}$	cub	3.4	6.8	1.70		
Hydromagnesite	$3MgCO_3 \cdot Mg(OH)_2 \cdot 3H_2O$	monocl	2.24	3.5	1.523	1.527	1.545
Illite	$KAl_4[Si_7AlO_{20}](OH)_4$	monocl	2.8	1.5	1.56	1.59	1.59
Ilmenite	$FeTiO_3$	rhomb	4.72	5.5			
Iodyrite	AgI	hex	5.68	1.5	2.21	2.22	
Jacobsite	$MnFe_2O_4$	cub	4.87	7.8	2.3		
Jadeite	$NaAlSi_2O_6$	monocl	3.34	6	1.649	1.654	1.663
Jamesonite	$Pb_4FeSb_6S_{14}$	monocl	5.63	2.5			
Jarosite	$KFe_3(SO_4)_2(OH)_6$	rhomb	3.09	3	1.715	1.820	
Kainite	$KMg(SO_4)Cl \cdot 3H_2O$	monocl	2.15	2.8	1.494	1.505	1.516
Kaliophylite	$KAlSiO_4$	hex	2.61	6	1.532	1.537	
Kaolinite	$Al_4Si_4O_{10}(OH)_8$	tricl	2.65	2.3	1.549	1.564	1.565
Kernite	$Na_2B_4O_7 \cdot 4H_2O$	monocl	1.95	2.5	1.454	1.472	1.488
Kieserite	$MgSO_4 \cdot H_2O$	monocl	2.57	3.5	1.520	1.533	1.584
Kyanite	Al_2OSiO_4	tricl	3.59	6.3	1.715	1.722	1.731
Lanarkite	$Pb_2(SO_4)O$	monocl	6.92	2.3	1.928	2.007	2.036
Lanthanite	$(La,Ce)_2(CO_3)_3 \cdot 8H_2O$	rhomb	2.72	2.8	1.52	1.587	1.613
Laumontite	$Ca_4[Al_8Si_{16}O_{48}] \cdot 16H_2O$	monocl	2.3	3.3	1.508	1.517	1.519
Laurionite	$Pb(OH)Cl$	rhomb	6.24	3.3	2.08	2.12	2.16
Lawsonite	$CaAl_2(OH)_2Si_2O_7 \cdot H_2O$	rhomb	3.08	6	1.655	1.675	1.685
Lazulite	$(Mg,Fe)Al_2(PO_4)_2(OH)_2$	monocl	3.23	5.8	1.615	1.64	1.650
Lazurite	$Na_4SSi_3Al_3O_{12}$	cub	2.42	5.3	1.500		
Leadhillite	$Pb_4(SO_4)(CO_3)_2(OH)_2$	monocl	6.55	2.8	1.87	2.00	2.01
Lepidocrocite	$FeO(OH)$	orth	4.26	5	1.94	2.20	2.51
Lepidolite	$K_2(Li,Al)_{5-6}[Si_{6-7}Al_{2-1}O_{20}](OH,F)_4$	monocl	2.85	3.3	1.536	1.565	1.566
Leucite	$KAlSi_2O_6$	tetr	2.49	5.8	1.510		
Levyne	$(Ca,Na_2)Al_2Si_4O_{12} \cdot 6H_2O$	rhomb	2.10	4.5	1.496	1.501	
Litharge	PbO	tetr	9.35	2	2.535	2.665	
Loellingite	$FeAs_2$	rhomb	7.40	5.3			
Maghemite	Fe_2O_3	cub	4.88	7.8	2.63		
Magnesite	$MgCO_3$	hex	3.05	4	1.536	1.741	
Magnetite	Fe_3O_4	cub	5.17	6	2.42		
Malachite	$Cu_2(OH)_2(CO_3)$	monocl	4.05	3.8	1.655	1.875	1.909
Manganite	$MnO(OH)$	monocl	≈4.3	4	2.25	2.25	2.53
Manganosite	MnO	cub	5.37	5.5			
Marcasite	FeS_2	cub	5.02	6.3			
Marialite	$Na_4Al_3Si_9O_{24}Cl$	tetr	2.56	5.5	1.541	1.548	
Marshite	CuI	cub	5.67	2.5	2.346		
Mascagnite	$(NH_4)_2SO_4$	orth	1.77	2.3	1.520	1.523	1.533
Matlockite	$PbClF$	tetr	7.05	2.8	2.006	2.145	

Name	Formula	Crystal system	Density g/cm³	Hardness	Index of refraction n_α	n_β	n_γ
Meionite	$Ca_4Al_6Si_6O_{24}CO_3$	tetr	2.78	5.5	1.559	1.595	
Melanterite	$FeSO_4 \cdot 7H_2O$	monocl	1.89	2	1.47	1.48	1.49
Melilite	$(Ca,Na)_2(Mg,Fe,Al,Si)_3O_7$	tetr	3.00	5.5	1.639	1.645	
Mellite	$Al_2C_{12}O_{12} \cdot 18H_2O$	tetr	1.64	2.3	1.511	1.539	
Mendipite	$Pb_3O_2Cl_2$	rhomb	7.24	2.5	2.24	2.27	2.31
Mesolite	$Na_2Ca_2(Al_2Si_3O_{10})_3 \cdot 8H_2O$	orth	2.26	5	1.506		
Metacinnabar	HgS	cub	7.70	3			
Microcline	$KAlSi_3O_8$	monocl	2.56	6.3	1.522	1.526	1.530
Miersite	AgI	hex	5.68	2.5	2.20		
Millerite	NiS	hex	5.5	3.3			
Mimetite	$Pb_5(AsO_4,PO_4)_3Cl$	hex	7.24	3.8	2.128	2.147	
Minium	Pb_3O_4	tetr	8.9	2.5			
Mirabilite	$Na_2SO_4 \cdot 10H_2O$	monocl	1.46	1.8	1.394	1.396	1.398
Moissanite	SiC	hex	3.16	9.5	2.648	2.691	
Molybdenite	MoS_2	hex	5.06	1.3			
Monazite	$(Ce,La,Th)PO_4$	monocl	5.2	5	1.787	1.789	1.840
Monetite	$CaHPO_4$	tricl	2.92	3.5	1.587	1.61	1.640
Monticellite	$Ca(Mg,Fe)SiO_4$	orth	3.18	5.5	1.647	1.655	1.664
Montmorillonite	$(0.5Ca,Na)_{0.7}(Al,Mg,Fe)_4[(Si,Al)_8O_{20}](OH)_4 \cdot nH_2O$	monocl	2.5	1.5	1.55	1.57	1.57
Montroydite	HgO	orth	11.14	2.5	2.37	2.50	2.65
Mordenite	$(Na,K,Ca)[Al_2Si_{10}O_{24}] \cdot 7H_2O$	orth	2.13	3.5	1.478	1.480	1.482
Muscovite	$KAl_2Si_3AlO_{10}(OH,F)_2$	monocl	2.83	2.8	1.563	1.596	1.602
Nantokite	$CuCl$	cub	4.14	2.5	1.930		
Natrolite	$Na_2Al_2Si_3O_{10} \cdot 2H_2O$	orth	2.23	5	1.478	1.481	1.491
Nepheline	$Na_3KAl_4Si_4O_{16}$	hex	2.61	5.8	1.534	1.538	
Newberyite	$MgHPO_4 \cdot 3H_2O$	orth	2.13	3.3	1.514	1.517	1.533
Niccolite	$NiAs$	hex	7.77	5.3			
Norbergite	$Mg(OH,F)_2 \cdot Mg_2SiO_4$	orth	3.21	6.5	1.565	1.573	1.592
Nosean	$Na_8Al_6Si_6O_{24}SO_4$	cub	2.35	5.5	1.495		
Oldhamite	CaS	cub	2.59	4	2.137		
Oligoclase	$([NaSi]_{0.9-0.7}[CaAl]_{0.1-0.3})AlSi_2O_8$	tricl	2.64	6.3	1.539	1.543	1.547
Olivenite	$Cu_2(AsO_4)(OH)$	rhomb	4.2	3	1.77	1.80	1.85
Olivine	$(Mg,Fe)SiO_4$	rhomb	3.81	6.8	1.73	1.76	1.78
Opal	$SiO_2 \cdot nH_2O$	amorp	1.9	5	1.44		
Orpiment	As_2S_3	monocl	3.46	1.8	2.40	2.81	3.02
Orthoclase	$KAlSi_3O_8$	monocl	2.56	6	1.523	1.527	1.531
Orthopyroxene	$(Mg,Fe)SiO_3$	rhomb	3.6	5.5	1.709	1.712	1.723
Paragonite	$NaAl_2Si_3AlO_{10}(OH)_2$	monocl	2.85	2.5	1.572	1.602	1.605
Parisite	$(Ce,La,Na)FCO_3 \cdot CaCO_3$	hex	4.42	4.5	1.672	1.771	
Pectolite	$Ca_2NaH(SiO_3)_3$	tricl	2.88	4.8	1.603	1.610	1.639
Penfieldite	$Pb_4Cl_6(OH)_2$	hex	6.6		2.13	2.21	
Pentlandite	$(Fe,Ni)_9S_8$	cub	4.8	3.8			
Percylite	$PbCuCl_2(OH)_2$	cub		2.5	2.05		
Periclase	MgO	cub	3.6	5.5	1.735		
Perovskite	$CaTiO_3$	cub	3.98	5.5	2.34		
Petalite	$LiAlSi_4O_{10}$	monocl	2.42	6.5	1.506	1.511	1.519
Pharmacosiderite	$Fe_3(AsO_4)_2(OH)_3 \cdot 5H_2O$	cub	2.80	2.5	1.690		
Phenakite	Be_2SiO_4	rhomb	2.98	7.5	1.654	1.670	
Phillipsite	$K(Ca_{0.5},Na)_2[Al_3Si_5O_{16}] \cdot 6H_2O$	monocl	2.2	4.3	1.494	1.497	1.505
Phlogopite	$KMg_3AlSi_3O_{10}(OH,F)_2$	monocl	2.83	2.3	1.560	1.597	1.598
Phosgenite	$Pb_2(CO_3)Cl_2$	tetr	6.13	2.5	2.118	2.145	
Piemontite	$Ca_2(Mn,Fe,Al)_3O(Si_2O_7)(SiO_4)(OH)$	monocl	3.49	6	1.762	1.773	1.796
Pigeonite	$(Mg,Fe,Ca)(Mg,Fe)Si_2O_6$	monocl	3.38	6	1.702	1.703	1.728
Pollucite	$CsAlSi_2O_6$	tetr	2.9	6.5	1.517		
Polybasite	$(Ag,Cu)_{16}Sb_2S_{11}$	monocl	6.1	2.5			
Powellite	$Ca(Mo,W)O_4$	tetr	4.35	3.8	1.971	1.980	
Prehnite	$Ca_2Al_2Si_3O_{10}(OH)_2$	rhomb	2.93	6.3	1.622	1.628	1.648
Proustite	Ag_3AsS_3	rhomb	5.57	2.3	2.792	3.088	
Pseudobrookite	Fe_2TiO_5	rhomb	4.36	6	2.38	2.39	2.42
Psilomelane	$BaMn_9O_{16}(OH)_4$	rhomb	4.71	5.5			

Name	Formula	Crystal system	Density g/cm³	Hardness	Index of refraction		
					n_α	n_β	n_γ
Pumpellyite	$Ca_2Al_2(Al,Fe,Mg)[Si_2(O,OH)_7](SiO_4)(OH,O)_3$	monocl	3.21	5.5	1.688	1.695	1.705
Pyrargyrite	Ag_3SbS_3	rhomb	5.85	2.5	2.88	3.08	
Pyrite	FeS_2	cub	5.02	6.3			
Pyrochlore	$NaCaNb_2O_6F$	cub	5.3	5.3			
Pyrochroite	$Mn(OH)_2$	hex	3.26	2.5	1.68	1.72	
Pyrolusite	MnO_2	tetr	5.08	6.3			
Pyromorphite	$Pb_5(PO_4,AsO_4)_3Cl$	hex	7.04	3.8	2.048	2.058	
Pyrope	$Mg_3Al_2Si_3O_{12}$	cub	3.58	6.8	1.714		
Pyrophyllite	$Al_2Si_4O_{10}(OH)_2$	monocl	2.78	1.5	1.545	1.579	1.599
Pyrrhotite	Fe_7S_8	hex	4.62	4			
Quartz	SiO_2	hex	2.65	7	1.544	1.553	
Rammelsbergite	$NiAs_2$	orth	7.1	5.8			
Raspite	$PbWO_4$	monocl	8.46	2.8	1.27	1.27	1.30
Realgar	As_4S_4	monocl	3.5	1.8	2.538	2.684	2.704
Rhodochrosite	$MnCO_3$	hex	3.70	3.8	1.597	1.816	
Rhodonite	$(Mn,Fe,Ca)SiO_3$	orth	3.48	6	1.725	1.729	1.737
Riebeckite	$Na_2Fe_5(Si_8O_{22})(OH)_2$	monocl	3.3	5	1.675	1.683	1.694
Rutile	TiO_2	tetr	4.23	6.2	2.609	2.900	
Safflorite	$(Co,Fe)As_2$	rhomb	7.3	4.8			
Samarskite	$(Y,Er,Ce,U,Ca,Fe,Pb,Th)(Nb,Ta,Ti,Sn)_2O_6$	rhomb	5.69	5.5	2.200		
Sapphirine	$(Mg,Fe)_2Al_6SiO_4$	monocl	3.49	7.5	1.709	1.712	1.715
Scapolite	$(Na,Ca)_4Al_3(Al,Si)_3Si_6O_{24}(Cl,F,OH,CO_3,SO_4)$	tetr	2.64	5.5	1.551	1.573	
Scheelite	$CaWO_4$	tetr	6.06	4.8	1.920	1.936	
Scolecite	$CaAl_2Si_3O_{10}\cdot3H_2O$	monocl	2.27	5	1.510	1.518	1.519
Scorodite	$Fe(AsO_4)\cdot2H_2O$	rhomb	3.28	3.8	1.784	1.795	1.814
Sellaite	MgF_2	tetr	3.15	5	1.378	1.390	
Senarmontite	Sb_2O_3	cub	5.58	2.3	2.087		
Serpentine	$Mg_3Si_2O_5(OH)_4$	monocl	2.55	3	1.55	1.56	1.56
Siderite	$FeCO_3$	hex	3.9	4.3	1.635	1.875	
Sillimanite	Al_2OSiO_4	rhomb	3.25	7	1.658	1.660	1.660
Skutterudite	$(Co,Ni)As_3$	cub	6.8	5.8			
Smithsonite	$ZnCO_3$	rhomb	4.4	4.3	1.621	1.848	
Sodalite	$Na_8Al_6Si_6O_{24}Cl_2$	cub	2.30	5.8	1.485		
Sperrylite	$PtAs_2$	cub	10.58	6.5			
Spessartite	$Mn_3Al_2Si_3O_{12}$	cub	4.19	6.8	1.800		
Sphalerite	ZnS	cub	4.0	3.8	2.369		
Sphene	$CaTiSiO_4(O,OH,F)$	monocl	3.50	5	1.90	1.95	2.03
Spinel	$MgAl_2O_4$	cub	3.55	7.8	1.719		
Spodumene	$LiAlSi_2O_6$	monocl	3.13	6.8	1.656	1.662	1.671
Stannite	Cu_2FeSn_4	tetr	4.4	4			
Staurolite	$(Fe,Mg,Zn)_2(Al,Fe,Ti)_9O_6[(Si,Al)O_4]_4(O,OH)_2$	monocl	3.79	7.5	1.743	1.747	1.755
Stercorite	$Na(NH_4)H(PO_4)\cdot4H_2O$	tricl	1.62	2	1.439	1.442	1.469
Stibiotantalite	$Sb(Ta,Nb)O_4$	rhomb	6.6	5.5	2.38	2.41	2.46
Stibnite	Sb_2S_3	orth	4.56	2			
Stilbite	$NaCa_2[Al_5Si_{13}O_{36}]\cdot14H_2O$	monocl	2.2	3.8	1.492	1.499	1.503
Stilpnomelane	$(K,Na,Ca)_{0.6}(Fe,Mg)_6Si_8Al(O,OH)_{27}\cdot2H_2O$	monocl	2.8	3.5	1.585	1.665	1.665
Stolzite	$PbWO_4$	tetr	8.2	2.8	2.19	2.27	
Strengite	$FePO_4\cdot2H_2O$	orth	2.87	4	1.707	1.719	1.741
Strontianite	$SrCO_3$	orth	3.5	3.5	1.518	1.666	1.668
Struvite	$Mg(NH_4)(PO_4)\cdot6H_2O$	rhomb	1.71	2	1.495	1.496	1.504
Sulfur	S	orth	2.07	2	1.958	2.038	2.245
Sylvanite	$(Ag,Au)Te_2$	monocl	8.16	1.8			
Sylvite	KCl	cub	1.99	2	1.490		
Talc	$Mg_3Si_4O_{10}(OH)_2$	monocl	2.71	1	1.545	1.592	1.595
Tantalite	$(Fe,Mn)(Ta,Nb)_2O_6$	rhomb	7.95	6.5	2.26	2.32	2.43
Tapiolite	$FeTa_2O_6$	tetr	7.9	6.3	2.27	2.42	
Tellurobismuthite	Bi_2Te_3	hex	7.74	1.8			
Terlinguaite	Hg_2OCl	monocl	8.73	2.5	2.35	2.64	2.66
Tetrahedrite	$(Cu,Fe)_{12}Sb_4S_{13}$	cub	4.9	3.8			
Thenardite	Na_2SO_4	orth	2.7	2.8	1.468	1.475	1.483

Name	Formula	Crystal system	Density g/cm³	Hardness	n_α	n_β	n_γ
Thermonatrite	$Na_2CO_3 \cdot H_2O$	orth	2.25	1.3	1.420	1.506	1.524
Thomsenolite	$NaCaAlF_6 \cdot H_2O$	monocl	2.98	2	1.407	1.414	1.415
Thorianite	ThO_2	cub	10.0	6.5	2.200		
Thorite	$ThSiO_4$	tetr	6.7	4.8	1.8		
Topaz	$Al_2SiO_4(OH,F)_2$	rhomb	3.53	8	1.618	1.620	1.627
Torbernite	$Cu(UO_2)_2(PO_4)_2 \cdot 8H_2O$	tetr	3.22	2.3	1.582	1.592	
Tourmaline	$Na(Mg,Fe,Mn,Li,Al)_3Al_6Si_6O_{18}(BO_3)_3$	rhomb	3.14	7	1.62	1.65	
Tremolite	$Ca_2Mg_5Si_8O_{22}(OH,F)_2$	monocl	3.0	5.5	1.599	1.612	1.622
Trevorite	$NiFe_2O_4$	cub	5.33	7.8	2.3		
Tridymite	SiO_2	hex	2.27	7	1.475	1.476	1.479
Triphyllite-Lithiophyllite	$Li(Fe,Mn)PO_4$	rhomb	3.46	4.5	1.68	1.68	1.69
Troegerite	$(UO_2)_3(AsO_4)_2 \cdot 12H_2O$	tetr		2.5	1.59	1.630	
Troilite	FeS	hex	4.7	4			
Trona	$Na_3H(CO_3)_2 \cdot 2H_2O$	monocl	2.14	2.8	1.412	1.492	1.540
Turquois	$Cu(Al,Fe)_6(PO_4)_4(OH)_8 \cdot 4H_2O$	tricl	2.9	5.3	1.70	1.73	1.75
Ullmannite	$NiSbS$	cub	6.65	5.3			
Uraninite	UO_2	cub	11.0	5.5			
Uvarovite	$Ca_3Cr_2Si_3O_{12}$	cub	3.83	6.8	1.865		
Valentinite	Sb_2O_3	orth	5.7	2.8	2.18	2.35	2.35
Vanadinite	$Pb_5(VO_4)_3Cl$	hex	6.8	2.9	2.350	2.416	
Variseite-Strengite	$(Al,Fe)(PO_4) \cdot 2H_2O$	rhomb	2.72	4	1.635	1.654	1.668
Vaterite	$CaCO_3$	hex	2.71		1.550	1.645	
Vermiculite	$(Mg,Ca)_{0.7}(Mg,Fe,Al)_6[(Al,Si)_8O_{20}](OH)_4 \cdot 8H_2O$	monocl	2.3	1.5	1.542	1.556	1.556
Vesuvianite	$Ca_{10}(Mg,Fe)_2Al_4(Si_2O_7)_2(SiO_4)_5(OH,F)_4$	tetr	3.33	6.5	1.72	1.73	
Villiaumite	NaF	cub	2.78	2.3	1.327		
Vivianite	$Fe_3(PO_4)_2 \cdot 8H_2O$	monocl	2.58	1.8	1.598	1.629	1.652
Wagnerite	$Mg_2(PO_4)F$	monocl	3.15	5.3	1.568	1.572	1.582
Wavellite	$Al_3(OH)_3(PO_4)_2 \cdot 5H_2O$	rhomb	2.36	3.6	1.527	1.535	1.553
Whewellite	$CaC_2O_4 \cdot H_2O$	cub	2.2	2.8	1.491	1.554	1.650
Willemite	Zn_2SiO_4	hex	4.1	5.5	1.691	1.719	
Witherite	$BaCO_3$	orth	4.29	3.5	1.529	1.676	1.677
Wolframite	$(Fe,Mn)WO_4$	monocl	7.3	4.3	2.26	2.32	2.42
Wollastonite	$CaSiO_3$	monocl	2.92	4.8	1.628	1.639	1.642
Wulfenite	$PbMoO_4$	tetr	6.7	2.9	2.283	2.403	
Wurtzite	ZnS	hex	4.09	3.8	2.356	2.378	
Xenotime	YPO_4	tetr	4.8	4.5	1.721	1.816	
Zeunerite	$Cu(UO_2)_2(AsO_4)_2 \cdot 10H_2O$	tetr			1.606		
Zincite	ZnO	hex	5.6	4	2.013	2.029	
Zircon	$ZrSiO_4$	tetr	4.6	7.5	1.94	1.99	
Zoisite	$Ca_2Al_3Si_3O_{12}(OH)$	rhomb	3.26	6	1.695	1.699	1.711

CRYSTALLOGRAPHIC DATA ON MINERALS

This table contains x-ray crystallographic data on about 400 common minerals, as well as selected crystalline elements. Entries are arranged alphabetically by mineral name. The columns are:

Name: Common name of the mineral.

Formula: Chemical formula for a typical sample of the mineral. Composition often varies considerably with the origin of the sample.

Crystal system: tricl = triclinic; monocl = monoclinic; orth = orthorhombic; tetr = tetragonal; hex = hexagonal; rhomb = rhombohedral; cubic = cubic.

Structure type: Prototype for the structural arrangement of the crystallographic cell.

Z: Number of formula units per the unit cell.

a, b, c: Lengths of the cell edges in Å (1Å = 10^{-8} cm).

α, β, γ: Angles between cell axes.

References

1. Robie, R.A., Bethke, P.M., and Beardsley, K.M., *U. S. Geological Survey Bulletin 1248*, U. S. Government Printing Office, Washington, D.C.
2. Donnay, J.D.H., and Ondik, H.M., *Crystal Data Determinative Tables, Third Edition, Volume 2, Inorganic Compounds*, Joint Committee on Powder Diffraction Standards, Swarthmore, PA, 1973.
3. Deer, W.A., Howie, R.A., and Zussman, J., *An Introduction to the Rock-Forming Minerals, 2nd Edition*, Longman Scientific & Technical, Harlow, Essex, 1992.

Name	Formula	Crystal system	Structure type	Z	a/Å	b/Å	c/Å	α	β	γ
Acanthite	Ag_2S	monocl		4	4.228	6.928	7.862		99.58°	
Acmite (Aegirine)	$NaFe(SiO_3)_2$	monocl	diopside	4	9.658	8.795	5.294		107.42°	
Akermanite	$Ca_2MgSi_2O_7$	tetr	melilite	2	7.8435		5.010			
Alabandite	MnS	cubic	rock salt	4	5.223					
Almandine (Almandite)	$Fe_3Al_2Si_3O_{12}$	cubic	garnet	8	11.526					
Altaite	PbTe	cubic	rock salt	4	6.4606					
Aluminum	Al	cubic	copper	4	4.049					
Alunite	$KAl_3(SO_4)_2(OH)_6$	rhomb		3	6.982		17.32			
Analcite	$NaAlSi_2O_6 \cdot H_2O$	cubic		16	13.733					
Anatase	TiO_2	tetr		4	3.785		9.514			
Andalusite	Al_2OSiO_4	orth		4	7.7959	7.8983	5.5583			
Andradite	$Ca_3Fe_2Si_3O_{12}$	cubic	garnet	8	12.048					
Anglesite	$PbSO_4$	orth	barite	4	8.480	5.398	6.958			
Anhydrite	$CaSO_4$	orth	anhydrite	4	6.991	6.996	6.238			
Annite	$KFe_3[AlSi_3O_{10}](OH)_2$	monocl	1M mica	2	10.29	9.33	5.39		105.1°	
Anorthite	$CaAl_2Si_2O_8$	tricl	primitive cell	8	8.177	12.877	14.169	93.17°	115.85°	91.22°
Anthophyllite	$Mg_7Si_8O_{22}(OH)_2$	orth		4	18.61	18.01	5.24			
Antimony	Sb	rhomb	arsenic	6	4.2996		11.2516			
Aragonite	$CaCO_3$	orth	aragonite	4	5.741	7.968	4.959			
Arcanite	K_2SO_4	orth	arcanite	4	5.772	10.072	7.483			
Argentite	Ag_2S	cubic		2	4.870					
Argentopyrite	$AgFe_2S_3$	orth		4	6.64	11.47	6.45			
Arsenic	As	rhomb	arsenic	6	3.760		10.555			
Arsenolite	As_2O_3	cubic	diamond	16	11.074					
Arsenopyrite	FeAsS	tricl		4	5.760	5.690	5.785	90.00°	112.23°	90.00°
Azurite	$Cu_3(OH)_2(CO_3)_2$	monocl		2	5.008	5.844	10.336		92.45°	
Baddeleyite	ZrO_2	monocl	baddeleyite	4	5.1454	5.2075	5.3107		99.23°	
Banalsite	$BaNa_2Al_4Si_4O_{16}$	orth		4	8.50	9.97	16.72			
Barite	$BaSO_4$	orth	barite	4	8.878	5.450	7.152			
Berlinite	$AlPO_4$	hex	α-quartz	3	4.942		10.97			
Beryl	$Be_3Al_2(SiO_3)_6$	hex	beryl	2	9.215		9.192			
Berzelianite	Cu_2Se	cubic		4	5.85					
Bismite	Bi_2O_3	monocl	pseudo-orth	4	7.48	8.14	5.83		112.9°	
Bismuth	Bi	rhomb	arsenic	6	4.5367		11.8383			
Bismuthinite	Bi_2S_3	orth	stibnite	4	11.150	11.300	3.981			
Bixbyite	Mn_2O_3	cubic	thallium trioxide	16	9.411					
Boehmite	AlO(OH)	orth	lepidocrocite	4	2.868	12.227	3.700			
Borax	$Na_2B_4O_7 \cdot 10H_2O$	monocl		4	11.858	10.674	12.197		106.68°	

Name	Formula	Crystal system	Structure type	Z	a/Å	b/Å	c/Å	α	β	γ
Bornite (metastable)	Cu_5FeS_4	cubic		8	10.94					
Breithauptite	NiSb	hex	niccolite	2	3.942		5.155			
Brochantite	$Cu_4SO_4(OH)_6$	monocl		4	13.066	9.85	6.022		103.27°	
Bromargyrite	AgBr	cubic	rock salt	4	5.7745					
Bromellite	BeO	hex	zincite	2	2.6979		4.3772			
Brookite	TiO_2	orth		8	5.456	9.182	5.143			
Brucite	$Mg(OH)_2$	hex	cadmium iodide	1	3.147		4.769			
Bunsenite	NiO	cubic	rock salt	4	4.177					
Bustamite	$CaMn(SiO_3)_2$	tricl		6	7.736	7.157	13.824	90.52°	94.58°	103.87°
Cadmium telluride	CdTe	cubic	sphalerite	4	6.4805					
Cadmoselite	CdSe	hex	zincite	2	4.2977		7.0021			
Calcite	$CaCO_3$	rhomb	calcite	6	4.9899		17.064			
Calomel	Hg_2Cl_2	tetr		4	4.478		10.910			
Carbonate-apatite	$Ca_{10}(PO_4)_6CO_3 \cdot H_2O$	hex	apatite	1	9.436		6.883			
Cassiterite	SnO_2	tetr	rutile	2	4.738		3.188			
Cattierite	CoS_2	cubic	pyrite	4	5.5345					
Celestite	$SrSO_4$	orth	barite	4	8.359	5.352	6.866			
Celsian	$BaAl_2Si_2O_8$	monocl		8	8.627	13.045	14.408		115.20°	
Cerianite	CeO_2	cubic	fluorite	4	5.4110					
Cerussite	$PbCO_3$	orth	aragonite	4	6.152	8.436	5.195			
Cervantite	Sb_2O_4	orth		4	5.424	11.76	4.804			
Chalcanthite	$CuSO_4 \cdot 5H_2O$	tricl		2	6.1045	10.72	5.949	97.57°	107.28°	77.43°
Chalcocite	Cu_2S	orth		96	11.881	27.323	13.491			
Chalcopyrite	$CuFeS_2$	tetr		4	5.2988		10.434			
Chlorapatite	$Ca_5(PO_4)_3Cl$	hex	apatite	2	9.629		6.777			
Chlorargyrite	AgCl	cubic	rock salt	4	5.5491					
Chloritoid	$FeAl_4O_2(SiO_4)_2(OH)_4$	monocl		8	9.48	5.48	18.18		101.77°	
Chloromagnesite	$MgCl_2$	rhomb		3	3.632		17.795			
Chondrodite	$2Mg_2SiO_4 \cdot MgF_2$	monocl		2	7.89	4.743	10.29		109.03°	
Chrysoberyl	$BeAl_2O_4$	orth	olivine	4	5.4756	9.4041	4.4267			
Cinnabar	HgS	hex	cinnabar	3	4.149		9.495			
Claudetite	As_2O_3	monocl		4	5.339	12.984	4.5405		94.27°	
Clausthalite	PbSe	cubic	rock salt	4	6.1255					
Clinoenstatite	$MgSiO_3$	monocl		8	9.620	8.825	5.188		108.33°	
Clinoferrosilite	$FeSiO_3$	monocl		8	9.7085	9.0872	5.2284		108.43°	
Clinohumite	$4Mg_2SiO_4 \cdot MgF_2$	monocl		2	13.68	4.75	10.27		100.83°	
Clinozoisite	$Ca_2Al_3(SiO_4)_3OH$	monocl		2	8.887	5.581	10.14		115.93°	
Cobalt olivine	Co_2SiO_4	orth	olivine	4	4.782	10.301	6.003			
Cobalt oxide	CoO	cubic	rock salt	4	4.260					
Cobalt sulfide	CoS	cubic	sphalerite	4	5.339					
Cobalt titanate	$CoTiO_3$	rhomb	ilmenite	6	5.066		13.918			
Cobalticalcite	$CoCO_3$	rhomb	calcite	6	4.6581		14.958			
Cobaltite	CoAsS	cubic	NiSbS	4	5.60					
Coesite	SiO_2	monocl		16	7.152	12.379	7.152		120.00°	
Coffinite	$USiO_4$	tetr	zircon	4	6.995		6.263			
Colemanite	$Ca_2B_6O_{11} \cdot 5H_2O$	monocl		4	8.743	11.264	6.102		110.12°	
Coloradoite	HgTe	cubic	sphalerite	4	6.4600					
Cooperite	PtS	tetr		2	3.4699		6.1098			
Copper	Cu	cubic	face-centered cubic	4	3.6150					
Corundum	Al_2O_3	rhomb	corundum	6	4.7591		12.9894			
Cotunnite	$PbCl_2$	orth		4	4.535	7.62	9.05			
Covellite	CuS	hex		6	3.792		16.34			
Cristobalite (α)	SiO_2	tetr		4	4.971		6.918			
Cristobalite (β)	SiO_2	cubic		8	7.1382					
Cryolite	Na_3AlF_6	monocl		2	5.40	5.60	7.776		90.18°	
Cubanite	$CuFe_2S_3$	orth		4	6.46	11.12	6.23			
Cummingtonite	$(Mg,Fe,Mn)_7(Si_4O_{11})_2(OH)_2$	monocl	tremolite	2	9.522	18.223	5.332		101.92°	
Cuprite	Cu_2O	cubic		2	4.2696					
Danburite	$CaB_2Si_2O_8$	orth		4	8.04	8.77	7.74			

Name	Formula	Crystal system	Structure type	Z	a/Å	b/Å	c/Å	α	β	γ
Datolite	$CaBSiO_4(OH)$	monocl		4	9.62	7.60	4.84		90.15°	
Daubreeite	$FeCr_2S_4$	cubic	spinel	8	9.966					
Diamond	C	cubic	diamond	8	3.5670					
Diaspore	$AlO(OH)$	orth		4	4.401	9.421	2.845			
Dickite	$Al_2Si_2O_5(OH)_4$	monocl		4	5.150	8.940	14.736		103.58°	
Digenite	$Cu_{1.79}S$	cubic	deformed fluorite	4	5.5695					
Diopside	$CaMg(SiO_3)_2$	monocl	diopside	4	9.743	8.923	5.251		105.93°	
Dioptase	$CuSiO_2(OH)_2$	rhomb	phenacite	18	14.61		7.80			
Dolerophanite	$Cu_2O(SO_4)$	monocl		4	8.334	6.312	7.628		108.4°	
Dolomite	$CaMg(CO_3)_2$	rhomb	calcite	3	4.8079		16.010			
Dravite	$NaMg_3Al_6B_3Si_6O_{27}(OH)_4$	rhomb	tourmaline	3	15.942		7.224			
Elbaite	$NaLiAl_{7.67}B_3Si_6O_{27}(OH)_4$	rhomb	tourmaline	3	15.842		7.009			
Enargite	Cu_3AsS_4	orth		2	6.426	7.422	6.144			
Enstatite	$MgSiO_3$	orth		16	8.829	18.22	5.192			
Epidote	$Ca_2Al_2(Al,Fe)OH(SiO_4)_3$	monocl		2	8.89	5.63	10.19		115.40°	
Epsomite	$MgSO_4·7H_2O$	orth		4	11.86	11.99	6.858			
Eskolaite	Cr_2O_3	rhomb	corundum	6	4.9607		13.599			
Eucairite	AgCuSe	orth		10	4.105	20.35	6.31			
Euclase	$AlBeSiO_4(OH)$	monocl		4	4.763	14.29	4.618		100.25°	
Famatimite	Cu_3SbS_4	tetr		2	5.384		10.770			
Fayalite	Fe_2SiO_4	orth	olivine	4	4.817	10.477	6.105			
Fe-Cordierite	$Fe_2Al_3(AlSi_5O_{18})$	orth	cordierite	4	9.726	17.065	9.287			
Fe-Gehlenite	$Ca_2Fe_2SiO_7$	tetr	melilite	2	7.54		4.855			
Fe-Indialite	$Fe_2Al_3(AlSi_5O_{18})$	hex	beryl	2	9.860		9.285			
Fe-Leucite	$KFeSi_2O_6$	tetr		16	13.205		13.970			
Fe-Microcline	$KFeSi_3O_8$	tricl		4	8.68	13.10	7.340	90.75°	116.05°	86.23°
Fe-Sanidine	$KFeSi_3O_8$	monocl		4	8.689	13.12	7.319		116.10°	
Fe-Skutterudite	$FeAs_{2.95}$	cubic		8	8.1814					
Ferberite	$FeWO_4$	monocl	wolframite	2	4.732	5.708	4.965		90.00°	
Ferriannite	$KFe_3[FeSi_3O_{10}](OH)_2$	monocl		2	5.430	9.404	10.341		100.07°	
Ferroselite	$FeSe_2$	orth	marcasite	2	4.801	5.778	3.587			
Ferrotremolite	$Ca_2Fe_5[Si_8O_{22}](OH)_2$	monocl	tremolite	2	9.97	18.34	5.30		104.50°	
Fluor-edenite	$NaCa_2Mg_5[AlSi_7O_{22}]F_2$	monocl	tremolite	2	9.847	18.00	5.282		104.83°	
Fluor-humite	$3Mg_2SiO_4·MgF_2$	orth		4	10.243	20.72	4.735			
Fluor-norbergite	$Mg_2SiO_4·MgF_2$	orth		4	8.727	10.271	4.709			
Fluor-phlogopite	$KMg_3[AlSi_3O_{10}]F_2$	monocl	1M mica	2	5.299	9.188	10.135		99.92°	
Fluor-richterite	$Na_2CaMg_5[Si_8O_{22}]F_2$	monocl	tremolite	2	9.823	17.96	5.268		104.33°	
Fluor-tremolite	$Ca_2Mg_5[Si_8O_{22}]F_2$	monocl	tremolite	2	9.781	18.01	5.267		104.52°	
Fluorapatite	$Ca_5(PO_4)_3F$	hex	apatite	2	9.3684		6.8841			
Fluorite	CaF_2	cubic	fluorite	4	5.4638					
Forsterite	Mg_2SiO_4	orth	olivine	4	4.758	10.214	5.984			
Frohbergite	$FeTe_2$	orth	marcasite	2	5.265	6.265	3.869			
Gahnite	$ZnAl_2O_4$	cubic	spinel	8	8.0848					
Galaxite	$MnAl_2O_4$	cubic	spinel	8	8.258					
Galena	PbS	cubic	rock salt	4	5.9360					
Gallium oxide	Ga_2O_3	rhomb	corundum	6	4.9793		13.429			
Gehlenite	$Ca_2Al_2SiO_7$	tetr	melilite	2	7.690		5.0675			
Geikielite	$MgTiO_3$	rhomb	ilmenite	6	5.054		13.898			
Gerhardite	$Cu_2(NO_3)(OH)_3$	orth		4	6.075	13.812	5.592			
Gersdorfite	NiAsS	cubic		4	5.693					
Gibbsite	$Al(OH)_3$	monocl		8	9.719	5.0705	8.6412		94.57°	
Glauchroite	$CaMnSiO_4$	orth	olivine	4	4.944	11.19	6.529			
Glaucodot	(Co,Fe)AsS	orth		24	6.64	28.39	5.64			
Glaucophane I	$Na_2Mg_3Al_2[Si_8O_{22}](OH)_2$	monocl	tremolite	2	9.748	17.915	5.273		102.78°	
Glaucophane II	$Na_2Mg_3Al_2[Si_8O_{22}](OH)_2$	monocl	tremolite	2	9.663	17.696	5.277		103.67°	
Goethite	$FeO(OH)$	orth		4	4.596	9.957	3.021			
Gold	Au	cubic	face-centered cubic	4	4.0786					
Goldmanite	$Ca_3V_2Si_3O_{12}$	cubic	garnet	8	12.070					
Goslarite	$ZnSO_4·7H_2O$	orth	epsomite	4	11.779	12.050	6.822			

Name	Formula	Crystal system	Structure type	Z	a/Å	b/Å	c/Å	α	β	γ
Graphite	C	hex	graphite	4	2.4612		6.7079			
Greenockite	CdS	hex	zincite	2	4.1354		6.7120			
Greigite	Fe_3S_4	cubic	spinel	8	9.876					
Grossularite	$Ca_3Al_2Si_3O_{12}$	cubic	garnet	8	11.851					
Grunerite	$Fe_7[Si_8O_{22}](OH)_2$	monocl	tremolite	2	9.572	18.44	5.342		101.77°	
Gudmundite	FeSbS	monocl		8	10.00	5.93	6.73		90.00°	
Gypsum	$CaSO_4 \cdot 2H_2O$	monocl		4	5.68	15.18	6.29		113.83°	
Hafnia	HfO_2	monocl	baddeleyite	4	5.1156	5.1722	5.2948		99.18°	
Halite	NaCl	cubic	rock salt	4	5.6402					
Hambergite	$Be_2(OH,F)BO_3$	orth		8	9.755	12.201	4.426			
Hardystonite	$Ca_2ZnSi_2O_7$	tetr	melilite	2	7.87		5.01			
Hauerite	MnS_2	cubic	pyrite	4	6.1014					
Hausmannite	Mn_3O_4	tetr		8	8.136		9.422			
Hawleyite	CdS	cubic	sphalerite	4	5.833					
Heazelwoodite	Ni_3S_2	rhomb		3	5.746		7.134			
Hedenbergite	$CaFe(SiO_3)_2$	monocl	diopside	4	9.854	9.024	5.263		104.23°	
Hematite	Fe_2O_3	rhomb	corundum	6	5.025		13.735			
Hemimorphite	$Zn_4(OH)_2Si_2O_7 \cdot H_2O$	orth		2	8.370	10.719	5.120			
Hercynite	$Fe(AlO_2)_2$	cubic	spinel	8	8.150					
Herzenbergite	SnS	orth	germanium sulfide	4	4.328	11.190	3.978			
Hessite	Ag_2Te	monocl		4	8.13	4.48	8.09		111.9°	
Hexahydrite	$MgSO_4 \cdot 6H_2O$	monocl		8	10.110	7.212	24.41		98.30°	
High albite (Analbite)	$NaAlSi_3O_8$	tricl		4	8.160	12.870	7.106	93.54°	116.36°	90.19°
High argentite	Ag_2S	cubic		4	6.269					
High bornite	Cu_5FeS_4	cubic		1	5.50					
High carnegeite	$NaAlSiO_4$	cubic		4	7.325					
High chalcocite	Cu_2S	hex		2	3.961		6.722			
High clinoenstatite	$MgSiO_3$	tricl		8	10.000	8.934	5.170	88.27°	70.03°	91.01°
High digenite	Cu_2S	cubic		4	5.725					
High germania	GeO_2	hex	α–quartz	3	4.987		5.652			
High leucite	$KAlSi_2O_6$	cubic		16	13.43					
High naumanite	Ag_2Se	cubic		2	4.993					
High sanidine	$KAlSi_3O_8$	monocl		4	8.615	13.031	7.177		115.98°	
Huebnerite	$MnWO_4$	monocl	wolframite	2	4.834	5.758	4.999		91.18°	
Huntite	$Mg_3Ca(CO_3)_4$	rhomb	calcite	3	9.498		7.816			
Hydroxylapatite	$Ca_5(PO_4)_3OH$	hex	apatite	2	9.418		6.883			
Ice	H_2O	hex		4	4.5212		7.3666			
Ilmenite	$FeTiO_3$	rhomb	ilmenite	6	5.093		14.055			
Indialite (Cordierite)	$Mg_2Al_3(AlSi_5O_{18})$	hex	beryl	2	9.7698		9.3517			
Iodargyrite	AgI	hex	zincite	2	4.5955		7.5005			
Iron (α)	Fe	cubic	body-centered cubic	2	2.8664					
Jacobsite	$MnFe_2O_4$	cubic	spinel	8	8.499					
Jadeite	$NaAl(SiO_3)_2$	monocl	diopside	4	9.409	8.564	5.220		107.50°	
Jalpaite	$Ag_{1.55}Cu_{0.45}S$	tetr		16	8.673		11.756			
Johannsenite	$CaMn(SiO_3)_2$	monocl	diopside	4	9.83	9.04	5.27		105.00°	
Kaliophilite	$KAlSiO_4$	hex		54	26.930		8.522			
Kalsilite	$KAlSiO_4$	hex		2	5.1597		8.7032			
Kaolinite	$Al_2Si_2O_5(OH)_4$	tricl		2	5.155	8.959	7.407	91.68°	104.87°	89.93°
Karelianite	V_2O_3	rhomb	corundum	6	4.952		14.002			
Keatite	SiO_2	tetr		12	7.456		8.604			
Kernite	$Na_2B_4O_7 \cdot 4H_2O$	monocl		4	7.022	9.151	15.676		108.83°	
Kerschsteinite	$CaFeSiO_4$	orth	olivine	4	4.886	11.146	6.434			
Klockmannite	CuSe	hex	deformed covellite	78	14.206		17.25			
Knebelite	$MnFeSiO_4$	orth	olivine	4	4.854	10.602	6.162			
Kyanite	Al_2OSiO_4	tricl		4	7.123	7.848	5.564	89.92°	101.25°	105.97°
Larnite	Ca_2SiO_4	monocl		4	5.48	6.76	9.28		94.55°	
Laurite	RuS_2	cubic	pyrite	4	5.60					

Name	Formula	Crystal system	Structure type	Z	a/Å	b/Å	c/Å	α	β	γ
Lawrencite	$FeCl_2$	rhomb		3	3.593		17.58			
Lawsonite	$CaAl_2Si_2O_7(OH)_2 \cdot H_2O$	orth		4	8.787	5.836	13.123			
Lead	Pb	cubic	face-centered cubic	4	4.9505					
Leonhardtite	$MgSO_4 \cdot 4H_2O$	monocl		4	5.922	13.604	7.905		90.85°	
Lepidocrocite	FeO(OH)	orth		4	3.868	12.525	3.066			
Lepidolite	$K_2Al_2Li_2AlSi_7O_{20}(OH)_4$	monocl	2M2 mica	2	9.2	5.3	20.0		98.00°	
Leucite	$KAlSi_2O_6$	tetr		16	13.074		13.738			
Lime	CaO	cubic	rock salt	4	4.8108					
Lime olivine	Ca_2SiO_4	orth	olivine	4	5.091	11.371	6.782			
Linnaeite	Co_3S_4	cubic	spinel	8	9.401					
Litharge	PbO	tetr		2	3.9759		5.023			
Loellingite	$FeAs_2$	orth	marcasite	2	5.300	5.981	2.882			
Low albite	$NaAlSi_3O_8$	tricl		4	8.139	12.788	7.160	94.27°	116.57°	87.68°
Low bornite	Cu_5FeS_4	tetr		16	10.94		21.88			
Low cordierite	$Mg_2Al_3(AlSi_5O_{18})$	orth		4	9.721	17.062	9.339			
Low germania	GeO_2	tetr	rutile	2	4.3963		2.8626			
Low nepheline	$NaAlSiO_4$	hex		8	9.986		8.330			
Luzonite	Cu_3AsS_4	tetr		2	5.289		10.440			
Mackinawite	FeS	tetr		2	3.675		5.030			
Magnesioriebeckite	$Na_2Mg_3Fe_2[Si_8O_{22}](OH)_2$	monocl	tremolite	2	9.733	17.946	5.299		103.30°	
Magnesite	$MgCO_3$	rhomb	calcite	6	4.6330		15.016			
Magnetite	Fe_3O_4	cubic	spinel	8	8.3940					
Malachite	$Cu_2(OH)_2CO_3$	monocl		4	9.502	11.974	3.240		98.75°	
Maldonite	Au_2Bi	cubic		8	7.958					
Manganese sulfide (γ)	MnS	hex	zincite	2	3.976		6.432			
Manganese sulfide (β)	MnS	cubic	sphalerite	4	5.611					
Manganosite	MnO	cubic	rock salt	4	4.4448					
Marcasite	FeS_2	orth	marcasite	2	4.443	5.423	3.3876			
Margarite	$CaAl_2[AlSi_2O_{10}](OH)_2$	monocl	2M mica	4	5.13	8.92	19.50		95.00°	
Marialite	$Na_4Al_3Si_9O_{24}Cl$	tetr		2	12.064		7.514			
Marshite	CuI	cubic	sphalerite	4	6.0507					
Mascagnite	$(NH_4)_2SO_4$	orth	arcanite	4	7.782	5.993	10.636			
Massicot	PbO	orth		4	5.489	4.755	5.891			
Matlockite	PbClF	tetr		2	4.106		7.23			
Maucherite	$Ni_{11}As_8$	tetr		4	6.870		21.81			
Meionite	$Ca_4Al_6Si_6O_{24}CO_3$	tetr		2	12.174		7.652			
Melanophlogite	SiO_2	cubic	clathrate type	46	13.402					
Melanterite	$FeSO_4 \cdot 7H_2O$	monocl		4	14.072	6.503	11.041		105.57°	
Melonite	$NiTe_2$	hex	cadmium iodide	1	3.869		5.308			
Metacinnabar	HgS	cubic	sphalerite	4	5.8517					
Miargyrite	$AgSbS_2$	monocl		8	12.862	4.111	13.220		98.63°	
Microcline	$KAlSi_3O_8$	tricl		4	8.582	12.964	7.222	90.62°	115.92°	87.68°
Miersite	AgI	cubic	sphalerite	4	6.4963					
Millerite	NiS	rhomb		9	9.616		3.152			
Minium	Pb_3O_4	tetr		4	8.815		6.565			
Minnesotaite	$Fe_3Si_4O_{10}(OH)_2$	monocl		4	5.4	9.42	19.4		100.00°	
Mirabilite	$Na_2SO_4 \cdot 10H_2O$	monocl		4	11.51	10.38	12.83		107.75°	
Mn-Indialite	$Mn_2Al_3(AlSi_5O_{18})$	hex	beryl	2	9.925		9.297			
Molybdenite	MoS_2	hex	molybdenite	2	3.1604		12.295			
Molybdenum	Mo	cubic		2	3.1653					
Molybdite	MoO_3	orth		4	3.962	13.858	3.697			
Monteponite	CdO	cubic	rock salt	4	4.6953					
Monticellite	$CaMgSiO_4$	orth	olivine	4	4.827	11.084	6.376			
Montroydite	HgO	orth		4	6.608	5.518	3.519			
Mullite (2:1)	$2Al_2O_3 \cdot SiO_2$	orth		6	7.5788	7.6909	2.8883			
Mullite (3:2)	$3Al_2O_3 \cdot 2SiO_2$	orth		3	7.557	7.6876	2.8842			
Muscovite	$KAl_2AlSi_3O_{10}(OH)_2$	monocl	2M2 mica	4	5.203	8.995	20.030		94.47°	

Name	Formula	Crystal system	Structure type	Z	a/Å	b/Å	c/Å	α	β	γ
Nacrite	$Al_2Si_2O_5(OH)_4$	monocl		4	8.909	5.146	15.697		113.70°	
Nantokite	CuCl	cubic	sphalerite	4	5.416					
Natroalunite	$NaAl_3(SO_4)_2(OH)_6$	rhomb		3	6.974		16.69			
Natrolite	$Na_2Al_2Si_3O_{10}\cdot2H_2O$	orth		8	18.30	18.63	6.60			
Neighborite	$NaMgF_3$	orth	perovskite	4	5.363	7.676	5.503			
Ni-Skutterudite	$NiAs_{2.95}$	cubic		8	8.3300					
Niccolite	NiAs	hex	niccolite	2	3.618		5.034			
Nickel	Ni	cubic	face-centered cubic	4	3.5238					
Nickel carbonate	$NiCO_3$	rhomb	calcite	6	4.5975		14.723			
Nickel olivine	Ni_2SiO_4	orth	olivine	4	4.727	10.121	5.915			
Nickel selenide	$NiSe_2$	cubic	pyrite	4	5.9604					
Niter	KNO_3	orth	aragonite	4	6.431	9.164	5.414			
Norsethite	$BaMg(CO_3)_2$	rhomb	calcite	3	5.020		16.75			
Oldhamite	CaS	cubic	rock salt	4	5.689					
Orpiment	As_2S_3	monocl		4	11.49	9.59	4.25		90.45°	
Orthoclase	$KAlSi_3O_8$	monocl		4	8.562	12.996	7.193		116.02°	
Orthoferrosilite	$FeSiO_3$	orth	enstatite	16	9.080	18.431	5.238			
Otavite	$CdCO_3$	rhomb	calcite	6	4.9204		16.298			
Paracelsian	$BaAl_2Si_2O_8$	monocl		4	8.58	9.583	9.08		90.00°	
Paragonite	$NaAl_2AlSi_3O_{10}(OH)_2$	monocl	2M1 mica	4	5.13	8.89	19.32		95.17°	
Pararammelsbergite	$NiAs_2$	orth		8	5.75	5.82	11.428			
Paratellurite	TeO_2	tetr		4	4.810		7.613			
Parawollastonite	$CaSiO_3$	monocl		12	15.417	7.321	7.066		95.40°	
Pectolite	$Ca_2NaH(SiO_3)_3$	tricl		2	7.99	7.04	7.02	90.05°	95.27°	102.47°
Pentlandite	$Fe_{5.25}Ni_{3.75}S_8$	cubic		4	10.196					
Pentlandite	$Fe_{4.75}Ni_{5.25}S_8$	cubic		4	10.095					
Periclase	MgO	cubic	rock salt	4	4.2117					
Perovskite	$CaTiO_3$	orth	perovskite	4	5.3670	7.6438	5.4439			
Petalite	$LiAlSi_4O_{10}$	monocl		2	11.32	5.14	7.62		105.90°	
Petzite	Ag_3AuTe_2	cubic		8	10.38					
Phenacite	Be_2SiO_4	rhomb	phenacite	18	12.472		8.252			
Phlogopite	$KMg_3AlSi_3O_{10}(OH)_2$	monocl	1M mica	2	5.326	9.210	10.311		100.17°	
Picrochromite	$MgCr_2O_4$	cubic	spinel	8	8.333					
Piemontite	$Ca_2Al_{1.5}Mn_{1.5}(SiO_4)_3OH$	monocl		2	8.95	5.70	9.41		115.70°	
Platinum	Pt	cubic	face-centered cubic	4	3.9231					
Polymidite	Ni_3S_4	cubic	spinel	8	9.480					
Portlandite	$Ca(OH)_2$	hex	cadmium iodide	1	3.5933		4.9086			
Powellite	$CaMoO_4$	tetr	scheelite	4	5.226		11.43			
Protoenstatite	$MgSiO_3$	orth		8	9.25	8.74	5.32			
Proustite	Ag_3AsS_3	rhomb		6	10.816		8.6948			
Pseudowollastonite	$CaSiO_3$	tricl		24	6.90	11.78	19.65	90.00°	90.80°	90.00°
Pyrargyrite	Ag_3SbS_3	rhomb		6	11.052		8.7177			
Pyrite	FeS_2	cubic	pyrite	4	5.4175					
Pyrolusite	MnO_2	tetr	rutile	2	4.388		2.865			
Pyrope	$Mg_3Al_2Si_3O_{12}$	cubic	garnet	8	11.459					
Pyrophanite	$MnTiO_3$	rhomb	ilmenite	6	5.155		14.18			
Pyrophyllite	$Al_2Si_4O_{10}(OH)_2$	monocl	2M1 mica	4	5.14	8.90	18.55		99.92°	
Pyroxmangite	$MnFe(SiO_3)_2$	tricl		7	7.56	17.45	6.67	84.00°	94.30°	113.70°
Pyrrhotite	$Fe_{0.980}S$	hex	defect niccolite	2	3.446		5.848			
Pyrrhotite	$Fe_{0.885}S$	hex	defect niccolite	2	3.440		5.709			
Quartz (α)	SiO_2	hex		3	4.9136		5.4051			
Quartz (β)	SiO_2	hex		3	4.999		5.4592			
Rammelsbergite	$NiAs_2$	orth	marcasite	2	4.757	5.797	3.542			
Realgar	As_4S_4	monocl		16	9.29	13.53	6.57		106.55°	
Retgersite	$NiSO_4\cdot4H_2O$	tetr		4	6.782		18.28			
Rhodochrosite	$MnCO_3$	rhomb	calcite	6	4.7771		15.664			

Name	Formula	Crystal system	Structure type	Z	a/Å	b/Å	c/Å	α	β	γ
Rhodonite	$MnSiO_3$	tricl		10	7.682	11.818	6.707	92.36°	93.95°	105.66°
Riebeckite	$Na_2Fe_5FSi_8O_{22}(OH)_2$	monocl	tremolite	2	9.729	18.065	5.334		103.31°	
Rutile	TiO_2	tetr		2	4.5937		2.9618			
Safflorite	$Co_{0.5}Fe_{0.5}As_2$	orth	marcasite	2	5.231	5.953	2.962			
Sanmartinite	$ZnWO_4$	monocl	wolframite	2	4.691	5.720	4.925		89.36°	
Sapphirine	$Mg_2Al_6O_6SiO_4$	monocl		8	9.96	28.60	9.85		110.5°	
Scacchite	$MnCl_2$	rhomb		3	3.711		17.59			
Scheelite	$CaWO_4$	tetr	scheelite	4	5.242		11.372			
Schorl	$NaFe_3Al_6B_3Si_6O_{27}(OH)_4$	rhomb	tourmaline	3	16.032		7.149			
Selenium	Se	hex		3	4.3642		4.9588			
Selenolite	SeO_2	tetr		8	8.35		5.05			
Sellaite	MgF_2	tetr	rutile	2	4.621		3.050			
Senarmontite	Sb_2O_3	cubic	arsenic trioxide	16	11.152					
Shandite	$Ni_3Pb_2S_2$	rhomb		3	5.576		13.658			
Shortite	$Na_2Ca_2(CO_3)_3$	orth		2	4.961	11.03	7.12			
Siderite	$FeCO_3$	rhomb	calcite	6	4.6887		15.373			
Silicon	Si	cubic	diamond	8	5.4305					
Sillimanite	Al_2OSiO_4	orth		4	7.4843	7.6730	5.7711			
Silver	Ag	cubic	face-centered cubic	4	4.0862					
Silver telluride I	Ag_2Te	cubic		2	5.29					
Silver telluride II	Ag_2Te	cubic		4	6.585					
Smithsonite	$ZnCO_3$	rhomb	calcite	6	4.6528		15.025			
Soda niter	$NaNO_3$	rhomb	calcite	6	5.0696		16.829			
Sodium melilite	$NaCaAlSi_2O_7$	tetr	melilite	2	8.511		4.809			
Sperrylite	$PtAs_2$	cubic	pyrite	4	5.968					
Spessartite	$Mn_3Al_2Si_3O_{12}$	cubic	garnet	8	11.621					
Sphalerite	ZnS	cubic	sphalerite	4	5.4093					
Sphene	$CaTiSiO_5$	monocl		4	7.07	8.72	6.56		113.95°	
Spinel	$MgAl_2O_4$	cubic	spinel	8	8.080					
Spodumene	$LiAl(SiO_3)_2$	monocl	diopside	4	9.451	8.387	5.208		110.07°	
Spodumene (β)	$LiAl(SiO_3)_2$	tetr		4	7.5332		9.1540			
Staurolite	$Fe_2Al_9Si_4O_{22}(OH)_2$	monocl		2	7.90	16.65	5.63		90.00°	
Sternbergite	$AgFe_2S_3$	orth		8	11.60	12.675	6.63			
Stibnite	Sb_2S_3	orth	stibnite	4	11.229	11.310	3.8389			
Stilleite	$ZnSe$	cubic	sphalerite	4	5.6685					
Stishovite	SiO_2	tetr	rutile	2	4.1790		2.6649			
Stolzite	$PbWO_4$	tetr	scheelite	4	5.4616		12.046			
Stromeyerite	$Ag_{0.93}Cu_{1.07}S$	orth		4	4.066	6.628	7.972			
Strontianite	$SrCO_3$	orth	aragonite	4	6.029	8.414	5.107			
Sulfur (monoclinic)	S	monocl	S8 ring molecules	48	11.04	10.98	10.92		96.73°	
Sulfur (orthorhombic)	S	orth	S8 ring molecules	128	10.4646	12.8660	24.4860			
Sulfur (rhombohedral)	S	rhomb	S6 ring molecules	18	10.818		4.280			
Sylvite	KCl	cubic	rock salt	4	6.2931					
Syngenite	$K_2Ca(SO_4)_2 \cdot H_2O$	monocl		2	9.775	7.156	6.251		104.00°	
Synthetic anorthite	$CaAl_2Si_2O_8$	hex		2	5.10		14.72			
Synthetic anorthite	$CaAl_2Si_2O_8$	orth		2	8.22	8.60	4.83			
Talc	$Mg_3Si_4O_{10}(OH)_2$	monocl	2M1 mica	4	5.287	9.158	18.95		99.50°	
Tantalum	Ta	cubic	tungsten	2	3.3058					
Teallite	$PbSnS_2$	orth	germanium sulfide	2	4.266	11.419	4.090			
Tellurite	TeO_2	orth	tellurite	8	5.607	12.034	5.463			
Tellurium	Te	hex	selenium	3	4.4570		5.9290			
Tellurobismuthite	Bi_2Te_3	rhomb		3	4.3835		30.487			
Tennantite	$Cu_{12}As_4S_{13}$	cubic	tetrahedrite	2	10.190					
Tenorite	CuO	monocl		4	4.684	3.425	5.129		99.47°	
Tephroite	Mn_2SiO_4	orth	olivine	4	4.871	10.636	6.232			

Name	Formula	Crystal system	Structure type	Z	a/Å	b/Å	c/Å	α	β	γ
Tetrahedrite	$Cu_{12}Sb_4S_{13}$	cubic	tetrahedrite	2	10.327					
Thenardite	Na_2SO_4	orth	thenardite	8	5.863	12.304	9.821			
Thorianite	ThO_2	cubic	fluorite	4	5.5952					
Thorite	$ThSiO_4$	tetr	zircon	4	7.143		6.327			
Tiemannite	HgSe	cubic	sphalerite	4	6.0853					
Tin	Sn	tetr		4	5.8315		3.1813			
Titanium	Ti	hex		2	2.953		4.729			
Titanium(III) oxide	Ti_2O_3	rhomb	corundum	6	5.149		13.642			
Topaz	$Al_2SiO_4(OH,F)_2$	orth		4	8.394	8.792	4.649			
Tremolite	$Ca_2Mg_5Si_8O_{22}(OH)_2$	monocl	tremolite	2	9.840	18.052	5.275		104.70°	
Trevorite	$NiFe_2O_4$	cubic	spinel	8	8.339					
Tridymite (β)	SiO_2	hex		4	5.0463		8.2563			
Trogtalite	$CoSe_2$	cubic	pyrite	4	5.8588					
Troilite	FeS	hex	niccolite	2	3.446		5.877			
Tschermakite	$CaAl_2SiO_6$	monocl	diopside	4	9.615	8.661	5.272		106.12°	
Tungsten	W	cubic		2	3.1653					
Tungstenite	WS_2	hex	molybdenite	2	3.154		12.362			
Turquois	$CuAl_6(PO_4)_4(OH)_8 \cdot 4H_2O$	tricl		1	7.424	7.629	9.910	68.61°	69.71°	65.08°
Umangite	Cu_3Se_2	tetr		2	6.402		4.276			
Uraninite	UO_2	cubic	fluorite	4	5.4682					
Ureyite	$NaCr(SiO_3)_2$	monocl	diopside	4	9.550	8.712	5.273		107.44°	
Uvarovite	$Ca_3Cr_2Si_3O_{12}$	cubic	garnet	8	11.999					
Uvite	$CaMg_4Al_5B_3Si_6O_{27}(OH)_4$	rhomb	tourmaline	3	15.86		7.19			
Vaesite	NiS_2	cubic	pyrite	4	5.6873					
Valentinite	Sb_2O_3	orth	antimony trioxide	4	4.914	12.468	5.421			
Vanthoffite	$MgSO_4 \cdot 3Na_2SO_4$	monocl		2	9.797	9.217	8.199		113.50°	
Vaterite	$CaCO_3$	hex		6	7.135		8.524			
Villiaumite	NaF	cubic	rock salt	4	4.6342					
Violarite	$FeNi_2S_4$	cubic	spinel	8	9.464					
Willemite	Zn_2SiO_4	rhomb	phenacite	18	13.94		9.309			
Witherite	$BaCO_3$	orth	aragonite	4	6.430	8.904	5.314			
Wolframite	$Fe_{0.5}Mn_{0.5}WO_4$	monocl	wolframite	2	4.782	5.731	4.982		90.57°	
Wollastonite	$CaSiO_3$	tricl		6	7.94	7.32	7.07	90.03°	95.37°	103.43°
Wulfenite	$PbMoO_4$	tetr	scheelite	4	5.435		12.110			
Wurtzite	ZnS	hex	zincite	2	3.8230		6.2565			
Wustite	$Fe_{0.953}O$	cubic	defect rock salt	4	4.3088					
Xenotime	YPO_4	tetr	zircon	4	6.885		5.982			
Zinc	Zn	hex	hexagonal close pack	2	2.665		4.947			
Zinc telluride	ZnTe	cubic	sphalerite	4	6.1020					
Zincite	ZnO	hex	zincite	2	3.2495		5.2069			
Zinkosite	$ZnSO_4$	orth	barite	4	8.588	6.740	4.770			
Zircon	$ZrSiO_4$	tetr	zircon	4	6.604		5.979			
Zoisite	$Ca_2Al_3(SiO_4)_3OH$	orth		4	16.15	5.581	10.06			

Section 5
Thermochemistry, Electrochemistry, and Solution Chemistry

CODATA KEY VALUES FOR THERMODYNAMICS

The Committee on Data for Science and Technology (CODATA) has conducted a project to establish internationally agreed values for the thermodynamic properties of key chemical substances. This table presents the final results of the project. Use of these recommended, internally consistent values is encouraged in the analysis of thermodynamic measurements, data reduction, and preparation of other thermodynamic tables.

The table includes the standard enthalpy of formation at 298.15 K, the entropy at 298.15 K, and the quantity $H°$ (298.15 K)$-H°$ (0). A value of 0 in the $\Delta_f H°$ column for an element indicates the reference state for that element. The standard state pressure is 100,000

Pa (1 bar). See the reference for information on the dependence of gas-phase entropy on the choice of standard state pressure.

Substances are listed in alphabetical order of their chemical formulas when written in the most common form.

The table is reprinted with permission of CODATA.

Reference

Cox, J. D., Wagman, D. D., and Medvedev, V. A., *CODATA Key Values for Thermodynamics*, Hemisphere Publishing Corp., New York, 1989.

Substance	State	$\Delta_f H°$ (298.15 K) kJ·mol^{-1}	$S°$ (298.15 K) J·K^{-1}·mol^{-1}	$H°$ (298.15 K) $- H°$ (0) kJ·mol^{-1}
Ag	cr	0	42.55 ± 0.20	5.745 ± 0.020
Ag	g	284.9 ± 0.8	172.997 ± 0.004	6.197 ± 0.001
Ag$^+$	aq	105.79 ± 0.08	73.45 ± 0.40	
AgCl	cr	−127.01 ± 0.05	96.25 ± 0.20	12.033 ± 0.020
Al	cr	0	28.30 ± 0.10	4.540 ± 0.020
Al	g	330.0 ± 4.0	164.554 ± 0.004	6.919 ± 0.001
Al^{+3}	aq	−538.4 ± 1.5	−325 ± 10	
AlF$_3$	cr	−1510.4 ± 1.3	66.5 ± 0.5	11.62 ± 0.04
Al$_2$O$_3$	cr, corundum	−1675.7 ± 1.3	50.92 ± 0.10	10.016 ± 0.020
Ar	g	0	154.846 ± 0.003	6.197 ± 0.001
B	cr, rhombic	0	5.90 ± 0.08	1.222 ± 0.008
B	g	565 ± 5	153.436 ± 0.015	6.316 ± 0.002
BF$_3$	g	−1136.0 ± 0.8	254.42 ± 0.20	11.650 ± 0.020
B$_2$O$_3$	cr	−1273.5 ± 1.4	53.97 ± 0.30	9.301 ± 0.040
Be	cr	0	9.50 ± 0.08	1.950 ± 0.020
Be	g	324 ± 5	136.275 ± 0.003	6.197 ± 0.001
BeO	cr	−609.4 ± 2.5	13.77 ± 0.04	2.837 ± 0.008
Br	g	111.87 ± 0.12	175.018 ± 0.004	6.197 ± 0.001
Br$^-$	aq	−121.41 ± 0.15	82.55 ± 0.20	
Br$_2$	l	0	152.21 ± 0.30	24.52 ± 0.01
Br$_2$	g	30.91 ± 0.11	245.468 ± 0.005	9.725 ± 0.001
C	cr, graphite	0	5.74 ± 0.10	1.050 ± 0.020
C	g	716.68 ± 0.45	158.100 ± 0.003	6.536 ± 0.001
CO	g	−110.53 ± 0.17	197.660 ± 0.004	8.671 ± 0.001
CO$_2$	g	−393.51 ± 0.13	213.785 ± 0.010	9.365 ± 0.003
CO$_2$	aq, undissoc.	−413.26 ± 0.20	119.36 ± 0.60	
CO$_3^{-2}$	aq	−675.23 ± 0.25	−50.0 ± 1.0	
Ca	cr	0	41.59 ± 0.40	5.736 ± 0.040
Ca	g	177.8 ± 0.8	154.887 ± 0.004	6.197 ± 0.001
Ca^{+2}	aq	−543.0 ± 1.0	−56.2 ± 1.0	
CaO	cr	−634.92 ± 0.90	38.1 ± 0.4	6.75 ± 0.06
Cd	cr	0	51.80 ± 0.15	6.247 ± 0.015
Cd	g	111.80 ± 0.20	167.749 ± 0.004	6.197 ± 0.001
Cd^{+2}	aq	−75.92 ± 0.60	−72.8 ± 1.5	
CdO	cr	−258.35 ± 0.40	54.8 ± 1.5	8.41 ± 0.08
CdSO$_4$·8/3H$_2$O	cr	−1729.30 ± 0.80	229.65 ± 0.40	35.56 ± 0.04
Cl	g	121.301 ± 0.008	165.190 ± 0.004	6.272 ± 0.001
Cl$^-$	aq	−167.080 ± 0.10	56.60 ± 0.20	
ClO$_4^-$	aq	−128.10 ± 0.40	184.0 ± 1.5	
Cl$_2$	g	0	223.081 ± 0.010	9.181 ± 0.001
Cs	cr	0	85.23 ± 0.40	7.711 ± 0.020
Cs	g	76.5 ± 1.0	175.601 ± 0.003	6.197 ± 0.001
Cs$^+$	aq	−258.00 ± 0.50	132.1 ± 0.5	
Cu	cr	0	33.15 ± 0.08	5.004 ± 0.008

Substance	State	$\dfrac{\Delta_f H° \,(298.15\ \text{K})}{\text{kJ} \cdot \text{mol}^{-1}}$	$\dfrac{S° \,(298.15\ \text{K})}{\text{J} \cdot \text{K}^{-1} \cdot \text{mol}^{-1}}$	$\dfrac{H° \,(298.15\ \text{K}) - H° \,(0)}{\text{kJ} \cdot \text{mol}^{-1}}$
Cu	g	337.4 ± 1.2	166.398 ± 0.004	6.197 ± 0.001
Cu^{+2}	aq	64.9 ± 1.0	−98 ± 4	
$CuSO_4$	cr	−771.4 ± 1.2	109.2 ± 0.4	16.86 ± 0.08
F	g	79.38 ± 0.30	158.751 ± 0.004	6.518 ± 0.001
F^-	aq	−335.35 ± 0.65	−13.8 ± 0.8	
F_2	g	0	202.791 ± 0.005	8.825 ± 0.001
Ge	cr	0	31.09 ± 0.15	4.636 ± 0.020
Ge	g	372 ± 3	167.904 ± 0.005	7.398 ± 0.001
GeF_4	g	−1190.20 ± 0.50	301.9 ± 1.0	17.29 ± 0.10
GeO_2	cr, tetragonal	−580.0 ± 1.0	39.71 ± 0.15	7.230 ± 0.020
H	g	217.998 ± 0.006	114.717 ± 0.002	6.197 ± 0.001
H^+	aq	0	0	
HBr	g	−36.29 ± 0.16	198.700 ± 0.004	8.648 ± 0.001
HCO_3^-	aq	−689.93 ± 0.20	98.4 ± 0.5	
HCl	g	−92.31 ± 0.10	186.902 ± 0.005	8.640 ± 0.001
HF	g	−273.30 ± 0.70	173.779 ± 0.003	8.599 ± 0.001
HI	g	26.50 ± 0.10	206.590 ± 0.004	8.657 ± 0.001
HPO_4^{-2}	aq	−1299.0 ± 1.5	−33.5 ± 1.5	
HS^-	aq	−16.3 ± 1.5	67 ± 5	
HSO_4^-	aq	−886.9 ± 1.0	131.7 ± 3.0	
H_2	g	0	130.680 ± 0.003	8.468 ± 0.001
H_2O	l	−285.830 ± 0.040	69.95 ± 0.03	13.273 ± 0.020
H_2O	g	−241.826 ± 0.040	188.835 ± 0.010	9.905 ± 0.005
$H_2PO_4^-$	aq	−1302.6 ± 1.5	92.5 ± 1.5	
H_2S	g	−20.6 ± 0.5	205.81 ± 0.05	9.957 ± 0.010
H_2S	aq, undissoc.	−38.6 ± 1.5	126 ± 5	
H_3BO_3	cr	−1094.8 ± 0.8	89.95 ± 0.60	13.52 ± 0.04
H_3BO_3	aq, undissoc.	−1072.8 ± 0.8	162.4 ± 0.6	
He	g	0	126.153 ± 0.002	6.197 ± 0.001
Hg	l	0	75.90 ± 0.12	9.342 ± 0.008
Hg	g	61.38 ± 0.04	174.971 ± 0.005	6.197 ± 0.001
Hg^{+2}	aq	170.21 ± 0.20	−36.19 ± 0.80	
HgO	cr, red	−90.79 ± 0.12	70.25 ± 0.30	9.117 ± 0.025
Hg_2^{+2}	aq	166.87 ± 0.50	65.74 ± 0.80	
Hg_2Cl_2	cr	−265.37 ± 0.40	191.6 ± 0.8	23.35 ± 0.20
Hg_2SO_4	cr	−743.09 ± 0.40	200.70 ± 0.20	26.070 ± 0.030
I	g	106.76 ± 0.04	180.787 ± 0.004	6.197 ± 0.001
I^-	aq	−56.78 ± 0.05	106.45 ± 0.30	
I_2	cr	0	116.14 ± 0.30	13.196 ± 0.040
I_2	g	62.42 ± 0.08	260.687 ± 0.005	10.116 ± 0.001
K	cr	0	64.68 ± 0.20	7.088 ± 0.020
K	g	89.0 ± 0.8	160.341 ± 0.003	6.197 ± 0.001
K^+	aq	−252.14 ± 0.08	101.20 ± 0.20	
Kr	g	0	164.085 ± 0.003	6.197 ± 0.001
Li	cr	0	29.12 ± 0.20	4.632 ± 0.040
Li	g	159.3 ± 1.0	138.782 ± 0.010	6.197 ± 0.001
Li^+	aq	−278.47 ± 0.08	12.24 ± 0.15	
Mg	cr	0	32.67 ± 0.10	4.998 ± 0.030
Mg	g	147.1 ± 0.8	148.648 ± 0.003	6.197 ± 0.001
Mg^{+2}	aq	−467.0 ± 0.6	−137 ± 4	
MgF_2	cr	−1124.2 ± 1.2	57.2 ± 0.5	9.91 ± 0.06
MgO	cr	−601.60 ± 0.30	26.95 ± 0.15	5.160 ± 0.020
N	g	472.68 ± 0.40	153.301 ± 0.003	6.197 ± 0.001
NH_3	g	−45.94 ± 0.35	192.77 ± 0.05	10.043 ± 0.010
NH_4^+	aq	−133.26 ± 0.25	111.17 ± 0.40	
NO_3^-	aq	−206.85 ± 0.40	146.70 ± 0.40	
N_2	g	0	191.609 ± 0.004	8.670 ± 0.001
Na	cr	0	51.30 ± 0.20	6.460 ± 0.020
Na	g	107.5 ± 0.7	153.718 ± 0.003	6.197 ± 0.001
Na^+	aq	−240.34 ± 0.06	58.45 ± 0.15	

Substance	State	$\Delta_f H°$ (298.15 K) $kJ \cdot mol^{-1}$	$S°$ (298.15 K) $J \cdot K^{-1} \cdot mol^{-1}$	$H°$ (298.15 K) $- H°$ (0) $kJ \cdot mol^{-1}$
Ne	g	0	146.328 ± 0.003	6.197 ± 0.001
O	g	249.18 ± 0.10	161.059 ± 0.003	6.725 ± 0.001
OH^-	aq	−230.015 ± 0.040	−10.90 ± 0.20	
O_2	g	0	205.152 ± 0.005	8.680 ± 0.002
P	cr, white	0	41.09 ± 0.25	5.360 ± 0.015
P	g	316.5 ± 1.0	163.199 ± 0.003	6.197 ± 0.001
P_2	g	144.0 ± 2.0	218.123 ± 0.004	8.904 ± 0.001
P_4	g	58.9 ± 0.3	280.01 ± 0.50	14.10 ± 0.20
Pb	cr	0	64.80 ± 0.30	6.870 ± 0.030
Pb	g	195.2 ± 0.8	175.375 ± 0.005	6.197 ± 0.001
Pb^{+2}	aq	0.92 ± 0.25	18.5 ± 1.0	
$PbSO_4$	cr	−919.97 ± 0.40	148.50 ± 0.60	20.050 ± 0.040
Rb	cr	0	76.78 ± 0.30	7.489 ± 0.020
Rb	g	80.9 ± 0.8	170.094 ± 0.003	6.197 ± 0.001
Rb^+	aq	−251.12 ± 0.10	121.75 ± 0.25	
S	cr, rhombic	0	32.054 ± 0.050	4.412 ± 0.006
S	g	277.17 ± 0.15	167.829 ± 0.006	6.657 ± 0.001
SO_2	g	−296.81 ± 0.20	248.223 ± 0.050	10.549 ± 0.010
SO_4^{-2}	aq	−909.34 ± 0.40	18.50 ± 0.40	
S_2	g	128.60 ± 0.30	228.167 ± 0.010	9.132 ± 0.002
Si	cr	0	18.81 ± 0.08	3.217 ± 0.008
Si	g	450 ± 8	167.981 ± 0.004	7.550 ± 0.001
SiF_4	g	−1615.0 ± 0.8	282.76 ± 0.50	15.36 ± 0.05
SiO_2	cr, alpha quartz	−910.7 ± 1.0	41.46 ± 0.20	6.916 ± 0.020
Sn	cr, white	0	51.18 ± 0.08	6.323 ± 0.008
Sn	g	301.2 ± 1.5	168.492 ± 0.004	6.215 ± 0.001
Sn^{+2}	aq	−8.9 ± 1.0	−16.7 ± 4.0	
SnO	cr, tetragonal	−280.71 ± 0.20	57.17 ± 0.30	8.736 ± 0.020
SnO_2	cr, tetragonal	−577.63 ± 0.20	49.04 ± 0.10	8.384 ± 0.020
Th	cr	0	51.8 ± 0.5	6.35 ± 0.05
Th	g	602 ± 6	190.17 ± 0.05	6.197 ± 0.003
ThO_2	cr	−1226.4 ± 3.5	65.23 ± 0.20	10.560 ± 0.020
Ti	cr	0	30.72 ± 0.10	4.824 ± 0.015
Ti	g	473 ± 3	180.298 ± 0.010	7.539 ± 0.002
$TiCl_4$	g	−763.2 ± 3.0	353.2 ± 4.0	21.5 ± 0.5
TiO_2	cr, rutile	−944.0 ± 0.8	50.62 ± 0.30	8.68 ± 0.05
U	cr	0	50.20 ± 0.20	6.364 ± 0.020
U	g	533 ± 8	199.79 ± 0.10	6.499 ± 0.020
UO_2	cr	−1085.0 ± 1.0	77.03 ± 0.20	11.280 ± 0.020
UO_2^{+2}	aq	−1019.0 ± 1.5	−98.2 ± 3.0	
UO_3	cr, gamma	−1223.8 ± 1.2	96.11 ± 0.40	14.585 ± 0.050
U_3O_8	cr	−3574.8 ± 2.5	282.55 ± 0.50	42.74 ± 0.10
Xe	g	0	169.685 ± 0.003	6.197 ± 0.001
Zn	cr	0	41.63 ± 0.15	5.657 ± 0.020
Zn	g	130.40 ± 0.40	160.990 ± 0.004	6.197 ± 0.001
Zn^{+2}	aq	−153.39 ± 0.20	−109.8 ± 0.5	
ZnO	cr	−350.46 ± 0.27	43.65 ± 0.40	6.933 ± 0.040

STANDARD THERMODYNAMIC PROPERTIES OF CHEMICAL SUBSTANCES

This table gives the standard state chemical thermodynamic properties of about 2500 individual substances in the crystalline, liquid, and gaseous states. Substances are listed by molecular formula in a modified Hill order; all substances not containing carbon appear first, followed by those that contain carbon. The properties tabulated are:

$\Delta_f H°$ Standard molar enthalpy (heat) of formation at 298.15 K in kJ/mol

$\Delta_f G°$ Standard molar Gibbs energy of formation at 298.15 K in kJ/mol

$S°$ Standard molar entropy at 298.15 K in J/mol K

C_p Molar heat capacity at constant pressure at 298.15 K in J/mol K

The standard state pressure is 100 kPa (1 bar). The standard states are defined for different phases by:

- The standard state of a pure gaseous substance is that of the substance as a (hypothetical) ideal gas at the standard state pressure.
- The standard state of a pure liquid substance is that of the liquid under the standard state pressure.
- The standard state of a pure crystalline substance is that of the crystalline substance under the standard state pressure.

An entry of 0.0 for $\Delta_f H°$ for an element indicates the reference state of that element. See References 1 and 2 for further information on reference states. A blank means no value is available.

The data are derived from the sources listed in the references, from other papers appearing in the *Journal of Physical and Chemical Reference Data*, and from the primary research literature. We are indebted to M. V. Korobov for providing data on fullerene compounds.

References

1. Cox, J. D., Wagman, D. D., and Medvedev, V. A., *CODATA Key Values for Thermodynamics*, Hemisphere Publishing Corp., New York, 1989.
2. Wagman, D. D., Evans, W. H., Parker, V. B., Schumm, R. H., Halow, I., Bailey, S. M., Churney, K. L., and Nuttall, R. L., *The NBS Tables of Chemical Thermodynamic Properties, J. Phys. Chem. Ref. Data*, Vol. 11, Suppl. 2, 1982.
3. Chase, M. W., Davies, C. A., Downey, J. R., Frurip, D. J., McDonald, R. A., and Syverud, A. N., *JANAF Thermochemical Tables, Third Edition, J. Phys. Chem. Ref. Data*, Vol. 14, Suppl. 1, 1985.
4. Chase, M. W., *NIST-JANAF Thermochemical Tables, Fourth Edition, J. Phys. Chem. Ref. Data*, Monograph 9, 1998.
5. Daubert, T. E., Danner, R. P., Sibul, H. M., and Stebbins, C. C., *Physical and Thermodynamic Properties of Pure Compounds: Data Compilation*, extant 1994 (core with 4 supplements), Taylor & Francis, Bristol, PA.
6. Pedley, J. B., Naylor, R. D., and Kirby, S. P., *Thermochemical Data of Organic Compounds, Second Edition*, Chapman & Hall, London, 1986.
7. Pedley, J. B., *Thermochemical Data and Structures of Organic Compounds*, Thermodynamic Research Center, Texas A & M University, College Station, TX, 1994.
8. Domalski, E. S., and Hearing, E. D., Heat Capacities and Entropies of Organic Compounds in the Condensed Phase, Volume III, *J. Phys. Chem. Ref. Data*, 25, 1–525, 1996.
9. Zabransky, M., Ruzicka , V., Majer, V., and Domalski, E. S., *Heat Capacity of Liquids, J. Phys. Chem. Ref. Data*, Monograph No. 6, 1996.
10. Gurvich, L. V., Veyts, I.V., and Alcock, C. B., *Thermodynamic Properties of Individual Substances, Fourth Edition, Vol. 1*, Hemisphere Publishing Corp., New York, 1989.
11. Gurvich, L. V., Veyts, I.V., and Alcock, C. B., *Thermodynamic Properties of Individual Substances, Vol. 3*, CRC Press, Boca Raton, FL, 1994.
12. *NIST Chemistry Webbook*, <webbook.nist.gov>

Molecular formula	Name	Crystal $\Delta_f H°$ kJ/mol	$\Delta_f G°$ kJ/mol	$S°$ J/mol K	C_p J/mol K	Liquid $\Delta_f H°$ kJ/mol	$\Delta_f G°$ kJ/mol	$S°$ J/mol K	C_p J/mol K	Gas $\Delta_f H°$ kJ/mol	$\Delta_f G°$ kJ/mol	$S°$ J/mol K	C_p J/mol K
Substances not containing carbon:													
Ac	Actinium	0.0		56.5	27.2					406.0	366.0	188.1	20.8
Ag	Silver	0.0		42.6	25.4					284.9	246.0	173.0	20.8
AgBr	Silver(I) bromide	-100.4	-96.9	107.1	52.4								
AgBrO₃	Silver(I) bromate	-10.5	71.3	151.9									
AgCl	Silver(I) chloride	-127.0	-109.8	96.3	50.8								
AgClO₃	Silver(I) chlorate	-30.3	64.5	142.0									
AgClO₄	Silver(I) perchlorate	-31.1											
AgF	Silver(I) fluoride	-204.6											
AgF₂	Silver(II) fluoride	-360.0											
AgI	Silver(I) iodide	-61.8	-66.2	115.5	56.8								
AgIO₃	Silver(I) iodate	-171.1	-93.7	149.4	102.9								
AgNO₃	Silver(I) nitrate	-124.4	-33.4	140.9	93.1								
Ag₂	Disilver									410.0	358.8	257.1	37.0
Ag₂CrO₄	Silver(I) chromate	-731.7	-641.8	217.6	142.3								
Ag₂O	Silver(I) oxide	-31.1	-11.2	121.3	65.9								
Ag₂O₂	Silver(II) oxide	-24.3	27.6	117.0	88.0								
Ag₂O₃	Silver(III) oxide	33.9	121.4	100.0									
Ag₂O₄S	Silver(I) sulfate	-715.9	-618.4	200.4	131.4								
Ag₂S	Silver(I) sulfide (argentite)	-32.6	-40.7	144.0	76.5								
Al	Aluminum	0.0		28.3	24.2					330.0	289.4	164.6	21.4
AlB₃H₁₂	Aluminum borohydride					-16.3	145.0	289.1	194.6	13.0	147.0	379.2	
AlBr	Aluminum monobromide									-4.0	-42.0	239.5	35.6

Molecular formula	Name	Crystal $\Delta_f H°$ kJ/mol	$\Delta_f G°$ kJ/mol	$S°$ J/mol K	C_p J/mol K	Liquid $\Delta_f H°$ kJ/mol	$\Delta_f G°$ kJ/mol	$S°$ J/mol K	C_p J/mol K	Gas $\Delta_f H°$ kJ/mol	$\Delta_f G°$ kJ/mol	$S°$ J/mol K	C_p J/mol K
$AlBr_3$	Aluminum bromide	-527.2		180.2	100.6					-425.1			
AlCl	Aluminum monochloride									-47.7	-74.1	228.1	35.0
$AlCl_2$	Aluminum dichloride									-331.0			
$AlCl_3$	Aluminum chloride	-704.2	-628.8	109.3	91.1					-583.2			
AlF	Aluminum monofluoride									-258.2	-283.7	215.0	31.9
AlF_3	Aluminum fluoride	-1510.4	-1431.1	66.5	75.1					-1204.6	-1188.2	277.1	62.6
AlF_4Na	Sodium tetrafluoroaluminate									-1869.0	-1827.5	345.7	105.9
AlH	Aluminum monohydride									259.2	231.2	187.9	29.4
AlH_3	Aluminum hydride	-46.0		30.0	40.2								
AlH_4K	Potassium aluminum hydride	-183.7											
AlH_4Li	Lithium aluminum hydride	-116.3	-44.7	78.7	83.2								
AlH_4Na	Sodium aluminum hydride	-115.5											
AlI	Aluminum monoiodide									65.5			36.0
AlI_3	Aluminum iodide	-302.9		195.9	98.7					-289.4		223.6	
AlN	Aluminum nitride	-318.0	-287.0	20.2	30.1								
AlO	Aluminum monoxide									91.2	65.3	218.4	30.9
AlO_4P	Aluminum phosphate	-1733.8	-1617.9	90.8	93.2								
AlP	Aluminum phosphide	-166.5											
AlS	Aluminum monosulfide									200.9	150.1	230.6	33.4
Al_2	Dialuminum									485.9	433.3	233.2	36.4
Al_2Br_6	Aluminum hexabromide									-970.7			
Al_2Cl_6	Aluminum hexachloride									-1290.8	-1220.4	490.0	
Al_2F_6	Aluminum hexafluoride									-2628.0			
Al_2I_6	Aluminum hexaiodide									-516.7			
Al_2O	Aluminum oxide (Al_2O)									-130.0	-159.0	259.4	45.7
Al_2O_3	Aluminum oxide (corundum)	-1675.7	-1582.3	50.9	79.0								
Al_2S_3	Aluminum sulfide	-724.0		116.9	105.1								
Am	Americium	0.0											
Ar	Argon									0.0		154.8	20.8
As	Arsenic (gray)	0.0		35.1	24.6					302.5	261.0	174.2	20.8
As	Arsenic (yellow)	14.6											
$AsBr_3$	Arsenic(III) bromide	-197.5								-130.0	-159.0	363.9	79.2
$AsCl_3$	Arsenic(III) chloride					-305.0	-259.4	216.3		-261.5	-248.9	327.2	75.7
AsF_3	Arsenic(III) fluoride					-821.3	-774.2	181.2	126.6	-785.8	-770.8	289.1	65.6
AsGa	Gallium arsenide	-71.0	-67.8	64.2	46.2								
AsH_3	Arsine									66.4	68.9	222.8	38.1
AsH_3O_4	Arsenic acid	-906.3											
AsI_3	Arsenic(III) iodide	-58.2	-59.4	213.1	105.8							388.3	80.6
AsIn	Indium arsenide	-58.6	-53.6	75.7	47.8								
AsO	Arsenic monoxide									70.0			
As_2	Diarsenic									222.2	171.9	239.4	35.0
As_2O_5	Arsenic(V) oxide	-924.9	-782.3	105.4	116.5								
As_2S_3	Arsenic(III) sulfide	-169.0	-168.6	163.6	116.3								
At	Astatine	0.0											
Au	Gold	0.0		47.4	25.4					366.1	326.3	180.5	20.8
AuBr	Gold(I) bromide	-14.0											
$AuBr_3$	Gold(III) bromide	-53.3											
AuCl	Gold(I) chloride	-34.7											
$AuCl_3$	Gold(III) chloride	-117.6											
AuF_3	Gold(III) fluoride	-363.6											
AuH	Gold hydride									295.0	265.7	211.2	29.2
AuI	Gold(I) iodide	0.0											
Au_2	Digold									515.1			36.9
B	Boron (β-rhombohedral)	0.0		5.9	11.1					565.0	521.0	153.4	20.8
BBr	Bromoborane(1)									238.1	195.4	225.0	32.9
BBr_3	Boron tribromide					-239.7	-238.5	229.7		-205.6	-232.5	324.2	67.8
BCl	Chloroborane(1)									149.5	120.9	213.2	31.7
BClO	Chloroxyborane									-314.0			
BCl_3	Boron trichloride					-427.2	-387.4	206.3	106.7	-403.8	-388.7	290.1	62.7
$BCsO_2$	Cesium metaborate	-972.0	-915.0	104.4	80.6								
BF	Fluoroborane(1)									-122.2	-149.8	200.5	29.6
BFO	Fluorooxyborane									-607.0			
BF_3	Boron trifluoride									-1136.0	-1119.4	254.4	
BF_3H_3N	Aminetrifluoroboron	-1353.9											
BF_3H_3P	Trihydro(phosphorus trifluoride)boron									-854.0			
BF_4Na	Sodium tetrafluoroborate	-1844.7	-1750.1	145.3	120.3								

Molecular formula	Name	Crystal				Liquid				Gas			
		$\Delta_f H°$ kJ/mol	$\Delta_f G°$ kJ/mol	$S°$ J/mol K	C_p J/mol K	$\Delta_f H°$ kJ/mol	$\Delta_f G°$ kJ/mol	$S°$ J/mol K	C_p J/mol K	$\Delta_f H°$ kJ/mol	$\Delta_f G°$ kJ/mol	$S°$ J/mol K	C_p J/mol K
BH	Borane(1)									442.7	412.7	171.8	29.2
BHO_2	Metaboric acid (β, monoclinic)	-794.3	-723.4	38.0						-561.9	-551.0	240.1	42.2
BH_3	Borane(3)									89.2	93.3	188.2	36.0
BH_3O_3	Boric acid	-1094.3	-968.9	90.0	86.1					-994.1			
BH_4K	Potassium borohydride	-227.4	-160.3	106.3	96.1								
BH_4Li	Lithium borohydride	-190.8	-125.0	75.9	82.6								
BH_4Na	Sodium borohydride	-188.6	-123.9	101.3	86.8								
BI_3	Boron triiodide									71.1	20.7	349.2	70.8
BKO_2	Potassium metaborate	-981.6	-923.4	80.0	66.7								
$BLiO_2$	Lithium metaborate	-1032.2	-976.1	51.5	59.8								
BN	Boron nitride	-254.4	-228.4	14.8	19.7					647.5	614.5	212.3	29.5
$BNaO_2$	Sodium metaborate	-977.0	-920.7	73.5	65.9								
BO	Boron monoxide									25.0	-4.0	203.5	29.2
BO_2	Boron dioxide									-300.4	-305.9	229.6	43.0
BO_2Rb	Rubidium metaborate	-971.0	-913.0	94.3	74.1								
BS	Boron monosulfide									342.0	288.8	216.2	30.0
B_2	Diboron									830.5	774.0	201.9	30.5
B_2Cl_4	Tetrachlorodiborane					-523.0	-464.8	262.3	137.7	-490.4	-460.6	357.4	95.4
B_2F_4	Tetrafluorodiborane									-1440.1	-1410.4	317.3	79.1
B_2H_6	Diborane									36.4	87.6	232.1	56.7
B_2O_2	Diboron dioxide									-454.8	-462.3	242.5	57.3
B_2O_3	Boron oxide	-1273.5	-1194.3	54.0	62.8					-843.8	-832.0	279.8	66.9
B_2S_3	Boron sulfide	-240.6		100.0	111.7					67.0			
$B_3H_6N_3$	Borazine					-541.0	-392.7	199.6					
B_4H_{10}	Tetraborane(10)									66.1	184.3	280.3	
$B_4Na_2O_7$	Sodium tetraborate	-3291.1	-3096.0	189.5	186.8								
B_5H_9	Pentaborane(9)					42.7	171.8	184.2	151.1	73.2	173.6	280.6	99.6
B_5H_{11}	Pentaborane(11)					73.2				103.3	230.6	321.0	130.3
B_6H_{10}	Hexaborane(10)					56.3				94.6	211.3	296.8	125.7
B_9H_{15}	Nonaborane(15)									158.4	357.5	364.9	187.0
$B_{10}H_{14}$	Decaborane(14)									47.3	232.8	350.7	186.1
Ba	Barium	0.0		62.5	28.1					180.0	146.0	170.2	20.8
$BaBr_2$	Barium bromide	-757.3	-736.8	146.0									
$BaCl_2$	Barium chloride	-855.0	-806.7	123.7	75.1								
$BaCl_2H_4O_2$	Barium chloride dihydrate	-1456.9	-1293.2	203.0									
BaF_2	Barium fluoride	-1207.1		-1156.8	96.4	71.2							
BaH_2	Barium hydride	-177.0	-138.2	63.0	46.0								
BaH_2O_2	Barium hydroxide	-944.7											
BaI_2	Barium iodide	-602.1											
BaN_2O_4	Barium nitrite	-768.2											
BaN_2O_6	Barium nitrate	-988.0	-792.6	214.0	151.4								
BaO	Barium oxide	-548.0	-520.3	72.1	47.3					-112.0			
BaO_4S	Barium sulfate	-1473.2	-1362.2	132.2	101.8								
BaS	Barium sulfide	-460.0	-456.0	78.2	49.4								
Be	Beryllium	0.0		9.5	16.4					324.0	286.6	136.3	20.8
$BeBr_2$	Beryllium bromide	-353.5		108.0	69.4								
$BeCl_2$	Beryllium chloride	-490.4	-445.6	75.8	62.4								
BeF_2	Beryllium fluoride	-1026.8	-979.4	53.4	51.8								
BeH_2O_2	Beryllium hydroxide	-902.5	-815.0	45.5	62.1								
BeI_2	Beryllium iodide	-192.5		121.0	71.1								
BeO	Beryllium oxide	-609.4	-580.1	13.8	25.6								
BeO_4S	Beryllium sulfate	-1205.2	-1093.8	77.9	85.7								
BeS	Beryllium sulfide	-234.3		34.0	34.0								
Bi	Bismuth	0.0		56.7	25.5					207.1	168.2	187.0	20.8
BiClO	Bismuth oxychloride	-366.9	-322.1	120.5									
$BiCl_3$	Bismuth trichloride	-379.1	-315.0	177.0	105.0					-265.7	-256.0	358.9	79.7
BiH_3O_3	Bismuth hydroxide	-711.3											
BiI_3	Bismuth triiodide		-175.3										
Bi_2	Dibismuth									219.7			36.9
Bi_2O_3	Bismuth oxide	-573.9	-493.7	151.5	113.5								
$Bi_2O_{12}S_3$	Bismuth sulfate	-2544.3											
Bi_2S_3	Bismuth sulfide	-143.1	-140.6	200.4	122.2								
Bk	Berkelium	0.0											
Br	Bromine (atomic)									111.9	82.4	175.0	20.8
BrCl	Bromine chloride									14.6	-1.0	240.1	35.0
$BrCl_3Si$	Bromotrichlorosilane											350.1	90.9

Molecular formula	Name	Crystal				Liquid				Gas			
		$\Delta_f H°$ kJ/mol	$\Delta_f G°$ kJ/mol	$S°$ J/mol K	C_p J/mol K	$\Delta_f H°$ kJ/mol	$\Delta_f G°$ kJ/mol	$S°$ J/mol K	C_p J/mol K	$\Delta_f H°$ kJ/mol	$\Delta_f G°$ kJ/mol	$S°$ J/mol K	C_p J/mol K
BrCs	Cesium bromide	-405.8	-391.4	113.1	52.9								
BrCu	Copper(I) bromide	-104.6	-100.8	96.1	54.7								
BrF	Bromine fluoride									-93.8	-109.2	229.0	33.0
BrF$_3$	Bromine trifluoride					-300.8	-240.5	178.2	124.6	-255.6	-229.4	292.5	66.6
BrF$_5$	Bromine pentafluoride					-458.6	-351.8	225.1		-428.9	-350.6	320.2	99.6
BrGe	Germanium monobromide									235.6			37.1
BrGeH$_3$	Bromogermane											274.8	56.4
BrH	Hydrogen bromide									-36.3	-53.4	198.7	29.1
BrHSi	Bromosilylene									-464.4			
BrH$_3$Si	Bromosilane											262.4	52.8
BrH$_4$N	Ammonium bromide	-270.8	-175.2	113.0	96.0								
BrI	Iodine bromide									40.8	3.7	258.8	36.4
BrIn	Indium(I) bromide	-175.3	-169.0	113.0						-56.9	-94.3	259.5	36.7
BrK	Potassium bromide	-393.8	-380.7	95.9	52.3								
BrKO$_3$	Potassium bromate	-360.2	-271.2	149.2	105.2								
BrKO$_4$	Potassium perbromate	-287.9	-174.4	170.1	120.2								
BrLi	Lithium bromide	-351.2	-342.0	74.3									
BrNO	Nitrosyl bromide									82.2	82.4	273.7	45.5
BrNa	Sodium bromide	-361.1	-349.0	86.8	51.4					-143.1	-177.1	241.2	36.3
BrNaO$_3$	Sodium bromate	-334.1	-242.6	128.9									
BrO	Bromine monoxide									125.8	109.6	233.0	34.2
BrO$_2$	Bromine dioxide									152.0	155.0	271.1	45.4
BrRb	Rubidium bromide	-394.6	-381.8	110.0	52.8								
BrSi	Bromosilyldyne									209.0			38.6
BrTl	Thallium(I) bromide	-173.2	-167.4	120.5						-37.7			
Br$_2$	Bromine					0.0		152.2	75.7	30.9	3.1	245.5	36.0
Br$_2$Ca	Calcium bromide	-682.8	-663.6	130.0									
Br$_2$Cd	Cadmium bromide	-316.2	-296.3	137.2	76.7								
Br$_2$Co	Cobalt(II) bromide	-220.9			79.5								
Br$_2$Cr	Chromium(II) bromide	-302.1											
Br$_2$Cu	Copper(II) bromide	-141.8											
Br$_2$Fe	Iron(II) bromide	-249.8	-238.1	140.6									
Br$_2$H$_2$Si	Dibromosilane											309.7	65.5
Br$_2$Hg	Mercury(II) bromide	-170.7	-153.1	172.0									
Br$_2$Hg$_2$	Mercury(I) bromide	-206.9	-181.1	218.0									
Br$_2$Mg	Magnesium bromide	-524.3	-503.8	117.2									
Br$_2$Mn	Manganese(II) bromide	-384.9											
Br$_2$Ni	Nickel(II) bromide	-212.1											
Br$_2$Pb	Lead(II) bromide	-278.7	-261.9	161.5	80.1								
Br$_2$Pt	Platinum(II) bromide	-82.0											
Br$_2$S$_2$	Sulfur bromide					-13.0							
Br$_2$Se	Selenium dibromide									-21.0			
Br$_2$Sn	Tin(II) bromide	-243.5											
Br$_2$Sr	Strontium bromide	-717.6	-697.1	135.1	75.3								
Br$_2$Ti	Titanium(II) bromide	-402.0											
Br$_2$Zn	Zinc bromide	-328.7	-312.1	138.5									
Br$_3$Ce	Cerium(III) bromide	-891.4											
Br$_3$ClSi	Tribromochlorosilane											377.1	95.3
Br$_3$Dy	Dysprosium(III) bromide	-836.2											
Br$_3$Fe	Iron(III) bromide	-268.2											
Br$_3$Ga	Gallium(III) bromide	-386.6	-359.8	180.0									
Br$_3$HSi	Tribromosilane					-355.6	-336.4	248.1		-317.6	-328.5	348.6	80.8
Br$_3$In	Indium(III) bromide	-428.9								-282.0			
Br$_3$OP	Phosphoric tribromide	-458.6										359.8	89.9
Br$_3$P	Phosphorus(III) bromide					-184.5	-175.7	240.2		-139.3	-162.8	348.1	76.0
Br$_3$Pt	Platinum(III) bromide	-120.9											
Br$_3$Re	Rhenium(III) bromide	-167.0											
Br$_3$Ru	Ruthenium(III) bromide	-138.0											
Br$_3$Sb	Antimony(III) bromide	-259.4	-239.3	207.1						-194.6	-223.9	372.9	80.2
Br$_3$Sc	Scandium bromide	-743.1											
Br$_3$Ti	Titanium(III) bromide	-548.5	-523.8	176.6	101.7								
Br$_4$Ge	Germanium(IV) bromide					-347.7	-331.4	280.7		-300.0	-318.0	396.2	101.8
Br$_4$Pa	Protactinium(IV) bromide	-824.0	-787.8	234.0									
Br$_4$Pt	Platinum(IV) bromide	-156.5											
Br$_4$Si	Tetrabromosilane					-457.3	-443.9	277.8		-415.5	-431.8	377.9	97.1
Br$_4$Sn	Tin(IV) bromide	-377.4	-350.2	264.4						-314.6	-331.4	411.9	103.4

Molecular formula	Name	Crystal				Liquid				Gas			
		$\Delta_f H°$ kJ/mol	$\Delta_f G°$ kJ/mol	$S°$ J/mol K	C_p J/mol K	$\Delta_f H°$ kJ/mol	$\Delta_f G°$ kJ/mol	$S°$ J/mol K	C_p J/mol K	$\Delta_f H°$ kJ/mol	$\Delta_f G°$ kJ/mol	$S°$ J/mol K	C_p J/mol K
Br_4Te	Tellurium tetrabromide	-190.4											
Br_4Ti	Titanium(IV) bromide	-616.7	-589.5	243.5	131.5					-549.4	-568.2	398.4	100.8
Br_4V	Vanadium(IV) bromide									-336.8			
Br_4Zr	Zirconium(IV) bromide	-760.7											
Br_5P	Phosphorus(V) bromide	-269.9											
Br_5Ta	Tantalum(V) bromide	-598.3											
Br_6W	Tungsten(VI) bromide	-348.5											
Ca	Calcium	0.0		41.6	25.9					177.8	144.0	154.9	
$CaCl_2$	Calcium chloride	-795.4	-748.8	108.4	72.9								
CaF_2	Calcium fluoride	-1228.0	-1175.6	68.5	67.0								
CaH_2	Calcium hydride	-181.5	-142.5	41.4	41.0								
CaH_2O_2	Calcium hydroxide	-985.2	-897.5	83.4	87.5								
CaI_2	Calcium iodide	-533.5	-528.9	142.0									
CaN_2O_6	Calcium nitrate	-938.2	-742.8	193.2	149.4								
CaO	Calcium oxide	-634.9	-603.3	38.1	42.0								
CaO_4S	Calcium sulfate	-1434.5	-1322.0	106.5	99.7								
CaS	Calcium sulfide	-482.4	-477.4	56.5	47.4								
$Ca_3O_8P_2$	Calcium phosphate	-4120.8	-3884.7	236.0	227.8								
Cd	Cadmium	0.0		51.8		26.0					111.8		167.7
$CdCl_2$	Cadmium chloride	-391.5	-343.9	115.3	74.7								
CdF_2	Cadmium fluoride	-700.4	-647.7	77.4									
CdH_2O_2	Cadmium hydroxide	-560.7	-473.6	96.0									
CdI_2	Cadmium iodide	-203.3	-201.4	161.1	80.0								
CdO	Cadmium oxide	-258.4	-228.7	54.8	43.4								
CdO_4S	Cadmium sulfate	-933.3	-822.7	123.0	99.6								
CdS	Cadmium sulfide	-161.9	-156.5	64.9									
$CdTe$	Cadmium telluride	-92.5	-92.0	100.0									
Ce	Cerium (γ, fcc)	0.0		72.0	26.9					423.0	385.0	191.8	23.1
$CeCl_3$	Cerium(III) chloride	-1060.5	-984.8	151.0	87.4								
CeI_3	Cerium(III) iodide	-669.3											
CeO_2	Cerium(IV) oxide	-1088.7	-1024.6	62.3	61.6								
CeS	Cerium(II) sulfide	-459.4	-451.5	78.2	50.0								
Ce_2O_3	Cerium(III) oxide	-1796.2	-1706.2	150.6	114.6								
Cf	Californium	0.0											
Cl	Chlorine (atomic)									121.3	105.3	165.2	21.8
$ClCs$	Cesium chloride	-443.0	-414.5	101.2	52.5								
$ClCsO_4$	Cesium perchlorate	-443.1	-314.3	175.1	108.3								
$ClCu$	Copper(I) chloride	-137.2	-119.9	86.2	48.5								
ClF	Chlorine fluoride									-50.3	-51.8	217.9	32.1
$ClFO_3$	Perchloryl fluoride									-23.8	48.2	279.0	64.9
ClF_3	Chlorine trifluoride					-189.5				-163.2	-123.0	281.6	63.9
ClF_5S	Sulfur chloride pentafluoride					-1065.7							
$ClGe$	Germanium monochloride									155.2	124.2	247.0	36.9
$ClGeH_3$	Chlorogermane											263.7	54.7
ClH	Hydrogen chloride									-92.3	-95.3	186.9	29.1
$ClHO$	Hypochlorous acid									-78.7	-66.1	236.7	37.2
$ClHO_4$	Perchloric acid					-40.6							
ClH_3Si	Chlorosilane											250.7	51.0
ClH_4N	Ammonium chloride	-314.4	-202.9	94.6	84.1								
ClH_4NO_4	Ammonium perchlorate	-295.3	-88.8	186.2									
ClH_4P	Phosphonium chloride	-145.2											
ClI	Iodine chloride					-23.9	-13.6	135.1		17.8	-5.5	247.6	35.6
$ClIn$	Indium(I) chloride	-186.2								-75.0			
ClK	Potassium chloride	-436.5	-408.5	82.6	51.3					-214.6	-233.3	239.1	36.5
$ClKO_3$	Potassium chlorate	-397.7	-296.3	143.1	100.3								
$ClKO_4$	Potassium perchlorate	-432.8	-303.1	151.0	112.4								
$ClLi$	Lithium chloride	-408.6	-384.4	59.3	48.0								
$ClLiO_4$	Lithium perchlorate	-381.0											
$ClNO$	Nitrosyl chloride									51.7	66.1	261.7	44.7
$ClNO_2$	Nitryl chloride									12.6	54.4	272.2	53.2
$ClNa$	Sodium chloride	-411.2	-384.1	72.1	50.5								
$ClNaO_2$	Sodium chlorite	-307.0											
$ClNaO_3$	Sodium chlorate	-365.8	-262.3	123.4									
$ClNaO_4$	Sodium perchlorate	-383.3	-254.9	142.3									
ClO	Chlorine oxide									101.8	98.1	226.6	31.5
$ClOV$	Vanadyl chloride	-607.0	-556.0	75.0									

Molecular formula	Name	Crystal				Liquid				Gas			
		$\Delta_f H°$ kJ/mol	$\Delta_f G°$ kJ/mol	$S°$ J/mol K	C_p J/mol K	$\Delta_f H°$ kJ/mol	$\Delta_f G°$ kJ/mol	$S°$ J/mol K	C_p J/mol K	$\Delta_f H°$ kJ/mol	$\Delta_f G°$ kJ/mol	$S°$ J/mol K	C_p J/mol K
ClO_2	Chlorine dioxide									102.5	120.5	256.8	42.0
ClO_2	Chlorine superoxide (ClOO)									89.1	105.0	263.7	46.0
ClO_4Rb	Rubidium perchlorate	-437.2	-306.9	161.1									
$ClRb$	Rubidium chloride	-435.4	-407.8	95.9	52.4								
$ClSi$	Chlorosilylidyne									189.9			36.9
$ClTl$	Thallium(I) chloride	-204.1	-184.9	111.3	50.9					-67.8			
Cl_2	Chlorine									0.0		223.1	33.9
Cl_2Co	Cobalt(II) chloride	-312.5	-269.8	109.2	78.5								
Cl_2Cr	Chromium(II) chloride	-395.4	-356.0	115.3	71.2								
Cl_2CrO_2	Chromyl chloride					-579.5	-510.8	221.8		-538.1	-501.6	329.8	84.5
Cl_2Cu	Copper(II) chloride	-220.1	-175.7	108.1	71.9								
Cl_2Fe	Iron(II) chloride	-341.8	-302.3	118.0	76.7								
Cl_2H_2Si	Dichlorosilane											285.7	60.5
Cl_2Hg	Mercury(II) chloride	-224.3	-178.6	146.0									
Cl_2Hg_2	Mercury(I) chloride	-265.4	-210.7	191.6									
Cl_2Mg	Magnesium chloride	-641.3	-591.8	89.6	71.4								
Cl_2Mn	Manganese(II) chloride	-481.3	-440.5	118.2	72.9								
Cl_2Ni	Nickel(II) chloride	-305.3	-259.0	97.7	71.7								
Cl_2O	Chlorine monoxide									80.3	97.9	266.2	45.4
Cl_2OS	Thionyl chloride					-245.6			121.0	-212.5	-198.3	309.8	66.5
Cl_2O_2S	Sulfuryl chloride					-394.1			134.0	-364.0	-320.0	311.9	77.0
Cl_2O_2U	Uranyl chloride	-1243.9	-1146.4	150.5	107.9								
Cl_2Pb	Lead(II) chloride	-359.4	-314.1	136.0									
Cl_2Pt	Platinum(II) chloride	-123.4											
Cl_2S	Sulfur dichloride					-50.0							
Cl_2S_2	Sulfur chloride					-59.4							
Cl_2Sn	Tin(II) chloride	-325.1											
Cl_2Sr	Strontium chloride	-828.9	-781.1	114.9	75.6								
Cl_2Ti	Titanium(II) chloride	-513.8	-464.4	87.4	69.8								
Cl_2Zn	Zinc chloride	-415.1	-369.4	111.5	71.3					-266.1			
Cl_2Zr	Zirconium(II) chloride	-502.0											
Cl_3Cr	Chromium(III) chloride	-556.5	-486.1	123.0	91.8								
Cl_3Dy	Dysprosium(III) chloride	-1000.0											
Cl_3Er	Erbium chloride	-998.7			100.0								
Cl_3Eu	Europium(III) chloride	-936.0											
Cl_3Fe	Iron(III) chloride	-399.5	-334.0	142.3	96.7								
Cl_3Ga	Gallium(III) chloride	-524.7	-454.8	142.0									
Cl_3Gd	Gadolinium(III) chloride	-1008.0			88.0								
Cl_3HSi	Trichlorosilane					-539.3	-482.5	227.6		-513.0	-482.0	313.9	75.8
Cl_3Ho	Holmium chloride	-1005.4			88.0								
Cl_3In	Indium(III) chloride	-537.2								-374.0			
Cl_3Ir	Iridium(III) chloride	-245.6											
Cl_3La	Lanthanum chloride	-1072.2			108.8								
Cl_3Lu	Lutetium chloride	-945.6								-649.0			
Cl_3N	Nitrogen trichloride						230.0						
Cl_3Nd	Neodymium chloride	-1041.0			113.0								
Cl_3OP	Phosphoric trichloride					-597.1	-520.8	222.5	138.8	-558.5	-512.9	325.5	84.9
Cl_3OV	Vanadyl trichloride					-734.7	-668.5	244.3		-695.6	-659.3	344.3	89.9
Cl_3Os	Osmium(III) chloride	-190.4											
Cl_3P	Phosphorus(III) chloride					-319.7	-272.3	217.1		-287.0	-267.8	311.8	71.8
Cl_3Pr	Praseodymium chloride	-1056.9			100.0								
Cl_3Pt	Platinum(III) chloride	-182.0											
Cl_3Re	Rhenium(III) chloride	-264.0	-188.0	123.8	92.4								
Cl_3Rh	Rhodium(III) chloride	-299.2											
Cl_3Ru	Ruthenium(III) chloride	-205.0											
Cl_3Sb	Antimony(III) chloride	-382.2	-323.7	184.1	107.9								
Cl_3Sc	Scandium chloride	-925.1											
Cl_3Sm	Samarium(III) chloride	-1025.9											
Cl_3Tb	Terbium chloride	-997.0											
Cl_3Ti	Titanium(III) chloride	-720.9	-653.5	139.7	97.2								
Cl_3Tl	Thallium(III) chloride	-315.1											
Cl_3Tm	Thulium chloride	-986.6											
Cl_3U	Uranium(III) chloride	-866.5	-799.1	159.0	102.5								
Cl_3V	Vanadium(III) chloride	-580.7	-511.2	131.0	93.2								
Cl_3Y	Yttrium chloride	-1000.0								-750.2			75.0
Cl_3Yb	Ytterbium(III) chloride	-959.8											

Molecular formula	Name	Crystal				Liquid				Gas			
		$\Delta_f H°$ kJ/mol	$\Delta_f G°$ kJ/mol	$S°$ J/mol K	C_p J/mol K	$\Delta_f H°$ kJ/mol	$\Delta_f G°$ kJ/mol	$S°$ J/mol K	C_p J/mol K	$\Delta_f H°$ kJ/mol	$\Delta_f G°$ kJ/mol	$S°$ J/mol K	C_p J/mol K
Cl_4Ge	Germanium(IV) chloride					-531.8	-462.7	245.6		-495.8	-457.3	347.7	96.1
Cl_4Hf	Hafnium(IV) chloride	-990.4	-901.3	190.8	120.5					-884.5			
Cl_4Pa	Protactinium(IV) chloride	-1043.0	-953.0	192.0									
Cl_4Pb	Lead(IV) chloride					-329.3							
Cl_4Pt	Platinum(IV) chloride	-231.8											
Cl_4Si	Tetrachlorosilane					-687.0	-619.8	239.7	145.3	-657.0	-617.0	330.7	90.3
Cl_4Sn	Tin(IV) chloride					-511.3	-440.1	258.6	165.3	-471.5	-432.2	365.8	98.3
Cl_4Te	Tellurium tetrachloride	-326.4			138.5								
Cl_4Th	Thorium(IV) chloride	-1186.2	-1094.1	190.4	120.3					-964.4	-932.0	390.7	107.5
Cl_4Ti	Titanium(IV) chloride					-804.2	-737.2	252.3	145.2	-763.2	-726.3	353.2	95.4
Cl_4U	Uranium(IV) chloride	-1019.2	-930.0	197.1	122.0					-809.6	-786.6	419.0	
Cl_4V	Vanadium(IV) chloride					-569.4	-503.7	255.0		-525.5	-492.0	362.4	96.2
Cl_4Zr	Zirconium(IV) chloride	-980.5	-889.9	181.6	119.8								
Cl_5Nb	Niobium(V) chloride	-797.5	-683.2	210.5	148.1					-703.7	-646.0	400.6	120.8
Cl_5P	Phosphorus(V) chloride	-443.5								-374.9	-305.0	364.6	112.8
Cl_5Pa	Protactinium(V) chloride	-1145.0	-1034.0	238.0									
Cl_5Ta	Tantalum(V) chloride	-859.0											
Cl_6U	Uranium(VI) chloride	-1092.0	-962.0	285.8	175.7					-1013.0	-928.0	431.0	
Cl_6W	Tungsten(VI) chloride	-602.5								-513.8			
Cm	Curium	0.0											
Co	Cobalt	0.0		30.0	24.8					424.7	380.3	179.5	23.0
CoF_2	Cobalt(II) fluoride	-692.0	-647.2	82.0	68.8								
CoH_2O_2	Cobalt(II) hydroxide	-539.7	-454.3	79.0									
CoI_2	Cobalt(II) iodide	-88.7											
CoN_2O_6	Cobalt(II) nitrate	-420.5											
CoO	Cobalt(II) oxide	-237.9	-214.2	53.0	55.2								
CoO_4S	Cobalt(II) sulfate	-888.3	-782.3	118.0									
CoS	Cobalt(II) sulfide	-82.8											
Co_2S_3	Cobalt(III) sulfide	-147.3											
Co_3O_4	Cobalt(II,III) oxide	-891.0	-774.0	102.5	123.4								
Cr	Chromium	0.0		23.8	23.4					396.6	351.8	174.5	20.8
CrF_2	Chromium(II) fluoride	778.0											
CrF_3	Chromium(III) fluoride	-1159.0	-1088.0	93.9	78.7								
CrI_2	Chromium(II) iodide	-156.9											
CrI_3	Chromium(III) iodide	-205.0											
CrO_2	Chromium(IV) oxide	-598.0											
CrO_3	Chromium(VI) oxide									-292.9		266.2	56.0
CrO_4Pb	Lead(II) chromate	-930.9											
Cr_2FeO_4	Chromium iron oxide	-1444.7	-1343.8	146.0	133.6								
Cr_2O_3	Chromium(III) oxide	-1139.7	-1058.1	81.2	118.7								
Cr_3O_4	Chromium(II,III) oxide	-1531.0											
Cs	Cesium	0.0		85.2	32.2					76.5	49.6	175.6	20.8
CsF	Cesium fluoride	-553.5	-525.5	92.8	51.1								
CsF_2H	Cesium hydrogen fluoride	-923.8	-858.9	135.2	87.3								
CsH	Cesium hydride	-54.2											
$CsHO$	Cesium hydroxide	-416.2	-371.8	104.2	69.9					-256.0	-256.5	254.8	49.7
$CsHO_4S$	Cesium hydrogen sulfate	-1158.1											
CsH_2N	Cesium amide	-118.4											
CsI	Cesium iodide	-346.6	-340.6	123.1	52.8								
$CsNO_3$	Cesium nitrate	-506.0	-406.5	155.2									
CsO_2	Cesium superoxide	-286.2											
Cs_2O	Cesium oxide	-345.8	-308.1	146.9	76.0								
Cs_2O_3S	Cesium sulfite	-1134.7											
Cs_2O_4S	Cesium sulfate	-1443.0	-1323.6	211.9	134.9								
Cs_2S	Cesium sulfide	-359.8											
Cu	Copper	0.0		33.2	24.4					337.4	297.7	166.4	20.8
CuF_2	Copper(II) fluoride	-542.7											
CuH_2O_2	Copper(II) hydroxide	-449.8											
CuI	Copper(I) iodide	-67.8	-69.5	96.7	54.1								
CuN_2O_6	Copper(II) nitrate	-302.9											
CuO	Copper(II) oxide	-157.3	-129.7	42.6	42.3								
CuO_4S	Copper(II) sulfate	-771.4	-662.2	109.2									
CuO_4W	Copper(II) tungstate	-1105.0											
CuS	Copper(II) sulfide	-53.1	-53.6	66.5	47.8								
$CuSe$	Copper(II) selenide	-39.5											
Cu_2	Dicopper									484.2	431.9	241.6	36.6

Molecular formula	Name	Crystal				Liquid				Gas			
		$\Delta_f H°$ kJ/mol	$\Delta_f G°$ kJ/mol	$S°$ J/mol K	C_p J/mol K	$\Delta_f H°$ kJ/mol	$\Delta_f G°$ kJ/mol	$S°$ J/mol K	C_p J/mol K	$\Delta_f H°$ kJ/mol	$\Delta_f G°$ kJ/mol	$S°$ J/mol K	C_p J/mol K
Cu_2O	Copper(I) oxide	-168.6	-146.0	93.1	63.6								
Cu_2S	Copper(I) sulfide	-79.5	-86.2	120.9	76.3								
Dy	Dysprosium	0.0		75.6	27.7					290.4	254.4	196.6	20.8
DyI_3	Dysprosium(III) iodide	-620.5											
Dy_2O_3	Dysprosium(III) oxide	-1863.1	-1771.5	149.8	116.3								
Er	Erbium	0.0		73.2	28.1					317.1	280.7	195.6	20.8
ErF_3	Erbium fluoride	-1711.0											
Er_2O_3	Erbium oxide	-1897.9	-1808.7	155.6	108.5								
Es	Einsteinium	0.0											
Eu	Europium	0.0		77.8	27.7					175.3	142.2	188.8	20.8
Eu_2O_3	Europium(III) oxide	-1651.4	-1556.8	146.0	122.2								
Eu_3O_4	Europium(II,III) oxide	-2272.0	-2142.0	205.0									
F	Fluorine (atomic)									79.4	62.3	158.8	22.7
FGa	Gallium monofluoride									-251.9			33.3
FGe	Germanium monofluoride									-33.4			34.7
$FGeH_3$	Fluorogermane											252.8	51.6
FH	Hydrogen fluoride					-299.8				-273.3	-275.4	173.8	
FH_3Si	Fluorosilane											238.4	47.4
FH_4N	Ammonium fluoride	-464.0	-348.7	72.0	65.3								
FI	Iodine fluoride									-95.7	-118.5	236.2	33.4
FIn	Indium(I) fluoride									-203.4			
FK	Potassium fluoride	-567.3	-537.8	66.6	49.0								
FLi	Lithium fluoride	-616.0	-587.7	35.7	41.6								
FNO	Nitrosyl fluoride									-66.5	-51.0	248.1	41.3
FNO_2	Nitryl fluoride											260.4	49.8
FNS	Thionitrosyl fluoride (NSF)											259.8	44.1
FNa	Sodium fluoride	-576.6	-546.3	51.1	46.9								
FO	Fluorine oxide									109.0	105.3	216.4	32.0
FO_2	Fluorine superoxide (FOO)									25.4	39.4	259.5	44.5
FRb	Rubidium fluoride	-557.7											
FSi	Fluorosilylidyne									7.1	-24.3	225.8	32.6
FTl	Thallium(I) fluoride	-324.7								-182.4			
F_2	Fluorine									0.0		202.8	31.3
F_2Fe	Iron(II) fluoride	-711.3	-668.6	87.0	68.1								
F_2HK	Potassium hydrogen fluoride	-927.7	-859.7	104.3	76.9								
F_2HN	Difluoramine											252.8	43.4
F_2HNa	Sodium hydrogen fluoride	-920.3	-852.2	90.9	75.0								
F_2HRb	Rubidium hydrogen fluoride	-922.6	-855.6	120.1	79.4								
F_2Mg	Magnesium fluoride	-1124.2	-1071.1	57.2	61.6								
F_2N	Difluoroamidogen									43.1	57.8	249.9	41.0
F_2N_2	cis-Difluorodiazine									69.5			
F_2N_2	trans-Difluorodiazine									82.0			
F_2Ni	Nickel(II) fluoride	-651.4	-604.1	73.6	64.1								
F_2O	Fluorine monoxide									24.5	41.8	247.5	43.3
F_2OS	Thionyl fluoride											278.7	56.8
F_2O_2	Fluorine dioxide									19.2	58.2	277.2	62.1
F_2O_2S	Sulfuryl fluoride											284.0	66.0
F_2O_2U	Uranyl fluoride	-1653.5	-1557.4	135.6	103.2								
F_2Pb	Lead(II) fluoride	-664.0	-617.1	110.5									
F_2Si	Difluorosilylene									-619.0	-628.0	252.7	43.9
F_2Sr	Strontium fluoride	-1216.3	-1164.8	82.1	70.0								
F_2Zn	Zinc fluoride	-764.4	-713.3	73.7	65.7								
F_3Ga	Gallium(III) fluoride	-1163.0	-1085.3	84.0									
F_3Gd	Gadolinium(III) fluoride									-1297.0			
F_3HSi	Trifluorosilane											271.9	60.5
F_3Ho	Holmium fluoride	-1707.0											
F_3N	Nitrogen trifluoride									-132.1	-90.6	260.8	53.4
F_3Nd	Neodymium fluoride	-1657.0											
F_3OP	Phosphoric trifluoride									-1254.3	-1205.8	285.4	68.8
F_3P	Phosphorus(III) fluoride									-958.4	-936.9	273.1	58.7
F_3Sb	Antimony(III) fluoride	-915.5											
F_3Sc	Scandium fluoride	-1629.2	-1555.6	92.0						-1247.0	-1234.0	300.5	67.8
F_3Sm	Samarium(III) fluoride	-1778.0											
F_3Th	Thorium(III) fluoride									-1166.1	-1160.6	339.2	73.3
F_3U	Uranium(III) fluoride	-1502.1	-1433.4	123.4	95.1					-1058.5	-1051.9	331.9	74.3
F_3Y	Yttrium fluoride	-1718.8	-1644.7	100.0						-1288.7	-1277.8	311.8	70.3

Standard Thermodynamic Properties of Chemical Substances

Molecular formula	Name	Crystal				Liquid				Gas			
		$\Delta_f H°$ kJ/mol	$\Delta_f G°$ kJ/mol	$S°$ J/mol K	C_p J/mol K	$\Delta_f H°$ kJ/mol	$\Delta_f G°$ kJ/mol	$S°$ J/mol K	C_p J/mol K	$\Delta_f H°$ kJ/mol	$\Delta_f G°$ kJ/mol	$S°$ J/mol K	C_p J/mol K
F_4Ge	Germanium(IV) fluoride									-1190.2	-1150.0	301.9	
F_4Hf	Hafnium fluoride	-1930.5	-1830.4	113.0						-1669.8			
F_4N_2	Tetrafluorohydrazine									-8.4	79.9	301.2	79.2
F_4Pb	Lead(IV) fluoride	-941.8											
F_4S	Sulfur tetrafluoride									-763.2	-722.0	299.6	77.6
F_4Si	Tetrafluorosilane									-1615.0	-1572.8	282.8	73.6
F_4Th	Thorium(IV) fluoride	-2097.8	-2003.4	142.0	110.7					-1759.0	-1724.0	341.7	93.0
F_4U	Uranium(IV) fluoride	-1914.2	-1823.3	151.7	116.0					-1598.7	-1572.7	368.0	91.2
F_4V	Vanadium(IV) fluoride	-1403.3											
F_4Xe	Xenon tetrafluoride	-261.5											
F_4Zr	Zirconium(IV) fluoride	-1911.3	-1809.9	104.6	103.7								
F_5I	Iodine pentafluoride					-864.8				-822.5	-751.7	327.7	99.2
F_5Nb	Niobium(V) fluoride	-1813.8	-1699.0	160.2	134.7					-1739.7	-1673.6	321.9	97.1
F_5P	Phosphorus(V) fluoride									-1594.4	-1520.7	300.8	84.8
F_5Ta	Tantalum(V) fluoride	-1903.6											
F_5V	Vanadium(V) fluoride					-1480.3	-1373.1	175.7		-1433.9	-1369.8	320.9	98.6
$F_6H_8N_2Si$	Ammonium hexafluorosilicate	-2681.7	-2365.3	280.2	228.1								
F_6Ir	Iridium(VI) fluoride	-579.7	-461.6	247.7						-544.0	-460.0	357.8	121.1
F_6K_2Si	Potassium hexafluorosilicate	-2956.0	-2798.6	226.0									
F_6Mo	Molybdenum(VI) fluoride					-1585.5	-1473.0	259.7	169.8	-1557.7	-1472.2	350.5	
F_6Na_2Si	Sodium hexafluorosilicate	-2909.6	-2754.2	207.1	187.1								
F_6Os	Osmium(VI) fluoride			246.0								358.1	120.8
F_6Pt	Platinum(VI) fluoride			235.6								348.3	122.8
F_6S	Sulfur hexafluoride									-1220.5	-1116.5	291.5	97.0
F_6Se	Selenium hexafluoride									-1117.0	-1017.0	313.9	110.5
F_6Si_2	Hexafluorodisilane	-2427.0	-2299.7	219.1	129.5					-2383.3	-2307.3	391.0	129.9
F_6Te	Tellurium hexafluoride									-1318.0			
F_6U	Uranium(VI) fluoride	-2197.0	-2068.5	227.6	166.8					-2147.4	-2063.7	377.9	129.6
F_6W	Tungsten(VI) fluoride					-1747.7	-1631.4	251.5		-1721.7	-1632.1	341.1	119.0
Fe	Iron	0.0		27.3	25.1					416.3	370.7	180.5	25.7
FeI_2	Iron(II) iodide	-113.0											
FeI_3	Iron(III) iodide									71.0			
$FeMoO_4$	Iron(II) molybdate	-1075.0	-975.0	129.3	118.5								
FeO	Iron(II) oxide	-272.0											
FeO_4S	Iron(II) sulfate	-928.4	-820.8	107.5	100.6								
FeO_4W	Iron(II) tungstate	-1155.0	-1054.0	131.8	114.6								
FeS	Iron(II) sulfide	-100.0	-100.4	60.3	50.5								
FeS_2	Iron disulfide	-178.2	-166.9	52.9	62.2								
Fe_2O_3	Iron(III) oxide	-824.2	-742.2	87.4	103.9								
Fe_2O_4Si	Iron(II) orthosilicate	-1479.9	-1379.0	145.2	132.9								
Fe_3O_4	Iron(II,III) oxide	-1118.4	-1015.4	146.4	143.4								
Fm	Fermium	0.0											
Fr	Francium	0.0		95.4									
Ga	Gallium	0.0	0.0	40.8	26.1	5.6				272.0	233.7	169.0	25.3
GaH_3O_3	Gallium(III) hydroxide	-964.4	-831.3	100.0									
GaI_3	Gallium(III) iodide	-238.9		205.0	100.0								
GaN	Gallium nitride	-110.5											
GaO	Gallium monoxide									279.5	253.5	231.1	32.1
GaP	Gallium phosphide	-88.0											
$GaSb$	Gallium antimonide	-41.8	-38.9	76.1	48.5								
Ga_2	Digallium									438.5			
Ga_2O	Gallium suboxide	-356.0											
Ga_2O_3	Gallium(III) oxide	-1089.1	-998.3	85.0	92.1								
Gd	Gadolinium	0.0		68.1	37.0					397.5	359.8	194.3	27.5
Gd_2O_3	Gadolinium(III) oxide	-1819.6			106.7								
Ge	Germanium	0.0		31.1	23.3					372.0	331.2	167.9	30.7
GeH_3I	Iodogermane											283.2	57.5
GeH_4	Germane									90.8	113.4	217.1	45.0
GeI_4	Germanium(IV) iodide	-141.8	-144.3	271.1						-56.9	-106.3	428.9	104.1
GeO	Germanium(II) oxide	-261.9	-237.2	50.0						-46.2	-73.2	224.3	30.9
GeO_2	Germanium(IV) oxide	-580.0	-521.4	39.7	52.1								
GeP	Germanium phosphide	-21.0	-17.0	63.0									
GeS	Germanium(II) sulfide	-69.0	-71.5	71.0						92.0	42.0	234.0	33.7
$GeTe$	Germanium(II) telluride	20.0											
Ge_2	Digermanium									473.1	416.3	252.8	35.6
Ge_2H_6	Digermane					137.3				162.3			

Molecular formula	Name	Crystal $\Delta_f H°$ kJ/mol	$\Delta_f G°$ kJ/mol	$S°$ J/mol K	C_p J/mol K	Liquid $\Delta_f H°$ kJ/mol	$\Delta_f G°$ kJ/mol	$S°$ J/mol K	C_p J/mol K	Gas $\Delta_f H°$ kJ/mol	$\Delta_f G°$ kJ/mol	$S°$ J/mol K	C_p J/mol K
Ge_3H_8	Trigermane					193.7				226.8			
H	Hydrogen (atomic)									218.0	203.3	114.7	20.8
HI	Hydrogen iodide									26.5	1.7	206.6	29.2
HIO_3	Iodic acid	-230.1											
HK	Potassium hydride	-57.7											
HKO	Potassium hydroxide	-424.6	-379.4	81.2	68.9					-232.0	-229.7	238.3	49.2
HKO_4S	Potassium hydrogen sulfate	-1160.6	-1031.3	138.1									
HLi	Lithium hydride	-90.5	-68.3	20.0	27.9								
HLiO	Lithium hydroxide	-487.5	-441.5	42.8	49.6					-229.0	-234.2	214.4	46.0
HN	Imidogen									351.5	345.6	181.2	29.2
HNO_2	Nitrous acid									-79.5	-46.0	254.1	45.6
HNO_3	Nitric acid					-174.1	-80.7	155.6	109.9	-133.9	-73.5	266.9	54.1
HN_3	Hydrazoic acid					264.0	327.3	140.6		294.1	328.1	239.0	43.7
HNa	Sodium hydride	-56.3	-33.5	40.0	36.4								
HNaO	Sodium hydroxide	-425.8	-379.7	64.4	59.5					-191.0	-193.9	229.0	48.0
$HNaO_4S$	Sodium hydrogen sulfate	-1125.5	-992.8	113.0									
HNa_2O_4P	Sodium hydrogen phosphate	-1748.1	-1608.2	150.5	135.3								
HO	Hydroxyl									39.0	34.2	183.7	29.9
HORb	Rubidium hydroxide	-418.8	-373.9	94.0	69.0					-238.0	-239.1	248.5	49.5
HOTl	Thallium(I) hydroxide	-238.9	-195.8	88.0									
HO_2	Hydroperoxy									10.5	22.6	229.0	34.9
HO_3P	Metaphosphoric acid	-948.5											
HO_4RbS	Rubidium hydrogen sulfate	-1159.0											
HO_4Re	Perrhenic acid	-762.3	-656.4	158.2									
HRb	Rubidium hydride	-52.3											
HS	Mercapto									142.7	113.3	195.7	32.3
HSi	Silylidyne									361.0			
HTa_2	Tantalum hydride	-32.6	-69.0	79.1	90.8								
H_2	Hydrogen									0.0		130.7	28.8
H_2KN	Potassium amide	-128.9											
H_2KO_4P	Potassium dihydrogen phosphate	-1568.3	-1415.9	134.9	116.6								
H_2LiN	Lithium amide	-179.5											
H_2Mg	Magnesium hydride	-75.3	-35.9	31.1	35.4								
H_2MgO_2	Magnesium hydroxide	-924.5	-833.5	63.2	77.0								
H_2N	Amidogen									184.9	194.6	195.0	33.9
H_2NNa	Sodium amide	-123.8	-64.0	76.9	66.2								
H_2NRb	Rubidium amide	-113.0											
$H_2N_2O_2$	Nitramide	-89.5											
H_2NiO_2	Nickel(II) hydroxide	-529.7	-447.2	88.0									
H_2O	Water					-285.8	-237.1	70.0	75.3	-241.8	-228.6	188.8	33.6
H_2O_2	Hydrogen peroxide					-187.8	-120.4	109.6	89.1	-136.3	-105.6	232.7	43.1
H_2O_2Sn	Tin(II) hydroxide	-561.1	-491.6	155.0									
H_2O_2Sr	Strontium hydroxide	-959.0											
H_2O_2Zn	Zinc hydroxide	-641.9	-553.5	81.2									
H_2O_3Si	Metasilicic acid	-1188.7	-1092.4	134.0									
H_2O_4S	Sulfuric acid					-814.0	-690.0	156.9	138.9				
H_2O_4Se	Selenic acid	-530.1											
H_2S	Hydrogen sulfide									-20.6	-33.4	205.8	34.2
H_2S_2	Hydrogen disulfide					-18.1			84.1	15.5			51.5
H_2Se	Hydrogen selenide									29.7	15.9	219.0	34.7
H_2Sr	Strontium hydride	-180.3											
H_2Te	Hydrogen telluride									99.6			
H_2Th	Thorium hydride	-139.7	-100.0	50.7	36.7								
H_2Zr	Zirconium(II) hydride	-169.0	-128.8	35.0	31.0								
H_3ISi	Iodosilane											270.9	54.4
H_3N	Ammonia									-45.9	-16.4	192.8	35.1
H_3NO	Hydroxylamine	-114.2											
H_3O_2P	Phosphinic acid	-604.6				-595.4							
H_3O_3P	Phosphonic acid	-964.4											
H_3O_4P	Phosphoric acid	-1284.4	-1124.3	110.5	106.1	-1271.7	-1123.6	150.8	145.0				
H_3P	Phosphine									5.4	13.5	210.2	37.1
H_3Sb	Stibine									145.1	147.8	232.8	41.1
H_3U	Uranium(III) hydride	-127.2	-72.8	63.7	49.3								
H_4IN	Ammonium iodide	-201.4	-112.5	117.0									
H_4N_2	Hydrazine					50.6	149.3	121.2	98.9	95.4	159.4	238.5	48.4
$H_4N_2O_2$	Ammonium nitrite	-256.5											

Molecular formula	Name	Crystal $\Delta_f H°$ kJ/mol	$\Delta_f G°$ kJ/mol	$S°$ J/mol K	C_p J/mol K	Liquid $\Delta_f H°$ kJ/mol	$\Delta_f G°$ kJ/mol	$S°$ J/mol K	C_p J/mol K	Gas $\Delta_f H°$ kJ/mol	$\Delta_f G°$ kJ/mol	$S°$ J/mol K	C_p J/mol K
$H_4N_2O_3$	Ammonium nitrate	-365.6	-183.9	151.1	139.3								
H_4N_4	Ammonium azide	115.5	274.2	112.5									
H_4O_4Si	Orthosilicic acid	-1481.1	-1332.9	192.0									
$H_4O_7P_2$	Diphosphoric acid	-2241.0				-2231.7							
H_4P_2	Diphosphine					-5.0				20.9			
H_4Si	Silane									34.3	56.9	204.6	42.8
H_4Sn	Stannane									162.8	188.3	227.7	49.0
H_5NO	Ammonium hydroxide					-361.2	-254.0	165.6	154.9				
H_5NO_3S	Ammonium hydrogen sulfite	-768.6											
H_5NO_4S	Ammonium hydrogen sulfate	-1027.0											
H_6Si_2	Disilane									80.3	127.3	272.7	80.8
$H_8N_2O_4S$	Ammonium sulfate	-1180.9	-901.7	220.1	187.5								
H_8Si_3	Trisilane					92.5				120.9			
$H_9N_2O_4P$	Ammonium hydrogen phosphate	-1566.9			188.0								
$H_{12}N_3O_4P$	Ammonium phosphate	-1671.9											
He	Helium									0.0		126.2	20.8
Hf	Hafnium	0.0		43.6	25.7					619.2	576.5	186.9	20.8
HfO_2	Hafnium oxide	-1144.7	-1088.2	59.3	60.3								
Hg	Mercury					0.0		75.9	28.0	61.4	31.8	175.0	20.8
HgI_2	Mercury(II) iodide	-105.4	-101.7	180.0									
HgO	Mercury(II) oxide	-90.8	-58.5	70.3	44.1								
HgO_4S	Mercury(II) sulfate	-707.5											
HgS	Mercury(II) sulfide (red)	-58.2	-50.6	82.4	48.4								
HgTe	Mercury(II) telluride	-42.0											
Hg_2	Dimercury									108.8	68.2	288.1	37.4
Hg_2I_2	Mercury(I) iodide	-121.3	-111.0	233.5									
Hg_2O_4S	Mercury(I) sulfate	-743.1	-625.8	200.7	132.0								
Ho	Holmium	0.0		75.3	27.2					300.8	264.8	195.6	20.8
Ho_2O_3	Holmium oxide	-1880.7	-1791.1	158.2	115.0								
I	Iodine (atomic)									106.8	70.2	180.8	20.8
IIn	Indium(I) iodide	-116.3	-120.5	130.0						7.5	-37.7	267.3	36.8
IK	Potassium iodide	-327.9	-324.9	106.3	52.9								
IKO_3	Potassium iodate	-501.4	-418.4	151.5	106.5								
IKO_4	Potassium periodate	-467.2	-361.4	175.7									
ILi	Lithium iodide	-270.4	-270.3	86.8	51.0								
INa	Sodium iodide	-287.8	-286.1	98.5	52.1								
$INaO_3$	Sodium iodate	-481.8			92.0								
$INaO_4$	Sodium periodate	-429.3	-323.0	163.0									
IO	Iodine monoxide									126.0	102.5	239.6	32.9
IRb	Rubidium iodide	-333.8	-328.9	118.4	53.2								
ITl	Thallium(I) iodide	-123.8	-125.4	127.6						7.1			
I_2	Iodine (rhombic)	0.0		116.1	54.4					62.4	19.3	260.7	36.9
I_2Mg	Magnesium iodide	-364.0	-358.2	129.7									
I_2Ni	Nickel(II) iodide	-78.2											
I_2Pb	Lead(II) iodide	-175.5	-173.6	174.9	77.4								
I_2Sn	Tin(II) iodide	-143.5											
I_2Sr	Strontium iodide	-558.1			81.6								
I_2Zn	Zinc iodide	-208.0	-209.0	161.1									
I_3In	Indium(III) iodide	-238.0								-120.5			
I_3La	Lanthanum iodide	-668.9											
I_3Lu	Lutetium iodide	-548.0											
I_3P	Phosphorus(III) iodide	-45.6										374.4	78.4
I_3Ru	Ruthenium(III) iodide	-65.7											
I_3Sb	Antimony(III) iodide	-100.4											
I_4Pt	Platinum(IV) iodide	-72.8											
I_4Si	Tetraiodosilane	-189.5											
I_4Sn	Tin(IV) iodide				84.9							446.1	105.4
I_4Ti	Titanium(IV) iodide	-375.7	-371.5	249.4	125.7					-277.8			
I_4V	Vanadium(IV) iodide									-122.6			
I_4Zr	Zirconium(IV) iodide	-481.6											
In	Indium	0.0		57.8	26.7					243.3	208.7	173.8	20.8
InO	Indium monoxide									387.0	364.4	236.5	32.6
InP	Indium phosphide	-88.7	-77.0	59.8	45.4								
InS	Indium(II) sulfide	-138.1	-131.8	67.0						238.0			
InSb	Indium antimonide	-30.5	-25.5	86.2	49.5					344.3			
In_2	Diindium									380.9			

Molecular formula	Name	Crystal				Liquid				Gas			
		$\Delta_f H°$ kJ/mol	$\Delta_f G°$ kJ/mol	$S°$ J/mol K	C_p J/mol K	$\Delta_f H°$ kJ/mol	$\Delta_f G°$ kJ/mol	$S°$ J/mol K	C_p J/mol K	$\Delta_f H°$ kJ/mol	$\Delta_f G°$ kJ/mol	$S°$ J/mol K	C_p J/mol K
In_2O_3	Indium(III) oxide	-925.8	-830.7	104.2	92.0								
In_2S_3	Indium(III) sulfide	-427.0	-412.5	163.6	118.0								
In_2Te_5	Indium(IV) telluride	-175.3											
Ir	Iridium	0.0		35.5	25.1					665.3	617.9	193.6	20.8
IrO_2	Iridium(IV) oxide	-274.1			57.3								
IrS_2	Iridium(IV) sulfide	-138.0											
Ir_2S_3	Iridium(III) sulfide	-234.0											
K	Potassium	0.0		64.7	29.6					89.0	60.5	160.3	20.8
$KMnO_4$	Potassium permanganate	-837.2	-737.6	171.7	117.6								
KNO_2	Potassium nitrite	-369.8	-306.6	152.1	107.4								
KNO_3	Potassium nitrate	-494.6	-394.9	133.1	96.4								
KNa	Potassium sodium					6.3							
KO_2	Potassium superoxide	-284.9	-239.4	116.7	77.5								
K_2	Dipotassium									123.7	87.5	249.7	37.9
K_2O	Potassium oxide	-361.5											
K_2O_2	Potassium peroxide	-494.1	-425.1	102.1									
K_2O_4S	Potassium sulfate	-1437.8	-1321.4	175.6	131.5								
K_2S	Potassium sulfide	-380.7	-364.0	105.0									
K_3O_4P	Potassium phosphate	-1950.2											
Kr	Krypton									0.0		164.1	20.8
La	Lanthanum	0.0		56.9	27.1					431.0	393.6	182.4	22.8
LaS	Lanthanum monosulfide	-456.0	-451.5	73.2	59.0								
La_2O_3	Lanthanum oxide	-1793.7	-1705.8	127.3	108.8								
Li	Lithium	0.0		29.1	24.8					159.3	126.6	138.8	20.8
$LiNO_2$	Lithium nitrite	-372.4	-302.0	96.0									
$LiNO_3$	Lithium nitrate	-483.1	-381.1	90.0									
Li_2	Dilithium									215.9	174.4	197.0	36.1
Li_2O	Lithium oxide	-597.9	-561.2	37.6	54.1								
Li_2O_2	Lithium peroxide	-634.3											
Li_2O_3Si	Lithium metasilicate	-1648.1	-1557.2	79.8	99.1								
Li_2O_4S	Lithium sulfate	-1436.5	-1321.7	115.1	117.6								
Li_2S	Lithium sulfide	-441.4											
Li_3O_4P	Lithium phosphate	-2095.8											
Lr	Lawrencium	0.0											
Lu	Lutetium	0.0		51.0	26.9					427.6	387.8	184.8	20.9
Lu_2O_3	Lutetium oxide	-1878.2	-1789.0	110.0	101.8								
Md	Mendelevium	0.0											
Mg	Magnesium	0.0		32.7	24.9					147.1	112.5	148.6	20.8
MgN_2O_6	Magnesium nitrate	-790.7	-589.4	164.0	141.9								
MgO	Magnesium oxide	-601.6	-569.3	27.0	37.2								
MgO_4S	Magnesium sulfate	-1284.9	-1170.6	91.6	96.5								
MgO_4Se	Magnesium selenate	-968.5											
MgS	Magnesium sulfide	-346.0	-341.8	50.3	45.6								
Mg_2	Dimagnesium									287.7			
Mg_2O_4Si	Magnesium orthosilicate	-2174.0	-2055.1	95.1	118.5								
Mn	Manganese	0.0		32.0	26.3					280.7	238.5	173.7	20.8
MnN_2O_6	Manganese(II) nitrate	-576.3											
$MnNaO_4$	Sodium permanganate	-1156.0											
MnO	Manganese(II) oxide	-385.2	-362.9	59.7	45.4								
MnO_2	Manganese(IV) oxide	-520.0	-465.1	53.1	54.1								
MnO_3Si	Manganese(II) metasilicate	-1320.9	-1240.5	89.1	86.4								
MnS	Manganese(II) sulfide (a form)	-214.2	-218.4	78.2	50.0								
MnSe	Manganese(II) selenide	-106.7	-111.7	90.8	51.0								
Mn_2O_3	Manganese(III) oxide	-959.0	-881.1	110.5	107.7								
Mn_2O_4Si	Manganese(II) orthosilicate	-1730.5	-1632.1	163.2	129.9								
Mn_3O_4	Manganese(II,III) oxide	-1387.8	-1283.2	155.6	139.7								
Mo	Molybdenum	0.0		28.7	24.1					658.1	612.5	182.0	20.8
$MoNa_2O_4$	Sodium molybdate	-1468.1	-1354.3	159.7	141.7								
MoO_2	Molybdenum(IV) oxide	-588.9	-533.0	46.3	56.0								
MoO_3	Molybdenum(VI) oxide	-745.1	-668.0	77.7	75.0								
MoO_4Pb	Lead(II) molybdate	-1051.9	-951.4	166.1	119.7								
MoS_2	Molybdenum(IV) sulfide	-235.1	-225.9	62.6	63.6								
Mo_3Si	Molybdenum silicide	-125.2	-125.7	106.3	93.1								
N	Nitrogen (atomic)									472.7	455.5	153.3	20.8
$NNaO_2$	Sodium nitrite	-358.7	-284.6	103.8									
$NNaO_3$	Sodium nitrate	-467.9	-367.0	116.5	92.9								

Molecular formula	Name	Crystal				Liquid				Gas			
		$\Delta_f H°$ kJ/mol	$\Delta_f G°$ kJ/mol	$S°$ J/mol K	C_p J/mol K	$\Delta_f H°$ kJ/mol	$\Delta_f G°$ kJ/mol	$S°$ J/mol K	C_p J/mol K	$\Delta_f H°$ kJ/mol	$\Delta_f G°$ kJ/mol	$S°$ J/mol K	C_p J/mol K
NO	Nitric oxide									91.3	87.6	210.8	29.9
NO_2	Nitrogen dioxide									33.2	51.3	240.1	37.2
NO_2Rb	Rubidium nitrite	-367.4	-306.2	172.0									
NO_3Rb	Rubidium nitrate	-495.1	-395.8	147.3	102.1								
NO_3Tl	Thallium(I) nitrate	-243.9	-152.4	160.7	99.5								
NP	Phosphorus nitride	-63.0								171.5	149.4	211.1	29.7
N_2	Nitrogen									0.0		191.6	29.1
N_2O	Nitrous oxide									81.6	103.7	220.0	38.6
N_2O_3	Nitrogen trioxide					50.3				86.6	142.4	314.7	72.7
N_2O_4	Nitrogen tetroxide					-19.5	97.5	209.2	142.7	11.1	99.8	304.4	79.2
N_2O_4Sr	Strontium nitrite	-762.3											
N_2O_5	Nitrogen pentoxide	-43.1	113.9	178.2	143.1					13.3	117.1	355.7	95.3
N_2O_6Pb	Lead(II) nitrate	-451.9											
N_2O_6Ra	Radium nitrate	-992.0	-796.1	222.0									
N_2O_6Sr	Strontium nitrate	-978.2	-780.0	194.6	149.9								
N_2O_6Zn	Zinc nitrate	-483.7											
N_3Na	Sodium azide	21.7	93.8	96.9	76.6								
N_4Si_3	Silicon nitride	-743.5	-642.6	101.3									
Na	Sodium	0.0		51.3	28.2					107.5	77.0	153.7	20.8
NaO_2	Sodium superoxide	-260.2	-218.4	115.9	72.1								
Na_2	Disodium									142.1	103.9	230.2	37.6
Na_2O	Sodium oxide	-414.2	-375.5	75.1	69.1								
Na_2O_2	Sodium peroxide	-510.9	-447.7	95.0	89.2								
Na_2O_3S	Sodium sulfite	-1100.8	-1012.5	145.9	120.3								
Na_2O_3Si	Sodium metasilicate	-1554.9	-1462.8	113.9									
Na_2O_4S	Sodium sulfate	-1387.1	-1270.2	149.6	128.2								
Na_2S	Sodium sulfide	-364.8	-349.8	83.7									
Nb	Niobium	0.0		36.4	24.6					725.9	681.1	186.3	30.2
NbO	Niobium(II) oxide	-405.8	-378.6	48.1	41.3								
NbO_2	Niobium(IV) oxide	-796.2	-740.5	54.5	57.5								
Nb_2O_5	Niobium(V) oxide	-1899.5	-1766.0	137.2	132.1								
Nd	Neodymium	0.0		71.5	27.5					327.6	292.4	189.4	22.1
Nd_2O_3	Neodymium oxide	-1807.9	-1720.8	158.6	111.3								
Ne	Neon									0.0		146.3	20.8
Ni	Nickel	0.0		29.9	26.1					429.7	384.5	182.2	23.4
NiO_4S	Nickel(II) sulfate	-872.9	-759.7	92.0	138.0								
NiS	Nickel(II) sulfide	-82.0	-79.5	53.0	47.1								
Ni_2O_3	Nickel(III) oxide	-489.5											
No	Nobelium	0.0											
O	Oxygen (atomic)									249.2	231.7	161.1	21.9
OP	Phosphorus monoxide									-28.5	-51.9	222.8	31.8
OPb	Lead(II) oxide (massicot)	-217.3	-187.9	68.7	45.8								
OPb	Lead(II) oxide (litharge)	-219.0	-188.9	66.5	45.8								
OPd	Palladium(II) oxide	-85.4			31.4					348.9	325.9	218.0	
ORa	Radium oxide	-523.0											
ORb_2	Rubidium oxide	-339.0											
ORh	Rhodium monoxide									385.0			
OS	Sulfur monoxide									6.3	-19.9	222.0	30.2
OSe	Selenium monoxide									53.4	26.8	234.0	31.3
OSi	Silicon monoxide									-99.6	-126.4	211.6	29.9
OSn	Tin(II) oxide	-280.7	-251.9	57.2	44.3					15.1	-8.4	232.1	31.6
OSr	Strontium oxide	-592.0	-561.9	54.4	45.0					1.5			
OTi	Titanium(II) oxide	-519.7	-495.0	50.0	40.0								
OTl_2	Thallium(I) oxide	-178.7	-147.3	126.0									
OU	Uranium(II) oxide									21.0			
OV	Vanadium(II) oxide	-431.8	-404.2	38.9	45.4								
OZn	Zinc oxide	-350.5	-320.5	43.7	40.3								
O_2	Oxygen									0.0		205.2	29.4
O_2P	Phosphorus dioxide									-279.9	-281.6	252.1	39.5
O_2Pb	Lead(IV) oxide	-277.4	-217.3	68.6	64.6								
O_2Rb	Rubidium superoxide	-278.7											
O_2Rb_2	Rubidium peroxide	-472.0											
O_2Ru	Ruthenium(IV) oxide	-305.0											
O_2S	Sulfur dioxide					-320.5				-296.8	-300.1	248.2	39.9
O_2Se	Selenium dioxide	-225.4											
O_2Si	Silicon dioxide (α-quartz)	-910.7	-856.3	41.5	44.4					-322.0			

Molecular formula	Name	Crystal				Liquid				Gas			
		$\Delta_f H°$ kJ/mol	$\Delta_f G°$ kJ/mol	$S°$ J/mol K	C_p J/mol K	$\Delta_f H°$ kJ/mol	$\Delta_f G°$ kJ/mol	$S°$ J/mol K	C_p J/mol K	$\Delta_f H°$ kJ/mol	$\Delta_f G°$ kJ/mol	$S°$ J/mol K	C_p J/mol K
O₂Sn	Tin(IV) oxide	-577.6	-515.8	49.0	52.6								
O₂Te	Tellurium dioxide	-322.6	-270.3	79.5									
O₂Th	Thorium(IV) oxide	-1226.4	-1169.2	65.2	61.8								
O₂Ti	Titanium(IV) oxide	-944.0	-888.8	50.6	55.0								
O₂U	Uranium(IV) oxide	-1085.0	-1031.8	77.0	63.6					-465.7	-471.5	274.6	51.4
O₂W	Tungsten(IV) oxide	-589.7	-533.9	50.5	56.1								
O₂Zr	Zirconium(IV) oxide	-1100.6	-1042.8	50.4	56.2								
O₃	Ozone									142.7	163.2	238.9	39.2
O₃PbS	Lead(II) sulfite	-669.9											
O₃PbSi	Lead(II) metasilicate	-1145.7	-1062.1	109.6	90.0								
O₃Pr₂	Praseodymium oxide	-1809.6			117.4								
O₃Rh₂	Rhodium(III) oxide	-343.0			103.8								
O₃S	Sulfur trioxide	-454.5	-374.2	70.7		-441.0	-373.8	113.8		-395.7	-371.1	256.8	50.7
O₃Sc₂	Scandium oxide	-1908.8	-1819.4	77.0	94.2								
O₃SiSr	Strontium metasilicate	-1633.9	-1549.7	96.7	88.5								
O₃Sm₂	Samarium(III) oxide	-1823.0	-1734.6	151.0	114.5								
O₃Tb₂	Terbium oxide	-1865.2			115.9								
O₃Ti₂	Titanium(III) oxide	-1520.9	-1434.2	78.8	97.4								
O₃Tm₂	Thulium oxide	-1888.7	-1794.5	139.7	116.7								
O₃U	Uranium(VI) oxide	-1223.8	-1145.7	96.1	81.7								
O₃V₂	Vanadium(III) oxide	-1218.8	-1139.3	98.3	103.2								
O₃W	Tungsten(VI) oxide	-842.9	-764.0	75.9	73.8								
O₃Y₂	Yttrium oxide	-1905.3	-1816.6	99.1	102.5								
O₃Yb₂	Ytterbium(III) oxide	-1814.6	-1726.7	133.1	115.4								
O₄Os	Osmium(VIII) oxide	-394.1	-304.9	143.9						-337.2	-292.8	293.8	74.1
O₄PbS	Lead(II) sulfate	-920.0	-813.0	148.5	103.2								
O₄PbSe	Lead(II) selenate	-609.2	-504.9	167.8									
O₄Pb₂Si	Lead(II) orthosilicate	-1363.1	-1252.6	186.6	137.2								
O₄Pb₃	Lead(II,II,IV) oxide	-718.4	-601.2	211.3	146.9								
O₄RaS	Radium sulfate	-1471.1	-1365.6	138.0									
O₄Rb₂S	Rubidium sulfate	-1435.6	-1316.9	197.4	134.1								
O₄Ru	Ruthenium(VIII) oxide	-239.3	-152.2	146.4									
O₄SSr	Strontium sulfate	-1453.1	-1340.9	117.0									
O₄STl₂	Thallium(I) sulfate	-931.8	-830.4	230.5									
O₄SZn	Zinc sulfate	-982.8	-871.5	110.5	99.2								
O₄SiSr₂	Strontium orthosilicate	-2304.5	-2191.1	153.1	134.3								
O₄SiZn₂	Zinc orthosilicate	-1636.7	-1523.2	131.4	123.3								
O₄SiZr	Zirconium(IV) orthosilicate	-2033.4	-1919.1	84.1	98.7								
O₄TiZr	Zirconium titanate	-2024.1	-1915.8	116.7	114.0								
O₅Sb₂	Antimony(V) oxide	-971.9	-829.2	125.1									
O₅Ta₂	Tantalum(V) oxide	-2046.0	-1911.2	143.1	135.1								
O₅Ti₃	Titanium(III,IV) oxide	-2459.4	-2317.4	129.3	154.8								
O₅V₂	Vanadium(V) oxide	-1550.6	-1419.5	131.0	127.7								
O₅V₃	Vanadium(III,IV) oxide	-1933.0	-1803.0	163.0									
O₇Re₂	Rhenium(VII) oxide	-1240.1	-1066.0	207.1	166.1					-1100.0	-994.0	452.0	
O₇U₃	Uranium(IV,VI) oxide	-3427.1	-3242.9	250.5	215.5								
O₈S₂Zr	Zirconium(IV) sulfate	-2217.1			172.0								
O₈U₃	Uranium(V,VI) oxide	-3574.8	-3369.5	282.6	238.4								
O₉U₄	Uranium(IV,V) oxide	-4510.4	-4275.1	334.1	293.3								
Os	Osmium	0.0		32.6	24.7					791.0	745.0	192.6	20.8
P	Phosphorus (white)	0.0		41.1	23.8					316.5	280.1	163.2	20.8
P	Phosphorus (red)	-17.6		22.8	21.2								
P	Phosphorus (black)	-39.3											
P₂	Diphosphorus									144.0	103.5	218.1	32.1
P₄	Tetraphosphorus									58.9	24.4	280.0	67.2
Pa	Protactinium	0.0		51.9						607.0	563.0	198.1	22.9
Pb	Lead	0.0		64.8	26.4					195.2	162.2	175.4	20.8
PbS	Lead(II) sulfide	-100.4	-98.7	91.2	49.5								
PbSe	Lead(II) selenide	-102.9	-101.7	102.5	50.2								
PbTe	Lead(II) telluride	-70.7	-69.5	110.0	50.5								
Pd	Palladium	0.0		37.6	26.0					378.2	339.7	167.1	20.8
PdS	Palladium(II) sulfide	-75.0	-67.0	46.0									
Pm	Promethium	0.0										187.1	24.3
Po	Polonium	0.0											
Pr	Praseodymium	0.0		73.2	27.2					355.6	320.9	189.8	21.4
Pt	Platinum	0.0		41.6	25.9					565.3	520.5	192.4	25.5

Standard Thermodynamic Properties of Chemical Substances

Molecular formula	Name	Crystal				Liquid				Gas			
		$\Delta_f H°$ kJ/mol	$\Delta_f G°$ kJ/mol	$S°$ J/mol K	C_p J/mol K	$\Delta_f H°$ kJ/mol	$\Delta_f G°$ kJ/mol	$S°$ J/mol K	C_p J/mol K	$\Delta_f H°$ kJ/mol	$\Delta_f G°$ kJ/mol	$S°$ J/mol K	C_p J/mol K
PtS	Platinum(II) sulfide	-81.6	-76.1	55.1	43.4								
PtS$_2$	Platinum(IV) sulfide	-108.8	-99.6	74.7	65.9								
Pu	Plutonium	0.0											
Ra	Radium	0.0		71.0						159.0	130.0	176.5	20.8
Rb	Rubidium	0.0		76.8	31.1					80.9	53.1	170.1	20.8
Re	Rhenium	0.0		36.9	25.5					769.9	724.6	188.9	20.8
Rh	Rhodium	0.0		31.5	25.0					556.9	510.8	185.8	21.0
Rn	Radon									0.0		176.2	20.8
Ru	Ruthenium	0.0		28.5	24.1					642.7	595.8	186.5	21.5
S	Sulfur (rhombic)	0.0		32.1	22.6					277.2	236.7	167.8	23.7
S	Sulfur (monoclinic)	0.3											
SSi	Silicon monosulfide									112.5	60.9	223.7	32.3
SSn	Tin(II) sulfide	-100.0	-98.3	77.0	49.3								
SSr	Strontium sulfide	-472.4	-467.8	68.2	48.7								
STl$_2$	Thallium(I) sulfide	-97.1	-93.7	151.0									
SZn	Zinc sulfide (wurtzite)	-192.6											
SZn	Zinc sulfide (sphalerite)	-206.0	-201.3	57.7	46.0								
S$_2$	Disulfur									128.6	79.7	228.2	32.5
Sb	Antimony	0.0		45.7	25.2					262.3	222.1	180.3	20.8
Sb$_2$	Diantimony									235.6	187.0	254.9	36.4
Sc	Scandium	0.0		34.6	25.5					377.8	336.0	174.8	22.1
Se	Selenium (gray)	0.0		42.4	25.4					227.1	187.0	176.7	20.8
Se	Selenium (α form)	6.7								227.1			
Se	Selenium (vitreous)	5.0								227.1			
SeSr	Strontium selenide	-385.8											
SeTl$_2$	Thallium(I) selenide	-59.0	-59.0	172.0									
SeZn	Zinc selenide	-163.0	-163.0	84.0									
Se$_2$	Diselenium									146.0	96.2	252.0	35.4
Si	Silicon	0.0		18.8	20.0					450.0	405.5	168.0	22.3
Si$_2$	Disilicon									594.0	536.0	229.9	34.4
Sm	Samarium	0.0		69.6	29.5					206.7	172.8	183.0	30.4
Sn	Tin (white)	0.0		51.2	27.0					301.2	266.2	168.5	21.3
Sn	Tin (gray)	-2.1	0.1	44.1	25.8								
Sr	Strontium	0.0		55.0	26.8					164.4	130.9	164.6	20.8
Ta	Tantalum	0.0		41.5	25.4					782.0	739.3	185.2	20.9
Tb	Terbium	0.0		73.2	28.9					388.7	349.7	203.6	24.6
Tc	Technetium	0.0								678.0		181.1	20.8
Te	Tellurium	0.0		49.7	25.7					196.7	157.1	182.7	20.8
Te$_2$	Ditellurium									168.2	118.0	268.1	36.7
Th	Thorium	0.0		51.8	27.3					602.0	560.7	190.2	20.8
Ti	Titanium	0.0		30.7	25.0					473.0	428.4	180.3	24.4
Tl	Thallium	0.0		64.2	26.3					182.2	147.4	181.0	20.8
Tm	Thulium	0.0		74.0	27.0					232.2	197.5	190.1	20.8
U	Uranium	0.0		50.2	27.7					533.0	488.4	199.8	23.7
V	Vanadium	0.0		28.9	24.9					514.2	754.4	182.3	26.0
W	Tungsten	0.0		32.6	24.3					849.4	807.1	174.0	21.3
Xe	Xenon									0.0		169.7	20.8
Y	Yttrium	0.0		44.4	26.5					421.3	381.1	179.5	25.9
Yb	Ytterbium	0.0		59.9	26.7					152.3	118.4	173.1	20.8
Zn	Zinc	0.0		41.6	25.4					130.4	94.8	161.0	20.8
Zr	Zirconium	0.0		39.0	25.4					608.8	566.5	181.4	26.7

Substances containing carbon:

Molecular formula	Name	Crystal				Liquid				Gas			
C	Carbon (graphite)	0.0		5.7	8.5					716.7	671.3	158.1	20.8
C	Carbon (diamond)	1.9	2.9	2.4	6.1								
CAgN	Silver(I) cyanide	146.0	156.9	107.2	66.7								
CAg$_2$O$_3$	Silver(I) carbonate	-505.8	-436.8	167.4	112.3								
CBaO$_3$	Barium carbonate	-1213.0	-1134.4	112.1	86.0								
CBeO$_3$	Beryllium carbonate	-1025.0		52.0	65.0								
CBrClF$_2$	Bromochlorodifluoromethane											318.5	74.6
CBrCl$_2$F	Bromodichlorofluoromethane											330.6	80.0
CBrCl$_3$	Bromotrichloromethane									-41.1			85.3
CBrF$_3$	Bromotrifluoromethane									-648.3			69.3
CBrN	Cyanogen bromide	140.5								186.2	165.3	248.3	46.9
CBrN$_3$O$_6$	Bromotrinitromethane								32.5	80.3			
CBr$_2$ClF	Dibromochlorofluoromethane											342.8	82.4

Molecular formula	Name	Crystal $\Delta_f H°$ kJ/mol	$\Delta_f G°$ kJ/mol	$S°$ J/mol K	C_p J/mol K	Liquid $\Delta_f H°$ kJ/mol	$\Delta_f G°$ kJ/mol	$S°$ J/mol K	C_p J/mol K	Gas $\Delta_f H°$ kJ/mol	$\Delta_f G°$ kJ/mol	$S°$ J/mol K	C_p J/mol K
CBr_2Cl_2	Dibromodichloromethane											347.8	87.1
CBr_2F_2	Dibromodifluoromethane											325.3	77.0
CBr_2O	Carbonyl bromide					-127.2				-96.2	-110.9	309.1	61.8
CBr_3Cl	Tribromochloromethane											357.8	89.4
CBr_3F	Tribromofluoromethane											345.9	84.4
CBr_4	Tetrabromomethane	29.4	47.7	212.5	144.3					83.9	67.0	358.1	91.2
$CCaO_3$	Calcium carbonate (calcite)	-1207.6	-1129.1	91.7	83.5								
$CCaO_3$	Calcium carbonate (aragonite)	-1207.8	-1128.2	88.0	82.3								
$CCdO_3$	Cadmium carbonate	-750.6	-669.4	92.5									
$CClFO$	Carbonyl chloride fluoride											276.7	52.4
$CClF_3$	Chlorotrifluoromethane									-706.3			66.9
$CClN$	Cyanogen chloride					112.1				138.0	131.0	236.2	45.0
$CClN_3O_6$	Chlorotrinitromethane					-27.1				18.4			
CCl_2F_2	Dichlorodifluoromethane									-477.4	-439.4	300.8	72.3
CCl_2O	Carbonyl chloride									-219.1	-204.9	283.5	57.7
CCl_3	Trichloromethyl									59.0			
CCl_3F	Trichlorofluoromethane					-301.3	-236.8	225.4	121.6	-268.3			78.1
CCl_4	Tetrachloromethane					-128.2			130.7	-95.7			83.3
$CCoO_3$	Cobalt(II) carbonate	-713.0											
CCs_2O_3	Cesium carbonate	-1139.7	-1054.3	204.5	123.9								
$CCuN$	Copper(I) cyanide	96.2	111.3	84.5									
CFN	Cyanogen fluoride											224.7	41.8
CF_2O	Carbonyl fluoride									-639.8			46.8
CF_3	Trifluoromethyl									-477.0	-464.0	264.5	49.6
CF_3I	Trifluoroiodomethane									-587.8		307.4	70.9
CF_4	Tetrafluoromethane									-933.6		261.6	61.1
$CFeO_3$	Iron(II) carbonate	-740.6	-666.7	92.9	82.1								
CFe_3	Iron carbide	25.1	20.1	104.6	105.9								
CH	Methylidyne									595.8			
$CHBrClF$	Bromochlorofluoromethane											304.3	63.2
$CHBrCl_2$	Bromodichloromethane											316.4	67.4
$CHBrF_2$	Bromodifluoromethane									-424.9		295.1	58.7
$CHBr_2Cl$	Chlorodibromomethane											327.7	69.2
$CHBr_2F$	Dibromofluoromethane											316.8	65.1
$CHBr_3$	Tribromomethane					-22.3	-5.0	220.9	130.7	23.8	8.0	330.9	71.2
$CHClF_2$	Chlorodifluoromethane									-482.6		280.9	55.9
$CHCl_2F$	Dichlorofluoromethane											293.1	60.9
$CHCl_3$	Trichloromethane					-134.1	-73.7	201.7	114.2	-102.7	6.0	295.7	65.7
$CHCsO_3$	Cesium hydrogen carbonate	-966.1											
$CHFO$	Formyl fluoride											246.6	39.9
CHF_3	Trifluoromethane									-695.4		259.7	51.0
CHI_3	Triiodomethane	-181.1								251.0		356.2	75.0
$CHKO_2$	Potassium formate	-679.7											
$CHKO_3$	Potassium hydrogen carbonate	-963.2	-863.5	115.5									
CHN	Hydrogen cyanide					108.9	125.0	112.8	70.6	135.1	124.7	201.8	35.9
$CHNO$	Isocyanic acid (HNCO)											238.0	44.9
$CHNS$	Isothiocyanic acid									127.6	113.0	247.8	46.9
CHN_3O_6	Trinitromethane					-32.8				-13.4		435.6	134.1
$CHNaO_2$	Sodium formate	-666.5	-599.9	103.8	82.7								
$CHNaO_3$	Sodium hydrogen carbonate	-950.8	-851.0	101.7	87.6								
CHO	Oxomethyl (HCO)									43.1	28.0	224.7	34.6
CH_2	Methylene									390.4	372.9	194.9	33.8
CH_2BrCl	Bromochloromethane											287.6	52.7
CH_2BrF	Bromofluoromethane											276.3	49.2
CH_2Br_2	Dibromomethane											293.2	54.7
CH_2ClF	Chlorofluoromethane											264.4	47.0
CH_2Cl_2	Dichloromethane					-124.2		177.8	101.2	-95.4		270.2	51.0
CH_2F_2	Difluoromethane									-452.3		246.7	42.9
CH_2I_2	Diiodomethane					68.5	90.4	174.1	134.0	119.5	95.8	309.7	57.7
CH_2N_2	Diazomethane											242.9	52.5
CH_2N_2	Cyanamide	58.8											
$CH_2N_2O_4$	Dinitromethane					-104.9				-61.5		358.1	86.4
CH_2O	Formaldehyde									-108.6	-102.5	218.8	35.4
$(CH_2O)_x$	Paraformaldehyde	-177.6											
CH_2O_2	Formic acid					-425.0	-361.4	129.0	99.0	-378.7			
CH_2S_3	Trithiocarbonic acid					24.0							

Molecular formula	Name	Crystal				Liquid				Gas			
		$\Delta_f H°$ kJ/mol	$\Delta_f G°$ kJ/mol	$S°$ J/mol K	C_p J/mol K	$\Delta_f H°$ kJ/mol	$\Delta_f G°$ kJ/mol	$S°$ J/mol K	C_p J/mol K	$\Delta_f H°$ kJ/mol	$\Delta_f G°$ kJ/mol	$S°$ J/mol K	C_p J/mol K
CH_3	Methyl									145.7	147.9	194.2	38.7
CH_3BO	Borane carbonyl									-111.2	-92.9	249.4	59.5
CH_3Br	Bromomethane					-59.8				-35.4	-26.3	246.4	42.4
CH_3Cl	Chloromethane									-81.9		234.6	40.8
CH_3Cl_3Si	Methyltrichlorosilane							262.8	163.1	-528.9		351.1	102.4
CH_3F	Fluoromethane											222.9	37.5
CH_3I	Iodomethane					-13.6		163.2	126.0	14.4		254.1	44.1
CH_3NO	Formamide					-254.0				-193.9			
CH_3NO_2	Nitromethane					-112.6	-14.4	171.8	106.6	-80.8		282.9	55.5
CH_3NO_2	Methyl nitrite									-66.1			
CH_3NO_3	Methyl nitrate					-156.3	-43.4	217.1	157.3	-122.0		305.8	76.6
CH_4	Methane									-74.6	-50.5	186.3	35.7
CH_4N_2	Ammonium cyanide	0.4			134.0								
CH_4N_2O	Urea	-333.1								-245.8			
CH_4N_2S	Thiourea	-89.1								22.9			
$CH_4N_4O_2$	Nitroguanidine	-92.4											
CH_4O	Methanol					-239.2	-166.6	126.8	81.1	-201.0	-162.3	239.9	44.1
CH_4S	Methanethiol					-46.7	-7.7	169.2	90.5	-22.9	-9.3	255.2	50.3
CH_5N	Methylamine					-47.3	35.7	150.2	102.1	-22.5	32.7	242.9	50.1
CH_5NO_3	Ammonium hydrogen carbonate	-849.4	-665.9	120.9									
CH_5N_3	Guanidine	-56.0											
CH_5N_3S	Hydrazinecarbothioamide	24.7											
$CH_5N_5O_2$	3-Amino-1-nitroguanidine	22.1											
CH_6ClN	Methylamine hydrochloride	-298.1											
CH_6N_2	Methylhydrazine					54.2	180.0	165.9	134.9	94.7	187.0	278.8	71.1
CH_6Si	Methylsilane											256.5	65.9
CHg_2O_3	Mercury(I) carbonate	-553.5	-468.1	180.0									
CIN	Cyanogen iodide	166.2	185.0	96.2						225.5	196.6	256.8	48.3
CI_4	Tetraiodomethane	-392.9								474.0		391.9	95.9
CKN	Potassium cyanide	-113.0	-101.9	128.5	66.3								
$CKNS$	Potassium thiocyanate	-200.2	-178.3	124.3	88.5								
CK_2O_3	Potassium carbonate	-1151.0	-1063.5	155.5	114.4								
CLi_2O_3	Lithium carbonate	-1215.9	-1132.1	90.4	99.1								
$CMgO_3$	Magnesium carbonate	-1095.8	-1012.1	65.7	75.5								
$CMnO_3$	Manganese(II) carbonate	-894.1	-816.7	85.8	81.5								
CN	Cyanide									437.6	407.5	202.6	29.2
$CNNa$	Sodium cyanide	-87.5	-76.4	115.6	70.4								
$CNNaO$	Sodium cyanate	-405.4	-358.1	96.7	86.6								
CN_4O_8	Tetranitromethane					38.4				82.4		503.7	176.1
CNa_2O_3	Sodium carbonate	-1130.7	-1044.4	135.0	112.3								
CO	Carbon monoxide									-110.5	-137.2	197.7	29.1
COS	Carbon oxysulfide									-142.0	-169.2	231.6	41.5
CO_2	Carbon dioxide									-393.5	-394.4	213.8	37.1
CO_3Pb	Lead(II) carbonate	-699.1	-625.5	131.0	87.4								
CO_3Rb_2	Rubidium carbonate	-1136.0	-1051.0	181.3	117.6								
CO_3Sr	Strontium carbonate	-1220.1	-1140.1	97.1	81.4								
CO_3Tl_2	Thallium(I) carbonate	-700.0	-614.6	155.2									
CO_3Zn	Zinc carbonate	-812.8	-731.5	82.4	79.7								
CS	Carbon monosulfide									280.3	228.8	210.6	29.8
CS_2	Carbon disulfide					89.0	64.6	151.3	76.4	116.7	67.1	237.8	45.4
CSe_2	Carbon diselenide					164.8							
CSi	Silicon carbide (cubic)	-65.3	-62.8	16.6	26.9								
CSi	Silicon carbide (hexagonal)	-62.8	-60.2	16.5	26.7								
C_2	Dicarbon									831.9	775.9	199.4	43.2
C_2BrF_5	Bromopentafluoroethane									-1064.4			
$C_2Br_2ClF_3$	1,2-Dibromo-1-chloro-1,2,2-trifluoroethane					-691.7				-656.6			
$C_2Br_2F_4$	1,2-Dibromotetrafluoroethane					-817.7				-789.1			
C_2Br_4	Tetrabromoethene											387.1	102.7
C_2Br_6	Hexabromoethane											441.9	139.3
C_2Ca	Calcium carbide	-59.8	-64.9	70.0	62.7								
C_2CaN_2	Calcium cyanide	-184.5											
C_2CaO_4	Calcium oxalate	-1360.6											
C_2ClF_3	Chlorotrifluoroethene					-522.7				-505.5	-523.8	322.1	83.9
C_2ClF_5	Chloropentafluoroethane									-1118.8			184.2
$C_2Cl_2F_4$	1,2-Dichloro-1,1,2,2-tetrafluoroethane					-960.2			111.7	-937.0			
$C_2Cl_2O_2$	Oxalyl chloride					-367.6				-335.8			

Standard Thermodynamic Properties of Chemical Substances

Molecular formula	Name	Crystal $\Delta_f H°$ kJ/mol	$\Delta_f G°$ kJ/mol	$S°$ J/mol K	C_p J/mol K	Liquid $\Delta_f H°$ kJ/mol	$\Delta_f G°$ kJ/mol	$S°$ J/mol K	C_p J/mol K	Gas $\Delta_f H°$ kJ/mol	$\Delta_f G°$ kJ/mol	$S°$ J/mol K	C_p J/mol K
C₂Cl₃F₃	1,1,2-Trichloro-1,2,2-trifluoroethane					-745.0			170.1	-716.8			
C₂Cl₃N	Trichloroacetonitrile											336.6	96.1
C₂Cl₄	Tetrachloroethene					-50.6	3.0	266.9	143.4	-10.9			
C₂Cl₄F₂	1,1,1,2-Tetrachloro-2,2-difluoroethane									-489.9	-407.0	382.9	123.4
C₂Cl₄F₂	1,1,2,2-Tetrachloro-1,2-difluoroethane								173.6				
C₂Cl₄O	Trichloroacetyl chloride					-280.8				-239.8			
C₂Cl₆	Hexachloroethane	-202.8	237.3	198.2						-143.6			
C₂F₃N	Trifluoroacetonitrile									-497.9		298.1	77.9
C₂F₄	Tetrafluoroethene	-820.5								-658.9		300.1	80.5
C₂F₆	Hexafluoroethane									-1344.2		332.3	106.7
C₂HBr	Bromoacetylene											253.7	55.7
C₂HBrClF₃	1-Bromo-2-chloro-1,1,2-trifluoroethane					-675.3				-644.8			
C₂HBrClF₃	2-Bromo-2-chloro-1,1,1-trifluoroethane					-720.0				-690.4			
C₂HCl	Chloroacetylene											242.0	54.3
C₂HClF₂	1-Chloro-2,2-difluoroethene									-315.5	-289.1	303.0	72.1
C₂HCl₂F	1,1-Dichloro-2-fluoroethene											313.9	76.5
C₂HCl₂F₃	2,2-Dichloro-1,1,1-trifluoroethane											352.8	102.5
C₂HCl₃	Trichloroethene					-43.6		228.4	124.4	-9.0		324.8	80.3
C₂HCl₃O	Trichloroacetaldehyde					-234.5			151.0	-196.6			
C₂HCl₃O	Dichloroacetyl chloride					-280.4				-241.0			
C₂HCl₃O₂	Trichloroacetic acid	-503.3											
C₂HCl₅	Pentachloroethane					-187.6			173.8	-142.0			
C₂HF	Fluoroacetylene											231.7	52.4
C₂HF₃	Trifluoroethene									-490.5			
C₂HF₃O₂	Trifluoroacetic acid					-1069.9				-1031.4			
C₂HF₅	Pentafluoroethane									-1100.4			
C₂H₂	Acetylene									227.4	209.9	200.9	44.0
C₂H₂BrF₃	2-Bromo-1,1,1-trifluoroethane									-694.5			
C₂H₂Br₂	cis-1,2-Dibromoethene											311.3	68.8
C₂H₂Br₂	trans-1,2-Dibromoethene											313.5	70.3
C₂H₂Br₂Cl₂	1,2-Dibromo-1,2-dichloroethane									-36.9			
C₂H₂Br₄	1,1,2,2-Tetrabromoethane								165.7				
C₂H₂ClF₃	2-Chloro-1,1,1-trifluoroethane											326.5	89.1
C₂H₂Cl₂	1,1-Dichloroethene					-23.9	24.1	201.5	111.3	2.8	25.4	289.0	67.1
C₂H₂Cl₂	cis-1,2-Dichloroethene					-26.4		198.4	116.4	4.6		289.6	65.1
C₂H₂Cl₂	trans-1,2-Dichloroethene					-24.3	27.3	195.9	116.8	5.0	28.6	290.0	66.7
C₂H₂Cl₂O	Chloroacetyl chloride					-283.7				-244.8			
C₂H₂Cl₂O₂	Dichloroacetic acid					-496.3							
C₂H₂Cl₃NO	2,2,2-Trichloroacetamide	-358.0											
C₂H₂Cl₄	1,1,1,2-Tetrachloroethane											356.0	102.7
C₂H₂Cl₄	1,1,2,2-Tetrachloroethane					-195.0		246.9	162.3	-149.2		362.8	100.8
C₂H₂F₂	1,1-Difluoroethene									-335.0		266.2	60.1
C₂H₂F₂	cis-1,2-Difluoroethene											268.3	58.2
C₂H₂F₃I	1,1,1-Trifluoro-2-iodoethane									-644.5			
C₂H₂I₂	cis-1,2-Diiodoethene									-207.4			
C₂H₂O	Ketene					-67.9				-47.5	-48.3	247.6	51.8
C₂H₂O₂	Glyoxal									-212.0	-189.7	272.5	60.6
C₂H₂O₄	Oxalic acid	-829.9	109.8	91.0						-731.8	-662.7	320.6	86.2
C₂H₂O₄Sr	Strontium formate	-1393.3											
C₂H₂S	Thiirene									300.0	275.8	255.3	54.7
C₂H₃Br	Bromoethene									79.2	81.8	275.8	55.5
C₂H₃BrO	Acetyl bromide					-223.5				-190.4			
C₂H₃BrO₂	Bromoacetic acid									-383.5	-338.3	337.0	80.5
C₂H₃Cl	Chloroethene	-94.1		59.4		14.6				37.2	53.6	264.0	53.7
C₂H₃ClF₂	1-Chloro-1,1-difluoroethane											307.2	82.5
C₂H₃ClO	Acetyl chloride					-272.9	-208.0	200.8	117.0	-242.8	-205.8	295.1	67.8
C₂H₃ClO₂	Chloroacetic acid	-509.7								-427.6	-368.5	325.9	78.8
C₂H₃Cl₂F	1,1-Dichloro-1-fluoroethane											320.2	88.7
C₂H₃Cl₃	1,1,1-Trichloroethane					-177.4		227.4	144.3	-144.4		323.1	93.3
C₂H₃Cl₃	1,1,2-Trichloroethane					-190.8		232.6	150.9	-151.3		337.2	89.0
C₂H₃F	Fluoroethene									-138.8			
C₂H₃FO	Acetyl fluoride					-467.2				-442.1			
C₂H₃F₃	1,1,1-Trifluoroethane									-744.6		279.9	78.2
C₂H₃F₃	1,1,2-Trifluoroethane									-730.7			
C₂H₃F₃O	2,2,2-Trifluoroethanol					-932.4				-888.4			
C₂H₃I	Iodoethene											285.0	57.9

Standard Thermodynamic Properties of Chemical Substances

Molecular formula	Name	Crystal				Liquid				Gas			
		$\Delta_f H°$ kJ/mol	$\Delta_f G°$ kJ/mol	$S°$ J/mol K	C_p J/mol K	$\Delta_f H°$ kJ/mol	$\Delta_f G°$ kJ/mol	$S°$ J/mol K	C_p J/mol K	$\Delta_f H°$ kJ/mol	$\Delta_f G°$ kJ/mol	$S°$ J/mol K	C_p J/mol K
C_2H_3IO	Acetyl iodide					-163.5				-126.4			
$C_2H_3KO_2$	Potassium acetate	-723.0											
C_2H_3N	Acetonitrile					40.6	86.5	149.6	91.5	74.0	91.9	243.4	52.2
C_2H_3N	Isocyanomethane					130.8	159.5	159.0		163.5	165.7	246.9	52.9
C_2H_3NO	Methyl isocyanate					-92.0							
$C_2H_3NO_2$	Nitroethene									33.3		300.5	73.7
$C_2H_3NO_3$	Oxamic acid	-661.2								-552.3			
C_2H_3NS	Methyl isothiocyanate	79.4											
$C_2H_3NaO_2$	Sodium acetate	-708.8	-607.2	123.0	79.9								
C_2H_4	Ethylene									52.4	68.4	219.3	42.9
C_2H_4BrCl	1-Bromo-2-chloroethane							130.1					
$C_2H_4Br_2$	1,1-Dibromoethane					-66.2						327.7	80.8
$C_2H_4Br_2$	1,2-Dibromoethane					-79.2		223.3	136.0	-37.5			
C_2H_4ClF	1-Chloro-1-fluoroethane									-313.4			
$C_2H_4Cl_2$	1,1-Dichloroethane					-158.4	-73.8	211.8	126.3	-127.7	-70.8	305.1	76.2
$C_2H_4Cl_2$	1,2-Dichloroethane					-166.8			128.4	-126.4		308.4	78.7
$C_2H_4F_2$	1,1-Difluoroethane									-497.0		282.5	67.8
$C_2H_4I_2$	1,2-Diiodoethane	9.3								75.0			
$C_2H_4N_2O_2$	Oxamide	-504.4								-387.1			
$C_2H_4N_2O_2$	Ethanedial dioxime	-90.5											
$C_2H_4N_2O_4$	1,1-Dinitroethane					-148.2							
$C_2H_4N_2O_4$	1,2-Dinitroethane					-165.2							
$C_2H_4N_2S_2$	Ethanedithioamide	-20.8								83.0			
$C_2H_4N_4$	1H-1,2,4-Triazol-3-amine	76.8											
C_2H_4O	Acetaldehyde					-192.2	-127.6	160.2	89.0	-166.2	-133.0	263.8	55.3
C_2H_4O	Oxirane					-78.0	-11.8	153.9	88.0	-52.6	-13.0	242.5	47.9
C_2H_4OS	Thioacetic acid					-216.9				-175.1			
$C_2H_4O_2$	Acetic acid					-484.3	-389.9	159.8	123.3	-432.2	-374.2	283.5	63.4
$C_2H_4O_2$	Methyl formate					-386.1			119.1	-357.4		285.3	64.4
$C_2H_4O_3$	Peroxyacetic acid								82.4				
$C_2H_4O_3$	Glycolic acid									-583.0	-504.9	318.6	87.1
C_2H_4S	Thiirane					51.6				82.0	96.8	255.2	53.3
C_2H_4Si	Ethynylsilane									269.4			72.6
C_2H_5Br	Bromoethane					-90.5	-25.8	198.7	100.8	-61.9	-23.9	286.7	64.5
C_2H_5Cl	Chloroethane					-136.8	-59.3	190.8	104.3	-112.1	-60.4	276.0	62.8
C_2H_5ClO	2-Chloroethanol					-295.4							
C_2H_5F	Fluoroethane									264.5			58.6
C_2H_5I	Iodoethane					-40.0	14.7	211.7	115.1	-8.1	19.2	306.0	66.9
C_2H_5N	Ethyleneimine					91.9				126.5			
C_2H_5NO	Acetamide	-317.0		115.0	91.3					-238.3			
C_2H_5NO	N-Methylformamide								123.8				
$C_2H_5NO_2$	Nitroethane					-143.9			134.4	-103.8		320.5	79.0
$C_2H_5NO_2$	Glycine	-528.5								-392.1			
$C_2H_5NO_3$	2-Nitroethanol					-350.7							
$C_2H_5NO_3$	Ethyl nitrate					-190.4				-154.1			
C_2H_5NS	Thioacetamide	-71.7								11.4			
C_2H_6	Ethane									-84.0	-32.0	229.2	52.5
C_2H_6Cd	Dimethyl cadmium					63.6	139.0	201.9	132.0	101.6	146.9	303.0	
C_2H_6Hg	Dimethyl mercury					59.8	140.3	209.0		94.4	146.1	306.0	83.3
$C_2H_6N_2O$	N-Methylurea	-332.8											
$C_2H_6N_4O_2$	1,2-Hydrazinedicarboxamide	-498.7											
$C_2H_6N_4O_2$	Oxalyl dihydrazide	-295.2											
C_2H_6O	Ethanol					-277.6	-174.8	160.7	112.3	-234.8	-167.9	281.6	65.6
C_2H_6O	Dimethyl ether					-203.3				-184.1	-112.6	266.4	64.4
C_2H_6OS	Dimethyl sulfoxide					-204.2	-99.9	188.3	153.0	-151.3			
$C_2H_6O_2$	Ethylene glycol					-460.0		163.2	148.6	-392.2		303.8	82.7
$C_2H_6O_2S$	Dimethyl sulfone	-450.1	-302.4	142.0						-373.1	-272.7	310.6	100.0
$C_2H_6O_3S$	Dimethyl sulfite					-523.6				-483.4			
$C_2H_6O_4S$	Dimethyl sulfate					-735.5				-687.0			
C_2H_6S	Ethanethiol					-73.6	-5.5	207.0	117.9	-46.1	-4.8	296.2	72.7
C_2H_6S	Dimethyl sulfide					-65.3		196.4	118.1	-37.4		286.0	74.1
$C_2H_6S_2$	1,2-Ethanedithiol					-54.3				-9.7			
$C_2H_6S_2$	Dimethyl disulfide					-62.6		235.4	146.1	-24.7			
C_2H_6Zn	Dimethyl zinc					23.4		201.6	129.2	53.0			
C_2H_7N	Ethylamine					-74.1			130.0	-47.5	36.3	283.8	71.5
C_2H_7N	Dimethylamine					-43.9	70.0	182.3	137.7	-18.8	68.5	273.1	70.7

Molecular formula	Name	Crystal				Liquid				Gas			
		$\Delta_f H°$ kJ/mol	$\Delta_f G°$ kJ/mol	$S°$ J/mol K	C_p J/mol K	$\Delta_f H°$ kJ/mol	$\Delta_f G°$ kJ/mol	$S°$ J/mol K	C_p J/mol K	$\Delta_f H°$ kJ/mol	$\Delta_f G°$ kJ/mol	$S°$ J/mol K	C_p J/mol K
C₂H₇NO	Ethanolamine								195.5				
C₂H₈ClN	Dimethylamine hydrochloride	-289.3											
C₂H₈N₂	1,2-Ethanediamine					-63.0			172.6	-18.0			
C₂H₈N₂	1,1-Dimethylhydrazine					48.9	206.4	198.0	164.1	84.1			
C₂H₈N₂	1,2-Dimethylhydrazine					52.7				92.2			
C₂H₈N₂O₄	Ammonium oxalate	-1123.0			226.0								
C₂HgO₄	Mercury(II) oxalate	-678.2											
C₂I₂	Diiodoacetylene									313.1			70.3
C₂I₄	Tetraiodoethene	305.0											
C₂K₂O₄	Potassium oxalate	-1346.0											
C₂MgO₄	Magnesium oxalate	-1269.0											
C₂N₂	Cyanogen					285.9				306.7		241.9	56.8
C₂N₄O₆	Trinitroacetonitrile					183.7							
C₂Na₂O₄	Sodium oxalate									-1318.0			
C₂O₄Pb	Lead(II) oxalate	-851.4	-750.1	146.0	105.4								
C₃F₈	Perfluoropropane									-1783.2			
C₃H₂N₂	Malononitrile	186.4								265.5			
C₃H₂O₂	2-Propynoic acid					-193.2							
C₃H₂O₃	1,3-Dioxol-2-one					-459.9				-418.6			
C₃H₃Cl₃	1,2,3-Trichloropropene					-101.8							
C₃H₃F₃	3,3,3-Trifluoropropene									-614.2			
C₃H₃N	Acrylonitrile					147.1				180.6			
C₃H₃NO	Oxazole					-48.0				-15.5			
C₃H₃NO	Isoxazole					42.1				78.6			
C₃H₄	Allene									190.5			
C₃H₄	Propyne									184.9			
C₃H₄	Cyclopropene									277.1			
C₃H₄Cl₂	2,3-Dichloropropene					-73.3							
C₃H₄Cl₄	1,1,1,3-Tetrachloropropane					-208.7							
C₃H₄Cl₄	1,2,2,3-Tetrachloropropane					-251.8							
C₃H₄F₄O	2,2,3,3-Tetrafluoro-1-propanol					-1114.9				-1061.3			
C₃H₄N₂	1H-Pyrazole	105.4			81.0					179.4			
C₃H₄N₂	Imidazole	49.8								132.9			
C₃H₄O	Acrolein												71.3
C₃H₄O₂	1,2-Propanedione					-309.1				-271.0			
C₃H₄O₂	Acrylic acid					-383.8			145.7				
C₃H₄O₂	2-Oxetanone					-329.9	175.3	122.1		-282.9			
C₃H₄O₃	Ethylene carbonate					-682.8			133.9	-508.4			
C₃H₅Br	cis-1-Bromopropene					7.9				40.8			
C₃H₅Br	3-Bromopropene					12.2				45.2			
C₃H₅BrO	Bromoacetone									-181.0			
C₃H₅Cl	2-Chloropropene									-21.0			
C₃H₅Cl	3-Chloropropene								125.1				
C₃H₅ClO	Epichlorohydrin					-148.4			131.6	-107.8			
C₃H₅ClO₂	2-Chloropropanoic acid					-522.5				-475.8			
C₃H₅ClO₂	3-Chloropropanoic acid	-549.3											
C₃H₅ClO₂	Ethyl chloroformate					-505.3				-462.9			
C₃H₅ClO₂	Methyl chloroacetate					-487.0				-444.0			
C₃H₅Cl₃	1,2,3-Trichloropropane					-230.6			183.6	-182.9			
C₃H₅I	3-Iodopropene					53.7				91.5			
C₃H₅IO	Iodoacetone									-130.5			
C₃H₅IO₂	3-Iodopropanoic acid	-460.0											
C₃H₅N	Propanenitrile					15.5			119.3	51.7			
C₃H₅N	2-Propyn-1-amine					205.7							
C₃H₅N	Ethyl isocyanide					108.6				141.7			
C₃H₅NO	Acrylamide	-212.1			110.6	-224.0				-130.2			
C₃H₅NO₃	Nitroacetone					-278.6							
C₃H₅NO₄	Methyl nitroacetate					-464.0							
C₃H₅N₃O₉	Trinitroglycerol					-370.9				-279.1		545.9	234.2
C₃H₆	Propene					4.0				20.0			
C₃H₆	Cyclopropane					35.2				53.3	104.5	237.5	55.6
C₃H₆Br₂	1,2-Dibromopropane					-113.6				-71.6			
C₃H₆Cl₂	1,2-Dichloropropane, (±)					-198.8			149.1	-162.8			
C₃H₆Cl₂	1,3-Dichloropropane					-199.9				-159.2			
C₃H₆Cl₂	2,2-Dichloropropane					-205.8				-173.2			
C₃H₆Cl₂O	2,3-Dichloro-1-propanol					-381.5				-316.3			

Molecular formula	Name	Crystal				Liquid				Gas			
		$\Delta_f H°$ kJ/mol	$\Delta_f G°$ kJ/mol	$S°$ J/mol K	C_p J/mol K	$\Delta_f H°$ kJ/mol	$\Delta_f G°$ kJ/mol	$S°$ J/mol K	C_p J/mol K	$\Delta_f H°$ kJ/mol	$\Delta_f G°$ kJ/mol	$S°$ J/mol K	C_p J/mol K
$C_3H_6Cl_2O$	1,3-Dichloro-2-propanol					-385.3				-318.4			
$C_3H_6I_2$	1,2-Diiodopropane									35.6			
$C_3H_6I_2$	1,3-Diiodopropane					-9.0							
$C_3H_6N_2O_2$	Propanediamide	-546.1											
$C_3H_6N_2O_2$	N-(Aminocarbonyl)acetamide	-544.2								-441.2			
$C_3H_6N_2O_4$	1,1-Dinitropropane					-163.2				-100.7			
$C_3H_6N_2O_4$	1,3-Dinitropropane					-207.1							
$C_3H_6N_2O_4$	2,2-Dinitropropane					-181.2							
$C_3H_6N_6O_6$	Hexahydro-1,3,5-trinitro-1,3,5-triazine									192.0		482.4	230.2
C_3H_6O	Allyl alcohol					-171.8			138.9	-124.5			
C_3H_6O	Propanal					-215.6				-185.6		304.5	80.7
C_3H_6O	Acetone					-248.4		199.8	126.3	-217.1	-152.7	295.3	74.5
C_3H_6O	Methyloxirane					-123.0		196.5	120.4	-94.7		286.9	72.6
C_3H_6O	Oxetane					-110.8				-80.5			
$C_3H_6O_2$	Propanoic acid					-510.7		191.0	152.8	-455.7			
$C_3H_6O_2$	Ethyl formate								149.3				
$C_3H_6O_2$	Methyl acetate					-445.9			141.9	-413.3		324.4	86.0
$C_3H_6O_2$	1,3-Dioxolane					-333.5			118.0	-298.0			
$C_3H_6O_2S$	Thiolactic acid					-468.4							
$C_3H_6O_3$	1,3,5-Trioxane	-522.5		133.0	111.4					-465.9			
C_3H_6S	Thietane					24.7		184.9		60.6	107.1	285.0	68.3
C_3H_6S	Methylthiirane					11.3				45.8			
$C_3H_6S_2$	1,2-Dithiolane									0.0	47.7	313.5	86.5
$C_3H_6S_2$	1,3-Dithiolane									10.0	54.7	323.3	84.7
$C_3H_6S_3$	1,3,5-Trithiane									80.0	130.4	336.4	111.3
C_3H_7Br	1-Bromopropane					-121.9				-87.0			
C_3H_7Br	2-Bromopropane					-130.5				-99.4			
C_3H_7Cl	1-Chloropropane					-160.5				-131.9			
C_3H_7Cl	2-Chloropropane					-172.3				-144.9			
$C_3H_7ClO_2$	3-Chloro-1,2-propanediol					-525.3							
$C_3H_7ClO_2$	2-Chloro-1,3-propanediol					-517.5							
C_3H_7F	1-Fluoropropane									-285.9			
C_3H_7F	2-Fluoropropane									-293.5			
C_3H_7I	1-Iodopropane					-66.0				-30.0			
C_3H_7I	2-Iodopropane					-74.8				-40.3			
C_3H_7N	Allylamine					-10.0							
C_3H_7N	Cyclopropylamine					45.8		187.7	147.1	77.0			
C_3H_7NO	N,N-Dimethylformamide					-239.3			150.6	-192.4			
C_3H_7NO	Propanamide	-338.2								-259.0			
$C_3H_7NO_2$	1-Nitropropane					-167.2				-124.3		350.0	104.1
$C_3H_7NO_2$	2-Nitropropane					-180.3			170.3	-138.9			
$C_3H_7NO_2$	Ethyl carbamate	-517.1		156.4		-497.3				-446.3			
$C_3H_7NO_2$	DL-Alanine	-563.6											
$C_3H_7NO_2$	D-Alanine	-561.2											
$C_3H_7NO_2$	L-Alanine	-604.0								-465.9			
$C_3H_7NO_2$	β-Alanine	-558.0								-424.0			
$C_3H_7NO_2$	Sarcosine	-513.3								-367.3			
$C_3H_7NO_2S$	L-Cysteine	-534.1											
$C_3H_7NO_3$	Propyl nitrate					-214.5				-174.1		362.6	123.2
$C_3H_7NO_3$	Isopropyl nitrate					-229.7				-191.0			
$C_3H_7NO_3$	DL-Serine	-739.0											
$C_3H_7NO_3$	L-Serine	-732.7											
C_3H_8	Propane					-120.9				-103.8	-23.4	270.3	73.6
$C_3H_8N_2O$	N-Ethylurea	-357.8											
$C_3H_8N_2O$	N,N-Dimethylurea	-319.1											
$C_3H_8N_2O$	N,N'-Dimethylurea	-312.1											
$C_3H_8N_2O_3$	Oxymethurea	-717.0											
C_3H_8O	1-Propanol					-302.6		193.6	143.9	-255.1		322.6	85.6
C_3H_8O	2-Propanol					-318.1		181.1	156.5	-272.6		309.2	89.3
C_3H_8O	Ethyl methyl ether									-216.4		309.2	93.3
$C_3H_8O_2$	1,2-Propylene glycol					-501.0			190.8	-429.8			
$C_3H_8O_2$	1,3-Propylene glycol					-480.8				-408.0			
$C_3H_8O_2$	Ethylene glycol monomethyl ether								171.1				
$C_3H_8O_2$	Dimethoxymethane					-377.8		244.0	162.0	-348.5			
$C_3H_8O_3$	Glycerol					-669.6		206.3	218.9	-577.9			
C_3H_8S	1-Propanethiol					-99.9		242.5	144.6	-67.8			

Molecular formula	Name	Crystal				Liquid				Gas			
		$\Delta_f H°$ kJ/mol	$\Delta_f G°$ kJ/mol	$S°$ J/mol K	C_p J/mol K	$\Delta_f H°$ kJ/mol	$\Delta_f G°$ kJ/mol	$S°$ J/mol K	C_p J/mol K	$\Delta_f H°$ kJ/mol	$\Delta_f G°$ kJ/mol	$S°$ J/mol K	C_p J/mol K
C_3H_8S	2-Propanethiol					-105.9		233.5	145.3	-76.2			
C_3H_8S	Ethyl methyl sulfide					-91.6		239.1	144.6	-59.6			
$C_3H_8S_2$	1,3-Propanedithiol					-79.4				-29.8			
C_3H_9Al	Trimethyl aluminum					-136.4	-9.9	209.4	155.6	-74.1			
C_3H_9B	Trimethylborane					-143.1	-32.1	238.9		-124.3	-35.9	314.7	88.5
$C_3H_9BO_3$	Trimethyl borate							189.9					
C_3H_9ClSi	Trimethylchlorosilane					-382.8	-246.4	278.2		-352.8	-243.5	369.1	
C_3H_9N	Propylamine					-101.5			164.1	-70.1	39.9	325.4	91.2
C_3H_9N	Isopropylamine					-112.3		218.3	163.8	-83.7	32.2	312.2	97.5
C_3H_9N	Trimethylamine					-45.7		208.5	137.9	-23.6		287.1	91.8
$C_3H_{10}ClN$	Propylamine hydrochloride	-354.7											
$C_3H_{10}ClN$	Trimethylamine hydrochloride	-282.9											
$C_3H_{10}N_2$	1,2-Propanediamine, (±)					-97.8				-53.6			
$C_3H_{10}Si$	Trimethylsilane											331.0	117.9
$C_3H_{12}BN$	Trimethylamine borane	-142.5	70.7	187.0									
$C_3H_{12}BN$	Aminetrimethylboron	-284.1	-79.3	218.0									
C_4Cl_6	Hexachloro-1,3-butadiene					-24.5							
C_4F_8	Perfluorocyclobutane									-1542.6			
C_4F_{10}	Perfluorobutane								127.2				
$C_4H_2N_2$	trans-2-Butenedinitrile	268.2								340.2			
$C_4H_2O_3$	Maleic anhydride	-469.8								-398.3			
$C_4H_2O_4$	2-Butynedioic acid	-577.3											
$C_4H_3NO_3$	2-Nitrofuran	-104.1								-28.8			
$C_4H_4BrNO_2$	N-Bromosuccinimide	-335.9											
$C_4H_4ClNO_2$	N-Chlorosuccinimide	-357.9											
$C_4H_4N_2$	Succinonitrile	139.7		191.6	145.6					209.7			
$C_4H_4N_2$	Pyrazine	139.8								196.1			
$C_4H_4N_2$	Pyrimidine					145.9				195.7			
$C_4H_4N_2$	Pyridazine					224.9				278.3			
$C_4H_4N_2O_2$	Uracil	-429.4			120.5					-302.9			
$C_4H_4N_2O_3$	Barbituric acid	-634.7											
C_4H_4O	Furan					-62.3		177.0	114.8	-34.8		267.2	65.4
$C_4H_4O_2$	Diketene					-233.1				-190.3			
$C_4H_4O_3$	Succinic anhydride	-608.6								-527.9			
$C_4H_4O_4$	Maleic acid	-789.4		160.8	137.0					-679.4			
$C_4H_4O_4$	Fumaric acid	-811.7		168.0	142.0					-675.8			
C_4H_4S	Thiophene					80.2		181.2	123.8	114.9	126.1	278.8	72.8
C_4H_5N	trans-2-Butenenitrile					95.1				134.3			
C_4H_5N	3-Butenenitrile					117.8				159.7			
C_4H_5N	2-Methylacrylonitrile								126.3				
C_4H_5N	Pyrrole					63.1		156.4	127.7	108.2			
C_4H_5N	Cyclopropanecarbonitrile					140.8				182.8			
$C_4H_5NO_2$	Succinimide	-459.0								-375.4			
C_4H_5NS	4-Methylthiazole					67.9				111.8			
$C_4H_5N_3O$	Cytosine	-221.3			132.6								
C_4H_6	1,2-Butadiene					138.6				162.3			
C_4H_6	1,3-Butadiene					88.5		199.0	123.6	110.0			
C_4H_6	1-Butyne					141.4				165.2			
C_4H_6	2-Butyne					119.1				145.7			
C_4H_6	Cyclobutene									156.7			
$C_4H_6N_2O_2$	2,5-Piperazinedione	-446.5											
C_4H_6O	Divinyl ether					-39.8				-13.6			
C_4H_6O	trans-2-Butenal					-138.7				-100.6			
$C_4H_6O_2$	trans-2-Butenoic acid												
$C_4H_6O_2$	Methacrylic acid								161.1				
$C_4H_6O_2$	Vinyl acetate					-349.2				-314.4			
$C_4H_6O_2$	Methyl acrylate					-362.2	239.5	158.8		-333.0			
$C_4H_6O_2$	γ-Butyrolactone					-420.9			141.4	-366.5			
$C_4H_6O_3$	Acetic anhydride					-624.4				-572.5			
$C_4H_6O_3$	Propylene carbonate					-613.2			218.6	-582.5			
$C_4H_6O_4$	Succinic acid	-940.5		167.3	153.1					-823.0			
$C_4H_6O_4$	Dimethyl oxalate	-756.3								-708.9			
C_4H_6S	2,3-Dihydrothiophene					52.9				90.7	133.5	303.5	79.8
C_4H_6S	2,5-Dihydrothiophene					47.0				86.9	131.6	297.1	83.3
C_4H_7ClO	2-Chloroethyl vinyl ether					-208.1				-170.1			
$C_4H_7ClO_2$	2-Chlorobutanoic acid					-575.5							

Molecular formula	Name	Crystal				Liquid				Gas			
		$\Delta_f H°$ kJ/mol	$\Delta_f G°$ kJ/mol	$S°$ J/mol K	C_p J/mol K	$\Delta_f H°$ kJ/mol	$\Delta_f G°$ kJ/mol	$S°$ J/mol K	C_p J/mol K	$\Delta_f H°$ kJ/mol	$\Delta_f G°$ kJ/mol	$S°$ J/mol K	C_p J/mol K
$C_4H_7ClO_2$	3-Chlorobutanoic acid					-556.3							
$C_4H_7ClO_2$	4-Chlorobutanoic acid					-566.3							
$C_4H_7ClO_2$	Propyl chlorocarbonate					-533.4				-492.7			
C_4H_7N	Butanenitrile					-5.8				33.6			
C_4H_7N	2-Methylpropanenitrile					-13.8				23.4			
C_4H_7NO	Acetone cyanohydrin					-120.9							
C_4H_7NO	2-Pyrrolidone					-286.2							
C_4H_7NO	2-Methyl-2-oxazoline					-169.5				-130.5			
$C_4H_7NO_4$	Iminodiacetic acid	-932.6											
$C_4H_7NO_4$	Ethyl nitroacetate					-487.1							
$C_4H_7NO_4$	L-Aspartic acid	-973.3											
$C_4H_7N_3O$	Creatinine	-238.5											
C_4H_8	1-Butene					-20.8		227.0	118.0	0.1			
C_4H_8	cis-2-Butene					-29.8		219.9	127.0	-7.1			
C_4H_8	trans-2-Butene					-33.3				-11.4			
C_4H_8	Isobutene					-37.5				-16.9			
C_4H_8	Cyclobutane					3.7				27.7			
C_4H_8	Methylcyclopropane					1.7							
$C_4H_8Br_2$	1,2-Dibromobutane					-142.1				-91.6			
$C_4H_8Br_2$	1,3-Dibromobutane					-148.0							
$C_4H_8Br_2$	1,4-Dibromobutane					-140.3				-87.8			
$C_4H_8Br_2$	2,3-Dibromobutane					-139.6				-102.0			
$C_4H_8Br_2$	1,2-Dibromo-2-methylpropane					-156.6				-113.3			
$C_4H_8Cl_2$	1,3-Dichlorobutane					-237.3				-195.0			
$C_4H_8Cl_2$	1,4-Dichlorobutane					-229.8				-183.4			
$C_4H_8Cl_2O$	Bis(2-chloroethyl) ether								220.9				
$C_4H_8I_2$	1,4-Diiodobutane					-30.0							
$C_4H_8N_2O_2$	Succinamide	-581.2											
$C_4H_8N_2O_2$	Dimethylglyoxime	-199.7											
$C_4H_8N_2O_3$	L-Asparagine	-789.4											
$C_4H_8N_2O_3$	N-Glycylglycine	-747.7											
$C_4H_8N_2O_4$	1,4-Dinitrobutane					-237.5							
$C_4H_8N_8O_8$	Cyclotetramethylenetetranitramine									187.9		568.8	275.5
C_4H_8O	Ethyl vinyl ether					-167.4				-140.8			
C_4H_8O	1,2-Epoxybutane					-168.9		230.9	147.0				
C_4H_8O	Butanal					-239.2		246.6	163.7	-204.8		343.7	103.4
C_4H_8O	Isobutanal					-247.3				-215.7			
C_4H_8O	2-Butanone					-273.3		239.1	158.7	-238.5		339.9	101.7
C_4H_8O	Tetrahydrofuran					-216.2		204.3	124.0	-184.1		302.4	76.3
C_4H_8OS	S-Ethyl thioacetate					-268.2				-228.1			
$C_4H_8O_2$	Butanoic acid					-533.8		222.2	178.6	-475.9			
$C_4H_8O_2$	2-Methylpropanoic acid								173.0				
$C_4H_8O_2$	Propyl formate					-500.3				-462.7			
$C_4H_8O_2$	Ethyl acetate					-479.3		257.7	170.7	-443.6			
$C_4H_8O_2$	Methyl propanoate								171.2				
$C_4H_8O_2$	1,3-Dioxane					-379.7			143.9	-340.6			
$C_4H_8O_2$	1,4-Dioxane					-353.9		270.2	152.1	-315.3			
$C_4H_8O_2$	2-Methyl-1,3-dioxolane					-386.9				-352.0			
$C_4H_8O_2S$	Sulfolane								180.0				
C_4H_8S	Tetrahydrothiophene					-72.9				-34.1	45.8	309.6	92.5
$C_4H_8S_2$	1,3-Dithiane									-10.0	72.4	333.5	110.4
$C_4H_8S_2$	1,4-Dithiane									0.0	84.5	326.2	109.7
C_4H_9Br	1-Bromobutane					-143.8				-107.1			
C_4H_9Br	2-Bromobutane, (±)					-154.9				-120.3			
C_4H_9Br	2-Bromo-2-methylpropane					-164.4				-132.4			
C_4H_9Cl	1-Chlorobutane					-188.1				-154.4			
C_4H_9Cl	2-Chlorobutane					-192.8				-161.1			
C_4H_9Cl	1-Chloro-2-methylpropane					-191.1				-159.3			
C_4H_9Cl	2-Chloro-2-methylpropane					-211.3				-182.2			
C_4H_9ClO	2-Chloroethyl ethyl ether					-335.6				-301.3			
C_4H_9I	1-Iodo-2-methylpropane								162.3				
C_4H_9I	2-Iodo-2-methylpropane					-107.5				-72.1			
C_4H_9N	Cyclobutanamine					5.6				41.2			
C_4H_9N	Pyrrolidine					-41.1		204.1	156.6	-3.6			
C_4H_9NO	Butanamide	-364.8								-282.0			
C_4H_9NO	N-Methylpropanamide								179.0				

Molecular formula	Name	Crystal				Liquid				Gas			
		$\Delta_f H°$ kJ/mol	$\Delta_f G°$ kJ/mol	$S°$ J/mol K	C_p J/mol K	$\Delta_f H°$ kJ/mol	$\Delta_f G°$ kJ/mol	$S°$ J/mol K	C_p J/mol K	$\Delta_f H°$ kJ/mol	$\Delta_f G°$ kJ/mol	$S°$ J/mol K	C_p J/mol K
C_4H_9NO	2-Methylpropanamide	-368.6								-282.6			
C_4H_9NO	N,N-Dimethylacetamide					-278.3			175.6	-228.0			
C_4H_9NO	Morpholine								164.8				
$C_4H_9NO_2$	1-Nitrobutane					-192.5				-143.9		369.9	115.1
$C_4H_9NO_2$	2-Nitroisobutane					-217.2				-177.1			
$C_4H_9NO_2$	Propyl carbamate	-552.6								-471.4			
$C_4H_9NO_2$	4-Aminobutanoic acid	-581.0								-441.0			
$C_4H_9NO_3$	3-Nitro-2-butanol					-390.0							
$C_4H_9NO_3$	2-Methyl-2-nitro-1-propanol	-410.1											
$C_4H_9NO_3$	DL-Threonine	-758.8											
$C_4H_9NO_3$	L-Threonine	-807.2											
$C_4H_9N_3O_2$	Creatine	-537.2											
C_4H_{10}	Butane					-147.3			140.9	-125.7			
C_4H_{10}	Isobutane					-154.2				-134.2			
$C_4H_{10}Hg$	Diethyl mercury					30.1			182.8	75.3			
$C_4H_{10}N_2$	Piperazine	-45.6											
$C_4H_{10}N_2O$	Trimethylurea	-330.5											
$C_4H_{10}N_2O_2$	N-Nitrodiethylamine					-106.2				-53.0			
$C_4H_{10}N_2O_4$	L-Asparagine, monohydrate	-1086.6											
$C_4H_{10}O$	1-Butanol					-327.3		225.8	177.2	-274.9			
$C_4H_{10}O$	2-Butanol					-342.6		214.9	196.9	-292.8		359.5	112.7
$C_4H_{10}O$	2-Methyl-1-propanol					-334.7		214.7	181.5	-283.8			
$C_4H_{10}O$	2-Methyl-2-propanol					-359.2		193.3	218.6	-312.5		326.7	113.6
$C_4H_{10}O$	Diethyl ether					-279.5		253.5	172.5	-252.1		342.7	119.5
$C_4H_{10}O$	Methyl propyl ether					-266.0		262.9	165.4	-238.1			
$C_4H_{10}O$	Isopropyl methyl ether					-278.8		253.8	161.9	-252.0			
$C_4H_{10}OS$	Diethyl sulfoxide					-268.0				-205.6			
$C_4H_{10}O_2$	1,2-Butanediol, (±)					-523.6							
$C_4H_{10}O_2$	1,3-Butanediol					-501.0				-433.2			
$C_4H_{10}O_2$	1,4-Butanediol					-505.3		223.4	200.1	-428.7			
$C_4H_{10}O_2$	2,3-Butanediol					-541.5			213.0	-482.3			
$C_4H_{10}O_2$	2-Methyl-1,2-propanediol					-539.7							
$C_4H_{10}O_2$	Ethylene glycol monoethyl ether								210.8				
$C_4H_{10}O_2$	Ethylene glycol dimethyl ether					-376.6			193.3				
$C_4H_{10}O_2$	Dimethylacetal					-420.6				-389.7			
$C_4H_{10}O_2$	tert-Butyl hydroperoxide					-293.6				-245.9			
$C_4H_{10}O_3$	Diethylene glycol					-628.5			244.8	-571.2			
$C_4H_{10}O_3S$	Diethyl sulfite					-600.7				-552.2			
$C_4H_{10}O_4S$	Diethyl sulfate					-813.2				-756.3			
$C_4H_{10}S$	1-Butanethiol					-124.7			171.2	-88.0			
$C_4H_{10}S$	2-Butanethiol					-131.0				-96.9			
$C_4H_{10}S$	2-Methyl-1-propanethiol					-132.0				-97.3			
$C_4H_{10}S$	2-Methyl-2-propanethiol					-140.5				-109.6			
$C_4H_{10}S$	Diethyl sulfide					-119.4		269.3	171.4	-83.5		368.1	117.0
$C_4H_{10}S$	Methyl propyl sulfide					-118.5		272.5	171.6	-82.2			
$C_4H_{10}S$	Isopropyl methyl sulfide					-124.7		263.1	172.4	-90.5			
$C_4H_{10}S_2$	1,4-Butanedithiol					-105.7				-50.6			
$C_4H_{10}S_2$	Diethyl disulfide					-120.1		269.3	171.4	-79.4			
$C_4H_{11}N$	Butylamine					-127.6			179.2	-91.9			
$C_4H_{11}N$	sec-Butylamine					-137.5				-104.6			
$C_4H_{11}N$	tert-Butylamine					-150.6			192.1	-121.0			
$C_4H_{11}N$	Isobutylamine					-132.6			183.2	-98.7			
$C_4H_{11}N$	Diethylamine					-103.7			169.2	-72.2			
$C_4H_{11}NO$	N,N-Dimethylethanolamine					-253.7				-203.6			
$C_4H_{11}NO_2$	Diethanolamine	-493.8		233.5						-397.1			
$C_4H_{11}NO_3$	Tris(hydroxymethyl)methylamine	-717.8											
$C_4H_{12}BrN$	Tetramethylammonium bromide	-251.0											
$C_4H_{12}ClN$	Diethylamine hydrochloride	-358.6											
$C_4H_{12}ClN$	Tetramethylammonium chloride	-276.4											
$C_4H_{12}IN$	Tetramethylammonium iodide	-203.9											
$C_4H_{12}N_2$	2-Methyl-1,2-propanediamine					-133.9				-90.3			
$C_4H_{12}Pb$	Tetramethyl lead					97.9				135.9			
$C_4H_{12}Si$	Tetramethylsilane					-264.0	-100.0	277.3	204.1	-239.1	-99.9	359.0	143.9
$C_4H_{12}Sn$	Tetramethylstannane					-52.3				-18.8			
$C_4H_{13}N_3$	Bis(2-aminoethyl)amine								254.0				
C_4N_2	2-Butynedinitrile					500.4				529.2			

Standard Thermodynamic Properties of Chemical Substances

Molecular formula	Name	Crystal				Liquid				Gas			
		$\Delta_f H°$ kJ/mol	$\Delta_f G°$ kJ/mol	$S°$ J/mol K	C_p J/mol K	$\Delta_f H°$ kJ/mol	$\Delta_f G°$ kJ/mol	$S°$ J/mol K	C_p J/mol K	$\Delta_f H°$ kJ/mol	$\Delta_f G°$ kJ/mol	$S°$ J/mol K	C_p J/mol K
C_4NiO_4	Nickel carbonyl					-633.0	-588.2	313.4	204.6	-602.9	-587.2	410.6	145.2
C_5FeO_5	Iron pentacarbonyl					-774.0	-705.3	338.1	240.6				
$C_5H_2F_6O_2$	Hexafluoroacetylacetone	-2286.7											
$C_5H_3NO_5$	5-Nitro-2-furancarboxylic acid	-516.8											
$C_5H_4N_4$	1H-Purine	169.4											
$C_5H_4N_4O$	Hypoxanthine	-110.8		145.6	134.5								
$C_5H_4N_4O_2$	Xanthine	-379.6		161.1	151.3								
$C_5H_4N_4O_3$	Uric acid	-618.8		173.2	166.1								
$C_5H_4O_2$	Furfural					-201.6			163.2	-151.0			
$C_5H_4O_3$	2-Furancarboxylic acid	-498.4								-390.0			
$C_5H_4O_3$	3-Methyl-2,5-furandione					-504.5				-447.2			
$C_5H_5F_3O_2$	1,1,1-Trifluoro-2,4-pentanedione					-1040.2				-993.3			
C_5H_5N	Pyridine					100.2			132.7	140.4			
C_5H_5NO	1H-Pyrrole-2-carboxaldehyde	-106.4											
$C_5H_5N_5$	Adenine	96.9		147.0						205.7			
$C_5H_5N_5O$	Guanine	-183.9											
C_5H_6	cis-3-Penten-1-yne					226.5							
C_5H_6	trans-3-Penten-1-yne					228.2							
C_5H_6	1,3-Cyclopentadiene					105.9				134.3			
$C_5H_6N_2O_2$	Thymine	-462.8		150.8						-328.7			
$C_5H_6O_2$	Furfuryl alcohol					-276.2			204.0	-211.8			
$C_5H_6O_4$	trans-1-Propene-1,2-dicarboxylic acid	-824.4											
C_5H_6S	2-Methylthiophene					44.6	218.5	149.8		83.5			
C_5H_6S	3-Methylthiophene					43.1				82.5			
C_5H_7N	trans-3-Pentenenitrile					80.9				125.7			
C_5H_7N	Cyclobutanecarbonitrile					103.0				147.4			
C_5H_7N	1-Methylpyrrole					62.4				103.1			
C_5H_7N	2-Methylpyrrole					23.3				74.0			
C_5H_7N	3-Methylpyrrole					20.5				70.2			
$C_5H_7NO_2$	Ethyl cyanoacetate								220.2				
C_5H_8	1,2-Pentadiene									140.7			
C_5H_8	cis-1,3-Pentadiene									81.4			
C_5H_8	trans-1,3-Pentadiene									76.1			
C_5H_8	1,4-Pentadiene									105.7			
C_5H_8	2,3-Pentadiene									133.1			
C_5H_8	3-Methyl-1,2-butadiene					101.2							
C_5H_8	2-Methyl-1,3-butadiene					48.2	229.3	152.6		75.5			
C_5H_8	Cyclopentene					4.3	201.2	122.4		34.0			
C_5H_8	Spiropentane					157.5	193.7	134.5		185.2			
C_5H_8	Methylenecyclobutane					93.8				121.6			
$C_5H_8N_4O_{12}$	Pentaerythritol tetranitrate	-538.6								-387.0		614.7	294.8
C_5H_8O	Cyclopentanone					-235.9				-192.1			
$C_5H_8O_2$	4-Pentenoic acid					-430.6							
$C_5H_8O_2$	Allyl acetate								184.1				
$C_5H_8O_2$	Ethyl acrylate					-370.6				-354.2			
$C_5H_8O_2$	Methyl trans-2-butenoate					-382.9				-341.9			
$C_5H_8O_2$	Methyl methacrylate								191.2				
$C_5H_8O_2$	2,4-Pentanedione					-423.8				-382.0			
$C_5H_8O_2$	Dihydro-4-methyl-2(3H)-furanone					-461.3				-406.5			
$C_5H_8O_2$	Tetrahydro-2H-pyran-2-one					-436.7				-379.6			
$C_5H_8O_3$	Methyl acetoacetate					-623.2							
$C_5H_8O_4$	Glutaric acid	-960.0											
$C_5H_9ClO_2$	Propyl chloroacetate					-515.5				-467.0			
C_5H_9N	Pentanenitrile					-33.1				10.5			
C_5H_9N	2,2-Dimethylpropanenitrile					-39.8	232.0	179.4		-2.3			
C_5H_9N	1,2,5,6-Tetrahydropyridine					33.5							
C_5H_9NO	2-Piperidinone	-306.6											
C_5H_9NO	N-Methyl-2-pyrrolidone					-262.2			307.8				
$C_5H_9NO_2$	L-Proline	-515.2								-366.2			
$C_5H_9NO_4$	D-Glutamic acid	-1005.3											
$C_5H_9NO_4$	L-Glutamic acid	-1009.7											
C_5H_{10}	1-Pentene					-46.9	262.6	154.0		-21.1			
C_5H_{10}	cis-2-Pentene					-53.7	258.6	151.7		-27.6			
C_5H_{10}	trans-2-Pentene					-58.2	256.5	157.0		-31.9			
C_5H_{10}	2-Methyl-1-butene					-61.1	254.0	157.2		-35.2			
C_5H_{10}	3-Methyl-1-butene					-51.5	253.3	156.1		-27.5			

Molecular formula	Name	Crystal $\Delta_f H°$ kJ/mol	$\Delta_f G°$ kJ/mol	$S°$ J/mol K	C_p J/mol K	Liquid $\Delta_f H°$ kJ/mol	$\Delta_f G°$ kJ/mol	$S°$ J/mol K	C_p J/mol K	Gas $\Delta_f H°$ kJ/mol	$\Delta_f G°$ kJ/mol	$S°$ J/mol K	C_p J/mol K
C_5H_{10}	2-Methyl-2-butene					-68.6		251.0	152.8	-41.7			
C_5H_{10}	Cyclopentane					-105.1		204.5	128.8	-76.4			
C_5H_{10}	Methylcyclobutane					-44.5							
C_5H_{10}	Ethylcyclopropane					-24.8							
C_5H_{10}	1,1-Dimethylcyclopropane					-33.3				-8.2			
C_5H_{10}	cis-1,2-Dimethylcyclopropane					-26.3							
C_5H_{10}	trans-1,2-Dimethylcyclopropane					-30.7							
$C_5H_{10}Br_2$	2,3-Dibromo-2-methylbutane									-137.6			
$C_5H_{10}N_2O$	N-Nitrosopiperidine					-31.1				16.6			
$C_5H_{10}N_2O_2$	N-Nitropiperidine					-93.0				-44.5			
$C_5H_{10}N_2O_3$	L-Glutamine	-826.4											
$C_5H_{10}O$	Cyclopentanol					-300.1		204.1	182.5	-242.5		362.9	
$C_5H_{10}O$	Pentanal					-267.2				-228.4			
$C_5H_{10}O$	2-Pentanone					-297.3			184.1	-258.8			
$C_5H_{10}O$	3-Pentanone					-296.5		266.0	190.9	-257.9			
$C_5H_{10}O$	3-Methyl-2-butanone					-299.5		268.5	179.9	-262.6			
$C_5H_{10}O$	3,3-Dimethyloxetane					-182.2				-148.2			
$C_5H_{10}O$	Tetrahydropyran					-258.3				-223.4			
$C_5H_{10}OS$	S-Propyl thioacetate					-294.5				-250.4			
$C_5H_{10}O_2$	Pentanoic acid					-559.4		259.8	210.3	-491.9			
$C_5H_{10}O_2$	2-Methylbutanoic acid					-554.5							
$C_5H_{10}O_2$	3-Methylbutanoic acid					-561.6				-510.0			
$C_5H_{10}O_2$	2,2-Dimethylpropanoic acid	-564.5								-491.3			
$C_5H_{10}O_2$	Butyl formate								200.2				
$C_5H_{10}O_2$	Propyl acetate								196.2				
$C_5H_{10}O_2$	Isopropyl acetate					-518.9			199.4	-481.6			
$C_5H_{10}O_2$	Ethyl propanoate					-502.7				-463.4			
$C_5H_{10}O_2$	Methyl butanoate								198.2				
$C_5H_{10}O_2$	(Ethoxymethyl)oxirane					-296.5							
$C_5H_{10}O_2$	4-Methyl-1,3-dioxane					-416.1				-376.9			
$C_5H_{10}O_2$	cis-1,2-Cyclopentanediol	-485.0											
$C_5H_{10}O_2$	trans-1,2-Cyclopentanediol	-490.1											
$C_5H_{10}O_2$	Tetrahydrofurfuryl alcohol					-435.7				-369.1			
$C_5H_{10}O_3$	Diethyl carbonate					-681.5				-637.9			
$C_5H_{10}O_3$	Ethylene glycol monomethyl ether acetate								310.0				
$C_5H_{10}O_3$	Ethyl lactate								254.0				
$C_5H_{10}O_4$	Glycerol 1-acetate, (DL)					-909.2							
$C_5H_{10}O_5$	D-Ribose	-1047.2											
$C_5H_{10}O_5$	D-Xylose	-1057.8											
$C_5H_{10}O_5$	α-D-Arabinopyranose	-1057.9											
$C_5H_{10}S$	Thiacyclohexane					-106.3		218.2	163.3	-63.5	53.1	323.0	109.7
$C_5H_{10}S$	Cyclopentanethiol					-89.5		256.9	165.2	-48.1			
$C_5H_{11}Br$	1-Bromopentane					-170.2				-128.9			
$C_5H_{11}Cl$	1-Chloropentane					-213.2				-174.9			
$C_5H_{11}Cl$	1-Chloro-3-methylbutane					-216.0				-179.7			
$C_5H_{11}Cl$	2-Chloro-2-methylbutane					-235.7				-202.2			
$C_5H_{11}Cl$	2-Chloro-3-methylbutane					-226.6				-185.1			
$C_5H_{11}N$	Cyclopentylamine					-95.1		241.0	181.2	-54.9			
$C_5H_{11}N$	Piperidine					-86.4		210.0	179.9	-47.1			
$C_5H_{11}NO$	Pentanamide	-379.5								-290.2			
$C_5H_{11}NO$	2,2-Dimethylpropanamide	-399.7								-313.1			
$C_5H_{11}NO_2$	1-Nitropentane					-215.4				-164.4		390.9	137.1
$C_5H_{11}NO_2$	DL-Valine	-628.9											
$C_5H_{11}NO_2$	L-Valine	-617.9								-455.1			
$C_5H_{11}NO_2$	5-Aminopentanoic acid	-604.1								-460.0			
$C_5H_{11}NO_2S$	L-Methionine	-577.5								-413.5			
$C_5H_{11}NO_4$	2-Ethyl-2-nitro-1,3-propanediol	-606.4											
C_5H_{12}	Pentane					-173.5			167.2	-146.9			
C_5H_{12}	Isopentane					-178.4		260.4	164.8	-153.6			
C_5H_{12}	Neopentane					-190.2				-168.0			
$C_5H_{12}N_2O$	Butylurea	-419.5											
$C_5H_{12}N_2O$	tert-Butylurea	-417.4											
$C_5H_{12}N_2O$	N,N-Diethylurea	-372.2											
$C_5H_{12}N_2O$	Tetramethylurea					-262.2							
$C_5H_{12}N_2S$	Tetramethylthiourea	-38.1								44.9			
$C_5H_{12}O$	1-Pentanol					-351.6			208.1	-294.6			

Molecular formula	Name	Crystal				Liquid				Gas			
		$\Delta_f H°$ kJ/mol	$\Delta_f G°$ kJ/mol	$S°$ J/mol K	C_p J/mol K	$\Delta_f H°$ kJ/mol	$\Delta_f G°$ kJ/mol	$S°$ J/mol K	C_p J/mol K	$\Delta_f H°$ kJ/mol	$\Delta_f G°$ kJ/mol	$S°$ J/mol K	C_p J/mol K
$C_5H_{12}O$	2-Pentanol					-365.2				-311.0			
$C_5H_{12}O$	3-Pentanol					-368.9			239.7	-314.9			
$C_5H_{12}O$	2-Methyl-1-butanol, (±)					-356.6				-301.4			
$C_5H_{12}O$	3-Methyl-1-butanol					-356.4				-300.7			
$C_5H_{12}O$	2-Methyl-2-butanol					-379.5			247.1	-329.3			
$C_5H_{12}O$	3-Methyl-2-butanol, (±)					-366.6				-313.5			
$C_5H_{12}O$	2,2-Dimethyl-1-propanol					-399.4							
$C_5H_{12}O$	Butyl methyl ether					-290.6		295.3	192.7	-258.1			
$C_5H_{12}O$	Methyl tert-butyl ether					-313.6		265.3	187.5	-283.7			
$C_5H_{12}O$	Ethyl propyl ether					-303.6		295.0	197.2	-272.0			
$C_5H_{12}O_2$	1,5-Pentanediol					-528.8				-450.8			
$C_5H_{12}O_2$	2,2-Dimethyl-1,3-propanediol	-551.2											
$C_5H_{12}O_2$	Diethoxymethane					-450.5				-414.7			
$C_5H_{12}O_2$	1,1-Dimethoxypropane					-443.6							
$C_5H_{12}O_2$	2,2-Dimethoxypropane					-459.4				-429.9			
$C_5H_{12}O_3$	Diethylene glycol monomethyl ether								271.1				
$C_5H_{12}O_3$	2-(Hydroxymethyl)-2-methyl-1,3-propanediol	-744.6											
$C_5H_{12}O_4$	Pentaerythritol	-920.6								-776.7			
$C_5H_{12}O_5$	Xylitol	-1118.5											
$C_5H_{12}S$	1-Pentanethiol					-151.3				-110.0			
$C_5H_{12}S$	2-Methyl-1-butanethiol, (+)					-154.4				-114.9			
$C_5H_{12}S$	3-Methyl-1-butanethiol					-154.4				-114.9			
$C_5H_{12}S$	2-Methyl-2-butanethiol					-162.8		290.1	198.1	-127.1			
$C_5H_{12}S$	3-Methyl-2-butanethiol					-158.8				-121.3			
$C_5H_{12}S$	2,2-Dimethyl-1-propanethiol					-165.4				-129.0			
$C_5H_{12}S$	Butyl methyl sulfide					-142.9		307.5	200.9	-102.4			
$C_5H_{12}S$	tert-Butyl methyl sulfide					-157.1		276.1	199.9	-121.3			
$C_5H_{12}S$	Ethyl propyl sulfide					-144.8		309.5	198.4	-104.8			
$C_5H_{12}S$	Ethyl isopropyl sulfide					-156.1				-118.3			
$C_5H_{13}N$	Pentylamine							218.0					
$C_5H_{14}N_2$	N,N,N',N'-Tetramethylmethanediamine					-51.1				-18.2			
C_6ClF_5	Chloropentafluorobenzene	-858.4								-809.3			
C_6Cl_6	Hexachlorobenzene	-127.6		260.2	201.2					-35.5			
C_6F_6	Hexafluorobenzene					-991.3		280.8	221.6	-955.4			
C_6F_{10}	Perfluorocyclohexene					-1963.5				-1932.7			
C_6F_{12}	Perfluorocyclohexane					-2406.3				-2370.4			
C_6HCl_5O	Pentachlorophenol	-292.5		253.2	202.0								
C_6HF_5	Pentafluorobenzene	-852.7								-806.5			
C_6HF_5O	Pentafluorophenol	-1024.1				-1007.7							
$C_6H_2F_4$	1,2,4,5-Tetrafluorobenzene					-683.8							
$C_6H_3Cl_3$	1,2,3-Trichlorobenzene	-70.8								3.8			
$C_6H_3Cl_3$	1,2,4-Trichlorobenzene					-63.1				-8.1			
$C_6H_3Cl_3$	1,3,5-Trichlorobenzene	-78.4								-13.4			
$C_6H_3N_3O_6$	1,3,5-Trinitrobenzene	-37.0			214.6								
$C_6H_3N_3O_7$	2,4,6-Trinitrophenol	-217.9			239.7								
$C_6H_3N_3O_8$	2,4,6-Trinitro-1,3-benzenediol	-467.5											
$C_6H_4ClNO_2$	1-Chloro-4-nitrobenzene	-48.7			250.2								
$C_6H_4Cl_2$	o-Dichlorobenzene					-17.5			162.4	30.2			
$C_6H_4Cl_2$	m-Dichlorobenzene					-20.7				25.7			
$C_6H_4Cl_2$	p-Dichlorobenzene	-42.3		175.4	147.8					22.5			
$C_6H_4Cl_2O$	2,4-Dichlorophenol	-226.4								-156.3			
$C_6H_4F_2$	o-Difluorobenzene					-330.0		222.6	159.0	-293.8			
$C_6H_4F_2$	m-Difluorobenzene					-343.9		223.8	159.1	-309.2			
$C_6H_4F_2$	p-Difluorobenzene					-342.3			157.5	-306.7			
$C_6H_4N_2O_4$	1,2-Dinitrobenzene	-2.0			200.4								
$C_6H_4N_2O_4$	1,3-Dinitrobenzene	-27.0			197.5	-36.0							
$C_6H_4N_2O_4$	1,4-Dinitrobenzene	-38.0			200.0								
$C_6H_4N_2O_5$	2,4-Dinitrophenol	-232.7								-128.1			
$C_6H_4O_2$	p-Benzoquinone	-185.7			129.0					-122.9			
C_6H_5Br	Bromobenzene					60.9		219.2	154.3				
C_6H_5Cl	Chlorobenzene					11.1			150.1	52.0			
C_6H_5ClO	2-Chlorophenol							188.7					
C_6H_5ClO	3-Chlorophenol	-206.4				-189.3							
C_6H_5ClO	4-Chlorophenol	-197.7				-181.3							
$C_6H_5Cl_2N$	3,4-Dichloroaniline	-89.1											
C_6H_5F	Fluorobenzene					-150.6		205.9	146.4	-115.9			

Molecular formula	Name	Crystal				Liquid				Gas			
		$\Delta_f H°$ kJ/mol	$\Delta_f G°$ kJ/mol	$S°$ J/mol K	C_p J/mol K	$\Delta_f H°$ kJ/mol	$\Delta_f G°$ kJ/mol	$S°$ J/mol K	C_p J/mol K	$\Delta_f H°$ kJ/mol	$\Delta_f G°$ kJ/mol	$S°$ J/mol K	C_p J/mol K
C_6H_5I	Iodobenzene					117.2		205.4	158.7	164.9			
$C_6H_5NO_2$	Nitrobenzene					12.5			185.8	68.5		348.8	120.4
$C_6H_5NO_2$	3-Pyridinecarboxylic acid	-344.9								-221.5			
$C_6H_5NO_3$	2-Nitrophenol	-202.4											
$C_6H_5N_3$	1H-Benzotriazole	236.5								335.5			
$C_6H_5N_3O_4$	2,3-Dinitroaniline	-11.7											
$C_6H_5N_3O_4$	2,4-Dinitroaniline	-67.8											
$C_6H_5N_3O_4$	2,5-Dinitroaniline	-44.3											
$C_6H_5N_3O_4$	2,6-Dinitroaniline	-50.6											
$C_6H_5N_3O_4$	3,5-Dinitroaniline	-38.9											
C_6H_6	1,5-Hexadiyne					384.2							
C_6H_6	Benzene					49.1	124.5	173.4	136.0	82.9	129.7	269.2	82.4
C_6H_6ClN	2-Chloroaniline					-4.6							
C_6H_6ClN	3-Chloroaniline					-20.3		198.7					
C_6H_6ClN	4-Chloroaniline	-33.3			147.3								
$C_6H_6N_2O_2$	2-Nitroaniline	-26.1			166.0	-9.4				63.8			
$C_6H_6N_2O_2$	3-Nitroaniline	-38.3			158.8	-14.4				58.4			
$C_6H_6N_2O_2$	4-Nitroaniline	-42.0			167.0	-20.7				58.8			
C_6H_6O	Phenol	-165.1		144.0	127.4					-96.4			
C_6H_6O	2-Vinylfuran					-10.3				27.8			
$C_6H_6O_2$	p-Hydroquinone	-364.5			136.0					-265.3			
$C_6H_6O_2$	Pyrocatechol	-354.1								-267.5			
$C_6H_6O_2$	Resorcinol	-368.0								-274.7			
$C_6H_6O_3$	1,2,3-Benzenetriol	-551.1								-434.2			
$C_6H_6O_3$	1,2,4-Benzenetriol	-563.8								-444.0			
$C_6H_6O_3$	1,3,5-Benzenetriol	-584.6								-452.9			
$C_6H_6O_3$	3,4-Dimethyl-2,5-furandione	-581.4											
$C_6H_6O_6$	cis-1-Propene-1,2,3-tricarboxylic acid	-1224.4											
$C_6H_6O_6$	trans-1-Propene-1,2,3-tricarboxylic acid	-1232.7											
C_6H_6S	Benzenethiol					63.7		222.8	173.2	111.3			
C_6H_7N	Aniline					31.6		191.9	87.5	-7.0	317.9	107.9	
C_6H_7N	2-Methylpyridine					56.7		158.6	99.2				
C_6H_7N	3-Methylpyridine					61.9	216.3	158.7	106.4				
C_6H_7N	4-Methylpyridine					59.2	209.1	159.0	103.8				
C_6H_7N	1-Cyclopentenecarbonitrile					111.5				156.5			
$C_6H_8N_2$	Adiponitrile					85.1		128.7		149.5			
$C_6H_8N_2$	1,2-Benzenediamine	-0.3											
$C_6H_8N_2$	1,3-Benzenediamine	-7.8		154.5	159.6								
$C_6H_8N_2$	1,4-Benzenediamine	3.0											
$C_6H_8N_2$	Phenylhydrazine					141.0		217.0		202.9			
$C_6H_8N_2S$	Bis(2-cyanoethyl) sulfide					96.3							
$C_6H_8O_4$	Dimethyl maleate							263.2					
$C_6H_8O_6$	L-Ascorbic acid	-1164.6											
$C_6H_8O_7$	Citric acid	-1543.8											
$C_6H_9Cl_3O_2$	Butyl trichloroacetate					-545.8				-492.3			
$C_6H_9Cl_3O_2$	Isobutyl trichloroacetate					-553.4				-500.2			
C_6H_9N	Cyclopentanecarbonitrile					0.7				44.1			
C_6H_9N	2,4-Dimethylpyrrole	-422.3											
C_6H_9N	2,5-Dimethylpyrrole					-16.7				39.8			
$C_6H_9NO_3$	Triacetamide					-610.5				-550.1			
$C_6H_9NO_6$	Nitrilotriacetic acid	-1311.9											
$C_6H_9N_3O_2$	L-Histidine	-466.7											
C_6H_{10}	1,5-Hexadiene					54.1				84.2			
C_6H_{10}	3,3-Dimethyl-1-butyne					78.4							
C_6H_{10}	Cyclohexene					-38.5	214.6	148.3		-5.0			
C_6H_{10}	1-Methylcyclopentene					-36.4				-3.8			
C_6H_{10}	3-Methylcyclopentene					-23.7				7.4			
C_6H_{10}	4-Methylcyclopentene					-17.6				14.6			
$C_6H_{10}Cl_2O_2$	Butyl dichloroacetate					-550.1				-497.8			
$C_6H_{10}O$	Cyclohexanone					-271.2		182.2		-226.1			
$C_6H_{10}O$	2-Methylcyclopentanone					-265.2							
$C_6H_{10}O$	Mesityl oxide							212.5					
$C_6H_{10}O_2$	Ethyl trans-2-butenoate					-420.0				-375.6			
$C_6H_{10}O_2$	Methyl cyclobutanecarboxylate					-395.0				-350.2			
$C_6H_{10}O_3$	Ethyl acetoacetate							248.0					
$C_6H_{10}O_3$	Propanoic anhydride					-679.1				-626.5			

Standard Thermodynamic Properties of Chemical Substances

Molecular formula	Name	Crystal				Liquid				Gas			
		$\Delta_f H°$ kJ/mol	$\Delta_f G°$ kJ/mol	$S°$ J/mol K	C_p J/mol K	$\Delta_f H°$ kJ/mol	$\Delta_f G°$ kJ/mol	$S°$ J/mol K	C_p J/mol K	$\Delta_f H°$ kJ/mol	$\Delta_f G°$ kJ/mol	$S°$ J/mol K	C_p J/mol K
$C_6H_{10}O_4$	Adipic acid	-994.3											
$C_6H_{10}O_4$	Diethyl oxalate					-805.5				-742.0			
$C_6H_{10}O_4$	Ethylene glycol diacetate								310.0				
$C_6H_{11}Cl$	Chlorocyclohexane					-207.2				-163.7			
$C_6H_{11}ClO_2$	Ethyl 4-chlorobutanoate					-566.5				-513.8			
$C_6H_{11}ClO_2$	Propyl 3-chloropropanoate					-537.6				-485.7			
$C_6H_{11}ClO_2$	Butyl chloroacetate					-538.4				-487.4			
$C_6H_{11}NO$	Caprolactam	-329.4			156.8					-239.6			
$C_6H_{11}NO$	1-Methyl-2-piperidinone					-293.0							
C_6H_{12}	1-Hexene					-74.2		295.2	183.3	-43.5			
C_6H_{12}	cis-2-Hexene					-83.9				-52.3			
C_6H_{12}	trans-2-Hexene					-85.5				-53.9			
C_6H_{12}	cis-3-Hexene					-78.9				-47.6			
C_6H_{12}	trans-3-Hexene					-86.1				-54.4			
C_6H_{12}	2-Methyl-1-pentene					-90.0				-59.4			
C_6H_{12}	3-Methyl-1-pentene					-78.2				-49.5			
C_6H_{12}	4-Methyl-1-pentene					-80.0				-51.3			
C_6H_{12}	2-Methyl-2-pentene					-98.5				-66.9			
C_6H_{12}	3-Methyl-cis-2-pentene					-94.5				-62.3			
C_6H_{12}	3-Methyl-trans-2-pentene					-94.6				-63.1			
C_6H_{12}	4-Methyl-cis-2-pentene					-87.0				-57.5			
C_6H_{12}	4-Methyl-trans-2-pentene					-91.6				-61.5			
C_6H_{12}	2-Ethyl-1-butene					-87.1				-56.0			
C_6H_{12}	2,3-Dimethyl-1-butene					-93.2				-62.4			
C_6H_{12}	3,3-Dimethyl-1-butene					-87.5				-60.3			
C_6H_{12}	2,3-Dimethyl-2-butene					-101.4		270.2	174.7	-68.1			
C_6H_{12}	Cyclohexane					-156.4			154.9	-123.4			
C_6H_{12}	Methylcyclopentane					-137.9				-106.2			
C_6H_{12}	Ethylcyclobutane					-59.0				-27.5			
C_6H_{12}	1,1,2-Trimethylcyclopropane					-96.2							
$C_6H_{12}N_2O_4S_2$	L-Cystine	-1032.7											
$C_6H_{12}N_2S_4$	Thiram	40.2			301.7								
$C_6H_{12}O$	Butyl vinyl ether					-218.8			232.0	-182.6			
$C_6H_{12}O$	Hexanal							280.3	210.4				
$C_6H_{12}O$	2-Hexanone					-322.0			213.3	-278.9			
$C_6H_{12}O$	3-Hexanone					-320.2		305.3	216.9	-277.6			
$C_6H_{12}O$	4-Methyl-2-pentanone								213.3				
$C_6H_{12}O$	2-Methyl-3-pentanone					-325.9				-286.0			
$C_6H_{12}O$	3,3-Dimethyl-2-butanone					-328.6				-290.6			
$C_6H_{12}O$	Cyclohexanol					-348.2			208.2	-286.2			
$C_6H_{12}O$	cis-2-Methylcyclopentanol					-345.5							
$C_6H_{12}O_2$	Hexanoic acid					-583.8				-511.9			
$C_6H_{12}O_2$	Butyl acetate					-529.2			227.8	-485.3			
$C_6H_{12}O_2$	tert-Butyl acetate					-554.5			231.0	-516.5			
$C_6H_{12}O_2$	Isobutyl acetate								233.8				
$C_6H_{12}O_2$	Ethyl butanoate								228.0				
$C_6H_{12}O_2$	Methyl pentanoate					-514.2			229.3	-471.1			
$C_6H_{12}O_2$	Methyl 2,2-dimethylpropanoate					-530.0			257.9	-491.2			
$C_6H_{12}O_2$	Diacetone alcohol								221.3				
$C_6H_{12}O_3$	Ethylene glycol monoethyl ether acetate								376.0				
$C_6H_{12}O_3$	Paraldehyde					-673.1				-631.7			
$C_6H_{12}O_6$	β-D-Fructose	-1265.6											
$C_6H_{12}O_6$	D-Galactose	-1286.3											
$C_6H_{12}O_6$	α-D-Glucose	-1273.3											
$C_6H_{12}O_6$	D-Mannose	-1263.0											
$C_6H_{12}O_6$	L-Sorbose	-1271.5											
$C_6H_{12}S$	Thiepane									-65.8	79.4	363.5	131.3
$C_6H_{12}S$	Cyclohexanethiol					-140.7		255.6	192.6	-96.2			
$C_6H_{12}S$	Cyclopentyl methyl sulfide					-109.8				-64.7			
$C_6H_{13}Br$	1-Bromohexane					-194.2		453.0	204.0	-148.3			
$C_6H_{13}Cl$	2-Chlorohexane					-246.1				-204.3			
$C_6H_{13}N$	Cyclohexylamine					-147.6				-104.0			
$C_6H_{13}N$	2-Methylpiperidine, (±)					-124.9				-84.4			
$C_6H_{13}NO$	Hexanamide	-423.0								-324.2			
$C_6H_{13}NO$	N-Butylacetamide					-380.9				-305.9			
$C_6H_{13}NO_2$	DL-Leucine	-640.6											

Molecular formula	Name	Crystal $\Delta_f H°$ kJ/mol	$\Delta_f G°$ kJ/mol	$S°$ J/mol K	C_p J/mol K	Liquid $\Delta_f H°$ kJ/mol	$\Delta_f G°$ kJ/mol	$S°$ J/mol K	C_p J/mol K	Gas $\Delta_f H°$ kJ/mol	$\Delta_f G°$ kJ/mol	$S°$ J/mol K	C_p J/mol K
$C_6H_{13}NO_2$	D-Leucine	-637.3											
$C_6H_{13}NO_2$	L-Leucine	-637.4		200.1						-486.8			
$C_6H_{13}NO_2$	DL-Isoleucine	-635.3											
$C_6H_{13}NO_2$	L-Isoleucine	-637.8											
$C_6H_{13}NO_2$	L-Norleucine	-639.1											
$C_6H_{13}NO_2$	6-Aminohexanoic acid	-637.3											
C_6H_{14}	Hexane					-198.7			195.6	-166.9			
C_6H_{14}	2-Methylpentane					-204.6		290.6	193.7	-174.6			
C_6H_{14}	3-Methylpentane					-202.4		292.5	190.7	-171.9			
C_6H_{14}	2,2-Dimethylbutane					-213.8		272.5	191.9	-185.9			
C_6H_{14}	2,3-Dimethylbutane					-207.4		287.8	189.7	-178.1			
$C_6H_{14}N_2$	Azopropane					11.5				51.3			
$C_6H_{14}N_2O_2$	DL-Lysine	-678.7											
$C_6H_{14}N_4O_2$	D-Arginine	-623.5		250.6	232.0								
$C_6H_{14}O$	1-Hexanol					-377.5		287.4	240.4	-315.9			
$C_6H_{14}O$	2-Hexanol					-392.0				-333.5			
$C_6H_{14}O$	3-Hexanol					-392.4			286.2				
$C_6H_{14}O$	2-Methyl-1-pentanol								248.0				
$C_6H_{14}O$	3-Methyl-2-pentanol								275.9				
$C_6H_{14}O$	4-Methyl-2-pentanol					-394.7			273.0				
$C_6H_{14}O$	2-Methyl-3-pentanol					-396.4							
$C_6H_{14}O$	3-Methyl-3-pentanol								293.4				
$C_6H_{14}O$	Dipropyl ether					-328.8		323.9	221.6	-293.0			
$C_6H_{14}O$	Diisopropyl ether					-351.5			216.8	-319.2			
$C_6H_{14}O$	Butyl ethyl ether								159.0				
$C_6H_{14}O$	tert-Butyl ethyl ether									-313.9			
$C_6H_{14}OS$	Dipropyl sulfoxide					-329.4				-254.9			
$C_6H_{14}O_2$	1,2-Hexanediol					-577.1				-490.1			
$C_6H_{14}O_2$	1,6-Hexanediol	-569.9				-548.6				-461.2			
$C_6H_{14}O_2$	2-Methyl-2,4-pentanediol								336.0				
$C_6H_{14}O_2$	Ethylene glycol monobutyl ether								281.0				
$C_6H_{14}O_2$	1,1-Diethoxyethane					-491.4				-453.5			
$C_6H_{14}O_2$	Ethylene glycol diethyl ether					-451.4			259.4	-408.1			
$C_6H_{14}O_3$	Diethylene glycol monoethyl ether								301.0				
$C_6H_{14}O_3$	Diethylene glycol dimethyl ether								274.1				
$C_6H_{14}O_3$	Trimethylolpropane	-750.9											
$C_6H_{14}O_4$	Triethylene glycol					-804.3				-725.0			
$C_6H_{14}O_4S$	Dipropyl sulfate					-859.0				-792.0			
$C_6H_{14}O_6$	Galactitol	-1317.0											
$C_6H_{14}O_6$	D-Mannitol	-1314.5											
$C_6H_{14}S$	1-Hexanethiol					-175.7				-129.9			
$C_6H_{14}S$	2-Methyl-2-pentanethiol					-188.3				-148.3			
$C_6H_{14}S$	2,3-Dimethyl-2-butanethiol					-187.1				-147.9			
$C_6H_{14}S$	Diisopropyl sulfide					-181.6		313.0	232.0	-142.0			
$C_6H_{14}S$	Butyl ethyl sulfide					-172.3				-127.8			
$C_6H_{14}S$	Methyl pentyl sulfide					-167.1				-121.8			
$C_6H_{14}S_2$	Dipropyl disulfide					-171.5				-118.3			
$C_6H_{15}B$	Triethylborane					-194.6	9.4	336.7	241.2	-157.7	16.1	437.8	
$C_6H_{15}N$	Dipropylamine					-156.1				-116.0			
$C_6H_{15}N$	Diisopropylamine					-178.5				-143.8			
$C_6H_{15}N$	Triethylamine					-127.7			219.9	-92.7			
$C_6H_{15}NO$	2-Diethylaminoethanol					-305.9							
$C_6H_{15}NO_3$	Triethanolamine	-664.2		389.0						-558.3			
$C_6H_{16}N_2$	1,6-Hexanediamine	-205.0											
$C_6H_{18}N_3OP$	Hexamethylphosphoric triamide								321.0				
$C_6H_{18}OSi_2$	Hexamethyldisiloxane					-815.0	-541.5	433.8	311.4	-777.7	-534.5	535.0	238.5
C_6MoO_6	Molybdenum hexacarbonyl	-982.8	-877.7	325.9	242.3					-912.1	-856.0	490.0	205.0
C_6N_4	Tetracyanoethene	623.8								705.0			
C_7F_8	Perfluorotoluene					-1311.1		355.5	262.3				
C_7F_{14}	Perfluoromethylcyclohexane					-2931.1			353.1	-2897.2			
C_7F_{16}	Perfluoroheptane					-3420.0		561.8	419.0	-3383.6			
$C_7H_3F_5$	2,3,4,5,6-Pentafluorotoluene					-883.8		306.4	225.8	-842.7			
$C_7H_4Cl_2O$	3-Chlorobenzoyl chloride					-189.7							
$C_7H_4N_2O_6$	3,5-Dinitrobenzoic acid	-409.8											
C_7H_5ClO	Benzoyl chloride					-158.0				-103.2			
$C_7H_5ClO_2$	2-Chlorobenzoic acid	-404.5								-325.0			

Molecular formula	Name	Crystal				Liquid				Gas			
		$\Delta_f H°$ kJ/mol	$\Delta_f G°$ kJ/mol	$S°$ J/mol K	C_p J/mol K	$\Delta_f H°$ kJ/mol	$\Delta_f G°$ kJ/mol	$S°$ J/mol K	C_p J/mol K	$\Delta_f H°$ kJ/mol	$\Delta_f G°$ kJ/mol	$S°$ J/mol K	C_p J/mol K
$C_7H_5ClO_2$	3-Chlorobenzoic acid	-424.3								-342.3			
$C_7H_5ClO_2$	4-Chlorobenzoic acid	-428.9		163.2						-341.0			
$C_7H_5F_3$	(Trifluoromethyl)benzene								188.4				
C_7H_5N	Benzonitrile					163.2		209.1	165.2	215.7			
C_7H_5NO	Benzoxazole	-24.2								44.8			
$C_7H_5NO_4$	2-Nitrobenzoic acid	-378.8											
$C_7H_5NO_4$	3-Nitrobenzoic acid	-394.7											
$C_7H_5NO_4$	4-Nitrobenzoic acid	-392.2											
$C_7H_5N_3O_6$	2,4,6-Trinitrotoluene	-63.2		243.3									
$C_7H_6N_2$	1H-Benzimidazole	79.5								181.7			
$C_7H_6N_2$	1H-Indazole	151.9								243.0			
$C_7H_6N_2O_4$	1-Methyl-2,4-dinitrobenzene	-66.4								33.2			
C_7H_6O	Benzaldehyde					-87.0		221.2	172.0	-36.7			
$C_7H_6O_2$	Benzoic acid	-385.2		167.6	146.8					-294.0			
$C_7H_6O_2$	Salicylaldehyde								222.0				
$C_7H_6O_2$	3-(2-Furanyl)-2-propenal	-182.0								-105.9			
$C_7H_6O_3$	2-Hydroxybenzoic acid	-589.9								-494.8			
C_7H_7Br	4-Bromotoluene					12.0							
C_7H_7Cl	2-Chlorotoluene								166.8				
C_7H_7Cl	(Chloromethyl)benzene					-32.5				18.9			
C_7H_7F	4-Fluorotoluene					-186.9			171.2	-147.4			
C_7H_7NO	Benzamide	-202.6								-100.9			
$C_7H_7NO_2$	Aniline-2-carboxylic acid	-401.1								-296.0			
$C_7H_7NO_2$	Aniline-3-carboxylic acid	-417.3								-283.6			
$C_7H_7NO_2$	Aniline-4-carboxylic acid	-410.0			177.8					-296.7			
$C_7H_7NO_2$	2-Nitrotoluene					-9.7							
$C_7H_7NO_2$	3-Nitrotoluene					-31.5							
$C_7H_7NO_2$	4-Nitrotoluene	-48.1		172.3						31.0			
$C_7H_7NO_2$	(Nitromethyl)benzene					-22.8				30.7			
$C_7H_7NO_2$	Salicylaldoxime	-183.7											
C_7H_8	Toluene					12.4			157.3	50.5			
$C_7H_8N_2O$	Phenylurea	-218.6											
C_7H_8O	o-Cresol	-204.6		165.4	154.6					-128.6			
C_7H_8O	m-Cresol					-194.0		212.6	224.9	-132.3			
C_7H_8O	p-Cresol	-199.3		167.3	150.2					-125.4			
C_7H_8O	Benzyl alcohol					-160.7		216.7	217.9	-100.4			
C_7H_8O	Anisole					-114.8				-67.9			
C_7H_9N	Benzylamine					34.2			207.2	94.4			
C_7H_9N	2-Methylaniline					-6.3				56.4	167.6	351.0	130.2
C_7H_9N	3-Methylaniline					-8.1				54.6	165.4	352.5	125.5
C_7H_9N	4-Methylaniline	-23.5								55.3	167.7	347.0	126.2
C_7H_9N	N-Methylaniline								207.1				
C_7H_9N	1-Cyclohexenecarbonitrile					48.1				101.6			
C_7H_9N	2,3-Dimethylpyridine					19.4		243.7	189.5	67.1			
C_7H_9N	2,4-Dimethylpyridine					16.1		248.5	184.8	63.6			
C_7H_9N	2,5-Dimethylpyridine					18.7		248.8	184.7	66.5			
C_7H_9N	2,6-Dimethylpyridine					12.7		244.2	185.2	58.1			
C_7H_9N	3,4-Dimethylpyridine					18.3		240.7	191.8	68.8			
C_7H_9N	3,5-Dimethylpyridine					22.5		241.7	184.5	72.0			
$C_7H_{10}O_2$	Ethyl 2-pentynoate					-301.8				-250.3			
$C_7H_{10}O_2$	Methyl 2-hexynoate					-242.7							
$C_7H_{11}Cl_3O_2$	Isopentyl trichloroacetate					-580.9				-523.1			
$C_7H_{11}N$	Cyclohexanecarbonitrile					-47.2				4.8			
C_7H_{12}	Bicyclo[2.2.1]heptane	-95.1		151.0						-54.8			
C_7H_{12}	1-Methylbicyclo(3,1,0)hexane					-33.2				1.7			
C_7H_{12}	Methylenecyclohexane					-61.3				-25.2			
C_7H_{12}	Vinylcyclopentane					-34.8							
C_7H_{12}	1-Ethylcyclopentene					-53.3				-19.8			
$C_7H_{12}O$	2-Methylenecyclohexanol					-277.6							
$C_7H_{12}O_2$	Butyl acrylate					-422.6			251.0	-375.3			
$C_7H_{12}O_4$	Diethyl malonate								285.0				
$C_7H_{13}ClO_2$	Butyl 2-chloropropanoate					-571.7				-517.3			
$C_7H_{13}ClO_2$	Isobutyl 2-chloropropanoate					-603.1				-549.6			
$C_7H_{13}ClO_2$	Butyl 3-chloropropanoate					-557.9				-502.3			
$C_7H_{13}ClO_2$	Isobutyl 3-chloropropanoate					-572.6				-517.3			
$C_7H_{13}ClO_2$	Propyl 2-chlorobutanoate					-630.7				-578.4			

Molecular formula	Name	Crystal				Liquid				Gas			
		$\Delta_f H°$ kJ/mol	$\Delta_f G°$ kJ/mol	$S°$ J/mol K	C_p J/mol K	$\Delta_f H°$ kJ/mol	$\Delta_f G°$ kJ/mol	$S°$ J/mol K	C_p J/mol K	$\Delta_f H°$ kJ/mol	$\Delta_f G°$ kJ/mol	$S°$ J/mol K	C_p J/mol K
$C_7H_{13}N$	Heptanenitrile					-82.8				-31.0			
C_7H_{14}	1-Heptene					-97.9		327.6	211.8	-62.3			
C_7H_{14}	cis-2-Heptene					-105.1							
C_7H_{14}	trans-2-Heptene					-109.5							
C_7H_{14}	cis-3-Heptene					-104.3							
C_7H_{14}	trans-3-Heptene					-109.3							
C_7H_{14}	5-Methyl-1-hexene					-100.0				-65.7			
C_7H_{14}	cis-3-Methyl-3-hexene					-115.9				-79.4			
C_7H_{14}	trans-3-Methyl-3-hexene					-112.7				-76.8			
C_7H_{14}	2,4-Dimethyl-1-pentene					-117.0				-83.8			
C_7H_{14}	4,4-Dimethyl-1-pentene					-110.6				-81.6			
C_7H_{14}	2,4-Dimethyl-2-pentene					-123.1				-88.7			
C_7H_{14}	cis-4,4-Dimethyl-2-pentene					-105.3				-72.6			
C_7H_{14}	trans-4,4-Dimethyl-2-pentene					-121.7				-88.8			
C_7H_{14}	2-Ethyl-3-methyl-1-butene					-114.1				-79.5			
C_7H_{14}	2,3,3-Trimethyl-1-butene					-117.7				-85.5			
C_7H_{14}	Cycloheptane					-156.6				-118.1			
C_7H_{14}	Methylcyclohexane					-190.1			184.8	-154.7			
C_7H_{14}	Ethylcyclopentane					-163.4		279.9		-126.9			
C_7H_{14}	1,1-Dimethylcyclopentane					-172.1				-138.2			
C_7H_{14}	cis-1,2-Dimethylcyclopentane					-165.3		269.2		-129.5			
C_7H_{14}	trans-1,2-Dimethylcyclopentane					-171.2				-136.6			
C_7H_{14}	cis-1,3-Dimethylcyclopentane					-170.1				-135.8			
C_7H_{14}	trans-1,3-Dimethylcyclopentane					-168.1				-133.6			
C_7H_{14}	1,1,2,2-Tetramethylcyclopropane					-119.8							
$C_7H_{14}Br_2$	1,2-Dibromoheptane					-212.3				-157.9			
$C_7H_{14}O$	1-Heptanal					-311.5		335.4	230.1	-263.8			
$C_7H_{14}O$	2-Heptanone								232.6				
$C_7H_{14}O$	3-Heptanone									-297.1			
$C_7H_{14}O$	4-Heptanone									-298.3			
$C_7H_{14}O$	2,2-Dimethyl-3-pentanone					-356.1				-313.6			
$C_7H_{14}O$	2,4-Dimethyl-3-pentanone					-352.9		318.0	233.7	-311.3			
$C_7H_{14}O$	cis-2-Methylcyclohexanol					-390.2				-327.0			
$C_7H_{14}O$	trans-2-Methylcyclohexanol, (±)					-415.7				-352.5			
$C_7H_{14}O$	cis-3-Methylcyclohexanol, (±)					-416.1				-350.9			
$C_7H_{14}O$	trans-3-Methylcyclohexanol, (±)					-394.4				-329.1			
$C_7H_{14}O$	cis-4-Methylcyclohexanol					-413.2				-347.5			
$C_7H_{14}O$	trans-4-Methylcyclohexanol					-433.3				-367.2			
$C_7H_{14}O_2$	Heptanoic acid					-610.2			265.4	-536.2			
$C_7H_{14}O_2$	Pentyl acetate								261.0				
$C_7H_{14}O_2$	Isopentyl acetate								248.5				
$C_7H_{14}O_2$	Ethyl pentanoate					-553.0				-505.9			
$C_7H_{14}O_2$	Ethyl 3-methylbutanoate					-571.0				-527.0			
$C_7H_{14}O_2$	Ethyl 2,2-dimethylpropanoate					-577.2				-536.0			
$C_7H_{14}O_2$	Methyl hexanoate					-540.2				-492.2			
$C_7H_{14}O_6$	α-Methylglucoside	-1233.3											
$C_7H_{15}Br$	1-Bromoheptane					-218.4				-167.8			
C_7H_{16}	Heptane					-224.2			224.7	-187.6			
C_7H_{16}	2-Methylhexane					-229.5		323.3	222.9	-194.5			
C_7H_{16}	3-Methylhexane					-226.4				-191.3			
C_7H_{16}	3-Ethylpentane					-224.9		314.5	219.6	-189.5			
C_7H_{16}	2,2-Dimethylpentane					-238.3		300.3	221.1	-205.7			
C_7H_{16}	2,3-Dimethylpentane					-233.1				-198.7			
C_7H_{16}	2,4-Dimethylpentane					-234.6		303.2	224.2	-201.6			
C_7H_{16}	3,3-Dimethylpentane					-234.2				-201.0			
C_7H_{16}	2,2,3-Trimethylbutane					-236.5		292.2	213.5	-204.4			
$C_7H_{16}O$	1-Heptanol					-403.3			272.1	-336.5			
$C_7H_{16}O$	tert-Butyl isopropyl ether					-392.8				-358.1			
$C_7H_{16}O_2$	1,7-Heptanediol					-574.2							
$C_7H_{16}O_2$	2,2-Diethoxypropane					-538.9				-506.9			
$C_7H_{16}S$	1-Heptanethiol					-200.5				-149.9			
$C_8H_4O_3$	Phthalic anhydride	-460.1		180.0	160.0					-371.4			
$C_8H_5NO_2$	1H-Indole-2,3-dione	-268.2											
$C_8H_6O_4$	Phthalic acid	-782.0		207.9	188.1								
$C_8H_6O_4$	Isophthalic acid	-803.0								-696.3			
$C_8H_6O_4$	Terephthalic acid	-816.1								-717.9			

Molecular formula	Name	Crystal				Liquid				Gas			
		$\Delta_f H°$ kJ/mol	$\Delta_f G°$ kJ/mol	$S°$ J/mol K	C_p J/mol K	$\Delta_f H°$ kJ/mol	$\Delta_f G°$ kJ/mol	$S°$ J/mol K	C_p J/mol K	$\Delta_f H°$ kJ/mol	$\Delta_f G°$ kJ/mol	$S°$ J/mol K	C_p J/mol K
C_8H_6S	Benzo[b]thiophene	100.6								166.3			
C_8H_7N	1H-Indole	86.6								156.5			
C_8H_8	Styrene					103.8			182.0	147.9			
C_8H_8O	Phenyl vinyl ether					-26.2				22.7			
C_8H_8O	Acetophenone					-142.5				-86.7			
$C_8H_8O_2$	o-Toluic acid	-416.5			174.9								
$C_8H_8O_2$	m-Toluic acid	-426.1			163.6								
$C_8H_8O_2$	p-Toluic acid	-429.2			169.0								
$C_8H_8O_2$	Methyl benzoate					-343.5			221.3	-287.9			
$C_8H_8O_3$	Methyl salicylate								249.0				
C_8H_9NO	Acetanilide	-209.4			179.3								
C_8H_{10}	1,7-Octadiyne					334.4							
C_8H_{10}	Ethylbenzene					-12.3			183.2	29.9			
C_8H_{10}	o-Xylene					-24.4			186.1	19.1			
C_8H_{10}	m-Xylene					-25.4			183.0	17.3			
C_8H_{10}	p-Xylene					-24.4			181.5	18.0			
$C_8H_{10}O$	2-Ethylphenol					-208.8				-145.2			
$C_8H_{10}O$	3-Ethylphenol					-214.3				-146.1			
$C_8H_{10}O$	4-Ethylphenol	-224.4			206.9					-144.1			
$C_8H_{10}O$	2,3-Xylenol	-241.1								-157.2			
$C_8H_{10}O$	2,4-Xylenol					-228.7				-163.8			
$C_8H_{10}O$	2,5-Xylenol	-246.6								-161.6			
$C_8H_{10}O$	2,6-Xylenol	-237.4								-162.1			
$C_8H_{10}O$	3,4-Xylenol	-242.3								-157.3			
$C_8H_{10}O$	3,5-Xylenol	-244.4								-162.4			
$C_8H_{10}O$	Benzeneethanol								252.6				
$C_8H_{10}O$	Ethoxybenzene					-152.6			228.5	-101.6			
$C_8H_{10}O_2$	1,2-Dimethoxybenzene					-290.3				-223.3			
$C_8H_{11}N$	N-Ethylaniline					8.2				56.3			
$C_8H_{11}N$	N,N-Dimethylaniline					46.0				100.5			
$C_8H_{11}N$	2,4-Dimethylaniline					-39.2							
$C_8H_{11}N$	2,5-Dimethylaniline					-38.9							
$C_8H_{11}N$	2,6-Dimethylaniline								238.9				
C_8H_{12}	1-Octen-3-yne					140.7							
C_8H_{12}	cis-1,2-Divinylcyclobutane					124.3				166.5			
C_8H_{12}	trans-1,2-Divinylcyclobutane					101.3				143.5			
$C_8H_{12}N_4$	2,2'-Azobis[isobutyronitrile]	246.0			237.6								
$C_8H_{12}O_2$	2,2,4,4-Tetramethyl-1,3-cyclobutanedione	-379.9								-307.6			
C_8H_{14}	Ethylidenecyclohexane					-103.5				-59.5			
C_8H_{14}	Allylcyclopentane					-64.5				-24.1			
$C_8H_{14}ClN_5$	Atrazine	-125.4											
$C_8H_{14}O_3$	Butanoic anhydride								283.7				
$C_8H_{15}ClO_2$	3-Methylbutyl 2-chloropropanoate					-627.3				-575.0			
$C_8H_{15}ClO_2$	3-Methylbutyl 3-chloropropanoate					-593.4				-539.4			
$C_8H_{15}N$	Octanenitrile					-107.3				-50.5			
C_8H_{16}	1-Octene					-124.5			241.0	-81.3			
C_8H_{16}	cis-2-Octene					-135.7			239.0				
C_8H_{16}	trans-2-Octene					-135.7			239.0				
C_8H_{16}	cis-2,2-Dimethyl-3-hexene					-126.4				-89.3			
C_8H_{16}	trans-2,2-Dimethyl-3-hexene					-144.9				-107.7			
C_8H_{16}	3-Ethyl-2-methyl-1-pentene					-137.9				-100.3			
C_8H_{16}	2,4,4-Trimethyl-1-pentene					-145.9				-110.5			
C_8H_{16}	2,4,4-Trimethyl-2-pentene					-142.4				-104.9			
C_8H_{16}	Cyclooctane					-167.7				-124.4			
C_8H_{16}	Ethylcyclohexane					-212.1		280.9	211.8	-171.5			
C_8H_{16}	1,1-Dimethylcyclohexane					-218.7		267.2	209.2	-180.9			
C_8H_{16}	cis-1,2-Dimethylcyclohexane					-211.8		274.1	210.2	-172.1			
C_8H_{16}	trans-1,2-Dimethylcyclohexane					-218.2		273.2	209.4	-179.9			
C_8H_{16}	cis-1,3-Dimethylcyclohexane					-222.9		272.6	209.4	-184.6			
C_8H_{16}	trans-1,3-Dimethylcyclohexane					-215.7		276.3	212.8	-176.5			
C_8H_{16}	cis-1,4-Dimethylcyclohexane					-215.6		271.1	212.1	-176.6			
C_8H_{16}	trans-1,4-Dimethylcyclohexane					-222.4		268.0	210.2	-184.5			
C_8H_{16}	Propylcyclopentane					-188.8		310.8	216.3	-147.7			
C_8H_{16}	1-Ethyl-1-methylcyclopentane					-193.8							
C_8H_{16}	cis-1-Ethyl-2-methylcyclopentane					-190.8							
C_8H_{16}	trans-1-Ethyl-2-methylcyclopentane					-195.1				-156.2			

Molecular formula	Name	Crystal				Liquid				Gas			
		$\Delta_f H°$ kJ/mol	$\Delta_f G°$ kJ/mol	$S°$ J/mol K	C_p J/mol K	$\Delta_f H°$ kJ/mol	$\Delta_f G°$ kJ/mol	$S°$ J/mol K	C_p J/mol K	$\Delta_f H°$ kJ/mol	$\Delta_f G°$ kJ/mol	$S°$ J/mol K	C_p J/mol K
C₈H₁₆	cis-1-Ethyl-3-methylcyclopentane					-194.4							
C₈H₁₆	trans-1-Ethyl-3-methylcyclopentane					-196.0							
C₈H₁₆O	Octanal									-291.9		365.4	
C₈H₁₆O	2-Ethylhexanal					-348.5				-299.6			
C₈H₁₆O	2-Octanone								273.3				
C₈H₁₆O	2,2,4-Trimethyl-3-pentanone					-381.6				-338.3			
C₈H₁₆O₂	Octanoic acid					-636.0			297.9	-554.3			
C₈H₁₆O₂	2-Ethylhexanoic acid					-635.1				-559.5			
C₈H₁₆O₂	Hexyl acetate								282.8				
C₈H₁₆O₂	Isobutyl isobutanoate					-587.4				-542.9			
C₈H₁₆O₂	Propyl pentanoate					-583.0				-533.6			
C₈H₁₆O₂	Isopropyl pentanoate					-592.2				-544.9			
C₈H₁₆O₂	Methyl heptanoate					-567.1			285.1	-515.5			
C₈H₁₇Br	1-Bromooctane					-245.1				-189.3			
C₈H₁₇Cl	1-Chlorooctane					-291.3				-238.9			
C₈H₁₇NO	Octanamide	-473.2								-362.7			
C₈H₁₈	Octane					-250.1			254.6	-208.5			
C₈H₁₈	2-Methylheptane					-255.0		356.4	252.0	-215.3			
C₈H₁₈	3-Methylheptane, (S)					-252.3		362.6	250.2	-212.5			
C₈H₁₈	4-Methylheptane					-251.6			251.1	-211.9			
C₈H₁₈	3-Ethylhexane					-250.4				-210.7			
C₈H₁₈	2,2-Dimethylhexane					-261.9				-224.5			
C₈H₁₈	2,3-Dimethylhexane					-252.6				-213.8			
C₈H₁₈	2,4-Dimethylhexane					-257.0				-219.2			
C₈H₁₈	2,5-Dimethylhexane					-260.4			249.2	-222.5			
C₈H₁₈	3,3-Dimethylhexane					-257.5			246.6	-219.9			
C₈H₁₈	3,4-Dimethylhexane					-251.8				-212.8			
C₈H₁₈	3-Ethyl-2-methylpentane					-249.6				-211.0			
C₈H₁₈	3-Ethyl-3-methylpentane					-252.8				-214.8			
C₈H₁₈	2,2,3-Trimethylpentane					-256.9				-220.0			
C₈H₁₈	2,2,4-Trimethylpentane					-259.2			239.1	-224.0			
C₈H₁₈	2,3,3-Trimethylpentane					-253.5			245.6	-216.3			
C₈H₁₈	2,3,4-Trimethylpentane					-255.0		329.3	247.3	-217.3			
C₈H₁₈	2,2,3,3-Tetramethylbutane	-269.0		273.7	239.2					-226.0			
C₈H₁₈N₂	Azobutane					-40.1				9.2			
C₈H₁₈O	1-Octanol					-426.5			305.2	-355.6			
C₈H₁₈O	2-Octanol								330.1				
C₈H₁₈O	2-Ethyl-1-hexanol					-432.8		347.0	317.5	-365.3			
C₈H₁₈O	Dibutyl ether					-377.9			278.2	-332.8			
C₈H₁₈O	Di-sec-butyl ether					-401.5				-360.6			
C₈H₁₈O	Di-tert-butyl ether					-399.6			276.1	-362.0			
C₈H₁₈O	tert-Butyl isobutyl ether					-409.1				-369.0			
C₈H₁₈O₂	1,8-Octanediol	-626.6											
C₈H₁₈O₂	2,5-Dimethyl-2,5-hexanediol	-681.7											
C₈H₁₈O₃	Diethylene glycol monobutyl ether								354.9				
C₈H₁₈O₃	Diethylene glycol diethyl ether								341.4				
C₈H₁₈O₃S	Dibutyl sulfite					-693.1				-625.3			
C₈H₁₈O₅	Tetraethylene glycol					-981.7			428.8	-883.0			
C₈H₁₈S	Dibutyl sulfide					-220.7		405.1	284.3	-167.7			
C₈H₁₈S	Di-sec-butyl sulfide					-220.7				-167.7			
C₈H₁₈S	Di-tert-butyl sulfide					-232.6				-188.8			
C₈H₁₈S	Diisobutyl sulfide					-229.2				-180.5			
C₈H₁₈S₂	Dibutyl disulfide					-222.9				-160.6			
C₈H₁₈S₂	Di-tert-butyl disulfide					-255.2				-201.0			
C₈H₁₉N	Dibutylamine					-206.0			292.9	-156.6			
C₈H₁₉N	Diisobutylamine					-218.5				-179.2			
C₈H₂₀BrN	Tetraethylammonium bromide	-342.7											
C₈H₂₀O₄Si	Ethyl silicate							533.1	364.4				
C₈H₂₀Pb	Tetraethyl lead					52.7		464.6	307.4	109.6			
C₈H₂₀Si	Tetraethylsilane								298.1				
C₉H₆N₂O₂	Toluene-2,4-diisocyanate								287.8				
C₉H₇N	Quinoline					141.2			200.5				
C₉H₇N	Isoquinoline					144.3		216.0	196.2	204.6			
C₉H₇NO	2-Quinolinol	-144.9								-25.5			
C₉H₇NO	8-Quinolinol	82.1											
C₉H₈	Indene					110.6		215.3	186.9	163.4			

Molecular formula	Name	Crystal				Liquid				Gas			
		$\Delta_f H°$ kJ/mol	$\Delta_f G°$ kJ/mol	$S°$ J/mol K	C_p J/mol K	$\Delta_f H°$ kJ/mol	$\Delta_f G°$ kJ/mol	$S°$ J/mol K	C_p J/mol K	$\Delta_f H°$ kJ/mol	$\Delta_f G°$ kJ/mol	$S°$ J/mol K	C_p J/mol K
$C_9H_8O_4$	2-(Acetyloxy)benzoic acid	-815.6											
C_9H_{10}	Cyclopropylbenzene					100.3				150.5			
C_9H_{10}	Indan					11.5		56.0	190.2	60.3			
$C_9H_{10}Cl_2N_2O$	Diuron	-329.0											
$C_9H_{10}N_2$	2,2'-Dipyrrolylmethane	126.2											
$C_9H_{10}O_2$	Ethyl benzoate								246.0				
$C_9H_{10}O_2$	Benzyl acetate								148.5				
$C_9H_{11}NO_2$	L-Phenylalanine	-466.9		213.6	203.0					-312.9			
$C_9H_{11}NO_3$	L-Tyrosine	-685.1		214.0	216.4								
C_9H_{12}	Propylbenzene					-38.3		287.8	214.7	7.9			
C_9H_{12}	Isopropylbenzene					-41.1			210.7	4.0			
C_9H_{12}	2-Ethyltoluene					-46.4				1.3			
C_9H_{12}	3-Ethyltoluene					-48.7				-1.8			
C_9H_{12}	4-Ethyltoluene					-49.8				-3.2			
C_9H_{12}	1,2,3-Trimethylbenzene					-58.5		267.9	216.4	-9.5			
C_9H_{12}	1,2,4-Trimethylbenzene					-61.8			215.0	-13.8			
C_9H_{12}	1,3,5-Trimethylbenzene					-63.4			209.3	-15.9			
$C_9H_{12}O$	2-Isopropylphenol					-233.7				-182.2			
$C_9H_{12}O$	3-Isopropylphenol					-252.5				-196.0			
$C_9H_{12}O$	4-Isopropylphenol	-270.0								-175.3			
$C_9H_{12}O_2$	Isopropylbenzene hydroperoxide					-148.3				-78.4			
$C_9H_{13}NO_2$	Ethyl 3,5-dimethylpyrrole-2-carboxylate	-474.5											
$C_9H_{13}NO_2$	Ethyl 2,4-dimethylpyrrole-3-carboxylate	-463.2											
$C_9H_{13}NO_2$	Ethyl 2,5-dimethylpyrrole-3-carboxylate	-478.7											
$C_9H_{13}NO_2$	Ethyl 4,5-dimethylpyrrole-3-carboxylate	-470.3											
$C_9H_{14}O$	Isophorone								253.5				
$C_9H_{14}O_6$	Triacetin					-1330.8		458.3	384.7	-1245.0			
$C_9H_{15}N$	3-Ethyl-2,4,5-trimethylpyrrole	-89.2											
C_9H_{16}	1-Nonyne					16.3				62.3			
$C_9H_{16}O_4$	Nonanedioic acid	-1054.3											
$C_9H_{17}NO$	2,2,6,6-Tetramethyl-4-piperidinone	-334.2								-273.4			
C_9H_{18}	Propylcyclohexane					-237.4		311.9	242.0	-192.3			
C_9H_{18}	1α,3α,5β-1,3,5-Trimethylcyclohexane									-212.1			
$C_9H_{18}O$	2-Nonanone					-397.2				-340.7			
$C_9H_{18}O$	5-Nonanone					-398.2		401.4	303.6	-344.9			
$C_9H_{18}O$	2,6-Dimethyl-4-heptanone					-408.5			297.3	-357.6			
$C_9H_{18}O_2$	Nonanoic acid					-659.7			362.4	-577.3			
$C_9H_{18}O_2$	Butyl pentanoate					-613.3				-560.2			
$C_9H_{18}O_2$	sec-Butyl pentanoate					-624.2				-573.2			
$C_9H_{18}O_2$	Isobutyl pentanoate					-620.0				-568.6			
$C_9H_{18}O_2$	Methyl octanoate					-590.3				-533.9			
$C_9H_{19}N$	N-Butylpiperidine					-171.8							
$C_9H_{19}N$	2,2,6,6-Tetramethylpiperidine					-206.9				-159.9			
C_9H_{20}	Nonane					-274.7			284.4	-228.2			
C_9H_{20}	2,2-Dimethylheptane					-288.1							
C_9H_{20}	2,2,3-Trimethylhexane					-282.7							
C_9H_{20}	2,2,4-Trimethylhexane					-282.8							
C_9H_{20}	2,2,5-Trimethylhexane					-293.3							
C_9H_{20}	2,3,3-Trimethylhexane					-281.1							
C_9H_{20}	2,3,5-Trimethylhexane					-284.0				-242.6			
C_9H_{20}	2,4,4-Trimethylhexane					-280.2							
C_9H_{20}	3,3,4-Trimethylhexane					-277.5							
C_9H_{20}	3,3-Diethylpentane					-275.4			278.2	-233.3			
C_9H_{20}	3-Ethyl-2,2-dimethylpentane					-272.7							
C_9H_{20}	3-Ethyl-2,4-dimethylpentane					-269.7							
C_9H_{20}	2,2,3,3-Tetramethylpentane					-278.3			271.5	-237.1			
C_9H_{20}	2,2,3,4-Tetramethylpentane					-277.7				-236.9			
C_9H_{20}	2,2,4,4-Tetramethylpentane					-280.0			266.3	-241.6			
C_9H_{20}	2,3,3,4-Tetramethylpentane					-277.9				-236.1			
$C_9H_{20}N_2O$	Tetraethylurea					-380.0				-316.4			
$C_9H_{20}O$	1-Nonanol					-453.4				-376.5			
$C_9H_{20}O_2$	1,9-Nonanediol	-657.6											
$C_9H_{21}N$	Tripropylamine					-207.1				-161.0			
$C_{10}H_6N_2$	2-Quinolinecarbonitrile	246.5											
$C_{10}H_6N_2$	3-Quinolinecarbonitrile	242.3											
$C_{10}H_6N_2O_4$	1,5-Dinitronaphthalene	29.8											

Molecular formula	Name	Crystal $\Delta_f H°$ kJ/mol	$\Delta_f G°$ kJ/mol	$S°$ J/mol K	C_p J/mol K	Liquid $\Delta_f H°$ kJ/mol	$\Delta_f G°$ kJ/mol	$S°$ J/mol K	C_p J/mol K	Gas $\Delta_f H°$ kJ/mol	$\Delta_f G°$ kJ/mol	$S°$ J/mol K	C_p J/mol K
$C_{10}H_6N_2O_4$	1,8-Dinitronaphthalene	39.7											
$C_{10}H_7Cl$	1-Chloronaphthalene					54.6			212.6	119.8			
$C_{10}H_7Cl$	2-Chloronaphthalene	55.4								137.4			
$C_{10}H_7I$	1-Iodonaphthalene					161.5				233.8			
$C_{10}H_7I$	2-Iodonaphthalene	144.3								235.1			
$C_{10}H_7NO_2$	1-Nitronaphthalene	42.6								111.2			
$C_{10}H_8$	Naphthalene	78.5	201.6	167.4	165.7					150.6	224.1	333.1	131.9
$C_{10}H_8$	Azulene	212.3								289.1			
$C_{10}H_8O$	1-Naphthol	-121.5			166.9					-30.4			149.4
$C_{10}H_8O$	2-Naphthol	-124.1		179.0	172.8					-29.9		366.6	147.8
$C_{10}H_9N$	1-Naphthylamine	67.8								132.8			
$C_{10}H_9N$	2-Naphthylamine	60.2								134.3			
$C_{10}H_{10}$	1,2-Dihydronaphthalene					71.6							
$C_{10}H_{10}$	1,4-Dihydronaphthalene					84.2							
$C_{10}H_{10}O$	1-Tetralone	-209.6											
$C_{10}H_{10}O_4$	Dimethyl phthalate								303.1				
$C_{10}H_{10}O_4$	Dimethyl isophthalate	-730.9											
$C_{10}H_{10}O_4$	Dimethyl terephthalate	-732.6			261.1								
$C_{10}H_{12}$	1,2,3,4-Tetrahydronaphthalene					-29.2			217.5	26.0			
$C_{10}H_{14}$	Butylbenzene					-63.2		321.2	243.4	-11.8			
$C_{10}H_{14}$	sec-Butylbenzene, (±)					-66.4				-18.4			
$C_{10}H_{14}$	tert-Butylbenzene					-71.9				-23.0			
$C_{10}H_{14}$	Isobutylbenzene					-69.8				-21.9			
$C_{10}H_{14}$	1-Isopropyl-2-methylbenzene					-73.3							
$C_{10}H_{14}$	1-Isopropyl-3-methylbenzene					-78.6							
$C_{10}H_{14}$	1-Isopropyl-4-methylbenzene					-78.0			236.4				
$C_{10}H_{14}$	o-Diethylbenzene					-68.5							
$C_{10}H_{14}$	m-Diethylbenzene					-73.5							
$C_{10}H_{14}$	p-Diethylbenzene					-72.8							
$C_{10}H_{14}$	3-Ethyl-1,2-dimethylbenzene					-80.5							
$C_{10}H_{14}$	4-Ethyl-1,2-dimethylbenzene					-86.0							
$C_{10}H_{14}$	2-Ethyl-1,3-dimethylbenzene					-80.1							
$C_{10}H_{14}$	2-Ethyl-1,4-dimethylbenzene					-84.8							
$C_{10}H_{14}$	1-Ethyl-2,4-dimethylbenzene					-84.1							
$C_{10}H_{14}$	1-Ethyl-3,5-dimethylbenzene					-87.8							
$C_{10}H_{14}$	1,2,4,5-Tetramethylbenzene	-119.9		245.6	215.1								
$C_{10}H_{14}O$	Thymol	-309.7								-218.5			
$C_{10}H_{16}$	Dipentene					-50.8			249.4	-2.6			
$C_{10}H_{16}$	d-Limonene					-54.5			249.0				
$C_{10}H_{16}$	α-Pinene					-16.4				28.3			
$C_{10}H_{16}$	β-Pinene					-7.7				38.7			
$C_{10}H_{16}$	α-Terpinene									-20.6			
$C_{10}H_{16}$	β-Myrcene					14.5							
$C_{10}H_{16}$	cis, cis-2,6-Dimethyl-2,4,6-octatriene					-24.0							
$C_{10}H_{16}N_2O_8$	Ethylenediaminetetraacetic acid	-1759.5											
$C_{10}H_{16}O$	Camphor, (±)	-319.4			271.2					-267.5			
$C_{10}H_{18}$	1,1'-Bicyclopentyl					-178.9							
$C_{10}H_{18}$	cis-Decahydronaphthalene					-219.4		265.0	232.0	-169.2			
$C_{10}H_{18}$	trans-Decahydronaphthalene					-230.6		264.9	228.5	-182.1			
$C_{10}H_{18}O_4$	Sebacic acid	-1082.6								-921.9			
$C_{10}H_{19}N$	Decanenitrile					-158.4				-91.5			
$C_{10}H_{20}$	1-Decene					-173.8		425.0	300.8	-123.3			
$C_{10}H_{20}$	cis-1,2-Di-tert-butylethene					-163.6							
$C_{10}H_{20}$	Butylcyclohexane					-263.1		345.0	271.0	-213.7			
$C_{10}H_{20}O_2$	Decanoic acid	-713.7				-684.3				-594.9			
$C_{10}H_{20}O_2$	Methyl nonanoate					-616.2				-554.2			
$C_{10}H_{21}NO_2$	1-Nitrodecane					-351.5							
$C_{10}H_{22}$	Decane					-300.9			314.4	-249.5			
$C_{10}H_{22}$	2-Methylnonane					-309.8		420.1	313.3	-260.2			
$C_{10}H_{22}$	5-Methylnonane					-307.9		423.8	314.4	-258.6			
$C_{10}H_{22}O$	1-Decanol					-478.1			370.6	-396.6			
$C_{10}H_{22}O$	Dipentyl ether								250.0				
$C_{10}H_{22}O$	Diisopentyl ether								379.0				
$C_{10}H_{22}O_2$	1,10-Decanediol	-678.9											
$C_{10}H_{22}O_2$	Ethylene glycol dibutyl ether								350.0				
$C_{10}H_{22}S$	1-Decanethiol	-309.9				-276.5		476.1	350.4	-211.5			

Standard Thermodynamic Properties of Chemical Substances

Molecular formula	Name	Crystal ΔfH° kJ/mol	ΔfG° kJ/mol	S° J/mol K	Cp J/mol K	Liquid ΔfH° kJ/mol	ΔfG° kJ/mol	S° J/mol K	Cp J/mol K	Gas ΔfH° kJ/mol	ΔfG° kJ/mol	S° J/mol K	Cp J/mol K
$C_{10}H_{22}S$	Dipentyl sulfide					-266.4				-204.9			
$C_{10}H_{22}S$	Diisopentyl sulfide					-281.8				-221.5			
$C_{10}H_{23}N$	Octyldimethylamine					-232.8							
$C_{11}H_8O_2$	1-Naphthalenecarboxylic acid	-333.5								-223.1			
$C_{11}H_8O_2$	2-Naphthalenecarboxylic acid	-346.1								-232.5			
$C_{11}H_{10}$	1-Methylnaphthalene					56.3		254.8	224.4				
$C_{11}H_{10}$	2-Methylnaphthalene	44.9		220.0	196.0					106.7			
$C_{11}H_{12}N_2O_2$	L-Tryptophan	-415.3		251.0	238.1								
$C_{11}H_{14}$	1,1-Dimethylindan					-53.6				-1.6			
$C_{11}H_{16}$	1-tert-Butyl-3-methylbenzene					-109.7							
$C_{11}H_{16}$	1-tert-Butyl-4-methylbenzene					-109.7				-57.0			
$C_{11}H_{16}$	Pentamethylbenzene	-144.6								-67.2			
$C_{11}H_{20}$	Spiro[5.5]undecane					-244.5				-188.3			
$C_{11}H_{22}$	1-Undecene								344.9				
$C_{11}H_{22}O_2$	Methyl decanoate					-640.5				-573.8			
$C_{11}H_{24}$	Undecane					-327.2			344.9	-270.8			
$C_{11}H_{24}O$	1-Undecanol					-504.8							
$C_{12}F_{27}N$	Tris(perfluorobutyl)amine								418.4				
$C_{12}H_8$	Acenaphthylene	186.7			166.4					259.7			
$C_{12}H_8N_2$	Phenazine	237.0								328.8			
$C_{12}H_8O$	Dibenzofuran	-5.3								83.4			
$C_{12}H_8S$	Dibenzothiophene	120.0								205.1			
$C_{12}H_8S_2$	Thianthrene	182.0								286.0			
$C_{12}H_9N$	Carbazole	101.7								200.7			
$C_{12}H_{10}$	Acenaphthene	70.3		188.9	190.4					156.0			
$C_{12}H_{10}$	Biphenyl	99.4		209.4	198.4					181.4			
$C_{12}H_{10}N_2O$	trans-Azoxybenzene	243.4								342.0			
$C_{12}H_{10}N_2O$	N-Nitrosodiphenylamine	227.2											
$C_{12}H_{10}O$	Diphenyl ether	-32.1		233.9	216.6	-14.9				52.0			
$C_{12}H_{10}O_2$	1-Naphthaleneacetic acid	-359.2											
$C_{12}H_{10}O_2$	2-Naphthaleneacetic acid	-371.9											
$C_{12}H_{11}N$	Diphenylamine	130.2								219.3			
$C_{12}H_{11}N$	2-Aminobiphenyl	93.8								184.4			
$C_{12}H_{11}N$	4-Aminobiphenyl	81.0											
$C_{12}H_{12}N_2$	p-Benzidine	70.7											
$C_{12}H_{14}O_4$	Diethyl phthalate					-776.6		425.1	366.1	-688.4			
$C_{12}H_{16}$	Cyclohexylbenzene					-76.6				-16.7			
$C_{12}H_{17}NO_4$	Diethyl 3,5-dimethylpyrrole-2,4-dicarboxylate	-916.7											
$C_{12}H_{18}$	3,9-Dodecadiyne					197.8							
$C_{12}H_{18}$	5,7-Dodecadiyne					181.5							
$C_{12}H_{18}$	1-tert-Butyl-3,5-dimethylbenzene					-146.5							
$C_{12}H_{18}$	Hexamethylbenzene	-162.4		306.3	245.6					-77.4			
$C_{12}H_{22}$	Cyclohexylcyclohexane					-273.7				-215.7			
$C_{12}H_{22}O_4$	Dodecanedioic acid	-1130.0								-976.9			
$C_{12}H_{22}O_{11}$	Sucrose	-2226.1											
$C_{12}H_{22}O_{11}$	β-D-Lactose	-2236.7											
$C_{12}H_{24}$	1-Dodecene					-226.2		484.8	360.7	-165.4			
$C_{12}H_{24}O_2$	Dodecanoic acid	-774.6			404.3	-737.9				-642.0			
$C_{12}H_{24}O_2$	Methyl undecanoate					-665.2				-593.8			
$C_{12}H_{24}O_{12}$	α-Lactose monohydrate	-2484.1											
$C_{12}H_{25}Br$	1-Bromododecane					-344.7				-269.9			
$C_{12}H_{25}Cl$	1-Chlorododecane					-392.3				-321.1			
$C_{12}H_{26}$	Dodecane					-350.9			375.8	-289.4			
$C_{12}H_{26}O$	1-Dodecanol					-528.5			438.1	-436.6			
$C_{12}H_{26}O_3$	Diethylene glycol dibutyl ether								452.0				
$C_{12}H_{27}N$	Tributylamine					-281.6							
$C_{12}H_{27}O_4P$	Tributyl phosphate								379.4				
$C_{13}H_8O_2$	Xanthone	-191.5											
$C_{13}H_9N$	Acridine	179.4								273.9			
$C_{13}H_9N$	Phenanthridine	141.9								240.5			
$C_{13}H_9N$	Benzo[f]quinoline	150.6								233.7			
$C_{13}H_{10}$	9H-Fluorene	89.9		207.3	203.1					175.0			173.1
$C_{13}H_{10}N_2$	9-Acridinamine	159.2											
$C_{13}H_{10}O$	Benzophenone	-34.5			224.8					54.9			
$C_{13}H_{11}N$	9-Methyl-9H-carbazole	105.5								201.0			
$C_{13}H_{12}$	Diphenylmethane	71.5		239.3		89.7				139.0			

Molecular formula	Name	Crystal				Liquid				Gas			
		$\Delta_f H°$ kJ/mol	$\Delta_f G°$ kJ/mol	$S°$ J/mol K	C_p J/mol K	$\Delta_f H°$ kJ/mol	$\Delta_f G°$ kJ/mol	$S°$ J/mol K	C_p J/mol K	$\Delta_f H°$ kJ/mol	$\Delta_f G°$ kJ/mol	$S°$ J/mol K	C_p J/mol K
$C_{13}H_{13}N$	N-Benzylaniline	101.4											
$C_{13}H_{14}N_2$	4,4'-Diaminodiphenylmethane			270.9									
$C_{13}H_{24}O_4$	Tridecanedioic acid	-1148.3											
$C_{13}H_{26}$	1-Tridecene								391.8				
$C_{13}H_{26}O_2$	Methyl dodecanoate					-693.0				-614.9			
$C_{13}H_{28}$	Tridecane								406.7				
$C_{13}H_{28}O$	1-Tridecanol	-599.4											
$C_{14}H_8O_2$	9,10-Anthracenedione	-188.5								-75.7			
$C_{14}H_8O_2$	9,10-Phenanthrenedione	-154.7								-46.6			
$C_{14}H_8O_4$	1,4-Dihydroxy-9,10-anthracenedione	-595.8								-471.7			
$C_{14}H_{10}$	Anthracene	129.2		207.5	210.5					230.9			
$C_{14}H_{10}$	Phenanthrene	116.2		215.1	220.6					207.5			
$C_{14}H_{10}$	Diphenylacetylene	312.4			225.9								
$C_{14}H_{10}O_2$	Benzil	-153.9								-55.5			
$C_{14}H_{10}O_4$	Benzoyl peroxide	-369.4								-281.7			
$C_{14}H_{12}$	cis-Stilbene					183.3				252.3			
$C_{14}H_{12}$	trans-Stilbene	136.9								236.1			
$C_{14}H_{14}$	1,1-Diphenylethane					48.7							
$C_{14}H_{14}$	1,2-Diphenylethane	51.5								142.9			
$C_{14}H_{22}$	1,3-Di-tert-butylbenzene					-188.8							
$C_{14}H_{22}$	1,4-Di-tert-butylbenzene	-212.0											
$C_{14}H_{23}N_3O_{10}$	Pentetic acid	-2225.2											
$C_{14}H_{27}N$	Tetradecanenitrile					-260.2				-174.9			
$C_{14}H_{28}O_2$	Tetradecanoic acid	-833.5			432.0	-788.8				-693.7			
$C_{14}H_{28}O_2$	Methyl tridecanoate					-717.9				-635.3			
$C_{14}H_{30}O$	1-Tetradecanol	-629.6			388.0	-580.6							
$C_{15}H_{16}O_2$	2,2-Bis(4-hydroxyphenyl)propane	-368.6											
$C_{15}H_{24}$	1,3-Di-tert-butyl-5-methylbenzene	-245.8											
$C_{15}H_{24}O$	2,6-Di-tert-butyl-4-methylphenol	-410.0								-296.9			
$C_{15}H_{30}$	Decylcyclopentane					-367.3							
$C_{15}H_{30}O_2$	Pentadecanoic acid	-861.7			443.3	-811.7				-699.0			
$C_{15}H_{30}O_2$	Methyl tetradecanoate					-743.9				-656.9			
$C_{15}H_{32}O$	1-Pentadecanol	-658.2											
$C_{16}H_{10}$	Fluoranthene	189.9		230.6	230.2					289.0			
$C_{16}H_{10}$	Pyrene	125.5		224.9	229.7					225.7			
$C_{16}H_{22}O_4$	Dibutyl phthalate					-842.6				-750.9			
$C_{16}H_{22}O_{11}$	α-D-Glucose pentaacetate	-2249.4											
$C_{16}H_{22}O_{11}$	β-D-Glucose pentaacetate	-2232.6											
$C_{16}H_{26}$	Decylbenzene					-218.3				-138.6			
$C_{16}H_{32}$	1-Hexadecene					-328.7		587.9	488.9	-248.4			
$C_{16}H_{32}O_2$	Hexadecanoic acid	-891.5		452.4	460.7	-838.1				-737.1			
$C_{16}H_{32}O_2$	Methyl pentadecanoate					-771.0				-680.0			
$C_{16}H_{33}Br$	1-Bromohexadecane					-444.5				-350.2			
$C_{16}H_{34}$	Hexadecane					-456.1			501.6	-374.8			
$C_{16}H_{34}O$	1-Hexadecanol	-686.5			422.0					-517.0			
$C_{16}H_{36}IN$	Tetrabutylammonium iodide	-498.6											
$C_{17}H_{34}O_2$	Heptadecanoic acid	-924.4			475.7	-865.6							
$C_{18}H_{12}$	Benz[a]anthracene	170.8								293.0			
$C_{18}H_{12}$	Chrysene	145.3								269.8			
$C_{18}H_{14}$	o-Terphenyl			298.8	274.8			337.1	369.1				
$C_{18}H_{14}$	p-Terphenyl	163.0		285.6	278.7					279.0			
$C_{18}H_{15}N$	Triphenylamine	234.7								326.8			
$C_{18}H_{15}O_4P$	Triphenyl phosphate			397.5	356.2								
$C_{18}H_{15}P$	Triphenylphosphine				312.5								
$C_{18}H_{30}$	1,3,5-Tri-tert-butylbenzene	-320.0											
$C_{18}H_{34}O_2$	Oleic acid								577.0				
$C_{18}H_{34}O_4$	Dibutyl sebacate								619.0				
$C_{18}H_{36}O_2$	Stearic acid	-947.7			501.5	-884.7				-781.2			
$C_{18}H_{37}Cl$	1-Chlorooctadecane					-544.1				-446.0			
$C_{18}H_{38}$	Octadecane	-567.4		480.2	485.6					-414.6			
$C_{18}H_{39}N$	Trihexylamine					-433.0							
$C_{19}H_{16}O$	Triphenylmethanol	-2.5											
$C_{19}H_{36}O_2$	Methyl oleate					-734.5				-649.9			
$C_{19}H_{36}O_2$	Methyl trans-9-octadecenoate					-737.0							
$C_{20}H_{12}$	Perylene	182.8		264.6	274.9								
$C_{20}H_{12}$	Benzo[a]pyrene												254.8

Molecular formula	Name	Crystal				Liquid				Gas			
		$\Delta_f H°$ kJ/mol	$\Delta_f G°$ kJ/mol	$S°$ J/mol K	C_p J/mol K	$\Delta_f H°$ kJ/mol	$\Delta_f G°$ kJ/mol	$S°$ J/mol K	C_p J/mol K	$\Delta_f H°$ kJ/mol	$\Delta_f G°$ kJ/mol	$S°$ J/mol K	C_p J/mol K
$C_{20}H_{14}O_4$	Diphenyl phthalate	-489.2											
$C_{20}H_{38}O_2$	Ethyl oleate					-775.8							
$C_{20}H_{38}O_2$	Ethyl *trans*-9-octadecenoate					-773.3							
$C_{20}H_{40}O_2$	Eicosanoic acid	-1011.9			545.1	-940.0				-812.4			
$C_{21}H_{21}O_4P$	Tri-*o*-cresyl phosphate			570.0	578.0								
$C_{22}H_{14}$	Dibenz[a,h]anthracene												283.9
$C_{22}H_{42}O_2$	*trans*-13-Docosenoic acid	-960.7											
$C_{22}H_{42}O_2$	Butyl oleate					-816.9							
$C_{22}H_{44}O_2$	Butyl stearate												
$C_{24}H_{38}O_4$	Bis(2-ethylhexyl) phthalate								704.7				
$C_{24}H_{51}N$	Trioctylamine					-585.0							
$C_{26}H_{18}$	9,10-Diphenylanthracene	308.7								465.6			
$C_{26}H_{54}$	5-Butyldocosane					-713.5				-587.6			
$C_{26}H_{54}$	11-Butyldocosane					-716.0				-593.4			
$C_{28}H_{18}$	9,9'-Bianthracene	326.2								454.3			
$C_{31}H_{64}$	11-Decylheneicosane					-848.0				-705.8			
$C_{32}H_{66}$	Dotriacontane	-968.3								-697.2			
C_{60}	Carbon (fullerene-C_{60})	2327.0	2302.0	426.0	520.0					2502.0	2442.0	544.0	512.0
C_{70}	Carbon (fullerene-C_{70})	2555.0	2537.0	464.0	650.0					2755.0	2692.0	614.0	585.0

THERMODYNAMIC PROPERTIES AS A FUNCTION OF TEMPERATURE

L. V. Gurvich, V. S. Iorish, V. S. Yungman, and O. V. Dorofeeva

The thermodynamic properties $C_p^\circ(T)$, $S^\circ(T)$, $H^\circ(T)-H^\circ(T_r)$, $-[G^\circ(T)-H^\circ(T_r)]/T$ and formation properties $\Delta_f H^\circ(T)$, $\Delta_f G^\circ(T)$, $\log K_f^\circ(T)$ are tabulated as functions of temperature in the range 298.15 to 1500 K for 80 substances in the standard state. The reference temperature, T_r, is equal to 298.15 K. The standard state pressure is taken as 1 bar (100,000 Pa). The tables are presented in the JANAF Thermochemical Tables format (Reference 2). The numerical data are extracted from IVTANTHERMO databases except for C_2H_4O, C_3H_6O, C_6H_6, C_6H_6O, $C_{10}H_8$, and CH_5N, which are based upon TRC Tables. See the references for information on standard states and other details.

References

1. Gurvich, L. V., Veyts, I. V., and Alcock, C. B., Eds., *Thermodynamic Properties of Individual Substances, 4th ed.*, Hemisphere Publishing Corp., New York, 1989.
2. Chase, M. W., et al., *JANAF Thermochemical Tables, 3rd ed., J. Phys. Chem. Ref. Data*, 14, Suppl. 1, 1985.

Order of Listing of Tables

No.	Formula	Name	State	No.	Formula	Name	State
1	Ar	Argon	g	42	$CuCl_2$	Copper dichloride	g
2	Br	Bromine	g	43	F	Fluorine	g
3	Br_2	Dibromine	g	44	F_2	Difluorine	g
4	BrH	Hydrogen bromide	g	45	FH	Hydrogen fluoride	g
5	C	Carbon (graphite)	cr	46	Ge	Germanium	cr, l
6	C	Carbon (diamond)	cr	47	Ge	Germanium	g
7	C_2	Dicarbon	g	48	GeO_2	Germanium dioxide	cr, l
8	C_3	Tricarbon	g	49	$GeCl_4$	Germanium tetrachloride	g
9	CO	Carbon oxide	g	50	H	Hydrogen	g
10	CO_2	Carbon dioxide	g	51	H_2	Dihydrogen	g
11	CH_4	Methane	g	52	HO	Hydroxyl	g
12	C_2H_2	Acetylene	g	53	H_2O	Water	l
13	C_2H_4	Ethylene	g	54	H_2O	Water	g
14	C_2H_6	Ethane	g	55	I	Iodine	g
15	C_3H_6	Cyclopropane	g	56	I_2	Diiodine	cr, l
16	C_3H_8	Propane	g	57	I_2	Diiodine	g
17	C_6H_6	Benzene	l	58	IH	Hydrogen iodide	g
18	C_6H_6	Benzene	g	59	K	Potassium	cr, l
19	$C_{10}H_8$	Naphthalene	cr, l	60	K	Potassium	g
20	$C_{10}H_8$	Naphthalene	g	61	K_2O	Dipotassium oxide	cr, l
21	CH_2O	Formaldehyde	g	62	KOH	Potassium hydroxide	cr, l
22	CH_4O	Methanol	g	63	KOH	Potassium hydroxide	g
23	C_2H_4O	Acetaldehyde	g	64	KCl	Potassium chloride	cr, l
24	C_2H_6O	Ethanol	g	65	KCl	Potassium chloride	g
25	$C_2H_4O_2$	Acetic acid	g	66	N_2	Dinitrogen	g
26	C_3H_6O	Acetone	g	67	NO	Nitric oxide	g
27	C_6H_6O	Phenol	g	68	NO_2	Nitrogen dioxide	g
28	CF_4	Carbon tetrafluoride	g	69	NH_3	Ammonia	g
29	CHF_3	Trifluoromethane	g	70	O	Oxygen	g
30	$CClF_3$	Chlorotrifluoromethane	g	71	O_2	Dioxygen	g
31	CCl_2F_2	Dichlorodifluoromethane	g	72	S	Sulfur	cr, l
32	$CHClF_2$	Chlorodifluoromethane	g	73	S	Sulfur	g
33	CH_5N	Methylamine	g	74	S_2	Disulfur	g
34	Cl	Chlorine	g	75	S_8	Octasulfur	g
35	Cl_2	Dichlorine	g	76	SO_2	Sulfur dioxide	g
36	ClH	Hydrogen chloride	g	77	Si	Silicon	cr
37	Cu	Copper	cr, l	78	Si	Silicon	g
38	Cu	Copper	g	79	SiO_2	Silicon dioxide	cr
39	CuO	Copper oxide	cr	80	$SiCl_4$	Silicon tetrachloride	g
40	Cu_2O	Dicopper oxide	cr				
41	$CuCl_2$	Copper dichloride	cr, l				

T/K	$C_p°$	J/K·mol $S°$	$-(G°-H°(T_r))/T$	$H°-H°(T_r)$	kJ/mol $\Delta_f H°$	$\Delta_f G°$	Log K_f
1. ARGON	**Ar (g)**						
298.15	20.786	154.845	154.845	0.000	0.000	0.000	0.000
300	20.786	154.973	154.845	0.038	0.000	0.000	0.000
400	20.786	160.953	155.660	2.117	0.000	0.000	0.000
500	20.786	165.591	157.200	4.196	0.000	0.000	0.000
600	20.786	169.381	158.924	6.274	0.000	0.000	0.000
700	20.786	172.585	160.653	8.353	0.000	0.000	0.000
800	20.786	175.361	162.322	10.431	0.000	0.000	0.000
900	20.786	177.809	163.909	12.510	0.000	0.000	0.000
1000	20.786	179.999	165.410	14.589	0.000	0.000	0.000
1100	20.786	181.980	166.828	16.667	0.000	0.000	0.000
1200	20.786	183.789	168.167	18.746	0.000	0.000	0.000
1300	20.786	185.453	169.434	20.824	0.000	0.000	0.000
1400	20.786	186.993	170.634	22.903	0.000	0.000	0.000
1500	20.786	188.427	171.773	24.982	0.000	0.000	0.000
2. BROMINE	**Br (g)**						
298.15	20.786	175.017	175.017	0.000	111.870	82.379	−14.432
300	20.786	175.146	175.018	0.038	111.838	82.196	−14.311
400	20.787	181.126	175.833	2.117	96.677	75.460	−9.854
500	20.798	185.765	177.373	4.196	96.910	70.129	−7.326
600	20.833	189.559	179.097	6.277	97.131	64.752	−5.637
700	20.908	192.776	180.827	8.364	97.348	59.338	−4.428
800	21.027	195.575	182.499	10.461	97.568	53.893	−3.519
900	21.184	198.061	184.093	12.571	97.796	48.420	−2.810
1000	21.365	200.302	185.604	14.698	98.036	42.921	−2.242
1100	21.559	202.347	187.034	16.844	98.291	37.397	−1.776
1200	21.752	204.231	188.390	19.010	98.560	31.850	−1.386
1300	21.937	205.980	189.676	21.195	98.844	26.279	−1.056
1400	22.107	207.612	190.900	23.397	99.141	20.686	−0.772
1500	22.258	209.142	192.065	25.615	99.449	15.072	−0.525
3. DIBROMINE	**Br$_2$ (g)**						
298.15	36.057	245.467	245.467	0.000	30.910	3.105	−0.544
300	36.074	245.690	245.468	0.067	30.836	2.933	−0.511
332.25	36.340	249.387	245.671	1.235		pressure = 1 bar	
400	36.729	256.169	246.892	3.711	0.000	0.000	0.000
500	37.082	264.406	249.600	7.403	0.000	0.000	0.000
600	37.305	271.188	252.650	11.123	0.000	0.000	0.000
700	37.464	276.951	255.720	14.862	0.000	0.000	0.000
800	37.590	281.962	258.694	18.615	0.000	0.000	0.000
900	37.697	286.396	261.530	22.379	0.000	0.000	0.000
1000	37.793	290.373	264.219	26.154	0.000	0.000	0.000
1100	37.883	293.979	266.763	29.938	0.000	0.000	0.000
1200	37.970	297.279	269.170	33.730	0.000	0.000	0.000
1300	38.060	300.322	271.451	37.532	0.000	0.000	0.000
1400	38.158	303.146	273.615	41.343	0.000	0.000	0.000
1500	38.264	305.782	275.673	45.164	0.000	0.000	0.000
4. HYDROGEN BROMIDE	**HBr (g)**						
298.15	29.141	198.697	198.697	0.000	−36.290	−53.360	9.348
300	29.141	198.878	198.698	0.054	−36.333	−53.466	9.309
400	29.220	207.269	199.842	2.971	−52.109	−55.940	7.305
500	29.454	213.811	202.005	5.903	−52.484	−56.854	5.939
600	29.872	219.216	204.436	8.868	−52.844	−57.694	5.023
700	30.431	223.861	206.886	11.882	−53.168	−58.476	4.363
800	31.063	227.965	209.269	14.957	−53.446	−59.214	3.866
900	31.709	231.661	211.555	18.095	−53.677	−59.921	3.478
1000	32.335	235.035	213.737	21.298	−53.864	−60.604	3.166
1100	32.919	238.145	215.816	24.561	−54.012	−61.271	2.909

	J/K·mol			kJ/mol			
T/K	C_p°	S°	$-(G^\circ-H^\circ(T_r))/T$	$H^\circ-H^\circ(T_r)$	$\Delta_f H^\circ$	$\Delta_f G^\circ$	Log K_f
1200	33.454	241.032	217.799	27.880	−54.129	−61.925	2.696
1300	33.938	243.729	219.691	31.250	−54.220	−62.571	2.514
1400	34.374	246.261	221.499	34.666	−54.291	−63.211	2.358
1500	34.766	248.646	223.230	38.123	−54.348	−63.846	2.223

5. CARBON (GRAPHITE)   C (cr; graphite)

298.15	8.536	5.740	5.740	0.000	0.000	0.000	0.000
300	8.610	5.793	5.740	0.016	0.000	0.000	0.000
400	11.974	8.757	6.122	1.054	0.000	0.000	0.000
500	14.537	11.715	6.946	2.385	0.000	0.000	0.000
600	16.607	14.555	7.979	3.945	0.000	0.000	0.000
700	18.306	17.247	9.113	5.694	0.000	0.000	0.000
800	19.699	19.785	10.290	7.596	0.000	0.000	0.000
900	20.832	22.173	11.479	9.625	0.000	0.000	0.000
1000	21.739	24.417	12.662	11.755	0.000	0.000	0.000
1100	22.452	26.524	13.827	13.966	0.000	0.000	0.000
1200	23.000	28.502	14.968	16.240	0.000	0.000	0.000
1300	23.409	30.360	16.082	18.562	0.000	0.000	0.000
1400	23.707	32.106	17.164	20.918	0.000	0.000	0.000
1500	23.919	33.749	18.216	23.300	0.000	0.000	0.000

6. CARBON (DIAMOND)   C (cr; diamond)

298.15	6.109	2.362	2.362	0.000	1.850	2.857	−0.501
300	6.201	2.400	2.362	0.011	1.846	2.863	−0.499
400	10.321	4.783	2.659	0.850	1.645	3.235	−0.422
500	13.404	7.431	3.347	2.042	1.507	3.649	−0.381
600	15.885	10.102	4.251	3.511	1.415	4.087	−0.356
700	17.930	12.709	5.274	5.205	1.361	4.537	−0.339
800	19.619	15.217	6.361	7.085	1.338	4.993	−0.326
900	21.006	17.611	7.479	9.118	1.343	5.450	−0.316
1000	22.129	19.884	8.607	11.277	1.372	5.905	−0.308
1100	23.020	22.037	9.731	13.536	1.420	6.356	−0.302
1200	23.709	24.071	10.842	15.874	1.484	6.802	−0.296
1300	24.222	25.990	11.934	18.272	1.561	7.242	−0.291
1400	24.585	27.799	13.003	20.714	1.646	7.675	−0.286
1500	24.824	29.504	14.047	23.185	1.735	8.103	−0.282

7. DICARBON   C$_2$ (g)

298.15	43.548	197.095	197.095	0.000	830.457	775.116	−135.795
300	43.575	197.365	197.096	0.081	830.506	774.772	−134.898
400	42.169	209.809	198.802	4.403	832.751	755.833	−98.700
500	39.529	218.924	201.959	8.483	834.170	736.423	−76.933
600	37.837	225.966	205.395	12.342	834.909	716.795	−62.402
700	36.984	231.726	208.758	16.078	835.148	697.085	−52.016
800	36.621	236.637	211.943	19.755	835.020	677.366	−44.227
900	36.524	240.943	214.931	23.411	834.618	657.681	−38.170
1000	36.569	244.793	217.728	27.065	834.012	638.052	−33.328
1100	36.696	248.284	220.349	30.728	833.252	618.492	−29.369
1200	36.874	251.484	222.812	34.406	832.383	599.006	−26.074
1300	37.089	254.444	225.133	38.104	831.437	579.596	−23.288
1400	37.329	257.201	227.326	41.824	830.445	560.261	−20.903
1500	37.589	259.785	229.405	45.570	829.427	540.997	−18.839

8. TRICARBON   C$_3$ (g)

298.15	42.202	237.611	237.611	0.000	839.958	774.249	−135.643
300	42.218	237.872	237.611	0.078	839.989	773.841	−134.736
400	43.383	250.164	239.280	4.354	841.149	751.592	−98.147
500	44.883	260.003	242.471	8.766	841.570	729.141	−76.172
600	46.406	268.322	246.104	13.331	841.453	706.659	−61.519
700	47.796	275.582	249.807	18.042	840.919	684.230	−51.057
800	48.997	282.045	253.440	22.884	840.053	661.901	−43.217

T/K	C_p°	S°	$-(G^\circ-H^\circ(T_r))/T$	$H^\circ-H^\circ(T_r)$	$\Delta_f H^\circ$	$\Delta_f G^\circ$	Log K_f
		J/K·mol			kJ/mol		
900	50.006	287.876	256.948	27.835	838.919	639.698	−37.127
1000	50.844	293.189	260.310	32.879	837.572	617.633	−32.261
1100	51.535	298.069	263.524	37.999	836.059	595.711	−28.288
1200	52.106	302.578	266.593	43.182	834.420	573.933	−24.982
1300	52.579	306.768	269.524	48.417	832.690	552.295	−22.191
1400	52.974	310.679	272.326	53.695	830.899	530.793	−19.804
1500	53.307	314.346	275.006	59.010	829.068	509.421	−17.739

9. CARBON OXIDE   CO (g)

T/K	C_p°	S°	$-(G^\circ-H^\circ(T_r))/T$	$H^\circ-H^\circ(T_r)$	$\Delta_f H^\circ$	$\Delta_f G^\circ$	Log K_f
298.15	29.141	197.658	197.658	0.000	−110.530	−137.168	24.031
300	29.142	197.838	197.659	0.054	−110.519	−137.333	23.912
400	29.340	206.243	198.803	2.976	−110.121	−146.341	19.110
500	29.792	212.834	200.973	5.930	−110.027	−155.412	16.236
600	30.440	218.321	203.419	8.941	−110.157	−164.480	14.319
700	31.170	223.067	205.895	12.021	−110.453	−173.513	12.948
800	31.898	227.277	208.309	15.175	−110.870	−182.494	11.915
900	32.573	231.074	210.631	18.399	−111.378	−191.417	11.109
1000	33.178	234.538	212.851	21.687	−111.952	−200.281	10.461
1100	33.709	237.726	214.969	25.032	−112.573	−209.084	9.928
1200	34.169	240.679	216.990	28.426	−113.228	−217.829	9.482
1300	34.568	243.430	218.920	31.864	−113.904	−226.518	9.101
1400	34.914	246.005	220.763	35.338	−114.594	−235.155	8.774
1500	35.213	248.424	222.527	38.845	−115.291	−243.742	8.488

10. CARBON DIOXIDE CO_2 (g)

T/K	C_p°	S°	$-(G^\circ-H^\circ(T_r))/T$	$H^\circ-H^\circ(T_r)$	$\Delta_f H^\circ$	$\Delta_f G^\circ$	Log K_f
298.15	37.135	213.783	213.783	0.000	−393.510	−394.373	69.092
300	37.220	214.013	213.784	0.069	−393.511	−394.379	68.667
400	41.328	225.305	215.296	4.004	−393.586	−394.656	51.536
500	44.627	234.895	218.280	8.307	−393.672	−394.914	41.256
600	47.327	243.278	221.762	12.909	−393.791	−395.152	34.401
700	49.569	250.747	225.379	17.758	−393.946	−395.367	29.502
800	51.442	257.492	228.978	22.811	−394.133	−395.558	25.827
900	53.008	263.644	232.493	28.036	−394.343	−395.724	22.967
1000	54.320	269.299	235.895	33.404	−394.568	−395.865	20.678
1100	55.423	274.529	239.172	38.893	−394.801	−395.984	18.803
1200	56.354	279.393	242.324	44.483	−395.035	−396.081	17.241
1300	57.144	283.936	245.352	50.159	−395.265	−396.159	15.918
1400	57.818	288.196	248.261	55.908	−395.488	−396.219	14.783
1500	58.397	292.205	251.059	61.719	−395.702	−396.264	13.799

11. METHANE CH_4 (g)

T/K	C_p°	S°	$-(G^\circ-H^\circ(T_r))/T$	$H^\circ-H^\circ(T_r)$	$\Delta_f H^\circ$	$\Delta_f G^\circ$	Log K_f
298.15	35.695	186.369	186.369	0.000	−74.600	−50.530	8.853
300	35.765	186.590	186.370	0.066	−74.656	−50.381	8.772
400	40.631	197.501	187.825	3.871	−77.703	−41.827	5.462
500	46.627	207.202	190.744	8.229	−80.520	−32.525	3.398
600	52.742	216.246	194.248	13.199	−82.969	−22.690	1.975
700	58.603	224.821	198.008	18.769	−85.023	−12.476	0.931
800	64.084	233.008	201.875	24.907	−86.693	−1.993	0.130
900	69.137	240.852	205.773	31.571	−88.006	8.677	−0.504
1000	73.746	248.379	209.660	38.719	−88.996	19.475	−1.017
1100	77.919	255.607	213.511	46.306	−89.698	30.358	−1.442
1200	81.682	262.551	217.310	54.289	−90.145	41.294	−1.797
1300	85.067	269.225	221.048	62.630	−90.367	52.258	−2.100
1400	88.112	275.643	224.720	71.291	−90.390	63.231	−2.359
1500	90.856	281.817	228.322	80.242	−90.237	74.200	−2.584

12. ACETYLENE C_2H_2 (g)

T/K	C_p°	S°	$-(G^\circ-H^\circ(T_r))/T$	$H^\circ-H^\circ(T_r)$	$\Delta_f H^\circ$	$\Delta_f G^\circ$	Log K_f
298.15	44.036	200.927	200.927	0.000	227.400	209.879	−36.769
300	44.174	201.199	200.927	0.082	227.397	209.770	−36.524
400	50.388	214.814	202.741	4.829	227.161	203.928	−26.630
500	54.751	226.552	206.357	10.097	226.846	198.154	−20.701

	J/K·mol			kJ/mol			
T/K	C_p°	S°	$-(G^\circ - H^\circ(T_r))/T$	$H^\circ - H^\circ(T_r)$	$\Delta_f H^\circ$	$\Delta_f G^\circ$	Log K_f
600	58.121	236.842	210.598	15.747	226.445	192.452	−16.754
700	60.970	246.021	215.014	21.704	225.968	186.823	−13.941
800	63.511	254.331	219.418	27.931	225.436	181.267	−11.835
900	65.831	261.947	223.726	34.399	224.873	175.779	−10.202
1000	67.960	268.995	227.905	41.090	224.300	170.355	−8.898
1100	69.909	275.565	231.942	47.985	223.734	164.988	−7.835
1200	71.686	281.725	235.837	55.067	223.189	159.672	−6.950
1300	73.299	287.528	239.592	62.317	222.676	154.400	−6.204
1400	74.758	293.014	243.214	69.721	222.203	149.166	−5.565
1500	76.077	298.218	246.709	77.264	221.774	143.964	−5.013

13. ETHYLENE C_2H_4 (g)

298.15	42.883	219.316	219.316	0.000	52.400	68.358	−11.976
300	43.059	219.582	219.317	0.079	52.341	68.457	−11.919
400	53.045	233.327	221.124	4.881	49.254	74.302	−9.703
500	62.479	246.198	224.864	10.667	46.533	80.887	−8.450
600	70.673	258.332	229.441	17.335	44.221	87.982	−7.659
700	77.733	269.770	234.393	24.764	42.278	95.434	−7.121
800	83.868	280.559	239.496	32.851	40.655	103.142	−6.734
900	89.234	290.754	244.630	41.512	39.310	111.036	−6.444
1000	93.939	300.405	249.730	50.675	38.205	119.067	−6.219
1100	98.061	309.556	254.756	60.280	37.310	127.198	−6.040
1200	101.670	318.247	259.688	70.271	36.596	135.402	−5.894
1300	104.829	326.512	264.513	80.599	36.041	143.660	−5.772
1400	107.594	334.384	269.225	91.223	35.623	151.955	−5.669
1500	110.018	341.892	273.821	102.107	35.327	160.275	−5.581

14. ETHANE C_2H_6 (g)

298.15	52.487	229.161	229.161	0.000	−84.000	−32.015	5.609
300	52.711	229.487	229.162	0.097	−84.094	−31.692	5.518
400	65.459	246.378	231.379	5.999	−88.988	−13.473	1.759
500	77.941	262.344	235.989	13.177	−93.238	5.912	−0.618
600	89.188	277.568	241.660	21.545	−96.779	26.086	−2.271
700	99.136	292.080	247.835	30.972	−99.663	46.800	−3.492
800	107.936	305.904	254.236	41.334	−101.963	67.887	−4.433
900	115.709	319.075	260.715	52.525	−103.754	89.231	−5.179
1000	122.552	331.628	267.183	64.445	−105.105	110.750	−5.785
1100	128.553	343.597	273.590	77.007	−106.082	132.385	−6.286
1200	133.804	355.012	279.904	90.131	−106.741	154.096	−6.708
1300	138.391	365.908	286.103	103.746	−107.131	175.850	−7.066
1400	142.399	376.314	292.178	117.790	−107.292	197.625	−7.373
1500	145.905	386.260	298.121	132.209	−107.260	219.404	−7.640

15. CYCLOPROPANE C_3H_6 (g)

298.15	55.571	237.488	237.488	0.000	53.300	104.514	−18.310
300	55.941	237.832	237.489	0.103	53.195	104.832	−18.253
400	76.052	256.695	239.924	6.708	47.967	122.857	−16.043
500	93.859	275.637	245.177	15.230	43.730	142.091	−14.844
600	108.542	294.092	251.801	25.374	40.405	162.089	−14.111
700	120.682	311.763	259.115	36.854	37.825	182.583	−13.624
800	130.910	328.564	266.755	49.447	35.854	203.404	−13.281
900	139.658	344.501	274.516	62.987	34.384	224.441	−13.026
1000	147.207	359.616	282.277	77.339	33.334	245.618	−12.830
1100	153.749	373.961	289.965	92.395	32.640	266.883	−12.673
1200	159.432	387.588	297.538	108.060	32.249	288.197	−12.545
1300	164.378	400.549	304.967	124.257	32.119	309.533	−12.437
1400	168.689	412.892	312.239	140.915	32.215	330.870	−12.345
1500	172.453	424.662	319.344	157.976	32.507	352.193	−12.264

16. PROPANE C_3H_8 (g)

298.15	73.597	270.313	270.313	0.000	−103.847	−23.458	4.110

T/K	C_p°	S°	$-(G^\circ - H^\circ(T_r))/T$	$H^\circ - H^\circ(T_r)$	$\Delta_f H^\circ$	$\Delta_f G^\circ$	$\text{Log } K_f$
		J/K·mol			kJ/mol		
300	73.931	270.769	270.314	0.136	-103.972	-22.959	3.997
400	94.014	294.739	273.447	8.517	-110.33	15.029	-0.657
500	112.591	317.768	280.025	18.872	-115.658	34.507	-3.605
600	128.700	339.753	288.162	30.955	-119.973	64.961	-5.655
700	142.674	360.668	297.039	44.540	-123.384	96.065	-7.168
800	154.766	380.528	306.245	59.427	-126.016	127.603	-8.331
900	165.352	399.381	315.555	75.444	-127.982	159.430	-9.253
1000	174.598	417.293	324.841	92.452	-129.380	191.444	-10.000
1100	182.673	434.321	334.026	110.325	-130.296	223.574	-10.617
1200	189.745	450.526	343.064	128.954	-130.802	255.770	-11.133
1300	195.853	465.961	351.929	148.241	-130.961	287.993	-11.572
1400	201.209	480.675	360.604	168.100	-130.829	320.217	-11.947
1500	205.895	494.721	369.080	188.460	-130.445	352.422	-12.272

17. BENZENE C_6H_6 (l)

T/K	C_p°	S°	$-(G^\circ - H^\circ(T_r))/T$	$H^\circ - H^\circ(T_r)$	$\Delta_f H^\circ$	$\Delta_f G^\circ$	$\text{Log } K_f$
298.15	135.950	173.450	173.450	0.000	49.080	124.521	-21.815
300	136.312	174.292	173.453	.252	49.077	124.989	-21.762
400	161.793	216.837	179.082	15.102	48.978	150.320	-19.630
500	207.599	257.048	190.639	33.204	50.330	175.559	-18.340

18. BENZENE C_6H_6 (g)

T/K	C_p°	S°	$-(G^\circ - H^\circ(T_r))/T$	$H^\circ - H^\circ(T_r)$	$\Delta_f H^\circ$	$\Delta_f G^\circ$	$\text{Log } K_f$
298.15	82.430	269.190	269.190	0.000	82.880	129.750	-22.731
300	83.020	269.700	269.190	0.153	82.780	130.040	-22.641
400	113.510	297.840	272.823	10.007	77.780	146.570	-19.140
500	139.340	326.050	280.658	22.696	73.740	164.260	-17.160
600	160.090	353.360	290.517	37.706	70.490	182.680	-15.903
700	176.790	379.330	301.360	54.579	67.910	201.590	-15.042
800	190.460	403.860	312.658	72.962	65.910	220.820	-14.418
900	201.840	426.970	324.084	92.597	64.410	240.280	-13.945
1000	211.430	448.740	335.473	113.267	63.340	259.890	-13.575
1100	219.580	469.280	346.710	134.827	62.620	277.640	-13.184
1200	226.540	488.690	357.743	157.137	62.200	299.320	-13.029
1300	232.520	507.070	368.534	180.097	62.000	319.090	-12.821
1400	237.680	524.490	379.056	203.607	61.990	338.870	-12.643
1500	242.140	541.040	389.302	227.607	62.110	358.640	-12.489

19. NAPHTHALENE $C_{10}H_8$ (cr, l)

T/K	C_p°	S°	$-(G^\circ - H^\circ(T_r))/T$	$H^\circ - H^\circ(T_r)$	$\Delta_f H^\circ$	$\Delta_f G^\circ$	$\text{Log } K_f$
298.15	165.720	167.390	167.390	0.000	78.530	201.585	-35.316
300	167.001	168.419	167.393	0.308	78.466	202.349	-35.232
353.43	208.722	198.948	169.833	10.290	96.099	224.543	-33.186

PHASE TRANSITION: $\Delta_{trs} H$ = 18.980 kJ/mol, $\Delta_{trs} S$ = 53.702 J/K·mol, cr–l

T/K	C_p°	S°	$-(G^\circ - H^\circ(T_r))/T$	$H^\circ - H^\circ(T_r)$	$\Delta_f H^\circ$	$\Delta_f G^\circ$	$\text{Log } K_f$
353.43	217.200	252.650	169.833	29.270	96.099	224.543	-33.186
400	241.577	280.916	181.124	39.917	96.067	241.475	-31.533
470	276.409	322.712	199.114	58.091	97.012	266.859	-29.658

20. NAPHTHALENE $C_{10}H_8$ (g)

T/K	C_p°	S°	$-(G^\circ - H^\circ(T_r))/T$	$H^\circ - H^\circ(T_r)$	$\Delta_f H^\circ$	$\Delta_f G^\circ$	$\text{Log } K_f$
298.15	131.920	333.150	333.150	0.000	150.580	224.100	-39.260
300	132.840	333.970	333.157	0.244	150.450	224.560	-39.098
400	180.070	378.800	338.950	15.940	144.190	250.270	-32.681
500	219.740	423.400	351.400	36.000	139.220	277.340	-28.973
600	251.530	466.380	367.007	59.624	135.350	305.330	-26.581
700	277.010	507.140	384.146	86.096	132.330	333.950	-24.919
800	297.730	545.520	401.935	114.868	130.050	362.920	-23.696
900	314.850	581.610	419.918	145.523	128.430	392.150	-22.759
1000	329.170	615.550	437.806	177.744	127.510	421.700	-22.027
1100	341.240	647.500	455.426	211.281	127.100	450.630	-21.398
1200	351.500	677.650	472.707	245.932	126.960	480.450	-20.913
1300	360.260	706.130	489.568	281.531	127.060	509.770	-20.482
1400	367.780	733.110	506.009	317.941	127.390	539.740	-20.137
1500	374.270	758.720	522.019	355.051	127.920	568.940	-19.812

T/K	C_p°	S°	$-(G^\circ-H^\circ(T_r))/T$	$H^\circ-H^\circ(T_r)$	$\Delta_f H^\circ$	$\Delta_f G^\circ$	Log K_f
	J/K·mol			kJ/mol			

21. FORMALDEHYDE   H$_2$CO (g)

T/K	C_p°	S°	$-(G^\circ-H^\circ(T_r))/T$	$H^\circ-H^\circ(T_r)$	$\Delta_f H^\circ$	$\Delta_f G^\circ$	Log K_f
298.15	35.387	218.760	218.760	0.000	−108.700	−102.667	17.987
300	35.443	218.979	218.761	0.066	−108.731	−102.630	17.869
400	39.240	229.665	220.192	3.789	−110.438	−100.340	13.103
500	43.736	238.900	223.028	7.936	−112.073	−97.623	10.198
600	48.181	247.270	226.381	12.534	−113.545	−94.592	8.235
700	52.280	255.011	229.924	17.560	−114.833	−91.328	6.815
800	55.941	262.236	233.517	22.975	−115.942	−87.893	5.739
900	59.156	269.014	237.088	28.734	−116.889	−84.328	4.894
1000	61.951	275.395	240.603	34.792	−117.696	−80.666	4.213
1100	64.368	281.416	244.042	41.111	−118.382	−76.929	3.653
1200	66.453	287.108	247.396	47.655	−118.966	−73.134	3.183
1300	68.251	292.500	250.660	54.392	−119.463	−69.294	2.784
1400	69.803	297.616	253.833	61.297	−119.887	−65.418	2.441
1500	71.146	302.479	256.915	68.346	−120.249	−61.514	2.142

22. METHANOL   CH$_3$OH (g)

T/K	C_p°	S°	$-(G^\circ-H^\circ(T_r))/T$	$H^\circ-H^\circ(T_r)$	$\Delta_f H^\circ$	$\Delta_f G^\circ$	Log K_f
298.15	44.101	239.865	239.865	0.000	−201.000	−162.298	28.434
300	44.219	240.139	239.866	0.082	−201.068	−162.057	28.216
400	51.713	253.845	241.685	4.864	−204.622	−148.509	19.393
500	59.800	266.257	245.374	10.442	−207.750	−134.109	14.010
600	67.294	277.835	249.830	16.803	−210.387	−119.125	10.371
700	73.958	288.719	254.616	23.873	−212.570	−103.737	7.741
800	79.838	298.987	259.526	31.569	−214.350	−88.063	5.750
900	85.025	308.696	264.455	39.817	−215.782	−72.188	4.190
1000	89.597	317.896	269.343	48.553	−216.916	−56.170	2.934
1100	93.624	326.629	274.158	57.718	−217.794	−40.050	1.902
1200	97.165	334.930	278.879	67.262	−218.457	−23.861	1.039
1300	100.277	342.833	283.497	77.137	−218.936	−7.624	0.306
1400	103.014	350.367	288.007	87.304	−219.261	8.644	−0.322
1500	105.422	357.558	292.405	97.729	−219.456	24.930	−0.868

23. ACETALDEHYDE   C$_2$H$_4$O (g)

T/K	C_p°	S°	$-(G^\circ-H^\circ(T_r))/T$	$H^\circ-H^\circ(T_r)$	$\Delta_f H^\circ$	$\Delta_f G^\circ$	Log K_f
298.15	55.318	263.840	263.840	0.000	−166.190	−133.010	23.302
300	55.510	264.180	263.837	0.103	−166.250	−132.800	23.122
400	66.282	281.620	266.147	6.189	−169.530	−121.130	15.818
500	76.675	297.540	270.850	13.345	−172.420	−108.700	11.356
600	85.942	312.360	276.550	21.486	−174.870	−95.720	8.334
700	94.035	326.230	282.667	30.494	−176.910	−82.350	6.145
800	101.070	339.260	288.938	40.258	−178.570	−68.730	4.487
900	107.190	351.520	295.189	50.698	−179.880	−54.920	3.187
1000	112.490	363.100	301.431	61.669	−180.850	−40.930	2.138
1100	117.080	374.040	307.537	73.153	−181.560	−27.010	1.283
1200	121.060	384.400	313.512	85.065	−182.070	−12.860	0.560
1300	124.500	394.230	319.350	97.344	−182.420	1.240	−0.050
1400	127.490	403.570	325.031	109.954	−182.640	15.470	−0.577
1500	130.090	412.460	330.571	122.834	−182.750	29.580	−1.030

24. ETHANOL   C$_2$H$_5$OH (g)

T/K	C_p°	S°	$-(G^\circ-H^\circ(T_r))/T$	$H^\circ-H^\circ(T_r)$	$\Delta_f H^\circ$	$\Delta_f G^\circ$	Log K_f
298.15	65.652	281.622	281.622	0.000	−234.800	−167.874	29.410
300	65.926	282.029	281.623	0.122	−234.897	−167.458	29.157
400	81.169	303.076	284.390	7.474	−239.826	−144.216	18.832
500	95.400	322.750	290.115	16.318	−243.940	−119.820	12.517
600	107.656	341.257	297.112	26.487	−247.260	−94.672	8.242
700	118.129	358.659	304.674	37.790	−249.895	−69.023	5.151
800	127.171	375.038	312.456	50.065	−251.951	−43.038	2.810
900	135.049	390.482	320.276	63.185	−253.515	−16.825	0.976
1000	141.934	405.075	328.033	77.042	−254.662	9.539	−0.498
1100	147.958	418.892	335.670	91.543	−255.454	36.000	−1.709
1200	153.232	431.997	343.156	106.609	−255.947	62.520	−2.721

	J/K·mol			kJ/mol			
T/K	C_p°	S°	$-(G^\circ-H^\circ(T_r))/T$	$H^\circ-H^\circ(T_r)$	$\Delta_f H^\circ$	$\Delta_f G^\circ$	$\log K_f$
1300	157.849	444.448	350.473	122.168	−256.184	89.070	−3.579
1400	161.896	456.298	357.612	138.160	−256.206	115.630	−4.314
1500	165.447	467.591	364.571	154.531	−256.044	142.185	−4.951

25. ACETIC ACID $C_2H_4O_2$ (g)

T/K	C_p°	S°	$-(G^\circ-H^\circ(T_r))/T$	$H^\circ-H^\circ(T_r)$	$\Delta_f H^\circ$	$\Delta_f G^\circ$	$\log K_f$
298.15	63.438	283.470	283.470	0.000	−432.249	−374.254	65.567
300	63.739	283.863	283.471	0.118	−432.324	−373.893	65.100
400	79.665	304.404	286.164	7.296	−436.006	−353.840	46.206
500	93.926	323.751	291.765	15.993	−438.875	−332.950	34.783
600	106.181	341.988	298.631	26.014	−440.993	−311.554	27.123
700	116.627	359.162	306.064	37.169	−442.466	−289.856	21.629
800	125.501	375.331	313.722	49.287	−443.395	−267.985	17.497
900	132.989	390.558	321.422	62.223	−443.873	−246.026	14.279
1000	139.257	404.904	329.060	75.844	−443.982	−224.034	11.702
1100	144.462	418.429	336.576	90.039	−443.798	−202.046	9.594
1200	148.760	431.189	343.933	104.707	−443.385	−180.086	7.839
1300	152.302	443.240	351.113	119.765	−442.795	−158.167	6.355
1400	155.220	454.637	358.105	135.146	−442.071	−136.299	5.085
1500	157.631	465.432	364.903	150.793	−441.247	−114.486	3.987

26. ACETONE C_3H_6O (g)

T/K	C_p°	S°	$-(G^\circ-H^\circ(T_r))/T$	$H^\circ-H^\circ(T_r)$	$\Delta_f H^\circ$	$\Delta_f G^\circ$	$\log K_f$
298.15	74.517	295.349	295.349	0.000	−217.150	−152.716	26.757
300	74.810	295.809	295.349	0.138	−217.233	−152.339	26.521
400	91.755	319.658	298.498	8.464	−222.212	−129.913	16.962
500	107.864	341.916	304.988	18.464	−226.522	−106.315	11.107
600	122.047	362.836	312.873	29.978	−230.120	−81.923	7.133
700	134.306	382.627	321.470	42.810	−233.049	−56.986	4.252
800	144.934	401.246	330.265	56.785	−235.350	−31.673	2.068
900	154.097	418.860	339.141	71.747	−237.149	−6.109	0.353
1000	162.046	435.513	347.950	87.563	−238.404	19.707	−1.030
1100	168.908	451.286	356.617	104.136	−239.283	45.396	−2.157
1200	174.891	466.265	365.155	121.332	−239.827	71.463	−3.110
1300	180.079	480.491	373.513	139.072	−240.120	97.362	−3.912
1400	184.556	493.963	381.596	157.314	−240.203	123.470	−4.607
1500	188.447	506.850	389.533	175.975	−240.120	149.369	−5.202

27. PHENOL C_6H_6O (g)

T/K	C_p°	S°	$-(G^\circ-H^\circ(T_r))/T$	$H^\circ-H^\circ(T_r)$	$\Delta_f H^\circ$	$\Delta_f G^\circ$	$\log K_f$
298.15	103.220	314.810	314.810	0.000	−96.400	−32.630	5.720
300	103.860	315.450	314.810	0.192	−96.490	−32.230	5.610
400	135.790	349.820	319.278	12.217	−100.870	−10.180	1.330
500	161.910	383.040	328.736	27.152	−104.240	12.970	−1.360
600	182.480	414.450	340.430	44.412	−106.810	36.650	−3.190
700	198.840	443.860	353.134	63.508	−108.800	60.750	−4.530
800	212.140	471.310	366.211	84.079	−110.300	85.020	−5.550
900	223.190	496.950	379.327	105.861	−111.370	109.590	−6.360
1000	232.490	520.960	392.302	128.658	−111.990	134.280	−7.010
1100	240.410	543.500	405.033	152.314	−112.280	158.620	−7.530
1200	247.200	564.720	417.468	176.703	−112.390	183.350	−7.980
1300	253.060	584.740	429.568	201.723	−112.330	208.070	−8.360
1400	258.120	603.680	441.331	227.288	−112.120	233.050	−8.700
1500	262.520	621.650	452.767	253.325	−111.780	257.540	−8.970

28. CARBON TETRAFLUORIDE CF_4 (g)

T/K	C_p°	S°	$-(G^\circ-H^\circ(T_r))/T$	$H^\circ-H^\circ(T_r)$	$\Delta_f H^\circ$	$\Delta_f G^\circ$	$\log K_f$
298.15	61.050	261.455	261.455	0.000	−933.200	−888.518	155.663
300	61.284	261.833	261.456	0.113	−933.219	−888.240	154.654
400	72.399	281.057	264.001	6.822	−933.986	−873.120	114.016
500	80.713	298.153	269.155	14.499	−934.372	−857.852	89.618
600	86.783	313.434	275.284	22.890	−934.490	−842.533	73.348
700	91.212	327.162	281.732	31.801	−934.431	−827.210	61.726
800	94.479	339.566	288.199	41.094	−934.261	−811.903	53.011
900	96.929	350.842	294.542	50.670	−934.024	−796.622	46.234

T/K	$C_p°$	$S°$	$-(G°-H°(T_r))/T$	$H°-H°(T_r)$	$\Delta_f H°$	$\Delta_f G°$	Log K_f
		J/K·mol			kJ/mol		
1000	98.798	361.156	300.695	60.460	−933.745	−781.369	40.814
1100	100.250	370.643	306.629	70.416	−933.442	−766.146	36.381
1200	101.396	379.417	312.334	80.500	−933.125	−750.952	32.688
1300	102.314	387.571	317.811	90.687	−932.800	−735.784	29.564
1400	103.059	395.181	323.069	100.957	−932.470	−720.641	26.887
1500	103.671	402.313	328.116	111.295	−932.137	−705.522	24.568

29. TRIFLUOROMETHANE CHF_3 (g)

T/K	$C_p°$	$S°$	$-(G°-H°(T_r))/T$	$H°-H°(T_r)$	$\Delta_f H°$	$\Delta_f G°$	Log K_f
298.15	51.069	259.675	259.675	0.000	−696.700	−662.237	116.020
300	51.258	259.991	259.676	0.095	−696.735	−662.023	115.267
400	61.148	276.113	261.807	5.722	−698.427	−650.186	84.905
500	69.631	290.700	266.149	12.275	−699.715	−637.969	66.647
600	76.453	304.022	271.368	19.593	−700.634	−625.528	54.456
700	81.868	316.230	276.917	27.519	−701.253	−612.957	45.739
800	86.201	327.455	282.542	35.930	−701.636	−600.315	39.196
900	89.719	337.818	288.116	44.732	−701.832	−587.636	34.105
1000	92.617	347.426	293.572	53.854	−701.879	−574.944	30.032
1100	95.038	356.370	298.879	63.240	−701.805	−562.253	26.699
1200	97.084	364.730	304.022	72.849	−701.629	−549.574	23.922
1300	98.833	372.571	308.997	82.647	−701.368	−536.913	21.573
1400	100.344	379.952	313.804	92.607	−701.033	−524.274	19.561
1500	101.660	386.921	318.449	102.709	−700.635	−511.662	17.817

30. CHLOROTRIFLUOROMETHANE $CClF_3$ (g)

T/K	$C_p°$	$S°$	$-(G°-H°(T_r))/T$	$H°-H°(T_r)$	$\Delta_f H°$	$\Delta_f G°$	Log K_f
298.15	66.886	285.419	285.419	0.000	−707.800	−667.238	116.896
300	67.111	285.834	285.421	0.124	−707.810	−666.986	116.131
400	77.528	306.646	288.187	7.383	−708.153	−653.316	85.313
500	85.013	324.797	293.734	15.532	−708.170	−639.599	66.818
600	90.329	340.794	300.271	24.314	−707.975	−625.901	54.489
700	94.132	355.020	307.096	33.547	−707.654	−612.246	45.686
800	96.899	367.780	313.897	43.106	−707.264	−598.642	39.087
900	98.951	379.317	320.536	52.903	−706.837	−585.090	33.957
1000	100.507	389.827	326.947	62.880	−706.396	−571.586	29.856
1100	101.708	399.465	333.108	72.993	−705.950	−558.126	26.503
1200	102.651	408.357	339.013	83.213	−705.505	−544.707	23.710
1300	103.404	416.604	344.668	93.517	−705.064	−531.326	21.349
1400	104.012	424.290	350.084	103.889	−704.628	−517.977	19.326
1500	104.512	431.484	355.273	114.316	−704.196	−504.660	17.574

31. DICHLORODIFLUOROMETHANE CCl_2F_2 (g)

T/K	$C_p°$	$S°$	$-(G°-H°(T_r))/T$	$H°-H°(T_r)$	$\Delta_f H°$	$\Delta_f G°$	Log K_f
298.15	72.476	300.903	300.903	0.000	−486.000	−447.030	78.317
300	72.691	301.352	300.905	0.134	−486.002	−446.788	77.792
400	82.408	323.682	303.883	7.919	−485.945	−433.716	56.637
500	89.063	342.833	309.804	16.514	−485.618	−420.692	43.949
600	93.635	359.500	316.729	25.663	−485.136	−407.751	35.497
700	96.832	374.189	323.909	35.196	−484.576	−394.897	29.467
800	99.121	387.276	331.027	44.999	−483.984	−382.126	24.950
900	100.801	399.053	337.942	55.000	−483.388	−369.429	21.441
1000	102.062	409.742	344.596	65.146	−482.800	−356.799	18.637
1100	103.030	419.517	350.969	75.402	−482.226	−344.227	16.346
1200	103.786	428.515	357.061	85.745	−481.667	−331.706	14.439
1300	104.388	436.847	362.882	96.154	−481.121	−319.232	12.827
1400	104.874	444.602	368.445	106.618	−480.588	−306.799	11.447
1500	105.270	451.851	373.767	117.126	−480.065	−294.404	10.252

32. CHLORODIFLUOROMETHANE $CHClF_2$ (g)

T/K	$C_p°$	$S°$	$-(G°-H°(T_r))/T$	$H°-H°(T_r)$	$\Delta_f H°$	$\Delta_f G°$	Log K_f
298.15	55.853	280.915	280.915	0.000	−475.000	−443.845	77.759
300	56.039	281.261	280.916	0.104	−475.028	−443.652	77.246
400	65.395	298.701	283.231	6.188	−476.390	−432.978	56.540
500	73.008	314.145	287.898	13.123	−477.398	−422.001	44.086
600	78.940	328.003	293.448	20.733	−478.103	−410.851	35.767

T/K	$C_p°$	$S°$	$-(G°-H°(T_r))/T$	$H°-H°(T_r)$	$\Delta_f H°$	$\Delta_f G°$	Log K_f
		J/K·mol			kJ/mol		
700	83.551	340.533	299.294	28.867	−478.574	−399.603	29.818
800	87.185	351.936	305.172	37.411	−478.870	−388.299	25.353
900	90.100	362.379	310.956	46.280	−479.031	−376.967	21.878
1000	92.475	371.999	316.586	55.413	−479.090	−365.622	19.098
1100	94.433	380.908	322.033	64.761	−479.068	−354.276	16.823
1200	96.066	389.196	327.289	74.289	−478.982	−342.935	14.927
1300	97.438	396.941	332.352	83.966	−478.843	−331.603	13.324
1400	98.601	404.206	337.228	93.769	−478.661	−320.283	11.950
1500	99.593	411.044	341.923	103.681	−478.443	−308.978	10.759

33. METHYLAMINE   CH$_5$N (g)

T/K	$C_p°$	$S°$	$-(G°-H°(T_r))/T$	$H°-H°(T_r)$	$\Delta_f H°$	$\Delta_f G°$	Log K_f
298.15	50.053	242.881	242.881	0.000	−22.529	32.734	−5.735
300	50.227	243.196	242.893	0.091	−22.614	33.077	−5.759
400	60.171	258.986	244.975	5.604	−26.846	52.294	−6.829
500	70.057	273.486	249.244	12.121	−30.431	72.510	−7.575
600	78.929	287.063	254.431	19.579	−33.364	93.382	−8.129
700	86.711	299.826	260.008	27.873	−35.712	114.702	−8.559
800	93.545	311.865	265.749	36.893	−37.548	136.316	−8.900
900	99.573	323.239	271.511	46.555	−38.949	158.138	−9.178
1000	104.886	334.006	277.220	56.786	−39.967	180.098	−9.407
1100	109.576	344.233	282.861	67.509	−40.681	201.822	−9.584
1200	113.708	353.944	288.374	78.685	−41.136	224.240	−9.761
1300	117.341	363.190	293.775	90.239	−41.376	246.364	−9.899
1400	120.542	372.012	299.061	102.131	−41.451	268.504	−10.018
1500	123.353	380.426	304.209	114.326	−41.381	290.639	−10.121

34. CHLORINE   Cl (g)

T/K	$C_p°$	$S°$	$-(G°-H°(T_r))/T$	$H°-H°(T_r)$	$\Delta_f H°$	$\Delta_f G°$	Log K_f
298.15	21.838	165.190	165.190	0.000	121.302	105.306	−18.449
300	21.852	165.325	165.190	0.040	121.311	105.207	−18.318
400	22.467	171.703	166.055	2.259	121.795	99.766	−13.028
500	22.744	176.752	167.708	4.522	122.272	94.203	−9.841
600	22.781	180.905	169.571	6.800	122.734	88.546	−7.709
700	22.692	184.411	171.448	9.074	123.172	82.813	−6.179
800	22.549	187.432	173.261	11.337	123.585	77.019	−5.029
900	22.389	190.079	174.986	13.584	123.971	71.175	−4.131
1000	22.233	192.430	176.615	15.815	124.334	65.289	−3.410
1100	22.089	194.542	178.150	18.031	124.675	59.368	−2.819
1200	21.959	196.458	179.597	20.233	124.996	53.416	−2.325
1300	21.843	198.211	180.963	22.423	125.299	47.439	−1.906
1400	21.742	199.826	182.253	24.602	125.587	41.439	−1.546
1500	21.652	201.323	183.475	26.772	125.861	35.418	−1.233

35. DICHLORINE   Cl$_2$ (g)

T/K	$C_p°$	$S°$	$-(G°-H°(T_r))/T$	$H°-H°(T_r)$	$\Delta_f H°$	$\Delta_f G°$	Log K_f
298.15	33.949	223.079	223.079	0.000	0.000	0.000	0.000
300	33.981	223.290	223.080	0.063	0.000	0.000	0.000
400	35.296	233.263	224.431	3.533	0.000	0.000	0.000
500	36.064	241.229	227.021	7.104	0.000	0.000	0.000
600	36.547	247.850	229.956	10.736	0.000	0.000	0.000
700	36.874	253.510	232.926	14.408	0.000	0.000	0.000
800	37.111	258.450	235.815	18.108	0.000	0.000	0.000
900	37.294	262.832	238.578	21.829	0.000	0.000	0.000
1000	37.442	266.769	241.203	25.566	0.000	0.000	0.000
1100	37.567	270.343	243.692	29.316	0.000	0.000	0.000
1200	37.678	273.617	246.052	33.079	0.000	0.000	0.000
1300	37.778	276.637	248.290	36.851	0.000	0.000	0.000
1400	37.872	279.440	250.416	40.634	0.000	0.000	0.000
1500	37.961	282.056	252.439	44.426	0.000	0.000	0.000

36. HYDROGEN CHLORIDE   HCl (g)

T/K	$C_p°$	$S°$	$-(G°-H°(T_r))/T$	$H°-H°(T_r)$	$\Delta_f H°$	$\Delta_f G°$	Log K_f
298.15	29.136	186.902	186.902	0.000	−92.310	−95.298	16.696
300	29.137	187.082	186.902	0.054	−92.314	−95.317	16.596

T/K	J/K·mol			kJ/mol			
	$C_p°$	$S°$	$-(G°-H°(T_r))/T$	$H°-H°(T_r)$	$\Delta_f H°$	$\Delta_f G°$	Log K_f
400	29.175	195.468	188.045	2.969	−92.587	−96.278	12.573
500	29.304	201.990	190.206	5.892	−92.911	−97.164	10.151
600	29.576	207.354	192.630	8.835	−93.249	−97.983	8.530
700	29.988	211.943	195.069	11.812	−93.577	−98.746	7.368
800	30.500	215.980	197.435	14.836	−93.879	−99.464	6.494
900	31.063	219.604	199.700	17.913	−94.149	−100.145	5.812
1000	31.639	222.907	201.858	21.049	−94.384	−100.798	5.265
1100	32.201	225.949	203.912	24.241	−94.587	−101.430	4.816
1200	32.734	228.774	205.867	27.488	−94.760	−102.044	4.442
1300	33.229	231.414	207.732	30.786	−94.908	−102.645	4.124
1400	33.684	233.893	209.513	34.132	−95.035	−103.235	3.852
1500	34.100	236.232	211.217	37.522	−95.146	−103.817	3.615

37. COPPER   Cu (cr, l)

T/K	$C_p°$	$S°$	$-(G°-H°(T_r))/T$	$H°-H°(T_r)$	$\Delta_f H°$	$\Delta_f G°$	Log K_f
298.15	24.440	33.150	33.150	0.000	0.000	0.000	0.000
300	24.460	33.301	33.150	0.045	0.000	0.000	0.000
400	25.339	40.467	34.122	2.538	0.000	0.000	0.000
500	25.966	46.192	35.982	5.105	0.000	0.000	0.000
600	26.479	50.973	38.093	7.728	0.000	0.000	0.000
700	26.953	55.090	40.234	10.399	0.000	0.000	0.000
800	27.448	58.721	42.322	13.119	0.000	0.000	0.000
900	28.014	61.986	44.328	15.891	0.000	0.000	0.000
1000	28.700	64.971	46.245	18.726	0.000	0.000	0.000
1100	29.553	67.745	48.075	21.637	0.000	0.000	0.000
1200	30.617	70.361	49.824	24.644	0.000	0.000	0.000
1300	31.940	72.862	51.501	27.769	0.000	0.000	0.000
1358	32.844	74.275	52.443	29.647	0.000	0.000	0.000

PHASE TRANSITION: $\Delta_{trs} H$ = 13.141 kJ/mol, $\Delta_{trs} S$ = 9.676 J/K·mol, cr–l

T/K	$C_p°$	$S°$	$-(G°-H°(T_r))/T$	$H°-H°(T_r)$	$\Delta_f H°$	$\Delta_f G°$	Log K_f
1358	32.800	83.951	52.443	42.788	0.000	0.000	0.000
1400	32.800	84.950	53.403	44.166	0.000	0.000	0.000
1500	32.800	87.213	55.583	47.446	0.000	0.000	0.000

38. COPPER   Cu (g)

T/K	$C_p°$	$S°$	$-(G°-H°(T_r))/T$	$H°-H°(T_r)$	$\Delta_f H°$	$\Delta_f G°$	Log K_f
298.15	20.786	166.397	166.397	0.000	337.600	297.873	−52.185
300	20.786	166.525	166.397	0.038	337.594	297.626	−51.821
400	20.786	172.505	167.213	2.117	337.179	284.364	−37.134
500	20.786	177.143	168.752	4.196	336.691	271.215	−28.333
600	20.786	180.933	170.476	6.274	336.147	258.170	−22.475
700	20.786	184.137	172.205	8.353	335.554	245.221	−18.298
800	20.786	186.913	173.874	10.431	334.913	232.359	−15.171
900	20.786	189.361	175.461	12.510	334.219	219.581	−12.744
1000	20.786	191.551	176.963	14.589	333.463	206.883	−10.806
1100	20.788	193.532	178.380	16.667	332.631	194.265	−9.225
1200	20.793	195.341	179.719	18.746	331.703	181.726	−7.910
1300	20.803	197.006	180.986	20.826	330.657	169.270	−6.801
1400	20.823	198.548	182.186	22.907	316.342	157.305	−5.869
1500	20.856	199.986	183.325	24.991	315.146	145.987	−5.084

39. COPPER OXIDE   CuO (cr)

T/K	$C_p°$	$S°$	$-(G°-H°(T_r))/T$	$H°-H°(T_r)$	$\Delta_f H°$	$\Delta_f G°$	Log K_f
298.15	42.300	42.740	42.740	0.000	−162.000	−134.277	23.524
300	42.417	43.002	42.741	0.078	−161.994	−134.105	23.349
400	46.783	55.878	44.467	4.564	−161.487	−124.876	16.307
500	49.190	66.596	47.852	9.372	−160.775	−115.803	12.098
600	50.827	75.717	51.755	14.377	−159.973	−106.883	9.305
700	52.099	83.651	55.757	19.526	−159.124	−98.102	7.320
800	53.178	90.680	59.691	24.791	−158.247	−89.444	5.840
900	54.144	97.000	63.491	30.158	−157.356	−80.897	4.695
1000	55.040	102.751	67.134	35.617	−156.462	−72.450	3.784
1100	55.890	108.037	70.615	41.164	−155.582	−64.091	3.043
1200	56.709	112.936	73.941	46.794	−154.733	−55.812	2.429

T/K	C_p°	S°	$-(G^\circ - H^\circ(T_r))/T$	$H^\circ - H^\circ(T_r)$	$\Delta_f H^\circ$	$\Delta_f G^\circ$	Log K_f
		J/K·mol			kJ/mol		
1300	57.507	117.507	77.118	52.505	−153.940	−47.601	1.913
1400	58.288	121.797	80.158	58.295	−166.354	−39.043	1.457
1500	59.057	125.845	83.070	64.163	−165.589	−29.975	1.044

40. DICOPPER OXIDE   Cu₂O (cr)

T/K	C_p°	S°	$-(G^\circ - H^\circ(T_r))/T$	$H^\circ - H^\circ(T_r)$	$\Delta_f H^\circ$	$\Delta_f G^\circ$	Log K_f
298.15	62.600	92.550	92.550	0.000	−173.100	−150.344	26.339
300	62.721	92.938	92.551	0.116	−173.102	−150.203	26.152
400	67.587	111.712	95.078	6.654	−173.036	−142.572	18.618
500	70.784	127.155	99.995	13.580	−172.772	−134.984	14.101
600	73.323	140.291	105.643	20.789	−172.389	−127.460	11.096
700	75.552	151.764	111.429	28.235	−171.914	−120.009	8.955
800	77.616	161.989	117.121	35.894	−171.363	−112.631	7.354
900	79.584	171.245	122.629	43.755	−170.750	−105.325	6.113
1000	81.492	179.729	127.920	51.809	−170.097	−98.091	5.124
1100	83.360	187.584	132.992	60.052	−169.431	−90.922	4.317
1200	85.202	194.917	137.850	68.480	−168.791	−83.814	3.648
1300	87.026	201.808	142.507	77.092	−168.223	−76.756	3.084
1400	88.836	208.324	146.978	85.885	−194.030	−68.926	2.572
1500	90.636	214.515	151.276	94.858	−193.438	−60.010	2.090

41. COPPER DICHLORIDE   CuCl₂ (cr, l)

T/K	C_p°	S°	$-(G^\circ - H^\circ(T_r))/T$	$H^\circ - H^\circ(T_r)$	$\Delta_f H^\circ$	$\Delta_f G^\circ$	Log K_f
298.15	71.880	108.070	108.070	0.000	−218.000	−173.826	30.453
300	71.998	108.515	108.071	0.133	−217.975	−173.552	30.218
400	76.338	129.899	110.957	7.577	−216.494	−158.962	20.758
500	78.654	147.204	116.532	15.336	−214.873	−144.765	15.123
600	80.175	161.687	122.884	23.282	−213.182	−130.901	11.396
675	81.056	171.183	127.732	29.329	−211.185	−120.693	9.340

PHASE TRANSITION: $\Delta_{trs} H$ = 0.700 kJ/mol, $\Delta_{trs} S$ = 1.037 J/K·mol, crII–crI

T/K	C_p°	S°	$-(G^\circ - H^\circ(T_r))/T$	$H^\circ - H^\circ(T_r)$	$\Delta_f H^\circ$	$\Delta_f G^\circ$	Log K_f
675	82.400	172.220	127.732	30.029	−211.185	−120.693	9.340
700	82.400	175.216	129.375	32.089	−210.719	−117.350	8.757
800	82.400	186.219	135.808	40.329	−208.898	−104.137	6.799
871	82.400	193.226	140.207	46.179	−192.649	−94.893	5.691

PHASE TRANSITION: $\Delta_{trs} H$ = 15.001 kJ/mol, $\Delta_{trs} S$ = 17.221 J/K·mol, crI–l

T/K	C_p°	S°	$-(G^\circ - H^\circ(T_r))/T$	$H^\circ - H^\circ(T_r)$	$\Delta_f H^\circ$	$\Delta_f G^\circ$	Log K_f
871	100.000	210.447	140.207	61.180	−192.649	−94.893	5.691
900	100.000	213.723	142.523	64.080	−191.640	−91.655	5.319
1000	100.000	224.259	150.179	74.080	−188.212	−80.730	4.217
1100	100.000	233.790	157.353	84.080	−184.873	−70.144	3.331
1130.75	100.000	236.547	159.470	87.155	−183.867	−66.951	3.093

42. COPPER DICHLORIDE   CuCl₂ (g)

T/K	C_p°	S°	$-(G^\circ - H^\circ(T_r))/T$	$H^\circ - H^\circ(T_r)$	$\Delta_f H^\circ$	$\Delta_f G^\circ$	Log K_f
298.15	56.814	278.418	278.418	0.000	−43.268	−49.883	8.739
300	56.869	278.769	278.419	0.105	−43.271	−49.924	8.692
400	58.992	295.456	280.679	5.911	−43.428	−52.119	6.806
500	60.111	308.752	285.010	11.871	−43.606	−54.271	5.670
600	60.761	319.774	289.911	17.918	−43.814	−56.385	4.909
700	61.168	329.173	294.865	24.015	−44.060	−58.462	4.362
800	61.439	337.360	299.677	30.147	−44.349	−60.500	3.950
900	61.630	344.608	304.274	36.301	−44.688	−62.499	3.627
1000	61.776	351.109	308.638	42.471	−45.088	−64.457	3.367
1100	61.900	357.003	312.771	48.655	−45.566	−66.372	3.152
1200	62.022	362.394	316.685	54.851	−46.139	−68.239	2.970
1300	62.159	367.364	320.395	61.060	−46.829	−70.053	2.815
1400	62.325	371.976	323.916	67.284	−60.784	−71.404	2.664
1500	62.531	376.283	327.265	73.526	−61.613	−72.133	2.512

43. FLUORINE   F (g)

T/K	C_p°	S°	$-(G^\circ - H^\circ(T_r))/T$	$H^\circ - H^\circ(T_r)$	$\Delta_f H^\circ$	$\Delta_f G^\circ$	Log K_f
298.15	22.746	158.750	158.750	0.000	79.380	62.280	−10.911
300	22.742	158.891	158.750	0.042	79.393	62.173	−10.825
400	22.432	165.394	159.639	2.302	80.043	56.332	−7.356
500	22.100	170.363	161.307	4.528	80.587	50.340	−5.259

T/K	J/K·mol			kJ/mol			
	$C_p°$	$S°$	$-(G°-H°(T_r))/T$	$H°-H°(T_r)$	$\Delta_f H°$	$\Delta_f G°$	Log K_f
600	21.832	174.368	163.161	6.724	81.046	44.246	-3.852
700	21.629	177.717	165.008	8.897	81.442	38.081	-2.842
800	21.475	180.595	166.780	11.052	81.792	31.862	-2.080
900	21.357	183.117	168.458	13.193	82.106	25.601	-1.486
1000	21.266	185.362	170.039	15.324	82.391	19.308	-1.009
1100	21.194	187.386	171.525	17.447	82.654	12.986	-0.617
1200	21.137	189.227	172.925	19.563	82.897	6.642	-0.289
1300	21.091	190.917	174.245	21.675	83.123	0.278	-0.011
1400	21.054	192.479	175.492	23.782	83.335	-6.103	0.228
1500	21.022	193.930	176.673	25.886	83.533	-12.498	0.435

44. DIFLUORINE F_2 (g)

T/K	$C_p°$	$S°$	$-(G°-H°(T_r))/T$	$H°-H°(T_r)$	$\Delta_f H°$	$\Delta_f G°$	Log K_f
298.15	31.304	202.790	202.790	0.000	0.000	0.000	0.000
300	31.337	202.984	202.790	0.058	0.000	0.000	0.000
400	32.995	212.233	204.040	3.277	0.000	0.000	0.000
500	34.258	219.739	206.453	6.643	0.000	0.000	0.000
600	35.171	226.070	209.208	10.117	0.000	0.000	0.000
700	35.839	231.545	212.017	13.669	0.000	0.000	0.000
800	36.343	236.365	214.765	17.279	0.000	0.000	0.000
900	36.740	240.669	217.409	20.934	0.000	0.000	0.000
1000	37.065	244.557	219.932	24.625	0.000	0.000	0.000
1100	37.342	248.103	222.334	28.346	0.000	0.000	0.000
1200	37.588	251.363	224.619	32.093	0.000	0.000	0.000
1300	37.811	254.381	226.794	35.863	0.000	0.000	0.000
1400	38.019	257.191	228.866	39.654	0.000	0.000	0.000
1500	38.214	259.820	230.843	43.466	0.000	0.000	0.000

45. HYDROGEN FLUORIDE   HF (g)

T/K	$C_p°$	$S°$	$-(G°-H°(T_r))/T$	$H°-H°(T_r)$	$\Delta_f H°$	$\Delta_f G°$	Log K_f
298.15	29.137	173.776	173.776	0.000	-273.300	-275.399	48.248
300	29.137	173.956	173.776	0.054	-273.302	-275.412	47.953
400	29.149	182.340	174.919	2.968	-273.450	-276.096	36.054
500	29.172	188.846	177.078	5.884	-273.679	-276.733	28.910
600	29.230	194.169	179.496	8.804	-273.961	-277.318	24.142
700	29.350	198.683	181.923	11.732	-274.277	-277.852	20.733
800	29.549	202.614	184.269	14.676	-274.614	-278.340	18.174
900	29.827	206.110	186.505	17.645	-274.961	-278.785	16.180
1000	30.169	209.270	188.626	20.644	-275.309	-279.191	14.583
1100	30.558	212.163	190.636	23.680	-275.652	-279.563	13.275
1200	30.974	214.840	192.543	26.756	-275.988	-279.904	12.184
1300	31.403	217.336	194.355	29.875	-276.315	-280.217	11.259
1400	31.831	219.679	196.081	33.037	-276.631	-280.505	10.466
1500	32.250	221.889	197.729	36.241	-276.937	-280.771	9.777

46. GERMANIUM   Ge (cr, l)

T/K	$C_p°$	$S°$	$-(G°-H°(T_r))/T$	$H°-H°(T_r)$	$\Delta_f H°$	$\Delta_f G°$	Log K_f
298.15	23.222	31.090	31.090	0.000	0.000	0.000	0.000
300	23.249	31.234	31.090	0.043	0.000	0.000	0.000
400	24.310	38.083	32.017	2.426	0.000	0.000	0.000
500	24.962	43.582	33.798	4.892	0.000	0.000	0.000
600	25.452	48.178	35.822	7.414	0.000	0.000	0.000
700	25.867	52.133	37.876	9.980	0.000	0.000	0.000
800	26.240	55.612	39.880	12.586	0.000	0.000	0.000
900	26.591	58.723	41.804	15.227	0.000	0.000	0.000
1000	26.926	61.542	43.639	17.903	0.000	0.000	0.000
1100	27.252	64.124	45.386	20.612	0.000	0.000	0.000
1200	27.571	66.509	47.048	23.353	0.000	0.000	0.000
1211.4	27.608	66.770	47.232	23.668	0.000	0.000	0.000

PHASE TRANSITION: $\Delta_{trs} H$ = 37.030 kJ/mol, $\Delta_{trs} S$ = 30.568 J/K·mol, cr–l

T/K	$C_p°$	$S°$	$-(G°-H°(T_r))/T$	$H°-H°(T_r)$	$\Delta_f H°$	$\Delta_f G°$	Log K_f
1211.4	27.600	97.338	47.232	60.698	0.000	0.000	0.000
1300	27.600	99.286	50.714	63.143	0.000	0.000	0.000
1400	27.600	101.331	54.258	65.903	0.000	0.000	0.000
1500	27.600	103.236	57.460	68.663	0.000	0.000	0.000

	J/K·mol			kJ/mol			
T/K	C_p°	S°	$-(G^\circ - H^\circ(T_r))/T$	$H^\circ - H^\circ(T_r)$	$\Delta_f H^\circ$	$\Delta_f G^\circ$	$\mathrm{Log}\, K_f$

47. GERMANIUM   Ge (g)

298.15	30.733	167.903	167.903	0.000	367.800	327.009	−57.290
300	30.757	168.094	167.904	0.057	367.814	326.756	−56.893
400	31.071	177.025	169.119	3.162	368.536	312.959	−40.868
500	30.360	183.893	171.415	6.239	369.147	298.991	−31.235
600	29.265	189.334	173.965	9.222	369.608	284.914	−24.804
700	28.102	193.758	176.487	12.090	369.910	270.773	−20.205
800	27.029	197.439	178.882	14.845	370.060	256.598	−16.754
900	26.108	200.567	181.122	17.501	370.073	242.414	−14.069
1000	25.349	203.277	183.205	20.072	369.969	228.234	−11.922
1100	24.741	205.664	185.141	22.575	369.763	214.069	−10.165
1200	24.264	207.795	186.941	25.025	369.471	199.928	−8.703
1300	23.898	209.722	188.621	27.432	332.088	188.521	−7.575
1400	23.624	211.483	190.192	29.807	331.704	177.492	−6.622
1500	23.426	213.105	191.666	32.159	331.296	166.491	−5.798

48. GERMANIUM DIOXIDE   GeO$_2$ (cr, l)

298.15	50.166	39.710	39.710	0.000	−580.200	−521.605	91.382
300	50.475	40.021	39.711	0.093	−580.204	−521.242	90.755
400	61.281	56.248	41.850	5.759	−579.893	−501.610	65.503
500	66.273	70.519	46.191	12.164	−579.013	−482.134	50.368
600	69.089	82.872	51.299	18.943	−577.915	−462.859	40.295
700	70.974	93.671	56.597	25.952	−576.729	−443.776	33.115
800	72.449	103.247	61.841	33.125	−575.498	−424.866	27.741
900	73.764	111.857	66.928	40.436	−574.235	−406.113	23.570
1000	75.049	119.696	71.819	47.877	−572.934	−387.502	20.241
1100	76.378	126.910	76.504	55.447	−571.582	−369.024	17.523
1200	77.796	133.616	80.987	63.155	−570.166	−350.671	15.264
1300	79.332	139.903	85.279	71.010	−605.685	−329.732	13.249
1308	79.460	140.390	85.615	71.646	−584.059	−328.034	13.100

PHASE TRANSITION: $\Delta_{trs} H$ = 21.500 kJ/mol, $\Delta_{trs} S$ = 16.437 J/K·mol, crII–crI

1308	80.075	156.827	85.615	93.146	−584.059	−328.034	13.100
1388	81.297	161.617	89.858	99.601	−565.504	−312.415	11.757

PHASE TRANSITION: $\Delta_{trs} H$ = 17.200 kJ/mol, $\Delta_{trs} S$ = 12.392 J/K·mol, crI–l

1388	78.500	174.009	89.858	116.801	−565.504	−312.415	11.757
1400	78.500	174.685	90.582	117.743	−565.328	−310.228	11.575
1500	78.500	180.100	96.372	125.593	−563.882	−292.057	10.170

49. GERMANIUM TETRACHLORIDE   GeCl$_4$ (g)

298.15	95.918	348.393	348.393	0.000	−500.000	−461.582	80.866
300	96.041	348.987	348.395	0.178	−499.991	−461.343	80.326
400	100.750	377.342	352.229	10.045	−499.447	−448.540	58.573
500	103.206	400.114	359.604	20.255	−498.845	−435.882	45.536
600	104.624	419.067	367.980	30.652	−498.234	−423.347	36.855
700	105.509	435.266	376.463	41.162	−497.634	−410.914	30.662
800	106.096	449.396	384.715	51.744	−497.057	−398.565	26.023
900	106.504	461.917	392.611	62.375	−496.509	−386.287	22.419
1000	106.799	473.155	400.113	73.041	−495.993	−374.068	19.539
1100	107.020	483.344	407.224	83.733	−495.512	−361.899	17.185
1200	107.189	492.664	413.961	94.444	−495.067	−349.772	15.225
1300	107.320	501.249	420.349	105.169	−531.677	−334.973	13.459
1400	107.425	509.206	426.416	115.907	−531.265	−319.857	11.934
1500	107.509	516.621	432.185	126.654	−530.861	−304.771	10.613

50. HYDROGEN   H (g)

298.15	20.786	114.716	114.716	0.000	217.998	203.276	−35.613
300	20.786	114.845	114.716	0.038	218.010	203.185	−35.377
400	20.786	120.824	115.532	2.117	218.635	198.149	−25.875
500	20.786	125.463	117.071	4.196	219.253	192.956	−20.158

Thermodynamic Properties as a Function of Temperature

T/K	$C_p°$	$S°$	$-(G°-H°(T_r))/T$	$H°-H°(T_r)$	$\Delta_f H°$	$\Delta_f G°$	Log K_f
	J/K·mol			kJ/mol			
600	20.786	129.252	118.795	6.274	219.867	187.639	−16.335
700	20.786	132.457	120.524	8.353	220.476	182.219	−13.597
800	20.786	135.232	122.193	10.431	221.079	176.712	−11.538
900	20.786	137.680	123.780	12.510	221.670	171.131	−9.932
1000	20.786	139.870	125.282	14.589	222.247	165.485	−8.644
1100	20.786	141.852	126.700	16.667	222.806	159.781	−7.587
1200	20.786	143.660	128.039	18.746	223.345	154.028	−6.705
1300	20.786	145.324	129.305	20.824	223.864	148.230	−5.956
1400	20.786	146.864	130.505	22.903	224.360	142.393	−5.313
1500	20.786	148.298	131.644	24.982	224.835	136.522	−4.754

51. DIHYDROGEN H_2 (g)

T/K	$C_p°$	$S°$	$-(G°-H°(T_r))/T$	$H°-H°(T_r)$	$\Delta_f H°$	$\Delta_f G°$	Log K_f
298.15	28.836	130.680	130.680	0.000	0.000	0.000	0.000
300	28.849	130.858	130.680	0.053	0.000	0.000	0.000
400	29.181	139.217	131.818	2.960	0.000	0.000	0.000
500	29.260	145.738	133.974	5.882	0.000	0.000	0.000
600	29.327	151.078	136.393	8.811	0.000	0.000	0.000
700	29.440	155.607	138.822	11.749	0.000	0.000	0.000
800	29.623	159.549	141.172	14.702	0.000	0.000	0.000
900	29.880	163.052	143.412	17.676	0.000	0.000	0.000
1000	30.204	166.217	145.537	20.680	0.000	0.000	0.000
1100	30.580	169.113	147.550	23.719	0.000	0.000	0.000
1200	30.991	171.791	149.460	26.797	0.000	0.000	0.000
1300	31.422	174.288	151.275	29.918	0.000	0.000	0.000
1400	31.860	176.633	153.003	33.082	0.000	0.000	0.000
1500	32.296	178.846	154.653	36.290	0.000	0.000	0.000

52. HYDROXYL   OH (g)

T/K	$C_p°$	$S°$	$-(G°-H°(T_r))/T$	$H°-H°(T_r)$	$\Delta_f H°$	$\Delta_f G°$	Log K_f
298.15	29.886	183.737	183.737	0.000	39.349	34.631	−6.067
300	29.879	183.922	183.738	0.055	39.350	34.602	−6.025
400	29.604	192.476	184.906	3.028	39.384	33.012	−4.311
500	29.495	199.067	187.104	5.982	39.347	31.422	−3.283
600	29.513	204.445	189.560	8.931	39.252	29.845	−2.598
700	29.655	209.003	192.020	11.888	39.113	28.287	−2.111
800	29.914	212.979	194.396	14.866	38.945	26.752	−1.747
900	30.265	216.522	196.661	17.874	38.763	25.239	−1.465
1000	30.682	219.731	198.810	20.921	38.577	23.746	−1.240
1100	31.135	222.677	200.848	24.012	38.393	22.272	−1.058
1200	31.603	225.406	202.782	27.149	38.215	20.814	−0.906
1300	32.069	227.954	204.621	30.332	38.046	19.371	−0.778
1400	32.522	230.347	206.374	33.562	37.886	17.941	−0.669
1500	32.956	232.606	208.048	36.836	37.735	16.521	−0.575

53. WATER H_2O (l)

T/K	$C_p°$	$S°$	$-(G°-H°(T_r))/T$	$H°-H°(T_r)$	$\Delta_f H°$	$\Delta_f G°$	Log K_f
298.15	75.300	69.950	69.950	0.000	−285.830	−237.141	41.546
300	75.281	70.416	69.951	0.139	−285.771	−236.839	41.237
373.21	76.079	86.896	71.715	5.666	−283.454	−225.160	31.513

54. WATER H_2O (g)

T/K	$C_p°$	$S°$	$-(G°-H°(T_r))/T$	$H°-H°(T_r)$	$\Delta_f H°$	$\Delta_f G°$	Log K_f
298.15	33.598	188.832	188.832	0.000	−241.826	−228.582	40.046
300	33.606	189.040	188.833	0.062	−241.844	−228.500	39.785
400	34.283	198.791	190.158	3.453	−242.845	−223.900	29.238
500	35.259	206.542	192.685	6.929	−243.822	−219.050	22.884
600	36.371	213.067	195.552	10.509	−244.751	−214.008	18.631
700	37.557	218.762	198.469	14.205	−245.620	−208.814	15.582
800	38.800	223.858	201.329	18.023	−246.424	−203.501	13.287
900	40.084	228.501	204.094	21.966	−247.158	−198.091	11.497
1000	41.385	232.792	206.752	26.040	−247.820	−192.603	10.060
1100	42.675	236.797	209.303	30.243	−248.410	−187.052	8.882
1200	43.932	240.565	211.753	34.574	−248.933	−181.450	7.898
1300	45.138	244.129	214.108	39.028	−249.392	−175.807	7.064

T/K	J/K·mol			kJ/mol			
	$C_p°$	$S°$	$-(G°-H°(T_r))/T$	$H°-H°(T_r)$	$\Delta_f H°$	$\Delta_f G°$	$\text{Log } K_f$
1400	46.281	247.516	216.374	43.599	−249.792	−170.132	6.348
1500	47.356	250.746	218.559	48.282	−250.139	−164.429	5.726

55. IODINE   I (g)

298.15	20.786	180.787	180.787	0.000	106.760	70.172	−12.294
300	20.786	180.915	180.787	0.038	106.748	69.945	−12.178
400	20.786	186.895	181.602	2.117	97.974	58.060	−7.582
500	20.786	191.533	183.142	4.196	75.988	50.202	−5.244
600	20.786	195.323	184.866	6.274	76.190	45.025	−3.920
700	20.786	198.527	186.594	8.353	76.385	39.816	−2.971
800	20.787	201.303	188.263	10.432	76.574	34.579	−2.258
900	20.789	203.751	189.851	12.510	76.757	29.319	−1.702
1000	20.795	205.942	191.352	14.589	76.936	24.038	−1.256
1100	20.806	207.924	192.770	16.669	77.109	18.740	−0.890
1200	20.824	209.735	194.110	18.751	77.277	13.426	−0.584
1300	20.851	211.403	195.377	20.835	77.440	8.098	−0.325
1400	20.889	212.950	196.577	22.921	77.596	2.758	−0.103
1500	20.936	214.392	197.717	25.013	77.745	−2.592	0.090

56. DIIODINE I_2 (cr, l)

298.15	54.440	116.139	116.139	0.000	0.000	0.000	0.000
300	54.518	116.476	116.140	0.101	0.000	0.000	0.000
386.75	61.531	131.039	117.884	5.088	0.000	0.000	0.000

PHASE TRANSITION: $\Delta_{trs} H$ = 15.665 kJ/mol, $\Delta_{trs} S$ = 40.504 J/K·mol, cr−l

386.75	79.555	171.543	117.884	20.753	0.000	0.000	0.000
400	79.555	174.223	119.706	21.807	0.000	0.000	0.000
457.67	79.555	184.938	127.266	26.395	0.000	0.000	0.000

57. DIIODINE I_2 (g)

298.15	36.887	260.685	260.685	0.000	62.420	19.324	−3.385
300	36.897	260.913	260.685	0.068	62.387	19.056	−3.318
400	37.256	271.584	262.138	3.778	44.391	5.447	−0.711
457.67	37.385	276.610	263.652	5.931	pressure = 1 bar		
500	37.464	279.921	264.891	7.515	0.000	0.000	0.000
600	37.613	286.765	267.983	11.269	0.000	0.000	0.000
700	37.735	292.573	271.092	15.037	0.000	0.000	0.000
800	37.847	297.619	274.099	18.816	0.000	0.000	0.000
900	37.956	302.083	276.965	22.606	0.000	0.000	0.000
1000	38.070	306.088	279.681	26.407	0.000	0.000	0.000
1100	38.196	309.722	282.249	30.220	0.000	0.000	0.000
1200	38.341	313.052	284.679	34.047	0.000	0.000	0.000
1300	38.514	316.127	286.981	37.890	0.000	0.000	0.000
1400	38.719	318.989	289.166	41.751	0.000	0.000	0.000
1500	38.959	321.668	291.245	45.635	0.000	0.000	0.000

58. HYDROGEN IODIDE   HI (g)

298.15	29.157	206.589	206.589	0.000	26.500	1.700	−0.298
300	29.158	206.769	206.589	0.054	26.477	1.546	−0.269
400	29.329	215.176	207.734	2.977	17.093	−6.289	0.821
500	29.738	221.760	209.904	5.928	−5.481	−9.946	1.039
600	30.351	227.233	212.348	8.931	−5.819	−10.806	0.941
700	31.070	231.965	214.820	12.002	−6.101	−11.614	0.867
800	31.807	236.162	217.230	15.145	−6.323	−12.386	0.809
900	32.511	239.950	219.548	18.362	−6.489	−13.133	0.762
1000	33.156	243.409	221.763	21.646	−6.608	−13.865	0.724
1100	33.735	246.597	223.878	24.991	−6.689	−14.586	0.693
1200	34.249	249.555	225.896	28.391	−6.741	−15.302	0.666
1300	34.703	252.314	227.823	31.839	−6.775	−16.014	0.643
1400	35.106	254.901	229.666	35.330	−6.797	−16.723	0.624
1500	35.463	257.336	231.430	38.858	−6.814	−17.432	0.607

	J/K·mol			kJ/mol			
T/K	$C_p°$	$S°$	$-(G°-H°(T_r))/T$	$H°-H°(T_r)$	$\Delta_f H°$	$\Delta_f G°$	Log K_f

59. POTASSIUM   K (cr, l)

298.15	29.600	64.680	64.680	0.000	0.000	0.000	0.000
300	29.671	64.863	64.681	0.055	0.000	0.000	0.000
336.86	32.130	68.422	64.896	1.188	0.000	0.000	0.000

PHASE TRANSITION: $\Delta_{trs} H$ = 2.321 kJ/mol, $\Delta_{trs} S$ = 6.891 J/K·mol, cr–l

336.86	32.129	75.313	64.896	3.509	0.000	0.000	0.000
400	31.552	80.784	66.986	5.519	0.000	0.000	0.000
500	30.741	87.734	70.469	8.632	0.000	0.000	0.000
600	30.158	93.283	73.824	11.675	0.000	0.000	0.000
700	29.851	97.905	76.943	14.673	0.000	0.000	0.000
800	29.838	101.887	79.818	17.655	0.000	0.000	0.000
900	30.130	105.415	82.470	20.651	0.000	0.000	0.000
1000	30.730	108.618	84.927	23.691	0.000	0.000	0.000
1039.4	31.053	109.812	85.847	24.908	0.000	0.000	0.000

60. POTASSIUM   K (g)

298.15	20.786	160.340	160.340	0.000	89.000	60.479	-10.596
300	20.786	160.468	160.340	0.038	88.984	60.302	-10.499
400	20.786	166.448	161.155	2.117	85.598	51.332	-6.703
500	20.786	171.086	162.695	4.196	84.563	42.887	-4.480
600	20.786	174.876	164.419	6.274	83.599	34.643	-3.016
700	20.786	178.080	166.148	8.353	82.680	26.557	-1.982
800	20.786	180.856	167.817	10.431	81.776	18.601	-1.215
900	20.786	183.304	169.404	12.510	80.859	10.759	-0.624
1000	20.786	185.494	170.905	14.589	79.897	3.021	-0.158
1039.4	20.786	186.297	171.474	15.408		pressure = 1 bar	
1100	20.786	187.475	172.323	16.667	0.000	0.000	0.000
1200	20.786	189.284	173.662	18.746	0.000	0.000	0.000
1300	20.789	190.948	174.929	20.825	0.000	0.000	0.000
1400	20.793	192.489	176.129	22.904	0.000	0.000	0.000
1500	20.801	193.923	177.268	24.983	0.000	0.000	0.000

61. DIPOTASSIUM OXIDE   K$_2$O (cr, l)

298.15	72.000	96.000	96.000	0.000	-361.700	-321.171	56.267
300	72.130	96.446	96.001	0.133	-361.704	-320.920	55.876
400	79.154	118.158	98.914	7.698	-366.554	-306.416	40.013
500	86.178	136.575	104.647	15.964	-366.043	-291.423	30.444
590	92.500	151.348	110.662	24.005	-364.204	-278.079	24.619

PHASE TRANSITION: $\Delta_{trs} H$ = 0.700 kJ/mol, $\Delta_{trs} S$ = 1.186 J/K·mol, crIII–crII

590	100.000	152.534	110.662	24.705	-364.204	-278.079	24.619
600	100.000	154.215	111.374	25.705	-363.968	-276.621	24.082
645	100.000	161.447	114.618	30.205	-358.901	-270.109	21.874

PHASE TRANSITION: $\Delta_{trs} H$ = 4.000 kJ/mol, $\Delta_{trs} S$ = 6.202 J/K·mol, crII–crI

645	100.000	167.649	114.618	34.205	-358.901	-270.109	21.874
700	100.000	175.832	119.111	39.705	-357.592	-262.592	19.595
800	100.000	189.185	127.054	49.705	-355.224	-249.183	16.270
900	100.000	200.963	134.625	59.705	-352.919	-236.067	13.701
1000	100.000	211.499	141.794	69.705	-350.732	-223.202	11.659
1013	100.000	212.791	142.697	71.005	-323.459	-221.546	11.424

PHASE TRANSITION: $\Delta_{trs} H$ = 27.000 kJ/mol, $\Delta_{trs} S$ =26.654 J/K·mol, crI–l

1013	100.000	239.444	142.697	98.005	-323.459	-221.546	11.424
1100	100.000	247.684	150.679	106.705	-479.439	-203.633	9.670
1200	100.000	256.385	159.131	116.705	-475.371	-178.740	7.780
1300	100.000	264.389	166.924	126.705	-471.321	-154.185	6.195
1400	100.000	271.800	174.154	136.705	-467.287	-129.941	4.848
1500	100.000	278.699	180.896	146.705	-463.268	-105.986	3.691

T/K	C_p°	S°	$-(G^\circ-H^\circ(T_r))/T$	$H^\circ-H^\circ(T_r)$	$\Delta_f H^\circ$	$\Delta_f G^\circ$	$\text{Log } K_f$
	J/K·mol			kJ/mol			

62. POTASSIUM HYDROXIDE   KOH (cr, l)

T/K	C_p°	S°	$-(G^\circ-H^\circ(T_r))/T$	$H^\circ-H^\circ(T_r)$	$\Delta_f H^\circ$	$\Delta_f G^\circ$	$\text{Log } K_f$
298.15	64.900	78.870	78.870	0.000	−424.580	−378.747	66.354
300	65.038	79.272	78.871	0.120	−424.569	−378.463	65.895
400	72.519	99.007	81.512	6.998	−426.094	−362.765	47.372
500	80.000	115.993	86.745	14.624	−424.572	−347.093	36.260
520	81.496	119.159	87.931	16.239	−417.725	−344.002	34.555

PHASE TRANSITION: $\Delta_{trs} H$ = 6.450 kJ/mol, $\Delta_{trs} S$ = 12.404 J/K·mol, crII–crI

520	79.000	131.563	87.931	22.689	−417.725	−344.002	34.555
600	79.000	142.868	94.520	29.009	−416.274	−332.766	28.969
678	79.000	152.523	100.649	35.171	−405.464	−321.998	24.807

PHASE TRANSITION: $\Delta_{trs} H$ = 9.400 kJ/mol, $\Delta_{trs} S$ = 13.865 J/K·mol, crI–l

678	83.000	166.388	100.649	44.571	−405.464	−321.998	24.807
700	83.000	169.038	102.757	46.397	−404.981	−319.297	23.826
800	83.000	180.121	111.750	54.697	−402.808	−307.206	20.058
900	83.000	189.897	119.901	62.997	−400.694	−295.383	17.143
1000	83.000	198.642	127.345	71.297	−398.668	−283.791	14.824
1100	83.000	206.553	134.192	79.597	−475.618	−267.780	12.716
1200	83.000	213.775	140.527	87.897	−472.711	−249.014	10.839
1300	83.000	220.418	146.421	96.197	−469.843	−230.490	9.261
1400	83.000	226.569	151.929	104.497	−467.011	−212.184	7.917
1500	83.000	232.296	157.098	112.797	−464.217	−194.080	6.758

63. POTASSIUM HYDROXIDE   KOH (g)

T/K	C_p°	S°	$-(G^\circ-H^\circ(T_r))/T$	$H^\circ-H^\circ(T_r)$	$\Delta_f H^\circ$	$\Delta_f G^\circ$	$\text{Log } K_f$
298.15	49.184	238.283	238.283	0.000	−227.989	−229.685	40.239
300	49.236	238.588	238.284	0.091	−228.007	−229.696	39.993
400	51.178	253.053	240.243	5.124	−231.377	−229.667	29.991
500	52.178	264.591	243.998	10.296	−232.309	−229.129	23.937
600	52.804	274.163	248.251	15.547	−233.145	−228.413	19.885
700	53.296	282.340	252.551	20.853	−233.934	−227.562	16.981
800	53.758	289.487	256.730	26.206	−234.708	−226.599	14.795
900	54.229	295.846	260.730	31.605	−235.495	−225.538	13.090
1000	54.713	301.585	264.533	37.052	−236.322	−224.388	11.721
1100	55.203	306.823	268.143	42.548	−316.077	−218.535	10.377
1200	55.686	311.647	271.570	48.092	−315.925	−209.674	9.127
1300	56.153	316.122	274.827	53.684	−315.764	−200.826	8.069
1400	56.598	320.300	277.927	59.322	−315.595	−191.991	7.163
1500	57.016	324.220	280.884	65.003	−315.420	−183.169	6.378

64. POTASSIUM CHLORIDE   KCl (cr, l)

T/K	C_p°	S°	$-(G^\circ-H^\circ(T_r))/T$	$H^\circ-H^\circ(T_r)$	$\Delta_f H^\circ$	$\Delta_f G^\circ$	$\text{Log } K_f$
298.15	51.300	82.570	82.570	0.000	−436.490	−408.568	71.579
300	51.333	82.887	82.571	0.095	−436.481	−408.395	71.107
400	52.977	97.886	84.605	5.312	−438.463	−398.651	52.058
500	54.448	109.867	88.498	10.685	−437.990	−388.749	40.612
600	55.885	119.921	92.919	16.201	−437.332	−378.960	32.991
700	57.425	128.649	97.413	21.865	−436.502	−369.295	27.557
800	59.205	136.430	101.812	27.694	−435.505	−359.760	23.490
900	61.361	143.523	106.058	33.719	−434.337	−350.360	20.334
1000	64.032	150.121	110.138	39.983	−432.981	−341.100	17.817
1044	65.405	152.908	111.882	42.830	−485.450	−336.720	16.847

PHASE TRANSITION: $\Delta_{trs} H$ = 26.320 kJ/mol, $\Delta_{trs} S$ = 25.210 J/K·mol, cr–l

1044	72.000	178.118	111.882	69.150	−485.450	−336.720	16.847
1100	72.000	181.880	115.351	73.182	−483.633	−328.790	15.613
1200	72.000	188.145	121.160	80.382	−480.393	−314.856	13.705
1300	72.000	193.908	126.537	87.582	−477.158	−301.192	12.102
1400	72.000	199.244	131.542	94.782	−473.928	−287.778	10.737
1500	72.000	204.211	136.223	101.982	−470.704	−274.594	9.562

T/K	$C_p°$	$S°$	$-(G°-H°(T_r))/T$	$H°-H°(T_r)$	$\Delta_f H°$	$\Delta_f G°$	Log K_f
	J/K·mol			kJ/mol			

65. POTASSIUM CHLORIDE   KCl (g)

T/K	$C_p°$	$S°$	$-(G°-H°(T_r))/T$	$H°-H°(T_r)$	$\Delta_f H°$	$\Delta_f G°$	Log K_f
298.15	36.505	239.091	239.091	0.000	−214.575	−233.320	40.876
300	36.518	239.317	239.092	0.068	−214.594	−233.436	40.644
400	37.066	249.904	240.532	3.749	−218.112	−239.107	31.224
500	37.384	258.212	243.267	7.473	−219.287	−244.219	25.513
600	37.597	265.048	246.344	11.222	−220.396	−249.100	21.686
700	37.769	270.857	249.441	14.991	−221.461	−253.799	18.938
800	37.907	275.910	252.441	18.775	−222.509	−258.347	16.868
900	38.041	280.382	255.302	22.572	−223.568	−262.764	15.250
1000	38.162	284.397	258.014	26.383	−224.667	−267.061	13.950
1100	38.279	288.039	260.581	30.205	−304.696	−266.627	12.661
1200	38.401	291.375	263.010	34.039	−304.821	−263.161	11.455
1300	38.518	294.454	265.312	37.885	−304.941	−259.684	10.434
1400	38.639	297.313	267.496	41.743	−305.053	−256.199	9.559
1500	38.761	299.983	269.574	45.613	−305.159	−252.706	8.800

66. DINITROGEN   N₂ (g)

T/K	$C_p°$	$S°$	$-(G°-H°(T_r))/T$	$H°-H°(T_r)$	$\Delta_f H°$	$\Delta_f G°$	Log K_f
298.15	29.124	191.608	191.608	0.000	0.000	0.000	0.000
300	29.125	191.788	191.608	0.054	0.000	0.000	0.000
400	29.249	200.180	192.752	2.971	0.000	0.000	0.000
500	29.580	206.738	194.916	5.911	0.000	0.000	0.000
600	30.109	212.175	197.352	8.894	0.000	0.000	0.000
700	30.754	216.864	199.812	11.936	0.000	0.000	0.000
800	31.433	221.015	202.208	15.046	0.000	0.000	0.000
900	32.090	224.756	204.509	18.222	0.000	0.000	0.000
1000	32.696	228.169	206.706	21.462	0.000	0.000	0.000
1100	33.241	231.311	208.802	24.759	0.000	0.000	0.000
1200	33.723	234.224	210.801	28.108	0.000	0.000	0.000
1300	34.147	236.941	212.708	31.502	0.000	0.000	0.000
1400	34.517	239.485	214.531	34.936	0.000	0.000	0.000
1500	34.842	241.878	216.275	38.404	0.000	0.000	0.000

67. NITRIC OXIDE   NO (g)

T/K	$C_p°$	$S°$	$-(G°-H°(T_r))/T$	$H°-H°(T_r)$	$\Delta_f H°$	$\Delta_f G°$	Log K_f
298.15	29.862	210.745	210.745	0.000	91.277	87.590	−15.345
300	29.858	210.930	210.746	0.055	91.278	87.567	−15.247
400	29.954	219.519	211.916	3.041	91.320	86.323	−11.272
500	30.493	226.255	214.133	6.061	91.340	85.071	−8.887
600	31.243	231.879	216.635	9.147	91.354	83.816	−7.297
700	32.031	236.754	219.168	12.310	91.369	82.558	−6.160
800	32.770	241.081	221.642	15.551	91.386	81.298	−5.308
900	33.425	244.979	224.022	18.862	91.405	80.036	−4.645
1000	33.990	248.531	226.298	22.233	91.426	78.772	−4.115
1100	34.473	251.794	228.469	25.657	91.445	77.505	−3.680
1200	34.883	254.811	230.540	29.125	91.464	76.237	−3.318
1300	35.234	257.618	232.516	32.632	91.481	74.967	−3.012
1400	35.533	260.240	234.404	36.170	91.495	73.697	−2.750
1500	35.792	262.700	236.209	39.737	91.506	72.425	−2.522

68. NITROGEN DIOXIDE   NO₂ (g)

T/K	$C_p°$	$S°$	$-(G°-H°(T_r))/T$	$H°-H°(T_r)$	$\Delta_f H°$	$\Delta_f G°$	Log K_f
298.15	37.178	240.166	240.166	0.000	34.193	52.316	−9.165
300	37.236	240.397	240.167	0.069	34.181	52.429	−9.129
400	40.513	251.554	241.666	3.955	33.637	58.600	−7.652
500	43.664	260.939	244.605	8.167	33.319	64.882	−6.778
600	46.383	269.147	248.026	12.673	33.174	71.211	−6.199
700	48.612	276.471	251.575	17.427	33.151	77.553	−5.787
800	50.405	283.083	255.107	22.381	33.213	83.893	−5.478
900	51.844	289.106	258.555	27.496	33.334	90.221	−5.236
1000	53.007	294.631	261.891	32.741	33.495	96.534	−5.042
1100	53.956	299.729	265.102	38.090	33.686	102.828	−4.883
1200	54.741	304.459	268.187	43.526	33.898	109.105	−4.749

T/K	C_p°	S°	$-(G^\circ - H^\circ(T_r))/T$	$H^\circ - H^\circ(T_r)$	$\Delta_f H^\circ$	$\Delta_f G^\circ$	$\text{Log } K_f$
		J/K·mol			kJ/mol		
1300	55.399	308.867	271.148	49.034	34.124	115.363	-4.635
1400	55.960	312.994	273.992	54.603	34.360	121.603	-4.537
1500	56.446	316.871	276.722	60.224	34.604	127.827	-4.451

69. AMMONIA　NH₃ (g)

T/K	C_p°	S°	$-(G^\circ - H^\circ(T_r))/T$	$H^\circ - H^\circ(T_r)$	$\Delta_f H^\circ$	$\Delta_f G^\circ$	$\text{Log } K_f$
298.15	35.630	192.768	192.768	0.000	-45.940	-16.407	2.874
300	35.678	192.989	192.769	0.066	-45.981	-16.223	2.825
400	38.674	203.647	194.202	3.778	-48.087	-5.980	0.781
500	41.994	212.633	197.011	7.811	-49.908	4.764	-0.498
600	45.229	220.578	200.289	12.174	-51.430	15.846	-1.379
700	48.269	227.781	203.709	16.850	-52.682	27.161	-2.027
800	51.112	234.414	207.138	21.821	-53.695	38.639	-2.523
900	53.769	240.589	210.516	27.066	-54.499	50.231	-2.915
1000	56.244	246.384	213.816	32.569	-55.122	61.903	-3.233
1100	58.535	251.854	217.027	38.309	-55.589	73.629	-3.496
1200	60.644	257.039	220.147	44.270	-55.920	85.392	-3.717
1300	62.576	261.970	223.176	50.432	-56.136	97.177	-3.905
1400	64.339	266.673	226.117	56.779	-56.251	108.975	-4.066
1500	65.945	271.168	228.971	63.295	-56.282	120.779	-4.206

70. OXYGEN　O (g)

T/K	C_p°	S°	$-(G^\circ - H^\circ(T_r))/T$	$H^\circ - H^\circ(T_r)$	$\Delta_f H^\circ$	$\Delta_f G^\circ$	$\text{Log } K_f$
298.15	21.911	161.058	161.058	0.000	249.180	231.743	-40.600
300	21.901	161.194	161.059	0.041	249.193	231.635	-40.331
400	21.482	167.430	161.912	2.207	249.874	225.677	-29.470
500	21.257	172.197	163.511	4.343	250.481	219.556	-22.937
600	21.124	176.060	165.290	6.462	251.019	213.319	-18.571
700	21.040	179.310	167.067	8.570	251.500	206.997	-15.446
800	20.984	182.115	168.777	10.671	251.932	200.610	-13.098
900	20.944	184.584	170.399	12.767	252.325	194.171	-11.269
1000	20.915	186.789	171.930	14.860	252.686	187.689	-9.804
1100	20.893	188.782	173.372	16.950	253.022	181.173	-8.603
1200	20.877	190.599	174.733	19.039	253.335	174.628	-7.601
1300	20.864	192.270	176.019	21.126	253.630	168.057	-6.753
1400	20.853	193.815	177.236	23.212	253.908	161.463	-6.024
1500	20.845	195.254	178.389	25.296	254.171	154.851	-5.392

71. DIOXYGEN　O₂ (g)

T/K	C_p°	S°	$-(G^\circ - H^\circ(T_r))/T$	$H^\circ - H^\circ(T_r)$	$\Delta_f H^\circ$	$\Delta_f G^\circ$	$\text{Log } K_f$
298.15	29.378	205.148	205.148	0.000	0.000	0.000	0.000
300	29.387	205.330	205.148	0.054	0.000	0.000	0.000
400	30.109	213.873	206.308	3.026	0.000	0.000	0.000
500	31.094	220.695	208.525	6.085	0.000	0.000	0.000
600	32.095	226.454	211.045	9.245	0.000	0.000	0.000
700	32.987	231.470	213.612	12.500	0.000	0.000	0.000
800	33.741	235.925	216.128	15.838	0.000	0.000	0.000
900	34.365	239.937	218.554	19.244	0.000	0.000	0.000
1000	34.881	243.585	220.878	22.707	0.000	0.000	0.000
1100	35.314	246.930	223.096	26.217	0.000	0.000	0.000
1200	35.683	250.019	225.213	29.768	0.000	0.000	0.000
1300	36.006	252.888	227.233	33.352	0.000	0.000	0.000
1400	36.297	255.568	229.162	36.968	0.000	0.000	0.000
1500	36.567	258.081	231.007	40.611	0.000	0.000	0.000

72. SULFUR　S (cr, l)

T/K	C_p°	S°	$-(G^\circ - H^\circ(T_r))/T$	$H^\circ - H^\circ(T_r)$	$\Delta_f H^\circ$	$\Delta_f G^\circ$	$\text{Log } K_f$
298.15	22.690	32.070	32.070	0.000	0.000	0.000	0.000
300	22.737	32.210	32.070	0.042	0.000	0.000	0.000
368.3	24.237	37.030	32.554	1.649	0.000	0.000	0.000

PHASE TRANSITION: $\Delta_{trs} H$ = 0.401 kJ/mol, $\Delta_{trs} S$ = 1.089 J/K·mol, crII–crI

T/K	C_p°	S°	$-(G^\circ - H^\circ(T_r))/T$	$H^\circ - H^\circ(T_r)$	$\Delta_f H^\circ$	$\Delta_f G^\circ$	$\text{Log } K_f$
368.3	24.773	38.119	32.553	2.050	0.000	0.000	0.000
388.36	25.180	39.444	32.875	2.551	0.000	0.000	0.000

PHASE TRANSITION: $\Delta_{trs} H$ = 1.722 kJ/mol, $\Delta_{trs} S$ = 4.431 J/K·mol, crI–l

T/K	$C_p°$	$S°$	$-(G°-H°(T_r))/T$	$H°-H°(T_r)$	$\Delta_f H°$	$\Delta_f G°$	Log K_f
		J/K·mol			kJ/mol		
388.36	31.710	43.875	32.872	4.273	0.000	0.000	0.000
400	32.369	44.824	33.206	4.647	0.000	0.000	0.000
500	38.026	53.578	36.411	8.584	0.000	0.000	0.000
600	34.371	60.116	39.842	12.164	0.000	0.000	0.000
700	32.451	65.278	43.120	15.511	0.000	0.000	0.000
800	32.000	69.557	46.163	18.715	0.000	0.000	0.000
882.38	32.000	72.693	48.496	21.351	0.000	0.000	0.000

73. SULFUR S (g)

T/K	$C_p°$	$S°$	$-(G°-H°(T_r))/T$	$H°-H°(T_r)$	$\Delta_f H°$	$\Delta_f G°$	Log K_f
298.15	23.673	167.828	167.828	0.000	277.180	236.704	-41.469
300	23.669	167.974	167.828	0.044	277.182	236.453	-41.170
400	23.233	174.730	168.752	2.391	274.924	222.962	-29.115
500	22.741	179.860	170.482	4.689	273.286	210.145	-21.953
600	22.338	183.969	172.398	6.942	271.958	197.646	-17.206
700	22.031	187.388	174.302	9.160	270.829	185.352	-13.831
800	21.800	190.314	176.125	11.351	269.816	173.210	-11.309
900	21.624	192.871	177.847	13.522	215.723	162.258	-9.417
1000	21.489	195.142	179.465	15.677	216.018	156.301	-8.164
1100	21.386	197.185	180.985	17.821	216.284	150.317	-7.138
1200	21.307	199.043	182.413	19.955	216.525	144.309	-6.282
1300	21.249	200.746	183.759	22.083	216.743	138.282	-5.556
1400	21.209	202.319	185.029	24.206	216.940	132.239	-4.934
1500	21.186	203.781	186.231	26.325	217.119	126.182	-4.394

74. DISULFUR S_2 (g)

T/K	$C_p°$	$S°$	$-(G°-H°(T_r))/T$	$H°-H°(T_r)$	$\Delta_f H°$	$\Delta_f G°$	Log K_f
298.15	32.505	228.165	228.165	0.000	128.600	79.696	-13.962
300	32.540	228.366	228.165	0.060	128.576	79.393	-13.823
400	34.108	237.956	229.462	3.398	122.703	63.380	-8.276
500	35.133	245.686	231.959	6.863	118.296	49.031	-5.122
600	35.815	252.156	234.800	10.413	114.685	35.530	-3.093
700	36.305	257.715	237.686	14.020	111.599	22.588	-1.685
800	36.697	262.589	240.501	17.671	108.841	10.060	-0.657
882.38	36.985	266.200	242.734	20.706	pressure = 1 bar		
900	37.045	266.932	243.201	21.358	0.000	0.000	0.000
1000	37.377	270.852	245.773	25.079	0.000	0.000	0.000
1100	37.704	274.430	248.218	28.833	0.000	0.000	0.000
1200	38.030	277.725	250.541	32.620	0.000	0.000	0.000
1300	38.353	280.781	252.751	36.439	0.000	0.000	0.000
1400	38.669	283.635	254.856	40.290	0.000	0.000	0.000
1500	38.976	286.314	256.865	44.173	0.000	0.000	0.000

75. OCTASULFUR S_8 (g)

T/K	$C_p°$	$S°$	$-(G°-H°(T_r))/T$	$H°-H°(T_r)$	$\Delta_f H°$	$\Delta_f G°$	Log K_f
298.15	156.500	432.536	432.536	0.000	101.277	48.810	-8.551
300	156.768	433.505	432.539	0.290	101.231	48.484	-8.442
400	167.125	480.190	438.834	16.542	80.642	32.003	-4.179
500	173.181	518.176	451.022	33.577	66.185	21.409	-2.237
600	177.936	550.180	464.951	51.137	55.101	13.549	-1.180
700	182.441	577.948	479.152	69.157	46.349	7.343	-0.548
800	186.764	602.596	493.071	87.620	39.177	2.263	-0.148
900	190.595	624.821	506.495	106.494	-392.062	6.554	-0.380
1000	193.618	645.067	519.355	125.712	-387.728	50.614	-2.644
1100	195.684	663.625	531.639	145.185	-383.272	94.233	-4.475
1200	196.825	680.707	543.359	164.817	-378.786	137.444	-5.983
1300	197.195	696.480	554.539	184.524	-374.356	180.283	-7.244
1400	196.988	711.089	565.206	204.237	-370.048	222.785	-8.312
1500	196.396	724.662	575.389	223.909	-365.905	264.984	-9.227

76. SULFUR DIOXIDE SO_2 (g)

T/K	$C_p°$	$S°$	$-(G°-H°(T_r))/T$	$H°-H°(T_r)$	$\Delta_f H°$	$\Delta_f G°$	Log K_f
298.15	39.842	248.219	248.219	0.000	-296.810	-300.090	52.574
300	39.909	248.466	248.220	0.074	-296.833	-300.110	52.253
400	43.427	260.435	249.828	4.243	-300.240	-300.935	39.298

T/K	J/K·mol			kJ/mol			
	$C_p°$	$S°$	$-(G°-H°(T_r))/T$	$H°-H°(T_r)$	$\Delta_f H°$	$\Delta_f G°$	Log K_f
500	46.490	270.465	252.978	8.744	−302.735	−300.831	31.427
600	48.938	279.167	256.634	13.520	−304.699	−300.258	26.139
700	50.829	286.859	260.413	18.513	−306.308	−299.386	22.340
800	52.282	293.746	264.157	23.671	−307.691	−298.302	19.477
900	53.407	299.971	267.796	28.958	−362.075	−295.987	17.178
1000	54.290	305.646	271.301	34.345	−362.012	−288.647	15.077
1100	54.993	310.855	274.664	39.810	−361.934	−281.314	13.358
1200	55.564	315.665	277.882	45.339	−361.849	−273.989	11.926
1300	56.033	320.131	280.963	50.920	−361.763	−266.671	10.715
1400	56.426	324.299	283.911	56.543	−361.680	−259.359	9.677
1500	56.759	328.203	286.735	62.203	−361.605	−252.053	8.777

77. SILICON   Si (cr)

T/K	$C_p°$	$S°$	$-(G°-H°(T_r))/T$	$H°-H°(T_r)$	$\Delta_f H°$	$\Delta_f G°$	Log K_f
298.15	19.789	18.810	18.810	0.000	0.000	0.000	0.000
300	19.855	18.933	18.810	0.037	0.000	0.000	0.000
400	22.301	25.023	19.624	2.160	0.000	0.000	0.000
500	23.610	30.152	21.231	4.461	0.000	0.000	0.000
600	24.472	34.537	23.092	6.867	0.000	0.000	0.000
700	25.124	38.361	25.006	9.348	0.000	0.000	0.000
800	25.662	41.752	26.891	11.888	0.000	0.000	0.000
900	26.135	44.802	28.715	14.478	0.000	0.000	0.000
1000	26.568	47.578	30.464	17.114	0.000	0.000	0.000
1100	26.974	50.130	32.138	19.791	0.000	0.000	0.000
1200	27.362	52.493	33.737	22.508	0.000	0.000	0.000
1300	27.737	54.698	35.265	25.263	0.000	0.000	0.000
1400	28.103	56.767	36.728	28.055	0.000	0.000	0.000
1500	28.462	58.719	38.130	30.883	0.000	0.000	0.000

78. SILICON   Si (g)

T/K	$C_p°$	$S°$	$-(G°-H°(T_r))/T$	$H°-H°(T_r)$	$\Delta_f H°$	$\Delta_f G°$	Log K_f
298.15	22.251	167.980	167.980	0.000	450.000	405.525	−71.045
300	22.234	168.117	167.980	0.041	450.004	405.249	−70.559
400	21.613	174.416	168.843	2.229	450.070	390.312	−50.969
500	21.316	179.204	170.456	4.374	449.913	375.388	−39.216
600	21.153	183.074	172.246	6.497	449.630	360.508	−31.385
700	21.057	186.327	174.032	8.607	449.259	345.682	−25.795
800	21.000	189.135	175.748	10.709	448.821	330.915	−21.606
900	20.971	191.606	177.375	12.808	448.329	316.205	−18.352
1000	20.968	193.815	178.911	14.904	447.791	301.553	−15.751
1100	20.989	195.815	180.358	17.002	447.211	286.957	−13.626
1200	21.033	197.643	181.723	19.103	446.595	272.416	−11.858
1300	21.099	199.329	183.014	21.209	445.946	257.927	−10.364
1400	21.183	200.895	184.236	23.323	445.268	243.489	−9.085
1500	21.282	202.360	185.396	25.446	444.563	229.101	−7.978

79. SILICON DIOXIDE SiO_2 (cr)

T/K	$C_p°$	$S°$	$-(G°-H°(T_r))/T$	$H°-H°(T_r)$	$\Delta_f H°$	$\Delta_f G°$	Log K_f
298.15	44.602	41.460	41.460	0.000	−910.700	−856.288	150.016
300	44.712	41.736	41.461	0.083	−910.708	−855.951	149.032
400	53.477	55.744	43.311	4.973	−910.912	−837.651	109.385
500	60.533	68.505	47.094	10.705	−910.540	−819.369	85.598
600	64.452	79.919	51.633	16.971	−909.841	−801.197	69.749
700	68.234	90.114	56.414	23.590	−908.958	−783.157	58.439
800	76.224	99.674	61.226	30.758	−907.668	−765.265	49.966
848	82.967	104.298	63.533	34.569	−906.310	−756.747	46.613

PHASE TRANSITION: $\Delta_{trs} H$ = 0.411 kJ/mol, $\Delta_{trs} S$ = 0.484 J/K·mol, crII–crII′

T/K	$C_p°$	$S°$	$-(G°-H°(T_r))/T$	$H°-H°(T_r)$	$\Delta_f H°$	$\Delta_f G°$	Log K_f
848	67.446	104.782	63.532	34.980	−906.310	−756.747	46.613
900	67.953	108.811	66.033	38.500	−905.922	−747.587	43.388
1000	68.941	116.021	70.676	45.345	−905.176	−730.034	38.133
1100	69.940	122.639	75.104	52.289	−904.420	−712.557	33.836
1200	70.947	128.768	79.323	59.333	−901.382	−695.148	30.259

PHASE TRANSITION: $\Delta_{trs} H$ = 2.261 kJ/mol, $\Delta_{trs} S$ = 1.883 J/K·mol, crII′–crI

T/K	C°_p	S°	−(G°−H°(T_r))/T	H°−H°(T_r)	Δ_fH°	Δ_fG°	Log K_f
		J/K·mol			**kJ/mol**		
1200	71.199	130.651	79.323	61.594	−901.382	−695.148	30.259
1300	71.743	136.372	83.494	68.742	−900.574	−677.994	27.242
1400	72.249	141.707	87.463	75.941	−899.782	−660.903	24.658
1500	72.739	146.709	91.248	83.191	−899.004	−643.867	22.421

80. SILICON TETRACHLORIDE   SiCl₄ (g)

T/K	C°_p	S°	−(G°−H°(T_r))/T	H°−H°(T_r)	Δ_fH°	Δ_fG°	Log K_f
298.15	90.404	331.446	331.446	0.000	−662.200	−622.390	109.039
300	90.562	332.006	331.448	0.167	−662.195	−622.143	108.323
400	96.893	359.019	335.088	9.572	−661.853	−608.841	79.505
500	100.449	381.058	342.147	19.456	−661.413	−595.637	62.225
600	102.587	399.576	350.216	29.616	−660.924	−582.527	50.713
700	103.954	415.500	358.432	39.948	−660.417	−569.501	42.496
800	104.875	429.445	366.455	50.392	−659.912	−556.548	36.338
900	105.523	441.837	374.155	60.914	−659.422	−543.657	31.553
1000	105.995	452.981	381.490	71.491	−658.954	−530.819	27.727
1100	106.349	463.101	388.456	82.109	−658.515	−518.027	24.599
1200	106.620	472.366	395.068	92.758	−658.107	−505.274	21.994
1300	106.834	480.909	401.347	103.431	−657.735	−492.553	19.791
1400	107.003	488.833	407.316	114.123	−657.400	−479.860	17.904
1500	107.141	496.220	413.000	124.830	−657.104	−467.189	16.269

THERMODYNAMIC PROPERTIES OF AQUEOUS IONS

This table contains standard state thermodynamic properties of positive and negative ions in aqueous solution. It includes enthalpy and Gibbs energy of formation, entropy, and heat capacity, and thus serves as a companion to the preceding table, "Standard Thermodynamic Properties of Chemical Substances". The standard state is the hypothetical ideal solution with molality $m = 1$ mol/kg (mean ionic molality $m_\pm$ in the case of a species which is assumed to dissociate at infinite dilution). Further details on conventions may be found in Reference 1.

All values refer to standard conditions of 25 °C and 100 kPa pressure.

References

1. Wagman, D. D., Evans, W. H., Parker, V. B., Schumm, R. H., Halow, I., Bailey, S. M., Churney, K. L., and Nuttall, R. L., *The NBS Tables of Chemical Thermodynamic Properties*, J. Phys. Chem. Ref. Data, Vol. 11, Suppl. 2, 1982.
2. Zemaitis, J. F., Clark, D. M., Rafal, M., and Scrivner, N. C., *Handbook of Aqueous Electrolyte Thermodynamics*, American Institute of Chemical Engineers, New York, 1986.

Species	$\Delta_f H°/$ kJ mol^{-1}	$\Delta_f G°/$ kJ mol^{-1}	$S°/$ J mol^{-1}K^{-1}	$C_p/$ J mol^{-1}K^{-1}
Cations				
Ag$^+$	105.6	77.1	72.7	21.8
Al^{+3}	−531.0	−485.0	−321.7	
AlOH^{+2}		−694.1		
Ba^{+2}	−537.6	−560.8	9.6	
BaOH$^+$		−730.5		
Be^{+2}	−382.8	−379.7	−129.7	
Bi^{+3}		82.8		
BiOH^{+2}		−146.4		
Ca^{+2}	−542.8	−553.6	−53.1	
CaOH$^+$		−718.4		
Cd^{+2}	−75.9	−77.6	−73.2	
CdOH$^+$		−261.1		
Ce^{+3}	−696.2	−672.0	−205.0	
Ce^{+4}	−537.2	−503.8	−301.0	
Co^{+2}	−58.2	−54.4	−113.0	
Co^{+3}	92.0	134.0	−305.0	
Cr^{+2}	−143.5			
Cs$^+$	−258.3	−292.0	133.1	−10.5
Cu$^+$	71.7	50.0	40.6	
Cu^{+2}	64.8	65.5	−99.6	
Dy^{+3}	−699.0	−665.0	−231.0	21.0
Er^{+3}	−705.4	−669.1	−244.3	21.0
Eu^{+2}	−527.0	−540.2	−8.0	
Eu^{+3}	−605.0	−574.1	−222.0	8.0
Fe^{+2}	−89.1	−78.9	−137.7	
Fe^{+3}	−48.5	−4.7	−315.9	
FeOH$^+$	−324.7	−277.4	−29.0	
FeOH^{+2}	−290.8	−229.4	−142.0	
Fe(OH)$_2$$^+$		−438.0		
Ga^{+2}		−88.0		
Ga^{+3}	−211.7	−159.0	−331.0	
GaOH^{+2}		−380.3		
Ga(OH)$_2$$^+$		−597.4		
Gd^{+3}	−686.0	−661.0	−205.9	
H$^+$	0	0	0	0
Hg^{+2}	171.1	164.4	−32.2	
Hg$_2$$^{+2}$	172.4	153.5	84.5	
HgOH$^+$	−84.5	−52.3	71.0	
Ho^{+3}	−705.0	−673.7	−226.8	17.0
In$^+$		−12.1		
In^{+2}		−50.7		
In^{+3}	−105.0	−98.0	−151.0	

Species	$\Delta_f H°/$ kJ mol^{-1}	$\Delta_f G°/$ kJ mol^{-1}	$S°/$ J mol^{-1}K^{-1}	$C_p/$ J mol^{-1}K^{-1}
InOH^{+2}	−370.3	−313.0	−88.0	
In(OH)$_2$$^+$	−619.0	−525.0	25.0	
K$^+$	−252.4	−283.3	102.5	21.8
La^{+3}	−707.1	−683.7	−217.6	−13.0
Li$^+$	−278.5	−293.3	13.4	68.6
Lu^{+3}	−665.0	−628.0	−264.0	25.0
LuF^{+2}		−931.4		
Mg^{+2}	−466.9	−454.8	−138.1	
MgOH$^+$		−626.7		
Mn^{+2}	−220.8	−228.1	−73.6	50.0
MnOH$^+$	−450.6	−405.0	−17.0	
NH$_4$$^+$	−132.5	−79.3	113.4	79.9
N$_2$H$_5$$^+$	−7.5	82.5	151.0	70.3
Na$^+$	−240.1	−261.9	59.0	46.4
Nd^{+3}	−696.2	−671.6	−206.7	−21.0
Ni^{+2}	−54.0	−45.6	−128.9	
NiOH$^+$	−287.9	−227.6	−71.0	
PH$_4$$^+$		92.1		
Pa^{+4}	−619.0			
Pb^{+2}	−1.7	−24.4	10.5	
PbOH$^+$		−226.3		
Pd^{+2}	149.0	176.5	−184.0	
Po^{+2}		71.0		
Po^{+4}		293.0		
Pr^{+3}	−704.6	−679.1	−209.0	−29.0
Pt^{+2}		254.8		
Ra^{+2}	−527.6	−561.5	54.0	
Rb$^+$	−251.2	−284.0	121.5	
Re$^+$		−33.0		
Sc^{+3}	−614.2	−586.6	−255.0	
ScOH^{+2}	−861.5	−801.2	−134.0	
Sm^{+2}		−497.5		
Sm^{+3}	−691.6	−666.6	−211.7	−21.0
Sn^{+2}	−8.8	−27.2	−17.0	
SnOH$^+$	−286.2	−254.8	50.0	
Sr^{+2}	−545.8	−559.5	−32.6	
SrOH$^+$		−721.3		
Tb^{+3}	−682.8	−651.9	−226.0	17.0
Te(OH)$_3$$^+$	−608.4	−496.1	111.7	
Th^{+4}	−769.0	−705.1	−422.6	
Th(OH)$^{+3}$	−1030.1	−920.5	−343.0	
Th(OH)$_2$$^{+2}$	−1282.4	−1140.9	−218.0	
Tl$^+$	5.4	−32.4	125.5	

Species	$\Delta_f H°/$ kJ mol^{-1}	$\Delta_f G°/$ kJ mol^{-1}	$S°/$ J mol^{-1}K^{-1}	$C_p/$ J mol^{-1}K^{-1}	Species	$\Delta_f H°/$ kJ mol^{-1}	$\Delta_f G°/$ kJ mol^{-1}	$S°/$ J mol^{-1}K^{-1}	$C_p/$ J mol^{-1}K^{-1}
Tl^{+3}	196.6	214.6	−192.0		HF$_2^-$	−649.9	−578.1	92.5	
TlOH^{+2}		−15.9			HPO$_3$F$^-$		−1198.2		
Tl(OH)$_2^+$		−244.7			HPO$_4^{-2}$	−1292.1	−1089.2	−33.5	
Tm^{+3}	−697.9	−662.0	−243.0	25.0	HP$_2$O$_7^{-3}$	−2274.8	−1972.2	46.0	
U^{+3}	−489.1	−476.2	−188.0		HS$^-$	−17.6	12.1	62.8	
U^{+4}	−591.2	−531.9	−410.0		HSO$_3^-$	−626.2	−527.7	139.7	
Y^{+3}	−723.4	−693.8	−251.0		HSO$_4^-$	−887.3	−755.9	131.8	−84.0
Y$_2$(OH)$_2^{+4}$		−1780.3			HS$_2$O$_4^-$		−614.5		
Yb^{+2}		−527.0			HSe$^-$	15.9	44.0	79.0	
Yb^{+3}	−674.5	−644.0	−238.0	25.0	HSeO$_3^-$	−514.6	−411.5	135.1	
Y(OH)$^{+2}$		−879.1			HSeO$_4^-$	−581.6	−452.2	149.4	
Zn^{+2}	−153.9	−147.1	−112.1	46.0	H$_2$AsO$_3^-$	−714.8	−587.1	110.5	
ZnOH$^+$		−330.1			H$_2$AsO$_4^-$	−909.6	−753.2	117.0	
Anions					H$_2$PO$_4^-$	−1296.3	−1130.2	90.4	
AlO$_2^-$	−930.9	−830.9	−36.8		H$_2$P$_2$O$_7^{-2}$	−2278.6	−2010.2	163.0	
Al(OH)$_4^-$	−1502.5	−1305.3	102.9		I$^-$	−55.2	−51.6	111.3	−142.3
AsO$_2^-$	−429.0	−350.0	40.6		IO$^-$	−107.5	−38.5	−5.4	
AsO$_4^{-3}$	−888.1	−648.4	−162.8		IO$_3^-$	−221.3	−128.0	118.4	
BF$_4^-$	−1574.9	−1486.9	180.0		IO$_4^-$	−151.5	−58.5	222.0	
BH$_4^-$	48.2	114.4	110.5		MnO$_4^-$	−541.4	−447.2	191.2	−82.0
BO$_2^-$	−772.4	−678.9	−37.2		MnO$_4^{-2}$	−653.0	−500.7	59.0	
B$_4$O$_7^{-2}$		−2604.8			MoO$_4^{-2}$	−997.9	−836.3	27.2	
BeO$_2^{-2}$	−790.8	−640.1	−159.0		NO$_2^-$	−104.6	−32.2	123.0	−97.5
Br$^-$	−121.6	−104.0	82.4	−141.8	NO$_3^-$	−207.4	−111.3	146.4	−86.6
BrO$^-$	−94.1	−33.4	42.0		N$_3^-$	275.1	348.2	107.9	
BrO$_3^-$	−67.1	18.6	161.7		OCN$^-$	−146.0	−97.4	106.7	
BrO$_4^-$	13.0	118.1	199.6		OH$^-$	−230.0	−157.2	−10.8	−148.5
CHOO$^-$	−425.6	−351.0	92.0	−87.9	PO$_4^{-3}$	−1277.4	−1018.7	−220.5	
CH$_3$COO$^-$	−486.0	−369.3	86.6	−6.3	P$_2$O$_7^{-4}$	−2271.1	−1919.0	−117.0	
C$_2$O$_4^{-2}$	−825.1	−673.9	45.6		Re$^-$	46.0	10.1	230.0	
C$_2$O$_4$H$^-$	−818.4	−698.3	149.4		S^{-2}	33.1	85.8	−14.6	
Cl$^-$	−167.2	−131.2	56.5	−136.4	SCN$^-$	76.4	92.7	144.3	−40.2
ClO$^-$	−107.1	−36.8	42.0		SO$_3^{-2}$	−635.5	−486.5	−29.0	
ClO$_2^-$	−66.5	17.2	101.3		SO$_4^{-2}$	−909.3	−744.5	20.1	−293.0
ClO$_3^-$	−104.0	−8.0	162.3		S$_2^{-2}$	30.1	79.5	28.5	
ClO$_4^-$	−129.3	−8.5	182.0		S$_2$O$_3^{-2}$	−652.3	−522.5	67.0	
CN$^-$	150.6	172.4	94.1		S$_2$O$_4^{-2}$	−753.5	−600.3	92.0	
CO$_3^{-2}$	−677.1	−527.8	−56.9		S$_2$O$_8^{-2}$	−1344.7	−1114.9	244.3	
CrO$_4^{-2}$	−881.2	−727.8	50.2		Se^{-2}			129.3	
Cr$_2$O$_7^{-2}$	−1490.3	−1301.1	261.9		SeO$_3^{-2}$	−509.2	−369.8	13.0	
F$^-$	−332.6	−278.8	−13.8	−106.7	SeO$_4^{-2}$	−599.1	−441.3	54.0	
Fe(CN)$_6^{-3}$	561.9	729.4	270.3		VO$_3^-$	−888.3	−783.6	50.0	
Fe(CN)$_6^{-4}$	455.6	695.1	95.0		VO$_4^{-3}$		−899.0		
HB$_4$O$_7^-$		−2685.1			WO$_4^{-2}$	−1075.7			
HCO$_3^-$	−692.0	−586.8	91.2						

HEAT OF COMBUSTION

The heat of combustion of a substance at 25 °C can be calculated from the enthalpy of formation ($\Delta_f H°$) data in the table "Standard Thermodynamic Properties of Chemical Substances" in this section. We can write the general combustion reaction as

$$X + O_2 \rightarrow CO_2(g) + H_2O(l) + \text{other products}$$

For a compound containing only carbon, hydrogen, and oxygen, the reaction is simply

$$C_aH_bO_c + \left(a + \frac{1}{4}b - \frac{1}{2}c\right)O_2 \rightarrow a\,CO_2(g) + \frac{1}{2}b\,H_2O(l)$$

and the standard heat of combustion $\Delta_c H°$, which is defined as the negative of the enthalpy change for the reaction (i.e., the heat released in the combustion process), is given by

$$\Delta_c H° = -a\Delta_f H°(CO_2, g) - \frac{1}{2}b\Delta_f H°(H_2O, l) + \Delta_f H°(C_aH_bO_c)$$

$$= 393.51a + 142.915b + \Delta_f H°(C_aH_bO_c)$$

This equation applies if the reactants start in their standard states (25 °C and one atmosphere pressure) and the products return to the same conditions. The same equation applies to a compound containing another element if that element ends in its standard reference state (e.g., nitrogen, if the product is N_2); in general, however, the exact products containing the other elements must be known in order to calculate the heat of combustion.

The following table gives the standard heat of combustion calculated in this manner for a few representative substances.

Molecular formula	Name	$\Delta_c H°$/kJ mol^{-1}
Inorganic substances		
C	Carbon (graphite)	393.5
CO	Carbon monoxide (g)	283.0
H_2	Hydrogen (g)	285.8
H_3N	Ammonia (g)	382.8
H_4N_2	Hydrazine (g)	667.1
N_2O	Nitrous oxide (g)	82.1
Hydrocarbons		
CH_4	Methane (g)	890.8
C_2H_2	Acetylene (g)	1301.1
C_2H_4	Ethylene (g)	1411.2
C_2H_6	Ethane (g)	1560.7
C_3H_6	Propylene (g)	2058.0
C_3H_6	Cyclopropane (g)	2091.3
C_3H_8	Propane (g)	2219.2
C_4H_6	1,3-Butadiene (g)	2541.5
C_4H_{10}	Butane (g)	2877.6
C_5H_{12}	Pentane (l)	3509.0
C_6H_6	Benzene (l)	3267.6
C_6H_{12}	Cyclohexane (l)	3919.6
C_6H_{14}	Hexane (l)	4163.2
C_7H_8	Toluene (l)	3910.3
C_7H_{16}	Heptane (l)	4817.0
$C_{10}H_8$	Naphthalene (s)	5156.3
Alcohols and ethers		
CH_4O	Methanol (l)	726.1
C_2H_6O	Ethanol (l)	1366.8
C_2H_6O	Dimethyl ether (g)	1460.4
$C_2H_6O_2$	Ethylene glycol (l)	1189.2
C_3H_8O	1-Propanol (l)	2021.3
$C_3H_8O_3$	Glycerol (l)	1655.4
$C_4H_{10}O$	Diethyl ether (l)	2723.9
$C_5H_{12}O$	1-Pentanol (l)	3330.9
C_6H_6O	Phenol (s)	3053.5
Carbonyl compounds		
CH_2O	Formaldehyde (g)	570.7
C_2H_2O	Ketene (g)	1025.4
C_2H_4O	Acetaldehyde (l)	1166.9
C_3H_6O	Acetone (l)	1789.9
C_3H_6O	Propanal (l)	1822.7
C_4H_8O	2-Butanone (l)	2444.1
Acids and esters		
CH_2O_2	Formic acid (l)	254.6
$C_2H_4O_2$	Acetic acid (l)	874.2
$C_2H_4O_2$	Methyl formate (l)	972.6
$C_3H_6O_2$	Methyl acetate (l)	1592.2
$C_4H_8O_2$	Ethyl acetate (l)	2238.1
$C_6H_5NO_2$	Nicotinic acid (s)	2731.1
$C_7H_6O_2$	Benzoic acid (s)	3228.2
Nitrogen compounds		
CHN	Hydrogen cyanide (g)	671.5
CH_3NO_2	Nitromethane (l)	709.2
CH_4N_2O	Urea (s)	632.7
CH_5N	Methylamine (g)	1085.6
C_2H_3N	Acetonitrile (l)	1247.2
C_2H_5NO	Acetamide (s)	1184.6
C_3H_9N	Trimethylamine (g)	2443.1
C_5H_5N	Pyridine (l)	2782.3
C_6H_7N	Aniline (l)	3392.8

ENERGY CONTENT OF FUELS

Several fuels are compared in this table with respect to their energy content per unit mass and the amount of CO_2 released per unit of available energy. The energy content is taken to be the negative of the standard enthalpy of combustion (see the table "Heat of Combustion" in this section for more details). The energy is assumed to be released by combustion with oxygen at normal atmospheric pressure, with products of gaseous CO_2 and liquid H_2O at room temperature. This quantity is often called the "gross heat of combustion" to distinguish it from the "net heat of combustion," for which the water remains in the gas state. The latter quantity is typically 5% to 10% less than the values given here.

The energy content is given both in SI units of MJ/kg and conventional units of BTU/lb. Values for the fossil fuels and other materials are typical; individual samples show wide variations.

The last column gives the grams of carbon released as carbon dioxide per megajoule of energy. Examination of the table shows that the minimum CO_2 release occurs for fuels that have a high ratio of hydrogen to carbon. Furthermore, fuels containing oxygen have a lower energy content and higher CO_2 release than hydrocarbons with the same number of carbon atoms.

References

1. Domalski, E. S., Jobe, T. L., and Milne, T. A., *Thermodynamic Data for Biomass Conversion and Waste Incineration*, SERI/SP-271-2839, Solar Technical Information Program, U. S. Department of Energy, September 1986; see also NBSIR 78-1479, National Bureau of Standards, August 1978.
2. Green, D. W., and Ackers, D. E, *Perry's Chemical Engineers' Handbook, Eighth Edition*, McGraw-Hill, New York, 2007.
3. *Transportation Energy Data Book*, U. S. Department of Energy, May 2007, http://cta.ornl.gov/data/index.shtml.
4. *Chemical Composition of Natural Gas*, Union Gas Limited, http://www.uniongas.com/aboutus/aboutng/composition.asp.

Fuel	Energy content MJ/kg	Energy content 10³ BTU/lb	g of C per MJ
Pure compounds			
Hydrogen	141.8	61.0	0.0
Methane	55.5	23.9	13.5
Ethane	51.9	22.3	15.4
Propane	50.3	21.7	16.2
Hexane	48.3	20.8	17.3
Heptane	48.1	20.7	17.5
Octane	47.9	20.6	17.6
Methanol	22.7	9.7	16.5
Ethanol	29.7	12.8	17.6
1-Propanol	33.6	14.5	17.8
1-Butanol	36.1	15.5	18.0
1-Octanol	40.7	17.5	18.1
Methyl *tert*-butyl ether	38.2	16.4	17.8

Fuel	Energy content MJ/kg	Energy content 10³ BTU/lb	g of C per MJ
Fossil fuels			
Natural gas[a]	54.0	23.2	13.9
Gasoline	46.5	20.0	17.6
Kerosene	46.4	20.0	18.5
Fuel oil	40.9	17.6	21.3
Coal, high bituminous	36.3	15.6	23.5
Coal, low bituminous	28.9	12.4	26.3
Coal, anthracite	34.6	14.9	27.3
Other materials			
Wood, oak	18.9	8.1	25.3
Wood, locust	19.7	8.5	25.7
Wood, Ponderosa pine	20.0	8.6	24.6
Wood, redwood	20.7	8.9	24.4
Charcoal, wood	34.7	14.9	26.8
Newsprint	18.6	8.0	26.5
Cellulose	17.3	7.5	25.6
Grass (lawn clippings)	19.3	8.3	24.9

[a] Assumed to be 95% methane, 2.5% ethane, and 2.5% inert compounds; however, the actual composition varies widely (see Reference 4).

IONIZATION CONSTANT OF WATER

Serguei N. Lvov and Allan H. Harvey

This table gives values of $pK_w = -\log_{10}(K_w)$, where K_w is the equilibrium constant of the reaction $2H_2O = H_3O^+(aq) + OH^-(aq)$. K_w is defined as $K_w = a_{H_3O^+} a_{OH^-} / a_{H_2O}^2$, where a_i is the dimensionless activity of species i. The activities are on the molality basis for ions and mole fraction basis for water molecules. It is assumed that the activity of $H_3O^+(aq)$ is the same as the activity of $H^+(aq)$, so that K_w is numerically equal to $a_{H^+} a_{OH^-} / a_{H_2O}^2$, the equilibrium constant for the ionization reaction of water, $H_2O = H^+(aq) + OH^-(aq)$, that is most commonly used in the literature. Values in the table are calculated using an analytical equation given in Refs. 1 and 2 where K_w is presented as a function of temperature and density from 0 °C to 800 °C and 0 g cm^{-3} to 1.25 g cm^{-3}.

References

1. International Association for the Properties of Water and Steam, *Release on the Ionization Constant of H₂O* (2007), available from www.iapws.org.
2. Bandura, A.V., and Lvov, S.N., *J. Phys. Chem. Ref. Data* 35, 15, 2006.

Pressure, MPa	Temperature, °C								
	0	25	50	75	100	150	200	250	300
0.1 or p_s	14.946	13.995	13.264	12.696	12.252	11.641	11.310	11.205	11.339
25	14.848	13.908	13.181	12.613	12.165	11.543	11.189	11.050	11.125
50	14.754	13.824	13.102	12.533	12.084	11.450	11.076	10.898	10.893
75	14.665	13.745	13.026	12.458	12.006	11.364	10.974	10.769	10.715
100	14.580	13.668	12.953	12.385	11.933	11.283	10.880	10.655	10.568
150	14.422	13.524	12.815	12.249	11.795	11.135	10.713	10.458	10.327
200	14.278	13.390	12.687	12.123	11.668	11.000	10.564	10.289	10.131
250	14.145	13.265	12.567	12.004	11.549	10.876	10.430	10.140	9.963
300	14.021	13.148	12.453	11.892	11.437	10.760	10.306	10.005	9.814
350	13.906	13.037	12.346	11.786	11.331	10.651	10.191	9.881	9.679
400	13.797	12.932	12.243	11.685	11.230	10.548	10.083	9.766	9.555
500	13.595	12.736	12.052	11.496	11.042	10.356	9.884	9.557	9.332
600	13.411	12.556	11.875	11.322	10.868	10.181	9.703	9.369	9.135
700	13.240	12.389	11.710	11.159	10.705	10.018	9.537	9.197	8.956
800	13.080	12.233	11.556	11.006	10.553	9.865	9.381	9.037	8.791
900	12.930	12.085	11.410	10.861	10.410	9.721	9.236	8.888	8.638
1000	12.788	11.946	11.272	10.725	10.273	9.585	9.098	8.748	8.495

Pressure, MPa	Temperature, °C						
	350	400	450	500	600	700	800
0.1 or p_s	11.920	47.961	47.873	47.638	46.384	43.925	40.785
25	11.551	16.566	18.135	18.758	19.425	19.829	20.113
50	11.076	11.557	12.710	14.195	15.621	16.279	16.693
75	10.802	11.045	11.491	12.162	13.507	14.301	14.791
100	10.600	10.744	11.005	11.381	12.296	13.040	13.544
150	10.295	10.345	10.464	10.642	11.117	11.613	12.032
200	10.062	10.063	10.119	10.220	10.513	10.853	11.171
250	9.869	9.839	9.859	9.917	10.112	10.360	10.609
300	9.702	9.651	9.646	9.677	9.810	9.998	10.199
350	9.554	9.487	9.465	9.476	9.567	9.712	9.877
400	9.420	9.341	9.305	9.302	9.361	9.475	9.613
500	9.182	9.086	9.031	9.007	9.024	9.094	9.191
600	8.974	8.866	8.798	8.761	8.749	8.790	8.861
700	8.787	8.670	8.593	8.546	8.514	8.536	8.587
800	8.616	8.493	8.409	8.354	8.308	8.314	8.352
900	8.458	8.330	8.240	8.180	8.122	8.117	8.144
1000	8.311	8.178	8.084	8.019	7.952	7.939	7.957

Note: Pressure for first row is 0.1 MPa at $t < 100$ °C and $t \geq 400$ °C, or p_s (saturated liquid) for 100 °C $\leq t \leq 350$ °C

IONIZATION CONSTANT OF NORMAL AND HEAVY WATER

Serguei N. Lvov and Allan H. Harvey

This table gives the ionization constant for liquid H_2O and D_2O at temperatures from 0 °C to 100 °C at the saturated vapor pressure. The quantity tabulated is $-\log_{10}(K_w)$, where $K_w = a_{H_3O^+} a_{OH^-} / a_{H_2O}^2$ for H_2O and $K_w = a_{D_3O^+} a_{OD^-} / a_{D_2O}^2$ for D_2O. Values in the table are calculated using analytical equations given in Refs. 1 and 2 for H_2O and Ref. 3 for D_2O.

References

1. International Association for the Properties of Water and Steam, *Release on the Ionization Constant of H₂O* (2007), available from www.iapws.org.
2. Bandura, A. V., and Lvov, S. N., *J. Phys. Chem. Ref. Data* 35, 15, 2006.
3. Mesmer, R. E., and Herting, D. L., *J. Solution Chem.* 7, 901, 1978.

	$-\log_{10} K_w$	
$t/°C$	H_2O	D_2O
0	14.947	15.972
5	14.734	15.743
10	14.534	15.527
15	14.344	15.324
20	14.165	15.132
25	13.995	14.951
30	13.833	14.779
35	13.680	14.616
40	13.535	14.462
45	13.396	14.316
50	13.265	14.176
55	13.140	14.044
60	13.020	13.918
65	12.907	13.798
70	12.799	13.683
75	12.696	13.574
80	12.598	13.470
85	12.505	13.371
90	12.417	13.276
95	12.332	13.186
100	12.252	13.099

ELECTRICAL CONDUCTIVITY OF WATER

This table gives the electrical conductivity of highly purified water over a range of temperature and pressure. The first column of conductivity data refers to water at its own vapor pressure. Equations for calculating the conductivity at any temperature and pressure may be found in the reference.

Reference

Marshall, W. L., *J. Chem. Eng. Data* 32, 221, 1987.

Conductivity in µS/cm at the Indicated Pressure

$t/°C$	Sat. vapor	50 MPa	100 MPa	200 MPa	400 MPa	600 MPa
0	0.0115	0.0150	0.0189	0.0275	0.0458	0.0667
25	0.0550	0.0686	0.0836	0.117	0.194	0.291
100	0.765	0.942	1.13	1.53	2.45	3.51
200	2.99	4.08	5.22	7.65	13.1	19.5
300	2.41	4.87	7.80	14.1	28.9	46.5
400		1.17	4.91	14.3	39.2	71.3
600			0.134	4.65	33.8	85.7

ELECTRICAL CONDUCTIVITY OF AQUEOUS SOLUTIONS

The following table gives the electrical conductivity of aqueous solutions of some acids, bases, and salts as a function of concentration. All values refer to 20 °C. The conductivity κ (often called specific conductance in older literature) is the reciprocal of the resistivity. The molar conductivity Λ is related to this by Λ = κ/c, where c is the amount-of-substance concentration of the electrolyte. Thus if κ has units of millisiemens per centimeter (mS/cm), as in this table, and c is expressed in mol/L, then Λ has units of S cm^2 mol^{-1}. For these electrolytes the concentration c corresponding to the mass percent values given here can be found in the table "Concentrative Properties of Aqueous Solutions" in Section 8.

References

1. *CRC Handbook of Chemistry, and Physics, 70th Edition*, Weast, R. C., Ed., CRC Press, Boca Raton, FL, 1989, p. D-221.
2. Wolf, A. V., *Aqueous Solutions and Body Fluids*, Harper and Row, New York, 1966.

Electrical Conductivity κ in mS/cm for the Indicated Concentration in Mass Percent

Name	Formula	0.5%	1%	2%	5%	10%	15%	20%	25%	30%	40%	50%
Acetic acid	CH_3COOH	0.3	0.6	0.8	1.2	1.5	1.7	1.7	1.6	1.4	1.1	0.8
Ammonia	NH_3	0.5	0.7	1.0	1.1	1.0	0.7	0.5	0.4			
Ammonium chloride	NH_4Cl	10.5	20.4	40.3	95.3	180						
Ammonium sulfate	$(NH_4)_2SO_4$	7.4	14.2	25.7	57.4	105	147	185	215			
Barium chloride	$BaCl_2$	4.7	9.1	17.4	40.4	76.7	109.0	137.0				
Calcium chloride	$CaCl_2$	8.1	15.7	29.4	67.0	117	157	177	183	172	106	
Cesium chloride	$CsCl$	3.8	7.4	13.8	32.9	65.8	102	142				
Citric acid	$H_3C(OH)(COO)_3$	1.2	2.1	3.0	4.7	6.2	7.0	7.2	7.1			
Copper(II) sulfate	$CuSO_4$	2.9	5.4	9.3	19.0	32.2	42.3					
Formic acid	$HCOOH$	1.4	2.4	3.5	5.6	7.8	9.0	9.9	10.4	10.5	9.9	8.6
Hydrogen chloride	HCl	45.1	92.9	183								
Lithium chloride	$LiCl$	10.1	19.0	34.9	76.4	127	155	170	165	146		
Magnesium chloride	$MgCl_2$	8.6	16.6	31.2	66.9	108	129	134	122	98		
Magnesium sulfate	$MgSO_4$	4.1	7.6	13.3	27.4	42.7	54.2	51.1	44.1			
Manganese(II) sulfate	$MnSO_4$		6.2	10.6	21.6	34.5	43.7	47.6				
Nitric acid	HNO_3	28.4	56.1	108								
Oxalic acid	$H_2C_2O_4$	14.0	21.8	35.3	65.6							
Phosphoric acid	H_3PO_4	5.5	10.1	16.2	31.5	59.4	88.4	118	146	173	209	
Potassium bromide	KBr	5.2	10.2	19.5	47.7	95.6	144	194				
Potassium carbonate	K_2CO_3	7.0	13.6	25.4	58.0	109	152	188	223			
Potassium chloride	KCl	8.2	15.7	29.5	71.9	143	208					
Potassium dihydrogen phosphate	KH_2PO_4	3.0	5.9	11.0	25.0	44.6						
Potassium hydrogen carbonate	$KHCO_3$	4.6	8.9	17.0	38.8	72.4	101	128				
Potassium hydrogen phosphate	K_2HPO_4	5.2	9.9	18.3	40.3							
Potassium hydroxide	KOH	20.0	38.5	75.0	178							
Potassium iodide	KI	3.8	7.5	14.2	35.2	71.8	110	188	224			
Potassium nitrate	KNO_3	5.5	10.7	20.1	47.0	87.3	124	157	182			
Potassium permanganate	$KMnO_4$	3.5	6.9	13.0	30.5							
Potassium sulfate	K_2SO_4	5.8	11.2	21.0	48.0	88.6						
Silver(I) nitrate	$AgNO_3$	3.1	6.1	12.0	26.7	49.8	72.0	92.8	112	129	162	
Sodium acetate	$NaCH_3COO$	3.9	7.6	14.4	30.9	53.4	64.1	69.3	69.2	64.3		
Sodium bromide	$NaBr$	5.0	9.7	18.4	44.0	84.6	122	157	191	216		
Sodium carbonate	Na_2CO_3	7.0	13.1	23.3	47.0	74.4	88.6					
Sodium chloride	$NaCl$	8.2	16.0	30.2	70.1	126	171	204	222			
Sodium citrate	$Na_3C_6H_5O_7$		7.4	12.8	26.2	42.1	52.0	57.1	57.3	53.5		
Sodium dihydrogen phosphate	NaH_2PO_4	2.2	4.4	9.1	21.0	33.2	43.3	49.6	53.1	54.0	46.1	
Sodium hydrogen carbonate	$NaHCO_3$	4.2	8.2	15.0	31.4							
Sodium hydrogen phosphate	Na_2HPO_4	4.6	8.7	15.6	31.4							
Sodium hydroxide	$NaOH$	24.8	48.6	93.1	206							
Sodium nitrate	$NaNO_3$	5.4	10.6	20.4	46.2	82.6	111	134	152	165	178	
Sodium phosphate	Na_3PO_4	7.3	14.1	22.7	43.5							
Sodium sulfate	Na_2SO_4	5.9	11.2	19.8	42.7	71.3	91.1	109				
Sodium thiosulfate	$Na_2S_2O_3$	5.7	10.7	19.5	43.3	76.7	104	123	134	136	118	
Strontium chloride	$SrCl_2$	5.9	11.4	22.0	49.1	91.5	127	153	168	178		
Sulfuric acid	H_2SO_4	24.3	47.8	92	211							
Trichloroacetic acid	CCl_3COOH	10.3	19.6	37.2	84.7	148	193	221				
Zinc sulfate	$ZnSO_4$	2.8	5.4	10.0	20.5	33.7	43.3					

STANDARD KCl SOLUTIONS FOR CALIBRATING CONDUCTIVITY CELLS

This table presents recommended electrolytic conductivity (κ) values for aqueous potassium chloride solutions with molalities of 0.01 mol/kg, 0.1 mol/kg and 1.0 mol/kg at temperatures from 0 °C to 50 °C. The values, which are based on measurements at the National Institute of Standards and Technology, provide primary standards for the calibration of conductivity cells. The measurements at 0.01 and 0.1 molal are described in Reference 1, while those at 1.0 molal are in Reference 2. Temperatures are given on the ITS-90 scale. The uncertainty in the conductivity is about 0.03% for the 0.01 molal values and about 0.04% for the 0.1 and 1.0 molal values. The conductivity of water saturated with atmospheric CO_2 is given in the last column. These values were sub-tracted from the original measurements to give the values in the second, third, and fourth columns. All κ values are given in units of 10^{-4} S/m (numerically equal to µS/cm).

The assistance of Kenneth W. Pratt is appreciated.

References

1. Wu, Y. C., Koch, W. F., and Pratt, K. W., *J. Res. Natl. Inst. Stand. Technol.* 96, 191, 1991.
2. Wu, Y. C., Koch, W. F., Feng, D., Holland, L. A., Juhasz, E., Arvay, E., and Tomek, A., *J. Res. Natl. Inst. Stand. Technol.* 99, 241, 1994.
3. Pratt, K. W., Koch, W. F., Wu, Y. C., and Berezansky, P. A., *Pure Appl. Chem.* 73, 1783, 2001.

	10^4 κ/S m^{-1}			
t/°C	0.01 m KCl	0.1 m KCl	1.0 m KCl	H_2O (CO_2 sat.)
0	772.92	7 116.85	63 488	0.58
5	890.96	8 183.70	72 030	0.68
10	1 013.95	9 291.72	80 844	0.79
15	1 141.45	10 437.1	89 900	0.89
18	1 219.93	11 140.6	—	0.95
20	1 273.03	11 615.9	99 170	0.99
25	1 408.23	12 824.6	108 620	1.10
30	1 546.63	14 059.2	118 240	1.20
35	1 687.79	15 316.0	127 970	1.30
40	1 831.27	16 591.0	137 810	1.40
45	1 976.62	17 880.6	147 720	1.51
50	2 123.43	19 180.9	157 670	1.61

MOLAR CONDUCTIVITY OF AQUEOUS HF, HCl, HBr, AND HI

The molar conductivity Λ of an electrolyte solution is defined as the conductivity divided by amount-of-substance concentration. The customary unit is S cm² mol⁻¹ (i.e., Ω^{-1} cm² mol⁻¹). The first part of this table gives the molar conductivity of the hydrohalogen acids at 25 °C as a function of the concentration in mol/L. The second part gives the temperature dependence of Λ for HCl and HBr. More extensive tables and mathematical representations may be found in the reference.

Reference

Hamer, W. J., and DeWane, H. J., *Electrolytic Conductance and the Conductances of the Hydrohalogen Acids in Water*, Natl. Stand. Ref. Data Sys.- Natl. Bur. Standards (U.S.), No. 33, 1970.

c/mol L⁻¹	HF	HCl	HBr	HI
Inf. dil.	405.1	426.1	427.7	426.4
0.0001		424.5	425.9	424.6
0.0005		422.6	424.3	423.0
0.001		421.2	422.9	421.7
0.005	128.1	415.7	417.6	416.4
0.01	96.1	411.9	413.7	412.8
0.05	50.1	398.9	400.4	400.8
0.10	39.1	391.1	391.9	394.0
0.5	26.3	360.7	361.9	369.8
1.0	24.3	332.2	334.5	343.9
1.5		305.8	307.6	316.4
2.0		281.4	281.7	288.9
2.5		258.9	257.8	262.5
3.0		237.6	236.8	237.9

c/mol L⁻¹	HF	HCl	HBr	HI
3.5		218.3	217.5	215.4
4.0		200.0	199.4	195.1
4.5		183.1	182.4	176.8
5.0		167.4	166.5	160.4
5.5		152.9	151.8	145.5
6.0		139.7	138.2	131.7
6.5		127.7	125.7	118.6
7.0		116.9	114.2	105.7
7.5		107.0	103.8	
8.0		98.2	94.4	
8.5		90.3	85.8	
9.0		83.1		
9.5		76.6		
10.0		70.7		

c/mol L⁻¹	−20 °C	−10 °C	0 °C	10 °C	20 °C	30 °C	40 °C	50 °C
HCl								
0.5			228.7	283.0	336.4	386.8	436.9	482.4
1.0			211.7	261.6	312.2	359.0	402.9	445.3
1.5			196.2	241.5	287.5	331.1	371.6	410.8
2.0			182.0	222.7	262.9	303.3	342.4	378.2
2.5		131.7	168.5	205.1	239.8	277.0	315.2	347.6
3.0		120.8	154.6	188.5	219.3	253.3	289.3	319.0
3.5	85.5	111.3	139.6	172.2	201.6	232.9	263.9	292.1
4.0	79.3	102.7	129.2	158.1	185.6	214.2	242.2	268.2
4.5	73.7	94.9	119.5	145.4	170.6	196.6	222.5	246.7
5.0	68.5	87.8	110.3	133.5	156.6	180.2	204.1	226.5
5.5	63.6	81.1	101.7	122.5	143.6	165.0	187.1	207.7
6.0	58.9	74.9	93.7	112.3	131.5	151.0	171.3	190.3
6.5	54.4	69.1	86.2	103.0	120.4	138.2	156.9	174.3
7.0	50.2	63.7	79.3	94.4	110.2	126.4	143.3	159.7
7.5	46.3	58.6	73.0	86.5	100.9	115.7	131.6	146.2
8.0	42.7	54.0	67.1	79.4	92.4	106.1	120.6	134.0
8.5	39.4	49.8	61.7	72.9	84.7	97.3	110.7	123.0
9.0	36.4	45.9	56.8	67.1	77.8	89.4	101.7	112.9
9.5	33.6	42.3	52.3	61.8	71.5	82.3	93.6	103.9
10.0	31.2	39.1	48.2	57.0	65.8	75.9	86.3	95.7
10.5	28.9	36.1	44.5	52.7	60.7	70.1	79.6	88.4
11.0	26.8	33.4	41.1	48.8	56.1	64.9	73.6	81.7
11.5	24.9	31.0	38.0	45.3	51.9	60.1	68.0	75.6
12.0	23.1	28.7	35.3	42.0	48.0	55.6	62.8	70.0
12.5	21.4	26.7	32.7	39.0	44.4	51.4	57.9	64.8

c/mol L⁻¹	−20 °C	−10 °C	0 °C	10 °C	20 °C	30 °C	40 °C	50 °C
HBr								
0.5			240.9	295.9	347.0	398.9	453.6	496.8
1.0			229.6	276.0	329.0	380.4	418.6	465.2
1.5			209.5	254.9	298.9	340.6	381.8	421.4
2.0		150.8	188.6	231.3	271.8	314.1	350.5	387.4
2.5		136.8	171.7	208.3	244.8	281.7	316.0	349.1
3.0		125.7	157.2	189.5	222.2	255.0	287.8	318.6
3.5		116.1	144.1	174.6	203.2	234.4	263.7	291.9
4.0	84.0	107.5	132.3	160.2	186.8	214.2	239.7	266.9
4.5	78.0	99.0	123.0	146.4	171.2	195.1	218.8	242.6
5.0	72.3	91.4	112.6	134.0	155.7	178.2	199.6	221.3
5.5	67.0	84.2	103.1	122.7	142.1	162.8	181.4	201.8
6.0	61.8	77.2	94.3	112.0	129.6	148.0	165.4	183.4
6.5	56.8	70.7	86.0	102.0	118.0	134.1	150.5	166.3
7.0	51.9	64.6	78.4	92.6	107.1	121.4	136.3	150.8

EQUIVALENT CONDUCTIVITY OF ELECTROLYTES IN AQUEOUS SOLUTION

Petr Vanýsek

This table gives the equivalent (molar) conductivity Λ at 25 °C for some common electrolytes in aqueous solution at concentrations up to 0.1 mol/L. The units of Λ are 10^{-4} m^2 S mol^{-1}.

For very dilute solutions, the equivalent conductivity for any electrolyte of concentration c can be approximately calculated using the Debye–Hückel–Onsager equation, which can be written for a symmetrical (equal charge on cation and anion) electrolyte as

$$\Lambda = \Lambda° - (A + B\Lambda°)c^{1/2}$$

For a solution at 25 °C and both cation and anion with charge $|1|$, the constants are $A = 60.20$ and $B = 0.229$. $\Lambda°$ can be found from the next table, "Ionic Conductivity and Diffusion at Infinite Dilution." The equation is reliable for $c < 0.001$ mol/L; with higher concentration the error increases.

Compound	Infinite dilution $\Lambda°$	Concentration (mol/L) Λ (10^{-4} m^2 S mol^{-1})						
		0.0005	0.001	0.005	0.01	0.02	0.05	0.1
AgNO₃	133.29	131.29	130.45	127.14	124.70	121.35	115.18	109.09
1/2BaCl₂	139.91	135.89	134.27	127.96	123.88	119.03	111.42	105.14
1/2CaCl₂	135.77	131.86	130.30	124.19	120.30	115.59	108.42	102.41
1/2Ca(OH)₂	258	—	—	233	226	214	—	—
CuSO₄	133.6	121.6	115.20	94.02	83.08	72.16	59.02	50.55
HCl	425.95	422.53	421.15	415.59	411.80	407.04	398.89	391.13
KBr	151.9	149.8	148.9	146.02	143.36	140.41	135.61	131.32
KCl	149.79	147.74	146.88	143.48	141.20	138.27	133.30	128.90
KClO₄	139.97	138.69	137.80	134.09	131.39	127.86	121.56	115.14
1/3K₃Fe(CN)₆	174.5	166.4	163.1	150.7	—	—	—	—
1/4K₄Fe(CN)₆	184	—	167.16	146.02	134.76	122.76	107.65	97.82
KHCO₃	117.94	116.04	115.28	112.18	110.03	107.17	—	—
KI	150.31	148.2	143.32	144.30	142.11	139.38	134.90	131.05
KIO₄	127.86	125.74	124.88	121.18	118.45	114.08	106.67	98.2
KNO₃	144.89	142.70	141.77	138.41	132.75	132.34	126.25	120.34
KMnO₄	134.8	132.7	131.9	—	126.5	—	—	113
KOH	271.5	—	234	230	228	—	219	213
KReO₄	128.20	126.03	125.12	121.31	118.49	114.49	106.40	97.40
1/3LaCl₃	145.9	139.6	137.0	127.5	121.8	115.3	106.2	99.1
LiCl	114.97	113.09	112.34	109.35	107.27	104.60	100.06	95.81
LiClO₄	105.93	104.13	103.39	100.52	98.56	96.13	92.15	88.52
1/2MgCl₂	129.34	125.55	124.15	118.25	114.49	109.99	103.03	97.05
NH₄Cl	149.6	147.5	146.7	143.9	141.21	138.25	133.22	128.69
NaCl	126.39	124.44	123.68	120.59	118.45	115.70	111.01	106.69
NaClO₄	117.42	115.58	114.82	111.70	109.54	106.91	102.35	98.38
NaI	126.88	125.30	124.19	121.19	119.18	116.64	112.73	108.73
NaOOCCH₃	91.0	89.2	88.5	85.68	83.72	81.20	76.88	72.76
NaOH	247.7	245.5	244.6	240.7	237.9	—	—	—
Na picrate	80.45	78.7	78.6	75.7	73.7	—	66.3	61.8
1/2Na₂SO₄	129.8	125.68	124.09	117.09	112.38	106.73	97.70	89.94
1/2SrCl₂	135.73	131.84	130.27	124.18	120.23	115.48	108.20	102.14
ZnSO₄	132.7	121.3	114.47	95.44	84.87	74.20	61.17	52.61

IONIC CONDUCTIVITY AND DIFFUSION AT INFINITE DILUTION

Petr Vanýsek

This table gives the molar (equivalent) conductivity λ for common ions at infinite dilution. All values refer to aqueous solutions at 25 °C. It also lists the diffusion coefficient D of the ion in dilute aqueous solution, which is related to λ through the equation

$$D = (RT/F^2)(\lambda/|z|)$$

where R is the molar gas constant, T the temperature, F the Faraday constant, and z the charge on the ion. The variation with temperature is fairly sharp; for typical ions, λ and D increase by 2 to 3% per degree as the temperature increases from 25 °C.

The diffusion coefficient for a salt, D_{salt}, may be calculated from the D_+ and D_- values of the constituent ions by the relation

$$D_{salt} = \frac{(z_+ + |z_-|)D_+D_-}{z_+D_+ + |z_-|D_-}$$

For solutions of simple, pure electrolytes (one positive and one negative ionic species), such as NaCl, equivalent ionic conductivity $\Lambda°$, which is the molar conductivity per unit concentration of charge, is defined as

$$\Lambda° = \Lambda_+ + \Lambda_-$$

where Λ_+ and Λ_- are equivalent ionic conductivities of the cation and anion. The more general formula is

$$\Lambda° = \nu_+\Lambda_+ + \nu_-\Lambda_-$$

where ν_+ and ν_- refer to the number of moles of cations and anions to which one mole of the electrolyte gives a rise in the solution.

References

1. Gray, D. E., Ed., *American Institute of Physics Handbook,* McGraw-Hill, New York, 1972, 2–226.
2. Robinson, R. A., and Stokes, R. H., *Electrolyte Solutions*, Butterworths, London, 1959.
3. Lobo, V. M. M., and Quaresma, J. L., *Handbook of Electrolyte Solutions*, Physical Science Data Series 41, Elsevier, Amsterdam, 1989.
4. Conway, B. E., *Electrochemical Data*, Elsevier, Amsterdam, 1952.
5. Milazzo, G., *Electrochemistry: Theoretical Principles and Practical Applications*, Elsevier, Amsterdam, 1963.

Ion	$\Lambda_\pm$ 10^{-4} m^2 S mol^{-1}	D 10^{-5} cm^2 s^{-1}
Inorganic Cations		
Ag$^+$	61.9	1.648
1/3Al^{3+}	61	0.541
1/2Ba^{2+}	63.6	0.847
1/2Be^{2+}	45	0.599
1/2Ca^{2+}	59.47	0.792
1/2Cd^{2+}	54	0.719
1/3Ce^{3+}	69.8	0.620
1/2Co^{2+}	55	0.732
1/3[Co(NH$_3$)$_6$]$^{3+}$	101.9	0.904
1/3[Co(en)$_3$]$^{3+}$	74.7	0.663
1/6[Co$_2$(trien)$_3$]$^{6+}$	69	0.306
1/3Cr^{3+}	67	0.595
Cs$^+$	77.2	2.056
1/2Cu^{2+}	53.6	0.714
D$^+$	249.9	6.655
1/3Dy^{3+}	65.6	0.582
1/3Er^{3+}	65.9	0.585
1/3Eu^{3+}	67.8	0.602
1/2Fe^{2+}	54	0.719
1/3Fe^{3+}	68	0.604
1/3Gd^{3+}	67.3	0.597
H$^+$	349.65	9.311
1/2Hg^{2+}	68.6	0.913
1/2Hg$_2^{2+}$	63.6	0.847
1/3Ho^{3+}	66.3	0.589
K$^+$	73.48	1.957
1/3La^{3+}	69.7	0.619
Li$^+$	38.66	1.029
1/2Mg^{2+}	53.0	0.706
1/2Mn^{2+}	53.5	0.712
NH$_4^+$	73.5	1.957
N$_2$H$_5^+$	59	1.571

Ion	$\Lambda_\pm$ 10^{-4} m^2 S mol^{-1}	D 10^{-5} cm^2 s^{-1}
Na$^+$	50.08	1.334
1/3Nd^{3+}	69.4	0.616
1/2Ni^{2+}	49.6	0.661
1/4[Ni$_2$(trien)$_3$]$^{4+}$	52	0.346
1/2Pb^{2+}	71	0.945
1/3Pr^{3+}	69.5	0.617
1/2Ra^{2+}	66.8	0.889
Rb$^+$	77.8	2.072
1/3Sc^{3+}	64.7	0.574
1/3Sm^{3+}	68.5	0.608
1/2Sr^{2+}	59.4	0.791
Tl$^+$	74.7	1.989
1/3Tm^{3+}	65.4	0.581
1/2UO$_2^{2+}$	32	0.426
1/3Y^{3+}	62	0.550
1/3Yb^{3+}	65.6	0.582
1/2Zn^{2+}	52.8	0.703
Inorganic Anions		
Au(CN)$_2^-$	50	1.331
Au(CN)$_4^-$	36	0.959
B(C$_6$H$_5$)$_4^-$	21	0.559
Br$^-$	78.1	2.080
Br$_3^-$	43	1.145
BrO$_3^-$	55.7	1.483
CN$^-$	78	2.077
CNO$^-$	64.6	1.720
1/2CO$_3^{2-}$	69.3	0.923
Cl$^-$	76.31	2.032
ClO$_2^-$	52	1.385
ClO$_3^-$	64.6	1.720
ClO$_4^-$	67.3	1.792
1/3[Co(CN)$_6$]$^{3-}$	98.9	0.878
1/2CrO$_4^{2-}$	85	1.132

Ion	$\Lambda_{\pm}$ $10^{-4} \, m^2 \, S \, mol^{-1}$	D $10^{-5} \, cm^2 \, s^{-1}$	Ion	$\Lambda_{\pm}$ $10^{-4} \, m^2 \, S \, mol^{-1}$	D $10^{-5} \, cm^2 \, s^{-1}$
F^-	55.4	1.475	Histidyl$^+$	23.0	0.612
$1/4[Fe(CN)_6]^{4-}$	110.4	0.735	Hydroxyethyltrimethylarsonium$^+$	39.4	1.049
$1/3[Fe(CN)_6]^{3-}$	100.9	0.896	Methylammonium$^+$	58.7	1.563
$H_2AsO_4^-$	34	0.905	Octadecylpyridinium$^+$	20	0.533
HCO_3^-	44.5	1.185	Octadecyltributylammonium$^+$	16.6	0.442
HF_2^-	75	1.997	Octadecyltriethylammonium$^+$	17.9	0.477
$1/2HPO_4^{2-}$	57	0.759	Octadecyltrimethylammonium$^+$	19.9	0.530
$H_2PO_4^-$	36	0.959	Octadecyltripropylammonium$^+$	17.2	0.458
$H_2PO_2^-$	46	1.225	Octyltrimethylammonium$^+$	26.5	0.706
HS^-	65	1.731	Pentylammonium$^+$	37	0.985
HSO_3^-	58	1.545	Piperidinium$^+$	37.2	0.991
HSO_4^-	52	1.385	Propylammonium$^+$	40.8	1.086
$H_2SbO_4^-$	31	0.825	Pyrilammonium$^+$	24.3	0.647
I^-	76.8	2.045	Tetrabutylammonium$^+$	19.5	0.519
IO_3^-	40.5	1.078	Tetradecyltrimethylammonium$^+$	21.5	0.573
IO_4^-	54.5	1.451	Tetraethylammonium$^+$	32.6	0.868
MnO_4^-	61.3	1.632	Tetramethylammonium$^+$	44.9	1.196
$1/2MoO_4^{2-}$	74.5	1.984	Tetraisopentylammonium$^+$	17.9	0.477
$N(CN)_2^-$	54.5	1.451	Tetrapentylammmonium$^+$	17.5	0.466
NO_2^-	71.8	1.912	Tetrapropylammonium$^+$	23.4	0.623
NO_3^-	71.42	1.902	Triethylammonium$^+$	34.3	0.913
$NH_2SO_3^-$	48.3	1.286	Triethylsulfonium$^+$	36.1	0.961
N_3^-	69	1.837	Trimethylammonium$^+$	47.23	1.258
OCN^-	64.6	1.720	Trimethylhexylammonium$^+$	34.6	0.921
OD^-	119	3.169	Trimethylsulfonium$^+$	51.4	1.369
OH^-	198	5.273	Tripropylammonium$^+$	26.1	0.695
PF_6^-	56.9	1.515			
$1/2PO_3F^{2-}$	63.3	0.843	***Organic Anions***		
$1/3PO_4^{3-}$	92.8	0.824	Acetate$^-$	40.9	1.089
$1/4P_2O_7^{4-}$	96	0.639	*p*-Anisate$^-$	29.0	0.772
$1/3P_3O_9^{3-}$	83.6	0.742	$1/2$Azelate^{2-}	40.6	0.541
$1/5P_3O_{10}^{5-}$	109	0.581	Benzoate$^-$	32.4	0.863
ReO_4^-	54.9	1.462	Bromoacetate$^-$	39.2	1.044
SCN^-	66	1.758	Bromobenzoate$^-$	30	0.799
$1/2SO_3^{2-}$	72	0.959	Butyrate$^-$	32.6	0.868
$1/2SO_4^{2-}$	80.0	1.065	Chloroacetate$^-$	39.8	1.060
$1/2S_2O_3^{2-}$	85.0	1.132	*m*-Chlorobenzoate$^-$	31	0.825
$1/2S_2O_4^{2-}$	66.5	0.885	*o*-Chlorobenzoate$^-$	30.2	0.804
$1/2S_2O_6^{2-}$	93	1.238	$1/3$Citrate^{3-}	70.2	0.623
$1/2S_2O_8^{2-}$	86	1.145	Crotonate$^-$	33.2	0.884
$Sb(OH)_6^-$	31.9	0.849	Cyanoacetate$^-$	43.4	1.156
$SeCN^-$	64.7	1.723	Cyclohexane carboxylate$^-$	28.7	0.764
$1/2SeO_4^{2-}$	75.7	1.008	$1/2$ 1,1-Cyclopropanedicarboxylate^{2-}	53.4	0.711
$1/2WO_4^{2-}$	69	0.919	Decylsulfate$^-$	26	0.692
			Dichloroacetate$^-$	38.3	1.020
Organic Cations			$1/2$Diethylbarbiturate^{2-}	26.3	0.350
Benzyltrimethylammonium$^+$	34.6	0.921	Dihydrogencitrate$^-$	30	0.799
Isobutylammonium$^+$	38	1.012	$1/2$Dimethylmalonate^{2-}	49.4	0.658
Butyltrimethylammonium$^+$	33.6	0.895	3,5-Dinitrobenzoate$^-$	28.3	0.754
Decylpyridinium$^+$	29.5	0.786	Dodecylsulfate$^-$	24	0.639
Decyltrimethylammonium$^+$	24.4	0.650	Ethylmalonate$^-$	49.3	1.313
Diethylammonium$^+$	42.0	1.118	Ethylsulfate$^-$	39.6	1.055
Dimethylammonium$^+$	51.8	1.379	Fluoroacetate$^-$	44.4	1.182
Dipropylammonium$^+$	30.1	0.802	Fluorobenzoate$^-$	33	0.879
Dodecylammonium$^+$	23.8	0.634	Formate$^-$	54.6	1.454
Dodecyltrimethylammonium$^+$	22.6	0.602	$1/2$Fumarate^{2-}	61.8	0.823
Ethanolammonium$^+$	42.2	1.124	$1/2$Glutarate^{2-}	52.6	0.700
Ethylammonium$^+$	47.2	1.257	Hydrogenoxalate$^-$	40.2	1.070
Ethyltrimethylammonium$^+$	40.5	1.078	Isovalerate$^-$	32.7	0.871
Hexadecyltrimethylammonium$^+$	20.9	0.557	Iodoacetate$^-$	40.6	1.081
Hexyltrimethylammonium$^+$	29.6	0.788	Lactate$^-$	38.8	1.033

Ion	$\Lambda_{\pm}$ 10^{-4} m^2 S mol^{-1}	D 10^{-5} cm^2 s^{-1}	Ion	$\Lambda_{\pm}$ 10^{-4} m^2 S mol^{-1}	D 10^{-5} cm^2 s^{-1}
1/2Malate^{2-}	58.8	0.783	Picrate$^-$	30.37	0.809
1/2Maleate^{2-}	61.9	0.824	Pivalate$^-$	31.9	0.849
1/2Malonate^{2-}	63.5	0.845	Propionate$^-$	35.8	0.953
Methylsulfate$^-$	48.8	1.299	Propylsulfate$^-$	37.1	0.988
Naphthylacetate$^-$	28.4	0.756	Salicylate$^-$	36	0.959
1/2Oxalate^{2-}	74.11	0.987	1/2Suberate^{2-}	36	0.479
Octylsulfate$^-$	29	0.772	1/2Succinate^{2-}	58.8	0.783
Phenylacetate$^-$	30.6	0.815	p-Sulfonate	29.3	0.780
1/2o-Phthalate^{2-}	52.3	0.696	1/2Tartarate^{2-}	59.6	0.794
1/2m-Phthalate^{2-}	54.7	0.728	Trichloroacetate$^-$	35	0.932

ELECTROCHEMICAL SERIES

Petr Vanýsek

There are three tables for this electrochemical series. Each table lists standard reduction potentials, $E°$ values, at 298.15 K (25 °C), and at a pressure of 101.325 kPa (1 atm). Table 1 is an alphabetical listing of the elements, according to the symbol of the elements. Thus, data for silver (Ag) precede those for aluminum (Al). Table 2 lists only those reduction reactions that have $E°$ values positive in respect to the standard hydrogen electrode. In Table 2, the reactions are listed in the order of increasing positive potential, and they range from 0.0000 V to + 3.4 V. Table 3 lists only those reduction potentials which have $E°$ negative with respect to the standard hydrogen electrode. In Table 3, the reactions are listed in the order of decreasing potential and range from 0.0000 V to −4.10 V. The reliability of the potentials is not the same for all the data. Typically, the values with fewer significant figures have lower reliability. The values of reduction potentials, in particular those of less common reactions, are not definite; they are subject to occasional revisions.

Abbreviations: ac = acetate; bipy = 2,2´-dipyridine, or bipyridine; en = ethylenediamine; phen = 1,10-phenanthroline.

References

1. Milazzo, G., Caroli, S., and Sharma, V. K. *Tables of Standard Electrode Potentials*, Wiley, Chichester, 1978.
2. Bard, A. J., Parsons, R., and Jordan, J. *Standard Potentials in Aqueous Solutions*, Marcel Dekker, New York, 1985.
3. Bratsch, S. G. *J. Phys. Chem. Ref. Data*, 18, 1–21, 1989.

TABLE 1. Alphabetical Listing

Reaction	$E°$/V	Reaction	$E°$/V
$Ac^{3+} + 3\,e \rightleftharpoons Ac$	−2.20	$As + 3\,H^+ + 3\,e \rightleftharpoons AsH_3$	−0.608
$Ag^+ + e \rightleftharpoons Ag$	0.7996	$As_2O_3 + 6\,H^+ + 6\,e \rightleftharpoons 2\,As + 3\,H_2O$	0.234
$Ag^{2+} + e \rightleftharpoons Ag^+$	1.980	$HAsO_2 + 3\,H^+ + 3\,e \rightleftharpoons As + 2\,H_2O$	0.248
$Ag(ac) + e \rightleftharpoons Ag + (ac)^-$	0.643	$AsO_2^- + 2\,H_2O + 3\,e \rightleftharpoons As + 4\,OH^-$	−0.68
$AgBr + e \rightleftharpoons Ag + Br^-$	0.07133	$H_3AsO_4 + 2\,H^+ + 2\,e^- \rightleftharpoons HAsO_2 + 2\,H_2O$	0.560
$AgBrO_3 + e \rightleftharpoons Ag + BrO_3^-$	0.546	$AsO_4^{3-} + 2\,H_2O + 2\,e \rightleftharpoons AsO_2^- + 4\,OH^-$	−0.71
$Ag_2C_2O_4 + 2\,e \rightleftharpoons 2\,Ag + C_2O_4^{2-}$	0.4647	$At_2 + 2\,e \rightleftharpoons 2\,At^-$	0.3
$AgCl + e \rightleftharpoons Ag + Cl^-$	0.22233	$Au^+ + e \rightleftharpoons Au$	1.692
$AgCN + e \rightleftharpoons Ag + CN^-$	−0.017	$Au^{3+} + 2\,e \rightleftharpoons Au^+$	1.401
$Ag_2CO_3 + 2\,e \rightleftharpoons 2\,Ag + CO_3^{2-}$	0.47	$Au^{3+} + 3\,e \rightleftharpoons Au$	1.498
$Ag_2CrO_4 + 2\,e \rightleftharpoons 2\,Ag + CrO_4^{2-}$	0.4470	$Au^{2+} + e^- \rightleftharpoons Au^+$	1.8
$AgF + e \rightleftharpoons Ag + F^-$	0.779	$AuOH^{2+} + H^+ + 2\,e \rightleftharpoons Au^+ + H_2O$	1.32
$Ag_4[Fe(CN)_6] + 4\,e \rightleftharpoons 4\,Ag + [Fe(CN)_6]^{4-}$	0.1478	$AuBr_2^- + e \rightleftharpoons Au + 2\,Br^-$	0.959
$AgI + e \rightleftharpoons Ag + I^-$	−0.15224	$AuBr_4^- + 3\,e \rightleftharpoons Au + 4\,Br^-$	0.854
$AgIO_3 + e \rightleftharpoons Ag + IO_3^-$	0.354	$AuCl_4^- + 3\,e \rightleftharpoons Au + 4\,Cl^-$	1.002
$Ag_2MoO_4 + 2\,e \rightleftharpoons 2\,Ag + MoO_4^{2-}$	0.4573	$Au(OH)_3 + 3\,H^+ + 3\,e \rightleftharpoons Au + 3\,H_2O$	1.45
$AgNO_2 + e \rightleftharpoons Ag + 2\,NO_2^-$	0.564	$H_2BO_3^- + 5\,H_2O + 8\,e \rightleftharpoons BH_4^- + 8\,OH^-$	−1.24
$Ag_2O + H_2O + 2\,e \rightleftharpoons 2\,Ag + 2\,OH^-$	0.342	$H_2BO_3^- + H_2O + 3\,e \rightleftharpoons B + 4\,OH^-$	−1.79
$Ag_2O_3 + H_2O + 2\,e \rightleftharpoons 2\,AgO + 2\,OH^-$	0.739	$H_3BO_3 + 3\,H^+ + 3\,e \rightleftharpoons B + 3\,H_2O$	−0.8698
$Ag^{3+} + 2\,e \rightleftharpoons Ag^+$	1.9	$B(OH)_3 + 7\,H^+ + 8\,e \rightleftharpoons BH_4^- + 3\,H_2O$	−0.481
$Ag^{3+} + e \rightleftharpoons Ag^{2+}$	1.8	$Ba^{2+} + 2\,e \rightleftharpoons Ba$	−2.912
$Ag_2O_2 + 4\,H^+ + e \rightleftharpoons 2\,Ag + 2\,H_2O$	1.802	$Ba^{2+} + 2\,e \rightleftharpoons Ba(Hg)$	−1.570
$2\,AgO + H_2O + 2\,e \rightleftharpoons Ag_2O + 2\,OH^-$	0.607	$Ba(OH)_2 + 2\,e \rightleftharpoons Ba + 2\,OH^-$	−2.99
$AgOCN + e \rightleftharpoons Ag + OCN^-$	0.41	$Be^{2+} + 2\,e \rightleftharpoons Be$	−1.847
$Ag_2S + 2\,e \rightleftharpoons 2\,Ag + S^{2-}$	−0.691	$Be_2O_3^{2-} + 3\,H_2O + 4\,e \rightleftharpoons 2\,Be + 6\,OH^-$	−2.63
$Ag_2S + 2\,H^+ + 2\,e \rightleftharpoons 2\,Ag + H_2S$	−0.0366	p–benzoquinone + 2 H$^+$ + 2 e $\rightleftharpoons$ hydroquinone	0.6992
$AgSCN + e \rightleftharpoons Ag + SCN^-$	0.08951	$Bi^+ + e \rightleftharpoons Bi$	0.5
$Ag_2SeO_3 + 2\,e \rightleftharpoons 2\,Ag + SeO_4^{2-}$	0.3629	$Bi^{3+} + 3\,e \rightleftharpoons Bi$	0.308
$Ag_2SO_4 + 2\,e \rightleftharpoons 2\,Ag + SO_4^{2-}$	0.654	$Bi^{3+} + 2\,e \rightleftharpoons Bi^+$	0.2
$Ag_2WO_4 + 2\,e \rightleftharpoons 2\,Ag + WO_4^{2-}$	0.4660	$Bi + 3\,H^+ + 3\,e \rightleftharpoons BiH_3$	−0.8
$Al^{3+} + 3\,e \rightleftharpoons Al$	−1.662	$BiCl_4^- + 3\,e \rightleftharpoons Bi + 4\,Cl^-$	0.16
$Al(OH)_3 + 3\,e \rightleftharpoons Al + 3\,OH^-$	−2.31	$Bi_2O_3 + 3\,H_2O + 6\,e \rightleftharpoons 2\,Bi + 6\,OH^-$	−0.46
$Al(OH)_4^- + 3\,e \rightleftharpoons Al + 4\,OH^-$	−2.328	$Bi_2O_4 + 4\,H^+ + 2\,e \rightleftharpoons 2\,BiO^+ + 2\,H_2O$	1.593
$H_2AlO_3^- + H_2O + 3\,e \rightleftharpoons Al + 4\,OH^-$	−2.33	$BiO^+ + 2\,H^+ + 3\,e \rightleftharpoons Bi + H_2O$	0.320
$AlF_6^{3-} + 3\,e \rightleftharpoons Al + 6\,F^-$	−2.069	$BiOCl + 2\,H^+ + 3\,e \rightleftharpoons Bi + Cl^- + H_2O$	0.1583
$Am^{4+} + e \rightleftharpoons Am^{3+}$	2.60	$Bk^{4+} + e \rightleftharpoons Bk^{3+}$	1.67
$Am^{2+} + 2\,e \rightleftharpoons Am$	−1.9	$Bk^{2+} + 2\,e \rightleftharpoons Bk$	−1.6
$Am^{3+} + 3\,e \rightleftharpoons Am$	−2.048	$Bk^{3+} + e \rightleftharpoons Bk^{2+}$	−2.8
$Am^{3+} + e \rightleftharpoons Am^{2+}$	−2.3		

Reaction	$E°/V$
$Br_2(aq) + 2e \rightleftharpoons 2 Br^-$	1.0873
$Br_2(l) + 2e \rightleftharpoons 2 Br^-$	1.066
$HBrO + H^+ + 2e \rightleftharpoons Br^- + H_2O$	1.331
$HBrO + H^+ + e \rightleftharpoons 1/2 Br_2(aq) + H_2O$	1.574
$HBrO + H^+ + e \rightleftharpoons 1/2 Br_2(l) + H_2O$	1.596
$BrO^- + H_2O + 2e \rightleftharpoons Br^- + 2 OH^-$	0.761
$BrO_3^- + 6 H^+ + 5e \rightleftharpoons 1/2 Br_2 + 3 H_2O$	1.482
$BrO_3^- + 6 H^+ + 6e \rightleftharpoons Br^- + 3 H_2O$	1.423
$BrO_3^- + 3 H_2O + 6e \rightleftharpoons Br^- + 6 OH^-$	0.61
$(CN)_2 + 2 H^+ + 2e \rightleftharpoons 2 HCN$	0.373
$2 HCNO + 2 H^+ + 2e \rightleftharpoons (CN)_2 + 2 H_2O$	0.330
$(CNS)_2 + 2e \rightleftharpoons 2 CNS^-$	0.77
$CO_2 + 2 H^+ + 2e \rightleftharpoons HCOOH$	−0.199
$Ca^+ + e \rightleftharpoons Ca$	−3.80
$Ca^{2+} + 2e \rightleftharpoons Ca$	−2.868
$Ca(OH)_2 + 2e \rightleftharpoons Ca + 2 OH^-$	−3.02
Calomel electrode, 1 molal KCl	0.2800
Calomel electrode, 1 molar KCl (NCE)	0.2801
Calomel electrode, 0.1 molar KCl	0.3337
Calomel electrode, saturated KCl (SCE)	0.2412
Calomel electrode, saturated NaCl (SSCE)	0.2360
$Cd^{2+} + 2e \rightleftharpoons Cd$	−0.4030
$Cd^{2+} + 2e \rightleftharpoons Cd(Hg)$	−0.3521
$Cd(OH)_2 + 2e \rightleftharpoons Cd(Hg) + 2 OH^-$	−0.809
$CdSO_4 + 2e \rightleftharpoons Cd + SO_4^{2-}$	−0.246
$Cd(OH)_4^{2-} + 2e \rightleftharpoons Cd + 4 OH^-$	−0.658
$CdO + H_2O + 2e \rightleftharpoons Cd + 2 OH^-$	−0.783
$Ce^{3+} + 3e \rightleftharpoons Ce$	−2.336
$Ce^{3+} + 3e \rightleftharpoons Ce(Hg)$	−1.4373
$Ce^{4+} + e \rightleftharpoons Ce^{3+}$	1.72
$CeOH^{3+} + H^+ + e \rightleftharpoons Ce^{3+} + H_2O$	1.715
$Cf^{4+} + e \rightleftharpoons Cf^{3+}$	3.3
$Cf^{3+} + e \rightleftharpoons Cf^{2+}$	−1.6
$Cf^{3+} + 3e \rightleftharpoons Cf$	−1.94
$Cf^{2+} + 2e \rightleftharpoons Cf$	−2.12
$Cl_2(g) + 2e \rightleftharpoons 2 Cl^-$	1.35827
$HClO + H^+ + e \rightleftharpoons 1/2 Cl_2 + H_2O$	1.611
$HClO + H^+ + 2e \rightleftharpoons Cl^- + H_2O$	1.482
$ClO^- + H_2O + 2e \rightleftharpoons Cl^- + 2 OH^-$	0.81
$ClO_2 + H^+ + e \rightleftharpoons HClO_2$	1.277
$HClO_2 + 2 H^+ + 2e \rightleftharpoons HClO + H_2O$	1.645
$HClO_2 + 3 H^+ + 3e \rightleftharpoons 1/2 Cl_2 + 2 H_2O$	1.628
$HClO_2 + 3 H^+ + 4e \rightleftharpoons Cl^- + 2 H_2O$	1.570
$ClO_2^- + H_2O + 2e \rightleftharpoons ClO^- + 2 OH^-$	0.66
$ClO_2^- + 2 H_2O + 4e \rightleftharpoons Cl^- + 4 OH^-$	0.76
$ClO_2(aq) + e \rightleftharpoons ClO_2^-$	0.954
$ClO_3^- + 2 H^+ + e \rightleftharpoons ClO_2 + H_2O$	1.152
$ClO_3^- + 3 H^+ + 2e \rightleftharpoons HClO_2 + H_2O$	1.214
$ClO_3^- + 6 H^+ + 5e \rightleftharpoons 1/2 Cl_2 + 3 H_2O$	1.47
$ClO_3^- + 6 H^+ + 6e \rightleftharpoons Cl^- + 3 H_2O$	1.451
$ClO_3^- + H_2O + 2e \rightleftharpoons ClO_2^- + 2 OH^-$	0.33
$ClO_3^- + 3 H_2O + 6e \rightleftharpoons Cl^- + 6 OH^-$	0.62
$ClO_4^- + 2 H^+ + 2e \rightleftharpoons ClO_3^- \cdot H_2O$	1.189
$ClO_4^- + 8 H^+ + 7e \rightleftharpoons 1/2 Cl_2 + 4 H_2O$	1.39
$ClO_4^- + 8 H^+ + 8e \rightleftharpoons Cl^- + 4 H_2O$	1.389
$ClO_4^- + H_2O + 2e \rightleftharpoons ClO_3^- + 2 OH^-$	0.36
$Cm^{4+} + e \rightleftharpoons Cm^{3+}$	3.0
$Cm^{3+} + 3e \rightleftharpoons Cm$	−2.04
$Co^{2+} + 2e \rightleftharpoons Co$	−0.28
$Co^{3+} + e \rightleftharpoons Co^{2+}$	1.92

Reaction	$E°/V$
$[Co(NH_3)_6]^{3+} + e \rightleftharpoons [Co(NH_3)_6]^{2+}$	0.108
$Co(OH)_2 + 2e \rightleftharpoons Co + 2 OH^-$	−0.73
$Co(OH)_3 + e \rightleftharpoons Co(OH)_2 + OH^-$	0.17
$Cr^{2+} + 2e \rightleftharpoons Cr$	−0.913
$Cr^{3+} + e \rightleftharpoons Cr^{2+}$	−0.407
$Cr^{3+} + 3e \rightleftharpoons Cr$	−0.744
$Cr_2O_7^{2-} + 14 H^+ + 6e \rightleftharpoons 2 Cr^{3+} + 7 H_2O$	1.36
$CrO_2^- + 2 H_2O + 3e \rightleftharpoons Cr + 4 OH^-$	−1.2
$HCrO_4^- + 7 H^+ + 3e \rightleftharpoons Cr^{3+} + 4 H_2O$	1.350
$CrO_2 + 4 H^+ + e \rightleftharpoons Cr^{3+} + 2H_2O$	1.48
$Cr(V) + e \rightleftharpoons Cr(IV)$	1.34
$CrO_4^{2-} + 4 H_2O + 3e \rightleftharpoons Cr(OH)_3 + 5 OH^-$	−0.13
$Cr(OH)_3 + 3e \rightleftharpoons Cr + 3 OH^-$	−1.48
$Cs^+ + e \rightleftharpoons Cs$	−3.026
$Cu^+ + e \rightleftharpoons Cu$	0.521
$Cu^{2+} + e \rightleftharpoons Cu^+$	0.153
$Cu^{2+} + 2e \rightleftharpoons Cu$	0.3419
$Cu^{2+} + 2e \rightleftharpoons Cu(Hg)$	0.345
$Cu^{3+} + e \rightleftharpoons Cu^{2+}$	2.4
$Cu_2O_3 + 6 H^+ + 2e \rightleftharpoons 2Cu^{2+} + 3 H_2O$	2.0
$Cu^{2+} + 2 CN^- + e \rightleftharpoons [Cu(CN)_2]^-$	1.103
$CuI_2^- + e \rightleftharpoons Cu + 2 I^-$	0.00
$Cu_2O + H_2O + 2e \rightleftharpoons 2 Cu + 2 OH^-$	−0.360
$Cu(OH)_2 + 2e \rightleftharpoons Cu + 2 OH^-$	−0.222
$2 Cu(OH)_2 + 2e \rightleftharpoons Cu_2O + 2 OH^- + H_2O$	−0.080
$2 D^+ + 2e \rightleftharpoons D_2$	−0.013
$Dy^{2+} + 2e \rightleftharpoons Dy$	−2.2
$Dy^{3+} + 3e \rightleftharpoons Dy$	−2.295
$Dy^{3+} + e \rightleftharpoons Dy^{2+}$	−2.6
$Er^{2+} + 2e \rightleftharpoons Er$	−2.0
$Er^{3+} + 3e \rightleftharpoons Er$	−2.331
$Er^{3+} + e \rightleftharpoons Er^{2+}$	−3.0
$Es^{3+} + e \rightleftharpoons Es^{2+}$	−1.3
$Es^{3+} + 3e \rightleftharpoons Es$	−1.91
$Es^{2+} + 2e \rightleftharpoons Es$	−2.23
$Eu^{2+} + 2e \rightleftharpoons Eu$	−2.812
$Eu^{3+} + 3e \rightleftharpoons Eu$	−1.991
$Eu^{3+} + e \rightleftharpoons Eu^{2+}$	−0.36
$F_2 + 2 H^+ + 2e \rightleftharpoons 2 HF$	3.053
$F_2 + 2e \rightleftharpoons 2 F^-$	2.866
$F_2O + 2 H^+ + 4e \rightleftharpoons H_2O + 2 F^-$	2.153
$Fe^{2+} + 2e \rightleftharpoons Fe$	−0.447
$Fe^{3+} + 3e \rightleftharpoons Fe$	−0.037
$Fe^{3+} + e \rightleftharpoons Fe^{2+}$	0.771
$2 HFeO_4^- + 8 H^+ + 6e \rightleftharpoons Fe_2O_3 + 5 H_2O$	2.09
$HFeO_4^- + 4 H^+ + 3e \rightleftharpoons FeOOH + 2 H_2O$	2.08
$HFeO_4^- + 7 H^+ + 3e \rightleftharpoons Fe^{3+} + 4 H_2O$	2.07
$Fe_2O_3 + 4 H^+ + 2e \rightleftharpoons 2 FeOH^+ + H_2O$	0.16
$[Fe(CN)_6]^{3-} + e \rightleftharpoons [Fe(CN)_6]^{4-}$	0.358
$FeO_4^{2-} + 8 H^+ + 3e \rightleftharpoons Fe^{3+} + 4 H_2O$	2.20
$[Fe(bipy)_2]^{3+} + e \rightleftharpoons Fe(bipy)_2]^{2+}$	0.78
$[Fe(bipy)_3]^{3+} + e \rightleftharpoons Fe(bipy)_3]^{2+}$	1.03
$Fe(OH)_3 + e \rightleftharpoons Fe(OH)_2 + OH^-$	−0.56
$[Fe(phen)_3]^{3+} + e \rightleftharpoons [Fe(phen)_3]^{2+}$	1.147
$[Fe(phen)_3]^{3+} + e \rightleftharpoons [Fe(phen)_3]^{2+}$ (1 molar H_2SO_4)	1.06
$[Ferricinium]^+ + e \rightleftharpoons$ ferrocene	0.400
$Fm^{3+} + e \rightleftharpoons Fm^{2+}$	−1.1
$Fm^{3+} + 3e \rightleftharpoons Fm$	−1.89
$Fm^{2+} + 2e \rightleftharpoons Fm$	−2.30

Reaction	$E°/V$	Reaction	$E°/V$
$Fr^+ + e \rightleftharpoons Fr$	-2.9	$La(OH)_3 + 3\,e \rightleftharpoons La + 3\,OH^-$	-2.90
$Ga^{3+} + 3\,e \rightleftharpoons Ga$	-0.549	$Li^+ + e \rightleftharpoons Li$	-3.0401
$Ga^+ + e \rightleftharpoons Ga$	-0.2	$Lr^{3+} + 3\,e \rightleftharpoons Lr$	-1.96
$GaOH^{2+} + H^+ + 3\,e \rightleftharpoons Ga + H_2O$	-0.498	$Lu^{3+} + 3\,e \rightleftharpoons Lu$	-2.28
$H_2GaO_3^- + H_2O + 3\,e \rightleftharpoons Ga + 4\,OH^-$	-1.219	$Md^{3+} + e \rightleftharpoons Md^{2+}$	-0.1
$Gd^{3+} + 3\,e \rightleftharpoons Gd$	-2.279	$Md^{3+} + 3\,e \rightleftharpoons Md$	-1.65
$Ge^{2+} + 2\,e \rightleftharpoons Ge$	0.24	$Md^{2+} + 2\,e \rightleftharpoons Md$	-2.40
$Ge^{4+} + 4\,e \rightleftharpoons Ge$	0.124	$Mg^+ + e \rightleftharpoons Mg$	-2.70
$Ge^{4+} + 2\,e \rightleftharpoons Ge^{2+}$	0.00	$Mg^{2+} + 2\,e \rightleftharpoons Mg$	-2.372
$GeO_2 + 2\,H^+ + 2\,e \rightleftharpoons GeO + H_2O$	-0.118	$Mg(OH)_2 + 2\,e \rightleftharpoons Mg + 2\,OH^-$	-2.690
$H_2GeO_3 + 4\,H^+ + 4\,e \rightleftharpoons Ge + 3\,H_2O$	-0.182	$Mn^{2+} + 2\,e \rightleftharpoons Mn$	-1.185
$2\,H^+ + 2\,e \rightleftharpoons H_2$	0.00000	$Mn^{3+} + e \rightleftharpoons Mn^{2+}$	1.5415
$H_2 + 2\,e \rightleftharpoons 2\,H^-$	-2.23	$MnO_2 + 4\,H^+ + 2\,e \rightleftharpoons Mn^{2+} + 2\,H_2O$	1.224
$HO_2 + H^+ + e \rightleftharpoons H_2O_2$	1.495	$MnO_4^- + e \rightleftharpoons MnO_4^{2-}$	0.558
$2\,H_2O + 2\,e \rightleftharpoons H_2 + 2\,OH^-$	-0.8277	$MnO_4^- + 4\,H^+ + 3\,e \rightleftharpoons MnO_2 + 2\,H_2O$	1.679
$H_2O_2 + 2\,H^+ + 2\,e \rightleftharpoons 2\,H_2O$	1.776	$MnO_4^- + 8\,H^+ + 5\,e \rightleftharpoons Mn^{2+} + 4\,H_2O$	1.507
$Hf^{4+} + 4\,e \rightleftharpoons Hf$	-1.55	$MnO_4^- + 2\,H_2O + 3\,e \rightleftharpoons MnO_2 + 4\,OH^-$	0.595
$HfO^{2+} + 2\,H^+ + 4\,e \rightleftharpoons Hf + H_2O$	-1.724	$MnO_4^{2-} + 2\,H_2O + 2\,e \rightleftharpoons MnO_2 + 4\,OH^-$	0.60
$HfO_2 + 4\,H^+ + 4\,e \rightleftharpoons Hf + 2\,H_2O$	-1.505	$Mn(OH)_2 + 2\,e \rightleftharpoons Mn + 2\,OH^-$	-1.56
$HfO(OH)_2 + H_2O + 4\,e \rightleftharpoons Hf + 4\,OH^-$	-2.50	$Mn(OH)_3 + e \rightleftharpoons Mn(OH)_2 + OH^-$	0.15
$Hg^{2+} + 2\,e \rightleftharpoons Hg$	0.851	$Mn_2O_3 + 6\,H^+ + e \rightleftharpoons 2\,Mn^{2+} + 3\,H_2O$	1.485
$2\,Hg^{2+} + 2\,e \rightleftharpoons Hg_2^{2+}$	0.920	$Mo^{3+} + 3\,e \rightleftharpoons Mo$	-0.200
$Hg_2^{2+} + 2\,e \rightleftharpoons 2\,Hg$	0.7973	$MoO_2 + 4\,H^+ + 4\,e \rightleftharpoons Mo + 4\,H_2O$	-0.152
$Hg_2(ac)_2 + 2\,e \rightleftharpoons 2\,Hg + 2(ac)^-$	0.51163	$H_3Mo_7O_{24}^{3-} + 45\,H^+ + 42\,e \rightleftharpoons 7\,Mo + 24\,H_2O$	0.082
$Hg_2Br_2 + 2\,e \rightleftharpoons 2\,Hg + 2\,Br^-$	0.13923	$MoO_3 + 6\,H^+ + 6\,e \rightleftharpoons Mo + 3\,H_2O$	0.075
$Hg_2Cl_2 + 2\,e \rightleftharpoons 2\,Hg + 2\,Cl^-$	0.26808	$N_2 + 2\,H_2O + 6\,H^+ + 6\,e \rightleftharpoons 2\,NH_4OH$	0.092
$Hg_2HPO_4 + 2\,e \rightleftharpoons 2\,Hg + HPO_4^{2-}$	0.6359	$3\,N_2 + 2\,H^+ + 2\,e \rightleftharpoons 2\,HN_3$	-3.09
$Hg_2I_2 + 2\,e \rightleftharpoons 2\,Hg + 2\,I^-$	-0.0405	$N_5^+ + 3\,H^+ + 2\,e \rightleftharpoons 2\,NH_4^+$	1.275
$Hg_2O + H_2O + 2\,e \rightleftharpoons 2\,Hg + 2\,OH^-$	0.123	$N_2O + 2\,H^+ + 2\,e \rightleftharpoons N_2 + H_2O$	1.766
$HgO + H_2O + 2\,e \rightleftharpoons Hg + 2\,OH^-$	0.0977	$H_2N_2O_2 + 2\,H^+ + 2\,e \rightleftharpoons N_2 + 2\,H_2O$	2.65
$Hg(OH)_2 + 2\,H^+ + 2\,e \rightleftharpoons Hg + 2\,H_2O$	1.034	$N_2O_4 + 2\,e \rightleftharpoons 2\,NO_2^-$	0.867
$Hg_2SO_4 + 2\,e \rightleftharpoons 2\,Hg + SO_4^{2-}$	0.6125	$N_2O_4 + 2\,H^+ + 2\,e \rightleftharpoons 2\,NHO_2$	1.065
$Ho^{2+} + 2\,e \rightleftharpoons Ho$	-2.1	$N_2O_4 + 4\,H^+ + 4\,e \rightleftharpoons 2\,NO + 2\,H_2O$	1.035
$Ho^{3+} + 3\,e \rightleftharpoons Ho$	-2.33	$2\,NH_3OH^+ + H^+ + 2\,e \rightleftharpoons N_2H_5^+ + 2\,H_2O$	1.42
$Ho^{3+} + e \rightleftharpoons Ho^{2+}$	-2.8	$2\,NO + 2\,H^+ + 2\,e \rightleftharpoons N_2O + H_2O$	1.591
$I_2 + 2\,e \rightleftharpoons 2\,I^-$	0.5355	$2\,NO + H_2O + 2\,e \rightleftharpoons N_2O + 2\,OH^-$	0.76
$I_3^- + 2\,e \rightleftharpoons 3\,I^-$	0.536	$HNO_2 + H^+ + e \rightleftharpoons NO + H_2O$	0.983
$H_3IO_6^{2-} + 2\,e \rightleftharpoons IO_3^- + 3\,OH^-$	0.7	$2\,HNO_2 + 4\,H^+ + 4\,e \rightleftharpoons H_2N_2O_2 + 2\,H_2O$	0.86
$H_5IO_6 + H^+ + 2\,e \rightleftharpoons IO_3^- + 3\,H_2O$	1.601	$2\,HNO_2 + 4\,H^+ + 4\,e \rightleftharpoons N_2O + 3\,H_2O$	1.297
$2\,HIO + 2\,H^+ + 2\,e \rightleftharpoons I_2 + 2\,H_2O$	1.439	$NO_2^- + H_2O + e \rightleftharpoons NO + 2\,OH^-$	-0.46
$HIO + H^+ + 2\,e \rightleftharpoons I^- + H_2O$	0.987	$2\,NO_2^- + 2\,H_2O + 4\,e \rightleftharpoons N_2O_2^{2-} + 4\,OH^-$	-0.18
$IO^- + H_2O + 2\,e \rightleftharpoons I^- + 2\,OH^-$	0.485	$2\,NO_2^- + 3\,H_2O + 4\,e \rightleftharpoons N_2O + 6\,OH^-$	0.15
$2\,IO_3^- + 12\,H^+ + 10\,e \rightleftharpoons I_2 + 6\,H_2O$	1.195	$NO_3^- + 3\,H^+ + 2\,e \rightleftharpoons HNO_2 + H_2O$	0.934
$IO_3^- + 6\,H^+ + 6\,e \rightleftharpoons I^- + 3\,H_2O$	1.085	$NO_3^- + 4\,H^+ + 3\,e \rightleftharpoons NO + 2\,H_2O$	0.957
$IO_3^- + 2\,H_2O + 4\,e \rightleftharpoons IO^- + 4\,OH^-$	0.15	$2\,NO_3^- + 4\,H^+ + 2\,e \rightleftharpoons N_2O_4 + 2\,H_2O$	0.803
$IO_3^- + 3\,H_2O + 6\,e \rightleftharpoons IO^- + 6\,OH^-$	0.26	$NO_3^- + H_2O + 2\,e \rightleftharpoons NO_2^- + 2\,OH^-$	0.01
$In^+ + e \rightleftharpoons In$	-0.14	$2\,NO_3^- + 2\,H_2O + 2\,e \rightleftharpoons N_2O_4 + 4\,OH^-$	-0.85
$In^{2+} + e \rightleftharpoons In^+$	-0.40	$Na^+ + e \rightleftharpoons Na$	-2.71
$In^{3+} + e \rightleftharpoons In^{2+}$	-0.49	$Nb^{3+} + 3\,e \rightleftharpoons Nb$	-1.099
$In^{3+} + 2\,e \rightleftharpoons In^+$	-0.443	$NbO_2 + 2\,H^+ + 2\,e \rightleftharpoons NbO + H_2O$	-0.646
$In^{3+} + 3\,e \rightleftharpoons In$	-0.3382	$NbO_2 + 4\,H^+ + 4\,e \rightleftharpoons Nb + 2\,H_2O$	-0.690
$In(OH)_3 + 3\,e \rightleftharpoons In + 3\,OH^-$	-0.99	$NbO + 2\,H^+ + 2\,e \rightleftharpoons Nb + H_2O$	-0.733
$In(OH)_4^- + 3\,e \rightleftharpoons In + 4\,OH^-$	-1.007	$Nb_2O_5 + 10\,H^+ + 10\,e \rightleftharpoons 2\,Nb + 5\,H_2O$	-0.644
$In_2O_3 + 3\,H_2O + 6\,e \rightleftharpoons 2\,In + 6\,OH^-$	-1.034	$Nd^{3+} + 3\,e \rightleftharpoons Nd$	-2.323
$Ir^{3+} + 3\,e \rightleftharpoons Ir$	1.156	$Nd^{2+} + 2\,e \rightleftharpoons Nd$	-2.1
$[IrCl_6]^{2-} + e \rightleftharpoons [IrCl_6]^{3-}$	0.8665	$Nd^{3+} + e \rightleftharpoons Nd^{2+}$	-2.7
$[IrCl_6]^{3-} + 3\,e \rightleftharpoons Ir + 6\,Cl^-$	0.77	$Ni^{2+} + 2\,e \rightleftharpoons Ni$	-0.257
$Ir_2O_3 + 3\,H_2O + 6\,e \rightleftharpoons 2\,Ir + 6\,OH^-$	0.098	$Ni(OH)_2 + 2\,e \rightleftharpoons Ni + 2\,OH^-$	-0.72
$K^+ + e \rightleftharpoons K$	-2.931	$NiO_2 + 4\,H^+ + 2\,e \rightleftharpoons Ni^{2+} + 2\,H_2O$	1.678
$La^{3+} + 3\,e \rightleftharpoons La$	-2.379	$NiO_2 + 2\,H_2O + 2\,e \rightleftharpoons Ni(OH)_2 + 2\,OH^-$	-0.490

Reaction	$E°/V$
$No^{3+} + e \rightleftharpoons No^{2+}$	1.4
$No^{3+} + 3 e \rightleftharpoons No$	−1.20
$No^{2+} + 2 e \rightleftharpoons No$	−2.50
$Np^{3+} + 3 e \rightleftharpoons Np$	−1.856
$Np^{4+} + e \rightleftharpoons Np^{3+}$	0.147
$NpO_2 + H_2O + H^+ + e \rightleftharpoons Np(OH)_3$	−0.962
$O_2 + 2 H^+ + 2 e \rightleftharpoons H_2O_2$	0.695
$O_2 + 4 H^+ + 4 e \rightleftharpoons 2 H_2O$	1.229
$O_2 + H_2O + 2 e \rightleftharpoons HO_2^- + OH^-$	−0.076
$O_2 + 2 H_2O + 2 e \rightleftharpoons H_2O_2 + 2 OH^-$	−0.146
$O_2 + 2 H_2O + 4 e \rightleftharpoons 4 OH^-$	0.401
$O_3 + 2 H^+ + 2 e \rightleftharpoons O_2 + H_2O$	2.076
$O_3 + H_2O + 2 e \rightleftharpoons O_2 + 2 OH^-$	1.24
$O(g) + 2 H^+ + 2 e \rightleftharpoons H_2O$	2.421
$OH + e \rightleftharpoons OH^-$	2.02
$HO_2^- + H_2O + 2 e \rightleftharpoons 3 OH^-$	0.878
$OsO_4 + 8 H^+ + 8 e \rightleftharpoons Os + 4 H_2O$	0.838
$OsO_4 + 4 H^+ + 4 e \rightleftharpoons OsO_2 + 2 H_2O$	1.02
$[Os(bipy)_2]^{3+} + e \rightleftharpoons [Os(bipy)_2]^{2+}$	0.81
$[Os(bipy)_3]^{3+} + e \rightleftharpoons [Os(bipy)_3]^{2+}$	0.80
$P(red) + 3 H^+ + 3 e \rightleftharpoons PH_3(g)$	−0.111
$P(white) + 3 H^+ + 3 e \rightleftharpoons PH_3(g)$	−0.063
$P + 3 H_2O + 3 e \rightleftharpoons PH_3(g) + 3 OH^-$	−0.87
$H_2P_2^- + e \rightleftharpoons P + 2 OH^-$	−1.82
$H_3PO_2 + H^+ + e \rightleftharpoons P + 2 H_2O$	−0.508
$H_3PO_3 + 2 H^+ + 2 e \rightleftharpoons H_3PO_2 + H_2O$	−0.499
$H_3PO_3 + 3 H^+ + 3 e \rightleftharpoons P + 3 H_2O$	−0.454
$HPO_3^{2-} + 2 H_2O + 2 e \rightleftharpoons H_2PO_2^- + 3 OH^-$	−1.65
$HPO_3^{2-} + 2 H_2O + 3 e \rightleftharpoons P + 5 OH^-$	−1.71
$H_3PO_4 + 2 H^+ + 2 e \rightleftharpoons H_3PO_3 + H_2O$	−0.276
$PO_4^{3-} + 2 H_2O + 2 e \rightleftharpoons HPO_3^{2-} + 3 OH^-$	−1.05
$Pa^{3+} + 3 e \rightleftharpoons Pa$	−1.34
$Pa^{4+} + 4 e \rightleftharpoons Pa$	−1.49
$Pa^{4+} + e \rightleftharpoons Pa^{3+}$	−1.9
$Pb^{2+} + 2 e \rightleftharpoons Pb$	−0.1262
$Pb^{2+} + 2 e \rightleftharpoons Pb(Hg)$	−0.1205
$PbBr_2 + 2 e \rightleftharpoons Pb + 2 Br^-$	−0.284
$PbCl_2 + 2 e \rightleftharpoons Pb + 2 Cl^-$	−0.2675
$PbF_2 + 2 e \rightleftharpoons Pb + 2 F^-$	−0.3444
$PbHPO_4 + 2 e \rightleftharpoons Pb + HPO_4^{2-}$	−0.465
$PbI_2 + 2 e \rightleftharpoons Pb + 2 I^-$	−0.365
$PbO + H_2O + 2 e \rightleftharpoons Pb + 2 OH^-$	−0.580
$PbO_2 + 4 H^+ + 2 e \rightleftharpoons Pb^{2+} + 2 H_2O$	1.455
$HPbO_2^- + H_2O + 2 e \rightleftharpoons Pb + 3 OH^-$	−0.537
$PbO_2 + H_2O + 2 e \rightleftharpoons PbO + 2 OH^-$	0.247
$PbO_2 + SO_4^{2-} + 4 H^+ + 2 e \rightleftharpoons PbSO_4 + 2 H_2O$	1.6913
$PbSO_4 + 2 e \rightleftharpoons Pb + SO_4^{2-}$	−0.3588
$PbSO_4 + 2 e \rightleftharpoons Pb(Hg) + SO_4^{2-}$	−0.3505
$Pd^{2+} + 2 e \rightleftharpoons Pd$	0.951
$[PdCl_4]^{2-} + 2 e \rightleftharpoons Pd + 4 Cl^-$	0.591
$[PdCl_6]^{2-} + 2 e \rightleftharpoons [PdCl_4]^{2-} + 2 Cl^-$	1.288
$Pd(OH)_2 + 2 e \rightleftharpoons Pd + 2 OH^-$	0.07
$Pm^{2+} + 2 e \rightleftharpoons Pm$	−2.2
$Pm^{3+} + 3 e \rightleftharpoons Pm$	−2.30
$Pm^{3+} + e \rightleftharpoons Pm^{2+}$	−2.6
$Po^{4+} + 2 e \rightleftharpoons Po^{2+}$	0.9
$Po^{4+} + 4 e \rightleftharpoons Po$	0.76
$Pr^{4+} + e \rightleftharpoons Pr^{3+}$	3.2
$Pr^{2+} + 2 e \rightleftharpoons Pr$	−2.0
$Pr^{3+} + 3 e \rightleftharpoons Pr$	−2.353

Reaction	$E°/V$
$Pr^{3+} + e \rightleftharpoons Pr^{2+}$	−3.1
$Pt^{2+} + 2 e \rightleftharpoons Pt$	1.18
$[PtCl_4]^{2-} + 2 e \rightleftharpoons Pt + 4 Cl^-$	0.755
$[PtCl_6]^{2-} + 2 e \rightleftharpoons [PtCl_4]^{2-} + 2 Cl^-$	0.68
$Pt(OH)_2 + 2 e \rightleftharpoons Pt + 2 OH^-$	0.14
$PtO_3 + 2 H^+ + 2 e \rightleftharpoons PtO_2 + H_2O$	1.7
$PtO_3 + 4 H^+ + 2 e \rightleftharpoons Pt(OH)_2^{2+} + H_2O$	1.5
$PtOH^+ + H^+ + 2 e \rightleftharpoons Pt + H_2O$	1.2
$PtO_2 + 2 H^+ + 2 e \rightleftharpoons PtO + H_2O$	1.01
$PtO_2 + 4 H^+ + 4 e \rightleftharpoons Pt + 2 H_2O$	1.00
$Pu^{3+} + 3 e \rightleftharpoons Pu$	−2.031
$Pu^{4+} + e \rightleftharpoons Pu^{3+}$	1.006
$Pu^{5+} + e \rightleftharpoons Pu^{4+}$	1.099
$PuO_2(OH)_2 + 2 H^+ + 2 e \rightleftharpoons Pu(OH)_4$	1.325
$PuO_2(OH)_2 + H^+ + e \rightleftharpoons PuO_2OH + H_2O$	1.062
$Ra^{2+} + 2 e \rightleftharpoons Ra$	−2.8
$Rb^+ + e \rightleftharpoons Rb$	−2.98
$Re^{3+} + 3 e \rightleftharpoons Re$	0.300
$ReO_4^- + 4 H^+ + 3 e \rightleftharpoons ReO_2 + 2 H_2O$	0.510
$ReO_2 + 4 H^+ + 4 e \rightleftharpoons Re + 2 H_2O$	0.2513
$ReO_4^- + 2 H^+ + e \rightleftharpoons ReO_3 + H_2O$	0.768
$ReO_4^- + 4 H_2O + 7 e \rightleftharpoons Re + 8 OH^-$	−0.584
$ReO_4^- + 8 H^+ + 7 e \rightarrow Re + 4 H_2O$	0.368
$Rh^+ + e \rightleftharpoons Rh$	0.600
$Rh^{3+} + 3 e \rightleftharpoons Rh$	0.758
$[RhCl_6]^{3-} + 3 e \rightleftharpoons Rh + 6 Cl^-$	0.431
$RhOH^{2+} + H^+ + 3 e \rightleftharpoons Rh + H_2O$	0.83
$Ru^{2+} + 2 e \rightleftharpoons Ru$	0.455
$Ru^{3+} + e \rightleftharpoons Ru^{2+}$	0.2487
$RuO_2 + 4 H^+ + 2 e \rightleftharpoons Ru^{2+} + 2 H_2O$	1.120
$RuO_4^- + e \rightleftharpoons RuO_4^{2-}$	0.59
$RuO_4 + e \rightleftharpoons RuO_4^-$	1.00
$RuO_4 + 6 H^+ + 4 e \rightleftharpoons Ru(OH)_2^{2+} + 2 H_2O$	1.40
$RuO_4 + 8 H^+ + 8 e \rightleftharpoons Ru + 4 H_2O$	1.038
$[Ru(bipy)_3]^{3+} + e^- \rightleftharpoons [Ru(bipy)_3]^{2+}$	1.24
$[Ru(H_2O)_6]^{3+} + e^- \rightleftharpoons [Ru(H_2O)_6]^{2+}$	0.23
$[Ru(NH_3)_6]^{3+} + e^- \rightleftharpoons [Ru(NH_3)_6]^{2+}$	0.10
$[Ru(en)_3]^{3+} + e^- \rightleftharpoons [Ru(en)_3]^{2+}$	0.210
$[Ru(CN)_6]^{3-} + e^- \rightleftharpoons [Ru(CN)_6]^{4-}$	0.86
$S + 2 e \rightleftharpoons S^{2-}$	−0.47627
$S + 2 H^+ + 2 e \rightleftharpoons H_2S(aq)$	0.142
$S + H_2O + 2 e \rightleftharpoons SH^- + OH^-$	−0.478
$2 S + 2 e \rightleftharpoons S_2^{2-}$	−0.42836
$S_2O_6^{2-} + 4 H^+ + 2 e \rightleftharpoons 2 H_2SO_3$	0.564
$S_2O_8^{2-} + 2 e \rightleftharpoons 2 SO_4^{2-}$	2.010
$S_2O_8^{2-} + 2 H^+ + 2 e \rightleftharpoons 2 HSO_4^-$	2.123
$S_4O_6^{2-} + 2 e \rightleftharpoons 2 S_2O_3^{2-}$	0.08
$2 H_2SO_3 + H^+ + 2 e \rightleftharpoons HS_2O_4^- + 2 H_2O$	−0.056
$H_2SO_3 + 4 H^+ + 4 e \rightleftharpoons S + 3 H_2O$	0.449
$2 SO_3^{2-} + 2 H_2O + 2 e \rightleftharpoons S_2O_4^{2-} + 4 OH^-$	−1.12
$2 SO_3^{2-} + 3 H_2O + 4 e \rightleftharpoons S_2O_3^{2-} + 6 OH^-$	−0.571
$SO_4^{2-} + 4 H^+ + 2 e \rightleftharpoons H_2SO_3 + H_2O$	0.172
$2 SO_4^{2-} + 4 H^+ + 2 e \rightleftharpoons S_2O_6^{2-} + H_2O$	−0.22
$SO_4^{2-} + H_2O + 2 e \rightleftharpoons SO_3^{2-} + 2 OH^-$	−0.93
$Sb + 3 H^+ + 3 e \rightleftharpoons SbH_3$	−0.510
$Sb_2O_3 + 6 H^+ + 6 e \rightleftharpoons 2 Sb + 3 H_2O$	0.152
Sb_2O_5 (senarmontite) $+ 4 H^+ + 4 e \rightleftharpoons Sb_2O_3 + 2 H_2O$	0.671
Sb_2O_5 (valentinite) $+ 4 H^+ + 4 e \rightleftharpoons Sb_2O_3 + 2 H_2O$	0.649

Reaction	$E°/V$
$Sb_2O_5 + 6 H^+ + 4 e \rightleftharpoons 2 SbO^+ + 3 H_2O$	0.581
$SbO^+ + 2 H^+ + 3 e \rightleftharpoons Sb + 2 H_2O$	0.212
$SbO_2^- + 2 H_2O + 3 e \rightleftharpoons Sb + 4 OH^-$	−0.66
$SbO_3^- + H_2O + 2 e \rightleftharpoons SbO_2^- + 2 OH^-$	−0.59
$Sc^{3+} + 3 e \rightleftharpoons Sc$	−2.077
$Se + 2 e \rightleftharpoons Se^{2-}$	−0.924
$Se + 2 H^+ + 2 e \rightleftharpoons H_2Se(aq)$	−0.399
$H_2SeO_3 + 4 H^+ + 4 e \rightleftharpoons Se + 3 H_2O$	0.74
$Se + 2 H^+ + 2 e \rightleftharpoons H_2Se$	−0.082
$SeO_3^{2-} + 3 H_2O + 4 e \rightleftharpoons Se + 6 OH^-$	−0.366
$SeO_4^{2-} + 4 H^+ + 2 e \rightleftharpoons H_2SeO_3 + H_2O$	1.151
$SeO_4^{2-} + H_2O + 2 e \rightleftharpoons SeO_3^{2-} + 2 OH^-$	0.05
$SiF_6^{2-} + 4 e \rightleftharpoons Si + 6 F^-$	−1.24
$SiO + 2 H^+ + 2 e \rightleftharpoons Si + H_2O$	−0.8
$SiO_2 \text{ (quartz)} + 4 H^+ + 4 e \rightleftharpoons Si + 2 H_2O$	0.857
$SiO_3^{2-} + 3 H_2O + 4 e \rightleftharpoons Si + 6 OH^-$	−1.697
$Sm^{3+} + e \rightleftharpoons Sm^{2+}$	−1.55
$Sm^{3+} + 3 e \rightleftharpoons Sm$	−2.304
$Sm^{2+} + 2 e \rightleftharpoons Sm$	−2.68
$Sn^{2+} + 2 e \rightleftharpoons Sn$	−0.1375
$Sn^{4+} + 2 e \rightarrow Sn^{2+}$	0.151
$Sn(OH)_3^+ + 3 H^+ + 2 e \rightleftharpoons Sn^{2+} + 3 H_2O$	0.142
$SnO_2 + 4 H^+ + 2 e^- \rightleftharpoons Sn^{2+} + 2 H_2O$	−0.094
$SnO_2 + 4 H^+ + 4 e \rightleftharpoons Sn + 2 H_2O$	−0.117
$SnO_2 + 3 H^+ + 2 e \rightleftharpoons SnOH^+ + H_2O$	−0.194
$SnO_2 + 2 H_2O + 4 e \rightleftharpoons Sn + 4 OH^-$	−0.945
$HSnO_2^- + H_2O + 2 e \rightleftharpoons Sn + 3 OH^-$	−0.909
$Sn(OH)_6^{2-} + 2 e \rightleftharpoons HSnO_2^- + 3 OH^- + H_2O$	−0.93
$Sr^+ + e \rightleftharpoons Sr$	−4.10
$Sr^{2+} + 2 e \rightleftharpoons Sr$	−2.899
$Sr^{2+} + 2 e \rightleftharpoons Sr(Hg)$	−1.793
$Sr(OH)_2 + 2 e \rightleftharpoons Sr + 2 OH^-$	−2.88
$Ta_2O_5 + 10 H^+ + 10 e \rightleftharpoons 2 Ta + 5 H_2O$	−0.750
$Ta^{3+} + 3 e \rightleftharpoons Ta$	−0.6
$Tc^{2+} + 2 e \rightleftharpoons Tc$	0.400
$TcO_4^- + 4 H^+ + 3 e \rightleftharpoons TcO_2 + 2 H_2O$	0.782
$Tc^{3+} + e \rightleftharpoons Tc^{2+}$	0.3
$TcO_4^- + 8 H^+ + 7 e \rightleftharpoons Tc + 4 H_2O$	0.472
$Tb^{4+} + e \rightleftharpoons Tb^{3+}$	3.1
$Tb^{3+} + 3 e \rightleftharpoons Tb$	−2.28
$Te + 2 e \rightleftharpoons Te^{2-}$	−1.143
$Te + 2 H^+ + 2 e \rightleftharpoons H_2Te$	−0.793
$Te^{4+} + 4 e \rightleftharpoons Te$	0.568
$TeO_2 + 4 H^+ + 4 e \rightleftharpoons Te + 2 H_2O$	0.593
$TeO_3^{2-} + 3 H_2O + 4 e \rightleftharpoons Te + 6 OH^-$	−0.57
$TeO_4^- + 8 H^+ + 7 e \rightleftharpoons Te + 4 H_2O$	0.472
$H_6TeO_6 + 2 H^+ + 2 e \rightleftharpoons TeO_2 + 4 H_2O$	1.02
$Th^{4+} + 4 e \rightleftharpoons Th$	−1.899
$ThO_2 + 4 H^+ + 4 e \rightleftharpoons Th + 2 H_2O$	−1.789
$Th(OH)_4 + 4 e \rightleftharpoons Th + 4 OH^-$	−2.48
$Ti^{2+} + 2 e \rightleftharpoons Ti$	−1.630
$Ti^{3+} + e \rightleftharpoons Ti^{2+}$	−0.9
$TiO_2 + 4 H^+ + 2 e \rightleftharpoons Ti^{2+} + 2 H_2O$	−0.502
$Ti^{3+} + 3 e \rightleftharpoons Ti$	−1.37
$TiOH^{3+} + H^+ + e \rightleftharpoons Ti^{3+} + H_2O$	−0.055

Reaction	$E°/V$
$Tl^+ + e \rightleftharpoons Tl$	−0.336
$Tl^+ + e \rightleftharpoons Tl(Hg)$	−0.3338
$Tl^{3+} + 2 e \rightleftharpoons Tl^+$	1.252
$Tl^{3+} + 3 e \rightleftharpoons Tl$	0.741
$TlBr + e \rightleftharpoons Tl + Br^-$	−0.658
$TlCl + e \rightleftharpoons Tl + Cl^-$	−0.5568
$TlI + e \rightleftharpoons Tl + I^-$	−0.752
$Tl_2O_3 + 3 H_2O + 4 e \rightleftharpoons 2 Tl^+ + 6 OH^-$	0.02
$TlOH + e \rightleftharpoons Tl + OH^-$	−0.34
$Tl(OH)_3 + 2 e \rightleftharpoons TlOH + 2 OH^-$	−0.05
$Tl_2SO_4 + 2 e \rightleftharpoons Tl + SO_4^{2-}$	−0.4360
$Tm^{3+} + e \rightleftharpoons Tm^{2+}$	−2.2
$Tm^{3+} + 3 e \rightleftharpoons Tm$	−2.319
$Tm^{2+} + 2 e \rightleftharpoons Tm$	−2.4
$U^{3+} + 3 e \rightleftharpoons U$	−1.798
$U^{4+} + e \rightleftharpoons U^{3+}$	−0.607
$UO_2^+ + 4 H^+ + e \rightleftharpoons U^{4+} + 2 H_2O$	0.612
$UO_2^{2+} + e \rightleftharpoons UO_2^+$	0.062
$UO_2^{2+} + 4 H^+ + 2 e \rightleftharpoons U^{4+} + 2 H_2O$	0.327
$UO_2^{2+} + 4 H^+ + 6 e \rightleftharpoons U + 2 H_2O$	−1.444
$V^{2+} + 2 e \rightleftharpoons V$	−1.175
$V^{3+} + e \rightleftharpoons V^{2+}$	−0.255
$VO^{2+} + 2 H^+ + e \rightleftharpoons V^{3+} + H_2O$	0.337
$VO_2^+ + 2 H^+ + e \rightleftharpoons VO^{2+} + H_2O$	0.991
$V_2O_5 + 6 H^+ + 2 e \rightleftharpoons 2 VO^{2+} + 3 H_2O$	0.957
$V_2O_5 + 10 H^+ + 10 e \rightleftharpoons 2 V + 5 H_2O$	−0.242
$V(OH)_4^+ + 2 H^+ + e \rightleftharpoons VO^{2+} + 3 H_2O$	1.00
$V(OH)_4^+ + 4 H^+ + 5 e \rightleftharpoons V + 4 H_2O$	−0.254
$[V(phen)_3]^{3+} + e \rightleftharpoons [V(phen)_3]^{2+}$	0.14
$W^{3+} + 3 e \rightleftharpoons W$	0.1
$W_2O_5 + 2 H^+ + 2 e \rightleftharpoons 2 WO_2 + H_2O$	−0.031
$WO_2 + 4 H^+ + 4 e \rightleftharpoons W + 2 H_2O$	−0.119
$WO_3 + 6 H^+ + 6 e \rightleftharpoons W + 3 H_2O$	−0.090
$WO_3 + 2 H^+ + 2 e \rightleftharpoons WO_2 + H_2O$	0.036
$2 WO_3 + 2 H^+ + 2 e \rightleftharpoons W_2O_5 + H_2O$	−0.029
$H_4XeO_6 + 2 H^+ + 2 e \rightleftharpoons XeO_3 + 3 H_2O$	2.42
$XeO_3 + 6 H^+ + 6 e \rightleftharpoons Xe + 3 H_2O$	2.10
$XeF + e \rightleftharpoons Xe + F^-$	3.4
$Y^{3+} + 3 e \rightleftharpoons Y$	−2.372
$Yb^{3+} + e \rightleftharpoons Yb^{2+}$	−1.05
$Yb^{3+} + 3 e \rightleftharpoons Yb$	−2.19
$Yb^{2+} + 2 e \rightleftharpoons Yb$	−2.76
$Zn^{2+} + 2 e \rightleftharpoons Zn$	−0.7618
$Zn^{2+} + 2 e \rightleftharpoons Zn(Hg)$	−0.7628
$ZnO_2^{2-} + 2 H_2O + 2 e \rightleftharpoons Zn + 4 OH^-$	−1.215
$ZnSO_4 \cdot 7 H_2O + 2 e = Zn(Hg) + SO_4^{2-} + 7$ $H_2O \text{ (Saturated } ZnSO_4)$	−0.7993
$ZnOH^+ + H^+ + 2 e \rightleftharpoons Zn + H_2O$	−0.497
$Zn(OH)_4^{2-} + 2 e \rightleftharpoons Zn + 4 OH^-$	−1.199
$Zn(OH)_2 + 2 e \rightleftharpoons Zn + 2 OH^-$	−1.249
$ZnO + H_2O + 2 e \rightleftharpoons Zn + 2 OH^-$	−1.260
$ZrO_2 + 4 H^+ + 4 e \rightleftharpoons Zr + 2 H_2O$	−1.553
$ZrO(OH)_2 + H_2O + 4 e \rightleftharpoons Zr + 4 OH^-$	−2.36
$Zr^{4+} + 4 e \rightleftharpoons Zr$	−1.45

TABLE 2. Reduction Reactions Having $E°$ Values More Positive than That of the Standard Hydrogen Electrode

Reaction	$E°/V$	Reaction	$E°/V$
$2 H^+ + 2 e \rightleftharpoons H_2$	0.00000	$Hg_2Cl_2 + 2 e \rightleftharpoons 2 Hg + 2 Cl^-$	0.26808
$CuI_2^- + e \rightleftharpoons Cu + 2 I^-$	0.00	Calomel electrode, 1 molal KCl	0.2800
$Ge^{4+} + 2 e \rightleftharpoons Ge^{2+}$	0.00	Calomel electrode, 1 molar KCl (NCE)	0.2801
$NO_3^- + H_2O + 2 e \rightleftharpoons NO_2^- + 2 OH^-$	0.01	$At_2 + 2 e \rightleftharpoons 2 At^-$	0.3
$Tl_2O_3 + 3 H_2O + 4 e \rightleftharpoons 2 Tl^+ + 6 OH^-$	0.02	$Re^{3+} + 3 e \rightleftharpoons Re$	0.300
$SeO_4^{2-} + H_2O + 2 e \rightleftharpoons SeO_3^{2-} + 2 OH^-$	0.05	$Tc^{3+} + e \rightleftharpoons Tc^{2+}$	0.3
$WO_3 + 2 H^+ + 2 e \rightleftharpoons WO_2 + H_2O$	0.036	$Bi^{3+} + 3 e \rightleftharpoons Bi$	0.308
$UO_2^{2+} + e \rightleftharpoons UO_2^+$	0.062	$BiO^+ + 2 H^+ + 3 e \rightleftharpoons Bi + H_2O$	0.320
$Pd(OH)_2 + 2 e \rightleftharpoons Pd + 2 OH^-$	0.07	$UO_2^{2+} + 4 H^+ + 2 e \rightleftharpoons U^{4+} + 2 H_2O$	0.327
$AgBr + e \rightleftharpoons Ag + Br^-$	0.07133	$ClO_3^- + H_2O + 2 e \rightleftharpoons ClO_2^- + 2 OH^-$	0.33
$MoO_3 + 6 H^+ + 6 e \rightleftharpoons Mo + 3 H_2O$	0.075	$2 HCNO + 2 H^+ + 2 e \rightleftharpoons (CN)_2 + 2 H_2O$	0.330
$S_4O_6^{2-} + 2 e \rightleftharpoons 2 S_2O_3^{2-}$	0.08	Calomel electrode, 0.1 molar KCl	0.3337
$H_3Mo_7O_{24}^{3-} + 45 H^+ + 42 e \rightleftharpoons 7 Mo + 24 H_2O$	0.082	$VO^{2+} + 2 H^+ + e \rightleftharpoons V^{3+} + H_2O$	0.337
$AgSCN + e \rightleftharpoons Ag + SCN^-$	0.8951	$Cu^{2+} + 2 e \rightleftharpoons Cu$	0.3419
$N_2 + 2 H_2O + 6 H^+ + 6 e \rightleftharpoons 2 NH_4OH$	0.092	$Ag_2O + H_2O + 2 e \rightleftharpoons 2 Ag + 2 OH^-$	0.342
$HgO + H_2O + 2 e \rightleftharpoons Hg + 2 OH^-$	0.0977	$Cu^{2+} + 2 e \rightleftharpoons Cu(Hg)$	0.345
$Ir_2O_3 + 3 H_2O + 6 e \rightleftharpoons 2 Ir + 6 OH^-$	0.098	$AgIO_3 + e \rightleftharpoons Ag + IO_3^-$	0.354
$2 NO + 2 e \rightleftharpoons N_2O_2^{2-}$	0.10	$[Fe(CN)_6]^{3-} + e \rightleftharpoons [Fe(CN)_6]^{4-}$	0.358
$[Ru(NH_3)_6]^{3+} + e \rightleftharpoons [Ru(NH_3)_6]^{2+}$	0.10	$ClO_4^- + H_2O + 2 e \rightleftharpoons ClO_3^- + 2 OH^-$	0.36
$W^{3+} + 3 e \rightleftharpoons W$	0.1	$Ag_2SeO_3 + 2 e \rightleftharpoons 2 Ag + SeO_3^{2-}$	0.3629
$[Co(NH_3)_6]^{3+} + e \rightleftharpoons [Co(NH_3)_6]^{2+}$	0.108	$ReO_4^- + 8 H^+ + 7 e \rightleftharpoons Re + 4 H_2O$	0.368
$Hg_2O + H_2O + 2 e \rightleftharpoons 2 Hg + 2 OH^-$	0.123	$(CN)_2 + 2 H^+ + 2 e \rightleftharpoons 2 HCN$	0.373
$Ge^{4+} + 4 e \rightleftharpoons Ge$	0.124	$[Ferricinium]^+ + e \rightleftharpoons ferrocene$	0.400
$Hg_2Br_2 + 2 e \rightleftharpoons 2 Hg + 2 Br^-$	0.13923	$Tc^{2+} + 2 e \rightleftharpoons Tc$	0.400
$Pt(OH)_2 + 2 e \rightleftharpoons Pt + 2 OH^-$	0.14	$O_2 + 2 H_2O + 4 e \rightleftharpoons 4 OH^-$	0.401
$[V(phen)_3]^{3+} + e \rightleftharpoons [V(phen)_3]^{2+}$	0.14	$AgOCN + e \rightleftharpoons Ag + OCN^-$	0.41
$S + 2 H^+ + 2 e \rightleftharpoons H_2S(aq)$	0.142	$[RhCl_6]^{3-} + 3 e \rightleftharpoons Rh + 6 Cl^-$	0.431
$Sn(OH)_3^+ + 3 H^+ + 2 e \rightleftharpoons Sn^{2+} + 3 H_2O$	0.142	$Ag_2CrO_4 + 2 e \rightleftharpoons 2 Ag + CrO_4^{2-}$	0.4470
$Np^{4+} + e \rightleftharpoons Np^{3+}$	0.147	$H_2SO_3 + 4 H^+ + 4 e \rightleftharpoons S + 3 H_2O$	0.449
$Ag_6[Fe(CN)_6] + 4 e \rightleftharpoons 4 Ag + [Fe(CN)_6]^{4-}$	0.1478	$Ru^{2+} + 2 e \rightleftharpoons Ru$	0.455
$IO_3^- + 2 H_2O + 4 e \rightleftharpoons IO^- + 4 OH^-$	0.15	$Ag_2MoO_4 + 2 e \rightleftharpoons 2 Ag + MoO_4^{2-}$	0.4573
$Mn(OH)_3 + e \rightleftharpoons Mn(OH)_2 + OH^-$	0.15	$Ag_2C_2O_4 + 2 e \rightleftharpoons 2 Ag + C_2O_4^{2-}$	0.4647
$2 NO_2^- + 3 H_2O + 4 e \rightleftharpoons N_2O + 6 OH^-$	0.15	$Ag_2WO_4 + 2 e \rightleftharpoons 2 Ag + WO_4^{2-}$	0.4660
$Sn^{4+} + 2 e \rightleftharpoons Sn^{2+}$	0.151	$Ag_2CO_3 + 2 e \rightleftharpoons 2 Ag + CO_3^{2-}$	0.47
$Sb_2O_3 + 6 H^+ + 6 e \rightleftharpoons 2 Sb + 3 H_2O$	0.152	$TcO_4^- + 8 H^+ + 7 e \rightleftharpoons Tc + 4 H_2O$	0.472
$Cu^{2+} + e \rightleftharpoons Cu^+$	0.153	$TeO_4^- + 8 H^+ + 7 e \rightleftharpoons Te + 4 H_2O$	0.472
$BiOCl + 2 H^+ + 3 e \rightleftharpoons Bi + Cl^- + H_2O$	0.1583	$IO^- + H_2O + 2 e \rightleftharpoons I^- + 2 OH^-$	0.485
$BiCl_4^- + 3 e \rightleftharpoons Bi + 4 Cl^-$	0.16	$NiO_2 + 2 H_2O + 2 e \rightleftharpoons Ni(OH)_2 + 2 OH^-$	0.490
$Fe_2O_3 + 4 H^+ + 2 e \rightleftharpoons 2 FeOH^+ + H_2O$	0.16	$Bi^+ + e \rightleftharpoons Bi$	0.5
$Co(OH)_3 + e \rightleftharpoons Co(OH)_2 + OH^-$	0.17	$ReO_4^- + 4 H^+ + 3 e \rightleftharpoons ReO_2 + 2 H_2O$	0.510
$SO_4^{2-} + 4 H^+ + 2 e \rightleftharpoons H_2SO_3 + H_2O$	0.172	$Hg_2(ac)_2 + 2 e \rightleftharpoons 2 Hg + 2(ac)^-$	0.51163
$Bi^{3+} + 2 e \rightleftharpoons Bi^+$	0.2	$Cu^+ + e \rightleftharpoons Cu$	0.521
$[Ru(en)_3]^{3+} + e \rightleftharpoons [Ru(en)_3]^{2+}$	0.210	$I_2 + 2 e \rightleftharpoons 2 I^-$	0.5355
$SbO^+ + 2 H^+ + 3 e \rightleftharpoons Sb + 2 H_2O$	0.212	$I_3^- + 2 e \rightleftharpoons 3 I^-$	0.536
$AgCl + e \rightleftharpoons Ag + Cl^-$	0.22233	$AgBrO_3 + e \rightleftharpoons Ag + BrO_3^-$	0.546
$[Ru(H_2O)_6]^{3+} + e \rightleftharpoons [Ru(H_2O)_6]^{2+}$	0.23	$MnO_4^- + e \rightleftharpoons MnO_4^-$	0.558
$As_2O_3 + 6 H^+ + 6 e \rightleftharpoons 2 As + 3 H_2O$	0.234	$H_3AsO_4 + 2 H^+ + 2 e \rightleftharpoons HAsO_2 + 2 H_2O$	0.560
Calomel electrode, saturated NaCl (SSCE)	0.2360	$S_2O_6^{2-} + 4 H^+ + 2 e \rightleftharpoons 2 H_2SO_3$	0.564
$Ge^{2+} + 2 e \rightleftharpoons Ge$	0.24	$AgNO_2 + e \rightleftharpoons Ag + NO_2^-$	0.564
$Ru^{3+} + e \rightleftharpoons Ru^{2+}$	0.24	$Te^{4+} + 4 e \rightleftharpoons Te$	0.568
Calomel electrode, saturated KCl	0.2412	$Sb_2O_5 + 6 H^+ + 4 e \rightleftharpoons 2 SbO^+ + 3 H_2O$	0.581
$PbO_2 + H_2O + 2 e \rightleftharpoons PbO + 2 OH^-$	0.247	$RuO_4^- + e \rightleftharpoons RuO_4^{2-}$	0.59
$HAsO_2 + 3 H^+ + 3 e \rightleftharpoons As + 2 H_2O$	0.248	$[PdCl_4]^{2-} + 2 e \rightleftharpoons Pd + 4 Cl^-$	0.591
$Ru^{3+} + e \rightarrow Ru^{2+}$	0.2487	$TeO_2 + 4 H^+ + 4 e \rightleftharpoons Te + 2 H_2O$	0.593
$ReO_2 + 4 H^+ + 4 e \rightleftharpoons Re + 2 H_2O$	0.2513	$MnO_4^- + 2 H_2O + 3 e \rightleftharpoons MnO_2 + 4 OH^-$	0.595
$IO_3^- + 3 H_2O + 6 e \rightleftharpoons I^- + OH^-$	0.26	$Rh^{2+} + 2 e \rightleftharpoons Rh$	0.600

Reaction	$E°/V$
$Rh^+ + e \rightleftharpoons Rh$	0.600
$MnO_4^{2-} + 2 H_2O + 2 e \rightleftharpoons MnO_2 + 4 OH^-$	0.60
$2 AgO + H_2O + 2 e \rightleftharpoons Ag_2O + 2 OH^-$	0.607
$BrO_3^- + 3 H_2O + 6 e \rightleftharpoons Br^- + 6 OH^-$	0.61
$UO_2^+ + 4 H^+ + e \rightleftharpoons U^{4+} + 2 H_2O$	0.612
$Hg_2SO_4 + 2 e \rightleftharpoons 2 Hg + SO_4^{2-}$	0.6125
$ClO_3^- + 3 H_2O + 6 e \rightleftharpoons Cl^- + 6 OH^-$	0.62
$Hg_2HPO_4 + 2 e \rightleftharpoons 2 Hg + HPO_4^{2-}$	0.6359
$Ag(ac) + e \rightleftharpoons Ag + (ac)^-$	0.643
$Sb_2O_5(valentinite) + 4 H^+ + 4 e \rightleftharpoons Sb_2O_3 + 2 H_2O$	0.649
$Ag_2SO_4 + 2 e \rightleftharpoons 2 Ag + SO_4^{2-}$	0.654
$ClO_2^- + H_2O + 2 e \rightleftharpoons ClO^- + 2 OH^-$	0.66
$Sb_2O_5(senarmontite) + 4 H^+ + 4 e \rightleftharpoons Sb_2O_5 + 2 H_2O$	0.671
$[PtCl_6]^{2-} + 2 e \rightleftharpoons [PtCl_4]^{2-} + 2 Cl^-$	0.68
$O_2 + 2 H^+ + 2 e \rightleftharpoons H_2O_2$	0.695
$p-benzoquinone + 2 H^+ + 2 e \rightleftharpoons hydroquinone$	0.6992
$H_3IO_6^{2-} + 2 e \rightleftharpoons IO_3^- + 3 OH^-$	0.7
$Ag_2O_3 + H_2O + 2 e \rightleftharpoons 2 AgO + 2 OH^-$	0.739
$Tl^{3+} + 3 e \rightleftharpoons Tl$	0.741
$[PtCl_4]^{2-} + 2 e \rightleftharpoons Pt + 4 Cl^-$	0.755
$Rh^{3+} + 3 e \rightleftharpoons Rh$	0.758
$ClO_2^- + 2 H_2O + 4 e \rightleftharpoons Cl^- + 4 OH^-$	0.76
$2 NO + H_2O + 2 e \rightleftharpoons N_2O + 2 OH^-$	0.76
$Po^{4+} + 4 e \rightleftharpoons Po$	0.76
$BrO^- + H_2O + 2 e \rightleftharpoons Br^- + 2 OH^-$	0.761
$ReO_4^- + 2 H^+ + e \rightleftharpoons ReO_3 + H_2O$	0.768
$(CNS)_2 + 2 e \rightleftharpoons 2 CNS^-$	0.77
$[IrCl_6]^{3-} + 3 e \rightleftharpoons Ir + 6 Cl^-$	0.77
$Fe^{3+} + e \rightleftharpoons Fe^{2+}$	0.771
$AgF + e \rightleftharpoons Ag + F^-$	0.779
$[Fe(bipy)_2]^{3+} + e \rightleftharpoons [Fe(bipy)_2]^{2+}$	0.78
$TcO_4^- + 4 H^+ + 3 e \rightleftharpoons TcO_2 + 2 H_2O$	0.782
$Hg_2^{2+} + 2 e \rightleftharpoons 2 Hg$	0.7973
$Ag^+ + e \rightleftharpoons Ag$	0.7996
$[Os(bipy)_3]^{3+} + e \rightleftharpoons [Os(bipy)_3]^{2+}$	0.80
$2 NO_3^- + 4 H^+ + 2 e \rightleftharpoons N_2O_4 + 2 H_2O$	0.803
$[Os(bipy)_2]^{3+} + e \rightleftharpoons [Os(bipy)_2]^{2+}$	0.81
$RhOH^{2+} + H + 3 e \rightleftharpoons Rh + H_2O$	0.83
$OsO_4 + 8 H^+ + 8 e \rightleftharpoons Os + 4 H_2O$	0.838
$ClO^- + H_2O + 2 e \rightleftharpoons Cl^- + 2 OH^-$	0.841
$Hg^{2+} + 2 e \rightleftharpoons Hg$	0.851
$AuBr_4^- + 3 e \rightleftharpoons Au + 4 Br^-$	0.854
$SiO_2(quartz) + 4 H^+ + 4 e \rightleftharpoons Si + 2 H_2O$	0.857
$2 HNO_2 + 4 H^+ + 4 e \rightleftharpoons H_2N_2O_2 + H_2O$	0.86
$[Ru(CN)_6]^{3-} + e^- \rightleftharpoons [Ru(CN)_6]^{4-}$	0.86
$[IrCl_6]^{2-} + e \rightleftharpoons [IrCl_6]^{3-}$	0.8665
$N_2O_4 + 2 e \rightleftharpoons 2 NO_2^-$	0.867
$HO_2^- + H_2O + 2 e \rightleftharpoons 3 OH^-$	0.878
$Po^{4+} + 2 e \rightleftharpoons Po^{2+}$	0.9
$2 Hg^{2+} + 2 e \rightleftharpoons Hg_2^{2+}$	0.920
$NO_3^- + 3 H^+ + 2 e \rightleftharpoons HNO_2 + H_2O$	0.934
$Pd^{2+} + 2 e \rightleftharpoons Pd$	0.951
$ClO_2(aq) + e \rightleftharpoons ClO_2^-$	0.954
$NO_3^- + 4 H^+ + 3 e \rightleftharpoons NO + 2 H_2O$	0.957
$V_2O_5 + 6 H^+ + 2 e \rightleftharpoons 2 VO^{2+} + 3 H_2O$	0.957
$AuBr_2^- + e \rightleftharpoons Au + 2 Br^-$	0.959
$HNO_2 + H^+ + e \rightleftharpoons NO + H_2O$	0.983

Reaction	$E°/V$
$HIO + H^+ + 2 e \rightleftharpoons I^- + H_2O$	0.987
$VO_2^+ + 2 H^+ + e \rightleftharpoons VO^{2+} + H_2O$	0.991
$PtO_2 + 4 H^+ + 4 e \rightleftharpoons Pt + 2 H_2O$	1.00
$RuO_4 + e \rightleftharpoons RuO_4^-$	1.00
$V(OH)_4^+ + 2 H^+ + e \rightleftharpoons VO^{2+} + 3 H_2O$	1.00
$AuCl_4^- + 3 e \rightleftharpoons Au + 4 Cl^-$	1.002
$Pu^{4+} + e \rightleftharpoons Pu^{3+}$	1.006
$PtO_2 + 2 H^+ + 2 e \rightleftharpoons PtO + H_2O$	1.01
$OsO_4 + 4 H + 4 e \rightleftharpoons OsO_2 + 2 H_2O$	1.02
$H_6TeO_6 + 2 H^+ + 2 e \rightleftharpoons TeO_2 + 4 H_2O$	1.02
$[Fe(bipy)_3]^{3+} + e \rightleftharpoons [Fe(bipy)_3]^{2+}$	1.03
$Hg(OH)_2 + 2 H^+ + 2 e \rightleftharpoons Hg + 2 H_2O$	1.034
$N_2O_4 + 4 H^+ + 4 e \rightleftharpoons 2 NO + 2 H_2O$	1.035
$RuO_4 + 8 H^+ + 8 e \rightleftharpoons Ru + 4 H_2O$	1.038
$[Fe(phen)_3]^{3+} + e \rightleftharpoons [Fe(phen)_3]^{2+} \text{ (1 molar } H_2SO_4)$	1.06
$PuO_2(OH)_2 + H^+ + e \rightleftharpoons PuO_2OH + H_2O$	1.062
$N_2O_4 + 2 H^+ + 2 e \rightleftharpoons 2 HNO_2$	1.065
$Br_2(l) + 2 e \rightleftharpoons 2Br^-$	1.066
$IO_3^- + 6 H^+ + 6 e \rightleftharpoons I^- + 3 H_2O$	1.085
$Br_2(aq) + 2 e \rightleftharpoons 2Br^-$	1.0873
$Pu^{5+} + e \rightleftharpoons Pu^{4+}$	1.099
$Cu^{2+} + 2 CN^- + e \rightleftharpoons [Cu(CN)_2]^-$	1.103
$RuO_2 + 4 H^+ + 2 e \rightleftharpoons Ru^{2+} + 2 H_2O$	1.120
$[Fe(phen)_3]^{3+} + e \rightleftharpoons [Fe(phen)_3]^{2+}$	1.147
$SeO_4^{2-} + 4 H^+ + 2 e \rightleftharpoons H_2SeO_3 + H_2O$	1.151
$ClO_3^- + 2 H^+ + e \rightleftharpoons ClO_2 + H_2O$	1.152
$Ir^{3+} + 3 e \rightleftharpoons Ir$	1.156
$Pt^{2+} + 2 e \rightleftharpoons Pt$	1.18
$ClO_4^- + 2 H^+ + 2 e \rightleftharpoons ClO_3^- + H_2O$	1.189
$2 IO_3^- + 12 H^+ + 10 e \rightleftharpoons I_2 + 6 H_2O$	1.195
$PtOH^+ + H^+ + 2 e \rightleftharpoons Pt + H_2O$	1.2
$ClO_3^- + 3 H^+ + 2 e \rightleftharpoons HClO_2 + H_2O$	1.214
$MnO_2 + 4 H^+ + 2 e \rightleftharpoons Mn^{2+} + 2 H_2O$	1.224
$O_2 + 4 H^+ + 4 e \rightleftharpoons 2 H_2O$	1.229
$O_3 + H_2O + 2 e \rightleftharpoons O_2 + 2 OH^-$	1.24
$[Ru(bipy)_3]^{3+} + e \rightleftharpoons [Ru(bipy)_3]^{2+}$	1.24
$Tl^{3+} + 2 e \rightleftharpoons Tl^+$	1.252
$N_2H_5^+ + 3 H^+ + 2 e \rightleftharpoons 2 NH_4^+$	1.275
$ClO_2 + H^+ + e \rightleftharpoons HClO_2$	1.277
$[PdCl_6]^{2-} + 2 e \rightleftharpoons [PdCl_4]^{2-} + 2 Cl^-$	1.288
$2 HNO_2 + 4 H^+ + 4 e \rightleftharpoons N_2O + 3 H_2O$	1.297
$AuOH^{2+} + H^+ + 2 e \rightleftharpoons Au^+ + H_2O$	1.32
$PuO_2(OH)_2 + 2 H^- + 2 e \rightleftharpoons Pu(OH)_4$	1.325
$HBrO + H^+ + 2 e \rightleftharpoons Br^- + H_2O$	1.331
$Cr(V) + e \rightleftharpoons Cr(IV)$	1.34
$HCrO_4^- + 7 H^+ + 3 e \rightleftharpoons Cr^{3+} + 4 H_2O$	1.350
$Cl_2(g) + 2 e \rightleftharpoons 2Cl^-$	1.35827
$Cr_2O_7^{2-} + 14 H^+ + 6 e \rightleftharpoons 2 Cr^{3+} + 7 H_2O$	1.36
$ClO_4^- + 8 H^+ + 8 e \rightleftharpoons Cl^- + 4 H_2O$	1.389
$ClO_4^- + 8 H^+ + 7 e \rightleftharpoons 1/2 Cl_2 + 4 H_2O$	1.39
$No^{3+} + e \rightleftharpoons No^{2+}$	1.4
$RuO_4 + 6 H^+ + 4 e \rightleftharpoons Ru(OH)_2^{2+} + 2 H_2O$	1.40
$Au^{3+} + 2 e \rightleftharpoons Au^+$	1.401
$2 NH_3OH^+ + H^+ + 2 e \rightleftharpoons N_2H_5^+ + 2 H_2O$	1.42
$BrO_3^- + 6 H^+ + 6 e \rightleftharpoons Br^- + 3 H_2O$	1.423
$2 HIO + 2 H^+ + 2 e \rightleftharpoons I_2 + 2 H_2O$	1.439
$Au(OH)_3 + 3 H^+ + 3 e \rightleftharpoons Au^- + 3 H_2O$	1.45

Reaction	$E°/V$
$3 IO_3^- + 6 H^+ + 6 e \rightleftharpoons Cl^- + 3 H_2O$	1.451
$PbO_2 + 4 H^+ + 2 e \rightleftharpoons Pb^{2+} + 2 H_2O$	1.455
$ClO_3^- + 6 H^+ + 5 e \rightleftharpoons 1/2 Cl_2 + 3 H_2O$	1.47
$CrO_2 + 4 H^+ + e \rightleftharpoons Cr^{3+} + 2 H_2O$	1.48
$BrO_3^- + 6 H^+ + 5 e \rightleftharpoons 1/2 Br_2 + 3 H_2O$	1.482
$HClO + H^+ + 2 e \rightleftharpoons Cl^- + H_2O$	1.482
$Mn_2O_3 + 6 H^+ + e \rightleftharpoons 2 Mn^{2+} + 3 H_2O$	1.485
$HO_2 + H^+ + e \rightleftharpoons H_2O_2$	1.495
$Au^{3+} + 3 e \rightleftharpoons Au$	1.498
$PtO_3 + 4 H^+ + 2 e \rightleftharpoons Pt(OH)_2^{2+} + H_2O$	1.5
$MnO_4^- + 8 H^+ + 5 e \rightleftharpoons Mn^{2+} + 4 H_2O$	1.507
$Mn^{3+} + e \rightleftharpoons Mn^{2-}$	1.5415
$HClO_2 + 3 H^+ + 4 e \rightleftharpoons Cl^- + 2 H_2O$	1.570
$HBrO + H^+ + e \rightleftharpoons 1/2 Br_2(aq) + H_2O$	1.574
$2 NO + 2 H^+ + 2 e \rightleftharpoons N_2O + H_2O$	1.591
$Bi_2O_4 + 4 H^+ + 2 e \rightleftharpoons 2 BiO^+ + 2 H_2O$	1.593
$HBrO + H^+ + e \rightleftharpoons 1/2 Br_2(l) + H_2O$	1.596
$H_5IO_6 + H^+ + 2 e \rightleftharpoons IO_3^- + 3 H_2O$	1.601
$HClO + H^+ + e \rightleftharpoons 1/2 Cl_2 + H_2O$	1.611
$HClO_2 + 3 H^+ + 3 e \rightleftharpoons 1/2 Cl_2 + 2 H_2O$	1.628
$HClO_2 + 2 H^+ + 2 e \rightleftharpoons HClO + H_2O$	1.645
$Bk^{4+} + e \rightleftharpoons Bk^{3+}$	1.67
$NiO_2 + 4 H^+ + 2 e \rightleftharpoons Ni^{2+} + 2 H_2O$	1.678
$MnO_4^- + 4 H^+ + 3 e \rightarrow MnO_2 + 2 H_2O$	1.679
$PbO_2 + SO_4^{2-} + 4 H^+ + 2 e \rightleftharpoons PbSO_4 + 2 H_2O$	1.6913
$Au^+ + e \rightleftharpoons Au$	1.692
$PtO_3 + 2 H^+ + 2 e \rightleftharpoons PtO_2 + H_2O$	1.7
$CeOH^{3+} + H^+ + e \rightleftharpoons Ce^{3+} + H_2O$	1.715
$Ce^{4+} + e \rightleftharpoons Ce^{3+}$	1.72
$N_2O + 2 H^+ + 2 e \rightleftharpoons N_2 + H_2O$	1.766
$H_2O_2 + 2 H^+ + 2 e \rightleftharpoons 2 H_2O$	1.776

Reaction	$E°/V$
$Ag^{3+} + e \rightleftharpoons Ag^{2+}$	1.8
$Au^{2+} + e^- \rightleftharpoons Au^+$	1.8
$Ag_2O_2 + 4 H^+ + e \rightleftharpoons 2 Ag + 2 H_2O$	1.802
$Co^{3+} + e \rightleftharpoons Co^{2-}(2 \text{ molar } H_2SO_4)$	1.83
$Ag^{3+} + 2 e \rightleftharpoons Ag^+$	1.9
$Co^{3+} + e \rightleftharpoons Co^{2+}$	1.92
$Ag^{2+} + e \rightleftharpoons Ag^+$	1.980
$Cu_2O_3 + 6 H^+ + 2 e \rightleftharpoons 2 Cu^{2+} + 3 H_2O$	2.0
$S_2O_8^{2-} + 2 e \rightleftharpoons 2 SO_4^{2-}$	2.010
$OH + e \rightleftharpoons OH^-$	2.02
$HFeO_4^- + 7 H^+ + 3 e \rightleftharpoons Fe^{3+} + 4 H_2O$	2.07
$O_3 + 2 H^+ + 2 e \rightleftharpoons O_2 + H_2O$	2.076
$HFeO_4^- + 4 H^+ + 3 e \rightleftharpoons FeOOH + 2 H_2O$	2.08
$2 HFeO_4^- + 8 H^+ + 6 e \rightleftharpoons Fe_2O_3 + 5 H_2O$	2.09
$XeO_3 + 6 H^+ + 6 e \rightleftharpoons Xe + 3 H_2O$	2.10
$S_2O_8^{2-} + 2 H^+ + 2 e \rightleftharpoons 2 HSO_4^-$	2.123
$F_2O + 2 H^+ + 4 e \rightleftharpoons H_2O + 2 F^-$	2.153
$FeO_4^{2-} + 8 H^+ + 3 e \rightleftharpoons Fe^{3+} + 4 H_2O$	2.20
$Cu^{3+} + e \rightleftharpoons Cu^{2+}$	2.4
$H_4XeO_6 + 2 H^+ + 2 e \rightleftharpoons XeO_3 + 3 H_2O$	2.42
$O(g) + 2 H^+ + 2 e \rightleftharpoons H_2O$	2.421
$Am^{4+} + e \rightleftharpoons Am^{3+}$	2.60
$H_2N_2O_2 + 2 H^+ + 2 e \rightleftharpoons N_2 + 2 H_2O$	2.65
$F_2 + 2 e \rightleftharpoons 2 F^-$	2.866
$Cm^{4+} + e \rightleftharpoons Cm^{3+}$	3.0
$F_2 + 2 H^+ + 2 e \rightleftharpoons 2 HF$	3.053
$Tb^{4+} + e \rightleftharpoons Tb^{3+}$	3.1
$Pr^{4+} + e \rightleftharpoons Pr^{3+}$	3.2
$Cf^{4+} + e \rightleftharpoons Cf^{3+}$	3.3
$XeF + e \rightleftharpoons Xe + F^-$	3.4

TABLE 3. Reduction Reactions Having $E°$ Values More Negative than That of the Standard Hydrogen Electrode

Reaction	$E°/V$
$2 H^+ + 2 e \rightleftharpoons H_2$	0.00000
$2 D^+ + 2 e \rightleftharpoons D_2$	−0.013
$AgCN + e \rightleftharpoons Ag + CN^-$	−0.017
$2 WO_3 + 2 H^+ + 2 e \rightleftharpoons W_2O_5 + H_2O$	−0.029
$W_2O_5 + 2 H^+ + 2 e \rightleftharpoons 2 WO_2 + H_2O$	−0.031
$Ag_2S + 2 H^+ + 2 e \rightleftharpoons 2 Ag + H_2S$	−0.0366
$Fe^{3+} + 3 e \rightleftharpoons Fe$	−0.037
$Hg_2I_2 + 2 e \rightleftharpoons 2 Hg + 2 I^-$	−0.0405
$Tl(OH)_3 + 2 e \rightleftharpoons TlOH + 2 OH^-$	−0.05
$TiOH^{3+} + H^+ + e \rightleftharpoons Ti^{3+} + H_2O$	−0.055
$2 H_2SO_3 + H^+ + 2 e \rightleftharpoons HS_2O_4^- + 2 H_2O$	−0.056
$P(\text{white}) + 3 H^+ + 3 e \rightleftharpoons PH_3(g)$	−0.063
$O_2 + H_2O + 2 e \rightleftharpoons HO_2^- + OH^-$	−0.076
$2 Cu(OH)_2 + 2 e \rightleftharpoons Cu_2O + 2 OH^- + H_2O$	−0.080
$Se + 2 H^+ + 2 e \rightleftharpoons H_2Se$	−0.082
$WO_3 + 6 H^+ + 6 e \rightleftharpoons W + 3 H_2O$	−0.090
$SnO_2 + 4 H^+ + 2 e \rightleftharpoons Sn^{2+} + 2 H_2O$	−0.094
$Md^{3+} + e \rightleftharpoons Md^{2+}$	−0.1
$P(\text{red}) + 3 H^+ + 3 e \rightleftharpoons PH_3(g)$	−0.111
$SnO_2 + 4 H^+ + 4 e \rightleftharpoons Sn + 2 H_2O$	−0.117
$GeO_2 + 2 H^+ + 2 e \rightleftharpoons GeO + H_2O$	−0.118

Reaction	$E°/V$
$WO_2 + 4 H^+ + 4 e \rightleftharpoons W + 2 H_2O$	−0.119
$Pb^{2+} + 2 e \rightleftharpoons Pb(Hg)$	−0.1205
$Pb^{2+} + 2 e \rightleftharpoons Pb$	−0.1262
$CrO_4^{2-} + 4 H_2O + 3 e \rightleftharpoons Cr(OH)_3 + 5 OH^-$	−0.13
$Sn^{2-} + 2 e \rightleftharpoons Sn$	−0.1375
$In^+ + e \rightleftharpoons In$	−0.14
$O_2 + 2 H_2O + 2 e \rightleftharpoons H_2O_2 + 2 OH^-$	−0.146
$MoO_2 + 4 H^+ + 4 e \rightleftharpoons Mo + 4 H_2O$	−0.152
$AgI + e \rightleftharpoons Ag + I^-$	−0.15224
$2 NO_2^- + 2 H_2O + 4 e \rightleftharpoons N_2O_2^{2-} + 4 OH^-$	−0.18
$H_2GeO_3 + 4 H^+ + 4 e \rightleftharpoons Ge + 3 H_2O$	−0.182
$SnO_2 + 3 H^+ + 2 e \rightleftharpoons SnOH^+ + H_2O$	−0.194
$CO_2 + 2 H^+ + 2 e \rightleftharpoons HCOOH$	−0.199
$Mo^{3+} + 3 e \rightleftharpoons Mo$	−0.200
$Ga^+ + e \rightleftharpoons Ga$	−0.2
$2 SO_3^{2-} + 4 H^+ + 2 e \rightleftharpoons S_2O_6^{2-} + H_2O$	−0.22
$Cu(OH)_2 + 2 e \rightleftharpoons Cu + 2 OH^-$	−0.222
$V_2O_5 + 10 H^+ + 10 e \rightleftharpoons 2 V + 5 H_2O$	−0.242
$CdSO_4 + 2 e \rightleftharpoons Cd + SO_4^{2-}$	−0.246
$V(OH)_4^+ + 4 H^+ + 5 e \rightleftharpoons V + 4 H_2O$	−0.254
$V^{3+} + e \rightleftharpoons V^{2+}$	−0.255

Reaction	$E°/V$
$Ni^{2+} + 2\,e \rightleftharpoons Ni$	-0.257
$PbCl_2 + 2\,e \rightleftharpoons Pb + 2\,Cl^-$	-0.2675
$H_3PO_4 + 2\,H^+ + 2\,e \rightleftharpoons H_3PO_3 + H_2O$	-0.276
$Co^{2+} + 2\,e \rightleftharpoons Co$	-0.28
$PbBr_2 + 2\,e \rightleftharpoons Pb + 2\,Br^-$	-0.284
$Tl^+ + e \rightleftharpoons Tl(Hg)$	-0.3338
$Tl^+ + e \rightleftharpoons Tl$	-0.336
$In^{3+} + 3\,e \rightleftharpoons In$	-0.3382
$TlOH + e \rightleftharpoons Tl + OH^-$	-0.34
$PbF_2 + 2\,e \rightleftharpoons Pb + 2\,F^-$	-0.3444
$PbSO_4 + 2\,e \rightleftharpoons Pb(Hg) + SO_4^{2-}$	-0.3505
$Cd^{2+} + 2\,e \rightleftharpoons Cd(Hg)$	-0.3521
$PbSO_4 + 2\,e \rightleftharpoons Pb + SO_4^{2-}$	-0.3588
$Cu_2O + H_2O + 2\,e \rightleftharpoons 2\,Cu + 2\,OH^-$	-0.360
$Eu^{3+} + e \rightleftharpoons Eu^{2+}$	-0.36
$PbI_2 + 2\,e \rightleftharpoons Pb + 2\,I^-$	-0.365
$SeO_3^{2-} + 3\,H_2O + 4\,e \rightleftharpoons Se + 6\,OH^-$	-0.366
$Se + 2\,H^+ + 2\,e \rightleftharpoons H_2Se(aq)$	-0.399
$In^{2+} + e \rightleftharpoons In^+$	-0.40
$Cd^{2+} + 2\,e \rightleftharpoons Cd$	-0.4030
$Cr^{3+} + e \rightleftharpoons Cr^{2+}$	-0.407
$2\,S + 2\,e \rightleftharpoons S_2^{2-}$	-0.42836
$Tl_2SO_4 + 2\,e \rightleftharpoons Tl + SO_4^{2-}$	-0.4360
$In^{3+} + 2\,e \rightleftharpoons In^+$	-0.443
$Fe^{2+} + 2\,e \rightleftharpoons Fe$	-0.447
$H_3PO_3 + 3\,H^+ + 3\,e \rightleftharpoons P + 3\,H_2O$	-0.454
$Bi_2O_3 + 3\,H_2O + 6\,e \rightleftharpoons 2\,Bi + 6\,OH^-$	-0.46
$NO_2^- + H_2O + e \rightleftharpoons NO + 2\,OH$	-0.46
$PbHPO_4 + 2\,e \rightleftharpoons Pb + HPO_4^{2-}$	-0.465
$S + 2\,e \rightleftharpoons S^{2-}$	-0.47627
$S + H_2O + 2\,e \rightleftharpoons HS^- + OH^-$	-0.478
$B(OH)_3 + 7\,H^+ + 8\,e \rightleftharpoons BH_4^- + 3\,H_2O$	-0.481
$In^{3+} + e \rightleftharpoons In^{2+}$	-0.49
$ZnOH^+ + H^+ + 2\,e \rightleftharpoons Zn + H_2O$	-0.497
$GaOH^{2+} + H^+ + 3\,e \rightleftharpoons Ga + H_2O$	-0.498
$H_3PO_3 + 2\,H^+ + 2\,e \rightleftharpoons H_3PO_2 + H_2O$	-0.499
$TiO_2 + 4\,H^+ + 2\,e \rightleftharpoons Ti^{2+} + 2\,H_2O$	-0.502
$H_3PO_2 + H^+ + e \rightleftharpoons P + 2\,H_2O$	-0.508
$Sb + 3\,H^+ + 3\,e \rightleftharpoons SbH_3$	-0.510
$HPbO_2^- + H_2O + 2\,e \rightleftharpoons Pb + 3\,OH^-$	-0.537
$Ga^{3+} + 3\,e \rightleftharpoons Ga$	-0.549
$TlCl + e \rightleftharpoons Tl + Cl^-$	-0.5568
$Fe(OH)_3 + e \rightleftharpoons Fe(OH)_2 + OH^-$	-0.56
$TeO_3^{2-} + 3\,H_2O + 4\,e \rightleftharpoons Te + 6\,OH^-$	-0.57
$2\,SO_3^{2-} + 3\,H_2O + 4\,e \rightleftharpoons S_2O_3^{2-} + 6\,OH^-$	-0.571
$PbO + H_2O + 2\,e \rightleftharpoons Pb + 2\,OH^-$	-0.580
$ReO_2^- + 4\,H_2O + 7\,e \rightleftharpoons Re + 8\,OH^-$	-0.584
$SbO_3^- + H_2O + 2\,e \rightleftharpoons SbO_2^- + 2\,OH^-$	-0.59
$Ta^{3+} + 3\,e \rightleftharpoons Ta$	-0.6
$U^{4+} + e \rightleftharpoons U^{3+}$	-0.607
$As + 3\,H^+ + 3\,e \rightleftharpoons AsH_3$	-0.608
$Nb_2O_5 + 10\,H^+ + 10\,e \rightleftharpoons 2\,Nb + 5\,H_2O$	-0.644
$NbO_2 + 2\,H^+ + 2\,e \rightleftharpoons NbO + H_2O$	-0.646
$Cd(OH)_4^{2-} + 2\,e \rightleftharpoons Cd + 4\,OH^-$	-0.658
$TlBr + e \rightleftharpoons Tl + Br^-$	-0.658
$SbO_2^- + 2\,H_2O + 3\,e \rightleftharpoons Sb + 4\,OH^-$	-0.66
$AsO_2^- + 2\,H_2O + 3\,e \rightleftharpoons As + 4\,OH^-$	-0.68

Reaction	$E°/V$
$NbO_2 + 4\,H^+ + 4\,e \rightleftharpoons Nb + 2\,H_2O$	-0.690
$Ag_2S + 2\,e \rightleftharpoons 2\,Ag + S^{2-}$	-0.691
$AsO_4^{3-} + 2\,H_2O + 2\,e \rightleftharpoons AsO_2^- + 4\,OH^-$	-0.71
$Ni(OH)_2 + 2\,e \rightleftharpoons Ni + 2\,OH^-$	-0.72
$Co(OH)_2 + 2\,e \rightleftharpoons Co + 2\,OH^-$	-0.73
$NbO + 2\,H^+ + 2\,e \rightleftharpoons Nb + H_2O$	-0.733
$H_2SeO_3 + 4\,H^+ + 4\,e \rightleftharpoons Se + 3\,H_2O$	-0.74
$Cr^{3+} + 3\,e \rightleftharpoons Cr$	-0.744
$Ta_2O_5 + 10\,H^+ + 10\,e \rightleftharpoons 2\,Ta + 5\,H_2O$	-0.750
$TlI + e \rightleftharpoons Tl + I^-$	-0.752
$Zn^{2+} + 2\,e \rightleftharpoons Zn$	-0.7618
$Zn^{2+} + 2\,e \rightleftharpoons Zn(Hg)$	-0.7628
$CdO + H_2O + 2\,e \rightleftharpoons Cd + 2\,OH^-$	-0.783
$Te + 2\,H^+ + 2\,e \rightleftharpoons H_2Te$	-0.793
$ZnSO_4 7H_2O + 2\,e \rightleftharpoons Zn(Hg) + SO_4^{2-} + 7\,H_2O$ (Saturated $ZnSO_4$)	-0.7993
$Bi + 3\,H^+ + 3\,e \rightleftharpoons BiH_3$	-0.8
$SiO + 2\,H^+ + 2\,e \rightleftharpoons Si + H_2O$	-0.8
$Cd(OH)_2 + 2\,e \rightleftharpoons Cd(Hg) + 2\,OH^-$	-0.809
$2\,H_2O + 2\,e \rightleftharpoons H_2 + 2\,OH^-$	-0.8277
$2\,NO_3^- + 2\,H_2O + 2\,e \rightleftharpoons N_2O_4 + 4\,OH^-$	-0.85
$H_3BO_3 + 3\,H^+ + 3\,e \rightleftharpoons B + 3\,H_2O$	-0.8698
$P + 3\,H_2O + 3\,e \rightleftharpoons PH_3(g) + 3\,OH^-$	-0.87
$Ti^{3+} + e \rightleftharpoons Ti^{2+}$	-0.9
$HSnO_2^- + H_2O + 2\,e \rightleftharpoons Sn + 3\,OH^-$	-0.909
$Cr^{2+} + 2\,e \rightleftharpoons Cr$	-0.913
$Se + 2\,e \rightleftharpoons Se^{2-}$	-0.924
$SO_4^{2-} + H_2O + 2\,e \rightleftharpoons SO_3^{2-} + 2\,OH^-$	-0.93
$Sn(OH)_6^{2-} + 2\,e \rightleftharpoons HSnO_2^- + 3\,OH^- + H_2O$	-0.93
$SnO_2 + 2\,H_2O + 4\,e \rightleftharpoons Sn + 4\,OH^-$	-0.945
$In(OH)_3 + 3\,e \rightleftharpoons In + 3\,OH^-$	-0.99
$NpO_2 + H_2O + H^+ + e \rightleftharpoons Np(OH)_3$	-0.962
$In(OH)_4^- + 3\,e \rightleftharpoons In + 4\,OH^-$	-1.007
$In_2O_3 + 3\,H_2O + 6\,e \rightleftharpoons 2\,In + 6\,OH^-$	-1.034
$PO_4^{3-} + 2\,H_2O + 2\,e \rightleftharpoons HPO_3^{2-} + 3\,OH^-$	-1.05
$Yb^{3+} + e \rightleftharpoons Yb^{2+}$	-1.05
$Nb^{3+} + 3\,e \rightleftharpoons Nb$	-1.099
$Fm^{3+} + e \rightleftharpoons Fm^{2+}$	-1.1
$2\,SO_3^{2-} + 2\,H_2O + 2\,e \rightleftharpoons S_2O_4^{2-} + 4\,OH^-$	-1.12
$Te + 2\,e \rightleftharpoons Te^{2-}$	-1.143
$V^{2+} + 2\,e \rightleftharpoons V$	-1.175
$Mn^{2+} + 2\,e \rightleftharpoons Mn$	-1.185
$Zn(OH)_4^{2-} + 2\,e \rightleftharpoons Zn + 4\,OH^-$	-1.199
$CrO_2 + 2\,H_2O + 3\,e \rightleftharpoons Cr + 4\,OH^-$	-1.2
$No^{3+} + 3\,e \rightleftharpoons No$	-1.20
$ZnO_2^- + 2\,H_2O + 2\,e \rightleftharpoons Zn + 4\,OH^-$	-1.215
$H_2GaO_3^- + H_2O + 3\,e \rightleftharpoons Ga + 4\,OH^-$	-1.219
$H_2BO_3^- + 5\,H_2O + 8\,e \rightleftharpoons BH_4^- + 8\,OH^-$	-1.24
$SiF_6^{2-} + 4\,e \rightleftharpoons Si + 6\,F^-$	-1.24
$Zn(OH)_2 + 2\,e \rightleftharpoons Zn + 2\,OH^-$	-1.249
$ZnO + H_2O + 2\,e \rightleftharpoons Zn + 2\,OH^-$	-1.260
$Es^{3+} + e \rightleftharpoons Es^{2+}$	-1.3
$Pa^{3+} + 3\,e \rightleftharpoons Pa$	-1.34
$Ti^{3+} + 3\,e \rightleftharpoons Ti$	-1.37
$Ce^{3+} + 3\,e \rightleftharpoons Ce(Hg)$	-1.4373
$UO_2^{2+} + 4\,H^+ + 6\,e \rightleftharpoons U + 2\,H_2O$	-1.444
$Zr^{4+} + 4\,e \rightleftharpoons Zr$	-1.45

Reaction	$E°/V$	Reaction	$E°/V$
$Cr(OH)_3 + 3e \rightleftharpoons Cr + 3OH^-$	−1.48	$Am^{3+} + e \rightleftharpoons Am^{2+}$	−2.3
$Pa^{4+} + 4e \rightleftharpoons Pa$	−1.49	$Fm^{2+} + 2e \rightleftharpoons Fm$	−2.30
$HfO_2 + 4H^+ + 4e \rightleftharpoons Hf + 2H_2O$	−1.505	$Pm^{3+} + 3e \rightleftharpoons Pm$	−2.30
$Hf^{4+} + 4e \rightleftharpoons Hf$	−1.55	$Sm^{3+} + 3e \rightleftharpoons Sm$	−2.304
$Sm^{3+} + e \rightleftharpoons Sm^{2+}$	−1.55	$Al(OH)_3 + 3e \rightleftharpoons Al + 3OH^-$	−2.31
$ZrO_2 + 4H^+ + 4e \rightleftharpoons Zr + 2H_2O$	−1.553	$Tm^{3+} + 3e \rightleftharpoons Tm$	−2.319
$Mn(OH)_2 + 2e \rightleftharpoons Mn + 2OH^-$	−1.56	$Nd^{3+} + 3e \rightleftharpoons Nd$	−2.323
$Ba^{2+} + 2e \rightleftharpoons Ba(Hg)$	−1.570	$Al(OH)^- + 3e \rightleftharpoons Al + 4OH^-$	−2.328
$Bk^{2+} + 2e \rightleftharpoons Bk$	−1.6	$H_2AlO_3^- + H_2O + 3e \rightleftharpoons Al + 4OH^-$	−2.33
$Cf^{3+} + e \rightleftharpoons Cf^{2+}$	−1.6	$Ho^{3+} + 3e \rightleftharpoons Ho$	−2.33
$Ti^{2+} + 2e \rightleftharpoons Ti$	−1.630	$Er^{3+} + 3e \rightleftharpoons Er$	−2.331
$Md^{3+} + 3e \rightleftharpoons Md$	−1.65	$Ce^{3+} + 3e \rightleftharpoons Ce$	−2.336
$HPO_3^{2-} + 2H_2O + 2e \rightleftharpoons H_2PO_2^- + 3OH^-$	−1.65	$Pr^{3+} + 3e \rightleftharpoons Pr$	−2.353
$Al^{3+} + 3e \rightleftharpoons Al$	−1.662	$ZrO(OH)_2 + H_2O + 4e \rightleftharpoons Zr + 4OH^-$	−2.36
$SiO_3^{2-} + H_2O + 4e \rightleftharpoons Si + 6OH^-$	−1.697	$Mg^{2+} + 2e \rightleftharpoons Mg$	−2.372
$HPO_3^{2-} + 2H_2O + 3e \rightleftharpoons P + 5OH^-$	−1.71	$Y^{3+} + 3e \rightleftharpoons Y$	−2.372
$HfO^{2+} + 2H^+ + 4e \rightleftharpoons Hf + H_2O$	−1.724	$La^{3+} + 3e \rightleftharpoons La$	−2.379
$ThO_2 + 4H^+ + 4e \rightleftharpoons Th + 2H_2O$	−1.789	$Tm^{2+} + 2e \rightleftharpoons Tm$	−2.4
$H_2BO_3^- + H_2O + 3e \rightleftharpoons B + 4OH^-$	−1.79	$Md^{2+} + 2e \rightleftharpoons Md$	−2.40
$Sr^{2+} + 2e \rightleftharpoons Sr(Hg)$	−1.793	$Th(OH)_4 + 4e \rightleftharpoons Th + 4OH^-$	−2.48
$U^{3+} + 3e \rightleftharpoons U$	−1.798	$HfO(OH)_2 + H_2O + 4e \rightleftharpoons Hf + 4OH^-$	−2.50
$H_2PO_2^- + e \rightleftharpoons P + 2OH^-$	−1.82	$No^{2+} + 2e \rightleftharpoons No$	−2.50
$Be^{2+} + 2e \rightleftharpoons Be$	−1.847	$Dy^{3+} + e \rightleftharpoons Dy^{2+}$	−2.6
$Np^{3+} + 3e \rightleftharpoons Np$	−1.856	$Pm^{3+} + e \rightleftharpoons Pm^{2+}$	−2.6
$Fm^{3+} + 3e \rightleftharpoons Fm$	−1.89	$Be_2O_3^{2-} + 3H_2O + 4e \rightleftharpoons 2Be + 6OH^-$	−2.63
$Th^{4+} + 4e \rightleftharpoons Th$	−1.899	$Sm^{2+} + 2e \rightleftharpoons Sm$	−2.68
$Am^{2+} + 2e \rightleftharpoons Am$	−1.9	$Mg(OH)_2 + 2e \rightleftharpoons Mg + 2OH^-$	−2.690
$Pa^{4+} + e \rightleftharpoons Pa^{3+}$	−1.9	$Nd^{3+} + e \rightleftharpoons Nd^{2+}$	−2.7
$Es^{3+} + 3e \rightleftharpoons Es$	−1.91	$Mg^+ + e \rightleftharpoons Mg$	−2.70
$Cf^{3+} + 3e \rightleftharpoons Cf$	−1.94	$Na^+ + e \rightleftharpoons Na$	−2.71
$Lr^{3+} + 3e \rightleftharpoons Lr$	−1.96	$Yb^{2+} + 2e \rightleftharpoons Yb$	−2.76
$Eu^{3+} + 3e \rightleftharpoons Eu$	−1.991	$Bk^{3+} + e \rightleftharpoons Bk^{2+}$	−2.8
$Er^{2+} + 2e \rightleftharpoons Er$	−2.0	$Ho^{3+} + e \rightleftharpoons Ho^{2+}$	−2.8
$Pr^{2+} + 2e \rightleftharpoons Pr$	−2.0	$Ra^{2+} + 2e \rightleftharpoons Ra$	−2.8
$Pu^{3+} + 3e \rightleftharpoons Pu$	−2.031	$Eu^{2+} + 2e \rightleftharpoons Eu$	−2.812
$Cm^{3+} + 3e \rightleftharpoons Cm$	−2.04	$Ca^{2+} + 2e \rightleftharpoons Ca$	−2.868
$Am^{3+} + 3e \rightleftharpoons Am$	−2.048	$Sr(OH)_2 + 2e \rightleftharpoons Sr + 2OH^-$	−2.88
$AlF_6^{3-} + 3e \rightleftharpoons Al + 6F^-$	−2.069	$Sr^{2+} + 2e \rightleftharpoons Sr$	−2.899
$Sc^{3+} + 3e \rightleftharpoons Sc$	−2.077	$Fr^+ + e \rightleftharpoons Fr$	−2.9
$Ho^{2+} + 2e \rightleftharpoons Ho$	−2.1	$La(OH)_3 + 3e \rightleftharpoons La + 3OH^-$	−2.90
$Nd^{2+} + 2e \rightleftharpoons Nd$	−2.1	$Ba^{2+} + 2e \rightleftharpoons Ba$	−2.912
$Cf^{2+} + 2e \rightleftharpoons Cf$	−2.12	$K^+ + e \rightleftharpoons K$	−2.931
$Yb^{3+} + 3e \rightleftharpoons Yb$	−2.19	$Rb^+ + e \rightleftharpoons Rb$	−2.98
$Ac^{3+} + 3e \rightleftharpoons Ac$	−2.20	$Ba(OH)_2 + 2e \rightleftharpoons Ba + 2OH^-$	−2.99
$Dy^{2+} + 2e \rightleftharpoons Dy$	−2.2	$Er^{3+} + e \rightleftharpoons Er^{2+}$	−3.0
$Tm^{3+} + e \rightleftharpoons Tm^{2+}$	−2.2	$Ca(OH)_2 + 2e \rightleftharpoons Ca + 2OH^-$	−3.02
$Pm^{2+} + 2e \rightleftharpoons Pm$	−2.2	$Cs^+ + e \rightleftharpoons Cs$	−3.026
$Es^{2+} + 2e \rightleftharpoons Es$	−2.23	$Li^+ + e \rightleftharpoons Li$	−3.0401
$H_2 + 2e \rightleftharpoons 2H^-$	−2.23	$3N_2 + 2H^+ + 2e \rightleftharpoons 2HN_3$	−3.09
$Gd^{3+} + 3e \rightleftharpoons Gd$	−2.279	$Pr^{3+} + e \rightleftharpoons Pr^{2+}$	−3.1
$Tb^{3+} + 3e \rightleftharpoons Tb$	−2.28	$Ca^+ + e \rightleftharpoons Ca$	−3.80
$Lu^{3+} + 3e \rightleftharpoons Lu$	−2.28	$Sr^+ + e \rightleftharpoons Sr$	−4.10
$Dy^{3+} + 3e \rightleftharpoons Dy$	−2.295		

REDUCTION AND OXIDATION POTENTIALS FOR CERTAIN ION RADICALS

Petr Vanýsek

There are two tables for ion radicals. The first table lists reduction potentials for organic compounds that produce anion radicals during reduction, a process described as $A + e^- \rightleftharpoons A^{-\cdot}$. The second table lists oxidation potentials for organic compounds that produce cation radicals during oxidation, a process described as $A \rightleftharpoons A^{+\cdot} + e^-$. To obtain reduction potential for a reverse reaction, the sign for the potential is changed.

Unlike the table of the Electrochemical Series, which lists *standard* potentials, values for radicals are experimental values with experimental conditions given in the second column. Since the measurements leading to potentials for ion radicals are very dependent on conditions, an attempt to report standard potentials for radicals would serve no useful purpose. For the same reason, the potentials are also reported as experimental values, usually a half-wave potential ($E_{1/2}$ in polarography) or a peak potential (E_p in cyclic voltammetry). Unless otherwise stated, the values are reported vs. SCE (saturated calomel electrode). To obtain a value vs.

normal hydrogen electrode, 0.241 V has to be added to the SCE values. All the ion radicals chosen for inclusion in the tables result from electrochemically reversible reactions. More detailed data on ion radicals can be found in the *Encyclopedia of Electrochemistry of Elements*, (A. J. Bard, Ed.), Vols. XI and XII in particular, Marcel Dekker, New York, 1978.

Abbreviations are: CV — cyclic voltammetry; DMF — *N,N*-Dimethylformamide; E swp — potential sweep; $E°$ — standard potential; E_p — peak potential; $E_{p/2}$ — half-peak potential; $E_{1/2}$ — half wave potential; M — mol/L; MeCN — acetonitrile; pol — polarography; rot Pt dsk — rotated Pt disk; SCE — saturated calomel electrode; $TBABF_4$ — tetrabutylammonium tetrafluoroborate; TBAI — tetrabutylammonium iodide; TBAP — tetrabutylammonium perchlorate; TEABr — tetraethylammonium bromide; TEAP — tetraethylammonium perchlorate; THF — tetrahydrofuran; $TPACF_3SO_3$ — tetrapropylammonium trifluoromethanesulfite; TPAP — tetrapropylammonium perchlorate; and wr — wire.

Reduction Potentials (Products Are Anion Radicals)

Substance	Conditions/electrode/technique	Potential V (vs. SCE)
Acetone	DMF, 0.1 M TEABr/Hg/pol	$E_{1/2} = -2.84$
1-Naphthphenylacetylene	DMF, 0.03 M TBAI/Hg/pol	$E_{1/2} = -1.91$
1-Naphthalenecarboxyaldehyde	-/Hg/pol	$E_{1/2} = -0.91$
2-Naphthalenecarboxyaldehyde	-/Hg/pol	$E_{1/2} = -0.96$
2-Phenanthrenecarboxaldehyde	-/Hg/pol	$E_{1/2} = -1.00$
3-Phenanthrenecarboxaldehyde	-/Hg/pol	$E_{1/2} = -0.94$
9-Phenanthrenecarboxaldehyde	-/Hg/pol	$E_{1/2} = -0.83$
1-Anthracenecarboxaldehyde	-/Hg/pol	$E_{1/2} = -0.75$
1-Pyrenecarboxaldehyde	-/Hg/pol	$E_{1/2} = -0.76$
2-Pyrenecarboxaldehyde	-/Hg/pol	$E_{1/2} = -1.00$
Anthracene	DMF, 0.1 M TBAP/Pt dsk/CV	$E_p = -2.00$
	DMF, 0.5 M $TBABF_4$/Hg/CV	$E_{1/2} = -1.93$
	MeCN, 0.1 M TEAP/Hg/CV	$E_{1/2} = -2.07$
	DMF, 0.1 M TBAI/Hg/pol	$E_{1/2} = -1.92$
9,10-Dimethylanthracene	DMF, 0.1 M TBAP/Pt/CV	$E_p = -2.08$
	MeCN, 0.1 M TBAP/Pt/CV	$E_p = -2.10$
1-Phenylanthracene	DMF, 0.5 M $TBABF_2$/Hg/CV	$E_{1/2} = -1.91$
	DMF, 0.1 M TBAI/Hg/pol	$E_{1/2} = -1.878$
2-Phenylanthracene	DMF, 0.1 M TBAI/Hg/pol	$E_{1/2} = -1.875$
8-Phenylanthracene	DMF, 0.5 M $TBABF_4$/Hg/CV	$E_{1/2} = -1.91$
9-Phenylanthracene	DMF, 0.5 M $TBABF_4$/Hg/CV	$E_{1/2} = -1.93$
	DMF, 0.1 M TBAI/Hg/pol	$E_{1/2} = -1.863$
1,8-Diphenylanthracene	DMF, 0.5 M $TBABF_4$/Hg/CV	$E_{1/2} = -1.88$
1,9-Diphenylanthracene	DMF, 0.1 M TBAI/Hg/pol	$E_{1/2} = -1.846$
1,10-Diphenylanthracene	DMF, 0.1 M TBAI/Hg/pol	$E_{1/2} = -1.786$
8,9-Diphenylanthracene	DMF, 0.5 M $TBABF_4$/Hg/CV	$E_{1/2} = -1.90$
9,10-Diphenylanthracene	MeCN, 0.1 M TBAP/rot Pt/E swp	$E_{1/2} = -1.83$
	DMF, 0.1 M TBAI/Hg/pol	$E_{1/2} = -1.835$
1,8,9-Triphenylanthracene	DMF, 0.5 M $TBABF_4$/Hg/CV	$E_{1/2} = -1.85$
1,8,10-Triphenylanthracene	DMF, 0.5 M $TBABF_4$/Hg/CV	$E_{1/2} = -1.81$
9,10-Dibiphenylanthracene	MeCN, 0.1 M TBAP/rot Pt/E swp	$E_{1/2} = -1.94$
Benz(a)anthracene	MeCN, 0.1 M TEAP/Hg/CV	$E_{1/2} = -2.11$
	MeCN, 0.1 M TEAP/Hg/pol	$E_{1/2} = -2.40$[a]
Azulene	DMF, 0.1 M TBAI/Hg/pol	$E_{1/2} = -1.10$[c]
Annulene	DMF, 0.5 M TBAP 0 °C/Hg/pol	$E_{1/2} = -1.23$
Benzaldehyde	DMF, 0.1 M TBAP/Hg/pol	$E_{1/2} = -1.67$
Benzil	DMSO, 0.1 M TBAP/Hg/pol	$E_{1/2} = -1.04$
Benzophenone	-/Hg/pol	$E_{1/2} = -1.80$
	DMF/Pt dsk/CV	$E° = -1.72$
Chrysene	MeCN, 0.1 M TEAP/Hg/pol	$E_{1/2} = -2.73$[a]
Fluoranthrene	DMF, 0.1 M TBAP/Pt dsk/CV	$E_p = -1.76$

Substance	Conditions/electrode/technique	Potential V (vs. SCE)
Cyclohexanone	DMF, 0.1 M TEABr/Hg/pol	$E_{1/2} = -2.79$
5,5-Dimethyl-3-phenyl-2-cyclohexen-1-one	DMF, 0.5 M/Hg/pol	$E_{1/2} = -1.71$
1,2,3-Indanetrione hydrate (ninhydrin)	DMF, 0.2 M NaNO$_3$/Hg/pol	$E_{1/2} = -0.039$
Naphthacene	DMF, 0.1 M TBAI/Hg/pol	$E_{1/2} = -1.53$
Naphthalene	DMF, 0.1 M TBAP/Pt dsk/CV	$E_p = -2.55$
	DMF, 0.5 M TBABF$_4$/Hg/CV	$E_{1/2} = -2.56$
	DMF, MeCN, 0.1 M TEAP/Hg/CV	$E_{1/2} = -2.63$
	DMF, 0.1 M TBAI/Hg/pol	$E_{1/2} = -2.50$
1-Phenylnaphthalene	DMF, 0.5 M TBABF$_4$/Hg/CV	$E_{1/2} = -2.36$
1,2-Diphenylnaphthalene	DMF, 0.5 M TBABF$_4$/Hg/CV	$E_{1/2} = -2.25$
Cyclopentanone	DMF, 0.1 M TEABr/Hg/pol	$E_{1/2} = -2.82$
Phenanthrene	MeCN, 0.1 M TBAP/Pt wr/CV	$E_{1/2} = -2.47$
	MeCN, 0.1 M TEAP/Hg/pol	$E_{1/2} = -2.88^a$
Pentacene	THF, 0.1 M TBAP/rot Pt dsk/E swp	$E_{1/2} = -1.40$
Perylene	MeCN, 0.1 M TEAP/Hg/CV	$E_{1/2} = -1.73$
1,3-Diphenyl-1,3-propanedione	DMSO, 0.2 M TBAP/Hg/CV	$E_{1/2} = -1.42$
2,2-Dimethyl-1,3-diphenyl-1,3 propanedione	DMSO, TBAP/Hg/CV	$E_{1/2} = -1.80$
Pyrene	DMF, 0.1 M TBAP/Pt/CV	$E_p = -2.14$
	MeCN, 0.1 M TEAP/Hg/pol	$E_{1/2} = -2.49^a$
Diphenylsulfone	DMF, TEABr	$E_{1/2} = -2.16$
Triphenylene	MeCN, 0.1 M TEAP/Hg/pol	$E_{1/2} = -2.87^a$
9,10-Anthraquinone	DMF, 0.5 M TBAP, 20°/Pt dsk/CV	$E_{1/2} = -1.01$
1,4-Benzoquinone	MeCN, 0.1 M TEAP/Pt/CV	$E_p = -0.54$
1,4-Naphthohydroquinone, dipotassium salt	DMF, 0.5 M TBAP, 20°/Pt dsk/CV	$E_{1/2} = -1.55$
Rubrene	DMF, 0.1 M TBAP/Pt dsk/CV	$E_p = -1.48$
	DMF, 0.1 M TBAI/Hg/pol	$E_{1/2} = -1.410$
Benzocyclooctatetraene	THF, 0.1 M TBAP/Hg/pol	$E_{1/2} = -2.13$
sym-Dibenzocyclooctatetraene	THF, 0.1 M TBAP/Hg/pol	$E_{1/2} = -2.29$
Ubiquinone-6	MeCN, 0.1 M TEAP/Pt/CV	$E_p = -1.05^e$
(9-Phenyl-fluorenyl)$^+$	10.2 M H$_2$SO$_4$/Hg/CV	$E_p = -0.01^b$
(Triphenylcyclopropenyl)$^+$	MeCN, 0.1 M TEAP/Hg/CV	$E_p = -1.87$
(Triphenylmethyl)$^+$	MeCN, 0.1 M TBAP/Hg/pol	$E_{1/2} = 0.27$
	H$_2$SO$_4$, 10.2 M/Hg/CV	$E_p = -0.58^b$
(Tribiphenylmethyl)$^+$	MeCN, 0.1 M TBAP/Hg/pol	$E_{1/2} = 0.19$
(Tri-4-t-butyl-5-phenylmethyl)$^+$	MeCN, 0.1 M TBAP/Hg/pol	$E_{1/2} = 0.13$
(Tri-4-isopropylphenylmethyl)$^+$	MeCN, 0.1 M TBAP/Hg/pol	$E_{1/2} = 0.07$
(Tri-4-methylphenylmethyl)$^+$	MeCN, 0.1 M TBAP/Hg/pol	$E_{1/2} = 0.05$
(Tri-4-cyclopropylphenylmethyl)$^+$	MeCN, 0.1 M TBAP/Hg/pol	$E_{1/2} = 0.01$
(Tropylium)$^+$	MeCN, 0.1 M TBAP/Hg/pol	$E_{1/2} = -0.17$
	DMF, 0.15 M TBAI/Hg/pol	$E_{1/2} = -1.55$
	DMF, 0.15 M TBAI/Hg/pol	$E_{1/2} = -1.55$
	DMF, 0.15 M TBAI/Hg/pol	$E_{1/2} = -1.57$
	DMF, 0.15 M TBAI/Hg/pol	$E_{1/2} = -1.60$
	DMF, 0.15 M TBAI/Hg/pol	$E_{1/2} = -1.87$
	DMF, 0.15 M TBAI/Hg/pol	$E_{1/2} = -1.96$
	DMF, 0.15 M TBAI/Hg/pol	$E_{1/2} = -2.05$

Oxidation Potentials (Products Are Cation Radicals)

Substance	Conditions/electrode/technique	Potential V (vs. SCE)
Anthracene	CH$_2$Cl$_2$, 0.2 M TBABF$_4$, −70 °C/Pt dsk/CV	$E_p = +0.73^d$
9,10-Dimethylanthracene	MeCN, 0.1 M LiClO$_4$/Pt wr/CV	$E_p = +1.0$
9,10-Dipropylanthracene	MeCN, 0.1 M TEAP/Pt/CV	$E_p = +1.08$
1,8-Diphenylanthracene	CH$_2$Cl$_2$, 0.2 M TPrACF$_3$SO$_3$/rot Pt wr/E swp	$E_{1/2} = +1.34$
8,9-Diphenylanthracene	CH$_2$Cl$_2$, 0.2 M TPrACF$_3$SO$_3$/rot Pt wr/E swp	$E_{1/2} = +1.30$
9,10-Diphenylanthracene	MeCN/Pt/CV	$E_p = +1.22$
Perylene	MeCN, 0.1 M TBAP/Pt/CV	$E_p = +1.34$
Pyrene	DMF, 0.1 M TBAP/Pt dsk/CV	$E_p = +1.25$
Rubrene	DMF, 0.1 M TBAP/Pt dsk/CV	$E_p = +1.10$
Tetracene	CH$_2$Cl$_2$, 0.2 M TBABF$_4$, −70 °C/Pt wr/CV	$E_p = +0.35^d$
1,4-Dithiabenzene	MeCN, 0.1 M TEAP/Pt dsk/rot	$E_{1/2} = +0.69$
1,4-Dithianaphthalene	MeCN, 0.1 M TEAP/Pt dsk/rot	$E_{1/2} = +0.80$
Thianthrene	0.1 M TPAP/Pt/CV	$E_{1/2} = +1.28$

[a] vs. 0.01 M Ag/AgClO$_4$
[b] vs. Hg/Hg$_2$SO$_4$, 17 M H$_2$SO$_4$
[c] vs. Hg pool
[d] vs. Ag/saturated AgNO$_3$
[e] vs. Ag/0.01 M Ag+

DISSOCIATION CONSTANTS OF INORGANIC ACIDS AND BASES

The data in this table are presented as values of pK_a, defined as the negative logarithm of the acid dissociation constant K_a for the reaction

$$BH \rightleftharpoons B^- + H^+$$

Thus $pK_a = -\log K_a$, and the hydrogen ion concentration $[H^+]$ can be calculated from

$$K_a = \frac{[H^+][B^-]}{[BH]}$$

In the case of bases, the entry in the table is for the conjugate acid; e.g., ammonium ion for ammonia. The OH^- concentration in the system

$$NH_3 + H_2O \rightleftharpoons NH_4^+ + OH^-$$

can be calculated from the equation

$$K_b = K_{water}/K_a = \frac{[OH^-][NH_4^+]}{[NH_3]}$$

where $K_{water} = 1.01 \times 10^{-14}$ at 25 °C. Note that $pK_a + pK_b = pK_{water}$.

All values refer to dilute aqueous solutions at zero ionic strength at the temperature indicated. The table is arranged alphabetically by compound name.

Reference

1. Perrin, D. D., *Ionization Constants of Inorganic Acids and Bases in Aqueous Solution, Second Edition*, Pergamon, Oxford, 1982.

Name	Formula	Step	t/°C	pK$_a$
Aluminum(III) ion	Al^{+3}		25	5.0
Ammonia	NH$_3$		25	9.25
Arsenic acid	H$_3$AsO$_4$	1	25	2.26
		2	25	6.76
		3	25	11.29
Arsenious acid	H$_2$AsO$_3$		25	9.29
Barium(II) ion	Ba^{+2}		25	13.4
Boric acid	H$_3$BO$_3$	1	20	9.27
		2	20	>14
Calcium(II) ion	Ca^{+2}		25	12.6
Carbonic acid	H$_2$CO$_3$	1	25	6.35
		2	25	10.33
Chlorous acid	HClO$_2$		25	1.94
Chromic acid	H$_2$CrO$_4$	1	25	0.74
		2	25	6.49
Cyanic acid	HCNO		25	3.46
Germanic acid	H$_2$GeO$_3$	1	25	9.01
		2	25	12.3
Hydrazine	N$_2$H$_4$		25	8.1
Hydrazoic acid	HN$_3$		25	4.6
Hydrocyanic acid	HCN		25	9.21
Hydrofluoric acid	HF		25	3.20
Hydrogen peroxide	H$_2$O$_2$		25	11.62
Hydrogen selenide	H$_2$Se	1	25	3.89
		2	25	11.0
Hydrogen sulfide	H$_2$S	1	25	7.05
		2	25	19
Hydrogen telluride	H$_2$Te	1	18	2.6
		2	25	11
Hydroxylamine	NH$_2$OH		25	5.94
Hypobromous acid	HBrO		25	8.55
Hypochlorous acid	HClO		25	7.40
Hypoiodous acid	HIO		25	10.5
Iodic acid	HIO$_3$		25	0.78
Lithium ion	Li$^+$		25	13.8
Magnesium(II) ion	Mg^{+2}		25	11.4
Nitrous acid	HNO$_2$		25	3.25
Perchloric acid	HClO$_4$		20	-1.6
Periodic acid	HIO$_4$		25	1.64
Phosphoric acid	H$_3$PO$_4$	1	25	2.16
		2	25	7.21

Name	Formula	Step	$t/°C$	pK_a
		3	25	12.32
Phosphorous acid	H_3PO_3	1	20	1.3
		2	20	6.70
Pyrophosphoric acid	$H_4P_2O_7$	1	25	0.91
		2	25	2.10
		3	25	6.70
		4	25	9.32
Selenic acid	H_2SeO_4	2	25	1.7
Selenious acid	H_2SeO_3	1	25	2.62
		2	25	8.32
Silicic acid	H_4SiO_4	1	30	9.9
		2	30	11.8
		3	30	12
		4	30	12
Sodium ion	Na^+		25	14.8
Strontium(II) ion	Sr^{+2}		25	13.2
Sulfamic acid	NH_2SO_3H		25	1.05
Sulfuric acid	H_2SO_4	2	25	1.99
Sulfurous acid	H_2SO_3	1	25	1.85
		2	25	7.2
Telluric acid	H_2TeO_4	1	18	7.68
		2	18	11.0
Tellurous acid	H_2TeO_3	1	25	6.27
		2	25	8.43
Tetrafluoroboric acid	HBF_4		25	0.5
Thiocyanic acid	$HSCN$		25	−1.8
Water	H_2O		25	13.995

DISSOCIATION CONSTANTS OF ORGANIC ACIDS AND BASES

This table lists the dissociation (ionization) constants of over 1070 organic acids, bases, and amphoteric compounds. All data apply to dilute aqueous solutions and are presented as values of pK_a, which is defined as the negative of the logarithm of the equilibrium constant K_a for the reaction

$$HA \rightleftharpoons H^+ + A^-$$

i.e.,

$$K_a = [H^+][A^-]/[HA]$$

where $[H^+]$, etc. represent the concentrations of the respective species in mol/L. It follows that $pK_a = pH + \log[HA] - \log[A^-]$, so that a solution with 50% dissociation has pH equal to the pK_a of the acid.

Data for bases are presented as pK_a values for the conjugate acid, i.e., for the reaction

$$BH^+ \rightleftharpoons H^+ + B$$

In older literature, an ionization constant K_b was used for the reaction $B + H_2O$ $\rightleftharpoons BH^+ + OH^-$. This is related to K_a by

$$pK_a + pK_b = pK_{water} = 14.00 \quad (at\ 25\ °C)$$

Compounds are listed by molecular formula in Hill order.

References

1. Perrin, D. D., *Dissociation Constants of Organic Bases in Aqueous Solution*, Butterworths, London, 1965; Supplement, 1972.
2. Serjeant, E. P., and Dempsey, B., *Ionization Constants of Organic Acids in Aqueous Solution*, Pergamon, Oxford, 1979.
3. Albert, A., "Ionization Constants of Heterocyclic Substances", in Katritzky, A. R., Ed., *Physical Methods in Heterocyclic Chemistry*, Academic Press, New York, 1963.
4. Sober, H.A., Ed., *CRC Handbook of Biochemistry*, CRC Press, Boca Raton, FL, 1968.
5. Perrin, D. D., Dempsey, B., and Serjeant, E. P., pK_a *Prediction for Organic Acids and Bases*, Chapman and Hall, London, 1981.
6. Albert, A., and Serjeant, E. P., *The Determination of Ionization Constants, Third Edition*, Chapman and Hall, London, 1984.
7. O'Neil, M.J., Ed., *The Merck Index, 14th Edition*, Merck & Co., Whitehouse Station, NJ, 2006.

Mol. form.	Name	Step	t/°C	pK_a
CHNO	Cyanic acid		25	3.7
CH$_2$N$_2$	Cyanamide		29	1.1
CH$_2$O	Formaldehyde		25	13.27
CH$_2$O$_2$	Formic acid		25	3.75
CH$_3$NO$_2$	Nitromethane		25	10.21
CH$_3$NS$_2$	Carbamodithioic acid		25	2.95
CH$_4$N$_2$O	Urea		25	0.10
CH$_4$N$_2$S	Thiourea		25	-1
CH$_4$O	Methanol		25	15.5
CH$_4$S	Methanethiol		25	10.33
CH$_5$N	Methylamine		25	10.66
CH$_5$NO	*O*-Methylhydroxylamine			12.5
CH$_5$N$_3$	Guanidine		25	13.6
C$_2$HCl$_3$O	Trichloroacetaldehyde		25	10.04
C$_2$HCl$_3$O$_2$	Trichloroacetic acid		20	0.66
C$_2$HF$_3$O$_2$	Trifluoroacetic acid		25	0.52
C$_2$H$_2$Cl$_2$O$_2$	Dichloroacetic acid		25	1.35
C$_2$H$_2$O$_3$	Glyoxylic acid		25	3.18
C$_2$H$_2$O$_4$	Oxalic acid	1	25	1.25
		2	25	3.81
C$_2$H$_3$BrO$_2$	Bromoacetic acid		25	2.90
C$_2$H$_3$ClO$_2$	Chloroacetic acid		25	2.87
C$_2$H$_3$Cl$_3$O	2,2,2-Trichloroethanol		25	12.24
C$_2$H$_3$FO$_2$	Fluoroacetic acid		25	2.59
C$_2$H$_3$F$_3$O	2,2,2-Trifluoroethanol		25	12.37
C$_2$H$_3$IO$_2$	Iodoacetic acid		25	3.18
C$_2$H$_3$NO$_4$	Nitroacetic acid		24	1.48
C$_2$H$_3$N$_3$	1H-1,2,3-Triazole		20	1.17
C$_2$H$_3$N$_3$	1H-1,2,4-Triazole		20	2.27
C$_2$H$_4$N$_2$	Aminoacetonitrile		25	5.34
C$_2$H$_4$O	Acetaldehyde		25	13.57
C$_2$H$_4$OS	Thioacetic acid		25	3.33
C$_2$H$_4$O$_2$	Acetic acid		25	4.756
C$_2$H$_4$O$_2$S	Thioglycolic acid		25	3.68
C$_2$H$_4$O$_3$	Glycolic acid		25	3.83
C$_2$H$_5$N	Ethyleneimine		25	8.04

Mol. form.	Name	Step	t/°C	pK_a
C$_2$H$_5$NO	Acetamide		25	15.1
C$_2$H$_5$NO$_2$	Acetohydroxamic acid			8.70
C$_2$H$_5$NO$_2$	Nitroethane		25	8.46
C$_2$H$_5$NO$_2$	Glycine	1	25	2.35
		2	25	9.78
C$_2$H$_6$N$_2$	Ethanimidamide		25	12.1
C$_2$H$_6$O	Ethanol		25	15.5
C$_2$H$_6$OS	2-Mercaptoethanol		25	9.72
C$_2$H$_6$O$_2$	Ethyleneglycol		25	15.1
C$_2$H$_7$AsO$_2$	Dimethylarsinic acid	1	25	1.57
		2	25	6.27
C$_2$H$_7$N	Ethylamine		25	10.65
C$_2$H$_7$N	Dimethylamine		25	10.73
C$_2$H$_7$NO	Ethanolamine		25	9.50
C$_2$H$_7$NO$_3$S	2-Aminoethanesulfonic acid	1	25	1.5
		2	25	9.06
C$_2$H$_7$NS	Cysteamine	1	25	8.27
		2	25	10.53
C$_2$H$_7$N$_5$	Biguanide	1		11.52
		2		2.93
C$_2$H$_8$N$_2$	1,2-Ethanediamine	1	25	9.92
		2	25	6.86
C$_2$H$_8$O$_7$P$_2$	1-Hydroxy-1,1-diphosphonoethane	1		1.35
		2		2.87
		3		7.03
		4		11.3
C$_3$H$_2$O$_2$	2-Propynoic acid		25	1.84
C$_3$H$_3$NO	Oxazole		33	0.8
C$_3$H$_3$NO	Isoxazole		25	-2.0
C$_3$H$_3$NO$_2$	Cyanoacetic acid		25	2.47
C$_3$H$_3$NS	Thiazole		25	2.52
C$_3$H$_3$N$_3$O$_3$	Cyanuric acid	1		6.88
		2		11.40
		3		13.5
C$_3$H$_4$N$_2$	1*H*-Pyrazole		25	2.49
C$_3$H$_4$N$_2$	Imidazole		25	6.99

Mol. form.	Name	Step	$t/°C$	pK_a
$C_3H_4N_2S$	2-Thiazolamine		20	5.36
C_3H_4O	Propargyl alcohol		25	13.6
$C_3H_4O_2$	Acrylic acid		25	4.25
$C_3H_4O_3$	Pyruvic acid		25	2.39
$C_3H_4O_4$	Malonic acid	1	25	2.85
		2	25	5.70
$C_3H_4O_5$	Hydroxypropanedioic acid	1		2.42
		2		4.54
$C_3H_5BrO_2$	3-Bromopropanoic acid		25	4.00
$C_3H_5ClO_2$	2-Chloropropanoic acid		25	2.83
$C_3H_5ClO_2$	3-Chloropropanoic acid		25	3.98
$C_3H_6N_2$	3-Aminopropanenitrile		20	7.80
$C_3H_6N_6$	1,3,5-Triazine-2,4,6-triamine		25	5.00
C_3H_6O	Allyl alcohol		25	15.5
$C_3H_6O_2$	Propanoic acid		25	4.87
$C_3H_6O_2S$	(Methylthio)acetic acid		25	3.66
$C_3H_6O_3$	Lactic acid		25	3.86
$C_3H_6O_3$	3-Hydroxypropanoic acid		25	4.51
$C_3H_6O_4$	Glyceric acid		25	3.52
C_3H_7N	Allylamine		25	9.49
C_3H_7N	Azetidine		25	11.29
C_3H_7NO	2-Propanone oxime		25	12.42
$C_3H_7NO_2$	L-Alanine	1	25	2.34
		2	25	9.87
$C_3H_7NO_2$	$β$-Alanine	1	25	3.55
		2	25	10.24
$C_3H_7NO_2$	Sarcosine	1	25	2.21
		2	25	10.1
$C_3H_7NO_2S$	L-Cysteine	1	25	1.5
		2	25	8.7
		3	25	10.2
$C_3H_7NO_3$	L-Serine	1	25	2.19
		2	25	9.21
$C_3H_7NO_5S$	DL-Cysteic acid	1	25	1.3
		2	25	1.9
		3	25	8.70
$C_3H_7N_3O_2$	Glycocyamine		25	2.82
$C_3H_8O_2$	Ethylene glycol monomethyl ether		25	14.8
$C_3H_8O_3$	Glycerol		25	14.15
C_3H_9N	Propylamine		25	10.54
C_3H_9N	Isopropylamine		25	10.63
C_3H_9N	Trimethylamine		25	9.80
C_3H_9NO	2-Methoxyethylamine		25	9.40
C_3H_9NO	Trimethylamine oxide		20	4.65
$C_3H_{10}N_2$	1,2-Propanediamine, (±)	1	25	9.82
		2	25	6.61
$C_3H_{10}N_2$	1,3-Propanediamine	1	25	10.55
		2	25	8.88
$C_3H_{10}N_2O$	1,3-Diamino-2-propanol	1	20	9.69
		2	20	7.93
$C_3H_{11}N_3$	1,2,3-Triaminopropane	1	20	9.59
		2	20	7.95
$C_4H_4FN_3O$	Flucytosine			3.26
$C_4H_4N_2$	Pyrazine		20	0.65
$C_4H_4N_2$	Pyrimidine		20	1.23
$C_4H_4N_2$	Pyridazine		20	2.24
$C_4H_4N_2O_2$	Uracil		25	9.45
$C_4H_4N_2O_3$	Barbituric acid		25	4.01
$C_4H_4N_2O_5$	Alloxanic acid		25	6.64

Mol. form.	Name	Step	$t/°C$	pK_a
$C_4H_4N_4O_2$	5-Nitropyrimidinamine		20	0.35
$C_4H_4O_2$	2-Butynoic acid		25	2.62
$C_4H_4O_4$	Maleic acid	1	25	1.92
		2	25	6.23
$C_4H_4O_4$	Fumaric acid	1	25	3.02
		2	25	4.38
$C_4H_4O_5$	Oxaloacetic acid	1	25	2.55
		2	25	4.37
		3	25	13.03
C_4H_5N	Pyrrole		25	-3.8
$C_4H_5NO_2$	Succinimide		25	9.62
$C_4H_5N_3$	2-Pyrimidinamine		20	3.45
$C_4H_5N_3$	4-Pyrimidinamine		20	5.71
$C_4H_5N_3O$	Cytosine	1		4.60
		2		12.16
$C_4H_5N_3O_2$	6-Methyl-1,2,4-triazine-3,5(2H,4H)-dione			7.6
$C_4H_6N_2$	1-Methylimidazol		25	6.95
$C_4H_6N_4O_3$	Allantoin		25	8.96
$C_4H_6N_4O_3S_2$	Acetazolamide			7.2
$C_4H_6O_2$	$trans$-Crotonic acid		25	4.69
$C_4H_6O_2$	3-Butenoic acid		25	4.34
$C_4H_6O_2$	Cyclopropanecarboxylic acid		25	4.83
$C_4H_6O_3$	2-Oxobutanoic acid		25	2.50
$C_4H_6O_3$	Acetoacetic acid		25	3.6
$C_4H_6O_4$	Succinic acid	1	25	4.21
		2	25	5.64
$C_4H_6O_4$	Methylmalonic acid	1	25	3.07
		2	25	5.76
$C_4H_6O_5$	Malic acid	1	25	3.40
		2	25	5.11
$C_4H_6O_6$	DL-Tartaric acid	1	25	3.03
		2	25	4.37
$C_4H_6O_6$	$meso$-Tartaric acid	1	25	3.17
		2	25	4.91
$C_4H_6O_6$	L-Tartaric acid	1	25	2.98
		2	25	4.34
$C_4H_6O_8$	Dihydroxytartaric acid		25	1.92
$C_4H_7ClO_2$	2-Chlorobutanoic acid			2.86
$C_4H_7ClO_2$	3-Chlorobutanoic acid			4.05
$C_4H_7ClO_2$	4-Chlorobutanoic acid			4.52
$C_4H_7NO_2$	4-Cyanobutanoic acid		25	2.42
$C_4H_7NO_3$	N-Acetylglycine		25	3.67
$C_4H_7NO_4$	Iminodiacetic acid	1		2.98
		2		9.89
$C_4H_7NO_4$	L-Aspartic acid	1	25	1.99
		2	25	3.90
		3	25	9.90
$C_4H_7N_3O$	Creatinine	1	25	4.8
		2		9.2
$C_4H_7N_5$	2,4,6-Pyrimidinetriamine		20	6.84
$C_4H_8N_2O_3$	L-Asparagine	1	20	2.1
		2	20	8.80
$C_4H_8N_2O_3$	N-Glycylglycine	1	25	3.14
		2		8.17
$C_4H_8O_2$	Butanoic acid		25	4.83
$C_4H_8O_2$	2-Methylpropanoic acid		20	4.84
$C_4H_8O_3$	3-Hydroxybutanoic acid, (±)		25	4.70
$C_4H_8O_3$	4-Hydroxybutanoic acid		25	4.72
$C_4H_8O_3$	Ethoxyacetic acid		18	3.65
C_4H_9N	Pyrrolidine		25	11.31

Mol. form.	Name	Step	$t/°C$	pK_a
C_4H_9NO	Morpholine		25	8.50
$C_4H_9NO_2$	2-Methylalanine	1	25	2.36
		2	25	10.21
$C_4H_9NO_2$	N,N-Dimethylglycine		25	9.89
$C_4H_9NO_2$	DL-2-Aminobutanoic acid	1	25	2.29
		2	25	9.83
$C_4H_9NO_2$	4-Aminobutanoic acid	1	25	4.031
		2	25	10.556
$C_4H_9NO_2S$	DL-Homocysteine	1	25	2.22
		2	25	8.87
		3	25	10.86
$C_4H_9NO_3$	L-Threonine	1	25	2.09
		2	25	9.10
$C_4H_9NO_3$	L-Homoserine	1	25	2.71
		2	25	9.62
$C_4H_9N_3O_2$	Creatine	1	25	2.63
		2	25	14.3
$C_4H_{10}N_2$	Piperazine	1	25	9.73
		2	25	5.33
$C_4H_{10}N_2O_2$	2,4-Diaminobutanoic acid	1	25	1.85
		2	25	8.24
		3	25	10.44
$C_4H_{10}O_4$	1,2,3,4-Butanetetrol			13.9
$C_4H_{11}N$	Butylamine		25	10.60
$C_4H_{11}N$	sec-Butylamine		25	10.56
$C_4H_{11}N$	tert-Butylamine		25	10.68
$C_4H_{11}N$	Diethylamine		25	10.84
$C_4H_{11}NO_3$	Tris(hydroxymethyl) methylamine		20	8.3
$C_4H_{12}N_2$	1,4-Butanediamine	1	25	10.80
		2	25	9.63
C_5H_4BrN	3-Bromopyridine		25	2.84
C_5H_4ClN	2-Chloropyridine		25	0.49
C_5H_4ClN	3-Chloropyridine		25	2.81
C_5H_4ClN	4-Chloropyridine		25	3.83
C_5H_4FN	2-Fluoropyridine		25	-0.44
$C_5H_4N_2O_2$	4-Nitropyridine		25	1.61
$C_5H_4N_4$	1H-Purine	1	20	2.30
		2	20	8.96
$C_5H_4N_4O$	Hypoxanthine		25	8.7
$C_5H_4N_4O$	Allopurinol			10.2
$C_5H_4N_4O_3$	Uric acid		12	3.89
$C_5H_4N_4S$	1,7-Dihydro-6H-purine-6-thione	1		7.77
		2		11.17
$C_5H_4O_2S$	2-Thiophenecarboxylic acid		25	3.49
$C_5H_4O_2S$	3-Thiophenecarboxylic acid		25	4.1
$C_5H_4O_3$	2-Furancarboxylic acid		25	3.16
$C_5H_4O_3$	3-Furancarboxylic acid		25	3.9
C_5H_5N	Pyridine		25	5.23
C_5H_5NO	2-Pyridinol	1	20	0.75
		2	20	11.65
C_5H_5NO	3-Pyridinol	1	20	4.79
		2	20	8.75
C_5H_5NO	4-Pyridinol	1	20	3.20
		2	20	11.12
C_5H_5NO	2(1H)-Pyridinone	1	20	0.75
		2	20	11.65
C_5H_5NO	Pyridine-1-oxide		24	0.79
$C_5H_5NO_2$	1H-Pyrrole-2-carboxylic acid		20	4.45

Mol. form.	Name	Step	$t/°C$	pK_a
$C_5H_5NO_2$	1H-Pyrrole-3-carboxylic acid		20	5.00
$C_5H_5N_3O$	Pyrazinecarboxamide			0.5
$C_5H_5N_5$	Adenine	1		4.3
		2		9.83
$C_5H_5N_5O$	Guanine		40	9.92
$C_5H_6N_2$	2-Pyridinamine		20	6.82
$C_5H_6N_2$	3-Pyridinamine		25	6.04
$C_5H_6N_2$	4-Pyridinamine		25	9.11
$C_5H_6N_2$	2-Methylpyrazine		27	1.45
$C_5H_6N_2O_2$	Thymine		25	9.94
$C_5H_6O_4$	1,1-Cyclopropanedi-carboxylic acid	1	25	1.82
		2	25	7.43
$C_5H_6O_4$	trans-1-Propene-1,2-dicarboxylic acid	1	25	3.09
		2	25	4.75
$C_5H_6O_4$	1-Propene-2,3-dicarboxylic acid	1	25	3.85
		2	25	5.45
$C_5H_6O_5$	2-Oxoglutaric acid	1	25	2.47
		2	25	4.68
$C_5H_7NO_3$	5,5-Dimethyl-2,4-oxazolidinedione		37	6.13
$C_5H_7NO_3$	L-Pyroglutamic acid		25	3.32
$C_5H_7N_3$	2,5-Pyridinediamine		20	6.48
$C_5H_7N_3$	Methylaminopyrazine		25	3.39
$C_5H_7N_3O_4$	Azaserine			8.55
$C_5H_8N_2$	2,4-Dimethylimidazole		25	8.36
$C_5H_8N_4O_3S_2$	Methazolamide			7.30
$C_5H_8O_2$	trans-3-Pentenoic acid		25	4.51
$C_5H_8O_4$	Dimethylmalonic acid		25	3.15
$C_5H_8O_4$	Glutaric acid	1	18	4.32
		2	25	5.42
$C_5H_8O_4$	Methylsuccinic acid	1	25	4.13
		2	25	5.64
$C_5H_9NO_2$	L-Proline	1	25	1.95
		2	25	10.64
$C_5H_9NO_3$	5-Amino-4-oxopentanoic acid	1	25	4.05
		2	25	8.90
$C_5H_9NO_3$	trans-4-Hydroxyproline	1	25	1.82
		2	25	9.66
$C_5H_9NO_4$	L-Glutamic acid	1	25	2.13
		2	25	4.31
		3		9.67
$C_5H_9N_3$	Histamine	1	25	6.04
		2	25	9.75
$C_5H_{10}N_2O_3$	Glycylalanine		25	3.15
$C_5H_{10}N_2O_3$	L-Glutamine	1	25	2.17
		2	25	9.13
$C_5H_{10}N_2O_4$	Glycylserine	1	25	2.98
		2	25	8.38
$C_5H_{10}O_2$	Pentanoic acid		20	4.83
$C_5H_{10}O_2$	2-Methylbutanoic acid		25	4.80
$C_5H_{10}O_2$	3-Methylbutanoic acid		25	4.77
$C_5H_{10}O_2$	2,2-Dimethylpropanoic acid		20	5.03
$C_5H_{10}O_4$	D-2-Deoxyribose		25	12.61
$C_5H_{10}O_5$	L-Ribose		25	12.22
$C_5H_{10}O_5$	D-Xylose		18	12.14
$C_5H_{11}N$	Piperidine		25	11.123
$C_5H_{11}N$	N-Methylpyrrolidine		25	10.46
$C_5H_{11}NO$	4-Methylmorpholine		25	7.38
$C_5H_{11}NO_2$	L-Valine	1	25	2.29
		2	25	9.74

Mol. form.	Name	Step	$t/°C$	pK_a	Mol. form.	Name	Step	$t/°C$	pK_a
$C_5H_{11}NO_2$	*DL*-Norvaline	1		2.36			3	20	9.31
		2		9.72	C_6H_6BrN	2-Bromoaniline		25	2.53
$C_5H_{11}NO_2$	*L*-Norvaline	1	25	2.32	C_6H_6BrN	3-Bromoaniline		25	3.53
		2	25	9.81	C_6H_6BrN	4-Bromoaniline		25	3.89
$C_5H_{11}NO_2$	*N*-Propylglycine	1	25	2.35	C_6H_6ClN	2-Chloroaniline		25	2.66
		2	25	10.19	C_6H_6ClN	3-Chloroaniline		25	3.52
$C_5H_{11}NO_2$	5-Aminopentanoic acid	1	25	4.27	C_6H_6ClN	4-Chloroaniline		25	3.98
		2	25	10.77	C_6H_6FN	2-Fluoroaniline		25	3.20
$C_5H_{11}NO_2$	Betaine		0	1.83	C_6H_6FN	3-Fluoroaniline		25	3.59
$C_5H_{11}NO_2S$	*L*-Methionine	1	25	2.13	C_6H_6FN	4-Fluoroaniline		25	4.65
		2	25	9.27	C_6H_6IN	2-Iodoaniline		25	2.54
$C_5H_{12}N_2O$	Tetramethylurea			2	C_6H_6IN	3-Iodoaniline		25	3.58
$C_5H_{12}N_2O_2$	*L*-Ornithine	1	25	1.71	C_6H_6IN	4-Iodoaniline		25	3.81
		2	25	8.69	$C_6H_6N_2O$	3-Pyridinecarboxamide		20	3.3
		3	25	10.76	$C_6H_6N_2O$	2-Pyridinecarbox-	1	20	3.59
$C_5H_{13}N$	Pentylamine		25	10.63		aldehyde oxime	2	20	10.18
$C_5H_{13}N$	3-Pentanamine		17	10.59	$C_6H_6N_2O_2$	2-Nitroaniline		25	-0.25
$C_5H_{13}N$	3-Methyl-1-butanamine		25	10.60	$C_6H_6N_2O_2$	3-Nitroaniline		25	2.46
$C_5H_{13}N$	2-Methyl-2-butanamine		19	10.85	$C_6H_6N_2O_2$	4-Nitroaniline		25	1.02
$C_5H_{13}N$	2,2-Dimethylpropylamine		25	10.15	C_6H_6O	Phenol		25	9.99
$C_5H_{13}N$	Diethylmethylamine		25	10.35	$C_6H_6O_2$	*p*-Hydroquinone	1	25	9.85
$C_5H_{14}NO$	Choline		25	13.9			2	25	11.4
$C_5H_{14}N_2$	1,5-Pentanediamine	1	25	10.05	$C_6H_6O_2$	Pyrocatechol	1	25	9.34
		2	25	10.93			2	25	12.6
$C_6H_3Cl_3N_2O_2$	4-Amino-3,5,6-trichloro-			3.6	$C_6H_6O_2$	Resorcinol	1	25	9.32
	2-pyridinecarboxlic acid						2	25	11.1
$C_6H_3N_3O_7$	2,4,6-Trinitrophenol		24	0.42	$C_6H_6O_2S$	Benzenesulfinic acid		20	1.3
$C_6H_4Cl_2O$	2,3-Dichlorophenol		25	7.44	$C_6H_6O_3S$	Benzenesulfonic acid		25	0.70
$C_6H_4N_2O_5$	2,4-Dinitrophenol		25	4.07	$C_6H_6O_4$	5-Hydroxy-2-(hydroxy-			7.9
$C_6H_4N_2O_5$	2,5-Dinitrophenol		15	5.15		methyl)-4H-pyran-4-one			
$C_6H_4N_4$	Pteridine		20	4.05	$C_6H_6O_4S$	3-Hydroxybenzene-		25	9.07
C_6H_5BrO	2-Bromophenol		25	8.45		sulfonic acid			
C_6H_5BrO	3-Bromophenol		25	9.03	$C_6H_6O_4S$	4-Hydroxybenzene-		25	9.11
C_6H_5BrO	4-Bromophenol		25	9.37		sulfonic acid			
$C_6H_5Br_2N$	3,5-Dibromoaniline		25	2.34	$C_6H_6O_6$	*cis*-1-Propene-1,2,3-		25	1.95
C_6H_5ClO	2-Chlorophenol		25	8.56		tricarboxylic acid			
C_6H_5ClO	3-Chlorophenol		25	9.12	$C_6H_6O_6$	*trans*-1-Propene-1,2,3-	1	25	2.80
C_6H_5ClO	4-Chlorophenol		25	9.41		tricarboxylic acid	2	25	4.46
$C_6H_5Cl_2N$	2,4-Dichloroaniline		22	2.05	C_6H_6S	Benzenethiol		25	6.62
C_6H_5FO	2-Fluorophenol		25	8.73	$C_6H_7BO_2$	Benzeneboronic acid			8.83
C_6H_5FO	3-Fluorophenol		25	9.29	C_6H_7N	Aniline		25	4.87
C_6H_5FO	4-Fluorophenol		25	9.89	C_6H_7N	2-Methylpyridine		25	6.00
C_6H_5IO	2-Iodophenol		25	8.51	C_6H_7N	3-Methylpyridine		25	5.70
C_6H_5IO	3-Iodophenol		25	9.03	C_6H_7N	4-Methylpyridine		25	5.99
C_6H_5IO	4-Iodophenol		25	9.33	C_6H_7NO	2-Aminophenol	1	20	4.78
C_6H_5NO	2-Pyridinecarboxaldehyde		25	12.68			2	20	9.97
C_6H_5NO	4-Pyridinecarboxaldehyde		30	12.05	C_6H_7NO	3-Aminophenol	1	20	4.37
$C_6H_5NO_2$	Nitrobenzene		0	3.98			2	20	9.82
$C_6H_5NO_2$	2-Pyridinecarboxylic acid	1	20	0.99	C_6H_7NO	4-Aminophenol	1	25	5.48
		2	20	5.39			2	25	10.30
$C_6H_5NO_2$	3-Pyridinecarboxylic acid	1	25	2.00	C_6H_7NO	2-Methoxypyridine		20	3.28
		2	25	4.82	C_6H_7NO	3-Methoxypyridine		25	4.78
$C_6H_5NO_2$	4-Pyridinecarboxylic acid	1	25	1.77	C_6H_7NO	4-Methoxypyridine		25	6.58
		2	25	4.84	$C_6H_7NO_3S$	2-Aminobenzenesulfonic		25	2.46
$C_6H_5NO_3$	2-Nitrophenol		25	7.23		acid			
$C_6H_5NO_3$	3-Nitrophenol		25	8.36	$C_6H_7NO_3S$	3-Aminobenzenesulfonic		25	3.74
$C_6H_5NO_3$	4-Nitrophenol		25	7.15		acid			
$C_6H_5N_3$	1*H*-Benzotriazole		20	1.6	$C_6H_7NO_3S$	4-Aminobenzenesulfonic		25	3.23
$C_6H_5N_5O$	2-Amino-4-	1	20	2.27		acid			
	hydroxypteridine	2	20	7.96	$C_6H_8N_2$	*N*-Methylpyridinamine		20	9.65
$C_6H_5N_5O_2$	Xanthopterin	2	20	6.59	$C_6H_8N_2$	*o*-Phenylenediamine	1	20	4.57

Mol. form.	Name	Step	t/°C	pK_a	Mol. form.	Name	Step	t/°C	pK_a
		2	20	0.80	$C_6H_{13}NO_2$	L-Leucine	1	25	2.33
$C_6H_8N_2$	m-Phenylenediamine	1	20	5.11			2	25	9.74
		2	20	2.50	$C_6H_{13}NO_2$	L-Isoleucine	1	25	2.32
$C_6H_8N_2$	p-Phenylenediamine	1	20	6.31			2	25	9.76
		2	20	2.97	$C_6H_{13}NO_2$	L-Norleucine	1	25	2.34
$C_6H_8N_2$	Phenylhydrazine		15	8.79			2	25	9.83
$C_6H_8O_2$	2,4-Hexadienoic acid		25	4.76	$C_6H_{13}NO_2$	6-Aminohexanoic acid	1	25	4.37
$C_6H_8O_2$	1,3-Cyclohexanedione		25	5.26			2	25	10.80
$C_6H_8O_4$	2,2-Dimethyl-1,3-dioxane-4,6-dione			5.1	$C_6H_{13}NO_4$	N,N-Bis(2-hydroxyethyl)glycine	2	20	8.35
$C_6H_8O_6$	L-Ascorbic acid	1	25	4.04	$C_6H_{13}N_3O_3$	Citrulline	1	25	2.43
		2	16	11.7			2	25	9.69
$C_6H_8O_7$	Citric acid	1	25	3.13	$C_6H_{14}N_2$	cis-1,2-Cyclohexanediamine	1	20	9.93
		2	25	4.76			2	20	6.13
		3	25	6.40	$C_6H_{14}N_2$	$trans$-1,2-Cyclohexanediamine	1	20	9.94
$C_6H_8O_7$	Isocitric acid	1	25	3.29			2	20	6.47
		2	25	4.71	$C_6H_{14}N_2$	cis-2,5-Dimethylpiperazine	1	25	9.66
		3	25	6.40			2	25	5.20
$C_6H_9NO_6$	Nitrilotriacetic acid	1	20	3.03	$C_6H_{14}N_2O_2$	L-Lysine	1	25	2.16
		2	20	3.07			2	25	9.06
		3	20	10.70			3	25	10.54
$C_6H_9NO_6$	L-γ-Carboxyglutamic acid	1	25	1.7	$C_6H_{14}N_4O_2$	L-Arginine	1	25	1.82
		2	25	3.2			2	25	8.99
		3	25	4.75			3	25	12.5
		4	25	9.9	$C_6H_{14}O_6$	D-Mannitol		18	13.5
$C_6H_9N_3$	4,6-Dimethylpyrimidinamine		20	4.82	$C_6H_{15}N$	Hexylamine		25	10.56
					$C_6H_{15}N$	Diisopropylamine		25	11.05
$C_6H_9N_3O_2$	L-Histidine	1	25	1.80	$C_6H_{15}N$	Triethylamine		25	10.75
		2	25	6.04	$C_6H_{15}NO_3$	Triethanolamine		25	7.76
		3	25	9.33	$C_6H_{16}N_2$	1,6-Hexanediamine	1	0	11.86
$C_6H_{10}O_2$	Cyclopentanecarboxylic acid		25	4.99			2	0	10.76
					$C_6H_{16}N_2$	N,N,N',N'-Tetramethyl-1,2-ethanediamine	1	25	10.40
$C_6H_{10}O_3$	Ethyl acetoacetate		25	10.68			2	25	8.26
$C_6H_{10}O_4$	3-Methylglutaric acid		25	4.24	$C_6H_{19}NSi_2$	Hexamethyldisilazane			7.55
$C_6H_{10}O_4$	Adipic acid	1	18	4.41	$C_7HF_5O_2$	Pentafluorobenzoic acid		25	1.75
		2	18	5.41	$C_7H_3Br_2NO$	3,5-Dibromo-4-hydroxybenzonitrile			4.06
$C_6H_{11}NO_2$	2-Piperidinecarboxylic acid	1	25	2.28					
		2	25	10.72	$C_7H_3N_3O_8$	2,4,6-Trinitrobenzoic acid		25	0.65
$C_6H_{11}NO_3$	Adipamic acid		25	4.63	$C_7H_4Cl_3NO_3$	Triclopyr			2.68
$C_6H_{11}NO_4$	2-Aminoadipic acid	1	25	2.14	$C_7H_4N_2O_6$	2,4-Dinitrobenzoic acid		25	1.43
		2	25	4.21	$C_7H_5BrO_2$	2-Bromobenzoic acid		25	2.85
		3	25	9.77	$C_7H_5BrO_2$	3-Bromobenzoic acid		25	3.81
$C_6H_{11}N_3O_4$	N-(N-Glycylglycyl)glycine	1	25	3.225	$C_7H_5BrO_2$	4-Bromobenzoic acid		25	3.96
		2	25	8.09	$C_7H_5ClO_2$	2-Chlorobenzoic acid		25	2.90
$C_6H_{11}N_3O_4$	Glycylasparagine	1	25	2.942	$C_7H_5ClO_2$	3-Chlorobenzoic acid		25	3.84
		2	18	8.44	$C_7H_5ClO_2$	4-Chlorobenzoic acid		25	4.00
$C_6H_{12}N_2$	Triethylenediamine	1		3.0	$C_7H_5FO_2$	2-Fluorobenzoic acid		25	3.27
		2		8.7	$C_7H_5FO_2$	3-Fluorobenzoic acid		25	3.86
$C_6H_{12}N_2O_4S_2$	L-Cystine	1		1	$C_7H_5FO_2$	4-Fluorobenzoic acid		25	4.15
		2		2.1	$C_7H_5F_3O$	2-(Trifluoromethyl)phenol		25	8.95
		3		8.02	$C_7H_5F_3O$	3-(Trifluoromethyl)phenol		25	8.68
		4		8.71	$C_7H_5IO_2$	2-Iodobenzoic acid		25	2.86
$C_6H_{12}O_2$	Hexanoic acid		25	4.85	$C_7H_5IO_2$	3-Iodobenzoic acid		25	3.87
$C_6H_{12}O_2$	4-Methylpentanoic acid		18	4.84	$C_7H_5IO_2$	4-Iodobenzoic acid		25	4.00
$C_6H_{12}O_6$	β-D-Fructose		25	12.27	C_7H_5NO	2-Hydroxybenzonitrile		25	6.86
$C_6H_{12}O_6$	α-D-Glucose		25	12.46	C_7H_5NO	3-Hydroxybenzonitrile		25	8.61
$C_6H_{12}O_6$	D-Mannose		25	12.08	C_7H_5NO	4-Hydroxybenzonitrile		25	7.97
$C_6H_{13}N$	Cyclohexylamine		25	10.64	$C_7H_5NO_3S$	Saccharin		18	11.68
$C_6H_{13}N$	1-Methylpiperidine		25	10.38	$C_7H_5NO_4$	2-Nitrobenzoic acid		25	2.17
$C_6H_{13}N$	1,2-Dimethylpyrrolidine		26	10.20	$C_7H_5NO_4$	3-Nitrobenzoic acid		25	3.46
$C_6H_{13}NO$	N-Ethylmorpholine		25	7.67	$C_7H_5NO_4$	4-Nitrobenzoic acid		25	3.43

Mol. form.	Name	Step	t/°C	pKa
$C_7H_5NO_4$	2,3-Pyridinedicarboxylic acid	1	25	2.43
		2	25	4.78
$C_7H_5NO_4$	2,4-Pyridinedicarboxylic acid	1	25	2.15
$C_7H_5NO_4$	2,6-Pyridinedicarboxylic acid	1	25	2.16
		2	25	4.76
$C_7H_5NO_4$	3,5-Pyridinedicarboxylic acid	1	25	2.80
	Chlorothiazide	1		6.85
$C_7H_6ClN_3O_4S_2$		2		9.45
$C_7H_6F_3N$	3-(Trifluoromethyl)aniline		25	3.49
$C_7H_6F_3N$	4-(Trifluoromethyl)aniline		25	2.45
$C_7H_6N_2$	1H-Benzimidazole		25	5.53
$C_7H_6N_2$	2-Aminobenzonitrile		25	0.77
$C_7H_6N_2$	3-Aminobenzonitrile		25	2.75
$C_7H_6N_2$	4-Aminobenzonitrile		25	1.74
C_7H_6O	Benzaldehyde		25	14.90
$C_7H_6O_2$	Benzoic acid		25	4.204
$C_7H_6O_2$	Salicylaldehyde		25	8.37
$C_7H_6O_2$	3-Hydroxybenzaldehyde		25	8.98
$C_7H_6O_2$	4-Hydroxybenzaldehyde		25	7.61
$C_7H_6O_3$	2-Hydroxybenzoic acid	1	20	2.98
		2	20	13.6
$C_7H_6O_3$	3-Hydroxybenzoic acid	1	25	4.08
		2	19	9.92
$C_7H_6O_3$	4-Hydroxybenzoic acid	1	25	4.57
		2	25	9.46
$C_7H_6O_4$	2,4-Dihydroxybenzoic acid	1	25	3.11
		2	25	8.55
		3	25	14.0
$C_7H_6O_4$	2,5-Dihydroxybenzoic acid	1	25	2.97
$C_7H_6O_4$	3,4-Dihydroxybenzoic acid	1	25	4.48
		2	25	8.83
		3	25	12.6
$C_7H_6O_4$	3,5-Dihydroxybenzoic acid	1	25	4.04
$C_7H_6O_5$	2,4,6-Trihydroxybenzoic acid		25	1.68
$C_7H_6O_5$	3,4,5-Trihydroxybenzoic acid		25	4.41
C_7H_7NO	Benzamide		25	¯13
$C_7H_7NO_2$	Aniline-2-carboxylic acid	1	25	2.17
		2	25	4.85
$C_7H_7NO_2$	Aniline-3-carboxylic acid	1	25	3.07
		2	25	4.79
$C_7H_7NO_2$	Aniline-4-carboxylic acid	1	25	2.50
		2	25	4.87
$C_7H_7NO_3$	4-Amino-2-hydroxybenzoic acid			3.25
$C_7H_8ClN_3O_4S_2$	Hydrochlorothiazide	1		7.9
		2		9.2
$C_7H_8N_4O_2$	Theobromine		18	7.89
$C_7H_8N_4O_2$	Theophylline	1	25	8.77
C_7H_8O	o-Cresol		25	10.29
C_7H_8O	m-Cresol		25	10.09
C_7H_8O	p-Cresol		25	10.26
C_7H_8OS	4-(Methylthio)phenol		25	9.53
$C_7H_8O_2$	2-Methoxyphenol		25	9.98
$C_7H_8O_2$	3-Methoxyphenol		25	9.65
$C_7H_8O_2$	4-Methoxyphenol		25	10.21
C_7H_8S	Benzenemethanethiol		25	9.43
C_7H_9N	Benzylamine		25	9.34

Mol. form.	Name	Step	t/°C	pKa
C_7H_9N	2-Methylaniline		25	4.45
C_7H_9N	3-Methylaniline		25	4.71
C_7H_9N	4-Methylaniline		25	5.08
C_7H_9N	N-Methylaniline		25	4.85
C_7H_9N	2-Ethylpyridine		25	5.89
C_7H_9N	2,3-Dimethylpyridine		25	6.57
C_7H_9N	2,4-Dimethylpyridine		25	6.99
C_7H_9N	2,5-Dimethylpyridine		25	6.40
C_7H_9N	2,6-Dimethylpyridine		25	6.65
C_7H_9N	3,4-Dimethylpyridine		25	6.46
C_7H_9N	3,5-Dimethylpyridine		25	6.15
C_7H_9NO	2-Methoxyaniline		25	4.53
C_7H_9NO	3-Methoxyaniline		25	4.20
C_7H_9NO	4-Methoxyaniline		25	5.36
C_7H_9NS	2-(Methylthio)aniline		25	3.45
C_7H_9NS	4-(Methylthio)aniline		25	4.35
$C_7H_9N_5$	2-Dimethylaminopurine	1	20	4.00
		2	20	10.24
$C_7H_{11}N_3O_2$	L-1-Methylhistidine	1	25	1.69
		2	25	6.48
		3	25	8.85
$C_7H_{11}N_3O_2$	L-3-Methylhistidine	1	25	1.92
		2	25	6.56
		3	25	8.73
$C_7H_{12}O_2$	Cyclohexanecarboxylic acid		25	4.91
$C_7H_{12}O_4$	Heptanedioic acid	1	25	4.71
		2	25	5.58
$C_7H_{12}O_4$	Butylpropanedioic acid	1	5	2.96
$C_7H_{13}NO_4$	α-Ethylglutamic acid	1	25	3.846
		2	25	7.838
$C_7H_{14}O_2$	Heptanoic acid		25	4.89
$C_7H_{14}O_6$	α-Methylglucoside		25	13.71
$C_7H_{15}N$	1-Ethylpiperidine		23	10.45
$C_7H_{15}N$	1,2-Dimethylpiperidine,(±)		25	10.22
$C_7H_{15}NO_3$	Carnitine		25	3.80
$C_7H_{17}N$	Heptylamine		25	10.67
$C_7H_{17}N$	2-Heptanamine		19	10.7
$C_8H_5NO_2$	3-Cyanobenzoic acid		25	3.60
$C_8H_5NO_2$	4-Cyanobenzoic acid		25	3.55
$C_8H_6N_2$	Cinnoline		20	2.37
$C_8H_6N_2$	Quinazoline		29	3.43
$C_8H_6N_2$	Quinoxaline		20	0.56
$C_8H_6N_2$	Phthalazine		20	3.47
$C_8H_6N_4O_5$	Nitrofurantoin			7.2
$C_8H_6O_3$	3-Formylbenzoic acid		25	3.84
$C_8H_6O_3$	4-Formylbenzoic acid		25	3.77
$C_8H_6O_4$	Phthalic acid	1	25	2.943
		2	25	5.432
$C_8H_6O_4$	Isophthalic acid	1	25	3.70
		2	25	4.60
$C_8H_6O_4$	Terephthalic acid	1	25	3.54
		2	25	4.34
$C_8H_7ClO_2$	2-Chlorobenzeneacetic acid		25	4.07
$C_8H_7ClO_2$	3-Chlorobenzeneacetic acid		25	4.14
$C_8H_7ClO_2$	4-Chlorobenzeneacetic acid		25	4.19
$C_8H_7ClO_3$	2-Chlorophenoxyacetic acid		25	3.05
$C_8H_7ClO_3$	3-Chlorophenoxyacetic acid		25	3.10
$C_8H_7NO_4$	2-Nitrobenzeneacetic acid		25	4.00
$C_8H_7NO_4$	3-Nitrobenzeneacetic acid		25	3.97
$C_8H_7NO_4$	4-Nitrobenzeneacetic acid		25	3.85
$C_8H_8F_3N_3O_4S_2$	Hydroflumethiazide	1		8.9

Dissociation Constants of Organic Acids and Bases

Mol. form.	Name	Step	t/°C	pKa
		2		9.7
$C_8H_8N_2$	2-Methyl-1H-benzimidazole	1	25	6.19
$C_8H_8O_2$	o-Toluic acid		25	3.91
$C_8H_8O_2$	m-Toluic acid		25	4.25
$C_8H_8O_2$	p-Toluic acid		25	4.37
$C_8H_8O_2$	Benzeneacetic acid		25	4.31
$C_8H_8O_2$	1-(2-Hydroxyphenyl)ethanone		25	10.06
$C_8H_8O_2$	1-(3-Hydroxyphenyl)ethanone		25	9.19
$C_8H_8O_2$	1-(4-Hydroxyphenyl)ethanone		25	8.05
$C_8H_8O_3$	2-Methoxybenzoic acid		25	4.08
$C_8H_8O_3$	3-Methoxybenzoic acid		25	4.10
$C_8H_8O_3$	4-Methoxybenzoic acid		25	4.50
$C_8H_8O_3$	Phenoxyacetic acid		25	3.17
$C_8H_8O_3$	Mandelic acid		25	3.37
$C_8H_8O_4$	2,5-Hydroxybenzeneacetic acid		25	4.40
C_8H_9NO	Acetanilide		25	0.5
$C_8H_9NO_2$	2-(Methylamino)benzoic acid		25	5.34
$C_8H_9NO_2$	3-(Methylamino)benzoic acid		25	5.10
$C_8H_9NO_2$	4-(Methylamino)benzoic acid		25	5.04
$C_8H_9NO_2$	N-Phenylglycine	1	25	1.83
		2		4.39
$C_8H_{10}BrN$	4-Bromo-N,N-dimethylaniline		25	4.23
$C_8H_{10}ClN$	3-Chloro-N,N-dimethylaniline		20	3.83
$C_8H_{10}ClN$	4-Chloro-N,N-dimethylaniline		20	4.39
$C_8H_{10}N_2O_2$	N,N-Dimethyl-3-nitroaniline		25	2.62
$C_8H_{11}N$	N-Ethylaniline		25	5.12
$C_8H_{11}N$	N,N-Dimethylaniline		25	5.07
$C_8H_{11}N$	2,6-Dimethylaniline		25	3.89
$C_8H_{11}N$	Benzeneethanamine		25	9.83
$C_8H_{11}N$	2,4,6-Trimethylpyridine		25	7.43
$C_8H_{11}NO$	2-Ethoxyaniline		28	4.43
$C_8H_{11}NO$	3-Ethoxyaniline		25	4.18
$C_8H_{11}NO$	4-Ethoxyaniline		28	5.20
$C_8H_{11}NO$	4-(2-Aminoethyl)phenol	1	25	9.74
		2	25	10.52
$C_8H_{11}NO$	2-(2-Methoxyethyl)pyridine			5.5
$C_8H_{11}NO_2$	Dopamine	1	25	8.9
		2	25	10.6
$C_8H_{11}NO_3$	Norepinephrine	1	25	8.64
		2	25	9.70
$C_8H_{11}N_3O_6$	6-Azauridine			6.70
$C_8H_{11}N_5$	Phenylbiguanide	1		10.76
		2		2.13
$C_8H_{12}N_2O_3$	Barbital		25	7.43
$C_8H_{12}O_2$	5,5-Dimethyl-1,3-cyclohexanedione		25	5.15
$C_8H_{13}NO_2$	Arecoline			6.84
$C_8H_{14}O_2S_2$	Thioctic acid			5.4
$C_8H_{14}O_4$	Octanedioic acid	1	25	4.52
$C_8H_{15}NO$	Tropine		15	3.80
$C_8H_{15}NO$	Pseudotropine		15	3.80
$C_8H_{16}N_2O_3$	N-Glycylleucine		25	3.18
$C_8H_{16}N_2O_3$	N-Leucylglycine	1	25	3.25

Mol. form.	Name	Step	t/°C	pKa
		2	25	8.2
$C_8H_{16}N_2O_4S_2$	Homocystine	1	25	1.59
		2	25	2.54
		3	25	8.52
		4	25	9.44
$C_8H_{16}O_2$	Octanoic acid		25	4.89
$C_8H_{16}O_2$	2-Propylpentanoic acid			4.6
$C_8H_{17}N$	2-Propylpiperidine,(S)			10.9
$C_8H_{17}N$	2,2,4-Trimethylpiperidine		30	11.04
$C_8H_{17}NO$	trans-6-Propyl-3-piperidinol,(3S)			10.3
$C_8H_{19}N$	Octylamine		25	10.65
$C_8H_{19}N$	N-Methyl-2-heptanamine		17	10.99
$C_8H_{19}N$	Dibutylamine		21	11.25
$C_8H_{20}N_2$	1,8-Octanediamine	1	20	11.00
		2	20	10.1
C_9H_6BrN	3-Bromoquinoline		25	2.69
$C_9H_7ClO_2$	trans-o-Chlorocinnamic acid		25	4.23
$C_9H_7ClO_2$	trans-m-Chlorocinnamic acid		25	4.29
$C_9H_7ClO_2$	trans-p-Chlorocinnamic acid		25	4.41
C_9H_7N	Quinoline		20	4.90
C_9H_7N	Isoquinoline		20	5.40
C_9H_7NO	2-Quinolinol	1	20	-0.31
		2	20	11.76
C_9H_7NO	3-Quinolinol	1	20	4.28
		2	20	8.08
C_9H_7NO	4-Quinolinol	1	20	2.23
		2	20	11.28
C_9H_7NO	6-Quinolinol	1	20	5.15
		2	20	8.90
C_9H_7NO	8-Quinolinol	1	25	4.91
		2	25	9.81
C_9H_7NO	7-Isoquinolinol	1	20	5.68
		2	20	8.90
$C_9H_7NO_3$	2-Cyanophenoxyacetic acid		25	2.98
$C_9H_7NO_3$	3-Cyanophenoxyacetic acid		25	3.03
$C_9H_7NO_3$	4-Cyanophenoxyacetic acid		25	2.93
$C_9H_7N_7O_2S$	Azathioprine			8.2
$C_9H_8N_2$	2-Quinolinamine		20	7.34
$C_9H_8N_2$	3-Quinolinamine		20	4.91
$C_9H_8N_2$	4-Quinolinamine		20	9.17
$C_9H_8N_2$	1-Isoquinolinamine		20	7.62
$C_9H_8N_2$	3-Isoquinolinamine		20	5.05
$C_9H_8O_2$	cis-Cinnamic acid		25	3.88
$C_9H_8O_2$	trans-Cinnamic acid		25	4.44
$C_9H_8O_2$	α-Methylenebenzene-acetic acid			4.35
$C_9H_8O_4$	2-(Acetyloxy)benzoic acid		25	3.48
$C_9H_9Br_2NO_3$	3,5-Dibromo-L-tyrosine	1		2.17
		2		6.45
		3		7.60
$C_9H_9ClO_2$	3-(2-Chlorophenyl)-propanoic acid		25	4.58
$C_9H_9ClO_2$	3-(3-Chlorophenyl)-propanoic acid		25	4.59
$C_9H_9ClO_2$	3-(4-Chlorophenyl)-propanoic acid		25	4.61
$C_9H_9I_2NO_3$	L-3,5-Diiodotyrosine	1	25	2.12

Mol. form.	Name	Step	$t/°C$	pK_a	Mol. form.	Name	Step	$t/°C$	pK_a
		2	25	5.32	$C_{10}H_8O$	1-Naphthol		25	9.39
		3	25	9.48	$C_{10}H_8O$	2-Naphthol		25	9.63
$C_9H_9NO_3$	N-Benzoylglycine		25	3.62	$C_{10}H_9N$	1-Naphthylamine		25	3.92
$C_9H_9NO_4$	3-(2-Nitrophenyl)- propanoic acid		25	4.50	$C_{10}H_9N$	2-Naphthylamine		25	4.16
					$C_{10}H_9N$	2-Methylquinoline		20	5.83
$C_9H_9NO_4$	3-(4-Nitrophenyl)- propanoic acid		25	4.47	$C_{10}H_9N$	4-Methylquinoline		20	5.67
					$C_{10}H_9N$	5-Methylquinoline		20	5.20
$C_9H_9N_3O_2$	Carbendazim			4.48	$C_{10}H_9NO$	5-Amino-1-naphthol		25	3.97
$C_9H_9N_3O_2S_2$	Sulfathiazole			7.2	$C_{10}H_9NO$	6-Methoxyquinoline		20	5.03
$C_9H_{10}INO_3$	L-3-Iodotyrosine	1	25	2.2	$C_{10}H_9NO_2$	1H-Indole-3-acetic acid			4.75
		2	25	8.7	$C_{10}H_{10}O_2$	o-Methylcinnamic acid		25	4.50
		3	25	9.1	$C_{10}H_{10}O_2$	m-Methylcinnamic acid		25	4.44
$C_9H_{10}N_2$	2-Ethylbenzimidazole		25	6.18	$C_{10}H_{10}O_2$	p-Methylcinnamic acid		25	4.56
$C_9H_{10}O_2$	3,5-Dimethylbenzoic acid		25	4.32	$C_{10}H_{12}N_2$	Tryptamine		25	10.2
$C_9H_{10}O_2$	Benzenepropanoic acid		25	4.66	$C_{10}H_{12}N_2O$	5-Hydroxytryptamine	1	25	9.8
$C_9H_{10}O_2$	α-Methylbenzeneacetic acid		25	4.64			2	25	11.1
$C_9H_{10}O_3$	α-Hydroxy-α-methyl- benezeneacetic acid		25	3.47	$C_{10}H_{12}N_2O_5$	Dinoseb			4.62
					$C_{10}H_{12}N_4O_3$	Dideoxyinosine			9.12
$C_9H_{11}Cl_2N_3O_4S_2$	Methylclothiazide			9.4	$C_{10}H_{12}O$	5,6,7,8-Tetrahydro-2- naphthalenol		25	10.48
$C_9H_{11}N$	N-Allylaniline		25	4.17					
$C_9H_{11}N$	1-Indanamine		22	9.21	$C_{10}H_{12}O_2$	Benzenebutanoic acid		25	4.76
$C_9H_{11}NO_2$	4-(Dimethylamino)- benzoic acid	1		6.03	$C_{10}H_{12}O_5$	Propyl 3,4,5-trihydroxy- benzoate			8.11
		2		11.49	$C_{10}H_{13}N_5O_4$	Adenosine	1	25	3.6
$C_9H_{11}NO_2$	Ethyl 4-aminobenzoate			2.5			2	25	12.4
$C_9H_{11}NO_2$	L-Phenylalanine	1	25	2.20	$C_{10}H_{14}N_2$	L-Nicotine	1		8.02
		2	25	9.31			2		3.12
$C_9H_{11}NO_3$	L-Tyrosine	1	25	2.20	$C_{10}H_{14}N_5O_7P$	5'-Adenylic acid	1		3.8
		2	25	9.11			2		6.2
		3	25	10.1	$C_{10}H_{14}O$	2-tert-Butylphenol		25	10.62
$C_9H_{11}NO_4$	Levodopa	1	25	2.32	$C_{10}H_{14}O$	3-tert-Butylphenol		25	10.12
		2	25	8.72	$C_{10}H_{14}O$	4-tert-Butylphenol		25	10.23
		3	25	9.96	$C_{10}H_{15}N$	N-tert-Butylaniline		25	7.00
		4	25	11.79	$C_{10}H_{15}N$	N,N-Diethylaniline		25	6.57
$C_9H_{12}N_2O_2$	Tyrosineamide		25	7.33	$C_{10}H_{15}NO$	d-Ephedrine		10	10.139
$C_9H_{13}N$	N-Isopropylaniline		25	5.77	$C_{10}H_{15}NO$	l-Ephedrine		10	9.958
$C_9H_{13}NO_3$	Epinephrine	1	25	8.66	$C_{10}H_{17}N_3O_6S$	l-Glutathione	1	25	2.12
		2	25	9.95			2	25	3.59
$C_9H_{13}N_2O_9P$	5'-Uridylic acid	1		6.4			3	25	8.75
		2		9.5			4	25	9.65
$C_9H_{13}N_3O_5$	Cytidine	1		4.22	$C_{10}H_{18}N_4O_5$	L-Argininosuccinic acid	1	25	1.62
		2		12.5			2	25	2.70
$C_9H_{14}ClNO$	Phenylpropanolamine hydrochloride			9.44			3	25	4.26
							4	25	9.58
$C_9H_{14}N_2O_3$	Metharbital			8.45	$C_{10}H_{18}O_4$	Sebacic acid	1	25	4.59
$C_9H_{14}N_3O_8P$	3'-Cytidylic acid	1		0.8			2	25	5.59
		2		4.28	$C_{10}H_{19}N$	Bornylamine		25	10.17
		3		6.0	$C_{10}H_{19}N$	Neobornylamine		25	10.01
$C_9H_{14}N_4O_3$	Carnosine	1	20	2.73	$C_{10}H_{21}N$	Butylcyclohexylamine		25	11.23
		2	20	6.87	$C_{10}H_{21}N$	1,2,2,6,6-Pentamethyl- piperidine		30	11.25
		3	20	9.73					
$C_9H_{15}NO_3S$	Captopril	1		3.7	$C_{10}H_{23}N$	Decylamine		25	10.64
		2		9.8	$C_{11}H_8N_2$	1H-Perimidine		20	6.35
$C_9H_{15}N_5O$	Minoxidil			4.61	$C_{11}H_8O_2$	1-Naphthalenecarboxylic acid		25	3.69
$C_9H_{16}O_4$	Nonanedioic acid	1	25	4.53					
		2	25	5.33	$C_{11}H_8O_2$	2-Naphthalenecarboxylic acid		25	4.16
$C_9H_{18}O_2$	Nonanoic acid		25	4.96					
$C_9H_{19}N$	N-Butylpiperidine		23	10.47	$C_{11}H_{11}N$	Methyl-1-naphthylamine		27	3.67
$C_9H_{19}N$	2,2,6,6-Tetramethyl- piperidine		25	11.07	$C_{11}H_{12}INO_2$	Iopanoic acid			4.8
					$C_{11}H_{12}N_2O_2$	L-Tryptophan	1	25	2.46
$C_9H_{21}N$	Nonylamine		25	10.64			2	25	9.41
$C_{10}H_7NO_2$	8-Quinolinecarboxylic acid		25	1.82					

Mol. form.	Name	Step	$t/°C$	pK_a
$C_{11}H_{12}N_4O_3S$	Sulfamethoxypyridazine			6.7
$C_{11}H_{13}F_3N_2O_3S$	Mefluidide			4.6
$C_{11}H_{13}NO_3$	Hydrastinine			11.38
$C_{11}H_{13}N_3O_3S$	Sulfisoxazole			5
$C_{11}H_{14}N_2O$	Cytisine	1		6.11
		2		13.08
$C_{11}H_{14}O_2$	2-*tert*-Butylbenzoic acid		25	3.54
$C_{11}H_{14}O_2$	3-*tert*-Butylbenzoic acid		25	4.20
$C_{11}H_{14}O_2$	4-*tert*-Butylbenzoic acid		25	4.38
$C_{11}H_{16}N_2O_2$	Pilocarpine	1	25	1.6
		2	25	6.9
$C_{11}H_{16}N_4O_4$	Pentostatin			5.2
$C_{11}H_{17}N$	N,N-Diethyl-2-methyl-aniline		25	7.24
$C_{11}H_{17}NO_3$	Isoproterenol			8.64
$C_{11}H_{17}N_3O_8$	Tetrodotoxin			8.76
$C_{11}H_{18}ClNO_3$	Methoxamine hydrochloride		25	9.2
$C_{11}H_{18}N_2O_3$	Amobarbital		25	8.0
$C_{11}H_{25}N$	Undecylamine		25	10.63
$C_{11}H_{26}NO_2PS$	Methylphosphonothioic acid S[2-[bis(1-isopropyl)amino]-ethyl], O-ethylester			7.9
$C_{12}H_6Cl_4O_2S$	Bithionol	1		4.82
		2		10.50
$C_{12}H_8N_2$	1,10-Phenanthroline		25	4.84
$C_{12}H_8N_2$	Phenazine		20	1.20
$C_{12}H_{10}O$	2-Hydroxybiphenyl		25	10.01
$C_{12}H_{10}O$	3-Hydroxybiphenyl		25	9.64
$C_{12}H_{10}O$	4-Hydroxybiphenyl		25	9.55
$C_{12}H_{11}N$	Diphenylamine		25	0.79
$C_{12}H_{11}N$	2-Aminobiphenyl		25	3.83
$C_{12}H_{11}N$	3-Aminobiphenyl		18	4.25
$C_{12}H_{11}N$	4-Aminobiphenyl		18	4.35
$C_{12}H_{11}N$	2-Benzylpyridine		25	5.13
$C_{12}H_{11}N_3$	4-Aminoazobenzene		25	2.82
$C_{12}H_{12}N_2$	p-Benzidine	1	20	4.65
		2	20	3.43
$C_{12}H_{12}N_2O_3$	Phenobarbital	1		7.3
		2		11.8
$C_{12}H_{13}I_3N_2O_3$	Iocetamic acid			4
$C_{12}H_{13}N$	N,N-Dimethyl-1-naphthylamine		25	4.83
$C_{12}H_{13}N$	N,N-Dimethyl-2-naphthylamine		25	4.566
$C_{12}H_{14}N_4O_2S$	Sulfamethazine	1		7.4
		2		2.65
$C_{12}H_{14}N_4O_3S$	Sulfacytine			6.9
$C_{12}H_{17}N_3O_4$	Agaritine	1		3.4
		2		8.86
$C_{12}H_{20}N_2O_2$	Aspergillic acid			5.5
$C_{12}H_{21}N_5O_2S_2$	Nizatidine	1		2.1
		2		6.8
$C_{12}H_{22}O_{11}$	Sucrose		25	12.7
$C_{12}H_{22}O_{11}$	α-Maltose		21	12.05
$C_{12}H_{23}N$	Dicyclohexylamine			10.4
$C_{12}H_{27}N$	Dodecylamine		25	10.63
$C_{13}H_9N$	Acridine		20	5.58
$C_{13}H_9N$	Phenanthridine		20	5.58
$C_{13}H_{10}N_2$	9-Acridinamine		20	9.99
$C_{13}H_{10}N_2$	2-Phenylbenzimidazole	1	25	5.23
		2	25	11.91

Mol. form.	Name	Step	$t/°C$	pK_a
$C_{13}H_{10}O_2$	2-Phenylbenzoic acid		25	3.46
$C_{13}H_{10}O_3$	2-Phenoxybenzoic acid		25	3.53
$C_{13}H_{10}O_3$	3-Phenoxybenzoic acid		25	3.95
$C_{13}H_{10}O_3$	4-Phenoxybenzoic acid		25	4.57
$C_{13}H_{11}N_3$	3,6-Acridinediamine		20	9.65
$C_{13}H_{12}Cl_2O_4$	Ethacrynic acid			3.50
$C_{13}H_{12}N_2O$	Harmine			7.70
$C_{13}H_{12}N_2O_3S$	Sulfabenzamide		25	4.57
$C_{13}H_{13}N$	4-Benzylaniline		25	2.17
$C_{13}H_{14}N_2O_{13}$	Harmaline			4.2
$C_{13}H_{15}N_3O_3$	Imazapyr	1		1.9
		2		3.6
$C_{13}H_{16}ClNO$	Ketamine			7.5
$C_{13}H_{19}NO_4S$	4-[(Dipropylamino)-sulfonyl]benzoic acid			5.8
$C_{13}H_{21}N$	2,6-Di-*tert*-butylpyridine			3.58
$C_{13}H_{29}N$	(Tridecyl)amine		25	10.63
$C_{14}H_{12}F_3NO_4S_2$	Perfluidone			2.5
$C_{14}H_{12}O_2$	α-Phenylbenzeneacetic acid		25	3.94
$C_{14}H_{12}O_3$	α-Hydroxy-α-phenyl-benzeneacetic acid		25	3.04
$C_{14}H_{18}N_4O_3$	Trimethoprim			6.6
$C_{14}H_{19}NO_2$	Methylphenidate			8.9
$C_{14}H_{21}N_3O_3S$	Tolazamide		25	3.6
$C_{14}H_{22}N_2O_3$	Atenolol			9.6
$C_{14}H_{31}N$	Tetradecylamine		25	10.62
$C_{15}H_{10}ClN_3O_3$	Clonazepam	1		1.5
		2		10.5
$C_{15}H_{11}I_4NO_4$	L-Thyroxine	1	25	2.2
		2	25	6.45
		3	25	10.1
$C_{15}H_{14}O_3$	Fenoprofen			4.5
$C_{15}H_{15}NO_2$	Mefenamic acid			4.2
$C_{15}H_{15}N_3O_2$	Methyl Red	1		2.5
		2		9.5
$C_{15}H_{17}ClN_4$	NeutralRed			6.7
$C_{15}H_{19}NO_2$	Tropacocaine		15	4.32
$C_{15}H_{19}N_3O_3$	Imazethapyr	1		2.1
		2		3.9
$C_{15}H_{21}N_3O_2$	Physostigmine	1		6.12
		2		12.24
$C_{15}H_{26}N_2$	Sparteine	1	20	2.24
		2	20	9.46
$C_{15}H_{33}N$	Pentadecylamine		25	10.61
$C_{16}H_{13}ClN_2O$	Valium			3.4
$C_{16}H_{14}ClN_3O$	Chlorodiazepoxide			4.8
$C_{16}H_{16}N_2O_2$	Lysergic acid	1		3.44
		2		7.68
$C_{16}H_{17}N_3O_4S$	Cephalexin	1		5.2
		2		7.3
$C_{16}H_{19}N_3O_4S$	Cephradine	1		2.63
		2		7.27
$C_{16}H_{22}N_2$	Lycodine	1		3.97
		2		8.08
$C_{16}H_{35}N$	Hexadecylamine		25	10.61
$C_{17}H_{17}NO_2$	Apomorphine	1		7.0
		2		8.92
$C_{17}H_{19}NO_3$	Piperine		18	12.22
$C_{17}H_{19}NO_3$	Morphine	1	25	8.21
		2	20	9.85
$C_{17}H_{20}N_4O_6$	Riboflavin	1		1.7

Mol. form.	Name	Step	t/°C	pKa	Mol. form.	Name	Step	t/°C	pKa
		2	25	9.69	$C_{21}H_{23}ClFNO_2$	Haloperidol			8.3
$C_{17}H_{20}O_6$	Mycophenolic acid			4.5	$C_{21}H_{31}NO_4$	Furethidine			7.48
$C_{17}H_{23}NO_3$	Hyoscyamine		21	9.7	$C_{21}H_{35}N_3O_7$	Lisinopril	1		2.5
$C_{17}H_{27}NO_4$	Nadolol			9.67			2		4.0
$C_{18}H_{19}ClN_4$	Clozapine	1		3.70			3		6.7
		2		7.60			4		10.1
$C_{18}H_{21}NO_3$	Codeine			8.21	$C_{22}H_{18}O_4$	o-Cresolphthalein			9.4
$C_{18}H_{21}N_3O$	Dibenzepin			8.25	$C_{22}H_{22}FN_3O_2$	Droperidol			7.64
$C_{18}H_{32}O_2$	Linoleic acid		25	4.77	$C_{22}H_{23}NO_7$	Noscapine			7.8
$C_{18}H_{33}ClN_2O_5S$	Clindamycin			7.6	$C_{22}H_{25}NO_6$	Colchicine		20	12.36
$C_{18}H_{39}N$	Octadecylamine		25	10.60	$C_{22}H_{25}N_3O$	Benzpiperylon	1		6.73
$C_{19}H_{10}Br_4O_5S$	Bromophenol Blue			4.0			2		9.13
$C_{19}H_{14}O_5S$	Phenol Red			7.9	$C_{22}H_{33}NO_2$	Atisine			12.2
$C_{19}H_{16}ClNO_4$	Indomethacin			4.5	$C_{23}H_{26}N_2O_4$	Brucine	1		6.04
$C_{19}H_{17}N_3O_4S_2$	Cephaloridine			3.2			2		11.07
$C_{19}H_{20}N_2O_2$	Phenylbutazone			4.5	$C_{24}H_{40}O_4$	Deoxycholic acid		20	5.15
$C_{19}H_{21}N$	Protriptyline			8.2	$C_{24}H_{40}O_5$	Cholic acid		20	4.98
$C_{19}H_{21}NO_3$	Thebaine		15	6.05	$C_{25}H_{29}I_2NO_3$	Amiodarone		25	6.56
$C_{19}H_{22}N_2O$	Cinchonine	1		5.85	$C_{25}H_{41}NO_9$	Aconine			9.52
		2		9.92	$C_{26}H_{43}NO_6$	Glycocholic acid			4.4
$C_{19}H_{22}N_2O$	Cinchonidine	1		5.80	$C_{26}H_{45}NO_7S$	Taurocholic acid			1.4
		2		10.03	$C_{27}H_{28}Br_2O_5S$	Bromothymol Blue			7.0
$C_{19}H_{22}N_2O_2$	Cupreine			6.57	$C_{27}H_{38}N_2O_4$	Verapamil			8.6
$C_{19}H_{22}O_6$	Gibberellic acid			4.0	$C_{29}H_{32}O_{13}$	Etoposide			9.8
$C_{19}H_{23}N_3O_2$	Ergometrinine			7.3	$C_{29}H_{40}N_2O_4$	Emetine	1		5.77
$C_{19}H_{23}N_3O_2$	Ergonovine			6.8			2		6.64
$C_{20}H_{14}O_4$	Phenolphthalein		25	9.7	$C_{30}H_{23}BrO_4$	Bromadiolone		21	4.04
$C_{20}H_{21}NO_4$	Papaverine			6.4	$C_{30}H_{48}O_3$	Oleanolic acid			2.52
$C_{20}H_{23}N$	Amitriptyline			9.4	$C_{31}H_{36}N_2O_{11}$	Novobiocin	1		4.3
$C_{20}H_{23}N_7O_7$	Folinic acid	1		3.1			2		9.1
		2		4.8	$C_{32}H_{32}O_{13}S$	Teniposide			10.13
		3		10.4	$C_{33}H_{40}N_2O_9$	Reserpine			6.6
$C_{20}H_{24}N_2O_2$	Quinine	1	25	8.52	$C_{34}H_{47}NO_{11}$	Aconitine			5.88
		2	25	4.13	$C_{36}H_{51}NO_{11}$	Veratridine			9.54
$C_{20}H_{24}N_2O_2$	Quinidine	1	20	5.4	$C_{37}H_{67}NO_{13}$	Erythromycin			8.8
		2	20	10.0	$C_{43}H_{58}N_4O_{12}$	Rifampin	1		1.7
$C_{20}H_{26}N_2O_2$	Hydroquinine			5.33			2		7.9
$C_{21}H_{14}Br_4O_5S$	Bromocresol Green			4.7	$C_{45}H_{73}NO_{15}$	Solanine		15	6.66
$C_{21}H_{16}Br_2O_5S$	Bromocresol Purple			6.3	$C_{46}H_{56}N_4O_{10}$	Vincristine			5.4
$C_{21}H_{18}O_5S$	CresolRed			8.3	$C_{46}H_{58}N_4O_9$	Vinblastine	1		5.4
$C_{21}H_{21}NO_6$	Hydrastine			7.8			2		7.4
$C_{21}H_{22}N_2O_2$	Strychnine		25	8.26					

ACTIVITY COEFFICIENTS OF ACIDS, BASES, AND SALTS

Petr Vanýsek

This table gives mean activity coefficients at 25 °C for molalities in the range 0.1 to 1.0. See the following table for definitions, references, and data over a wider concentration range.

	0.1	0.2	0.3	0.4	0.5	0.6	0.7	0.8	0.9	1.0
$AgNO_3$	0.734	0.657	0.606	0.567	0.536	0.509	0.485	0.464	0.446	0.429
$AlCl_3$	0.337	0.305	0.302	0.313	0.331	0.356	0.388	0.429	0.479	0.539
$Al_2(SO_4)_3$	0.035	0.0225	0.0176	0.0153	0.0143	0.014	0.0142	0.0149	0.0159	0.0175
$BaCl_2$	0.500	0.444	0.419	0.405	0.397	0.391	0.391	0.391	0.392	0.395
$BeSO_4$	0.150	0.109	0.0885	0.0769	0.0692	0.0639	0.0600	0.0570	0.0546	0.0530
$CaCl_2$	0.518	0.472	0.455	0.448	0.448	0.453	0.460	0.470	0.484	0.500
$CdCl_2$	0.2280	0.1638	0.1329	0.1139	0.1006	0.0905	0.0827	0.0765	0.0713	0.0669
$Cd(NO_3)_2$	0.513	0.464	0.442	0.430	0.425	0.423	0.423	0.425	0.428	0.433
$CdSO_4$	0.150	0.103	0.0822	0.0699	0.0615	0.0553	0.0505	0.0468	0.0438	0.0415
$CoCl_2$	0.522	0.479	0.463	0.459	0.462	0.470	0.479	0.492	0.511	0.531
$CrCl_3$	0.331	0.298	0.294	0.300	0.314	0.335	0.362	0.397	0.436	0.481
$Cr(NO_3)_3$	0.319	0.285	0.279	0.281	0.291	0.304	0.322	0.344	0.371	0.401
$Cr_2(SO_4)_3$	0.0458	0.0300	0.0238	0.0207	0.0190	0.0182	0.0181	0.0185	0.0194	0.0208
$CsBr$	0.754	0.694	0.654	0.626	0.603	0.586	0.571	0.558	0.547	0.538
$CsCl$	0.756	0.694	0.656	0.628	0.606	0.589	0.575	0.563	0.553	0.544
CsI	0.754	0.692	0.651	0.621	0.599	0.581	0.567	0.554	0.543	0.533
$CsNO_3$	0.733	0.655	0.602	0.561	0.528	0.501	0.478	0.458	0.439	0.422
$CsOH$	0.795	0.761	0.744	0.739	0.739	0.742	0.748	0.754	0.762	0.771
$CsOAc$	0.799	0.771	0.761	0.759	0.762	0.768	0.776	0.783	0.792	0.802
Cs_2SO_4	0.456	0.382	0.338	0.311	0.291	0.274	0.262	0.251	0.242	0.235
$CuCl_2$	0.508	0.455	0.429	0.417	0.411	0.409	0.409	0.410	0.413	0.417
$Cu(NO_3)_2$	0.511	0.460	0.439	0.429	0.426	0.427	0.431	0.437	0.445	0.455
$CuSO_4$	0.150	0.104	0.0829	0.0704	0.0620	0.0559	0.0512	0.0475	0.0446	0.0423
$FeCl_2$	0.5185	0.473	0.454	0.448	0.450	0.454	0.463	0.473	0.488	0.506
HBr	0.805	0.782	0.777	0.781	0.789	0.801	0.815	0.832	0.850	0.871
HCl	0.796	0.767	0.756	0.755	0.757	0.763	0.772	0.783	0.795	0.809
$HClO_4$	0.803	0.778	0.768	0.766	0.769	0.776	0.785	0.795	0.808	0.823
HI	0.818	0.807	0.811	0.823	0.839	0.860	0.883	0.908	0.935	0.963
HNO_3	0.791	0.754	0.735	0.725	0.720	0.717	0.717	0.718	0.721	0.724
H_2SO_4	0.2655	0.2090	0.1826	—	0.1557	—	0.1417	—	—	0.1316
KBr	0.772	0.722	0.693	0.673	0.657	0.646	0.636	0.629	0.622	0.617
KCl	0.770	0.718	0.688	0.666	0.649	0.637	0.626	0.618	0.610	0.604
$KClO_3$	0.749	0.681	0.635	0.599	0.568	0.541	0.518	—	—	—
K_2CrO_4	0.456	0.382	0.340	0.313	0.292	0.276	0.263	0.253	0.243	0.235
KF	0.775	0.727	0.700	0.682	0.670	0.661	0.654	0.650	0.646	0.645
$K_3Fe(CN)_6$	0.268	0.212	0.184	0.167	0.155	0.146	0.140	0.135	0.131	0.128
$K_4Fe(CN)_6$	0.139	0.0993	0.0808	0.0693	0.0614	0.0556	0.0512	0.0479	0.0454	—
KH_2PO_4	0.731	0.653	0.602	0.561	0.529	0.501	0.477	0.456	0.438	0.421
KI	0.778	0.733	0.707	0.689	0.676	0.667	0.660	0.654	0.649	0.645
KNO_3	0.739	0.663	0.614	0.576	0.545	0.519	0.496	0.476	0.459	0.443
$KOAc$	0.796	0.766	0.754	0.750	0.751	0.754	0.759	0.766	0.774	0.783
KOH	0.798	0.760	0.742	0.734	0.732	0.733	0.736	0.742	0.749	0.756
$KSCN$	0.769	0.716	0.685	0.663	0.646	0.633	0.623	0.614	0.606	0.599
K_2SO_4	0.441	0.360	0.316	0.286	0.264	0.246	0.232	—	—	—
$LiBr$	0.796	0.766	0.756	0.752	0.753	0.758	0.767	0.777	0.789	0.803
$LiCl$	0.790	0.757	0.744	0.740	0.739	0.743	0.748	0.755	0.764	0.774
$LiClO_4$	0.812	0.794	0.792	0.798	0.808	0.820	0.834	0.852	0.869	0.887
LiI	0.815	0.802	0.804	0.813	0.824	0.838	0.852	0.870	0.888	0.910
$LiNO_3$	0.788	0.752	0.736	0.728	0.726	0.727	0.729	0.733	0.737	0.743
$LiOH$	0.760	0.702	0.665	0.638	0.617	0.599	0.585	0.573	0.563	0.554
$LiOAc$	0.784	0.742	0.721	0.709	0.700	0.691	0.689	0.688	0.688	0.689
Li_2SO_4	0.468	0.398	0.361	0.337	0.319	0.307	0.297	0.289	0.282	0.277
$MgCl_2$	0.529	0.489	0.477	0.475	0.481	0.491	0.506	0.522	0.544	0.570

	0.1	0.2	0.3	0.4	0.5	0.6	0.7	0.8	0.9	1.0
$MgSO_4$	0.150	0.107	0.0874	0.0756	0.0675	0.0616	0.0571	0.0536	0.0508	0.0485
$MnCl_2$	0.516	0.469	0.450	0.442	0.440	0.443	0.448	0.455	0.466	0.479
$MnSO_4$	0.150	0.105	0.0848	0.0725	0.0640	0.0578	0.0530	0.0493	0.0463	0.0439
NH_4Cl	0.770	0.718	0.687	0.665	0.649	0.636	0.625	0.617	0.609	0.603
NH_4NO_3	0.740	0.677	0.636	0.606	0.582	0.562	0.545	0.530	0.516	0.504
$(NH_4)_2SO_4$	0.439	0.356	0.311	0.280	0.257	0.240	0.226	0.214	0.205	0.196
$NaBr$	0.782	0.741	0.719	0.704	0.697	0.692	0.689	0.687	0.687	0.687
$NaCl$	0.778	0.735	0.710	0.693	0.681	0.673	0.667	0.662	0.659	0.657
$NaClO_3$	0.772	0.720	0.688	0.664	0.645	0.630	0.617	0.606	0.597	0.589
$NaClO_4$	0.775	0.729	0.701	0.683	0.668	0.656	0.648	0.641	0.635	0.629
Na_2CrO_4	0.464	0.394	0.353	0.327	0.307	0.292	0.280	0.269	0.261	0.253
NaF	0.765	0.710	0.676	0.651	0.632	0.616	0.603	0.592	0.582	0.573
NaH_2PO_4	0.744	0.675	0.629	0.593	0.563	0.539	0.517	0.499	0.483	0.468
NaI	0.787	0.751	0.735	0.727	0.723	0.723	0.724	0.727	0.731	0.736
$NaNO_3$	0.762	0.703	0.666	0.638	0.617	0.599	0.583	0.570	0.558	0.548
$NaOAc$	0.791	0.757	0.744	0.737	0.735	0.736	0.740	0.745	0.752	0.757
$NaOH$	0.766	0.727	0.708	0.697	0.690	0.685	0.681	0.679	0.678	0.678
$NaSCN$	0.787	0.750	—	0.720	0.715	0.712	0.710	0.710	0.711	0.712
Na_2SO_4	0.445	0.365	0.320	0.289	0.266	0.248	0.233	0.221	0.210	0.201
$NiCl_2$	0.522	0.479	0.463	0.460	0.464	0.471	0.482	0.496	0.515	0.563
$NiSO_4$	0.150	0.105	0.0841	0.0713	0.0627	0.0562	0.0515	0.0478	0.0448	0.0425
$Pb(NO_3)_2$	0.395	0.308	0.260	0.228	0.205	0.187	0.172	0.160	0.150	0.141
$RbBr$	0.763	0.706	0.673	0.650	0.632	0.617	0.605	0.595	0.586	0.578
$RbCl$	0.764	0.709	0.675	0.652	0.634	0.620	0.608	0.599	0.590	0.583
RbI	0.762	0.705	0.671	0.647	0.629	0.614	0.602	0.591	0.583	0.575
$RbNO_3$	0.734	0.658	0.606	0.565	0.534	0.508	0.485	0.465	0.446	0.430
$RbOAc$	0.796	0.767	0.756	0.753	0.755	0.759	0.766	0.773	0.782	0.792
Rb_2SO_4	0.451	0.374	0.331	0.301	0.279	0.263	0.249	0.238	0.228	0.219
$SrCl_2$	0.511	0.462	0.442	0.433	0.430	0.431	0.434	0.441	0.449	0.461
$TlClO_4$	0.730	0.652	0.599	0.559	0.527	—	—	—	—	—
$TlNO_3$	0.702	0.606	0.545	0.500	—	—	—	—	—	—
UO_2Cl_2	0.544	0.510	0.520	0.505	0.517	0.532	0.549	0.571	0.595	0.620
UO_2SO_4	0.150	0.102	0.0807	0.0689	0.0611	0.0566	0.0515	0.0483	0.0458	0.0439
$ZnCl_2$	0.515	0.462	0.432	0.411	0.394	0.380	0.369	0.357	0.348	0.339
$Zn(NO_3)_2$	0.531	0.489	0.474	0.469	0.473	0.480	0.489	0.501	0.518	0.535
$ZnSO_4$	0.150	0.10	0.0835	0.0714	0.0630	0.0569	0.0523	0.0487	0.0458	0.0435

MEAN ACTIVITY COEFFICIENTS OF ELECTROLYTES AS A FUNCTION OF CONCENTRATION

The mean activity coefficient γ of an electrolyte X_aY_b is defined as

$$\gamma = (\gamma_+^a \gamma_-^b)^{1/(a+b)}$$

where γ_+ and γ_- are activity coefficients of the individual ions (which cannot be directly measured). This table gives the mean activity coefficients of about 100 electrolytes in aqueous solution as a function of concentration, expressed in molality terms. All values refer to a temperature of 25 °C. Substances are arranged in alphabetical order by formula.

References

1. Hamer, W. J., and Wu, Y. C., *J. Phys. Chem. Ref. Data*, 1, 1047, 1972.
2. Staples, B. R., *J. Phys. Chem. Ref. Data*, 6, 385, 1977; 10, 767, 1981; 10, 779, 1981.
3. Goldberg, R. N. et al., *J. Phys. Chem. Ref. Data*, 7, 263, 1978; 8, 923, 1979; 8, 1005, 1979; 10, 1, 1981; 10, 671, 1981.

Mean Activity Coefficient at 25 °C

m/mol kg^{-1}	AgNO$_3$	BaBr$_2$	BaCl$_2$	BaI$_2$	CaBr$_2$	CaCl$_2$	CaI$_2$
0.001	0.964	0.881	0.887	0.890	0.890	0.888	0.890
0.002	0.950	0.850	0.849	0.853	0.853	0.851	0.853
0.005	0.924	0.785	0.782	0.792	0.791	0.787	0.791
0.010	0.896	0.727	0.721	0.737	0.735	0.727	0.736
0.020	0.859	0.661	0.653	0.678	0.674	0.664	0.677
0.050	0.794	0.573	0.559	0.600	0.594	0.577	0.600
0.100	0.732	0.517	0.492	0.551	0.540	0.517	0.552
0.200	0.656	0.463	0.436	0.520	0.502	0.469	0.524
0.500	0.536	0.435	0.391	0.536	0.500	0.444	0.554
1.000	0.430	0.470	0.393	0.664	0.604	0.495	0.729
2.000	0.316	0.654		1.242	1.125	0.784	
5.000	0.181				18.7	5.907	
10.000	0.108					43.1	
15.000	0.085						

m/mol kg^{-1}	Cd(NO$_2$)$_2$	Cd(NO$_3$)$_2$	CoBr$_2$	CoCl$_2$	CoI$_2$	Co(NO$_3$)$_2$	CsBr
0.001	0.881	0.888	0.890	0.889	0.887	0.888	0.965
0.002	0.837	0.851	0.854	0.852	0.849	0.850	0.951
0.005	0.759	0.787	0.794	0.789	0.783	0.786	0.925
0.010	0.681	0.728	0.740	0.732	0.724	0.728	0.898
0.020	0.589	0.664	0.681	0.670	0.661	0.663	0.864
0.050	0.451	0.576	0.605	0.586	0.582	0.576	0.806
0.100	0.344	0.515	0.556	0.528	0.540	0.516	0.752
0.200	0.247	0.465	0.523	0.483	0.527	0.469	0.691
0.500	0.148	0.428	0.538	0.465	0.596	0.446	0.605
1.000	0.098	0.437	0.685	0.532	0.845	0.492	0.540
2.000	0.069	0.517	1.421	0.864	2.287	0.722	0.485
5.000	0.054		13.9		55.3	3.338	0.454
10.000					196		

m/mol kg^{-1}	CsCl	CsF	CsI	CsNO$_3$	CsOH	Cs$_2$SO$_4$	CuBr$_2$
0.001	0.965	0.965	0.965	0.964	0.966	0.885	0.889
0.002	0.951	0.952	0.951	0.951	0.953	0.845	0.853
0.005	0.925	0.929	0.925	0.924	0.930	0.775	0.791
0.010	0.898	0.905	0.898	0.897	0.906	0.709	0.735
0.020	0.864	0.876	0.863	0.860	0.878	0.634	0.674
0.050	0.805	0.830	0.804	0.796	0.836	0.526	0.594
0.100	0.751	0.792	0.749	0.733	0.802	0.444	0.541
0.200	0.691	0.755	0.688	0.655	0.772	0.369	0.504
0.500	0.607	0.721	0.601	0.529	0.755	0.285	0.503
1.000	0.546	0.726	0.534	0.421	0.782	0.233	0.591
2.000	0.496	0.803	0.470				0.859
5.000	0.474						
10.000	0.508						

Mean Activity Coefficients of Electrolytes as a Function of Concentration

m/mol kg^{-1}	CuCl$_2$	Cu(ClO$_4$)$_2$	Cu(NO$_3$)$_2$	FeCl$_2$	HBr	HCl	HClO$_4$
0.001	0.887	0.890	0.888	0.888	0.966	0.965	0.966
0.002	0.849	0.854	0.851	0.850	0.953	0.952	0.953
0.005	0.783	0.795	0.787	0.785	0.930	0.929	0.929
0.010	0.722	0.741	0.729	0.725	0.907	0.905	0.906
0.020	0.654	0.685	0.664	0.659	0.879	0.876	0.878
0.050	0.561	0.613	0.577	0.570	0.837	0.832	0.836
0.100	0.495	0.572	0.516	0.509	0.806	0.797	0.803
0.200	0.441	0.553	0.466	0.462	0.783	0.768	0.776
0.500	0.401	0.617	0.431	0.443	0.790	0.759	0.769
1.000	0.405	0.892	0.456	0.500	0.872	0.811	0.826
2.000	0.453	2.445	0.615	0.782	1.167	1.009	1.055
5.000	0.601		2.083		3.800	2.380	3.100
10.000					33.4	10.4	30.8
15.000							323

m/mol kg^{-1}	HF	HI	HNO$_3$	H$_2$SO$_4$	KBr	KCNS	KCl
0.001	0.551	0.966	0.965	0.804	0.965	0.965	0.965
0.002	0.429	0.953	0.952	0.740	0.952	0.951	0.951
0.005	0.302	0.931	0.929	0.634	0.927	0.927	0.927
0.010	0.225	0.909	0.905	0.542	0.902	0.901	0.901
0.020	0.163	0.884	0.875	0.445	0.870	0.869	0.869
0.050	0.106	0.847	0.829	0.325	0.817	0.815	0.816
0.100	0.0766	0.823	0.792	0.251	0.771	0.768	0.768
0.200	0.0550	0.811	0.756	0.195	0.722	0.716	0.717
0.500	0.0352	0.845	0.725	0.146	0.658	0.647	0.649
1.000	0.0249	0.969	0.730	0.125	0.617	0.598	0.604
2.000	0.0175	1.363	0.788	0.119	0.593	0.556	0.573
5.000	0.0110	4.760	1.063	0.197	0.626	0.525	0.593
10.000	0.0085	49.100	1.644	0.527			
15.000	0.0077		2.212	1.077			
20.000	0.0075		2.607	1.701			

m/mol kg^{-1}	KClO$_3$	K$_2$CrO$_4$	KF	KH$_2$PO$_4$*	K$_2$HPO$_4$**	KI	KNO$_3$
0.001	0.965	0.886	0.965	0.964	0.886	0.965	0.964
0.002	0.951	0.847	0.952	0.950	0.847	0.952	0.950
0.005	0.926	0.779	0.927	0.924	0.779	0.927	0.924
0.010	0.899	0.715	0.902	0.896	0.715	0.902	0.896
0.020	0.865	0.643	0.870	0.859	0.643	0.871	0.860
0.050	0.805	0.539	0.818	0.793	0.538	0.820	0.797
0.100	0.749	0.460	0.773	0.730	0.457	0.776	0.735
0.200	0.681	0.385	0.726	0.652	0.379	0.731	0.662
0.500	0.569	0.296	0.670	0.529	0.283	0.676	0.546
1.000		0.239	0.645	0.422		0.646	0.444
2.000		0.199	0.658			0.638	0.332
5.000			0.871				
10.000			1.715				
15.000			3.120				

m/mol kg^{-1}	KOH	K$_2$SO$_4$	LiBr	LiCl	LiClO$_4$	LiI	LiNO$_3$
0.001	0.965	0.885	0.965	0.965	0.966	0.966	0.965
0.002	0.952	0.844	0.952	0.952	0.953	0.953	0.952
0.005	0.927	0.772	0.929	0.928	0.931	0.930	0.928
0.010	0.902	0.704	0.905	0.904	0.908	0.908	0.904
0.020	0.871	0.625	0.877	0.874	0.882	0.882	0.874
0.050	0.821	0.511	0.832	0.827	0.843	0.843	0.827
0.100	0.779	0.424	0.797	0.789	0.815	0.817	0.788
0.200	0.740	0.343	0.767	0.756	0.795	0.802	0.753
0.500	0.710	0.251	0.754	0.739	0.806	0.824	0.726
1.000	0.733		0.803	0.775	0.887	0.912	0.743
2.000	0.860		1.012	0.924	1.161	1.197	0.837
5.000	1.697		2.696	2.000			1.298

m/mol kg^{-1}	KOH	K$_2$SO$_4$	LiBr	LiCl	LiClO$_4$	LiI	LiNO$_3$
10.000	6.110		20.0	9.600			2.500
15.000	19.9		147	30.9			3.960
20.000	46.4		486				4.970

m/mol kg^{-1}	LiOH	Li$_2$SO$_4$	MgBr$_2$	MgCl$_2$	MgI$_2$	MnBr$_2$	MnCl$_2$
0.001	0.964	0.887	0.889	0.889	0.889	0.889	0.888
0.002	0.950	0.847	0.852	0.852	0.853	0.853	0.850
0.005	0.923	0.780	0.790	0.790	0.791	0.791	0.786
0.010	0.895	0.716	0.733	0.734	0.736	0.735	0.727
0.020	0.858	0.645	0.672	0.672	0.677	0.674	0.662
0.050	0.794	0.544	0.593	0.590	0.602	0.595	0.574
0.100	0.735	0.469	0.543	0.535	0.556	0.543	0.513
0.200	0.668	0.400	0.512	0.493	0.535	0.508	0.464
0.500	0.579	0.325	0.540	0.485	0.594	0.519	0.437
1.000	0.522	0.284	0.715	0.577	0.858	0.650	0.477
2.000	0.484	0.270	1.590	1.065	2.326	1.224	0.661
5.000	0.493		36.1	14.40	109.8	6.697	1.539

m/mol kg^{-1}	Mn(ClO$_4$)$_2$	NH$_4$Cl	NH$_4$ClO$_4$	(NH$_4$)$_2$HPO$_4$**	NH$_4$NO$_3$	NaBr	NaBrO$_3$
0.001	0.892	0.965	0.964	0.882	0.964	0.965	0.965
0.002	0.858	0.952	0.950	0.839	0.951	0.952	0.951
0.005	0.801	0.927	0.924	0.763	0.925	0.928	0.926
0.010	0.752	0.901	0.895	0.688	0.897	0.903	0.900
0.020	0.700	0.869	0.859	0.600	0.862	0.873	0.867
0.050	0.637	0.816	0.794	0.469	0.801	0.824	0.811
0.100	0.604	0.769	0.734	0.367	0.744	0.783	0.759
0.200	0.596	0.718	0.663	0.273	0.678	0.742	0.698
0.500	0.686	0.649	0.560	0.171	0.582	0.697	0.605
1.000	1.030	0.603	0.479	0.114	0.502	0.687	0.528
2.000	3.072	0.569	0.399	0.074	0.419	0.730	0.449
5.000		0.563			0.303	1.083	
10.000					0.220		
15.000					0.179		
20.000					0.154		

m/mol kg^{-1}	Na$_2$CO$_3$	NaCl	NaClO$_3$	NaClO$_4$	Na$_2$CrO$_4$	NaF	Na$_2$HPO$_4$*
0.001	0.887	0.965	0.965	0.965	0.887	0.965	0.887
0.002	0.847	0.952	0.952	0.952	0.849	0.951	0.848
0.005	0.780	0.928	0.927	0.928	0.783	0.926	0.780
0.010	0.716	0.903	0.902	0.903	0.722	0.901	0.717
0.020	0.644	0.872	0.870	0.872	0.653	0.868	0.644
0.050	0.541	0.822	0.818	0.821	0.554	0.813	0.539
0.100	0.462	0.779	0.771	0.777	0.479	0.764	0.456
0.200	0.385	0.734	0.719	0.729	0.406	0.710	0.373
0.500	0.292	0.681	0.646	0.668	0.318	0.633	0.266
1.000	0.229	0.657	0.590	0.630	0.261	0.573	0.191
2.000	0.182	0.668	0.537	0.608	0.231		0.133
5.000		0.874		0.648			

m/mol kg^{-1}	NaI	NaNO$_3$	NaOH	Na$_2$SO$_3$	Na$_2$SO$_4$	Na$_2$WO$_4$	NiBr$_2$
0.001	0.965	0.965	0.965	0.887	0.886	0.886	0.889
0.002	0.952	0.951	0.952	0.847	0.846	0.846	0.853
0.005	0.928	0.926	0.927	0.779	0.777	0.777	0.791
0.010	0.904	0.900	0.902	0.716	0.712	0.712	0.735
0.020	0.874	0.866	0.870	0.644	0.637	0.638	0.675
0.050	0.827	0.810	0.819	0.540	0.529	0.534	0.596
0.100	0.789	0.759	0.775	0.462	0.446	0.457	0.546
0.200	0.753	0.701	0.731	0.386	0.366	0.388	0.514
0.500	0.722	0.617	0.685	0.296	0.268	0.320	0.535
1.000	0.734	0.550	0.674	0.237	0.204	0.291	0.692
2.000	0.823	0.480	0.714	0.196	0.155	0.291	1.476

m/mol kg^{-1}	NaI	NaNO$_3$	NaOH	Na$_2$SO$_3$	Na$_2$SO$_4$	Na$_2$WO$_4$	NiBr$_2$
5.000	1.402	0.388	1.076				
10.000	4.011	0.329	3.258				
15.000			9.796				
20.000			19.410				

m/mol kg^{-1}	NiCl$_2$	Ni(ClO$_4$)$_2$	Ni(NO$_3$)$_2$	Pb(ClO$_4$)$_2$	Pb(NO$_3$)$_2$	RbBr	RbCl
0.001	0.889	0.891	0.889	0.889	0.882	0.965	0.965
0.002	0.852	0.855	0.851	0.851	0.840	0.951	0.951
0.005	0.789	0.797	0.787	0.787	0.764	0.926	0.926
0.010	0.732	0.745	0.730	0.729	0.690	0.900	0.900
0.020	0.669	0.690	0.666	0.666	0.604	0.866	0.867
0.050	0.584	0.621	0.581	0.580	0.476	0.811	0.811
0.100	0.527	0.582	0.524	0.522	0.379	0.760	0.761
0.200	0.482	0.567	0.481	0.476	0.291	0.705	0.707
0.500	0.465	0.639	0.467	0.458	0.195	0.630	0.633
1.000	0.538	0.946	0.528	0.516	0.136	0.578	0.583
2.000	0.915	2.812	0.797	0.799		0.535	0.546
5.000	4.785			4.043		0.514	0.544
10.000				33.8			

m/mol kg^{-1}	RbF	RbI	RbNO$_3$	Rb$_2$SO$_4$	SrBr$_2$	SrCl$_2$	SrI$_2$
0.001	0.965	0.965	0.964	0.886	0.889	0.888	0.890
0.002	0.952	0.951	0.950	0.845	0.852	0.850	0.854
0.005	0.927	0.926	0.924	0.776	0.790	0.785	0.793
0.010	0.902	0.900	0.896	0.710	0.734	0.725	0.740
0.020	0.871	0.866	0.859	0.635	0.673	0.659	0.681
0.050	0.821	0.810	0.795	0.526	0.591	0.569	0.606
0.100	0.780	0.759	0.733	0.443	0.535	0.506	0.557
0.200	0.739	0.703	0.657	0.365	0.492	0.455	0.526
0.500	0.701	0.627	0.536	0.274	0.476	0.421	0.542
1.000	0.697	0.574	0.430	0.217	0.545	0.451	0.686
2.000	0.724	0.532	0.320		0.921	0.650	
5.000		0.517					

m/mol kg^{-1}	UO$_2$Cl$_2$	UO$_2$(NO$_3$)$_2$	ZnBr$_2$	ZnCl$_2$	ZnI$_2$
0.001	0.888	0.888	0.890	0.887	0.893
0.002	0.851	0.849	0.854	0.847	0.859
0.005	0.787	0.784	0.794	0.781	0.804
0.010	0.729	0.726	0.741	0.719	0.757
0.020	0.666	0.663	0.683	0.652	0.708
0.050	0.583	0.583	0.606	0.561	0.644
0.100	0.529	0.535	0.553	0.499	0.601
0.200	0.493	0.509	0.515	0.447	0.574
0.500	0.501	0.532	0.516	0.384	0.635
1.000	0.601	0.673	0.558	0.330	0.836
2.000	0.948	1.223	0.578	0.283	1.062
5.000		3.020	0.788	0.342	1.546
10.000			2.317	0.876	4.698
15.000			5.381	1.914	
20.000			7.965	2.968	

* The anion is H$_2$PO$_4^-$.

** The anion is HPO$_4^{-2}$.

ENTHALPY OF DILUTION OF ACIDS

The quantity given in this table is $-\Delta_{dil}H$, the negative of the enthalpy (heat) of dilution to infinite dilution for aqueous solutions of several common acids; i.e., the negative of the enthalpy change when a solution of molality m at a temperature of 25 °C is diluted with an infinite amount of water. The tabulated numbers thus represent the heat produced (or, if the value is negative, the heat absorbed) when the acid is diluted. The initial molality m is given in the first column. The second column gives the dilution ratio, which is the number of moles of water that must be added to one mole of the acid to produce a solution of the molality in the first column.

Reference

Parker, V. B., *Thermal Properties of Aqueous Uni-Univalent Electrolytes*, Natl. Stand. Ref. Data Ser. - Natl. Bur. Stand. (U.S.) 2, U.S. Government Printing Office, 1965.

$-\Delta_{dil}H$ in kJ/mol at 25 °C

m	Dil. ratio	HF	HCl	$HClO_4$	HBr	HI	HNO_3	CH_2O_2	$C_2H_4O_2$
55.506	1.0		45.61		48.83		19.73	0.046	2.167
20	2.775	14.88	19.87	13.81	19.92	21.71	9.498	0.038	2.075
15	3.700	14.34	15.40	7.920	14.29	14.02	6.883	0.109	1.962
10	5.551	13.87	10.24	2.013	8.694	7.615	3.933	0.205	1.824
9	6.167	13.81	9.213	1.280	7.719	6.569	3.368	0.230	1.782
8	6.938	13.77	8.201	0.611	6.786	5.607	2.791	0.255	1.724
7	7.929	13.73	7.217	0.046	5.925	4.728	2.251	0.272	1.648
6	9.251	13.69	6.268	-0.351	5.004	3.975	1.749	0.280	1.540
5.5506	10	13.66	5.841	-0.490	4.590	3.577	1.540	0.285	1.477
5	11.10	13.62	5.318	-0.628	4.113	3.197	1.310	0.289	1.393
4.5	12.33	13.58	4.899	-0.732	3.711	2.828	1.109	0.289	1.310
4	13.88	13.53	4.402	-0.787	3.330	2.460	0.958	0.289	1.218
3.5	15.86	13.47	3.958	-0.820	2.966	2.105	0.791	0.289	1.121
3	18.50	13.45	3.506	-0.782	2.611	1.787	0.665	0.289	1.025
2.5	22.20	13.43	3.063	-0.724	2.301	1.527	0.582	0.285	0.912
2	27.75	13.40	2.623	-0.623	1.996	1.318	0.527	0.276	0.803
1.5	37.00	13.36	2.167	-0.431	1.665	1.125	0.506	0.259	0.678
1	55.51	13.30	1.695	-0.201	1.314	0.933	0.506	0.226	0.544
0.5551	100	13.22	1.234	0.050	0.983	0.736	0.502	0.184	0.423
0.5	111.0	13.20	1.172	0.075	0.941	0.711	0.498	0.176	0.406
0.2	277.5	13.09	0.761	0.247	0.649	0.536	0.439	0.146	0.331
0.1	555.1	12.80	0.556	0.272	0.498	0.439	0.372	0.134	0.289
0.0925	600	12.79	0.540	0.272	0.481	0.427	0.368	0.134	0.285
0.0793	700	12.70	0.502	0.272	0.452	0.402	0.351	0.134	0.285
0.0694	800	12.61	0.473	0.268	0.427	0.385	0.339	0.130	0.280
0.0617	900	12.50	0.448	0.264	0.406	0.368	0.326	0.126	0.276
0.05551	1000	12.42	0.427	0.259	0.385	0.351	0.318	0.121	0.272
0.05	1110	12.24	0.406	0.259	0.372	0.339	0.305	0.121	0.272
0.02775	2000	11.29	0.310	0.226	0.285	0.264	0.247	0.117	0.264
0.01850	3000	10.66	0.251	0.197	0.234	0.218	0.213	0.117	0.259
0.01388	4000	10.25	0.226	0.180	0.205	0.192	0.192	0.113	0.259
0.01110	5000	9.874	0.197	0.167	0.184	0.172	0.176	0.109	0.255
0.00555	10000	8.912	0.142	0.126	0.130	0.121	0.130	0.105	0.243
0.00278	20000	7.531	0.105	0.092	0.092	0.084	0.096	0.096	0.230
0.00111	50000	5.439	0.067	0.059	0.054	0.050	0.063	0.084	0.222
0.000555	100000	3.766	0.042	0.042	0.038	0.038	0.046	0.054	0.209
0.000111	500000	1.255	0.021	0.021	0.021	0.021	0.021	0.038	0.167
0	∞	0	0	0	0	0	0	0	0

ENTHALPY OF SOLUTION OF ELECTROLYTES

This table gives the molar enthalpy (heat) of solution at infinite dilution for some common uni-univalent electrolytes. This is the enthalpy change when 1 mol of solute in its standard state is dissolved in an infinite amount of water. Values are given in kilojoules per mole at 25 °C.

Reference

Parker, V. B., *Thermal Properties of Uni-Univalent Electrolytes*, Natl. Stand. Ref. Data Series — Natl. Bur. Stand.(U.S.), No.2, 1965.

Solute	State	$\Delta_{sol}H°$ kJ/mol	Solute	State	$\Delta_{sol}H°$ kJ/mol	Solute	State	$\Delta_{sol}H°$ kJ/mol
HF	g	−61.50	$LiBr \cdot 2H_2O$	c	−9.41	$KClO_3$	c	41.38
HCl	g	−74.84	$LiBrO_3$	c	1.42	$KClO_4$	c	51.04
$HClO_4$	l	−88.76	LiI	c	−63.30	KBr	c	19.87
$HClO_4 \cdot H_2O$	c	−32.95	$LiI \cdot H_2O$	c	−29.66	$KBrO_3$	c	41.13
HBr	g	−85.14	$LiI \cdot 2H_2O$	c	−14.77	KI	c	20.33
HI	g	−81.67	$LiI \cdot 3H_2O$	c	0.59	KIO_3	c	27.74
HIO_3	c	8.79	$LiNO_2$	c	−11.00	KNO_2	c	13.35
HNO_3	l	−33.28	$LiNO_2 \cdot H_2O$	c	7.03	KNO_3	c	34.89
HCOOH	l	−0.86	$LiNO_3$	c	−2.51	$KC_2H_3O_2$	c	−15.33
CH_3COOH	l	−1.51				KCN	c	11.72
			NaOH	c	−44.51	KCNO	c	20.25
NH_3	g	−30.50	$NaOH \cdot H_2O$	c	−21.41	KCNS	c	24.23
NH_4Cl	c	14.78	NaF	c	0.91	$KMnO_4$	c	43.56
NH_4ClO_4	c	33.47	NaCl	c	3.88			
NH_4Br	c	16.78	$NaClO_2$	c	0.33	RbOH	c	−62.34
NH_4I	c	13.72	$NaClO_2 \cdot 3H_2O$	c	28.58	$RbOH \cdot H_2O$	c	−17.99
NH_4IO_3	c	31.80	$NaClO_3$	c	21.72	$RbOH \cdot 2H_2O$	c	0.88
NH_4NO_2	c	19.25	$NaClO_4$	c	13.88	RbF	c	−26.11
NH_4NO_3	c	25.69	$NaClO_4 \cdot H_2O$	c	22.51	$RbF \cdot H_2O$	c	−0.42
$NH_4C_2H_3O_2$	c	−2.38	NaBr	c	−0.60	$RbF \cdot 1.5H_2O$	c	1.34
NH_4CN	c	17.57	$NaBr \cdot 2H_2O$	c	18.64	RbCl	c	17.28
NH_4CNS	c	22.59	$NaBrO_3$	c	26.90	$RbClO_3$	c	47.74
CH_3NH_3Cl	c	5.77	NaI	c	−7.53	$RbClO_4$	c	56.74
$(CH_3)_3NHCl$	c	1.46	$NaI \cdot 2H_2O$	c	16.13	RbBr	c	21.88
$N(CH_3)_4Cl$	c	4.08	$NaIO_3$	c	20.29	$RbBrO_3$	c	48.95
$N(CH_3)_4Br$	c	24.27	$NaNO_2$	c	13.89	RbI	c	25.10
$N(CH_3)_4I$	c	42.07	$NaNO_3$	c	20.50	$RbNO_3$	c	36.48
			$NaC_2H_3O_2$	c	−17.32			
$AgClO_4$	c	7.36	$NaC_2H_3O_2 \cdot 3H_2O$	c	19.66	CsOH	c	−71.55
$AgNO_2$	c	36.94	NaCN	c	1.21	$CsOH \cdot H_2O$	c	−20.50
$AgNO_3$	c	22.59	$NaCN \cdot 0.5H_2O$	c	3.31	CsF	c	−36.86
			$NaCN \cdot 2H_2O$	c	18.58	$CsF \cdot H_2O$	c	−10.46
LiOH	c	−23.56	NaCNO	c	19.20	$CsF \cdot 1.5H_2O$	c	−5.44
$LiOH \cdot H_2O$	c	−6.69	NaCNS	c	6.83	CsCl	c	17.78
LiF	c	4.73				$CsClO_4$	c	55.44
LiCl	c	−37.03	KOH	c	−57.61	CsBr	c	25.98
$LiCl \cdot H_2O$	c	−19.08	$KOH \cdot H_2O$	c	−14.64	$CsBrO_3$	c	50.46
$LiClO_4$	c	−26.55	$KOH \cdot 1.5H_2O$	c	−10.46	CsI	c	33.35
$LiClO_4 \cdot 3H_2O$	c	32.61	KF	c	−17.73	$CsNO_3$	c	40.00
LiBr	c	−48.83	$KF \cdot 2H_2O$	c	6.97			
$LiBr \cdot H_2O$	c	−23.26	KCl	c	17.22			

ENTHALPY OF HYDRATION OF GASES

The molar enthalpy of hydration $\Delta_{hyd}H^{\infty}$ is defined as the enthalpy change when one mole of an ideal gas is dissolved in an infinite amount of water. Another term for this quantity is enthalpy (heat) of solvation in water at infinite dilution. The enthalpy of hydration influences the distribution of a volatile compound between the aqueous solution phase and air and is thus important in fields such as environmental science, geochemistry, and chemical engineering. It is related to the enthalpy of solution of the liquid or solid phase, $\Delta_{sol}H^{\infty}$, by

$$\Delta_{hyd}H^{\infty} = \Delta_{sol}H^{\infty} - \Delta_{vap}H^{0}$$

where $\Delta_{vap}H^{0}$ is the molar enthalpy of vaporization. This table gives the molar enthalpy of hydration for a number of common substances.

References

1. Plyasunov, A. V., and Shock, E. L., *J. Chem. Eng. Data* [Hydrocarbons, alcohols, and ketones] 46, 1016, 2001.
2. Plyasunov, A. V., Plyasunova, N. V., and Shock, E. L., *J. Chem. Eng. Data* [Esters] 49, 1152, 2004.
3. Plyasunov, A. V., Plyasunova, N. V., and Shock, E. L., *J. Chem. Eng. Data* [Ethers, diethers, and polyethers] 51, 276, 2006.
4. Plyasunova, N. V., Plyasunov, A. V., and Shock, E. L., *J. Chem. Eng. Data* [Thiols, sulfides, and polysulfides] 50, 246, 2005.
5. Plyasunov, A. V., Plyasunova, N. V., and Shock, E. L., *J. Chem. Eng. Data* [Nitriles and dinitriles] 51, 1481, 2006.
6. Plyasunov, A. V., and Shock, E. L., *Geochim. Cosmochim. Acta* [Inorganics and halogen compounds; temperature dependence] 67, 4981, 2003.
7. Kühne, R., Ebert, R-U, and Schüürmann, G., *Environ. Sci. Technol.* [Compilation; temperature dependence] 39, 6705, 2005.
8. Mintz, C., Clark, M., Acree, W. E., and Abraham, M. H., *J. Chem. Inf. Model.* [Compilation and modelling] 47, 115, 2007.

Name	Mol. Form.	$\Delta_{hyd}H^{\infty}$/kJ mol^{-1} at 298.15 K	Name	Mol. Form.	$\Delta_{hyd}H^{\infty}$/kJ mol^{-1} at 298.15 K
Acenaphthene	$C_{12}H_{10}$	−52.1	*sec*-Butyl acetate	$C_6H_{12}O_2$	−51.9
Acetic acid	$C_2H_4O_2$	−52.8	*tert*-Butyl acetate	$C_6H_{12}O_2$	−46.2
Acetone	C_3H_6O	−39.7	Butylamine	$C_4H_{11}N$	−59.2
Acetonitrile	C_2H_3N	−34.9	*sec*-Butylamine	$C_4H_{11}N$	−57.1
Acetophenone	C_8H_8O	−53.3	*tert*-Butylamine	$C_4H_{11}N$	−59.0
Ammonia	H_3N	−35.4	Butylbenzene	$C_{10}H_{14}$	−38.5
Aniline	C_6H_7N	−56.5	Butyl butanoate	$C_8H_{16}O_2$	−63.5
Anisole	C_7H_8O	−41.4	Butyl ethyl ether	$C_6H_{14}O$	−48.4
Argon	Ar	−12.2	*tert*-Butyl ethyl ether	$C_6H_{14}O$	−53.4
Benzaldehyde	C_7H_6O	−42.1	4-*tert*-Butylphenol	$C_{10}H_{14}O$	−63.8
Benzene	C_6H_6	−28.1	Butyl propanoate	$C_7H_{14}O_2$	−57.8
Benzonitrile	C_7H_5N	−48.5	1-Butyne	C_4H_6	−13.5
Benzyl alcohol	C_7H_8O	−66.9	Carbon dioxide	CO_2	−17.9
Biphenyl	$C_{12}H_{10}$	−47.2	Carbon monoxide	CO	−11.1
Bromobenzene	C_6H_5Br	−33.5	Chlorine	Cl_2	−23.4
Bromodichloromethane	$CHBrCl_2$	−28.9	Chlorine dioxide	ClO_2	−27.8
Bromoethane	C_2H_5Br	−29.5	Chlorobenzene	C_6H_5Cl	−30.6
Bromomethane	CH_3Br	−23.8	2-Chloro-1,1′-biphenyl	$C_{12}H_9Cl$	−42.8
2-Bromo-2-methylpropane	C_4H_9Br	−25.4	1-Chlorobutane	C_4H_9Cl	−28.2
1,3-Butadiene	C_4H_6	−31.4	2-Chlorobutane	C_4H_9Cl	−34.6
Butane	C_4H_{10}	−24.8	Chlorodibromomethane	$CHBr_2Cl$	−33.3
1,4-Butanediamine	$C_4H_{12}N_2$	−91.6	Chlorodifluoromethane	$CHClF_2$	−22.8
1,4-Butanediol	$C_4H_{10}O_2$	−89.6	Chloroethane	C_2H_5Cl	−22.0
Butanenitrile	C_4H_7N	−42.1	Chlorofluoromethane	CH_2ClF	−21.7
1,2,3,4-Butanetetrol	$C_4H_{10}O_4$	−114	1-Chlorohexane	$C_6H_{13}Cl$	−34.5
1-Butanethiol	$C_4H_{10}S$	−36.3	Chloromethane	CH_3Cl	−20.2
Butanoic acid	$C_4H_8O_2$	−59.5	1-Chloropentane	$C_5H_{11}Cl$	−34.1
1-Butanol	$C_4H_{10}O$	−61.9	3-Chlorophenol	C_6H_5ClO	−50.3
2-Butanol	$C_4H_{10}O$	−62.7	1-Chloropropane	C_3H_7Cl	−27.0
2-Butanone	C_4H_8O	−41.9	2-Chloropyridine	C_5H_4ClN	−42.6
1-Butene	C_4H_8	−24.1	3-Chloropyridine	C_5H_4ClN	−46.2
2-Butoxyethanol	$C_6H_{14}O_2$	−73.6	2-Chlorotoluene	C_7H_7Cl	−38.3
Butyl acetate	$C_6H_{12}O_2$	−52.7	3-Chlorotoluene	C_7H_7Cl	−37.0

Name	Mol. Form.	$\Delta_{hyd}H^{\infty}$/kJ mol^{-1} at 298.15 K	Name	Mol. Form.	$\Delta_{hyd}H^{\infty}$/kJ mol^{-1} at 298.15 K
4-Chlorotoluene	C_7H_7Cl	−33.3	Dimethylamine	C_2H_7N	−53.1
o-Cresol	C_7H_8O	−64.8	2,4-Dimethylaniline	$C_8H_{11}N$	−58.7
m-Cresol	C_7H_8O	−58.7	2,5-Dimethylaniline	$C_8H_{11}N$	−61.5
p-Cresol	C_7H_8O	−61.3	2,6-Dimethylaniline	$C_8H_{11}N$	−60.5
Cycloheptanol	$C_7H_{14}O$	−74.6	N,N-Dimethylaniline	$C_8H_{11}N$	−49.6
Cyclohexane	C_6H_{12}	−30.0	2,3-Dimethylbutane	C_6H_{14}	−32.4
cis-1,2-Cyclohexanediol	$C_6H_{12}O_2$	−82.4	3,3-Dimethyl-2-butanone	$C_6H_{12}O$	−47.5
Cyclohexanol	$C_6H_{12}O$	−70.7	cis-1,2-Dimethylcyclohexane	C_8H_{16}	−38.3
Cyclohexanone	$C_6H_{10}O$	−49.8	trans-1,2-Dimethylcyclohexane	C_8H_{16}	−36.1
Cyclohexene	C_6H_{10}	−27.3	Dimethyl ether	C_2H_6O	−34.0
Cyclooctane	C_8H_{16}	−39.0	N,N-Dimethylformamide	C_3H_7NO	−62.9
cis-Cyclooctene	C_8H_{14}	−45.5	2,4-Dimethyl-3-pentanone	$C_7H_{14}O$	−54.0
Cyclopentane	C_5H_{10}	−30.3	2,3-Dimethylpyridine	C_7H_9N	−57.7
Cyclopentanol	$C_5H_{10}O$	−58.5	2,4-Dimethylpyridine	C_7H_9N	−60.7
Cyclopentanone	C_5H_8O	−44.3	2,5-Dimethylpyridine	C_7H_9N	−54.9
Cyclopropane	C_3H_6	−15.4	2,6-Dimethylpyridine	C_7H_9N	−52.3
Dibromomethane	CH_2Br_2	−33.0	3,4-Dimethylpyridine	C_7H_9N	−50.5
Dibutylamine	$C_8H_{19}N$	−59.3	3,5-Dimethylpyridine	C_7H_9N	−51.3
Dibutyl ether	$C_8H_{18}O$	−55.8	Dimethyl sulfide	C_2H_6S	−31.5
o-Dichlorobenzene	$C_6H_4Cl_2$	−37.3	Dimethyl sulfoxide	C_2H_6OS	−71.9
m-Dichlorobenzene	$C_6H_4Cl_2$	−35.3	2,5-Dimethyltetrahydrofuran	$C_6H_{12}O$	−56.3
p-Dichlorobenzene	$C_6H_4Cl_2$	−28.4	1,4-Dioxane	$C_4H_8O_2$	−48.4
2,3-Dichloro-1,1'-biphenyl	$C_{12}H_8Cl_2$	−45.6	1,2-Dipropoxyethane	$C_8H_{18}O_2$	−76.8
2,4-Dichloro-1,1'-biphenyl	$C_{12}H_8Cl_2$	−43.0	Dipropylamine	$C_6H_{15}N$	−65.2
2,4'-Dichloro-1,1'-biphenyl	$C_{12}H_8Cl_2$	−44.2	Dipropyl ether	$C_6H_{14}O$	−49.9
2,5-Dichlorobiphenyl	$C_{12}H_8Cl_2$	−45.6	Dipropyl sulfide	$C_6H_{14}S$	−47.7
Dichlorodifluoromethane	CCl_2F_2	−26.0	1-Dodecanol	$C_{12}H_{26}O$	−81.9
2,2-Dichloro-1,1-difluoro-1-methoxyethane	$C_3H_4Cl_2F_2O$	−30.4	Ethane	C_2H_6	−17.9
			1,2-Ethanediamine	$C_2H_8N_2$	−76.1
1,1-Dichloroethane	$C_2H_4Cl_2$	−30.3	1,2-Ethanediol	$C_2H_6O_2$	−77.3
1,2-Dichloroethane	$C_2H_4Cl_2$	−27.9	Ethanethiol	C_2H_6S	−28.9
1,1-Dichloroethene	$C_2H_2Cl_2$	−28.5	Ethanol	C_2H_6O	−50.6
cis-1,2-Dichloroethene	$C_2H_2Cl_2$	−26.9	2-Ethoxyethanol	$C_4H_{10}O_2$	−66.4
trans-1,2-Dichloroethene	$C_2H_2Cl_2$	−29.3	1-Ethoxy-2-methoxyethane	$C_5H_{12}O_2$	−66.1
Dichloromethane	CH_2Cl_2	−30.3	Ethyl acetate	$C_4H_8O_2$	−45.3
1,2-Dichloropropane	$C_3H_6Cl_2$	−31.1	Ethylamine	C_2H_7N	−53.7
1,3-Dichloropropane	$C_3H_6Cl_2$	−29.7	2-Ethylaniline	$C_8H_{11}N$	−59.7
1,2-Dichloro-1,1,2,2-tetrafluoroethane	$C_2Cl_2F_4$	−20.2	4-Ethylaniline	$C_8H_{11}N$	−65.0
			Ethylbenzene	C_8H_{10}	−39.4
1,2-Diethoxyethane	$C_6H_{14}O_2$	−71.9	Ethyl butanoate	$C_6H_{12}O_2$	−52.7
Diethylamine	$C_4H_{11}N$	−64.3	Ethylcyclohexane	C_8H_{16}	−36.8
N,N-Diethylaniline	$C_{10}H_{15}N$	−45.7	Ethyl 2,2-dimethylpropanoate	$C_7H_{14}O_2$	−50.3
p-Diethylbenzene	$C_{10}H_{14}$	−46.4	Ethylene	C_2H_4	−13.7
Diethylene glycol dimethyl ether	$C_6H_{14}O_3$	−96.2	Ethyl formate	$C_3H_6O_2$	−38.1
Diethyl ether	$C_4H_{10}O$	−46.4	Ethyl hexanoate	$C_8H_{16}O_2$	−60.2
Diethyl sulfide	$C_4H_{10}S$	−40.2	Ethyl 3-methylbutanoate	$C_7H_{14}O_2$	−56.0
1,1-Difluoroethane	$C_2H_4F_2$	−20.7	Ethyl 2-methylbutanoate	$C_7H_{14}O_2$	−55.4
Difluoromethane	CH_2F_2	−17.2	Ethyl 2-methylpropanoate	$C_6H_{12}O_2$	−51.3
Diiodomethane	CH_2I_2	−41.6	Ethyl pentanoate	$C_7H_{14}O_2$	−56.5
Diisopropyl ether	$C_6H_{14}O$	−51.7	Ethyl propanoate	$C_5H_{10}O_2$	−49.5
1,2-Dimethoxyethane	$C_4H_{10}O_2$	−59.3	2-Ethylpyridine	C_7H_9N	−55.7

Name	Mol. Form.	$\Delta_{hyd}H^\infty$/kJ mol^{-1} at 298.15 K	Name	Mol. Form.	$\Delta_{hyd}H^\infty$/kJ mol^{-1} at 298.15 K
3-Ethylpyridine	C_7H_9N	−53.5	Isopropyl formate	$C_4H_8O_2$	−43.0
4-Ethylpyridine	C_7H_9N	−52.2	1-Isopropyl-4-methylbenzene	$C_{10}H_{14}$	−34.6
9H-Fluorene	$C_{13}H_{10}$	−42.7	Krypton	Kr	−15.6
Fluorobenzene	C_6H_5F	−29.3	Methane	CH_4	−12.0
Fluoromethane	CH_3F	−16.1	Methanethiol	CH_4S	−24.4
Glycerol	$C_3H_8O_3$	−103.5	Methanol	CH_4O	−52.0
Helium	He	−0.67	2-Methoxyethanol	$C_3H_8O_2$	−60.4
1,1,1,2,3,3,3-Heptafluoropropane	C_3HF_7	−24.8	2-Methoxy-2-methylbutane	$C_6H_{14}O$	−52.5
Heptanal	$C_7H_{14}O$	−56.6	2-Methoxyphenol	$C_7H_8O_2$	−62.6
Heptane	C_7H_{16}	−34.0	Methyl acetate	$C_3H_6O_2$	−40.1
1-Heptanol	$C_7H_{16}O$	−72.1	Methylamine	CH_5N	−45.3
2-Heptanol	$C_7H_{16}O$	−72.6	Methyl benzoate	$C_8H_8O_2$	−50.3
4-Heptanol	$C_7H_{16}O$	−75.3	Methyl butanoate	$C_5H_{10}O_2$	−47.5
2-Heptanone	$C_7H_{14}O$	−54.9	3-Methyl-1-butanol	$C_5H_{12}O$	−66.0
4-Heptanone	$C_7H_{14}O$	−58.1	2-Methyl-2-butanol	$C_5H_{12}O$	−68.4
1,1,1,3,3,3-Hexafluoro-2-propanol	$C_3H_2F_6O$	−57.1	3-Methyl-2-butanone	$C_5H_{10}O$	−57.6
Hexanal	$C_6H_{12}O$	−55.2	2-Methyl-2-butene	C_5H_{10}	−26.6
Hexane	C_6H_{14}	−31.9	Methyl tert-butyl ether	$C_5H_{12}O$	−48.7
Hexanedinitrile	$C_6H_8N_2$	−66.6	Methyl 2,2-dimethylpropanoate	$C_6H_{12}O_2$	−46.2
1-Hexanol	$C_6H_{14}O$	−67.4	Methyl formate	$C_2H_4O_2$	−32.0
3-Hexanol	$C_6H_{14}O$	−69.6	Methyl hexanoate	$C_7H_{14}O_2$	−54.7
2-Hexanone	$C_6H_{12}O$	−48.9	Methyl isobutanoate	$C_5H_{10}O_2$	−46.0
3-Hexanone	$C_6H_{12}O$	−46.0	4-Methylmorpholine	$C_5H_{11}NO$	−68.7
1-Hexene	C_6H_{12}	−30.4	1-Methylnaphthalene	$C_{11}H_{10}$	−45.0
Hexyl acetate	$C_8H_{16}O_2$	−60.8	2-Methylnaphthalene	$C_{11}H_{10}$	−44.9
Hexylamine	$C_6H_{15}N$	−65.9	2-Methylpentane	C_6H_{14}	−30.5
Hexylbenzene	$C_{12}H_{18}$	−52.7	3-Methylpentane	C_6H_{14}	−36.8
Hydrogen	H_2	−0.402	Methyl pentanoate	$C_6H_{12}O_2$	−50.4
Hydrogen selenide	H_2Se	−15.7	4-Methyl-2-pentanol	$C_6H_{14}O$	−69.9
Hydrogen sulfide	H_2S	−18.0	4-Methyl-2-pentanone	$C_6H_{12}O$	−44.6
3-Hydroxybenzaldehyde	$C_7H_6O_2$	−70.7	1-Methylpiperidine	$C_6H_{13}N$	−65.8
3-Hydroxybenzonitrile	C_7H_5NO	−70.7	2-Methylpropanenitrile	C_4H_7N	−40.0
4-Hydroxybenzonitrile	C_7H_5NO	−70.3	Methyl propanoate	$C_4H_8O_2$	−44.5
Iodoethane	C_2H_5I	−31.7	2-Methyl-1-propanol	$C_4H_{10}O$	−60.2
Iodomethane	CH_3I	−28.2	2-Methyl-2-propanol	$C_4H_{10}O$	−62.9
1-Iodopropane	C_3H_7I	−35.3	Methyl propyl ether	$C_4H_{10}O$	−38.0
2-Iodopropane	C_3H_7I	−36.6	2-Methylpyridine	C_6H_7N	−50.3
Isobutanal	C_4H_8O	−40.0	3-Methylpyridine	C_6H_7N	−50.3
Isobutane	C_4H_{10}	−21.7	4-Methylpyridine	C_6H_7N	−51.8
Isobutene	C_4H_8	−22.7	N-Methylpyrrolidine	$C_5H_{11}N$	−63.4
Isobutyl acetate	$C_6H_{12}O_2$	−51.8	2-Methyltetrahydrofuran	$C_5H_{10}O$	−51.4
Isobutyl formate	$C_5H_{10}O_2$	−43.0	Morpholine	C_4H_9NO	−69.5
Isobutyl isobutanoate	$C_8H_{16}O_2$	−55.3	Naphthalene	$C_{10}H_8$	−42.8
Isobutyl propanoate	$C_7H_{14}O_2$	−54.7	Neon	Ne	−3.90
Isoflurane	$C_3H_2ClF_5O$	−35.3	Neopentane	C_5H_{12}	−23.4
Isopentyl acetate	$C_7H_{14}O_2$	−53.8	Nitric oxide	NO	−11.9
Isopentyl formate	$C_6H_{12}O_2$	−47.7	Nitrobenzene	$C_6H_5NO_2$	−43.8
Isophorone	$C_9H_{14}O$	−59.1	Nitroethane	$C_2H_5NO_2$	−32.5
Isopropyl acetate	$C_5H_{10}O_2$	−46.8	Nitrogen	N_2	−1.04
Isopropylamine	C_3H_9N	−55.0	Nitromethane	CH_3NO_2	−35.7
Isopropylbenzene	C_9H_{12}	−33.7	2-Nitrophenol	$C_6H_5NO_3$	−49.8

Name	Mol. Form.	$\Delta_{hyd}H^\circ$/kJ mol^{-1} at 298.15 K	Name	Mol. Form.	$\Delta_{hyd}H^\circ$/kJ mol^{-1} at 298.15 K
3-Nitrophenol	$C_6H_5NO_3$	−67.7	Propylbenzene	C_9H_{12}	−36.4
4-Nitrophenol	$C_6H_5NO_3$	−68.6	Propyl butanoate	$C_7H_{14}O_2$	−54.9
1-Nitropropane	$C_3H_7NO_2$	−34.4	Propyl formate	$C_4H_8O_2$	−40.5
2-Nitropropane	$C_3H_7NO_2$	−34.1	Propyl propanoate	$C_6H_{12}O_2$	−51.2
2-Nitrotoluene	$C_7H_7NO_2$	−46.4	Propyne	C_3H_4	−15.6
3-Nitrotoluene	$C_7H_7NO_2$	−38.5	Pyridine	C_5H_5N	−42.1
Nitrous oxide	N_2O	−19.8	Quinoline	C_9H_7N	−58.2
5-Nonanone	$C_9H_{18}O$	−62.8	Radon	Rn	−24.0
2-Nonanone	$C_9H_{18}O$	−65.3	Styrene	C_8H_8	−28.4
Octanal	$C_8H_{16}O$	−48.8	Succinonitrile	$C_4H_4N_2$	−58.2
Octane	C_8H_{18}	−36.0	Sulfur hexafluoride	F_6S	−20.7
1-Octanol	$C_8H_{18}O$	−74.1	1,2,3,4-Tetrachlorobenzene	$C_6H_2Cl_4$	−35.0
2-Octanone	$C_8H_{16}O$	−58.3	1,1,2,2-Tetrachloroethane	$C_2H_2Cl_4$	−34.8
1-Octene	C_8H_{16}	−39.2	1,1,1,2-Tetrachloroethane	$C_2H_2Cl_4$	−36.2
Octylamine	$C_8H_{19}N$	−52.3	Tetrachloroethene	C_2Cl_4	−41.5
Oxygen	O_2	−1.20	Tetrachloromethane	CCl_4	−30.5
Pentachlorobenzene	C_6HCl_5	−39.9	Tetraethylene glycol dimethyl ether	$C_{10}H_{22}O_5$	−126
Pentafluoroethane	C_2HF_5	−21.5	1,1,1,2-Tetrafluoroethane	$C_2H_2F_4$	−22.2
2,2,3,3,3-Pentafluoro-1-propanol	$C_3H_3F_5O$	−51.9	Tetrafluoroethene	C_2F_4	−15.1
Pentanal	$C_5H_{10}O$	−42.9	Tetrafluoromethane	CF_4	−13.5
Pentane	C_5H_{12}	−28.3	2,2,3,3-Tetrafluoro-1-propanol	$C_3H_4F_4O$	−57.9
1,5-Pentanediamine	$C_5H_{14}N_2$	−95.1	Tetrahydrofuran	C_4H_8O	−47.3
Pentanedinitrile	$C_5H_6N_2$	−63.5	Tetrahydropyran	$C_5H_{10}O$	−48.9
Pentanenitrile	C_5H_9N	−45.6	Thiophene	C_4H_4S	−29.9
1-Pentanol	$C_5H_{12}O$	−61.9	Toluene	C_7H_8	−32.4
2-Pentanol	$C_5H_{12}O$	−63.3	Tribromomethane	$CHBr_3$	−35.8
3-Pentanol	$C_5H_{12}O$	−59.6	1,2,3-Trichlorobenzene	$C_6H_3Cl_3$	−32.6
2-Pentanone	$C_5H_{10}O$	−45.3	1,3,5-Trichlorobenzene	$C_6H_3Cl_3$	−34.2
3-Pentanone	$C_5H_{10}O$	−49.6	1,1,1-Trichloroethane	$C_2H_3Cl_3$	−28.7
Pentyl acetate	$C_7H_{14}O_2$	−55.3	1,1,2-Trichloroethane	$C_2H_3Cl_3$	−32.5
Pentylamine	$C_5H_{13}N$	−62.1	Trichloroethene	C_2HCl_3	−32.2
Pentylbenzene	$C_{11}H_{16}$	−49.5	Trichlorofluoromethane	CCl_3F	−19.8
Pentyl formate	$C_6H_{12}O_2$	−48.1	Trichloromethane	$CHCl_3$	−33.5
Perfluoropropene	C_3F_6	−17.4	1,1,2-Trichloro-1,2,2-trifluoroethane	$C_2Cl_3F_3$	−28.8
Phenol	C_6H_6O	−57.7	Triethylamine	$C_6H_{15}N$	−69.7
1-Phenyl-1-propanone	$C_9H_{10}O$	−61.9	Triethylene glycol dimethyl ether	$C_8H_{18}O_4$	−102.4
Piperidine	$C_5H_{11}N$	−65.4	2,2,2-Trifluoroethanol	$C_2H_3F_3O$	−50.2
Propanal	C_3H_6O	−39.4	Trifluoromethane	CHF_3	−22.6
Propanamide	C_3H_7NO	−73.4	1,1,1-Trifluoro-2-propanol	$C_3H_5F_3O$	−53.5
Propane	C_3H_8	−20.4	Trimethylamine	C_3H_9N	−52.7
1,3-Propanediamine	$C_3H_{10}N_2$	−85.6	1,2,3-Trimethylbenzene	C_9H_{12}	−37.4
1,3-Propanediol	$C_3H_8O_2$	−81.1	1,2,4-Trimethylbenzene	C_9H_{12}	−36.6
Propanenitrile	C_3H_5N	−39.5	1,3,5-Trimethylbenzene	C_9H_{12}	−39.1
1-Propanethiol	C_3H_8S	−30.2	2,2,4-Trimethylpentane	C_8H_{18}	−31.0
Propanoic acid	$C_3H_6O_2$	−56.5	2,3,4-Trimethylpentane	C_8H_{18}	−38.5
1-Propanol	C_3H_8O	−59.9	Xenon	Xe	−19.4
2-Propanol	C_3H_8O	−58.2	o-Xylene	C_8H_{10}	−37.7
Propene	C_3H_6	−21.6	m-Xylene	C_8H_{10}	−38.6
2-Propoxyethanol	$C_5H_{12}O_2$	−69.6	p-Xylene	C_8H_{10}	−34.8
Propyl acetate	$C_5H_{10}O_2$	−48.7			
Propylamine	C_3H_9N	−56.0			

pH SCALE FOR AQUEOUS SOLUTIONS

A.K. Covington

A Working Party of IUPAC, after extensive considerations over five years, has produced a report (1) which sets pH firmly within the International System of Units (SI). A summary of these important developments is given below.

The concept of pH is unique amongst the commonly encountered physicochemical quantities in that, in terms of its definition,

$$pH = - \lg a_H \qquad (1)$$

it involves a single ion quantity, the activity of the hydrogen ion, which is immeasurable by any thermodynamically valid method and requires a convention for its evaluation.

pH was originally defined by Sørensen (2) in terms of the concentration of hydrogen ions (in modern nomenclature) as pH = $- \lg (c_H/c^\circ)$ where c_H is the hydrogen ion concentration in mol dm^{-3}, and $c^\circ = 1$ mol dm^{-3} is the standard amount concentration. Subsequently (3), it was accepted as more satisfactory to define pH in terms of the relative activity of hydrogen ions in solution

$$pH = - \lg a_H = - \lg (m_H \gamma_H/m^\circ) \qquad (2)$$

where a_H is the relative (molality basis) activity and γ_H is the molal activity coefficient of the hydrogen ion H$^+$ at the molality m_H, and m° the standard molality. The quantity pH is intended to be a measure of the activity of hydrogen ions in solution. However, since it is defined in terms of a quantity that cannot be measured by a thermodynamically valid method, eqn. (2) can only be considered a *notional definition* of pH.

pH being a single ion quantity, it is not determinable in terms of a fundamental (or base) unit of any measurement system, and there is difficulty providing a proper basis for the traceability of pH measurements. A satisfactory approach is now available in that pH determinations can be incorporated into the International System (SI) if they can be traced to measurements made using a method that fulfils the definition of a 'primary method of measurement' (4).

The essential feature of a primary method is that it must operate according to a well-defined measurement equation in which all of the variables can be determined experimentally in terms of SI units. Any limitation in the determination of the experimental variables, or in the theory, must be included within the estimated uncertainty of the method if traceability to the SI is to be established. If a convention were used without an estimate of its uncertainty, true traceability to SI would not be established. The electrochemical cell without liquid junction, known as the Harned cell (5), fulfills the definition of a primary method for the measurement of the acidity function, $p(a_H\gamma_{Cl})$, and subsequently of the pH of buffer solutions.

The Harned cell is written as

$$Pt \mid H_2 \mid buffer\ S,\ Cl^- \mid AgCl \mid Ag \qquad (Cell\ I)$$

and contains a standard buffer, S, with chloride ions, as potassium or sodium chloride, added in order to use the silver–silver chloride electrode as reference electrode. The application of the Nernst equation to the spontaneous cell reaction of Cell I:

$$½ H_2 + AgCl \rightarrow Ag(s) + H^+ + Cl^-$$

yields the potential difference E_I of the cell (corrected to 1 atm (101.325 kPa), the partial pressure of hydrogen gas used in electrochemistry in preference to 100 kPa) as

$$E_I = E^\circ - (RT/F)\ln 10\ \lg\ [(m_H\gamma_H/m^\circ)(m_{Cl}\gamma_{Cl}/m^\circ)] \qquad (3)$$

which can be rearranged, since $a_H = m_H\gamma_H/m^\circ$, to give the acidity function

$$p(a_H\gamma_{Cl}) = - \lg(a_{HyCl}) = (E_I - E^\circ)/[(RT/F)\ln10] + \lg(m_{Cl}/m^\circ) \quad (4)$$

where E° is the standard potential difference of the cell, and hence of the silver–silver chloride electrode, and γ_{Cl} is the activity coefficient of the chloride ion.

The standard potential difference of the silver–silver chloride electrode, E°, is determined from a Harned cell in which only HCl is present at a fixed molality (e.g., $m = 0.01$ mol kg^{-1})

$$Pt \mid H_2 \mid HCl\ (m) \mid AgCl \mid Ag \qquad (Cell\ Ia)$$

The application of the Nernst equation to the HCl cell (Ia) gives

$$E_{Ia} = E^\circ - (2RT/F)\ln 10\ \lg[(m_{HCl}/m^\circ)(\gamma_{\pm HCl})] \qquad (5)$$

where E_{Ia} has been corrected to 1 atmosphere partial pressure of hydrogen gas (101.325 kPa) and $\gamma_{\pm HCl}$ is the mean ionic activity coefficient of HCl.

Values of the activity coefficient ($\gamma_{\pm HCl}$) at molality 0.01 mol kg^{-1} and various temperatures were given by Bates and Robinson (6). The standard potential difference depends on the method of preparation of the electrodes, but individual determinations of the activity coefficient of HCl at 0.01 mol kg^{-1} are more uniform than values of E°. Hence the practical determination of the potential difference of the cell with HCl at 0.01 mol kg^{-1} is recommended at 298.15 K at which the mean ionic activity coefficient is 0.904. (It is unnecessary to repeat the measurement of E° at other temperatures but simply to correct published smoothed values by the observed difference in E^0 at 298.15 K.)

In national metrology institutes (NMIs), measurements of Cells I and Ia are often done simultaneously in a thermostat bath. Subtracting eqn. (5) from (3) gives

$$\Delta E = E_I - E_{Ia} = - (RT/F)\ln 10\{\lg[(m_H\gamma_H/m^\circ)(m_{Cl}\gamma_{Cl}/m^\circ)] - \lg[(m_{HCl}/m^\circ)^2\gamma^2_{\pm HCl}]\} \qquad (6)$$

which is independent of the standard potential difference. Therefore, the subsequently calculated pH does not depend on the standard potential difference and hence does not depend on the assumption that the standard potential of the hydrogen electrode is zero at all temperatures. Therefore, the Harned cell gives an exact comparison between hydrogen ion activities at different temperatures.

The quantity $p(a_H\gamma_{Cl}) = - \lg (a_H\gamma_{Cl})$, on the left-hand side of (4), is called the acidity function (5). To obtain the quantity pH according to eqn. (2) from the acidity function, it is necessary to evaluate $\lg \gamma_{Cl}$ independently. This is done in two steps: (i) the value of $\lg (a_H\gamma_{Cl})$ at zero chloride molality, $\lg (a_H\gamma_{Cl})^\circ$, is evaluated and (ii) a value for the activity of the chloride ion γ°_{Cl}, at zero chloride molality (sometimes referred to as the limiting or 'trace' activity coefficient) is calculated using the Bates-Guggenheim convention (7). The value of $\lg (a_H\gamma_{Cl})^\circ$ corresponding to zero chloride molality is determined by linear extrapolation of measurements using Harned cells with at least three added molalities of sodium or potassium chloride ($I < 0.1$ mol kg^{-1}).

The value of $\lg(a_H\gamma_{Cl})^\circ$ corresponding to zero chloride molality is determined by linear extrapolation of measurements using Harned cells with at least three added molalities of sodium or potassium chloride ($I < 0.1$ mol kg^{-1}) in accord with eqn. (7):

$$-\lg(a_H\gamma_{Cl}) = -\lg(a_H\gamma_{Cl})^\circ + Sm_{Cl} \quad (7)$$

where S is an empirical, temperature dependent, constant.

The Bates-Guggenheim convention (7) assumes that the trace activity coefficient of the chloride ion γ°_{Cl} is given by

$$\lg\gamma^\circ_{Cl} = -A\,I^{1/2}/(1 + Ba\,I^{1/2}) \quad (8)$$

where A is the Debye-Hückel temperature dependent constant (limiting slope), a is the *mean* distance of closest approach of the ions (ion size parameter), Ba is set equal to 1.5 (mol kg^{-1})$^{-1/2}$ at all temperatures in the range 5–50 °C, and I is the ionic strength of the buffer (which for its evaluation requires knowledge of appropriate acid dissociation constants).

The various stages in the assignment of primary standard pH values are combined in eqn. (9), which is derived from eqns. (4), (5) and (8)

$$\text{pH(PS)} = \lim m_{Cl\to 0}\{(E_t - E^\circ)/[(RT/F)\ln 10] + \lg(m_{Cl}/m^\circ)\} \\ - AI^{1/2}/[1 + 1.5\,(I/m^\circ)^{1/2}] \quad (9)$$

In order for a particular buffer solution to be considered a primary buffer solution, it must be of the "highest metrological" quality (4) in accordance with the definition of a primary standard. It is recommended that it have the following attributes (9):

1. High buffer value in the range 0.016–0.07 (mol OH$^-$)/pH.
2. Small dilution value at half concentration (change in pH with change in buffer concentration) in the range 0.01–0.20.
3. Small dependence of pH on temperature less than ± 0.01 K^{-1}.
4. Low residual liquid junction potential < 0.01 in pH.
5. Ionic strength ≤0.1 mol kg^{-1} to permit applicability of Bates-Guggenheim convention.
6. NMI certificate for specific batch.
7. Reproducible purity of preparation (lot to lot differences of $|\Delta\text{pH(PS)}| < 0.003$).
8. Long-term stability of stored solid material.

Values for the above and other important parameters for the primary and secondary buffer materials are given in Table 1.

Primary Standard Buffers

As there can be significant variations in the purity of samples of a buffer of the same nominal chemical composition, it is essential that the primary buffer material used has been certified with values that have been measured with Cell I. The Harned cell is used by many national metrological institutes for accurate measurements of pH of buffer solutions.

Typical values of the pH(PS) of the seven solutions from the six accepted primary standard reference buffers, which meet the conditions stated above, are listed in Table 2. Batch-to-batch variations in purity can result in changes in the pH value of samples of at most 0.003. The typical values in Table 2 should not be used in place of the certified value (from a Harned cell measurement) for a specific batch of buffer material.

The required attributes listed above effectively limit the range of primary buffers available to between pH 3 and 10 (at 25 °C). Calcium

hydroxide and potassium tetraoxalate are excluded because the contribution of hydroxide or hydrogen ions to the ionic strength is significant. Also excluded are the nitrogen bases of the type BH$^+$ (such as tris(hydroxymethyl)aminomethane and piperazine phosphate) and the zwitterionic buffers (e.g., HEPES and MOPS (10)). These do not comply because either the Bates-Guggenheim convention is not applicable, or the liquid junction potentials are high. This means the choice of primary standards is restricted to buffers derived from oxy-carbon, -phosphorus, -boron and mono, di- and tri-protic carboxylic acids. The uncertainties (11) associated with Harned cell measurements are calculated (1) to be 0.004 in pH at NMIs, with typical variation between batches of primary standard buffers of 0.003.

Secondary Standards

Substances that do not fulfill all the criteria for primary standards, but to which pH values can be assigned using Cell I are considered to be secondary standards (Table 3). Reasons for their exclusion as primary standards include difficulties in achieving consistent and suitable chemical quality (e.g. acetic acid is a liquid), suspected high liquid junction potential, or inappropriateness of the Bates-Guggenheim convention (e.g., other charge-type buffers). The uncertainty is higher (e.g., 0.01) for biological buffers. Certain other substances, which cannot be used in cells containing hydrogen gas electrodes, are also classed as secondary standards.

Calibration Procedures

(a) One-point calibration
A single point calibration is insufficient to determine both slope and one-point parameters. The theoretical value for the slope can be assumed but the practical slope may be up to 5% lower. Alternatively, a value for the practical slope can be assumed from the manufacturer's prior calibration. The one-point calibration therefore yields only an estimate of pH(X). Since both parameters may change with age of the electrodes, this is not a reliable procedure.

(b) Two-point calibration [target uncertainty: 0.02–0.03 at 25 °C]
In the majority of practical applications, glass electrodes cells are calibrated by a two-point calibration, or bracketing, procedure using two standard buffer solutions, with pH values, pH(S$_1$) and pH(S$_2$), bracketing the unknown pH(X). Bracketing is often taken to mean that the pH(S$_1$) and pH(S$_2$) buffers selected from Table 2 should be those that are immediately above and below pH(X). This may not be appropriate in all situations and choice of a wider range may be better.

(c) Multi-point calibration [target uncertainty: 0.01–0.03 at 25 °C].
Multi-point calibration is carried out using up to five standard buffers. The use of more than five points yields no significant improvement in the statistical information obtainable.
Details of uncertainty computations (11) have been given (1).

Measurement of pH and choice of pH Standard Solutions

1a) If pH is not required to better than ±0.05 any pH standard solution may be selected.

1b) If pH is required to ±0.002 and interpretation in terms of hydrogen ion concentration or activity is desired, choose a standard solution, pH(PS), to match X as closely as possible in terms of pH, composition and ionic strength.

2) Alternatively, a bracketing procedure may be adopted whereby two standard solutions are chosen whose pH values, pH(S1), pH(S2) are on either side of pH(X). Then if the corresponding potential difference measurements are $E(S1)$, $E(S2)$, $E(X)$, then pH(X) is obtained from

$$pH(X) = pH(S1) + [E(X) - E(S1)]/\%k$$

where $\%k = 100[E(S2) - E(S1)]/[pH(S2) - pH(S1)]$ is the apparent percentage slope. This procedure is very easily done on some pH meters simply by adjusting downwards the slope factor control with the electrodes in S2. The purpose of the bracketing procedure is to compensate for deficiencies in the electrodes and measuring system.

Information to be given about the measurement of pH(X)

The standard solutions selected for calibration of the pH meter system should be reported with the measurement as follows:

System calibrated with pH(S) = ... at ... K.

System calibrated with two primary standards, pH(PS1) = ... and pH(PS2) = ... at .... K.

System calibrated with n standards, pH(S1) = ..., pH(S2) = ... etc. at .... K.

Interpretation of pH(X) in terms of hydrogen ion concentration

The defined pH has no simple interpretation in terms of hydrogen ion concentration but the mean ionic activity coefficient of a typical 1:1 electrolyte can be used to obtain hydrogen ion concentration subject to an uncertainty of 3.9% in concentration, corresponding to 0.02 in pH.

References

1. Buck, R.P., Rondinini, S., Covington, A.K., Baucke, F.G.K., Brett, C.M.A., Camoes, M.F.C., Milton, M.J.T., Mussini, T., Naumann, R., Pratt, K.W., Spitzer, P., and Wilson, G.S. *Pure Appl. Chem.*, 74, 2105, 2002.
2. Sørensen, S.P.L. *Comp. Rend. Trav. Lab. Carlsberg*, 8, 1, 1909.
3. Sørensen, S.P.L., and Linderstrøm-Lang, K.L., *Comp. Rend. Trav. Lab. Carlsberg*, 15, 1924.
4. BIPM, *Com. Cons. Quantité de Matière* 4, 1998. See also M.J.T. Milton and T.J. Quinn, *Metrologia* 38, 289, 2001.
5. Harned H.S., and Owen, B.B., *The Physical Chemistry of Electrolytic Solutions*, Ch 14, Reinhold, New York, 1958.
6. Bates R.G., and Robinson, R.A., *J. Soln. Chem.* 9, 455, 1980.
7. Bates R.G., and Guggenheim, E.A., *Pure Appl. Chem.* 1, 163, 1960.
8. *International Vocabulary of Basic and General Terms in Metrology* (VIM), Beuth, Berlin, 2nd Edn. 1994.
9. Bates, R.G. *Determination of pH*, Wiley, New York, 1973.
10. Good, N.E. et al., *Biochem. J.* 5, 467, 1966.
11. *Guide to the Expression of Uncertainty* (GUM), BIPM, IEC, IFCC, ISO, IUPAC, IUPAP, OIML, 1993.

TABLE 1. Summary of Useful Properties of Some Primary and Secondary Standard Buffer Substances and Solutions

Salt or solid substance	Formula	Molality/ mol kg^{-1}	Molar mass/ g mol^{-1}	Density/ g/mL	Amount conc. at 20 °C/ mol dm^{-3}	Mass/g to make 1 dm^3	Dilution value ΔpH$_{1/2}$	Buffer value (β)/ mol OH$^-$ dm^{-3}	pH Temperature coefficient/ K^{-1}
Potassium tetroxalate dihydrate	KH$_3$C$_4$O$_8$·2H$_2$O	0.1	254.191	1.0091	0.09875	25.101			
Potassium tetraoxalate dihydrate	KH$_3$C$_4$O$_8$·2H$_2$O	0.05	254.191	1.0032	0.04965	12.620	0.186	0.070	0.001
Potassium hydrogen tartrate (sat at 25 °C)	KHC$_4$H$_4$O$_6$	0.0341	188.18	1.0036	0.034	6.4	0.049	0.027	−0.0014
Potassium dihydrogen citrate	KH$_2$C$_6$H$_5$O$_7$	0.05	230.22	1.0029	0.04958	11.41	0.024	0.034	−0.022
Potassium hydrogen phthalate	KHC$_8$H$_4$O$_4$	0.05	204.44	1.0017	0.04958	10.12	0.052	0.016	0.00012
Disodium hydrogen orthophosphate +	Na$_2$HPO$_4$	0.025	141.958	1.0038	0.02492	3.5379	0.080	0.029	−0.0028
potassium dihydrogen orthophosphate	KH$_2$PO$_4$	0.025	136.085			3.3912			
Disodium hydrogen orthophosphate +	Na$_2$HPO$_4$	0.03043	141.959	1.0020	0.08665	4.302	0.07	0.016	−0.0028
potassium dihydrogen orthophosphate	KH$_2$PO$_4$	0.00869	136.085		0.03032	1.179			
Disodium tetraborate decahydrate	Na$_2$B$_4$O$_7$·10H$_2$O	0.05	381.367	1.0075	0.04985	19.012			
Disodium tetraborate decahydrate	Na$_2$B$_4$O$_7$·10H$_2$O	0.01	381.367	1.0001	0.00998	3.806	0.01	0.020	−0.0082
Sodium hydrogen carbonate +	NaHCO$_3$	0.025	84.01	1.0013	0.02492	2.092	0.079	0.029	−0.0096
sodium carbonate	Na$_2$CO$_3$	0.025	105.99			2.640			
Calcium hydroxide (sat. at 25 °C)	Ca(OH)$_2$	0.0203	74.09	0.9991	0.02025	1.5	−0.28	0.09	−0.033

TABLE 2. Typical Values of pH(PS) for Primary Standards at 0–50 °C

Primary standards (PS)	Temperature in °C										
	0	5	10	15	20	25	30	35	37	40	50
Sat. potassium hydrogen tartrate (at 25 °C)						3.557	3.552	3.549	3.548	3.547	3.549
0.05 mol kg⁻¹ potassium dihydrogen citrate	3.863	3.840	3.820	3.802	3.788	3.776	3.766	3.759	3.756	3.754	3.749
0.05 mol kg⁻¹ potassium hydrogen phthalate	4.000	3.998	3.997	3.998	4.000	4.005	4.011	4.018	4.022	4.027	4.050
0.025 mol kg⁻¹ disodium hydrogen phosphate + 0.025 mol kg⁻¹ potassium dihydrogen phosphate	6.984	6.951	6.923	6.900	6.881	6.865	6.853	6.844	6.841	6.838	6.833
0.03043 mol kg⁻¹ disodium hydrogen phosphate + 0.008695 mol kg⁻¹ potassium dihydrogen phosphate	7.534	7.500	7.472	7.448	7.429	7.413	7.400	7.389	7.386	7.380	7.367
0.01 mol kg⁻¹ disodium tetraborate	9.464	9.395	9.332	9.276	9.225	9.180	9.139	9.102	9.088	9.068	9.011
0.025 mol kg⁻¹ sodium hydrogen carbonate + 0.025 mol kg⁻¹ sodium carbonate	10.317	10.245	10.179	10.118	10.062	10.012	9.966	9.926	9.910	9.889	9.828

TABLE 3. Values of pH(SS) of Some Secondary Standards from Harned Cell I Measurements

Secondary standards	Temperature in °C									
	0	5	10	15	20	25	30	37	40	50
0.05 mol kg⁻¹ potassium tetroxalate[a]	1.67	1.67	1.67	1.67	1.68	1.68	1.68	1.69	1.69	1.71
0.05 mol kg⁻¹ sodium hydrogen diglycolate[b]		3.47	3.47	3.48	3.48	3.49	3.50	3.52	3.53	3.56
0.1 mol dm⁻³ acetic acid + 0.1 mol dm⁻³ sodium acetate	4.68	4.67	4.67	4.66	4.66	4.65	4.65	4.66	4.66	4.68
0.01 mol dm⁻³ acetic acid + 0.1 mol dm⁻³ sodium acetate	4.74	4.73	4.73	4.72	4.72	4.72	4.72	4.73	4.73	4.75
0.02 mol kg⁻¹ piperazine phosphate[c]	6.58	6.51	6.45	6.39	6.34	6.29	6.24	6.16	6.14	6.06
0.05 mol kg⁻¹ tris hydrochloride + 0.01667 mol kg⁻¹ tris[c]	8.47	8.30	8.14	7.99	7.84	7.70	7.56	7.38	7.31	7.07
0.05 mol kg⁻¹ disodium tetraborate	9.51	9.43	9.36	9.30	9.25	9.19	9.15	9.09	9.07	9.01
Saturated (at 25 °C) calcium hydroxide	13.42	13.21	13.00	12.81	12.63	12.45	12.29	12.07	11.98	11.71

[a] Potassium trihydrogen dioxalate ($KH_3C_4O_8$)
[b] Sodium hydrogen 2,2 -oxydiacetate
[c] 2-Amino-2-(hydroxymethyl)-1,3 propanediol or tris(hydroxymethyl)aminomethane

PRACTICAL pH MEASUREMENTS ON NATURAL WATERS

A. K. Covington and W. Davison

(1) Dilute solutions and freshwater including 'acid-rain' samples ($I < 0.02$ mol kg^{-1})

Major problems could be encountered due to errors associated with the liquid junction. It is recommended that either a free diffusion junction is used or it is verified that the junction is working correctly using dilute solutions as follows. For commercial electrodes calibrated with IUPAC aqueous RVS or PS standards, the pH(X) of dilute solutions should be within ±0.02 of those given in Table 1. The difference in determined pH(X) between a stirred and unstirred dilute solution should be < 0.02. The characteristics of glass electrodes are such that below pH 5 the readings should be stable within 2 min, but for pH 5 to 8.8 or so minutes may be necessary to attain stability. Interpretation of pH(X) measured in this way in terms of activity of hydrogen ion, a_{H+} is subject[1] to an uncertainty of ±0.02 in pH.

(2) Seawater

Measurements made by calibration of electrodes with IUPAC aqueous RVS or PS standards to obtain pH(X) are perfectly valid. However, the interpretation of pH(X) in terms of the activity of hydrogen ion is complicated by the nonzero residual liquid junction potential as well as by systematic differences between electrode pairs, principally attributable to the reference electrode. For 35‰ salinity seawater ($S = 0.035$) a_{H+} calculated from pH(X) is typically 12% too low. Special seawater pH scales have been devised to overcome this problem:

(i) The total hydrogen ion scale, pH$_T$, is defined in terms of the sum of free and complexed (total) hydrogen ion concentrations, where

$$^T C_H = [H^+] + [HSO_4^-] + [HF].$$

$$\text{So, pH}_T = - \log {}^T C_H$$

Calibration of the electrodes with a buffer having a composition similar to that of seawater, to which pH$_T$ has been assigned, results in values of pHT(X) (Tables 2, 3) which are accurately interpretable in terms of $^T C_H$.

(ii) The free hydrogen ion scale, pH$_F$, is defined, and fully interpretable, in terms of the concentration of free hydrogen ions.

$$\text{pH}_F = - \log [H^+]$$

Values of pH$_F$ as a function of temperature have been assigned to the same set of pH$_T$ seawater buffers, and so alternatively can be used for calibration (Tables 2, 3).[2,3]

(3) Estuarine water

Prescriptions for seawater scale buffers are available for a range of salinities. Reliable estuarine pH measurements can be made by calibrating with a buffer of the same salinity as the sample. However, these buffers are difficult to prepare and their use presumes prior knowledge of salinity of the sample. Interpretable measurements of estuarine pH can be made by calibration with IUPAC aqueous RVS or PS standards if the electrode pair is additionally calibrated using a 20‰ salinity seawater buffer.[4] The difference between the assigned pH$_{SWS}$ of the seawater buffer and its measured pH(X) value using RVS or PS standards is

$$\Delta pH = pH_{SWS} - pH(X)$$

Values of ΔpH should be in the range of 0.08 to 0.18. It empirically corrects for differences between the two pH scales and for measurement errors associated with the electrode pair. The pH(X) of samples measured using IUPAC aqueous buffers, can be converted to pH$_T$ or pH$_F$ using the appropriate measured ΔpH:

$$pH_T = pH(X) - \Delta pH$$

$$\text{or } pH_F = pH(X) - \Delta pH$$

This simple procedure is appropriate to pH measurement at salinities from 2‰ to 35‰. For salinities lower than 2‰ the procedures for freshwaters should be adopted.

References

1. Davison, W. and Harbinson, T. R., *Analyst*, 113, 709, 1988.
2. Culberson, C. H., in *Marine Electrochemistry*, Whitfield, M. and Jagner, D., Eds., Wiley, 1981.
3. Millero, F. J., *Limnol. Oceanogr.*, 31, 839, 1986.
4. Covington, A. K., Whalley, P. D., Davison, W., and Whitfield, M., in *The Determination of Trace Metals in Natural Waters*, West, T. S. and Nurnberg, H. W., Eds., Blackwell, Oxford, 1988.
5. Koch, W. F., Marinenko, G., and Paule, R. C., *J. Res. NBS*, 91, 33, 1986.

TABLE 1. pH of Dilute Solutions at 25 °C, Degassed and Equilibrated with Air, Suitable as Quality Control Standards

	Ionic strength mmol kg^{-1}	Concentration(x) mmol kg^{-1}	pH $p_{CO_2} = 0$	pH $p_{CO_2} = $ air
Potassium hydrogen phthalate	10.7	10	4.12	4.12
	1.1	1	4.33	4.33
xKH$_2$PO$_4$ + xNa$_2$HPO$_4$	9.9	2.5	7.07	7.05
xKH$_2$PO$_4$ + 3.5xNa$_2$HPO$_4$	10	0.87	7.61	7.58
Na$_2$B$_4$O$_7$ · 10H$_2$O	10	5	9.20	—
HCl	0.1	0.1	4.03	4.03
SRM2694-I[a]	—	—	4.30	—
SRM2694-II[a]	—	—	3.59	—

Note: The pH of solutions near to pH 4 is virtually independent of temperature over the range of 5 to 30 °C.

[a] Simulated rainwater samples are available (Reference 5) from NIST containing sulfate, nitrate, chloride, fluoride, sodium, potassium, calcium and magnesium.

TABLE 2. Composition of Seawater Buffer of Salinity S = 35‰ at 25 °C (Reference 3)

Solute	mol dm^{-3}	mol kg^{-1}	g kg^{-1}	g dm^{-3}
NaCl	0.3666	0.3493	20.416	20.946
Na$_2$SO$_4$	0.02926	0.02788	3.96	4.063
KCl	0.01058	0.01008	0.752	0.772
CaCl$_2$	0.01077	0.01026	1.139	1.169
MgCl$_2$	0.05518	0.05258	5.006	5.139
Tris	0.06	0.05717	6.926	7.106
Tris · HCl	0.06	0.05717	9.010	9.244

Tris = tris(hydroxymethyl)aminomethane (HOCH$_2$)$_3$CNH$_2$.
A 20‰ buffer is made by diluting the 35‰ in the ratio 20:35.

TABLE 3. Assigned Values of 20‰ and 35‰ Buffers on Free and Total Hydrogen Ion Scales. Calculated from Equations Provided by Millero (Reference 3)

Temp (°C)	pH$_T$ $S = 20‰$	pH$_T$ $S = 35‰$	pH$_F$ $S = 20‰$	pH$_F$ $S = 35‰$
5	8.683	8.718	8.759	8.81
10	8.513	8.542	8.597	8.647
15	8.351	8.374	8.442	8.491
20	8.195	8.212	8.292	8.341
25	8.045	8.057	8.149	8.197
30	7.901	7.908	8.011	8.059
35	7.762	7.764	7.879	7.926

BUFFER SOLUTIONS GIVING ROUND VALUES OF pH AT 25 °C

A pH	A x	B pH	B x	C pH	C x	D pH	D x	E pH	E x
1.00	67.0	2.20	49.5	4.10	1.3	5.80	3.6	7.00	46.6
1.10	52.8	2.30	45.8	4.20	3.0	5.90	4.6	7.10	45.7
1.20	42.5	2.40	42.2	4.30	4.7	6.00	5.6	7.20	44.7
1.30	33.6	2.50	38.8	4.40	6.6	6.10	6.8	7.30	43.4
1.40	26.6	2.60	35.4	4.50	8.7	6.20	8.1	7.40	42.0
1.50	20.7	2.70	32.1	4.60	11.1	6.30	9.7	7.50	40.3
1.60	16.2	2.80	28.9	4.70	13.6	6.40	11.6	7.60	38.5
1.70	13.0	2.90	25.7	4.80	16.5	6.50	13.9	7.70	36.6
1.80	10.2	3.00	22.3	4.90	19.4	6.60	16.4	7.80	34.5
1.90	8.1	3.10	18.8	5.00	22.6	6.70	19.3	7.90	32.0
2.00	6.5	3.20	15.7	5.10	25.5	6.80	22.4	8.00	29.2
2.10	5.10	3.30	12.9	5.20	28.8	6.90	25.9	8.10	26.2
2.20	3.9	3.40	10.4	5.30	31.6	7.00	29.1	8.20	22.9
		3.50	8.2	5.40	34.1	7.10	32.1	8.30	19.9
		3.60	6.3	5.50	36.6	7.20	34.7	8.40	17.2
		3.70	4.5	5.60	38.8	7.30	37.0	8.50	14.7
		3.80	2.9	5.70	40.6	7.40	39.1	8.60	12.2
		3.90	1.4	5.80	42.3	7.50	40.9	8.70	10.3
		4.00	0.1	5.90	43.7	7.60	42.4	8.80	8.5
						7.70	43.5	8.90	7.0
						7.80	44.5	9.00	5.7
						7.90	45.3		
						8.00	46.1		

F pH	F x	G pH	G x	H pH	H x	I pH	I x	J pH	J x
8.00	20.5	9.20	0.9	9.60	5.0	10.90	3.3	12.00	6.0
8.10	19.7	9.30	3.6	9.70	6.2	11.00	4.1	12.10	8.0
8.20	18.8	9.40	6.2	9.80	7.6	11.10	5.1	12.20	10.2
8.30	17.7	9.50	8.8	9.90	9.1	11.20	6.3	12.30	12.8
8.40	16.6	9.60	11.1	10.00	10.7	11.30	7.6	12.40	16.2
8.50	15.2	9.70	13.1	10.10	12.2	11.40	9.1	12.50	20.4
8.60	13.5	9.80	15.0	10.20	13.8	11.50	11.1	12.60	25.6
8.70	11.6	9.90	16.7	10.30	15.2	11.60	13.5	12.70	32.2
8.80	9.6	10.00	18.3	10.40	16.5	11.70	16.2	12.80	41.2
8.90	7.1	10.10	19.5	10.50	17.8	11.80	19.4	12.90	53.0
9.00	4.6	10.20	20.5	10.60	19.1	11.90	23.0	13.00	66.0
9.10	2.0	10.30	21.3	10.70	20.2	12.00	26.9		
		10.40	22.1	10.80	21.2				
		10.50	22.7	10.90	22.0				
		10.60	23.3	11.00	22.7				
		10.70	23.8						
		10.80	24.25						

A. 25 ml of 0.2 molar KCl + x ml of 0.2 molar HCl.
B. 50 ml of 0.1 molar potassium hydrogen phthalate + x ml of 0.1 molar HCl.
C. 50 ml of 0.1 molar potassium hydrogen phthalate + x ml of 0.1 molar NaOH.
D. 50 ml of 0.1 molar potassium dihydrogen phosphate + x ml of 0.1 molar NaOH.
E. 50 ml of 0.1 molar tris(hydroxymethyl)aminomethane + x ml of 0.1 M HCl.
F. 50 ml of 0.025 molar borax + x ml of 0.1 molar HCl.
G. 50 ml of 0.025 molar borax + x ml of 0.1 molar NaOH.
H. 50 ml of 0.05 molar sodium bicarbonate + x ml of 0.1 molar NaOH.
I. 50 ml of 0.05 molar disodium hydrogen phosphate + x ml of 0.1 molar NaOH.
J. 25 ml of 0.2 molar KCl + x ml of 0.2 molar NaOH.

Final volume of mixtures = 100 ml.

References

1. Bower, V. E., and Bates, R. G., *J. Res. Natl. Bur. Stand.*, 55, 197, 1955 (A–D).
2. Bates, R. G., and Bower, V. E., *Anal. Chem.*, 28, 1322, 1956 (E–J).

CONCENTRATIVE PROPERTIES OF AQUEOUS SOLUTIONS: DENSITY, REFRACTIVE INDEX, FREEZING POINT DEPRESSION, AND VISCOSITY

This table gives properties of aqueous solutions of 66 substances as a function of concentration. All data refer to a temperature of 20 °C. The properties are:

Mass %: Mass of solute divided by total mass of solution, expressed as percent.

m Molality (moles of solute per kg of water).

c Molarity (moles of solute per liter of solution).

ρ Density of solution in g/cm³.

n Index of refraction, relative to air, at a wavelength of 589 nm (sodium D line); the index of pure water at 20 °C is 1.3330.

Δ Freezing point depression in °C relative to pure water.

η Absolute (dynamic) viscosity in mPa s (equal to centipoise, cP); the viscosity of pure water at 20 °C is 1.002 mPa s.

Density data for aqueous solutions over a wider range of temperatures and pressures (and for other compounds) may be found in Reference 2. Solutes are listed in the following order:

Acetic acid
Acetone
Ammonia
Ammonium chloride
Ammonium sulfate
Barium chloride
Calcium chloride
Cesium chloride
Citric acid
Copper sulfate
Disodium ethylenediamine tetraacetate (EDTA sodium)
Ethanol
Ethylene glycol
Ferric chloride
Formic acid
D-Fructose
D-Glucose
Glycerol
Hydrochloric acid
Lactic acid
Lactose

Lithium chloride
Magnesium chloride
Magnesium sulfate
Maltose
Manganese(II) sulfate
D-Mannitol
Methanol
Nitric acid
Oxalic acid
Phosphoric acid
Potassium bicarbonate
Potassium bromide
Potassium carbonate
Potassium chloride
Potassium hydroxide
Potassium iodide
Potassium nitrate
Potassium permanganate
Potassium hydrogen phosphate
Potassium dihydrogen phosphate
Potassium sulfate
1-Propanol

2-Propanol
Silver nitrate
Sodium acetate
Sodium bicarbonate
Sodium bromide
Sodium carbonate
Sodium chloride
Sodium citrate
Sodium hydroxide
Sodium nitrate
Sodium phosphate
Sodium hydrogen phosphate
Sodium dihydrogen phosphate
Sodium sulfate
Sodium thiosulfate
Strontium chloride
Sucrose
Sulfuric acid
Trichloroacetic acid
Tris(hydroxymethyl)methylamine
Urea
Zinc sulfate

References

1. Wolf, A. V., *Aqueous Solutions and Body Fluids*, Hoeber, 1966.
2. Söhnel, O., and Novotny, P., *Densities of Aqueous Solutions of Inorganic Substances*, Elsevier, Amsterdam, 1985.

Solute	Mass %	m/mol kg⁻¹	c/mol L⁻¹	ρ/g cm⁻³	n	Δ/°C	η/mPa s
Acetic acid	0.5	0.084	0.083	0.9989	1.3334	0.16	1.012
CH₃COOH	1.0	0.168	0.166	0.9996	1.3337	0.32	1.022
	2.0	0.340	0.333	1.0011	1.3345	0.63	1.042
	3.0	0.515	0.501	1.0025	1.3352	0.94	1.063
	4.0	0.694	0.669	1.0038	1.3359	1.26	1.084
	5.0	0.876	0.837	1.0052	1.3366	1.58	1.105
	6.0	1.063	1.006	1.0066	1.3373	1.90	1.125
	7.0	1.253	1.175	1.0080	1.3381	2.23	1.143
	8.0	1.448	1.345	1.0093	1.3388	2.56	1.162
	9.0	1.647	1.515	1.0107	1.3395	2.89	1.186
	10.0	1.850	1.685	1.0121	1.3402	3.23	1.210
	12.0	2.271	2.028	1.0147	1.3416	3.91	1.253
	14.0	2.711	2.372	1.0174	1.3430	4.61	1.298
	16.0	3.172	2.718	1.0200	1.3444	5.33	1.341
	18.0	3.655	3.065	1.0225	1.3458	6.06	1.380
	20.0	4.163	3.414	1.0250	1.3472	6.81	1.431
	22.0	4.697	3.764	1.0275	1.3485	7.57	1.478
	24.0	5.259	4.116	1.0299	1.3498	8.36	1.525

Solute	Mass %	m/mol kg^{-1}	c/mol L^{-1}	ρ/g cm^{-3}	n	Δ/°C	η/mPa s
	26.0	5.851	4.470	1.0323	1.3512	9.17	1.572
	28.0	6.476	4.824	1.0346	1.3525	10.00	1.613
	30.0	7.137	5.180	1.0369	1.3537	10.84	1.669
	32.0	7.837	5.537	1.0391	1.3550	11.70	1.715
	34.0	8.579	5.896	1.0413	1.3562	12.55	1.762
	36.0	9.367	6.255	1.0434	1.3574	13.38	1.812
	38.0	10.207	6.615	1.0454	1.3586		1.852
	40.0	11.102	6.977	1.0474	1.3598		1.912
	50.0	16.653	8.794	1.0562	1.3653		2.158
	60.0	24.979	10.620	1.0629	1.3700		2.409
	70.0	38.857	12.441	1.0673	1.3738		2.629
	80.0	66.611	14.228	1.0680	1.3767		2.720
	90.0	149.875	15.953	1.0644	1.3771		2.386
	92.0	191.507	16.284	1.0629	1.3766		2.240
	94.0	260.894	16.602	1.0606	1.3759		2.036
	96.0	399.667	16.911	1.0578	1.3748		1.813
	98.0	815.987	17.198	1.0538	1.3734		1.535
	100.0		17.447	1.0477	1.3716		1.223
Acetone	0.5	0.087	0.086	0.9975	1.3334	0.16	1.013
$(CH_3)_2CO$	1.0	0.174	0.172	0.9968	1.3337	0.32	1.024
	2.0	0.351	0.343	0.9954	1.3344	0.65	1.047
	3.0	0.533	0.513	0.9940	1.3352	0.97	1.072
	4.0	0.717	0.684	0.9926	1.3359	1.30	1.099
	5.0	0.906	0.853	0.9912	1.3366	1.63	1.125
	6.0	1.099	1.023	0.9899	1.3373	1.96	1.150
	7.0	1.296	1.191	0.9886	1.3381	2.29	1.174
	8.0	1.497	1.360	0.9874	1.3388	2.62	1.198
	9.0	1.703	1.528	0.9861	1.3395	2.95	1.221
	10.0	1.913	1.696	0.9849	1.3402	3.29	1.244
Ammonia	0.5	0.295	0.292	0.9960	1.3332	0.55	1.009
NH_3	1.0	0.593	0.584	0.9938	1.3335	1.14	1.015
	2.0	1.198	1.162	0.9895	1.3339	2.32	1.029
	3.0	1.816	1.736	0.9853	1.3344	3.53	1.043
	4.0	2.447	2.304	0.9811	1.3349	4.78	1.057
	5.0	3.090	2.868	0.9770	1.3354	6.08	1.071
	6.0	3.748	3.428	0.9730	1.3359	7.43	1.085
	7.0	4.420	3.983	0.9690	1.3365	8.95	1.099
	8.0	5.106	4.533	0.9651	1.3370	10.34	1.113
	9.0	5.807	5.080	0.9613	1.3376	11.90	1.127
	10.0	6.524	5.622	0.9575	1.3381	13.55	1.141
	12.0	8.007	6.695	0.9502	1.3393	17.13	1.169
	14.0	9.558	7.753	0.9431	1.3404	21.13	1.195
	16.0	11.184	8.794	0.9361	1.3416	25.63	1.218
	18.0	12.889	9.823	0.9294	1.3428	30.70	1.237
	20.0	14.679	10.837	0.9228	1.3440	36.42	1.254
	22.0	16.561	11.838	0.9164	1.3453	43.36	1.268
	24.0	18.542	12.826	0.9102	1.3465	51.38	1.280
	26.0	20.630	13.801	0.9040	1.3477	60.77	1.288
	28.0	22.834	14.764	0.8980	1.3490	71.66	
	30.0	25.164	15.713	0.8920	1.3502	84.06	
Ammonium	0.5	0.094	0.093	0.9998	1.3340	0.32	0.999
chloride	1.0	0.189	0.187	1.0014	1.3349	0.64	0.996
NH_4Cl	2.0	0.382	0.376	1.0045	1.3369	1.27	0.992
	3.0	0.578	0.565	1.0076	1.3388	1.91	0.988
	4.0	0.779	0.756	1.0107	1.3407	2.57	0.985
	5.0	0.984	0.948	1.0138	1.3426	3.25	0.982
	6.0	1.193	1.141	1.0168	1.3445	3.94	0.979
	7.0	1.407	1.335	1.0198	1.3464	4.66	0.976

Solute	Mass %	$m/\text{mol kg}^{-1}$	$c/\text{mol L}^{-1}$	$\rho/\text{g cm}^{-3}$	n	$\Delta/°\text{C}$	$\eta/\text{mPa s}$
	8.0	1.626	1.529	1.0227	1.3483	5.40	0.974
	9.0	1.849	1.726	1.0257	1.3502	6.16	0.972
	10.0	2.077	1.923	1.0286	1.3521	6.95	0.970
	12.0	2.549	2.320	1.0344	1.3559	8.60	0.969
	14.0	3.043	2.722	1.0401	1.3596		0.969
	16.0	3.561	3.128	1.0457	1.3634		0.971
	18.0	4.104	3.537	1.0512	1.3671		0.973
	20.0	4.674	3.951	1.0567	1.3708		0.978
	22.0	5.273	4.368	1.0621	1.3745		0.986
	24.0	5.903	4.789	1.0674	1.3782		0.996
Ammonium	0.5	0.038	0.038	1.0012	1.3338	0.17	1.008
sulfate	1.0	0.076	0.076	1.0042	1.3346	0.33	1.014
$(NH_4)_2SO_4$	2.0	0.154	0.153	1.0101	1.3363	0.63	1.027
	3.0	0.234	0.231	1.0160	1.3379	0.92	1.041
	4.0	0.315	0.309	1.0220	1.3395	1.21	1.057
	5.0	0.398	0.389	1.0279	1.3411	1.49	1.073
	6.0	0.483	0.469	1.0338	1.3428	1.77	1.090
	7.0	0.570	0.551	1.0397	1.3444	2.05	1.108
	8.0	0.658	0.633	1.0456	1.3460	2.33	1.127
	9.0	0.748	0.716	1.0515	1.3476	2.61	1.147
	10.0	0.841	0.800	1.0574	1.3492	2.89	1.168
	12.0	1.032	0.971	1.0691	1.3523	3.47	1.210
	14.0	1.232	1.145	1.0808	1.3555	4.07	1.256
	16.0	1.441	1.323	1.0924	1.3586	4.69	1.305
	18.0	1.661	1.504	1.1039	1.3616		1.359
	20.0	1.892	1.688	1.1154	1.3647		1.421
	22.0	2.134	1.876	1.1269	1.3677		1.490
	24.0	2.390	2.067	1.1383	1.3707		1.566
	26.0	2.659	2.262	1.1496	1.3737		1.650
	28.0	2.943	2.460	1.1609	1.3766		1.743
	30.0	3.243	2.661	1.1721	1.3795		1.847
	32.0	3.561	2.866	1.1833	1.3824		1.961
	34.0	3.898	3.073	1.1945	1.3853		2.086
	36.0	4.257	3.284	1.2056	1.3881		2.222
	38.0	4.638	3.499	1.2166	1.3909		2.371
	40.0	5.045	3.716	1.2277	1.3938		2.530
Barium	0.5	0.024	0.024	1.0026	1.3337	0.12	1.009
chloride	1.0	0.049	0.048	1.0070	1.3345	0.23	1.016
$BaCl_2$	2.0	0.098	0.098	1.0159	1.3360	0.46	1.026
	3.0	0.149	0.148	1.0249	1.3375	0.69	1.037
	4.0	0.200	0.199	1.0341	1.3391	0.93	1.049
	5.0	0.253	0.251	1.0434	1.3406	1.18	1.062
	6.0	0.307	0.303	1.0528	1.3422	1.44	1.075
	7.0	0.361	0.357	1.0624	1.3438	1.70	1.087
	8.0	0.418	0.412	1.0721	1.3454	1.98	1.101
	9.0	0.475	0.468	1.0820	1.3470	2.27	1.114
	10.0	0.534	0.524	1.0921	1.3487	2.58	1.129
	12.0	0.655	0.641	1.1128	1.3520	3.22	1.161
	14.0	0.782	0.763	1.1342	1.3555	3.92	1.195
	16.0	0.915	0.889	1.1564	1.3591	4.69	1.234
	18.0	1.054	1.019	1.1793	1.3627		1.277
	20.0	1.201	1.156	1.2031	1.3664		1.325
	22.0	1.355	1.297	1.2277	1.3703		1.378
	24.0	1.517	1.444	1.2531	1.3741		1.437
	26.0	1.687	1.597	1.2793	1.3781		1.503
Calcium	0.5	0.045	0.045	1.0024	1.3342	0.22	1.015
chloride	1.0	0.091	0.091	1.0065	1.3354	0.44	1.028
$CaCl_2$	2.0	0.184	0.183	1.0148	1.3378	0.88	1.050

Solute	Mass %	m/mol kg^{-1}	c/mol L^{-1}	ρ/g cm^{-3}	n	Δ/°C	η/mPa s
	3.0	0.279	0.277	1.0232	1.3402	1.33	1.078
	4.0	0.375	0.372	1.0316	1.3426	1.82	1.110
	5.0	0.474	0.469	1.0401	1.3451	2.35	1.143
	6.0	0.575	0.567	1.0486	1.3475	2.93	1.175
	7.0	0.678	0.667	1.0572	1.3500	3.57	1.208
	8.0	0.784	0.768	1.0659	1.3525	4.28	1.242
	9.0	0.891	0.872	1.0747	1.3549	5.04	1.279
	10.0	1.001	0.976	1.0835	1.3575	5.86	1.319
	12.0	1.229	1.191	1.1014	1.3625	7.70	1.408
	14.0	1.467	1.413	1.1198	1.3677	9.83	1.508
	16.0	1.716	1.641	1.1386	1.3730	12.28	1.625
	18.0	1.978	1.878	1.1579	1.3784	15.11	1.764
	20.0	2.253	2.122	1.1775	1.3839	18.30	1.930
	22.0	2.541	2.374	1.1976	1.3895	21.70	2.127
	24.0	2.845	2.634	1.2180	1.3951	25.30	2.356
	26.0	3.166	2.902	1.2388	1.4008	29.70	2.645
	28.0	3.504	3.179	1.2600	1.4066	34.70	3.000
	30.0	3.862	3.464	1.2816	1.4124	41.00	3.467
	32.0	4.240	3.759	1.3036	1.4183	49.70	4.035
	34.0	4.642	4.062	1.3260	1.4242		4.820
	36.0	5.068	4.375	1.3488	1.4301		5.807
	38.0	5.522	4.698	1.3720	1.4361		7.321
	40.0	6.007	5.030	1.3957	1.4420		8.997
Cesium	0.5	0.030	0.030	1.0020	1.3334	0.10	1.000
chloride	1.0	0.060	0.060	1.0058	1.3337	0.20	0.997
CsCl	2.0	0.121	0.120	1.0135	1.3345	0.40	0.992
	3.0	0.184	0.182	1.0214	1.3353	0.61	0.988
	4.0	0.247	0.245	1.0293	1.3361	0.81	0.984
	5.0	0.313	0.308	1.0374	1.3369	1.02	0.980
	6.0	0.379	0.373	1.0456	1.3377	1.22	0.977
	7.0	0.447	0.438	1.0540	1.3386	1.43	0.974
	8.0	0.516	0.505	1.0625	1.3394	1.64	0.971
	9.0	0.587	0.573	1.0711	1.3403	1.85	0.969
	10.0	0.660	0.641	1.0798	1.3412	2.06	0.966
	12.0	0.810	0.782	1.0978	1.3430	2.51	0.961
	14.0	0.967	0.928	1.1163	1.3448	2.97	0.955
	16.0	1.131	1.079	1.1355	1.3468	3.46	0.950
	18.0	1.304	1.235	1.1552	1.3487	3.96	0.945
	20.0	1.485	1.397	1.1756	1.3507	4.49	0.939
	22.0	1.675	1.564	1.1967	1.3528		0.934
	24.0	1.876	1.737	1.2185	1.3550		0.930
	26.0	2.087	1.917	1.2411	1.3572		0.926
	28.0	2.310	2.103	1.2644	1.3594		0.924
	30.0	2.546	2.296	1.2885	1.3617		0.922
	32.0	2.795	2.497	1.3135	1.3641		0.922
	34.0	3.060	2.705	1.3393	1.3666		0.924
	36.0	3.341	2.921	1.3661	1.3691		0.926
	38.0	3.640	3.146	1.3938	1.3717		0.930
	40.0	3.960	3.380	1.4226	1.3744		0.934
	42.0	4.301	3.624	1.4525	1.3771		0.940
	44.0	4.667	3.877	1.4835	1.3800		0.947
	46.0	5.060	4.142	1.5158	1.3829		0.956
	48.0	5.483	4.418	1.5495	1.3860		0.967
	50.0	5.940	4.706	1.5846	1.3892		0.981
	60.0	8.910	6.368	1.7868	1.4076		1.120
	64.0	10.560	7.163	1.8842	1.4167		1.238
Citric acid	0.5	0.026	0.026	1.0002	1.3336	0.05	1.013
$(HO)C(COOH)_3$	1.0	0.053	0.052	1.0022	1.3343	0.11	1.024
	2.0	0.106	0.105	1.0063	1.3356	0.21	1.048

Solute	Mass %	m/mol kg^{-1}	c/mol L^{-1}	ρ/g cm^{-3}	n	Δ/°C	η/mPa s
	3.0	0.161	0.158	1.0105	1.3368	0.32	1.073
	4.0	0.217	0.211	1.0147	1.3381	0.43	1.098
	5.0	0.274	0.265	1.0189	1.3394	0.54	1.125
	6.0	0.332	0.320	1.0232	1.3407	0.65	1.153
	7.0	0.392	0.374	1.0274	1.3420	0.76	1.183
	8.0	0.453	0.430	1.0316	1.3433	0.88	1.214
	9.0	0.515	0.485	1.0359	1.3446	1.00	1.247
	10.0	0.578	0.541	1.0402	1.3459	1.12	1.283
	12.0	0.710	0.655	1.0490	1.3486	1.38	1.357
	14.0	0.847	0.771	1.0580	1.3514	1.66	1.436
	16.0	0.991	0.889	1.0672	1.3541	1.95	1.525
	18.0	1.143	1.008	1.0764	1.3569	2.26	1.625
	20.0	1.301	1.130	1.0858	1.3598	2.57	1.740
	22.0	1.468	1.254	1.0953	1.3626	2.88	1.872
	24.0	1.644	1.380	1.1049	1.3655	3.21	2.017
	26.0	1.829	1.508	1.1147	1.3684	3.55	2.178
	28.0	2.024	1.639	1.1246	1.3714	3.89	2.356
	30.0	2.231	1.772	1.1346	1.3744	4.25	2.549
Copper	0.5	0.031	0.031	1.0033	1.3339	0.08	1.017
sulfate	1.0	0.063	0.063	1.0085	1.3348	0.14	1.036
CuSO$_4$	2.0	0.128	0.128	1.0190	1.3367	0.26	1.084
	3.0	0.194	0.194	1.0296	1.3386	0.37	1.129
	4.0	0.261	0.261	1.0403	1.3405	0.48	1.173
	5.0	0.330	0.329	1.0511	1.3424	0.59	1.221
	6.0	0.400	0.399	1.0620	1.3443	0.70	1.276
	7.0	0.472	0.471	1.0730	1.3462	0.82	1.336
	8.0	0.545	0.543	1.0842	1.3481	0.93	1.400
	9.0	0.620	0.618	1.0955	1.3501	1.05	1.469
	10.0	0.696	0.694	1.1070	1.3520	1.18	1.543
	12.0	0.854	0.850	1.1304	1.3560	1.45	1.701
	14.0	1.020	1.013	1.1545	1.3601	1.75	1.889
	16.0	1.193	1.182	1.1796	1.3644		2.136
	18.0	1.375	1.360	1.2059	1.3689		2.449
Disodium	0.5	0.015	0.015	1.0009	1.3339	0.07	1.017
ethylenediamine	1.0	0.030	0.030	1.0036	1.3348	0.14	1.032
tetraacetate	1.5	0.045	0.045	1.0062	1.3356	0.21	1.046
(EDTA sodium)	2.0	0.061	0.060	1.0089	1.3365	0.27	1.062
Na$_2$C$_{10}$H$_{14}$N$_2$O$_8$	2.5	0.076	0.075	1.0115	1.3374	0.33	1.077
	3.0	0.092	0.090	1.0142	1.3383	0.40	1.093
	3.5	0.108	0.106	1.0169	1.3392	0.46	1.109
	4.0	0.124	0.121	1.0196	1.3400	0.52	1.125
	4.5	0.140	0.137	1.0223	1.3409	0.58	1.142
	5.0	0.157	0.152	1.0250	1.3418	0.65	1.160
	5.5	0.173	0.168	1.0277	1.3427	0.71	1.178
	6.0	0.190	0.184	1.0305	1.3436	0.77	1.197
Ethanol	0.5	0.109	0.108	0.9973	1.3333	0.20	1.023
CH$_3$CH$_2$OH	1.0	0.219	0.216	0.9963	1.3336	0.40	1.046
	2.0	0.443	0.432	0.9945	1.3342	0.81	1.095
	3.0	0.671	0.646	0.9927	1.3348	1.23	1.140
	4.0	0.904	0.860	0.9910	1.3354	1.65	1.183
	5.0	1.142	1.074	0.9893	1.3360	2.09	1.228
	6.0	1.385	1.286	0.9878	1.3367	2.54	1.279
	7.0	1.634	1.498	0.9862	1.3374	2.99	1.331
	8.0	1.887	1.710	0.9847	1.3381	3.47	1.385
	9.0	2.147	1.921	0.9833	1.3388	3.96	1.442
	10.0	2.412	2.131	0.9819	1.3395	4.47	1.501
	12.0	2.960	2.551	0.9792	1.3410	5.56	1.627
	14.0	3.534	2.967	0.9765	1.3425	6.73	1.761

Solute	Mass %	m/mol kg^{-1}	c/mol L^{-1}	ρ/g cm^{-3}	n	Δ/°C	η/mPa s
	16.0	4.134	3.382	0.9739	1.3440	8.01	1.890
	18.0	4.765	3.795	0.9713	1.3455	9.40	2.019
	20.0	5.427	4.205	0.9687	1.3469	10.92	2.142
	22.0	6.122	4.613	0.9660	1.3484	12.60	2.259
	24.0	6.855	5.018	0.9632	1.3498	14.47	2.370
	26.0	7.626	5.419	0.9602	1.3511	16.41	2.476
	28.0	8.441	5.817	0.9571	1.3524	18.43	2.581
	30.0	9.303	6.212	0.9539	1.3535	20.47	2.667
	32.0	10.215	6.601	0.9504	1.3546	22.44	2.726
	34.0	11.182	6.987	0.9468	1.3557	24.27	2.768
	36.0	12.210	7.370	0.9431	1.3566	25.98	2.803
	38.0	13.304	7.747	0.9392	1.3575	27.62	2.829
	40.0	14.471	8.120	0.9352	1.3583	29.26	2.846
	42.0	15.718	8.488	0.9311	1.3590	30.98	2.852
	44.0	17.055	8.853	0.9269	1.3598	32.68	2.850
	46.0	18.490	9.213	0.9227	1.3604	34.36	2.843
	48.0	20.036	9.568	0.9183	1.3610	36.04	2.832
	50.0	21.706	9.919	0.9139	1.3616	37.67	2.813
	60.0	32.559	11.605	0.8911	1.3638	44.93	2.547
	70.0	50.648	13.183	0.8676	1.3652		2.214
	80.0	86.824	14.649	0.8436	1.3658		1.881
	90.0	195.355	15.980	0.8180	1.3650		1.542
	92.0	249.620	16.225	0.8125	1.3646		1.475
	94.0	340.062	16.466	0.8070	1.3642		1.407
	96.0	520.946	16.697	0.8013	1.3636		1.342
	98.0		16.920	0.7954	1.3630		1.273
	100.0		17.133	0.7893	1.3614		1.203
Ethylene	0.5	0.081	0.080	0.9988	1.3335	0.15	1.010
glycol	1.0	0.163	0.161	0.9995	1.3339	0.30	1.020
$(CH_2OH)_2$	2.0	0.329	0.322	1.0007	1.3348	0.61	1.048
	3.0	0.498	0.484	1.0019	1.3358	0.92	1.074
	4.0	0.671	0.646	1.0032	1.3367	1.24	1.099
	5.0	0.848	0.809	1.0044	1.3377	1.58	1.125
	6.0	1.028	0.972	1.0057	1.3386	1.91	1.153
	7.0	1.213	1.136	1.0070	1.3396	2.26	1.182
	8.0	1.401	1.299	1.0082	1.3405	2.62	1.212
	9.0	1.593	1.464	1.0095	1.3415	2.99	1.243
	10.0	1.790	1.628	1.0108	1.3425	3.37	1.277
	12.0	2.197	1.959	1.0134	1.3444	4.16	1.348
	14.0	2.623	2.292	1.0161	1.3464	5.01	1.424
	16.0	3.069	2.626	1.0188	1.3484	5.91	1.500
	18.0	3.537	2.962	1.0214	1.3503	6.89	1.578
	20.0	4.028	3.300	1.0241	1.3523	7.93	1.661
	24.0	5.088	3.981	1.0296	1.3564	10.28	1.843
	28.0	6.265	4.669	1.0350	1.3605	13.03	2.047
	32.0	7.582	5.364	1.0405	1.3646	16.23	2.280
	36.0	9.062	6.067	1.0460	1.3687	19.82	2.537
	40.0	10.741	6.776	1.0514	1.3728	23.84	2.832
	44.0	12.659	7.491	1.0567	1.3769	28.32	3.166
	48.0	14.872	8.212	1.0619	1.3811	33.30	3.544
	52.0	17.453	8.939	1.0670	1.3851	38.81	3.981
	56.0	20.505	9.671	1.0719	1.3892	44.83	4.475
	60.0	24.166	10.406	1.0765	1.3931	51.23	5.026
Ferric	0.5	0.031	0.031	1.0025	1.3344	0.21	1.024
chloride	1.0	0.062	0.062	1.0068	1.3358	0.39	1.047
$FeCl_3$	2.0	0.126	0.125	1.0153	1.3386	0.75	1.093
	3.0	0.191	0.189	1.0238	1.3413	1.15	1.139
	4.0	0.257	0.255	1.0323	1.3441	1.56	1.187
	5.0	0.324	0.321	1.0408	1.3468	2.00	1.238

Solute	Mass %	m/mol kg^{-1}	c/mol L^{-1}	ρ/g cm^{-3}	n	Δ/°C	η/mPa s
	6.0	0.394	0.388	1.0493	1.3496	2.48	1.292
	7.0	0.464	0.457	1.0580	1.3524	2.99	1.350
	8.0	0.536	0.526	1.0668	1.3552	3.57	1.412
	9.0	0.610	0.597	1.0760	1.3581	4.19	1.480
	10.0	0.685	0.669	1.0853	1.3611	4.85	1.553
	12.0	0.841	0.817	1.1040	1.3670	6.38	1.707
	14.0	1.004	0.969	1.1228	1.3730	8.22	1.879
	16.0	1.174	1.126	1.1420		10.45	2.080
	18.0	1.353	1.289	1.1615		13.08	2.311
	20.0	1.541	1.457	1.1816		16.14	2.570
	24.0	1.947	1.810	1.2234		23.79	3.178
	28.0	2.398	2.189	1.2679		33.61	4.038
	32.0	2.901	2.595	1.3153		49.16	5.274
	36.0	3.468	3.030	1.3654			7.130
	40.0	4.110	3.496	1.4176			9.674
Formic acid	0.5	0.109	0.109	0.9994	1.3333	0.21	1.006
HCOOH	1.0	0.219	0.217	1.0006	1.3336	0.42	1.011
	2.0	0.443	0.436	1.0029	1.3342	0.82	1.017
	3.0	0.672	0.655	1.0053	1.3348	1.24	1.195
	4.0	0.905	0.876	1.0077	1.3354	1.67	1.032
	5.0	1.143	1.097	1.0102	1.3359	2.10	1.039
	6.0	1.387	1.320	1.0126	1.3365	2.53	1.046
	7.0	1.635	1.544	1.0150	1.3371	2.97	1.052
	8.0	1.889	1.768	1.0175	1.3376	3.40	1.058
	9.0	2.149	1.994	1.0199	1.3382	3.84	1.064
	10.0	2.414	2.221	1.0224	1.3387	4.27	1.070
	12.0	2.962	2.678	1.0273	1.3397	5.19	1.082
	14.0	3.537	3.139	1.0322	1.3408	6.11	1.094
	16.0	4.138	3.605	1.0371	1.3418	7.06	1.106
	18.0	4.769	4.074	1.0419	1.3428	8.08	1.119
	20.0	5.431	4.548	1.0467	1.3437	9.11	1.132
	28.0	8.449	6.481	1.0654	1.3475	13.10	1.179
	36.0	12.220	8.477	1.0839	1.3511	17.65	1.227
	44.0	17.070	10.529	1.1015	1.3547	22.93	1.281
	52.0	23.535	12.633	1.1183	1.3581	29.69	1.340
	60.0	32.587	14.813	1.1364	1.3612	38.26	1.410
	68.0	46.166	17.054	1.1544	1.3641		1.490
D-Fructose	0.5	0.028	0.028	1.0002	1.3337	0.05	1.015
C$_6$H$_{12}$O$_6$	1.0	0.056	0.056	1.0021	1.3344	0.10	1.028
	2.0	0.113	0.112	1.0061	1.3358	0.21	1.054
	3.0	0.172	0.168	1.0101	1.3373	0.32	1.080
	4.0	0.231	0.225	1.0140	1.3387	0.43	1.106
	5.0	0.292	0.283	1.0181	1.3402	0.54	1.134
	6.0	0.354	0.340	1.0221	1.3417	0.66	1.165
	7.0	0.418	0.399	1.0262	1.3431	0.78	1.198
	8.0	0.483	0.458	1.0303	1.3446	0.90	1.232
	9.0	0.549	0.517	1.0344	1.3461	1.03	1.270
	10.0	0.617	0.576	1.0385	1.3476	1.16	1.309
	12.0	0.757	0.697	1.0469	1.3507	1.43	1.391
	14.0	0.904	0.820	1.0554	1.3538	1.71	1.483
	16.0	1.057	0.945	1.0640	1.3569	2.01	1.587
	18.0	1.218	1.072	1.0728	1.3601	2.32	1.703
	20.0	1.388	1.201	1.0816	1.3634	2.64	1.837
	22.0	1.566	1.332	1.0906	1.3667	3.05	1.986
	24.0	1.753	1.465	1.0996	1.3700	3.43	2.154
	26.0	1.950	1.600	1.1089	1.3734	3.82	2.348
	28.0	2.159	1.738	1.1182	1.3768	4.20	2.562
	30.0	2.379	1.878	1.1276	1.3803		2.817
	32.0	2.612	2.020	1.1372	1.3839		3.112

Solute	Mass %	m/mol kg^{-1}	c/mol L^{-1}	ρ/g cm^{-3}	n	Δ/°C	η/mPa s
	34.0	2.859	2.164	1.1469	1.3874		3.462
	36.0	3.122	2.312	1.1568	1.3911		3.899
	38.0	3.402	2.461	1.1668	1.3948		4.418
	40.0	3.700	2.613	1.1769	1.3985		5.046
	42.0	4.019	2.767	1.1871	1.4023		5.773
	44.0	4.361	2.925	1.1975	1.4062		6.644
	46.0	4.728	3.084	1.2080	1.4101		7.753
	48.0	5.124	3.247	1.2187	1.4141		9.060
D-Glucose	0.5	0.028	0.028	1.0001	1.3337	0.05	1.010
$C_6H_{12}O_6$	1.0	0.056	0.056	1.0020	1.3344	0.11	1.021
	2.0	0.113	0.112	1.0058	1.3358	0.21	1.052
	3.0	0.172	0.168	1.0097	1.3373	0.32	1.083
	4.0	0.231	0.225	1.0136	1.3387	0.43	1.113
	5.0	0.292	0.282	1.0175	1.3402	0.55	1.145
	6.0	0.354	0.340	1.0214	1.3417	0.67	1.179
	7.0	0.418	0.398	1.0254	1.3432	0.79	1.214
	8.0	0.483	0.457	1.0294	1.3447	0.91	1.250
	9.0	0.549	0.516	1.0334	1.3462	1.04	1.289
	10.0	0.617	0.576	1.0375	1.3477	1.17	1.330
	12.0	0.757	0.697	1.0457	1.3508	1.44	1.416
	14.0	0.904	0.819	1.0540	1.3539	1.73	1.512
	16.0	1.057	0.944	1.0624	1.3571	2.03	1.625
	18.0	1.218	1.070	1.0710	1.3603	2.35	1.757
	20.0	1.388	1.199	1.0797	1.3635	2.70	1.904
	22.0	1.566	1.329	1.0884	1.3668	3.07	2.063
	24.0	1.753	1.462	1.0973	1.3702	3.48	2.242
	26.0	1.950	1.597	1.1063	1.3736	3.90	2.458
	28.0	2.159	1.734	1.1154	1.3770	4.34	2.707
	30.0	2.379	1.873	1.1246	1.3805	4.79	2.998
	32.0	2.612	2.014	1.1340	1.3840		3.324
	34.0	2.859	2.158	1.1434	1.3876		3.704
	36.0	3.122	2.304	1.1529	1.3912		4.193
	38.0	3.402	2.452	1.1626	1.3949		4.786
	40.0	3.700	2.603	1.1724	1.3986		5.493
	42.0	4.019	2.756	1.1823	1.4024		6.288
	44.0	4.361	2.912	1.1924	1.4062		7.235
	46.0	4.728	3.071	1.2026	1.4101		8.454
	48.0	5.124	3.232	1.2130	1.4141		9.883
	50.0	5.551	3.396	1.2235	1.4181		11.884
	52.0	6.013	3.562	1.2342	1.4222		14.489
	54.0	6.516	3.732	1.2451	1.4263		17.916
	56.0	7.064	3.905	1.2562	1.4306		22.886
	58.0	7.665	4.081	1.2676	1.4349		29.389
	60.0	8.326	4.261	1.2793	1.4394		37.445
Glycerol	0.5	0.055	0.054	0.9994	1.3336	0.07	1.022
$CH_2OHCHOHCH_2OH$	1.0	0.110	0.109	1.0005	1.3342	0.18	1.034
	2.0	0.222	0.218	1.0028	1.3353	0.41	1.060
	3.0	0.336	0.327	1.0051	1.3365	0.63	1.088
	4.0	0.452	0.438	1.0074	1.3376	0.85	1.116
	5.0	0.572	0.548	1.0097	1.3388	1.08	1.145
	6.0	0.693	0.659	1.0120	1.3400	1.32	1.176
	7.0	0.817	0.771	1.0144	1.3412	1.56	1.207
	8.0	0.944	0.883	1.0167	1.3424	1.81	1.240
	9.0	1.074	0.996	1.0191	1.3436	2.06	1.275
	10.0	1.207	1.109	1.0215	1.3448	2.32	1.310
	12.0	1.481	1.337	1.0262	1.3472	2.88	1.386
	14.0	1.768	1.568	1.0311	1.3496	3.47	1.469
	16.0	2.068	1.800	1.0360	1.3521	4.09	1.560
	18.0	2.384	2.035	1.0409	1.3547	4.76	1.658

Solute	Mass %	m/mol kg^{-1}	c/mol L^{-1}	ρ/g cm^{-3}	n	Δ/°C	η/mPa s
	20.0	2.715	2.271	1.0459	1.3572	5.46	1.766
	24.0	3.429	2.752	1.0561	1.3624	7.01	2.01
	28.0	4.223	3.242	1.0664	1.3676	8.77	2.32
	32.0	5.110	3.742	1.0770	1.3730	10.74	2.69
	36.0	6.108	4.252	1.0876	1.3785	12.96	3.15
	40.0	7.239	4.771	1.0984	1.3841	15.50	3.73
	44.0	8.532	5.300	1.1092	1.3897		4.48
	48.0	10.024	5.838	1.1200	1.3954		5.45
	52.0	11.764	6.385	1.1308	1.4011		6.73
	56.0	13.820	6.944	1.1419	1.4069		8.47
	60.0	16.288	7.512	1.1530	1.4129		10.9
	64.0	19.305	8.092	1.1643	1.4189		14.3
	68.0	23.075	8.680	1.1755	1.4249		19.4
	72.0	27.923	9.277	1.1866	1.4310		27.2
	76.0	34.387	9.884	1.1976	1.4370		39.6
	80.0	43.436	10.498	1.2085	1.4431		60.6
	84.0	57.009	11.121	1.2192	1.4492		98
	88.0	79.632	11.753	1.2299	1.4553		170
	92.0	124.878	12.392	1.2404	1.4613		319
	96.0	260.615	13.039	1.2508	1.4674		648
	100.0		13.694	1.2611	1.4735		1460
Hydrochloric acid HCl	0.5	0.138	0.137	1.0007	1.3341	0.49	1.008
	1.0	0.277	0.275	1.0031	1.3353	0.99	1.015
	2.0	0.560	0.553	1.0081	1.3376	2.08	1.029
	3.0	0.848	0.833	1.0130	1.3399	3.28	1.044
	4.0	1.143	1.117	1.0179	1.3422	4.58	1.059
	5.0	1.444	1.403	1.0228	1.3445	5.98	1.075
	6.0	1.751	1.691	1.0278	1.3468	7.52	1.091
	7.0	2.064	1.983	1.0327	1.3491	9.22	1.108
	8.0	2.385	2.277	1.0377	1.3515	11.10	1.125
	9.0	2.713	2.574	1.0426	1.3538	13.15	1.143
	10.0	3.047	2.873	1.0476	1.3561	15.40	1.161
	12.0	3.740	3.481	1.0576	1.3607	20.51	1.199
	14.0	4.465	4.099	1.0676	1.3653		1.239
	16.0	5.224	4.729	1.0777	1.3700		1.282
	18.0	6.020	5.370	1.0878	1.3746		1.326
	20.0	6.857	6.023	1.0980	1.3792		1.374
	22.0	7.736	6.687	1.1083	1.3838		1.426
	24.0	8.661	7.362	1.1185	1.3884		1.483
	26.0	9.636	8.049	1.1288	1.3930		1.547
	28.0	10.666	8.748	1.1391	1.3976		1.620
	30.0	11.754	9.456	1.1492	1.4020		1.705
	32.0	12.907	10.175	1.1594	1.4066		1.799
	34.0	14.129	10.904	1.1693	1.4112		1.900
	36.0	15.427	11.642	1.1791	1.4158		2.002
	38.0	16.810	12.388	1.1886	1.4204		2.105
	40.0	18.284	13.140	1.1977	1.4250		
Lactic acid CH$_3$CHOHCOOH	0.5	0.056	0.055	0.9992	1.3335	0.10	1.014
	1.0	0.112	0.111	1.0002	1.3340	0.19	1.027
	2.0	0.227	0.223	1.0023	1.3350	0.38	1.056
	3.0	0.343	0.334	1.0043	1.3360	0.57	1.084
	4.0	0.463	0.447	1.0065	1.3370	0.76	1.110
	5.0	0.584	0.560	1.0086	1.3380	0.95	1.138
	6.0	0.709	0.673	1.0108	1.3390	1.16	1.167
	7.0	0.836	0.787	1.0131	1.3400	1.36	1.198
	8.0	0.965	0.902	1.0153	1.3410	1.57	1.229
	9.0	1.098	1.017	1.0176	1.3420	1.79	1.262
	10.0	1.233	1.132	1.0199	1.3430	2.02	1.296
	12.0	1.514	1.365	1.0246	1.3450	2.49	1.366

Solute	Mass %	m/mol kg^{-1}	c/mol L^{-1}	ρ/g cm^{-3}	n	Δ/°C	η/mPa s
	14.0	1.807	1.600	1.0294	1.3470	2.99	1.441
	16.0	2.115	1.837	1.0342	1.3491	3.48	1.522
	18.0	2.437	2.076	1.0390	1.3511	3.96	1.607
	20.0	2.775	2.318	1.0439	1.3532	4.44	1.699
	24.0	3.506	2.807	1.0536	1.3573		1.902
	28.0	4.317	3.305	1.0632	1.3615		2.136
	32.0	5.224	3.811	1.0728	1.3657		2.414
	36.0	6.244	4.325	1.0822	1.3700		2.730
	40.0	7.401	4.847	1.0915	1.3743		3.114
	44.0	8.722	5.377	1.1008	1.3786		3.566
	48.0	10.247	5.917	1.1105	1.3828		4.106
	52.0	12.026	6.466	1.1201	1.3871		4.789
	56.0	14.129	7.023	1.1297	1.3914		5.579
	60.0	16.652	7.588	1.1392	1.3958		6.679
	64.0	19.736	8.161	1.1486	1.4001		8.024
	68.0	23.590	8.741	1.1579	1.4045		9.863
	72.0	28.546	9.328	1.1670	1.4088		12.866
	76.0	35.154	9.922	1.1760	1.4131		16.974
	80.0	44.405	10.522	1.1848	1.4173		22.164
Lactose	0.5	0.015	0.015	1.0002	1.3337	0.03	1.013
$C_{12}H_{22}O_{11}$	1.0	0.030	0.029	1.0021	1.3345	0.06	1.026
	2.0	0.060	0.059	1.0061	1.3359	0.11	1.058
	3.0	0.090	0.089	1.0102	1.3375	0.17	1.089
	4.0	0.122	0.119	1.0143	1.3390	0.23	1.120
	5.0	0.154	0.149	1.0184	1.3406	0.29	1.154
	6.0	0.186	0.179	1.0225	1.3421	0.35	1.191
	7.0	0.220	0.210	1.0267	1.3437	0.42	1.232
	8.0	0.254	0.241	1.0308	1.3453	0.50	1.276
	9.0	0.289	0.272	1.0349	1.3468		1.321
	10.0	0.325	0.304	1.0390	1.3484		1.370
	12.0	0.398	0.367	1.0473	1.3515		1.476
	14.0	0.476	0.432	1.0558	1.3548		1.593
	16.0	0.556	0.498	1.0648	1.3582		1.724
	18.0	0.641	0.565	1.0746	1.3619		1.869
Lithium	0.5	0.119	0.118	1.0012	1.3341	0.42	1.019
chloride	1.0	0.238	0.237	1.0041	1.3351	0.84	1.037
LiCl	2.0	0.481	0.476	1.0099	1.3373	1.72	1.072
	3.0	0.730	0.719	1.0157	1.3394	2.68	1.108
	4.0	0.983	0.964	1.0215	1.3415	3.73	1.146
	5.0	1.241	1.211	1.0272	1.3436	4.86	1.185
	6.0	1.506	1.462	1.0330	1.3457	6.14	1.226
	7.0	1.775	1.715	1.0387	1.3478	7.56	1.269
	8.0	2.051	1.971	1.0444	1.3499	9.11	1.313
	9.0	2.333	2.230	1.0502	1.3520	10.79	1.360
	10.0	2.621	2.491	1.0560	1.3541	12.61	1.411
	12.0	3.217	3.022	1.0675	1.3583	16.59	1.522
	14.0	3.840	3.564	1.0792	1.3625	21.04	1.647
	16.0	4.493	4.118	1.0910	1.3668		1.787
	18.0	5.178	4.683	1.1029	1.3711		1.942
	20.0	5.897	5.260	1.1150	1.3755		2.128
	22.0	6.653	5.851	1.1274	1.3799		2.341
	24.0	7.449	6.453	1.1399	1.3844		2.600
	26.0	8.288	7.069	1.1527	1.3890		2.925
	28.0	9.173	7.700	1.1658	1.3936		3.318
	30.0	10.109	8.344	1.1791	1.3983		3.785
Magnesium	0.5	0.053	0.053	1.0022	1.3343	0.26	1.024
chloride	1.0	0.106	0.106	1.0062	1.3356	0.52	1.046
$MgCl_2$	2.0	0.214	0.213	1.0144	1.3381	1.06	1.091

Solute	Mass %	m/mol kg^{-1}	c/mol L^{-1}	ρ/g cm^{-3}	n	Δ/°C	η/mPa s
	3.0	0.325	0.322	1.0226	1.3406	1.65	1.139
	4.0	0.438	0.433	1.0309	1.3432	2.30	1.188
	5.0	0.553	0.546	1.0394	1.3457	3.01	1.241
	6.0	0.670	0.660	1.0479	1.3483		1.298
	7.0	0.791	0.777	1.0564	1.3508		1.358
	8.0	0.913	0.895	1.0651	1.3534		1.423
	9.0	1.039	1.015	1.0738	1.3560		1.493
	10.0	1.167	1.137	1.0826	1.3587		1.570
	12.0	1.432	1.387	1.1005	1.3641		1.745
	14.0	1.710	1.645	1.1189	1.3695		1.956
	16.0	2.001	1.911	1.1372	1.3749		2.207
	18.0	2.306	2.184	1.1553	1.3804		2.507
	20.0	2.626	2.467	1.1742	1.3859		2.867
	22.0	2.962	2.758	1.1938	1.3915		3.323
	24.0	3.317	3.060	1.2140	1.3972		3.917
	26.0	3.690	3.371	1.2346	1.4030		4.694
	28.0	4.085	3.692	1.2555	1.4089		5.709
	30.0	4.501	4.022	1.2763	1.4148		7.017
Magnesium sulfate $MgSO_4$	0.5	0.042	0.042	1.0033	1.3340	0.10	1.027
	1.0	0.084	0.084	1.0084	1.3350	0.19	1.054
	2.0	0.170	0.169	1.0186	1.3371	0.36	1.112
	3.0	0.257	0.256	1.0289	1.3391	0.52	1.177
	4.0	0.346	0.345	1.0392	1.3411	0.69	1.249
	5.0	0.437	0.436	1.0497	1.3431	0.87	1.328
	6.0	0.530	0.528	1.0602	1.3451	1.05	1.411
	7.0	0.625	0.623	1.0708	1.3471	1.24	1.498
	8.0	0.722	0.719	1.0816	1.3492	1.43	1.593
	9.0	0.822	0.817	1.0924	1.3512	1.64	1.702
	10.0	0.923	0.917	1.1034	1.3532	1.85	1.829
	12.0	1.133	1.122	1.1257	1.3572	2.31	2.104
	14.0	1.352	1.336	1.1484	1.3613	2.86	2.412
	16.0	1.582	1.557	1.1717	1.3654	3.67	2.809
	18.0	1.824	1.788	1.1955	1.3694		3.360
	20.0	2.077	2.027	1.2198	1.3735		4.147
	22.0	2.343	2.275	1.2447	1.3776		5.199
	24.0	2.624	2.532	1.2701	1.3817		6.498
	26.0	2.919	2.800	1.2961	1.3858		8.066
Maltose $C_{12}H_{22}O_{11}$	0.5	0.015	0.015	1.0003	1.3337	0.03	1.016
	1.0	0.030	0.029	1.0023	1.3345	0.06	1.030
	2.0	0.060	0.059	1.0063	1.3359	0.11	1.060
	3.0	0.090	0.089	1.0104	1.3374	0.17	1.092
	4.0	0.122	0.119	1.0144	1.3389	0.23	1.126
	5.0	0.154	0.149	1.0184	1.3404	0.29	1.162
	6.0	0.186	0.179	1.0224	1.3420	0.35	1.200
	7.0	0.220	0.210	1.0265	1.3435	0.42	1.239
	8.0	0.254	0.241	1.0305	1.3450	0.48	1.281
	9.0	0.289	0.272	1.0345	1.3466	0.55	1.325
	10.0	0.325	0.303	1.0385	1.3482	0.62	1.372
	12.0	0.398	0.367	1.0465	1.3513	0.77	1.474
	14.0	0.476	0.431	1.0545	1.3546	0.92	1.588
	16.0	0.556	0.497	1.0629	1.3578	1.08	1.715
	18.0	0.641	0.564	1.0716	1.3612	1.25	1.859
	20.0	0.730	0.631	1.0801	1.3644	1.43	2.021
	22.0	0.824	0.700	1.0894	1.3678	1.64	2.216
	24.0	0.923	0.770	1.0984	1.3714	1.85	2.463
	26.0	1.026	0.842	1.1080	1.3749	2.08	2.753
	28.0	1.136	0.914	1.1171	1.3785	2.34	3.066
	30.0	1.252	0.988	1.1269	1.3821	2.62	3.427
	40.0	1.948	1.375	1.1769	1.4013	4.41	6.926

Solute	Mass %	m/mol kg^{-1}	c/mol L^{-1}	ρ/g cm^{-3}	n	Δ/°C	η/mPa s
	50.0	2.921	1.797	1.2304	1.4217		17.786
	52.0	3.165	1.886	1.2416	1.4260		22.034
	54.0	3.429	1.976	1.2528	1.4308		28.757
	56.0	3.718	2.068	1.2638	1.4350		38.226
	58.0	4.034	2.159	1.2740	1.4394		49.298
	60.0	4.382	2.253	1.2855	1.4440		
Manganese(II)	1.0	0.067	0.067	1.0080	1.3348	0.16	1.046
sulfate	2.0	0.135	0.135	1.0178	1.3366	0.31	1.090
MnSO$_4$	3.0	0.205	0.204	1.0277	1.3384	0.44	1.137
	4.0	0.276	0.275	1.0378	1.3402	0.57	1.187
	5.0	0.349	0.347	1.0480	1.3420	0.70	1.242
	6.0	0.423	0.421	1.0583	1.3438	0.84	1.301
	7.0	0.498	0.495	1.0688	1.3457	0.98	1.363
	8.0	0.576	0.572	1.0794	1.3475	1.12	1.431
	9.0	0.655	0.650	1.0902	1.3494	1.28	1.505
	10.0	0.736	0.729	1.1012	1.3513	1.44	1.587
	12.0	0.903	0.893	1.1236	1.3551	1.80	1.779
	14.0	1.078	1.063	1.1467	1.3589	2.21	2.005
	16.0	1.261	1.240	1.1705	1.3629	2.67	2.272
	18.0	1.454	1.424	1.1950	1.3668	3.19	2.580
	20.0	1.656	1.616	1.2203	1.3708	3.80	2.938
D-Mannitol	0.5	0.028	0.027	1.0000	1.3337	0.05	1.019
CH$_2$(CHOH)$_4$CH$_2$OH	1.0	0.055	0.055	1.0017	1.3345	0.10	1.032
	2.0	0.112	0.110	1.0053	1.3359	0.21	1.057
	3.0	0.170	0.166	1.0088	1.3374	0.32	1.081
	4.0	0.229	0.222	1.0124	1.3389	0.43	1.107
	5.0	0.289	0.279	1.0159	1.3403	0.54	1.135
	6.0	0.350	0.336	1.0195	1.3418	0.66	1.166
	7.0	0.413	0.393	1.0230	1.3433	0.77	1.200
	8.0	0.477	0.451	1.0266	1.3447	0.90	1.236
	9.0	0.543	0.509	1.0302	1.3462	1.02	1.275
	10.0	0.610	0.567	1.0338	1.3477	1.15	1.314
	11.0	0.678	0.626	1.0375	1.3491	1.28	1.355
	12.0	0.749	0.686	1.0412	1.3506	1.41	1.398
	13.0	0.820	0.746	1.0450	1.3521	1.55	1.443
	14.0	0.894	0.806	1.0489	1.3536	1.69	1.489
	15.0	0.969	0.867	1.0529	1.3552	1.84	1.537
Methanol	0.5	0.157	0.156	0.9973	1.3331	0.28	1.022
CH$_3$OH	1.0	0.315	0.311	0.9964	1.3332	0.56	1.040
	2.0	0.637	0.621	0.9947	1.3334	1.14	1.070
	3.0	0.965	0.930	0.9930	1.3336	1.75	1.100
	4.0	1.300	1.238	0.9913	1.3339	2.37	1.131
	5.0	1.643	1.544	0.9896	1.3341	3.02	1.163
	6.0	1.992	1.850	0.9880	1.3343	3.71	1.196
	7.0	2.349	2.155	0.9864	1.3346	4.41	1.229
	8.0	2.714	2.459	0.9848	1.3348	5.13	1.264
	9.0	3.087	2.762	0.9832	1.3351	5.85	1.297
	10.0	3.468	3.064	0.9816	1.3354	6.60	1.329
	12.0	4.256	3.665	0.9785	1.3359	8.14	1.389
	14.0	5.081	4.262	0.9755	1.3365	9.72	1.446
	16.0	5.945	4.856	0.9725	1.3370	11.36	1.501
	18.0	6.851	5.447	0.9695	1.3376	13.13	1.554
	20.0	7.803	6.034	0.9666	1.3381	15.02	1.604
	22.0	8.803	6.616	0.9636	1.3387	16.98	1.652
	24.0	9.856	7.196	0.9606	1.3392	19.04	1.697
	26.0	10.966	7.771	0.9576	1.3397	21.23	1.735
	28.0	12.138	8.341	0.9545	1.3402	23.59	1.769
	30.0	13.376	8.908	0.9514	1.3407	25.91	1.795

Solute	Mass %	$m/\text{mol kg}^{-1}$	$c/\text{mol L}^{-1}$	$\rho/\text{g cm}^{-3}$	n	$\Delta/°C$	$\eta/\text{mPa s}$
	32.0	14.688	9.470	0.9482	1.3411	28.15	1.814
	34.0	16.078	10.028	0.9450	1.3415	30.48	1.827
	36.0	17.556	10.580	0.9416	1.3419	32.97	1.835
	38.0	19.129	11.127	0.9382	1.3422	35.60	1.839
	40.0	20.807	11.669	0.9347	1.3425	38.60	1.837
	50.0	31.211	14.288	0.9156	1.3431	54.50	1.761
	60.0	46.816	16.749	0.8944	1.3426	74.50	1.600
	70.0	72.826	19.040	0.8715	1.3411		1.368
	80.0	124.844	21.144	0.8468	1.3385		1.128
	90.0	280.899	23.045	0.8204	1.3348		0.861
	100.0		24.710	0.7917	1.3290		0.586
Nitric acid	0.5	0.080	0.079	1.0009	1.3336	0.28	1.004
HNO_3	1.0	0.160	0.159	1.0037	1.3343	0.56	1.005
	2.0	0.324	0.320	1.0091	1.3356	1.12	1.007
	3.0	0.491	0.483	1.0146	1.3368	1.70	1.010
	4.0	0.661	0.648	1.0202	1.3381	2.32	1.014
	5.0	0.835	0.814	1.0257	1.3394	2.96	1.018
	6.0	1.013	0.982	1.0314	1.3407	3.63	1.022
	7.0	1.194	1.152	1.0370	1.3421	4.33	1.027
	8.0	1.380	1.324	1.0427	1.3434	5.05	1.032
	9.0	1.570	1.498	1.0485	1.3447	5.81	1.038
	10.0	1.763	1.673	1.0543	1.3460	6.60	1.044
	12.0	2.164	2.030	1.0660	1.3487	8.27	1.058
	14.0	2.583	2.395	1.0780	1.3514	10.08	1.075
	16.0	3.023	2.768	1.0901	1.3541	12.04	1.094
	18.0	3.484	3.149	1.1025	1.3569	14.16	1.116
	20.0	3.967	3.539	1.1150	1.3596		1.141
	22.0	4.476	3.937	1.1277	1.3624		1.169
	24.0	5.011	4.344	1.1406	1.3652		1.199
	26.0	5.576	4.760	1.1536	1.3680		1.233
	28.0	6.172	5.185	1.1668	1.3708		1.271
	30.0	6.801	5.618	1.1801	1.3736		1.311
	32.0	7.468	6.060	1.1934	1.3763		1.354
	34.0	8.175	6.512	1.2068	1.3790		1.400
	36.0	8.927	6.971	1.2202	1.3817		1.450
	38.0	9.727	7.439	1.2335	1.3842		1.504
	40.0	10.580	7.913	1.2466	1.3867		1.561
Oxalic acid	0.5	0.056	0.056	1.0006	1.3336	0.16	1.013
$(COOH)_2$	1.0	0.112	0.111	1.0030	1.3342	0.30	1.023
	1.5	0.169	0.167	1.0054	1.3347	0.44	1.033
	2.0	0.227	0.224	1.0079	1.3353	0.57	1.044
	2.5	0.285	0.281	1.0103	1.3359	0.71	1.055
	3.0	0.343	0.337	1.0126	1.3364	0.84	1.065
	3.5	0.403	0.395	1.0150	1.3370	0.97	1.076
	4.0	0.463	0.452	1.0174	1.3375	1.09	1.086
	4.5	0.523	0.510	1.0197	1.3381		1.097
	5.0	0.585	0.568	1.0220	1.3386		1.108
	6.0	0.709	0.684	1.0265	1.3397		1.129
	7.0	0.836	0.802	1.0310	1.3407		1.150
	8.0	0.966	0.920	1.0355	1.3418		1.172
Phosphoric	0.5	0.051	0.051	1.0010	1.3335	0.12	1.010
acid	1.0	0.103	0.102	1.0038	1.3340	0.24	1.020
H_3PO_4	2.0	0.208	0.206	1.0092	1.3349	0.46	1.050
	3.0	0.316	0.311	1.0146	1.3358	0.69	1.079
	4.0	0.425	0.416	1.0200	1.3367	0.93	1.108
	5.0	0.537	0.523	1.0254	1.3376	1.16	1.138
	6.0	0.651	0.631	1.0309	1.3385	1.38	1.169
	7.0	0.768	0.740	1.0363	1.3394	1.62	1.200

Solute	Mass %	m/mol kg^{-1}	c/mol L^{-1}	ρ/g cm^{-3}	n	Δ/°C	η/mPa s
	8.0	0.887	0.850	1.0418	1.3403	1.88	1.232
	9.0	1.009	0.962	1.0474	1.3413	2.16	1.267
	10.0	1.134	1.075	1.0531	1.3422	2.45	1.303
	12.0	1.392	1.304	1.0647	1.3441	3.01	1.382
	14.0	1.661	1.538	1.0765	1.3460	3.76	1.469
	16.0	1.944	1.777	1.0885	1.3480	4.45	1.565
	18.0	2.240	2.022	1.1009	1.3500	5.25	1.671
	20.0	2.551	2.273	1.1135	1.3520	6.23	1.788
	22.0	2.878	2.529	1.1263	1.3540	7.38	1.914
	24.0	3.223	2.791	1.1395	1.3561	8.69	2.049
	26.0	3.585	3.059	1.1528	1.3582	10.12	2.198
	28.0	3.968	3.333	1.1665	1.3604	11.64	2.365
	30.0	4.373	3.614	1.1804	1.3625	13.23	2.553
	32.0	4.802	3.901	1.1945	1.3647	14.94	2.766
	34.0	5.257	4.194	1.2089	1.3669	16.81	3.001
	36.0	5.740	4.495	1.2236	1.3691	18.85	3.260
	38.0	6.254	4.803	1.2385	1.3713	21.09	3.544
	40.0	6.803	5.117	1.2536	1.3735	23.58	3.856
Potassium	0.5	0.050	0.050	1.0014	1.3335	0.18	1.009
bicarbonate	1.0	0.101	0.100	1.0046	1.3341	0.34	1.015
KHCO$_3$	2.0	0.204	0.202	1.0114	1.3353	0.67	1.027
	3.0	0.309	0.305	1.0181	1.3365	0.98	1.040
	4.0	0.416	0.409	1.0247	1.3376	1.29	1.053
	5.0	0.526	0.515	1.0310	1.3386	1.60	1.067
	6.0	0.638	0.622	1.0379	1.3397	1.91	1.081
	7.0	0.752	0.730	1.0446	1.3409	2.22	1.096
	8.0	0.869	0.840	1.0514	1.3419	2.53	1.112
	9.0	0.988	0.951	1.0581	1.3430	2.84	1.128
	10.0	1.110	1.064	1.0650	1.3441	3.16	1.145
	12.0	1.362	1.293	1.0788	1.3462	3.79	1.183
	14.0	1.626	1.528	1.0929	1.3484	4.41	1.224
	16.0	1.903	1.770	1.1073	1.3506		1.270
	18.0	2.193	2.017	1.1221	1.3528		1.319
	20.0	2.497	2.272	1.1372	1.3550		1.373
	22.0	2.817	2.533	1.1527	1.3572		1.432
	24.0	3.154	2.801	1.1685	1.3595		1.497
Potassium	0.5	0.042	0.042	1.0018	1.3336	0.15	1.000
bromide	1.0	0.085	0.084	1.0054	1.3342	0.29	0.998
KBr	2.0	0.171	0.170	1.0127	1.3354	0.59	0.994
	3.0	0.260	0.257	1.0200	1.3366	0.88	0.990
	4.0	0.350	0.345	1.0275	1.3379	1.18	0.985
	5.0	0.442	0.435	1.0350	1.3391	1.48	0.981
	6.0	0.536	0.526	1.0426	1.3403	1.78	0.977
	7.0	0.633	0.618	1.0503	1.3416	2.10	0.974
	8.0	0.731	0.711	1.0581	1.3429	2.42	0.970
	9.0	0.831	0.806	1.0660	1.3441	2.74	0.967
	10.0	0.934	0.903	1.0740	1.3454	3.07	0.964
	12.0	1.146	1.099	1.0903	1.3481	3.76	0.958
	14.0	1.368	1.302	1.1070	1.3507	4.49	0.953
	16.0	1.601	1.512	1.1242	1.3535	5.25	0.949
	18.0	1.845	1.727	1.1419	1.3562	6.04	0.946
	20.0	2.101	1.950	1.1601	1.3591	6.88	0.944
	22.0	2.370	2.179	1.1788	1.3620	7.76	0.943
	24.0	2.654	2.416	1.1980	1.3650	8.70	0.943
	26.0	2.952	2.661	1.2179	1.3680	9.68	0.944
	28.0	3.268	2.914	1.2383	1.3711	10.72	0.947
	30.0	3.601	3.175	1.2593	1.3743	11.82	0.952
	32.0	3.954	3.445	1.2810	1.3776	12.98	0.959
	34.0	4.329	3.724	1.3033	1.3809		0.968

Solute	Mass %	m/mol kg^{-1}	c/mol L^{-1}	ρ/g cm^{-3}	n	Δ/°C	η/mPa s
	36.0	4.727	4.012	1.3263	1.3843		0.979
	38.0	5.150	4.311	1.3501	1.3878		0.993
	40.0	5.602	4.620	1.3746	1.3914		1.010
Potassium	0.5	0.036	0.036	1.0027	1.3339	0.18	1.013
carbonate	1.0	0.073	0.073	1.0072	1.3347	0.34	1.025
K_2CO_3	2.0	0.148	0.147	1.0163	1.3365	0.66	1.048
	3.0	0.224	0.223	1.0254	1.3382	0.99	1.071
	4.0	0.301	0.299	1.0345	1.3399	1.32	1.094
	5.0	0.381	0.378	1.0437	1.3416	1.67	1.119
	6.0	0.462	0.457	1.0529	1.3433	2.03	1.146
	7.0	0.545	0.538	1.0622	1.3450	2.40	1.174
	8.0	0.629	0.620	1.0715	1.3467	2.77	1.204
	9.0	0.716	0.704	1.0809	1.3484	3.17	1.235
	10.0	0.804	0.789	1.0904	1.3501	3.57	1.269
	12.0	0.987	0.963	1.1095	1.3535	4.45	1.339
	14.0	1.178	1.144	1.1291	1.3569	5.39	1.414
	16.0	1.378	1.330	1.1490	1.3603	6.42	1.497
	18.0	1.588	1.523	1.1692	1.3637	7.55	1.594
	20.0	1.809	1.722	1.1898	1.3671	8.82	1.707
	24.0	2.285	2.139	1.2320	1.3739	11.96	1.978
	28.0	2.814	2.584	1.2755	1.3807	16.01	2.331
	32.0	3.405	3.057	1.3204	1.3874	21.46	2.834
	36.0	4.070	3.559	1.3665	1.3940	28.58	3.503
	40.0	4.824	4.093	1.4142	1.4006	37.55	4.360
	50.0	7.236	5.573	1.5404	1.4168		9.369
Potassium	0.5	0.067	0.067	1.0014	1.3337	0.23	1.000
chloride	1.0	0.135	0.135	1.0046	1.3343	0.46	0.999
KCl	2.0	0.274	0.271	1.0110	1.3357	0.92	0.999
	3.0	0.415	0.409	1.0174	1.3371	1.38	0.998
	4.0	0.559	0.549	1.0239	1.3384	1.85	0.997
	5.0	0.706	0.691	1.0304	1.3398	2.32	0.996
	6.0	0.856	0.835	1.0369	1.3411	2.80	0.994
	7.0	1.010	0.980	1.0434	1.3425	3.29	0.992
	8.0	1.166	1.127	1.0500	1.3438	3.80	0.990
	9.0	1.327	1.276	1.0566	1.3452	4.30	0.989
	10.0	1.490	1.426	1.0633	1.3466	4.81	0.988
	12.0	1.829	1.733	1.0768	1.3493	5.88	0.990
	14.0	2.184	2.048	1.0905	1.3521		0.994
	16.0	2.555	2.370	1.1043	1.3549		0.999
	18.0	2.944	2.701	1.1185	1.3577		1.004
	20.0	3.353	3.039	1.1328	1.3606		1.012
	22.0	3.783	3.386	1.1474	1.3635		1.024
	24.0	4.236	3.742	1.1623	1.3665		1.040
Potassium	0.5	0.090	0.089	1.0025	1.3340	0.30	1.010
hydroxide	1.0	0.180	0.179	1.0068	1.3350	0.61	1.019
KOH	2.0	0.364	0.362	1.0155	1.3369	1.24	1.038
	3.0	0.551	0.548	1.0242	1.3388	1.89	1.058
	4.0	0.743	0.736	1.0330	1.3408	2.57	1.079
	5.0	0.938	0.929	1.0419	1.3427	3.36	1.102
	6.0	1.138	1.124	1.0509	1.3445	4.14	1.126
	7.0	1.342	1.322	1.0599	1.3464	4.92	1.151
	8.0	1.550	1.524	1.0690	1.3483		1.177
	9.0	1.763	1.729	1.0781	1.3502		1.205
	10.0	1.980	1.938	1.0873	1.3520		1.233
	12.0	2.431	2.365	1.1059	1.3558		1.294
	14.0	2.902	2.806	1.1246	1.3595		1.361
	16.0	3.395	3.261	1.1435	1.3632		1.436
	18.0	3.913	3.730	1.1626	1.3670		1.521

Solute	Mass %	m/mol kg^{-1}	c/mol L^{-1}	ρ/g cm^{-3}	n	Δ/°C	η/mPa s
	20.0	4.456	4.213	1.1818	1.3707		1.619
	22.0	5.027	4.711	1.2014	1.3744		1.732
	24.0	5.629	5.223	1.2210	1.3781		1.861
	26.0	6.262	5.750	1.2408	1.3818		2.006
	28.0	6.931	6.293	1.2609	1.3854		2.170
	30.0	7.639	6.851	1.2813	1.3889		2.357
	40.0	11.882	9.896	1.3881	1.4068		3.879
	50.0	17.824	13.389	1.5024	1.4247		7.892
Potassium	0.5	0.030	0.030	1.0019	1.3337	0.11	0.999
iodide	1.0	0.061	0.061	1.0056	1.3343	0.22	0.997
KI	2.0	0.123	0.122	1.0131	1.3357	0.43	0.991
	3.0	0.186	0.184	1.0206	1.3370	0.64	0.986
	4.0	0.251	0.248	1.0282	1.3384	0.86	0.981
	5.0	0.317	0.312	1.0360	1.3397	1.08	0.976
	6.0	0.385	0.377	1.0438	1.3411	1.30	0.969
	7.0	0.453	0.443	1.0517	1.3425	1.53	0.963
	8.0	0.524	0.511	1.0598	1.3440	1.77	0.957
	9.0	0.596	0.579	1.0679	1.3454	2.01	0.951
	10.0	0.669	0.648	1.0762	1.3469	2.26	0.946
	12.0	0.821	0.790	1.0931	1.3498	2.77	0.937
	14.0	0.981	0.937	1.1105	1.3529	3.30	0.929
	16.0	1.147	1.088	1.1284	1.3560	3.87	0.921
	18.0	1.322	1.244	1.1469	1.3593	4.46	0.915
	20.0	1.506	1.405	1.1659	1.3626	5.09	0.910
	22.0	1.699	1.571	1.1856	1.3661	5.76	0.905
	24.0	1.902	1.744	1.2060	1.3696	6.46	0.901
	26.0	2.117	1.922	1.2270	1.3733	7.21	0.898
	28.0	2.343	2.106	1.2487	1.3771	8.01	0.895
	30.0	2.582	2.297	1.2712	1.3810	8.86	0.892
	32.0	2.835	2.495	1.2944	1.3851	9.76	0.891
	34.0	3.103	2.700	1.3185	1.3893	10.72	0.890
	36.0	3.388	2.913	1.3434	1.3936	11.73	0.890
	38.0	3.692	3.134	1.3692	1.3981	12.81	0.893
	40.0	4.016	3.364	1.3959	1.4027	13.97	0.897
Potassium	0.5	0.050	0.050	1.0014	1.3335	0.17	0.999
nitrate	1.0	0.100	0.099	1.0045	1.3339	0.33	0.996
KNO$_3$	2.0	0.202	0.200	1.0108	1.3349	0.64	0.990
	3.0	0.306	0.302	1.0171	1.3358	0.94	0.986
	4.0	0.412	0.405	1.0234	1.3368	1.22	0.983
	5.0	0.521	0.509	1.0298	1.3377	1.50	0.980
	6.0	0.631	0.615	1.0363	1.3386	1.76	0.977
	7.0	0.744	0.722	1.0428	1.3396	2.02	0.975
	8.0	0.860	0.830	1.0494	1.3405	2.27	0.973
	9.0	0.978	0.940	1.0560	1.3415	2.52	0.971
	10.0	1.099	1.051	1.0627	1.3425	2.75	0.970
	12.0	1.349	1.277	1.0762	1.3444		0.970
	14.0	1.610	1.509	1.0899	1.3463		0.972
	16.0	1.884	1.747	1.1039	1.3482		0.976
	18.0	2.171	1.991	1.1181	1.3502		0.982
	20.0	2.473	2.240	1.1326	1.3521		0.990
	22.0	2.790	2.497	1.1473	1.3541		0.999
	24.0	3.123	2.759	1.1623	1.3561		1.010
Potassium	0.5	0.032	0.032	1.0017		0.11	1.001
permanganate	1.0	0.064	0.064	1.0051		0.22	1.000
KMnO$_4$	1.5	0.096	0.096	1.0085		0.32	0.999
	2.0	0.129	0.128	1.0118		0.43	0.998
	3.0	0.196	0.193	1.0186			0.995
	4.0	0.264	0.260	1.0254			0.992

Solute	Mass %	m/mol kg^{-1}	c/mol L^{-1}	ρ/g cm^{-3}	n	Δ/°C	η/mPa s
	5.0	0.333	0.327	1.0322			0.989
	6.0	0.404	0.394	1.0390			0.985
Potassium	0.5	0.029	0.029	1.0025	1.3338	0.13	1.013
hydrogen	1.0	0.058	0.058	1.0068	1.3345	0.25	1.023
phosphate	1.5	0.087	0.087	1.0110	1.3353	0.37	1.034
K_2HPO_4	2.0	0.117	0.117	1.0153	1.3361	0.49	1.046
	2.5	0.147	0.146	1.0195	1.3368	0.61	1.057
	3.0	0.178	0.176	1.0238	1.3376	0.73	1.069
	3.5	0.208	0.207	1.0281	1.3384	0.86	1.081
	4.0	0.239	0.237	1.0324	1.3392	0.97	1.094
	4.5	0.271	0.268	1.0368	1.3399	1.10	1.107
	5.0	0.302	0.299	1.0412	1.3407	1.22	1.120
	6.0	0.366	0.362	1.0500	1.3422	1.46	1.147
	7.0	0.432	0.426	1.0590	1.3438	1.70	1.177
	8.0	0.499	0.491	1.0680	1.3453	1.95	1.209
Potassium	0.5	0.037	0.037	1.0018	1.3336	0.13	1.010
dihydrogen	1.0	0.074	0.074	1.0053	1.3342	0.25	1.019
phosphate	1.5	0.112	0.111	1.0089	1.3348	0.37	1.028
KH_2PO_4	2.0	0.150	0.149	1.0125	1.3354	0.49	1.038
	3.0	0.227	0.225	1.0197	1.3365	0.72	1.060
	4.0	0.306	0.302	1.0269	1.3377	0.96	1.083
	5.0	0.387	0.380	1.0342	1.3388	1.19	1.108
	6.0	0.469	0.459	1.0414	1.3400	1.41	1.133
	7.0	0.553	0.539	1.0486	1.3411	1.63	1.160
	8.0	0.639	0.621	1.0558	1.3422	1.84	1.187
	9.0	0.727	0.703	1.0630	1.3434	2.04	1.215
	10.0	0.816	0.786	1.0703	1.3445	2.23	1.245
Potassium	0.5	0.029	0.029	1.0022	1.3336	0.14	1.006
sulfate	1.0	0.058	0.058	1.0062	1.3343	0.26	1.011
K_2SO_4	2.0	0.117	0.116	1.0143	1.3355	0.50	1.021
	3.0	0.177	0.176	1.0224	1.3368	0.73	1.033
	4.0	0.239	0.237	1.0306	1.3380	0.95	1.045
	5.0	0.302	0.298	1.0388	1.3393	1.17	1.058
	6.0	0.366	0.360	1.0470	1.3405		1.072
	7.0	0.432	0.424	1.0553	1.3417		1.087
	8.0	0.499	0.488	1.0637	1.3428		1.102
	9.0	0.568	0.554	1.0721	1.3440		1.117
	10.0	0.638	0.620	1.0806	1.3452		1.132
1-Propanol	1.0	0.168	0.166	0.9963	1.3339	0.31	1.051
$CH_3CH_2CH_2OH$	2.0	0.340	0.331	0.9946	1.3348	0.61	1.100
	3.0	0.515	0.496	0.9928	1.3357	0.93	1.152
	4.0	0.693	0.660	0.9911	1.3366	1.24	1.208
	5.0	0.876	0.823	0.9896	1.3376	1.57	1.267
	6.0	1.062	0.987	0.9882	1.3385	1.91	1.325
	7.0	1.252	1.149	0.9868	1.3394	2.26	1.387
	8.0	1.447	1.312	0.9855	1.3404	2.61	1.449
	9.0	1.646	1.474	0.9842	1.3414	2.99	1.514
	10.0	1.849	1.635	0.9829	1.3423	3.36	1.577
	12.0	2.269	1.958	0.9804	1.3442	4.09	1.710
	14.0	2.709	2.278	0.9779	1.3460	4.91	1.849
	16.0	3.169	2.595	0.9749	1.3477	5.78	1.986
	18.0	3.652	2.911	0.9719	1.3494	6.67	2.106
	20.0	4.160	3.223	0.9686	1.3510	7.76	2.218
	24.0	5.254	3.838	0.9612	1.3539	9.12	2.432
	28.0	6.471	4.441	0.9533	1.3566	10.17	2.612
	32.0	7.830	5.033	0.9452	1.3592	10.66	2.765
	36.0	9.359	5.613	0.9370	1.3614		2.900

Solute	Mass %	m/mol kg^{-1}	c/mol L^{-1}	ρ/g cm^{-3}	n	$\Delta/°C$	η/mPa s
	40.0	11.093	6.182	0.9288	1.3635		3.010
	60.0	24.958	8.860	0.8875	1.3734		3.186
	80.0	66.556	11.275	0.8470	1.3812		2.822
	100.0		13.368	0.8034	1.3852		2.227
2-Propanol	1.0	0.168	0.166	0.9960	1.3338	0.30	1.056
CH$_3$CHOHCH$_3$	2.0	0.340	0.331	0.9939	1.3346	0.60	1.112
	3.0	0.515	0.495	0.9920	1.3355	0.93	1.166
	4.0	0.693	0.659	0.9902	1.3364	1.26	1.225
	5.0	0.876	0.822	0.9884	1.3373	1.61	1.287
	6.0	1.062	0.985	0.9871	1.3382	1.96	1.352
	7.0	1.252	1.148	0.9855	1.3392	2.32	1.417
	8.0	1.447	1.310	0.9843	1.3400	2.68	1.485
	9.0	1.646	1.472	0.9831	1.3410	3.06	1.553
	10.0	1.849	1.633	0.9816	1.3420	3.48	1.629
	12.0	2.269	1.955	0.9793	1.3439	4.43	1.794
	14.0	2.709	2.276	0.9772	1.3459	5.29	1.970
	16.0	3.169	2.596	0.9751	1.3478	6.36	2.160
	18.0	3.652	2.913	0.9725	1.3496	7.40	2.352
	20.0	4.160	3.227	0.9696	1.3514	8.52	2.550
	40.0	11.093	6.191	0.9302	1.3642		
	60.0	24.958	8.809	0.8824	1.3717		
	80.0	66.556	11.103	0.8341	1.3742		
	100.0		13.058	0.7848	1.3776		
Silver	0.5	0.030	0.030	1.0027	1.3336	0.10	1.003
nitrate	1.0	0.059	0.059	1.0070	1.3342	0.20	1.005
AgNO$_3$	2.0	0.120	0.120	1.0154	1.3352	0.40	1.009
	3.0	0.182	0.181	1.0239	1.3363	0.59	1.013
	4.0	0.245	0.243	1.0327	1.3374	0.78	1.016
	5.0	0.310	0.307	1.0417	1.3385	0.96	1.020
	6.0	0.376	0.371	1.0506	1.3396	1.15	1.024
	7.0	0.443	0.437	1.0597	1.3407	1.33	1.027
	8.0	0.512	0.503	1.0690	1.3419	1.51	1.031
	9.0	0.582	0.571	1.0785	1.3431	1.69	1.035
	10.0	0.654	0.641	1.0882	1.3443	1.87	1.039
	12.0	0.803	0.783	1.1079	1.3467	2.21	1.049
	14.0	0.958	0.930	1.1284	1.3493	2.55	1.060
	16.0	1.121	1.083	1.1496	1.3519	2.86	1.072
	18.0	1.292	1.241	1.1715	1.3546		1.086
	20.0	1.472	1.406	1.1942	1.3574		1.101
	22.0	1.660	1.577	1.2177	1.3602		1.117
	24.0	1.859	1.755	1.2420	1.3632		1.135
	26.0	2.068	1.940	1.2672	1.3662		1.154
	28.0	2.289	2.132	1.2933	1.3694		1.176
	30.0	2.523	2.332	1.3204	1.3726		1.200
	32.0	2.770	2.541	1.3487	1.3760		1.227
	34.0	3.033	2.758	1.3780	1.3795		1.257
	36.0	3.311	2.985	1.4087	1.3832		1.290
	38.0	3.608	3.223	1.4407	1.3871		1.326
	40.0	3.925	3.472	1.4743	1.3911		1.366
Sodium	0.5	0.061	0.061	1.0008	1.3337	0.22	1.021
acetate	1.0	0.123	0.122	1.0034	1.3344	0.43	1.040
CH$_3$COONa	2.0	0.249	0.246	1.0085	1.3358	0.88	1.080
	3.0	0.377	0.371	1.0135	1.3372	1.34	1.124
	4.0	0.508	0.497	1.0184	1.3386	1.82	1.171
	5.0	0.642	0.624	1.0234	1.3400	2.32	1.222
	6.0	0.778	0.752	1.0283	1.3414	2.85	1.278
	7.0	0.918	0.882	1.0334	1.3428	3.40	1.337
	8.0	1.060	1.013	1.0386	1.3442	3.98	1.401

Solute	Mass %	m/mol kg^{-1}	c/mol L^{-1}	ρ/g cm^{-3}	n	Δ/°C	η/mPa s
	9.0	1.206	1.145	1.0440	1.3456	4.57	1.468
	10.0	1.354	1.279	1.0495	1.3470		1.539
	12.0	1.662	1.552	1.0607	1.3498		1.688
	14.0	1.984	1.829	1.0718	1.3526		1.855
	16.0	2.322	2.112	1.0830	1.3554		2.054
	18.0	2.676	2.400	1.0940	1.3583		2.284
	20.0	3.047	2.694	1.1050	1.3611		2.567
	22.0	3.438	2.993	1.1159	1.3639		2.948
	24.0	3.849	3.297	1.1268	1.3666		3.400
	26.0	4.283	3.606	1.1377	1.3693		3.877
	28.0	4.741	3.921	1.1488	1.3720		4.388
	30.0	5.224	4.243	1.1602	1.3748		4.940
Sodium	0.5	0.060	0.060	1.0018	1.3337	0.20	1.015
bicarbonate	1.0	0.120	0.120	1.0054	1.3344	0.40	1.028
NaHCO$_3$	1.5	0.181	0.180	1.0089	1.3351	0.59	1.042
	2.0	0.243	0.241	1.0125	1.3357	0.78	1.057
	2.5	0.305	0.302	1.0160	1.3364	0.98	1.071
	3.0	0.368	0.364	1.0196	1.3370	1.16	1.086
	3.5	0.432	0.426	1.0231	1.3377	1.35	1.102
	4.0	0.496	0.489	1.0266	1.3383	1.54	1.118
	4.5	0.561	0.552	1.0301	1.3390	1.72	1.134
	5.0	0.627	0.615	1.0337	1.3396	1.90	1.151
	5.5	0.693	0.679	1.0372	1.3403	2.08	1.168
	6.0	0.760	0.743	1.0408	1.3409	2.26	1.185
Sodium	0.5	0.049	0.049	1.0021	1.3337	0.17	1.004
bromide	1.0	0.098	0.098	1.0060	1.3344	0.34	1.007
NaBr	2.0	0.198	0.197	1.0139	1.3358	0.69	1.012
	3.0	0.301	0.298	1.0218	1.3372	1.04	1.017
	4.0	0.405	0.400	1.0298	1.3386	1.39	1.022
	5.0	0.512	0.504	1.0380	1.3401	1.76	1.028
	6.0	0.620	0.610	1.0462	1.3415	2.14	1.034
	7.0	0.732	0.717	1.0546	1.3430	2.53	1.040
	8.0	0.845	0.826	1.0630	1.3445	2.93	1.046
	9.0	0.961	0.937	1.0716	1.3460	3.34	1.053
	10.0	1.080	1.050	1.0803	1.3475	3.77	1.060
	12.0	1.325	1.281	1.0981	1.3506	4.67	1.077
	14.0	1.582	1.519	1.1164	1.3538	5.65	1.096
	16.0	1.851	1.765	1.1352	1.3570	6.74	1.119
	18.0	2.133	2.020	1.1546	1.3604		1.144
	20.0	2.430	2.283	1.1745	1.3638		1.174
	22.0	2.741	2.555	1.1951	1.3673		1.207
	24.0	3.069	2.837	1.2163	1.3708		1.244
	26.0	3.415	3.129	1.2382	1.3745		1.287
	28.0	3.780	3.431	1.2608	1.3783		1.336
	30.0	4.165	3.744	1.2842	1.3822		1.395
	32.0	4.574	4.069	1.3083	1.3862		1.465
	34.0	5.007	4.406	1.3333	1.3903		1.546
	36.0	5.467	4.755	1.3592	1.3946		1.639
	38.0	5.957	5.119	1.3860	1.3990		1.745
	40.0	6.479	5.496	1.4138	1.4035		1.866
Sodium	0.5	0.047	0.047	1.0034	1.3341	0.22	1.025
carbonate	1.0	0.095	0.095	1.0086	1.3352	0.43	1.049
Na$_2$CO$_3$	2.0	0.193	0.192	1.0190	1.3375	0.75	1.102
	3.0	0.292	0.291	1.0294	1.3397	1.08	1.159
	4.0	0.393	0.392	1.0398	1.3419	1.42	1.222
	5.0	0.497	0.495	1.0502	1.3440	1.77	1.292
	6.0	0.602	0.600	1.0606	1.3462	2.13	1.367
	7.0	0.710	0.707	1.0711	1.3483		1.448

Solute	Mass %	m/mol kg^{-1}	c/mol L^{-1}	ρ/g cm^{-3}	n	Δ/°C	η/mPa s
	8.0	0.820	0.816	1.0816	1.3504		1.538
	9.0	0.933	0.927	1.0922	1.3525		1.638
	10.0	1.048	1.041	1.1029	1.3547		1.754
	11.0	1.166	1.156	1.1136	1.3568		1.884
	12.0	1.287	1.273	1.1244	1.3589		2.028
	13.0	1.410	1.392	1.1353	1.3610		2.186
	14.0	1.536	1.514	1.1463	1.3631		2.361
	15.0	1.665	1.638	1.1574	1.3652		2.551
Sodium	0.5	0.086	0.086	1.0018	1.3339	0.30	1.011
chloride	1.0	0.173	0.172	1.0053	1.3347	0.59	1.020
NaCl	2.0	0.349	0.346	1.0125	1.3365	1.19	1.036
	3.0	0.529	0.523	1.0196	1.3383	1.79	1.052
	4.0	0.713	0.703	1.0268	1.3400	2.41	1.068
	5.0	0.901	0.885	1.0340	1.3418	3.05	1.085
	6.0	1.092	1.069	1.0413	1.3435	3.70	1.104
	7.0	1.288	1.256	1.0486	1.3453	4.38	1.124
	8.0	1.488	1.445	1.0559	1.3470	5.08	1.145
	9.0	1.692	1.637	1.0633	1.3488	5.81	1.168
	10.0	1.901	1.832	1.0707	1.3505	6.56	1.193
	12.0	2.333	2.229	1.0857	1.3541	8.18	1.250
	14.0	2.785	2.637	1.1008	1.3576	9.94	1.317
	16.0	3.259	3.056	1.1162	1.3612	11.89	1.388
	18.0	3.756	3.486	1.1319	1.3648	14.04	1.463
	20.0	4.278	3.928	1.1478	1.3684	16.46	1.557
	22.0	4.826	4.382	1.1640	1.3721	19.18	1.676
	24.0	5.403	4.847	1.1804	1.3757		1.821
	26.0	6.012	5.326	1.1972	1.3795		1.990
Sodium	1.0	0.039	0.039	1.0049	1.3348	0.20	1.043
citrate	2.0	0.079	0.078	1.0120	1.3366	0.39	1.081
(HO)C(COONa)$_3$	3.0	0.120	0.118	1.0186	1.3383	0.59	1.122
	4.0	0.161	0.159	1.0260	1.3401	0.79	1.166
	5.0	0.204	0.200	1.0331	1.3419	0.97	1.210
	6.0	0.247	0.242	1.0405	1.3437	1.17	1.263
	7.0	0.292	0.284	1.0482	1.3455	1.36	1.314
	8.0	0.337	0.327	1.0557	1.3473	1.57	1.371
	9.0	0.383	0.371	1.0632	1.3491	1.77	1.427
	10.0	0.431	0.415	1.0708	1.3509	1.96	1.499
	12.0	0.528	0.505	1.0861	1.3546	2.38	1.649
	14.0	0.631	0.598	1.1019	1.3583	2.82	1.832
	16.0	0.738	0.693	1.1173	1.3618	3.27	2.045
	18.0	0.851	0.790	1.1327	1.3656	3.82	2.290
	20.0	0.969	0.891	1.1492	1.3693	4.39	2.596
	24.0	1.224	1.099	1.1813	1.3767		3.409
	28.0	1.507	1.318	1.2151	1.3845		4.586
	32.0	1.823	1.548	1.2487	1.3923		6.541
	36.0	2.180	1.792	1.2843	1.4001		9.788
Sodium	0.5	0.126	0.125	1.0039	1.3344	0.43	1.027
hydroxide	1.0	0.253	0.252	1.0095	1.3358	0.86	1.054
NaOH	2.0	0.510	0.510	1.0207	1.3386	1.74	1.112
	3.0	0.773	0.774	1.0318	1.3414	2.64	1.176
	4.0	1.042	1.043	1.0428	1.3441	3.59	1.248
	5.0	1.316	1.317	1.0538	1.3467	4.57	1.329
	6.0	1.596	1.597	1.0648	1.3494	5.60	1.416
	7.0	1.882	1.883	1.0758	1.3520	6.69	1.510
	8.0	2.174	2.174	1.0869	1.3546	7.87	1.616
	9.0	2.473	2.470	1.0979	1.3572	9.12	1.737
	10.0	2.778	2.772	1.1089	1.3597	10.47	1.882
	12.0	3.409	3.393	1.1309	1.3648	13.42	2.201

Solute	Mass %	m/mol kg^{-1}	c/mol L^{-1}	ρ/g cm^{-3}	n	Δ/°C	η/mPa s
	14.0	4.070	4.036	1.1530	1.3697	16.76	2.568
	15.0	4.412	4.365	1.1640	1.3722		2.789
	16.0	4.762	4.701	1.1751	1.3746		3.043
	18.0	5.488	5.387	1.1971	1.3793		3.698
	20.0	6.250	6.096	1.2192	1.3840		4.619
	22.0	7.052	6.827	1.2412	1.3885		5.765
	24.0	7.895	7.579	1.2631	1.3929		7.100
	26.0	8.784	8.352	1.2848	1.3971		8.744
	28.0	9.723	9.145	1.3064	1.4012		10.832
	30.0	10.715	9.958	1.3277	1.4051		13.517
	32.0	11.766	10.791	1.3488	1.4088		16.844
	34.0	12.880	11.643	1.3697	1.4123		20.751
	36.0	14.064	12.512	1.3901	1.4156		25.290
	38.0	15.324	13.398	1.4102	1.4186		30.461
	40.0	16.668	14.300	1.4299	1.4215		36.312
Sodium	0.5	0.059	0.059	1.0016	1.3336	0.20	1.004
nitrate	1.0	0.119	0.118	1.0050	1.3341	0.40	1.007
$NaNO_3$	2.0	0.240	0.238	1.0117	1.3353	0.79	1.012
	3.0	0.364	0.359	1.0185	1.3364	1.18	1.018
	4.0	0.490	0.483	1.0254	1.3375	1.56	1.025
	5.0	0.619	0.607	1.0322	1.3387	1.94	1.032
	6.0	0.751	0.734	1.0392	1.3398	2.32	1.040
	7.0	0.886	0.862	1.0462	1.3409	2.70	1.049
	8.0	1.023	0.991	1.0532	1.3421	3.08	1.059
	9.0	1.164	1.123	1.0603	1.3432	3.46	1.069
	10.0	1.307	1.256	1.0674	1.3443	3.84	1.081
	12.0	1.604	1.527	1.0819	1.3466	4.60	1.107
	14.0	1.915	1.806	1.0967	1.3489	5.37	1.138
	18.0	2.583	2.387	1.1272	1.3536	6.98	1.215
	20.0	2.941	2.689	1.1429	1.3559	7.81	1.263
	30.0	5.042	4.326	1.2256	1.3678		1.609
	40.0	7.844	6.200	1.3175	1.3802		2.226
Sodium	0.5	0.031	0.031	1.0042	1.3343	0.19	1.033
phosphate	1.0	0.062	0.062	1.0100	1.3356	0.37	1.064
Na_3PO_4	1.5	0.093	0.093	1.0158	1.3369	0.53	1.094
	2.0	0.124	0.125	1.0216	1.3381	0.67	1.126
	2.5	0.156	0.157	1.0275	1.3394	0.79	1.161
	3.0	0.189	0.189	1.0335	1.3406		1.198
	3.5	0.221	0.222	1.0395	1.3419		1.238
	4.0	0.254	0.255	1.0456	1.3432		1.281
	4.5	0.287	0.289	1.0517	1.3444		1.327
	5.0	0.321	0.323	1.0579	1.3457		1.375
	5.5	0.355	0.357	1.0642	1.3470		1.426
	6.0	0.389	0.392	1.0705	1.3482		1.480
	6.5	0.424	0.427	1.0768	1.3495		1.538
	7.0	0.459	0.462	1.0832	1.3507		1.598
	7.5	0.495	0.498	1.0896	1.3519		1.662
	8.0	0.530	0.535	1.0961	1.3532		1.729
Sodium hydrogen	0.5	0.035	0.035	1.0032	1.3340	0.17	1.021
phosphate	1.0	0.071	0.071	1.0082	1.3349	0.32	1.042
Na_2HPO_4	1.5	0.107	0.107	1.0131	1.3358	0.46	1.064
	2.0	0.144	0.143	1.0180	1.3368		1.088
	2.5	0.181	0.180	1.0229	1.3377		1.113
	3.0	0.218	0.217	1.0279	1.3386		1.138
	3.5	0.255	0.255	1.0328	1.3396		1.165
	4.0	0.293	0.292	1.0378	1.3405		1.193
	4.5	0.332	0.331	1.0428	1.3414		1.223
	5.0	0.371	0.369	1.0478	1.3424		1.254

Solute	Mass %	m/mol kg^{-1}	c/mol L^{-1}	ρ/g cm^{-3}	n	Δ/°C	η/mPa s
	5.5	0.410	0.408	1.0528	1.3433		1.286
Sodium	0.5	0.042	0.042	1.0019	1.3336	0.14	1.018
dihydrogen	1.0	0.084	0.084	1.0056	1.3343	0.28	1.035
phosphate	1.5	0.127	0.126	1.0094	1.3349	0.42	1.051
NaH$_2$PO$_4$	2.0	0.170	0.169	1.0131	1.3356	0.56	1.068
	2.5	0.214	0.212	1.0168	1.3362	0.70	1.085
	3.0	0.258	0.255	1.0206	1.3369	0.84	1.103
	3.5	0.302	0.299	1.0244	1.3375	0.98	1.121
	4.0	0.347	0.343	1.0281	1.3382	1.12	1.140
	4.5	0.393	0.387	1.0319	1.3388	1.25	1.160
	5.0	0.439	0.432	1.0358	1.3395	1.39	1.180
	6.0	0.532	0.522	1.0434	1.3408	1.65	1.223
	7.0	0.627	0.613	1.0511	1.3421	1.89	1.270
	8.0	0.725	0.706	1.0589	1.3434	2.12	1.319
	9.0	0.824	0.800	1.0668	1.3447	2.35	1.371
	10.0	0.926	0.896	1.0747	1.3460	2.58	1.428
	12.0	1.137	1.091	1.0907	1.3486	3.06	1.552
	14.0	1.357	1.292	1.1070	1.3512	3.53	1.694
	16.0	1.588	1.499	1.1236	1.3538	4.03	1.861
	18.0	1.830	1.711	1.1404	1.3565	4.55	2.050
	20.0	2.084	1.930	1.1576	1.3592	5.10	2.283
	22.0	2.351	2.155	1.1752	1.3618		2.550
	24.0	2.632	2.387	1.1931	1.3646		2.850
	26.0	2.929	2.625	1.2113	1.3673		3.214
	28.0	3.242	2.870	1.2299	1.3700		3.682
	30.0	3.572	3.123	1.2488	1.3728		4.300
	32.0	3.923	3.383	1.2682	1.3756		5.079
	34.0	4.294	3.650	1.2879	1.3784		6.008
	36.0	4.689	3.925	1.3080	1.3812		7.098
	38.0	5.109	4.208	1.3285	1.3840		8.363
	40.0	5.557	4.499	1.3493	1.3869		9.814
Sodium	0.5	0.035	0.035	1.0027	1.3338	0.17	1.013
sulfate	1.0	0.071	0.071	1.0071	1.3345	0.32	1.026
Na$_2$SO$_4$	2.0	0.144	0.143	1.0161	1.3360	0.61	1.058
	3.0	0.218	0.217	1.0252	1.3376	0.87	1.091
	4.0	0.293	0.291	1.0343	1.3391	1.13	1.126
	5.0	0.371	0.367	1.0436	1.3406	1.36	1.163
	6.0	0.449	0.445	1.0526	1.3420	1.56	1.202
	7.0	0.530	0.523	1.0619	1.3435		1.244
	8.0	0.612	0.603	1.0713	1.3449		1.289
	9.0	0.696	0.685	1.0808	1.3464		1.337
	10.0	0.782	0.768	1.0905	1.3479		1.390
	12.0	0.960	0.938	1.1101	1.3509		1.508
	14.0	1.146	1.114	1.1301	1.3539		1.646
	16.0	1.341	1.296	1.1503	1.3567		1.812
	18.0	1.545	1.483	1.1705	1.3595		2.005
	20.0	1.760	1.677	1.1907	1.3620		2.227
	22.0	1.986	1.875	1.2106	1.3643		2.481
Sodium	0.5	0.032	0.032	1.0024	1.3340	0.14	1.012
thiosulfate	1.0	0.064	0.064	1.0065	1.3351	0.28	1.023
Na$_2$S$_2$O$_3$	2.0	0.129	0.128	1.0148	1.3371	0.57	1.044
	3.0	0.196	0.194	1.0231	1.3392	0.84	1.066
	4.0	0.264	0.261	1.0315	1.3413	1.09	1.090
	5.0	0.333	0.329	1.0399	1.3434	1.34	1.115
	6.0	0.404	0.398	1.0483	1.3454	1.59	1.141
	7.0	0.476	0.468	1.0568	1.3475	1.83	1.169
	8.0	0.550	0.539	1.0654	1.3496	2.06	1.199
	9.0	0.626	0.611	1.0740	1.3517	2.30	1.231

Solute	Mass %	m/mol kg^{-1}	c/mol L^{-1}	ρ/g cm^{-3}	n	Δ/°C	η/mPa s
	10.0	0.703	0.685	1.0827	1.3538	2.55	1.267
	12.0	0.862	0.835	1.1003	1.3581	3.06	1.345
	14.0	1.030	0.990	1.1182	1.3624	3.60	1.435
	16.0	1.205	1.150	1.1365	1.3667	4.17	1.537
	18.0	1.388	1.315	1.1551	1.3711	4.76	1.657
	20.0	1.581	1.485	1.1740	1.3756	5.37	1.798
	30.0	2.711	2.417	1.2739	1.3987		2.903
	40.0	4.216	3.498	1.3827	1.4229		5.758
Strontium	0.5	0.032	0.032	1.0027	1.3339	0.16	1.012
chloride	1.0	0.064	0.064	1.0071	1.3348	0.31	1.021
SrCl$_2$	2.0	0.129	0.128	1.0161	1.3366	0.62	1.039
	3.0	0.195	0.194	1.0252	1.3384	0.93	1.057
	4.0	0.263	0.261	1.0344	1.3402	1.26	1.076
	5.0	0.332	0.329	1.0437	1.3421	1.61	1.096
	6.0	0.403	0.399	1.0532	1.3440	1.98	1.116
	7.0	0.475	0.469	1.0628	1.3459	2.38	1.136
	8.0	0.549	0.541	1.0726	1.3478	2.80	1.157
	9.0	0.624	0.615	1.0825	1.3498	3.25	1.180
	10.0	0.701	0.689	1.0925	1.3518	3.74	1.204
	12.0	0.860	0.843	1.1131	1.3558	4.81	1.258
	14.0	1.027	1.002	1.1342	1.3599	6.03	1.317
	16.0	1.202	1.167	1.1558	1.3641	7.41	1.383
	18.0	1.385	1.338	1.1780	1.3684	8.98	1.460
	20.0	1.577	1.515	1.2008	1.3728	10.74	1.549
	22.0	1.779	1.699	1.2241	1.3772	12.74	1.650
	24.0	1.992	1.890	1.2481	1.3817	14.99	1.765
	26.0	2.216	2.087	1.2728	1.3864		1.897
	28.0	2.453	2.293	1.2983	1.3911		2.056
	30.0	2.703	2.507	1.3248	1.3961		2.245
	32.0	2.968	2.730	1.3523	1.4013		2.527
	34.0	3.250	2.962	1.3811	1.4067		2.846
	36.0	3.548	3.205	1.4114	1.4124		3.206
Sucrose	0.5	0.015	0.015	1.0002	1.3337	0.03	1.015
C$_{12}$H$_{22}$O$_{11}$	1.0	0.030	0.029	1.0021	1.3344	0.06	1.028
	2.0	0.060	0.059	1.0060	1.3359	0.11	1.055
	3.0	0.090	0.089	1.0099	1.3373	0.17	1.084
	4.0	0.122	0.118	1.0139	1.3388	0.23	1.114
	5.0	0.154	0.149	1.0178	1.3403	0.29	1.146
	6.0	0.186	0.179	1.0218	1.3418	0.35	1.179
	7.0	0.220	0.210	1.0259	1.3433	0.42	1.215
	8.0	0.254	0.241	1.0299	1.3448	0.49	1.254
	9.0	0.289	0.272	1.0340	1.3463	0.55	1.294
	10.0	0.325	0.303	1.0381	1.3478	0.63	1.336
	12.0	0.398	0.367	1.0465	1.3509	0.77	1.429
	14.0	0.476	0.431	1.0549	1.3541	0.93	1.534
	16.0	0.556	0.497	1.0635	1.3573	1.10	1.653
	18.0	0.641	0.564	1.0722	1.3606	1.27	1.790
	20.0	0.730	0.632	1.0810	1.3639	1.47	1.945
	22.0	0.824	0.700	1.0899	1.3672	1.67	2.124
	24.0	0.923	0.771	1.0990	1.3706	1.89	2.331
	26.0	1.026	0.842	1.1082	1.3741	2.12	2.573
	28.0	1.136	0.914	1.1175	1.3776	2.37	2.855
	30.0	1.252	0.988	1.1270	1.3812	2.64	3.187
	32.0	1.375	1.063	1.1366	1.3848	2.94	3.762
	34.0	1.505	1.139	1.1464	1.3885	3.27	4.052
	36.0	1.643	1.216	1.1562	1.3922	3.63	4.621
	38.0	1.791	1.295	1.1663	1.3960	4.02	5.315
	40.0	1.948	1.375	1.1765	1.3999	4.45	6.162
	42.0	2.116	1.456	1.1868	1.4038	4.93	7.234

Solute	Mass %	m/mol kg^{-1}	c/mol L^{-1}	ρ/g cm^{-3}	n	Δ/°C	η/mPa s
	44.0	2.295	1.539	1.1972	1.4078		8.596
	46.0	2.489	1.623	1.2079	1.4118		10.301
	48.0	2.697	1.709	1.2186	1.4159		12.515
	50.0	2.921	1.796	1.2295	1.4201		15.431
	60.0	4.382	2.255	1.2864	1.4419		58.487
	70.0	6.817	2.755	1.3472	1.4654		481.561
	80.0	11.686	3.299	1.4117	1.4906		
Sulfuric acid	0.5	0.051	0.051	1.0016	1.3336	0.21	1.010
H$_2$SO$_4$	1.0	0.103	0.102	1.0049	1.3342	0.42	1.019
	2.0	0.208	0.206	1.0116	1.3355	0.80	1.036
	3.0	0.315	0.311	1.0183	1.3367	1.17	1.059
	4.0	0.425	0.418	1.0250	1.3379	1.60	1.085
	5.0	0.537	0.526	1.0318	1.3391	2.05	1.112
	6.0	0.651	0.635	1.0385	1.3403	2.50	1.136
	7.0	0.767	0.746	1.0453	1.3415	2.95	1.159
	8.0	0.887	0.858	1.0522	1.3427	3.49	1.182
	9.0	1.008	0.972	1.0591	1.3439	4.08	1.206
	10.0	1.133	1.087	1.0661	1.3451	4.64	1.230
	12.0	1.390	1.322	1.0802	1.3475	5.93	1.282
	14.0	1.660	1.563	1.0947	1.3500	7.49	1.337
	16.0	1.942	1.810	1.1094	1.3525	9.26	1.399
	18.0	2.238	2.064	1.1245	1.3551	11.29	1.470
	20.0	2.549	2.324	1.1398	1.3576	13.64	1.546
	22.0	2.876	2.592	1.1554	1.3602	16.48	1.624
	24.0	3.220	2.866	1.1714	1.3628	19.85	1.706
	26.0	3.582	3.147	1.1872	1.3653	24.29	1.797
	28.0	3.965	3.435	1.2031	1.3677	29.65	1.894
	30.0	4.370	3.729	1.2191	1.3701	36.21	2.001
	32.0	4.798	4.030	1.2353	1.3725	44.76	2.122
	34.0	5.252	4.339	1.2518	1.3749	55.28	2.255
	36.0	5.735	4.656	1.2685	1.3773		2.392
	38.0	6.249	4.981	1.2855	1.3797		2.533
	40.0	6.797	5.313	1.3028	1.3821		2.690
	42.0	7.383	5.655	1.3205	1.3846		2.872
	44.0	8.011	6.005	1.3386	1.3870		3.073
	46.0	8.685	6.364	1.3570	1.3895		3.299
	48.0	9.411	6.734	1.3759	1.3920		3.546
	50.0	10.196	7.113	1.3952	1.3945		3.826
	52.0	11.045	7.502	1.4149	1.3971		4.142
	54.0	11.969	7.901	1.4351	1.3997		4.499
	56.0	12.976	8.312	1.4558	1.4024		4.906
	58.0	14.080	8.734	1.4770	1.4050		5.354
	60.0	15.294	9.168	1.4987	1.4077		5.917
	70.0	23.790	11.494	1.6105			
	80.0	40.783	14.088	1.7272			
	90.0	91.762	16.649	1.8144			
	92.0	117.251	17.109	1.8240			
	94.0	159.734	17.550	1.8312			
	96.0	244.698	17.966	1.8355			
	98.0	499.592	18.346	1.8361			
	100.0		18.663	1.8305			
Trichloroacetic	0.5	0.031	0.031	1.0008	1.3337	0.11	1.011
acid	1.0	0.062	0.061	1.0034	1.3343	0.21	1.021
CCl$_3$COOH	2.0	0.125	0.123	1.0083	1.3356	0.42	1.044
	3.0	0.189	0.186	1.0133	1.3369	0.64	1.069
	4.0	0.255	0.249	1.0182	1.3381	0.86	1.096
	5.0	0.322	0.313	1.0230	1.3394	1.08	1.123
	6.0	0.391	0.377	1.0279	1.3406	1.30	1.150
	7.0	0.461	0.442	1.0328	1.3418	1.53	1.177

Solute	Mass %	m/mol kg^{-1}	c/mol L^{-1}	ρ/g cm^{-3}	n	Δ/°C	η/mPa s
	8.0	0.532	0.508	1.0378	1.3431	1.76	1.204
	9.0	0.605	0.574	1.0428	1.3444	1.99	1.233
	10.0	0.680	0.641	1.0479	1.3456	2.23	1.263
	12.0	0.835	0.777	1.0583	1.3483	2.73	1.326
	14.0	0.996	0.916	1.0692	1.3510	3.26	1.393
	16.0	1.166	1.058	1.0806	1.3539	3.82	1.462
	18.0	1.343	1.203	1.0921	1.3568		1.533
	20.0	1.530	1.351	1.1035	1.3597		1.608
	24.0	1.933	1.654	1.1260	1.3652		1.768
	28.0	2.380	1.968	1.1485	1.3705		1.935
	32.0	2.880	2.294	1.1713	1.3759		2.118
	36.0	3.443	2.632	1.1947	1.3813		2.320
	40.0	4.080	2.984	1.2188	1.3868		1.543
	44.0	4.809	3.349	1.2435	1.3923		2.797
	48.0	5.650	3.726	1.2682	1.3977		3.076
Tris	0.5	0.041	0.041	0.9994	1.3337	0.08	1.014
(hydroxymethyl)-	1.0	0.083	0.083	1.0006	1.3344	0.16	1.027
methylamine	2.0	0.168	0.166	1.0030	1.3359	0.31	1.054
$H_2NC(CH_2OH)_3$	3.0	0.255	0.249	1.0054	1.3374	0.47	1.083
	4.0	0.344	0.333	1.0078	1.3388	0.64	1.115
	5.0	0.434	0.417	1.0103	1.3403	0.80	1.148
	6.0	0.527	0.502	1.0128	1.3418	0.97	1.182
	7.0	0.621	0.587	1.0153	1.3433	1.15	1.218
	8.0	0.718	0.672	1.0179	1.3448	1.33	1.256
	9.0	0.816	0.758	1.0204	1.3463	1.51	1.295
	10.0	0.917	0.844	1.0230	1.3478	1.70	1.337
	12.0	1.126	1.019	1.0282	1.3508	2.08	1.427
	14.0	1.344	1.194	1.0335	1.3539	2.47	1.527
	16.0	1.572	1.372	1.0389	1.3570	2.90	1.642
	18.0	1.812	1.552	1.0443	1.3601	3.36	1.772
	20.0	2.064	1.733	1.0498	1.3633	3.85	1.920
	30.0	3.538	2.670	1.0781	1.3797		2.998
	40.0	5.503	3.657	1.1076	1.3970		5.208
Urea	0.5	0.084	0.083	0.9995	1.3337	0.16	1.007
$(NH_2)_2CO$	1.0	0.168	0.167	1.0007	1.3344	0.31	1.010
	2.0	0.340	0.334	1.0033	1.3358	0.62	1.012
	3.0	0.515	0.502	1.0058	1.3372	0.93	1.017
	4.0	0.694	0.672	1.0085	1.3387	1.24	1.025
	5.0	0.876	0.842	1.0111	1.3401	1.55	1.033
	6.0	1.063	1.013	1.0138	1.3416	1.88	1.041
	7.0	1.253	1.185	1.0165	1.3431	2.22	1.049
	8.0	1.448	1.358	1.0192	1.3446	2.56	1.057
	9.0	1.647	1.531	1.0220	1.3461	2.91	1.065
	10.0	1.850	1.706	1.0248	1.3476	3.26	1.074
	12.0	2.270	2.059	1.0304	1.3506	3.95	1.091
	14.0	2.710	2.415	1.0360	1.3537	4.66	1.109
	16.0	3.171	2.775	1.0417	1.3568	5.40	1.130
	18.0	3.655	3.139	1.0473	1.3599	6.19	1.153
	20.0	4.163	3.506	1.0530	1.3629	7.00	1.178
	22.0	4.696	3.878	1.0586	1.3661	7.81	1.205
	24.0	5.258	4.253	1.0643	1.3692	8.64	1.235
	26.0	5.850	4.632	1.0699	1.3723	9.52	1.266
	28.0	6.475	5.014	1.0756	1.3754	10.45	1.298
	30.0	7.136	5.401	1.0812	1.3785	11.40	1.332
	32.0	7.835	5.791	1.0869	1.3817	12.34	1.371
	34.0	8.577	6.185	1.0926	1.3848	13.27	1.413
	36.0	9.366	6.584	1.0984	1.3881	14.20	1.459
	38.0	10.205	6.988	1.1044	1.3913	15.11	1.509
	40.0	11.100	7.397	1.1106	1.3947	15.99	1.565

Solute	Mass %	m/mol kg^{-1}	c/mol L^{-1}	ρ/g cm^{-3}	n	Δ/°C	η/mPa s
	42.0	12.057	7.812	1.1171	1.3982	16.83	1.629
	44.0	13.082	8.234	1.1239	1.4018	17.62	1.700
	46.0	14.183	8.665	1.1313	1.4056		1.780
Zinc sulfate	0.5	0.031	0.031	1.0034	1.3339	0.08	1.021
ZnSO$_4$	1.0	0.063	0.062	1.0085	1.3348	0.15	1.040
	2.0	0.126	0.126	1.0190	1.3366	0.28	1.081
	3.0	0.192	0.191	1.0296	1.3384	0.41	1.126
	4.0	0.258	0.258	1.0403	1.3403	0.53	1.175
	5.0	0.326	0.326	1.0511	1.3421	0.65	1.227
	6.0	0.395	0.395	1.0620	1.3439	0.77	1.283
	7.0	0.466	0.465	1.0730	1.3457	0.89	1.341
	8.0	0.539	0.537	1.0842	1.3475	1.01	1.403
	9.0	0.613	0.611	1.0956	1.3494	1.14	1.470
	10.0	0.688	0.686	1.1071	1.3513	1.27	1.545
	12.0	0.845	0.840	1.1308	1.3551	1.55	1.716
	14.0	1.008	1.002	1.1553	1.3590	1.89	1.918
	16.0	1.180	1.170	1.1806	1.3630	2.31	2.152

SOLUBILITY OF SELECTED GASES IN WATER

L. H. Gevantman

The values in this table are taken almost exclusively from the International Union of Pure and Applied Chemistry "Solubility Data Series." Unless noted, they comprise evaluated data fitted to a smoothing equation. The data at each temperature are then derived from the smoothing equation which expresses the mole fraction solubility X_1 of the gas in solution as:

$$\ln X_1 = A + B/T^* + C \ln T^*$$

where

$$T^* = T/100 \text{ K}$$

All values refer to a partial pressure of the gas of 101.325 kPa (one atmosphere).

The equation constants, the standard deviation for $\ln X_1$ (except where noted), and the temperature range over which the equation applies are given in the column headed Equation constants. There are two exceptions. The equation for methane has an added term, DT^*. The equation for H_2Se and H_2S takes the form,

$$\ln X_1 = A + B/T + C \ln T + DT$$

where T is the temperature in kelvin.

Solubilities given for those gases that react with water, namely ozone, nitrogen oxides, chlorine and its oxides, carbon dioxide, hydrogen sulfide, hydrogen selenide and sulfur dioxide, are recorded as bulk solubilities; i.e., all chemical species of the gas and its reaction products with water are included.

Gas	T/K	Solubility (X_1)	Equation constants	Ref.
Hydrogen (H_2)	288.15	1.510×10^{-5}	$A = -48.1611$	1
$M_r = 2.01588$	293.15	1.455×10^{-5}	$B = 55.2845$	
	298.15	1.411×10^{-5}	$C = 16.8893$	
	303.15	1.377×10^{-5}	Std. dev. = ± 0.54%	
	308.15	1.350×10^{-5}	Temp. range = 273.15—353.15	
Deuterium (D_2)	283.15	1.675×10^{-5} ± 0.57%	Averaged experimental values	1
$M_r = 4.0282$	288.15	1.595×10^{-5} ± 0.57%		
	293.15	1.512×10^{-5} ± 0.78%	Temp. range = 278.15—303.15	
	298.15	1.460×10^{-5} ± 0.52%		
	303.15	1.395×10^{-5} ± 0.37%		
Helium (He)	288.15	7.123×10^{-6}	$A = -41.4611$	2
$A_r = 4.0026$	293.15	7.044×10^{-6}	$B = 42.5962$	
	298.15	6.997×10^{-6}	$C = 14.0094$	
	303.15	6.978×10^{-6}	Std. dev. = ±0.54%	
	308.15	6.987×10^{-6}	Temp. range = 273.15—348.15	
Neon (Ne)	288.15	8.702×10^{-6}	$A = -52.8573$	2
$A_r = 20.1797$	293.15	8.395×10^{-6}	$B = 61.0494$	
	298.15	8.152×10^{-6}	$C = 18.9157$	
	303.15	7.966×10^{-6}	Std. dev. = ±0.47%	
	308.15	7.829×10^{-6}	Temp. range = 273.15—348.15	
Argon (Ar)	288.15	3.025×10^{-5}	$A = -57.6661$	3
$A_r = 39.948$	293.15	2.748×10^{-5}	$B = 74.7627$	
	298.15	2.519×10^{-5}	$C = 20.1398$	
	303.15	2.328×10^{-5}	Std. dev. = ±0.26%	
	308.15	2.169×10^{-5}	Temp. range = 273.15—348.15	
Krypton (Kr)	288.15	5.696×10^{-5}	$A = -66.9928$	4
$A_r = 83.80$	293.15	5.041×10^{-5}	$B = 91.0166$	
	298.15	4.512×10^{-5}	$C = 24.2207$	
	303.15	4.079×10^{-5}	Std. dev. = ±0.32%	
	308.15	3.725×10^{-5}	Temp. range = 273.15—353.15	
Xenon (Xe)	288.15	10.519×10^{-5}	$A = -74.7398$	4
$A_r = 131.29$	293.15	9.051×10^{-5}	$B = 105.210$	
	298.15	7.890×10^{-5}	$C = 27.4664$	
	303.15	6.961×10^{-5}	Std. dev. = ±0.35%	
	308.15	6.212×10^{-5}	Temp. range = 273.15—348.15	

Gas	T/K	Solubility (X_1)	Equation constants	Ref.
Radon-222(^{222}Rn)	288.15	2.299×10^{-4}	$A = -90.5481$	
$A_r = 222$	293.15	1.945×10^{-4}	$B = 130.026$	
	298.15	1.671×10^{-4}	$C = 35.0047$	
	303.15	1.457×10^{-4}	Std. dev. = ±1.02%	
	308.15	1.288×10^{-4}	Temp. range = 273.15—373.15	
Oxygen (O_2)	288.15	2.756×10^{-5}	$A = -66.7354$	5
$M_r = 31.9988$	293.15	2.501×10^{-5}	$B = 87.4755$	
	298.15	2.293×10^{-5}	$C = 24.4526$	
	303.15	2.122×10^{-5}	Std. dev. = ±0.36%	
	308.15	1.982×10^{-5}	Temp. range = 273.15—348.15	
Ozone (O_3)	293.15	$1.885 \times 10^{-6} \pm 10\%$	Experimental value derived	5
$M_r = 47.9982$		pH = 7.0	from Henry's Law Constant Equation	
Nitrogen (N_2)	288.15	1.386×10^{-5}	$A = -67.3877$	6
$M_r = 28.0134$	293.15	1.274×10^{-5}	$B = 86.3213$	
	298.15	1.183×10^{-5}	$C = 24.7981$	
	303.15	1.108×10^{-5}	Std. dev. = ±0.72%	
	308.15	1.047×10^{-5}	Temp. range = 273.15—348.15	
Nitrous oxide (N_2O)	288.15	5.948×10^{-4}	$A = -60.7467$	7
$M_r = 44.0129$	293.15	5.068×10^{-4}	$B = 88.8280$	
	298.15	4.367×10^{-4}	$C = 21.2531$	
	303.15	3.805×10^{-4}	Std. dev. = ±1.2%	
	308.15	3.348×10^{-4}	Temp. range = 273.15—313.15	
Nitric oxide (NO)	288.15	4.163×10^{-5}	$A = -62.8086$	7
$M_r = 30.0061$	293.15	3.786×10^{-5}	$B = 82.3420$	
	298.15	3.477×10^{-5}	$C = 22.8155$	
	303.15	3.222×10^{-5}	Std. dev. = ±0.76%	
	308.15	3.012×10^{-5}	Temp. range = 273.15—358.15	
Carbon monoxide (CO)	288.15	2.095×10^{-5}	Derived from Henry's	8
$M_r = 28.0104$	293.15	1.918×10^{-5}	Law Constant Equation	
	298.15	1.774×10^{-5}	Std. dev. = ±0.043%	
	303.15	1.657×10^{-5}	Temp. range = 273.15—328.15	
	308.15	1.562×10^{-5}		
Carbon dioxide (CO_2)	288.15	8.21×10^{-4}	Derived from Henry's	9
$M_r = 44.0098$	293.15	7.07×10^{-4}	Law Constant Equation	
	298.15	6.15×10^{-4}	Std. dev. = ±1.1%	
	303.15	5.41×10^{-4}	Temp. range = 273.15—353.15	
	308.15	4.80×10^{-4}		
Hydrogen selenide (H_2Se)	288.15	1.80×10^{-3}	$A = 9.15$	10
$M_r = 80.976$	298.15	1.49×10^{-3}	$B = 974$	
	308.15	1.24×10^{-3}	$C = -3.542$	
			$D = 0.0042$	
			Std. dev. = ±2.3 × 10^{-5}	
			Temp. range = 288.15—343.15	
Hydrogen sulfide (H_2S)	288.15	2.335×10^{-3}	$A = -24.912$	10
$M_r = 34.082$	293.15	2.075×10^{-3}	$B = 3477$	
	298.15	1.85×10^{-3}	$C = 0.3993$	
	303.15	1.66×10^{-3}	$D = 0.0157$	
	308.15	1.51×10^{-3}	Std. dev. = ±6.5 × 10^{-5}	
			Temp. range = 283.15—603.15	
Sulfur dioxide (SO_2)	288.15	3.45×10^{-2}	$A = -25.2629$	11
$M_r = 64.0648$	293.15	2.90×10^{-2}	$B = 45.7552$	

Gas	T/K	Solubility (X_1)	Equation constants	Ref.
	298.15	2.46×10^{-2}	$C = 5.6855$	
	303.15	2.10×10^{-2}	Std. dev. = ±1.8%	
	308.15	1.80×10^{-2}	Temp. range = 278.15—328.15	
Chlorine (Cl_2)	283.15	$2.48 \times 10^{-3} \pm 2\%$	Experimental data	11
$M_r = 70.9054$	293.15	$1.88 \times 10^{-3} \pm 2\%$	Temp. range = 283.15—333.15	
	303.15	$1.50 \times 10^{-3} \pm 2\%$		
	313.15	$1.23 \times 10^{-3} \pm 2\%$		
Chlorine monoxide (Cl_2O)	273.15	$5.25 \times 10^{-1} \pm 1\%$	Experimental data	11
$M_r = 86.9048$	276.61	$4.54 \times 10^{-1} \pm 1\%$	Temp. range = 273.15—293.15	
	283.15	$4.273 \times 10^{-1} \pm 1\%$		
	293.15	$3.353 \times 10^{-1} \pm 1\%$		
Chlorine dioxide (ClO_2)	288.15	2.67×10^{-2}	$A = 7.9163$	11
$M_r = 67.4515$	293.15	2.20×10^{-2}	$B = 0.4791$	
	298.15	1.823×10^{-2}	$C = 11.0593$	
	303.15	1.513×10^{-2}	Std. dev. = ±4.6%	
	308.15	1.259×10^{-2}	Temp. range = 283.15—333.15	
Methane (CH_4)	288.15	3.122×10^{-5}	$A = -115.6477$	12
$M_r = 16.0428$	293.15	2.806×10^{-5}	$B = 155.5756$	
	298.15	2.552×10^{-5}	$C = 65.2553$	
	303.15	2.346×10^{-5}	$D = -6.1698$	
	308.15	2.180×10^{-5}	Std. dev. = ±0.056%	
			Temp. range = 273.15—328.15	
Ethane (C_2H_6)	288.15	4.556×10^{-5}	$A = -90.8225$	13
$M_r = 30.0696$	293.15	3.907×10^{-5}	$B = 126.9559$	
	298.15	3.401×10^{-5}	$C = 34.7413$	
	303.15	3.002×10^{-5}	Std. dev. = ±0.13%	
	308.15	2.686×10^{-5}	Temp. range = 273.15—323.15	
Propane (C_3H_8)	288.15	3.813×10^{-5}	$A = -102.044$	14
$M_r = 44.097$	293.15	3.200×10^{-5}	$B = 144.345$	
	298.15	2.732×10^{-5}	$C = 39.4740$	
	303.15	2.370×10^{-5}	Std. dev. = ±0.012%	
	308.15	2.088×10^{-5}	Temp. range = 273.15—347.15	
Butane (C_4H_{10})	288.15	3.274×10^{-5}	$A = -102.029$	14
$M_r = 58.123$	293.15	2.687×10^{-5}	$B = 146.040$	
	298.15	2.244×10^{-5}	$C = 38.7599$	
	303.15	1.906×10^{-5}	Std. dev. = ±0.026%	
	308.15	1.645×10^{-5}	Temp. range = 273.15—349.15	
2-Methyl propane (Isobutane)	288.15	2.333×10^{-5}	$A = -129.714$	14
(C_4H_{10})	293.15	1.947×10^{-5}	$B = 183.044$	
$M_r = 58.123$	298.15	1.659×10^{-5}	$C = 53.4651$	
	303.15	1.443×10^{-5}	Std. dev. = ±0.034%	
	308.15	1.278×10^{-5}	Temp. range = 278.15—318.15	

References

1. C. L. Young, Ed., *IUPAC Solubility Data Series*, Vol. 5/6, Hydrogen and Deuterium, Pergamon Press, Oxford, England, 1981.
2. H. L. Clever, Ed., *IUPAC Solubility Data Series*, Vol. 1, Helium and Neon, Pergamon Press, Oxford, England, 1979.
3. H. L. Clever, Ed., *IUPAC Solubility Data Series*, Vol. 4, Argon, Pergamon Press, Oxford, England, 1980.
4. H. L. Clever, Ed., *IUPAC Solubility Data Series*, Vol. 2, Krypton, Xenon and Radon, Pergamon Press, Oxford, England, 1979.
5. R. Battino, Ed., *IUPAC Solubility Data Series*, Vol. 7, Oxygen and Ozone, Pergamon Press, Oxford, England, 1981.
6. R. Battino, Ed., *IUPAC Solubility Data Series*, Vol. 10, Nitrogen and Air, Pergamon Press, Oxford, England, 1982.
7. C. L. Young, Ed., *IUPAC Solubility Data Series*, Vol. 8, Oxides of Nitrogen, Pergamon Press, Oxford, England, 1981.
8. R. W. Cargill, Ed., *IUPAC Solubility Data Series*, Vol. 43, Carbon Monoxide, Pergamon Press, Oxford, England, 1990.
9. R. Crovetto, Evaluation of Solubility Data for the System CO2-H2O, *J. Phys. Chem. Ref. Data*, 20, 575, 1991.

10. P. G. T. Fogg and C. L. Young, Eds., *IUPAC Solubility Data Series*, Vol. 32, Hydrogen Sulfide, Deuterium Sulfide, and Hydrogen Selenide, Pergamon Press, Oxford, England, 1988.

11. C. L. Young, Ed., *IUPAC Solubility Data Series*, Vol. 12, Sulfur Dioxide, Chlorine, Fluorine and Chlorine Oxides, Pergamon Press, Oxford, England, 1983.

12. H. L. Clever and C. L. Young, Eds., *IUPAC Solubility Data Series*, Vol. 27/28, Methane, Pergamon Press, Oxford, England, 1987.

13. W. Hayduk, Ed., *IUPAC Solubility Data Series*, Vol. 9, Ethane, Pergamon Press, Oxford, England, 1982.

14. W. Hayduk, Ed., *IUPAC Solubility Data Series*, Vol. 24, Propane, Butane and 2-Methylpropane, Pergamon Press, Oxford, England, 1986.

SOLUBILITY OF CARBON DIOXIDE IN WATER AT VARIOUS TEMPERATURES AND PRESSURES

The solubility of CO_2 in water, expressed as mole fraction of CO_2 in the liquid phase, is given for pressures up to atmospheric and temperatures of 0 to 100 °C. Note that 1 standard atmosphere equals 101.325 kPa. The references give data over a wider range of temperature and pressure. The estimated uncertainty is about 2%.

References

1. Carroll, J. J., Slupsky, J. D., and Mather, A. E., *J. Phys. Chem. Ref. Data*, 20, 1201, 1991.
2. Fernandez-Prini, R. and Crovetto, R., *J. Phys. Chem. Ref. Data*, 18, 1231, 1989.
3. Crovetto, R., *J. Phys. Chem. Ref. Data*, 20, 575, 1991.

	1000 × mole fraction of CO_2 in liquid phase						
	Partial pressure of CO_2 in kPa						
$t/°C$	5	10	20	30	40	50	100
0	0.067	0.135	0.269	0.404	0.538	0.671	1.337
5	0.056	0.113	0.226	0.338	0.451	0.564	1.123
10	0.048	0.096	0.191	0.287	0.382	0.477	0.950
15	0.041	0.082	0.164	0.245	0.327	0.409	0.814
20	0.035	0.071	0.141	0.212	0.283	0.353	0.704
25	0.031	0.062	0.123	0.185	0.247	0.308	0.614
30	0.027	0.054	0.109	0.163	0.218	0.271	0.541
35	0.024	0.048	0.097	0.145	0.193	0.242	0.481
40	0.022	0.043	0.087	0.130	0.173	0.216	0.431
45	0.020	0.039	0.078	0.117	0.156	0.196	0.389
50	0.018	0.036	0.071	0.107	0.142	0.178	0.354
55	0.016	0.033	0.065	0.098	0.131	0.163	0.325
60	0.015	0.030	0.060	0.090	0.121	0.150	0.300
65	0.014	0.028	0.056	0.084	0.112	0.140	0.279
70	0.013	0.026	0.052	0.079	0.105	0.131	0.261
75	0.012	0.025	0.049	0.074	0.099	0.123	0.245
80	0.012	0.023	0.047	0.070	0.093	0.116	0.232
85	0.011	0.022	0.044	0.067	0.089	0.111	0.221
90	0.011	0.021	0.042	0.064	0.085	0.106	0.211
95	0.010	0.020	0.041	0.061	0.082	0.102	0.203
100	0.010	0.020	0.039	0.059	0.079	0.098	0.196

AQUEOUS SOLUBILITY AND HENRY'S LAW CONSTANTS
OF ORGANIC COMPOUNDS

The solubility in water of about 1300 organic compounds, including many compounds of environmental interest, is tabulated here. When data are available, values are given at several temperatures between 0 °C and 100 °C. Solids, liquids, and gases are included; additional data on gases can be found in the table "Solubility of Selected Gases in Water" in Section 8.

Solubility of solids is defined as the concentration of the compound in a solution that is in equilibrium with the solid phase at the specified temperature and one atmosphere pressure. For liquids whose water mixtures separate into two phases, the solubility given here is the concentration of the specified compound in the water-rich phase at equilibrium. In the case of gases (i.e., compounds whose vapor pressure at the specified temperature exceeds one atmosphere) the solubility is defined here as the concentration in the water phase when the partial pressure of the compound above the solution is 101.325 kPa (1 atm). Values for gases are marked with an asterisk.

The solubility values in this table are expressed as mass percent of solute, $s = 100w_2$, where the mass fraction w_2 is defined as

$$w_2 = m_2/(m_1 + m_2)$$

where m_2 is the mass of solute and m_1 the mass of water. For convenience, the solubility expressed in grams of solute that will dissolve in 1 kilogram of water is tabulated in the adjacent column to mass percent. For compounds with low solubility (e.g., $s < 1\%$), that column is, to a high approximation, numerically identical to the solubility expressed in grams of solute per liter of solution.

The mass fraction w_2 is related to other common measures of solubility as follows:

Molality:	$m_2 = 1000 \, w_2/M_2(1 - w_2)$
Molarity:	$c_2 = 1000\rho w_2/M_2$
Mole fraction:	$x_2 = (w_2/M_2)/\{(w_2/M_2) + (1 - w_2)/M_1\}$
Mass of solute per 100 g of H_2O:	$100w_2/(1 - w_2)$
Mass of solute per liter of solution:	$1000\rho w_2$

Here, M_2 is the molar mass of the solute, $M_1 = 18.015$ g/mol is the molar mass of water, and ρ is the density of the solution in g/mL.

Data have been selected from evaluated sources wherever possible, in particular the *IUPAC Solubility Data Series*. Many values come from experimental measurements reported in the *Journal of Chemical and Engineering Data* and the *Journal of Chemical Thermodynamics*, as well as critical review papers in the *Journal of Physical and Chemical Reference Data*. The primary source for each value is listed in the column following the solubility values; additional references of interest are sometimes given. Many of the references contain solubility data at other temperatures and pH values and in the presence of other compounds. The user is cautioned that wide variations of data are found in the literature for the lower solubility compounds. The references should be consulted for more information on these compounds.

The table also contains values of the Henry's Law constant k_H, which provides a measure of the partition of a substance between the atmosphere and the aqueous phase. Here, k_H is defined as the limit of p_2/c_2 as the concentration approaches zero, where p_2 is the partial pressure of the solute above the solution and c_2 is the concentration in the solution at equilibrium (other formulations of Henry's Law are often used; see Reference 5). The values of k_H listed here are based on direct experimental measurement whenever available, but many of them are simply calculated as the ratio of the pure compound vapor pressure to the solubility. This approximation is reliable only for compounds of very low solubility. In fact, values of k_H found in the literature frequently differ by a factor of two or three, and variations over an order of magnitude are not unusual (Reference 5). Therefore, the data given here should be taken only as a rough indication of the true Henry's Law constant, which is difficult to measure precisely.

All values of k_H refer to 25 °C. If the vapor pressure of the compound at 25 °C is greater than one atmosphere, it can be assumed that the k_H value has been calculated as $101.325/c_2$. The source of the Henry's Law data is given in the last column. The air-water partition coefficient (i.e., ratio of air concentration to water concentration when both are expressed in the same units) is equal to k_H/RT or $k_H/2.48$ in the units used here.

Compounds are listed by systematic name. To locate a compound by molecular formula or CAS Registry Number, use the indexes to the table "Physical Constants of Organic Compounds" in Section 3, which point to the entry in that table from which the name can be determined.

References

1. *Solubility Data Series, International Union of Pure and Applied Chemistry, Vol. 15*, Pergamon Press, Oxford, 1982.
2. *Solubility Data Series, International Union of Pure and Applied Chemistry, Vol. 20*, Pergamon Press, Oxford, 1985.
3. *Solubility Data Series, International Union of Pure and Applied Chemistry, Vol. 37*, Pergamon Press, Oxford, 1988.
4. *Solubility Data Series, International Union of Pure and Applied Chemistry, Vol. 38*, Pergamon Press, Oxford, 1988.
5. Mackay, D., and Shiu, W. Y., *J. Phys. Chem. Ref. Data*, 10, 1175, 1981.
6. Pearlman, R. S., and Yalkowsky, S. H., *J. Phys. Chem. Ref. Data*, 13, 975, 1984.
7. Shiu, W. Y., and Mackay, D., *J. Phys. Chem. Ref. Data*, 15, 911, 1986.
8. Varhanickova, D., Lee, S. C., Shiu, W. Y., and Mackay, D., *J. Chem. Eng. Data*, 40, 620, 1995.
9. Miller, M. M., Ghodbane, S., Wasik, S. P., Tewari, Y. B., and Martire, D. E., *J. Chem. Eng. Data*, 29, 184, 1984.
10. Riddick, J. A., Bunger, W. B., and Sakano, T. K., *Organic Solvents, Fourth Edition*, John Wiley & Sons, New York, 1986.
11. Mackay, D., Shiu, W. Y., and Ma, K. C., *Illustrated Handbook of Physical-Chemical Properties and Environmental Fate for Organic Chemicals, Vol. I*, Lewis Publishers/CRC Press, Boca Raton, FL, 1992.
12. Mackay, D., Shiu, W. Y., and Ma, K. C., *Illustrated Handbook of Physical-Chemical Properties and Environmental Fate for Organic Chemicals, Vol. II*, Lewis Publishers/CRC Press, Boca Raton, FL, 1992.
13. Mackay, D., Shiu, W. Y., and Ma, K. C., *Illustrated Handbook of Physical-Chemical Properties and Environmental Fate for Organic Chemicals, Vol. III*, Lewis Publishers/CRC Press, Boca Raton, FL, 1993.
14. Horvath, A. L., *Halogenated Hydrocarbons*, Marcel Dekker, New York, 1982.
15. Howard, P. H., *Handbook of Environmental Fate and Exposure Data for Organic Chemicals, Vol. I*, Lewis Publishers/CRC Press, Boca Raton, FL, 1989.

---

* Indicates a value of s for a gas at a partial pressure of 101.325 kPa (1 atm) in equilibrium with the solution.

16. Howard, P. H., *Handbook of Environmental Fate and Exposure Data for Organic Chemicals, Vol. II*, Lewis Publishers/CRC Press, Boca Raton, FL, 1990.

17. Banergee, S., Yalkowsky, S. H., and Valvani, S. C., *Environ. Sci. Technol.*, 14, 1227, 1980.

18. Gevantman, L. H., in *CRC Handbook of Chemistry and Physics, 90th Edition*, p. 8–80, CRC Press, Boca Raton, FL, 2009.

19. Wilhelm, E., Battino, R., and Wilcock, R. J., *Chem. Rev.* 77, 219, 1977.

20. Stephenson, R. M., *J. Chem. Eng. Data*, 37, 80, 1992.

21. Stephenson, R. M., Stuart, J., and Tabak, M., *J. Chem. Eng. Data*, 29, 287, 1984.

22. Shiu, W.-Y., and Ma, K.-C, *J. Phys. Chem. Ref. Data*, 29, 41, 2000.

23. Lun, R., Varhanickova, D., Shiu, W.-Y., and Mackay, D., *J. Chem. Eng. Data*, 42, 951 (1997).

24. Huang, G.-L., Xiao, H., Chi, J., Shiu, W.-Y., and Mackay, D., *J. Chem. Eng. Data*, 45, 411, 2000.

25. Horvath, A. L., Getzen, F. W., and Maczynska, Z., *J. Phys. Chem. Ref. Data*, 28, 395, 2000 [IUPAC No. 67].

26. Dawson, R. M. C., Elliott, D. C., Elliott, W. H., and Jones, K. M., *Data for Biochemical Research*, Third Edition, Clarendon Press, Oxford, 1986.

27. Stephen, H., and Stephen, T., *Solubilities of Organic and Inorganic Compounds*, MacMillan, New York, 1963.

28. Shiu, W.-Y., and Mackay, D., *J. Chem. Eng. Data* 42, 27, 1997.

29. Hinz, H.-J., ed., *Thermodynamic Data for Biochemistry and Biotechnology*, Springer-Verlag, Berlin, 1986.

30. Budavari, S., ed., *The Merck Index, Twelfth Edition*, Merck & Co., Rahway, NJ, 1996.

31. Bamford, H. A., Poster, D. L., and Baker, J. E., *J. Chem. Eng. Data*, 45, 1069, 2000.

32. Lide, D. R., and Milne, G. W. A., *Handbook of Data on Organic Compounds, Third Edition*, CRC Press, Boca Raton, FL, 1994.

33. Apelblat, A., and Manzurola, E., *J. Chem. Thermodynamics* 21, 1005, 1989.

34. Apelblat, A., and Manzurola, E., *J. Chem. Thermodynamics* 22, 289, 1990.

35. Horvath, A. L., and Getzen, F. W., *J. Phys. Chem. Ref. Data* 28, 649, 1999 [IUPAC No. 68].

36. Sazonov, V. P., Marsh, K. N., and Hefter, G. T., *J. Phys. Chem. Ref. Data* 29, 1165, 2000 [IUPAC No. 71].

37. Verbruggen, E. M. J., Hermens, J. L. M., and Tolls, J., *J. Phys. Chem. Ref. Data* 29, 1435, 2000.

38. Sazonov, V. P., Shaw, D. G., and Marsh, K. N., *J. Phys. Chem. Ref. Data* 31, 1, 2002 [IUPAC No. 77].

39. Sazonov, V. P., and Shaw, D. G., *J. Phys. Chem. Ref. Data* 31, 989, 2002 [IUPAC No. 78].

40. Yalkowsky, S. H., and He, Y., *Handbook of Aqueous Solubility Data*, CRC Press, Boca Raton, FL, 2003.

41. Shiu, W.-Y., and Ma, K.-C., *J. Phys. Chem. Ref. Data* 29, 387, 2000.

42. Shaw, D. G., and Maczynski, A., *J. Phys. Chem. Ref. Data* 35, 687, 2006 [IUPAC No. 81, Part 11].

43. Nordstrom, F. L., and Rasmuson, A. C., *J. Chem. Eng. Data* 51, 1668, 2006.

44. Nordstrom, F. L., and Rasmuson, A. C., *J. Chem. Eng. Data* 51, 1775, 2006.

45. Sapoundjiev, D., Lorenz, H,. and Seidel-Morgenstern, A., *J. Chem. Eng. Data* 51, 1562, 2006.

46. Marche, C., Ferronato, C., and Jose, J., *J. Chem. Eng. Data* 48, 967, 2003.

47. Lu, J., Wang, X., Yang, X., and Ching, C., *J. Chem. Eng. Data* 51, 1593, 2006.

48. Achard, C., Jaoui, M., Schwing, M., and Rogalski, M., *J. Chem. Eng. Data* 41, 504, 1996.

49. Shareef, A., et al., *J. Chem. Eng. Data* 51, 879, 2006.

50. Clever, H. L., et al., *J. Phys. Chem. Ref. Data* 34, 201, 2005 [IUPAC No. 80].

51. Jaoui, M., Achard, C., and Rogalski, M., *J. Chem. Eng. Data* 47, 297, 2002.

52. Fichan, I., Larroche, C., and Gros, J. B., *J. Chem. Eng. Data* 44, 56, 1999.

53. Freire, M. G., et al., *J. Chem. Eng. Data* 50, 237, 2005.

54. Domanska, U., and Kozlowska, M. K., *J. Chem. Eng. Data* 47, 456, 2002.

55. Phelan, J. M., and Barnett, J. L., *J. Chem. Eng. Data* 46, 375, 2001.

56. Long, B-W., Wang, L-S., and Wu, J-S., *J. Chem. Eng. Data* 50, 136, 2005.

57. Marche, C., Ferronato, C., and Jose, J., *J. Chem. Eng. Data* 49, 937, 2004.

58. Oleszek-Kudlak, S., Shibata, E., and Nakamura, T., *J. Chem. Eng. Data* 49, 570, 2004.

59. Lynch, J. C., et al., *J. Chem. Eng. Data* 46, 1549, 2001.

60. Xiao, H., Li, N., and Wania, F., *J. Chem. Eng. Data* 49, 173, 2004.

61. Ma, J. H. Y., Hung, H., Shiu, W-Y., and Mackay, D., *J. Chem. Eng. Data* 46, 619, 2001.

62. Carta, R., and Tola, G., *J. Chem. Eng. Data* 41, 414, 1996; 44, 563, 1999.

63. Kao, H. D., et al., *Pharm. Res.* 17, 978, 2000.

64. Heric, E. L., and Langford, R. E., *J. Chem. Eng. Data* 17, 471, 1972.

65. Marche, C., Delépine, H., Ferronato, C., and Jose, J., *J. Chem. Eng. Data* 48, 398, 2003.

66. Wang, L-C, and Wang, F-A, *J. Chem. Eng. Data* 49, 155, 2004.

67. Shen, L, and Wania, F., *J. Chem. Eng. Data* 50, 742, 2005.

68. Oleszek-Kudlak, S., Shibata, E., and Nakamura, T., *J. Chem. Eng. Data* 52, 1824, 2007.

69. Zhao, H-K., Li, R-R, Ji, H-Z, Zhang, D-S., Tang, C., and Yang, L-Q., *J. Chem. Eng. Data* 52, 2072, 2007.

70. Yang, X., Wang, X., and Ching, C. B., *J. Chem. Eng. Data* 53, 1133, 2008.

71. Liu, L., and Chen, J., *J. Chem. Eng. Data* 53, 1649, 2008.

72. Szterner, P., *J. Chem. Eng. Data* 53, 1738, 2008.

73. Kong, M-Z., Shi, X-H, Cao, Y-C., and Zhou, C-R., *J. Chem. Eng. Data* 53, 615, 2008.

74. Daneshfar, A., Ghaziaskar, H. S., and Homayoun, N., *J. Chem. Eng. Data* 53, 776, 2008.

75. Manzurola, E., and Apelblat, A., *J. Chem. Thermodynamics* 34, 1127, 2002.

76. Apelblat, A., Manzurola, E., and Balal, N. A., *J. Chem. Thermodynamics* 38, 565, 2006.

77. Apelblat, A., and Mishelevich, A., *J. Chem. Thermodynamics* 40, 897, 2008.

78. Góral, M., Wiśniewska-Goclowska, B., and Mączyński, A., *J. Phys. Chem. Ref. Data* 35, 1391, 2006.

79. Mączyński, A., Shaw, D. G., Góral, M., and Wiśniewska-Goclowska, B., *J. Phys. Chem. Ref. Data* 37, 1119, 2008.

80. Mączyński, A., Shaw, D. G., Góral, M., and Wiśniewska-Goclowska, B., *J. Phys. Chem. Ref. Data* 37, 1147, 2008.

81. Mączyński, A., Shaw, D. G., Góral, M., and Wiśniewska-Goclowska, B., *J. Phys. Chem. Ref. Data* 37, 1169, 2008.

82. Mączyński, A., Shaw, D. G., Góral, M., and Wiśniewska-Goclowska, B., *J. Phys. Chem. Ref. Data* 37, 1517, 2008.

83. Mączyński, A., Shaw, D. G., Góral, M., and Wiśniewska-Goclowska, B., *J. Phys. Chem. Ref. Data* 37, 1575, 2008.

84. Mączyński, A., Shaw, D. G., Góral, M., and Wiśniewska-Goclowska, B., *J. Phys. Chem. Ref. Data* 37, 1611, 2008.

85. Luning Prak, D. J., and O'Sullivam, D. W., *J. Chem. Eng. Data* 51, 448, 2006.

86. Shiu, W.-Y., Wania, F., Hung, H., and Mackay, D., *J. Chem. Eng. Data* 42, 293, 1997.

87. Mączyński, A., and Shaw, D. G., *J. Phys. Chem. Ref. Data* 36, 59, 2007.

88. Mączyński, A., and Shaw, D. G., *J. Phys. Chem. Ref. Data* 36, 133, 2007.

89. Góral, M., Wiśniewska-Goclowska, B., and Mączyński, A., *J. Phys. Chem. Ref. Data* 33, 1159, 2004.

90. Heric, E. L., and Langford, R. E., *J. Chem. Eng. Data* 17, 209, 1972.

Name	Mol. Form.	Mol. Wt.	$t/°C$	Solubility, s 100 w_2 (mass%)	g per kg H_2O	Ref.	Henry Const., k_H kPa m^3mol^{-1}	Ref.
Acenaphthene	$C_{12}H_{10}$	154.207	0	0.00015	0.0015	4		
			25	0.000380	0.00380	22	0.01217	22
			50	0.00092	0.0092	4		
Acenaphthylene	$C_{12}H_8$	152.192	20	0.0016	0.016	28	0.012	28
Acephate	$C_4H_{10}NO_3PS$	183.166	20	≈28	≈390	40		
Acetamide	C_2H_5NO	59.067	20	40.8	689	10		
Acetanilide	C_8H_9NO	135.163	20	0.52	5.2	27		
			70	2.7	28	27		
Acetazolamide	$C_4H_6N_4O_3S_2$	222.246	30	0.10	1.0	40		
Acetohexamide	$C_{15}H_{20}N_2O_4S$	324.396	37	0.0013	0.013	40		
Acetonitrile	C_2H_3N	41.052	−3	40.5	681	39		
			−10	31.7	464	39		
Acetophenone	C_8H_8O	120.149	20	0.67	6.7	84	0.00108	28
			50	0.81	8.2	84	0.00108	28
			80	1.16	11.7	84	0.00108	28
Acetylene	C_2H_2	26.037	25	0.108*	1.08*	19		
2-(Acetyloxy)benzoic acid	$C_9H_8O_4$	180.158		0.25	2.5	27		
2-(Acetyloxy)-5-bromobenzoic acid	$C_9H_7BrO_4$	259.054		0.07	0.7	30		
Acridine	$C_{13}H_9N$	179.217	25	0.00466	0.0466	6		
Acrolein	C_3H_4O	56.063	20	20.8	263	10		
Acrylamide	C_3H_5NO	71.078	20	≈27	≈370	40		
Acrylonitrile	C_3H_3N	53.063	20	7.35	79.3	10		
Adenine	$C_5H_5N_5$	135.128	25	0.104	1.04	29		
Adenosine	$C_{10}H_{13}N_5O_4$	267.242	25	0.51	5.1	29		
Alachlor	$C_{14}H_{20}ClNO_2$	269.768	23	0.024	0.24	40		
L-Alanine	$C_3H_7NO_2$	89.094	25	14.30	167	26		
β-Alanine	$C_3H_7NO_2$	89.094	25	47.1	890	26		
Aldicarb	$C_7H_{14}N_2O_2S$	190.263	20	0.60	6.0	40		
Aldrin	$C_{12}H_8Cl_6$	364.910	25	0.00002	0.0002	67		
Allopurinol	$C_5H_4N_4O$	136.112	25	0.057	0.57	40		
Ametryn	$C_9H_{17}N_5S$	227.330	20	0.0190	0.190	40		
2-Amino-9,10-anthracenedione	$C_{14}H_9NO_2$	223.227	25	0.000016	0.00016	40		
4-Aminoazobenzene	$C_{12}H_{11}N_3$	197.235	25	0.0030	0.030	40		
			97	0.068	0.68	40		
4-Aminobenzenesulfonamide	$C_6H_8N_2O_2S$	172.205	20	0.71	7.2	40		
4-Aminobenzenesulfonic acid	$C_6H_7NO_3S$	173.190	7	0.59	5.9	27		
DL-2-Aminobutanoic acid	$C_4H_9NO_2$	103.120	25	17.4	211	26		
DL-3-Aminobutanoic acid	$C_4H_9NO_2$	103.120	25	55.6	1250	26		
4-Amino-N-[(butylamino)carbonyl]benzenesulfonamide	$C_{11}H_{17}N_3O_3S$	271.336	37	0.053	0.53	40		
3-Amino-2,5-dichlorobenzoic acid	$C_7H_5Cl_2NO_2$	206.027	25	0.070	0.70	40		
6-Amino-1,3-dihydro-2H-purin-2-one	$C_5H_5N_5O$	151.127	25	0.006	0.06	26		
4-(2-Aminoethyl)phenol	$C_8H_{11}NO$	137.179	15	1.03	10.4	40		
6-Aminohexanoic acid	$C_6H_{13}NO_2$	131.173	25	46	852	29		
4-Amino-2-hydroxybenzoic acid	$C_7H_7NO_3$	153.136	20	0.20	2.0	40		
2-Amino-2-methylpropanoic acid	$C_4H_9NO_2$	103.120	25	12.1	138	26		
4-Amino-5-methyl-2(1H)-pyrimidinone	$C_5H_7N_3O$	125.129	25	0.45	4.5	26		
2-Aminophenol	C_6H_7NO	109.126	20	1.92	19.6	40		
3-Aminophenol	C_6H_7NO	109.126	20	2.56	26.3	40		

Name	Mol. Form.	Mol. Wt.	$t/°C$	Solubility, s		Ref.	Henry Const., k_H	Ref.
				$100\,w_2$ (mass%)	g per kg H_2O		kPa m^3mol^{-1}	
			70	≈24	≈320	40		
4-Aminophenol	C_6H_7NO	109.126	20	1.55	15.7	40		
Aminopyrine	$C_{13}H_{17}N_3O$	231.293	25	4.8	50	40		
Amitriptyline	$C_{20}H_{23}N$	277.404	24	0.00097	0.0097	40		
Amobarbital	$C_{11}H_{18}N_2O_3$	226.272	25	0.06	0.6	40		
Anilazine	$C_9H_5Cl_3N_4$	275.522	20	0.001	0.01	40		
Aniline	C_6H_7N	93.127	25	3.38	35.0	10	14	15
Aniline-2-carboxylic acid	$C_7H_7NO_2$	137.137	20	0.349	3.49	40		
Aniline-4-carboxylic acid	$C_7H_7NO_2$	137.137	25	0.54	5.4	40		
Aniline hydrochloride	C_6H_8ClN	129.588	15	15.1	178	27		
Anisole	C_7H_8O	108.138	20	0.203	2.03	20	0.025	13
			40	0.184	1.84	20	0.025	13
			81	0.294	2.95	20	0.025	13
Anthracene	$C_{14}H_{10}$	178.229	0	0.0000022	0.000022	42,4		
			25	0.0000044	0.000044	42,22	0.00396	22
			50	0.000029	0.00029	42		
9,10-Anthracenedione	$C_{14}H_8O_2$	208.213	25	0.00014	0.0014	40		
Apomorphine	$C_{17}H_{17}NO_2$	267.323	25	2.0	20	40		
L-Arginine	$C_6H_{14}N_4O_2$	174.201	25	15.44	183	26		
L-Ascorbic acid	$C_6H_8O_6$	176.124	25	25.2	337	33		
			50	41.0	695	33		
L-Asparagine	$C_4H_8N_2O_3$	132.118	25	2.45	25.1	26		
L-Aspartic acid	$C_4H_7NO_4$	133.104	10	0.29	2.9	77		
			25	0.49	4.9	77		
			50	1.31	13.3	77		
Atrazine	$C_8H_{14}ClN_5$	215.684	25	0.007	0.07	26		
Atropine	$C_{17}H_{23}NO_3$	289.370	20	0.3	3	40		
Azinphos-methyl	$C_{10}H_{12}N_3O_3PS_2$	317.324	20	0.00209	0.0209	40		
trans-Azobenzene	$C_{12}H_{10}N_2$	182.220	20	0.03	0.3	27		
Bayleton	$C_{14}H_{16}ClN_3O_2$	293.749	20	0.026	0.26	40		
Bendiocarb	$C_{11}H_{13}NO_4$	223.226	25	0.004	0.04	40		
Bentazon	$C_{10}H_{12}N_2O_3S$	240.278	20	0.050	0.50	40		
Benzaldehyde	C_7H_6O	106.122	20	0.3	3	10		
Benzamide	C_7H_7NO	121.137	12	0.577	5.77	27		
Benz[a]anthracene	$C_{18}H_{12}$	228.288	10	0.00000038	0.0000038	42		
			25	0.00000093	0.0000093	42,22	0.00058	22
Benzene	C_6H_6	78.112	10	0.174	1.74	22		
			20	0.177	1.77	22		
			30	0.183	1.83	22		
			40	0.192	1.92	22		
			50	0.206	2.06	22		
			70	0.249	2.50	65		
			101	0.398	4.00	65		
Benzeneacetic acid	$C_8H_8O_2$	136.149	25	1.71	17.4	27		
1,2-Benzenediamine	$C_6H_8N_2$	108.141	20	3.02	31.1	40		
1,3-Benzenediamine	$C_6H_8N_2$	108.141	20	3.48	36.1	40		
1,4-Benzenediamine	$C_6H_8N_2$	108.141	24	3.45	35.7	40		
1,2-Benzenedicarboxamide	$C_8H_8N_2O_2$	164.162	30	0.59	5.9	40		
Benzeneethanol	$C_8H_{10}O$	122.164	25	1.72	17.5	40		

Name	Mol. Form.	Mol. Wt.	$t/°C$	Solubility, s 100 w_2 (mass%)	g per kg H_2O	Ref.	Henry Const., k_H kPa m^3mol^{-1}	Ref.
Benzenehexacarboxylic acid	$C_{12}H_6O_{12}$	342.169	25	49.3	972	76		
Benzenepentacarboxylic acid	$C_{11}H_6O_{10}$	298.160	10	11.9	135	76		
			25	21.1	267	76		
			50	36.2	567	76		
1,2,3,4-Benzenetetracarboxylic acid	$C_{10}H_6O_8$	254.150	10	11.0	124	76		
			25	20.9	264	76		
			50	39.5	653	76		
1,2,3,5-Benzenetetracarboxylic acid	$C_{10}H_6O_8$	254.150	10	7.50	81.1	76		
			25	10.1	112	76		
			50	15.8	188	76		
1,2,4,5-Benzenetetracarboxylic acid	$C_{10}H_6O_8$	254.150	10	0.51	5.1	76		
			25	1.06	10.7	76		
			50	3.82	39.7	76		
1,2,3-Benzenetricarboxylic acid	$C_9H_6O_6$	210.140	10	2.39	24.5	76		
			25	4.78	50.2	76		
			50	17.4	211	76		
1,2,4-Benzenetricarboxylic acid	$C_9H_6O_6$	210.140	10	1.02	10.3	76		
			25	1.92	19.6	76		
			50	5.45	57.6	76		
1,3,5-Benzenetricarboxylic acid	$C_9H_6O_6$	210.140	10	0.110	1.10	76		
			25	0.207	2.07	76		
			50	0.598	6.02	76		
1,2,3-Benzenetriol	$C_6H_6O_3$	126.110	25	38.5	626	27		
1,3,5-Benzenetriol	$C_6H_6O_3$	126.110	20	1.12	11.3	27		
p-Benzidine	$C_{12}H_{12}N_2$	184.236	24	0.0360	0.360	40		
1H-Benzimidazole	$C_7H_6N_2$	118.136	15	0.33	3.3	54		
			20	0.201	2.01	6		
1,3-Benzodioxole-5-carboxaldehyde	$C_8H_6O_3$	150.132	20	0.35	3.5	40		
Benzo[b]fluoranthene	$C_{20}H_{12}$	252.309	20	0.0000002	0.000002	40		
Benzo[k]fluoranthene	$C_{20}H_{12}$	252.309		0.00000008	0.0000008	40		
11H-Benzo[a]fluorene	$C_{17}H_{12}$	216.277	25	0.0000045	0.000045	42,4		
11H-Benzo[b]fluorene	$C_{17}H_{12}$	216.277	25	0.0000002	0.000002	42,4		
Benzoic acid	$C_7H_6O_2$	122.122	10	0.209	2.09	76		
			25	0.343	3.44	76		
			50	0.842	8.49	76		
Benzoin	$C_{14}H_{12}O_2$	212.244	25	0.03	0.3	40		
Benzonitrile	C_7H_5N	103.122	25	0.2	2	10		
Benzo[ghi]perylene	$C_{22}H_{12}$	276.330	25	0.000000026	0.00000026	42,4	0.000075	12
Benzophenone	$C_{13}H_{10}O$	182.217	20	0.0075	0.075	40		
2H-1-Benzopyran-2-one	$C_9H_6O_2$	146.143	20	0.190	1.90	40		
			60	0.69	6.9	40		
Benzo[a]pyrene	$C_{20}H_{12}$	252.309	25	0.00000043	0.0000043	42,22	0.0000465	22
Benzo[e]pyrene	$C_{20}H_{12}$	252.309	8	0.00000032	0.0000032	42		
			17	0.00000044	0.0000044	42,22	0.0000467	22
			25	0.00000048	0.0000048	42		
Benzo[f]quinoline	$C_{13}H_9N$	179.217	25	0.0079	0.079	6		
p-Benzoquinone	$C_6H_4O_2$	108.095	25	1.36	13.8	27		
Benzo[b]thiophene	C_8H_6S	134.199	20	0.0130	0.130	6		
Benzo[b]triphenylene	$C_{22}H_{14}$	278.346	25	0.0000027	0.000027	4		

Name	Mol. Form.	Mol. Wt.	$t/°C$	Solubility, s 100 w_2 (mass%)	g per kg H$_2$O	Ref.	Henry Const., k_H kPa m^{3}mol^{-1}	Ref.
Benzoxazole	C$_7$H$_5$NO	119.121	20	0.834	8.34	6		
N-Benzoylglycine	C$_9$H$_9$NO$_3$	179.172	25	0.37	3.7	29		
Benzoyl peroxide	C$_{14}$H$_{10}$O$_4$	242.227	20	0.000016	0.00016	40		
N-Benzoyl-L-phenylalanine	C$_{16}$H$_{15}$NO$_3$	269.295	25	0.085	0.85	29		
Benzyl acetate	C$_9$H$_{10}$O$_2$	150.174	25	0.150	1.50	40		
Benzyl alcohol	C$_7$H$_8$O	108.138	20	0.08	0.8	10		
Benzyl formate	C$_8$H$_8$O$_2$	136.149	20	1.07	10.8	20		
			80	1.43	14.5	20		
Bifenthrin	C$_{23}$H$_{22}$ClF$_3$O$_2$	422.868	25	0.00001	0.0001	32		
Biotin	C$_{10}$H$_{16}$N$_2$O$_3$S	244.310	25	0.035	0.35	40		
Biphenyl	C$_{12}$H$_{10}$	154.207	0	0.000272	0.00272	4		
			25	0.00054	0.0054	58,22	0.0280	22
			50	0.0022	0.022	4		
2,2'-Bipyridine	C$_{10}$H$_8$N$_2$	156.184	25	0.61	6.1	40		
2,2'-Biquinoline	C$_{18}$H$_{12}$N$_2$	256.301	24	0.000102	0.00102	6		
Bis(4-aminophenyl) sulfone	C$_{12}$H$_{12}$N$_2$O$_2$S	248.300	25	0.016	0.16	40		
Bis(2-chloroethyl) ether	C$_4$H$_8$Cl$_2$O	143.012	20	1.04	10.5	20	0.003	13
			81	1.26	12.8	20		
1,1-Bis(4-chlorophenyl)-2,2,2-trichloroethanol	C$_{14}$H$_9$Cl$_5$O	370.485	25	0.00013	0.0013	40		
Bis(2-ethylhexyl) phthalate	C$_{24}$H$_{38}$O$_4$	390.557	25	0.000027	0.00027	40		
2,2-Bis(4-hydroxyphenyl)propane	C$_{15}$H$_{16}$O$_2$	228.287	25	0.0300	0.30	49		
1,3-Bis(trifluoromethyl)benzene	C$_8$H$_4$F$_6$	214.108	25	0.0041	0.041	2		
Borneol	C$_{10}$H$_{18}$O	154.249	25	0.046	0.46	52		
Bromacil	C$_9$H$_{13}$BrN$_2$O$_2$	261.115	25	0.082	0.82	40		
Bromobenzene	C$_6$H$_5$Br	157.008	10	0.0387	0.387	2		
			25	0.0445	0.445	2	0.250	28
			40	0.0516	0.516	2		
2-Bromobenzoic acid	C$_7$H$_5$BrO$_2$	201.018	25	0.185	1.85	27		
3-Bromobenzoic acid	C$_7$H$_5$BrO$_2$	201.018	25	0.040	0.40	27		
4-Bromobenzoic acid	C$_7$H$_5$BrO$_2$	201.018	25	0.0056	0.056	27		
1-Bromobutane	C$_4$H$_9$Br	137.018	25	0.087	0.87	35	1.2	13
4-Bromo-1-butene	C$_4$H$_7$Br	135.003	25	0.076	0.76	35		
1-Bromo-2-chlorobenzene	C$_6$H$_4$BrCl	191.453	25	0.0124	0.124	2		
1-Bromo-3-chlorobenzene	C$_6$H$_4$BrCl	191.453	25	0.0118	0.118	2		
1-Bromo-4-chlorobenzene	C$_6$H$_4$BrCl	191.453	25	0.00442	0.0442	2		
1-Bromo-2-chloroethane	C$_2$H$_4$BrCl	143.410	30	0.683	6.83	25		
Bromochloromethane	CH$_2$BrCl	129.384	25	1.7	17	10	0.18	13
1-Bromo-3-chloropropane	C$_3$H$_6$BrCl	157.437	25	0.223	2.23	35		
2-Bromo-2-chloro-1,1,1-trifluoroethane	C$_2$HBrClF$_3$	197.381	10	0.52	5.2	25		
			25	0.41	4.1	25		
			40	0.40	4.0	25		
Bromodichloromethane	CHBrCl$_2$	163.829	30	0.300	3.00	40		
Bromoethane	C$_2$H$_5$Br	108.965	0	1.05	10.6	25		
			25	0.90	9.0	25	1.23	13
1-Bromoheptane	C$_7$H$_{15}$Br	179.098	25	0.00067	0.0067	35		
1-Bromohexane	C$_6$H$_{13}$Br	165.071	25	0.00258	0.0258	35		
1-Bromo-4-iodobenzene	C$_6$H$_4$BrI	282.904	25	0.000794	0.00794	2		
Bromomethane	CH$_3$Br	94.939	20	1.80*	18.3*	5	0.63	13
1-Bromo-3-methylbutane	C$_5$H$_{11}$Br	151.045	16	0.020	0.20	35		

Name	Mol. Form.	Mol. Wt.	$t/°C$	Solubility, s			Henry Const., k_H	
				100 w_2 (mass%)	g per kg H_2O	Ref.	kPa m^3mol^{-1}	Ref.
1-Bromo-2-methylpropane	C_4H_9Br	137.018	18	0.051	0.51	35		
1-Bromooctane	$C_8H_{17}Br$	193.125	25	0.000167	0.00167	35		
1-Bromopentane	$C_5H_{11}Br$	151.045	25	0.0127	0.127	35		
4-Bromophenol	C_6H_5BrO	173.007	25	1.86	19.0	2		
1-Bromopropane	C_3H_7Br	122.992	0	0.298	2.98	35		
			25	0.234	2.34	35	3.8	13
2-Bromopropane	C_3H_7Br	122.992	20	0.32	3.2	35	1.27	13
3-Bromopropene	C_3H_5Br	120.976	25	0.38	3.8	35		
4-Bromotoluene	C_7H_7Br	171.035	25	0.011	0.11	2		
Bromotrifluoromethane	$CBrF_3$	148.910	25	0.032*	0.32*	14		
5-Bromouracil	$C_4H_3BrN_2O_2$	190.983	25	0.288	2.89	72		
Brucine	$C_{23}H_{26}N_2O_4$	394.463	20	0.012	0.12	27		
1,3-Butadiene	C_4H_6	54.091	25	0.0735*	0.735*	5	20.7	13
Butanal	C_4H_8O	72.106	25	7.1	76	10		
Butanamide	C_4H_9NO	87.120	25	≈19	≈230	40		
Butane	C_4H_{10}	58.122	25	0.00724*	0.0724*	18	95.9	5
2,3-Butanedione	$C_4H_6O_2$	86.090	20	31.7	464	20		
			80	21.8	279	20		
Butanenitrile	C_4H_7N	69.106	20	3.3	34	10		
1,2,3,4-Butanetetrol	$C_4H_{10}O_4$	122.120	20	38.0	613	27		
1-Butanethiol	$C_4H_{10}S$	90.187	20	0.0597	0.597	10		
1-Butanol	$C_4H_{10}O$	74.121	0	10.5	117	78,1		
			25	7.3	79	78,1		
			50	6.4	68	78,1		
			100	8.8	96	78		
2-Butanol	$C_4H_{10}O$	74.121	10	23.9	314	1,87		
			25	18.1	221	1,87		
			50	14.0	163	1,87		
2-Butanone	C_4H_8O	72.106	0	35.9	560	82		
			25	25.6	344	82		
			40	21.5	274	82		
			70	18.1	221	20		
			100	19.3	239	82		
trans-2-Butenal	C_4H_6O	70.090	20	15.6	185	10		
1-Butene	C_4H_8	56.107	25	0.0222*	0.222*	5	25.6	13
trans-2-Butenoic acid	$C_4H_6O_2$	86.090	20	7.1	76	26		
cis-2-Buten-1-ol	C_4H_8O	72.106	20	16.6	199	10		
3-Buten-2-one	C_4H_6O	70.090	28	54.3	1190	82		
			50	35.6	553	82		
			80	37.6	603	82		
Butyl acetate	$C_6H_{12}O_2$	116.158	20	0.68	6.8	10		
sec-Butyl acetate	$C_6H_{12}O_2$	116.158	20	0.62	6.2	10		
Butyl 4-aminobenzoate	$C_{11}H_{15}NO_2$	193.243	25	0.018	0.18	40		
Butylbenzene	$C_{10}H_{14}$	134.218	25	0.00138	0.0138	22,89	1.33	22
sec-Butylbenzene	$C_{10}H_{14}$	134.218	25	0.0014	0.014	4,89	1.89	11
tert-Butylbenzene	$C_{10}H_{14}$	134.218	25	0.0032	0.032	4	1.28	11
Butyl ethyl ether	$C_6H_{14}O$	102.174	20	0.65	6.5	20		
			70	0.39	3.9	20		
Butyl 4-hydroxybenzoate	$C_{11}H_{14}O_3$	194.227	25	0.020	0.20	40		

Name	Mol. Form.	Mol. Wt.	$t/°C$	100 w_2 (mass%)	g per kg H_2O	Ref.	Henry Const., k_H kPa m³mol⁻¹	Ref.
Butyl methyl ether	$C_5H_{12}O$	88.148	0	2.51	25.7	79		
			25	0.89	9.0	79		
4-*tert*-Butylphenol	$C_{10}H_{14}O$	150.217	25	0.058	0.58	40		
Butyl propanoate	$C_7H_{14}O_2$	130.185	22	0.572	5.72	27		
Butyl stearate	$C_{22}H_{44}O_2$	340.583	25	0.2	2	10		
Butyl vinyl ether	$C_6H_{12}O$	100.158	20	0.3	3	10		
1-Butyne	C_4H_6	54.091	25	0.287*	2.87*	5	1.91	5
Caffeine	$C_8H_{10}N_4O_2$	194.191	25	2.12	21.7	29		
Camphor, (+)	$C_{10}H_{16}O$	152.233	20	0.01	0.1	10		
trans-Camphoric acid	$C_{10}H_{16}O_4$	200.232	25	0.8	8	27		
Cantharidin	$C_{10}H_{12}O_4$	196.200	25	0.003	0.03	40		
Caprolactam	$C_6H_{11}NO$	113.157	25	84.0	5250	10		
Captafol	$C_{10}H_9Cl_4NO_2S$	349.061	20	0.000142	0.00142	40		
Captan	$C_9H_8Cl_3NO_2S$	300.590	20	0.00005	0.0005	40		
Carbaryl	$C_{12}H_{11}NO_2$	201.221	20	0.0102	0.102	40		
Carbazole	$C_{12}H_9N$	167.206	22	0.000120	0.00120	6		
Carbofuran	$C_{12}H_{15}NO_3$	221.252	20	0.032	0.32	40		
Carbon dioxide	CO_2	44.010	25	0.150*	1.50*	18		
Carbon disulfide	CS_2	76.141	20	0.210	2.10	10		
Carbon monoxide	CO	28.010	25	0.00276*	0.0276*	18		
Carboxin	$C_{12}H_{13}NO_2S$	235.302	25	0.017	0.17	40		
Carminic acid	$C_{22}H_{20}O_{13}$	492.386	20	0.13	1.3	40		
Carnosine	$C_9H_{14}N_4O_3$	226.232	25	24.4	323	26		
Carvenol	$C_{10}H_{16}O$	152.233	25	0.29	2.9	52		
Carvenone, (*S*)-	$C_{10}H_{16}O$	152.233	15	0.22	2.2	27		
Carvone	$C_{10}H_{14}O$	150.217	15	0.13	1.3	27		
(*S*)-Carvone	$C_{10}H_{14}O$	150.217	25	0.13	1.3	52		
Cephalexin	$C_{16}H_{17}N_3O_4S$	347.389	25	1.2	12	40		
Chloramphenicol	$C_{11}H_{12}Cl_2N_2O_5$	323.129	25	0.38	3.8	40		
Chlordane	$C_{10}H_6Cl_8$	409.779	25	0.00006	0.0006	67		
2-Chloroaniline	C_6H_6ClN	127.572	25	0.876	8.76	10		
3-Chloroaniline	C_6H_6ClN	127.572	20	0.54	5.4	40		
4-Chloroaniline	C_6H_6ClN	127.572	20	0.275	2.75	40		
Chlorobenzene	C_6H_5Cl	112.557	5	0.050	0.50	61		
			25	0.050	0.50	61		
			45	0.055	0.55	61,2		
Chlorobenzilate	$C_{16}H_{14}Cl_2O_3$	325.186	20	0.001	0.01	32		
2-Chlorobenzoic acid	$C_7H_5ClO_2$	156.567	25	0.209	2.09	27		
3-Chlorobenzoic acid	$C_7H_5ClO_2$	156.567	25	0.040	0.40	27		
4-Chlorobenzoic acid	$C_7H_5ClO_2$	156.567	25	0.072	0.72	27		
2-Chlorobiphenyl	$C_{12}H_9Cl$	188.652	25	0.00055	0.0055	7	0.0701	7
1-Chlorobutane	C_4H_9Cl	92.567	1	0.062	0.62	35		
			25	0.087	0.87	35	1.54	13
2-Chlorobutane	C_4H_9Cl	92.567	0	0.107	1.07	35		
			25	0.092	0.92	35		
3-Chloro-2-butanone	C_4H_7ClO	106.551	19	2.80	28.8	20		
			92	3.38	35.0	20		
Chlorodiazepoxide	$C_{16}H_{14}ClN_3O$	299.754	20	0.2	2	40		
Chlorodibromomethane	$CHBr_2Cl$	208.280	30	0.251	2.51	40		

Name	Mol. Form.	Mol. Wt.	$t/°C$	Solubility, s			Henry Const., k_H	
				100 w_2 (mass%)	g per kg H_2O	Ref.	kPa m^3mol^{-1}	Ref.
Chlorodifluoromethane	$CHClF_2$	86.469	25	0.30*	3.0*	10	3.0	13
4-Chloro-2,5-dimethylphenol	C_8H_9ClO	156.609	25	0.89	8.9	2		
4-Chloro-2,6-dimethylphenol	C_8H_9ClO	156.609	25	0.52	5.2	2		
4-Chloro-3,5-dimethylphenol	C_8H_9ClO	156.609	25	0.34	3.4	2		
1-Chloro-2,4-dinitrobenzene	$C_6H_3ClN_2O_4$	202.552	25	0.00092	0.0092	40		
Chloroethane	C_2H_5Cl	64.514	0	0.45	4.5	25		
			25	0.67*	6.7*	25	1.02	13
Chloroethene	C_2H_3Cl	62.498	25	0.27*	2.7*	5	2.68	13
1-Chloro-2-fluorobenzene	C_6H_4ClF	130.547	25	0.0502	0.502	40		
Chlorofluoromethane	CH_2ClF	68.478	25	1.05*	10.6*	14		
1-Chloroheptane	$C_7H_{15}Cl$	134.647	25	0.00136	0.0136	35		
1-Chlorohexane	$C_6H_{13}Cl$	120.620	5	0.0047	0.047	35		
			25	0.0064	0.064	35		
2-Chloro-4-hydroxy-5-methoxybenzaldehyde	$C_8H_7ClO_3$	186.593	25	0.013	0.13	8		
3-Chloro-4-hydroxy-5-methoxybenzaldehyde	$C_8H_7ClO_3$	186.593	25	0.093	0.93	8		
1-Chloro-2-iodobenzene	C_6H_4ClI	238.453	25	0.00689	0.0689	2		
1-Chloro-3-iodobenzene	C_6H_4ClI	238.453	25	0.00674	0.0674	2		
1-Chloro-4-iodobenzene	C_6H_4ClI	238.453	25	0.00311	0.0311	2		
Chloromethane	CH_3Cl	50.488	25	0.535*	5.35*	5	0.98	13
1-Chloro-2-methoxyethane	C_3H_7ClO	94.540	20	7.79	84.5	20		
			70	6.31	67.3	20		
(Chloromethyl)benzene	C_7H_7Cl	126.584	20	0.0493	0.493	10		
3-(Chloromethyl)heptane	$C_8H_{17}Cl$	148.674	20	0.01	0.1	10		
2-Chloro-6-methylphenol	C_7H_7ClO	142.583	25	0.36	3.6	2		
4-Chloro-2-methylphenol	C_7H_7ClO	142.583	25	0.68	6.8	2		
4-Chloro-3-methylphenol	C_7H_7ClO	142.583	25	0.40	4.0	2		
(4-Chloro-2-methylphenoxy)acetic acid	$C_9H_9ClO_3$	200.618	25	0.117	1.17	40		
1-Chloro-2-methylpropane	C_4H_9Cl	92.567	25	0.92	9.2	35		
2-Chloro-2-methylpropane	C_4H_9Cl	92.567	15	0.29	2.9	35		
1-Chloro-2-methylpropene	C_4H_7Cl	90.552	25	0.916	9.16	5	0.12	5
1-Chloronaphthalene	$C_{10}H_7Cl$	162.616	25	0.00224	0.0224	5	0.0363	28
2-Chloronaphthalene	$C_{10}H_7Cl$	162.616	25	0.00117	0.0117	5	0.0335	28
1-Chloro-2-nitrobenzene	$C_6H_4ClNO_2$	157.555	20	0.0441	0.441	40		
1-Chloro-3-nitrobenzene	$C_6H_4ClNO_2$	157.555	20	0.0273	0.273	40		
1-Chloro-4-nitrobenzene	$C_6H_4ClNO_2$	157.555	20	0.0453	0.453	40		
3-Chloro-2-nitrobenzoic acid	$C_7H_4ClNO_4$	201.565	25	0.047	0.47	27		
5-Chloro-2-nitrobenzoic acid	$C_7H_4ClNO_4$	201.565	25	0.96	9.6	27		
1-Chlorooctane	$C_8H_{17}Cl$	148.674	25	0.0345	0.345	35		
Chloropentafluoroethane	C_2ClF_5	154.466	25	0.006*	0.06*	10	260	13
1-Chloropentane	$C_5H_{11}Cl$	106.594	5	0.020	0.20	35		
			25	0.0201	0.201	35	2.37	13
3-Chloropentane	$C_5H_{11}Cl$	106.594	25	0.025	0.25	35		
5-Chloro-2-pentanone	C_5H_9ClO	120.577	22	4.7	49	20		
			71	13.5	156	20		
2-Chlorophenol	C_6H_5ClO	128.556	25	2.27	23.2	48,51,2		
3-Chlorophenol	C_6H_5ClO	128.556	25	2.2	22	2		
4-Chlorophenol	C_6H_5ClO	128.556	25	2.55	26.2	48,51,2		
N'-(4-Chlorophenyl)-N,N-dimethylurea	$C_9H_{11}ClN_2O$	198.648	25	0.023	0.23	26		
1-Chloropropane	C_3H_7Cl	78.541	25	0.250	2.50	35	1.41	13

Name	Mol. Form.	Mol. Wt.	$t/°C$	Solubility, s			Henry Const., k_H	
				$100\,w_2$ (mass%)	g per kg H_2O	Ref.	kPa $m^3 mol^{-1}$	Ref.
2-Chloropropane	C_3H_7Cl	78.541	0	0.44	4.4	35		
			20	0.30	3.0	35		
3-Chloropropene	C_3H_5Cl	76.525	25	0.40	4.0	35	1.10	5
			50	0.13	1.3	35		
Chloropropham	$C_{10}H_{12}ClNO_2$	213.661	25	0.0080	0.080	40		
1-Chlorotetradecane	$C_{14}H_{29}Cl$	232.833	25	0.0232	0.232	35		
Chlorothalonil	$C_8Cl_4N_2$	265.911	25	0.00006	0.0006	40		
Chlorothiazide	$C_7H_6ClN_3O_4S_2$	295.724	25	0.0283	0.283	40		
2-Chlorotoluene	C_7H_7Cl	126.584	25	0.0117	0.117	61		
3-Chlorotoluene	C_7H_7Cl	126.584	25	0.0117	0.117	61		
4-Chlorotoluene	C_7H_7Cl	126.584	25	0.0123	0.123	61		
Chlorotrifluoromethane	$CClF_3$	104.459	25	0.009*	0.09*	10	6.9	13
3-Chloro-1,1,1-trifluoropropane	$C_3H_4ClF_3$	132.512	20	0.133	1.33	35		
2-Chloro-1,3,5-trinitrobenzene	$C_6H_2ClN_3O_6$	247.549	15	0.018	0.18	40		
5-Chlorouracil	$C_4H_3ClN_2O_2$	146.532	25	0.250	2.51	72		
Chlorpyrifos	$C_9H_{11}Cl_3NO_3PS$	350.586	20	0.000073	0.00073	40		
Chlorsulfuron	$C_{12}H_{12}ClN_5O_4S$	357.773	25	2.71	27.9	32		
Cholic acid	$C_{24}H_{40}O_5$	408.572	20	0.028	0.28	26		
Chrysene	$C_{18}H_{12}$	228.288	7	0.00000007	0.0000007	42		
			25	0.00000019	0.0000019	42,22	0.000065	22
trans-Cinnamaldehyde	C_9H_8O	132.159	25	0.135	1.35	40		
trans-Cinnamic acid	$C_9H_8O_2$	148.159	20	0.1	1	26		
			98	0.59	5.9	26		
Citric acid	$C_6H_8O_7$	192.124	20	59	1440	26		
Clopyralid	$C_6H_3Cl_2NO_2$	192.000	20	0.1	1	40		
Clorophene	$C_{13}H_{11}ClO$	218.678	20	0.42	4.2	40		
Cocaine	$C_{17}H_{21}NO_4$	303.354	25	0.17	1.7	27		
Codeine	$C_{18}H_{21}NO_3$	299.365	25	0.79	7.9	27		
Colchicine	$C_{22}H_{25}NO_6$	399.437	20	4	42	26		
Coronene	$C_{24}H_{12}$	300.352	25	0.000000014	0.00000014	42,4		
Creatine	$C_4H_9N_3O_2$	131.133	25	1.6	16	26		
o-Cresol	C_7H_8O	108.138	40	3.08	31.8	10		
m-Cresol	C_7H_8O	108.138	40	2.51	25.7	10		
p-Cresol	C_7H_8O	108.138	40	2.26	23.1	10		
Crufomate	$C_{12}H_{19}ClNO_3P$	291.711	20	0.50	5.0	40		
Cyanazine	$C_9H_{13}ClN_6$	240.692	25	0.0171	0.171	40		
2-Cyanoacetamide	$C_3H_4N_2O$	84.076	20	11.5	130	40		
Cyanogen	C_2N_2	52.034	25	0.8*	8*	30		
Cyanogen chloride	$CClN$	61.471	0	5.7	60	40		
Cyanoguanidine	$C_2H_4N_4$	84.080	25	3.8	40	40		
Cyanuric acid	$C_3H_3N_3O_3$	129.074	25	0.259	2.59	40		
Cycloheptane	C_7H_{14}	98.186	25	0.0030	0.030	3	9.59	13
Cycloheptanone	$C_7H_{12}O$	112.169	20	3.61	37.5	20		
			92	2.82	29.0	20		
1,3,5-Cycloheptatriene	C_7H_8	92.139	25	0.064	0.64	3	0.47	13
Cycloheptene	C_7H_{12}	96.170	25	0.0066	0.066	3	4.9	13
1,4-Cyclohexadiene	C_6H_8	80.128	25	0.08	0.8	3	1.03	13
Cyclohexane	C_6H_{12}	84.159	25	0.0058	0.058	3	19.4	13
			70	0.0092	0.092	65		

| Name | Mol. Form. | Mol. Wt. | $t/°C$ | Solubility, s | | Ref. | Henry Const., k_H | Ref. |
				$100\,w_2$ (mass%)	g per kg H_2O		kPa m^3mol^{-1}	
			100	0.0163	0.163	65		
Cyclohexanecarboxylic acid	$C_7H_{12}O_2$	128.169	15	0.201	2.01	27		
Cyclohexanol	$C_6H_{12}O$	100.158	10	4.62	48.4	1		
			25	3.8	40	1		
			40	3.30	34.1	1		
Cyclohexanone	$C_6H_{10}O$	98.142	10	12.2	139	83		
			25	9.5	105	83		
			50	7.6	82	83		
			80	6.8	73	20		
Cyclohexanone oxime	$C_6H_{11}NO$	113.157	25	1.57	16.0	40		
Cyclohexene	C_6H_{10}	82.143	25	0.016	0.16	3	4.57	13
Cyclohexyl butanoate	$C_{10}H_{18}O_2$	170.249	20	0.11	1.1	20		
			90	0.09	0.90	20		
Cyclooctane	C_8H_{16}	112.213	25	0.00079	0.0079	4	10.7	13
1,3-Cyclopentadiene	C_5H_6	66.102	25	0.068	0.68	3		
Cyclopentane	C_5H_{10}	70.133	25	0.0157	0.157	3	19.1	13
Cyclopentanol	$C_5H_{10}O$	86.132	19	10.6	119	88		
			50	8.3	91	88		
			90	9.2	101	88		
Cyclopentanone	C_5H_8O	84.117	0	37.7	605	20		
			20	31.0	449	20		
			80	24.8	330	20		
Cyclopentene	C_5H_8	68.118	25	0.054	0.54	3	6.56	13
Cyclopropane	C_3H_6	42.080	25	0.0484*	0.484*	19		
Cyfluthrin	$C_{22}H_{18}Cl_2FNO_3$	434.287	20	0.0000002	0.000002	32		
Cygon	$C_5H_{12}NO_3PS_2$	229.258	20	2.6	27	40		
Cyhalothrin	$C_{23}H_{19}ClF_3NO_3$	449.850	20	0.0000005	0.000005	32		
Cypermethrin	$C_{22}H_{19}Cl_2NO_3$	416.297	20	0.000001	0.00001	32		
L-Cystine	$C_6H_{12}N_2O_4S_2$	240.300	25	0.0166	0.166	62		
Cytisine	$C_{11}H_{14}N_2O$	190.241	16	≈30	≈430	40		
Cytosine	$C_4H_5N_3O$	111.102	25	0.73	7.3	29		
Daminozide	$C_6H_{12}N_2O_3$	160.170	25	9.1	100	40		
Dazomet	$C_5H_{10}N_2S_2$	162.276	25	0.12	1.2	40		
Decabromobiphenyl ether	$C_{12}Br_{10}O$	959.167	25	0.0000025	0.000025	40		
Decachlorobiphenyl	$C_{12}Cl_{10}$	498.658	25	0.00000000012	0.0000000012	7	0.0208	7
cis-Decahydronaphthalene	$C_{10}H_{18}$	138.250	25	0.000089	0.00089	37		
trans-Decahydronaphthalene	$C_{10}H_{18}$	138.250	25	0.000089	0.00089	4	3	13
Decane	$C_{10}H_{22}$	142.282	0	0.0000015	0.000015	4	479	13
Decanedioic acid	$C_{10}H_{18}O_4$	202.248	20	0.10	1.0	40		
Decanoic acid	$C_{10}H_{20}O_2$	172.265	20	0.015	0.15	26		
1-Decanol	$C_{10}H_{22}O$	158.281	25	0.0037	0.037	1		
2-Decanone	$C_{10}H_{20}O$	156.265	25	0.0079	0.079	84		
4-Decanone	$C_{10}H_{20}O$	156.265	20	0.0238	0.238	20		
			80	0.0064	0.064	20		
1-Decene	$C_{10}H_{20}$	140.266	25	0.00057	0.0057	4		
2'-Deoxyadenosine	$C_{10}H_{13}N_5O_3$	251.242	25	0.67	6.7	29		
Dexamethasone	$C_{22}H_{29}FO_5$	392.460	25	0.009	0.09	40		
Dibenz[a,j]acridine	$C_{21}H_{13}N$	279.335	25	0.000016	0.00016	6		
Dibenz[a,h]anthracene	$C_{22}H_{14}$	278.346	25	0.00000005	0.0000005	42,4		

Name	Mol. Form.	Mol. Wt.	$t/°C$	Solubility, s			Henry Const., k_H	
				$100\,w_2$ (mass%)	g per kg H_2O	Ref.	kPa m^3mol^{-1}	Ref.
Dibenz[a,j]anthracene	$C_{22}H_{14}$	278.346	27	0.0000012	0.000012	42,4		
13H-Dibenzo[a,i]carbazole	$C_{20}H_{13}N$	267.324	24	0.00000104	0.0000104	6		
Dibenzo[b,e][1,4]dioxin	$C_{12}H_8O_2$	184.191	25	0.000126	0.00126	68		
Dibenzofuran	$C_{12}H_8O$	168.191	25	0.000475	0.00475	41	0.011	12
Dibenzothiophene	$C_{12}H_8S$	184.257	25	0.000103	0.00103	6		
Dibenzyl ether	$C_{14}H_{14}O$	198.260	35	0.0040	0.040	10		
o-Dibromobenzene	$C_6H_4Br_2$	235.904	25	0.00748	0.0748	2		
m-Dibromobenzene	$C_6H_4Br_2$	235.904	25	0.0064	0.064	2		
p-Dibromobenzene	$C_6H_4Br_2$	235.904	25	0.0020	0.020	2		
1,4-Dibromobutane	$C_4H_8Br_2$	215.915	25	0.035	0.35	35		
1,2-Dibromo-1-chloroethane	$C_2H_3Br_2Cl$	222.306	20	0.060	0.60	25		
1,2-Dibromo-3-chloropropane	$C_3H_5Br_2Cl$	236.333	20	0.123	1.23	35		
1,2-Dibromo-1,2-dichloroethane	$C_2H_2Br_2Cl_2$	256.751	20	0.070	0.70	25		
1,2-Dibromoethane	$C_2H_4Br_2$	187.861	20	0.412	4.14	20		
			50	0.493	4.95	20	0.066	13
			80	0.572	5.75	20		
1,2-Dibromo-1,1,2,3,3,3-hexafluoropropane	$C_3Br_2F_6$	309.830	21	0.0068	0.068	35		
3,5-Dibromo-4-hydroxybenzonitrile	$C_7H_3Br_2NO$	276.913	25	0.013	0.13	40		
Dibromomethane	CH_2Br_2	173.835	20	1.28	13.0	20	0.086	13
			90	1.51	15.3	20		
2,4-Dibromophenol	$C_6H_4Br_2O$	251.903	25	0.2	2	2		
1,2-Dibromopropane	$C_3H_6Br_2$	201.888	25	0.143	1.43	10		
1,3-Dibromopropane	$C_3H_6Br_2$	201.888	25	0.169	1.69	35		
1,2-Dibromotetrafluoroethane	$C_2Br_2F_4$	259.823	25	0.00030	0.0030	25		
Dibutylamine	$C_8H_{19}N$	129.244	20	0.47	4.7	10		
Dibutyl ether	$C_8H_{18}O$	130.228	0	0.040	0.40	20	0.48	13
			20	0.023	0.23	20	0.48	13
			90	0.010	0.10	20		
Dibutyl phthalate	$C_{16}H_{22}O_4$	278.344	25	0.00112	0.0112	15		
Dibutyl sebacate	$C_{18}H_{34}O_4$	314.461	20	0.004	0.04	10		
o-Dichlorobenzene	$C_6H_4Cl_2$	147.002	5	0.012	0.12	61,58,2		
			25	0.015	0.15	61,58,2		
			45	0.020	0.20	61,58,2		
m-Dichlorobenzene	$C_6H_4Cl_2$	147.002	10	0.0103	0.103	41,2		
			25	0.0120	0.120	41,2	0.376	11
			45	0.0141	0.141	61,2		
p-Dichlorobenzene	$C_6H_4Cl_2$	147.002	10	0.00512	0.0512	2		
			25	0.0080	0.080	41	0.244	28
			50	0.0167	0.167	2		
3,5-Dichloro-1,2-benzenediol	$C_6H_4Cl_2O_2$	179.001	25	0.78	7.8	8		
4,5-Dichloro-1,2-benzenediol	$C_6H_4Cl_2O_2$	179.001	25	1.19	12.0	8		
3,3'-Dichloro-p-benzidine	$C_{12}H_{10}Cl_2N_2$	253.126	25	0.00031	0.0031	40		
2,5-Dichlorobiphenyl	$C_{12}H_8Cl_2$	223.098	25	0.0002	0.002	7	0.0201	7
2,6-Dichlorobiphenyl	$C_{12}H_8Cl_2$	223.098	25	0.00014	0.0014	7		
1,1-Dichloro-2,2-bis(p-chlorophenyl)ethane	$C_{14}H_{10}Cl_4$	320.041	25	0.000009	0.00009	40		
			45	0.000024	0.00024	40		
1,1-Dichlorobutane	$C_4H_8Cl_2$	127.013	25	0.050	0.50	35		
1,4-Dichlorobutane	$C_4H_8Cl_2$	127.013	25	0.16	1.6	35		
2,3-Dichlorobutane	$C_4H_8Cl_2$	127.013	20	0.056	0.56	35		

| Name | Mol. Form. | Mol. Wt. | $t/°C$ | Solubility, s | | Ref. | Henry Const., k_H | Ref. |
				100 w_2 (mass%)	g per kg H$_2$O		kPa m^{3}mol^{-1}	
2,7-Dichlorodibenzo-*p*-dioxin	C$_{12}$H$_6$Cl$_2$O$_2$	253.081	25	0.00000041	0.0000041	68		
1,2-Dichloro-1,1-difluoroethane	C$_2$H$_2$Cl$_2$F$_2$	134.940	24	0.49	4.9	25		
Dichlorodifluoromethane	CCl$_2$F$_2$	120.914	20	0.028*	0.28*	5	41	13
1,3-Dichloro-5,5-dimethyl hydantoin	C$_5$H$_6$Cl$_2$N$_2$O$_2$	197.019	20	0.050	0.50	40		
1,1-Dichloroethane	C$_2$H$_4$Cl$_2$	98.959	0	0.62	6.2	25		
			25	0.50	5.0	25	0.63	13
			50	0.50	5.0	25		
1,2-Dichloroethane	C$_2$H$_4$Cl$_2$	98.959	0	0.92	9.2	25		
			25	0.86	8.6	25	0.14	13
			50	1.05	10.6	25		
			100	2.17	22.2	25		
1,1-Dichloroethene	C$_2$H$_2$Cl$_2$	96.943	5	0.310	3.10	25		
			25	0.242	2.42	25	2.62	13
			50	0.225	2.25	25		
			90	0.355	3.55	25		
cis-1,2-Dichloroethene	C$_2$H$_2$Cl$_2$	96.943	10	0.76	7.6	25		
			25	0.64	6.4	25	0.46	13
			40	0.66	6.6	25		
trans-1,2-Dichloroethene	C$_2$H$_2$Cl$_2$	96.943	10	0.53	5.3	25		
			25	0.45	4.5	25	0.96	13
			40	0.41	4.1	25		
1,1-Dichloro-1-fluoroethane	C$_2$H$_3$Cl$_2$F	116.949	25	0.042	0.42	25		
Dichlorofluoromethane	CHCl$_2$F	102.923	25	0.95*	9.5*	10		
1,2-Dichloro-1,1,2,3,3,3-hexafluoropropane	C$_3$Cl$_2$F$_6$	220.928	21	0.0096	0.096	35		
1,4-Dichloro-5-isopropyl-2-methylbenzene	C$_{10}$H$_{12}$Cl$_2$	203.108	25	0.00049	0.0049	23		
Dichloromethane	CH$_2$Cl$_2$	84.933	25	1.73	17.6	20	0.30	13
3,6-Dichloro-2-methoxybenzoic acid	C$_8$H$_6$Cl$_2$O$_3$	221.038	25	0.45	4.5	40		
(Dichloromethyl)benzene	C$_7$H$_6$Cl$_2$	161.029	30	0.025	0.25	10		
2,3-Dichloro-2-methylbutane	C$_5$H$_{10}$Cl$_2$	141.038	25	0.029	0.29	35		
2,4-Dichloro-6-methylphenol	C$_7$H$_6$Cl$_2$O	177.028	25	0.0283	0.283	2		
2,6-Dichloro-4-methylphenol	C$_7$H$_6$Cl$_2$O	177.028	25	0.0673	0.673	2		
2,3-Dichloro-1,4-naphthalenedione	C$_{10}$H$_4$Cl$_2$O$_2$	227.044	25	0.00001	0.0001	40		
1,2-Dichloro-4-nitrobenzene	C$_6$H$_3$Cl$_2$NO$_2$	192.000	20	0.0121	0.121	40		
1,2-Dichloropentane	C$_5$H$_{10}$Cl$_2$	141.038	25	0.029	0.29	35		
1,5-Dichloropentane	C$_5$H$_{10}$Cl$_2$	141.038	19	0.02	0.2	35		
2,3-Dichloropentane	C$_5$H$_{10}$Cl$_2$	141.038	25	0.029	0.29	35		
Dichlorophene	C$_{13}$H$_{10}$Cl$_2$O$_2$	269.123	25	0.003	0.03	40		
2,3-Dichlorophenol	C$_6$H$_4$Cl$_2$O	163.001	25	0.82	8.3	40		
2,4-Dichlorophenol	C$_6$H$_4$Cl$_2$O	163.001	25	0.55	5.5	48,51,24		
2,6-Dichlorophenol	C$_6$H$_4$Cl$_2$O	163.001	25	0.262	2.62	40		
(2,4-Dichlorophenoxy)acetic acid	C$_8$H$_6$Cl$_2$O$_3$	221.038	25	0.07	0.7	40		
4-(2,4-Dichlorophenoxy)butanoic acid	C$_{10}$H$_{10}$Cl$_2$O$_3$	249.090	25	0.0046	0.046	40		
2-(2,4-Dichlorophenoxy)propanoic acid	C$_9$H$_8$Cl$_2$O$_3$	235.064	25	0.083	0.83	40		
1,2-Dichloropropane	C$_3$H$_6$Cl$_2$	112.986	5	0.270	2.70	35		
			25	0.274	2.74	35	0.29	13
			40	0.297	2.97	35		
1,3-Dichloropropane	C$_3$H$_6$Cl$_2$	112.986	5	0.218	2.18	35		
			25	0.280	2.80	35		
cis-1,3-Dichloropropene	C$_3$H$_4$Cl$_2$	110.970	20	0.27	2.7	5	0.24	5

Name	Mol. Form.	Mol. Wt.	$t/°C$	Solubility, s			Henry Const., k_H	
				$100\ w_2$ (mass%)	g per kg H_2O	Ref.	kPa m^3mol^{-1}	Ref.
trans-1,3-Dichloropropene	$C_3H_4Cl_2$	110.970	20	0.28	2.8	5	0.18	5
2,3-Dichloropropene	$C_3H_4Cl_2$	110.970	25	0.215*	2.15	5	0.36	5
1,2-Dichloro-1,1,2,2-tetrafluoroethane	$C_2Cl_2F_4$	170.921	25	0.013*	0.13*	10	127	13
2,4-Dichlorotoluene	$C_7H_6Cl_2$	161.029	25	0.00260	0.0260	61		
2,6-Dichlorotoluene	$C_7H_6Cl_2$	161.029	25	0.00233	0.0233	61		
2,2-Dichloro-1,1,1-trifluoroethane	$C_2HCl_2F_3$	152.930	25	0.46	4.6	25		
Diclofop-methyl	$C_{16}H_{14}Cl_2O_4$	341.186	20	0.0003	0.003	32		
Dieldrin	$C_{12}H_8Cl_6O$	380.909	25	0.000020	0.00020	67		
Diethanolamine	$C_4H_{11}NO_2$	105.136	20	95.4	20700	10		
1,1-Diethoxyethane	$C_6H_{14}O_2$	118.174	25	5	53	10		
1,2-Diethoxyethane	$C_6H_{14}O_2$	118.174	20	21.0	266	10		
2-(Diethylamino)-N-(2,6-dimethylphenyl) acetamide	$C_{14}H_{22}N_2O$	234.337	25	0.38	3.8	40		
o-Diethylbenzene	$C_{10}H_{14}$	134.218	20	0.0071	0.071	40		
p-Diethylbenzene	$C_{10}H_{14}$	134.218	20	0.0025	0.025	40		
Diethyl carbonate	$C_5H_{10}O_3$	118.131	20	1.8	18	40		
Diethyl ether	$C_4H_{10}O$	74.121	0	12.5	143	79	0.088	13
			25	5.9	63	79	0.088	13
			38	4.6	48	79	0.088	13
			82	3.1	32	79	0.088	13
Diethyl glutarate	$C_9H_{16}O_4$	188.221	30	1.20	12.1	20		
			91	0.91	9.2	20		
Diethyl maleate	$C_8H_{12}O_4$	172.179	20	1.56	15.8	20		
			91	1.75	17.8	20		
Diethyl malonate	$C_7H_{12}O_4$	160.168	20	2.26	23.1	20		
			91	2.47	25.3	20		
Diethyl phthalate	$C_{12}H_{14}O_4$	222.237	25	0.12	1.2	40		
trans-Diethylstilbestrol	$C_{18}H_{20}O_2$	268.351	20	0.01	0.1	40		
Diethyl succinate	$C_8H_{14}O_4$	174.195	20	0.19	1.9	40		
Diethyl sulfide	$C_4H_{10}S$	90.187	25	0.307	3.07	40		
Diflubenzuron	$C_{14}H_9ClF_2N_2O_2$	310.683	20	0.00002	0.0002	40		
o-Difluorobenzene	$C_6H_4F_2$	114.093	25	0.114	1.14	2		
m-Difluorobenzene	$C_6H_4F_2$	114.093	25	0.114	1.14	2		
p-Difluorobenzene	$C_6H_4F_2$	114.093	25	0.122	1.22	2		
1,1-Difluoroethane	$C_2H_4F_2$	66.050	20	0.29*	2.9*	50		
Digitoxin	$C_{41}H_{64}O_{13}$	764.939	25	0.0004	0.004	40		
Diglycolic acid	$C_4H_6O_5$	134.088	24	40.0	667	34		
			50	59.9	1490	34		
Digoxin	$C_{41}H_{64}O_{14}$	780.939	25	0.0059	0.059	40		
Dihexyl ether	$C_{12}H_{26}O$	186.333	20	0.019	0.19	20		
			90	0.019	0.19	20		
1,2-Dihydrobenz[j]aceanthrylene	$C_{20}H_{14}$	254.325	27	0.00000035	0.0000035	42,6		
1,3-Dihydro-2H-benzimidazol-2-one	$C_7H_6N_2O$	134.135	24	0.37	3.7	54		
1,2-Dihydro-3-methylbenz[j]aceanthrylene	$C_{21}H_{16}$	268.352	25	0.00000022	0.0000022	42,6		
			27	0.00000028	0.0000028	42		
2,3-Dihydro-6-propyl-2-thioxo-4(1H)-pyrimidinone	$C_7H_{10}N_2OS$	170.231	25	0.120	1.20	40		
1,7-Dihydro-6H-purine-6-thione	$C_5H_4N_4S$	152.178	25	0.0124	0.124	40		
3,4-Dihydro-2H-pyran	C_5H_8O	84.117	20	1.04	10.5	20		
			82	2.26	23.1	20		

| Name | Mol. Form. | Mol. Wt. | $t/°C$ | Solubility, s | | | Henry Const., k_H | |
				$100 w_2$ (mass%)	g per kg H_2O	Ref.	kPa m^3mol^{-1}	Ref.
1,4-Dihydroxy-9,10-anthracenedione	$C_{14}H_8O_4$	240.212	25	0.0000096	0.000096	40		
3,4-Dihydroxybenzoic acid	$C_7H_6O_4$	154.121	14	1.8	18	26		
			80	21.3	271	26		
3,12-Dihydroxycholan-24-oic acid, (3α,5β,12α)	$C_{24}H_{40}O_4$	392.573	20	0.001	0.01	40		
17,21-Dihydroxypregna-1,4-diene-3,11,20-trione	$C_{21}H_{26}O_5$	358.428	25	0.012	0.12	40		
17,21-Dihydroxypregn-4-ene-3,11,20-trione	$C_{21}H_{28}O_5$	360.444	25	0.028	0.28	30		
o-Diiodobenzene	$C_6H_4I_2$	329.905	25	0.00192	0.0192	2		
m-Diiodobenzene	$C_6H_4I_2$	329.905	25	0.000185	0.00185	2		
p-Diiodobenzene	$C_6H_4I_2$	329.905	25	0.000893	0.00893	2		
cis-1,2-Diiodoethene	$C_2H_2I_2$	279.846	25	0.046	0.46	25		
trans-1,2-Diiodoethene	$C_2H_2I_2$	279.846	25	0.015	0.15	25		
Diiodomethane	CH_2I_2	267.836	30	0.124	1.24	10	0.032	13
3,5-Diiodo-L-tyrosine	$C_9H_9I_2NO_3$	432.981	25	0.062	0.62	26		
Diisopentyl ether	$C_{10}H_{22}O$	158.281	20	0.02	0.2	10		
Diisopropyl ether	$C_6H_{14}O$	102.174	20	0.79	8.0	20	0.26	13
			61	0.22	2.2	20		
1,2-Dimethoxybenzene	$C_8H_{10}O_2$	138.164	20	0.716	7.21	20		
			92	1.073	10.85*	20		
3,3'-Dimethoxybenzidine	$C_{14}H_{16}N_2O_2$	244.289	25	0.006	0.06	40		
Dimethoxymethane	$C_3H_8O_2$	76.095	16	24.4	323	10		
4-(Dimethylamino)azobenzene	$C_{14}H_{15}N_3$	225.289	20	0.00014	0.0014	40		
2',3-Dimethyl-4-aminoazobenzene	$C_{14}H_{15}N_3$	225.289	37	0.0007	0.007	40		
2,5-Dimethylaniline	$C_8H_{11}N$	121.180	20	0.66	6.6	27		
N,N-Dimethylaniline	$C_8H_{11}N$	121.180	25	0.111	1.11	40		
9,10-Dimethylanthracene	$C_{16}H_{14}$	206.282	25	0.0000056	0.000056	42,4		
Dimethylarsinic acid	$C_2H_7AsO_2$	137.998	25	≈41	≈700	40		
7,12-Dimethylbenz[a]anthracene	$C_{20}H_{16}$	256.341	25	0.0000061	0.000061	42		
2,2-Dimethylbutane	C_6H_{14}	86.175	25	0.0021	0.021	3	199	13
2,3-Dimethylbutane	C_6H_{14}	86.175	25	0.0021	0.021	3	144	13
2,2-Dimethyl-1-butanol	$C_6H_{14}O$	102.174	25	0.78	7.9	78,1		
2,3-Dimethyl-2-butanol	$C_6H_{14}O$	102.174	25	4.2	44	1		
3,3-Dimethyl-2-butanol	$C_6H_{14}O$	102.174	25	2.4	25	1		
3,3-Dimethyl-2-butanone	$C_6H_{12}O$	100.158	0	2.92	30.1	83		
			19	1.97	20.1	20		
			25	1.85	18.8	83		
			50	1.46	14.8	83		
			90	1.14	11.5	20		
2,3-Dimethyl-1-butene	C_6H_{12}	84.159	30	0.046	0.46	3		
cis-1,2-Dimethylcyclohexane	C_8H_{16}	112.213	25	0.00060	0.0060	4	36	5
trans-1,2-Dimethylcyclohexane	C_8H_{16}	112.213	30	0.00050	0.0050	57,4	88.2	5
			100	0.00293	0.0293	57,4		
Dimethyl ether	C_2H_6O	46.068	25	35.3*	546	79		
			50	29.2*	412	79		
Dimethylglyoxime	$C_4H_8N_2O_2$	116.119	20	0.06	0.6	40		
3,5-Dimethyl-4-heptanol	$C_9H_{20}O$	144.254	15	0.072	0.72	1		
2,6-Dimethyl-4-heptanone	$C_9H_{18}O$	142.238	21	0.045	0.45	20		
			91	0.037	0.37	20		
1,2-Dimethyl-1H-imidazole	$C_5H_8N_2$	96.131	19	94.3	16500	54		
Dimethyl maleate	$C_6H_8O_4$	144.126	25	8.0	87	10		

Name	Mol. Form.	Mol. Wt.	$t/°C$	Solubility, s			Henry Const., k_H	
				$100 w_2$ (mass%)	g per kg H_2O	Ref.	kPa m^3mol^{-1}	Ref.
Dimethyl malonate	$C_5H_8O_4$	132.116	19	14.9	175	20		
			90	29.8	425	20		
1,3-Dimethylnaphthalene	$C_{12}H_{12}$	156.223	25	0.0008	0.008	4		
1,4-Dimethylnaphthalene	$C_{12}H_{12}$	156.223	25	0.00114	0.0114	4		
1,5-Dimethylnaphthalene	$C_{12}H_{12}$	156.223	25	0.00031	0.0031	4	0.036	28
2,3-Dimethylnaphthalene	$C_{12}H_{12}$	156.223	25	0.00025	0.0025	4		
2,6-Dimethylnaphthalene	$C_{12}H_{12}$	156.223	25	0.00017	0.0017	4		
Dimethyl oxalate	$C_4H_6O_4$	118.089	20	5.82	61.8	27		
2,2-Dimethylpentane	C_7H_{16}	100.202	25	0.00044	0.0044	3	318	5
2,3-Dimethylpentane	C_7H_{16}	100.202	25	0.00052	0.0052	3	175	5
2,4-Dimethylpentane	C_7H_{16}	100.202	25	0.00042	0.0042	3	323	13
3,3-Dimethylpentane	C_7H_{16}	100.202	25	0.00059	0.0059	3	186	5
2,3-Dimethyl-2-pentanol	$C_7H_{16}O$	116.201	25	1.5	15	1		
2,4-Dimethyl-2-pentanol	$C_7H_{16}O$	116.201	25	1.3	13	1		
2,2-Dimethyl-3-pentanol	$C_7H_{16}O$	116.201	25	0.82	8.2	1		
2,3-Dimethyl-3-pentanol	$C_7H_{16}O$	116.201	25	1.6	16	1		
2,4-Dimethyl-3-pentanol	$C_7H_{16}O$	116.201	25	0.70	7.0	1		
2,4-Dimethyl-3-pentanone	$C_7H_{14}O$	114.185	20	0.52	5.2	20		
			90	0.30	3.0	20		
N,N-Dimethyl-N'-phenylurea	$C_9H_{12}N_2O$	164.203	25	0.32	3.2	40		
Dimethyl phthalate	$C_{10}H_{10}O_4$	194.184	25	0.40	4.0	15		
2,2-Dimethyl-1-propanol	$C_5H_{12}O$	88.148	12	3.87	40.3	78,1		
			25	3.26	33.7	78,1		
			80	2.84	29.2	78,1		
4-(1,1-Dimethylpropyl)phenol	$C_{11}H_{16}O$	164.244	25	0.017	0.17	40		
Dimethyl succinate	$C_6H_{10}O_4$	146.141	21	12.4	142	20		
			92	17.1	206	20		
Dimethyl sulfate	$C_2H_6O_4S$	126.132	18	2.7	28	27		
Dimethyl sulfide	C_2H_6S	62.134	25	2	20	10		
Dimethyl sulfoxide	C_2H_6OS	78.133	25	25.3	339	10		
Dimethyl terephthalate	$C_{10}H_{10}O_4$	194.184	25	0.00328	0.0328	40		
Dimethyl tetrachloroterephthalate	$C_{10}H_6Cl_4O_4$	331.965	25	0.00005	0.0005	40		
N,N-Dimethyl-N'-[3-(trifluoromethyl)phenyl]urea	$C_{10}H_{11}F_3N_2O$	232.201	20	0.0105	0.105	40		
2,4-Dinitroaniline	$C_6H_5N_3O_4$	183.122	25	0.0078	0.078	40		
1,2-Dinitrobenzene	$C_6H_4N_2O_4$	168.107	20	0.21	2.1	27		
1,3-Dinitrobenzene	$C_6H_4N_2O_4$	168.107	20	2.09	21.3	27		
1,4-Dinitrobenzene	$C_6H_4N_2O_4$	168.107	20	1.30	13.2	27		
3,5-Dinitrobenzoic acid	$C_7H_4N_2O_6$	212.116	25	0.134	1.34	27		
2,4-Dinitrophenol	$C_6H_4N_2O_5$	184.106	25	0.069	0.69	48,51		
			35	0.098	0.98	48,51		
Dipentyl ether	$C_{10}H_{22}O$	158.281	25	0.11	1.1	81		
Diphenamid	$C_{16}H_{17}NO$	239.312	27	0.026	0.26	32		
Diphenylamine	$C_{12}H_{11}N$	169.222	20	0.0055	0.055	40		
			50	0.0058	0.058	40		
1,2-Diphenylethane	$C_{14}H_{14}$	182.261	25	0.00044	0.0044	6	0.017	12
Diphenyl ether	$C_{12}H_{10}O$	170.206	25	0.0018	0.0180	6	0.027	13
Diphenylmethane	$C_{13}H_{12}$	168.234	25	0.00014	0.0014	42,4	0.001	12
Diphenyl phthalate	$C_{20}H_{14}O_4$	318.323	24	0.000008	0.00008	40		
1,3-Diphenyl-1-triazene	$C_{12}H_{11}N_3$	197.235	20	0.050	0.50	40		

Name	Mol. Form.	Mol. Wt.	$t/°C$	Solubility, s			Henry Const., k_H	
				$100\,w_2$ (mass%)	g per kg H_2O	Ref.	kPa m^3mol^{-1}	Ref.
N,N'-Diphenylurea	$C_{13}H_{12}N_2O$	212.246	20	0.015	0.15	40		
Dipropylamine	$C_6H_{15}N$	101.190	20	2.5	26	10		
Dipropyl ether	$C_6H_{14}O$	102.174	0	2.67	27.4	80	0.26	13
			25	0.91	9.2	80	0.26	13
Diuron	$C_9H_{10}Cl_2N_2O$	233.093	25	0.0042	0.042	40		
Docosane	$C_{22}H_{46}$	310.600	22	0.0000006	0.000006	37		
Dodecane	$C_{12}H_{26}$	170.334	25	0.00000037	0.0000037	4	750	5
Dodecanedioic acid	$C_{12}H_{22}O_4$	230.301	20	0.004	0.04	40		
Dodecanoic acid	$C_{12}H_{24}O_2$	200.318	20	0.0055	0.055	26		
1-Dodecanol	$C_{12}H_{26}O$	186.333	25	0.0004	0.004	1		
Droperidol	$C_{22}H_{22}FN_3O_2$	379.427	30	0.00041	0.0041	40		
Eicosane	$C_{20}H_{42}$	282.547	25	0.00000019	0.0000019	42,4		
Emetine	$C_{29}H_{40}N_2O_4$	480.639	15	0.096	0.96	40		
Endrin	$C_{12}H_8Cl_6O$	380.909	25	0.000025	0.00025	67		
l-Ephedrine	$C_{10}H_{15}NO$	165.232	25	0.57	5.7	40		
Epichlorohydrin	C_3H_5ClO	92.524	20	6.58	70.4	10	0.003	13
			65	7.2	78	40		
Epinephrine	$C_9H_{13}NO_3$	183.204	20	0.018	0.18	40		
1,2-Epoxy-4-(epoxyethyl)cyclohexane	$C_8H_{12}O_2$	140.180	20	13.4	155	40		
2,3-Epoxy-α-pinane	$C_{10}H_{16}O$	152.233	25	0.039	0.39	52		
Erythromycin	$C_{37}H_{67}NO_{13}$	733.927	30	0.12	1.2	40		
			80	0.04	0.4	40		
Estra-1,3,5(10)-triene-3,17-diol (17β)	$C_{18}H_{24}O_2$	272.383	25	0.000151	0.00151	49		
Estrone	$C_{18}H_{22}O_2$	270.367	25	0.000130	0.00130	49		
Ethane	C_2H_6	30.069	25	0.00568*	0.0568*	18	50.6	5
1,2-Ethanediol, diacetate	$C_6H_{10}O_4$	146.141	25	13.3	153	40		
Ethinylestradiol	$C_{20}H_{24}O_2$	296.404	25	0.000921	0.00921	49		
Ethoxybenzene	$C_8H_{10}O$	122.164	25	0.12	1.2	10		
2-Ethoxyethyl acetate	$C_6H_{12}O_3$	132.157		14	163	30		
N-(4-Ethoxyphenyl)acetamide	$C_{10}H_{13}NO_2$	179.216	25	0.0502	0.502	40		
Ethyl acetate	$C_4H_8O_2$	88.106	25	8.08	87.9	10		
Ethyl acetoacetate	$C_6H_{10}O_3$	130.141	25	12	136	10		
Ethyl acrylate	$C_5H_8O_2$	100.117	25	1.50	15.2	10		
Ethylbenzene	C_8H_{10}	106.165	0	0.020	0.20	4,89		
			25	0.0161	0.161	22,89	0.843	22
			40	0.0200	0.200	4,89		
Ethyl benzoate	$C_9H_{10}O_2$	150.174	25	0.083	0.83	20		
Ethyl butanoate	$C_6H_{12}O_2$	116.158	20	0.49	4.9	10		
2-Ethyl-1-butanol	$C_6H_{14}O$	102.174	20	0.92	9.3	78		
			50	0.80	8.1	78		
Ethyl carbamate	$C_3H_7NO_2$	89.094	15	48	920	27		
Ethyl cyanoacetate	$C_5H_7NO_2$	113.116	20	25.9	350	10		
Ethylcyclohexane	C_8H_{16}	112.213	30	0.00061	0.0061	57,4		
			100	0.00212	0.0212	57,4		
Ethylcyclopentane	C_7H_{14}	98.186	20	0.012	0.12	3		
Ethyl decanoate	$C_{12}H_{24}O_2$	200.318	20	0.0015	0.015	27		
Ethylene	C_2H_4	28.053	25	0.01336*	0.1336*	19	21.7	5
Ethyleneimine	C_2H_5N	43.068	20	0.90	9.1	40		
Ethyl formate	$C_3H_6O_2$	74.079	25	11.8	134	10		

Name	Mol. Form.	Mol. Wt.	t/°C	Solubility, s 100 w_2 (mass%)	g per kg H_2O	Ref.	Henry Const., k_H kPa m^3mol^{-1}	Ref.
Ethyl heptanoate	$C_9H_{18}O_2$	158.238	20	0.029	0.29	27		
Ethyl hexanoate	$C_8H_{16}O_2$	144.212	20	0.063	0.63	27		
2-Ethyl-1-hexanol	$C_8H_{18}O$	130.228	25	0.071	0.71	78		
			50	0.074	0.74	78		
2-Ethylhexylamine	$C_8H_{19}N$	129.244	20	0.25	2.5	10		
Ethyl 4-hydroxybenzoate	$C_9H_{10}O_3$	166.173	25	0.0080	0.080	40		
Ethyl isopropyl ether	$C_5H_{12}O$	88.148	25	0.52	5.2	79		
Ethyl 2-methylbutanoate, (+)	$C_7H_{14}O_2$	130.185	19	0.257	2.58	20		
			91	0.151	1.51	20		
Ethyl 3-methylbutanoate	$C_7H_{14}O_2$	130.185	20	0.2	2	10		
Ethyl N-methylcarbamate	$C_4H_9NO_2$	103.120	15	69	2230	27		
1-Ethylnaphthalene	$C_{12}H_{12}$	156.223	25	0.00101	0.0101	4	0.039	12
2-Ethylnaphthalene	$C_{12}H_{12}$	156.223	25	0.00080	0.0080	4	0.078	12
O-Ethyl O-p-nitrophenyl benzenethiophosphonate	$C_{14}H_{14}NO_4PS$	323.304	22	0.00031	0.0031	40		
N-Ethyl-N-nitrosourea	$C_3H_7N_3O_2$	117.107	20	1.3	13	40		
Ethyl nonanoate	$C_{11}H_{22}O_2$	186.292	20	0.003	0.03	27		
Ethyl octanoate	$C_{10}H_{20}O_2$	172.265	20	0.007	0.07	27		
Ethyl pentanoate	$C_7H_{14}O_2$	130.185	25	0.3	3	27		
3-Ethyl-3-pentanol	$C_7H_{16}O$	116.201	25	1.7	17	1		
4-Ethylphenol	$C_8H_{10}O$	122.164	20	0.59	5.9	40		
Ethyl propanoate	$C_5H_{10}O_2$	102.132	20	1.92	19.6	10		
Ethyl N-propylcarbamate	$C_6H_{13}NO_2$	131.173	15	7.70	83.4	27		
Ethyl propyl ether	$C_5H_{12}O$	88.148	25	1.87	19.2	79		
2-Ethyltoluene	C_9H_{12}	120.191	25	0.0075	0.075	89,5	0.529	13
4-Ethyltoluene	C_9H_{12}	120.191	25	0.0094	0.094	5	0.500	13
Ethyl vinyl ether	C_4H_8O	72.106	20	0.9	9	10		
Etoposide	$C_{29}H_{32}O_{13}$	588.556	20	0.02	0.2	40		
Eucalyptol	$C_{10}H_{18}O$	154.249	21	0.379	3.79	40		
			50	0.170	1.70	40		
Fenamiphos	$C_{13}H_{22}NO_3PS$	303.358	20	0.0329	0.329	40		
Fenbutatin oxide	$C_{60}H_{78}OSn_2$	1052.68	23	0.0000005	0.000005	32		
α-Fenchol, (+)-	$C_{10}H_{18}O$	154.249	25	0.083	0.83	52		
Fenchone	$C_{10}H_{16}O$	152.233	20	0.2	2	84		
Fenoxycarb	$C_{17}H_{19}NO_4$	301.338	20	0.0006	0.006	32		
Ferbam	$C_9H_{18}FeN_3S_6$	416.494	20	0.013	0.13	40		
Fluoranthene	$C_{16}H_{10}$	202.250	20	0.000017	0.00017	42		
			25	0.000021	0.00021	42,22	0.00096	22
9H-Fluorene	$C_{13}H_{10}$	166.218	0	0.00007	0.0007	42,4		
			25	0.00019	0.0019	42,22	0.00787	22
			50	0.00063	0.0063	42,4		
Fluorescein	$C_{20}H_{12}O_5$	332.306	20	0.005	0.05	27		
Fluorobenzene	C_6H_5F	96.102	19	0.170	1.70	20	0.70	11
			80	0.188	1.88	20	0.70	11
2-Fluorobenzoic acid	$C_7H_5FO_2$	140.112	25	0.72	7.2	27		
3-Fluorobenzoic acid	$C_7H_5FO_2$	140.112	25	0.15	1.5	27		
4-Fluorobenzoic acid	$C_7H_5FO_2$	140.112	25	0.12	1.2	27		
Fluoroethane	C_2H_5F	48.059	25	0.216*	2.16*	14		
Fluoromethane	CH_3F	34.033	0	0.420*	4.20*	50		
			25	0.201*	2.01*	50		

| Name | Mol. Form. | Mol. Wt. | $t/°C$ | Solubility, s | | | Henry Const., k_H | |
				$100\ w_2$ (mass%)	g per kg H_2O	Ref.	kPa m^3mol^{-1}	Ref.
			80	0.082*	0.82*	50		
1-Fluoropropane	C_3H_7F	62.086	14	0.386*	3.86*	14		
2-Fluoropropane	C_3H_7F	62.086	15	0.366*	3.66*	14		
5-Fluorouracil	$C_4H_3FN_2O_2$	130.077	25	1.77	18.0	72		
Folic acid	$C_{19}H_{19}N_7O_6$	441.397	0	0.001	0.01	26		
			100	0.05	0.5	26		
Folpet	$C_9H_4Cl_3NO_2S$	296.558	20	0.00010	0.0010	40		
β-D-Fructose	$C_6H_{12}O_6$	180.155	20	≈31	≈450	40		
Furan	C_4H_4O	68.074	25	1	10	10	0.54	13
2-Furancarboxylic acid	$C_5H_4O_3$	112.084	25	4.76	50.0	33		
			50	25.2	337	33		
Furfural	$C_5H_4O_2$	96.085	20	8.2	89	10		
Galactaric acid	$C_6H_{10}O_8$	210.138	14	0.33	3.3	40		
D-Galactose	$C_6H_{12}O_6$	180.155	20	40.6	684	27		
D-Glucitol	$C_6H_{14}O_6$	182.171	20	≈41	≈700	40		
α-D-Glucose	$C_6H_{12}O_6$	180.155	15	45.0	818	27		
			30	54.6	1200	27		
			80	81.5	4400	27		
DL-Glutamic acid	$C_5H_9NO_4$	147.130	25	2.30	23.5	29		
L-Glutamic acid	$C_5H_9NO_4$	147.130	10	0.444	4.46	75		
			25	0.824	8.31	75		
			50	2.13	21.8	75		
L-Glutamine	$C_5H_{10}N_2O_3$	146.144	25	4.0	42	26		
Glycerol triacetate	$C_9H_{14}O_6$	218.203	25	5.8	62	10		
Glycine	$C_2H_5NO_2$	75.067	25	18.5	227	70		
			36	22.1	284	70		
			50	26.1	353	70		
Glycolic acid	$C_2H_4O_3$	76.051	25	71.2	2470	34		
			55	77.9	3520	34		
N-Glycylglycine	$C_4H_8N_2O_3$	132.118	25	18.8	232	47,29		
Glyphosate	$C_3H_8NO_5P$	169.074	25	1.2	12	32		
Guanidinoacetic acid	$C_3H_7N_3O_2$	117.107	25	0.5	5	26		
Guanine	$C_5H_5N_5O$	151.127	25	0.0068	0.068	29		
Guanosine	$C_{10}H_{13}N_5O_5$	283.241	25	0.0500	0.500	29		
Haloperidol	$C_{21}H_{23}ClFNO_2$	375.865	30	0.0003	0.003	40		
Heptachlor	$C_{10}H_5Cl_7$	373.318	25	0.000018	0.00018	67		
2,2',3,3',4,4',6-Heptachlorobiphenyl	$C_{12}H_3Cl_7$	395.323	25	0.0000002	0.000002	7	0.0054	7
Heptadecanoic acid	$C_{17}H_{34}O_2$	270.451	20	0.00042	0.0042	26		
1,6-Heptadiyne	C_7H_8	92.139	25	0.125	1.25	3		
Heptanal	$C_7H_{14}O$	114.185	11	0.124	1.24	27		
Heptane	C_7H_{16}	100.202	25	0.000242	0.00242	46		
			50	0.000341	0.00341	46	209	13
			75	0.000570	0.00570	46		
			100	0.00108	0.0108	46		
Heptanedioic acid	$C_7H_{12}O_4$	160.168	25	6.347	67.77	33		
			50	42.80	748	33		
Heptanoic acid	$C_7H_{14}O_2$	130.185	15	0.24	2.4	27		
1-Heptanol	$C_7H_{16}O$	116.201	0	0.236	2.37	78		
			25	0.164	1.64	78,1	0.00562	28

Name	Mol. Form.	Mol. Wt.	$t/°C$	Solubility, s 100 w_2 (mass%)	g per kg H_2O	Ref.	Henry Const., k_H kPa m^3mol^{-1}	Ref.
			50	0.164	1.64	78,1		
			90	0.245	2.46	78		
2-Heptanol	$C_7H_{16}O$	116.201	30	0.33	3.3	1		
3-Heptanol, (S)-	$C_7H_{16}O$	116.201	25	0.43	4.3	1		
4-Heptanol	$C_7H_{16}O$	116.201	25	0.47	4.7	1		
2-Heptanone	$C_7H_{14}O$	114.185	25	0.435	4.37	20	0.0171	28
			90	0.353	3.53	20	0.0171	28
3-Heptanone	$C_7H_{14}O$	114.185	20	0.479	4.81	20		
			90	0.309	3.10	20		
4-Heptanone	$C_7H_{14}O$	114.185	20	0.457	4.57	20		
			90	0.316	3.16	20		
1-Heptene	C_7H_{14}	98.186	25	0.032	0.32	3	40.3	13
trans-2-Heptene	C_7H_{14}	98.186	25	0.015	0.15	3	42.2	13
Heptyl butanoate	$C_{11}H_{22}O_2$	186.292	20	0.028	0.28	20		
			80	0.020	0.20	20		
1-Heptyne	C_7H_{12}	96.170	25	0.0094	0.094	3	4.47	13
Hesperetin	$C_{16}H_{14}O_6$	302.278	15	0.00004	0.0004	71		
			25	0.00014	0.0014	71		
			35	0.00052	0.0052	71		
Hexachlorobenzene	C_6Cl_6	284.782	25	0.00000096	0.0000096	58	0.131	11
			35	0.0000018	0.000018	58		
			55	0.0000038	0.000038	58		
2,2',3,3',4,4'-Hexachlorobiphenyl	$C_{12}H_4Cl_6$	360.878	25	0.00000006	0.0000006	7	0.0354	31
2,2',4,4',6,6'-Hexachlorobiphenyl	$C_{12}H_4Cl_6$	360.878	25	0.0000003	0.000003	41	0.818	7
2,2',3,3',6,6'-Hexachlorobiphenyl	$C_{12}H_4Cl_6$	360.878	25	0.0000004	0.000004	41		
Hexachloro-1,3-butadiene	C_4Cl_6	260.761	25	0.41	4.1	35		
1,2,3,4,5,6-Hexachlorocyclohexane, (1α,2α,3β,4α,5α,6β)	$C_6H_6Cl_6$	290.830	25	0.00078	0.0078	60		
			45	0.0015	0.015	60		
1,2,3,4,5,6-Hexachlorocyclohexane, (1α,2α,3β,4α,5β,6β)	$C_6H_6Cl_6$	290.830	25	0.00018	0.0018	60		
1,2,3,4,5,6-Hexachlorocyclohexane, (1α,2β,3α,4β,5α,6β)	$C_6H_6Cl_6$	290.830	25	0.00002	0.0002	60		
Hexachloroethane	C_2Cl_6	236.739	25	0.005	0.05	25	0.85	13
Hexachloropropene	C_3Cl_6	248.750	20	0.00118	0.0118	35		
Hexacosafluorododecane	$C_{12}F_{26}$	638.086	20	0.00000096	0.0000096	35		
Hexacosane	$C_{26}H_{54}$	366.707	25	0.00000017	0.0000017	42,37		
Hexadecane	$C_{16}H_{34}$	226.441	25	0.0000004	0.000004	42,37		
Hexadecanoic acid	$C_{16}H_{32}O_2$	256.424	20	0.00072	0.0072	26		
1-Hexadecanol	$C_{16}H_{34}O$	242.440	25	0.000003	0.00003	1		
1,5-Hexadiene	C_6H_{10}	82.143	25	0.017	0.17	3		
Hexafluorobenzene	C_6F_6	186.054	8	0.0778	0.778	53		
			28	0.0616	0.616	53		
			67	0.0636	0.636	53		
Hexahydro-1,3,5-trinitro-1,3,5-triazine	$C_3H_6N_6O_6$	222.116	3	0.0014	0.014	59		
			20	0.0037	0.037	59		
			25	0.0060	0.060	17		
			34	0.0086	0.086	59		
Hexamethylenetetramine	$C_6H_{12}N_4$	140.186	12	44.8	812	27		
Hexane	C_6H_{14}	86.175	25	0.00098	0.0098	46		

Name	Mol. Form.	Mol. Wt.	$t/°C$	Solubility, s			Henry Const., k_H	
				$100\,w_2$ (mass%)	g per kg H_2O	Ref.	kPa m^3mol^{-1}	Ref.
			50	0.00114	0.0114	46		
			75	0.00167	0.0167	46	183	13
			100	0.00291	0.0291	46		
1,6-Hexanediamine	$C_6H_{16}N_2$	116.204	5	≈42	≈720	40		
Hexanedinitrile	$C_6H_8N_2$	108.141	20	0.80	8.0	16		
1,6-Hexanedioic acid	$C_6H_{10}O_4$	146.141	15	1.48	15.0	26		
			100	61.5	1600	26		
Hexanoic acid	$C_6H_{12}O_2$	116.158	25	1.01	10.2	64		
			35	1.09	11.0	64		
			60	1.16	11.7	26		
1-Hexanol	$C_6H_{14}O$	102.174	0	0.79	7.9	1		
			10	0.70	7.0	78		
			25	0.59	5.9	78,1		
			50	0.55	5.5	78,1		
2-Hexanol	$C_6H_{14}O$	102.174	25	1.4	14	1		
3-Hexanol	$C_6H_{14}O$	102.174	25	1.6	16	1		
2-Hexanone	$C_6H_{12}O$	100.158	10	1.91	19.5	83		
			25	1.49	15.1	83		
			50	1.17	11.8	83		
3-Hexanone	$C_6H_{12}O$	100.158	25	1.47	14.9	83		
Hexatriacontane	$C_{36}H_{74}$	506.973	25	0.00000017	0.0000017	42,37		
Hexazinone	$C_{12}H_{20}N_4O_2$	252.313	25	3.2	33	40		
1-Hexene	C_6H_{12}	84.159	25	0.0053	0.053	3	41.8	5
trans-2-Hexene	C_6H_{12}	84.159	25	0.0067	0.067	3		
1-Hexen-3-ol	$C_6H_{12}O$	100.158	25	2.52	25.9	1		
4-Hexen-2-ol	$C_6H_{12}O$	100.158	25	3.81	39.6	1		
Hexyl acetate	$C_8H_{16}O_2$	144.212	20	0.02	0.2	10		
sec-Hexyl acetate	$C_8H_{16}O_2$	144.212	20	0.13	1.3	10		
Hexylbenzene	$C_{12}H_{18}$	162.271	25	0.00021	0.0021	4		
4-Hexyl-1,3-benzenediol	$C_{12}H_{18}O_2$	194.270	18	0.05	0.5	40		
Hexyl butanoate	$C_{10}H_{20}O_2$	172.265	29	0.021	0.21	20		
1-Hexyne	C_6H_{10}	82.143	25	0.036	0.36	3	4.14	13
L-Histidine	$C_6H_9N_3O_2$	155.154	25	4.17	43.5	26		
Homocystine	$C_8H_{16}N_2O_4S_2$	268.354	25	0.02	0.2	26		
L-Homoserine	$C_4H_9NO_3$	119.119	25	52.4	1100	26		
Hydramethylnon	$C_{25}H_{24}F_6N_4$	494.476	20	0.0000006	0.000006	32		
Hydrochlorothiazide	$C_7H_8ClN_3O_4S_2$	297.740	25	0.007	0.07	40		
Hydrocortisone	$C_{21}H_{30}O_5$	362.460	25	0.029	0.29	40		
Hydroflumethiazide	$C_8H_8F_3N_3O_4S_2$	331.293	37	0.068	0.68	40		
p-Hydroquinone	$C_6H_6O_2$	110.111	25	7.42	80.1	27		
17-Hydroxyandrost-4-en-3-one, (17β)	$C_{19}H_{28}O_2$	288.424	25	0.0024	0.024	40		
4-Hydroxybenzaldehyde	$C_7H_6O_2$	122.122	30	1.27	12.9	40		
2-Hydroxybenzamide	$C_7H_7NO_2$	137.137	10	0.122	1.22	44		
			25	0.241	2.42	44		
			50	0.737	7.42	44		
α-Hydroxybenzeneacetic acid	$C_8H_8O_3$	152.148	25	11.3	127	27		
2-Hydroxybenzoic acid	$C_7H_6O_3$	138.121	10	0.119	1.19	43,33		
			25	0.189	1.89	43,33		
			50	0.521	5.24	43,33		

Name	Mol. Form.	Mol. Wt.	$t/°C$	$100\,w_2$ (mass%)	g per kg H_2O	Ref.	kPa m^3mol^{-1}	Ref.
				Solubility, s			Henry Const., k_H	
4-Hydroxybenzoic acid	$C_7H_6O_3$	138.121	15	0.8	8	26		
			75	2.5	26	27		
2-Hydroxybiphenyl	$C_{12}H_{10}O$	170.206	25	0.07	0.7	40		
4-Hydroxybiphenyl	$C_{12}H_{10}O$	170.206	25	0.0056	0.056	40		
4-Hydroxy-3-methoxybenzaldehyde	$C_8H_8O_3$	152.148	25	0.247	2.47	8		
3-Hydroxy-4-oxo-4H-pyran-2,6-dicarboxylic acid	$C_7H_4O_7$	200.103	25	0.84	8.4	27		
N-(4-Hydroxyphenyl)acetamide	$C_8H_9NO_2$	151.163	25	1.3	13	40		
trans-4-Hydroxy-L-proline	$C_5H_9NO_3$	131.130	25	26.5	361	26		
Hyoscyamine	$C_{17}H_{23}NO_3$	289.370	20	0.36	3.6	40		
Hypoxanthine	$C_5H_4N_4O$	136.112	25	0.070	0.70	29		
Ibuprofen	$C_{13}H_{18}O_2$	206.281	25	0.0011	0.011	40		
			60	0.0048	0.048	40		
Imazaquin	$C_{17}H_{17}N_3O_3$	311.335	20	0.009	0.09	32		
Imidacloprid	$C_9H_{10}ClN_5O_2$	255.66	30	0.038	0.38	73		
			51	0.117	1.17	73		
Imidazole	$C_3H_4N_2$	68.077	19	67.3	2060	54		
2,4-Imidazolidinedione	$C_3H_4N_2O_2$	100.076	25	3.93	40.9	29		
Imidodicarbonic diamide	$C_2H_5N_3O_2$	103.080	15	1.5	15	40		
Iminodiacetic acid	$C_4H_7NO_4$	133.104	5	2.32	23.8	40		
Indan	C_9H_{10}	118.175	25	0.010	0.10	4		
1H-Indazole	$C_7H_6N_2$	118.136	20	0.0827	0.827	6		
Indeno[1,2,3-cd]pyrene	$C_{22}H_{12}$	276.330	20	0.00000002	0.0000002	40		
1H-Indole	C_8H_7N	117.149	20	0.187	1.87	6		
Indomethacin	$C_{19}H_{16}ClNO_4$	357.788	25	0.001	0.01	40		
Inosine	$C_{10}H_{12}N_4O_5$	268.226	20	1.6	16	29		
Iodobenzene	C_6H_5I	204.008	10	0.0193	0.193	2		
			25	0.0226	0.226	2	0.078	11
			45	0.0279	0.279	2		
2-Iodobenzoic acid	$C_7H_5IO_2$	248.018	25	0.095	0.95	27		
3-Iodobenzoic acid	$C_7H_5IO_2$	248.018	25	0.016	0.16	27		
4-Iodobenzoic acid	$C_7H_5IO_2$	248.018	25	0.0027	0.027	27		
1-Iodobutane	C_4H_9I	184.018	17	0.021	0.21	10	1.87	13
Iodoethane	C_2H_5I	155.965	0	0.44	4.4	25		
			25	0.40	4.0	25	0.52	13
1-Iodoheptane	$C_7H_{15}I$	226.098	25	0.00035	0.0035	35		
Iodomethane	CH_3I	141.939	20	1.4	14	10	0.54	13
1-Iodopropane	C_3H_7I	169.992	0	0.114	1.14	35		
			20	0.100	1.00	35	0.93	13
2-Iodopropane	C_3H_7I	169.992	0	0.167	1.67	35		
			20	0.140	1.40	35		
5-Iodouracil	$C_4H_3IN_2O_2$	237.983	25	0.49	4.9	72		
trans-β-Ionone	$C_{13}H_{20}O$	192.297	25	0.017	0.17	52		
Iopanoic acid	$C_{11}H_{12}I_3NO_2$	570.932	37	0.034	0.34	40		
Iprodione	$C_{13}H_{13}Cl_2N_3O_3$	330.166	20	0.0013	0.013	40		
Isobutanal	C_4H_8O	72.106	20	9.1	100	10		
Isobutane	C_4H_{10}	58.122	25	0.00535*	0.0535*	18	120	5
Isobutene	C_4H_8	56.107	25	0.0263*	0.263*	5	21.6	13
Isobutyl acetate	$C_6H_{12}O_2$	116.158	20	0.63	6.3	10		
Isobutylbenzene	$C_{10}H_{14}$	134.218	25	0.0010	0.010	4	3.32	11

Name	Mol. Form.	Mol. Wt.	$t/°C$	Solubility, s 100 w_2 (mass%)	g per kg H_2O	Ref.	Henry Const., k_H kPa m^3mol^{-1}	Ref.
Isobutyl formate	$C_5H_{10}O_2$	102.132	22	1.0	10	10		
Isobutyl isobutanoate	$C_8H_{16}O_2$	144.212	20	0.5	5	10		
Isobutyl propanoate	$C_7H_{14}O_2$	130.185	19	0.225	2.26	20		
			91	0.142	1.42	20		
$1H$-Isoindole-1,3(2H)-dione	$C_8H_5NO_2$	147.132	25	0.036	0.36	40		
L-Isoleucine	$C_6H_{13}NO_2$	131.173	25	3.31	34.2	26		
Isoniazid	$C_6H_7N_3O$	137.139	25	11.0	124	40		
Isopentane	C_5H_{12}	72.149	25	0.00485	0.0485	3	479	13
Isopentyl acetate	$C_7H_{14}O_2$	130.185	20	0.2	2	10		
Isopentyl formate	$C_6H_{12}O_2$	116.158	22	0.3	3	27		
Isophorone	$C_9H_{14}O$	138.206	20	1.57	16.0	20		
			80	1.27	12.9	20		
Isophthalic acid	$C_8H_6O_4$	166.132	10	0.0062	0.062	76		
			25	0.0154	0.154	56		
			50	0.0395	0.395	56		
			80	0.123	1.23	56		
Isopropenylbenzene	C_9H_{10}	118.175	20	0.0116	0.116	40		
Isopropyl acetate	$C_5H_{10}O_2$	102.132	20	2.9	30	10		
Isopropylbenzene	C_9H_{12}	120.191	25	0.0050	0.050	22	1.466	22
1-Isopropyl-2-methylbenzene	$C_{10}H_{14}$	134.218	25	0.00482	0.0482	23		
1-Isopropyl-3-methylbenzene	$C_{10}H_{14}$	134.218	25	0.00425	0.0425	23		
1-Isopropyl-4-methylbenzene	$C_{10}H_{14}$	134.218	25	0.0051	0.051	23	0.80	5
Isopropyl phenylcarbamate	$C_{10}H_{13}NO_2$	179.216	20	0.01	0.1	40		
Isoquinoline	C_9H_7N	129.159	20	0.452	4.52	6		
Isosorbide dinitrate	$C_6H_8N_2O_8$	236.136	25	0.055	0.55	40		
Kepone	$C_{10}Cl_{10}O$	490.636	100	0.4	4	40		
L-Lanthionine	$C_6H_{12}N_2O_4S$	208.235	25	0.15	1.5	26		
Lasiocarpine	$C_{21}H_{33}NO_7$	411.490	20	0.67	6.7	40		
L-Leucine	$C_6H_{13}NO_2$	131.173	25	2.32	23.8	62		
Levodopa	$C_9H_{11}NO_4$	197.188	20	0.165	1.65	63		
d-Limonene	$C_{10}H_{16}$	136.234	0	0.001	0.01	4		
			25	0.0020	0.020	52		
Linalol	$C_{10}H_{18}O$	154.249	25	0.156	1.56	52		
Linuron	$C_9H_{10}Cl_2N_2O_2$	249.093	25	0.0075	0.075	40		
L-Lysine	$C_6H_{14}N_2O_2$	146.187	25	0.58	5.8	26		
Maleic acid	$C_4H_4O_4$	116.073	25	44.1	789	26		
Malic acid	$C_4H_6O_5$	134.088	26	59	1440	26		
Malonic acid	$C_3H_4O_4$	104.062	0	37.9	610	26		
			20	42.4	736	26		
			50	48.1	927	26		
Malononitrile	$C_3H_2N_2$	66.061	20	10.6	119	40		
α-Maltose	$C_{12}H_{22}O_{11}$	342.296	20	51.9	1080	27		
D-Mannitol	$C_6H_{14}O_6$	182.171	25	17.7	215	27		
Mefenamic acid	$C_{15}H_{15}NO_2$	241.286	20	0.0026	0.026	40		
Melphalan	$C_{13}H_{18}Cl_2N_2O_2$	305.200	30	0.44	4.4	40		
Mercury(II) phenyl acetate	$C_8H_8HgO_2$	336.74	20	0.2	2	30		
Mesityl oxide	$C_6H_{10}O$	98.142	20	2.8	29	83		
Methacrylic acid	$C_4H_6O_2$	86.090	20	8.9	98	10		
Methane	CH_4	16.043	25	0.00227*	0.0227*	18	67.4	5

Name	Mol. Form.	Mol. Wt.	$t/°C$	Solubility, s		Ref.	Henry Const., k_H	Ref.
				$100\,w_2$ (mass%)	g per kg H_2O		kPa m^3mol^{-1}	
Methazolamide	$C_5H_8N_4O_3S_2$	236.273	15	0.0472	0.472	40		
Methazole	$C_9H_6Cl_2N_2O_3$	261.061	24	0.00015	0.0015	40		
Methidathion	$C_6H_{11}N_2O_4PS_3$	302.330	20	0.0187	0.187	40		
L-Methionine	$C_5H_{11}NO_2S$	149.212	25	5.3	56	26		
Methomyl	$C_5H_{10}N_2O_2S$	162.210	25	5.5	58	40		
Methoxsalen	$C_{12}H_8O_4$	216.190	30	0.0048	0.048	40		
2-Methoxyaniline	C_7H_9NO	123.152	25	1.24	12.6	40		
4-Methoxyaniline	C_7H_9NO	123.152	20	1.14	11.5	40		
4-Methoxybenzaldehyde	$C_8H_8O_2$	136.149	25	0.429	4.29	40		
4-Methoxybenzoic acid	$C_8H_8O_3$	152.148	25	0.023	0.23	27		
Methoxychlor	$C_{16}H_{15}Cl_3O_2$	345.648	25	0.000005	0.00005	40		
2-Methoxy-2-methylbutane	$C_6H_{14}O$	102.174	20	1.10	11.1	20		
			79	0.36	3.6	20		
4-Methoxyphenol	$C_7H_8O_2$	124.138	20	2.51	25.7	40		
Methyclothiazide	$C_9H_{11}Cl_2N_3O_4S_2$	360.237	20	0.005	0.05	40		
Methyl acetate	$C_3H_6O_2$	74.079	20	24.5	325	10		
Methyl acrylate	$C_4H_6O_2$	86.090	25	4.94	52.0	10		
2-Methylacrylonitrile	C_4H_5N	67.090	20	2.57	26.4	10		
2-Methylaniline	C_7H_9N	107.153	20	1.66	16.9	10		
4-Methylaniline	C_7H_9N	107.153	21	7.35	79.3	10		
N-Methylaniline	C_7H_9N	107.153	25	0.56	5.6	40		
2-Methylanthracene	$C_{15}H_{12}$	192.256	6	0.0000007	0.000007	42		
			25	0.0000021	0.000021	42,22		
9-Methylanthracene	$C_{15}H_{12}$	192.256	25	0.000026	0.00026	42,4		
9-Methylbenz[a]anthracene	$C_{19}H_{14}$	242.314	27	0.0000066	0.000066	42,4		
10-Methylbenz[a]anthracene	$C_{19}H_{14}$	242.314	27	0.0000055	0.000055	42,4		
2-Methylbenzenesulfonamide	$C_7H_9NO_2S$	171.217	25	0.162	1.62	27		
3-Methylbenzenesulfonamide	$C_7H_9NO_2S$	171.217	25	0.78	7.8	27		
4-Methylbenzenesulfonamide	$C_7H_9NO_2S$	171.217	25	0.316	3.16	27		
2-Methyl-1H-benzimidazole	$C_8H_8N_2$	132.163	20	0.145	1.45	6		
Methyl benzoate	$C_8H_8O_2$	136.149	20	0.21	2.1	10		
2-Methyl-1,3-butadiene	C_5H_8	68.118	25	0.061	0.61	3	7.78	5
			50	0.076*	0.76*	3		
Methyl butanoate	$C_5H_{10}O_2$	102.132		1.6	16	30		
3-Methylbutanoic acid	$C_5H_{10}O_2$	102.132	20	4.0	42	26		
2-Methyl-1-butanol	$C_5H_{12}O$	88.148	10	3.38	35.0	78		
			25	2.75	28.3	78		
			50	2.35	24.1	78		
3-Methyl-1-butanol	$C_5H_{12}O$	88.148	10	3.17	32.7	78,1		
			25	2.59	26.6	78,1		
			70	2.24	22.9	78,1		
2-Methyl-2-butanol	$C_5H_{12}O$	88.148	25	11.0	124	88,1		
			60	6.6	71	88,1		
3-Methyl-2-butanol	$C_5H_{12}O$	88.148	25	5.6	59	1		
3-Methyl-2-butanone	$C_5H_{10}O$	86.132	0	9.4	104	82		
			25	6.1	65	82		
			40	5.2	55	82		
3-Methyl-1-butene	C_5H_{10}	70.133	25	0.013*	0.13*	3	54.7	5

Name	Mol. Form.	Mol. Wt.	t/°C	Solubility, s		Ref.	Henry Const., k_H	
				$100\,w_2$ (mass%)	g per kg H_2O		kPa m³mol⁻¹	Ref.
2-Methyl-2-butene	C_5H_{10}	70.133	25	0.041	0.41	3		
2-Methyl-3-buten-2-ol	$C_5H_{10}O$	86.132	18	27.4	377	88		
			29	18.4	225	88		
Methyl *tert*-butyl ether	$C_5H_{12}O$	88.148	0	7.72	83.7	79		
			25	3.25	33.6	79		
			35	2.56	26.3	79		
			70	1.64	16.7	79		
Methyl carbamate	$C_2H_5NO_2$	75.067	15	69	2230	27		
5-Methylchrysene	$C_{19}H_{14}$	242.314	27	0.0000062	0.000062	42,4		
Methylcyclohexane	C_7H_{14}	98.186	26	0.00161	0.0161	3	43.3	13
			100	0.00548	0.0548	3		
2-Methylcyclohexanone	$C_7H_{12}O$	112.169	0	2.93	30.2	84		
			20	1.98	20.2	20		
			31	1.72	17.5	84		
			60	1.44	14.6	84		
			90	1.54	15.6	20		
4-Methylcyclohexanone	$C_7H_{12}O$	112.169	20	2.43	24.9	20		
			80	1.95	19.9	20		
1-Methylcyclohexene	C_7H_{12}	96.170	25	0.0052	0.052	3		
Methylcyclopentane	C_6H_{12}	84.159	25	0.0043	0.043	3	36.7	5
1-Methyl-2,4-dinitrobenzene	$C_7H_6N_2O_4$	182.134	12	0.0130	0.130	55		
			32	0.0270	0.270	85		
			62	0.098	0.98	85		
2-Methyl-4,6-dinitrophenol	$C_7H_6N_2O_5$	198.133		0.0130	0.130	40		
Methyl formate	$C_2H_4O_2$	60.052	25	23	300	10		
3-Methylheptane	C_8H_{18}	114.229	25	0.000079	0.00079	4	376	5
2-Methyl-2-heptanol	$C_8H_{18}O$	130.228	30	0.25	2.5	1		
5-Methyl-3-heptanone	$C_8H_{16}O$	128.212	20	0.192	1.92	20		
			90	0.131	1.31	20		
2-Methylhexane	C_7H_{16}	100.202	25	0.00025	0.0025	3	346	5
3-Methylhexane	C_7H_{16}	100.202	25	0.00026	0.0026	3	249	13
2-Methyl-2-hexanol	$C_7H_{16}O$	116.201	25	1.0	10	1		
5-Methyl-2-hexanol	$C_7H_{16}O$	116.201	25	0.49	4.9	1		
3-Methyl-3-hexanol	$C_7H_{16}O$	116.201	25	1.2	12	1		
5-Methyl-2-hexanone	$C_7H_{14}O$	114.185	19	0.537	5.40	20		
			90	0.417	4.19	20		
5-Methyl-3-hexanone	$C_7H_{14}O$	114.185	20	0.47	4.7	20		
			81	0.32	3.2	20		
Methyl 4-hydroxybenzoate	$C_8H_8O_3$	152.148	25	0.24	2.4	40		
2-Methyl-1*H*-imidazole	$C_4H_6N_2$	82.104	18	23.2	302	54		
3-Methyl-1*H*-indole	C_9H_9N	131.174	20	0.050	0.50	6		
3-Methylisoquinoline	$C_{10}H_9N$	143.185	20	0.092	0.92	6		
Methyl isothiocyanate	C_2H_3NS	73.117	20	0.75	7.6	40		
Methylmalonic acid	$C_4H_6O_4$	118.089	0	30.1	431	26		
			20	40	670	26		
Methyl methacrylate	$C_5H_8O_2$	100.117	20	1.56	15.8	10		
2-Methyl-3-(2-methylphenyl)-4(3*H*)-quinazolinone	$C_{16}H_{14}N_2O$	250.294	23	0.03	0.3	40		
1-Methylnaphthalene	$C_{11}H_{10}$	142.197	25	0.00281	0.0281	22	0.045	22

Name	Mol. Form.	Mol. Wt.	$t/°C$	Solubility, s		Ref.	Henry Const., k_H	Ref.
				$100\,w_2$ (mass%)	g per kg H_2O		$kPa\ m^3mol^{-1}$	
2-Methylnaphthalene	$C_{11}H_{10}$	142.197	25	0.0025	0.025	4	0.051	12
2-Methyl-1,4-naphthalenedione	$C_{11}H_8O_2$	172.181	25	0.016	0.16	40		
N-Methyl-N-nitrosourea	$C_2H_5N_3O_2$	103.080	14	2.3	24	40		
4-Methyloctane	C_9H_{20}	128.255	25	0.0000115	0.000115	4	1000	5
Methyloxirane	C_3H_6O	58.079	20	40.5	681	10	0.0087	13
Methyl parathion	$C_8H_{10}NO_5PS$	263.208	10	0.00218	0.0218	40		
			20	0.00380	0.0380	40		
			30	0.0059	0.059	40		
2-Methylpentane	C_6H_{14}	86.175	25	0.00137	0.0137	3	176	13
3-Methylpentane	C_6H_{14}	86.175	25	0.00129	0.0129	3	170	13
2-Methyl-1-pentanol	$C_6H_{14}O$	102.174	25	0.76	7.7	78,1		
			50	0.70	7.0	78		
4-Methyl-1-pentanol	$C_6H_{14}O$	102.174	25	0.76	7.6	1		
2-Methyl-2-pentanol	$C_6H_{14}O$	102.174	25	3.2	33	1		
3-Methyl-2-pentanol	$C_6H_{14}O$	102.174	25	1.9	19	1		
4-Methyl-2-pentanol	$C_6H_{14}O$	102.174	27	1.5	15	1		
2-Methyl-3-pentanol	$C_6H_{14}O$	102.174	25	2.0	20	1		
3-Methyl-3-pentanol	$C_6H_{14}O$	102.174	25	4.3	45	1		
4-Methyl-2-pentanone	$C_6H_{12}O$	100.158	0	2.92	30.1	83		
			25	1.85	18.8	83		
			50	1.46	14.8	83		
2-Methyl-3-pentanone	$C_6H_{12}O$	100.158	25	1.5	15	83		
2-Methyl-1-pentene	C_6H_{12}	84.159	25	0.0078	0.078	3	28.1	5
4-Methyl-1-pentene	C_6H_{12}	84.159	25	0.0048	0.048	3	63.2	5
1-Methylphenanthrene	$C_{15}H_{12}$	192.256	7	0.0000095	0.000095	42		
			25	0.0000269	0.000269	42,4		
Methylprednisolone	$C_{22}H_{30}O_5$	374.470	25	0.012	0.12	40		
Methyl propanoate	$C_4H_8O_2$	88.106		6	60	30		
2-Methylpropanoic acid	$C_4H_8O_2$	88.106	20	22.8	295	10		
2-Methyl-1-propanol	$C_4H_{10}O$	74.121	0	12.2	139	78,1		
			25	8.1	88	78,1	0.00273	28
			50	7.0	70	78,1		
Methyl propyl ether	$C_4H_{10}O$	74.121	0	5.4	57	79		
			25	3.0	31	79		
2-Methyl-2-propyl-1,3-propanediol dicarbamate	$C_9H_{18}N_2O_4$	218.250	25	0.33	3.3	40		
Methyl salicylate	$C_8H_8O_3$	152.148	30	0.74	7.4	10		
17-Methyltestosterone	$C_{20}H_{30}O_2$	302.451	25	0.0033	0.033	40		
2-Methyltetrahydrofuran	$C_5H_{10}O$	86.132	19	14.4	168	20	0.67	13
			71	6.0	64	20		
N-Methyl-N,2,4,6-tetranitroaniline	$C_7H_5N_5O_8$	287.144	20	0.0074	0.074	40		
Methylthiouracil	$C_5H_6N_2OS$	142.179	25	0.0533	0.533	40		
1-Methyl-2,3,4-trinitrobenzene	$C_7H_5N_3O_6$	227.131	14	0.0091	0.091	85,59		
			23	0.0116	0.116	85,59		
			61	0.0643	0.643	85,59		
Metronidazole	$C_6H_9N_3O_3$	171.153	20	0.93	9.4	40		
Mirex	$C_{10}Cl_{12}$	545.543	25	0.0000085	0.000085	40		
Morphine	$C_{17}H_{19}NO_3$	285.338	20	0.015	0.15	27		
β-Myrcene	$C_{10}H_{16}$	136.234	25	0.030	0.30	52		
Naphtacene	$C_{18}H_{12}$	228.288	25	0.00000007	0.0000007	42,4	0.000004	12

Name	Mol. Form.	Mol. Wt.	t/°C	Solubility, s			Henry Const., k_H	
				$100\ w_2$ (mass%)	g per kg H_2O	Ref.	kPa m^3mol^{-1}	Ref.
Naphthalene	$C_{10}H_8$	128.171	10	0.0019	0.019	4		
			25	0.00316	0.0316	22	0.043	22
			50	0.0082	0.082	4		
1-Naphthaleneacetic acid	$C_{12}H_{10}O_2$	186.206	25	0.0415	0.415	40		
1-Naphthalenecarboxylic acid	$C_{11}H_8O_2$	172.181	25	0.0058	0.058	27		
1-Naphthalenylthiourea	$C_{11}H_{10}N_2S$	202.275	20	0.06	0.6	40		
1-Naphthol	$C_{10}H_8O$	144.170	20	0.111	1.11	40		
2-Naphthol	$C_{10}H_8O$	144.170	20	0.064	0.64	40		
			80	0.67	6.7	40		
1-Naphthylamine	$C_{10}H_9N$	143.185	20	0.17	1.7	40		
2-Naphthylamine	$C_{10}H_9N$	143.185	20	0.0189	0.189	40		
Narceine	$C_{23}H_{27}NO_8$	445.462	13	0.078	0.78	27		
Neopentane	C_5H_{12}	72.149	25	0.00332*	0.0332*	3	220	13
Nitrapyrin	$C_6H_3Cl_4N$	230.907	20	0.0040	0.040	40		
2-Nitroaniline	$C_6H_6N_2O_2$	138.124	30	1.47	14.9	27		
3-Nitroaniline	$C_6H_6N_2O_2$	138.124	30	0.121	1.21	27		
4-Nitroaniline	$C_6H_6N_2O_2$	138.124	30	0.073	0.73	27		
2-Nitroanisole	$C_7H_7NO_3$	153.136	30	0.169	1.69	10		
4-Nitroanisole	$C_7H_7NO_3$	153.136	30	0.059	0.59	27		
3-Nitrobenzaldehyde	$C_7H_5NO_3$	151.120	25	0.16	1.6	27		
4-Nitrobenzaldehyde	$C_7H_5NO_3$	151.120	25	0.23	2.3	27		
Nitrobenzene	$C_6H_5NO_2$	123.110	25	0.21	2.1	17		
3-Nitro-1,2-benzenedicarboxylic acid	$C_8H_5NO_6$	211.129	25	1.63	16.6	69		
			40	2.97	30.6	69		
4-Nitro-1,2-benzenedicarboxylic acid	$C_8H_5NO_6$	211.129	25	60.9	1560	69		
			40	68.0	2125	69		
2-Nitrobenzoic acid	$C_7H_5NO_4$	167.120	25	0.55	5.5	40		
3-Nitrobenzoic acid	$C_7H_5NO_4$	167.120	10	0.197	1.97	75		
			25	0.313	3.14	75		
			50	0.90	9.1	75		
4-Nitrobenzoic acid	$C_7H_5NO_4$	167.120	25	0.0422	0.422	40		
Nitroethane	$C_2H_5NO_2$	75.067	25	4.4	46	38		
			50	5.3	56	38		
Nitrofen	$C_{12}H_7Cl_2NO_3$	284.095	22	0.00095	0.0095	40		
Nitrofurantoin	$C_8H_6N_4O_5$	238.158	30	0.011	0.11	40		
Nitrofurazone	$C_6H_6N_4O_4$	198.137	20	0.0238	0.238	40		
Nitroguanidine	$CH_4N_4O_2$	104.069	25	1.2	12	40		
Nitromethane	CH_3NO_2	61.041	0	9.2	101	36		
			25	11.0	124	36		
			50	14.8	174	36		
1-Nitronaphthalene	$C_{10}H_7NO_2$	173.169	18	0.005	0.05	40		
2-Nitrophenol	$C_6H_5NO_3$	139.109	25	0.170	1.70	48,51		
3-Nitrophenol	$C_6H_5NO_3$	139.109	20	2.14	21.9	27		
4-Nitrophenol	$C_6H_5NO_3$	139.109	20	1.56	15.8	48,51		
1-Nitropropane	$C_3H_7NO_2$	89.094	25	1.54	15.6	38		
			90	2.29	23.4	20		
2-Nitropropane	$C_3H_7NO_2$	89.094	25	1.75	17.8	38		
			90	2.36	24.2	20		
N-Nitrosodiethylamine	$C_4H_{10}N_2O$	102.134	24	9.6	106	40		

| Name | Mol. Form. | Mol. Wt. | t/°C | Solubility, s | | Ref. | Henry Const., k_H | Ref. |
				100 w_2 (mass%)	g per kg H_2O		kPa m^3mol^{-1}	
N-Nitrosodiphenylamine	$C_{12}H_{10}N_2O$	198.219	25	0.0035	0.035	17		
2-Nitrotoluene	$C_7H_7NO_2$	137.137	30	0.065	0.65	27		
3-Nitrotoluene	$C_7H_7NO_2$	137.137	30	0.050	0.50	27		
4-Nitrotoluene	$C_7H_7NO_2$	137.137	30	0.044	0.44	27		
2,2',3,3',4,5,5',6,6'-Nonachlorobiphenyl	$C_{12}HCl_9$	464.213	25	0.0000000018	0.000000018	7		
1,8-Nonadiyne	C_9H_{12}	120.191	25	0.0125	0.125	4		
Nonane	C_9H_{20}	128.255	25	0.000017	0.00017	4	333	13
			50	0.000022	0.00022	4		
Nonanedioic acid	$C_9H_{16}O_4$	188.221	25	0.1780	1.780	34		
			65	1.322	13.40	34		
Nonanoic acid	$C_9H_{18}O_2$	158.238	20	0.0284	0.284	26		
1-Nonanol	$C_9H_{20}O$	144.254	25	0.0129	0.129	78,1		
			90	0.0291	0.291	78		
2-Nonanol	$C_9H_{20}O$	144.254	15	0.026	0.26	1		
3-Nonanol	$C_9H_{20}O$	144.254	15	0.032	0.32	1		
4-Nonanol	$C_9H_{20}O$	144.254	15	0.0026	0.026	1		
5-Nonanol	$C_9H_{20}O$	144.254	15	0.0032	0.032	1		
2-Nonanone	$C_9H_{18}O$	142.238	20	0.038	0.38	20		
			70	0.034	0.34	20		
3-Nonanone	$C_9H_{18}O$	142.238	30	0.056	0.56	20		
			80	0.046	0.46	20		
5-Nonanone	$C_9H_{18}O$	142.238	20	0.054	0.54	20		
			80	0.029	0.29	20		
1-Nonene	C_9H_{18}	126.239	25	0.000112	0.00112	40		
Nonyl formate	$C_{10}H_{20}O_2$	172.265	10	0.012	0.12	20		
			90	0.039	0.39	20		
4-Nonylphenol	$C_{15}H_{24}O$	220.351	25	0.000636	0.00636	40		
1-Nonyne	C_9H_{16}	124.223	25	0.00072	0.0072	4		
Norethisterone	$C_{20}H_{26}O_2$	298.419	25	0.00063	0.0063	40		
Norflurazon	$C_{12}H_9ClF_3N_3O$	303.666	25	0.0028	0.028	40		
L-Norleucine	$C_6H_{13}NO_2$	131.173	25	1.5	15	26		
L-Norvaline	$C_5H_{11}NO_2$	117.147	25	9.7	107	26		
Noscapine	$C_{22}H_{23}NO_7$	413.421	25	0.03	0.3	40		
2,2',3,3',5,5',6,6'-Octachlorobiphenyl	$C_{12}H_2Cl_8$	429.768	25	0.00000015	0.0000015	41	0.0381	7
Octachlorodibenzo-p-dioxin	$C_{12}Cl_8O_2$	459.751	25	$2.3 \cdot 10^{-11}$	$2.3 \cdot 10^{-10}$	68		
Octachloro-1,3-pentadiene	C_5Cl_8	343.678	20	0.000020	0.00020	35		
Octacosane	$C_{28}H_{58}$	394.761	22	0.0000006	0.000006	37		
Octadecane	$C_{18}H_{38}$	254.495	25	0.00000021	0.0000021	42,37		
1-Octadecanol	$C_{18}H_{38}O$	270.494	34	0.000011	0.00011	1		
Octane	C_8H_{18}	114.229	25	0.000073	0.00073	46	311	13
			50	0.000102	0.00102	47		
			75	0.000179	0.00179	46		
			100	0.000377	0.00377	46		
Octanedioic acid	$C_8H_{14}O_4$	174.195	25	0.242	2.43	34		
			50	0.557	5.570	34		
Octanoic acid	$C_8H_{16}O_2$	144.212	25	0.080	0.80	26		
1-Octanol	$C_8H_{18}O$	130.228	25	0.0460	0.460	78		
			60	0.0536	0.536	78		
2-Octanol	$C_8H_{18}O$	130.228	25	0.4	4	1		

Name	Mol. Form.	Mol. Wt.	$t/°C$	Solubility, s			Henry Const., k_H	
				$100\,w_2$ (mass%)	g per kg H_2O	Ref.	kPa m^3mol^{-1}	Ref.
2-Octanone	$C_8H_{16}O$	128.212	20	0.134	1.34	84		
			50	0.098	0.98	84		
			80	0.091	0.91	84		
3-Octanone	$C_8H_{16}O$	128.212	20	0.137	1.37	20		
			91	0.106	1.06	20		
1-Octene	C_8H_{16}	112.213	25	0.00027	0.0027	4	96.3	13
Octyl acetate	$C_{10}H_{20}O_2$	172.265	19	0.020	0.20	20		
			92	0.012	0.12	20		
1-Octyne	C_8H_{14}	110.197	25	0.0024	0.024	4	7.87	13
Orotic acid	$C_5H_4N_2O_4$	156.097	18	0.18	1.8	26		
Oryzalin	$C_{12}H_{18}N_4O_6S$	346.359	25	0.00024	0.0024	40		
Ouabain	$C_{29}H_{44}O_{12}$	584.652	25	1.3	13	40		
Oxalic acid	$C_2H_2O_4$	90.035	20	8.69	95.2	27		
			80	45.8	845	27		
Oxamyl	$C_7H_{13}N_3O_3S$	219.261	25	≈21	≈270	40		
4-Oxopentanoic acid	$C_5H_8O_3$	116.116	10	63.6	1750	34		
			25	84.0	5250	34		
4-Oxo-4H-pyran-2,6-dicarboxylic acid	$C_7H_4O_6$	184.103	25	1.45	14.7	27		
Papaverine	$C_{20}H_{21}NO_4$	339.386	37	0.0037	0.037	40		
Paraldehyde	$C_6H_{12}O_3$	132.157	25	11	124	30		
Parathion	$C_{10}H_{14}NO_5PS$	291.261	20	0.00129	0.0129	40		
Pendimethalin	$C_{13}H_{19}N_3O_4$	281.308	20	0.00003	0.0003	40		
Pentachlorobenzene	C_6HCl_5	250.337	25	0.000050	0.00050	41	0.085	11
2,3,4,5,6-Pentachlorobiphenyl	$C_{12}H_5Cl_5$	326.433	25	0.0000008	0.000008	7		
2,2',4,5,5'-Pentachlorobiphenyl	$C_{12}H_5Cl_5$	326.433	25	0.000001	0.00001	7	0.0421	31
Pentachloroethane	C_2HCl_5	202.294	25	0.049	0.49	25	0.25	13
Pentachloronitrobenzene	$C_6Cl_5NO_2$	295.335	20	0.000044	0.00044	40		
Pentachlorophenol	C_6HCl_5O	266.336	25	0.0021	0.021	48,51,24		
2,3,4,5,6-Pentachlorotoluene	$C_7H_3Cl_5$	264.364	25	0.0000028	0.000028	61		
Pentadecanoic acid	$C_{15}H_{30}O_2$	242.398	20	0.0012	0.012	26		
1-Pentadecanol	$C_{15}H_{32}O$	228.414	25	0.000010	0.00010	1		
1,4-Pentadiene	C_5H_8	68.118	25	0.056	0.56	3	12	5
Pentaerythritol	$C_5H_{12}O_4$	136.147	15	5.3	56	30		
Pentaerythritol tetranitrate	$C_5H_8N_4O_{12}$	316.138	20	0.0002	0.002	40		
Pentanal	$C_5H_{10}O$	86.132	25	1.2	12	40		
Pentane	C_5H_{12}	72.149	25	0.0041	0.041	3	128	13
Pentanedioic acid	$C_5H_8O_4$	132.116	25	58.3	1400	33		
			50	78.1	3570	33		
2,4-Pentanedione	$C_5H_8O_2$	100.117	20	16.1	192	20		
			80	32.2	475	20		
Pentanoic acid	$C_5H_{10}O_2$	102.132	16	3.6	37	26		
			25	4.32	45.2	90		
			35	5.26	55.5	90		
1-Pentanol	$C_5H_{12}O$	88.148	0	3.23	33.4	78,1		
			25	2.14	21.9	78,1		
			50	1.83	18.6	78,1		
			90	2.12	21.7	78		
2-Pentanol	$C_5H_{12}O$	88.148	25	4.3	45	21		
3-Pentanol	$C_5H_{12}O$	88.148	25	5.6	59	21		

Name	Mol. Form.	Mol. Wt.	$t/°C$	Solubility, s 100 w_2 (mass%)	g per kg H_2O	Ref.	Henry Const., k_H kPa m^3mol^{-1}	Ref.
2-Pentanone	$C_5H_{10}O$	86.132	0	8.7	95	20	0.00847	28
			25	5.5	58	20	0.00847	28
			80	3.8	40	20	0.00847	28
3-Pentanone	$C_5H_{10}O$	86.132	0	7.6	82	82		
			25	4.9	52	82		
			80	3.6	37	82		
1-Pentene	C_5H_{10}	70.133	25	0.0148	0.148	3	40.3	5
cis-2-Pentene	C_5H_{10}	70.133	25	0.0203	0.203	3	22.8	5
Pentyl acetate	$C_7H_{14}O_2$	130.185	20	0.17	1.7	10		
sec-Pentyl acetate (S)-	$C_7H_{14}O_2$	130.185	25	0.2	2	27		
Pentylbenzene	$C_{11}H_{16}$	148.245	25	0.00043	0.0043	89,5	1.69	11
Pentylcyclopentane	$C_{10}H_{20}$	140.266	25	0.0000115	0.000115	4	185	5
Pentyl propanoate	$C_8H_{16}O_2$	144.212	20	0.1	1	27		
1-Pentyne	C_5H_8	68.118	25	0.157	1.57	3	2.5	5
Perfluorocyclobutane	C_4F_8	200.030	5	0.00638*	0.0638*	50		
			25	0.00247*	0.0247*	50		
			45	0.00158*	0.0158*	50		
Perfluorodecane	$C_{10}F_{22}$	538.072	20	0.000031	0.00031	35		
Perfluoroheptane	C_7F_{16}	388.049	25	0.0000013	0.000013	35		
Perfluorohexane	C_6F_{14}	338.042	25	0.0000098	0.000098	35		
Perfluoro-2-methylpentane	C_6F_{14}	338.042	25	0.000017	0.00017	35		
Perfluorooctane	C_8F_{18}	438.057	25	0.00000017	0.0000017	35		
Perfluoropentane	C_5F_{12}	288.035	25	0.00012	0.0012	35		
Perfluoropropane	C_3F_8	188.019	15	0.0015*	0.015*	14		
Perfluoropropene	C_3F_6	150.022	25	0.0194*	0.194*	14		
Permethrin	$C_{21}H_{20}Cl_2O_3$	391.288	20	0.00002	0.0002	32		
Perylene	$C_{20}H_{12}$	252.309	25	0.00000004	0.0000004	42,4	0.000003	12
Phenanthrene	$C_{14}H_{10}$	178.229	0	0.000039	0.00039	42		
			10	0.000047	0.00047	42,4		
			25	0.00012	0.0012	42,22	0.00324	22
			50	0.00042	0.0042	42,4		
Phenmedipham	$C_{16}H_{16}N_2O_4$	300.309	25	0.00047	0.0047	32		
Phenobarbital	$C_{12}H_{12}N_2O_3$	232.234	25	0.12	1.2	40		
			45	0.26	2.6	40		
Phenol	C_6H_6O	94.111	15	7.60	82.3	48,51		
			25	8.40	91.7	48,51		
			35	9.31	102.7	48,51		
Phenolphthalein	$C_{20}H_{14}O_4$	318.323	20	0.018	0.18	27		
10H-Phenothiazine	$C_{12}H_9NS$	199.271	25	0.00016	0.0016	40		
2-Phenoxyethanol	$C_8H_{10}O_2$	138.164	20	2.53	26.0	40		
Phenyl acetate	$C_8H_8O_2$	136.149	20	0.59	5.9	20		
			91	0.91	9.2	20		
DL-Phenylalanine	$C_9H_{11}NO_2$	165.189	25	1.40	14.2	29		
L-Phenylalanine	$C_9H_{11}NO_2$	165.189	25	2.71	27.9	26		
Phenylbutazone	$C_{19}H_{20}N_2O_2$	308.374	25	0.0034	0.034	40		
1-Phenyl-1-propanone	$C_9H_{10}O$	134.174	19	0.32	3.2	20		
			80	0.24	2.4	20		
Phenylthiourea	$C_7H_8N_2S$	152.217	25	2.55	26.2	27		
Phenytoin	$C_{15}H_{12}N_2O_2$	252.268	37	0.0038	0.038	40		

Name	Mol. Form.	Mol. Wt.	t/°C	Solubility, s 100 w_2 (mass%)	g per kg H_2O	Ref.	Henry Const., k_H kPa m³mol⁻¹	Ref.
Phosalone	$C_{12}H_{15}ClNO_4PS_2$	367.808	20	0.00026	0.0026	40		
Phosmet	$C_{11}H_{12}NO_4PS_2$	317.321	25	0.0025	0.025	40		
Phthalic acid	$C_8H_6O_4$	166.132	10	0.464	4.66	76		
			25	0.719	7.24	76		
			50	1.76	17.9	76		
			65	3.57	37.0	33		
Phthalic anhydride	$C_8H_4O_3$	148.116	27	0.62	6.20	40		
Picene	$C_{22}H_{14}$	278.346	27	0.00000025	0.0000025	42,4		
α-Pinene, (-)	$C_{10}H_{16}$	136.234	25	0.00050	0.0050	52		
β-Pinene, (1S)-	$C_{10}H_{16}$	136.234	25	0.00110	0.0110	52		
2,5-Piperazinedione	$C_4H_6N_2O_2$	114.103	25	1.64	16.7	29		
2-Pivaloyl-1,3-indandione	$C_{14}H_{14}O_3$	230.259	25	0.0018	0.018	40		
Prednisolone	$C_{21}H_{28}O_5$	360.444	25	0.03	0.3	40		
Progesterone	$C_{21}H_{30}O_2$	314.462	25	0.00088	0.0088	40		
			41	0.00206	0.0206	40		
L-Proline	$C_5H_9NO_2$	115.131	25	61.9	1625	26		
Prometone	$C_{10}H_{19}N_5O$	225.291	20	0.075	0.75	40		
Prometryn	$C_{10}H_{19}N_5S$	241.357	20	0.0048	0.048	32		
Propachlor	$C_{11}H_{14}ClNO$	211.688	20	0.07	0.7	40		
Propanal	C_3H_6O	58.079	25	30.6	441	10		
Propane	C_3H_8	44.096	25	0.00669*	0.0669*	18	71.6	5
Propanenitrile	C_3H_5N	55.079	25	10.3	115	10		
Propanil	$C_9H_9Cl_2NO$	218.079	20	0.013	0.13	40		
Propazine	$C_9H_{16}ClN_5$	229.710	20	0.00086	0.0086	40		
Propene	C_3H_6	42.080	25	0.0200*	0.200*	5	21.3	5
1-Propene-2,3-dicarboxylic acid	$C_5H_6O_4$	130.100	20	7.7	83	26		
trans-1-Propene-1,2,3-tricarboxylic acid	$C_6H_6O_6$	174.108	25	20.9	264	26		
			90	52.5	1105	26		
Propoxur	$C_{11}H_{15}NO_3$	209.242	20	0.193	1.93	40		
Propyl acetate	$C_5H_{10}O_2$	102.132	20	2.3	24	10		
Propylbenzene	C_9H_{12}	120.191	25	0.0052	0.052	22	1.041	22
Propyl butanoate	$C_7H_{14}O_2$	130.185	17	0.162	1.62	27		
Propylcyclopentane	C_8H_{16}	112.213	25	0.00020	0.0020	4	90.2	5
Propyl formate	$C_4H_8O_2$	88.106	22	2.05	20.9	10		
Propyl 4-hydroxybenzoate	$C_{10}H_{12}O_3$	180.200	25	0.04	0.4	40		
Propyl propanoate	$C_6H_{12}O_2$	116.158	25	0.6	6	27		
Propyne	C_3H_4	40.064	25	0.364*	3.64*	5	1.11	5
Propyzamide	$C_{12}H_{11}Cl_2NO$	256.127	25	0.0015	0.015	32		
Pyrene	$C_{16}H_{10}$	202.250	0	0.0000049	0.000049	42		
			15	0.0000069	0.000069	42		
			25	0.0000139	0.000139	42,22	0.00092	22
			50	0.000053	0.00053	42,4		
			75	0.000231	0.00231	42		
3-Pyridinecarboxamide	$C_6H_6N_2O$	122.124	20	≈33	≈490	40		
3-Pyridinecarboxylic acid	$C_6H_5NO_2$	123.110	24	1.63	16.6	66		
			52	3.40	35.2	66		
			72	5.20	54.9	66		
Pyrocatechol	$C_6H_6O_2$	110.111	20	31.1	451	27		
Pyrrole	C_4H_5N	67.090	25	4.5	47	10		

Name	Mol. Form.	Mol. Wt.	$t/°C$	Solubility, s			Henry Const., k_H	
				$100\,w_2$ (mass%)	g per kg H_2O	Ref.	kPa m^3mol^{-1}	Ref.
Quinic acid	$C_7H_{12}O_6$	192.166	9	29	410	26		
Quinidine	$C_{20}H_{24}N_2O_2$	324.417	20	0.020	0.20	27		
Quinine	$C_{20}H_{24}N_2O_2$	324.417	25	0.057	0.57	27		
Quinoline	C_9H_7N	129.159	20	0.633	6.33	6		
8-Quinolinol	C_9H_7NO	145.158	25	0.065	0.65	40		
Quinoxaline	$C_8H_6N_2$	130.147	50	54	1170	6		
Raffinose	$C_{18}H_{32}O_{16}$	504.437	20	12.5	143	27		
Reserpine	$C_{33}H_{40}N_2O_9$	608.679	30	0.0073	0.073	40		
Resorcinol	$C_6H_6O_2$	110.111	20	63.7	1750	27		
Riboflavin	$C_{17}H_{20}N_4O_6$	376.364	25	0.0075	0.075	40		
Ronnel	$C_8H_8Cl_3O_3PS$	321.546	20	0.00011	0.0011	40		
Rotenone	$C_{23}H_{22}O_6$	394.417	25	0.000017	0.00017	40		
Saccharin	$C_7H_5NO_3S$	183.185	25	0.40	4.0	27		
			100	4.0	42	27		
Salicylaldehyde	$C_7H_6O_2$	122.122	86	1.68	17.1	10		
Sarcosine	$C_3H_7NO_2$	89.094	25	30.0	429	26		
L-Serine	$C_3H_7NO_3$	105.093	25	20	250	26		
Shikimic acid	$C_7H_{10}O_5$	174.151		15	176	26		
Silvex	$C_9H_7Cl_3O_3$	269.509	25	0.014	0.14	40		
Solanine	$C_{45}H_{73}NO_{15}$	868.060	15	0.0026	0.026	40		
L-Sorbose	$C_6H_{12}O_6$	180.155	17	≈26	≈350	40		
Stearic acid	$C_{18}H_{36}O_2$	284.478	20	0.00029	0.0029	26		
trans-Stilbene	$C_{14}H_{12}$	180.245	25	0.000029	0.00029	42,4	0.040	12
Streptozotocin	$C_8H_{15}N_3O_7$	265.221	25	0.50	5.0	40		
Strychnine	$C_{21}H_{22}N_2O_2$	334.412	20	0.013	0.13	27		
Styrene	C_8H_8	104.150	25	0.032	0.32	22	0.286	22
			50	0.046	0.46	4,89	0.30	13
Succinamide	$C_4H_8N_2O_2$	116.119	50	18.4	225	27		
Succinic acid	$C_4H_6O_4$	118.089	25	7.71	83.5	27		
			100	55	1220	27		
Succinonitrile	$C_4H_4N_2$	80.088	25	11.5	130	10		
Sucrose	$C_{12}H_{22}O_{11}$	342.296	20	67.1	2040	27		
			50	72.3	2610	27		
			100	83.0	4880	27		
Sulfamethazine	$C_{12}H_{14}N_4O_2S$	278.330	20	0.053	0.53	40		
Sulfamethoxazole	$C_{10}H_{11}N_3O_3S$	253.277	25	0.0281	0.281	40		
Sulfathiazole	$C_9H_9N_3O_2S_2$	255.316	20	0.048	0.48	40		
Sulfisoxazole	$C_{11}H_{13}N_3O_3S$	267.304	37	0.03	0.3	40		
DL-Tartaric acid	$C_4H_6O_6$	150.087	0	8.95	98.3	26		
			20	17.1	206	26		
			100	65	1860	26		
L-Tartaric acid	$C_4H_6O_6$	150.087	20	58	1380	26		
			100	77	3350	26		
Tebuthiuron	$C_9H_{16}N_4OS$	228.314	20	0.23	2.3	40		
Terbacil	$C_9H_{13}ClN_2O_2$	216.664	25	0.071	0.71	40		
Terephthalic acid	$C_8H_6O_4$	166.132	10	0.0082	0.082	76		
			25	0.0065	0.065	76		
			50	0.0074	0.074	76		
o-Terphenyl	$C_{18}H_{14}$	230.304	25	0.000124	0.00124	42,40		

| Name | Mol. Form. | Mol. Wt. | $t/°C$ | Solubility, s | | | Henry Const., k_H | |
				$100\,w_2$ (mass%)	g per kg H_2O	Ref.	kPa m^3mol^{-1}	Ref.
m-Terphenyl	$C_{18}H_{14}$	230.304	25	0.000152	0.00152	42,40		
p-Terphenyl	$C_{18}H_{14}$	230.304	25	0.00000180	0.000018	42,40		
α-Terpineol	$C_{10}H_{18}O$	154.249	25	0.189	1.89	52		
1,2,4,5-Tetrabromobenzene	$C_6H_2Br_4$	393.696	25	0.00000434	0.0000434	2		
1,1,2,2-Tetrabromoethane	$C_2H_2Br_4$	345.653	0	0.052	0.52	25		
			25	0.068	0.68	25		
			50	0.106	1.06	25		
			100	0.307	3.07	25		
Tetrabromomethane	CBr_4	331.627	30	0.024	0.24	14		
1,2,3,4-Tetrachlorobenzene	$C_6H_2Cl_4$	215.892	25	0.0007	0.007	41	0.144	11
1,2,3,5-Tetrachlorobenzene	$C_6H_2Cl_4$	215.892	25	0.00035	0.0035	41	0.59	11
1,2,4,5-Tetrachlorobenzene	$C_6H_2Cl_4$	215.892	25	0.000060	0.00060	41	0.122	11
3,4,5,6-Tetrachloro-1,2-benzenediol	$C_6H_2Cl_4O_2$	247.891	25	0.071	0.71	8		
2,2',4',5-Tetrachlorobiphenyl	$C_{12}H_6Cl_4$	291.988	25	0.0000016	0.000016	9		
2,3,4,5-Tetrachlorobiphenyl	$C_{12}H_6Cl_4$	291.988	25	0.000002	0.00002	7		
2,3,5,6-Tetrachloro-2,5-cyclohexadiene-1,4-dione	$C_6Cl_4O_2$	245.875	20	0.025	0.25	40		
2,3,7,8-Tetrachlorodibenzo-p-dioxin	$C_{12}H_4Cl_4O_2$	321.971	22	0.0000000019	0.000000019	40		
1,1,2,2-Tetrachloro-1,2-difluoroethane	$C_2Cl_4F_2$	203.830	27	0.016	0.16	25		
1,1,1,2-Tetrachloroethane	$C_2H_2Cl_4$	167.849	0	0.120	1.20	25		
			25	0.107	1.07	25	0.24	13
			50	0.123	1.23	25		
1,1,2,2-Tetrachloroethane	$C_2H_2Cl_4$	167.849	5	0.302	3.02	25		
			25	0.283	2.83	25	0.026	13
			50	0.318	3.18	25		
Tetrachloroethene	C_2Cl_4	165.833	0	0.0273	0.273	20		
			20	0.0286	0.286	20	1.73	13
			80	0.0380	0.380	20		
Tetrachloromethane	CCl_4	153.823	25	0.065	0.65	20	2.99	13
			75	0.115	1.15	20	2.99	13
2,3,4,6-Tetrachloro-5-methylphenol	$C_7H_4Cl_4O$	245.918	25	0.00061	0.0061	2		
2,3,4,6-Tetrachlorophenol	$C_6H_2Cl_4O$	231.891	25	0.017	0.17	24		
1,1,1,3-Tetrachloro-2,2,3,3-tetrafluoropropane	$C_3Cl_4F_4$	253.838	21	0.0052	0.052	35		
Tetracosane	$C_{24}H_{50}$	338.654	22	0.0000004	0.000004	37		
Tetradecane	$C_{14}H_{30}$	198.388	25	0.00000023	0.0000023	42,5		
Tetradecanoic acid	$C_{14}H_{28}O_2$	228.371	20	0.0020	0.020	26		
1-Tetradecanol	$C_{14}H_{30}O$	214.387	25	0.000031	0.00031	1		
Tetraethylsilane	$C_8H_{20}Si$	144.331	25	0.0000325	0.000325	10		
Tetrafluoroethene	C_2F_4	100.015	25	0.0158*	0.158*	19,50		
			70	0.0090*	0.090*	50		
Tetrafluoromethane	CF_4	88.005	0	0.00390*	0.0390*	50		
			25	0.00185*	0.0185*	50,19		
			50	0.00134*	0.0134*	50		
Tetrahydro-2,5-dimethoxyfuran	$C_6H_{12}O_3$	132.157	21	32	470	20		
			90	19	235	20		
1,2,3,4-Tetrahydronaphthalene	$C_{10}H_{12}$	132.202	20	0.0045	0.045	40		
Tetrahydropyran	$C_5H_{10}O$	86.132	20	8.57	93.7	20		
			81	4.29	44.8	20		
1,2,4,5-Tetramethylbenzene	$C_{10}H_{14}$	134.218	25	0.000348	0.00348	4	2.55	11
N,N,N',N'-Tetramethyl-4,4'-diaminobenzophenone	$C_{17}H_{20}N_2O$	268.353	20	0.04	0.4	40		

Name	Mol. Form.	Mol. Wt.	$t/°C$	Solubility, s			Henry Const., k_H	
				$100\,w_2$ (mass%)	g per kg H_2O	Ref.	kPa m^3mol^{-1}	Ref.
Tetramethylsilane	$C_4H_{12}Si$	88.224	25	0.00196	0.0196	10		
Theophylline	$C_7H_8N_4O_2$	180.165	20	0.52	5.2	29		
Thioacetamide	C_2H_5NS	75.133	25	12.3	140	40		
Thiourea	CH_4N_2S	76.121	20	10.6	119	40		
			80	≈37	≈590	40		
2-Thioxo-4-thiazolidinone	$C_3H_3NOS_2$	133.192	25	0.225	2.25	40		
Thiram	$C_6H_{12}N_2S_4$	240.432	20	0.003	0.03	40		
DL-Threonine	$C_4H_9NO_3$	119.119	10	14.34	167	45		
			20	15.69	186	45		
			40	19.84	248	45		
L-Threonine	$C_4H_9NO_3$	119.119	10	7.34	79.2	45		
			20	8.31	90.6	45		
			40	10.78	121	45		
Thymidine	$C_{10}H_{14}N_2O_5$	242.228	25	5.1	54	29		
Thymine	$C_5H_6N_2O_2$	126.114	25	0.35	3.5	29		
Thymol	$C_{10}H_{14}O$	150.217		0.1	1	30		
Tolazamide	$C_{14}H_{21}N_3O_3S$	311.400	30	0.0065	0.065	40		
Tolbutamide	$C_{12}H_{18}N_2O_3S$	270.347	25	0.011	0.11	40		
o-Tolidine	$C_{14}H_{16}N_2$	212.290	25	0.13	1.3	40		
Toluene	C_7H_8	92.139	5	0.054	0.54	61		
			25	0.0519	0.519	61,22	0.660	22
			45	0.063	0.63	61		
			90	0.12	1.2	22		
p-Toluenesulfonic acid	$C_7H_8O_3S$	172.202	40	≈33	≈490	40		
o-Toluic acid	$C_8H_8O_2$	136.149	25	0.118	1.18	27		
m-Toluic acid	$C_8H_8O_2$	136.149	25	0.098	0.98	27		
p-Toluic acid	$C_8H_8O_2$	136.149	10	0.030	0.30	75		
			25	0.036	0.36	75		
			50	0.089	0.89	75		
1,3,5-Triazine-2,4,6-triamine	$C_3H_6N_6$	126.120	20	0.323	3.23	40		
			95	4.2	44	40		
1H-1,2,4-Triazol-3-amine	$C_2H_4N_4$	84.080	23	22	280	26		
1,2,4-Tribromobenzene	$C_6H_3Br_3$	314.800	25	0.0010	0.010	2		
1,3,5-Tribromobenzene	$C_6H_3Br_3$	314.800	25	0.0000789	0.000789	2		
1,1,2-Tribromoethane	$C_2H_3Br_3$	266.757	20	0.050	0.50	25		
Tribromofluoromethane	CBr_3F	270.721	25	0.040	0.40	14		
Tribromomethane	$CHBr_3$	252.731	25	0.30	3.0	5	0.047	13
2,4,6-Tribromophenol	$C_6H_3Br_3O$	330.799	15	0.0007	0.007	2		
Tributylamine	$C_{12}H_{27}N$	185.349	25	0.0142	0.142	40		
Tributyl phosphate	$C_{12}H_{27}O_4P$	266.313	25	0.039	0.39	10		
Tributyrin	$C_{15}H_{26}O_6$	302.363	20	0.010	0.10	40		
Trichloroacetaldehyde	C_2HCl_3O	147.387	25	≈39	≈640	40		
Trichloroacetic acid	$C_2HCl_3O_2$	163.387	25	92.3	11990	27		
1,2,3-Trichlorobenzene	$C_6H_3Cl_3$	181.447	25	0.0021	0.021	41	0.242	11
1,2,4-Trichlorobenzene	$C_6H_3Cl_3$	181.447	15	0.0029	0.029	61		
			25	0.0037	0.037	61,41	0.277	11
			45	0.0047	0.047	61		
1,3,5-Trichlorobenzene	$C_6H_3Cl_3$	181.447	25	0.0008	0.008	41	1.1	11
3,4,5-Trichloro-1,2-benzenediol	$C_6H_3Cl_3O_2$	213.446	25	0.051	0.51	8		

| Name | Mol. Form. | Mol. Wt. | $t/°C$ | Solubility, s | | Ref. | Henry Const., k_H | Ref. |
				$100\,w_2$ (mass%)	g per kg H_2O		kPa m^3mol^{-1}	
2,4,5-Trichlorobiphenyl	$C_{12}H_7Cl_3$	257.543	25	0.000014	0.00014	7	0.0379	31
2,4,6-Trichlorobiphenyl	$C_{12}H_7Cl_3$	257.543	25	0.00002	0.0002	7	0.0495	7
1,1,1-Trichloro-2,2-bis(4-chlorophenyl)ethane	$C_{14}H_9Cl_5$	354.486	25	0.0000004	0.000004	67		
2,4,6-Trichloro-3,5-dimethylphenol	$C_8H_7Cl_3O$	225.500	25	0.00050	0.0050	2		
1,1,1-Trichloroethane	$C_2H_3Cl_3$	133.404	0	0.134	1.34	25		
			25	0.129	1.29	25	1.76	13
			50	0.138	1.38	25		
1,1,2-Trichloroethane	$C_2H_3Cl_3$	133.404	0	0.425	4.25	25		
			25	0.459	4.59	25	0.092	13
			50	0.536	5.36	25		
Trichloroethene	C_2HCl_3	131.388	0	0.145	1.45	25		
			25	0.128	1.28	25	1.03	13
			60	0.133	1.33	25		
Trichlorofluoromethane	CCl_3F	137.368	20	0.11	1.1	5	10.2	13
Trichloromethane	$CHCl_3$	119.378	25	0.80	8.0	20	0.43	13
			59	0.79	7.9	20	0.43	13
1,2,4-Trichloro-5-methylbenzene	$C_7H_5Cl_3$	195.474	25	0.00023	0.0023	61		
(Trichloromethyl)benzene	$C_7H_5Cl_3$	195.474	5	0.0053	0.053	10		
2,4,6-Trichloro-3-methylphenol	$C_7H_5Cl_3O$	211.473	25	0.0112	0.112	2		
Trichloronitromethane	CCl_3NO_2	164.376	0	0.227	2.27	40		
			25	0.162	1.62	40		
1,1,1-Trichloro-2,2,3,3,3-pentafluoropropane	$C_3Cl_3F_5$	237.383	21	0.0058	0.058	35		
2,4,5-Trichlorophenol	$C_6H_3Cl_3O$	197.446	25	0.1	1	2		
2,4,6-Trichlorophenol	$C_6H_3Cl_3O$	197.446	25	0.069	0.69	48,51,24		
2,4,5-Trichlorophenoxyacetic acid	$C_8H_5Cl_3O_3$	255.483	25	0.028	0.28	40		
1,2,3-Trichloropropane	$C_3H_5Cl_3$	147.431	10	0.14	1.4	35		
			25	0.20	2.0	35	0.038	13
1,1,2-Trichloro-1,2,2-trifluoroethane	$C_2Cl_3F_3$	187.375	25	0.017	0.17	25	32	13
Tri-p-cresyl phosphate	$C_{21}H_{21}O_4P$	368.363	25	0.00004	0.0004	40		
Tridecane	$C_{13}H_{28}$	184.361	25	0.000000033	0.00000033	37		
Tridecanoic acid	$C_{13}H_{26}O_2$	214.344	20	0.0033	0.033	26		
Triethylamine	$C_6H_{15}N$	101.190	20	5.5	58	10		
Triethylamine hydrochloride	$C_6H_{16}ClN$	137.651	25	57.8	1370	27		
Trifluoromethane	CHF_3	70.014	25	0.15*	1.5*	50,14		
3,4,5-Trihydroxybenzoic acid	$C_7H_6O_5$	170.120	25	1.52	15.4	74		
			50	3.82	39.7	74		
			100	25.0	333	27		
Triiodomethane	CHI_3	393.732	25	0.012	0.12	14		
Trimethoprim	$C_{14}H_{18}N_4O_3$	290.318	25	0.04	0.4	40		
1,2,3-Trimethylbenzene	C_9H_{12}	120.191	25	0.0070	0.070	22	0.343	22
1,2,4-Trimethylbenzene	C_9H_{12}	120.191	25	0.0057	0.057	22	0.569	22
1,3,5-Trimethylbenzene	C_9H_{12}	120.191	25	0.0050	0.050	22	0.781	22
2,3,3-Trimethyl-2-butanol	$C_7H_{16}O$	116.201	40	2.2	22	1		
1,1,3-Trimethylcyclohexane	C_9H_{18}	126.239	25	0.000177	0.00177	4	105	13
1,1,3-Trimethylcyclopentane	C_8H_{16}	112.213	25	0.00037	0.0037	4	159	5
2,2,5-Trimethylhexane	C_9H_{20}	128.255	25	0.00008	0.0008	4	246	13
1,4,5-Trimethylnaphthalene	$C_{13}H_{14}$	170.250	25	0.00021	0.0021	42,4		
2,6,8-Trimethyl-4-nonanone	$C_{12}H_{24}O$	184.318	10	0.012	0.12	20		
			80	0.014	0.14	20		

| Name | Mol. Form. | Mol. Wt. | $t/°C$ | Solubility, s | | Ref. | Henry Const., k_H | Ref. |
				$100\,w_2$ (mass%)	g per kg H_2O		kPa m^3mol^{-1}	
2,2,4-Trimethylpentane	C_8H_{18}	114.229	25	0.00022	0.0022	4	307	13
2,3,4-Trimethylpentane	C_8H_{18}	114.229	25	0.00018	0.0018	4	206	13
Trimethyl phosphate	$C_3H_9O_4P$	140.074	25	≈33	≈490	40		
1,3,5-Trinitrobenzene	$C_6H_3N_3O_6$	213.104	15	0.028	0.28	40		
2,4,6-Trinitrobenzoic acid	$C_7H_3N_3O_8$	257.114	23	1.97	20.1	40		
Trinitroglycerol	$C_3H_5N_3O_9$	227.087	25	0.13	1.3	40		
			80	0.34	3.4	40		
2,4,6-Trinitrophenol	$C_6H_3N_3O_7$	229.104	25	1.25	12.7	40		
			90	4.9	52	40		
2,4,6-Trinitrotoluene	$C_7H_5N_3O_6$	227.131	20	0.012	0.12	40		
			100	0.15	1.5	40		
2,4,6-Trinitro-N-(2,4,6-trinitrophenyl)aniline	$C_{12}H_5N_7O_{12}$	439.208	17	0.0060	0.060	40		
1,3,5-Trioxane	$C_3H_6O_3$	90.078	25	17.4	211	30		
Triphenylene	$C_{18}H_{12}$	228.288	25	0.0000043	0.000043	42,4	0.00001	12
Triphenyl phosphate	$C_{18}H_{15}O_4P$	326.283	24	0.000073	0.00073	40		
Triphenyltin hydroxide	$C_{18}H_{16}OSn$	367.029	20	0.0001	0.001	32		
Tris(hydroxymethyl)methylamine	$C_4H_{11}NO_3$	121.135	25	≈41	≈700	40		
L-Tryptophan	$C_{11}H_{12}N_2O_2$	204.225	25	1.30	13.2	26		
DL-Tyrosine	$C_9H_{11}NO_3$	181.188	25	0.35	3.5	30		
L-Tyrosine	$C_9H_{11}NO_3$	181.188	25	0.0507	0.507	62		
Undecane	$C_{11}H_{24}$	156.309	25	0.0000004	0.000004	37		
Uracil	$C_4H_4N_2O_2$	112.087	25	0.460	4.62	72		
Urea	CH_4N_2O	60.055	5	44	790	26		
			25	54.4	1200	26		
Uric acid	$C_5H_4N_4O_3$	168.111	20	0.002	0.02	26		
L-Valine	$C_5H_{11}NO_2$	117.147	25	8.13	88.5	26		
Valium	$C_{16}H_{13}ClN_2O$	284.739	25	0.005	0.05	40		
Vidarabine	$C_{10}H_{15}N_5O_5$	285.257	20	0.051	0.51	40		
Vinclozolin	$C_{12}H_9Cl_2NO_3$	286.110	20	0.1	1	32		
Vinyl acetate	$C_4H_6O_2$	86.090	20	2.0	20	10		
4-Vinylcyclohexene	C_8H_{12}	108.181	25	0.005	0.05	4		
Warfarin	$C_{19}H_{16}O_4$	308.328	20	0.004	0.04	40		
Xanthine	$C_5H_4N_4O_2$	152.112	20	0.05	0.5	26		
o-Xylene	C_8H_{10}	106.165	25	0.0171	0.171	22	0.551	22
			45	0.021	0.21	4		
m-Xylene	C_8H_{10}	106.165	0	0.0203	0.203	4		
			25	0.0161	0.161	22	0.730	22
			40	0.022	0.22	4		
p-Xylene	C_8H_{10}	106.165	0	0.0160	0.160	4		
			25	0.0181	0.181	22	0.690	22
			40	0.022	0.22	4		
2,3-Xylenol	$C_8H_{10}O$	122.164	25	0.457	4.57	40		
2,4-Xylenol	$C_8H_{10}O$	122.164	25	0.787	7.87	10		
2,5-Xylenol	$C_8H_{10}O$	122.164	25	0.354	3.54	40		
2,6-Xylenol	$C_8H_{10}O$	122.164	25	0.60	6.0	40		
3,4-Xylenol	$C_8H_{10}O$	122.164	25	0.477	4.77	40		
3,5-Xylenol	$C_8H_{10}O$	122.164	29	0.62	6.2	10		
D-Xylose	$C_5H_{10}O_5$	150.130	25	≈30	≈430	40		
Ziram	$C_6H_{12}N_2S_4Zn$	305.841	20	0.0065	0.065	40		

AQUEOUS SOLUBILITY OF INORGANIC COMPOUNDS AT VARIOUS TEMPERATURES

The solubility of over 300 common inorganic compounds in water is tabulated here as a function of temperature. Solubility is defined as the concentration of the compound in a solution that is in equilibrium with a solid phase at the specified temperature. In this table the solid phase is generally the most stable crystalline phase at the temperature in question. An asterisk * on solubility values in adjacent columns indicates that the solid phase changes between those two temperatures (usually from one hydrated phase to another or from a hydrate to the anhydrous solid). In such cases the slope of the solubility vs. temperature curve may show a discontinuity.

All solubility values are expressed as mass percent of solute, $100 \cdot w_2$, where

$$w_2 = m_2/(m_1 + m_2)$$

and m_2 is the mass of solute and m_1 the mass of water. This quantity is related to other common measures of solubility as follows:

Molarity: $c_2 = 1000 \, \rho w_2/M_2$
Molality: $m_2 = 1000 w_2/M_2(1-w_2)$
Mole fraction: $x_2 = (w_2/M_2)/\{(w_2/M_2) + (1-w_2)/M_1\}$
Mass of solute per 100 g of H_2O: $r_2 = 100 w_2/(1-w_2)$

Here M_2 is the molar mass of the solute and $M_1 = 18.015$ g/mol is the molar mass of water. ρ is the density of the solution in g cm^{-3}.

The data in the table have been derived from the references indicated; in many cases the data have been refitted or interpolated in order to present solubility at rounded values of temperature. Where available, values were taken from the IUPAC *Solubility Data Series* (Reference 1) or the related papers in the *Journal of Physical and Chemical Reference Data* (References 2 to 5), which present carefully evaluated data.

The solubility of sparingly soluble compounds that do not appear in this table may be calculated from the data in the table "Solubility Product Constants." Solubility of inorganic gases may be found in the table "Solubility of Selected Gases in Water."

Compounds are listed alphabetically by chemical formula in the most commonly used form (e.g., NaCl, NH$_4$NO$_3$, etc.).

References

1. *Solubility Data Series*, International Union of Pure and Applied Chemistry. Volumes 1 to 53 were published by Pergamon Press, Oxford, from 1979 to 1994; subsequent volumes were published by Oxford University Press, Oxford. The number following the colon is the volume number in the series.
2. Clever, H. L., and Johnston, F. J., *J. Phys. Chem. Ref. Data*, 9, 751, 1980.
3. Marcus, Y., *J. Phys. Chem. Ref. Data*, 9, 1307, 1980.
4. Clever, H. L., Johnson, S. A., and Derrick, M. E., *J. Phys. Chem. Ref. Data*, 14, 631, 1985.
5. Clever, H. L., Johnson, S. A., and Derrick, M. E., *J. Phys. Chem. Ref. Data*, 21, 941, 1992.
6. Söhnel, O., and Novotny, P., *Densities of Aqueous Solutions of Inorganic Substances*, Elsevier, Amsterdam, 1985.
7. Krumgalz, B.S., *Mineral Solubility in Water at Various Temperatures*, Israel Oceanographic and Limnological Research Ltd., Haifa, 1994.
8. Potter, R. W., and Clynne, M. A., *J. Research U.S. Geological Survey*, 6, 701, 1978; Clynne, M. A., and Potter, R. W., *J. Chem. Eng. Data*, 24, 338, 1979.
9. Marshal, W. L., and Slusher, R., *J. Phys. Chem.*, 70, 4015, 1966; Knacke, O., and Gans, W., *Zeit. Phys. Chem.*, NF, 104, 41, 1977.
10. Stephen, H., and Stephen, T., *Solubilities of Inorganic and Organic Compounds, Vol. 1*, Macmillan, New York, 1963.

Compound	0 °C	10 °C	20 °C	25 °C	30 °C	40 °C	50 °C	60 °C	70 °C	80 °C	90 °C	100 °C	Ref.
AgBrO$_3$				0.193							1.32		7
AgClO$_2$	0.17	0.31	0.47	0.55	0.64	0.82	1.02	1.22	1.44	1.66	1.88	2.11	7
AgClO$_3$				15									7
AgClO$_4$	81.6	83.0	84.2	84.8	85.3	86.3	86.9	87.5	87.9	88.3	88.6	88.8	6
AgNO$_2$	0.155			0.413									7
AgNO$_3$	55.9	62.3	67.8	70.1	72.3	76.1	79.2	81.7	83.8	85.4	86.7	87.8	6
Ag$_2$SO$_4$	0.56	0.67	0.78	0.83	0.88	0.97	1.05	1.13	1.20	1.26	1.32	1.39	7
AlCl$_3$	30.84	30.91	31.03	31.10	31.18	31.37	31.60	31.87	32.17	32.51	32.90	33.32	7
Al(ClO$_4$)$_3$	54.9										64.4		7
AlF$_3$	0.25	0.34	0.44	0.50	0.56	0.68	0.81	0.96	1.11	1.28	1.45	1.64	7
Al(NO$_3$)$_3$	37.0	38.2	39.9	40.8	42.0	44.5	47.3	50.4	53.8*			61.5*	6
Al$_2$(SO$_4$)$_3$	27.5			27.8	28.2	29.2	30.7	32.6	34.9	37.6	40.7	44.2	7
As$_2$O$_3$	1.19	1.48	1.80	2.01	2.27	2.86	3.43	4.11	4.89	5.77	6.72	7.71	10
BaBr$_2$	47.6	48.5	49.5	50.0	50.4	51.4	52.5	53.5	54.5	55.5	56.6	57.6	6
Ba(BrO$_3$)$_2$	0.285	0.442	0.656	0.788	0.935	1.30	1.74	2.27	2.90	3.61	4.40	5.25	1:14
Ba(C$_2$H$_3$O$_2$)$_2$	37.0			44.2									7
BaCl$_2$	23.30	24.88	26.33	27.03	27.70	29.00	30.27	31.53	32.81	34.14	35.54	37.05	8
Ba(ClO$_2$)$_2$	30.5			31.3								44.7	7
Ba(ClO$_3$)$_2$	16.90	21.23	23.66	27.50	29.43	33.16	36.69	40.05	43.04	45.90	48.70	51.17	1:14
Ba(ClO$_4$)$_2$	67.30	70.96	74.30	75.75	77.05	79.23	80.92	82.21	83.16	83.88	84.43	84.90	7
BaF$_2$		0.158		0.161									7
BaI$_2$	62.5	64.7	67.3	68.8	69.1	69.5	70.1	70.7	71.3	72.0	72.7	73.4	6

Compound	0 °C	10 °C	20 °C	25 °C	30 °C	40 °C	50 °C	60 °C	70 °C	80 °C	90 °C	100 °C	Ref.
$Ba(IO_3)_2$	0.0182	0.0262	0.0342	0.0396	0.045*	0.058*	0.073	0.090	0.109	0.131	0.156	0.182	1:14
$Ba(NO_2)_2$	31.1	36.6	41.8	44.3	46.8	51.6	56.2	60.5	64.6	68.5	72.1	75.6	10
$Ba(NO_3)_2$	4.7	6.3	8.2	9.3	10.2	12.4	14.7	17.0	19.3	21.5	23.5	25.5	6
$Ba(OH)_2$	1.67			4.68	8.4	19	33	52	74	100			7
BaS	2.79	4.78	6.97	8.21	9.58	12.67	16.18	20.05	24.19	28.55	33.04	37.61	7
$Ba(SCN)_2$				62.6									7
$BaSO_3$				0.0011									1:26
$BeCl_2$	40.5			41.7									7
$Be(ClO_4)_2$				59.5									7
$BeSO_4$	26.69	27.58	28.61	29.22	29.90	31.51	33.39	35.50	37.78	40.21	42.72	45.28	7
$CaBr_2$	55	56	59	61	63	68	71	73					10
$CaCl_2$	36.70	39.19	42.13	44.83*	49.12*	52.85*	56.05*	56.73	57.44	58.21	59.04	59.94	8
$Ca(ClO_3)_2$	63.2	64.2	65.5	66.3	67.2	69.0	71.0	73.2	75.5*	77.4*	77.7	78.0	1:14
$Ca(ClO_4)_2$				65.3									7
CaF_2	0.0013			0.0016									10
CaI_2	64.6	66.0	67.6	68.3	69.0	70.8	72.4	74.0	76.0	78.0	79.6	81.0	7
$Ca(IO_3)_2$	0.082	0.155	0.243	0.305	0.384*	0.517*	0.590	0.652	0.811*	0.665*	0.668		1:14
$Ca(NO_2)_2$	38.6	39.5	44.5	48.6									7
$Ca(NO_3)_2$	50.1	53.1	56.7	59.0	60.9	65.4	77.8	78.1	78.2	78.3	78.4	78.5	6
$CaSO_3$			0.0059	0.0054	0.0049	0.0041	0.0035	0.0030	0.0026	0.0023	0.0020	0.0019	1:26
$CaSO_4$	0.174	0.191	0.202	0.205	0.208	0.210	0.207	0.201	0.193	0.184	0.173	0.163	9
$CdBr_2$	36.0	43.0	49.9	53.4	56.4	60.3*	60.3*	60.5	60.7	60.9	61.3	61.6	6
CdC_2O_4				0.0060									5
$CdCl_2$	47.2	50.1	53.2	54.6	56.3*	57.3*	57.5	57.8	58.1	58.51	58.98	59.5	6
$Cd(ClO_4)_2$				58.7								66.9	7
CdF_2		5.82	4.65	4.18	3.76								5
CdI_2	44.1	44.9	45.8	46.3	46.8	47.9	49.0	50.2	51.5	52.7	54.1	55.4	6
$Cd(IO_3)_2$				0.091									5
$Cd(NO_3)_2$	55.4	57.1	59.6	61.0	62.8	66.5	70.6	86.1	86.5	86.8	87.1	87.4	6
$CdSO_4$	43.1	43.1	43.2	43.4	43.6	44.1	43.5	42.5	41.4	40.2	38.5	36.7	6
$CdSeO_4$	42.04	40.59	39.02	38.18	37.29	35.35	33.15	30.65	27.84	24.69	21.24	17.49	5
$Ce(NO_3)_3$	57.99	59.80	61.89	63.05	64.31*	67.0*	68.6	71.1*	74.9*	79.2	80.9	83.1	1:13
$CoCl_2$	30.30	32.60	34.87	35.99	37.10	39.27	41.38	43.46	45.50	47.51	49.51	51.50	7
$Co(ClO_4)_2$	50.0			53.0									7
CoF_2				1.4									7
CoI_2	58.00	61.78	65.35	66.99	68.51	71.17	73.41	75.29	76.89	78.28	79.52	80.70	7
$Co(NO_2)_2$	0.076			0.49									7
$Co(NO_3)_2$	45.5	47.0	49.4	50.8	52.4	56.0	60.1	62.6	64.9	67.7			6
$CoSO_4$	19.9	23.0	26.1	27.7	29.2	32.3	34.4	35.9	35.5	33.2	30.6	27.8	6
$Co(SCN)_2$				50.7									7
CrO_3	62.2	62.3	62.6	62.8	63.0	63.5	64.1	64.7	65.5	66.2	67.1	67.9	6
$CsBr$				55.2									7
$CsBrO_3$	1.16	1.93	3.01	3.69	4.46	6.32	8.60	11.32	14.45	17.96	21.83	25.98	1:30
$CsCl$	61.83	63.48	64.96	65.64	66.29	67.50	68.60	69.61	70.54	71.40	72.21	72.96	1:47
$CsClO_3$	2.40	3.87	5.94	7.22	8.69	12.15	16.33	21.14	26.45	32.10	37.89	43.42	1:30
$CsClO_4$	0.79	1.01	1.51	1.96	2.57	4.28	6.55	9.29	12.41	15.80	19.39	23.07	7
CsI	30.9	37.2	43.2	45.9	48.6	53.3	57.3	60.7	63.6	65.9	67.7	69.2	6
$CsIO_3$	1.08	1.58	2.21	2.59	3.02	3.96	5.06	6.29	7.70	9.20	10.79	12.45	1:30
$CsNO_3$	8.46	13.0	18.6	21.8	25.1	32.0	39.0	45.7	51.9	57.3	62.1	66.2	6
$CsOH$				75									7
Cs_2SO_4	62.6	63.4	64.1	64.5	64.8	65.5	66.1	66.7	67.3	67.8	68.3	68.8	6
$CuBr_2$				55.8									7
$CuCl_2$	40.8	41.7	42.6	43.1	43.7	44.8	46.0	47.2	48.5	49.9	51.3	52.7	6
$Cu(ClO_4)_2$	54.3			59.3									7
CuF_2				0.075									7
$Cu(NO_3)_2$	45.2	49.8	56.3	59.2	61.1	62.0	63.1	64.5	65.9	67.5	69.2	71.0	6
$CuSO_4$	12.4	14.4	16.7	18.0	19.3	22.2	25.4	28.8	32.4	36.3	40.3	43.5	6
$CuSeO_4$	10.6			16.0									7
$Dy(NO_3)_3$	58.79	59.99	61.49	62.35	63.29	65.43	68.04	71.58					1:13
$Er(NO_3)_3$	61.58	63.15	64.84	65.75	66.69	68.70	70.96	73.64	77.75				1:13
$Eu(NO_3)_3$	55.2	56.7	58.5	59.4	60.4	62.5	64.6						1:13

Compound	0 °C	10 °C	20 °C	25 °C	30 °C	40 °C	50 °C	60 °C	70 °C	80 °C	90 °C	100 °C	Ref.
$FeBr_2$				54.6								64.8*	7
$FeCl_2$	33.2*			39.4*								48.7*	7
$FeCl_3$	42.7	44.9	47.9	47.7	51.6	74.8	76.7	84.6	84.3	84.3	84.4	84.7	6
$Fe(ClO_4)_2$	63.39			67.76									7
FeF_3				5.59									7
$Fe(NO_3)_3$	40.15			46.57									7
$Fe(NO_3)_2$	41.44			46.67									7
$FeSO_4$	13.5	17.0	20.8	22.8	24.8	28.8	32.8	35.5	33.6	30.4	27.1	24.0	6
$Gd(NO_3)_3$	56.3	57.7	59.2	60.1	61.0	62.9	65.2	67.9	71.5				1:13
HIO_3	73.45	74.10	74.98	75.48	76.03	77.20	78.46	79.78	81.13	82.48	83.82	85.14	1:30
H_3BO_3	2.61	3.57	4.77	5.48	6.27	8.10	10.3	12.9	15.9	19.3	23.1	27.3	6
$HgBr_2$	0.26	0.37	0.52	0.61	0.72	0.96	1.26	1.63	2.08	2.61	3.23	3.95	4
$Hg(CN)_2$	6.57	7.83	9.33	10.2	11.1	13.1	15.5	18.2	21.2	24.6	28.3	32.3	6
$HgCl_2$	4.24	5.05	6.17	6.81	7.62	9.53	12.02	15.18	19.16	24.06	29.90	36.62	4
HgI_2			0.0041	0.0055	0.0072	0.0122	0.0199						4
$Hg(SCN)_2$				0.070									4
Hg_2Cl_2				0.0004									3
$Hg_2(ClO_4)_2$	73.8			79.8*								85.3*	7
Hg_2SO_4	0.038	0.043	0.048	0.051	0.054	0.059	0.065	0.070	0.076	0.082	0.088	0.093	4
$Ho(NO_3)_3$				63.8									1:13
KBF_4	0.28	0.34	0.45	0.55	0.75	1.38	2.09	2.82	3.58	4.34	5.12	5.90	10
KBr	35.0	37.3	39.4	40.4	41.4	43.2	44.8	46.2	47.6	48.8	49.8	50.8	6
$KBrO_3$	2.97	4.48	6.42	7.55	8.79	11.57	14.71	18.14	21.79	25.57	29.42	33.28	1:30
$KC_2H_3O_2$	68.40	70.29	72.09	72.92	73.70	75.08	76.27	77.31	78.22	79.04	79.80	80.55	7
KCl	21.74	23.61	25.39	26.22	27.04	28.59	30.04	31.40	32.66	33.86	34.99	36.05	1:47
$KClO_3$	3.03	4.67	6.74	7.93	9.21	12.06	15.26	18.78	22.65	26.88	31.53	36.65	1:30
$KClO_4$	0.70	1.10	1.67	2.04	2.47	3.54	4.94	6.74	8.99	11.71	14.94	18.67	6
KF	30.90	39.8	47.3	50.41	53.2				60.0				7
$KHCO_3$	18.62	21.73	24.92	26.6	28.13	31.32	34.46	37.51	40.45				6
$KHSO_4$	27.1	29.7	32.3	33.6	35.0	37.8	40.5	43.4	46.2	49.02	51.82	54.6	6
KH_2PO_4	11.74	14.91	18.25	19.97	21.77	25.28	28.95	32.76	36.75	40.96	45.41	50.12	1:31
KI	56.0	57.6	59.0	59.7	60.4	61.6	62.8	63.8	64.8	65.7	66.6	67.4	6
KIO_3	4.53	5.96	7.57	8.44	9.34	11.09	13.22	15.29	17.41	19.58	21.78	24.03	1:30
KIO_4	0.16	0.22	0.37	0.51	0.70	1.24	1.96	2.83	3.82	4.89	6.02	7.17	7
$KMnO_4$	2.74	4.12	5.96	7.06	8.28	11.11	14.42	18.16					6
KNO_2	73.7	74.6	75.3	75.7	76.0	76.7	77.4	78.0	78.5	79.1	79.6	80.1	6
KNO_3	12.0	17.6	24.2	27.7	31.3	38.6	45.7	52.2	58.0	63.0	67.3	70.8	6
KOH	48.7	50.8	53.2	54.7	56.1	57.9	58.6	59.5	60.6	61.8	63.1	64.6	6
$KSCN$	63.8	66.4	69.1	70.4	71.6	74.1	76.5	78.9	81.1	83.3	85.3	87.3	6
K_2CO_3	51.3	51.7	52.3	52.7	53.1	54.0	54.9	56.0	57.2	58.4	59.6	61.0	6
K_2CrO_4	37.1	38.1	38.9	39.4	39.8	40.5	41.3	41.9	42.6	43.2	43.8	44.3	6
$K_2Cr_2O_7$	4.30	7.12	10.9	13.1	15.5	20.8	26.3	31.7	36.9	41.5	45.5	48.9	6
K_2HAsO_4	48.5*			63.6*								79.8*	7
K_2HPO_4	57.0	59.1	61.5	62.7	64.1	67.7*		72.7*					1:31
K_2MoO_4				64.7							66.5		7
K_2SO_3	51.30	51.39	51.49	51.55	51.62	51.76	51.93	52.11	52.32	52.54	52.79	53.06	1:26
K_2SO_4	7.11	8.46	9.95	10.7	11.4	12.9	14.2	15.5	16.7	17.7	18.6	19.3	6
$K_2S_2O_3$	49.0*			62.3*							75.7*		7
$K_2S_2O_5$	22.1	26.7	31.1	33.1	35.2	39.0	42.6	46.0	49.1	52.0	54.6		1:26
K_2SeO_3	68.4*			68.5*								68.5*	7
K_2SeO_4	52.70	52.93	53.17	53.30	53.43	53.70	53.99	54.30	54.61	54.94	55.26	55.60	7
K_3AsO_4	51.5*			55.6*								73*	7
$K_3Fe(CN)_6$	23.9	27.6	31.1	32.8	34.3	37.2	39.6	41.7	43.5	45.0	46.1	47.0	6
K_3PO_4	44.3			51.4									7
$K_4Fe(CN)_6$	12.5	17.3	22.0	23.9	25.6	29.2	32.5	35.5	38.2	40.6	41.4	43.1	6
$LaCl_3$	49.0	48.5	48.6	48.9	49.3	50.5	52.1	54.0	56.3	58.9	61.7		6
$La(NO_3)_3$	55.0	56.9	58.9	60.0	61.1	63.6	66.3	69.9*	74.1*				1:13
$LiBr$	58.4	60.1	62.7	64.4	65.9	67.8	68.3	69.0	69.8	70.7	71.7	72.8	6
$LiBrO_3$	61.03	62.62	64.44	65.44	66.51	68.90	71.68*	73.24*	74.43	75.66	76.93	78.32	1:30
$LiC_2H_3O_2$	23.76	26.49	29.42	31.02	32.72	36.48	40.65	45.15	49.93	54.91	60.04	65.26	7
$LiCl$	40.45	42.46*	45.29*	45.81	46.25	47.30	48.47	49.78	51.27	52.98	54.98*	56.34*	1:47

Compound	0 °C	10 °C	20 °C	25 °C	30 °C	40 °C	50 °C	60 °C	70 °C	80 °C	90 °C	100 °C	Ref.
LiClO$_3$	73.2	75.6*	80.8*	82.1	83.4	85.9*	87.1*	88.2	89.6	91.3	93.4	95.7	1:30
LiClO$_4$	30.1	32.6	35.5	37.0	38.6	41.9	45.5	49.2	53.2	57.2	61.3	71.4	6
LiF	0.120	0.126	0.131	0.134									7
LiH$_2$PO$_4$	55.8												7
LiI	59.4	60.5	61.7	62.3	63.0	64.3	65.8	67.3	68.8	81.3	81.7	82.6	6
LiIO$_3$				43.8									1:30
LiNO$_2$	41	45	49	51	53	56	60	63	66	68			10
LiNO$_3$	34.8	37.6	42.7	50.5	57.9	60.1	62.2	64.0	65.7	67.2	68.5	69.7	6
LiOH	10.8	10.8	11.0	11.1	11.3	11.7	12.2	12.7	13.4	14.2	15.1	16.1	6
LiSCN				54.5									7
Li$_2$CO$_3$	1.54	1.43	1.33	1.28	1.24	1.15	1.07	0.99	0.92	0.85	0.78	0.72	7
Li$_2$C$_2$O$_4$				5.87									7
Li$_2$HPO$_3$	9.07	8.40	7.77	7.47	7.18	6.64	6.16	5.71	5.30	4.91	4.53	4.16	7
Li$_2$SO$_4$	26.3	25.9	25.6	25.5	25.3	25.0	24.8	24.5	24.3	24.0	23.8	23.6	6
Li$_3$PO$_4$				0.027									1:31
Lu(NO$_3$)$_3$				71.1									1:13
MgBr$_2$	49.3	49.8	50.3	50.6	50.9	51.5	52.1	52.8	53.5	54.2	55.0	55.7	6
Mg(BrO$_3$)$_2$	43.0	45.2	48.0	49.4	51.0	54.3	57.9	61.6	65.3	69.0*	70.9*	71.7	1:14
Mg(C$_2$H$_3$O$_2$)$_2$	36.18	37.55	38.92	39.61									7
MgC$_2$O$_4$				0.038									7
MgCl$_2$	33.96	34.85	35.58	35.90	36.20	36.77	37.34	37.97	38.71	39.62	40.75	42.15	8
Mg(ClO$_3$)$_2$	53.35	54.40	56.81	58.66	60.91*	65.46*	67.33	69.27	71.01	72.44	73.48		1:14
Mg(ClO$_4$)$_2$	47.8	48.7	49.6	50.1	50.5	51.3	52.1						6
MgCrO$_4$	32.06*			35.39*									7
MgCr$_2$O$_7$				58.9						67.0			7
MgF$_2$				0.013									7
MgI$_2$	54.7	56.1	58.2	59.4	60.8	63.9	65.0	65.0	65.0	65.0	65.1	65.2	6
Mg(IO$_3$)$_2$	3.19*	6.70*	7.92	8.52	9.11	10.45	11.99	13.7	15.6	17.6	19.6		1:14
Mg(NO$_2$)$_2$				47									7
Mg(NO$_3$)$_2$	38.4	39.5	40.8	41.6	42.4	44.1	45.9	47.9	50.0	52.2	70.6	72.0	6
MgSO$_3$	0.32	0.37	0.46	0.52	0.61	0.87*	0.85*	0.76	0.69	0.64	0.62	0.60	1:26
MgSO$_4$	18.2	21.7	25.1	26.3	28.2	30.9	33.4	35.6	36.9	35.9	34.7	33.3	6
MgS$_2$O$_3$	30.7			34.1									7
MgSeO$_4$	31.4*			35.7*								47*	7
MnBr$_2$	56.00	57.72	59.39	60.19	60.96	62.41	63.75	65.01	66.19	67.32	68.42	69.50	7
MnCl$_2$	38.7	40.6	42.5	43.6	44.7	47.0	49.4	54.1	54.7	55.2	55.7	56.1	6
MnF$_2$	0.80*			1.01*								0.48	7
Mn(IO$_3$)$_2$				0.27							0.34		7
Mn(NO$_3$)$_2$	50.5			61.7									7
MnSO$_4$	34.6	37.3	38.6	38.9	38.9	37.7	36.3	34.6	32.8	30.8	28.8	26.7	6
NH$_4$Br	37.5	40.2	42.7	43.9	45.1	47.3	49.4	51.3	53.0	54.6	56.1	57.4	7
NH$_4$Cl	22.92	25.12	27.27	28.34	29.39	31.46	33.50	35.49	37.46	39.40	41.33	43.24	1:47
NH$_4$ClO$_4$	10.8	14.1	17.8	19.7	21.7	25.8	29.8	33.6	37.3	40.7	43.8	46.6	6
NH$_4$F	41.7	43.2	44.7	45.5	46.3	47.8	49.3	50.9	52.5	54.1			7
NH$_4$HCO$_3$	10.6	13.7	17.6	19.9	22.4	27.9	34.2	41.4	49.3	58.1	67.6	78.0	7
NH$_4$H$_2$AsO$_4$	25.2	29.0	32.7	34.5	36.3	39.7	43.1	46.2	49.3	52.2	55.0		7
NH$_4$H$_2$PO$_4$	17.8	22.0	26.4	28.8	31.2	36.2	41.6	47.2	53.0	59.2	65.7	72.4	7
NH$_4$I	60.7	62.1	63.4	64.0	64.6	65.8	66.8	67.8	68.7	69.6	70.4	71.1	6
NH$_4$IO$_3$				3.70	4.20	5.64	7.63						1:30
NH$_4$NO$_2$	55.7	59.0	64.9	68.8									7
NH$_4$NO$_3$	54.0	60.1	65.5	68.0	70.3	74.3	77.7	80.8	83.4	85.8	88.2	90.3	6
NH$_4$SCN				64.4						81.1			7
(NH$_4$)$_2$C$_2$O$_4$	2.31	3.11	4.25	4.94	5.73	7.56	9.73	12.2	15.1	18.3	21.8	25.7	7
(NH$_4$)$_2$HPO$_4$	36.4	38.2	40.0	41.0	42.0	44.1	46.2	48.5	50.9	53.3	55.9	58.6	7
(NH$_4$)$_2$S$_2$O$_5$	65.5	67.9	69.8	70.5	71.3	72.3	72.9	73.1					1:26
(NH$_4$)$_2$S$_2$O$_8$	37.00	40.45	43.84	45.49	47.11	50.25	53.28	56.23	59.13	62.00			7
(NH$_4$)$_2$SO$_3$	32.2	34.9	37.7	39.1	40.6	43.7	47.0	50.6	54.5	58.9			1:26
(NH$_4$)$_2$SO$_4$	41.3	42.1	42.9	43.3	43.8	44.7	45.6	46.6	47.5	48.5	49.5	50.5	6
(NH$_4$)$_2$SeO$_3$	49.0	51.1	53.4	54.7	56.0	58.9	62.0	65.4	69.1				7
(NH$_4$)$_2$SeO$_4$				54.02									7
(NH$_4$)$_3$PO$_4$				15.5									7

Compound	0 °C	10 °C	20 °C	25 °C	30 °C	40 °C	50 °C	60 °C	70 °C	80 °C	90 °C	100 °C	Ref.
NaBr	44.4	45.9	47.7	48.6	49.6	51.6	53.7	54.1	54.3	54.5	54.7	54.9	6
NaBrO$_3$	20.0	23.22	26.65	28.28	29.86	32.83	35.55	38.05	40.37	42.52			1:30
NaCHO$_2$	30.8	37.9	45.7	48.7	50.6	52.0	53.5	55.0					6
NaC$_2$H$_3$O$_2$	26.5	28.8	31.8	33.5	35.5	39.9	45.1	58.3	59.3	60.5	61.7	62.9	6
NaCl	26.28	26.32	26.41	26.45	26.52	26.67	26.84	27.03	27.25	27.50	27.78	28.05	1:47
NaClO	22.7			44.4									7
NaClO$_2$				97.0*				95.3*					7
NaClO$_3$	44.27	46.67	49.3	50.1	51.2	53.6	55.5	57.0	58.5	60.5	63.3	67.1	1:30
NaClO$_4$	61.9	64.1	66.2	67.2	68.3	70.4	72.5	74.1	74.7	75.4	76.1	76.7	6
NaF	3.52	3.72	3.89	3.97	4.05	4.20	4.34	4.46	4.57	4.66	4.75	4.82	6
NaHCO$_3$	6.48	7.59	8.73	9.32	9.91	11.13	12.40	13.70	15.02	16.37	17.73	19.10	7
NaHSO$_4$				22.2								33.3	10
NaH$_2$PO$_4$	36.54	41.07	46.00	48.68	51.54	57.89*	61.7*	62.3*	65.9	68.7			1:31
NaI	61.2	62.4	63.9	64.8	65.7	67.7	69.8	72.0	74.7	74.8	74.9	75.1	6
NaIO$_3$	2.43	4.40	7.78*	8.65*	9.60	11.67	13.99	16.52	19.25*	21.1*	22.9	24.7	1:30
NaIO$_4$				12.62									7
NaNO$_2$	41.9	43.4	45.1	45.9	46.8	48.7	50.7	52.8	55.0	57.2	59.5	61.8	6
NaNO$_3$	42.2	44.4	46.6	47.7	48.8	51.0	53.2	55.3	57.5	59.6	61.7	63.8	6
NaOH	30	39	46	50	53	58	63	67	71	74	76	79	10
NaSCN		52.9	57.1	60.2	62.7	63.5	64.2	65.0	65.9	66.9	67.9	69.0	6
Na$_2$B$_4$O$_7$	1.23	1.71	2.50	3.07	3.82	6.02	9.7	14.9	17.1	19.9	23.5	28.0	6
Na$_2$CO$_3$	6.44	10.8	17.9	23.5	28.7	32.8	32.2	31.7	31.3	31.1	30.9	30.9	6
Na$_2$C$_2$O$_4$	2.62	2.95	3.30	3.48	3.65	4.00	4.36	4.71	5.06	5.41	5.75	6.08	6
Na$_2$CrO$_4$	22.6	32.3	44.6	46.7	46.9	48.9	51.0	53.4	55.3	55.5	55.8	56.1	6
Na$_2$Cr$_2$O$_7$	62.1	63.1	64.4	65.2	66.1	68.0	70.1	72.3	74.6	77.0	79.6	80.7	6
Na$_2$HAsO$_4$	5.6*			29.3*								67*	7
Na$_2$HPO$_4$	1.66	4.19	7.51	10.55	16.34*	35.17*	44.64*	45.20	46.81	48.78	50.52	51.53	1:31
Na$_2$MoO$_4$	30.6	38.8	39.4	39.4	39.8	40.3	41.0	41.7	42.6	43.5	44.5	45.5	6
Na$_2$S	11.1	13.2	15.7	17.1	18.6	22.1	26.7	28.1	30.2	33.0	36.4	41.0	6
Na$_2$SO$_3$	12.0	16.1	20.9	23.5	26.3*	27.3*	25.9	24.8	23.7	22.8	22.1	21.5	1:26
Na$_2$SO$_4$			16.13	21.94	29.22*	32.35*	31.55	30.90	30.39	30.02	29.79	29.67	8
Na$_2$S$_2$O$_3$	33.1	36.3	40.6	43.3	45.9	52.0	62.3	65.7	68.8	69.4	70.1	71.0	6
Na$_2$S$_2$O$_5$		38.4	39.5	40.0	40.6	41.8	43.0	44.2	45.5	46.8	48.1	49.5	1:26
Na$_2$SeO$_3$				47.3*								45*	7
Na$_2$SeO$_4$	11.7			36.9*								42.1*	7
Na$_2$WO$_4$	41.6	41.9	42.3	42.6	42.9	43.6	44.4	45.3	46.2	47.3	48.4	49.5	6
Na$_3$PO$_4$	4.28	7.30	10.8	12.6	14.1	16.6	22.9	28.4	32.4	37.6	40.4	43.5	6
Na$_4$P$_2$O$_7$	2.23	3.28	4.81	6.62	7.00	10.10	14.38	20.07	27.31	36.03	32.37	30.67	6
NdCl$_3$	49.0	49.3	49.7	50.0	50.4	51.2	52.2	53.3	54.5	55.8	57.1	58.5	6
Nd(NO$_3$)$_3$	55.76	57.49	59.37	60.38	61.43	63.69	66.27	69.47					1:13
NiCl$_2$	34.7	36.1	38.5	40.3	41.7	42.1	43.2	45.0	46.1	46.2	46.4	46.6	6
Ni(ClO$_4$)$_2$	51.1			52.8									7
NiF$_2$				2.50							2.52		7
NiI$_2$	55.40	57.68	59.78	60.69	61.50	62.80	63.73	64.38	64.80	65.09	65.30		7
Ni(NO$_3$)$_2$	44.1	46.0	48.4	49.8	51.3	54.6	58.3	61.0	63.1	65.6	67.9	69.0	6
NiSO$_4$	21.4	24.4	27.4	28.8	30.3*	32.0*	34.1	35.8	37.7	39.9	42.3	44.8	6
Ni(SCN)$_2$				35.48									7
NiSeO$_4$	21.6		26.2*									45.6*	7
PbBr$_2$	0.449	0.620	0.841	0.966	1.118	1.46	1.89						2
PbCl$_2$	0.66	0.81	0.98	1.07	1.17	1.39	1.64	1.93	2.24	2.60	2.99	3.42	2
Pb(ClO$_4$)$_2$				81.5									7
PbF$_2$		0.0603	0.0649	0.0670	0.0693								2
PbI$_2$	0.041	0.052	0.067	0.076	0.086	0.112	0.144	0.187	0.243	0.315			2
Pb(IO$_3$)$_2$				0.0025									7
Pb(NO$_3$)$_2$	28.46	32.13	35.67	37.38	39.05	42.22	45.17	47.90	50.42	52.72	54.82	56.75	2
PbSO$_4$	0.0033	0.0038	0.0042	0.0044	0.0047	0.0052	0.0058						2
PrCl$_3$	48.0	48.1	48.6	49.0	49.5	50.8	52.3	54.1	56.1	58.3			6
Pr(NO$_3$)$_3$	57.50	59.20	61.16	62.24	63.40*	65.7*	67.8	70.2	73.4				1:13
RbBr	47.4	50.1	52.6	53.8	54.9	57.0	58.8	60.6	62.1	63.5	64.8	65.9	6
RbBrO$_3$	0.97	1.55	2.36	2.87	3.45	4.87	6.64	8.78	11.29	14.15	17.32	20.76	1:30
RbCl	43.58	45.65	47.53	48.42	49.27	50.86	52.34	53.67	54.92	56.08	57.16	58.15	1:47

Compound	0 °C	10 °C	20 °C	25 °C	30 °C	40 °C	50 °C	60 °C	70 °C	80 °C	90 °C	100 °C	Ref.
$RbClO_3$	2.10	3.38	5.14	6.22	7.45	10.35	13.85	17.93	22.53	27.57	32.96	38.60	1:30
$RbClO_4$	1			1.5								17	7
RbF			75										7
$RbHCO_3$			53.7										7
RbI	55.8	58.6	61.1	62.3	63.4	65.4	67.2	68.8	70.3	71.6	72.7	73.8	6
$RbIO_3$	1.09	1.53	2.07	2.38	2.74	3.52	4.41	5.42	6.52	7.74	9.00	10.36	1:30
$RbNO_3$	16.4	25.0	34.6	39.4	44.2	53.1	60.8	67.2	72.2	76.1	79.0	81.2	6
$RbOH$					63.4								7
Rb_2CrO_4	38.27			43.26									7
Rb_2SO_4	27.3	30.0	32.5	33.7	34.8	36.9	38.7	40.3	41.8	43.0	44.1	44.9	6
$SbCl_3$	85.7			90.8									7
SbF_3	79.4			83.1									7
$Sc(NO_3)_3$	57.0	59.3	61.6	62.8	63.9	66.2	68.5						1:13
$Sm(NO_3)_3$	54.83	56.33	58.08	59.05	60.08	62.38	65.05*	68.1*	70.8	74.2			1:13
$SmCl_3$		48.0	48.2	48.4	48.6	49.2	50.0						6
$SnCl_2$	46	64											7
SnI_2			0.97									3.87	7
$SrBr_2$	46.0	48.3	50.6	51.7	52.9	55.2	57.6	59.9	62.3	64.6	66.8	69.0	6
$Sr(BrO_3)_2$	18.53	22.00	25.39	27.02	28.59	31.55	34.21	36.57	38.64*	40.2*	40.8	41.0	1:14
$SrCl_2$	31.94	32.93	34.43	35.37	36.43	38.93	41.94	45.44*	46.81*	47.69	48.70	49.87	8
$Sr(ClO_2)_2$	13.0	13.6	14.1	14.3	14.5	14.9	15.3	15.6	15.9				7
$Sr(ClO_3)_2$	63.29	63.42	63.64	63.77	63.93	64.29	64.70	65.16	65.65	66.18	66.74	67.31	1:14
$Sr(ClO_4)_2$	70.04*			75.35*		78.44*							7
SrF_2	0.011			0.021									7
SrI_2	62.5	62.8	63.5	63.9	64.5	65.8	67.3	69.0	70.8	72.7	74.7	79.2	6
$Sr(IO_3)_2$	0.102	0.126	0.152	0.165	0.179	0.206	0.233	0.259	0.284	0.307	0.328	0.346	1:14
$Sr(MnO_4)_2$	2.5												7
$Sr(NO_2)_2$					41.9	44.3						58.6	7
$Sr(NO_3)_2$	28.2	34.6	41.0	44.5	47.0	47.4	47.9	48.4	48.9	49.5	50.1	50.7	6
$Sr(OH)_2$	0.9			2.2									7
$SrSO_3$				0.0015									1:26
$SrSO_4$				0.0135									7
SrS_2O_3	8.8	13.2	17.7	20.0	22.2	26.8							7
$Tb(NO_3)_3$			60.6	61.02									1:13
Tl_2SO_4	2.65	3.56	4.61	5.19	5.80	7.09	8.46	9.89	11.33	12.77	14.18	15.53	6
$Tm(NO_3)_3$				67.9									1:13
$UO_2(NO_3)_2$	49.52	51.82	54.42	55.85	57.55	61.59	67.07						1:55
$Y(NO_3)_3$	55.57	56.93	58.75	59.86	61.11*	63.3*	64.9	67.9	72.5				1:13
$Yb(NO_3)_3$				70.5									1:13
$ZnBr_2$	79.3	80.1	81.8	83.0	84.1	85.6	85.8	86.1	86.3	86.6	86.8	87.1	6
ZnC_2O_4		0.0010	0.0019	0.0026									5
$ZnCl_2$		76.6	79.0	80.3	81.4	81.8	82.4	83.0	83.7	84.4	85.2	86.0	6
$Zn(ClO_4)_2$	44.29*			46.27*		48.70							7
ZnF_2				1.53									5
ZnI_2	81.1	81.2	81.3	81.4	81.5	81.7	82.0	82.3	82.6	83.0	83.3	83.7	6
$Zn(IO_3)_2$			0.58	0.64	0.69	0.77	0.82						5
$Zn(NO_3)_2$	47.8	50.8	54.4	54.6	58.5	79.1	80.1	87.5	89.9				6
$ZnSO_3$			0.1786	0.1790	0.1794	0.1803	0.1812						5
$ZnSO_4$	29.1	32.0	35.0	36.6	38.2	41.3	43.0	42.1	41.0	39.9	38.8	37.6	6
$ZnSeO_4$	33.06	34.98	37.38	38.79	40.34								5

SOLUBILITY PRODUCT CONSTANTS

The solubility product constant K_{sp} is a useful parameter for calculating the aqueous solubility of sparingly soluble compounds under various conditions. It may be determined by direct measurement or calculated from the standard Gibbs energies of formation $\Delta_f G°$ of the species involved at their standard states. Thus if K_{sp} = $[M^+]^m$, $[A^-]^n$ is the equilibrium constant for the reaction

$$M_m A_n(s) \rightleftharpoons mM^+(aq) + nA^-(aq),$$

where $M_m A_n$ is the slightly soluble substance and M^+ and A^- are the ions produced in solution by the dissociation of $M_m A_n$, then the Gibbs energy change is

$$\Delta G° = m\,\Delta_f G°\,(M^+,aq) + n\,\Delta_f G°\,(A^-,aq) - \Delta_f G°\,(M_m A_n, s)$$

The solubility product constant is calculated from the equation

$$\ln K_{sp} = -\Delta G°/RT$$

The first table below gives selected values of K_{sp} at 25 °C. Many of these have been calculated from standard state thermodynamic data in References 1 and 2; other values are taken from publications of the IUPAC Solubility Data Project (References 3 to 7).

The above formulation is not convenient for treating sulfides because the S^{-2} ion is usually not present in significant concentrations (see Reference 8). This is due to the hydrolysis reaction

$$S^{-2} + H_2O \rightleftharpoons HS^- + OH^-$$

which is strongly shifted to the right except in very basic solutions. Furthermore, the equilibrium constant for this reaction, which depends on the second ionization constant of H_2S, is poorly known. Therefore it is more useful in the case of sulfides to define a different solubility product K_{spa} based on the reaction

$$M_m S_n(s) + 2H^+ \rightleftharpoons mM^+ + nH_2S\,(aq)$$

Values of K_{spa}, taken from Reference 8, are given for several sulfides in the auxiliary table following the main table. Additional discussion of sulfide equilibria may be found in References 7 and 9.

References

1. Wagman, D. D., Evans, W. H., Parker, V. B., Schumm, R. H., Halow, I., Bailey, S. M., Churney, K. L., and Nuttall, R. L., *The NBS Tables of Chemical Thermodynamic Properties*, J. Phys. Chem. Ref. Data, Vol. 11, Suppl. 2, 1982.
2. Garvin, D., Parker, V. B., and White, H. J., *CODATA Thermodynamic Tables*, Hemisphere, New York, 1987.
3. *Solubility Data Series* (53 Volumes), International Union of Pure and Applied Chemistry, Pergamon Press, Oxford, 1979–1992.
4. Clever, H. L., and Johnston, F. J., *J. Phys. Chem. Ref. Data*, 9, 751, 1980.
5. Marcus, Y., *J. Phys. Chem. Ref. Data*, 9, 1307, 1980.
6. Clever, H. L., Johnson, S. A., and Derrick, M. E., *J. Phys. Chem. Ref. Data*, 14, 631, 1985.
7. Clever, H. L., Johnson, S. A., and Derrick, M. E., *J. Phys. Chem. Ref. Data*, 21, 941, 1992.
8. Myers, R. J., *J. Chem. Educ.*, 63, 687, 1986.
9. Licht, S., *J. Electrochem. Soc.*, 135, 2971, 1988.

Compound	Formula	K_{sp}
Aluminum phosphate	$AlPO_4$	$9.84 \cdot 10^{-21}$
Barium bromate	$Ba(BrO_3)_2$	$2.43 \cdot 10^{-4}$
Barium carbonate	$BaCO_3$	$2.58 \cdot 10^{-9}$
Barium chromate	$BaCrO_4$	$1.17 \cdot 10^{-10}$
Barium fluoride	BaF_2	$1.84 \cdot 10^{-7}$
Barium hydroxide octahydrate	$Ba(OH)_2 \cdot 8H_2O$	$2.55 \cdot 10^{-4}$
Barium iodate	$Ba(IO_3)_2$	$4.01 \cdot 10^{-9}$
Barium iodate monohydrate	$Ba(IO_3)_2 \cdot H_2O$	$1.67 \cdot 10^{-9}$
Barium molybdate	$BaMoO_4$	$3.54 \cdot 10^{-8}$
Barium selenate	$BaSeO_4$	$3.40 \cdot 10^{-8}$
Barium sulfate	$BaSO_4$	$1.08 \cdot 10^{-10}$
Barium sulfite	$BaSO_3$	$5.0 \cdot 10^{-10}$
Beryllium hydroxide	$Be(OH)_2$	$6.92 \cdot 10^{-22}$
Bismuth arsenate	$BiAsO_4$	$4.43 \cdot 10^{-10}$
Bismuth iodide	BiI_3	$7.71 \cdot 10^{-19}$
Cadmium arsenate	$Cd_3(AsO_4)_2$	$2.2 \cdot 10^{-33}$
Cadmium carbonate	$CdCO_3$	$1.0 \cdot 10^{-12}$
Cadmium fluoride	CdF_2	$6.44 \cdot 10^{-3}$
Cadmium hydroxide	$Cd(OH)_2$	$7.2 \cdot 10^{-15}$
Cadmium iodate	$Cd(IO_3)_2$	$2.5 \cdot 10^{-8}$
Cadmium oxalate trihydrate	$CdC_2O_4 \cdot 3H_2O$	$1.42 \cdot 10^{-8}$
Cadmium phosphate	$Cd_3(PO_4)_2$	$2.53 \cdot 10^{-33}$
Calcium carbonate (calcite)	$CaCO_3$	$3.36 \cdot 10^{-9}$
Calcium fluoride	CaF_2	$3.45 \cdot 10^{-11}$
Calcium hydroxide	$Ca(OH)_2$	$5.02 \cdot 10^{-6}$
Calcium iodate	$Ca(IO_3)_2$	$6.47 \cdot 10^{-6}$
Calcium iodate hexahydrate	$Ca(IO_3)_2 \cdot 6H_2O$	$7.10 \cdot 10^{-7}$
Calcium molybdate	$CaMoO_4$	$1.46 \cdot 10^{-8}$
Calcium oxalate monohydrate	$CaC_2O_4 \cdot H_2O$	$2.32 \cdot 10^{-9}$
Calcium phosphate	$Ca_3(PO_4)_2$	$2.07 \cdot 10^{-33}$
Calcium sulfate	$CaSO_4$	$4.93 \cdot 10^{-5}$
Calcium sulfate dihydrate	$CaSO_4 \cdot 2H_2O$	$3.14 \cdot 10^{-5}$
Calcium sulfite hemihydrate	$CaSO_3 \cdot 0.5H_2O$	$3.1 \cdot 10^{-7}$
Cesium perchlorate	$CsClO_4$	$3.95 \cdot 10^{-3}$
Cesium periodate	$CsIO_4$	$5.16 \cdot 10^{-6}$
Cobalt(II) arsenate	$Co_3(AsO_4)_2$	$6.80 \cdot 10^{-29}$
Cobalt(II) hydroxide (blue)	$Co(OH)_2$	$5.92 \cdot 10^{-15}$
Cobalt(II) iodate dihydrate	$Co(IO_3)_2 \cdot 2H_2O$	$1.21 \cdot 10^{-2}$
Cobalt(II) phosphate	$Co_3(PO_4)_2$	$2.05 \cdot 10^{-35}$
Copper(I) bromide	$CuBr$	$6.27 \cdot 10^{-9}$
Copper(I) chloride	$CuCl$	$1.72 \cdot 10^{-7}$
Copper(I) cyanide	$CuCN$	$3.47 \cdot 10^{-20}$
Copper(I) iodide	CuI	$1.27 \cdot 10^{-12}$
Copper(I) thiocyanate	$CuSCN$	$1.77 \cdot 10^{-13}$
Copper(II) arsenate	$Cu_3(AsO_4)_2$	$7.95 \cdot 10^{-36}$
Copper(II) iodate monohydrate	$Cu(IO_3)_2 \cdot H_2O$	$6.94 \cdot 10^{-8}$
Copper(II) oxalate	CuC_2O_4	$4.43 \cdot 10^{-10}$
Copper(II) phosphate	$Cu_3(PO_4)_2$	$1.40 \cdot 10^{-37}$
Europium(III) hydroxide	$Eu(OH)_3$	$9.38 \cdot 10^{-27}$
Gallium(III) hydroxide	$Ga(OH)_3$	$7.28 \cdot 10^{-36}$
Iron(II) carbonate	$FeCO_3$	$3.13 \cdot 10^{-11}$
Iron(II) fluoride	FeF_2	$2.36 \cdot 10^{-6}$

Compound	Formula	K_{sp}
Iron(II) hydroxide	$Fe(OH)_2$	$4.87 \cdot 10^{-17}$
Iron(III) hydroxide	$Fe(OH)_3$	$2.79 \cdot 10^{-39}$
Iron(III) phosphate dihydrate	$FePO_4 \cdot 2H_2O$	$9.91 \cdot 10^{-16}$
Lanthanum iodate	$La(IO_3)_3$	$7.50 \cdot 10^{-12}$
Lead(II) bromide	$PbBr_2$	$6.60 \cdot 10^{-6}$
Lead(II) carbonate	$PbCO_3$	$7.40 \cdot 10^{-14}$
Lead(II) chloride	$PbCl_2$	$1.70 \cdot 10^{-5}$
Lead(II) fluoride	PbF_2	$3.3 \cdot 10^{-8}$
Lead(II) hydroxide	$Pb(OH)_2$	$1.43 \cdot 10^{-20}$
Lead(II) iodate	$Pb(IO_3)_2$	$3.69 \cdot 10^{-13}$
Lead(II) iodide	PbI_2	$9.8 \cdot 10^{-9}$
Lead(II) selenate	$PbSeO_4$	$1.37 \cdot 10^{-7}$
Lead(II) sulfate	$PbSO_4$	$2.53 \cdot 10^{-8}$
Lithium carbonate	Li_2CO_3	$8.15 \cdot 10^{-4}$
Lithium fluoride	LiF	$1.84 \cdot 10^{-3}$
Lithium phosphate	Li_3PO_4	$2.37 \cdot 10^{-11}$
Magnesium carbonate	$MgCO_3$	$6.82 \cdot 10^{-6}$
Magnesium carbonate trihydrate	$MgCO_3 \cdot 3H_2O$	$2.38 \cdot 10^{-6}$
Magnesium carbonate pentahydrate	$MgCO_3 \cdot 5H_2O$	$3.79 \cdot 10^{-6}$
Magnesium fluoride	MgF_2	$5.16 \cdot 10^{-11}$
Magnesium hydroxide	$Mg(OH)_2$	$5.61 \cdot 10^{-12}$
Magnesium oxalate dihydrate	$MgC_2O_4 \cdot 2H_2O$	$4.83 \cdot 10^{-6}$
Magnesium phosphate	$Mg_3(PO_4)_2$	$1.04 \cdot 10^{-24}$
Manganese(II) carbonate	$MnCO_3$	$2.24 \cdot 10^{-11}$
Manganese(II) iodate	$Mn(IO_3)_2$	$4.37 \cdot 10^{-7}$
Manganese(II) oxalate dihydrate	$MnC_2O_4 \cdot 2H_2O$	$1.70 \cdot 10^{-7}$
Mercury(I) bromide	Hg_2Br_2	$6.40 \cdot 10^{-23}$
Mercury(I) carbonate	Hg_2CO_3	$3.6 \cdot 10^{-17}$
Mercury(I) chloride	Hg_2Cl_2	$1.43 \cdot 10^{-18}$
Mercury(I) fluoride	Hg_2F_2	$3.10 \cdot 10^{-6}$
Mercury(I) iodide	Hg_2I_2	$5.2 \cdot 10^{-29}$
Mercury(I) oxalate	$Hg_2C_2O_4$	$1.75 \cdot 10^{-13}$
Mercury(I) sulfate	Hg_2SO_4	$6.5 \cdot 10^{-7}$
Mercury(I) thiocyanate	$Hg_2(SCN)_2$	$3.2 \cdot 10^{-20}$
Mercury(II) bromide	$HgBr_2$	$6.2 \cdot 10^{-20}$
Mercury(II) iodide	HgI_2	$2.9 \cdot 10^{-29}$
Neodymium carbonate	$Nd_2(CO_3)_3$	$1.08 \cdot 10^{-33}$
Nickel(II) carbonate	$NiCO_3$	$1.42 \cdot 10^{-7}$
Nickel(II) hydroxide	$Ni(OH)_2$	$5.48 \cdot 10^{-16}$
Nickel(II) iodate	$Ni(IO_3)_2$	$4.71 \cdot 10^{-5}$
Nickel(II) phosphate	$Ni_3(PO_4)_2$	$4.74 \cdot 10^{-32}$
Palladium(II) thiocyanate	$Pd(SCN)_2$	$4.39 \cdot 10^{-23}$
Potassium hexachloroplatinate	K_2PtCl_6	$7.48 \cdot 10^{-6}$
Potassium perchlorate	$KClO_4$	$1.05 \cdot 10^{-2}$
Potassium periodate	KIO_4	$3.71 \cdot 10^{-4}$
Praseodymium hydroxide	$Pr(OH)_3$	$3.39 \cdot 10^{-24}$
Radium iodate	$Ra(IO_3)_2$	$1.16 \cdot 10^{-9}$
Radium sulfate	$RaSO_4$	$3.66 \cdot 10^{-11}$
Rubidium perchlorate	$RbClO_4$	$3.00 \cdot 10^{-3}$
Scandium fluoride	ScF_3	$5.81 \cdot 10^{-24}$
Scandium hydroxide	$Sc(OH)_3$	$2.22 \cdot 10^{-31}$
Silver(I) acetate	$AgCH_3COO$	$1.94 \cdot 10^{-3}$
Silver(I) arsenate	Ag_3AsO_4	$1.03 \cdot 10^{-22}$
Silver(I) bromate	$AgBrO_3$	$5.38 \cdot 10^{-5}$
Silver(I) bromide	$AgBr$	$5.35 \cdot 10^{-13}$
Silver(I) carbonate	Ag_2CO_3	$8.46 \cdot 10^{-12}$
Silver(I) chloride	$AgCl$	$1.77 \cdot 10^{-10}$
Silver(I) chromate	Ag_2CrO_4	$1.12 \cdot 10^{-12}$
Silver(I) cyanide	$AgCN$	$5.97 \cdot 10^{-17}$
Silver(I) iodate	$AgIO_3$	$3.17 \cdot 10^{-8}$
Silver(I) iodide	AgI	$8.52 \cdot 10^{-17}$
Silver(I) oxalate	$Ag_2C_2O_4$	$5.40 \cdot 10^{-12}$
Silver(I) phosphate	Ag_3PO_4	$8.89 \cdot 10^{-17}$
Silver(I) sulfate	Ag_2SO_4	$1.20 \cdot 10^{-5}$
Silver(I) sulfite	Ag_2SO_3	$1.50 \cdot 10^{-14}$
Silver(I) thiocyanate	$AgSCN$	$1.03 \cdot 10^{-12}$
Strontium arsenate	$Sr_3(AsO_4)_2$	$4.29 \cdot 10^{-19}$
Strontium carbonate	$SrCO_3$	$5.60 \cdot 10^{-10}$
Strontium fluoride	SrF_2	$4.33 \cdot 10^{-9}$
Strontium iodate	$Sr(IO_3)_2$	$1.14 \cdot 10^{-7}$
Strontium iodate monohydrate	$Sr(IO_3)_2 \cdot H_2O$	$3.77 \cdot 10^{-7}$
Strontium iodate hexahydrate	$Sr(IO_3)_2 \cdot 6H_2O$	$4.55 \cdot 10^{-7}$
Strontium sulfate	$SrSO_4$	$3.44 \cdot 10^{-7}$
Thallium(I) bromate	$TlBrO_3$	$1.10 \cdot 10^{-4}$
Thallium(I) bromide	$TlBr$	$3.71 \cdot 10^{-6}$
Thallium(I) chloride	$TlCl$	$1.86 \cdot 10^{-4}$
Thallium(I) chromate	Tl_2CrO_4	$8.67 \cdot 10^{-13}$
Thallium(I) iodate	$TlIO_3$	$3.12 \cdot 10^{-6}$
Thallium(I) iodide	TlI	$5.54 \cdot 10^{-8}$
Thallium(I) thiocyanate	$TlSCN$	$1.57 \cdot 10^{-4}$
Thallium(III) hydroxide	$Tl(OH)_3$	$1.68 \cdot 10^{-44}$
Tin(II) hydroxide	$Sn(OH)_2$	$5.45 \cdot 10^{-27}$
Yttrium carbonate	$Y_2(CO_3)_3$	$1.03 \cdot 10^{-31}$
Yttrium fluoride	YF_3	$8.62 \cdot 10^{-21}$
Yttrium hydroxide	$Y(OH)_3$	$1.00 \cdot 10^{-22}$
Yttrium iodate	$Y(IO_3)_3$	$1.12 \cdot 10^{-10}$
Zinc arsenate	$Zn_3(AsO_4)_2$	$2.8 \cdot 10^{-28}$
Zinc carbonate	$ZnCO_3$	$1.46 \cdot 10^{-10}$
Zinc carbonate monohydrate	$ZnCO_3 \cdot H_2O$	$5.42 \cdot 10^{-11}$
Zinc fluoride	ZnF_2	$3.04 \cdot 10^{-2}$
Zinc hydroxide	$Zn(OH)_2$	$3 \cdot 10^{-17}$
Zinc iodate dihydrate	$Zn(IO_3)_2 \cdot 2H_2O$	$4.1 \cdot 10^{-6}$
Zinc oxalate dihydrate	$ZnC_2O_4 \cdot 2H_2O$	$1.38 \cdot 10^{-9}$
Zinc selenide	$ZnSe$	$3.6 \cdot 10^{-26}$
Zinc selenite monohydrate	$ZnSeO_3 \cdot H_2O$	$1.59 \cdot 10^{-7}$

Sulfides

Compound	Formula	K_{spa}
Cadmium sulfide	CdS	$8 \cdot 10^{-7}$
Copper(II) sulfide	CuS	$6 \cdot 10^{-16}$
Iron(II) sulfide	FeS	$6 \cdot 10^{2}$
Lead(II) sulfide	PbS	$3 \cdot 10^{-7}$
Manganese(II) sulfide (green)	MnS	$3 \cdot 10^{7}$
Mercury(II) sulfide (red)	HgS	$4 \cdot 10^{-33}$
Mercury(II) sulfide (black)	HgS	$2 \cdot 10^{-32}$
Silver(I) sulfide	Ag$_2$S	$6 \cdot 10^{-30}$
Tin(II) sulfide	SnS	$1 \cdot 10^{-5}$
Zinc sulfide (sphalerite)	ZnS	$2 \cdot 10^{-4}$
Zinc sulfide (wurtzite)	ZnS	$3 \cdot 10^{-2}$

SOLUBILITY OF COMMON SALTS AT AMBIENT TEMPERATURES

This table gives the aqueous solubility of selected salts at temperatures from 10 °C to 40 °C. Values are given in molality terms.

References

1. Apelblat, A., *J. Chem. Thermodynamics*, 24, 619, 1992.
2. Apelblat, A., *J. Chem. Thermodynamics*, 25, 63, 1993.
3. Apelblat, A., *J. Chem. Thermodynamics*, 25, 1513, 1993.
4. Apelblat, A. and Korin, E., *J. Chem. Thermodynamics*, 30, 59, 1998.

Salt	10 °C	15 °C	20 °C	25 °C	30 °C	35 °C	40 °C	Ref.
$BaCl_2$	1.603	1.659	1.716	1.774	1.834	1.895	1.958	1
$Ca(NO_3)_2$	6.896	7.398	7.986	8.675	9.480	10.421		1
$CuSO_4$	1.055	1.153	1.260	1.376	1.502	1.639		3
$FeSO_4$	1.352	1.533	1.729	1.940	2.165	2.405		3
KBr	5.002	5.237	5.471	5.703	5.932	6.157		3
KIO_3	0.291	0.333	0.378	0.426	0.478	0.534	0.593	4
K_2CO_3	7.756	7.846	7.948	8.063	8.191	8.331	8.483	1
$LiCl$	19.296	19.456	19.670	19.935				2
$Mg(NO_3)_2$	4.403	4.523	4.656	4.800	4.958	5.130	5.314	1
$MnCl_2$	5.421	5.644	5.884	6.143	6.422	6.721		3
NH_4Cl	6.199	6.566	6.943	7.331				2
NH_4NO_3	18.809	21.163	23.721	26.496				2
$(NH_4)_2SO_4$	5.494	5.589	5.688	5.790	5.896	6.005		3
$NaBr$	8.258	8.546	8.856	9.191	9.550	9.937	10.351	4
$NaCl$	6.110	6.121	6.136	6.153	6.174	6.197	6.222	4
$NaNO_2$	11.111	11.484	11.883	12.310	12.766	13.253	13.772	4
$NaNO_3$	9.395	9.819	10.261	10.723	11.204	11.706	12.230	4
$RbCl$	6.911	7.180	7.449	7.717	7.986	8.253	8.520	4
$ZnSO_4$	2.911	3.116	3.336	3.573	3.827	4.099	4.194	1

SOLUBILITY OF HYDROCARBONS IN SEAWATER

Concern about pollution of the oceans has stimulated measurements of the solubility of organic compounds in seawater. This table gives the solubility of several hydrocarbons in seawater. The data are derived from a review in the IUPAC Solubility Data Series (Reference 1).

Solubility is expressed in this table as parts per million by mass, i.e.,

$$S/\text{ppm(mass)} = 10^6 \times w_2 = 10^6 \times m_2/(m_1 + m_2)$$

where m_1 and m_2 are the masses of solvent (seawater) and solute, respectively, under saturation conditions, and w_2 is the mass fraction. Since the solubilities in this table are very low, the value of S is effectively the mass of hydrocarbon in grams per 1000 kg of seawater.

The temperature and salinity of each measurement are given in the table. Salinity is a standardized measure of the concentration of dissolved salts, as explained in the table "Properties of Seawater" in Section 14. Salinity values in the open oceans at mid-latitude typically fall between 34 and 36.

Reference 1 gives details of the method of measurement and an indication of the reliability of the measurements.

Reference

1. Shaw, David G., and Maczynski, A., IUPAC-NIST Solubility Data Series 81. Hydrocarbons with Water and Seawater — Revised and Updated. Part 12. C_5-C_{26} Hydrocarbons with Seawater, *J. Phys. Chem. Ref. Data* 35, 785, 2006.

Name	Mol. Form.	Salinity	$t/°C$	S/ppm (mass)
Acenaphthene	$C_{12}H_{10}$	35	15	0.21
Acenaphthene	$C_{12}H_{10}$	35	25	1.8
Anthracene	$C_{14}H_{10}$	35	25	0.031
Benz[a]anthracene	$C_{18}H_{12}$	35	25	0.0056
Benzene	C_6H_6	34.4	0	1320
Benzene	C_6H_6	35	25	1360
Benzo[ghi]perylene	$C_{22}H_{12}$	6	25	0.00021
Benzo[a]pyrene	$C_{20}H_{12}$	6	25	0.00013
Benzo[e]pyrene	$C_{20}H_{12}$	30	25	0.0033
Benzo[b]triphenylene	$C_{22}H_{14}$	6	25	0.027
Biphenyl	$C_{12}H_{10}$	35	25	4.76
Butylbenzene	$C_{10}H_{14}$	34.5	25	7.1
sec-Butylbenzene	$C_{10}H_{14}$	34.5	25	12
tert-Butylbenzene	$C_{10}H_{14}$	34.5	25	21
Chrysene	$C_{18}H_{12}$	35	25	0.0011
Dibenz[a,h]anthracene	$C_{22}H_{14}$	6	25	0.021
Dibenz[a,j]anthracene	$C_{22}H_{14}$	6	25	0.010
Dodecane	$C_{12}H_{26}$	35	25	0.0029
Eicosane	$C_{20}H_{42}$	35	25	0.0008
Ethylbenzene	C_8H_{10}	34.4	0	140
Ethylbenzene	C_8H_{10}	34.4	10	129
Ethylbenzene	C_8H_{10}	34.4	25	111
Fluoranthene	$C_{16}H_{10}$	35	25	0.124
9*H*-Fluorene	$C_{13}H_{10}$	35	25	1.2
Heptane	C_7H_{16}	6	25	10.3
Hexacosane	$C_{26}H_{54}$	35	25	0.0001
Hexadecane	$C_{16}H_{34}$	35	25	0.0004
Hexane	C_6H_{14}	35.3	25	7.9
Isopropylbenzene	C_9H_{12}	34.5	25	43
2-Methylanthracene	$C_{15}H_{12}$	35	25	0.013
Methylcyclopentane	C_6H_{12}	34.5	25	29
1-Methylnaphthalene	$C_{11}H_{10}$	30	25	23
1-Methylphenanthrene	$C_{15}H_{12}$	35	25	0.20
Naphthalene	$C_{10}H_8$	35	25	22.8
Nonane	C_9H_{20}	6	25	0.43
Octadecane	$C_{18}H_{38}$	35	25	0.0008
Pentane	C_5H_{12}	34.5	25	28
Phenanthrene	$C_{14}H_{10}$	34	25	0.69

Name	Mol. Form.	Salinity	$t/°C$	S/ppm (mass)
Pyrene	$C_{16}H_{10}$	35	25	0.086
Tetradecane	$C_{14}H_{30}$	35	25	0.0017
Toluene	C_7H_8	34.4	0	450
Toluene	C_7H_8	35	25	387
1,2,3-Trimethylbenzene	C_9H_{12}	34.5	25	49
1,2,4-Trimethylbenzene	C_9H_{12}	34.5	25	40
1,3,5-Trimethylbenzene	C_9H_{12}	34.5	25	31
Undecane	$C_{11}H_{24}$	6	25	0.01
o-Xylene	C_8H_{10}	34.5	25	130
m-Xylene	C_8H_{10}	34.5	25	106
p-Xylene	C_8H_{10}	34.5	25	111

SOLUBILITY OF ORGANIC COMPOUNDS IN PRESSURIZED HOT WATER

Liquid water at elevated temperatures and pressures, but still in the subcritical region, is of interest as a solvent in various laboratory and industrial processes. In effect, this means water at a temperature between about 100 °C and 373 °C, the critical temperature, and at pressures up to 400 bar or greater. Since the dielectric constant of water decreases with increasing temperature, the solubility of many compounds, especially non-polar compounds, increases dramatically at higher temperature. The fact that solubility can be fine-tuned by controlling temperature and pressure makes pressurized hot water a useful tool in various extraction and reaction processes.

This table gives a sample of the variations of solubility with temperature and pressure for several compounds, mostly hydrocarbons. The solubility is expressed in both mole fraction of solute, x_2, and mass percent, $100w_2$, where w_2 is the mass fraction. More information is available in the references.

References

1. *Solubility Data Series, International Union of Pure and Applied Chemistry, Vol. 38*, Pergamon Press, Oxford, 1988.
2. Shaw, D. G., and Maczynski, A., *J. Phys. Chem. Ref. Data* 35, 687, 2006.
3. Stephenson, R. M., *J. Chem. Eng. Data* 37, 80, 1992.
4. Lun, R., Varhanickova, D., Shiu, W.-Y., and Mackay, D., *J. Chem. Eng. Data* 42, 951, 1997.
5. Miller, D. J., et al., *J. Chem. Eng. Data* 43, 1043, 1998.
6. Miller, D. J., and Hawthorne, S. B., *J. Chem. Eng. Data* 45, 78, 2000.
7. Ma, J. H. Y., Hung, H., Shiu, W-Y., and Mackay, D., *J. Chem. Eng. Data* 46, 619, 2001.
8. Marche, C., Ferronato, C., and Jose, J., *J. Chem. Eng. Data* 48, 967, 2003.
9. Oleszek-Kudlak, S., Shibata, E., and Nakamura, T., *J. Chem. Eng. Data* 49, 570, 2004.
10. Marche, C., Ferronato, C., and Jose, J., *J. Chem. Eng. Data* 49, 937, 2004.
11. Andersson, T. A., Hartonen, K. M., and Riekkola, M-L., *J. Chem. Eng. Data* 50, 1177, 2005.
12. Karasek, P., Planeta, J., and Roth, M., *J. Chem. Eng. Data* 51, 616, 2006.
13. Shiu, W.-Y., and Ma, K.-C, *J. Phys. Chem. Ref. Data* 29, 41, 2000.

Name	Mol. Form.	$t/°C$	p/bar	$10^3 x_2$	Mass%	Ref.
Acenaphthene	$C_{12}H_{10}$	25	1	0.000444	0.000380	13
		250	50	1.25	1.06	11
Anthracene	$C_{14}H_{10}$	25	1	0.0000074	0.0000044	2
		50	50	0.000017	0.000017	5
		100	45	0.00032	0.00032	5
		100	39	0.000457	0.00045	12
		150	50	0.0102	0.0101	11
		200	77	0.13	0.13	12
		250	50	0.497	0.49	11
		300	100	3.78	3.62	11
Benz[a]anthracene	$C_{18}H_{12}$	25	1	0.00000073	0.00000093	2
		60	50	0.00000846	0.0000107	12
		100	50	0.000113	0.000143	12
		120	52	0.000418	0.00053	12
		150	49	0.00296	0.00375	12
Benzene	C_6H_6	25	1	0.40	0.178	13
		25	65	0.40	0.173	6
		25	400	0.33	0.143	6
		50	65	0.47	0.203	6
		100	65	0.89	0.38	6
		150	65	2.2	0.95	6
		200	65	5.0	2.13	6
		200	400	4.1	1.75	6
Carbazole	$C_{12}H_9N$	25	1	0.00013	0.00012	5
		25	54	0.00011	0.000102	5
		50	56	0.00045	0.00042	5
		100	54	0.0099	0.0092	5
		150	54	0.162	0.150	5
		200	52	1.9	1.74	5
Chrysene	$C_{18}H_{12}$	25	1	0.00000016	0.00000019	2

Name	Mol. Form.	$t/°C$	p/bar	Solubility		Ref.
				$10^3 x_2$	Mass%	
		25	32	0.00000063	0.0000008	5
		50	36	0.000001	0.0000013	5
		100	38	0.000013	0.000016	5
		150	43	0.00060	0.00076	5
		200	45	0.0158	0.020	5
		225	62	0.0758	0.096	5
o-Dichlorobenzene	$C_6H_4Cl_2$	25	1	0.018	0.0094	9
		50	65	0.023	0.019	6
		100	65	0.055	0.045	6
		150	65	0.18	0.15	6
		200	65	0.57	0.46	6
trans-1,2-Dimethylcyclohexane	C_8H_{16}	25	1	0.008	0.00050	10
		101	7	0.0047	0.0029	10
		131	7	0.0108	0.0067	10
		151	7	0.0223	0.0139	10
		170	7	0.0356	0.0222	10
Ethylcyclohexane	C_8H_{16}	25	1	0.00098	0.00061	10
		100	7	0.00340	0.00212	10
		131	7	0.0085	0.0053	10
		151	7	0.01665	0.0104	10
		171	7	0.0334	0.0208	10
Heptane	C_7H_{16}	25	1	0.0004352	0.000242	8
		50	7	0.000613	0.00034096	8
		100	7	0.001938	0.00108	8
		125	7	0.00400	0.00222	8
		150	7	0.00878	0.00488	8
		170	7	0.01701	0.00946	8
Hexane	C_6H_{14}	25	1	0.002045	0.00098	8
		100	7	0.006074	0.0029	8
		125	7	0.01192	0.0057	8
		150	7	0.02555	0.0122	8
		170	7	0.04935	0.0236	8
1-Isopropyl-4-methylbenzene	$C_{10}H_{14}$	25	1	0.0030	0.0051	4
		50	60	0.0040	0.0030	6
		100	60	0.011	0.0082	6
		150	60	0.043	0.032	6
		200	60	0.20	0.15	6
Methylcyclohexane	C_7H_{14}	25	1	0.00293	0.00151	10
		100	7	0.01006	0.0055	10
		131	7	0.0244	0.0133	10
		151	7	0.0423	0.0231	10
		171	7	0.0708	0.0386	10
Naphthalene	$C_{10}H_8$	25	1	0.00444	0.00316	13
		40	50	0.00692	0.0049	12
		50	50	0.0114	0.0081	12
		65	50	0.0264	0.0188	12
		75	50	0.0435	0.0309	12
Octane	C_8H_{18}	25	1	0.0001158	0.000073	8
		100	7	0.0005943	0.000377	8
		125	7	0.0014163	0.000898	8
		150	7	0.0036957	0.00234	8
		170	7	0.0083483	0.00529	8
		200	65	0.029	0.018	6
Perylene	$C_{20}H_{12}$	25	1	0.00000003	0.00000004	2

Name	Mol. Form.	$t/°C$	p/bar	Solubility		Ref.
				$10^3 x_2$	Mass%	
		50	50	0.00000029	0.0000004	5
		100	45	0.00000210	0.00000294	5
		150	47	0.000120	0.000168	5
		200	48	0.0050	0.0070	5
Pyrene	$C_{16}H_{10}$	25	1	0.000012	0.0000139	2
		100	50	0.000637	0.00072	11
		100	200	0.00078	0.00087	5
		140	50	0.0054	0.0061	11
		200	50	0.0492	0.055	11
		250	50	0.205	0.23	11
		300	50	1.41	1.56	11
p-Terphenyl	$C_{18}H_{14}$	25	1	0.00000141	0.00000180	2
		100	49	0.0000219	0.000028	12
		140	51	0.000372	0.000476	12
		180	55	0.00626	0.0080	12
		200	53	0.0241	0.0308	12
		210	54	0.0393	0.0502	12
Tetrachloroethene	C_2Cl_4	25	1	0.0285	0.0286	3
		50	65	0.027	0.025	6
		100	65	0.059	0.054	6
		150	65	0.18	0.17	6
		200	65	0.59	0.54	6
Toluene	C_7H_8	25	1	0.107	0.0519	7,13
		50	50	0.125	0.064	6
		100	50	0.27	0.138	6
		150	50	0.66	0.337	6
		200	50	1.9	0.96	6
2,2,4-Trimethylpentane	C_8H_{18}	25	1	0.00035	0.00022	1
		50	65	0.00052	0.00033	6
		100	65	0.0020	0.00127	6
		150	65	0.0102	0.0065	6
		200	65	0.061	0.0387	6
Triphenylene	$C_{18}H_{12}$	25	1	0.0000034	0.0000043	2
		100	51	0.0000899	0.000114	12
		140	50	0.00126	0.00160	12
		180	64	0.0123	0.0156	12
		195	60	0.0283	0.0359	12
m-Xylene	C_8H_{10}	25	1	0.028	0.0161	13
		50	60	0.036	0.021	6
		100	60	0.085	0.050	6
		150	60	0.27	0.159	6
		200	60	0.88	0.516	6

SOLUBILITY CHART

Abbreviations: **W**, soluble in water; **A**, insoluble in water but soluble in acids; **w**, sparingly soluble in water but soluble in acids; **a**, insoluble in water and only sparingly soluble in acids; **I**, insoluble in water and acids; **d**, decomposes in water. * Indicates two modifications of the salt.

No.		Al	NH₄	Sb	Ba	Bi	Cd	Ca	Cr	Co	Cu	Au (I)	Au (II)	H	Fe (II)	Fe (III)
1	Acetate	W	W		W	W	W	W	W	W	W	W	W	W	W	W
	—$(C_2H_3O_2)$	$Al(—)_3$	$NH_4(—)$		$Ba(—)_2$	$Bi(—)_3$	$Cd(—)_2$	$Ca(—)_2$	$Cr(—)_3$	$Co(—)_2$	$Cu(—)_2$			$C_2H_4O_2$	$Fe(—)_2$	$Fe_2(—)_6$
2	Arsenate	a	W	A	w	A	A	w		A	A			W	A	A
	—(AsO_4)	$Al(—)$	$(NH_4)_3(—)$	$Sb(—)$	$Ba_3(—)_2$	$Bi(—)$	$Cd_3(—)_2$	$Ca_5(—)_2$		$Co_3(—)_2$	$Cu_3(—)_2$			H_3AsO_4	$Fe_3(—)_2$	$Fe(—)$
3	Arsenite		W	A				w		A	A					
	—(AsO_3)		NH_4AsO_2	$Sb(—)$				$Ca_3(—)_2$		$Co_3H_6(—)_4$	$CuH(—)$					
4	Benzoate		W		W	A	W	W		W	w			W	W	A
	—$(C_7H_5O_2)$		$NH_4(—)$		$Ba(—)_2$	$Bi(—)_3$	$Cd(—)_2$	$Ca(—)_2$		$Co(—)_2$	$Cu(—)_2$			$C_7H_6O_2$	$Fe(—)_2$	$Fe_2(—)_6$
5	Bromide	W	W	d	W	d	W	W	W(I)*	W	W	W	W	W	W	W
		$AlBr_3$	NH_4Br	$SbBr_3$	$BaBr_2$	$BiBr_3$	$CdBr_2$	$CaBr_2$	$CrBr_3$	$CoBr_2$	$CuBr_2$	$AuBr$	$AuBr_3$	HBr	$FeBr_2$	$FeBr_3$
6	Carbonate		W		w	A	A	A	W	A						
			$(NH_4)_2CO_3$		$BaCO_3$		$CdCO_3$	$CaCO_3$	$CrCO_3$	$CoCO_3$					$FeCO_3$	
7	Chlorate	W	W		W	W	W	W		W	W			W	W	W
	—(ClO_3)	$Al(—)_3$	$NH_4(—)$		$Ba(—)_2$	$Bi(—)_3$	$Cd(—)_2$	$Ca(—)_2$		$Co(—)_2$	$Cu(—)_2$			$HClO_3$	$Fe(—)_2$	$Fe(—)_3$
8	Chloride	W	W	W	W	d	W	W	I	W	W	w	W	W	W	W
		$AlCl_3$	NH_4Cl	$SbCl_3$	$BaCl_2$	$BiCl_3$	$CdCl_2$	$CaCl_2$	$CrCl_3$	$CoCl_2$	$CuCl_2$	$AuCl$	$AuCl_3$	HCl	$FeCl_2$	$FeCl_3$
9	Chromate		W		A		A	W		A						A
	—(CrO_4)		$(NH_4)_2(—)$		$Ba(—)$		$Cd(—)$	$Ca(—)$		$Co(—)$						$Fe_2(—)_3$
10	Citrate	W	W		w	A	A	w		W				W		W
	—$(C_6H_5O_7)$	$Al(—)$	$(NH_4)_3(—)$		$Ba_3(—)_2$	$Bi(—)$	$Cd_3(—)_2$	$Ca_3(—)_2$		$Co_3(—)_2$				$C_6H_8O_7$		$Fe(—)$
11	Cyanide		W		W	w	W	W	A	A	A	w	W	W	a	
			NH_4CN		$Ba(CN)_2$	$Bi(CN)_3$	$Cd(CN)_2$	$Ca(CN)_2$	$Cr(CN)_3$	$Co(CN)_2$	$Cu(CN)_2$	$AuCN$	$Au(CN)_3$	HCN	$Fe(CN)_2$	
12	Ferricy'de		W		w		A	W		I	I			W	I	
	—$(Fe(CN)_6)$		$(NH_4)_3(—)$		$Ba_3(—)_2$		$Cd_3(—)_2$	$Ca_3(—)_2$		$Co_3(—)_2$	$Cu_3(—)_2$			$H_3(—)$	$Fe_3(—)_2$	
13	Ferrocy'de	w	W		W	A	W	W		I	I			W	I	a
	—$(Fe(CN)_6)$	$Al_4(—)_3$	$(NH_4)_4(—)$		$Ba_2(—)$		$Cd_2(—)$	$Ca_2(—)$		$Co_2(—)$	$Cu_2(—)$			$H_4(—)$	$Fe_2(—)$	$Fe_4(—)_3$
14	Fluoride	W	W	W	W	W	W	W	W(a)*	W	w			W	w	w
		AlF_3	NH_4F	SbF_3	BaF_2	BiF_3	CdF_2	CaF_2	CrF_3	CoF_2	CuF_2			HF	FeF_2	FeF_3
15	Formate	W	W		W	W	W	W		W	W			W	W	W
	—(CHO_2)	$Al(—)_3$	$NH_4(—)$		$Ba(—)_2$	$Bi(—)_3$	$Cd(—)_2$	$Ca(—)_2$		$Co(—)_2$	$Cu(—)_2$			CH_2O_2	$Fe(—)_2$	$Fe(—)_3$
16	Hydroxide	A	W		W	A	A	W	A	A	A	W	A		A	A
		$Al(OH)_3$	NH_4OH		$Ba(OH)_2$	$Bi(OH)_3$	$Cd(OH)_2$	$Ca(OH)_2$	$Cr(OH)_3$	$Co(OH)_2$	$Cu(OH)_2$	$AuOH$	$Au(OH)_3$		$Fe(OH)_2$	$Fe(OH)_3$
17	Iodide	W	W	d	W	A	W	W	W	W	W	a	a	W	W	W
		AlI_3	NH_4I	SbI_3	BaI_2	BiI_3	CdI_2	CaI_2	CrI_3	CoI_2	CuI	AuI	AuI_3	HI	FeI_2	FeI_3
18	Nitrate	W	W	W	W	d	W	W	W	W	W			W	W	W
		$Al(NO_3)_3$	NH_4NO_3		$Ba(NO_3)_2$	$Bi(NO_3)_3$	$Cd(NO_3)_2$	$Ca(NO_3)_2$	$Cr(NO_3)_3$	$Co(NO_3)_2$	$Cu(NO_3)_2$			HNO_3	$Fe(NO_3)_2$	$Fe(NO_3)_3$
19	Oxalate	A	W	w	w	A	w	A	W	A	A			W	A	W
	—(C_2O_4)	$Al_2(—)_3$	$(NH_4)_2(—)$		$Ba(—)$	$Bi_2(—)_3$	$Cd(—)$	$Ca(—)$	$Cr(—)$	$Co(—)$	$Cu(—)$			$C_2H_2O_4$	$Fe(—)$	$Fe_2(—)_3$
20	Oxide	a		w	W	A	A	w	a	A	A	A	A	W	A	A
		Al_2O_3		Sb_2O_3	BaO	Bi_2O_3	CdO	CaO	Cr_2O_3	CoO	CuO	Au_2O	Au_2O_3	H_2O_2	FeO	Fe_2O_3
21	Phosphate	A	W		A	A	A	w	w	A	A			W	A	w
		$AlPO_4$	$NH_4H_2PO_4$		$Ba_3(PO_4)_2$	$BiPO_4$	$Cd_3(PO_4)_2$	$Ca_3(PO_4)_2$	$Cr_2(PO_4)_3$	$Co_3(PO_4)_2$	$Cu_3(PO_4)_2$			H_3PO_4	$Fe_3(PO_4)_2$	$FePO_4$
22	Silicate,	I			W		A	w						I		
	—(SiO_3)	$Al_2(—)_3$			$Ba(—)$		$Cd(—)$	$Ca(—)$		Co_2SiO_4	$Cu(—)$			H_2SiO_3		
23	Sulfate	W	W	A	a	d	W	W	w	W(I)*	W			W	W	w
		$Al_2(SO_4)_3$	$(NH_4)_2SO_4$	$Sb_2(SO_4)_3$	$BaSO_4$	$Bi_2(SO_4)_3$	$CdSO_4$	$CaSO_4$	$Cr_2(SO_4)_3$	$CoSO_4$	$CuSO_4$			H_2SO_4	$FeSO_4$	$Fe_2(SO_4)_3$
24	Sulfide	d	W	A	d	A	A	w	d	A	A	I	I	W	A	d
		Al_2S_3	$(NH_4)_2S$	Sb_2S_3	BaS	Bi_2S_3	CdS	CaS	Cr_2S_3	CoS	CuS	Au_2S	Au_2S_3	H_2S	FeS	Fe_2S_3
25	Tartrate	w	W	W	A	A	W	A		W	W	A	I	W	A	d
	—$(C_4H_4O_6)$	$Al_2(—)_3$	$(NH_4)_2(—)$	$Sb_2(—)_3$	$Ba(—)$	$Bi_2(—)_3$	$Cd(—)$	$Ca(—)$		$Co(—)$	$Cu(—)$			$C_4H_6O_6$	$Fe(—)$	$Fe_2(—)_3$
26	Thiocy'te		W		W			W		W	d			W	W	W
			NH_4CNS		$Ba(CNS)_2$			$Ca(CNS)$		$Co(CNS)_2$	$CuCNS$			$CNSH$	$Fe(CNS)_2$	$Fe(CNS)_3$

Solubility Chart

No.		Pb	Mg	Mn	Hg (I)	Hg (II)	Ni	K	Pt	Ag	Na	Sn (IV)	Sn (II)	Sr	Zn
1	Acetate	W	W	W	w	W	W	W		w	W	W	d	W	W
	—$(C_2H_3O_2)$	$Pb(—)_2$	$Mg(—)_2$	$Mn(—)_2$	$Hg(—)$	$Hg(—)_2$	$Ni(—)_2$	$K(—)$		$Ag(—)$	$Na(—)$	$Sn(—)_4$	$Sn(—)_2$	$Sr(—)_2$	$Zn(—)_2$
2	Arsenate	A	A	w	A	w	A	W		A	W			w	A
	—(AsO_4)	$PbH(—)$	$Mg_3(—)$	$MnH(—)$	$Hg_3(—)$	$Hg_3(—)_2$	$Ni_3(—)_2$	$K_3(—)$		$Ag_3(—)$	$Na_3(—)$			$SrH(—)$	$Zn_3(—)_2$
3	Arsenite		W	A	A	A	A	W		A	W		A	w	
	—(AsO_3)		$Mg_3(—)_2$	$Mn_3H_6(—)_4$	$Hg_3(—)$	$Hg_5(—)$	$Ni_3H_6(—)_4$	K_3AsO_3		$Ag_3(—)$	$Na_2H(—)$		$Sn_3(—)_2$	$Sr_3(—)_2$	
4	Benzoate	w	W	W	W	A	w	W		w	W				W
	—$(C_7H_5O_2)$	$Pb(—)_2$	$Mg(—)_2$	$Mn(—)_2$	$Hg_2(—)_2$	$Hg(—)_2$	$Ni(—)_2$	$K(—)$		$Ag(—)$	$Na(—)$				$Zn(—)_2$
5	Bromide	W	W	W	A	W	W	W	w	a	W	W	W	W	W
		$PbBr_2$	$MgBr_2$	$MnBr_2$	$HgBr$	$HgBr_2$	$NiBr_2$	KBr	$PtBr_4$	$AgBr$	$NaBr$	$SnBr_4$	$SnBr_2$	$SrBr_2$	$ZnBr_2$
6	Carbonate	A	w	w	A		w	W		A	W			w	w
		$PbCO_3$	$MgCO_3$	$MnCO_3$	Hg_2CO_3		$NiCO_3$	K_2CO_3		Ag_2CO_3	Na_2CO_3			$SrCO_3$	$ZnCO_3$
7	Chlorate	W	W	W	W	W	W	W		W	W		W	W	W
	—(ClO_3)	$Pb(—)_2$	$Mg(—)_2$	$Mn(—)_2$	$Hg(—)$	$Hg(—)_2$	$Ni(—)_2$	$K(—)$		$Ag(—)$	$Na(—)$		$Sn(—)_2$	$Sr(—)_2$	$Zn(—)_2$
8	Chloride	W	W	W	a	W	W	W	W	a	W	W	W	W	W
		$PbCl_2$	$MgCl_2$	$MnCl_2$	$HgCl$	$HgCl_2$	$NiCl_2$	KCl	$PtCl_4$	$AgCl$	$NaCl$	$SnCl_4$	$SnCl_2$	$SrCl_2$	$ZnCl_2$
9	Chromate	A	W		w	w	A	W		w	W	W	A	w	w
	—(CrO_4)	$Pb(—)$	$Mg(—)$		$Hg_2(—)$	$Hg(—)$	$Ni(—)$	$K_2(—)$		$Ag_2(—)$	$Ma_2(—)$	$Sn(—)_2$	$Sn(—)$	$Sr(—)$	$Zn(—)$
10	Citrate	W	W	w	w		W	W		w	W			A	A
	—$(C_6H_5O_7)$	$Pb_3(—)_2$	$Mg_3(—)_2$	$MnH(—)$	$Hg_3(—)$		$Ni_3(—)_2$	$K_3(—)$		$Ag_3(—)$	$Na_3(—)$			$SrH(—)$	$Zn_3(—)_2$
11	Cyanide	w	W		A	W	a	W	I	a	W			W	A
		$Pb(CN)_2$	$Mg(CN)_2$		$HgCN$	$Hg(CN)_2$	$Ni(CN)_2$	KCN	$Pt(CN)_2$	$AgCN$	$NaCN$			$Sr(CN)_2$	$Zn(CN)_2$
12	Ferricy'de	w	W			A	I	W		A	W		A	W	A
	—$Fe(CN)_6$	$Pb_3(—)_2$	$Mg_3(—)_2$			$Hg_3(—)_2$	$Ni_3(—)_2$	$K_3(—)$		$Ag_3(—)$	$Na_3(—)$		$Sn_3(—)_2$	$Sr_3(—)_2$	$Zn_3(—)_2$
13	Ferrocy'de	a	W	A	I	I	I	W		I	W		a	W	I
	—$Fe(CN)_6$	$Pb_2(—)$	$Mg_2(—)$	$Mn_2(—)$		$Hg_2(—)$	$Ni_2(—)$	$K_4(—)$		$Ag_4(—)$	$Na_4(—)$		$Sn_2(—)$	$Sr_2(—)$	$Zn_2(—)$
14	Fluoride	w	w	A	d	d	w	W	W	W	W	W	W	w	w
		PbF_2	MgF_2	MnF_2	HgF	HgF_2	NiF_2	KF	PtF_4	AgF	NaF	SnF_4	SnF_2	SrF_2	ZnF_2
15	Formate	W	W	W	w	W	W	W		W	W			W	W
	—(CHO_2)	$Pb(—)_2$	$Mg(—)_2$	$Mn(—)_2$	$Hg(—)$	$Hg(—)_2$	$Ni(—)_2$	$K(—)$		$Ag(—)$	$Na(—)$			$Sr(—)_2$	$Zn(—)_2$
16	Hydroxide	w	A	A		A	w	W	A		W	w	A	W	A
		$Pb(OH)_2$	$Mg(OH)_2$	$Mn(OH)_2$		$Hg(OH)_2$	$Ni(OH)_2$	KOH	$Pt(OH)_4$		$NaOH$	$Sn(OH)_4$	$Sn(OH)_2$	$Sr(OH)_2$	$Zn(OH)_2$
17	Iodide	w	W	W	A	w	W	W	I	I	W	d	W	W	W
		PbI_2	MgI_2	MnI_2	HgI	HgI_2	NiI_2	KI	PtI_2	AgI	NaI	SnI_4	SnI_2	SrI_2	ZnI_2
18	Nitrate	W	W	W	W	W	W	W	W	W	W		d	W	W
		$Pb(NO_3)_2$	$Mg(NO_3)_2$	$Mn(NO_3)_2$	$HgNO_3$	$Hg(NO_3)_2$	$Ni(NO_3)_2$	KNO_3	$Pt(NO_3)_4$	$AgNO_3$	$NaNO_3$		$Sn(NO_3)_2$	$Sr(NO_3)_2$	$Zn(NO_3)_2$
19	Oxalate	A	w	w	a	A	A	W		a	W		A	A	A
	—(C_2O_4)	$Pb(—)$	$Mg(—)$	$Mn(—)$	$Hg_2(—)$	$Hg(—)$	$Ni(—)$	$K_2(—)$		$Ag_2(—)$	$Na_2(—)$		$Sn(—)$	$Sr(—)$	$Zn(—)$
20	Oxide	w	A	A	A	A	w	W	A	w	d	A	A	W	A
		PbO	MgO	MnO	Hg_2O	HgO	NiO	K_2O	PtO	Ag_2O	Na_2O	SnO_2	SnO	SrO	ZnO
21	Phosphate	A	w	w	A	A	A	W		A	W	A		A	A
		$Pb_3(PO_4)_2$	$Mg_3(PO_4)_2$	$Mn_3(PO_4)_2$	Hg_3PO_4	$Hg_3(PO_4)_2$	$Ni_3(PO_4)_2$	K_3PO_4		Ag_3PO_4	Na_3PO_4	$Sn_3(PO_4)$		$Sr_3(PO_4)_2$	$Zn_3(PO_4)_2$
22	Silicate	A	A	I				W			W			A	A
	—(SiO_3)	$Pb(—)$	$Mg(—)$	$Mn(—)$				$K_2(—)$			$Na_2(—)$			$Sr(—)$	$Zn(—)$
23	Sulfate	w	W	W	w	d	W	W	W	w	W	W	W	w	W
		$PbSO_4$	$MgSO_4$	$MnSO_4$	Hg_2SO_4	$HgSO_4$	$NiSO_4$	K_2SO_4	$Pt(SO_4)_2$	Ag_2SO_4	Na_2SO_4	$Sn(SO_4)_2$	$SnSO_4$	$SrSO_4$	$ZnSO_4$
24	Sulfide	A	d	A	I	I	A	W	I	A	W	I	A	W	A
		PbS	MgS	MnS	Hg_2S	HgS	NiS	K_2S	PtS	Ag_2S	Na_2S	SnS_2	SnS	SrS	ZnS
25	Tartrate	A	w	w	I	I	A	W	I	w	W	A	A	W	A
	—$(C_4H_4O_6)$	$Pb(—)$	$Mg(—)$	$Mn(—)$	$Hg_2(—)$		$Ni(—)$	$K_2(—)$		$Ag_2(—)$	$Na_2(—)$	$Sn(—)$		$Sr(—)$	$Zn(—)$
26	Thiocy'te	w	W	W	A	w		W		I	W			W	W
		$Pb(CNS)_2$	$Mg(CNS)_2$	$Mn(CNS)_2$	$HgCNS$	$Hg(CNS)_2$		$KCNS$		$AgCNS$	$NaCNS$			$Sr(CNS)_2$	$Zn(CNS)_2$

Section 6
Fluid Properties

THERMOPHYSICAL PROPERTIES OF WATER AND STEAM

Eric W. Lemmon

These tables summarize the thermophysical properties of water and steam at equilibrium as accepted by the International Association for the Properties of Water and Steam (www.iapws. org) for general and scientific use. The thermodynamic properties are calculated from the equation of state of Wagner and Pruss (Ref. 6). The reference state for these tables is the liquid at the triple point, at which the internal energy and entropy are taken as zero. These tables refer to states at 1 bar (100 kPa) pressure with temperatures in °C in the first section; liquid and gaseous states at equilibrium as a function of temperature in the second section; and properties along isobars in the third section. The tabulated properties are pressure (P), density (ρ), enthalpy (H), entropy (S), isochoric heat capacity (C_v), isobaric heat capacity (C_p), speed of sound (u), viscosity (η), thermal conductivity (λ), and static dielectric constant (D). In the saturation tables, the first line of identical temperatures is for the liquid state and the second line is for the vapor state. A duplicate entry in the isobar section indicates a phase transition (liquid-vapor) at that temperature; property values are then given for both phases. These are identified by the high densities in the liquid and the low densities in the vapor. The temperature scale is ITS-90. Additional calculations at state points not listed below can be obtained by using the NIST Standard Reference Data program REFPROP (Ref. 5) or the water-specific program Steam (Ref. 2).

References

1. Fernández, D.P., Goodwin, A.R.H., Lemmon, E.W., Levelt Sengers, J.M.H., and Williams, R.C., A Formulation for the Static Permittivity of Water and Steam at Temperatures from 238 K to 873 K at Pressures up to 1200 MPa, Including Derivatives and Debye-Hückel Coefficients, *J. Phys. Chem. Ref. Data* 26, 1125, 1997.

2. Harvey, A.H., Peskin, A.P., and Klein, S.A., NIST Standard Reference Database 10: NIST/ASME Steam Properties, Version 2.22, National Institute of Standards and Technology, Standard Reference Data Program, Gaithersburg, Maryland, 2008 (www.nist.gov/srd/nist10. cfm).

3. Huber, M.L., Perkins, R.A., Laesecke, A., Friend, D.G., Sengers, J.V., Assael, M.J., Metaxa, I.M., Vogel, E., Mares, R., and Miyagawa, K., New International Formulation for the Viscosity of Water, *J. Phys. Chem. Ref. Data* 38, 101, 2009.

4. Kestin, J., Sengers, J.V., Kamgar-Parsi, B., and Levelt Sengers, J.M.H., Thermophysical Properties of Fluid H₂O, *J. Phys. Chem. Ref. Data* 13, 175, 1984.

5. Lemmon, E.W., Huber, M.L., and McLinden, M.O., NIST Standard Reference Database 23: Reference Fluid Thermodynamic and Transport Properties-REFPROP, Version 9.0, National Institute of Standards and Technology, Standard Reference Data Program, Gaithersburg, Maryland, 2010 (www.nist.gov/srd/nist23.cfm).

6. Wagner, W. and Pruss, A., The IAPWS Formulation 1995 for the Thermodynamic Properties of Ordinary Water Substance for General and Scientific Use, *J. Phys. Chem. Ref. Data* 31, 387, 2002.

T	P	ρ	H	S	C_v	C_p	u		η	λ
°C	MPa	kg m⁻³	kJ kg⁻¹	kJ kg⁻¹ K⁻¹	kJ kg⁻¹ K⁻¹	kJ kg⁻¹ K⁻¹	m s⁻¹	D	μPa s	mW m⁻¹ K⁻¹
P = 0.1 MPa (1 bar)										
0.01	0.1	999.84	0.10186	0.000007	4.2170	4.2194	1402.4	87.899	1791.1	561.09
10	0.1	999.70	42.118	0.15108	4.1906	4.1952	1447.3	83.974	1305.9	580.05
20	0.1	998.21	84.006	0.29646	4.1567	4.1841	1482.3	80.223	1001.6	598.46
25	0.1	997.05	104.92	0.36720	4.1376	4.1813	1496.7	78.408	890.02	607.19
30	0.1	995.65	125.82	0.43673	4.1172	4.1798	1509.2	76.634	797.22	615.50
40	0.1	992.22	167.62	0.57237	4.0734	4.1794	1528.9	73.201	652.73	630.63
50	0.1	988.03	209.42	0.70377	4.0262	4.1813	1542.6	69.916	546.52	643.59
60	0.1	983.20	251.25	0.83125	3.9765	4.1850	1551.0	66.774	466.03	654.39
70	0.1	977.76	293.12	0.95509	3.9251	4.1901	1554.7	63.770	403.55	663.13
80	0.1	971.79	335.05	1.0755	3.8728	4.1968	1554.4	60.898	354.05	670.01
90	0.1	965.31	377.06	1.1928	3.8204	4.2052	1550.4	58.152	314.17	675.27
99.606	0.1	958.63	417.50	1.3028	3.7702	4.2152	1543.5	55.628	282.75	678.97
99.606	0.1	0.59034	2674.9	7.3588	1.5548	2.0784	471.99	1.0058	12.218	25.053
100	0.1	0.58967	2675.8	7.3610	1.5535	2.0766	472.28	1.0058	12.234	25.079
Saturation										
0.01	0.000612	999.79	0.000612	0.0	4.2174	4.2199	1402.3	87.895	1791.4	561.04
0.01	0.000612	0.0048546	2500.9	9.1555	1.4184	1.8844	409.00	1.00006	8.9458	17.071
10	0.0012282	999.65	42.021	0.15109	4.1910	4.1955	1447.1	83.971	1306.0	580.00
10	0.0012282	0.0094071	2519.2	8.8998	1.4269	1.8947	416.17	1.00012	9.2384	17.621
20	0.0023393	998.16	83.914	0.29648	4.1570	4.1844	1482.2	80.219	1001.6	598.42
20	0.0023393	0.017314	2537.4	8.6660	1.4359	1.9059	423.18	1.00021	9.5441	18.227
25	0.0031699	997.00	104.83	0.36722	4.1379	4.1816	1496.5	78.405	890.04	607.15
25	0.0031699	0.023075	2546.5	8.5566	1.4405	1.9118	426.63	1.00028	9.7009	18.550
30	0.0042470	995.61	125.73	0.43675	4.1175	4.1801	1509.0	76.630	797.22	615.46
30	0.0042470	0.030415	2555.5	8.4520	1.4452	1.9180	430.03	1.00036	9.8602	18.887
40	0.0073849	992.18	167.53	0.57240	4.0737	4.1796	1528.7	73.197	652.72	630.58

T	P	ρ	H	S	C_v	C_p	u		η	λ
°C	MPa	kg m⁻³	kJ kg⁻¹	kJ kg⁻¹ K⁻¹	kJ kg⁻¹ K⁻¹	kJ kg⁻¹ K⁻¹	m s⁻¹	D	μPa s	mW m⁻¹ K⁻¹
40	0.0073849	0.051242	2573.5	8.2555	1.4552	1.9314	436.71	1.00059	10.185	19.599
50	0.012352	988.00	209.34	0.70381	4.0264	4.1815	1542.4	69.913	546.50	643.55
50	0.012352	0.083147	2591.3	8.0748	1.4663	1.9468	443.21	1.00094	10.516	20.365
60	0.019946	983.16	251.18	0.83129	3.9767	4.1851	1550.8	66.772	466.02	654.35
60	0.019946	0.13043	2608.8	7.9081	1.4789	1.9648	449.50	1.0014	10.854	21.187
70	0.031201	977.73	293.07	0.95513	3.9252	4.1902	1554.6	63.768	403.53	663.09
70	0.031201	0.19843	2626.1	7.7540	1.4937	1.9862	455.57	1.0021	11.195	22.068
80	0.047414	971.77	335.01	1.0756	3.8729	4.1969	1554.3	60.896	354.04	669.99
80	0.047414	0.29367	2643.0	7.6111	1.5111	2.0120	461.39	1.0030	11.539	23.011
90	0.070182	965.30	377.04	1.1929	3.8204	4.2053	1550.4	58.151	314.17	675.25
90	0.070182	0.42390	2659.5	7.4781	1.5316	2.0429	466.94	1.0043	11.885	24.019
100	0.10142	958.35	419.17	1.3072	3.7682	4.2157	1543.2	55.527	281.58	679.09
100	0.10142	0.59817	2675.6	7.3541	1.5558	2.0800	472.20	1.0059	12.232	25.096
100	0.10142	958.35	419.17	1.3072	3.7682	4.2157	1543.2	55.527	281.58	679.09
100	0.10142	0.59817	2675.6	7.3541	1.5558	2.0800	472.20	1.0059	12.232	25.096
120	0.19867	943.11	503.81	1.5279	3.6662	4.2435	1519.9	50.620	232.03	683.19
120	0.19867	1.1221	2705.9	7.1291	1.6177	2.1770	481.73	1.0105	12.927	27.467
140	0.36154	926.13	589.16	1.7392	3.5694	4.2826	1486.2	46.131	196.64	683.30
140	0.36154	1.9667	2733.4	6.9293	1.7002	2.3109	489.82	1.0177	13.618	30.140
160	0.61823	907.45	675.47	1.9426	3.4788	4.3354	1443.2	42.018	170.43	679.96
160	0.61823	3.2596	2757.4	6.7491	1.8044	2.4883	496.29	1.0282	14.304	33.131
180	1.0028	887.00	763.05	2.1392	3.3949	4.4050	1391.7	38.235	150.38	673.32
180	1.0028	5.1588	2777.2	6.5840	1.9279	2.7129	501.04	1.0431	14.985	36.449
200	1.5549	864.66	852.27	2.3305	3.3179	4.4958	1332.1	34.742	134.58	663.31
200	1.5549	7.8610	2792.0	6.4302	2.0666	2.9895	503.92	1.0636	15.666	40.113
220	2.3196	840.22	943.58	2.5177	3.2479	4.6146	1264.5	31.495	121.77	649.65
220	2.3196	11.615	2800.9	6.2840	2.2172	3.3289	504.77	1.0915	16.354	44.170
240	3.3469	813.37	1037.6	2.7020	3.1850	4.7719	1189.0	28.455	111.06	631.85
240	3.3469	16.749	2803.0	6.1423	2.3794	3.7537	503.32	1.1292	17.062	48.729
260	4.6923	783.63	1135.0	2.8849	3.1301	4.9856	1105.3	25.580	101.81	609.24
260	4.6923	23.712	2796.6	6.0016	2.5555	4.3075	499.21	1.1802	17.810	54.035
280	6.4166	750.28	1236.9	3.0685	3.0849	5.2889	1012.6	22.824	93.550	581.15
280	6.4166	33.165	2779.9	5.8579	2.7503	5.0731	491.93	1.2505	18.630	60.622
300	8.5879	712.14	1345.0	3.2552	3.0530	5.7504	909.40	20.135	85.855	547.43
300	8.5879	46.168	2749.6	5.7059	2.9708	6.2197	480.73	1.3504	19.580	69.667
320	11.284	667.09	1462.2	3.4494	3.0428	6.5373	793.16	17.440	78.310	509.21
320	11.284	64.638	2700.6	5.5372	3.2276	8.1589	464.43	1.5012	20.773	83.944
340	14.601	610.67	1594.5	3.6601	3.0781	8.2080	658.27	14.606	70.331	468.55
340	14.601	92.759	2621.8	5.3356	3.5430	12.236	440.72	1.7555	22.477	111.00
360	18.666	527.59	1761.7	3.9167	3.2972	15.004	479.74	11.225	60.306	425.77
360	18.666	143.90	2481.5	5.0536	4.0068	27.356	402.37	2.3096	25.638	181.77
373.95	22.064	322.00	2084.3	4.4070				5.3606	79.791	

T	P	ρ	H	S	C_v	C_p	u		η	λ
K	MPa	kg m⁻³	kJ kg⁻¹	kJ kg⁻¹ K⁻¹	kJ kg⁻¹ K⁻¹	kJ kg⁻¹ K⁻¹	m s⁻¹	D	μPa s	mW m⁻¹ K⁻¹
P = 0.1 MPa (1 bar)										
273.16	0.1	999.84	0.10186	0.000007	4.2170	4.2194	1402.4	87.899	1791.1	561.09
280	0.1	999.91	28.894	0.10411	4.1998	4.2009	1434.3	85.192	1433.6	574.09
300	0.1	996.56	112.65	0.39306	4.1302	4.1806	1501.5	77.747	853.74	610.32
320	0.1	989.43	196.25	0.66281	4.0414	4.1805	1538.9	70.935	576.73	639.75
340	0.1	979.54	279.93	0.91646	3.9414	4.1883	1554.0	64.702	421.63	660.58
360	0.1	967.40	363.82	1.1562	3.8369	4.2023	1552.1	59.004	325.86	673.78
372.756	0.1	958.63	417.50	1.3028	3.7702	4.2152	1543.5	55.628	282.75	678.97
372.756	0.1	0.59034	2674.9	7.3588	1.5548	2.0784	471.99	1.0058	12.218	25.053
380	0.1	0.57824	2689.9	7.3986	1.5356	2.0507	477.08	1.0056	12.498	25.546

T	P	ρ	H	S	C_v	C_p	u	D	η	λ
K	MPa	kg m⁻³	kJ kg⁻¹	kJ kg⁻¹ K⁻¹	kJ kg⁻¹ K⁻¹	kJ kg⁻¹ K⁻¹	m s⁻¹		μPa s	mW m⁻¹ K⁻¹
400	0.1	0.54761	2730.4	7.5025	1.5082	2.0078	490.31	1.0051	13.278	27.008
450	0.1	0.48458	2829.7	7.7365	1.4943	1.9752	520.60	1.0040	15.267	31.168
500	0.1	0.43514	2928.6	7.9447	1.5082	1.9813	548.31	1.0033	17.299	35.861
550	0.1	0.39507	3028.1	8.1344	1.5319	2.0010	574.19	1.0027	19.356	40.953
600	0.1	0.36185	3128.8	8.3096	1.5600	2.0268	598.61	1.0023	21.425	46.367
650	0.1	0.33384	3230.8	8.4730	1.5903	2.0557	621.79	1.0020	23.496	52.049
700	0.1	0.30988	3334.4	8.6264	1.6222	2.0867	643.92	1.0017	25.562	57.964
750	0.1	0.28915	3439.5	8.7715	1.6553	2.1191	665.11	1.0015	27.617	64.083
800	0.1	0.27102	3546.3	8.9093	1.6892	2.1525	685.47	1.0013	29.657	70.385
900	0.1	0.24085	3765.0	9.1668	1.7589	2.2216	724.03	1.0011	33.680	83.466
1000	0.1	0.21673	3990.7	9.4045	1.8297	2.2921	760.17	1.00088	37.615	97.085
1100	0.1	0.19701	4223.4	9.6263	1.9000	2.3621	794.33	1.00074	41.453	111.15
1200	0.1	0.18058	4463.0	9.8348	1.9682	2.4302	826.85	1.00063	45.192	125.58

P = 1 MPa (10 bar)

T	P	ρ	H	S	C_v	C_p	u	D	η	λ
273.16	1.0	1000.3	1.0180	0.000066	4.2127	4.2150	1403.9	87.937	1789.1	561.59
280	1.0	1000.3	29.783	0.10407	4.1960	4.1973	1435.7	85.228	1432.5	574.54
300	1.0	996.96	113.48	0.39281	4.1272	4.1781	1503.0	77.781	853.66	610.73
320	1.0	989.82	197.03	0.66242	4.0390	4.1784	1540.5	70.967	576.89	640.16
340	1.0	979.93	280.67	0.91594	3.9395	4.1863	1555.7	64.734	421.86	661.02
360	1.0	967.81	364.52	1.1556	3.8353	4.2004	1553.9	59.035	326.10	674.24
380	1.0	953.74	448.73	1.3832	3.7315	4.2220	1538.4	53.827	262.82	681.49
400	1.0	937.87	533.47	1.6005	3.6315	4.2535	1511.3	49.065	218.82	684.10
450	1.0	890.39	749.20	2.1086	3.4076	4.3924	1400.6	38.814	153.23	674.64
453.028	1.0	887.13	762.52	2.1381	3.3954	4.4045	1392.0	38.258	150.49	673.37
453.028	1.0	5.1450	2777.1	6.5850	1.9271	2.7114	501.02	1.0430	14.981	36.428
500	1.0	4.5323	2891.2	6.8250	1.6699	2.2795	535.74	1.0344	17.054	38.799
550	1.0	4.0581	3001.8	7.0359	1.6159	2.1647	565.75	1.0282	19.215	42.798
600	1.0	3.6871	3109.0	7.2224	1.6098	2.1292	592.58	1.0236	21.349	47.636
650	1.0	3.3843	3215.2	7.3925	1.6227	2.1254	617.34	1.0202	23.462	53.002
700	1.0	3.1305	3321.7	7.5504	1.6447	2.1368	640.55	1.0174	25.555	58.735
750	1.0	2.9140	3429.0	7.6984	1.6715	2.1566	662.53	1.0153	27.628	64.745
800	1.0	2.7265	3537.5	7.8384	1.7014	2.1816	683.48	1.0135	29.680	70.983
900	1.0	2.4174	3758.5	8.0986	1.7663	2.2402	722.85	1.0108	33.716	84.000
1000	1.0	2.1723	3985.7	8.3380	1.8346	2.3048	759.50	1.0088	37.655	97.573
1100	1.0	1.9729	4219.5	8.5608	1.9034	2.3713	794.01	1.0074	41.494	111.57
1200	1.0	1.8074	4460.0	8.7699	1.9708	2.4371	826.77	1.0063	45.231	125.89

P = 10 MPa (100 bar)

T	P	ρ	H	S	C_v	C_p	u	D	η	λ
273.16	10.0	1004.8	10.111	0.00049	4.1721	4.1726	1418.4	88.311	1770.0	566.56
280	10.0	1004.7	38.613	0.10355	4.1593	4.1622	1450.3	85.588	1422.2	579.06
300	10.0	1001.0	121.73	0.39029	4.0984	4.1536	1518.2	78.113	852.99	614.81
320	10.0	993.70	204.84	0.65846	4.0157	4.1580	1556.4	71.284	578.57	644.31
340	10.0	983.84	288.08	0.91079	3.9205	4.1672	1572.6	65.044	424.17	665.40
360	10.0	971.85	371.56	1.1493	3.8198	4.1810	1572.0	59.345	328.53	678.93
380	10.0	957.99	455.37	1.3759	3.7188	4.2013	1558.0	54.140	265.22	686.53
400	10.0	942.42	539.67	1.5921	3.6210	4.2302	1532.7	49.385	221.17	689.57
450	10.0	896.16	753.94	2.0967	3.4010	4.3553	1428.4	39.172	155.48	681.51
500	10.0	838.02	977.18	2.5669	3.2211	4.6022	1271.3	30.794	119.83	651.64
550	10.0	761.82	1218.8	3.0270	3.0865	5.1407	1054.6	23.531	96.080	592.51
584.147	10.0	688.42	1408.1	3.3606	3.0438	6.1237	847.33	18.660	81.718	526.84
584.147	10.0	55.463	2725.5	5.6160	3.1065	7.1408	472.51	1.4248	20.194	76.569
600	10.0	49.773	2820.0	5.7756	2.6239	5.1365	503.34	1.3649	21.017	71.115
650	10.0	40.479	3022.6	6.1009	2.1103	3.3968	562.10	1.2672	23.472	67.341
700	10.0	35.355	3177.4	6.3305	1.9338	2.8741	602.20	1.2145	25.773	69.297
750	10.0	31.810	3314.6	6.5200	1.8625	2.6452	634.58	1.1793	27.973	73.331

T	P	ρ	H	S	C_v	C_p	u		η	λ
K	MPa	kg m⁻³	kJ kg⁻¹	kJ kg⁻¹ K⁻¹	kJ kg⁻¹ K⁻¹	kJ kg⁻¹ K⁻¹	m s⁻¹	D	μPa s	mW m⁻¹ K⁻¹
800	10.0	29.107	3443.7	6.6867	1.8367	2.5313	662.61	1.1536	30.101	78.476
900	10.0	25.123	3691.6	6.9787	1.8439	2.4458	710.98	1.1182	34.201	90.516
1000	10.0	22.241	3935.5	7.2357	1.8843	2.4397	753.03	1.0948	38.144	103.50
1100	10.0	20.017	4180.6	7.4693	1.9377	2.4661	791.02	1.0782	41.960	116.73
1200	10.0	18.230	4429.2	7.6855	1.9957	2.5070	826.16	1.0659	45.663	130.00
P = 100 MPa (1000 bar)										
273.16	100.0	1045.3	95.444	−0.0083717	3.8761	3.9053	1575.5	91.834	1660.1	612.20
280	100.0	1043.6	122.26	0.088571	3.8869	3.9328	1603.8	88.976	1367.4	621.51
300	100.0	1037.2	201.44	0.36171	3.8751	3.9798	1667.9	81.216	859.19	654.50
320	100.0	1028.9	281.30	0.61941	3.8289	4.0043	1707.7	74.222	599.18	684.86
340	100.0	1019.0	361.55	0.86265	3.7637	4.0194	1728.7	67.891	448.03	707.90
360	100.0	1007.8	442.05	1.0927	3.6883	4.0309	1735.4	62.150	352.49	724.05
380	100.0	995.37	522.79	1.3110	3.6089	4.0427	1730.9	56.937	288.37	734.91
400	100.0	981.82	603.78	1.5187	3.5293	4.0569	1717.3	52.201	243.33	741.80
450	100.0	943.51	807.84	1.9993	3.3430	4.1105	1652.8	42.149	175.71	745.40
500	100.0	899.21	1015.4	2.4366	3.1820	4.1968	1555.7	34.149	139.57	730.42
550	100.0	848.78	1228.2	2.8421	3.0452	4.3234	1435.5	27.667	117.36	696.87
600	100.0	791.49	1448.6	3.2256	2.9295	4.5019	1300.4	22.290	101.85	645.83
650	100.0	726.21	1679.5	3.5952	2.8329	4.7503	1158.6	17.717	89.647	580.63
700	100.0	651.77	1925.0	3.9589	2.7538	5.0832	1020.0	13.754	79.123	510.19
750	100.0	568.52	2188.5	4.3223	2.6866	5.4492	898.84	10.336	69.732	433.71
800	100.0	482.23	2466.5	4.6811	2.6169	5.6108	813.97	7.5622	61.842	351.53
900	100.0	343.61	3000.1	5.3104	2.4386	4.8879	765.30	4.2835	52.771	257.05
1000	100.0	265.45	3440.1	5.7749	2.2950	3.9788	792.50	2.9559	50.506	232.07
1100	100.0	220.62	3809.8	6.1276	2.2276	3.4715	832.67	2.3472	51.089	223.70
1200	100.0	191.53	4142.5	6.4172	2.2098	3.2098	872.28	2.0111	52.802	219.07

VAPOR PRESSURE AND OTHER SATURATION PROPERTIES OF WATER

Eric W. Lemmon

This table summarizes the vapor pressure, enthalpy (heat) of vaporization, and surface tension of water as accepted by the International Association for the Properties of Water and Steam (www.iapws.org) for general and scientific use. The vapor pressure and heat of vaporization are calculated from the equation of state of Wagner and Pruss (Ref. 1). The temperature scale is ITS-90. Additional calculations at state points not listed below can be obtained by using the NIST Standard Reference Data program REFPROP (www.nist.gov/srd/nist23.htm) or the water-specific program Steam (www.nist.gov/srd/nist10.htm).

References

1. Wagner, W. and Pruss, A., The IAPWS Formulation 1995 for the Thermodynamic Properties of Ordinary Water Substance for General and Scientific Use, *J. Phys. Chem. Ref. Data*, 31, 387, 2002.
2. International Association for the Properties of Water and Steam, Release on the surface tension of ordinary water substance, Physical Chemistry of Aqueous Systems: Proceedings of the 12th International Conference on the Properties of Water and Steam, Orlando, Florida, September 11–16, 1994, pp. A139–A142.

t °C	P kPa	$\Delta_{vap}H$ kJ kg^{-1}	Surf. Ten. mN m^{-1}	t °C	P kPa	$\Delta_{vap}H$ kJ kg^{-1}	Surf. Ten. mN m^{-1}
0.01	0.61165	2500.9	75.65	80	47.414	2308.0	62.67
2	0.70599	2496.2	75.37	82	51.387	2302.9	62.31
4	0.81355	2491.4	75.08	84	55.635	2297.9	61.94
6	0.93536	2486.7	74.80	86	60.173	2292.8	61.56
8	1.0730	2481.9	74.51	88	65.017	2287.6	61.19
10	1.2282	2477.2	74.22	90	70.182	2282.5	60.82
12	1.4028	2472.5	73.93	92	75.684	2277.3	60.44
14	1.5990	2467.7	73.63	94	81.541	2272.1	60.06
16	1.8188	2463.0	73.34	96	87.771	2266.9	59.68
18	2.0647	2458.3	73.04	98	94.390	2261.7	59.30
20	2.3393	2453.5	72.74	100	101.42	2256.4	58.91
22	2.6453	2448.8	72.43	102	108.87	2251.1	58.53
24	2.9858	2444.0	72.13	104	116.78	2245.8	58.14
25	3.1699	2441.7	71.97	106	125.15	2240.4	57.75
26	3.3639	2439.3	71.82	108	134.01	2235.1	57.36
28	3.7831	2434.6	71.51	110	143.38	2229.6	56.96
30	4.2470	2429.8	71.19	112	153.28	2224.2	56.57
32	4.7596	2425.1	70.88	114	163.74	2218.7	56.17
34	5.3251	2420.3	70.56	116	174.77	2213.2	55.77
36	5.9479	2415.5	70.24	118	186.41	2207.7	55.37
38	6.6328	2410.8	69.92	120	198.67	2202.1	54.97
40	7.3849	2406.0	69.60	122	211.59	2196.5	54.56
42	8.2096	2401.2	69.27	124	225.18	2190.9	54.16
44	9.1124	2396.4	68.94	126	239.47	2185.2	53.75
46	10.099	2391.6	68.61	128	254.50	2179.5	53.34
48	11.177	2386.8	68.28	130	270.28	2173.7	52.93
50	12.352	2381.9	67.94	132	286.85	2167.9	52.52
52	13.631	2377.1	67.61	134	304.23	2162.1	52.11
54	15.022	2372.3	67.27	136	322.45	2156.2	51.69
56	16.533	2367.4	66.93	138	341.54	2150.3	51.27
58	18.171	2362.5	66.58	140	361.54	2144.3	50.86
60	19.946	2357.7	66.24	142	382.47	2138.3	50.44
62	21.867	2352.8	65.89	144	404.37	2132.2	50.01
64	23.943	2347.8	65.54	146	427.26	2126.1	49.59
66	26.183	2342.9	65.19	148	451.18	2119.9	49.17
68	28.599	2338.0	64.84	150	476.16	2113.7	48.74
70	31.201	2333.0	64.48	152	502.25	2107.5	48.31
72	34.000	2328.1	64.12	154	529.46	2101.2	47.89
74	37.009	2323.1	63.76	156	557.84	2094.8	47.46
76	40.239	2318.1	63.40	158	587.42	2088.4	47.02
78	43.703	2313.0	63.04	160	618.23	2082.0	46.59

t °C	P kPa	$\Delta_{vap}H$ kJ kg^{-1}	Surf. Ten. mN m^{-1}		t °C	P kPa	$\Delta_{vap}H$ kJ kg^{-1}	Surf. Ten. mN m^{-1}
162	650.33	2075.5	46.16		270	5503.0	1604.4	21.34
164	683.73	2068.9	45.72		272	5677.2	1592.5	20.87
166	718.48	2062.3	45.28		274	5855.6	1580.4	20.40
168	754.62	2055.6	44.85		276	6038.3	1568.1	19.93
170	792.19	2048.8	44.41		278	6225.2	1555.6	19.46
172	831.22	2042.0	43.97		280	6416.6	1543.0	18.99
174	871.76	2035.1	43.52		282	6612.4	1530.1	18.53
176	913.84	2028.2	43.08		284	6812.8	1517.1	18.06
178	957.51	2021.2	42.64		286	7017.7	1503.8	17.59
180	1002.8	2014.2	42.19		288	7227.4	1490.4	17.13
182	1049.8	2007.0	41.74		290	7441.8	1476.7	16.66
184	1098.5	1999.8	41.30		292	7661.0	1462.7	16.20
186	1148.9	1992.6	40.85		294	7885.2	1448.6	15.74
188	1201.1	1985.3	40.40		296	8114.3	1434.2	15.28
190	1255.2	1977.9	39.95		298	8348.5	1419.5	14.82
192	1311.2	1970.4	39.49		300	8587.9	1404.6	14.36
194	1369.1	1962.8	39.04		302	8832.5	1389.4	13.90
196	1429.0	1955.2	38.59		304	9082.4	1374.0	13.45
198	1490.9	1947.5	38.13		306	9337.8	1358.2	12.99
200	1554.9	1939.7	37.67		308	9598.6	1342.1	12.54
202	1621.0	1931.9	37.22		310	9865.1	1325.7	12.09
204	1689.3	1923.9	36.76		312	10137.	1309.0	11.64
206	1759.8	1915.9	36.30		314	10415.	1291.9	11.19
208	1832.6	1907.8	35.84		316	10699.	1274.5	10.75
210	1907.7	1899.6	35.38		318	10989.	1256.6	10.30
212	1985.1	1891.4	34.92		320	11284.	1238.4	9.86
214	2065.0	1883.0	34.46		322	11586.	1219.7	9.43
216	2147.3	1874.6	33.99		324	11895.	1200.6	8.99
218	2232.2	1866.0	33.53		326	12209.	1180.9	8.56
220	2319.6	1857.4	33.07		328	12530.	1160.8	8.13
222	2409.6	1848.6	32.60		330	12858.	1140.2	7.70
224	2502.3	1839.8	32.14		332	13193.	1118.9	7.28
226	2597.8	1830.9	31.67		334	13534.	1097.1	6.86
228	2696.0	1821.8	31.20		336	13882.	1074.6	6.44
230	2797.1	1812.7	30.74		338	14238.	1051.3	6.03
232	2901.0	1803.5	30.27		340	14601.	1027.3	5.63
234	3008.0	1794.1	29.80		342	14971.	1002.5	5.22
236	3117.9	1784.7	29.33		344	15349.	976.7	4.83
238	3230.8	1775.1	28.86		346	15734.	949.9	4.43
240	3346.9	1765.4	28.39		348	16128.	922.0	4.05
242	3466.2	1755.6	27.92		350	16529.	892.7	3.67
244	3588.7	1745.7	27.45		352	16939.	862.1	3.29
246	3714.5	1735.6	26.98		354	17358.	829.8	2.93
248	3843.6	1725.5	26.51		356	17785.	795.5	2.57
250	3976.2	1715.2	26.04		358	18221.	759.0	2.22
252	4112.2	1704.7	25.57		360	18666.	719.8	1.88
254	4251.8	1694.2	25.10		362	19121.	677.3	1.55
256	4394.9	1683.5	24.63		364	19585.	630.5	1.23
258	4541.7	1672.6	24.16		366	20060.	578.2	0.93
260	4692.3	1661.6	23.69		368	20546.	517.8	0.65
262	4846.6	1650.5	23.22		370	21044.	443.8	0.39
264	5004.7	1639.2	22.75		372	21554.	340.3	0.16
266	5166.8	1627.8	22.28		373.95	22064.	0.0	0.0
268	5332.9	1616.2	21.81					

STANDARD DENSITY OF WATER

This table gives the density ρ of water in the temperature range from 0 °C to 100 °C at a pressure of 101325 Pa (one standard atmosphere). Temperatures are given on the ITS-90 scale. From 0 °C to 40 °C the values are taken from the publication in Reference 1 and refer to standard mean ocean water (SMOW), free from dissolved salts and gases. SMOW is a standard water sample of high purity and known isotopic composition. Methods of correcting for different isotopic compositions are discussed in Ref. 2. The remaining values are calculated from the NIST REFPROP program, Ref. 3, which obtains thermodynamic properties from the equation of state of Wagner and Pruss given in Ref. 4.

References

1. Tanaka, M., Girard, G., Davis, R., Peuto, A., and Bignell, N., *Metrologia* 38, 301, 2001.
2. Marsh, K. N., Ed., *Recommended Reference Materials for the Realization of Physicochemical Properties*, Blackwell Scientific Publications, Oxford, 1987.
3. Lemmon, E.W., Huber, M.L., and McLinden, M.O., NIST Standard Reference Database 23: Reference Fluid Thermodynamic and Transport Properties-REFPROP, Version 9.0, National Institute of Standards and Technology, Standard Reference Data Program, Gaithersburg, Maryland, 2010 (www.nist.gov/srd/nist23.cfm).
4. Wagner, W., and Pruss, A., *J. Phys. Chem. Ref. Data* 31, 387, 2002.

t/°C	ρ/g cm^{-3}	t/°C	ρ/g cm^{-3}	t/°C	ρ/g cm^{-3}	t/°C	ρ/g cm^{-3}	t/°C	ρ/g cm^{-3}
0.1	0.9998495	4.1	0.9999748	8.1	0.9998452	12.1	0.9994890	16.1	0.9989296
0.2	0.9998560	4.2	0.9999746	8.2	0.9998389	12.2	0.9994774	16.2	0.9989132
0.3	0.9998624	4.3	0.9999742	8.3	0.9998325	12.3	0.9994657	16.3	0.9988967
0.4	0.9998685	4.4	0.9999736	8.4	0.9998260	12.4	0.9994539	16.4	0.9988800
0.5	0.9998745	4.5	0.9999728	8.5	0.9998193	12.5	0.9994419	16.5	0.9988633
0.6	0.9998803	4.6	0.9999719	8.6	0.9998125	12.6	0.9994298	16.6	0.9988464
0.7	0.9998859	4.7	0.9999709	8.7	0.9998056	12.7	0.9994176	16.7	0.9988294
0.8	0.9998913	4.8	0.9999697	8.8	0.9997985	12.8	0.9994052	16.8	0.9988123
0.9	0.9998966	4.9	0.9999683	8.9	0.9997912	12.9	0.9993927	16.9	0.9987951
1.0	0.9999017	5.0	0.9999668	9.0	0.9997839	13.0	0.9993801	17.0	0.9987778
1.1	0.9999066	5.1	0.9999651	9.1	0.9997764	13.1	0.9993674	17.1	0.9987603
1.2	0.9999113	5.2	0.9999633	9.2	0.9997687	13.2	0.9993546	17.2	0.9987428
1.3	0.9999158	5.3	0.9999613	9.3	0.9997610	13.3	0.9993416	17.3	0.9987251
1.4	0.9999202	5.4	0.9999592	9.4	0.9997530	13.4	0.9993285	17.4	0.9987073
1.5	0.9999244	5.5	0.9999569	9.5	0.9997450	13.5	0.9993153	17.5	0.9986895
1.6	0.9999285	5.6	0.9999544	9.6	0.9997368	13.6	0.9993020	17.6	0.9986715
1.7	0.9999323	5.7	0.9999518	9.7	0.9997285	13.7	0.9992885	17.7	0.9986534
1.8	0.9999360	5.8	0.9999491	9.8	0.9997200	13.8	0.9992749	17.8	0.9986351
1.9	0.9999396	5.9	0.9999462	9.9	0.9997114	13.9	0.9992612	17.9	0.9986168
2.0	0.9999429	6.0	0.9999431	10.0	0.9997027	14.0	0.9992474	18.0	0.9985984
2.1	0.9999461	6.1	0.9999400	10.1	0.9996938	14.1	0.9992335	18.1	0.9985798
2.2	0.9999491	6.2	0.9999366	10.2	0.9996848	14.2	0.9992194	18.2	0.9985611
2.3	0.9999519	6.3	0.9999331	10.3	0.9996757	14.3	0.9992052	18.3	0.9985424
2.4	0.9999546	6.4	0.9999295	10.4	0.9996665	14.4	0.9991909	18.4	0.9985235
2.5	0.9999571	6.5	0.9999257	10.5	0.9996571	14.5	0.9991765	18.5	0.9985045
2.6	0.9999595	6.6	0.9999217	10.6	0.9996475	14.6	0.9991619	18.6	0.9984854
2.7	0.9999616	6.7	0.9999176	10.7	0.9996379	14.7	0.9991473	18.7	0.9984662
2.8	0.9999636	6.8	0.9999134	10.8	0.9996281	14.8	0.9991325	18.8	0.9984469
2.9	0.9999655	6.9	0.9999090	10.9	0.9996182	14.9	0.9991176	18.9	0.9984275
3.0	0.9999672	7.0	0.9999045	11.0	0.9996081	15.0	0.9991026	19.0	0.9984079
3.1	0.9999687	7.1	0.9998998	11.1	0.9995979	15.1	0.9990874	19.1	0.9983883
3.2	0.9999700	7.2	0.9998950	11.2	0.9995876	15.2	0.9990722	19.2	0.9983686
3.3	0.9999712	7.3	0.9998900	11.3	0.9995772	15.3	0.9990568	19.3	0.9983487
3.4	0.9999722	7.4	0.9998849	11.4	0.9995666	15.4	0.9990413	19.4	0.9983287
3.5	0.9999731	7.5	0.9998797	11.5	0.9995559	15.5	0.9990257	19.5	0.9983087
3.6	0.9999738	7.6	0.9998743	11.6	0.9995451	15.6	0.9990100	19.6	0.9982885
3.7	0.9999743	7.7	0.9998687	11.7	0.9995341	15.7	0.9989942	19.7	0.9982682
3.8	0.9999747	7.8	0.9998631	11.8	0.9995230	15.8	0.9989782	19.8	0.9982478
3.9	0.9999749	7.9	0.9998572	11.9	0.9995118	15.9	0.9989621	19.9	0.9982273
4.0	0.9999749	8.0	0.9998513	12.0	0.9995005	16.0	0.9989459	20.0	0.9982067

$t/°C$	$\rho/\text{g cm}^{-3}$	$t/°C$	$\rho/\text{g cm}^{-3}$	$t/°C$	$\rho/\text{g cm}^{-3}$	$t/°C$	$\rho/\text{g cm}^{-3}$	$t/°C$	$\rho/\text{g cm}^{-3}$
20.1	0.9981860	25.3	0.9969696	30.5	0.9954967	35.7	0.9937899	49.0	0.98848
20.2	0.9981652	25.4	0.9969436	30.6	0.9954660	35.8	0.9937549	50.0	0.98804
20.3	0.9981443	25.5	0.9969176	30.7	0.9954352	35.9	0.9937199	51.0	0.98758
20.4	0.9981233	25.6	0.9968914	30.8	0.9954044	36.0	0.9936847	52.0	0.98712
20.5	0.9981022	25.7	0.9968651	30.9	0.9953734	36.1	0.9936495	53.0	0.98665
20.6	0.9980810	25.8	0.9968387	31.0	0.9953424	36.2	0.9936142	54.0	0.98617
20.7	0.9980596	25.9	0.9968123	31.1	0.9953113	36.3	0.9935788	55.0	0.98569
20.8	0.9980382	26.0	0.9967857	31.2	0.9952801	36.4	0.9935434	56.0	0.98521
20.9	0.9980167	26.1	0.9967591	31.3	0.9952488	36.5	0.9935078	57.0	0.98471
21.0	0.9979950	26.2	0.9967324	31.4	0.9952175	36.6	0.9934722	58.0	0.98421
21.1	0.9979733	26.3	0.9967055	31.5	0.9951860	36.7	0.9934365	59.0	0.98371
21.2	0.9979514	26.4	0.9966786	31.6	0.9951545	36.8	0.9934007	60.0	0.98320
21.3	0.9979295	26.5	0.9966516	31.7	0.9951228	36.9	0.9933649	61.0	0.98268
21.4	0.9979074	26.6	0.9966245	31.8	0.9950911	37.0	0.9933290	62.0	0.98216
21.5	0.9978853	26.7	0.9965973	31.9	0.9950593	37.1	0.9932929	63.0	0.98163
21.6	0.9978630	26.8	0.9965700	32.0	0.9950275	37.2	0.9932569	64.0	0.98109
21.7	0.9978407	26.9	0.9965426	32.1	0.9949955	37.3	0.9932207	65.0	0.98055
21.8	0.9978182	27.0	0.9965151	32.2	0.9949635	37.4	0.9931844	66.0	0.98000
21.9	0.9977956	27.1	0.9964875	32.3	0.9949313	37.5	0.9931481	67.0	0.97945
22.0	0.9977730	27.2	0.9964599	32.4	0.9948991	37.6	0.9931117	68.0	0.97890
22.1	0.9977502	27.3	0.9964321	32.5	0.9948668	37.7	0.9930753	69.0	0.97833
22.2	0.9977273	27.4	0.9964043	32.6	0.9948344	37.8	0.9930387	70.0	0.97776
22.3	0.9977044	27.5	0.9963763	32.7	0.9948020	37.9	0.9930021	71.0	0.97719
22.4	0.9976813	27.6	0.9963483	32.8	0.9947694	38.0	0.9929654	72.0	0.97661
22.5	0.9976582	27.7	0.9963202	32.9	0.9947368	38.1	0.9929286	73.0	0.97603
22.6	0.9976349	27.8	0.9962920	33.0	0.9947041	38.2	0.9928917	74.0	0.97544
22.7	0.9976115	27.9	0.9962637	33.1	0.9946713	38.3	0.9928548	75.0	0.97484
22.8	0.9975881	28.0	0.9962353	33.2	0.9946384	38.4	0.9928178	76.0	0.97424
22.9	0.9975645	28.1	0.9962068	33.3	0.9946055	38.5	0.9927807	77.0	0.97364
23.0	0.9975408	28.2	0.9961783	33.4	0.9945724	38.6	0.9927435	78.0	0.97303
23.1	0.9975171	28.3	0.9961496	33.5	0.9945393	38.7	0.9927063	79.0	0.97241
23.2	0.9974932	28.4	0.9961208	33.6	0.9945061	38.8	0.9926689	80.0	0.97179
23.3	0.9974692	28.5	0.9960920	33.7	0.9944728	38.9	0.9926316	81.0	0.97116
23.4	0.9974452	28.6	0.9960631	33.8	0.9944394	39.0	0.9925941	82.0	0.97053
23.5	0.9974210	28.7	0.9960341	33.9	0.9944060	39.1	0.9925565	83.0	0.96990
23.6	0.9973968	28.8	0.9960050	34.0	0.9943724	39.2	0.9925189	84.0	0.96926
23.7	0.9973724	28.9	0.9959758	34.1	0.9943388	39.3	0.9924812	85.0	0.96861
23.8	0.9973480	29.0	0.9959465	34.2	0.9943051	39.4	0.9924434	86.0	0.96796
23.9	0.9973234	29.1	0.9959171	34.3	0.9942713	39.5	0.9924056	87.0	0.96731
24.0	0.9972988	29.2	0.9958876	34.4	0.9942375	39.6	0.9923677	88.0	0.96664
24.1	0.9972740	29.3	0.9958581	34.5	0.9942035	39.7	0.9923297	89.0	0.96598
24.2	0.9972492	29.4	0.9958285	34.6	0.9941695	39.8	0.9922916	90.0	0.96531
24.3	0.9972243	29.5	0.9957987	34.7	0.9941354	39.9	0.9922534	91.0	0.96463
24.4	0.9971992	29.6	0.9957689	34.8	0.9941012	40.0	0.9922152	92.0	0.96396
24.5	0.9971741	29.7	0.9957390	34.9	0.9940669	41.0	0.99183	93.0	0.96327
24.6	0.9971489	29.8	0.9957090	35.0	0.9940326	42.0	0.99144	94.0	0.96258
24.7	0.9971236	29.9	0.9956790	35.1	0.9939982	43.0	0.99104	95.0	0.96189
24.8	0.9970981	30.0	0.9956488	35.2	0.9939637	44.0	0.99063	96.0	0.96119
24.9	0.9970726	30.1	0.9956185	35.3	0.9939291	45.0	0.99021	97.0	0.96049
25.0	0.9970470	30.2	0.9955882	35.4	0.9938944	46.0	0.98979	98.0	0.95978
25.1	0.9970213	30.3	0.9955578	35.5	0.9938597	47.0	0.98936	99.0	0.95907
25.2	0.9969955	30.4	0.9955273	35.6	0.9938248	48.0	0.98893	99.974	0.95837

FIXED-POINT PROPERTIES OF H_2O AND D_2O

Allan H. Harvey

Temperatures are given on the ITS-90 scale.

References

1. International Association for the Properties of Water and Steam (IAPWS), *Guideline on the Use of Fundamental Physical Constants and Basic Constants of Water* (2001; 2008 update), available from http://www.iapws.org.
2. IAPWS, *Revised Release on the Pressure along the Melting and Sublimation Curves of Ordinary Water Substance* (2008), available from http://www.iapws.org.
3. IAPWS, *IAPWS Release on the Values of Temperature, Pressure, and Density of Ordinary and Heavy Water Substances at their Respective Critical Points* (1992), available from http://www.iapws.org.
4. Wagner, W., and Pruß, A., *J. Phys. Chem. Ref. Data* 31, 387, 2002.
5. Hill, P. G., MacMillan, R. D. C., and Lee, V., *J. Phys. Chem. Ref. Data* 11, 1, 1982.
6. Guildner, L. A., Johnson, D. P., and Jones, F. E., *J. Res. Nat. Bur. Stand.* 80A, 505, 1976.
7. Harvey, A. H., and Lemmon, E. W., *J. Phys. Chem. Ref. Data* 31, 173, 2002.
8. Harvey, A. H., Peskin, A. P., and Klein, S. A., NIST Standard Reference Database 10: *NIST/ASME Steam Properties*, Version 2.22, National Institute of Standards and Technology, Standard Reference Data Program, Gaithersburg, Maryland, 2008 (http://www.nist.gov/srd/nist10.cfm).
9. Lemmon, E. W., Huber, M. L., and McLinden, M. O., NIST Standard Reference Database 23: *Reference Fluid Thermodynamic and Transport Properties-REFPROP*, Version 9.0, National Institute of Standards and Technology, Standard Reference Data Program, Gaithersburg, Maryland, 2010 (http://www.nist.gov/srd/nist23.cfm).

	Unit	H_2O	D_2O
Molar mass	g mol^{-1}	18.015268	20.02751
Melting point (101.325 kPa)	°C	0.0025	3.81
Boiling point (101.325 kPa)	°C	99.974	101.40
Triple-point temperature	°C	0.01 (exact)	3.82
Triple-point pressure	Pa	611.657	661
Triple-point density (liq)	g cm^{-3}	0.99979	1.1055
Critical temperature	°C	373.946	370.697
Critical pressure	MPa	22.064	21.671
Critical density	g cm^{-3}	0.322	0.356
Maximum density (101.325 kPa)	g cm^{-3}	0.999975	1.1060
Temperature of maximum density	°C	3.98	11.2

PROPERTIES OF SATURATED LIQUID D_2O

Allan H. Harvey

Properties of saturated liquid heavy water, D_2O, are given in this table as a function of temperature from the melting point to the critical point. The vapor pressure was calculated from the formulation of Harvey and Lemmon (Ref. 2). The other properties were generated from the NIST REFPROP program (Ref. 1) and are consistent with formulations adopted for general and scientific use by The International Association for the Properties of Water and Steam (IAPWS). The background for the equation of state used for density and heat capacity is given by Hill et al. (Ref. 3), and the background for the transport property correlations is given by Matsunaga and Nagashima (Ref. 4). The unpublished surface tension correlation and the other IAPWS formulations may be found on the IAPWS Web site (http://www.iapws.org). The temperature scale is ITS-90. Additional calculations at state points not listed below can be obtained by using the REFPROP program (http://www.nist.gov/srd/nist23.htm). The properties are

P: vapor pressure
ρ: density
C_p: isobaric heat capacity

η: viscosity
λ: thermal conductivity
σ: surface tension

References

1. Lemmon, E.W., Huber, M.L., and McLinden, M.O., NIST Standard Reference Database 23: Reference Fluid Thermodynamic and Transport Properties-REFPROP, Version 9.0, National Institute of Standards and Technology, Standard Reference Data Program, Gaithersburg, Maryland, 2010 (www.nist.gov/srd/nist23.cfm).
2. Harvey, A. H. and Lemmon, E. W., *J. Phys. Chem. Ref. Data* 31, 173, 2002.
3. Hill, P. G., MacMillan, R. D. C., and Lee, V., *J. Phys. Chem. Ref. Data* 11, 1, 1982.
4. Matsunaga, N. and Nagashima, A., *J. Phys. Chem. Ref. Data* 12, 933, 1983.

$t/°C$	P/kPa	$\rho/kg\ m^{-3}$	$C_p/kJ\ kg^{-1}\ K^{-1}$	$\eta/mPa\ s$	$\lambda/W\ m^{-1}\ K^{-1}$	$\sigma/mN\ m^{-1}$
3.82	0.661	1105.5	4.211	2.086	0.565	74.93
10	1.026	1106.0	4.231	1.679	0.575	74.06
20	1.999	1105.3	4.243	1.247	0.589	72.61
30	3.702	1103.2	4.240	0.972	0.600	71.09
40	6.550	1099.9	4.232	0.785	0.610	69.52
50	11.12	1095.6	4.220	0.651	0.618	67.89
60	18.20	1090.5	4.207	0.552	0.625	66.21
70	28.81	1084.6	4.194	0.476	0.629	64.47
80	44.24	1078.1	4.181	0.416	0.633	62.67
90	66.09	1071.0	4.170	0.368	0.635	60.82
100	96.30	1063.3	4.161	0.329	0.636	58.93
110	137.1	1055.1	4.155	0.296	0.636	56.98
120	191.3	1046.4	4.153	0.269	0.635	54.99
130	261.8	1037.2	4.155	0.247	0.632	52.95
140	352.0	1027.5	4.162	0.227	0.629	50.87
150	465.8	1017.2	4.174	0.210	0.625	48.75
160	607.3	1006.5	4.190	0.195	0.620	46.59
170	781.1	995.2	4.212	0.182	0.614	44.39
180	992.0	983.4	4.240	0.171	0.607	42.16
190	1246	971.1	4.273	0.161	0.600	39.90
200	1547	958.2	4.313	0.152	0.592	37.61
210	1903	944.7	4.360	0.144	0.583	35.29
220	2319	930.5	4.415	0.136	0.574	32.95
230	2802	915.7	4.479	0.130	0.563	30.59
240	3359	900.1	4.554	0.123	0.553	28.22
250	3998	883.7	4.643	0.118	0.541	25.84
260	4725	866.4	4.750	0.112	0.529	23.45
270	5550	848.0	4.880	0.107	0.516	21.07
280	6480	828.4	5.038	0.103	0.502	18.69
290	7525	807.5	5.237	0.098	0.488	16.33
300	8694	784.8	5.490	0.094	0.473	13.99
310	9998	760.1	5.823	0.089	0.458	11.68

$t/°C$	P/kPa	$\rho/kg\ m^{-3}$	$C_p/kJ\ kg^{-1}\ K^{-1}$	$\eta/mPa\ s$	$\lambda/W\ m^{-1}\ K^{-1}$	$\sigma/mN\ m^{-1}$
320	11449	732.7	6.281	0.085	0.442	9.428
330	13059	702.0	6.952	0.080	0.425	7.238
340	14845	666.5	8.041	0.075	0.408	5.141
350	16824	623.3	10.17	0.069	0.391	3.173
360	19024	565.1	16.41	0.062	0.382	1.405
370	21487	430.7	268.8	0.046	0.548	0.0467
370.697	21671	356.0				0

PROPERTIES OF ICE AND SUPERCOOLED WATER

Allan H. Harvey

The common form of ice at ambient pressure is hexagonal ice, designated as ice Ih (see phase diagram in Section 12). The data given here refer to that form, at standard atmospheric pressure (101.325 kPa). Data have been taken from the references indicated, which in most cases are formulations based on critical evaluation of available experimental data. Most properties are sensitive to the method of preparation of the sample, because air and other gases are sometimes occluded. For this reason, there is often disagreement among values in the literature. For all properties except the dielectric constant of ice, the cited reference contains information on the uncertainty of the property.

References

1. Wagner, W., and Pruß, A., *J. Phys. Chem. Ref. Data* 31, 387, 2002.
2. Feistel, R., and Wagner, W., *J. Phys. Chem. Ref. Data* 35, 1021, 2006.
3. International Association for the Properties of Water and Steam (IAPWS), *Revised Release on the Equation of State 2006 for H₂O Ice Ih* (2009), available from http://www.iapws.org.
4. Andersson, O., and Inaba, A., *Phys. Chem. Chem. Phys.* 7, 1441, 2005.
5. Slack, G.A., *Phys. Rev. B* 22, 3065, 1980.
6. Wörz, O., and Cole, R.H., *J. Chem. Phys.* 51, 1546, 1969.

Density of supercooled water (Ref. 1)

$t/°C$	$\rho/\text{g cm}^{-3}$
0	0.9998
−5	0.9993
−10	0.9981
−15	0.9963
−20	0.9936
−25	0.9896
−30	0.9838

Phase Transition Properties (Ref. 2)

$\Delta_{fus}H(0\ °C) = 333.4\ \text{J g}^{-1}$
$\Delta_{subl}H(0\ °C) = 2834\ \text{J g}^{-1}$

Thermophysical Properties of Ice I$_h$

ρ: mass density (Refs. 2, 3)
α_V: cubic expansion coefficient, $\alpha_V = -(1/V)(\partial V/\partial T)_p$ (Refs. 2, 3)
κ_s: isentropic compressibility, $\kappa_s = -(1/V)(\partial V/\partial p)_s$ (Refs. 2, 3)
c_p: specific heat capacity at constant pressure (Refs. 2, 3)
λ: thermal conductivity (Refs. 4, 5)
ε: static dielectric constant (relative permittivity) (Ref. 6)

$t/°C$	$\rho/\text{g cm}^{-3}$	$10^3\alpha_V/\text{K}^{-1}$	κ_s/GPa^{-1}	$c_p/\text{J g}^{-1}\text{K}^{-1}$	$\lambda/\text{W m}^{-1}\text{K}^{-1}$	ε
0	0.9167	0.160	0.114	2.10	2.16	91.2
−10	0.9182	0.155	0.113	2.02	2.26	95.1
−20	0.9196	0.150	0.111	1.95	2.38	99.4
−30	0.9209	0.144	0.110	1.88	2.50	104.1
−40	0.9222	0.138	0.108	1.80	2.63	109.2
−50	0.9235	0.132	0.107	1.73	2.77	115.0
−60	0.9247	0.126	0.106	1.66	2.93	121.4
−80	0.9269	0.111	0.103	1.52	3.27	136.6
−100	0.9288	0.095	0.101	1.38	3.69	
−120	0.9304	0.078	0.099	1.25	4.2	
−140	0.9317	0.060	0.097	1.11	4.9	
−160	0.9326	0.041	0.096	0.97	5.7	
−180	0.9332	0.025	0.095	0.82	7.0	
−200	0.9336	0.013	0.095	0.65	8.9	
−220	0.9337	0.0050	0.095	0.47	12.2	
−240	0.9338	0.0012	0.095	0.27	20	
−260	0.9338	0.00008	0.095	0.036		

VAPOR PRESSURE OF ICE

The values of the vapor (sublimation) pressure of ice Ih were calculated from the equation recommended by the International Association for the Properties of Water and Steam (IAPWS) in 2008. See Refs. 1 and 2 for details on the uncertainty in different temperature ranges. The first entry in the table is the triple point of water, whose pressure has an uncertainty of 0.010 Pa.

References

1. International Association for the Properties of Water and Steam (IAPWS), *Revised Release on the Pressure along the Melting and Sublimation Curves of Ordinary Water Substance* (2008), available from http://www.iapws.org.
2. Wagner, W., Riethmann, T., Feistel, R., and Harvey, A. H., New Equations for the Melting Pressure and Sublimation Pressure of H_2O Ice Ih, *J. Phys. Chem. Ref. Data*, to be submitted.

t/°C	p/Pa	t/°C	p/Pa	t/°C	p/Pa	t/°C	p/Pa	t/°C	p/Pa
0.01	611.657	−10	259.87	−22	85.08	−50	3.938	−130	8.75×10^{-7}
0	611.15	−11	237.71	−24	69.89	−55	2.094	−140	3.62×10^{-8}
−1	562.66	−12	217.29	−26	57.24	−60	1.081	−150	9.00×10^{-10}
−2	517.70	−13	198.49	−28	46.72	−65	0.541	−160	1.18×10^{-11}
−3	476.04	−14	181.19	−30	38.01	−70	0.262	−170	6.80×10^{-14}
−4	437.45	−15	165.27	−32	30.81	−75	0.122	−180	1.32×10^{-16}
−5	401.74	−16	150.65	−34	24.89	−80	5.48×10^{-2}	−190	5.89×10^{-20}
−6	368.71	−17	137.22	−36	20.04	−90	9.68×10^{-3}	−200	3.31×10^{-24}
−7	338.17	−18	124.90	−38	16.07	−100	1.40×10^{-3}	−210	8.86×10^{-30}
−8	309.95	−19	113.60	−40	12.84	−110	1.61×10^{-4}	−220	2.06×10^{-37}
−9	283.91	−20	103.24	−45	7.203	−120	1.41×10^{-5}		

T/K	p/Pa	T/K	p/Pa	T/K	p/Pa	T/K	p/Pa	T/K	p/Pa
273.16	611.657	265	305.91	250	76.01	215	1.386	130	1.20×10^{-8}
273.15	611.15	264	280.18	248	62.33	210	0.702	120	2.48×10^{-10}
273	603.65	263	256.43	246	50.95	205	0.344	110	2.57×10^{-12}
272	555.70	262	234.54	244	41.51	200	0.163	100	1.09×10^{-14}
271	511.25	261	214.37	242	33.70	190	3.24×10^{-2}	90	1.39×10^{-17}
270	470.06	260	195.80	240	27.27	180	5.39×10^{-3}	80	3.50×10^{-21}
269	431.92	258	163.00	235	15.81	170	7.30×10^{-4}	70	8.58×10^{-26}
268	396.62	256	135.30	230	8.947	160	7.73×10^{-5}	60	6.51×10^{-32}
267	363.97	254	111.98	225	4.939	150	6.10×10^{-6}	50	1.93×10^{-40}
266	333.79	252	92.40	220	2.654	140	3.37×10^{-7}		

MELTING POINT OF ICE AS A FUNCTION OF PRESSURE

This table gives the melting temperature of ice at various pressures, calculated from the equation for the ice Ih – liquid water phase boundary recommended by the International Association for the Properties of Water and Steam (IAPWS) in 2008. See Refs. 1 and 2 for information on the solid/liquid transitions for high-pressure forms of ice. IAPWS gives the following locations for the triple points where equilibrium exists among two ice forms and liquid water:

References

1. International Association for the Properties of Water and Steam (IAPWS), *Revised Release on the Pressure along the Melting and Sublimation Curves of Ordinary Water Substance* (2008), available from http://www.iapws.org.
2. Wagner, W., Riethmann, T., Feistel, R., and Harvey, A. H., New Equations for the Melting Pressure and Sublimation Pressure of H_2O Ice Ih, *J. Phys. Chem. Ref. Data*, to be submitted.

ice I – ice III	208.566 MPa	−21.985 °C
ice III – ice V	350.1 MPa	−16.986 °C
ice V – ice VI	632.4 MPa	0.16 °C
ice VI – ice VII	2216 MPa	81.85 °C

p/MPa	t/°C	p/MPa	t/°C	p/MPa	t/°C
0.000 612	0.01	20	−1.54	90	−7.91
0.1	0.0026	30	−2.36	100	−8.94
1	−0.064	40	−3.21	120	−11.09
2	−0.14	50	−4.09	140	−13.35
5	−0.37	60	−5.00	160	−15.73
10	−0.75	70	−5.94	180	−18.22
15	−1.14	80	−6.91	200	−20.83

PERMITTIVITY (DIELECTRIC CONSTANT) OF WATER AT VARIOUS FREQUENCIES

The permittivity of liquid water in the radiofrequency and microwave regions can be represented by the Debye equation (References 1 and 2):

$$\varepsilon' = \varepsilon_\infty + \frac{\varepsilon_s - \varepsilon_\infty}{1 + \omega^2 \tau^2}$$

$$\varepsilon'' = \frac{(\varepsilon_s - \varepsilon_\infty)\omega\tau}{1 + \omega^2 \tau^2}$$

where $\varepsilon = \varepsilon' + i\,\varepsilon''$ is the (complex) relative permittivity (i.e., the absolute permittivity divided by the permittivity of free space $\varepsilon_0 = 8.854 \cdot 10^{-12}$ F m^{-1}). Here ε_s is the static permittivity (see Reference 3 and the table "Thermophysical Properties of Water and Steam" in this Section); ε_∞ is a parameter describing the permittivity in the high frequency limit; τ is the relaxation time for molecular orientation; and $\omega = 2\pi f$ is the angular frequency. The values in this table have been calculated from parameters given in Reference 2:

	0 °C	25 °C	50 °C
ε_∞	5.7	5.2	4.0
τ/ps	17.67	8.27	4.75

Other useful quantities that can be calculated from the values in the table are the loss tangent:

$$\tan\delta = \varepsilon''/\varepsilon'$$

and the absorption coefficient α which describes the power attenuation per unit length ($P = P_0\,e^{-\alpha l}$):

$$\alpha = \frac{\pi f \varepsilon''}{c\sqrt{\varepsilon'}}$$

and c is the speed of light. The last equation is valid when $\varepsilon''/\varepsilon' \ll 1$.

References

1. Fernández, D. P., Mulev, Y., Goodwin, A. R. H., and Levelt Sengers, J. M. H., *J. Phys. Chem. Ref. Data*, 24, 33, 1995.
2. Kaatze, U., *J. Chem. Eng. Data*, 34, 371, 1989.
3. Archer, D. G., and Wang, P., *J. Phys. Chem. Ref. Data*, 12, 817, 1983.

Frequency	0 °C ε'	0 °C ε''	25 °C ε'	25 °C ε''	50 °C ε'	50 °C ε''
0	87.90	0.00	78.36	0.00	69.88	0.00
1 kHz	87.90	0.00	78.36	0.00	69.88	0.00
1 MHz	87.90	0.01	78.36	0.00	69.88	0.00
10 MHz	87.90	0.09	78.36	0.04	69.88	0.02
100 MHz	87.89	0.91	78.36	0.38	69.88	0.20
200 MHz	87.86	1.82	78.35	0.76	69.88	0.39
500 MHz	87.65	4.55	78.31	1.90	69.87	0.98
1 GHz	86.90	9.01	78.16	3.79	69.82	1.96
2 GHz	84.04	17.39	77.58	7.52	69.65	3.92
3 GHz	79.69	24.64	76.62	11.13	69.36	5.85
4 GHz	74.36	30.49	75.33	14.58	68.95	7.75
5 GHz	68.54	34.88	73.73	17.81	68.45	9.62
10 GHz	42.52	40.88	62.81	29.93	64.49	18.05
20 GHz	19.56	30.78	40.37	36.55	52.57	28.99
30 GHz	12.50	22.64	26.53	33.25	40.57	32.74
40 GHz	9.67	17.62	18.95	28.58	31.17	32.43
50 GHz	8.28	14.34	14.64	24.53	24.42	30.47

THERMOPHYSICAL PROPERTIES OF AIR

Eric W. Lemmon

These tables summarize the thermophysical properties of air in the liquid and gaseous states as calculated from the pseudo-pure fluid equation of state of Lemmon et al. (2000). The first table refers to liquid and gaseous air at equilibrium as a function of temperature. The tabulated properties are the bubble-point pressure (i.e., pressure at which boiling begins as the pressure of the liquid is lowered); the dew-point pressure (i.e., pressure at which condensation begins as the pressure of the gas is raised); density (ρ); enthalpy (H); entropy (S); isochoric heat capacity (C_v); isobaric heat capacity (C_p); speed of sound (u); viscosity (η); and thermal conductivity (λ). The first line of identical temperatures is the bubble-point (liquid) and the second line is the dew point (vapor). The normal boiling point of air, i.e., the temperature at which the bubble-point pressure reaches 1 standard atmosphere (1.01325 bar), is 78.90 K (−194.25 °C).

The second table gives the properties of air along various isobars. An entry with non-integer temperatures in the isobar section indicates a phase transition (liquid–vapor) at these temperatures; property values are then given for both phases. These are identi-fied by the high densities in the liquid and the low densities in the vapor. Additional calculations at state points not listed below can be obtained by using the NIST program REFPROP (http://www.nist.gov/srd/nist23.htm).

References

1. Lemmon, E.W., Jacobsen, R.T, Penoncello, S.G., and Friend, D.G., Thermodynamic Properties of Air and Mixtures of Nitrogen, Argon, and Oxygen from 60 to 2000 K at Pressures to 2000 MPa, *J. Phys. Chem. Ref. Data*, 29, 331, 2000.
2. Lemmon, E.W. and Jacobsen, R.T, Viscosity and Thermal Conductivity Equations for Nitrogen, Oxygen, Argon, and Air, *Int. J. Thermophys.*, 25, 21, 2004.
3. Lemmon, E.W., Huber, M.L., McLinden, M.O., NIST Standard Reference Database 23: Reference Fluid Thermodynamic and Transport Properties-REFPROP, Version 9.0, National Institute of Standards and Technology, Standard Reference Data Program, Gaithersburg, Maryland, 2010 (www.nist.gov/srd/nist23.cfm).

Thermophysical Properties of Air along the Boiling and Condensation Curves

T K	P MPa	ρ kg m^{-3}	H kJ kg^{-1}	S kJ kg^{-1} K^{-1}	C_v kJ kg^{-1} K^{-1}	C_p kJ kg^{-1} K^{-1}	u m s^{-1}	η μPa s	λ mW m^{-1} K^{-1}
59.75	0.005265	957.6	−36.66	−0.5306	1.174	1.901	1030.	376.6	171.4
59.75	0.002432	0.1421	185.5	3.340	0.7184	1.009	154.8	4.220	5.294
60	0.005546	956.5	−36.19	−0.5226	1.173	1.901	1028.	371.9	171.0
60	0.002584	0.1504	185.8	3.326	0.7186	1.009	155.1	4.238	5.320
62	0.008270	948.2	−32.38	−0.4603	1.157	1.901	1012.	336.9	167.8
62	0.004111	0.2318	187.7	3.225	0.7198	1.012	157.6	4.386	5.529
64	0.01200	939.9	−28.58	−0.3999	1.143	1.902	995.8	306.3	164.5
64	0.006325	0.3460	189.6	3.132	0.7212	1.015	160.0	4.532	5.739
66	0.01699	931.5	−24.77	−0.3414	1.129	1.903	979.1	279.4	161.3
66	0.009442	0.5018	191.5	3.047	0.7230	1.019	162.3	4.679	5.950
68	0.02352	923.0	−20.95	−0.2846	1.115	1.906	962.2	255.7	158.0
68	0.01371	0.7089	193.4	2.968	0.7252	1.024	164.5	4.825	6.162
70	0.03191	914.4	−17.13	−0.2293	1.102	1.908	945.1	234.8	154.7
70	0.01943	0.9785	195.2	2.896	0.7277	1.030	166.7	4.970	6.376
72	0.04250	905.7	−13.31	−0.1756	1.090	1.912	927.7	216.3	151.4
72	0.02692	1.322	197.0	2.828	0.7305	1.037	168.7	5.115	6.592
74	0.05566	897.0	−9.468	−0.1232	1.078	1.917	910.0	199.9	148.1
74	0.03655	1.753	198.7	2.766	0.7338	1.046	170.6	5.260	6.810
76	0.07179	888.1	−5.617	−0.07209	1.067	1.923	892.1	185.2	144.8
76	0.04870	2.285	200.4	2.708	0.7375	1.055	172.5	5.405	7.031
78	0.09129	879.1	−1.751	−0.02217	1.056	1.930	873.9	172.1	141.5
78	0.06381	2.933	202.0	2.653	0.7416	1.066	174.2	5.549	7.256
80	0.1146	870.0	2.132	0.02665	1.045	1.938	855.4	160.4	138.2
80	0.08232	3.711	203.6	2.602	0.7460	1.078	175.8	5.694	7.485
82	0.1422	860.7	6.036	0.07444	1.035	1.948	836.7	149.8	134.8
82	0.1047	4.635	205.1	2.554	0.7510	1.092	177.4	5.839	7.719
84	0.1745	851.3	9.962	0.1213	1.025	1.959	817.6	140.2	131.4
84	0.1315	5.724	206.5	2.509	0.7563	1.108	178.8	5.984	7.959
86	0.2121	841.7	13.91	0.1673	1.016	1.972	798.2	131.5	128.1
86	0.1631	6.993	207.8	2.466	0.7620	1.125	180.0	6.131	8.206
88	0.2553	832.0	17.90	0.2125	1.007	1.986	778.6	123.6	124.8
88	0.2002	8.464	209.1	2.425	0.7682	1.144	181.2	6.278	8.461

T K	P MPa	ρ kg m⁻³	H kJ kg⁻¹	S kJ kg⁻¹ K⁻¹	C_v kJ kg⁻¹ K⁻¹	C_p kJ kg⁻¹ K⁻¹	u m s⁻¹	η μPa s	λ mW m⁻¹ K⁻¹
90	0.3048	822.0	21.91	0.2569	0.9984	2.003	758.5	116.4	121.4
90	0.2432	10.16	210.3	2.386	0.7748	1.166	182.2	6.427	8.725
92	0.3609	811.8	25.97	0.3007	0.9902	2.022	738.2	109.7	118.0
92	0.2927	12.09	211.4	2.349	0.7817	1.190	183.1	6.578	9.001
94	0.4243	801.4	30.06	0.3439	0.9825	2.044	717.5	103.6	114.6
94	0.3493	14.29	212.3	2.313	0.7891	1.217	183.8	6.732	9.289
96	0.4954	790.7	34.21	0.3866	0.9752	2.069	696.5	97.88	111.2
96	0.4136	16.78	213.2	2.279	0.7969	1.248	184.5	6.889	9.593
98	0.5749	779.7	38.41	0.4288	0.9684	2.098	675.0	92.57	107.8
98	0.4861	19.60	214.0	2.246	0.8052	1.282	184.9	7.050	9.915
100	0.6631	768.4	42.66	0.4707	0.9619	2.131	653.3	87.61	104.4
100	0.5674	22.76	214.6	2.213	0.8138	1.320	185.3	7.215	10.26
102	0.7608	756.7	46.98	0.5122	0.9560	2.168	631.1	82.94	101.0
102	0.6582	26.32	215.1	2.182	0.8230	1.363	185.5	7.387	10.63
104	0.8684	744.6	51.38	0.5535	0.9505	2.212	608.5	78.53	97.62
104	0.7590	30.31	215.5	2.151	0.8326	1.413	185.6	7.566	11.02
106	0.9864	732.1	55.86	0.5947	0.9456	2.262	585.5	74.35	94.25
106	0.8706	34.78	215.7	2.120	0.8429	1.470	185.5	7.754	11.46
108	1.116	719.1	60.44	0.6358	0.9412	2.321	562.1	70.36	90.89
108	0.9934	39.79	215.8	2.089	0.8537	1.536	185.2	7.952	11.94
110	1.256	705.5	65.12	0.6769	0.9375	2.390	538.2	66.54	87.55
110	1.128	45.41	215.6	2.059	0.8653	1.614	184.9	8.163	12.47
112	1.409	691.2	69.93	0.7182	0.9345	2.472	513.9	62.87	84.24
112	1.276	51.73	215.2	2.028	0.8777	1.708	184.3	8.391	13.07
114	1.575	676.2	74.87	0.7598	0.9324	2.571	489.0	59.31	80.96
114	1.437	58.84	214.6	1.997	0.8912	1.821	183.6	8.637	13.76
116	1.755	660.3	79.98	0.8019	0.9312	2.693	463.5	55.85	77.72
116	1.612	66.88	213.8	1.965	0.9059	1.961	182.7	8.909	14.56
118	1.948	643.4	85.29	0.8447	0.9312	2.847	437.3	52.47	74.52
118	1.801	76.04	212.6	1.932	0.9220	2.139	181.7	9.210	15.50
120	2.156	625.1	90.83	0.8885	0.9327	3.048	410.2	49.13	71.36
120	2.007	86.55	211.0	1.898	0.9402	2.374	180.4	9.552	16.63
122	2.379	605.3	96.66	0.9338	0.9363	3.323	382.0	45.81	68.24
122	2.229	98.76	208.9	1.861	0.9608	2.694	179.1	9.946	18.04
124	2.617	583.3	102.9	0.9811	0.9427	3.723	352.3	42.46	65.17
124	2.468	113.2	206.3	1.821	0.9847	3.157	177.5	10.41	19.85
126	2.872	558.3	109.7	1.032	0.9537	4.367	320.4	39.01	62.18
126	2.727	130.6	202.9	1.777	1.013	3.882	175.8	10.98	22.29
128	3.143	528.3	117.3	1.088	0.9728	5.589	285.0	35.33	59.44
128	3.006	152.6	198.3	1.725	1.049	5.166	174.0	11.72	25.84
130	3.429	488.3	126.7	1.157	1.010	8.849	243.7	31.07	58.05
130	3.308	182.7	191.7	1.660	1.096	8.033	171.9	12.77	31.81
132	3.723	411.2	142.6	1.273	1.117	35.04	189.1	24.47	67.80
132	3.646	235.4	179.7	1.556	1.168	20.65	169.4	14.80	47.00
132.63	3.785	302.6	164.5	1.437				17.83	

Thermophysical Properties of Air along Various Isobars

T K	ρ kg m⁻³	H kJ kg⁻¹	S kJ kg⁻¹ K⁻¹	C_v kJ kg⁻¹ K⁻¹	C_p kJ kg⁻¹ K⁻¹	u m s⁻¹	η μPa s	λ mW m⁻¹ K⁻¹
P = 0.1 MPa (1 bar)								
60	956.7	−36.11	−0.5230	1.173	1.901	1029.	372.4	171.1
78.79	875.5	−0.2237	−0.002818	1.051	1.933	866.7	167.4	140.2
81.61	4.442	204.8	2.563	0.7500	1.089	177.1	5.811	7.673
100	3.557	224.3	2.779	0.7282	1.040	198.2	7.107	9.469

T	ρ	H	S	C_v	C_p	u	η	λ
K	kg m⁻³	kJ kg⁻¹	kJ kg⁻¹ K⁻¹	kJ kg⁻¹ K⁻¹	kJ kg⁻¹ K⁻¹	m s⁻¹	μPa s	mW m⁻¹ K⁻¹
120	2.938	244.9	2.966	0.7211	1.022	218.3	8.457	11.38
140	2.507	265.2	3.123	0.7184	1.014	236.4	9.750	13.24
160	2.188	285.5	3.258	0.7172	1.011	253.2	10.99	15.05
180	1.942	305.6	3.377	0.7166	1.008	268.8	12.18	16.80
200	1.746	325.8	3.483	0.7163	1.007	283.5	13.33	18.50
220	1.586	345.9	3.579	0.7163	1.006	297.4	14.44	20.16
240	1.453	366.0	3.667	0.7164	1.006	310.7	15.51	21.77
260	1.341	386.2	3.747	0.7168	1.006	323.4	16.55	23.35
280	1.245	406.3	3.822	0.7173	1.006	335.6	17.56	24.88
300	1.161	426.4	3.891	0.7181	1.007	347.4	18.54	26.38
320	1.089	446.5	3.956	0.7192	1.007	358.7	19.49	27.85
340	1.024	466.7	4.018	0.7206	1.009	369.6	20.41	29.29
360	0.9674	486.9	4.075	0.7223	1.010	380.3	21.32	30.71
380	0.9164	507.1	4.130	0.7243	1.012	390.5	22.20	32.09
400	0.8706	527.4	4.182	0.7266	1.014	400.5	23.06	33.45
500	0.6964	629.5	4.410	0.7426	1.030	446.4	27.09	39.94
600	0.5803	733.6	4.599	0.7641	1.051	487.1	30.77	46.01
700	0.4974	839.9	4.763	0.7879	1.075	523.9	34.18	51.76
800	0.4352	948.6	4.908	0.8117	1.099	557.8	37.37	57.25
900	0.3869	1060.	5.039	0.8340	1.121	589.6	40.39	62.54
1000	0.3482	1173.	5.158	0.8540	1.141	619.6	43.28	67.68

P = 0.5 MPa (5 bar)

T	ρ	H	S	C_v	C_p	u	η	λ
60	957.3	−35.80	−0.5248	1.173	1.900	1031.	374.6	171.4
80	870.9	2.387	0.02430	1.046	1.934	858.5	161.4	138.6
96.12	790.0	34.46	0.3892	0.9748	2.071	695.2	97.55	111.0
98.36	20.14	214.1	2.240	0.8067	1.288	185.0	7.079	9.974
100	19.65	216.2	2.261	0.7967	1.261	187.4	7.192	10.10
120	15.48	239.6	2.475	0.7461	1.115	212.4	8.542	11.81
140	12.94	261.4	2.643	0.7311	1.068	232.8	9.834	13.58
160	11.17	282.5	2.784	0.7245	1.045	250.9	11.07	15.32
180	9.842	303.3	2.906	0.7213	1.032	267.4	12.26	17.04
200	8.811	323.8	3.014	0.7195	1.025	282.6	13.41	18.71
220	7.981	344.3	3.112	0.7186	1.020	297.0	14.51	20.34
240	7.297	364.6	3.200	0.7182	1.017	310.6	15.58	21.94
260	6.724	385.0	3.282	0.7182	1.015	323.5	16.62	23.50
280	6.235	405.2	3.357	0.7185	1.014	335.9	17.62	25.02
300	5.813	425.5	3.427	0.7191	1.013	347.8	18.60	26.51
320	5.446	445.8	3.492	0.7200	1.013	359.3	19.54	27.97
340	5.123	466.0	3.553	0.7213	1.013	370.3	20.47	29.41
360	4.836	486.3	3.611	0.7229	1.014	381.0	21.37	30.81
380	4.580	506.6	3.666	0.7248	1.016	391.3	22.24	32.19
400	4.350	526.9	3.718	0.7271	1.018	401.3	23.10	33.55
500	3.477	629.3	3.947	0.7429	1.032	447.3	27.13	40.02
600	2.897	733.5	4.137	0.7643	1.053	488.0	30.80	46.07
700	2.483	840.0	4.301	0.7881	1.076	524.8	34.20	51.80
800	2.173	948.8	4.446	0.8119	1.100	558.7	37.40	57.29
900	1.932	1060.	4.577	0.8341	1.122	590.5	40.42	62.58
1000	1.739	1173.	4.696	0.8542	1.142	620.4	43.30	67.71

P = 1 MPa (10 bar)

T	ρ	H	S	C_v	C_p	u	η	λ
60	958.0	−35.42	−0.5271	1.174	1.898	1033.	377.2	171.7
80	872.2	2.720	0.02129	1.047	1.930	862.5	162.8	139.1
100	770.1	42.76	0.4673	0.9623	2.119	658.2	88.33	105.0
106.22	730.7	56.36	0.5992	0.9451	2.268	583.0	73.90	93.88
108.10	40.07	215.8	2.088	0.8543	1.540	185.2	7.963	11.96

T K	ρ kg m^{-3}	H kJ kg^{-1}	S kJ kg^{-1} K^{-1}	C_v kJ kg^{-1} K^{-1}	C_p kJ kg^{-1} K^{-1}	u m s^{-1}	η μPa s	λ mW m^{-1} K^{-1}
120	33.48	232.3	2.233	0.7844	1.285	204.0	8.718	12.59
140	27.02	256.3	2.419	0.7481	1.148	228.2	9.978	14.11
160	22.94	278.7	2.568	0.7341	1.093	248.1	11.20	15.74
180	20.03	300.2	2.695	0.7273	1.065	265.7	12.38	17.38
200	17.83	321.3	2.806	0.7236	1.048	281.7	13.51	19.00
220	16.09	342.2	2.906	0.7216	1.038	296.6	14.61	20.60
240	14.68	362.9	2.996	0.7204	1.031	310.6	15.67	22.17
260	13.50	383.5	3.078	0.7199	1.026	323.8	16.70	23.70
280	12.50	403.9	3.154	0.7199	1.023	336.4	17.70	25.21
300	11.64	424.4	3.224	0.7203	1.021	348.4	18.67	26.68
320	10.90	444.8	3.290	0.7211	1.020	360.0	19.62	28.13
340	10.25	465.2	3.352	0.7222	1.019	371.1	20.54	29.55
360	9.668	485.6	3.410	0.7237	1.019	381.9	21.43	30.95
380	9.153	506.0	3.465	0.7255	1.020	392.3	22.31	32.32
400	8.690	526.4	3.518	0.7277	1.022	402.3	23.16	33.67
500	6.943	629.1	3.747	0.7434	1.034	448.5	27.18	40.11
600	5.784	733.5	3.937	0.7646	1.054	489.2	30.84	46.15
700	4.957	840.0	4.101	0.7884	1.077	526.0	34.24	51.87
800	4.338	948.9	4.247	0.8121	1.100	559.9	37.43	57.35
900	3.857	1060.	4.378	0.8343	1.122	591.5	40.45	62.63
1000	3.472	1173.	4.497	0.8543	1.142	621.5	43.33	67.75
P = 2 MPa (20 bar)								
60.11[a]	959.1	−34.44	−0.5282	1.175	1.895	1037.	380.5	172.2
80	874.6	3.390	0.01535	1.048	1.921	870.2	165.5	140.1
100	775.0	43.09	0.4576	0.9636	2.086	672.5	90.43	106.6
118.52	638.8	86.69	0.8559	0.9314	2.894	430.4	51.60	73.71
119.94	86.20	211.0	1.899	0.9396	2.365	180.5	9.540	16.59
120	86.03	211.2	1.900	0.9382	2.354	180.7	9.542	16.57
140	59.88	244.8	2.161	0.7883	1.387	218.3	10.44	15.68
160	48.61	270.5	2.333	0.7545	1.213	242.7	11.55	16.80
180	41.54	294.0	2.471	0.7396	1.139	262.8	12.67	18.21
200	36.52	316.3	2.589	0.7319	1.100	280.3	13.77	19.69
220	32.70	338.0	2.692	0.7275	1.076	296.2	14.84	21.19
240	29.66	359.4	2.785	0.7249	1.060	310.8	15.88	22.68
260	27.18	380.5	2.870	0.7235	1.050	324.6	16.89	24.16
280	25.11	401.4	2.947	0.7228	1.042	337.6	17.88	25.62
300	23.34	422.2	3.019	0.7227	1.037	349.9	18.84	27.06
320	21.82	442.9	3.086	0.7231	1.033	361.7	19.77	28.47
340	20.49	463.5	3.148	0.7240	1.031	373.0	20.68	29.87
360	19.32	484.1	3.207	0.7253	1.030	383.9	21.57	31.24
380	18.27	504.7	3.263	0.7269	1.029	394.4	22.44	32.59
400	17.34	525.3	3.315	0.7290	1.029	404.5	23.29	33.93
500	13.84	628.6	3.546	0.7442	1.039	450.8	27.28	40.31
600	11.52	733.4	3.737	0.7653	1.057	491.5	30.93	46.30
700	9.878	840.2	3.902	0.7889	1.079	528.3	34.32	52.00
800	8.646	949.3	4.047	0.8125	1.102	562.1	37.49	57.46
900	7.689	1061.	4.178	0.8346	1.123	593.7	40.50	62.73
1000	6.923	1174.	4.298	0.8546	1.143	623.6	43.38	67.84
P = 5 MPa (50 bar)								
60.64[a]	961.4	−31.10	−0.5246	1.176	1.886	1048.	386.2	173.3
80	881.7	5.437	−0.001766	1.054	1.898	892.1	173.5	143.0
100	788.3	44.27	0.4311	0.9681	2.009	710.6	96.44	111.1
120	665.1	87.67	0.8256	0.9194	2.429	496.8	57.06	79.02
140	321.3	172.5	1.467	1.049	8.515	199.5	19.25	43.18

T	ρ	H	S	C_v	C_p	u	η	λ
K	kg m^{-3}	kJ kg^{-1}	kJ kg^{-1} K^{-1}	kJ kg^{-1} K^{-1}	kJ kg^{-1} K^{-1}	m s^{-1}	µPa s	mW m^{-1} K^{-1}
160	151.4	240.7	1.930	0.8269	1.916	231.8	13.80	22.96
180	116.7	273.4	2.123	0.7785	1.453	258.2	14.14	22.02
200	97.93	300.6	2.267	0.7569	1.288	279.4	14.90	22.54
220	85.43	325.5	2.385	0.7451	1.205	297.7	15.77	23.49
240	76.25	349.0	2.488	0.7381	1.155	314.0	16.68	24.62
260	69.12	371.8	2.579	0.7338	1.123	328.8	17.60	25.83
280	63.36	394.0	2.661	0.7312	1.101	342.7	18.52	27.09
300	58.59	415.8	2.736	0.7297	1.085	355.6	19.42	28.39
320	54.55	437.4	2.806	0.7291	1.074	367.9	20.31	29.69
340	51.07	458.8	2.871	0.7292	1.065	379.6	21.18	30.98
360	48.05	480.1	2.932	0.7298	1.059	390.7	22.04	32.27
380	45.38	501.2	2.989	0.7310	1.055	401.4	22.88	33.55
400	43.01	522.3	3.043	0.7327	1.052	411.7	23.70	34.81
500	34.21	627.4	3.277	0.7466	1.052	458.3	27.61	40.97
600	28.48	733.2	3.470	0.7671	1.066	498.9	31.20	46.83
700	24.42	840.8	3.636	0.7903	1.085	535.4	34.55	52.43
800	21.39	950.4	3.782	0.8136	1.106	569.0	37.69	57.83
900	19.03	1062.	3.914	0.8356	1.127	600.3	40.68	63.04
1000	17.14	1176.	4.034	0.8554	1.146	629.9	43.54	68.12

P = 10 MPa (100 bar)

T	ρ	H	S	C_v	C_p	u	η	λ
61.52[a]	965.2	−25.55	−0.5187	1.177	1.871	1064.	395.1	175.0
80	892.4	8.950	−0.02831	1.063	1.868	925.2	186.8	147.5
100	806.9	46.73	0.3930	0.9767	1.924	763.5	105.8	117.8
120	706.1	86.65	0.7565	0.9192	2.094	591.1	67.00	89.01
140	573.7	132.1	1.106	0.8916	2.517	418.1	41.82	63.61
160	397.0	188.0	1.479	0.8787	2.882	297.6	24.99	43.38
180	273.3	238.0	1.774	0.8267	2.098	283.1	19.31	32.89
200	214.1	274.8	1.968	0.7908	1.646	296.3	18.16	29.63
220	179.9	305.4	2.114	0.7701	1.434	312.4	18.17	28.80
240	157.0	332.8	2.233	0.7574	1.317	328.0	18.60	28.89
260	140.2	358.3	2.336	0.7492	1.245	342.7	19.20	29.38
280	127.3	382.7	2.426	0.7439	1.196	356.5	19.90	30.15
300	116.9	406.3	2.507	0.7404	1.162	369.5	20.64	31.12
320	108.3	429.3	2.582	0.7383	1.138	381.8	21.40	32.15
340	101.0	451.8	2.650	0.7372	1.120	393.5	22.17	33.22
360	94.80	474.1	2.714	0.7369	1.106	404.6	22.95	34.33
380	89.36	496.1	2.773	0.7374	1.096	415.3	23.72	35.45
400	84.57	517.9	2.829	0.7384	1.088	425.6	24.49	36.58
500	67.06	625.7	3.070	0.7505	1.073	471.8	28.19	42.26
600	55.82	733.3	3.266	0.7699	1.080	511.9	31.67	47.84
700	47.90	842.0	3.433	0.7926	1.095	547.8	34.94	53.26
800	42.00	952.4	3.581	0.8155	1.113	580.8	38.04	58.52
900	37.42	1065.	3.713	0.8372	1.132	611.6	40.99	63.64
1000	33.75	1179.	3.833	0.8568	1.150	640.7	43.81	68.64

P = 20 MPa (200 bar)

T	ρ	H	S	C_v	C_p	u	η	λ
63.24[a]	972.3	−14.49	−0.5069	1.180	1.847	1094.	411.4	178.4
80	911.3	16.25	−0.07568	1.081	1.825	982.1	213.1	155.4
100	836.2	52.71	0.3311	0.9948	1.827	846.0	123.1	128.7
120	755.9	89.57	0.6670	0.9334	1.864	711.9	82.28	103.9
140	668.7	127.4	0.9589	0.8906	1.926	589.6	58.52	82.08
160	575.8	166.5	1.220	0.8606	1.977	491.8	43.00	65.18
180	484.9	206.0	1.452	0.8365	1.946	428.0	33.22	53.43
200	407.6	243.6	1.651	0.8143	1.808	397.2	27.80	46.02
220	348.5	278.1	1.815	0.7956	1.642	387.6	25.12	41.70

T K	ρ kg m^{-3}	H kJ kg^{-1}	S kJ kg^{-1} K^{-1}	C_v kJ kg^{-1} K^{-1}	C_p kJ kg^{-1} K^{-1}	u m s^{-1}	η μPa s	λ mW m^{-1} K^{-1}
240	304.6	309.5	1.952	0.7812	1.506	388.7	23.92	39.30
260	271.5	338.6	2.068	0.7705	1.405	394.7	23.48	37.98
280	245.7	365.9	2.169	0.7627	1.331	402.8	23.47	37.46
300	225.0	392.0	2.259	0.7572	1.277	412.0	23.70	37.57
320	208.0	417.1	2.340	0.7532	1.236	421.5	24.08	37.94
340	193.7	441.5	2.414	0.7506	1.204	431.1	24.56	38.47
360	181.5	465.3	2.482	0.7491	1.180	440.6	25.10	39.13
380	170.9	488.7	2.546	0.7485	1.161	450.0	25.68	39.88
400	161.7	511.7	2.605	0.7486	1.146	459.2	26.29	40.69
500	128.3	624.0	2.856	0.7575	1.108	501.8	29.49	45.27
600	107.0	734.4	3.057	0.7753	1.103	539.5	32.69	50.19
700	92.10	845.1	3.227	0.7969	1.112	573.7	35.78	55.18
800	80.99	957.0	3.377	0.8191	1.126	605.2	38.75	60.13
900	72.34	1070.	3.510	0.8402	1.142	634.8	41.61	65.02
1000	65.41	1185.	3.631	0.8594	1.157	662.8	44.36	69.84

P = 50 MPa (500 bar)

T K	ρ kg m^{-3}	H kJ kg^{-1}	S kJ kg^{-1} K^{-1}	C_v kJ kg^{-1} K^{-1}	C_p kJ kg^{-1} K^{-1}	u m s^{-1}	η μPa s	λ mW m^{-1} K^{-1}
68.21[a]	991.5	18.30	−0.4728	1.192	1.793	1173.	450.9	187.5
80	955.5	39.24	−0.1896	1.128	1.761	1112.	292.7	174.1
100	895.9	73.98	0.1982	1.043	1.715	1012.	171.6	152.8
120	837.8	107.9	0.5077	0.9811	1.680	920.9	118.9	133.2
140	781.1	141.2	0.7643	0.9342	1.649	839.4	90.60	115.9
160	726.2	173.9	0.9826	0.8983	1.619	770.2	72.86	101.0
180	673.9	205.9	1.171	0.8706	1.586	714.1	60.73	88.73
200	624.9	237.3	1.337	0.8488	1.549	670.6	52.14	79.05
220	579.8	267.9	1.482	0.8313	1.509	638.1	45.98	71.65
240	539.0	297.7	1.612	0.8171	1.467	614.7	41.60	66.11
260	502.4	326.6	1.728	0.8057	1.425	598.5	38.49	61.97
280	470.0	354.7	1.832	0.7964	1.385	588.0	36.32	58.99
300	441.2	382.0	1.926	0.7890	1.349	581.7	34.82	57.07
320	415.6	408.7	2.012	0.7832	1.316	578.4	33.83	55.77
340	392.9	434.7	2.091	0.7787	1.287	577.4	33.19	54.93
360	372.7	460.2	2.164	0.7754	1.262	578.1	32.82	54.41
380	354.6	485.2	2.232	0.7731	1.241	580.0	32.66	54.15
400	338.2	509.8	2.295	0.7717	1.223	582.9	32.64	54.07
500	276.1	629.0	2.561	0.7749	1.168	604.1	33.85	55.45
600	234.6	744.6	2.772	0.7890	1.148	629.5	36.00	58.34
700	204.6	859.2	2.948	0.8080	1.147	655.5	38.46	61.93
800	181.8	974.2	3.102	0.8285	1.154	681.0	41.00	65.86
900	163.8	1090.	3.238	0.8483	1.164	705.9	43.55	69.98
1000	149.2	1207.	3.362	0.8665	1.175	730.1	46.07	74.20

[a] Freezing point for the liquid state.

THERMOPHYSICAL PROPERTIES OF FLUIDS

Eric W. Lemmon

These tables give thermodynamic and transport properties of a variety of fluids, as generated from the equations of state presented in the references below. The properties tabulated are pressure (P), density (ρ), enthalpy (H), entropy (S), isochoric heat capacity (C_v), isobaric heat capacity (C_p), speed of sound (u), viscosity (η), thermal conductivity (λ), and static dielectric constant (D). All extensive properties are given on a mass basis. Not all properties are included for every substance. The references should be consulted for information on the uncertainties.

Values are given first along the saturation line. The first two points are the properties at the triple point. The final line gives the properties at the critical point. Two lines are given for each temperature (except at the critical point); the first line gives the values of the liquid phase (note the high density) and the second line gives the values of the vapor phase (at low densities). Following the saturation tables, values are given as a function of temperature for several isobars. A duplicate entry in the isobar section indicates a phase transition (liquid–vapor) at that temperature; property values are then given for both phases. The phase can be determined by noting the sharp decrease in density between two successive temperature entries; all lines above this point refer to the liquid phase, and all lines below refer to the gas phase. If there is no sharp discontinuity in density, all data in the table refer to the supercritical region (*i.e.*, the isobar is above the critical pressure). If the first temperature in the isobars is not an integer, this state point refers to the properties of the liquid at the melting line.

All temperatures are given on ITS-90, except those for oxygen (Ref. 16) and helium (Ref. 13), where the source equations of state still use the IPTS-68 temperature scale. Nitrogen, oxygen, and argon use a reference state based on zero enthalpy in the gas phase at 0 K; parahydrogen, helium, methane, and ethane use a reference state of zero enthalpy and entropy at the saturated liquid state at the normal boiling point; and propane and carbon dioxide use a reference state of 200 kJ kg^{-1} and 1 kJ kg^{-1} K^{-1}, respectively, at −40 °C. Additional calculations at state points not listed below and for fluids not contained here can be obtained at the NIST Chemistry WebBook website (webbook.nist.gov/chemistry/fluid/), or by using the NIST Standard Reference Data program REFPROP (Ref. 9).

References

1. Arp, V.D., McCarty, R.D., and Friend, D.G., Thermophysical Properties of Helium-4 from 0.8 to 1500 K with Pressures to 2000 MPa, NIST Technical Note 1334 (revised), 1998.
2. Buecker, D. and Wagner, W., A Reference Equation of State for the Thermodynamic Properties of Ethane for Temperatures from the Melting Line to 675 K and Pressures up to 900 MPa, *J. Phys. Chem. Ref. Data*, 35, 205, 2006.
3. Friend, D.G., Ely, J.F., and Ingham, H., Tables for the Thermophysical Properties of Methane, NIST Technical Note 1325, 1989.
4. Friend, D.G., Ingham, H., and Ely, J.F., Thermophysical Properties of Ethane, *J. Phys. Chem. Ref. Data*, 20, 275, 1991.
5. Hands, B.A. and Arp, V.D., A Correlation of Thermal Conductivity Data for Helium, *Cryogenics*, 21, 697, 1981.
6. Harvey, A.H. and Lemmon, E.W., Method for Estimating the Dielectric Constant of Natural Gas Mixtures, *Int. J. Thermophys.*, 26, 31, 2005.
7. Leachman, J.W., Jacobsen, R.T, Lemmon, E.W., and Penoncello, S.G., Fundamental Equations of State for Parahydrogen, Normal Hydrogen, and Orthohydrogen, submitted to *J. Phys. Chem. Ref. Data*, 38, 721, 2009.
8. Lemmon, E.W. and Jacobsen, R.T, Viscosity and Thermal Conductivity Equations for Nitrogen, Oxygen, Argon, and Air, *Int. J. Thermophys.*, 25, 21, 2004.
9. Lemmon, E.W., Huber, M.L., and McLinden, M.O., NIST Standard Reference Database 23: Reference Fluid Thermodynamic and Transport Properties-REFPROP, Version 9.0, National Institute of Standards and Technology, Standard Reference Data Program, Gaithersburg, Maryland, 2010 (www.nist.gov/srd/nist23.cfm).
10. Lemmon, E.W., McLinden, M.O., and Friend, D.G., Thermophysical Properties of Fluid Systems, NIST Chemistry WebBook, NIST Standard Reference Database Number 69, Linstrom, P.J. and Mallard, W.G., Eds., National Institute of Standards and Technology, Gaithersburg, Maryland.
11. Lemmon, E.W., McLinden, M.O., and Wagner, W., Thermodynamic Properties of Propane. III. A Reference Equation of State for Temperatures from the Melting Line to 650 K and Pressures up to 1000 MPa, submitted to *J. Chem. Eng. Data*, 2009.
12. Marsh, K., Perkins, R., and Ramires, M.L.V., Measurement and Correlation of the Thermal Conductivity of Propane from 86 to 600 K at Pressures to 70 MPa, *J. Chem. Eng. Data*, 47, 932, 2002.
13. McCarty, R.D. and Arp, V.D., A New Wide Range Equation of State for Helium, *Adv. Cryo. Eng.*, 35, 1465, 1990.
14. McCarty, R.D. and Weber, L.A., Thermophysical Properties of Parahydrogen from the Freezing Liquid Line to 5000 R for Pressures to 10,000 psia, Natl. Bur. Stand., Tech. Note 617, 1972.
15. Quinones-Cisneros, S.E., Huber, M.L., and Deiters, U.K., Reference Correlation for the Viscosity of Methane, for submission to *J. Phys. Chem. Ref. Data*, 2009.
16. Schmidt, R. and Wagner, W., A New Form of the Equation of State for Pure Substances and its Application to Oxygen, *Fluid Phase Equilib.*, 19, 175, 1985.
17. Setzmann, U. and Wagner, W., A New Equation of State and Tables of Thermodynamic Properties for Methane Covering the Range from the Melting Line to 625 K at Pressures up to 1000 MPa, *J. Phys. Chem. Ref. Data*, 20, 1061, 1991.
18. Span, R. and Wagner, W., A New Equation of State for Carbon Dioxide Covering the Fluid Region from the Triple-Point Temperature to 1100 K at Pressures up to 800 MPa, *J. Phys. Chem. Ref. Data*, 25, 1509, 1996.
19. Span, R., Lemmon, E.W., Jacobsen, R.T, Wagner, W., and Yokozeki, A., A Reference Equation of State for the Thermodynamic Properties of Nitrogen for Temperatures from 63.151 to 1000 K and Pressures to 2200 MPa, *J. Phys. Chem. Ref. Data*, 29, 1361, 2000.
20. Tegeler, Ch., Span, R., and Wagner, W., A New Equation of State for Argon Covering the Fluid Region for Temperatures from the Melting Line to 700 K at Pressures up to 1000 MPa, *J. Phys. Chem. Ref. Data*, 28, 779, 1999.
21. Vogel, E., Kuechenmeister, C., Bich, E., and Laesecke, A., Reference Correlation of the Viscosity of Propane, *J. Phys. Chem. Ref. Data*, 27, 947, 1998.

Nitrogen (N₂)

T K	P MPa	ρ kg m⁻³	H kJ kg⁻¹	S kJ kg⁻¹ K⁻¹	C_v kJ kg⁻¹ K⁻¹	C_p kJ kg⁻¹ K⁻¹	u m s⁻¹	D	η μPa s	λ mW m⁻¹ K⁻¹
Saturation										
63.15	0.01252	867.2	−150.7	2.426	1.176	2.000	995.3	1.47003	311.6	173.2
63.15	0.01252	0.6743	64.78	5.838	0.7499	1.058	161.1	1.00032	4.376	5.621
70	0.03854	838.5	−137.0	2.632	1.130	2.014	925.7	1.45241	220.2	159.5
70	0.03854	1.896	71.10	5.605	0.7580	1.082	168.4	1.00089	4.883	6.355
80	0.1369	793.9	−116.6	2.903	1.069	2.056	824.4	1.42541	145.1	139.5
80	0.1369	6.089	79.10	5.349	0.7773	1.145	176.7	1.00286	5.652	7.506
90	0.3605	745.0	−95.52	3.147	1.020	2.141	719.0	1.39622	102.8	119.8
90	0.3605	15.08	84.97	5.153	0.8078	1.266	181.8	1.00710	6.482	8.868
100	0.7783	689.4	−73.21	3.376	0.9832	2.318	605.2	1.36351	75.76	100.1
100	0.7783	31.96	87.77	4.986	0.8548	1.503	183.3	1.01510	7.429	10.73
110	1.466	621.5	−48.49	3.601	0.9667	2.743	476.4	1.32430	55.99	80.44
110	1.466	62.58	85.84	4.823	0.9284	2.062	180.8	1.02974	8.626	13.83
120	2.511	523.4	−17.87	3.851	1.011	4.508	317.3	1.26895	38.43	61.01
120	2.511	125.1	74.17	4.618	1.099	4.631	172.6	1.06012	10.62	21.72
126.19	3.396	313.3	29.23	4.215				1.15559	18.30	
P = 0.1 MPa (1 bar)										
63.17	0.1	867.3	−150.6	2.426	1.176	2.000	995.6	1.47007	311.6	173.3
77.24	0.1	806.6	−122.2	2.831	1.085	2.041	852.5	1.43304	161.4	145.0
77.24	0.1	4.556	77.07	5.412	0.7710	1.123	174.7	1.00214	5.435	7.174
80	0.1	4.379	80.15	5.451	0.7666	1.112	178.3	1.00206	5.623	7.443
100	0.1	3.437	101.9	5.694	0.7514	1.071	201.6	1.00162	6.958	9.381
120	0.1	2.840	123.2	5.888	0.7466	1.057	222.0	1.00133	8.244	11.27
140	0.1	2.424	144.2	6.050	0.7447	1.050	240.4	1.00114	9.480	13.11
160	0.1	2.116	165.2	6.190	0.7438	1.047	257.3	1.00099	10.67	14.89
180	0.1	1.878	186.1	6.313	0.7433	1.045	273.2	1.00088	11.81	16.61
200	0.1	1.688	207.0	6.423	0.7430	1.043	288.1	1.00079	12.91	18.28
220	0.1	1.534	227.9	6.523	0.7429	1.043	302.3	1.00072	13.97	19.90
240	0.1	1.405	248.7	6.613	0.7428	1.042	315.8	1.00066	15.00	21.48
260	0.1	1.297	269.5	6.697	0.7428	1.042	328.7	1.00061	15.99	23.01
280	0.1	1.204	290.4	6.774	0.7429	1.041	341.2	1.00057	16.96	24.51
300	0.1	1.123	311.2	6.846	0.7432	1.041	353.2	1.00053	17.89	25.97
320	0.1	1.053	332.0	6.913	0.7436	1.042	364.7	1.00049	18.80	27.39
340	0.1	0.9909	352.9	6.976	0.7441	1.042	375.9	1.00047	19.68	28.79
360	0.1	0.9357	373.7	7.036	0.7450	1.043	386.8	1.00044	20.55	30.15
400	0.1	0.8421	415.5	7.146	0.7475	1.045	407.5	1.00040	22.21	32.81
500	0.1	0.6736	520.5	7.380	0.7592	1.056	454.6	1.00032	26.06	39.04
600	0.1	0.5613	627.0	7.574	0.7781	1.075	496.3	1.00026	29.58	44.84
700	0.1	0.4811	735.6	7.741	0.8011	1.098	533.9	1.00023	32.83	50.31
800	0.1	0.4210	846.6	7.890	0.8254	1.122	568.4	1.00020	35.89	55.51
900	0.1	0.3742	960.0	8.023	0.8488	1.146	600.7	1.00018	38.78	60.52
1000	0.1	0.3368	1076.	8.145	0.8705	1.167	631.1	1.00016	41.54	65.36
P = 1 MPa (10 bar)										
63.37	1.0	868.0	−149.5	2.427	1.177	1.995	998.9	1.47049	311.8	173.6
80	1.0	796.3	−116.0	2.896	1.071	2.044	832.3	1.42683	147.3	140.5
100	1.0	690.8	−73.18	3.373	0.9833	2.305	609.4	1.36432	76.26	100.6
103.75	1.0	665.8	−64.33	3.460	0.9739	2.431	559.2	1.34984	67.78	92.74
103.75	1.0	41.33	87.73	4.926	0.8786	1.652	182.8	1.01956	7.835	11.67
120	1.0	32.10	110.9	5.134	0.7980	1.297	208.9	1.01517	8.714	12.49
140	1.0	26.01	135.5	5.324	0.7701	1.177	232.8	1.01227	9.844	14.00
160	1.0	22.11	158.5	5.477	0.7588	1.127	252.9	1.01043	10.96	15.60
180	1.0	19.33	180.7	5.608	0.7531	1.101	270.6	1.00911	12.06	17.21
200	1.0	17.21	202.6	5.723	0.7500	1.085	286.9	1.00811	13.13	18.81
220	1.0	15.54	224.2	5.826	0.7480	1.075	301.9	1.00732	14.16	20.37
240	1.0	14.17	245.6	5.919	0.7468	1.067	316.1	1.00667	15.17	21.90
260	1.0	13.04	266.9	6.005	0.7460	1.062	329.5	1.00614	16.14	23.39

T K	P MPa	ρ kg m^{-3}	H kJ kg^{-1}	S kJ kg^{-1} K^{-1}	C_v kJ kg^{-1} K^{-1}	C_p kJ kg^{-1} K^{-1}	u m s^{-1}	D	η μPa s	λ mW m^{-1} K^{-1}
280	1.0	12.07	288.1	6.083	0.7456	1.059	342.4	1.00569	17.09	24.86
300	1.0	11.25	309.2	6.156	0.7454	1.056	354.6	1.00530	18.01	26.29
320	1.0	10.53	330.3	6.224	0.7455	1.054	366.4	1.00496	18.91	27.69
340	1.0	9.901	351.4	6.288	0.7458	1.053	377.8	1.00466	19.79	29.07
360	1.0	9.343	372.4	6.348	0.7464	1.052	388.8	1.00440	20.64	30.42
400	1.0	8.399	414.5	6.459	0.7486	1.052	409.7	1.00395	22.29	33.04
500	1.0	6.711	520.1	6.695	0.7600	1.061	457.0	1.00316	26.13	39.23
600	1.0	5.592	627.0	6.889	0.7786	1.078	498.7	1.00263	29.63	44.99
700	1.0	4.793	735.8	7.057	0.8015	1.100	536.2	1.00226	32.87	50.43
800	1.0	4.195	847.0	7.206	0.8257	1.124	570.7	1.00197	35.92	55.63
900	1.0	3.730	960.6	7.339	0.8492	1.147	602.9	1.00176	38.81	60.62
1000	1.0	3.358	1076.	7.461	0.8708	1.168	633.2	1.00158	41.57	65.45

P = 10 MPa (100 bar)

T K	P MPa	ρ kg m^{-3}	H kJ kg^{-1}	S kJ kg^{-1} K^{-1}	C_v kJ kg^{-1} K^{-1}	C_p kJ kg^{-1} K^{-1}	u m s^{-1}	D	η μPa s	λ mW m^{-1} K^{-1}
65.32	10.0	875.0	−138.3	2.441	1.186	1.955	1031.	1.47447	314.0	177.1
80	10.0	818.4	−109.6	2.837	1.093	1.960	904.1	1.43984	169.0	150.4
100	10.0	733.6	−69.89	3.280	0.9997	2.022	734.2	1.38918	93.65	115.9
120	10.0	632.9	−27.82	3.663	0.9401	2.218	559.6	1.33060	58.57	84.11
140	10.0	499.8	20.66	4.036	0.9127	2.676	392.0	1.25566	36.11	57.60
160	10.0	344.2	76.44	4.408	0.8849	2.668	300.8	1.17162	22.40	40.20
180	10.0	248.7	122.6	4.681	0.8407	1.990	294.7	1.12203	18.50	32.06
200	10.0	199.4	158.4	4.870	0.8107	1.628	307.7	1.09701	17.70	29.38
220	10.0	169.4	188.9	5.015	0.7927	1.445	323.2	1.08198	17.74	28.72
240	10.0	148.8	216.7	5.136	0.7813	1.340	338.4	1.07172	18.13	28.87
260	10.0	133.4	242.8	5.240	0.7737	1.273	352.9	1.06415	18.67	29.39
280	10.0	121.4	267.7	5.333	0.7685	1.227	366.6	1.05827	19.29	30.21
300	10.0	111.7	291.9	5.417	0.7649	1.195	379.5	1.05351	19.96	31.14
320	10.0	103.6	315.6	5.493	0.7623	1.171	391.8	1.04958	20.66	32.13
340	10.0	96.80	338.8	5.563	0.7606	1.153	403.6	1.04625	21.37	33.16
360	10.0	90.89	361.7	5.629	0.7596	1.138	414.8	1.04338	22.08	34.21
400	10.0	81.19	406.8	5.748	0.7593	1.119	436.0	1.03869	23.51	36.35
500	10.0	64.50	517.5	5.995	0.7672	1.100	483.1	1.03065	26.99	41.73
600	10.0	53.74	627.6	6.195	0.7841	1.104	523.9	1.02549	30.28	47.00
700	10.0	46.15	738.7	6.367	0.8059	1.118	560.4	1.02187	33.40	52.10
800	10.0	40.49	851.4	6.517	0.8294	1.137	593.8	1.01917	36.35	57.04
900	10.0	36.08	966.1	6.652	0.8523	1.157	625.0	1.01708	39.17	61.85
1000	10.0	32.56	1083.	6.775	0.8735	1.176	654.4	1.01540	41.88	66.54

Oxygen (O$_2$)

T K	P MPa	ρ kg m^{-3}	H kJ kg^{-1}	S kJ kg^{-1} K^{-1}	C_v kJ kg^{-1} K^{-1}	C_p kJ kg^{-1} K^{-1}	u m s^{-1}	D	η μPa s	λ mW m^{-1} K^{-1}
Saturation										
54.36	0.0001463	1306.	−193.6	2.092	1.195	1.673	1123	1.56799	773.6	201.9
54.36	0.0001463	0.01036	49.11	6.557	0.6638	0.9260	140.3	1.00000	4.096	4.420
60	0.0007258	1282.	−184.2	2.257	1.089	1.673	1127.	1.55615	578.1	193.9
60	0.0007258	0.04659	54.19	6.230	0.6817	0.9475	147.0	1.00002	4.553	4.984
70	0.006262	1237.	−167.4	2.516	1.017	1.678	1066.	1.53399	371.8	179.7
70	0.006262	0.3457	63.09	5.809	0.7052	0.9780	158.1	1.00013	5.356	5.992
80	0.03012	1190.	−150.6	2.740	0.9697	1.682	987.4	1.51120	261.2	165.4
80	0.03012	1.468	71.69	5.519	0.6950	0.9743	168.4	1.00054	6.149	7.028
90	0.09935	1142.	−133.7	2.938	0.9296	1.699	905.9	1.48766	195.6	151.0
90	0.09935	4.387	79.55	5.308	0.6758	0.9705	177.3	1.00163	6.936	8.124
100	0.2540	1091.	−116.4	3.118	0.8949	1.738	822.2	1.46295	152.6	136.6
100	0.2540	10.42	86.16	5.144	0.6752	1.006	184.1	1.00387	7.728	9.336
110	0.5434	1035.	−98.64	3.286	0.8658	1.807	734.8	1.43650	121.5	121.9
110	0.5434	21.28	91.05	5.010	0.6988	1.101	188.1	1.00791	8.547	10.75
120	1.022	973.9	−79.90	3.444	0.8430	1.927	641.5	1.40746	97.43	107.2
120	1.022	39.31	93.75	4.892	0.7415	1.276	189.4	1.01465	9.427	12.51

T	P	ρ	H	S	C_v	C_p	u	D	η	λ
K	MPa	kg m^{-3}	kJ kg^{-1}	kJ kg^{-1} K^{-1}	kJ kg^{-1} K^{-1}	kJ kg^{-1} K^{-1}	m s^{-1}		μPa s	mW m^{-1} K^{-1}
130	1.749	902.5	−59.66	3.600	0.8293	2.153	539.5	1.37432	77.57	92.63
130	1.749	68.37	93.47	4.778	0.8002	1.600	187.8	1.02558	10.45	14.94
140	2.788	813.2	−36.70	3.761	0.8323	2.691	423.1	1.33363	60.22	78.22
140	2.788	116.8	88.47	4.655	0.8834	2.370	182.8	1.04395	11.82	18.98
150	4.219	675.5	−6.671	3.955	0.9057	5.464	273.8	1.27248	42.90	64.19
150	4.219	214.9	72.56	4.483	1.049	6.625	172.8	1.08192	14.72	29.67
154.58	5.043	436.1	32.42	4.201				1.17084	24.84	

P = 0.1 MPa (1 bar)

54.37	0.1	1306.	−193.5	2.092	1.195	1.673	1124.	1.56802	773.8	201.9
60	0.1	1282.	−184.1	2.257	1.089	1.673	1128.	1.55621	578.5	194.0
80	0.1	1191.	−150.6	2.739	0.9699	1.681	987.7	1.51126	261.4	165.5
90.06	0.1	1142.	−133.6	2.939	0.9293	1.699	905.4	1.48751	195.3	151.0
90.06	0.1	4.413	79.60	5.307	0.6757	0.9705	177.4	1.00164	6.940	8.131
100	0.1	3.941	88.99	5.405	0.6527	0.9352	188.4	1.00146	7.712	9.085
120	0.1	3.252	107.6	5.575	0.6543	0.9280	207.3	1.00121	9.219	10.99
140	0.1	2.774	126.1	5.718	0.6530	0.9218	224.6	1.00103	10.67	12.86
160	0.1	2.420	144.5	5.841	0.6519	0.9180	240.5	1.00090	12.07	14.69
180	0.1	2.147	162.8	5.949	0.6513	0.9158	255.4	1.00080	13.41	16.48
200	0.1	1.930	181.1	6.045	0.6512	0.9146	269.3	1.00072	14.72	18.24
220	0.1	1.753	199.4	6.132	0.6516	0.9142	282.6	1.00065	15.98	19.95
240	0.1	1.606	217.7	6.212	0.6525	0.9146	295.2	1.00060	17.20	21.64
260	0.1	1.482	236.0	6.285	0.6539	0.9156	307.2	1.00055	18.38	23.28
280	0.1	1.376	254.3	6.353	0.6560	0.9174	318.7	1.00051	19.53	24.90
300	0.1	1.284	272.7	6.416	0.6587	0.9199	329.7	1.00048	20.65	26.49
320	0.1	1.203	291.1	6.476	0.6621	0.9231	340.3	1.00045	21.74	28.04
340	0.1	1.132	309.6	6.532	0.6661	0.9269	350.5	1.00042	22.80	29.58
360	0.1	1.069	328.2	6.585	0.6707	0.9313	360.4	1.00040	23.84	31.08
400	0.1	0.9622	365.7	6.684	0.6812	0.9416	379.0	1.00036	25.84	34.03
500	0.1	0.7696	461.3	6.897	0.7119	0.9722	421.3	1.00029	30.49	41.05
600	0.1	0.6413	560.1	7.077	0.7432	1.003	458.9	1.00024	34.73	47.66
700	0.1	0.5497	661.9	7.234	0.7710	1.031	493.3	1.00020	38.65	53.97
800	0.1	0.4809	766.2	7.373	0.7945	1.055	525.4	1.00018	42.33	60.02
900	0.1	0.4275	872.6	7.498	0.8140	1.074	555.6	1.00016	45.81	65.87
1000	0.1	0.3848	980.9	7.612	0.8301	1.090	584.3	1.00014	49.12	71.55

P = 1 MPa (10 bar)

54.47	1.0	1307.	−192.8	2.093	1.190	1.669	1127.	1.56831	775.0	202.1
60	1.0	1283.	−183.6	2.254	1.089	1.671	1130.	1.55674	582.7	194.3
80	1.0	1192.	−150.1	2.736	0.9716	1.679	991.1	1.51204	263.6	166.0
100	1.0	1093.	−116.1	3.115	0.8964	1.731	826.8	1.46397	153.9	137.2
119.62	1.0	976.3	−80.64	3.438	0.8438	1.921	645.2	1.40862	98.25	107.8
119.62	1.0	38.46	93.70	4.896	0.7396	1.268	189.4	1.01433	9.392	12.43
120	1.0	38.25	94.18	4.900	0.7355	1.257	190.0	1.01425	9.421	12.45
140	1.0	30.39	116.8	5.075	0.6851	1.065	214.2	1.01131	10.89	13.90
160	1.0	25.65	137.5	5.213	0.6708	1.005	233.7	1.00954	12.29	15.54
180	1.0	22.33	157.2	5.329	0.6631	0.9743	250.8	1.00831	13.64	17.21
200	1.0	19.84	176.5	5.431	0.6590	0.9568	266.3	1.00738	14.94	18.88
220	1.0	17.88	195.5	5.521	0.6571	0.9462	280.6	1.00665	16.20	20.53
240	1.0	16.30	214.4	5.603	0.6566	0.9397	294.0	1.00606	17.41	22.15
260	1.0	14.98	233.1	5.678	0.6572	0.9359	306.6	1.00557	18.59	23.76
280	1.0	13.86	251.8	5.748	0.6586	0.9342	318.5	1.00515	19.73	25.34
300	1.0	12.91	270.5	5.812	0.6609	0.9340	329.9	1.00480	20.85	26.89
320	1.0	12.08	289.2	5.872	0.6640	0.9351	340.8	1.00449	21.93	28.43
340	1.0	11.35	307.9	5.929	0.6677	0.9373	351.2	1.00422	22.99	29.93
360	1.0	10.71	326.7	5.983	0.6720	0.9404	361.2	1.00398	24.02	31.42
400	1.0	9.623	364.5	6.082	0.6823	0.9487	380.1	1.00358	26.01	34.33
500	1.0	7.683	460.7	6.297	0.7126	0.9763	422.7	1.00286	30.63	41.29
600	1.0	6.398	559.8	6.478	0.7436	1.006	460.4	1.00238	34.85	47.86

T K	P MPa	ρ kg m^{-3}	H kJ kg^{-1}	S kJ kg^{-1} K^{-1}	C_v kJ kg^{-1} K^{-1}	C_p kJ kg^{-1} K^{-1}	u m s^{-1}	D	η μPa s	λ mW m^{-1} K^{-1}
700	1.0	5.483	661.8	6.635	0.7713	1.033	494.9	1.00204	38.77	54.14
800	1.0	4.798	766.2	6.774	0.7948	1.056	526.9	1.00179	42.43	60.17
900	1.0	4.265	872.8	6.900	0.8142	1.075	557.1	1.00159	45.90	66.00
1000	1.0	3.839	981.1	7.014	0.8303	1.091	585.8	1.00143	49.20	71.67

P = 10 MPa (100 bar)

T K	P MPa	ρ kg m^{-3}	H kJ kg^{-1}	S kJ kg^{-1} K^{-1}	C_v kJ kg^{-1} K^{-1}	C_p kJ kg^{-1} K^{-1}	u m s^{-1}	D	η μPa s	λ mW m^{-1} K^{-1}
55.50	10.0	1312.	−185.4	2.103	1.149	1.640	1158.	1.57109	786.0	203.9
60	10.0	1294.	−178.0	2.231	1.092	1.653	1155.	1.56193	625.3	197.9
80	10.0	1207.	−144.8	2.708	0.9885	1.654	1023.	1.51940	285.2	171.5
100	10.0	1116.	−111.6	3.078	0.9136	1.672	877.1	1.47515	169.5	144.8
120	10.0	1015.	−77.51	3.389	0.8571	1.755	727.0	1.42679	112.4	118.1
140	10.0	892.5	−40.64	3.673	0.8191	1.966	564.5	1.36967	75.70	91.70
160	10.0	716.0	4.057	3.970	0.8137	2.659	379.7	1.29022	47.96	66.52
180	10.0	423.3	70.35	4.360	0.8161	3.312	250.9	1.16550	26.07	42.24
200	10.0	277.6	119.7	4.621	0.7506	1.894	259.0	1.10662	20.74	31.77
220	10.0	219.9	152.6	4.778	0.7152	1.462	278.2	1.08384	20.24	29.46
240	10.0	186.7	179.8	4.896	0.6973	1.280	295.8	1.07090	20.67	29.19
260	10.0	164.2	204.3	4.995	0.6878	1.183	311.6	1.06220	21.40	29.67
280	10.0	147.6	227.3	5.080	0.6830	1.124	326.1	1.05578	22.25	30.48
300	10.0	134.6	249.4	5.156	0.6809	1.086	339.4	1.05078	23.15	31.47
320	10.0	124.0	270.8	5.225	0.6809	1.060	351.7	1.04675	24.07	32.56
340	10.0	115.2	291.9	5.289	0.6823	1.042	363.3	1.04339	25.00	33.74
360	10.0	107.8	312.6	5.348	0.6848	1.030	374.2	1.04054	25.92	34.96
400	10.0	95.67	353.5	5.456	0.6923	1.017	394.5	1.03595	27.74	37.45
500	10.0	75.32	454.8	5.682	0.7187	1.015	438.7	1.02825	32.07	43.71
600	10.0	62.44	557.1	5.869	0.7477	1.031	476.8	1.02339	36.10	49.86
700	10.0	53.46	661.2	6.029	0.7743	1.050	511.2	1.02002	39.87	55.84
800	10.0	46.79	767.1	6.170	0.7970	1.069	543.1	1.01752	43.43	61.66
900	10.0	41.63	874.8	6.297	0.8159	1.085	572.9	1.01558	46.80	67.32
1000	10.0	37.51	984.0	6.412	0.8316	1.098	601.2	1.01404	50.03	72.85

Parahydrogen (H$_2$)

T K	P MPa	ρ kg m^{-3}	H kJ kg^{-1}	S kJ kg^{-1} K^{-1}	C_v kJ kg^{-1} K^{-1}	C_p kJ kg^{-1} K^{-1}	u m s^{-1}	D	η μPa s	λ mW m^{-1} K^{-1}
Saturation										
13.80	0.007041	76.98	−53.74	−3.084	5.131	6.924	1263.	1.25267	25.90	75.27
13.80	0.007041	0.1255	396.3	29.52	6.226	10.53	305.6	1.00038	0.6507	10.46
14	0.007884	76.82	−52.36	−2.986	5.158	6.981	1257.	1.25208	25.21	76.86
14	0.007884	0.1388	398.1	29.19	6.236	10.56	307.6	1.00042	0.6669	10.65
16	0.02155	75.13	−37.60	−2.013	5.334	7.659	1210.	1.24585	19.81	90.02
16	0.02155	0.3377	415.8	26.33	6.323	10.87	325.4	1.00102	0.8106	12.60
18	0.04815	73.25	−21.18	−1.068	5.472	8.513	1168.	1.23903	16.22	98.39
18	0.04815	0.6880	431.5	24.08	6.387	11.29	340.7	1.00207	0.9355	14.60
20	0.09341	71.14	−2.692	−.1281	5.637	9.569	1119.	1.23147	13.63	103.0
20	0.09341	1.244	444.5	22.23	6.450	11.92	353.5	1.00375	1.057	16.71
22	0.1635	68.74	18.32	0.8244	5.811	10.86	1059.	1.22301	11.65	104.9
22	0.1635	2.071	454.2	20.64	6.541	12.88	363.7	1.00625	1.182	18.98
24	0.2648	66.01	42.37	1.805	5.974	12.52	989.9	1.21345	10.06	104.7
24	0.2648	3.255	459.7	19.19	6.681	14.40	371.3	1.00984	1.315	21.54
26	0.4038	62.83	70.23	2.832	6.120	14.83	909.3	1.20242	8.725	102.6
26	0.4038	4.924	460.0	17.82	6.890	16.90	376.1	1.01491	1.463	24.53
28	0.5875	58.98	103.2	3.941	6.266	18.57	813.3	1.18922	7.533	98.79
28	0.5875	7.300	453.4	16.45	7.188	21.53	378.2	1.02216	1.634	28.26
30	0.8232	53.98	144.2	5.211	6.472	26.65	693.0	1.17223	6.390	92.78
30	0.8232	10.87	435.7	14.93	7.625	32.58	377.2	1.03313	1.865	33.47
32	1.120	45.90	204.1	6.945	7.010	68.19	522.6	1.14520	5.076	82.53
32	1.120	17.49	392.3	12.83	8.336	92.39	372.0	1.05368	2.337	43.17
32.94	1.286	31.32	295.6	9.625				1.09756	3.512	

T K	P MPa	ρ kg m^{-3}	H kJ kg^{-1}	S kJ kg^{-1} K^{-1}	C_v kJ kg^{-1} K^{-1}	C_p kJ kg^{-1} K^{-1}	u m s^{-1}	D	η µPa s	λ mW m^{-1} K^{-1}
P = 0.1 MPa (1 bar)										
20	0.1	71.14	−2.629	−.1296	5.637	9.566	1119.	1.23150	13.63	103.1
20.23	0.1	70.88	−.4432	−.02096	5.657	9.702	1112.	1.23055	13.38	103.4
20.23	0.1	1.323	445.8	22.04	6.458	12.01	354.8	1.00399	1.071	16.95
40	0.1	0.6157	661.5	29.52	6.218	10.57	521.4	1.00186	2.039	31.85
60	0.1	0.4059	873.1	33.81	6.503	10.72	635.7	1.00122	2.865	45.53
80	0.1	0.3035	1096.	37.02	7.598	11.77	713.9	1.00091	3.566	61.36
100	0.1	0.2425	1348.	39.82	9.277	13.43	772.5	1.00073	4.190	80.43
120	0.1	0.2020	1633.	42.41	10.83	14.98	827.4	1.00061	4.763	101.4
140	0.1	0.1731	1943.	44.80	11.82	15.96	883.4	1.00052	5.301	119.7
160	0.1	0.1514	2267.	46.96	12.21	16.35	940.4	1.00046	5.814	134.8
180	0.1	0.1346	2594.	48.89	12.19	16.32	997.7	1.00041	6.306	146.5
200	0.1	0.1211	2918.	50.60	11.95	16.08	1054.	1.00037	6.780	155.6
220	0.1	0.1101	3237.	52.11	11.63	15.76	1110.	1.00033	7.239	163.4
240	0.1	0.1010	3549.	53.47	11.33	15.46	1163.	1.00030	7.684	170.6
260	0.1	0.09319	3856.	54.70	11.07	15.20	1214.	1.00028	8.118	177.7
280	0.1	0.08654	4157.	55.82	10.87	15.00	1263.	1.00026	8.540	184.9
300	0.1	0.08077	4456.	56.85	10.72	14.85	1310.	1.00024	8.953	192.3
320	0.1	0.07572	4752.	57.80	10.61	14.74	1355.	1.00023	9.357	200.0
340	0.1	0.07127	5045.	58.69	10.53	14.66	1398.	1.00022	9.752	208.0
360	0.1	0.06731	5338.	59.53	10.48	14.61	1439.	1.00020	10.14	216.4
400	0.1	0.06058	5921.	61.06	10.43	14.55	1518.	1.00018	10.89	233.1
500	0.1	0.04847	7374.	64.31	10.40	14.53	1698.	1.00015	12.67	275.0
600	0.1	0.04040	8828.	66.96	10.42	14.55	1859.	1.00012	14.34	317.6
700	0.1	0.03463	10290	69.20	10.48	14.60	2006.	1.00011	15.92	360.8
800	0.1	0.03030	11750	71.16	10.57	14.70	2142.	1.00009	17.43	404.7
900	0.1	0.02693	13230	72.90	10.71	14.83	2268.	1.00008	18.89	449.2
1000	0.1	0.02424	14720	74.47	10.87	14.99	2386.	1.00007	20.30	494.3
P = 1 MPa (10 bar)										
20	1.0	72.29	6.046	−.3233	5.634	9.187	1164.	1.23552	14.53	105.1
31.24	1.0	49.65	177.4	6.181	6.718	40.77	596.5	1.15769	5.627	87.27
31.24	1.0	14.31	414.3	13.76	8.017	52.90	374.7	1.04376	2.096	38.44
40	1.0	7.270	590.6	18.85	6.455	14.12	497.2	1.02207	2.228	37.28
60	1.0	4.228	838.4	23.90	6.588	11.70	634.2	1.01279	2.976	48.40
80	1.0	3.072	1075.	27.31	7.645	12.25	717.9	1.00929	3.648	63.68
100	1.0	2.429	1335.	30.19	9.307	13.72	778.4	1.00734	4.255	82.51
120	1.0	2.015	1624.	32.83	10.85	15.17	834.2	1.00609	4.817	103.1
140	1.0	1.723	1938.	35.24	11.84	16.09	890.7	1.00520	5.348	121.2
160	1.0	1.506	2264.	37.42	12.23	16.45	948.1	1.00455	5.854	136.1
180	1.0	1.338	2593.	39.36	12.20	16.40	1006.	1.00404	6.341	147.6
200	1.0	1.204	2918.	41.07	11.96	16.14	1062.	1.00364	6.812	156.7
220	1.0	1.095	3238.	42.60	11.64	15.81	1117.	1.00331	7.268	164.3
240	1.0	1.004	3551.	43.96	11.34	15.50	1171.	1.00303	7.711	171.5
260	1.0	0.9267	3858.	45.19	11.08	15.24	1222.	1.00280	8.143	178.6
280	1.0	0.8607	4161.	46.31	10.88	15.03	1271.	1.00260	8.563	185.7
300	1.0	0.8035	4460.	47.34	10.73	14.87	1317.	1.00243	8.975	193.0
320	1.0	0.7534	4756.	48.30	10.62	14.76	1362.	1.00228	9.377	200.7
340	1.0	0.7093	5050.	49.19	10.54	14.68	1405.	1.00215	9.771	208.7
360	1.0	0.6700	5343.	50.03	10.49	14.62	1446.	1.00203	10.16	217.0
400	1.0	0.6033	5927.	51.56	10.43	14.56	1525.	1.00183	10.91	233.7
500	1.0	0.4830	7381.	54.81	10.41	14.53	1704.	1.00146	12.68	275.5
600	1.0	0.4027	8835.	57.46	10.43	14.55	1865.	1.00122	14.35	318.0
700	1.0	0.3454	10290	59.71	10.48	14.61	2011.	1.00105	15.93	361.2
800	1.0	0.3023	11760	61.66	10.58	14.70	2147.	1.00092	17.44	405.0
900	1.0	0.2688	13230	63.40	10.71	14.83	2272.	1.00082	18.90	449.5
1000	1.0	0.2420	14720	64.97	10.87	15.00	2390.	1.00074	20.30	494.6

T K	P MPa	ρ kg m^{-3}	H kJ kg^{-1}	S kJ kg^{-1} K^{-1}	C_v kJ kg^{-1} K^{-1}	C_p kJ kg^{-1} K^{-1}	u m s^{-1}	D	η μPa s	λ mW m^{-1} K^{-1}
P = 10 MPa (100 bar)										
20	10.0	79.91	98.92	−1.567	5.529	7.566	1467.	1.26244	23.45	120.3
40	10.0	63.19	310.5	5.532	6.665	13.45	1154.	1.20345	8.791	121.8
60	10.0	42.86	617.0	11.71	6.960	16.19	923.2	1.13508	5.594	100.6
80	10.0	29.87	933.6	16.27	7.952	15.47	884.9	1.09289	5.034	96.74
100	10.0	23.01	1244.	19.73	9.546	15.76	904.3	1.07104	5.169	106.3
120	10.0	18.87	1567.	22.67	11.05	16.56	942.4	1.05803	5.497	121.6
140	10.0	16.08	1905.	25.27	12.00	17.11	990.1	1.04932	5.888	136.6
160	10.0	14.06	2249.	27.57	12.37	17.22	1042.	1.04303	6.303	149.4
180	10.0	12.51	2592.	29.59	12.32	17.01	1096.	1.03824	6.727	159.4
200	10.0	11.28	2928.	31.36	12.07	16.63	1150.	1.03445	7.151	167.3
220	10.0	10.28	3257.	32.93	11.74	16.22	1203.	1.03138	7.571	173.9
240	10.0	9.453	3577.	34.32	11.43	15.84	1254.	1.02883	7.985	180.4
260	10.0	8.751	3891.	35.58	11.17	15.52	1303.	1.02667	8.393	186.8
280	10.0	8.149	4199.	36.72	10.96	15.27	1350.	1.02483	8.795	193.3
300	10.0	7.626	4502.	37.77	10.80	15.08	1394.	1.02323	9.189	200.2
320	10.0	7.167	4802.	38.73	10.68	14.94	1437.	1.02183	9.577	207.4
340	10.0	6.762	5100.	39.64	10.60	14.84	1478.	1.02059	9.959	215.0
360	10.0	6.400	5396.	40.48	10.55	14.77	1518.	1.01948	10.33	223.0
400	10.0	5.784	5985.	42.03	10.49	14.68	1593.	1.01761	11.07	239.2
500	10.0	4.665	7447.	45.30	10.45	14.60	1764.	1.01420	12.81	280.0
600	10.0	3.911	8907.	47.96	10.46	14.59	1919.	1.01191	14.46	321.9
700	10.0	3.368	10370	50.21	10.51	14.63	2062.	1.01026	16.02	364.7
800	10.0	2.957	11830	52.17	10.60	14.72	2193.	1.00902	17.52	408.2
900	10.0	2.636	13310	53.91	10.73	14.84	2315.	1.00805	18.97	452.4
1000	10.0	2.378	14800	55.48	10.89	15.00	2430.	1.00727	20.37	497.3

Helium (He−4)

T K	P MPa	ρ kg m^{-3}	H kJ kg^{-1}	S kJ kg^{-1} K^{-1}	C_v kJ kg^{-1} K^{-1}	C_p kJ kg^{-1} K^{-1}	u m s^{-1}	D	η μPa s	λ mW m^{-1} K^{-1}
Saturation										
2.177	0.004856	146.2	−7.501	−2.177	6.287	6.318	216.8	1.07842	3.596	13.52
2.177	0.004856	1.146	15.73	8.494	3.470	6.061	83.22	1.00045	0.5376	3.977
2.5	0.01000	145.0	−6.180	−1.620	2.521	2.700	216.0	1.07491	3.746	14.89
2.5	0.01000	2.116	16.97	7.641	3.486	6.258	87.62	1.00082	0.6382	4.765
3	0.02373	141.3	−4.869	−1.177	2.061	2.598	214.3	1.06975	3.694	16.55
3	0.02373	4.428	18.66	6.666	3.452	6.592	93.04	1.00173	0.7961	5.893
3.5	0.04663	136.1	−3.258	−.7324	2.332	3.414	202.4	1.06471	3.511	17.75
3.5	0.04663	8.022	19.95	5.898	3.378	7.122	96.94	1.00315	0.9640	7.060
4	0.08100	128.9	−1.164	−.2439	2.504	4.523	185.7	1.05941	3.278	18.47
4	0.08100	13.41	20.68	5.217	3.284	8.179	99.47	1.00529	1.151	8.356
4.5	0.1292	118.8	1.586	0.3106	2.601	6.776	164.0	1.05319	3.010	18.77
4.5	0.1292	21.76	20.52	4.519	3.179	10.96	100.8	1.00865	1.372	10.04
5	0.1945	100.8	5.854	1.081	2.709	20.24	133.0	1.04367	2.634	19.28
5	0.1945	38.21	18.18	3.547	3.053	29.09	101.6	1.01543	1.687	13.60
5.195	0.2275	69.64	11.87	2.183				1.02904	2.129	
P = 0.1 MPa (1 bar)										
20	0.1	2.408	108.5	13.94	3.121	5.250	264.3	1.00093	3.582	26.20
40	0.1	1.200	212.9	17.56	3.119	5.206	373.4	1.00047	5.542	40.44
60	0.1	0.8006	317.0	19.67	3.118	5.198	456.9	1.00031	7.116	52.55
80	0.1	0.6007	420.9	21.16	3.117	5.195	527.2	1.00023	8.503	63.52
100	0.1	0.4807	524.8	22.32	3.117	5.194	589.2	1.00019	9.778	73.71
120	0.1	0.4007	628.7	23.27	3.117	5.194	645.3	1.00016	10.79	83.33
140	0.1	0.3435	732.6	24.07	3.116	5.193	696.9	1.00013	11.94	92.50
160	0.1	0.3006	836.4	24.76	3.116	5.193	744.9	1.00012	13.05	101.3
180	0.1	0.2672	940.3	25.37	3.116	5.193	790.0	1.00010	14.11	109.8
200	0.1	0.2405	1044.	25.92	3.116	5.193	832.7	1.00009	15.14	118.0

T K	P MPa	ρ kg m^{-3}	H kJ kg^{-1}	S kJ kg^{-1} K^{-1}	C_v kJ kg^{-1} K^{-1}	C_p kJ kg^{-1} K^{-1}	u m s^{-1}	D	η μPa s	λ mW m^{-1} K^{-1}
220	0.1	0.2187	1148.	26.42	3.116	5.193	873.3	1.00008	16.14	126.0
240	0.1	0.2005	1252.	26.87	3.116	5.193	912.0	1.00008	17.12	133.7
260	0.1	0.1851	1356.	27.28	3.116	5.193	949.2	1.00007	18.08	141.3
280	0.1	0.1718	1460.	27.67	3.116	5.193	985.0	1.00007	19.01	148.7
300	0.1	0.1604	1563.	28.03	3.116	5.193	1020.	1.00006	19.93	156.0
320	0.1	0.1504	1667.	28.36	3.116	5.193	1053.	1.00006	20.83	163.1
340	0.1	0.1415	1771.	28.68	3.116	5.193	1085.	1.00005	21.72	170.1
360	0.1	0.1337	1875.	28.97	3.116	5.193	1117.	1.00005	22.59	177.0
400	0.1	0.1203	2083.	29.52	3.116	5.193	1177.	1.00005	24.29	190.4
500	0.1	0.09626	2602.	30.68	3.116	5.193	1316.	1.00004	28.36	222.3
600	0.1	0.08022	3121.	31.63	3.116	5.193	1442.	1.00003	32.22	252.4
700	0.1	0.06876	3641.	32.43	3.116	5.193	1557.	1.00003	35.89	281.1
800	0.1	0.06017	4160.	33.12	3.116	5.193	1664.	1.00002	39.43	308.5
900	0.1	0.05348	4679.	33.73	3.116	5.193	1765.	1.00002	42.85	335.0
1000	0.1	0.04814	5199.	34.28	3.116	5.193	1861.	1.00002	46.16	360.6
P = 1 MPa (10 bar)										
20	1.0	24.02	104.1	8.936	3.156	5.728	276.4	1.00938	3.933	29.04
40	1.0	11.72	213.2	12.73	3.143	5.317	384.9	1.00455	5.740	42.01
60	1.0	7.850	318.6	14.87	3.133	5.242	466.8	1.00305	7.269	53.72
80	1.0	5.912	423.1	16.37	3.128	5.217	535.8	1.00229	8.633	64.50
100	1.0	4.745	527.4	17.54	3.125	5.206	596.9	1.00184	9.892	74.60
120	1.0	3.963	631.4	18.48	3.123	5.200	652.2	1.00154	10.90	84.16
140	1.0	3.403	735.4	19.29	3.121	5.197	703.2	1.00132	12.04	93.29
160	1.0	2.982	839.3	19.98	3.120	5.195	750.7	1.00116	13.13	102.1
180	1.0	2.653	943.2	20.59	3.120	5.194	795.4	1.00103	14.19	110.5
200	1.0	2.390	1047.	21.14	3.119	5.194	837.7	1.00093	15.21	118.7
220	1.0	2.174	1151.	21.63	3.119	5.193	878.0	1.00084	16.20	126.7
240	1.0	1.994	1255.	22.09	3.118	5.193	916.5	1.00077	17.17	134.4
260	1.0	1.842	1359.	22.50	3.118	5.193	953.5	1.00071	18.12	142.0
280	1.0	1.711	1463.	22.89	3.118	5.192	989.1	1.00066	19.05	149.4
300	1.0	1.597	1566.	23.24	3.118	5.192	1023.	1.00062	19.96	156.6
320	1.0	1.498	1670.	23.58	3.117	5.192	1057.	1.00058	20.86	163.8
340	1.0	1.410	1774.	23.89	3.117	5.192	1089.	1.00055	21.74	170.7
360	1.0	1.332	1878.	24.19	3.117	5.192	1120.	1.00052	22.61	177.6
400	1.0	1.200	2086.	24.74	3.117	5.192	1180.	1.00047	24.32	191.0
500	1.0	0.9604	2605.	25.90	3.117	5.192	1319.	1.00037	28.38	222.9
600	1.0	0.8007	3124.	26.84	3.116	5.192	1444.	1.00031	32.23	253.0
700	1.0	0.6866	3643.	27.64	3.116	5.192	1559.	1.00027	35.91	281.6
800	1.0	0.6009	4163.	28.34	3.116	5.192	1666.	1.00023	39.44	309.1
900	1.0	0.5342	4682.	28.95	3.116	5.193	1767.	1.00021	42.86	335.5
1000	1.0	0.4809	5201.	29.50	3.116	5.193	1863.	1.00019	46.17	361.1
P = 10 MPa (100 bar)										
20	10.0	147.2	111.0	3.756	3.286	5.413	497.6	1.06056	6.928	58.72
40	10.0	91.33	225.2	7.703	3.305	5.721	512.2	1.03614	7.483	59.90
60	10.0	65.36	337.4	9.980	3.251	5.506	566.3	1.02560	8.638	68.12
80	10.0	51.09	446.0	11.54	3.215	5.376	620.9	1.01993	9.802	76.83
100	10.0	42.05	552.8	12.74	3.192	5.303	672.3	1.01637	10.93	85.42
120	10.0	35.77	658.4	13.70	3.176	5.261	720.4	1.01391	11.82	93.82
140	10.0	31.14	763.3	14.51	3.165	5.235	765.6	1.01210	12.88	102.0
160	10.0	27.58	867.8	15.21	3.157	5.219	808.5	1.01071	13.89	110.1
180	10.0	24.75	972.1	15.82	3.151	5.208	849.2	1.00961	14.87	118.0
200	10.0	22.46	1076.	16.37	3.146	5.201	888.2	1.00872	15.82	125.7
220	10.0	20.55	1180.	16.86	3.142	5.196	925.6	1.00798	16.75	133.3
240	10.0	18.94	1284.	17.31	3.139	5.193	961.6	1.00735	17.65	140.7
260	10.0	17.57	1388.	17.73	3.137	5.190	996.4	1.00682	18.54	148.0
280	10.0	16.38	1492.	18.11	3.135	5.188	1030.	1.00635	19.41	155.2
300	10.0	15.34	1595.	18.47	3.133	5.187	1063.	1.00595	20.26	162.3

T	P	ρ	H	S	C_v	C_p	u	D	η	λ
K	MPa	kg m^{-3}	kJ kg^{-1}	kJ kg^{-1} K^{-1}	kJ kg^{-1} K^{-1}	kJ kg^{-1} K^{-1}	m s^{-1}		μPa s	mW m^{-1} K^{-1}
320	10.0	14.43	1699.	18.81	3.132	5.186	1094.	1.00560	21.14	169.2
340	10.0	13.62	1803.	19.12	3.130	5.186	1125.	1.00528	22.01	176.0
360	10.0	12.89	1907.	19.42	3.129	5.185	1155.	1.00500	22.86	182.8
400	10.0	11.65	2114.	19.96	3.127	5.185	1213.	1.00452	24.54	196.0
500	10.0	9.387	2632.	21.12	3.124	5.185	1347.	1.00364	28.56	227.6
600	10.0	7.860	3151.	22.07	3.122	5.186	1469.	1.00305	32.38	257.4
700	10.0	6.760	3670.	22.87	3.121	5.186	1582.	1.00262	36.04	285.9
800	10.0	5.930	4188.	23.56	3.120	5.187	1687.	1.00230	39.56	313.2
900	10.0	5.281	4707.	24.17	3.119	5.188	1786.	1.00205	42.96	339.6
1000	10.0	4.760	5226.	24.72	3.119	5.188	1880.	1.00185	46.26	365.1

Argon (Ar)

T	P	ρ	H	S	C_v	C_p	u	D	η	λ
K	MPa	kg m^{-3}	kJ kg^{-1}	kJ kg^{-1} K^{-1}	kJ kg^{-1} K^{-1}	kJ kg^{-1} K^{-1}	m s^{-1}		μPa s	mW m^{-1} K^{-1}
Saturation										
83.81	0.06889	1417.	−121.4	1.329	0.5496	1.116	862.4	1.51232	290.2	133.6
83.81	0.06889	4.055	42.28	3.283	0.3247	0.5550	168.1	1.00126	6.856	5.359
90	0.1335	1379.	−114.5	1.409	0.5268	1.121	819.5	1.49650	240.0	124.5
90	0.1335	7.436	44.57	3.176	0.3309	0.5757	172.8	1.00231	7.413	5.835
100	0.3238	1314.	−103.1	1.528	0.4976	1.154	746.9	1.46993	181.3	110.2
100	0.3238	16.86	47.40	3.032	0.3445	0.6269	178.9	1.00525	8.349	6.689
110	0.6653	1243.	−91.13	1.639	0.4747	1.218	669.2	1.44136	140.4	96.41
110	0.6653	33.29	48.84	2.911	0.3633	0.7122	183.0	1.01039	9.366	7.732
120	1.213	1163.	−78.35	1.746	0.4576	1.332	584.2	1.40965	110.2	83.13
120	1.213	60.14	48.41	2.802	0.3893	0.8627	185.1	1.01884	10.54	9.154
130	2.025	1068.	−64.16	1.854	0.4492	1.564	487.9	1.37273	85.86	70.43
130	2.025	103.6	45.30	2.696	0.4275	1.172	184.8	1.03259	12.03	11.45
140	3.168	943.7	−47.16	1.971	0.4598	2.225	371.6	1.32519	63.62	58.06
140	3.168	178.9	37.47	2.576	0.4940	2.104	181.5	1.05677	14.32	16.39
150	4.735	680.4	−17.88	2.159	0.7060	23.58	174.7	1.22816	36.78	57.63
150	4.735	394.5	11.52	2.355	0.8218	35.47	157.0	1.12827	21.18	55.88
150.69	4.863	535.6	−4.332	2.248				1.17684	27.63	
P = 0.1 MPa (1 bar)										
83.81	0.1	1417.	−121.4	1.329	0.5496	1.116	862.5	1.51233	290.2	133.6
87.18	0.1	1396.	−117.7	1.373	0.5366	1.117	839.2	1.50375	261.3	128.6
87.18	0.1	5.704	43.57	3.223	0.3279	0.5654	170.8	1.00178	7.157	5.614
100	0.1	4.915	50.69	3.299	0.3206	0.5470	184.2	1.00153	8.234	6.450
120	0.1	4.058	61.49	3.398	0.3161	0.5347	202.8	1.00126	9.878	7.735
140	0.1	3.461	72.12	3.480	0.3144	0.5293	219.6	1.00108	11.48	8.987
160	0.1	3.020	82.68	3.550	0.3135	0.5264	235.1	1.00094	13.03	10.21
180	0.1	2.680	93.19	3.612	0.3131	0.5247	249.6	1.00083	14.53	11.39
200	0.1	2.409	103.7	3.667	0.3128	0.5236	263.2	1.00075	16.00	12.54
220	0.1	2.189	114.1	3.717	0.3127	0.5229	276.2	1.00068	17.42	13.66
240	0.1	2.005	124.6	3.762	0.3125	0.5224	288.5	1.00062	18.80	14.74
260	0.1	1.850	135.0	3.804	0.3125	0.5220	300.3	1.00058	20.15	15.80
280	0.1	1.717	145.5	3.843	0.3124	0.5217	311.7	1.00053	21.46	16.83
300	0.1	1.603	155.9	3.879	0.3124	0.5215	322.7	1.00050	22.74	17.84
320	0.1	1.502	166.3	3.913	0.3124	0.5214	333.3	1.00047	23.99	18.82
340	0.1	1.414	176.8	3.944	0.3123	0.5212	343.5	1.00044	25.21	19.77
360	0.1	1.335	187.2	3.974	0.3123	0.5211	353.5	1.00042	26.40	20.71
400	0.1	1.201	208.0	4.029	0.3123	0.5209	372.7	1.00037	28.70	22.52
500	0.1	0.9608	260.1	4.145	0.3123	0.5207	416.6	1.00030	34.08	26.73
600	0.1	0.8006	312.2	4.240	0.3122	0.5206	456.4	1.00025	39.00	30.57
700	0.1	0.6862	364.2	4.320	0.3122	0.5205	493.0	1.00021	43.56	34.13
800	0.1	0.6004	416.3	4.390	0.3122	0.5205	527.0	1.00019	47.82	37.46
900	0.1	0.5337	468.3	4.451	0.3122	0.5204	558.9	1.00017	51.85	40.60
1000	0.1	0.4803	520.3	4.506	0.3122	0.5204	589.1	1.00015	55.69	43.58

T K	P MPa	ρ kg m^{-3}	H kJ kg^{-1}	S kJ kg^{-1} K^{-1}	C_v kJ kg^{-1} K^{-1}	C_p kJ kg^{-1} K^{-1}	u m s^{-1}	D	η μPa s	λ mW m^{-1} K^{-1}
P = 1 MPa (10 bar)										
84.04	1.0	1418.	−120.8	1.330	0.5498	1.113	865.3	1.51276	291.1	134.0
100	1.0	1316.	−102.8	1.525	0.4984	1.148	751.6	1.47104	183.1	110.9
116.60	1.0	1191.	−82.82	1.710	0.4627	1.285	614.1	1.42088	119.7	87.58
116.60	1.0	49.55	48.81	2.839	0.3795	0.8007	184.6	1.01550	10.12	8.608
120	1.0	47.20	51.45	2.861	0.3682	0.7559	189.3	1.01476	10.37	8.732
140	1.0	37.75	65.11	2.967	0.3375	0.6350	212.1	1.01179	11.87	9.704
160	1.0	31.93	77.31	3.048	0.3266	0.5907	230.5	1.00997	13.36	10.80
180	1.0	27.83	88.89	3.116	0.3214	0.5686	246.8	1.00868	14.83	11.90
200	1.0	24.74	100.1	3.176	0.3185	0.5556	261.6	1.00772	16.26	13.00
220	1.0	22.31	111.1	3.228	0.3168	0.5473	275.3	1.00696	17.65	14.07
240	1.0	20.33	122.0	3.275	0.3157	0.5417	288.3	1.00634	19.01	15.12
260	1.0	18.69	132.8	3.319	0.3149	0.5376	300.5	1.00583	20.34	16.15
280	1.0	17.30	143.5	3.358	0.3144	0.5346	312.2	1.00539	21.64	17.16
300	1.0	16.11	154.2	3.395	0.3140	0.5324	323.4	1.00502	22.90	18.14
320	1.0	15.08	164.8	3.429	0.3137	0.5306	334.2	1.00470	24.14	19.10
340	1.0	14.17	175.4	3.462	0.3135	0.5292	344.7	1.00441	25.35	20.04
360	1.0	13.37	186.0	3.492	0.3133	0.5280	354.7	1.00416	26.53	20.96
400	1.0	12.01	207.1	3.547	0.3131	0.5263	374.1	1.00374	28.82	22.75
500	1.0	9.593	259.6	3.664	0.3127	0.5239	418.3	1.00299	34.16	26.92
600	1.0	7.988	311.9	3.760	0.3126	0.5227	458.1	1.00249	39.06	30.73
700	1.0	6.846	364.1	3.840	0.3125	0.5220	494.6	1.00213	43.61	34.27
800	1.0	5.990	416.3	3.910	0.3124	0.5215	528.6	1.00186	47.87	37.58
900	1.0	5.325	468.4	3.971	0.3124	0.5212	560.5	1.00166	51.90	40.71
1000	1.0	4.793	520.6	4.026	0.3124	0.5210	590.7	1.00149	55.72	43.68
P = 10 MPa (100 bar)										
86.27	10.0	1428.	−114.2	1.333	0.5519	1.085	891.3	1.51685	300.0	137.1
100	10.0	1349.	−99.27	1.493	0.5089	1.093	805.9	1.48430	205.8	118.6
120	10.0	1222.	−76.82	1.698	0.4650	1.163	674.3	1.43296	130.0	93.55
140	10.0	1066.	−52.01	1.888	0.4374	1.349	527.4	1.37161	85.36	70.81
160	10.0	833.6	−20.15	2.100	0.4356	1.970	357.8	1.28395	50.84	49.64
180	10.0	491.9	26.15	2.373	0.4254	2.091	259.3	1.16159	27.88	30.35
200	10.0	337.7	57.55	2.539	0.3815	1.215	267.7	1.10910	23.43	23.57
220	10.0	270.9	78.52	2.639	0.3585	0.9257	283.7	1.08686	22.90	21.77
240	10.0	231.2	95.59	2.713	0.3457	0.7960	299.0	1.07382	23.22	21.31
260	10.0	204.0	110.7	2.774	0.3379	0.7240	313.1	1.06491	23.87	21.39
280	10.0	183.6	124.7	2.826	0.3327	0.6789	326.1	1.05829	24.68	21.73
300	10.0	167.6	138.0	2.872	0.3290	0.6481	338.4	1.05312	25.58	22.19
320	10.0	154.6	150.7	2.913	0.3264	0.6261	350.0	1.04892	26.53	22.78
340	10.0	143.7	163.1	2.950	0.3243	0.6095	361.0	1.04542	27.51	23.43
360	10.0	134.4	175.1	2.985	0.3228	0.5967	371.6	1.04245	28.49	24.11
400	10.0	119.4	198.6	3.046	0.3205	0.5784	391.5	1.03765	30.48	25.50
500	10.0	94.09	255.0	3.173	0.3174	0.5540	436.1	1.02957	35.35	29.04
600	10.0	78.03	309.8	3.272	0.3159	0.5423	475.7	1.02448	39.97	32.48
700	10.0	66.80	363.7	3.355	0.3151	0.5357	511.7	1.02093	44.34	35.76
800	10.0	58.47	417.0	3.427	0.3145	0.5317	545.1	1.01830	48.47	38.88
900	10.0	52.03	470.0	3.489	0.3142	0.5290	576.4	1.01628	52.40	41.86
1000	10.0	46.88	522.8	3.545	0.3139	0.5271	606.0	1.01466	56.16	44.71

Methane (CH$_4$)

T K	P MPa	ρ kg m^{-3}	H kJ kg^{-1}	S kJ kg^{-1} K^{-1}	C_v kJ kg^{-1} K^{-1}	C_p kJ kg^{-1} K^{-1}	u m s^{-1}	D	η μPa s	λ mW m^{-1} K^{-1}
Saturation										
90.69	0.01170	451.5	−71.82	−.7099	2.168	3.368	1539.	1.67721	193.4	211.2
90.69	0.01170	0.2507	472.4	5.291	1.574	2.110	249.1	1.00031	3.585	8.776
100	0.03438	438.9	−40.27	−.3793	2.114	3.408	1452.	1.65478	151.1	199.6
100	0.03438	0.6746	490.2	4.925	1.589	2.146	260.1	1.00082	3.914	9.910

T	P	ρ	H	S	C_v	C_p	u	D	η	λ
K	MPa	kg m⁻³	kJ kg⁻¹	kJ kg⁻¹ K⁻¹	kJ kg⁻¹ K⁻¹	kJ kg⁻¹ K⁻¹	m s⁻¹		μPa s	mW m⁻¹ K⁻¹
110	0.08813	424.8	−5.813	−.05217	2.064	3.469	1355.	1.62995	121.0	186.1
110	0.08813	1.598	508.0	4.619	1.611	2.205	270.0	1.00196	4.273	11.22
120	0.1914	409.9	29.41	0.2521	2.020	3.549	1253.	1.60408	98.68	172.0
120	0.1914	3.262	524.0	4.374	1.639	2.293	277.8	1.00399	4.636	12.66
130	0.3673	394.0	65.63	0.5385	1.980	3.658	1148.	1.57683	81.29	157.7
130	0.3673	5.980	537.7	4.170	1.674	2.421	283.1	1.00733	5.008	14.28
140	0.6412	376.9	103.2	0.8116	1.945	3.813	1038.	1.54771	67.35	143.5
140	0.6412	10.15	548.3	3.991	1.717	2.611	285.9	1.01248	5.398	16.15
150	1.040	357.9	142.6	1.076	1.919	4.047	920.8	1.51599	55.99	129.2
150	1.040	16.33	555.2	3.827	1.773	2.908	286.0	1.02013	5.827	18.38
160	1.592	336.3	184.8	1.338	1.904	4.435	795.4	1.48045	46.55	115.0
160	1.592	25.38	557.1	3.664	1.847	3.419	283.0	1.03144	6.335	21.23
170	2.328	310.5	231.2	1.605	1.910	5.187	657.5	1.43870	38.45	100.5
170	2.328	38.97	551.5	3.490	1.956	4.459	276.7	1.04861	7.012	25.34
180	3.285	276.2	285.9	1.899	1.967	7.292	497.0	1.38453	30.92	85.50
180	3.285	61.38	532.8	3.271	2.140	7.574	266.0	1.07739	8.127	33.50
190	4.519	200.8	378.3	2.369	2.602	94.01	250.3	1.27031	20.34	94.07
190	4.519	125.2	459.0	2.794	2.855	140.8	238.5	1.16274	12.24	120.5
190.56	4.599	162.7	415.6	2.562				1.21521	15.91	

P = 0.1 MPa (1 bar)

T	P	ρ	H	S	C_v	C_p	u	D	η	λ
90.72	0.1	451.5	−71.60	−.7097	2.168	3.367	1539.	1.67726	193.6	211.2
100	0.1	438.9	−40.17	−.3798	2.114	3.408	1453.	1.65487	151.2	199.6
111.51	0.1	422.6	−.5573	−.004967	2.057	3.480	1340.	1.62612	117.2	184.0
111.51	0.1	1.795	510.6	4.579	1.615	2.216	271.3	1.00220	4.327	11.43
120	0.1	1.655	529.2	4.740	1.594	2.174	282.8	1.00202	4.668	12.35
140	0.1	1.403	572.1	5.071	1.574	2.129	307.6	1.00172	5.449	14.65
160	0.1	1.221	614.5	5.354	1.568	2.110	330.0	1.00149	6.213	16.98
180	0.1	1.081	656.6	5.602	1.568	2.104	350.7	1.00132	6.963	19.32
200	0.1	0.9709	698.7	5.824	1.574	2.106	370.0	1.00119	7.697	21.65
220	0.1	0.8812	740.9	6.025	1.587	2.115	388.0	1.00108	8.416	23.99
240	0.1	0.8069	783.4	6.209	1.607	2.133	404.9	1.00099	9.117	26.39
260	0.1	0.7442	826.3	6.381	1.634	2.159	420.8	1.00091	9.803	28.88
280	0.1	0.6906	869.8	6.542	1.670	2.194	435.7	1.00085	10.47	31.47
300	0.1	0.6443	914.1	6.695	1.713	2.236	449.7	1.00079	11.13	34.19
320	0.1	0.6038	959.3	6.841	1.763	2.285	463.1	1.00074	11.77	37.04
340	0.1	0.5681	1006.	6.981	1.819	2.340	475.7	1.00070	12.39	40.03
360	0.1	0.5364	1053.	7.117	1.880	2.401	487.8	1.00066	13.00	43.15
400	0.1	0.4826	1152.	7.376	2.013	2.534	510.6	1.00059	14.18	49.80
500	0.1	0.3859	1423.	7.981	2.381	2.901	561.9	1.00047	16.92	68.34
600	0.1	0.3215	1732.	8.543	2.754	3.273	608.0	1.00039	19.41	88.80
700	0.1	0.2756	2077.	9.074	3.110	3.629	650.8	1.00034	21.68	110.4
800	0.1	0.2411	2457.	9.581	3.441	3.959	690.9	1.00030	23.76	132.5
900	0.1	0.2143	2868.	10.06	3.744	4.263	729.0	1.00026	25.70	154.7
1000	0.1	0.1929	3308.	10.53	4.020	4.538	765.2	1.00024	27.49	176.7

P = 1 MPa (10 bar)

T	P	ρ	H	S	C_v	C_p	u	D	η	λ
90.95	1.0	451.8	−69.37	−.7070	2.169	3.363	1543.	1.67772	195.7	211.7
100	1.0	439.6	−38.76	−.3862	2.116	3.401	1460.	1.65604	153.4	200.5
120	1.0	410.8	30.49	0.2447	2.022	3.536	1262.	1.60560	99.86	173.0
140	1.0	377.5	103.5	0.8069	1.946	3.798	1044.	1.54879	67.81	144.0
149.14	1.0	359.6	139.2	1.054	1.920	4.023	931.2	1.51885	56.88	130.4
149.14	1.0	15.70	554.8	3.841	1.767	2.876	286.1	1.01935	5.788	18.17
160	1.0	13.97	584.2	4.031	1.682	2.588	305.2	1.01720	6.287	18.79
180	1.0	11.81	633.6	4.322	1.631	2.380	333.7	1.01453	7.089	20.65
200	1.0	10.33	680.2	4.568	1.613	2.289	357.8	1.01269	7.838	22.74
220	1.0	9.217	725.5	4.784	1.613	2.248	379.1	1.01133	8.558	24.90
240	1.0	8.344	770.3	4.978	1.626	2.234	398.2	1.01025	9.258	27.19
260	1.0	7.634	815.0	5.157	1.648	2.238	415.8	1.00938	9.939	29.60

T	P	ρ	H	S	C_v	C_p	u	D	η	λ
K	MPa	kg m^{-3}	kJ kg^{-1}	kJ kg^{-1} K^{-1}	kJ kg^{-1} K^{-1}	kJ kg^{-1} K^{-1}	m s^{-1}		μPa s	mW m^{-1} K^{-1}
280	1.0	7.043	859.9	5.324	1.681	2.258	432.0	1.00865	10.60	32.12
300	1.0	6.542	905.4	5.480	1.722	2.289	447.0	1.00803	11.25	34.79
320	1.0	6.110	951.5	5.629	1.770	2.330	461.1	1.00750	11.89	37.59
340	1.0	5.733	998.6	5.772	1.825	2.379	474.4	1.00704	12.51	40.54
360	1.0	5.402	1047.	5.910	1.885	2.434	487.0	1.00663	13.11	43.64
400	1.0	4.846	1147.	6.172	2.017	2.559	510.6	1.00595	14.28	50.23
500	1.0	3.860	1420.	6.781	2.383	2.915	563.0	1.00474	17.01	68.68
600	1.0	3.210	1730.	7.345	2.755	3.282	609.7	1.00394	19.48	89.08
700	1.0	2.750	2076.	7.878	3.110	3.635	652.8	1.00338	21.74	110.6
800	1.0	2.405	2456.	8.385	3.441	3.964	693.1	1.00296	23.82	132.7
900	1.0	2.137	2868.	8.870	3.744	4.266	731.2	1.00263	25.75	154.9
1000	1.0	1.924	3308.	9.334	4.020	4.541	767.5	1.00237	27.54	176.8

P = 10 MPa (100 bar)

T	P	ρ	H	S	C_v	C_p	u	D	η	λ
93.22	10.0	454.5	−47.07	−.6806	2.176	3.325	1582.	1.68223	212.8	216.9
100	10.0	446.0	−24.48	−.4466	2.139	3.344	1526.	1.66701	174.8	209.0
120	10.0	419.8	43.14	0.1696	2.047	3.424	1352.	1.62085	112.1	183.3
140	10.0	391.2	112.8	0.7062	1.969	3.552	1171.	1.57167	78.30	157.0
160	10.0	358.8	185.8	1.194	1.909	3.777	981.8	1.51718	56.70	131.5
180	10.0	319.6	265.3	1.661	1.873	4.226	780.8	1.45298	41.52	107.0
200	10.0	266.2	358.9	2.153	1.878	5.304	567.9	1.36870	29.55	83.93
220	10.0	187.6	482.7	2.742	1.903	6.713	404.4	1.25087	19.37	63.29
240	10.0	128.4	599.6	3.252	1.847	4.853	383.6	1.16707	14.89	49.74
260	10.0	101.0	683.8	3.589	1.803	3.730	402.8	1.12974	13.80	44.85
280	10.0	85.51	752.9	3.846	1.793	3.242	424.5	1.10902	13.61	43.81
300	10.0	75.18	815.1	4.060	1.807	3.002	444.5	1.09539	13.75	44.37
320	10.0	67.61	873.8	4.250	1.838	2.878	462.8	1.08549	14.04	45.79
340	10.0	61.75	930.6	4.422	1.880	2.817	479.5	1.07786	14.41	47.75
360	10.0	57.01	986.7	4.582	1.931	2.795	494.8	1.07173	14.82	50.09
400	10.0	49.74	1099.	4.877	2.051	2.819	522.6	1.06238	15.72	55.61
500	10.0	38.32	1391.	5.529	2.401	3.055	581.0	1.04783	18.06	72.56
600	10.0	31.47	1712.	6.113	2.766	3.370	630.6	1.03918	20.32	92.15
700	10.0	26.82	2066.	6.657	3.117	3.696	675.2	1.03334	22.44	113.2
800	10.0	23.41	2451.	7.171	3.445	4.008	716.3	1.02907	24.43	134.9
900	10.0	20.79	2866.	7.660	3.747	4.300	754.7	1.02581	26.28	156.8
1000	10.0	18.72	3310.	8.128	4.022	4.568	791.1	1.02323	28.02	178.5

Ethane (C$_2$H$_6$)

T	P	ρ	H	S	C_v	C_p	u	D	η	λ
K	MPa	kg m^{-3}	kJ kg^{-1}	kJ kg^{-1} K^{-1}	kJ kg^{-1} K^{-1}	kJ kg^{-1} K^{-1}	m s^{-1}		μPa s	mW m^{-1} K^{-1}
Saturation										
90.37	0.0000011	651.5	−219.2	−1.655	1.605	2.326	2009.	1.94483	1281.	255.6
90.37	0.0000011	0.0000457	375.6	4.927	0.8916	1.168	180.9	1.00000	3.043	2.908
100	0.0000111	640.9	−197.0	−1.422	1.541	2.283	1938.	1.92579	873.2	247.8
100	0.0000111	0.0004007	386.9	4.418	0.9107	1.187	189.9	1.00000	3.316	3.456
120	0.0003523	618.9	−151.5	−1.007	1.478	2.280	1794.	1.88657	485.8	229.9
120	0.0003523	0.01062	411.0	3.681	0.9528	1.230	206.9	1.00001	3.896	4.657
140	0.003814	596.6	−105.6	−.6532	1.450	2.311	1649.	1.84725	319.8	210.6
140	0.003814	0.09880	435.6	3.213	1.003	1.284	222.0	1.00011	4.496	5.966
160	0.02141	573.6	−58.95	−.3418	1.436	2.357	1501.	1.80737	231.2	190.8
160	0.02141	0.4890	460.3	2.904	1.048	1.338	235.1	1.00054	5.110	7.422
180	0.07864	549.5	−11.13	−.06082	1.434	2.421	1350.	1.76641	176.2	171.3
180	0.07864	1.625	484.2	2.691	1.098	1.409	245.5	1.00181	5.739	9.079
200	0.2172	524.0	38.30	0.1981	1.444	2.512	1196.	1.72364	138.3	152.6
200	0.2172	4.170	506.2	2.538	1.179	1.537	252.3	1.00465	6.391	11.01
220	0.4920	496.3	90.01	0.4419	1.468	2.645	1037.	1.67809	110.1	134.6
220	0.4920	9.017	525.5	2.421	1.280	1.720	254.6	1.01008	7.089	13.32
240	0.9668	465.3	145.0	0.6767	1.507	2.847	873.3	1.62822	87.80	117.5

T K	P MPa	ρ kg m^{-3}	H kJ kg^{-1}	S kJ kg^{-1} K^{-1}	C_v kJ kg^{-1} K^{-1}	C_p kJ kg^{-1} K^{-1}	u m s^{-1}	D	η μPa s	λ mW m^{-1} K^{-1}
240	0.9668	17.43	540.9	2.326	1.388	1.976	252.1	1.01957	7.881	16.18
260	1.712	429.1	204.9	0.9095	1.566	3.195	700.5	1.57121	69.30	101.2
260	1.712	31.58	550.2	2.238	1.516	2.418	243.8	1.03567	8.876	20.02
280	2.807	382.7	273.1	1.152	1.654	3.987	512.4	1.50037	52.84	85.43
280	2.807	56.37	548.6	2.136	1.696	3.522	228.1	1.06440	10.37	26.33
300	4.357	303.5	364.4	1.451	1.912	10.02	274.9	1.38458	34.97	71.49
300	4.357	114.5	514.1	1.950	2.089	13.30	200.5	1.13418	14.02	47.46
305.32	4.872	206.2	439.0	1.690				1.25115	21.80	

P = 0.1 MPa (1 bar)

T K	P MPa	ρ kg m^{-3}	H kJ kg^{-1}	S kJ kg^{-1} K^{-1}	C_v kJ kg^{-1} K^{-1}	C_p kJ kg^{-1} K^{-1}	u m s^{-1}	D	η μPa s	λ mW m^{-1} K^{-1}
90.38	0.1	651.5	−219.0	−1.655	1.605	2.326	2009.	1.94485	1281.	255.6
100	0.1	641.0	−196.9	−1.422	1.541	2.283	1939.	1.92586	873.9	247.9
120	0.1	619.0	−151.4	−1.007	1.478	2.279	1795.	1.88665	486.1	230.0
140	0.1	596.6	−105.5	−.6535	1.450	2.311	1650.	1.84735	320.0	210.6
160	0.1	573.6	−58.85	−.3420	1.436	2.357	1502.	1.80747	231.3	190.9
180	0.1	549.5	−11.10	−.06091	1.434	2.421	1351.	1.76644	176.2	171.4
184.33	0.1	544.1	−.5954	−.003215	1.435	2.438	1317.	1.75734	166.9	167.2
184.33	0.1	2.030	489.1	2.654	1.113	1.432	247.3	1.00226	5.877	9.471
200	0.1	1.856	511.8	2.771	1.149	1.457	257.9	1.00207	6.368	10.72
220	0.1	1.676	541.3	2.912	1.200	1.499	270.3	1.00187	6.991	12.47
240	0.1	1.529	571.8	3.045	1.260	1.554	281.8	1.00170	7.608	14.39
260	0.1	1.407	603.5	3.172	1.328	1.617	292.5	1.00157	8.217	16.48
280	0.1	1.303	636.5	3.294	1.401	1.688	302.6	1.00145	8.817	18.76
300	0.1	1.214	671.1	3.413	1.479	1.764	312.2	1.00135	9.408	21.22
320	0.1	1.137	707.1	3.530	1.560	1.844	321.4	1.00127	9.989	23.85
340	0.1	1.069	744.8	3.644	1.644	1.927	330.3	1.00119	10.56	26.65
360	0.1	1.009	784.2	3.756	1.729	2.011	338.8	1.00112	11.12	29.61
400	0.1	0.9067	868.1	3.977	1.901	2.181	355.2	1.00101	12.21	35.96
500	0.1	0.7242	1107.	4.509	2.315	2.594	393.1	1.00081	14.78	53.77
600	0.1	0.6031	1386.	5.015	2.689	2.967	427.6	1.00067	17.13	73.34
700	0.1	0.5167	1699.	5.498	3.021	3.298	459.6	1.00058	19.32	93.86
800	0.1	0.4520	2044.	5.958	3.314	3.591	489.6	1.00050	21.36	114.8
900	0.1	0.4018	2416.	6.396	3.572	3.849	517.9	1.00045	23.28	135.9
1000	0.1	0.3616	2813.	6.814	3.799	4.076	544.8	1.00040	25.10	156.8

P = 1 MPa (10 bar)

T K	P MPa	ρ kg m^{-3}	H kJ kg^{-1}	S kJ kg^{-1} K^{-1}	C_v kJ kg^{-1} K^{-1}	C_p kJ kg^{-1} K^{-1}	u m s^{-1}	D	η μPa s	λ mW m^{-1} K^{-1}
90.53	1.0	651.7	−217.5	−1.653	1.605	2.324	2012.	1.94510	1283.	255.9
100	1.0	641.3	−195.7	−1.424	1.542	2.282	1942.	1.92644	880.2	248.3
120	1.0	619.4	−150.2	−1.010	1.480	2.278	1799.	1.88737	489.1	230.5
140	1.0	597.2	−104.4	−.6563	1.452	2.309	1654.	1.84822	322.0	211.2
160	1.0	574.3	−57.80	−.3452	1.437	2.354	1508.	1.80855	232.8	191.6
180	1.0	550.4	−10.13	−.06458	1.435	2.416	1358.	1.76780	177.5	172.1
200	1.0	524.9	39.03	0.1943	1.445	2.505	1204.	1.72518	139.3	153.3
220	1.0	497.1	90.36	0.4388	1.468	2.637	1044.	1.67944	110.7	135.2
240	1.0	465.4	145.0	0.6764	1.507	2.846	873.9	1.62834	87.85	117.6
241.10	1.0	463.5	148.1	0.6894	1.510	2.861	864.0	1.62531	86.70	116.6
241.10	1.0	18.04	541.6	2.321	1.395	1.994	251.8	1.02025	7.929	16.36
260	1.0	15.93	577.9	2.466	1.400	1.880	268.9	1.01786	8.485	17.92
280	1.0	14.32	615.3	2.605	1.449	1.869	283.8	1.01604	9.072	19.91
300	1.0	13.07	652.9	2.735	1.513	1.898	296.9	1.01463	9.652	22.18
320	1.0	12.05	691.3	2.859	1.585	1.948	308.7	1.01349	10.22	24.67
340	1.0	11.21	730.9	2.979	1.663	2.010	319.6	1.01254	10.79	27.36
360	1.0	10.49	771.8	3.095	1.745	2.079	329.9	1.01173	11.34	30.24
400	1.0	9.312	857.9	3.322	1.911	2.230	348.8	1.01041	12.42	36.46
500	1.0	7.323	1101.	3.862	2.321	2.620	390.2	1.00818	14.95	54.08
600	1.0	6.058	1381.	4.372	2.693	2.983	426.6	1.00677	17.29	73.56
700	1.0	5.173	1696.	4.857	3.023	3.309	459.7	1.00578	19.46	94.03
800	1.0	4.518	2042.	5.318	3.315	3.598	490.3	1.00505	21.48	115.0

T K	P MPa	ρ kg m^{-3}	H kJ kg^{-1}	S kJ kg^{-1} K^{-1}	C_v kJ kg^{-1} K^{-1}	C_p kJ kg^{-1} K^{-1}	u m s^{-1}	D	η μPa s	λ mW m^{-1} K^{-1}
900	1.0	4.011	2415.	5.757	3.573	3.855	519.0	1.00448	23.39	136.0
1000	1.0	3.608	2812.	6.175	3.800	4.081	546.2	1.00403	25.20	156.9

P = 10 MPa (100 bar)

T K	P MPa	ρ kg m^{-3}	H kJ kg^{-1}	S kJ kg^{-1} K^{-1}	C_v kJ kg^{-1} K^{-1}	C_p kJ kg^{-1} K^{-1}	u m s^{-1}	D	η μPa s	λ mW m^{-1} K^{-1}
91.96	10.0	653.3	−202.5	−1.640	1.605	2.310	2034.	1.94751	1310.	258.6
100	10.0	644.8	−184.1	−1.448	1.553	2.275	1976.	1.93215	947.3	252.5
120	10.0	623.6	−138.8	−1.035	1.490	2.266	1839.	1.89429	520.6	235.4
140	10.0	602.2	−93.25	−.6839	1.463	2.290	1701.	1.85661	342.0	216.9
160	10.0	580.4	−47.10	−.3758	1.449	2.326	1564.	1.81879	248.1	197.9
180	10.0	558.1	−.1206	−.09918	1.447	2.374	1425.	1.78049	190.5	179.3
200	10.0	534.8	47.98	0.1542	1.456	2.439	1287.	1.74125	151.3	161.4
220	10.0	510.2	97.60	0.3905	1.478	2.527	1147.	1.70050	122.4	144.4
240	10.0	483.8	149.3	0.6152	1.511	2.646	1008.	1.65744	99.96	128.5
260	10.0	454.6	203.7	0.8331	1.557	2.809	866.9	1.61088	81.74	113.6
280	10.0	421.4	262.1	1.049	1.616	3.044	724.4	1.55888	66.38	99.57
300	10.0	381.3	326.4	1.271	1.689	3.419	579.3	1.49786	52.87	86.39
320	10.0	328.8	400.9	1.511	1.783	4.109	434.8	1.42048	40.36	73.81
340	10.0	255.9	493.3	1.791	1.888	5.070	319.9	1.31781	28.77	62.21
360	10.0	187.0	591.3	2.071	1.941	4.452	286.8	1.22573	21.62	53.36
400	10.0	126.2	738.6	2.460	2.025	3.205	310.4	1.14850	17.94	48.70
500	10.0	80.57	1035.	3.123	2.369	2.933	377.8	1.09293	17.95	59.52
600	10.0	62.41	1338.	3.675	2.722	3.150	427.4	1.07144	19.50	76.88
700	10.0	51.79	1666.	4.180	3.043	3.415	467.8	1.05903	21.26	96.40
800	10.0	44.59	2021.	4.653	3.330	3.672	502.9	1.05068	23.01	116.8
900	10.0	39.30	2400.	5.100	3.584	3.909	534.3	1.04457	24.73	137.5
1000	10.0	35.20	2802.	5.523	3.809	4.122	563.2	1.03988	26.40	158.2

Propane (C$_3$H$_8$)

T K	P MPa	ρ kg m^{-3}	H kJ kg^{-1}	S kJ kg^{-1} K^{-1}	C_v kJ kg^{-1} K^{-1}	C_p kJ kg^{-1} K^{-1}	u m s^{-1}	D	η μPa s	λ mW m^{-1} K^{-1}
Saturation										
85.53	0.17×10^{-9}	733.1	−196.6	−1.396	1.355	1.916	2136.	2.08838	10780	207.9
85.53	0.17×10^{-9}	0.11×10^{-7}	366.3	5.186	0.6907	0.8792	143.3	1.00000	2.641	1.706
100	0.25×10^{-7}	718.1	−168.8	−1.095	1.341	1.930	2038.	2.05908	3774.	203.2
100	0.25×10^{-7}	0.0000013	379.4	4.387	0.7475	0.9361	153.7	1.00000	2.979	2.417
120	0.0000030	697.8	−129.9	−.7411	1.335	1.957	1901.	2.01930	1500.	194.4
120	0.0000030	0.0001310	398.9	3.666	0.8210	1.010	166.8	1.00000	3.467	3.500
140	0.0000790	677.6	−90.51	−.4372	1.335	1.988	1767.	1.98025	822.2	183.9
140	0.0000790	0.002993	419.7	3.207	0.8878	1.076	178.9	1.00000	3.970	4.699
160	0.0008502	657.3	−50.38	−.1693	1.342	2.025	1633.	1.94163	535.0	172.3
160	0.0008502	0.02821	441.8	2.907	0.9514	1.141	190.0	1.00003	4.483	6.013
180	0.005068	636.6	−9.433	0.07174	1.356	2.070	1499.	1.90310	382.2	160.2
180	0.005068	0.1499	464.9	2.707	1.017	1.209	200.1	1.00016	5.000	7.439
200	0.02019	615.4	32.53	0.2926	1.383	2.127	1366.	1.86434	288.0	147.8
200	0.02019	0.5417	488.6	2.573	1.088	1.287	208.7	1.00059	5.515	8.971
220	0.06057	593.4	75.80	0.4984	1.421	2.199	1233.	1.82499	224.2	135.7
220	0.06057	1.499	512.7	2.484	1.169	1.381	215.6	1.00164	6.026	10.61
240	0.1480	570.4	120.7	0.6932	1.469	2.289	1102.	1.78460	178.4	124.0
240	0.1480	3.438	536.6	2.426	1.259	1.494	220.1	1.00376	6.540	12.38
260	0.3107	545.8	167.7	0.8801	1.528	2.403	971.7	1.74256	143.9	112.9
260	0.3107	6.903	560.1	2.389	1.357	1.630	221.9	1.00755	7.073	14.32
280	0.5817	519.2	217.3	1.062	1.596	2.547	840.3	1.69800	117.1	102.5
280	0.5817	12.62	582.3	2.365	1.466	1.803	220.4	1.01383	7.655	16.54
300	0.9977	489.4	270.2	1.241	1.675	2.740	706.8	1.64961	95.28	92.86
300	0.9977	21.63	602.6	2.349	1.588	2.041	214.8	1.02381	8.340	19.24
320	1.599	454.9	327.3	1.421	1.764	3.028	569.1	1.59504	76.78	83.94
320	1.599	35.74	619.5	2.334	1.729	2.416	204.4	1.03957	9.230	22.78
340	2.431	411.8	390.9	1.608	1.872	3.585	422.5	1.52907	59.98	75.57

T K	P MPa	ρ kg m⁻³	H kJ kg⁻¹	S kJ kg⁻¹ K⁻¹	C_v kJ kg⁻¹ K⁻¹	C_p kJ kg⁻¹ K⁻¹	u m s⁻¹	D	η μPa s	λ mW m⁻¹ K⁻¹
340	2.431	58.88	629.8	2.311	1.898	3.197	187.4	1.06579	10.57	28.20
360	3.555	345.6	468.2	1.820	2.049	5.984	251.0	1.43242	42.27	67.72
360	3.555	105.4	622.4	2.249	2.183	7.111	161.1	1.11995	13.38	41.36
369.89	4.251	220.5	555.2	2.052				1.26277	22.89	

P = 0.1 MPa (1 bar)

T K	P MPa	ρ kg m⁻³	H kJ kg⁻¹	S kJ kg⁻¹ K⁻¹	C_v kJ kg⁻¹ K⁻¹	C_p kJ kg⁻¹ K⁻¹	u m s⁻¹	D	η μPa s	λ mW m⁻¹ K⁻¹
85.53	0.1	733.1	−196.5	−1.396	1.355	1.916	2137.	2.08840	10790	207.9
100	0.1	718.2	−168.7	−1.096	1.341	1.930	2038.	2.05912	3778.	203.2
120	0.1	697.8	−129.8	−.7413	1.335	1.957	1902.	2.01936	1502.	194.5
140	0.1	677.6	−90.39	−.4374	1.335	1.988	1767.	1.98032	822.9	184.0
160	0.1	657.3	−50.27	−.1695	1.342	2.025	1633.	1.94172	535.5	172.4
180	0.1	636.7	−9.328	0.07149	1.357	2.070	1500.	1.90320	382.5	160.2
200	0.1	615.5	32.62	0.2924	1.383	2.127	1366.	1.86444	288.2	147.9
220	0.1	593.5	75.84	0.4983	1.421	2.198	1234.	1.82505	224.3	135.7
230.74	0.1	581.2	99.69	0.6042	1.446	2.245	1163.	1.80347	197.9	129.4
230.74	0.1	2.387	525.6	2.450	1.216	1.439	218.3	1.00261	6.301	11.54
240	0.1	2.284	539.0	2.507	1.249	1.467	222.9	1.00249	6.559	12.40
260	0.1	2.092	569.0	2.627	1.324	1.535	232.3	1.00228	7.110	14.32
280	0.1	1.932	600.5	2.744	1.405	1.610	241.1	1.00211	7.656	16.36
300	0.1	1.796	633.5	2.857	1.490	1.692	249.4	1.00196	8.196	18.51
320	0.1	1.679	668.2	2.969	1.578	1.777	257.3	1.00183	8.732	20.78
340	0.1	1.576	704.6	3.080	1.668	1.865	264.9	1.00172	9.262	23.16
360	0.1	1.486	742.8	3.189	1.758	1.954	272.3	1.00162	9.787	25.66
400	0.1	1.334	824.5	3.404	1.937	2.130	286.2	1.00145	10.82	30.99
500	0.1	1.064	1059.	3.925	2.356	2.547	318.3	1.00116	13.29	46.36
600	0.1	0.8852	1332.	4.422	2.722	2.912	347.4	1.00096	15.59	64.63
700	0.1	0.7581	1640.	4.896	3.038	3.228	374.2	1.00082	17.70	85.81
800	0.1	0.6631	1976.	5.345	3.313	3.502	399.3	1.00072	19.62	109.9
900	0.1	0.5892	2339.	5.772	3.553	3.743	422.8	1.00064	21.36	136.9
1000	0.1	0.5302	2724.	6.177	3.764	3.953	445.1	1.00058	22.93	166.8

P = 1 MPa (10 bar)

T K	P MPa	ρ kg m⁻³	H kJ kg⁻¹	S kJ kg⁻¹ K⁻¹	C_v kJ kg⁻¹ K⁻¹	C_p kJ kg⁻¹ K⁻¹	u m s⁻¹	D	η μPa s	λ mW m⁻¹ K⁻¹
85.62	1.0	733.3	−195.3	−1.396	1.356	1.916	2139.	2.08856	10840	208.1
100	1.0	718.5	−167.6	−1.097	1.342	1.930	2041.	2.05953	3816.	203.5
120	1.0	698.2	−128.8	−.7432	1.336	1.956	1905.	2.01987	1515.	194.8
140	1.0	678.1	−89.34	−.4394	1.336	1.987	1771.	1.98096	829.3	184.3
160	1.0	657.8	−49.24	−.1717	1.343	2.024	1638.	1.94249	539.4	172.8
180	1.0	637.3	−8.331	0.06918	1.358	2.068	1505.	1.90415	385.3	160.7
200	1.0	616.2	33.57	0.2899	1.384	2.124	1372.	1.86560	290.4	148.4
220	1.0	594.4	76.72	0.4955	1.422	2.194	1241.	1.82648	226.2	136.3
240	1.0	571.5	121.5	0.6901	1.470	2.284	1111.	1.78630	180.0	124.6
260	1.0	547.0	168.2	0.8771	1.528	2.395	980.1	1.74433	145.2	113.5
280	1.0	520.1	217.5	1.060	1.596	2.539	846.7	1.69945	117.9	102.9
300	1.0	489.5	270.2	1.241	1.675	2.740	706.9	1.64962	95.28	92.86
300.09	1.0	489.3	270.4	1.242	1.675	2.741	706.2	1.64937	95.19	92.82
300.09	1.0	21.68	602.7	2.349	1.588	2.042	214.8	1.02386	8.344	19.25
320	1.0	19.34	642.9	2.479	1.646	2.009	229.9	1.02125	8.877	21.49
340	1.0	17.61	683.2	2.601	1.717	2.032	242.4	1.01931	9.414	23.91
360	1.0	16.25	724.3	2.719	1.795	2.081	253.5	1.01779	9.948	26.46
400	1.0	14.17	810.1	2.944	1.958	2.211	272.6	1.01549	11.00	31.92
500	1.0	10.93	1050.	3.478	2.363	2.583	311.6	1.01191	13.48	47.56
600	1.0	8.964	1326.	3.980	2.725	2.933	344.1	1.00976	15.76	66.04
700	1.0	7.623	1635.	4.456	3.040	3.242	372.9	1.00830	17.86	87.38
800	1.0	6.641	1973.	4.907	3.315	3.512	399.1	1.00723	19.76	111.6
900	1.0	5.889	2336.	5.334	3.555	3.750	423.4	1.00641	21.49	138.7
1000	1.0	5.292	2722.	5.741	3.765	3.959	446.3	1.00576	23.04	168.7

P = 10 MPa (100 bar)

T K	P MPa	ρ kg m⁻³	H kJ kg⁻¹	S kJ kg⁻¹ K⁻¹	C_v kJ kg⁻¹ K⁻¹	C_p kJ kg⁻¹ K⁻¹	u m s⁻¹	D	η μPa s	λ mW m⁻¹ K⁻¹
86.45	10.0	735.2	−182.9	−1.394	1.362	1.914	2159.	2.09013	11310	209.8
100	10.0	721.5	−156.9	−1.115	1.350	1.927	2070.	2.06349	4212.	205.7

T K	P MPa	ρ kg m^{-3}	H kJ kg^{-1}	S kJ kg^{-1} K^{-1}	C_v kJ kg^{-1} K^{-1}	C_p kJ kg^{-1} K^{-1}	u m s^{-1}	D	η μPa s	λ mW m^{-1} K^{-1}
120	10.0	701.8	−118.1	−.7615	1.344	1.951	1939.	2.02490	1649.	197.6
140	10.0	682.2	−78.80	−.4586	1.345	1.979	1809.	1.98713	894.7	187.7
160	10.0	662.7	−38.90	−.1922	1.352	2.012	1682.	1.94997	579.3	176.7
180	10.0	643.1	1.726	0.04696	1.367	2.052	1556.	1.91318	413.7	165.2
200	10.0	623.2	43.23	0.2656	1.393	2.101	1432.	1.87653	312.8	153.5
220	10.0	602.8	85.84	0.4686	1.430	2.162	1310.	1.83978	245.3	142.0
240	10.0	581.9	129.8	0.6598	1.478	2.236	1192.	1.80267	197.1	130.9
260	10.0	560.1	175.4	0.8422	1.535	2.325	1077.	1.76484	161.3	120.5
280	10.0	537.1	222.9	1.018	1.600	2.429	963.4	1.72590	133.7	110.7
300	10.0	512.7	272.7	1.190	1.673	2.550	852.7	1.68529	111.7	101.8
320	10.0	486.1	325.0	1.359	1.751	2.692	744.0	1.64228	93.65	93.68
340	10.0	456.7	380.5	1.527	1.833	2.865	637.1	1.59580	78.30	86.37
360	10.0	423.2	439.9	1.697	1.921	3.085	532.1	1.54422	64.77	79.79
400	10.0	334.5	576.0	2.054	2.111	3.790	339.0	1.41464	41.09	68.25
500	10.0	143.3	947.1	2.887	2.438	3.221	275.9	1.16393	19.83	58.85
600	10.0	99.02	1263.	3.462	2.755	3.183	330.8	1.11103	19.67	73.95
700	10.0	78.82	1591.	3.967	3.058	3.386	372.8	1.08759	20.88	95.32
800	10.0	66.46	1940.	4.434	3.328	3.610	407.1	1.07350	22.27	120.1
900	10.0	57.89	2312.	4.871	3.565	3.821	436.7	1.06382	23.62	148.0
1000	10.0	51.49	2704.	5.284	3.774	4.013	463.1	1.05665	24.89	178.8

Carbon Dioxide (CO$_2$)

T K	P MPa	ρ kg m^{-3}	H kJ kg^{-1}	S kJ kg^{-1} K^{-1}	C_v kJ kg^{-1} K^{-1}	C_p kJ kg^{-1} K^{-1}	u m s^{-1}	D	η μPa s	λ mW m^{-1} K^{-1}
Saturation										
216.59	0.5180	1178.	80.04	0.5213	0.9747	1.953	975.8	1.75696	256.7	180.6
216.59	0.5180	13.76	430.4	2.139	0.6292	0.9087	222.8	1.00694	10.95	11.01
220	0.5991	1166.	86.73	0.5517	0.9698	1.962	951.2	1.74853	242.0	176.2
220	0.5991	15.82	431.6	2.119	0.6389	0.9303	223.1	1.00798	11.14	11.30
230	0.8929	1129.	106.6	0.6387	0.9567	1.997	879.1	1.72239	204.2	163.3
230	0.8929	23.27	434.6	2.065	0.6700	1.005	223.6	1.01178	11.69	12.22
240	1.282	1089.	126.8	0.7235	0.9454	2.051	806.4	1.69398	173.0	150.7
240	1.282	33.30	436.5	2.014	0.7053	1.103	223.0	1.01692	12.27	13.30
250	1.785	1046.	147.7	0.8068	0.9364	2.132	731.8	1.66298	146.7	138.5
250	1.785	46.64	437.0	1.964	0.7459	1.237	221.2	1.02382	12.90	14.61
260	2.419	998.9	169.4	0.8895	0.9323	2.255	652.6	1.62876	124.4	126.3
260	2.419	64.42	435.9	1.914	0.7943	1.429	218.2	1.03309	13.61	16.31
270	3.203	945.8	192.4	0.9732	0.9396	2.453	565.5	1.59026	105.0	114.3
270	3.203	88.37	432.6	1.863	0.8517	1.731	213.8	1.04574	14.47	18.69
280	4.161	883.6	217.3	1.060	0.9605	2.814	471.5	1.54547	87.73	102.0
280	4.161	121.7	425.9	1.805	0.9232	2.277	207.7	1.06363	15.60	22.47
290	5.318	804.7	245.6	1.154	0.9937	3.676	371.9	1.48961	71.41	89.55
290	5.318	172.0	413.8	1.734	1.026	3.614	199.4	1.09112	17.36	29.82
300	6.713	679.2	283.4	1.276	1.120	8.698	245.7	1.40346	53.11	80.59
300	6.713	268.6	387.1	1.622	1.248	11.92	185.3	1.14588	21.31	53.69
304.13	7.377	467.6	332.2	1.434				1.26600	33.04	
P = 0.1 MPa (1 bar)										
220	0.1	2.439	442.2	2.492	0.5791	0.7807	233.4	1.00122	11.06	10.90
240	0.1	2.228	458.0	2.561	0.5981	0.7962	243.2	1.00112	12.07	12.24
260	0.1	2.052	474.1	2.625	0.6184	0.8142	252.3	1.00103	13.06	13.68
280	0.1	1.902	490.6	2.686	0.6390	0.8333	261.0	1.00095	14.05	15.20
300	0.1	1.773	507.4	2.745	0.6593	0.8525	269.4	1.00089	15.02	16.79
320	0.1	1.661	524.7	2.800	0.6791	0.8715	277.4	1.00083	15.98	18.42
340	0.1	1.562	542.3	2.854	0.6982	0.8900	285.2	1.00078	16.93	20.09
360	0.1	1.474	560.3	2.905	0.7165	0.9079	292.8	1.00074	17.87	21.77
400	0.1	1.326	597.3	3.002	0.7510	0.9417	307.3	1.00066	19.70	25.14
500	0.1	1.059	695.2	3.221	0.8255	1.015	340.6	1.00053	24.02	33.49
600	0.1	0.8824	799.9	3.411	0.8867	1.076	370.8	1.00044	28.00	41.55

T K	P MPa	ρ kg m^{-3}	H kJ kg^{-1}	S kJ kg^{-1} K^{-1}	C_v kJ kg^{-1} K^{-1}	C_p kJ kg^{-1} K^{-1}	u m s^{-1}	D	η μPa s	λ mW m^{-1} K^{-1}
700	0.1	0.7562	910.2	3.581	0.9375	1.127	398.7	1.00038	31.68	49.30
800	0.1	0.6616	1025.	3.735	0.9800	1.169	424.7	1.00033	35.09	56.71
900	0.1	0.5880	1144.	3.874	1.015	1.205	449.2	1.00029	38.27	63.80
1000	0.1	0.5292	1266.	4.003	1.045	1.234	472.4	1.00027	41.26	70.57
P = 1 MPa (10 bar)										
216.70	1.0	1179.	80.37	0.5210	0.9751	1.950	977.8	1.75735	257.2	180.9
220	1.0	1167.	86.83	0.5506	0.9703	1.959	953.6	1.74911	242.8	176.5
233.03	1.0	1117.	112.7	0.6646	0.9530	2.011	857.2	1.71404	194.1	159.5
233.03	1.0	26.01	435.3	2.049	0.6803	1.032	223.5	1.01318	11.86	12.53
240	1.0	24.86	442.4	2.079	0.6733	0.9991	228.5	1.01259	12.20	12.94
260	1.0	22.21	461.7	2.157	0.6656	0.9450	241.2	1.01123	13.18	14.26
280	1.0	20.20	480.4	2.226	0.6709	0.9252	252.3	1.01019	14.15	15.71
300	1.0	18.58	498.8	2.289	0.6822	0.9209	262.4	1.00937	15.11	17.25
320	1.0	17.23	517.3	2.349	0.6960	0.9243	271.8	1.00868	16.07	18.84
340	1.0	16.09	535.8	2.405	0.7111	0.9320	280.6	1.00810	17.01	20.47
360	1.0	15.11	554.6	2.459	0.7266	0.9421	289.0	1.00760	17.94	22.12
400	1.0	13.48	592.7	2.559	0.7575	0.9655	304.7	1.00677	19.76	25.46
500	1.0	10.66	692.4	2.781	0.8282	1.027	339.8	1.00535	24.06	33.74
600	1.0	8.845	798.0	2.974	0.8881	1.083	370.9	1.00444	28.04	41.76
700	1.0	7.564	908.8	3.144	0.9384	1.132	399.2	1.00379	31.71	49.47
800	1.0	6.610	1024.	3.298	0.9805	1.173	425.5	1.00332	35.12	56.86
900	1.0	5.872	1143.	3.438	1.016	1.207	450.2	1.00295	38.30	63.93
1000	1.0	5.283	1265.	3.567	1.046	1.236	473.6	1.00265	41.28	70.69
P = 10 MPa (100 bar)										
218.60	10.0	1190.	86.78	0.5155	0.9827	1.902	1012.	1.76433	266.3	185.4
220	10.0	1186.	89.44	0.5277	0.9806	1.904	1003.	1.76113	260.2	183.6
240	10.0	1115.	127.9	0.6949	0.9536	1.949	870.9	1.71160	189.4	159.2
260	10.0	1035.	167.8	0.8545	0.9340	2.052	735.5	1.65372	139.5	136.0
280	10.0	938.2	210.8	1.014	0.9263	2.280	588.7	1.58337	102.2	113.1
300	10.0	801.6	261.8	1.189	0.9496	2.991	414.3	1.48612	71.03	88.91
320	10.0	448.3	362.9	1.514	1.058	7.617	219.1	1.25290	32.39	60.44
340	10.0	258.6	443.4	1.759	0.9025	2.402	238.1	1.13908	22.80	35.70
360	10.0	208.2	483.0	1.873	0.8526	1.707	257.8	1.11030	21.83	31.88
400	10.0	161.5	542.1	2.029	0.8287	1.333	286.9	1.08425	22.23	31.58
500	10.0	113.1	663.8	2.301	0.8552	1.162	337.4	1.05801	25.39	37.26
600	10.0	89.94	779.3	2.511	0.9017	1.156	375.4	1.04580	28.94	44.39
700	10.0	75.49	895.9	2.691	0.9465	1.178	407.4	1.03828	32.39	51.59
800	10.0	65.35	1015.	2.850	0.9859	1.205	435.8	1.03305	35.66	58.65
900	10.0	57.76	1137.	2.993	1.020	1.231	461.8	1.02916	38.75	65.48
1000	10.0	51.82	1261.	3.124	1.048	1.255	485.9	1.02613	41.66	72.05

THERMOPHYSICAL PROPERTIES OF SELECTED FLUIDS AT SATURATION

Eric W. Lemmon

These tables give thermodynamic and transport properties of a variety of fluids, as generated from the equations of state presented in the references below. The properties tabulated are pressure (P), density (ρ), enthalpy (H), entropy (S), isochoric heat capacity (C_v), isobaric heat capacity (C_p), speed of sound (u), viscosity (η), thermal conductivity (λ), and static dielectric constant (D). All extensive properties are given on a mass basis. Not all properties are included for every substance. The references should be consulted for information on the uncertainties. Values are given along the saturation line. The first two lines give the properties at the triple point (except for R-1234yf). The final line gives the properties at the critical point. Two lines are given for each temperature (except at the critical point); the first line gives the values of the liquid phase (note the high density), and the second line gives the values of the vapor phase (at low densities). Ammonia uses a reference state based on zero energy and entropy at the triple point; benzene, carbon monoxide, ethylene, hydrogen sulfide, sulfur dioxide, and toluene use a reference state of zero enthalpy and entropy at the saturated liquid state at the normal boiling point; butane, isobutane, ethanol, propylene, R-134a, R-1234yf, and sulfur hexafluoride use a reference state of 200 kJ kg^{-1} and 1 kJ kg^{-1} K^{-1} at -40 °C. Additional calculations at state points not listed below and for fluids not contained here can be obtained at the NIST Chemistry WebBook website (http://webbook.nist.gov/chemistry/fluid/), or by using the NIST Standard Reference Data program REFPROP (http://www.nist.gov/srd/nist23.cfm).

References

1. Bücker, D., and Wagner, W., Reference Equations of State for the Thermodynamic Properties of Fluid Phase n-Butane and Isobutane, *J. Phys. Chem. Ref. Data* 35, 929, 2006.
2. Dillon, H. E., and Penoncello, S. G., A Fundamental Equation for Calculation of the Thermodynamic Properties of Ethanol, *Int. J. Thermophys.* 25, 321, 2004.
3. Fenghour, A., Wakeham, W. A., Vesovic, V., Watson, J. T. R., Millat, J., and Vogel, E., The Viscosity of Ammonia, *J. Phys. Chem. Ref. Data* 24, 1649, 1995.
4. Guder, C., and Wagner, W., A Reference Equation of State for the Thermodynamic Properties of Sulfur Hexafluoride (SF$_6$) for Temperatures from the Melting Line to 625 K and Pressures up to 150 MPa, *J. Phys. Chem. Ref. Data* 38, 33, 2009.
5. Harvey, A. H., and Lemmon, E. W., Method for Estimating the Dielectric Constant of Natural Gas Mixtures, *Int. J. Thermophys.* 26, 31, 2005.
6. Holland, P. M., Eaton, B. E., and Hanley, H. J. M., A Correlation of the Viscosity and Thermal Conductivity Data of Gaseous and Liquid Ethylene, *J. Phys. Chem. Ref. Data* 12, 917, 1983.
7. Huber, M. L., Lasecke, A., and Perkins, R. A., Model for the Viscosity and Thermal Conductivity of Refrigerants, Including a New Correlation for the Viscosity of R134a, *Ind. Eng. Chem. Res.* 42, 3163, 2003.
8. Kiselev, S. B., Ely, J. F., Abdulagatov, I. M., and Huber, M. L., Generalized SAFT-DFT/DMT Model for the Thermodynamic, Interfacial, and Transport Properties of Associating Fluids: Application for n-Alkanols, *Ind. Eng. Chem. Res.* 44, 6916, 2005.
9. Lemmon, E. W., and Span, R., Short Fundamental Equations of State for 20 Industrial Fluids, *J. Chem. Eng. Data* 51, 785, 2006.
10. Overhoff, U., Development of a New Equation of State for the Fluid Region of Propene for Temperatures from the Melting Line to 575 K with Pressures to 1000 MPa as Well as Software for the Computation of Thermodynamic Properties of Fluids, Ph.D. Dissertation, Ruhr University, Bochum, Germany, 2006.
11. Perkins, R. A, Ramires, M. L. V., Nieto de Castro, C. A., and Cusco, L., Measurement and Correlation of the Thermal Conductivity of Butane from 135 K to 600 K at Pressures to 70 MPa, *J. Chem. Eng. Data* 47, 1263, 2002.
12. Perkins, R. A., Laesecke, A., Howley, J., Ramires, M. L. V., Gurova, A. N., and Cusco, L., Experimental Thermal Conductivity Values for the IUPAC Round-Robin Sample of 1,1,1,2-Tetrafluoroethane (R134a), NISTIR, 2000.
13. Perkins, R. A., Measurement and Correlation of the Thermal Conductivity of Isobutane from 114 K to 600 K at Pressures to 70 MPa, *J. Chem. Eng. Data* 47, 1272, 2002.
14. Polt, A., Platzer, B., and Maurer, G., Parameter der thermischen Zustandsgleichung von Bender fuer 14 mehratomige reine Stoffe, *Chem. Tech.* (*Leipzig*) 44, 216, 1992.
15. Schmidt, K. A. G., Carroll, J. J., Quiñones-Cisneros, S. E., and Kvamme, B., Hydrogen Sulphide Viscosity Model, Proceedings of the 86th Annual GPA Convention, San Antonio, Texas, 2007.
16. Smukala, J., Span, R., and Wagner, W., New Equation of State for Ethylene Covering the Fluid Region for Temperatures from the Melting Line to 450 K at Pressures up to 300 MPa, *J. Phys. Chem. Ref. Data* 29, 1053, 2000.
17. Tillner-Roth, R. and Baehr, H. D., An International Standard Formulation for the Thermodynamic Properties of 1,1,1,2-Tetrafluoroethane (HFC-134a) for Temperatures from 170 K to 455 K and Pressures up to 70 MPa, *J. Phys. Chem. Ref. Data* 23, 657, 1994.
18. Tillner-Roth, R., Harms-Watzenberg, F., and Baehr, H. D., A New Fundamental Equation for Ammonia, *DKV-Tagungsbericht* 20, 167, 1993.
19. Tufeu, R., Ivanov, D. Y., Garrabos, Y., and Le Neindre, B., Thermal Conductivity of Ammonia in a Large Temperature and Pressure Range Including the Critical Region, *Ber. Bunsenges. Phys. Chem.* 88, 422, 1984.
20. Vogel, E., Kuechenmeister, C., and Bich, E., Viscosity Correlation for Isobutane over Wide Ranges of the Fluid Region, *Int. J. Thermophys.* 21, 343, 2000.
21. Vogel, E., Kuechenmeister, C., and Bich, E., Viscosity Correlation for n-Butane in the Fluid Region, *High Temp.-High Press.* 31, 173, 1999.
22. Richter, M., McLinden, M. O., and Lemmon, E. W., Thermodynamic Properties of 2,3,3,3-Tetrafluoroprop-1-ene (R1234yf): p-ρ-T Measurements and an Equation of State, submitted to *J. Chem. Eng. Data*, 2011.

Ammonia (NH$_3$)

T	P	ρ	H	S	C_v	C_p	u	η	λ
K	MPa	kg m^{-3}	kJ kg^{-1}	kJ kg^{-1} K^{-1}	kJ kg^{-1} K^{-1}	kJ kg^{-1} K^{-1}	m s^{-1}	μPa s	mW m^{-1} K^{-1}
195.50	0.006091	732.9	0.008311	0.	2.934	4.202	2124.	559.6	819.0
195.50	0.006091	0.06409	1484.	7.593	1.557	2.063	354.1	6.840	19.64
200.00	0.008651	728.1	19.00	0.09601	2.926	4.227	2080.	507.3	803.1

T K	P MPa	ρ kg m⁻³	H kJ kg⁻¹	S kJ kg⁻¹ K⁻¹	C_v kJ kg⁻¹ K⁻¹	C_p kJ kg⁻¹ K⁻¹	u m s⁻¹	η μPa s	λ mW m⁻¹ K⁻¹
200.00	0.008651	0.08908	1493.	7.465	1.565	2.075	357.9	6.952	19.68
220.00	0.03379	705.8	104.7	0.5042	2.889	4.342	1914.	346.7	733.2
220.00	0.03379	0.3188	1529.	6.978	1.620	2.160	373.4	7.485	20.13
240.00	0.1022	681.8	192.7	0.8866	2.855	4.449	1767.	254.9	665.1
240.00	0.1022	0.8969	1562.	6.591	1.705	2.298	386.3	8.059	20.98
260.00	0.2553	656.2	282.8	1.246	2.821	4.548	1625.	197.3	600.1
260.00	0.2553	2.115	1590.	6.274	1.823	2.503	396.2	8.656	22.26
280.00	0.5509	629.1	375.0	1.586	2.790	4.656	1481.	158.1	538.5
280.00	0.5509	4.382	1612.	6.005	1.972	2.788	402.6	9.266	24.03
300.00	1.062	600.0	469.7	1.910	2.762	4.800	1333.	129.3	480.3
300.00	1.062	8.251	1628.	5.770	2.148	3.177	405.0	9.894	26.41
320.00	1.873	568.2	567.9	2.222	2.743	5.018	1178.	106.9	424.8
320.00	1.873	14.51	1634.	5.554	2.349	3.718	402.7	10.56	29.57
340.00	3.080	532.4	671.3	2.529	2.739	5.385	1012.	88.55	371.5
340.00	3.080	24.40	1629.	5.346	2.574	4.530	395.1	11.33	33.94
360.00	4.793	490.3	783.0	2.838	2.764	6.082	830.6	72.80	319.2
360.00	4.793	40.19	1608.	5.129	2.832	5.954	380.8	12.35	40.75
380.00	7.140	436.1	910.3	3.169	2.853	7.818	628.8	58.31	266.6
380.00	7.140	67.37	1558.	4.874	3.147	9.395	358.0	14.02	54.56
400.00	10.30	344.6	1085.	3.595	3.177	22.73	384.6	41.80	216.0
400.00	10.30	131.1	1432.	4.462	3.598	34.92	318.2	18.53	113.5
405.40	11.33	225.0	1262.	4.026				27.04	

Benzene (C₆H₆)

T K	P MPa	ρ kg m⁻³	H kJ kg⁻¹	S kJ kg⁻¹ K⁻¹	C_v kJ kg⁻¹ K⁻¹	C_p kJ kg⁻¹ K⁻¹	u m s⁻¹
278.70	0.004799	893.6	−131.6	−0.4170	1.226	1.703	1369.
278.70	0.004799	0.1623	315.2	1.186	0.8603	0.9684	182.1
280.00	0.005147	892.3	−129.4	−0.4091	1.219	1.699	1366.
280.00	0.005147	0.1733	316.4	1.183	0.8659	0.9741	182.5
300.00	0.01381	871.7	−95.46	−0.2922	1.191	1.706	1296.
300.00	0.01381	0.4357	336.2	1.147	0.9505	1.061	187.4
320.00	0.03203	850.3	−60.78	−0.1804	1.237	1.766	1204.
320.00	0.03203	0.9537	357.3	1.126	1.034	1.148	191.7
340.00	0.06613	828.3	−24.71	−0.07116	1.306	1.840	1108.
340.00	0.06613	1.872	379.5	1.118	1.116	1.236	195.3
360.00	0.1242	805.8	12.89	0.03606	1.377	1.915	1015.
360.00	0.1242	3.370	402.5	1.118	1.198	1.326	197.9
380.00	0.2161	782.7	51.99	0.1414	1.446	1.990	925.2
380.00	0.2161	5.662	426.3	1.127	1.280	1.421	199.5
400.00	0.3528	758.6	92.59	0.2451	1.509	2.064	839.4
400.00	0.3528	9.003	450.6	1.140	1.363	1.521	199.7
450.00	0.9724	692.4	201.1	0.4984	1.657	2.279	633.9
450.00	0.9724	24.23	511.9	1.189	1.573	1.816	193.2
500.00	2.165	609.2	322.2	0.7495	1.813	2.637	425.3
500.00	2.165	57.26	569.5	1.244	1.800	2.314	172.5
550.00	4.215	466.9	469.8	1.023	2.053	4.765	179.1
550.00	4.215	150.3	600.4	1.260	2.104	5.869	125.6
562.05	4.894	309.0	547.3	1.159			

Butane (C₄H₁₀)

T K	P MPa	ρ kg m⁻³	H kJ kg⁻¹	S kJ kg⁻¹ K⁻¹	C_v kJ kg⁻¹ K⁻¹	C_p kJ kg⁻¹ K⁻¹	u m s⁻¹	D	η μPa s	λ mW m⁻¹ K⁻¹
134.90	0.6657·10⁻⁶	735.0	−89.82	−0.4651	1.441	1.973	1827.	2.03981	2304.	176.6
134.90	0.6657·10⁻⁶	0.3450E·10⁻⁴	406.1	3.211	0.9634	1.106	148.9	1.00000	3.321	4.855

T K	P MPa	ρ kg m^{-3}	H kJ kg^{-1}	S kJ kg^{-1} K^{-1}	C_v kJ kg^{-1} K^{-1}	C_p kJ kg^{-1} K^{-1}	u m s^{-1}	D	η µPa s	λ mW m^{-1} K^{-1}
150.00	0.8573·10^{-5}	720.9	−59.93	−0.2550	1.442	1.986	1730.	2.01487	1370.	171.2
150.00	0.8573·10^{-5}	0.3995E·10^{-3}	423.2	2.966	1.015	1.158	156.5	1.00000	3.709	5.579
200.00	0.001939	674.0	41.02	0.3252	1.473	2.062	1438.	1.93301	496.5	149.4
200.00	0.001939	0.06792	484.8	2.544	1.174	1.319	178.9	1.00007	4.979	8.497
220.00	0.007805	654.8	82.75	0.5240	1.505	2.113	1326.	1.90007	373.4	139.9
220.00	0.007805	0.2496	511.3	2.472	1.244	1.392	186.4	1.00027	5.477	9.884
240.00	0.02409	635.1	125.6	0.7105	1.548	2.176	1216.	1.86666	290.9	130.4
240.00	0.02409	0.7120	538.6	2.431	1.322	1.476	192.9	1.00076	5.966	11.39
260.00	0.06098	614.6	170.0	0.8875	1.601	2.254	1106.	1.83250	232.3	121.2
260.00	0.06098	1.688	566.6	2.413	1.408	1.573	197.9	1.00180	6.450	13.03
280.00	0.1328	593.3	216.0	1.058	1.662	2.345	998.3	1.79722	189.0	112.3
280.00	0.1328	3.490	595.0	2.411	1.501	1.684	201.1	1.00373	6.935	14.82
300.00	0.2576	570.7	264.0	1.222	1.729	2.451	890.9	1.76038	155.6	103.9
300.00	0.2576	6.516	623.6	2.421	1.602	1.811	202.2	1.00697	7.436	16.78
320.00	0.4562	546.4	314.3	1.384	1.803	2.577	783.6	1.72136	129.1	96.10
320.00	0.4562	11.29	651.9	2.439	1.708	1.960	200.7	1.01209	7.978	19.00
340.00	0.7520	519.7	367.4	1.543	1.881	2.729	675.4	1.67929	107.2	88.88
340.00	0.7520	18.52	679.5	2.461	1.820	2.142	196.2	1.01990	8.601	21.58
360.00	1.170	489.6	423.8	1.702	1.965	2.926	564.3	1.63273	88.54	82.27
360.00	1.170	29.35	705.6	2.484	1.935	2.379	187.9	1.03167	9.369	24.72
380.00	1.740	454.2	484.3	1.862	2.059	3.221	447.1	1.57904	71.86	76.22
380.00	1.740	45.85	728.7	2.505	2.058	2.754	174.7	1.04981	10.41	28.81
400.00	2.495	408.5	551.2	2.029	2.173	3.838	318.4	1.51174	55.96	70.60
400.00	2.495	73.08	745.2	2.514	2.210	3.623	154.8	1.08025	12.03	35.03
420.00	3.490	327.8	635.2	2.227	2.357	8.852	165.6	1.39801	37.39	67.19
420.00	3.490	135.0	739.9	2.476	2.444	10.72	124.5	1.15196	15.96	53.10
425.13	3.796	228.0	693.9	2.363				1.26627	23.90	

Carbon monoxide (CO)

T K	P MPa	ρ kg m^{-3}	H kJ kg^{-1}	S kJ kg^{-1} K^{-1}	C_v kJ kg^{-1} K^{-1}	C_p kJ kg^{-1} K^{-1}	u m s^{-1}	η µPa s	λ mW m^{-1} K^{-1}
68.16	0.01554	849.5	−28.96	−0.3863	1.262	2.157	998.2	274.2	180.3
68.16	0.01554	0.7761	203.0	3.017	0.7529	1.063	167.2	4.637	6.687
70.00	0.02105	842.1	−24.99	−0.3290	1.243	2.150	980.5	252.2	175.5
70.00	0.02105	1.027	204.7	2.953	0.7553	1.069	169.2	4.777	6.884
80.00	0.08374	800.3	−3.524	−0.04334	1.154	2.142	883.4	178.0	151.4
80.00	0.08374	3.658	213.4	2.668	0.7747	1.120	178.4	5.608	7.982
90.00	0.2385	755.4	18.18	0.2099	1.088	2.188	783.6	139.5	129.2
90.00	0.2385	9.660	220.4	2.457	0.8073	1.216	184.7	6.618	9.129
100.00	0.5444	705.4	40.71	0.4428	1.037	2.306	678.7	114.0	108.3
100.00	0.5444	21.20	224.9	2.285	0.8544	1.389	187.5	7.924	10.37
110.00	1.067	647.4	64.89	0.6657	1.002	2.558	564.7	93.40	88.95
110.00	1.067	41.74	226.0	2.130	0.9190	1.725	186.7	9.722	11.83
120.00	1.877	574.6	92.30	0.8924	0.9951	3.202	433.1	73.95	71.30
120.00	1.877	78.77	221.2	1.967	1.016	2.601	182.0	12.51	14.10
130.00	3.065	456.2	129.3	1.169	1.110	8.070	254.0	51.35	53.11
130.00	3.065	164.8	201.1	1.721	1.235	9.853	171.9	18.94	21.85
132.86	3.494	303.9	164.7	1.429				32.12	

Ethanol (C$_2$H$_6$O)

T K	P MPa	ρ kg m^{-3}	H kJ kg^{-1}	S kJ kg^{-1} K^{-1}	C_v kJ kg^{-1} K^{-1}	C_p kJ kg^{-1} K^{-1}	u m s^{-1}	η µPa s	λ mW m^{-1} K^{-1}
250.00	0.2701·10^{-3}	825.1	150.4	0.8103	1.664	2.032	1325.	3141.	178.1
250.00	0.2701·10^{-3}	0.005988	1110.	4.647	1.278	1.459	226.9	7.272	14.94
260.00	0.6105·10^{-3}	816.9	171.1	0.8917	1.764	2.124	1281.	2454.	175.0

T K	P MPa	ρ kg m⁻³	H kJ kg⁻¹	S kJ kg⁻¹ K⁻¹	C_v kJ kg⁻¹ K⁻¹	C_p kJ kg⁻¹ K⁻¹	u m s⁻¹	η μPa s	λ mW m⁻¹ K⁻¹
260.00	0.6105·10⁻³	0.01302	1124.	4.558	1.306	1.487	231.0	7.586	15.46
280.00	0.002582	800.5	215.8	1.057	1.989	2.351	1203.	1564.	169.6
280.00	0.002582	0.05120	1154.	4.409	1.362	1.544	238.9	8.211	16.61
300.00	0.008841	783.8	265.3	1.228	2.213	2.597	1132.	1047.	164.7
300.00	0.008841	0.1641	1185.	4.293	1.420	1.605	246.2	8.829	17.90
320.00	0.02546	766.4	319.7	1.403	2.412	2.838	1065.	728.9	160.1
320.00	0.02546	0.4457	1216.	4.204	1.481	1.673	252.6	9.439	19.34
340.00	0.06354	747.7	378.7	1.582	2.582	3.064	997.9	524.9	155.7
340.00	0.06354	1.058	1247.	4.136	1.548	1.753	257.9	10.04	20.97
360.00	0.1409	727.5	442.2	1.763	2.719	3.274	930.2	388.5	151.3
360.00	0.1409	2.255	1277.	4.082	1.625	1.852	261.7	10.63	22.83
380.00	0.2833	705.3	509.8	1.945	2.827	3.471	860.3	293.8	146.9
380.00	0.2833	4.400	1305.	4.039	1.714	1.980	263.8	11.22	25.04
400.00	0.5245	680.6	581.2	2.127	2.910	3.665	787.2	225.9	142.5
400.00	0.5245	8.009	1331.	4.002	1.818	2.148	263.8	11.82	27.74
420.00	0.9063	653.0	656.6	2.310	2.971	3.868	709.9	175.7	138.1
420.00	0.9063	13.80	1353.	3.969	1.939	2.375	261.4	12.44	31.21
440.00	1.477	621.3	736.1	2.493	3.016	4.109	627.0	137.5	133.8
440.00	1.477	22.81	1371.	3.935	2.079	2.697	256.2	13.13	35.92
460.00	2.292	584.0	820.7	2.678	3.048	4.451	535.8	107.6	129.9
460.00	2.292	36.68	1382.	3.897	2.241	3.197	247.6	13.96	42.73
480.00	3.406	537.1	912.7	2.869	3.075	5.091	431.4	83.25	126.8
480.00	3.406	58.46	1383.	3.848	2.431	4.128	234.1	15.13	53.47
500.00	4.883	467.9	1020.	3.081	3.109	7.255	302.5	61.12	127.6
500.00	4.883	96.58	1365.	3.771	2.669	6.943	211.9	17.23	73.49
513.90	6.148	276.0	1192.	3.413				30.79	

Ethylene (C₂H₄)

T K	P MPa	ρ kg m⁻³	H kJ kg⁻¹	S kJ kg⁻¹ K⁻¹	C_v kJ kg⁻¹ K⁻¹	C_p kJ kg⁻¹ K⁻¹	u m s⁻¹	D	η μPa s	λ mW m⁻¹ K⁻¹
103.99	0.1220·10⁻³	654.6	−158.1	−1.179	1.622	2.429	1767.	2.00849	685.7	270.6
103.99	0.1220·10⁻³	0.003958	409.4	4.279	0.8901	1.187	202.7	1.00000	0.7727	6.801
120.00	0.001368	634.2	−119.2	−0.8308	1.553	2.427	1660.	1.96600	427.5	248.8
120.00	0.001368	0.03852	428.3	3.731	0.8937	1.192	217.5	1.00004	3.339	6.780
140.00	0.01185	608.0	−70.81	−0.4581	1.465	2.408	1521.	1.91418	277.6	222.3
140.00	0.01185	0.2876	451.3	3.271	0.9064	1.212	233.9	1.00033	4.841	7.892
160.00	0.05623	580.9	−22.66	−0.1372	1.395	2.407	1377.	1.86255	200.1	197.5
160.00	0.05623	1.212	473.0	2.961	0.9336	1.260	247.4	1.00139	5.700	9.115
180.00	0.1818	552.2	25.87	0.1473	1.346	2.441	1227.	1.80986	153.0	174.6
180.00	0.1818	3.589	492.3	2.739	0.9783	1.347	257.0	1.00413	6.401	10.36
200.00	0.4555	521.2	75.68	0.4070	1.321	2.529	1070.	1.75467	120.5	153.7
200.00	0.4555	8.494	508.1	2.569	1.043	1.492	261.9	1.00979	7.133	11.86
220.00	0.9566	486.7	128.0	0.6516	1.320	2.700	903.5	1.69488	95.68	134.5
220.00	0.9566	17.45	519.1	2.429	1.132	1.738	261.5	1.02023	8.007	13.94
240.00	1.773	446.1	184.9	0.8911	1.344	3.043	724.4	1.62679	74.96	116.1
240.00	1.773	33.07	522.7	2.299	1.254	2.211	254.9	1.03865	9.162	17.14
260.00	3.003	393.5	250.4	1.141	1.407	3.946	524.1	1.54141	55.95	97.24
260.00	3.003	61.54	513.5	2.153	1.439	3.512	240.5	1.07303	10.96	23.22
265.00	3.390	376.5	269.4	1.210	1.437	4.473	467.2	1.51452	51.11	92.17
265.00	3.390	72.62	507.3	2.108	1.506	4.308	235.3	1.08667	11.63	26.05
270.00	3.812	356.4	290.3	1.284	1.484	5.409	405.0	1.48318	46.03	86.86
270.00	3.812	86.80	498.3	2.054	1.592	5.766	229.2	1.10433	12.51	30.45
275.00	4.275	330.8	314.6	1.368	1.566	7.588	334.7	1.44393	40.40	81.58
275.00	4.275	106.5	484.3	1.985	1.716	9.293	221.6	1.12922	13.76	39.17
280.00	4.784	290.7	347.8	1.481	1.778	19.56	246.7	1.38366	33.08	83.42

T K	P MPa	ρ kg m^{-3}	H kJ kg^{-1}	S kJ kg^{-1} K^{-1}	C_v kJ kg^{-1} K^{-1}	C_p kJ kg^{-1} K^{-1}	u m s^{-1}	D	η μPa s	λ mW m^{-1} K^{-1}
280.00	4.784	140.7	457.5	1.873	1.981	29.26	208.9	1.17365	16.18	71.72
282.35	5.042	214.2	399.4	1.661				1.27358	22.88	

Hydrogen sulfide (H$_2$S)

T K	P MPa	ρ kg m^{-3}	H kJ kg^{-1}	S kJ kg^{-1} K^{-1}	C_v kJ kg^{-1} K^{-1}	C_p kJ kg^{-1} K^{-1}	u m s^{-1}		η μPa s	λ mW m^{-1} K^{-1}
187.70	0.02326	992.3	−50.47	−0.2520	1.302	2.020	1438.		543.7	249.9
187.70	0.02326	0.5120	524.5	2.811	0.7437	0.9976	245.8		7.673	6.240
200.00	0.05034	971.5	−25.72	−0.1244	1.263	2.003	1373.		378.5	236.7
200.00	0.05034	1.046	535.5	2.682	0.7507	1.012	252.8		8.188	7.174
220.00	0.1437	936.5	14.28	0.06577	1.209	1.995	1268.		265.5	215.8
220.00	0.1437	2.758	552.2	2.511	0.7671	1.047	262.6		9.055	8.740
240.00	0.3377	899.8	54.38	0.2393	1.164	2.009	1162.		211.3	195.6
240.00	0.3377	6.093	566.9	2.375	0.7898	1.102	270.1		9.954	10.42
260.00	0.6875	860.6	95.09	0.4006	1.128	2.050	1054.		175.0	176.1
260.00	0.6875	11.87	579.1	2.262	0.8190	1.183	275.0		10.88	12.32
280.00	1.256	818.0	137.0	0.5533	1.099	2.127	942.1		146.2	157.6
280.00	1.256	21.16	588.0	2.164	0.8549	1.302	277.1		11.83	14.58
300.00	2.110	770.5	180.8	0.7008	1.078	2.262	825.0		121.9	139.7
300.00	2.110	35.48	592.7	2.074	0.8981	1.488	276.1		12.80	17.48
300.00	2.110	770.5	180.8	0.7008	1.078	2.262	825.0		121.9	139.7
300.00	2.110	35.48	592.7	2.074	0.8981	1.488	276.1		12.80	17.48
310.00	2.667	744.3	203.8	0.7738	1.070	2.365	763.8		110.9	131.0
310.00	2.667	45.26	592.9	2.029	0.9230	1.626	274.3		13.31	19.32
320.00	3.323	715.7	227.8	0.8472	1.064	2.509	700.2		100.6	122.5
320.00	3.323	57.40	591.4	1.983	0.9509	1.816	271.6		13.85	21.59
330.00	4.089	684.1	253.2	0.9218	1.062	2.719	633.5		90.67	114.0
330.00	4.089	72.67	587.6	1.935	0.9825	2.095	268.0		14.44	24.50
340.00	4.976	648.3	280.4	0.9989	1.064	3.055	562.6		81.07	105.6
340.00	4.976	92.34	580.8	1.883	1.020	2.548	263.4		15.13	28.43
350.00	5.997	605.8	310.4	1.081	1.074	3.678	485.6		71.43	97.33
350.00	5.997	118.9	569.3	1.821	1.065	3.409	257.7		16.04	34.18
360.00	7.171	551.1	345.6	1.175	1.099	5.271	398.9		61.01	89.85
360.00	7.171	158.3	549.8	1.742	1.126	5.653	250.8		17.54	43.94
370.00	8.529	457.9	397.2	1.309	1.176	18.59	292.8		46.39	88.51
370.00	8.529	238.3	506.6	1.604	1.225	23.66	242.8		21.88	72.03
373.10	9.000	347.3	450.5	1.449					31.95	

Isobutane (C$_4$H$_{10}$)

T K	P MPa	ρ kg m^{-3}	H kJ kg^{-1}	S kJ kg^{-1} K^{-1}	C_v kJ kg^{-1} K^{-1}	C_p kJ kg^{-1} K^{-1}	u m s^{-1}	D	η μPa s	λ mW m^{-1} K^{-1}
113.73	0.2289·10^{-7}	740.3	−112.4	−0.6798	1.174	1.689	2000.	2.10871	8767.	157.9
113.73	0.2289·10^{-7}	0.1407·10^{-5}	368.3	3.547	0.7366	0.8796	139.4	1.00000	2.848	2.272
150.00	0.2388·10^{-4}	706.0	−49.02	−0.1971	1.251	1.805	1714.	2.03196	1748.	148.4
150.00	0.2388·10^{-4}	0.001113	403.1	2.817	0.8944	1.037	157.8	1.00000	3.799	4.461
200.00	0.003814	657.7	45.22	0.3440	1.363	1.968	1389.	1.93554	547.1	129.1
200.00	0.003814	0.1338	459.6	2.416	1.095	1.241	179.4	1.00015	5.080	8.027
220.00	0.01402	637.7	85.31	0.5349	1.415	2.041	1269.	1.89795	395.7	120.6
220.00	0.01402	0.4500	484.5	2.349	1.180	1.330	186.5	1.00049	5.576	9.620
240.00	0.04022	616.9	127.0	0.7159	1.475	2.123	1151.	1.86024	298.3	112.1
240.00	0.04022	1.197	510.3	2.313	1.270	1.428	192.3	1.00130	6.063	11.30
260.00	0.09588	595.4	170.4	0.8893	1.541	2.215	1036.	1.82195	231.6	103.9
260.00	0.09588	2.685	536.7	2.298	1.366	1.538	196.3	1.00291	6.545	13.07
280.00	0.1988	572.6	215.8	1.057	1.613	2.320	922.8	1.78256	183.8	95.99
280.00	0.1988	5.316	563.5	2.299	1.469	1.664	198.3	1.00575	7.033	14.96

T K	P MPa	ρ kg m^{-3}	H kJ kg^{-1}	S kJ kg^{-1} K^{-1}	C_v kJ kg^{-1} K^{-1}	C_p kJ kg^{-1} K^{-1}	u m s^{-1}	D	η μPa s	λ mW m^{-1} K^{-1}
300.00	0.3700	548.3	263.5	1.220	1.690	2.442	810.3	1.74142	148.2	88.60
300.00	0.3700	9.610	590.4	2.310	1.578	1.810	197.7	1.01041	7.546	17.02
320.00	0.6333	521.8	313.9	1.381	1.773	2.589	697.4	1.69761	120.7	81.78
320.00	0.6333	16.27	616.7	2.328	1.694	1.985	194.2	1.01765	8.118	19.37
340.00	1.015	492.1	367.4	1.541	1.862	2.777	582.7	1.64968	98.39	75.56
340.00	1.015	26.36	642.0	2.349	1.814	2.209	186.9	1.02871	8.813	22.21
360.00	1.543	457.2	425.0	1.703	1.958	3.052	463.6	1.59499	79.37	69.94
360.00	1.543	41.79	664.7	2.368	1.939	2.548	175.0	1.04579	9.758	25.97
380.00	2.252	412.8	488.6	1.870	2.069	3.588	335.1	1.52752	61.79	64.93
380.00	2.252	66.87	681.9	2.379	2.101	3.298	156.7	1.07403	11.28	31.85
400.00	3.186	341.0	565.4	2.060	2.250	6.349	184.4	1.42315	42.50	61.90
400.00	3.186	118.4	683.3	2.355	2.354	7.555	128.9	1.13392	14.76	46.45
407.81	3.629	225.5	633.9	2.226				1.26683	24.42	

Propylene (C$_3$H$_6$)

T K	P MPa	ρ kg m^{-3}	H kJ kg^{-1}	S kJ kg^{-1} K^{-1}	C_v kJ kg^{-1} K^{-1}	C_p kJ kg^{-1} K^{-1}	u m s^{-1}	η μPa s	λ mW m^{-1} K^{-1}
87.95	0.7421·10^{-9}	768.3	−200.6	−1.433	1.598	2.280	2130.		
87.95	0.7421·10^{-9}	0.4271E·10^{-7}	386.1	5.238	0.7005	0.8981	149.3	2.589	3.100
100.00	0.4108·10^{-7}	754.1	−173.8	−1.146	1.509	2.184	2027.	3817.	191.4
100.00	0.4108·10^{-7}	0.2079E·10^{-5}	397.1	4.562	0.7310	0.9286	158.4	2.900	3.578
150.00	0.4107·10^{-3}	697.6	−68.84	−0.2936	1.355	2.060	1669.	595.0	179.0
150.00	0.4107·10^{-3}	0.01386	446.6	3.142	0.8573	1.056	190.9	4.250	5.894
200.00	0.02676	640.8	35.10	0.3038	1.345	2.117	1345.	265.1	160.5
200.00	0.02676	0.6864	501.0	2.633	1.007	1.217	215.6	5.647	8.829
220.00	0.07823	616.9	78.03	0.5079	1.366	2.174	1216.	208.4	151.9
220.00	0.07823	1.855	523.0	2.531	1.079	1.305	222.4	6.204	10.20
240.00	0.1872	591.8	122.3	0.6998	1.397	2.250	1087.	167.9	142.6
240.00	0.1872	4.178	544.6	2.459	1.159	1.413	226.7	6.773	11.74
260.00	0.3862	565.0	168.4	0.8829	1.439	2.352	957.0	137.5	132.8
260.00	0.3862	8.275	565.1	2.408	1.249	1.549	228.1	7.386	13.57
280.00	0.7128	535.8	216.9	1.060	1.489	2.488	825.1	113.8	122.6
280.00	0.7128	14.99	583.8	2.371	1.350	1.728	226.0	8.097	15.88
300.00	1.209	503.1	268.5	1.235	1.550	2.684	690.5	94.34	112.0
300.00	1.209	25.57	599.8	2.339	1.457	1.976	219.8	9.000	19.07
320.00	1.919	464.5	324.6	1.411	1.623	3.006	549.0	77.47	101.0
320.00	1.919	42.32	611.3	2.307	1.596	2.420	208.3	10.27	23.88
340.00	2.897	414.7	387.9	1.596	1.726	3.746	396.0	61.42	89.47
340.00	2.897	70.68	614.1	2.261	1.764	3.484	190.3	12.37	32.30
360.00	4.219	325.4	472.8	1.828	1.993	10.80	200.4	41.56	80.42
360.00	4.219	139.2	586.6	2.144	2.158	15.87	161.1	17.86	59.05
364.21	4.555	230.1	529.6	1.981				27.37	

R−134a (1,1,1,2−tetrafluoroethane, CF$_3$CH$_2$F)

T K	P MPa	ρ kg m^{-3}	H kJ kg^{-1}	S kJ kg^{-1} K^{-1}	C_v kJ kg^{-1} K^{-1}	C_p kJ kg^{-1} K^{-1}	u m s^{-1}	η μPa s	λ mW m^{-1} K^{-1}
169.85	0.3896·10^{-3}	1591.	71.46	0.4126	0.7922	1.184	1120.	2154.	145.2
169.85	0.3896·10^{-3}	0.02817	334.9	1.964	0.5030	0.5853	126.8	6.829	3.080
180.00	0.001128	1564.	83.48	0.4814	0.7912	1.187	1068.	1479.	139.1
180.00	0.001128	0.07701	340.9	1.911	0.5267	0.6097	130.1	7.232	3.893
200.00	0.006313	1510.	107.4	0.6073	0.8016	1.206	967.6	867.3	127.7
200.00	0.006313	0.3898	353.1	1.836	0.5732	0.6586	136.0	8.015	5.498
220.00	0.02443	1455.	131.8	0.7235	0.8193	1.233	869.9	582.2	117.2
220.00	0.02443	1.385	365.7	1.787	0.6204	0.7109	141.0	8.779	7.108
240.00	0.07248	1398.	156.8	0.8321	0.8403	1.267	775.0	420.2	107.3

T	P	ρ	H	S	C_v	C_p	u	η	λ
K	MPa	kg m^{-3}	kJ kg^{-1}	kJ kg^{-1} K^{-1}	kJ kg^{-1} K^{-1}	kJ kg^{-1} K^{-1}	m s^{-1}	µPa s	mW m^{-1} K^{-1}
240.00	0.07248	3.837	378.3	1.755	0.6700	0.7705	144.7	9.521	8.732
260.00	0.1768	1337.	182.6	0.9349	0.8631	1.308	682.1	316.6	97.92
260.00	0.1768	8.905	390.8	1.736	0.7234	0.8418	146.8	10.25	10.39
280.00	0.3727	1272.	209.3	1.033	0.8877	1.361	590.2	244.3	88.99
280.00	0.3727	18.23	402.5	1.724	0.7810	0.9296	146.6	10.98	12.12
300.00	0.7028	1200.	237.2	1.129	0.9144	1.432	497.9	190.5	80.34
300.00	0.7028	34.19	413.3	1.716	0.8426	1.044	143.9	11.77	14.01
320.00	1.217	1117.	266.8	1.223	0.9443	1.543	404.0	147.8	71.78
320.00	1.217	60.71	422.3	1.709	0.9093	1.211	137.9	12.73	16.30
340.00	1.972	1015.	298.9	1.318	0.9802	1.751	306.4	111.8	63.08
340.00	1.972	105.7	428.2	1.698	0.9852	1.524	127.6	14.16	19.71
360.00	3.040	870.1	336.1	1.421	1.039	2.437	196.0	78.15	54.06
360.00	3.040	193.6	427.1	1.674	1.085	2.606	111.2	17.14	27.37
374.21	4.059	511.9	389.6	1.562				34.69	

R-1234yf (2,3,3,3-Tetrafluoroprop-1-ene, CF$_3$CF=CH$_2$)

T	P	ρ	H	S	C_v	C_p	u
K	MPa	kg m^{-3}	kJ kg^{-1}	kJ kg^{-1} K^{-1}	kJ kg^{-1} K^{-1}	kJ kg^{-1} K^{-1}	m s^{-1}
240.00	0.08603	1273.	159.1	0.8412	0.7990	1.178	704.6
240.00	0.08603	5.129	341.2	1.600	0.7096	0.8004	134.5
250.00	0.1327	1245.	171.0	0.8898	0.8205	1.210	659.9
250.00	0.1327	7.713	347.9	1.597	0.7374	0.8352	135.4
260.00	0.1972	1216.	183.3	0.9378	0.8411	1.243	616.0
260.00	0.1972	11.23	354.6	1.597	0.7656	0.8725	135.8
270.00	0.2834	1186.	195.9	0.9852	0.8607	1.278	572.8
270.00	0.2834	15.89	361.2	1.597	0.7940	0.9127	135.6
280.00	0.3959	1154.	208.9	1.032	0.8794	1.315	529.7
280.00	0.3959	22.00	367.7	1.599	0.8227	0.9569	134.7
290.00	0.5393	1121.	222.3	1.079	0.8973	1.356	486.6
290.00	0.5393	29.88	374.0	1.602	0.8514	1.006	133.2
300.00	0.7187	1085.	236.1	1.125	0.9147	1.401	443.1
300.00	0.7187	39.99	379.9	1.604	0.8808	1.065	130.9
310.00	0.9394	1047.	250.3	1.171	0.9321	1.454	399.2
310.00	0.9394	52.95	385.5	1.607	0.9122	1.139	127.7
320.00	1.207	1005.	265.1	1.217	0.9496	1.521	354.6
320.00	1.207	69.65	390.5	1.609	0.9467	1.238	123.6
330.00	1.528	957.6	280.5	1.263	0.9667	1.617	308.2
330.00	1.528	91.48	394.8	1.610	0.9847	1.382	118.3
340.00	1.910	902.8	296.8	1.311	0.9855	1.768	257.2
340.00	1.910	120.8	397.9	1.608	1.027	1.614	111.7
350.00	2.361	834.9	314.5	1.360	1.015	2.061	200.6
350.00	2.361	162.8	399.2	1.602	1.077	2.087	103.4
360.00	2.893	738.9	335.0	1.416	1.074	3.029	138.4
360.00	2.893	232.4	396.3	1.587	1.146	3.724	92.64
367.85	3.382	475.6	369.6	1.509			

Sulfur hexafluoride (SF$_6$)

T	P	ρ	H	S	C_v	C_p	u
K	MPa	kg m^{-3}	kJ kg^{-1}	kJ kg^{-1} K^{-1}	kJ kg^{-1} K^{-1}	kJ kg^{-1} K^{-1}	m s^{-1}
223.56	0.2314	1845.	154.1	0.8175	0.5275	0.8371	552.3
223.56	0.2314	19.56	264.7	1.312	0.4833	0.5631	112.8
230.00	0.3013	1813.	159.5	0.8414	0.5407	0.8573	521.5
230.00	0.3013	25.14	267.4	1.310	0.4996	0.5851	112.8
240.00	0.4401	1761.	168.3	0.8784	0.5604	0.8902	474.7
240.00	0.4401	36.21	271.6	1.309	0.5249	0.6219	112.3
250.00	0.6220	1705.	177.4	0.9151	0.5796	0.9264	428.6
250.00	0.6220	50.87	275.6	1.308	0.5506	0.6636	111.0
260.00	0.8546	1646.	186.9	0.9517	0.5986	0.9681	382.8

T K	P MPa	ρ kg m^{-3}	H kJ kg^{-1}	S kJ kg^{-1} K^{-1}	C_v kJ kg^{-1} K^{-1}	C_p kJ kg^{-1} K^{-1}	u m s^{-1}
260.00	0.8546	70.10	279.5	1.308	0.5762	0.7121	109.0
270.00	1.146	1580.	196.8	0.9884	0.6176	1.019	336.9
270.00	1.146	95.32	283.0	1.308	0.6026	0.7739	106.1
280.00	1.506	1507.	207.2	1.025	0.6372	1.087	290.2
280.00	1.506	128.7	286.1	1.307	0.6327	0.8623	102.1
290.00	1.942	1423.	218.3	1.063	0.6589	1.190	241.6
290.00	1.942	173.9	288.4	1.305	0.6650	0.9997	96.92
300.00	2.468	1319.	230.3	1.103	0.6863	1.380	189.8
300.00	2.468	238.4	289.6	1.300	0.7029	1.272	90.15
310.00	3.098	1175.	244.3	1.147	0.7308	1.961	131.5
310.00	3.098	344.1	288.2	1.289	0.7618	2.195	81.33
318.72	3.755	742.3	269.5	1.225			

Sulfur dioxide (SO$_2$)

T K	P MPa	ρ kg m^{-3}	H kJ kg^{-1}	S kJ kg^{-1} K^{-1}	C_v kJ kg^{-1} K^{-1}	C_p kJ kg^{-1} K^{-1}	u m s^{-1}
197.70	0.001660	1620.	−89.30	−0.3900	0.8776	1.376	1362.
197.70	0.001660	0.06483	355.0	1.857	0.4424	0.5746	182.2
200.00	0.002026	1615.	−86.14	−0.3741	0.8756	1.374	1350.
200.00	0.002026	0.07823	356.3	1.838	0.4443	0.5768	183.2
220.00	0.009334	1567.	−58.76	−0.2437	0.8578	1.364	1253.
220.00	0.009334	0.3289	367.1	1.692	0.4638	0.6006	191.1
240.00	0.03199	1519.	−31.52	−0.1252	0.8406	1.360	1160.
240.00	0.03199	1.041	377.7	1.580	0.4885	0.6330	198.1
260.00	0.08791	1469.	−4.277	−0.01632	0.8248	1.363	1070.
260.00	0.08791	2.677	387.6	1.491	0.5180	0.6750	203.9
280.00	0.2043	1417.	23.14	0.08495	0.8112	1.375	980.3
280.00	0.2043	5.897	396.6	1.419	0.5514	0.7274	208.4
300.00	0.4172	1362.	50.92	0.1802	0.7999	1.398	891.0
300.00	0.4172	11.57	404.6	1.359	0.5875	0.7915	211.3
320.00	0.7702	1304.	79.31	0.2710	0.7910	1.436	801.0
320.00	0.7702	20.83	411.1	1.308	0.6253	0.8712	212.4
340.00	1.313	1240.	108.6	0.3586	0.7849	1.495	709.5
340.00	1.313	35.23	415.7	1.262	0.6648	0.9756	211.5
360.00	2.099	1168.	139.3	0.4445	0.7819	1.590	615.4
360.00	2.099	57.07	417.8	1.218	0.7065	1.126	208.2
380.00	3.190	1086.	172.2	0.5306	0.7832	1.759	516.7
380.00	3.190	90.35	416.1	1.173	0.7529	1.380	202.2
400.00	4.656	983.7	208.7	0.6205	0.7922	2.134	409.4
400.00	4.656	143.8	408.2	1.119	0.8091	1.942	192.9
420.00	6.590	832.0	254.3	0.7264	0.8262	3.867	279.4
420.00	6.590	248.6	385.7	1.039	0.8916	4.570	179.3
430.64	7.884	525.0	320.7	0.8778			

Toluene (CH$_3$–C$_6$H$_5$)

T K	P MPa	ρ kg m^{-3}	H kJ kg^{-1}	S kJ kg^{-1} K^{-1}	C_v kJ kg^{-1} K^{-1}	C_p kJ kg^{-1} K^{-1}	u m s^{-1}
178.00	0.3939·10^{-7}	974.8	−344.9	−1.261	1.024	1.472	1888.
178.00	0.3939·10^{-7}	0.2453·10^{-5}	147.1	1.503	0.5940	0.6843	136.0
180.00	0.5534·10^{-7}	972.9	−342.0	−1.244	1.024	1.472	1877.
180.00	0.5534·10^{-7}	0.3407·10^{-5}	148.5	1.480	0.6004	0.6907	136.7
200.00	0.1083·10^{-5}	953.5	−312.5	−1.089	1.039	1.479	1768.
200.00	0.1083·10^{-5}	0.6002·10^{-4}	163.0	1.288	0.6668	0.7570	143.1
220.00	0.1148·10^{-4}	934.6	−282.7	−0.9471	1.068	1.504	1665.
220.00	0.1148·10^{-4}	0.5783·10^{-3}	178.8	1.151	0.7373	0.8275	149.3

T K	P MPa	ρ kg m^{-3}	H kJ kg^{-1}	S kJ kg^{-1} K^{-1}	C_v kJ kg^{-1} K^{-1}	C_p kJ kg^{-1} K^{-1}	u m s^{-1}
240.00	$0.7754 \cdot 10^{-4}$	916.0	−252.2	−0.8147	1.108	1.542	1566.
240.00	$0.7754 \cdot 10^{-4}$	0.003581	196.1	1.053	0.8111	0.9014	155.1
260.00	$0.3731 \cdot 10^{-3}$	897.5	−220.9	−0.6895	1.156	1.590	1472.
260.00	$0.3731 \cdot 10^{-3}$	0.01591	214.8	0.9865	0.8876	0.9781	160.7
280.00	0.001383	879.0	−188.6	−0.5697	1.210	1.646	1381.
280.00	0.001383	0.05483	235.0	0.9433	0.9660	1.057	166.0
300.00	0.004177	860.4	−155.1	−0.4541	1.268	1.707	1294.
300.00	0.004177	0.1549	256.7	0.9184	1.046	1.138	171.0
320.00	0.01073	841.7	−120.3	−0.3418	1.329	1.773	1211.
320.00	0.01073	0.3745	279.7	0.9080	1.126	1.221	175.5
340.00	0.02417	822.6	−84.12	−0.2323	1.391	1.842	1130.
340.00	0.02417	0.7995	303.9	0.9090	1.208	1.305	179.4
360.00	0.04898	803.1	−46.54	−0.1250	1.455	1.915	1051.
360.00	0.04898	1.545	329.3	0.9190	1.289	1.391	182.7
380.00	0.09099	783.0	−7.483	−0.01956	1.518	1.989	974.1
380.00	0.09099	2.757	355.7	0.9361	1.370	1.479	185.1
400.00	0.1573	762.2	33.10	0.08428	1.581	2.066	898.1
400.00	0.1573	4.612	382.9	0.9587	1.450	1.570	186.4
450.00	0.4862	705.7	141.6	0.3385	1.737	2.272	709.6
450.00	0.4862	13.62	453.2	1.031	1.649	1.814	184.2
500.00	1.177	638.1	261.2	0.5882	1.887	2.529	516.5
500.00	1.177	33.64	523.7	1.113	1.846	2.132	170.9
550.00	2.428	544.8	395.8	0.8406	2.043	3.029	306.4
550.00	2.428	80.11	585.4	1.185	2.057	2.873	140.1
591.75	4.126	292.0	565.8	1.130			

VIRIAL COEFFICIENTS OF SELECTED GASES

Henry V. Kehiaian

This table gives second virial coefficients of about 110 inorganic and organic gases as a function of temperature. Selected data from the literature have been fitted by least squares to the equation

$$B/\mathrm{cm}^3\mathrm{mol}^{-1} = \sum_{i=1}^{n} a(i)[(T_0/T)-1]^{i-1}$$

where $T_0 = 298.15$ K. The table gives the coefficients $a(i)$ and values of B at fixed temperature increments, as calculated from this smoothing equation.

The equation may be used with the tabulated coefficients for interpolation within the indicated temperature range. It should not be used for extrapolation beyond this range.

Compounds are listed in the modified Hill order (see Introduction), with carbon-containing compounds following those compounds not containing carbon.

A useful compilation of virial coefficient data from the literature may be found in the reference.

Reference

J. H. Dymond and E. B. Smith, *The Virial Coefficients of Pure Gases and Mixtures, A Critical Compilation*, Oxford University Press, Oxford, 1980.

Compounds Not Containing Carbon

Mol. form.	Name		T/K	B/cm³ mol⁻¹
Ar	Argon		100	−184
			120	−131
			140	−98
		$a(1) = -16$	160	−76
		$a(2) = -60$	180	−60
		$a(3) = -9.7$	200	−48
		$a(4) = -1.5$	300	−16
			400	−1
			500	7
			600	12
			700	15
			800	18
			900	20
			1000	22
BF₃	Boron trifluoride		200	−338
			240	−202
			280	−129
		$a(1) = -106$	320	−85
		$a(2) = -330$	360	−56
		$a(3) = -251$	400	−37
		$a(4) = -80$	440	−23
ClH	Hydrogen chloride		190	−451
			230	−269
			270	−181
		$a(1) = -144$	310	−132
		$a(2) = -325$	350	−102
		$a(3) = -277$	390	−81
		$a(4) = -170$	430	−66
			470	−54
Cl₂	Chlorine		210	−508
			220	−483
			230	−457
		$a(1) = -303$	240	−432
		$a(2) = -555$	250	−407
		$a(3) = 9$	260	−383
		$a(4) = 329$	270	−360
		$a(5) = 68$	280	−339
			290	−318
			300	−299
			350	−221
			400	−166
			450	−126

Mol. form.	Name		T/K	B/cm³ mol⁻¹
			500	−97
			600	−59
			700	−36
			800	−22
			900	−12
F₂	Fluorine		80	−378
			110	−165
			140	−109
		$a(1) = 8.5$	170	−79
		$a(2) = -163.2$	200	−55
		$a(3) = 84.0$	230	−33
		$a(4) = -27.9$	260	−14
F₄Si	Silicon tetrafluoride		210	−268
			240	−213
			270	−170
		$a(1) = -138$	300	−136
		$a(2) = -312$	330	−108
			360	−84
			390	−64
			420	−47
			450	−32
F₅I	Iodine pentafluoride		320	−2540
			330	−2344
			340	−2172
		$a(1) = -3077$	350	−2021
		$a(2) = -8474$	360	−1890
		$a(3) = -9116$	370	−1775
			380	−1674
			390	−1587
			400	−1510
			410	−1443
F₅P	Phosphorus pentafluoride		320	−162
			340	−143
			360	−127
		$a(1) = -186$	380	−112
		$a(2) = -345$	400	−98
			420	−86
			440	−75
			460	−64
F₆Mo	Molybdenum hexafluoride		300	−896
			310	−810
			320	−737

Mol. form.	Name		T/K	B/cm³ mol⁻¹
		$a(1) = -914$	330	-677
		$a(2) = -2922$	340	-627
		$a(3) = -4778$	350	-586
			360	-553
			370	-527
			380	-506
			390	-491
F_6S	Sulfur hexafluoride		200	-685
			250	-416
			300	-275
		$a(1) = -279$	350	-190
		$a(2) = -647$	400	-135
		$a(3) = -335$	450	-96
		$a(4) = -72$	500	-68
F_6U	Uranium hexafluoride		320	-1030
			340	-905
			360	-805
		$a(1) = -1204$	380	-724
		$a(2) = -2690$	400	-658
		$a(3) = -2144$	420	-604
			440	-560
F_6W	Tungsten hexafluoride		320	-641
			340	-578
			360	-523
		$a(1) = -719$	380	-473
		$a(2) = -1143$	400	-428
			420	-387
			440	-350
			460	-317
H_2	Hydrogen		15	-230
			20	-151
			25	-108
		$a(1) = 15.4$	30	-82
		$a(2) = -9.0$	35	-64
		$a(3) = -0.21$	40	-52
			45	-42
			50	-35
			60	-24
			70	-16
			80	-11
			90	-7
			100	-3
			200	11
			300	15
			400	18
H_2O	Water		300	-1126
			320	-850
			340	-660
		$a(1) = -1158$	360	-526
		$a(2) = -5157$	380	-428
		$a(3) = -10301$	400	-356
		$a(4) = -10597$	420	-301
		$a(5) = -4415$	440	-258
			460	-224
			480	-197
			500	-175
			600	-104
			700	-67
			800	-44
			900	-30
			1000	-20

Mol. form.	Name		T/K	B/cm³ mol⁻¹
			1100	-14
			1200	-11
H_3N	Ammonia		290	-302
			300	-265
			310	-236
		$a(1) = -271$	320	-213
		$a(2) = -1022$	330	-194
		$a(3) = -2715$	340	-179
		$a(4) = -4189$	350	-166
			360	-154
			370	-144
			380	-135
			400	-118
			420	-101
H_3P	Phosphine		190	-457
			200	-404
			210	-364
		$a(1) = -146$	220	-332
		$a(2) = -733$	230	-305
		$a(3) = 1022$	240	-281
		$a(4) = -1220$	250	-258
			260	-235
			270	-213
			280	-190
			290	-166
He	Helium		2	-172
			6	-48
			10	-24
		$a(1) = 12.44$	14	-13
		$a(2) = -1.25$	18	-7
			22	-3
			26	-1
			30	1
			50	6
			70	8
			90	10
			110	10
			150	11
			250	12
			650	13
			700	13
Kr	Krypton		110	-363
			120	-307
			130	-263
		$a(1) = -51$	140	-229
		$a(2) = -118$	150	-201
		$a(3) = -29$	160	-178
		$a(4) = -5$	170	-159
			180	-143
			190	-129
			200	-117
			250	-75
			300	-51
			400	-23
			500	-8
			600	2
			700	8
NO	Nitric oxide		120	-232
			130	-176
			140	-138
		$a(1) = -12$	150	-113

Mol. form.	Name	T/K	B/cm³ mol⁻¹
$a(2) = -119$		160	-96
$a(3) = 89$		170	-83
$a(4) = -73$		180	-73
		190	-65
		200	-58
		210	-52
		230	-42
		250	-32
		270	-24
N_2	Nitrogen	75	-274
		100	-161
		125	-104
$a(1) = -4.3$		150	-71
$a(2) = -55.7$		175	-49
$a(3) = -11.8$		200	-34
		225	-24
		250	-15
		300	-4
		400	9
		500	16
		600	21
		700	24
N_2O	Nitrous oxide	240	-219
		260	-181
		280	-151
$a(1) = -130$		300	-128
$a(2) = -307$		320	-110
$a(3) = -248$		340	-96
		360	-85
		380	-76
		400	-68
Ne	Neon	60	-25
		80	-13
		100	-6
$a(1) = 10.8$		120	-1
$a(2) = -7.5$		140	2
$a(3) = -0.4$		160	4
		180	6
		200	7
		300	11
		400	13
		500	14
		600	15
O_2	Oxygen	90	-241
		110	-161
		130	-117
$a(1) = -16$		150	-88
$a(2) = -62$		170	-69
$a(3) = -8$		190	-55
$a(4) = -3$		210	-44
		230	-36
		250	-29
		270	-23
		290	-18
		310	-14
		330	-10
		350	-7
		400	-1
O_2S	Sulfur dioxide	290	-465
		320	-354
		350	-276

Mol. form.	Name	T/K	B/cm³ mol⁻¹
$a(1) = -430$		380	-221
$a(2) = -1193$		410	-181
$a(3) = -1029$		440	-153
		470	-132
Xe	Xenon	160	-421
		170	-377
		180	-340
$a(1) = -130$		190	-307
$a(2) = -262$		200	-280
$a(3) = -87$		210	-255
		220	-234
		230	-215
		240	-199
		250	-184
		300	-129
		350	-93
		400	-69
		500	-39
		600	-21
		650	-14

Compounds Containing Carbon

Mol. form.	Name	T/K	B/cm³ mol⁻¹
$CClF_3$	Chlorotrifluoromethane	240	-369
		290	-237
		340	-165
$a(1) = -223$		390	-119
$a(2) = -504$		440	-86
$a(3) = -340$		490	-60
$a(4) = -291$		540	-39
CCl_2F_2	Dichlorodifluoromethane	250	-769
		280	-570
		310	-441
$a(1) = -486$		340	-353
$a(2) = -1217$		370	-289
$a(3) = -1188$		400	-241
$a(4) = -698$		430	-204
		460	-174
CCl_3F	Trichlorofluoromethane	240	-1140
		280	-879
		320	-689
$a(1) = -786$		360	-545
$a(2) = -1428$		400	-431
$a(3) = -142$		440	-340
		480	-265
CCl_4	Tetrachloromethane	320	-1345
		340	-1171
		360	-1040
$a(1) = -1600$		380	-942
$a(2) = -4059$		400	-868
$a(3) = -4653$		420	-814
CF_4	Tetrafluoromethane	250	-137
		300	-87
		350	-55
$a(1) = -88$		400	-32
$a(2) = -238$		450	-16
$a(3) = -70$		500	-4
		600	14
		700	25

Mol. form.	Name	T/K	$B/cm^3\ mol^{-1}$	Mol. form.	Name	T/K	$B/cm^3\ mol^{-1}$
		800	33	CH_3Cl	Chloromethane	280	−466
$CHClF_2$	Chlorodifluoromethane	300	−343			300	−402
		325	−298			320	−348
		350	−257		$a(1) = -407$	340	−304
	$a(1) = -347$	375	−221		$a(2) = -887$	360	−266
	$a(2) = -575$	400	−188		$a(3) = -385$	380	−234
	$a(3) = 187$	425	−158			400	−206
$CHCl_2F$	Dichlorofluoromethane	250	−728			420	−182
		275	−634			440	−161
		300	−557			460	−142
	$a(1) = -562$	325	−491			480	−126
	$a(2) = -862$	350	−434			500	−112
		375	−385			600	−58
		400	−343	CH_3F	Fluoromethane	280	−244
		425	−305			300	−205
		450	−271			320	−174
$CHCl_3$	Trichloromethane	320	−1001		$a(1) = -209$	340	−150
		330	−926		$a(2) = -525$	360	−129
		340	−858		$a(3) = -365$	380	−112
	$a(1) = -1193$	350	−797			400	−99
	$a(2) = -2936$	360	−740			420	−87
	$a(3) = -1751$	370	−689	CH_3I	Iodomethane	310	−725
		380	−642			320	−646
		390	−599			330	−582
		400	−559		$a(1) = -844$	340	−531
CHF_3	Trifluoromethane	200	−433		$a(2) = -3353$	350	−492
		220	−350		$a(3) = -6590$	360	−462
		240	−288			370	−441
	$a(1) = -177$	260	−241			380	−427
	$a(2) = -399$	280	−204	CH_4	Methane	110	−328
	$a(3) = -250$	300	−174			120	−276
		320	−151			130	−237
		340	−132		$a(1) = -43$	140	−206
		360	−116		$a(2) = -114$	150	−181
		380	−103		$a(3) = -19$	160	−160
		400	−91		$a(4) = -7$	170	−143
CH_2Cl_2	Dichloromethane	320	−706			180	−128
		330	−634			190	−116
		340	−574			200	−105
	$a(1) = -913$	350	−524			250	−66
	$a(2) = -3371$	360	−482			300	−43
	$a(3) = -5013$	370	−447			350	−27
		380	−420			400	−16
		400	−380			500	0
		420	−357			600	10
CH_2F_2	Difluoromethane	280	−375	CH_4O	Methanol	320	−1431
		290	−343			330	−1299
		300	−316			340	−1174
	$a(1) = -321$	310	−294			350	−1056
	$a(2) = -754$	320	−275		$a(1) = -1752$	360	−945
	$a(3) = -1300$	330	−260		$a(2) = -4694$	370	−840
		340	−248			380	−741
		350	−238			390	−646
CH_3Br	Bromomethane	280	−645			400	−557
		290	−596	CH_5N	Methylamine	300	−451
		300	−551			325	−367
	$a(1) = -559$	310	−509			350	−304
	$a(2) = -1324$	320	−469		$a(1) = -459$	375	−257
		340	−396		$a(2) = -1191$	400	−220
		360	−332		$a(3) = -995$	425	−192
		380	−274			450	−170

Mol. form.	Name	T/K	B/cm³ mol⁻¹
		500	−140
		550	−122
CO	Carbon monoxide	210	−36
		240	−24
		270	−15
	$a(1) = -9$	300	−8
	$a(2) = -58$	330	−3
	$a(3) = -18$	360	1
		420	7
		480	11
CO₂	Carbon dioxide	220	−244
		240	−204
		260	−172
	$a(1) = -127$	280	−146
	$a(2) = -288$	300	−126
	$a(3) = -118$	320	−108
		340	−94
		360	−81
		380	−71
		400	−62
		500	−30
		600	−13
		700	−1
		800	7
		900	12
		1000	16
		1100	19
CS₂	Carbon disulfide	280	−932
		310	−740
		340	−603
	$a(1) = -807$	370	−504
	$a(2) = -1829$	400	−431
	$a(3) = -1371$	430	−375
C₂Cl₂F₄	1,2-Dichloro-1,1,2,2-tetrafluoroethane	300	−801
		320	−695
		340	−608
	$a(1) = -812$	360	−536
	$a(2) = -1773$	380	−475
	$a(3) = -963$	400	−423
		420	−379
		440	−341
		460	−307
		480	−279
		500	−253
C₂Cl₃F₃	1,1,2-Trichloro-1,2,2-trifluoroethane	290	−1041
		310	−943
		330	−856
		350	−780
	$a(1) = -999$	370	−712
	$a(2) = -1479$	390	−651
		410	−596
		430	−546
		450	−500
C₂H₂	Ethyne	200	−573
		210	−500
		220	−440
	$a(1) = -216$	230	−390
	$a(2) = -375$	240	−349
	$a(3) = -716$	250	−315
		260	−287

Mol. form.	Name	T/K	B/cm³ mol⁻¹
		270	−263
C₂H₃N	Ethanenitrile	330	−3468
		340	−2971
		350	−2563
	$a(1) = -5840$	360	−2233
	$a(2) = -29175$	370	−1970
	$a(3) = -47611$	380	−1765
		390	−1610
		400	−1499
		410	−1425
C₂H₄	Ethene	240	−218
		270	−172
		300	−139
	$a(1) = -140$	330	−113
	$a(2) = -296$	360	−92
	$a(3) = -101$	390	−76
		420	−63
		450	−52
C₂H₄Cl₂	1,2-Dichloroethane	370	−812
		390	−716
		410	−635
	$a(1) = -1362$	430	−566
	$a(2) = -3240$	450	−508
	$a(3) = -2100$	470	−458
		490	−416
		510	−379
		530	−347
		550	−319
		570	−295
C₂H₄O	Ethanal	290	−1352
		320	−927
		350	−654
	$a(1) = -1217$	380	−482
	$a(2) = -4647$	410	−375
	$a(3) = -5725$	440	−314
		470	−283
C₂H₄O₂	Methyl methanoate	320	−821
		330	−744
		340	−677
	$a(1) = -1035$	350	−620
	$a(2) = -3425$	360	−571
	$a(3) = -4203$	370	−528
		380	−492
		390	−461
		400	−435
C₂H₅Cl	Chloroethane	320	−634
		360	−450
		400	−330
	$a(1) = -777$	440	−249
	$a(2) = -2205$	480	−195
	$a(3) = -1764$	520	−157
		560	−131
		600	−114
C₂H₆	Ethane	200	−409
		220	−337
		240	−284
		260	−242
	$a(1) = -184$	280	−209
	$a(2) = -376$	300	−181
	$a(3) = -143$	320	−159
	$a(4) = -54$	340	−140

Mol. form.	Name	T/K	B/cm^3 mol^{-1}	Mol. form.	Name	T/K	B/cm^3 mol^{-1}
		360	−123			440	−146
		380	−109			460	−131
		400	−96			480	−118
		500	−52			500	−106
		600	−24	C_3H_6O	2-Propanone	300	−1996
C_2H_6O	Ethanol	320	−2710			320	−1522
		330	−2135			340	−1198
		340	−1676	$a(1) = -2051$		360	−971
$a(1) = -4475$		350	−1317	$a(2) = -8903$		380	−806
$a(2) = -29719$		360	−1043	$a(3) = -18056$		400	−683
$a(3) = -56716$		370	−843	$a(4) = -16448$		420	−586
		380	−705			440	−506
		390	−622			460	−437
C_2H_6O	Dimethyl ether	275	−536			480	−375
		280	−517	C_3H_6O	Ethyl methanoate	330	−1003
		285	−499			340	−916
$a(1) = -455$		290	−482			350	−839
$a(2) = -965$		295	−465	$a(1) = -1371$		360	−771
		300	−449	$a(2) = -4231$		370	−712
		305	−433	$a(3) = -4312$		380	−660
		310	−418			390	−614
C_2H_7N	Dimethylamine	310	−606	C_3H_6O	Methyl ethanoate	320	−1320
		320	−563			330	−1186
		330	−523			340	−1074
$a(1) = -662$		340	−487	$a(1) = -1709$		350	−980
$a(2) = -1504$		350	−454	$a(2) = -6348$		360	−903
$a(3) = -667$		360	−423	$a(3) = -9650$		370	−840
		370	−395			380	−789
		380	−369			390	−749
		390	−345	C_3H_7Cl	1-Chloropropane	310	−1001
		400	−322			340	−772
C_2H_7N	Ethylamine	300	−773			370	−614
		310	−710	$a(1) = -1121$		400	−501
		320	−654	$a(2) = -3271$		430	−417
$a(1) = -785$		330	−604	$a(3) = -3786$		460	−352
$a(2) = -2012$		340	−558	$a(4) = -1974$		490	−302
$a(3) = -1397$		350	−517			520	−261
		360	−480			550	−227
		370	−447			580	−198
		380	−416	C_3H_8	Propane	240	−641
		390	−389			260	−527
		400	−363			280	−444
C_3H_6	Cyclopropane	300	−383	$a(1) = -386$		300	−381
		310	−356	$a(2) = -844$		320	−331
		320	−332	$a(3) = -720$		340	−292
$a(1) = -388$		330	−310	$a(4) = -574$		360	−259
$a(2) = -861$		340	−290			380	−232
$a(3) = -538$		350	−272			400	−208
		360	−256			440	−169
		370	−241			480	−138
		380	−227			520	−112
		390	−215			560	−90
		400	−204	C_3H_8O	1-Propanol	380	−873
C_3H_6	Propene	280	−395			385	−826
		300	−342			390	−783
		320	−299			395	−744
$a(1) = -347$		340	−262	$a(1) = -2690$		400	−709
$a(2) = -727$		360	−232	$a(2) = -12040$		405	−679
$a(3) = -325$		380	−205	$a(3) = -16738$		410	−651
		400	−183			415	−627
		420	−163			420	−606

Mol. form.	Name	T/K	B/cm³ mol⁻¹	Mol. form.	Name	T/K	B/cm³ mol⁻¹
C_3H_8O	2-Propanol	380	−821				
		385	−766	C_4H_{10}	Butane	250	−1170
		390	−717			280	−863
	$a(1) = -3165$	395	−674			310	−668
	$a(2) = -16092$	400	−636		$a(1) = -735$	340	−536
	$a(3) = -24197$	405	−604		$a(2) = -1835$	370	−442
		410	−576		$a(3) = -1922$	400	−371
		415	−552		$a(4) = -1330$	430	−315
		420	−533			460	−270
C_3H_9N	Trimethylamine	310	−675			490	−232
		320	−628			520	−199
		330	−585			550	−171
	$a(1) = -737$	340	−547	C_4H_{10}	2-Methylpropane	270	−900
	$a(2) = -1669$	350	−512			300	−697
	$a(3) = -986$	360	−480			330	−553
		370	−450		$a(1) = -707$	360	−450
C_4H_8	1-Butene	300	−624		$a(2) = -1719$	390	−374
		320	−539		$a(3) = -1282$	420	−317
		340	−470			450	−273
	$a(1) = -633$	360	−413			480	−240
	$a(2) = -1442$	380	−366			510	−215
	$a(3) = -932$	400	−327	$C_4H_{10}O$	1-Butanol	350	−1693
		420	−294			360	−1544
C_4H_8O	2-Butanone	310	−2056			370	−1402
		320	−1878		$a(1) = -2629$	380	−1268
		330	−1712		$a(2) = -6315$	390	−1141
	$a(1) = -2282$	340	−1555			400	−1021
	$a(2) = -5907$	350	−1407			420	−796
		360	−1267			440	−593
		370	−1135	$C_4H_{10}O$	2-Methyl-1-propanol	390	−1076
$C_4H_8O_2$	Propyl methanoate	330	−1496			400	−979
		340	−1354			410	−887
		350	−1231		$a(1) = -2269$	420	−800
	$a(1) = -2118$	360	−1126		$a(2) = -5065$	430	−716
	$a(2) = -7299$	370	−1035			440	−636
	$a(3) = -8851$	380	−957	$C_4H_{10}O$	2-Butanol	380	−1110
		390	−890			390	−1005
		400	−834			400	−906
$C_4H_8O_2$	Ethyl ethanoate	330	−1543		$a(1) = -2232$	410	−811
		340	−1385		$a(2) = -5209$	420	−721
		350	−1254	$C_4H_{10}O$	2-Methyl-2-propanol	380	−924
	$a(1) = -2272$	360	−1144			390	−827
	$a(2) = -8818$	370	−1055			400	−736
	$a(3) = -13130$	380	−982		$a(1) = -1952$	410	−649
		390	−923		$a(2) = -4775$	420	−567
		400	−878	$C_4H_{10}O$	Diethyl ether	280	−1550
$C_4H_8O_2$	Methyl propanoate	330	−1588			300	−1199
		340	−1444			320	−954
		350	−1319		$a(1) = -1226$	340	−776
	$a(1) = -2216$	360	−1211		$a(2) = -4458$	360	−638
	$a(2) = -7339$	370	−1117		$a(3) = -7746$	380	−525
	$a(3) = -8658$	380	−1037		$a(4) = -10005$	400	−428
		390	−968			420	−340
		400	−908	$C_4H_{11}N$	Diethylamine	320	−1228
C_4H_9Cl	1-Chlorobutane	330	−1224			330	−1134
		370	−898			340	−1056
		410	−691		$a(1) = -1522$	350	−988
	$a(1) = -1643$	450	−551		$a(2) = -5204$	360	−926
	$a(2) = -4897$	490	−449		$a(3) = -15047$	370	−868
	$a(3) = -6178$	530	−371		$a(4) = -28835$	380	−812
	$a(4) = -3718$	570	−309			390	−755

Mol. form.	Name	T/K	$B/cm^3\ mol^{-1}$	Mol. form.	Name	T/K	$B/cm^3\ mol^{-1}$
		400	−697			390	−492
C_5H_5N	Pyridine	350	−1257			400	−464
		360	−1176			450	−357
		370	−1099			500	−279
$a(1) = -1765$		380	−1026			550	−218
$a(2) = -3431$		390	−957	C_6H_6	Benzene	290	−1588
		400	−892			300	−1454
		420	−770			310	−1335
		440	−659	$a(1) = -1477$		320	−1231
C_5H_{10}	Cyclopentane	300	−1049	$a(2) = -3851$		330	−1139
		305	−1015	$a(3) = -3683$		340	−1056
		310	−981	$a(4) = -1423$		350	−983
$a(1) = -1062$		315	−949			400	−712
$a(2) = -2116$		320	−918			450	−542
C_5H_{10}	1-Pentene	310	−966			500	−429
		320	−898			550	−349
		330	−836			600	−291
$a(1) = -1055$		340	−780	C_6H_7N	2-Methylpyridine	360	−1656
$a(2) = -2377$		350	−729			370	−1523
$a(3) = -1189$		360	−681			380	−1404
		370	−638	$a(1) = -2940$		390	−1297
		380	−598	$a(2) = -8813$		400	−1202
		390	−561	$a(3) = -7809$		410	−1117
		400	−527			420	−1040
		410	−495			430	−972
$C_5H_{10}O$	2-Pentanone	330	−2850	C_6H_7N	3-Methylpyridine	380	−1819
		340	−2420			390	−1612
		350	−2076			400	−1448
$a(1) = -4962$		360	−1804	$a(1) = -6304$		410	−1322
$a(2) = -26372$		370	−1595	$a(2) = -30415$		420	−1230
$a(3) = -46537$		380	−1440	$a(3) = -44549$		430	−1166
		390	−1332	C_6H_7N	4-Methylpyridine	380	−1787
C_5H_{12}	Pentane	300	−1234			390	−1578
		310	−1130			400	−1417
		320	−1038	$a(1) = -6553$		410	−1297
$a(1) = -1254$		330	−957	$a(2) = -32873$		420	−1214
$a(2) = -3345$		340	−884	$a(3) = -49874$		430	−1163
$a(3) = -2726$		350	−818	C_6H_{12}	Cyclohexane	300	−1698
		400	−579			320	−1391
		450	−436			340	−1170
		500	−348	$a(1) = -1733$		360	−1007
		550	−294	$a(2) = -5618$		380	−883
C_5H_{12}	2-Methylbutane	280	−1263	$a(3) = -9486$		400	−786
		290	−1166	$a(4) = -7936$		420	−707
		300	−1079			440	−641
$a(1) = -1095$		310	−1001			460	−584
$a(2) = -2503$		320	−931			480	−534
$a(3) = -1534$		330	−867			500	−488
		340	−810			520	−446
		350	−757			540	−406
		400	−557			560	−368
		450	−424	C_6H_{12}	Methylcyclopentane	305	−1447
C_5H_{12}	2,2-Dimethylpropane	300	−916			315	−1357
		310	−843			325	−1272
		320	−780	$a(1) = -1512$		335	−1192
$a(1) = -931$		330	−724	$a(2) = -2910$		345	−1117
$a(2) = -2387$		340	−674	C_6H_{14}	Hexane	300	−1920
$a(3) = -2641$		350	−629			310	−1724
$a(4) = -1810$		360	−590			320	−1561
		370	−554	$a(1) = -1961$		330	−1424
		380	−521	$a(2) = -6691$		340	−1309

Mol. form.	Name	T/K	B/cm^3 mol^{-1}	Mol. form.	Name	T/K	B/cm^3 mol^{-1}
	$a(3) = -13167$	350	−1209			500	−702
	$a(4) = -15273$	360	−1123			540	−583
		370	−1046			580	−490
		380	−978			620	−416
		390	−916			660	−355
		400	−859			700	−304
		410	−806	C$_8$H$_{10}$	1,2-Dimethylbenzene	380	−2046
		430	−707			390	−1848
		450	−616			400	−1681
C$_6$H$_{15}$N	Triethylamine	330	−1562		$a(1) = -5632$	410	−1543
		340	−1444		$a(2) = -22873$	420	−1428
		350	−1340		$a(3) = -28900$	430	−1335
	$a(1) = -2061$	360	−1249			440	−1261
	$a(2) = -5735$	370	−1169	C$_8$H$_{10}$	1,3-Dimethylbenzene	380	−2082
	$a(3) = -5899$	380	−1099			390	−1865
		390	−1037			400	−1679
		400	−983		$a(1) = -5808$	410	−1521
C$_7$H$_8$	Toluene	350	−1641		$a(2) = -23244$	420	−1388
		360	−1511		$a(3) = -27607$	430	−1276
		370	−1394			440	−1184
	$a(1) = -2620$	380	−1289	C$_8$H$_{10}$	1,4-Dimethylbenzene	380	−2043
	$a(2) = -7548$	390	−1195			390	−1851
	$a(3) = -6349$	400	−1110			400	−1680
		410	−1034		$a(1) = -4921$	410	−1529
		420	−965		$a(2) = -16843$	420	−1395
		430	−903		$a(3) = -16159$	430	−1276
C$_7$H$_{14}$	1-Heptene	340	−1781			440	−1171
		350	−1651	C$_8$H$_{16}$	1-Octene	360	−2147
		360	−1532			370	−2000
	$a(1) = -2491$	370	−1424			380	−1861
	$a(2) = -6230$	380	−1324		$a(1) = -3273$	390	−1729
	$a(3) = -3780$	390	−1233		$a(2) = -6557$	400	−1604
		400	−1150			410	−1485
		410	−1073	C$_8$H$_{18}$	Octane	300	−4042
C$_7$H$_{16}$	Heptane	300	−2782			350	−2511
		320	−2297			400	−1704
		340	−1928			450	−1234
	$a(1) = -2834$	360	−1641		$a(1) = -4123$	500	−936
	$a(2) = -8523$	380	−1415		$a(2) = -13120$	550	−732
	$a(3) = -10068$	400	−1233		$a(3) = -16408$	600	−583
	$a(4) = -5051$	420	−1085		$a(4) = -8580$	650	−468
		440	−963			700	−375
		460	−862				
		480	−775				

VAN DER WAALS CONSTANTS FOR GASES

The van der Waals equation of state for a real gas is

$$(P + n^2 a/V^2)(V - nb) = nRT$$

where P is the pressure, V the volume, T the temperature, n the amount of substance (in moles), and R the gas constant. The van der Waals constants a and b are characteristic of the substance and are independent of temperature. They are related to the critical temperature and pressure, T_c and P_c, by

$$a = 27R^2T_c^2/64P_c \quad b = RT_c/8P_c$$

This table gives values of a and b for some common gases. Most of the values have been calculated from the critical temperature and pressure values given in the table "Critical Constants" in this section. Van der Waals constants for other gases may easily be calculated from the data in that table.

To convert the van der Waals constants to SI units, note that 1 bar L²/mol² = 0.1 Pa m⁶/mol² and 1 L/mol = 0.001 m³/mol.

Reference

Reid, R. C, Prausnitz, J. M., and Poling, B. E., *The Properties of Gases and Liquids, Fourth Edition*, McGraw-Hill, New York, 1987.

Substance	*a* bar L²/mol²	*b* L/mol
Acetic acid	17.71	0.1065
Acetone	16.02	0.1124
Acetylene	4.516	0.0522
Ammonia	4.225	0.0371
Aniline	29.14	0.1486
Argon	1.355	0.0320
Benzene	18.82	0.1193
Bromine	9.75	0.0591
Butane	13.89	0.1164
1-Butanol	20.94	0.1326
2-Butanone	19.97	0.1326
Carbon dioxide	3.658	0.0429
Carbon disulfide	11.25	0.0726
Carbon monoxide	1.472	0.0395
Chlorine	6.343	0.0542
Chlorobenzene	25.80	0.1454
Chloroethane	11.66	0.0903
Chloromethane	7.566	0.0648
Cyclohexane	21.92	0.1411
Cyclopropane	8.34	0.0747
Decane	52.74	0.3043
1-Decanol	59.51	0.3086
Diethyl ether	17.46	0.1333
Dimethyl ether	8.690	0.0774
Dodecane	69.38	0.3758
1-Dodecanol	75.70	0.3750
Ethane	5.580	0.0651
Ethanol	12.56	0.0871
Ethylene	4.612	0.0582
Fluorine	1.171	0.0290
Furan	12.74	0.0926
Helium	0.0346	0.0238
Heptane	31.06	0.2049
1-Heptanol	38.17	0.2150
Hexane	24.84	0.1744
1-Hexanol	31.79	0.1856
Hydrazine	8.46	0.0462
Hydrogen	0.2452	0.0265
Hydrogen bromide	4.500	0.0442
Hydrogen chloride	3.700	0.0406
Hydrogen cyanide	11.29	0.0881
Hydrogen fluoride	9.565	0.0739
Hydrogen iodide	6.309	0.0530

Substance	*a* bar L²/mol²	*b* L/mol
Hydrogen sulfide	4.544	0.0434
Isobutane	13.32	0.1164
Krypton	5.193	0.0106
Methane	2.303	0.0431
Methanol	9.476	0.0659
Methylamine	7.106	0.0588
Neon	0.208	0.0167
Neopentane	17.17	0.1411
Nitric oxide	1.46	0.0289
Nitrogen	1.370	0.0387
Nitrogen dioxide	5.36	0.0443
Nitrogen trifluoride	3.58	0.0545
Nitrous oxide	3.852	0.0444
Octane	37.88	0.2374
1-Octanol	44.71	0.2442
Oxygen	1.382	0.0319
Ozone	3.570	0.0487
Pentane	19.09	0.1449
1-Pentanol	25.88	0.1568
Phenol	22.93	0.1177
Propane	9.39	0.0905
1-Propanol	16.26	0.1079
2-Propanol	15.82	0.1109
Propene	8.442	0.0824
Pyridine	19.77	0.1137
Pyrrole	18.82	0.1049
Silane	4.38	0.0579
Sulfur dioxide	6.865	0.0568
Sulfur hexafluoride	7.857	0.0879
Tetrachloromethane	20.01	0.1281
Tetrachlorosilane	20.96	0.1470
Tetrafluoroethylene	6.954	0.0809
Tetrafluoromethane	4.040	0.0633
Tetrafluorosilane	5.259	0.0724
Tetrahydrofuran	16.39	0.1082
Thiophene	17.21	0.1058
Toluene	24.86	0.1497
1,1,1-Trichloroethane	20.15	0.1317
Trichloromethane	15.34	0.1019
Trifluoromethane	5.378	0.0640
Trimethylamine	13.37	0.1101
Water	5.537	0.0305
Xenon	4.192	0.0516

MEAN FREE PATH AND RELATED PROPERTIES OF GASES

In the simplest version of the kinetic theory of gases, molecules are treated as hard spheres of diameter d which make binary collisions only. In this approximation the mean distance traveled by a molecule between successive collisions, the mean free path l, is related to the collision diameter by:

$$l = \frac{kT}{\pi\sqrt{2}Pd^2}$$

where P is the pressure, T the absolute temperature, and k the Boltzmann constant. At standard conditions ($P = 100,000$ Pa and $T = 298.15$ K) this relation becomes:

$$l = \frac{9.27 \cdot 10^{27}}{d^2}$$

where l and d are in meters.

Using the same model and the same standard pressure, the collision diameter can be calculated from the viscosity η by the kinetic theory relation:

$$\eta = \frac{2.67 \cdot 10^{-20}(MT)^{1/2}}{d^2}$$

where η is in units of μPa s and M is the molar mass in g/mol. Kinetic theory also gives a relation for the mean velocity v of molecules of mass m:

$$\bar{v} = \left(\frac{8kT}{\pi m}\right)^{1/2} = 145.5(T/M)^{1/2}\,\text{m/s}$$

Finally, the mean time τ between collisions can be calculated from the relation $\tau\bar{v} = l$.

The table below gives values of l, $\bar{v}$, and τ for some common gases at 25 °C and atmospheric pressure, as well as the value of d, all calculated from measured gas viscosities (see References 2 and 3 and the table "Viscosity of Gases" in this section). It is seen from the above equations that the mean free path varies directly with T and inversely with P, while the mean velocity varies as the square root of T and, in this approximation, is independent of P.

A more accurate model, in which molecular interactions are described by a Lennard-Jones potential, gives mean free path values about 5% lower than this table (see Reference 4).

References

1. Reid, R. C., Prausnitz, J. M., and Poling, B. E., *The Properties of Gases and Liquids, Fourth Edition*, McGraw-Hill, New York, 1987.
2. Lide, D. R., and Kehiaian, H. V., *CRC Handbook of Thermophysical and Thermochemical Data*, CRC Press, Boca Raton, FL, 1994.
3. Vargaftik, N. B., *Tables of Thermophysical Properties of Liquids and Gases, Second Edition*, John Wiley, New York, 1975.
4. Kaye, G. W. C., and Laby, T. H., *Tables of Physical and Chemical Constants, 15th Edition*, Longman, London, 1986.

Gas	d/m	l/m	$\bar{v}$/m s^{-1}	τ/ps
Air	$3.66 \cdot 10^{-10}$	$6.91 \cdot 10^{-8}$	467	148
Ar	3.58	7.22	397	182
CO_2	4.53	4.51	379	119
H_2	2.71	12.6	1769	71
He	2.15	20.0	1256	159
Kr	4.08	5.58	274	203
N_2	3.70	6.76	475	142
NH_3	4.32	4.97	609	82
Ne	2.54	14.3	559	256
O_2	3.55	7.36	444	166
Xe	4.78	4.05	219	185

INFLUENCE OF PRESSURE ON FREEZING POINTS

This table illustrates the variation of the freezing point of representative types of liquids with pressure. Substances are listed in alphabetical order. Note that 1 MPa = 0.01 kbar = 9.87 atm.

References

1. Isaacs, N. S., *Liquid Phase High Pressure Chemistry*, John Wiley, New York, 1981.
2. Merrill, L., *J. Phys. Chem. Ref. Data*, 6, 1205, 1977; 11, 1005, 1982.

Substance	Molecular formula	Freezing point in °C at:		
		0.1 MPa	100 MPa	1000 MPa
Acetic acid	$C_2H_4O_2$	16.6	37	
Acetophenone	C_8H_8O	20.0	41.2	
Aniline	C_6H_7N	−6.0	13.5	140
Benzene	C_6H_6	5.5	33.4	
Benzonitrile	C_7H_5N	−12.8	7.6	
Benzyl alcohol	C_7H_8O	−15.2	0.2	
Bromobenzene	C_6H_5Br	−30.6	−12	108
Bromoethane	C_2H_5Br	−118.6	−108	
1-Bromonaphthalene	$C_{10}H_7Br$	−1.8	6.1	
1-Bromopropane	C_3H_7Br	−110	−98	
p-Bromotoluene	C_7H_7Br	28.0	56.7	
Butanoic acid	$C_4H_8O_2$	−5.7	13.8	
1-Butanol	$C_4H_{10}O$	−89.8	−77.2	
Carbon disulfide	CS_2	−111.5	−98	
Chlorobenzene	C_6H_5Cl	−45.2	−28	84
p-Chlorotoluene	C_7H_7Cl	6.9	33.1	
o-Cresol	C_7H_8O	29.8	47.7	
m-Cresol	C_7H_8O	11.8	25.6	
p-Cresol	C_7H_8O	35.8	56.2	
Cyclohexane	C_6H_{12}	6.6	32.5	
Cyclohexanol	$C_6H_{12}O$	25.5	62.3	
1,2-Dibromoethane	$C_2H_4Br_2$	9.9	34.0	
p-Dichlorobenzene	$C_6H_4Cl_2$	52.7	79.1	
Dichloromethane	CH_2Cl_2	−95.1	−83	
N,N-Dimethylaniline	$C_8H_{11}N$	2.5	26.3	
1,4-Dioxane	$C_4H_8O_2$	11	23	
Ethanol	C_2H_6O	−114.1	−108	
Formamide	CH_3NO	−15.5	10.8	
Formic acid	CH_2O_2	8.3	20.6	
Furan	C_4H_4O	−85.6	−73	
Hexamethyldisiloxane	$C_6H_{18}OSi_2$	−66	−37	
Menthol	$C_{10}H_{20}O$	42	60	
Methyl benzoate	$C_8H_8O_2$	−15	31.8	
2-Methyl-2-butanol	$C_5H_{12}O$	−8.8	13.4	
2-Methyl-2-propanol	$C_4H_{10}O$	25.4	58.1	
Naphthalene	$C_{10}H_8$	78.2	115.7	
Nitrobenzene	$C_6H_5NO_2$	5.7	13.5	
m-Nitrotoluene	$C_7H_7NO_2$	15.5	40.6	
Pentachloroethane	C_2HCl_5	−29.0	−6.3	
Potassium	K	63.7	78	170
Potassium chloride	ClK	771		945
Propanoic acid	$C_3H_6O_2$	−20.7	−1.2	
Silver chloride	AgCl	455		545
Sodium	Na	97.8	106	167
Sodium chloride	ClNa	800.7		997
Sodium fluoride	FNa	996		1115
Tetrachloromethane	CCl_4	−23.0	14.2	
Tribromomethane	$CHBr_3$	8.1	31.5	
Trichloromethane	$CHCl_3$	−63.6	−45.2	
Water	H_2O	0.0	−9.0	
o-Xylene	C_8H_{10}	−25.2	−3.5	
m-Xylene	C_8H_{10}	−47.8	−25.2	
p-Xylene	C_8H_{10}	13.2	46.0	

CRITICAL CONSTANTS OF ORGANIC COMPOUNDS

Chris D. Muzny, Vladimir Diky, Andrei Kazakov, Robert D. Chirico, and Michael Frenkel

The parameters of the liquid-gas critical point are important constants in determining the behavior of fluids. This table lists the critical temperature, pressure, and molar volume, as well as the normal boiling point, for over 850 organic substances. The properties and their units are:

T_b: Normal boiling point in K at a pressure of 101.325 kPa (1 atmosphere); an "s" following the value indicates a sublimation point (temperature at which the solid is in equilibrium with the gas at a pressure of 101.325 kPa)

T_c: Critical temperature in K

P_c: Critical pressure in MPa

V_c: Critical molar volume in cm³ mol⁻¹

The listed values of the critical constants are critically evaluated using the NIST ThermoData Engine, TDE (Ref. 1), designed to implement the dynamic data evaluation concept (Refs. 2–5). This concept requires large electronic databases capable of storing essentially all relevant experimental data known to date with detailed descriptions of metadata and uncertainties. The combination of these electronic databases with expert-system software, designed to automatically generate recommended property values based on available experimental and predicted data, leads to the ability to produce critically evaluated data dynamically or "to order." The evaluated data have been generated only for compounds for which experimental data for critical properties are available. Group contribution methods such as Joback-Reed (Ref. 6), Constantinou-Gani (Ref. 7), Marrero-Pardillo (Ref. 8), and Wilson-Jasperson (Ref. 9) as well as quantitative structure-property relationship (QSPR) methods (Ref. 5) were used within the TDE environment to validate available experimental data. Each recommended value in the table is characterized with a combined expanded uncertainty (Ref. 10) (level of confidence, approximately 95%) listed in parentheses. Only references to original experimental data actually used by TDE to generate critically evaluated data are indicated for each compound. Compounds are listed alphabetically by name.

The values of the normal boiling temperatures provided in the table along with the combined expanded uncertainties listed in parentheses have also been critically evaluated using TDE. Additional details on the determination of the normal boiling temperatures using TDE can be found in the Physical Constants of Organic Compounds table in Section 3. The remaining values of the normal boiling temperatures (without uncertainties) are taken from the compilation presented in the 91st Edition of the CRC *Handbook of Chemistry and Physics*.

Name	Mol. Form.	T_b/K	T_c/K	P_c/MPa	V_c/cm³ mol⁻¹	Ref.
Acetaldehyde	C_2H_4O	293.9(0.6)	462(8)	7.5(1)	154(5)	11–13
Acetic acid	$C_2H_4O_2$	391.0(0.2)	593(2)	5.79(0.03)	171(2)	14–21
Acetic anhydride	$C_4H_6O_3$	412.6(0.3)	606(1)	4.00(0.08)	294(12)	22
Acetone	C_3H_6O	329.23(0.07)	508.1(0.2)	4.7(0.1)	221(20)	18, 23–31
Acetonitrile	C_2H_3N	354.8(0.2)	545.47(0.07)	4.88(0.01)	173(59)	32–40
Acetophenone	C_8H_8O	475.2(0.2)	709.5(0.7)	4.01(0.05)	373(40)	11, 41
Acetylene	C_2H_2	188.45 s	308.4(0.4)	6.24(0.04)	119(11)	42–48
Acrylonitrile	C_3H_3N	350.3(0.2)	540(2)	4.6(0.1)	211(10)	49
Allene	C_3H_4	238.3(0.3)	394(4)	6.5(0.7)	167(8)	50
Allyl alcohol	C_3H_6O	370.0(0.5)	539.8(0.6)	5.76(0.04)	222(9)	51
Allylamine	C_3H_7N	327(2)	540.0(0.7)	4.83(0.03)	217(11)	52
Allyl ethyl ether	$C_5H_{10}O$	338(4)	518(10)	3(2)	320(10)	21
2-Aminobiphenyl	$C_{12}H_{11}N$	571.4(0.2)	838(2)	3.52(0.03)	548(99)	53
2-Aminoethanol	C_2H_7NO	443.4(0.4)	671(3)	8.0(0.5)	207(13)	41
2-(2-Aminoethoxy)ethanol	$C_4H_{11}NO_2$	496.2(0.1)	721(4)	4.88(0.1)	333(19)	54
N-(2-Aminoethyl)ethanolamine	$C_4H_{12}N_2O$	515(5)	739(2)	4.53(0.09)	340(17)	55
Aminoimidazole ribotide	$C_8H_{14}N_3O_7P$		132.5(0.6)			56
Amyl orthosilicate	$C_{20}H_{44}O_4Si$		714(14)			57
Aniline	C_6H_7N	457.2(0.4)	704(7)	5.3(0.1)	291(3)	39, 40, 58–60
Anisole	C_7H_8O	426.8(0.2)	646.1(0.2)	4.2(0.1)	355(12)	25, 39, 40, 49
Benzene	C_6H_6	353.23(0.07)	562.0(0.1)	4.90(0.02)	257(11)	19, 27, 28, 34, 61–97
Benzeneacetic acid	$C_8H_8O_2$	541(2)	766(8)	3.9(0.3)	372(16)	98
Benzenebutanoic acid	$C_{10}H_{12}O_2$	569(2)	783(8)	3.2(0.2)	493(18)	98
Benzeneethanol	$C_8H_{10}O$	493(3)	724(4)	4.0(0.2)	390(15)	99
Benzeneheptanoic acid	$C_{13}H_{18}O_2$	585(27)	798(8)	2.5(0.3)	662(21)	98
Benzenehexanoic acid	$C_{12}H_{16}O_2$	574(27)	794(8)	2.6(0.3)	611(20)	98
Benzenepentanoic acid	$C_{11}H_{14}O_2$	583(1)	790(8)	3.1(0.2)	526(19)	98
Benzenepropanoic acid	$C_9H_{10}O_2$	557(2)	776(8)	3.5(0.2)	440(17)	98
Benzo[b]thiophene	C_8H_6S	494.0(0.4)	764(2)	4.68(0.04)	359(59)	100
Benzonitrile	C_7H_5N	464(1)	691(9)	4.22(0.04)	348(7)	39, 40, 101, 102
Benzophenone	$C_{13}H_{10}O$	579.0(0.2)	830(2)	3.0(0.1)	568(44)	101
Benzyl alcohol	C_7H_8O	478.4(0.2)	715(3)	4.3(0.2)	333(13)	103
[1,1'-Bicyclohexyl]-2-one	$C_{12}H_{20}O$	537	787(70)	3(7)	584(20)	104

Name	Mol. Form.	T_b/K	T_c/K	P_c/MPa	V_c/cm^3 mol^{-1}	Ref.
1,1'-Bicyclopentyl	C$_{10}$H$_{18}$	463.61(0.03)	690(2)	3.27(0.03)	497(41)	105
Biphenyl	C$_{12}$H$_{10}$	528.3(0.3)	773(5)	3.43(0.06)	481(69)	39, 106–109
Bis(2-aminoethyl)amine	C$_4$H$_{13}$N$_3$	479.6(0.3)	710(2)	4.43(0.07)	350(21)	55
1,1-Bis(difluoromethoxy)-1,2,2,2-tetrafluoroethane	C$_4$H$_2$F$_8$O$_2$	319.78(0.08)	450(1)	2.40(0.08)	410(4)	110
Bis(difluoromethyl) ether	C$_2$H$_2$F$_4$O	278.6(0.4)	420.2(0.1)	4.16(0.06)	223(13)	111
Bis(2-ethylhexyl) phthalate	C$_{24}$H$_{38}$O$_4$	657	835(9)	1.1(0.2)	1495(27)	112
Bis(2-hydroxyethyl)methylamine	C$_5$H$_{13}$NO$_2$	518(1)	742(4)	4.2(0.4)	404(17)	54
Bis(2,2,2-trifluoroethyl) ether	C$_4$H$_4$F$_6$O	336.91	476.31(0.09)	2.78(0.01)	365(1)	113
Bis(trimethylsilyl)methane	C$_7$H$_{20}$Si$_2$	406	573.9(0.3)			114
Bromochlorodifluoromethane	CBrClF$_2$	269.2(0.7)	428(12)	4.31(0.01)	229(16)	115
Bromodifluoromethane	CHBrF$_2$	257.5(0.5)	412.0(0.3)	5.2(0.1)	173(18)	116
Bromoethane	C$_2$H$_5$Br	311.3(0.6)	503.9(0.4)	6.2(0.1)	214(10)	117, 118
1-Bromo-2-fluorobenzene	C$_6$H$_4$BrF	427	669.6(0.6)	4.3(0.6)	342(18)	119
1-Bromo-3-fluorobenzene	C$_6$H$_4$BrF	423	652.0(0.4)	4.2(0.6)	337(18)	119
1-Bromo-4-fluorobenzene	C$_6$H$_4$BrF	423(2)	654.8(0.4)	4.2(0.2)	338(18)	119
1-Bromopropane	C$_3$H$_7$Br	343.9(0.2)	536.9(0.1)	4.33(0.06)	271(6)	120
Bromotrifluoromethane	CBrF$_3$	215.3(0.4)	340.06(0.05)	3.96(0.01)	199(6)	121
1-Bromo-2-(trifluoromethyl)benzene	C$_7$H$_4$BrF$_3$	440.7	656.5(0.4)	3.3(0.8)	415(24)	119
1-Bromo-3-(trifluoromethyl)benzene	C$_7$H$_4$BrF$_3$	424.7	627.1(0.4)	3.2(0.7)	413(24)	119
1-Bromo-4-(trifluoromethyl)benzene	C$_7$H$_4$BrF$_3$	433	629.8(0.4)	3.2(0.8)	413(24)	119
1,3-Butadiene	C$_4$H$_6$	268.5(0.2)	425(1)	4.35(0.07)	221(23)	122, 123
Butanal	C$_4$H$_8$O	347.9(0.2)	537(2)	4.41(0.1)	258(9)	104, 124
Butane	C$_4$H$_{10}$	272.6(0.5)	425.2(0.1)	3.79(0.01)	257(4)	29, 125–139
1,4-Butanediamine	C$_4$H$_{12}$N$_2$	429(10)	651(7)	4.5(0.5)	317(14)	140
1,2-Butanediol	C$_4$H$_{10}$O$_2$	469.57(0.06)	680(2)	5.4(0.1)	298(12)	141
1,3-Butanediol	C$_4$H$_{10}$O$_2$	481.3(0.1)	679(17)	4.7(0.1)	302(67)	54, 141
1,4-Butanediol	C$_4$H$_{10}$O$_2$	502.6(0.4)	724(4)	5.5(0.2)	307(14)	49, 55
Butanenitrile	C$_4$H$_7$N	390.8(0.4)	585.40(0.07)	3.82(0.05)	265(6)	35, 39, 40, 142
1-Butanethiol	C$_4$H$_{10}$S	371.5(0.5)	570.1(0.6)	4.01(0.02)	324(12)	143, 144
Butanoic acid	C$_4$H$_8$O$_2$	436.8(0.1)	623(6)	4.0(0.3)	292(10)	145–147
1-Butanol	C$_4$H$_{10}$O	390.8(0.2)	563.0(0.4)	4.43(0.07)	280(14)	34, 80, 148–155
2-Butanol	C$_4$H$_{10}$O	372.5(0.2)	535(4)	4.2(0.1)	269(4)	148, 149, 153, 156
2-Butanone	C$_4$H$_8$O	352.8(0.2)	537(1)	4.18(0.02)	274(30)	25, 30, 33, 34, 38, 157
1-Butene	C$_4$H$_8$	276.87(0.08)	419.3(0.1)	4.00(0.05)	236(14)	123, 158–161
cis-2-Butene	C$_4$H$_8$	274.03(0.09)	435.7(0.2)	4.23(0.02)	235(4)	86, 123, 158
trans-2-Butene	C$_4$H$_8$	266.8(0.2)	428.6(0.1)	4.03(0.02)	238(4)	86, 123, 158
2-Butoxyethanol	C$_6$H$_{14}$O$_2$	444(2)	633.9(1)	3.3(0.1)	424(15)	11, 41
1-tert-Butoxy-2-ethoxyethane	C$_8$H$_{18}$O$_2$	421.2	585(3)	2.5(0.4)	546(14)	162
2-Butoxyethyl acetate	C$_8$H$_{16}$O$_3$	464.2(0.9)	640(2)	2.7(0.2)	551(21)	144, 163
1-tert-Butoxy-2-methoxyethane	C$_7$H$_{16}$O$_2$	404(15)	574(1)	2.8(0.7)	480(13)	162
1-Butoxy-2-propanol	C$_7$H$_{16}$O$_2$	445(3)	625(1)	2.7(0.1)	479(20)	32
Butyl acetate	C$_6$H$_{12}$O$_2$	381(4)	578(10)	3.16(0.06)	403(6)	162, 164–167
sec-Butyl acetate	C$_6$H$_{12}$O$_2$	371(1)	571.1(0.5)	3.01(0.1)	398(14)	164, 165
tert-Butyl acetate	C$_6$H$_{12}$O$_2$	399.1(0.1)	541(4)	3.0(0.1)	399(7)	54
Butyl acrylate	C$_7$H$_{12}$O$_2$	419.8(0.6)	597.4(0.6)	2.76(0.03)	445(7)	51
Butylamine	C$_4$H$_{11}$N	335.86(0.08)	531.9(0.2)	4.20(0.04)	291(13)	168
sec-Butylamine	C$_4$H$_{11}$N	317.17(0.07)	514.3(0.2)	4.0(0.2)	284(11)	168
tert-Butylamine	C$_4$H$_{11}$N	350.1(0.2)	483.7(0.6)	3.85(0.06)	293(23)	75
Butylbenzene	C$_{10}$H$_{14}$	446.4(0.4)	660.5(0.1)	2.89(0.03)	498(18)	79, 86, 88
sec-Butylbenzene	C$_{10}$H$_{14}$	442.2(0.3)	652(1)	2.94(0.03)	488(39)	58
tert-Butylbenzene	C$_{10}$H$_{14}$	456.4(0.3)	648(1)	3.00(0.03)	474(30)	58, 64
Butyl benzoate	C$_{11}$H$_{14}$O$_2$	522(3)	725(14)	2.4(0.3)	594(10)	169
Butyl butanoate	C$_8$H$_{16}$O$_2$	438.1(0.1)	612(3)	2.4(0.2)	550(9)	162
Butylcyclohexane	C$_{10}$H$_{20}$	444.8(0.4)	653.1(0.4)	2.57(0.07)	547(14)	170, 171
tert-Butylcyclohexane	C$_{10}$H$_{20}$	454.0(0.6)	652.0(0.4)	2.82(0.09)	537(15)	170
tert-Butyl ethyl ether	C$_6$H$_{14}$O	345.8(0.1)	509(2)	3.0(0.2)	394(4)	172
Butyl methyl ether	C$_5$H$_{12}$O	343.2(0.3)	512.7(0.1)	3.37(0.02)	340(2)	25, 173, 174
Butyl propanoate	C$_7$H$_{14}$O$_2$	418.2(0.1)	594(1)	2.8(0.2)	464(10)	162
Butyl vinyl ether	C$_6$H$_{12}$O	367(1)	540(1)	3.12(0.01)	379(10)	175
γ-Butyrolactone	C$_4$H$_6$O$_2$	477.8(0.4)	731(1)	5(1)	246(18)	49

Name	Mol. Form.	T_b/K	T_c/K	P_c/MPa	V_c/cm³mol⁻¹	Ref.
Chlorobenzene	C_6H_5Cl	404.8(0.2)	632.4(0.1)	4.5(0.1)	303(79)	19, 33, 34
1-Chlorobutane	C_4H_9Cl	351.5(0.2)	539.2(0.6)	4.1(0.2)	303(11)	119
2-Chlorobutane	C_4H_9Cl	341.4	518.6(0.6)	3.4(0.2)	307(14)	119
1-Chloro-2,4-difluorobenzene	$C_6H_3ClF_2$	400	609.6(0.4)	4.0(0.7)	333(16)	119
1-Chloro-2,5-difluorobenzene	$C_6H_3ClF_2$	401	612.5(0.4)	4.0(0.7)	333(16)	119
1-Chloro-3,4-difluorobenzene	$C_6H_3ClF_2$	400	609.2(0.4)	4.0(0.6)	333(16)	119
1-Chloro-3,5-difluorobenzene	$C_6H_3ClF_2$	391.7	592.0(0.4)	3.9(0.7)	327(16)	119
1-Chloro-1,1-difluoroethane	$C_2H_3ClF_2$	264.03(0.07)	410.31(0.05)	4.06(0.03)	230(6)	113, 176–178
1-Chloro-2,2-difluoroethene	C_2HClF_2	254.3(0.5)	400.5(0.7)	4.54(0.07)	197(6)	179
Chlorodifluoromethane	$CHClF_2$	232.3(0.5)	369.30(0.05)	4.98(0.01)	165(2)	180–191
2-Chloro-2-(difluoromethoxy)-1,1,1-trifluoroethane	$C_3H_2ClF_5O$	322.4(0.1)	467.8(0.6)	3.05(0.03)	316(24)	192
Chloroethane	C_2H_5Cl	285.4(0.2)	460.3(0.4)	5.24(0.04)	198(11)	193
Chloroethene	C_2H_3Cl	259.3(0.3)	425(5)	5.60(0.03)	171(9)	194
1-Chloro-2-fluorobenzene	C_6H_4ClF	410.8	633.8(0.4)	4.3(0.6)	319(21)	119
1-Chloro-3-fluorobenzene	C_6H_4ClF	401(25)	615.9(0.4)	4.2(0.6)	324(21)	119
1-Chloro-4-fluorobenzene	C_6H_4ClF	403	620.1(0.4)	4.2(0.4)	322(18)	119
1-Chloroheptane	$C_7H_{15}Cl$	432(2)	614(8)	3.1(0.6)	492(14)	119
1-Chlorohexane	$C_6H_{13}Cl$	408.1(0.5)	599(3)	3.3(0.3)	422(12)	119
Chloromethane	CH_3Cl	249.0(0.3)	416.24(0.04)	6.72(0.03)	136(2)	195, 196
2-Chloro-2-methylbutane	$C_5H_{11}Cl$	358(1)	509.1(0.6)	3.2(0.5)	397(15)	119
3-Chloro-3-methylpentane	$C_6H_{13}Cl$	389	528(3)	3(1)	414(14)	119
2-Chloro-2-methylpropane	C_4H_9Cl	324.0(0.5)	497.8(0.1)	3.7(0.4)	308(13)	34
1-Chlorooctane	$C_8H_{17}Cl$	456(3)	643(2)	2.5(0.4)	543(13)	119
Chloropentafluoroacetone	C_3ClF_5O	280.9(0.9)	410.6(0.1)	2.89(0.01)	277(21)	197
Chloropentafluorobenzene	C_6ClF_5	391.11	570(1)	3.2(0.2)	367(25)	198
Chloropentafluoroethane	C_2ClF_5	233.9(0.2)	353.0(0.2)	3.141(0.01)	255(4)	199, 200
1-Chloropentane	$C_5H_{11}Cl$	381.0(0.3)	571.2(0.4)	3.3(0.2)	361(12)	119
1-Chloropropane	C_3H_7Cl	319.3(0.5)	503.3(0.4)	4.56(0.04)	268(24)	118, 193, 201, 202
2-Chloropropane	C_3H_7Cl	308.1(0.6)	482.4(0.4)	4.25(0.04)	245(16)	201, 202
1-Chloro-1,2,2,2-tetrafluoroethane	C_2HClF_4	261.19(0.09)	395.43(0.06)	3.62(0.01)	244(4)	203, 204
4-Chlorotoluene	C_7H_7Cl	435.0(0.2)	615.9(0.5)	2.33(0.09)	377(16)	88
2-Chloro-1,1,1-trifluoroethane	$C_2H_2ClF_3$	279.1(0.6)	425.0(0.2)	4.02(0.02)	232(6)	205
Chlorotrifluoroethene	C_2ClF_3	244.8(0.3)	380.1(0.1)	3.95(0.03)	214(12)	206, 207
2-Chloro-1,1,2-trifluoroethyl difluoromethyl ether	$C_3H_2ClF_5O$	329.9(0.5)	475.0(0.6)	2.98(0.03)	343(25)	192
Chlorotrifluoromethane	$CClF_3$	191.8	301.9(0.2)	3.89(0.01)	180.3(1)	191, 208–216
m-Cresol	C_7H_8O	475.3(0.1)	705.8(0.4)	4.4(0.2)	337(12)	39, 40, 217, 218
o-Cresol	C_7H_8O	464.1(0.1)	697.6(0.2)	4.2(0.2)	336(12)	217, 218
p-Cresol	C_7H_8O	475.0(0.1)	704.6(0.3)	4.1(0.2)	349(13)	217, 218
Cyanogen	C_2N_2	252.1	397(3)	6.2(0.4)	149(8)	219
Cycloheptane	C_7H_{14}	391.9(0.2)	604.2(0.1)	3.85(0.04)	361(12)	34, 220, 221
Cyclohexane	C_6H_{12}	353.8(0.7)	553.4(0.3)	4.07(0.01)	307(12)	19, 34, 78, 82, 84, 88, 90, 144, 163, 170, 171, 220, 222–231
Cyclohexanol	$C_6H_{12}O$	434.0(0.2)	647.1(0.3)	4.3(0.1)	334(33)	49, 232, 233
Cyclohexanone	$C_6H_{10}O$	428.5(0.1)	665(1)	4.61(0.09)	354(12)	162, 166
Cyclohexene	C_6H_{10}	356.0(0.2)	560.45(0.02)	4.43(0.08)	290(20)	86, 201, 202
Cyclohexylamine	$C_6H_{13}N$	406.8(0.5)	626.8(0.9)	3.9(0.6)	349(15)	170
Cyclooctane	C_8H_{16}	424.2(0.1)	647.2(0.4)	3.55(0.06)	417(28)	34, 170, 220, 221
Cyclopentane	C_5H_{10}	322.3(0.1)	511.7(0.2)	4.51(0.07)	264(10)	34, 78, 90, 234–236
Cyclopentanol	$C_5H_{10}O$	413.5(0.2)	619(1)	4.9(0.1)	288(19)	233
Cyclopentanone	C_5H_8O	403.6(0.2)	624(2)	4.59(0.05)	276(17)	233
Cyclopentene	C_5H_8	317.3(0.2)	506.1(0.2)	4.78(0.05)	252(16)	90, 144, 163, 201, 202
Cyclopropane	C_3H_6	242(2)	398.2(0.4)	5.58(0.02)	164(9)	237, 238
Decafluorobiphenyl	$C_{12}F_{10}$	480(2)	640(4)	2.3(0.3)	641(40)	239
1,1,1,2,2,3,4,5,5,5-Decafluoro-3,4-bis(trifluoromethyl)-pentane	C_7F_{16}				632(5)	240
1,1,1,2,2,3,4,5,5,5-Decafluoro-3-[1,2,2,2-tetrafluoro-1-(trifluoromethyl)ethyl]-4-(trifluoromethyl)pentane	C_9F_{20}	399(24)	581(43)	1(2)	800(7)	240
cis-Decahydronaphthalene	$C_{10}H_{18}$	469.0(0.3)	702(1)	3.2(0.3)	492(19)	241

Name	Mol. Form.	T_b/K	T_c/K	P_c/MPa	V_c/cm³ mol⁻¹	Ref.
trans-Decahydronaphthalene	C₁₀H₁₈	460.4(0.2)	687(1)	3.1(0.1)	499(19)	241
Decamethylcyclopentasiloxane	C₁₀H₃₀O₅Si₅	486(3)	617.4(0.3)	1.04(0.02)	1201(2)	34, 114
Decanal	C₁₀H₂₀O	485(3)	674(1)	2.6(0.3)	601(14)	163, 242
Decane	C₁₀H₂₂	447.2(0.1)	618.1(0.9)	2.10(0.03)	621(35)	16, 33, 65, 78, 86, 131, 243–252
1,10-Decanediamine	C₁₀H₂₄N₂	535(12)	736(8)	2.4(0.3)	654(28)	140
Decanedioic acid	C₁₀H₁₈O₄	647(5)	845(13)	2.5(0.1)	724(21)	253
Decanoic acid	C₁₀H₂₀O₂	543(1)	724(5)	1.9(0.8)	638(24)	145, 146
1-Decanol	C₁₀H₂₂O	502(3)	690(10)	2.3(0.1)	624(87)	149, 152, 254, 255
2-Decanol	C₁₀H₂₂O	484	668.5(0.3)	2.3(0.5)	646(13)	255
3-Decanol	C₁₀H₂₂O	490(7)	666.1(0.3)	2.3(0.3)	643(13)	255
4-Decanol	C₁₀H₂₂O	487(3)	663.7(0.3)	2.3(0.1)	643(13)	255
5-Decanol	C₁₀H₂₂O	489(5)	663.2(0.4)	2.3(0.4)	646(13)	255
2-Decanone	C₁₀H₂₀O	484(3)	671.8(0.5)	2.2(0.3)	625(25)	256
3-Decanone	C₁₀H₂₀O	485(4)	668(1)	2.2(0.2)	628(15)	256
4-Decanone	C₁₀H₂₀O	479.7	662.9(0.5)	2.2(0.2)	636(18)	256
5-Decanone	C₁₀H₂₀O	477	661.0(0.4)	2.2(0.2)	628(25)	256
1-Decene	C₁₀H₂₀	444(1)	616.0(0.3)	2.157(0.01)	594(3)	257
Decylbenzene	C₁₆H₂₆	571(1)	752(8)	1.72(0.1)	879(29)	258
Dibenzofuran	C₁₂H₈O	558.3(0.3)	824(2)	3.37(0.03)	494(32)	259
Dibenzothiophene	C₁₂H₈S	604.8(0.4)	897(2)	3.9(0.2)	506(108)	260
1,2-Dibromo-1-chloro-1,2,2-trifluoroethane	C₂Br₂ClF₃	365.9(0.2)	560.6(0.2)	3.61(0.02)	368(4)	261
1,4-Dibromooctafluorobutane	C₄Br₂F₈	371(25)	532(2)	2.4(0.3)	452(29)	198
Dibutylamine	C₈H₁₉N	435(2)	607.5(0.2)	3.11(0.03)	532(21)	168
1,4-Di-*tert*-butylbenzene	C₁₄H₂₂	510.4(0.5)	708(2)	2.23(0.01)	732(70)	232
Dibutyl ether	C₈H₁₈O	414.8(0.3)	584.1(0.2)	2.4(0.2)	521(12)	262
Dibutyl phthalate	C₁₆H₂₂O₄	611(9)	797(9)	1.6(0.3)	954(18)	112
m-Dichlorobenzene	C₆H₄Cl₂	445(2)	685.7(0.4)	4.2(0.2)	366(22)	119
1,4-Dichlorobenzene	C₆H₄Cl₂	447.0(0.2)	669(5)	3.54(0.07)	364(22)	263
Dichlorodiethylsilane	C₄H₁₀Cl₂Si	403(2)	595.7(0.6)	3.06(0.03)	455(4)	264
Dichlorodifluoromethane	CCl₂F₂	243.3(0.1)	384.9(0.2)	4.12(0.01)	218(36)	191, 216, 265
Dichlorodimethylsilane	C₂H₆Cl₂Si	343.6(0.5)	520.3(0.6)	3.49(0.03)	350(5)	266
1,1-Dichloroethane	C₂H₄Cl₂	329.4(0.7)	523.4(0.1)	5.1(0.5)	248(12)	267
1,2-Dichloroethane	C₂H₄Cl₂	356.5(0.1)	561.5(0.4)	5.4(0.1)	225(8)	33, 34, 268, 269
cis-1,2-Dichloroethene	C₂H₂Cl₂	333(2)	535.8(0.4)	5.4(0.3)	220(15)	34
trans-1,2-Dichloroethene	C₂H₂Cl₂	320.79(0.08)	515.5(0.2)	5.3(0.2)	216(14)	33, 34
1,1-Dichloro-1-fluoroethane	C₂H₃Cl₂F	305.20(0.09)	477.3(0.1)	4.20(0.02)	253.7(0.6)	178, 270
Dichlorofluoromethane	CHCl₂F	282.1	451.6(0.4)	5.20(0.01)	196(1)	271
1,2-Dichlorohexafluoropropane	C₃Cl₂F₆	307(2)	451.8(0.1)	2.63(0.07)	365(48)	272
Dichloromethane	CH₂Cl₂	312.9(0.3)	508.0(0.2)	6.35(0.05)	177(13)	273
1,2-Dichloropropane	C₃H₆Cl₂	369.6	578(2)	4.63(0.06)	292(6)	119, 232
1,3-Dichloropropane	C₃H₆Cl₂	393.9(0.3)	615(3)	4.7(0.5)	299(16)	119
1,1-Dichloro-1,2,2,2-tetrafluoroethane	C₂Cl₂F₄	276(1)	418.6(0.8)	3.31(0.03)	294(8)	179
1,2-Dichloro-1,1,2,2-tetrafluoroethane	C₂Cl₂F₄	276.8(0.5)	418.74(0.06)	3.25(0.02)	295(3)	191, 274
1,2-Dichloro-1,1,2-trifluoroethane	C₂HCl₂F₃	303.1(0.1)	461.6(0.1)	3.77(0.08)	283.2(0.5)	178
2,2-Dichloro-1,1,1-trifluoroethane	C₂HCl₂F₃	300.9(0.6)	456.8(0.2)	3.67(0.01)	278(2)	182, 275–280
Didecyl phthalate	C₂₈H₄₆O₄	736(4)	870(10)	0.94(0.05)	1807(27)	112
1,1-Diethoxyethane	C₆H₁₄O₂	375(2)	539.7(0.4)	3.22(0.08)	426(12)	166
1,2-Diethoxyethane	C₆H₁₄O₂	393.8(0.7)	542(3)	2.14(0.02)	432(11)	162
Diethoxymethane	C₅H₁₂O₂	359(2)	532(1)	3.4(0.5)	370(10)	162
Diethylamine	C₄H₁₁N	328.5(0.1)	499.5(0.4)	3.75(0.02)	304(32)	38, 117, 133, 281, 282
p-Diethylbenzene	C₁₀H₁₄	457(1)	657.90(0.03)	2.80(0.08)	494(12)	79, 86
Diethylene glycol	C₄H₁₀O₃	518.6(0.2)	753(4)	4.8(0.2)	325(19)	283
Diethylene glycol diethyl ether	C₈H₁₈O₃	458(4)	612(10)	2.4(0.7)	587(18)	162
Diethylene glycol dimethyl ether	C₆H₁₄O₃	435(2)	617(4)	3.0(0.6)	450(16)	162
Diethylene glycol monobutyl ether	C₈H₁₈O₃	505(4)	692(3)	2.8(0.6)	546(26)	41
Diethylene glycol monobutyl ether acetate	C₁₀H₂₀O₄	521(2)	694(2)	2.15(0.05)	627(20)	55
Diethylene glycol monoethyl ether	C₆H₁₄O₃	475(3)	670(4)	3.2(0.1)	427(23)	49
Diethylene glycol monoethyl ether acetate	C₈H₁₆O₄	491(1)	670(12)	2.50(0.06)	524(18)	49, 55
Diethylene glycol monomethyl ether	C₅H₁₂O₃	467(2)	672(2)	3.7(0.2)	378(25)	49
Diethylene glycol monopropyl ether	C₇H₁₆O₃	488.0(0.4)	680(2)	3.05(0.07)	495(18)	41, 104

Name	Mol. Form.	T_b/K	T_c/K	P_c/MPa	V_c/cm^3 mol^{-1}	Ref.
Diethyl ether	$C_4H_{10}O$	307.5(0.5)	466.8(0.3)	3.64(0.01)	280(5)	19, 77, 95, 155, 196, 284–301
Diethyl oxalate	$C_6H_{10}O_4$	459(1)	618(2)	2.14(0.02)	464(37)	101
Diethyl phthalate	$C_{12}H_{14}O_4$	571(2)	776(9)	2.2(0.2)	687(15)	112
Diethyl succinate	$C_8H_{14}O_4$	490(1)	663(30)	2.26(0.02)	567(44)	101, 302
Diethyl sulfide	$C_4H_{10}S$	365.2(0.2)	557.5(1)	4.0(0.1)	322(8)	32, 303, 304
m-Difluorobenzene	$C_6H_4F_2$	356.1(0.5)	548.4(0.4)	4.20(0.01)	289(21)	119
o-Difluorobenzene	$C_6H_4F_2$	367.0(0.5)	566.0(0.4)	4.28(0.01)	290(21)	119
p-Difluorobenzene	$C_6H_4F_2$	362.0(0.3)	556.9(0.4)	4.28(0.07)	297(22)	119
1,1-Difluoroethane	$C_2H_4F_2$	249.1	386.4(0.1)	4.52(0.01)	178(2)	178, 180, 185, 305–307
1,1-Difluoroethene	$C_2H_2F_2$	187.6(0.8)	302.9(0.6)	4.48(0.05)	155(4)	179, 215
2,2-Difluoroethylbis(trifluoromethyl)amine	$C_4H_3F_8N$	324.5	460.20(0.09)	2.64(0.01)	375(1)	308
Difluoromethane	CH_2F_2	221.50(0.07)	351.28(0.03)	5.79(0.01)	121(4)	306, 309–316
3-Difluoromethoxy-1,1,1,2,2-pentafluoropropane	$C_4H_3F_7O$	319.09	455.1(0.1)	2.77(0.02)	363(1)	113
2-(Difluoromethoxy)-1,1,1-trifluoroethane	$C_3H_3F_5O$	302.3(0.2)	444.9(0.3)	3.43(0.01)	291(19)	113
2,4-Difluorotoluene	$C_7H_6F_2$	390	581.4(0.4)	3.7(0.4)	340(21)	119
2,5-Difluorotoluene	$C_7H_6F_2$	391	587.8(0.4)	3.8(0.5)	341(21)	119
2,6-Difluorotoluene	$C_7H_6F_2$	385	581.8(0.4)	3.7(0.4)	341(21)	119
3,4-Difluorotoluene	$C_7H_6F_2$	385	598.5(0.5)	3.8(0.6)	342(22)	119
Diheptyl phthalate	$C_{22}H_{34}O_4$	633	830(9)	1.24(0.08)	1153(22)	112
Dihexyl phthalate	$C_{20}H_{30}O_4$	652(5)	817(9)	1.3(0.1)	1061(22)	112
3,4-Dihydro-2H-pyran	C_5H_8O	358.6(0.2)	561(2)	4.63(0.08)	268(34)	75
Diisobutylamine	$C_8H_{19}N$	412.8	584.4(0.2)	3.20(0.06)	518(22)	168
Diisopropylamine	$C_6H_{15}N$	357(3)	523.1(0.2)	3.02(0.02)	407(18)	168
1,4-Diisopropylbenzene	$C_{12}H_{18}$	483.4(0.2)	675(1)	2.30(0.04)	610(65)	317
Diisopropyl ether	$C_6H_{14}O$	341.5(0.2)	500.2(0.7)	2.85(0.04)	386(5)	25, 290, 318, 319
1,2-Dimethoxyethane	$C_4H_{10}O_2$	358.1(0.1)	539(4)	3.91(0.05)	305(9)	162, 166, 175, 290
Dimethoxymethane	$C_3H_8O_2$	315.4(0.2)	488(11)	4.0(0.2)	259(10)	75, 162, 320
1,2-Dimethoxypropane	$C_5H_{12}O_2$	369	543(1)	3.4(0.6)	356(12)	162
2,2-Dimethoxypropane	$C_5H_{12}O_2$	350.5(0.7)	510(3)	4(1)	360(13)	162
Dimethyl adipate	$C_8H_{14}O_4$	504(3)	692(14)	2.5(0.5)	561(13)	321
Dimethylamine	C_2H_7N	280.4(0.4)	437.5(0.4)	5.34(0.05)	188(13)	193, 322, 323
N,N-Dimethylaniline	$C_8H_{11}N$	466(1)	687.7(0.6)	3.63(0.09)	407(15)	39, 40
2,2-Dimethylbutane	C_6H_{14}	322.8(0.2)	489.1(0.5)	3.10(0.01)	363.74(0.02)	86, 324–328
2,3-Dimethylbutane	C_6H_{14}	331.1(0.3)	500.2(0.3)	3.13(0.01)	358(1)	19, 86, 252, 324–330
3,3-Dimethyl-2-butanone	$C_6H_{12}O$	379.2(0.2)	570.9(0.3)	3.67(0.03)	383(6)	164, 165
2,3-Dimethyl-1-butene	C_6H_{12}	328.74(0.04)	497.7(0.9)	3.31(0.01)	346(9)	170
3,3-Dimethyl-1-butene	C_6H_{12}	346.34(0.06)	477.4(0.9)	3.18(0.02)	348(10)	170
2,3-Dimethyl-2-butene	C_6H_{12}	314.39(0.04)	521.0(0.9)	3.4(0.1)	344(7)	170
Dimethyl carbonate	$C_3H_6O_3$	363.26(0.09)	557(1)	4.8(0.2)	251(51)	331, 332
cis-1,3-Dimethylcyclohexane	C_8H_{16}	397.5(0.6)	587.7(0.5)	2.88(0.01)	429(10)	64
cis-1,4-Dimethylcyclohexane	C_8H_{16}	392.4(0.5)	603.2(0.3)	3.44(0.02)	434(7)	164, 165
$trans$-1,4-Dimethylcyclohexane	C_8H_{16}	397.4(0.7)	588(2)	3.04(0.01)	439(18)	74
Dimethyl disulfide	$C_2H_6S_2$	382.87(0.08)	608(4)	5.1(0.1)	266(8)	54
Dimethyl ether	C_2H_6O	248.3(0.2)	400.1(0.8)	5.31(0.03)	171(3)	125, 174, 186, 333–343
N,N-Dimethylformamide	C_3H_7NO	426.0(0.5)	649.6(0.8)	4.4(0.1)	262(9)	11, 344
Dimethyl glutarate	$C_7H_{12}O_4$	489(4)	682(14)	2.8(0.4)	488(14)	321
2,2-Dimethylheptane	C_9H_{20}	406(1)	576.7(0.5)	2.35(0.07)	546(12)	345
2,2-Dimethylhexane	C_8H_{18}	379.9(0.4)	549.9(0.4)	2.53(0.03)	481(10)	346
2,3-Dimethylhexane	C_8H_{18}	388.8(0.5)	563.5(0.4)	2.63(0.02)	466(16)	346
2,4-Dimethylhexane	C_8H_{18}	382.5(0.4)	553(3)	2.55(0.02)	480(39)	330, 346
2,5-Dimethylhexane	C_8H_{18}	382.2(0.7)	550.0(0.3)	2.49(0.02)	485(20)	19, 346
3,3-Dimethylhexane	C_8H_{18}	385.0(0.6)	562.0(0.4)	2.65(0.02)	450(17)	346
3,4-Dimethylhexane	C_8H_{18}	390.8(0.4)	568.8(0.4)	2.69(0.02)	467(21)	346
Dimethyl malonate	$C_5H_8O_4$	454.2(0.6)	647(1)	3.5(0.1)	368(13)	317, 321
2,7-Dimethylnaphthalene	$C_{12}H_{12}$	535.5(0.3)	775(2)	3.02(0.03)	515(45)	347
Dimethyl octanedioate	$C_{10}H_{18}O_4$	532(8)	723(14)	2.3(0.3)	672(14)	321
Dimethyl oxalate	$C_4H_6O_4$	436.5(0.5)	632(14)	4.0(0.3)	315(12)	321, 348
2,2-Dimethyloxirane	C_4H_8O	324(2)	500(4)	4.4(0.1)	327(31)	54
2,2-Dimethylpentane	C_7H_{16}	352.3(0.3)	520.6(0.6)	2.77(0.04)	409(12)	346, 349
2,3-Dimethylpentane	C_7H_{16}	362.9(0.6)	537.5(0.6)	2.92(0.08)	394(23)	346, 349, 350

Name	Mol. Form.	T_b/K	T_c/K	P_c/MPa	V_c/cm^3 mol^{-1}	Ref.
2,4-Dimethylpentane	C$_7$H$_{16}$	353.5(0.5)	520.0(0.7)	2.74(0.06)	415(19)	346, 349
3,3-Dimethylpentane	C$_7$H$_{16}$	359.1(0.6)	536.4(0.4)	2.94(0.02)	413(24)	346
2,3-Dimethyl-1-pentene	C$_7$H$_{14}$	357(1)	534(4)	2.9(0.8)	400(11)	170
4,4-Dimethyl-1-pentene	C$_7$H$_{14}$	345.6(0.2)	516(4)	2.91(0.01)	406(11)	170
Dimethyl phthalate	C$_{10}$H$_{10}$O$_4$	555.8(0.2)	772(9)	2.76(0.08)	557(17)	112
Dimethyl pimelate	C$_9$H$_{16}$O$_4$	518(2)	711(14)	2.4(0.1)	608(14)	321
2,3-Dimethylpyridine	C$_7$H$_9$N	434.2(0.4)	655.5(0.3)	4.03(0.01)	337(28)	86, 351
2,4-Dimethylpyridine	C$_7$H$_9$N	431.5(0.3)	647.1(0.9)	3.83(0.02)	363(34)	90, 352
2,5-Dimethylpyridine	C$_7$H$_9$N	430.15(0.05)	644.2(0.3)	4.11(0.04)	371(45)	86
2,6-Dimethylpyridine	C$_7$H$_9$N	417.1(0.1)	623.8(0.2)	3.80(0.04)	353(43)	90
3,4-Dimethylpyridine	C$_7$H$_9$N	452.2(0.3)	683.8(0.4)	4.06(0.01)	353(13)	86, 351, 352
3,5-Dimethylpyridine	C$_7$H$_9$N	445.0(0.1)	667.3(0.3)	3.84(0.02)	369(12)	86, 351
2,6-Dimethylquinoline	C$_{11}$H$_{11}$N	541.2(0.4)	786(2)	3.27(0.02)	505(24)	529
Dimethyl sebacate	C$_{12}$H$_{22}$O$_4$	562(3)	742(14)	2.1(0.2)	695(14)	321
Dimethyl succinate	C$_6$H$_{10}$O$_4$	470(1)	662(14)	3.5(0.2)	426(16)	321
Dimethyl sulfide	C$_2$H$_6$S	310.47(0.05)	503.0(0.3)	5.40(0.06)	201(9)	52, 117, 304
Dimethyl sulfoxide	C$_2$H$_6$OS	465.0(0.9)	707(1)	4.6(0.7)	228(7)	23
1,3-Dimethyl-1,1,3,3-tetraphenyldisiloxane	C$_{26}$H$_{26}$OSi$_2$	701(2)	893(9)	1.38(0.1)	1300(89)	353
1,3-Dimethyltricyclo[3.3.1.1^{3,7}]decane	C$_{12}$H$_{20}$	476.53	708(2)	2.86(0.01)	595(115)	141
Dinonyl phthalate	C$_{26}$H$_{42}$O$_4$	686	858(9)	1.0(0.3)	1652(26)	112
Dioctyl phthalate	C$_{24}$H$_{38}$O$_4$	688(4)	840(9)	1.1(0.1)	1510(22)	112
1,4-Dioxane	C$_4$H$_8$O$_2$	374.3(0.3)	587.3(0.1)	5.2(0.2)	251(5)	34, 290, 354
Dipentyl phthalate	C$_{18}$H$_{26}$O$_4$	614(40)	811(9)	1.4(0.7)	957(21)	112
Diphenyl ether	C$_{12}$H$_{10}$O	531.1(0.1)	766.9(0.8)	3.10(0.04)	526(23)	25, 355
Diphenylmethane	C$_{13}$H$_{12}$	537.3(0.3)	776(9)	3.02(0.07)	546(144)	39, 302, 356, 357
1,3-Diphenyltetramethyldisiloxane	C$_{16}$H$_{22}$OSi$_2$		750(8)			353
Dipropylamine	C$_6$H$_{15}$N	380.6(0.9)	555.8(0.1)	3.6(0.1)	414(15)	168
Dipropylene glycol	C$_6$H$_{14}$O$_3$	504(2)	705(4)	3.4(0.1)	444(18)	54
Dipropyl ether	C$_6$H$_{14}$O	363.2(0.3)	531(2)	2.92(0.05)	402(8)	25, 52
Dipropyl phthalate	C$_{14}$H$_{18}$O$_4$	592(2)	784(9)	1.9(0.1)	816(20)	112
Diundecyl phthalate	C$_{30}$H$_{50}$O$_4$	711(25)	886(10)	0.89(0.1)	1590(25)	112
Docosane	C$_{22}$H$_{46}$	642(5)	786(6)	1.0(0.1)	1434(50)	62, 358, 359
Docosanoic acid	C$_{22}$H$_{44}$O$_2$	693(3)	837(8)	1.11(0.08)	1485(29)	360
1-Docosanol	C$_{22}$H$_{46}$O	680(12)	827(8)	1.0(0.6)	1243(20)	361
1,2,2,3,3,4,4,5,5,6,6,7-Dodecafluoro-1-heptanol	C$_7$H$_4$F$_{12}$O	444.7(0.7)	589(5)	2.0(0.2)	620(34)	362
Dodecane	C$_{12}$H$_{26}$	489.4(0.2)	658.8(0.9)	1.80(0.09)	747(18)	34, 86, 118, 243, 244, 363, 364
1,12-Dodecanediamine	C$_{12}$H$_{28}$N$_2$	572(13)	767(8)	2.0(0.3)	765(37)	140
Dodecanedioic acid	C$_{12}$H$_{22}$O$_4$	621(10)	859(13)	2.1(0.2)	730(19)	253
1-Dodecanethiol	C$_{12}$H$_{26}$S	550(3)	734(4)	1.81(0.1)	726(25)	99
Dodecanoic acid	C$_{12}$H$_{24}$O$_2$	572(1)	743(7)	1.9(0.2)	787(19)	360
1-Dodecanol	C$_{12}$H$_{26}$O	537.2(0.3)	719.4(0.6)	2.02(0.05)	805(15)	149
2-Dodecanone	C$_{12}$H$_{24}$O	520(6)	702(4)	1.9(0.9)	742(20)	256
3-Dodecanone	C$_{12}$H$_{24}$O	523(3)	701(2)	1.9(0.2)	680(19)	256
4-Dodecanone	C$_{12}$H$_{24}$O	528(5)	697(2)	1.9(0.4)	672(19)	256
5-Dodecanone	C$_{12}$H$_{24}$O	521(4)	695(5)	1.9(0.4)	678(19)	256
6-Dodecanone	C$_{12}$H$_{24}$O	522(4)	694(2)	1.9(0.3)	677(19)	256
1-Dodecene	C$_{12}$H$_{24}$	486.5(0.9)	657.6(0.6)	1.88(0.01)	710(15)	257
Eicosane	C$_{20}$H$_{42}$	617.2(0.9)	768(6)	1.08(0.05)	1325(48)	358, 359
Eicosanoic acid	C$_{20}$H$_{40}$O$_2$	673(6)	820(8)	1.2(0.1)	1346(27)	360
1-Eicosanol	C$_{20}$H$_{42}$O	629	808(8)	1.1(0.2)	1130(18)	361
1-Eicosene	C$_{20}$H$_{40}$	620(15)	772(15)	1.1(0.3)	1213(36)	365
Ethane	C$_2$H$_6$	184.5(0.4)	305.36(0.04)	4.88(0.01)	146(3)	44, 48, 92, 131, 132, 134, 214, 215, 229, 246, 366–393
1,2-Ethanediamine	C$_2$H$_8$N$_2$	390.0(0.5)	613.1(0.3)	6.71(0.04)	204(13)	49
1,2-Ethanediol	C$_2$H$_6$O$_2$	470.6(0.1)	719(5)	8.1(0.4)	180(11)	11, 32, 41, 394
1,1-Ethanediol, diacetate	C$_6$H$_{10}$O$_4$	441(3)	618(4)	2.9(0.1)	457(13)	54
Ethanethiol	C$_2$H$_6$S	308.1(0.1)	498.7(0.3)	5.53(0.08)	208(10)	304

Name	Mol. Form.	T_b/K	T_c/K	P_c/MPa	V_c/cm³ mol⁻¹	Ref.
Ethanol	C_2H_6O	351.39(0.09)	515(1)	6.25(0.04)	169(4)	19, 25, 34, 80, 85, 128, 148–150, 152, 222, 395–409
Ethoxybenzene	$C_8H_{10}O$	443.0(0.2)	647(2)	3.45(0.05)	407(16)	39, 40
2-Ethoxyethyl acetate	$C_6H_{12}O_3$	429.8(0.4)	609(2)	3.07(0.03)	443(38)	141, 144, 163
2-Ethoxy-2-methylbutane	$C_7H_{16}O$	374.6(0.4)	546(2)	2.83(0.09)	448(29)	410
1-Ethoxy-1,1,2,2,3,3,4,4,4-nonafluorobutane	$C_6H_5F_9O$	350.04	482.0(0.1)	1.98(0.01)	518(2)	113
Ethyl acetate	$C_4H_8O_2$	350.2(0.2)	523.27(0.07)	3.88(0.02)	288(19)	19, 23, 400, 411–413
Ethylamine	C_2H_7N	289.8(0.2)	456.5(0.9)	5.6(0.1)	183(22)	193, 414
Ethylbenzene	C_8H_{10}	409.3(0.4)	617.1(0.1)	3.61(0.01)	365(61)	68, 78, 79, 86, 88, 172, 345
Ethyl benzoate	$C_9H_{10}O_2$	485.6(0.2)	700(14)	3.01(0.05)	470(12)	169
Ethyl butanoate	$C_6H_{12}O_2$	394.2(0.4)	566.1(0.1)	3.2(0.3)	421(30)	415, 416
Ethyl trans-2-butenoate	$C_6H_{10}O_2$	413(5)	599(10)	3(2)	382(7)	21
Ethylcyclohexane	C_8H_{16}	404.9(0.4)	606.9(0.4)	3.27(0.04)	431(12)	170, 171
Ethylcyclopentane	C_7H_{14}	376.6(0.6)	569.48(0.05)	3.40(0.07)	377(3)	236
Ethyl 2,2-dimethylpropanoate	$C_7H_{14}O_2$	391.4(0.4)	566(2)	2.88(0.02)	461(8)	417
Ethylene	C_2H_4	169.3(0.3)	282.35(0.03)	5.06(0.01)	130.9(0.2)	61, 246, 371, 418–436
Ethyl 3-ethoxypropanoate	$C_7H_{14}O_3$	441(2)	621(3)	2.7(0.1)	478(19)	11, 41
Ethyl formate	$C_3H_6O_2$	327.2(0.1)	508.5(0.5)	4.78(0.03)	223(47)	413, 415
Ethyl heptanoate	$C_9H_{18}O_2$	461(2)	634(1)	2.3(0.2)	587(8)	162
3-Ethylhexane	C_8H_{18}	391.6(0.5)	565.5(0.4)	2.61(0.02)	450(15)	346
Ethyl hexanoate	$C_8H_{16}O_2$	438(1)	615(1)	2.6(0.2)	528(8)	162
2-Ethylhexanoic acid	$C_8H_{16}O_2$	500.6(0.1)	674(1)	2.75(0.06)	543(8)	144, 163
2-Ethyl-1-hexanol	$C_8H_{18}O$	459.3(0.2)	640.2(0.3)	3.0(0.2)	508(29)	437
2-Ethylhexyl acetate	$C_{10}H_{20}O_2$	473(1)	642(2)	2.02(0.01)	644(10)	410
Ethyl 3-methylbutanoate	$C_7H_{14}O_2$	408(3)	584(6)	3(1)	463(9)	147, 162
Ethyl methyl ether	C_3H_8O	279(2)	437.8(0.2)	4.39(0.06)	219(5)	174, 304, 438
3-Ethyl-2-methylpentane	C_8H_{18}	388.8(0.6)	567.1(0.4)	2.70(0.02)	442(22)	346
3-Ethyl-3-methylpentane	C_8H_{18}	391.3(0.9)	576.5(0.4)	2.77(0.01)	463(13)	346
Ethyl 2-methylpropanoate	$C_6H_{12}O_2$	384(2)	554(4)	3.1(0.3)	421(76)	415
Ethyl methyl sulfide	C_3H_8S	339.8(0.3)	533(10)	4.62(0.03)	260(6)	303
Ethyl nonanoate	$C_{11}H_{22}O_2$	497(5)	664(1)	2.0(0.4)	715(8)	162
Ethyl octanoate	$C_{10}H_{20}O_2$	479(1)	652(12)	2(2)	657(8)	147, 162
3-Ethylpentane	C_7H_{16}	366.5(0.4)	540.7(0.4)	2.90(0.03)	412(14)	346, 349
Ethyl pentanoate	$C_7H_{14}O_2$	415(3)	593(1)	2.8(0.4)	466(10)	162
2-Ethylphenol	$C_8H_{10}O$	477.6(0.1)	703(1)	3.7(0.3)	388(15)	218
3-Ethylphenol	$C_8H_{10}O$	491.5(0.1)	716(1)	3.8(0.3)	393(15)	218
4-Ethylphenol	$C_8H_{10}O$	491.12(0.06)	716(1)	3.1(0.6)	395(15)	218
Ethyl propanoate	$C_5H_{10}O_2$	372.0(0.2)	547(1)	3.37(0.05)	343(23)	19, 64, 162, 413, 415
Ethyl propyl ether	$C_5H_{12}O$	336(3)	500.2(0.4)	3.37(0.01)	343(44)	25, 304
S-Ethyl thioacetate	C_4H_8OS	387(3)	590.5(0.2)	4.1(0.1)	320(10)	49
4-Ethyltoluene	C_9H_{12}	435.1(0.6)	640.2(0.5)	3.23(0.04)	446(12)	439, 440
Ethyl vinyl ether	C_4H_8O	309(2)	475(2)	4.06(0.04)	262(8)	290
Fluorobenzene	C_6H_5F	357.8(0.3)	560.10(0.07)	4.55(0.01)	272(9)	86, 441
Fluoroethane	C_2H_5F	235.4(0.3)	375.2(0.2)	5.02(0.01)	164(3)	206, 442, 443
Fluoromethane	CH_3F	194.8	317.42(0.01)	5.88(0.01)	112.41(0.01)	444, 445
2-Fluorotoluene	C_7H_7F	387(2)	591.2(0.4)	3.9(0.3)	323(47)	119
3-Fluorotoluene	C_7H_7F	389(2)	591.8(0.4)	3.9(0.3)	332(17)	119
4-Fluorotoluene	C_7H_7F	389.8(0.4)	592.1(0.8)	3.85(0.01)	332(17)	119
Formic acid	CH_2O_2	374	588(10)		115.88(0.08)	146
Furan	C_4H_4O	304.4(0.2)	490.2(0.2)	5.43(0.08)	218(3)	241, 290
Glycerol	$C_3H_8O_3$	562(3)	850(9)	7.6(0.8)	251(15)	394
Heneicosane	$C_{21}H_{44}$	632(6)	778(8)	1.0(0.1)	1366(48)	359
Heptadecane	$C_{17}H_{36}$	576(2)	736(1)	1.33(0.07)	1081(28)	359
Heptadecanoic acid	$C_{17}H_{34}O_2$	635(4)	792(8)	1.4(0.1)	1130(24)	360
1-Heptadecanol	$C_{17}H_{36}O$	597	780(8)	1.4(0.1)	1097(18)	361
1-Heptadecene	$C_{17}H_{34}$	574(3)	734(7)	1.34(0.08)	1053(35)	365
2,2,3,3,5,5,6-Heptafluoro-1,4-dioxane	$C_4HF_7O_2$	312.5(0.1)	453(1)	2.86(0.06)	359(4)	110
1,1,1,2,2,3,3-Heptafluoropentan-4-one	$C_5H_3F_7O$	337.4	476.55(0.08)	2.57(0.01)	394.2(0.7)	308

Name	Mol. Form.	T_b/K	T_c/K	P_c/MPa	V_c/cm^3 mol^{-1}	Ref.
1,1,1,2,3,3,3-Heptafluoropropane	C$_3$HF$_7$	257.65	375.0(0.1)	2.93(0.01)	299(7)	446–449
1,1,1,2,4,4,4-Heptafluoro-2-trifluoromethoxybutane	C$_5$H$_2$F$_{10}$O	322.73(0.09)	447(1)	2.15(0.06)	465(5)	110
1,1,1,2,2,3,3-Heptafluoro-3-(trifluoromethoxy)propane	C$_4$F$_{10}$O	280.0(0.4)	391.7(0.7)	1.89(0.05)	431(31)	450
2,2,4,6,8,8-Heptamethylnonane	C$_{16}$H$_{34}$	519(1)	692(4)	1.53(0.01)	957(27)	451
1,1,1,3,5,5,5-Heptamethyltrisiloxane	C$_7$H$_{22}$O$_2$Si$_3$	416(1)	553.4(0.6)	1.48(0.02)	828(3)	452
Heptanal	C$_7$H$_{14}$O	426(3)	616.8(0.4)	3.2(0.2)	434(7)	242
2-Heptanamine	C$_7$H$_{17}$N	414(4)	598(2)	2.9(0.3)	455(18)	170
Heptane	C$_7$H$_{16}$	371.53(0.07)	540.1(0.2)	2.74(0.01)	428(15)	34, 36, 62, 75, 78, 86, 131, 133–135, 157, 243, 244, 254, 286, 324, 346, 359, 394, 432, 453–463
Heptanedioic acid	C$_7$H$_{12}$O$_4$	615.1	842(13)	3.3(0.2)	463(15)	253
Heptanoic acid	C$_7$H$_{14}$O$_2$	495(2)	678(2)	3.0(0.3)	476(5)	145, 146, 464
1-Heptanol	C$_7$H$_{16}$O	451(1)	632.4(0.6)	3.1(0.2)	430(9)	149, 254, 465
2-Heptanol	C$_7$H$_{16}$O	432	608.4(0.6)	3.0(0.1)	442(2)	149, 465
3-Heptanol	C$_7$H$_{16}$O	436(2)	605.4(0.3)	3.1(0.4)	451(3)	465
4-Heptanol	C$_7$H$_{16}$O	434(2)	602.6(0.3)	3.1(0.6)	455(4)	465
2-Heptanone	C$_7$H$_{14}$O	424.1(0.3)	611.4(0.2)	2.98(0.04)	436(4)	25, 172, 466
3-Heptanone	C$_7$H$_{14}$O	419(2)	606.6(0.2)	3.0(0.1)	433(5)	466
4-Heptanone	C$_7$H$_{14}$O	417(1)	602.0(0.2)	3.0(0.3)	434(5)	466
1-Heptene	C$_7$H$_{14}$	370(2)	537.3(0.3)	2.85(0.02)	409(2)	51, 52, 86, 124, 257, 467
cis-2-Heptene	C$_7$H$_{14}$	371(2)	548.5(0.6)	3.0(0.3)	410(7)	170
trans-2-Heptene	C$_7$H$_{14}$	369(2)	542.8(0.4)	3.0(0.2)	410(7)	170
trans-3-Heptene	C$_7$H$_{14}$	367(1)	538.6(0.7)	3.0(0.2)	411(7)	170
Heptylbenzene	C$_{13}$H$_{20}$	515(4)	708(7)	2.1(0.2)	680(16)	258
Heptyl orthosilicate	C$_{28}$H$_{60}$O$_4$Si		778(16)			57
Hexacosane	C$_{26}$H$_{54}$	688(11)	816(8)	0.8(0.2)	1740(59)	358
Hexadecane	C$_{16}$H$_{34}$	560.0(0.7)	722.2(0.8)	1.4(0.2)	1009(53)	34, 244, 245
Hexadecanoic acid	C$_{16}$H$_{32}$O$_2$	624(6)	785(8)	1.5(0.2)	1059(23)	360
1-Hexadecanol	C$_{16}$H$_{34}$O	598(2)	770(8)	1.47(0.1)	1019(17)	361
1-Hexadecene	C$_{16}$H$_{32}$	558(1)	718(7)	1.4(0.2)	986(33)	365
Hexaethyldisiloxane	C$_{12}$H$_{30}$OSi$_2$	525(7)	692.9(0.1)	1.7(0.7)	955(2)	34
Hexafluoroacetylacetone	C$_5$H$_2$F$_6$O$_2$	342(2)	485.1(0.5)	2.9(0.2)	313(49)	468
Hexafluorobenzene	C$_6$F$_6$	353.3(0.2)	516.4(0.5)	3.28(0.01)	337(4)	74, 198, 286, 469–473
2,2,4,5,5,5-Hexafluoro-1,3-dioxolane	C$_3$F$_6$O$_2$	251.0(0.2)	368.1(0.7)	2.72(0.04)	293(30)	116
Hexafluoroethane	C$_2$F$_6$	195.0(0.1)	292.9(0.2)	3.03(0.01)	223(3)	172, 474–477
1,1,1,3,3,3-Hexafluoro-2-methoxy-2-(trifluoromethyl)propane	C$_5$H$_3$F$_9$O	327(1)	463(1)	2.37(0.07)	448(5)	110
1,1,1,2,3,3-Hexafluoro-3-(2,2,3,3,3-pentafluoropropoxy)propane	C$_6$H$_3$F$_{11}$O	360.64	486.48(0.07)	1.95(0.01)	529(2)	113
1,1,1,2,3,3-Hexafluoropropane	C$_3$H$_2$F$_6$	277.65	412.40(0.06)	3.42(0.01)	270(5)	203, 447, 448, 478, 479
1,1,1,3,3,3-Hexafluoropropane	C$_3$H$_2$F$_6$	271.8(0.2)	398.07(0.06)	3.18(0.01)	262(18)	203
1,1,1,2,3,3-Hexafluoro-3-(2,2,3,3-tetrafluoropropoxy)propane	C$_6$H$_4$F$_{10}$O	379.07	516.2(0.3)	2.2(0.2)	543(34)	113
1,1,1,2,3,3-Hexafluoro-3-(2,2,2-trifluoroethoxy)propane	C$_5$H$_3$F$_9$O	345.87	475.74(0.09)	2.23(0.02)	455(2)	113
Hexamethylbenzene	C$_{12}$H$_{18}$	541(3)	758(2)	2.6(0.4)	581(15)	25
1,1,1,5,5,5-Hexamethyl-3,3-bis[(trimethylsilyl)oxy]trisiloxane	C$_{12}$H$_{36}$O$_4$Si$_5$	494.6(0.2)	622.6(0.2)	1.03(0.02)	1323(90)	34
Hexamethyldisiloxane	C$_6$H$_{18}$OSi$_2$	373.6(0.3)	518.7(0.6)	1.95(0.02)	629(15)	480
2,6,10,15,19,23-Hexamethyltetracosane	C$_{30}$H$_{62}$	693(6)	796(2)	0.60(0.04)	2060(70)	481
Hexanal	C$_6$H$_{12}$O	402.8(0.4)	592(3)	3.4(0.2)	378(7)	163, 242
Hexane	C$_6$H$_{14}$	341.87(0.06)	507.5(0.1)	3.03(0.01)	366.0(0.8)	19, 29, 33, 34, 38, 52, 77, 78, 84, 86, 96, 106, 124, 131, 133, 134, 243, 244, 254, 281, 286, 318, 324–328, 401, 456, 458, 470, 482–488
1,6-Hexanediamine	C$_6$H$_{16}$N$_2$	470(2)	685(7)	3.6(0.5)	446(17)	140

Name	Mol. Form.	T_b/K	T_c/K	P_c/MPa	V_c/cm³ mol⁻¹	Ref.
1,6-Hexanedioic acid	$C_6H_{10}O_4$	610.5	841(13)	3.8(0.3)	449(17)	253
1,6-Hexanediol	$C_6H_{14}O_2$	481	741(10)	4.1(0.1)	404(13)	489
Hexanenitrile	$C_6H_{11}N$	436.6(0.3)	633.8(0.2)	2.99(0.06)	378(8)	35
Hexanoic acid	$C_6H_{12}O_2$	478.0(0.6)	661(7)		413(15)	145, 146, 360, 464
1-Hexanol	$C_6H_{14}O$	430.0(0.7)	611.0(0.4)	3.40(0.09)	381(30)	34, 149, 151, 152, 254, 255, 456
2-Hexanol	$C_6H_{14}O$	413	585.9(0.5)	3.3(0.3)	406(8)	149, 255, 437
3-Hexanol	$C_6H_{14}O$	416(2)	582.4(0.4)	3.3(0.1)	378(14)	64, 254, 255
2-Hexanone	$C_6H_{12}O$	400.8(0.1)	586.7(0.5)	3.31(0.04)	377(4)	25, 172, 466
3-Hexanone	$C_6H_{12}O$	396.6(0.3)	583.1(0.5)	3.32(0.01)	378(4)	25, 466
Hexatriacontane	$C_{36}H_{74}$	777(7)	872(9)	0.47(0.07)	2711(2)	358
1-Hexene	C_6H_{12}	336.5(0.1)	504.1(0.9)	3.20(0.03)	381(9)	86, 201, 202, 252, 257, 490
cis-2-Hexene	C_6H_{12}	342.0(0.5)	513.4(0.9)	3.34(0.06)	347(7)	170
trans-2-Hexene	C_6H_{12}	341.00(0.09)	509.0(0.7)	3.16(0.01)	353(7)	170
cis-3-Hexene	C_6H_{12}	339.5(0.5)	510(1)	3.29(0.01)	351(7)	170
trans-3-Hexene	C_6H_{12}	340.21(0.09)	507(2)	3.18(0.01)	352(7)	170
5-Hexen-2-one	$C_6H_{10}O$	402.2(0.5)	593.5(0.6)	3.51(0.04)	359(10)	51
Hexyl acetate	$C_8H_{16}O_2$	444.2(0.7)	618(1)	2.5(0.1)	526(8)	162
Hexylamine	$C_6H_{15}N$	405(1)	592.3(0.7)	3.4(0.3)	402(15)	170
Hexylbenzene	$C_{12}H_{18}$	499(2)	695(7)	2.4(0.2)	620(14)	258
Hexyl benzoate	$C_{13}H_{18}O_2$	550(26)	748(14)	2.0(0.3)	658(14)	169
Indan	C_9H_{10}	451.0(0.4)	684.8(0.4)	3.95(0.03)	385(22)	25
Isobutanal	C_4H_8O	337.2(0.2)	543.6(0.6)	5.12(0.08)	283(10)	164, 165
Isobutane	C_4H_{10}	261.4(0.5)	407.84(0.07)	3.64(0.02)	256(7)	125, 132, 491, 492
Isobutene	C_4H_8	266.1(0.2)	418.0(0.3)	4.00(0.04)	240(3)	123, 158, 493, 494
Isobutyl acetate	$C_6H_{12}O_2$	390.0(0.6)	562(2)	2.97(0.06)	369(37)	166, 201, 202, 415
Isobutylbenzene	$C_{10}H_{14}$	445.8(0.4)	650(3)	3.0(0.2)	493(15)	96
Isobutyl butanoate	$C_8H_{16}O_2$	430(1)	611(6)	2.5(0.3)	524(9)	147
Isobutylcyclohexane	$C_{10}H_{20}$	444.5	642.1(0.6)	2.61(0.07)	550(14)	170
Isobutyl formate	$C_5H_{10}O_2$	371.5(0.3)	551(4)	3.9(0.4)	359(25)	415
Isobutyl isobutanoate	$C_8H_{16}O_2$	421(3)	602(6)	2.5(0.8)	530(9)	147
Isobutyl 3-methylbutanoate	$C_9H_{18}O_2$	442(3)	621(6)	2.3(0.7)	581(9)	147
Isobutyl propanoate	$C_7H_{14}O_2$	409(2)	586(8)	3(1)	462(9)	162, 167
Isopentane	C_5H_{12}	300.98(0.06)	460.37(0.09)	3.35(0.06)	313(17)	19, 86, 96, 127, 495
Isopentyl acetate	$C_7H_{14}O_2$	414.8(0.7)	586.1(0.4)	2.76(0.07)	464(9)	166
Isopentyl butanoate	$C_9H_{18}O_2$	458.0(0.3)	619(6)	3(1)	595(10)	147
Isopentyl nitrite	$C_5H_{11}NO_2$	372(3)	626(16)	5.07(0.04)	386.2(0.1)	52
Isopentyl propanoate	$C_8H_{16}O_2$	446(4)	611(6)	2.5(0.9)	523(9)	147
Isopropyl acetate	$C_5H_{10}O_2$	361.8(0.2)	531.1(0.6)	3.31(0.04)	343(4)	64, 166, 172, 411, 490
Isopropylamine	C_3H_9N	304.9(0.2)	472.2(0.9)	4.55(0.07)	231(5)	75, 170
Isopropylbenzene	C_9H_{12}	425.5(0.2)	631(1)	3.2(0.1)	423(5)	79, 90, 96, 172
Isopropylcyclohexane	C_9H_{18}	427.5(0.4)	632.2(0.4)	3.1(0.2)	484(14)	170
Isopropyl formate	$C_4H_8O_2$	341(2)	534.6(0.5)	3.95(0.03)	294(11)	164, 165
1-Isopropyl-4-methylbenzene	$C_{10}H_{14}$	450(2)	654(8)	2.8(0.1)	495(16)	96, 147
(1S,2R,5S)-2-Isopropyl-5-methylcyclohexanol	$C_{10}H_{20}O$	489(3)	694(5)	2.7(0.6)	539(14)	302
Isopropyl methyl ether	$C_4H_{10}O$	303.9(0.5)	464.4(0.2)	3.76(0.01)	287(11)	25
Isoquinoline	C_9H_7N	516.3(0.6)	803(8)	5.07(0.03)	380(17)	218
d-Limonene	$C_{10}H_{16}$	450.8(0.5)	653(2)	2.81(0.02)	498(11)	496
Mesityl oxide	$C_6H_{10}O$	402.8(0.4)	605(2)	3.85(0.02)	353(27)	497
Methane	CH_4	111.6(0.2)	190.56(0.02)	4.60(0.01)	99(3)	132, 134, 498–508
Methane-d_4	CD_4		189.2(0.6)		98(3)	507
Methanethiol	CH_4S	279.1(0.1)	469.9(0.3)	7.24(0.09)	148(4)	304
Methanol	CH_4O	337.6(0.7)	512.7(0.6)	8.01(0.03)	117(4)	19, 38, 67, 80, 85, 87, 131, 148, 152, 155, 196, 405, 458, 482, 509–516
1-Methoxy-2,4-dimethylbenzene	$C_9H_{12}O$	465	682(4)	3.2(0.7)	451(18)	162
2-Methoxy-1,4-dimethylbenzene	$C_9H_{12}O$	467	677(1)	3.2(0.7)	451(15)	162
2-Methoxyethanol	$C_3H_8O_2$	397.4(0.1)	598(1)	5.28(0.08)	263(6)	49
2-Methoxyethyl acetate	$C_5H_{10}O_3$	415(3)	603(3)	3.6(0.5)	368(15)	162

Name	Mol. Form.	T_b/K	T_c/K	P_c/MPa	V_c/cm³ mol⁻¹	Ref.
4-Methoxy-1,1,1,2,2,3,3-heptafluorobutane	$C_5H_5F_7O$	344.13	481.5(0.2)	2.38(0.01)	431(2)	113
1-Methoxy-1,1,2,2,3,3-hexafluoropropane	$C_4H_4F_6O$	341.02	487.0(0.3)	2.9(0.1)	370(25)	113
2-Methoxy-2-methylbutane	$C_6H_{14}O$	359.5(0.1)	536(2)	3.23(0.09)	372(19)	172, 410, 517
5-Methoxy-1,1,2,2,3,3,4,4-octafluoropentane	$C_6H_6F_8O$	395.83	546.1(0.3)	2.40(0.07)	493(30)	113
1-Methoxy-2-propanol	$C_4H_{10}O_2$	393.1(0.6)	579.8(0.3)	4.11(0.04)	304(12)	49
2-Methoxypropene	C_4H_8O	308.8(0.3)	478.5(0.6)	4.2(0.3)	257(13)	518
Methyl acetate	$C_3H_6O_2$	329.8(0.2)	506.7(0.4)	4.73(0.07)	227(22)	19, 411–413
Methylamine	CH_5N	266.8(0.3)	430.6(0.6)	7.61(0.09)	139(1)	130, 193, 322, 323
N-Methylaniline	C_7H_9N	470(1)	702(5)	5.2(0.6)	347(16)	117
2-Methylaniline	C_7H_9N	473.1(0.4)	710(1)	3.6(0.1)	377(52)	519
3-Methylaniline	C_7H_9N	476.4(0.5)	709(10)	4.6(0.5)	346(17)	106
4-Methylaniline	C_7H_9N	474(1)	667(10)	3.3(0.7)	334(16)	106
2-Methylanisole	$C_8H_{10}O$	446(2)	662(1)	3.6(0.3)	397(18)	162
3-Methylanisole	$C_8H_{10}O$	450(2)	665(1)	3.6(0.3)	449(22)	162
4-Methylanisole	$C_8H_{10}O$	448(2)	667(1)	3.6(0.4)	396(15)	162
α-Methylbenzenemethanol	$C_8H_{10}N$	478(4)	699(5)	3.8(0.7)	399(16)	32
Methyl benzoate	$C_8H_8O_2$	472(2)	702(1)	3.8(0.1)	408(11)	517
2-Methylbutanal	$C_5H_{10}O$	363(2)	531.6(0.1)	4.04(0.03)	318(10)	124
Methyl butanoate	$C_5H_{10}O_2$	375.0(0.1)	554.4(0.1)	3.49(0.08)	341(20)	19, 413, 416
3-Methylbutanoic acid	$C_5H_{10}O_2$	449.6(0.2)	629(1)	3.4(0.2)	355(10)	146
2-Methyl-1-butanol	$C_5H_{12}O$	402.1(0.4)	575.4(0.5)	3.9(0.1)	342(8)	254
2-Methyl-2-butanol	$C_5H_{12}O$	375.6	544(1)	3.71(0.05)	326(9)	147, 254
3-Methyl-1-butanol	$C_5H_{12}O$	403.9(0.3)	579(2)	3.9(0.3)	335(7)	21, 91, 222, 254, 511, 520
3-Methyl-2-butanol	$C_5H_{12}O$	386.8(0.4)	556.1(0.5)	3.9(0.4)	336(9)	254
3-Methyl-2-butanone	$C_5H_{10}O$	367.3(0.2)	553.1(0.3)	3.83(0.1)	321(33)	157, 166
2-Methyl-2-butene	C_5H_{10}	311.6(0.4)	470(1)	3.4(0.1)	299(7)	521
3-Methyl-1-butene	C_5H_{10}	293.2(0.2)	452.7(0.5)	3.51(0.04)	305(8)	490
Methyl tert-butyl ether	$C_5H_{12}O$	328.2(0.1)	497.0(0.6)	3.41(0.05)	335(10)	25, 440
Methylcyclohexane	C_7H_{14}	374.0(0.1)	572.3(0.2)	3.48(0.09)	368(3)	34, 74, 78, 86, 88, 171, 235, 236
Methylcyclopentane	C_6H_{12}	344.9(0.2)	532.78(0.05)	3.79(0.05)	322(2)	78, 235, 236
2-Methylcyclopentanone	$C_6H_{10}O$	413(3)	631(2)	4.0(0.6)	328(17)	162
2-Methyl-N,N-dimethylaniline	$C_9H_{13}N$	458(2)	668.0(0.7)	3.12(0.08)	466(15)	39, 40
Methyl dodecanoate	$C_{13}H_{26}O_2$	540(2)	712(5)	1.4(0.4)	842(9)	218
1,1'-Methylenebis[(1-methylethyl)benzene]	$C_{19}H_{24}$	592(36)	795(8)	1.6(0.1)	871(30)	394
Methyl formate	$C_2H_4O_2$	304.8(0.3)	487.16(0.1)	6.01(0.01)	172(6)	19, 412, 413
2-Methylfuran	C_5H_6O	337.0(0.2)	528(3)	4.77(0.08)	252(3)	290
2-Methylheptane	C_8H_{18}	390.8(0.9)	559.6(0.1)	2.50(0.02)	487(12)	86, 346, 522
3-Methylheptane	C_8H_{18}	392.0(0.6)	563.7(0.4)	2.54(0.02)	463(12)	346
4-Methylheptane	C_8H_{18}	390.8(0.5)	561.7(0.4)	2.54(0.02)	480(14)	346
Methyl heptanoate	$C_8H_{16}O_2$	442.8(0.4)	628(2)	2.6(0.4)	521(8)	162
4-Methyl-3-heptanol	$C_8H_{18}O$	430(2)	623.5(0.7)	2.8(0.4)	505(13)	437
5-Methyl-3-heptanol	$C_8H_{18}O$	427(2)	621.2(0.3)	2.8(0.3)	493(13)	437
2-Methyl-3-heptanone	$C_8H_{16}O$	431	615(1)	2.7(0.3)	487(11)	162
5-Methyl-3-heptanone	$C_8H_{16}O$	432(4)	619(4)	2.7(0.7)	484(11)	162
2-Methyl-1-heptene	C_8H_{16}	392(2)	567.5(0.9)	2.6(0.2)	466(10)	170
2-Methyl-2-heptene	C_8H_{16}	395(2)	569(1)	2.6(0.4)	465(11)	170
2-Methylhexane	C_7H_{16}	363.1(0.8)	530.4(0.1)	2.73(0.03)	420(15)	86, 346, 349, 522
3-Methylhexane	C_7H_{16}	365.0(0.1)	535.4(0.5)	2.82(0.06)	405(19)	346, 349
2-Methyl-3-hexanone	$C_7H_{14}O$	407(3)	593(1)	2.9(0.4)	428(13)	162
5-Methyl-2-hexanone	$C_7H_{14}O$	412(2)	604(1)	2.9(0.3)	434(13)	162
2-Methyl-1-hexene	C_7H_{14}	365(2)	542(1)	2.9(0.3)	407(8)	170
5-Methyl-1-hexene	C_7H_{14}	358(1)	528.7(0.4)	2.9(0.2)	410(9)	170
N-Methylhexylamine	$C_7H_{17}N$	418(8)	592(1)	2.8(0.8)	458(20)	170
Methyl isobutanoate	$C_5H_{10}O_2$	365(1)	540.7(0.5)	3.43(0.01)	341(42)	19, 413
Methyl methacrylate	$C_5H_8O_2$	373.8(0.2)	540.3(0.6)	2.97(0.06)	320(6)	51
1-Methylnaphthalene	$C_{11}H_{10}$	517.5(0.9)	771(5)	3.56(0.07)	479(22)	86, 218, 523
2-Methylnaphthalene	$C_{11}H_{10}$	514.2(0.3)	761(3)	3.37(0.06)	464(20)	218
2-Methyloctane	C_9H_{20}	416(1)	582.8(0.2)	2.30(0.02)	547(17)	345, 522
Methyloxirane	C_3H_6O	308	488.11(0.08)	5.44(0.02)	197(42)	290, 524

Name	Mol. Form.	T_b/K	T_c/K	P_c/MPa	V_c/cm³ mol⁻¹	Ref.
Methyl pentafluoroethyl ether	$C_3H_3F_5O$	278.8(0.9)	406.81(0.05)	2.89(0.01)	301(5)	176, 448, 525
2-Methylpentane	C_6H_{14}	333.36(0.09)	497.9(0.2)	3.03(0.01)	371(2)	86, 127, 252, 325–328, 487, 522
3-Methylpentane	C_6H_{14}	336.4(0.5)	504.6(0.2)	3.12(0.01)	368.7(0.3)	252, 324–328, 526
Methyl pentanoate	$C_6H_{12}O_2$	400.51(0.06)	588.9(0.3)	3.20(0.05)	398(6)	164, 165
2-Methyl-1-pentanol	$C_6H_{14}O$	430(6)	604.4(0.5)	3.4(0.2)	410(8)	254
2-Methyl-2-pentanol	$C_6H_{14}O$	394(1)	559.5(0.7)	3.6(0.4)	410(11)	437
2-Methyl-3-pentanol	$C_6H_{14}O$	401.0(0.2)	576(1)	3.5(0.1)	380(9)	254
3-Methyl-3-pentanol	$C_6H_{14}O$	402(4)	575.6(0.6)	3.5(0.2)	376(10)	254
4-Methyl-1-pentanol	$C_6H_{14}O$	424(2)	603.5(0.7)	3.4(0.4)	406(7)	437
4-Methyl-2-pentanol	$C_6H_{14}O$	405.1(0.5)	574.4(0.5)	4(2)	389(9)	437
4-Methyl-2-pentanone	$C_6H_{12}O$	388.8(0.2)	575.4(1)	3.4(0.1)	378(12)	166, 527
2-Methyl-2-pentene	C_6H_{12}	340.4(0.5)	509.3(0.5)	3.26(0.01)	348(7)	170
4-Methyl-1-pentene	C_6H_{12}	327(2)	493.1(0.5)	3.18(0.07)	348(77)	170, 497
4-Methyl-cis-2-pentene	C_6H_{12}	329.5(0.1)	496.3(0.7)	3.24(0.01)	350(7)	170
Methyl pentyl ether	$C_6H_{14}O$	372(3)	546.5(0.2)	3.04(0.1)	395(14)	173, 174
2-Methyl-1,3-propanediol	$C_4H_{10}O_2$	494(4)	708(2)	5.4(0.4)	300(12)	55
Methyl propanoate	$C_4H_8O_2$	351.8(0.2)	530.57(0.1)	4.0(0.2)	280(98)	19, 412, 413, 415
2-Methylpropanoic acid	$C_4H_8O_2$	427.5(0.2)	605(2)	3.7(0.3)	296(10)	146
2-Methyl-1-propanol	$C_4H_{10}O$	380.99(0.07)	548(2)	4.30(0.04)	274(17)	25, 91, 148, 153, 155
2-Methyl-2-propanol	$C_4H_{10}O$	355.4(0.1)	506.2(0.1)	3.98(0.07)	283(4)	153
Methyl propyl ether	$C_4H_{10}O$	312(1)	476.2(0.2)	3.80(0.01)	281(7)	25
2-Methylpyridine	C_6H_7N	402.5(0.2)	622(1)	4.62(0.04)	306(119)	75, 528
3-Methylpyridine	C_6H_7N	417.2(0.1)	644.8(0.6)	4.63(0.03)	302(30)	90, 528
4-Methylpyridine	C_6H_7N	418.4(0.1)	645.8(0.5)	4.68(0.04)	316(67)	75, 90
N-Methyl-2-pyrrolidinone	C_5H_9NO	477.3(0.3)	721.7(0.4)	4.5(0.4)	330(12)	144, 242
2-Methylquinoline	$C_{10}H_9N$	520.5(0.4)	778(2)	3.91(0.02)	447(49)	530
8-Methylquinoline	$C_{10}H_9N$	520.5(0.7)	787(2)	4.22(0.02)	426(124)	530
Methyl salicylate	$C_8H_8O_3$	495.8(0.5)	709(30)	4.4(0.7)	436(17)	302
2-Methyltetrahydrofuran	$C_5H_{10}O$	353(1)	537(2)	3.74(0.06)	292(4)	290
(Methylthio)benzene	C_7H_8S	467.4(0.2)	706(4)	4.1(0.1)	374(13)	99
Methyl trifluoromethyl ether	$C_2H_3F_3O$	247.9(0.7)	377.92(0.06)	3.64(0.03)	219(2)	447, 525
Methyltris(trimethylsiloxy)silane	$C_{10}H_{30}O_3Si_4$	464.3(0.2)	597.4(0.2)	1.23(0.02)	1089(74)	34
4-Morpholinecarboxaldehyde	$C_5H_9NO_2$	511(1)	779(4)	5.0(0.4)	326(14)	54
Naphthalene	$C_{10}H_8$	491.1(0.1)	748.3(0.4)	4.06(0.04)	408(21)	79, 82, 86, 241, 251, 355, 451, 531
Neopentane	C_5H_{12}	282.65(0.06)	433.71(0.01)	3.20(0.01)	311.6(0.7)	532
Nitromethane	CH_3NO_2	374.3(0.1)	588(3)	6.0(0.2)	175(2)	533, 534
Nonadecane	$C_{19}H_{40}$	603(3)	756(5)	1.16(0.07)	1216(43)	358, 359
1-Nonadecene	$C_{19}H_{38}$	604(17)	755(8)	1.2(0.2)	1196(36)	365
1,1,1,2,2,3,3,4,4-Nonafluorohexan-5-one	$C_6H_3F_9O$	360.47	498.97(0.08)	2.20(0.02)	504(2)	308
Nonanal	$C_9H_{18}O$	468(3)	658(2)	2.7(0.1)	546(10)	242
Nonane	C_9H_{20}	424.0(0.2)	594.2(0.5)	2.29(0.05)	547(23)	34, 36, 78, 86, 131, 243–245, 247, 249, 251, 454, 456, 535, 536
1,9-Nonanediamine	$C_9H_{22}N_2$	531.8	726(7)	2.6(0.3)	600(23)	140
Nonanedioic acid	$C_9H_{16}O_4$	630.2	844(13)	2.7(0.2)	586(17)	253
Nonanoic acid	$C_9H_{18}O_2$	529(1)	712(3)	2.3(0.8)	592(16)	146
1-Nonanol	$C_9H_{20}O$	486.8(0.4)	670.6(0.5)	2.54(0.07)	555(75)	149, 152, 254, 255
2-Nonanol	$C_9H_{20}O$	466.7	649(1)	2.53(0.1)	575(11)	149, 255
3-Nonanol	$C_9H_{20}O$	468	648.0(0.3)	2.5(0.3)	577(12)	255
4-Nonanol	$C_9H_{20}O$	465.7	645.1(0.3)	2.5(0.3)	577(12)	255
2-Nonanone	$C_9H_{18}O$	467(1)	652.1(0.7)	2.5(0.1)	560(8)	49, 466
3-Nonanone	$C_9H_{18}O$	460(4)	648(4)	2.4(0.6)	560(7)	466
4-Nonanone	$C_9H_{18}O$	461(4)	643.7(0.3)	2.4(0.3)	560(7)	466
5-Nonanone	$C_9H_{18}O$	461.5(0.3)	641.4(0.3)	2.35(0.02)	560(7)	466
1-Nonene	C_9H_{18}	420.0(0.6)	594(1)	2.38(0.01)	529(2)	257
Octacosane	$C_{28}H_{58}$	705(6)	824(8)	0.8(0.1)	1916(65)	358
Octadecane	$C_{18}H_{38}$	589(2)	748(1)	1.3(0.1)	1167(41)	244
1-Octadecanol	$C_{18}H_{38}O$	624(2)	790(8)	1.28(0.1)	1157(18)	361
1-Octadecene	$C_{18}H_{36}$	588.7(1)	748(8)	1.3(0.1)	1119(34)	365

Name	Mol. Form.	T_b/K	T_c/K	P_c/MPa	V_c/cm³ mol⁻¹	Ref.
1,1,1,2,2,3,3,4-Octafluorobutane	$C_4H_2F_8$	300.62(0.02)	432.0(0.1)	2.80(0.02)	360(22)	537
1,2,2,3,3,4,4,5-Octafluoro-1-pentanol	$C_5H_4F_8O$	413.1(0.3)	571(1)	2.9(0.1)	440(24)	362
Octafluorotetrahydrofuran	C_4F_8O	272.3(0.5)	399.6(0.7)	2.68(0.09)	350(30)	450
Octamethylcyclotetrasiloxane	$C_8H_{24}O_4Si_4$	448.5(0.9)	585.8(0.9)	1.33(0.01)	1006(67)	114, 538
Octamethyltrisiloxane	$C_8H_{24}O_2Si_3$	425.6(0.8)	564.1(0.2)	1.42(0.01)	882(16)	539
Octanal	$C_8H_{16}O$	447(3)	639.3(0.3)	3.0(0.3)	489(7)	163, 242
Octane	C_8H_{18}	398.8(0.1)	568.7(0.1)	2.48(0.01)	490(22)	11, 19, 33, 34, 36, 78, 83, 86, 91, 96, 131, 133, 134, 243–245, 247, 249, 318, 319, 324, 346, 465, 469, 535, 540, 541
1,8-Octanediamine	$C_8H_{20}N_2$	498.7	712(7)	2.8(0.3)	547(20)	140
Octanedioic acid	$C_8H_{14}O_4$	618.6	843(13)	3.0(0.2)	520(16)	253
Octanenitrile	$C_8H_{15}N$	475(3)	674.4(0.4)	2.85(0.03)	494(10)	35
Octanoic acid	$C_8H_{16}O_2$	513(1)	694(1)	2.9(0.3)	522(19)	145, 146, 360, 464
1-Octanol	$C_8H_{18}O$	467.8(0.8)	651(2)	2.80(0.07)	490(47)	25, 149, 152, 254, 255
2-Octanol	$C_8H_{18}O$	452.5	629.5(0.9)	2.75(0.04)	519(10)	149, 254, 255
3-Octanol	$C_8H_{18}O$	457(6)	628.4(0.3)	2.8(0.4)	515(10)	255
4-Octanol	$C_8H_{18}O$	449.5	625.1(0.3)	2.8(0.3)	516(10)	255
2-Octanone	$C_8H_{16}O$	446(3)	632.7(0.2)	2.7(0.5)	497(6)	466
3-Octanone	$C_8H_{16}O$	439(4)	627.7(0.2)	2.7(0.3)	497(6)	466
4-Octanone	$C_8H_{16}O$	439(3)	623.8(0.2)	2.7(0.3)	497(6)	466
1-Octene	C_8H_{16}	398.0(0.5)	566.58(0.05)	2.68(0.02)	464(2)	86, 257
trans-2-Octene	C_8H_{16}	395.5(0.5)	569.8(0.4)	2.58(0.09)	471(9)	170
trans-4-Octene	C_8H_{16}	394.4(0.2)	566(1)	2.55(0.06)	472(9)	170
Octylamine	$C_8H_{19}N$	451.8(0.2)	641(1)	2.82(0.03)	494(41)	542
Octylbenzene	$C_{14}H_{22}$	536(2)	725(7)	2.0(0.2)	746(17)	258
Octyl orthosilicate	$C_{32}H_{68}O_4Si$		812(16)			57
Oxazole	C_3H_3NO	342.6(0.2)	551(4)	6.8(0.2)	185(23)	54
Oxirane	C_2H_4O	283.5(0.1)	469(1)	7.2(0.2)	138(4)	543, 544
Paraldehyde	$C_6H_{12}O_3$	397(2)	563(10)	4(3)	410(15)	12
Pentacene	$C_{22}H_{14}$		1115(47)		806(23)	545
1*H*-Pentadecafluoroheptane	C_7HF_{15}	368(2)	495.8(0.7)	1.7(0.5)	644(38)	546
Pentadecane	$C_{15}H_{32}$	543.8(0.4)	707(2)	1.54(0.09)	938(36)	243, 244, 245, 247, 249, 363, 547
Pentadecanoic acid	$C_{15}H_{30}O_2$	612(4)	777(8)	1.6(0.2)	1002(22)	360
1-Pentadecanol	$C_{15}H_{32}O$	591(2)	757(8)	1.6(0.2)	961(16)	361
1-Pentadecene	$C_{15}H_{30}$	541.5(0.4)	705(7)	1.56(0.05)	933(30)	365
Pentafluorobenzene	C_6HF_5	358(3)	530.93(0.03)	3.53(0.01)	322(22)	472, 548
3,3,4,4,4-Pentafluoro-2-butanone	$C_4H_3F_5O$	314.36(0.04)	453(1)	2.90(0.06)	333(4)	110
Pentafluoroethane	C_2HF_5	224.65	339.2(0.2)	3.63(0.01)	210(3)	180, 309, 312, 315, 447, 448, 549–554
1,1,1,2,2-Pentafluoropentan-3-one	$C_5H_5F_5O$	335.24	475.5(0.1)	2.64(0.01)	356(1)	308
1,1,1,2,2-Pentafluoropropane	$C_3H_3F_5$	255.1(0.3)	380.1(0.4)	3.14(0.02)	273(3)	555
1,1,1,3,3-Pentafluoropropane	$C_3H_3F_5$	288.5	427.20(0.07)	3.66(0.02)	262(14)	203
1,1,2,2,3-Pentafluoropropane	$C_3H_3F_5$	298.2	447.57(0.06)	3.96(0.02)	258(13)	203
1,1,1,2,2-Pentafluoro-3-(1,1,2,2-tetrafluoroethoxy)propane	$C_5H_3F_9O$	343.4	473.0(0.1)	2.24(0.01)	457(2)	113
Pentafluoro(trifluoromethoxy)ethane	C_3F_8O	249.5(0.3)	356.8(0.1)	2.4(0.5)	319(5)	442
Pentafluoro(trifluoromethyl)sulfur	CF_8S	252.4(0.2)	381.2(0.1)	3.4(0.1)	284(4)	442
Pentanal	$C_5H_{10}O$	376(2)	567(3)	3.1(0.3)	313(11)	104, 124
Pentane	C_5H_{12}	309.21(0.07)	469.7(0.1)	3.37(0.01)	310(1)	19, 34, 36, 52, 64, 77, 84–86, 127, 131, 133, 134, 201, 202, 243, 244, 254, 286, 324, 359, 394, 456, 490, 556–562
Pentanedioic acid	$C_5H_8O_4$	546(10)	840(13)	4.3(0.5)	343(13)	253
Pentanenitrile	C_5H_9N	413(1)	610.3(0.2)	3.58(0.05)	320(8)	35
Pentanoic acid	$C_5H_{10}O_2$	459.2(0.3)	639(2)	3.6(0.1)	347(15)	11, 41, 145, 146, 464
1-Pentanol	$C_5H_{12}O$	410.8(0.4)	587.9(0.4)	3.9(0.3)	331(9)	25, 34, 149, 329, 456, 465, 563

Name	Mol. Form.	T_b/K	T_c/K	P_c/MPa	V_c/cm³mol⁻¹	Ref.
2-Pentanol	$C_5H_{12}O$	392.2(0.5)	560.4(0.2)	4.2(0.4)	340(3)	149, 254, 465
3-Pentanol	$C_5H_{12}O$	396(2)	559.6(0.3)	4.9(0.9)	325(2)	465
2-Pentanone	$C_5H_{10}O$	375.3(0.1)	561.0(0.2)	3.70(0.06)	324(4)	25, 172
3-Pentanone	$C_5H_{10}O$	375.0(0.1)	561.4(0.2)	3.73(0.07)	319(11)	25
1-Pentene	C_5H_{10}	310.0(0.2)	464.74(0.04)	3.55(0.02)	301.0(0.1)	86, 470, 559
cis-2-Pentene	C_5H_{10}	303.1(0.3)	474.9(0.4)	3.69(0.02)	301(7)	564
Pentyl acetate	$C_7H_{14}O_2$	422.5(0.3)	600(2)	2.79(0.03)	466(123)	141, 162, 166
Pentylbenzene	$C_{11}H_{16}$	476(3)	675(7)	2.6(0.3)	559(12)	258
Pentyl benzoate	$C_{12}H_{16}O_2$	533(3)	736(14)	2.2(0.2)	661(14)	169
Pentyl formate	$C_6H_{12}O_2$	399(3)	576(4)	3.5(0.8)	453(35)	415
Perfluoroacetone	C_3F_6O	245.7(0.4)	357.2(0.1)	2.85(0.01)	329(7)	197, 470
Perfluorobutane	C_4F_{10}	271.0(0.8)	386.3(0.2)	2.33(0.02)	380(18)	565, 566
Perfluorocyclobutane	C_4F_8	267.3	388.4(0.1)	2.78(0.01)	316(8)	265, 470, 567, 568
Perfluorocyclohexane	C_6F_{12}	325.95 s	457.1(0.5)	2.24(0.03)	424(32)	569
Perfluorocyclohexene	C_6F_{10}	324.8(0.1)	461.7(0.7)	2.6(0.1)	434(28)	546
Perfluorodecane	$C_{10}F_{22}$	408(3)	542.4(0.4)	1.45(0.03)	892(7)	240, 570
Perfluorodimethoxymethane	$C_3F_8O_2$	263.1(0.7)	372.4(0.2)	2.34(0.01)	370(30)	116, 571
Perfluoro-2,3-dimethylbutane	C_6F_{14}	332.9(0.3)	463.0(0.1)	1.95(0.03)	523(18)	240, 572
Perfluoroethyl ethyl ether	$C_4H_5F_5O$	301(3)	431.23(0.08)	2.53(0.01)	366(3)	176
Perfluoroethyl 2,2,2-trifluoroethyl ether	$C_4H_2F_8O$	301.04	421.68(0.08)	2.33(0.01)	409(3)	176
Perfluoroheptane	C_7F_{16}	355.6(0.2)	477(3)	1.63(0.01)	603(14)	232, 566, 570, 573–576
Perfluoro-1-heptene	C_7F_{14}	354(4)	478.2(0.7)	1.7(0.3)	555(34)	546
1H-Perfluorohexane	C_6HF_{13}	345(4)	471.8(0.7)	2.0(0.3)	504(32)	546
Perfluorohexane	C_6F_{14}	330.3(0.2)	451(3)	1.88(0.02)	552(77)	198, 470, 546, 570, 572, 573, 577
Perfluoro-1-hexene	C_6F_{12}	330.2	454.3(0.7)	1.9(0.7)	462(31)	546
Perfluoroisobutane	C_4F_{10}	273	395.4(0.7)		396(22)	578
Perfluoroisopentane	C_5F_{12}	303.26(0.03)	423(19)	2.12(0.02)	465(8)	240
Perfluoroisopropyl methyl ether	$C_4H_3F_7O$	302(1)	433.30(0.08)	2.55(0.01)	369(3)	176
Perfluoromethylcyclohexane	C_7F_{14}	349.4(0.2)	486.5(1)	2.02(0.01)	561(4)	325, 326, 327, 566, 569
Perfluoromethylcyclopentane	C_6F_{12}	321.58(0.01)	451.43(0.04)	2.17(0.01)	419(37)	579
Perfluoro-2-methylpentane	C_6F_{14}	330.8(0.3)	454.6(0.2)	1.87(0.02)	585(10)	572, 580
Perfluoro-3-methylpentane	C_6F_{14}	331(9)	450(1)	1.69(0.01)	511(36)	572
Perfluoronaphthalene	$C_{10}F_8$	473(8)	673(1)	2.9(0.6)	464(29)	546
Perfluorononane	C_9F_{20}	390(3)	524.0(0.1)	1.56(0.04)	846(48)	570
Perfluorooctane	C_8F_{18}	378(2)	502.3(0.1)	1.66(0.02)	738(17)	34, 240, 570, 573
Perfluorooxetane	C_3F_6O	244.5(0.6)	361.8(0.5)	3.10(0.02)	274(24)	116, 571
1H-Perfluoropentane	C_5HF_{11}	319(2)	443.9(0.7)	2.2(0.3)	413(28)	546
Perfluoropentane	C_5F_{12}	302.3(0.2)	421.8(0.1)	2.04(0.02)	463(7)	570, 573
Perfluoropropane	C_3F_8	236.3(0.3)	345.03(0.08)	2.67(0.01)	301(12)	187, 305, 470, 581
Perfluoropropyl methyl ether	$C_4H_3F_7O$	307(1)	437.7(0.1)	2.48(0.01)	382(3)	176
Perfluorotoluene	C_7F_8	377.8(0.3)	534.4(0.2)	2.70(0.02)	423(26)	25
Perfluorotributylamine	$C_{12}F_{27}N$	451(2)	566(4)	1.24(0.09)	1196(66)	582
Perfluorovaleric acid	$C_5HF_9O_2$	415.3(0.3)	545.5(0.2)	2.10(0.05)	496(29)	583
Phenol	C_6H_6O	455.0(0.1)	694.3(0.1)	5.5(0.2)	283(10)	218, 584
Phenyl acetate	$C_8H_8O_2$	468(1)	686(2)	3.60(0.06)	407(12)	55
4-Phenyl-1-butanol	$C_{10}H_{14}O$	537(3)	746(14)	3.1(0.2)	493(19)	584
Phenyl isocyanate	C_7H_5NO	439.4(0.4)	657(10)	3.6(0.1)	342(15)	54
3-Phenyl-1-propanol	$C_9H_{12}O$	514(4)	732(14)	3.4(0.5)	450(14)	584
(1S)-(-)-α-Pinene	$C_{10}H_{16}$	429.1(0.4)	644(2)	3.4(0.2)	472(10)	496
Piperazine	$C_4H_{10}N_2$	421.78(0.05)	660(5)	5.4(0.4)	283(16)	331, 489
Piperidine	$C_5H_{11}N$	379.34(0.09)	594.14(0.02)	4.7(0.1)	294(10)	39, 218
Propanal	C_3H_6O	321.1(0.2)	503.7(0.8)	5.04(0.03)	218(9)	104, 124, 163, 242
Propane	C_3H_8	231.04(0.09)	369.9(0.1)	4.25(0.01)	199(6)	29, 125, 127, 131–134, 252, 322, 323, 336, 371, 387, 470, 486, 509, 540, 585–596
1,3-Propanediamine	$C_3H_{10}N_2$	412.3(0.7)	632(7)	5.7(0.6)	257(13)	140
1,2-Propanediol	$C_3H_8O_2$	460.4(0.2)	676(1)	5.9(0.2)	237(10)	32
1,3-Propanediol	$C_3H_8O_2$	487.8(0.3)	718(2)	6.7(0.2)	255(12)	55

Name	Mol. Form.	T_b/K	T_c/K	P_c/MPa	V_c/cm³mol⁻¹	Ref.
Propanenitrile	C_3H_5N	370.4(0.4)	561.3(0.2)	4.26(0.07)	211(7)	35
1-Propanethiol	C_3H_8S	340.8(0.1)	536.6(0.6)	4.7(0.1)	286(11)	143, 144
Propanoic acid	$C_3H_6O_2$	414.6(0.2)	603(3)	4.5(0.7)	232(12)	39, 145, 146, 493, 597, 598
1-Propanol	C_3H_8O	370.19(0.09)	536.8(0.2)	5.1(0.1)	220(22)	19, 34, 80, 148–150, 152, 153, 599, 600
2-Propanol	C_3H_8O	355.36(0.09)	508.3(0.2)	4.7(0.1)	226(1)	25, 91, 148, 149, 153, 324, 440, 520, 601, 602
Propene	C_3H_6	225.5(0.1)	364.9(0.5)	4.59(0.02)	184(11)	67, 123, 172, 181, 427, 603–610
2-Propoxyethanol	$C_5H_{12}O_2$	425(3)	614.7(0.7)	3.65(0.09)	364(13)	11, 41
1-Propoxy-2-propanol	$C_6H_{14}O_2$	423.3(0.7)	605(1)	3.1(0.1)	417(16)	32
Propyl acetate	$C_5H_{10}O_2$	374.1(0.2)	549.69(0.08)	3.37(0.07)	346(24)	19, 411, 413, 415
Propylamine	C_3H_9N	320.36(0.08)	499.2(0.4)	4.77(0.06)	230(12)	170, 193
Propylbenzene	C_9H_{12}	432.3(0.5)	638.3(0.1)	3.20(0.02)	441(6)	79, 86, 88
Propyl benzoate	$C_{10}H_{12}O_2$	504(2)	710(14)	2.6(0.3)	530(10)	169
Propyl butanoate	$C_7H_{14}O_2$	417(2)	593(1)	2.72(0.06)	463(10)	162, 166
Propylcyclohexane	C_9H_{18}	429.8(0.3)	630.8(0.9)	2.87(0.04)	489(13)	170, 171
Propylene carbonate	$C_4H_6O_3$	514.8(0.7)	763(2)	4.1(0.2)	256.5(0.1)	55
1,2-Propylene glycol 1-*tert*-butyl ether	$C_7H_{16}O_2$	425.2	601(4)	2.7(0.1)	468(15)	99
1,2-Propylene glycol monomethyl ether acetate	$C_6H_{12}O_3$	419.1(0.4)	598(1)	3.1(0.2)	432(16)	41, 144
Propyl formate	$C_4H_8O_2$	353.8(0.2)	538.1(0.1)	4.07(0.02)	281(31)	19, 412, 413, 415
Propyl isobutanoate	$C_7H_{14}O_2$	407(4)	582(11)	3(1)	463(9)	162, 167
Propyl 3-methylbutanoate	$C_8H_{16}O_2$	428(3)	609(6)	2.5(0.8)	523(9)	147
Propyl propanoate	$C_6H_{12}O_2$	395.3(0.1)	569(3)	3.1(0.1)	403(6)	162, 166
Propyne	C_3H_4	250	402(2)	5.63(0.06)	160(10)	605, 611
Pyrazine	$C_4H_4N_2$	389.4(0.1)	627(1)	6.49(0.03)	225(16)	612
Pyridine	C_5H_5N	388.3(0.1)	619(2)	5.63(0.07)	248(12)	16, 17, 60, 90, 246, 290, 520, 613
Pyrrole	C_4H_5N	402.89(0.04)	639.7(0.2)	8.0(0.2)	222(15)	241
Pyrrolidine	C_4H_9N	359.8(0.1)	568.6(0.2)	5.69(0.08)	259(3)	241, 290
Quinoline	C_9H_7N	510.2(0.5)	782(3)	4.75(0.1)	382(17)	218
Resorcinol	$C_6H_6O_2$	553(2)	836(10)	6.3(0.3)	292(10)	489
Stearic acid	$C_{18}H_{36}O_2$	644(3)	803(8)	1.3(0.2)	1251(27)	360
Styrene	C_8H_8	418.4(0.6)	635(2)	3.9(0.2)	357(15)	481
Succinic acid	$C_4H_6O_4$	507(3)	851(20)		308(21)	253
m-Terphenyl	$C_{18}H_{14}$	648(1)	883(7)	2.2(0.2)	747(37)	108, 358
o-Terphenyl	$C_{18}H_{14}$	610(5)	857(6)	2.9(0.1)	737(37)	108
p-Terphenyl	$C_{18}H_{14}$	649	913(22)	2.5(0.5)	713(37)	108, 614
Tetrabutyl silicate	$C_{16}H_{36}O_4Si$		682(14)			57
Tetrachloromethane	CCl_4	349.8(0.2)	556.5(0.3)	4.57(0.07)	276(9)	19, 27, 469, 615–619
Tetracosane	$C_{24}H_{50}$	664(5)	800(5)	0.9(0.1)	1585(55)	358, 359
Tetradecamethylcycloheptasiloxane	$C_{14}H_{42}O_7Si_7$	548.4(0.2)	683.2(0.2)	0.99(0.02)	1634(110)	34
Tetradecane	$C_{14}H_{30}$	526.6(0.4)	693(1)	1.56(0.08)	870(49)	34, 244
Tetradecanedioic acid	$C_{14}H_{26}O_4$	639(10)		1.9(0.2)		253
Tetradecanoic acid	$C_{14}H_{28}O_2$	599(1)	763(8)	1.6(0.2)	921(20)	360
1-Tetradecanol	$C_{14}H_{30}O$	569.0(0.4)	743(7)	1.70(0.04)	887(15)	361
2-Tetradecanone	$C_{14}H_{28}O$	562(6)	728(9)	1.6(0.5)	896(26)	256
3-Tetradecanone	$C_{14}H_{28}O$	552(7)	727(6)	1.6(0.5)	896(26)	256
4-Tetradecanone	$C_{14}H_{28}O$	552(7)	725(6)	1.6(0.5)	900(27)	256
7-Tetradecanone	$C_{14}H_{28}O$	552(8)	723(8)	1.6(0.6)	904(27)	256
1-Tetradecene	$C_{14}H_{28}$	524.2(0.4)	691(7)	1.58(0.07)	851(24)	365
Tetradecyl orthosilicate	$C_{40}H_{84}O_4Si$		849(16)			57
Tetraethoxysilane	$C_8H_{20}O_4Si$	441(1)	587(12)	2.0(0.4)	701.3(0.2)	57
Tetraethylene glycol	$C_8H_{18}O_5$	588(7)	800(30)	2.8(0.7)	608(31)	283
Tetraethylsilane	$C_8H_{20}Si$	426.5(0.7)	606(2)	2.297(0.01)	596.4(0.2)	620
1,2,3,4-Tetrafluorobenzene	$C_6H_2F_4$	367.4(0.8)	550.8(0.2)	3.791(0.01)	312(22)	25
1,2,3,5-Tetrafluorobenzene	$C_6H_2F_4$	357.4(0.8)	535.2(0.2)	3.75(0.01)	311(22)	25
1,2,4,5-Tetrafluorobenzene	$C_6H_2F_4$	363.3(0.3)	543.3(0.2)	3.80(0.01)	309(22)	25
1,1,2,2-Tetrafluoro-2-(2,2-difluoromethoxy) ethane	$C_4H_4F_6O$	352.13	501.08(0.08)	3.09(0.02)	356(1)	113

Name	Mol. Form.	T_b/K	T_c/K	P_c/MPa	$V_c/cm^3\,mol^{-1}$	Ref.
1,1,1,2-Tetrafluoroethane	$C_2H_2F_4$	247.0(0.1)	374.2(0.2)	4.06(0.01)	200(2)	113, 182, 183, 188, 278–280, 305, 313, 315, 447, 448, 478, 621–631
1,1,2,2-Tetrafluoroethane	$C_2H_2F_4$	253(1)	391.75(0.08)	4.61(0.01)	192(1)	178, 632
Tetrafluoroethene	C_2F_4	197(1)	307(1)	3.94(0.05)	183(3)	633, 634
1,2,2,2-Tetrafluoroethyl difluoromethyl ether	$C_3H_2F_6O$	296(2)	428.95(0.08)	3.05(0.01)	315(2)	176
1,1,2,2-Tetrafluoroethyl 1,1,1-trifluoroethyl ether	$C_4H_3F_7O$	329.37	463.89(0.07)	2.71(0.01)	373(1)	113
Tetrafluoromethane	CF_4	145.2(0.1)	227.54(0.03)	3.73(0.03)	140(1)	506, 635
1,1,2,2-Tetrafluoro-3-methoxypropane	$C_4H_6F_4O$	347.4(0.1)	505.4(0.1)	3.28(0.01)	331(1)	113
1,2,2,3-Tetrafluoro-1-propanol	$C_3H_4F_4O$	386.4(0.4)	554(2)	3.3(0.2)	280(15)	362
1,1,2,2-Tetrafluoro-3-(1,1,2,2-tetrafluoroethoxy)propane	$C_5H_4F_8O$	366.32	510.07(0.08)	2.58(0.01)	440(2)	113
1,1,1,2-Tetrafluoro-2-(trifluoromethoxy)ethane	C_3HF_7O	264(2)	377.26(0.06)	2.62(0.01)	321(2)	525
3,4,4,4-Tetrafluoro-3-(trifluoromethyl)-2-butanone	$C_5H_3F_7O$	328.76(0.05)	468(1)	2.50(0.06)	409(5)	110
4,5,5,5-Tetrafluoro-2-(trifluoromethyl)-1,3-dioxolane	$C_4HF_7O_2$	304.6(0.1)	435(1)	2.62(0.07)	376(4)	110
Tetrahexoxysilane	$C_{24}H_{52}O_4Si$		757(16)			57
Tetrahydrofuran	C_4H_8O	339.1(0.1)	540(1)	5.29(0.06)	223(2)	23, 241, 290
1,2,3,4-Tetrahydronaphthalene	$C_{10}H_{12}$	480.3(0.3)	720(1)	3.6(0.1)	431(40)	144, 242, 636
Tetrahydropyran	$C_5H_{10}O$	361.1(0.4)	572.0(0.3)	4.8(0.2)	278(18)	75
Tetrahydrothiophene	C_4H_8S	394.2(0.2)	632.0(0.2)	5.4(0.6)	276(19)	143, 241
1,2,4,5-Tetraisopropylbenzene	$C_{18}H_{30}$	532	703(1)	1.65(0.02)	983(83)	317
Tetramethoxysilane	$C_4H_{12}O_4Si$	393.2(0.7)	558(12)	2.8(0.6)	464(22)	57
1,2,4,5-Tetramethylbenzene	$C_{10}H_{14}$	470(1)	676(2)	2.9(0.3)	489(11)	39
2,2,3,3-Tetramethylhexane	$C_{10}H_{22}$	433(2)	623.0(0.5)	2.51(0.06)	574(14)	536
2,2,5,5-Tetramethylhexane	$C_{10}H_{22}$	410(2)	581.4(0.5)	2.19(0.01)	600(14)	536
2,2,3,3-Tetramethylpentane	C_9H_{20}	413.3(0.4)	607.5(0.5)	2.74(0.03)	514(15)	536
2,2,3,4-Tetramethylpentane	C_9H_{20}	406.1(0.8)	592.6(0.5)	2.60(0.03)	517(17)	536
2,2,4,4-Tetramethylpentane	C_9H_{20}	395(1)	574.6(0.5)	2.49(0.01)	532(16)	536
2,3,3,4-Tetramethylpentane	C_9H_{20}	414.6(0.7)	607.5(0.5)	2.72(0.04)	517(17)	536
Tetramethylsilane	$C_4H_{12}Si$	299.8(0.5)	449(2)	2.82(0.01)	362(7)	637–639
Tetramethylstannane	$C_4H_{12}Sn$	350(1)	521.77(0.02)	2.98(0.01)	109(109)	640, 641
Tetranonoxysilane	$C_{36}H_{76}O_4Si$		830(16)			57
Tetrapropyl silicate	$C_{12}H_{28}O_4Si$		649(12)			57
Thiacyclohexane	$C_5H_{10}S$	414.88(0.04)	684(44)	6.50(0.07)	284(11)	104
Thiobis(trifluoromethane)	C_2F_6S	251.3(0.3)	376.8(0.1)	3.2(0.5)	216(19)	442
Thiophene	C_4H_4S	357.2(0.1)	579.4(0.2)	5.7(0.2)	230(3)	241, 290
Thymol	$C_{10}H_{14}O$	506(3)	698(10)	3(2)	528(21)	302
Toluene	C_7H_8	383.75(0.07)	591.9(0.2)	4.13(0.02)	314(7)	34, 63, 68, 74, 78, 79, 84, 86, 88, 90, 96, 163, 172, 258, 481, 636, 642–646
Triacontane	$C_{30}H_{62}$	724(7)	843(8)	0.6(0.1)	2055(69)	358
Tribromomethane	$CHBr_3$	422.3(0.5)	682(1)	5.8(0.2)	261(12)	647
Trichloroacetyl chloride	C_2Cl_4O	391.3(0.3)	604(2)	4.21(0.03)	331(54)	410
Trichloroethylsilane	$C_2H_5Cl_3Si$	371.8(0.7)	559.9(0.6)	3.34(0.04)	403(5)	266
Trichlorofluoromethane	CCl_3F	296.8(0.6)	471.1(0.2)	4.40(0.03)	248.0(0.9)	33, 34, 191, 271, 571
Trichloromethane	$CHCl_3$	334.3(0.1)	536.0(0.4)	5.5(0.2)	237(7)	18, 27, 28, 117
Trichloromethylsilane	CH_3Cl_3Si	339(2)	517.7(0.3)	3.52(0.03)	329(9)	264, 648
1,3,5-Trichloro-2,4,6-trifluorobenzene	$C_6Cl_3F_3$	472(27)	684.7(0.4)	3.3(0.1)	443(27)	25
1,1,2-Trichloro-1,2,2-trifluoroethane	$C_2Cl_3F_3$	320.8(0.2)	487.4(0.2)	3.40(0.02)	325(1)	191, 265, 271, 649
Tricosane	$C_{23}H_{48}$	654(9)	790(8)	0.9(0.1)	1527(53)	359
Tridecane	$C_{13}H_{28}$	508.5(0,4)	676(1)	1.68(0.04)	824(30)	34, 244, 249
1-Tridecanol	$C_{13}H_{28}O$	560(8)	732(7)	1.8(0.2)	828(14)	361
2-Tridecanone	$C_{13}H_{26}O$	541(1)	717(6)	1.8(0.2)	820(24)	256
3-Tridecanone	$C_{13}H_{26}O$	539(7)	716(5)	1.7(0.5)	823(24)	256
4-Tridecanone	$C_{13}H_{26}O$	539(7)	712(6)	1.7(0.5)	823(24)	256
5-Tridecanone	$C_{13}H_{26}O$	539(8)	710(8)	1.7(0.7)	826(17)	256
6-Tridecanone	$C_{13}H_{26}O$	539(7)	709(5)	1.7(0.5)	826(24)	256
7-Tridecanone	$C_{13}H_{26}O$	539(9)	708(5)	1.7(0.5)	830(24)	256

Name	Mol. Form.	T_b/K	T_c/K	P_c/MPa	V_c/cm^3 mol^{-1}	Ref.
1-Tridecene	$C_{13}H_{26}$	506.0(0.7)	673(7)	1.74(0.05)	770(17)	365
Tridecylbenzene	$C_{19}H_{32}$	613(4)	790(8)	1.5(0.1)	1079(43)	258
Triethylamine	$C_6H_{15}N$	361.9(0.2)	535.6(0.3)	3.1(0.3)	392(25)	117, 286
1,3,5-Triethylbenzene	$C_{12}H_{18}$	489.0(0.9)	679(2)	2.32(0.01)	624(60)	332
Triethylene glycol	$C_6H_{14}O_4$	561.8(0.2)	775(30)	3.3(0.2)	454(25)	283
Trifluoroacetonitrile	C_2F_3N	204.3(0.8)	311.1(0.4)	3.61(0.04)	202(4)	470
1,2,3-Trifluorobenzene	$C_6H_3F_3$	368	560.3(0.4)	4.1(0.4)	296(20)	119
1,2,4-Trifluorobenzene	$C_6H_3F_3$	363	551.1(0.4)	4.1(0.6)	297(20)	119
1,3,5-Trifluorobenzene	$C_6H_3F_3$	350.1(0.5)	530.9(0.4)	3.8(0.2)	300(20)	119
1,1,1-Trifluoroethane	$C_2H_3F_3$	225.9(0.1)	345.89(0.07)	3.77(0.01)	195(15)	179, 203, 309, 478, 479, 549, 624, 650–653
2,2,2-Trifluoroethanol	$C_2H_3F_3O$	346.9(0.3)	498.57(0.05)	4.81(0.01)	211(12)	654, 655
2,2,2-Trifluoroethyl methyl ether	$C_3H_5F_3O$	304.77	448.98(0.08)	3.51(0.06)	277(3)	176
Trifluoroiodomethane	CF_3I	251.3(0.6)	396.44(0.06)	3.95(0.01)	231(3)	656–659
Trifluoromethane	CHF_3	191.1(0.1)	299.00(0.02)	4.82(0.01)	133(1)	310, 606, 660–664
Trifluoromethyl difluoromethyl ether	C_2HF_5O	238.1(0.2)	354.49(0.06)	3.36(0.02)	226(20)	203
Trifluoromethyl 1,1,2,2-tetrafluoroethyl ether	C_3HF_7O	270(1)	387.8(0.5)	2.65(0.01)	341(22)	116, 571
3,3,3-Trifluoropropene	$C_3H_3F_3$	246(4)	378.6(0.5)	3.61(0.08)	229(14)	440
Trimethylamine	C_3H_9N	275.9(0.2)	433.0(0.6)	4.08(0.04)	254(6)	322, 323, 665
1,2,3-Trimethylbenzene	C_9H_{12}	449.1(0.4)	664.4(0.1)	3.45(0.03)	423(11)	79
1,2,4-Trimethylbenzene	C_9H_{12}	442.5(0.3)	649.1(0.1)	3.3(0.1)	436(12)	79, 86, 96, 251
1,3,5-Trimethylbenzene	C_9H_{12}	437.8(0.3)	637.31(0.1)	3.13(0.05)	435(12)	79
3,7,7-Trimethyl-bicyclo[4.1.0]hept-3-ene	$C_{10}H_{16}$	445(2)	658(2)	2.9(0.5)	487(10)	496
2,2,3-Trimethylbutane	C_7H_{16}	353.9(0.1)	531.3(0.5)	2.96(0.03)	401(13)	346, 349
Trimethylchlorosilane	C_3H_9ClSi	330.8(0.4)	497.7(0.6)	3.20(0.03)	366(6)	266
1α,3α,5β-1,3,5-Trimethylcyclohexane	C_9H_{18}	414(2)	602(2)	2.6(0.3)	494(14)	74
3,3,5-Trimethylheptane	$C_{10}H_{22}$	430(3)	609.5(0.5)	2.32(0.05)	583(18)	536
2,2,5-Trimethylhexane	C_9H_{20}	397(2)	570(2)	2.46(0.03)	547(18)	350
2,2,3-Trimethylpentane	C_8H_{18}	382.9(0.4)	563.5(0.4)	2.73(0.02)	442(16)	346
2,2,4-Trimethylpentane	C_8H_{18}	372.3(0.2)	543.9(0.4)	2.57(0.02)	475(20)	33, 34, 86, 159, 346, 666, 667
2,3,3-Trimethylpentane	C_8H_{18}	387.8(0.3)	573.5(0.4)	2.82(0.03)	454(14)	346
2,3,4-Trimethylpentane	C_8H_{18}	386.5(0.3)	566.4(0.4)	2.72(0.02)	462(12)	346
cis-Tri(methylphenyl)trisiloxane	$C_{21}H_{24}O_3Si_3$		824(8)			353
trans-2,4,6-Trimethyl-2,4,6-triphenylcyclotrisiloxane	$C_{21}H_{24}O_3Si_3$		839(8)			353
Undecafluorocyclohexane	C_6HF_{11}	335.2	477.7(0.7)			546
Undecane	$C_{11}H_{24}$	469.0(0.3)	638.8(0.2)	2.01(0.03)	683(20)	34, 86, 131, 243, 249, 363, 535
Undecanoic acid	$C_{11}H_{22}O_2$	553	728(7)	2.1(0.2)	741(20)	360
1-Undecanol	$C_{11}H_{24}O$	519(2)	703.0(0.6)	2.15(0.07)	707(12)	149
2-Undecanone	$C_{11}H_{22}O$	506.2(0.3)	688(2)	2.08(0.01)	692(20)	256
3-Undecanone	$C_{11}H_{22}O$	500	685(2)	2.0(0.4)	692(20)	256
4-Undecanone	$C_{11}H_{22}O$	501(3)	681(2)	2.0(0.2)	692(20)	256
5-Undecanone	$C_{11}H_{22}O$	500	679(2)	2.0(0.2)	692(20)	256
6-Undecanone	$C_{11}H_{22}O$	500.5(0.5)	678(2)	2.02(0.01)	692(20)	256
Undecylbenzene	$C_{17}H_{28}$	585(3)	763(8)	1.6(0.1)	946(35)	258
Vinyl acetate	$C_4H_6O_2$	345.8(0.3)	519.2(0.2)	4.17(0.03)	269(7)	156, 440
m-Xylene	C_8H_{10}	412.2(0.4)	616.9(0.3)	3.54(0.01)	377(7)	34, 68, 79, 88, 90, 106, 668
o-Xylene	C_8H_{10}	417.5(0.4)	630.26(0.1)	3.74(0.01)	372(40)	34, 68, 78, 79, 88, 669
p-Xylene	C_8H_{10}	411.4(0.5)	616.17(0.09)	3.55(0.02)	372(35)	34, 68, 74, 79, 88, 90, 669
2,3-Xylenol	$C_8H_{10}O$	490.03(0.05)	723(1)	4.1(0.3)	397(15)	218
2,4-Xylenol	$C_8H_{10}O$	484.09(0.03)	708(1)	3.5(0.3)	389(15)	218
2,5-Xylenol	$C_8H_{10}O$	484.29(0.08)	707(1)	3.9(0.1)	397(15)	25
2,6-Xylenol	$C_8H_{10}O$	474.18(0.05)	701(1)	3.8(0.1)	396(15)	218
3,4-Xylenol	$C_8H_{10}O$	500.46(0.05)	730(1)	4.9(0.5)	388(15)	218
3,5-Xylenol	$C_8H_{10}O$	494.86(0.05)	716(1)	3.8(0.2)	396(15)	218

References

1. Frenkel, M., Chirico, R. D., Diky, V. V., Kazakov, A., and Muzny, C. D., *ThermoData Engine*. NIST Standard Reference Database 103b, Version 4.0 (Pure Compounds, Binary Mixtures, and Chemical Reactions, TDE-SOURCE Version 4.3), National Institute of Standards and Technology, Gaithersburg, MD – Boulder, CO, 2009, http://www.nist.gov/srd/nist103b.cfm.

2. Frenkel, M., Chirico, R. D., Diky, V., Yan, X., Dong, Q., and Muzny, C., *J. Chem. Inf. Model.* 45, 816, 2005.

3. Diky, V., Muzny, C. D., Lemmon, E. W., Chirico, R. D., and Frenkel, M., *J. Chem. Inf. Model.* 47, 1713, 2007.

4. Diky, V., Chirico, R. D., Kazakov, A. F., Muzny, C., and Frenkel, M., *J. Chem. Inf. Model.* 49, 503, 2009.

5. Diky, V., Chirico, R. D., Kazakov, A. F., Muzny, C., and Frenkel, M., *J. Chem. Inf. Model.* 49, 2883, 2009.

6. Joback, K. G., and Reid, R. C., *Chem. Eng. Commun.* 57, 233, 1987.

7. Constantinou, L., and Gani, R., *AIChE J.* 40, 1697, 1994.

8. Marrero-Morejon, J., and Pardillo-Fontdevila, E., *AIChE J.* 45, 615, 1999.

9. Wilson, G. M., and Jasperson, L. V., AIChE Meeting, New Orleans, LA, 1996.

10. Chirico, R. D., Frenkel, M., Diky, V. V., Marsh, K. N., and Wilhoit, R. C., *J. Chem. Eng. Data* 48, 1344, 2003.

11. Teja, A. S., and Anselme, M. J., *AIChE Symp. Ser.* 86 (279), 115, 1990.

12. Hollmann, R., *Z. Phys. Chem., Stoechiom. Verwandtschaftsl.* 43, 129, 1903.

13. Van der Waals, J. D., *Continuity of Gas and Liquid Data*, 1st edition, Leipzig, p. 168, 1881.

14. Vandana, V., and Teja, A. S., *Fluid Phase Equilib.* 103, 113, 1995.

15. Ambrose, D., Ellender, J. H., Sprake, C. H. S., and Townsend, R., *J. Chem. Thermodyn.* 9, 735, 1977.

16. Kreglewski, A., *Rocz. Chem.* 31, 1001, 1957.

17. Swietoslawski, W., and Kreglewski, A., *Bull. Acad. Pol. Sci., Cl. 3* 2, 77, 1954.

18. Swietoslawski, W., and Kreglewski, A., *Bull. Acad. Pol. Sci., Cl. 3* 2, 187, 1954.

19. Young, S., *Sci. Proc. R. Dublin Soc.* 12, 374, 1910.

20. Young, S., *J. Chem. Soc.* 59, 903, 1891.

21. Pawlewski, B., *Ber. Dtsch. Chem. Ges.* 16, 2633, 1883.

22. Ambrose, D., and Ghiassee, N. B., *J. Chem. Thermodyn.* 19, 911, 1987.

23. Sassa, Y., Konishi, R., and Katayama, T., *J. Chem. Eng. Data* 19, 44, 1974.

24. Ambrose, D., Sprake, C. H. S., and Townsend, R., *J. Chem. Thermodyn.* 6, 693, 1974.

25. Ambrose, D., Broderick, B. E., and Townsend, R., *J. Appl. Chem. Biotechnol.* 24, 359, 1974.

26. Campbell, A. N., and Musbally, G. M., *Can. J. Chem.* 48, 3173, 1970.

27. Campbell, A. N., and Chatterjee, R. M., *Can. J. Chem.* 47, 3893, 1969.

28. Campbell, A. N., and Chatterjee, R. M., *Can. J. Chem.* 46, 575, 1968.

29. Kay, W. B., *J. Phys. Chem.* 68, 827, 1964.

30. Rosenbaum, M., M.S. Thesis, Univ. Texas, Austin, TX, 1951.

31. Kuenen, J. P., and Robson, W. G., *Philos. Mag.* 3, 622, 1902.

32. VonNiederhausern, D. M., Wilson, L. C., Giles, N. F., and Wilson, G. M., *J. Chem. Eng. Data* 45, 154, 2000.

33. Christou, G., Young, C. L., and Svejda, P., *Ber. Bunsen-Ges. Phys. Chem.* 95, 510, 1991.

34. Christou, G., Ph.D. Dissertation, Univ. Melbourne, 1988.

35. Castillo-Lopez, N., and Trejo Rodriguez, A., *J. Chem. Thermodyn.* 19, 671, 1987.

36. Trejo Rodriguez, A., and McLure, I. A., *Fluid Phase Equilib.* 12, 297, 1983.

37. Trejo Rodriguez, A., and McLure, I. A., *J. Chem. Thermodyn.* 11, 1113, 1979.

38. Khera, R., Ph.D. Thesis, Ohio State Univ., Columbus, OH, 1968.

39. Guye, P. A., and Mallet, E., *Arch. Sci. Phys. Nat.* 13, 274, 1902.

40. Guye, P. A., and Mallet, E., *C. R. Hebd. Seances Acad. Sci.* 133, 168, 1902.

41. Teja, A. S., and Rosenthal, D. J., *Experimental Results for Phase Equilibria and Pure Component Properties*, DIPPR DATA Series No. 1, p. 96, 1991.

42. Goloborod'ko, N. P., and Khodeeva, S. M., *Russ. J. Phys. Chem. (Engl. Transl.)* 46, 235, 1972.

43. Mislavskaya, V. S., and Khodeeva, S. M., *Zh. Fiz. Khim.* 43, 2367, 1969.

44. Khodeeva, S. M., *Russ. J. Phys. Chem. (Engl. Transl.)* 40, 1061, 1966.

45. Ambrose, D., and Townsend, R., *Trans. Faraday Soc.* 60, 1025, 1964.

46. Ambrose, D., *Trans. Faraday Soc.* 52, 772, 1956.

47. McIntosh, D., *J. Phys. Chem.* 11, 306, 1907.

48. Kuenen, J. P., *Philos. Mag.* 44, 174, 1897.

49. Wilson, L. C., Wilson, H. L., Wilding, W. V., and Wilson, G. M., *J. Chem. Eng. Data* 41, 1252, 1996.

50. Lespieau, R., and Chavanne, G., *C. R. Hebd. Seances Acad. Sci.* 140, 1035, 1905.

51. Wang, X., Jia, Q., Gao, J., Xia, S., and Ma, P. S., *J. Chem. Ind. Eng. (China)* 56, 1385, 2005.

52. Liang, Y.-H., Ma, P. S., and Zhang, H., *J. Chem. Ind. Eng. (China)* 51, 243, 2000.

53. Steele, W. V., Chirico, R. D., Knipmeyer, S. E., and Nguyen, A., *J. Chem. Thermodyn.* 23, 957, 1991.

54. VonNiederhausern, D. M., Wilson, G. M., and Giles, N. F., *J. Chem. Eng. Data* 51, 1990, 2006.

55. Wilson, G. M., VonNiederhausern, D. M., and Giles, N. F., *J. Chem. Eng. Data* 47, 761, 2002.

56. Chashkin, Yu. R., Gorbunova, V. G., and Voronel, A. V., *Zh. Exp. Teor. Fiz.* 49, 432, 1965.

57. Nikitin, E. D., and Popov, A. P., *J. Chem. Eng. Data* 53, 1371, 2008.

58. Steele, W. V., Chirico, R. D., Knipmeyer, S. E., and Nguyen, A., *J. Chem. Eng. Data* 47, 648–666, 2002.

59. Lagutkin, O. D., and Kuropatkin, E. I., *Zh. Fiz. Khim.* 55, 1329, 1981.

60. Livingston, J., Morgan, R., and Higgins, E., *Z. Phys. Chem., Stoechiom. Verwandtschaftsl.* 64, 170, 1908.

61. Liu, T., Fu, J., Wang, K., Gao, Y., and Yuan, W., *J. Chem. Eng. Data* 46, 809, 2001.

62. Nikitin, E. D., Pavlov, P. A., and Skutin, M., *Fluid Phase Equilib.* 161, 119, 1999.

63. Chirico, R. D., and Steele, W. V., *Ind. Eng. Chem. Res.* 33, 157, 1994.

64. Zhang, J., Zhao, X., and Ma, P., *Huagong Xuebao* 43, 105, 1992.

65. Knipmeyer, S. E., Archer, D. G., Chirico, R. D., Gammon, B. E., Hossenlopp, I. A., Nguyen, A., Smith, N. K., Steele, W. V., and Strube, M. M., *Fluid Phase Equilib.* 52, 185, 1989.

66. Goodwin, R. D., *J. Phys. Chem. Ref. Data* 17, 1541, 1988.

67. Brunner, E., *J. Chem. Thermodyn.* 20, 1397, 1988.

68. Ambrose, D., *J. Chem. Thermodyn.* 19, 1007, 1987.

69. Kay, W. B., and Kreglewski, A., *Fluid Phase Equilib.* 11, 251, 1983.

70. Hales, J. L., and Gundry, H. A., *J. Phys. E* 16, 91, 1983.

71. Hugill, J. A., and McGlashan, M. L., *J. Chem. Thermodyn.* 13, 429, 1981.

72. Ewing, M. B., McGlashan, M. L., and Tzias, P., *J. Chem. Thermodyn.* 13, 527, 1981.

73. Akhundov, T. S., and Abdullaev, F. G., *Izv. Vyssh. Uchebn. Zaved., Neft Gaz* 20, 73, 1977.

74. Powell, R. J., Swinton, F. L., and Young, C. L., *J. Chem. Thermodyn.* 2, 105, 1970.

75. Kobe, K. A., and Mathews, J. F., *J. Chem. Eng. Data* 15, 182, 1970.

76. Artyukhovskaya, L. M., Shimanskaya, E. T., and Shimanskii, Yu. I., *Ukr. Fiz. Zh. (Ukr. Ed.)* 15, 1974, 1970.

77. Skripov, V. P., and Sinitsyn, E. N., *Zh. Fiz. Khim.* 42, 309, 1968.

78. Kay, W. B., and Hissong, D. W., *Proc. - Am. Pet. Inst., Div. Refin.* 47, 653, 1967.

79. Ambrose, D., Broderick, B. E., and Townsend, R., *J. Chem. Soc. A* 633, 1967.

80. Skaates, J. M., and Kay, W. B., *Chem. Eng. Sci.* 19, 431, 1964.

81. Makhan'ko, I. G., and Nozdrev, V. F., *Akust. Zh.* 10, 249, 1964.

82. Cheng, D. C. H., *Chem. Eng. Sci.* 18, 715, 1963.

83. Connolly, J. F., and Kandalic, G. A., *J. Chem. Eng. Data* 7, 137, 1962.

84. Partington, E. J., Rowlinson, J. S., and Weston, J. F., *Trans. Faraday Soc.* 56, 479, 1960.

85. McCracken, P. G., Storvick, T. S., and Smith, J. M., *J. Chem. Eng. Data* 5, 130, 1960.

86. Ambrose, D., Cox, J. D., and Townsend, R., *Trans. Faraday Soc.* 56, 1452, 1960.

87. Krichevskii, I. R., Khazanova, N. E., and Linshits, L. R., *Tr. GIAP* No. 9, 40, 1959.
88. Simon, M., *Bull. Soc. Chim. Belg.* 66, 375, 1957.
89. Krichevskii, I. R., Khazanova, N. E., and Linshits, L. R., *Zh. Fiz. Khim.* 31, 2711, 1957.
90. Ambrose, D., and Grant, D. G., *Trans. Faraday Soc.* 53, 771, 1957.
91. Kreglewski, A., *Rocz. Chem.* 29, 754, 1955.
92. Kay, W. B., and Nevens, T. D., *Chem. Eng. Prog., Symp. Ser.* 48, 108, 1952.
93. Bender, P., Furukawa, G. T., and Hyndman, J. R., *Ind. Eng. Chem.* 44, 387, 1952.
94. Gornowski, E. J., Amick, E. H., and Hixson, A. N., *Ind. Eng. Chem.* 39, 1348, 1947.
95. Schamhardt, H. O., Thesis, Amsterdam, The Netherlands, 1908.
96. Altschul, M., *Z. Phys. Chem., Stoechiom. Verwandtschaftsl.* 11, 577, 1893.
97. Young, S., *J. Chem. Soc., Trans.* 55, 486, 1889.
98. Nikitin, E. D., Popov, A. P., and Yatluk, Y. G., *J. Chem. Eng. Data* 51, 1335, 2006.
99. VonNiederhausern, D. M., Wilson, G. M., and Giles, N. F., *J. Chem. Eng. Data* 51, 1982, 2006.
100. Chirico, R. D., Knipmeyer, S. E., Nguyen, A., and Steele, W. V., *J. Chem. Thermodyn.* 23, 759, 1991.
101. Steele, W. V., Chirico, R. D., Hossenlopp, I. A., Knipmeyer, S. E., Nguyen, A., and Smith, N. K., *Experimental Results for DIPPR 1990– 91 Projects on Phase Equilibria and Pure Component Properties*, DIPPR Data Ser. No. 2, p. 188, 1994.
102. Guseinov, S. O., Naziev, Y. M., Farzaliev, B. I., and Movsunov, T. G., *Izv. Vyssh. Uchebn. Zaved., Neft Gaz* 21, 48, 1978.
103. Ambrose, D., and Ghiassee, N. B., *J. Chem. Thermodyn.* 22, 307, 1990.
104. Anselme, M. J., and Teja, A. S., *AIChE Symp. Ser.* 86 (279), 128, 1990.
105. Chirico, R. D., and Steele, W. V., *J. Chem. Thermodyn.* 36, 633, 2004.
106. Glaser, F., and Ruland, H., *Chem.-Ing.-Tech.* 29, 772, 1957.
107. Chirico, R. D., Knipmeyer, S. E., Nguyen, A., and Steele, W. V., *J. Chem. Thermodyn.* 21, 1307, 1989.
108. Reiter, R. W., *NASA Document* N63, 1963.
109. Ellard, J. A., and Yanko, W. H., *U. S. A. E. C. Rep.* IDO-11008, 1963.
110. Sako, T., Yasumoto, M., Nakazawa, N., and Kamizawa, C., *J. Chem. Eng. Data* 46, 1078, 2001.
111. Defibaugh, D. R., Gillis, K. A., Moldover, M. R., Morrison, G., and Schmidt, J. W., *Fluid Phase Equilib.* 81, 285, 1992.
112. Nikitin, E. D., Popov, A. P., and Yatluk, Y. G., *J. Chem. Eng. Data* 51, 1326, 2006.
113. Yasumoto, M., Yamada, Y., Murata, J., Urata, S., and Otake, K., *J. Chem. Eng. Data* 48, 1368, 2003.
114. McLure, I. A., and Neville, J. F., *J. Chem. Thermodyn.* 14, 385, 1982.
115. Badylkes, S., *Kholod. Tekh.* 43, 18, 1966.
116. Salvi-Narkhede, M., Wang, B. -H., Adcock, J. I., and Van Hook, W. A., *J. Chem. Thermodyn.* 24, 1065, 1992.
117. Herz, W., and Neukirch, E., *Z. Phys. Chem., Stoechiom. Verwandtschaftsl.* 104, 433, 1923.
118. Adamenko, I. I., and Chernyavskaya, I. A., *Ukr. Fiz. Zh.* 11, 336, 1966.
119. Morton, D. W., Lui, M. P. W., Tran, C. A., and Young, C. L., *J. Chem. Eng. Data* 45, 437, 2000.
120. Li, Y., Ma, P., and Ruan, Y., *Shiyou Huagong* 22, 322, 1993.
121. Higashi, Y., Uematsu, M., and Watanabe, K., *Bull. JSME* 28, 2660, 1985.
122. Scott, R. B., Meyers, C. H., Rands, R. D., Brickwedde, F. G., and Bekkedahl, N., *J. Res. Natl. Bur. Stand. (U. S.)* 35, 39, 1945.
123. Cragoe, C. S., *Natl. Bur. Stand. (U. S.)* LC-736, 1943.
124. Ma, P. S., Gao, J., and Xia, S., *Chin. J. Chem. Eng.* 10, 473, 2002.
125. Yasumoto, M., Uchida, Y., Ochi, K., Furuya, T., and Otake, K., *J. Chem. Eng. Data* 50, 596, 2005.
126. Warowny, W., *J. Chem. Eng. Data* 41, 689, 1996.
127. Holcomb, C. D., Magee, J. W., and Haynes, W. M., *Research Report RR-147*, Gas Processors Association Project 916, Tulsa, OK, 1995.
128. Deak, A., Victorov, A. I., and De Loos, T. W., *Fluid Phase Equilib.* 107, 277, 1995.
129. Vasserman, A. A., Khasilev, I. P., and Cymarnyi, V. A., *Deposited Doc. VNIIKI*, Doc. No. 604-kk, 1989.

130. Li, L., and Kiran, E., *J. Chem. Eng. Data* 33, 342, 1988.
131. Brunner, E., *J. Chem. Thermodyn.* 20, 273, 1988.
132. Younglove, B. A., and Ely, J. F., *J. Phys. Chem. Ref. Data* 16, 577, 1987.
133. Kreglewski, A., and Kay, W. B., *J. Phys. Chem.* 73, 3359, 1969.
134. Golubev, I. F., and Agaev, N. A., *Dokl. Akad. Nauk SSSR* 151(4), 875, 1963.
135. Kay, W. B., *Ind. Eng. Chem.* 33, 590, 1941.
136. Sage, B. H., Hicks, B. L., and Lacey, W. N., *Ind. Eng. Chem.* 32, 1085, 1940.
137. Kay, W. B., *Ind. Eng. Chem.* 32, 353, 1940.
138. Beattie, J. A., Simard, G. L., and Su, G.-J., *J. Am. Chem. Soc.* 61, 24, 1939.
139. Kuenen, J. P., *Commun. Kamerlingh Onnes Lab., Univ. Leiden*, No. 125, 1, 1911.
140. Nikitin, E. D., Popov, A. P., and Yatluk, Y. G., *J. Chem. Eng. Data* 51, 609, 2006.
141. Steele, W. V., Chirico, R. D., Knipmeyer, S. E., and Nguyen, A., *J. Chem. Eng. Data* 41, 1255, 1996.
142. Eliosa, G., Murrieta-Guevara, F., Reza, J., and Trejo Rodriguez, A., *Fluid Phase Equilib.* 61, 99, 1990.
143. Tsonopoulos, C., and Ambrose, D., *J. Chem. Eng. Data* 46, 480, 2001.
144. Teja, A. S., and Anselme, M. J., *AIChE Symp. Ser.* 86 (279), 122, 1990.
145. Gude, M. T., Mendez-Santiago, J., and Teja, A. S., *J. Chem. Eng. Data* 42, 278, 1997.
146. Ambrose, D., and Ghiassee, N. B., *J. Chem. Thermodyn.* 19, 505, 1987.
147. Brown, J. C., *J. Chem. Soc., Trans.* 89, 311, 1906.
148. Lydersen, A. L., and Tsochev, V., *Chem. Eng. Technol.* 13, 125, 1990.
149. Rosenthal, D. J., and Teja, A. S., *Ind. Eng. Chem. Res.* 28, 1693, 1989.
150. Christou, G., and Young, C. L., *Int. DATA Ser.*, Sel. Data Mixtures, Ser. A, 14(4), 245, 1986.
151. Naumova, A. A., Tyvina, T. N., and Fokina, V. V., *Zh. Prikl. Khim.* 1980, 1667, 1980.
152. Efremov, Yu. V., *Zh. Fiz. Khim.* 40, 1240, 1966.
153. Ambrose, D., and Townsend, R., *J. Chem. Soc.* 54, 3614, 1963.
154. Singh, R., and Shemilt, L. W., *J. Chem. Phys.* 23, 1370, 1955.
155. Kay, W. B., and Donham, W. E., *Chem. Eng. Sci.* 4, 1, 1955.
156. Stevens, R. M. M., Van Roermund, J. C., Jager, M. D., De Loos, T. W., and De Swaan Arons, J., *Fluid Phase Equilib.*, 138, 159, 1997.
157. Kobe, K. A., Crawford, H. R., and Stephenson, R. W., *Ind. Eng. Chem.* 47, 1767, 1955.
158. Ihmels, E. C., Fischer, K., and Gmehling, J., *Fluid Phase Equilib.* 228– 229, 155, 2005.
159. Li, J., Qin, Z., Wang, G., Dong, M., and Wang, J., *J. Chem. Eng. Data* 52, 1736, 2007.
160. Beattie, J. A., and Marple, S., *J. Am. Chem. Soc.* 72, 1449, 1950.
161. Olds, R. H., Sage, B. H., and Lacey, W. N., *Ind. Eng. Chem.* 38, 301, 1946.
162. Morton, D. W., Lui, M., and Young, C. L., *J. Chem. Thermodyn.* 31, 675, 1999.
163. Teja, A. S., and Rosenthal, D. J., *AIChE Symp. Ser.* 86 (279), 133, 1990.
164. Ma, F., Wang, J., and Ruan, Y., *J. Chem. Eng. Chin. Univ.* 9, 62, 1995.
165. Ma, P., and Ruan, Y., *Gaoxiao Huaxue Gongcheng Xuebao* 9, 62, 1995.
166. Quadri, S. K., and Kudchadker, A. P., *J. Chem. Thermodyn.* 23, 129, 1991.
167. Pawlewski, B., *Ber. Dtsch. Chem. Ges.* 15, 2460, 1882.
168. Toczylkin. L. S., and Young, C. L., *J. Chem. Thermodyn.* 12, 365, 1980.
169. Nikitin, E. D., and Popov, A. P., *J. Chem. Eng. Data* 52, 1336, 2007.
170. Morton, D. W., Lui, M. P. W., Tran, C. A., and Young, C. L., *J. Chem. Eng. Data* 49, 283, 2004.
171. Nikitin, E. D., Popov, A. P., and Bogatishcheva, N. S., *J. Chem. Eng. Data* 48, 1137, 2003.
172. Wilson, L. C., Wilding, W. V., Wilson, H. L., and Wilson, G. M., *J. Chem. Eng. Data* 40, 765, 1995.
173. Zawisza, A. C., and Glowka, S., *Bull. Acad. Pol. Sci., Ser. Sci. Chim.* 19, 191, 1971.
174. Osipiuk, B., and Stryjek, R., *Bull. Acad. Pol. Sci., Ser. Sci. Chim.* 18, 289, 1970.
175. Steele, W. V., Chirico, R. D., Knipmeyer, S. E., Nguyen, A., and Smith, N. K., *J. Chem. Eng. Data* 41, 1285, 1996.

176. Sako, T., Sato, M., Nakazawa, N., Oowa, M., Yasumoto, M., Ito, H., and Yamashita, S., *J. Chem. Eng. Data* 41, 802, 1996.

177. Tanikawa, S., Tatoh, J., Maezawa, Y., Sato, H., and Watanabe, K., *J. Chem. Eng. Data* 37, 74, 1992.

178. Chae, H. B., Schmidt, J. W., and Moldover, M. R., *J. Phys. Chem.* 94, 8840, 1990.

179. Mears, W. H., Stahl, R. F., Orfeo, S. R., Shair, R. C., Kells, L. F., Thompson, W., and McCann, H., *Ind. Eng. Chem.* 47, 1449, 1955.

180. Yata, J., Hori, M., Kawakatsu, H., and Minamiyama, T., *Int. J. Thermophys.* 17, 65, 1996.

181. Zhao, X., and Ma, P. S., *Chin. J. Chem. Eng.* 3, 233, 1995.

182. Nishiumi, H., Kohmatsu, S., Yokoyama, T., and Konda, A., *Fluid Phase Equilib.* 104, 131, 1995.

183. Economou, I. G., Peters, C. J., Florusse, L. J., and De Swaan Arons, J., *Fluid Phase Equilib.* 111, 239, 1995.

184. Nishiumi, H., Komatsu, M., Yokoyama, T., and Kohmatsu, S., *Fluid Phase Equilib.* 83, 109, 1993.

185. Wang, J., Liu, Z. G., Tan, L. C., and Yin, J. M., *Fluid Phase Equilib.* 80, 203, 1992.

186. Noles, J. R., and Zollweg, J. A., *J. Chem. Eng. Data* 37, 306, 1992.

187. Leu, A. D., and Robinson, D. B., *J. Chem. Eng. Data* 37, 7, 1992.

188. Goodwin, A. R. H., Defibaugh, D. R., and Weber, L. A., *Int. J. Thermophys.* 13, 837, 1992.

189. Zhimai, H., and Jianfen, H., *Gongcheng Rewuli Xuebao* 10, 233, 1989.

190. He, Z., Zhang, Y., and Hong, J., *The Second Asian Thermophysical Properties Conference*, Hunan University of Science and Technology, Guangzhou, China, p. 519, 1989.

191. *Chemicals and Plastics Physical Properties*, Union Carbide Corp. (Product Bulletin), 1968.

192. Ambrose, D., and Ghiassee, N. B., *J. Chem. Thermodyn.* 20, 765, 1988.

193. Berthoud, A., *J. Chim. Phys. Phys.-Chim. Biol.* 15, 3, 1917.

194. Cullick, A. S., and Ely, J. F., *J. Chem. Eng. Data* 27, 276, 1982.

195. Mansoorian, H., Hall, K. R., Holste, J. C., and Eubank, P. T., *J. Chem. Thermodyn.* 13, 1001, 1981.

196. Centnerszwer, M., *Z. Phys. Chem., Stoechiom. Verwandtschaftsl.* 49, 199, 1904.

197. Murphy, K. P., *J. Chem. Eng. Data* 9, 259, 1964.

198. Skripov, V. P., and Muratov, G. N., *Russ. J. Phys. Chem. (Engl. Transl.)* 51, 806, 1977.

199. Yada, N., Uematsu, M., and Watanabe, K., *Nippon Kikai Gakkai Ronbunshu, B-hen* 55, 2426, 1989.

200. Mears, W. H., Rosenthal, E., and Sinka, J. V., *J. Chem. Eng. Data* 11, 338, 1966.

201. Ma, P., Fang, Z., Zhang, J., and Ruan, Y., *J. Chem. Eng. Chin. Univ.* 6, 112, 1992.

202. Ma, P., Fang, Z., Zhang, J., and Ruan, Y., *Gaoxiao Huaxue Gongcheng Xuebao* 6, 112, 1992.

203. Schmidt, J. W., Carrillo-Nava, E., and Moldover, M. R., *Fluid Phase Equilib.* 122, 187, 1996.

204. Fukushima, M., and Watanabe, N., *Nippon Reito Kyokai Ronbunshu* 10, 75, 1993.

205. Liu, Z., Liang, D., He, M., Ju, B., and Yin, J., *Gongchengrewuli Xuebao* 18, 261, 1997.

206. Booth, H. S., and Swinehart, C. F., *J. Am. Chem. Soc.* 57, 1337, 1935.

207. Oliver, G. D., Grisard, J. W., and Cunningham, C. W., *J. Am. Chem. Soc.* 73, 5719, 1951.

208. Weber, L. A., *J. Chem. Eng. Data* 34, 171, 1989.

209. Shavandrin, A. M., and Li, S. A., *Inzh.-Fiz. Zh.* 37, 830, 1979.

210. Vitkalov, V. S., Kolpakov, Y. D., and Skripov, V. P., *Zh. Fiz. Khim.* 50, 2336, 1976.

211. Oguchi, K., Tanishita, I., Watanabe, K., Yamaguchi, T., and Sasayama, A., *Bull. JSME* 18, 1456, 1975.

212. Muratov, G. N., and Skripov, V. P., *Zh. Fiz. Khim.* 49, 2148, 1975.

213. Levelt Sengers, J. M. H., Straub, J., and Vincentini-Missoni, M., *J. Chem. Phys.* 54, 5034, 1971.

214. Tsiklis, D. S., and Prokhorov, V. M., *Dokl. Akad. Nauk SSSR* 174, 470, 1967.

215. Tsiklis, D. S., and Prokhorov, V. M., *Zh. Fiz. Khim.* 41, 2195, 1967.

216. Michels, A., Wassenaar, T., Wolkers, G. J., Prins, Chr., and van de Klundert, L., *J. Chem. Eng. Data* 11, 449, 1966.

217. Delaunois, C., *Ann. Mines Belg.* No. 1, 9, 1968.

218. Ambrose, D., *Trans. Faraday Soc.* 59, 1988, 1963.

219. Dewar, J., *Philos. Mag.* 18, 210, 1884.

220. Young, C. L., *Aust. J. Chem.* 25, 1625, 1972.

221. Hicks, C. P., and Young, C. L., *Trans. Faraday Soc.* 67, 1605, 1971.

222. Fischer, R., and Reichel, T., *Mikrochem. Ver. Mikrochim. Acta* 31, 102, 1943.

223. Zhang, R., Qin, Z., Wang, G., Dong, M., Hou, X., and Wang, J., *J. Chem. Eng. Data* 50, 1414, 2005.

224. Hugill, J. A., and McGlashan, M. L., *J. Chem. Thermodyn.* 10, 95, 1978.

225. Naziev, Y. M., Abasov, A. A., Nurberdiev, A. A., and Shakhverdiev, A. N., *Zh. Fiz. Khim.* 68, 434, 1974.

226. Krichevskii, I. R., and Sorina, G. A., *Zh. Fiz. Khim.* 34, 1420, 1960.

227. Richardson, M. J., and Rowlinson, J. S., *Trans. Faraday Soc.* 53, 1586, 1959.

228. Reamer, H. H., Sage, B. H., and Lacey, W. N., *Chem. Eng. Data Ser.* 3, 240, 1958.

229. Kay, W. B., and Albert, R. E., *Ind. Eng. Chem.* 48, 422, 1956.

230. Rotinyantz, L., and Nagornov, N. N., *Z. Phys. Chem., Abt. A* 169, 20, 1934.

231. Young, S., and Fortey, E. C., *J. Chem. Soc., Trans.* 75, 873, 1899.

232. Steele, W. V., Chirico, R. D., Knipmeyer, S. E., and Nguyen, A., *J. Chem. Eng. Data* 42, 1021, 1997.

233. Ambrose, D., and Ghiassee, N. B., *J. Chem. Thermodyn.* 19, 903, 1987.

234. Alekhin, O. D., Krupskii, N. P., and Minchenko, Y. B., *Ukr. Fiz. Zh. (Ukr. Ed.)* 15, 509, 1970.

235. Kudchadker, A. P., Alani, G. H., and Zwolinski, B. J., *Chem. Rev.* 68, 659, 1968.

236. Kay, W. B., *J. Am. Chem. Soc.* 69, 1273, 1947.

237. Lin, D. C. K., Silberberg, I. H., and McKetta, J. J., *J. Chem. Eng. Data* 15, 483, 1970.

238. Booth, H. S., and Morris, W. C., *J. Phys. Chem.* 62, 875, 1958.

239. Grzyll, L. R., Ramos, C., and Back, D. D., *J. Chem. Eng. Data* 41, 446, 1996.

240. Ermakov, G. V., and Skripov, V. P., *Zh. Fiz. Khim.* 43, 1308, 1969.

241. Cheng, D. C. H., McCoubrey, J. C., and Phillips, D. G., *Trans. Faraday Soc.* 58, 224, 1962.

242. Gude, M. T., and Teja, A. S., *Experimental Results for DIPPR 1990-91 Projects on Phase Equilibria and Pure Component Properties*, DIPPR Data Series No. 2, p. 174, 1994.

243. Anselme, M. J., Gude, M., and Teja, A. S., *Fluid Phase Equilib.* 57, 317, 1990.

244. Rosenthal, D. J., and Teja, A. S., *AIChE J.* 35, 1829, 1989.

245. Smith, R. L., Teja, A. S., and Kay, W. B., *AIChE J.* 33, 232, 1987.

246. Brunner, E., *J. Chem. Thermodyn.* 19, 823, 1987.

247. Smith, R. L., Anselme, M., and Teja, A. S., *Proc. World Congress III Chem. Eng.*, Tokyo, Vol. II, p. 135, 1986.

248. Gehrig, M., and Lentz, H., *J. Chem. Thermodyn.* 15, 1159, 1983.

249. Mogollon, E., Kay, W. B., and Teja, A. S., *Ind. Eng. Chem. Fundam.* 21, 173, 1982.

250. Cholpan, P. F., Sperkach, V. S., and Garkusha, L. N., *Fiz. Zhidk. Sostoyaniya* 9, 79, 1981.

251. Kay, W. B., and Pak, S. C., *J. Chem. Thermodyn.* 12, 673, 1980.

252. Chun, S. W., Ph.D. Thesis, Ohio State Univ., Columbus, OH, 1964.

253. Nikitin, E. D., Popov, A. P., Bogatishcheva, N. S., and Yatluk, Y. G., *J. Chem. Eng. Data* 49, 1515, 2004.

254. Quadri, S. K., Khilar, K. C., Kudchadker, A. P., and Patni, M. J., *J. Chem. Thermodyn.* 23, 67, 1991.

255. Anselme, M. J., and Teja, A. S., *Fluid Phase Equilib.* 40, 127, 1988.

256. Pulliam, M. K., Gude, M. T., and Teja, A. S., *J. Chem. Eng. Data* 40, 455, 1995.

257. Gude, M. T., Rosenthal, D. J., and Teja, A. S., *Fluid Phase Equilib.* 70, 55, 1991.

258. Nikitin, E. D., Popov, A. P., Bogatishcheva, N. S., and Yatluk, Y. G., *J. Chem. Eng. Data* 47, 1012, 2002.

259. Chirico, R. D., Gammon, B. E., Knipmeyer, S. E., Nguyen, A., Strube, M. M., Tsonopoulos, C., and Steele, W. V., *J. Chem. Thermodyn.* 22, 1075, 1990.

260. Chirico, R. D., Knipmeyer, S. E., Nguyen, A., and Steele, W. V., *J. Chem. Thermodyn.* 23, 431, 1991.

261. Nisel'son, L. A., Tret'yakova, K. V., Yatko, M. E., Tsirut, E. K., and Antonova, N. P., *Thermophysical Properties of Matter and Substances*, Vol. 4, Rabinovich, V. A., Ed., Amerind Pub., New Delhi, p. 132, 1975.

262. Toczylkin. L. S., and Young, C. L., *J. Chem. Thermodyn.* 12, 355, 1980.

263. Golik, A. Z., and Ravikovich, S. D., *Zh. Fiz. Khim.* 23, 86, 1949.

264. Stepanov, N. G., *Russ. J. Phys. Chem. (Engl. Transl.)* 46, 464, 1972.

265. Krauss, R., and Stephan, K., *J. Phys. Chem. Ref. Data* 18, 43, 1989.

266. Stepanov, N. G., and Nozdrev, V. F., *Russ. J. Phys. Chem. (Engl. Transl.)* 42, 1300, 1968.

267. Garcia-Sanchez, F., and Trejo Rodriguez, A., *J. Chem. Thermodyn.* 19, 359, 1987.

268. Garcia-Sanchez, F., and Trejo Rodriguez, A., *J. Chem. Thermodyn.* 17, 981, 1985.

269. Hojendahl, K., *Mat.-Fys. Medd. - K. Dan. Vidensk. Selsk.* 24, 1, 1946.

270. Duarte-Garza, H. A., Hwang, C.-A., Kellerman, S. A., Miller, R. C., Hall, K. R., Holste, J. C., Marsh, K. N., and Gammon, B. E., *J. Chem. Eng. Data* 42, 497, 1997.

271. Benning, A. F., and McHarness, R. C., *Ind. Eng. Chem.* 32, 814, 1940.

272. Gorchakovskii, V. K., Zadov, V. E., and Podvezennyi, V. N., *Inzh.-Fiz. Zh.* 59, 122, 1990.

273. Garcia-Sanchez, F., Romero-Martinez, A., and Trejo Rodriguez, A., *J. Chem. Thermodyn.* 21, 823, 1989.

274. Higashi, Y., Uematsu, M., and Watanabe, K., *Bull. JSME* 28, 2968, 1985.

275. Piao, C. C., Sato, H., and Watanabe, K., *J. Chem. Eng. Data* 36, 398, 1991.

276. Weber, L. A., and Levelt Sengers, J. M. H., *Fluid Phase Equilib.* 55, 241, 1990.

277. Tanikawa, S., Kabata, Y., Sato, H., and Watanabe, K., *J. Chem. Eng. Data* 35, 381, 1990.

278. Fukushima, M., Watanabe, N., and Kamimura, T., *Nippon Reito Kyokai Ronbunshu* 7, 243, 1990.

279. Fukushima, M., Watanabe, N., and Kamimura, T., *Nippon Reito Kyokai Ronbunshu* 7, 189, 1990.

280. Yamashita, T., Kubota, H., Tanaka, Y., Makita, T., and Kashiwagi, H., *Proc. 10th Symp. Thermophys. Prop.*, Japan, pp. 75–78, 1989.

281. Mandlekar, A. V., Kay, W. B., Smith, R. L., and Teja, A. S., *Fluid Phase Equilib.* 23, 79, 1985.

282. Herz, W., *Z. Anorg. Allg. Chem.* 149, 230, 1925.

283. Nikitin, E. D., Pavlov, P. A., and Popov, A. P., *J. Chem. Thermodyn.* 27, 43, 1995.

284. Ratzsh, M. T., *Z. Phys. Chem. Leipzig* 243, 212, 1970.

285. Schmidt, G. C., *Justus Liebigs Ann. Chem.* 266, 266, 1891.

286. Young, C. L., *Int. DATA Ser., Sel. Data Mixtures, Ser. A*, No. 1, 66, 1975.

287. Ambrose, D., Sprake, C. H. S., and Townsend, R., *J. Chem. Thermodyn.* 4, 247, 1972.

288. Zawisza, A. C., *Bull. Acad. Pol. Sci., Ser. Sci. Chim.* 15, 291, 1967.

289. Stryjek, R., and Kreglewski, A., *Bull. Acad. Pol. Sci., Ser. Sci. Chim.* 13, 201, 1965.

290. Kobe, K. A., Ravicz, A. E., and Vohra, S. P., *J. Chem. Eng. Data* 1, 50, 1956.

291. Schroeer, E., *Z. Phys. Chem., Abt. A* 140, 379, 1929.

292. Schroeer, E., *Z. Phys. Chem., Abt. A* 140, 241, 1929.

293. Wilip, J., *Eesti Vabariigi Tartu Ulik. Toim. A* 6 (2), 1924.

294. Audant, *C. R. Hebd. Seances Acad. Sci.* 170, 1573, 1920.

295. Prins, A., and Scheffer, F. E. C., *J. Phys. Chem.* 84, 827, 1913.

296. Travers, M. W., and Usher, F. L., *Z. Phys. Chem., Stoechiom. Verwandtschaftsl.* 57, 365, 1906.

297. Centerszwer, M., and Pakalneet, A., *Z. Phys. Chem., Stoechiom. Verwandtschaftsl.* 55, 303, 1906.

298. Smits, A., *Z. Phys. Chem., Stoechiom. Verwandtschaftsl.* 52, 587, 1905.

299. Galitzine, B., and Wilip, J., *Bull. Acad. Pet.* 11, No. 3, 117, 1901.

300. De Vries, E. C., *Arch. Neerl. Sci. Exactes Nat.* 28, 215, 1895.

301. Ramsay, W., and Young, S., *Philos. Trans. R. Soc. London, A* 178, 57, 1887.

302. Radice, G., Ph.D. Thesis, Univ. of Geneve, 1899.

303. Vespigniani, G. R., *Gazz. Chim. Ital.* 33, 73, 1903.

304. Berthoud, A., and Brum, R., *J. Chim. Phys. Phys.-Chim. Biol.* 21, 143, 1924.

305. Grebenkov, A. J., Zhelezny, V. P., Klepatsky, P. M., Beljajeva, O. V., Chernjak, Y. A., Kotelevsky, Y. G., and Timofejev, B. D., *Int. J. Thermophys.* 17, 535, 1996.

306. Holcomb, C. D., Niesen, V. G., Van Poolen, L. J., and Outcalt, S. L., *Fluid Phase Equilib.* 91, 145, 1993.

307. Higashi, Y., Ashizawa, M., Kabata, Y., Majima, T., Uematsu, M., and Watanabe, K., *JSME Int. J.* 30, 1106, 1987.

308. Otake, K., Yasumoto, M., Yamada, Y., Murata, J., and Urata, S., *J. Chem. Eng. Data* 48, 1380, 2003.

309. Pitschmann, M., and Straub, J., *Int. J. Thermophys.* 23, 877, 2002.

310. Diefenbacher, A., and Tuerk, M., *J. Chem. Thermodyn.* 31, 905, 1999.

311. Shi, L., Zhu, M., Han, L., Duan, Y., Sun, L;, and Fu, Y.-D., *Science in China, Ser. E* 41, 435, 1998.

312. Kuwabara, S., Aoyama, H., Sato, H., and Watanabe, K., *J. Chem. Eng. Data* 40, 112, 1995.

313. Higashi, Y., *Int. J. Thermophys.* 16, 1175, 1995.

314. Fu, Y. D., Han, L.-Z., and Zhu, M.-S., *Fluid Phase Equilib.* 111, 273, 1995.

315. Higashi, Y., *Int. J. Refrig.* 17, 524, 1994.

316. Malbrunot, P. F., Meunier, P. A., Scatena, G. M., Mears, W. H., Murphy, K. P., and Sinka, J. V., *J. Chem. Eng. Data* 13, 16, 1968.

317. Steele, W. V., Chirico, R. D., Cowell, A. B., Knipmeyer, S. E., and Nguyen, A., *J. Chem. Eng. Data* 47, 725, 2002.

318. Young, C. L., *Int. DATA Ser., Sel. Data Mixtures, Ser. A*, No. 1, 159, 1975.

319. Durig, J. R., and Li, Y. S., *J. Chem. Phys.* 63, 4110, 1975.

320. Bourgou, A., *Bull. Soc. Chim. Belg.* 33, 101, 1924.

321. Verevkin, S. P., Kozlova, S. A., Emel'yanenko, V. N., Nikitin, E. D., Popov, A. P., and Krasnykh, E. L., *J. Chem. Eng. Data* 51, 1896, 2006.

322. Kay, W. B., and Young, C. L., *Int. DATA Ser., Sel. Data Mixtures, Ser. A*, No. 2, 154, 1974.

323. Weaver, D. L., M.S. Thesis, Ohio State Univ., Columbus, OH, 1973.

324. Young, C. L., *Int. DATA Ser., Sel. Data Mixtures, Ser. A*, No. 1, 47, 1974.

325. Genco, J. M., Teja, A. S., and Kay, W. B., *J. Chem. Eng. Data* 25, 350, 1980.

326. Kay, W. B., and Young, C. L., *Int. DATA Ser., Sel. Data Mixtures, Ser. A*, No. 1, 52, 1975.

327. Genco, J. M., Ph.D. Thesis, Ohio State Univ., Columbus, OH, 1965.

328. Kay, W. B., *J. Am. Chem. Soc.* 68, 1336, 1946.

329. Quadri, S. K., and Kudchadker, A. P., *J. Chem. Thermodyn.* 24, 473, 1992.

330. Young, S., and Fortey, E. C., *J. Chem. Soc.* 35, 1126, 1879.

331. Steele, W. V., Chirico, R. D., Knipmeyer, S. E., Nguyen, A., and Smith, N. K., *J. Chem. Eng. Data* 42, 1037, 1997.

332. Steele, W. V., Chirico, R. D., Knipmeyer, S. E., and Nguyen, A., *J. Chem. Eng. Data* 42, 1008, 1997.

333. Ihmels, E. C. C., and Lemmon, E. W., *Fluid Phase Equilib.* 260, 36, 2007.

334. Wu, J., Liu, Z., Wang, B., and Pan, J., *J. Chem. Eng. Data* 49, 704, 2004.

335. Noles, J. R., and Zollweg, J. A., *Fluid Phase Equilib.* 66, 275, 1991.

336. Glowka, S., *Bull. Acad. Pol. Sci., Ser. Sci. Chim.* 20, 163, 1972.

337. Zawisza, A. C., and Glowka, S., *Bull. Acad. Pol. Sci., Ser. Sci. Chim.* 18, 549, 1970.

338. Edwards, J., and Maass, O., *Can. J. Res., Sect. A* 12, 357, 1935.

339. Tapp, J. S., Steacie, E. W. R., and Maass, O., *Can. J. Res.* 9, 217, 1933.

340. Cardoso, E., and Coppola, A. A., *J. Chim. Phys. Phys.-Chim. Biol.* 20, 337, 1923.

341. Cardoso, E., and Bruno, A., *J. Chim. Phys. Phys.-Chim. Biol.* 20, 347, 1923.

342. Briner, E., and Cardoso, E., *J. Chim. Phys. Phys.-Chim. Biol.* 6, 641, 1908.

343. Briner, E., and Cardoso, E., *C. R. Hebd. Seances Acad. Sci.* 144, 911, 1907.

344. Bogoslovskii, V. E., Mikhalyuk, G. I., and Shamolin, A. I., *Zh. Prikl. Khim. (Leningrad)* 45, 1154, 1972.

345. Kay, W. B., and Hissong, D. W., *Proc. - Am. Pet. Inst., Div. Refin.* 49, 13, 1969.

346. McMicking, J. H., and Kay, W. B., *Proc., Am. Pet. Inst., Sect. 3* 45, 75, 1965.

347. Chirico, R. D., Knipmeyer, S. E., Nguyen, A., and Steele, W. V., *J. Chem. Thermodyn.* 25, 1461, 1993.

348. Stern, S. A., and Kay, W. B., *J. Phys. Chem.* 61, 374, 1957.

349. Edgar, G., and Calingaert, G., *J. Am. Chem. Soc.* 51, 1540, 1929.

350. Francis, A. W., *Ind. Eng. Chem.* 49, 1779, 1957.

351. Cox, J. D., *Trans. Faraday Soc.* 56, 959, 1960.

352. Steele, W. V., Chirico, R. D., Nguyen, A., and Knipmeyer, S. E., *J. Chem. Thermodyn.* 27, 311, 1995.

353. Nikitin, E. D., Pavlov, P. A., and Popov, A. P., *J. Chem. Thermodyn.* 26, 1047, 1994.

354. Cristou, G., Young, C. L., and Svejda, P., *Fluid Phase Equilib.* 67, 45, 1991.

355. Zhuravlev, D. I., *Zh. Fiz. Khim.* 9, 875, 1937.

356. Chirico, R. D., and Steele, W. V., *J. Chem. Eng. Data* 50, 1052, 2005.

357. Smith, R. L., Ph.D. Dissertation, Georgia Institute of Technology 1985.

358. Nikitin, E. D., Pavlov, P. A., and Popov, A. P., *Fluid Phase Equilib.* 141, 155, 1997.

359. Nikitin, E. D., Pavlov, P. A., and Bessonova, N. V., *J. Chem. Thermodyn.* 26, 177, 1994.

360. Nikitin, E. D., Pavlov, P. A., and Popov, A. P., *Fluid Phase Equilib.* 189, 151, 2001.

361. Nikitin, E. D., Pavlov, P. A., and Popov, A. P., *Fluid Phase Equilib.* 149, 223, 1998.

362. Sinicyn, E. N., Mikhalevich, L. A., and Yankovskaya, O. P., *Deposited Doc. VINITI*, Doc. No. 2510-V90, 1990.

363. Teja, A. S., Gude, M., and Rosenthal, D. J., *Fluid Phase Equilib.* 52, 193, 1989.

364. Beale, E. S., and Docksey, P., *J. Inst. Pet.* 21, 860, 1935.

365. Nikitin, E. D., and Popov, A. P., *Fluid Phase Equilib.* 166, 237, 1999.

366. Horstman, S., Fischer, K., Gmehling, J., and Kolar, P., *J. Chem. Thermodyn.* 32, 451, 2000.

367. Colgate, S. O., Sivaraman, A., and Dejsupa, C., *Fluid Phase Equilib.* 76, 175, 1992.

368. Friend, D. G., Ingham, H., and Ely, J. F., *J. Phys. Chem. Ref. Data* 20, 275, 1991.

369. Jangkamolkulchai, A., and Luks, K. D., *J. Chem. Eng. Data* 34, 92, 1989.

370. Calado, J. C. G., Chang, E., Clancy, P., and Streett, W. B., *J. Phys. Chem.* 91, 3914, 1987.

371. Brunner, E., *J. Chem. Thermodyn.* 17, 871, 1985.

372. Morrison, G., and Kincaid, J. M., *AIChE J.* 30, 257, 1984.

373. Sychev, V. V., Vasserman, A. A., Kozlov, A. D., Zagoruchenko, V. A., Spiridonov, G. A., and Tsymarny, V. A., *Thermodynamic Properties of Ethane*, Standards Publishing House, Moscow, 1982.

374. Morrison, G., *J. Phys. Chem.* 85, 759, 1981.

375. Bulavin, L. A., and Shimanskii, Yu. I., *Zh. Eksp. Teor. Fiz.* 29, 482, 1979.

376. Strumpf, H. J., Collings, A. F., and Pings, C. J., *J. Chem. Phys.* 60, 3109, 1974.

377. Burton, M., and Balzarini, D., *Can. J. Phys.* 52, 2011, 1974.

378. Douslin, D. R., and Harrison, R. H., *J. Chem. Thermodyn.* 5, 491, 1973.

379. Berestov, A. T., Giterman, M. S., and Shmakov, N. G., *Sov. Phys. - JETP (Engl. Transl.)* 37, 1128, 1973.

380. Efremova, G. D., and Shvarts, A. V., *Russ. J. Phys. Chem. (Engl. Transl.)* 46, 237, 1972.

381. Miniovich, V. M., and Sorina, G. A., *Russ. J. Phys. Chem. (Engl. Transl.)* 45, 306, 1971.

382. Khazanova, N. E., and Sominskaya, E. E., *Russ. J. Phys. Chem. (Engl. Transl.)* 45, 88, 1971.

383. Bulavin, L. A., Ostanevich, Yu M., Simkina, A. P., and Stelkov, A. V., *Ukr. Fiz. Zh. (Ukr. Ed.)* 16, 90, 1971.

384. Chashkin, Yu. R., Smirnov, V. A., and Voronel, A. V., *Teplofiz. Svoistva Veshchestv Mater.* 2, 139, 1970.

385. Sliwinski, P., *Z. Phys. Chem. (Munich)* 63, 263, 1969.

386. Khazanova, N. E., Lesnevskaya, L. S., and Zakharova, A. V., *Khim. Prom-st. (Moscow)* 42, 364, 1966.

387. Matschke, D. E., and Thodos, G., *J. Chem. Eng. Data* 7, 232, 1962.

388. Schmidt, E., and Thomas, W., *Forsch. Geb. Ingenieurw.* 20B, 161, 1954.

389. Kay, W. B., and Brice, D. B., *Ind. Eng. Chem.* 45, 615, 1953.

390. Murray, F. E., and Mason, S. G., *Can. J. Chem.* 30, 550, 1952.

391. Mason, S. G., Naldrett, S. N., and Maass, O., *Can. J. Res., Sect. B* 18, 103, 1940.

392. Beattie, J. A., Su, G.-J., and Simard, G. L., *J. Am. Chem. Soc.* 61, 924, 1939.

393. Price, T. W., *J. Chem. Soc.* 107, 188, 1915.

394. Nikitin, E. D., Pavlov, P. A., and Skripov, P. V., *J. Chem. Thermodyn.* 25, 869, 1993.

395. Marshall, W. L., and Jones, E. V., *J. Inorg. Nucl. Chem.* 36, 2319, 1974.

396. Sajotschewsky, W., *Beibl. Ann. Phys.* 3, 741, 1879.

397. Mocharnyuk, R. F., *Zh. Obshch. Khim.* 30, 1098, 1960.

398. Golik, A. Z., Ravikovich, S. D., and Orishchenko, A. V., *Ukr. Khim. Zh. (Russ. Ed.)* 21, 167, 1955.

399. Polikhronidi, N. G., Abdulagatov, I. M., Stepanov, G. V., and Batyrova, R. G., *J. Supercrit. Fluids* 43, 1, 2007.

400. Hu, T., Qin, Z., Wang, G., Hou, X., and Wang, J., *J. Chem. Eng. Data* 49, 1809, 2004.

401. Sauermann, P., Holzapfel, K., Oprzynski, J., Kohler, F., Poot, W., and De Loos, T. W., *Fluid Phase Equilib.* 112, 249, 1995.

402. Mousa, A. H. N., *J. Chem. Eng. Jpn.* 20, 635, 1987.

403. Wilson, K. S., Lindley, D. D., Kay, W. B., and Hershey, H. C., *J. Chem. Eng. Data* 29, 243, 1984.

404. Hentze, G., *Thermochim. Acta* 20, 27, 1977.

405. Nozdrev, V. F., *Akust. Zh.* 2, 209, 1956.

406. Griswold, J., Haney, J. D., and Klein, V. A., *Ind. Eng. Chem.* 35, 701, 1943.

407. Battelli, A., *Mem. Torino, Ser. 2* 44, 57, 1893.

408. Ramsay, W., and Young, S., *Philos. Trans. R. Soc. London* 177, 123, 1886.

409. Strauss, O, *Beibl. Ann. Phys.* 6, 282, 1882.

410. Steele, W. V., Chirico, R. D., Knipmeyer, S. E., and Nguyen, A., *Experimental Results for DIPPR 1990–91 Projects on Phase Equilibria and Pure Component Properties*, DIPPR Data Series No. 2, 154, 1994.

411. Ambrose, D., Ellender, J. H., Gundry, H. A., Lee, D. A., and Townsend, R., *J. Chem. Thermodyn.* 13, 795, 1981.

412. Lambert, J. D., Clarke, J. S., Duke, J. F., Hicks, C. L., Lawrence, S. D., Morris, D. M., and Shone, M. G. T., *Proc. R. Soc. London, A* 249, 414, 1959.

413. Young, S., and Thomas, G. L., *J. Chem. Soc.* 63, 1191, 1893.

414. Pohland, E., and Mehl, W., *Z. Phys. Chem., Abt. A* 164, 48, 1933.

415. Nadezhdin, A., *Rep. Phys.* 23, 708, 1887.

416. Guseinov, K. D., and Zhabbarov, O., *Izv. Vyssh. Uchebn. Zaved. Neft Gaz* 2, 76, 1975.

417. Steele, W. V., Chirico, R. D., Cowell, A. B., Knipmeyer, S. E., and Nguyen, A., *J. Chem. Eng. Data* 47, 700, 2002.

418. Nowak, P., Kleinrahm, R., and Wagner, W., *J. Chem. Thermodyn.* 28, 1441, 1996.

419. Hasch, B. M., and McHugh, M. A., *Fluid Phase Equilib.* 64, 251, 1991.

420. Jahangiri, M., Jacobsen, R. T., Stewart, R. B., and McCarty, R. D., *J. Phys. Chem. Ref. Data* 15, 593, 1986.

421. Younglove, B. A., *J. Phys. Chem. Ref. Data*, Vol. 11, Suppl. No. 1, Am. Chem. Soc., Washington, DC, 1982.

422. McCarty, R. D., and Jacobsen, R. T., *NBS Tech. Note (U.S.)* 1045, 1981.

423. Thomas, W., and Zander, M., *Int. J. Thermophys* 1, 383, 1980.

424. Hastings, J. R., Levelt Sengers, J. M. H., and Balfour, F. W., *J. Chem. Thermodyn.* 12, 1009, 1980.

425. Hastings, J. R., and Levelt Sengers, J. M. H., *Proc. 7th Symp. Thermophys. Prop.*, Cezairliyan, A., Ed., ASME, New York, p. 794, 1977.

426. Douslin, D. R., and Harrison, R. H., *J. Chem. Thermodyn.* 8, 301, 1976.

427. Bender, E., *Cryogenics* 15, 667, 1975.

428. Moldover, M. R., *J. Chem. Phys.* 61, 1766, 1974.

429. Angus, S., Armstrong, B., and de Reuck, K. M., *International Thermodynamic Tables of the Fluid State - 2 Ethylene*, Butterworths, London, 1974.

430. Zernov, V. S., Kogan, V. B., and Lyubetskii, S. G., *Zh. Prikl. Khim. (Leningrad)* 44, 1819, 1971.

431. Shim, J., and Kohn, J. P., *J. Chem. Eng. Data* 9, 1, 1964.

432. Kay, W. B., *Ind. Eng. Chem.* 40, 1459, 1948.
433. Diepen, G. A. M., and Scheffer, F. E. C., *J. Am. Chem. Soc.* 70, 4081, 1948.
434. Naldrett, S. N., and Maass, O., *Can. J. Res., Sect. B* 18, 118, 1940.
435. Dacey, J. R., McIntosh, R. L., and Maass, O., *Can. J. Res., Sect. B* 17, 206, 1939.
436. Maass, O., and Geddes, A. L., *Philos. Trans. R. Soc. London, A* 236, 303, 1937.
437. Lawrenson, I. J., and Lee, D. A., *J. Chem. Thermodyn.* 10, 1111, 1978.
438. Zawisza, A. C., and Glowka, S., *Bull. Acad. Pol. Sci., Ser. Sci. Chim.* 18, 555, 1970.
439. Lyons, R. L., M.S. Thesis, Pennsylvania State Univ., University Park, PA, 1985.
440. Daubert, T. E., Jalowka, J. W., and Goren, V., *AIChE Symp. Ser.* 83 (256), 128, 1987.
441. Douslin, D. R., Moore, R. T., Dawson, J. P., and Waddington, G., *J. Am. Chem. Soc.* 80, 2031, 1958.
442. Beyerlein, A. L., DesMarteau, D. D., Kul, I., and Zhao, G., *Fluid Phase Equilib.* 150, 287, 1998.
443. Parthasarathy, S., *Proc. - Indian Acad. Sci., Sect. A* 2, 497, 1935.
444. Bominaar, S. A. R. C., Trappeniers, N. J., and Biswas, S. N., *J. Phys. Chem.* 94, 1097, 1990.
445. Bominaar, S. A. R. C., Biswas, S. N., Trappeniers, N. J., and Ten Seldam, C. A., *J. Chem. Thermodyn.* 19, 959, 1987.
446. Froba, A. P., Botero, C., and Leipertz, A., *Int. J. Thermophys.* 27, 1609, 2006.
447. Uchida, Y., Yasumoto, M., Yamada, Y., Ochi, K., Furuya, T., and Otake, K., *J. Chem. Eng. Data* 49, 1615, 2004.
448. Otake, K., Uchida, Y., Yasumoto, M., Yamada, Y., Furuya, T., and Ochi, K., *J. Chem. Eng. Data* 49, 1643, 2004.
449. Hu, P., and Chen, Z. S., *Fluid Phase Equilib.* 221, 7, 2004.
450. Salvi-Narkhede, M., Adcock, J. L., Gakh, A., and Van Hook, W. A., *J. Chem. Thermodyn.* 25, 643, 1993.
451. Ambrose, D., and Ghiassee, N. B., *J. Chem. Thermodyn.* 20, 1231, 1988.
452. Myers, J. E., Hershey, H. C., and Kay, W. B., *J. Chem. Thermodyn.* 11, 1019, 1979.
453. Golik, A. Z., and Adamenko, I. I., *Ukr. Fiz. Zh. (Ukr. Ed.)* 10, 443, 1965.
454. Golik, A. Z., and Ivanova, I. I., *Zh. Fiz. Khim.* 36, 1768, 1962.
455. Kay, W. B., *Ind. Eng. Chem.* 30, 459, 1938.
456. Christou, G., Sadus, R. J., and Young, C. L., *Fluid Phase Equilib.* 67, 259, 1991.
457. Kurumov, D. S., Grigor'ev, B. A., and Vasil'ev, Yu. L., *Teplofiz. Svoistva Veshchestv Mater.* No. 27, 101, 1989.
458. De Loos, T. W., Poot, W., and De Swaan Arons, J., *Fluid Phase Equilib.* 42, 209, 1988.
459. Artyukhovskaya, L. M., Shimanskaya, E. T., and Shimanskii, Yu I., *Opt. Spektrosk.* 37, 935, 1974.
460. Artyukhovskaya, L. M., Shimanskaya, E. T., and Shimanskii, Yu I., *Sov. Phys. - JETP (Engl. Transl.)* 37, 848, 1973.
461. Artyukhovskaya, L. M., Shimanskaya, E. T., and Shimanskii, Yu I., *Zh. Eksp. Teor. Fiz.* 63, 2159, 1972.
462. Smith, L. B., Beattie, J. A., and Kay, W. C., *J. Am. Chem. Soc.* 59, 1587, 1937.
463. Beattie, J. A., and Kay, W. C., *J. Am. Chem. Soc.* 59, 1586, 1937.
464. Rosenthal, D. J., Gude, M. T., Teja, A. S., and Mendez-Santiago, J., *Fluid Phase Equilib.* 135, 89, 1997.
465. Smith, R. L., Anselme, M. J., and Teja, A. S., *Fluid Phase Equilib.* 31, 161, 1986.
466. Pulliam, M. K., Gude, M. T., and Teja, A. S., *Experimental Results for DIPPR 1990–91 Projects on Phase Equilibria and Pure Component Properties*, DIPPR Data Ser. No. 2, p. 184, 1994.
467. Naziev, Y. M., and Abasov, A. A., *Izv. Vyssh. Uchebn. Zaved., Neft Gaz* 12, 81, 1969.
468. Mousa, A. H. N., *J. Chem. Eng. Data* 26, 248, 1981.
469. Hicks, C. P., and Young, C. L., *Chem. Rev.* 75, 119, 1975.
470. Mousa, A. H. N., Kay, W. B., and Kreglewski, A., *J. Chem. Thermodyn.* 4, 301, 1972.
471. Douslin, D. R., Harrison, R. H., and Moore, R. T., *J. Chem. Thermodyn.* 1, 305, 1969.

472. Evans, F. D., and Tiley, P. F., *J. Chem. Soc. B* 134, 1966.
473. Counsell, J. F., Green, J. H. S., Hales, J. L., and Martin, J. F., *Trans. Faraday Soc.* 61, 212, 1965.
474. Saikawa, K., Kijima, J., Uematsu, M., and Watanabe, K., *J. Chem. Eng. Data* 24, 165, 1979.
475. Kijima, J., Saikawa, K., Watanabe, K., Oguchi, K., and Tanishita, I., *Proc. 7th Symp. Thermophys. Prop.*, Cezairliyan, A., Ed., ASME, New York, p. 480, 1977.
476. Kim, K. Y., Ph.D. Dissertation, Univ. Michigan, Ann Arbor, MI, 1974.
477. Swarts, F., *Bull. Soc. Chim. Belg.* 42, 114, 1933.
478. Aoyama, H., Kishizawa, G., Sato, H., and Watanabe, K., *J. Chem. Eng. Data* 41, 1046, 1996.
479. Aoyama, H., Sato, H., and Watanabe, K., *Sixteenth Japan Symposium on Thermophysical Properties*, Hiroshima, p. 173, 1995.
480. McLure, I. A., and Dickinson, E., *J. Chem. Thermodyn.* 8, 93, 1976.
481. VonNiederhausern, D. M., Wilson, G. M., and Giles, N. F., *J. Chem. Eng. Data* 45, 157, 2000.
482. Liu, J., Qin, Z., Wang, G., Hou, X., and Wang, J., *J. Chem. Eng. Data* 48, 1610, 2003.
483. Gude, M. T., and Teja, A. S., *Fluid Phase Equilib.* 83, 139, 1993.
484. Grigor'ev, B. A., Rastorguev, Yu. L., Gerasimov, A. A., Kurumov, D. S., and Plotnikov, S. A., *Int. J. Thermophys.* 9, 439, 1988.
485. Zawisza, A., *J. Chem. Thermodyn.* 17, 941, 1985.
486. Mousa, A. H. N., *J. Chem. Thermodyn.* 9, 1063, 1977.
487. Young, C. L., *Int. DATA Ser., Sel. Data Mixtures, Ser. A*, No. 1, 157, 1975.
488. Nichols, W. B., Reamer, H. H., and Sage, B. H., *AIChE J.* 3, 262, 1957.
489. VonNiederhausern, D. M., Wilson, G. M., and Giles, N. F., *J. Chem. Eng. Data* 51, 1986, 2006.
490. Ma, P., Ma, Y., and Zhang, J., *Gaoxiao Huaxue Gongcheng Xuebao* 5, 175, 1991.
491. Masui, G., Honda, Y., and Uematsu, M., *J. Chem. Thermodyn.* 38, 1711, 2006.
492. Goodwin, R. D., and Haynes, W. M., *NBS Tech. Note (U.S.)* No. 1051, 1982.
493. Pryanikova, R. O., Plenkina, R. M., Kuzyakina, N. V., and Markina, I. A., *Khim. Prom-st. (Moscow)*, 13. 1987.
494. Beattie, J. A., Ingersoll, H. G., and Stockmayer, W. H., *J. Am. Chem. Soc.* 64, 546, 1942.
495. Vohra, S. P., and Kobe, K. A., *J. Chem. Eng. Data* 4, 329, 1959.
496. Smith, R. L., Negishi, E., Arai, K., and Saito, S., *J. Chem. Eng. Jpn.* 23, 99, 1990.
497. Steele, W. V., Chirico, R. D., Cowell, A. B., Knipmeyer, S. E., and Nguyen, A., *J. Chem. Eng. Data* 42, 1053, 1997.
498. Setzmann, U., and Wagner, W., *J. Phys. Chem. Ref. Data* 20, 1061, 1991.
499. Friend, D. G., Ely, J. F., and Ingham, H., *J. Phys. Chem. Ref. Data* 18, 583, 1989.
500. Kleinrahm, R., and Wagner, W., *J. Chem. Thermodyn.* 18, 739, 1986.
501. Calado, J. C. G., Dieters, U., and Strett, W. B., *J. Chem. Soc., Faraday Trans. 1* 77, 2503, 1981.
502. Angus, S., Armstrong, B., and de Reuck, K. M., *International Thermodynamic Tables of the Fluid State - 5 Methane*, Pergamon, Oxford, 1978.
503. Goodwin, R. D., *NBS Tech. Note (U.S.)* No. 653, 1974.
504. Gielen, H., Jansoone, F., and Verbeke, O. B., *J. Chem. Phys.* 59, 5763, 1973.
505. Jansoone, V., Gielen, H., De Boelpaep, J., and Verbeke, O. B., *Physica (Amsterdam)* 46, 213, 1970.
506. Terry, M. J., Lynch, J. T., Bunclark, M., Mansell, K. R., and Staveley, L. A. K., *J. Chem. Thermodyn.* 1, 413, 1969.
507. Grigor, A. F., and Steele, W. A., *J. Chem. Phys.* 48, 1032, 1968.
508. Keyes, F. G., Taylor, R. S., and Smith, L. B., *J. Math. Phys. (Cambridge, Mass.)* 1, 211, 1922.
509. Kuenen, J. P., *Philos. Mag.* 6, 637, 1903.
510. Crismer, L., *Bull. Soc. Chim. Belg.* 18, 18, 1904.
511. Schmidt, G. C., *Z. Phys. Chem., Stoechiom. Verwandtschaftsl.* 8, 628, 1891.
512. Polikhronidi, N. G., Radzhabova, L. M., Rasulov, A. R., and Stepanov, G. V., *High Temp. (Engl. Transl.)* 44, 512, 2006.
513. Francesconi, Artur Zaghini Lentz, H., and Franck, E. U., *J. Phys. Chem.* 85, 3303, 1981.

514. Swami, D. R., Kumarkrishna Rao, V. N., and Narasinga Rao, N., *Trans., Indian Inst. Chem. Eng.* 9, 32, 1956.

515. Salzwedel, E., *Ann. Phys. (Leipzig)* 5, 853, 1930.

516. Ramsay, W., and Young, S., *Philos. Trans. R. Soc. London, A* 178, 313, 1887.

517. Steele, W. V., Chirico, R. D., Cowell, A. B., Knipmeyer, S. E., and Nguyen, A., *J. Chem. Eng. Data* 47, 667, 2002.

518. Gurarii, L. L., Kuleshov, G. G., Baglai, A. K., and Petrashkevich, R. I., *Khim.-Farm. Zh.* 21, 247, 1987.

519. Guseinov, S. O., Farzaliev, B. I., and Naziev, Y. M., *Izv. Vyssh. Uchebn. Zaved., Neft Gaz* 22, 52, 1979.

520. Kreglewski, A., *Bull. Acad. Pol. Sci., Cl. 3* 2, 191, 1954.

521. Kiyama, R., Suzuki, K., and Ikegami, T., *Rev. Phys. Chem. Jpn.* 21, 50, 1951.

522. Abara, J. A., Jennings, D. W., Kay, W. B., and Teja, A. S., *J. Chem. Eng. Data* 33, 242, 1988.

523. Wilson, G. M., Johnston, R. H., Hwang, S.-C., and Tsonopoulos, C., *Ind. Eng. Chem. Process Des. Dev.* 20, 94, 1981.

524. Rutenberg, O. L., and Shakhova, S. F., *Russ. J. Phys. Chem. (Engl. Transl.)* 47, 124, 1973.

525. Yasumoto, M., Uchida, Y., Ochi, K., Furuya, T., Shono, A., and Otake, K., *J. Chem. Eng. Data* 52, 1726, 2007.

526. Day, H. O., and Felsing, W. A., *J. Am. Chem. Soc.* 74, 1951, 1952.

527. Ambrose, D., and Ghiassee, N. B., *J. Chem. Thermodyn.* 20, 767, 1988.

528. Chirico, R. D., Knipmeyer, S. E., Nguyen, A., and Steele, W. V., *J. Chem. Thermodyn.* 31, 339, 1999.

529. Chirico, R. D., Johnson, R. D. I., and Steele, W. V., *J. Chem. Thermodyn.* 39, 698, 2007.

530. Chirico, R. D., and Steele, W. V., *J. Chem. Eng. Data* 50, 697, 2005.

531. Schroeer, E., *Z. Phys. Chem., Abt. B* 49, 271, 1941.

532. Dawson, P. P., Silberberg, I. H., and McKetta, J. J., *J. Chem. Eng. Data* 18, 7, 1973.

533. Ambrose, D., Counsell, J. F., and Hicks, C. P., *J. Chem. Thermodyn.* 10, 771, 1978.

534. Griffin, D. N., *J. Am. Chem. Soc.* 71, 1423, 1949.

535. Matzik, I., and Schneider, G. M., *Ber. Bunsen-Ges. Phys. Chem.* 89, 551, 1985.

536. Ambrose, D., and Townsend, R., *Trans. Faraday Soc.* 64, 2622, 1968.

537. Defibaugh, D. R., Carrillo-Nava, E., Hurly, J. J., Moldover, M. R., Schmidt, J. W., and Weber, L. A., *J. Chem. Eng. Data* 42, 488, 1997.

538. Young, C. L., *J. Chem. Thermodyn.* 4, 65, 1972.

539. Lindley, D. D., and Hershey, H. C., *Fluid Phase Equilib.* 55, 109, 1990.

540. Kreglewski, A., *Bull. Acad. Pol. Sci., Cl. 3* 5, 323, 1957.

541. Young, S., *J. Chem. Soc.* 77, 1145, 1900.

542. Steele, W. V., Chirico, R. D., Knipmeyer, S. E., Nguyen, A., Smith, N. K., and Tasker, I. R., *J. Chem. Eng. Data* 41, 1269, 1996.

543. Post, R. G., *Unpublished Rep., Chem. Eng. No. 362*, Univ. Texas, Austin, TX, 1950.

544. Hess, L. G., and Tilton, V. V., *Ind. Eng. Chem.* 42, 1251, 1950.

545. Thodos, G., *AIChE J.* 3, 428, 1957.

546. Cheng, D. C. H., and McCoubrey, J. C., *J. Chem. Soc.* 4993, 1963.

547. Teja, A. S., and Smith, R. L., *AIChE J.* 33, 1560, 1987.

548. Ambrose, D., and Sprake, C. H. S., *J. Chem. Soc. A* 1263, 1971.

549. Yata, J., Hori, M., Kohno, K., and Minamiyama, T., *High Temp. - High Press.* 29, 19, 1997.

550. Duarte-Garza, H. A., Stouffer, C. E., Hall, K. R., Hall, K. R., Holste, J. C., Marsh, K. N., and Gammon, B. E., *J. Chem. Eng. Data* 42, 745, 1997.

551. Ye, F., Sato, H., and Watanabe, K., *J. Chem. Eng. Data* 40, 148, 1995.

552. Tsvetkov, O. B., Kletskii, A. V., Laptev, Yu. A., Asambaev, A. J., and Zausaev, I. A., *Int. J. Thermophys.* 16, 1185, 1995.

553. Sagawa, T., Sato, H., and Watanabe, K., *High Temp. - High Press.* 26, 193, 1994.

554. Wilson, L. C., Wilding, W. V., Wilson, G. M., Rowley, R. L., Felix, V. M., and Chisolm-Carter, T., *Fluid Phase Equilib.* 80, 167, 1992.

555. Shank, R. L., *J. Chem. Eng. Data* 12, 474, 1967.

556. Gude, M. T., and Teja, A. S., *AIChE Symp. Ser.* 90 (298), 14, 1994.

557. Grigor'ev, B. A., Rastorguev, Yu. L., Kurumov, D. S., Gerasimov, A. A., Kharin, V. E., and Plotnikov, S. A., *Int. J. Thermophys.* 11, 487, 1990.

558. Kratzke, H., *AIChE J.* 31, 693, 1985.

559. Wolfe, D., Kay, W. B., and Teja, A. S., *J. Chem. Eng. Data* 28, 319, 1983.

560. Artyukhovskaya, L. M., Shimanskaya, E. T., and Shimanskii, Yu I., *Sov. Phys. - JETP (Engl. Transl.)* 59, 375, 1970.

561. Beattie, J. A., Levine, S. W., and Douslin, D. R., *J. Am. Chem. Soc.* 73, 4431, 1951.

562. Sage, B. H., and Lacey, W. N., *Ind. Eng. Chem.* 32, 992, 1940.

563. Quadri, S. K., and Kudchadker, A. P., *AIChE Symp. Ser.* 298, 1, 1994.

564. Lenoir, J. M., Rebert, C. J., and Hipkin, H. G., *J. Chem. Eng. Data* 16, 401, 1971.

565. Brown, J. A., and Mears, W. H., *J. Phys. Chem.* 62, 960, 1958.

566. Fowler, R. D., Hamilton, J. M., Kasper, J. S., Weber, C. E., Burford, W. B., and Anderson, H. C., *Ind. Eng. Chem.* 39, 375, 1947.

567. Martin, J. J., *J. Chem. Eng. Data* 7, 68, 1962.

568. Douslin, D. R., Moore, R. T., and Waddington, G., *J. Phys. Chem.* 63, 1959, 1959.

569. Rowlinson, J. S., and Thacker, R., *Trans. Faraday Soc.* 53, 1, 1957.

570. Ermakov, G. V., and Skripov, V. P., *Russ. J. Phys. Chem. (Engl. Transl.)* 41, 39, 1967.

571. Wang, B. -H., Adcock, J. L., Mathur, S. B., and Van Hook, W. A., *J. Chem. Thermodyn.* 23, 699, 1991.

572. Taylor, Z. L., and Reed, T. M., *AIChE J.* 16, 738, 1970.

573. Vandana, V., Rosenthal, D. J., and Teja, A. S., *Fluid Phase Equilib.* 99, 209, 1994.

574. Milton, H. T., and Oliver, G. D., *J. Am. Chem. Soc.* 74, 3951, 1952.

575. Oliver, G. D., and Grisard, J. W., *J. Am. Chem. Soc.* 73, 1688, 1951.

576. Oliver, G. D., Blumkin, S., and Cunningham, C. W., *J. Am. Chem. Soc.* 73, 5722, 1951.

577. Mousa, A. H. N., *J. Chem. Eng. Data* 23, 133, 1978.

578. McLure, I. A., Trejo Rodriguez, A., and Soares, V. A. M., *J. Chem. Thermodyn.* 14, 402, 1982.

579. Ewing, M. B., and Sanchez Ochoa, J. C., *J. Chem. Thermodyn.* 30, 189, 1998.

580. Ernst, G., Gurtner, J., and Wirbser, H., *J. Chem. Thermodyn.* 29, 1125, 1997.

581. Brown, J. A., *J. Chem. Eng. Data* 8, 106, 1963.

582. Young, C. L., *Int. DATA Ser., Sel. Data Mixtures, Ser. A*, No. 4, 291, 1985.

583. Sinitsyn, E. N., Mikhalevich, L. A., Biryukova, L. V., Danilov, N. N., Muratov, G. N., and Fedorov, A. P., *Deposited Doc. VINITI*, Doc. No. 1516-80, 1980.

584. Nikitin, E. D., Popov, A. P., and Yatluk, Y. G., *J. Chem. Eng. Data* 52, 315, 2007.

585. Honda, Y., Sato, T., and Uematsu, M., *J. Chem. Thermodyn.* 40, 208, 2008.

586. Jou, F.-Y., Carroll, J. J., and Mather, A. E., *Fluid Phase Equilib.* 109, 235, 1995.

587. Sychev, V. V., Vasserman, A. A., Kozlov, A. D., and Tsymarny, V. A., *Thermodynamic Properties of Propane*, Standards Publishing House, Moscow, 1989.

588. Goodwin, R. D., and Haynes, W. M., *NBS Monogr. (U.S.)* No. 170, 1982.

589. Barber, J. R., Kay, W. B., and Teja, A. S., *AIChE J.* 28, 134, 1982.

590. Yesavage, V. F., Katz, D. L., and Powers, J. E., *J. Chem. Eng. Data* 14, 197, 1969.

591. Clegg, H. P., and Rowlinson, J. S., *Trans. Faraday Soc.* 51, 1333, 1955.

592. Kay, W. B., and Rambosek, G. M., *Ind. Eng. Chem.* 45, 221, 1953.

593. Meyers, C. H., *J. Res. Natl. Bur. Stand. (U.S.)* 29, 157, 1942.

594. Meyer, R. E., Ph.D. Thesis, Pennsylvania State Univ., University Park, PA, 1941.

595. Deschner, W. W., and Brown, G. G., *Ind. Eng. Chem.* 32, 836, 1940.

596. Beattie, J. A., Poffenberger, N., and Hadlock, C., *J. Chem. Phys.* 3, 96, 1935.

597. Efremova, G. D., and Sokolova, E. S., *Russ. J. Phys. Chem. (Engl. Transl.)* 46, 1084, 1972.

598. Anonymous, B., *International Critical Tables of Numerical Data, Phys., Chem. Technol.*, Vol. III, Washburn, E. W., Ed., McGraw-Hill, New York, 1928.

599. Kuenen, J. P., and Robson, W. G., *Philos. Mag.* 4, 116, 1902.

600. Ramsay, W., and Young, S., *Philos. Trans. R. Soc. London, A* 180, 137, 1889.

601. Oh, B. C., Lee, S., Seo, J., and Kim, H., *J. Chem. Eng. Data* 49, 221, 2004.
602. Ambrose, D., Counsell, J. F., Lawrenson, I. J., and Lewis, G. B., *J. Chem. Thermodyn.* 10, 1033, 1978.
603. Seibert, F. M., and Burrell, G. A., *J. Am. Chem. Soc.* 37, 2683, 1915.
604. Lu, H., Newitt, D. M., and Ruhemann, M., *Proc. R. Soc. London, A* 178, 506, 1941.
605. Maass, O., and Wright, C. H., *J. Am. Chem. Soc.* 43, 1098, 1921.
606. Ohgaki, K., Umezono, S., and Katayama, T., *J. Supercrit. Fluids* 3, 78, 1990.
607. Marchman, H., Prengle, H. W., and Motard, R. L., *Ind. Eng. Chem.* 41, 2658, 1949.
608. Farrington, P. S., and Sage, B. H., *Ind. Eng. Chem.* 41, 1734, 1949.
609. Vaughan, W. E., and Graves, N. R., *Ind. Eng. Chem.* 32, 1252, 1940.
610. Winkler, C. A., and Maass, O., *Can. J. Res.* 9, 613, 1933.
611. Vohra, S. P., Kang, T.-L., Kobe, K. A., and McKetta, J. J., *J. Chem. Eng. Data* 7, 150, 1962.
612. Steele, W. V., Chirico, R. D., Knipmeyer, S. E., and Nguyen, A., *J. Chem. Eng. Data* 47, 689, 2002.
613. Chirico, R. D., Steele, W. V., Nguyen, A., Klots, T. D., and Knipmeyer, S. E., *J. Chem. Thermodyn.* 28, 797, 1996.
614. Mandel, H., and Ewbank, N., *Atomics International* NAA-S-R-5129, 1960.
615. Gallant, R. W., *Hydrocarbon Process.* 45, 161, 1966.
616. Altunin, V. V., Geller, V. Z., Kremenvskaya, E. A., Perel'shtein, I. I., and Petrov, E. K., *Thermophysical Properties of Freons, Methane Ser.*, Part 2, Vol. 9, NSRDS-USSR, Selover, T. B., Ed., Hemisphere, New York, 1987.
617. Toczylkin. L. S., and Young, C. L., *Aust. J. Chem.* 30, 1591, 1977.
618. Kordes, E., *Z. Elektrochem.* 58, 76, 1954.
619. Lewis, D. T., *J. Appl. Chem.* 3, 154, 1953.
620. Steele, W. V., Chirico, R. D., Nguyen, A., Hossenlopp, I. A., and Smith, N. K., *DIPPR Data Ser.* 1, 101, 1991.
621. Poot, W., and De Loos, T. W., *Fluid Phase Equilib.* 222–223, 255, 2004.
622. Poot, W., and De Loos, T. W., *Fluid Phase Equilib.* 210, 69, 2003.
623. Yata, J., Hori, M., Niki, M., Isono, Y., and Yanagitani, Y., *Fluid Phase Equilib.* 174, 221, 2000.
624. Fujiwara, K., Nakamura, S., and Noguchi, M., *J. Chem. Eng. Data* 43, 55, 1998.
625. Morrison, G., and Ward, D., *Fluid Phase Equilib.* 62, 65, 1991.
626. Piao, C. C., Sato, H., and Watanabe, K., *ASHRAE Trans.* 96, 132, 1990.
627. Piao, C. C., Sato, H., and Watanabe, K., *ASHRAE Trans.* 41, 132, 1989.
628. Kubota, H., Yamashita, T., Tanaka, Y., and Makita, T., *Int. J. Thermophys.* 10, 629, 1989.
629. Kabata, Y., Tanikawa, S., Uematsu, M., and Watanabe, K., *Int. J. Thermophys.* 10, 605, 1989.
630. Basu, R. S., and Wilson, D. P., *Int. J. Thermophys.* 10, 591, 1989.
631. Wilson, D. P., and Basu, R. S., *ASHRAE Trans.* 94, 2095, 1988.
632. Tatoh, J., Kuwabara, S., Sato, H., and Watanabe, K., *J. Chem. Eng. Data* 38, 116, 1993.
633. Lebedeva, E. S., and Khodeeva, S. M., *Zh. Fiz. Khim.* 41, 2081, 1967.
634. Renfrew, M. M., and Lewis, E. E., *Ind. Eng. Chem.* 38, 870, 1946.
635. Chari, N. C., Ph.D. Dissertation, Univ. Michigan, Ann Arbor, MI, 1960.
636. Steele, W. V., Chirico, R. D., Knipmeyer, S. E., and Smith, N. K., *Report*, NIPPR-360, NTIS Order No. DE89000709, Dec. 1988.
637. McGlashan, M. L., and McKinnon, I. R., *J. Chem. Thermodyn.* 9, 1205, 1977.
638. Cipollint, N. E., and Allen, A. O., *J. Chem. Phys.* 67, 131, 1977.
639. Hicks, C. P., and Young, C. L., *J. Chem. Soc., Faraday Trans. 1* 72, 122, 1976.
640. Hugill, J. A., and McGlashan, M. L., *J. Chem. Thermodyn.* 10, 85, 1978.
641. Bendtsen, J., *J. Raman Spectrosc.* 6, 306, 1977.
642. Abdulagatov, I. M., Polikhronidi, N. G., Bruno, T. J., Batyrova, R. G., and Stepanov, G. V., *Fluid Phase Equilib.* 263, 71, 2008.
643. Polikhronidi, N. G., Abdulagatov, I. M., Magee, J. W., and Batyrova, R. G., *J. Chem. Eng. Data* 46, 1064, 2001.
644. Goodwin, R. D., *J. Phys. Chem. Ref. Data* 18, 1565, 1989.
645. Akhundov, T. S., and Abdullaev, F. G., *Izv. Vyssh. Uchebn. Zaved., Neft Gaz* 12, 44, 1969.
646. Krase, N. W., and Goodman, J. B., *Ind. Eng. Chem.* 22, 13, 1930.
647. Buchowski, H., Janaszewski, B., and Teperek, J., *Bull. Acad. Pol. Sci., Ser. Sci. Chim.* 14, 403, 1966.
648. Sokolova, T. D., Prokof'eva, N. K., and Nisel'son, L. A., *Russ. J. Phys. Chem. (Engl. Transl.)* 47, 154, 1973.
649. Mastroianni, M. J., Stahl, R. F., and Sheldon, P. N., *J. Chem. Eng. Data* 23, 113, 1978.
650. Weber, L. A., and Defibaugh, D. R., *J. Chem. Eng. Data* 41, 1477, 1996.
651. Higashi, Y., and Ikeda, T., *Fluid Phase Equilib.* 125, 139, 1996.
652. Wang, H., Ma, Y., Lu, C., and Tian, Y., *J. Eng. Thermophys.* 14, 122, 1993.
653. Fukushima, M., *Nippon Reito Kyokai Ronbunshu* 10, 87, 1993.
654. Bier, K., Turk, M., and Zhai, J., *Vapour Pressure of Trifluoroethanol*, Insitut für Technische Thermodynamik und Kaltetechnik, Universität Karlsruhe (TH), D 7500 Karlsruhe l, FRG, 1991.
655. Bier, K., Tuerk, M., and Zhai, J., *Proc. Int. Inst. Ref., Comm. B1 Meet.*, Herzlia, Israel, pp. 129–139, 1990.
656. Zhang, C., Duan, Y. -Y., Shi, L., Zhu, M. -S., and Han, L. -Z., *J. Tsinghua Univ. (Sci. & Technol.)* 40, 77, 2000.
657. Duan, Y. -Y., Shi, L., Zhu, M. -S., and Han, L. -Z., *J. Tsinghua Univ. (Sci. & Technol.)* 40, 60, 2000.
658. Duan, Y. -Y., Shi, L., Sun, L. -Q., Zhu, M. -S., and Han, L. -Z., *Int. J. Thermophys.* 21, 393, 2000.
659. Duan, Y. -Y., Shi, L., Zhu, M. -S., and Han, L. -Z., *J. Chem. Eng. Data* 44, 501, 1999.
660. Khodeeva, S. M., and Gubochkin, I. V., *Russ. J. Phys. Chem. (Engl. Transl.)* 51, 998, 1977.
661. Diefenbacher, A., Crone, M., and Turk, M., *J. Chem. Thermodyn.* 30, 481, 1998.
662. Hori, K., Okazaki, S., Uematsu, M., and Watanabe, K., *Proc. 8th Symp. Thermophys. Prop.*, Vol. II, Sengers, J. V., Ed., ASME, New York, pp. 370–376, 1982.
663. Wagner, W., *Kaeltetech.-Klim.* 20, 238, 1968.
664. Hou, Y.-C., and Martin, J. J., *AIChE J.* 5, 125, 1959.
665. Day, H. O., and Felsing, W. A., *J. Am. Chem. Soc.* 72, 1698, 1950.
666. Kay, W. B., and Warzel, F. M., *Ind. Eng. Chem.* 43, 1150, 1951.
667. Beattie, J. A., and Edwards, D. G., *J. Am. Chem. Soc.* 70, 3382, 1948.
668. Akhundov, T. S., and Asadullaeva, N. N., *Izv. Vyssh. Uchebn. Zaved., Neft Gaz* 11, 83, 1968.
669. Akhundov, T. S., and Imanov, Sh. Yu., *Teplofiz. Svoistva Zhidk.* 48, 1970.

CRITICAL CONSTANTS OF INORGANIC COMPOUNDS

The parameters of the liquid-gas critical point are important constants in determining the behavior of fluids. This table lists the critical temperature, pressure, and molar volume, as well as the normal boiling point, for over 140 inorganic substances. The properties and their units are:

T_b: Normal boiling point in K at a pressure of 101.325 kPa (1 atmosphere); an "s" following the value indicates a sublimation point (temperature at which the solid is in equilibrium with the gas at a pressure of 101.325 kPa)

T_c: Critical temperature in K

P_c: Critical pressure in MPa

V_c: Critical molar volume in cm^3 mol^{-1}

The number of digits given for T_b, T_c, and P_c indicates the estimated accuracy of these quantities; however, values of T_c greater than 750 K may be in error by 10 K or more. Although most V_c values are given to three figures, they cannot be assumed accurate to better than a few percent. All values are experimentally determined except for a few values, indicated by an asterisk*, that are based on extrapolations. Methods of measurement are described and critiqued in Reference 1. Compounds are listed alphabetically by name.

References

1. Ambrose, D., and Young, C. L., *J. Chem. Eng. Data* 40, 345, 1995.
2. Morel, V., Bultel, A., and Chéron, B. G., *Int. J. Thermophys.* 30, 1853, 2009.
3. Ambrose, D., "Vapor-Liquid Constants of Fluids," in *Handbook of the Thermodynamics of Organic Compounds*, Stevenson, R. M., and Malanowski, S., Eds., Elsevier, New York, 1987.
4. Sato, M., Masui, G., and Uematsu, M., *J. Chem. Thermodyn.* 37, 931, 2005.
5. Velasco, S., Roman, F. L., White, J. A., and Mulero, A., *Fluid Phase Equilib.* 244, 11, 2006.
6. Nowak, P., Tielkes, T., Kleinrahm, R., and Wagner, W., *J. Chem. Thermodyn.* 29, 885, 1997.
7. Lemmon, E. W., and Span, R., *J. Chem. Eng. Data* 51, 785, 2006.
8. Goodwin, R. D., *J. Phys. Chem. Ref. Data* 14, 849, 1985.
9. Vargaftik, N. B., *Int. J. Thermophys.* 11, 467, 1990.
10. Nikitin, E. D., Pavlov, P. A., Popov, A. P., and Nikitina, H. E., *J. Chem. Thermodyn.* 27, 945, 1995.
11. Dillon, I. G., Nelson, P. A., and Swanson, B. S., *J. Chem. Phys.* 44, 4229, 1966.
12. Huber, M. L., Laesecke, A., and Friend, D. G., *The Vapor Pressure of Mercury*, NISTIR 6643, National Institute of Standards and Technology, Boulder, CO, March 2006; *Ind. Eng. Chem. Res.* 45, 7351, 2006.
13. Rau, H., Kutty, T. R. N., and Guedes de Carvalho, J. R. F., *J. Chem. Thermodyn.* 5, 291, 1973.
14. Funke, M., Kleinrahm, R., and Wagner, W., *J. Chem. Thermodyn.* 34, 717, 2002.
15. Sifner, O., and Klomfar, J., *J. Phys. Chem. Ref. Data* 23, 63, 1994.

Name	Formula	T_b/K	T_c/K	P_c/MPa	V_c/cm^3 mol^{-1}	Ref.
Aluminum	Al	2792	6700*			2
Aluminum bromide	AlBr$_3$	528	763	2.89	310	3
Aluminum chloride	AlCl$_3$	453 s	620	2.63	257	3
Aluminum iodide	AlI$_3$	655	983		408	3
Ammonia	NH$_3$	239.82	405.56	11.357	69.8	3,4
Ammonium chloride	NH$_4$Cl	611 s	1155	163.5		3
Antimony(III) bromide	SbBr$_3$	561	904		300	3
Antimony(III) chloride	SbCl$_3$	493.5	794		272	3
Antimony(III) iodide	SbI$_3$	673	1102			3
Argon	Ar	87.302	150.687	4.863	75	3
Arsenic	As	889 s	1673	22.3	35	3
Arsenic(III) chloride	AsCl$_3$	403	654		252	3
Arsine	AsH$_3$	210.7	373.1			3
Beryllium	Be	2741	5205*			5
Bismuth	Bi	1837	4620*			5
Bismuth tribromide	BiBr$_3$	735	1220		301	3
Bismuth trichloride	BiCl$_3$	714	1179	12.0	261	3
Boron tribromide	BBr$_3$	364.4	581		272	3
Boron trichloride	BCl$_3$	285.80	455	3.87	239	3
Boron trifluoride	BF$_3$	173.3	260.8	4.98	115	3
Boron triiodide	BI$_3$	482.7	773		356	3
Bromine	Br$_2$	332.0	588	10.34	127	3
Carbon dioxide	CO$_2$	194.6 s	304.13	7.375	94	6
Carbon disulfide	CS$_2$	319	552	7.90	173	3
Carbon monoxide	CO	81.7	132.86	3.494	93	3,7,8
Carbon oxysulfide	COS	223	375	5.88	137	3,7
Cesium	Cs	944	1938	9.4	341	9
Chlorine	Cl$_2$	239.11	417.0	7.991	123	3
Chlorine pentafluoride	ClF$_5$	260.1	416	5.27	233	3

Name	Formula	T_b/K	T_c/K	P_c/MPa	V_c/cm³ mol⁻¹	Ref.
Chlorotrifluorosilane	SiClF₃	203.2	307.7	3.46		3
Diborane	B₂H₆	180.8	289.8	4.05		3
Dichlorodifluorosilane	SiCl₂F₂	241	369.0	3.5		3
Difluoramine	NHF₂	250	403			3
cis-Difluorodiazine	N₂F₂	167.40	272	7.09		3
trans-Difluorodiazine	N₂F₂	161.70	260	5.57		3
Fluorine	F₂	85.04	144.41	5.1724	66	3
Fluorine monoxide	F₂O	128.8	215			3
Gallium(III) bromide	GaBr₃	552	806.7		303	3
Gallium(III) chloride	GaCl₃	474	694		263	3
Gallium(III) iodide	GaI₃	613	951		395	3
Germane	GeH₄	185.1	312.2	4.95	147	3
Germanium	Ge	3106	9802*			5
Germanium(IV) bromide	GeBr₄	459.50	718		392	3
Germanium(IV) chloride	GeCl₄	359.70	553.2	3.861	330	3
Germanium(IV) iodide	GeI₄	621	973		500	3
Hafnium(IV) bromide	HfBr₄	596 s	746		415	3
Hafnium(IV) chloride	HfCl₄	590 s	725.7	5.42	314	3
Hafnium(IV) iodide	HfI₄	667 s	916		528	3
Helium	He	4.222	5.1953	0.22746	57	3
Hydrazine	N₂H₄	386.70	653	14.7		3
Hydrogen	H₂	20.388	33.14	1.2964	65	3
Hydrogen bromide	HBr	206.77	363.2	8.55		3
Hydrogen chloride	HCl	188	324.7	8.31	81	3
Hydrogen fluoride	HF	293	461	6.48	69	3
Hydrogen iodide	HI	237.60	424.0	8.31		3
Hydrogen peroxide	H₂O₂	423.4	728*	22*		10
Hydrogen selenide	H₂Se	231.90	411	8.92		3
Hydrogen sulfide	H₂S	213.60	373.1	9.00	99	3,7
Iodine	I₂	457.6	819		155	3
Iodine bromide	IBr	389	719		139	3
Iron	Fe	3134	9340*			5
Krypton	Kr	119.735	209.48	5.525	91	3,7
Lithium	Li	1615	3223*	67*	66*	11
Manganese	Mn	2334	4325*			5
Mercury	Hg	629.769	1764	167	43	3,12
Mercury(II) bromide	HgBr₂	591	1012			3
Mercury(II) chloride	HgCl₂	577	973		174	3
Mercury(II) iodide	HgI₂	624	1072			3
Molybdenum(V) chloride	MoCl₅	541	850		369	3
Molybdenum(VI) fluoride	MoF₆	307.2	473	4.75	226	3
Neon	Ne	27.097	44.49	2.6786	42	3
Niobium(V) chloride	NbCl₅	520.6	803.5	4.88	397	3
Niobium(V) fluoride	NbF₅	507	737	6.28	155	3
Nitric oxide	NO	121.41	180	6.48	58	3
Nitrogen	N₂	77.355	126.192	3.39	90	3
Nitrogen chloride difluoride	NClF₂	206	337.5	5.15		3
Nitrogen tetroxide	N₂O₄	294.30	431	10.1	167	3
Nitrogen trifluoride	NF₃	144.40	234.0	4.46	126	3
Nitrosyl chloride	NOCl	267.7	440			3
Nitrous oxide	N₂O	184.67	309.52	7.245	97	3,7
Nitryl fluoride	NO₂F	200.8	349.5			3
Osmium(VIII) oxide	OsO₄	404.4	678			3
Oxygen	O₂	90.188	154.581	5.043	73	3
Ozone	O₃	161.80	261.1	5.57	89	3
Perchloryl fluoride	ClO₃F	226.40	368.4	5.37	161	3
Phosphine	PH₃	185.40	324.5	6.54		3
Phosphonium chloride	PH₄Cl	246 s	322.3	7.37		3
Phosphorothioc chloride difluoride	PSClF₂	279.5	439.2	4.14		3
Phosphorothioc trifluoride	PSF₃	220.90	346.0	3.82		3

Name	Formula	T_b/K	T_c/K	P_c/MPa	V_c/cm³ mol⁻¹	Ref.
Phosphorus	P	553.7	994			3
Phosphorus(III) bromide	PBr_3	446.4	711		300	3
Phosphorus(III) chloride	PCl_3	349.3	563		264	3
Phosphorus(V) chloride	PCl_5	433 s	646			3
Phosphorus(III) chloride difluoride	$PClF_2$	225.9	362.4	4.52		3
Phosphorus(III) dichloride fluoride	PCl_2F	287.00	463.0	4.96		3
Phosphorus(III) fluoride	PF_3	171.4	271.2	4.33		3
Potassium	K	1032	2223*	16*	209*	11
Radon	Rn	211.5	377	6.28		3
Rhenium(VII) oxide	Re_2O_7	633	942		334	3
Rhenium(VI) oxytetrachloride	$ReOCl_4$	496	781		362	3
Rubidium	Rb	961	2093*	16*	247*	11
Selenium	Se	958	1766	27.2		3
Selenium hexafluoride	SeF_6	226.55 s	345.5			3
Selenium oxychloride	$SeOCl_2$	450	730	7.09	235	3
Silver	Ag	2435	6410*			5
Sodium	Na	1156	2573*	35*	116*	11
Sulfur	S	717.76	1314	20.7	57.0	3,13
Sulfur chloride pentafluoride	SF_5Cl	254.10	390.9			3
Sulfur dioxide	SO_2	263.10	430.64	7.884	122	3,7
Sulfur hexafluoride	SF_6	209.35 s	318.723	3.77	197	3,14
Sulfur tetrafluoride	SF_4	232.70	364			3
Sulfur trioxide	SO_3	317.7	491.0	8.2	127	3
Tantalum(V) bromide	$TaBr_5$	622	974		461	3
Tantalum(V) chloride	$TaCl_5$	512	767		402	3
Tellurium	Te	1261	2329*			5
Tellurium hexafluoride	TeF_6	234.25 s	356			3
Tellurium tetrachloride	$TeCl_4$	660	1002	8.56	310	3
Tetrabromosilane	$SiBr_4$	427	663		382	3
Tetrachlorosilane	$SiCl_4$	330.80	508.1	3.593	326	3
Tetrafluorohydrazine	N_2F_4	199	309	3.75		3
Tetrafluorosilane	SiF_4	187	259.0	3.72		3
Tetraiodosilane	SiI_4	560.50	944		558	3
Tin(IV) bromide	$SnBr_4$	478	744		417	3
Tin(IV) chloride	$SnCl_4$	387.30	591.9	3.75	351	3
Tin(IV) iodide	SnI_4	637.50	968		531	3
Titanium(IV) bromide	$TiBr_4$	506.7	795.7		391	3
Titanium(IV) chloride	$TiCl_4$	409.60	638	4.66	339	3
Titanium(IV) iodide	TiI_4	650	1040		505	3
Tribromosilane	$SiHBr_3$	382	610.0		305	3
Trichlorofluorosilane	$SiCl_3F$	285.40	438.6	3.58		3
Trichlorosilane	$SiHCl_3$	306	479		268	3
Trifluoramine oxide	NOF_3	185.7	303	6.43	147	3
Tungsten(VI) chloride	WCl_6	610	923		422	3
Tungsten(VI) fluoride	WF_6	290.3	444	4.34	233	3
Tungsten(VI) oxytetrachloride	$WOCl_4$	503	782		338	3
Uranium(VI) fluoride	UF_6	329.65 s	505.8	4.66	250	3
Vanadyl chloride	VOCl	400	636		171	3
Water	H_2O	373.12	647.10	22.06	56	3
Xenon	Xe	165.051	289.733	5.842	118	7,15
Xenon difluoride	XeF_2	387.50 s	631	9.32	148	3
Xenon tetrafluoride	XeF_4	388.90 s	612	7.04	188	3
Zirconium(IV) bromide	$ZrBr_4$	633 s	805		424	3
Zirconium(IV) chloride	$ZrCl_4$	604 s	778	5.77	319	3
Zirconium(IV) iodide	ZrI_4	704 s	960		530	3

SUBLIMATION PRESSURE OF SOLIDS

This table gives the sublimation (vapor) pressure of some representative solids as a function of temperature. Entries include simple inorganic and organic substances in their solid phase below room temperature, as well as polycyclic organic compounds which show measurable sublimation pressure only at elevated temperatures. Substances are listed by molecular formula in the Hill order. Values marked by * represent the solid–liquid–gas triple point. Note that some pressure values are in pascals (Pa) and others are in kilopascals (kPa). For conversion, 1 kPa = 7.506 mmHg = 0.0098692 atm.

References

1. Lide, D. R. and Kehiaian, H. V., *CRC Handbook of Thermophysical and Thermochemical Data*, CRC Press, Boca Raton, FL, 1994.
2. *TRC Thermodynamic Tables*, Thermodynamic Research Center, Texas A&M University, College Station, TX.
3. Oja, V. and Suuberg, E. M., *J. Chem. Eng. Data*, 43, 486, 1998.

Ar	T/K	55	60	65	70	75	80	83.81*	
Argon	p/kPa	0.2	0.8	2.8	7.7	18.7	40.7	68.8*	
BrH	T/K	135	140	150	160	170	180	185.1*	
Hydrogen bromide	p/kPa	0.1	0.3	1.1	3.3	8.7	20.1	27.4*	
Br_2	T/K	170	180	190	200	210	220	230	240*
Bromine	p/Pa	0.069	0.416	2.04	8.45	30.3	96.0	273	710*
ClH	T/K	120	130	140	150	155	159.0*		
Hydrogen chloride	p/kPa	0.1	0.5	1.9	5.8	9.5	13.5*		
Cl_2	T/K	120	130	140	150	160	170*		
Chlorine	p/Pa	0.144	1.52	11.2	63.1	283	1054*		
F_4Si	T/K	130	140	150	160	170	175	180	186.3*
Tetrafluorosilane	p/kPa	0.2	0.9	3.9	14.0	43.8	74.2	122.4	220.8*
F_6S	T/K	150	165	180	190	200	210	220	223.1*
Sulfur hexafluoride	p/kPa	0.4	2.6	11.3	25.9	54.5	106.1	195.1	232.7*
HI	T/K	160	170	180	190	200	210	220	222.4*
Hydrogen iodide	p/kPa	0.2	0.8	2.2	5.3	11.7	23.6	44.1	49.3*
H_2O	T/K	190	210	225	240	250	260	270	273.16*
Water	p/Pa	0.032	0.702	4.942	27.28	76.04	195.8	470.1	611.66*
H_2S	T/K	140	150	160	165	170	175	180	187.6*
Hydrogen sulfide	p/kPa	0.2	0.6	1.9	3.2	5.2	8.3	12.7	22.7*
H_3N	T/K	160	170	180	190	195	195.4*		
Ammonia	p/kPa	0.1	0.4	1.2	3.5	5.8	6.12*		
I_2	T/K	240	250	260	270	280	290	300	310*
Iodine	p/Pa	0.081	0.297	0.971	2.89	7.92	20.1	47.9	107*
Kr	T/K	80	90	95	100	105	110	115.8*	
Krypton	p/kPa	0.4	2.7	6.0	12.1	22.8	40.4	73.1*	
NO	T/K	85	90	95	100	105	109.5*		
Nitric oxide	p/kPa	0.1	0.4	1.3	3.8	10.0	21.9*		
Xe	T/K	110	120	130	140	150	155	160	161.4*
Xenon	p/kPa	0.3	1.5	4.9	14.0	34.2	51.1	74.2	81.7*
CHN	T/K	200	210	220	230	240	250	255	259.83*
Hydrogen cyanide	p/kPa	0.2	0.4	1.0	2.2	4.8	9.7	13.6	18.62*
CH_4	T/K	65	70	75	80	85	90.69*		
Methane	p/kPa	0.1	0.3	0.8	2.1	4.9	11.70*		
CO	T/K	50	55	60	65	68.13*			
Carbon monoxide	p/kPa	0.1	0.6	2.6	8.2	15.4*			
CO_2	T/K	130	140	155	170	185	194.7	205	216.58*
Carbon dioxide	p/kPa	0.032	0.187	1.674	9.987	44.02	101.3	227.1	518.0*
C_2Cl_6	T/K	275	300	325	350	375	400	425	459.9*
Hexachloroethane	p/Pa	0.004	0.056	0.383	1.62	5.30	14.8	36.4	107.4*
C_2H_2	T/K	130	140	150	160	170	180	190	192.4*
Acetylene	p/kPa	0.2	0.7	2.6	7.8	20.6	49.0	106.3	126.0*
$C_2H_4O_2$	T/K	250	260	270	280	289.7*			
Acetic acid	p/kPa	0.092	0.199	0.406	0.79	1.29*			
C_5H_{12}	T/K	200	210	220	230	240	250	255	256.58*
Neopentane	p/kPa	0.7	1.6	3.6	7.3	13.9	24.8	32.4	35.8*

$C_6H_6Cl_6$	T/K	300	320	330	340	350	360	370	380
1,2,3,4,5,6–Hexa–chlorocyclohexane (Lindane)	p/Pa	0.01	0.13	0.39	1.04	2.66	6.42	14.8	32.7
$C_6H_6O_2$	T/K	330	340	350	360	370	380		
Resorcinol	p/Pa	1.03	2.78	7.09	17.2	39.6	87.6		
$C_6H_6O_2$	T/K	350	360	370	380	390	400		
p–Hydroquinone	p/Pa	1.20	3.18	7.96	19.0	43.4	95.1		
$C_{10}H_8$	T/K	250	270	280	290	300	310	330	353.43*
Naphthalene	p/Pa	0.036	0.514	1.662	4.918	13.43	34.15	182.9	999.6*
$C_{12}H_8N_2$	T/K	290	300	310	320				
Phenazine	p/Pa	0.0013	0.0046	0.0150	0.0448				
$C_{12}H_8O$	T/K	300	310	320	330	340	350		
Dibenzofuran	p/Pa	0.408	1.21	3.35	8.71	21.4	50.0		
$C_{12}H_9N$	T/K	350	355	360					
Carbazole	p/Pa	0.086	0.140	0.245					
$C_{13}H_7NO_2$	T/K	330	340	350	360	370	380		
Benz[g]isoquinoline–5,10–dione	p/Pa	0.006	0.018	0.053	0.148	0.394	0.994		
$C_{13}H_8O$	T/K	330	340	350					
1H–Phenalen–1–one	p/Pa	0.040	0.113	0.302					
$C_{13}H_8O_2$	T/K	400	410	420	430				
3–Hydroxy–1H–phenalen–1–one	p/Pa	0.006	0.018	0.053	0.144				
$C_{13}H_9N$	T/K	290	300	310	320				
Acridine	p/Pa	0.0024	0.0085	0.0278	0.0845				
$C_{13}H_9N$	T/K	310	320	330	340				
Phenanthridine	p/Pa	0.020	0.066	0.206	0.603				
$C_{14}H_{10}$	T/K	320	330	340	350	360	370	380	390
Anthracene	p/Pa	0.014	0.043	0.125	0.342	1.01	2.38	5.35	11.5
$C_{14}H_{10}$	T/K	300	310	320	330	340	350	360	
Phenanthrene	p/Pa	0.025	0.085	0.270	0.796	2.02	4.89	11.2	
$C_{16}H_{10}$	T/K	320	330	340	350	360	370	380	390
Pyrene	p/Pa	0.008	0.024	0.073	0.208	0.556	1.32	2.86	6.30
$C_{16}H_{10}O$	T/K	360	370	380	390	400			
1–Pyrenol	p/Pa	0.005	0.016	0.047	0.135	0.364			
$C_{16}H_{12}S$	T/K	330	340	350	360	370	380	390	
Benzo[b]naphtho–(2,1–d)thiophene	p/Pa	0.001	0.004	0.012	0.036	0.098	0.255	0.631	
$C_{17}H_{12}$	T/K	340	350	360	370	380	390	400	
11H–Benzo[b]fluorene	p/Pa	0.003	0.009	0.029	0.085	0.235	0.619	1.55	
$C_{18}H_{10}O_4$	T/K	420	430	440	450				
6,11–Dihydroxy–5,12–naphthacenedione	p/Pa	0.008	0.022	0.055	0.131				
$C_{18}H_{12}$	T/K	390	400	410	420				
Chrysene	p/Pa	0.087	0.221	0.539	1.26				
$C_{18}H_{12}$	T/K	390	400	410	420	430	440	450	460
Naphthacene	p/Pa	0.005	0.014	0.035	0.084	0.194	0.432	0.928	1.929
$C_{20}H_{12}$	T/K	390	400	410	420	430			
Perylene	p/Pa	0.006	0.015	0.040	0.102	0.246			
$C_{22}H_{14}$	T/K	450	460	470	480	490			
Pentacene	p/Pa	0.002	0.006	0.013	0.031	0.069			
$C_{24}H_{12}$	T/K	430	440	450	460	470	480	490	500
Coronene	p/Pa	0.004	0.010	0.021	0.046	0.097	0.197	0.389	0.747

VAPOR PRESSURE

This table gives vapor pressure data for about 1800 inorganic and organic substances. In order to accommodate elements and compounds ranging from refractory to highly volatile in a single table, the temperature at which the vapor pressure reaches specified pressure values is listed. The pressure values run in decade steps from 1 Pa (about 7.5 μm Hg) to 100 kPa (about 750 mm Hg). All temperatures are given in °C.

The data used in preparing the table came from a large number of sources; the main references used for each substance are indicated in the last column. Since the data were refit in most cases, values appearing in this table may not be identical with values in the source cited. The temperature entry in the 100 kPa column is close to, but not identical with, the normal boiling point (which is defined as the temperature at which the vapor pressure reaches 101.325 kPa). Although some temperatures are quoted to 0.1 °C, uncertainties of several degrees should generally be assumed. Values followed by an "e" were obtained by extrapolating (usually with an Antoine equation) beyond the region for which experimental measurements were available and are thus subject to even greater uncertainty.

Compounds are listed by molecular formula following the Hill convention. Substances not containing carbon are listed first, followed by those that contain carbon. To locate an organic compound by name or CAS Registry Number when the molecular formula is not known, use the table "Physical Constants of Organic Compounds" in Section 3 and its indexes to determine the molecular formula. The indexes to "Physical Constants of Inorganic Compounds" in Section 4 can be used in a similar way.

More extensive and detailed vapor pressure data on selected important substances appear in other tables in this section of the *Handbook*. These substances are flagged by a symbol following the name as follows:

*   See "Vapor Pressure of Fluids at Temperatures below 300 K"
**   See "IUPAC Recommended Data for Vapor Pressure Calibration"
*** See "Vapor Pressure of Ice" and "Vapor Pressure and other Saturation Properties of Water"

The following notations appear after individual temperature entries:

s — Indicates the substance is a solid at this temperature.
e — Indicates an extrapolation beyond the region where experimental measurements exist.
i — Indicates the value was calculated from ideal gas thermodynamic functions, such as those in the *JANAF Thermochemical Tables* (see Reference 8).

References

1. Lide, D.R., and Kehiaian, H.V., CRC Handbook of *Thermophysical and Thermochemical Data*, CRC Press, Boca Raton, FL, 1994.
2. Stull, D., in *American Institute of Physics Handbook, Third Edition*, Gray, D.E., Ed., McGraw Hill, New York, 1972.
3. Hultgren, R., Desai, P.D., Hawkins, D.T., Gleiser, M., Kelley, K.K., and Wagman, D.D., *Selected Values of Thermodynamic Properties of the Elements*, American Society for Metals, Metals Park, OH, 1973.
4. Stull, D., *Ind. Eng. Chem.*, 39, 517, 1947.
5. TRCVP, *Vapor Pressure Database, Version 2.2P*, Thermodynamic Research Center, Texas A&M University, College Station, TX.
6. *TRC Thermodynamic Tables*, Thermodynamic Research Center, Texas A&M University, College Station, TX.
7. Ohe, S., *Computer Aided Data Book of Vapor Pressure*, Data Book Publishing Co., Tokyo, 1976.
8. Chase, M.W., Davies, C.A., Downey, J.R., Frurip, D.J., McDonald, R.A., and Syverud, A.N., *JANAF Thermochemical Tables, Third Edition*, J. Phys. Chem. Ref. Data, Vol. 14, Suppl. 1, 1985.
9. Barin, I., *Thermochemical Data of Pure Substances*, VCH Publishers, New York, 1993.
10. Jacobsen, R.T., et al, *International Thermodynamic Tables of the Fluid State, No. 10. Ethylene*, Blackwell Scientific Publications, Oxford, 1988.
11. Wakeham, W.A., *International Thermodynamic Tables of the Fluid State, No. 12. Methanol*, Blackwell Scientific Publications, Oxford, 1993.
12. Janz, G.J., *Molten Salts Handbook*, Academic Press, New York, 1967.
13. Ohse, R.W. *Handbook of Thermodynamic and Transport Properties of Alkali Metals*, Blackwell Scientific Publications, Oxford, 1994.
14. Gschneidner, K.A., in *CRC Handbook of Chemistry and Physics, 77th Edition*, p. 4–112, CRC Press, Boca Raton, FL, 1996.
15. Leider, H.R., Krikorian, O.H., and Young, D.A., *Carbon*, 11, 555, 1973.
16. Ruzicka, K., and Majer, V., *J. Phys. Chem. Ref. Data*, 23, 1, 1994.
17. Tillner–Roth, R., and Baehr, H.D., *J. Phys. Chem. Ref. Data*, 23, 657, 1994.
18. Younglove, B.A., and McLinden, M.O., *J. Phys. Chem. Ref. Data*, 23, 731, 1994.
19. Outcalt, S.L., and McLinden, M.O., *J. Phys. Chem. Ref. Data*, 25, 605, 1996.
20. Weber, L.A., and Defibaugh, D.R., *J. Chem. Eng. Data*, 41, 382, 1996.
21. Rodrigues, M.F., and Bernardo-Gil, M.G., *J. Chem. Eng. Data*, 41, 581, 1996.
22. Piacente, V., Gigli, G., Scardala, P., and Giustini, A., *J. Phys. Chem.*, 100, 9815, 1996.
23. Barton, J.L., and Bloom, H., *J. Phys. Chem.*, 60, 1413, 1956.
24. Sense, K.A., Alexander, C.A., Bowman, R.E., and Filbert, R.B., *J. Phys. Chem.*, 61, 337, 1957.
25. Ewing, C.T., and Stern, K.H., *J. Phys. Chem.* 78, 1998, 1974.
26. Cady, G.H., and Hargreaves, G.B., *J. Chem. Soc.*, 1563, 1961; 1568, 1961.
27. Skudlarski, K., Dudek, J., and Kapala, J., *J. Chem. Thermodynamics*, 19, 857, 1987.
28. Wagner, W., and de Reuck, K.M., *International Thermodynamic Tables of the Fluid State, No. 9. Oxygen*, Blackwell Scientific Publications, Oxford, 1987.
29. Marsh, K.N., Ed., *Recommended Reference Materials for the Realization of Physicochemical Properties*, Blackwell Scientific Publications, Oxford, 1987.
30. Alcock, C.B., Itkin, V.P., and Horrigan, M.K., *Canadian Metallurgical Quarterly*, 23, 309, 1984.
31. Stewart, R.B., and Jacobsen, R.T., *J. Phys. Chem. Ref. Data*, 18, 639, 1989.
32. Sifner, O., and Klomfar, J., *J. Phys. Chem. Ref. Data*, 23, 63, 1994.
33. Bah, A., and Dupont–Pavlovsky, N., *J. Chem. Eng. Data*, 40, 869, 1995.
34. Behrens, R.G., and Rosenblatt, G., *J. Chem. Thermodynamics*, 4, 175, 1972.
35. Behrens, R.G., and Rosenblatt, G., *J. Chem. Thermodynamics*, 5, 173, 1973.
36. Haar, L., Gallagher, J.S., and Kell, G.S., *NBS/NRC Steam Tables*, Hemisphere Publishing Corp., New York, 1984.
37. Wagner, W., Saul, A., and Pruss, A., *J. Phys. Chem. Ref. Data*, 23, 515, 1994.
38. Behrens, R.G., Lemons, R.S., and Rosenblatt, G., *J. Chem. Thermodynamics*, 6, 457, 1974.
39. Boublik, T., Fried, V., and Hala, E., *The Vapor Pressure of Pure Substances, Second Edition*, Elsevier, Amsterdam, 1984.
40. Goodwin, R.D., *J. Phys. Chem. Ref. Data*, 14, 849, 1985.
41. Younglove, B.A., and Ely, J.F., *J. Phys. Chem. Ref. Data*, 16, 577, 1987.

Mol. form.	Name	Temperature in °C for the indicated pressure						Ref.
		1 Pa	10 Pa	100 Pa	1 kPa	10 kPa	100 kPa	

Substances not containing carbon:

Mol. form.	Name	1 Pa	10 Pa	100 Pa	1 kPa	10 kPa	100 kPa	Ref.
Ag	Silver	1010	1140	1302	1509	1782	2160	2
AgBr	Silver(I) bromide	569 i	656 i	765 i	905 i	1093 i	1359 i	9
AgCl	Silver(I) chloride	670	769	873	1052	1264	1561	4
AgI	Silver(I) iodide	594	686	803	959	1177	1503	4
Al	Aluminum	1209	1359	1544	1781	2091	2517	2
AlB_3H_{12}	Aluminum borohydride				−46.8	−9.4	45.5	4
$AlCl_3$	Aluminum trichloride	58.4 s	76.5 s	97.1 s	120.7 s	148.2 s	180.5 s	4
AlF_3	Aluminum trifluoride	744 s	819 s	906 s	1008 s	1130 s	1276 s	8
AlI_3	Aluminum triiodide				218	285	385	4
Al_2O_3	Aluminum oxide			2122	2351	2629	2975	4
Ar	Argon*		−226.4 s	−220.3 s	−212.4 s	−201.7 s	−186.0	1,5,31
As	Arsenic	280 s	323 s	373 s	433 s	508 s	601 s	3
$AsCl_3$	Arsenic(III) chloride			−8 e	21.3	63.1	129.4	1
AsF_3	Arsenic(III) fluoride					8.1	56.0	4
AsI_3	Arsenic(III) iodide				187	261	367 e	7
As_2O_3	Arsenic(III) oxide (arsenolite)	133.7 s	163.0 s	196.8 s	236.2 s	283.0		34
At	Astatine	88 s	119 s	156 s	202 s	258 s	334	2
Au	Gold	1373	1541	1748	2008	2347	2805	2
B	Boron	2075	2289	2549	2868	3272	3799	2
BBr_3	Boron tribromide			−45 e	−15 e	27.5	90.4	1
BCl_3	Boron trichloride*			−94.0	−70.5	−37.4	12.3	4
BF_3	Boron trifluoride*	−173.9 s	−166.0 s	−156.0 s	−143.0 s	−125.9	−101.1	4
B_2F_4	Tetrafluorodiborane						−34	1
B_2H_6	Diborane			−162 e	−147.0	−125.8	−92.6	1
B_5H_9	Pentaborane(9)				−34.8	3.8	57.6	4
Ba	Barium	638 s	765	912	1115	1413	1897	9
Be	Beryllium	1189 s	1335	1518	1750	2054	2469	2
$BeBr_2$	Beryllium bromide	203 s	240 s	283 s	335 s	397 s	473 s	4
$BeCl_2$	Beryllium chloride	196 s	237 s	284 s	339 s	402 s	487	4
BeF_2	Beryllium fluoride		686 e	767 e	869	999	1172 e	7
BeI_2	Beryllium iodide	188 s	229 s	276 s	333 s	402 s	487	4
Bi	Bismuth	668	768	892	1052	1265	1562	2
$BiBr_3$	Bismuth tribromide			217 s	273 i	348 i	455 i	4,9
$BiCl_3$	Bismuth trichloride				248.9	328.6	438.7	1,4
BrCs	Cesium bromide	531 s	601 s	701 i	834 i	1019 i	1293 e	9
BrH	Hydrogen bromide*		−153.3 s	−140.4 s	−123.8 s	−101.5 s	−67.0	5
BrH_3Si	Bromosilane				−81.0	−47.3	2.2	4
BrH_4N	Ammonium bromide	121 s	154 s	195 s	246 s	310.4 s	395.1 s	5
BrK	Potassium bromide	597 s	674 s	773				25
BrLi	Lithium bromide		630	733	868	1049	1308	4
BrNa	Sodium bromide			791	931	1120	1389	4
BrRb	Rubidium bromide			766	903	1087	1350	4
BrTl	Thallium(I) bromide				509	635	817	4
Br_2	Bromine*	−87.7 s	−71.8 s	−52.7 s	−29.3 s	2.5	58.4	1
Br_2Cd	Cadmium bromide	373 s	435 s	509 s				27
Br_2Hg	Mercury(II) bromide	71 s	98 s	132 s	174 s	227 s	318	4
Br_2OS	Thionyl bromide	−49 e	−29 e	−5 e	27.8	72.9	139.6	5
Br_2Pb	Lead(II) bromide	374	431	502	597	726	914	4
Br_2S_2	Sulfur bromide	−7 e	15 e	42 e	78.4	128.1	200.9	5
Br_3In	Indium(III) bromide			304.6 s	328.7 s	364.8 s		1
Br_3OP	Phosphorus(V) oxybromide				64 e	115.5	191.4	5
Br_3P	Phosphorus(III) bromide		−23 e	5 e	42.3	94.6	172.6	5
Br_3Sb	Antimony(III) bromide				136.5	196.9	286.5	1
Br_4Ge	Germanium(IV) bromide				51	105	188	4
Br_4Sn	Tin(IV) bromide				67	122	204	4
Br_4Zr	Zirconium(IV) bromide	136 s	167 s	203 s	245 s	295 s	356 s	4
Br_5P	Phosphorus(V) bromide		−19 s	4 s	31 s	65.5 s	110.1	5
Ca	Calcium	591 s	683 s	798 s	954	1170	1482	2
Cd	Cadmium	257 s	310 s	381	472	594	767	2

		Temperature in °C for the indicated pressure						
Mol. form.	Name	1 Pa	10 Pa	100 Pa	1 kPa	10 kPa	100 kPa	Ref.
$CdCl_2$	Cadmium chloride	412 s	471 s	541 s	634	768	959	23, 27
CdF_2	Cadmium fluoride				1257	1461	1742	4
CdI_2	Cadmium iodide	296 s	344 s	406	498	622	795	4,27
CdO	Cadmium oxide	770 s	866 s	983 s	1128 s	1314 s	1558 s	4
Ce	Cerium	1719	1921	2169	2481	2886	3432	14
$ClCs$	Cesium chloride			730	864	1043	1297	4
$ClCu$	Copper(I) chloride		459	543	675	914	1477	4
ClF	Chlorine fluoride*				−144.4	−122.6	−90.2	5
ClF_2P	Phosphorus(III) chloride difluoride				−119.5	−91.1	−47.6	5
ClF_3	Chlorine trifluoride				−63.7	−33.0	11.4	5
ClF_5	Chlorine pentafluoride				−88 e	−59	−14	7
ClH	Hydrogen chloride*				−138.2 s	−118.0	−85.2	1,5
$ClHO_3S$	Chlorosulfonic acid	−40 e	−20 e	5 e	38.7	85.0	153.6	5
ClH_4N	Ammonium chloride	91 s	121 s	159 s	204.7 s	263.1 s	339.5 s	5
ClK	Potassium chloride	625 s	704 s	804	945	1137	1411	23,25
$ClLi$	Lithium chloride		649 i	761 i	905 i	1101 i	1381 i	8
$ClNO$	Nitrosyl chloride		−116 s	−100 s	−78.7 s	−50.2	−5.7	5
$ClNO_2$	Nitryl chloride	−121 e	−113 e	−102 e	−86.1	−60.9	−15.7	5
$ClNa$	Sodium chloride	653 s	733 s	835	987	1182	1461	23,25
ClO_2	Chlorine dioxide*				−34.3	10.5		5
$ClRb$	Rubidium chloride			777	916	1105	1379	4
$ClTl$	Thallium(I) chloride				504	626	806	4
Cl_2	Chlorine*	−145 s	−133.7 s	−120.2 s	−103.6 s	−76.1	−34.2	1
Cl_2Co	Cobalt(II) chloride					818	1048	4
Cl_2FP	Phosphorus(III) dichloride fluoride				−71.1	−37.4	13.5	5
Cl_2F_3P	Phosphorus(V) dichloride trifluoride		−120 e	−101 e	−77.1	−44.3	3 e	7
Cl_2Fe	Iron(II) chloride				685	821	1025	4
Cl_2Hg	Mercury(II) chloride	64.4 s	94.7 s	130.8 s	174.5 s	228.5 s	304.0	4
Cl_2Mg	Magnesium chloride			762	908	1111	1414	4
Cl_2Mn	Manganese(II) chloride				760	933	1189	4
Cl_2Ni	Nickel(II) chloride	534 s	592 s	662 s	747 s	852 s	985 s	4
Cl_2OS	Thionyl chloride	−99 e	−81 e	−58 e	−27.1	14.6	75.2	5
Cl_2O_2S	Sulfuryl chloride				−27 e	11.8	69.0	5
Cl_2Pb	Lead(II) chloride			541 e	637	765	949	23
Cl_2S	Sulfur dichloride	−76 e	−61 e	−41 e	−16.7	15.3	58.7	5
Cl_2S_2	Sulfur chloride	−55 e	−36 e	−12 e	21.0	67.2	137.1	5
Cl_2Sn	Tin(II) chloride		253	308	381	479	622	4
Cl_2Zn	Zinc chloride	305 i	356 i	419 i	497 i	596 i	726 i	4,9,12
Cl_3Fe	Iron(III) chloride	118 s	153 s	190 s	229 s	268 s	319	4
Cl_3HSi	Trichlorosilane			−81 e	−56 e	−21 e	31.6	7
Cl_3N	Nitrogen trichloride				−25 e	13.2	70.6	5
Cl_3OP	Phosphorus(V) oxychloride					39.9	105.0	5
Cl_3P	Phosphorus(III) chloride	−93 e	−77 e	−55 e	−26.0	14.5	75.7	5
Cl_4Po	Polonium(IV) chloride					300.6	389.4	5
Cl_4Se	Selenium tetrachloride	23 s	45 s	71 s	102 s	141.4 s	191.1 s	5
Cl_4Si	Tetrachlorosilane*				−39 e	0 e	57.3	1
Cl_4Te	Tellurium tetrachloride				237 e	299.4	387.8	5
Cl_4Zr	Zirconium(IV) chloride	117 s	146 s	181 s	222 s	272 s	336 z	9
Cl_5P	Phosphorus(V) chloride	−2 s	19 s	44 s	74 s	111.4 s	158.9 s	5
Co	Cobalt	1517	1687	1892	2150	2482	2925	2
Cr	Chromium	1383 s	1534 s	1718 s	1950	2257	2669	2
Cs	Cesium	144.5	195.6	260.9	350.0	477.1	667.0	13,30
CsF	Cesium fluoride				825	999	1249	4
CsI	Cesium iodide	523 s	595 s	692	854	1029	1278	4,25
Cu	Copper	1236	1388	1577	1816	2131	2563	2
CuI	Copper(I) iodide				636	864	1331	4
Dy	Dysprosium	1105 s	1250 s	1431 i	1681 i	2031 i	2558 i	3
Er	Erbium	1231 s	1390 s	1612 i	1890 i	2279 i	2859 i	3

Mol. form.	Name	Temperature in °C for the indicated pressure						Ref.
		1 Pa	10 Pa	100 Pa	1 kPa	10 kPa	100 kPa	
Eu	Europium	590 s	684 s	799 s	961	1179	1523	14
FH	Hydrogen fluoride*				−71.1	−33.7	19.2	1,5
FHO$_3$S	Fluorosulfonic acid	−14 e	4 e	28 e	59.1	101.3	162.2	5
FK	Potassium fluoride			869	1017	1216	1499	4
FLi	Lithium fluoride	801 s	896	1024	1188	1395	1672	4,12,25
FNO	Nitrosyl fluoride			−131 e	−116.1	−94.3	−60.1	5
FNO$_2$	Nitryl fluoride		−156 e	−144 e	−128.1	−106.0	−72.6	5
FNO$_3$	Fluorine nitrate	−160 e	−149 e	−135 e	−115.1	−87.4	−45.0	5
FNa	Sodium fluoride		920 s	1058	1218	1426	1702	4,12,24
FRb	Rubidium fluoride			910	1001	1145	1409	4,12
F$_2$	Fluorine*	−235 s	−229.5 s	−222.9 s	−214.8	−204.3	−188.3	1,5
F$_2$O	Fluorine monoxide*	−211.7	−204.7	−195.9	−184.2	−168.2	−144.9	5
F$_2$OS	Thionyl fluoride			−124 e	−106.5	−81.5	−44.1	5
F$_2$O$_2$Re	Rhenium(VI) dioxydifluoride				89.2	131.9	185 e	26
F$_2$Pb	Lead(II) fluoride				865	1054	1292	4
F$_2$Xe	Xenon difluoride			2.9 s	31.8 s	67.9 s	114 s	1,5
F$_2$Zn	Zinc fluoride	731 s	813 s	911 i	1048 i	1237 i	1503 i	9
F$_3$N	Nitrogen trifluoride*	−201 e	−194 e	−185 e	−172.8	−155.5	−129.2	5
F$_3$OP	Phosphorus(V) oxyfluoride	−124 s	−113 s	−100 s	−83.7 s	−64.1 s	−39.7 s	5
F$_3$P	Phosphorus(III) fluoride*				−152 e	−132.6	−101.4	5
F$_4$MoO	Molybdenum(VI) oxytetrafluoride	−21 s	3 s	33 s	69.3 s	117.3	184.1	26
F$_4$ORe	Rhenium(VI) oxytetrafluoride	5 s	26 s	50.7 s	80.1 s	117.1	171.2	26
F$_4$OW	Tungsten(VI) oxytetrafluoride	2 s	25 s	52.1 s	84.3 s	126.7	185.4	26
F$_4$S	Sulfur tetrafluoride				−110.0	−82.1	−40.3	5
F$_4$Se	Selenium tetrafluoride				13.6	51.6	104.7	5
F$_4$Si	Tetrafluorosilane*	−166 s	−157 s	−145.6 s	−132.3 s	−115.7 s	−94.9 s	4,7
F$_5$Mo	Molybdenum(V) fluoride				86.6	140.3	213 e	26
F$_5$Nb	Niobium(V) fluoride				80	140	224	4
F$_5$ORe	Rhenium(VII) oxypentafluoride	−103 s	−84 s	−59 s	−28 s	13.7 s	72.8	26
F$_5$Os	Osmium(V) fluoride			74.1	113.2	162.3	226 e	26
F$_5$P	Phosphorus(V) fluoride	−157 s	−148 s	−137 s	−124.5 s	−108.6 s	−84.8 s	5
F$_5$Re	Rhenium(V) fluoride			58.8	99.5	152 e	221 e	26
F$_5$Ta	Tantalum(V) fluoride					119	229	4
F$_6$Ir	Iridium(VI) fluoride	−88 s	−71 s	−51 s	−27 s	3.8 s	53.1	26
F$_6$Mo	Molybdenum(VI) fluoride	−98 s	−82 s	−64 s	−41.2 s	−13.4 s	33.5	26
F$_6$Os	Osmium(VI) fluoride	−89 s	−73 s	−54 s	−30.6 s	−1.7 s	47.4	26
F$_6$Re	Rhenium(VI) fluoride	−97 s	−82 s	−63 s	−40.2 s	−11.9 s	33.4	26
F$_6$S	Sulfur hexafluoride*	−158 s	−147 s	−133.6 s	−116.6 s	−94.4 s	−64.1 s	5
F$_6$Se	Selenium hexafluoride	−143 s	−132 s	−118 s	−100.7 s	−77.8 s	−46.5 s	5
F$_6$Te	Tellurium hexafluoride	−142 s	−130 s	−115 s	−96 s	−71.8 s	−39.1 s	5
F$_6$W	Tungsten(VI) fluoride	−107 s	−92 s	−74 s	−52.1 s	−24.8 s	16.9	26
F$_{10}$S$_2$	Sulfur decafluoride					−22.0	28.5	5
Fe	Iron	1455 s	1617	1818	2073	2406	2859	2
Fr	Francium	131 e	181 e	246 e	335 e	465 e	673 e	2
Ga	Gallium	1037	1175	1347	1565	1852	2245	2
Gd	Gadolinium	1563 i	1755 i	1994 i	2300 i	2703 i	3262 i	3
Ge	Germanium	1371	1541	1750	2014	2360	2831	2
HI	Hydrogen iodide*	−146 s	−135.2 s	−120.8 s	−101.9 s	−75.9 s	−35.9	5
HKO	Potassium hydroxide	520 e	601 e	704	842	1035	1325	4
HNO$_3$	Nitric acid				−37 e	−9 e	28.4	82.2	5
HN$_3$	Hydrazoic acid				−79 e	−54 e	−18.0	35.7	5
HNaO	Sodium hydroxide	513	605	722	874	1080	1377	4
H$_2$	Hydrogen*					−258.6	−252.8	1
H$_2$I$_2$Si	Diiodosilane				11.8	70.5	149.4	4
H$_2$O	Water***	−60.7 s	−42.2 s	−20.3 s	7.0	45.8	99.6	36,37
H$_2$O$_2$	Hydrogen peroxide			13 e	45 e	89.0	149.8	5
H$_2$O$_4$S	Sulfuric acid	72	103	140	187	248	330	4
H$_2$S	Hydrogen sulfide*		−149 s	−136 s	−118.9 s	−95.9 s	−60.5	1,5
H$_2$S$_2$	Hydrogen disulfide				−27 e	12.2	70.7	5

Mol. form.	Name	Temperature in °C for the indicated pressure						Ref.
		1 Pa	10 Pa	100 Pa	1 kPa	10 kPa	100 kPa	
H_2Se	Hydrogen selenide	−145 s	−134 s	−120 s	−102.8 s	−78.9 s	−41.5	5
H_2Te	Hydrogen telluride					−46.6	−2.3	5
H_3ISi	Iodosilane				−47.7	−10.1	45.2	4
H_3N	Ammonia*	−139 s	−127 s	−112 s	−94.5 s	−71.3	−33.6	1,5,6
H_3NO	Hydroxylamine				43.7	73.3	109.8	4
H_3P	Phosphine*	−182 s	−173 s	−161 s	−145 s	−122.7	−88.0	5
H_4IN	Ammonium iodide	125 s	159 s	201 s	253 s	318.4 s	405.2 s	5
H_4N_2	Hydrazine				14.7	55.6	113 e	5
H_4Si	Silane*			−181	−165.4	−143.7	−111.8	4
He	Helium*					−270.6	−268.9	2
Hf	Hafnium	2416	2681	3004	3406	3921	4603	9
Hg	Mercury**	42.0	76.6	120.0	175.6	250.3	355.9	29,30
HgI_2	Mercury(II) iodide	85.1 s	115.6 s	152.4 s	197.8 s	255.1 s	353.6	4
Ho	Holmium	1159 s	1311 s	1502 i	1767 i	2137 i	2691 i	3
IK	Potassium iodide			731	866	1052	1322	4
ILi	Lithium iodide	545	619	710	824	972	1170	4
INa	Sodium iodide			753	883	1058	1301	4
IRb	Rubidium iodide			733	866	1045	1302	4
ITl	Thallium(I) iodide				520	644	821	4
I_2	Iodine (rhombic)	−12.8 s	9.3 s	35.9 s	68.7 s	108 s	184.0	1,2
I_2Pb	Lead(II) iodide			470	558	682	869	4
I_2Zn	Zinc iodide	301 s	351 s	409 s	488 i	598 i	750 i	9
I_3Sb	Antimony(III) iodide				214.9	292.0	401.2	4
I_4Sn	Tin(IV) iodide				167.1	242.7	347.7	4
I_4Zr	Zirconium(IV) iodide	187 s	220 s	259 s	305 s	361 s	430 s	4
In	Indium	923	1052	1212	1417	1689	2067	2
Ir	Iridium	2440 s	2684	2979	3341	3796	4386	2
K	Potassium	200.2	256.5	328	424	559	756.2	13,30
Kr	Krypton*	−214.0 s	−208.0 s	−199.4 s	−188.9 s	−174.6 s	−153.6	5
La	Lanthanum	1732 i	1935 i	2185 i	2499 i	2905 i	3453 i	3
Li	Lithium	524.3	612.3	722.1	871.2	1064.3	1337.1	13,30
Lu	Lutetium	1633 s	1829.8	2072.8	2380 i	2799 i	3390 i	3
Mg	Magnesium	428 s	500 s	588 s	698	859	1088	2
Mn	Manganese	955 s	1074 s	1220 s	1418	1682	2060	2
Mo	Molybdenum	2469 s	2721	3039	3434	3939	4606	2
MoO_3	Molybdenum(VI) oxide				801	935	1151	4
NO	Nitric oxide*	−201 s	−195 s	−188 s	−179.3 s	−168.1 s	−151.9	5
N_2	Nitrogen*	−236 s	−232 s	−226.8 s	−220.2 s	−211.1 s	−195.9	1,5
N_2O	Nitrous oxide*	−167 s	−157 s	−145.4 s	−131.1 s	−112.9 s	−88.7	5
N_2O_4	Nitrogen tetroxide	−92 s	−78 s	−61 s	−41.1 s	−16.6 s	28.7	5
N_2O_5	Nitrogen pentoxide	−71 s	−56 s	−40 s	−19.9 s	3.9 s	33.2	5
Na	Sodium	280.6	344.2	424.3	529	673	880.2	13,30
Nb	Niobium	2669	2934	3251	3637	4120	4740	2
Nd	Neodymium	1322.3	1501.2	1725.3	2023 i	2442 i	3063 i	3
Ne	Neon*	−261 s	−260 s	−258 s	−255 s	−252 s	−246.1	2
Ni	Nickel	1510	1677	1881	2137	2468	2911	2
OPb	Lead(II) oxide	724	816	928	1065	1241	1471	4
OSr	Strontium oxide	1789 s	1903 s	2047 s	2235 s	2488 s		4
O_2	Oxygen*				−211.9	−200.5	−183.1	1,28
O_2S	Sulfur dioxide*			−98 s	−80 s	−52.2	−10.3	1,5
O_2Se	Selenium dioxide	124.5 s	153.9 s	188 s	228 s	275 s	315 s	38
O_2Si	Silicon dioxide	1966 i	2149 i	2368 i				8
O_3	Ozone*	−189 e	−182 e	−172 e	−158 e	−139.7	−111.5	5
O_3P_2	Phosphorus(III) oxide				47.3	100.3	172.8	4
O_3S	Sulfur trioxide				−20 s	6.6 s	44.5	5
O_3Sb_2	Antimony(III) oxide (valentinite)	426.1 s	478 s	539 s	610 s	907	1420	4,35
O_5P_2	Phosphorus(V) oxide	285 s	328 s	377.5 s	434.4 s	500.5 s	591	4
O_7Re_2	Rhenium(VII) oxide	147 s	176 s	208 s	244 s	284 s	362	4
Os	Osmium	2887 s	3150	3478	3875	4365	4983	2

		Temperature in °C for the indicated pressure						
Mol. form.	Name	1 Pa	10 Pa	100 Pa	1 kPa	10 kPa	100 kPa	Ref.
P	Phosphorus (white)	6 s	34 s	69	115	180	276	3,9
P	Phosphorus (red)	182 s	216 s	256 s	303 s	362 s	431 s	2,3
Pb	Lead	705	815	956	1139	1387	1754	2
PbS	Lead(II) sulfide	656 s	741 s	838 s	953 s	1088 s	1280	4
Pd	Palladium	1448 s	1624	1844	2122	2480	2961	2
Po	Polonium				573 e	730.2	963.3	5
Pr	Praseodymium	1497.7	1699.4	1954 i	2298 i	2781 i	3506 i	3
Pt	Platinum	2057	2277 e	2542	2870	3283	3821	2
Pu	Plutonium	1483	1680	1925	2238	2653	3226	2
Ra	Radium	546 s	633 s	764	936	1173	1526	2
Rb	Rubidium	160.4	212.5	278.9	368	496.1	685.3	13,30
Re	Rhenium	3030 s	3341	3736	4227	4854	5681	2
Rh	Rhodium	2015	2223	2476	2790	3132	3724	2
Rn	Radon*	−163 s	−152 s	−139 s	−121.4 s	−97.6 s	−62.3	5
Ru	Ruthenium	2315 s	2538	2814	3151	3572	4115	2
S	Sulfur	102 s	135	176	235	318	444	3
Sb	Antimony	534 s	603 s	738	946	1218	1585	2,3
Sc	Scandium	1372 s	1531 s	1733 i	1993 i	2340 i	2828 i	3
Se	Selenium	227	279	344	431	540	685	3
Si	Silicon	1635	1829	2066	2363	2748	3264	2
Sm	Samarium	728 s	833 s	967 s	1148 i	1402 i	1788 i	3
Sn	Tin	1224	1384	1582	1834	2165	2620	2
Sr	Strontium	523 s	609 s	717 s	866	1072	1373	2
Ta	Tantalum	3024	3324	3684	4122	4666	5361	2
Tb	Terbium	1516.1	1706.1	1928 i	2232 i	2640 i	3218 i	3
Tc	Technetium	2454 e	2725 e	3051 e	3453 e	3961 e	4621 e	2
Te	Tellurium			502 e	615 e	768.8	992.4	5
Th	Thorium	2360	2634	2975	3410	3986	4782	2
Ti	Titanium	1709	1898	2130 e	2419	2791	3285	2
Tl	Thallium	609	704	824	979	1188	1485	2
Tm	Thulium	844 s	962 s	1108 s	1297 s	1548 i	1944 i	3
U	Uranium	2052	2291	2586	2961	3454	4129	2
V	Vanadium	1828 s	2016	2250	2541	2914	3406	2
W	Tungsten	3204 s	3500	3864	4306	4854	5550	2
Xe	Xenon*	−190 s	−181 s	−170 s	−155.8 s	−136.6 s	−108.4	5,32
Y	Yttrium	1610.1	1802.3	2047 i	2354 i	2763 i	3334 i	3
Yb	Ytterbium	463 s	540 s	637 s	774 s	993 i	1192 i	3
Zn	Zinc	337 s	397 s	477	579	717	912 e	2
Zr	Zirconium	2366	2618	2924	3302	3780	4405	2

Substances containing carbon:

Mol. form.	Name	1 Pa	10 Pa	100 Pa	1 kPa	10 kPa	100 kPa	Ref.
C	Carbon (graphite)		2566 s	2775 s	3016 s	3299 s	3635 s	15
CBrClF$_2$	Bromochloro-difluoromethane	−136 e	−123 e	−106 e	−83.4	−51.8	−4.3	1
CBrCl$_3$	Bromotrichloromethane				−6 e	38.9	104.4	5
CBrF$_3$	Bromotrifluoromethane*	−168 e	−156 e	−142 e	−122.8	−96.6	−58.1	5
CBrN	Cyanogen bromide				−13 s	17.7 s	61.0	1
CBr$_2$F$_2$	Dibromodifluoromethane		−110 e	−91 e	−66 e	−30 e	22.5	1
CBr$_4$	Tetrabromomethane			25.6 s	65.8 s	111.6	188.9	5
CClF$_3$	Chlorotrifluoromethane	−176 e	−167 e	−155 e	−139 e	−116 e	−81.7	5
CClN	Cyanogen chloride		−94.6 s	−78.1 s	−57 s	−29 s	13.0	5
CCl$_2$F$_2$	Dichlorodifluoromethane*	−150 e	−138 e	−122 e	−101.8	−73.1	−30.0	5
CCl$_2$O	Carbonyl chloride	−127 e	−113 e	−96 e	−73 e	−40.6	7.2	5
CCl$_3$F	Trichlorofluoromethane*		−107 e	−89 e	−63 e	−28.5	23.3	1,5
CCl$_3$NO$_2$	Trichloronitromethane		−59 e	−30 e	4.4	47.8	112.0	5
CCl$_4$	Tetrachloromethane*	−79.4 s	−70.8 s	−53.5 s	−24.4 s	15.8	76.2	1,5
CFN	Cyanogen fluoride		−135 s	−121.2 s	−104.1 s	−82.8 s	−46.2	1,5
CF$_4$	Tetrafluoromethane*	−199.9 s	−193 s	−183.9 s	−171.6	−153.9	−128.3	1,5
CHBrF$_2$	Bromodifluoromethane		−128 s	−111.4 s	−89.7 s	−59.7 s	−16 s	5
CHBr$_3$	Tribromomethane				30.5	78.3	148.8	1
CHClF$_2$	Chlorodifluoromethane*	−152 e	−141 e	−126 e	−107.1	−80.5	−41.1	5
CHCl$_2$F	Dichlorofluoromethane	−76 e	−70 e	−61 e	−49 e	−28.7	8.6	1

Mol. form.	Name	Temperature in °C for the indicated pressure						Ref.
		1 Pa	10 Pa	100 Pa	1 kPa	10 kPa	100 kPa	
$CHCl_3$	Trichloromethane*			−61 e	−34 e	4.3	60.8	1
CHF_3	Trifluoromethane*			−152 e	−136 e	−114.4	−82.3	1
CHI_3	Triiodomethane	51.1 s	82.7 s	121 e			218.0	5
CHN	Hydrogen cyanide*			−77 s	−52.6 s	−22.7 s	25.4	1,5
$CHNO$	Cyanic acid			−81.1	−56.8	−23.9	23 e	5
CH_2BrCl	Bromochloromethane	−83 e	−69 e	−50 e	−25 e	11.4	67.7	1
CH_2Br_2	Dibromomethane			−37 e	−7 e	35.2	96.5	5
CH_2ClF	Chlorofluoromethane		−124 e	−108 e	−86.2	−55.7	−9.4	5
CH_2Cl_2	Dichloromethane*		−92 e	−73 e	−48 e	−12.5	39.3	1
CH_2F_2	Difluoromethane*	−156.7	−145.8	−131.9	−113.6	−88.6	−51.9	1
CH_2I_2	Diiodomethane			17 e	55 e	106.1	181.6	5
CH_2O	Formaldehyde*			−91 e	−61.7	−19.3		1
CH_2O_2	Formic acid	−56 s	−40.4 s	−22.3 s	−0.8 s	37.0	100.2	1,5
CH_3AsF_2	Methyldifluoroarsine			−15 e	22.1	76.1		5
CH_3BO	Borane carbonyl			−124	−99	−64		4
CH_3Br	Bromomethane			−77 e	−44.3	3.3		1
CH_3Cl	Chloromethane*	−140.2 s	−128.6 s	−114.7 s	−96 e	−67.1	−24.4	1,33
CH_3Cl_3Si	Methyltrichlorosilane		−83 e	−61 e	−33 e	7 e	65.7	1
CH_3F	Fluoromethane*			−130 e	−111 e	−78.6		1
CH_3I	Iodomethane			−49 e	−12.4	42.1		1
CH_3NO	Formamide		22 e	53 e	93 e	145.0	218 e	5
CH_3NO_2	Nitromethane			−2 e	40 e	100.8		1
CH_3NO_3	Methyl nitrate		−75 e	−55 e	−27 e	9.8	63 e	5
CH_4	Methane*	−220 s	−214.2 s	−206.8 s	−197 s	−183.6 s	−161.7	5,41
CH_4Cl_2Si	Dichloromethylsilane			−77 e	−51 e	−14 e	40.5	1
CH_4O	Methanol*	−87 e	−69 e	−47.5	−20.4	15.2	64.2	11
CH_4S	Methanethiol		−115 e	−97 e	−74 e	−41.7	5.7	1
CH_5ClSi	Chloromethylsilane	−129 e	−115 e	−97.9	−74.4	−41.5	8.3	5
CH_5N	Methylamine			−76.7	−48.1	−6.6		1
CH_6N_2	Methylhydrazine			−31 e	−4.7	32.9	91 e	1
CH_6OSi	Methyl silyl ether			−90.2	−61.8	−18 e		1
CH_6Si	Methylsilane			−144 e	−124.6	−97.5	−57.5	5
CIN	Cyanogen iodide						153.8	5
$CNNa$	Sodium cyanide		672 e	798	961	1182	1497	4
CN_4O_8	Tetranitromethane				18.0	61.8	124 e	5
CO	Carbon monoxide*			−223 s	−216.5 s	−207.2 s	−191.7	40
COS	Carbon oxysulfide*			−136 e	−117 e	−90.0	−50.4	1
$COSe$	Carbon oxyselenide			−120	−98	−67	−22	4
CO_2	Carbon dioxide*	−159.1 s	−148.9 s	−136.7 s	−121.6 s	−103.1 s	−78.6 s	5
CS_2	Carbon disulfide		−96 e	−76 e	−49 e	−10.9	45.9	1
CSe_2	Carbon diselenide			−24 e	9.4	56.2	127 e	1
$C_2Br_2ClF_3$	1,2-Dibromo-1-chloro-1,2,2-trifluoroethane						92.3	5
$C_2Br_2F_4$	1,2-Dibromotetrafluoroethane		−97 e	−75 e	−46 e	−7.2	47.1	5
C_2Br_4	Tetrabromoethylene		−54.5 s	−31.7 s	−3.5 s	32.2 s	226.0	5
C_2ClF_3	Chlorotrifluoroethylene	−146 e	−134 e	−119 e	−99 e	−71 e	−28.4	1
C_2ClF_5	Chloropentafluoroethane					−80.3	−39.4	1
$C_2Cl_2F_4$	1,1-Dichlorotetrafluoroethane					−45.4	2.7	1
$C_2Cl_2F_4$	1,2-Dichlorotetrafluoroethane				−76.8	−44.9	3.2	5
$C_2Cl_3F_3$	1,1,1-Trichlorotrifluoroethane						45.6	1,5
$C_2Cl_3F_3$	1,1,2-Trichlorotrifluoroethane					−8.2	47.3	1,5
C_2Cl_3N	Trichloroacetonitrile				−16 e	25.3	85.1	1
C_2Cl_4	Tetrachloroethylene			−22 e	10 e	54.4	120.7	1
$C_2Cl_4F_2$	1,1,1,2-Tetrachloro-2,2-difluoroethane				−7 e	31.0	91.1	5
$C_2Cl_4F_2$	1,1,2,2-Tetrachloro-1,2-difluoroethane					32.3	92.5	1
C_2Cl_4O	Trichloroacetyl chloride			−25 e	7 e	51.7	117.8	1,5
C_2Cl_6	Hexachloroethane	−7.6 s	9.9 s	33.6 s	67.7 s	116.9 s	184.2 s	5
C_2F_3N	Trifluoroacetonitrile				−126.1	−102.5	−67.8	1
C_2F_4	Tetrafluoroethylene				−132.3	−109.7	−75.8	1
$C_2F_4N_2O_4$	1,1,2,2-Tetrafluoro-1,2-dinitroethane				−30 e	6.4	59.5	5

Mol. form.	Name	\multicolumn{6}{c}{Temperature in °C for the indicated pressure}	Ref.
		1 Pa	10 Pa	100 Pa	1 kPa	10 kPa	100 kPa	
C_2F_6	Hexafluoroethane**			−155.2 s	−137.5 s	−113.4 s	−78.4 s	1,5
$C_2HBrClF_3$	2-Bromo-2-chloro-1,1,1-trifluoroethane				−41.4	−4.8	49.8	1
C_2HBr_3O	Tribromoacetaldehyde			15.0	52.7	103.0	173.5	5
C_2HClF_4	1-Chloro-1,1,2,2-tetrafluoroethane			−110 e	−87.6	−57.0	−12.1	5
$C_2HCl_2F_3$	2,2-Dichloro-1,1,1-trifluoroethane		−101.0	−82.2	−57.4	−23.3	26.7	18
C_2HCl_3	Trichloroethylene	−74 e	−59 e	−39 e	−12 e	26.7	86.8	1
C_2HCl_3O	Trichloroacetaldehyde			−41.6	−9.8	33.8	97.4	5
$C_2HCl_3O_2$	Trichloroacetic acid				83.8	130.0	197.2	1,5
C_2HCl_5	Pentachloroethane		−23 e	3 e	37.4	86.0	159.4	1
$C_2HF_3O_2$	Trifluoroacetic acid					16.8	71.4	1,5
C_2HF_5O	Trifluoromethyl difluoromethyl ether	−147 e	−136 e	−121 e	−102 e	−75.0	−35.4	20
C_2H_2	Acetylene*			−146.6 s	−130.7 s	−110.6 s	−84.8 s	5
$C_2H_2Br_2$	cis-1,2-Dibromoethylene		−45 e	−21 e	10 e	52.2	114.8	1
$C_2H_2Br_2$	trans-1,2–Dibromoethylene			−4 e	42.2	107.4	5	
$C_2H_2Br_2Cl_2$	1,2-Dibromo-1,1-dichloroethane					103.6	177.8	5
$C_2H_2Br_2Cl_2$	1,2-Dibromo-1,2-dichloroethane		−11 e	22 e	64.1	119 e	193 e	5
$C_2H_2Br_4$	1,1,2,2-Tetrabromoethane	14 e	38 e	69 e	109 e	163.7	242.9	5
$C_2H_2Cl_2$	1,1-Dichloroethylene	−116 e	−101 e	−82 e	−57 e	−21.4	31.2	1
$C_2H_2Cl_2$	cis-1,2-Dichloroethylene			−62 e	−34 e	3.8	60.3	1
$C_2H_2Cl_2$	trans-1,2-Dichloroethylene			−44 e	−7.5	47.3	1	
$C_2H_2Cl_2F_2$	1,2-Dichloro-1,1-difluoroethane	−101 e	−87 e	−68 e	−42.2	−6.8	46.3	5
$C_2H_2Cl_2O$	Chloroacetyl chloride			−23.7	5.6	46.1	105.6	5
$C_2H_2Cl_4$	1,1,1,2-Tetrachloroethane	−58 e	−40 e	−15 e	17 e	62.2	129.7	1
$C_2H_2Cl_4$	1,1,2,2-Tetrachloroethane		−22 e	1 e	32.4	76.9	144.7	1
$C_2H_2F_4$	1,1,1,2-Tetrafluoroethane				−94.3	−66.8	−26.4	17
$C_2H_2F_4$	1,1,2,2-Tetrafluoroethane				−96.0	−66.9	−23.3	5
C_2H_2O	Ketene		−151 e	−135 e	−115 e	−88.2	−50.0	1
C_2H_3Br	Bromoethylene	−124 e	−110 e	−92 e	−68 e	−34.5	15.4	5
C_2H_3BrO	Acetyl bromide	−78 e	−65 e	−49 e	−25 e	13.9	84 e	5
$C_2H_3Br_3$	1,1,2-Tribromoethane	−18 e	4 e	32 e	68 e	117.1	188.4	5
C_2H_3Cl	Chloroethylene	−139 e	−127 e	−110 e	−89 e	−59.0	−14.1	1
$C_2H_3ClF_2$	1-Chloro-1,1-difluoroethane		−123 e	−107 e	−85.3	−55.4	−10.5	5
C_2H_3ClO	Acetyl chloride	−100 e	−85 e	−66 e	−40 e	−3.6	50.4	1
$C_2H_3ClO_2$	Chloroacetic acid				78.4	123.9	188.9	1
$C_2H_3Cl_2F$	1,1-Dichloro-1-fluoroethane		−101 e	−83 e	−57.9	−22.7	31.4	5
$C_2H_3Cl_2F$	1,2-Dichloro-1-fluoroethane			−50 e	−23.8	14.1	73.4	5
$C_2H_3Cl_3$	1,1,1-Trichloroethane				−25.3	14.2	73.7	5
$C_2H_3Cl_3$	1,1,2-Trichloroethane			−23 e	7 e	49.9	113.4	1
C_2H_3F	Fluoroethylene			−153.3	−135.2	−109.9	−72.2	5
C_2H_3FO	Acetyl fluoride					−64.1	17.0	5
$C_2H_3F_3$	1,1,1-Trifluoroethane				−113 e	−86.6	−47.8	1
$C_2H_3F_3O$	2,2,2-Trifluoroethanol			−33 e	−8 e	26.0	74 e	5
C_2H_3I	Iodoethylene				−41 e	−3 e	55.6	5
C_2H_3IO	Acetyl iodide				−0.6	47 e	107.0	5
C_2H_3N	Acetonitrile				−20 e	21.4	81.2	1
C_2H_3NO	Methylisocyanate				−43.5	−10.2	38.8	1
C_2H_3NS	Methyl thiocyanate			−18.4	16.2	63.5	132.5	5
C_2H_4	Ethylene*				−155.6	−135.1	−104.0	1,10
C_2H_4BrCl	1-Bromo-2-chloroethane				−0.4	41.7	105.7	6
$C_2H_4Br_2$	1,1-Dibromoethane		−49 e	−26 e	5 e	46.4	107.6	5
$C_2H_4Br_2$	1,2-Dibromoethane				18 e	62.2	130.9	1
C_2H_4ClF	1-Chloro-1-fluoroethane				−69.9	−36.1	15.8	5
$C_2H_4Cl_2$	1,1-Dichloroethane		−84 e	−64 e	−36.7	1.0	56.9	1
$C_2H_4Cl_2$	1,2-Dichloroethane				−16.4	23.7	83.1	1
$C_2H_4F_2$	1,1-Difluoroethane			−115.2	−94.6	−66.1	−24.3	19
$C_2H_4N_2O_6$	Ethylene glycol dinitrate	4 e	25.6	51.0	81 e	117 e	162 e	5
C_2H_4O	Acetaldehyde		−105 e	−87 e	−62.8	−29.4	20.0	5
C_2H_4O	Ethylene oxide		−111 e	−93 e	−70 e	−37.0	10.2	1
$C_2H_4O_2$	Acetic acid	−42.8 s	−26.7 s	−8 s	14.2 s	55.9	117.5	1,5
$C_2H_4O_2$	Methyl formate		−95 e	−76 e	−51.8	−18.1	31.4	5

Mol. form.	Name	Temperature in °C for the indicated pressure						Ref.
		1 Pa	10 Pa	100 Pa	1 kPa	10 kPa	100 kPa	
$C_2H_4O_3$	Peroxyacetic acid				14.4	55.3	109.7	5
$C_2H_4O_3$	Glycolic acid						99.9	5
$C_2H_5AsF_2$	Ethyldifluoroarsine			−36 e	−6.0	35.0	93.1	5
C_2H_5Br	Bromoethane	−111 e	−96 e	−77 e	−51.3	−15.5	38.0	5
C_2H_5Cl	Chloroethane	−126 e	−112 e	−94 e	−70 e	−37.0	12.0	1
C_2H_5ClO	2-Chloroethanol	−61 e	−39 e	−12 e	23 e	67.1	127.3	5
C_2H_5ClO	Chloromethyl methyl ether	−96 e	−80 e	−59 e	−32 e	6 e	61 e	5
$C_2H_5Cl_3OSi$	Trichloroethoxysilane	−78 e	−60 e	−36.0	−4.6	38.7	102.0	5
$C_2H_5Cl_3Si$	Trichloroethylsilane	−79 e	−61 e	−38 e	−8 e	34.9	98.7	5
C_2H_5F	Fluoroethane		−142 e	−127 e	−106.3	−78.7	−37.9	1
C_2H_5FO	2-Fluoroethanol			−22 e	8.3	47.5	99 e	5
C_2H_5I	Iodoethane	−94 e	−78 e	−56 e	−27.9	11.9	71.9	5
C_2H_5N	Ethyleneimine		−74 e	−55 e	−30 e	4.1	55 e	5
C_2H_5NO	Acetamide	16.7 s	39.1 s	65.2 s	102.8	150.8	218.2	5
C_2H_5NO	N-Methylformamide		13 e	41 e	78 e	127.9	199.1	1
$C_2H_5NO_2$	Nitroethane	−61 e	−44 e	−21 e	8.3	50.1	113.5	5
$C_2H_5NO_3$	Ethyl nitrate	−81 e	−63 e	−41 e	−12 e	28.2	87 e	1
C_2H_6	Ethane*	−183.3 s	−173.2	−161.3	−145.3	−122.8	−88.8	41
$C_2H_6Cl_2Si$	Dichlorodimethylsilane					11.1	70.1	5
C_2H_6Hg	Dimethyl mercury				−13.5	29.0	92.1	5
$C_2H_6N_2O$	N-Nitrosodimethylamine				30.7	80.5	149.8	5
C_2H_6O	Ethanol	−73 e	−56 e	−34 e	−7 e	29.2	78.0	1,5
C_2H_6O	Dimethyl ether*		−135 e	−118 e	−96.8	−67.6	−25.1	1,5
C_2H_6OS	Dimethyl sulfoxide			27.4	65.0	115.9	188.6	1
$C_2H_6O_2$	Ethylene glycol	2 e	24 e	51.1	86.1	132.5	196.9	1
$C_2H_6O_2$	Ethyl hydroperoxide	−70 e	−49 e	−25 e	6.8	47.0	101 e	5
$C_2H_6O_2S$	Dimethyl sulfone				109 e	166.8	248.9	5
C_2H_6S	Ethanethiol	−112 e	−97 e	−78 e	−53 e	−18 e	34.7	1
C_2H_6S	Dimethyl sulfide		−96 e	−77 e	−51.2	−16.0	37.0	1,5
$C_2H_6S_2$	Dimethyl disulfide	−71 e	−53 e	−29 e	1.7	45.0	109.3	5
$C_2H_7BO_2$	Dimethoxyborane	−116 e	−101.9	−83.5	−59.2	−25.4	25 e	5
C_2H_7N	Ethylamine			−71 e	−53 e	−27 e	16.4	1
C_2H_7N	Dimethylamine			−88 e	−66.9	−37.2	6.6	1
C_2H_7NO	Ethanolamine		11 e	35 e	66.2	109.0	170.6	1
$C_2H_8N_2$	1,2-Ethanediamine				17.0	57.5	116.6	1,5
$C_2H_8N_2$	1,1-Dimethylhydrazine			−52 e	−25.6	10.5	63 e	5
$C_2H_8N_2$	1,2-Dimethylhydrazine		−49 e	−33 e	−9 e	26.4	88 e	1
C_2N_2	Cyanogen	−127 s	−114.1 s	−98.5 s	−79.2 s	−54.9 s	−21.4	5
C_3ClF_5O	Chloropentafluoroacetone	−122 e	−109 e	−93 e	−71 e	−39.4	7.4	5
C_3Cl_6	Hexachloropropene	−12 e	11 e	40 e	79 e	132.8	213.6	5
C_3F_6	Perfluoropropene	−150 e	−138 e	−122 e	−101 e	−72 e	−30.6	5
C_3F_6O	Perfluoroacetone			−113 e	−94 e	−67.8	−27.6	5
C_3F_8	Perfluoropropane		−139 e	−124 e	−105 e	−77.5	−37.0	1
C_3HN	Cyanoacetylene			−58.7 s	−35.6 s	−7 s	42.0	5
$C_3H_2F_6O$	1,1,1,3,3,3-Hexafluoro-2-propanol					12.7	57.1	5
$C_3H_3F_5$	1,1,1,2,2–Pentafluoropropane					−60 e	−17.9	5
C_3H_3N	2-Propenenitrile			−72 e	−50 e	−22 e	17.7	77.0	1
C_3H_3NS	Thiazole					54.4	117.8	5
C_3H_4	Allene*		−129 e	−118 e	−101.4	−76.7	−34.7	5
C_3H_4	Propyne				−94 e	−65.3	−23.2	1
$C_3H_4ClF_3$	3-Chloro-1,1,1-trifluoropropane	−102 e	−87 e	−68 e	−43 e	−8 e	45.3	5
$C_3H_4Cl_2O$	1,1-Dichloroacetone				1 e	47.8	118.0	5
$C_3H_4Cl_2O_2$	Methyl dichloroacetate	−44 e	−25 e	0 e	33 e	77.7	142.3	5
$C_3H_4Cl_4$	1,1,1,2-Tetrachloropropane	−48 e	−28 e	−2 e	32 e	79.1	149.5	5
$C_3H_4F_4O$	2,2,3,3-Tetrafluoro-1-propanol			−10 e	17 e	53.9	107.2	5
C_3H_4O	Acrolein		−87 e	−67 e	−40 e	−3.0	52.8	1
$C_3H_4O_2$	Propenoic acid				35 e	78.0	140.7	1
$C_3H_4O_2$	Vinyl formate			−58 e	−34 e	−1.6	46.2	1
$C_3H_4O_2$	2-Oxetanone		−21 e	8 e	45.5	93.8	159.3	5
$C_3H_4O_3$	Ethylene carbonate	12.7 s	37 e				247	5

		Temperature in °C for the indicated pressure						
Mol. form.	Name	1 Pa	10 Pa	100 Pa	1 kPa	10 kPa	100 kPa	Ref.
C$_3$H$_5$Br	cis-1-Bromopropene	−100 e	−84 e	−64 e	−37 e	1.0	57.4	5
C$_3$H$_5$Br	2-Bromopropene	−112 e	−95 e	−75 e	−47 e	−9 e	48.0	5
C$_3$H$_5$Br	3-Bromopropene	−98 e	−80 e	−58 e	−28 e	12 e	69.6	5
C$_3$H$_5$Cl	cis-1-Chloropropene	−114 e	−100 e	−81 e	−55 e	−20.1	32.4	5
C$_3$H$_5$Cl	trans-1-Chloropropene		−97 e	−77 e	−52 e	−16.2	37.0	5
C$_3$H$_5$Cl	2-Chloropropene	−120 e	−106 e	−87 e	−63 e	−28.7	22.3	5
C$_3$H$_5$Cl	3-Chloropropene	−107 e	−92 e	−72.4	−46.3	−9.8	44.6	5
C$_3$H$_5$ClO	Epichlorohydrin			−21 e	11 e	53.8	115.5	5
C$_3$H$_5$ClO$_2$	Methyl chloroacetate		−28 e	−5 e	25 e	66.9	129.1	5
C$_3$H$_5$Cl$_3$	1,1,3-Trichloropropane	−51 e	−31 e	−5 e	28 e	75.3	145.1	5
C$_3$H$_5$Cl$_3$	1,2,3-Trichloropropane			2 e	37 e	84.9	156.3	5
C$_3$H$_5$Cl$_3$Si	Trichloro-2-propenylsilane					53.0	116.5	5
C$_3$H$_5$I	3-Iodopropene	−80 e	−62 e	−39 e	−8 e	36 e	101.5	5
C$_3$H$_5$N	Propanenitrile	−69.4	−55.3	−36.0	−7.9	35.2	97.4	1,5
C$_3$H$_5$NO	Acrylamide			109.6	161 e			5
C$_3$H$_5$NO	3-Hydroxypropanenitrile	−11 e	18 e	53 e	96.1	150.3	220.8	5
C$_3$H$_5$NS	Ethyl thiocyanate	−39 e	−20 e	4 e	35 e	79.1	143.4	5
C$_3$H$_5$NS	Ethyl isothiocyanate			17.4	66 e	136 e		5
C$_3$H$_5$N$_3$O$_9$	Trinitroglycerol	48.6	75.7	118 e	191 e	353 e	1007 e	5
C$_3$H$_6$	Propene*	−160.6	−149.0	−134.3	−114.9	−88.2	−47.9	1,5
C$_3$H$_6$	Cyclopropane			−124 e	−104 e	−75.7	−33.1	1
C$_3$H$_6$BrCl	1-Bromo-3-chloropropane	−51 e	−31 e	−6 e	28 e	74.1	142.9	5
C$_3$H$_6$Br$_2$	1,2-Dibromopropane	−46 e	−26 e	−2 e	31 e	75.3	139.5	5
C$_3$H$_6$Br$_2$	1,3-Dibromopropane	−30 e	−9 e	17 e	52 e	98.7	166.8	5
C$_3$H$_6$Cl$_2$	1,1-Dichloropropane				−14 e	27.0	87.7	5
C$_3$H$_6$Cl$_2$	1,2-Dichloropropane	−78 e	−61 e	−38.1	−8.1	33.7	95.9	5
C$_3$H$_6$Cl$_2$	1,3-Dichloropropane	−65 e	−46 e	−22 e	10 e	54.0	119.9	5
C$_3$H$_6$Cl$_2$	2,2-Dichloropropane				−28 e	10.8	68.9	5
C$_3$H$_6$Cl$_2$O	1,3-Dichloro-2-propanol			21.8	59.0	107.6	173.9	5
C$_3$H$_6$N$_2$O$_4$	1,1-Dinitropropane	−9 e	12 e	39 e	73.2	120 e	187 e	5
C$_3$H$_6$O	Allyl alcohol	−63 e	−48 e	−21.9	6.8	44.5	96.2	5
C$_3$H$_6$O	Methyl vinyl ether			−114 e	−89 e	−52.7	4.6	1
C$_3$H$_6$O	Propanal			−69 e	−42 e	−6 e	47.7	1
C$_3$H$_6$O	Acetone	−95	−81.8	−62.8	−35.6	1.3	55.7	1,5
C$_3$H$_6$O	Methyloxirane	−109 e	−95 e	−76 e	−51.5	−17.2	33.9	5
C$_3$H$_6$O$_2$	Propanoic acid			0 e	35.1	79.9	140.8	1,5
C$_3$H$_6$O$_2$	Ethyl formate		−80 e	−61 e	−35 e	1 e	54.0	1
C$_3$H$_6$O$_2$	Methyl acetate	−95 e	−79 e	−59 e	−33 e	3.3	56.6	1
C$_3$H$_6$O$_2$	1,3-Dioxolane		−72 e	−50 e	−22 e	17.0	75.3	1
C$_3$H$_6$O$_3$	1,3,5-Trioxane					53 e	113.7	1
C$_3$H$_6$S	Thietane		−62 e	−40 e	−9 e	32.5	94.5	5
C$_3$H$_7$Br	1-Bromopropane	−95 e	−78 e	−57 e	−28 e	11.6	70.6	1
C$_3$H$_7$Br	2-Bromopropane		−84 e	−65 e	−39.6	−1.7	59.1	1,5
C$_3$H$_7$Cl	1-Chloropropane	−106 e	−90 e	−71 e	−44.5	−8.1	46.2	1
C$_3$H$_7$Cl	2-Chloropropane		−91 e	−74 e	−51.1	−17.8	35.4	1,5
C$_3$H$_7$ClO	2-Chloro-1-propanol				23 e	63.8	125.7	5
C$_3$H$_7$F	1-Fluoropropane	−133 e	−120 e	−103 e	−80.7	−49.4	−2.8	5
C$_3$H$_7$I	1-Iodopropane	−78 e	−60 e	−37 e	−6 e	36.9	102.0	5
C$_3$H$_7$I	2-Iodopropane	−89 e	−71 e	−47 e	−16.3	26.5	89.2	5
C$_3$H$_7$N	Allylamine		−88 e	−65 e	−37 e	0.4	52 e	5
C$_3$H$_7$NO	N,N-Dimethylformamide	−39 e	−20 e	5 e	38.0	83.9	152.6	1
C$_3$H$_7$NO	N-Methylacetamide	−13.3 s	13 s	43 e	83.8	136.1	206.3	5
C$_3$H$_7$NO$_2$	1-Nitropropane	−56 e	−37 e	−13 e	20 e	64.8	130.8	1
C$_3$H$_7$NO$_2$	2-Nitropropane		−48 e	−22 e	10.7	55.6	119.8	1
C$_3$H$_7$NO$_3$	Propyl nitrate			−23.9	6.1	48.1	111 e	5
C$_3$H$_8$	Propane*	−156.9	−145.6	−130.9	−111.4	−83.8	−42.3	1,41
C$_3$H$_8$O	1-Propanol	−54 e	−38 e	−16 e	10 e	47 e	96.9	1,5
C$_3$H$_8$O	2-Propanol	−65 e	−49 e	−28 e	−1.3	33.6	82.0	1,5
C$_3$H$_8$O	Ethyl methyl ether	−98 e	−89 e	−77 e	−60 e	−34.8	7.0	5
C$_3$H$_8$O$_2$	1,2-Propylene glycol	−11 e	13 e	42 e	78 e	125.0	187.2	5

Mol. form.	Name	1 Pa	10 Pa	100 Pa	1 kPa	10 kPa	100 kPa	Ref.
$C_3H_8O_2$	1,3-Propylene glycol	4 e	30 e	62 e	101 e	149.9	214.0	5
$C_3H_8O_2$	Ethylene glycol monomethyl ether	−57 e	−37 e	−12 e	21 e	63.8	124.3	1
$C_3H_8O_2$	Dimethoxymethane	−93 e	−81 e	−64 e	−42 e	−9.3	41.7	5
$C_3H_8O_3$	Glycerol	96 e	113 e	136 e	168 e	213.4	287 e	1
C_3H_8S	1-Propanethiol	−94 e	−78 e	−57 e	−29.1	9.6	67.4	1,5
C_3H_8S	2-Propanethiol	−102 e	−87 e	−67 e	−41 e	−3 e	52.2	1
C_3H_8S	Ethyl methyl sulfide	−94 e	−78 e	−57 e	−29.7	8.8	66.3	1
$C_3H_8S_2$	1,3–Propanedithiol	−53 e	−28 e	3 e	43 e	97 e	172.4	5
C_3H_9As	Trimethylarsine			−74 e	−45 e	−5.4	52.0	5
$C_3H_9BO_3$	Trimethyl borate			−14 e	15.6	67.9		5
C_3H_9BS	Methyl dimethylthioborane			−62 e	−30.4	11.4	70.7	5
C_3H_9ClSi	Trimethylchlorosilane			−37.8	0.4	57.3		5
C_3H_9N	Propylamine		−81 e	−63 e	−38.3	−4.1	46.9	1,5
C_3H_9N	Isopropylamine		−91 e	−74 e	−50.4	−17.6	31.5	1,5
C_3H_9N	Trimethylamine		−114 e	−97 e	−75.0	−43.8	2.6	1,5
C_3H_9NO	1-Amino-2-propanol			18 e	53.2	98.2	157.9	5
$C_3H_9O_4P$	Trimethyl phosphate	−31 e	−7 e	23.6	62.8	116.0	192.0	5
C_3H_9P	Trimethylphosphine			−81 e	−53 e	−15.0	37.1	5
C_3H_9Sb	Trimethylstibine			−56 e	−23.8	19 e	80 e	5
$C_3H_{10}N_2$	1,2-Propanediamine		−35.4	−12.0	18.8	61 e	119 e	5
C_3N_2O	Carbonyl dicyanide			−21.7	15.3	65.2		5
C_4Cl_6	Hexachloro-1,3-butadiene	−1 e	22 e	50 e	86.7	137.0	209.7	5
$C_4F_6O_3$	Trifluoroacetic acid anhydride			−63 e	−39 e	−7.1	38.8	5
C_4F_8	Perfluorocyclobutane						−6.2	1
C_4F_{10}	Perfluorobutane		−122 e	−105 e	−82 e	−49.8	−2.5	1,5
$C_4H_2Cl_2O_2$	trans-2-Butenedioyl dichloride			8.0	45.6	94.3	159.8	5
$C_4H_2Cl_2S$	2,5-Dichlorothiophene			−20 e	22 e	81.4	171 e	5
$C_4H_2O_3$	Maleic anhydride			73.7	127.9	201.7		5
C_4H_3ClS	2-Chlorothiophene		−62 e	−35 e	2 e	51.8	123 e	5
C_4H_3IS	2-Iodothiophene			−25 e	23 e	94.9	181.0	5
C_4H_4	1-Buten-3-yne			−96.1	−73.4	−41.8	4.9	5
$C_4H_4N_2$	Succinonitrile	24.8 s					266.0	5
C_4H_4O	Furan			−78 e	−54 e	−20 e	31.0	1
$C_4H_4O_2$	Diketene				19.3	63.3	126 e	5
$C_4H_4O_3$	Succinic anhydride				121 e	180.8	260.8	5
$C_4H_4O_4$	Fumaric acid	123.9 s	150 s	180 s				5
C_4H_4S	Thiophene				−17 e	23.7	83.7	5
C_4H_5Cl	2-Chloro-1,3-butadiene	−113 e	−95 e	−71 e	−41 e	0.3	59.0	5
C_4H_5ClO	2-Methyl-2-propenoyl chloride		−57 e	−35 e	−5 e	36.4	98.2	5
$C_4H_5Cl_3O_2$	Ethyl trichloroacetate			15.3	51.9	100.1	166.6	5
C_4H_5N	3-Butenenitrile	−67 e	−48 e	−23.1	9.3	53.7	118.4	5
C_4H_5N	Methylacrylonitrile			−12 e	29.0	89.8		5
C_4H_5N	Pyrrole			−8 e	24 e	66.7	129.4	1
$C_4H_5NO_2$	Methyl cyanoacetate	−3 e	19 e	48 e	84 e	134.0	204.6	5
C_4H_5NS	Allyl isothiocyanate	−45 e	−27 e	−3 e	32.1	89 e	198 e	5
C_4H_5NS	4-Methylthiazole					67.0		5
C_4H_6	1,2-Butadiene	−132 e	−117 e	−98 e	−72.8	−38.9	10.5	5
C_4H_6	1,3-Butadiene*			−106 e	−83 e	−51.9	−4.7	1
C_4H_6	1-Butyne	−125 e	−111 e	−94 e	−71.2	−39.4	7.8	1
C_4H_6	2-Butyne		−89.2 s	−73.8 s	−53.5 s	−23.9	26.6	5
$C_4H_6Cl_2O_2$	Ethyl dichloroacetate			2.6	40.1	89.1	156.3	5
C_4H_6O	Divinyl ether		−99 e	−80 e	−56 e	−22.1	28.0	5
C_4H_6O	trans-2-Butenal	−74 e	−56 e	−33 e	−3 e	39.7	102.4	5
C_4H_6O	3-Buten-2-one					21 e	81.0	5
C_4H_6O	Cyclobutanone			−34 e	−4 e	37.1	97 e	5
$C_4H_6O_2$	cis-Crotonic acid			30 e	63 e	106.7	168.9	5
$C_4H_6O_2$	trans-Crotonic acid				74 e	120.8	184.9	5
$C_4H_6O_2$	3-Butenoic acid	−19 e	2 e	27 e	61 e	105.6	168.6	5
$C_4H_6O_2$	Methacrylic acid			22 e	56 e	99.9	161.5	5
$C_4H_6O_2$	Vinyl acetate	−88 e	−71 e	−50 e	−22 e	16.2	72.2	1

		Temperature in °C for the indicated pressure						
Mol. form.	Name	1 Pa	10 Pa	100 Pa	1 kPa	10 kPa	100 kPa	Ref.
C$_4$H$_6$O$_2$	Methyl acrylate		−71 e	−48 e	−18 e	22 e	79.9	5
C$_4$H$_6$O$_2$	2,3-Butanedione					30.7	84.8	5
C$_4$H$_6$O$_2$	gamma-Butyrolactone		−17 e	24 e	72 e	130.2	203 e	5
C$_4$H$_6$O$_3$	Acetic anhydride	−44 e	−25 e	−1 e	31 e	75.1	139.7	1
C$_4$H$_6$O$_3$	Propylene carbonate	−40 e	−5 e	43 e	112 e	220 e	410 e	5
C$_4$H$_6$O$_4$	Dimethyl oxalate				50.5	98.1	163.0	5
C$_4$H$_7$Br	trans-1-Bromo-1-butene	−87 e	−68 e	−43.3	−11.4	31.9	94.4	5
C$_4$H$_7$Br	2-Bromo-1-butene	−87 e	−70 e	−48 e	−20 e	20.7	80.6	5
C$_4$H$_7$Br	cis-2-Bromo-2-butene	−90 e	−72 e	−49.0	−18.5	23.5	85.2	5
C$_4$H$_7$Br	trans-2-Bromo-2-butene	−86 e	−67 e	−43.4	−12.0	31.0	93.5	5
C$_4$H$_7$Br$_3$	1,2,3-Tribromobutane	0 e	23 e	53 e	91 e	143.7	219.5	5
C$_4$H$_7$Br$_3$	1,2,4-Tribromobutane	−3 e	20 e	49 e	87 e	139.4	214.5	5
C$_4$H$_7$Cl	3-Chloro-1-butene			−64 e	−36 e	4 e	63.6	5
C$_4$H$_7$Cl	cis-2-Chloro-2-butene	−100 e	−83 e	−62 e	−34 e	6 e	66.4	5
C$_4$H$_7$Cl	trans-2-Chloro-2-butene	−102 e	−86 e	−65 e	−37 e	3 e	62.2	5
C$_4$H$_7$Cl	3-Chloro-2-methylpropene		−75 e	−54 e	−25 e	13.8	71.5	5
C$_4$H$_7$ClO$_2$	Ethyl chloroacetate			−2.6	32.6	79.1	143.8	5
C$_4$H$_7$N	Butanenitrile	−67 e	−48 e	−24 e	8 e	52.3	117.2	1
C$_4$H$_8$	1-Butene	−139.0	−125.2	−107.8	−85.3	−53.7	−6.6	1,5
C$_4$H$_8$	cis-2-Butene	−131.2	−117.4	−99.8	−76.7	−44.8	3.4	1,5
C$_4$H$_8$	trans-2-Butene			−102 e	−80 e	−47.6	0.6	1
C$_4$H$_8$	Isobutene	−139.1	−125.5	−108.2	−85.5	−54.5	−7.3	1,5
C$_4$H$_8$	Cyclobutane				−71.8	−38.1	12.1	5
C$_4$H$_8$	Methylcyclopropane	−130 e	−116 e	−99.3	−76.3	−44.2	4.2	5
C$_4$H$_8$Br$_2$	1,2-Dibromobutane	−54 e	−30 e	0.4	39.6	92.1	166.1	5
C$_4$H$_8$Br$_2$	1,4-Dibromobutane	−13 e	9 e	37 e	74 e	124.0	196.5	5
C$_4$H$_8$Cl$_2$	1,1-Dichlorobutane			−25 e	6 e	49.3	113.4	5
C$_4$H$_8$Cl$_2$	1,2-Dichlorobutane			−28.4	5.8	53.1	123.1	5
C$_4$H$_8$Cl$_2$	1,4-Dichlorobutane			−26 e	0 e	35 e	82.4	153.4	5
C$_4$H$_8$Cl$_2$	2,2-Dichlorobutane			−58 e	−35 e	−5 e	37.8	102.1	5
C$_4$H$_8$Cl$_2$O	Bis(2-chloroethyl) ether	−32 e	−9 e	19.8	56.9	106.9	177.9	5
C$_4$H$_8$O	Ethyl vinyl ether		−102 e	−81 e	−53.1	−16.5	34.7	5
C$_4$H$_8$O	1,2-Epoxybutane	−135 e	−114 e	−87 e	−53 e	−5.5	62.1	5
C$_4$H$_8$O	Butanal	−88 e	−72 e	−50 e	−22 e	16.6	74.5	1,5
C$_4$H$_8$O	Isobutanal			−56 e	−29 e	8 e	63.8	1
C$_4$H$_8$O	2-Butanone	−85 e	−68 e	−46 e	−18.1	21.2	79.2	1
C$_4$H$_8$O	Tetrahydrofuran	−94 e	−78 e	−57.3	−29.8	9 e	65.6	1
C$_4$H$_8$O$_2$	Butanoic acid			12.9	52.2	101.4	163.3	1,5
C$_4$H$_8$O$_2$	2-Methylpropanoic acid	−30.1	−8.2	18.1	50.5	92.9	154.0	5
C$_4$H$_8$O$_2$	Propyl formate	−78 e	−62 e	−42 e	−15.1	23.0	80.4	1,5
C$_4$H$_8$O$_2$	Isopropyl formate	−80 e	−65 e	−47 e	−22.2	13.2	67.7	5
C$_4$H$_8$O$_2$	Ethyl acetate	−83 e	−66 e	−45 e	−18 e	20.4	76.8	1
C$_4$H$_8$O$_2$	Methyl propanoate	−80 e	−64 e	−43 e	−15.8	22.2	79.0	1
C$_4$H$_8$O$_2$	cis-2-Butene-1,4-diol	17 e	44 e	77 e	117.4	168.5	234.9	5
C$_4$H$_8$O$_2$	1,3-Dioxane			−37 e	−3 e	43.4	106.0	5
C$_4$H$_8$O$_2$	1,4-Dioxane					39.6	101.0	1
C$_4$H$_8$O$_2$S	Sulfolane		49 e	87 e	135 e	198.0	283.5	5
C$_4$H$_8$S	Tetrahydrothiophene	−66 e	−47 e	−23 e	9.4	54.1	120.5	1
C$_4$H$_9$Br	1-Bromobutane	−68.4	−53.9	−34.1	−5.4	37.6	101.1	1,5
C$_4$H$_9$Br	2-Bromobutane	−86 e	−68 e	−46 e	−16 e	26.6	90.7	5
C$_4$H$_9$Br	1-Bromo-2-methylpropane	−85 e	−68 e	−46 e	−16 e	26.8	91.1	5
C$_4$H$_9$Br	2-Bromo-2-methylpropane					11.7	72.4	1,5
C$_4$H$_9$Cl	1-Chlorobutane	−87 e	−71 e	−49 e	−21 e	18.4	78.1	1
C$_4$H$_9$Cl	2-Chlorobutane	−96 e	−80 e	−59 e	−31.0	8.5	67.9	1
C$_4$H$_9$Cl	1-Chloro-2-methylpropane	−94 e	−78 e	−56.6	−28.7	10.2	68.5	5
C$_4$H$_9$Cl	2-Chloro-2-methylpropane					−4.2	50.3	5
C$_4$H$_9$Cl$_3$Si	Butyltrichlorosilane					77.2	148.4	5
C$_4$H$_9$F	1-Fluorobutane	−114 e	−99 e	−80 e	−55 e	−20.0	32.1	5
C$_4$H$_9$F	2-Fluorobutane	−117 e	−103 e	−85 e	−60.7	−26.7	24.7	5
C$_4$H$_9$I	1-Iodobutane	−62 e	−43 e	−19 e	14 e	60.5	130.0	5

Mol. form.	Name	Temperature in °C for the indicated pressure						Ref.
		1 Pa	10 Pa	100 Pa	1 kPa	10 kPa	100 kPa	
C_4H_9I	2-Iodobutane	−70 e	−51 e	−27 e	5 e	50 e	119.5	5
C_4H_9I	1-Iodo-2-methylpropane		−47 e	−21.4	12.0	56.8	120.0	5
C_4H_9I	2-Iodo-2-methylpropane	−75.1 s	−58.8 s	−39.5 s	-5.2	41 e	100.0	5
C_4H_9N	Pyrrolidine		−59 e	−38 e	−10 e	28.5	86.2	1
C_4H_9NO	N-Methylpropanamide				81.1	105 e		5
C_4H_9NO	N,N-Dimethylacetamide	−8 e	8 e	28.0	56.4	98.2	165.7	1
C_4H_9NO	2-Butanone oxime		−18 e	7 e	38.9	81.9	142.9	5
C_4H_9NO	Morpholine				21 e	64.5	128.5	1
$C_4H_9NO_3$	Isobutyl nitrate			−18 e	15.1	59.2	123.0	5
C_4H_{10}	Butane*	−134.3	−121.0	−103.9	−81.1	−49.1	−0.8	1,41
C_4H_{10}	Isobutane*		−129.0	−113.0	−90.9	−59.4	−12.0	1,41
$C_4H_{10}O$	1-Butanol	−37 e	−20 e	0 e	28 e	64 e	117.4	1
$C_4H_{10}O$	2-Butanol	−50 e	−34 e	−14 e	12.6	48.2	99.2	1,5
$C_4H_{10}O$	2-Methyl-1-propanol	−39 e	−24 e	−5 e	20.9	56.0	107.6	1,5
$C_4H_{10}O$	2-Methyl-2-propanol					34.4	82.1	1,5
$C_4H_{10}O$	Diethyl ether	−111 e	−96 e	−77 e	−52.6	−17.8	34.1	1
$C_4H_{10}O$	Methyl propyl ether				−40 e	−11.3	38.7	5
$C_4H_{10}O$	Isopropyl methyl ether				−56 e	−21.2	30.4	5
$C_4H_{10}O_2$	1,3-Butanediol	−4 e	23 e	55 e	94 e	142.9	206.1	5
$C_4H_{10}O_2$	1,4-Butanediol		45 e	77 e	116 e	164.7	227.6	5
$C_4H_{10}O_2$	2,3-Butanediol		15 e	43 e	77 e	121.2	180.3	5
$C_4H_{10}O_2$	Ethylene glycol monoethyl ether	−49 e	−29 e	−3 e	30 e	73.6	135.3	1
$C_4H_{10}O_2$	Ethylene glycol dimethyl ether			−44 e	−15 e	25.2	85.2	1
$C_4H_{10}O_2$	Dimethylacetal	−89 e	−74 e	−55 e	−29 e	7.7	64.1	5
$C_4H_{10}O_2$	Diethylperoxide				−39 e	3.6	65.0	5
$C_4H_{10}O_2S$	Bis(2-hydroxyethyl) sulfide			31 e	114.2		282.0	5
$C_4H_{10}O_3$	Diethylene glycol	35 e	58 e	86 e	123 e	173.6	245.2	1
$C_4H_{10}O_4S$	Diethyl sulfate		3 e	36 e	79 e	134 e	208.3	5
$C_4H_{10}S$	1-Butanethiol	−77 e	−59 e	−37 e	−6 e	35.4	98.0	5
$C_4H_{10}S$	2-Butanethiol	−86 e	−69 e	−47 e	−17 e	23.4	84.5	5
$C_4H_{10}S$	2-Methyl-1-propanethiol		−66 e	−44 e	−15 e	26.5	88.1	5
$C_4H_{10}S$	2-Methyl-2-propanethiol					5.8	63.8	5
$C_4H_{10}S$	Diethyl sulfide	−80 e	−62 e	−40 e	−10.8	30.3	91.7	1
$C_4H_{10}S$	Methyl propyl sulfide	−78 e	−61 e	−38 e	−8 e	33.1	95.1	5
$C_4H_{10}S$	Isopropyl methyl sulfide	−85 e	−68 e	−46 e	−17 e	23.4	84.3	5
$C_4H_{10}S_2$	1,4-Butanedithiol	−17 e	5 e	32 e	69.1	119.9	195.1	5
$C_4H_{10}S_2$	Diethyl disulfide	−46 e	−26 e	0 e	35 e	82.4	153.5	5
$C_4H_{11}N$	Butylamine			−46 e	−18.1	20.0	75.9	5
$C_4H_{11}N$	sec-Butylamine			−55 e	−29.1	7.5	62.3	5
$C_4H_{11}N$	tert-Butylamine			−67 e	−42.4	−8.1	43.7	5
$C_4H_{11}N$	Isobutylamine	−85 e	−70 e	−50 e	−24.5	12.0	67.3	5
$C_4H_{11}N$	Diethylamine			−46 e	−26 e	5 e	55.2	1
$C_4H_{11}NO$	N,N-Dimethylethanolamine	−52 e	−31 e	−6 e	27 e	70.9	133 e	5
$C_4H_{11}NO_2$	Diethanolamine	53 e	77 e	107 e	146 e	197.3	268 e	5
$C_4H_{12}BN$	(Dimethylamino)dimethyl-borane		−81 e	−60.1	−31.9	7.0	64.2	5
$C_4H_{12}Cl_2OSi_2$	1,3-Dichloro-1,1,3,3-tetramethyldisiloxane		−33 e	−9 e	23.8	69.1	136.5	5
$C_4H_{12}O_4Si$	Tetramethyl silicate				14.4	59.3	119.7	5
$C_4H_{12}Si$	Tetramethylsilane			−83 e	−59 e	−25 e	26.7	5
$C_4H_{12}Sn$	Tetramethylstannane			−55.0	−25.6	16.6	77.7	5
$C_4H_{13}N_3$	Diethylenetriamine	−10 e	13 e	43 e	80 e	129.6	198 e	5
C_4NiO_4	Nickel carbonyl					−12	42	4
C_5F_{12}	Perfluoropentane				−54.7	−20.9	28.6	5
C_5FeO_5	Iron pentacarbonyl				0	44	105	4
C_5H_4ClN	2-Chloropyridine			7.4	45.8	97.3	169.9	5
$C_5H_4O_2$	Furfural	−26 e	−8 e	16 e	47 e	92.4	161.4	1
C_5H_5N	Pyridine			−23 e	8 e	51.0	114.9	1
C_5H_6	1,3-Cyclopentadiene			−77 e	−51 e	−14 e	39.8	5
$C_5H_6N_2$	Pentanedinitrile	24.1	52 e	85 e	126 e	178 e	245 e	5
C_5H_6O	2-Methylfuran			−66 e	−35 e	6 e	64.5	1

Mol. form.	Name	Temperature in °C for the indicated pressure						Ref.
		1 Pa	10 Pa	100 Pa	1 kPa	10 kPa	100 kPa	
C$_5$H$_6$O$_2$	Furfuryl alcohol	−30 e	−5 e	25 e	62.6	109.3	169.7	5
C$_5$H$_6$S	2-Methylthiophene		−58 e	−32 e	2 e	47.9	112.2	1
C$_5$H$_6$S	3-Methylthiophene		−53 e	−28 e	6 e	50.6	115.1	1
C$_5$H$_7$N	1-Methylpyrrole				8 e	49.9	112.3	5
C$_5$H$_7$NO$_2$	Ethyl cyanoacetate	16 e	39 e	67.0	102.1	146.7	205.6	5
C$_5$H$_8$	1,2-Pentadiene	−109 e	−93 e	−73 e	−46.1	−9.7	44.5	5
C$_5$H$_8$	cis-1,3-Pentadiene	−109 e	−93 e	−73 e	−47.0	−10.5	43.7	1,5
C$_5$H$_8$	trans-1,3-Pentadiene			−75 e	−49.0	−13 e	42 e	1
C$_5$H$_8$	1,4-Pentadiene	−120 e	−105 e	−86 e	−60.9	−26.2	25.6	5
C$_5$H$_8$	2,3-Pentadiene	−106 e	−90 e	−70 e	−42.9	−6.3	47.9	5
C$_5$H$_8$	3-Methyl-1,2-butadiene	−111 e	−95 e	−75 e	−49.2	−13.1	40.4	5
C$_5$H$_8$	2-Methyl-1,3-butadiene	−115 e	−100 e	−81 e	−55.4	−19.7	33.7	1,5
C$_5$H$_8$	1-Pentyne			−75 e	−49.1	−13.5	39.9	5
C$_5$H$_8$	2-Pentyne	−100 e	−85 e	−65 e	−37.9	−0.5	55.7	5
C$_5$H$_8$	3-Methyl-1-butyne			−82 e	−57.5	−23.1	28.6	5
C$_5$H$_8$	Cyclopentene	−109 e	−94 e	−74 e	−48 e	−11.1	43.8	5
C$_5$H$_8$	Spiropentane	−110 e	−95 e	−76 e	−51 e	−15 e	38.6	5
C$_5$H$_8$O	3-Methyl-3-buten-2-one			−35 e	−5 e	36.0	97.3	5
C$_5$H$_8$O	Cyclopropyl methyl ketone		−57 e	−31 e	3 e	49 e	112 e	5
C$_5$H$_8$O	Cyclopentanone		−39 e	−14 e	19 e	64 e	130.3	1
C$_5$H$_8$O	3,4-Dihydro-2H-pyran			−22 e	22.0	84.9		5
C$_5$H$_8$O$_2$	4-Pentenoic acid	0 e	19 e	44 e	77 e	122.0	187.5	5
C$_5$H$_8$O$_2$	Vinyl propanoate				31.2	94 e		5
C$_5$H$_8$O$_2$	Ethyl acrylate		−55 e	−32.7	−2.8	38.5	99.2	5
C$_5$H$_8$O$_2$	Methyl methacrylate			−31 e	−1 e	39.7	100.0	1
C$_5$H$_8$O$_2$	2,4-Pentanedione			−5 e	24.7	67.8	137.4	1
C$_5$H$_8$O$_2$	Tetrahydro-2H-pyran-2-one		5 e	35.1	74.4	128.3	207.0	5
C$_5$H$_8$O$_3$	Methyl acetoacetate				50.1	101.1	171.3	5
C$_5$H$_8$O$_4$	Glutaric acid		121 e	153.2	191.9	240.3	302.5	5
C$_5$H$_8$O$_4$	Dimethyl malonate	−22 e	1 e	30.0	66.7	114.7	180.2	5
C$_5$H$_9$ClO$_2$	Ethyl 2-chloropropanoate			1.4	36.4	82.5	146.0	5
C$_5$H$_9$ClO$_2$	Isopropyl chloroacetate			−2 e	35.0	83.3	148.1	5
C$_5$H$_9$N	Pentanenitrile	−54 e	−34 e	−8 e	26 e	72.2	140.9	1
C$_5$H$_9$N	2,2-Dimethylpropanenitrile					41.1	104.8	5
C$_5$H$_9$NO	N-Methyl-2-pyrrolidone	1 e	24 e	53.1	92.3	147.2	229 e	5
C$_5$H$_{10}$	1-Pentene	−118.9	−103.4	−84.0	−58.8	−23.3	29.6	1,5
C$_5$H$_{10}$	cis-2-Pentene	−113.8	−98.1	−78.4	−52.7	−16.8	36.6	1,5
C$_5$H$_{10}$	trans-2-Pentene	−114.5	−98.9	−79.1	−53.3	−17.5	36.0	1,5
C$_5$H$_{10}$	2-Methyl-1-butene	−117.7	−102.2	−82.7	−57.2	−21.9	30.8	1,5
C$_5$H$_{10}$	3-Methyl-1-butene	−125.0	−110.1	−91.2	−66.7	−32.1	19.7	1,5
C$_5$H$_{10}$	2-Methyl-2-butene	−113.4	−97.6	−77.7	−51.6	−15.8	38.2	1,5
C$_5$H$_{10}$	Cyclopentane			−77.0	−45.4	−7.1	48.8	5
C$_5$H$_{10}$	Ethylcyclopropane	−118 e	−102 e	−83 e	−57 e	−20 e	35.5	5
C$_5$H$_{10}$	cis-1,2-Dimethylcyclo-propane	−118 e	−103 e	−83 e	−57 e	−20 e	36.6	5
C$_5$H$_{10}$	trans-1,2-Dimethylcyclo-propane	−122 e	−108 e	−89 e	−63 e	−27 e	27.8	5
C$_5$H$_{10}$Br$_2$	1,5-Dibromopentane	1 e	25 e	54 e	93 e	145.6	221.8	5
C$_5$H$_{10}$Cl$_2$	1,2-Dichloropentane				30 e	77.4	147.8	5
C$_5$H$_{10}$Cl$_2$	1,5-Dichloropentane	−31 e	−10 e	17 e	54 e	104.1	178.9	5
C$_5$H$_{10}$N$_2$	3-(Dimethylamino)-propanenitrile				51.1	101.8	171.4	5
C$_5$H$_{10}$O	Cyclopentanol		−13 e	11.5	42.2	82.5	140.0	5
C$_5$H$_{10}$O	Allyl ethyl ether			−56 e	−28.7	9.8	67.2	5
C$_5$H$_{10}$O	Pentanal	−71 e	−53 e	−31 e	−1 e	40.8	102.6	5
C$_5$H$_{10}$O	2-Pentanone				−1 e	40.3	101.9	1,5
C$_5$H$_{10}$O	3-Pentanone			−31 e	−1 e	40 e	101.6	1
C$_5$H$_{10}$O	3-Methyl-2-butanone	−69 e	−54 e	−34 e	−6.9	32.2	94.0	1,5
C$_5$H$_{10}$O	Tetrahydropyran				−15 e	26.0	88 e	5
C$_5$H$_{10}$O	2-Methyltetrahydrofuran				−20 e	19.7	79.8	5
C$_5$H$_{10}$O$_2$	Pentanoic acid	−7.4	15.3	42.7	76.3	122.1	185.7	5
C$_5$H$_{10}$O$_2$	2-Methylbutanoic acid	−10 e	10 e	36 e	69 e	112.8	175.2	5
C$_5$H$_{10}$O$_2$	3-Methylbutanoic acid	−15.8	4 e	30.0	64.7	110.6	176.1	5

Mol. form.	Name	Temperature in °C for the indicated pressure						Ref.
		1 Pa	10 Pa	100 Pa	1 kPa	10 kPa	100 kPa	
$C_5H_{10}O_2$	Butyl formate			−29 e	2 e	44.4	105.7	5
$C_5H_{10}O_2$	Isobutyl formate	−69 e	−53 e	−31 e	−3 e	37.4	97.6	5
$C_5H_{10}O_2$	Propyl acetate	−69 e	−51 e	−29 e	0 e	40.9	101.2	1
$C_5H_{10}O_2$	Isopropyl acetate		−61 e	−40 e	−11 e	29.8	88.2	5
$C_5H_{10}O_2$	Ethyl propanoate	−69 e	−52 e	−30 e	−1 e	38.9	98.7	1
$C_5H_{10}O_2$	Methyl butanoate	−68 e	−50 e	−28 e	0.9	41.7	102.3	5
$C_5H_{10}O_2$	Methyl isobutanoate	−83 e	−65 e	−41 e	−11 e	31 e	92.1	5
$C_5H_{10}O_2$	Tetrahydrofurfuryl alcohol	−40 e	−16 e	15 e	55 e	106 e	176.8	5
$C_5H_{10}O_3$	Diethyl carbonate		−42 e	−17 e	17 e	61.6	125.9	5
$C_5H_{10}O_3$	Ethylene glycol monomethyl ether acetate	−47 e	−26 e	0 e	34 e	79.4	144.1	5
$C_5H_{10}S$	Thiacyclohexane				24 e	71.1	141.2	5
$C_5H_{10}S$	Cyclopentanethiol				18 e	64 e	131.7	5
$C_5H_{11}Br$	1-Bromopentane	−60 e	−41 e	−16 e	16 e	61.5	129.1	5
$C_5H_{11}Br$	2-Bromopentane	−69 e	−51 e	−27 e	5 e	49.7	116.9	5
$C_5H_{11}Br$	3-Bromopentane	−68 e	−50 e	−26 e	6 e	50.8	118.1	5
$C_5H_{11}Br$	1-Bromo-3-methylbutane	−67 e	−49 e	−25 e	8 e	52.4	119.9	5
$C_5H_{11}Cl$	1-Chloropentane	−73 e	−55 e	−32 e	−1 e	42.5	107.9	5
$C_5H_{11}Cl$	2-Chloropentane	−80 e	−62 e	−39 e	−9 e	33.2	96.1	5
$C_5H_{11}Cl$	3-Chloropentane	−77 e	−60 e	−37 e	−7 e	34.9	97.3	5
$C_5H_{11}Cl$	2-Chloro-2-methylbutane			−52 e	−21 e	21.8	85.2	5
$C_5H_{11}Cl$	1-Chloro-2,2-dimethyl-propane				−17 e	23.5	83.9	5
$C_5H_{11}F$	1-Fluoropentane	−97 e	−80 e	−60 e	−32 e	5.7	62.4	5
$C_5H_{11}I$	1-Iodopentane	−47 e	−27 e	−1 e	34 e	83.0	156.5	5
$C_5H_{11}I$	1-Iodo-3-methylbutane		−34 e	−6.6	28.8	77.3	147.8	5
$C_5H_{11}N$	Cyclopentylamine	−66 e	−48 e	−26 e	4 e	45.8	108 e	5
$C_5H_{11}N$	Piperidine				2 e	43.3	105.8	5
$C_5H_{11}N$	N-Methylpyrrolidine				−23 e	18.5	78 e	5
$C_5H_{11}NO_3$	3-Methylbutyl nitrate		−26 e	1.0	35.5	81.7	147.0	5
C_5H_{12}	Pentane**	−115.5	−99.8	−80.0	−54.0	−18.1	35.7	16
C_5H_{12}	Isopentane	−119 e	−105 e	−86 e	−61 e	−26 e	27.5	1
C_5H_{12}	Neopentane*		−107.5 s	−90.8 s	−68.8 s	−38.5 s	9.2	1,5
$C_5H_{12}N_2O$	Tetramethylurea			20.7	58.0	106.7	179.5	5
$C_5H_{12}O$	1-Pentanol	−27 e	−10 e	12 e	41 e	79.8	137.4	5
$C_5H_{12}O$	2-Pentanol	−35 e	−19 e	1 e	28.0	64.9	118.7	1
$C_5H_{12}O$	3-Pentanol	−41 e	−25 e	−4 e	24 e	61.1	114.9	5
$C_5H_{12}O$	2-Methyl-1-butanol	−27 e	−11 e	9 e	36.2	73.4	128.3	1
$C_5H_{12}O$	3-Methyl-1-butanol	−22 e	−7 e	13 e	39.1	75.7	130.1	5
$C_5H_{12}O$	2-Methyl-2-butanol			−5 e	17.7	50.6	101.7	1,5
$C_5H_{12}O$	3-Methyl-2-butanol			−3 e	22.7	58.2	111.1	5
$C_5H_{12}O$	2,2-Dimethyl-1-propanol					59.2	112.7	5
$C_5H_{12}O$	Butyl methyl ether			−54 e	−27 e	12 e	69.8	1
$C_5H_{12}O$	Methyl tert-butyl ether			−66 e	−39 e	−2 e	54.8	1
$C_5H_{12}O$	Ethyl propyl ether	−92 e	−77 e	−57 e	−30.5	6.7	63.4	1,5
$C_5H_{12}O_2$	1,5-Pentanediol	25 e	52 e	85 e	125 e	175.1	238.9	5
$C_5H_{12}O_2$	Ethylene glycol monopropyl ether				40 e	85.6	149.3	5
$C_5H_{12}O_2$	Diethoxymethane		−65 e	−43 e	−14 e	27.3	87.7	5
$C_5H_{12}O_3$	Diethylene glycol monomethyl ether		12 e	40 e	76 e	124.2	193.7	1
$C_5H_{12}S$	1-Pentanethiol	−60 e	−41 e	−17 e	15 e	60 e	126.2	1
$C_5H_{12}S$	2-Pentanethiol	−70 e	−52 e	−28 e	3 e	46.6	111.9	5
$C_5H_{12}S$	3-Pentanethiol	−70 e	−51 e	−28 e	4 e	47.7	113.4	5
$C_5H_{12}S$	2-Methyl-1-butanethiol				8.0	52.3	118.5	5
$C_5H_{12}S$	3-Methyl-1-butanethiol				7.8	51.9	117.9	5
$C_5H_{12}S$	2-Methyl-2-butanethiol				−8.0	34.6	98.7	5
$C_5H_{12}S$	Butyl methyl sulfide		−43 e	−19 e	13 e	57 e	123.0	1
$C_5H_{12}S$	tert-Butyl methyl sulfide				−7.8	34.7	98.4	5
$C_5H_{12}S$	Ethyl propyl sulfide	−64 e	−46 e	−23 e	9 e	52.7	118.0	5
$C_5H_{12}S$	Ethyl isopropyl sulfide	−72 e	−54 e	−31 e	0 e	42.7	106.9	5
$C_5H_{13}N$	Pentylamine		−52 e	−29 e	1 e	42.8	104.0	5
C_6BrF_5	Bromopentafluorobenzene			−10 e	23 e	68 e	136.0	5
C_6ClF_5	Chloropentafluorobenzene		−44 e	−21 e	11 e	53.8	117.6	1

Mol. form.	Name	1 Pa	10 Pa	100 Pa	1 kPa	10 kPa	100 kPa	Ref.
					Temperature in °C for the indicated pressure			
C$_6$Cl$_3$F$_3$	1,3,5-Trichloro-2,4,6-trifluorobenzene	−19 e	4 e	32 e	70 e	121.7	197.9	1
C$_6$F$_6$	Hexafluorobenzene		−56.9 s	−36 s	−11.5 s	22.6	79.9	1,5
C$_6$F$_{12}$	Perfluorocyclohexane				−46.2 s	−7.6 s	48.9 s	5
C$_6$F$_{14}$	Perfluorohexane		−75 e	−57 e	−32 e	2.8	56.8	5
C$_6$F$_{14}$	Perfluoro-2-methylpentane				−33 e	2.9	57.1	5
C$_6$F$_{14}$	Perfluoro-3-methylpentane	−95 e	−80 e	−60 e	−34 e	2.8	57.9	5
C$_6$F$_{14}$	Perfluoro-2,3-dimethylbutane					4.3	59.3	5
C$_6$HF$_5$	Pentafluorobenzene			−41 e	−13 e	27 e	85.3	5
C$_6$HF$_5$O	Pentafluorophenol				39 e	82 e	145.2	5
C$_6$H$_2$F$_4$	1,2,3,4-Tetrafluorobenzene			−36 e	−7 e	33.8	94.0	1
C$_6$H$_2$F$_4$	1,2,3,5-Tetrafluorobenzene			−43 e	−14 e	25.5	84.1	1
C$_6$H$_2$F$_4$	1,2,4,5-Tetrafluorobenzene					30.7	89.9	1
C$_6$H$_3$Cl$_3$O	2,4,6-Trichlorophenol			71.8	114.0	169.5	245.7	5
C$_6$H$_3$F$_3$	1,3,5-Trifluorobenzene					18.2	75.0	5
C$_6$H$_4$Br$_2$	m-Dibromobenzene	−7 e	16 e	44 e	83 e	137.0	218.2	5
C$_6$H$_4$ClNO$_2$	1-Chloro-4-nitrobenzene	15.4 s	35.8 s		97 e	156.0	238 e	5
C$_6$H$_4$Cl$_2$	o-Dichlorobenzene		−13 e	16.3	53.9	104.6	180.0	1,5
C$_6$H$_4$Cl$_2$	m-Dichlorobenzene		−22 e	8.0	46.7	97.8	172.5	1,5
C$_6$H$_4$Cl$_2$	p-Dichlorobenzene	−45.5 s	−21.8 s	8 s	46.7 s	99.0	173.6	1,5
C$_6$H$_4$O$_2$	p-Benzoquinone	−4.1 s	17.8 s	43.5 s	74.3 s	111.6 s		5
C$_6$H$_5$AsCl$_2$	Dichlorophenylarsine	6.9	35.2	70 e	113 e	170 e	245 e	5
C$_6$H$_5$Br	Bromobenzene		−25 e	1 e	34.9	83.1	155.4	1
C$_6$H$_5$Cl	Chlorobenzene		−43 e	−17 e	16.8	62.9	131.3	1,5
C$_6$H$_5$ClO	o-Chlorophenol				45.8	97.9	173.9	5
C$_6$H$_5$ClO	m-Chlorophenol			39.7	80.2	135.1	213.4	5
C$_6$H$_5$ClO	p-Chlorophenol			45.0	86.5	142.0	219.9	5
C$_6$H$_5$Cl$_3$Si	Trichlorophenylsilane			33 e	70.2	122.6	201 e	5
C$_6$H$_5$F	Fluorobenzene				−16.9	24.2	84.4	1
C$_6$H$_5$I	Iodobenzene	−30 e	−7 e	20.9	58.5	110.6	187.8	1
C$_6$H$_5$NO$_2$	Nitrobenzene		10 e	40 e	78 e	132 e	210.3	1
C$_6$H$_5$NO$_3$	p-Nitrophenol	72.6 s	97.4 s					5
C$_6$H$_6$	1,5-Hexadien-3-yne	−82 e	−66 e	−44.3	−16.0	23.7	83.6	5
C$_6$H$_6$	Benzene**			−40 s	−15.1 s	20.0	79.7	1,5
C$_6$H$_6$ClN	o-Chloroaniline		10 e	39.0	75.2	131.4	208.3	5
C$_6$H$_6$ClN	m-Chloroaniline	−5 e	19.7	49.4	94.2	162 e	1069 e	5
C$_6$H$_6$N$_2$O$_2$	p-Nitroaniline	87.8 s			192.0	252.6	331.2	5
C$_6$H$_6$O	Phenol	−9.7 s	9.6 s	34.1 s	68.9	113.7	181.4	1,5
C$_6$H$_6$O$_3$	1,2,3-Benzenetriol				162.0	222.8	308.3	5
C$_6$H$_6$S	Benzenethiol		−15 e	12 e	47 e	96.0	168.6	5
C$_6$H$_7$N	Aniline		−2.5	26.7	63.5	112.5	183.5	1,5
C$_6$H$_7$N	2-Methylpyridine	−56.5	−37.8	−13.9	18.3	62.9	129.0	1,5
C$_6$H$_7$N	3-Methylpyridine			−5 e	28.8	75.2	143.7	1
C$_6$H$_7$N	4-Methylpyridine	−58.2 s	−43.1 s	−3.9 s	29.6	76.1	144.9	1,5
C$_6$H$_8$	cis-1,3,5-Hexatriene					21 e	78 e	5
C$_6$H$_8$	1,3-Cyclohexadiene	−88 e	−71 e	−50 e	−21 e	19 e	79.9	5
C$_6$H$_8$	1,4-Cyclohexadiene				−15 e	27.3	85.0	5
C$_6$H$_8$N$_2$	Adiponitrile	30 e	61 e	100 e	148.6	211.8	297 e	5
C$_6$H$_8$N$_2$	m-Phenylenediamine			94.5	140.2	200.8	285.0	5
C$_6$H$_8$N$_2$	Phenylhydrazine		38 e	69 e	109 e	163.9	242.5	5
C$_6$H$_8$O$_4$	Dimethyl maleate		5 e	36 e	76 e	127.3	197 e	5
C$_6$H$_8$S	2,5-Dimethylthiophene		−43 e	−16 e	20 e	67.5	134.8	5
C$_6$H$_{10}$	trans-1,3-Hexadiene	−86 e	−70 e	−51 e	−24 e	14 e	72 e	5
C$_6$H$_{10}$	trans-1,4-Hexadiene	−98 e	−81 e	−60 e	−33 e	7 e	65 e	5
C$_6$H$_{10}$	1,5-Hexadiene	−99 e	−84 e	−64 e	−37 e	0.9	59.2	5
C$_6$H$_{10}$	cis,cis-2,4-Hexadiene					18 e	79.6	5
C$_6$H$_{10}$	trans,cis-2,4-Hexadiene	−89 e	−73 e	−52 e	−23 e	18 e	79.6	5
C$_6$H$_{10}$	trans,trans-2,4-Hexadiene				−23 e	18 e	79.6	5
C$_6$H$_{10}$	trans-2-Methyl-1,3-pentadiene	−92 e	−75 e	−54 e	−26 e	14 e	75.6	5
C$_6$H$_{10}$	2,3-Dimethyl-1,3-butadiene			−59 e	−30 e	9.7	68.1	5
C$_6$H$_{10}$	1-Hexyne	−91 e	−75 e	−54 e	−26 e	12.8	71.0	5

Mol. form.	Name	Temperature in °C for the indicated pressure						Ref.
		1 Pa	10 Pa	100 Pa	1 kPa	10 kPa	100 kPa	
C_6H_{10}	2-Hexyne	−84 e	−67 e	−46 e	−17 e	23.6	84.1	5
C_6H_{10}	3-Hexyne	−86 e	−69 e	−48 e	−19.1	21.0	81.0	1,5
C_6H_{10}	4-Methyl-1-pentyne	−97 e	−81 e	−61 e	−34 e	4.1	60.7	5
C_6H_{10}	4-Methyl-2-pentyne	−91 e	−74 e	−54 e	−26 e	13.8	72.7	5
C_6H_{10}	Cyclohexene	−87 e	−70 e	−49 e	−19 e	21 e	82.6	1
$C_6H_{10}Cl_2$	1,1-Dichlorocyclohexane	−39 e	−19 e	8 e	43 e	93.5	170.5	5
$C_6H_{10}Cl_2$	cis-1,2-Dichlorocyclohexane			27 e	69 e	125.7	206.2	5
$C_6H_{10}O$	4-Methyl-4-penten-2-one	−59 e	−41 e	−17 e	14 e	57.0	121.0	5
$C_6H_{10}O$	Cyclohexanone		−25 e	1 e	36 e	84 e	155.2	1
$C_6H_{10}O$	Mesityl oxide	−56 e	−37 e	−13 e	19 e	63.5	129.3	5
$C_6H_{10}O_2$	Vinyl butanoate				53 e		114.5	5
$C_6H_{10}O_2$	Ethyl methacrylate				8 e	53.2	116.8	5
$C_6H_{10}O_2$	Allyl glycidyl ether				40.1	85.7	152.8	5
$C_6H_{10}O_3$	Ethyl acetoacetate	−25 e	−3 e	25.7	62.3	111.3	180.2	5
$C_6H_{10}O_3$	Propanoic anhydride	−32 e	−15 e	6 e	36 e	77.6	142.9	5
$C_6H_{10}O_4$	Diethyl oxalate	−5 e	18 e	44.9	79.4	124.3	185.2	5
$C_6H_{10}O_4$	Dimethyl succinate			30 e	70.4	123.3	195.4	5
$C_6H_{10}O_4$	Ethylene glycol diacetate	−17 e	6 e	35.0	71.9	121.1	190.0	5
$C_6H_{10}S$	Diallylsulfide	−58 e	−38 e	−12.4	21.7	68.8	138.1	5
$C_6H_{11}Cl$	Chlorocyclohexane		−35 e	−9 e	25 e	71.6	142.1	5
$C_6H_{11}N$	Hexanenitrile	−40 e	−19 e	8 e	43 e	91.5	163.2	1,5
$C_6H_{11}N$	4-Methylpentanenitrile		−50 e	−20 e	20 e	75.2	155.2	5
$C_6H_{11}NO$	Caprolactam	36.8 s	58.9 s	86.6 s			270	5
C_6H_{12}	1-Hexene	−99.8	−82.8	−61.4	−33.7	5.2	63.1	1,5
C_6H_{12}	cis-2-Hexene	−97 e	−80 e	−58 e	−30 e	9.9	68.5	5
C_6H_{12}	trans-2-Hexene	−94 e	−78 e	−57 e	−30 e	9.3	67.5	5
C_6H_{12}	cis-3-Hexene	−96 e	−79 e	−59 e	−30.8	7.9	66.0	5
C_6H_{12}	trans-3-Hexene	−95 e	−79 e	−58 e	−30.0	8.8	66.7	5
C_6H_{12}	2-Methyl-1-pentene	−98 e	−82 e	−62 e	−34.2	4.1	61.7	5
C_6H_{12}	3-Methyl-1-pentene	−104 e	−88 e	−68 e	−41.5	−3.6	53.8	5
C_6H_{12}	4-Methyl-1-pentene	−105 e	−89 e	−69 e	−41.6	−3.6	53.5	5
C_6H_{12}	2-Methyl-2-pentene	−95 e	−78 e	−58 e	−30 e	9.0	66.9	5
C_6H_{12}	3-Methyl-cis-2-pentene	−95 e	−79 e	−58 e	−30 e	8.9	67.3	5
C_6H_{12}	3-Methyl-trans-2-pentene	−93 e	−77 e	−55 e	−27.4	11.7	70.0	5
C_6H_{12}	4-Methyl-cis-2-pentene	−102 e	−86 e	−66 e	−38.7	−0.9	56.0	5
C_6H_{12}	4-Methyl-trans-2-pentene	−100 e	−84 e	−64 e	−36.8	1.2	58.2	5
C_6H_{12}	2-Ethyl-1-butene	−98 e	−81 e	−60 e	−32 e	6.6	64.3	5
C_6H_{12}	2,3-Dimethyl-1-butene	−103 e	−87 e	−67 e	−39.9	−1.9	55.2	5
C_6H_{12}	3,3-Dimethyl-1-butene	−110 e	−95 e	−76 e	−50.1	−14.5	40.8	5
C_6H_{12}	2,3-Dimethyl-2-butene		−75 e	−54 e	−25 e	14 e	72.9	1
C_6H_{12}	Cyclohexane	−85.6 s	−68.9 s	−47.6 s	−19.8 s	19.3	80.4	1,5
C_6H_{12}	Methylcyclopentane	−97 e	−80 e	−58 e	−28.8	11.6	71.4	1,5
C_6H_{12}	Ethylcyclobutane	−99 e	−82 e	−61 e	−32 e	9 e	70.2	5
C_6H_{12}	Isopropylcyclopropane	−104 e	−88 e	−68 e	−40 e	−1 e	57.9	5
C_6H_{12}	1-Ethyl-1-methylcyclopropane	−105 e	−89 e	−69 e	−41 e	−3 e	56.3	5
C_6H_{12}	1,1,2-Trimethylcyclopropane	−109 e	−94 e	−73 e	−46 e	−7 e	52.0	5
$C_6H_{12}Cl_2$	1,2-Dichlorohexane				49 e	98.1	171.7	5
$C_6H_{12}Cl_2O$	2,2′-Dichlorodiisopropyl ether		−1 e	27.3	63.4	112.3	182.1	5
$C_6H_{12}O$	Butyl vinyl ether	−87 e	−67 e	−42 e	−9.3	33.6	93.2	5
$C_6H_{12}O$	Isobutyl vinyl ether	−87 e	−68 e	−44 e	−13 e	26.5	80.7	5
$C_6H_{12}O$	Hexanal	−56 e	−37 e	−13 e	19 e	62.6	127.8	5
$C_6H_{12}O$	2-Hexanone	−43 e	−21 e	4.2	34.5	61.9	127.2	1,5
$C_6H_{12}O$	3-Hexanone		−40 e	−16 e	15 e	58.5	123.1	1
$C_6H_{12}O$	3-Methyl-2-pentanone				8.5	52.7	117.0	5
$C_6H_{12}O$	4-Methyl-2-pentanone	−61 e	−43 e	−21 e	9 e	51.5	116.1	5
$C_6H_{12}O$	2-Methyl-3-pentanone					50.2	113.0	5
$C_6H_{12}O$	3,3-Dimethyl-2-butanone			−30 e	0 e	42.5	105.7	1
$C_6H_{12}O$	Cyclohexanol			34 e	61 e	99.2	160.7	1
$C_6H_{12}O_2$	Hexanoic acid		33 e	59 e	93 e	139.3	204.5	1
$C_6H_{12}O_2$	4-Methylpentanoic acid	36 e	49 e	67.1	92.9	133.6	206.8	5

		Temperature in °C for the indicated pressure						
Mol. form.	Name	1 Pa	10 Pa	100 Pa	1 kPa	10 kPa	100 kPa	Ref.
$C_6H_{12}O_2$	Diethylacetic acid	−9 e	16 e	46 e	83 e	130.7	192.5	5
$C_6H_{12}O_2$	Isopentyl formate	−60 e	−41 e	−17 e	15 e	59.1	124 e	5
$C_6H_{12}O_2$	Butyl acetate	−63 e	−43 e	−19 e	14 e	61.0	125.6	1,5
$C_6H_{12}O_2$	Isobutyl acetate	−63 e	−45 e	−21 e	10 e	53.4	116 e	5
$C_6H_{12}O_2$	Propyl propanoate	−62 e	−42 e	−18 e	14 e	58.3	122.0	5
$C_6H_{12}O_2$	Ethyl butanoate	−49 e	−34 e	−14 e	14.3	55.2	121.1	5
$C_6H_{12}O_2$	Ethyl 2-methylpropanoate	−65 e	−47 e	−24.6	5.4	47.3	109.8	5
$C_6H_{12}O_2$	Methyl pentanoate				19.2	63.7	127.4	5
$C_6H_{12}O_2$	Methyl isopentanoate					53.3	116.3	5
$C_6H_{12}O_2$	Diacetone alcohol	−41 e	−17 e	13 e	50.1	98.5	164 e	5
$C_6H_{12}O_3$	Ethylene glycol monoethyl ether acetate	−25 e	−8 e	14 e	44.6	88.0	155.6	5
$C_6H_{12}O_3$	Paraldehyde				17 e	62.2	124 e	5
$C_6H_{12}S$	Cyclohexanethiol					84.8	158.3	5
$C_6H_{12}S$	cis-Tetrahydro-2,5-dimethylthiophene	−53 e	−34 e	−8 e	25 e	72.0	142.1	5
$C_6H_{12}S$	Tetrahydro-3-methyl-2H-thiopyran	−48 e	−27 e	0 e	35 e	84.1	157.5	5
$C_6H_{13}Br$	1-Bromohexane	−45 e	−25 e	2 e	36 e	83.7	154.8	5
$C_6H_{13}Cl$	1-Chlorohexane	−55 e	−36 e	−11 e	21 e	66.7	134.6	5
$C_6H_{13}F$	1-Fluorohexane	−80 e	−62 e	−40 e	−11 e	30.4	91.1	5
$C_6H_{13}I$	1-Iodohexane	−33 e	−11 e	16 e	53 e	104.0	180.8	5
$C_6H_{13}N$	Cyclohexylamine			−9 e	22 e	66.6	133.5	1
C_6H_{14}	Hexane	−96.4 s	−79.2	−57.6	−29.3	9.8	68.3	16
C_6H_{14}	2-Methylpentane	−100 e	−84 e	−64 e	−36 e	2 e	59.9	1
C_6H_{14}	3-Methylpentane	−99 e	−83 e	−62 e	−34.3	4.6	62.9	1
C_6H_{14}	2,2-Dimethylbutane		−90 e	−71.5	−45.5	−7.7	49.4	1
C_6H_{14}	2,3-Dimethylbutane	−103 e	−87 e	−66 e	−39.0	−0.4	57.6	1
$C_6H_{14}O$	1-Hexanol		5 e	28 e	56.8	97.3	157.1	1
$C_6H_{14}O$	2-Hexanol	−28 e	−10 e	12 e	41.4	81.5	139.6	1
$C_6H_{14}O$	3-Hexanol	−43 e	−23 e	1 e	33 e	75.4	135.1	1
$C_6H_{14}O$	2-Methyl-1-pentanol			14 e	45.9	88.3	147.6	5
$C_6H_{14}O$	4-Methyl-1-pentanol			24 e	53 e	92.4	151.4	5
$C_6H_{14}O$	2-Methyl-2-pentanol	−29 e	−15 e	3 e	27.1	63.0	120.9	5
$C_6H_{14}O$	3-Methyl-2-pentanol				36.5	76.1	133.8	5
$C_6H_{14}O$	4-Methyl-2-pentanol	−43 e	−24 e	0 e	30 e	71.9	131.3	5
$C_6H_{14}O$	2-Methyl-3-pentanol				29.8	68.8	126.0	5
$C_6H_{14}O$	3-Methyl-3-pentanol		−23 e	−4 e	22.9	61.1	121.1	5
$C_6H_{14}O$	2-Ethyl-1-butanol		−5 e	17 e	46 e	85.7	146.1	5
$C_6H_{14}O$	3,3-Dimethyl-1-butanol	−37 e	−16 e	9 e	42 e	84.3	142.5	5
$C_6H_{14}O$	2,3-Dimethyl-2-butanol		−5 e	23 e	61.3	118.2		5
$C_6H_{14}O$	Dipropyl ether	−80 e	−63 e	−41 e	−12 e	28.8	89.7	1
$C_6H_{14}O$	Diisopropyl ether		−76 e	−55 e	−28 e	11 e	68.1	1
$C_6H_{14}O$	Butyl ethyl ether	−78 e	−61 e	−39 e	−10 e	31.0	91.9	1
$C_6H_{14}O$	tert-Butyl ethyl ether	−90 e	−74 e	−53 e	−24.6	14.4	72.6	5
$C_6H_{14}O_2$	2-Methyl-2,4-pentanediol	−8 e	17 e	48 e	86 e	134.4	197.5	5
$C_6H_{14}O_2$	Ethylene glycol monobutyl ether	−31 e	−8 e	20 e	55 e	103.2	170.2	5
$C_6H_{14}O_2$	1,1-Diethoxyethane	−68 e	−49 e	−26 e	3.7	44.2	101.9	5
$C_6H_{14}O_2$	Ethylene glycol diethyl ether		−59 e	−35.3	−2.8	44.4	118.8	5
$C_6H_{14}O_3$	1,2,6-Hexanetriol	92 e	114.8	146.0	191 e			5
$C_6H_{14}O_3$	Dipropylene glycol				110 e	162.6	231.4	5
$C_6H_{14}O_3$	Diethylene glycol monoethyl ether			40 e	80.3	132.4	201.4	5
$C_6H_{14}O_3$	Diethylene glycol dimethyl ether	−42 e	−20 e	8.3	44.3	92.3	159.4	5
$C_6H_{14}O_3$	Trimethylolpropane	73 e	98 e	128 e	167.8	220.5	295 e	5
$C_6H_{14}O_4$	Triethylene glycol	44 e	74 e	109.0	152.6	207.2	277.9	5
$C_6H_{14}S$	1-Hexanethiol	−45 e	−25 e	1 e	35 e	81.7	152.2	5
$C_6H_{14}S$	2-Hexanethiol	−50 e	−32 e	−8 e	25 e	69.9	138.4	5
$C_6H_{14}S$	Dipropyl sulfide	−50 e	−30 e	−6 e	28 e	73.6	142.4	5
$C_6H_{14}S$	Diisopropyl sulfide	−65 e	−47 e	−23 e	9 e	53.1	119.6	5
$C_6H_{14}S$	Isopropyl propyl sulfide				18.5	63.8	131.6	5
$C_6H_{14}S$	Butyl ethyl sulfide	−49 e	−30 e	−5 e	29 e	74.8	143.8	5
$C_6H_{15}N$	Hexylamine			−10 e	22 e	66.0	130.6	5
$C_6H_{15}N$	Butylethylamine				6.1	47.7	107.0	5

		Temperature in °C for the indicated pressure						
Mol. form.	Name	1 Pa	10 Pa	100 Pa	1 kPa	10 kPa	100 kPa	Ref.
$C_6H_{15}N$	Dipropylamine		−48 e	−25 e	6 e	47.5	108.8	5
$C_6H_{15}N$	Diisopropylamine			−47 e	−17.5	23.5	84.0	5
$C_6H_{15}N$	Triethylamine	−58 e	−45 e	−29 e	−5 e	29.9	88.5	1
$C_6H_{15}NO$	2-Diethylaminoethanol					97 e	160.6	5
$C_6H_{15}NO_3$	Triethanolamine	75 e	108 e	148 e	196 e	256.7	334 e	5
$C_6H_{15}O_4P$	Triethyl phosphate			34	76	132	211	4
$C_6H_{16}N_2$	Hexamethylenediamine				76.0	128.2	199.0	5
$C_6H_{16}O_2Si$	Diethoxydimethylsilane	−62 e	−44 e	−21.2	9.1	51.0	113.0	5
$C_6H_{18}Cl_2O_2Si_3$	1,5-Dichloro-1,1,3,3,5,5-hexamethyltrisiloxane	−29 e	−7 e	22.2	59.7	110.5	183.4	5
$C_6H_{18}OSi_2$	Hexamethyldisiloxane		−56 e	−34 e	−5 e	37.1	100.1	5
C_6MoO_6	Molybdenum hexacarbonyl		17.4 s	42.8 s	73.1 s	109.9 s	155.4 s	5
C_7F_{14}	Perfluoromethylcyclohexane				−21 e	18 e	75.9	1
C_7F_{16}	Perfluoroheptane		−62 e	−41 e	−14 e	24.7	82.1	1
C_7HF_{15}	1H-Pentadecafluoroheptane				−7 e	35.9	96.0	5
$C_7H_3ClF_3NO_2$	1-Chloro-2-nitro-4-trifluoromethyl)benzene	3 e	26 e	55 e	92.8	145.2	222.0	5
$C_7H_3F_5$	2,3,4,5,6-Pentafluorotoluene			−20 e	11 e	53.6	117.0	5
$C_7H_4ClF_3$	1-Chloro-2-(trifluoromethyl) benzene			1 e	34.5	81.8	151.8	5
$C_7H_4ClF_3$	1-Chloro-3-(trifluoromethyl) benzene	−53 e	−34 e	−9 e	24.2	69.8	137.2	5
$C_7H_4ClF_3$	1-Chloro-4-(trifluoromethyl) benzene			−9 e	24.2	70.4	138.1	5
$C_7H_4Cl_2O$	o-Chlorobenzoyl chloride			93 e	149 e	237.0		5
$C_7H_4Cl_2O$	m-Chlorobenzoyl chloride			87.8	147 e	225.0		5
$C_7H_4F_3NO_2$	1-Nitro-3-(trifluoromethyl) benzene		11 e	39 e	76.2	127.3	202.2	5
$C_7H_4F_4$	1-Fluoro-4-(trifluoromethyl) benzene			−38 e	−6 e	38.6	102.3	5
C_7H_5BrO	Benzoyl bromide	−15 e	11 e	42.6	83.9	139.5	218.0	5
C_7H_5ClO	Benzoyl chloride			27.5	67.0	120.4	196.7	5
$C_7H_5Cl_3$	(Trichloromethyl)benzene		9 e	40.6	81.5	136.2	213.0	5
$C_7H_5F_3$	(Trifluoromethyl)benzene			−3 e	39 e	101.6		5
C_7H_5N	Benzonitrile		−6 e	23.9	63.1	115.7	190.0	5
C_7H_5NS	Phenyl isothiocyanate			79.4	105 e	117 e		5
$C_7H_6Cl_2$	2,4-Dichlorotoluene		6 e	33 e	68.3	119.5	199.1	5
$C_7H_6Cl_2$	3,4-Dichlorotoluene	−13 e	9 e	38 e	76 e	129.3	208.4	5
$C_7H_6Cl_2$	(Dichloromethyl)benzene			31	72	130	213	4
C_7H_6O	Benzaldehyde		−9 e	19 e	54.6	104.6	178.3	1
$C_7H_6O_2$	Salicylaldehyde		−1 e	29 e	68 e	120.7	196.2	5
C_7H_7Br	o-Bromotoluene		−10 e	17 e	54 e	104.8	181.1	5
C_7H_7Br	m-Bromotoluene	−34 e	−11 e	19.4	58.1	109.9	183.1	5
C_7H_7Br	p-Bromotoluene				57 e	107.8	183.8	5
C_7H_7Br	(Bromomethyl)benzene			25.4	66.8	121.7	198.3	5
C_7H_7Cl	o-Chlorotoluene		−24 e	3 e	38 e	86.3	158.7	1,5
C_7H_7Cl	m-Chlorotoluene	−41 e	−21 e	6 e	41 e	89 e	161.8	5
C_7H_7Cl	p-Chlorotoluene				40 e	88.9	161.5	1,5
C_7H_7Cl	(Chloromethyl)benzene	−34 e	−11 e	17.7	55.4	106.3	178.9	5
C_7H_7ClO	1-Chloro-2-methoxy-benzene	−22 e	2 e	33 e	72 e	125.2	201.0	5
C_7H_7F	o-Fluorotoluene		−50 e	−26 e	5 e	49.0	113.9	5
C_7H_7F	m-Fluorotoluene	−67 e	−48 e	−25 e	7 e	51.0	116.1	5
C_7H_7F	p-Fluorotoluene		−48 e	−24 e	7 e	51 e	116.2	5
$C_7H_7NO_2$	o-Nitrotoluene	23 e	40 e	62 e	94 e	141.9	221.9	5
$C_7H_7NO_2$	m-Nitrotoluene			45 e	89.7	148.7	231.3	5
$C_7H_7NO_3$	2-Nitroanisole	15 e	45 e	82 e	129 e	189.4	271.8	5
C_7H_8	Toluene	−78.1	−57.1	−31.3	1.5	45.2	110.1	5
C_7H_8	Bicyclo[2.2.1]hepta-2,5-diene				−15 e	27.4	91 e	5
$C_7H_8Cl_2Si$	Dichloromethylphenylsilane			32.4	71.8	126.0	205.0	5
C_7H_8O	o-Cresol	−6.4 s	12.8 s	40.2	72.3	120.3	190.5	1,5
C_7H_8O	m-Cresol	20.8	33.6	52.4	82.6	130.6	201.8	1,5
C_7H_8O	p-Cresol	−0.2 s	20.7 s	52.7	83.1	130.7	201.5	1,5
C_7H_8O	Benzyl alcohol	8 e	28 e	54 e	88 e	134.7	204.9	1
C_7H_8O	Anisole		−21 e	4 e	38 e	84 e	153.2	1,5
C_7H_8S	3-Methylbenzenethiol		0 e	29 e	66 e	117.9	194.6	5
C_7H_9N	Benzylamine			25.6	62.6	112.7	183.9	5

Mol. form.	Name	Temperature in °C for the indicated pressure						Ref.
		1 Pa	10 Pa	100 Pa	1 kPa	10 kPa	100 kPa	
C$_7$H$_9$N	o-Methylaniline	1.0	18.8	42.6	76.1	125.6	199.9	1,5
C$_7$H$_9$N	m-Methylaniline	3.8	22.0	46.2	80.1	128.8	202.9	1,5
C$_7$H$_9$N	p-Methylaniline				77.1	126.2	199.9	5
C$_7$H$_9$N	N-Methylaniline	−16 e	6 e	34 e	70.3	121.1	195.8	1
C$_7$H$_9$N	2-Ethylpyridine	−46 e	−26 e	−1 e	33 e	79.3	149.0	5
C$_7$H$_9$N	3-Ethylpyridine	−38 e	−17 e	9 e	44 e	92.7	166.5	5
C$_7$H$_9$N	4-Ethylpyridine	−35 e	−15 e	11 e	46 e	94.4	168.6	5
C$_7$H$_9$N	2,3-Dimethylpyridine				42 e	89.9	160.6	5
C$_7$H$_9$N	2,4-Dimethylpyridine		−25 e	3.7	40.0	87.5	157.9	1,5
C$_7$H$_9$N	2,5-Dimethylpyridine			4 e	39 e	86.2	156.6	1
C$_7$H$_9$N	2,6-Dimethylpyridine			−3 e	29.9	75.8	143.6	1
C$_7$H$_9$N	3,4-Dimethylpyridine		−9 e	19 e	55 e	104.8	178.6	5
C$_7$H$_9$N	3,5-Dimethylpyridine			11 e	48 e	98 e	171.5	1
C$_7$H$_{10}$N$_2$	Toluene-2,4-diamine			100.4	145.3	202.9	279.5	5
C$_7$H$_{12}$	1-Heptyne	−75 e	−57 e	−35 e	−5 e	37.1	99.5	5
C$_7$H$_{12}$	2-Heptyne		−51 e	−27 e	4 e	46.9	111.5	5
C$_7$H$_{12}$	3-Heptyne	−71 e	−53 e	−31 e	0 e	42.7	106.4	5
C$_7$H$_{12}$	5-Methyl-1-hexyne	−80 e	−62 e	−40 e	−11 e	30.1	91.4	5
C$_7$H$_{12}$	5-Methyl-2-hexyne	−75 e	−57 e	−34 e	−4 e	38.6	102.0	5
C$_7$H$_{12}$	2-Methyl-3-hexyne	−78 e	−61 e	−39 e	−9 e	32.6	94.8	5
C$_7$H$_{12}$	4,4-Dimethyl-1-pentyne		−73 e	−52 e	−24 e	15.9	75.6	5
C$_7$H$_{12}$	4,4-Dimethyl-2-pentyne		−70 e	−48 e	−19 e	21.4	82.6	5
C$_7$H$_{12}$	Bicyclo[4.1.0]heptane					49.9	116.3	5
C$_7$H$_{12}$	Cycloheptene			−30.0	3.4	47.5	108 e	5
C$_7$H$_{12}$	1-Methylbicyclo(3,1,0)hexane					29.8	92.6	5
C$_7$H$_{12}$	Methylenecyclohexane	−76 e	−58 e	−35 e	−5 e	38 e	103.0	5
C$_7$H$_{12}$	1-Methylcyclohexene	−72 e	−53 e	−30 e	1 e	45 e	109.8	5
C$_7$H$_{12}$	4-Methylcyclohexene	−76 e	−59 e	−36 e	−5 e	37.9	102.3	5
C$_7$H$_{12}$	1-Ethylcyclopentene	−75 e	−57 e	−34 e	−3 e	40.7	105.8	5
C$_7$H$_{12}$	1,2-Dimethylcyclopentene	−75 e	−57 e	−34 e	−3 e	40.2	105.3	5
C$_7$H$_{12}$	1,5-Dimethylcyclopentene	−77 e	−59 e	−36 e	−5.5	37.3	101.5	5
C$_7$H$_{12}$O	Cycloheptanone			18 e	53.7	104.0	178.7	5
C$_7$H$_{12}$O$_2$	Butyl acrylate	−52 e	−31 e	−4.5	30.4	78.0	146.9	5
C$_7$H$_{12}$O$_2$	Propyl methacrylate				26 e	73.8	139.7	5
C$_7$H$_{12}$O$_3$	Ethyl levulinate		17 e	45.3	82.6	133.2	205.7	5
C$_7$H$_{12}$O$_4$	Diethyl malonate	−23 e	4 e	36.0	76.4	128.5	198.3	5
C$_7$H$_{12}$O$_4$	Dimethyl glutarate	−11 e	15 e	47 e	87.7	139.8	209.5	5
C$_7$H$_{13}$ClO	Heptanoyl chloride	−17 e	4 e	29.4	59.7	96.9	144.0	5
C$_7$H$_{14}$	1-Heptene	−82.1	−63.8	−40.6	−10.7	31.1	93.2	1,5
C$_7$H$_{14}$	cis-2-Heptene	−79 e	−61 e	−38 e	−8 e	34.3	98.0	5
C$_7$H$_{14}$	trans-2-Heptene	−79 e	−61 e	−39 e	−8 e	34.0	97.5	5
C$_7$H$_{14}$	cis-3-Heptene	−80 e	−62 e	−40 e	−10 e	32.3	95.3	5
C$_7$H$_{14}$	trans-3-Heptene	−80 e	−62 e	−40 e	−10 e	32.2	95.2	5
C$_7$H$_{14}$	2-Methyl-1-hexene	−81 e	−64 e	−42 e	−12 e	29.3	91.6	5
C$_7$H$_{14}$	4-Methyl-1-hexene	−84 e	−67 e	−45 e	−16 e	25.3	86.3	5
C$_7$H$_{14}$	2-Methyl-2-hexene	−80 e	−63 e	−40 e	−10 e	32.0	95.0	5
C$_7$H$_{14}$	cis-3-Methyl-2-hexene	−79 e	−62 e	−39 e	−9 e	33.4	96.8	5
C$_7$H$_{14}$	trans-4-Methyl-2-hexene	−83 e	−66 e	−44 e	−15 e	25.9	87.1	5
C$_7$H$_{14}$	trans-5-Methyl-2-hexene	−83 e	−66 e	−44 e	−15 e	26.3	87.7	5
C$_7$H$_{14}$	trans-2-Methyl-3-hexene	−84 e	−67 e	−45 e	−16 e	24.6	85.5	5
C$_7$H$_{14}$	3-Ethyl-1-pentene	−85 e	−68 e	−46 e	−17 e	23.2	83.7	5
C$_7$H$_{14}$	2,3-Dimethyl-1-pentene	−85 e	−68 e	−46 e	−17 e	23.4	83.8	5
C$_7$H$_{14}$	2,4-Dimethyl-1-pentene	−88 e	−71 e	−50 e	−21 e	20.0	81.2	5
C$_7$H$_{14}$	3,3-Dimethyl-1-pentene	−87 e	−71 e	−50 e	−21 e	18.1	77.1	5
C$_7$H$_{14}$	4,4-Dimethyl-1-pentene	−94 e	−78 e	−57 e	−28 e	11.5	72.1	5
C$_7$H$_{14}$	2,3-Dimethyl-2-pentene	−79 e	−62 e	−39 e	−9 e	33.5	96.9	5
C$_7$H$_{14}$	2,4-Dimethyl-2-pentene	−84 e	−68 e	−46 e	−18 e	22.6	82.9	5
C$_7$H$_{14}$	cis-3,4-Dimethyl-2-pentene	−83 e	−65 e	−43 e	−14 e	27.2	88.8	5
C$_7$H$_{14}$	trans-3,4-Dimethyl-2-pentene	−82 e	−64 e	−42 e	−13 e	29.0	91.1	5
C$_7$H$_{14}$	cis-4,4-Dimethyl-2-pentene	−90 e	−73 e	−51 e	−22 e	18.6	80.0	5

Mol. form.	Name	Temperature in °C for the indicated pressure						Ref.
		1 Pa	10 Pa	100 Pa	1 kPa	10 kPa	100 kPa	
C_7H_{14}	trans-4,4-Dimethyl-2-pentene	−90 e	−73 e	−52 e	−23 e	16.6	76.3	5
C_7H_{14}	2,3,3-Trimethyl-1-butene	−91 e	−75 e	−53 e	−24.2	16.3	77.5	5
C_7H_{14}	Cycloheptane				6 e	51.1	118.4	1
C_7H_{14}	Methylcyclohexane	−79 e	−62 e	−39 e	−7.9	35.5	100.5	1
C_7H_{14}	Ethylcyclopentane	−76 e	−59 e	−35 e	−5 e	38.4	103.0	5
C_7H_{14}	1,1-Dimethylcyclopentane		−69 e	−47 e	−17 e	24.8	87.4	5
C_7H_{14}	cis-1,2-Dimethylcyclopentane			−38 e	−8 e	34.9	99.0	5
C_7H_{14}	trans-1,2-Dimethylcyclopentane	−83 e	−66 e	−43 e	−13 e	28.4	91.4	5
C_7H_{14}	cis-1,3-Dimethylcyclopentane	−84 e	−66 e	−44 e	−14 e	28.2	91.1	5
C_7H_{14}	trans-1,3-Dimethylcyclopentane	−84 e	−67 e	−44 e	−14 e	27.4	90.3	5
$C_7H_{14}O$	1-Heptanal	−41 e	−21 e	4 e	37 e	83.7	152.3	5
$C_7H_{14}O$	2-Heptanone		−22 e	3 e	36 e	82.2	150.6	1
$C_7H_{14}O$	3-Heptanone		−28 e	0 e	36 e	83.2	147.0	5
$C_7H_{14}O$	4-Heptanone	−27 e	−6 e	18.8	50.2	90.3	143.4	5
$C_7H_{14}O$	5-Methyl-2-hexanone		−27 e	−2 e	31.0	76.6	144.4	5
$C_7H_{14}O$	2,4-Dimethyl-3-pentanone	−61 e	−42 e	−18 e	14 e	58.5	124.8	1
$C_7H_{14}O_2$	Heptanoic acid	24 e	46 e	72 e	107 e	154.6	222.6	5
$C_7H_{14}O_2$	Pentyl acetate	−58 e	−39 e	−14 e	20 e	70.1	149 e	5
$C_7H_{14}O_2$	Isopentyl acetate	−51 e	−30 e	−4 e	30.3	76.2	141.4	5
$C_7H_{14}O_2$	Isobutyl propanoate	−35 e	−19 e	2 e	31 e	72.0	136.1	5
$C_7H_{14}O_2$	Propyl butanoate	−35 e	−19 e	3 e	32.0	74.9	142.8	5
$C_7H_{14}O_2$	Propyl isobutanoate		−28 e	−5.7	24.5	67.5	133.3	5
$C_7H_{14}O_2$	Isopropyl isobutanoate		−44 e	−19.7	12.2	56.0	120.1	5
$C_7H_{14}O_2$	Ethyl 3-methylbutanoate	−57 e	−36 e	−10 e	23.9	69.5	134.4	5
$C_7H_{14}O_2$	Methyl hexanoate	−47 e	−26 e	2 e	36.6	83.3	149 e	5
$C_7H_{14}O_2$	4-Methoxy-4-methyl-2-pentanone				43 e	89.8	160 e	5
$C_7H_{15}Br$	1-Bromoheptane	−30 e	−9 e	18 e	54 e	104.4	178.4	5
$C_7H_{15}Cl$	1-Chloroheptane	−39 e	−19 e	7 e	41 e	88.6	159.9	5
$C_7H_{15}F$	1-Fluoroheptane	−64 e	−45 e	−22 e	10 e	53.3	117.4	5
$C_7H_{15}I$	1-Iodoheptane	−19 e	3 e	32 e	71 e	123.8	203.4	5
C_7H_{16}	Heptane	−78.6	−60.2	−37.0	−6.6	35.4	98.0	16
C_7H_{16}	2-Methylhexane	−82 e	−65 e	−43 e	−13 e	27.8	89.7	1
C_7H_{16}	3-Methylhexane	−81 e	−64 e	−42 e	−12 e	29.2	91.5	1
C_7H_{16}	3-Ethylpentane	−81 e	−63 e	−41 e	−11 e	30.5	93.1	1
C_7H_{16}	2,2-Dimethylpentane	−90 e	−73 e	−52 e	−22.9	17.6	78.8	1
C_7H_{16}	2,3-Dimethylpentane	−87 e	−68.4	−45.3	−14.9	26.8	89.3	1
C_7H_{16}	2,4-Dimethylpentane	−89 e	−72 e	−50 e	−21.3	19.2	80.1	1
C_7H_{16}	3,3-Dimethylpentane	−88 e	−71 e	−49 e	−18.8	22.9	85.6	1
C_7H_{16}	2,2,3-Trimethylbutane				−23.2	18.1	80.4	5
$C_7H_{16}O$	1-Heptanol		17 e	40 e	70.1	112.5	176 e	1
$C_7H_{16}O$	2-Heptanol	−9 e	7 e	27 e	55.0	95.2	158.7	5
$C_7H_{16}O$	3-Heptanol	−8 e	7 e	27 e	54.5	93.9	156.3	5
$C_7H_{16}O$	4-Heptanol	−16 e	1 e	22 e	51 e	91.9	154.6	5
$C_7H_{16}O$	2,2-Dimethyl-3-pentanol			9 e	35 e	73.1	135.5	5
$C_7H_{16}S$	1-Heptanethiol	−30 e	−9 e	18 e	53 e	102.7	176.4	5
$C_7H_{17}N$	Heptylamine			5 e	39 e	86.7	156.4	5
$C_7H_{18}N_2$	N,N-Diethyl-1,3-propanediamine				50.1	99.9	167.7	5
C_8F_{18}	Perfluorooctane				5 e	45.0	105.6	5
$C_8H_4O_3$	Phthalic anhydride	48.2 s	72.4 s			192.7	284.2	5
C_8H_6O	Benzofuran		−16 e	12 e	47.9	97.7	170.7	5
C_8H_7Cl	o-Chlorostyrene	−33 e	−10 e	20 e	58 e	110.8	188 e	5
C_8H_7N	2-Methylbenzonitrile		1 e	32.1	72.2	126.6	204.7	5
C_8H_7N	4-Methylbenzonitrile			40.1	78.7	134.3	221.3	5
C_8H_7N	Benzeneacetonitrile	−3 e	23 e	55.3	97.4	153.7	233.1	5
C_8H_7N	Indole	20.6 s	44.5 s				254.0	5
$C_8H_7NO_4$	Methyl 2-nitrobenzoate	17 e	49 e	89 e	140 e	208 e	302 e	5
C_8H_8	Styrene		−31 e	−5 e	28.6	75.4	144.7	1
C_8H_8	1,3,5,7-Cyclooctatetraene				24.3	71.0	140.1	5
C_8H_8O	Acetophenone			36 e	73 e	125.3	201.5	5
$C_8H_8O_2$	Phenyl acetate		3 e	33.1	72.2	123.9	195.5	5

Mol. form.	Name	Temperature in °C for the indicated pressure						Ref.
		1 Pa	10 Pa	100 Pa	1 kPa	10 kPa	100 kPa	
$C_8H_8O_2$	Methyl benzoate		−1 e	29 e	68 e	121.2	198.9	5
$C_8H_8O_2$	4-Methoxybenzaldehyde	9 e	35 e	68.1	110.8	167.9	248.5	5
$C_8H_8O_3$	Methyl salicylate	−1 e	22 e	51 e	88.8	141.8	219.9	5
C_8H_9Cl	1-Chloro-2-ethylbenzene	−30 e	−9 e	18 e	54 e	103.7	177.9	5
C_8H_9Cl	1-Chloro-4-ethylbenzene	−27 e	−6 e	22 e	58 e	108.7	183.9	5
$C_8H_9NO_2$	1-Ethyl-4-nitrobenzene	10 e	36 e	69 e	111.6	168 e	245 e	5
C_8H_{10}	Ethylbenzene	−56.2	−36.8	−12.0	21.1	67.1	135.7	1
C_8H_{10}	o-Xylene		−7 e	27 e		74.2	143.9	1
C_8H_{10}	m-Xylene		−35 e	−10 e	23.4	69.8	138.7	1
C_8H_{10}	p-Xylene				22.4	68.9	137.9	1
$C_8H_{10}O$	o-Ethylphenol		16.9	44.5	81.1	130.9	204.0	5
$C_8H_{10}O$	m-Ethylphenol	5.6	29.2	57.5	91.9	144.8	217.9	5
$C_8H_{10}O$	p-Ethylphenol			60 e	95.5	144.6	217.5	5
$C_8H_{10}O$	2,3-Xylenol	14.3 s	34.3 s	57.2 s	91.4	141.7	216.4	1,5
$C_8H_{10}O$	2,4-Xylenol			50.2	85.5	137.2	210.5	1,5
$C_8H_{10}O$	2,5-Xylenol	13.4 s	33.2 s	55.9 s	87.4	137.0	210.6	5
$C_8H_{10}O$	2,6-Xylenol	−3.1 s	16.7 s	39.6 s	75.3	125.9	200.6	1,5
$C_8H_{10}O$	3,4-Xylenol	19.7 s	40.2 s	63.7 s	102.1	152.3	226.4	1,5
$C_8H_{10}O$	3,5-Xylenol	16.5 s	37.2 s	61.1 s	98.0	147.9	221.3	1,5
$C_8H_{10}O$	Benzeneethanol	2 e	25 e	54 e	92 e	143.6	217.7	5
$C_8H_{10}O$	Phenetole		−9 e	17 e	51 e	99 e	169.3	5
$C_8H_{10}O_2$	2-Phenoxyethanol	21 e	46 e	75.9	115.4	168.7	244.8	5
$C_8H_{10}O_2$	1,3-Dimethoxybenzene	18 e	34 e	56 e	86.7	135.5	223 e	5
$C_8H_{11}N$	p-Ethylaniline	−2 e	21 e	49 e	87 e	139.4	216.7	5
$C_8H_{11}N$	N-Ethylaniline	−15 e	8 e	38 e	76.4	128.8	204.2	5
$C_8H_{11}N$	N,N-Dimethylaniline			28 e	66 e	118.1	193.6	1
$C_8H_{11}N$	2,4-Xylidine	−2 e	21 e	51 e	88 e	139.1	210.9	5
$C_8H_{11}N$	2,6-Xylidine			37 e	80 e	137.7	217.7	5
$C_8H_{11}N$	5-Ethyl-2-picoline	−33 e	−9.3	20 e			178.0	5
$C_8H_{11}NO$	o-Phenetidine	0 e	27 e	60 e	102.2	156.0	228.1	5
C_8H_{12}	1,5-Cyclooctadiene		−37 e	−8 e	30 e	80.2	150 e	5
C_8H_{12}	4-Vinylcyclohexene	−62 e	−43 e	−19 e	14.1	59.9	129 e	5
$C_8H_{12}O_4$	Diethyl maleate	−6 e	20 e	52.2	93.5	148.4	224.8	5
C_8H_{14}	2,5-Dimethyl-1,5-hexadiene	−38 e	−26 e	−10 e	14 e	50.8	115.1	5
C_8H_{14}	1-Octyne	−59 e	−40 e	−16 e	16 e	60.3	125.8	1
C_8H_{14}	2-Octyne	−52 e	−33 e	−8 e	25 e	70.6	137.8	1
C_8H_{14}	3-Octyne	−55 e	−35 e	−11 e	22 e	66.8	132.8	1
C_8H_{14}	4-Octyne	−56 e	−36 e	−12 e	21 e	65.6	131.4	1
C_8H_{14}	1-Ethylcyclohexene	−55 e	−35 e	−11 e	22 e	68 e	136.5	5
$C_8H_{14}O_2$	Cyclohexyl acetate					103.1	172.9	5
$C_8H_{14}O_2$	Butyl methacrylate				47 e	93.3	159.0	5
$C_8H_{14}O_3$	Butanoic anhydride	−28 e	−2 e	30 e	71 e	123.8	196.5	5
$C_8H_{14}O_4$	Ethyl succinate	−6 e	20 e	51.0	91.1	143.7	216.1	5
$C_8H_{14}O_4$	Dipropyl oxalate	−4 e	20 e	49.9	88.6	140.4	213.0	5
$C_8H_{14}O_4$	Dimethyl adipate		28 e	61 e	103 e	156.1	227.3	5
$C_8H_{15}Br$	(2-Bromoethyl)cyclohexane	−14 e	8 e	36.9	75.3	129.7	212.5	5
$C_8H_{15}ClO$	Octanoyl chloride	1 e	22 e	46 e	74.7	109 e	150 e	5
$C_8H_{15}N$	Octanenitrile	−15 e	8 e	37 e	75 e	127.7	204.4	5
C_8H_{16}	1-Octene	−65.7	−46.1	−21.4	10.5	54.9	120.9	1,5
C_8H_{16}	cis-2-Octene	−59 e	−41 e	−17 e	15 e	59 e	125.2	5
C_8H_{16}	trans-2-Octene	−59 e	−41 e	−17 e	14 e	59 e	124.5	5
C_8H_{16}	cis-3-Octene	−65 e	−46 e	−22 e	10 e	55.1	122.4	5
C_8H_{16}	trans-3-Octene	−61 e	−43 e	−19 e	13 e	57 e	122.8	5
C_8H_{16}	cis-4-Octene	−63 e	−44 e	−20 e	11 e	56 e	122.1	5
C_8H_{16}	trans-4-Octene	−65 e	−46 e	−22 e	10 e	54.6	121.8	5
C_8H_{16}	2-Methyl-1-heptene	−66 e	−48 e	−24 e	8 e	52.3	118.7	5
C_8H_{16}	2,2-Dimethyl-cis-3-hexene	−74 e	−56 e	−33 e	−3 e	40.1	105.0	5
C_8H_{16}	2,3-Dimethyl-2-hexene	−65 e	−47 e	−23 e	10 e	54.3	121.3	5
C_8H_{16}	2,3,3-Trimethyl-1-pentene		−53 e	−30 e	1 e	43.8	107.9	5
C_8H_{16}	2,4,4-Trimethyl-1-pentene	−79 e	−61 e	−38 e	−7 e	36.2	101.0	5

Mol. form.	Name	Temperature in °C for the indicated pressure						Ref.
		1 Pa	10 Pa	100 Pa	1 kPa	10 kPa	100 kPa	
C_8H_{16}	2,3,4-Trimethyl-2-pentene	−68 e	−49 e	−26 e	6 e	50.0	115.8	5
C_8H_{16}	2,4,4-Trimethyl-2-pentene	−73 e	−56 e	−33 e	−2 e	40.4	104.5	5
C_8H_{16}	Cyclooctane				30 e	78 e	150.7	1
C_8H_{16}	Ethylcyclohexane	−61 e	−42 e	−17 e	15.8	61.9	131.3	5
C_8H_{16}	1,1-Dimethylcyclohexane			−27 e	5 e	50.6	119.1	5
C_8H_{16}	cis-1,2-Dimethylcyclohexane		−44 e	−20 e	14 e	59.7	129.2	5
C_8H_{16}	trans-1,2-Dimethylcyclohexane	−68 e	−49 e	−25 e	8 e	53.9	122.9	5
C_8H_{16}	cis-1,3-Dimethylcyclohexane	−68 e	−48 e	−23 e	10 e	55.6	123.1	5
C_8H_{16}	trans-1,3-Dimethylcyclohexane	−62 e	−45 e	−23 e	8 e	51.5	120.9	5
C_8H_{16}	cis-1,4-Dimethylcyclohexane	−66 e	−47 e	−23 e	10 e	55.3	123.8	5
C_8H_{16}	trans-1,4-Dimethylcyclohexane			−27 e	5 e	50.6	118.9	5
C_8H_{16}	Propylcyclopentane	−60 e	−41 e	−16 e	16.5	62.1	130.5	5
C_8H_{16}	Isopropylcyclopentane	−65 e	−46 e	−21 e	12 e	57.3	125.9	5
C_8H_{16}	1-Ethyl-1-methylcyclopentane	−67 e	−49 e	−24 e	8 e	53.2	121.0	5
C_8H_{16}	cis-1-Ethyl-2-methylcyclopentane	−63 e	−44 e	−19 e	13.3	59.1	127.6	5
C_8H_{16}	1,1,2-Trimethylcyclopentane				2 e	46.2	113.2	5
C_8H_{16}	1,1,3-Trimethylcyclopentane	−77 e	−59 e	−36 e	−5 e	38.7	104.4	5
C_8H_{16}	1′,2′,4a-1,2,4-Trimethylcyclo-pentane	−70 e	−52 e	−28 e	4 e	48.9	116.2	5
C_8H_{16}	1′,2a,4′-1,2,4-Trimethylcyclo-pentane	−74 e	−56 e	−33 e	−1 e	42.8	108.8	5
$C_8H_{16}O$	1-Propylcyclopentanol	9 e	24 e	43 e	69.0	108.4	173.5	5
$C_8H_{16}O$	Octanal			6 e	45.7	97.8	170.2	5
$C_8H_{16}O$	2-Octanone		−3 e	23 e	57 e	103.8	172.1	5
$C_8H_{16}O$	3-Octanone			8 e	47.7	97 e	161 e	5
$C_8H_{16}O$	2,2,4-Trimethyl-3-pentanone			11.3	42.1	81.7	134.6	5
$C_8H_{16}O_2$	Octanoic acid	37 e	58 e	85 e	120 e	165.5	238.4	1,5
$C_8H_{16}O_2$	2-Ethylhexanoic acid				108 e	159.6	226.6	5
$C_8H_{16}O_2$	Hexyl acetate	−37 e	−13 e	16 e	52.8	100.4	164 e	5
$C_8H_{16}O_2$	Isopentyl propanoate			3.1	40.7	90.6	159.8	5
$C_8H_{16}O_2$	Isobutyl isobutanoate	−47 e	−26 e	0.4	34.8	81.1	147.0	5
$C_8H_{16}O_2$	Propyl 3-methylbutanoate			1.8	38.9	87.9	155.6	5
$C_8H_{16}O_2$	Ethyl hexanoate	−31 e	−9 e	18.7	53.9	100.7	166.2	5
$C_8H_{16}O_2$	Methyl heptanoate	−30 e	−9 e	19 e	54.2	102.4	172 e	5
$C_8H_{16}O_4$	Diethylene glycol monoethyl ether acetate	−16 e	10.6	43.9	86.2	141.3	216.6	5
$C_8H_{17}Br$	1-Bromooctane	−17 e	6 e	34 e	72 e	123.8	200.3	5
$C_8H_{17}Cl$	1-Chlorooctane	−25 e	−4 e	23 e	59 e	108.8	182.9	5
$C_8H_{17}Cl$	3-(Chloromethyl)heptane					100.3	172.4	5
$C_8H_{17}F$	1-Fluorooctane				29 e	74.6	141.8	5
$C_8H_{17}I$	1-Iodooctane	−6 e	18 e	48 e	87 e	142.5	224.5	5
C_8H_{18}	Octane		−42.6	−17.9	14.4	58.9	125.3	16
C_8H_{18}	2-Methylheptane	−69 e	−49.1	−24.5	7.6	51.6	117.2	1,5
C_8H_{18}	3-Methylheptane	−67 e	−48.1	−23.6	8.5	52.7	118.5	1,5
C_8H_{18}	4-Methylheptane	−65 e	−47 e	−24 e	7.8	51.6	117.2	5
C_8H_{18}	3-Ethylhexane				8 e	52.1	118.1	5
C_8H_{18}	2,2-Dimethylhexane	−73 e	−55 e	−32 e	−1.5	41.6	106.4	5
C_8H_{18}	2,3-Dimethylhexane				5 e	49.2	115.1	5
C_8H_{18}	2,4-Dimethylhexane				0.6	43.9	109.0	5
C_8H_{18}	2,5-Dimethylhexane	−71 e	−53 e	−30 e	0.7	43.8	108.6	5
C_8H_{18}	3,3-Dimethylhexane	−72 e	−54 e	−30 e	1.4	45.4	111.5	5
C_8H_{18}	3,4-Dimethylhexane				7 e	50.9	117.3	5
C_8H_{18}	3-Ethyl-2-methylpentane	−69 e	−50 e	−27 e	5 e	48.9	115.2	5
C_8H_{18}	3-Ethyl-3-methylpentane	−70 e	−51 e	−27 e	5 e	50.2	117.8	5
C_8H_{18}	2,2,3-Trimethylpentane	−74 e	−56 e	−32 e	−0.8	43.1	109.4	5
C_8H_{18}	2,2,4-Trimethylpentane	−81.9	−63.4	−39.8	−8.9	34.0	98.8	5
C_8H_{18}	2,3,3-Trimethylpentane	−72 e	−54 e	−30 e	2.1	46.9	114.3	5
C_8H_{18}	2,3,4-Trimethylpentane	−74 e	−54.5	−30.0	2.2	46.7	113.1	1,5
C_8H_{18}	2,2,3,3-Tetramethylbutane	−62.5 s	−44 s	−20.9 s	8.9 s	48.8 s	105.8	5
$C_8H_{18}O$	1-Octanol	12 e	30 e	53 e	84 e	128.2	194.8	1,39
$C_8H_{18}O$	2-Octanol			40 e	69.9	112.5	179.4	1,39
$C_8H_{18}O$	3-Octanol	12 e	24 e	40 e	64 e	102.8	174.1	1

		Temperature in °C for the indicated pressure						
Mol. form.	Name	1 Pa	10 Pa	100 Pa	1 kPa	10 kPa	100 kPa	Ref.
$C_8H_{18}O$	4-Octanol			40 e	66.9	107.3	176.0	1,39
$C_8H_{18}O$	4-Methyl-3-heptanol	−52 e	−28 e	1 e	39 e	87.6	155.0	5
$C_8H_{18}O$	5-Methyl-3-heptanol	−35 e	−16 e	8 e	40 e	84.8	153.0	5
$C_8H_{18}O$	4-Methyl-4-heptanol	−17 e	1 e	24 e	55 e	97.2	160.7	5
$C_8H_{18}O$	2-Ethyl-1-hexanol			45 e	75 e	118.3	184.2	1
$C_8H_{18}O$	2-Ethyl-2-hexanol	−13 e	4 e	26 e	55 e	96.3	160.3	5
$C_8H_{18}O$	2,4,4-Trimethyl-2-pentanol		−7 e	13 e	40 e	79.8	146.1	5
$C_8H_{18}O$	2,2,4-Trimethyl-3-pentanol	−2 e	9 e	24 e	47 e	82.6	150.4	5
$C_8H_{18}O$	Dibutyl ether	−55 e	−35 e	−8 e	26 e	73.0	141.2	5
$C_8H_{18}O$	Di-sec-butyl ether			−19 e	12.1	55.4	120.6	5
$C_8H_{18}O$	Di-tert-butyl ether			−33 e	−2 e	41.7	106.8	1
$C_8H_{18}O_2$	Ethylene glycol monohexyl ether	−13 e	14 e	46 e	86 e	137.7	206.9	5
$C_8H_{18}O_2$	1,2-Dipropoxyethane			−44.2	−2.0	63.6	179.2	5
$C_8H_{18}O_2$	Di-tert-butyl peroxide			−26 e	4.3	46.6	110.5	5
$C_8H_{18}O_3$	Diethylene glycol monobutyl ether	14 e	37 e	66.8	104.9	153 e	230.4	5
$C_8H_{18}O_3$	Diethylene glycol diethyl ether	−32 e	−7 e	25 e	64.9	117.1	189 e	5
$C_8H_{18}O_5$	Tetraethylene glycol	89 e	117 e	151.1	192.2	242.9	307.3	5
$C_8H_{18}S$	1-Octanethiol	−15 e	6 e	34 e	71 e	122.1	198.5	5
$C_8H_{18}S$	Dibutyl sulfide	−22 e	0 e	27 e	63 e	113.5	188.4	5
$C_8H_{19}N$	Dibutylamine	−37 e	−16 e	10 e	44 e	90.8	159.1	5
$C_8H_{19}N$	Diisobutylamine	−57 e	−36 e	−9.0	25.5	72.2	139.0	5
$C_8H_{20}O_4Si$	Ethyl silicate	−77 e	−52 e	−21 e	21.6	80.5	164.1	5
$C_8H_{20}Si$	Tetraethylsilane			−6.5	30.5	80.6	152.6	5
C_9F_{20}	Perfluorononane					40 e	114.7	5
$C_9H_6N_2O_2$	Toluene-2,4-diisocyanate		39 e	72 e	113.9	169.7	247 e	5
C_9H_7N	Quinoline	−1.3	23.7	55.4	96.8	153.4	236.5	1,5
C_9H_7N	Isoquinoline		30.2	60.7	101.3	157.9	242.7	1,5
C_9H_8	Indene			12 e	53.0	106.8	181.0	5
C_9H_{10}	cis-1-Propenylbenzene	−38 e	−15.4	13.3	51.4	103.7	178.4	5
C_9H_{10}	trans-1-Propenylbenzene		−16 e	13.3	51.6	103.7	178.4	5
C_9H_{10}	Isopropenylbenzene			3.2	41.5	92.8	164.9	5
C_9H_{10}	Indan	−33 e	−12 e	16 e	52 e	102.3	177.5	1
$C_9H_{10}O$	2,4-Dimethylbenzaldehyde	−3 e	23 e	54 e	93.2	144.6	214.5	5
$C_9H_{10}O_2$	Ethyl benzoate	−18 e	8 e	39 e	80.1	135.1	212.8	5
$C_9H_{10}O_2$	Benzyl acetate	−11 e	15 e	46.6	86.9	139.5	211 e	5
$C_9H_{11}Br$	1-Bromo-4-isopropylbenzene	−8 e	15 e	45 e	84 e	138.1	218.5	5
$C_9H_{11}Cl$	1-Chloro-2-isopropylbenzene	−23 e	−1 e	27 e	64 e	114.6	190.5	5
$C_9H_{11}Cl$	1-Chloro-4-isopropylbenzene		3 e	31 e	69 e	120.5	197.8	5
C_9H_{12}	Propylbenzene	−43 e	−23 e	4 e	38 e	86.7	158.8	1
C_9H_{12}	Isopropylbenzene	−46 e	−26 e	−1 e	33 e	80.9	152.0	1
C_9H_{12}	o-Ethyltoluene	−40 e	−19 e	8 e	43 e	92.1	164.7	5
C_9H_{12}	m-Ethyltoluene	−42 e	−21 e	5 e	40.4	88.9	160.8	5
C_9H_{12}	p-Ethyltoluene	−41 e	−21 e	6 e	41 e	89.2	161.5	5
C_9H_{12}	1,2,3-Trimethylbenzene		−12 e	15 e	52 e	101.5	175.6	1
C_9H_{12}	1,2,4-Trimethylbenzene	−37 e	−16 e	11 e	47 e	95.9	168.9	1
C_9H_{12}	1,3,5-Trimethylbenzene	−39 e	−18 e	9 e	43.7	92.4	164.3	1
$C_9H_{12}O$	Benzyl ethyl ether		−10 e	20.4	59.3	111.3	184.5	5
$C_9H_{12}O$	Phenyl propyl ether		−10 e	21 e	61 e	113.9	189.3	5
$C_9H_{12}O$	Phenyl isopropyl ether	−20 e	−1 e	23 e	56 e	103.7	176.9	5
$C_9H_{13}N$	2,4,6-Trimethylaniline	12 e	36 e	66 e	104.1	154.9	226 e	5
$C_9H_{13}N$	N,N-Dimethyl-o-toluidine	−25 e	−3 e	24.4	60.6	110.7	184.5	5
$C_9H_{13}N$	Amphetamine			33 e	70.1	118 e	202.0	5
$C_9H_{14}O$	Isophorone		1 e	33.1	75.1	132.4	215.1	5
$C_9H_{14}O_6$	Triacetin	37.6	62 e	90 e	124 e	165 e	214 e	5
$C_9H_{16}O_4$	Diethyl glutarate	−1 e	26 e	60.2	103.3	159.6	236.5	5
$C_9H_{17}N$	Nonanenitrile	−3 e	21 e	50.9	90.7	145.4	225.1	5
C_9H_{18}	1-Nonene	−50.1	−29.4	−3.3	30.4	77.1	146.4	1,5
C_9H_{18}	2-Methyl-1-octene	−53 e	−34 e	−9 e	25 e	72 e	144.1	5
C_9H_{18}	Butylcyclopentane	−45 e	−24 e	1 e	36 e	84 e	156.1	5
C_9H_{18}	Propylcyclohexane	−46 e	−26 e	0 e	35.1	83.6	156.2	5

Mol. form.	Name	1 Pa	10 Pa	100 Pa	1 kPa	10 kPa	100 kPa	Ref.
		\ Temperature in °C for the indicated pressure							
C_9H_{18}	Isopropylcyclohexane	−48 e	−28 e	−2 e	33 e	81.3	154.0	5
C_9H_{18}	trans-1-Ethyl-4-methylcyclo-hexane	−53 e	−33 e	−8 e	25 e	71.8	141.5	5
C_9H_{18}	1,1,2-Trimethylcyclohexane			−12 e	23 e	71.5	145.5	5
C_9H_{18}	1,1,3-Trimethylcyclohexane	−60 e	−41 e	−16 e	18 e	65.2	136.1	5
C_9H_{18}	1′,2a,4a-1,2,4-Trimethylcyclo-hexane	−71 e	−50 e	−22 e	15 e	65.7	140.7	5
C_9H_{18}	1′,3′,5′-1,3,5-Trimethylcyclo-hexane	−72 e	−50 e	−22 e	14 e	65.1	140.0	5
C_9H_{18}	Isobutylcyclopentane	−105 e	−88 e	−64 e	−28 e	31 e	147.0	5
C_9H_{18}	cis-1-Methyl-2-propylcyclo-pentane	−52 e	−33 e	−7 e	28 e	77 e	152.0	5
C_9H_{18}	trans-1-Methyl-2-propylcyclo-pentane	−56 e	−36 e	−11 e	23 e	72 e	145.8	5
C_9H_{18}	1,1,3,3-Tetramethylcyclo-pentane	−72 e	−54 e	−30 e	2 e	47 e	117.4	5
$C_9H_{18}O$	Nonanal		−3 e	27.4	65.5	115.6	184.6	5
$C_9H_{18}O$	2-Nonanone		8 e	35 e	71 e	121.0	194.0	5
$C_9H_{18}O$	5-Nonanone			−1 e	39.1	94 e	188 e	5
$C_9H_{18}O$	2,6–Dimethyl-4-heptanone	−32 e	−12 e	14 e	48 e	96.2	167.7	5
$C_9H_{18}O_2$	Nonanoic acid	48 e	69 e	97 e	133 e	182.7	255.1	5
$C_9H_{18}O_2$	Heptyl acetate	−16 e	6 e	34 e	70 e	119.9	191.9	5
$C_9H_{18}O_2$	Isopentyl butanoate				55 e	105.6	178.4	5
$C_9H_{18}O_2$	Isobutyl 3-methylbutanoate			11.3	48.3	97.9	168.3	5
$C_9H_{18}O_2$	Propyl hexanoate	−26 e	−2 e	28 e	65.1	113.4	178 e	5
$C_9H_{18}O_2$	Methyl octanoate	−26 e	−9 e	13 e	40 e	76 e	127.9	5
$C_9H_{19}Cl$	1-Chlorononane	−11 e	11 e	39 e	76 e	127.8	204.7	5
C_9H_{20}	Nonane	−46.8	−26.0	0.0	34.0	80.8	150.3	16
C_9H_{20}	2-Methyloctane	−49 e	−30 e	−5 e	28 e	73.9	142.8	5
C_9H_{20}	3-Methyloctane	−49 e	−29 e	−5 e	29 e	74.7	143.7	5
C_9H_{20}	4-Methyloctane	−50 e	−30 e	−6 e	27 e	73.2	141.9	5
C_9H_{20}	2,2-Dimethylheptane	−58 e	−39 e	−15 e	18 e	63.6	132.3	5
C_9H_{20}	2,3-Dimethylheptane	−53 e	−33 e	−9 e	25 e	70.8	140.0	5
C_9H_{20}	2,6-Dimethylheptane	−55 e	−36 e	−12 e	21 e	66.4	134.7	5
C_9H_{20}	3-Ethyl-4-methylhexane			−9 e	24 e	70.6	139.9	5
C_9H_{20}	2,2,4-Trimethylhexane	−66.1	−46.4	−21.3	11.8	57.7	126.0	5
C_9H_{20}	2,2,5-Trimethylhexane	−65.1	−45.8	−21.2	11.2	56.2	123.7	1,5
C_9H_{20}	2,3,3-Trimethylhexane	−58 e	−38 e	−13 e	20 e	66.7	137.2	5
C_9H_{20}	2,3,5-Trimethylhexane	−60 e	−41 e	−16 e	17 e	62.3	130.9	5
C_9H_{20}	2,4,4-Trimethylhexane	−62 e	−43 e	−18 e	15 e	61.0	130.2	5
C_9H_{20}	3,3,4-Trimethylhexane	−53 e	−33 e	−7 e	28 e	76.3	148.9	5
C_9H_{20}	3,3-Diethylpentane			−9 e	26 e	73.7	145.7	1
C_9H_{20}	3-Ethyl-2,4-dimethylpentane	−58 e	−38 e	−13 e	20 e	66.7	136.2	5
C_9H_{20}	2,2,3,3-Tetramethylpentane				21 e	68.5	139.8	1
C_9H_{20}	2,2,3,4-Tetramethylpentane	−61 e	−42 e	−17 e	16 e	62.5	132.6	1
C_9H_{20}	2,2,4,4-Tetramethylpentane		−49 e	−25 e	8 e	53.2	121.8	1
C_9H_{20}	2,3,3,4-Tetramethylpentane	−57 e	−37 e	−12 e	22 e	69.7	141.1	1
$C_9H_{20}O$	1-Nonanol		40 e	64 e	96.9	141.0	213.0	5,39
$C_9H_{20}O$	3-Nonanol		24 e	47 e	78 e	123.0	194.2	5
$C_9H_{20}O$	4-Nonanol			45 e	76.4	121.3	192.0	5
$C_9H_{20}O$	5-Nonanol	13 e	31 e	54 e	84.5	128.1	194.7	5
$C_9H_{20}O$	2,2,4,4-Tetramethyl-3-pentanol				58	100	167	5
$C_9H_{20}S$	1-Nonanethiol	−2 e	21 e	49 e	87 e	140.4	219.2	5
$C_9H_{21}BO_3$	Triisopropyl borate				73.1		139.0	5
$C_9H_{21}N$	Nonylamine		9 e	37 e	75 e	126.2	202.1	5
$C_9H_{21}N$	Tripropylamine	−39 e	−18 e	8 e	42 e	88.2	156.0	5
$C_{10}F_8$	Perfluoronaphthalene	5.2 s	25.1 s	48.1 s				5
$C_{10}F_{22}$	Perfluorodecane					52 e	132.9	5
$C_{10}H_7Br$	1-Bromonaphthalene	17 e	45 e	80.3	126.7	189.8	280.5	5
$C_{10}H_7Cl$	1-Chloronaphthalene	14 e	39 e	70.5	112.8	171.6	258.6	5
$C_{10}H_8$	Naphthalene**	3.2 s	24.1 s	49.3 s	80.7	135.6	217.5	1,5
$C_{10}H_8$	Azulene	24.1 s	46 s	71.5 s	103.3	162.6	244.0	5
$C_{10}H_8O$	1-Naphthol				137.2	196.7	281.8	5
$C_{10}H_8O$	2-Naphthol				140.7	200.5	286.8	5
$C_{10}H_9N$	1-Naphthalenamine		62 e	99.0	146.9	210.7	300.1	5
$C_{10}H_9N$	2-Naphthalenamine	36.3 s	65.9 s	103 s	150.9	215.1	305.5	5

		Temperature in °C for the indicated pressure						
Mol. form.	Name	1 Pa	10 Pa	100 Pa	1 kPa	10 kPa	100 kPa	Ref.
$C_{10}H_9N$	2-Methylquinoline	5.3	31.9	63.8	102.9	165.8	247.2	5
$C_{10}H_9N$	4-Methylquinoline	29 e	54 e	85 e	127 e	183.0	265.1	5
$C_{10}H_9N$	6-Methylquinoline	27 e	51 e	81 e	122 e	179.2	264.5	5
$C_{10}H_9N$	8-Methylquinoline	15 e	40 e	70 e	111 e	166.1	247.3	5
$C_{10}H_{10}$	m-Divinylbenzene	−29 e	−4 e	27.1	67.6	122.1	199 e	5
$C_{10}H_{10}O_4$	Dimethyl phthalate	27 e	56 e	92.7	137.8	195.8	272.7	5
$C_{10}H_{10}O_4$	Dimethyl isophthalate			85 e	129.5	189.2	273 e	5
$C_{10}H_{10}O_4$	Dimethyl terephthalate	56.6 s	79.4 s	106.1 s	137.9 s	197.9	282 e	5
$C_{10}H_{12}$	1,2,3,4-Tetrahydronaphthalene	−21 e	3 e	33.2	74.1	127.4	207.8	5
$C_{10}H_{12}$	2-Ethylstyrene	−31 e	−8 e	21 e	60 e	111.7	187 e	5
$C_{10}H_{12}$	3-Ethylstyrene	−28 e	−5.3	24.1	62.6	116 e	193 e	5
$C_{10}H_{12}$	4-Ethylstyrene	−31 e	−8.2	21.3	60.5	115 e	196 e	5
$C_{10}H_{12}O$	Estragole			48.5	88.0	140.7	214.6	5
$C_{10}H_{12}O$	4-Isopropylbenzaldehyde			54.1	96.0	152.2	231.5	5
$C_{10}H_{12}O_2$	4-Allyl-2-methoxyphenol	9 e	37 e	72 e	115.9	173.8	252.9	5
$C_{10}H_{12}O_2$	2-Phenylethyl acetate	−4 e	22 e	54 e	96 e	152.3	232.0	5
$C_{10}H_{12}O_2$	Propyl benzoate	−8 e	18 e	50.2	92.3	149.2	230.5	5
$C_{10}H_{12}O_2$	Ethyl phenylacetate	−9 e	19 e	52 e	95 e	150.2	225 e	5
$C_{10}H_{12}O_2$	Isoeugenol				125 e	185.3	267.1	5
$C_{10}H_{14}$	Butylbenzene	−28 e	−7 e	21 e	56.9	107.6	182.8	1,5
$C_{10}H_{14}$	sec-Butylbenzene	−35 e	−14 e	13 e	48 e	98.3	172.8	5
$C_{10}H_{14}$	tert-Butylbenzene	−37 e	−16 e	10 e	46 e	94.9	168.6	5
$C_{10}H_{14}$	Isobutylbenzene	−36 e	−15 e	12 e	47.9	97.8	172.3	5
$C_{10}H_{14}$	o-Cymene	−39 e	−16 e	13 e	51 e	103.1	177.8	5
$C_{10}H_{14}$	m-Cymene	−34 e	−13 e	14 e	50 e	99.9	174.6	5
$C_{10}H_{14}$	p-Cymene	−33 e	−12 e	16 e	52 e	102.2	176.6	5
$C_{10}H_{14}$	o-Diethylbenzene	−28 e	−6 e	21 e	58 e	107.9	182.9	5
$C_{10}H_{14}$	m-Diethylbenzene	−28 e	−7 e	20 e	56 e	106.2	180.6	5
$C_{10}H_{14}$	p-Diethylbenzene	−28 e	−6 e	21 e	57 e	108.1	183.3	5
$C_{10}H_{14}$	3-Ethyl-1,2-dimethylbenzene	−22 e	0 e	28 e	66 e	117.2	193.4	5
$C_{10}H_{14}$	4-Ethyl-1,2-dimethylbenzene	−24 e	−2 e	26 e	63 e	113.6	189.2	5
$C_{10}H_{14}$	2-Ethyl-1,3-dimethylbenzene		−2 e	26 e	63 e	113.7	189.5	5
$C_{10}H_{14}$	2-Ethyl-1,4-dimethylbenzene	−27 e	−5 e	23 e	60 e	110.6	186.4	5
$C_{10}H_{14}$	1-Ethyl-2,4-dimethylbenzene	−25 e	−4 e	24 e	61 e	112.2	187.9	5
$C_{10}H_{14}$	1-Ethyl-3,5-dimethylbenzene	−28 e	−6 e	21 e	58 e	108.3	183.2	5
$C_{10}H_{14}$	1-Methyl-2-propylbenzene	−27 e	−6 e	22 e	58.2	108.9	184.3	5
$C_{10}H_{14}$	1-Methyl-3-propylbenzene	−29 e	−8 e	20 e	56.1	106.5	181.3	5
$C_{10}H_{14}$	1-Methyl-4-propylbenzene	−29 e	−7 e	20 e	56.6	107.4	182.8	5
$C_{10}H_{14}$	1,2,3,4-Tetramethylbenzene		7 e	36 e	74 e	126.6	204.5	5
$C_{10}H_{14}$	1,2,3,5-Tetramethylbenzene	−19 e	3 e	32 e	69 e	120.9	197.5	5
$C_{10}H_{14}$	1,2,4,5-Tetramethylbenzene					119.9	196.3	5
$C_{10}H_{14}O$	2-Butylphenol	7 e	31 e	61 e	101 e	155.2	234.4	5
$C_{10}H_{14}O$	Butyl phenyl ether	−16 e	8 e	38 e	77 e	131.3	209.7	5
$C_{10}H_{14}O$	Thymol	18.9 s	37.9 s	59.5	101.2	155.0	230.4	5
$C_{10}H_{15}N$	2-Methyl-5-isopropylaniline	19 e	43 e	72 e	107.4	150 e	204 e	5
$C_{10}H_{15}N$	N-Butylaniline	11 e	35 e	66 e	106 e	160.9	241.0	5
$C_{10}H_{15}N$	N,N-Diethylaniline	−11 e	14 e	44.3	84.2	138.4	216.3	5
$C_{10}H_{16}$	Dipentene	−42 e	−19 e	10.6	48.7	100.2	173.9	5
$C_{10}H_{16}$	d-Limonene	−45 e	−21 e	9.1	48.0	100.4	174.5	5
$C_{10}H_{16}$	l-Limonene	−33 e	−12 e	16 e	52.0	102.3	177.0	21
$C_{10}H_{16}$	β-Myrcene			9.4	47.3	98.3	171.0	5
$C_{10}H_{16}$	α-Pinene	−48 e	−27 e	−1 e	33.6	82.2	155.1	21
$C_{10}H_{16}$	β-Pinene	−43 e	−22 e	5.0	40.6	90.5	165.5	21
$C_{10}H_{16}$	Camphene					90.7	160.1	4
$C_{10}H_{16}$	Terpinolene			26.5	64.9	115.4	184.6	5
$C_{10}H_{16}$	β-Phellandrene			16 e	53.2	104 e	171.0	5
$C_{10}H_{16}O$	(+)-Camphor	−15.8 s	10 s	41.5 s	80.8 s	131.4 s	207.6	5
$C_{10}H_{16}O$	Pulegone	37 e	49.1	66.4	92.2	135.1	220.2	5
$C_{10}H_{18}$	1-Decyne	−34 e	−13 e	14 e	51 e	100.3	173.5	5
$C_{10}H_{18}$	cis-Decahydronaphthalene	−26 e	−4 e	24 e	62.4	115.5	195.3	1

Mol. form.	Name	Temperature in °C for the indicated pressure						Ref.
		1 Pa	10 Pa	100 Pa	1 kPa	10 kPa	100 kPa	
$C_{10}H_{18}$	*trans*-Decahydronaphthalene		−10 e	18 e	55.3	107.9	186.8	1
$C_{10}H_{18}O$	α-Terpineol			48	89	142	217	4
$C_{10}H_{18}O$	Eucalyptol			10.6	48.5	100.3	175.4	5
$C_{10}H_{18}O$	*trans*-Geraniol	4 e	31 e	63.2	104.3	157.7	229.6	5
$C_{10}H_{18}O_4$	Sebacic acid	125.9 s						5
$C_{10}H_{18}O_4$	Dipropyl succinate	11 e	38 e	72.1	115.4	172.3	250.4	5
$C_{10}H_{18}O_4$	Diethyl adipate	4 e	35 e	72 e	116.6	171.2	239.5	5
$C_{10}H_{19}N$	Decanenitrile	13 e	36 e	66 e	105.8	160.6	241.6	5
$C_{10}H_{20}$	1-Decene	−35.5	−13.7	13.7	49.0	97.9	170.1	1,5
$C_{10}H_{20}$	Cyclodecane			29 e	68 e	121.3	201.8	1
$C_{10}H_{20}$	Butylcyclohexane	−31 e	−9 e	18 e	54 e	104.7	180.4	5
$C_{10}H_{20}$	Isobutylcyclohexane	−37 e	−16 e	10 e	46 e	95.9	170.8	5
$C_{10}H_{20}$	*tert*-Butylcyclohexane	−39 e	−18 e	9 e	45 e	95.3	171.1	5
$C_{10}H_{20}O$	Decanal			16 e	47.2	86.3	137.7	208.0	5
$C_{10}H_{20}O_2$	Decanoic acid	58 e	80 e	108 e	145 e	195.2	269.5	5
$C_{10}H_{20}O_2$	Octyl acetate	−26 e	−3 e	27 e	66.3	120.0	198.2	5
$C_{10}H_{20}O_2$	2-Ethylhexyl acetate	−11 e	5 e	26 e	57.6	107.1	197.2	5
$C_{10}H_{20}O_2$	Isopentyl isopentanoate			22 e	62.8	116.9	193.6	5
$C_{10}H_{20}O_2$	Ethyl octanoate	−17 e	9 e	41 e	81.4	133.2	203 e	5
$C_{10}H_{20}O_4$	Diethylene glycol monobutyl ether acetate	6 e	34 e	69 e	112.6	169.2	245.4	5
$C_{10}H_{21}Br$	1-Bromodecane	9 e	33 e	63 e	104 e	159.2	240.0	5
$C_{10}H_{21}Cl$	1-Chlorodecane	2 e	25 e	54 e	92 e	145.7	225.3	5
$C_{10}H_{21}F$	1-Fluorodecane	−22 e	0 e	27 e	64 e	113.3	185.7	5
$C_{10}H_{22}$	Decane		−10.6	16.7	52.3	101.1	173.7	16
$C_{10}H_{22}$	2-Methylnonane	−34 e	−14 e	12 e	47 e	94.8	166.5	5
$C_{10}H_{22}$	3-Methylnonane	−34 e	−14 e	12 e	47 e	95.1	167.3	5
$C_{10}H_{22}$	4-Methylnonane	−36 e	−16 e	10 e	45 e	93.1	165.2	5
$C_{10}H_{22}$	5-Methylnonane	−36 e	−16 e	10 e	45 e	92.6	164.6	5
$C_{10}H_{22}$	2,4-Dimethyloctane				38 e	84.9	155.4	5
$C_{10}H_{22}$	2,7-Dimethyloctane	−39 e	−19 e	7 e	41 e	88.4	159.4	5
$C_{10}H_{22}$	2,2,6-Trimethylheptane	−46 e	−27 e	−2 e	32 e	78.5	148.4	5
$C_{10}H_{22}$	3,3,5-Trimethylheptane			0 e	35 e	82.7	155.2	5
$C_{10}H_{22}$	2,2,3,3-Tetramethylhexane	−46 e	−25 e	1 e	36 e	85.6	159.8	5
$C_{10}H_{22}$	2,2,5,5-Tetramethylhexane			−10 e	22 e	68.3	137.0	5
$C_{10}H_{22}$	2,4-Dimethyl-3-isopropylpentane	−46 e	−26 e	0 e	35 e	83.2	156.5	5
$C_{10}H_{22}$	2,2,3,3,4-Pentamethylpentane		−24 e	3 e	39 e	89.1	165.5	5
$C_{10}H_{22}$	2,2,3,4,4-Pentamethylpentane		−29 e	−3 e	33 e	82.8	158.7	5
$C_{10}H_{22}O$	1-Decanol	30 e	50 e	75 e	109 e	157.3	230.6	1,39
$C_{10}H_{22}O$	4-Decanol	18 e	37 e	61 e	93 e	139 e	210 e	5
$C_{10}H_{22}O$	Dipentyl ether	−31 e	−8 e	22 e	60 e	111.6	186.2	5
$C_{10}H_{22}O$	Diisopentyl ether			14.0	51.5	101.8	172.8	5
$C_{10}H_{22}O_2$	Ethylene glycol dibutyl ether	0 e	20 e	44 e	78.4	127.1	202.9	5
$C_{10}H_{22}O_5$	Tetraethylene glycol dimethyl ether				138 e	200.9	275.3	5
$C_{10}H_{22}S$	1-Decanethiol	11 e	34 e	64 e	103 e	157.5	238.6	5
$C_{10}H_{22}S$	Diisopentylsulfide			7 e	82 e	118 e	139 e	5
$C_{10}H_{23}N$	Dipentylamine				77 e	127.7	202.0	5
$C_{10}H_{30}O_3Si_4$	Decamethyltetrasiloxane	−31 e	−6 e	26 e	66.8	118.8	193.9	5
$C_{10}H_{30}O_5Si_5$	Decamethylcyclopentasiloxane	−2 e	19 e	46 e	82 e	132.9	210.4	5
$C_{11}H_8O_2$	1-Naphthalenecarboxylic acid				191.9	239.3	299.6	5
$C_{11}H_8O_2$	2-Naphthalenecarboxylic acid				197.9	246.0	308.1	5
$C_{11}H_{10}$	1-Methylnaphthalene	5 e	29 e	60 e	102 e	159.1	244.1	1
$C_{11}H_{10}$	2-Methylnaphthalene			57 e	99 e	156.0	240.5	1
$C_{11}H_{12}O_2$	Ethyl *trans*-cinnamate			79	125	187	271	4
$C_{11}H_{12}O_3$	Myristicin	23 e	53 e	88.9	135.2	196.0	279.4	5
$C_{11}H_{14}$	4-Isopropylstyrene	−25 e	−1 e	30.2	70.3	124.5	202.1	5
$C_{11}H_{14}$	1,2,3,4-Tetrahydro-5-methylnaphthalene	9 e	31 e	60 e	99 e	153.1	233.8	5
$C_{11}H_{14}$	1,2,3,4-Tetrahydro-6-methylnaphthalene	17 e	36 e	62 e	97 e	147.8	228.5	5
$C_{11}H_{14}O_2$	Butyl benzoate	6 e	34 e	67.9	110.3	165 e	237 e	5
$C_{11}H_{16}$	Pentylbenzene	−14 e	8 e	37 e	74 e	126.7	204.9	5

		Temperature in °C for the indicated pressure						
Mol. form.	Name	1 Pa	10 Pa	100 Pa	1 kPa	10 kPa	100 kPa	Ref.
$C_{11}H_{16}$	*p-tert*-Butyltoluene	−24 e	−2 e	27 e	64.1	115.5	190.8	5
$C_{11}H_{16}$	1,3-Diethyl-5-methylbenzene	−26 e	−1 e	29.5	69.5	123.5	200.2	5
$C_{11}H_{16}$	2-Ethyl-1,3,5-trimethylbenzene		6 e	36 e	75.7	129.6	207.6	5
$C_{11}H_{16}$	1-Ethyl-2,4,5-trimethylbenzene	−13 e	11 e	40 e	79.4	132.1	207.7	5
$C_{11}H_{20}$	1-Undecyne	−22 e	0 e	29 e	67 e	118.5	194.5	5
$C_{11}H_{20}$	2-Undecyne	−17 e	6 e	35 e	74 e	127.4	205.4	5
$C_{11}H_{20}O_2$	10-Undecenoic acid	35 e	67 e	105 e	150.0	205.4	274.5	5
$C_{11}H_{20}O_4$	Ethyl diethylmalonate			74 e	105 e	149.4	219 e	5
$C_{11}H_{21}N$	Undecanenitrile			78.6	120.3	177.3	259.9	5
$C_{11}H_{22}$	1-Undecene	−21.6	1.2	29.7	66.4	117.1	192.2	5
$C_{11}H_{22}$	*cis*-2-Undecene	−14 e	7 e	34 e	70.2	120.6	196 e	5
$C_{11}H_{22}$	*trans*-2-Undecene	−14 e	7 e	33 e	69.3	119.6	195 e	5
$C_{11}H_{22}$	*cis*-4-Undecene	−19 e	3 e	30 e	66.6	117.1	192 e	5
$C_{11}H_{22}$	*trans*-4-Undecene	−17 e	4 e	31 e	67.1	117.4	193 e	5
$C_{11}H_{22}$	*cis*-5-Undecene	−19 e	2 e	30 e	66.2	116.7	191 e	5
$C_{11}H_{22}$	*trans*-5-Undecene	−18 e	3 e	31 e	67.0	117.4	192 e	5
$C_{11}H_{22}$	Pentylcyclohexane	−17 e	6 e	34 e	72 e	124.2	202.7	5
$C_{11}H_{22}$	Hexylcyclopentane	−15 e	7 e	36 e	73 e	125.0	202.5	5
$C_{11}H_{22}O$	2-Undecanone	17 e	37 e	64.3	103.0	153.6	232.6	1,5
$C_{11}H_{22}O$	6-Undecanone		28 e	57 e	95 e	148.4	226.9	1
$C_{11}H_{22}O_2$	Undecanoic acid	68 e	90 e	118 e	156 e	207.2	283.6	5
$C_{11}H_{22}O_2$	Heptyl butanoate	2 e	29 e	62 e	102.6	155.1	224.7	5
$C_{11}H_{22}O_2$	Propyl octanoate	−2 e	23 e	55 e	94.0	145.2	215 e	5
$C_{11}H_{22}O_2$	Methyl decanoate	10 e	33 e	62 e	100.9	154.0	232 e	5
$C_{11}H_{24}$	Undecane	−18.4	4.3	32.6	69.5	120.2	195.4	16
$C_{11}H_{24}$	2-Methyldecane	−20 e	1 e	28 e	64 e	114.0	188.7	5
$C_{11}H_{24}$	3-Methyldecane	−35 e	−10 e	22 e	61.9	115.6	190.4	5
$C_{11}H_{24}$	4-Methyldecane	−38 e	−12 e	20 e	60.8	113.9	186.4	5
$C_{11}H_{24}$	2,4,7-Trimethyloctane				43 e	94 e	170.4	5
$C_{11}H_{24}O$	1-Undecanol	52.2	80.0	82 e	118 e	167.6	244.1	5
$C_{11}H_{24}S$	1-Undecanethiol	23 e	47 e	77 e	118 e	173.6	256.8	5
$C_{12}F_{27}N$	Trinonafluorobutylamine		3 e	29.0	63.3	109.9	176.8	5
$C_{12}H_8$	Acenaphthylene	24 s	49.8 s	80.6 s				5
$C_{12}H_9N$	Carbazole					254.7	354.0	5
$C_{12}H_{10}$	Acenaphthene			126.2	187 e	276 e		1
$C_{12}H_{10}$	Biphenyl			69.0	111.1	169.5	254.7	1
$C_{12}H_{10}N_2$	Azobenzene			98.1	144.8	206.7	292.7	4
$C_{12}H_{10}O$	Diphenyl ether		44 e	75 e	116 e	173 e	257.4	5
$C_{12}H_{10}O$	1-Acetonaphthone	37 e	69 e	107.0	154.6	215.2	294.9	5
$C_{12}H_{10}O$	2-Acetonaphthone	48.3 s		118.7	163.0	221.1	300.3	5
$C_{12}H_{10}S$	Diphenyl sulfide	20 e	51 e	88.7	137.5	202.2	291.8	5
$C_{12}H_{11}N$	Diphenylamine	48 s		102.8	150.5	213.7	301.4	5
$C_{12}H_{12}$	1-Ethylnaphthalene	16 e	41 e	72 e	114 e	171.8	257.7	5
$C_{12}H_{12}$	2-Ethylnaphthalene	14 e	39 e	71 e	113 e	171.2	257.3	5
$C_{12}H_{12}$	1,2-Dimethylnaphthalene	26 e	51 e	82 e	123 e	180.5	265.7	5
$C_{12}H_{12}$	2,7-Dimethylnaphthalene	31.5 s	53.1 s	78.8 s	115.9	175 e	260 e	5
$C_{12}H_{14}O_4$	Diethyl phthalate	12 e	51 e	96 e	150.5	215.9	296.2	5
$C_{12}H_{16}$	*p*-Isopropenylisopropylbenzene	−11 e	15 e	46 e	87 e	142.4	221 e	5
$C_{12}H_{16}$	Cyclohexylbenzene		28 e	58 e	98 e	154.7	239.5	5
$C_{12}H_{16}O_2$	3-Methylbutyl benzoate			66 e	115.0	177.7	261.4	5
$C_{12}H_{18}$	Hexylbenzene	−2 e	22 e	51 e	90 e	144.5	225.5	5
$C_{12}H_{18}$	1,2-Diisopropylbenzene	−14 e	9 e	37 e	74 e	125.9	203.2	5
$C_{12}H_{18}$	1,3-Diisopropylbenzene	−14 e	8 e	36 e	74 e	125.5	202.6	5
$C_{12}H_{18}$	1,4-Diisopropylbenzene	−6 e	18 e	49 e	90 e	148.8	238 e	5
$C_{12}H_{18}$	Hexamethylbenzene	46.3 s	72.5 s	81.7 s	121.8 s	178.3	263.7	5
$C_{12}H_{18}$	1,5,9-Cyclododecatriene	−14 e	11 e	44 e	87 e	145.0	229.8	5
$C_{12}H_{20}O_2$	Geranyl acetate			67.7	110.8	166.9	242.9	5
$C_{12}H_{20}O_4$	Dibutyl maleate	12.3	50.4	94.0	144.2	203 e	272 e	5
$C_{12}H_{22}$	1-Dodecyne	−11 e	13 e	43 e	82 e	135.8	214.4	5
$C_{12}H_{22}$	Cyclohexylcyclohexane		20 e	53.1	96.0	154.1	237.2	5

Mol. form.	Name	Temperature in °C for the indicated pressure						Ref.
		1 Pa	10 Pa	100 Pa	1 kPa	10 kPa	100 kPa	
$C_{12}H_{22}O_2$	Methyl 10-undecenoate	10 e	38 e	73 e	116 e	172.2	247.1	5
$C_{12}H_{22}O_4$	Dimethyl sebacate		53 e	97	150	214	293	4
$C_{12}H_{23}N$	Dodecanenitrile	36 e	60 e	92 e	133 e	190.5	275.5	5
$C_{12}H_{24}$	1-Dodecene	−8.3	15.2	44.8	82.9	135.4	212.8	5
$C_{12}H_{24}$	Hexylcyclohexane	−3 e	20 e	50 e	89 e	143.1	224.2	5
$C_{12}H_{24}$	Heptylcyclopentane	−1 e	22 e	51 e	90 e	143.5	223.5	5
$C_{12}H_{24}O$	Dodecanal			70 e	116.2	175.9	256.6	5
$C_{12}H_{24}O_2$	Dodecanoic acid	78 e	100 e	128 e	166 e	219.1	298.1	5
$C_{12}H_{24}O_2$	Decyl acetate	12 e	40 e	74 e	115.1	168.1	238 e	5
$C_{12}H_{24}O_2$	Ethyl decanoate	8 e	35 e	69 e	111.8	166.1	238 e	5
$C_{12}H_{25}Br$	1-Bromododecane	31 e	57 e	90 e	132 e	190.8	275.3	5
$C_{12}H_{25}Cl$	1-Chlorododecane	27 e	51 e	81 e	122 e	178.7	262.6	5
$C_{12}H_{26}$	Dodecane	−5.4	18.2	47.6	85.8	138.2	215.8	16
$C_{12}H_{26}O$	1-Dodecanol				133 e	185.0	264.1	1
$C_{12}H_{26}O_3$	Diethylene glycol dibutyl ether	5 e	34.4	70.2	115.3	174.1	253.8	5
$C_{12}H_{27}N$	Tributylamine	−26 e	1 e	35 e	77.7	134.5	213.4	5
$C_{12}H_{27}N$	Triisobutylamine		1 e	28.9	64.9	112.5	178.5	5
$C_{12}H_{27}O_4P$	Tributyl phosphate					205 e	288.3	5
$C_{12}H_{36}O_6Si_6$	Dodecamethylcyclohexasiloxane	18 e	41 e	69 e	108 e	162.2	244.7	5
$C_{13}H_9N$	Acridine			124.4	176.2	246.0	345.4	5
$C_{13}H_9N$	Phenanthridine	79 s						5
$C_{13}H_{10}$	Fluorene	48.4 s			137.4	205.4	295 e	5
$C_{13}H_{10}O_2$	Phenyl benzoate			102.3	151.4	217.9	313.3	5
$C_{13}H_{10}O_3$	Phenyl salicylate				166.0	224.8	312.4	5
$C_{13}H_{12}$	Diphenylmethane		45 e	77 e	119.3	177.7	263.6	1,5
$C_{13}H_{13}N$	Methyldiphenylamine	35 e	63 e	98.4	143.1	201.6	281.6	5
$C_{13}H_{14}$	1-Isopropylnaphthalene	27 e	51 e	82 e	123.2	180.8	267.3	5
$C_{13}H_{20}$	Heptylbenzene	12 e	36 e	66 e	107 e	162.7	246.2	5
$C_{13}H_{24}O_2$	Ethyl 10-undecenoate	32 e	55 e	86 e	125.2	179.5	258.4	5
$C_{13}H_{26}$	1-Tridecene	4.1	28.5	59.0	98.3	152.5	232.3	5
$C_{13}H_{26}$	Heptylcyclohexane	11 e	34 e	65 e	105 e	160.9	244.3	5
$C_{13}H_{26}$	Octylcyclopentane	13 e	36 e	66 e	106 e	160.9	243.1	5
$C_{13}H_{26}O_2$	Tridecanoic acid	87 e	109 e	138 e	176 e	230.3	311.5	5
$C_{13}H_{26}O_2$	Methyl dodecanoate	38 e	61 e	90 e	130 e	184.9	269 e	5
$C_{13}H_{28}$	Tridecane	7.2	31.5	61.8	101.1	155.1	234.9	16
$C_{13}H_{28}O$	1-Tridecanol	71.6	101.0	103 e	140 e	192.3	273.1	5
$C_{14}H_{10}$	Anthracene	89.2 s	125.9 s	151.5 s	165 s	238.8	340.2	1,5
$C_{14}H_{10}$	Phenanthrene	53 s	83 s	120.8	170.4	238.4	337.7	5
$C_{14}H_{10}O_2$	Benzil			123	175	246	346	4
$C_{14}H_{12}$	cis-Stilbene	26 e	54 e	88 e	130.4	183 e	253 e	5
$C_{14}H_{12}$	trans-Stilbene				155.6	218.1	305.8	5
$C_{14}H_{12}O_2$	Benzoin				181	248	342	4
$C_{14}H_{14}$	1,1-Diphenylethane	19 e	47 e	82.0	125.3	181 e	254 e	5
$C_{14}H_{15}N$	Dibenzylamine	48 e	77 e	113.1	158.9	218.5	299.4	5
$C_{14}H_{16}$	1-Butylnaphthalene	67 e	82 e	103 e	135 e	186.7	288.6	5
$C_{14}H_{16}$	2-Butylnaphthalene	44 e	67 e	98 e	139 e	197.5	287.4	5
$C_{14}H_{22}$	Octylbenzene	20.1	46.2	79.1	121.9	178.1	263.8	5
$C_{14}H_{26}O_4$	Diethyl sebacate		83 e	120	166	225	305	4
$C_{14}H_{27}N$	Tetradecanenitrile	52 e	79 e	114.0	159.0	219.7	306.3	5
$C_{14}H_{28}$	1-Tetradecene	16.1	41.3	72.7	113.2	168.7	250.6	5
$C_{14}H_{28}$	Octylcyclohexane	16.9	44.3	77.8	120.0	177.6	263.2	5
$C_{14}H_{28}$	Nonylcyclopentane	25 e	49 e	80 e	120 e	177.2	261.5	5
$C_{14}H_{28}O_2$	Tetradecanoic acid	96 e	118 e	147 e	186 e	241.3	325.6	5
$C_{14}H_{30}$	Tetradecane	19.1	44.1	75.3	115.7	171.1	253.0	16
$C_{14}H_{30}O$	1-Tetradecanol	80.0	110.5	149.6	152 e	205.3	286.7	5
$C_{14}H_{31}N$	Tetradecylamine			104 e	147 e	206.1	290.9	5
$C_{14}H_{42}O_5Si_6$	Tetradecamethylhexasiloxane	6 e	36 e	72 e	117 e	176.0	259.1	5
$C_{15}H_{18}$	1-Pentylnaphthalene	34 e	62 e	96 e	141.3	202.2	289 e	5
$C_{15}H_{24}$	Nonylbenzene	33.0	58.9	92.0	135.4	193.7	281.4	5
$C_{15}H_{30}$	Nonylcyclohexane	35 e	60 e	92 e	134 e	193.4	280.9	5

Mol. form.	Name	Temperature in °C for the indicated pressure						Ref.
		1 Pa	10 Pa	100 Pa	1 kPa	10 kPa	100 kPa	
$C_{15}H_{30}$	Decylcyclopentane	37 e	61 e	93 e	134 e	192.5	278.8	5
$C_{15}H_{30}O_2$	Methyl tetradecanoate		75 e	110	155	214	295	4
$C_{15}H_{32}$	Pentadecane	30.5	56.1	88.1	129.6	186.3	270.1	16
$C_{16}H_{22}O_4$	Dibutyl phthalate		104.0	142.7	191.5	254.5	339.4	4
$C_{16}H_{32}$	1-Hexadecene	38.4	65.0	98.1	140.5	198.8	284.3	5
$C_{16}H_{32}O_2$	Hexadecanoic acid		136 e	165 e	205 e	261.9	350.2	5
$C_{16}H_{34}$	Hexadecane	41.1	67.4	100.3	142.7	200.7	286.3	16
$C_{16}H_{34}O$	1-Hexadecanol	99.5	130.6	171.9	175 e	229.0	311.7	5
$C_{16}H_{35}N$	Hexadecylamine	63 e	91 e	126 e	171 e	232.6	320.5	5
$C_{17}H_{10}O$	Benzanthrone		184 e	229.3	290.3	377.2	511 e	5
$C_{17}H_{34}O_2$	Methyl hexadecanoate	65 e	93	129	177			4
$C_{17}H_{36}$	Heptadecane	51.5	78.5	112.0	155.3	214.5	302 e	16
$C_{17}H_{36}O$	1-Heptadecanol	94 e	117 e	146 e	185 e	240.1	323.3	5
$C_{18}H_{14}$	o-Terphenyl	66 e	94 e	129 e	176 e	241.3	336.3	5
$C_{18}H_{14}$	m-Terphenyl	87 e	118 e	156 e	206.6	275.3	374.6	5
$C_{18}H_{14}$	p-Terphenyl	127.1 s	154.7 s		217.2	284.0	383.0	5
$C_{18}H_{30}$	Hexaethylbenzene				144.1	206.8	297.5	5
$C_{18}H_{34}O_2$	Oleic acid	94 e	126 e	165.5	214.5	277.0	359.7	5
$C_{18}H_{34}O_2$	Elaidic acid		124 e	166	216	280	361	4
$C_{18}H_{36}O$	Stearaldehyde			142 e	186 e	246.9	336.7	5
$C_{18}H_{36}O_2$	Stearic acid		153 e	183 e	223 e	281.6	374.5	5
$C_{18}H_{38}$	Octadecane	61.5	89.0	123.1	167.3	227.6	316 e	16
$C_{18}H_{38}O$	1-Octadecanol	106 e	130 e	160 e	200.5	257.3	343.0	5
$C_{19}H_{16}$	Triphenylmethane	81 s		112 e	175 e	254.6	360.0	5
$C_{19}H_{36}O_2$	Methyl oleate	85 e	114 e	149.7	195.6	256 e	340 e	5
$C_{19}H_{40}$	Nonadecane	71.1	99.1	133.8	178.8	240.1	330 e	16
$C_{20}H_{42}$	Eicosane	80.4	108.9	144.2	189.8	252.1	344 e	16
$C_{20}H_{42}O$	1-Eicosanol	119 e	143 e	173 e	213 e	270.0	355.1	5
$C_{20}H_{60}O_8Si_9$	Eicosamethylnonasiloxane			141 e	183.1	236.7	307.1	5
$C_{21}H_{21}O_4P$	Tri-o-cresyl phosphate	119.0	156.1	201.0	256.3	326.3	418 e	5
$C_{21}H_{21}O_4P$	Tri-m-cresyl phosphate	147.8	177.3	211.4	251.3	298 e	355 e	5
$C_{21}H_{21}O_4P$	Tri-p-cresyl phosphate	140.6	174 e	214 e	262 e	320 e	392 e	5
$C_{21}H_{44}$	Heneicosane	82.3	113.5	152.2	201.6	263.8	355.9	5
$C_{22}H_{42}O_2$	Brassidic acid	134 e	166 e	203.6	249.8	307.6	382.0	5
$C_{22}H_{42}O_2$	Erucic acid	126 e	160 e	199.4	247.4	306.5	381.1	5
$C_{22}H_{42}O_2$	Butyl oleate	95.5	124.2	158 e	198 e	245 e	304 e	5
$C_{22}H_{44}O_2$	Behenic acid	145.4	176.5	213.7	259.3	316.2	390 e	5
$C_{22}H_{44}O_2$	Butyl stearate	99.6	128 e	162 e	201 e	249 e	307 e	5
$C_{22}H_{46}$	Docosane	83.5	115.0	154.0	203.6	274.8	368.0	5
$C_{23}H_{48}$	Tricosane	102.9	135.1	174.8	221 e	285.3	379.5	5
$C_{24}H_{38}O_4$	Dioctyl phthalate	130 e	163.7	203.8	252 e	311 e	385 e	5
$C_{24}H_{38}O_4$	Bis(2-ethylhexyl) phthalate	122.0	153.2	189.2	231.3	281.1	341.1	5
$C_{24}H_{50}$	Tetracosane	115.0	148.1	188.5	239.1	295.4	390.6	5
$C_{25}H_{52}$	Pentacosane	119.7	152.7	193.2	244.4	305.0	401.1	5
$C_{26}H_{54}$	Hexacosane	125.1	158.8	200.1	252.1	314.3	411.3	5
$C_{27}H_{56}$	Heptacosane	136.7	168.8	206.5	255.8	323.3	421.2	5
$C_{28}H_{58}$	Octacosane	136.5	169.8	210.9	263.1	332.0	430.6	5
$C_{29}H_{60}$	Nonacosane	148.2	182.8	221.2	271.5	340.2	439.7	5
$C_{30}H_{62}$	Squalane	66 e	84 e	105.8	131.9	163.7	203.2	5
C_{70}	Carbon (fullerene-C_{70})	598 s	662 s					22

VAPOR PRESSURE OF FLUIDS AT TEMPERATURES BELOW 300 K

This table gives vapor pressures of 67 important fluids in the temperature range 2 to 300 K. Helium (^{4}He), hydrogen (H$_2$), and neon (Ne) are covered on this page. The remaining fluids are listed on subsequent pages by molecular formula in the Hill order (see Introduction). The data have been taken from evaluated sources; references are listed at the end of the table.

Pressures are given in kilopascals (kPa). Note that:
- 1 kPa = 7.50062 Torr
- 100 kPa = 1 bar
- 101.325 kPa = 1 atm

"s" following an entry indicates that the compound is solid at that temperature.

Helium		Hydrogen		Neon			Helium		Hydrogen		Neon	
T/K	P/kPa	T/K	P/kPa	T/K	P/kPa		T/K	P/kPa	T/K	P/kPa	T/K	P/kPa
2.2	5.3	14.0	7.90	25.0	51.3		4.2	99.0	24.0	264.2		
2.3	6.7	14.5	10.38	26.0	71.8		4.3	108.7	24.5	295.1		
2.4	8.3	15.0	13.43	27.0	98.5		4.4	119.0	25.0	328.5		
2.5	10.2	15.5	17.12	28.0	132.1		4.5	129.9	25.5	364.3		
2.6	12.4	16.0	21.53	29.0	173.5		4.6	141.6	26.0	402.9		
2.7	14.8	16.5	26.74	30.0	223.8		4.7	153.9	26.5	444.3		
2.8	17.5	17.0	32.84	31.0	284.0		4.8	167.0	27.0	488.5		
2.9	20.6	17.5	39.92	32.0	355.2		4.9	180.8	27.5	535.7		
3.0	24.0	18.0	48.08	33.0	438.6		5.0	195.4	28.0	586.1		
3.1	27.8	18.5	57.39	34.0	535.2		5.1	210.9	28.5	639.7		
3.2	32.0	19.0	67.96	35.0	646.2				29.0	696.7		
3.3	36.5	19.5	79.89	36.0	772.8				29.5	757.3		
3.4	41.5	20.0	93.26	37.0	916.4				30.0	821.4		
3.5	47.0	20.5	108.2	38.0	1078				30.5	889.5		
3.6	52.9	21.0	124.7	39.0	1260				31.0	961.5		
3.7	59.3	21.5	143.1	40.0	1462				31.5	1038.0		
3.8	66.1	22.0	163.2	41.0	1688				32.0	1119.0		
3.9	73.5	22.5	185.3	42.0	1939				32.5	1204.0		
4.0	81.5	23.0	209.4	43.0	2216		Ref.	17,18		1		13
4.1	90.0	23.5	235.7	44.0	2522							

T/K	Ar Argon		BCl$_3$ Boron trichloride	BF$_3$ Boron trifluoride	BrH Hydrogen bromide		Br$_2$ Bromine	ClF Chlorine fluoride	ClH Hydrogen chloride	
50	0.1	s								
55	0.2	s								
60	0.8	s								
65	2.8	s								
70	7.7	s								
75	18.7	s								
80	40.7	s								
85	79.0									
90	134									
95	213									
100	324									
105	473									
110	666									
115	910							0.1		
120	1214							0.3	0.1	s
125	1584							0.6	0.3	s
130	2027							1.2	0.5	s
135	2553				0.1	s		2.1	1.0	s
140	3170				0.3	s		3.6	1.9	s
145	3892			7.7	0.6	s		6.0	3.4	s
150	4736			13.4	1.1	s		9.5	5.8	s
155				22.3	1.9	s		14.6	9.5	s
160				35.2	3.3	s		21.8	14.7	
165				53.7	5.4	s		31.7	22.0	
170				79.1	8.7	s		44.8	31.9	
175				113	13.4	s		62.0	45.1	

T/K	Ar Argon	BCl₃ Boron trichloride	BF₃ Boron trifluoride	BrH Hydrogen bromide	Br₂ Bromine	ClF Chlorine fluoride	ClH Hydrogen chloride
180		0.1	157	20.1 s		84.2	62.5
185		0.2	214	29.5 s		112	84.7
190		0.3	285	37.9		147	113
195		0.5	372	51.8		190	148
200		0.8	479	69.5		242	190
205		1.2	608	91.8		304	242
210		1.8	762	119		378	304
215		2.6	944	153		464	377
220		3.8	1160	194	0.1 s	564	463
225		5.2	1413	242	0.2 s	680	563
230		7.2	1709	299	0.3 s	812	678
235		9.7	2056	366	0.4 s	961	811
240		12.9	2460	443	0.7 s	1130	961
245		17.0	2913	532	1.1 s	1319	1132
250		22.0	3481	633	1.7 s	1529	1325
255		28.1	4123	748	2.6 s	1762	1542
260		35.6	4874	878	3.8 s	2019	1784
265		44.5		1023	5.5 s	2301	2054
270		55.1		1185	7.3	2608	2354
275		67.6		1364	9.5	2941	2686
280		82.2		1562	12.3	3303	3053
285		99.1		1780	15.6	3693	3457
290		119		2018	19.7	4111	3901
295		141		2278	24.6	4560	4388
300		166		2561	30.5	5039	4921
Ref.	8,15	12	12	12	12	12	12

T/K	ClO₂ Chlorine dioxide	Cl₂ Chlorine	Cl₄Si Silicon tetrachloride	FH Hydrogen fluoride	F₂ Fluorine	F₂O Difluorine oxide	F₃N Nitrogen trifluoride
50							
55					0.4		
60					1.5		
65					4.8		
70					12.3		
75					27.6	0.1	
80					55.3	0.2	
85					101	0.5	0.1
90					172	1.2	0.2
95					276	2.6	0.4
100					420	5.3	0.9
105					615	10.1	2.0
110					870	18.0	4.0
115					1196	30.5	7.3
120					1605	49.3	12.8
125					2108	76.7	21.1
130					2721	115	33.5
135					3458	168	51.1
140					4339	237	75.4
145						328	108
150						444	150
155						588	205
160						766	273
165						981	357
170						1238	459
175		1.8				1541	581
180		2.8				1895	726
185		4.2				2303	896
190		6.1		0.3		2771	1092
195	0.1	8.7		0.5		3302	1319
200	0.3	12.3		0.8		3899	1578
205	0.5	16.9		1.2		4567	1871
210	0.9	22.9	0.1	1.7		5308	2203

Vapor Pressure of Fluids at Temperatures below 300 K

T/K	ClO$_2$ Chlorine dioxide	Cl$_2$ Chlorine	Cl$_4$Si Silicon tetrachloride	FH Hydrogen fluoride	F$_2$ Fluorine	F$_2$O Difluorine oxide	F$_3$N Nitrogen trifluoride
215	1.4	30.5	0.2	2.3			2577
220	2.3	40.1	0.3	3.2			2995
225	3.5	51.9	0.5	4.4			3464
230	5.3	66.4	0.7	5.9			3991
235	7.6	84.0	1.0	7.9			
240	10.8	105	1.5	10.3			
245	14.9	130	2.0	13.4			
250	20.1	160	2.8	17.2			
255	26.6	194	3.8	21.8			
260	34.6	234	5.0	27.4			
265	44.4	280	6.6	34.2			
270	56.1	332	8.6	42.2			
275	69.9	392	11.1	51.8			
280	86.2	459	14.2	63.1			
285	105	535	17.9	76.3			
290	127	619	22.3	91.7			
295	151	714	27.7	110			
300	179	818	34.0	130			
Ref.	12	5	12	12	12	12	1

T/K	F$_3$P Phosphorus trifluoride	F$_4$Si Silicon tetrafluoride		F$_6$S Sulfur hexafluoride		HI Hydrogen iodide		H$_2$S Hydrogen sulfide		H$_3$N Ammonia		H$_3$P Phosphine
105	0.1											
110	0.2											0.1
115	0.5											0.2
120	1.0											0.4
125	1.9	0.1	s									0.7
130	3.5	0.2	s									1.3
135	5.9	0.4	s					0.1	s			2.3
140	9.5	0.9	s	0.1	s			0.2	s			3.9
145	14.9	1.9	s	0.2	s			0.3	s			6.2
150	22.5	3.8	s	0.4	s			0.6	s			9.6
155	33.1	7.5	s	0.8	s	0.1	s	1.1	s			14.5
160	47.3	14.0	s	1.5	s	0.2	s	1.9	s	0.1	s	21.1
165	66.0	25.2	s	2.6	s	0.4	s	3.2	s	0.2	s	30.0
170	90.1	43.8	s	4.4	s	0.8	s	5.2	s	0.3	s	41.6
175	121	74.2	s	7.1	s	1.3	s	8.3	s	0.6	s	56.6
180	159	122	s	11.3	s	2.2	s	12.7	s	1.2	s	75.6
185	206	197	s	17.3	s	3.4	s	18.9	s	2.1	s	99.2
190	262	280		25.9	s	5.3	s	26.6		3.5	s	128
195	330	376		38.0	s	8.0	s	36.7		5.8	s	163
200	410	488		54.4	s	11.7	s	49.8		8.7		205
205	503	618		76.6	s	16.8	s	66.4		12.6		254
210	611	766		106	s	23.6	s	87.1		17.9		312
215	736	932		145	s	32.5	s	113		24.9		379
220	877	1117		195	s	44.0	s	144		34.1		456
225	1037	1324		249		56.2		182		45.9		544
230	1217	1555		305		71.4		227		60.8		644
235	1418	1816		371		89.7		281		79.6		756
240	1640	2111		448		112		344		103		881
245	1885	2449		536		137		416		131		1019
250	2154	2841		636		168		500		165		1172
255	2448	3301		750		203		597		207		1341
260	2767			878		244		706		256		1525
265	3112			1021		290		830		313		1725
270				1181		343		969		381		1942
275				1358		404		1124		460		2176
280				1554		472		1297		552		2428
285				1768		548		1488		655		2699
290				2003		633		1698		774		2987
295				2258		727		1929		909		3295
300				2534		831		2181		1062		3621
Ref.	12	12		12,15		12		12,15		11		12

T/K	H₄Si Silane	Kr Krypton	NO Nitric oxide	N₂ Nitrogen	N₂O Nitrous oxide	O₂ Oxygen	O₂S Sulfur dioxide
50				0.4 s			
55				1.8 s		0.2	
60				6.3 s		0.7	
65				17.4		2.3	
70				38.6		6.3	
75		0.1 s		76.1		14.5	
80		0.4 s		137		30.1	
85		1.1 s	0.1 s	229		56.8	
90		2.7 s	0.4 s	361		99.3	
95	0.1	6.0 s	1.3 s	541		163	
100	0.2	12.1 s	3.8 s	779		254	
105	0.4	22.8 s	10.0 s	1084		379	
110	1.0	40.4 s	23.5	1467		543	
115	1.9	68.0 s	46.8	1939	0.1	756	
120	3.5	103	86.5	2513	0.1	1022	
125	6.1	150	151	3209	0.3	1351	
130	10.0	211	248		0.7	1749	
135	15.8	290	391		1.3	2225	
140	24.1	388	592		2.5	2788	
145	35.3	509	867		4.3	3448	
150	50.3	655	1231		7.1	4219	
155	69.8	830	1703		11.4		
160	94.6	1037	2302		17.6		
165	126	1278	3050		26.4		
170	164	1557	3971		38.5		0.1
175	210	1877	5089		54.7		0.2
180	265	2241	6433		75.9		0.3
185	331	2655			103		0.5
190	408	3120			138		0.8
195	498	3641			181		1.3
200	602	4223			234		2.0
205	722	4870			298		3.0
210	859				374		4.4
215	1017				465		6.3
220	1196				571		9.0
225	1398				694		12.6
230	1628				835		17.3
235	1888				996		23.3
240	2180				1179		31.1
245	2509				1385		40.9
250	2880				1615		53.2
255	3296				1870		68.3
260	3763				2152		86.7
265	4288				2462		109
270					2802		136
275					3172		168
280					3573		205
285					4006		249
290					4473		300
295					4973		359
300					5508		426
Ref.	12	13, 15	12, 15	1	12	3	12

T/K	O₃ Ozone	Rn Radon	Xe Xenon	CBrF₃ Bromotri-fluoromethane	CClF₃ Chlorotri-fluoromethane	CCl₂F₂ Dichlorodi-fluoromethane	CCl₃F Trichloro-fluoromethane
100	0.1		0.1 s				
105	0.2		0.1 s				
110	0.4		0.3 s				
115	1.0		0.7 s		0.1		
120	2.0		1.5 s		0.2		
125	3.8		2.7 s		0.3		
130	6.8	0.1	4.9 s		0.6		

T/K	O₃ Ozone	Rn Radon	Xe Xenon		CBrF₃ Bromotrifluoromethane	CClF₃ Chlorotrifluoromethane	CCl₂F₂ Dichlorodifluoromethane	CCl₃F Trichlorofluoromethane
135	11.5	0.3	8.5	s	0.1	1.1		
140	18.7	0.5	14.0	s	0.3	2.0		
145	29.1	0.9	22.2	s	0.5	3.3		
150	43.7	1.5	34.2	s	0.9	5.3		
155	63.6	2.4	51.1	s	1.5	8.3	0.1	
160	89.9	3.8	74.2	s	2.5	12.6	0.3	
165	124	5.8	101		3.9	18.6	0.5	
170	168	8.6	134		5.9	26.8	0.8	
175	222	12.5	173		8.8	37.6	1.3	
180	289	17.7	222		12.8	51.7	2.1	
185	367	24.5	280		18.1	69.7	3.2	
190	468	33.2	348		25.1	92.3	4.8	0.2
195	584	44.4	428		34.1	120	6.9	0.3
200	721	58.2	521		45.6	155	9.9	0.4
205	881	75.3	628		60.0	196	13.7	0.6
210	1068	96	750		77.8	246	18.8	1.0
215	1285	121	889		99.5	304	25.2	1.4
220	1536	151	1045		126	372	33.3	2.0
225	1824	185	1220		157	451	43.3	2.9
230	2155		1416		194	542	55.5	4.1
235	2534		1633		237	646	70.4	5.6
240	2968		1872		287	763	88.1	7.6
245	3464		2136		344	896	109	10.1
250	4031		2425		410	1044	134	13.3
255	4678		2742		485	1210	163	17.2
260	5417		3087		570	1394	196	22.1
265			3462		665	1598	234	28.0
270			3869		771	1823	278	35.1
275			4310		889	2071	327	43.7
280			4786		1021	2343	383	53.8
285			5299		1166	2641	445	65.7
290					1325	2968	515	79.6
295					1501	3325	593	95.6
300					1692	3716	679	114.1
Ref.	12	15	12,13		12	12	12	12

T/K	CCl₄ Tetrachloromethane	CF₄ Tetrafluoromethane	CO Carbon monoxide		COS Carbon oxysulfide	CO₂ Carbon dioxide		CHClF₂ Chlorodifluoromethane	CHCl₃ Trichloromethane
50			0.1	s					
55			0.6	s					
60			2.6	s					
65			8.2	s					
70			21.0						
75			44.4						
80			83.7						
85			147						
90		0.1	239						
95		0.3	371						
100		0.8	545						
105		1.7	771						
110		3.4	1067						
115		6.5	1428						
120		11.5	1877						
125		19.3	2400						
130		30.8	3064						
135		47.4				0.1	s		
140		70.2			0.1	0.2	s		
145		101			0.2	0.4	s		
150		141			0.4	0.8	s	0.1	
155		191			0.8	1.7	s	0.3	

T/K	CCl$_4$ Tetrachloro-methane	CF$_4$ Tetrafluoro-methane	CO Carbon monoxide	COS Carbon oxysulfide	CO$_2$ Carbon dioxide		CHClF$_2$ Chlorodifluoro-methane	CHCl$_3$ Trichloro-methane
160		254		1.3	3.1	s	0.5	
165		332		2.2	5.7	s	0.8	
170		425		3.4	9.9	s	1.4	
175		537		5.2	16.8	s	2.3	
180		669		7.8	27.6	s	3.6	
185		824		11.3	44.0	s	5.5	
190		1005		15.9	68.4	s	8.1	
195		1216		22.1	104	s	11.8	
200		1460		30.0	155	s	16.7	
205		1743		40.1	227	s	23.1	
210		2073		52.7	327	s	31.5	
215		2457		68.2	465	s	42.1	0.1
220		2907		87.2	599		55.3	0.2
225		3438		110	734		71.7	0.3
230				137	893		91.6	0.4
235				169	1075		116	0.7
240				207	1283		144	1.0
245				250	1519		178	1.4
250				301	1785		218	2.0
255	1.5			358	2085		264	2.7
260	2.1			423	2419		317	3.7
265	2.8			497	2790		377	5.0
270	3.7			580	3203		446	6.6
275	4.9			673	3658		525	8.7
280	6.4			777	4161		613	11.3
285	8.2			892	4714		711	14.4
290	10.5			1019	5318		821	18.3
295	13.2			1159	5984		944	22.9
300	16.5			1313	6713		1080	28.5
Ref.	12	12	9	12	6, 19		12	12

T/K	CHF$_3$ Trifluoro-methane	CHN Hydrogen cyanide		CH$_2$Cl$_2$ Dichloro-methane	CH$_2$F$_2$ Difluoro-methane	CH$_2$O Formaldehyde	CH$_3$Cl Chloromethane	CH$_3$F Fluoromethane
120	0.1							
125	0.2							
130	0.4							
135	0.7							0.6
140	1.4				0.1			1.2
145	2.5				0.2			2.1
150	4.3				0.3			3.6
155	7.1				0.6			5.9
160	11.1				1.0			9.3
165	17.0				1.7			14.1
170	25.3				2.8			20.9
175	36.5				4.4			29.9
180	51.4				6.8			42.0
185	70.9				10.2	1.3	2.1	57.6
190	95.8				14.8	2.0	3.1	77.4
195	127				21.2	3.0	4.6	102
200	166	0.1	s	0.1	29.5	4.4	6.7	133
205	214	0.2	s	0.2	40.5	6.4	9.5	171
210	271	0.4	s	0.3	54.5	9.1	13.1	216
215	340	0.6	s	0.4	72.1	12.7	17.9	270
220	421	1	s	0.6	94.1	17.4	24.0	333
225	516	1.5	s	0.9	121	23.4	31.8	408
230	626	2.2	s	1.4	154	31.0	41.4	495
235	754	3.3	s	2.0	193	40.6	53.3	595
240	900	4.7	s	2.8	240	52.5	67.7	711
245	1067	6.8	s	3.8	295	67.0	85.1	843
250	1257	9.7	s	5.3	360	84.6	106	993
255	1472	13.6	s	7.1	434	106	131	1163
260	1713	18.8		9.5	521	131	159	1355

T/K	CHF$_3$ Trifluoro-methane	CHN Hydrogen cyanide	CH$_2$Cl$_2$ Dichloro-methane	CH$_2$F$_2$ Difluoro-methane	CH$_2$O Formaldehyde	CH$_3$Cl Chloromethane	CH$_3$F Fluoromethane
265	1984	24.1	12.4	620	161	193	1571
270	2287	30.5	16.1	732	196	232	1813
275	2624	38.3	20.7	860	236	277	2084
280	3000	47.7	26.3	1004	283	327	2387
285	3418	58.8	33.0	1165	337	385	2724
290	3881	72.1	41.1	1346	399	450	3099
295	4393	87.6	50.8	1547	470	524	3516
300		105.9	62.1	1770	549	606	3978
Ref.	12	12,16	12	12	12	12	12

T/K	CH$_4$ Methane	CH$_4$O Methanol	C$_2$H$_2$ Acetylene	C$_2$H$_4$ Ethylene	C$_2$H$_6$ Ethane	C$_2$H$_6$O Dimethyl ether	C$_3$H$_4$ Propadiene
65	0.1						
70	0.3						
75	0.8						
80	2.1						
85	4.9						
90	10.6						
95	20.0						
100	34.5						
105	57.0						
110	88.4			0.3			
115	133			0.8	0.1		
120	192			1.4	0.4		
125	269			2.7	0.7		
130	368		0.1 s	4.5	1.3		
135	491		0.3 s	7.7	2.2		
140	642		0.7 s	11.9	3.8		
145	824		1.3 s	18.3	6.0		
150	1041		2.6 s	27.5	9.7		0.1
155	1297		4.6 s	39.9	15.0	0.1	0.2
160	1594		7.8 s	56.4	21.5	0.2	0.3
165	1937		12.8 s	77.9	31.0	0.3	0.6
170	2331		20.6 s	105	42.9	0.5	1.0
175	2779		32.2 s	140	59.0	0.9	1.7
180	3288		49.0 s	182	78.7	1.4	2.7
185	3865		72.9 s	234	104	2.1	4.1
190	4520		106 s	296	135	3.2	6.1
195			146	369	172	4.7	8.9
200			190	456	217	6.8	12.5
205			244	557	271	9.6	17.4
210			309	673	334	13.3	23.7
215			385	806	407	18.1	31.6
220			475	958	492	24.3	41.4
225			579	1128	590	32.1	53.5
230		0.1	699	1321	700	41.9	68.2
235		0.2	837	1535	826	53.9	85.8
240		0.4	993	1774	967	68.6	107
245		0.5	1170	2039	1125	86.3	131
250		0.8	1370	2331	1301	108	160
255		1.2	1593	2652	1496	133	193
260		1.7	1843	3005	1712	162	230
265		2.4	2121	3391	1949	197	273
270		3.3	2429	3813	2210	237	322
275		4.5	2771	4275	2495	283	376
280		6.2	3150		2806	335	438
285		8.3	3567		3146	395	506
290		11	4028		3515	463	582
295		14.4	4535		3917	538	666
300		18.7	5093		4355	623	759
Ref.	2,16	12	12,16	4	2	12	12

T/K	C_3H_6 Propylene	C_3H_8 Propane	C_4H_6 Buta-1,3-diene	C_4H_{10} Butane	C_4H_{10} Isobutane	C_5H_{12} Pentane	C_5H_{12} Neopentane	
140	0.1							
145	0.2							
150	0.4							
155	0.7							
160	1.2	0.8			0.1			
165	2.0	1.4			0.1			
170	3.1	2.2	0.1	0.1	0.3			
175	4.7	3.3	0.2	0.2	0.4			
180	7.0	5.0	0.4	0.3	0.7			
185	10.1	7.3	0.6	0.5	1.1		0.1	s
190	14.2	10.5	1.0	0.8	1.7		0.2	s
195	19.7	15.0	1.5	1.3	2.5		0.4	s
200	26.9	20.1	2.3	1.9	3.7		0.7	s
205	35.9	27.0	3.4	2.8	5.3		1.1	s
210	47.3	36.0	4.8	4.0	7.4		1.6	s
215	61.3	47.0	6.7	5.7	10.2		2.4	s
220	78.5	60.0	9.2	7.8	13.8	1.0	3.6	s
225	99.2	77.0	12.5	10.6	18.3	1.5	5.2	s
230	124	97.0	16.7	14.1	24.0	2.1	7.3	s
235	153	120	21.9	18.5	31.1	3.0	10.2	s
240	188	148	28.4	24.1	39.8	4.2	13.9	s
245	228	180	36.3	30.9	50.3	5.7	18.7	s
250	274	218	46.0	39.1	62.9	7.6	24.8	s
255	327	261	57.6	49.1	77.8	10.0	32.4	s
260	387	311	71.3	61.0	95.4	13.0	41.6	
265	456	367	87.6	75.0	116	16.6	51.4	
270	533	431	107	91.5	140	21.1	63.0	
275	619	502	129	111	167	26.6	76.6	
280	715	582	154	133	198	33.1	92.3	
285	822	671	184	159	234	40.8	111	
290	940	769	217	188	274	50.0	131	
295	1069	878	255	221	319	60.7	155	
300	1212	998	297	258	370	73.2	182	
Ref.	7	2	12	2	2	14	12,16	

References

1. B. A. Younglove, Thermophysical properties of fluids. I. Ethylene, parahydrogen, nitrogen trifluoride, and oxygen, *J. Phys. Chem. Ref. Data*, 11, Supp. 1, 1982.
2. B. A. Younglove and J. F. Ely, Thermophysical properties of fluids. II. Methane, ethane, propane, isobutane, and normal butane, *J. Phys. Chem. Ref. Data*, 16, 577, 1987.
3. W. Wagner, et al., *International Tables for the Fluid State: Oxygen*, Blackwell Scientific Publications, Oxford, 1987.
4. R. T. Jacobsen, et al., *International Tables for the Fluid State: Ethylene*, Blackwell Scientific Publications, Oxford, 1988.
5. S. Angus, et al., *International Tables for the Fluid State: Chlorine*, Pergamon Press, Oxford, 1985.
6. S. Angus, et al., *International Tables for the Fluid State: Carbon Dioxide*, Pergamon Press, Oxford, 1976.
7. S. Angus, et al., *International Tables for the Fluid State: Propylene*, Pergamon Press, Oxford, 1980.
8. R. B. Stewart and R. T. Jacobsen, Thermophysical properties of argon, *J. Phys. Chem. Ref. Data*, 18, 639, 1989.
9. R. D. Goodwin, Carbon monoxide thermophysical properties, *J. Phys. Chem. Ref. Data*, 14, 849, 1985.
10. R. D. Goodwin, Methanol thermophysical properties, *J. Phys. Chem. Ref. Data*, 16, 799, 1987.
11. L. Haar, Thermodynamic properties of ammonia, *J. Phys. Chem. Ref. Data*, 7, 635, 1978.
12. DIPPR Data Compilation of Pure Compound Properties, Design Institute for Physical Properties Data, American Institute of Chemical Engineers, 1987.
13. V. A. Rabinovich, et al., *Thermophysical Properties of Neon, Argon, Krypton, and Xenon*, Hemisphere Publishing Corp., New York, 1987.
14. K. N. Marsh, *Recommended Reference Methods for the Realization of Physicochemical Properties*, Blackwell Scientific Publications, Oxford, 1987.
15. TRC Thermodynamic Tables: Non-Hydrocarbons, Thermodynamic Research Center, Texas A & M University, College Station, Texas, 1985.
16. R. M. Stevenson and S. Malanowski, *Handbook of the Thermodynamics of Organic Compounds*, Elsevier, New York, 1987.
17. S. Angus and K. M. de Reuck, *International Tables of the Fluid State: Helium-4*, Pergamon Press, Oxford, 1977.
18. R. D. McCarty, *J. Phys. Chem. Ref. Data*, 2, 923, 1973.
19. R. Span and W. Wagner, *J. Phys. Chem. Ref. Data*, 25, 1509, 1996.

VAPOR PRESSURE OF SATURATED SALT SOLUTIONS

This table gives the vapor pressure of water above saturated solutions of some common salts at ambient temperatures. Data on pure water are given on the last line for comparison.

The references provide additional information on water activity, osmotic coefficient, and enthalpy of vaporization.

References

1. Apelblat, A., *J. Chem. Thermodynamics*, 24, 619, 1992.
2. Apelblat, A., *J. Chem. Thermodynamics*, 25, 63, 1993.
3. Apelblat, A., *J. Chem. Thermodynamics*, 25, 1513, 1993.
4. Apelblat, A. and Korin, E., *J. Chem. Thermodynamics*, 30, 59, 1998.

Vapor Pressure in kPa

Salt	10 °C	15 °C	20 °C	25 °C	30 °C	35 °C	40 °C	Ref.
$BaCl_2$	0.971	1.443	2.073	2.887	3.903	5.133	6.576	1
$Ca(NO_3)_2$	0.701	1.015	1.381	1.772	2.154	2.487		1
$CuSO_4$	1.113	1.574	2.189	2.996	4.037	5.363		3
$FeSO_4$	0.978	1.516	2.208	3.035	3.950	4.884		3
KBr	0.953	1.338	1.853	2.533	3.419	4.563		3
KIO_3	1.100	1.564	2.177	2.970	3.979	5.236	6.778	4
K_2CO_3	0.541	0.802	1.134	1.536	1.997	2.499	3.016	1
$LiCl$	0.128	0.193	0.279	0.384				2
$Mg(NO_3)_2$	0.726	0.999	1.339	1.749	2.231	2.782	3.397	1
$MnCl_2$	0.697	1.064	1.515	2.020	2.535	3.002		3
NH_4Cl	0.971	1.328	1.836	2.481				2
NH_4NO_3	0.853	1.152	1.524	1.972				2
$(NH_4)_2SO_4$	0.901	1.319	1.871	2.573	3.439	4.474		3
$NaBr$	0.722	1.004	1.376	1.858	2.475	3.255	4.229	4
$NaCl$	0.921	1.285	1.768	2.401	3.218	4.262	5.581	4
$NaNO_2$	0.703	0.994	1.381	1.888	2.540	3.368	4.403	4
$NaNO_3$	0.884	1.244	1.719	2.335	3.121	4.109	5.333	4
$RbCl$	0.862	1.215	1.684	2.298	3.088	4.089	5.343	4
$ZnSO_4$	0.945	1.401	1.986	2.698	3.523	4.431	5.382	1
Water	1.228	1.706	2.339	3.169	4.246	5.627	7.381	

RECOMMENDED DATA FOR VAPOR-PRESSURE CALIBRATION

Eric W. Lemmon

These precise vapor pressure values are recommended as secondary standards. Values are given in kPa (1 kPa = 0.0098692 atm = 7.5006 Torr). References for water and CO_2 are given in the *Thermophysical Properties of Fluids* table.

References

1. Huber, M.L., Laesecke, A., and Friend, D.G., Correlation for the Vapor Pressure of Mercury, *Ind. Eng. Chem. Res.* 45, 7351, 2006.
2. Ruzicka, K., Fulem, M., and Ruzicka, V., Recommended Vapor Pressure of Solid Naphthalene, *J. Chem. Eng. Data* 50, 1956, 2005.
3. Span, R., and Wagner, W., Equations of State for Technical Applications. II. Results for Nonpolar Fluids, *Int. J. Thermophys.* 24, 41, 2003. [pentane]
4. Thol, M., Lemmon, E.W., and Span, R., Equation of State for Benzene for Temperatures from the Melting Line up to 750 K and Pressures up to 500 MPa, private communication, 2010.

T/K	CO_2(s)	H_2O(s)	$C_{10}H_8$(s)	C_5H_{12}	C_6H_6	H_2O	Hg
180	27.56		$3.04 \cdot 10^{-11}$				
190	68.34		$4.33 \cdot 10^{-10}$				
200	155.03	0.00016	$4.69 \cdot 10^{-9}$				
210	327.09	0.0007	$4.02 \cdot 10^{-8}$				
220		0.0027	$2.82 \cdot 10^{-7}$				
230		0.0089	$1.66 \cdot 10^{-6}$				
240		0.0273	$8.37 \cdot 10^{-6}$				
250		0.0760	$3.69 \cdot 10^{-5}$	7.59			
260		0.1958	$1.446 \cdot 10^{-4}$	12.96			
270		0.4701	$5.09 \cdot 10^{-4}$	21.14			
280			0.00163	33.10	5.139	0.992	
290			0.004798	50.00	8.602	1.920	
300			0.01308	73.17	13.82	3.537	
310			0.03328	104.07	21.40	6.231	
320			0.07956	144.3	32.07	10.546	
330			0.1797	195.7	46.68	17.213	
340			0.3854	260.2	66.19	27.188	
350			0.7884	339.6	91.67	41.682	
360				436.1	124.3	62.194	
370				551.8	165.3	90.535	
380				689.2	216.1	128.85	
390				850.5	278.0	179.64	
400				1038	352.6	245.77	0.139
410				1256	441.3	330.45	0.216
420				1505	545.9	437.30	0.330
430				1790	668.0	570.26	0.494
440				2115	809.2	733.67	0.726
450				2484	971.5	932.20	1.048
460				2903	1156	1170.9	1.489
470					1366	1455.1	2.082
480					1603	1790.5	2.870
490					1868	2183.1	3.905
500					2165	2639.2	5.245
510					2495	3165.5	6.962
520					2863	3769.0	9.139
530					3270	4456.9	11.871
540					3721	5236.9	15.267
550					4223	6117.2	19.452
560					4784	7106.2	24.564
570						8213.2	30.76
580						9448.0	38.22
590						10821	47.12
600						12345	57.69

ENTHALPY OF VAPORIZATION

The molar enthalpy (heat) of vaporization $\Delta_{vap}H$, which is defined as the enthalpy change in the conversion of one mole of liquid to gas at constant temperature, is tabulated here for about 950 inorganic and organic compounds. Values are given, when available, both at the normal boiling point t_b, referred to a pressure of 101.325 kPa (760 mmHg), and at 25 °C.

The values in this table were measured either by calorimetric techniques or by application of the Claperyon equation to the variation of vapor pressure with temperature. See Reference 1 for a discussion of the accuracy of different experimental techniques and methods of estimating enthalpy of vaporization at other temperatures. Several of the references present empirical techniques for correlating enthalpy of vaporization with molecular structure.

Compounds are listed by systematic name, with compounds not containing carbon preceding those that do contain carbon. To locate a compound by molecular formula or CAS Registry Number, use the indexes to the table "Physical Constants of Organic Compounds" in Section 3, which point to the entry in that table from which the name can be determined.

References

1. Majer, V., and Svoboda, V., *Enthalpies of Vaporization of Organic Compounds*, Blackwell Scientific Publications, Oxford, 1985.

2. Chase, M. W., Davies, C. A., Downey, J. R., Frurip, D. J., McDonald, R. A., and Syverud, A. N., *JANAF Thermochemical Tables, Third Edition, J. Phys. Chem. Ref. Data*, 14, Suppl. 1, 1985.
3. *Landolt-Börnstein, Numerical Data and Functional Relationships in Science and Technology, Sixth Edition*, II/4, *Caloric Quantities of State*, Springer-Verlag, Heidelberg, 1961.
4. Daubert, T. E., Danner, R. P., Sibul, H. M., and Stebbins, C. C., *Physical and Thermodynamic Properties of Pure Compounds: Data Compilation*, extant 1994 (core with 4 supplements), Taylor & Francis, Bristol, PA.
5. Ruzicka, K., and Majer, V., Simultaneous Treatment of Vapor Pressures and Related Thermal Data Between the Triple and Normal Boiling Temperatures for *n*-Alkanes C5–C20, *J. Phys. Chem. Ref. Data*, 23, 1, 1994.
6. Verevkin, S. P., Thermochemistry of Amines: Experimental Standard Molar Enthalpies of Formation of Some Aliphatic and Aromatic Amines, *J. Chem. Thermodynamics*, 29, 891, 1997.
7. Cady, G. H., and Hargreaves, G. B., The Vapor Pressure of Some Heavy Transition Metal Hexafluorides, *J. Chem. Soc.*, 1563, 1578, 1961.
8. Steele, W. V., Chirico, R. D., Knipmeyer, S. E., and Nguyen, A., *J. Chem. Eng. Data* 41, 1255, 1996.
9. Nichols, G., et al., *J. Chem. Eng. Data* 51, 475, 2006.
10. Dias, A. M. A., et al., *J. Chem. Eng. Data* 50, 1328, 2005.
11. Umnahanant, P., et al., *J. Chem. Eng. Data* 51, 2246, 2006.
12. Raganov, G. N., Pisarev, P. N., and Emel'yanenko, V. N., *J. Chem. Eng. Data* 50, 1114, 2005.
13. Verevkin, S. P., et al., *J. Chem. Eng. Data* 51, 1896, 2006.
14. Verevkin, S. P., *J. Chem. Thermodynamics* 38, 1111, 2006.

Name	Mol. Form.	t_b °C	$\Delta_{vap}H(t_b)$ kJ/mol	$\Delta_{vap}H(25\ °C)$ kJ/mol
Compounds not containing carbon				
Aluminum	Al	2519	294	
Aluminum borohydride	AlB_3H_{12}	44.5	30	
Aluminum bromide	$AlBr_3$	255	23.5	
Aluminum iodide	AlI_3	382	32.2	
Ammonia	H_3N	−33.33	23.33	19.86
Antimony(III) bromide	Br_3Sb	288	59	
Antimony(III) chloride	Cl_3Sb	220.3	45.19	
Antimony(III) iodide	I_3Sb	400	68.6	
Argon	Ar	−185.85	6.43	
Arsenic(III) bromide	$AsBr_3$	221	41.8	
Arsenic(III) chloride	$AsCl_3$	130	35.01	
Arsenic(III) fluoride	AsF_3	57.13	29.7	
Arsenic(V) fluoride	AsF_5	−52.8	20.8	
Arsenic(III) iodide	AsI_3	424	59.3	
Arsine	AsH_3	−62.5	16.69	
Barium	Ba	1845	140	
Beryllium chloride	$BeCl_2$	482	105	
Beryllium iodide	BeI_2	590	70.5	
Bismuth	Bi	1564	151	
Bismuth tribromide	$BiBr_3$	462	75.4	
Bismuth trichloride	$BiCl_3$	441	72.61	
Boron	B	4000	480	
Boron tribromide	BBr_3	91.3	30.5	
Boron trichloride	BCl_3	12.5	23.77	23.1
Boron trifluoride	BF_3	−99.9	19.33	
Boron triiodide	BI_3	209.5	40.5	

Name	Mol. Form.	t_b °C	$\Delta_{vap}H(t_b)$ kJ/mol	$\Delta_{vap}H(25\,°C)$ kJ/mol
Bromine	Br_2	58.8	29.96	30.91
Bromine fluoride	BrF	20	25.1	
Bromine pentafluoride	BrF_5	41.3	30.6	
Bromine trifluoride	BrF_3	125.8	47.57	
Bromosilane	BrH_3Si	1.9	24.4	
Cadmium	Cd	767	99.87	
Cadmium bromide	Br_2Cd	863	115	
Cadmium chloride	$CdCl_2$	964	124.3	
Cadmium fluoride	CdF_2	1750	214	
Cadmium iodide	CdI_2	744	115	
Chlorine	Cl_2	−34.04	20.41	17.65
Chlorine dioxide	ClO_2	11	30	
Chlorine fluoride	ClF	−101.1	24	
Chlorine monoxide	Cl_2O	2.2	25.9	
Chlorine trifluoride	ClF_3	11.75	27.53	
Chlorosilane	ClH_3Si	−30.4	21	
Chlorotrifluorosilane	ClF_3Si	−70.0	18.7	
Chromium(II) chloride	Cl_2Cr	1120	197	
Chromium(VI) dichloride dioxide	Cl_2CrO_2	117	35.1	
Diborane	B_2H_6	−92.49	14.28	
Dibromosilane	Br_2H_2Si	66	31	
Dichlorodifluorosilane	Cl_2F_2Si	−32	21.2	
Dichlorosilane	Cl_2H_2Si	8.3	25	24.2
Difluorine dioxide	F_2O_2	−57	19.1	
Difluorosilane	F_2H_2Si	−77.8	16.3	
Digermane	Ge_2H_6	29	25.1	
Diphosphine	H_4P_2	63.5	28.8	
Disilane	H_6Si_2	−14.8	21.2	
Fluorine	F_2	−188.11	6.62	
Fluorine monoxide	F_2O	−144.3	11.09	
Fluorosilane	FH_3Si	−98.6	18.8	
Gallium	Ga	2229	254	
Gallium(III) bromide	Br_3Ga	279	38.9	
Gallium(III) chloride	Cl_3Ga	201	23.9	
Gallium(III) iodide	GaI_3	340	56.5	
Germane	GeH_4	−88.1	14.06	
Germanium	Ge	2833	334	
Germanium(IV) bromide	Br_4Ge	186.35	41.4	
Germanium(IV) chloride	Cl_4Ge	86.55	27.9	
Gold	Au	2836	324	
Helium	He	−268.928	0.08	
Hydrazine	H_4N_2	113.55	41.8	44.7
Hydrazoic acid	HN_3	35.7	30.5	
Hydrogen	H_2	−252.76	0.90	
Hydrogen bromide	BrH	−66.38		12.69
Hydrogen chloride	ClH	−85	16.15	9.08
Hydrogen disulfide	H_2S_2	70.7		33.78
Hydrogen iodide	HI	−35.55	19.76	17.36
Hydrogen peroxide	H_2O_2	150.2		51.6
Hydrogen selenide	H_2Se	−41.25	19.7	
Hydrogen sulfide	H_2S	−59.55	18.67	14.08
Hydrogen telluride	H_2Te	−2	19.2	
Indium(I) bromide	BrIn	656	92	
Indium(I) iodide	IIn	712	90.8	
Iodine	I_2	184.4	41.57	

Name	Mol. Form.	t_b °C	$\Delta_{vap}H(t_b)$ kJ/mol	$\Delta_{vap}H(25\ °C)$ kJ/mol
Iodine pentafluoride	F_5I	100.5	41.3	
Iridium(VI) fluoride	F_6Ir	53.6	30.9	
Krypton	Kr	−153.415	9.08	
Lead	Pb	1749	179.5	
Lead(II) bromide	Br_2Pb	892	133	
Lead(II) chloride	Cl_2Pb	951	127	
Lead(II) fluoride	F_2Pb	1293	160.4	
Lead(II) iodide	I_2Pb	872	104	
Lithium fluoride	FLi	1673	147	
Lithium hydroxide	HLiO	1626	188	
Mercury	Hg	356.619	59.11	
Mercury(II) bromide	Br_2Hg	318	58.89	
Mercury(II) chloride	Cl_2Hg	304	58.9	
Mercury(II) iodide	HgI_2	351	59.2	
Molybdenum(V) chloride	Cl_5Mo	268	62.8	
Molybdenum(V) fluoride	F_5Mo	213.6	51.8	
Molybdenum(VI) fluoride	F_6Mo	34.0	29.0	
Molybdenum(VI) oxide	MoO_3	1155	138	
Molybdenum(VI) oxytetrafluoride	F_4MoO	186.0	50.6	
Neon	Ne	−246.05	1.71	
Niobium(V) chloride	Cl_5Nb	247.4	52.7	
Niobium(V) fluoride	F_5Nb	234	52.3	
Nitric acid	HNO_3	83		39.1
Nitric oxide	NO	−151.74	13.83	
Nitrogen	N_2	−195.79	5.57	
Nitrogen tetroxide	N_2O_4	21.15	38.12	
Nitrogen trifluoride	F_3N	−128.75	11.56	
Nitrosyl chloride	ClNO	−5.5	25.78	
Nitrosyl fluoride	FNO	−59.9	19.28	
Nitrous oxide	N_2O	−88.48	16.53	
Nitryl chloride	$ClNO_2$	−15	25.7	
Nitryl fluoride	FNO_2	−72.4	18.05	
Osmium(V) fluoride	F_5Os	233	65.6	
Osmium(VI) fluoride	F_6Os	47.5	28.1	
Oxygen	O_2	−182.95	6.82	
Pentaborane(11)	B_5H_{11}	65	31.8	
Perchloryl fluoride	$ClFO_3$	−46.75	19.33	
Phosphine	H_3P	−87.75	14.6	
Phosphorothioic trifluoride	F_3PS	−52.25	19.6	
Phosphorus	P	280.5	12.4	14.2
Phosphorus(III) bromide	Br_3P	173.2	38.8	
Phosphorus(III) chloride	Cl_3P	76	30.5	32.1
Phosphorus(III) chloride difluoride	ClF_2P	−47.3	17.6	
Phosphorus(III) dichloride fluoride	Cl_2FP	13.85	24.9	
Phosphorus(III) fluoride	F_3P	−101.8	16.5	
Phosphorus(V) fluoride	F_5P	−84.6	17.2	
Phosphorus(III) iodide	I_3P	227	43.9	
Phosphoryl bromide	Br_3OP	191.7	38	
Phosphoryl chloride	Cl_3OP	105.5	34.35	38.6
Rhenium(VII) dioxytrifluoride	F_3O_2Re	185.4	65.7	
Rhenium(V) fluoride	F_5Re	221.3	58.1	
Rhenium(VI) fluoride	F_6Re	33.8	28.7	
Rhenium(VI) oxytetrafluoride	F_4ORe	171.7	61.0	
Selenium	Se	685	95.48	
Selenium tetrafluoride	F_4Se	101.6	47.2	

Name	Mol. Form.	t_b °C	$\Delta_{vap}H(t_b)$ kJ/mol	$\Delta_{vap}H(25\ °C)$ kJ/mol
Silane	H_4Si	−111.9	12.1	
Silver(I) bromide	AgBr	1502	198	
Silver(I) chloride	AgCl	1547	199	
Silver(I) iodide	AgI	1506	143.9	
Sodium hydroxide	HNaO	1388	175	
Stannane	H_4Sn	−51.8	19.05	
Stibine	H_3Sb	−17	21.3	
Sulfur	S	444.61	45	
Sulfur dioxide	O_2S	−10.05	24.94	22.92
Sulfur hexafluoride	F_6S			8.99
Sulfur tetrafluoride	F_4S	−40.45	26.44	
Sulfur trioxide	O_3S	44.5	40.69	43.14
Sulfuryl chloride	Cl_2O_2S	69.4	31.4	30.1
Tantalum(V) bromide	Br_5Ta	348.8	62.3	
Tantalum(V) chloride	Cl_5Ta	239	54.8	
Tantalum(V) fluoride	F_5Ta	229.5	56.9	
Tellurium	Te	988	114.1	
Tellurium tetrachloride	Cl_4Te	387	77	
Tetraborane(10)	B_4H_{10}	18	27.1	
Tetrabromosilane	Br_4Si	154	37.9	
Tetrachlorosilane	Cl_4Si	57.65	28.7	29.7
Tetrafluorodiborane	B_2F_4	−34.0	28	
Tetrafluorohydrazine	F_4N_2	−74	13.27	
Tetraiodosilane	I_4Si	287.35	50.2	
Thallium(I) bromide	BrTl	819	99.56	
Thallium(I) chloride	ClTl	720	102.2	
Thallium(I) iodide	ITl	824	104.7	
Thallium(I) sulfide	STl_2	1367	154	
Thionitrosyl fluoride (NSF)	FNS	4.8	22.2	
Thionyl chloride	Cl_2OS	75.6	31.7	31
Thionyl fluoride	F_2OS	−43.8	21.8	
Thorium(IV) chloride	Cl_4Th	921	146.4	
Thorium(IV) fluoride	F_4Th	1680	258	
Tin(II) bromide	Br_2Sn	639	102	
Tin(IV) bromide	Br_4Sn	205	43.5	
Tin(II) chloride	Cl_2Sn	623	86.8	
Tin(IV) chloride	Cl_4Sn	114.15	34.9	
Tin(II) iodide	I_2Sn	714	105	
Tin(IV) iodide	I_4Sn	364.35	56.9	
Titanium(IV) bromide	Br_4Ti	233.5	44.37	
Titanium(II) chloride	Cl_2Ti	1500	232	
Titanium(III) chloride	Cl_3Ti	960	124	
Titanium(IV) chloride	Cl_4Ti	136.45	36.2	
Titanium(IV) iodide	I_4Ti	377	58.4	
Tribromosilane	Br_3HSi	109	34.8	
Trichlorosilane	Cl_3HSi	33		25.7
Trifluorosilane	F_3HSi	−95	16.2	
Trigermane	Ge_3H_8	110.5	32.2	
Trisilane	H_8Si_3	52.9	28.5	
Tungsten(VI) chloride	Cl_6W	337	52.7	
Tungsten(VI) fluoride	F_6W	17.1	26.5	
Tungsten(VI) oxytetrachloride	Cl_4OW	230	67.8	
Tungsten(VI) oxytetrafluoride	F_4OW	185.9	59.5	
Vanadium(IV) chloride	Cl_4V	151	41.4	42.5
Vanadium(V) fluoride	F_5V	48.3	44.52	

Name	Mol. Form.	t_b °C	$\Delta_{vap}H(t_b)$ kJ/mol	$\Delta_{vap}H(25\,°C)$ kJ/mol
Vanadyl trichloride	Cl$_3$OV	127	36.78	
Water	H$_2$O	99.97	40.65	43.98
Xenon	Xe	−108.09	12.57	
Zinc bromide	Br$_2$Zn	670	118	
Zinc chloride	Cl$_2$Zn	732	126	
Zinc fluoride	F$_2$Zn	1500	190.1	
Compounds containing carbon				
Acetaldehyde	C$_2$H$_4$O	20.1	25.76	25.47
Acetic acid	C$_2$H$_4$O$_2$	117.9	23.70	23.36
Acetic anhydride	C$_4$H$_6$O$_3$	139.5	38.2	
Acetone	C$_3$H$_6$O	56.05	29.10	30.99
Acetonitrile	C$_2$H$_3$N	81.65	29.75	32.94
Acetophenone	C$_8$H$_8$O	202	43.98	55.40
Acrolein	C$_3$H$_4$O	52.6	28.3	
Acrylonitrile	C$_3$H$_3$N	77.3	32.6	
Allyl acetate	C$_5$H$_8$O$_2$	103.5	36.3	
Allyl alcohol	C$_3$H$_6$O	97.4	40.0	
2-Amino-2-methyl-1-propanol	C$_4$H$_{11}$NO	165.5	50.6	
Aniline	C$_6$H$_7$N	184.17	42.44	55.83
Anisole	C$_7$H$_8$O	153.7	38.97	46.90
Azobutane	C$_8$H$_{18}$N$_2$			49.31
Azopropane	C$_6$H$_{14}$N$_2$	114		39.88
Benzaldehyde	C$_7$H$_6$O	178.8	42.5	
Benzene	C$_6$H$_6$	80.09	30.72	33.83
Benzenethiol	C$_6$H$_6$S	169.1	39.93	47.56
Benzonitrile	C$_7$H$_5$N	191.1	45.9	
Benzyl acetate	C$_9$H$_{10}$O$_2$	213	49.4	
Benzyl alcohol	C$_7$H$_8$O	205.31	50.48	
Benzylamine	C$_7$H$_9$N	185		60.16
N-Benzylaniline	C$_{13}$H$_{13}$N	306.5		79.6
Benzyl benzoate	C$_{14}$H$_{12}$O$_2$	323.5	53.6	
Bis(2-chloroethyl) ether	C$_4$H$_8$Cl$_2$O	178.5	45.2	
Bis(ethoxymethyl) ether	C$_6$H$_{14}$O$_3$	140.6	36.17	44.69
Bromobenzene	C$_6$H$_5$Br	156.06		44.54
1-Bromobutane	C$_4$H$_9$Br	101.6	32.51	36.64
2-Bromobutane	C$_4$H$_9$Br	91.3	30.77	34.41
Bromochloromethane	CH$_2$BrCl	68.0	30.0	
2-Bromo-2-chloro-1,1,1-trifluoroethane	C$_2$HBrClF$_3$	50.2	28.08	29.61
Bromoethane	C$_2$H$_5$Br	38.5	27.04	28.03
Bromoethene	C$_2$H$_3$Br	15.8	23.4	
1-Bromoheptane	C$_7$H$_{15}$Br	178.9		50.60
1-Bromohexane	C$_6$H$_{13}$Br	155.3		45.89
Bromomethane	CH$_3$Br	3.5	23.91	22.81
1-Bromo-2-methylpropane	C$_4$H$_9$Br	91.1	31.33	34.82
2-Bromo-2-methylpropane	C$_4$H$_9$Br	73.3	29.23	31.81
1-Bromonaphthalene	C$_{10}$H$_7$Br	281	39.3	
1-Bromooctane	C$_8$H$_{17}$Br	200.8		55.77
1-Bromopentane	C$_5$H$_{11}$Br	129.8	35.01	41.28
1-Bromopropane	C$_3$H$_7$Br	71.1	29.84	32.01
2-Bromopropane	C$_3$H$_7$Br	59.5	28.33	30.17
3-Bromopropene	C$_3$H$_5$Br	70.1	30.24	32.73
1,2-Butadiene	C$_4$H$_6$	10.9	24.02	23.21
1,3-Butadiene	C$_4$H$_6$	−4.41	22.47	20.86
Butanal	C$_4$H$_8$O	74.8	31.5	

Name	Mol. Form.	t_b °C	$\Delta_{vap}H(t_b)$ kJ/mol	$\Delta_{vap}H(25\ °C)$ kJ/mol
Butane	C_4H_{10}	−0.5	22.44	21.02
1,2-Butanediol	$C_4H_{10}O_2$	190.5	52.84	71.55
1,3-Butanediol	$C_4H_{10}O_2$	207.5	54.31	74.46
1,4-Butanediol	$C_4H_{10}O_2$	235		77.1
1,4-Butanedithiol	$C_4H_{10}S_2$	195.5		55.10
Butanenitrile	C_4H_7N	117.6	33.68	39.33
1-Butanethiol	$C_4H_{10}S$	98.5	32.23	36.63
2-Butanethiol	$C_4H_{10}S$	85.0	30.59	33.99
Butanoic acid	$C_4H_8O_2$	163.75		40.45
Butanoic anhydride	$C_8H_{14}O_3$	200	50.0	
1-Butanol	$C_4H_{10}O$	117.73	43.29	52.35
2-Butanol	$C_4H_{10}O$	99.51	40.75	49.72
2-Butanone	C_4H_8O	79.59	31.30	34.79
1-Butene	C_4H_8	−6.26	22.07	20.22
cis-2-Butene	C_4H_8	3.71	23.34	22.16
trans-2-Butene	C_4H_8	0.88	22.72	21.40
2-Butoxyethanol	$C_6H_{14}O_2$	168.4		56.59
Butyl acetate	$C_6H_{12}O_2$	126.1	36.28	43.86
tert-Butyl acetate	$C_6H_{12}O_2$	95.1	33.07	38.03
Butylamine	$C_4H_{11}N$	77.00	31.81	35.72
sec-Butylamine	$C_4H_{11}N$	62.73	29.92	32.85
tert-Butylamine	$C_4H_{11}N$	44.04	28.27	29.64
Butylbenzene	$C_{10}H_{14}$	183.31	38.87	50.8
sec-Butylbenzene	$C_{10}H_{14}$	173.3		48.1
tert-Butylbenzene	$C_{10}H_{14}$	169.1		47.6
Butylcyclohexane	$C_{10}H_{20}$	180		49.36
Butylcyclopentane	C_9H_{18}	156.6	36.16	45.89
Butylethylamine	$C_6H_{15}N$	107.5	33.97	40.15
Butyl ethyl ether	$C_6H_{14}O$	92.3	31.63	36.32
Butyl ethyl sulfide	$C_6H_{14}S$	144.3	37.01	44.51
Butyl formate	$C_5H_{10}O_2$	106.1	36.58	41.11
tert-Butyl isobutyl ether	$C_8H_{18}O$	112.0	33.11	40.5
Butyl methyl ether	$C_5H_{12}O$	70.16	29.55	32.37
sec-Butyl methyl ether	$C_5H_{12}O$	59.1	28.09	30.23
Butyl methyl sulfide	$C_5H_{12}S$	123.4	34.47	40.46
tert-Butyl methyl sulfide	$C_5H_{12}S$	98.9	31.47	35.84
Butyl propyl ether	$C_7H_{16}O$	118.1	33.72	40.22
Butyl vinyl ether	$C_6H_{12}O$	94	31.58	36.17
1-Butyne	C_4H_6	8.08	24.52	23.35
2-Butyne-1,4-diol	$C_4H_6O_2$	238		81.5
γ-Butyrolactone	$C_4H_6O_2$	204	52.2	
Camphor, (+)	$C_{10}H_{16}O$	207.4	59.5	
Carbon disulfide	CS_2	46	26.74	27.51
Carbon monoxide	CO	−191.5	6.04	
2-Chloroaniline	C_6H_6ClN	208.8	44.4	
Chlorobenzene	C_6H_5Cl	131.72	35.19	40.97
1-Chlorobutane	C_4H_9Cl	78.4	30.39	33.51
2-Chlorobutane	C_4H_9Cl	68.2	29.17	31.53
Chlorodifluoromethane	$CHClF_2$	−40.7	20.2	
Chloroethane	C_2H_5Cl	12.3	24.65	
2-Chloroethanol	C_2H_5ClO	128.6	41.4	
Chloroethene	C_2H_3Cl	−13.8	20.8	
1-Chloroheptane	$C_7H_{15}Cl$	160.4		47.66
1-Chlorohexane	$C_6H_{13}Cl$	135.1	35.67	42.83
Chloromethane	CH_3Cl	−24.09	21.40	18.92

Name	Mol. Form.	t_b °C	$\Delta_{vap}H(t_b)$ kJ/mol	$\Delta_{vap}H(25\ °C)$ kJ/mol
1-Chloro-3-methylbutane	$C_5H_{11}Cl$	98.9	32.02	36.24
1-Chloro-2-methylpropane	C_4H_9Cl	68.5	29.22	31.67
2-Chloro-2-methylpropane	C_4H_9Cl	50.9	27.55	28.98
1-Chloronaphthalene	$C_{10}H_7Cl$	259	52.1	
1-Chlorooctane	$C_8H_{17}Cl$	183.5		52.42
Chloropentafluorobenzene	C_6ClF_5	117.96	34.76	41.07
Chloropentafluoroethane	C_2ClF_5	−39.1	19.41	
1-Chloropentane	$C_5H_{11}Cl$	108.4	33.15	38.24
2-Chloropentane	$C_5H_{11}Cl$	97.0	31.79	36.03
1-Chloropropane	C_3H_7Cl	46.5	27.18	28.35
2-Chloropropane	C_3H_7Cl	35.7	26.30	26.90
3-Chloropropene	C_3H_5Cl	45.1	29.0	
2-Chlorotoluene	C_7H_7Cl	159.0	37.5	
4-Chlorotoluene	C_7H_7Cl	162.4	38.7	
Chlorotrifluoromethane	$CClF_3$	−81.37	15.8	
Cholesterol	$C_{27}H_{46}O$	360 dec		148.0
o-Cresol	C_7H_8O	191.04	45.19	
m-Cresol	C_7H_8O	202.27	47.40	61.71
p-Cresol	C_7H_8O	201.98	47.45	
Cyanogen	C_2N_2	−21.1	23.33	19.75
Cyclobutane	C_4H_8	12.6	24.19	23.51
Cyclobutanecarbonitrile	C_5H_7N	149.6	36.88	44.34
Cyclohexane	C_6H_{12}	80.73	29.97	33.01
Cyclohexanecarbonitrile	$C_7H_{11}N$	184		51.92
Cyclohexanethiol	$C_6H_{12}S$	158.8	37.06	44.57
Cyclohexanol	$C_6H_{12}O$	160.84		62.01
Cyclohexanone	$C_6H_{10}O$	155.43		45.06
Cyclohexene	C_6H_{10}	82.98	30.46	33.47
1-Cyclohexenecarbonitrile	C_7H_9N			53.55
Cyclohexylamine	$C_6H_{13}N$	134	36.14	43.67
Cyclohexylbenzene	$C_{12}H_{16}$	240.1		60.8
Cyclohexylcyclohexane	$C_{12}H_{22}$	238		57.98
Cyclopentane	C_5H_{10}	49.3	27.30	28.52
Cyclopentanecarbonitrile	C_6H_9N	170		43.43
Cyclopentanethiol	$C_5H_{10}S$	132.1	35.32	41.42
Cyclopentanol	$C_5H_{10}O$	140.42		57.05
Cyclopentanone	C_5H_8O	130.57	36.35	42.72
1-Cyclopentenecarbonitrile	C_6H_7N			44.98
Cyclopropane	C_3H_6	−32.81	20.05	16.93
Cyclopropanecarbonitrile	C_4H_5N	135.1	35.55	41.94
Cyclopropylbenzene	C_9H_{10}	173.6		50.22
Cyclopropyl methyl ketone	C_5H_8O	111.3	34.07	39.41
cis-Decahydronaphthalene	$C_{10}H_{18}$	195.8	41.0	
trans-Decahydronaphthalene	$C_{10}H_{18}$	187.3	40.2	
Decane	$C_{10}H_{22}$	174.15	39.58	51.42
1,10-Decanediol	$C_{10}H_{22}O_2$			120.0
Decanenitrile	$C_{10}H_{19}N$	243		66.84
1-Decanethiol	$C_{10}H_{22}S$	240.6		65.48
1-Decanol	$C_{10}H_{22}O$	231.1		81.50
1-Decene	$C_{10}H_{20}$	170.5		50.43
Decylbenzene	$C_{16}H_{26}$	293		78.2
1,4-Dibromobutane	$C_4H_8Br_2$	197		53.09
1,2-Dibromo-1-chloro-1,2,2-trifluoroethane	$C_2Br_2ClF_3$	93	31.17	35.04
1,2-Dibromoethane	$C_2H_4Br_2$	131.6	34.77	41.73
Dibromomethane	CH_2Br_2	97	32.92	36.97

Name	Mol. Form.	t_b °C	$\Delta_{vap}H(t_b)$ kJ/mol	$\Delta_{vap}H(25\ °C)$ kJ/mol
1,2-Dibromopropane	$C_3H_6Br_2$	141.9	35.61	41.67
1,3-Dibromopropane	$C_3H_6Br_2$	167.3		47.45
1,2-Dibromotetrafluoroethane	$C_2Br_2F_4$	47.35	27.03	28.39
Dibutylamine	$C_8H_{19}N$	159.6	38.44	49.45
Dibutyl ether	$C_8H_{18}O$	140.28	36.49	44.97
Di-sec-butyl ether	$C_8H_{18}O$	121.1	34.06	40.84
Di-tert-butyl ether	$C_8H_{18}O$	107.23	32.15	37.61
Dibutyl phthalate	$C_{16}H_{22}O_4$	340	79.2	
Dibutyl sulfide	$C_8H_{18}S$	185		52.96
Di-tert-butyl sulfide	$C_8H_{18}S$	149.1	33.26	43.76
o-Dichlorobenzene	$C_6H_4Cl_2$	180	39.66	50.21
m-Dichlorobenzene	$C_6H_4Cl_2$	173	38.62	48.58
p-Dichlorobenzene	$C_6H_4Cl_2$	174	38.79	49.0
1,2-Dichlorobutane	$C_4H_8Cl_2$	124.1	33.90	39.58
1,4-Dichlorobutane	$C_4H_8Cl_2$	161		46.36
Dichlorodifluoromethane	CCl_2F_2	−29.8	20.1	
1,1-Dichloroethane	$C_2H_4Cl_2$	57.3	28.85	30.62
1,2-Dichloroethane	$C_2H_4Cl_2$	83.5	31.98	35.16
1,1-Dichloroethene	$C_2H_2Cl_2$	31.6	26.14	26.48
cis-1,2-Dichloroethene	$C_2H_2Cl_2$	60.1	30.2	
trans-1,2-Dichloroethene	$C_2H_2Cl_2$	48.7	28.9	
1,1-Dichloro-1-fluoroethane	$C_2H_3Cl_2F$	32.0	26.06	26.48
Dichlorofluoromethane	$CHCl_2F$	8.9	25.2	
1,2-Dichloro-1,1,2,3,3,3-hexafluoropropane	$C_3Cl_2F_6$	34.1	26.28	26.93
1,2-Dichlorohexane	$C_6H_{12}Cl_2$	173		48.16
Dichloromethane	CH_2Cl_2	40	28.06	28.82
1,2-Dichloropentane	$C_5H_{10}Cl_2$	148.3	36.45	43.89
1,5-Dichloropentane	$C_5H_{10}Cl_2$	179		50.71
1,3-Dichloropropane	$C_3H_6Cl_2$	120.9	35.18	40.75
1,2-Dichloro-1,1,2,2-tetrafluoroethane	$C_2Cl_2F_4$	3.5	23.3	
Dicyclopropyl ketone	$C_7H_{10}O$	161		53.70
Diethanolamine	$C_4H_{11}NO_2$	268.8	65.2	
1,1-Diethoxyethane	$C_6H_{14}O_2$	102.25	36.28	43.20
1,2-Diethoxyethane	$C_6H_{14}O_2$	121.2	36.28	43.20
Diethoxymethane	$C_5H_{12}O_2$	88	31.33	35.65
Diethylamine	$C_4H_{11}N$	55.5	29.06	31.31
Diethyl carbonate	$C_5H_{10}O_3$	126		43.60
Diethyl disulfide	$C_4H_{10}S_2$	154.0	37.58	45.18
Diethylene glycol	$C_4H_{10}O_3$	245.8	52.3	
Diethylene glycol diethyl ether	$C_8H_{18}O_3$	188		58.40
Diethylene glycol dimethyl ether	$C_6H_{14}O_3$	162	36.17	44.69
Diethylene glycol monoethyl ether	$C_6H_{14}O_3$	196	47.5	
Diethylene glycol monomethyl ether	$C_5H_{12}O_3$	193	46.6	
Diethyl ether	$C_4H_{10}O$	34.5	26.52	27.10
Diethyl malonate	$C_7H_{12}O_4$	200	54.8	
Diethyl oxalate	$C_6H_{10}O_4$	185.7	42.0	
3,3-Diethylpentane	C_9H_{20}	146.3	34.61	42.0
Diethyl sulfide	$C_4H_{10}S$	92.1	31.77	35.80
o-Difluorobenzene	$C_6H_4F_2$	94	32.21	36.18
m-Difluorobenzene	$C_6H_4F_2$	82.6	31.10	34.59
p-Difluorobenzene	$C_6H_4F_2$	89	31.77	35.54
1,1-Difluoroethane	$C_2H_4F_2$	−24.05	21.56	19.08
2,3-Dihydrothiophene	C_4H_6S	112.1	33.24	37.74
2,5-Dihydrothiophene	C_4H_6S	122.4	34.83	39.95
Diiodomethane	CH_2I_2	182	42.5	

Name	Mol. Form.	t_b °C	$\Delta_{vap}H(t_b)$ kJ/mol	$\Delta_{vap}H(25\ °C)$ kJ/mol
Diisobutyl sulfide	$C_8H_{18}S$	171		48.71
Diisopentyl ether	$C_{10}H_{22}O$	172.5	35.1	
Diisopropylamine	$C_6H_{15}N$	83.9	30.40	34.61
Diisopropyl ether	$C_6H_{14}O$	68.4	29.10	32.12
Diisopropyl sulfide	$C_6H_{14}S$	120.0	33.80	39.60
Diketene	$C_4H_4O_2$	126.1	36.80	42.89
1,2-Dimethoxyethane	$C_4H_{10}O_2$	84.5	32.42	36.39
N,N-Dimethylacetamide	C_4H_9NO	165		50.24
Dimethylamine	C_2H_7N	6.88	26.40	25.05
2,4-Dimethylaniline	$C_8H_{11}N$	214		61.3
2,5-Dimethylaniline	$C_8H_{11}N$	214		61.7
N,N-Dimethylaniline	$C_8H_{11}N$	194.15		52.83
2,2-Dimethylbutane	C_6H_{14}	49.73	26.31	27.68
2,3-Dimethylbutane	C_6H_{14}	57.93	27.38	29.12
2,3-Dimethyl-2-butanethiol	$C_6H_{14}S$	126.1		39.3
3,3-Dimethyl-2-butanone	$C_6H_{12}O$	106.1	33.39	37.91
2,3-Dimethyl-1-butene	C_6H_{12}	55.6		29.18
3,3-Dimethyl-1-butene	C_6H_{12}	41.2		26.61
2,3-Dimethyl-2-butene	C_6H_{12}	73.3	29.64	32.51
1,1-Dimethylcyclohexane	C_8H_{16}	119.6	32.51	37.92
cis-1,2-Dimethylcyclohexane	C_8H_{16}	129.8	33.47	39.70
trans-1,2-Dimethylcyclohexane	C_8H_{16}	123.5	32.96	38.36
cis-1,3-Dimethylcyclohexane	C_8H_{16}	120.1	32.91	38.26
trans-1,3-Dimethylcyclohexane	C_8H_{16}	124.5	33.39	39.16
cis-1,4-Dimethylcyclohexane	C_8H_{16}	124.4	33.28	39.02
trans-1,4-Dimethylcyclohexane	C_8H_{16}	119.4	32.56	37.90
cis-1,3-Dimethylcyclopentane	C_7H_{14}	90.8	30.40	34.20
Dimethyl decanedioate	$C_{12}H_{22}O_4$			86.4
Dimethyl disulfide	$C_2H_6S_2$	109.74	33.78	37.86
Dimethyl ether	C_2H_6O	−24.8	21.51	18.51
N,N-Dimethylformamide	C_3H_7NO	153		46.89
Dimethyl glutarate	$C_7H_{12}O_4$	214		65.7
Dimethyl heptanedioate	$C_9H_{16}O_4$			73.5
2,6-Dimethyl-4-heptanol	$C_9H_{20}O$	174.5		65.17
2,6-Dimethyl-4-heptanone	$C_9H_{18}O$	169.4		50.92
2,2-Dimethylhexane	C_8H_{18}	106.86	32.07	37.28
2,3-Dimethylhexane	C_8H_{18}	115.62	33.17	38.78
2,4-Dimethylhexane	C_8H_{18}	109.5	32.51	37.76
2,5-Dimethylhexane	C_8H_{18}	109.12	32.54	37.85
3,3-Dimethylhexane	C_8H_{18}	111.97	32.31	37.53
3,4-Dimethylhexane	C_8H_{18}	117.73	33.24	38.97
Dimethyl 1,6-hexanedioate	$C_8H_{14}O_4$			69.0
2,5-Dimethyl-2,5-hexanediol	$C_8H_{18}O_2$	214		85.2
cis-2,2-Dimethyl-3-hexene	C_8H_{16}	105.5		36.86
trans-2,2-Dimethyl-3-hexene	C_8H_{16}	100.8		37.03
2,5-Dimethyl-3-hexyne-2,5-diol	$C_8H_{14}O_2$	205		82.8
1,1-Dimethylhydrazine	$C_2H_8N_2$	63.9	32.55	35.0
Dimethyl malonate	$C_5H_8O_4$	181.4		57.5
Dimethyl nonanedioate	$C_{11}H_{20}O_4$			82.3
2,4-Dimethyloctane	$C_{10}H_{22}$	156	36.47	47.13
Dimethyl octanedioate	$C_{10}H_{18}O_4$	268		78.1
Dimethyl oxalate	$C_4H_6O_4$	163.5		54.7
3,3-Dimethyloxetane	$C_5H_{10}O$	80.6	30.85	33.94
2,2-Dimethylpentane	C_7H_{16}	79.2	29.23	32.42
2,3-Dimethylpentane	C_7H_{16}	89.78	30.46	34.26

Name	Mol. Form.	t_b °C	$\Delta_{vap}H(t_b)$ kJ/mol	$\Delta_{vap}H(25\,°C)$ kJ/mol
2,4-Dimethylpentane	C_7H_{16}	80.49	29.55	32.88
3,3-Dimethylpentane	C_7H_{16}	86.06	29.62	33.03
2,2-Dimethyl-3-pentanone	$C_7H_{14}O$	125.6	36.09	42.34
2,4-Dimethyl-3-pentanone	$C_7H_{14}O$	125.4	34.64	41.51
2,4-Dimethyl-1-pentene	C_7H_{14}	81.6		33.03
4,4-Dimethyl-1-pentene	C_7H_{14}	72.5		31.13
2,4-Dimethyl-2-pentene	C_7H_{14}	83.4		34.19
cis-4,4-Dimethyl-2-pentene	C_7H_{14}	80.4		32.56
trans-4,4-Dimethyl-2-pentene	C_7H_{14}	76.7		32.81
2,2-Dimethylpropanenitrile	C_5H_9N	106.1	32.40	37.35
2,2-Dimethyl-1-propanethiol	$C_5H_{12}S$	103.7		36.4
2,3-Dimethylpyridine	C_7H_9N	161.12	39.08	47.82
2,4-Dimethylpyridine	C_7H_9N	158.38	38.53	47.49
2,5-Dimethylpyridine	C_7H_9N	156.98	38.68	47.04
2,6-Dimethylpyridine	C_7H_9N	144.01	37.46	45.34
3,4-Dimethylpyridine	C_7H_9N	179.10	39.99	50.50
3,5-Dimethylpyridine	C_7H_9N	171.84	39.46	49.33
Dimethyl succinate	$C_6H_{10}O_4$	196.4		61.0
Dimethyl sulfide	C_2H_6S	37.33	27.0	27.65
Dimethyl sulfoxide	C_2H_6OS	189	43.1	
1,3-Dioxane	$C_4H_8O_2$	106.1	34.37	39.09
1,4-Dioxane	$C_4H_8O_2$	101.5	34.16	38.60
Diphenyl ether	$C_{12}H_{10}O$	258.0	48.2	
1,2-Dipropoxyethane	$C_8H_{18}O_2$	163.2		50.62
Dipropylamine	$C_6H_{15}N$	109.3	33.47	40.04
Dipropyl ether	$C_6H_{14}O$	90.08	31.31	35.69
Dipropyl sulfide	$C_6H_{14}S$	142.9	36.60	44.21
1-Docosanol	$C_{22}H_{46}O$			135.9
Dodecane	$C_{12}H_{26}$	216.32	44.09	61.52
1,12-Dodecanediol	$C_{12}H_{26}O_2$			135
Dodecanenitrile	$C_{12}H_{23}N$	277		76.12
1-Dodecanol	$C_{12}H_{26}O$	260		90.8
1-Dodecene	$C_{12}H_{24}$	213.8		60.78
Dodecylbenzene	$C_{18}H_{30}$	328		86.6
Eicosane	$C_{20}H_{42}$	343	58.49	101.81
1-Eicosanol	$C_{20}H_{42}O$	356		125.9
1,2-Epoxybutane	C_4H_8O	63.4	30.3	
Ethane	C_2H_6	−88.6	14.69	5.16
1,2-Ethanediamine	$C_2H_8N_2$	117	37.98	44.98
1,2-Ethanediol	$C_2H_6O_2$	197.3	50.5	63.9
1,2-Ethanediol, diacetate	$C_6H_{10}O_4$	190		61.44
1,2-Ethanedithiol	$C_2H_6S_2$	146.1	37.93	44.68
Ethanethiol	C_2H_6S	35.0	26.79	27.30
Ethanol	C_2H_6O	78.29	38.56	42.32
Ethanolamine	C_2H_7NO	171	49.83	
Ethoxybenzene	$C_8H_{10}O$	169.81		51.04
2-Ethoxyethanol	$C_4H_{10}O_2$	135	39.22	48.21
2-Ethoxyethyl acetate	$C_6H_{12}O_3$	156.4	40.76	52.61
1-Ethoxy-2-methoxyethane	$C_5H_{12}O_2$	103.5	34.33	39.83
N-Ethylacetamide	C_4H_9NO	205		64.89
Ethyl acetate	$C_4H_8O_2$	77.11	31.94	35.60
Ethyl acrylate	$C_5H_8O_2$	99.4	34.7	
N-Ethylaniline	$C_8H_{11}N$	203.0		58.3
Ethylbenzene	C_8H_{10}	136.16	35.57	42.24
Ethyl butanoate	$C_6H_{12}O_2$	121.3	35.47	42.68

Name	Mol. Form.	t_b °C	$\Delta_{vap}H(t_b)$ kJ/mol	$\Delta_{vap}H(25\ °C)$ kJ/mol
2-Ethyl-1-butanol	$C_6H_{14}O$	147	43.2	
2-Ethyl-1-butene	C_6H_{12}	64.7		31.13
Ethyl chloroacetate	$C_4H_7ClO_2$	144.3	40.43	49.47
Ethylcyclobutane	C_6H_{12}	70.8	28.67	31.24
Ethylcyclohexane	C_8H_{16}	131.9	34.04	40.56
Ethylcyclopentane	C_7H_{14}	103.5	31.96	36.40
Ethyl dichloroacetate	$C_4H_6Cl_2O_2$	155		50.60
Ethyl 2,2-dimethylpropanoate	$C_7H_{14}O_2$	118.4	34.51	41.25
Ethylene	C_2H_4	−103.77	13.53	
N-Ethylformamide	C_3H_7NO	198		58.44
Ethyl formate	$C_3H_6O_2$	54.4	29.91	31.96
3-Ethylhexane	C_8H_{18}	118.6	33.59	39.64
Ethyl hexanoate	$C_8H_{16}O_2$	167		51.72
2-Ethylhexanoic acid	$C_8H_{16}O_2$	228		75.60
2-Ethyl-1-hexanol	$C_8H_{18}O$	184.6	54.2	68.51
2-Ethylhexyl acetate	$C_{10}H_{20}O_2$	199	43.5	
2-Ethylhexylamine	$C_8H_{19}N$	169.2	40.0	
Ethylisopropylamine	$C_5H_{13}N$	69.6	29.94	33.13
Ethyl isopropyl ether	$C_5H_{12}O$	54.1	28.21	30.08
Ethyl isopropyl sulfide	$C_5H_{12}S$	107.5	32.74	37.78
Ethyl 3-methylbutanoate	$C_7H_{14}O_2$	135.0	37.0	
2-Ethyl-3-methyl-1-butene	C_7H_{14}	89		34.35
1-Ethyl-1-methylcyclopentane	C_8H_{16}	121.6	33.20	38.85
3-Ethyl-2-methylpentane	C_8H_{18}	115.66	32.93	38.52
3-Ethyl-3-methylpentane	C_8H_{18}	118.27	32.78	37.99
3-Ethyl-2-methyl-1-pentene	C_8H_{16}	109.5		37.27
Ethyl 2-methylpropanoate	$C_6H_{12}O_2$	110.1	33.67	39.83
Ethyl methyl sulfide	C_3H_8S	66.7	29.53	31.85
3-Ethylpentane	C_7H_{16}	93.5	31.12	35.22
Ethyl pentanoate	$C_7H_{14}O_2$	146.1	36.96	47.01
3-Ethyl-3-pentanol	$C_7H_{16}O$	142		57.34
Ethyl pentyl ether	$C_7H_{16}O$	117.6	34.41	41.01
Ethyl propanoate	$C_5H_{10}O_2$	99.1	33.88	39.21
Ethyl propyl ether	$C_5H_{12}O$	63.21	28.94	31.43
Ethyl propyl sulfide	$C_5H_{12}S$	118.6	34.24	39.97
Ethyl trichloroacetate	$C_4H_5Cl_3O_2$	167.5		50.97
Ethyl vinyl ether	C_4H_8O	35.5	26.2	
Fluorobenzene	C_6H_5F	84.73	31.19	34.58
1-Fluorooctane	$C_8H_{17}F$	142.3	40.43	49.65
2-Fluorotoluene	C_7H_7F	115	35.4	
4-Fluorotoluene	C_7H_7F	116.6	34.08	39.42
Formamide	CH_3NO	220		60.15
Formic acid	CH_2O_2	101	22.69	20.10
Furan	C_4H_4O	31.5	27.10	27.45
Furfural	$C_5H_4O_2$	161.7	43.2	
Furfuryl alcohol	$C_5H_6O_2$	171	53.6	
Glycerol	$C_3H_8O_3$	290	61.0	
Glycerol triacetate	$C_9H_{14}O_6$	259		85.74
Heptadecane	$C_{17}H_{36}$	302.0	53.58	86.47
1-Heptadecanol	$C_{17}H_{36}O$	324		112.5
6-Heptadecanol	$C_{17}H_{36}O$			108.6
7-Heptadecanol	$C_{17}H_{36}O$			108.2
9-Heptadecanol	$C_{17}H_{36}O$			108.5
Heptane	C_7H_{16}	98.4	31.77	36.57
1,7-Heptanediol	$C_7H_{16}O_2$	262		97.9

Name	Mol. Form.	t_b °C	$\Delta_{vap}H(t_b)$ kJ/mol	$\Delta_{vap}H(25\ °C)$ kJ/mol
1-Heptanol	$C_7H_{16}O$	176.45		66.81
3-Heptanol	$C_7H_{16}O$	157	42.5	
2-Heptanone	$C_7H_{14}O$	151.05		47.24
1-Heptene	C_7H_{14}	93.64		35.49
cis-2-Heptene	C_7H_{14}	98.4		36.26
trans-2-Heptene	C_7H_{14}	98		36.27
cis-3-Heptene	C_7H_{14}	95.8		35.81
trans-3-Heptene	C_7H_{14}	95.7		35.84
Heptylamine	$C_7H_{17}N$	156		49.96
Heptylbenzene	$C_{13}H_{20}$	240		64.2
1-Hexacosanol	$C_{26}H_{54}O$			153.7
Hexadecane	$C_{16}H_{34}$	286.86	51.84	81.35
1,16-Hexadecanediol	$C_{16}H_{34}O_2$			163
1-Hexadecanol	$C_{16}H_{34}O$	312		107.7
1-Hexadecene	$C_{16}H_{32}$	284.9		80.25
Hexadecylbenzene	$C_{22}H_{38}$	385		104.8
Hexafluoroacetylacetone	$C_5H_2F_6O_2$	54.15	27.05	30.58
Hexafluorobenzene	C_6F_6	80.32	31.66	35.71
Hexafluoroethane	C_2F_6	−78.1	16.15	
Hexane	C_6H_{14}	68.73	28.85	31.56
1,6-Hexanediol	$C_6H_{14}O_2$	208		90.2
Hexanenitrile	$C_6H_{11}N$	163.65		47.91
1-Hexanol	$C_6H_{14}O$	157.6	44.50	61.61
2-Hexanol	$C_6H_{14}O$	140	41.01	58.46
2-Hexanone	$C_6H_{12}O$	127.6	36.35	43.14
3-Hexanone	$C_6H_{12}O$	123.5	35.36	42.47
1-Hexene	C_6H_{12}	63.48		30.61
cis-2-Hexene	C_6H_{12}	68.8		32.19
trans-2-Hexene	C_6H_{12}	67.9		31.60
cis-3-Hexene	C_6H_{12}	66.4		31.23
trans-3-Hexene	C_6H_{12}	67.1		31.55
Hexylamine	$C_6H_{15}N$	132.8	36.54	45.10
Hexylbenzene	$C_{12}H_{18}$	226.1		60.4
Hexyl methyl ether	$C_7H_{16}O$	126.1	34.93	42.07
Indan	C_9H_{10}	177.97	39.63	48.79
Iodobenzene	C_6H_5I	188.4	39.5	
1-Iodobutane	C_4H_9I	130.5	34.66	40.63
2-Iodobutane	C_4H_9I	120.1	33.27	38.46
Iodoethane	C_2H_5I	72.3	29.44	31.93
1-Iodohexane	$C_6H_{13}I$	181.3		49.75
Iodomethane	CH_3I	42.43	27.34	27.97
1-Iodo-2-methylpropane	C_4H_9I	121.1	33.54	38.83
2-Iodo-2-methylpropane	C_4H_9I	100.1	31.43	35.41
1-Iodopentane	$C_5H_{11}I$	157.0		45.27
1-Iodopropane	C_3H_7I	102.5	32.08	36.25
2-Iodopropane	C_3H_7I	89.5	30.68	34.06
Isobutane	C_4H_{10}	−11.73	21.30	19.23
Isobutyl acetate	$C_6H_{12}O_2$	116.5	35.9	
Isobutylamine	$C_4H_{11}N$	67.75	30.61	33.85
Isobutylbenzene	$C_{10}H_{14}$	172.79		48.0
Isobutyl formate	$C_5H_{10}O_2$	98.2	33.6	
Isobutyl isobutanoate	$C_8H_{16}O_2$	148.6	38.2	
Isobutyl methyl ether	$C_5H_{12}O$	58.6	28.02	30.13
Isopentane	C_5H_{12}	27.88	24.69	24.85
Isopentyl acetate	$C_7H_{14}O_2$	142.5	37.5	

Name	Mol. Form.	t_b °C	$\Delta_{vap}H(t_b)$ kJ/mol	$\Delta_{vap}H(25\,°C)$ kJ/mol
Isopentyl isopentanoate	$C_{10}H_{20}O_2$	190.4	45.9	
Isopropyl acetate	$C_5H_{10}O_2$	88.7	32.93	37.20
Isopropylamine	C_3H_9N	31.76	27.83	28.36
Isopropylbenzene	C_9H_{12}	152.41		45.13
Isopropylcyclohexane	C_9H_{18}	154.8		44.02
Isopropylcyclopentane	C_8H_{16}	126.5	33.56	39.44
Isopropylmethylamine	$C_4H_{11}N$	50.4	28.71	30.69
1-Isopropyl-4-methylbenzene	$C_{10}H_{14}$	177.1	38.2	
Isopropyl methyl ether	$C_4H_{10}O$	30.77	26.05	26.41
Isopropyl methyl sulfide	$C_4H_{10}S$	84.8	30.71	34.15
Isopropylpropylamine	$C_6H_{15}N$	96.9	32.14	37.23
Isopropyl propyl sulfide	$C_6H_{14}S$	132.1	35.11	41.78
Isoquinoline	C_9H_7N	243.22	49.0	60.26
Mesityl oxide	$C_6H_{10}O$	130	36.1	
Methane	CH_4	−161.48	8.19	
Methanol	CH_4O	64.6	35.21	37.43
2-Methoxyethanol	$C_3H_8O_2$	124.1	37.54	45.17
2-Methoxyethyl acetate	$C_5H_{10}O_3$	143	43.9	
Methyl acetate	$C_3H_6O_2$	56.87	30.32	32.29
Methyl acrylate	$C_4H_6O_2$	80.7	33.1	
2-Methylacrylonitrile	C_4H_5N	90.3	31.8	
Methylamine	CH_5N	−6.32	25.60	23.37
2-Methylaniline	C_7H_9N	200.3	44.6	
3-Methylaniline	C_7H_9N	203.3	44.9	
4-Methylaniline	C_7H_9N	200.4	44.3	
Methyl benzoate	$C_8H_8O_2$	199		55.57
1-Methylbicyclo[3,1,0]hexane	C_7H_{12}	93.1	31.07	34.77
3-Methylbutanenitrile	C_5H_9N	127.5	35.10	41.64
2-Methyl-1-butanethiol	$C_5H_{12}S$	119.1	33.79	39.45
2-Methyl-2-butanethiol	$C_5H_{12}S$	99.1	31.37	35.67
3-Methyl-2-butanethiol	$C_5H_{12}S$	109.8		37.5
Methyl butanoate	$C_5H_{10}O_2$	102.8	33.79	39.28
2-Methylbutanoic acid	$C_5H_{10}O_2$	177		46.91
2-Methyl-1-butanol	$C_5H_{12}O$	127.5		55.16
3-Methyl-1-butanol	$C_5H_{12}O$	131.1	44.07	55.61
2-Methyl-2-butanol	$C_5H_{12}O$	102.4	39.04	50.10
3-Methyl-2-butanol	$C_5H_{12}O$	112.9		53.0
3-Methyl-2-butanone	$C_5H_{10}O$	94.33	32.35	36.78
2-Methyl-1-butene	C_5H_{10}	31.2	25.50	25.92
3-Methyl-1-butene	C_5H_{10}	20.1		23.77
2-Methyl-2-butene	C_5H_{10}	38.56	26.31	27.06
(1-Methylbutyl)benzene	$C_{11}H_{16}$	199		53.0
Methyl tert-butyl ether	$C_5H_{12}O$	55.0	27.94	29.82
Methyl chloroacetate	$C_3H_5ClO_2$	129.5	39.23	46.73
Methyl cyanoacetate	$C_4H_5NO_2$	200.5	48.2	
Methyl cyclobutanecarboxylate	$C_6H_{10}O_2$	135.5	37.13	44.72
Methylcyclohexane	C_7H_{14}	100.93	31.27	35.36
1-Methylcyclohexanol	$C_7H_{14}O$	155	79.0	
cis-2-Methylcyclohexanol	$C_7H_{14}O$	165	48.5	
trans-2-Methylcyclohexanol	$C_7H_{14}O$	167.5	53.0	
Methylcyclopentane	C_6H_{12}	71.8	29.08	31.64
Methyl cyclopropanecarboxylate	$C_5H_8O_2$	114.9	35.25	41.27
2-Methyldecane	$C_{11}H_{24}$	189.3	40.25	54.28
4-Methyldecane	$C_{11}H_{24}$	187	40.70	53.76
Methyl dichloroacetate	$C_3H_4Cl_2O_2$	142.9	39.28	47.72

Name	Mol. Form.	t_b °C	$\Delta_{vap}H(t_b)$ kJ/mol	$\Delta_{vap}H(25\,°C)$ kJ/mol
Methyl 2,2-dimethylpropanoate	$C_6H_{12}O_2$	101.1	33.42	38.76
Methyl dodecanoate	$C_{13}H_{26}O_2$	267		77.17
N-Methylformamide	C_2H_5NO	199.51		56.19
Methyl formate	$C_2H_4O_2$	31.7	27.92	28.35
2-Methylheptane	C_8H_{18}	117.66	33.26	39.67
3-Methylheptane	C_8H_{18}	118.9	33.66	39.83
4-Methylheptane	C_8H_{18}	117.72	33.35	39.69
Methyl heptanoate	$C_8H_{16}O_2$	174		51.62
2-Methyl-2-heptanol	$C_8H_{18}O$	156		62.87
2-Methylhexane	C_7H_{16}	90.04	30.62	34.87
3-Methylhexane	C_7H_{16}	92	30.9	
Methyl hexanoate	$C_7H_{14}O_2$	149.5	38.55	48.04
2-Methyl-2-hexanol	$C_7H_{16}O$	143		58.57
5-Methyl-3-hexanol	$C_7H_{16}O$	147		59.82
cis-3-Methyl-3-hexene	C_7H_{14}	95.4		36.31
trans-3-Methyl-3-hexene	C_7H_{14}	93.5		35.70
Methylhydrazine	CH_6N_2	87.5	36.12	40.37
Methyl isobutanoate	$C_5H_{10}O_2$	92.5	32.61	37.32
Methyl methacrylate	$C_5H_8O_2$	100.5	36.0	
1-Methylnaphthalene	$C_{11}H_{10}$	244.7	45.5	
2-Methylnonane	$C_{10}H_{22}$	167.1	38.23	49.63
3-Methylnonane	$C_{10}H_{22}$	167.9	38.26	49.71
5-Methylnonane	$C_{10}H_{22}$	165.1	38.14	49.34
Methyl octanoate	$C_9H_{18}O_2$	192.9		56.41
Methyloxirane	C_3H_6O	35	27.35	27.89
2-Methylpentane	C_6H_{14}	60.26	27.79	29.89
3-Methylpentane	C_6H_{14}	63.27	28.06	30.28
2-Methyl-2,4-pentanediol	$C_6H_{14}O_2$	197.1	57.3	
Methyl pentanoate	$C_6H_{12}O_2$	127.4	35.36	43.10
2-Methyl-1-pentanol	$C_6H_{14}O$	149	50.2	
4-Methyl-1-pentanol	$C_6H_{14}O$	151.9	44.46	60.47
2-Methyl-2-pentanol	$C_6H_{14}O$	121.1	39.59	54.77
4-Methyl-2-pentanol	$C_6H_{14}O$	131.6	44.2	
3-Methyl-2-pentanone	$C_6H_{12}O$	117.5	34.16	40.53
4-Methyl-2-pentanone	$C_6H_{12}O$	116.5	34.49	40.61
2-Methyl-3-pentanone	$C_6H_{12}O$	113.5	33.84	39.79
2-Methyl-1-pentene	C_6H_{12}	62.1		30.48
3-Methyl-1-pentene	C_6H_{12}	54.2		28.62
4-Methyl-1-pentene	C_6H_{12}	53.9		28.71
2-Methyl-2-pentene	C_6H_{12}	67.3		31.60
3-Methyl-cis-2-pentene	C_6H_{12}	67.7		32.09
3-Methyl-trans-2-pentene	C_6H_{12}	70.4		31.35
4-Methyl-cis-2-pentene	C_6H_{12}	56.3		29.48
4-Methyl-trans-2-pentene	C_6H_{12}	58.6		29.97
Methyl pentyl ether	$C_6H_{14}O$	99	32.02	36.85
Methyl pentyl sulfide	$C_6H_{14}S$	145.1	37.41	45.24
2-Methylpropanenitrile	C_4H_7N	103.9	32.39	37.13
2-Methyl-1-propanethiol	$C_4H_{10}S$	88.5	31.01	34.63
2-Methyl-2-propanethiol	$C_4H_{10}S$	64.2	28.45	30.78
Methyl propanoate	$C_4H_8O_2$	79.8	32.24	35.85
2-Methylpropanoic acid	$C_4H_8O_2$	154.45		35.30
2-Methyl-1-propanol	$C_4H_{10}O$	107.89	41.82	50.82
2-Methyl-2-propanol	$C_4H_{10}O$	82.4	39.07	46.69
Methyl propyl ether	$C_4H_{10}O$	39.1	26.75	27.60
Methyl propyl sulfide	$C_4H_{10}S$	95.6	32.08	36.24

Name	Mol. Form.	t_b °C	$\Delta_{vap}H(t_b)$ kJ/mol	$\Delta_{vap}H(25\ °C)$ kJ/mol
2-Methylpyridine	C_6H_7N	129.38	36.17	42.48
3-Methylpyridine	C_6H_7N	144.14	37.35	44.44
4-Methylpyridine	C_6H_7N	145.36	37.51	44.56
2-Methylquinoline	$C_{10}H_9N$	246.5		66.1
4-Methylquinoline	$C_{10}H_9N$	262		67.6
6-Methylquinoline	$C_{10}H_9N$	258.6		67.7
8-Methylquinoline	$C_{10}H_9N$	247.5		65.7
Methyl salicylate	$C_8H_8O_3$	222.9	46.7	
4-Methylthiazole	C_4H_5NS	133.3	37.58	43.85
2-Methylthiophene	C_5H_6S	112.6	33.90	38.87
3-Methylthiophene	C_5H_6S	115.5	34.24	39.43
Methyl trichloroacetate	$C_3H_3Cl_3O_2$	153.8		48.33
Morpholine	C_4H_9NO	128	37.1	
Naphthalene	$C_{10}H_8$	217.9	43.2	
Neopentane	C_5H_{12}	9.48	22.74	21.84
Nitrobenzene	$C_6H_5NO_2$	210.8		55.01
Nitroethane	$C_2H_5NO_2$	114.0	38.0	
Nitromethane	CH_3NO_2	101.19	33.99	38.27
1-Nitropropane	$C_3H_7NO_2$	131.1	38.5	
2-Nitropropane	$C_3H_7NO_2$	120.2	36.8	
Nonadecane	$C_{19}H_{40}$	329.9	56.93	96.4
Nonane	C_9H_{20}	150.82	37.18	46.55
1,9-Nonanediol	$C_9H_{20}O_2$			112.5
1-Nonanol	$C_9H_{20}O$	213.37		76.86
2-Nonanone	$C_9H_{18}O$	195.3		56.44
5-Nonanone	$C_9H_{18}O$	188.45		53.30
Nonylbenzene	$C_{15}H_{24}$	280.5		74.1
Octadecane	$C_{18}H_{38}$	316.3	55.23	91.44
1-Octadecanol	$C_{18}H_{38}O$	335		116.8
cis-9-Octadecenoic acid	$C_{18}H_{34}O_2$	360	67.4	
Octane	C_8H_{18}	125.67	34.41	41.49
1,8-Octanediol	$C_8H_{18}O_2$			104.9
Octanenitrile	$C_8H_{15}N$	205.25		56.80
Octanoic acid	$C_8H_{16}O_2$	239	58.5	
1-Octanol	$C_8H_{18}O$	195.16		70.98
2-Octanol	$C_8H_{18}O$	179.3	44.4	
1-Octene	C_8H_{16}	121.29	34.07	40.34
Octylbenzene	$C_{14}H_{22}$	264		69.1
1-Octyne	C_8H_{14}	126.3	35.83	42.30
2-Octyne	C_8H_{14}	137.6	37.26	44.49
3-Octyne	C_8H_{14}	133.1	36.94	43.92
4-Octyne	C_8H_{14}	131.6	36.0	42.73
Oxetane	C_3H_6O	47.6	28.67	29.85
2-Oxetanone	$C_3H_4O_2$	162		47.03
Oxirane	C_2H_4O	10.6	25.54	24.75
Pentachloroethane	C_2HCl_5	162.0	36.9	
Pentadecane	$C_{15}H_{32}$	270.6	50.08	76.77
1,15-Pentadecanediol	$C_{15}H_{32}O_2$			139
1-Pentadecanol	$C_{15}H_{32}O$	300		103.5
Pentadecylbenzene	$C_{21}H_{36}$	373		100.3
Pentafluorobenzene	C_6HF_5	85.74	32.15	36.27
2,3,4,5,6-Pentafluorotoluene	$C_7H_3F_5$	117.5	34.75	41.12
2,2,4,6,6-Pentamethylheptane	$C_{12}H_{26}$	177.8		48.97
Pentane	C_5H_{12}	36.06	25.79	26.43
1,5-Pentanediol	$C_5H_{12}O_2$	239	60.7	83.0

Name	Mol. Form.	t_b °C	$\Delta_{vap}H(t_b)$ kJ/mol	$\Delta_{vap}H(25\,°C)$ kJ/mol
2,4-Pentanedione	$C_5H_8O_2$	138	34.30	41.77
Pentanenitrile	C_5H_9N	141.3	36.09	43.60
1-Pentanethiol	$C_5H_{12}S$	126.6	34.88	41.24
Pentanoic acid	$C_5H_{10}O_2$	186.1	44.1	
1-Pentanol	$C_5H_{12}O$	137.98	44.36	57.02
2-Pentanol	$C_5H_{12}O$	119.3	41.40	54.21
3-Pentanol	$C_5H_{12}O$	116.25		54.0
2-Pentanone	$C_5H_{10}O$	102.26	33.44	38.40
3-Pentanone	$C_5H_{10}O$	101.7	33.45	38.52
1-Pentene	C_5H_{10}	29.96	25.20	25.47
cis-2-Pentene	C_5H_{10}	36.93		26.86
trans-2-Pentene	C_5H_{10}	36.34		26.76
trans-3-Pentenenitrile	C_5H_7N	144	37.09	44.77
Pentyl acetate	$C_7H_{14}O_2$	149.2	38.42	48.56
Pentylamine	$C_5H_{13}N$	104.3	34.01	40.08
Pentylbenzene	$C_{11}H_{16}$	205.4		55.1
Pentylcyclohexane	$C_{11}H_{22}$	203.7		53.88
Perfluorobutane	C_4F_{10}	-1.9	22.9	
Perfluorocyclobutane	C_4F_8	-5.91	23.2	
Perfluorodecalin	$C_{10}F_{18}$	142.02		41.54
Perfluorohexane	C_6F_{14}	57.14		32.47
Perfluorononane	C_9F_{20}	117.61		45.27
Perfluorooctane	C_8F_{18}	105.9	33.38	41.13
Perfluorotoluene	C_7F_8	103.55		40.52
Phenanthrene	$C_{14}H_{10}$	340		75.50
Phenol	C_6H_6O	181.87	45.69	57.82
Piperidine	$C_5H_{11}N$	106.22		39.29
Propanal	C_3H_6O	48	28.31	29.62
Propane	C_3H_8	-42.1	19.04	14.79
1,3-Propanediamine	$C_3H_{10}N_2$	139.8	40.85	50.16
1,2-Propanediol	$C_3H_8O_2$	187.6	52.4	
1,3-Propanediol	$C_3H_8O_2$	214.4	57.9	69.8
1,3-Propanedithiol	$C_3H_8S_2$	172.9		49.66
Propanenitrile	C_3H_5N	97.14	31.81	36.03
1-Propanethiol	C_3H_8S	67.8	29.54	31.89
2-Propanethiol	C_3H_8S	52.6	27.91	29.45
Propanoic acid	$C_3H_6O_2$	141.15		32.14
Propanoic anhydride	$C_6H_{10}O_3$	170	41.7	
1-Propanol	C_3H_8O	97.2	41.44	47.45
2-Propanol	C_3H_8O	82.3	39.85	45.39
Propene	C_3H_6	-47.69	18.42	14.24
2-Propoxyethanol	$C_5H_{12}O_2$	149.8	41.40	52.12
Propyl acetate	$C_5H_{10}O_2$	101.3	33.92	39.72
Propylamine	C_3H_9N	47.22	29.55	31.27
Propylbenzene	C_9H_{12}	159.24		46.22
Propylcyclohexane	C_9H_{18}	156		45.08
Propylcyclopentane	C_8H_{16}	131	34.70	41.08
Propyl formate	$C_4H_8O_2$	80.9	33.61	37.53
Propyl propanoate	$C_6H_{12}O_2$	122.5	35.54	43.45
Pyridazine	$C_4H_4N_2$	208		53.47
Pyridine	C_5H_5N	115.23	35.09	40.21
Pyrimidine	$C_4H_4N_2$	123.8	43.09	49.79
Pyrrole	C_4H_5N	129.79	38.75	45.09
Pyrrolidine	C_4H_9N	86.56	33.01	37.52
Quinoline	C_9H_7N	237.16	49.7	59.30

Name	Mol. Form.	t_b °C	$\Delta_{vap}H(t_b)$ kJ/mol	$\Delta_{vap}H(25\ °C)$ kJ/mol
Salicylaldehyde	$C_7H_6O_2$	197	38.2	
Spiro[2.2]pentane	C_5H_8	39	26.76	27.49
Styrene	C_8H_8	145	38.7	
Succinonitrile	$C_4H_4N_2$	266	48.5	
1,1,2,2-Tetrabromoethane	$C_2H_2Br_4$	243.5	48.7	
1,1,2,2-Tetrachloroethane	$C_2H_2Cl_4$	145.2	37.64	45.71
Tetrachloroethene	C_2Cl_4	121.3	34.68	39.68
Tetrachloromethane	CCl_4	76.8	29.82	32.43
Tetradecane	$C_{14}H_{30}$	253.58	48.16	71.73
1,14-Tetradecanediol	$C_{14}H_{30}O_2$			149.7
Tetradecanenitrile	$C_{14}H_{27}N$			85.29
1-Tetradecanol	$C_{14}H_{30}O$	287		98.9
Tetradecylbenzene	$C_{20}H_{34}$	359		95.8
Tetrahydrofuran	C_4H_8O	65	29.81	31.99
Tetrahydrofurfuryl alcohol	$C_5H_{10}O_2$	178	45.2	
1,2,3,4-Tetrahydronaphthalene	$C_{10}H_{12}$	207.6	43.9	
Tetrahydropyran	$C_5H_{10}O$	88	31.17	34.58
Tetrahydrothiophene	C_4H_8S	121.1	34.66	39.43
2,2,3,3-Tetramethylbutane	C_8H_{18}	106.45		42.90
2,2,4,4-Tetramethylpentane	C_9H_{20}	122.29	32.51	38.49
Tetranitromethane	CN_4O_8	126.1	40.74	49.93
Thiacyclohexane	$C_5H_{10}S$	141.8	35.96	42.58
Thietane	C_3H_6S	95.0	32.32	35.97
Thiophene	C_4H_4S	84.0	31.48	34.70
Toluene	C_7H_8	110.63	33.18	38.01
Triacetamide	$C_6H_9NO_3$			60.41
Tribromomethane	$CHBr_3$	149.1	39.66	46.05
Tributylamine	$C_{12}H_{27}N$	216.5	46.9	
Tributyl borate	$C_{12}H_{27}BO_3$	234	56.1	
1,1,1-Trichloroethane	$C_2H_3Cl_3$	74.09	29.86	32.50
1,1,2-Trichloroethane	$C_2H_3Cl_3$	113.8	34.82	40.24
Trichloroethene	C_2HCl_3	87.21	31.40	34.54
Trichlorofluoromethane	CCl_3F	23.7	25.1	
Trichloromethane	$CHCl_3$	61.17	29.24	31.28
1,2,3-Trichloropropane	$C_3H_5Cl_3$	157	37.1	
1,1,1-Trichloro-2,2,2-trifluoroethane	$C_2Cl_3F_3$	45.5	26.85	28.08
1,1,2-Trichloro-1,2,2-trifluoroethane	$C_2Cl_3F_3$	47.7	27.04	28.40
Tridecane	$C_{13}H_{28}$	235.47	46.20	66.68
1,13-Tridecanediol	$C_{13}H_{28}O_2$			133
1-Tridecanol	$C_{13}H_{28}O$	274		94.7
Tridecylbenzene	$C_{19}H_{32}$	346		91.8
Triethylamine	$C_6H_{15}N$	89	31.01	34.84
Triethylene glycol	$C_6H_{14}O_4$	285	71.4	
Trifluoroacetic acid	$C_2HF_3O_2$	73	33.3	
1,1,1-Trifluoroethane	$C_2H_3F_3$	-47.25	18.99	
(Trifluoromethyl)benzene	$C_7H_5F_3$	102.1	32.63	37.60
Trimethylamine	C_3H_9N	2.87	22.94	21.66
1,2,3-Trimethylbenzene	C_9H_{12}	176.12		49.05
1,2,4-Trimethylbenzene	C_9H_{12}	169.38		47.93
1,3,5-Trimethylbenzene	C_9H_{12}	164.74		47.50
2,2,3-Trimethylbutane	C_7H_{16}	80.86	28.90	32.05
2,3,3-Trimethyl-1-butene	C_7H_{14}	77.9		32.09
2,2,5-Trimethylhexane	C_9H_{20}	124.09	33.65	40.16
2,3,5-Trimethylhexane	C_9H_{20}	131.4	34.43	41.41
3,5,5-Trimethyl-1-hexanol	$C_9H_{20}O$	194		67.86

Name	Mol. Form.	t_b °C	$\Delta_{vap}H(t_b)$ kJ/mol	$\Delta_{vap}H(25\,°C)$ kJ/mol
2,4,7-Trimethyloctane	$C_{11}H_{24}$	168.1	38.22	49.91
2,2,3-Trimethylpentane	C_8H_{18}	110	31.94	36.91
2,2,4-Trimethylpentane	C_8H_{18}	99.22	30.79	35.14
2,3,3-Trimethylpentane	C_8H_{18}	114.8	32.12	37.27
2,3,4-Trimethylpentane	C_8H_{18}	113.5	32.36	37.75
2,2,4-Trimethyl-3-pentanone	$C_8H_{16}O$	135.1	35.64	43.30
2,4,4-Trimethyl-1-pentene	C_8H_{16}	101.4		35.59
2,4,4-Trimethyl-2-pentene	C_8H_{16}	104.9		37.23
2,3,6-Trimethylpyridine	$C_8H_{11}N$	171.6	39.95	50.61
2,4,6-Trimethylpyridine	$C_8H_{11}N$	170.6	39.87	50.33
Tris(perfluorobutyl)amine	$C_{12}F_{27}N$	178	46.4	
Undecane	$C_{11}H_{24}$	195.9	41.91	56.58
1,11-Undecanediol	$C_{11}H_{24}O_2$			132
Undecanenitrile	$C_{11}H_{21}N$	253		71.14
1-Undecanol	$C_{11}H_{24}O$	245		85.8
Undecylbenzene	$C_{17}H_{28}$	316		82.4
Vinyl acetate	$C_4H_6O_2$	72.8	34.6	
o-Xylene	C_8H_{10}	144.5	36.24	43.43
m-Xylene	C_8H_{10}	139.07	35.66	42.65
p-Xylene	C_8H_{10}	138.23	35.67	42.40
2,4-Xylenol	$C_8H_{10}O$	210.98		64.96
2,5-Xylenol	$C_8H_{10}O$	211.1	46.9	
2,6-Xylenol	$C_8H_{10}O$	201.07		75.31
3,4-Xylenol	$C_8H_{10}O$	227		85.03
3,5-Xylenol	$C_8H_{10}O$	221.74		82.01

ENTHALPY OF FUSION

This table lists the molar enthalpy (heat) of fusion, $\Delta_{fus}H$, of over 1100 inorganic and organic compounds. All values refer to the enthalpy change at equilibrium between the liquid phase and the most stable solid phase at the phase transition temperature. Most values of $\Delta_{fus}H$ are given at the normal melting point t_m. However, a "t" following the entry in the melting point column indicates a triple–point temperature, where the solid, liquid, and gas phases are in equilibrium. Temperatures are given on the ITS–90 scale.

A * following an entry indicates that the value includes the enthalpy of transition between crystalline phases whose transformation occurs within 1 °C of the melting point.

Substances are listed by name, either an IUPAC systematic name or, in the case of drugs and other complex compounds, a common synonym. Inorganic compounds, including metal salts of organic acids, are listed first, followed by organic compounds. The molecular formula in the Hill convention is included.

References

1. Chase, M. W., Davies, C. A., Downey, J. R., Frurip, D. J., McDonald, R. A., and Syverud, A. N., *JANAF Thermochemical Tables, Third Edition, J. Phys. Chem. Ref. Data*, Vol. 14, Suppl. 1, 1985.
2. Chase, M. W., *NIST–JANAF Thermochemical Tables, Fourth Edition, J. Phys. Chem. Ref. Data*, Monograph No. 9, 1998.
3. Gurvich, L. V., Veyts, I. V., and Alcock, C. B., *Thermodynamic Properties of Individual Substances, Fourth Edition*; Vol. 2, Hemisphere Publishing Corp., New York, 1991; Vol. 3, CRC Press, Boca Raton, FL, 1994.
4. Dinsdale, A. T., "SGTE Data for Pure Elements", *CALPHAD*, 15, 317–425, 1991.
5. *Landolt–Börnstein, Numerical Data and Functional Relationships in Science and Technology*, New Series, IV/8A, "Enthalpies of Fusion and Transition of Organic Compounds," Springer–Verlag, Heidelberg, 1995.
6. *Landolt–Börnstein, Numerical Data and Functional Relationships in Science and Technology*, New Series, IV/19A, "Thermodynamic Properties of Inorganic Materials compiled by SGTE," Springer–Verlag, Heidelberg; Part 1, 1999; Part 2; 1999; Part 3, 2000; Part 4, 2001.
7. Janz, G. J., et al., *Physical Properties Data Compilations Relevant to Energy Storage. II. Molten Salts*, Nat. Stand. Ref. Data Sys.– Nat. Bur. Standards (U.S.), No. 61, Part 2, 1979.
8. Dirand, M., Bouroukba, M., Chevallier, V., Petitjean, D., Behar, E., and Ruffier–Meray, V., "Normal Alkanes, Multialkane Synthetic Model Mixtures, and Real Petroleum Waxes: Crystallographic Structures, Thermodynamic Properties, and Crystallization," *J. Chem. Eng. Data*, 47, 115–143, 2002.
9. Linstrom, P. J., and Mallard, W. G., Editors, *NIST Chemistry WebBook*, NIST Standard Reference Database No. 69, June 2005, National Institute of Standards and Technology, Gaithersburg, MD, <http://webbook.nist.gov>.
10. Thermodynamic Research Center, National Institute of Standards and Technology, *TRC Thermodynamic Tables*, <http://trc.nist.gov>.
11. Sangster, J., "Phase Diagrams and Thermodynamic Properties of Binary Systems of Drugs," *J. Phys. Chem. Ref. Data* 28, 889, 1999.

Name	Molecular formula	t_m/°C	$\Delta_{fus}H$/ kJ mol^{-1}
Inorganic compounds (including salts of organic acids)			
Actinium	Ac	1050	12.0
Aluminum	Al	660.323	10.71
Aluminum bromide	AlBr$_3$	97.5	11.25
Aluminum chloride	AlCl$_3$	192.6	35.35
Aluminum fluoride	AlF$_3$	2250 t	98
Aluminum iodide	AlI$_3$	188.28	15.90
Aluminum oxide (α)	Al$_2$O$_3$	2054	111.1
Aluminum sulfide	Al$_2$S$_3$	1100	66
Americium	Am	1176	14.39
Ammonia	H$_3$N	−77.73	5.66
Ammonium chloride	ClH$_4$N	520.1	10.6
Ammonium fluoride	FH$_4$N	238	12.6
Ammonium iodide	H$_4$IN	551	21
Ammonium nitrate	H$_4$N$_2$O$_3$	169.7	5.86
Antimony (gray)	Sb	630.628	19.79
Antimony(III) bromide	Br$_3$Sb	97	14.6
Antimony(III) chloride	Cl$_3$Sb	73.4	12.97
Antimony(III) fluoride	F$_3$Sb	287	22.8
Antimony(III) iodide	I$_3$Sb	171	22.8
Antimony(III) oxide (valentinite)	O$_3$Sb$_2$	655	54
Antimony(III) sulfide	S$_3$Sb$_2$	550	47.9
Argon	Ar	−189.34	1.18
Arsenic (gray)	As	817	24.44
Arsenic(III) bromide	AsBr$_3$	31.1	11.7
Arsenic(III) chloride	AsCl$_3$	−16	10.1
Arsenic(III) fluoride	AsF$_3$	−5.9	10.4
Arsenic(III) iodide	AsI$_3$	141	21.8
Arsenic(III) oxide (claudetite)	As$_2$O$_3$	314	18
Arsenic(V) oxide	As$_2$O$_5$	730	60
Arsenic(III) selenide	As$_2$Se$_3$	377	40.8
Arsenic(III) sulfide	As$_2$S$_3$	312	28.7
Arsenic sulfide	As$_4$S$_4$	307	25.4
Arsenic(III) telluride	As$_2$Te$_3$	375	46.0
Barium	Ba	727	7.12
Barium bromide	BaBr$_2$	857	32.2
Barium carbonate	CBaO$_3$	1555 (high pres.)	40
Barium chloride	BaCl$_2$	961	15.85
Barium fluoride	BaF$_2$	1368	23.36
Barium hydride	BaH$_2$	1200	25
Barium hydroxide	BaH$_2$O$_2$	408	16
Barium iodide	BaI$_2$	711	26.5
Barium oxide	BaO	1973	46
Barium sulfate	BaO$_4$S	1580	40
Barium sulfide	BaS	2227	63
Beryllium	Be	1287	7.895
Beryllium bromide	BeBr$_2$	508	18
Beryllium carbide	CBe$_2$	2127	75.3
Beryllium chloride	BeCl$_2$	415	8.66
Beryllium fluoride	BeF$_2$	552	4.77
Beryllium iodide	BeI$_2$	480	20.92
Beryllium nitride	Be$_3$N$_2$	2200	111
Beryllium oxide	BeO	2578	86
Beryllium sulfate	BeO$_4$S	1127	6
Bismuth	Bi	271.406	11.106
Bismuth oxide	Bi$_2$O$_3$	825	14.7

Name	Molecular formula	t_m/°C	$\Delta_{fus}H$/ kJ mol^{-1}	Name	Molecular formula	t_m/°C	$\Delta_{fus}H$/ kJ mol^{-1}
Bismuth sulfide	Bi_2S_3	777	78.2	Chromium(II) fluoride	CrF_2	894	34
Bismuth tribromide	$BiBr_3$	219	21.7	Chromium(III) fluoride	CrF_3	1425	66
Bismuth trichloride	$BiCl_3$	234	23.6	Chromium(II) iodide	CrI_2	867	46
Bismuth trifluoride	BiF_3	649	21.6	Chromium(III) iodide	CrI_3	857	61
Bismuth triiodide	BiI_3	408.6	39.1	Chromium(III) oxide	Cr_2O_3	2432	125
Boric acid	BH_3O_3	170.9	22.3	Chromium(VI) oxide	CrO_3	197	14.2
Boron	B	2077	50.2	Chromium(II) sulfide	CrS	1567	25.5
Boron nitride	BN	2967	81	Cobalt	Co	1495	16.20
Boron oxide	B_2O_3	450	24.56	Cobalt(II) bromide	Br_2Co	678	43
Boron sulfide	B_2S_3	563	48.12	Cobalt(II) chloride	Cl_2Co	737	46.0
Boron trichloride	BCl_3	−107.3	2.10	Cobalt(II) fluoride	CoF_2	1127	58.1
Boron trifluoride	BF_3	−126.8	4.20	Cobalt(II) iodide	CoI_2	520	35
Bromine	Br_2	−7.2	10.57	Cobalt(II) selenite	CoO_3Se	659	16.3
Bromine pentafluoride	BrF_5	−60.5	5.67	Cobalt(II) sulfide	CoS	1117	30
Cadmium	Cd	321.069	6.21	Copper	Cu	1084.62	13.26
Cadmium bromide	Br_2Cd	568	33.35	Copper(I) bromide	BrCu	483	5.1
Cadmium chloride	$CdCl_2$	568	48.58	Copper(I) chloride	ClCu	423	7.08
Cadmium fluoride	CdF_2	1075	22.6	Copper(II) chloride	Cl_2Cu	598	15.0
Cadmium iodide	CdI_2	388	15.3	Copper(II) fluoride	CuF_2	836	55
Cadmium nitrate	CdN_2O_6	360	18.3	Copper(I) iodide	CuI	591	7.93
Calcium	Ca	842	8.54	Copper(I) oxide	Cu_2O	1244	65.6
Calcium bromide	Br_2Ca	742	29.1	Copper(II) oxide	CuO	1227	49
Calcium carbonate (calcite)	$CCaO_3$	800	36	Copper(I) sulfide	Cu_2S	1129	9.62
Calcium chloride	$CaCl_2$	775	28.05	Curium	Cm	1345	14.64
Calcium fluoride	CaF_2	1418	30	Decaborane(14)	$B_{10}H_{14}$	98.78	21.97
Calcium hydride	CaH_2	1000	6.7	Dysprosium	Dy	1412	11.35
Calcium iodide	CaI_2	783	41.8	Dysprosium(III) fluoride	DyF_3	1157	58.6
Calcium nitrate	CaN_2O_6	561	23.4	Dysprosium(III) oxide	Dy_2O_3	2408	120
Calcium oxide	CaO	2613	80	Einsteinium	Es	860	9.41
Calcium sulfate	CaO_4S	1460	28	Erbium	Er	1529	19.90
Calcium sulfide	CaS	2524	70	Erbium chloride	Cl_3Er	776	32.6
Carbon (graphite)	C	4489	117.4	Erbium fluoride	ErF_3	1146	28.2
Cerium	Ce	799	5.460	Erbium oxide	Er_2O_3	2418	130
Cerium(III) bromide	Br_3Ce	732	51.9	Europium	Eu	822	9.21
Cerium(III) chloride	$CeCl_3$	807	53.1	Europium(II) bromide	Br_2Eu	683	25.1
Cerium(III) fluoride	CeF_3	1430	55.6	Europium(III) chloride	Cl_3Eu	623	33.1
Cerium(III) iodide	CeI_3	760	51.0	Europium(III) fluoride	EuF_3	647	6.40
Cerium(III) oxide	Ce_2O_3	2250	120	Europium (II) oxide	EuO	1967	40
Cerium(IV) oxide	CeO_2	2480	80	Europium(III) oxide	Eu_2O_3	2350	117
Cesium	Cs	28.5	2.09	Fluorine	F_2	−219.67	0.51
Cesium carbonate	CCs_2O_3	793	31	Gadolinium	Gd	1313	9.67
Cesium chloride	ClCs	646	20.4	Gadolinium(III) bromide	Br_3Gd	785	38.1
Cesium chromate	$CrCs_2O_4$	963	35.3	Gadolinium(III) chloride	Cl_3Gd	602	40.6
Cesium fluoride	CsF	703	21.7	Gadolinium(III) fluoride	F_3Gd	1232	52.4
Cesium hydride	CsH	528	15	Gadolinium(III) iodide	GdI_3	930	54.0
Cesium hydroxide	CsHO	342.3	7.78	Gadolinium(III) oxide	Gd_2O_3	2425	60
Cesium iodide	CsI	632	25.7	Gallium	Ga	29.7646	5.585
Cesium metaborate	$BCsO_2$	732	27	Gallium antimonide	GaSb	712	25.1
Cesium molybdate	Cs_2MoO_4	956.3	31.8	Gallium arsenide	AsGa	1238	87.64
Cesium nitrate	$CsNO_3$	409	13.8	Gallium(III) bromide	Br_3Ga	123	11.7
Cesium nitrite	$CsNO_2$	406	10.9	Gallium(III) chloride	Cl_3Ga	77.9	11.51
Cesium oxide	Cs_2O	495	20	Gallium(III) iodide	GaI_3	212	12.9
Cesium peroxide	Cs_2O_2	594	22	Gallium(III) oxide	Ga_2O_3	1807	100
Cesium sulfate	Cs_2O_4S	1005	35.7	Germanium	Ge	938.25	36.94
Chlorine	Cl_2	−101.5	6.40	Germanium(IV) bromide	Br_4Ge	26.1	12
Chromium	Cr	1907	21.00	Germanium(II) iodide	GeI_2	428	33.3
Chromium(II) bromide	Br_2Cr	842	45	Germanium(IV) iodide	GeI_4	146	19.1
Chromium(III) bromide	Br_3Cr	812	60	Germanium(IV) oxide	GeO_2	1116	12.6
Chromium(II) chloride	Cl_2Cr	824	45.0	Germanium(II) selenide	GeSe	675	24.7
Chromium(III) chloride	Cl_3Cr	827	60	Germanium(II) sulfide	GeS	658	21.3

Name	Molecular formula	$t_m/°C$	$\Delta_{fus}H/$ kJ mol^{-1}	Name	Molecular formula	$t_m/°C$	$\Delta_{fus}H/$ kJ mol^{-1}
Germanium(IV) sulfide	GeS$_2$	840	16.3	Lead(II) oxide (massicot)	OPb	887	25.6
Germanium(II) telluride	GeTe	724	47.3	Lead(II) sulfate	O$_4$PbS	1087	40.2
Gold	Au	1064.18	12.55	Lead(II) sulfide	PbS	1113	49.4
Hafnium	Hf	2233	27.20	Lithium	Li	180.50	3.00
Hafnium nitride	HfN	3310	62.8	Lithium aluminate	AlLiO$_2$	1610	87.9
Hafnium(IV) oxide	HfO$_2$	2800	96	Lithium bromide	BrLi	550	17.66
Holmium	Ho	1472	11.76	Lithium carbonate	CLi$_2$O$_3$	732	44.8
Holmium bromide	Br$_3$Ho	919	50.1	Lithium chloride	ClLi	610	19.8
Holmium chloride	Cl$_3$Ho	720	30.5	Lithium chromate	CrLi$_2$O$_4$	482	30.5
Holmium fluoride	F$_3$Ho	1143	56.3	Lithium fluoride	FLi	848.2	27.09
Holmium oxide	Ho$_2$O$_3$	2415	130	Lithium hexafluoroaluminate	AlF$_6$Li$_3$	785	86.19
Hydrazine	H$_4$N$_2$	1.54	12.66	Lithium hydride	HLi	692	21.8
Hydrogen	H$_2$	−259.16	0.12	Lithium hydride-d	DLi	694	22
Hydrogen bromide	BrH	−86.80	2.41	Lithium hydroxide	HLiO	473	20.9
Hydrogen chloride	ClH	−114.17	2.00	Lithium iodide	ILi	469	14.6
Hydrogen fluoride	FH	−83.36	4.58	Lithium metasilicate	Li$_2$O$_3$Si	1201	28
Hydrogen iodide	HI	−50.76	2.87	Lithium nitrate	LiNO$_3$	253	26.7
Hydrogen peroxide	H$_2$O$_2$	−0.43	12.50	Lithium nitrite	LiNO$_2$	222	9.2
Hydrogen sulfide	H$_2$S	−85.5	2.38	Lithium oxide	Li$_2$O	1438	35.6
Indium	In	156.60	3.291	Lithium perchlorate	ClLiO$_4$	236	29.3
Indium antimonide	InSb	524	47.7	Lithium sulfate	Li$_2$O$_4$S	860	9.00
Indium arsenide	AsIn	942	77.0	Lutetium	Lu	1663	18.65
Indium(I) bromide	BrIn	285	24.3	Lutetium oxide	Lu$_2$O$_3$	2490	133
Indium(III) bromide	Br$_3$In	420	26	Magnesium	Mg	650	8.48
Indium(I) chloride	ClIn	225	9.20	Magnesium bromide	Br$_2$Mg	711	39.3
Indium(III) chloride	Cl$_3$In	583	27	Magnesium carbonate	CMgO$_3$	990	59
Indium(III) fluoride	F$_3$In	1172	64	Magnesium chloride	Cl$_2$Mg	714	43.1
Indium(I) iodide	IIn	364.4	17.26	Magnesium fluoride	F$_2$Mg	1263	58.7
Indium(II) iodide	I$_2$In	155	1.29	Magnesium hydride	H$_2$Mg	327	14
Indium(III) iodide	I$_3$In	207	18.48	Magnesium iodide	I$_2$Mg	634	26
Indium(III) oxide	In$_2$O$_3$	1912	105	Magnesium orthosilicate	Mg$_2$O$_4$Si	1897	71
Indium(II) sulfide	InS	692	36.0	Magnesium oxide	MgO	2825	77
Iodine	I$_2$	113.7	15.52	Magnesium phosphate	Mg$_3$O$_8$P$_2$	1348	121
Iodine chloride	ClI	27.38	11.6	Magnesium sulfate	MgO$_4$S	1137	14.6
Iridium	Ir	2446	41.12	Magnesium sulfide	MgS	2226	63
Iridium(VI) fluoride	F$_6$Ir	44	8.40	Magnesium tetraboride	B$_4$Mg	727	0.0
Iron	Fe	1538	13.81	Manganese	Mn	1246	12.91
Iron boride (FeB)	BFe	1658	62.66	Manganese(II) bromide	Br$_2$Mn	698	33.5
Iron(II) bromide	Br$_2$Fe	691	43.0	Manganese(II) chloride	Cl$_2$Mn	650	30.7
Iron(II) chloride	Cl$_2$Fe	677	42.83	Manganese(II) fluoride	F$_2$Mn	900	30
Iron(III) chloride	Cl$_3$Fe	307.6	40	Manganese(II) iodide	I$_2$Mn	638	41.8
Iron(II) fluoride	F$_2$Fe	1100	50	Manganese(II) oxide	MnO	1842	43.9
Iron(III) fluoride	F$_3$Fe	367	0.58	Manganese(II) sulfide (α form)	MnS	1530	26.1
Iron(II) iodide	FeI$_2$	594	39	Mercury	Hg	−38.829	2.295
Iron(II) oxide	FeO	1377	24.1	Mercury(II) bromide	Br$_2$Hg	241	17.9
Iron(II,III) oxide	Fe$_3$O$_4$	1597	138	Mercury(II) chloride	Cl$_2$Hg	277	19.41
Iron(III) oxide	Fe$_2$O$_3$	1539	87	Mercury(II) fluoride	F$_2$Hg	645	23.0
Iron sodium oxide	FeNaO$_2$	1347	49.4	Mercury(I) iodide	Hg$_2$I$_2$	290	31.4
Iron(II) sulfide	FeS	1188	31.5	Mercury(II) iodide (yellow)	HgI$_2$	256	15.6
Krypton	Kr	−157.37	1.64	Mercury(II) sulfide (black)	HgS	820	40
Lanthanum	La	920	6.20	Metaboric acid (γ form)	BHO$_2$	236	14.3
Lanthanum bromide	Br$_3$La	788	54.0	Molybdenum	Mo	2622	37.48
Lanthanum chloride	Cl$_3$La	858	54.4	Molybdenum boride (Mo$_2$B$_5$)	B$_5$Mo$_2$	2210	226
Lanthanum fluoride	F$_3$La	1493	50.2	Molybdenum(IV) chloride	Cl$_4$Mo	317	16.7
Lanthanum iodide	I$_3$La	778	56.1	Molybdenum(V) chloride	Cl$_5$Mo	194	19
Lead	Pb	327.462	4.774	Molybdenum(VI) dioxydichloride	Cl$_2$MoO$_2$	176	17.0
Lead(II) bromide	Br$_2$Pb	371	16.44	Molybdenum(V) fluoride	F$_5$Mo	45.67	6.1
Lead(II) chloride	Cl$_2$Pb	501	21.88	Molybdenum(VI) fluoride	F$_6$Mo	17.5	4.33
Lead(II) fluoride	F$_2$Pb	830	14.7	Molybdenum monoboride	BMo	2600	55.23
Lead(II) iodide	I$_2$Pb	410	23.4	Molybdenum(VI) oxide	MoO$_3$	802	48.7

Name	Molecular formula	$t_m/°C$	$\Delta_{fus}H/$ kJ mol^{-1}	Name	Molecular formula	$t_m/°C$	$\Delta_{fus}H/$ kJ mol^{-1}
Molybdenum(VI) oxytetrachloride	Cl$_4$MoO	105	14.3	Plutonium(III) iodide	I$_3$Pu	777	50.2
Molybdenum(VI) oxytetrafluoride	F$_4$MoO	97.2	4	Plutonium(III) oxide	O$_3$Pu$_2$	2085	113
Molybdenum(V) oxytrichloride	Cl$_3$MoO	310	22	Plutonium(IV) oxide	O$_2$Pu	2390	67
Molybdenum(III) sulfide	Mo$_2$S$_3$	1807	0.13	Polonium	Po	254	10.0
Neodymium	Nd	1016	7.14	Potassium	K	63.5	2.335
Neodymium(III) bromide	Br$_3$Nd	682	45.3	Potassium aluminate	AlKO$_2$	1713	82
Neodymium(III) chloride	Cl$_3$Nd	759	48.5	Potassium bromide	BrK	734	25.52
Neodymium(III) fluoride	F$_3$Nd	1377	54.8	Potassium carbonate	CK$_2$O$_3$	899	27.6
Neodymium(III) iodide	I$_3$Nd	787	41.5	Potassium chloride	ClK	771	26.28
Neon	Ne	−248.59	0.328	Potassium chromate	CrK$_2$O$_4$	974	33.0
Neptunium	Np	644	3.20	Potassium cyanide	CKN	622	14.6
Nickel	Ni	1455	17.48	Potassium fluoride	FK	858	27.2
Nickel boride (Ni$_2$B)	BNi$_2$	1125	42.15	Potassium fluoroborate	BF$_4$K	570	17.66
Nickel boride (Ni$_3$B)	BNi$_3$	1166	72.28	Potassium hydride	HK	619	21
Nickel(II) bromide	Br$_2$Ni	963	56	Potassium hydrogen fluoride	F$_2$HK	238.8	6.62
Nickel(II) chloride	Cl$_2$Ni	1031	77.9	Potassium hydroxide	HKO	406	7.90
Nickel(II) fluoride	F$_2$Ni	1380	69	Potassium iodide	IK	681	24.0
Nickel(II) iodide	I$_2$Ni	800	48	Potassium metaborate	BKO$_2$	947	31.38
Nickel(II) oxide	NiO	1957	50.7	Potassium nitrate	KNO$_3$	334	9.6
Nickel(II) sulfide	NiS	976	30.1	Potassium nitrite	KNO$_2$	438	16.7
Nickel disulfide	NiS$_2$	1007	65.7	Potassium oxide	K$_2$O	740	27
Nickel subsulfide	Ni$_3$S$_2$	789	19.7	Potassium peroxide	K$_2$O$_2$	545	20.5
Niobium	Nb	2477	30	Potassium sulfate	K$_2$O$_4$S	1069	36.6
Niobium(V) bromide	Br$_5$Nb	254	24.0	Potassium sulfide	K$_2$S	948	16.15
Niobium(V) chloride	Cl$_5$Nb	205.8	33.9	Potassium superoxide	KO$_2$	535	20.6
Niobium(V) fluoride	F$_5$Nb	80	12.2	Praseodymium	Pr	931	6.89
Niobium(V) iodide	I$_5$Nb	327	37.7	Praseodymium(III) bromide	Br$_3$Pr	693	47.3
Niobium nitride	NNb	2050	46.0	Praseodymium(III) chloride	Cl$_3$Pr	786	50.6
Niobium(II) oxide	NbO	1937	85.4	Praseodymium(III) fluoride	F$_3$Pr	1399	57.3
Niobium(IV) oxide	NbO$_2$	1901	92	Praseodymium(III) iodide	I$_3$Pr	738	53.1
Niobium(V) oxide	Nb$_2$O$_5$	1512	104.3	Protactinium	Pa	1572	12.34
Nitric acid	HNO$_3$	−41.6	10.5	Radium	Ra	696	7.7
Nitric oxide	NO	−163.6	2.30	Rhenium	Re	3185	34.08
Nitrogen	N$_2$	−210.0	0.71	Rhenium(VII) oxide	O$_7$Re$_2$	327	65.7
Nitrogen tetroxide	N$_2$O$_4$	−9.3	14.65	Rhodium	Rh	1963	26.59
Nitrous oxide	N$_2$O	−90.8	6.54	Rubidium	Rb	39.30	2.19
Osmium	Os	3033	57.85	Rubidium bromide	BrRb	692	23.3
Osmium(VIII) oxide	O$_4$Os	40.6	14.3	Rubidium carbonate	CO$_3$Rb$_2$	873	30
Oxygen	O$_2$	−218.79	0.44	Rubidium chloride	ClRb	724	24.4
Palladium	Pd	1554.8	16.74	Rubidium fluoride	FRb	795	25.8
Palladium(II) chloride	Cl$_2$Pd	679	18.41	Rubidium hydride	HRb	585	22
Phosphinic acid	H$_3$O$_2$P	26.5	9.7	Rubidium hydroxide	HORb	385	8.0
Phosphonic acid	H$_3$O$_3$P	74.4	12.8	Rubidium iodide	IRb	656	22.1
Phosphoric acid	H$_3$O$_4$P	42.4	13.4	Rubidium metaborate	BO$_2$Rb	860	31
Phosphorus (white)	P	44.15	0.659	Rubidium nitrate	NO$_3$Rb	310	4.6
Phosphorus (red)	P	579.2	18.54	Rubidium nitrite	NO$_2$Rb	422	12.1
Phosphorus(III) chloride	Cl$_3$P	−93	7.10	Rubidium oxide	ORb$_2$	505	20
Phosphorus heptasulfide	P$_4$S$_7$	308	36.6	Rubidium peroxide	O$_2$Rb$_2$	570	21
Phosphorus(V) oxide	O$_5$P$_2$	562	27.2	Rubidium sulfate	O$_4$Rb$_2$S	1066	37.3
Phosphorus sesquisulfide	P$_4$S$_3$	173	20.1	Rubidium superoxide	O$_2$Rb	540	21
Phosphoryl chloride	Cl$_3$OP	1.18	13.1	Ruthenium	Ru	2333	38.59
Platinum	Pt	1768.2	22.175	Ruthenium(V) fluoride	F$_5$Ru	101	74.5
Plutonium	Pu	640	2.824	Samarium	Sm	1072	8.62
Plutonium(III) bromide	Br$_3$Pu	681	58.6	Samarium(III) oxide	O$_3$Sm$_2$	2335	119
Plutonium(III) chloride	Cl$_3$Pu	760	63.6	Scandium	Sc	1541	14.10
Plutonium(III) fluoride	F$_3$Pu	1396	59.8	Scandium chloride	Cl$_3$Sc	967	67.4
Plutonium(IV) fluoride	F$_4$Pu	1037	42.7	Scandium fluoride	F$_3$Sc	1552	62.6
Plutonium(VI) fluoride	F$_6$Pu	51.6	18.6	Scandium oxide	O$_3$Sc$_2$	2489	127
				Selenium (gray)	Se	220.8	6.69
				Selenium dioxide	O$_2$Se	360	17.6

Name	Molecular formula	$t_m/°C$	$\Delta_{fus}H/$ kJ mol^{-1}	Name	Molecular formula	$t_m/°C$	$\Delta_{fus}H/$ kJ mol^{-1}
Silicon	Si	1414	50.21	Tantalum(V) oxide	O_5Ta_2	1875	120
Silicon dioxide (cristobalite)	O_2Si	1722	9.6	Technetium	Tc	2157	33.29
Silicon monosulfide	SSi	1090	31	Tellurium	Te	449.51	17.38
Silver	Ag	961.78	11.30	Tellurium dioxide	O_2Te	733	28.9
Silver(I) bromide	AgBr	430	9.163	Tellurium tetrabromide	Br_4Te	380	24.7
Silver(I) chloride	AgCl	455	13.054	Tellurium tetrachloride	Cl_4Te	224	18.9
Silver(I) iodide	AgI	558	9.414	Terbium	Tb	1359	10.15
Silver(I) nitrate	$AgNO_3$	210	11.72	Terbium(III) bromide	Br_3Tb	830	31.5
Silver(I) oxide	Ag_2O	827	15	Terbium(III) chloride	Cl_3Tb	582	19.5
Silver(I) sulfate	Ag_2O_4S	660	17.99	Tetrachlorosilane	Cl_4Si	−68.74	7.60
Silver(I) sulfide	Ag_2S	836	7.9	Tetraiodosilane	I_4Si	120.5	19.7
Sodium	Na	97.794	2.60	Thallium	Tl	304	4.142
Sodium bromate	$BrNaO_3$	381	28.11	Thallium(I) bromide	BrTl	460	16.4
Sodium bromide	BrNa	747	26.23	Thallium(I) carbonate	CO_3Tl_2	273	18
Sodium carbonate	CNa_2O_3	856	29.7	Thallium(I) chloride	ClTl	431	15.56
Sodium chlorate	$ClNaO_3$	248	22.6	Thallium(I) fluoride	FTl	326	13.87
Sodium chloride	ClNa	800.7	28.16	Thallium(I) formate	CHO_2Tl	101	10.9
Sodium chromate	$CrNa_2O_4$	794	24.7	Thallium(I) iodide	ITl	441.7	14.7
Sodium cyanide	CNNa	562	8.79	Thallium(I) nitrate	NO_3Tl	206	9.6
Sodium fluoride	FNa	996	33.35	Thallium(I) oxide	OTl_2	579	30.3
Sodium formate	$CHNaO_2$	257.3	17.7	Thallium(III) oxide	O_3Tl_2	834	53
Sodium hexafluoroaluminate	AlF_6Na_3	1013	114.4	Thallium(I) sulfate	O_4STl_2	632	23.8
Sodium hexafluorosilicate	F_6Na_2Si	847	99.6	Thallium(I) sulfide	STl_2	457	23.0
Sodium hydride	HNa	638	26	Thorium	Th	1750	13.81
Sodium hydroxide	HNaO	323	6.60	Thorium(IV) bromide	Br_4Th	679	54.4
Sodium iodate	$INaO_3$	422	35.1	Thorium(IV) chloride	Cl_4Th	770	43.9
Sodium iodide	INa	661	23.7	Thorium(IV) fluoride	F_4Th	1110	41.8
Sodium metaborate	$BNaO_2$	966	36.2	Thorium(IV) iodide	I_4Th	566	48.1
Sodium metasilicate	Na_2O_3Si	1089	51.8	Thorium(IV) oxide	O_2Th	3350	90
Sodium nitrate	$NNaO_3$	306.5	15.5	Thulium	Tm	1545	16.84
Sodium nitrite	$NNaO_2$	284	14.9	Thulium(III) chloride	Cl_3Tm	845	34.9
Sodium oxide	Na_2O	1134	47.7	Thulium(III) fluoride	F_3Tm	1158	28.9
Sodium peroxide	Na_2O_2	675	24.5	Tin (white)	Sn	231.928	7.15
Sodium sulfate	Na_2O_4S	884	23.85	Tin(II) bromide	Br_2Sn	232	18.0
Sodium sulfide	Na_2S	1172	19	Tin(IV) bromide	Br_4Sn	29.1	12.2
Sodium sulfite	Na_2O_3S	911	25.9	Tin(II) chloride	Cl_2Sn	247.0	14.52
Strontium	Sr	777	7.43	Tin(IV) chloride	Cl_4Sn	−34.07	9.20
Strontium bromide	Br_2Sr	657	10.5	Tin(II) fluoride	F_2Sn	215	10.5
Strontium carbonate	CO_3Sr	1494	40	Tin(IV) fluoride	F_4Sn	442	27.6
Strontium chloride	Cl_2Sr	874	16.22	Tin(II) iodide	I_2Sn	320	18.0
Strontium fluoride	F_2Sr	1477	29.7	Tin(IV) iodide	I_4Sn	402	0.16
Strontium hydride	H_2Sr	1050	23	Tin(II) oxide	OSn	977	27.7
Strontium hydroxide	H_2O_2Sr	535	23	Tin(IV) oxide	O_2Sn	1630	23.4
Strontium iodide	I_2Sr	538	19.7	Tin(II) sulfide	SSn	881	31.6
Strontium nitrate	N_2O_6Sr	570	44.6	Tin(II) telluride	SnTe	806	45.2
Strontium oxide	OSr	2531	81	Titanium	Ti	1670	14.15
Strontium sulfate	O_4SSr	1606	36	Titanium boride	B_2Ti	2920	100.4
Strontium sulfide	SSr	2226	63	Titanium(IV) bromide	Br_4Ti	38.3	12.9
Sulfur (monoclinic)	S	115.21	1.721	Titanium(II) chloride	Cl_2Ti	1035	34.3
Sulfur hexafluoride	F_6S	−49.596	5.02	Titanium(IV) chloride	Cl_4Ti	−24.12	9.97
Sulfuric acid	H_2O_4S	10.31	10.71	Titanium(IV) fluoride	F_4Ti	377	41
Sulfur trioxide (γ-form)	O_3S	16.8	8.60	Titanium(IV) iodide	I_4Ti	155	19.8
Tantalum	Ta	3017	36.57	Titanium nitride	NTi	2947	66.9
Tantalum boride (TaB$_2$)	B_2Ta	3100	83.68	Titanium(III) oxide	O_3Ti_2	1842	104.6
Tantalum(V) bromide	Br_5Ta	240	37.7	Titanium(IV) oxide (rutile)	O_2Ti	1912	68
Tantalum(V) chloride	Cl_5Ta	216.6	35.1	Titanium(II) sulfide	STi	1927	32
Tantalum(V) fluoride	F_5Ta	96.9	12	Tungsten	W	3414	52.31
Tantalum(V) iodide	I_5Ta	496	7.74	Tungsten boride (WB)	BW	2800	80
Tantalum nitride (TaN)	NTa	3090	6.7	Tungsten boride (W_2B)	BW_2	2740	117
Tantalum nitride (Ta$_2$N)	NTa_2	2727	92.0	Tungsten boride (W_2B_5)	B_5W_2	2370	240

Name	Molecular formula	t_m/°C	$\Delta_{fus}H$/ kJ mol^{-1}	Name	Molecular formula	t_m/°C	$\Delta_{fus}H$/ kJ mol^{-1}
Tungsten(V) bromide	Br_5W	286	17.2	Zirconium(II) iodide	I_2Zr	827	28
Tungsten(V) chloride	Cl_5W	253	20.6	Zirconium(III) iodide	I_3Zr	727	33
Tungsten(VI) chloride	Cl_6W	282	6.69	Zirconium(IV) iodide	I_4Zr	500	32
Tungsten(VI) fluoride	F_6W	1.9	4.10	Zirconium nitride	NZr	2952	67.4
Tungsten(VI) oxide	O_3W	1473	73	Zirconium(IV) oxide	O_2Zr	2710	90
Tungsten(VI) oxytetrachloride	Cl_4OW	210	18.8	Zirconium(IV) sulfide	S_2Zr	1550	45
Tungsten(VI) oxytetrafluoride	F_4OW	105	6				
Uranium	U	1135	9.14	*Organic compounds*			
Uranium(III) bromide	Br_3U	727	43.9	Acenaphthene	$C_{12}H_{10}$	93.4	21.49
Uranium(IV) bromide	Br_4U	519	55.2	Acenaphthylene	$C_{12}H_8$	91.8	6.9
Uranium(IV) chloride	Cl_4U	590	44.8	Acetaldehyde	C_2H_4O	−123.37	2.31
Uranium(III) fluoride	F_3U	1495	36.8	Acetamide	C_2H_5NO	80.16	15.59
Uranium(IV) fluoride	F_4U	1036	47	Acetaminophen	$C_8H_9NO_2$	169.3	30.5
Uranium(V) fluoride	F_5U	348	35	Acetanilide	C_8H_9NO	114.3	21.3
Uranium(VI) fluoride	F_6U	64.06	19.2	Acetic acid	$C_2H_4O_2$	16.64	11.73
Uranium(IV) iodide	I_4U	506	42.1	Acetic anhydride	$C_4H_6O_3$	−74.1	10.5
Uranium(IV) oxide	O_2U	2847	74.2	Acetone	C_3H_6O	−94.7	5.77
Uranyl chloride	Cl_2O_2U	577	44.06	Acetonitrile	C_2H_3N	−43.82	8.16
Vanadium	V	1910	21.5	Acrylic acid	$C_3H_4O_2$	12.5	9.51
Vanadium(II) chloride	Cl_2V	1350	35.0	Acrylonitrile	C_3H_3N	−83.48	6.23
Vanadium(IV) chloride	Cl_4V	−28	2.30	Allene	C_3H_4	−136.6	4.40
Vanadium(II) fluoride	F_2V	1490	44	Allobarbital	$C_{10}H_{12}N_2O_3$	172	32.3
Vanadium(III) fluoride	F_3V	1395	57	2-Aminobenzoic acid	$C_7H_7NO_2$	146	20.5
Vanadium(V) fluoride	F_5V	19.5	49.96	4-Aminobenzoic acid	$C_7H_7NO_2$	188.2	22.5
Vanadium(II) oxide	OV	1790	50	3-Amino-1-propanol	C_3H_9NO	12.4	19.7
Vanadium(III) oxide	O_3V_2	1957	140	Aminopyrine	$C_{13}H_{17}N_3O$	107.5	27.6
Vanadium(IV) oxide	O_2V	1545	56.0	Ampyrone	$C_{11}H_{13}N_3O$	109	24.9
Vanadium(V) oxide	O_5V_2	681	64	Aniline	C_6H_7N	−6.02	10.54
Water	H_2O	0.00	6.01	Anisole	C_7H_8O	−37.13	12.9
Xenon	Xe	−111.75	2.27	Anthracene	$C_{14}H_{10}$	215.76	29.4
Xenon difluoride	F_2Xe	129.03	16.8	Antipyrine	$C_{11}H_{12}N_2O$	112	27.3
Xenon tetrafluoride	F_4Xe	117.1	16.3	*trans*-Azobenzene	$C_{12}H_{10}N_2$	67.88	22.52
Xenon hexafluoride	F_6Xe	49.48	5.74	*trans*-Azoxybenzene	$C_{12}H_{10}N_2O$	34.6	17.9
Ytterbium	Yb	824	7.66	Barbital	$C_8H_{12}N_2O_3$	190	24.7
Ytterbium(III) chloride	Cl_3Yb	854	35.4	Benzaldehyde	C_7H_6O	−57.1	9.32
Yttrium	Y	1522	11.39	Benzamide	C_7H_7NO	127.3	19.5
Yttrium chloride	Cl_3Y	721	31.5	Benz[a]anthracene	$C_{18}H_{12}$	160.5	21.4
Yttrium fluoride	F_3Y	1155	27.9	Benzene	C_6H_6	5.49	9.87
Yttrium oxide	O_3Y_2	2439	81	Benzeneacetic acid	$C_8H_8O_2$	76.5	16.3
Zinc	Zn	419.527	7.068	1,2-Benzenediamine	$C_6H_8N_2$	102.1	23.1
Zinc bromide	Br_2Zn	402	15.7	1,3-Benzenediamine	$C_6H_8N_2$	66.0	15.57
Zinc chloride	Cl_2Zn	325	10.30	1,4-Benzenediamine	$C_6H_8N_2$	141.1	23.8
Zinc fluoride	F_2Zn	872	40	Benzenethiol	C_6H_6S	−14.93	11.48
Zinc iodide	I_2Zn	450	17	*p*-Benzidine	$C_{12}H_{12}N_2$	127	19.1
Zinc oxide	OZn	1974	70	Benzil	$C_{14}H_{10}O_2$	94.87	23.5
Zinc phosphide (ZnP$_2$)	P_2Zn	980	92.9	Benzocaine	$C_9H_{11}NO_2$	89.7	22.3
Zinc selenite	O_3SeZn	621	46.4	Benzoic acid	$C_7H_6O_2$	122.35	18.02
Zinc sulfide (wurtzite)	SZn	1827	30	Benzonitrile	C_7H_5N	−13.99	9.1
Zinc telluride	$TeZn$	1295	63	Benzo[c]phenanthrene	$C_{18}H_{12}$	68	16.3
Zirconium	Zr	1854	21.00	Benzophenone	$C_{13}H_{10}O$	47.9	18.19
Zirconium boride	B_2Zr	3050	104.6	Benzo[a]pyrene	$C_{20}H_{12}$	181.1	17.3
Zirconium(II) bromide	Br_2Zr	827	28	Benzo[e]pyrene	$C_{20}H_{12}$	181.4	16.6
Zirconium(III) bromide	Br_3Zr	727	33	*p*-Benzoquinone	$C_6H_4O_2$	115	18.5
Zirconium(IV) bromide	Br_4Zr	450		Benzoyl chloride	C_7H_5ClO	−0.4	19.2
Zirconium(II) chloride	Cl_2Zr	722	27.0	Benzyl alcohol	C_7H_8O	−15.4	8.97
Zirconium(III) chloride	Cl_3Zr	627	30	2,2'-Binaphthalene	$C_{20}H_{14}$	187.9	38.9
Zirconium(IV) chloride	Cl_4Zr	437	29	Biphenyl	$C_{12}H_{10}$	68.93	18.57
Zirconium(II) fluoride	F_2Zr	902	37.7	Bromobenzene	C_6H_5Br	−30.72	10.70
Zirconium(III) fluoride	F_3Zr	927	50	1-Bromobutane	C_4H_9Br	−112.6	9.23
Zirconium(IV) fluoride	F_4Zr	910	61	2-Bromobutane	C_4H_9Br	−112.65	6.89

Name	Molecular formula	$t_m/°C$	$\Delta_{fus}H/$ kJ mol^{-1}	Name	Molecular formula	$t_m/°C$	$\Delta_{fus}H/$ kJ mol^{-1}
Bromoethane	C_2H_5Br	−118.6	7.47	2-Chlorophenol	C_6H_5ClO	9.4	13.0
Bromoethene	C_2H_3Br	−139.54	5.12	3-Chlorophenol	C_6H_5ClO	32.6	14.9
1-Bromoheptane	$C_7H_{15}Br$	−56.1	21.8	4-Chlorophenol	C_6H_5ClO	42.8	14.1
1-Bromohexane	$C_6H_{13}Br$	−83.7	18.1	1-Chloropropane	C_3H_7Cl	−122.9	5.54
Bromomethane	CH_3Br	−93.68	5.98	2-Chloropropane	C_3H_7Cl	−117.18	7.39
1-Bromonaphthalene	$C_{10}H_7Br$	6.1	15.2	2-Chlorotoluene	C_7H_7Cl	−35.8	9.6
2-Bromonaphthalene	$C_{10}H_7Br$	55.9	14.4	Chlorotrifluoroethene	C_2ClF_3	−158.2	5.55
1-Bromooctane	$C_8H_{17}Br$	−55.0	24.7	Chrysene	$C_{18}H_{12}$	255.5	26.2
1-Bromopentane	$C_5H_{11}Br$	−88.0	14.37	Coronene	$C_{24}H_{12}$	437.4	19.2
1-Bromopropane	C_3H_7Br	−110.3	6.44	o-Cresol	C_7H_8O	31.03	15.82
2-Bromopropane	C_3H_7Br	−89.0	6.53	m-Cresol	C_7H_8O	12.24	10.71
Bromotrichloromethane	$CBrCl_3$	−5.65	2.53	p-Cresol	C_7H_8O	34.77	12.71
1,2-Butadiene	C_4H_6	−136.2	6.96	Cyanamide	CH_2N_2	45.56	7.27
1,3-Butadiene	C_4H_6	−108.91	7.98	Cyanogen	C_2N_2	−27.83	8.11
Butanal	C_4H_8O	−96.86	10.77	Cyclobutane	C_4H_8	−90.7	1.09
Butane	C_4H_{10}	−138.3	4.66	Cycloheptane	C_7H_{14}	−8.46	1.88
1,4-Butanediol	$C_4H_{10}O_2$	20.4	18.70	Cycloheptanol	$C_7H_{14}O$	7.2	1.60
1-Butanethiol	$C_4H_{10}S$	−115.7	10.46	Cyclohexane	C_6H_{12}	6.59	2.68
Butanoic acid	$C_4H_8O_2$	−5.1	11.59	Cyclohexanol	$C_6H_{12}O$	25.93	1.78
1-Butanol	$C_4H_{10}O$	−88.6	9.37	Cyclohexanone	$C_6H_{10}O$	−27.9	1.328
2-Butanol	$C_4H_{10}O$	−88.5	5.97	Cyclohexene	C_6H_{10}	−103.5	3.29
2-Butanone	C_4H_8O	−86.64	8.39	Cyclohexylamine	$C_6H_{13}N$	−17.8	17.5
1-Butene	C_4H_8	−185.34	3.96	Cyclohexylbenzene	$C_{12}H_{16}$	7.07	15.6
cis-2-Butene	C_4H_8	−138.88	7.31	Cyclooctane	C_8H_{16}	14.59	2.41
trans-2-Butene	C_4H_8	−105.52	9.76	Cyclopentane	C_5H_{10}	−93.4	0.61
cis-2-Butenoic acid	$C_4H_6O_2$	15	12.6	Cyclopentanol	$C_5H_{10}O$	−17.5	1.535
trans-2-Butenoic acid	$C_4H_6O_2$	71.5	13.0	Cyclopentene	C_5H_8	−135.0	3.36
tert-Butylamine	$C_4H_{11}N$	−66.94	0.882	Cyclopentylamine	$C_5H_{11}N$	−82.7	8.31
Butylbenzene	$C_{10}H_{14}$	−87.85	11.22	Cyclopropane	C_3H_6	−127.58	5.44
Butylcyclohexane	$C_{10}H_{20}$	−74.73	14.16	Cyclopropylamine	C_3H_7N	−35.39	13.18
Butyl methyl ether	$C_5H_{12}O$	−115.7	10.85	cis-Decahydronaphthalene	$C_{10}H_{18}$	−42.9	9.49
1-Butyne	C_4H_6	−125.7	6.03	trans-Decahydronaphthalene	$C_{10}H_{18}$	−30.4	14.41
2-Butyne	C_4H_6	−32.2	9.23	Decanal	$C_{10}H_{20}O$	−4.0	34.5
γ-Butyrolactone	$C_4H_6O_2$	−43.61	9.57	Decane	$C_{10}H_{22}$	−29.6	28.72
Caffeine	$C_8H_{10}N_4O_2$	236.3	22.0	Decanoic acid	$C_{10}H_{20}O_2$	31.4	27.8
Carbazole	$C_{12}H_9N$	246.3	24.1	1-Decanol	$C_{10}H_{22}O$	6.9	43
Carbon dioxide	CO_2	−56.558	9.02	1-Decene	$C_{10}H_{20}$	−66.3	13.81
Carbon diselenide	CSe_2	−43.7	6.36	1,2-Dibromoethane	$C_2H_4Br_2$	9.84	10.89
Carbon disulfide	CS_2	−112.1	4.39	1,2-Dibromopropane	$C_3H_6Br_2$	−55.49	8.94
Carbon monoxide	CO	−205.02	0.833	1,3-Dibromopropane	$C_3H_6Br_2$	−34.5	14.6
Carbon oxysulfide	COS	−138.8	4.73	1,2-Dibromotetrafluoroethane	$C_2Br_2F_4$	−110.32	7.04
Carbonyl chloride	CCl_2O	−127.78	5.74	o-Dichlorobenzene	$C_6H_4Cl_2$	−17.0	12.4
Chloroacetic acid	$C_2H_3ClO_2$	63	12.28	m-Dichlorobenzene	$C_6H_4Cl_2$	−24.8	12.6
2-Chloroaniline	C_6H_6ClN	−1.9	11.9	p-Dichlorobenzene	$C_6H_4Cl_2$	53.09	18.19
3-Chloroaniline	C_6H_6ClN	−10.28	10.15	1,1-Dichloroethane	$C_2H_4Cl_2$	−96.9	7.87
4-Chloroaniline	C_6H_6ClN	70.5	20.0	1,2-Dichloroethane	$C_2H_4Cl_2$	−35.7	8.84
Chlorobenzene	C_6H_5Cl	−45.31	9.6	1,1-Dichloroethene	$C_2H_2Cl_2$	−122.56	6.51
2-Chlorobenzoic acid	$C_7H_5ClO_2$	140.2	25.6	cis-1,2-Dichloroethene	$C_2H_2Cl_2$	−80.0	7.2
Chlorocyclohexane	$C_6H_{11}Cl$	−43.81	2.043	Dichloromethane	CH_2Cl_2	−97.2	4.60
Chlorodifluoromethane	$CHClF_2$	−157.42	4.12	1,2-Dichloropropane	$C_3H_6Cl_2$	−100.53	6.40
Chloroethane	C_2H_5Cl	−138.4	4.45	2,2-Dichloropropane	$C_3H_6Cl_2$	−33.9	2.30
Chloroethene	C_2H_3Cl	−153.84	4.92	1,2-Dichloro-1,1,2,2-tetrafluoroethane	$C_2Cl_2F_4$	−92.53	1.51
Chloromethane	CH_3Cl	−97.7	6.43	Diethyl ether	$C_4H_{10}O$	−116.2	7.19
2-Chloro-2-methylpropane	C_4H_9Cl	−25.60	2.07	3,3-Diethylpentane	C_9H_{20}	−33.1	10.09
1-Chloronaphthalene	$C_{10}H_7Cl$	−2.5	12.9	Diethyl sulfide	$C_4H_{10}S$	−103.91	10.90
2-Chloronaphthalene	$C_{10}H_7Cl$	58.0	14.0	o-Difluorobenzene	$C_6H_4F_2$	−47.1	11.05
1-Chloro-2-nitrobenzene	$C_6H_4ClNO_2$	32.1	17.9	m-Difluorobenzene	$C_6H_4F_2$	−69.12	8.58
1-Chloro-3-nitrobenzene	$C_6H_4ClNO_2$	44.4	19.4	Diisopropyl ether	$C_6H_{14}O$	−85.4	12.04
1-Chloro-4-nitrobenzene	$C_6H_4ClNO_2$	82	14.1	1,2-Dimethoxyethane	$C_4H_{10}O_2$	−69.20	12.6
Chloropentafluoroethane	C_2ClF_5	−99.4	1.86	Dimethoxymethane	$C_3H_8O_2$	−105.1	8.33

Name	Molecular formula	$t_m/°C$	$\Delta_{fus}H/$ kJ mol^{-1}	Name	Molecular formula	$t_m/°C$	$\Delta_{fus}H/$ kJ mol^{-1}
Dimethylamine	C_2H_7N	−92.18	5.94	Formamide	CH_3NO	2.49	8.44
2,2-Dimethylbutane	C_6H_{14}	−98.8	0.58	Formic acid	CH_2O_2	8.3	12.68
2,3-Dimethylbutane	C_6H_{14}	−128.10	0.79	Furan	C_4H_4O	−85.61	3.80
2,3-Dimethyl-2-butene	C_6H_{12}	−74.19	6.45	Furfural	$C_5H_4O_2$	−38.1	14.37
1,1-Dimethylcyclohexane	C_8H_{16}	−33.3	2.07	Furfuryl alcohol	$C_5H_6O_2$	−14.6	13.13
cis-1,2-Dimethylcyclohexane	C_8H_{16}	−49.8	1.64	Glycerol	$C_3H_8O_3$	18.1	18.3
trans-1,2-Dimethylcyclohexane	C_8H_{16}	−88.15	10.49	Heneicosane	$C_{21}H_{44}$	40.01	45.21
cis-1,3-Dimethylcyclohexane	C_8H_{16}	−75.53	10.82	Heptacosane	$C_{27}H_{56}$	59.23	61.9
trans-1,3-Dimethylcyclohexane	C_8H_{16}	−90.07	9.87	Heptadecane	$C_{17}H_{36}$	22.0	40.16
cis-1,4-Dimethylcyclohexane	C_8H_{16}	−87.39	9.31	Heptanal	$C_7H_{14}O$	−43.4	23.2
trans-1,4-Dimethylcyclohexane	C_8H_{16}	−36.93	12.33	Heptane	C_7H_{16}	−90.55	14.03
Dimethyl disulfide	$C_2H_6S_2$	−84.67	9.19	Heptanoic acid	$C_7H_{14}O_2$	−7.17	15.13
Dimethyl ether	C_2H_6O	−141.5	4.94	1-Heptanol	$C_7H_{16}O$	−33.2	18.17
N,N-Dimethylformamide	C_3H_7NO	−60.48	7.90	1-Heptene	C_7H_{14}	−118.9	12.41
1,1-Dimethylhydrazine	$C_2H_8N_2$	−57.20	10.07	Hexachlorobenzene	C_6Cl_6	228.83	25.2
1,2-Dimethylhydrazine	$C_2H_8N_2$	−8.9	13.64	Hexachloroethane	C_2Cl_6	186.8t	9.75
Dimethyl oxalate	$C_4H_6O_4$	54.8	21.1	Hexacontane	$C_{60}H_{122}$	99.3	193.2
2,2-Dimethylpentane	C_7H_{16}	−123.7	5.82	Hexacosane	$C_{26}H_{54}$	56.1	60.0
2,4-Dimethylpentane	C_7H_{16}	−119.2	6.85	Hexadecane	$C_{16}H_{34}$	18.12	53.36
3,3-Dimethylpentane	C_7H_{16}	−134.4	6.85	Hexadecanoic acid	$C_{16}H_{32}O_2$	62.5	53.7
Dimethyl sulfide	C_2H_6S	−98.24	7.99	1-Hexadecanol	$C_{16}H_{34}O$	49.2	33.6
Dimethyl sulfone	$C_2H_6O_2S$	108.9	18.30	Hexafluorobenzene	C_6F_6	5.03	11.59
Dimethyl sulfoxide	C_2H_6OS	17.89	14.37	Hexafluoroethane	C_2F_6	−100.05	2.69
N,N-Dimethylurea	$C_3H_8N_2O$	182.1	23.0	Hexamethylbenzene	$C_{12}H_{18}$	165.5	20.6
N,N'-Dimethylurea	$C_3H_8N_2O$	106.6	13.0	Hexanal	$C_6H_{12}O$	−56	13.3
Dimethyl zinc	C_2H_6Zn	−43.0	6.83	Hexane	C_6H_{14}	−95.35	13.08
1,4-Dioxane	$C_4H_8O_2$	11.85	12.84	1,6-Hexanedioic acid	$C_6H_{10}O_4$	152.5	36.3
1,3-Dioxolane	$C_3H_6O_2$	−97.22	6.57	1,6-Hexanediol	$C_6H_{14}O_2$	41.5	22.2
Diphenylamine	$C_{12}H_{11}N$	53.2	18.5	1-Hexanol	$C_6H_{14}O$	−47.4	15.38
Diphenyl ether	$C_{12}H_{10}O$	26.864	17.22	2-Hexanone	$C_6H_{12}O$	−55.5	14.9
Diphenylmethane	$C_{13}H_{12}$	25.4	18.6	3-Hexanone	$C_6H_{12}O$	−55.4	13.49
Dipropyl ether	$C_6H_{14}O$	−114.8	10.8	Hexatetracontane	$C_{46}H_{94}$	87.6	176.0
Divinyl ether	C_4H_6O	−100.6	7.9	Hexatriacontane	$C_{36}H_{74}$	75.8	87.7
Docosane	$C_{22}H_{46}$	43.6	48.8	1-Hexene	C_6H_{12}	−139.76	9.35
Dodecane	$C_{12}H_{26}$	−9.57	36.8	cis-2-Hexene	C_6H_{12}	−141.11	8.88
Dodecanoic acid	$C_{12}H_{24}O_2$	43.8	36.3	Hydrogen cyanide	CHN	−13.29	8.41
1-Dodecanol	$C_{12}H_{26}O$	23.9	40.2	p-Hydroquinone	$C_6H_6O_2$	172.4	26.8
1-Dodecene	$C_{12}H_{24}$	−35.2	19.9	2-Hydroxybenzoic acid	$C_7H_6O_3$	159.0	14.2
Dotriacontane	$C_{32}H_{66}$	69.4	75.8	Imidazole	$C_3H_4N_2$	89.5	12.82
Eicosane	$C_{20}H_{42}$	36.6	69.9	Indan	C_9H_{10}	−51.38	8.60
1-Eicosanol	$C_{20}H_{42}O$	65.4	42	Indene	C_9H_8	−1.5	10.20
Estradiol benzoate	$C_{25}H_{28}O_3$	193	41.8	Indomethacin	$C_{19}H_{16}ClNO_4$	160	36.9
Ethane	C_2H_6	−182.79	2.72*	Iodobenzene	C_6H_5I	−31.3	9.75
1,2-Ethanediamine	$C_2H_8N_2$	11.14	22.58	Isobutane	C_4H_{10}	−159.4	4.54
1,2-Ethanediol	$C_2H_6O_2$	−12.69	9.96	Isobutene	C_4H_8	−140.7	5.92
Ethanethiol	C_2H_6S	−147.88	4.98	Isopentane	C_5H_{12}	−159.77	5.15
Ethanol	C_2H_6O	−114.14	4.931	Isopropylamine	C_3H_9N	−95.13	7.33
Ethinylestradiol	$C_{20}H_{24}O_2$	183.5	27.9	Isopropylbenzene	C_9H_{12}	−96.02	7.33
Ethyl acetate	$C_4H_8O_2$	−83.8	10.48	1-Isopropyl-4-methylbenzene	$C_{10}H_{14}$	−67.94	9.66
Ethylbenzene	C_8H_{10}	−94.96	9.18	Isoquinoline	C_9H_7N	26.47	13.54
Ethylcyclohexane	C_8H_{16}	−111.3	8.33	Khellin	$C_{14}H_{12}O_5$	154	32.3
Ethylene	C_2H_4	−169.15	3.35	Maleic anhydride	$C_4H_2O_3$	52.56	13.60
Ethyl methyl sulfide	C_3H_8S	−105.93	9.76	Methane	CH_4	−182.47	0.94
3-Ethylpentane	C_7H_{16}	−118.55	9.55	Methanethiol	CH_4S	−123	5.91
2-Ethyltoluene	C_9H_{12}	−79.83	9.96	Methanol	CH_4O	−97.53	3.215
3-Ethyltoluene	C_9H_{12}	−95.6	7.6	Methyl acetate	$C_3H_6O_2$	−98.25	7.49
4-Ethyltoluene	C_9H_{12}	−62.35	12.7	Methylamine	CH_5N	−93.5	6.13
Fluoranthene	$C_{16}H_{10}$	110.19	18.69	2-Methylaniline	C_7H_9N	−14.41	11.66
9H-Fluorene	$C_{13}H_{10}$	114.77	19.58	3-Methylaniline	C_7H_9N	−31.3	7.9
Fluorobenzene	C_6H_5F	−42.18	11.31	4-Methylaniline	C_7H_9N	43.6	18.9

Name	Molecular formula	$t_m/°C$	$\Delta_{fus}H/$ kJ mol^{-1}	Name	Molecular formula	$t_m/°C$	$\Delta_{fus}H/$ kJ mol^{-1}
Methyl benzoate	$C_8H_8O_2$	−12.4	9.74	1-Octene	C_8H_{16}	−101.7	15.31
2-Methyl-1,3-butadiene	C_5H_8	−145.9	4.93	2-Oxepanone	$C_6H_{10}O_2$	−1.0	13.83
2-Methyl-2-butanol	$C_5H_{12}O$	−9.1	4.46	Oxetane	C_3H_6O	−97	6.5
3-Methyl-2-butanone	$C_5H_{10}O$	−93.1	9.34	Oxirane	C_2H_4O	−112.5	5.17
2-Methyl-1-butene	C_5H_{10}	−137.53	7.91	4-Oxopentanoic acid	$C_5H_8O_3$	33	9.22
3-Methyl-1-butene	C_5H_{10}	−168.43	5.36	Paraldehyde	$C_6H_{12}O_3$	12.6	13.5
2-Methyl-2-butene	C_5H_{10}	−133.72	7.60	Pentachloroethane	C_2HCl_5	−28.78	11.3
Methyl *tert*-butyl ether	$C_5H_{12}O$	−108.6	7.60	Pentacontane	$C_{50}H_{102}$	92.1	162.4
Methylcyclohexane	C_7H_{14}	−126.6	6.75	Pentacosane	$C_{25}H_{52}$	53.93	56.9
Methylcyclopentane	C_6H_{12}	−142.42	6.93	Pentadecane	$C_{15}H_{32}$	9.95	34.6
Methylcyclopropane	C_4H_8	−177.6	2.8	*cis*-1,3-Pentadiene	C_5H_8	−140.8	5.64
2-Methylfuran	C_5H_6O	−91.3	8.55	*trans*-1,3-Pentadiene	C_5H_8	−87.4	7.14
2-Methylheptane	C_8H_{18}	−109.02	11.92	1,4-Pentadiene	C_5H_8	−148.2	6.12
3-Methylheptane	C_8H_{18}	−120.48	11.69	Pentaerythritol	$C_5H_{12}O_4$	258	4.8
4-Methylheptane	C_8H_{18}	−121.0	10.8	Pentafluorobenzene	C_6HF_5	−47.4	10.87
2-Methylhexane	C_7H_{16}	−118.2	9.19	Pentafluorophenol	C_6HF_5O	37.5	16.41
Methylhydrazine	CH_6N_2	−52.36	10.42	2,3,4,5,6-Pentafluorotoluene	$C_7H_3F_5$	−29.78	13.1
Methyl methacrylate	$C_5H_8O_2$	−47.55	14.4	Pentane	C_5H_{12}	−129.67	8.40
1-Methylnaphthalene	$C_{11}H_{10}$	−30.43	6.95	Pentanedioic acid	$C_5H_8O_4$	97.8	20.3
2-Methylnaphthalene	$C_{11}H_{10}$	34.6	12.13	Pentanenitrile	C_5H_9N	−96.2	9
Methyl nitrate	CH_3NO_3	−83.0	8.24	1-Pentanethiol	$C_5H_{12}S$	−75.65	17.53
Methyloxirane	C_3H_6O	−111.9	6.53	Pentanoic acid	$C_5H_{10}O_2$	−33.6	14.16
2-Methylpentane	C_6H_{14}	−153.6	6.27	1-Pentanol	$C_5H_{12}O$	−77.6	10.50
3-Methylpentane	C_6H_{14}	−162.90	5.30	2-Pentanone	$C_5H_{10}O$	−76.8	10.63
2-Methyl-1-propanol	$C_4H_{10}O$	−101.9	6.32	3-Pentanone	$C_5H_{10}O$	−39	11.59
2-Methyl-2-propanol	$C_4H_{10}O$	25.69	6.70	Pentatriacontane	$C_{35}H_{72}$	74.6	86.3
2-Methylpyridine	C_6H_7N	−66.68	9.72	1-Pentene	C_5H_{10}	−165.12	5.94
3-Methylpyridine	C_6H_7N	−18.14	14.18	*cis*-2-Pentene	C_5H_{10}	−151.36	7.11
4-Methylpyridine	C_6H_7N	3.67	12.58	*trans*-2-Pentene	C_5H_{10}	−140.21	8.35
N-Methylurea	$C_2H_6N_2O$	104.9	14.0	Perfluoroacetone	C_3F_6O	−125.45	8.38
Morpholine	C_4H_9NO	−4.8	14.5	Perfluorobutane	C_4F_{10}	−129.1	7.66
Naphthalene	$C_{10}H_8$	80.26	19.01	Perfluorocyclobutane	C_4F_8	−40.19	2.77
1-Naphthol	$C_{10}H_8O$	95.0	23.1	Perfluoroheptane	C_7F_{16}	−51.2	6.95
2-Naphthol	$C_{10}H_8O$	121.5	18.1	Perfluorohexane	C_6F_{14}	−88.2	6.84
Neopentane	C_5H_{12}	−16.4	3.10	Perfluoropropane	C_3F_8	−147.70	0.477
Niacinamide	$C_6H_6N_2O$	130	23.2	Perfluorotoluene	C_7F_8	−65.49	11.54
2-Nitroaniline	$C_6H_6N_2O_2$	71.0	16.1	Perylene	$C_{20}H_{12}$	277.76	31.9
3-Nitroaniline	$C_6H_6N_2O_2$	113.4	23.6	Phenacetin	$C_{10}H_{13}NO_2$	134	33.0
4-Nitroaniline	$C_6H_6N_2O_2$	147.5	21.2	Phenanthrene	$C_{14}H_{10}$	99.24	16.46
Nitrobenzene	$C_6H_5NO_2$	5.7	12.12	Phenobarbital	$C_{12}H_{12}N_2O_3$	174.0	27.8
Nitroethane	$C_2H_5NO_2$	−89.5	9.85	Phenol	C_6H_6O	40.89	11.51
Nitromethane	CH_3NO_2	−28.38	9.70	α-Phenylbenzeneacetic acid	$C_{14}H_{12}O_2$	147.29	31.3
2-Nitrophenol	$C_6H_5NO_3$	44.8	17.7	Phenylbutazone	$C_{19}H_{20}N_2O_2$	105	27.7
3-Nitrophenol	$C_6H_5NO_3$	96.8	20.6	Phenylhydrazine	$C_6H_8N_2$	20.6	14.05
4-Nitrophenol	$C_6H_5NO_3$	113.6	18.8	Piperidine	$C_5H_{11}N$	−11.02	14.85
Nitrosobenzene	C_6H_5NO	67	31.0	Potassium acetate	$C_2H_3KO_2$	309	7.65
4-Nitrotoluene	$C_7H_7NO_2$	51.63	16.81	Propane	C_3H_8	−187.63	3.50
Nonacosane	$C_{29}H_{60}$	63.7	66.9	1,3-Propanediol	$C_3H_8O_2$	−27.7	7.1
Nonadecane	$C_{19}H_{40}$	32.0	45.8	Propanenitrile	C_3H_5N	−92.78	5.03
Nonanal	$C_9H_{18}O$	−19.3	30.5	1-Propanethiol	C_3H_8S	−113.13	5.48
Nonane	C_9H_{20}	−53.46	15.47	2-Propanethiol	C_3H_8S	−130.5	5.74
Nonanoic acid	$C_9H_{18}O_2$	12.4	19.82	Propanoic acid	$C_3H_6O_2$	−20.5	10.66
5-Nonanone	$C_9H_{18}O$	−3.8	24.93	1-Propanol	C_3H_8O	−124.39	5.37
Octacosane	$C_{28}H_{58}$	61.1	65.1	2-Propanol	C_3H_8O	−87.9	5.41
Octadecane	$C_{18}H_{38}$	28.2	61.7	Propene	C_3H_6	−185.24	3.003
1-Octadecanol	$C_{18}H_{38}O$	57.9	45	Propylamine	C_3H_9N	−84.75	10.97
Octane	C_8H_{18}	−56.82	20.73	Propylbenzene	C_9H_{12}	−99.6	9.27
Octanoic acid	$C_8H_{16}O_2$	16.5	21.35	Propylcyclohexane	C_9H_{18}	−94.9	10.37
1-Octanol	$C_8H_{18}O$	−14.8	23.7	Pyrazine	$C_4H_4N_2$	51.0	12.9
Octatriacontane	$C_{38}H_{78}$	78.6	133.2	1*H*-Pyrazole	$C_3H_4N_2$	70.7	14.0

Name	Molecular formula	$t_m/°C$	$\Delta_{fus}H/$ kJ mol^{-1}	Name	Molecular formula	$t_m/°C$	$\Delta_{fus}H/$ kJ mol^{-1}
Pyrene	$C_{16}H_{10}$	150.62	17.36	Thiazole	C_3H_3NS	−33.62	9.57
Pyridine	C_5H_5N	−41.70	8.28	Thietane	C_3H_6S	−73.24	8.25
Pyrocatechol	$C_6H_6O_2$	104.6	22.8	Thiophene	C_4H_4S	−38.21	5.07
Pyrrole	C_4H_5N	−23.39	7.91	Thiourea	CH_4N_2S	178	14.0
Pyrrolidine	C_4H_9N	−57.79	8.58	Thymol	$C_{10}H_{14}O$	49.5	21.3
Quinoline	C_9H_7N	−14.78	10.66	Toluene	C_7H_8	−94.95	6.64
Resorcinol	$C_6H_6O_2$	109.4	20.4	o-Toluic acid	$C_8H_8O_2$	103.5	19.5
Sebacic acid	$C_{10}H_{18}O_4$	130.9	40.8	m-Toluic acid	$C_8H_8O_2$	109.9	15.7
Sodium acetate	$C_2H_3NaO_2$	328.2	17.9	p-Toluic acid	$C_8H_8O_2$	179.6	22.7
Sodium hydrogen carbonate	$CHNaO_3$	527	25	Triacontane	$C_{30}H_{62}$	65.1	68.3
Spiro[2.2]pentane	C_5H_8	−107.0	6.43	1,3,5-Triazine	$C_3H_3N_3$	80.3	14.56
Stearic acid	$C_{18}H_{36}O_2$	69.3	61.2	Tribromomethane	$CHBr_3$	8.69	11.05
trans-Stilbene	$C_{14}H_{12}$	124.2	27.7	Trichloroacetic acid	$C_2HCl_3O_2$	59.2	5.90
Styrene	C_8H_8	−30.65	10.9	1,2,3-Trichlorobenzene	$C_6H_3Cl_3$	51.3	17.9
Succinic acid	$C_4H_6O_4$	187.9	32.4	1,2,4-Trichlorobenzene	$C_6H_3Cl_3$	16.92	16.4
Succinic anhydride	$C_4H_4O_3$	119	20.4	1,3,5-Trichlorobenzene	$C_6H_3Cl_3$	62.8	18.1
Succinonitrile	$C_4H_4N_2$	58.06	3.70	1,1,1-Trichloroethane	$C_2H_3Cl_3$	−30.01	2.35
Sulfacetamide	$C_8H_{10}N_2O_3S$	183	22.4	1,1,2-Trichloroethane	$C_2H_3Cl_3$	−36.3	11.46
Sulfadiazine	$C_{10}H_{10}N_4O_2S$	258	42.6	Trichloroethene	C_2HCl_3	−84.7	8.45
Sulfamerazine	$C_{11}H_{12}N_4O_2S$	236	38.7	Trichlorofluoromethane	CCl_3F	−110.44	6.89
Sulfamethoxazole	$C_{10}H_{11}N_3O_3S$	170	32.2	Trichloromethane	$CHCl_3$	−63.41	9.5
Sulfamethoxypyridazine	$C_{11}H_{12}N_4O_3S$	182.5	31.3	1,1,2-Trichloro-1,2,2-trifluoroethane	$C_2Cl_3F_3$	−36.22	2.47
Sulfapyridine	$C_{11}H_{11}N_3O_2S$	192	34.4				
Sulfathiazole	$C_9H_9N_3O_2S_2$	202	26.4	Tricosane	$C_{23}H_{48}$	47.76	50.86
Sulfisoxazole	$C_{11}H_{13}N_3O_3S$	196	30.2	Tridecane	$C_{13}H_{28}$	−5.4	28.50
o-Terphenyl	$C_{18}H_{14}$	56.20	17.19	1-Tridecanol	$C_{13}H_{28}O$	31.7	41.4
p-Terphenyl	$C_{18}H_{14}$	213.9	35.3	1,1,1-Trifluoroethane	$C_2H_3F_3$	−111.3	6.19
Tetrabromomethane	CBr_4	92.3	3.76	Trifluoromethane	CHF_3	−155.2	4.06
1,1,2,2-Tetrachloro-1,2-difluoroethane	$C_2Cl_4F_2$	24.8	3.67	Triiodomethane	CHI_3	121.2	16.44
				Trimethoprim	$C_{14}H_{18}N_4O_3$	199	49.4
1,1,2,2-Tetrachloroethane	$C_2H_2Cl_4$	−42.4	9.17	Trimethylamine	C_3H_9N	−117.1	7
Tetrachloroethene	C_2Cl_4	−22.3	10.88	1,2,3-Trimethylbenzene	C_9H_{12}	−25.4	8.18
Tetrachloromethane	CCl_4	−22.62	2.56	1,2,4-Trimethylbenzene	C_9H_{12}	−43.77	13.19
Tetracontane	$C_{40}H_{82}$	81.5	135.5	1,3,5-Trimethylbenzene	C_9H_{12}	−44.72	9.51
Tetracosane	$C_{24}H_{50}$	50.4	54.4	2,2,3-Trimethylbutane	C_7H_{16}	−24.6	2.26
Tetradecane	$C_{14}H_{30}$	5.82	45.07	2,2,4-Trimethylpentane	C_8H_{18}	−107.3	9.20
Tetradecanoic acid	$C_{14}H_{28}O_2$	54.2	45.1	1,3,5-Trinitrobenzene	$C_6H_3N_3O_6$	122.9	15.4
1-Tetradecanol	$C_{14}H_{30}O$	38.2	25.1*	Trinitroglycerol	$C_3H_5N_3O_9$	13.5	21.87
1,2,3,5-Tetrafluorobenzene	$C_6H_2F_4$	−46.25	6.36	2,4,6-Trinitrotoluene	$C_7H_5N_3O_6$	80.5	22.9
1,2,4,5-Tetrafluorobenzene	$C_6H_2F_4$	3.88	15.05	1,3,5-Trioxane	$C_3H_6O_3$	60.29	15.11
Tetrafluoroethene	C_2F_4	−131.15	7.72	Triphenylamine	$C_{18}H_{15}N$	126.5	24.9
Tetrafluoromethane	CF_4	−183.60	0.704	Triphenylene	$C_{18}H_{12}$	197.8	24.74
Tetrahydrofuran	C_4H_8O	−108.44	8.54	Tritriacontane	$C_{33}H_{68}$	71.2	79.5
Tetrahydropyran	$C_5H_{10}O$	−49.1	1.8	Undecane	$C_{11}H_{24}$	−25.5	22.2
Tetrahydrothiophene	C_4H_8S	−96.2	7.35	Urea	CH_4N_2O	133.3	13.9
1,2,4,5-Tetramethylbenzene	$C_{10}H_{14}$	79.3	21	o-Xylene	C_8H_{10}	−25.2	13.6
Tetramethyl lead	$C_4H_{12}Pb$	−30.2	10.80	m-Xylene	C_8H_{10}	−47.8	11.6
2,2,3,3-Tetramethylpentane	C_9H_{20}	−9.75	2.33	p-Xylene	C_8H_{10}	13.25	17.12
2,2,4,4-Tetramethylpentane	C_9H_{20}	−66.54	9.74	2,3-Xylenol	$C_8H_{10}O$	72.5	21.0
Tetramethylsilane	$C_4H_{12}Si$	−99.06	6.87	2,5-Xylenol	$C_8H_{10}O$	74.8	23.4
Tetramethylstannane	$C_4H_{12}Sn$	−55.1	9.30	2,6-Xylenol	$C_8H_{10}O$	45.8	18.9
Tetratetracontane	$C_{44}H_{90}$	85.6	149.6	3,4-Xylenol	$C_8H_{10}O$	65.1	18.1
Tetratriacontane	$C_{34}H_{70}$	72.5	79.4	3,5-Xylenol	$C_8H_{10}O$	63.4	17.4
1H-Tetrazole	CH_2N_4	157.3	18.2				

COMPRESSIBILITY AND EXPANSION COEFFICIENTS OF LIQUIDS

This table gives data on the variation of the density of some common liquids with pressure and temperature. The pressure dependence is described to first order by the isothermal compressibility coefficient κ defined as

$$\kappa_T = -(1/V)\,(\partial V/\partial P)_T$$

where V is the volume, and the temperature dependence by the cubic expansion coefficient α,

$$\alpha_V = (1/V)\,(\partial V/\partial T)_P$$

Substances are listed by molecular formula in the Hill order. More precise data on the variation of density with temperature over a wide temperature range can be found in Reference 1.

References

1. Lide, D. R., and Kehiaian, H. V., *CRC Handbook of Thermophysical and Thermochemical Data*, CRC Press, Boca Raton, FL, 1994.
2. Le Neindre, B., *Effets des Hautes et Très Hautes Pressions*, in *Techniques de l'Ingénieur*, Paris, 1991.
3. *Landolt-Börnstein, Numerical Data and Functional Relationships in Science and Technology, New Series, IV/4, High-Pressure Properties of Matter*, Springer-Verlag, Heidelberg, 1980.
4. Riddick, J.A., Bunger, W.B., and Sakano, T.K., *Organic Solvents, Fourth Edition*, John Wiley & Sons, New York, 1986.
5. Isaacs, N. S., *Liquid Phase High Pressure Chemistry*, John Wiley, New York, 1981.

Molecular formula	Name	Isothermal compressibility		Cubic expansion coefficient	
		$t/°C$	$\kappa_T \times 10^4/\text{MPa}^{-1}$	$t/°C$	$\alpha_V \times 10^3/°C^{-1}$
Cl_3P	Phosphorus trichloride	20	9.45	20	1.9
H_2O	Water	20	4.591	20	0.206
		25	4.524	25	0.256
		30	4.475	30	0.302
Hg	Mercury	20	0.401	20	0.1811
CCl_4	Tetrachloromethane	20	10.50	20	1.14
		40	12.20	40	1.21
		70	15.6	70	1.33
$CHBr_3$	Tribromomethane	50	8.76	25	0.91
$CHCl_3$	Trichloromethane	20	9.96	20	1.21
		50	12.9	50	1.33
CH_2Br_2	Dibromomethane	27	6.85		
CH_2Cl_2	Dichloromethane	25	10.3	25	1.39
CH_3I	Iodomethane	27	10.3	25	1.26
CH_4O	Methanol	20	12.14	20	1.49
		40	13.83	40	1.59
CS_2	Carbon disulfide	20	9.38	20	1.12
		40	10.6	35	1.16
C_2Cl_4	Tetrachloroethylene	25	7.56	25	1.02
C_2HCl_3	Trichloroethylene	25	8.57	25	1.17
$C_2H_2Cl_2$	*trans*-1,2-Dichloroethylene	25	11.2	25	1.36
$C_2H_4Cl_2$	1,1-Dichloroethane	20	7.97	25	0.93
$C_2H_4Cl_2$	1,2-Dichloroethane	30	8.46	20	1.14
$C_2H_4O_2$	Acetic acid	20	9.08	20	1.08
		80	13.7	80	1.38
C_2H_5Br	Bromoethane	20	11.53	20	1.31
C_2H_5I	Iodoethane	20	9.82	25	1.17
C_2H_6O	Ethanol	20	11.19	20	1.40
		70	15.93	70	1.67
$C_2H_6O_2$	Ethylene glycol	20	3.64	20	0.626
C_3H_6O	Acetone	20	12.62	20	1.46
		40	15.6	40	1.57
C_3H_7Br	1-Bromopropane	0	10.22	25	1.2
C_3H_7Cl	1-Chloropropane	0	12.09	20	1.4
C_3H_7I	1-Iodopropane	0	10.22	25	1.09
C_3H_8O	1-Propanol	0	8.43	0	1.22
C_3H_8O	2-Propanol	40	13.32	40	1.55
$C_3H_8O_2$	1,2-Propanediol	0	4.45	20	0.695
$C_3H_8O_2$	1,3-Propanediol	0	4.09	20	0.61
$C_3H_8O_3$	Glycerol	0	2.54	20	0.520

Molecular formula	Name	Isothermal compressibility		Cubic expansion coefficient	
		$t/°C$	$\kappa_T \times 10^4/MPa^{-1}$	$t/°C$	$\alpha_V \times 10^3/°C^{-1}$
$C_4H_8O_2$	Ethyl acetate	20	11.32	20	1.35
		60	16.2	60	1.54
C_4H_9Br	1-Bromobutane	25	10.26	20	1.13
C_4H_9I	1-Iodobutane	0	7.73	25	1.02
$C_4H_{10}O$	1-Butanol	0	8.10	0	1.12
$C_4H_{10}O$	Diethyl ether	20	18.65	20	1.65
		30	20.85	30	1.72
$C_4H_{10}O_3$	Diethylene glycol	0	3.34	20	0.635
C_5H_{10}	Cyclopentane	20	13.31	20	1.35
$C_5H_{11}Br$	1-Bromopentane	0	8.42	25	1.04
$C_5H_{11}I$	1-Iodopentane	0	7.56		
C_5H_{12}	Pentane	25	21.80	25	1.64
$C_5H_{12}O$	1-Pentanol	0	7.71	0	1.02
C_6H_5Br	Bromobenzene	20	6.46	20	0.86
C_6H_5Cl	Chlorobenzene	20	7.45	20	0.94
$C_6H_5NO_2$	Nitrobenzene	20	4.93	25	0.833
C_6H_6	Benzene	25	9.66	25	1.14
		45	11.28	45	1.21
C_6H_6O	Phenol	60	6.05	60	0.82
C_6H_7N	Aniline	20	4.53	20	0.81
		80	6.32	80	0.91
C_6H_{12}	Cyclohexane	20	11.30	20	1.15
		60	15.2	60	1.29
C_6H_{14}	Hexane	25	16.69	25	1.41
		45	20.27	45	1.52
C_6H_{14}	2-Methylpentane	0	13.97	25	1.43
C_6H_{14}	3-Methylpentane	0	14.57	25	1.40
C_6H_{14}	2,3-Dimethylbutane	20	17.97	25	1.39
$C_6H_{14}O$	1-Hexanol	25	8.24	25	1.03
$C_6H_{15}NO_3$	Triethanolamine	0	3.61	55	0.53
C_7H_8	Toluene	20	8.96	20	1.05
		50	11.0	50	1.13
C_7H_8O	Anisole	20	6.60	20	0.951
C_7H_{14}	Cycloheptane	20	9.22		
C_7H_{16}	Heptane	25	14.38	25	1.26
C_8H_{10}	o-Xylene	25	8.10	25	0.96
C_8H_{10}	m-Xylene	20	8.46	20	0.99
C_8H_{10}	p-Xylene	25	8.59	25	1.00
C_8H_{16}	Cyclooctane	20	8.03		
C_8H_{18}	Octane	25	12.82	25	1.16
		45	15.06	45	1.23
$C_8H_{18}O$	1-Octanol	25	7.64	25	0.827
C_9H_{12}	Mesitylene	25	8.14	25	0.94
$C_9H_{14}O_6$	Triacetin	0	4.49	25	0.94
C_9H_{20}	Nonane	25	11.75	25	1.08
$C_{10}H_{22}$	Decane	25	10.94	25	1.02
$C_{11}H_{24}$	Undecane	25	10.31	25	0.97
$C_{12}H_{26}$	Dodecane	25	9.88	25	0.93
$C_{13}H_{28}$	Tridecane	25	9.48	25	0.90
$C_{14}H_{30}$	Tetradecane	25	9.10	25	0.87
$C_{15}H_{32}$	Pentadecane	25	8.82		
$C_{16}H_{22}O_4$	Butyl phthalate	0	5.0	25	0.86
$C_{16}H_{34}$	Hexadecane	25	8.57		
		45	9.78		
$C_{19}H_{36}O_2$	Methyl oleate	0	6.18	60	0.85

TEMPERATURE AND PRESSURE DEPENDENCE OF LIQUID DENSITY

Ivan Cibulka

This table records parameters of the Tait equation (Refs. 1,2) that gives the ratio between the density at pressure P, $\rho(T,P)$, relative to the density at a reference pressure, $\rho(T,P_{\text{ref}})$, at the same temperature T:

$$\frac{\rho(T,P)}{\rho(T,P_{\text{ref}})} = \frac{1}{1 - C(T)\ln\left\{\dfrac{B(T)+P}{B(T)+P_{\text{ref}}}\right\}},$$

$$C(T) = a_1 + b_1(T/\text{K}) + c_1(T/\text{K})^2, \tag{1}$$

$$B(T)/\text{MPa} = a_2 + a_3(T/\text{K}) + a_4(T/\text{K})^2 + a_5(T/\text{K})^3 + a_6(T/\text{K})^4$$

Parameters a_i ($i = 1, \ldots, 6$) and b_1, c_1 were adjusted to selected experimental data using the weighted least-squares method. Parameters b_1 and c_1 are zero for most substances, and therefore their values are given in footnotes for those few substances where the statistical significance was justified. The reference pressure is $P_{\text{ref}} = 0.101325$ MPa at temperatures either at or below the normal boiling point temperature (T_{nbp}) and $P_{\text{ref}} = P_{\text{sat}}(T)$ (saturated vapor pressure) at temperatures $T > T_{\text{nbp}}$. Ranges of validity of the equation (T_{min}, T_{max}, P_{max}) are derived from ranges of experimental data; the minimal pressure of validity is taken as P_{ref}, i.e., interpolation between P_{ref} and lowest experimental pressure is allowed. The upper limit of application is the freezing line (if not limited by the ranges of validity). To avoid any large-scale extrapolation, the validity ranges are rectangular areas $(T_{\text{max}} - T_{\text{min}})P_{\text{max}}$. If in particular temperature interval(s) the maximum experimental pressure exceeded the given value of P_{max}, then the maximum pressure given in the table is denoted by (r) which means that the validity range given in the table is a rectangular subset of the non-rectangular experimental T, P range. In a few cases P_{max} is given as a ratio where the first value corresponds to T_{min} and the second one to T_{max}, i.e., the validity range has approximately a trapezoidal shape. Values of parameters were taken from the papers (Refs. 3–10) where detailed information on the fits, experimental data, and application ranges is available. The numerical values of the parameters are different from those reported in papers (Refs. 3–10) since the forms of polynomials $C(T)$ and $B(T)$ differ. Besides, the parameters recorded in the table below must not necessarily correspond to those in Refs. 3–10 since some fits were updated using newly published experimental data.

In the second line for each substance, the parameters of the smoothing function for the density at reference pressure $\rho(T,P_{\text{ref}})$ are given. The functions are either the polynomial expansion,

$$\rho(T,P_{\text{ref}})/\left(\text{kg m}^{-3}\right) = \sum_{i=1}^{N_p} a_i(T/\text{K})^{(i-1)} \tag{2}$$

or the expansion,

$$\rho(T,P_{\text{ref}}) = \rho_c\left[1 + \sum_{i=1}^{N_p} a_i(1-T_r)^{(i/3)}\right], \quad T_r = T/T_c \tag{3}$$

where N_p is the number of adjustable parameters. Values of the critical density ρ_c and the critical temperature T_c used for the fits

using Eq. 3 are also recorded in the table. Parameters were mostly taken from Refs. 3–10, those for 1-alkanols C_1 to C_{10} and n-alkanes C_5 to C_{16} are from Ref. 11. Data used for the fits were predominantly recommended values published in the TRC Thermodynamic Tables (Refs. 12,13), sometimes combined with the original experimental data or, in a few cases, the original experimental data were correlated.

$RMSD$ is a relative root-mean-square deviation (in per cent) between experimental values of density and those calculated from the particular function (Tait Eq. 1 or Eqs. 2, 3);

$$RMSD/\% = 100\left\{\frac{1}{N}\sum_{i=1}^{N}\left(\frac{\rho_{\text{exp}} - \rho_{\text{calc}}}{\rho_{\text{exp}}}\right)^2\right\}^{1/2}$$

where N is the number of experimental values included in the fit.

If the maximum temperature T_{max} of validity of the Tait equation is greater than the normal boiling point temperature T_{nbp}, then parameters of the Wagner equation in the form of either

$$P_{\text{sat}}(T) =$$

$$P_c \exp\left[\frac{a_1(1-T_r) + a_2(1-T_r)^{1.5} + a_3(1-T_r)^{2.5} + a_4(1-T_r)^5}{T_r}\right] \tag{4}$$

or

$$P_{\text{sat}}(T) =$$

$$P_c \exp\left[\frac{a_1(1-T_r) + a_2(1-T_r)^{1.5} + a_3(1-T_r)^3 + a_4(1-T_r)^6}{T_r}\right] \tag{5}$$

where $T_r = T/T_c$ are recorded in the third line for each substance. Values of the critical pressure P_c and critical temperature T_c used in Eqs. 4 and 5 are also recorded in the table (values of the critical temperature may differ a little from those recorded for the function, Eq. 3). Parameters of Eqs. 4 and 5 were taken mostly from the papers by McGarry (Ref. 14) and Ambrose and Walton (Ref. 15); in a few cases, the fits were performed using original experimental data or in combination with the recommended values from the TRC Thermodynamic Tables (Refs. 12,13).

The two right-hand-most columns gives values of the isothermal compressibility coefficient, $\kappa_T = -(1/V)(\partial V/\partial P)_T = (1/\rho)(\partial \rho/\partial P)_T$, and the isobaric cubic expansion coefficient, $\alpha_P = (1/V)(\partial V/\partial T)_P = -(1/\rho)(\partial \rho/\partial T)_P$, calculated for $T = 298.15$ K and $P = 0.101325$ MPa from the Tait Eq. 1 and from the $\rho(T, P_{\text{ref}})$ equation, respectively. In a very few cases when the lower temperature limit of the Tait equation T_{min} is greater than 298.15 K, the extrapolated values of isothermal compressibility are given.

References

1. Tait, P. G., in *Physics and Chemistry of the Voyage of H.M.S. Challenger*, Vol. II, Part IV, Thomson, C. W., and Murray, J., Eds., H.M.S.O., London, 1889.
2. Tamman, G., *Z. Phys. Chem.* 17, 620, 1895.

3. Cibulka, I., and Ziková, M., *J. Chem. Eng. Data* 39, 876, 1994.

4. Cibulka, I., and Hnědkovský, L., *J. Chem. Eng. Data* 41, 657, 1996.

5. Cibulka, I., Hnědkovský, L., and Takagi, T., *J. Chem. Eng. Data* 42, 2, 1997.

6. Cibulka, I., Hnědkovský, L., and Takagi, T., *J. Chem. Eng. Data* 42, 415, 1997.

7. Cibulka, I., and Takagi, T., *J. Chem. Eng. Data* 44, 411, 1999.

8. Cibulka, I., and Takagi, T., *J. Chem. Eng. Data* 44, 1105, 1999.

9. Cibulka, I., Takagi, T., and Růžička, K., *J. Chem. Eng. Data* 46, 2, 2001.

10. Cibulka, I., and Takagi, T., *J. Chem. Eng. Data* 47, 1037, 2002.

11. Cibulka, I., *Fluid Phase Equilib.* 89, 1, 1993.

12. TRC Thermodynamic Tables, Hydrocarbons, Thermodynamics Research Center (TRC), NIST, Thermophysical Properties Division, Boulder, Colorado.

13. TRC Thermodynamic Tables, Non-Hydrocarbons, Thermodynamics Research Center (TRC), NIST, Thermophysical Properties Division, Boulder, Colorado.

14. McGarry, J., *Ind. Eng. Chem., Process Des. Develop.* 22, 313, 1983.

15. Ambrose, D., and Walton, J., *Pure Appl. Chem.* 61, 1395, 1989.

Eq.	a_1	a_2	a_3	a_4	a_5	a_6	T_{min}/T_{max} K	P_{max} MPa	T_c K	P_c MPa	ρ_c kg m⁻³	RMSD %	T_{nbp} K	κ_T GPa⁻¹	α_P kK⁻¹
CCl_4 Tetrachloromethane (Ref. 9)															
1	$9.33340 \cdot 10^{-2}$	$1.11363 \cdot 10^{3}$	-8.68453	$2.80698 \cdot 10^{-2}$	$-4.22880 \cdot 10^{-5}$	$2.37923 \cdot 10^{-8}$	273/413	51/388				0.040	349.9	1.074	1.209
3	1.58994	2.51946	-5.82313	6.96793	-2.51359		253/554		556.40		557.33	0.042			
5	-7.07139	1.71497	-2.89930	-2.49466			250/556		556.40	4.551					
$CHBr_3$ Tribromomethane (Ref. 9)															
1	$1.03492 \cdot 10^{-1}$	$2.64208 \cdot 10^{2}$	$-4.57399 \cdot 10^{-1}$				323/368	150/343				0.058	422.3	0.809	0.907
2	$3.55953 \cdot 10^{3}$	-1.96212	$-1.08712 \cdot 10^{-3}$				283/403					0.001			
$CHCl_3$ Trichloromethane (Ref. 9)															
1	$9.57210 \cdot 10^{-2}$	$4.79593 \cdot 10^{2}$	-1.84011	$1.81340 \cdot 10^{-3}$			273/348	100(r)				0.031	334.4	1.037	1.274
3	3.56339	-3.86051	3.35636				213/333		536.40		499.49	0.043			
5	-6.95546	1.16625	-2.13970	-3.44421			215/536		536.40	5.366					
CH_2Cl_2 Dichloromethane (Ref. 9)															
1	$9.76370 \cdot 10^{-2}$	$5.24365 \cdot 10^{2}$	-2.06633	$2.09494 \cdot 10^{-3}$			293/423	100(r)				0.091	313.4	1.032	1.428
3	3.00368	-2.19763	2.34269				178/383		510.00		440.07	0.014			
5	-7.35739	2.17546	-4.07038	3.50701			233/510		510.00	6.300					
CH_3I Iodomethane (Ref. 9)															
1	$9.54770 \cdot 10^{-2}$	$5.36810 \cdot 10^{2}$	-2.25115	$2.53188 \cdot 10^{-3}$			253/313	160				0.038	315.6	1.052	1.255
2	$3.48981 \cdot 10^{3}$	-7.47709	$1.83592 \cdot 10^{-2}$	$-2.36742 \cdot 10^{-5}$			213/313					0.011			
CH_4O Methanol (Ref. 3)															
1[a]	$1.15068 \cdot 10^{-1}$	$6.49718 \cdot 10^{2}$	-4.34583	$1.30722 \cdot 10^{-2}$	$-2.00292 \cdot 10^{-5}$	$1.20566 \cdot 10^{-8}$	183/483	104(r)				0.059	337.7	1.231	1.201
3	2.62781	-4.04742	$1.58343 \cdot 10$	$-2.25066 \cdot 10$	$1.09160 \cdot 10$	$3.04774 \cdot 10^{-1}$	175/509		512.60		272.00	0.180			
4	-8.63571	1.17982	-2.47900	-1.02400			175/513		512.64	8.092					
C_2Cl_4 Tetrachloroethylene (Ref. 9)															
1	$1.01727 \cdot 10^{-1}$	$1.36610 \cdot 10^{2}$					298/298	101				0.007	394.5	0.744	1.012
3	1.10232	1.45142					298/313		620.20		571.84	0.019			
C_2HCl_3 Trichloroethylene (Ref. 9)															
1	$1.00987 \cdot 10^{-1}$	$1.16005 \cdot 10^{2}$					298/298	101				0.010	360.4	0.870	1.142
3	1.30631	1.33512					291/315		571.00		513.24	0.008			
$C_2H_4Cl_2$ 1,1-Dichloroethane (Ref. 9)															
1	$9.79930 \cdot 10^{-2}$	$4.40906 \cdot 10^{2}$	-1.67041	$1.60492 \cdot 10^{-3}$			298/398	101(r)				0.028	330.4	1.144	1.320
3	2.81773	-2.41755	2.40906				263/398		523.00		419.32	0.084			
4	-9.51254	7.49332	-7.87028	$6.32540 \cdot 10^{-1}$			320/450		523.00	5.070					
$C_2H_4Cl_2$ 1,2-Dichloroethane (Ref. 9)															
1	$9.60440 \cdot 10^{-2}$	$5.39476 \cdot 10^{2}$	-1.94358	$1.79758 \cdot 10^{-3}$			278/398	101(r)				0.027	356.7	0.801	1.158
3	2.77660	-2.33774	2.31530				263/398		561.60		439.82	0.063			
5	-7.36864	1.76727	-3.34295	-1.43530			260/566		566.00	5.362					
$C_2H_4O_2$ Acetic acid (Ref. 5)															
1	$9.44550 \cdot 10^{-2}$	$7.22131 \cdot 10^{2}$	-3.33202	$4.23588 \cdot 10^{-3}$			293/328	25/253				0.040	391.1	0.897	1.085
3	4.74861	$-1.76644 \cdot 10$	$4.44807 \cdot 10$	$-4.82535 \cdot 10$	$1.98308 \cdot 10$		293/493		592.71		351.19	0.003			
C_2H_5Br Bromoethane (Ref. 9)															
1	$9.43490 \cdot 10^{-2}$	$7.70215 \cdot 10^{2}$	-4.06470	$5.75736 \cdot 10^{-3}$			253/313	157(r)				0.055	311.6	1.344	1.475
3	1.50516	1.35063					253/313		503.90		506.82	0.058			
5	-9.14807	5.49831	-6.68657	6.27287			301/504		503.80	6.232					
C_2H_6O Ethanol (Ref. 3)															
1[b]	$1.09012 \cdot 10^{-1}$	$5.22523 \cdot 10^{2}$	-2.99367	$8.25136 \cdot 10^{-3}$	$-1.31014 \cdot 10^{-5}$	$8.74008 \cdot 10^{-9}$	193/480	200(r)				0.055	351.4	1.117	1.097
3[c]	$-9.92640 \cdot 10^{-1}$	$3.80287 \cdot 10$	$-1.81117 \cdot 10^{2}$	$4.45905 \cdot 10^{2}$	$-5.88518 \cdot 10^{2}$	$3.93920 \cdot 10^{2}$	159/508		513.88		276.00	0.160			
4	-8.68587	1.17831	-4.87620	1.58800			159/514		513.92	6.132					
$C_2H_6O_2$ Ethylene glycol (Ref. 6)															
1	$9.50140 \cdot 10^{-2}$	$6.74427 \cdot 10^{2}$	-1.77594	$1.27583 \cdot 10^{-3}$			298/378	100(r)				0.031	470.5	0.368	0.639
3	1.77482	1.11208					298/378		790.00		333.70	0.022			
C_3H_6O Acetone (Ref. 5)															
1	$9.92390 \cdot 10^{-2}$	$6.22264 \cdot 10^{2}$	-2.94315	$3.73329 \cdot 10^{-3}$			278/323	392				0.032	329.2	1.293	1.457
3	1.83805	$5.82830 \cdot 10^{-1}$	$-1.00291 \cdot 10^{-1}$	$5.70410 \cdot 10^{-1}$			179/506		508.10		277.90	0.025			

Temperature and Pressure Dependence of Liquid Density

Eq.	a_1	a_2	a_3	a_4	a_5	a_6	T_{min}/T_{max} K	P_{max} MPa	T_c K	P_c MPa	ρ_c kg m⁻³	RMSD %	T_{nbp} K	κ_T GPa⁻¹	α_P kK⁻¹
C_3H_7Br 1-Bromopropane (Ref. 9)															
1	$8.88710\cdot10^{-2}$	$4.24716\cdot10^2$	-1.81352	$2.55049\cdot10^{-3}$	$-1.23227\cdot10^{-6}$		280/500	100				0.094	344.0	1.137	1.251
3	1.65835	1.17536					280/340		536.00		455.53	0.009			
4	-9.68919	7.16793	-7.05282	1.00727			300/536		536.00	4.750					
C_3H_7Cl 1-Chloropropane (Ref. 9)															
1	$1.07245\cdot10^{-1}$	$2.78606\cdot10^2$	$-5.17553\cdot10^{-2}$	$-3.83546\cdot10^{-3}$	$5.85824\cdot10^{-6}$		273/368	98(r)				0.078	319.6	1.382	1.462
3	4.15272	-5.73954	4.74431				253/343		503.00		309.22	0.015			
5	-7.55764	2.60153	-5.06041	3.31163			248/503		503.00	4.580					
C_3H_7I 1-Iodopropane (Ref. 9)															
1	$1.01044\cdot10^{-1}$	$3.01114\cdot10^2$	$-6.61560\cdot10^{-1}$	$8.74342\cdot10^{-5}$			273/368	1177				0.070	375.7	0.904	1.072
2	$2.22323\cdot10^3$	-1.38867	$-7.96985\cdot10^{-4}$				288/361					0.196			
C_3H_8O 1-Propanol (Ref. 3)															
1[d]	$9.60290\cdot10^{-2}$	$3.10861\cdot10^2$	$-8.35531\cdot10^{-1}$	$4.27987\cdot10^{-5}$	$7.78234\cdot10^{-7}$		170/524	49(r)				0.109	370.3	1.007	0.999
3	$9.40534\cdot10^1$	$1.29442\cdot10$	$-5.39519\cdot10$	$1.13639\cdot10^2$	$-1.13866\cdot10^2$	$4.33832\cdot10$	153/530		536.74		274.00	0.160			
4	-8.53706	1.96214	-7.69180	2.94500			147/537		536.78	5.168					
C_3H_8O 2-Propanol (Ref. 6)															
1	$8.90020\cdot10^{-2}$	$1.87411\cdot10^2$	$-2.36391\cdot10^{-1}$	$-4.24787\cdot10^{-4}$			273/400	50(r)				0.054	355.4	1.123	1.099
3	$6.24542\cdot10^{-1}$	6.18938	-6.79039	2.49009			243/430		508.30		273.16	0.093			
5	-8.16927	$-9.43213\cdot10^{-2}$	-8.10040	7.85000			250/508		508.30	4.742					
$C_3H_8O_2$ 1,2-Propanediol (Ref. 6)															
1	$8.38940\cdot10^{-2}$	$9.09931\cdot10^2$	-4.03075	$5.24295\cdot10^{-3}$			273/368	200				0.019	460.2	0.481	0.714
2	$1.18071\cdot10^3$	$-2.56645\cdot10^{-1}$	$-8.05917\cdot10^{-4}$				283/363					0.017			
$C_3H_8O_2$ 1,3-Propanediol (Ref. 6)															
1	$9.10310\cdot10^{-2}$	$5.95160\cdot10^2$	-1.64714	$1.36994\cdot10^{-3}$			273/368	200				0.067	487.6	0.403	0.593
3	3.64841	-2.06696	1.22738				283/363		685.70		318.39	0.002			
$C_3H_8O_3$ Glycerol (Ref. 6)															
1	$1.14255\cdot10^{-1}$	$9.25959\cdot10^2$	-1.58140	$4.48909\cdot10^{-4}$			223/368	686				0.061	563.1	0.231	0.486
3	6.94960	-9.25236	5.43535				214/364		800.00		351.51	0.069			
$C_4H_8O_2$ Ethyl acetate (Ref. 5)															
1	$8.94090\cdot10^{-2}$	$1.05394\cdot10^3$	-7.58700	$1.98689\cdot10^{-2}$	$-1.82587\cdot10^{-5}$		253/343	49(r)				0.106	350.3	1.204	1.425
3	1.69043	2.64150	-4.70476	3.52368			253/473		523.20		308.06	0.015			
C_4H_9Br 1-Bromobutane (Ref. 9)															
1	$9.76600\cdot10^{-2}$	$4.97301\cdot10^2$	-2.02541	$2.24791\cdot10^{-3}$			273/368	1177				0.074	374.8	1.046	1.119
3	1.59525	1.23746					293/323		572.00		421.60	0.003			
$C_4H_{10}O$ 1-Butanol (Ref. 3)															
1[e]	$1.01542\cdot10^{-1}$	$5.01888\cdot10^2$	-2.41576	$5.35877\cdot10^{-3}$	$-7.31982\cdot10^{-6}$	$4.60628\cdot10^{-9}$	195/524	49(r)				0.053	390.9	0.914	0.949
3	-3.57379	$5.02450\cdot10$	$-1.75934\cdot10^2$	$3.08588\cdot10^2$	$-2.65628\cdot10^2$	$8.94997\cdot10$	186/559		563.01		271.00	0.352			
4	-8.40615	2.23010	-8.24860	$-7.11000\cdot10^{-1}$			185/563		563.05	4.424					
$C_4H_{10}O$ Diethyl ether (Ref. 5)															
1	$9.70250\cdot10^{-2}$	$3.95413\cdot10^2$	-1.68339	$1.82398\cdot10^{-3}$			293/353	981				0.239	307.6	1.740	1.657
3	3.31808	-8.55179	$2.09100\cdot10$	$-2.10672\cdot10$	8.26525		140/430		466.74		264.72	0.027			
5	-7.29916	1.24828	-2.91931	-3.36740			250/467		466.74	3.646					
C_5H_{10} Cyclopentane (Ref. 8)															
1	$8.85580\cdot10^{-2}$	$8.94968\cdot10^2$	-7.76576	$3.01636\cdot10^{-2}$	$-5.82645\cdot10^{-5}$	$4.43996\cdot10^{-8}$	193/353	48/196				0.050	322.4	1.308	1.325
3	1.68431	$6.19856\cdot10^{-1}$	$4.90670\cdot10^{-2}$	$2.19121\cdot10^{-1}$			179/508		511.70		275.04	0.031			
5	-6.51809	$3.84422\cdot10^{-1}$	-1.11706	-4.50275			289/512		511.60	4.509					
$C_5H_{11}Br$ 1-Bromopentane (Ref. 9)															
1	$8.98660\cdot10^{-2}$	$5.28042\cdot10^2$	-2.41620	$3.92466\cdot10^{-3}$	$-2.31231\cdot10^{-6}$		283/523	98				0.061	402.7	0.943	1.076
2	$1.60209\cdot10^3$	-1.30486					283/363					0.006			
4	-8.59341	3.46288	-3.06661	-4.07997			294/606		605.90	3.900					
C_5H_{12} Pentane (Ref. 4)															
1[f]	$1.00620\cdot10^{-1}$	$6.52014\cdot10^2$	-4.94167	$1.54854\cdot10^{-2}$	$-2.36811\cdot10^{-5}$	$1.44530\cdot10^{-8}$	173/373	87/285				0.119	309.2	2.133	1.610
3	1.17756	3.89157	-5.50896	3.29181			143/444		469.80		232.00	0.154			
4	-7.30698	1.75845	-2.16290	-2.91300			143/470		469.80	3.375					
$C_5H_{12}O$ 1-Pentanol (Ref. 3)															
1	$9.52000\cdot10^{-2}$	$3.00019\cdot10^2$	$-5.62481\cdot10^{-1}$	$-5.98193\cdot10^{-4}$	$1.15744\cdot10^{-6}$		233/550	59(r)				0.149	411.2	0.866	0.905
3	-2.61129	$4.23375\cdot10$	$-1.49640\cdot10^2$	$2.62210\cdot10^2$	$-2.23090\cdot10^2$	$7.37879\cdot10$	195/585		588.11		270.00	0.353			
4	-8.98005	3.91624	-9.90810	-2.19100			196/588		588.15	3.909					
C_6H_5Br Bromobenzene (Ref. 9)															
1	$9.73880\cdot10^{-2}$	$4.87590\cdot10^2$	-1.49313	$1.16283\cdot10^{-3}$			278/358	100(r)				0.043	429.2	0.668	0.901
3	2.84948	-2.30055	2.31394				298/358		670.00		484.60	0.002			
C_6H_5Cl Chlorobenzene (Ref. 9)															
1[g]	$1.06896\cdot10^{-1}$	$5.95178\cdot10^2$	-2.52143	$4.02563\cdot10^{-3}$	$-3.10524\cdot10^{-6}$	$1.06813\cdot10^{-9}$	278/583	50(r)				0.057	404.9	0.752	0.969
3	$1.17137\cdot10^{-1}$	$1.23712\cdot10$	$-2.84464\cdot10$	$2.94048\cdot10$	$-1.07075\cdot10$		253/630		632.40		365.45	0.008			
5	-7.58700	2.26551	-4.09118	$1.70377\cdot10^{-1}$			335/632		632.40	4.519					

Eq.	a_1	a_2	a_3	a_4	a_5	a_6	T_{min}/T_{max} K	P_{max} MPa	T_c K	P_c MPa	ρ_c kg m⁻³	RMSD %	T_{nbp} K	κ_T GPa⁻¹	α_P kK⁻¹
$C_6H_5NO_2$ Nitrobenzene (Ref. 10)															
1	$9.31940 \cdot 10^{-2}$	$6.10698 \cdot 10^2$	-1.89353	$1.56305 \cdot 10^{-3}$			293/358	100				0.005	484.0	0.503	0.826
3	4.47929	-5.72846	4.39478				273/373		718.00		362.09	0.027			
C_6H_6 Benzene (Ref. 7)															
1	$9.36560 \cdot 10^{-2}$	$3.28066 \cdot 10^2$	$2.79270 \cdot 10^{-2}$	$-6.54288 \cdot 10^{-3}$	$1.66249 \cdot 10^{-5}$	$-1.24527 \cdot 10^{-8}$	283/499	58(r)				0.079	353.2	0.965	1.222
3	1.81679	$-1.44828 \cdot 10^{-1}$	3.65168	-5.85231	3.50125		279/561		562.16		301.60	0.024			
4	-7.01433	1.55256	-1.84790	-3.71300			288/562		562.16	4.898					
C_6H_7N Aniline (Ref. 10)															
1	$9.43630 \cdot 10^{-2}$	$7.27479 \cdot 10^2$	-2.43017	$2.23750 \cdot 10^{-3}$			298/358	100(r)				0.043	457.3	0.467	0.830
3	2.71043	-3.14236	$1.48381 \cdot 10$	$-3.89610 \cdot 10$	$4.62777 \cdot 10$	$-1.91216 \cdot 10$	263/699		699.00		332.60	0.093			
C_6H_{12} Cyclohexane (Ref. 8)															
1	$8.51590 \cdot 10^{-2}$	$6.06614 \cdot 10^2$	-3.64438	$9.15077 \cdot 10^{-3}$	$-1.15323 \cdot 10^{-5}$	$5.95736 \cdot 10^{-9}$	287/523	81(r)				0.087	353.9	1.135	1.207
3	1.62642	$9.95043 \cdot 10^{-1}$	$-2.55766 \cdot 10^{-1}$	$-8.79980 \cdot 10^{-2}$	$4.74534 \cdot 10^{-1}$		273/553		553.50		273.25	0.038			
5	-6.96009	1.31328	-2.75683	-2.45491			293/554		553.64	4.075					
C_6H_{14} Hexane (Ref. 4)															
1[h]	$1.05863 \cdot 10^{-1}$	$4.87649 \cdot 10^2$	-2.71631	$5.75951 \cdot 10^{-3}$	$-5.74075 \cdot 10^{-6}$	$2.33450 \cdot 10^{-9}$	223/498	200(r)				0.112	341.9	1.645	1.387
3	1.59756	1.84266	-1.72631	$4.94308 \cdot 10^{-1}$	$6.46314 \cdot 10^{-1}$		183/507		507.90		234.00	0.124			
4	-7.53998	1.83759	-2.54380	-3.16300			178/508		507.90	3.035					
C_6H_{14} 2-Methylpentane (Ref. 8)															
1	$8.95370 \cdot 10^{-2}$	$3.60920 \cdot 10^2$	-1.70650	$2.61411 \cdot 10^{-3}$	$-1.30788 \cdot 10^{-6}$		273/473	32(r)				0.272	333.4	1.793	1.430
3	5.72227	$-2.82931 \cdot 10$	$7.80587 \cdot 10$	$-9.08956 \cdot 10$	$3.90010 \cdot 10$		273/473		497.50		234.82	0.090			
5	-7.28750	1.29015	-2.97853	-2.17234			240/498		498.10	3.033					
C_6H_{14} 3-Methylpentane (Ref. 8)															
1	$8.79150 \cdot 10^{-2}$	$2.68648 \cdot 10^2$	-1.00644	$9.29350 \cdot 10^{-4}$			293/473	32(r)				0.086	336.4	1.714	1.350
3	4.03149	$-1.18248 \cdot 10$	$2.42559 \cdot 10$	$-1.83295 \cdot 10$	4.33405		293/473		504.50		234.82	0.143			
5	-7.27084	1.26113	-2.81741	-2.17642			235/504		504.40	3.122					
C_6H_{14} 2,3-Dimethylbutane (Ref. 8)															
1[i]	$6.09940 \cdot 10^{-2}$	$2.88312 \cdot 10^2$	-1.38880	$2.23422 \cdot 10^{-3}$	$-1.24468 \cdot 10^{-6}$		208/473	32(r)				0.072	331.2	1.831	1.364
3	4.23996	$-1.47401 \cdot 10$	$3.56742 \cdot 10$	$-3.61641 \cdot 10$	$1.37668 \cdot 10$		208/463		499.98		240.72	0.078			
5	-7.27870	1.56349	-3.05387	-1.57752			235/500		500.30	3.146					
$C_6H_{14}O$ 1-Hexanol (Ref. 3)															
1	$9.45430 \cdot 10^{-2}$	$3.56654 \cdot 10^2$	$-9.63938 \cdot 10^{-1}$	$3.19317 \cdot 10^{-4}$	$7.27932 \cdot 10^{-7}$	$-3.68450 \cdot 10^{-10}$	298/503	400				0.045	430.5	0.828	0.878
3	$-1.55309 \cdot 10^{-1}$	$1.77262 \cdot 10$	$-5.76007 \cdot 10$	$9.99026 \cdot 10$	$-8.65883 \cdot 10$	$2.96281 \cdot 10$	223/607		610.70		268.00	0.191			
4	-9.49034	5.13288	$-1.05817 \cdot 10$	-5.15400			226/611		610.70	3.470					
C_7H_8 Toluene (Ref. 7)															
1[j]	$1.08655 \cdot 10^{-1}$	$6.58594 \cdot 10^2$	-3.36063	$6.79417 \cdot 10^{-3}$	$-6.73894 \cdot 10^{-6}$	$2.80015 \cdot 10^{-9}$	179/583	50(r)				0.052	383.8	0.900	1.080
3	2.33057	-3.10784	$1.04839 \cdot 10$	$-1.28168 \cdot 10$	6.29458	$-2.64550 \cdot 10^{-1}$	178/588		591.79		291.59	0.041			
5	-7.28607	1.38091	-2.83433	-2.79168			309/592		591.72	4.106					
C_7H_8O Anisole (Ref. 5)															
1	$9.57370 \cdot 10^{-2}$	$3.76997 \cdot 10^2$	$-7.76199 \cdot 10^{-1}$				298/353	196				0.032	426.9	0.657	0.949
3	2.32023	-1.28589	1.68888				273/353		645.60		335.84	0.031			
C_7H_{14} Cycloheptane (Ref. 8)															
1	$9.09350 \cdot 10^{-2}$	$3.25145 \cdot 10^2$	$-8.81409 \cdot 10^{-1}$	$4.29576 \cdot 10^{-4}$			294/393	40(r)				0.041	391.6	0.904	1.062
2	$1.06207 \cdot 10^3$	$-8.56800 \cdot 10^{-1}$					298/353					0.003			
C_7H_{16} Heptane (Ref. 4)															
1[k]	$8.14360 \cdot 10^{-2}$	$5.31891 \cdot 10^2$	-3.38490	$9.42976 \cdot 10^{-3}$	$-1.36183 \cdot 10^{-5}$	$8.10257 \cdot 10^{-9}$	198/511	150(r)				0.153	371.6	1.429	1.251
3	1.33159	3.30092	-4.50961	2.76549			183/538		540.11		236.00	0.377			
4	-7.77404	1.85614	-2.82980	-3.50700			183/540		540.15	2.735					
C_8H_{10} o-Xylene (Ref. 7)															
1	$8.02720 \cdot 10^{-2}$	$4.52200 \cdot 10^2$	-1.81150	$2.45174 \cdot 10^{-3}$	$-1.15349 \cdot 10^{-6}$		257/598	51(r)				0.082	417.6	0.806	0.944
3	1.93903	$6.93214 \cdot 10^{-1}$	$-5.52518 \cdot 10^{-1}$	$7.42481 \cdot 10^{-1}$			248/628		630.30		287.72	0.006			
5	-7.53357	1.40968	-3.10985	-2.85992			337/630		630.25	3.733					
C_8H_{10} m-Xylene (Ref. 7)															
1	$8.13550 \cdot 10^{-2}$	$5.95204 \cdot 10^2$	-3.17132	$6.99286 \cdot 10^{-3}$	$-7.69418 \cdot 10^{-6}$	$3.47267 \cdot 10^{-9}$	230/598	20(r)				0.054	412.3	0.857	0.987
3	1.96653	$7.70424 \cdot 10^{-1}$	$-8.14622 \cdot 10^{-1}$	$9.44937 \cdot 10^{-1}$			225/613		617.05		282.36	0.007			
5	-7.59222	1.39441	-3.22746	-2.40376			332/617		616.97	3.537					
C_8H_{10} p-Xylene (Ref. 7)															
1	$8.45420 \cdot 10^{-2}$	$4.38870 \cdot 10^2$	-1.77395	$2.41453 \cdot 10^{-3}$	$-1.13719 \cdot 10^{-6}$		288/598	51(r)				0.076	411.5	0.894	1.003
3	2.01680	$6.60504 \cdot 10^{-1}$	$-7.70676 \cdot 10^{-1}$	$9.97453 \cdot 10^{-1}$			286/603		616.20		280.13	0.005			
5	-7.63495	1.50724	-3.19678	-2.78710			331/616		616.15	3.513					
C_8H_{16} Cyclooctane (Ref. 8)															
1	$8.66040 \cdot 10^{-2}$	$3.85135 \cdot 10^2$	-1.19668	$9.00890 \cdot 10^{-4}$			314/394	40				0.026	422.2	0.798	0.960
3	1.25843	1.53566					293/394		647.20		273.70	0.040			

Eq.	a_1	a_2	a_3	a_4	a_5	a_6	T_{min}/T_{max} K	P_{max} MPa	T_c K	P_c MPa	ρ_c kg m^{-3}	RMSD %	T_{nbp} K	κ_T GPa^{-1}	α_p kK^{-1}
C_8H_{18} Octane (Ref. 4)															
1[i]	$1.09943 \cdot 10^{-1}$	$1.17639 \cdot 10^3$	$-1.02441 \cdot 10$	$3.73444 \cdot 10^{-2}$	$-6.42362 \cdot 10^{-5}$	$4.23381 \cdot 10^{-8}$	248/393	108(r)				0.069	398.8	1.255	1.158
3	1.96977	−1.10062	6.36417	−8.69348	4.42005		216/563		568.91		237.00	0.221			
4	−8.04937	2.03865	−3.31200	−3.64800			216/569		568.95	2.490					
$C_8H_{18}O$ 1-Octanol (Ref. 3)															
1	$9.39730 \cdot 10^{-2}$	$5.22871 \cdot 10^2$	−1.99322	$2.56178 \cdot 10^{-3}$	$-1.13255 \cdot 10^{-6}$		283/623	79(r)				0.177	468.3	0.743	0.844
3	−3.53373	$5.48581 \cdot 10$	$-1.99653 \cdot 10^2$	$3.45797 \cdot 10^2$	$-2.85030 \cdot 10^2$	$9.05445 \cdot 10$	258/643		652.50		266.00	0.262			
4	$-1.00144 \cdot 10$	5.90629	$-1.04026 \cdot 10$	−9.04800			258/653		652.50	2.860					
C_9H_{12} Mesitylene (Ref. 7)															
1	$8.84360 \cdot 10^{-2}$	$4.00368 \cdot 10^2$	−1.18616	$7.38886 \cdot 10^{-4}$			238/362	200				0.056	437.9	0.786	0.944
3	$1.23001 \cdot 10$	$-3.23171 \cdot 10$	$3.37700 \cdot 10$	$-1.06994 \cdot 10$			238/353		637.25		277.59	0.012			
C_9H_{20} Nonane (Ref. 4)															
1[m]	$1.04176 \cdot 10^{-1}$	$6.60874 \cdot 10^2$	−3.94793	$9.86964 \cdot 10^{-3}$	$-1.23014 \cdot 10^{-5}$	$6.27200 \cdot 10^{-9}$	248/511	65(r)				0.064	424.0	1.125	1.088
3	1.92778	$9.30219 \cdot 10^{-1}$	−1.33413	1.39282			223/511		594.90		238.00	0.078			
4	−8.32886	2.25707	−3.82570	−3.73200			220/595		594.90	2.290					
$C_{10}H_{22}$ Decane (Ref. 4)															
1[n]	$7.76110 \cdot 10^{-2}$	$5.12895 \cdot 10^2$	−2.98749	$7.58178 \cdot 10^{-3}$	$-9.71758 \cdot 10^{-6}$	$5.01670 \cdot 10^{-9}$	248/503	200(r)				0.079	447.3	1.096	1.042
3	$3.29139 \cdot 10^{-1}$	7.36434	−9.98510	5.28361			243/511		617.61		239.00	0.108			
4	−8.60643	2.44659	−4.29250	−3.90800			244/618		617.65	2.105					
$C_{11}H_{24}$ Undecane (Ref. 4)															
1[o]	$9.36240 \cdot 10^{-2}$	$9.78153 \cdot 10^3$	$-1.13416 \cdot 10^2$	$4.97036 \cdot 10^{-1}$	$-9.65143 \cdot 10^{-4}$	$6.97765 \cdot 10^{-7}$	258/423	50/500				0.044	469.1	1.078	0.998
3	$-2.75532 \cdot 10^{-1}$	9.28503	$-1.19923 \cdot 10$	5.96957			258/473		638.81		240.00	0.069			
$C_{12}H_{26}$ Dodecane (Ref. 4)															
1	$9.05450 \cdot 10^{-2}$	$9.00366 \cdot 10^3$	$-1.03134 \cdot 10^2$	$4.49084 \cdot 10^{-1}$	$-8.70464 \cdot 10^{-4}$	$6.31007 \cdot 10^{-7}$	268/393	20/442				0.083	489.5	0.998	0.977
3	$-3.04946 \cdot 10^{-2}$	8.32535	$-1.08268 \cdot 10$	5.55178			263/483		658.60		240.00	0.055			
$C_{13}H_{28}$ Tridecane (Ref. 4)															
1	$8.79880 \cdot 10^{-2}$	$3.53129 \cdot 10^2$	−1.17372	$1.00334 \cdot 10^{-3}$			303/473	500				0.080	508.6	0.951	0.947
3	2.52779	−1.64864	2.10099				265/473		676.00		240.00	0.139			
$C_{14}H_{30}$ Tetradecane (Ref. 4)															
1	$9.01310 \cdot 10^{-2}$	$2.44902 \cdot 10^2$	$-4.92510 \cdot 10^{-1}$				293/358	50/367				0.040	526.7	0.918	0.924
3	3.50992	−4.09768	3.61064				279/372		693.00		241.00	0.050			
$C`_{15}H_{32}$ Pentadecane (Ref. 4)															
1	$8.85030 \cdot 10^{-2}$	$4.04942 \cdot 10^2$	−1.37693	$1.25370 \cdot 10^{-3}$			311/408	69/655				0.048	543.8	0.835	0.882
3	1.08801	1.82524					288/408		708.00		241.00	0.038			
$C_{16}H_{34}$ Hexadecane (Ref. 4)															
1	$9.05120 \cdot 10^{-2}$	$-6.87292 \cdot 10^2$	8.07128	$-2.59370 \cdot 10^{-2}$	$2.60661 \cdot 10^{-5}$		293/393	10/451				0.047	560.0	0.866	0.889
3	2.61306	−1.99964	2.40011				293/490		722.00		241.00	0.035			

[a] $b_1 = -5.322 \cdot 10^{-5}$

[b] $b_1 = -4.588 \cdot 10^{-5}$

[c] Additional term $a_7(1 - T_r)^{(7/3)}$ is included in Eq. 3 with $a_7 = -1.04344 \; 10^2$

[d] $b_1 = -3.058 \cdot 10^{-5}$

[e] $b_1 = -3.268 \cdot 10^{-5}$

[f] $b_1 = -3.788 \cdot 10^{-5}$

[g] $b_1 = -3.690 \cdot 10^{-5}$

[h] $b_1 = -4.522 \cdot 10^{-5}$

[i] $b_1 = 4.085 \cdot 10^{-5}$

[j] $b_1 = -5.004 \cdot 10^{-5}$

[k] $b_1 = 7.924 \cdot 10^{-5}, c_1 = -1.513 \cdot 10^{-7}$

[l] $b_1 = -5.741 \cdot 10^{-5}$

[m] $b_1 = -2.960 \cdot 10^{-5}$

[n] $b_1 = 2.778 \cdot 10^{-5}$

[o] $b_1 = -1.105 \cdot 10^{-5}$

VOLUMETRIC PROPERTIES OF AQUEOUS SODIUM CHLORIDE SOLUTIONS

This table gives the following properties of aqueous solutions of NaCl as a function of temperature and concentration:

Specific volume v (reciprocal of density) in cm³/g
Isothermal compressibility $\kappa_T = -(1/v)(\partial v/\partial P)_T$ in GPa⁻¹
Cubic expansion coefficient $\alpha_v = (1/v)(\partial v/\partial T)_P$ in kK⁻¹

All data refer to a pressure of 100 kPa (1 bar). The reference gives properties over a wider range of temperature and pressure.

Reference

Rogers, P. S. Z., and Pitzer, K. S., *J. Phys. Chem. Ref. Data*, 11, 15, 1982.

					Molality in mol/kg				
t/°C	0.100	0.250	0.500	0.750	1.000	2.000	3.000	4.000	5.000
Specific volume v in cm³/g									
0	0.995732	0.989259	0.978889	0.968991	0.959525	0.925426	0.896292	0.870996	0.848646
10	0.995998	0.989781	0.979804	0.970256	0.961101	0.927905	0.899262	0.874201	0.851958
20	0.997620	0.991564	0.981833	0.972505	0.963544	0.930909	0.902565	0.877643	0.855469
25	0.998834	0.992832	0.983185	0.973932	0.965038	0.932590	0.904339	0.879457	0.857301
30	1.000279	0.994319	0.984735	0.975539	0.966694	0.934382	0.906194	0.881334	0.859185
40	1.003796	0.997883	0.988374	0.979243	0.970455	0.938287	0.910145	0.885276	0.863108
50	1.008064	1.002161	0.992668	0.983551	0.974772	0.942603	0.914411	0.889473	0.867241
60	1.0130	1.0071	0.9976	0.9885	0.9797	0.9474	0.9191	0.8940	0.8716
70	1.0186	1.0127	1.0031	0.9939	0.9851	0.9526	0.9240	0.8987	0.8762
80	1.0249	1.0188	1.0092	0.9999	0.9909	0.9581	0.9293	0.9037	0.8809
90	1.0317	1.0256	1.0157	1.0063	0.9972	0.9640	0.9348	0.9089	0.8858
100	1.0391	1.0329	1.0228	1.0133	1.0040	0.9703	0.9406	0.9144	0.8910
Isothermal Compressibility κ_T in GPa⁻¹									
0	0.503	0.492	0.475	0.459	0.443	0.389	0.346	0.315	0.294
10	0.472	0.463	0.449	0.436	0.423	0.377	0.341	0.313	0.294
20	0.453	0.446	0.433	0.422	0.411	0.371	0.338	0.313	0.294
25	0.447	0.440	0.428	0.417	0.407	0.369	0.337	0.313	0.294
30	0.443	0.436	0.425	0.414	0.404	0.367	0.337	0.313	0.294
40	0.438	0.432	0.421	0.411	0.401	0.367	0.338	0.315	0.296
50	0.438	0.431	0.421	0.411	0.402	0.369	0.340	0.317	0.299
60	0.44	0.44	0.43	0.42	0.41	0.38	0.35	0.32	0.30
70	0.45	0.44	0.43	0.42	0.42	0.38	0.36	0.33	0.31
80	0.46	0.45	0.44	0.43	0.43	0.39	0.37	0.34	0.32
90	0.47	0.47	0.46	0.45	0.44	0.41	0.38	0.35	0.33
100	0.49	0.48	0.47	0.46	0.45	0.42	0.39	0.37	0.34
Cubic expansion coefficient α_v in kK⁻¹									
0	−0.058	−0.026	0.024	0.069	0.110	0.237	0.313	0.355	
10	0.102	0.123	0.156	0.186	0.213	0.297	0.349	0.380	
20	0.218	0.232	0.254	0.274	0.292	0.349	0.384	0.406	
25	0.267	0.278	0.296	0.312	0.327	0.373	0.401	0.420	
30	0.311	0.320	0.334	0.347	0.359	0.395	0.418	0.433	
40	0.389	0.394	0.402	0.410	0.417	0.438	0.451	0.460	
50	0.458	0.460	0.464	0.467	0.470	0.479	0.484	0.486	
60	0.52	0.52	0.52	0.52	0.52	0.52	0.52	0.52	
70	0.58	0.58	0.58	0.57	0.57	0.56	0.55	0.54	
80	0.64	0.63	0.63	0.62	0.61	0.60	0.58	0.56	
90	0.69	0.68	0.67	0.67	0.66	0.63	0.61	0.59	
100	0.74	0.73	0.72	0.71	0.70	0.66	0.64	0.61	

PROPERTIES OF CRYOGENIC FLUIDS

Eric W. Lemmon

This table gives physical and thermodynamic properties of ten cryogenic fluids. The properties are:

M	Molar mass
T_t	Triple-point temperature
P_t	Triple-point pressure
ρ_t (l)	Liquid density at the triple point
$\Delta_{fus}H @ T_t$	Enthalpy of fusion at the triple point
T_b	Normal boiling point at a pressure of 101.325 kPa (760 mmHg)
$\Delta_{vap}H @ T_b$	Enthalpy of vaporization at the normal boiling point
$\rho @ T_b$	Density at the normal boiling point for the liquid (l) or vapor (v)
$C_p @ T_b$	Heat capacity at constant pressure at the normal boiling point for the liquid (l) or vapor (v)
$u @ T_b$	Speed of sound at the normal boiling point for the liquid (l) or vapor (v)
T_c	Critical temperature
P_c	Critical pressure
ρ_c	Critical density

All properties except the heat of fusion are calculated from the REFPROP program, see Reference 4. Temperatures are listed on the ITS-90 scale, except those values for neon and oxygen that are obtained from older equations of state based on the IPTS-68 temperature scale. The references for all fluids except air, helium, neon, krypton, and xenon are given in the *Thermophysical Properties of Fluids* table. The properties of hydrogen are given for the para form of the molecule (see the Leachman et al. reference for details).

The triple-point temperature of air is the solidification temperature of the liquid (see Reference 2 for details). The boiling-point temperature for air is the bubble-point temperature (i.e., the temperature at which boiling begins as the pressure of the liquid is lowered). The dew-point (vapor) properties of air at 101.325 kPa are calculated at a temperature of 81.72 K; the liquid and vapor properties of these two state points are not in equilibrium. The triple-point properties of helium are given at the temperature of the lambda line (change from normal-to-superfluid helium) for the saturated-liquid state.

References

1. Katti, R., Jacobsen, R. T, Stewart, R. B., and Jahangiri, M., Thermodynamic Properties of Neon for Temperatures from the Triple Point to 700 K at Pressures to 700 MPa, *Adv. Cryo. Eng.* 31, 1189, 1986.
2. Lemmon, E. W., Jacobsen, R. T, Penoncello, S. G., and Friend, D. G., Thermodynamic Properties of Air and Mixtures of Nitrogen, Argon, and Oxygen from 60 to 2000 K at Pressures to 2000 MPa, *J. Phys. Chem. Ref. Data* 29, 331, 2000.
3. Lemmon, E. W., and Span, R., Short Fundamental Equations of State for 20 Industrial Fluids, *J. Chem. Eng. Data* 51, 785, 2006.
4. Lemmon, E. W., Huber, M. L., and McLinden, M. O., NIST Standard Reference Database 23: Reference Fluid Thermodynamic and Transport Properties-REFPROP, Version 9.0, National Institute of Standards and Technology, Standard Reference Data Program, Gaithersburg, Maryland, 2010 (www.nist.gov/srd/nist23.cfm).
5. Ortiz-Vega, D. O., Hall, K. R., Arp, V. D., and Lemmon, E. W., Interim Helium Equation, to be published in *Int. J. Thermophys.*, 2011.

Property	Units	Air	N₂	O₂	H₂	He	Ne	Ar	Kr	Xe	CH₄
M	g mol⁻¹	28.9655	28.0134	31.9988	2.01588	4.0026	20.1797	39.948	83.798	131.293	16.0428
T_t	K	59.75	63.151	54.361	13.8033	2.1768	24.556	83.8058	115.775	161.405	90.6941
P_t	kPa	5.265	12.52	0.1463	7.041	5.043	43.37	68.89	73.53	81.77	11.70
ρ_t(l)	g cm⁻³	0.9578	0.8672	1.306	0.07698	0.1459	1.252	1.417	2.447	2.966	0.4515
$\Delta_{fus}H @ T_t$	J g⁻¹		25.3	13.7	59.5		16.8	28.0	16.3	13.8	58.41
T_b	K	78.903	77.355	90.188	20.271	4.222	27.104	87.302	119.735	165.051	111.667
$\Delta_{vap}H @ T_b$	J g⁻¹	204.8	199.2	213.1	446.1	20.91	85.75	161.1	107.1	95.59	510.8
ρ (l) @ T_b	g cm⁻³	0.8752	0.8061	1.141	0.07083	0.1250	1.207	1.395	2.417	2.942	0.4224
ρ (v) @ T_b	g dm⁻³	4.497	4.612	4.467	1.339	16.70	9.577	5.774	8.818	10.01	1.816
C_p(l) @ T_b	J g⁻¹ K⁻¹	1.933	2.041	1.699	9.729	5.105	1.862	1.117	0.5198	0.3393	3.481
C_p(v) @ T_b	J g⁻¹ K⁻¹	1.090	1.124	0.9707	12.03	9.327	1.415	0.5658	0.2748	0.1751	2.218
u(l) @ T_b	m s⁻¹	865.5	851.4	904.3	1111	177.3	594.1	838.3	682.5	643.1	1338
u(v) @ T_b	m s⁻¹	177.1	174.8	177.5	355.0	101.5	126.6	170.9	137.9	129.4	271.5
T_c	K	132.5306	126.192	154.581	32.938	5.1953	44.4918	150.687	209.48	289.733	190.564
P_c	MPa	3.7860	3.3958	5.0430	1.2858	0.22746	2.6786	4.8630	5.5250	5.8420	4.5992
ρ_c	g cm⁻³	0.3426	0.3133	0.4361	0.03132	0.06958	0.4819	0.5356	0.9092	1.103	0.1627

PROPERTIES OF LIQUID HELIUM

The following data were obtained by a critical evaluation of all existing experimental measurements on liquid helium, using a fitting procedure described in the reference. All values refer to liquid helium at saturated vapor pressure; temperatures are on the ITS-90 scale. Several properties show a singularity at the lambda point (2.1768 K).

p: vapor pressure
ρ: density
C_s: molar heat capacity
$\Delta_{vap}H$: molar enthalpy of vaporization

ε: relative permittivity (dielectric constant)
σ: surface tension
α_v: cubic expansion coefficient
η: viscosity
λ: thermal conductivity

Reference

Donnelly, R. J., and Barenghi, C. F., *J. Phys. Chem. Ref. Data*, 27, 1217, 1998.

T/K	p/kPa	$\rho/g\,cm^{-3}$	$C_s/J\,mol^{-1}\,K^{-1}$	$\Delta_{vap}H/J\,mol^{-1}$	ε	$\sigma/mN\,m^{-1}$	$10^3\alpha_v/K^{-1}$	$\eta/\mu Pa\,s$	$\lambda/W\,cm^{-1}\,K^{-1}$
0.0		0.1451397	0	59.83	1.057255		0.000		
0.5		0.1451377	0.010	70.24	1.057254	0.3530	0.107		
1.0	0.01558	0.1451183	0.415	80.33	1.057246	0.3471	0.309	3.873	
1.5	0.4715	0.1451646	4.468	89.35	1.057265	0.3322	−2.36	1.346	
2.0	3.130	0.1456217	21.28	93.07	1.057449	0.3021	−12.2	1.468	
2.5	10.23	0.1448402	9.083	92.50	1.057135	0.2623	39.4	3.259	0.1497
3.0	24.05	0.1412269	9.944	94.11	1.055683	0.2161	61.5	3.517	0.1717
3.5	47.05	0.1360736	12.37	92.84	1.053615	0.1626	88.7	3.509	0.1868
4.0	81.62	0.1289745	15.96	87.00	1.050770	0.1095	129	3.319	0.1965
4.5	130.3	0.1188552	21.8	75.86	1.046725	0.0609	211		
5.0	196.0		44.7	47.67		0.0157			

PROPERTIES OF REFRIGERANTS

This table gives physical properties of compounds that have been used as working fluids in traditional refrigeration systems or are under consideration as replacements in newer systems. Some are also used as solvents and blowing agents. Many of the compounds listed are believed to be less harmful to the environment than the traditional halocarbon refrigerants.

Compounds are listed by their ASHRAE standard refrigerant designations (Reference 1), which appear in the first column. These codes are often prefixed by symbols such as CFC- (for chlorofluorocarbon), HCFC- (for hydrochlorofluorocarbon), or simply R- (for refrigerant). The "R" number assigned to refrigerants is specified by ANSI/ASHRAE Standard 34. This system is most useful for the hydrocarbons and halocarbons with one to three carbons; for such molecules the chemical composition can be determined from the number and vice versa. The first digit on the far right is the number of fluorine atoms in the compound. The second digit from the right is one more than the number of hydrogen atoms. The third digit from the right is one less than the number of carbon atoms; for single-carbon compounds, this digit is omitted. The fourth digit from the right is equal to the number of unsaturated carbon–carbon bonds; for saturated compounds, this digit is omitted. The number of bromine and iodine atoms is indicated, if needed, by appending "Bn" or "In" to the digits specified by the above rules, where "n" is the number of bromine or iodine atoms. All atoms not specified by the above are assumed to be chlorine. Appended lowercase letter(s) designate different isomers. Additional rules are used to specify cyclic compounds, ethers, inorganic fluids (R700- and R7000-series), miscellaneous organic compounds (R600-series), and blends (R400- and R500-series).

The properties tabulated are

t_m normal melting point in °C
t_b normal boiling point in °C (at 101.325 kPa or 760 mmHg)
t_c critical temperature in °C
TLV   Threshold Limit Value, which is the maximum safe concentration in air in the workplace, expressed as the time-weighted average (TWA) in parts per million by volume, over an 8-hr workday and 40-hr workweek.

Many of the critical temperatures and normal boiling points have been calculated from the equations of state described in References 9 and 10. These values differ slightly in some cases (generally no more than about 0.1 °C) from values elsewhere in this book, but the differences are probably within the experimental uncertainty. Further references and additional data on the critical properties may be found in the table "Critical Constants" in this Section.

References

1. ASHRAE (2007). ANSI/ASHRAE Standard 34-2007 Designation and Safety Classification of Refrigerants, American Society of Heating, Refrigerating and Air-Conditioning Engineers, Atlanta, GA.
2. *ASHRAE Fundamentals Handbook 2001*, Chapter 19, Refrigerants, American Society of Heating, Refrigerating, and Air-Conditioning Engineers, Atlanta, GA, 2001.
3. Platzer, B., Polt, A., and Mauer, G., *Thermophysical Properties of Refrigerants*, Springer, Berlin, 1990.
4. Sako, T., Sato, M., Nakazawa, N., Oowa, M., Yasumoto, M., Ito, H., and Yamashita, S., *J. Chem. Eng. Data* 41, 802, 1996.
5. Schmidt, J. W., Carrillo-Nava, E., and Moldover, M. R., *Fluid Phase Equilib.* 122, 187, 1996.
6. Salvi-Narkhede, M., Wang, B-H., Adcock, J. L., and Van Hook, W. A., *J. Chem. Thermodyn.* 24, 1065, 1992.
7. Fialho, P. S., and Nieto de Castro, C. A., *Int. J. Thermophys.* 21, 385, 2000.
8. Daubert, T. E., Danner, R. P., Sibul, H. M., and Stebbins, C. C., *Physical and Thermodynamic Properties of Pure Compounds: Data Compilation*, extant 2002 (core with supplements), Taylor & Francis, Bristol, PA.
9. McLinden, M.O., Lemmon, E.W., and Huber, M.L., The REFPROP Database for the Thermophysical Properties of Refrigerants, 21st International Congress of Refrigeration, Washington D.C., International Institute of Refrigeration, Paper ICR0443, 2003.
10. Lemmon, E.W., Huber, M.L., and McLinden, M.O., NIST Standard Reference Database 23: Reference Fluid Thermodynamic and Transport Properties - REFPROP, Version 9.0, National Institute of Standards and Technology, Standard Reference Data Program, Gaithersburg, MD, 2010 (www.nist.gov/srd/nist23.cfm).

Code	Name	Molecular Formula	CAS Reg. No.	t_m/°C	t_b/°C	t_c/°C	TLV/ ppm
10	Tetrachloromethane	CCl_4	56-23-5	−22.62	76.8	283.4	5
11	Trichlorofluoromethane	CCl_3F	75-69-4	−110.47	23.71	197.96	1000
12	Dichlorodifluoromethane	CCl_2F_2	75-71-8	−157.05	−29.75	111.97	1000
12B1	Bromochlorodifluoromethane	$CBrClF_2$	353-59-3	−159.5	−3.7	153.73	
12B2	Dibromodifluoromethane	CBr_2F_2	75-61-6	−110.1	22.76	198.1	100
13	Chlorotrifluoromethane	$CClF_3$	75-72-9	−181.15	−81.48	28.85	
13B1	Bromotrifluoromethane	$CBrF_3$	75-63-8	−172	−57.8	67.0	1000
14	Tetrafluoromethane	CF_4	75-73-0	−183.61	−128.05	−45.64	
20	Trichloromethane	$CHCl_3$	67-66-3	−63.41	61.17	263.2	10
21	Dichlorofluoromethane	$CHCl_2F$	75-43-4	−130.35	8.86	178.33	10
22	Chlorodifluoromethane	$CHClF_2$	75-45-6	−157.42	−40.81	96.15	1000
22B1	Bromodifluoromethane	$CHBrF_2$	1511-62-2	−145	−14.6	138.83	
23	Trifluoromethane	CHF_3	75-46-7	−155.13	−82.02	26.14	
30	Dichloromethane	CH_2Cl_2	75-09-2	−97.2	40	237	50

Code	Name	Molecular Formula	CAS Reg. No.	$t_m/°C$	$t_b/°C$	$t_c/°C$	TLV/ ppm
31	Chlorofluoromethane	CH_2ClF	593-70-4	−135.1	−9.1	154	
32	Difluoromethane	CH_2F_2	75-10-5	−136.81	−51.65	78.11	
40	Chloromethane	CH_3Cl	74-87-3	−97.7	−24.09	143.10	50
41	Fluoromethane	CH_3F	593-53-3	−143.33	−78.31	44.13	
50	Methane	CH_4	74-82-8	−182.43	−161.48	−82.59	1000
110	Hexachloroethane	C_2Cl_6	67-72-1	186.8	184.7 sp	422	1
111	Pentachlorofluoroethane	C_2Cl_5F	354-56-3	101.3	138		
112	1,1,2,2-Tetrachloro-1,2-difluoroethane	$C_2Cl_4F_2$	76-12-0	24.8	92.8	278	50
112a	1,1,1,2-Tetrachloro-2,2-difluoroethane	$C_2Cl_4F_2$	76-11-9	41.0	92.8		100
113	1,1,2-Trichloro-1,2,2-trifluoroethane	$C_2Cl_3F_3$	76-13-1	−36.22	47.59	214.06	1000
113a	1,1,1-Trichloro-2,2,2-trifluoroethane	$C_2Cl_3F_3$	354-58-5	14.37	45.5	209.7	
114	1,2-Dichloro-1,1,2,2-tetrafluoroethane	$C_2Cl_2F_4$	76-14-2	−92.52	3.59	145.68	1000
114a	1,1-Dichloro-1,2,2,2-tetrafluoroethane	$C_2Cl_2F_4$	374-07-2	−56.6	3.4	145.4	
114B2	1,2-Dibromotetrafluoroethane	$C_2Br_2F_4$	124-73-2	−110.32	47.35	214.6	
115	Chloropentafluoroethane	C_2ClF_5	76-15-3	−99.4	−39.22	79.95	1000
116	Hexafluoroethane	C_2F_6	76-16-4	−100.05	−78.09	19.88	
120	Pentachloroethane	C_2HCl_5	76-01-7	−28.78	162.0		
121	1,1,2,2-Tetrachloro-1-fluoroethane	C_2HCl_4F	354-14-3	−82.6	116.7		
121a	1,1,1,2-Tetrachloro-2-fluoroethane	C_2HCl_4F	354-11-0	−95.3	117.1		
122	1,2,2-Trichloro-1,1-difluoroethane	$C_2HCl_3F_2$	354-21-2	−140	71.9		
122a	1,2,2-Trichloro-1,2-difluoroethane	$C_2HCl_3F_2$	354-15-4	−174	72.5		
122b	1,1,1-Trichloro-2,2-difluoroethane	$C_2HCl_3F_2$	354-12-1		73		
123	2,2-Dichloro-1,1,1-trifluoroethane	$C_2HCl_2F_3$	306-83-2	−107.15	27.82	183.68	
123a	1,2-Dichloro-1,1,2-trifluoroethane	$C_2HCl_2F_3$	354-23-4	−78	29.5	188.4	
124	1-Chloro-1,2,2,2-tetrafluoroethane	C_2HClF_4	2837-89-0	−199.15	−11.96	122.28	
124a	1-Chloro-1,1,2,2-tetrafluoroethane	C_2HClF_4	354-25-6	−117	−11.7	126.7	
125	Pentafluoroethane	C_2HF_5	354-33-6	−100.63	−48.09	66.02	
E125	Trifluoromethyl difluoromethyl ether	C_2HF_5O	3822-68-2	−157	−38	80.8	
130	1,1,2,2-Tetrachloroethane	$C_2H_2Cl_4$	79-34-5	−42.4	145.2	388.00	1
131	1,1,2-Trichloro-2-fluoroethane	$C_2H_2Cl_3F$	359-28-4		102.4		
132	1,2-Dichloro-1,2-difluoroethane	$C_2H_2Cl_2F_2$	431-06-1	−101.2	59.6		
132b	1,2-Dichloro-1,1-difluoroethane	$C_2H_2Cl_2F_2$	1649-08-7	−101.2	46.2		
133	1-Chloro-1,2,2-trifluoroethane	$C_2H_2ClF_3$	431-07-2		17.3		
133a	2-Chloro-1,1,1-trifluoroethane	$C_2H_2ClF_3$	75-88-7	−105.5	6.1	151.86	
133b	1-Chloro-1,1,2-trifluoroethane	$C_2H_2ClF_3$	421-04-5		12		
134	1,1,2,2-Tetrafluoroethane	$C_2H_2F_4$	359-35-3	−89	−19.9	118.59	
134a	1,1,1,2-Tetrafluoroethane	$C_2H_2F_4$	811-97-2	−103.3	−26.07	101.06	
E134	Bis(difluoromethyl) ether	$C_2H_2F_4O$	1691-17-4		2	147.10	
140	1,1,2-Trichloroethane	$C_2H_3Cl_3$	79-00-5	−36.3	113.8	329	10
140a	1,1,1-Trichloroethane	$C_2H_3Cl_3$	71-55-6	−30.01	74.09	272	350
141	1,2-Dichloro-1-fluoroethane	$C_2H_3Cl_2F$	430-57-9	−60	73.8		
141b	1,1-Dichloro-1-fluoroethane	$C_2H_3Cl_2F$	1717-00-6	−103.47	32.05	204.35	
142	1-Chloro-2,2-difluoroethane	$C_2H_3ClF_2$	338-65-8		35.1		
142b	1-Chloro-1,1-difluoroethane	$C_2H_3ClF_2$	75-68-3	−130.43	−9.12	137.11	
143	1,1,2-Trifluoroethane	$C_2H_3F_3$	430-66-0	−84	3.7	156.6	
143a	1,1,1-Trifluoroethane	$C_2H_3F_3$	420-46-2	−111.81	−47.24	72.71	
143m	Methyl trifluoromethyl ether	$C_2H_3F_3O$	421-14-7	−149	−23.66	104.77	
E143a	2,2,2-Trifluoroethyl methyl ether	$C_3H_5F_3O$	460-43-5		31.62	175.83	
150	1,2-Dichloroethane	$C_2H_4Cl_2$	107-06-2	−35.7	83.5	288	10
150a	1,1-Dichloroethane	$C_2H_4Cl_2$	75-34-3	−96.9	57.3	250	100
151	1-Chloro-2-fluoroethane	C_2H_4ClF	762-50-5		52.8		

Code	Name	Molecular Formula	CAS Reg. No.	$t_m/°C$	$t_b/°C$	$t_c/°C$	TLV/ ppm
151a	1-Chloro-1-fluoroethane	C_2H_4ClF	1615-75-4		16.2		
152	1,2-Difluoroethane	$C_2H_4F_2$	624-72-6		26		
152a	1,1-Difluoroethane	$C_2H_4F_2$	75-37-6	−118.59	−24.02	113.26	
160	Chloroethane	C_2H_5Cl	75-00-3	−138.4	12.3	187.2	100
161	Fluoroethane	C_2H_5F	353-36-6	−143.2	−37.7	102.16	
170	Ethane	C_2H_6	74-84-0	−182.77	−88.58	32.17	1000
E170	Dimethyl ether	C_2H_6O	115-10-6	−141.50	−24.81	127.15	
216ca	1,3-Dichloro-1,1,2,2,3,3-hexafluoropropane	$C_3Cl_2F_6$	662-01-1	−125.4	35.7	180	
218	Perfluoropropane	C_3F_8	76-19-7	−147.70	−36.79	71.87	
227ca2	Trifluoromethyl 1,1,2,2-tetrafluoroethyl ether	C_3HF_7O	2356-61-8	−141	−3	114.63	
227ea	1,1,1,2,3,3,3-Heptafluoropropane	C_3HF_7	431-89-0	−126.80	−16.34	101.75	
227me	Trifluoromethyl 1,2,2,2-tetrafluoroethyl ether	C_3HF_7O	2356-62-9		−9.6		
236ea	1,1,1,2,3,3-Hexafluoropropane	$C_3H_2F_6$	431-63-0		6.20	139.29	
236fa	1,1,1,3,3,3-Hexafluoropropane	$C_3H_2F_6$	690-39-1	−93.63	−1.44	124.92	
236me	1,2,2,2-Tetrafluoroethyl difluoromethyl ether	$C_3H_2F_6O$	57041-67-5		23.35	155.80	
245ca	1,1,2,2,3-Pentafluoropropane	$C_3H_3F_5$	679-86-7		25.13	174.42	
245cb	1,1,1,2,2-Pentafluoropropane	$C_3H_3F_5$	1814-88-6		−17.4	106.96	
245fa	1,1,1,3,3-Pentafluoropropane	$C_3H_3F_5$	460-73-1	−102.10	15.14	154.01	
245mc	Methyl pentafluoroethyl ether	$C_3H_3F_5O$	22410-44-2		5.59	133.65	
245mf	Difluoromethyl 2,2,2-trifluoroethyl ether	$C_3H_3F_5O$	1885-48-9		29.24	170.84	
245qc	Difluoromethyl 1,1,2-trifluoroethyl ether	$C_3H_3F_5O$	69948-24-9		43.1		
254pc	Methyl 1,1,2,2-tetrafluoroethyl ether	$C_3H_4F_4O$	425-88-7	−107	37.1		
290	Propane	C_3H_8	74-98-6	−187.62	−42.11	96.74	1000
C316	1,2-Dichloro-1,2,3,3,4,4-hexafluorocyclobutane	$C_4Cl_2F_6$	356-18-3	−24.2	59.5	224	
C317	1-Chloro-1,2,2,3,3,4,4-heptafluorocyclobutane	C_4ClF_7	377-41-3	−39.1	25		
C318	Perfluorocyclobutane	C_4F_8	115-25-3	−40.19	−5.9	115.31	
347mcc	Perfluoropropyl methyl ether	$C_4H_3F_7O$	375-03-1		34.23	164.55	
347mmy	Perfluoroisopropyl methyl ether	$C_4H_3F_7O$	22052-84-2		29.34	160.15	
600	Butane	C_4H_{10}	106-97-8	−138.24	−0.49	151.98	1000
600a	Isobutane	C_4H_{10}	75-28-5	−159.38	−11.75	134.66	1000
610	Diethyl ether	$C_4H_{10}O$	60-29-7	−116.2	34.5	193.5	400
611	Methyl formate	$C_2H_4O_2$	107-31-3	−99	31.7	214.0	100
717	Ammonia	H_3N	7664-41-7	−77.65	−33.33	132.25	25
744	Carbon dioxide	CO_2	124-38-9	−56.56	−78.46 sp	30.98	5000
764	Sulfur dioxide	O_2S	7446-09-5	−75.45	−10.02	157.49	2
1112a	1,1-Dichloro-2,2-difluoroethene	$C_2Cl_2F_2$	79-35-6	−116	19		
1113	Chlorotrifluoroethene	C_2ClF_3	79-38-9	−158.2	−27.8	106	
1114	Tetrafluoroethene	C_2F_4	116-14-3	−131.15	−75.9	33.3	2
1120	Trichloroethene	C_2HCl_3	79-01-6	−84.7	87.21	271.0	10
1130	trans-1,2-Dichloroethene	$C_2H_2Cl_2$	156-60-5	−49.8	48.7	243.3	200
1132a	1,1-Difluoroethene	$C_2H_2F_2$	75-38-7	−144	−85.7	29.7	500
1140	Chloroethene	C_2H_3Cl	75-01-4	−153.84	−13.8	159	1
1141	Fluoroethene	C_2H_3F	75-02-5	−160.5	−72	54.7	1
1150	Ethylene	C_2H_4	74-85-1	−169.15	−103.77	9.20	200
1234yf	2,3,3,3-Tetrafluoroprop-l-ene	$C_3H_2F_4$	754-12-1		−29.5	94.7	
1270	Propene	C_3H_6	115-07-1	−185.19	−47.62	91.06	500

PROPERTIES OF GAS CLATHRATE HYDRATES

Carolyn A. Koh and E. Dendy Sloan

Gas clathrate hydrates (also known as gas hydrates) are crystalline inclusion compounds composed of hydrogen-bonded water cavities (host) which encage small gas (guest) molecules. Generally, a maximum of one guest molecule occupies each water cavity. Typical guest molecules that form gas hydrates are methane, ethane, carbon dioxide, and propane (see gas hydrate phase equilibria data in Table II). The structural and physical properties of gas hydrates are given in Tables Ia and Ib. Data have been taken from the references indicated.

Table Ia. Gas Hydrate Structural Properties (Ref. 1)

Structure	sI		sII		sH		
Crystal system	Cubic		Cubic		Hexagonal		
Space group	$Pm3n$ (No. 223)[b]		$Fd3m$ (No. 227)[b]		$P6/mmm$ (No. 191)[b]		
Lattice description	Primitive		Face centered		Hexagonal		
Lattice parameters[a]	$a = 12$ Å		$a = 17.3$ Å		$a = 12.2$ Å, $c = 10.1$ Å		
	$\alpha = \beta = \gamma = 90°$		$\alpha = \beta = \gamma = 90°$		$\alpha = \beta = 90°, \gamma = 120°$		
Ideal unit cell formula	$6(5^{12}6^2) \cdot 2(5^{12}) \cdot 46H_2O$		$8(5^{12}6^4) \cdot 16(5^{12}) \cdot 136H_2O$		$1(5^{12}6^8) \cdot 3(5^{12}) \cdot 2(4^35^66^3) \cdot 34H_2O$		
Cavity	Small	Large	Small	Large	Small	Medium	Large
Description	5^{12}	$5^{12}6^2$	5^{12}	$5^{12}6^4$	5^{12}	$4^35^66^3$	$5^{12}6^8$
Number of cavities/unit cell	2	6	16	8	3	2	1
Average cavity radius[c] (Å)	3.95	4.33	3.91	4.73	3.94[d]	4.04[d]	5.79[d]
H_2O molecules/cavity[e]	20	24	20	28	20	20	36

[a] Lattice parameters are a function of temperature, pressure, and guest composition. Typical average values given.
[b] Space group reference numbers from the International Tables of Crystallography.
[c] The average cavity radius will vary with temperature, pressure, and guest composition.
[d] From the atomic coordinates measured using single crystal x-ray diffraction on 2,2-dimethylpentane·5(Xe,H_2S)·34H_2O at 173 K (Ref. 2). The Rietveld refinement package, GSAS was used to determine the atomic distances for each cage oxygen to the cage center.
[e] Number of oxygen atoms at the periphery of each cavity.

Table Ib. Physical Properties of sI, sII Hydrates Compared to Ice, Ih (Ref. 1,3,4,5)

Property	Ice	sI	sII
Dielectric constant at 273 K	94	~58	~58
H_2O reorientation time at 273 K (µs)	21	~10	~10
H_2O diffusion jump time (µs)	2.7	>200	>200
Isothermal Young's modulus at 268 K (10^9 Pa)	9.5	8.4[est]	8.2[est]
Poisson's ratio	0.3301[f]	0.31403[f]	0.31119[f]
Bulk modulus (GPa)	9.097[f]	8.762[f]	8.482[f]
Shear modulus (GPa)	3.488[f]	3.574[f]	3.6663[f]
Compressional velocity, V_p (m/s)	3870.1[f]	3778[f]	3821.8[f]
Shear velocity, V_s (m/s)	1949[f]	1963.6	2001.14[g]
Linear thermal expansion at 200 K (K^{-1})	56 x 10^{-6}	77 x 10^{-6}	52 x 10^{-6}
Thermal conductivity ($W\ m^{-1}K^{-1}$) at 263 K	2.18±0.01[h]	0.51±0.01[h]	0.50±0.01[h]
Adiabatic bulk compression at 273 K (GPa)	12	14[est]	14[est]
Heat capacity ($J\ kg^{-1}K^{-1}$)	1700±200[h]	2080	2130±40[h]
Refractive index (632.8 nm, −3 °C)	1.3082 (Ref. 9)	1.346 (Ref. 9)	1.350 (Ref. 9)
Density (g/cm^3)	0.91[j]	0.94	1.291[k]

[f] At 253–268 K, 22.4–32.8 MPa (ice, Ih), 258–288 K, 27.1–62.1 MPa (CH_4, sI), 258–288 K, 30.5–91.6 MPa (CH_4–C_2H_6, sII), Ref. 6.
[g] At 258–288 K, 26.6–62.1 MPa, Ref. 7.
[h] At 248–268 K (ice, Ih), 253–288 K (CH_4, sI), 248–265.5 K (THF, sII), Ref. 8.
[j] Fractional occupancy (calculated from a theoretical model) in small (S) and large (L) cavities: sI = CH_4: 0.87 (S) and CH_4: 0.973 (L); sII = CH_4: 0.672 (S), 0.057 (L); C_2H_6: 0.096 (L) only; C_3H_8: 0.84 (L) only.
[k] Calculated for 2,2-dimethylpentane·5(Xe,H_2S)·34H_2O, Ref. 2; est = estimated.

References for Table I

1. Sloan, E.D. and Koh, C.A., *Clathrate Hydrates of Natural Gases*, 3rd Edition, CRC Press, 2008.
2. Udachin, K.A., Ratcliffe, C.I., Enright, G.D., and Ripmeester, J.A., *Supramol. Chem.*, 8, 173, 1997.
3. Davidson, D.W., *Natural Gas Hydrates* (Cox, J.L., Ed.) Butterworths, Boston, 1, 1983.
4. Davidson, D.W., Handa, Y.P., and Ripmeester, J.A., *J. Phys. Chem.*, 90, 6549, 1986.
5. Ripmeester, J.A., Ratcliffe, C.I., Klug, D.D., and Tse, J.S., in *Proc. First International Conference on Natural Gas Hydrates*, (Sloan, E.D.,

Happel, J., and Hnatow, M.A., eds.) *Annals of the New York Academy of Sciences*, 715, 161, 1994.
6. Helgerud, M.B., Circone, S., Stern, L., Kirby, S., and Lorenson, T.D., in *Proc. Fourth International Conference on Gas Hydrates*, Yokohama May 19–23, 2002, 716, 2002.
7. Helgerud, M.B., Waite, W.F., Kirby, S.H., and Nur, A., *Can. J. Phys.*, 81, 47, 2003.
8. Waite, W.F., Gilbert, L.Y., Winters, W.J., and Mason, D.H., in *Proc. Fifth International Conference on Gas Hydrates*, Trondheim, Norway, June 13–16, Paper 5042, 2005.
9. Bylov, M. and Rasmussen, P., *Chem. Eng. Sci.*, 52, 3295, 1997.

Table II: Phase Equilibria Data of Gas Clathrate Hydrates

This table gives measured phase equilibria data of sI and sII gas clathrate hydrates (see Table I for gas hydrate structure and physical property data). The temperature and pressure conditions at which gas hydrates are stable are listed here for typical guest molecules (Tables IIa–d). For example, data for methane hydrate show that at 277.1 K methane hydrate will dissociate at pressures below 3.81 MPa.

Table IIa. Methane Hydrate (Ref. 1)

I–H–V

T (K)	P (MPa)	T (K)	P (MPa)	T (K)	P (MPa)	T (K)	P (MPa)
262.4	1.79	266.5	2.08	268.6	2.22	270.9	2.39
264.2	1.90						

L_w–H–V

T (K)	P (MPa)	T (K)	P (MPa)	T (K)	P (MPa)	T (K)	P (MPa)
273.7	2.77	275.9	3.43	280.4	5.35	282.6	6.77
274.3	2.90	277.1	3.81	280.9	5.71	284.3	8.12
275.4	3.24	279.3	4.77	281.5	6.06	285.9	9.78
275.9	3.42						Ref. 2

L_w-H-V

T (K)	P (MPa)	T (K)	P (MPa)	T (K)	P (MPa)	T (K)	P (MPa)	
295.7	33.99	295.9	35.30	301.0	64.81	302.0	77.50	Ref. 3

L_w–H–V

T (K)	P (MPa)	T (K)	P (MPa)	T (K)	P (MPa)	T (K)	P (MPa)
285.7	9.62	285.7	9.62	295.9	34.75	300.9	62.40
286.3	10.31	289.0	13.96	298.7	48.68	301.6	68.09
286.1	10.10	292.1	21.13				Ref. 4

L_w-H-V

T (K)	P (MPa)	T (K)	P (MPa)	T (K)	P (MPa)	T (K)	P (MPa)
275.4	2.87	277.2	3.90	279.2	4.90	281.2	6.10
276.2	3.37	278.2	4.50				Ref. 5

I–H–V

T (K)	P (MPa)	T (K)	P (MPa)	T (K)	P (MPa)	T (K)	P (MPa)
190.2	0.08251	208.2	0.222	243.2	0.9550	262.4	1.798
198.2	0.1314	218.2	0.3571				Ref. 6

Table IIb. Ethane Hydrate (Ref. 1)

T (K)	P (kPa)	Phases	T (K)	P (kPa)	Phases	
260.8	294	I–H–V	285.8	2537	L_w–H–V	
260.9	290	I–H–V	287.0	3054	L_w–H–V	
269.3	441	I–H–V	287.7	4909	L_w–H–L_E	
273.4	545	L_w–H–V	287.8	3413	L_w–H–L_E	
275.4	669	L_w–H–V	287.8	4289	L_w–H–L_E	
277.6	876	L_w–H–V	288.1	3716	L_w–H–L_E	
279.1	1048	L_w–H–V	288.1	6840	L_w–H–L_E	
219.7	1131	L_w–H–V	288.2	4944	L_w–H–L_E	
281.1	1317	L_w–H–V	288.2	5082	L_w–H–L_E	
282.8	1641	L_w–H–V	288.3	4358	L_w–H–L_E	
284.4	2137	L_w–H–V	288.4	6840	L_w–H–L_E	Ref. 7
284.6	2055	L_w–H–V				

I–H–V

T (K)	P (kPa)	T (K)	P (kPa)	T (K)	P (kPa)	T (K)	P (kPa)
263.6	313	266.5	357	269.3	405	272.0	457

L_w–H–V

T (K)	P (kPa)	T (K)	P (kPa)	T (K)	P (kPa)	T (K)	P (kPa)	
273.7	510	278.7	931	280.4	1165	283.2	1689	
273.7	503	278.7	931	280.9	1255	284.3	1986	
274.8	579	279.3	1007	281.5	1345	285.4	2303	
275.9	662	279.8	1083	282.1	1448	285.4	2310	
277.6	814	280.4	1165	282.6	1558	286.5	2730	Ref. 2

L_w–H–V

T (K)	P (kPa)	T (K)	P (kPa)	T (K)	P (kPa)	T (K)	P (kPa)	
277.5	780	279.9	1040	283.3	1660	286.5	2620	
278.1	840	281.5	1380	284.5	2100			Ref. 8

Table IIc. Propane Hydrate (Ref. 1)

I–H–V

T (K)	P (kPa)	T (K)	P (kPa)	T (K)	P (kPa)	T (K)	P (kPa)
261.2	100	267.4	132	269.8	149	272.9	172
264.2	115	267.6	135	272.2	167		

L_w–H–V

T (K)	P (kPa)	T (K)	P (kPa)	T (K)	P (kPa)	T (K)	P (kPa)	
273.7	183	274.8	232	275.9	301	277.1	386	
273.7	183	275.4	270					Ref. 2

I–H–V

T (K)	P (kPa)	T (K)	P (kPa)	T (K)	P (kPa)	T (K)	P (kPa)	
247.9	48.2	251.6	58.3	258.2	81.1	260.9	94.5	
251.4	58.3	255.4	69.6	260.8	90.5	262.1	99.4	Ref. 9

Feed composition: $x_{H_2O} = 0.9503$, $x_{C_3H_8} = 0.0407$
Q_2 at $T = 278.62$, $P = 0.6$ MPa

L_w–H–V		L_w–H–L $x_{C_3H_8}$	
T (K)	P (MPa)	T (K)	P (MPa)
276.77	0.368	278.71	0.643
277.01	0.377	278.75	0.893
277.22	0.405	278.75	1.393
277.36	0.425	278.75	1.891
277.44	0.433	278.78	1.893
277.87	0.473	278.80	2.391
278.01	0.527	278.80	2.891
278.22	0.483	278.79	2.893
278.55	0.547	278.75	3.891
		278.77	3.391
		278.81	4.391
		278.79	5.892
		278.86	6.392
		278.88	6.892
		278.80	8.393
		278.84	8.893
		278.89	9.893 Ref. 10

Table IId. Carbon Dioxide Hydrate (Ref. 1)

		L_w–H–V			
T (K)	P (MPa)	T (K)	P (MPa)	T (K)	P (MPa)
279.6	2.74	282.1	4.01	282.8	4.36

		L_w–H–L_{CO_2}					
T (K)	P (MPa)	T (K)	P (MPa)	T (K)	P (MPa)	T (K)	P (MPa)
282.9	5.03	283.1	6.47	283.6	11.98	283.9	14.36
282.9	5.62	283.2	9.01				Ref. 11

Overall feed composition:
$x_{H_2O} = 0.8668$, $x_{CO_2} = 0.1332$
Q_2 at 283.27 K and 4.48 MPa

L_w–H–V		L_w–H–L_{CO_2}	
T (K)	P (MPa)	T (K)	P (MPa)
276.52	1.82	283.33	5.97
277.85	1.95	283.36	7.35
278.52	2.21		
279.49	2.62		
280.44	2.88		
281.49	3.35		
281.97	3.68		
282.00	3.69		
282.45	3.85		
282.50	4.01		Ref. 12

References for Table II

1. Sloan, E.D. and Koh, C.A., *Clathrate Hydrates of Natural Gases*, 3rd Edition, CRC Press, 2008.
2. Deaton, W.M. and Frost, E.M., Jr., *Gas Hydrates and Their Relation to the Operation of Natural-Gas Pipe Lines*, U.S. Bureau of Mines Monograph 8, p. 101, 1946.
3. Kobayashi, R. and Katz, D.L., *Trans AIME*, **186**, 66, 1949.
4. McLeod, H.O. and Campbell, J.M., *J. Petl Tech.*, **222**, 590, 1961.
5. Thakore, J.L. and Holder, G.D., *Ind. Eng Chem. Res.*, **26**, 462, 1987.
6. Makogon, T.Y. and Sloan, E.D., *J. Chem. Eng. Data*, **39**, 351, 1994.
7. Roberts, O.L., Brownscombe, E.R., and Howe, L.S., *Oil Gas J.*, **39**, 37, 1940.
8. Holder, G.D. and Grigoriou, G.C., *J. Chem. Thermodyn.*, **12**, 1093, 1980.
9. Holder, G.D. and Godbole, S.P., *AIChE J.*, **28**, 930, 1982.
10. Mooijer-van den Heuvel, M.M., Peters, C.J., and de Swaan Arons, J., *Fluid Phase Equilib.*, **193**, 245, 2002.
11. Ng, H.-J. and Robinson, D.B., *Fluid Phase Equilib.*, **21**, 145, 1985.
12. Mooijer-van den Heuvel, M.M., Witteman, R., and Peters, C.J., *Fluid Phase Equilib.*, **182**, 97, 2001.

IONIC LIQUIDS

Ionic liquids are a class of organic salts with relatively low melting points. The term usually implies a melting point of 100 °C or lower, and many are liquid at room temperature. They offer several advantages as solvents, such as very low vapor pressure, good thermal stability, and nonflammable behavior. For these reasons they are attractive as constituents of environmentally friendly chemical processes.

This table lists some of the ionic liquids that have been studied. The following properties are given:

Mol. Form. — molecular formula in the Hill convention
CASRN — Chemical Abstracts Service Registry Number
Mol. Wt. — molecular weight (relative molar mass)
t_m — normal melting point in °C; the notation "gl" indicates a glass–liquid transition, rather than a crystal–liquid transition
ρ — density in g/cm^3. The superscript indicates the temperature in °C; if there is no superscript, room temperature can be assumed.
η — viscosity in mPa s. The superscript indicates the temperature in °C; if there is no superscript, room temperature can be assumed.

The phase behavior of ionic liquids can be complicated. Some are crystalline at low temperatures and show a sharp transition from crystal to liquid state (a true melting point) as the temperature is raised, but others exist as a glass at low temperatures and convert to a liquid at the glass–liquid transition temperature, denoted by a small change in heat capacity. Still others are glasses at very low temperatures, transform to crystals as the temperature is raised, and finally become liquid at a still higher temperature. See Reference 3 for a discussion of the types of phase behavior.

References

1. *Ionic Liquids Database-(ILThermo)*, NIST Standard Reference Database #147, 2006, National Institute of Standards and Technology, Gaithersburg, MD, <ilthermo.boulder.nist.gov>.
2. *Ionic Liquids — New Materials for New Applications*, 2006, < ildb. merck.de/ionicliquids/en/startpage.htm >.
3. Crosthwaite, J. M., Muldoon, M. J., Dixon, J. K., Anderson, J. L., and Brennecke, J. F., *J. Chem. Thermodynamics* 37, 559, 2005.
4. Sun, J., Forsyth, M., and MacFarlane, D. R., *J. Phys. Chem.* B 102, 8858, 1998.
5. Fredlake, C. P., Crosthwaite, J. M., Hert, D. G., Aki, S. N. V. K., and Brennecke, J. F., *J. Chem. Eng. Data* 49, 954, 2004.

Name	Mol. Form.	CASRN	Mol. Wt	t_m/°C	ρ/g cm^{-3}	η/mPa s
1-Benzyl-3-methylimidazolium tetrafluoroborate	$C_{11}H_{13}BF_4N_2$	500996-04-3	260.039	63		
1-Butyl-4-(dimethylamino)pyridinium bromide	$C_{11}H_{19}BrN_2$		259.186	222		
1-Butyl-2,3-dimethylimidazolium chloride	$C_9H_{17}ClN_2$	98892-75-2	188.697	99		
1-Butyl-2,3-dimethylimidazolium hexafluorophosphate	$C_9H_{17}F_6N_2P$	227617-70-1	298.208	−58 gl	1.2416^{23}	
1-Butyl-2,3-dimethylimidazolium iodide	$C_9H_{17}IN_2$	108203-70-9	280.148	97		
1-Butyl-2,3-dimethylimidazolium octylsulfate	$C_{17}H_{34}N_2O_4S$		362.528	90		
1-Butyl-2,3-dimethylimidazolium tetrafluoroborate	$C_9H_{17}N_2BF_4$	402846-78-0	240.049	37	1.0762^{40}	
1-Butyl-2,3-dimethylimidazolium trifluoromethanesulfonate	$C_{10}H_{18}F_3N_2O_3S$		303.321	41		
1-Butyl-3,5-dimethylpyridinium bromide	$C_{11}H_{18}BrN$		244.172	95		
1-Butyl-3,4-dimethylpyridinium chloride	$C_{11}H_{18}ClN$		199.721	72		
1-Butyl-3,5-dimethylpyridinium chloride	$C_{11}H_{18}ClN$		199.721	100		
1-Butyl-3-methylimidazolium acetate	$C_{10}H_{18}N_2O_2$		198.262			440^{25}
1-Butyl-3-methylimidazolium bis(trifluoromethylsulfonyl)imide	$C_{10}H_{15}F_6N_3O_4S_2$	174899-83-3	419.364	−2	1.4370^{25}	70^{25}
1-Butyl-3-methylimidazolium bromide	$C_8H_{15}BrN_2$	85100-77-2	219.122	78		
1-Butyl-3-methylimidazolium chloride	$C_8H_{15}ClN_2$	79917-90-1	174.671	67	1.08^{25}	
1-Butyl-3-methylimidazolium dicyanamide	$C_{10}H_{15}N_5$	448245-52-1	205.260	−6	1.0580^{24}	
1-Butyl-3-methylimidazolium (diethylene glycol monomethyl ether)sulfate	$C_{13}H_{27}N_2O_6S$		339.427	−62 gl		1033^{25}
1-Butyl-3-methylimidazolium hexafluorophosphate	$C_8H_{15}F_6N_2P$	174501-64-5	284.182	11	1.367^{20}	382^{20}
1-Butyl-3-methylimidazolium iodide	$C_8H_{15}IN_2$	65039-05-6	266.122	−72	1.44^{25}	1100^{25}
1-Butyl-3-methylimidazolium methide	$C_{12}H_{15}F_9N_2O_6S_2$	731774-32-6	550.437		1.57^{25}	
1-Butyl-3-methylimidazolium methylsulfate	$C_9H_{18}N_2O_4S$	401788-98-5	250.315	13	1.212^{25}	
1-Butyl-3-methylimidazolium nitrate	$C_8H_{15}N_3O_3$	179075-88-8	201.223		1.15^{40}	67^{35}
1-Butyl-3-methylimidazolium octylsulfate	$C_{16}H_{32}N_2O_4S$	445473-58-5	348.501	31	1.07	
1-Butyl-3-methylimidazolium tetrafluoroborate	$C_8H_{15}BF_4N_2$	174501-65-6	226.023	−85 gl	1.2048^{22}	120^{25}
1-Butyl-3-methylimidazolium tosylate	$C_{15}H_{22}N_2O_3S$	410522-18-8	310.412	72		
1-Butyl-3-methylimidazolium trifluoroacetate	$C_{10}H_{15}F_3N_2O_2$	174899-94-6	252.233			70^{25}
1-Butyl-3-methylimidazolium trifluoromethanesulfonate	$C_9H_{15}F_3N_2O_3S$	174899-66-2	288.286	13	1.3013^{23}	
1-Butyl-3-methylpyridinium bis(trifluoromethylsulfonyl)imide	$C_{12}H_{16}F_6N_2O_4S_2$		430.386	−84 gl		63^{25}
1-Butyl-3-methylpyridinium bromide	$C_{10}H_{16}BrN$		230.145	−36 gl		
1-Butyl-3-methylpyridinium chloride	$C_{10}H_{16}ClN$	125652-55-3	185.694	117		
1-Butyl-3-methylpyridinium hexafluorophosphate	$C_{10}H_{16}F_6NP$		295.205	46		

Name	Mol. Form.	CASRN	Mol. Wt	t_m/°C	ρ/g cm^{-3}	η/mPa s
1-Butyl-4-methylpyridinium hexafluorophosphate	$C_{10}H_{16}F_6NP$	401788-99-6	295.205	44		
1-Butyl-3-methylpyridinium methylsulfate	$C_{11}H_{19}NO_4S$		261.339	<−50	1.19	
1-Butyl-4-methylpyridinium tetrafluorborate	$C_{10}H_{16}BF_4N$	343952-33-0	237.046		1.1842[25]	
1-Butyl-3-methylpyridinium tetrafluoroborate	$C_{10}H_{16}BF_4N$		237.046	−76 gl		177[25]
1-Butyl-1-methylpyrrolidinium bis(trifluoromethylsulfonyl)imide	$C_{11}H_{20}F_6N_2O_4S_2$	223437-11-4	422.408		1.394[25]	
1-Butyl-1-methylpyrrolidinium dicyanamide	$C_{11}H_{20}N_4$	370865-80-8	208.304	<−50	1.02	
1-Butyl-1-methylpyrrolidinium hexafluorophosphate	$C_9H_{20}F_6NP$	330671-29-9	287.226	85		
1-Butyl-1-methylpyrrolidinium methylsulfate	$C_{10}H_{23}NO_4S$		253.360	10	1.17	
1-Butyl-1-methylpyrrolidinium trifluoroacetate	$C_{11}H_{20}F_3NO_2$		255.278	31		
1-Butyl-1-methylpyrrolidinium trifluoromethanesulfonate	$C_{10}H_{20}F_3NO_3S$	367522-96-1	291.331	3	1.25	
1-Butyl-1-methylpyrrolidinium tris(pentafluoroethyl)trifluorophosphate	$C_{15}H_{20}F_{18}NP$	851856-47-8	587.272	4	1.59	
1-Butylnicotinic acid	$C_{16}H_{22}F_6N_2O_6S_2$		516.475	15		531[25]
1-Butylpyridinium bromide	$C_9H_{14}BrN$	874-80-6	216.118	105		
1-Butylpyridinium chloride	$C_9H_{14}ClN$	1124-64-7	171.667	127		
1-Butylpyridinium hexafluorophosphate	$C_9H_{14}F_6NP$	186088-50-6	281.178	75		
1-Butylpyridinium methylsulfate	$C_{10}H_{17}NO_4S$		247.312	<−50	1.22	
1-Butylpyridinium tetrafluoroborate	$C_9H_{14}BF_4N$	203389-28-0	223.019		1.214[25]	
1-Butylpyridinium trifluoromethanesulfonate	$C_{10}H_{14}F_3NO_3S$	390423-43-5	285.283	35		
Cocosalky pentaethoxi methylammonium methylsulfate	$C_{23}H_{51}NO_9S$		517.718			2800[25]
1-Decyl-3-methylimidazolium bromide	$C_{14}H_{27}BrN_2$	188589-32-4	303.281	16	112[20]	
1-Decyl-3-methylimidazolium chloride	$C_{14}H_{27}ClN_2$	171058-18-7	258.830	38		
1,1-Dibutylpyrrolidinium bis(trifluoromethylsulfonyl)imide	$C_{14}H_{26}F_6N_2O_4S_2$		464.487	41		
N,N-Diethyl-N-(1-methylethyl)-2-propanaminium bis(trifluoromethylsulfonyl)imide	$C_{12}H_{24}F_6N_2O_4S_2$	210230-41-4	438.450	148		
1,3-Dimethylimidazolium bis(trifluoromethylsulfonyl)imide	$C_7H_9F_6N_3O_4S_2$	174899-81-1	377.284		1.570[25]	
1,3-Dimethylimidazolium chloride	$C_5H_9ClN_2$	79917-88-7	132.591	126		
1,3-Dimethylimidazolium dimethylphosphate	$C_7H_{15}N_2O_4P$	654058-04-5	222.178		1.253[30]	
1,3-Dimethylimidazolium methoxyethylsulfate	$C_8H_{16}N_2O_5S$	790663-78-4	252.288		1.314[25]	
1,3-Dimethylimidazolium methylsulfate	$C_6H_{12}N_2O_4S$	97345-90-9	208.235			
1,3-Dimethylimidazolium trifluoromethylsulfonate	$C_6H_9F_3N_2O_3S$	121091-30-3	246.206	43		
1,2-Dimethyl-3-propylimidazolium bis(trifluoromethylsulfonyl)imide	$C_{10}H_{15}F_6N_3O_4S_2$	169051-76-7	419.364	11	1.457[22]	90.0[25]
1,1-Dimethylpyrrolidinium tris(pentafluoroethyl)trifluorophosphate	$C_{12}H_{14}F_{18}NP$		545.191	107	1.81	
1-Dodecyl-3-methylimidazolium chloride	$C_{16}H_{31}ClN_2$	114569-84-5	286.883	97		
1-Dodecyl-3-methylimidazolium hexafluorophosphate	$C_{16}H_{31}F_6N_2P$	219947-93-0	396.394			
1-Dodecyl-3-methylimidazolium tetrafluoroborate	$C_{16}H_{31}BF_4N_2$	244193-59-7	338.235			
N-Ethyl-N,N-bis(1-methylethyl)-1-heptanaminium bis(trifluoromethylsulfonyl)imide	$C_{17}H_{34}F_6N_2O_4S_2$	210230-53-8	508.583	−82 gl	1.27[20]	362[25]
1-Ethyl-2,3-dimethylimidazolium bis(trifluoromethylsulfonyl)imide	$C_9H_{13}F_6N_3O_4S_2$	174899-90-2	405.337	25	1.4913[23]	
1-Ethyl-2,3-dimethylimidazolium bromide	$C_7H_{13}BrN_2$	98892-76-3	205.095	138		
1-Ethyl-2,3-dimethylimidazolium methylsulfate	$C_8H_{16}N_2O_4S$		236.289	46		
1-Ethyl-2,3-dimethylimidazolium tetrafluoroborate	$C_7H_{13}BF_4N_2$	307492-75-7	211.996	94		
1-Ethyl-2,3-dimethylimidazolium trifluoromethanesulfonate	$C_8H_{13}F_3N_2O_3S$	174899-72-0	274.260	110		
1-Ethyl-3-methylimidazolium bis(pentafluroethyl)phosphinate	$C_{10}H_{11}F_{10}N_2O_2P$		412.164	20	1.53	
1-Ethyl-3-methylimidazolium bis(trifluoromethylsulfonyl)imide	$C_8H_{11}F_6N_3O_4S_2$	174899-82-2	391.311	−17	1.5213[23]	32[25]
1-Ethyl-3-methylimidazolium bromide	$C_6H_{11}BrN_2$	65039-08-9	191.068	77		
1-Ethyl-3-methylimidazolium chloride	$C_6H_{11}ClN_2$	65039-09-0	146.617	85		
1-Ethyl-3-methylimidazolium ethylsulfate	$C_8H_{16}N_2O_4S$	342573-75-5	236.289		1.2388[25]	100[25]
1-Ethyl-3-methylimidazolium hexafluorophosphate	$C_6H_{11}F_6N_2P$	155371-19-0	256.128	59.6		
1-Ethyl-3-methylimidazolium methanesulfonate	$C_7H_{14}N_2O_3S$	145022-45-3	206.262		1.2437[25]	
1-Ethyl-3-methylimidazolium octyl sulfate	$C_{14}H_{28}N_2O_4S$	790663-79-5	320.448			
1-Ethyl-3-methylimidazolium tetrafluoroborate	$C_6H_{11}BF_4N_2$	143314-16-3	197.969	14	1.2526[60]	36.1[25]
1-Ethyl-3-methylimidazolium thiocyanate	$C_7H_{11}N_3S$	331717-63-6	169.247	<−50	1.11	
1-Ethyl-3-methylimidazolium tosylate	$C_{13}H_{18}N_2O_3S$	328090-25-1	282.358	56		
1-Ethyl-3-methylimidazolium trifluoromethanesulfonate	$C_7H_{11}F_3N_2O_3S$	145022-44-2	260.233		1.385[25]	
N-Ethyl-N-methyl-N-(1-methylethyl)-2-propanaminium bis(trifluoromethylsulfonyl)imide	$C_{11}H_{22}F_6N_2O_4S_2$	210230-42-5	424.424	140		

Name	Mol. Form.	CASRN	Mol. Wt	$t_m/°C$	$\rho/g\,cm^{-3}$	$\eta/mPa\,s$
1-Ethyl-3-methylpyridinium ethylsulfate	$C_{10}H_{17}NO_4S$		247.312	−71 gl		150^{25}
1-Ethyl-1-methylpyrrolidinium methylsulfate	$C_8H_{19}NO_4S$		225.307	23	1.23	
1-Ethyl-1-methylpyrrolidinium tetrafluoroborate	$C_7H_{16}BF_4N$	15302-90-6	201.014	69		
1-Ethylnicotinic acid ethyl ester ethylsulfate	$C_{12}H_{19}NO_6S$		305.347	−44 gl		3200^{25}
1-Ethylpyridinium bromide	$C_7H_{10}BrN$	1906-79-2	188.065	122		
1-Ethylpyridinium chloride	$C_7H_{10}ClN$	2294-38-4	143.614	119		
1-Ethylpyridinium ethylsulfate	$C_9H_{15}NO_4S$		233.285			137^{25}
1-Ethylpyridinium tetrafluoroborate	$C_7H_{10}BF_4N$	350-48-1	194.966		1.302^{20}	
1-Ethylpyridinium trifluoroacetate	$C_9H_{10}F_3NO_2$	474461-33-1	221.176		1.273^{20}	
1-Ethylpyridinium bis(trifluoromethylsulfonyl)imide	$C_9H_{10}F_6N_2O_4S_2$	712354-97-7	388.306		1.536^{25}	
1-Heptyl-3-methylimidazolium hexafluorophosphate	$C_{11}H_{21}F_6N_2P$	357915-04-9	326.262	−84 gl	1.274^{21}	
1-Hexadecyl-2,3-dimethylimidazolium chloride	$C_{21}H_{41}ClN_2$		357.017	210		
1-Hexadecyl-3-methylimidazolium chloride	$C_{20}H_{39}ClN_2$	61546-01-8	342.990	222		
1-Hexyl-4-(dimethylamino)pyridinium bis(trifluoromethylsulfonyl)imide	$C_{15}H_{23}F_6N_3O_4S_2$		487.482	−69		111^{25}
1-Hexyl-4-(dimethylamino)pyridinium bromide	$C_{13}H_{23}BrN_2$		287.239	196		
1-Hexyl-2,3-dimethylimidazolium bis(trifluoromethylsulfonyl)imide	$C_{13}H_{21}F_6N_3O_4S_2$	384347-22-2	461.444	−5	1.361^{25}	131^{25}
1-Hexyl-2,3-dimethylimidazolium chloride	$C_{11}H_{21}ClN_2$	455270-59-4	216.751	46		
1-Hexyl-2,3-dimethylimidazolium tetrafluoroborate	$C_{11}H_{21}BF_4N_2$	384347-21-1	268.103	14	1.15	
1-Hexyl-3,5-dimethylpyridinium bis(trifluoromethylsulfonyl)imide	$C_{15}H_{22}F_6N_2O_4S_2$		472.467	10		104^{25}
1-Hexyl-2-ethyl-3,5-dimethylpyridinium bis(trifluoromethylsulfonyl)imide	$C_{17}H_{26}F_6N_2O_4S_2$		500.519	−66 gl		245^{25}
1-Hexyl-3-methyl-4-(dimethylamino)pyridinium bis(trifluoromethylsulfonyl)imide	$C_{16}H_{25}F_6N_3O_4S_2$		501.508	−2		112^{25}
1-Hexyl-3-methyl-4-(dimethylamino)pyridinium bromide	$C_{14}H_{25}BrN_2$		301.266	119		
1-Hexyl-3-methylimidazolium bis(trifluoromethylsulfonyl)imide	$C_{12}H_{19}F_6N_3O_4S_2$	382150-50-7	447.417	−7	1.3708^{25}	68^{25}
1-Hexyl-3-methylimidazolium bromide	$C_{10}H_{19}BrN_2$		247.175	−49 gl		
1-Hexyl-3-methylimidazolium chloride	$C_{10}H_{19}ClN_2$	171058-17-6	202.724	−75 gl	1.0400^{25}	3400000^{40}
1-Hexyl-3-methylimidazolium hexafluorophosphate	$C_{10}H_{19}F_6N_2P$	304680-35-1	312.235	−79 gl	1.294^{25}	585^{25}
1-Hexyl-3-methylimidazolium tetrafluoroborate	$C_{10}H_{19}BF_4N_2$	244193-50-8	254.076	−79 gl	1.136^{37}	
1-Hexyl-3-methylimidazolium trifluoromethanesulfonate	$C_{11}H_{19}F_3N_2O_3S$	460345-16-8	316.340	28	1.24	
1-Hexyl-4-(4-methylpiperidino)pyridinium bis(trifluoromethylsulfonyl)imide	$C_{19}H_{29}F_6N_3O_4S_2$		541.571	37		
1-Hexyl-4-(4-methylpiperidino)pyridinium bromide	$C_{17}H_{29}BrN_2$		341.329	33 gl		
1-Hexyl-3-methylpyridinium bis(trifluoromethylsulfonyl)imide	$C_{14}H_{20}F_6N_2O_4S_2$		458.440	−82 gl		85^{25}
1-Hexyl-3-methylpyridinium bromide	$C_{12}H_{20}BrN$		258.198	−37 gl		
1-Hexyl-2-propyl-3,5-diethylpyridinium bis(trifluoromethylsulfonyl)imide	$C_{20}H_{32}F_6N_2O_4S_2$		542.599	−67 gl		206^{25}
1-Hexylpyridinium bis(trifluoromethylsulfonyl)imide	$C_{13}H_{18}F_6N_2O_4S_2$		444.413	0		80^{25}
1-Hexylpyridinium bromide	$C_{11}H_{18}BrN$	74440-81-6	244.172	46		
1-Hexylpyridinium hexafluorophosphate	$C_{11}H_{18}F_6NP$		309.232	48		
1-Hexylpyridinium trifluoromethanesulfonate	$C_{12}H_{18}F_3NO_3S$		313.336	62		
1-Methylimidazolium tetrafluoroborate	$C_4H_7BF_4N_2$	151200-14-5	169.917	63		
1-Methylimidazolium tosylate	$C_{11}H_{14}N_2O_3S$	63458-90-2	254.305	89		
1-Methyl-3-octadecylimidazolium bis(trifluoromethylsulfonyl)imide	$C_{24}H_{43}F_6N_3O_4S_2$	404001-51-0	615.736	52		
1-Methyl-3-octadecylimidazolium tris(pentafluoroethyl)trifluorophosphate	$C_{28}H_{43}F_{18}N_2P$		780.599	43		
1-Methyl-3-octylimidazolium bromide	$C_{12}H_{23}BrN_2$	61545-99-1	275.228			
1-Methyl-3-octylimidazolium chloride	$C_{12}H_{23}ClN_2$	64697-40-1	230.777	−87 gl	1.0088^{25}	4100000^{40}
1-Methyl-3-octylimidazolium 2-(2-methoxyethoxy)ethyl sulfate	$C_{17}H_{34}N_2O_6S$	595565-55-2	394.526			
1-Methyl-3-octylimidazolium nitrate	$C_{12}H_{23}N_3O_3$	203389-27-9	257.329			1240^{20}
1-Methyl-3-octylimidazolium octylsulfate	$C_{20}H_{40}N_2O_4S$		404.608	76		
1-Methyl-3-octylimidazolium tetrafluoroborate	$C_{12}H_{23}BF_4N_2$	244193-52-0	282.129	−81 gl	1.104^{25}	341^{25}
1-Methyl-3-octylimidazolium trifluoromethanesulfonate	$C_{13}H_{23}F_3N_2O_3S$	403842-84-2	344.393	14	1.19	
1-Methyl-3-octylpyridinium bis(trifluoromethylsulfonyl)imide	$C_{16}H_{24}F_6N_2O_4S_2$		486.493	−80		112^{25}
1-Methyl-3-octylpyrrolidinium bis(trifluoromethylsulfonyl)imide	$C_{15}H_{28}F_6N_2O_4S_2$		478.514	−12	1.29	
1-Methyl-3-pentylimidazolium hexafluorophosphate	$C_9H_{17}F_6N_2P$	280779-52-4	298.208	−80 gl	1.333^{21}	
1-Methyl-3-propylimidazolium chloride	$C_7H_{13}ClN_2$	79917-89-8	160.644	62		

Name	Mol. Form.	CASRN	Mol. Wt	t_m/°C	ρ/g cm^{-3}	η/mPa s
1-Methyl-1-propylpyrrolidinium bis(trifluoromethylsulfonyl)imide	$C_{10}H_{18}F_6N_2O_4S_2$	223437-05-6	408.381			
3-Methyl-1-tetradecylimidazolium chloride	$C_{18}H_{35}ClN_2$	171058-21-2	314.937	195		
1-Methyl-3-tetradecylimidazolium tetrafluoroborate	$C_{18}H_{35}BF_4N_2$	244193-61-1	366.289	116		
Methyltrioctylammonium trifluoroacetate	$C_{27}H_{54}F_3NO_2$	121107-16-2	481.719	<−50	0.97	
Methyltrioctylammonium trifluoromethanesulfonate	$C_{26}H_{54}F_3NO_3S$	121107-18-4	517.772	56		
1-Nonyl-3-methylimidazolium hexafluorophosphate	$C_{13}H_{25}F_6N_2P$	343952-29-4	354.315	14	1.20[21]	
1-Octyl-3-methylimidazolium bis(trifluoromethylsulfonyl)imide	$C_{14}H_{23}F_6N_3O_4S_2$	178631-04-4	475.471	−84 gl	1.325[25]	
1-Octyl-3-methylimidazolium hexafluorophosphate	$C_{12}H_{23}F_6N_2P$	304680-36-2	340.288	−82 gl	1.237[25]	734[25]
1-Octylpyridinium chloride	$C_{13}H_{22}ClN$	4086-73-1	227.774	46		
1-Pentyl-3-methylimidazolium bis(trifluoromethylsulfonyl)imide	$C_{10}H_{15}F_6N_3O_4S_2$	280779-53-5	419.364			
1-(3,4,5,6-Perfluorohexyl)-3-methylimidazolium bis(trifluoromethyl-sulfonyl)imide	$C_{12}H_{10}F_{15}N_3O<s$		609.331	−56		
1-Propyl-3-methylimidazolium bromide	$C_7H_{13}BrN_2$		205.095	37		
1-Propyl-3-methylimidazolium tetrafluoroborate	$C_7H_{13}BF_4N_2$		211.996	−17	1.240[25]	103[25]
Pyridinium ethoxyethylsulfate	$C_9H_{15}NO_5S$	630393-27-0	249.284		1.281[25]	
Tetrabutylammonium bis(trifluoromethylsulfonyl)imide	$C_{18}H_{35}F_6N_2O_4S_2$	210230-40-3	522.610	92		
Tetrabutylammonium docusate	$C_{36}H_{73}NO_7S$		664.033	−62 gl		12000[25]
Tetrabutylammonium tris(pentafluoroethyl)trifluorophosphate	$C_{22}H_{36}F_{18}NP$		687.473	58		
Tetrabutylammonium tris(trifluoromethylsulfonyl)methide	$C_{20}H_{36}F_9NO_6S_3$	196958-57-3	653.684	96		
Tetramethylammonium tris(pentafluoroethyl)trifluorophosphate	$C_{14}H_{20}F_{18}NP$	394692-80-9	575.261	97		
N,N,N-Tributyl-1-heptanaminium bis(trifluoromethylsulfonyl)imide	$C_{21}H_{42}F_6N_2O_4S_2$	210230-50-5	564.689	−67 gl	1.17[20]	606[25]
N,N,N-Tributyl-1-heptanaminium trifluoromethanesulfonate	$C_{20}H_{42}F_3NO_3S$	210230-54-9	433.612	−55 gl		
N,N,N-Tributyl-1-hexanaminium bis(trifluoromethylsulfonyl)imide	$C_{20}H_{40}F_6N_2O_4S_2$	210230-49-2	550.663	26	1.15[20]	595[25]
Tributylmethylammonium bis(trifluoromethylsulfonyl)imide	$C_{15}H_{30}F_6N_2O_4S_2$	405514-94-5	480.530		1.266[24]	
N,N,N-Tributyl-1-octanaminium bis(trifluoromethylsulfonyl)imide	$C_{22}H_{44}F_6N_2O_4S_2$	210230-51-6	578.715	−63 gl	1.12[20]	574[25]
N,N,N-Tributyl-1-octanaminium trifluoromethanesulfonate	$C_{21}H_{44}F_3NO_3S$	210230-58-3	447.639	−57 gl	1.02[20]	2000[25]
N,N,N-Triethylethanaminium bis(trifluoromethylsulfonyl)imide	$C_{10}H_{20}F_6N_2O_4S_2$	161401-26-9	410.397			
N,N,N-Triethyl-1-heptanaminium bis(trifluoromethylsulfonyl)imide	$C_{15}H_{30}F_6N_2O_4S_2$	210230-47-0	480.530	−79 gl	1.26[20]	76[25]
N,N,N-Triethyl-1-hexanaminium bis(trifluoromethylsulfonyl)imide	$C_{14}H_{28}F_6N_2O_4S_2$	210230-46-9	466.503	20	1.27[20]	167[25]
N,N,N-Triethyl-1-octanaminium bis(trifluoromethylsulfonyl)imide	$C_{16}H_{32}F_6N_2O_4S_2$	210230-48-1	494.556	−74 gl	1.25[20]	202[25]
Trihexyl(tetradecyl)phosphonium acetate	$C_{34}H_{71}O_2P$	460092-04-0	542.901		0.890[25]	
Trihexyl(tetradecyl)phosphonium bis(trifluoromethylsulfonyl)imide	$C_{34}H_{68}F_6NO_4PS_2$	460092-03-9	764.002		1.067[25]	
Trihexyl(tetradecyl)phosphonium chloride	$C_{32}H_{68}ClP$	258864-54-9	519.309		0.89[25]	
Trihexyltetradecylphosphonium hexafluorophosphate	$C_{32}H_{68}F_6P_2$	374683-44-0	628.820	39		
Trihexyltetradecylphosphonium tetrafluoroborate	$C_{32}H_{68}BF_4P$	374683-55-3	570.661	25	0.94	
Trihexyl(tetradecyl)phosphonium tris(pentafluoroethyl)trifluoro-phosphate	$C_{38}H_{68}F_{18}P_2$		928.866	<−50	1.18	
Triisobutylmethylphosphonium p-toluenesulfonate	$C_{20}H_{37}O_3PS$	344774-05-6	388.545		1.069[25]	
N,N,N-Trimethyl-1-heptanaminium bis(trifluoromethylsulfonyl)imide	$C_{12}H_{24}F_6N_2O_4S_2$	210230-44-7	438.450	−73 gl	1.28[20]	153[25]
N,N,N-Trimethyl-1-hexanaminium bis(trifluoromethylsulfonyl)imide	$C_{11}H_{22}F_6N_2O_4S_2$	210230-43-6	424.424	−74 gl	1.33[20]	153[25]
N,N,N-Trimethylmethanaminium bis(trifluoromethylsulfonyl)imide	$C_6H_{12}F_6N_2O_4S_2$	161401-25-8	354.290	133		
N,N,N-Trimethyl-1-octanaminium bis(trifluoromethylsulfonyl)imide	$C_{13}H_{26}F_6N_2O_4S_2$	210230-45-8	452.476	−73 gl	1.27[20]	181[25]
N,N,N-Tripropyl-1-propanaminium bis(trifluoromethylsulfonyl)imide	$C_{14}H_{28}F_6N_2O_4S_2$	210230-39-0	466.503	105		

DENSITY AND SPECIFIC VOLUME OF MERCURY

The data in this table have been adjusted to the ITS-90 temperature scale. The uncertainty in density values is 0.0003 g/mL between −20 and −10 °C; 0.0001 or less between −10 and 200 °C; and 0.0002 between 200 and 300 °C.

Reference

Ambrose, D., *Metrologia*, 27, 245, 1990.

$t/°C$	$\rho/(g/mL)$	$v/(mL/kg)$	$t/°C$	$\rho/(g/mL)$	$v/(mL/kg)$	$t/°C$	$\rho/(g/mL)$	$v/(mL/kg)$
−20	13.64461	73.2890	30	13.52134	73.9572	80	13.39971	74.6285
−19	13.64212	73.3024	31	13.51889	73.9705	81	13.39729	74.6420
−18	13.63964	73.3157	32	13.51645	73.9839	82	13.39487	74.6554
−17	13.63716	73.3291	33	13.51400	73.9973	83	13.39245	74.6689
−16	13.63468	73.3424	34	13.51156	74.0107	84	13.39003	74.6824
−15	13.63220	73.3558	35	13.50911	74.0241	85	13.38762	74.6959
−14	13.62972	73.3691	36	13.50667	74.0375	86	13.38520	74.7094
−13	13.62724	73.3824	37	13.50422	74.0509	87	13.38278	74.7229
−12	13.62476	73.3958	38	13.50178	74.0643	88	13.38037	74.7364
−11	13.62228	73.4091	39	13.49934	74.0777	89	13.37795	74.7498
−10	13.61981	73.4225	40	13.49690	74.0911	90	13.37554	74.7633
−9	13.61733	73.4358	41	13.49446	74.1045	91	13.37313	74.7768
−8	13.61485	73.4492	42	13.49202	74.1179	92	13.37071	74.7903
−7	13.61238	73.4625	43	13.48958	74.1313	93	13.36830	74.8038
−6	13.60991	73.4759	44	13.48714	74.1447	94	13.36589	74.8173
−5	13.60743	73.4892	45	13.48470	74.1581	95	13.36347	74.8308
−4	13.60496	73.5026	46	13.48226	74.1715	96	13.36106	74.8443
−3	13.60249	73.5160	47	13.47982	74.1850	97	13.35865	74.8579
−2	13.60002	73.5293	48	13.47739	74.1984	98	13.35624	74.8714
−1	13.59755	73.5427	49	13.47495	74.2118	99	13.35383	74.8849
0	13.59508	73.5560	50	13.47251	74.2252	100	13.35142	74.8984
1	13.59261	73.5694	51	13.47008	74.2386	110	13.3273	75.0337
2	13.59014	73.5827	52	13.46765	74.2520	120	13.3033	75.1693
3	13.58768	73.5961	53	13.46521	74.2655	130	13.2793	75.3052
4	13.58521	73.6095	54	13.46278	74.2789	140	13.2553	75.4413
5	13.58275	73.6228	55	13.46035	74.2923	150	13.2314	75.5778
6	13.58028	73.6362	56	13.45791	74.3057	160	13.2075	75.7147
7	13.57782	73.6495	57	13.45548	74.3192	170	13.1836	75.8519
8	13.57535	73.6629	58	13.45305	74.3326	180	13.1597	75.9895
9	13.57289	73.6763	59	13.45062	74.3460	190	13.1359	76.1274
10	13.57043	73.6896	60	13.44819	74.3594	200	13.1120	76.2659
11	13.56797	73.7030	61	13.44576	74.3729	210	13.0882	76.4047
12	13.56551	73.7164	62	13.44333	74.3863	220	13.0644	76.5440
13	13.56305	73.7297	63	13.44090	74.3998	230	13.0406	76.6838
14	13.56059	73.7431	64	13.43848	74.4132	240	13.0167	76.8241
15	13.55813	73.7565	65	13.43605	74.4266	250	12.9929	76.9650
16	13.55567	73.7698	66	13.43362	74.4401	260	12.9691	77.1064
17	13.55322	73.7832	67	13.43120	74.4535	270	12.9453	77.2484
18	13.55076	73.7966	68	13.42877	74.4670	280	12.9214	77.3909
19	13.54831	73.8100	69	13.42635	74.4804	290	12.8975	77.5341
20	13.54585	73.8233	70	13.42392	74.4939	300	12.8736	77.6779
21	13.54340	73.8367	71	13.42150	74.5073			
22	13.54094	73.8501	72	13.41908	74.5208			
23	13.53849	73.8635	73	13.41665	74.5342			
24	13.53604	73.8769	74	13.41423	74.5477			
25	13.53359	73.8902	75	13.41181	74.5612			
26	13.53114	73.9036	76	13.40939	74.5746			
27	13.52869	73.9170	77	13.40697	74.5881			
28	13.52624	73.9304	78	13.40455	74.6016			
29	13.52379	73.9438	79	13.40213	74.6150			

THERMAL PROPERTIES OF MERCURY

Lev R. Fokin

The first of these tables gives the molar heat capacity at constant pressure of liquid and gaseous mercury as a function of temperature. To convert to specific heat in units of J/g K, divide these values by 200.59, the atomic weight of mercury.

Reference

Douglas, T. B., Ball, A. T., and Ginnings, D. C., *J. Res. Natl. Bur. Stands.*, 46, 334, 1951.

$t/°C$	C_p/(J/mol K) Liquid	C_p/(J/mol K) Gas	$t/°C$	C_p/(J/mol K) Liquid	C_p/(J/mol K) Gas	$t/°C$	C_p/(J/mol K) Liquid	C_p/(J/mol K) Gas
−38.84	28.2746	20.786	140	27.3675	20.786	340	27.1500	20.836
−20	28.1466	20.786	160	27.3090	20.786	356.73	27.1677	20.849
0	28.0190	20.786	180	27.2588	20.790	360	27.1709	20.853
20	27.9002	20.786	200	27.2169	20.790	380	27.1981	20.870
25	27.8717	20.786	220	27.1834	20.794	400	27.2324	20.891
40	27.7897	20.786	240	27.1583	20.794	420	27.2738	20.916
60	27.6880	20.786	260	27.1412	20.799	440	27.3207	20.941
80	27.5952	20.786	280	27.1320	20.807	460	27.3742	20.974
100	27.5106	20.786	300	27.1303	20.815	480	27.4332	21.008
120	27.4349	20.786	320	27.1366	20.824	500	27.4985	21.046

The second table gives the molar heat capacity of solid mercury in its rhombohedral (α-mercury) form.

References

1. Busey and Giaque, *J. Am. Chem. Soc.*, 75, 806, 1953.
2. Amitin, Lebedeva, and Paukov, *Rus. J. Phys. Chem.*, 2666, 1979.

$t/°C$	C_p/(J/mol K)	$t/°C$	C_p/(J/mol K)	$t/°C$	C_p/(J/mol K)	$t/°C$	C_p/(J/mol K)
−268.99	0.99*	−248.15	12.74	−193.15	23.16	−93.15	26.69
−268.99	0.97**	−243.15	14.78	−183.15	23.76	−73.15	27.28
−268.15	1.6	−233.15	17.90	−173.15	24.24	−53.15	27.96
−263.15	4.6	−223.15	19.94	−153.15	25.00	−38.87	28.5
−258.15	7.6	−213.15	21.40	−133.15	25.61		
−253.15	10.33	−203.15	22.42	−113.15	26.15		

* Superconducting state
**Normal state

The final table gives the cubic thermal expansion coefficient α, the isothermal compressibility coefficient κ_T, and the speed of sound u for liquid mercury as a function of temperature. These properties are defined as follows:

$$\alpha = \frac{1}{v}\left(\frac{\partial v}{\partial T}\right)_p \qquad \kappa_T = -\frac{1}{v}\left(\frac{\partial v}{\partial P}\right)_T \qquad u^2 = \left(\frac{\partial P}{\partial \rho}\right)_s \qquad \rho = v^{-1}$$

where v is the specific volume (given in the table on the preceding page).

Reference

Vukalovich, M. P., et al., *Thermophysical Properties of Mercury*, Moscow Standard Press, 1971.

$t/°C$	$\alpha \times 10^4/K^{-1}$	$\kappa_T \times 10^6/bar^{-1}$ At 1 bar	$\kappa_T \times 10^6/bar^{-1}$ At 1000 bar	$u/m\ s^{-1}$	$t/°C$	$\alpha \times 10^4/K^{-1}$	$\kappa_T \times 10^6/bar^{-1}$ At 1 bar	$\kappa_T \times 10^6/bar^{-1}$ At 1000 bar	$u/m\ s^{-1}$
−20	1.818	3.83		1470	120	1.8058	4.513	4.33	1404.7
0	1.8144	3.918	3.78	1460.8	140	1.8074	4.622		1395.4
20	1.8110	4.013	3.87	1451.4	160	1.8100	4.731	4.53	1386.1
40	1.8083	4.109	3.96	1442.0	180	1.8136	4.844		1376.7
60	1.8064	4.207		1432.7	200	1.818	4.96		1367
80	1.8053	4.308	4.14	1423.4	250	1.834	5.26		1344
100	1.8051	4.410		1414.1	300	1.856	5.59		1321

MELTING CURVE OF MERCURY

The solid–liquid phase boundary of mercury provides a convenient means of calibrating pressures up to 1 GPa in the neighborhood of room temperature. The best representation of this curve is given by:

$$p/\text{GPa} = 1.932845 \cdot 10^{-2}d + 1.8333 \cdot 10^{-6}d^2 + 5.9791 \cdot 10^{-8}d^3$$

where $d = t/°\text{C} - 38.8344$. Temperature is on the ITS-90 scale, and the relation is valid for pressures up to about 1.2 GPa. The following table is calculated from this equation.

Reference

Molinar, G. F., Bean, V., Houck, J., and Welch, B., *Metrologia* 16, 21, 1980; 28, 353, 1991.

p vs. t		t vs. p	
t/°C	p/GPa	p/GPa	t/°C
−38.83	0.000	0.010	−38.32
−35.00	0.074	0.020	−37.80
−30.00	0.171	0.050	−36.25
−25.00	0.268	0.100	−33.66
−20.00	0.365	0.200	−28.50
−15.00	0.463	0.300	−23.35
−10.00	0.560	0.400	−18.21
−5.00	0.658	0.500	−13.08
0.00	0.757	0.600	−7.97
5.00	0.856	0.700	−2.88
10.00	0.955	0.800	2.18
15.00	1.055	0.900	7.23
20.00	1.156	1.000	12.24
25.00	1.257	1.100	17.23
30.00	1.359	1.200	22.19

VAPOR PRESSURE OF MERCURY

The following table gives the vapor pressure of mercury in kilopascals (100 kPa = 1 bar) from the triple point (234.3156 K) to the critical point (1764 K). The data are generated from the formulation of Huber, Laesecke, and Friend in Reference 1, which is based on a critical evaluation of all the published data on mercury vapor pressure and related thermodynamic properties. The estimated uncertainty in the vapor pressure is:

−38 to −10 °C	3%
0 to 130 °C	1%
140 to 350 °C	0.15%
360 to 620 °C	0.5%
630 to 1491 °C	5%

Most of the entries in this table carry one significant figure beyond the estimated accuracy.

Note that the table refers to mercury vapor in equilibrium with liquid mercury, in the absence of air or other gases.

References

1. Huber, M. L., Laesecke, A., and Friend, D. G., *The Vapor Pressure of Mercury*, NISTIR 6643, National Institute of Standards and Technology, Boulder, CO, March 2006.
2. Huber, M. L., Laesecke, A., and Friend, D. G., *Ind. Eng. Chem. Res.* 45, 7351, 2006.
3. Vargaftik, N. B., Vinogradov, Y. K., and Yargin, V. S., *Handbook of Physical Properties of Liquids and Gases, Third Edition*, Begell House, New York, 1996.

$t/°C$	p/kPa	$t/°C$	p/kPa	$t/°C$	p/kPa	$t/°C$	p/kPa
−38.83	2.985×10^{-7}	38	0.0007350	300	32.965	720	6254
−30	9.451×10^{-7}	39	0.0007929	310	40.856	730	6718
−20	3.160×10^{-6}	40	0.0008551	320	50.260	740	7205
−10	9.625×10^{-6}	41	0.0009216	330	61.396	750	7718
0	2.699×10^{-5}	42	0.0009928	340	74.498	760	8258
1	2.979×10^{-5}	43	0.001069	350	89.823	770	8824
2	3.287×10^{-5}	44	0.001151	360	107.65	780	9417
3	3.623×10^{-5}	45	0.001238	370	128.26	790	10040
4	3.991×10^{-5}	46	0.001331	380	151.99	800	10690
5	4.393×10^{-5}	47	0.001430	390	179.17	810	11370
6	4.833×10^{-5}	48	0.001537	400	210.15	820	12080
7	5.312×10^{-5}	49	0.001650	410	245.32	830	12820
8	5.836×10^{-5}	50	0.001771	420	285.07	840	13600
9	6.406×10^{-5}	55	0.002506	430	329.82	850	14410
10	7.028×10^{-5}	60	0.003508	440	380.00	860	15250
11	7.705×10^{-5}	65	0.004862	450	436.07	870	16120
12	8.441×10^{-5}	70	0.006673	460	498.51	880	17030
13	9.242×10^{-5}	75	0.009075	470	567.81	890	17980
14	0.0001011	80	0.01223	480	644.46	900	18960
15	0.0001106	85	0.01635	490	729.01	910	19980
16	0.0001208	90	0.02167	500	821.99	920	21040
17	0.0001320	95	0.02850	510	923.96	930	22140
18	0.0001440	100	0.03721	520	1035.5	940	23270
19	0.0001571	110	0.06209	530	1157.2	950	24450
20	0.0001713	120	0.1009	540	1289.6	960	25670
21	0.0001866	130	0.1599	550	1433.3	970	26930
22	0.0002032	140	0.2478	560	1589.1	980	28230
23	0.0002211	150	0.3759	570	1757.4	990	29580
24	0.0002404	160	0.5592	580	1939	1000	30970
25	0.0002613	170	0.8168	590	2135	1050	38600
26	0.0002839	180	1.1728	600	2345	1100	47450
27	0.0003082	190	1.6573	610	2570	1150	57590
28	0.0003344	200	2.3071	620	2811	1200	69100
29	0.0003627	210	3.1670	630	3069	1250	82100
30	0.0003931	220	4.2906	640	3344	1300	96600
31	0.0004259	230	5.7414	650	3637	1350	112700
32	0.0004611	240	7.5939	660	3949	1400	130000
33	0.0004990	250	9.9347	670	4281	1450	150000
34	0.0005398	260	12.863	680	4632	1491	167000
35	0.0005835	270	16.494	690	5005		
36	0.0006305	280	20.955	700	5399		
37	0.0006809	290	26.392	710	5815		

SURFACE TENSION OF COMMON LIQUIDS

The surface tension γ of about 200 liquids is tabulated here as a function of temperature. Values of γ are given in units of milli-newtons per meter (mN/m), which is equivalent to dyn/cm in c.g.s. units. The values refer to a nominal pressure of one atmosphere (about 100 kPa) except in cases where the indicated temperature is above the normal boiling point of the substance; in those cases, the applicable pressure is the saturation vapor pressure at the temperature in question.

The uncertainty of the values is 0.1 to 0.2 mN/m or less in most cases. Values at temperatures between the points tabulated can be obtained by linear interpolation to a good approximation.

Substances are listed by molecular formula in the modified Hill order, with substances not containing carbon appearing before those that do contain carbon. A more extensive compilation of surface tension may be found in Reference 1.

References

1. Jasper, J. J., *J. Phys. Chem. Ref. Data*, 1, 841, 1972.
2. Kahl, H., Wadewitz, T., and Winkelmann, J., *J. Chem. Eng. Data*, 48, 580, 2003.

| Mol. formula | Name | γ in mN/m | | | | |
		10 °C	25 °C	50 °C	75 °C	100 °C
Br_2	Bromine	43.68	40.95	36.40		
Cl_2O_2S	Sulfuryl chloride		28.78			
Cl_3OP	Phosphoryl chloride		32.03	28.85	25.66	
Cl_3P	Phosphorus trichloride		27.98	24.81		
Cl_4Si	Silicon tetrachloride	19.78	18.29	15.80		
H_2O	Water	74.23	71.99	67.94	63.57	58.91
H_4N_2	Hydrazine		66.39			
Hg	Mercury	488.55	485.48	480.36	475.23	470.11
CCl_4	Tetrachloromethane		26.43	23.37	20.31	17.25
CS_2	Carbon disulfide	33.81	31.58	27.87		
$CHBr_3$	Tribromomethane		44.87	41.60	38.33	
$CHCl_3$	Trichloromethane		26.67	23.44	20.20	
CH_2Br_2	Dibromomethane		39.05	35.33	31.61	
CH_2Cl_2	Dichloromethane		27.20			
CH_2O_2	Formic acid		37.13	34.38	31.64	
CH_3I	Iodomethane	32.19	30.34			
CH_3NO	Formamide		57.03	54.92	52.82	50.71
CH_3NO_2	Nitromethane	39.04	36.53	32.33		
CH_4O	Methanol	23.23	22.07	20.14		
CH_5N	Methylamine		19.15			
C_2HCl_5	Pentachloroethane		34.15	31.20	28.26	
$C_2HF_3O_2$	Trifluoroacetic acid		13.53	11.42		
$C_2H_2Cl_4$	1,1,2,2-Tetrachloroethane		35.58	32.41	29.24	26.07
$C_2H_3Cl_3$	1,1,1-Trichloroethane		25.18	22.07		
$C_2H_3Cl_3$	1,1,2-Trichloroethane		34.02	30.65	27.27	23.89
C_2H_3N	Acetonitrile		28.66	25.51		
$C_2H_4Br_2$	1,2-Dibromoethane		39.55	36.25	32.95	
$C_2H_4Cl_2$	1,1-Dichloroethane		24.07			
$C_2H_4Cl_2$	1,2-Dichloroethane		31.86	28.29	24.72	
C_2H_4O	Acetaldehyde	22.54	20.50	17.10		
$C_2H_4O_2$	Acetic acid		27.10	24.61	22.13	
$C_2H_4O_2$	Methyl formate	26.72	24.36	20.43	16.50	12.57
C_2H_5Br	Bromoethane	25.36	23.62			
C_2H_5I	Iodoethane	30.38	28.46	25.24		
$C_2H_5NO_2$	Nitroethane	34.02	32.13	29.00		
C_2H_6O	Ethanol	23.22	21.97	19.89		
C_2H_6OS	Dimethyl sulfoxide		42.92	40.06		
$C_2H_6O_2$	Ethylene glycol		47.99	45.76	43.54	41.31
C_2H_6S	Dimethyl sulfide	25.27	24.06			
C_2H_6S	Ethanethiol		23.08			
$C_2H_6S_2$	Dimethyl disulfide		33.39	30.04		
C_2H_7N	Dimethylamine		26.34			
C_2H_7N	Ethylamine		19.20			
C_2H_7NO	Ethanolamine		48.32	45.53	42.73	
C_3H_5Br	3-Bromopropene		26.31	23.17		

Mol. formula	Name	γ in mN/m				
		10 °C	25 °C	50 °C	75 °C	100 °C
C_3H_5Cl	3-Chloropropene		23.14			
C_3H_5ClO	Epichlorohydrin	38.40	36.36	32.96	29.56	26.16
C_3H_5N	Propanenitrile		26.75	23.87		
$C_3H_6Cl_2$	1,2-Dichloropropane		28.32	25.22	22.12	
C_3H_6O	Acetone	24.57	22.72	19.65		
C_3H_6O	Allyl alcohol	26.63	25.28	23.02	20.77	
$C_3H_6O_2$	Ethyl formate	25.16	23.18			
$C_3H_6O_2$	Methyl acetate	26.66	24.73	21.51		
$C_3H_6O_2$	Propanoic acid		26.20	23.72	21.23	
C_3H_7Br	1-Bromopropane	27.08	25.26	22.21		
C_3H_7Br	2-Bromopropane	25.03	23.25	20.30		
C_3H_7Cl	1-Chloropropane	23.16	21.30			
C_3H_7Cl	2-Chloropropane	20.49	19.16			
C_3H_7NO	N,N-Dimethylformamide	37.56	35.74	32.70	29.66	26.62
$C_3H_7NO_2$	2-Nitropropane	31.02	29.29	26.39		
C_3H_8O	1-Propanol	24.48	23.32	21.38	19.43	
C_3H_8O	2-Propanol	22.11	20.93	18.96	16.98	
$C_3H_8O_2$	2-Methoxyethanol	32.32	30.84	28.38	25.92	23.46
C_3H_8S	1-Propanethiol		24.20	21.02		
C_3H_8S	2-Propanethiol		21.33	18.39		
C_3H_9N	Propylamine		21.75			
C_3H_9N	Trimethylamine		13.41			
$C_4H_4N_2$	Pyridazine	49.51	47.96	45.37	42.78	40.19
$C_4H_4N_2$	Pyrimidine		30.33	27.80	25.28	22.75
C_4H_4S	Thiophene		30.68	27.36		
C_4H_5N	Pyrrole	38.71	37.06	34.31		
$C_4H_6O_3$	Acetic anhydride	34.08	31.93	28.34	24.75	21.16
C_4H_7N	Butanenitrile		26.92	24.33	21.73	
C_4H_8O	2-Butanone		23.97	21.16		
$C_4H_8O_2$	1,4-Dioxane		32.75	29.28	25.80	22.32
$C_4H_8O_2$	Ethyl acetate	25.13	23.39	20.49	17.58	14.68
$C_4H_8O_2$	Methyl propanoate	26.32	24.44	21.29		
$C_4H_8O_2$	Butanoic acid		26.05	23.75	21.45	
C_4H_9Br	1-Bromobutane	27.58	25.90	23.08	20.27	17.45
C_4H_9Cl	1-Chlorobutane	24.85	23.18	20.39		
C_4H_9I	1-Iodobutane	29.79	28.24	25.67	23.09	20.51
C_4H_9N	Pyrrolidine	30.58	29.23	26.98		
$C_4H_{10}O$	1-Butanol	26.28	24.93	22.69	20.44	18.20
$C_4H_{10}O$	2-Butanol	23.74	22.54	20.56	18.57	16.58
$C_4H_{10}O$	2-Methyl-2-propanol		19.96	17.71		
$C_4H_{10}O$	Diethyl ether		16.65			
$C_4H_{10}O_2$	2-Ethoxyethanol		28.35	26.11	23.86	21.62
$C_4H_{10}O_3$	Diethylene glycol		44.77	42.57	40.37	38.17
$C_4H_{10}S$	Diethyl sulfide	26.22	24.57	21.80		
$C_4H_{11}N$	Butylamine		23.44	20.63		
$C_4H_{11}N$	Isobutylamine		21.75	19.02		
$C_4H_{11}N$	tert-Butylamine		16.87			
$C_4H_{11}N$	Diethylamine		19.85			
$C_5H_4O_2$	Furfural	45.08	43.09	39.78	36.46	33.14
C_5H_5N	Pyridine		36.56	33.29	30.03	
C_5H_8	Cyclopentene	24.45	22.20			
C_5H_8O	Cyclopentanone	34.45	32.80	30.05	27.30	24.55
C_5H_9NO	N-Methyl-2-pyrrolidinone	41.94	40.21	37.33	34.45	31.57
C_5H_{10}	1-Pentene	17.10	15.45			
C_5H_{10}	2-Methyl-2-butene	18.61	17.15			
C_5H_{10}	Cyclopentane	24.07	21.88	18.22		
$C_5H_{10}O$	2-Pentanone		23.25	21.62		
$C_5H_{10}O$	3-Pentanone		24.74	22.13		
$C_5H_{10}O$	Pentanal	26.95	25.44	22.91		
$C_5H_{10}O_2$	Butyl formate	26.05	24.52	21.95	19.39	16.82

Mol. formula	Name	γ in mN/m				
		10 °C	25 °C	50 °C	75 °C	100 °C
$C_5H_{10}O_2$	Propyl acetate	25.48	23.80	21.00	18.20	15.40
$C_5H_{10}O_2$	Isopropyl acetate	23.37	21.76	19.08	16.40	
$C_5H_{10}O_2$	Ethyl propanoate	25.55	23.80	20.88	17.96	
$C_5H_{10}O_2$	Methyl butanoate	26.34	24.62	21.76	18.89	16.03
$C_5H_{11}Cl$	1-Chloropentane	26.01	24.40	21.71	19.02	16.33
$C_5H_{11}N$	Piperidine	30.64	28.91	26.03	23.14	20.26
C_5H_{12}	Pentane	17.15	15.49			
$C_5H_{12}O$	1-Pentanol	26.67	25.36	23.17	20.99	18.80
$C_5H_{12}O$	2-Pentanol	24.96	23.45	20.94	18.43	15.92
$C_5H_{12}O$	3-Methyl-1-butanol	24.94	23.71	21.66	19.61	17.56
$C_5H_{13}N$	Pentylamine		24.69	22.14	19.58	
$C_6H_4Cl_2$	m-Dichlorobenzene	37.15	35.43	32.57	29.70	26.83
C_6H_5Br	Bromobenzene	36.98	35.24	32.34	29.44	26.54
C_6H_5Cl	Chlorobenzene	34.78	32.99	30.02	27.04	24.06
C_6H_5ClO	o-Chlorophenol		39.70	36.89	34.09	31.28
C_6H_5ClO	m-Chlorophenol		41.18	38.66	36.13	33.61
C_6H_5F	Fluorobenzene	28.47	26.66	23.65	20.64	
C_6H_5I	Iodobenzene	40.40	38.71	35.91	33.10	30.29
$C_6H_5NO_2$	Nitrobenzene			40.56	37.66	34.77
C_6H_6	Benzene		28.22	25.00	21.77	
C_6H_6O	Phenol			38.20	35.53	32.86
C_6H_7N	Aniline		42.12	39.41	36.69	
C_6H_7N	2-Methylpyridine		33.00	29.90	26.79	
$C_6H_8N_2$	Adiponitrile		45.45	43.02	40.58	
C_6H_{10}	Cyclohexene	28.01	26.17	23.12		
$C_6H_{10}O$	Cyclohexanone	36.43	34.57	31.46	28.36	25.25
$C_6H_{11}N$	Hexanenitrile		27.37	25.11	22.84	
C_6H_{12}	Cyclohexane	25.91	24.16	21.26	15.44	
C_6H_{12}	Methylcyclopentane	23.47	21.72	18.82		
C_6H_{12}	1-Hexene	19.44	17.90	15.33		
$C_6H_{12}O$	Cyclohexanol		32.92	30.50	28.09	25.67
$C_6H_{12}O$	2-Hexanone		25.45	22.72		
$C_6H_{12}O_2$	Butyl acetate	26.48	24.88	22.21	19.54	16.87
$C_6H_{12}O_2$	Isobutyl acetate	24.58	23.06	20.53	17.99	15.46
$C_6H_{12}O_2$	Ethyl butanoate	25.51	23.94	21.33	18.71	16.10
$C_6H_{12}O_3$	Paraldehyde	27.22	25.63	22.97	20.32	17.66
$C_6H_{13}Cl$	1-Chlorohexane	27.28	25.73	23.13	20.54	17.94
$C_6H_{13}N$	Cyclohexylamine		31.22	28.25	25.28	
C_6H_{14}	Hexane	19.42	17.89	15.33		
C_6H_{14}	2-Methylpentane	18.37	16.88	14.39		
C_6H_{14}	3-Methylpentane	19.20	17.61	14.96		
$C_6H_{14}O$	Diisopropyl ether		17.27	14.65		
$C_6H_{14}O$	1-Hexanol		25.81	23.81	21.80	19.80
$C_6H_{14}O_2$	1,1-Diethoxyethane		20.89	18.31	15.74	
$C_6H_{14}O_2$	2-Butoxyethanol	27.36	26.14	24.10	22.06	20.02
$C_6H_{15}N$	Triethylamine		20.22	17.74		
$C_6H_{15}N$	Dipropylamine		22.31	19.75	17.20	
$C_6H_{15}N$	Diisopropylamine		19.14	16.45		
C_7H_5N	Benzonitrile		38.79	35.90	33.00	
C_7H_6O	Benzaldehyde	39.63	38.00	35.27	32.55	29.82
C_7H_8	Toluene	29.46	27.73	24.85	21.98	19.10
C_7H_8O	o-Cresol		36.90	34.38	31.85	29.32
C_7H_8O	m-Cresol		35.69	33.38	31.07	28.76
C_7H_8O	Benzyl alcohol				27.89	24.44
C_7H_8O	Anisole		35.10	32.09	29.08	
C_7H_9N	N-Methylaniline		36.90	34.47	32.05	
C_7H_9N	2,3-Dimethylpyridine		32.71	30.04	27.36	
C_7H_9N	Benzylamine		39.30	36.27	33.23	
C_7H_{14}	Methylcyclohexane	24.98	23.29	20.46		
C_7H_{14}	1-Heptene	21.29	19.80	17.33	14.85	

Mol. formula	Name	γ in mN/m				
		10 °C	25 °C	50 °C	75 °C	100 °C
$C_7H_{14}O$	2-Heptanone		26.12	23.48		
$C_7H_{14}O_2$	Pentyl acetate	26.67	25.17	22.69	20.20	17.72
$C_7H_{14}O_2$	Heptanoic acid		27.76	25.64		
C_7H_{16}	Heptane	21.14	19.66	17.19	14.73	
C_7H_{16}	3-Methylhexane	20.76	19.31	16.88	14.46	
C_8H_8O	Acetophenone		39.04	36.15	33.27	
$C_8H_8O_2$	Methyl benzoate		37.17	34.25	31.32	
$C_8H_8O_3$	Methyl salicylate	40.98	39.22	36.28	33.35	30.41
C_8H_{10}	Ethylbenzene	30.39	28.75	26.01	23.28	20.54
C_8H_{10}	o-Xylene	31.41	29.76	27.01	24.25	21.50
C_8H_{10}	m-Xylene	30.13	28.47	25.71	22.95	20.19
C_8H_{10}	p-Xylene		28.01	25.32	22.64	19.95
$C_8H_{10}O$	Phenetole		32.41	29.65	26.89	
$C_8H_{11}N$	N,N-Dimethylaniline		35.52	32.90	30.27	
$C_8H_{11}N$	N-Ethylaniline		36.33	33.65	30.98	
C_8H_{16}	Ethylcyclohexane	26.73	25.15	22.51		
C_8H_{18}	Octane	22.57	21.14	18.77	16.39	14.01
C_8H_{18}	2,5-Dimethylhexane	20.77	19.40	17.12	14.84	12.56
$C_8H_{18}O$	1-Octanol	28.30	27.10	25.12		
$C_8H_{19}N$	Dibutylamine		24.12	21.74	19.36	
$C_8H_{19}N$	Diisobutylamine		21.72	19.44	17.16	
C_9H_7N	Quinoline	44.19	42.59	39.94	37.28	34.62
C_9H_{12}	Cumene	29.27	27.69	25.05	22.42	19.78
C_9H_{12}	1,2,4-Trimethylbenzene	30.74	29.20	26.64	24.07	21.51
C_9H_{12}	Mesitylene	28.89	27.55	25.31	23.07	20.82
$C_9H_{18}O$	5-Nonanone		26.28	23.85		
C_9H_{20}	Nonane	23.79	22.38	20.05	17.71	15.37
$C_9H_{20}O$	1-Nonanol	29.03	27.89	26.00	24.10	22.20
$C_{10}H_{12}$	1,2,3,4-Tetrahydronaphthalene		33.17	30.78	28.40	
$C_{10}H_{22}$	Decane	24.75	23.37	21.07	18.77	16.47
$C_{10}H_{22}O$	1-Decanol	29.61	28.51	26.68	24.85	23.02
$C_{11}H_{24}$	Undecane	25.56	24.21	21.96	19.70	17.45
$C_{12}H_{10}O$	Diphenyl ether		26.75	24.80		
$C_{12}H_{27}N$	Tributylamine		24.39	22.32	20.24	
$C_{13}H_{28}$	Tridecane	26.86	25.55	23.37	21.19	19.01
$C_{14}H_{12}O_2$	Benzyl benzoate	44.47	42.82	40.06	37.31	34.55
$C_{14}H_{30}$	Tetradecane	27.43	26.13	23.96	21.78	19.61
$C_{16}H_{34}$	Hexadecane		27.05	24.91	22.78	20.64
$C_{18}H_{38}$	Octadecane		27.87	25.77	23.66	21.55

SURFACE TENSION OF AQUEOUS MIXTURES

The composition dependence of the surface tension of binary mixtures of several compounds with water is given in this table. The data are tabulated as a function of the mass percent of the non-aqueous component. Data for methanol, ethanol, 1-propanol, and 2-propanol are taken from Reference 1, which also gives values at other temperatures.

References

1. Vazquez, G., Alvarez, E., and Navaza, J. M., *J. Chem. Eng. Data*, 40, 611, 1995.
2. *Landolt-Börnstein, Numerical Data and Functional Relationships in Science and Technology, New Series*, IV/16, *Surface Tension*, Springer-Verlag, Heidelberg, 1997.

Surface Tension in mN/m for the Specified Mass %

Compound	t/°C	0%	10%	20%	30%	40%	50%	60%	70%	80%	90%	100%
Acetic acid	30	71.2	51.4	43.3	41.2	38.2	37.4	36.1	33.5	31.5	30.2	26.3
Acetone	25	72.0	44.9	40.5	36.7	33.0	30.1	29.4	29.4	27.6	24.5	23.1
Acetonitrile	20	72.8	48.5	40.2	34.1	31.6	30.6	30.0	29.6	29.1	28.7	28.4
1,2-Butanediol	25	72.0	66.1	60.4	55.1	50.1	45.6	43.3	41.9	40.8	39.2	35.8
1,3-Butanediol	30	71.2	58.1	51.6	48.7	45.8	43.9	42.4	41.2	40.0	39.0	37.0
1,4-Butanediol	30	71.2	61.2	56.9	54.2	52.0	50.7	49.5	47.9	46.6	45.2	43.8
Butanoic acid	30	71.2	42.4	37.5	35.5	34.8	32.2	30.8	29.2	27.4	26.3	25.5
2-Butanone	20	72.8	41.6	32.2				25.2				24.6
γ-Butyrolactone	30	71.2	64	58	53	50	48	46	45	44	42.8	42.7
Chloroacetic acid	25	72.0	59.8	53.6	51.3	49.7	48.3	47.5	46.1			
Diethanolamine	25	72.0	66.8	63.2	60.7	58.8	57.2	55.7	54.3	52.7	50.6	47.2
N,N-Dimethylacetamide	25	72.0	72.0	72.0	72.4	73.5	74.9	75.4	73.0	65.7	54.7	36.4
N,N-Dimethylformamide	25	72.0	65.4	59.2	53.8	49.6	47.3	46.9	44.9	42.3	38.4	35.2
1,4-Dioxane	25	72.0					41.2	39.6	37.9	36.2	34.5	33.7
Ethanol	25	72.01	47.53	37.97	32.98	30.16	27.96	26.23	25.01	23.82	22.72	21.82
Ethylene glycol	20	72.8	68.5	64.9	61.9		57.0					48.2
Formic acid	20	72.8	66	60	55.7	52.2	50.3	48.8	47.1	44.7	40.9	38.0
Glycerol	25	72.0	70.5	69.5	68.5	67.9	67.4	66.9	66.5	65.7	64.5	62.5
Methanol	25	72.01	56.18	47.21	41.09	36.51	32.86	29.83	27.48	25.54	23.93	22.51
Morpholine	20	72.8	65.1	60.7	58.9	56.7	53.0	49.6	47.0	43.7	41.8	38.7
Nitric acid	20	72.8	71.9	70.7	68.9	66.6	63.8	60.6	56.8	52.6	47.9	42.6
Propanoic acid	30	71.2	46.6	42.2	37.7	35.6	33.1	31.7	30.2	28.2	27.4	25.8
1-Propanol	25	72.01	34.32	27.84	25.98	25.26	24.80	24.49	24.08	23.86	23.59	23.28
2-Propanol	25	72.01	40.42	30.57	26.82	25.27	24.26	23.51	22.68	22.14	21.69	21.22
1,2-Propylene glycol	30	71.2	60.5	54.9	50.7	47.2	44.5	41.5	38.6	37.6	36.3	35.5
1,3-Propylene glycol	30	71.2	62.6	58.8	55.7	53.8	52.8	51.7	50.8	49.6	48.2	47.0
Pyridine	25	72.0	52.8	51.2	48.0	46.8	46.6	45.8	45.0	43.6	40.9	37.0
Sulfolane	20	72.8					62.5	61.6	59.6	57.1	54.9	50.9
Sulfuric acid	50	67.9	73.5	75.1	73.6	71.2	68.0	64.1	60.0	56.4	53.6	51.7
Trichloroacetaldehyde	25	72.0	56.7	51.0	46.7	44.1	43.0	42.5	41.5	38.9	34.7	29.4
Trichloroacetic acid	25	72.0	55.8	46.5	42.8	41.6	40.6	39.4	38.3	37.4	36.5	

PERMITTIVITY (DIELECTRIC CONSTANT) OF LIQUIDS

Christian Wohlfarth

The permittivity of a substance (often called the dielectric constant) is the ratio of the electric displacement D to the electric field strength E when an external field is applied to the substance. The quantity tabulated here is the relative permittivity, which is the ratio of the actual permittivity to the permittivity of a vacuum; it is a dimensionless number.

The table gives the static relative permittivity ε_r, i.e., the relative permittivity measured in static fields or at low frequencies where no relaxation effects occur. The fourth column of the table lists the value of ε_r at the temperature specified in the third column, usually 293.15 or 298.15 K. Otherwise, the temperature closest to 293.15 K was chosen, or (as it is the case for many of the substances included here) ε_r is given at the only temperature for which data are available.

The static permittivity refers to nominal atmospheric pressure as long as the corresponding temperature is below the normal boiling point. Otherwise, at temperatures above the normal boiling point, the pressure is understood to be the saturated vapor pressure of the substance considered.

For substances where information on the temperature dependence of the permittivity is available, the table gives the coefficients of a simple polynomial fitting of permittivity to temperature with an equation of the form

$$\varepsilon_r(T) = a + bT + cT^2 + dT^3$$

where T is the absolute temperature in K. Since the parameter d was used in only a few cases where the quadratic fit was not satisfactory, only a, b, and c are listed as columns in the table, while the d values are given at the end of this introduction. For all other substances, $d = 0$. The temperature range of the fit is given in the last column. The coefficients of the fitting equation can be used to calculate dielectric constants within the fitted temperature range but should not be used for extrapolation outside this range. The user who needs dielectric constant data with more accuracy than can be provided by this equation is referred to Reference 1, which gives the original data together with their literature source.

Substances are listed by molecular formula in modified Hill order, with substances not containing carbon preceding those that do contain carbon.

* Indicates that the isomer was not specified in the original reference.

** Indicates a compound for which the cubic term is needed:

Ethanol	$d = -0.15512E-05$
N-Methylacetamide	$d = -0.12998E-04$
1,2-Propylene glycol	$d = -0.32544E-05$
1-Butanol	$d = -0.48841E-06$
2-Butanol	$d = -0.89512E-06$
2-Methyl-1-propanol	$d = -0.45229E-06$
2-Methyl-2-propanol	$d = -0.25968E-05$
N-Butylacetamide	$d = -0.48716E-05$

References

1. Wohlfarth, Ch., Static Dielectric Constants of Pure Liquids and Binary Liquid Mixtures, *Landolt-Börnstein, Numerical Data and Functional Relationships in Science and Technology, New Series*, Editor in Chief, O. Madelung, Group IV, Macroscopic and Technical Properties of Matter, Volume 6, Springer-Verlag, Berlin, 1991.
2. Marsh, K. N., Ed., *Recommended Reference Materials for the Realization of Physicochemical Properties*, Blackwell Scientific Publications, Oxford, 1987.

Mol. form.	Name	T/K	ε_r	a	b	c	Range/K
$AlBr_3$	Aluminum tribromide	373.2	3.38				
Ar	Argon	140.00	1.3247	0.12408E+01	0.68755E-02	−0.45344E-04	87-149
AsH_3	Arsine	200.9	2.40	0.37674E+01	−0.97454E-02	0.14537E-04	157-201
BBr_3	Boron tribromide	273.2	2.58				
B_2H_6	Diborane	180.66	1.8725	0.23848E+01	−0.29501E-02	0.64189E-06	108-181
B_5H_9	Pentaborane(9)	298.2	21.1	0.40952E+03	−0.24414E+01	0.38225E-02	226-298
BrF_3	Bromine trifluoride	298.2	106.8				
BrF_5	Bromine pentafluoride	297.7	7.91	0.11428E+02	−0.11822E-01		262-298
BrH	Hydrogen bromide	186.8	8.23				
BrNO	Nitrosyl bromide	288.4	13.4				
Br_2	Bromine	297.9	3.1484	0.32701E+01	−0.12535E-03		273-327
Br_2OS	Thionyl bromide	293.2	9.06				
Br_3OV	Vanadyl tribromide	298.2	3.6	0.61112E+01	−0.84211E-02		203-298
Br_4Ge	Germanium(IV) bromide	299.9	2.955	0.34450E+01	−0.16083E-02		300-316
Br_4Sn	Tin(IV) bromide	303.45	3.169	0.50001E+01	−0.60383E-02		304-316
$ClFO_3$	Perchloryl fluoride	150.2	2.194	0.23808E+01	−0.38629E-03	−0.57143E-05	125-150
ClF_3	Chlorine trifluoride	293.2	4.394	0.96716E+01	−0.18000E-01		273-313
ClF_5	Chlorine pentafluoride	193.2	4.28	0.78192E+01	−0.20860E-01	0.13132E-04	193-256
ClH	Hydrogen chloride	158.9	14.3	0.47316E+02	−0.28455E+00	0.48650E-03	159-258
ClNO	Nitrosyl chloride	285.2	18.2				
Cl_2	Chlorine	208.0	2.147	0.29440E+01	−0.44649E-02	0.30388E-05	208-240
Cl_2F_3P	Phosphorus(V) dichloride trifluoride	228.63	2.8129	0.46501E+01	−0.80358E-02		172-229
Cl_2OS	Thionyl chloride	298.2	8.675				
Cl_2OSe	Selenium oxychloride	293.2	46.2				
Cl_2O_2S	Sulfuryl chloride	293.2	9.1				

Mol. form.	Name	T/K	ε_r	a	b	c	Range/K
Cl_2S	Sulfur dichloride	298.2	2.915				
Cl_2S_2	Sulfur chloride	288.2	4.79				
Cl_3F_2P	Phosphorus(V) trichloride difluoride	268.0	2.3752	0.28905E+01	−0.19228E−02		215-268
Cl_3OP	Phosphorus(V) oxychloride	293.2	14.1				
Cl_3OV	Vanadyl trichloride	298.2	3.4				
Cl_3P	Phosphorus(III) chloride	290.2	3.498	0.59098E+01	−0.83322E−02		290-333
Cl_3PS	Phosphorus(V) sulfide trichloride	298.2	4.94				
Cl_4FP	Phosphorus(V) tetrachloride fluoride	272.64	2.6499	0.33503E+01	−0.29651E−02		244-273
Cl_4Ge	Germanium(IV) chloride	273.2	2.463	−0.55078E+01	0.64881E−01	−0.13091E−03	246-273
Cl_4Pb	Lead(IV) chloride	293.2	2.78				
Cl_4Si	Tetrachlorosilane	273.2	2.248	0.58041E+01	−0.27129E−01	0.51678E−04	207-273
Cl_4Sn	Tin(IV) chloride	273.2	3.014	0.43951E+01	−0.48805E−02		234-273
Cl_4Ti	Titanium(IV) chloride	257.4	2.843	0.33668E+01	−0.19675E−02		237-257
Cl_4V	Vanadium(IV) chloride	298.2	3.05				
Cl_5P	Phosphorus(V) chloride	433.2	2.85				
Cl_5Sb	Antimony(V) chloride	293.0	3.222	0.45413E+01	−0.45078E−02		276-320
FH	Hydrogen fluoride	273.2	83.6	0.50352E+03	−0.19297E+01	0.14372E−02	200-273
F_2	Fluorine	53.48	1.4913	0.14144E+01	0.26387E−02	−0.28356E−04	54-144
F_5I	Iodine pentafluoride	293.2	37.13	0.95184E+02	−0.19800E+00		273-313
F_6S	Sulfur hexafluoride	223.2	1.81				
F_6Xe	Xenon hexafluoride	328.2	4.10				
F_7I	Iodine heptafluoride	298.2	1.75				
$F_{10}S_2$	Sulfur decafluoride	293.2	2.0202				
HI	Hydrogen iodide	220.2	3.87	0.51557E+03	−0.44552E+01	0.96795E−02	220-236
H_2	Hydrogen	13.52	1.2792	0.13327E+01	−0.51946E−02		14-19
H_2O	Water	293.2	80.100	0.24921E+03	−0.79069E+00	0.72997E−03	273-372
H_2O_2	Hydrogen peroxide	290.2	74.6	0.48511E+03	−0.23145E+01	0.31020E−02	233-303
H_2S	Hydrogen sulfide	283.2	5.93	0.14736E+02	−0.33675E−01	0.96740E−05	212-363
H_3N	Ammonia	293.2	16.61	0.66756E+02	−0.24696E+00	0.25913E−03	238-323
H_4N_2	Hydrazine	298.2	51.7	0.22061E+03	−0.89633E+00	0.11066E−02	278-323
He	Helium	2.055	1.0555	0.10640E+01	−0.35584E−02		2-4
I_2	Iodine	391.25	11.08	0.64730E+02	−0.29266E+00	0.39759E−03	391-441
Kr	Krypton	119.80	1.664				
Mn_2O_7	Manganese(VII) oxide	293.2	3.28	0.37655E+01	−0.16463E−02		283-312
NO	Nitric oxide	124	2.00				
N_2	Nitrogen	63.15	1.4680	0.12550E+01	0.67949E−02	−0.56704E−04	63-126
N_2O_3	Nitrogen trioxide	203.2	31.13	0.92287E+02	−0.43306E+00	0.65000E−03	203-243
N_2O_4	Nitrogen tetroxide	293.2	2.44	0.28212E+01	−0.13000E−02		253-293
Ne	Neon	26.11	1.1907	0.12667E+01	−0.29064E−02		26-29
O_2	Oxygen	54.478	1.5684	0.15434E+01	0.14615E−02	−0.21964E−04	55-154
O_2S	Sulfur dioxide	298.2	16.3	0.52045E+02	−0.16125E+00	0.11042E−03	213-449
O_3	Ozone	90.2	4.75	0.86344E+01	−0.54807E−01	0.12596E−03	90-185
O_3S	Sulfur trioxide	291.2	3.11				
P	Phosphorus	307.2	4.096	0.79018E+00	0.23911E−01	−0.42826E−04	307-358
S	Sulfur	407.2	3.4991	0.51651E+01	−0.77381E−02	0.89120E−05	407-479
Se	Selenium	510.65	5.44	0.67569E+01	−0.25829E−02		511-575
Xe	Xenon	161.35	1.880				
$CBrClF_2$	Bromochlorodifluoromethane	123.2	3.920	0.52442E+01	−0.11000E−01		123-223
$CBrCl_3$	Bromotrichloromethane	293.2	2.405	0.29249E+01	−0.17650E−02		273-333
$CBrF_3$	Bromotrifluoromethane	123.2	3.730	0.54154E+01	−0.13680E−01		123-173
CBr_2Cl_2	Dibromodichloromethane	298.2	2.542	0.32330E+01	−0.23162E−02		298-333
CBr_2F_2	Dibromodifluoromethane	273.2	2.939	0.67296E+01	−0.22133E−01	0.30213E−04	139-273
CBr_3Cl	Tribromochloromethane	333.2	2.601				
CBr_3F	Tribromofluoromethane	293.2	3.00	0.53203E+01	−0.11061E−01	0.10688E−04	206-323
CBr_3NO_2	Tribromonitromethane	298.2	9.034	0.16079E+02	−0.23630E−01		298-328
$CClF_3$	Chlorotrifluoromethane	123.2	3.010	0.43677E+01	−0.11020E−01		123-173
CCl_2F_2	Dichlorodifluoromethane	123.2	3.500	0.46984E+01	−0.97600E−02		123-223
CCl_2O	Carbonyl chloride	295.2	4.30				
CCl_3D	Trichloromethane-d	298.2	4.67				
CCl_3F	Trichlorofluoromethane	293.2	3.00	0.53203E+01	−0.11061E−01	0.10688E−04	206-323
CCl_3NO_2	Trichloronitromethane	293.2	7.319	0.14403E+02	−0.24178E−01		276-333
CCl_4	Tetrachloromethane	293.2	2.2379	0.28280E+01	−0.20339E−02	0.71795E−07	283-333
CF_4	Tetrafluoromethane	126.3	1.685	0.20350E+01	−0.27616E−02		126-142

Mol. form.	Name	T/K	ε_r	a	b	c	Range/K
CHBr$_3$	Tribromomethane	283.2	4.404	0.71707E+01	−0.98000E-02		283-343
CHCl$_3$	Trichloromethane	293.2	4.8069	0.15115E+02	−0.51830E-01	0.56803E-04	218-323
CHF$_3$	Trifluoromethane	294.0	5.2	0.11442E+03	−0.75600E+00	0.13562E-02	130-263
CHN	Hydrogen cyanide	293.2	114.9	0.37331E+04	−0.23180E+02	0.36963E-01	258-299
CH$_2$Br$_2$	Dibromomethane	283.2	7.77	0.18060E+02	−0.36333E-01		283-313
CH$_2$Cl$_2$	Dichloromethane	298.0	8.93	0.40452E+02	−0.17748E+00	0.23942E-03	184-306
CH$_2$F$_2$	Difluoromethane	152.2	53.74	0.19428E+03	−0.12939E+01	0.24280E-02	152-224
CH$_2$I$_2$	Diiodomethane	298.2	5.32				
CH$_2$O$_2$	Formic acid	298.2	51.1	0.14040E+03	−0.24673E+00	−0.17151E-03	287-358
CH$_3$Br	Bromomethane	275.7	9.71	0.40580E+02	−0.18418E+00	0.26219E-03	195-276
CH$_3$Cl	Chloromethane	295.2	10.0	0.42775E+02	−0.16175E+00	0.17108E-03	190-392
CH$_3$ClO$_2$S	Methanesulfonyl chloride	293.2	34.0	0.10384E+03	−0.33838E+00	0.34156E-03	293-373
CH$_3$DO	Methan-d_1-ol	297.5	31.68	0.20839E+03	−0.10318E+01	0.14740E-02	176-298
CH$_3$F	Fluoromethane	131.0	51.0	0.11338E+03	−0.63979E+00	0.96983E-03	150-299
CH$_3$I	Iodomethane	293.2	6.97	0.24264E+02	−0.93914E-01	0.11926E-03	223-303
CH$_3$NO	Formamide	293.2	111.0	0.26076E+03	−0.61145E+00	0.34296E-03	278-333
CH$_3$NO$_2$	Nitromethane	293.2	37.27	0.11227E+03	−0.35591E+00	0.34206E-03	288-343
CH$_3$NO$_2$	Methyl nitrite	200.0	20.77	0.11071E+03	−0.73428E+00	0.14054E-02	110-260
CH$_3$NO$_3$	Methyl nitrate	293.2	23.9				
CH$_4$	Methane	91.0	1.6761	0.15996E+01	0.27434E-02	−0.22086E-04	91-184
CH$_4$O	Methanol	293.2	33.0	0.19341E+03	−0.92211E+00	0.12839E-02	177-293
CH$_5$N	Methylamine	215.2	16.7	0.34398E+02	−0.73630E-01	−0.41279E-04	198-258
CN$_4$O$_8$	Tetranitromethane	293.2	2.317				
COS	Carbon oxysulfide	185.0	4.47	0.84702E+01	−0.21488E-01		143-185
COSe	Carbon oxyselenide	283.2	3.47	0.48740E+01	−0.49425E-02		219-283
CO$_2$	Carbon dioxide	295.0	1.4492	0.79062E+00	0.10639E-01	−0.28510E-04	220-300
CS$_2$	Carbon disulfide	293.2	2.6320	0.45024E+01	−0.12054E-01	0.19147E-04	154-319
C$_2$Br$_2$F$_4$	1,2-Dibromotetrafluoroethane	298.2	2.34				
C$_2$Cl$_2$F$_4$	1,2-Dichlorotetrafluoroethane	273.2	2.4842	0.36663E+01	−0.42271E-02	−0.36255E-06	193-273
C$_2$Cl$_2$O$_2$	Oxalyl chloride	294.35	3.470				
C$_2$Cl$_3$N	Trichloroacetonitrile	292.2	7.85				
C$_2$Cl$_4$	Tetrachloroethylene	303.2	2.268				
C$_2$Cl$_4$F$_2$	1,1,2,2-Tetrachloro-1,2-difluoroethane	308.2	2.52				
C$_2$HBr$_3$O	Tribromoacetaldehyde	293.2	7.6				
C$_2$HCl$_3$	Trichloroethylene	301.5	3.390	0.58319E+01	−0.80828E-02		302-338
C$_2$HCl$_3$F$_2$	1,2,2-Trichloro-1,1-difluoroethane	303.2	4.01	0.75423E+01	−0.11667E-01		303-333
C$_2$HCl$_3$O	Trichloroacetaldehyde	298.2	6.8				
C$_2$HCl$_3$O$_2$	Trichloroacetic acid	333.2	4.34	0.13412E+01	0.90000E-02	−0.24130E-14	333-393
C$_2$HCl$_5$	Pentachloroethane	298.2	3.716	0.65972E+01	−0.96800E-02		298-338
C$_2$HF$_3$O$_2$	Trifluoroacetic acid	293.2	8.42	0.21652E+02	−0.68146E-01	0.78571E-04	263-323
C$_2$H$_2$	Acetylene	195.0	2.4841				
C$_2$H$_2$Br$_2$	cis-1,2-Dibromoethylene	298.2	7.08				
C$_2$H$_2$Br$_2$	trans-1,2-Dibromoethylene	298.2	2.88				
C$_2$H$_2$Br$_4$	1,1,2,2-Tetrabromoethane	303.2	6.72	0.16246E+02	−0.31500E-01		303-333
C$_2$H$_2$Cl$_2$	1,1-Dichloroethylene	293.2	4.60				
C$_2$H$_2$Cl$_2$	cis-1,2-Dichloroethylene	298.2	9.20				
C$_2$H$_2$Cl$_2$	trans-1,2-Dichloroethylene	293.2	2.14				
C$_2$H$_2$Cl$_2$O$_2$	Dichloroacetic acid	293.2	8.33	0.11014E+02	−0.10859E-01	0.49242E-05	284-363
C$_2$H$_2$Cl$_4$	1,1,1,2-Tetrachloroethane	207.2	9.22	0.19606E+02	−0.49847E-01		207-233
C$_2$H$_2$Cl$_4$	1,1,2,2-Tetrachloroethane	293.2	8.50				
C$_2$H$_2$I$_2$	cis-1,2-Diiodoethylene	345.65	4.46				
C$_2$H$_3$ClO	Acetyl chloride	295.2	15.8				
C$_2$H$_3$ClO$_2$	Chloroacetic acid	338.2	12.35	0.17310E+02	−0.14674E-01		338-393
C$_2$H$_3$Cl$_2$NO$_2$	1,1-Dichloro-1-nitroethane	303.2	16.3	0.37576E+02	−0.70400E-01		303-333
C$_2$H$_3$Cl$_3$	1,1,1-Trichloroethane	293.2	7.243	0.27705E+02	−0.10621E+00	0.12424E-03	258-318
C$_2$H$_3$Cl$_3$	1,1,2-Trichloroethane	298.2	7.1937	0.17147E+02	−0.33371E-01		288-318
C$_2$H$_3$F$_3$O	2,2,2-Trifluoroethanol	293.2	27.68	0.90593E+02	−0.21421E+00		293-318
C$_2$H$_3$N	Acetonitrile	293.2	36.64	0.29724E+03	−0.15508E+01	0.22591E-02	288-333
C$_2$H$_3$NO	Methyl isocyanate	288.7	21.75				
C$_2$H$_4$	Ethylene	270.0	1.4833	0.13546E+01	0.62614E-02	−0.21374E-04	200-270
C$_2$H$_4$BrCl	1-Bromo-2-chloroethane	283.2	7.41	0.19493E+02	−0.59054E-01	0.58036E-04	263-363
C$_2$H$_4$Br$_2$	1,2-Dibromoethane	293.2	4.9612	0.67142E+01	−0.59800E-02		293-313
C$_2$H$_4$Cl$_2$	1,1-Dichloroethane	298.2	10.10	0.24429E+02	−0.48000E-01		288-318

Mol. form.	Name	T/K	ε_r	a	b	c	Range/K
$C_2H_4Cl_2$	1,2-Dichloroethane	293.2	10.42	0.24404E+02	−0.47892E-01		293-343
$C_2H_4Cl_2O$	Bis(chloromethyl) ether	293.2	3.51				
$C_2H_4N_2O_6$	Ethylene glycol dinitrate	293.2	28.26				
C_2H_4O	Acetaldehyde	291.2	21.0				
C_2H_4O	Ethylene oxide	293.2	12.42	0.52661E+02	−0.21337E+00	0.25947E-03	293-243
C_2H_4OS	Thioacetic acid	298.2	14.30				
$C_2H_4O_2$	Acetic acid	293.2	6.20	−0.15731E+02	0.12662E+00	−0.17738E-03	293-363
$C_2H_4O_2$	Methyl formate	288.2	9.20	0.19699E+02	−0.36429E-01		288-302
$C_2H_4O_3S$	Ethylene glycol sulfite	298.2	39.6	0.85483E+02	−0.15400E+00		298-328
C_2H_5Br	Bromoethane	298.2	9.01	0.28473E+02	−0.85495E-01	0.67971E-04	243-308
C_2H_5Cl	Chloroethane	293.2	9.45	0.60693E+02	−0.31290E+00	0.47154E-03	237-293
C_2H_5ClO	2-Chloroethanol	293.2	25.80	0.11155E+03	−0.30149E+00		140-175
C_2H_5I	Iodoethane	293.2	7.82	0.25598E+02	−0.94367E-01	0.11424E-03	183-343
C_2H_5N	Ethyleneimine	298.2	18.3	0.61405E+02	−0.14474E+00		273-298
C_2H_5NO	Acetamide	363.7	67.6	−0.20055E+03	0.15515E+01	−0.22392E-02	364-448
C_2H_5NO	N-Methylformamide	293.2	189.0	0.10383E+04	−0.43165E+01	0.48398E-02	276-353
C_2H_5NO	Acetaldoxime	298.2	4.70				
$C_2H_5NO_2$	Nitroethane	288.2	29.11	0.57406E+02	−0.97657E-01		276-333
$C_2H_5NO_2$	Methyl carbamate	328.2	18.48	0.36773E+02	−0.55700E-01		328-368
$C_2H_5NO_3$	Ethyl nitrate	293.2	19.7				
C_2H_6	Ethane	95.0	1.9356	0.20185E+01	−0.51493E-03	−0.48148E-05	95-295
C_2H_6O	Ethanol**	293.2	25.3	0.15145E+03	−0.87020E+00	0.19570E-02	163-523
C_2H_6O	Dimethyl ether	258.0	6.18	0.22389E+02	−0.86524E-01	0.91291E-04	155-258
C_2H_6OS	Dimethyl sulfoxide	293.2	47.24	0.38478E+02	0.16939E+00	−0.47423E-03	288-343
$C_2H_6O_2$	Ethylene glycol	293.2	41.4	0.14355E+03	−0.48573E+00	0.46703E-03	293-423
$C_2H_6O_2S$	Dimethyl sulfone	383.2	47.39	0.10830E+03	−0.15900E+00		383-398
$C_2H_6O_4S$	Dimethyl sulfate	298.2	55.0				
C_2H_6S	Ethanethiol	298.2	6.667				
C_2H_6S	Dimethyl sulfide	294.2	6.70				
$C_2H_6S_2$	1,2-Ethanedithiol	293.2	7.26	0.11228E+02	−0.13500E-01		293-333
$C_2H_6S_2$	Dimethyl disulfide	298.2	9.6	0.19109E+02	−0.32000E-01		298-323
C_2H_7N	Ethylamine	273.2	8.7	0.30163E+02	−0.79000E-01		233-273
C_2H_7NO	Ethanolamine	293.2	31.94	0.14890E+03	−0.62491E+00	0.77143E-03	253-293
$C_2H_8N_2$	1,2-Ethanediamine	293.2	13.82	0.48922E+02	−0.17021E+00	0.17262E-03	273-333
C_3Cl_6O	Hexachloroacetone	291.9	3.925	0.76423E+01	−0.15838E-01	0.10618E-04	269-303
C_3F_6O	Perfluoroacetone	202.2	2.104	0.34809E+01	−0.92883E-02	0.12282E-04	151-238
C_3HN	Cyanoacetylene	291.9	72.3	0.91803E+03	−0.49149E+01	0.69104E-02	281-314
$C_3H_2F_6O$	1,1,1,3,3,3-Hexafluoro-2-propanol	293.2	16.70				
$C_3H_3ClO_3$	4-Chloro-1,3-dioxolan-2-one	313.2	62.0				
C_3H_3N	Acrylonitrile	293.2	33.0	0.11109E+03	−0.36806E+00	0.34879E-03	233-413
$C_3H_3NO_2$	Cyanoacetic acid	277.2	33.4				
C_3H_4	Allene	269.0	2.025	0.26049E+01	−0.44147E-03	−0.63420E-05	156-269
C_3H_4	Propyne	246.0	3.218	0.60871E+01	−0.11730E-01		185-246
$C_3H_4ClF_3$	3-Chloro-1,1,1-trifluoropropane	295.2	7.32	0.22361E+02	−0.68840E-01	0.60594E-04	275-313
C_3H_4ClNO	2-Chloroethyl isocyanate	288.2	29.1	0.64311E+02	−0.12217E+00		288-403
$C_3H_4Cl_2O$	1,1-Dichloroacetone	293.2	14.6				
$C_3H_4F_4O$	2,2,3,3-Tetrafluoro-1-propanol	298.2	21.03				
C_3H_4O	Propargyl alcohol	293.2	20.8	0.99895E+02	−0.38911E+00	0.40776E-03	213-293
$C_3H_4O_3$	Ethylene carbonate	313.2	89.78	0.20746E+03	−0.37610E+00		313-343
C_3H_5Br	3-Bromopropene	293.2	7.0				
$C_3H_5BrO_2$	2-Bromopropanoic acid	294.2	11.0				
$C_3H_5Br_3$	1,2,3-Tribromopropane	303.2	6.00	0.11024E+02	−0.16596E-01		303-358
C_3H_5Cl	2-Chloropropene	299.25	8.92				
C_3H_5Cl	3-Chloropropene	293.2	8.2				
$C_3H_5ClN_2O_6$	3-Chloro-1,2-propanediol dinitrate	293.2	17.50				
C_3H_5ClO	Epichlorohydrin	293.2	22.6				
$C_3H_5ClO_2$	Ethyl chloroformate	308.7	9.736	0.15356E+02	−0.18250E-01		309-349
$C_3H_5ClO_2$	Methyl chloroacetate	293.2	12.0				
$C_3H_5Cl_3$	1,2,3-Trichloropropane	293.2	7.5				
C_3H_5I	3-Iodopropene	292.2	6.1				
C_3H_5N	Propanenitrile	293.2	29.7	0.82222E+02	−0.22937E+00	0.17424E-03	213-473
C_3H_5NO	Ethyl isocyanate	293.2	19.7				
C_3H_5NS	Ethyl isothiocyanate	293.2	19.6				

Mol. form.	Name	T/K	ε_r	a	b	c	Range/K
$C_3H_5N_3O_9$	Trinitroglycerol	293.2	19.25				
C_3H_6	Propene	220.0	2.1365	0.29623E+01	−0.37564E−02		220-250
$C_3H_6Br_2$	1,2-Dibromopropane	283.2	4.60	0.54973E+01	−0.31695E−02		283-333
$C_3H_6Br_2$	1,3-Dibromopropane	293.2	9.482	0.29193E+02	−0.94450E−01	0.92800E−04	293-368
$C_3H_6ClNO_2$	2-Chloro-2-nitropropane	250.4	31.90				
$C_3H_6Cl_2$	1,2-Dichloropropane	293.2	8.37	0.18915E+02	−0.35907E−01		281-323
$C_3H_6Cl_2$	1,3-Dichloropropane	303.2	10.27	0.21609E+02	−0.37333E−01		303-333
$C_3H_6Cl_2$	2,2-Dichloropropane	293.2	11.37	0.32421E+02	−0.72188E−01		245-293
$C_3H_6N_2O_4$	2,2-Dinitropropane	325.1	42.4				
C_3H_6O	Allyl alcohol	293.2	19.7	0.62714E+02	−0.14771E+00	0.37879E−05	213-303
C_3H_6O	Propanal	290.2	18.5				
C_3H_6O	Acetone	293.2	21.01	0.88157E+02	−0.34300E+00	0.38925E−03	273-323
$C_3H_6O_2$	Propanoic acid	298.2	3.44	0.18793E+01	0.46841E−02	0.19983E−05	289-408
$C_3H_6O_2$	Ethyl formate	288.2	8.57	0.15884E+02	−0.25333E−01		288-318
$C_3H_6O_2$	Methyl acetate	288.2	7.07	0.13190E+02	−0.21226E−01		276-318
$C_3H_6O_3$	3-Hydroxypropanoic acid	296.2	30.0				
$C_3H_6O_3$	Dimethyl carbonate	298.2	3.087				
$C_3H_6O_3$	1,3,5-Trioxane	338.2	15.55				
C_3H_7Br	1-Bromopropane	293.2	8.09	0.17769E+02	−0.32599E−01		274-328
C_3H_7Br	2-Bromopropane	293.2	9.46	0.26195E+02	−0.72995E−01	0.55454E−04	186-328
C_3H_7Cl	1-Chloropropane	293.2	8.588	0.21214E+02	−0.43130E−01		273-313
C_3H_7ClO	3-Chloro-1-propanol	215.2	36.0	0.12436E+03	−0.60841E+00	0.92060E−03	145-215
C_3H_7ClO	1-Chloro-2-propanol	153.2	59.0	−0.19169E+02	0.13605E+01	−0.55567E−02	153-177
$C_3H_7ClO_2$	3-Chloro-1,2-propanediol	293.2	31.0				
C_3H_7I	1-Iodopropane	293.2	7.07	0.13744E+02	−0.22745E−01		293-323
C_3H_7I	2-Iodopropane	298.2	8.19				
C_3H_7NO	N-Ethylformamide	298.2	102.7	0.64764E+03	−0.28499E+01	0.34286E−02	298-338
C_3H_7NO	N,N-Dimethylformamide	293.2	38.25	0.15364E+03	−0.60367E+00	0.71505E−03	213-353
C_3H_7NO	N-Methylacetamide**	303.2	179.0	0.15975E+04	−0.90451E+01	0.18345E−01	303-473
$C_3H_7NO_2$	1-Nitropropane	288.2	24.70	0.94999E+02	−0.38358E+00	0.48480E−03	276-333
$C_3H_7NO_2$	2-Nitropropane	288.2	26.74	0.60138E+02	−0.11566E+00		276-303
$C_3H_7NO_2$	Propyl nitrite	250.0	12.35	0.70552E+02	−0.40362E+00	0.66687E−03	110-310
$C_3H_7NO_2$	Isopropyl nitrite	260.0	13.92	0.74578E+02	−0.38283E+00	0.57071E−03	150-300
$C_3H_7NO_2$	Ethyl carbamate	328.2	14.14	0.32431E+02	−0.65097E−01	0.28571E−04	328-368
C_3H_8	Propane	293.19	1.6678	0.22883E+01	−0.23276E−02	0.84710E−06	90-300
C_3H_8O	1-Propanol	293.2	20.8	0.98045E+02	−0.36860E+00	0.36422E−03	193-493
C_3H_8O	2-Propanol	293.2	20.18	0.10416E+03	−0.41011E+00	0.42049E−03	193-493
$C_3H_8O_2$	1,2-Propylene glycol**	303.2	27.5	0.24546E+03	−0.15738E+01	0.38068E−02	193-403
$C_3H_8O_2$	1,3-Propylene glycol	293.2	35.1	0.11365E+03	−0.36680E+00	0.33766E−03	288-328
$C_3H_8O_2$	Ethylene glycol monomethyl ether	298.2	17.2	0.11803E+03	−0.58000E+00	0.81001E−03	254-318
$C_3H_8O_2$	Dimethoxymethane	293.2	2.644	0.25877E+01	−0.93019E−03	0.38472E−05	171-293
$C_3H_8O_3$	Glycerol	293.2	46.53	0.77503E+02	−0.37984E−01	−0.23107E−03	288-343
C_3H_8S	1-Propanethiol	288.2	5.937	0.11602E+02	−0.19580E−01		273-318
C_3H_8S	2-Propanethiol	298.2	5.952				
$C_3H_8S_2$	1,2-Propanedithiol	293.2	7.24	0.14667E+02	−0.32660E−01	0.25000E−04	293-333
$C_3H_8S_2$	1,3-Propanedithiol	303.2	8.11	0.66607E+01	0.31310E−01	−0.87500E−04	303-343
$C_3H_9BO_3$	Trimethyl borate	293.2	2.2762				
C_3H_9ClSi	Trimethylchlorosilane	273.2	10.21	−0.19492E+02	0.29806E+00	−0.69284E−03	223-273
C_3H_9N	Propylamine	296.2	5.08	0.17719E+02	−0.59022E−01	0.54780E−04	204-296
C_3H_9N	Isopropylamine	293.2	5.6268	0.40429E+02	−0.21441E+00	0.32634E−03	213-298
C_3H_9N	Trimethylamine	298.2	2.440	0.39745E+01	−0.51331E−02		273-298
$C_3H_9O_4P$	Trimethyl phosphate	293.2	20.6				
C_4Cl_6	Hexachloro-1,3-butadiene	293.2	2.55				
$C_4Cl_6O_3$	Trichloroacetic anhydride	298.2	5.0				
$C_4F_6O_3$	Trifluoroacetic acid anhydride	298.2	2.7				
$C_4H_2Cl_4O_3$	Dichloroacetic anhydride	298.2	15.8				
$C_4H_2O_3$	Maleic anhydride	326.2	52.75				
$C_4H_3F_7O$	2,2,3,3,4,4,4-Heptafluoro-1-butanol	298.2	14.4				
$C_4H_4N_2$	Succinonitrile	298.2	62.6	0.17724E+03	−0.54654E+00	0.54046E−03	236-351
$C_4H_4N_2$	Pyrazine	323.2	2.80				
C_4H_4O	Furan	277.1	2.88	0.13636E+01	0.12864E−01	−0.22701E−04	188-277
C_4H_4S	Thiophene	293.2	2.739	0.32941E+01	−0.19019E−02		253-293
C_4H_5Cl	2-Chloro-1,3-butadiene	293.2	4.914				

Mol. form.	Name	T/K	ε_r	a	b	c	Range/K
$C_4H_5Cl_3O_2$	Ethyl trichloroacetate	293.2	8.428				
C_4H_5N	Pyrrole	293.0	8.00	0.12672E+02	−0.14075E−01	−0.62671E−05	293-357
C_4H_5NO	Allyl isocynate	288.2	15.15	0.34299E+02	−0.66444E−01		288-333
C_4H_6	1,3-Butadiene	265.0	2.050	0.27674E+01	−0.26738E−02		185-265
C_4H_6O	Divinyl ether	288.2	3.94				
C_4H_6O	Ethoxyacetylene	298.2	8.05				
C_4H_6O	Cyclobutanone	298.2	14.27	0.43974E+02	−0.15712E+00	0.19264E−03	220-317
$C_4H_6O_2$	Methyl acrylate	303.2	7.03	0.11968E+02	−0.16500E−01		303-333
$C_4H_6O_2$	2,3-Butanedione	298.2	4.04	0.46907E+01	−0.22302E−02		278-348
$C_4H_6O_2$	γ-Butyrolactone	293.2	39.0				
$C_4H_6O_3$	Acetic anhydride	293.2	22.45				
$C_4H_6O_3$	Propylene carbonate	293.0	66.14	0.15940E+03	−0.39530E+00	0.26284E−03	273-333
C_4H_7Br	cis-2-Bromo-2-butene	293.2	5.38				
C_4H_7Br	trans-2-Bromo-2-butene	293.2	6.76				
$C_4H_7BrO_2$	2-Bromobutanoic acid	293.2	7.2				
$C_4H_7BrO_2$	Ethyl bromoacetate	303.2	9.75	0.15627E+02	−0.19600E−01		303-333
$C_4H_7BrO_2$	Methyl 3-bromopropanoate	303.2	5.81	0.36001E+01	0.72500E−02		303-343
$C_4H_7ClO_2$	Propyl chlorocarbonate	293.2	11.2				
$C_4H_7ClO_2$	Methyl 2-chloropropanoate	303.2	11.45	0.22449E+02	−0.36250E−01		303-343
C_4H_7N	Butanenitrile	293.2	24.83	0.53884E+02	−0.99257E−01		293-333
C_4H_7N	2-Methylpropanenitrile	293.2	24.42	0.52554E+02	−0.96000E−01		293-313
C_4H_7NO	2-Pyrrolidone	298.2	28.18	0.11054E+03	−0.47945E+00	0.68182E−03	298-338
C_4H_8	1-Butene	220.0	2.2195	0.29354E+01	−0.32580E−02		220-250
C_4H_8	cis-2-Butene	296.0	1.960	0.28802E+01	−0.31064E−02		197-296
C_4H_8	Isobutene	288.7	2.1225	0.33701E+01	−0.43295E−02		220-289
$C_4H_8Br_2$	1,2-Dibromobutane	293.2	4.74	0.11199E+03	−0.63334E+00	0.91250E−03	293-333
$C_4H_8Br_2$	1,3-Dibromobutane	293.2	9.14	0.34031E+02	−0.13254E+00	0.16250E−03	293-333
$C_4H_8Br_2$	1,4-Dibromobutane	303.2	8.68	0.20944E+02	−0.55620E−01	0.50000E−04	303-333
$C_4H_8Br_2$	2,3-Dibromobutane	298.2	6.245	0.23849E+02	−0.96300E−01	0.12500E−03	293-333
$C_4H_8Br_2$	1,2-Dibromo-2-methylpropane	293.2	4.1				
$C_4H_8Cl_2$	1,2-Dichlorobutane	293.2	7.74	0.31925E+02	−0.13232E+00	0.17007E−03	293-356
$C_4H_8Cl_2$	1,4-Dichlorobutane	308.2	9.30	0.59766E+01	0.49300E−01	−0.12500E−03	308-338
$C_4H_8Cl_2$	1,2-Dichloro-2-methylpropane	296.0	7.15	0.39429E+02	−0.20028E+00	0.30917E−03	165-296
$C_4H_8Cl_2O$	Bis(2-chloroethyl) ether	293.2	21.20				
C_4H_8O	Butanal	298.2	13.45				
C_4H_8O	2-Butanone	293.2	18.56	0.15457E+02	0.90152E−01	−0.27100E−03	293-333
C_4H_8O	Tetrahydrofuran	295.2	7.52	0.30739E+02	−0.12946E+00	0.17195E−03	224-295
$C_4H_8O_2$	Butanoic acid	287.2	2.98	0.15010E+01	0.50046E−02		287-403
$C_4H_8O_2$	2-Methylpropanoic acid	293.2	2.58				
$C_4H_8O_2$	Propyl formate	303.2	6.92				
$C_4H_8O_2$	Ethyl acetate	293.2	6.0814	0.15646E+02	−0.44066E−01	0.39137E−04	293-433
$C_4H_8O_2$	Methyl propanoate	293.2	6.200	0.12798E+02	−0.22540E−01		293-333
$C_4H_8O_2$	1,4-Dioxane	293.2	2.2189	0.27299E+01	−0.17440E−02		293-313
$C_4H_8O_3$	2-Hydroxybutanoic acid	296.2	37.7				
$C_4H_8O_3$	3-Hydroxybutanoic acid	296.2	31.5				
$C_4H_8O_3$	Ethyl methyl carbonate	293.2	2.985				
$C_4H_8O_3$	Ethylene glycol monoacetate	303.2	12.95				
C_4H_9Br	1-Bromobutane	283.2	7.315	0.22542E+02	−0.79306E−01	0.89867E−04	183-363
C_4H_9Br	2-Bromobutane	298.2	8.64	0.18461E+02	−0.32933E−01		274-328
C_4H_9Br	1-Bromo-2-methylpropane	273.2	7.70	0.37558E+02	−0.20571E+00	0.35496E−03	112-273
C_4H_9Br	2-Bromo-2-methylpropane	293.0	10.98	0.35085E+02	−0.14075E+00	0.19960E−03	258-293
C_4H_9Cl	1-Chlorobutane	293.2	7.276	0.13565E+02	−0.10161E−01	−0.38750E−04	273-323
C_4H_9Cl	2-Chlorobutane	293.2	8.564	0.30376E+02	−0.11377E+00	0.13429E−03	273-323
C_4H_9Cl	1-Chloro-2-methylpropane	293.2	7.027	0.14945E+02	−0.33747E−01	0.23036E−04	273-323
C_4H_9Cl	2-Chloro-2-methylpropane	293.2	9.663	0.35077E+02	−0.12867E+00	0.14304E−03	273-323
C_4H_9I	1-Iodobutane	293.2	6.27	0.16493E+02	−0.50262E−01	0.52485E−04	293-323
C_4H_9I	2-Iodobutane	293.2	7.873	0.10883E+02	−0.14680E−02	−0.30000E−04	293-323
C_4H_9I	2-Iodo-2-methylpropane	283.2	6.65	0.76780E+01	0.69900E−02	−0.37500E−04	283-323
C_4H_9N	Pyrrolidine	293.0	8.30	0.38191E+02	−0.15462E+00	0.17941E−03	274-333
C_4H_9NO	N-Methylpropanamide	293.2	170.0				
C_4H_9NO	N-Ethylacetamide	293.2	135.0	0.74494E+03	−0.31400E+01	0.36131E−02	213-353
C_4H_9NO	N,N-Dimethylacetamide	294.2	38.85	0.15420E+03	−0.57506E+00	0.61911E−03	294-433
C_4H_9NO	2-Butanone oxime	293.2	3.4				

Mol. form.	Name	T/K	ε_r	a	b	c	Range/K
C$_4$H$_9$NO	Morpholine	298.2	7.42				
C$_4$H$_9$NO$_2$	*tert*-Butyl nitrite	298.2	11.47				
C$_4$H$_9$NO$_2$	Propyl carbamate	338.2	12.06	0.24356E+02	−0.36400E−01		338-378
C$_4$H$_9$NO$_2$	Ethyl-*N*-methyl carbamate	298.2	21.10	0.11477E+03	−0.47568E+00	0.54127E−03	298-373
C$_4$H$_9$NO$_2$	*N*-Acetylethanolamine	298.2	96.6	0.37016E+03	−0.13113E+01	0.13214E−02	298-348
C$_4$H$_9$NO$_3$	Butyl nitrate	293.2	13.10				
C$_4$H$_{10}$	Butane	295.0	1.7697	0.22379E+01	−0.13884E−02	−0.66711E−06	135-303
C$_4$H$_{10}$	Isobutane	295.0	1.7518	0.23295E+01	−0.19953E−02	0.14197E−06	115-303
C$_4$H$_{10}$O	1-Butanol**	293.2	17.84	0.10578E+03	−0.50587E+00	0.84733E−03	193-553
C$_4$H$_{10}$O	2-Butanol**	293.2	17.26	0.13850E+03	−0.75146E+00	0.14086E−02	172-533
C$_4$H$_{10}$O	2-Methyl-1-propanol**	293.2	17.93	0.10762E+03	−0.51398E+00	0.83702E−03	173-533
C$_4$H$_{10}$O	2-Methyl-2-propanol**	298.2	12.47	0.22541E+03	−0.14990E+01	0.34050E−02	298-503
C$_4$H$_{10}$O	Diethyl ether	293.2	4.2666	0.79725E+01	−0.12519E−01		283-301
C$_4$H$_{10}$O$_2$	1,2-Butanediol	298.2	22.4	0.63702E+02	−0.13807E+00		278-323
C$_4$H$_{10}$O$_2$	1,3-Butanediol	298.2	28.8	0.72883E+02	−0.14770E+00		278-323
C$_4$H$_{10}$O$_2$	1,4-Butanediol	298.2	31.9	0.13079E+03	−0.46985E+00	0.46320E−03	288-328
C$_4$H$_{10}$O$_2$	Ethylene glycol monoethyl ether	298.2	13.38				
C$_4$H$_{10}$O$_2$	Ethylene glycol dimethyl ether	296.7	7.30	0.48832E+02	−0.24218E+00	0.34413E−03	256-318
C$_4$H$_{10}$O$_2$S	Bis(2-hydroxyethyl) sulfide	293.2	28.61	0.13128E+03	−0.52719E+00	0.60465E−03	253-333
C$_4$H$_{10}$O$_3$	Diethylene glycol	293.2	31.82	0.13973E+03	−0.54725E+00	0.61149E−03	288-343
C$_4$H$_{10}$O$_3$S	Diethyl sulfite	293.2	15.6				
C$_4$H$_{10}$O$_4$	1,2,3,4-Butanetetrol	393.2	28.2				
C$_4$H$_{10}$O$_4$S	Diethyl sulfate	293.2	29.2				
C$_4$H$_{10}$S	1-Butanethiol	288.2	5.204	0.11201E+02	−0.20767E−01		273-318
C$_4$H$_{10}$S	2-Butanethiol	288.2	5.645	0.10866E+02	−0.17993E−01		273-318
C$_4$H$_{10}$S	2-Methyl-1-propanethiol	298.2	4.961				
C$_4$H$_{10}$S	2-Methyl-2-propanethiol	293.2	5.475	0.10597E+02	−0.17500E−01		283-313
C$_4$H$_{10}$S	Diethyl sulfide	298.2	5.723				
C$_4$H$_{11}$N	Butylamine	293.2	4.71	0.13322E+02	−0.44176E−01	0.50250E−04	223-333
C$_4$H$_{11}$N	Diethylamine	293.2	3.680	0.26462E+02	−0.13750E+00	0.20373E−03	243-323
C$_4$H$_{11}$NO$_2$	Diethanolamine	293.2	25.75	0.73435E+02	−0.21377E+00	0.17500E−03	273-323
C$_4$H$_{12}$O$_2$Si	Dimethoxydimethylsilane	298.2	3.663				
C$_4$H$_{12}$O$_3$Si	Trimethoxymethylsilane	298.2	4.9				
C$_4$H$_{12}$O$_4$Si	Tetramethyl silicate	293.2	6.0				
C$_4$H$_{12}$Si	Diethylsilane	293.2	2.544				
C$_4$H$_{12}$Si	Tetramethylsilane	293.2	1.921				
C$_4$H$_{13}$N$_3$	Diethylenetriamine	293.2	12.62	0.57840E+02	−0.23873E+00	0.28841E−03	213-333
C$_5$FeO$_5$	Iron pentacarbonyl	293.2	2.602				
C$_5$H$_4$BrN	2-Bromopyridine	298.2	23.18	0.73391E+02	−0.23678E+00	0.22930E−03	298-398
C$_5$H$_4$ClN	2-Chloropyridine	298.2	27.32	0.98702E+02	−0.34237E+00	0.34502E−03	298-398
C$_5$H$_4$F$_8$O	2,2,3,3,4,4,5,5-Octafluoro-1-pentanol	298.2	15.30				
C$_5$H$_4$O$_2$	Furfural	293.2	42.1				
C$_5$H$_5$N	Pyridine	293.2	13.260	0.43991E+02	−0.15150E+00	0.15925E−03	293-323
C$_5$H$_5$NO	Pyridine-1-oxide	343.0	35.94	0.20878E+02	0.16450E+00	−0.35269E−03	343-398
C$_5$H$_6$O	2-Methylfuran	293.2	2.76				
C$_5$H$_6$O$_2$	Furfuryl alcohol	298.2	16.85				
C$_5$H$_7$Cl$_3$O$_2$	Propyl trichloroacetate	298.2	8.32				
C$_5$H$_7$NO$_2$	Ethyl cyanoacetate	263.2	31.62				
C$_5$H$_8$	1,3-Pentadiene*	298.2	2.319				
C$_5$H$_8$	1,4-Pentadiene	294.0	2.054	0.29994E+01	−0.34578E−02	0.85300E−06	178-294
C$_5$H$_8$	2-Methyl-1,3-butadiene	293.2	2.098	0.28170E+01	−0.23147E−02	−0.43975E−06	198-293
C$_5$H$_8$	Cyclopentene	295.0	2.083	0.28177E+01	−0.27597E−02	0.89346E−06	171-319
C$_5$H$_8$O	Cyclopentanone	298.2	13.58	0.24083E+02	−0.30286E−01	−0.16802E−04	219-298
C$_5$H$_8$O$_2$	Ethyl acrylate	303.2	6.05	0.47827E+02	−0.24394E+00	0.35000E−03	303-343
C$_5$H$_8$O$_2$	Methyl *trans*-2-butenoate	293.2	6.6645				
C$_5$H$_8$O$_2$	Methyl methacrylate	303.2	6.32	0.32098E+02	−0.14568E+00	0.20000E−03	303-343
C$_5$H$_8$O$_2$	2,4-Pentanedione	303.2	26.524				
C$_5$H$_8$O$_4$	Dimethyl malonate	293.2	9.82	0.26470E+02	−0.76656E−01	0.67888E−04	293-433
C$_5$H$_9$BrO$_2$	Ethyl 2-bromopropanoate	293.2	9.4				
C$_5$H$_9$ClO$_2$	Isobutyl chlorocarbonate	293.2	9.1				
C$_5$H$_9$ClO$_2$	Ethyl 2-chloropropanoate	303.2	11.95	0.25965E+02	−0.46250E−01		303-343
C$_5$H$_9$ClO$_2$	Ethyl 3-chloropropanoate	303.2	10.19	0.21951E+02	−0.38750E−01		303-343
C$_5$H$_9$ClO$_2$	Methyl 4-chlorobutanoate	303.2	9.51	0.17127E+02	−0.25000E−01		303-343

Mol. form.	Name	T/K	ε_r	a	b	c	Range/K
C_5H_9N	Pentanenitrile	293.2	20.04	0.55793E+02	−0.15750E+00	0.12432E−03	183-333
C_5H_9N	2,2-Dimethylpropanenitrile	293.2	21.1	0.58418E+02	−0.16884E+00	0.14131E−03	293-453
C_5H_9NO	Isobutyl isocyanate	293.2	11.638	0.38026E+02	−0.12714E+00	0.12679E−03	293-353
C_5H_9NO	N-Methyl-2-pyrrolidone	293.2	32.55				
C_5H_{10}	1-Pentene	293.2	2.011	−0.11438E+01	0.25420E−01	−0.50000E−04	273-293
C_5H_{10}	2-Methyl-1-butene	293.2	2.180				
C_5H_{10}	2-Methyl-2-butene	296.0	1.979	0.26064E+01	−0.19578E−02	−0.53908E−06	225-296
C_5H_{10}	Cyclopentane	293.2	1.9687	0.24287E+01	−0.15304E−02	−0.13095E−06	278-313
C_5H_{10}	Ethylcyclopropane	293.2	1.933				
$C_5H_{10}Br_2$	1,2-Dibromopentane	298.2	4.39				
$C_5H_{10}Br_2$	1,4-Dibromopentane	293.2	9.05	0.26443E+02	−0.88640E−01	0.10000E−03	293-333
$C_5H_{10}Br_2$	1,5-Dibromopentane	303.2	9.14	0.38192E+02	−0.15648E+00	0.20000E−03	303-333
$C_5H_{10}Cl_2$	1,2-Dichloropentane	293.2	6.89	0.19016E+02	−0.57954E−01	0.56801E−04	293-356
$C_5H_{10}Cl_2$	1,5-Dichloropentane	298.2	9.92				
$C_5H_{10}O$	Cyclopentanol	288.2	18.5	0.10565E+03	−0.44244E+00	0.48657E−03	258-323
$C_5H_{10}O$	Pentanal	293.2	10.00				
$C_5H_{10}O$	2,2-Dimethylpropanal	293.2	9.051	0.18645E+02	−0.32395E−01	−0.16157E−05	280-333
$C_5H_{10}O$	2-Pentanone	293.2	15.45	0.40893E+02	−0.10423E+00	0.60557E−04	204-353
$C_5H_{10}O$	3-Pentanone	293.2	17.00	0.12690E+02	0.95177E−01	−0.27321E−03	233-353
$C_5H_{10}O$	3-Methyl-2-butanone	293.2	10.37	0.30695E+02	−0.10962E+00	0.13810E−03	293-328
$C_5H_{10}O$	Tetrahydropyran	293.2	5.66	0.19793E+02	−0.76071E−01	0.94852E−04	234-333
$C_5H_{10}O$	2-Methyltetrahydrofuran	298.2	6.97				
$C_5H_{10}O_2$	Pentanoic acid	294.4	2.661	0.33491E+01	−0.75156E−02	0.17820E−04	250-344
$C_5H_{10}O_2$	Butyl formate	303.2	6.10	0.21532E+02	−0.84106E−01	0.10952E−03	288-323
$C_5H_{10}O_2$	Isobutyl formate	293.2	6.41				
$C_5H_{10}O_2$	Propyl acetate	293.2	5.62	0.17677E+02	−0.61404E−01	0.69196E−04	253-353
$C_5H_{10}O_2$	Ethyl propanoate	293.2	5.76				
$C_5H_{10}O_2$	Methyl butanoate	301.2	5.48	0.38604E+02	−0.19171E+00	0.27128E−03	301-343
$C_5H_{10}O_2$	Tetrahydrofurfuryl alcohol	303.2	13.48				
$C_5H_{10}O_2S$	3-Methyl sulfolane	298.2	29.4	0.53158E+02	−0.93730E−01	0.47275E−04	298-398
$C_5H_{10}O_3$	Diethyl carbonate	297.2	2.820				
$C_5H_{10}O_3$	Ethyl lactate	303.2	15.4	0.31225E+02	−0.43531E−01	−0.28571E−04	273-373
$C_5H_{10}O_4$	1,2,3-Propanetriol-1-acetate	242.2	38.57	0.10653E+03	−0.26439E+00	−0.62371E−04	215-242
$C_5H_{11}Br$	2-Bromo-2-methylbutane	298.2	9.21				
$C_5H_{11}Br$	1-Bromopentane	299.2	6.31	0.20954E+02	−0.78743E−01	0.98908E−04	183-328
$C_5H_{11}Br$	3-Bromopentane	298.2	8.37				
$C_5H_{11}Br$	1-Bromo-3-methylbutane	291.5	6.33	0.27743E+02	−0.13927E+00	0.22627E−03	123-292
$C_5H_{11}Cl$	1-Chloropentane	293.2	6.654	0.18626E+02	−0.54719E−01	0.47143E−04	273-323
$C_5H_{11}Cl$	1-Chloro-3-methylbutane	292.0	6.10	0.22228E+02	−0.93189E−01	0.12991E−03	171-297
$C_5H_{11}Cl$	2-Chloro-2-methylbutane	222.75	12.31	0.55104E+02	−0.29866E+00	0.47840E−03	201-223
$C_5H_{11}F$	1-Fluoropentane	293.2	3.931				
$C_5H_{11}I$	1-Iodopentane	293.2	5.78	0.15753E+02	−0.50543E−01	0.56401E−04	293-323
$C_5H_{11}I$	3-Iodopentane	293.2	7.432				
$C_5H_{11}I$	1-Iodo-3-methylbutane	292.2	5.6				
$C_5H_{11}I$	2-Iodo-2-methylbutane	293.2	8.192				
$C_5H_{11}N$	Piperidine	293.0	4.33	0.82317E+01	−0.11229E−01	−0.71429E−05	293-333
$C_5H_{11}N$	N-Methylpyrrolidine	298.2	32.2				
$C_5H_{11}NO$	2,2-Dimethylpropanamide	298.2	20.13	0.10400E+03	−0.46017E+00	0.60000E−03	298-328
$C_5H_{11}NO$	N,N-Diethylformamide	293.2	29.6				
$C_5H_{11}NO$	2-Pentanone oxime	293.2	3.3				
$C_5H_{11}NO_2$	Pentyl nitrite	298.2	7.21				
C_5H_{12}	Pentane	293.2	1.8371				
C_5H_{12}	Isopentane	293.2	1.845	0.22384E+01	−0.12985E−02	−0.16182E−06	143-293
C_5H_{12}	Neopentane	296.0	1.769	0.10949E+02	−0.63057E−01	0.10835E−03	251-296
$C_5H_{12}N_2O$	Tetramethylurea	293.2	23.10				
$C_5H_{12}O$	1-Pentanol	298.2	15.13	0.73397E+02	−0.28165E+00	0.28427E−03	213-513
$C_5H_{12}O$	2-Pentanol	298.2	13.71	0.16437E+03	−0.86506E+00	0.11955E−02	273-323
$C_5H_{12}O$	3-Pentanol	298.2	13.35	0.12838E+03	−0.60980E+00	0.75000E−03	288-318
$C_5H_{12}O$	2-Methyl-1-butanol	298.2	15.63	0.14020E+02	0.13948E+00	−0.45000E−03	288-318
$C_5H_{12}O$	3-Methyl-1-butanol	293.2	15.63	0.79733E+02	−0.31272E+00	0.32014E−03	173-513
$C_5H_{12}O$	2-Methyl-2-butanol	298.2	5.78	0.11662E+03	−0.69756E+00	0.10920E−02	268-318
$C_5H_{12}O$	3-Methyl-2-butanol	298.2	12.1				
$C_5H_{12}O$	2,2-Dimethyl-1-propanol	333.2	8.35	0.92350E+02	−0.41870E+00	0.50000E−03	333-373

Mol. form.	Name	T/K	ε_r	a	b	c	Range/K
C₅H₁₂O₂	1,2-Pentanediol	296.8	17.31	0.18436E+03	−0.10682E+01	0.17037E-02	197-297
C₅H₁₂O₂	1,4-Pentanediol	295.7	26.74	0.13568E+03	−0.59198E+00	0.75398E-03	193-318
C₅H₁₂O₂	1,5-Pentanediol	293.2	26.2	0.11858E+03	−0.45920E+00	0.49341E-03	243-343
C₅H₁₂O₂	2,3-Pentanediol	296.9	17.37	0.95876E+02	−0.46463E+00	0.67434E-03	238-297
C₅H₁₂O₂	2,4-Pentanediol	294.2	24.69	0.11914E+03	−0.52569E+00	0.69607E-03	224-294
C₅H₁₂O₂	Diethoxymethane	293.2	2.527	0.25294E+01	0.73988E-04	−0.28331E-06	227-293
C₅H₁₂O₄	Tetramethoxymethane	293.2	2.40				
C₅H₁₂O₅	Xylitol	293.2	40.0				
C₅H₁₂S	1-Pentanethiol	293.2	4.847	0.71131E+01	−0.30228E-02	−0.16414E-04	273-333
C₅H₁₂S	2-Methyl-2-butanethiol	293.2	5.087	0.15116E+02	−0.50700E-01	0.56250E-04	273-333
C₅H₁₂S₄	Tetrakis(methylthio)methane	343.2	2.818				
C₅H₁₃N	Pentylamine	293.2	4.27	0.11274E+02	−0.34965E-01	0.37706E-04	223-353
C₅H₁₃N₃	1,1,3,3-Tetramethylguanidine	298.2	11.5				
C₅H₁₄OSi	Ethoxytrimethylsilane	298.2	3.013				
C₆F₆	Hexafluorobenzene	298.2	2.029	0.24041E+01	−0.83086E-03	−0.14286E-05	298-338
C₆F₁₄	Perfluorohexane	298.2	1.76				
C₆H₃N₃O₇	2,4,6-Trinitrophenol	294.2	4.0				
C₆H₄BrF	1-Bromo-2-fluorobenzene	298.2	4.72				
C₆H₄BrF	1-Bromo-3-fluorobenzene	298.2	4.85				
C₆H₄BrF	1-Bromo-4-fluorobenzene	298.2	2.60				
C₆H₄BrNO₂	1-Bromo-3-nitrobenzene	328.2	20.2	0.81413E+02	−0.27645E+00	0.27367E-03	328-413
C₆H₄Br₂	o-Dibromobenzene	293.2	7.86	−0.81849E-02	0.62671E-01	−0.12222E-03	293-353
C₆H₄Br₂	m-Dibromobenzene	293.2	4.81	0.93214E+01	−0.20273E-01	0.16667E-04	293-353
C₆H₄Br₂	p-Dibromobenzene	368.2	2.57				
C₆H₄ClF	1-Chloro-2-fluorobenzene	298.2	6.10				
C₆H₄ClF	1-Chloro-3-fluorobenzene	298.2	4.96				
C₆H₄ClF	1-Chloro-4-fluorobenzene	298.2	3.34				
C₆H₄ClNO₂	1-Chloro-2-nitrobenzene	323.2	37.7	0.16800E+03	−0.59708E+00	0.59957E-03	323-436
C₆H₄ClNO₂	1-Chloro-3-nitrobenzene	323.2	20.9	0.77193E+02	−0.25118E+00	0.23798E-03	323-433
C₆H₄ClNO₂	1-Chloro-4-nitrobenzene	393.2	8.09				
C₆H₄Cl₂	o-Dichlorobenzene	293.2	10.12	0.13629E+02	0.10622E-02	−0.44444E-04	293-353
C₆H₄Cl₂	m-Dichlorobenzene	293.2	5.02	0.77565E+01	−0.93333E-02	−0.26880E-14	293-353
C₆H₄Cl₂	p-Dichlorobenzene	328.2	2.3943	0.26999E+01	−0.35325E-03	−0.17619E-05	328-363
C₆H₄FI	1-Fluoro-2-iodobenzene	298.2	8.22				
C₆H₄FI	1-Fluoro-4-iodobenzene	298.2	3.12				
C₆H₄F₂	o-Difluorobenzene	301.2	13.38	0.59107E+02	−0.23611E+00	0.27987E-03	273-323
C₆H₄F₂	m-Difluorobenzene	301.2	5.01	0.14448E+02	−0.46982E-01	0.51948E-04	273-323
C₆H₄I₂	o-Diiodobenzene	323.2	5.41	0.31150E+02	−0.14428E+00	0.20000E-03	323-353
C₆H₄I₂	m-Diiodobenzene	323.2	4.11				
C₆H₄I₂	p-Diiodobenzene	393.2	2.88				
C₆H₄N₂	2-Pyridinecarbonitrile	303.2	93.77	0.45596E+03	−0.17746E+01	0.19105E-02	303-398
C₆H₄N₂	3-Pyridinecarbonitrile	323.2	20.54	0.60484E+02	−0.17280E+00	0.15218E-03	323-398
C₆H₄N₂	4-Pyridinecarbonitrile	353.2	5.23	0.12533E+02	−0.30115E-01	0.26674E-04	353-398
C₆H₄N₂O₄	1,3-Dinitrobenzene	365.2	22.9	0.10406E+03	−0.34133E+00	0.32609E-03	365-413
C₆H₅Br	Bromobenzene	293.2	5.45	0.94100E+01	−0.12537E-01	−0.31127E-05	234-333
C₆H₅Cl	Chlorobenzene	293.2	5.6895	0.19471E+02	−0.70786E-01	0.82466E-04	293-430
C₆H₅ClO	o-Chlorophenol	296.2	7.40	0.29755E+02	−0.11256E+00	0.12390E-03	296-448
C₆H₅ClO	m-Chlorophenol	293.2	6.255				
C₆H₅ClO	p-Chlorophenol	314.2	11.18	0.31997E+02	−0.94241E-01	0.88392E-04	314-453
C₆H₅ClO₂S	Benzenesulfonyl chloride	323.2	28.90	0.83886E+02	−0.23405E+00	0.19713E-03	323-473
C₆H₅ClS	4-Chlorobenzenethiol	338.2	3.59				
C₆H₅F	Fluorobenzene	293.2	5.465				
C₆H₅I	Iodobenzene	293.2	4.59	0.89442E+01	−0.20008E-01	0.17641E-04	243-323
C₆H₅NOS	N-Sulfinylaniline	298.2	6.97				
C₆H₅NO₂	Nitrobenzene	293.0	35.6	0.11212E+03	−0.35211E+00	0.31128E-03	279-533
C₆H₅NO₃	o-Nitrophenol	323.2	16.50	0.33827E+02	−0.62123E-01	0.26774E-04	323-453
C₆H₅NO₃	m-Nitrophenol	373.2	35.45	0.18967E+03	−0.66144E+00	0.66532E-03	373-458
C₆H₅NO₃	p-Nitrophenol	393.2	42.20	0.22901E+03	−0.74264E+00	0.68006E-03	393-463
C₆H₆	Benzene	293.2	2.2825	0.26706E+01	−0.91648E-03	−0.14257E-05	293-513
C₆H₆BrN	m-Bromoaniline	293.2	13.0				
C₆H₆ClN	o-Chloroaniline	293.2	13.40				
C₆H₆ClN	m-Chloroaniline	293.2	13.3				
C₆H₆N₂O₂	o-Nitroaniline	353.0	47.3	0.18900E+03	−0.56977E+00	0.47484E-03	353-468

Mol. form.	Name	T/K	ε_r	a	b	c	Range/K
$C_6H_6N_2O_2$	*m*-Nitroaniline	398.0	35.6	0.20352E+03	−0.66582E+00	0.61310E-03	398-468
$C_6H_6N_2O_2$	*p*-Nitroaniline	428.0	78.5	0.48673E+03	−0.15040E+01	0.12857E-02	428-468
C_6H_6O	Phenol	303.2	12.40	0.63391E+02	−0.24988E+00	0.26930E-03	303-433
$C_6H_6O_2$	Pyrocatechol	388.2	17.57	0.74930E+02	−0.22142E+00	0.18919E-03	388-463
$C_6H_6O_2$	Resorcinol	393.2	13.55	0.30252E+02	−0.56443E-01	0.35578E-04	393-463
C_6H_6S	Benzenethiol	303.2	4.26	0.57155E+01	−0.70336E-02	0.73617E-05	303-358
C_6H_7N	Aniline	293.2	7.06	0.89534E+01	0.38990E-02	−0.36310E-04	293-413
C_6H_7N	2-Methylpyridine	293.2	10.18	0.34560E+02	−0.11980E+00	0.12500E-03	293-333
C_6H_7N	3-Methylpyridine	303.0	11.10	0.19643E+03	−0.11167E+01	0.16667E-02	303-333
C_6H_7N	4-Methylpyridine	293.0	12.2	0.33765E+02	−0.10113E+00	0.93860E-04	274-333
C_6H_7NO	2-Methylpyridine-1-oxide	323.2	36.4	0.11705E+03	−0.35301E+00	0.32000E-03	323-398
C_6H_7NO	3-Methylpyridine-1-oxide	318.2	28.26	0.59851E+02	−0.12682E+00	0.86622E-04	318-398
C_6H_8	1,3-Cyclohexadiene	184.2	2.68				
C_6H_8	1,4-Cyclohexadiene	296.0	2.211	0.27459E+01	−0.16975E-02	−0.36461E-06	232-356
$C_6H_8N_2$	Phenylhydrazine	293.2	7.15				
$C_6H_8N_2$	2,5-Dimethylpyrazine	293.2	2.436				
$C_6H_8N_2$	2,6-Dimethylpyrazine	308.2	2.653				
$C_6H_8O_2$	1,4-Cyclohexanedione	351.2	4.40				
$C_6H_9Cl_3O_2$	Butyl trichloroacetate	293.2	7.480				
$C_6H_9Cl_3O_2$	Isobutyl trichloroacetate	293.2	7.667				
C_6H_9N	Cyclopentanecarbonitrile	293.2	22.68	0.69830E+02	−0.25303E+00	0.31491E-03	201-293
C_6H_{10}	1,5-Hexadiene	294.0	2.125	0.30014E+01	−0.28668E-02	−0.31026E-06	151-294
C_6H_{10}	*cis,cis*-2,4-Hexadiene	297.0	2.163	0.27284E+01	−0.17178E-02	−0.62926E-06	234-351
C_6H_{10}	*trans,trans*-2,4-Hexadiene	297.0	2.123	0.26774E+01	−0.16977E-02	−0.55637E-06	232-353
C_6H_{10}	2-Methyl-1,3-pentadiene*	298.2	2.422				
C_6H_{10}	3-Methyl-1,3-pentadiene	298.2	2.426				
C_6H_{10}	4-Methyl-1,3-pentadiene	293.2	2.599	0.51328E+01	−0.12774E-01	0.14215E-04	198-323
C_6H_{10}	2,3-Dimethyl-1,3-butadiene	293.2	2.102	0.26258E+01	−0.17990E-02	0.12035E-06	223-323
C_6H_{10}	1-Hexyne	296.0	2.621	0.58591E+01	−0.17099E-01	0.20856E-04	184-296
C_6H_{10}	Cyclohexene	293.2	2.2176	0.30598E+01	−0.39841E-02	0.37554E-05	141-313
$C_6H_{10}O$	Butoxyacetylene	298.2	6.62				
$C_6H_{10}O$	Cyclohexanone	293.0	16.1	0.41577E+02	−0.11463E+00	0.92454E-04	253-423
$C_6H_{10}O$	Mesityl oxide	273.2	15.6				
$C_6H_{10}O_2$	Ethyl 2-butenoate	293.2	5.4				
$C_6H_{10}O_2$	Ethyl methacrylate	303.2	5.68	0.40962E+02	−0.20520E+00	0.29286E-03	303-343
$C_6H_{10}O_3$	Ethyl acetoacetate	293.2	14.0				
$C_6H_{10}O_3$	Propanoic anhydride	293.2	18.30				
$C_6H_{10}O_4$	Monomethyl glutarate	293.2	8.37	0.16779E+02	−0.39839E-01	0.38095E-04	293-363
$C_6H_{10}O_4$	Diethyl oxalate	293.2	8.266	0.21938E+02	−0.66226E-01	0.66800E-04	293-368
$C_6H_{10}O_4$	Dimethyl succinate	293.2	7.19	0.13551E+02	−0.23109E-01	0.55440E-05	293-433
$C_6H_{10}O_4$	Ethylene glycol diacetate	290.2	7.7	0.25093E+02	−0.95171E-01	0.12224E-03	223-290
$C_6H_{11}Br$	Bromocyclohexane	303.2	8.0026				
$C_6H_{11}BrO_2$	Ethyl 2-bromobutanoate	303.2	8.57	0.49005E+02	−0.23193E+00	0.32500E-03	303-333
$C_6H_{11}BrO_2$	Ethyl 2-bromo-2-methylpropanoate	303.2	8.55	0.77044E+02	−0.40784E+00	0.60000E-03	303-333
$C_6H_{11}Cl$	Chlorocyclohexane	303.2	7.9505				
$C_6H_{11}N$	Hexanenitrile	298.2	17.26				
$C_6H_{11}N$	4-Methylpentanenitrile	295.2	17.5				
$C_6H_{11}NO$	Cyclohexanone oxime	362.2	3.04				
C_6H_{12}	1-Hexene	294.0	2.077	0.31476E+01	−0.50003E-02	0.46673E-05	149-294
C_6H_{12}	*trans*-2-Hexene	295.0	1.978	0.24338E+01	−0.11323E-02	−0.13720E-05	157-295
C_6H_{12}	*cis*-3-Hexene	296.0	2.069	0.30691E+01	−0.45458E-02	0.39898E-05	155-296
C_6H_{12}	*trans*-3-Hexene	293.2	1.954				
C_6H_{12}	Cyclohexane	293.2	2.0243	0.24293E+01	−0.12095E-02	−0.58741E-06	283-333
C_6H_{12}	Methylcyclopentane	293.2	1.9853	0.21587E+01	−0.22450E-03	−0.12500E-05	293-323
C_6H_{12}	Ethylcyclobutane	293.2	1.965				
$C_6H_{12}Br_2$	1,6-Dibromohexane	298.2	8.52	−0.55185E+01	0.11746E+00	−0.23658E-03	274-328
$C_6H_{12}Br_2$	3,4-Dibromohexane	298.2	6.732				
$C_6H_{12}Cl_2$	1,6-Dichlorohexane	308.2	8.60	0.11277E+02	0.67200E-02	−0.50000E-04	308-338
$C_6H_{12}O$	1-Methylcyclopentanol	310.1	7.11	0.75444E+02	−0.36617E+00	0.47021E-03	310-333
$C_6H_{12}O$	Isobutyl vinyl ether	293.2	3.34	0.48060E+01	−0.50000E-02	−0.41495E-14	293-323
$C_6H_{12}O$	2-Hexanone	293.2	14.56	0.70378E+02	−0.29385E+00	0.35289E-03	243-293
$C_6H_{12}O$	4-Methyl-2-pentanone	293.2	13.11	0.36341E+02	−0.97119E-01	0.61896E-04	204-373
$C_6H_{12}O$	3,3-Dimethyl-2-butanone	293.2	12.73	0.66857E+02	−0.28552E+00	0.34422E-03	243-293

Mol. form.	Name	T/K	ε_r	a	b	c	Range/K
$C_6H_{12}O$	Cyclohexanol	293.2	16.40	0.10173E+03	−0.43072E+00	0.47926E-03	293-423
$C_6H_{12}O_2$	Hexanoic acid	298.2	2.600	0.21730E+01	0.14840E-02	−0.16526E-06	298-433
$C_6H_{12}O_2$	2-Ethylbutanoic acid	296.2	2.72				
$C_6H_{12}O_2$	*tert*-Butylacetic acid	296.2	2.85				
$C_6H_{12}O_2$	Pentyl formate	292.2	5.7				
$C_6H_{12}O_2$	Isopentyl formate	288.2	5.44	0.29257E+02	−0.14028E+00	0.20000E-03	288-323
$C_6H_{12}O_2$	Butyl acetate	293.2	5.07	0.13825E+02	−0.43994E-01	0.48214E-04	253-353
$C_6H_{12}O_2$	*sec*-Butyl acetate	293.2	5.135	0.12427E+02	−0.32035E-01	0.24286E-04	273-323
$C_6H_{12}O_2$	*tert*-Butyl acetate	293.2	5.672	0.55435E+02	−0.30494E+00	0.46107E-03	273-323
$C_6H_{12}O_2$	Isobutyl acetate	293.2	5.068	0.14323E+02	−0.46048E-01	0.49286E-04	273-323
$C_6H_{12}O_2$	Propyl propanoate	293.2	5.249				
$C_6H_{12}O_2$	Ethyl butanoate	301.2	5.18	0.48698E+02	−0.25660E+00	0.37237E-03	301-343
$C_6H_{12}O_2$	Methyl pentanoate	293.2	4.992				
$C_6H_{12}O_2$	Diacetone alcohol	298.2	18.2				
$C_6H_{12}O_3$	Ethylene glycol monoethyl ether acetate	303.2	7.567	0.23290E+02	−0.71566E-01	0.65000E-04	303-323
$C_6H_{12}S$	Cyclohexanethiol	298.2	5.420				
$C_6H_{13}Br$	1-Bromohexane	298.2	5.82	0.15233E+02	−0.44385E-01	0.43039E-04	274-328
$C_6H_{13}Cl$	1-Chlorohexane	293.2	6.104	0.15994E+02	−0.43647E-01	0.33393E-04	273-323
$C_6H_{13}ClO$	6-Chloro-1-hexanol	242.2	21.6	−0.73364E+01	0.46377E+00	−0.14202E-02	195-242
$C_6H_{13}I$	1-Iodohexane	293.3	5.35	0.16685E+02	−0.61309E-01	0.77262E-04	293-323
$C_6H_{13}N$	Cyclohexylamine	293.2	4.547				
$C_6H_{13}NO$	*N*-Propylpropanamide	298.2	118.1	0.58846E+03	−0.22012E+01	0.20870E-02	298-328
$C_6H_{13}NO$	*N*-Butylacetamide**	293.2	104.0	0.70739E+03	−0.37369E+01	0.71585E-02	253-493
$C_6H_{13}NO$	*N,N*-Diethylacetamide	293.2	32.1				
C_6H_{14}	Hexane	293.2	1.8865	0.19768E+01	0.70933E-03	−0.34470E-05	293-473
C_6H_{14}	2-Methylpentane	293.2	1.886	0.20745E+01	0.50871E-03	−0.39286E-05	273-323
C_6H_{14}	3-Methylpentane	293.2	1.886	0.24739E+01	−0.23190E-02	0.10714E-05	273-323
C_6H_{14}	2,2-Dimethylbutane	293.2	1.869	0.22740E+01	−0.96229E-03	−0.14286E-05	273-313
C_6H_{14}	2,3-Dimethylbutane	293.2	1.889	0.24305E+01	−0.20081E-02	0.53571E-06	273-323
$C_6H_{14}O$	1-Hexanol	293.2	13.03	0.62744E+02	−0.24214E+00	0.24704E-03	233-513
$C_6H_{14}O$	2-Hexanol	298.2	11.06				
$C_6H_{14}O$	3-Hexanol	298.2	9.66				
$C_6H_{14}O$	3-Methyl-1-pentanol	298.2	15.2				
$C_6H_{14}O$	3-Methyl-3-pentanol	293.2	4.322				
$C_6H_{14}O$	2-Ethyl-1-butanol	362.2	6.19				
$C_6H_{14}O$	2,2-Dimethyl-1-butanol	293.2	10.5	0.14054E+03	−0.72925E+00	0.97821E-03	243-393
$C_6H_{14}O$	Dipropyl ether	297.0	3.38	0.14600E+02	−0.72670E-01	0.11742E-03	161-297
$C_6H_{14}O$	Diisopropyl ether	303.2	3.805				
$C_6H_{14}OS$	Dipropyl sulfoxide	303.2	30.37	0.84868E+02	−0.23486E+00	0.18198E-03	303-373
$C_6H_{14}O_2$	2-Methyl-2,4-pentanediol	293.2	25.86	0.14531E+03	−0.65285E+00	0.83503E-03	203-333
$C_6H_{14}O_2$	Ethylene glycol diethyl ether	293.2	3.90	0.99099E+01	−0.33403E-01	0.44048E-04	223-303
$C_6H_{14}O_2S$	Dipropyl sulfone	303.2	32.62	0.70195E+02	−0.15008E+00	0.86506E-04	303-398
$C_6H_{14}O_3$	1,2,6-Hexanetriol	285.3	31.5	0.26127E+03	−0.14552E+01	0.22765E-02	261-285
$C_6H_{14}O_3$	Diethylene glycol dimethyl ether	298.2	7.23	0.28291E+02	−0.11236E+00	0.14000E-03	298-333
$C_6H_{14}O_4$	Triethylene glycol	293.2	23.69	0.91845E+02	−0.33827E+00	0.36062E-03	253-333
$C_6H_{14}O_6$	*D*-Glucitol	353.2	35.5				
$C_6H_{14}O_6$	*D*-Mannitol	443.2	24.6				
$C_6H_{14}S$	1-Hexanethiol	293.2	4.436	0.11774E+02	−0.37298E-01	0.41875E-04	273-333
$C_6H_{15}B$	Triethylborane	293.2	1.974				
$C_6H_{15}N$	Hexylamine	293.2	4.08	0.80244E+01	−0.16627E-01	0.10874E-04	253-373
$C_6H_{15}N$	Dipropylamine	293.2	2.923	0.11376E+02	−0.49796E-01	0.71792E-04	243-323
$C_6H_{15}N$	Triethylamine	293.2	2.418	0.29205E+01	−0.14007E-02	−0.13469E-05	233-323
$C_6H_{15}OP$	Triethylphosphine oxide	323.2	35.5				
$C_6H_{15}O_4P$	Triethyl phosphate	298.2	13.20	0.61230E+02	−0.26047E+00	0.33333E-03	298-333
$C_6H_{15}PS$	Triethylphosphine sulfide	371.2	39.0				
$C_6H_{16}O_2Si$	Diethoxydimethylsilane	298.2	3.216				
$C_6H_{16}Si$	Triethylsilane	293.2	2.323				
$C_6H_{18}N_3OP$	Hexamethylphosphoric triamide	293.2	31.3	0.95666E+02	−0.29769E+00	0.26407E-03	283-363
$C_6H_{18}N_4$	*N,N'*-Bis(2-aminoethyl)-1,2-ethanediamine	293.2	10.76	0.50699E+02	−0.21730E+00	0.27582E-03	213-333
$C_6H_{18}OSi_2$	Hexamethyldisiloxane	293.2	2.179	0.34537E+01	−0.61530E-02	0.61544E-05	213-313
$C_6H_{18}O_3Si_3$	Hexamethylcyclotrisiloxane	343.2	2.139				

Mol. form.	Name	T/K	ε_r	a	b	c	Range/K
$C_6H_{19}NSi_2$	Hexamethyldisilazane	294.2	2.273	0.23358E+01	0.16127E-02	−0.62078E-05	294-333
C_7F_{14}	Perfluoromethylcyclohexane	298.2	1.82				
C_7F_{16}	Perfluoroheptane	289.2	1.847				
$C_7H_3Cl_5$	2,3,4,5,6-Pentachlorotoluene	293.2	4.8				
C_7H_4ClNO	4-Chlorophenyl isocyanate	288.2	3.177	0.40896E+01	−0.31667E-02		288-348
C_7H_5BrO	Benzoyl bromide	293.2	21.33	0.84231E+02	−0.31089E+00	0.32857E-03	283-313
C_7H_5ClO	Benzoyl chloride	293.2	23.0				
C_7H_5FO	Benzoyl fluoride	293.2	22.7				
$C_7H_5F_3$	(Trifluoromethyl)benzene	298.2	9.22				
C_7H_5N	Benzonitrile	293.2	25.9	0.57605E+02	−0.13354E+00	0.87767E-04	273-453
C_7H_5NO	Phenyl isocyanate	293.2	8.940	0.17541E+02	−0.29790E-01	0.15476E-05	293-353
$C_7H_6ClNO_2$	4-Chloro-3-nitrotoluene	301.2	28.07				
$C_7H_6Cl_2$	2,4-Dichlorotoluene	301.2	5.68				
$C_7H_6Cl_2$	2,6-Dichlorotoluene	301.2	3.36				
$C_7H_6Cl_2$	3,4-Dichlorotoluene	301.2	9.39				
$C_7H_6Cl_2$	(Dichloromethyl)benzene	293.2	6.9				
C_7H_6O	Benzaldehyde	293.2	17.85	0.35046E+02	−0.61271E-01	0.16222E-04	301-346
$C_7H_6O_2$	Salicylaldehyde	293.2	18.35	0.51315E+02	−0.15379E+00	0.14111E-03	289-453
C_7H_7Br	o-Bromotoluene	293.2	4.641	0.10229E+02	−0.25050E-01	0.20357E-04	273-323
C_7H_7Br	m-Bromotoluene	293.2	5.566	0.11522E+02	−0.24946E-01	0.15714E-04	273-323
C_7H_7Br	p-Bromotoluene	293.2	5.503	0.10014E+02	−0.13918E-01	−0.50000E-05	273-293
C_7H_7Br	(Bromomethyl)benzene	293.2	6.658	0.18482E+02	−0.57207E-01	0.57321E-04	273-323
C_7H_7BrO	o-Bromoanisole	303.2	8.96	0.12023E+02	−0.59116E-02	−0.13787E-04	303-358
C_7H_7BrO	p-Bromoanisole	303.2	7.40	0.74367E+01	0.12648E-01	−0.42128E-04	303-358
C_7H_7Cl	o-Chlorotoluene	293.2	4.721	0.11507E+02	−0.31148E-01	0.27143E-04	273-323
C_7H_7Cl	m-Chlorotoluene	293.2	5.763	0.13921E+02	−0.37186E-01	0.31786E-04	273-323
C_7H_7Cl	p-Chlorotoluene	293.2	6.25	0.20265E+01	0.40060E-01	−0.87500E-04	293-333
C_7H_7Cl	(Chloromethyl)benzene	293.2	6.854	0.17108E+02	−0.45285E-01	0.35000E-04	273-323
C_7H_7ClO	p-Chloroanisole	293.2	7.84	0.64019E+01	0.30560E-01	−0.87500E-04	293-333
$C_7H_7ClO_2S$	p-Toluenesulfonyl chloride	343.2	22.6				
$C_7H_7ClO_3S$	4-Methoxybenzenesulfonyl chloride	314.2	27.2				
C_7H_7F	o-Fluorotoluene	298.2	4.23				
C_7H_7F	m-Fluorotoluene	298.2	5.41				
C_7H_7F	p-Fluorotoluene	298.2	5.88				
C_7H_7I	p-Iodotoluene	308.2	4.4				
C_7H_7N	2-Vinylpyridine	293.2	9.126				
C_7H_7N	4-Vinylpyridine	293.2	10.50				
$C_7H_7NO_2$	Benzyl nitrite	298.2	7.78				
$C_7H_7NO_2$	o-Nitrotoluene	293.0	26.26	0.10420E+03	−0.41726E+00	0.51607E-03	273-323
$C_7H_7NO_2$	m-Nitrotoluene	303.2	24.95	0.62492E+02	−0.16235E+00	0.12844E-03	303-403
$C_7H_7NO_2$	p-Nitrotoluene	331.2	22.2				
$C_7H_7NO_2S$	4-Nitrothioanisole	346.0	21.7				
$C_7H_7NO_3$	2-Nitroanisole	293.2	45.75	0.16684E+03	−0.58196E+00	0.57382E-03	293-423
$C_7H_7NO_3$	3-Nitroanisole	318.2	25.7	0.65402E+02	−0.16460E+00	0.12560E-03	318-443
$C_7H_7NO_3$	4-Nitroanisole	338.2	26.95	0.59811E+02	−0.10955E+00	0.36042E-04	338-443
C_7H_8	Toluene	296.35	2.379	0.32584E+01	−0.34410E-02	0.15937E-05	207-316
C_7H_8O	o-Cresol	298.2	6.76	0.21633E+02	−0.71069E-01	0.70590E-04	298-453
C_7H_8O	m-Cresol	298.2	12.44	0.81716E+02	−0.35039E+00	0.39878E-03	274-463
C_7H_8O	p-Cresol	298.2	13.05	0.70253E+02	−0.28870E+00	0.31979E-03	298-453
C_7H_8O	Benzyl alcohol	303.2	11.916	0.13661E+03	−0.72127E+00	0.10225E-02	303-333
C_7H_8O	Anisole	294.2	4.30	0.10887E+02	−0.32372E-01	0.33629E-04	294-413
$C_7H_8O_2$	2-Methoxyphenol	298.2	11.95	0.31751E+02	−0.88173E-01	0.72953E-04	291-448
$C_7H_8O_2$	3-Methoxyphenol	298.2	11.59	0.37279E+02	−0.12113E+00	0.11698E-03	298-433
$C_7H_8O_2$	4-Methoxyphenol	333.7	11.05	0.39483E+02	−0.12142E+00	0.10841E-03	334-453
$C_7H_8O_2S$	Ethyl thiophene-2-carboxylate	293.2	6.18				
$C_7H_8O_2S$	Methyl phenyl sulfone	373.2	37.9				
C_7H_8S	Benzenemethanethiol	298.2	4.705	0.16628E+02	−0.68276E-01	0.94636E-04	298-358
C_7H_8S	4-Methylbenzenethiol	323.2	4.74	0.87052E+01	−0.15347E-01	0.95238E-05	323-358
C_7H_8S	(Methylthio)benzene	303.2	4.88	0.21841E+02	−0.97630E-01	0.13750E-03	303-343
C_7H_9N	Benzylamine	293.2	5.18				
C_7H_9N	o-Methylaniline	298.2	6.138	0.10988E+02	−0.18976E-01	0.91958E-05	298-398
C_7H_9N	m-Methylaniline	298.2	5.816	0.13477E+02	−0.35551E-01	0.33135E-04	298-398
C_7H_9N	p-Methylaniline	333.2	5.058	0.78897E+01	−0.10196E-01	0.51190E-05	333-403

Mol. form.	Name	T/K	ε_r	a	b	c	Range/K
C_7H_9N	*N*-Methylaniline	293.2	5.96				
C_7H_9N	2-Ethylpyridine	293.2	8.33	0.36397E+02	−0.15070E+00	0.18750E-03	293-333
C_7H_9N	4-Ethylpyridine	293.2	10.98	−0.73831E+01	0.14326E+00	−0.27500E-03	293-333
C_7H_9N	2,4-Dimethylpyridine	293.2	9.60	0.25895E+02	−0.73900E-01	0.62500E-04	293-333
C_7H_9N	2,6-Dimethylpyridine	293.2	7.33	0.17714E+02	−0.39080E-01	0.12500E-04	293-333
C_7H_9NO	2,6-Dimethylpyridine-1-oxide	298.2	46.11	0.22765E+03	−0.90760E+00	0.10011E-02	298-398
C_7H_9NO	*o*-Methoxyaniline	303.2	5.230	0.79911E+01	−0.92183E-02	0.37879E-06	303-393
C_7H_9NO	*m*-Methoxyaniline	298.2	8.76	0.28179E+02	−0.97840E-01	0.11027E-03	289-393
C_7H_9NO	*p*-Methoxyaniline	333.2	7.85	0.30149E+02	−0.10523E+00	0.11467E-03	333-453
$C_7H_{10}N_2$	1-Methyl-1-phenylhydrazine	292.2	7.3				
$C_7H_{11}Cl_3O_2$	Isopentyl trichloroacetate	293.2	7.287				
C_7H_{12}	1,6-Heptadiene	293.0	2.161	0.30815E+01	−0.36095E-02	0.16354E-05	184-293
C_7H_{12}	Cycloheptene	295.0	2.265	0.32309E+01	−0.42373E-02	0.32572E-05	227-363
$C_7H_{12}O$	Cycloheptanone	298.2	13.16	0.17511E+03	−0.11221E+01	0.19417E-02	258-298
$C_7H_{12}O$	2-Methylcyclohexanone	293.2	14.0				
$C_7H_{12}O$	3-Methylcyclohexanone	293.2	12.4				
$C_7H_{12}O$	4-Methylcyclohexanone	293.2	12.35				
$C_7H_{12}O_2$	Cyclohexanecarboxylic acid	304.2	2.67				
$C_7H_{12}O_2$	Cyclohexyl formate	293.2	6.47				
$C_7H_{12}O_2$	Butyl acrylate	301.2	5.25	0.38296E+02	−0.19109E+00	0.27006E-03	301-343
$C_7H_{12}O_4$	Monomethyl adipate	293.2	6.69	0.11962E+02	−0.23973E-01	0.20608E-04	293-433
$C_7H_{12}O_4$	Diethyl malonate	304.2	7.550	0.14809E+02	−0.31207E-01	0.24066E-04	304-393
$C_7H_{12}O_4$	Dimethyl glutarate	293.2	7.87	0.20697E+02	−0.57794E-01	0.48405E-04	293-433
$C_7H_{12}O_5$	1,2,3-Propanetriol-1,3-diacetate	288.2	9.80	0.28321E+02	−0.89073E-01	0.86891E-04	258-374
C_7H_{14}	1-Heptene	293.2	2.092	0.21755E+01	0.13896E-02	−0.57049E-05	273-323
C_7H_{14}	2-Methyl-2-hexene	293.2	2.962				
C_7H_{14}	3-Ethyl-2-pentene	293.2	2.051				
C_7H_{14}	Cycloheptane	293.2	2.0784	0.25136E+01	−0.15089E-02	0.84915E-07	278-333
C_7H_{14}	Methylcyclohexane	293.2	2.024				
$C_7H_{14}Br_2$	1,2-Dibromoheptane	298.2	3.77				
$C_7H_{14}Br_2$	2,3-Dibromoheptane	298.2	5.08				
$C_7H_{14}Br_2$	3,4-Dibromoheptane	298.2	4.70				
$C_7H_{14}Cl_2$	1,7-Dichloroheptane	298.2	8.34				
$C_7H_{14}O$	1-Heptanal	295.2	9.07				
$C_7H_{14}O$	2-Heptanone	293.2	11.95	0.38348E+02	−0.12531E+00	0.12005E-03	253-413
$C_7H_{14}O$	3-Heptanone	293.2	12.7				
$C_7H_{14}O$	4-Heptanone	293.2	12.60	0.41520E+02	−0.13839E+00	0.13497E-03	253-393
$C_7H_{14}O$	5-Methyl-2-hexanone	293.2	13.53	0.52353E+02	−0.17695E+00	0.15195E-03	293-333
$C_7H_{14}O$	Cyclohexanemethanol	333.2	9.70	0.10164E+03	−0.45839E+00	0.54762E-03	333-368
$C_7H_{14}O$	2-Methylcyclohexanol*	293.2	9.375	0.17315E+03	−0.98794E+00	0.14634E-02	273-323
$C_7H_{14}O$	3-Methylcyclohexanol*	293.2	13.79	0.65896E+02	−0.21954E+00	0.14107E-03	273-323
$C_7H_{14}O$	4-Methylcyclohexanol*	293.2	13.45	0.65021E+02	−0.22896E+00	0.17946E-03	273-323
$C_7H_{14}O_2$	Heptanoic acid	288.2	3.04	0.36423E+01	−0.31996E-02	0.39362E-05	288-423
$C_7H_{14}O_2$	Pentyl acetate	293.2	4.79	0.12091E+02	−0.36536E-01	0.39732E-04	253-353
$C_7H_{14}O_2$	Isopentyl acetate	293.2	4.72				
$C_7H_{14}O_2$	Butyl propanoate	293.2	4.838				
$C_7H_{14}O_2$	Propyl butanoate	293.2	4.3				
$C_7H_{14}O_2$	Ethyl pentanoate	291.2	4.71				
$C_7H_{14}O_2$	Ethyl 3-methylbutanoate	293.2	4.71				
$C_7H_{14}O_2$	Methyl hexanoate	293.2	4.615				
$C_7H_{15}Br$	1-Bromoheptane	303.2	5.255	0.15289E+02	−0.50621E-01	0.57753E-04	203-343
$C_7H_{15}Br$	2-Bromoheptane	295.2	6.46				
$C_7H_{15}Br$	4-Bromoheptane	295.2	6.81				
$C_7H_{15}Cl$	1-Chloroheptane	293.2	5.521	0.14279E+02	−0.39431E-01	0.32321E-04	273-323
$C_7H_{15}Cl$	2-Chloroheptane	295.2	6.52				
$C_7H_{15}Cl$	3-Chloroheptane	295.2	6.70				
$C_7H_{15}Cl$	4-Chloroheptane	295.2	6.54				
$C_7H_{15}I$	1-Iodoheptane	298.2	4.92	0.11856E+02	−0.33493E-01	0.34368E-04	294-323
$C_7H_{15}I$	3-Iodoheptane	295.2	6.39				
C_7H_{16}	Heptane	293.2	1.9209	0.24740E+01	−0.22577E-02	0.12428E-05	273-373
C_7H_{16}	2-Methylhexane	293.2	1.9221	0.24759E+01	−0.22535E-02	0.12500E-05	293-323
C_7H_{16}	3-Methylhexane	293.2	1.920	0.27089E+01	−0.37908E-02	0.37500E-05	273-323
C_7H_{16}	3-Ethylpentane	293.2	1.942	0.23771E+01	−0.15140E-02	0.10093E-06	163-363

Mol. form.	Name	T/K	ε_r	a	b	c	Range/K
C$_7$H$_{16}$	2,2-Dimethylpentane	293.2	1.915	0.23414E+01	−0.14362E-02	−0.51322E-07	153-353
C$_7$H$_{16}$	2,3-Dimethylpentane	293.2	1.929	0.25637E+01	−0.26328E-02	0.16071E-05	273-323
C$_7$H$_{16}$	2,4-Dimethylpentane	293.2	1.902	0.23979E+01	−0.17436E-02	0.17857E-06	273-323
C$_7$H$_{16}$	3,3-Dimethylpentane	291.3	1.9419	0.24007E+01	−0.16802E-02	0.36069E-06	291-322
C$_7$H$_{16}$	2,2,3-Trimethylbutane	293.2	1.930				
C$_7$H$_{16}$O	1-Heptanol	293.2	11.75	0.60662E+02	−0.24049E+00	0.25155E-03	239-513
C$_7$H$_{16}$O	2-Heptanol	293.7	9.72	0.10050E+03	−0.49793E+00	0.64504E-03	207-365
C$_7$H$_{16}$O	3-Heptanol	296.1	7.07	0.19586E+03	−0.11465E+01	0.17175E-02	248-349
C$_7$H$_{16}$O	4-Heptanol	296.2	6.18	0.28995E+03	−0.18499E+01	0.30109E-02	270-301
C$_7$H$_{16}$O	2-Methyl-2-hexanol	297.0	3.257				
C$_7$H$_{16}$O	3-Methyl-2-hexanol	297.2	4.990	0.59724E+02	−0.32417E+00	0.47058E-03	244-372
C$_7$H$_{16}$O	3-Methyl-3-hexanol	298.2	3.248				
C$_7$H$_{16}$O	3-Ethyl-3-pentanol	293.2	3.158				
C$_7$H$_{16}$O	2,2-Dimethyl-1-pentanol	293.2	6.020	0.37318E+02	−0.17095E+00	0.22022E-03	283-393
C$_7$H$_{16}$O	Ethyl pentyl ether	296.2	3.6				
C$_7$H$_{16}$O	Ethyl isopentyl ether	293.2	3.955	0.66541E+01	−0.55450E-02	−0.12500E-04	293-323
C$_7$H$_{16}$O$_3$	Triethoxymethane	293.2	4.779				
C$_7$H$_{16}$S	1-Heptanethiol	293.2	4.194	0.71333E+01	−0.97320E-02	−0.12500E-05	273-333
C$_7$H$_{17}$N	Heptylamine	293.2	3.81	0.87794E+01	−0.24363E-01	0.25325E-04	253-373
C$_7$H$_{18}$O$_3$Si	Triethoxymethylsilane	298.2	3.845				
C$_8$H$_4$F$_6$	1,3-Bis(trifluoromethyl)benzene	303.2	5.98				
C$_8$H$_6$	Phenylacetylene	298.2	2.98				
C$_8$H$_6$Cl$_2$	2,5-Dichlorostyrene	298.2	2.58				
C$_8$H$_6$Cl$_4$	1,2,3,4-Tetrachloro-5,6-dimethylbenzene	293.2	8.0				
C$_8$H$_6$Cl$_4$	1,2,3,5-Tetrachloro-4,6-dimethylbenzene	293.2	5.4				
C$_8$H$_6$O	Phenoxyacetylene	298.2	4.76				
C$_8$H$_7$N	Benzeneacetonitrile	299.2	17.87	0.82175E+02	−0.37416E+00	0.53220E-03	299-343
C$_8$H$_7$NO$_2$	4-Methoxyphenyl isocyanate	333.2	10.26	0.20780E+02	−0.31571E-01		333-403
C$_8$H$_7$NO$_4$	Methyl 2-nitrobenzoate	300.1	27.76				
C$_8$H$_8$	Styrene	293.2	2.4737	0.44473E+01	−0.11422E-01	0.16000E-04	293-313
C$_8$H$_8$O	Acetophenone	298.2	17.44	0.26099E+02	0.64048E-02	−0.11905E-03	298-333
C$_8$H$_8$O$_2$	Benzeneacetic acid	353.2	3.47	0.24104E+01	0.30000E-02		353-393
C$_8$H$_8$O$_2$	Benzyl formate	303.2	6.34	0.26162E+02	−0.11026E+00	0.14787E-03	303-358
C$_8$H$_8$O$_2$	Phenyl acetate	298.2	5.403	0.11327E+02	−0.26707E-01	0.22938E-04	298-404
C$_8$H$_8$O$_2$	Methyl benzoate	302.7	6.642	0.17486E+02	−0.51027E-01	0.50222E-04	303-393
C$_8$H$_8$O$_2$	(Hydroxyacetyl)benzene	298.2	21.33	0.42286E+02	−0.69215E-01	−0.35714E-05	298-368
C$_8$H$_8$O$_2$	4-Methoxybenzaldehyde	303.2	22.0				
C$_8$H$_8$O$_3$	Methyl salicylate	314.4	8.80	0.20501E+02	−0.39045E-01	0.68298E-05	223-398
C$_8$H$_9$Br	1-Bromo-2-ethylbenzene	298.2	5.55				
C$_8$H$_9$Br	1-Bromo-3-ethylbenzene	298.2	5.56				
C$_8$H$_9$Br	1-Bromo-4-ethylbenzene	298.2	5.42				
C$_8$H$_9$BrO	1-Bromo-2-ethoxybenzene	313.2	7.04	0.23146E+02	−0.75753E-01	0.77778E-04	313-358
C$_8$H$_9$Cl	1-Chloro-2-ethylbenzene	298.2	4.36				
C$_8$H$_9$Cl	1-Chloro-3-ethylbenzene	298.2	5.18				
C$_8$H$_9$Cl	1-Chloro-4-ethylbenzene	298.2	5.16				
C$_8$H$_9$NO$_2$	1-Ethyl-2-nitrobenzene	273.4	21.9				
C$_8$H$_9$NO$_2$	Methyl 2-aminobenzoate	298.2	21.9				
C$_8$H$_9$NO$_2$	Ethyl 4-pyridinecarboxylate	293.2	8.95				
C$_8$H$_{10}$	Ethylbenzene	293.2	2.4463	0.35969E+01	−0.53169E-02	0.47500E-05	293-323
C$_8$H$_{10}$	o-Xylene	293.2	2.562	0.36163E+01	−0.40177E-02	0.14286E-05	273-323
C$_8$H$_{10}$	m-Xylene	293.2	2.359	0.28421E+01	−0.10191E-02	−0.21429E-05	273-323
C$_8$H$_{10}$	p-Xylene	293.2	2.2735	0.23140E+01	0.97221E-03	−0.37500E-05	293-363
C$_8$H$_{10}$O	2,3-Xylenol	343.2	4.81	0.14399E+02	−0.41438E-01	0.39244E-04	343-433
C$_8$H$_{10}$O	2,4-Xylenol	303.2	5.060	0.22125E+02	−0.85543E-01	0.96548E-04	303-363
C$_8$H$_{10}$O	2,5-Xylenol	338.2	5.36	0.18049E+02	−0.54991E-01	0.51656E-04	338-455
C$_8$H$_{10}$O	2,6-Xylenol	313.2	4.90	0.12284E+02	−0.32996E-01	0.29867E-04	313-453
C$_8$H$_{10}$O	3,4-Xylenol	333.2	9.02	0.54423E+02	−0.21153E+00	0.22508E-03	333-453
C$_8$H$_{10}$O	3,5-Xylenol	323.2	9.06	0.54251E+02	−0.21647E+00	0.23542E-03	323-453
C$_8$H$_{10}$O	Benzeneethanol	293.2	12.31	0.12170E+03	−0.63124E+00	0.87776E-03	278-333
C$_8$H$_{10}$O	1-Phenylethanol	293.2	8.77	0.32971E+02	−0.12042E+00	0.12809E-03	293-423
C$_8$H$_{10}$O	Phenetole	293.2	4.216	−0.15043E+02	0.13752E+00	−0.24500E-03	293-313

Mol. form.	Name	T/K	ε_r	a	b	c	Range/K
$C_8H_{10}O$	2-Methylanisole	293.2	3.502	0.50825E+01	−0.62297E-02	0.28571E-05	293-333
$C_8H_{10}O$	3-Methylanisole	293.2	3.967	0.12830E+02	−0.49701E-01	0.66429E-04	293-333
$C_8H_{10}O$	4-Methylanisole	293.2	3.914	0.86608E+01	−0.23510E-01	0.25000E-04	293-333
$C_8H_{10}O_2$	1,2-Dimethoxybenzene	293.2	4.45	0.74604E+01	−0.13445E-01	0.10737E-04	293-443
$C_8H_{10}O_2$	1,3-Dimethoxybenzene	298.2	5.363	0.11911E+02	−0.30804E-01	0.29643E-04	298-358
$C_8H_{10}O_2$	1,4-Dimethoxybenzene	333.7	5.60	0.11289E+02	−0.20765E-01	0.11987E-04	334-463
$C_8H_{10}O_2S$	Ethyl phenyl sulfone	348.2	39.0				
$C_8H_{10}S$	(Ethylthio)benzene	298.2	4.95				
$C_8H_{11}N$	p-Ethylaniline	298.2	4.84				
$C_8H_{11}N$	N-Ethylaniline	293.2	5.87				
$C_8H_{11}N$	N,N-Dimethylaniline	298.2	4.90	0.84052E+01	−0.13549E-01	0.62835E-05	289-453
$C_8H_{11}N$	2,4,6-Trimethylpyridine	298.2	7.807	0.20990E+02	−0.57419E-01	0.44286E-04	298-358
$C_8H_{11}NO$	4-Ethoxyaniline	298.2	7.43				
$C_8H_{12}N_2O_2$	Hexamethylene diisocyanate	288.2	14.41	0.26715E+02	−0.42696E-01		288-403
$C_8H_{12}O_4$	Diethyl maleate	298.2	7.560	0.13953E+02	−0.21969E-01	0.17817E-05	298-343
$C_8H_{12}O_4$	Diethyl fumarate	296.2	6.56				
C_8H_{14}	1,7-Octadiene	293.0	2.186	0.28376E+01	−0.17442E-02	−0.16141E-05	214-293
C_8H_{14}	cis-Cyclooctene	296.0	2.306	0.31115E+01	−0.32058E-02	0.16713E-05	269-406
C_8H_{14}	1,2-Dimethylcyclohexene	296.0	2.144	0.26443E+01	−0.17973E-02	0.35815E-06	211-374
C_8H_{14}	1,3-Dimethylcyclohexene	296.0	2.182	0.29951E+01	−0.34615E-02	0.24026E-05	213-373
$C_8H_{14}O_2$	Methyl cyclohexanecarboxylate	293.2	4.87				
$C_8H_{14}O_2$	Cyclohexyl acetate	293.2	5.08				
$C_8H_{14}O_3$	Butanoic anhydride	293.2	12.8				
$C_8H_{14}O_3$	2-Methylpropanoic anhydride	292.2	13.6				
$C_8H_{14}O_4$	Diisopropyl oxalate	293.2	6.403	0.10709E+02	−0.16328E-01	0.56000E-05	293-368
$C_8H_{14}O_4$	Diethyl succinate	293.2	6.098	0.80213E+01	0.11810E-02	−0.26400E-04	293-343
$C_8H_{14}O_4$	Dimethyl adipate	293.2	6.84	0.11739E+02	−0.17281E-01	0.11447E-05	293-433
$C_8H_{15}N$	Octanenitrile	293.2	13.90				
C_8H_{16}	1-Octene	293.2	2.113	0.24348E+01	0.34200E-03	−0.50000E-05	273-323
C_8H_{16}	cis-3-Octene	298.2	2.062				
C_8H_{16}	trans-3-Octene	298.2	2.002				
C_8H_{16}	cis-4-Octene	298.2	2.053				
C_8H_{16}	trans-4-Octene	298.2	2.004				
C_8H_{16}	3-Methyl-2-heptene*	293.2	2.436				
C_8H_{16}	2,5-Dimethyl-2-hexene	293.2	2.431				
C_8H_{16}	2,4,4-Trimethyl-1-pentene	298.2	2.0908				
C_8H_{16}	Cyclooctane	295.0	2.116	0.25036E+01	−0.12460E-02	−0.23175E-06	295-411
$C_8H_{16}Br_2$	1,8-Dibromooctane	298.2	7.43	0.94117E+00	0.61520E-01	−0.13333E-03	298-328
$C_8H_{16}Cl_2$	1,8-Dichlorooctane	298.2	7.64				
$C_8H_{16}O$	2-Octanone	293.2	9.51	−0.16219E+02	0.18799E+00	−0.34156E-03	293-333
$C_8H_{16}O$	3-Octanone	303.2	10.50				
$C_8H_{16}O_2$	Octanoic acid	288.2	2.85	0.29391E+01	−0.38721E-03		288-423
$C_8H_{16}O_2$	2-Ethylhexanoic acid	296.2	2.64				
$C_8H_{16}O_2$	Hexyl acetate	293.2	4.42				
$C_8H_{16}O_2$	Pentyl propanoate	293.2	4.552				
$C_8H_{16}O_2$	Isopentyl propanoate	273.2	5.21	0.17665E+02	−0.71718E-01	0.95635E-04	273-373
$C_8H_{16}O_2$	Butyl butanoate	298.2	4.39	0.79684E+01	−0.12000E-01	0.15266E-13	298-318
$C_8H_{16}O_2$	Propyl pentanoate	292.2	4.0				
$C_8H_{16}O_2$	Ethyl hexanoate	293.2	4.45	0.11007E+02	−0.32800E-01	0.35714E-04	253-353
$C_8H_{16}O_2$	Methyl heptanoate	293.2	4.355				
$C_8H_{16}O_3$	Isopentyl lactate	273.2	11.2	0.48649E+02	−0.21253E+00	0.27619E-03	273-373
$C_8H_{17}Br$	1-Bromooctane	293.2	5.0957	0.12404E+02	−0.35050E-01	0.34542E-04	283-353
$C_8H_{17}Br$	2-Bromooctane	293.2	5.44				
$C_8H_{17}Cl$	1-Chlorooctane	298.2	5.05	0.11346E+02	−0.25120E-01	0.13450E-04	274-328
$C_8H_{17}Cl$	2-Chlorooctane	293.2	5.42				
$C_8H_{17}F$	1-Fluorooctane	293.2	3.89				
$C_8H_{17}I$	1-Iodooctane	293.2	4.67	0.12452E+02	−0.41229E-01	0.50108E-04	233-313
$C_8H_{17}NO_2$	1-Nitrooctane	293.2	11.46				
C_8H_{18}	Octane	293.2	1.948	0.22590E+01	−0.84212E-03	−0.75758E-06	233-393
C_8H_{18}	2-Methylheptane	293.2	1.9519				
C_8H_{18}	3-Ethylhexane	293.2	1.9617				
C_8H_{18}	2,2-Dimethylhexane	293.2	1.9498				
C_8H_{18}	2,5-Dimethylhexane	293.95	1.9619	0.25821E+01	−0.26804E-02	0.19404E-05	294-324

Mol. form.	Name	T/K	ε_r	a	b	c	Range/K
C_8H_{18}	3,3-Dimethylhexane	293.2	1.9645				
C_8H_{18}	3,4-Dimethylhexane	292.1	1.9814	0.26849E+01	−0.33712E-02	0.32949E-05	292-324
C_8H_{18}	3-Ethyl-3-methylpentane	291.49	1.9869	0.25983E+01	−0.28027E-02	0.24195E-05	292-324
C_8H_{18}	2,2,3-Trimethylpentane	293.2	1.960				
C_8H_{18}	2,2,4-Trimethylpentane	293.2	1.943	0.23677E+01	−0.14768E-02	0.94261E-07	173-373
C_8H_{18}	2,3,3-Trimethylpentane	293.2	1.9780				
C_8H_{18}	2,3,4-Trimethylpentane	293.2	1.9738				
$C_8H_{18}O$	1-Octanol	293.2	10.30	0.51647E+02	−0.20371E+00	0.21320E-03	258-513
$C_8H_{18}O$	2-Octanol	293.2	8.13	0.63760E+02	−0.27643E+00	0.31075E-03	213-513
$C_8H_{18}O$	3-Octanol	293.2	5.55	0.12505E+03	−0.70646E+00	0.10245E-02	223-383
$C_8H_{18}O$	4-Octanol	293.2	4.48	0.51049E+02	−0.26664E+00	0.37280E-03	243-403
$C_8H_{18}O$	2-Methyl-1-heptanol	293.1	5.16	0.61698E+02	−0.33647E+00	0.49066E-03	236-328
$C_8H_{18}O$	3-Methyl-1-heptanol	290.3	2.884	0.84687E+01	−0.33712E-01	0.49793E-04	241-316
$C_8H_{18}O$	4-Methyl-1-heptanol	290.6	4.63	0.48612E+02	−0.26773E+00	0.39972E-03	237-332
$C_8H_{18}O$	5-Methyl-1-heptanol	290.4	7.68	0.54581E+02	−0.24772E+00	0.29734E-03	235-328
$C_8H_{18}O$	6-Methyl-1-heptanol	290.3	10.54	0.57997E+02	−0.23517E+00	0.24663E-03	265-328
$C_8H_{18}O$	2-Methyl-2-heptanol	292.2	3.43				
$C_8H_{18}O$	3-Methyl-2-heptanol	289.6	7.47	0.39178E+02	−0.17976E+00	0.24218E-03	229-329
$C_8H_{18}O$	4-Methyl-2-heptanol	290.0	3.59	0.39715E+02	−0.23115E+00	0.36771E-03	240-333
$C_8H_{18}O$	5-Methyl-2-heptanol	278.5	7.5	0.68568E+02	−0.40706E+00	0.67433E-03	230-279
$C_8H_{18}O$	6-Methyl-2-heptanol	290.1	6.41	0.77520E+02	−0.41724E+00	0.59448E-03	239-329
$C_8H_{18}O$	2-Methyl-3-heptanol	293.2	3.260	−0.59739E+01	0.56700E-01	−0.83125E-04	343-403
$C_8H_{18}O$	3-Methyl-3-heptanol	293.2	3.013	−0.38440E+01	0.42327E-01	−0.61250E-04	343-403
$C_8H_{18}O$	4-Methyl-3-heptanol	293.2	3.312	−0.48003E+01	0.50740E-01	−0.75000E-04	343-403
$C_8H_{18}O$	5-Methyl-3-heptanol	293.2	3.832	0.61967E+01	−0.63750E-02		343-383
$C_8H_{18}O$	6-Methyl-3-heptanol	293.2	4.992	0.23037E+02	−0.98029E-01	0.12479E-03	283-383
$C_8H_{18}O$	2-Methyl-4-heptanol	296.3	3.338	0.42102E+00	0.10427E-01	−0.20438E-05	230-333
$C_8H_{18}O$	3-Methyl-4-heptanol	290.0	7.46	0.33354E+02	−0.14077E+00	0.17750E-03	230-330
$C_8H_{18}O$	4-Methyl-4-heptanol	296.2	2.902				
$C_8H_{18}O$	2-Ethyl-1-hexanol	298.2	7.58	0.86074E+02	−0.42636E+00	0.55078E-03	208-318
$C_8H_{18}O$	2,2-Dimethyl-1-hexanol	293.2	4.50	0.91244E+01	−0.21785E-01	0.21018E-04	283-393
$C_8H_{18}O$	Dibutyl ether	293.2	3.0830	0.65383E+01	−0.16172E-01	0.14969E-04	293-314
$C_8H_{18}OS$	Dibutyl sulfoxide	313.2	24.73	0.67156E+02	−0.16448E+00	0.92275E-04	313-393
$C_8H_{18}O_2$	2-Ethyl-1,3-hexanediol	293.2	18.73	0.57919E+02	−0.17128E+00	0.12949E-03	233-333
$C_8H_{18}O_2S$	Dibutyl sulfone	323.2	25.72	0.66248E+02	−0.16417E+00	0.12001E-03	323-398
$C_8H_{18}O_4$	Triethylene glycol dimethyl ether	298.2	7.62				
$C_8H_{18}O_5$	Tetraethylene glycol	293.2	20.44	0.83547E+02	−0.31691E+00	0.34689E-03	253-333
$C_8H_{18}S$	1-Octanethiol	293.2	3.949	0.63667E+01	−0.87920E-02	0.18750E-05	273-333
$C_8H_{18}S$	Dibutyl sulfide	298.2	4.29				
$C_8H_{19}N$	Octylamine	293.2	3.58	0.77931E+01	−0.20015E-01	0.19347E-04	273-373
$C_8H_{19}N$	Dibutylamine	293.2	2.765	0.52504E+01	−0.10538E-01	0.71485E-05	243-323
$C_8H_{20}O_4Si$	Ethyl silicate	293.2	2.50				
$C_8H_{20}Si$	Tetraethylsilane	293.2	2.090				
$C_8H_{20}Sn$	Tetraethylstannane	293.2	2.241				
$C_8H_{23}N_5$	Tetraethylenepentamine	293.2	9.40	0.40553E+02	−0.16681E+00	0.20659E-03	213-333
$C_8H_{24}O_4Si_4$	Octamethylcyclotetrasiloxane	296.2	2.390	0.36286E+01	−0.56885E-02	0.50874E-05	296-333
$C_9H_6N_2O_2$	Toluene-2,4-diisocyanate	293.2	8.433	0.22174E+02	−0.66982E-01	0.68571E-04	293-353
$C_9H_6O_2$	2H-1-Benzopyran-2-one	343.2	34.04	0.11311E+03	−0.33804E+00	0.31324E-03	343-423
C_9H_7N	Quinoline	293.2	9.16	0.33432E+02	−0.13497E+00	0.17788E-03	258-323
C_9H_7N	Isoquinoline	298.2	11.0	0.14412E+03	−0.79935E+00	0.11839E-02	298-323
C_9H_8O	Cinnamaldehyde	305.8	17.72	0.41837E+02	−0.11060E+00	0.10401E-03	306-354
$C_9H_8O_4$	2-(Acetyloxy)benzoic acid	333.2	6.55	0.69994E+01	−0.14553E-02		333-416
C_9H_{10}	1-Propenylbenzene	293.2	2.73				
C_9H_{10}	Allylbenzene	293.2	2.63				
C_9H_{10}	Isopropenylbenzene	293.2	2.28				
$C_9H_{10}OS$	4-Acetylthioanisole	355.2	11.34				
$C_9H_{10}O_2$	Ethyl benzoate	293.2	6.20	0.18216E+02	−0.62361E-01	0.72884E-04	288-343
$C_9H_{10}O_2$	Methyl 4-methylbenzoate	306.2	4.3				
$C_9H_{10}O_2$	Benzyl acetate	303.2	5.34	0.11727E+02	−0.30869E-01	0.32340E-04	303-358
$C_9H_{10}O_2$	Phenyl propanoate	293.2	4.77				
$C_9H_{10}O_2$	4-Acetylanisole	313.2	17.3				
$C_9H_{10}O_3$	Ethyl salicylate	308.2	8.48	0.18910E+02	−0.35623E-01	0.46529E-05	225-321
$C_9H_{10}O_3$	Methyl 2-methoxybenzoate	294.2	7.7				

Mol. form.	Name	T/K	ε_r	a	b	c	Range/K
C₉H₁₁Br	(3-Bromopropyl)benzene	302.2	5.41	0.11360E+02	−0.27471E−01	0.25775E−04	302-358
C₉H₁₁NO	N-Ethylbenzamide	352.7	42.6	−0.20109E+03	0.17866E+01	−0.31065E−02	353-389
C₉H₁₁NO	N,N-Dimethylbenzamide	318.2	20.77	0.76725E+02	−0.26908E+00	0.29409E−03	318-443
C₉H₁₁NO₂	Ethyl 2-aminobenzoate	298.2	4.14				
C₉H₁₂	Propylbenzene	293.2	2.370	0.26933E+01	0.21679E−03	−0.44643E−05	273-323
C₉H₁₂	Isopropylbenzene	293.2	2.381	0.31149E+01	−0.30801E−02	0.19643E−05	273-323
C₉H₁₂	o-Ethyltoluene	293.2	2.595				
C₉H₁₂	m-Ethyltoluene	293.2	2.365				
C₉H₁₂	p-Ethyltoluene	293.2	2.265				
C₉H₁₂	1,2,3-Trimethylbenzene	293.2	2.656	0.76006E+01	−0.29118E−01	0.41786E−04	273-323
C₉H₁₂	1,2,4-Trimethylbenzene	293.2	2.377	0.31517E+01	−0.30634E−02	0.14286E−05	273-323
C₉H₁₂	1,3,5-Trimethylbenzene	293.2	2.279	0.38998E+01	−0.88072E−02	0.11149E−04	288-358
C₉H₁₂O	Benzenepropanol	293.2	11.97	0.94482E+02	−0.45540E+00	0.59307E−03	213-303
C₉H₁₂O	α-Ethylbenzenemethanol	293.2	6.68	0.44520E+02	−0.21505E+00	0.29443E−03	233-373
C₉H₁₂O	α,α-Dimethylbenzenemethanol	303.2	5.61	0.57072E+01	0.86568E−02	−0.29580E−04	303-373
C₉H₁₂O	1-Phenyl-2-propanol	293.2	9.35	0.10762E+03	−0.56026E+00	0.76915E−03	233-373
C₉H₁₂O	Benzyl ethyl ether	298.2	3.90				
C₉H₁₂O	2,6-Dimethylanisole	293.2	3.780	0.76700E+01	−0.18298E−01	0.17143E−04	293-333
C₉H₁₂O	3,5-Dimethylanisole	293.2	3.711	0.54981E+01	−0.56651E−02	−0.14286E−05	293-333
C₉H₁₂O₂S	Butyl thiophene-2-carboxylate	293.2	6.40				
C₉H₁₂S	Benzenepropanethiol	303.2	4.36	0.82411E+01	−0.15034E−01	0.73617E−05	303-358
C₉H₁₃N	Benzylethylamine	293.2	4.3				
C₉H₁₃N	N-Propylaniline	293.2	5.48				
C₉H₁₃N	2-Methyl-N,N-dimethylaniline	293.2	3.4				
C₉H₁₃N	4-Methyl-N,N-dimethylaniline	293.2	3.9				
C₉H₁₄OSi	Trimethylphenoxysilane	298.2	3.3953				
C₉H₁₄O₆	Triacetin	293.6	7.11	0.17819E+02	−0.53656E−01	0.57759E−04	219-304
C₉H₁₄Si	Trimethylphenylsilane	298.2	2.3533	0.21463E+01	0.32711E−02	−0.86264E−05	288-323
C₉H₁₆O₂	2-Nonenoic acid	296.2	2.5				
C₉H₁₆O₂	Cyclohexyl propanoate	293.2	4.82				
C₉H₁₆O₂	Ethyl cyclohexanecarboxylate	293.2	4.64				
C₉H₁₆O₄	Diethyl glutarate	303.2	6.659				
C₉H₁₇N	Nonanenitrile	293.2	12.08				
C₉H₁₈	1-Nonene	293.2	2.180	0.22710E+01	0.15797E−02	−0.64286E−05	273-323
C₉H₁₈Br₂	1,9-Dibromononane	293.2	7.153	0.18931E+02	−0.57764E−01	0.60000E−04	293-343
C₉H₁₈O	2-Nonanone	295.2	9.14				
C₉H₁₈O	5-Nonanone	293.2	10.6				
C₉H₁₈O	Di-tert-butyl ketone	287.65	10.0				
C₉H₁₈O	2,6-Dimethyl-4-heptanone	293.2	9.91	0.33178E+02	−0.11290E+00	0.11454E−03	273-393
C₉H₁₈O₂	Nonanoic acid	294.9	2.475	0.25039E+01	0.67274E−03	−0.24180E−05	295-365
C₉H₁₈O₂	2-Methyloctanoic acid	293.2	2.39				
C₉H₁₈O₂	2-Ethylheptanoic acid	293.2	1.98				
C₉H₁₈O₂	Heptyl acetate	293.2	4.2				
C₉H₁₈O₂	Pentyl butanoate	301.2	4.08	0.59029E+01	−0.49905E−02	−0.34292E−05	301-343
C₉H₁₈O₂	Isopentyl butanoate	293.2	4.0				
C₉H₁₈O₂	Isobutyl pentanoate	292.2	3.8				
C₉H₁₈O₂	Methyl octanoate	293.2	4.101				
C₉H₁₉Br	1-Bromononane	298.2	4.74	0.79870E+01	−0.10488E−01	−0.13450E−05	274-328
C₉H₁₉Cl	1-Chlorononane	293.2	4.803	0.95528E+01	−0.16200E−01	−0.16365E−13	293-323
C₉H₁₉NO	N,N-Dibutylformamide	293.2	18.4				
C₉H₂₀	Nonane	293.2	1.9722	0.23894E+01	−0.14830E−02	0.14881E−06	253-393
C₉H₂₀	2-Methyloctane	293.2	1.967				
C₉H₂₀	4-Methyloctane	293.2	1.967				
C₉H₂₀	2,4-Dimethylheptane	293.2	1.89				
C₉H₂₀	2,5-Dimethylheptane	293.2	1.89				
C₉H₂₀	2,6-Dimethylheptane	293.2	1.987				
C₉H₂₀N₂O	Tetraethylurea	296.8	14.29	0.52820E+02	−0.18790E+00	0.19580E−03	205-411
C₉H₂₀O	1-Nonanol	293.2	8.83	0.97467E+02	−0.51103E+00	0.71429E−03	288-343
C₉H₂₀O	2-Nonanol	298.2	6.66	0.10136E+03	−0.55612E+00	0.80000E−03	288-308
C₉H₂₀O	3-Nonanol	298.2	4.49	0.55214E+02	−0.31920E+00	0.50000E−03	288-308
C₉H₂₀O	4-Nonanol	298.2	3.69	0.27954E+01	0.30000E−02	−0.52375E−13	288-308
C₉H₂₀O	5-Nonanol	298.2	3.54	−0.25463E+01	0.35320E−01	−0.50000E−04	288-308
C₉H₂₁B	Tripropylborane	293.2	2.026				

Mol. form.	Name	T/K	ε_r	a	b	c	Range/K
C$_9$H$_{21}$N	Nonylamine	293.2	3.42	0.53575E+01	−0.71982E-02	0.19481E-05	293-373
C$_9$H$_{21}$N	Tripropylamine	293.2	2.380	0.33380E+01	−0.86332E-02	0.18322E-04	243-293
C$_9$H$_{21}$O$_4$P	Tripropyl phosphate	293.2	10.93	0.33166E+02	−0.10514E+00	0.10000E-03	293-373
C$_{10}$H$_7$Br	1-Bromonaphthalene	298.2	4.768	0.10561E+02	−0.27671E-01	0.27655E-04	293-323
C$_{10}$H$_7$Cl	1-Chloronaphthalene	298.2	5.04	0.84861E+01	−0.12357E-01	0.26899E-05	274-328
C$_{10}$H$_7$NO$_2$	1-Nitronaphthalene	333.2	19.68	0.36267E+02	−0.41283E-01	−0.25595E-04	333-403
C$_{10}$H$_8$	Naphthalene	363.2	2.54				
C$_{10}$H$_8$O	1-Naphthol	373.0	5.03	0.16489E+02	−0.46700E-01	0.42857E-04	373-453
C$_{10}$H$_8$O	2-Naphthol	413.0	4.95	0.92865E+01	−0.10500E-01	0.42501E-15	413-453
C$_{10}$H$_9$N	1-Naphthylamine	333.2	5.20	0.10577E+02	−0.22114E-01	0.17857E-04	333-453
C$_{10}$H$_9$N	2-Naphthylamine	393.0	5.26	0.19722E+02	−0.60679E-01	0.60714E-04	393-473
C$_{10}$H$_9$N	2-Methylquinoline	293.2	7.24	0.11688E+02	−0.78400E-02	−0.25000E-04	293-333
C$_{10}$H$_9$N	4-Methylquinoline	293.2	9.31	0.17788E+02	−0.32580E-01	0.12500E-04	293-333
C$_{10}$H$_9$N	6-Methylquinoline	293.2	8.48	0.21696E+02	−0.63400E-01	0.62500E-04	293-333
C$_{10}$H$_9$N	8-Methylquinoline	293.2	6.58	0.19356E+02	−0.61900E-01	0.62500E-04	293-333
C$_{10}$H$_{10}$O$_4$	Methyl 2-(acetyloxy)benzoate	328.9	5.31	0.19579E+02	−0.69970E-01	0.80889E-04	329-371
C$_{10}$H$_{10}$O$_4$	Dimethyl phthalate	293.2	8.66				
C$_{10}$H$_{12}$	1,2,3,4-Tetrahydronaphthalene	298.2	2.771	0.29172E+01	0.12832E-02	−0.59453E-05	298-343
C$_{10}$H$_{12}$	4-Ethylstyrene	298.2	3.350				
C$_{10}$H$_{12}$	Dicyclopentadiene	313.2	2.43	0.30564E+01	−0.20000E-02	0.82443E-15	313-373
C$_{10}$H$_{12}$O	Tetrahydro-2-naphthol*	293.2	11.70	0.98978E+02	−0.48267E+00	0.63008E-03	293-363
C$_{10}$H$_{12}$O	4-Isopropylbenzaldehyde	288.2	10.68				
C$_{10}$H$_{12}$O$_2$	4-Allyl-2-methoxyphenol	293.2	9.55	0.52377E+02	−0.24380E+00	0.33333E-03	273-323
C$_{10}$H$_{12}$O$_2$	2-Phenylethyl acetate	297.2	4.93				
C$_{10}$H$_{12}$O$_2$	Benzyl propanoate	303.0	5.11	0.42301E+01	0.13962E-01	−0.36426E-04	303-358
C$_{10}$H$_{12}$O$_2$	Phenyl butanoate	293.2	4.48				
C$_{10}$H$_{12}$O$_2$	Propyl benzoate	303.2	5.78	0.10927E+02	−0.20535E-01	0.11745E-04	303-358
C$_{10}$H$_{12}$O$_2$	Ethyl phenylacetate	293.2	5.320				
C$_{10}$H$_{14}$	Butylbenzene	293.2	2.359				
C$_{10}$H$_{14}$	sec-Butylbenzene	293.2	2.357	0.28348E+01	−0.68586E-03	−0.32143E-05	273-323
C$_{10}$H$_{14}$	tert-Butylbenzene	293.2	2.359	0.27924E+01	−0.38350E-03	−0.37500E-05	273-323
C$_{10}$H$_{14}$	Isobutylbenzene	293.2	2.318	0.28055E+01	−0.92614E-03	−0.25000E-05	273-323
C$_{10}$H$_{14}$	1-Isopropyl-4-methylbenzene	298.2	2.2322	0.25266E+01	−0.25121E-03	−0.24867E-05	277-333
C$_{10}$H$_{14}$	o-Diethylbenzene	293.2	2.594				
C$_{10}$H$_{14}$	m-Diethylbenzene	293.2	2.369				
C$_{10}$H$_{14}$	p-Diethylbenzene	293.2	2.259				
C$_{10}$H$_{14}$	1-Ethyl-3,5-dimethylbenzene	293.2	2.275				
C$_{10}$H$_{14}$	1,2,3,4-Tetramethylbenzene	296.0	2.538	0.33822E+01	−0.33630E-02	0.17475E-05	273-412
C$_{10}$H$_{14}$	1,2,4,5-Tetramethylbenzene	356.0	2.223	0.26834E+01	−0.10327E-02	−0.73533E-06	356-430
C$_{10}$H$_{14}$N$_2$	L-Nicotine	293.2	8.937	0.21347E+02	−0.57177E-01	0.50655E-04	293-363
C$_{10}$H$_{14}$O	1-Phenyl-2-methyl-2-propanol	298.2	5.71	0.21922E+02	−0.84231E-01	0.99475E-04	298-423
C$_{10}$H$_{14}$O	Butyl phenyl ether	293.2	3.734				
C$_{10}$H$_{14}$O	Thymol	333.2	4.259				
C$_{10}$H$_{15}$N	N,N-Diethylaniline	303.2	5.15	0.50773E+01	0.15399E-01	−0.50000E-04	303-328
C$_{10}$H$_{16}$	γ-Terpinene	298.2	2.2738				
C$_{10}$H$_{16}$	d-Limonene	298.2	2.3746				
C$_{10}$H$_{16}$	l-Limonene	298.2	2.3738				
C$_{10}$H$_{16}$	Terpinolene	298.2	2.2918				
C$_{10}$H$_{16}$	α-Pinene	298.2	2.1787				
C$_{10}$H$_{16}$	β-Pinene	298.2	2.4970				
C$_{10}$H$_{16}$	α-Terpinene	298.2	2.4526				
C$_{10}$H$_{16}$	β-Myrcene	298.2	2.3				
C$_{10}$H$_{16}$O	Carvenone	293.2	18.8				
C$_{10}$H$_{16}$O	d-Fenchone	294.2	12.8				
C$_{10}$H$_{17}$Cl	2-Chlorobornane	368.2	5.21				
C$_{10}$H$_{18}$	Pinane	298.2	2.1456				
C$_{10}$H$_{18}$	cis-Decahydronaphthalene	293.2	2.219	0.25410E+01	−0.11420E-02	0.15092E-06	293-373
C$_{10}$H$_{18}$	trans-Decahydronaphthalene	293.2	2.184	0.26615E+01	−0.21241E-02	0.16864E-05	293-373
C$_{10}$H$_{18}$O	Eucalyptol	298.2	4.57				
C$_{10}$H$_{18}$O$_2$	Cyclohexyl butanoate	293.2	4.58				
C$_{10}$H$_{18}$O$_4$	Diethyl adipate	293.2	6.109	0.14824E+02	−0.40749E-01	0.37600E-04	293-343
C$_{10}$H$_{20}$	1-Decene	293.2	2.136	0.19091E+01	0.33442E-02	−0.87500E-05	273-323
C$_{10}$H$_{20}$	cis-5-Decene	298.2	2.071				

Mol. form.	Name	T/K	ε_r	a	b	c	Range/K
$C_{10}H_{20}$	trans-5-Decene	298.2	2.030				
$C_{10}H_{20}$	5-Methyl-4-nonene	293.2	2.175				
$C_{10}H_{20}$	2,4,6-Trimethyl-3-heptene	293.2	2.293				
$C_{10}H_{20}Br_2$	1,10-Dibromodecane	303.2	6.56	0.17350E+02	−0.50328E-01	0.48633E-04	303-368
$C_{10}H_{20}Cl_2$	1,10-Dichlorodecane	308.2	6.68	−0.57423E+01	0.94220E-01	−0.17500E-03	308-338
$C_{10}H_{20}O$	2-Decanone	287.2	8.3				
$C_{10}H_{20}O$	Menthol	309.3	3.90	0.68202E+01	−0.15894E-01	0.20837E-04	309-358
$C_{10}H_{20}O_2$	2,2-Dimethyloctanoic acid	296.2	2.8				
$C_{10}H_{20}O_2$	Octyl acetate	288.2	4.18	−0.34691E+01	0.58106E-01	−0.10952E-03	288-323
$C_{10}H_{20}O_2$	2-Methylheptyl acetate	288.2	4.27	0.23285E+02	−0.11538E+00	0.17143E-03	288-323
$C_{10}H_{20}O_2$	Pentyl pentanoate	305.6	4.076	0.77641E+01	−0.14335E-01	0.73740E-05	306-393
$C_{10}H_{20}O_2$	Isopentyl pentanoate	292.2	3.6				
$C_{10}H_{20}O_2$	Isopentyl isopentanoate	288.2	4.39	0.14698E+02	−0.57726E-01	0.76190E-04	288-323
$C_{10}H_{20}O_2$	Methyl nonanoate	293.2	3.943				
$C_{10}H_{21}Br$	1-Bromodecane	298.2	4.44	0.11202E+02	−0.33491E-01	0.36314E-04	274-328
$C_{10}H_{21}Cl$	1-Chlorodecane	293.2	4.581	0.68741E+01	−0.12210E-02	−0.22500E-04	293-323
$C_{10}H_{21}NO$	N,N-Dibutylacetamide	293.2	19.1				
$C_{10}H_{22}$	Decane	293.2	1.9853	0.24054E+01	−0.15445E-02	0.44643E-06	253-393
$C_{10}H_{22}$	2,7-Dimethyloctane	293.2	1.98				
$C_{10}H_{22}$	4-Propylheptane	293.2	1.9955				
$C_{10}H_{22}O$	1-Decanol	293.2	7.93	0.47195E+02	−0.20740E+00	0.24942E-03	293-343
$C_{10}H_{22}O$	2-Decanol	298.2	5.82	0.13621E+03	−0.81000E+00	0.12500E-02	288-308
$C_{10}H_{22}O$	3-Decanol	298.2	4.05	0.52090E+02	−0.31020E+00	0.50000E-03	288-308
$C_{10}H_{22}O$	4-Decanol	298.2	3.42	−0.11260E+02	0.93960E-01	−0.15000E-03	288-308
$C_{10}H_{22}O$	5-Decanol	298.2	3.24	−0.25832E+01	0.31456E-01	−0.40000E-04	288-308
$C_{10}H_{22}O$	2,2-Dimethyl-1-octanol	293.2	7.86	0.69536E+02	−0.34596E+00	0.46250E-03	293-333
$C_{10}H_{22}O$	Dipentyl ether	298.2	2.798				
$C_{10}H_{22}O$	Diisopentyl ether	293.2	2.817	0.44690E+01	−0.63710E-02	0.25000E-05	293-323
$C_{10}H_{22}OS$	Dipentyl sulfoxide	348.2	18.8				
$C_{10}H_{22}O_5$	Tetraethylene glycol dimethyl ether	298.2	7.68				
$C_{10}H_{22}S$	Dipentyl sulfide	298.2	3.826				
$C_{10}H_{23}N$	Decylamine	293.2	3.31	0.61497E+01	−0.12801E-01	0.10606E-04	293-373
$C_{10}H_{30}O_3Si_4$	Decamethyltetrasiloxane	293.2	2.370				
$C_{10}H_{30}O_5Si_5$	Decamethylcyclopentasiloxane	293.2	2.50				
$C_{11}H_{10}$	1-Methylnaphthalene	293.2	2.915	0.45126E+01	−0.76480E-02	0.75000E-05	293-333
$C_{11}H_{10}$	2-Methylnaphthalene	313.2	2.747				
$C_{11}H_{10}O$	1-Methoxynaphthalene	293.2	4.020	0.71885E+01	−0.14838E-01	0.13750E-04	293-333
$C_{11}H_{10}O$	2-Methoxynaphthalene	353.2	3.563	0.56702E+01	−0.69754E-02	0.28571E-05	353-373
$C_{11}H_{12}O_2$	Ethyl trans-cinnamate	293.2	5.63				
$C_{11}H_{12}O_3$	Ethyl benzoylacetate	303.2	13.50	0.93644E+01	0.74280E-01	−0.20000E-03	303-323
$C_{11}H_{14}O_2$	Benzyl butanoate	301.2	4.55				
$C_{11}H_{14}O_2$	Phenyl pentanoate	293.2	4.30				
$C_{11}H_{14}O_2$	Butyl benzoate	303.2	5.52	0.77854E+01	−0.34972E-02	−0.13149E-04	303-358
$C_{11}H_{14}O_2$	Isobutyl benzoate	291.2	5.39				
$C_{11}H_{16}$	1,3-Diethyl-5-methylbenzene	293.2	2.264				
$C_{11}H_{16}$	Pentamethylbenzene	334.0	2.358	0.30196E+01	−0.22619E-02	0.83831E-06	334-413
$C_{11}H_{22}$	1-Undecene	293.2	2.137	0.22132E+01	0.13121E-02	−0.53571E-05	273-323
$C_{11}H_{22}O$	2-Undecanone	285.3	8.3				
$C_{11}H_{22}O_2$	Nonyl acetate	293.2	3.87				
$C_{11}H_{22}O_2$	Pentyl hexanoate	288.2	4.22	0.83503E+01	−0.18449E-01	0.14286E-04	288-323
$C_{11}H_{23}Br$	1-Bromoundecane	272.6	4.61				
$C_{11}H_{24}$	Undecane	293.2	1.9972	0.23637E+01	−0.12500E-02	−0.85869E-16	283-363
$C_{11}H_{24}O$	1-Undecanol	313.2	5.98				
$C_{11}H_{25}N$	Undecylamine	293.2	3.25	0.54945E+01	−0.96161E-02	0.66017E-05	293-373
$C_{12}F_{27}N$	Tris(perfluorobutyl)amine	293.2	2.15				
$C_{12}H_8O$	Dibenzofuran	373.2	3.00				
$C_{12}H_{10}$	Biphenyl	348.2	2.53	0.26869E+01	0.63072E-03	−0.30995E-05	348-428
$C_{12}H_{10}N_2O$	trans-Azoxybenzene	311.2	5.2				
$C_{12}H_{10}O$	Diphenyl ether	283.2	3.726				
$C_{12}H_{10}O$	2-Acetonaphthone	333.2	13.03	0.14538E+03	−0.73040E+00	0.10000E-02	333-363
$C_{12}H_{10}OS$	Diphenyl sulfoxide	344.7	16.6				
$C_{12}H_{10}O_2S$	Diphenyl sulfone	406.2	21.1				
$C_{12}H_{10}S$	Diphenyl sulfide	298.2	5.43				

Mol. form.	Name	T/K	ε_r	a	b	c	Range/K
$C_{12}H_{11}N$	Diphenylamine	323.2	3.73				
$C_{12}H_{11}NO$	N-1-Naphthylenylacetamide	433.2	24.3	0.84739E+02	−0.12391E+00	−0.35714E-04	433-533
$C_{12}H_{12}$	1,6-Dimethylnaphthalene	293.2	2.7250				
$C_{12}H_{12}O$	1-Ethoxynaphthalene	292.2	3.3				
$C_{12}H_{14}O_2$	Propyl cinnamate	293.2	5.45				
$C_{12}H_{14}O_4$	Diethyl phthalate	293.2	7.86				
$C_{12}H_{16}O$	2-Cyclohexylphenol	328.2	3.97				
$C_{12}H_{16}O$	4-Cyclohexylphenol	404.2	4.42				
$C_{12}H_{16}O_2$	Pentyl benzoate	293.2	5.07				
$C_{12}H_{16}O_3$	Pentyl salicylate	301.2	6.25				
$C_{12}H_{16}O_3$	Isopentyl salicylate	293.12	7.26	0.13129E+02	−0.19190E-01	−0.36060E-05	225-397
$C_{12}H_{17}NO$	N-Butyl-N-phenylacetamide	298.2	11.66				
$C_{12}H_{18}$	Hexylbenzene	293.2	2.3				
$C_{12}H_{18}$	1,3,5-Triethylbenzene	293.2	2.256				
$C_{12}H_{18}$	Hexamethylbenzene	449.0	2.172	0.35710E+01	−0.46912E-02	0.35088E-05	449-489
$C_{12}H_{20}O_2$	l-Bornyl acetate	303.2	4.46	0.60791E+01	0.98200E-02	−0.50000E-04	303-323
$C_{12}H_{22}O$	Dicyclohexyl ether	293.2	3.45	0.95324E+01	−0.31740E-01	0.37500E-04	293-333
$C_{12}H_{22}O$	Cyclododecanone	303.2	11.4	0.39327E+02	−0.13248E+00	0.13298E-03	303-423
$C_{12}H_{22}O_6$	Dibutyl tartrate	314.2	9.4				
$C_{12}H_{24}$	1-Dodecene	293.2	2.152	0.22581E+01	0.11106E-02	−0.50000E-05	273-323
$C_{12}H_{24}O_2$	Decyl acetate	293.2	3.75				
$C_{12}H_{24}O_2$	Ethyl decanoate	293.2	3.75	0.70969E+01	−0.15080E-01	0.12500E-04	293-353
$C_{12}H_{24}O_2$	Methyl undecanoate	293.2	3.671				
$C_{12}H_{25}Br$	1-Bromododecane	298.2	4.07	0.86103E+01	−0.20891E-01	0.18994E-04	274-328
$C_{12}H_{25}Cl$	1-Chlorododecane	298.2	4.17	0.10002E+02	−0.27798E-01	0.27559E-04	274-328
$C_{12}H_{25}I$	1-Iodododecane	298.2	3.91	0.34641E+01	0.97404E-02	−0.27602E-04	293-323
$C_{12}H_{26}$	Dodecane	293.2	2.0120	0.23697E+01	−0.12200E-02	−0.36375E-16	283-363
$C_{12}H_{26}O$	1-Dodecanol	303.2	5.82	0.18518E+02	−0.44859E-01	0.99900E-05	303-358
$C_{12}H_{26}O$	2-Butyl-1-octanol	363.2	3.28				
$C_{12}H_{27}BO_3$	Tributyl borate	293.2	2.23				
$C_{12}H_{27}N$	Dodecylamine	303.2	3.07	0.27999E+01	0.44810E-02	−0.11905E-04	303-373
$C_{12}H_{27}N$	Tributylamine	293.2	2.340	0.19846E+01	0.28108E-02	−0.54545E-05	233-293
$C_{12}H_{27}O_4P$	Tributyl phosphate	293.2	8.34	0.26304E+02	−0.88480E-01	0.92857E-04	293-373
$C_{12}H_{28}O_4Si$	Tetrapropoxysilane	298.2	3.21				
$C_{12}H_{28}Sn$	Tetrapropylstannane	293.2	2.267				
$C_{12}H_{30}OSi_2$	Hexaethyldisiloxane	298.2	2.259	0.36559E+01	−0.72406E-02	0.85714E-05	298-333
$C_{13}H_{10}O$	Benzophenone	300.2	12.62	0.34130E+02	−0.10249E+00	0.10268E-03	300-420
$C_{13}H_{10}O_3$	Phenyl salicylate	290.2	6.92	0.26545E+02	−0.11180E+00	0.15220E-03	290-358
$C_{13}H_{12}$	Diphenylmethane	303.2	2.540	0.30638E+01	−0.17286E-02		303-333
$C_{13}H_{12}O$	Benzyl phenyl ether	313.2	3.748				
$C_{13}H_{18}O_2$	Hexyl benzoate	293.2	4.80				
$C_{13}H_{20}$	Heptylbenzene	293.2	2.26				
$C_{13}H_{20}O$	α-Ionone*	292.4	10.78				
$C_{13}H_{20}O$	β-Ionone*	297.65	11.66				
$C_{13}H_{24}O_4$	Diethyl nonanedioate	303.2	5.133				
$C_{13}H_{26}$	1-Tridecene	293.2	2.139	0.14154E+01	0.66514E-02	−0.14286E-04	273-323
$C_{13}H_{26}O$	7-Tridecanone	303.2	7.6				
$C_{13}H_{26}O_2$	Ethyl undecanoate	293.2	3.55				
$C_{13}H_{26}O_2$	Methyl dodecanoate	293.2	3.539				
$C_{13}H_{27}Br$	1-Bromotridecane	281.15	4.19				
$C_{13}H_{28}$	Tridecane	293.2	2.0213	0.23731E+01	−0.12000E-02	−0.21841E-15	283-363
$C_{13}H_{28}$	5-Butylnonane	293.2	2.0319				
$C_{13}H_{28}O$	1-Tridecanol	333.2	4.02				
$C_{14}H_{10}$	Anthracene	502.0	2.649	0.20571E+02	−0.69169E-01	0.66667E-04	502-516
$C_{14}H_{10}$	Phenanthrene	383.2	2.72				
$C_{14}H_{10}O_2$	Benzil	368.2	13.04	−0.23599E+02	0.22715E+00	−0.34667E-03	368-393
$C_{14}H_{12}O_2$	Benzyl benzoate	303.2	5.26	0.76856E+01	−0.80000E-02	−0.80361E-15	303-358
$C_{14}H_{12}O_3$	Benzyl salicylate	301.2	4.12				
$C_{14}H_{14}$	1,2-Diphenylethane	331.2	2.47	0.31178E+01	−0.21572E-02	0.59800E-06	331-451
$C_{14}H_{14}O$	Dibenzyl ether	293.2	3.821	0.80154E+01	−0.20536E-01	0.21250E-04	293-333
$C_{14}H_{15}N$	Dibenzylamine	293.2	3.446				
$C_{14}H_{16}O_2Si$	Dimethyldiphenoxysilane	298.2	3.500	0.51669E+01	−0.77001E-02	0.70156E-05	283-353
$C_{14}H_{18}O_2$	Pentyl cinnamate	293.2	4.89				

Mol. form.	Name	T/K	ε_r	a	b	c	Range/K
$C_{14}H_{22}$	Octylbenzene	293.2	2.26				
$C_{14}H_{26}O_4$	Diisobutyl adipate	293.2	5.19				
$C_{14}H_{26}O_4$	Diethyl sebacate	303.2	4.995	0.39143E+02	−0.20965E+00	0.32000E-03	303-313
$C_{14}H_{28}O_2$	Dodecyl acetate	293.2	3.6				
$C_{14}H_{28}O_2$	Ethyl laurate	273.2	3.94				
$C_{14}H_{28}O_2$	Methyl tridecanoate	293.2	3.442				
$C_{14}H_{29}Br$	1-Bromotetradecane	293.2	3.84	0.10058E+02	−0.33905E-01	0.43528E-04	274-328
$C_{14}H_{30}$	Tetradecane	293.2	2.0343	0.23832E+01	−0.11900E-02	−0.51229E-16	283-363
$C_{14}H_{30}O$	1-Tetradecanol	318.2	4.42	0.12272E+02	−0.24667E-01	−0.13168E-13	318-358
$C_{14}H_{31}N$	Tetradecylamine	312.55	2.90				
$C_{15}H_{12}O_4$	Phenyl 2-(acetyloxy)benzoate	384.2	4.33				
$C_{15}H_{26}O_6$	Tributyrin	282.8	5.72	0.13152E+02	−0.36684E-01	0.36795E-04	199-283
$C_{15}H_{30}O_2$	Methyl tetradecanoate	293.2	3.352				
$C_{15}H_{31}Br$	1-Bromopentadecane	293.35	3.88				
$C_{15}H_{32}$	Pentadecane	293.2	2.0391	0.23792E+01	−0.11600E-02	−0.71069E-16	283-363
$C_{15}H_{32}O$	1-Pentadecanol	333.2	3.70				
$C_{15}H_{33}N$	Pentadecylamine	313.25	2.85				
$C_{15}H_{33}N$	Triisopentylamine	294.2	2.29				
$C_{16}H_{22}O_4$	Dibutyl phthalate	293.2	6.58	0.12444E+02	−0.20000E-01		293-333
$C_{16}H_{32}O_2$	Hexadecanoic acid	338.2	2.417				
$C_{16}H_{32}O_2$	Ethyl myristate	293.2	3.50	0.52642E+01	−0.60000E-02	−0.47358E-15	293-353
$C_{16}H_{32}O_2$	Methyl pentadecanoate	293.2	3.296				
$C_{16}H_{33}Br$	1-Bromohexadecane	298.2	3.68	0.58668E+01	−0.73333E-02	−0.52666E-14	298-328
$C_{16}H_{33}I$	1-Iodohexadecane	293.2	3.57	0.79531E+01	−0.22859E-01	0.26955E-04	293-323
$C_{16}H_{34}$	Hexadecane	293.2	2.0460	0.23861E+01	−0.11600E-02	0.25555E-15	293-363
$C_{16}H_{34}O$	1-Hexadecanol	333.2	3.69	0.85935E+01	−0.14714E-01	−0.45533E-13	333-363
$C_{16}H_{35}N$	Hexadecylamine	328.35	2.71				
$C_{16}H_{36}Sn$	Tetrabutylstannane	293.2	9.74	0.56115E+02	−0.24812E+00	0.30682E-03	293-313
$C_{17}H_{12}O_3$	2-Naphthyl salicylate	293.0	6.30	0.11229E+02	−0.18857E-01	0.70332E-05	293-353
$C_{17}H_{34}O$	9-Heptadecanone	328.2	5.43	0.44176E+02	−0.21183E+00	0.28571E-03	328-363
$C_{17}H_{34}O_2$	Methyl palmitate	313.2	3.124				
$C_{17}H_{36}$	Heptadecane	293.2	2.0578	0.23627E+01	−0.10400E-02	−0.10397E-12	293-308
$C_{17}H_{36}O$	1-Heptadecanol	333.2	3.41				
$C_{18}H_{26}O_4$	Dipentyl phthalate	293.2	6.00				
$C_{18}H_{28}O_2$	Phenyl laurate	293.2	3.28				
$C_{18}H_{30}O_2$	Linolenic acid	293.2	2.825	0.33867E+01	−0.19181E-02		274-368
$C_{18}H_{30}O_4$	Dicyclohexyl adipate	308.2	4.84				
$C_{18}H_{32}O_2$	Linoleic acid	293.2	2.754	0.32073E+01	−0.15477E-02		275-368
$C_{18}H_{34}O_2$	Oleic acid	293.2	2.336	0.25385E+01	−0.69448E-03		275-368
$C_{18}H_{34}O_4$	Dibutyl sebacate	293.2	4.54				
$C_{18}H_{36}O_2$	Stearic acid	293.2	2.314	0.27159E+01	−0.13300E-02		293-373
$C_{18}H_{36}O_2$	Hexadecyl acetate	308.2	3.19	0.47310E+01	−0.50000E-02	0.41338E-14	308-348
$C_{18}H_{36}O_2$	Ethyl palmitate	303.2	3.07	0.57938E+01	−0.12294E-01	0.10919E-04	303-455
$C_{18}H_{36}O_2$	Methyl heptadecanoate	313.2	3.07				
$C_{18}H_{37}Br$	1-Bromooctadecane	303.35	3.53	0.46790E+01	−0.30355E-02	−0.24798E-05	303-332
$C_{18}H_{38}O$	1-Octadecanol	333.2	3.38	0.73784E+01	−0.12000E-01	−0.22871E-13	333-363
$C_{18}H_{39}BO_3$	Trihexyl borate	293.2	2.22				
$C_{18}H_{39}N$	Octadecylamine	326.35	2.67				
$C_{19}H_{16}$	Triphenylmethane	367.2	2.46	0.40201E+01	−0.66507E-02	0.65329E-05	367-448
$C_{19}H_{18}O_3Si$	Methyltriphenoxysilane	298.2	3.628				
$C_{19}H_{32}O_2$	Methyl linolenate	293.2	3.355				
$C_{19}H_{34}O_2$	Methyl linoleate	293.2	3.466				
$C_{19}H_{36}O_2$	Methyl oleate	293.2	3.211				
$C_{19}H_{38}O$	10-Nonadecanone	353.2	5.37				
$C_{19}H_{38}O_2$	Methyl stearate	313.2	3.021				
$C_{19}H_{40}$	Nonadecane	293.2	2.0706				
$C_{20}H_{30}O_4$	Dihexyl phthalate	293.2	5.62				
$C_{20}H_{38}O_2$	Ethyl oleate	301.2	3.17	0.57033E+01	−0.11223E-01	0.93447E-05	301-423
$C_{20}H_{40}O_2$	Octadecyl acetate	308.2	3.07	0.44569E+01	−0.45000E-02	0.33923E-14	308-348
$C_{20}H_{40}O_2$	Ethyl stearate	313.2	2.958	0.70930E+01	−0.19081E-01	0.19555E-04	331-440
$C_{20}H_{40}O_2$	Methyl nonadecanoate	313.2	2.982				
$C_{20}H_{42}O$	1-Eicosanol	338.2	3.13	0.21700E+01	0.12497E-01	−0.28571E-04	338-363
$C_{20}H_{42}O$	Didecyl ether	293.2	2.644	0.41465E+01	−0.62240E-02	0.37500E-05	293-333

Mol. form.	Name	T/K	ε_r	a	b	c	Range/K
$C_{20}H_{60}O_8Si_9$	Eicosamethylnonasiloxane	293.2	2.645	0.57840E+01	−0.16568E-01	0.20000E-04	293-323
$C_{21}H_{21}O_4P$	Tricresyl phosphate*	298.2	6.7				
$C_{21}H_{38}O_6$	1,2,3-Propanetriyl hexanoate	293.2	4.476				
$C_{22}H_{42}O_2$	Butyl oleate	298.2	4.00				
$C_{22}H_{44}O_2$	Butyl stearate	298.2	3.120	0.73894E+02	−0.46261E+00	0.75500E-03	298-343
$C_{22}H_{46}$	Docosane	293.2	2.0840				
$C_{22}H_{46}O$	1-Docosanol	348.2	2.94	0.82062E+01	−0.25069E-01	0.28571E-04	348-373
$C_{24}H_{20}O_4Si$	Tetraphenoxysilane	333.2	3.4915				
$C_{24}H_{38}O_4$	Dioctyl phthalate	293.2	5.22				
$C_{26}H_{50}O_4$	Dioctyl sebacate	299.2	4.01				
$C_{27}H_{50}O_6$	1,2,3-Propanetriyl octanoate	293.2	3.931				
$C_{30}H_{58}O_4$	Ethylene glycol ditetradecanoate	343.2	2.98				
$C_{30}H_{62}$	Triacontane	373.2	1.9112				
$C_{30}H_{62}$	2,6,10,15,19,23-Hexamethyltetracosane	373.2	1.9106				
$C_{34}H_{66}O_4$	Ethylene glycol dipalmitate	348.2	2.89				
$C_{34}H_{68}O_2$	Hexadecyl stearate	333.2	2.61				
$C_{38}H_{74}O_4$	Ethylene glycol distearate	353.2	2.79				
$C_{39}H_{74}O_6$	Glycerol trilaurate	313.2	3.287				
$C_{51}H_{98}O_6$	Glycerol tripalmitate	328.2	2.901	−0.29131E+01	0.32206E-01	−0.44154E-04	328-393
$C_{57}H_{104}O_6$	Glycerol trioleate	293.2	3.109				
$C_{57}H_{104}O_6$	Glycerol trielaidate	313.2	2.980				
$C_{57}H_{110}O_6$	Glycerol tristearate	353.2	2.740				

* Isomer was not specified in the original reference.

** Cubic term is needed; see introduction.

PERMITTIVITY (DIELECTRIC CONSTANT) OF GASES

This table gives the relative permittivity ε (often called the dielectric constant) of some common gases at a temperature of 20 °C and pressure of one atmosphere (101.325 kPa). Values of the permanent dipole moment μ in Debye Units (1 D = 3.33564 × 10⁻³⁰ C m) are also included.

The density dependence of the permittivity is given by the equation

$$\frac{\varepsilon-1}{\varepsilon-2} = \rho_m \left(\frac{4\pi N\alpha}{3} + \frac{4\pi N\mu^2}{9kT} \right)$$

where ρ_m is the molar density, N is Avogadro's number, k is the Boltzmann constant, T is the temperature, and α is the molecular polarizability. Therefore, in regions where the gas can be considered ideal, $\varepsilon - 1$ is approximately proportional to the pressure at constant temperature. For nonpolar gases ($\mu = 0$), $\varepsilon - 1$ is inversely proportional to temperature at constant pressure.

The number of significant figures indicates the accuracy of the values given. The values of ε for air, Ar, H₂, He, N₂, O₂, and CO₂ are recommended as reference values; these are accurate to 1 ppm or better.

The second part of the table gives the permittivity of water vapor in equilibrium with liquid water as a function of temperature (derived from Reference 4).

References

1. A. A. Maryott and F. Buckley, *Table of Dielectric Constants and Electric Dipole Moments of Substances in the Gaseous State*, National Bureau of Standards Circular 537, 1953.
2. Harvey, A. H., and Lemmon, E. W., *Int. J. Thermophys.* 26, 31, 2005 [for nonpolar gases and light hydrocarbons]
3. *Landolt-Börnstein, Numerical Data and Functional Relationships in Science and Technology*, New Series, Group IV, Vol. 4, Springer-Verlag, Heidelberg, 1980 (data at high pressures).
4. Fernández, D. P., Goodwin, A. R. H., Lemmon, E. W., Levelt Sengers, J. M. H., and Williams, R. C., *J. Phys. Chem. Ref. Data* 26, 1125, 1997 [for water vapor]

Mol. form.	Name	ε	μ/D
Compounds not containing carbon			
	Air (dry, CO₂ free)	1.0005360	
Ar	Argon	1.0005169	0
BF₃	Boron trifluoride	1.0011	0
BrH	Hydrogen bromide	1.00279	0.827
ClH	Hydrogen chloride	1.00390	1.109
F₃N	Nitrogen trifluoride	1.0013	0.235
F₆S	Sulfur hexafluoride	1.00200	0
HI	Hydrogen iodide	1.00214	0.448
H₂	Hydrogen	1.0002532	0
H₂S	Hydrogen sulfide	1.00344	0.97
H₃N	Ammonia	1.00622	1.471
He	Helium	1.0000645	0
Kr	Krypton	1.000784	0
NO	Nitric oxide	1.00060	0.159
N₂	Nitrogen	1.0005474	0
N₂O	Nitrous oxide	1.00104	0.161
Ne	Neon	1.000124	0
O₂	Oxygen	1.0004941	0
O₂S	Sulfur dioxide	1.00825	1.633
O₃	Ozone	1.0017	0.534
Xe	Xenon	1.00127	0

Mol. form.	Name	ε	μ/D
Compounds containing carbon			
CF₄	Tetrafluoromethane	1.00121	0
CO	Carbon monoxide	1.00065	0.110
CO₂	Carbon dioxide	1.0009217	0
CH₃Br	Bromomethane	1.01028	1.822
CH₃Cl	Chloromethane	1.01080	1.892
CH₃F	Fluoromethane	1.00973	1.858
CH₃I	Iodomethane	1.00914	1.62
CH₄	Methane	1.0008181	0
C₂H₂	Acetylene	1.00124	0
C₂H₃Cl	Chloroethylene	1.0075	1.45
C₂H₄	Ethylene	1.00135	0
C₂H₅Cl	Chloroethane	1.01325	2.05
C₂H₆	Ethane	1.001403	0
C₂H₆O	Dimethyl ether	1.0062	1.30
C₃H₆	Propene	1.00228	0.366
C₃H₆	Cyclopropane	1.00178	0
C₃H₈	Propane	1.002032	0.084
C₄H₁₀	Butane	1.00266	0.05
C₄H₁₀	Isobutane	1.00268	0.132

Permittivity of Saturated Water Vapor

t/°C	ε	t/°C	ε
0	1.000064	60	1.00143
10	1.000120	70	1.00211
20	1.000214	80	1.00304
30	1.000363	90	1.00428
40	1.000594	100	1.00589
50	1.000935		

AZEOTROPIC DATA FOR BINARY MIXTURES

J. Gmehling, J. Menke, J. Krafczyk, K. Fischer, J.-C. Fontaine, and H. V. Kehiaian

Binary homogeneous (single-phase) liquid mixtures having an extremum (maximum or minimum) vapor pressure P at constant temperature T, as a function of composition, are called azeotropic mixtures, or simply azeotropes. The composition is usually expressed as mole fractions, where x_1 for component 1 in the liquid phase and y_1 for component 1 in the vapor phase are identical. Mixtures that do not show a maximum or minimum are called zeotropic. A maximum (minimum) of the $P(x_1)$ or $P(y_1)$ curves corresponds to a minimum (maximum) of the boiling temperature T at constant P, plotted as a function of x_1 or y_1 [see $T(x_1)$ and $T(y_1)$ curves, Types I and III, in Fig. 1]. Azeotropes in which the pressure is a maximum (temperature is a minimum) are often called positive azeotropes, while pressure-minimum (temperature-maximum) azeotropes are called negative azeotropes. The coordinates of an azeotropic point are the azeotropic temperature T_{Az}, pressure P_{Az}, and the vapor-phase composition $y_{1,Az}$, which is the same as the liquid-phase composition $x_{1,Az}$.

In the two-phase liquid-liquid region of partially miscible (heterogeneous) mixtures, the vapor pressure at constant T (or the boiling temperature at constant P) is independent of the global composition x_1 of the two coexisting liquid phases between the equilibrium compositions x_1' and x_1'' ($x_1' < x_1''$).

The constant vapor pressure (boiling temperature) above the two-phase region of certain partially miscible mixtures is usually larger (smaller) than the vapor pressure (boiling temperature) at any other liquid-phase composition in the homogeneous region. In this case, the vapor-phase composition is inside the miscibility gap. Mixtures of this type are called heteroazeotropic mixtures, or simply heteroazeotropes. (Fig. 1, Type II), as opposed to the other types of azeotropes, called homoazeotropes.

Only in a few cases partially miscible mixtures present a positive or negative azeotropic point in the single-phase region, outside the miscibility gap, similar to the azeotropic points of homogeneous mixtures (Fig. 1, Types IV and VI).

A few binary mixtures, for example the system perfluorobenzene + benzene, may present two azeotropic points at constant temperature (pressure), a positive and a negative one. They are called double azeotropic mixtures, or simply double azeotropes. (Fig. 1, Type V).

The knowledge of the occurrence of azeotropic points in binary and higher systems is of special importance for the design of distillation processes. The number of theoretical stages of a distillation column required for the separation depends on the separation factor α_{12}, i.e., the ratio of the K_i-factors ($K_i = y_i/x_i$) of the components i ($i = 1, 2$). The required separation factor can be calculated with the following simplified relation (Reference 1):

$$\alpha_{12} = K_1/K_2 = (y_1/x_1)/(y_2/x_2) = (\gamma_1 P_1^s)/(\gamma_2 P_2^s) \qquad (1)$$

where γ_i is the activity coefficient of component i in the liquid phase and P_i^s is the vapor pressure of the pure component i.

In distillation processes, only the difference between the separation factor and unity ($\alpha_{12} - 1$) can be exploited for the separation. If the separation factor is close to unity, a large number of theoretical stages is required for the separation. If the binary system to be separated shows an azeotropic point ($\alpha_{12} = 1$), the separation is impossible by ordinary distillation, even with an infinitely large number of stages.

Following eq. (1) azeotropic behavior will always occur in homogeneous binary systems when the vapor pressure ratio P_1^s/P_2^s is equal to the ratio of the activity coefficients γ_2/γ_1.

Various thermodynamic methods based on g^E-models (Wilson, NRTL, UNIQUAC) or group contribution methods (UNIFAC, modified UNIFAC, ASOG, PSRK) can be used for either calculating or predicting the required activity coefficients for the components under given conditions of temperature and composition (Reference 2).

Because of the importance of azeotropic data for the design of distillation processes, compilations have been available in book form for quite some time (References 3-7). The most recent printed data collection was published in 1994 (Reference 8). A revised and extended version appeared in 2004 (Reference 9).

A collection of approximately 47,400 zeotropic and azeotropic data sets, compiled from 6600 references, are stored in a comprehensive computerized data bank (Reference 10). The references from the above-mentioned compilations and from the vapor-liquid equilibrium part of the Dortmund Data Bank (Reference 11) were supplemented by references found from CAS online searches, private communications, data from industry, etc.. Over 24,000 zeotropic data and over 20,000 azeotropic data are available for binary systems. Nearly 90% of the binary azeotropic data show a pressure maximum. In most cases (ca. 90%) these are homogeneous azeotropes, and in approximately 7–8% of the cases heterogeneous azeotropes are reported. Less than 10% of the data stored show a pressure minimum. Approximately 21,000 of the data sets stored were published after 1970.

The table below provides information about azeotropes for 808 selected binary systems. Compounds are listed in the modified Hill order, with carbon-containing compounds following those compounds not containing carbon. In columns 1 and 2 are the molecular formulas of components 1 and 2 written in the Hill convention. In column 3 the names of the components are given, either a systematic IUPAC name or a name in ubiquitous use. Columns 4, 5, and 6 contain the azeotropic coordinates of the mixtures: temperature T_{Az}, pressure P_{Az}, and vapor-phase composition $y_{1,Az}$. The explanation of the type of azeotrope (column 7) is given by the following codes:

O: homogeneous azeotrope in a completely miscible system
L: homogeneous azeotrope in a partially miscible system
E: heterogeneous azeotrope
X: pressure maximum
N: pressure minimum
D: double azeotrope
C: system contains a supercritical compound

References

1. Gmehling, J. and Brehm, A., *Grundoperationen*, Thieme-Verlag, Stuttgart, 1996.
2. Gmehling, J. and Kolbe, B., *Thermodynamik*, VCH-Verlag, Weinheim, 1992.
3. Lecat, M., *Doctoral Dissertation*, 1908.
4. Lecat, M., *L'Azeotropisme*, Monograph, L'Auteur, Brussel, 1918.
5. Lecat, M., *Tables Azeotropiques*, Monograph, Lamertin, Brussel 1949.

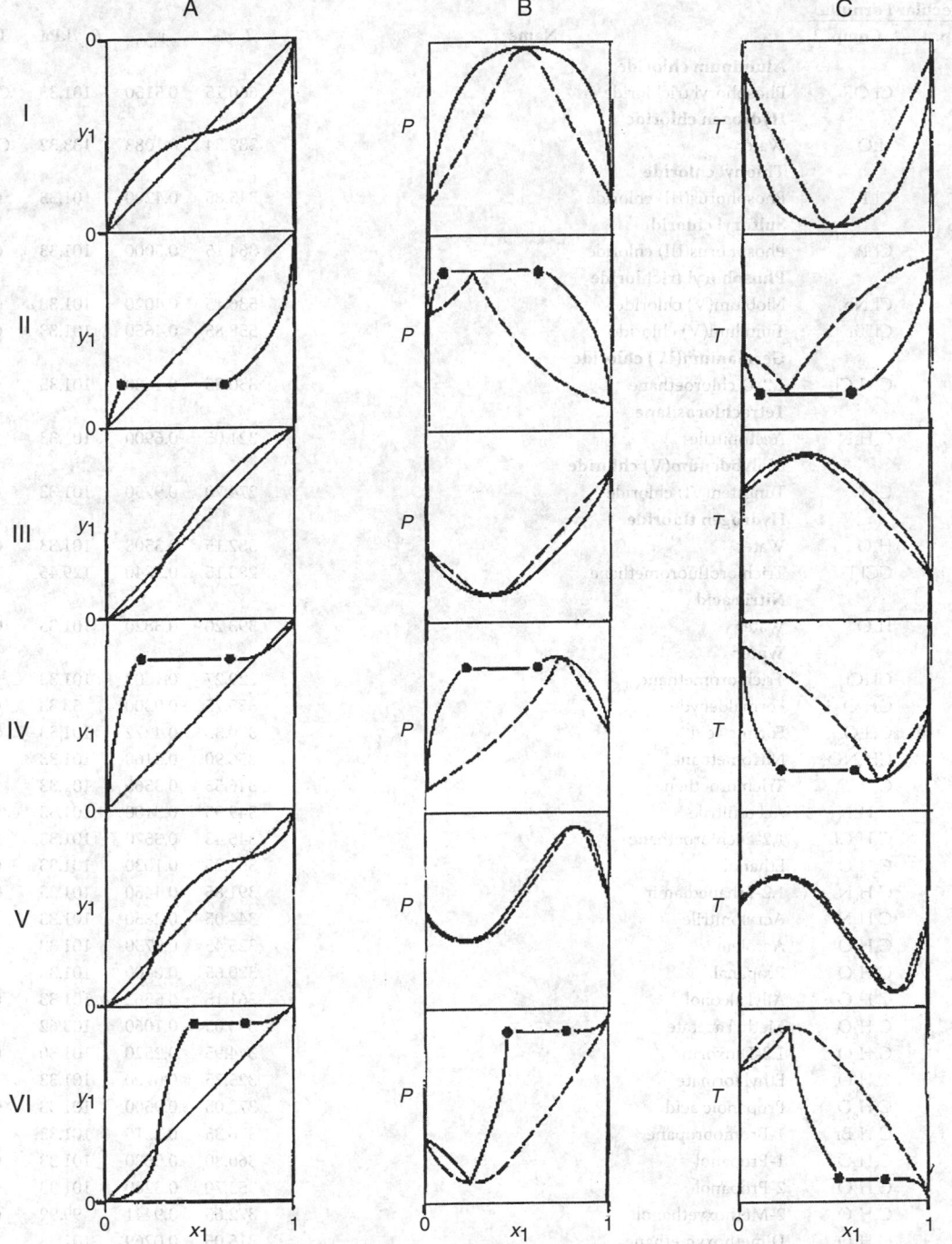

Figure 1 Different types of binary azeotropic systems: I — homogeneous pressure-maximum azeotrope in a completely miscible system (OX); II — heterogeneous pressure-maximum azeotrope (EX); III — homogeneous pressure-minimum azeotrope in a completely miscible system (ON); IV — homogeneous pressure-maximum azeotrope in a partially miscible system (LX); V–D: double azeotrope (OND, OXD); VI — homogeneous pressure-minimum azeotrope in a partially miscible system (LN). A — $y_1(x_1)$; B — $P(x_1)$ and $P(y_1)$; C — $T(x_1)$ and $T(y_1)$. Continuous line — (x_1); Dashed line — (y_1).

6. Ogorodnikov, S. K., Lesteva, T. M., and Kogan V. B., *Azeotropic Mixtures*, Khimia, Leningrad, 1971.

7. Horsley, L. H., *Azeotropic Data III*, American Chemical Society, Washington, 1973.

8. Gmehling, J., Menke, J., Krafczyk, J., and Fischer, K., *Azeotropic Data*, 2 Volumes, VCH Verlag, Weinheim, 1994.

9. Gmehling, J., Menke, J., Krafczyk, J., and Fischer, K., *Azeotropic Data*, 2nd Ed., 3 Volumes, VCH Verlag, Weinheim, 2004.

10. Gmehling, J., Menke, J., Krafczyk, J., and Fischer, K., *A Data Bank for Azeotropic Data, Status and Applications, Fluid Phase Equilib.* 103, 51, 1995.

11. Dortmund Data Bank, www.ddbst.de

| Molecular Formula | | | | | | |
Comp. 1	Comp. 2	Name	T_{Az}/K	$y_{1,Az}$	P_{Az}/kPa	Type
AlCl₃		**Aluminum chloride**				
	Cl₃OP	Phosphoryl trichloride	660.15	0.5150	101.33	ONC
ClH		**Hydrogen chloride**				
	H₂O	Water	389.34	0.1083	133.32	ONC
Cl₂OS		**Thionyl chloride**				
	Cl₃P	Phosphorus(III) chloride	345.85	0.4200	101.33	OX
Cl₂O₂S		**Sulfuryl chloride**				
	Cl₃P	Phosphorus(III) chloride	364.15	0.5000	101.33	ON
Cl₃OP		**Phosphoryl trichloride**				
	Cl₅Nb	Niobium(V) chloride	536.15	0.4020	101.33	ON
	Cl₅Ta	Tantalum(V) chloride	558.85	0.4650	101.33	ON
Cl₄Ge		**Germanium(IV) chloride**				
	C₂H₄Cl₂	1,2-Dichloroethane	350.75	0.4630	101.33	OX
Cl₄Si		**Tetrachlorosilane**				
	C₂H₃N	Acetonitrile	321.05	0.6900	101.33	EX
Cl₅Mo		**Molybdenum(V) chloride**				
	Cl₆W	Tungsten(VI) chloride	274.70	0.9750	101.33	OX
FH		**Hydrogen fluoride**				
	H₂O	Water	382.15	0.3508	101.33	ON
	CCl₃F	Trichlorofluoromethane	283.15	0.7840	129.45	EX
HNO₃		**Nitric acid**				
	H₂O	Water	393.20	0.3820	101.33	ON
H₂O		**Water**				
	CHCl₃	Trichloromethane	329.27	0.1603	101.33	EX
	CH₂O	Formaldehyde	355.75	0.9300	53.33	OX
	CH₂O₂	Formic acid	380.35	0.4272	101.33	ON
	CH₃NO₂	Nitromethane	356.90	0.5160	101.33	EX
	C₂HCl₃	Trichloroethene	346.55	0.3560	101.33	EX
	C₂H₃N	Acetonitrile	349.95	0.3100	101.33	OX
	C₂H₄Cl₂	1,2-Dichloroethane	345.43	0.3570	101.33	EX
	C₂H₆O	Ethanol	351.25	0.1030	101.33	OX
	C₂H₈N₂	1,2-Ethanediamine	391.85	0.4450	101.33	ON
	C₃H₃N	Acrylonitrile	344.05	0.2850	101.33	EX
	C₃H₄O	Acrolein	325.45	0.0730	101.33	LX
	C₃H₆O	Propanal	320.65	0.0600	101.33	LX
	C₃H₆O	Allyl alcohol	361.15	0.5562	101.33	OX
	C₃H₆O₂	Methyl acetate	330.05	0.1060	103.62	LX
	C₃H₆O₂	1,3-Dioxolane	344.95	0.2520	101.30	OX
	C₃H₆O₂	Ethyl formate	325.75	0.0700	101.33	EX
	C₃H₆O₂	Propanoic acid	373.05	0.9500	101.33	OX
	C₃H₇Br	1-Bromopropane	336.35	0.2210	101.33	EX
	C₃H₈O	1-Propanol	360.80	0.5680	101.33	OX
	C₃H₈O	2-Propanol	353.70	0.3260	101.33	OX
	C₃H₈O₂	2-Methoxyethanol	372.65	0.9441	99.99	OX
	C₃H₈O₂	Dimethoxymethane	315.05	0.0269	101.38	LX
	C₄H₅N	cis-2-Butenenitrile	358.45	0.3832	101.33	EX
	C₄H₅N	trans-2-Butenenitrile	363.05	0.6843	101.33	EX
	C₄H₅N	Pyrrole	348.15	0.7514	50.13	EX
	C₄H₆O₂	Methacrylic acid	372.25	0.9464	98.93	OX
	C₄H₈O	2-Butanone	346.54	0.3480	101.33	LX
	C₄H₈O	Tetrahydrofuran	336.67	0.1828	101.33	OX
	C₄H₈O	Isobutanal	332.80	0.1698	100.99	EX
	C₄H₈O₂	Ethyl acetate	343.55	0.2990	101.33	EX
	C₄H₈O₂	Butanoic acid	372.95	0.9559	101.33	OX
	C₄H₈O₂	1,4-Dioxane	360.65	0.5280	101.33	OX
	C₄H₈O₂	Propyl formate	344.85	0.3090	101.33	EX
	C₄H₈O₂	Methyl propanoate	344.75	0.3050	101.33	EX
	C₄H₉Br	1-Bromobutane	353.95	0.4950	101.33	EX
	C₄H₉Br	1-Bromo-2-methylpropane	348.45	0.3730	101.33	EX

Molecular Formula					
Comp. 1	Comp. 2	Name	T_{Az}/K	$y_{1,Az}$	P_{Az}/kPa	Type
	C_4H_9Cl	1-Chloro-2-methylpropane	333.95	0.1970	101.33	LX
	$C_4H_{10}O$	1-Butanol	365.45	0.7540	101.33	EX
	$C_4H_{10}O$	2-Butanol	360.50	0.6200	101.33	LX
	$C_4H_{10}O$	2-Methyl-2-propanol	353.00	0.4011	101.33	OX
	$C_4H_{11}N$	Butylamine	349.85	0.0700	101.33	OX
	C_5H_5N	Pyridine	367.30	0.7500	101.33	OX
	C_5H_8	2-Methyl-1,3-butadiene	305.85	0.0520	101.33	EX
	C_5H_8	Methylenecyclobutane	313.15	0.0212	101.30	EX
	C_5H_8O	Cyclopropyl methyl ketone	361.65	0.7060	101.19	EX
	$C_5H_8O_2$	Methyl methacrylate	354.45	0.4996	101.33	EX
	C_5H_{10}	2-Methyl-2-butene	309.75	0.0650	101.33	EX
	$C_5H_{10}O$	3-Methyl-2-buten-1-ol	369.55	0.9141	101.33	EX
	$C_5H_{10}O$	3-Methyl-3-buten-1-ol	333.15	0.8680	101.33	EX
	$C_5H_{10}O$	2-Methyl-3-buten-2-ol	359.25	0.5770	101.33	LX
	$C_5H_{10}O$	3-Pentanone	356.05	0.4750	101.33	EX
	$C_5H_{10}O_2$	Isopropyl acetate	349.75	0.3960	101.33	EX
	$C_5H_{10}O_2$	Propyl acetate	355.91	0.5228	101.33	EX
	$C_5H_{10}O_2$	Butyl formate	356.95	0.5360	101.33	EX
	$C_5H_{10}O_2$	Isobutyl formate	352.75	0.4460	101.33	EX
	$C_5H_{12}O$	3-Methyl-1-butanol	367.97	0.8265	101.33	EX
	$C_5H_{12}O$	2-Methyl-2-butanol	360.85	0.6355	101.75	EX
	$C_5H_{12}O$	1-Pentanol	369.08	0.8633	101.33	EX
	$C_5H_{12}O$	2-Pentanol	363.15	0.7550	92.49	EX
	C_6H_6	Benzene	342.35	0.2980	101.33	EX
	C_6H_7N	Aniline	372.55	0.9580	101.33	EX
	C_6H_7N	4-Methylpyridine	370.50	0.8972	101.33	OX
	C_6H_{10}	Cyclohexene	343.95	0.3090	101.33	EX
	$C_6H_{10}O$	Cyclohexanone	369.45	0.8694	101.33	EX
	$C_6H_{10}O$	Methyldihydropyran (unspecified isomer)	360.75	0.5841	100.93	EX
	$C_6H_{10}O_2$	4-Vinyl-1,3-dioxane	367.65	0.8955	101.33	EX
	C_6H_{12}	1-Hexene	318.15	0.1510	63.35	EX
	$C_6H_{12}O_2$	Butyl acetate	363.35	0.7013	101.33	EX
	$C_6H_{12}O_2$	Isobutyl acetate	361.05	0.6440	101.33	EX
	$C_6H_{12}O_2$	4,4-Dimethyl-1,3-dioxane	366.00	0.7779	101.33	EX
	$C_6H_{12}O_2$	4,5-Dimethyl-1,3-dioxane (unspecified isomer)	365.05	0.7966	101.50	EX
	$C_6H_{12}O_2$	4-Ethyl-1,3-dioxane	365.75	0.7257	101.30	EX
	$C_6H_{12}O_2$	Diacetone alcohol	370.00	0.9900	90.79	OX
	$C_6H_{12}O_2$	Propyl propanoate	362.05	0.6600	101.33	EX
	$C_6H_{13}N$	Cyclohexylamine	369.55	0.8692	101.33	OX
	C_6H_{14}	Hexane	334.75	0.2110	101.33	EX
	$C_6H_{14}O$	Butyl ethyl ether	349.85	0.4070	101.33	EX
	$C_6H_{14}O$	1-Hexanol	367.89	0.9432	101.33	EX
	$C_6H_{14}O_3$	Di(ethylene glycol) dimethyl ether	372.70	0.9679	101.33	OX
	$C_6H_{15}N$	Diisopropylamine	347.25	0.3654	101.33	EX
	$C_6H_{15}N$	Dipropylamine	359.00	0.6046	101.33	EX
	C_7H_8	Toluene	357.25	0.5230	101.33	EX
	C_7H_8O	Benzyl alcohol	373.05	0.9840	101.33	EX
	C_7H_9N	2,6-Dimethylpyridine	369.17	0.8647	101.33	EX
	$C_7H_{12}O_4$	1,2-Propanediol diacetate	358.15	0.9740	59.41	EX
	C_7H_{14}	1-Heptene	350.20	0.4100	101.33	EX
	$C_7H_{14}O_2$	Isopentyl acetate	367.05	0.7990	101.46	EX
	$C_7H_{14}O_2$	Butyl propanoate	367.95	0.8340	101.33	EX
	C_7H_{16}	Heptane	352.35	0.4510	101.33	EX
	$C_7H_{16}O$	1-Heptanol	371.99	0.9703	101.33	EX
	C_8H_8	Styrene	367.15	0.8000	101.33	EX
	C_8H_8O	Acetophenone	371.15	0.9675	101.19	EX
	C_8H_{10}	m-Xylene	365.15	0.7667	101.33	EX
	C_8H_{10}	p-Xylene	365.15	0.7450	101.33	EX
	C_8H_{10}	Ethylbenzene	364.15	0.7221	101.33	EX

Molecular Formula						
Comp. 1	Comp. 2	Name	T_{Az}/K	$y_{1,Az}$	P_{Az}/kPa	Type
	$C_8H_{16}O_2$	Butyl butanoate	369.85	0.9110	101.33	EX
	C_8H_{18}	Octane	362.75	0.6850	101.33	EX
	C_8H_{18}	2,2,4-Trimethylpentane	351.95	0.4420	101.33	EX
	$C_8H_{18}O$	Dibutyl ether	368.65	0.7628	101.33	EX
	$C_8H_{18}O$	1-Octanol	372.75	0.9820	101.33	EX
	$C_8H_{19}N$	Dibutylamine	370.05	0.8850	101.33	EX
	C_9H_{10}	Isopropenylbenzene	369.95	0.8880	101.33	EX
	C_9H_{12}	Isopropylbenzene	368.15	0.8340	101.33	EX
	$C_9H_{12}O$	2-Phenyl-2-propanol	371.25	0.9718	101.33	EX
	C_9H_{20}	Nonane	367.95	0.8280	101.33	EX
	$C_9H_{20}O$	1-Nonanol	373.00	0.9846	101.33	EX
	$C_{10}H_{22}$	Decane	370.75	0.9180	101.33	EX
	$C_{10}H_{22}O$	1-Decanol	373.13	0.9865	101.33	EX
	$C_{12}H_{27}N$	Tributylamine	372.80	0.9762	101.46	EX
CCl_4		**Tetrachloromethane**				
	C_2H_6O	Ethanol	338.19	0.6140	101.33	OX
	C_3H_6O	Acetone	341.25	0.0337	149.93	OX
	C_3H_8O	1-Propanol	346.28	0.8032	101.33	OX
	C_3H_8O	2-Propanol	341.83	0.6686	101.33	OX
	C_4H_6O	2-Butenal	348.15	0.6500	97.86	OX
	C_4H_6O	2-Methylpropenal	339.15	0.6000	97.86	OX
	C_4H_8O	2-Butanone	346.99	0.6630	101.33	OX
	$C_4H_8O_2$	Ethyl acetate	347.95	0.5700	101.33	OX
	$C_4H_{10}O$	1-Butanol	349.71	0.9500	101.33	OX
	$C_4H_{10}O$	2-Methyl-1-propanol	348.95	0.9080	101.33	OX
	$C_5H_{10}O$	2-Methyl-3-buten-2-ol	348.45	0.9009	101.06	OX
CS_2		**Carbon disulfide**				
	CH_4O	Methanol	310.65	0.7000	101.33	LX
$CHCl_3$		**Trichloromethane**				
	CH_4O	Methanol	328.15	0.6480	107.99	OX
	C_2H_6O	Ethanol	332.45	0.8410	101.33	OX
	C_3H_6O	Acetone	337.58	0.6398	101.33	ON
	$C_3H_6O_2$	Methyl acetate	337.51	0.6760	101.33	ON
	C_3H_8O	2-Propanol	334.15	0.9500	101.33	OX
	C_4H_6O	2-Butenal	329.15	0.9950	97.86	OX
	C_6H_{12}	2-Methyl-1-pentene	333.95	0.6235	101.19	OX
	C_6H_{14}	Hexane	333.45	0.7840	101.33	OX
CHN		**Hydrogen cyanide**				
	C_3H_5Cl	3-Chloropropene	296.45	0.8016	101.33	OX
CH_2Cl_2		**Dichloromethane**				
	C_2H_6O	Ethanol	312.05	0.9600	101.33	OX
CH_2O_2		**Formic acid**				
	$C_2H_4Cl_2$	1,2-Dichloroethane	350.17	0.4275	101.33	OX
	$C_5H_{10}O_2$	Butyl formate	372.15	0.8700	101.33	OX
	C_8H_{10}	m-Xylene	365.95	0.8545	101.33	EX
CH_3NO_2		**Nitromethane**				
	C_2H_6O	Ethanol	333.15	0.2850	53.61	OX
	C_3H_7Br	1-Bromopropane	343.25	0.1020	99.82	OX
	$C_4H_8O_2$	1,4-Dioxane	373.25	0.4101	101.48	OX
	C_5H_{10}	2-Methyl-2-butene	311.15	0.0570	101.33	LX
	C_7H_{14}	Methylcyclohexane	354.85	0.5123	101.33	EX
	C_7H_{16}	Heptane	353.25	0.4790	101.33	EX
	C_8H_{18}	Octane	363.38	0.6964	99.73	EX
	C_9H_{20}	Nonane	369.29	0.8403	99.73	EX
	$C_{10}H_{22}$	Decane	371.96	0.9239	99.73	EX
	$C_{11}H_{24}$	Undecane	373.16	0.9619	99.73	EX
	$C_{12}H_{26}$	Dodecane	373.75	0.9846	99.73	EX
CH_4O		**Methanol**				
	$C_2HBrClF_3$	2-Bromo-2-chloro-1,1,1-trifluoroethane	317.25	0.1890	93.33	OX

Molecular Formula						
Comp. 1	Comp. 2	Name	T_{Az}/K	$y_{1,Az}$	P_{Az}/kPa	Type
	C_2H_5Br	Bromoethane	308.05	0.1610	101.33	OX
	C_3H_5Cl	3-Chloropropene	312.15	0.2570	100.39	OX
	C_3H_6O	Acetone	328.29	0.2400	101.33	OX
	$C_3H_6O_2$	Methyl acetate	328.15	0.3480	107.19	OX
	$C_3H_6O_2$	1,3-Dioxolane	334.66	0.6910	101.30	OX
	$C_3H_6O_2$	Ethyl formate	318.15	0.3000	81.34	OX
	$C_3H_6O_3$	Dimethyl carbonate	337.25	0.8504	102.52	OX
	C_3H_7Cl	1-Chloropropane	313.35	0.2500	101.59	OX
	$C_4H_4F_6O$	Bis(2,2,2-trifluoroethyl) ether	326.28	0.4450	101.30	OX
	$C_4H_6O_2$	Vinyl acetate	332.05	0.6182	101.33	OX
	C_4H_8O	2-Butanone	323.15	0.8020	58.80	OX
	C_4H_8O	Tetrahydrofuran	332.75	0.5040	101.33	OX
	$C_4H_8O_2$	Ethyl acetate	335.66	0.7120	101.33	OX
	$C_4H_{10}O$	Diethyl ether	305.15	0.0500	93.33	OX
	$C_4H_{10}O_2$	Dimethylacetal	330.35	0.4700	101.33	OX
	$C_5H_3F_9O$	1,1,1,2,3,3-Hexafluoro-3-(2,2,2-trifluoroethoxy)propane	330.67	0.5600	101.30	OX
	C_5H_6	1,3-Cyclopentadiene	309.05	0.2120	101.33	OX
	C_5H_8	2-Methyl-1,3-butadiene	303.55	0.1670	101.33	OX
	C_5H_8	Methylenecyclobutane	309.05	0.2190	101.33	OX
	C_5H_8	1-Methylcyclobutene	304.85	0.1900	101.33	OX
	C_5H_8	cis-1,3-Pentadiene	311.10	0.2300	101.33	OX
	C_5H_8	trans-1,3-Pentadiene	309.65	0.2110	101.33	OX
	C_5H_{10}	2-Methyl-1-butene	300.55	0.1720	101.33	OX
	C_5H_{10}	3-Methyl-1-butene	291.05	0.0890	101.33	OX
	C_5H_{10}	2-Methyl-2-butene	306.25	0.2160	101.33	OX
	C_5H_{10}	1-Pentene	300.05	0.1469	102.47	OX
	$C_5H_{10}O$	2,3-Epoxy-2-methylbutane	334.95	0.6590	101.33	OX
	C_5H_{12}	Isopentane	297.05	0.0930	101.33	OX
	C_5H_{12}	Pentane	303.20	0.1930	101.30	OX
	$C_5H_{12}O$	Butyl methyl ether	330.00	0.5515	100.08	OX
	$C_5H_{12}O$	Methyl tert-butyl ether	325.00	0.3140	103.15	OX
	$C_5H_{12}O$	Ethyl propyl ether	330.00	0.4050	112.25	OX
	$C_5H_{12}O_2$	Diethoxymethane	336.03	0.8127	101.52	OX
	$C_5H_{12}O_2$	2,2-Dimethoxypropane	334.15	0.7250	100.00	OX
	$C_5H_{14}N_2$	N,N,N',N'-Tetramethylmethanediamine	335.15	0.7670	101.33	OX
	C_6F_6	Hexafluorobenzene	318.15	0.6100	61.73	OX
	C_6H_5F	Fluorobenzene	333.35	0.6625	101.62	OX
	C_6H_6	Benzene	331.56	0.6090	101.33	OX
	C_6H_{12}	Cyclohexane	328.75	0.6090	106.66	OX
	C_6H_{12}	2-Methyl-1-pentene	330.00	0.4517	141.80	OX
	C_6H_{14}	2,3-Dimethylbutane	313.15	0.3620	85.50	OX
	C_6H_{14}	Hexane	333.15	0.5160	149.64	OX
	$C_6H_{14}O$	tert-Butyl ethyl ether	330.95	0.6002	101.54	OX
	$C_6H_{14}O$	Diisopropyl ether	330.00	0.5390	101.61	OX
	$C_6H_{14}O$	Butyl ethyl ether	335.00	0.8010	98.84	OX
	$C_6H_{14}O$	2-Methoxy-2-methylbutane	335.55	0.7735	101.69	OX
	C_7H_8	Toluene	336.65	0.8820	101.33	OX
	C_7H_{14}	Methylcyclohexane	333.15	0.7520	102.87	EX
	C_7H_{16}	Heptane	331.95	0.7279	101.33	OX
	$C_7H_{16}O$	2-Ethoxy-2-methylbutane	335.15	0.8736	97.28	OX
	C_8H_{18}	Octane	335.55	0.8830	101.33	LX
	C_9H_{20}	Nonane	337.25	0.9526	101.33	OX
$C_2Cl_3F_3$		**1,1,2-Trichloro-1,2,2-trifluoroethane**				
	$C_2H_3F_3O$	2,2,2-Trifluoroethanol	316.58	0.7770	101.33	EX
	C_2H_6O	Ethanol	317.75	0.8456	101.42	OX
	C_3H_8O	2-Propanol	319.35	0.9159	100.95	OX
	$C_4H_{10}O$	2-Methyl-2-propanol	319.95	0.9426	101.09	OX
C_2Cl_4		**Tetrachloroethene**				
	$C_2H_3Cl_3$	1,1,2-Trichloroethane	385.95	0.2115	101.33	OX

Molecular Formula						
Comp. 1	Comp. 2	Name	T_{Az}/K	$y_{1,Az}$	P_{Az}/kPa	Type
	C_8H_{16}	1-Octene	393.15	0.5900	101.33	OX
	C_8H_{16}	*cis*-4-Octene	393.65	0.7100	101.33	OX
	C_8H_{16}	*trans*-4-Octene	393.45	0.6700	101.33	OX
	C_8H_{18}	Octane	371.90	0.8781	53.44	OX
$C_2Cl_4F_2$		**1,1,2,2-Tetrachloro-1,2-difluoroethane**				
	$C_2H_4Cl_2$	1,2-Dichloroethane	353.80	0.2700	101.33	OX
$C_2HBrClF_3$		**2-Bromo-2-chloro-1,1,1-trifluoroethane**				
	$C_4H_{10}O$	Diethyl ether	323.65	0.7200	93.33	ON
C_2HCl_3		**Trichloroethene**				
	$C_2H_4Cl_2$	1,2-Dichloroethane	355.35	0.3324	101.36	OX
	C_2H_6O	Ethanol	343.85	0.4741	101.33	OX
	C_4H_6O	2-Butenal	360.15	0.9000	97.86	OX
	C_6H_{12}	Cyclohexane	353.40	0.0975	101.32	OX
$C_2H_2Cl_2$		***trans*-1,2-Dichloroethene**				
	$C_5H_3F_9O$	1,1,1,2,3,3-Hexafluoro-3-(2,2,2-trifluoroethoxy)propane	318.50	0.8390	101.30	OX
C_2H_3N		**Acetonitrile**				
	C_3H_8O	2-Propanol	348.15	0.5287	100.81	OX
	$C_4H_6O_2$	Vinyl acetate	344.65	0.1948	98.33	OX
	C_4H_8O	2-Butanone	352.15	0.3195	101.15	OX
	C_4H_8O	Tetrahydrofuran	338.95	0.0784	101.13	OX
	$C_4H_{10}O$	2-Methyl-2-propanol	333.15	0.6200	56.93	OX
	C_5H_8	2-Methyl-1,3-butadiene	306.75	0.0410	101.33	OX
	C_5H_8	Methylenecyclobutane	312.45	0.1450	101.33	OX
	$C_5H_8O_2$	Methyl methacrylate	355.25	0.9866	102.07	OX
	C_5H_{10}	2-Methyl-2-butene	308.95	0.1320	101.33	OX
	C_5H_{10}	1-Pentene	301.85	0.0830	101.33	OX
	C_5H_{12}	Isopentane	298.45	0.1040	101.33	EX
	C_6H_6	Benzene	328.15	0.4560	54.65	OX
	$C_6H_{14}O$	2-Methoxy-2-methylbutane	346.13	0.5835	100.56	OX
	$C_7H_{16}O$	2-Ethoxy-2-methylbutane	348.85	0.7219	98.99	OX
	$C_{10}H_{20}$	1-Decene	354.55	0.9924	100.51	OX
$C_2H_4Cl_2$		**1,1-Dichloroethane**				
	C_3H_8O	2-Propanol	329.55	0.8928	101.60	OX
	C_6H_{14}	Hexane	329.30	0.8025	101.21	OX
$C_2H_4Cl_2$		**1,2-Dichloroethane**				
	C_3H_8O	2-Propanol	347.25	0.5258	100.32	OX
	$C_4H_{10}O$	2-Methyl-1-propanol	356.05	0.9173	101.26	OX
	$C_4H_{10}O$	2-Methyl-2-propanol	349.45	0.5336	101.43	OX
	C_7H_{14}	Methylcyclohexane	354.65	0.8036	101.21	OX
	C_8H_{18}	2,2,4-Trimethylpentane	343.15	0.7600	73.13	OX
C_2H_4O		**Acetaldehyde**				
	C_4H_6	1,3-Butadiene	268.15	0.0520	101.33	OX
	C_5H_8	2-Methyl-1,3-butadiene	292.23	0.8140	101.33	OX
$C_2H_4O_2$		**Acetic acid**				
	C_5H_5N	Pyridine	411.25	0.5780	101.33	ON
	$C_5H_{12}O$	3-Methyl-2-butanol	392.65	0.7210	101.33	ON
	C_6H_7N	2-Methylpyridine	417.27	0.5120	101.33	ON
	$C_6H_{10}O_2$	Vinyl butanoate	386.45	0.5750	101.33	OX
	C_6H_{14}	Hexane	341.40	0.0839	101.33	OX
	C_7H_9N	2,4-Dimethylpyridine	435.45	0.3022	101.33	ON
	C_7H_{16}	Heptane	364.95	0.4490	101.33	OX
	C_8H_{10}	*o*-Xylene	389.75	0.8640	101.33	OX
	C_8H_{10}	*p*-Xylene	388.40	0.8200	101.33	OX
	C_8H_{18}	Octane	378.85	0.6870	101.33	OX
	C_9H_{20}	Nonane	386.05	0.8250	101.33	OX
	$C_{10}H_{22}$	Decane	390.05	0.9250	101.33	OX
	$C_{11}H_{24}$	Undecane	391.15	0.9720	101.33	OX
$C_2H_4O_2$		**Methyl formate**				
	C_2H_5Br	Bromoethane	303.05	0.7360	101.33	OX

Molecular Formula						
Comp. 1	Comp. 2	Name	T_{Az}/K	$y_{1,Az}$	P_{Az}/kPa	Type
	$C_4H_{10}O$	Diethyl ether	301.55	0.6030	101.33	OX
	C_5H_8	2-Methyl-1,3-butadiene	298.90	0.5150	101.33	OX
	C_5H_{10}	2-Methyl-2-butene	297.75	0.5760	101.33	OX
	C_5H_{12}	Isopentane	291.55	0.4920	101.33	OX
	C_5H_{12}	Pentane	294.85	0.5740	101.33	OX
	C_6H_{14}	Hexane	302.65	0.8490	101.33	OX
C_2H_5Br		**Bromoethane**				
	C_5H_{10}	2-Methyl-2-butene	308.55	0.5110	101.33	OX
	C_5H_{12}	Isopentane	300.55	0.2180	101.33	OX
$C_2H_5NO_2$		**Nitroethane**				
	$C_4H_{10}O$	2-Methyl-1-propanol	375.81	0.4080	101.33	OX
	C_7H_{16}	Heptane	362.95	0.3520	101.33	OX
C_2H_6O		**Ethanol**				
	C_3H_3N	Acrylonitrile	343.95	0.4440	101.33	OX
	$C_3H_6O_2$	Methyl acetate	329.79	0.0362	101.33	OX
	$C_4H_3F_7O$	1,1,2,2-Tetrafluoroethyl 1,1,1-trifluoroethyl ether	326.67	0.2000	101.30	OX
	$C_4H_4F_6O$	Bis(2,2,2-trifluoroethyl) ether	331.90	0.2840	101.30	OX
	C_4H_8O	Butanal	345.45	0.3690	101.33	OX
	C_4H_8O	2-Butanone	347.15	0.5080	101.33	OX
	C_4H_8O	Tetrahydrofuran	344.95	0.1290	125.00	OX
	$C_4H_8O_2$	Ethyl acetate	344.85	0.4590	101.33	OX
	$C_4H_8O_2$	1,4-Dioxane	351.33	0.9480	101.33	OX
	$C_4H_8O_2$	Methyl propanoate	346.30	0.5140	103.91	OX
	$C_4H_{11}N$	Butylamine	354.99	0.5900	101.33	ON
	$C_5H_3F_9O$	1,1,1,2,3,3-Hexafluoro-3-(2,2,2-trifluoroethoxy)propane	337.88	0.3980	101.30	OX
	C_5H_8	2-Methyl-1,3-butadiene	305.95	0.1500	101.33	OX
	C_5H_8	Cyclopentene	323.40	0.1440	134.00	OX
	C_5H_{10}	2-Methyl-2-butene	309.79	0.0795	101.33	OX
	C_5H_{10}	Cyclopentane	323.44	0.1800	121.00	OX
	$C_5H_{10}O$	2,3-Epoxy-2-methylbutane	343.45	0.2930	101.33	OX
	$C_5H_{10}O$	3-Methyl-2-butanone	350.85	0.8250	101.33	OX
	$C_5H_{10}O$	2-Pentanone	351.15	0.9779	100.50	OX
	$C_5H_{10}O$	3-Pentanone	351.33	0.9590	101.33	OX
	$C_5H_{10}O_2$	Isopropyl acetate	349.85	0.7010	101.33	OX
	$C_5H_{10}O_2$	Methyl butanoate	346.30	0.8800	83.88	OX
	C_5H_{12}	Isopentane	299.95	0.0540	101.33	OX
	C_5H_{12}	Pentane	307.15	0.0537	101.33	OX
	$C_5H_{12}O$	Methyl *tert*-butyl ether	327.75	0.0380	101.33	OX
	$C_5H_{12}O_2$	Diethoxymethane	348.30	0.6497	102.35	OX
	C_6H_5F	Fluorobenzene	343.85	0.4752	101.54	OX
	C_6H_6	Benzene	341.25	0.4600	101.33	OX
	C_6H_{12}	Cyclohexane	337.95	0.4540	102.26	OX
	C_6H_{14}	Hexane	331.65	0.3410	101.33	OX
	$C_6H_{14}O$	*tert*-Butyl ethyl ether	339.95	0.3728	101.72	OX
	$C_6H_{14}O$	2-Methoxy-2-methylbutane	346.81	0.5820	101.32	OX
	C_7H_8	Toluene	349.75	0.8152	101.33	OX
	$C_7H_{16}O$	2-Ethoxy-2-methylbutane	349.35	0.7644	101.54	OX
	C_8H_{18}	Octane	349.85	0.8250	101.33	OX
	C_8H_{18}	2,2,4-Trimethylpentane	344.42	0.6450	101.33	OX
	C_9H_{20}	Nonane	351.35	0.9400	101.33	OX
$C_2H_6O_2$		**1,2-Ethanediol**				
	$C_5H_{12}O_3$	Di(ethylene glycol) monomethyl ether	463.95	0.4388	101.33	OX
	$C_6H_{14}O_3$	Di(ethylene glycol) monoethyl ether	467.15	0.6480	101.33	OX
	C_7H_8O	*o*-Cresol	462.67	0.3797	101.33	OX
	$C_7H_{16}O_3$	Di(ethylene glycol) monoisopropyl ether	466.35	0.6964	101.33	OX
	$C_7H_{16}O_3$	Di(ethylene glycol) monopropyl ether	468.55	0.8448	101.33	OX
	$C_7H_{16}O_3$	Di(propylene glycol) monomethyl ether (unspecified isomer)	457.65	0.3500	101.33	OX
	$C_8H_{11}N$	2,4,6-Trimethylpyridine	443.65	0.1734	101.33	OX
	$C_8H_{18}O_3$	Di(ethylene glycol) monobutyl ether	469.15	0.9102	101.33	OX

Molecular Formula						
Comp. 1	Comp. 2	Name	T_{Az}/K	$y_{1,Az}$	P_{Az}/kPa	Type
	$C_8H_{18}O_3$	Di(ethylene glycol) monoisobutyl ether	467.55	0.8355	101.33	OX
	$C_8H_{18}O_3$	Di(propylene glycol) monoethyl ether (unspecified isomer)	458.65	0.4800	101.33	OX
	$C_9H_{20}O_3$	Di(propylene glycol) monopropyl ether (unspecified isomer)	463.15	0.6590	101.33	OX
	$C_{10}H_{22}O_3$	Di(propylene glycol) monobutyl ether (unspecified isomer)	465.75	0.8130	101.33	OX
C_3H_3N		**Acrylonitrile**				
	C_5H_8	Methylenecyclobutane	313.80	0.1275	101.33	OX
	C_6H_6	Benzene	347.45	0.5575	101.46	OX
	C_6H_{12}	Cyclohexane	337.75	0.4836	101.94	OX
	C_6H_{14}	Hexane	330.90	0.4048	101.05	OX
C_3H_4O		**Acrolein**				
	C_5H_8	2-Methyl-1,3-butadiene	306.45	0.1980	101.33	OX
C_3H_6O		**Propanal**				
	C_5H_8	2-Methyl-1,3-butadiene	306.35	0.1700	101.33	OX
	C_5H_8	Methylenecyclobutane	311.30	0.3600	101.33	OX
C_3H_6O		**Acetone**				
	$C_3H_6O_2$	Methyl acetate	328.85	0.6470	101.33	OX
	C_3H_7Br	1-Bromopropane	328.75	0.9915	99.75	OX
	C_4H_8O	Tetrahydrofuran	328.85	0.9603	100.35	OX
	C_4H_9Cl	2-Chloro-2-methylpropane	322.05	0.1944	102.11	OX
	C_5H_8	2-Methyl-1,3-butadiene	306.95	0.0610	101.33	OX
	C_5H_8	Methylenecyclobutane	311.25	0.2800	101.33	OX
	C_5H_8	1-Methylcyclobutene	307.75	0.2220	101.33	OX
	C_5H_{10}	2-Methyl-1-butene	303.25	0.1400	101.33	OX
	C_5H_{10}	2-Methyl-2-butene	308.75	0.2440	101.33	OX
	C_5H_{12}	Isopentane	298.75	0.1730	101.33	OX
	$C_5H_{12}O$	Methyl *tert*-butyl ether	324.35	0.4824	102.19	OX
	C_6H_{12}	Cyclohexane	330.05	0.7590	109.32	OX
	C_6H_{12}	1-Hexene	323.35	0.5973	101.40	OX
	C_6H_{12}	2-Methyl-1-pentene	333.40	0.5793	140.60	OX
	C_6H_{14}	Hexane	322.95	0.6480	101.33	OX
	$C_6H_{14}O$	Diisopropyl ether	327.10	0.7424	100.17	OX
	$C_6H_{15}N$	Triethylamine	318.15	0.9800	68.13	OX
	C_7H_{14}	Methylcyclohexane	318.15	0.9500	68.66	OX
C_3H_6O		**Allyl alcohol**				
	$C_5H_{10}O_2$	Ethyl propanoate	367.65	0.5597	99.79	OX
	C_6H_6	Benzene	349.90	0.2203	101.33	OX
	C_6H_{12}	Cyclohexane	333.15	0.2790	63.98	OX
$C_3H_6O_2$		**Methyl acetate**				
	C_3H_7Br	1-Bromopropane	329.60	0.9727	99.56	OX
	C_6H_{10}	Cyclohexene	330.35	0.9121	102.87	OX
	C_6H_{12}	Cyclohexane	328.65	0.8000	101.33	OX
	C_6H_{12}	Methylcyclopentane	325.85	0.6917	99.50	OX
	C_6H_{12}	1-Hexene	323.15	0.6340	92.08	OX
	C_6H_{12}	2-Methyl-1-pentene	325.15	0.5931	100.38	OX
	C_6H_{14}	Hexane	326.65	0.6590	106.66	OX
	C_7H_{16}	Heptane	323.15	0.9570	79.48	OX
$C_3H_6O_2$		**Ethyl formate**				
	C_3H_7Br	2-Bromopropane	326.15	0.7090	101.33	OX
	C_6H_{12}	Cyclohexane	323.15	0.8210	91.46	OX
$C_3H_6O_2$		**Propanoic acid**				
	C_5H_5N	Pyridine	421.75	0.6860	101.33	ON
$C_3H_6O_3$		**Dimethyl carbonate**				
	$C_5H_{12}O_2$	Diethoxymethane	358.71	0.4437	100.42	OX
	C_6H_6	Benzene	353.50	0.1366	100.48	OX
	C_6H_{12}	Cyclohexane	346.95	0.3780	101.49	OX
	C_6H_{12}	Methylcyclopentane	342.35	0.2680	103.46	OX
	C_6H_{14}	Hexane	338.15	0.2540	98.46	OX
	$C_6H_{14}O$	Dipropyl ether	356.45	0.5044	100.73	OX
	C_7H_{16}	Heptane	355.15	0.5930	99.67	OX

Molecular Formula						
Comp. 1	Comp. 2	Name	T_{Az}/K	$y_{1,Az}$	P_{Az}/kPa	Type
C_3H_7Br		**1-Bromopropane**				
	C_3H_8O	2-Propanol	339.15	0.7349	99.97	OX
	C_6H_{12}	Cyclohexane	343.35	0.9219	98.84	OX
C_3H_7NO		**N,N-Dimethylformamide**				
	C_7H_{16}	Heptane	370.15	0.0800	101.33	OX
	$C_{10}H_{16}$	1,4-Dimethyl-4-vinylcyclohexene	415.65	0.5880	101.33	OX
	$C_{10}H_{16}$	1-Methyl-3-(1-methylethylidene)cyclohexene	419.05	0.7250	101.33	OX
$C_3H_7NO_2$		**1-Nitropropane**				
	C_7H_{16}	Heptane	369.25	0.1630	101.33	OX
$C_3H_7NO_2$		**2-Nitropropane**				
	C_7H_{16}	Heptane	367.55	0.2920	101.33	OX
C_3H_8O		**1-Propanol**				
	$C_4H_3F_7O$	1,1,2,2-Tetrafluoroethyl 1,1,1-trifluoroethyl ether	329.23	0.0350	101.30	OX
	$C_4H_4F_6O$	Bis(2,2,2-trifluoroethyl) ether	336.22	0.1100	101.30	OX
	$C_4H_6O_2$	2,3-Butanedione	359.30	0.3600	100.67	OX
	$C_4H_8O_2$	1,4-Dioxane	365.30	0.6418	101.30	OX
	$C_5H_{10}O_2$	Propyl acetate	367.88	0.6190	101.33	OX
	$C_5H_{12}O_2$	Diethoxymethane	359.01	0.2320	99.43	OX
	C_6H_6	Benzene	350.20	0.2060	101.33	OX
	C_6H_{12}	Cyclohexane	347.68	0.2490	101.33	OX
	C_6H_{12}	Methylcyclopentane	340.85	0.1729	101.19	OX
	$C_6H_{12}O_2$	4,4-Dimethyl-1,3-dioxane	368.20	0.9597	101.30	OX
	C_6H_{14}	Hexane	348.15	0.1900	137.23	OX
	C_7H_8	Toluene	365.35	0.6770	101.33	OX
	C_7H_{16}	Heptane	357.65	0.4830	101.33	OX
	C_8H_8	Styrene	369.08	0.9884	98.13	OX
	C_8H_{10}	o-Xylene	369.85	0.9886	98.66	OX
	C_8H_{10}	m-Xylene	369.90	0.9531	99.06	OX
	C_8H_{10}	p-Xylene	369.60	0.9531	99.99	OX
	C_8H_{14}	1-Octyne	369.00	0.8600	101.33	OX
	C_8H_{18}	Octane	366.85	0.7483	101.33	OX
	C_8H_{18}	2,2,4-Trimethylpentane	357.89	0.4580	101.30	OX
	C_9H_{20}	Nonane	369.95	0.9225	101.33	OX
C_3H_8O		**2-Propanol**				
	$C_4H_4F_6O$	Bis(2,2,2-trifluoroethyl) ether	334.16	0.2230	101.30	OX
	$C_4H_6O_2$	2,3-Butanedione	350.85	0.6454	100.95	OX
	C_4H_8O	2-Butanone	350.55	0.3830	101.33	OX
	$C_4H_{10}O$	2-Methyl-2-propanol	343.05	0.5551	60.27	ON
	$C_5H_3F_9O$	1,1,1,2,3,3-Hexafluoro-3-(2,2,2-trifluoroethoxy)propane	341.23	0.3420	101.30	OX
	C_5H_8	2-Methyl-1,3-butadiene	307.05	0.0150	101.33	OX
	C_5H_{10}	2-Methyl-2-butene	310.95	0.0460	101.33	OX
	$C_5H_{10}O$	2,3-Epoxy-2-methylbutane	346.10	0.1400	101.33	OX
	$C_5H_{10}O$	3-Methyl-2-butanone	354.75	0.8500	101.33	OX
	C_5H_{12}	Isopentane	298.15	0.1370	101.33	OX
	$C_5H_{12}O_2$	Diethoxymethane	351.45	0.6107	98.61	OX
	C_6H_5F	Fluorobenzene	347.75	0.4666	101.25	OX
	C_6H_6	Benzene	345.03	0.3960	101.33	OX
	C_6H_{10}	Cyclohexene	344.65	0.4271	101.40	OX
	C_6H_{12}	Cyclohexane	342.75	0.4050	101.33	OX
	C_6H_{12}	Methylcyclopentane	336.45	0.2900	98.14	OX
	C_6H_{14}	Hexane	338.15	0.2900	112.66	OX
	$C_6H_{14}O$	Diisopropyl ether	340.00	0.2050	103.36	OX
	$C_6H_{15}N$	Diisopropylamine	352.94	0.4890	101.33	OX
	C_7H_8	Toluene	354.65	0.8370	101.33	OX
	C_7H_{14}	Methylcyclohexane	350.85	0.6530	101.33	OX
	C_7H_{16}	Heptane	349.55	0.6023	101.33	OX
	$C_7H_{16}O$	tert-Butyl isopropyl ether	349.95	0.5306	102.70	OX
	C_8H_{18}	Octane	354.63	0.8990	101.33	OX
	C_8H_{18}	2,2,4-Trimethylpentane	349.58	0.6350	101.30	OX

Molecular Formula						
Comp. 1	Comp. 2	Name	T_{Az}/K	$y_{1,Az}$	P_{Az}/kPa	Type
$C_3H_8O_2$		**2-Methoxyethanol**				
	C_8H_8	Styrene	393.95	0.7787	98.93	OX
	C_8H_{10}	o-Xylene	392.65	0.7127	98.79	OX
	C_8H_{10}	m-Xylene	392.15	0.6397	99.73	OX
	C_8H_{10}	p-Xylene	392.65	0.6303	99.99	OX
	C_8H_{16}	1-Octene	380.75	0.4700	101.33	OX
	C_8H_{16}	cis-4-Octene	381.25	0.4900	101.33	OX
	C_8H_{16}	trans-4-Octene	381.05	0.4900	101.33	OX
$C_3H_8O_2$		**Dimethoxymethane**				
	C_5H_6	1,3-Cyclopentadiene	313.65	0.3350	101.33	OX
	C_5H_8	2-Methyl-1,3-butadiene	306.80	0.0160	101.33	OX
	C_5H_8	Methylenecyclobutane	310.35	0.4630	101.33	OX
	C_5H_8	1-Methylcyclobutene	309.05	0.2900	101.33	OX
$C_3H_8O_2$		**1,2-Propanediol**				
	$C_7H_{16}O_3$	Di(propylene glycol) monomethyl ether (unspecified isomer)	456.85	0.5691	101.33	OX
	$C_8H_{18}O_3$	Di(propylene glycol) monoethyl ether (unspecified isomer)	458.75	0.7778	101.33	OX
	$C_9H_{20}O_3$	Di(propylene glycol) monoisopropyl ether (unspecified isomer)	458.95	0.8130	101.33	OX
	$C_9H_{20}O_3$	Di(propylene glycol) monopropyl ether (unspecified isomer)	458.95	0.9010	101.33	OX
	$C_{10}H_{22}O_3$	Di(propylene glycol) monobutyl ether (unspecified isomer)	459.65	0.9721	101.33	OX
	$C_{10}H_{22}O_3$	Di(propylene glycol) monoisobutyl ether (unspecified isomer)	459.05	0.9255	101.33	OX
$C_3H_8O_2$		**1,3-Propanediol**				
	$C_5H_{12}O_3$	Di(ethylene glycol) monomethyl ether	455.25	0.6300	101.33	OX
	$C_6H_{14}O_3$	Di(ethylene glycol) monoethyl ether	459.25	0.9350	101.33	OX
C_4H_6		**1,3-Butadiene**				
	C_4H_8	2-Butene (unspecified isomer)	267.59	0.7650	101.33	OX
C_4H_6O		**2-Butenal**				
	C_7H_8	Toluene	374.15	0.5950	97.86	OX
	C_8H_{18}	Octane	353.15	0.4950	97.86	OX
$C_4H_6O_2$		**Vinyl acetate**				
	C_6H_{12}	Cyclohexane	340.45	0.6200	101.33	OX
	C_6H_{14}	Hexane	335.25	0.4450	101.33	OX
$C_4H_6O_2$		**2,3-Butanedione**				
	C_7H_8	Toluene	362.70	0.9513	101.34	OX
$C_4H_6O_3$		**Acetic anhydride**				
	C_8H_{16}	1-Octene	367.53	0.2840	53.88	OX
	C_8H_{18}	Octane	397.65	0.3500	129.80	OX
C_4H_8O		**Butanal**				
	C_6H_{12}	2-Methyl-1-pentene	334.15	0.2293	101.48	OX
C_4H_8O		**2-Butanone**				
	$C_4H_8O_2$	Ethyl acetate	349.55	0.1700	101.33	OX
	C_6H_6	Benzene	351.53	0.4790	101.33	OX
	C_6H_{10}	Cyclohexene	343.29	0.5110	89.35	OX
	C_6H_{12}	1-Hexene	334.75	0.1760	100.58	OX
	C_6H_{14}	Hexane	337.15	0.3280	101.33	OX
	$C_6H_{14}O$	Diisopropyl ether	340.55	0.1938	101.56	OX
	$C_6H_{14}O$	Dipropyl ether	351.40	0.7785	100.88	OX
	C_7H_{14}	Methylcyclohexane	350.50	0.7984	98.93	OX
	C_7H_{16}	Heptane	350.15	0.7670	101.33	OX
C_4H_8O		**Tetrahydrofuran**				
	C_6H_{12}	2-Methyl-1-pentene	334.65	0.2867	101.29	OX
	C_6H_{14}	Hexane	323.15	0.5900	65.83	OX
$C_4H_8O_2$		**Ethyl acetate**				
	$C_4H_{10}O$	2-Methyl-2-propanol	349.75	0.7778	101.28	OX
	C_6H_6	Benzene	350.55	0.9453	102.45	OX
	C_6H_{10}	Cyclohexene	347.45	0.6183	100.87	OX
	C_6H_{12}	Cyclohexane	345.00	0.5390	102.45	OX
	C_6H_{12}	1-Hexene	333.15	0.1230	91.47	OX
	C_6H_{14}	Hexane	338.00	0.3430	101.32	OX
	C_7H_{14}	Methylcyclohexane	349.90	0.9001	101.83	OX

Molecular Formula						
Comp. 1	Comp. 2	Name	T_{Az}/K	$y_{1,Az}$	P_{Az}/kPa	Type
$C_4H_8O_2$		**Butanoic acid**				
	C_5H_5N	Pyridine	436.35	0.9117	101.33	ON
	$C_8H_{16}O_2$	Butyl butanoate	434.60	0.6532	93.33	OXD
	$C_8H_{16}O_2$	Butyl butanoate	434.78	0.8639	93.33	OND
	$C_{11}H_{24}$	Undecane	435.55	0.9060	101.33	OX
$C_4H_8O_2$		**1,4-Dioxane**				
	$C_4H_{10}O$	2-Butanol	371.75	0.4732	100.77	OX
	$C_5H_{10}O_2$	Propyl acetate	373.35		0.6334	101.13
	$C_5H_{12}O$	2-Methyl-2-butanol	373.75	0.8119	99.62	OX
	C_6H_{10}	Cyclohexene	355.75	0.1065	101.44	OX
	C_6H_{12}	Methylcyclopentane	343.85	0.0538	99.79	OX
	$C_6H_{15}N$	Triethylamine	343.15	0.2500	56.80	OX
	C_7H_{16}	Heptane	364.30	0.4868	101.06	OX
	$C_7H_{16}O$	2-Ethoxy-2-methylbutane	369.15	0.5452	100.27	OX
$C_4H_8O_2$		**Propyl formate**				
	C_6H_6	Benzene	343.15	0.3770	76.08	OX
$C_4H_8O_2$		**Methyl propanoate**				
	C_7H_{14}	Methylcyclohexane	352.45	0.8956	101.33	OX
C_4H_9Cl		**1-Chlorobutane**				
	C_6H_{12}	Cyclohexane	348.31	0.5800	95.85	OX
C_4H_9NO		*N,N*-**Dimethylacetamide**				
	C_8H_{10}	*o*-Xylene	416.95	0.0591	103.40	OX
	C_8H_{10}	Ethylbenzene	408.95	0.0037	101.70	OX
$C_4H_{10}O$		**1-Butanol**				
	C_5H_5N	Pyridine	392.00	0.7050	101.33	ON
	$C_5H_{10}O_3$	Diethyl carbonate	370.85	0.6346	53.20	OX
	C_6H_5Cl	Chlorobenzene	388.25	0.6950	101.33	OX
	C_6H_{12}	Cyclohexane	352.68	0.0787	101.33	OX
	$C_6H_{12}O_2$	Butyl acetate	389.97	0.7700	101.33	OX
	$C_6H_{12}O_2$	Isobutyl acetate	387.15	0.5980	101.33	OX
	C_6H_{14}	Hexane	341.35	0.0370	101.33	OX
	C_7H_8	Toluene	378.85	0.3320	101.33	OX
	C_7H_{12}	3-Ethylcyclopentene	367.65	0.1900	101.33	OX
	C_7H_{16}	Heptane	366.55	0.2272	101.38	OX
	C_8H_8	Styrene	388.71	0.8923	98.39	OX
	C_8H_{10}	*o*-Xylene	388.05	0.8671	100.13	OX
	C_8H_{10}	*m*-Xylene	387.75	0.7865	101.46	OX
	C_8H_{10}	*p*-Xylene	387.85	0.7823	99.73	OX
	C_8H_{14}	1-Octyne	386.50	0.6200	101.33	OX
	C_8H_{14}	2-Octyne	398.30	0.7910	101.33	OX
	C_8H_{16}	1-Octene	363.45	0.4530	53.33	OX
	C_8H_{16}	*cis*-4-Octene	382.35	0.5300	101.33	OX
	C_8H_{16}	*trans*-4-Octene	382.15	0.5310	101.33	OX
	C_8H_{18}	Octane	383.15	0.5500	102.79	OX
	$C_8H_{18}O$	Dibutyl ether	390.59	0.8754	101.33	OX
	C_9H_{16}	1-Butylcyclopentene	356.70	0.8450	79.99	OX
	C_9H_{16}	1-Nonyne	390.60	0.9400	101.33	OX
	C_9H_{20}	Nonane	389.05	0.8128	101.33	OX
$C_4H_{10}O$		**2-Butanol**				
	$C_5H_{10}O$	3-Pentanone	370.50	0.6075	99.98	OX
	C_6H_{10}	Cyclohexene	352.75	0.2046	101.25	OX
	C_6H_{12}	Cyclohexane	349.90	0.1892	101.02	OX
	C_6H_{14}	Hexane	348.15	0.1010	128.66	OX
	$C_6H_{14}O$	2-Methoxy-2-methylbutane	359.15	0.0991	102.12	OX
	C_7H_8	Toluene	353.44	0.5550	56.67	OX
	C_7H_{16}	Heptane	361.95	0.4116	102.70	OX
	$C_7H_{16}O$	2-Ethoxy-2-methylbutane	367.75	0.4931	102.89	OX
	C_8H_{10}	*m*-Xylene	369.85	0.9717	101.06	OX
	C_8H_{10}	*p*-Xylene	369.55	0.9646	101.46	OX

Molecular Formula						
Comp. 1	**Comp. 2**	**Name**	T_{Az}/K	$y_{1,Az}$	P_{Az}/kPa	**Type**
	C_8H_{18}	Octane	371.05	0.8001	101.30	OX
$C_4H_{10}O$		**Diethyl ether**				
	C_5H_{12}	Pentane	306.85	0.5500	101.33	OX
$C_4H_{10}O$		**2-Methyl-1-propanol**				
	$C_5H_{10}O_2$	Isobutyl formate	370.90	0.1930	101.33	OX
	C_6H_6	Benzene	352.45	0.0780	101.33	OX
	C_6H_{10}	Cyclohexene	353.75	0.1363	100.31	OX
	C_6H_{12}	Cyclohexane	351.35	0.1325	101.45	OX
	C_6H_{12}	Methylcyclopentane	343.15	0.0567	100.35	OX
	C_7H_8	Toluene	374.35	0.4941	101.33	OX
	C_8H_{10}	m-Xylene	380.35	0.9300	101.33	OX
	C_8H_{10}	p-Xylene	380.30	0.9200	101.33	OX
	C_8H_{18}	Octane	376.58	0.6700	101.30	OX
$C_4H_{10}O$		**2-Methyl-2-propanol**				
	C_5H_8	Methylenecyclobutane	314.65	0.0150	101.33	OX
	C_6H_{10}	Cyclohexene	346.00	0.4172	99.61	OX
	C_6H_{12}	Methylcyclopentane	339.35	0.2559	99.93	OX
	C_6H_{12}	1-Hexene	333.25	0.2650	101.30	OX
	C_6H_{14}	Hexane	337.70	0.2502	101.30	OX
	$C_6H_{14}O$	tert-Butyl ethyl ether	342.85	0.2512	101.44	OX
	$C_6H_{14}O$	Diisopropyl ether	340.45	0.1058	101.72	OX
	$C_6H_{14}O$	2-Methoxy-2-methylbutane	353.20	0.5617	101.80	OX
	C_7H_8	Toluene	353.44	0.9200	93.61	OX
	$C_7H_{16}O$	tert-Butyl isopropyl ether	350.90	0.5390	102.94	OX
	C_8H_{18}	Octane	343.15	0.9680	61.18	OX
	C_8H_{18}	2,2,4-Trimethylpentane	339.28	0.6040	59.49	OX
$C_4H_{10}O_2$		**1,4-Butanediol**				
	$C_{15}H_{32}O$	1-Pentadecanol	502.75	0.9980	101.33	OX
$C_4H_{10}O_2$		**1,2-Dimethoxyethane**				
	C_7H_{14}	Methylcyclohexane	350.00	0.8190	79.42	OX
$C_4H_{10}O_2$		**2-Ethoxyethanol**				
	C_8H_8	Styrene	405.75	0.6438	101.33	OX
	C_8H_{10}	o-Xylene	404.95	0.5965	101.36	OX
	C_8H_{10}	m-Xylene	401.75	0.5159	101.33	OX
	C_8H_{10}	p-Xylene	402.55	0.5042	102.19	OX
	C_8H_{10}	Ethylbenzene	401.05	0.4632	100.94	OX
$C_4H_{10}O_3$		**Di(ethylene glycol)**				
	$C_9H_{20}O$	1-Nonanol	486.65	0.0095	101.33	OX
$C_4H_{11}N$		**Butylamine**				
	C_6H_6	Benzene	343.15	0.7000	80.89	OX
C_5H_5N		**Pyridine**				
	C_7H_8	Toluene	383.19	0.2250	101.33	OX
	C_7H_{16}	Heptane	368.61	0.3002	101.33	OX
	C_9H_{20}	Nonane	388.15	0.9350	101.33	OX
C_5H_6		**2-Methyl-1-buten-3-yne**				
	C_5H_8	2-Methyl-1,3-butadiene	305.88	0.7210	101.33	OX
	C_5H_{10}	2-Methyl-1-butene	303.15	0.3450	101.33	OX
	C_5H_{12}	Isopentane	299.35	0.3620	101.33	OX
C_5H_6		**1,3-Cyclopentadiene**				
	C_5H_{10}	2-Methyl-2-butene	310.85	0.3000	101.33	OX
	C_5H_{12}	Pentane	307.75	0.1959	101.30	OX
C_5H_8		**2-Methyl-1,3-butadiene**				
	C_5H_{12}	Pentane	310.55	0.7421	114.66	OX
	$C_6F_{15}N$	Tris(perfluoroethyl)amine	303.35	0.8200	101.33	EX
C_5H_8		**3-Methyl-1-butyne**				
	C_5H_{12}	Isopentane	297.15	0.5650	101.33	OX
C_5H_8		**1-Pentyne**				
	C_5H_{10}	2-Methyl-2-butene	310.95	0.3300	101.33	OX
	C_5H_{12}	Pentane	307.55	0.3050	101.33	OX

| Molecular Formula | | | | | | |
Comp. 1	Comp. 2	Name	T_{Az}/K	$y_{1,Az}$	P_{Az}/kPa	Type
C_5H_8O		**Cyclopentanone**				
	$C_5H_{12}O$	3-Methyl-1-butanol	402.02	0.5944	101.33	OX
	$C_5H_{12}O$	1-Pentanol	403.84	0.9196	101.33	OX
$C_5H_8O_2$		**Methyl methacrylate**				
	C_7H_{16}	Heptane	366.35	0.4597	99.94	OX
	C_8H_{18}	Octane	373.70	0.9651	100.16	OX
C_5H_{10}		**2-Methyl-1-butene**				
	$C_6F_{15}N$	Tris(perfluoroethyl)amine	301.95	0.8450	101.33	OX
C_5H_{10}		**2-Methyl-2-butene**				
	$C_6F_{15}N$	Tris(perfluoroethyl)amine	307.65	0.8170	101.33	OX
$C_5H_{10}O$		**3-Methyl-3-buten-1-ol**				
	C_6H_{12}	Cyclohexane	352.65	0.0215	101.10	OX
	$C_6H_{12}O_2$	4,4-Dimethyl-1,3-dioxane	403.05	0.7590	102.26	OX
	C_7H_8	Toluene	381.55	0.2391	101.60	OX
	C_7H_{16}	Heptane	370.00	0.2100	101.30	OX
$C_5H_{10}O$		**2-Methyl-3-buten-2-ol**				
	C_6H_{12}	Cyclohexane	350.15	0.1904	101.20	OX
	C_6H_{12}	1-Hexene	336.55	0.0479	101.30	OX
	C_7H_8	Toluene	366.55	0.7788	101.20	OX
$C_5H_{10}O$		**3-Pentanone**				
	C_7H_{14}	Methylcyclohexane	366.95	0.4441	99.82	OX
	$C_7H_{16}O$	2-Ethoxy-2-methylbutane	371.15	0.4764	100.21	OX
$C_5H_{10}O_2$		**Propyl acetate**				
	C_6H_{12}	Cyclohexane	353.15	0.0598	100.43	OX
	C_7H_{14}	Methylcyclohexane	368.40	0.4746	100.90	OX
	C_7H_{16}	Heptane	366.75	0.4215	101.38	OX
	$C_7H_{16}O$	2-Ethoxy-2-methylbutane	370.95	0.6529	100.03	OX
C_5H_{12}		**Isopentane**				
	$C_6F_{15}N$	Tris(perfluoroethyl)amine	299.65	0.9020	101.33	OX
$C_5H_{12}O$		**2-Methyl-1-butanol**				
	C_8H_{10}	*o*-Xylene	402.05	0.7417	101.87	OX
	C_8H_{10}	*m*-Xylene	400.65	0.6316	101.85	OX
	C_8H_{10}	*p*-Xylene	400.15	0.6273	101.07	OX
	C_8H_{10}	Ethylbenzene	398.75	0.5657	99.46	OX
$C_5H_{12}O$		**3-Methyl-1-butanol**				
	$C_6H_{10}O$	Cyclohexanone	404.87	0.9094	101.33	OX
	C_7H_8	Toluene	383.15	0.1250	101.33	OX
	$C_7H_{14}O_2$	Isopentyl acetate	403.95	0.9900	101.33	OX
	C_7H_{16}	Heptane	368.15	0.1016	95.06	OX
$C_5H_{12}O$		**2-Methyl-2-butanol**				
	C_6H_6	Benzene	352.35	0.1500	101.33	OX
	C_6H_{12}	Cyclohexane	351.95	0.1100	101.33	OX
	C_6H_{12}	Methylcyclopentane	344.75	0.0551	101.80	OX
	C_6H_{14}	Hexane	339.06	0.0436	93.55	OX
	C_7H_{14}	Methylcyclohexane	366.60	0.3965	99.87	OX
	C_7H_{16}	Heptane	348.15	0.3140	56.83	OX
	$C_7H_{16}O$	2-Ethoxy-2-methylbutane	369.85	0.3904	100.52	OX
$C_5H_{12}O$		**1-Pentanol**				
	$C_6H_{10}O$	Cyclohexanone	392.37	0.9748	53.32	OX
	$C_7H_{14}O_2$	Isopentyl acetate	407.45	0.6000	101.33	OX
	C_7H_{16}	Heptane	371.45	0.0576	101.33	OX
	C_8H_{18}	Octane	393.15	0.2847	101.33	OX
	C_9H_{20}	Nonane	404.45	0.6242	101.33	OX
	$C_{10}H_{22}$	Decane	410.65	0.9221	101.33	OX
$C_5H_{12}O$		**3-Pentanol**				
	C_7H_{16}	Heptane	368.15	0.2001	98.62	OX
$C_5H_{12}O_2$		**Diethoxymethane**				
	C_6H_{12}	Cyclohexane	353.21	0.1774	101.39	OX
	C_6H_{14}	Hexane	361.27	0.9101	102.30	OX

Molecular Formula						
Comp. 1	Comp. 2	Name	T_{Az}/K	$y_{1,Az}$	P_{Az}/kPa	Type
C_6F_6		**Hexafluorobenzene**				
	C_6H_6	Benzene	353.60	0.7600	101.33	OND
	C_6H_6	Benzene	352.50	0.1832	101.33	OXD
$C_6F_{15}N$		**Tris(perfluoroethyl)amine**				
	C_6H_6	Benzene	329.95	0.5900	101.33	EX
	C_6H_{12}	Cyclohexane	329.35	0.5690	101.33	EX
	C_6H_{14}	Hexane	327.65	0.4840	101.33	OX
C_6H_5Br		**Bromobenzene**				
	$C_6H_{12}O$	Cyclohexanol	403.15	0.7390	52.45	OX
C_6H_6		**Benzene**				
	C_6H_{12}	Cyclohexane	353.15	0.5460	109.18	OX
	C_6H_{12}	Methylcyclopentane	333.15	0.1390	69.93	OX
	C_6H_{14}	Hexane	341.45	0.0500	101.33	OX
	C_7H_{16}	Heptane	353.25	0.9922	101.32	OX
	C_8H_{18}	2,2,4-Trimethylpentane	353.25	0.9751	101.32	OX
C_6H_6O		**Phenol**				
	C_6H_7N	Aniline	459.09	0.3884	101.33	ON
	C_6H_7N	2-Methylpyridine	458.33	0.7852	101.32	ON
	C_6H_7N	3-Methylpyridine	462.93	0.6918	101.32	ON
	C_7H_5N	Benzonitrile	465.11	0.2345	101.33	ON
	C_7H_6O	Benzaldehyde	447.00	0.6001	73.00	ON
	C_7H_9N	2,6-Dimethylpyridine	459.32	0.7539	101.32	ON
	C_8H_{18}	Octane	398.17	0.0690	101.32	OX
	C_9H_{12}	Propylbenzene	428.15	0.1150	91.85	OX
	C_9H_{12}	1,2,3-Trimethylbenzene	443.45	0.3936	101.33	OX
	C_9H_{12}	1,2,4-Trimethylbenzene	440.65	0.2409	101.33	OX
	C_9H_{12}	1,3,5-Trimethylbenzene	436.95	0.1828	101.33	OX
	C_9H_{18}	1-Nonene	413.15	0.1297	86.57	OX
	C_9H_{20}	Nonane	419.18	0.2180	101.32	OX
	$C_{10}H_{12}$	1,2,3,4-Tetrahydronaphthalene	448.15	0.9031	84.25	OX
	$C_{10}H_{14}$	Butylbenzene	447.05	0.5535	101.33	OX
	$C_{10}H_{14}$	sec-Butylbenzene	441.15	0.3129	101.33	OX
	$C_{10}H_{14}$	tert-Butylbenzene	439.95	0.2773	101.33	OX
	$C_{10}H_{14}$	Diethylbenzene (unspecified isomer)	446.45	0.4705	101.33	OX
	$C_{10}H_{14}$	o-Diethylbenzene	447.15	0.5565	101.33	OX
	$C_{10}H_{14}$	m-Diethylbenzene	445.95	0.5152	101.33	OX
	$C_{10}H_{14}$	1-Ethyl-3,5-dimethylbenzene	447.95	0.5840	101.33	OX
	$C_{10}H_{14}$	2-Ethyl-1,4-dimethylbenzene	447.95	0.5840	101.33	OX
	$C_{10}H_{14}$	Isobutylbenzene	441.75	0.3522	101.33	OX
	$C_{10}H_{14}$	1-Isopropyl-2-methylbenzene	443.75	0.4643	101.33	OX
	$C_{10}H_{14}$	1-Isopropyl-3-methylbenzene	443.05	0.4027	101.33	OX
	$C_{10}H_{14}$	1-Isopropyl-4-methylbenzene	443.65	0.4430	101.33	OXD
	$C_{10}H_{14}$	1-Methyl-2-propylbenzene	447.75	0.5801	101.33	OX
	$C_{10}H_{14}$	1-Methyl-3-propylbenzene	446.35	0.5264	101.33	OX
	$C_{10}H_{14}$	1-Methyl-4-propylbenzene	447.15	0.5575	101.33	OX
	$C_{10}H_{14}$	1,2,3,5-Tetramethylbenzene	454.25	0.7957	101.33	OX
	$C_{10}H_{14}$	1,2,4,5-Tetramethylbenzene	453.36	0.7857	101.33	OX
	$C_{10}H_{18}$	trans-Decahydronaphthalene	443.15	0.5419	99.85	OX
	$C_{10}H_{22}$	Decane	434.15	0.4450	101.32	OX
	$C_{11}H_{16}$	1-Butyl-2-methylbenzene	455.55	0.8504	101.33	OX
	$C_{11}H_{16}$	1-Butyl-3-methylbenzene	454.15	0.8099	101.33	OX
	$C_{11}H_{16}$	1-Butyl-4-methylbenzene	454.55	0.8230	101.33	OX
	$C_{11}H_{22}$	1-Undecene	443.15	0.6426	92.31	OX
	$C_{12}H_{26}$	Dodecane	450.73	0.7900	101.32	OX
	$C_{14}H_{30}$	Tetradecane	452.48	0.9650	101.32	OX
C_6H_7N		**Aniline**				
	C_7H_8O	o-Cresol	464.29	0.0953	101.33	ON
	C_9H_{12}	1,2,3-Trimethylbenzene	444.65	0.3331	101.33	OX
	C_9H_{12}	1,2,4-Trimethylbenzene	441.80	0.1850	101.33	OX

Molecular Formula						
Comp. 1	Comp. 2	Name	T_{Az}/K	$y_{1,Az}$	P_{Az}/kPa	Type
	C_9H_{12}	1,3,5-Trimethylbenzene	437.68	0.1071	101.33	OX
	C_9H_{20}	Nonane	422.35	0.1770	101.33	OX
	$C_{10}H_{14}$	Butylbenzene	448.65	0.4993	101.33	OX
	$C_{10}H_{14}$	sec-Butylbenzene	443.15	0.3021	101.33	OX
	$C_{10}H_{14}$	tert-Butylbenzene	438.25	0.2104	101.33	OX
	$C_{10}H_{14}$	o-Diethylbenzene	448.75	0.5024	101.33	OX
	$C_{10}H_{14}$	m-Diethylbenzene	447.45	0.4584	101.33	OX
	$C_{10}H_{14}$	p-Diethylbenzene	448.85	0.5086	101.33	OX
	$C_{10}H_{14}$	1-Ethyl-3,5-dimethylbenzene	449.65	0.5310	101.33	OX
	$C_{10}H_{14}$	2-Ethyl-1,4-dimethylbenzene	449.65	0.5310	101.33	OX
	$C_{10}H_{14}$	Isobutylbenzene	442.75	0.2890	101.33	OX
	$C_{10}H_{14}$	1-Isopropyl-2-methylbenzene	445.95	0.4052	101.33	OX
	$C_{10}H_{14}$	1-Isopropyl-3-methylbenzene	444.15	0.3419	101.33	OX
	$C_{10}H_{14}$	1-Isopropyl-4-methylbenzene	445.35	0.3829	101.33	OX
	$C_{10}H_{14}$	1-Methyl-2-propylbenzene	449.45	0.5270	101.33	OX
	$C_{10}H_{14}$	1-Methyl-3-propylbenzene	447.85	0.4711	101.33	OX
	$C_{10}H_{14}$	1-Methyl-4-propylbenzene	448.75	0.5035	101.33	OX
	$C_{10}H_{14}$	1,2,3,5-Tetramethylbenzene	456.55	0.7504	101.33	OX
	$C_{10}H_{14}$	1,2,4,5-Tetramethylbenzene	455.36	0.7349	101.33	OX
	$C_{10}H_{22}$	Decane	440.43	0.4660	101.33	OX
	$C_{11}H_{16}$	1-Butyl-2-methylbenzene	458.15	0.8091	101.33	OX
	$C_{11}H_{16}$	1-Butyl-3-methylbenzene	456.55	0.7661	101.33	OX
	$C_{11}H_{16}$	1-Butyl-4-methylbenzene	457.05	0.7807	101.33	OX
	$C_{11}H_{24}$	Undecane	449.05	0.6970	101.33	OX
	$C_{12}H_{26}$	Dodecane	453.52	0.8220	101.33	OX
	$C_{13}H_{28}$	Tridecane	456.22	0.9300	101.33	OX
	$C_{14}H_{30}$	Tetradecane	457.05	0.9770	101.33	OX
C_6H_7N		**2-Methylpyridine**				
	C_8H_{18}	Octane	394.27	0.4610	101.33	OX
	C_9H_{20}	Nonane	402.35	0.8790	101.33	OX
C_6H_7N		**3-Methylpyridine**				
	C_7H_8O	m-Cresol	477.01	0.1556	101.32	ON
	C_7H_9N	2,6-Dimethylpyridine	416.64	0.2940	101.33	OX
C_6H_7N		**4-Methylpyridine**				
	C_7H_8O	m-Cresol	477.74	0.1822	101.32	ON
	C_7H_9N	2,6-Dimethylpyridine	417.08	0.2000	101.32	OX
$C_6H_{10}O$		**Methyldihydropyran (unspecified isomer)**				
	C_7H_8	Toluene	381.85	0.0207	101.30	OX
$C_6H_{10}O$		**4-Methylenetetrahydropyran**				
	C_7H_8	Toluene	381.15	0.5253	101.20	OX
$C_6H_{12}O$		**Cyclohexanol**				
	C_8H_{10}	o-Xylene	415.95	0.1426	101.33	OX
	C_8H_{10}	m-Xylene	411.85	0.0503	101.33	OX
	C_8H_{10}	p-Xylene	410.95	0.0505	101.33	OX
	C_9H_{20}	Nonane	410.20	0.3350	79.99	OX
$C_6H_{12}O_2$		**Butyl acetate**				
	C_8H_{16}	1-Octene	393.00	0.3030	101.33	OX
$C_6H_{12}O_2$		**4,4-Dimethyl-1,3-dioxane**				
	C_8H_{10}	o-Xylene	404.65	0.9662	101.30	OX
	C_8H_{18}	Octane	393.95	0.3343	101.20	OX
	C_9H_{20}	Nonane	402.15	0.8864	101.30	OX
	$C_{10}H_{22}$	Decane	405.35	0.9999	100.60	OX
$C_6H_{14}O$		**1-Hexanol**				
	C_8H_{18}	Octane	398.55	0.0886	101.33	OX
	C_9H_{20}	Nonane	416.95	0.3649	101.33	OX
	$C_{10}H_{22}$	Decane	427.05	0.7123	101.33	OX
$C_6H_{14}O_2$		**2,2-Dimethoxybutane**				
	C_7H_8	Toluene	380.15	0.9180	101.44	OX
C_7F_{16}		**Perfluoroheptane**				

Molecular Formula						
Comp. 1	Comp. 2	Name	T_{Az}/K	$y_{1,Az}$	P_{Az}/kPa	Type
	C_7H_{16}	Heptane	328.16	0.6100	53.60	OX
C_7H_5N		**Benzonitrile**				
	C_7H_8O	o-Cresol	468.91	0.5100	101.33	ON
	C_7H_8O	m-Cresol	476.10	0.1441	101.33	ON
	C_7H_8O	p-Cresol	476.95	0.0898	101.33	ON
	$C_8H_{10}O$	2,6-Xylenol	477.15	0.0807	101.33	ON
C_7H_8O		**o-Cresol**				
	$C_8H_{11}N$	2,4,6-Trimethylpyridine	470.35	0.6561	101.33	ON
	$C_{10}H_{14}$	sec-Butylbenzene	444.65	0.0938	101.33	OX
	$C_{10}H_{14}$	Diethylbenzene (unspecified isomer)	453.10	0.2694	101.33	OX
	$C_{10}H_{14}$	1,2,4,5-Tetramethylbenzene	462.37	0.6273	101.33	OX
	$C_{10}H_{22}$	Decane	433.15	0.3100	78.71	OX
	$C_{11}H_{22}$	1-Undecene	448.15	0.5516	83.07	OX
	$C_{11}H_{24}$	Undecane	433.15	0.5800	56.40	OX
	$C_{12}H_{26}$	Dodecane	458.15	0.8466	93.55	OX
C_7H_8O		**m-Cresol**				
	C_7H_9N	2,6-Dimethylpyridine	475.66	0.9869	101.32	ON
	C_9H_7N	Quinoline	511.20	0.0356	101.33	ON
	C_9H_{20}	Nonane	413.15	0.0400	76.54	OX
	$C_{10}H_8$	Naphthalene	474.65	0.9680	101.33	OX
	$C_{10}H_{12}$	1,2,3,4-Tetrahydronaphthalene	468.45	0.5900	93.10	OX
	$C_{10}H_{14}$	sec-Butylbenzene	445.85	0.0136	101.33	OX
	$C_{10}H_{14}$	Diethylbenzene (unspecified isomer)	454.10	0.1010	101.33	OX
	$C_{10}H_{14}$	1,2,4,5-Tetramethylbenzene	466.87	0.3591	101.33	OX
	$C_{10}H_{22}$	Decane	433.15	0.2170	75.85	OX
C_7H_8O		**p-Cresol**				
	$C_{10}H_8$	Naphthalene	474.55	0.9414	101.33	OX
	$C_{10}H_{14}$	sec-Butylbenzene	446.05	0.0186	101.33	OX
	$C_{10}H_{14}$	Diethylbenzene (unspecified isomer)	454.50	0.1105	101.33	OX
C_7H_8O		**Benzyl alcohol**				
	$C_{10}H_{22}$	Decane	445.75	0.2490	101.33	OX
C_7H_9N		**2-Methylaniline**				
	$C_{10}H_{22}$	Decane	446.91	0.1770	101.33	OX
	$C_{11}H_{24}$	Undecane	461.40	0.4930	101.33	OX
	$C_{12}H_{26}$	Dodecane	468.90	0.7650	101.33	OX
	$C_{13}H_{28}$	Tridecane	472.55	0.9070	101.33	OX
$C_7H_{14}O_2$		**Pentyl acetate**				
	C_9H_{20}	Nonane	419.20	0.5380	101.32	OX
$C_7H_{16}O$		**1-Heptanol**				
	C_9H_{20}	Nonane	423.45	0.1071	101.33	OX
	$C_{10}H_{22}$	Decane	438.75	0.4308	101.33	OX
	$C_{11}H_{24}$	Undecane	447.85	0.8014	101.33	OX
C_8H_{10}		**o-Xylene**				
	C_9H_{20}	Nonane	417.40	0.8498	101.33	OX
$C_8H_{10}O$		**2,6-Xylenol**				
	C_9H_7N	Quinoline	511.00	0.0890	101.33	ON
	$C_{10}H_8$	Naphthalene	475.70	0.9381	101.33	OX
	$C_{10}H_{14}$	1,2,4,5-Tetramethylbenzene	468.85	0.3480	101.33	OX
$C_8H_{10}O$		**2,3-Xylenol**				
	C_9H_7N	Quinoline	513.30	0.2684	101.33	ON
	$C_{10}H_8$	Naphthalene	485.45	0.4123	101.33	OX
$C_8H_{10}O$		**2,4-Xylenol**				
	C_9H_7N	Quinoline	512.30	0.1717	101.33	ON
	$C_{10}H_8$	Naphthalene	481.25	0.6435	101.33	OX
	$C_{10}H_{14}$	1,2,4,5-Tetramethylbenzene	474.05	0.1869	101.33	OX
$C_8H_{10}O$		**2,5-Xylenol**				
	C_9H_7N	Quinoline	512.30	0.1717	101.33	ON
	$C_{10}H_8$	Naphthalene	481.25	0.6435	101.33	OX
	$C_{10}H_{14}$	1,2,4,5-Tetramethylbenzene	474.35	0.1763	101.33	OX

Molecular Formula						
Comp. 1	Comp. 2	Name	T_{Az}/K	$y_{1,Az}$	P_{Az}/kPa	Type
$C_8H_{10}O$		**3,4-Xylenol**				
	C_9H_7N	Isoquinoline	519.75	0.2955	101.33	ON
	C_9H_7N	Quinoline	514.77	0.3907	101.33	ON
	$C_{10}H_8$	Naphthalene	490.95	0.1158	101.33	OX
	$C_{10}H_9N$	3-Methylisoquinoline	524.35	0.0811	101.33	ON
	$C_{10}H_9N$	2-Methylquinoline	521.17	0.1647	101.33	ON
	$C_{10}H_9N$	3-Methylquinoline	523.60	0.1152	101.33	ON
	$C_{10}H_9N$	7-Methylquinoline	525.85	0.0466	101.33	ON
	$C_{11}H_{11}N$	2,3-Dimethylquinoline	521.60	0.2113	101.33	ON
$C_8H_{10}O$		**3,5-Xylenol**				
	C_9H_7N	Isoquinoline	518.05	0.1915	101.33	ON
	C_9H_7N	Quinoline	513.58	0.3287	101.33	ON
	$C_{10}H_8$	Naphthalene	489.33	0.2601	101.33	OX
	$C_{10}H_9N$	2-Methylquinoline	520.65	0.0094	101.33	ON
	$C_{11}H_{11}N$	2,3-Dimethylquinoline	520.70	0.0530	101.33	ON
$C_8H_{10}O$		**2-Ethylphenol**				
	C_9H_7N	Quinoline	511.75	0.1041	101.33	ON
	$C_{10}H_8$	Naphthalene	478.35	0.8005	101.33	OX
	$C_{10}H_{14}$	1,2,4,5-Tetramethylbenzene	471.45	0.3707	101.33	OX
$C_8H_{10}O$		**3-Ethylphenol**				
	C_9H_7N	Quinoline	512.70	0.2089	101.33	ON
	$C_{10}H_8$	Naphthalene	483.45	0.5551	101.33	OX
	$C_{10}H_{14}$	1,2,4,5-Tetramethylbenzene	475.95	0.1249	101.33	OX
C_{8H10O}		**4-Ethylphenol**				
	C_9H_7N	Quinoline	513.45	0.2832	101.33	ON
	$C_{10}H_8$	Naphthalene	486.10	0.3762	101.33	OX
$C_8H_{11}N$		**2,4-Dimethylaniline**				
	$C_{11}H_{24}$	Undecane	468.13	0.1490	101.33	OX
	$C_{12}H_{26}$	Dodecane	482.95	0.4520	101.33	OX
	$C_{13}H_{28}$	Tridecane	488.43	0.7880	101.33	OX
	$C_{14}H_{30}$	Tetradecane	490.53	0.9840	101.33	OX
$C_8H_{18}O$		**1-Octanol**				
	$C_{10}H_{22}$	Decane	446.45	0.1029	101.33	OX
	$C_{11}H_{24}$	Undecane	460.05	0.4772	101.33	OX
	$C_{12}H_{26}$	Dodecane	466.95	0.8836	101.33	OX
C_9H_7N		**Isoquinoline**				
	$C_{11}H_{10}$	2-Methylnaphthalene	513.90	0.2074	101.33	OX
C_9H_7N		**Quinoline**				
	$C_9H_{12}O$	3-Isopropylphenol	514.70	0.6109	101.33	ON
	$C_9H_{12}O$	2-Isopropylphenol	512.75	0.8015	101.33	ON
	$C_9H_{12}O$	2-Propylphenol	513.60	0.7243	101.33	ON
	$C_9H_{12}O$	3-Propylphenol	514.70	0.6109	101.33	ON
	$C_9H_{12}O$	4-Propylphenol	515.35	0.5451	101.33	ON
	$C_{10}H_{14}O$	2-Butylphenol	515.70	0.5350	101.33	ON
	$C_{10}H_{14}O$	2-*tert*-Butylphenol	513.70	0.7299	101.33	ON
	C10	3-*tert*-Butylphenol	517.05	0.4315	101.33	ON
	$C_{10}H_{14}O$	4-Isobutylphenol	515.95	0.5061	101.33	ON
	$C_{10}H_{14}O$	2-*sec*-Butylphenol	514.70	0.6339	101.33	ON
	$C_{10}H_{14}O$	4-*sec*-Butylphenol	516.45	0.4551	101.33	ON
	$C_{11}H_{10}$	2-Methylnaphthalene	511.05	0.9213	101.33	OX
	$C_{11}H_{16}O$	2-*tert*-Butyl-5-methylphenol	515.45	0.5854	101.33	ON
	$C_{11}H_{16}O$	2-*sec*-Butyl-4-methylphenol	516.10	0.5139	101.33	ON
C_9H_{12}		**1,2,3-Trimethylbenzene**				
	$C_{10}H_{22}$	Decane	433.35	0.4010	72.54	OX
C_9H_{12}		**1,2,4-Trimethylbenzene**				
	$C_{10}H_{22}$	Decane	433.35	0.8600	80.25	OX
$C_9H_{12}O$		**2-Ethyl-4-methylphenol**				
	$C_{10}H_8$	Naphthalene	488.20	0.2218	101.33	OX
$C_9H_{12}O$		**2-Ethyl-5-methylphenol**				

Molecular Formula						
Comp. 1	Comp. 2	Name	T_{Az}/K	$y_{1,Az}$	P_{Az}/kPa	Type
	$C_{10}H_8$	Naphthalene	489.45	0.1710	101.33	OX
$C_9H_{12}O$		**2-Isopropylphenol**				
	$C_{10}H_8$	Naphthalene	483.15	0.5102	101.33	OX
	$C_{10}H_{14}$	1,2,4,5-Tetramethylbenzene	476.25	0.1036	101.33	OX
$C_9H_{12}O$		**2,4,6-Trimethylphenol**				
	$C_{10}H_8$	Naphthalene	486.70	0.3161	101.33	OX
$C_9H_{20}O$		**1-Nonanol**				
	$C_{11}H_{24}$	Undecane	468.45	0.0925	101.33	OX
	$C_{12}H_{26}$	Dodecane	480.65	0.5235	101.33	OX
$C_{10}H_{22}O$		**1-Decanol**				
	$C_{12}H_{26}$	Dodecane	489.25	0.1068	101.33	OX

VISCOSITY OF GASES

Marcia L. Huber and Allan H. Harvey

The following table gives the viscosity of some common gases as a function of temperature. Unless otherwise noted, the viscosity values refer to a pressure of 100 kPa (1 bar). The notation $P = 0$ indicates that the low-pressure limiting value is given. The difference between the viscosity at 100 kPa and the limiting value is generally less than 2%. Uncertainties for the viscosities of gases in this table are generally less than 3%; uncertainty information on specific fluids can be found in the references. Viscosity is given in units of µPa s; note that 1 µPa s = 10^{-5} poise. Substances are listed in the modified Hill order (see Introduction).

		Viscosity in µPa s						
		100 K	200 K	300 K	400 K	500 K	600 K	Ref.
	Air	7.1	13.3	18.5	23.1	27.1	30.8	1
Ar	Argon ($P = 0$)	8.1	15.9	22.7	28.6	33.9	38.8	2, 3*, 4*
BF_3	Boron trifluoride		12.3	17.1	21.7	26.1	30.2	5
ClH	Hydrogen chloride			14.6	19.7	24.3		5
F_6S	Sulfur hexafluoride ($P = 0$)			15.3	19.7	23.8	27.6	6
H_2	Normal hydrogen ($P = 0$)	4.1	6.8	8.9	10.9	12.8	14.5	3*, 7
D_2	Deuterium ($P = 0$)	5.9	9.6	12.6	15.4	17.9	20.3	8
H_2O	Water ($P = 0$)			9.8	13.4	17.3	21.4	9
D_2O	Deuterium oxide ($P = 0$)			10.2	13.7	17.8	22.0	10
H_2S	Hydrogen sulfide			12.5	16.9	21.2	25.4	11
H_3N	Ammonia			10.2	14.0	17.9	21.7	12
He	Helium ($P = 0$)	9.6	15.1	19.9	24.3	28.3	32.2	13
Kr	Krypton ($P = 0$)		17.4	25.5	32.9	39.6	45.8	14
NO	Nitric oxide		13.8	19.2	23.8	28.0	31.9	5
N_2	Nitrogen	7.0	12.9	17.9	22.2	26.1	29.6	1, 15*
N_2O	Nitrous oxide ($P = 0$)		10.0	15.0	19.8	24.1	27.9	16
Ne	Neon ($P = 0$)	14.4	24.1	31.9	38.6	44.8	50.6	17
O_2	Oxygen	7.7	14.7	20.7	25.8	30.5	34.7	1
O_2S	Sulfur dioxide		8.6	12.9	17.5	21.7		5
Xe	Xenon ($P = 0$)		15.7	23.2	30.5	37.2	43.5	3*, 14
CO	Carbon monoxide	6.7	12.9	17.8	22.1	25.8	29.1	5
CO_2	Carbon dioxide		10.1	15.0	19.7	24.0	28.0	18
$CHCl_3$	Chloroform			10.2	13.7	16.9	20.1	5
CH_4	Methane ($P = 0$)	3.9	7.7	11.1	14.2	17.0	19.5	3*, 19
CH_4O	Methanol ($P = 0$)		6.6	9.7	13.0	16.4	19.8	20
C_2H_2	Acetylene			10.4	13.5	16.5		5
C_2H_4	Ethylene		7.0	10.4	13.6	16.5	19.2	21
C_2H_6	Ethane		6.4	9.4	12.2	14.8	17.1	22
C_2H_6O	Ethanol			11.6	14.5	17.0		5
C_3H_8	Propane			8.2	10.8	13.3	15.6	23
C_4H_{10}	Butane			7.5	9.9	12.2	14.5	24
C_4H_{10}	Isobutane			7.5	9.9	12.2	14.4	25
$C_4H_{10}O$	Diethyl ether			7.6	10.1	12.4		5
C_5H_{12}	Pentane			6.7	9.2	11.4	13.4	5
C_6H_{14}	Hexane				8.6	10.8	12.8	5

* More accurate data covering a restricted temperature range.

References

1. Lemmon, E. W., and Jacobsen, R. T., Viscosity and Thermal Conductivity Equations for Nitrogen, Oxygen, Argon, and Air, *Int. J. Thermophys.* 25, 21, 2004.
2. Vogel, E., Jäger, B., Hellmann, R., and Bich, E., *Ab initio* Pair Potential Energy Curve for the Argon Atom Pair and Thermophysical Properties for the Dilute Argon Gas. II. Thermophysical Properties for Low-Density Argon, *Mol. Phys.* 108, 3335, 2010.
3. May, E. F., Berg, R. F., and Moldover, M. R., Reference Viscosities of H_2, CH_4, Ar, and Xe at Low Densities, *Int. J. Thermophys.* 28, 1085, 2007.
4. Vogel, E., Reference Viscosity of Argon at Low Density in the Temperature Range from 290 K to 680 K, *Int. J. Thermophys.* 31, 447, 2010.
5. Ho, C. Y., Ed., *Properties of Inorganic and Organic Fluids*, CINDAS Data Series on Materials Properties, Vol. V-1, Hemisphere Publishing Corp., New York, 1988.
6. Strehlow, T., and Vogel, E., Temperature Dependence and Initial Density Dependence of the Viscosity of Sulfur Hexafluoride, *Physica A* 161, 101, 1989.
7. Mehl, J. B., Huber, M. L., and Harvey, A. H., *Ab Initio* Transport Coefficients of Gaseous Hydrogen, *Int. J. Thermophys.* 31, 740, 2010.

8. Assael, M. J., Mixafendi, M., and Wakeham, W. A., The Viscosity of Normal Deuterium in the Limit of Zero Density, *J. Phys. Chem. Ref. Data* 16, 189, 1987.

9. Huber, M. L., Perkins, R. A., Laesecke, A., Friend, D. G., Sengers, J. V., Assael, M. J., Metaxa, I. M., Vogel, E., Mares, R., and Miyagawa, K., New International Formulation for the Viscosity of Water, *J. Phys. Chem. Ref. Data* 38, 101, 2009.

10. Matsunaga, N., and Nagashima, A., Transport Properties of Liquid and Gaseous D_2O over a Wide Range of Temperature and Pressure, *J. Phys. Chem. Ref. Dat*, 12, 933, 1983.

11. Schmidt, K. A. G., Quinones-Cisneros, S. E., Carroll, J. J., and Kvamme, B., Hydrogen Sulfide Viscosity Modeling, *Energy & Fuels* 22, 3424, 2008.

12. Fenghour, A., Wakeham, W. A., Vesovic, V., Watson, J. T. R., Millat, J., and Vogel, E., The Viscosity of Ammonia, *J. Phys. Chem. Ref. Data* 24, 1649, 1995.

13. Cencek, W., Komasa, J., Przybytek, M., Mehl, J. B., Jeziorski, B., and Szalewicz, K., Effects of Aadiabatic, Relativistic, and Quantum Electrodynamics Interactions in Helium Dimer on Thermophysical Properties of Helium, *J. Chem. Phys.*, to be submitted (2011).

14. Bich, E., Millat, J., and Vogel, E., The Viscosity and Thermal Conductivity of Pure Monatomic Gases from Their Normal Boiling Point up to 5000 K in the Limit of Zero Density and at 0.101325 MPa, *J. Phys. Chem. Ref. Data* 19, 1289, 1990.

15. Seibt, D., Herrmann, S., Vogel, E., Bich, E., and Hassel, E., Simultaneous Measurements on Helium and Nitrogen with a Newly Designed Viscometer-Densimeter over a Wide Range of Temperature and Pressure, *J. Chem. Eng. Data* 54, 2626, 2009.

16. Millat, J., Vesovic, V., and Wakeham, W. A., The Viscosity of Nitrous Oxide and Tetrafluoromethane in the Limit of Zero Density, *Int. J. Thermophys.* 12, 265, 1991.

17. Bich, E., Hellmann, R., and Vogel, E., *Ab initio* Potential Energy Curve for the Neon Atom Pair and Thermophysical Properties for the Dilute Neon Gas. II. Thermophysical Properties for Low-Density Neon, *Mol. Phys.* 106, 813, 2008.

18. Fenghour, A., Wakeham, W. A., and Vesovic, V., The Viscosity of Carbon Dioxide, *J. Phys. Chem. Ref. Data* 27, 31, 1998.

19. Hellmann, R., Bich, E., Vogel, E., Dickinson, A. S., and Vesovic, V., Calculation of the Transport and Relaxation Properties of Methane. I. Shear Viscosity, Viscomagnetic Effects, and Self-Diffusion, *J. Chem. Phys.* 129, 064302, 2008.

20. Xiang, H.-W., Huber, M. L., and Laesecke, A., A New Reference Correlation for the Viscosity of Methanol, *J. Phys. Chem. Ref. Data* 35, 1597, 2006.

21. Holland, P. M., Eaton, B. E., and Hanley, H. J. M., A Correlation of the Viscosity and Thermal Conductivity Data of Gaseous and Liquid Ethylene, *J. Phys. Chem. Ref. Data* 12, 917, 1983.

22. Friend, D. G., Ingham, H., and Ely, J. F., Thermophysical Properties of Ethane, *J. Phys. Chem. Ref. Data* 20, 275, 1991.

23. Vogel, E., Küchenmeister, C., Bich, E., and Laesecke, A., Reference Correlation of the Viscosity of Propane, *J. Phys. Chem. Ref. Data* 27, 947, 1998.

24. Vogel, E., Küchenmeister, C., and Bich, E., Viscosity for n-Butane in the Fluid Region, *High Temp. – High Press.* 31, 173, 1999.

25. Vogel, E., Küchenmeister, C., and Bich, E., Viscosity Correlation for Isobutane over Wide Ranges of the Fluid Region, *Int. J. Thermophys.* 21, 343, 2000.

VISCOSITY OF LIQUIDS

The absolute viscosity of some common liquids at temperatures between −25 and 100 °C is given in this table. Values were derived by fitting experimental data to suitable expressions for the temperature dependence. The substances are arranged by molecular formula in the modified Hill order (see Preface). All values are given in units of millipascal seconds (mPa s); this unit is identical to centipoise (cp).

Viscosity values correspond to a nominal pressure of 1 atmosphere. If a value is given at a temperature above the normal boiling point, the applicable pressure is understood to be the vapor pressure of the liquid at that temperature. A few values are given at a temperature slightly below the normal freezing point; these refer to the supercooled liquid.

The accuracy ranges from 1% in the best cases to 5 to 10% in the worst cases. Additional significant figures are included in the table to facilitate interpolation.

References

1. Viswanath, D. S. and Natarajan, G., *Data Book on the Viscosity of Liquids*, Hemisphere Publishing Corp., New York, 1989.
2. Daubert, T. E., Danner, R. P., Sibul, H. M., and Stebbins, C. C., *Physical and Thermodynamic Properties of Pure Compounds: Data Compilation*, extant 1994 (core with 4 supplements), Taylor & Francis, Bristol, PA (also available as database).
3. Ho, C. Y., Ed., *CINDAS Data Series on Material Properties*, Vol. V-1, *Properties of Inorganic and Organic Fluids*, Hemisphere Publishing Corp., New York, 1988.
4. Stephan, K. and Lucas, K., *Viscosity of Dense Fluids*, Plenum Press, New York, 1979.
5. Vargaftik, N. B., *Tables of Thermophysical Properties of Liquids and Gases*, 2nd ed., John Wiley, New York, 1975.

Molecular formula	Name	Viscosity in mPa s					
		−25 °C	0 °C	25 °C	50 °C	75 °C	100 °C
Compounds not containing carbon							
Br_2	Bromine		1.252	0.944	0.746		
Cl_3HSi	Trichlorosilane		0.415	0.326			
Cl_3P	Phosphorous trichloride	0.870	0.662	0.529	0.439		
Cl_4Si	Tetrachlorosilane			99.4	96.2		
H_2O	Water		1.793	0.890	0.547	0.378	0.282
H_4N_2	Hydrazine			0.876	0.628	0.480	0.384
Hg	Mercury			1.526	1.402	1.312	1.245
NO_2	Nitrogen dioxide		0.532	0.402			
Compounds containing carbon							
CCl_3F	Trichlorofluoromethane	0.740	0.539	0.421			
CCl_4	Tetrachloromethane		1.321	0.908	0.656	0.494	
CS_2	Carbon disulfide		0.429	0.352			
$CHBr_3$	Tribromomethane			1.857	1.367	1.029	
$CHCl_3$	Trichloromethane	0.988	0.706	0.537	0.427		
CHN	Hydrogen cyanide		0.235	0.183			
CH_2Br_2	Dibromomethane	1.948	1.320	0.980	0.779	0.652	
CH_2Cl_2	Dichloromethane	0.727	0.533	0.413			
CH_2O_2	Formic acid			1.607	1.030	0.724	0.545
CH_3I	Iodomethane		0.594	0.469			
CH_3NO	Formamide		7.114	3.343	1.833		
CH_3NO_2	Nitromethane	1.311	0.875	0.630	0.481	0.383	0.317
CH_4O	Methanol	1.258	0.793	0.544			
CH_5N	Methylamine	0.319	0.231				
$C_2Cl_3F_3$	1,1,2-Trichlorotrifluoro-ethane	1.465	0.945	0.656	0.481		
C_2Cl_4	Tetrachloroethylene		1.114	0.844	0.663	0.535	0.442
C_2HCl_3	Trichloroethylene		0.703	0.545	0.444	0.376	
C_2HCl_5	Pentachloroethane		3.761	2.254	1.491	1.061	
$C_2HF_3O_2$	Trifluoroacetic acid			0.808	0.571		
$C_2H_2Cl_2$	*cis*-1,2-Dichloroethylene	0.786	0.575	0.445			
$C_2H_2Cl_2$	*trans*-1,2-Dichloroethylene	0.522	0.398	0.317	0.261		
$C_2H_2Cl_4$	1,1,1,2-Tetrachloroethane	3.660	2.200	1.437	1.006	0.741	0.570
$C_2H_3ClF_2$	1-Chloro-1,1-difluoro-ethane	0.477	0.376				
C_2H_3ClO	Acetyl chloride			0.368	0.294		
$C_2H_3Cl_3$	1,1,1-Trichloroethane	1.847	1.161	0.793	0.578	0.428	
C_2H_3N	Acetonitrile		0.400	0.369	0.284	0.234	
$C_2H_4Br_2$	1,2-Dibromoethane			1.595	1.116	0.837	0.661

Molecular formula	Name	Viscosity in mPa s					
		−25 °C	0 °C	25 °C	50 °C	75 °C	100 °C
$C_2H_4Cl_2$	1,1-Dichloroethane			0.464	0.362		
$C_2H_4Cl_2$	1,2-Dichloroethane		1.125	0.779	0.576	0.447	
$C_2H_4O_2$	Acetic acid			1.056	0.786	0.599	0.464
$C_2H_4O_2$	Methyl formate		0.424	0.325			
C_2H_5Br	Bromoethane	0.635	0.477	0.374			
C_2H_5Cl	Chloroethane	0.416	0.319				
C_2H_5I	Iodoethane		0.723	0.556	0.444	0.365	
C_2H_5NO	N-Methylformamide		2.549	1.678	1.155	0.824	0.606
$C_2H_5NO_2$	Nitroethane	1.354	0.940	0.688	0.526	0.415	0.337
C_2H_6O	Ethanol	3.262	1.786	1.074	0.694	0.476	
C_2H_6OS	Dimethyl sulfoxide			1.987	1.290		
$C_2H_6O_2$	Ethylene glycol			16.1	6.554	3.340	1.975
C_2H_6S	Dimethyl sulfide		0.356	0.284			
C_2H_6S	Ethanethiol		0.364	0.287			
C_2H_7N	Dimethylamine	0.300	0.232				
C_2H_7NO	Ethanolamine			21.1	8.560	3.935	1.998
C_3H_5Br	3-Bromopropene		0.620	0.471	0.373		
C_3H_5Cl	3-Chloropropene		0.408	0.314			
C_3H_5ClO	Epichlorohydrin	2.492	1.570	1.073	0.781	0.597	0.474
C_3H_5N	Propanenitrile			0.294	0.240	0.202	
C_3H_6O	Acetone	0.540	0.395	0.306	0.247		
C_3H_6O	Allyl alcohol			1.218	0.759	0.505	
C_3H_6O	Propanal			0.321	0.249		
$C_3H_6O_2$	Ethyl formate		0.506	0.380	0.300		
$C_3H_6O_2$	Methyl acetate		0.477	0.364	0.284		
$C_3H_6O_2$	Propanoic acid		1.499	1.030	0.749	0.569	0.449
C_3H_7Br	1-Bromopropane		0.645	0.489	0.387		
C_3H_7Br	2-Bromopropane		0.612	0.458	0.359		
C_3H_7Cl	1-Chloropropane		0.436	0.334			
C_3H_7Cl	2-Chloropropane		0.401	0.303			
C_3H_7I	1-Iodopropane		0.970	0.703	0.541	0.436	0.363
C_3H_7I	2-Iodopropane		0.883	0.653	0.506	0.407	
C_3H_7NO	N,N-Dimethylformamide		1.176	0.794	0.624		
$C_3H_7NO_2$	1-Nitropropane	1.851	1.160	0.798	0.589	0.460	0.374
C_3H_8O	1-Propanol	8.645	3.815	1.945	1.107	0.685	
C_3H_8O	2-Propanol		4.619	2.038	1.028	0.576	
$C_3H_8O_2$	1,2-Propylene glycol		248	40.4	11.3	4.770	2.750
$C_3H_8O_3$	Glycerol			934	152	39.8	14.8
C_3H_8S	1-Propanethiol		0.503	0.385			
C_3H_8S	2-Propanethiol		0.477	0.357	0.280		
C_3H_9N	Propylamine			0.376			
C_3H_9N	Isopropylamine		0.454	0.325			
C_4H_4O	Furan	0.661	0.475	0.361			
C_4H_5N	Pyrrole		2.085	1.225	0.828	0.612	
$C_4H_6O_3$	Acetic anhydride		1.241	0.843	0.614	0.472	0.377
C_4H_7N	Butanenitrile			0.553	0.418	0.330	0.268
C_4H_8O	2-Butanone	0.720	0.533	0.405	0.315	0.249	
C_4H_8O	Tetrahydrofuran	0.849	0.605	0.456	0.359		
$C_4H_8O_2$	1,4-Dioxane			1.177	0.787	0.569	
$C_4H_8O_2$	Ethyl acetate		0.578	0.423	0.325	0.259	
$C_4H_8O_2$	Methyl propionate		0.581	0.431	0.333	0.266	
$C_4H_8O_2$	Propyl formate		0.669	0.485	0.370	0.293	
$C_4H_8O_2$	Butanoic acid		2.215	1.426	0.982	0.714	0.542
$C_4H_8O_2$	2-Methylpropanoic acid		1.857	1.226	0.863	0.639	0.492
$C_4H_8O_2S$	Sulfolane				6.280	3.818	2.559
C_4H_8S	Tetrahydrothiophene			0.973	0.912		
C_4H_9Br	1-Bromobutane		0.815	0.606	0.471	0.379	
C_4H_9Cl	1-Chlorobutane		0.556	0.422	0.329	0.261	
C_4H_9N	Pyrrolidine	1.914	1.071	0.704	0.512		
C_4H_9NO	N,N-Dimethylacetamide			1.927			

Molecular formula	Name	Viscosity in mPa s					
		−25 °C	0 °C	25 °C	50 °C	75 °C	100 °C
C_4H_9NO	Morpholine			2.021	1.247	0.850	0.627
$C_4H_{10}O$	1-Butanol	12.19	5.185	2.544	1.394	0.833	0.533
$C_4H_{10}O$	2-Butanol			3.096	1.332	0.698	0.419
$C_4H_{10}O$	2-Methyl-2-propanol			4.312	1.421	0.678	
$C_4H_{10}O$	Diethyl ether		0.283	0.224			
$C_4H_{10}O_3$	Diethylene glycol			30.200	11.130	4.917	2.505
$C_4H_{10}S$	Diethyl sulfide		0.558	0.422	0.331	0.267	
$C_4H_{11}N$	Butylamine		0.830	0.574	0.409	0.298	
$C_4H_{11}N$	Isobutylamine		0.770	0.571	0.367		
$C_4H_{11}N$	Diethylamine			0.319	0.239		
$C_4H_{11}NO_2$	Diethanolamine				109.5	28.7	9.100
$C_5H_4O_2$	Furfural		2.501	1.587	1.143	0.906	0.772
C_5H_5N	Pyridine		1.361	0.879	0.637	0.497	0.409
C_5H_{10}	1-Pentene	0.313	0.241	0.195			
C_5H_{10}	2-Methyl-2-butene		0.255	0.203			
C_5H_{10}	Cyclopentane		0.555	0.413	0.321		
$C_5H_{10}O$	Mesityl oxide	1.291	0.838	0.602	0.465	0.381	0.326
$C_5H_{10}O$	2-Pentanone		0.641	0.470	0.362	0.289	0.238
$C_5H_{10}O$	3-Pentanone		0.592	0.444	0.345	0.276	0.227
$C_5H_{10}O_2$	Butyl formate		0.937	0.644	0.472	0.362	0.289
$C_5H_{10}O_2$	Propyl acetate		0.768	0.544	0.406	0.316	0.255
$C_5H_{10}O_2$	Ethyl propanoate		0.691	0.501	0.380	0.299	0.242
$C_5H_{10}O_2$	Methyl butanoate		0.759	0.541	0.406	0.318	0.257
$C_5H_{10}O_2$	Methyl isobutanoate		0.672	0.488	0.373	0.296	
$C_5H_{11}N$	Piperidine			1.573	0.958	0.649	0.474
C_5H_{12}	Pentane	0.351	0.274	0.224			
C_5H_{12}	Isopentane	0.376	0.277	0.214			
$C_5H_{12}O$	1-Pentanol	25.4	8.512	3.619	1.820	1.035	0.646
$C_5H_{12}O$	2-Pentanol			3.470	1.447	0.761	0.465
$C_5H_{12}O$	3-Pentanol			4.149	1.473	0.727	0.436
$C_5H_{12}O$	2-Methyl-1-butanol			4.453	1.963	1.031	0.612
$C_5H_{12}O$	3-Methyl-1-butanol		8.627	3.692	1.842	1.031	0.631
$C_5H_{13}N$	Pentylamine		1.030	0.702	0.493	0.356	
C_6F_6	Hexafluorobenzene			2.789	1.730	1.151	
$C_6H_4Cl_2$	o-Dichlorobenzene		1.958	1.324	0.962	0.739	0.593
$C_6H_4Cl_2$	m-Dichlorobenzene		1.492	1.044	0.787	0.628	0.525
C_6H_5Br	Bromobenzene		1.560	1.074	0.798	0.627	0.512
C_6H_5Cl	Chlorobenzene	1.703	1.058	0.753	0.575	0.456	0.369
C_6H_5ClO	o-Chlorophenol			3.589	1.835	1.131	0.786
C_6H_5ClO	m-Chlorophenol				4.041		
C_6H_5F	Fluorobenzene		0.749	0.550	0.423	0.338	
C_6H_5I	Iodobenzene		2.354	1.554	1.117	0.854	0.683
$C_6H_5NO_2$	Nitrobenzene		3.036	1.863	1.262	0.918	0.704
C_6H_6	Benzene			0.604	0.436	0.335	
C_6H_6ClN	o-Chloroaniline			3.316	1.913	1.248	0.887
C_6H_6O	Phenol				3.437	1.784	1.099
C_6H_7N	Aniline			3.847	2.029	1.247	0.850
$C_6H_8N_2$	Phenylhydrazine			13.0	4.553	1.850	0.848
C_6H_{10}	Cyclohexene		0.882	0.625	0.467	0.364	
$C_6H_{10}O$	Cyclohexanone			2.017	1.321	0.919	0.671
$C_6H_{11}N$	Hexanenitrile			0.912	0.650	0.488	0.382
C_6H_{12}	Cyclohexane			0.894	0.615	0.447	
C_6H_{12}	Methylcyclopentane	0.927	0.653	0.479	0.364		
C_6H_{12}	1-Hexene	0.441	0.326	0.252	0.202		
$C_6H_{12}O$	Cyclohexanol			57.5	12.3	4.274	1.982
$C_6H_{12}O$	2-Hexanone	1.300	0.840	0.583	0.429	0.329	0.262
$C_6H_{12}O$	4-Methyl-2-pentanone			0.545	0.406		
$C_6H_{12}O_2$	Butyl acetate		1.002	0.685	0.500	0.383	0.305
$C_6H_{12}O_2$	Isobutyl acetate			0.676	0.493	0.370	0.286
$C_6H_{12}O_2$	Ethyl butanoate			0.639	0.453		

Molecular formula	Name	Viscosity in mPa s					
		−25 °C	0 °C	25 °C	50 °C	75 °C	100 °C
$C_6H_{12}O_2$	Diacetone alcohol	28.7	6.621	2.798	1.829	1.648	
$C_6H_{12}O_3$	Paraldehyde			1.079	0.692	0.485	0.362
$C_6H_{13}N$	Cyclohexylamine			1.944	1.169	0.782	0.565
C_6H_{14}	Hexane		0.405	0.300	0.240		
C_6H_{14}	2-Methylpentane		0.372	0.286	0.226		
C_6H_{14}	3-Methylpentane		0.395	0.306			
$C_6H_{14}O$	Dipropyl ether		0.542	0.396	0.304	0.242	
$C_6H_{14}O$	1-Hexanol			4.578	2.271	1.270	0.781
$C_6H_{15}N$	Triethylamine		0.455	0.347	0.273	0.221	
$C_6H_{15}N$	Dipropylamine		0.751	0.517	0.377	0.288	0.228
$C_6H_{15}N$	Diisopropylamine			0.393	0.300	0.237	
$C_6H_{15}NO_3$	Triethanolamine			609	114	31.5	11.7
C_7H_5N	Benzonitrile			1.267	0.883	0.662	0.524
C_7H_7Cl	o-Chlorotoluene		1.390	0.964	0.710	0.547	0.437
C_7H_7Cl	m-Chlorotoluene		1.165	0.823	0.616	0.482	0.391
C_7H_7Cl	p-Chlorotoluene			0.837	0.621	0.483	0.390
C_7H_8	Toluene	1.165	0.778	0.560	0.424	0.333	0.270
C_7H_8O	o-Cresol				3.035	1.562	0.961
C_7H_8O	m-Cresol			12.9	4.417	2.093	1.207
C_7H_8O	Benzyl alcohol			5.474	2.760	1.618	1.055
C_7H_8O	Anisole			1.056	0.747	0.554	0.427
C_7H_9N	N-Methylaniline		4.120	2.042	1.222	0.825	0.606
C_7H_9N	o-Methyl aniline		10.3	3.823	1.936	1.198	0.839
C_7H_9N	m-Methyl aniline		8.180	3.306	1.679	1.014	0.699
C_7H_9N	Benzylamine			1.624	1.080	0.769	0.577
C_7H_{14}	Methylcyclohexane		0.991	0.679	0.501	0.390	0.316
C_7H_{14}	1-Heptene		0.441	0.340	0.273	0.226	
$C_7H_{14}O$	2-Heptanone			0.714	0.407	0.297	
$C_7H_{14}O_2$	Heptanoic acid			3.840	2.282	1.488	1.041
C_7H_{16}	Heptane	0.757	0.523	0.387	0.301	0.243	
C_7H_{16}	3-Methylhexane			0.350			
$C_7H_{16}O$	1-Heptanol			5.810	2.603	1.389	0.849
$C_7H_{16}O$	2-Heptanol			3.955	1.799	0.987	0.615
$C_7H_{16}O$	3-Heptanol				1.957	0.976	0.584
$C_7H_{16}O$	4-Heptanol			4.207	1.695	0.882	0.539
$C_7H_{17}N$	Heptylamine			1.314	0.865	0.600	0.434
C_8H_8	Styrene		1.050	0.695	0.507	0.390	0.310
C_8H_8O	Acetophenone			1.681			0.634
$C_8H_8O_2$	Methyl benzoate			1.857			
$C_8H_8O_3$	Methyl salicylate					1.102	0.815
C_8H_{10}	Ethylbenzene		0.872	0.631	0.482	0.380	0.304
C_8H_{10}	o-Xylene		1.084	0.760	0.561	0.432	0.345
C_8H_{10}	m-Xylene		0.795	0.581	0.445	0.353	0.289
C_8H_{10}	p-Xylene			0.603	0.457	0.359	0.290
$C_8H_{10}O$	Phenetole			1.197	0.817	0.594	0.453
$C_8H_{11}N$	N,N-Dimethylaniline		1.996	1.300	0.911	0.675	0.523
$C_8H_{11}N$	N-Ethylaniline		3.981	2.047	1.231	0.825	0.596
C_8H_{16}	Ethylcyclohexane		1.139	0.784	0.579		
$C_8H_{16}O_2$	Octanoic acid			5.020	2.656	1.654	1.147
C_8H_{18}	Octane		0.700	0.508	0.385	0.302	0.243
$C_8H_{18}O$	1-Octanol			7.288	3.232	1.681	0.991
$C_8H_{18}O$	4-Methyl-3-heptanol		1.904	1.085	0.702	0.497	0.375
$C_8H_{18}O$	5-Methyl-3-heptanol		2.052	1.178	0.762	0.536	0.401
$C_8H_{18}O$	2-Ethyl-1-hexanol		20.7	6.271	2.631	1.360	0.810
$C_8H_{18}O$	Dibutyl ether	1.417	0.918	0.637	0.466	0.356	0.281
$C_8H_{19}N$	Dibutylamine		1.509	0.918	0.619	0.449	0.345
$C_8H_{19}N$	Diisobutylamine		1.115	0.723	0.511	0.384	0.303
C_9H_7N	Quinoline			3.337	1.892	1.201	0.833
C_9H_{10}	Indane		2.230	1.357	0.931	0.692	0.545
C_9H_{12}	Cumene		1.075	0.737	0.547		

Molecular formula	Name	Viscosity in mPa s					
		−25 °C	0 °C	25 °C	50 °C	75 °C	100 °C
$C_9H_{14}O$	Isophorone		4.201	2.329	1.415	0.923	0.638
$C_9H_{18}O$	5-Nonanone			1.199	0.834	0.619	0.484
$C_9H_{18}O_2$	Nonanoic acid			7.011	3.712	2.234	1.475
C_9H_{20}	Nonane		0.964	0.665	0.488	0.375	0.300
$C_9H_{20}O$	1-Nonanol			9.123	4.032		
$C_{10}H_{10}O_4$	Dimethyl phthalate		63.2	14.4	5.309	2.824	1.980
$C_{10}H_{14}$	Butylbenzene			0.950	0.683	0.515	
$C_{10}H_{18}$	cis-Decahydronaphthalene	12.8	5.645	3.042	1.875	1.271	0.924
$C_{10}H_{18}$	trans-Decahydronaphthalene	6.192	3.243	1.948	1.289	0.917	0.689
$C_{10}H_{20}O_2$	Decanoic acid				4.327	2.651	
$C_{10}H_{22}$	Decane	2.188	1.277	0.838	0.598	0.453	0.359
$C_{10}H_{22}O$	1-Decanol			10.9	4.590		
$C_{11}H_{24}$	Undecane		1.707	1.098	0.763	0.562	0.433
$C_{12}H_{10}O$	Diphenyl ether				2.130	1.407	1.023
$C_{12}H_{26}$	Dodecane		2.277	1.383	0.930	0.673	0.514
$C_{13}H_{12}$	Diphenylmethane					1.265	0.929
$C_{13}H_{28}$	Tridecane		2.909	1.724	1.129	0.796	0.594
$C_{14}H_{30}$	Tetradecane			2.128	1.376	0.953	0.697
$C_{16}H_{22}O_4$	Dibutyl phthalate	483	66.4	16.6	6.470	3.495	2.425
$C_{16}H_{34}$	Hexadecane			3.032	1.879	1.260	0.899
$C_{18}H_{38}$	Octadecane				2.487	1.609	1.132

VISCOSITY OF CARBON DIOXIDE ALONG THE SATURATION LINE

The table below gives the viscosity of gas and liquid CO_2 along the liquid–vapor saturation line.

References

1. Fenghour, A., Wakeham, W. A., and Vesovic, V., *J. Phys. Chem. Ref. Data*, 27, 31, 1998.
2. Angus, S., et al., *International Tables for the Fluid State: Carbon Dioxide*, Pergamon Press, Oxford, 1976.

T/K	P/kPa	Gas $\eta/\mu Pa\ s$	Liquid $\eta/\mu Pa\ s$
205	227	10.33	
210	327	10.60	
215	465	10.87	
220	600	11.13	241.68
225	735	11.41	221.72
230	894	11.69	203.75
235	1075	11.98	187.48
240	1283	12.27	172.67
245	1519	12.58	159.13
250	1786	12.90	146.69
255	2085	13.24	135.20
260	2419	13.61	124.30
265	2790	14.02	114.63
270	3203	14.47	105.21
275	3658	14.99	96.44
280	4160	15.61	87.89
285	4712	16.37	79.64
290	5315	17.36	71.47
295	5984	18.79	63.01
300	6710	21.29	53.33
302	6997	23.52	48.30

VISCOSITY AND DENSITY OF AQUEOUS HYDROXIDE SOLUTIONS

The viscosity and density of aqueous hydroxide solutions at 25 °C are tabulated here as a function of concentration. Viscosity is given in millipascal second, which is equal to the c.g.s. unit centipoise (cP). The last entry in each column refers to the saturated solution.

Reference

Sipos, P. M., Hefter, G., and May, P. M., *J. Chem. Eng. Data*, 45, 613, 2000.

Viscosity in mPa s

c/mol L^{-1}	LiOH	NaOH	KOH	CsOH	(CH$_3$)$_4$NOH
0.5	1.017	0.997	0.937	0.91	1.017
1.0	1.169	1.116	0.990	0.94	1.186
1.5	1.340	1.248	1.050	0.97	1.430
2.0	1.537	1.396	1.116	1.03	1.762
3.0	2.050	1.754	1.269	1.19	3.031
4.0	2.734	2.228	1.448	1.41	7.238
5.0		2.867	1.657	1.67	
6.0		3.727	1.902	1.98	
7.0		4.869	2.196	2.40	
8.0		6.351	2.554	3.09	
9.0		8.230	3.005	4.31	
10.0		10.554	3.581	6.46	
11.0		13.362	4.328		
12.0		16.677	5.303		
13.0		20.503	6.577		
14.0		24.826	8.235		
15.0		29.604			
16.0		34.767			
17.0		40.212			
18.0		45.800			
19.0		51.354			
Sat.	3.311	51.911	8.526		8.850

Density in g/cm^3

c/mol L^{-1}	LiOH	NaOH	KOH	CsOH	(CH$_3$)$_4$NOH
0.5	1.012	1.019	1.022	1.063	0.999
1.0	1.025	1.040	1.045	1.128	1.002
1.5	1.038	1.059	1.068	1.193	1.005
2.0	1.050	1.078	1.090	1.257	1.009
3.0	1.072	1.115	1.133	1.383	1.019
4.0	1.093	1.149	1.174	1.508	1.030
5.0		1.182	1.214	1.632	
6.0		1.213	1.253	1.755	
7.0		1.243	1.290	1.876	
8.0		1.271	1.326	1.997	
9.0		1.299	1.362	2.117	
10.0		1.325	1.396	2.236	
11.0		1.350	1.429	2.354	
12.0		1.374	1.462	2.471	
13.0		1.397	1.494	2.587	
14.0		1.419	1.524	2.703	
15.0		1.441			
16.0		1.461			
17.0		1.481			
18.0		1.499			
19.0		1.517			
Sat.	1.109	1.519	1.529	2.800	1.032

VISCOSITY OF LIQUID METALS

This table gives the viscosity of several liquid metals as a function of temperature. Experimental data from some of the references was smoothed to produce the table. Viscosity is given in millipascal second (mPa s), which equals the c.g.s. unit centipoise (cP).

References

1. Shpil'rain, E. E., Yakimovich, K. A., Fomin, V. A., Skovorodjko, S. N., and Mozgovoi, A. G., in *Handbook of Thermodynamic and Transport Properties of the Alkali Metals*, Ohse, R. H., Ed., Blackwell Scientific Publishers, Oxford, 1985. [Li, Na, K, Rb, Cs]
2. Culpin, M. F., *Proc. Phys. Soc.* 70, 1079, 1957. [Ca]
3. *Landolt-Börnstein, Numerical Data and Functional Relationships in Science and Technology, Sixth Edition*, II/5a, *Transport Phenomena I (Viscosity and Diffusion)*, Springer-Verlag, Heidelberg, 1961. [Co, Au, Mg, Ni, Ag]
4. Spells, K. E., *Proc. Phys. Soc.* 48, 299, 1936. [Ga]
5. Walsdorfer, H., Arpshofen, I., and Predel, B., *Z. Met.* 79, 503, 1988. [In]
6. Assael, M. J., Kakosimos, K., Banish, R. M., Brillo, J., Egry, I., Brooks, R., Quested, P. N., Mills, K. C., Nagashima, A., Sato, Y., and Wakeham, W. A., *J. Phys. Chem. Ref. Data* 35, 285, 2006. [Al, Fe]
7. Assael, M. J., Kalyva, A. E., Antoniadis, K. D., Banish, R. M., Egry, I., Wu, J., Kaschnitz, E., and Wakeham, W. A., *J. Phys. Chem. Ref. Data* 39, 033105-1, 2010. [Cu, Sn]

				Viscosity in mPa s				
$t/°C$	Lithium	Sodium	Potassium	Rubidium	Cesium	Gallium	Calcium	Magnesium
50				0.542	0.598	1.921		
100		0.687	0.441	0.435	0.469	1.608		
150		0.542	0.358	0.365	0.389	1.397		
200	0.566	0.451	0.303	0.316	0.334	1.245		
250	0.503	0.387	0.263	0.280	0.294	1.130		
300	0.453	0.341	0.234	0.252	0.264	1.040		
350	0.412	0.306	0.211	0.230	0.240	0.968		
400	0.379	0.278	0.193	0.212	0.221	0.909		
450	0.352	0.255	0.178	0.197	0.206	0.859		
500	0.328	0.237	0.166	0.185	0.192	0.817		
550	0.308	0.221	0.155	0.174	0.181	0.781		
600	0.290	0.208	0.146	0.165	0.171	0.750		
650	0.275	0.196	0.138	0.157	0.163	0.722		
700	0.261	0.186	0.132	0.150	0.156	0.698		1.10
750	0.249	0.177	0.126	0.143	0.149	0.677		0.96
800	0.238	0.170	0.120	0.138	0.143	0.657		0.84
850	0.228	0.163	0.115	0.133	0.138	0.640	1.107	0.74
900	0.219	0.156	0.111	0.128	0.134	0.624	0.959	0.67
950	0.211	0.151	0.107	0.124	0.129	0.610		
1000	0.204	0.146	0.104	0.120	0.125	0.597		
1050	0.197	0.141	0.101	0.117	0.122	0.585		
1100	0.191	0.137	0.098	0.114	0.119	0.574		
1150	0.185	0.133	0.095	0.111	0.116			
1200	0.180	0.129	0.092	0.108	0.113			
1250	0.175	0.126	0.090	0.105	0.110			
1300	0.170	0.123	0.088	0.103	0.108			
1350	0.166	0.120	0.086	0.101	0.106			
1400	0.162	0.117	0.084	0.099	0.104			
1450	0.158	0.115	0.082	0.097	0.102			
1500	0.155	0.113	0.081	0.095	0.100			
1550	0.151	0.110	0.079	0.093	0.098			
1600	0.148	0.108	0.078	0.092	0.097			
1650	0.145	0.106	0.076	0.090	0.095			
1700	0.142	0.105	0.075		0.094			
1750	0.139	0.103	0.074		0.092			
1800	0.137	0.101			0.091			
1850	0.135	0.100			0.090			
1900	0.132	0.098			0.089			
1950	0.130	0.097			0.088			
2000	0.128	0.096			0.086			

				Viscosity in mPa s					
$t/°C$	Aluminum	Cobalt	Copper	Gold	Indium	Iron	Nickel	Silver	Tin
250					1.35				1.77
300					1.22				1.55
350					1.12				1.39
400					1.04				1.27
450					0.98				1.17
500									1.09
550									1.02
600									0.97
650									0.92
700	1.24								0.88
750	1.13								0.85
800	1.04								0.82
850	0.96								0.79
900	0.90								0.77
950	0.84								0.75
1000								3.80	0.73
1050								3.56	
1100			3.92	5.130				3.31	
1150			3.61	4.874				3.06	
1200			3.34	4.640				2.82	
1250			3.11	4.429				2.61	
1300			2.91	4.240				2.42	
1350			2.73					2.28	
1400			2.58					2.20	
1450			2.44					2.19	
1500		4.15	2.31				4.35		
1550		3.89	2.20				4.09		
1600		3.64	2.10			5.22	3.87		
1650		3.41	2.01			4.79	3.67		
1700		3.20	1.92			4.41	3.49		
1750		2.99				4.08	3.32		
1800						3.79			
1850						3.54			
1900						3.31			
1950						3.10			
2000						2.92			
2050						2.75			
2100						2.60			
2150						2.46			
2200						2.34			
2250						2.22			

THERMAL CONDUCTIVITY OF GASES

Marcia L. Huber and Allan H. Harvey

The following table gives the thermal conductivity of some common gases as a function of temperature. Unless otherwise noted, the thermal conductivity values refer to a pressure of 100 kPa (1 bar) or to the saturation vapor pressure if that is less than 100 kPa. The notation $P = 0$ indicates that the low-pressure limiting value is given. The difference between the thermal conductivity at 100 kPa and the limiting value is generally less than 1%. Uncertainties for the thermal conductivities of gases in this table are generally less than 3%; uncertainty information on specific fluids can be found in the references. Thermal conductivity is given in units of mW m^{-1} K^{-1}. Substances are listed in the modified Hill order.

		Thermal conductivity in mW m^{-1} K^{-1}						
		100 K	200 K	300 K	400 K	500 K	600 K	Ref.
	Air	9.5	18.5	26.4	33.5	39.9	46.0	1
Ar	Argon ($P = 0$)	6.3	12.4	17.7	22.4	26.5	30.3	2, 3*
BF$_3$	Boron trifluoride			19.0	24.6			4
ClH	Hydrogen chloride		9.2	14.5	19.5	24.0	28.1	4
F$_6$S	Sulfur hexafluoride ($P = 0$)			13.0	20.6	27.5	33.8	5
H$_2$	Normal hydrogen ($P = 0$)	68.2	132.8	186.6	230.9	270.9	309.1	6
H$_2$O	Water ($P = 0$)			18.6	26.1	35.6	46.2	7
D$_2$O	Deuterium oxide ($P = 0$)			18.2	26.6	36.3	47.6	8
H$_2$S	Hydrogen sulfide			14.6	20.5	26.4	32.4	4
H$_3$N	Ammonia			25.1	37.2	53.1	68.6	9
He	Helium ($P = 0$)	74.7	118.3	155.7	189.6	221.4	251.6	10
Kr	Krypton ($P = 0$)		6.5	9.5	12.3	14.8	17.1	11
NO	Nitric oxide		17.8	25.9	33.1	39.6	46.2	4
N$_2$	Nitrogen	9.4	18.3	26.0	32.8	39.0	44.8	1
N$_2$O	Nitrous oxide		9.8	17.4	26.0	34.1	41.8	4
Ne	Neon ($P = 0$)	22.3	37.4	49.4	59.9	69.5	78.5	12
O$_2$	Oxygen	9.1	18.2	26.5	34.0	41.0	47.7	1
O$_2$S	Sulfur dioxide			9.6	14.3	20.0	25.6	4
Xe	Xenon ($P = 0$)		3.7	5.5	7.2	8.8	10.3	3*, 11
CCl$_2$F$_2$	Dichlorodifluoromethane			9.9	15.0	20.1	25.2	13
CF$_4$	Tetrafluoromethane ($P = 0$)			16.0	24.1	32.2	39.9	5
CO	Carbon monoxide ($P = 0$)			25.0	32.3	39.2	45.7	14
CO$_2$	Carbon dioxide		9.6	16.8	25.2	33.5	41.6	15
CHCl$_3$	Trichloromethane			7.5	11.1	15.1		4
CH$_4$	Methane ($P = 0$)	10.4	21.8	34.4	50.0	68.4	88.6	16
CH$_4$O	Methanol				26.2	38.6	53.0	4
C$_2$Cl$_2$F$_4$	1,2-Dichloro-1,1,2,2-tetrafluoroethane			10.3	15.7	21.1		13
C$_2$Cl$_3$F$_3$	1,1,2-Trichloro-1,2,2-trifluoroethane			9.0	13.6	18.3		13
C$_2$H$_2$	Acetylene			21.4	33.3	45.4	56.8	4
C$_2$H$_4$	Ethylene		11.3	20.6	34.7	49.9	68.6	17
C$_2$H$_6$	Ethane		10.7	21.2	36.0	53.8	73.3	18
C$_2$H$_6$O	Ethanol			14.4	25.8	38.4	53.2	4
C$_3$H$_6$O	Acetone			11.5	20.2	30.6	42.7	4
C$_3$H$_8$	Propane			18.5	31.0	46.4	64.6	19
C$_4$F$_8$	Perfluorocyclobutane			12.5	19.5			13
C$_4$H$_{10}$	Butane			16.7	28.3	43.0	60.9	20
C$_4$H$_{10}$	Isobutane			17.1	28.9	43.2	60.2	21
C$_4$H$_{10}$O	Diethyl ether			15.1	25.0	37.1		4
C$_5$H$_{12}$	Pentane				24.9	37.8	52.7	4
C$_6$H$_{14}$	Hexane				23.4	35.4	48.7	4

* More accurate data covering a restricted temperature range.

References

1. Lemmon, E. W., and Jacobsen, R. T, Viscosity and Thermal Conductivity Equations for Nitrogen, Oxygen, Argon, and Air, *Int. J. Thermophys.* 25, 21, 2004.
2. Vogel, E., Jäger, B., Hellmann, R., and Bich, E., *Ab initio* Pair Potential Energy Curve for the Argon Atom Pair and Thermophysical Properties for the Dilute Argon Gas. II. Thermophysical Properties for Low-Density Argon, *Mol. Phys.* 108, 3335, 2010.
3. May, E. F., Berg, R. F., and Moldover, M. R., Reference Viscosities of H₂, CH₄, Ar, and Xe at Low Densities, *Int. J. Thermophys.* 28, 1085, 2007.
4. Ho, C. Y., Ed., *Properties of Inorganic and Organic Fluids, CINDAS Data Series on Materials Properties*, Vol. V-1, Hemisphere Publishing Corp., New York, 1988.
5. Uribe, F. J., Mason, E. A., and Kestin, J., Thermal Conductivity of Nine Polyatomic Gases at Low Density, *J. Phys. Chem. Ref. Data* 19, 1123, 1990.
6. Mehl, J. B., Huber, M. L., and Harvey, A. H., *Ab Initio* Transport Coefficients of Gaseous Hydrogen, *Int. J. Thermophys.* 31, 740, 2010.
7. Sengers, J. V., and Watson, J. T. R., Improved International Formulations for the Viscosity and Thermal Conductivity of Water Substance, *J. Phys. Chem. Ref. Data* 15, 1291, 1986.
8. Matsunaga, N., and Nagashima, A., Transport Properties of Liquid and Gaseous D₂O over a Wide Range of Temperature and Pressure, *J. Phys. Chem. Ref. Data* 12, 933, 1983.
9. Tufeu, R., Ivanov, D. Y., Garrabos, Y., and Le Neindre, B., Thermal Conductivity of Ammonia in a Large Temperature and Pressure Range Including the Critical Region, *Ber. Bunsenges. Phys. Chem.* 88, 422, 1984.
10. Cencek, W., Komasa, J., Przybytek, M., Mehl, J. B., Jeziorski, B., and Szalewicz, K., Effects of Adiabatic, Relativistic, and Quantum Electrodynamics Interactions in Helium Dimer on Thermophysical Properties of Helium, *J. Chem. Phys.*, to be submitted (2011).
11. Bich, E., Millat, J., and Vogel, E., The Viscosity and Thermal Conductivity of Pure Monatomic Gases from Their Normal Boiling Point up to 5000 K in the Limit of Zero Density and at 0.101325 MPa, *J. Phys. Chem. Ref. Data* 19, 1289, 1990.
12. Bich, E., Hellmann, R., and Vogel, E., *Ab initio* Potential Energy Curve for the Neon Atom Pair and Thermophysical Properties for the Dilute Neon Gas. II. Thermophysical Properties for Low-Density Neon, *Mol. Phys.* 106, 813, 2008.
13. Krauss, R., and Stephan, K., Thermal Conductivity of Refrigerants in a Wide Range of Temperature and Pressure, *J. Phys. Chem. Ref. Data* 18, 43, 1989.
14. Millat J., and Wakeham, W. A., The Thermal Conductivity of Nitrogen and Carbon Monoxide in the Limit of Zero Density, *J. Phys. Chem. Ref. Data* 18, 565, 1989.
15. Vesovic, V., Wakeham, W. A., Olchowy, G. A., Sengers, J. V., Watson, J. T. R., and Millat, J., The Transport Properties of Carbon Dioxide, *J. Phys. Chem. Ref. Data*, 19, 763, 1990.
16. Hellmann, R., Bich, E., Vogel, E., Dickinson, A. S., and Vesovic, V., Calculation of the Transport and Relaxation Properties of Methane. II. Thermal Conductivity, Thermomagnetic Effects, Volume Viscosity, and Nuclear-Spin Relaxation, *J. Chem. Phys.* 130, 124309, 2009.
17. Holland, P. M., Eaton, B. E., and Hanley, H. J. M., A Correlation of the Viscosity and Thermal Conductivity Data of Gaseous and Liquid Ethylene, *J. Phys. Chem. Ref. Data* 12, 917, 1983.
18. Friend, D. G., Ingham, H., and Ely, J. F., Thermophysical Properties of Ethane, *J. Phys. Chem. Ref. Data* 20, 275, 1991.
19. Marsh, K., Perkins, R. A., and Ramires, M. L. V., Measurement and Correlation of the Thermal Conductivity of Propane from 86 to 600 K at Pressures to 70 MPa, *J. Chem. Eng. Data* 47, 932, 2002.
20. Perkins, R. A., Ramires, M. L. V., Nieto de Castro, C. A., and Cusco, L., Measurement and Correlation of the Thermal Conductivity of Butane from 135 K to 600 K at Pressures to 70 MPa, *J. Chem. Eng. Data* 47, 1263, 2002.
21. Perkins, R. A., Measurement and Correlation of the Thermal Conductivity of Isobutane from 114 K to 600 K at Pressures to 70 MPa, *J. Chem. Eng. Data* 47, 1272, 2002.

THERMAL CONDUCTIVITY OF LIQUIDS

This table gives the thermal conductivity of about 275 liquids at temperatures between −25 and 100 °C. Values refer to nominal atmospheric pressure; when an entry is given for a temperature above the normal boiling point of the liquid, the pressure is understood to be the saturation vapor pressure at that temperature. Reference 1 contains data on many of these liquids at high pressures. Data on halocarbon refrigerants over a wide range of temperature and pressure may be found in Reference 6.

Values given to three decimal places (i.e., to 0.001 W/m K) have an uncertainty of 2% to 5%. Values given to 0.0001 W/m K should be accurate to 1% or better.

Substances are arranged by molecular formula in Hill order, except that compounds not containing carbon precede those that do contain carbon.

References

1. Vargaftik, N. B., Filippov, L. P., Tarzimanov, A. A., and Totskii, E. E., *Handbook of Thermal Conductivity of Liquids and Gases*, CRC Press, Boca Raton FL, 1994.

2. Daubert, T. E., Danner, R. P., Sibul, H. M., and Stebbins, C. C., *Physical and Thermodynamic Properties of Pure Compounds: Data Compilation*, extant 1994 (core with four supplements), Taylor and Francis, Bristol, PA (also available as a database).

3. Watanabe, H., *J. Chem. Eng. Data* 48, 124, 2003.

4. Watanabe, H., and Seong, D. J., *Int. J. Thermophys.* 23, 337, 2002.

5. Nieto de Castro, C. A., Li, S. F. Y., Nagashima, A., Trengove, R. D., and Wakeham, W. A., *J. Phys. Chem. Ref. Data* 15, 1073, 1986.

6. Krauss, R., and Stephan, K., *J. Phys. Chem. Ref. Data* 18, 43, 1989.

7. Assael, M. J., Ramires, M. L. V., Nieto de Castro, C. A., and Wakeham, W. A., *J. Phys. Chem. Ref. Data* 19, 113, 1990.

8. Ramires, M. L. V., Nieto de Castro, C. A., Nagasaka, Y., Nagashima, A., Assael, M. J., and Wakeham, W. A., *J. Phys. Chem. Ref. Data* 24, 1377, 1995.

9. Ramires, M. L. V., Nieto de Castro, C. A., Perkins, R. A., Nagasaka, Y., Nagashima, A., Assael, M. J., and Wakeham, W. A., *J. Phys. Chem. Ref. Data* 29, 133, 2000.

10. Marsh, K. N., Ed., *Recommended Reference Materials for the Realization of Physicochemical Properties*, Blackwell Scientific Publications, Oxford, 1987.

11. Beaton, C. F., and Hewitt, G. F., *Physical Property Data for the Design Engineer*, Hemisphere Publishing Corp., New York, 1989.

Molecular Formula	Name	Thermal Conductivity in W/m K						
		−25 °C	0 °C	25 °C	50 °C	75 °C	100 °C	Ref.
Cl_4Ge	Germanium(IV) chloride	0.111	0.105	0.100	0.095	0.090	0.084	1
Cl_4Si	Tetrachlorosilane		0.099	0.096				2
Cl_4Sn	Tin(IV) chloride	0.123	0.117	0.112	0.106	0.101	0.095	1
Cl_4Ti	Titanium(IV) chloride		0.143	0.138	0.134	0.129	0.124	1
H_2O	Water		0.5562	0.6062	0.6423	0.6643	0.6729	8
Hg	Mercury	7.85	8.175	8.514	8.842	9.161	9.475	11
CCl_3F	Trichlorofluoromethane	0.102	0.096	0.089	0.083	0.076	0.070	1
CCl_4	Tetrachloromethane		0.109	0.103	0.098	0.092	0.087	1
$CHCl_3$	Trichloromethane	0.127	0.122	0.117	0.112	0.107	0.102	2
CH_2Br_2	Dibromomethane	0.120	0.114	0.108	0.103	0.097		2
CH_2Cl_2	Dichloromethane	0.158	0.149	0.140	0.133	0.128	0.127	1
CH_2I_2	Diiodomethane		0.098	0.093	0.088	0.083		1
CH_2O_2	Formic acid		0.267	0.265	0.263	0.261		1
CH_3NO_2	Nitromethane	0.226	0.215	0.204	0.193	0.182	0.171	1
CH_4O	Methanol	0.218	0.210	0.202	0.195	0.189	0.182	1
CS_2	Carbon disulfide		0.154	0.149				2
$C_2Br_2F_4$	1,2-Dibromotetrafluoroethane	0.071	0.066	0.061	0.057	0.053	0.049	1
$C_2Cl_3F_3$	1,1,2-Trichloro-1,2,2-trifluoroethane	0.0847	0.0790	0.0736	0.0683			6
C_2Cl_4	Tetrachloroethene		0.117	0.110	0.104	0.098	0.093	1
$C_2Cl_4F_2$	1,1,2,2-Tetrachloro-1,2-difluoroethane		0.082	0.078	0.074	0.069		1
C_2HCl_3	Trichloroethene	0.128	0.121	0.114	0.106	0.098	0.090	1
$C_2H_2Cl_4$	1,1,2,2-Tetrachloroethane	0.124	0.118	0.111	0.104	0.098	0.091	1
$C_2H_3Cl_3$	1,1,1-Trichloroethane		0.106	0.101	0.096			2
C_2H_3N	Acetonitrile	0.208	0.198	0.188	0.178	0.168		2
$C_2H_4Br_2$	1,2-Dibromoethane		0.100	0.096	0.092	0.088		1
$C_2H_4Cl_2$	1,2-Dichloroethane	0.144	0.139	0.133	0.128	0.122	0.117	1
$C_2H_4O_2$	Acetic acid			0.158	0.153	0.149	0.144	2
$C_2H_4O_2$	Methyl formate		0.194	0.187				1
C_2H_5Br	Bromoethane	0.107	0.104	0.101				1
C_2H_5Cl	Chloroethane	0.145	0.132	0.119	0.106	0.093		2
C_2H_5I	Iodoethane		0.091	0.087	0.083	0.079		1
C_2H_5NO	*N*-Methylformamide			0.203	0.201	0.199	0.196	2
$C_2H_5NO_2$	Nitroethane			0.173	0.161	0.149		1
C_2H_6O	Ethanol	0.181	0.174	0.167	0.160	0.153	0.148	1
$C_2H_6O_2$	1,2-Ethanediol		0.248	0.254	0.258	0.261	0.261	1
C_2H_7NO	Ethanolamine			0.240	0.238	0.236		1

Molecular Formula	Name	Thermal Conductivity in W/m K						
		−25 °C	0 °C	25 °C	50 °C	75 °C	100 °C	Ref.
C_3F_8	Perfluoropropane	0.062	0.056	0.051	0.046	0.041	0.035	1
C_3H_3N	Acrylonitrile	0.186	0.176	0.166	0.156	0.146	0.136	1
C_3H_5ClO	Epichlorohydrin	0.142	0.137	0.131	0.125	0.119	0.114	2
C_3H_6O	Allyl alcohol			0.162				1
C_3H_6O	Acetone		0.169	0.161				2
C_3H_6O	Methyloxirane		0.181	0.171				1
$C_3H_6O_2$	Propanoic acid		0.147	0.144	0.141	0.139	0.136	1
$C_3H_6O_2$	Ethyl formate	0.181	0.171	0.160	0.149	0.138		1
$C_3H_6O_2$	Methyl acetate	0.174	0.164	0.153	0.143	0.133	0.122	2
C_3H_7Br	1-Bromopropane	0.108	0.104	0.099	0.094			1
C_3H_7Cl	1-Chloropropane	0.129	0.123	0.116	0.110	0.104	0.098	1
C_3H_7I	1-Iodopropane	0.096	0.092	0.087	0.083	0.078	0.074	1
C_3H_7I	2-Iodopropane	0.089	0.085	0.082	0.078	0.074	0.071	1
C_3H_7NO	N,N-Dimethylformamide			0.183	0.175	0.167	0.159	1
$C_3H_7NO_2$	1-Nitropropane			0.152	0.144	0.137		1
C_3H_8O	1-Propanol	0.162	0.158	0.154	0.149	0.145	0.141	2
C_3H_8O	2-Propanol	0.146	0.141	0.135	0.129	0.124	0.118	2
$C_3H_8O_2$	1,2-Propanediol	0.199	0.200	0.200	0.200	0.199	0.197	1
$C_3H_8O_2$	2-Methoxyethanol			0.190	0.180	0.170		1
$C_3H_8O_3$	Glycerol			0.285	0.288	0.292	0.296	1
C_3H_9N	Trimethylamine	0.143	0.133					2
C_4F_8	Perfluorocyclobutane	0.082	0.072	0.063	0.053	0.044	0.034	1
C_4H_4O	Furan	0.142	0.134	0.126				2
C_4H_4S	Thiophene			0.199	0.195	0.191	0.186	2
C_4H_6	1,2-Butadiene	0.147	0.134					1
C_4H_6	2-Butyne	0.137	0.129	0.121				2
$C_4H_6O_2$	Vinyl acetate			0.151	0.141	0.131	0.120	1
$C_4H_6O_3$	Acetic anhydride		0.170	0.164	0.158	0.152	0.146	1
C_4H_8O	Butanal		0.155	0.147	0.140	0.132		1
C_4H_8O	2-Butanone	0.158	0.151	0.145	0.139	0.133		2
C_4H_8O	Tetrahydrofuran	0.132	0.126	0.120	0.114			2
$C_4H_8O_2$	Propyl formate		0.151	0.144	0.137	0.130		1
$C_4H_8O_2$	Ethyl acetate		0.151	0.144	0.136			1
$C_4H_8O_2$	Methyl propanoate			0.141	0.137			1
$C_4H_8O_2$	1,4-Dioxane			0.159	0.147	0.135	0.123	2
C_4H_9Br	1-Bromobutane	0.112	0.107	0.103	0.098	0.093	0.088	1
C_4H_9I	1-Iodobutane		0.094	0.090	0.085	0.081	0.077	1
C_4H_9NO	N,N-Dimethylacetamide			0.175	0.172	0.168		1
$C_4H_{10}O$	1-Butanol		0.158	0.153	0.147	0.142	0.137	1
$C_4H_{10}O$	2-Methyl-2-propanol			0.112	0.110	0.109	0.108	1
$C_4H_{10}O$	Diethyl ether	0.150	0.140	0.130	0.120	0.110	0.100	2
$C_4H_{10}O_2$	2-Ethoxyethanol			0.190	0.182	0.174	0.165	1
C_5H_5N	Pyridine		0.171	0.166	0.162	0.157	0.153	1
$C_5H_6O_2$	Furfuryl alcohol			0.179				1
C_5H_8	2-Methyl-1,3-butadiene	0.141	0.130	0.119				1
C_5H_8	1-Pentyne	0.144	0.136	0.127	0.119			1
C_5H_8	Cyclopentene	0.143	0.136	0.129				2
$C_5H_8O_2$	Methyl methacrylate		0.156	0.147	0.137	0.127	0.117	1
$C_5H_8O_2$	2,4-Pentanedione			0.154	0.150	0.146	0.143	1
C_5H_9NO	N-Methyl-2-pyrrolidone			0.167	0.162	0.157		1
C_5H_{10}	1-Pentene	0.131	0.124	0.116				2
C_5H_{10}	Cyclopentane	0.140	0.133	0.126				2
$C_5H_{10}O$	Pentanal		0.146	0.139	0.133	0.127	0.121	1
$C_5H_{10}O$	2-Pentanone		0.149	0.142	0.135	0.128	0.121	1
$C_5H_{10}O$	3-Pentanone		0.151	0.144	0.137	0.129	0.122	1
$C_5H_{10}O_2$	Pentanoic acid			0.140	0.137	0.133	0.130	1
$C_5H_{10}O_2$	Butyl formate			0.136	0.130	0.123	0.117	1
$C_5H_{10}O_2$	Propyl acetate		0.146	0.140	0.135	0.130	0.124	1
$C_5H_{10}O_2$	Ethyl propanoate			0.133	0.121			1
$C_5H_{10}O_2$	Methyl butanoate			0.140				1

Molecular Formula	Name	Thermal Conductivity in W/m K						Ref.
		−25 °C	0 °C	25 °C	50 °C	75 °C	100 °C	
C$_5$H$_{11}$Br	1-Bromopentane	0.113	0.109	0.105	0.101	0.097	0.093	1
C$_5$H$_{11}$Cl	1-Chloropentane		0.125	0.120	0.115	0.109		1
C$_5$H$_{11}$I	1-Iodopentane		0.096	0.092	0.088	0.084	0.081	1
C$_5$H$_{12}$	Pentane	0.130	0.1207	0.1113	0.1018	0.0923	0.083	4
C$_5$H$_{12}$	Isopentane			0.111				1
C$_5$H$_{12}$O	1-Pentanol	0.159	0.155	0.150	0.145	0.141	0.136	1
C$_5$H$_{12}$O	2-Methyl-2-butanol		0.119	0.116	0.113	0.109	0.106	1
C$_5$H$_{12}$O$_2$	1,5-Pentanediol		0.221	0.222				1
C$_5$H$_{12}$O$_3$	Diethylene glycol monomethyl ether			0.190	0.185	0.180	0.175	1
C$_6$F$_6$	Hexafluorobenzene			0.083				1
C$_6$F$_{14}$	Perfluorohexane		0.067	0.065	0.064			1
C$_6$H$_3$Cl$_3$	1,2,3-Trichlorobenzene			0.110	0.108	0.106		1
C$_6$H$_3$Cl$_3$	1,2,4-Trichlorobenzene			0.112	0.109	0.106		1
C$_6$H$_4$Cl$_2$	o-Dichlorobenzene		0.125	0.121	0.117	0.113	0.109	1
C$_6$H$_4$Cl$_2$	m-Dichlorobenzene		0.120	0.116	0.113	0.109		1
C$_6$H$_4$Cl$_2$	p-Dichlorobenzene			0.112	0.108	0.105		1
C$_6$H$_5$Br	Bromobenzene	0.119	0.115	0.111	0.107	0.103	0.099	1
C$_6$H$_5$Cl	Chlorobenzene	0.137	0.132	0.127	0.123	0.118	0.113	1
C$_6$H$_5$F	Fluorobenzene			0.136	0.131	0.126		1
C$_6$H$_5$I	Iodobenzene	0.106	0.103	0.101	0.098	0.095	0.092	1
C$_6$H$_5$NO$_2$	Nitrobenzene			0.149	0.145	0.142	0.139	1
C$_6$H$_6$	Benzene			0.1411	0.1329	0.1247		7
C$_6$H$_6$ClN	2-Chloroaniline			0.148				1
C$_6$H$_6$O	Phenol				0.153	0.149	0.147	1
C$_6$H$_7$N	Aniline		0.175					1
C$_6$H$_8$N$_2$	Hexanedinitrile			0.174	0.168			1
C$_6$H$_{10}$	Cyclohexene	0.142	0.136	0.130	0.124	0.118		2
C$_6$H$_{10}$O	Cyclohexanone			0.138	0.134	0.130	0.126	1
C$_6$H$_{10}$O	Mesityl oxide	0.170	0.163	0.156	0.149	0.142	0.134	2
C$_6$H$_{10}$O$_3$	Ethyl acetoacetate			0.155	0.152	0.148	0.144	1
C$_6$H$_{10}$O$_4$	Diethyl oxalate			0.157				1
C$_6$H$_{12}$	1-Hexene	0.138	0.129	0.121	0.113			1
C$_6$H$_{12}$	Cyclohexane			0.123	0.117	0.111		2
C$_6$H$_{12}$O	2-Hexanone		0.156	0.145	0.134	0.124	0.115	1
C$_6$H$_{12}$O	Cyclohexanol			0.138	0.134	0.130	0.126	1
C$_6$H$_{12}$O$_2$	Hexanoic acid		0.148	0.142	0.137	0.131		1
C$_6$H$_{12}$O$_2$	Butyl acetate		0.143	0.136	0.130	0.123	0.116	1
C$_6$H$_{12}$O$_2$	Propyl propanoate			0.133				1
C$_6$H$_{12}$O$_2$	Ethyl butanoate		0.143	0.137	0.131	0.126		1
C$_6$H$_{12}$O$_2$	Methyl pentanoate		0.143	0.138	0.132	0.127		1
C$_6$H$_{12}$O$_3$	Paraldehyde			0.130				1
C$_6$H$_{13}$Br	1-Bromohexane	0.115	0.111	0.108	0.104	0.101	0.097	1
C$_6$H$_{13}$I	1-Iodohexane		0.098	0.095	0.091	0.088	0.084	1
C$_6$H$_{14}$	Hexane	0.133	0.1250	0.1167	0.1083	0.0999	0.092	4
C$_6$H$_{14}$	2-Methylpentane	0.120	0.1127	0.1050	0.0972	0.0894	0.082	3
C$_6$H$_{14}$	3-Methylpentane	0.122	0.1142	0.1064	0.0986	0.0909	0.083	3
C$_6$H$_{14}$	2,2-Dimethylbutane	0.108	0.1006	0.0934	0.0861	0.0788	0.072	3
C$_6$H$_{14}$	2,3-Dimethylbutane	0.115	0.1076	0.1003	0.0930	0.0857	0.078	3
C$_6$H$_{14}$O	1-Hexanol	0.161	0.157	0.152	0.147	0.142	0.137	1
C$_6$H$_{14}$O	Dipropyl ether		0.137	0.130	0.123	0.117		1
C$_6$H$_{14}$O$_2$	1,2-Diethoxyethane			0.140	0.133	0.125		1
C$_6$H$_{14}$O$_3$	Diethylene glycol monoethyl ether			0.188	0.184	0.180		1
C$_6$H$_{14}$O$_4$	Triethylene glycol		0.193	0.195	0.196	0.196	0.196	1
C$_6$H$_{15}$N	Triethylamine	0.146	0.139	0.132	0.125	0.118	0.111	1
C$_7$F$_{16}$	Perfluoroheptane	0.068	0.064	0.060	0.056	0.053		1
C$_7$H$_5$N	Benzonitrile			0.148	0.142	0.136	0.130	1
C$_7$H$_6$O	Benzaldehyde			0.153	0.148	0.143	0.139	1
C$_7$H$_8$	Toluene	0.1455	0.1385	0.1310	0.1235	0.1162	0.1095	9
C$_7$H$_8$O	o-Cresol							1
C$_7$H$_8$O	m-Cresol			0.149	0.147	0.145		1

Molecular Formula	Name	Thermal Conductivity in W/m K						
		−25 °C	0 °C	25 °C	50 °C	75 °C	100 °C	Ref.
C_7H_8O	Benzyl alcohol			0.159	0.158	0.156	0.154	1
C_7H_8O	Anisole			0.145	0.142	0.139	0.136	1
C_7H_9N	2-Methylaniline			0.162				1
C_7H_9N	3-Methylaniline			0.161				1
C_7H_{14}	1-Heptene	0.139	0.132	0.125	0.118	0.111		1
C_7H_{14}	Cycloheptane			0.123	0.118	0.112	0.108	1
$C_7H_{14}O$	Heptanal			0.140				1
$C_7H_{14}O$	3-Heptanone		0.143	0.137	0.131	0.125	0.119	1
$C_7H_{14}O$	4-Heptanone			0.136	0.131	0.125	0.120	1
$C_7H_{14}O_2$	Hexyl formate			0.141	0.133	0.126	0.119	1
$C_7H_{14}O_2$	Heptanoic acid			0.140	0.137	0.133		1
$C_7H_{14}O_2$	Pentyl acetate		0.141	0.134	0.126	0.120	0.113	1
$C_7H_{14}O_2$	Butyl propanoate			0.139	0.133	0.126	0.121	1
$C_7H_{14}O_2$	Ethyl pentanoate			0.132				1
$C_7H_{14}O_2$	Methyl hexanoate			0.136	0.131	0.126	0.121	1
C_7H_{16}	Heptane	0.1378	0.1303	0.1228	0.1152	0.1077		5
C_7H_{16}	2-Methylhexane	0.125	0.1177	0.1105	0.1033	0.0961	0.089	3
C_7H_{16}	3-Methylhexane	0.126	0.1184	0.1112	0.1040	0.0968	0.090	3
C_7H_{16}	3-Ethylpentane	0.128	0.1203	0.1128	0.1053	0.0978	0.090	3
C_7H_{16}	2,2-Dimethylpentane	0.111	0.1046	0.0980	0.0913	0.0847	0.078	3
C_7H_{16}	2,3-Dimethylpentane	0.120	0.1127	0.1059	0.0990	0.0922	0.085	3
C_7H_{16}	2,4-Dimethylpentane	0.116	0.1089	0.1020	0.0951	0.0882	0.081	3
C_7H_{16}	3,3-Dimethylpentane	0.113	0.1068	0.1001	0.0934	0.0867	0.080	3
C_7H_{16}	2,2,3-Trimethylbutane	0.107	0.1011	0.0950	0.0889	0.0828	0.077	3
$C_7H_{16}O$	1-Heptanol	0.160	0.158	0.153	0.149	0.144	0.139	1
C_8F_{18}	Perfluorooctane		0.066	0.062	0.059	0.055	0.052	1
C_8H_8	Styrene	0.148	0.142	0.137	0.131	0.126	0.120	2
C_8H_8O	Acetophenone			0.147	0.146	0.144	0.142	1
$C_8H_8O_2$	Methyl benzoate			0.147				1
C_8H_{10}	Ethylbenzene	0.143	0.137	0.130	0.123	0.116	0.110	1
C_8H_{10}	o-Xylene			0.131	0.126	0.120	0.114	2
C_8H_{10}	m-Xylene			0.130	0.124	0.118	0.113	2
C_8H_{10}	p-Xylene			0.130	0.124	0.118	0.112	2
$C_8H_{10}O$	Ethoxybenzene	0.151	0.145	0.140	0.135	0.130		1
$C_8H_{10}O_2$	2-Phenoxyethanol			0.169	0.168	0.166	0.165	1
$C_8H_{11}N$	N-Ethylaniline			0.150				1
$C_8H_{11}N$	N,N-Dimethylaniline			0.122	0.119	0.115		1
C_8H_{16}	1-Octene	0.139	0.133	0.126	0.120	0.114	0.107	1
$C_8H_{16}O$	2-Octanone		0.141	0.135	0.129	0.124	0.118	1
$C_8H_{16}O_2$	Heptyl formate		0.141	0.137	0.132	0.128	0.123	1
$C_8H_{16}O_2$	Octanoic acid			0.146	0.143	0.139	0.135	1
$C_8H_{16}O_2$	Hexyl acetate			0.135	0.129	0.123	0.118	1
$C_8H_{16}O_2$	Pentyl propanoate			0.138	0.132			1
$C_8H_{16}O_2$	Ethyl hexanoate		0.142	0.137	0.133	0.128	0.123	1
$C_8H_{17}Cl$	1-Chlorooctane		0.130	0.127	0.124	0.121	0.119	1
C_8H_{18}	Octane	0.139	0.1317	0.1244	0.1171	0.1097	0.102	4
C_8H_{18}	2-Methylheptane	0.127	0.1206	0.1139	0.1072	0.1005	0.094	3
C_8H_{18}	3-Methylheptane	0.128	0.1216	0.1149	0.1081	0.1014	0.095	3
C_8H_{18}	2,2,4-Trimethylpentane	0.107	0.1007	0.0948	0.0888	0.0829	0.077	3
C_8H_{18}	2,3,4-Trimethylpentane	0.115	0.1093	0.1035	0.0976	0.0918	0.086	3
$C_8H_{18}O$	Ethyl hexyl ether		0.131	0.126	0.120	0.114	0.109	1
$C_8H_{18}O$	1-Octanol		0.162	0.158	0.153	0.148	0.143	1
$C_8H_{18}O$	Dibutyl ether		0.139	0.132	0.125	0.118	0.112	1
$C_8H_{18}O_3$	Diethylene glycol monobutyl ether			0.163	0.158	0.153	0.148	1
$C_8H_{18}O_4$	Triethylene glycol dimethyl ether			0.169	0.158	0.147		1
$C_8H_{18}O_5$	Tetraethylene glycol			0.191	0.192			1
C_9H_7N	Quinoline			0.147	0.144	0.141	0.138	1
C_9H_{10}	Indan			0.135				1
$C_9H_{10}O_2$	Ethyl benzoate			0.141				1
C_9H_{12}	Propylbenzene	0.134	0.130	0.125	0.120	0.115	0.109	1

Molecular Formula	Name	Thermal Conductivity in W/m K						
		−25 °C	0 °C	25 °C	50 °C	75 °C	100 °C	Ref.
C_9H_{12}	Isopropylbenzene	0.132	0.128	0.123	0.118	0.112	0.107	1
C_9H_{12}	1,2,4-Trimethylbenzene			0.129	0.124	0.118	0.114	1
C_9H_{12}	1,3,5-Trimethylbenzene	0.143	0.139	0.134	0.129	0.123	0.117	1
C_9H_{18}	1-Nonene	0.136	0.130	0.123	0.116	0.110	0.104	1
$C_9H_{18}O_2$	Nonanoic acid			0.150	0.146	0.142	0.138	1
$C_9H_{18}O_2$	Heptyl acetate			0.135	0.128	0.122	0.116	1
$C_9H_{19}Br$	1-Bromononane		0.116	0.112	0.109	0.106	0.103	1
$C_9H_{19}Cl$	1-Chlorononane		0.132	0.128	0.124	0.120	0.115	1
$C_9H_{19}I$	1-Iodononane		0.105	0.102	0.099	0.095	0.092	1
C_9H_{20}	Nonane	0.141	0.1337	0.1269	0.1201	0.1133	0.106	4
$C_9H_{20}O$	1-Nonanol		0.164	0.159	0.155	0.150	0.145	1
$C_{10}H_7Br$	1-Bromonaphthalene			0.110	0.109	0.108	0.106	1
$C_{10}H_7Cl$	1-Chloronaphthalene			0.126				1
$C_{10}H_{10}O_4$	Dimethyl phthalate			0.1473	0.1443	0.1409	0.1373	10
$C_{10}H_{12}$	1,2,3,4-Tetrahydronaphthalene			0.131	0.129	0.128	0.126	1
$C_{10}H_{14}$	Butylbenzene			0.126	0.121	0.116	0.111	1
$C_{10}H_{14}$	sec-Butylbenzene, (±)-		0.129	0.124	0.119	0.114	0.108	1
$C_{10}H_{14}$	tert-Butylbenzene			0.117	0.114	0.110	0.106	1
$C_{10}H_{14}$	1-Isopropyl-4-methylbenzene	0.132	0.127	0.122	0.117	0.112	0.107	2
$C_{10}H_{14}$	o-Diethylbenzene		0.133	0.127	0.122	0.116	0.111	1
$C_{10}H_{18}$	trans-Decahydronaphthalene			0.113				1
$C_{10}H_{20}$	1-Decene	0.138	0.132	0.126	0.120	0.114	0.109	1
$C_{10}H_{20}O$	Decanal		0.149	0.144	0.139	0.134	0.129	1
$C_{10}H_{20}O_2$	Heptyl propanoate			0.137	0.132	0.127	0.122	1
$C_{10}H_{20}O_2$	Hexyl butanoate			0.137	0.132	0.127	0.121	1
$C_{10}H_{20}O_2$	Decanoic acid				0.148	0.144	0.140	1
$C_{10}H_{22}$	Decane	0.142	0.1360	0.1296	0.1232	0.1167	0.110	4
$C_{10}H_{22}O$	1-Decanol			0.162	0.159	0.155	0.151	1
$C_{10}H_{22}O$	Dipentyl ether			0.131	0.125	0.121	0.116	1
$C_{10}H_{22}O_2$	1,2-Dibutoxyethane			0.140	0.134	0.127	0.120	1
$C_{11}H_{16}$	Pentylbenzene		0.135	0.130	0.125	0.120	0.115	1
$C_{11}H_{22}$	1-Undecene			0.126	0.118	0.114	0.108	1
$C_{11}H_{22}O$	6-Undecanone			0.137	0.132	0.127		1
$C_{11}H_{22}O_2$	Undecanoic acid				0.153	0.149		1
$C_{11}H_{22}O_2$	Octyl propanoate			0.135	0.130	0.125	0.120	1
$C_{11}H_{22}O_2$	Heptyl butanoate			0.139	0.134	0.129	0.123	1
$C_{11}H_{24}$	Undecane			0.136	0.128	0.122	0.116	1
$C_{11}H_{24}O$	1-Undecanol			0.169	0.165	0.161	0.158	1
$C_{12}H_{10}O$	Diphenyl ether				0.139	0.135	0.131	2
$C_{12}H_{14}O_4$	Diethyl phthalate			0.172	0.169	0.166		1
$C_{12}H_{16}$	Cyclohexylbenzene			0.121	0.119	0.117		1
$C_{12}H_{18}$	Hexylbenzene		0.141	0.137	0.132	0.128	0.124	1
$C_{12}H_{24}O_2$	Decyl acetate			0.146	0.136	0.126		1
$C_{12}H_{24}O_2$	Octyl butanoate			0.139	0.134	0.129	0.125	1
$C_{12}H_{26}$	Dodecane			0.135	0.130	0.124	0.119	1
$C_{12}H_{26}O$	1-Dodecanol			0.167	0.163	0.159		1
$C_{12}H_{26}O_3$	Diethylene glycol dibutyl ether		0.150	0.146	0.143	0.139	0.135	1
$C_{12}H_{27}N$	Tributylamine			0.129				1
$C_{13}H_{26}$	1-Tridecene			0.130	0.125	0.120	0.115	1
$C_{13}H_{28}$	Tridecane			0.130	0.125	0.120	0.115	1
$C_{14}H_{28}$	1-Tetradecene			0.136	0.131	0.126	0.121	1
$C_{14}H_{30}$	Tetradecane			0.139	0.134	0.129	0.124	1
$C_{14}H_{30}O$	1-Tetradecanol				0.167	0.162	0.157	2
$C_{16}H_{22}O_4$	Dibutyl phthalate		0.139	0.136	0.134	0.131	0.129	1
$C_{16}H_{34}$	Hexadecane			0.140	0.135	0.130	0.125	2
$C_{18}H_{38}$	Octadecane				0.146	0.142	0.137	2
$C_{20}H_{40}O_2$	Butyl palmitate			0.151	0.148	0.144	0.140	1
$C_{22}H_{42}O_2$	Butyl oleate			0.157	0.153	0.149	0.145	1
$C_{22}H_{42}O_4$	Dioctyl hexanedioate			0.157	0.153	0.149	0.145	1

DIFFUSION IN GASES

This table gives binary diffusion coefficients D_{12} for a number of common gases as a function of temperature. Values refer to atmospheric pressure. The diffusion coefficient is inversely proportional to pressure as long as the gas is in a regime where binary collisions dominate. See Reference 1 for a discussion of the dependence of D_{12} on temperature and composition.

The first part of the table gives data for several gases in the presence of a large excess of air. The remainder applies to equimolar mixtures of gases. Each gas pair is ordered alphabetically according to the most common way of writing the formula. The listing of pairs then follows alphabetical order by the first constituent.

References

1. Marrero, T. R., and Mason, E. A., *J. Phys. Chem. Ref. Data*, 1, 1, 1972.
2. Kestin, J., et al., *J. Phys. Chem. Ref. Data*, 13, 229, 1984.

$D_{12}/cm^2 s^{-1}$ for p = 101.325 kPa and the Specified T/K

System	200	273.15	293.15	373.15	473.15	573.15	673.15
Large Excess of Air							
Ar-air		0.167	0.189	0.289	0.437	0.612	0.810
CH$_4$-air			0.210	0.321	0.485	0.678	0.899
CO-air			0.208	0.315	0.475	0.662	0.875
CO$_2$-air			0.160	0.252	0.390	0.549	0.728
H$_2$-air		0.668	0.756	1.153	1.747	2.444	3.238
H$_2$O-air			0.242	0.399	0.638	0.873	1.135
He-air		0.617	0.697	1.057	1.594	2.221	2.933
SF$_6$-air				0.150	0.233	0.329	0.438
Equimolar Mixture							
Ar-CH$_4$				0.306	0.467	0.657	0.876
Ar-CO		0.168	0.190	0.290	0.439	0.615	0.815
Ar-CO$_2$		0.129	0.148	0.235	0.365	0.517	0.689
Ar-H$_2$		0.698	0.794	1.228	1.876	2.634	3.496
Ar-He	0.381	0.645	0.726	1.088	1.617	2.226	2.911
Ar-Kr	0.064	0.117	0.134	0.210	0.323	0.456	0.605
Ar-N$_2$		0.168	0.190	0.290	0.439	0.615	0.815
Ar-Ne	0.160	0.277	0.313	0.475	0.710	0.979	1.283
Ar-O$_2$		0.166	0.187	0.285	0.430	0.600	0.793
Ar-SF$_6$				0.128	0.202	0.290	0.389
Ar-Xe	0.052	0.095	0.108	0.171	0.264	0.374	0.498
CH$_4$-H$_2$			0.708	1.084	1.648	2.311	3.070
CH$_4$-He			0.650	0.992	1.502	2.101	2.784
CH$_4$-N$_2$			0.208	0.317	0.480	0.671	0.890
CH$_4$-O$_2$			0.220	0.341	0.523	0.736	0.978
CH$_4$-SF$_6$				0.167	0.257	0.363	0.482
CO-CO$_2$			0.162	0.250	0.384		
CO-H$_2$	0.408	0.686	0.772	1.162	1.743	2.423	3.196
CO-He	0.365	0.619	0.698	1.052	1.577	2.188	2.882
CO-Kr		0.131	0.149	0.227	0.346	0.485	0.645
CO-N$_2$	0.133	0.208	0.231	0.336	0.491	0.673	0.878
CO-O$_2$			0.202	0.307	0.462	0.643	0.849
CO-SF$_6$				0.144	0.226	0.323	0.432
CO$_2$-C$_3$H$_8$			0.084	0.133	0.209		
CO$_2$-H$_2$	0.315	0.552	0.627	0.964	1.470	2.066	2.745
CO$_2$-H$_2$O			0.162	0.292	0.496	0.741	1.021
CO$_2$-He	0.300	0.513	0.580	0.878	1.321		
CO$_2$-N$_2$			0.160	0.253	0.392	0.553	0.733
CO$_2$-N$_2$O	0.055	0.099	0.113	0.177	0.276		
CO$_2$-Ne	0.131	0.227	0.258	0.395	0.603	0.847	
CO$_2$-O$_2$			0.159	0.248	0.380	0.535	0.710
CO$_2$-SF$_6$				0.099	0.155		
D$_2$-H$_2$	0.631	1.079	1.219	1.846	2.778	3.866	5.103
H$_2$-He	0.775	1.320	1.490	2.255	3.394	4.726	6.242
H$_2$-Kr	0.340	0.601	0.682	1.053	1.607	2.258	2.999

System	200	273.15	293.15	373.15	473.15	573.15	673.15
H_2-N_2	0.408	0.686	0.772	1.162	1.743	2.423	3.196
H_2-Ne	0.572	0.982	1.109	1.684	2.541	3.541	4.677
H_2-O_2		0.692	0.782	1.188	1.792	2.497	3.299
H_2-SF_6			0.412	0.649	0.998	1.400	1.851
H_2-Xe		0.513	0.581	0.890	1.349	1.885	2.493
H_2O-N_2			0.242	0.399			
H_2O-O_2			0.244	0.403	0.645	0.882	1.147
He-Kr	0.330	0.559	0.629	0.942	1.404	1.942	2.550
He-N_2	0.365	0.619	0.698	1.052	1.577	2.188	2.882
He-Ne	0.563	0.948	1.066	1.592	2.362	3.254	4.262
He-O_2		0.641	0.723	1.092	1.640	2.276	2.996
He-SF_6			0.400	0.592	0.871	1.190	1.545
He-Xe	0.282	0.478	0.538	0.807	1.201	1.655	2.168
Kr-N_2		0.131	0.149	0.227	0.346	0.485	0.645
Kr-Ne	0.131	0.228	0.258	0.392	0.587	0.812	1.063
Kr-Xe	0.035	0.064	0.073	0.116	0.181	0.257	0.344
N_2-Ne			0.317	0.483	0.731	1.021	1.351
N_2-O_2			0.202	0.307	0.462	0.643	0.849
N_2-SF_6				0.148	0.231	0.328	0.436
N_2-Xe		0.107	0.122	0.188	0.287	0.404	0.539
Ne-Xe	0.111	0.193	0.219	0.332	0.498	0.688	0.901
O_2-SF_6			0.097	0.154	0.238	0.334	0.441

DIFFUSION OF GASES IN WATER

This table gives values of the diffusion coefficient, D, for diffusion of several common gases in water at various temperatures. For simple one-dimensional transport, the diffusion coefficient describes the time–rate of change of concentration, dc/dt, through the equation

$$dc/dt = D \, d^2c/dx^2$$

where x is, for example, the perpendicular distance from a gas–liquid interface. The values below have been selected from the references indicated; in some cases data have been refitted to permit interpolation in temperature.

Gas–liquid diffusion coefficients are difficult to measure, and large differences are found between values obtained by differ-

ent authors and through different experimental methods. See References 1 and 2 for a discussion of measurement techniques.

References

1. Jähne, B., Heinz, G., and Dietrich, W., *J. Geophys. Res.*, 92, 10767, 1987.
2. Himmelblau, D. M., *Chem. Rev.*, 64, 527, 1964.
3. Boerboom, A. J. H., and Kleyn, G., *J. Chem. Phys.*, 50, 1086, 1969.
4. O'Brien, R. N., and Hyslop, W. F., *Can. J. Chem.*, 55, 1415, 1977.
5. Maharajh, D. M., and Walkley, J., *Can. J. Chem.*, 51, 944, 1973.
6. *Landolt-Börnstein, Numerical Data and Functional Relationships in Science and Technology, Sixth Edition*, II/5a, *Transport Phenomena I (Viscosity and Diffusion)*, Springer-Verlag, Heidelberg, 1969.

$$D/10^{-5} \, cm^2 \, s^{-1}$$

	10°C	15°C	20°C	25°C	30°C	35°C	Ref.
Ar				2.5			3,4
CHCl$_2$F				1.80			5
CH$_3$Br				1.35			5
CH$_3$Cl				1.40			5
CH$_4$	1.24	1.43	1.62	1.84	2.08	2.35	1
CO$_2$	1.26	1.45	1.67	1.91	2.17	2.47	1
C$_2$H$_2$	1.43	1.59	1.78	1.99	2.23		2
Cl$_2$		1.13	1.5	1.89			2,6
HBr				3.15			6
HCl				3.07			6
H$_2$	3.62	4.08	4.58	5.11	5.69	6.31	1
H$_2$S				1.36			2,6
He	5.67	6.18	6.71	7.28	7.87	8.48	1,3
Kr	1.20	1.39	1.60	1.84	2.11	2.40	1,3
NH$_3$		1.3	1.5				2
NO$_2$			1.23	1.4	1.59		2,6
N$_2$				2.0			2
N$_2$O		1.62	2.11	2.57			2,6
Ne	2.93	3.27	3.64	4.03	4.45	4.89	1,3
O$_2$		1.67	2.01	2.42			2,6
Rn	0.81	0.96	1.13	1.33	1.55	1.80	1
SO$_2$			1.62	1.83	2.07	2.32	2
Xe	0.93	1.08	1.27	1.47	1.70	1.95	1,3

DIFFUSION COEFFICIENTS IN LIQUIDS AT INFINITE DILUTION

This table lists diffusion coefficients D_{AB} at infinite dilution for some binary liquid mixtures. Values are given at 25 °C when available; it should be noted that the diffusion coefficient generally increases by 10% to 20% for a 10 °C increase above ambient temperature.

Solvents are listed in alphabetical order, as are the solutes within each solvent group.

References

1. Safi, A., Nicolas, C., Neau, E., and Chevalier, J.-L., *J. Chem. Eng. Data* 52, 977, 2007.
2. Safi, A., Nicolas, C., Neau, E., and Chevalier, J.-L., *J. Chem. Eng. Data* 52, 126, 2007.
3. Sanni, S. A., Fell, C. J. D., and Hutchison, H. P., *J. Chem. Eng. Data* 16, 424, 1971.
4. Fan, Y., Qian, R., Shi, M., and Shi, J., *J. Chem. Eng. Data* 40, 1053, 1995.
5. *Landolt-Börnstein Numerical Data and Functional Relationships in Science and Technology*, Sixth Edition, Vol. II/5a, Springer-Verlag, Heidelberg, 1969.

Solute	Solvent	t/°C	$D_{AB}/$ 10^{-5} cm^{2}s^{-1}	Ref.
Acetic acid	Acetone	25	3.31	5
Benzene	Acetone	25	4.25	2
Benzoic acid	Acetone	25	2.62	5
Formic acid	Acetone	25	3.77	5
Nitrobenzene	Acetone	20	2.94	5
Tetrachloromethane	Acetone	25	3.29	5
Trichloromethane	Acetone	25	3.64	5
Water	Acetone	25	4.56	5
Acetic acid	Benzene	25	2.09	5
Aniline	Benzene	25	1.96	5
Benzoic acid	Benzene	25	1.38	5
Bromobenzene	Benzene	8	1.45	5
2-Butanone	Benzene	30	2.09	5
Chloroethene	Benzene	8	1.77	5
Cyclohexane	Benzene	25	2.09	3
Ethanol	Benzene	25	3.02	5
Formic acid	Benzene	25	2.28	5
Heptane	Benzene	25	1.79	3
Methanol	Benzene	25	3.80	5
Toluene	Benzene	25	1.85	3
1,2,4-Trichlorobenzene	Benzene	8	1.34	5
Trichloromethane	Benzene	25	2.26	5
Benzene	1-Butanol	25	1.00	5
Biphenyl	1-Butanol	25	0.63	5
Butanoic acid	1-Butanol	30	0.51	5
p-Dichlorobenzene	1-Butanol	25	0.82	5
1,6-Hexanedioic acid	1-Butanol	30	0.40	5
Methanol	1-Butanol	30	0.59	5
cis-9-Octadecenoic acid	1-Butanol	30	0.25	5
Propane	1-Butanol	25	1.57	5
Water	1-Butanol	25	0.56	5
Benzene	Cyclohexane	25	1.92	1
Chlorobenzene	Cyclohexane	25	1.34	1
p-Chlorotoluene	Cyclohexane	25	1.28	1
Ethylbenzene	Cyclohexane	25	1.36	1
Naphthalene	Cyclohexane	25	1.18	1
Perylene	Cyclohexane	25	0.79	1
Pyrene	Cyclohexane	25	0.95	1
Tetrachloromethane	Cyclohexane	25	1.49	3
Toluene	Cyclohexane	25	1.66	1
Benzene	Decane	25	2.16	1
Chlorobenzene	Decane	25	1.98	1
p-Chlorotoluene	Decane	25	1.80	1
Ethylbenzene	Decane	25	1.79	1
Naphthalene	Decane	25	1.65	1
Perylene	Decane	25	1.08	1
Pyrene	Decane	25	1.23	1
Toluene	Decane	25	1.93	1
Trichloromethane	Diethyl ether	25	4.48	3
Allyl alcohol	Ethanol	20	0.98	5
Benzene	Ethanol	25	1.88	2
Iodine	Ethanol	25	1.32	5
Iodobenzene	Ethanol	20	1.00	5
3-Methyl-1-butanol	Ethanol	20	0.81	5
Pyridine	Ethanol	20	1.10	5
Tetrachloromethane	Ethanol	25	1.50	5
Water	Ethanol	25	1.24	5
Acetic acid	Ethyl acetate	20	2.18	5
Acetone	Ethyl acetate	20	3.18	5
2-Butanone	Ethyl acetate	30	2.93	5
Ethyl benzoate	Ethyl acetate	20	1.85	5
Nitrobenzene	Ethyl acetate	20	2.25	5
Water	Ethyl acetate	25	3.20	5
Benzene	Heptane	25	3.75	1
Chlorobenzene	Heptane	25	3.42	1
p-Chlorotoluene	Heptane	25	3.11	1
Ethylbenzene	Heptane	25	3.15	1
Naphthalene	Heptane	25	2.81	1
Perylene	Heptane	25	1.89	1
Pyrene	Heptane	25	2.16	1
Toluene	Heptane	25	3.42	1
Benzene	Hexane	25	4.70	1
Bromobenzene	Hexane	8	2.60	5

Solute	Solvent	t/°C	$D_{AB}/10^{-5}\ cm^2s^{-1}$	Ref.	Solute	Solvent	t/°C	$D_{AB}/10^{-5}\ cm^2s^{-1}$	Ref.
2-Butanone	Hexane	30	3.74	5	Ethylbenzene	2,2,4-Trimethylpentane	30	2.63	4
Chlorobenzene	Hexane	25	4.16	1	Toluene	2,2,4-Trimethylpentane	30	3.15	4
p-Chlorotoluene	Hexane	25	3.74	1	1,3,5-Trimethylbenzene	2,2,4-Trimethylpentane	30	2.26	4
Dodecane	Hexane	25	2.73	5	o-Xylene	2,2,4-Trimethylpentane	30	2.62	4
Ethylbenzene	Hexane	25	3.73	1	p-Xylene	2,2,4-Trimethylpentane	30	3.03	4
Iodine	Hexane	25	4.45	5	Acetic acid	Water	25	1.29	5
Methane	Hexane	25	0.09	5	Acetone	Water	25	1.28	5
Naphthalene	Hexane	25	3.55	1	Acetonitrile	Water	15	1.26	5
Perylene	Hexane	25	2.30	1	L-Alanine	Water	25	0.91	5
Propane	Hexane	25	4.87	5	Allyl alcohol	Water	15	0.90	5
Pyrene	Hexane	25	2.58	1	Aniline	Water	20	0.92	5
Tetrachloromethane	Hexane	25	3.70	5	DL-Arabinose	Water	20	0.69	5
Toluene	Hexane	25	4.12	1	Benzene	Water	20	1.02	5
Benzene	Octane	25	3.19	1	1-Butanol	Water	25	0.56	5
Chlorobenzene	Octane	25	2.89	1	Caprolactam	Water	25	0.87	5
p-Chlorotoluene	Octane	25	2.62	1	Chloroethene	Water	25	1.34	5
Ethylbenzene	Octane	25	2.58	1	Cyclohexane	Water	20	0.84	5
Naphthalene	Octane	25	2.35	1	Diethylamine	Water	20	0.97	5
Perylene	Octane	25	1.58	1	1,2-Ethanediol	Water	25	1.16	5
Pyrene	Octane	25	1.81	1	Ethanol	Water	25	1.24	5
Toluene	Octane	25	2.83	1	Ethanolamine	Water	25	1.08	5
1,3,5-Trimethylbenzene	Octane	30	2.21	4	Ethyl acetate	Water	20	1.00	5
o-Xylene	Octane	30	2.65	4	Ethylbenzene	Water	20	0.81	5
p-Xylene	Octane	30	2.82	4	Ethyl carbamate	Water	15	0.80	5
Acetone	Tetrachloromethane	25	1.75	5	α-D-Glucose	Water	25	0.67	5
Benzene	Tetrachloromethane	25	1.42	5	Glycerol	Water	25	1.06	5
Cyclohexane	Tetrachloromethane	25	1.30	5	Glycine	Water	25	1.05	5
Ethanol	Tetrachloromethane	25	1.90	5	β-D-Lactose	Water	15	0.38	5
Iodine	Tetrachloromethane	30	1.63	5	α-Maltose	Water	15	0.38	5
Trichloromethane	Tetrachloromethane	25	1.66	5	D-Mannitol	Water	15	0.50	5
Acetic acid	Toluene	25	2.26	5	Methane	Water	25	1.49	5
Benzene	Toluene	25	2.54	3	Methanol	Water	15	1.28	5
Benzoic acid	Toluene	25	1.49	5	3-Methyl-1-butanol	Water	10	0.69	5
Cyclohexane	Toluene	25	2.42	3	Methylcyclopentane	Water	20	0.85	5
Formic acid	Toluene	25	2.65	5	Phenol	Water	20	0.89	5
Water	Toluene	25	6.19	5	1-Propanol	Water	15	0.87	5
Acetone	Trichloromethane	25	2.55	5	Propene	Water	25	1.44	5
Benzene	Trichloromethane	25	2.89	5	Pyridine	Water	25	0.58	5
2-Butanone	Trichloromethane	25	2.13	5	Raffinose	Water	15	0.33	5
Butyl acetate	Trichloromethane	25	1.71	5	Sucrose	Water	25	0.52	5
Cyclohexane	Trichloromethane	25	1.28	3	Toluene	Water	20	0.85	5
Diethyl ether	Trichloromethane	25	2.13	3	Urea	Water	25	1.38	5
Ethanol	Trichloromethane	15	2.20	5					
Ethyl acetate	Trichloromethane	25	2.02	5					
Benzene	2,2,4-Trimethylpentane	30	3.46	4					

Section 7
Biochemistry

PROPERTIES OF AMINO ACIDS

This table gives selected properties of some important amino acids and closely related compounds. The first part of the table lists the 20 "standard" amino acids that are the basic constituents of proteins. The second part includes other amino acids and related compounds of biochemical importance. Within each part of the table the compounds are listed by name in alphabetical order. Structures are given in the following table.

Symbol: Three-letter symbol for the standard amino acids
M_r: Molecular weight
t_m: Melting point
pK_a, pK_b, pK_c, pK_d: Negative of the logarithm of the acid dissociation constants for the COOH and NH_2 groups (and, in some cases, other groups) in the molecule (at 25 °C)
pI: pH at the isoelectric point
S: Solubility in water in units of grams of compound per kilogram of water; a temperature of 25 °C is understood unless otherwise stated in a superscript. When quantitative data are not available, the notations sl.s. (for slightly soluble), s. (for soluble), and v.s. (for very soluble) are used.
V_2^0: Partial molar volume in aqueous solution at infinite dilution (at 25 °C)

Data on the enthalpy of formation of many of these compounds are included in the table "Standard Thermodynamic Properties of Chemical Substances" in Section 5 of this *Handbook*. Absorption spectra and optical rotation data can be found in Reference 3. Partial molar volume is taken from Reference 5; other thermodynamic properties, including solubility as a function of temperature, are given in References 3 and 5. Most of the pK values come from References 1, 6, and 7.

References

1. Dawson, R. M. C., Elliott, D. C., Elliott, W. H., and Jones, K. M., *Data for Biochemical Research*, Third Edition, Clarendon Press, Oxford, 1986.
2. O'Neil, Maryadele J., Ed., *The Merck Index, Fourteenth Edition*, Merck & Co., Rahway, NJ, 2006.
3. Sober, H. A., Ed., *CRC Handbook of Biochemistry. Selected Data for Molecular Biology*, CRC Press, Boca Raton, FL, 1968.
4. Voet, D., and Voet, J. G., *Biochemistry, Second Edition*, John Wiley & Sons, New York, 1995.
5. Hinz, H. J., Ed., *Thermodynamic Data for Biochemistry and Biotechnology*, Springer-Verlag, Heidelberg, 1986.
6. Fasman, G. D., Ed. *Practical Handbook of Biochemistry and Molecular Biology*, CRC Press, Boca Raton, FL, 1989.
7. Smith, R. M., and Martell, A. E., *NIST Standard Reference Database 46: Critically Selected Stability Constants of Metal Complexes Database*, Version 3.0, National Institute of Standards and Technology, Gaithersburg, MD, 1997.
8. Ramasami, P., *J. Chem. Eng. Data*, 47, 1164, 2002.

Symbol	Name	Mol. form.	M_r	t_m/°C	pK_a	pK_b	pK_c	pK_d	pI	S/g kg^{-1}	V_2^0/ cm^3 mol^{-1}
Ala	L-Alanine	$C_3H_7NO_2$	89.09	297	2.33	9.71			6.00	166.9	60.54
Arg	L-Arginine	$C_6H_{14}N_4O_2$	174.20	244	2.03	9.00	12.10		10.76	182.6	127.42
Asn	L-Asparagine	$C_4H_8N_2O_3$	132.12	235	2.16	8.73			5.41	25.1	78.0
Asp	L-Aspartic acid	$C_4H_7NO_4$	133.10	270	1.95	9.66	3.71		2.77	5.04	74.8
Cys	L-Cysteine	$C_3H_7NO_2S$	121.16	240	1.91	10.28	8.14		5.07	v.s.	73.45
Gln	L-Glutamine	$C_5H_{10}N_2O_3$	146.14	185	2.18	9.00			5.65	42	
Glu	L-Glutamic acid	$C_5H_9NO_4$	147.13	160	2.16	9.58	4.15		3.22	8.6	89.85
Gly	Glycine	$C_2H_5NO_2$	75.07	290	2.34	9.58			5.97	239	43.26
His	L-Histidine	$C_6H_9N_3O_2$	155.15	287	1.70	9.09	6.04		7.59	43.5	98.3
Ile	L-Isoleucine	$C_6H_{13}NO_2$	131.17	284	2.26	9.60			6.02	34.2	105.80
Leu	L-Leucine	$C_6H_{13}NO_2$	131.17	293	2.32	9.58			5.98	23.8	107.77
Lys	L-Lysine	$C_6H_{14}N_2O_2$	146.19	224	2.15	9.16	10.67		9.74	5.8	108.5
Met	L-Methionine	$C_5H_{11}NO_2S$	149.21	281	2.16	9.08			5.74	56	105.57
Phe	L-Phenylalanine	$C_9H_{11}NO_2$	165.19	283	2.18	9.09			5.48	27.9	121.5
Pro	L-Proline	$C_5H_9NO_2$	115.13	221	1.95	10.47			6.30	1625	82.76
Ser	L-Serine	$C_3H_7NO_3$	105.09	228	2.13	9.05			5.68	250	60.62
Thr	L-Threonine	$C_4H_9NO_3$	119.12	256	2.20	8.96			5.60	90.6	76.90
Trp	L-Tryptophan	$C_{11}H_{12}N_2O_2$	204.23	289	2.38	9.34			5.89	13.2	143.8
Tyr	L-Tyrosine	$C_9H_{11}NO_3$	181.19	343	2.24	9.04	10.10		5.66	0.51	
Val	L-Valine	$C_5H_{11}NO_2$	117.15	315	2.27	9.52			5.96	88	90.75

Name	Mol. form.	M_r	t_m/°C	pK_a	pK_b	pK_c	pK_d	pI	S/g kg^{-1}	V_2^0/ cm^3 mol^{-1}
N-Acetylglutamic acid	$C_7H_{11}NO_5$	189.17	199						s.	
*N*6-Acetyl-L-lysine	$C_8H_{16}N_2O_3$	188.22	265	2.12	9.51					
β-Alanine	$C_3H_7NO_2$	89.09	200	3.51	10.08				890	58.28
2-Aminoadipic acid	$C_6H_{11}NO_4$	161.16	207	2.14	4.21	9.77		3.18	2.2^{40}	
DL-2-Aminobutanoic acid	$C_4H_9NO_2$	103.12	304	2.30	9.63			6.06	210	75.6
DL-3-Aminobutanoic acid	$C_4H_9NO_2$	103.12	194.3	3.43	10.05			7.30	1250	76.3
4-Aminobutanoic acid	$C_4H_9NO_2$	103.12	203	4.02	10.35				971	73.2
10-Aminodecanoic acid	$C_{10}H_{21}NO_2$	187.28	188.5							167.3

Name	Mol. form.	M_r	t_m/°C	pK_a	pK_b	pK_c	pK_d	pI	S/g kg^{-1}	V_2^0/cm^3 mol^{-1}
7-Aminoheptanoic acid	C$_7$H$_{15}$NO$_2$	145.20	195						v.s.	120.0
6-Aminohexanoic acid	C$_6$H$_{13}$NO$_2$	131.17	205					7.29	850	104.2
L-3-Amino-2-methylpropanoic acid	C$_4$H$_9$NO$_2$	103.12	185						s.	
2-Amino-2-methylpropanoic acid	C$_4$H$_9$NO$_2$	103.12	335	2.36	10.21			5.72	137	77.55
9-Aminononanoic acid	C$_9$H$_{19}$NO$_2$	173.26	191							151.3
8-Aminooctanoic acid	C$_8$H$_{17}$NO$_2$	159.23	192							136.1
5-Amino-4-oxopentanoic acid	C$_5$H$_9$NO$_3$	131.13	118	4.05	8.90					
5-Aminopentanoic acid	C$_5$H$_{11}$NO$_2$	117.15	157 dec						s.	87.6
o-Anthranilic acid	C$_7$H$_7$NO$_2$	137.14	146	2.05	4.95				3.5[14]	
Azaserine	C$_5$H$_7$N$_3$O$_4$	173.13	150		8.55				v.s.	
Canavanine	C$_5$H$_{12}$N$_4$O$_3$	176.17	172	2.50	6.60	9.25		7.93	v.s.	
L-γ-Carboxyglutamic acid	C$_6$H$_9$NO$_6$	191.14	167	1.70	9.90	4.75	3.20			
Carnosine	C$_9$H$_{14}$N$_4$O$_3$	226.23	260	2.51	9.35	6.76			322	
Citrulline	C$_6$H$_{13}$N$_3$O$_3$	175.19	222	2.32	9.30			5.92	s.	
Creatine	C$_4$H$_9$N$_3$O$_2$	131.13	303	2.63	14.30				16	
L-Cysteic acid	C$_3$H$_7$NO$_5$S	169.16	260	1.89	8.70	1.30			v.s.	
L-Cystine	C$_6$H$_{12}$N$_2$O$_4$S$_2$	240.30	260	1.50	8.80	2.05	8.03		0.17	
2,4-Diaminobutanoic acid	C$_4$H$_{10}$N$_2$O$_2$	118.13	118.1	1.85	8.24	10.44		9.27	s.	
3,5-Dibromo-L-tyrosine	C$_9$H$_9$Br$_2$NO$_3$	338.98	245						2.72	
3,5-Dichloro-L-tyrosine	C$_9$H$_9$Cl$_2$NO$_3$	250.08	247						1.97	
3,5-Diiodo-L-tyrosine	C$_9$H$_9$I$_2$NO$_3$	432.98	213	2.12	9.10	6.16			0.62	
Dopamine	C$_8$H$_{11}$NO$_2$	153.18			10.36	8.88			s.	
L-Ethionine	C$_6$H$_{13}$NO$_2$S	163.24	273	2.18	9.05	13.10				
N-Glycylglycine	C$_4$H$_8$N$_2$O$_3$	132.12	263	3.13	8.10				231	
Guanidinoacetic acid	C$_3$H$_7$N$_3$O$_2$	117.11	282	2.82					5	
Histamine	C$_5$H$_9$N$_3$	111.15	83		9.83	6.11			v.s.	
L-Homocysteine	C$_4$H$_9$NO$_2$S	135.19	232	2.15	8.57	10.38		5.55	s.	
Homocystine	C$_8$H$_{16}$N$_2$O$_4$S$_2$	268.35	264	1.59	9.44	2.54	8.52		0.2	
L-Homoserine	C$_4$H$_9$NO$_3$	119.12	203	2.27	9.28			6.17	1100	
3-Hydroxy-DL-glutamic acid	C$_5$H$_9$NO$_5$	163.13	209					3.28		
5-Hydroxylysine	C$_6$H$_{14}$N$_2$O$_3$	162.19		2.13	8.85	9.83		9.15		
trans-4-Hydroxy-L-proline	C$_5$H$_9$NO$_3$	131.13	274	1.82	9.47			5.74	361	84.49
L-3-Iodotyrosine	C$_9$H$_{10}$INO$_3$	307.08	205	2.20	9.10	8.70			sl.s.	
L-Kynurenine	C$_{10}$H$_{12}$N$_2$O$_3$	208.21	194						sl.s.	
L-Lanthionine	C$_6$H$_{12}$N$_2$O$_4$S	208.24	294						1.5	
Levodopa	C$_9$H$_{11}$NO$_4$	197.19	277	2.32	8.72	9.96	11.79		1.65[20]	
L-1-Methylhistidine	C$_7$H$_{11}$N$_3$O$_2$	169.18	249	1.69	8.85	6.48			200	
L-Norleucine	C$_6$H$_{13}$NO$_2$	131.17	301	2.31	9.68			6.09	15	107.7
L-Norvaline	C$_5$H$_{11}$NO$_2$	117.15	307	2.31	9.65				107	91.8
L-Ornithine	C$_5$H$_{12}$N$_2$O$_2$	132.16	140	1.94	8.78	10.52		9.73	v.s.	
O-Phosphoserine	C$_3$H$_8$NO$_6$P	185.07	166	2.14	9.80	5.70				
L-Pyroglutamic acid	C$_5$H$_7$NO$_3$	129.12	162	3.32						
Sarcosine	C$_3$H$_7$NO$_2$	89.09	212	2.18	9.97				428	
Taurine	C$_2$H$_7$NO$_3$S	125.15	328	-0.3	9.06				105	
L-Thyroxine	C$_{15}$H$_{11}$I$_4$NO$_4$	776.87	235	2.20	10.01	6.45			sl.s.	

STRUCTURES OF COMMON AMINO ACIDS

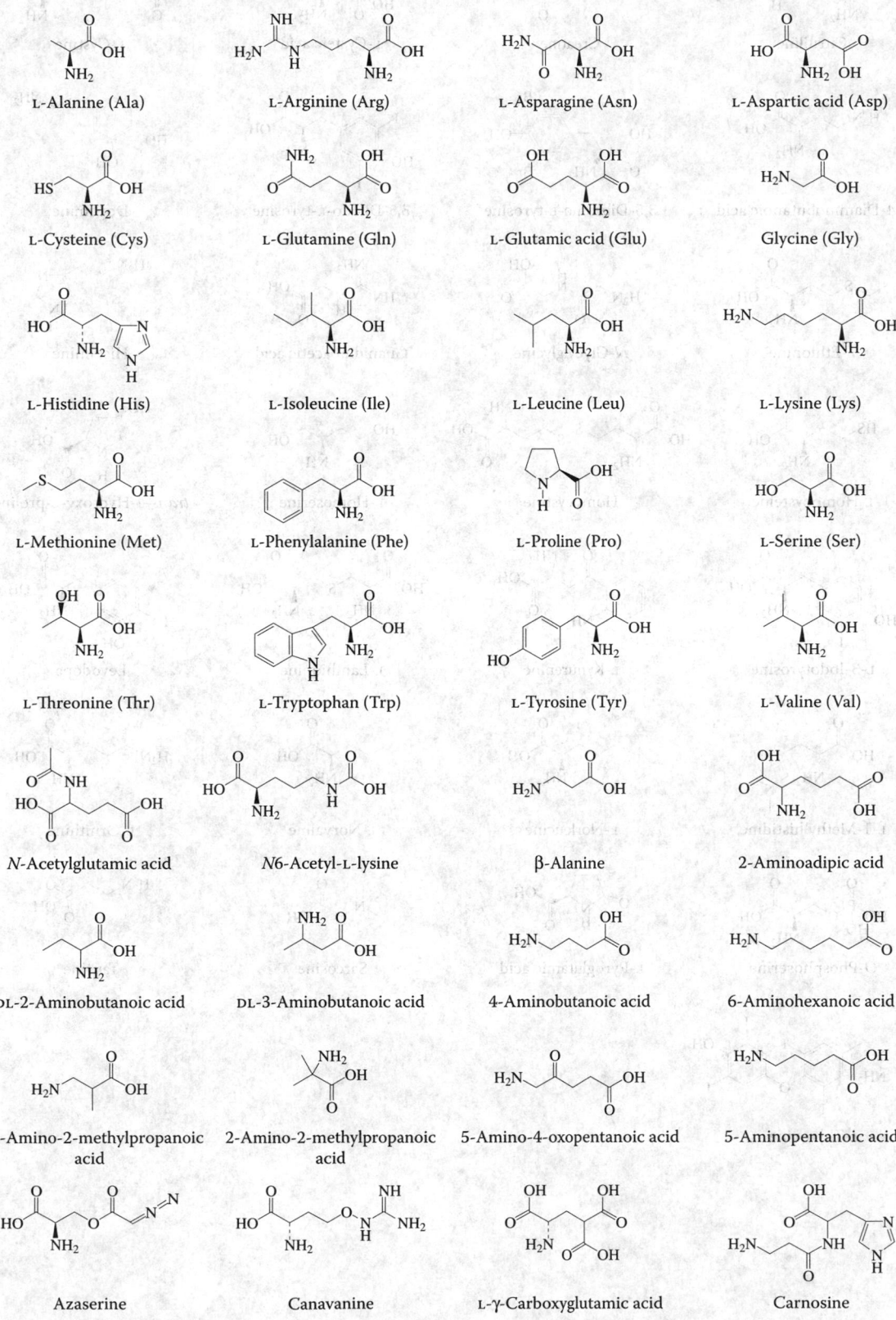

L-Alanine (Ala)

L-Arginine (Arg)

L-Asparagine (Asn)

L-Aspartic acid (Asp)

L-Cysteine (Cys)

L-Glutamine (Gln)

L-Glutamic acid (Glu)

Glycine (Gly)

L-Histidine (His)

L-Isoleucine (Ile)

L-Leucine (Leu)

L-Lysine (Lys)

L-Methionine (Met)

L-Phenylalanine (Phe)

L-Proline (Pro)

L-Serine (Ser)

L-Threonine (Thr)

L-Tryptophan (Trp)

L-Tyrosine (Tyr)

L-Valine (Val)

N-Acetylglutamic acid

*N*6-Acetyl-L-lysine

β-Alanine

2-Aminoadipic acid

DL-2-Aminobutanoic acid

DL-3-Aminobutanoic acid

4-Aminobutanoic acid

6-Aminohexanoic acid

L-3-Amino-2-methylpropanoic
acid

2-Amino-2-methylpropanoic
acid

5-Amino-4-oxopentanoic acid

5-Aminopentanoic acid

Azaserine

Canavanine

L-γ-Carboxyglutamic acid

Carnosine

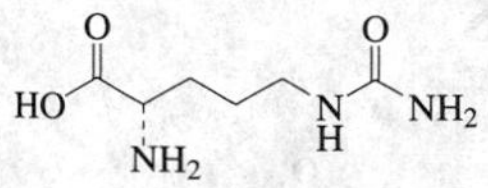

Citrulline

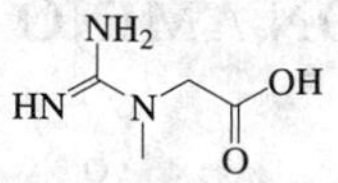

Creatine

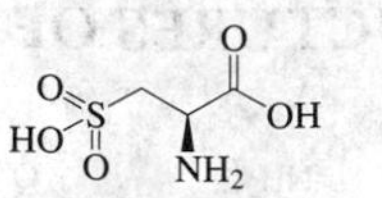

L-Cysteic acid

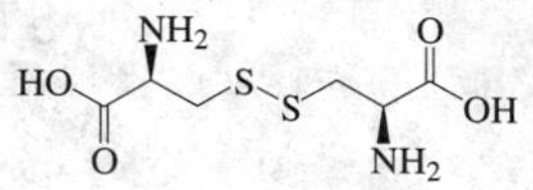

L-Cystine

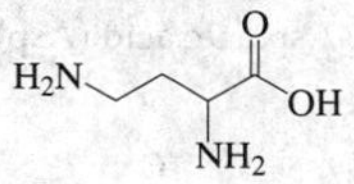

2,4-Diaminobutanoic acid

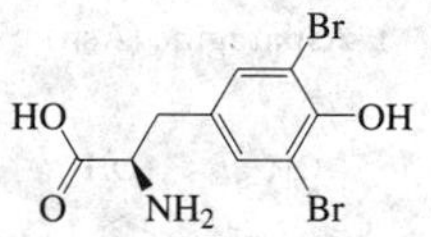

3,5-Dibromo-L-tyrosine

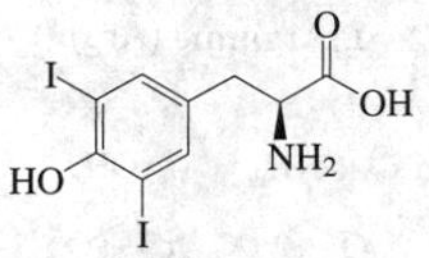

3,5-Diiodo-L-tyrosine

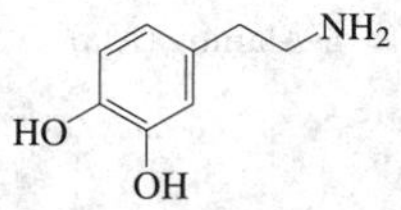

Dopamine

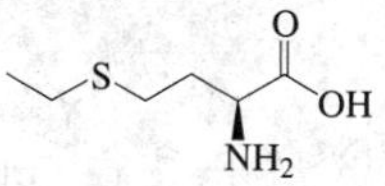

L-Ethionine

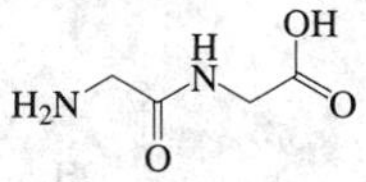

N-Glycylglycine

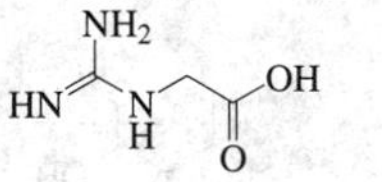

Guanidinoacetic acid

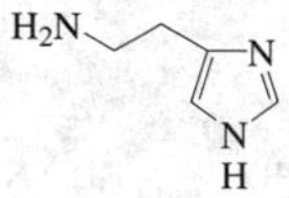

Histamine

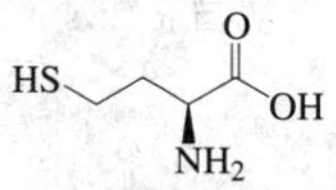

L-Homocysteine

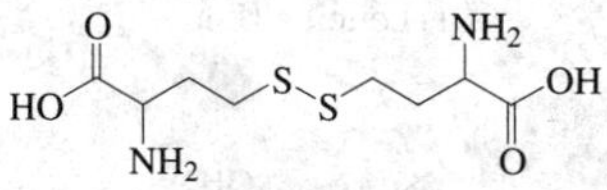

Homocystine

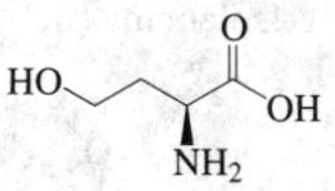

L-Homoserine

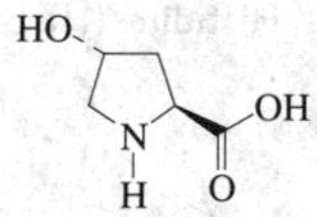

trans-4-Hydroxy-L-proline

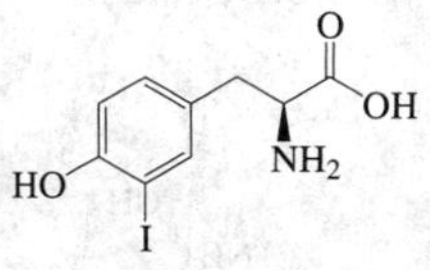

L-3-Iodotyrosine

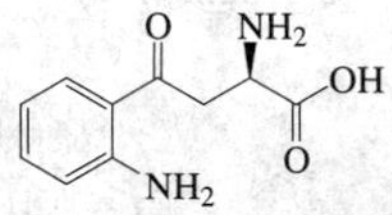

L-Kynurenine

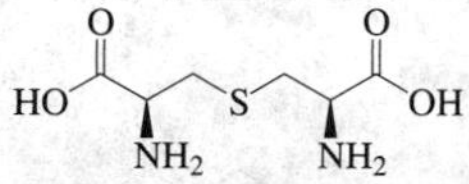

L-Lanthionine

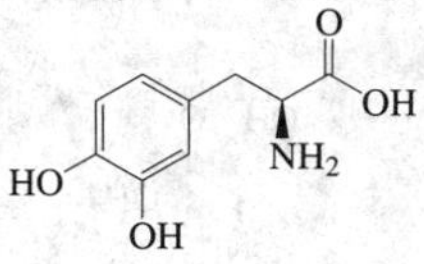

Levodopa

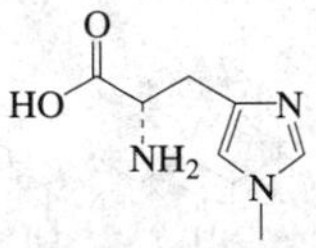

L-1-Methylhistidine

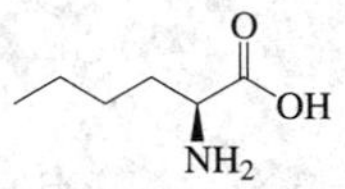

L-Norleucine

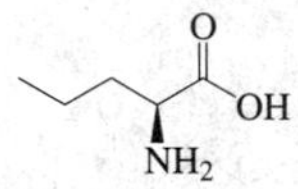

L-Norvaline

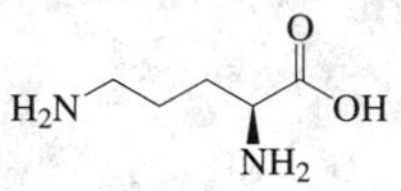

L-Ornithine

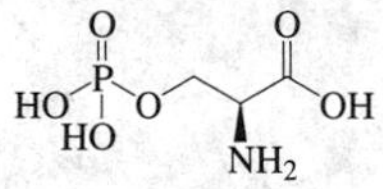

O-Phosphoserine

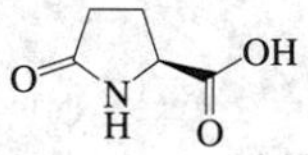

L-Pyroglutamic acid

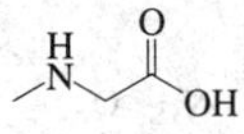

Sarcosine

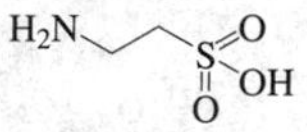

Taurine

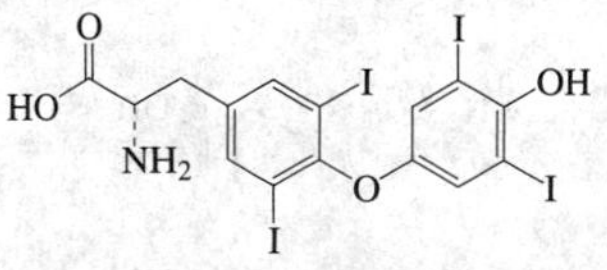

L-Thyroxine

PROPERTIES OF PURINE AND PYRIMIDINE BASES

This table lists some of the important purine and pyrimidine bases that occur in nucleic acids. The pK_a values (negative logarithm of the acid dissociation constant) are given for each ionization stage. The last column gives the aqueous solubility S at the indicated temperature in units of grams per 100 grams of solution.

The numbering system in the rings is:

Purine   Pyrimidine

References

1. Dawson, R. M. C., et al., *Data for Biochemical Research*, 3rd ed., Clarendon Press, Oxford, 1986.
2. O'Neil, M. J., Ed., *The Merck Index*, 13th ed., Merck and Co., Rahway, NJ, 2001.

Common name	Systematic name	Mol. form.	Mol. wt.	pK_a values			S/mass % (temp.)
Pyrimidines							
Cytosine	4-Amino-2-hydroxypyrimidine	$C_4H_5N_3O$	111.10	4.60	12.16		0.73 (25 °C)
5-Methylcytosine	4-Amino-2-hydroxy-5-methylpyrimidine	$C_5H_7N_3O$	125.13	4.6	12.4		0.45 (25 °C)
5-Hydroxymethylcytosine	4-Amino-2-hydroxy-5-hydroxymethylpyrimidine	$C_5H_7N_3O_2$	141.13	4.3	13		
Uracil	2,4-Dihydroxypyrimidine	$C_4H_4N_2O_2$	112.09	0.5	9.5	>13	0.27 (25 °C)
Thymine	5-Methyluracil	$C_5H_6N_2O_2$	126.11	9.94	>13		0.35 (25 °C)
Orotic acid	Uracil-6-carboxylic acid	$C_5H_4N_2O_4$	156.10	2.4	9.5	>13	0.18 (18 °C)
Purines							
Adenine	6-Aminopurine	$C_5H_5N_5$	135.14	<1	4.3	9.83	0.104 (25 °C)
Guanine	2-Amino-6-hydroxypurine	$C_5H_5N_5O$	151.13	3.3	9.2	12.3	0.0068 (40 °C)
7-Methylguanine	7-Methyl-2-amino-6-hydroxypurine	$C_6H_7N_5O$	165.16	3.5	9.9		
Isoguanine	6-Amino-2-hydroxypurine	$C_5H_5N_5O$	151.13	4.5	9.0		0.006 (25 °C)
Xanthine	2,6-Dioxopurine	$C_5H_4N_4O_2$	152.11	0.8	7.4	11.1	0.05 (20 °C)
Hypoxanthine	6-Hydroxypurine	$C_5H_4N_4O$	136.11	2.0	8.9	12.1	0.07 (19 °C)
Uric acid	2,6,8-Trihydroxypurine	$C_5H_4N_4O_3$	168.11	5.4	11.3		0.002 (20 °C)

THE GENETIC CODE

This table gives the correspondence between a messenger RNA codon and the amino acid that it specifies. The symbols for bases in the codon are:

U: uracil
C: cytosine
A: adenine
G: guanine

The amino acid symbols are given in the table entitled "Structures of Common Amino Acids." A chain-initiating codon is indicated by **init** and a chain-terminating codon by **term**.

Example: UCA codes for **Ser**, UAC codes for **Tyr**, etc.

First position	Second position				Third position
	U	C	A	G	
U	Phe	Ser	Tyr	Cys	U
	Phe	Ser	Tyr	Cys	C
	Leu	Ser	**term**	**term**	A
	Leu	Ser	**term**	Trp	G
C	Leu	Pro	His	Arg	U
	Leu	Pro	His	Arg	C
	Leu	Pro	Gln	Arg	A
	Leu	Pro	Gln	Arg	G
A	Ile	Thr	Asn	Ser	U
	Ile	Thr	Asn	Ser	C
	Ile	Thr	Lys	Arg	A
	Met (**init**)	Thr	Lys	Arg	G
G	Val	Ala	Asp	Gly	U
	Val	Ala	Asp	Gly	C
	Val	Ala	Glu	Gly	A
	Val (**init**)	Ala	Glu	Gly	G

PROPERTIES OF FATTY ACIDS AND THEIR METHYL ESTERS

This table gives the names and selected properties of some important fatty acids and their methyl esters. It includes most of the acids that are significant constituents of naturally occurring oils and fats. Compounds are listed first by number of carbon atoms and, secondly, by the degree of unsaturation. Both the systematic name and the common or trivial name are given, as well as the Chemical Abstracts Service Registry Number and the shorthand acid code that is frequently used. The first number in this code gives the number of carbon atoms; the number following the colon is the number of unsaturated centers (mainly double bonds). The location and orientation of the unsaturated centers follow. The symbols used are: c = *cis*; t = *trans*; a = acetylenic center; e = ethylenic center at end of chain; ep = *epoxy*. Thus 9c,11t indicates a double bond with *cis* orientation at the No. 9 carbon and another with *trans* orientation at the No. 11 carbon. More details on the codes can be found in Reference 1.

The table gives the molecular weight and melting point of the acid and the melting and boiling points of the methyl ester of the acid when available. A superscript on the boiling point indicates the pressure in mmHg (torr); if there is no superscript, the value refers to one atmosphere (760 mmHg). The references cover many other fatty acids beyond those listed here and give additional properties.

We are indebted to Frank D. Gunstone for advice on the content of the table.

References

1. Gunstone, F. D., Harwood, J. L., and Dijkstra, A. J., eds., *The Lipid Handbook, Third Edition,* CRC Press, Boca Raton, FL, 2006.
2. Gunstone, F. D., and Adlof, R. O., *Common (non-systematic) Names for Fatty Acids,* www.aocs.org/member/division/analytic/fanames. asp, 2003.
3. Firestone, D., *Physical and Chemical Characteristics of Oils, Fats, and Waxes,* 2nd Edition, AOCS Press, Urbana, IL, 2006.
4. Dawson, R. M. C., Elliott, D. C., Elliott, W. H., and Jones, K. M., *Data for Biochemical Research,* Third Edition, Clarendon Press, Oxford, 1986.
5. Altman, P. L., and Dittmer, D. S., eds., *Biology Data Book,* Second Edition, Vol. 1, Federation of American Societies for Experimental Biology, Bethesda, MD, 1972.
6. Fasman, G. D., Ed., *Practical Handbook of Biochemistry and Molecular Biology,* CRC Press, Boca Raton, FL, 1989.

Systematic name	Common name	Mol. form.	Acid code	CAS RN	Mol. weight	mp/°C	Methyl ester mp/°C	Methyl ester bp/°C
Butanoic acid	Butyric acid	$C_4H_8O_2$	4:0	107-92-6	88.106	−5.1	−85.8	102.8
Pentanoic acid	Valeric acid	$C_5H_{10}O_2$	5:0	109-52-4	102.132	−33.6		127.4
3-Methylbutanoic acid	Isovaleric acid	$C_5H_{10}O_2$	4:0 3-Me	503-74-2	102.132	−29.3		116.5
Hexanoic acid	Caproic acid	$C_6H_{12}O_2$	6:0	142-62-1	116.158	−3	−71	149.5
Heptanoic acid	Enanthic acid	$C_7H_{14}O_2$	7:0	111-14-8	130.185	−7.2	−56	174
Octanoic acid	Caprylic acid	$C_8H_{16}O_2$	8:0	124-07-2	144.212	16.5	−40	192.9
Nonanoic acid	Pelargonic acid	$C_9H_{18}O_2$	9:0	112-05-0	158.238	12.4		213.5
Decanoic acid	Capric acid	$C_{10}H_{20}O_2$	10:0	334-48-5	172.265	31.4	−18	224
9-Decenoic acid	Caproleic acid	$C_{10}H_{18}O_2$	10:1 9e	14436-32-9	170.249	26.5		120[20]
Undecanoic acid		$C_{11}H_{22}O_2$	11:0	112-37-8	186.292	28.6		123[10]
Dodecanoic acid	Lauric acid	$C_{12}H_{24}O_2$	12:0	143-07-7	200.318	43.8	5.2	267
cis-9-Dodecenoic acid	Lauroleic acid	$C_{12}H_{22}O_2$	12:1 9c	2382-40-3	198.302			
Tridecanoic acid		$C_{13}H_{26}O_2$	13:0	638-53-9	214.344	41.5	6.5	92[1]
Tetradecanoic acid	Myristic acid	$C_{14}H_{28}O_2$	14:0	544-63-8	228.371	54.2	19	295
cis-9-Tetradecenoic acid	Myristoleic acid	$C_{14}H_{26}O_2$	14:1 9c	13147-06-3	226.355	−4		
Pentadecanoic acid		$C_{15}H_{30}O_2$	15:0	1002-84-2	242.398	52.3	18.5	153.5
Hexadecanoic acid	Palmitic acid	$C_{16}H_{32}O_2$	16:0	57-10-3	256.424	62.5	30	417
cis-9-Hexadecenoic acid	Palmitoleic acid	$C_{16}H_{30}O_2$	16:1 9c	373-49-9	254.408	0.5		140[5]
Heptadecanoic acid	Margaric acid	$C_{17}H_{34}O_2$	17:0	506-12-7	270.451	61.3	30	185[9]
Octadecanoic acid	Stearic acid	$C_{18}H_{36}O_2$	18:0	57-11-4	284.478	69.3	39.1	443
cis-6-Octadecenoic acid	Petroselinic acid	$C_{18}H_{34}O_2$	18:1 6c	593-39-5	282.462	29.8		
cis-9-Octadecenoic acid	Oleic acid	$C_{18}H_{34}O_2$	18:1 9c	112-80-1	282.462	13.4	−19.9	218.5[20]
trans-9-Octadecenoic acid	Elaidic acid	$C_{18}H_{34}O_2$	18:1 9t	112-79-8	282.462	45	13.5	218[24]
cis-11-Octadecenoic acid	*cis*-Vaccenic acid	$C_{18}H_{34}O_2$	18:1 11c	506-17-2	282.462	15		163[0.1]
trans-11-Octadecenoic acid	Vaccenic acid	$C_{18}H_{34}O_2$	18:1 11t	693-72-1	282.462	44		172[3]
cis-12,13-Epoxy-*cis*-9-octadecenoic acid	Vernolic acid	$C_{18}H_{32}O_3$	18:1 12,13-ep,9c	503-07-1	296.445	32.5		
12-Hydroxy-*cis*-9-octadecenoic acid	Ricinoleic acid	$C_{18}H_{34}O_3$	18:1 12-OH,9c	141-22-0	298.461	5.5		226[15]
cis,trans-9,11-Octadecadienoic acid	Rumenic (CLA)	$C_{18}H_{32}O_2$	18:2 9c,11t	1839-11-8	280.446	20		
cis,cis-9,12-Octadecadienoic acid	Linoleic acid	$C_{18}H_{32}O_2$	18:2 9c,12c	60-33-3	280.446	−7	−35	215[20]

Systematic name	Common name	Mol. form.	Acid code	CAS RN	Mol. weight	mp/°C	Methyl ester mp/°C	Methyl ester bp/°C
trans,cis-10,12-Octadecadienoic acid	(CLA)	$C_{18}H_{32}O_2$	18:2 10t,12c	22880-03-1	280.446	23	−12	
cis-9-Octadecen-12-ynoic acid	Crepenynic acid	$C_{18}H_{30}O_2$	18:2 9c,12a	2277-31-8	278.430			
cis,cis,cis-5,9,12-Octadecatrienoic acid	Pinolenic acid	$C_{18}H_{30}O_2$	18:3 5c,9c,12c	27213-43-0	278.430			
trans,cis,cis-5,9,12-Octadecatrienoic acid	Columbinic acid	$C_{18}H_{30}O_2$	18:3 5t,9c,12c	2441-53-4	278.430			
cis,cis,cis-6,9,12-Octadecatrienoic acid	γ-Linolenic acid	$C_{18}H_{30}O_2$	18:3 6c,9c,12c	506-26-3	278.430			162[0.5]
trans,trans,cis-8,10,12-Octadecatrienoic acid	Calendic acid	$C_{18}H_{30}O_2$	18:3 8t,10t,12c	28872-28-8	278.430	40		
cis,trans,cis-9,11,13-Octadecatrienoic acid	Punicic acid	$C_{18}H_{30}O_2$	18:3 9c,11t,13c	544-72-9	278.430	45		
cis,trans,trans-9,11,13-Octadecatrienoic acid	α-Eleostearic acid	$C_{18}H_{30}O_2$	18:3 9c,11t,13t	506-23-0	278.430	49		148[1]
trans,trans,cis-9,11,13-Octadecatrienoic acid	Catalpic acid	$C_{18}H_{30}O_2$	18:3 9t,11t,13c	4337-71-7	278.430	32		
trans,trans,trans-9,11,13-Octadecatrienoic acid	β-Eleostearic acid	$C_{18}H_{30}O_2$	18:3 9t,11t,13t	544-73-0	278.430	71.5	13	162[1]
cis,cis,cis-9,12,15-Octadecatrienoic acid	α-Linolenic acid	$C_{18}H_{30}O_2$	18:3 9c,12c,15c	463-40-1	278.430	−11.3	−52	109[0.018]
6,9,12,15-Octadecatetraenoic acid, all *cis*	Stearidonic acid	$C_{18}H_{28}O_2$	18:4 6c,9c,12c,15c	20290-75-9	276.414	−57		
cis,trans,trans-9,11,13,15-Octadecatetraenoic acid	Parinaric acid	$C_{18}H_{28}O_2$	18:4 9c,11t,13t,15c	593-38-4	276.414	86		
Nonadecanoic acid		$C_{19}H_{38}O_2$	19:0	646-30-0	298.504	69.4	41.3	190[4]
Eicosanoic acid	Arachidic acid	$C_{20}H_{40}O_2$	20:0	506-30-9	312.531	76.5	54.5	215[10]
3,7,11,15-Tetramethylhexadecanoic acid	Phytanic acid	$C_{20}H_{40}O_2$	16:0 3,7,11,15-tetramethyl	14721-66-5	312.531	−65		
cis-5-Eicosenoic acid		$C_{20}H_{38}O_2$	20:1 5c	7050-07-9	310.515	27		
cis-9-Eicosenoic acid	Gadoleic acid	$C_{20}H_{38}O_2$	20:1 9c	29204-02-2	310.515	24.5		
cis-11-Eicosenoic acid	Gondoic acid	$C_{20}H_{38}O_2$	20:1 11c	2462-94-4	310.515	24		
cis,cis,cis-8,11,14-Eicosatrienoic acid	Dihomo-γ-linolenic acid	$C_{20}H_{34}O_2$	20:3 8c,11c,14c	1783-84-2				
5,8,11,14-Eicosatetraenoic acid, all *cis*	Arachidonic acid	$C_{20}H_{32}O_2$	20:4 5c,8c,11c,14c	506-32-1	304.467	−49.5		195[0.7]
5,8,11,14,17-Eicosapentaenoic acid, all *cis*	Timnodonic acid, EPA	$C_{20}H_{30}O_2$	20:5 5c,8c,11c,14c,17c	10417-94-4	302.451	−54		
Heneicosanoic acid		$C_{21}H_{42}O_2$	21:0	2363-71-5	326.557	82	49	207[4]
Docosanoic acid	Behenic acid	$C_{22}H_{44}O_2$	22:0	112-85-6	340.583	81.5	54	
cis-11-Docosenoic acid	Cetolic acid	$C_{22}H_{42}O_2$	22:1 11c	506-36-5	338.567	33		
cis-13-Docosenoic acid	Erucic acid	$C_{22}H_{42}O_2$	22:1 13c	112-86-7	338.567	34.7		221[5]
trans-13-Docosenoic acid	Brassidic acid	$C_{22}H_{42}O_2$	22:1 13t	506-33-2	338.567	61.9	35	
cis,cis-5,13-Docosadienoic acid		$C_{22}H_{40}O_2$	22:2 5c,13c	676-39-1	336.552	−4		
7,10,13,16,19-Docosapentaenoic acid, all *cis*		$C_{22}H_{34}O_2$	22:5 7c,10c,13c,16c,19c					
4,7,10,13,16,19-Docosahexaenoic acid, all *cis*	Cervonic acid, DHA	$C_{22}H_{32}O_2$	22:6 4c,7c,10c,13c,16c,19c	2091-24-9		−45		
Tricosanoic acid		$C_{23}H_{46}O_2$	23:0	2433-96-7		79.6	53.4	
Tetracosanoic acid	Lignoceric acid	$C_{24}H_{48}O_2$	24:0	557-59-5	368.637	87.5	60	
cis-15-Tetracosenoic acid	Nervonic acid	$C_{24}H_{46}O_2$	24:1 15c	506-37-6	366.621	43	15	165[0.02]
Pentacosanoic acid		$C_{25}H_{50}O_2$	25:0	506-38-7	382.664	77.5	62	
Hexacosanoic acid	Cerotic acid	$C_{26}H_{52}O_2$	26:0	506-46-7	396.690	88.5	63.8	286[15]
Heptacosanoic acid		$C_{27}H_{54}O_2$	27:0	7138-40-1		87.6	64	
Octacosanoic acid	Montanic acid	$C_{28}H_{56}O_2$	28:0	506-48-9	424.744	90.9	67	
Nonacosanoic acid		$C_{29}H_{58}O_2$	29:0	4250-38-8	438.770	90.3	69	
Triacontanoic acid	Melissic acid	$C_{30}H_{60}O_2$	30:0	506-50-3	452.796	93.6	72	
Hentriacontanoic acid		$C_{31}H_{62}O_2$	31:0	38232-01-8	466.823	93.1		
Dotriacontanoic acid	Lacceric acid	$C_{32}H_{64}O_2$	32:0	3625-52-3	480.849	96.2		192[0.01]

COMPOSITION AND PROPERTIES OF COMMON OILS AND FATS

This table lists some of the most common naturally occurring oils and fats. The list is separated into those of plant origin, fish and other marine life origin, and land animal origin. The oils and fats consist mainly of esters of glycerol (i.e., triglycerides) with fatty acids of 10 to 22 carbon atoms. The four fatty acids with the highest concentration are given for each oil; concentrations are given in weight percent. Because there is often a wide variation in composition depending on the source of the oil sample, a range (or sometimes an average) is generally given. More complete data on composition, including minor fatty acids, sterols, and tocopherols, can be found in the references.

The acids are labeled by the codes described in the previous table, "Properties of Fatty Acids and Their Methyl Esters," which gives the systematic and common names of the acids. Thus 18:2 9c,12c indicates a C_{18} acid with two double bonds in the 9 and 12 positions, both with a *cis* configuration (*cis,cis*-9,12-octadecadienoic acid, or linoleic acid).

The density and refractive index of the oils are typical values; superscripts indicate the temperature in °C.

Notes:
- The composition figure given for oleic acid (18:1 9c) often includes low levels of other 18:1 isomers.

- In some oils where a concentration is given for 18:2 9c,12c (linoleic acid), other isomers of 18:2 may be included.
- Likewise, where a concentration is given for 18:3 9c,12c,15c (α-linolenic acid), other isomers of 18:3 may be included.
- The acid 20:5 6c,9c,12c,15c,17c, which is prevalent in many fish oils, is often abbreviated as 20:5 ω-3 or 20:5 n-3.

The assistance of Frank D. Gunstone in preparing this table is gratefully acknowledged.

References

1. Firestone, D., *Physical and Chemical Characteristics of Oils, Fats, and Waxes*, 2nd Edition, AOCS Press, Urbana, IL, 2006.
2. Gunstone, F. D., Harwood, J. L., and Dijkstra, A. J., eds., *The Lipid Handbook*, Third Edition, CRC Press, Boca Raton, FL, 2006.
3. Dawson, R. M. C., Elliott, D. C., Elliott, W. H., and Jones, K. M., *Data for Biochemical Research*, Third Edition, Clarendon Press, Oxford, 1986.
4. Altman, P. L., and Dittmer, D. S., eds., *Biology Data Book*, Second Edition, Vol. 1, Federation of American Societies for Experimental Biology, Bethesda, MD, 1972.

Type of oil	Principal fatty acid components in weight %				mp/ °C	Density/ g cm⁻³	Refractive index	Iodine value	Saponification value
Plants									
Almond kernel oil	18:1 9c	43–70%	18:2 9c,12c	24–30%		0.910^{25}	1.467^{26}	89–101	188–200
	16:0	4–13%	18:0	1–10%					
Apricot kernel oil	18:1 9c	58–66%	18:2 9c,12c	29–33%		0.910^{25}	1.469^{25}	97–110	185–199
	16:0	4.6–6%	18:0	1%					
Argan seed oil	18:1 9c	42–55%	18:2 9c,12c	30–34%		0.912^{20}	1.467^{20}	92–102	189–195
	16:0	12–16%	18:0	2–7%					
Avocado pulp oil	18:1 9c	56–74%	18:2 9c,12c	10–17%		0.912^{25}	1.466^{25}	85–90	177–198
	16:0	9–18%	16:1 9c	3–9%					
Babassu palm oil	12:0	40–55%	14:0	11–27%	24	0.914^{25}	1.450^{40}	10–18	245–256
	18:1 9c	9–20%	16:0	5.2–11%					
Blackcurrant oil	18:2 9c,12c	45–50%	18:3 6c,9c,12c	14–20%		0.923^{20}	1.480^{20}	173–182	185–195
	18:3 9c,12c,15c	12–15%	18:1 9c	9–13%					
Borage (star-flower) oil	18:2 9c,12c	36–40%	18:3 6c,9c,12c	17–25%				141–160	189–192
	18:1 9c	14–21%	16:0	9.4–12%					
Borneo tallow	18:0	39–43%	18:1 9c	34–37%	38	0.855^{100}	1.456^{40}	29–38	189–200
	16:0	18–21%	20:0	1.0%					
Cameline oil	18:3 9c,12c,15c	33–38%	18:2 9c,12c	15–16%		0.924^{15}	1.477^{20}	127–155	180–190
	20:1 total	14–16%	18:1 9c	12–24%					
Canola (rapeseed) oil (low linolenic)	18:1 9c	59–66%	18:2 9c,12c	24–29%	–10			91	
	16:0	4–5%	18:3 9c,12c,15c	2–3%					
Canola (rapeseed) oil (low erucic)	18:1 9c	52–67%	18:2 9c,12c	16–25%	–10	0.915^{20}	1.466^{40}	110–126	182–193
	18:3 9c,12c,15c	6–14%	16:0	3.3–6.0%					
Caraway seed oil	18:1 9c	40%	18:2 9c,12c	30%			1.471^{35}	128	178
	18:1 6c	26%	16:0	3%					
Cashew nut oil	18:1 9c	57–80%	18:2 9c,12c	16–22%		0.914^{15}	1.463^{40}	79–89	180–196
	16:0	4–17%	18:0	2–12%					

Type of oil	Principal fatty acid components in weight %				mp/ °C	Density/ g cm⁻³	Refractive index	Iodine value	Saponification value
Castor oil	18:1 12-OH,9c	88%	18:2 9c,12c	3–5%	–18	0.952²⁵	1.475²⁵	81–91	176–187
	18:1 9c	2.9–6%	22:0	2.1%					
Cherry kernel oil	18:2 9c,12c	42–45%	18:1 9c	35–49%		0.918²⁵	1.468⁴⁰	110–118	190–198
	16:0	4–9%	18:3 9c,11t,13t	3–10%					
Chinese vegetable tallow	16:0	58–72%	18:1 9c	20–35%	44	0.887²⁵	1.456⁴⁰	16–29	200–218
	18:0	1–8%	14:0	0.5–3.7%					
Cocoa butter	18:0	31–37%	18:1 9c	31–35%	34	0.974²⁵	1.457⁴⁰	32–40	192–200
	16:0	25–27%	18:2 9c,12c	2.8–4.0%					
Coconut oil	12:0	45–51%	14:0	17–21%	25	0.913⁴⁰	1.449⁴⁰	5–13	248–265
	16:0	7.7–10.2%	18:1 9c	5.4–9.9%					
Cohune nut oil	12:0	44–48%	14:0	16–17%		0.914²⁵	1.450⁴⁰	9–14	251–260
	18:1 9c	8–10%	16:0	7–10%					
Coriander seed oil	18:1 6c	53%	18:1 9c	32%		0.908²⁵	1.464²⁵	86–100	182–191
	18:2 9c,12c	7–14%	16:0	3–8%					
Corn oil	18:2 9c,12c	40–66%	18:1 9c	20–42%	–20	0.919²⁰	1.472²⁵	107–135	187–195
	16:0	9–16%	18:0	0–3%					
Cottonseed oil	18:2 9c,12c	47–58%	16:0	18–26%	–1	0.920²⁰	1.462⁴⁰	96–115	189–198
	18:1 9c	14–22%	18:0	2.1–3.3%					
Crambe oil	22:1 13c	55–60%	18:1 9c	12–15%		0.906²⁵	1.470²⁵	87–113	
	18:2 9c,12c	8–10%	18:3 9c,12c,15c	6–7%					
Cuphea seed oil (caprylic acid rich)	8:0	65–78%	10:0	19–24%					
	18:2 9c,12c	1–4%	16:0	0.6–3%					
Euphorbia lagascae seed oil	18:1 12,13-ep,9c	64%	18:1 other	19%		0.952²⁵	1.473²⁵	102	
	18:2 9c,12c	9%	16:0	4%					
Evening primrose oil	18:2 9c,12c	65–80%	18:3 6c,9c,12c	8–14%			1.479²⁰	147–155	193–198
	16:0	6–10%	18:1 9c	5–12%					
Grape seed oil	18:2 9c,12c	58–78%	18:1 9c	12–28%		0.923²⁰	1.475⁴⁰	130–138	188–194
	16:0	5.5–11%	18:0	3–6%					
Hazelnut oil (Chilean)	18:1 9c	39%	16:1 11c	22.7%					
	20:1 total	9.7%	22:1 total	9.5%					
Hazelnut oil (Filbert)	18:1 9c	72–84%	18:2 9c,12c	5.7–22%		0.909²⁵	1.473²⁵	83–90	188–197
	16:0	4.1–7.2%	18:0	1.5–2.4%					
Hempseed oil	18:2 9c,12c	45–60%	18:3 9c,12c,15c	15–30%		0.921²⁵	1.472⁴⁰	145–166	190–195
	18:1 9c	11–16%	16:0	6–12%					
Illipe (mowrah) butter	18:1 9c	34%	16:0	23%	27	0.862¹⁰⁰	1.460⁴⁰	53–70	188–207
	18:0	23%	18:2 9c,12c	14%					
Jojoba oil[a]	20:1 total	66–74%	22:1 undefined	9–19%					
	18:1 9c	5–12%	24:1 15c	1–5%					
Kapok seed oil[b]	18:1 9c	45–65%	16:0	10–28%	30	0.926¹⁵	1.469²⁵	86–110	189–197
	18:2 9c,12c	7–35%	18:0	2–9%					
Kokum butter	18:0	49–56%	18:1 9c	39–49%	41		1.456⁴⁰	33–37	192
	16:0	2–5%	18:2 9c,12c	1–2%					
Kusum oil	18:1 9c	57–62%	20:0	20–25%			1.461⁴⁰	48–58	220–230
	16:0	5–8%	18:0	2–6%					
Linola oil	18:2 9c,12c	72%	18:1 9c	16%				142	
	16:0	5.6%	18:0	4.0%					
Linseed oil	18:3 9c,12c,15c	52–58%	18:1 9c	18–20%	–24	0.924²⁵	1.480²⁵	170–203	188–196
	18:2 9c,12c	17%	18:2 9c,12c	16%					
Macadamia nut oil	18:1 9c	56–59%	16:1 9c	21–22%					
	16:0	8–9%	18:0	2–4%					

Type of oil	Principal fatty acid components in weight %				mp/ °C	Density/ g cm⁻³	Refractive index	Iodine value	Saponification value
Mango seed oil	18:1 9c	38–50%	18:0	31–49%		0.912^{15}	1.461^{25}	39–48	188–195
	18:2 9c,12c	3–6%	20:0	2–6%					
Meadowfoam seed oil	20:1 5c	58–77%	22:1 total	8–24%			1.464^{40}	86–91	168
	22:2 5c,13c	7–15%	18:1 9c	1–3%					
Melon oil	18:2 9c,12c	67% (av.)	18:1 9c	12% (av.)					
	16:0	11% (av.)	18:0	9% (av.)					
Moringa peregrina seed oil	18:1 9c	70%	16:0	9%		0.903^{24}	1.460^{40}	70	185
	18:0	3.8%	22:0	2.4%					
Mustard seed oil	22:1 13c	43%	22:1 13c	22–50%		0.913^{20}	1.465^{40}	92–125	170–184
	18:3 9c,12c,15c	12%	18:2 9c,12c	10–24%					
Neem oil	18:1 9c	49–62%	18:0	14–24%	−3	0.912^{30}	1.462^{40}	68–71	195–205
	16:0	13–18%	18:2 9c,12c	7–15%					
Niger seed oil	18:2 9c,12c	52–78%	16:0	5–12%		0.924^{15}	1.468^{40}	126–135	188–193
	18:1 9c	4–10%	18:0	2–12%					
Nutmeg butter	14:0	76–83%	18:1 9c	5–11%	45		1.468^{40}	48–85	170–190
	16:0	4–10%	12:0	3–6%					
Oat oil	18:2 9c,12c	24–48%	18:1 9c	18–53%		0.917^{25}	1.467^{40}	105–116	190–199
	16:0	13–39%	18:0	0.5–4%					
Oiticica oil	18:3 9c,11t,13t, 4-oxo	70–80%	16:0	7%		0.972^{20}	1.514^{25}	140–150	188–193
	18:0	5%	18:1 9c	4–7%					
Olive oil	18:1 9c	55–83%	18:2 9c,12c	9%	−6	0.911^{20}	1.469^{20}	75–94	184–196
	16:0	7.5–20%	18:2 9c,12c	3.5–21%					
Palm kernel oil	12:0	40–55%	14:0	14–18%	24	0.922^{15}	1.450^{40}	14–21	230–250
	18:1 9c	12–21%	16:0	6.5–10%					
Palm oil	16:0	40–48%	18:1 9c	36–44%	35	0.914^{15}	1.455^{40}	49–55	190–209
	18:2 9c,12c	6.5–12%	18:0	3.5–6.5%					
Palm olein	18:1 9c	40–44%	16:0	38–43%		0.91^{40}	1.459^{40}	>56	194–202
	18:2 9c,12c	10–13%	18:0	3.7–4.8%					
Palm stearin	16:0	48–74%	18:1 9c	16–36%		0.884^{60}	1.449^{40}	<48	193–205
	18:0	3.9–5.6%	18:2 9c,12c	3.2–9.8%					
Parsley seed oil	18:1 6c	69–76%	18:1 9c	12–15%			1.4800^{40}	110–120	
	18:2 9c,12c	6–14%	16:0	2%					
Peanut oil	18:1 9c	36–67%	18:2 9c,12c	14–43%	3	0.914^{20}	1.463^{40}	86–107	187–196
	16:0	8.3–14%	22:0	2.1–4.4%					
Perilla oil	18:3 9c,12c,15c	59%	18:2 9c,12c	14–18%		0.924^{25}	1.477^{25}	192–208	188–197
	18:1 9c	11–13%	16:0	6–9%					
Phulwara butter	16:0	57–61%	18:1 9c	30–36%	43	0.862^{100}	1.458^{40}	40–51	188–200
	18:2 9c,12c	3–4%	18:0	3–4%					
Pine nut oil	18:2 9c,12c	47–51%	18:1 9c	36–39%		0.919^{15}		118–121	193–197
	16:0	6–8%	18:0	2–3%					
Poppy seed oil	18:2 9c,12c	62–73%	18:1 9c	16–30%	−15	0.916^{25}	1.469^{40}	132–146	188–196
	16:0	7–11%	18:0	1–4%					
Rice bran oil	18:1 9c	38–48%	18:2 9c,12c	16–36%		0.916^{25}	1.472^{25}	92–108	181–189
	16:0	16–28%	18:0	2–4%					
Safflower seed oil	18:2 9c,12c	68–83%	18:1 9c	8.4–30%		0.924^{15}	1.474^{25}	136–148	186–198
	16:0	5.3–8.0%	18:0	1.9–2.9%					
Safflower seed oil (high oleic)	18:1 9c	74–80%	18:2 9c,12c	13–18%		0.921^{20}	1.470^{25}	91–95	
	16:0	5–6%	18:0	1.5–2.0%					
Sal fat	18:0	33–57%	18:1 9c	31–52%	33		1.456^{40}	31–45	175–192
	16:0	6–23%	20:0	1–8%					

Type of oil	Principal fatty acid components in weight %				mp/ °C	Density/ g cm⁻³	Refractive index	Iodine value	Saponification value
Sesame seed oil	18:2 9c,12c	40–51%	18:1 9c	33–44%	−6	0.917²⁰	1.467⁴⁰	104–120	187–195
	16:0	7.9–10.2%	18:0	4.4–6.7%					
Sheanut butter	18:1 9c	45–50%	18:0	36–41%	38	0.863¹⁰⁰	1.465⁴⁰	52–66	178–198
	16:0	4–8%	18:2 9c,12c	4–8%					
Soybean oil	18:2 9c,12c	50–57%	18:1 9c	18–28%	−16	0.920²⁰	1.468⁴⁰	118–139	189–195
	16:0	9–13%	18:3 9c,12c,15c	5.5–9.5%					
Stillingia seed kernel oil[c]	18:3 total	41–54%	18:2 9c,12c	24–30%		0.937²⁵	1.483²⁵	169–191	202–212
	18:1 9c	7–10%	16:0	6–9%					
Sunflower seed oil	18:2 9c,12c	48–74%	18:1 9c	13–40%	−17	0.919²⁰	1.474²⁵	118–145	188–194
	16:0	5–8%	18:0	2.5–7.0%					
Sunflower oil, high-oleic (HO)	18:1 9c	80%	18:2 9c,12c	10%		0.911²⁵	1.468²⁵	81	
	18:0	4.4%	16:0	3.5%					
Sunflower oil, mid-Oleic (NuSun oil)	18:1 9c	65%	18:2, 18:3	25%					
	16:0, 18:0	10%							
Tall oil	18:2 9c,12c	41–52%	18:1 9c	41–48%		0.969²⁵	1.494²⁵	140–180	154–180
	16:0	5–6%	18:0	2–3%					
Tung oil	18:3 9c,11t,13t	71–82%	18:2 9c,12c	8–15%	−2	0.912²⁵	1.517²⁵	160–175	189–195
	18:1 9c	4–10%	18:0	3%					
Ucuhuba butter oil	14:0	64–73%	12:0	13–15%		0.870¹⁰⁰	1.451⁵⁰	11–17	221–229
	18:1 9c	6–8%	16:0	3–9%					
Vernonia seed oil	18:1 12,13-ep,9c	62–72%	18:2 9c,12c	9–17%		0.901³⁰	1.486³²	55	176
	16:0	3–7%	18:0	2–6%					
Walnut oil	18:2 9c,12c	56–60%	18:1 9c	17–19%		0.921²⁵	1.474²⁵	138–162	189–197
	18:3 9c,12c,15c	13–14%	16:0	6–8%					
Wheatgerm oil	18:2 9c,12c	50–59%	18:1 9c	13–23%		0.926²⁵	1.479²⁵	100–128	179–217
	16:0	12–20%	18:3 9c,12c,15c	2–9%					
Marine animals									
Anchovy oil	20:5 6c,9c,12c,15c,17c	22%	16:0	17%				163–169	191–194
	16:1 undefined	13%	18:1 undefined	10%					
Capelin oil[d]	20:1 undefined	17%	22:1 undefined	15%			1.463⁵⁰	94–164	185–202
	18:1 undefined	14%	16:0	10%					
Cod liver oil	18:1 undefined	24%	20:1 undefined	13%		0.924¹⁵	1.482²⁵	142–176	180–192
	22:6 4c,7c,10c,13c, 16c,19c	11%	16:0	10%					
Herring oil	22:1 undefined	19%	16:0	17%		0.914²⁰	1.474²⁵	115–160	161–192
	20:1 undefined	15%	18:1 undefined	14%					
Mackerel oil	22:1 undefined	15%	16:0	14%		0.929¹⁵	1.481²⁰	136–167	
	18:1 undefined	13%	20:1 undefined	12%					
Menhaden oil	16:0	19%	20:5 6c,9c,12c,15c, 17c	14%		0.920¹⁵		150–200	192–199
	16:1 undefined	12%	18:1 undefined	11%					
Salmon oil	22:6 4c,7c,10c,13c, 16c,19c	18%	20:5 6c,9c,12c,15c, 17c	13%		0.924¹⁵	1.475²⁵	130–160	183–186
	16:0	9.8%	16:1	4.8%					
Sardine oil	16:0	18%	20:5 6c,9c,12c,15c, 17c	16%		0.915²⁵	1.464⁶⁵	159–192	188–199
	18:1 undefined	13%	16:1 undefined	10%					
Seal blubber oil, harp	18:1 9c	21%	20:1	12%					
	22:6 4c,7c,10c,13c, 16c,19c	7.6%	20:5 6c,9c,12c,15c, 17c	6.4%					

Type of oil	Principal fatty acid components in weight %				mp/°C	Density/g cm^{-3}	Refractive index	Iodine value	Saponification value
Shark liver oil	18:1 undefined	45%	16:0	21%		0.917^{25}	1.476^{25}	150–300	170–190
	20:1	12%	22:1	9%					
Tuna oil	22:6 4c,7c,10c,13c, 16c,19c	22%	16:0	22%					
	18:1 undefined	21%	20:5 6c,9c,12c,15c, 17c	6%					
Trout lipids	16:0	21–24%	18:1 undefined	18–31%					
	18:2	7–16%	16:1	4–10%					
Whale oil, minke	18:1 undefined	18%	20:1	17%					
	22:1	11%	16:1	9%					
Land animals									
Beef tallow	18:1 undefined	31–50%	18:0	25–40%	47	0.902^{25}	1.454^{40}	33–47	190–200
	16:0	20–37%	14:0	1–6%					
Butterfat	16:0	28.1% (av.)	18:1 9c	20.8% (av.)	32	0.934^{15}	1.455^{40}	26–40	210–232
	14:0	10.8% (av.)	18:0	10.6% (av.)					
Chicken egg lipids, yolk	16:0	28%	18:1 9c	25%					
	18:0	17%	18:2 9c,12c	16%					
Chicken fat	18:1 undefined	37%	16:0	22%		0.918^{15}	1.456^{40}	76–80	
	18:2	20%	18:0	6%					
Milk fats, cow	16:0	28.2% (av.)	18:1 9c	21.4% (av.)					
	18:0	12.6% (av.)	14:0	10.6% (av.)					
Milk fats, human	18:1 9c	31.1% (av.)	16:0	21.6% (av.)					
	18:2 9c,12c	11.7% (av.)	14:0	6.6% (av.)					
Mutton tallow	18:1 undefined	30–42%	18:0	22–34%	48	0.946^{15}	1.455^{40}	35–46	
	16:0	20–27%	14:0	2–4%					
Pork lard	18:1 undefined	35–62%	16:0	20–32%	30	0.898^{20}			
	18:0	5–24%	18:2	3–16%					

[a] Jojoba oil consists primarily of wax esters of the acids listed here and long-chain alcohols.
[b] Kapok oil also contains up to 15% cyclopropene acids.
[c] Stillingia oil also contains 5–10% *trans,cis*-2,4-decadienoic acid (stillingic acid, 10:2 2t,4c).
[d] Capelin oil also contains about 10% 16:1.

CARBOHYDRATE NAMES AND SYMBOLS

The following table lists the systematic names and symbols for selected carbohydrates and some of their derivatives. The symbols for monosaccharide residues and derivatives are recommended by IUPAC for use in describing the structures of oligosaccharide chains. A more complete list can be found in the reference.

Reference

McNaught, A. D., *Pure Appl. Chem.*, 68, 1919–2008, 1996.

Common name	Symbol	Systematic name
Abequose	Abe	3,6-Dideoxy-D-*xylo*-hexose
N-Acetyl-2-deoxyneur-2-enaminic acid	Neu2en5Ac	
N-Acetylgalactosamine	GalNAc	
N-Acetylglucosamine	GlcNAc	
N-Acetylneuraminic acid	Neu5Ac	
Allose	All	*allo*-Hexose
Altrose	Alt	*altro*-Hexose
Apiose	Api	3-*C*-(Hydroxymethyl)-*glycero*-tetrose
Arabinitol	Ara-ol	Arabinitol
Arabinose	Ara	*arabino*-Pentose
Arcanose		2,6-Dideoxy-3-*C*-methyl-3-*O*-methyl-*xylo*-hexose
Ascarylose		3,6-Dideoxy-L-*arabino*-hexose
Boivinose		2,6-Dideoxy-D-gulose
Chalcose		4,6-Dideoxy-3-*O*-methyl-D-*xylo*-hexose
Cladinose		2,6-Dideoxy-3-*C*-methyl-3-*O*-methyl-L-*ribo*-hexose
Colitose		3,6-Dideoxy-L-*xylo*-hexose
Cymarose		6-Deoxy-3-*O*-methyl-*ribo*-hexose
3-Deoxy-D-*manno*-oct-2-ulosonic acid	Kdo	
2-Deoxyribose	dRib	2-Deoxy-*erythro*-pentose
2,3-Diamino-2,3-dideoxy-D-glucose	GlcN3N	
Diginose		2,6-Dideoxy-3-*O*-methyl-*lyxo*-hexose
Digitalose		6-Deoxy-3-*O*-methyl-D-galactose
Digitoxose		2,6-Dideoxy-D-*ribo*-hexose
3,4-Di-*O*-methylrhamnose	Rha3,4Me$_2$	
Ethyl glucopyranuronate	Glc*p*A6Et	
Evalose		6-Deoxy-3-*C*-methyl-D-mannose
Fructose	Fru	*arabino*-Hex-2-ulose
Fucitol	Fuc-ol	6-Deoxy-D-galactitol
Fucose	Fuc	6-Deoxygalactose
β-D-Galactopyranose 4-sulfate	β-D-Gal*p*4S	
Galactosamine	GalN	2-Amino-2-deoxygalactose
Galactose	Gal	*galacto*-Hexose
Glucitol	Glc-ol	
Glucosamine	GlcN	2-Amino-2-deoxyglucose
Glucose	Glc	*gluco*-Hexose
Glucuronic acid	GlcA	
N-Glycoloylneuraminic acid	Neu5Gc	
Gulose	Gul	*gulo*-Hexose
Hamamelose		2-*C*-(Hydroxymethyl)-D-ribose
Idose	Ido	*ido*-Hexose
Iduronic acid	IdoA	
Lactose	Lac	β-D-Galactopyranosyl-(1→4)-D-glucose
Lyxose	Lyx	*lyxo*-Pentose
Maltose		α-D-Glucopyranosyl-(1→4)-D-glucose
Mannose	Man	*manno*-Hexose
2-*C*-Methylxylose	Xyl2CMe	
Muramic acid	Mur	2-Amino-3-*O*-[(R)-1-carboxyethyl]-2-deoxy-D-glucose
Mycarose		2,6-Dideoxy-3-*C*-methyl-L-*ribo*-hexose
Mycinose		6-Deoxy-2,3-di-*O*-methyl-D-allose
Neuraminic acid	Neu	5-Amino-3,5-dideoxy-D-*glycero*-D-*galacto*-non-2-ulosonic acid

Common name	Symbol	Systematic name
Panose		α-D-Glucopyranosyl-(1→6)-α-D-glucopyranosyl-(1→4)-D-glucose
Paratose		3,6-Dideoxy-D-*ribo*-hexose
Primeverose		β-D-Xylopyranosyl-(1→6)-D-glucose
Psicose	Psi	*ribo*-Hex-2-ulose
Quinovose	Qui	6-Deoxyglucose
Raffinose		β-D-Fructofuranosyl-α-D-galactopyranosyl-(1→6)-α-D-glucopyranoside
Rhamnose	Rha	6-Deoxymannose
Rhodinose		2,3,6-Trideoxy-L-*threo*-hexose
Ribose	Rib	*ribo*-Pentose
Ribose 5-phosphate	Rib5*P*	
Ribulose	Ribulo (Rul)	*erythro*-Pent-2-ulose
Rutinose		α-L-Rhamnopyranosyl-(1→6)-D-glucose
Sarmentose		2,6-Dideoxy-3-*O*-methyl-D-*xylo*-hexose
Sedoheptulose		D-*altro*-Hept-2-ulose
Sorbose	Sor	*xylo*-Hex-2-ulose
Streptose		5-Deoxy-3-*C*-formyl-L-lyxose
Sucrose		β-D-Fructofuranosyl-α-D-glucopyranoside
Tagatose	Tag	*lyxo*-Hex-2-ulose
Talose	Tal	*talo*-Hexose
Turanose		α-D-Glucopyranosyl-(1→3)-D-fructose
Tyvelose	Tyv	3,6-Dideoxy-D-*arabino*-hexose
Xylose	Xyl	*xylo*-Pentose
Xylulose	Xylulo (Xul)	*threo*-Pent-2-ulose

STANDARD TRANSFORMED GIBBS ENERGIES OF FORMATION FOR BIOCHEMICAL REACTANTS

Robert N. Goldberg and Robert A. Alberty

This table contains values of the standard transformed Gibbs energies of formation $\Delta_f G'^\circ$ for 130 biochemical reactants. Values of $\Delta_f G'^\circ$ are given at pH 7.0, the temperature 298.15 K, and the pressure 100 kPa for three ionic strengths: $I = 0$, $I = 0.1$ mol/L and $I = 0.25$ mol/L. The table can be used for calculating apparent equilibrium constants K' and standard apparent reduction potentials E'° for biochemical reactions. Such a listing is more compact than tabulating the actual apparent equilibrium constants or standard apparent reduction potentials, which would require a very large number of reactant–product combinations. In the table, all reactants are in aqueous solution unless indicated otherwise.

A biochemical reactant is a sum of species. For example, ATP consists of an equilibrium mixture of the aqueous species ATP^{4-}, $HATP^{3-}$, H_2ATP^{2-}, $MgATP^{2-}$, etc. Similarly, phosphate refers to the equilibrium mixture of the aqueous species PO_4^{3-}, HPO_4^{2-}, $H_2PO_4^-$, H_3PO_4, $MgHPO_4$, etc. Biochemical reactions are written using biochemical reactants in terms of an apparent equilibrium constant K', which is distinct from the standard equilibrium constant K. This subject is discussed in an IUPAC report (see Reference 1 below).

The apparent equilibrium constant K' and the standard transformed Gibbs energy change $\Delta_r G'^\circ$ for a biochemical reaction can be calculated from the $\Delta_f G'^\circ$ values by using the relationship

$$-RT \ln K' = \Delta_r G'^\circ = \Sigma v'_i \Delta_f G'^\circ,$$

where the summation is over all of the biochemical reactants. The quantity v'_i is the stoichiometric number of reactant i (v'_i is positive for reactants on the right side of the equation and negative for reactants on the left side); R is the gas constant. As an example, the hydrolysis reaction of ATP is

$$ATP + H_2O(l) = ADP + phosphate.$$

At pH 7.00 and $I = 0.25$ M, $\Delta_r G'^\circ$ and K' are calculated as follows:

$$\Delta_r G'^\circ = \{-1424.70 - 1059.49 - (-2292.50 - 155.66)\} \cdot (kJ\ mol^{-1}) = -36.03\ kJ\ mol^{-1}$$

$$K' = \exp[-(-36030\ J\ mol^{-1})/\{(8.3145\ J\ mol^{-1}\ K^{-1}) \cdot (298.15\ K)\} = 2.05 \cdot 10^6$$

An example involving a biochemical half-cell reaction is

$$acetaldehyde(aq) + 2\ e^- = ethanol(aq).$$

At 298.15 K, pH 7.00, and $I = 0$, the standard apparent reduction potential E'° can be calculated as follows

$$E'^\circ = -(1/nF) \cdot \{\Delta_f G'^\circ(ethanol) - \Delta_f G'^\circ(acetaldehyde)\},$$

where n is the number of electrons in the half-cell reaction and F is the Faraday constant. Then,

$$E'^\circ = [-1/(2 \cdot 9.6485 \cdot 10^4\ C\ mol^{-1})] \cdot (58.10 \cdot 10^3\ J\ mol^{-1} - 20.83 \cdot 10^3\ J\ mol^{-1}) = -0.193\ V$$

References

1. Alberty, R.A., Cornish-Bowden, A., Gibson, Q.H., Goldberg, R.N., Hammes, G., Jencks, W., Tipton, K.F, Veech, R., Westerhoff, H.V., and Webb, E.C. *Pure Appl. Chem.* 66, 1641-1666, 1994.
2. Alberty, R.A., *Arch. Biochem. Biophys.*, 353, 116-130, 1998; 358, 25-39, 1998.
3. Alberty, R.A., *Thermodynamics of Biochemical Reactions*, Wiley-Interscience, New York, 2003.
4. Alberty, R.A., *BasicBiochemData2: Data and Programs for Biochemical Thermodynamics*, <http://library.wolfram.com/infocenter/MathSource/797>.

Reactant	$\Delta_f G'^\circ(I = 0)$ kJ mol^{-1}	$\Delta_f G'^\circ(I = 0.1\ M)$ kJ mol^{-1}	$\Delta_f G'^\circ(I = 0.25\ M)$ kJ mol^{-1}
Acetaldehyde	20.83	23.27	24.06
Acetate	-249.46	-248.23	-247.83
Acetone	80.04	83.71	84.90
Acetyl Coenzyme A	-60.49	-58.65	-58.06
Acetylphosphate	-1109.34	-1107.57	-1107.02
cis-Aconitate	-797.26	-800.93	-802.12
Adenine	510.45	513.51	514.50
Adenosine	324.93	332.89	335.46
Adenosine 5'-diphosphate (ADP)	-1428.93	-1425.55	-1424.70
Adenosine 5'-monophosphate (AMP)	-562.04	-556.53	-554.83
Adenosine 5'-triphosphate (ATP)	-2292.61	-2292.16	-2292.50
D-Alanine	-91.31	-87.02	-85.64
Ammonia	80.50	82.34	82.93
D-Arabinose	-342.67	-336.55	-334.57
L-Asparagine	-206.28	-201.38	-199.80
L-Aspartate	-456.14	-453.08	-452.09
1,3-Biphosphoglycerate	-2202.06	-2205.69	-2207.30
Butanoate	-72.94	-69.26	-68.08
1-Butanol	227.72	233.84	235.82

Reactant	$\Delta_f G'^\circ(I = 0)$ kJ mol^{-1}	$\Delta_f G'^\circ(I = 0.1\ M)$ kJ mol^{-1}	$\Delta_f G'^\circ(I = 0.25\ M)$ kJ mol^{-1}
Citrate	-963.46	-965.49	-966.23
Isocitrate	-956.82	-958.84	-959.58
Coenzyme A (CoA)	-7.98	-7.43	-7.26
CO(aq)	-119.90	-119.90	-119.90
CO(g)	-137.17	-137.17	-137.17
CO$_2$(aq)[total]	-547.33	-547.15	-547.10
CO$_2$(g)	-394.36	-394.36	-394.36
Creatine	100.41	105.92	107.69
Creatinine	256.55	260.84	262.22
L-Cysteine	-59.23	-55.01	-53.65
L-Cystine	-187.03	-179.69	-177.32
Cytochrome c [oxidized]	0.00	-5.51	-7.29
Cytochrome c [reduced]	-24.51	-26.96	-27.75
Dihydroxyacetone phosphate	-1096.60	-1095.91	-1095.70
Ethanol	58.10	61.77	62.96
Ethyl acetate	-18.00	-13.10	-11.52
Ferredoxin [oxidized]	0.00	-0.61	-0.81
Ferredoxin [reduced]	38.07	38.07	38.07
Flavine adenine dinucleotide (FAD) [oxidized]	1238.65	1255.17	1260.51
Flavine adenine dinucleotide (FAD) [reduced]	1279.68	1297.43	1303.16
Flavin adenine dinucleotide-enzyme (FADenz) [oxidized]	1238.65	1255.17	1260.51
Flavin adenine dinucleotide-enzyme (FADenz) [reduced]	1229.96	1247.71	1253.44
Flavin mononucleotide (FMN) [oxidized]	759.17	768.35	771.32
Flavin mononucleotide (FMN) [reduced]	800.20	810.61	813.97
Formate	-311.04	-311.04	-311.04
D-Fructose	-436.03	-428.69	-426.32
D-Fructose 1,6-diphosphate	-2202.84	-2205.66	-2206.78
D-Fructose 6-phosphate	-1321.71	-1317.16	-1315.74
Fumarate	-521.97	-523.19	-523.58
D-Galactose	-429.45	-422.11	-419.74
α-D-Galactose 1-phosphate	-1317.50	-1313.01	-1311.60
D-Glucose	-436.42	-429.08	-426.71
α-D-Glucose 1-phosphate	-1318.03	-1313.34	-1311.89
D-Glucose 6-phosphate	-1325.00	-1320.37	-1318.92
Glutamate	-377.82	-373.54	-372.16
D-Glutamine	-128.46	-122.34	-120.36
Glutathione [oxidized]	1198.69	1214.60	1219.74
Glutathione [reduced]	625.75	634.76	637.62
Glutathione-coenzyme A	563.49	572.06	574.83
D-Glyceraldehyde 3-phosphate	-1088.94	-1088.25	-1088.04
Glycerol	-177.83	-172.93	-171.35
sn-Glycerol 3-phosphate	-1080.22	-1077.83	-1077.13
Glycine	-180.13	-177.07	-176.08
Glycolate	-411.08	-409.86	-409.46
Glycylglycine	-200.55	-195.65	-194.07
Glyoxylate	-428.64	-428.64	-428.64
H$_2$(aq)	97.51	98.74	99.13
H$_2$(g)	79.91	81.14	81.53
H$_2$O(l)	-157.28	-156.05	-155.66
H$_2$O$_2$(aq)	-54.12	-52.89	-52.50
3-Hydroxypropanoate	-318.62	-316.17	-315.38
Hypoxanthine	249.33	251.77	252.56
Indole	503.49	507.78	509.16
Lactate	-316.94	-314.49	-313.70
Lactose	-688.29	-674.83	-670.48
L-Leucine	167.18	175.14	177.71
L-Isoleucine	175.53	183.49	186.06
D-Lyxose	-349.58	-343.46	-341.48
Malate	-682.88	-682.85	-682.85
Maltose	-695.65	-682.19	-677.84

Reactant	$\Delta_f G'^\circ(I = 0)$	$\Delta_f G'^\circ(I = 0.1\ M)$	$\Delta_f G'^\circ(I = 0.25\ M)$
	kJ mol^{-1}	kJ mol^{-1}	kJ mol^{-1}
D-Mannitol	-383.22	-374.65	-371.89
Mannose	-430.52	-423.18	-420.81
Methane(aq)	125.50	127.94	128.73
Methane(g)	109.11	111.55	112.34
Methanol	-15.48	-13.04	-12.25
L-Methionine	-63.40	-56.67	-54.49
N$_2$(aq)	18.70	18.70	18.70
N$_2$(g)	0.00	0.00	0.00
Nicotinamide Adenine Dinucleotide (NAD) [oxidized]	1038.86	1054.17	1059.11
Nicotinamide Adenine Dinucleotide (NAD) [reduced]	1101.47	1115.55	1120.09
Nicotinamide Adenine Dinucleotide Phosphate (NADP) [oxidized]	163.73	173.52	176.68
Nicotinamide Adenine Dinucleotide Phosphate (NADP) [reduced]	229.67	235.79	237.77
O$_2$(aq)	16.40	16.40	16.40
O$_2$(g)	0.00	0.00	0.00
Oxalate	-673.90	-676.35	-677.14
Oxaloacetate	-713.38	-714.60	-715.00
Oxalosuccinate	-979.05	-979.05	-979.05
2-Oxoglutarate	-633.58	-633.58	-633.58
Palmitate	979.25	997.61	1003.54
L-Phenylalanine	232.42	239.15	241.33
Phosphate	-1058.56	-1059.17	-1059.49
2-Phospho-*D*-glycerate	-1340.72	-1341.32	-1341.79
3-Phospho-*D*-glycerate	-1346.38	-1347.19	-1347.73
Phospho*enol*pyruvate	-1185.46	-1188.53	-1189.73
1-Propanol	143.84	148.74	150.32
2-Propanol	134.42	139.32	140.90
Pyrophosphate	-1934.95	-1939.13	-1940.66
Pyruvate	-352.40	-351.18	-350.78
Retinal	1118.78	1135.91	1141.45
Retinol	1170.78	1189.14	1195.07
Ribose	-339.23	-333.11	-331.13
Ribose 1-phosphate	-1215.87	-1212.24	-1211.14
Ribose 5-phosphate	-1223.95	-1220.32	-1219.22
Ribulose	-336.38	-330.26	-328.28
L-Serine	-231.18	-226.89	-225.51
Sorbose	-432.47	-425.13	-422.76
Succinate	-530.72	-530.65	-530.64
Succinyl Coenzyme A	-349.90	-348.06	-347.47
Sucrose	-685.66	-672.20	-667.85
Thioredoxin [oxidized]	0.00	0.00	0.00
Thioredoxin [reduced]	54.32	55.41	55.74
L-Tryptophan	364.78	372.12	374.49
L-Tyrosine	68.82	75.55	77.73
Ubiquinone [oxidized]	3596.07	3651.15	3668.94
Ubiquinone [reduced]	3586.06	3642.37	3660.55
Urate	-206.03	-204.81	-204.41
Urea	-42.97	-40.53	-39.74
Uric acid	-197.07	-194.63	-193.84
L-Valine	80.87	87.60	89.78
D-Xylose	-350.93	-344.81	-342.83
D-Xylulose	-346.59	-340.47	-338.49

APPARENT EQUILIBRIUM CONSTANTS FOR ENZYME-CATALYZED REACTIONS

Robert N. Goldberg

This table contains values of apparent equilibrium constants K' for selected enzyme-catalyzed reactions at specified temperatures T and pHs. In those cases where the ionic strength I and/or the pMg (pMg = $-\log_{10}[Mg^{2+}]$) have been reported, the values of these quantities are given. The Enzyme Commission numbers [Ref. 1] of the enzymes that were used to catalyze the reactions are also given.

There are two fundamentally different types of equilibrium constants. This is illustrated by the following example for the hydrolysis of adenosine 5′-triphosphate (ATP) to adenosine 5′-diphosphate (ADP) and phosphate:

$$ATP + H_2O = ADP + phosphate \qquad (1)$$

The apparent equilibrium constant for the overall biochemical reaction (1) is

$$K' = [ADP][phosphate]/([ATP]c^o) \qquad (2)$$

The biochemical reactants ATP, ADP, and phosphate each exist in several different ionized and metal bound forms. For example, ATP is an equilibrium mixture of the species ATP^{4-}, $HATP^{3-}$, H_2ATP^{2-}, $MgATP^{2-}$, $MgHATP^-$, Mg_2ATP^0. Additional species would also have to be considered if Ca^{2+} were present. Thus, ATP has often been denoted in the literature as ΣATP or as $(ATP)_{tot}$. When it is clear that one is dealing with total amounts of substances, it is not necessary to use either the Σ or "tot." Thus, these designations are not used in this table. In the above equation, $c^o = 1$ mol dm^{-3}; it is included to make K′ dimensionless. The standard transformed Gibbs energy of reaction $\Delta_r G'^o$ at specified conditions of temperature T, pressure P, ionic strength I, pH, and pMg can be calculated from K':

$$\Delta_r G'^o = -RT \ln K' \qquad (3)$$

The molar gas constant, R, is equal to 8.314 472 J K^{-1} mol^{-1}. $\Delta_r G'^o$ and the apparent equilibrium constant, K', can be used to calculate the position of equilibrium of overall biochemical reactions.

It is also possible to choose a chemical reference reaction that involves selected solute species:

$$ATP^{4-} + H_2O = ADP^{3-} + HPO_4^{2-} + H^+ \qquad (4)$$

The equilibrium constant for this reference reaction is

$$K = [ADP^{3-}][HPO_4^{2-}][H^+]/\{[ATP^{4-}](c^o)^2\} \qquad (5)$$

Equations and algorithms that relate these two different types of equilibrium constants have been published [Refs. 2–4]. To calculate the equilibrium constant K for the reference reaction from the apparent equilibrium constant K', or vice versa, one needs the equilibrium constants for the binding of H^+ and for the relevant metal ions to ATP^{4-}, ADP^{3-}, and HPO_4^{2-}.

To avoid confusion between the two different types of equilibrium constants (K' and K) and to avoid ambiguity about whether specific species or sums of species are intended, the word "ammonia," for example, rather than NH_3 or NH_4^+, is used for total ammonia, and chemical formulas are used for specific chemical species. Other substances such as carbon dioxide (CO_2, HCO_3^-, and CO_3^{2-}), and phosphate ($H_2PO_4^-$, HPO_4^{2-}, and PO_4^{3-}) are treated in the same manner. Exceptions are made for water, which is always written as H_2O, and for gaseous hydrogen and oxygen, which are written as $H_2(g)$ and $O_2(g)$, respectively.

For symmetrical reactions, there is no concern about the units used to calculate the value of an equilibrium constant. However, care must be exercised for reactions that are not symmetrical. In such cases, the units "mol dm^{-3}" have been used for all concentrations. As stated above, a c^o (1 mol dm^{-3}) is then used to make all equilibrium constants dimensionless.

All substances are assumed to be in aqueous solutions unless specified otherwise.

Values of $\Delta_r G'^o$ and K' can also be calculated for many biochemical reactions by using the table "Standard Transformed Gibbs Energies of Formation for Biochemical Reactants" in Section 7 of this Handbook.

Abbreviations

ADP	adenosine 5′-diphosphate
AMP	adenosine 5′-monophosphate
ATP	adenosine 5′-triphosphate
CoA	coenzyme A
GDP	guanosine 5′-diphosphate
GMP	guanosine 5′-monophosphate
GTP	guanosine 5′-triphosphate
IDP	inosine 5′-diphosphate
IMP	inosine 5′-monophosphate
ITP	inosine 5′-triphosphate
NAD_{ox}	β-nicotinamide-adenine dinucleotide, oxidized form
NAD_{red}	β-nicotinamide-adenine dinucleotide, reduced form
$NADP_{ox}$	β-nicotinamide-adenine dinucleotide phosphate, oxidized form
$NADP_{red}$	β-nicotinamide-adenine dinucleotide phosphate, reduced form
UDP	uridine 5′-diphosphate
UTP	uridine 5′-triphosphate

References

1. Webb, E.C., *Enzyme Nomenclature 1992*, Academic Press, New York, 1992. See also <www.chem.qmul.ac.uk/iubmb/enzyme/>.
2. Akers, D.L., and Goldberg, R.N., *Mathematica J.*, 8, 86–113 (2001).
3. Alberty, R.A., *J. Biol. Chem.*, 243, 1337–1343, 1969.
4. Alberty, R.A., *Thermodynamics of Biochemical Reactions*, Wiley-Interscience, Hoboken, NJ, 2003.
5. Goldberg, R.N., Tewari, Y.B., Bell, D., Fazio, K., and Anderson, E., *J. Phys. Chem. Ref. Data*, 22, 515-582 1993.
6. Goldberg, R.N., and Tewari, Y.B., *J. Phys. Chem. Ref. Data*, 23, 547-617, 1994.
7. Goldberg, R.N., and Tewari, Y.B., *J. Phys. Chem. Ref. Data*, 23, 1035-1103, 1994.
8. Goldberg, R.N., and Tewari, Y.B., *J. Phys. Chem. Ref. Data*, 24, 1669-1698, 1995.
9. Goldberg, R.N., and Tewari, Y.B., *J. Phys. Chem. Ref. Data*, 24, 1765-1801, 1995.
10. Goldberg, R.N., *J. Phys. Chem. Ref. Data*, 28, 931–965, 1999.
11. Goldberg, R.N., Tewari, Y.B., and Bhat, T.N., *Bioinformatics*, 20, 2874–2877, 2004; <xpdb.nist.gov/enzyme_thermodynamics/>.
12. Goldberg, R.N., Tewari, Y.B., and Bhat, T.N., *J. Phys. Chem. Ref. Data*, 36, 1347–1397, 2007.

Apparent Equilibrium Constants for Enzyme-Catalyzed Reactions

Reaction	K'	Enzyme Commission Number	T K	pH	I mol dm^{-3}	pMg
benzyl alcohol + NAD$_{ox}$ = benzaldehyde + NAD$_{red}$	$9.8 \cdot 10^{-4}$	1.1.1.1	298.15	7.5		
1-butanol + NAD$_{ox}$ = butanal + NAD$_{red}$	$1.8 \cdot 10^{-3}$	1.1.1.1	298.15	8.3		
cyclohexanol + NAD$_{ox}$ = cyclohexanone + NADH$_{red}$	0.090	1.1.1.1	298.15	7.2		
1-hexanol + NAD$_{ox}$ = hexanal + NAD$_{red}$	$2.87 \cdot 10^{-3}$	1.1.1.1	298.15	8.3		
1-octanol + NAD$_{ox}$ = octanal + NAD$_{red}$	$1.1 \cdot 10^{-3}$	1.1.1.1	298.15	8.3		
L-homoserine + NADP$_{ox}$ = L-aspartate 4-semialdehyde + NADP$_{red}$	$6.3 \cdot 10^{-4}$	1.1.1.3	298.15	7.9		
xylitol + NAD$_{ox}$ = L-xylulose + NAD$_{red}$	$2.97 \cdot 10^{-4}$	1.1.1.10	298.15	7.00		
D-sorbitol + NAD$_{ox}$ = D-fructose + NAD$_{red}$	0.032	1.1.1.14	298.15	7.0		
quinate + NAD$_{ox}$ = 5-dehydroquinate + NAD$_{red}$	$4.61 \cdot 10^{-3}$	1.1.1.24	305.15	7.2		
shikimate + NADP$_{ox}$ = 5-dehydroshikimate + NADP$_{red}$	0.036	1.1.1.25	303.15	7.0		
2-hydroxybutanoate + NAD$_{ox}$ = 2-oxobutanoate + NAD$_{red}$	$3.0 \cdot 10^{-3}$	1.1.1.27	298.65	8.0		
(R)-3-hydroxybutanoate + NAD$_{ox}$ = 3-oxobutanoate + NAD$_{red}$	$1.9 \cdot 10^{-3}$	1.1.1.30	298.15	7.0		
D-glucose 6-phosphate + NADP$_{ox}$ = D-glucono-1,5-lactone 6-phosphate + NADP$_{red}$	1.50	1.1.1.49	301.15	6.40		
5α-androstane-3α-ol-17-one + NAD$_{ox}$ = 5α-androstane-3,17-dione + NAD$_{red}$	0.058	1.1.1.50	298.15	7.0		
5α-pregnane-3α,17α,21-triol-20-one + NAD$_{ox}$ = 5α-pregnane-17α,21-diol-3,20-dione + NAD$_{red}$	0.0113	1.1.1.50	298.15	7.0		
5α-androstane-3β,17α-diol + NAD$_{ox}$ = 5α-androstane-17α-ol-3-one + NAD$_{red}$	0.0211	1.1.1.51	298.15	7.0		
4-androstene-17β-ol-3-one + NAD$_{ox}$ = 4-androstene-3,17-dione + NAD$_{red}$	0.378	1.1.1.51	298.15	7.0		
1,2-propanediol + NADP$_{ox}$ = L-lactaldehyde + NADP$_{red}$	$6.0 \cdot 10^{-5}$	1.1.1.55	298.15	8.4		
ribitol + NAD$_{ox}$ = D-ribulose + NAD$_{red}$	$3.1 \cdot 10^{-3}$	1.1.1.56	310.15	7.4		
3-hydroxypropanoate + NAD$_{ox}$ = 3-oxopropanoate + NAD$_{red}$	$9.0 \cdot 10^{-3}$	1.1.1.59	298.15	9.0		
estradiol-17β + NAD$_{ox}$ = estrone + NAD$_{red}$	0.18	1.1.1.62	298.15	7.00		
benzyl alcohol + NAD$_{ox}$ = benzaldehyde + NAD$_{red}$	0.097	1.1.1.90	300.15	9.5		
L-carnitine + NAD$_{ox}$ = 3-dehydrocarnitine + NAD$_{red}$	$1.3 \cdot 10^{-4}$	1.1.1.108	303.15	7.0		
L-threonate + NAD$_{ox}$ = 3-oxo-L-threonate + NAD$_{red}$	$3.42 \cdot 10^{-4}$	1.1.1.129	298.15	7.0		
prostaglandin E$_1$ + NAD$_{ox}$ = 15-oxo-prostaglandin E$_1$ + NAD$_{red}$	0.65	1.1.1.141	298.15	7.0		
7,8-dihydrobiopterin + NADP$_{ox}$ = sepiapterin + NADP$_{red}$	0.045	1.1.1.153	298.15	8.0		
glycine + acetaldehyde = L-threonine	56	2.1.2.1	310.15	7.6		
sedoheptulose 7-phosphate + D-glyceraldehyde 3-phosphate = D-ribose 5-phosphate + D-xylulose 5-phosphate	0.48	2.2.1.1	311.15	7.0	0.25	3.0
acetyl-CoA + choline = CoA + O-acetylcholine	1.60	2.3.1.7	298.15	7.0	0.25	
acetyl-CoA + acyl-carrier protein = CoA + acetyl-[acyl-carrier protein]	2.09	2.3.1.38	311.15	6.5		
UDPglucose + D-fructose = UDP + sucrose	6.7	2.4.1.13	298.15	7.5		
cellobiose + orthophosphate = D-glucose + α-D-glucose 1-phosphate	0.23	2.4.1.20	310.15	7.0		
laminaritriose + orthophosphate = laminaribiose + α-D-glucose 1-phosphate	0.26	2.4.1.31	310.15	6.5		
α,α-trehalose + orthophosphate = D-glucose + β-D-glucose 1-phosphate	0.24	2.4.1.64	310.15	7.0		
UDPglucose + sinapate = UDP + 1-sinapoyl-D-glucose	0.21	2.4.1.120	303.15	6.0		
inosine + orthophosphate = hypoxanthine + α-D-ribose 1-phosphate	0.0164	2.4.2.1	311.15	7.0	0.25	3.0
xanthosine + orthophosphate = xanthine + α-D-ribose 1-phosphate	0.0156	2.4.2.1	311.15	7.0	0.25	3.0
uridine + orthophosphate = uracil + α-D-ribose 1-phosphate	0.44	2.4.2.2	310.15	7.0		
adenine + 5-phospho-α-D-ribose 1-diphosphate = AMP + pyrophosphate	$2 \cdot 10^{3}$	2.4.2.7	311.15	7.4	0.25	3.0
GMP + hypoxanthine = IMP + guanine	0.38	2.4.2.8	310.15	7.4		
guanine + 5-phospho-α-D-ribose 1-diphosphate = GMP + pyrophosphate	$1 \cdot 10^{5}$	2.4.2.8	311.15	7.4	0.25	3.0
hypoxanthine + 5-phospho-α-D-ribose 1-diphosphate = IMP + pyrophosphate	$1 \cdot 10^{5}$	2.4.2.8	311.15	7.4	0.25	3.0
ATP + ammonium carbamate = ADP + carbamoyl phosphate	0.042	2.7.2.2	283.15	9.4		
ATP + creatine = ADP + phosphocreatine	$5.78 \cdot 10^{-3}$	2.7.3.2	310.15	7.11	0.25	2.47
ATP + L-arginine = ADP + Nω-phospho-L-arginine	0.10	2.7.3.3	285.15	7.25		
ATP + sulfate = adenosine 5'-phosphosulfate + pyrophosphate	$4 \cdot 10^{-8}$	2.7.7.4	303.15	7.5		
UTP + α-D-glucose 1-phosphate = pyrophosphate + UDPglucose	0.48	2.7.7.9	310.15	8.0		
succinyl-CoA + acetoacetate = succinate + acetoacetyl-CoA	$2.8 \cdot 10^{-3}$	2.8.3.5	303.15	7.0		

Reaction	K'	Enzyme Commission Number	T K	pH	I mol dm^{-3}	pMg
acetylcholine + H$_2$O = acetate + choline	$5.38 \cdot 10^2$	3.1.1.7	296.15	5.1		
IMP + H$_2$O = inosine + orthophosphate	$1.58 \cdot 10^2$	3.1.3.1	298.15	8.55	1.53	4.44
phosphorylcholine + H$_2$O = choline + orthophosphate	49.9	3.1.3.1	311.15	6.90		
L-O-phosphoserine + H$_2$O = L-serine + orthophosphate	56	3.1.3.1	308.15	7.0		
cytidine 2′:3′-(cyclic)phosphate + H$_2$O = cytidine 3′-monophosphate	$1.06 \cdot 10^3$	3.1.27.5	298.15	6.0		
isomaltose + H$_2$O = 2 D-glucose	17.2	3.2.1.3	298.15	5.65		
β-gentiobiose + H$_2$O = 2 D-glucose	17.7	3.2.1.21	298.15	5.65		
3-O-β-D-galactopyranosyl-D-arabinose + H$_2$O = D-galactose + D-arabinose	$1.04 \cdot 10^2$	3.2.1.23	298.15	5.65		
lactulose + H$_2$O = D-galactose + D-fructose	$1.28 \cdot 10^2$	3.2.1.23	298.15	5.65		
4′,5′-anhydroadenosine + H$_2$O = adenosine	0.48	3.3.1.1	310.15	7.0		
pteroylglutamate + H$_2$O = pteroate + L-glutamate	15.6	3.4.19.9	310.15	7.3		
N-acetyl-L-phenylalanine methyl ester + H$_2$O = N-acetyl-L-phenylalanine + methanol	$5.88 \cdot 10^2$	3.4.21.1	293.15	5.5		
hippurylanilide + H$_2$O = hippuric acid + aniline	11	3.4.22.2	312.15	5.0		
ammonium carbamate + H$_2$O = 2 ammonia + carbon dioxide	$1.92 \cdot 10^3$	3.5.1.5	293.15	6.5		
ampicillin + H$_2$O = 6-aminopenicillanic acid + D(−)-α-aminophenylacetic acid	0.013	3.5.1.11	298.15	5.0		
cephalexin + H$_2$O = 7-aminodeacetoxycephalosporanic acid + D(−)-α-aminophenylacetic acid	0.044	3.5.1.11	298.15	5.8		
cephaloridine + H$_2$O = 2-thienylacetic acid + 7-amino-3-(1-pyridyl-methyl)-3-cephem-4-carboxylic acid	0.015	3.5.1.11	298.15	5.0		
penicillin G + H$_2$O = 6-aminopenicillanic acid + phenylacetic acid	0.445	3.5.1.11	298.15	6.71		
N-acetyl-L-alanine + H$_2$O = acetate + L-alanine	7	3.5.1.14	298.15	6.0		
ampicillin + H$_2$O = ampicillinoic acid	95	3.5.2.6	282.35	5.55		
penicillin G + H$_2$O = penicillinoic acid	2.9	3.5.2.6	298.15	6.01		
cytidine + H$_2$O = uridine + ammonia	$1.03 \cdot 10^4$	3.5.4.5	298.15	7.00		
N^8-methylcytidine + H$_2$O = uridine + methylamine	$4.88 \cdot 10^2$	3.5.4.5	298.15	7.50		
5,10-methenyltetrahydrofolate + H$_2$O = 10-formyltetrahydrofolate	50	3.5.4.9	298.15	7.0		
ITP + oxaloacetate + H$_2$O = IDP + phospho$enol$pyruvate + carbon dioxide	12	4.1.1.32	303.15	7.6		
2-deoxy-D-ribose 5-phosphate = D-glyceraldehyde 3-phosphate + acetaldehyde	$2.5 \cdot 10^{-4}$	4.1.2.4	295.15	7.5		
6-phospho-2-dehydro-3-deoxy-D-gluconate = pyruvate + D-glyceraldehyde 3-phosphate	$1.2 \cdot 10^{-3}$	4.1.2.14	298.15	8.0	0.37	
L-fuculose 1-phosphate = glycerone phosphate + (S)-lactaldehyde	$4.6 \cdot 10^{-4}$	4.1.2.17	310.15	7.2		
L-rhamnulose 1-phosphate = glycerone phosphate + (S)-lactaldehyde	0.083	4.1.2.19	310.15	7.5		
isocitrate = succinate + glyoxylate	$2.3 \cdot 10^{-3}$	4.1.3.1	303.15	7.7		
(S)-2-methylmalate = acetate + pyruvate	0.151	4.1.3.22	298.15	7.4	0.845	
isocitrate = citrate	14.7	4.2.1.3	298.15	7.4		
3-dehydroquinate = 3-dehydroshikimate + H$_2$O	15	4.2.1.10	302.15	7.4		
(3R)-3-hydroxybutanoyl-CoA = cis-but-2-enoyl-CoA + H$_2$O	0.18	4.2.1.17	298.15	7.5		
indole + D-glyceraldehyde 3-phosphate = 1-(indol-3-yl)glycerol 3-phosphate	$1.2 \cdot 10^4$	4.2.1.20	298.15	7.54		
(R)-malate = maleate + H$_2$O	$4.88 \cdot 10^{-4}$	4.2.1.31	298.15	7.00	0.10	
(R)-2-methylmalate = 2-methylmaleate + H$_2$O	0.0962	4.2.1.35	298.15	7.0	0.10	
D-glutamate = 5-oxo-D-proline + H$_2$O	24.3	4.2.1.48	293.4	7.9		
L-$threo$-3-methylaspartate = 2-methylfumarate + ammonia	0.238	4.3.1.2	298.15	7.9		
L-histidine = urocanate + ammonia	3.01	4.3.1.3	298.25	8.41	0.167	
L-phenylalanine = $trans$-cinnamate + ammonia	2.47	4.3.1.5	298.05	7.69		
ATP = adenosine 3′:5′-(cyclic)phosphate + diphosphate	0.065	4.6.1.1	298.15	7.0		
L,L-2,6-diaminoheptanedioate = $meso$-diaminoheptanedioate	1.9	5.1.1.7	310.15	7.0		
D-ribulose 5-phosphate = D-xylulose 5-phosphate	1.82	5.1.3.1	311.15	7.0	0.25	3.0
UDPglucose = UDPgalactose	0.33	5.1.3.2	298.15	8.7		
GDPmannose = GDP-L-galactose	0.52	5.1.3.18	310.15	8.0		
all-$trans$-retinal = 11-cis-retinal	0.05	5.2.1.3	309.15	7.0		

Reaction	K'	Enzyme Commission Number	T K	pH	I mol dm^{-3}	pMg
9-*cis*,12-*cis*-octadecadienoate = 9-*cis*,11-*trans*-octadecadienoate	61	5.2.1.5	308.15	7.0		
D-erythrose = D-erythrulose	2.3	5.3.1.2	308.15	5.8		
D-arabinose = D-ribulose	0.146	5.3.1.3	320.25	7.4		
L-fucose = L-fuculose	0.12	5.3.1.3	310.15	8.0		
L-arabinose = L-ribulose	0.11	5.3.1.4	298.15	7.0		
D-psicose = β-D-allose	2.15	5.3.1.4	317.25	7.4		
D-ribose 5-phosphate = D-ribulose 5-phosphate	0.83	5.3.1.6	311.15	7.0	0.25	3.0
D-rhamnose = D-rhamnulose	0.58	5.3.1.7	303.15	7.4		
D-mannose 6-phosphate = D-fructose 6-phosphate	0.99	5.3.1.8	298.15	8.50		
6-amino-D-glucose 6-phosphate = 6-amino-D-fructose 6-phosphate	0.202	5.3.1.9	278.85	8.7		
D-glucosamine 6-phosphate + H_2O = D-fructose 6-phosphate + ammonia	0.15	5.3.1.10	310.15	8.4		
D-lyxose = D-xylulose	0.23	5.3.1.15	298.15	7.0		
D-ribose = D-ribulose	0.391	5.3.1.20	313.15	7.4		
keto-phenylpyruvate = *enol*-phenylpyruvate	0.1	5.3.2.1	298.15	7.8		
L-lysine = (3*S*)-3,6-diaminohexanoate	5.3	5.4.3.2	303.15	7.7		
(*R*)-methylmalonyl-CoA = succinyl-CoA	23.1	5.4.99.2	298.15	7.4		
(−)-4-carboxymethyl-Δ$^\alpha$-but-2-en-4-olide = *cis,trans*-hexadienedioate	4.0	5.5.1.1	303.15	8.0		
ATP + heptanoate + CoA = AMP + diphosphate + *n*-heptanoyl-CoA	1.11	6.2.1.2	311.15	8.0		
GTP + succinate + CoA = GDP + phosphate + succinyl-CoA	1.68	6.2.1.4	298.15	7.15	0.25	2.91
GTP + IMP + L-aspartate = GDP + phosphate + adenylosuccinate	2.9	6.3.4.4	310.15	8.0		
ATP + L-citrulline + L-aspartate = AMP + diphosphate + L-arginosuccinate	2.14	6.3.4.5	311.15	6.91		
ATP + propanoyl-CoA + carbon dioxide = ADP + phosphate + (*S*)-methylmalonyl-CoA	$8.1 \cdot 10^{-3}$	6.4.1.3	310.15	8.15		

THERMODYNAMIC QUANTITIES FOR THE IONIZATION REACTIONS OF BUFFERS IN WATER

Robert N. Goldberg, Nand Kishore, and Rebecca M. Lennen

This table contains selected values for the pK, standard molar enthalpy of reaction $\Delta_r H°$, and standard molar heat-capacity change $\Delta_r C_p°$ for the ionization reactions of 64 buffers many of which are relevant to biochemistry and to biology.[1] The values pertain to the temperature $T = 298.15$ K and the pressure $p = 0.1$ MPa. The standard state is the hypothetical ideal solution of unit molality. These data permit one to calculate values of the pK and of $\Delta_r H°$ at temperatures in the vicinity $\{T \approx (274$ K to 350 K$)\}$ of the reference temperature $\theta = 298.15$ K by using the following equations[2]

$$\Delta_r G°_T = -RT \ln K_T = \ln(10) \cdot RT \cdot pK_T, \qquad (1)$$

$$R\ln K_T = -(\Delta_r G°_\theta / \theta) + \Delta_r H°_\theta \{(1/\theta) - (1/T)\} + \Delta_r C°_{p\theta} \{(\theta/T) - 1 + \ln(T/\theta)\}, \qquad (2)$$

$$\Delta_r H°_T = \Delta_r H°_\theta + \Delta_r C°_{p\theta}(T - \theta). \qquad (3)$$

Here, $\Delta_r G°$ is the standard molar Gibbs energy change and K is the equilibrium constant for a reaction; R is the gas constant (8.314 472 J K^{-1} mol^{-1}). The subscripts T and θ denote the temperature to which a quantity pertains, the subscript p denotes constant pressure, and the subscript r denotes that the quantity refers to a reaction. Combination of equations (1) and (2) yields the following equation that gives pK as a function of temperature:

$$pK_T = -\{R \cdot \ln(10)\}^{-1}[-\{\ln(10) \cdot RT \cdot pK_\theta / \theta\} + \Delta_r H°_\theta \{(1/\theta) - (1/T)\} + \Delta_r C°_{p\theta} \{(\theta/T) - 1 + \ln(T/\theta)\}]. \qquad (4)$$

The above equations neglect higher order terms that involve temperature derivatives of $\Delta_r C_p°$. Also, it is important to recognize that the values of pK and $\Delta_r H°$ effectively pertain to ionic strength $I = 0$. However, the values of pK and $\Delta_r H°$ are almost always dependent on the ionic strength and the actual composition of the solution. These issues are discussed in Reference 1, which also gives an approximate method for making appropriate corrections.

References

1. Goldberg, R. N., Kishore, N., and Lennen, R. M., "Thermodynamic Quantities for the Ionization Reactions of Buffers," *J. Phys. Chem. Ref. Data*, 31, 231, 2002.
2. Clarke, E. C. W., and Glew, D. N., *Trans. Faraday Soc.*, 62, 539-547, 1966.

Selected Values of Thermodynamic Quantities for the Ionization Reactions of Buffers in Water at $T = 298.15$ K and $p = 0.1$ MPa

Buffer	Reaction	pK	$\Delta_r H°$ kJ mol^{-1}	$\Delta_r C_p°$ J mol^{-1} K^{-1}
ACES	$HL^\pm = H^+ + L^-$, $(HL = C_4H_{10}N_2O_4S)$	6.847	30.43	−49
Acetate	$HL = H^+ + L^-$, $(HL = C_2H_4O_2)$	4.756	−0.41	−142
ADA	$H_3L^+ = H^+ + H_2L^\pm$, $(H_2L = C_6H_{10}N_2O_5)$	1.59		
	$H_2L^\pm = H^+ + HL^-$	2.48	16.7	
	$HL^- = H^+ + L^{2-}$	6.844	12.23	−144
2-Amino-2-methyl-1,3-propanediol	$HL^+ = H^+ + L$, $(L = C_4H_{11}NO_2)$	8.801	49.85	−44
2-Amino-2-methyl-1-propanol	$HL^+ = H^+ + L$, $(L = C_4H_{11}NO)$	9.694	54.05	≈−21
3-Amino-1-propanesulfonic acid	$HL = H^+ + L^-$, $(HL = C_3H_9NO_3S)$	10.2		
Ammonia	$NH_4^+ = H^+ + NH_3$	9.245	51.95	8
AMPSO	$HL^\pm = H^+ + L^-$, $(HL = C_7H_{17}NO_5S)$	9.138	43.19	−61
Arsenate	$H_3AsO_4 = H^+ + H_2AsO_4^-$	2.31	−7.8	
	$H_2AsO_4^- = H^+ + HAsO^{2-}$	7.05	1.7	
	$HAsO_4^{2-} = H^+ + AsO^{3-\,4}$	11.9	15.9	
Barbital	$H_2L = H^+ + HL^-$, $(H_2L = C_8H_{12}N_2O_3)$	7.980	24.27	−135
	$HL^- = H^+ + L^{2-}$	12.8		
BES	$HL^\pm = H^+ + L^-$, $(HL = C_6H_{15}NO_5S)$	7.187	24.25	−2
Bicine	$H_2L^+ = H^+ + HL^\pm$, $(HL = C_6H_{13}NO_4)$	2.0		
	$HL^\pm = H^+ + L^-$	8.334	26.34	0
Bis-tris	$H_3L^+ = H^+ + H_2L^\pm$, $(H_2L = C_8H_{19}NO_5)$	6.484	28.4	27
Bis-tris propane	$H_2L^{2+} = H^+ + HL^+$, $(L = C_{11}H_{26}N_2O_6)$	6.65		
	$HL^+ = H^+ + L$	9.10		
Borate	$H_3BO_3 = H^+ + H_2BO_3^-$	9.237	13.8	≈−240
Cacodylate	$H_2L^+ = H^+ + HL$, $(HL = C_2H_6AsO_2)$	1.78	−3.5	
	$HL = H^+ + L^-$	6.28	−3.0	−86
CAPS	$HL^\pm = H^+ + L^-$, $(HL = C_9H_{19}NO_3S)$	10.499	48.1	57
CAPSO	$HL^\pm = H^+ + L^-$, $(HL = C_9H_{19}NO_4S)$	9.825	46.67	21
Carbonate	$H_2CO_3 = H^+ + HCO_3^-$	6.351	9.15	−371
	$HCO_3^- = H^+ + CO^{2-}$	10.329	14.70	−249
CHES	$HL^\pm = H^+ + L^-$, $(HL = C_8H_{17}NO_3S)$	9.394	39.55	9

Buffer	Reaction	pK	$\Delta_r H°$ kJ mol⁻¹	$\Delta_r C_p°$ J mol⁻¹ K⁻¹
Citrate	$H_3L = H^+ + H_2L^-$, $(H_3L = C_6H_8O_7)$	3.128	4.07	−131
	$H_2L^- = H^+ + HL^{2-}$	4.761	2.23	−178
	$HL^{2-} = H^+ + L^{3-}$	6.396	−3.38	−254
L-Cysteine	$H_3L^+ = H^+ + H_2L$, $(H_2L = C_3H_7NO_2S)$	1.71	≈−0.6	
	$H_2L = H^+ + HL^-$	8.36	36.1	≈−66
	$HL^- = H^+ + L^{2-}$	10.75	34.1	≈−204
Diethanolamine	$HL^+ = H^+ + L$, $(L = C_4H_{11}NO_2)$	8.883	42.08	36
Diglycolate	$H_2L = H^+ + HL^-$, $(H_2L = C_4H_6O_5)$	3.05	−0.1	≈−142
	$HL^- = H^+ + L^{2-}$	4.37	−7.2	≈−138
3,3-Dimethylglutarate	$H_2L = H^+ + HL^-$, $(H_2L = C_7H_{12}O_4)$	3.70		
	$HL^- = H^+ + L^{2-}$	6.34		
DIPSO	$HL^± = H^+ + L^-$, $(HL = C_7H_{17}NO_5S)$	7.576	30.18	42
Ethanolamine	$HL^+ = H^+ + L$, $(L = C_2H_7NO)$	9.498	50.52	26
N-Ethylmorpholine	$HL^+ = H^+ + L$, $(L = C_6H_{13}NO)$	7.77	27.4	
Glycerol 2-phosphate	$H_2L = H^+ + HL^-$, $(H_2L = C_3H_9NO_6P)$	1.329	−12.2	−330
	$HL^- = H^+ + L^{2-}$	6.650	−1.85	−212
Glycine	$H_2L^+ = H^+ + HL^±$, $(HL = C_2H_5NO_2)$	2.351	4.00	−139
	$HL^± = H^+ + L^-$	9.780	44.2	−57
Glycine amide	$HL^+ = H^+ + L$, $(L = C_2H_6N_2O)$	8.04	42.9	
Glycylglycine	$H_2L^+ = H^+ + HL^±$, $(HL = C_4H_8N_2O_3)$	3.140	0.11	−128
	$HL^± = H^+ + L^-$	8.265	43.4	−16
Glycylglycylglycine	$H_2L^+ = H^+ + HL^±$, $(HL = C_6H_{11}N_3O_4)$	3.224	0.84	
	$HL^± = H^+ + L^-$	8.090	41.7	
HEPES	$H_2L^+ = H^+ + HL^±$, $(HL = C_8H_{18}N_2O_4S)$	≈3.0		
	$HL^± = H^+ + L^-$	7.564	20.4	47
HEPPS	$HL^± = H^+ + L^-$, $(HL = C_6H_{20}N_2O_4S)$	7.957	21.3	48
HEPPSO	$HL^± = H^+ + L^-$, $(HL = C_9H_{20}N_2O_5S)$	8.042	23.70	47
L-Histidine	$H_3L^{2+} = H^+ + H_2L^+$, $(HL = C_6H_9N_3O_2)$	1.5₄	3.6	
	$H_2L^+ = H^+ + HL$	6.07	29.5	176
	$HL = H^+ + L^-$	9.34	43.8	−233
Hydrazine	$H_2L^{2+} = H^+ + HL^+$, $(L = H_4N_2)$	−0.99	38.1	
	$HL^+ = H^+ + L$	8.02	41.7	
Imidazole	$HL^+ = H^+ + L$, $(L = C_3H_4N_2)$	6.993	36.64	−9
Maleate	$H_2L = H^+ + HL^-$, $(H_2L = C_4H_4O_4)$	1.92	1.1	≈−21
	$HL^- = H^+ + L^{2-}$	6.27	−3.6	≈−31
2-Mercaptoethanol	$HL = H^+ + L^-$, $(HL = C_2H_6OS)$	9.7₅	26.2	
MES	$HL^± = H^+ + L^-$, $(HL = C_6H_{13}NO_4S)$	6.270	14.8	5
Methylamine	$HL^+ = H^+ + L$, $(L = CH_5N)$	10.645	55.34	33
2-Methylimidazole	$HL^+ = H^+ + L$, $(L = C_4H_6N_2)$	8.0₁	36.8	
MOPS	$HL^± = H^+ + L^-$, $(HL = C_7H_{15}NO_4S)$	7.184	21.1	25
MOPSO	$H_2L^+ = H^+ + HL^±$, $(HL = C_7H_{15}NO_5S)$	0.060		
	$HL^± = H^+ + L^-$	6.90	25.0	≈38
Oxalate	$H_2L = H^+ + HL^-$, $(H_2L = C_2H_2O_4)$	1.27	−3.9	≈−231
	$HL^- = H^+ + L^{2-}$	4.266	7.00	−231
Phosphate	$H_3PO_4 = H^+ + H_2PO_4^-$	2.148	−8.0	−141
	$H_2PO_4^- = H^+ + HPO_4^{2-}$	7.198	3.6	−230
	$HPO_4^{2-} = H^+ + PO_4^{3-}$	12.35	16.0	−242
Phthalate	$H_2L = H^+ + HL^-$, $(H_2L = C_8H_6O_4)$	2.950	−2.70	−91
	$HL^- = H^+ + L^{2-}$	5.408	−2.17	−295
Piperazine	$H_2L^{2+} = H^+ + HL^+$, $(L = C_4H_{10}N_2)$	5.333	31.11	86
	$HL^+ = H^+ + L$	9.731	42.89	75
PIPES	$HL^± = H^+ + L^-$, $(HL = C_8H_{18}N_2O_6S_2)$	7.141	11.2	22
POPSO	$HL^± = H^+ + L^-$, $(HL = C_{10}H_{22}N_2O_8S_2)$	≈8.0		
Pyrophosphate	$H_4P_2O_7 = H^+ + H_3P_2O_7^-$	0.83	−9.2	≈−90
	$H_3P_2O_7^- = H^+ + H_2P_2O_7^{2-}$	2.26	−5.0	≈−130
	$H_2P_2O_7^{2-} = H^+ + HP_2O_7^{3-}$	6.72	0.5	−136
	$HP_2O_7^{3-} = H^+ + P_2O_7^{4-}$	9.46	1.4	−141
Succinate	$H_2L = H^+ + HL^-$, $(H_2L = C_4H_6O_4)$	4.207	3.0	−121
	$HL^- = H^+ + L^{2-}$	5.636	−0.5	−217
Sulfate	$HSO_4^- = H^+ + SO_4^{2-}$	1.987	−22.4	−258

Buffer	Reaction	pK	$\Delta_r H°$ kJ mol^{-1}	$\Delta_r C_p°$ J mol^{-1} K^{-1}
Sulfite	$H_2SO_3 = H^+ + HSO_3^-$	1.857	−17.80	−272
	$HSO_3^- = H^+ + SO_3^{2-}$	7.172	−3.65	−262
TAPS	$HL^± = H^+ + L^-$, (HL = $C_7H_{17}NO_6S$)	8.44	40.4	15
TAPSO	$HL^± = H^+ + L^-$, (HL = $C_7H_{17}NO_7S$)	7.635	39.09	−16
L(+)-Tartaric acid	$H_2L = H^+ + HL^-$, ($H_2L = C_4H_6O_6$)	3.036	3.19	−147
	$HL^- = H^+ + L^{2-}$	4.366	0.93	−218
TES	$HL^± = H^+ + L^-$, (HL = $C_6H_{15}NO_6S$)	7.550	32.13	0
Tricine	$H_2L^+ = H^+ + HL^±$, (HL = $C_6H_{13}NO_5$)	2.023	5.85	−196
	$HL^± = H^+ + L^-$	8.135	31.37	−53
Triethanolamine	$HL^+ = H^+ + L$, (L = $C_6H_{15}NO_3$)	7.762	33.6	50
Triethylamine	$HL^+ = H^+ + L$, (L = $C_6H_{15}N$)	10.72	43.13	151
Tris	$HL^+ = H^+ + L$, (L = $C_4H_{11}NO_3$)	8.072	47.45	−59

BIOLOGICAL BUFFERS

This table of frequently used buffers gives the pK_a value at 25 °C and the useful pH range of each buffer. The buffers are listed in order of increasing pH.

The table is reprinted with permission of Sigma Chemical Company, St. Louis, MO.

Acronym	Name	Mol. wt.	pK_a	Useful pH range
MES	2-(N-Morpholino)ethanesulfonic acid	195.2	6.1	5.5–6.7
BIS TRIS	Bis(2-hydroxyethyl)iminotris(hydroxymethyl)methane	209.2	6.5	5.8–7.2
ADA	N-(2-Acetamido)-2-iminodiacetic acid	190.2	6.6	6.0–7.2
ACES	2-[(2-Amino-2-oxoethyl)amino]ethanesulfonic acid	182.2	6.8	6.1–7.5
PIPES	Piperazine-N,N´-bis(2-ethanesulfonic acid)	302.4	6.8	6.1–7.5
MOPSO	3-(N-Morpholino)-2-hydroxypropanesulfonic acid	225.3	6.9	6.2–7.6
BIS TRISPROPANE	1,3-Bis[tris(hydroxymethyl)methylamino]propane	282.3	6.8[a]	6.3–9.5
BES	N,N-Bis(2-hydroxyethyl)-2-aminoethanesulfonic acid	213.2	7.1	6.4–7.8
MOPS	3-(N-Morpholino)propanesulfonic acid	209.3	7.2	6.5–7.9
HEPES	N-(2-Hydroxyethyl)piperazine-N´-(2-ethanesulfonic acid)	238.3	7.5	6.8–8.2
TES	N-Tris(hydroxymethyl)methyl-2-aminoethanesulfonic acid	229.2	7.5	6.8–8.2
DIPSO	3-[N,N-Bis(2-hydroxyethyl)amino]-2-hydroxypropanesulfonic acid	243.3	7.6	7.0–8.2
TAPSO	3-[N-Tris(hydroxymethyl)methylamino)-2-hydroxypropanesulfonic acid	259.3	7.6	7.0–8.2
TRIZMA	Tris(hydroxymethyl)aminomethane	121.1	8.1	7.0–9.1
HEPPSO	N-(2-hydroxyethyl)piperazine-N´-(2-hydroxypropanesulfonic acid)	268.3	7.8	7.1–8.5
POPSO	Piperazine-N,N´-bis(2-hydroxypropanesulfonic acid)	362.4	7.8	7.2–8.5
EPPS	N-(2-Hydroxyethyl)piperazine-N´-(3-propanesulfonic acid)	252.3	8.0	7.3–8.7
TEA	Triethanolamine	149.2	7.8	7.3–8.3
TRICINE	N-Tris(hydroxymethyl)methylglycine	179.2	8.1	7.4–8.8
BICINE	N,N-Bis(2-hydroxyethyl)glycine	163.2	8.3	7.6–9.0
TAPS	N-Tris(hydroxymethyl)methyl-3-aminopropanesulfonic acid	243.3	8.4	7.7–9.1
AMPSO	3-[(1,1-Dimethyl-2-hydroxyethyl)amino]-2-hydroxypropanesulfonic acid	227.3	9.0	8.3–9.7
CHES	2-(N-Cyclohexylamino)ethanesulfonic acid	207.3	9.3	8.6–10.0
CAPSO	3-(Cyclohexylamino)-2-hydroxy-1-propanesulfonic acid	237.3	9.6	8.9–10.3
AMP	2-Amino-2-methyl-1-propanol	89.1	9.7	9.0–10.5
CAPS	3-(Cyclohexylamino)-1-propanesulfonic acid	221.3	10.4	9.7–11.1

[a] pK_a = 9.0 for the second dissociation stage.

TYPICAL pH VALUES OF BIOLOGICAL MATERIALS AND FOODS

This table gives typical pH ranges for various biological fluids and common foods. All values refer to 25 °C.

Biological Materials

Blood, human	7.35–7.45
Blood, dog	6.9–7.2
Spinal fluid, human	7.3–7.5
Saliva, human	6.5–7.5
Gastric contents, human	1.0–3.0
Duodenal contents, human	4.8–8.2
Feces, human	4.6–8.4
Urine, human	4.8–8.4
Milk, human	6.6–7.6
Bile, human	6.8–7.0

Foods

Apples	2.9–3.3
Apricots	3.6–4.0
Asparagus	5.4–5.8
Bananas	4.5–4.7
Beans	5.0–6.0
Beers	4.0–5.0
Beets	4.9–5.5
Blackberries	3.2–3.6
Bread, white	5.0–6.0
Butter	6.1–6.4
Cabbage	5.2–5.4
Carrots	4.9–5.3
Cheese	4.8–6.4
Cherries	3.2–4.0
Cider	2.9–3.3
Corn	6.0–6.5
Crackers	6.5–8.5
Dates	6.2–6.4
Eggs, fresh white	7.6–8.0
Flour, wheat	5.5–6.5
Gooseberries	2.8–3.0
Grapefruit	3.0–3.3
Grapes	3.5–4.5

Hominy (lye)	6.8–8.0
Jams, fruit	3.5–4.0
Jellies, fruit	2.8–3.4
Lemons	2.2–2.4
Limes	1.8–2.0
Maple syrup	6.5–7.0
Milk, cows	6.3–6.6
Olives	3.6–3.8
Oranges	3.0–4.0
Oysters	6.1–6.6
Peaches	3.4–3.6
Pears	3.6–4.0
Peas	5.8–6.4
Pickles, dill	3.2–3.6
Pickles, sour	3.0–3.4
Pimento	4.6–5.2
Plums	2.8–3.0
Potatoes	5.6–6.0
Pumpkin	4.8–5.2
Raspberries	3.2–3.6
Rhubarb	3.1–3.2
Salmon	6.1–6.3
Sauerkraut	3.4–3.6
Shrimp	6.8–7.0
Soft drinks	2.0–4.0
Spinach	5.1–5.7
Squash	5.0–5.4
Strawberries	3.0–3.5
Sweet potatoes	5.3–5.6
Tomatoes	4.0–4.4
Tuna	5.9–6.1
Turnips	5.2–5.6
Vinegar	2.4–3.4
Water, drinking	6.5–8.0
Wines	2.8–3.8

STRUCTURE AND FUNCTIONS OF SOME COMMON DRUGS

This table lists the names, categories, therapeutic uses, and chemical structures of selected drugs. The generic (chemical) name of each drug is given, along with some of the trade names under which it is sold. When available, physical properties are given in italics in the fourth column. The structure given refers to the active drug, but many of these are packaged as salts or other derivatives. The drugs have been selected to represent a variety of categories; most are widely used throughout the world.

The list is divided into therapeutic categories; within each category the listing is alphabetical by generic name. The index that follows the table can be used to locate a drug by either generic or trade name.

References

1. *The Combined Chemical Dictionary on DVD, Version 12:1*, CRC Press, Boca Raton, FL, June 2008; also available on the Internet at www.chemnetbase.com.
2. Milne, G. W. A., *Drugs: Synonyms and Properties*, Ashgate Publishing, Aldershot, Hampshire, UK, 2000.
3. Corey, E. J., Czakó, B., and Kürti, L., *Molecules and Medicine*, John Wiley & Sons, Hoboken, NJ, 2007.
4. *Physicians' Desk Reference, 61st Edition*, Thomson PDR, Montvale, NJ, 2007.
5. O'Neil, M. J., Editor, *The Merck Index, 14th Edition*, Merck & Co., Whitehouse Station, NJ, 2006.

	Generic Name	Trade Names	Category and Properties	Applications	Structure
			Antiallergic Agents		
1	Albuterol	Proventil; Ventolin; Volmax	β_2-Andrenergic receptor agonist	Treatment of troubled breathing caused by asthma, emphysema, and other lung diseases	
2	Budesonide	Budeson; Budamax; Rhinocort; Pulmicort; Inflammide	Glucocorticoid *mp 226 °C*	Management of asthma and treatment of inflammatory bowel disease	
3	Cetirizine	Zyrtec	Histamine H_1-receptor antagonist	Treatment of seasonal allergies and hives	
4	Fexofenadine	Carboxyterfenadine; Allegra; Telfast	Histamine H_1-receptor antagonist	Treatment of allergic rhinitis	
5	Fluticasone	Flovent (as propanoate); Flonase (as propanoate); Advair (with salmeterol)	Anti-inflammatory glucocorticoid	Treatment of asthma & rhinitis	
6	Loratadine	Claritin; Claratyne; Alavert	Long-acting antihistamine *mp 132 °C*	Relief of allergy symptoms	

	Generic Name	Trade Names	Category and Properties	Applications	Structure
7	Montelukast	Singulair	Leukotriene LTD$_4$ receptor antagonist	Control of asthma and relief of seasonal allergies	
8	Salmeterol	Serevent	β$_2$-Adrenergic receptor agonist	Treatment of asthma & chronic obstructive pulmonary disease	
9	Tiotropium bromide	Spiriva	Long-acting antimuscarinic bronchodilator	Treatment of chronic obstructive pulmonary disease	

Antibiotics

	Generic Name	Trade Names	Category and Properties	Applications	Structure
10	Amikacin	Amikin; Biclin; Chemacin; Flexilite; Negasin	Aminoglycoside antibiotic	Treatment of serious infections resistant to other antibiotics	
11	Amoxicillin	Amoxil; Isimoxin; Ospamox	β-Lactam/penicillin	Treatment of a broad spectrum of bacterial infections	
12	Azithromycin	Zithromax; Vinzam; Zmax; Azitrocin	Azalide/macrolide antibiotic *mp* 155 °C	Treatment of bacterial skin, ear, and respiratory infections	
13	Cefaclor	Ceclor; Kefolar; Panacef; Panoral	β-Lactam/cephalosporin	Treatment of bacterial infections, pneumonia, and urinary tract infections	
14	Ciprofloxacin	Cipro; Ciproxin; Ciprobay; Flociprin; Uniflox	Fluoroquinoline/broad-spectrum antibiotic *mp* 256 °C	Treatment of urinary & respiratory tract infections, anthrax, and sexually-transmitted diseases	

Generic Name	Trade Names	Category and Properties	Applications	Structure
15 Doxycycline	Vibramycin; Adoxa; Doryx; Liviatin; Deoxymykoin	Tetracycline/broad-spectrum antibiotic	Treatment of urinary tract, respiratory tract, and eye infections; anthrax, syphilis, cholera, etc.	
16 Erythromycin	E-Mycin; Erythrocin; Ilosone	Macrolide antibiotic *mp 191 °C*	Treatment of bacterial infections, including diphtheria, pertussis, rheumatic fever, venereal disease, etc.	
17 Isoniazid	Laniazid	Antimycobacterial agent *mp 171 °C*	Treatment of tuberculosis; reduction of tremors from multiple sclerosis	
18 Linezolid	Zyvox	Oxazolidinone *mp 182 °C*	Treatment of serious Gram-positive infections resistant to other antibiotics	
19 Trimethoprim	Triprim; Proloprim; Monotrim	Dihydrofolate reductase inhibitor *mp 199 °C*	Treatment of urinary tract infections, diarrhea, and ear infections	

Antidiabetic Drugs

Generic Name	Trade Names	Category and Properties	Applications	Structure
20 Glipizide	Glucotrol; Glydiazinamide; Glibenese; Minodiab	Potassium channel blocker/sulfonylurea *mp 208 °C*	Treatment of type 2 diabetes by stimulating insulin secretion in pancreas β-cells	
21 Metformin	Glucophage; Diabex; Diaformin; Fortamet	Antidiabetic biguanide	Treatment of type 2 diabetes by enhancing transport of glucose into muscle cells	$Me_2NCNHCNH_2$ with $\|$ NH $\|$ NH
22 Pioglitazone	Actos	Peroxisome proliferator-activated receptor *mp 174 °C*	Treatment of type 2 diabetes by increasing glucose metabolism and insulin sensitivity	

	Generic Name	Trade Names	Category and Properties	Applications	Structure
23	Rosiglitazone	Avandia	Thiazolidinedione	Treatment of type 2 diabetes by increasing insulin sensitivity. Possible adverse effects on patients with heart problems.	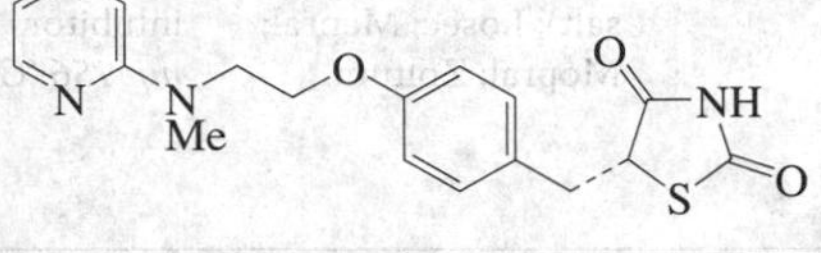
24	Sitagliptin	Januvia	Dipeptidyl peptidase IV inhibitor	Treatment of type 2 diabetes by enhancing the body's ability to lower elevated glucose levels	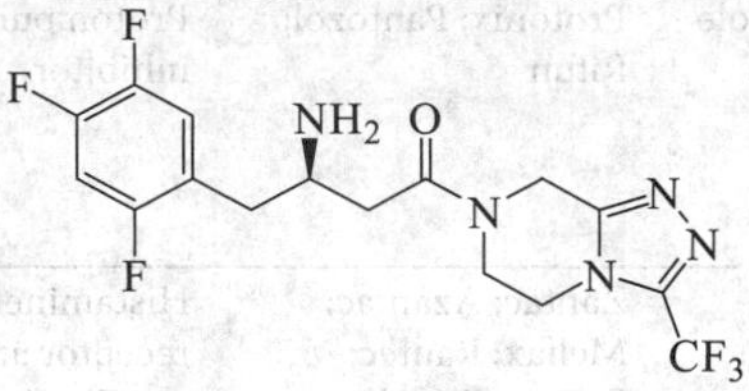

Anti-Inflammatory Agents

	Generic Name	Trade Names	Category and Properties	Applications	Structure
25	Acetylsalicylic acid	Aspirin	NSAID	Pain and fever relief; anticlotting agent	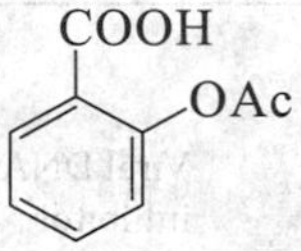
26	Celecoxib	Celebrex; Onsenal	NSAID (COX-2 inhibitor) *mp158 °C*	Treatment of osteoarthritis and rheumatoid arthritis	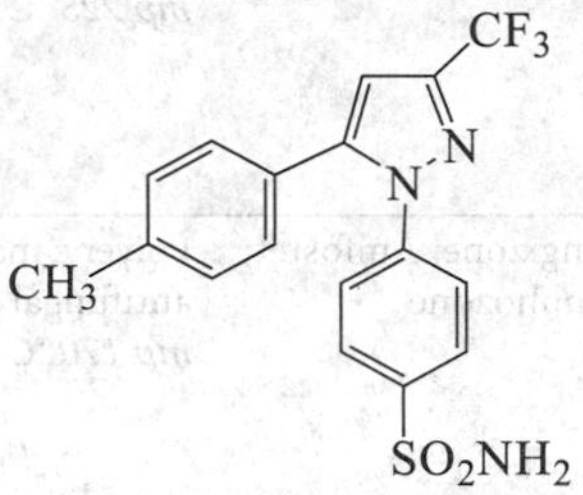
27	Ibuprofen	Advil	NSAID (cyclooxygenase inhibitor) *mp 76 °C*	Relief of inflammation and pain	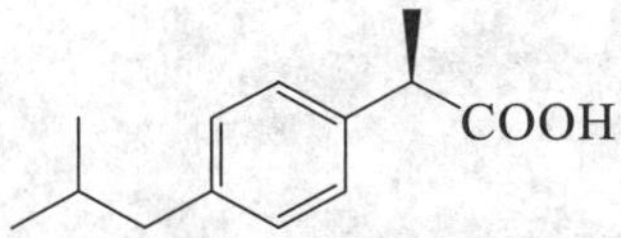
28	Meloxicam	Mobic; Metacam; Metacain	NSAID (cyclooxygenase inhibitor) *mp 254 °C*	Treatment of osteoarthritis and rheumatoid arthritis	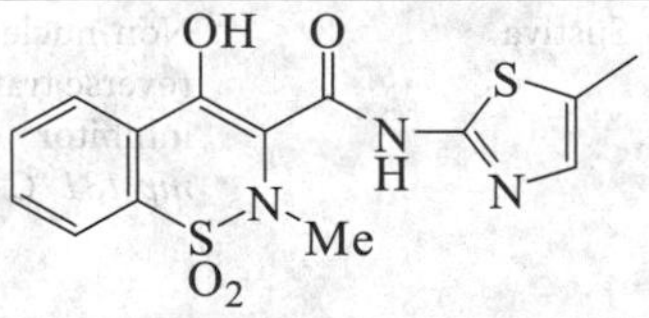
29	Naproxen	Aleve; Naprelan; Anaprox; Naprogesic	NSAID *mp 155 °C*	Relief of inflammation and pain	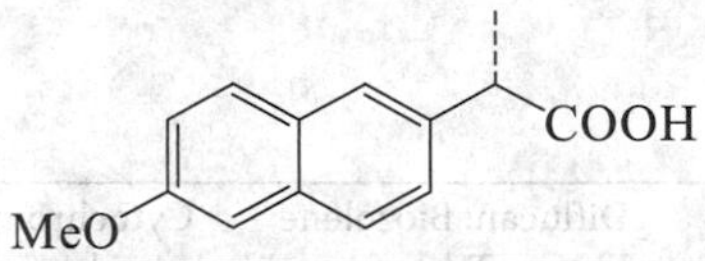
30	Prednisone	Meticorten; Deltasone	Adrenocortical steroid; anti-inflammatory agent; immunosuppressant	Treatment of asthma and other inflammatory diseases	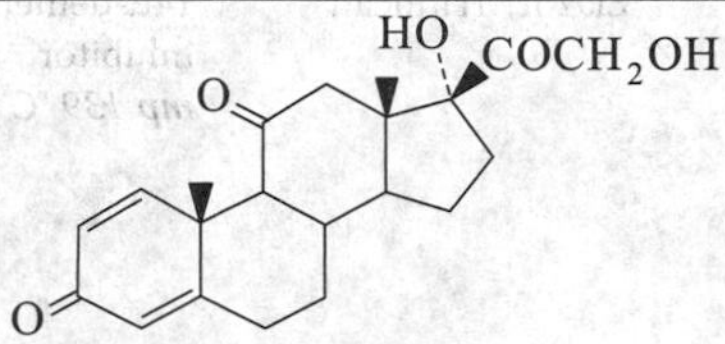

Anti-Ulcer Drugs

	Generic Name	Trade Names	Category and Properties	Applications	Structure
31	Cimetidine	Tagamet; Cimetimax; Gastromet; Peptimax	Histamine H_2-receptor antagonist *mp 142 °C*	Treatment of peptic ulcer, gastrointestinal bleeding, and gastroesophageal reflux disease	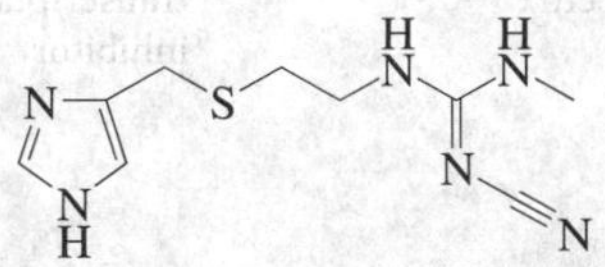
32	Lansoprazole	Prevacid; SoluTab; Prevpac; Zoton; Prezal	Proton pump inhibitor *mp 180 °C*	Treatment of duodenal ulcers & gastroesophageal reflux disease	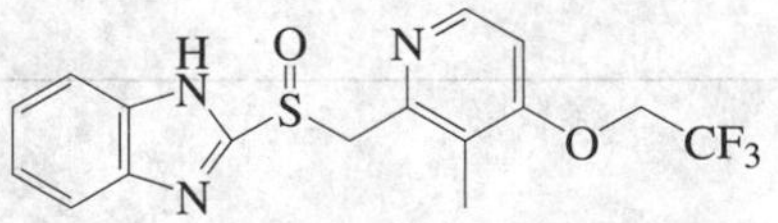

	Generic Name	Trade Names	Category and Properties	Applications	Structure
33	Omeprazole	Prilosec; Nexium (Mg salt); Losec; Mepral; Mopral; Zoltum	Proton pump inhibitor *mp 156 °C*	Treatment of peptic ulcer, dyspepsia, and gastroesophageal reflux disease	
34	Pantoprazole	Protonix; Pantozol; Rifun	Proton pump inhibitor	Treatment of gastric acid-related conditions	
35	Ranitidine	Zantac; Azantac; Melfax; Rantec; Sostril; Taural	Histamine H_2-receptor antagonist *mp 70 °C*	Treatment of peptic ulcer, gastrointestinal bleeding, and gastroesophageal reflux disease	

Antiviral and Antifungal Agents

	Generic Name	Trade Names	Category and Properties	Applications	Structure
36	Acyclovir	Zovirax; Zovir; Avirax; Mirolex	Viral DNA synthesis inhibitor *mp 225 °C*	Treatment of cold sores, genital herpes, chicken pox, etc.	
37	Amphotericin	Fungizone; Amfostet; Amphozone	Polyene macrolide antifungal agent *mp 170 °C*	Intravenous treatment of systemic fungal infections	
38	Efavirenz	Sustiva	Non-nucleoside HIV reverse transcriptase inhibitor *mp 131 °C*	Treatment of HIV-1 infections (as part of combination therapy)	
39	Fluconazole	Diflucan; Biozolene; Elazor; Triflucan	Cytochrome P450 14α-demethylase inhibitor *mp 139 °C*	Treatment and prevention of superficial & systemic fungal infections	
40	Lamivudine	Epivir; Heptodin; Zeffix	HIV reverse transcriptase inhibitor	Treatment of hepatitis B and human immunodeficiency virus	

	Generic Name	Trade Names	Category and Properties	Applications	Structure
41	Nevirapine	Viramune	Non-nucleoside HIV reverse transcriptase inhibitor *mp 250 °C*	Treatment of HIV infections (as part of combination therapy)	
42	Oseltamivir	Tamiflu	Neuraminidase inhibitor *Foam*	Prevention & treatment of influenza A and B	
43	Ribavirin	Virazole; Rebetol; Copegus; Ribasphere; Viratek; Cotronak	Nucleoside antimetabolite *mp 175 °C*	Treatment of hepatitis C	
44	Terbinafine	Lamisil	Squalene epoxidase inhibitor	Treatment of fungal infections of the skin & nails	
45	Zalcitabine	Hivid	Pyrimidine nucleoside reverse transcriptase inhibitor	Treatment of HIV infection & AIDS	
46	Zidovudine	Retrovir; AZT; Azidothymidine; Zidovir	Pyrimidine nucleoside reverse transcriptase inhibitor *mp 121 °C*	Treatment of HIV infection & prevention of mother-to-child transmission	

Cancer Chemotherapy Drugs

	Generic Name	Trade Names	Category and Properties	Applications	Structure
47	Anastrozole	Arimidex	Aromatase inhibitor; antineoplastic agent *mp 81 °C*	Treatment of breast cancer in postmenopausal women	
48	Bicalutamide	Casodex	Antiandrogen *mp 180 °C*	Treatment of advanced prostate cancer	

	Generic Name	Trade Names	Category and Properties	Applications	Structure
49	Bortezomib	Velcade	Proteasome inhibitor	Treatment of lymphomas and multiple myeloma	
50	Capecitabine	Xeloda	Antimetabolite	Treatment of breast & colon cancer	
51	Carboplatin	Paraplatin; Carboplat; Erbakar; Nonoplat	Platinum-based anticancer agent	Treatment of ovarian cancer	
52	Cyclo-phosphamide	Cytoxan; Neosar; Cytophosphan; Endoxan; Clafen	Oxazaphosphorine alkylating agent ("Nitrogen mustard") mp 143 °C	Treatment of lymphomas & leukemias, multiple myeloma, and other cancers	
53	Docetaxel	Taxotere	Antineoplastic agent mp 232 °C	Treatment of ovarian, breast and bronchial carcinomas	
54	Fluorouracil	Efudex; Carac; Fluoroplex; Adrucil	Thymidylate synthesis inhibitor mp 243 °C	Constituent of several antineoplastic combinations	
55	Gemcitabine	Gemzar (as hydrochloride)	Antineoplastic and antiviral agent mp 290 °C	Treatment of lung and pancreatic tumors	
56	Imatinib	Gleevec; Glivec	Protein kinase inhibitor mp 212 °C	Treatment of myelogenous leukemia and gastrointestinal tumors	

	Generic Name	Trade Names	Category and Properties	Applications	Structure
57	Irinotecan	Camptosar; Camptetin; Topotecin	Topoisomerase I inhibitor	Treatment of colorectal cancer	
58	Paclitaxel	Taxol; Abraxane; Yewtaxan	Microtubule-stabilizing agent *mp 214 °C*	Treatment of ovarian, breast, and lung cancer	
59	Sunitinib	Sutent	Tyrosine kineases inhibitor	Treatment of gastrointestinal & kidney tumors	
60	Tamoxifen	Nolvadex; Soltamox; Tamaxin; Tamoplex; Valodex	Selective estrogen receptor modulator *mp 97 °C*	Treatment of breast cancer	

Cardiovascular Agents

	Generic Name	Trade Names	Category and Properties	Applications	Structure
61	Aliskiren	Tekturna; Rasilex	Renin inhibitor	Treatment of hypertension	
62	Amlodipine	Norvasc	Calcium channel blocker	Treatment of hypertension, atrial fibrillation, and angina	
63	Atenolol	Tenormin	β-blocker *mp 147 °C*	Treatment of hypertension and excessive heart rate	
64	Carvedilol	Coreg; Dilatrend; Eucardic	Calcium channel blocker; β-adrenoceptor blocker	Treatment of heart failure and hypertension	
65	Clopidogrel	Plavix; Isocover; Meilax; Tipidyl	Antiplatelet agent	Prevention of blood clots after stroke or myocardial infarction	

	Generic Name	Trade Names	Category and Properties	Applications	Structure
66	Digoxin	Lanoxin; Davoxin; Digacin; Dilanacin; Rougoxin; Digosin; Cordioxil	Cardiac glycoside mp 249 °C	Treatment of congestive heart failure	
67	Enalaprilat	Vasotec	Angiotensin-converting enzyme inhibitor; antihypertensive agent mp 150 °C	Treatment of hypertension, atherosclerosis, and congestive heart failure	
68	Glycerol trinitrate	Nitroglycerin	Cardiac stimulant and vasodilator	Treatment of angina and congestive heart failure	CH_2ONO_2 $CHONO_2$ CH_2ONO_2
69	Irbesartan	Aprovel; Avapro; Avalide (in combination with hydrochlorothiazide	Angiotensin II AT_1-receptor antagonist mp 180 °C	Treatment of hypertension and diabetes-related kidney disease	
70	Lisinopril	Acecomb; Alapril; Carace; Novatec; Novazyd; Vivatec; Zestoretic; Zestril	Angiotensin-converting enzyme inhibitor; antihypertensive agent mp 159 °C	Treatment of hypertension	
71	Losartan	Cozaar; Hyzaar (with hydrochlorothiazide)	Angiotensin II AT_1-receptor antagonist mp 184 °C	Treatment of congestive heart failure and hypertension	
72	Metoprolol	Lopressor; Toprol XL	β-Adrenergic blocker	Treatment of angina and hypertension	
73	Telmisartan	Micardis	Angiotensin II (AT_1) receptor antagonist mp 262 °C	Treatment of hypertension	

	Generic Name	Trade Names	Category and Properties	Applications	Structure
74	Terazosin	Hytrin; Itrin; Hytrinex; Magnurol; Teraprost; Vasocard; Uroflo; etc.	Antihypertensive agent (α_1-adrenoceptor antagonist)	Treatment of hypertension and benign prostatic hyperplasia	
75	Warfarin	Coumadin	Anticlotting agent mp 161 °C	Reduction of possibility of stroke or coronary	

Cholesterol-Lowering Drugs

	Generic Name	Trade Names	Category and Properties	Applications	Structure
76	Atorvastatin	Lipitor; Caduet (in combination with amlodipine)	HMG-CoA reductase inhibitor (statin)	Reduction of LDL cholesterol levels by inhibiting cholesterol biosynthesis	
77	Ezetimibe	Zetia	Selective cholesterol absorption inhibitor mp 165 °C	Reduction of LDL cholesterol levels by inhibiting dietary cholesterol absorption	
78	Nicotinic acid	Niacin	Lipoprotein synthesis inhibitor	Reduction of LDL cholesterol levels	
79	Pravastatin	Pravochol	HMG-CoA reductase inhibitor (statin)	Reduction of LDL cholesterol levels by inhibiting cholesterol biosynthesis	
80	Simvastatin	Zocor; Vytorin (combination with Ezetimibe)	HMG-CoA reductase inhibitor (statin) mp 136 °C	Reduction of LDL cholesterol levels by inhibiting cholesterol biosynthesis	

Depression and Anxiety Drugs

	Generic Name	Trade Names	Category and Properties	Applications	Structure
81	Bupropion	Wellbutrin; Amfebutamone; Zyban	Dopamine reuptake inhibitor Pale yellow oil	Antidepressant; smoking cessation aid	
82	Diazepam	Valium; Stesolid; Seduxen; Antenex; Calmpose; Livotensin	Benzodiazepine central nervous system depressant mp 125 °C	Treatment of anxiety, seizures, muscular spasms, and insomnia	

	Generic Name	Trade Names	Category and Properties	Applications	Structure
83	Donepezil	Aricept	Acetylcholine esterase inhibitor	Treatment of Alzheimer's disease	
84	Fluoxetine	Prozac; Sarafem; Adofen; Fontex; Lorien	Selective serotonin reuptake inhibitor	Treatment of depression, panic attacks, and obsessive-compulsive disorder	
85	Levodopa	L-Dopa; Larodopa; Bendopa; Veldopa	Precursor to the neurotransmitter dopamine *mp 277 °C*	Treatment of Parkinson's disease	
86	Paroxetine	Paxil; Seroxat; Tagonis; Aropax; Motivan	Selective serotonin reuptake inhibitor	Treatment of depression, panic attacks, and obsessive-compulsive disorder	
87	Phenobarbital	Luminal; Fenemal; Gardenal; Barbivis	Anticonvulsant *mp 174 °C*	Epilepsy control; also used as a sedative	
88	Sertraline	Zoloft; Lustral; Serad; Serlain; Tatig	Selective serotonin reuptake inhibitor	Treatment of depression, panic attacks, and obsessive-compulsive disorder	
89	Tiagabine	Gabitril	GABA reuptake inhibitor; anticonvulsant	Treatment of epileptic seizures	
90	Venlafaxine	Effexor; Trewilor; Vandral; Dobupal	Selective serotonin & norepinephrine reuptake inhibitor	Treatment of anxiety and panic disorders	
			Osteoporosis Drugs		
91	Alendronic acid	Fosamax; Adronat; Alendros; Dronal	Farnesyl pyrophosphate synthase inhibitor	Prevention and treatment of osteoporosis and Paget's disease	

	Generic Name	Trade Names	Category and Properties	Applications	Structure
92	Calcitriol	Rocaltrol; Calcijex	Calcium and phosphate metabolism regulator *mp 170 °C*	Treatment of rickets and osteoporosis	
93	Raloxifene	Evista (as hydrochloride)	Selective estrogen receptor modulator	Prevention and treatment of osteoporosis	
94	Risedronic acid	Actonel (as Na salt); Optinate	Calcium regulator	Treatment of osteoporosis & Paget's disease	

Pain Relief Drugs

	Generic Name	Trade Names	Category and Properties	Applications	Structure
95	Acetamino-phen	Tylenol; APAP; Paracetamol; Hedrex; Tramil	Analgesic/antipyretic *mp 170 °C*	Relief of musculoskeletal, neuralgic, and other types of pain	
96	Fentanyl	Duragesic; Actiq; Fentora; Sublimaze	Opioid µ-receptor agonist *mp 87.5 °C*	Treatment of severe pain; spinal and epidural anesthesia	
97	Gabapentin	Neurotin; Aclonium	Anticonvulsant/ analgesic; CNS depressant *mp 164 °C*	Treatment of neuralgia, pain from shingles, migraine, and epilepsy	
98	Lidocaine	Xylocaine; Xylocard; Lidamantle	Aminoamide anesthetic/ antiarrhythmic agent	Local anesthetic for dental procedures	
99	Morphine	Avinza; Contin; Kadian; Roxanol; Meconium; Morfine	Opioid analgesic *mp 255 °C*	Pain management, especially in malignant diseases	
100	Thiopental	Pentothal; Penthiobarbital; Thiopentone	Short-acting barbiturate/ anesthetic	Induction of presurgical anesthesia	

Generic Name	Trade Names	Category and Properties	Applications	Structure
101 Sumatriptan	Imitrex; Imigrane; Megrelan; Permicran; Sumadol	Selective 5-HT$_{1D}$ receptor agonist	Treatment of severe migraine headaches	

Reproductive and Urinary System Drugs

Generic Name	Trade Names	Category and Properties	Applications	Structure
102 Finasteride	Proscar; Propecia; Andozac; Finastid; Procure; Urprosan	5α Reductase inhibitor *mp 252 °C*	Treatment of benign prostatic hyperplasia and male hair loss	
103 Mestranol	Devocin; Norinyl; Ovastol; Tranel	Estrogen, used in combination as oral contraceptive *mp 151 °C*	Prevention of unplanned pregnancy	
104 Mifepristone	Mifeprex; Mifegyne; Corlux; RU 486	Progesterone receptor modulator *mp 150 °C*	Termination of pregnancy	
105 Sildenafil	Viagra; Revatio	Cyclic GMP phosphodiesterase inhibitor	Treatment of erectile dysfunction and pulmonary hypertension	
106 Tamsulosin	Flomax; Amsulosin; Harnal; Omix	Prostate selective α$_1$-adrenoceptor antagonist	Treatment of benign prostatic hyperplasia	
107 Testosterone	Androderm; Androgel; Testrim; Striant	Anabolic steroid hormone	Treatment of male hypogonadism	

Index

The index below lists the trade and generic names for the drugs in this table; the generic names are in bold face. Each entry is referred to by its generic name and the identification number in the table. An asterisk* beside the trade name indicates a product that is a combination of two drugs.

Name	Generic Name	Name	Generic Name
Adoxa	Doxycycline (15)	**Budesonide**	Budesonide (2)
Adronat	Alendronic acid (91)	**Bupropion**	Bupropion (81)
Adrucil	Fluorouracil (54)	Caduet*	Atorvastatin (76)
Advair*	Fluticasone (5)	Calcijex	Calcitriol (92)
Advil	Ibuprofen (27)	**Calcitriol**	Calcitriol (92)
Alapril	Lisinopril (70)	Calmpose	Diazepam (82)
Alavert	Loratadine (6)	Camptetin	Irinotecan (57)
Albuterol	Albuterol (1)	Camptosar	Irinotecan (57)
Alendronic acid	Alendronic acid (91)	Capecitabine	Capecitabine (50)
Alendros	Alendronic acid (91)	Carac	Fluorouracil (54)
Aleve	Naproxen (29)	Carace	Lisinopril (70)
Aliskiren	Aliskiren (61)	Carboplat	Carboplatin (51)
Allegra	Fexofenadine (4)	**Carboplatin**	Carboplatin (51)
Amfebutamone	Bupropion (81)	Carboxyterfenadine	Fexofenadine (4)
Amfostet	Amphotericin (37)	**Carvedilol**	Carvedilol (64)
Amikacin	Amikacin (10)	Casodex	Bicalutamide (48)
Amikin	Amikacin (10)	Ceclor	Cefaclor (13)
Amlodipine	Amlodipine (62)	**Cefaclor**	Cefaclor (13)
Amoxicillin	Amoxicillin (11)	Celebrex	Celecoxib (26)
Amoxil	Amoxicillin (11)	**Celecoxib**	Celecoxib (26)
Amphotericin	Amphotericin (37)	**Cetirizine**	Cetirizine (3)
Amphozone	Amphotericin (37)	Chemacin	Amikacin (10)
Amsulosin	Tamsulosin (106)	**Cimetidine**	Cimetidine (31)
Anaprox	Naproxen (29)	Cimetimax	Cimetidine (31)
Anastrozole	Anastrozole (47)	Cipro	Ciprofloxacin (14)
Andozac	Finasteride (102)	Ciprobay	Ciprofloxacin (14)
Androderm	Testosterone (107)	**Ciprofloxacin**	Ciprofloxacin (14)
Androgel	Testosterone (107)	Ciproxin	Ciprofloxacin (14)
Antenex	Diazepam (82)	Clafen	Cyclophosphamide (52)
APAP	Acetaminophen (95)	Claratyne	Loratadine (6)
Aprovel	Irbesartan (69)	Claritin	Loratadine (6)
Aricept	Donepezil (83)	**Clopidogrel**	Clopidogrel (65)
Arimidex	Anastrozole (47)	Contin	Morphine (99)
Aropax	Paroxetine (86)	Copegus	Ribavirin (43)
Aspirin	Acetylsalicylic acid (25)	Cordioxil	Digoxin (66)
Atenolol	Atenolol (63)	Coreg	Carvedilol (64)
Atorvastatin	Atorvastatin (76)	Corlux	Mifepristone (104)
Avalide*	Irbesartan (69)	Cotronak	Ribavirin (43)
Avandia	Rosiglitazone (23)	Coumadin	Warfarin (75)
Avapro	Irbesartan (69)	Cozaar	Losartan (71)
Avinza	Morphine (99)	**Cyclophosphamide**	Cyclophosphamide (52)
Avirax	Acyclovir (36)	Cytophosphan	Cyclophosphamide (52)
Azantac	Ranitidine (35)	Cytoxan	Cyclophosphamide (52)
Azidothymidine	Zidovudine (46)	**Davoxin**	Digoxin (66)
Azithromycin	Azithromycin (12)	Deltasone	Prednisone (30)
Azitrocin	Azithromycin (12)	Deoxymykoin	Doxycycline (15)
AZT	Zidovudine (46)	Devocin	Mestranol (103)
Barbivis	Phenobarbital (87)	Diabex	Metformin (21)
Bendopa	Levodopa (85)	Diaformin	Metformin (21)
Bicalutamide	Bicalutamide (48)	**Diazepam**	Diazepam (82)
Biclin	Amikacin (10)	Diflucan	Fluconazole (39)
Biozolene	Fluconazole (39)	Digacin	Digoxin (66)
Bortezomib	Bortezomib (49)	Digosin	Digoxin (66)
Budamax	Budesonide (2)	**Digoxin**	Digoxin (66)
Budeson	Budesonide (2)	Dilanacin	Digoxin (66)

Name	Generic Name
Dilatrend	Carvedilol (64)
Dobupal	Venlafaxine (90)
Docetaxel	Docetaxel (53)
Donepezil	Donepezil (83)
Doryx	Doxycycline (15)
Doxycycline	Doxycycline (15)
Dronal	Alendronic acid (91)
Duragesic	Fentanyl (96)
E-Mycin	Erythromycin (16)
Efavirenz	Efavirenz (38)
Effexor	Venlafaxine (90)
Efudex	Fluorouracil (54)
Elazor	Fluconazole (39)
Enalaprilat	Enalaprilat (67)
Endoxan	Cyclophosphamide (52)
Epivir	Lamivudine (40)
Erbakar	Carboplatin (51)
Erythrocin	Erythromycin (16)
Erythromycin	Erythromycin (16)
Eucardic	Carvedilol (64)
Evista	Raloxifene (93)
Ezetimibe	Ezetimibe (77)
Fenemal	Phenobarbital (87)
Fentanyl	Fentanyl (96)
Fentora	Fentanyl (96)
Fexofenadine	Fexofenadine (4)
Finasteride	Finasteride (102)
Finastid	Finasteride (102)
Flexilite	Amikacin (10)
Flociprin	Ciprofloxacin (14)
Flomax	Tamsulosin (106)
Flonase	Fluticasone (5)
Flovent	Fluticasone (5)
Fluconazole	Fluconazole (39)
Fluoroplex	Fluorouracil (54)
Fluorouracil	Fluorouracil (54)
Fluoxetine	Fluoxetine (84)
Fluticasone	Fluticasone (5)
Fontex	Fluoxetine (84)
Fortamet	Metformin (21)
Fosamax	Alendronic acid (91)
Fungizone	Amphotericin (37)
Gabapentin	Gabapentin (97)
Gabitril	Tiagabine (89)
Gardenal	Phenobarbital (87)
Gastromet	Cimetidine (31)
Gemcitabine	Gemcitabine (55)
Gemzar	Gemcitabine (55)
Gleevec	Imatinib (56)
Glibenese	Glipizide (20)
Glipizide	Glipizide (20)
Glivec	Imatinib (56)
Glucophage	Metformin (21)
Glucotrol	Glipizide (20)
Glycerol trinitrate	Glycerol trinitrate (68)

Name	Generic Name
Glydiazinamide	Glipizide (20)
Harnal	Tamsulosin (106)
Hedrex	Acetaminophen (95)
Heptodin	Lamivudine (40)
Hivid	Zalcitabine (45)
Hytrin	Terazosin (74)
Hytrinex	Terazosin (74)
Hyzaar*	Losartan (71)
Ibuprofen	Ibuprofen (27)
Ilosone	Erythromycin (16)
Imatinib	Imatinib (56)
Imigrane	Sumatriptan (101)
Imitrex	Sumatriptan (101)
Inflammide	Budesonide (2)
Irbesartan	Irbesartan (69)
Irinotecan	Irinotecan (57)
Isimoxin	Amoxicillin (11)
Isocover	Clopidogrel (65)
Isoniazid	Isoniazid (17)
Itrin	Terazosin (74)
Januvia	Sitagliptin (24)
Kadian	Morphine (99)
Kefolar	Cefaclor (13)
L-Dopa	Levodopa (85)
Lamisil	Terbinafine (44)
Lamivudine	Lamivudine (40)
Laniazid	Isoniazid (17)
Lanoxin	Digoxin (66)
Lansoprazole	Lansoprazole (32)
Larodopa	Levodopa (85)
Levodopa	Levodopa (85)
Lidamantle	Lidocaine (98)
Lidocaine	Lidocaine (98)
Linezolid	Linezolid (18)
Lipitor	Atorvastatin (76)
Lisinopril	Lisinopril (70)
Liviatin	Doxycycline (15)
Livotensin	Diazepam (82)
Lopressor	Metoprolol (72)
Loratadine	Loratadine (6)
Lorien	Fluoxetine (84)
Losartan	Losartan (71)
Losec	Omeprazole (33)
Luminal	Phenobarbital (87)
Lustral	Sertraline (88)
Magnurol	Terazosin (74)
Meconium	Morphine (99)
Megrelan	Sumatriptan (101)
Meilax	Clopidogrel (65)
Melfax	Ranitidine (35)
Meloxicam	Meloxicam (28)
Mepral	Omeprazole (33)
Mestranol	Mestranol (103)
Metacain	Meloxicam (28)
Metacam	Meloxicam (28)

Name	Generic Name	Name	Generic Name
Metformin	Metformin (21)	Plavix	Clopidogrel (65)
Meticorten	Prednisone (30)	**Pravastatin**	Pravastatin (79)
Metoprolol	Metoprolol (72)	Pravochol	Pravastatin (79)
Micardis	Telmisartan (73)	**Prednisone**	Prednisone (30)
Mifegyne	Mifepristone (104)	Prevacid	Lansoprazole (32)
Mifeprex	Mifepristone (104)	Prevpac	Lansoprazole (32)
Mifepristone	Mifepristone (104)	Prezal	Lansoprazole (32)
Minodiab	Glipizide (20)	Prilosec	Omeprazole (33)
Mirolex	Acyclovir (36)	Procure	Finasteride (102)
Mobic	Meloxicam (28)	Proloprim	Trimethoprim (19)
Monotrim	Trimethoprim (19)	Propecia	Finasteride (102)
Montelukast	Montelukast (7)	Proscar	Finasteride (102)
Mopral	Omeprazole (33)	Protonix	Pantoprazole (34)
Morfine	Morphine (99)	Proventil	Albuterol (1)
Morphine	Morphine (99)	Prozac	Fluoxetine (84)
Motivan	Paroxetine (86)	Pulmicort	Budesonide (2)
Naprelan	Naproxen (29)	**Raloxifene**	Raloxifene (93)
Naprogesic	Naproxen (29)	**Ranitidine**	Ranitidine (35)
Naproxen	Naproxen (29)	Rantec	Ranitidine (35)
Negasin	Amikacin (10)	Rasilex	Aliskiren (61)
Neosar	Cyclophosphamide (52)	Rebetol	Ribavirin (43)
Neurotin	Gabapentin (97)	Retrovir	Zidovudine (46)
Nevirapine	Nevirapine (41)	Revatio	Sildenafil (105)
Nexium	Omeprazole (33)	Rhinocort	Budesonide (2)
Niacin	Nicotinic acid (78)	Ribasphere	Ribavirin (43)
Nicotinic acid	Nicotinic acid (78)	**Ribavirin**	Ribavirin (43)
Nitroglycerin	Glycerol trinitrate (68)	Rifun	Pantoprazole (34)
Nolvadex	Tamoxifen (60)	**Risedronic acid**	Risedronic acid (94)
Nonoplat	Carboplatin (51)	Rocaltrol	Calcitriol (92)
Norinyl	Mestranol (103)	**Rosiglitazone**	Rosiglitazone (23)
Norvasc	Amlodipine (62)	Rougoxin	Digoxin (66)
Novatec	Lisinopril (70)	Roxanol	Morphine (99)
Novazyd	Lisinopril (70)	RU 486	Mifepristone (104)
Omeprazole	Omeprazole (33)	**Salmeterol**	Salmeterol (8)
Omix	Tamsulosin (106)	Sarafem	Fluoxetine (84)
Onsenal	Celecoxib (26)	Seduxen	Diazepam (82)
Optinate	Risedronic acid (94)	Serad	Sertraline (88)
Oseltamivir	Oseltamivir (42)	Serevent	Salmeterol (8)
Ospamox	Amoxicillin (11)	Serlain	Sertraline (88)
Ovastol	Mestranol (103)	Seroxat	Paroxetine (86)
Paclitaxel	Paclitaxel (58)	**Sertraline**	Sertraline (88)
Panacef	Cefaclor (13)	**Sildenafil**	Sildenafil (105)
Panoral	Cefaclor (13)	**Simvastatin**	Simvastatin (80)
Pantoprazole	Pantoprazole (34)	Singulair	Montelukast (7)
Pantozol	Pantoprazole (34)	**Sitagliptin**	Sitagliptin (24)
Paracetamol	Acetaminophen (95)	Soltamox	Tamoxifen (60)
Paraplatin	Carboplatin (51)	SoluTab	Lansoprazole (32)
Paroxetine	Paroxetine (86)	Sostril	Ranitidine (35)
Paxil	Paroxetine (86)	Spiriva	Tiotropium bromide (9)
Penthiobarbital	Thiopental (100)	Stesolid	Diazepam (82)
Pentothal	Thiopental (100)	Striant	Testosterone (107)
Peptimax	Cimetidine (31)	Sublimaze	Fentanyl (96)
Permicran	Sumatriptan (101)	Sumadol	Sumatriptan (101)
Phenobarbital	Phenobarbital (87)	**Sumatriptan**	Sumatriptan (101)
Pioglitazone	Pioglitazone (22)	**Sunitinib**	Sunitinib (59)

Name	Generic Name
Sustiva	Efavirenz (38)
Sutent	Sunitinib (59)
Tagamet	Cimetidine (31)
Tagonis	Paroxetine (86)
Tamaxin	Tamoxifen (60)
Tamiflu	Oseltamivir (42)
Tamoplex	Tamoxifen (60)
Tamoxifen	Tamoxifen (60)
Tamsulosin	Tamsulosin (106)
Tatig	Sertraline (88)
Taural	Ranitidine (35)
Taxol	Paclitaxel (58)
Taxotere	Docetaxel (53)
Tekturna	Aliskiren (61)
Telfast	Fexofenadine (4)
Telmisartan	Telmisartan (73)
Tenormin	Atenolol (63)
Teraprost	Terazosin (74)
Terazosin	Terazosin (74)
Terbinafine	Terbinafine (44)
Testosterone	Testosterone (107)
Testrim	Testosterone (107)
Thiopental	Thiopental (100)
Thiopentone	Thiopental (100)
Tiagabine	Tiagabine (89)
Tiotropium bromide	Tiotropium bromide (9)
Tipidyl	Clopidogrel (65)
Topotecin	Irinotecan (57)
Toprol XL	Metoprolol (72)
Tramil	Acetaminophen (95)
Tranel	Mestranol (103)
Trewilor	Venlafaxine (90)
Triflucan	Fluconazole (39)
Trimethoprim	Trimethoprim (19)
Triprim	Trimethoprim (19)
Tylenol	Acetaminophen (95)
Uniflox	Ciprofloxacin (14)
Uroflo	Terazosin (74)
Urprosan	Finasteride (102)
Valium	Diazepam (82)
Valodex	Tamoxifen (60)

Name	Generic Name
Vandral	Venlafaxine (90)
Vasocard	Terazosin (74)
Vasotec	Enalaprilat (67)
Velcade	Bortezomib (49)
Veldopa	Levodopa (85)
Venlafaxine	Venlafaxine (90)
Ventolin	Albuterol (1)
Viagra	Sildenafil (105)
Vibramycin	Doxycycline (15)
Vinzam	Azithromycin (12)
Viramune	Nevirapine (41)
Viratek	Ribavirin (43)
Virazole	Ribavirin (43)
Vivatec	Lisinopril (70)
Volmax	Albuterol (1)
Vytorin*	Simvastatin (80)
Warfarin	Warfarin (75)
Wellbutrin	Bupropion (81)
Xeloda	Capecitabine (50)
Xylocaine	Lidocaine (98)
Xylocard	Lidocaine (98)
Yewtaxan	Paclitaxel (58)
Zalcitabine	Zalcitabine (45)
Zantac	Ranitidine (35)
Zeffix	Lamivudine (40)
Zestoretic	Lisinopril (70)
Zestril	Lisinopril (70)
Zetia	Ezetimibe (77)
Zidovir	Zidovudine (46)
Zidovudine	Zidovudine (46)
Zithromax	Azithromycin (12)
Zmax	Azithromycin (12)
Zocor	Simvastatin (80)
Zoloft	Sertraline (88)
Zoltum	Omeprazole (33)
Zoton	Lansoprazole (32)
Zovir	Acyclovir (36)
Zovirax	Acyclovir (36)
Zyban	Bupropion (81)
Zyrtec	Cetirizine (3)
Zyvox	Linezolid (18)

CHEMICAL CONSTITUENTS OF HUMAN BLOOD

This table lists typical concentrations of some of the chemical constituents of human blood. The table covers elements and compounds of relatively low molecular weight. References 1 and 4 give extensive information on enzymes, hormones, vitamins, and other blood constituents.

The values given for the normal range refer to healthy adults who have not been exposed to unusual environmental agents. In keeping with IUPAC practice, all values refer to a volume of one liter, and thus are stated in units of g/L, mg/L, μg/L or mmol/L. Many clinical test results, especially in the United States, are reported on a deciliter (dL) rather than a liter basis; thus the values in this table should be divided by 10 to place them on a dL basis. The symbols S (for serum), P (plasma), and WB (whole blood) in the second column indicate the nature of the blood sample to which the values apply. In some cases only a single mean value has been reported, rather than a range; these are given in italics.

The total volume of blood in a 100 kg (220 lb) adult is 7.5 L for a male and 6.7 L for a female. The corresponding volume of plasma is 4.4 L and 4.3 L, respectively (Reference 1).

Values from Reference 1 are so-called "reference values" against which clinical tests of blood chemistry are compared. In these cases the "normal range" is understood to include about 95% of the population. The remaining 5% may show values outside the normal range without necessarily implying a medical problem. Note that these reference values may vary slightly from one testing laboratory to another, depending on the detailed test procedure.

Accurate measurements on trace elements are very difficult to make, and wide variations can be found in the literature. Preferred measurement methods are discussed in References 2 and 6. Values for the trace elements can also vary from one country to another, depending on dietary or environmental factors. Thus cadmium levels tend to be higher in Japan because of the prevalence of seafood in the diet, and lead levels are higher in regions where lead additives are still used in gasoline. Variations with gender, age, geography, and occurrence of diseases are reviewed in Reference 6.

The Critical Values column gives levels that deviate far enough from the normal range to suggest a probable medical issue. Such values from Reference 3 are the Biological Exposure Indexes (BEI) that are specified by the American Council of Government Industrial Hygienists (ACGIH) as danger signals for the levels of pollutants in the workplace.

References

1. Wallach, J., *Interpretation of Diagnostic Tests, Eight Edition*, Wolters Kluwer, Philadelphia, 2007.
2. IUPAC Commission on Toxicology, "Sample Collection Guidelines for Trace Elements in Blood and Urine," *Pure & Appl. Chem.*, 67, 1575, 1995.
3. *2008 TLV's and BEI's*, American Conference of Governmental Industrial Hygienists, 1330 Kemper Meadow Drive, Cincinnati, OH 45240–1634, 2008 (www.acgih.org).
4. Altman, P. L., and Dittmer, D. S., Eds., *Biology Data Book, Second Edition, Vol. III*, Federation of American Societies for Experimental Biology, Bethesda, MD, 1974.
5. Bowen, H. J. M., *Trace Elements in Biochemistry*, Academic Press, New York, 1966.
6. Versieck, J., and Cornelis, R., *Trace Elements in Human Plasma or Serum*, CRC Press, Boca Raton, FL, 1989.

Component	Unit		Normal Range Low	Normal Range High	Critical Values	Ref.
			Inorganic			
Aluminum	S	μg/L	1	10	>60	6,2
Ammonia	P	μg/L	190	600	>700	1
Antimony	S,P	μg/L		*1*		6
Arsenic	S	μg/L	0.5	5		6,2
Barium	S,P	μg/L		*79*		4,5
Beryllium	S,P	μg/L		*<4*		4,5
Bicarbonate (HCO$_3^-$)	WB	mmol/L	22	28	<10 or >40	1
Bromine	S,P	mg/L	2	11		6,4
Cadmium	S	μg/L	0.1	1	>5	6,2,3
Calcium, total	S	mg/L	90	105	<65 or >140	1
Calcium ion (Ca^{++})	WB	mg/L	30	45		1
Carbon dioxide	P	mmol/L	21	30	<11 or >40	1
Carbon monoxide*	WB	%CO-Hb	0	5%	30%	1
Cesium	S,P	μg/L	0.5	2.0		6
Chloride (Cl$^-$)	S	mmol/L	98	106	<80 or >115	1
Chromium	S	μg/L	0.1	0.4		6,2
Cobalt	S	μg/L	0.05	0.35	>1	6,2,3
Copper	S	mg/L	0.7	1.4		1,2,6
Fluorine	S,P	μg/L	33	236		6
Hydrogen ion (H$^+$)	WB	pH	7.38	7.44	<7.10 or >7.59	1
Iodine (total)	S,P	μg/L	59	76		4
Iron	S	mg/L	0.5	1.7		1
Lead	S	μg/L	5	100	>300	1,3,6

Component		Unit	Normal Range		Critical Values	Ref.
			Low	High		
Lithium	S,P	μg/L		8		6
Magnesium	S	mg/L	18	30	<10 or >47	1
Manganese	S	μg/L	0.3	1.0		6,2
Mercury	S	μg/L	0.5	3	>15	2,3
Molybdenum	S,P	μg/L	0.3	1.3		6
Nickel	S	μg/L	0.1	1.3		6,2
Oxygen (arterial)	WB	% saturation	96%	100%		1
Oxygen (venous)	WB	% saturation	60%	85%		1
Phosphorus (inorganic)	S	mg/L	30	45	<11	1
Potassium	S	mmol/L	3.5	5.0	<2.8 or >6.2	1
		mg/L	137	196		
Rubidium	S,P	μg/L	100	300		6
Selenium	S,P	μg/L	40	160		2,6
Silver	S,P	μg/L		1		6
Sodium	S	mmol/L	135	145	<120 or >160	1
		g/L	3.11	3.34		
Strontium	S,P	μg/L		57		4,5
Sulfur (total)	S,P	mg/L		780		4
Tellurium	S,P	μg/L		30		4,5
Titanium	S,P	μg/L		33		4,5
Tin	S,P	μg/L		1		4,5
Vanadium	S,P	μg/L	0.02	1.0		6
Zinc	S,P	mg/L	0.5	1.2		6,2,4
Zirconium	S,P	μg/L		400		4,5

Organic

Component		Unit	Normal Range		Critical Values	Ref.
			Low	High		
Acetoacetate ion	P	mg/L		<10		1
Acetone	S,P	mg/L	3	20		1
Alanine	S,P	mg/L	30	37		4
Arginine	S,P	mg/L	12	19		4
Asparagine	S,P	mg/L	5.4	6.5		4
Cholesterol, total	P	mg/L	1000	2000**	>2400	1,4
HDL Cholesterol	P	mg/L	400	600		1
LDL Cholesterol	P	mg/L	0	1000	>1900	1
Citrulline	S,P	mg/L	2.1	9.7		4
Creatine	S,P	mg/L	2.8	6.2		4
Creatinine	S	mg/L	5	15	>50	1
Fructose	WB	mg/L	5	50		4
Glucosamine	S,P	mg/L	760	1110		4
Glucose (fasting)	S	mg/L	600	1000	<450 or >1300	1
Glutamic acid	S,P	mg/L	4.3	11.5		4
Glutamine	S,P	mg/L	61	102		4
Glycine	S,P	mg/L	13.4	17.3		4
Histidine	S,P	mg/L	7.9	14.8		4
Homocysteine	P	mg/L	0.54	1.62		1
Isoleucine	S,P	mg/L	6.9	12.8		4
Lactate (venous)	P	mg/L	50	150		1
Leucine	S,P	mg/L	14	23		4
Lysine	S,P	mg/L	25	30		4
Methionine	S,P	mg/L	3.3	4.3		4
Ornithine	S,P	mg/L	6.2	8.0		4
Phenylalanine	S,P	mg/L	5.8	14.0		1
Proline	S,P	mg/L	20	33		4
Serine	S,P	mg/L	10.1	12.5		4

Component		Unit	Normal Range		Critical Values	Ref.
			Low	High		
Taurine	S,P	mg/L	4.1	8.2		4
Threonine	S,P	mg/L	12	17		4
Triglyceride	S	mg/L	250	1750		1
Tyrosine	S,P	mg/L	8.1	14.5		4
Urea	S	mmol/L	3.5	7.0	<0.7 or >28	1
Urea nitrogen (BUN)	S	mg/L (of N)	100	200	<20 or >800	1
Uric acid (males)	S	mg/L	25	80		1
Uric acid (females)	S	mg/L	13	60		1
Valine	S,P	mg/L	24	37		4

* Measured as the percent of hemoglobin bound to CO. Typical value for heavy smokers is 5%–10%. Major symptoms begin around 30%, and respiratory failure sets in at >60%.

** This is the desirable upper limit. Values between 2000 and 2400 mg/L are considered borderline high.

CHEMICAL COMPOSITION OF THE HUMAN BODY

The elemental composition of the "standard man" of mass 70 kg is given below.

References

1. Padikal, T. N., and Fivozinsky, S. P., *Medical Physics Data Book, National Bureau of Standards Handbook 138*, U. S. Government Printing Office, Washington, DC, 1981.
2. Snyde, W. S., et al., *Reference Man: Anatomical, Physiological, and Metabolic Characteristics*, Pergamon, New York, 1975.

Element	Amount (g)	Percent of total body mass
Oxygen	43,000	61
Carbon	16,000	23
Hydrogen	7000	10
Nitrogen	1800	2.6
Calcium	1000	1.4
Phosphorus	780	1.1
Sulfur	140	0.20
Potassium	140	0.20
Sodium	100	0.14
Chlorine	95	0.12
Magnesium	19	0.027
Silicon	18	0.026
Iron	4.2	0.006
Fluorine	2.6	0.0037
Zinc	2.3	0.0033
Rubidium	0.32	0.00046
Strontium	0.32	0.00046
Bromine	0.20	0.00029
Lead	0.12	0.00017
Copper	0.072	0.00010
Aluminum	0.061	0.00009
Cadmium	0.050	0.00007
Boron	<0.048	0.00007
Barium	0.022	0.00003
Tin	<0.017	0.00002
Manganese	0.012	0.00002
Iodine	0.013	0.00002
Nickel	0.010	0.00001
Gold	<0.010	0.00001
Molybdenum	<0.0093	0.00001
Chromium	<0.0018	0.000003
Cesium	0.0015	0.000002
Cobalt	0.0015	0.000002
Uranium	0.00009	0.0000001
Beryllium	0.000036	
Radium	$3.1 \cdot 10^{-11}$	

NUTRIENT VALUES OF FOODS

The U. S. Department of Agriculture maintains the USDA National Nutrient Database for Standard Reference, which contains over 7000 food items with data on the energy content, minerals, vitamins, and other properties of nutritional interest. The table here includes about 600 common foods extracted from that database. The properties listed are the energy content (in effect, the enthalpy of combustion); the content of carbohydrates, proteins, and lipids (fats); the cholesterol content; and the amount of sodium, potassium, calcium, magnesium, iron, copper, zinc, manganese, phosphorus, and selenium. All values are given for a 100 gram sample of the food.

To conform with common practice in nutritional science, the energy content is given in kilocalories. For conversion to kilojoules, this number should be multiplied by 4.184. For conversion to the avoirdupois units frequently used in the United States, note that:

1 oz = 28.35 g (thus the 100 g basis in this table is approximately 3.5 oz)

1 lb = 453.6 g

The full USDA database covers specific proprietary brands of many processed foods, and it includes vitamin content as well as the minerals given in this table.

Reference

U.S. Department of Agriculture, Agricultural Research Service. 2005. USDA National Nutrient Database for Standard Reference, Release 18. Nutrient Data Laboratory Home Page, http://www.ars.usda.gov/ba/bhnrc/ndl

Food description	Energy kcal/ 100 g	Carb. g/ 100 g	Protein g/ 100 g	Fat g/ 100 g	Chol. mg/ 100 g	Na mg/ 100 g	K mg/ 100 g	Ca mg/ 100g	Mg mg/ 100 g	Fe mg/ 100 g	Cu mg/ 100 g	Zn mg/ 100 g	Mn mg/ 100 g	P mg/ 100 g	Se μg/ 100 g
Alcoholic beverage, beer, light	29	1.6	0.2	0.0	0	4	21	4	5	0.03	0.006	0.01	0.006	12	0.4
Alcoholic beverage, beer, regular	43	3.6	0.5	0.0	0	4	27	4	6	0.02	0.005	0.01	0.008	14	0.6
Alcoholic beverage, dessert wine, sweet	160	13.7	0.2	0.0	0	9	92	8	9	0.24	0.045	0.07	0.119	9	0.5
Alcoholic beverage, distilled (gin, rum, vodka, whiskey), 80 proof	231	0.0	0.0	0.0	0	1	2	0	0	0.04	0.021	0.04	0.018	4	0.0
Alcoholic beverage, distilled (gin, rum, vodka, whiskey), 90 proof	263	0.0	0.0	0.0	0	1	2	0	0	0.04	0.021	0.04	0.018	4	0.0
Alcoholic beverage, distilled, all, 100 proof	295	0.0	0.0	0.0	0	1	2	0	0	0.04	0.021	0.04	0.018	4	0.0
Alcoholic beverage, liqueur, coffee, 63 proof	308	32.2	0.1	0.3	0	8	30	1	3	0.06	0.040	0.03	0.017	6	0.3
Alcoholic beverage, table wine, red	85	2.6	0.1	0.0	0	4	127	8	12	0.46	0.011	0.14	0.132	23	0.2
Alcoholic beverage, table wine, white	83	2.6	0.1	0.0	0	5	71	9	10	0.27	0.004	0.12	0.117	18	0.1
Almonds, dry roasted, w/o salt	597	19.3	22.1	52.8	0	1	746	266	286	4.51	1.170	3.54	2.620	489	2.8
Almonds, oil roasted, w/o salt	607	17.7	21.2	55.2	0	1	699	291	274	3.68	0.955	3.07	2.460	466	2.8
Apple juice, canned or bottled, unsweetened, w/o vitamin C	47	11.7	0.1	0.1	0	3	119	7	3	0.37	0.022	0.03	0.113	7	0.1
Apples, raw, w/o skin	48	12.8	0.3	0.1	0	0	90	5	4	0.07	0.031	0.05	0.038	11	0.0
Apples, raw, with skin	52	13.8	0.3	0.2	0	1	107	6	5	0.12	0.027	0.04	0.035	11	0.0
Apricots, dried, sulfured, stewed, w. sugar	113	29.3	1.2	0.2	0	3	443	15	15	1.52	0.138	0.24	0.088	38	
Apricots, dried, sulfured, stewed, w/o sugar	85	22.2	1.2	0.2	0	4	411	19	11	0.94	0.121	0.14	0.083	25	0.8
Artichokes, (globe or French), boiled, w/o salt	50	11.2	3.5	0.2	0	95	354	45	60	1.29	0.233	0.49	0.259	86	0.2
Arugula, raw	25	3.7	2.6	0.7	0	27	369	160	47	1.46	0.076	0.47	0.321	52	0.3
Asparagus, boiled	22	4.1	2.4	0.2	0	14	224	23	14	0.91	0.165	0.60	0.154	54	6.1
Asparagus, raw	20	3.9	2.2	0.1	0	2	202	24	14	2.14	0.189	0.54	0.158	52	2.3
Avocados, raw, all commercial varieties	160	8.5	2.0	14.7	0	7	485	12	29	0.55	0.190	0.64	0.142	52	0.4
Bagel, plain, toasted, enriched w. calcium propanoate	257	50.5	10.0	1.6	0	448	75	89	22	6.05	0.130	1.90	0.515	87	22.8
Bagels, cinnamon-raisin, toasted	294	59.3	10.6	1.8	0	346	163	20	23	4.09	0.160	0.81	0.328	83	33.3
Baked beans, canned, no salt	105	20.6	4.8	0.4	0	1	296	50	32	0.29	0.206	1.40		104	4.5
Bamboo shoots, boiled, w/o salt	12	1.9	1.5	0.2	0	4	533	12	3	0.24	0.082	0.47	0.113	20	0.4
Bananas, raw	89	22.8	1.1	0.3	0	1	358	5	27	0.26	0.078	0.15	0.270	22	1.0
Bass, freshwater, mixed species, cooked, dry heat	146	0.0	24.2	4.7	87	90	456	103	38	1.91	0.119	0.83	1.140	256	16.2
Bass, striped, cooked, dry heat	124	0.0	22.7	3.0	103	88	328	19	51	1.08	0.040	0.51	0.019	254	46.8
Beans, French, mature seeds, boiled, w/o salt	129	24.0	7.1	0.8	0	6	370	63	56	1.08	0.115	0.64	0.382	102	1.2
Beans, kidney, all types, mature seeds, boiled, w/o salt	127	22.8	8.7	0.5	0	1	405	35	42	2.22	0.216	1.00	0.430	138	1.1
Beans, lima, immature seeds, canned	71	13.3	4.1	0.3	0	252	285	28	34	1.61	0.162	0.64	0.700	71	1.1
Beans, navy, mature seeds, canned	113	20.5	7.5	0.4	0	448	288	47	47	1.85	0.208	0.77	0.375	134	5.8
Beans, pinto, mature seeds, boiled, w/o salt	143	26.2	9.0	0.7	0	1	436	46	50	2.09	0.219	0.98	0.453	147	6.2

Food description	Energy kcal/ 100 g	Carb. g/ 100 g	Protein g/ 100 g	Fat g/ 100 g	Chol. mg/ 100 g	Na mg/ 100 g	K mg/ 100 g	Ca mg/ 100g	Mg mg/ 100g	Fe mg/ 100 g	Cu mg/ 100 g	Zn mg/ 100 g	Mn mg/ 100 g	P mg/ 100 g	Se µg/ 100 g
Beans, snap, green, boiled, w/o salt	35	7.9	1.9	0.3	0	1	146	44	18	0.65	0.057	0.25	0.285	29	0.2
Beef, bottom sirloin, tri-tip roast, lean & fat, 0" fat, choice, roasted	218	0.0	25.7	12.4	94	50	308	17	20	1.70	0.083	4.52	0.009	189	27.3
Beef, brisket, flat half, lean & fat, 1/8" fat, select, braised	280	0.0	29.0	17.4	71	49	237	17	19	2.40	0.090	6.92	0.010	180	28.1
Beef, chuck, arm pot roast, lean, 1/8" fat, choice, braised	224	0.0	34.7	8.4	81	56	275	15	23	3.04	0.129	8.20	0.012	213	34.7
Beef, ground, 75% lean meat / 25% fat, patty, broiled	278	0.0	25.6	18.7	89	78	289	30	20	2.37	0.075	6.19	0.010	189	21.4
Beef, ground, 95% lean meat / 5% fat, patty, broiled	171	0.0	26.3	6.6	76	65	348	7	22	2.83	0.096	6.43	0.014	206	21.7
Beef, loin, porterhouse steak, lean & fat, 1/4" fat, all grades, broiled	329	0.0	22.5	25.8	72	62	255	8	20	2.68	0.118	4.13	0.014	177	19.3
Beef, loin, T-bone steak, lean & fat, 1/4" fat, all grades, broiled	306	0.0	23.5	22.8	65	67	282	7	22	3.09	0.122	4.30	0.014	185	11.6
Beef, rib, eye, small end (ribs 10-12), lean, 0" fat, choice, broiled	205	0.0	28.9	9.0	91	60	363	16	25	1.98	0.092	5.49	0.011	227	33.6
Beef, rib, large end (ribs 6-9), lean & fat, 1/4" fat, choice, roasted	383	0.0	22.3	32.0	85	63	283	10	19	2.27	0.086	5.58	0.013	168	21.9
Beef, round, bottom round roast, lean, 0" fat, select, roasted	169	0.0	28.3	5.3	72	38	238	6	19	2.35	0.079	4.92	0.010	183	35.7
Beef, round, bottom round, lean & fat, 1/8" fat, all grades, roasted	218	0.0	26.4	11.6	75	35	214	6	17	2.16	0.063	4.43	0.009	164	27.0
Beef, round, eye of round, lean & fat, 1/8" fat, all grades, roasted	208	0.0	28.3	9.7	62	37	227	7	18	2.29	0.067	4.70	0.010	174	28.7
Beef, tenderloin, lean & fat, 1/8" fat, all grades, broiled	267	0.0	26.5	17.1	90	54	329	19	22	1.69	0.079	4.76	0.009	205	28.6
Beef, top sirloin, lean & fat, 1/8" fat, select, broiled	230	0.0	27.1	12.7	67	57	345	22	23	1.66	0.074	4.93	0.009	217	29.3
Beets, boiled	44	10.0	1.7	0.2	0	77	305	16	23	0.79	0.074	0.35	0.326	38	0.7
Blackberries, raw	43	9.6	1.4	0.5	0	1	162	29	20	0.62	0.165	0.53	0.646	22	0.4
Blueberries, raw	57	14.5	0.7	0.3	0	1	77	6	6	0.28	0.057	0.16	0.336	12	0.1
Bluefish, cooked, dry heat	159	0.0	25.7	5.4	76	77	477	9	42	0.62	0.068	1.04	0.027	291	46.8
Bologna, beef	314	4.0	10.3	28.2	56	1080	172	31	14	1.10	0.068	9.10	0.044	172	0.0
Bologna, pork	247	0.7	15.3	19.9	59	1184	281	11	14	0.77	0.080	2.03	0.036	139	12.7
Bratwurst, beef & pork, smoked	297	2.0	12.2	26.3	78	848	283	7	15	1.00	0.080	2.47	0.041	130	14.1
Bratwurst, pork, cooked	333	2.9	13.7	29.2	74	846	259	28	21	0.53	0.079	2.49	0.012	225	26.9
Brazil nuts, dried, unblanched	656	12.3	14.3	66.4	0	3	659	160	376	2.43	1.743	4.06	1.223	725	1917.0
Bread, cracked-wheat	260	49.5	8.7	3.9	0	538	177	43	52	2.81	0.222	1.24	1.371	153	25.3
Bread, Italian	271	50.0	8.8	3.5	0	584	110	78	27	2.94	0.191	0.86	0.464	103	27.2
Bread, mixed-grain, toasted (includes whole-grain, 7-grain)	272	50.4	10.9	4.1	0	530	222	99	58	3.77	0.276	1.38	1.615	191	32.1
Bread, oatmeal	269	48.5	8.4	4.4	0	599	142	66	37	2.70	0.209	1.02	0.940	126	24.6
Bread, pita, white, enriched	275	55.7	9.1	1.2	0	536	120	86	26	2.62	0.168	0.84	0.481	97	27.1
Bread, pumpernickel	250	47.5	8.7	3.1	0	671	208	68	54	2.87	0.287	1.48	1.305	178	24.5
Bread, rye	259	48.3	8.5	3.3	0	660	166	73	40	2.83	0.186	1.14	0.824	125	30.9
Bread, white, commercially prepared	266	50.6	7.6	3.3	0	681	100	151	23	3.74	0.253	0.74	0.478	99	17.3
Bread, whole-wheat, commercially prepared	246	46.1	9.7	4.2	0	527	252	72	86	3.30	0.284	1.94	2.324	229	36.6
Broccoli, boiled, w/o salt	35	7.2	2.4	0.4	0	41	293	40	21	0.67	0.061	0.45	0.194	67	1.6
Broccoli, raw	34	6.6	2.8	0.4	0	33	316	47	21	0.73	0.049	0.41	0.210	66	2.5
Brussels sprouts, boiled, w/o salt	36	7.1	2.6	0.5	0	21	317	36	20	1.20	0.083	0.33	0.227	56	1.5
Butter, w. salt	717	0.1	0.9	81.1	215	576	24	24	2	0.02	0.000	0.09	0.000	24	1.0
Butter, w/o salt	717	0.1	0.9	81.1	215	11	24	24	2	0.02	0.016	0.09	0.004	24	1.0
Cabbage, boiled, w/o salt	22	4.5	1.0	0.4	0	8	97	31	8	0.17	0.012	0.09	0.117	15	0.6
Cabbage, raw	24	5.6	1.4	0.1	0	18	246	47	15	0.59	0.023	0.18	0.159	23	0.9
Cake, angel food, commercially prepared	258	57.8	5.9	0.8	0	749	93	140	12	0.52	0.078	0.07	0.085	32	7.3
Cake, carrot, dry mix, pudding-type	415	79.2	5.1	9.8	0	567	169	172	8	1.80	0.050	0.20	0.528	247	14.9
Cake, fruitcake, commercially prepared	324	61.6	2.9	9.1	5	270	153	33	16	2.07	0.050	0.27	0.220	52	2.0
Cake, pound, commercially prepared, butter	388	48.8	5.5	19.9	221	398	119	35	11	1.38	0.035	0.46	0.090	137	8.8
Cake, white, dry mix	426	78.0	4.5	10.9	0	664	117	192	11	1.39	0.081	0.46	0.205	337	8.6
Cake, yellow, dry mix, enriched	432	78.1	4.4	11.6	2	657	82	135	10	1.50	0.072	0.27	0.191	310	3.0
Candies, fudge, chocolate, prepared-from-recipe	411	76.5	2.4	10.4	14	47	131	45	36	1.77	0.333	1.10	0.423	69	2.5

Food description	Energy kcal/ 100 g	Carb. g/ 100 g	Protein g/ 100 g	Fat g/ 100 g	Chol. mg/ 100 g	Na mg/ 100 g	K mg/ 100 g	Ca mg/ 100g	Mg mg/ 100 g	Fe mg/ 100 g	Cu mg/ 100 g	Zn mg/ 100 g	Mn mg/ 100 g	P mg/ 100 g	Se µg/ 100 g
Candies, gumdrops, starch jelly pieces	396	98.9	0.0	0.0	0	44	5	3	1	0.40	0.012	0.00	0.010	1	0.8
Candies, hard	394	98.0	0.0	0.2	0	38	5	3	3	0.30	0.029	0.01	0.010	3	0.6
Candies, marshmallows	318	81.3	1.8	0.2	0	80	5	3	2	0.23	0.097	0.04	0.008	8	1.7
Candies, milk chocolate	535	59.4	7.7	29.7	23	79	372	189	63	2.35	0.491	2.01	0.471	208	4.5
Candies, milk chocolate w. almond bites	550	51.0	9.8	35.7	19	74	471	220	59	1.50	0.200	1.34	0.010	227	3.4
Candies, peanut brittle, prepared-from-recipe	486	71.2	7.6	19.0	12	445	168	27	42	1.22	0.254	0.87	0.593	106	2.6
Candies, semisweet chocolate	479	63.1	4.2	30.0	0	11	365	32	115	3.13	0.700	1.62	0.800	132	4.2
Candies, sweet chocolate	505	59.6	3.9	34.2	0	16	290	24	113	2.76	0.574	1.50	0.494	147	2.8
Candies, white chocolate	539	59.2	5.9	32.1	14	90	286	199	12	0.24	0.060	0.74	0.008	176	4.5
Capers, canned	23	4.9	2.4	0.9	0	2964	40	40	33	1.67	0.374	0.32	0.078	10	1.2
Carbonated beverage, cola, contains caffeine	37	9.6	0.1	0.0	0	4	2	2	0	0.11	0.001	0.02	0.002	10	0.1
Carbonated beverage, ginger ale	34	8.8	0.0	0.0	0	7	1	3	1	0.18	0.018	0.05	0.013	0	0.1
Carbonated beverage, orange	48	12.3	0.0	0.0	0	12	2	5	1	0.06	0.015	0.10	0.013	1	0.0
Carbonated beverage, root beer	41	10.6	0.0	0.0	0	13	1	5	1	0.05	0.007	0.07	0.013	0	0.1
Carbonated beverage, tonic water	34	8.8	0.0	0.0	0	12	0	1	0	0.01	0.006	0.10	0.001	0	0.0
Carrots, boiled, w/o salt	35	8.2	0.8	0.2	0	58	235	30	10	0.34	0.017	0.20	0.155	30	0.7
Carrots, raw	41	9.6	0.9	0.2	0	69	320	33	12	0.30	0.045	0.24	0.143	35	0.1
Cashew nuts, dry roasted, w. salt	574	32.7	15.3	46.4	0	640	565	45	260	6.00	2.220	5.60	0.826	490	11.7
Cashew nuts, oil roasted, w. salt	581	30.2	16.8	47.8	0	308	632	43	273	6.05	2.043	5.35	1.668	531	20.3
Cashew nuts, raw	553	30.2	18.2	43.9	0	12	660	37	292	6.68	2.195	5.78	1.655	593	19.9
Catfish, channel, farmed, cooked, dry heat	152	0.0	18.7	8.0	64	80	321	9	26	0.82	0.122	1.05	0.020	245	14.5
Catsup	97	25.1	1.7	0.4	0	1114	382	18	19	0.51	0.181	0.26	0.128	33	0.3
Cauliflower, boiled, w/o salt	23	4.1	1.8	0.5	0	15	142	16	9	0.33	0.027	0.18	0.138	32	0.5
Cauliflower, raw	25	5.3	2.0	0.1	0	30	303	22	15	0.44	0.042	0.28	0.156	44	0.6
Celery, raw	14	3.0	0.7	0.2	0	80	260	40	11	0.20	0.035	0.13	0.103	24	0.4
Cereals, corn grits, white, enriched, cooked w. water, w/o salt	59	12.9	1.4	0.2	0	2	21	3	5	0.60	0.018	0.07	0.018	11	3.1
Chard, Swiss, boiled, w/o salt	20	4.1	1.9	0.1	0	179	549	58	86	2.26	0.163	0.33	0.334	33	0.9
Chard, Swiss, raw	19	3.7	1.8	0.2	0	213	379	51	81	1.80	0.179	0.36	0.366	46	0.9
Cheese, American cheddar, imitation	239	11.6	16.7	14.0	36	1345	242	562	29	0.33	0.033	2.59		712	15.2
Cheese, blue	353	2.3	21.4	28.7	75	1395	256	528	23	0.31	0.040	2.66	0.009	387	14.5
Cheese, brie	334	0.5	20.8	27.7	100	629	152	184	20	0.50	0.019	2.38	0.034	188	14.5
Cheese, camembert	300	0.5	19.8	24.3	72	842	187	388	20	0.33	0.021	2.38	0.038	347	14.5
Cheese, cheddar	403	1.3	24.9	33.1	105	621	98	721	28	0.68	0.031	3.11	0.010	512	13.9
Cheese, cheshire	387	4.8	23.4	30.6	103	700	95	643	21	0.21	0.042	2.79	0.012	464	14.5
Cheese, colby	394	2.6	23.8	32.1	95	604	127	685	26	0.76	0.042	3.07	0.012	457	14.5
Cheese, cottage, creamed, large or small curd	103	2.7	12.5	4.5	15	405	84	60	5	0.14	0.028	0.37	0.003	132	9.0
Cheese, cottage, low fat, 1% milkfat	72	2.7	12.4	1.0	4	406	86	61	5	0.14	0.028	0.38	0.003	134	9.0
Cheese, cream	349	2.7	7.6	34.9	110	296	119	80	6	1.20	0.016	0.54	0.004	104	2.4
Cheese, cream, low fat	231	7.0	10.6	17.6	56	296	167	112	8	1.68	0.022	0.76		146	4.0
Cheese, edam	357	1.4	25.0	27.8	89	965	188	731	30	0.44	0.036	3.75	0.011	536	14.5
Cheese, feta	264	4.1	14.2	21.3	89	1116	62	493	19	0.65	0.032	2.88	0.028	337	15.0
Cheese, gouda	356	2.2	24.9	27.4	114	819	121	700	29	0.24	0.036	3.90	0.011	546	14.5
Cheese, gruyere	413	0.4	29.8	32.3	110	336	81	1011	36	0.17	0.032	3.90	0.017	605	14.5
Cheese, mozzarella, whole milk	300	2.2	22.2	22.4	79	627	76	505	20	0.44	0.011	2.92	0.030	354	17.0
Cheese, muenster	368	1.1	23.4	30.0	96	628	134	717	27	0.41	0.031	2.81	0.008	468	14.5
Cheese, parmesan, grated	431	4.1	38.5	28.6	88	1529	125	1109	38	0.90	0.238	3.87	0.085	729	17.7
Cheese, pimento	375	1.7	22.1	31.2	94	1428	162	614	22	0.42	0.033	2.98	0.016	744	14.5
Cheese, provolone	351	2.1	25.6	26.6	69	876	138	756	28	0.52	0.026	3.23	0.010	496	14.5
Cheese, Swiss	380	5.4	26.9	27.8	92	192	77	791	38	0.20	0.043	4.36	0.005	567	18.2
Cheese, tilsit	340	1.9	24.4	26.0	102	753	65	700	13	0.23	0.026	3.50	0.013	500	14.5
Cheesecake, commercially prepared	321	25.5	5.5	22.5	55	207	90	51	11	0.63	0.020	0.51	0.140	93	5.2
Cherries, sour, red, raw	50	12.2	1.0	0.3	0	3	173	16	9	0.32	0.104	0.10	0.112	15	0.0
Cherries, sweet, raw	63	16.0	1.1	0.2	0	0	222	13	11	0.36	0.060	0.07	0.070	21	0.0

Food description	Energy kcal/ 100 g	Carb. g/ 100 g	Protein g/ 100 g	Fat g/ 100 g	Chol. mg/ 100 g	Na mg/ 100 g	K mg/ 100 g	Ca mg/ 100g	Mg mg/ 100 g	Fe mg/ 100 g	Cu mg/ 100 g	Zn mg/ 100 g	Mn mg/ 100 g	P mg/ 100 g	Se µg/ 100 g
Chestnuts, European, boiled & steamed	131	27.8	2.0	1.4	0	27	715	46	54	1.73	0.472	0.25	0.854	99	
Chestnuts, European, roasted	245	53.0	3.2	2.2	0	2	592	29	33	0.91	0.507	0.57	1.180	107	1.2
Chicken, broilers or fryers, breast, meat & skin, fried	222	1.6	31.8	8.9	89	76	259	16	30	1.19	0.057	1.10	0.026	233	23.9
Chicken, broilers or fryers, breast, meat & skin, roasted	197	0.0	29.8	7.8	84	71	245	14	27	1.07	0.050	1.02	0.018	214	24.7
Chicken, broilers or fryers, breast, meat only, fried	187	0.5	33.4	4.7	91	79	276	16	31	1.14	0.054	1.08	0.021	246	26.2
Chicken, broilers or fryers, breast, meat only, roasted	165	0.0	31.0	3.6	85	74	256	15	29	1.04	0.049	1.00	0.017	228	27.6
Chicken, broilers or fryers, dark meat, meat only, roasted	205	0.0	27.4	9.7	93	93	240	15	23	1.33	0.080	2.80	0.021	179	18.0
Chicken, broilers or fryers, drumstick, meat & skin, fried	245	1.6	27.0	13.7	90	89	229	12	23	1.34	0.080	2.89	0.028	176	18.4
Chicken, broilers or fryers, thigh, meat & skin, roasted	247	0.0	25.1	15.5	93	84	222	12	22	1.34	0.078	2.36	0.021	174	19.5
Chicken, broilers or fryers, thigh, meat only, roasted	209	0.0	25.9	10.9	95	88	238	12	24	1.31	0.081	2.57	0.021	183	29.0
Chicken, broilers or fryers, wing, meat & skin, fried	321	2.4	26.1	22.2	81	77	177	15	19	1.25	0.061	1.76	0.028	150	21.3
Chicken, Cornish game hens, meat & skin, roasted	260	0.0	22.3	18.2	131	64	245	13	18	0.91	0.061	1.49	0.015	146	15.5
Chicken, Cornish game hens, meat only, roasted	134	0.0	23.3	3.9	106	63	250	13	19	0.77	0.059	1.53	0.015	149	20.8
Chickpeas, mature seeds, boiled, w/o salt	164	27.4	8.9	2.6	0	7	291	49	48	2.89	0.352	1.53	1.030	168	3.7
Chicory greens, raw	23	4.7	1.7	0.3	0	45	420	100	30	0.90	0.295	0.42	0.429	47	0.3
Chocolate syrup	279	65.1	2.1	1.1	0	72	224	14	65	2.11	0.512	0.73	0.382	129	2.7
Clam, mixed species, breaded & fried	202	10.3	14.2	11.2	61	364	326	63	14	13.91	0.356	1.46	0.540	188	28.9
Clam, mixed species, raw	74	2.6	12.8	1.0	34	56	314	46	9	13.98	0.344	1.37	0.500	169	24.3
Cocoa, dry powder, unsweetened	229	54.3	19.6	13.7	0	21	1524	128	499	13.86	3.788	6.81	3.837	734	14.3
Coconut meat, raw	354	15.2	3.3	33.5	0	20	356	14	32	2.43	0.435	1.10	1.500	113	10.1
Cod, Atlantic, cooked, dry heat	105	0.0	22.8	0.9	55	78	244	14	42	0.49	0.036	0.58	0.020	138	37.6
Cod, Atlantic, dried & salted	290	0.0	62.8	2.4	152	7027	1458	160	133	2.50	0.176	1.59	0.050	950	147.8
Coffee, brewed from grounds, prep w. tap h2o	1	0.0	0.1	0.0	0	2	49	2	3	0.01	0.002	0.02	0.023	3	0.0
Coffee, brewed, espresso, rest-prep	2	0.0	0.1	0.2	0	14	115	2	80	0.13	0.050	0.05	0.050	7	0.0
Coleslaw, home-prepared	69	12.4	1.3	2.6	8	23	181	45	10	0.59	0.023	0.20	0.097	32	0.7
Collards, boiled, w/o salt	26	4.9	2.1	0.4	0	16	116	140	20	1.16	0.038	0.23	0.436	30	0.5
Cookies, brownies, commercially prepared	405	63.9	4.8	16.3	17	312	149	29	31	2.25	0.224	0.72	0.128	101	6.3
Cookies, chocolate chip, commercially prepared, higher fat, unenriched	481	66.8	5.4	22.6	0	315	135	25	31	1.00	0.212	0.64	0.449	108	
Cookies, chocolate sandwich, w/creme filling	466	71.6	5.3	19.1	0	483	187	21	48	3.93	0.431	0.98	0.629	92	8.1
Cookies, fig bars	348	70.9	3.7	7.3	0	350	207	64	27	2.90	0.147	0.39	0.343	62	3.3
Cookies, gingersnaps	416	76.9	5.6	9.8	0	654	346	77	49	6.40	0.305	0.55	1.555	83	5.1
Cookies, graham crackers, plain or honey	423	76.8	6.9	10.1	0	605	135	24	30	3.73	0.202	0.81	0.804	104	10.2
Cookies, oatmeal, commercially prepared	450	68.7	6.2	18.1	0	383	142	37	33	2.58	0.134	0.79	0.839	138	9.8
Cookies, peanut butter sandwich, regular	478	65.6	8.8	21.1	0	368	192	53	49	2.60	0.237	1.06	0.912	188	7.7
Cookies, shortbread, commercially prepared, plain	502	64.5	6.1	24.1	20	455	100	35	17	2.74	0.144	0.53	0.428	108	7.3
Cookies, vanilla wafers, higher fat	473	71.1	4.3	19.4	0	306	107	25	12	2.21	0.124	0.33	0.384	64	11.3
Cookies, vanilla wafers, lower fat	441	73.6	5.0	15.2	51	312	97	48	14	2.38	0.100	0.36	0.262	104	11.3
Corn, sweet, white, boiled, w. salt	108	25.1	3.3	1.3	0	253	249	2	32	0.61	0.053	0.48	0.194	103	0.8
Corn, sweet, yellow, boiled, w. salt	108	25.1	3.3	1.3	0	253	249	2	32	0.61	0.053	0.48	0.194	103	0.2
Couscous, cooked	112	23.2	3.8	0.2	0	5	58	8	8	0.38	0.041	0.26	0.084	22	27.5
Cowpeas (black-eyed), immature seeds, boiled, w. salt	97	20.3	3.2	0.4	0	240	418	128	52	1.12	0.133	1.03	0.572	51	2.5
Crab, Alaska king, cooked, moist heat	97	0.0	19.4	1.5	53	1072	262	59	63	0.76	1.182	7.62	0.040	280	40.0
Crab, Alaska king, raw	84	0.0	18.3	0.6	42	836	204	46	49	0.59	0.922	5.95	0.035	219	36.4
Crab, blue, cooked, moist heat	102	0.0	20.2	1.8	100	279	324	104	33	0.91	0.645	4.22	0.190	206	40.2
Crab, blue, raw	87	0.0	18.1	1.1	78	293	329	89	34	0.74	0.669	3.54	0.150	229	37.4
Crab, Dungeness, cooked, moist heat	110	1.0	22.3	1.2	76	378	408	59	58	0.43	0.734	5.47	0.097	175	47.6
Crackers, cheese, regular	503	58.2	10.1	25.3	13	995	145	151	36	4.77	0.210	1.13	0.629	218	8.6
Crackers, matzo, plain	395	83.7	10.0	1.4	0	2	112	13	25	3.16	0.060	0.68	0.650	89	36.9
Crackers, saltines (includes oyster, soda, soup)	428	70.9	9.2	11.4	0	1072	154	68	22	5.64	0.298	0.83	0.653	101	10.3

Food description	Energy kcal/ 100 g	Carb. g/ 100 g	Protein g/ 100 g	Fat g/ 100 g	Chol. mg/ 100 g	Na mg/ 100 g	K mg/ 100 g	Ca mg/ 100g	Mg mg/ 100 g	Fe mg/ 100 g	Cu mg/ 100 g	Zn mg/ 100 g	Mn mg/ 100 g	P mg/ 100 g	Se µg/ 100 g
Crackers, wheat, regular	473	64.9	8.6	20.6	0	795	183	49	62	4.40	0.318	1.60	1.781	220	6.3
Cranberries, raw	46	12.2	0.4	0.1	0	2	85	8	6	0.25	0.061	0.10	0.360	13	0.1
Cranberry juice, unsweetened	46	12.2	0.4	0.1	0	2	77	8	6	0.25	0.055	0.10		13	0.1
Crayfish, mixed species, farmed, cooked, moist heat	87	0.0	17.5	1.3	137	97	238	51	33	1.11	0.580	1.48	0.217	241	34.2
Cream, fluid, half and half	130	4.3	3.0	11.5	37	41	130	105	10	0.07	0.010	0.51	0.001	95	1.8
Cream, half & half, fat free	59	9.0	2.6	1.4	5	144	206	96	16	0.00	0.016	0.81	0.002	151	2.9
Cream, sour, cultured	214	4.3	3.2	21.0	44	53	144	116	11	0.06	0.019	0.27	0.003	85	2.2
Cream, whipped, cream topping, pressurized	257	12.5	3.2	22.2	76	130	147	101	11	0.05	0.010	0.37	0.001	89	1.4
Croissants, butter	406	45.8	8.2	21.0	67	744	118	37	16	2.03	0.080	0.75	0.330	105	22.7
Croutons, plain	407	73.5	11.9	6.6	0	698	124	76	31	4.08	0.163	0.89	0.500	115	37.5
Cucumber, peeled, raw	12	2.2	0.6	0.2	0	2	136	14	12	0.22	0.071	0.17	0.073	21	0.1
Cucumber, with peel, raw	15	3.6	0.7	0.1	0	2	147	16	13	0.28	0.041	0.20	0.079	24	0.3
Curry powder	325	58.2	12.7	13.8	0	52	1543	478	254	29.59	0.815	4.05	4.289	349	17.1
Dandelion greens, boiled, w/o salt	33	6.4	2.0	0.6	0	44	232	140	24	1.80	0.115	0.28	0.230	42	0.3
Dandelion greens, raw	45	9.2	2.7	0.7	0	76	397	187	36	3.10	0.171	0.41	0.342	66	0.5
Duck, domesticated, meat & skin, roasted	337	0.0	19.0	28.4	84	59	204	11	16	2.70	0.227	1.86	0.019	156	20.0
Duck, domesticated, meat only, roasted	201	0.0	23.5	11.2	89	65	252	12	20	2.70	0.231	2.60	0.019	203	22.4
Egg, white, raw, fresh	52	0.7	10.9	0.2	0	166	163	7	11	0.08	0.023	0.03	0.011	15	20.0
Egg, whole, fried	201	0.9	13.6	15.3	457	204	147	59	13	1.98	0.111	1.20	0.041	208	34.2
Egg, whole, hard-boiled	155	1.1	12.6	10.6	424	124	126	50	10	1.19	0.013	1.05	0.026	172	30.8
Egg, whole, poached	147	0.8	12.5	9.9	422	294	133	53	12	1.83	0.102	1.10	0.039	190	31.6
Egg, whole, scrambled	166	2.2	11.1	12.2	352	280	138	71	12	1.20	0.014	1.00	0.022	170	22.5
Egg, yolk, raw, fresh	322	3.6	15.9	26.5	1234	48	109	129	5	2.73	0.077	2.30	0.055	390	56.0
Eggplant, boiled, w/o salt	35	8.7	0.8	0.2	0	1	123	6	11	0.25	0.059	0.12	0.113	15	0.1
Eggplant, raw	24	5.7	1.0	0.2	0	2	230	9	14	0.24	0.082	0.16	0.250	25	0.3
Endive, raw	17	3.4	1.3	0.2	0	22	314	52	15	0.83	0.099	0.79	0.420	28	0.2
English muffins, whole-wheat, toasted	221	44.1	9.6	2.3	0	692	228	288	77	2.66	0.225	1.74	1.946	307	43.8
English, muffins, plain, toasted, enriched, w. calcium propanoate (includes sourdough)	270	52.7	10.3	2.0	0	477	129	197	28	4.65	0.160	1.40	0.630	107	26.3
Fat, beef tallow	902	0.0	0.0	100.0	109	0	0	0	0	0.00	0.000	0.00		0	0.2
Fat, chicken	900	0.0	0.0	99.8	85	0	0	0	0	0.00	0.000	0.00		0	0.2
Fat, duck	900	0.0	0.0	99.8	100	0	0	0	0	0.00	0.000	0.00		0	0.2
Fat, goose	900	0.0	0.0	99.8	100	0	0	0	0	0.00		0.00		0	0.2
Fat, turkey	900	0.0	0.0	99.8	102	0	0	0	0	0.00	0.000	0.00		0	0.2
Fennel, bulb, raw	31	7.3	1.2	0.2	0	52	414	49	17	0.73	0.066	0.20	0.191	50	0.7
Figs, dried, stewed	107	27.6	1.4	0.4	0	4	294	70	29	0.88	0.124	0.24	0.220	29	0.2
Figs, raw	74	19.2	0.8	0.3	0	1	232	35	17	0.37	0.070	0.15	0.128	14	0.2
Fish oil, cod liver	902	0.0	0.0	100.0	570	0	0	0	0	0.00	0.000	0.00	0.000	0	0.0
Frankfurter beef	330	4.1	11.2	29.6	53	1140	156	14	14	1.51	0.184	2.46	0.082	160	8.2
Frankfurter, chick	257	6.8	12.9	19.5	101	1370	84	95	10	2.00	0.050	1.04	0.015	107	18.4
Frankfurter, pork	269	0.3	12.8	23.7	66	816	264	267	15	3.70	0.074	2.09	0.016	171	27.8
Frankfurter, turkey	226	1.5	14.3	17.7	107	1426	179	106	14	1.84	0.100	3.11	0.016	134	15.4
Game meat, beaver, roasted	212	0.0	34.9	7.0	117	59	403	22	29	10.00	0.189	2.27		292	43.1
Game meat, beefalo, roasted	188	0.0	30.7	6.3	58	82	459	24		3.05		6.40		250	13.1
Game meat, bison, lean, roasted	143	0.0	28.4	2.4	82	57	361	8	26	3.42	0.107	3.68	0.008	209	35.5
Game meat, boar, wild, roasted	160	0.0	28.3	4.4	77	60	396	16	27	1.12	0.056	3.01		134	13.0
Game meat, deer, roasted	158	0.0	30.2	3.2	112	54	335	7	24	4.47	0.300	2.75	0.046	226	12.9
Game meat, elk, roasted	146	0.0	30.2	1.9	73	61	328	5	24	3.63	0.142	3.16	0.013	180	13.0
Game meat, moose, roasted	134	0.0	29.3	1.0	78	69	334	6	24	4.22	0.079	3.68	0.009	176	12.8
Game meat, rabbit, domesticated, roasted	197	0.0	29.1	8.1	82	47	383	19	21	2.27	0.189	2.27	0.032	263	38.5
Game meat, raccoon, roasted	255	0.0	29.2	14.5	97	79	398	14	30	7.10	0.189	2.27		261	18.0
Game meat, squirrel, roasted	173	0.0	30.8	4.7	121	119	352	3	28	6.81	0.148	1.78	0.032	211	15.1
Goose, domesticated, meat & skin, roasted	305	0.0	25.2	21.9	91	70	329	13	22	2.83	0.264	2.62	0.023	270	21.8
Goose, domesticated, meat only, roasted	238	0.0	29.0	12.7	96	76	388	14	25	2.87	0.276	3.17	0.024	309	25.5

Food description	Energy kcal/ 100 g	Carb. g/ 100 g	Protein g/ 100 g	Fat g/ 100 g	Chol. mg/ 100 g	Na mg/ 100 g	K mg/ 100 g	Ca mg/ 100g	Mg mg/ 100 g	Fe mg/ 100 g	Cu mg/ 100 g	Zn mg/ 100 g	Mn mg/ 100 g	P mg/ 100 g	Se µg/ 100 g
Goose, liver, raw	133	6.3	16.4	4.3	515	140	230	43	24	30.53	7.522	3.07	0.000	261	68.1
Gooseberries, raw	44	10.2	0.9	0.6	0	1	198	25	10	0.31	0.070	0.12	0.144	27	0.6
Grape juice, canned or bottled, unsweetened, w/o vitamin C	61	15.0	0.6	0.1	0	3	132	9	10	0.24	0.028	0.05	0.360	11	0.1
Grapefruit juice, white, canned, unsweetened	38	9.0	0.5	0.1	0	1	153	7	10	0.20	0.038	0.09	0.020	11	0.1
Grapefruit, raw, pink & red, all areas	42	10.7	0.8	0.1	0	0	135	22	9	0.08	0.032	0.07	0.022	18	0.1
Grapefruit, raw, white, all areas	33	8.4	0.7	0.1	0	0	148	12	9	0.06	0.050	0.07	0.013	8	1.4
Grapes, American type (slip skin), raw	67	17.2	0.6	0.4	0	2	191	14	5	0.29	0.040	0.04	0.718	10	0.1
Grouper, mixed species, cooked, dry heat	118	0.0	24.8	1.3	47	53	475	21	37	1.14	0.045	0.51	0.012	143	46.8
Haddock, cooked, dry heat	112	0.0	24.2	0.9	74	87	399	42	50	1.35	0.033	0.48	0.030	241	40.5
Haddock, smoked	116	0.0	25.2	1.0	77	763	415	49	54	1.40	0.042	0.50	0.030	251	42.9
Halibut, Atlantic & Pacific, cooked, dry heat	140	0.0	26.7	2.9	41	69	576	60	107	1.07	0.035	0.53	0.020	285	46.8
Ham, sliced, regular (approx 11% fat)	163	3.8	16.6	8.6	57	1304	287	24	22	1.02	0.089	1.35	0.557	153	20.7
Herring, Atlantic, kippered	217	0.0	24.6	12.4	82	918	447	84	46	1.51	0.135	1.36	0.050	325	52.6
Hominy, canned, white or yellow	72	14.3	1.5	0.9	0	210	9	10	16	0.62	0.030	1.05	0.070	35	3.0
Hummus, commercial	166	14.3	7.9	9.6	0	379	228	38	71	2.44	0.527	1.83	0.773	176	2.6
Ice creams, chocolate	216	28.2	3.8	11.0	34	76	249	109	29	0.93	0.135	0.58	0.140	107	2.5
Ice creams, French vanilla, soft-serve	222	22.2	4.1	13.0	91	61	177	131	12	0.21	0.030	0.52	0.005	116	3.0
Ice creams, strawberry	192	27.6	3.2	8.4	29	60	188	120	14	0.21	0.037	0.34	0.078	100	1.9
Ice creams, vanilla	201	23.6	3.5	11.0	44	80	199	128	14	0.09	0.023	0.69	0.008	105	1.8
Jams and preserves	278	68.9	0.4	0.1	0	32	77	20	4	0.49	0.100	0.06	0.040	19	2.0
Jellies	266	70.0	0.2	0.0	0	30	54	7	6	0.19	0.011	0.03	0.132	6	0.4
Kale, boiled, w/o salt	28	5.6	1.9	0.4	0	23	228	72	18	0.90	0.156	0.24	0.416	28	0.9
Kale, raw	50	10.0	3.3	0.7	0	43	447	135	34	1.70	0.290	0.44	0.774	56	0.9
Kiwi fruit, (Chinese gooseberries), fresh, raw	61	14.7	1.1	0.5	0	3	312	34	17	0.31	0.130	0.14	0.098	34	0.2
Knockwurst, pork or beef	307	3.2	11.1	27.7	60	930	199	11	11	0.66	0.060	1.66	0.021	98	13.5
Kumquats, raw	71	15.9	1.9	0.9	0	10	186	62	20	0.86	0.095	0.17	0.135	19	0.0
Lamb, domestic, composite of retail cuts, lean & fat, 1/4" fat, choice, cooked	294	0.0	24.5	20.9	97	72	310	17	23	1.88	0.119	4.46	0.022	188	26.4
Lamb, domestic, composite of retail cuts, lean & fat, 1/8" fat, choice, cooked	271	0.0	25.5	18.0	96	72	318	16	24	1.93	0.121	4.74	0.024	193	27.2
Lamb, domestic, composite of retail cuts, lean, 1/4" fat, choice, cooked	206	0.0	28.2	9.5	92	76	344	15	26	2.05	0.128	5.27	0.028	210	26.1
Lamb, domestic, leg, shank half, lean & fat, 1/8" fat, choice, roasted	217	0.0	26.7	11.4	90	65	329	9	25	1.99	0.118	4.72	0.026	200	29.7
Lamb, domestic, loin, lean & fat, 1/4" fat, choice, roasted	309	0.0	22.6	23.6	95	64	246	18	23	2.12	0.119	3.41	0.020	180	24.6
Lamb, domestic, rib, lean & fat, 1/8" fat, choice, roasted	341	0.0	21.8	27.5	96	74	277	22	20	1.62	0.117	3.62	0.021	170	22.3
Lamb, ground, broiled	283	0.0	24.8	19.7	97	81	339	22	24	1.79	0.128	4.67	0.024	201	27.7
Lard	902	0.0	0.0	100.0	95	0	0	0	0	0.00	0.000	0.11	0.000	0	0.2
Leeks, (bulb & lower leaf-portion), boiled, w/o salt	31	7.6	0.8	0.2	0	10	87	30	14	1.10	0.062	0.06	0.247	17	0.5
Lemon juice, canned or bottled	21	6.5	0.4	0.3	0	21	102	11	8	0.13	0.037	0.06	0.020	9	0.1
Lemons, raw, with peel	20	10.7	1.2	0.3	0	3	145	61	12	0.70	0.260	0.10		15	
Lemons, raw, without peel	29	9.3	1.1	0.3	0	2	138	26	8	0.60	0.037	0.06	0.030	16	0.4
Lentils, mature seeds, boiled, w. salt	116	20.1	9.0	0.4	0	238	369	19	36	3.33	0.251	1.27	0.494	180	2.8
Lentils, sprouted, stir-fried, w. salt	101	21.3	8.8	0.5	0	246	284	14	35	3.10	0.337	1.60	0.502	153	0.6
Lettuce, iceberg (includes crisp head types), raw	14	3.0	0.9	0.1	0	10	141	18	7	0.41	0.025	0.15	0.125	20	0.1
Lettuce, romaine, raw	17	3.3	1.2	0.3	0	8	247	33	14	0.97	0.048	0.23	0.155	30	0.4
Lima beans, large, mature seeds, boiled, w/o salt	115	20.9	7.8	0.4	0	2	508	17	43	2.39	0.235	0.95	0.516	111	4.5
Lima beans, thin seeded (baby), mature seeds, boiled, w/o salt	126	23.3	8.0	0.4	0	3	401	29	53	2.40	0.215	1.03	0.585	127	4.9
Limes, raw	30	10.5	0.7	0.2	0	2	102	33	6	0.60	0.065	0.11	0.008	18	0.4
Liverwurst spread	305	5.9	12.4	25.5	118	700	170	22	12	8.85	0.240	2.30	0.155	230	58.0
Lobster, northern, cooked, moist heat	98	1.3	20.5	0.6	72	380	352	61	35	0.39	1.940	2.92	0.061	185	42.7
Macadamia nuts, dry roasted, w/o salt	718	13.4	7.8	76.1	0	4	363	70	118	2.65	0.570	1.29	3.036	198	3.6
Macaroni, cooked, enriched	158	30.9	5.8	0.9	0	1	45	7	18	1.33	0.103	0.50	0.317	58	26.4
Mackerel, Atlantic, cooked, dry heat	262	0.0	23.9	17.8	75	83	401	15	97	1.57	0.094	0.94	0.020	278	51.6

Food description	Energy kcal/ 100 g	Carb. g/ 100 g	Protein g/ 100 g	Fat g/ 100 g	Chol. mg/ 100 g	Na mg/ 100 g	K mg/ 100 g	Ca mg/ 100g	Mg mg/ 100 g	Fe mg/ 100 g	Cu mg/ 100 g	Zn mg/ 100 g	Mn mg/ 100 g	P mg/ 100 g	Se µg/ 100 g
Mackerel, king, cooked, dry heat	134	0.0	26.0	2.6	68	203	558	40	41	2.28	0.033	0.72	0.006	318	46.8
Mackerel, Pacific & jack, mixed species, cooked, dry heat	201	0.0	25.7	10.1	60	110	521	29	36	1.49	0.119	0.86	0.019	160	46.8
Mackerel, salted	305	0.0	18.5	25.1	95	4450	520	66	60	1.40	0.100	1.10		254	73.4
Mackerel, Spanish, cooked, dry heat	158	0.0	23.6	6.3	73	66	554	13	38	0.74	0.065	0.62	0.012	271	40.6
Mangos, raw	65	17.0	0.5	0.3	0	2	156	10	9	0.13	0.110	0.04	0.027	11	0.6
Margarine, regular, hard, corn (hydrogenated)	719	0.9	0.9	80.5	0	943	42	30	3	0.00				23	0.0
Margarine, regular, hard, soybean (hydrogenated) & palm (hydrogenated)	719	0.9	0.9	80.5	0	943	42	30	3	0.00		0.00		23	0.0
Margarine, regular, stick, unsalted, 80% fat	719	0.9	0.9	80.5	0	2	25	17	2	0.00	0.000	0.00		13	0.0
Margarine, regular, tub, unsalted, 80% fat	716	0.5	0.8	80.4	0	28	38	26	2	0.00	0.000	0.00		20	0.0
Marmalade, orange	246	66.3	0.3	0.0	0	56	37	38	2	0.15	0.090	0.04	0.020	4	0.6
Mayonnaise dressing, no cholesterol	688	0.3	0.0	77.8	0	486	14	7	1	0.23	0.000	0.13		25	1.6
Mayonnaise, low sodium, diet	231	16.0	0.3	19.2	24	110	10	0	0	0.00	0.000	0.11	0.000	0	1.6
Melons, honeydew, raw	36	9.1	0.5	0.1	0	18	228	6	10	0.17	0.024	0.09	0.027	11	0.7
Milk, buttermilk, fluid, cultured, reduced fat	56	5.3	4.1	2.0	8	86	180	143	13	0.06	0.008	0.24		82	2.3
Milk, goat, fluid	69	4.5	3.6	4.1	11	50	204	134	14	0.05	0.046	0.30	0.018	111	1.4
Milk, low fat, fluid, 1% milkfat, w. vitamin A	42	5.0	3.4	1.0	5	44	150	119	11	0.03	0.010	0.42	0.003	95	3.3
Milk, low sodium, fluid	61	4.5	3.1	3.5	14	3	253	101	5	0.05	0.010	0.38	0.004	86	2.0
Milk, reduced fat, fluid, 2% milkfat, w. added vitamin A	50	4.7	3.3	2.0	8	41	150	117	11	0.03	0.012	0.43	0.003	94	2.5
Muffins, blueberry, commercially prepared	277	48.0	5.5	6.5	30	447	123	57	16	1.61	0.074	0.49	0.440	197	11.2
Muffins, corn, commercially prepared	305	50.9	5.9	8.4	26	521	69	74	32	2.81	0.299	0.54	0.355	284	15.2
Muffins, oat bran	270	48.3	7.0	7.4	0	393	507	63	157	4.20	0.330	1.84	2.630	376	11.0
Mullet, striped, cooked, dry heat	150	0.0	24.8	4.9	63	71	458	31	33	1.41	0.141	0.88	0.022	244	46.8
Mushrooms, boiled, w/o salt	28	5.3	2.2	0.5	0	2	356	6	12	1.74	0.504	0.87	0.115	87	11.9
Mushrooms, oyster, raw	35	6.4	3.3	0.4	0	18	420	3	18	1.33	0.244	0.77	0.113	120	2.6
Mushrooms, portabella, grilled	35	5.1	4.1	0.8	0	10	521	4	15	0.56	0.499	0.73	0.075	150	17.7
Mushrooms, raw	22	3.3	3.1	0.3	0	5	318	3	9	0.50	0.318	0.52	0.047	86	9.3
Mushrooms, shiitake, cooked, w/o salt	56	14.4	1.6	0.2	0	4	117	3	14	0.44	0.896	1.33	0.204	29	24.8
Mussels, blue, cooked, moist heat	172	7.4	23.8	4.5	56	369	268	33	37	6.72	0.149	2.67	6.800	285	89.6
Mussels, blue, raw	86	3.7	11.9	2.2	28	286	320	26	34	3.95	0.094	1.60	3.400	197	44.8
Mustard greens, boiled, w/o salt	15	2.1	2.3	0.2	0	16	202	74	15	0.70	0.084	0.11	0.274	41	0.6
Nectarines, raw	44	10.6	1.1	0.3	0	0	201	6	9	0.28	0.086	0.17	0.054	26	0.0
Noodles, egg, cooked, enriched, w. salt	138	25.2	4.5	2.1	29	165	38	12	21	1.47	0.098	0.65	0.315	76	23.9
Noodles, Japanese, soba, cooked	99	21.4	5.1	0.1	0	60	35	4	9	0.48	0.008	0.12	0.374	25	
Nutmeg, ground	525	49.3	5.8	36.3	0	16	350	184	183	3.04	1.027	2.15	2.900	213	1.6
Ocean perch, Atlantic, cooked, dry heat	121	0.0	23.9	2.1	54	96	350	137	39	1.18	0.033	0.61	0.020	277	55.5
Oil, avocado	884	0.0	0.0	100.0	0	0	0	0	0	0.00	0.000	0.00	0.000	0	0.0
Oil, canola	884	0.0	0.0	100.0	0	0	0	0	0	0.00	0.000	0.00	0.000	0	0.0
Oil, canola and soybean	884	0.0	0.0	100.0	0	0	0	0	0	0.00	0.000	0.00		0	0.0
Oil, coconut	862	0.0	0.0	100.0	0	0	0	0	0	0.04	0.000	0.00	0.000	0	0.0
Oil, corn and canola	884	0.0	0.0	100.0	0	0	0	0	0	0.00	0.000	0.00	0.000	0	0.0
Oil, mustard	884	0.0	0.0	100.0	0	0	0	0	0	0.00	0.000	0.00	0.000	0	0.0
Oil, olive, salad or cooking	884	0.0	0.0	100.0	0	2	1	1	0	0.56	0.000	0.00	0.000	0	0.0
Oil, palm	884	0.0	0.0	100.0	0	0	0	0	0	0.01	0.000	0.00		0	0.0
Oil, palm kernel	862	0.0	0.0	100.0	0	0	0	0	0	0.00	0.000	0.00	0.000	0	0.0
Oil, peanut, salad or cooking	884	0.0	0.0	100.0	0	0	0	0	0	0.03	0.000	0.01		0	0.0
Oil, safflower, salad or cooking, linoleic >70%	884	0.0	0.0	100.0	0	0	0	0	0	0.00	0.000	0.00	0.000	0	0.0
Oil, safflower, salad or cooking, oleic >70%	884	0.0	0.0	100.0	0	0	0	0	0	0.00	0.000	0.00	0.000	0	0.0
Oil, sesame, salad or cooking	884	0.0	0.0	100.0	0	0	0	0	0	0.00	0.000	0.00		0	0.0
Oil, soybean lecithin	763	0.0	0.0	100.0	0	0	0	0	0	0.00	0.000	0.00	0.000	0	0.0
Oil, soybean, salad or cooking	884	0.0	0.0	100.0	0	0	0	0	0	0.02	0.000	0.00		0	0.0
Oil, sunflower, linoleic, approx. 65%	884	0.0	0.0	100.0	0	0	0	0	0	0.00	0.000	0.00		0	0.0
Okra, boiled, w/o salt	22	4.6	1.9	0.2	0	6	135	77	36	0.28	0.085	0.43	0.294	32	0.4

Food description	Energy kcal/ 100 g	Carb. g/ 100 g	Protein g/ 100 g	Fat g/ 100 g	Chol. mg/ 100 g	Na mg/ 100 g	K mg/ 100 g	Ca mg/ 100g	Mg mg/ 100 g	Fe mg/ 100 g	Cu mg/ 100 g	Zn mg/ 100 g	Mn mg/ 100 g	P mg/ 100 g	Se µg/ 100 g
Olives, pickled, canned or bottled, green	145	3.8	1.0	15.3	0	1556	42	52	11	0.49	0.120	0.04		4	0.9
Olives, ripe, canned (small-extra large)	115	6.3	0.8	10.7	0	872	8	88	4	3.30	0.251	0.22	0.020	3	0.9
Onions, boiled, w/o salt	44	10.2	1.4	0.2	0	3	166	22	11	0.24	0.067	0.21	0.153	35	0.6
Onions, raw	42	10.1	0.9	0.1	0	3	144	22	10	0.19	0.038	0.16	0.132	27	0.5
Orange juice, includes from concentrate	44	10.1	0.8	0.3	0	1	190	10	11	0.17	0.040	0.04	0.023	11	0.1
Oranges, raw, all commercial varieties	47	11.8	0.9	0.1	0	0	181	40	10	0.10	0.045	0.07	0.025	14	0.5
Oranges, raw, California, Valencias	49	11.9	1.0	0.3	0	0	179	40	10	0.09	0.037	0.06	0.023	17	
Oranges, raw, Florida	46	11.5	0.7	0.2	0	0	169	43	10	0.09	0.039	0.08	0.024	12	0.5
Oranges, raw, navels	49	12.5	0.9	0.2	0	1	166	43	11	0.13	0.039	0.08	0.029	23	0.0
Ostrich, ground, cooked, pan-broiled	175	0.0	26.2	7.1	83	80	323	8	23	3.43	0.136	4.33	0.017	224	33.5
Ostrich, inside leg, cooked	141	0.0	29.0	1.9	73	83	352	6	25	3.12	0.148	4.71	0.018	244	36.5
Ostrich, top loin, cooked	155	0.0	28.1	3.9	93	77	353	6	25	3.31	0.148	4.72	0.018	245	36.6
Oyster, eastern, breaded & fried	197	11.6	8.8	12.6	81	417	244	62	58	6.95	4.294	87.13	0.490	159	66.5
Oyster, eastern, farmed, raw	59	5.5	5.2	1.6	25	178	124	44	33	5.78	0.738	37.92	0.394	93	63.7
Oyster, eastern, wild, raw	68	3.9	7.1	2.5	53	211	156	45	47	6.66	4.452	90.81	0.367	135	63.7
Oyster, Pacific, cooked, moist heat	163	9.9	18.9	4.6	100	212	302	16	44	9.20	2.679	33.24	1.222	243	154.0
Oyster, Pacific, raw	81	5.0	9.5	2.3	50	106	168	8	22	5.11	1.576	16.62	0.643	162	77.0
Palm hearts, raw	115	25.6	2.7	0.2	0	14	1806	18	10	1.69	0.644	3.73		140	0.7
Papayas, raw	39	9.8	0.6	0.1	0	3	257	24	10	0.10	0.016	0.07	0.011	5	0.6
Parsley, raw	36	6.3	3.0	0.8	0	56	554	138	50	6.20	0.149	1.07	0.160	58	0.1
Pastrami beef 98% fat-free	95	1.5	19.6	1.2	47	1010	228	9	18	2.78	0.079	4.26	0.013	150	10.4
Pate de foie gras, canned, smoked	462	4.7	11.4	43.8	150	697	138	70	13	5.50	0.400	0.92	0.120	200	44.0
Pate, liver, not specified, canned	319	1.5	14.2	28.0	255	697	138	70	13	5.50	0.400	2.85	0.120	200	41.6
Peaches, raw	39	9.5	0.9	0.3	0	0	190	6	9	0.25	0.068	0.17	0.061	20	0.1
Peanut butter, chunk style, w/o salt	589	21.6	24.1	49.9	0	17	745	45	160	1.90	0.578	2.79	1.800	319	8.2
Peanut butter, smooth style, w/o salt	588	19.6	25.1	50.4	0	17	649	43	154	1.87	0.473	2.91	1.466	358	5.6
Peanuts, all types, dry-roasted, w. salt	585	21.5	23.7	49.7	0	813	658	54	176	2.26	0.671	3.31	2.083	358	7.5
Peanuts, all types, oil-roasted, w. salt	599	15.3	28.0	52.5	0	320	726	61	176	1.52	0.533	3.28	1.845	397	3.3
Pears, raw	58	15.5	0.4	0.1	0	1	119	9	7	0.17	0.082	0.10	0.049	11	0.1
Peas, edible-podded, boiled, w. salt	42	7.1	3.3	0.2	0	240	240	42	26	1.97	0.077	0.37	0.168	55	0.7
Peas, edible-podded, raw	42	7.6	2.8	0.2	0	4	200	43	24	2.08	0.079	0.27	0.244	53	0.7
Peas, green, boiled, w/o salt	84	15.6	5.4	0.2	0	3	271	27	39	1.54	0.173	1.19	0.525	117	1.9
Pecans, dry roasted, w. salt	710	13.6	9.5	74.3	0	383	424	72	132	2.80	1.167	5.07	3.933	293	4.0
Pecans, oil roasted, w. salt	715	13.0	9.2	75.2	0	393	392	67	121	2.47	1.200	4.47	3.700	263	6.0
Peppers, hot chili, green, raw	40	9.5	2.0	0.2	0	7	340	18	25	1.20	0.174	0.30	0.237	46	0.5
Peppers, hot chili, red, raw	40	8.8	1.9	0.4	0	9	322	14	23	1.03	0.129	0.26	0.187	43	0.5
Peppers, jalapeno, raw	30	5.9	1.4	0.6	0	1	215	10	19	0.70	0.133	0.23	0.250	31	0.3
Peppers, sweet, green, boiled, w/o salt	28	6.7	0.9	0.2	0	2	166	9	10	0.46	0.065	0.12	0.115	18	0.3
Peppers, sweet, green, raw	20	4.6	0.9	0.2	0	3	175	10	10	0.34	0.066	0.13	0.122	20	0.0
Peppers, sweet, red, boiled, w/o salt	28	6.7	0.9	0.2	0	2	166	9	10	0.46	0.065	0.12	0.115	18	0.3
Peppers, sweet, red, raw	26	6.0	1.0	0.3	0	2	211	7	12	0.43	0.017	0.25	0.112	26	0.1
Peppers, sweet, yellow, raw	27	6.3	1.0	0.2	0	2	212	11	12	0.46	0.107	0.17	0.117	24	0.3
Persimmons, native, raw	127	33.5	0.8	0.4	0	1	310	27		2.50				26	
Pheasant, cooked, total edible	247	0.0	32.4	12.1	89	43	271	16	22	1.43	0.084	1.37		242	20.7
Pickle, cucumber, sour	11	2.3	0.3	0.2	0	1208	23	0	4	0.40	0.085	0.02	0.011	14	0.0
Pickle, cucumber, sweet	117	31.8	0.4	0.3	0	939	32	4	4	0.59	0.105	0.08	0.015	12	0.0
Pickles, cucumber, dill	18	4.1	0.6	0.2	0	1282	116	9	11	0.53	0.079	0.14	0.015	21	0.0
Pie, apple, commercially prepared, enriched flour	237	34.0	1.9	11.0	0	266	65	11	7	0.45	0.046	0.16	0.182	24	1.0
Pie, blueberry, commercially prepared	232	34.9	1.8	10.0	0	325	50	8	5	0.30	0.046	0.16	0.176	23	1.4
Pie, cherry, commercially prepared	260	39.8	2.0	11.0	0	246	81	12	8	0.48	0.040	0.18	0.140	29	1.2
Pie, chocolate creme, commercially prepared	304	33.6	2.6	19.4	5	136	127	36	21	1.07	0.050	0.23	0.200	68	7.5
Pie, coconut creme, commercially prepared	298	37.2	2.1	16.6	0	255	65	29	20	0.80	0.068	0.47	0.438	85	5.3
Pie, egg custard, commercially prepared	210	20.8	5.5	11.6	33	240	106	80	11	0.58	0.024	0.52	0.060	112	7.1
Pie, lemon meringue, commercially prepared	268	47.2	1.5	8.7	45	146	89	56	15	0.61	0.001	0.49	0.060	105	3.0

Food description	Energy kcal/ 100 g	Carb. g/ 100 g	Protein g/ 100 g	Fat g/ 100 g	Chol. mg/ 100 g	Na mg/ 100 g	K mg/ 100 g	Ca mg/ 100g	Mg mg/ 100 g	Fe mg/ 100 g	Cu mg/ 100 g	Zn mg/ 100 g	Mn mg/ 100 g	P mg/ 100 g	Se µg/ 100 g
Pie, mince, prepared from recipe	289	48.0	2.6	10.8	0	254	203	22	14	1.49	0.113	0.22	0.263	42	6.6
Pie, peach	223	32.9	1.9	10.0	0	270	125	8	6	0.50	0.053	0.09	0.152	22	1.3
Pie, pecan, commercially prepared	400	57.2	4.0	18.5	32	424	74	17	18	1.04	0.195	0.57	0.789	77	5.0
Pie, pumpkin, commercially prepared	210	27.3	3.9	9.5	20	282	154	60	15	0.79	0.048	0.45	0.240	71	2.6
Pike, northern, cooked, dry heat	113	0.0	24.7	0.9	50	49	331	73	40	0.71	0.065	0.86	0.310	282	16.2
Pike, walleye, raw	93	0.0	19.1	1.2	86	51	389	110	30	1.30	0.178	0.62	0.800	210	12.6
Pimento, canned	23	5.1	1.1	0.3	0	14	158	6	6	1.68	0.049	0.19	0.092	17	0.2
Pine nuts, pinyon, dried	629	19.3	11.6	61.0	0	72	628	8	234	3.06	1.035	4.28	4.333	35	
Pineapple juice, canned, unsweetened, w/o vitamin C	53	12.9	0.4	0.1	0	2	130	13	12	0.31	0.069	0.11	0.504	8	0.1
Pineapple, raw, all variety	48	12.6	0.5	0.1	0	1	115	13	12	0.28	0.099	0.10	1.177	8	0.1
Pineapple, raw, traditional variety	45	11.8	0.6	0.1	0	1	125	13	12	0.25	0.081	0.08	1.593	9	0.0
Pistachio nuts, dry roasted, w/o salt	571	27.7	21.4	46.0	0	10	1042	110	120	4.20	1.325	2.30	1.275	485	9.3
Plantains, cooked	116	31.2	0.8	0.2	0	5	465	2	32	0.58	0.066	0.13		28	1.4
Plums, dried (prunes), uncooked	240	63.9	2.2	0.4	0	2	732	43	41	0.93	0.281	0.44	0.299	69	0.3
Plums, raw	46	11.4	0.7	0.3	0	0	157	6	7	0.17	0.057	0.10	0.052	16	0.0
Pollock, Atlantic, cooked, dry heat	118	0.0	24.9	1.3	91	110	456	77	86	0.59	0.064	0.60	0.019	283	46.8
Pollock, walleye, cooked, dry heat	113	0.0	23.5	1.1	96	116	387	6	73	0.28	0.055	0.60	0.020	482	43.4
Pomegranates, raw	68	17.2	1.0	0.3	0	3	259	3	3	0.30	0.070	0.12		8	0.6
Pompano, Florida, cooked, dry heat	211	0.0	23.7	12.1	64	76	636	43	31	0.67	0.078	0.69	0.025	341	46.8
Popcorn, air-popped	387	77.8	12.9	4.5	0	8	329	7	144	3.19	0.262	3.08	1.113	358	0.0
Pork sausage, fresh, cooked	339	0.0	19.4	28.4	84	749	294	13	17	1.36	0.086	2.08	0.005	163	0.0
Pork, cured, bacon, broiled, pan-fried or roasted	541	1.4	37.0	41.8	110	2310	565	11	33	1.44	0.164	3.50	0.022	533	62.0
Pork, cured, breakfast strips, cooked	459	1.1	29.0	36.7	105	2099	466	14	26	1.97	0.153	3.68	0.044	265	24.7
Pork, cured, Canadian-style bacon, grilled	185	1.4	24.2	8.4	58	1546	390	10	21	0.82	0.054	1.70	0.027	296	24.7
Pork, cured, ham, boneless, extra lean (approx 5% fat), roasted	145	1.5	20.9	5.5	53	1203	287	8	14	1.48	0.079	2.88	0.054	196	19.5
Pork, cured, ham, boneless, regular (approx 11% fat), roasted	178	0.0	22.6	9.0	59	1500	409	8	22	1.34	0.145	2.47	0.041	281	19.8
Pork, cured, ham, whole, lean & fat, roasted	243	0.0	21.6	16.8	62	1187	286	7	19	0.87	0.083	2.32	0.014	214	22.7
Pork, cured, salt pork, raw	748	0.0	5.1	80.5	86	1424	66	6	7	0.44	0.050	0.90	0.005	52	5.8
Pork, fresh, composite of retail cuts (leg, loin, & shoulder), lean, cooked	212	0.0	29.3	9.7	86	59	375	21	26	1.10	0.061	2.97	0.018	237	45.0
Pork, fresh, leg (ham), rump half, lean & fat, roasted	252	0.0	28.9	14.3	96	62	374	12	27	1.05	0.103	2.82	0.023	272	46.8
Pork, fresh, leg (ham), shank half, lean & fat, roasted	289	0.0	25.3	20.1	92	59	338	15	22	0.98	0.098	3.06	0.028	257	43.3
Pork, fresh, loin, blade (chops), bone-in, lean & fat, broiled	320	0.0	22.5	24.9	86	70	344	29	22	0.93	0.083	3.37	0.008	212	37.1
Pork, fresh, loin, center loin (chops), bone-in, lean & fat, broiled	240	0.0	28.7	13.1	82	58	358	33	25	0.80	0.046	2.26	0.003	232	44.3
Pork, fresh, loin, center rib (chops), bone-in, lean, broiled	219	0.0	30.8	9.7	81	65	420	31	28	0.82	0.070	2.38	0.020	245	47.3
Pork, fresh, loin, center rib (chops), boneless, lean & fat, broiled	260	0.0	27.6	15.8	82	62	401	28	26	0.77	0.068	2.26	0.018	237	44.0
Pork, fresh, loin, center rib (roasts), boneless, lean & fat, roasted	252	0.0	27.0	15.2	81	48	346	6	22	0.93	0.016	2.64	0.010	214	40.3
Pork, fresh, loin, country-style ribs, lean & fat, braised	296	0.0	23.9	21.5	87	59	328	29	17	1.22	0.093	3.56	0.012	167	39.7
Pork, fresh, loin, sirloin (chops), bone-in, lean & fat, broiled	259	0.0	26.7	16.1	86	68	383	17	29	0.99	0.058	2.57	0.010	246	47.7
Pork, fresh, loin, sirloin (chops), boneless, lean & fat, broiled	208	0.0	30.5	8.6	91	56	372	18	27	1.21	0.053	2.62	0.003	243	50.5
Pork, fresh, loin, tenderloin, lean & fat, broiled	201	0.0	29.9	8.1	94	64	444	5	35	1.39	0.067	2.89	0.012	290	47.7
Pork, fresh, spareribs, lean & fat, braised	397	0.0	29.1	30.3	121	93	320	47	24	1.85	0.142	4.60	0.014	261	37.4
Potato chips, plain, salted	547	49.7	6.6	37.5	0	525	1642	24	70	1.61	0.398	2.39	0.664	155	8.1
Potatoes, baked, flesh & skin, w. salt	93	21.2	2.5	0.1	0	244	535	15	28	1.08	0.118	0.36	0.219	70	0.4
Potatoes, boiled in skin, flesh, w. salt	87	20.1	1.9	0.1	0	240	379	5	22	0.31	0.188	0.30	0.138	44	0.3
Potatoes, boiled w/o skin, flesh, w. salt	86	20.0	1.7	0.1	0	241	328	8	20	0.31	0.167	0.27	0.140	40	0.3
Potatoes, French fried, all types, frozen, oven-heated	172	28.7	2.7	5.2	0	32	451	12	26	0.74	0.135	0.38	0.210	97	0.2

Food description	Energy kcal/ 100 g	Carb. g/ 100 g	Protein g/ 100 g	Fat g/ 100 g	Chol. mg/ 100 g	Na mg/ 100 g	K mg/ 100 g	Ca mg/ 100g	Mg mg/ 100 g	Fe mg/ 100 g	Cu mg/ 100 g	Zn mg/ 100 g	Mn mg/ 100 g	P mg/ 100 g	Se µg/ 100 g
Potatoes, hashed brown, home-prepared	265	35.1	3.0	12.5	0	342	576	14	35	0.55	0.293	0.47	0.247	70	0.5
Potatoes, mashed, home-prepared, whole milk & margarine added	113	16.9	2.0	4.2	1	333	328	22	19	0.26	0.153	0.30	0.112	49	0.8
Potatoes, red, flesh & skin, baked	89	19.6	2.3	0.2	0	8	545	9	28	0.70	0.174	0.40	0.173	72	0.5
Potatoes, white, flesh & skin, baked	94	21.1	2.1	0.2	0	7	544	10	27	0.64	0.127	0.35	0.189	75	0.5
Pretzels, hard, plain, salted	380	79.8	10.3	2.6	0	1357	146	18	29	5.20	0.264	1.43	1.789	113	6.0
Prunes, dehydrated (low-moisture), stewed	113	29.7	1.2	0.2	0	2	353	24	21	1.17	0.204	0.25	0.104	37	
Pumpkin, boiled, w/o salt	20	4.9	0.7	0.1	0	1	230	15	9	0.57	0.091	0.23	0.089	30	0.2
Pumpkin, raw	26	6.5	1.0	0.1	0	1	340	21	12	0.80	0.127	0.32	0.125	44	0.3
Quail, cooked, total edible	234	0.0	25.1	14.1	86	52	216	15	22	4.43	0.592	3.10		279	21.8
Quinces, raw	57	15.3	0.4	0.1	0	4	197	11	8	0.70	0.130	0.04		17	0.6
Radicchio, raw	23	4.5	1.4	0.3	0	22	302	19	13	0.57	0.341	0.62	0.138	40	0.9
Radishes, raw	16	3.4	0.7	0.1	0	39	233	25	10	0.34	0.050	0.28	0.069	20	0.6
Raisins, golden seedless	302	79.5	3.4	0.5	0	12	746	53	35	1.79	0.363	0.32	0.308	115	0.7
Raisins, seeded	296	78.5	2.5	0.5	0	28	825	28	30	2.59	0.302	0.18	0.267	75	0.6
Raisins, seedless	299	79.2	3.1	0.5	0	11	749	50	32	1.88	0.318	0.22	0.299	101	0.6
Raspberries, raw	52	11.9	1.2	0.7	0	1	151	25	22	0.69	0.090	0.42	0.670	29	0.2
Rhubarb, frozen, cooked, w/sugar	116	31.2	0.4	0.1	0	1	96	145	12	0.21	0.027	0.08	0.073	8	0.9
Rhubarb, raw	21	4.5	0.9	0.2	0	4	288	86	12	0.22	0.021	0.10	0.196	14	1.1
Rice, brown, long-grain, cooked	111	23.0	2.6	0.9	0	5	43	10	43	0.42	0.100	0.63	0.905	83	9.8
Rice, brown, medium-grain, cooked	112	23.5	2.3	0.8	0	1	79	10	44	0.53	0.081	0.62	1.097	77	
Rice, white, glutinous, cooked	97	21.1	2.0	0.2	0	5	10	2	5	0.14	0.049	0.41	0.262	8	5.6
Rice, white, long-grain, regular, cooked	130	28.2	2.7	0.3	0	1	35	10	12	1.20	0.069	0.49	0.472	43	7.5
Rice, white, medium-grain, cooked	130	28.6	2.4	0.2	0	0	29	3	13	1.49	0.038	0.42	0.377	37	7.5
Rice, white, short-grain, cooked	130	28.7	2.4	0.2	0	0	26	1	8	1.46	0.072	0.40	0.357	33	7.5
Rockfish, Pacific, mixed species, cooked, dry heat	121	0.0	24.0	2.0	44	77	520	12	34	0.53	0.037	0.53	0.020	228	46.8
Rolls, hamburger or hotdog, plain	279	49.5	9.5	4.3	0	479	94	138	21	3.32	0.220	0.66	0.272	62	19.5
Rutabagas, boiled, w/o salt	39	8.7	1.3	0.2	0	20	326	48	23	0.53	0.041	0.35	0.174	56	0.7
Sablefish, cooked, dry heat	250	0.0	17.2	19.6	63	72	459	45	71	1.64	0.028	0.41	0.019	215	46.8
Sablefish, smoked	257	0.0	17.7	20.1	64	737	471	50	74	1.69	0.036	0.43	0.020	222	50.2
Salami, cooked, turkey	152	0.4	15.3	9.4	76	1004	216	40	22	1.25	0.190	2.32	0.020	266	26.4
Salami, dry or hard, pork	407	1.6	22.6	33.7	79	2260	378	13	22	1.30	0.160	4.20	0.070	229	25.4
Salami, dry or hard, pork, beef	385	3.8	23.2	30.1	100	2010	378	8	17	1.51	0.080	3.23		142	26.1
Salami, Italian, pork	425	1.2	21.7	37.0	80	1890	340	10	22	1.52	0.160	4.20	0.070	229	25.4
Salmon, Atlantic, farmed, cooked, dry heat	206	0.0	22.1	12.4	63	61	384	15	30	0.34	0.049	0.43	0.016	252	41.4
Salmon, Atlantic, farmed, raw	183	0.0	19.9	10.9	59	59	362	12	28	0.36	0.049	0.40	0.015	233	36.5
Salmon, Atlantic, wild, cooked, dry heat	182	0.0	25.4	8.1	71	56	628	15	37	1.03	0.321	0.82	0.021	256	46.8
Salmon, Atlantic, wild, raw	142	0.0	19.8	6.3	55	44	490	12	29	0.80	0.250	0.64	0.016	200	36.5
Salmon, Chinook, cooked, dry heat	231	0.0	25.7	13.4	85	60	505	28	122	0.91	0.053	0.56	0.019	371	46.8
Salmon, Chinook, raw	179	0.0	19.9	10.4	50	47	394	26	95	0.25	0.041	0.44	0.015	289	36.5
Salmon, Chinook, smoked	117	0.0	18.3	4.3	23	784	175	11	18	0.85	0.230	0.31	0.017	164	32.4
Salmon, Chinook, smoked (lox)	117	0.0	18.3	4.3	23	2000	175	11	18	0.85	0.230	0.31	0.017	164	38.1
Salmon, chum, cooked, dry heat	154	0.0	25.8	4.8	95	64	550	14	28	0.71	0.071	0.60	0.019	363	46.8
Salmon, coho, farmed, cooked, dry heat	178	0.0	24.3	8.2	63	52	460	12	34	0.39	0.089	0.47	0.021	332	14.1
Salmon, coho, farmed, raw	160	0.0	21.3	7.7	51	47	450	12	31	0.34	0.048	0.43	0.012	292	12.6
Salmon, coho, wild, cooked, dry heat	139	0.0	23.5	4.3	55	58	434	45	33	0.61	0.071	0.56	0.019	322	38.0
Salmon, coho, wild, raw	146	0.0	21.6	5.9	45	46	423	36	31	0.56	0.051	0.41	0.014	262	36.5
Salmon, pink, cooked, dry heat	149	0.0	25.6	4.4	67	86	414	17	33	0.99	0.099	0.71	0.019	295	57.2
Salmon, pink, raw	116	0.0	19.9	3.5	52	67	323	13	26	0.77	0.077	0.55	0.015	230	44.6
Salmon, sockeye, cooked, dry heat	216	0.0	27.3	11.0	87	66	375	7	31	0.55	0.067	0.51	0.020	276	37.8
Salmon, sockeye, raw	168	0.0	21.3	8.6	62	47	391	6	24	0.47	0.052	0.54	0.014	215	33.7
Sausage, chicken, beef, pork, skinless, smoked	216	8.1	13.6	14.3	120	1034	246	100	14	4.80	0.063	2.68	0.015	132	20.2
Sausage, Italian, pork, cooked	344	4.3	19.1	27.3	57	1207	304	21	18	1.43	0.080	2.39		170	22.0
Sausage, Italian, sweet, links	149	2.1	16.1	8.4	30	570	194	25	12	1.19	0.040	1.52	0.007	103	10.8

Food description	Energy kcal/ 100 g	Carb. g/ 100 g	Protein g/ 100 g	Fat g/ 100 g	Chol. mg/ 100 g	Na mg/ 100 g	K mg/ 100 g	Ca mg/ 100g	Mg mg/ 100 g	Fe mg/ 100 g	Cu mg/ 100 g	Zn mg/ 100 g	Mn mg/ 100 g	P mg/ 100 g	Se µg/ 100 g
Sausage, Polish, beef w. chicken, hot	259	3.6	17.6	19.4	66	1540	237	12	14	0.88	0.090	1.93	0.049	136	17.7
Sausage, smoked link sausage, pork & beef	320	2.4	12.0	28.7	58	911	179	12	13	0.75	0.077	1.26	0.048	121	0.0
Sausage, turkey, breakfast links, mild	235	1.6	15.4	18.1	60	585	197	32	25	1.07	0.111	2.13	0.066	185	22.2
Scallops, (bay & sea), steamed	112	0.0	23.2	1.4	53	265	476	115	55	3.00	0.300	3.00		338	27.9
Scrapple, pork	213	14.1	8.1	13.9	49	659	158	7	13	1.89	0.212	1.06		76	17.4
Sea bass, mixed species, cooked, dry heat	124	0.0	23.6	2.6	53	87	328	13	53	0.37	0.024	0.52	0.020	248	46.8
Shad, American, cooked, dry heat	252	0.0	21.7	17.7	96	65	492	60	38	1.24	0.082	0.47	0.054	349	46.8
Shark, mixed species, batter-dipped & fried	228	6.4	18.6	13.8	59	122	155	50	43	1.11	0.042	0.48	0.050	194	34.0
Shortening, frying (heavy duty), beef tallow & cottonseed	900	0.0	0.0	100.0	100	0	0	0	0	0.00		0.00		0	0.0
Shortening, household, soybean (hydrogenated) & palm	884	0.0	0.0	100.0	0	0	0	0	0	0.00	0.000	0.00	0.000	0	0.0
Shrimp, mixed species, cooked, moist heat	99	0.0	20.9	1.1	195	224	182	39	34	3.09	0.193	1.56	0.034	137	39.6
Shrimp, mixed species, imitation, made from surimi	101	9.1	12.4	1.5	36	705	89	19	43	0.60	0.032	0.33	0.011	282	22.9
Shrimp, mixed species, raw	106	0.9	20.3	1.7	152	148	185	52	37	2.41	0.264	1.11	0.050	205	38.0
Smelt, rainbow, cooked, dry heat	124	0.0	22.6	3.1	90	77	372	77	38	1.15	0.178	2.12	0.900	295	46.8
Snails, raw	90	2.0	16.1	1.4	50	70	382	10	250	3.50	0.400	1.00		272	27.4
Snapper, mixed species, cooked, dry heat	128	0.0	26.3	1.7	47	57	522	40	37	0.24	0.046	0.44	0.017	201	49.0
Sour cream, light	136	7.1	3.5	10.6	35	71	212	141	10	0.07	0.016	0.50		71	3.1
Sour cream, reduced fat	181	7.0	7.0	14.1	35	70	211	141	11	0.06	0.010	0.27		85	4.1
Soybeans, green, boiled, w/o salt	141	11.1	12.4	6.4	0	14	539	145	60	2.50	0.117	0.91	0.502	158	1.4
Soybeans, green, raw	147	11.1	13.0	6.8	0	15	620	197	65	3.55	0.128	0.99	0.547	194	1.5
Soybeans, mature cooked, boiled, w/o salt	173	9.9	16.6	9.0	0	1	515	102	86	5.14	0.407	1.15	0.824	245	7.3
Spaghetti, cooked, enriched, w. salt	157	30.6	5.8	0.9	0	128	45	7	18	1.33	0.103	0.50	0.317	58	26.4
Spaghetti, whole-wheat, cooked	124	26.5	5.3	0.5	0	3	44	15	30	1.06	0.167	0.81	1.379	89	25.9
Spinach, boiled, w/o salt	23	3.8	3.0	0.3	0	70	466	136	87	3.57	0.174	0.76	0.935	56	1.5
Spinach, raw	23	3.6	2.9	0.4	0	79	558	99	79	2.71	0.130	0.53	0.897	49	1.0
Squash, summer, all varieties, boiled, w/o salt	20	4.3	0.9	0.3	0	1	192	27	24	0.36	0.103	0.39	0.213	39	0.2
Squash, summer, all varieties, raw	16	3.4	1.2	0.2	0	2	262	15	17	0.35	0.051	0.29	0.175	38	0.2
Squash, summer, zucchini, includes skin, boiled, w/o salt	16	3.9	0.6	0.1	0	3	253	13	22	0.35	0.086	0.18	0.178	40	0.2
Squash, summer, zucchini, includes skin, raw	16	3.4	1.2	0.2	0	10	262	15	17	0.35	0.051	0.29	0.175	38	0.2
Squash, winter, acorn, baked, w/o salt	56	14.6	1.1	0.1	0	4	437	44	43	0.93	0.086	0.17	0.242	45	0.7
Squash, winter, butternut, baked, w/o salt	40	10.5	0.9	0.1	0	4	284	41	29	0.60	0.065	0.13	0.172	27	0.5
Squash, winter, Hubbard, baked, w/o salt	50	10.8	2.5	0.6	0	8	358	17	22	0.47	0.045	0.15	0.170	23	0.6
Squash, zucchini, baby, raw	21	3.1	2.7	0.4	0	3	459	21	33	0.79	0.097	0.83	0.196	93	0.3
Strawberries, raw	32	7.7	0.7	0.3	0	1	153	16	13	0.42	0.048	0.14	0.386	24	0.4
Sturgeon, mixed species, cooked, dry heat	135	0.0	20.7	5.2	77	69	364	17	45	0.90	0.053	0.54	0.030	271	16.2
Sturgeon, mixed species, raw	105	0.0	16.1	4.0	60	54	284	13	35	0.70	0.041	0.42	0.025	211	12.6
Sturgeon, mixed species, smoked	173	0.0	31.2	4.4	80	739	379	17	47	0.93	0.050	0.56	0.030	281	20.1
Sugars, brown	377	97.3	0.0	0.0	0	39	346	85	29	1.91	0.298	0.18	0.320	22	1.2
Sugars, granulated	387	100.0	0.0	0.0	0	0	2	1	0	0.01	0.000	0.00	0.000	0	0.6
Sugars, maple	354	90.9	0.1	0.2	0	11	274	90	19	1.61	0.099	6.06	4.422	3	0.8
Sweet potato, baked in skin, w/o salt	90	20.7	2.0	0.2	0	36	475	38	27	0.69	0.161	0.32	0.497	54	0.2
Sweet potato, boiled, w/o skin	76	17.7	1.4	0.1	0	27	230	27	18	0.72	0.094	0.20	0.266	32	0.2
Swordfish, cooked, dry heat	155	0.0	25.4	5.1	50	115	369	6	34	1.04	0.162	1.47	0.020	337	61.7
Swordfish, raw	121	0.0	19.8	4.0	39	90	288	4	27	0.81	0.126	1.15	0.019	263	48.1
Syrups, corn, dark	286	77.6	0.0	0.0	0	155	44	18	8	0.37	0.053	0.04	0.100	11	2.9
Syrups, corn, high-fructose	281	76.0	0.0	0.0	0	2	0	0	0	0.03	0.029	0.02	0.094	0	0.7
Syrups, maple	261	67.1	0.0	0.2	0	9	204	67	14	1.20	0.074	4.16	3.298	2	0.6
Syrups, table blends, cane & 15% maple	278	69.5	0.0	0.0	0	104	53	12	2	0.19	0.017	0.63	0.495	0	0.5
Tangerines, (mandarin oranges), raw	53	13.3	0.8	0.3	0	2	166	37	12	0.15	0.042	0.07	0.039	20	0.1
Tempeh, cooked	196	9.4	18.2	11.4		14	401	96	77	2.13	0.540	1.57	1.285	253	0.0
Tilefish, cooked, dry heat	147	0.0	24.5	4.7	64	59	512	26	33	0.31	0.052	0.53	0.015	236	51.5
Tilefish, raw	96	0.0	17.5	2.3	50	53	433	26	28	0.25	0.041	0.37	0.010	187	36.5

Food description	Energy kcal/ 100 g	Carb. g/ 100 g	Protein g/ 100 g	Fat g/ 100 g	Chol. mg/ 100 g	Na mg/ 100 g	K mg/ 100 g	Ca mg/ 100g	Mg mg/ 100 g	Fe mg/ 100 g	Cu mg/ 100 g	Zn mg/ 100 g	Mn mg/ 100 g	P mg/ 100 g	Se µg/ 100 g
Tofu, fried	271	10.5	17.2	20.2	0	16	146	372	60	4.87	0.398	1.99	1.495	287	28.5
Tomato juice, canned, w. salt	17	4.2	0.8	0.1	0	269	229	10	11	0.43	0.061	0.15	0.070	18	0.3
Tomatoes, green, raw	23	5.1	1.2	0.2	0	13	204	13	10	0.51	0.090	0.07	0.100	28	0.4
Tomatoes, orange, raw	16	3.2	1.2	0.2	0	42	212	5	8	0.47	0.062	0.14	0.088	29	0.4
Tomatoes, red, ripe, cooked	18	4.0	1.0	0.1	0	11	218	11	9	0.68	0.075	0.14	0.105	28	0.5
Tomatoes, red, ripe, raw, year-round average	18	3.9	0.9	0.2	0	5	237	10	11	0.27	0.059	0.17	0.114	24	0.0
Tomatoes, sun-dried	258	55.8	14.1	3.0	0	2095	3427	110	194	9.09	1.423	1.99	1.846	356	5.5
Tomatoes, yellow, raw	15	3.0	1.0	0.3	0	23	258	11	12	0.49	0.101	0.28	0.120	36	0.4
Trout, mixed species, cooked, dry heat	190	0.0	26.6	8.5	74	67	463	55	28	1.92	0.241	0.85	1.091	314	16.2
Trout, rainbow, farmed, cooked, dry heat	169	0.0	24.3	7.2	68	42	441	86	32	0.33	0.061	0.49	0.020	266	15.0
Trout, rainbow, wild, cooked, dry heat	150	0.0	22.9	5.8	69	56	448	86	31	0.38	0.058	0.51	0.021	269	13.2
Tuna, fresh, bluefin, cooked, dry heat	184	0.0	29.9	6.3	49	50	323	10	64	1.31	0.110	0.77	0.020	326	46.8
Tuna, fresh, bluefin, raw	144	0.0	23.3	4.9	38	39	252	8	50	1.02	0.086	0.60	0.015	254	36.5
Tuna, fresh, skipjack, raw	103	0.0	22.0	1.0	47	37	407	29	34	1.25	0.086	0.82	0.015	222	36.5
Tuna, fresh, yellowfin, raw	108	0.0	23.4	1.0	45	37	444	16	50	0.73	0.064	0.52	0.015	191	36.5
Tuna, skipjack, fresh, cooked, dry heat	132	0.0	28.2	1.3	60	47	522	37	44	1.60	0.110	1.05	0.019	285	46.8
Tuna, white, canned in oil	186	0.0	26.5	8.1	31	396	333	4	34	0.65	0.130	0.47	0.016	267	60.1
Tuna, white, canned in water	128	0.0	23.6	3.0	42	377	237	14	33	0.97	0.039	0.48	0.019	217	65.7
Tuna, yellowfin, fresh, cooked, dry heat	139	0.0	30.0	1.2	58	47	569	21	64	0.94	0.082	0.67	0.019	245	46.8
Turbot, European, cooked, dry heat	122	0.0	20.6	3.8	62	192	305	23	65	0.46	0.047	0.28	0.022	165	46.8
Turkey breast meat	104	4.2	17.1	1.7	43	1015	302	8	21	1.44	0.057	1.33	0.018	162	22.8
Turkey, all classes, breast, meat & skin, roasted	189	0.0	28.7	7.4	74	63	288	21	27	1.40	0.047	2.03	0.020	210	29.1
Turkey, all classes, dark meat, meat & skin, raw	160	0.0	18.9	8.8	72	71	261	17	20	1.69	0.137	2.95	0.021	170	26.4
Turkey, all classes, leg, meat & skin, roasted	208	0.0	27.9	9.8	85	77	280	32	23	2.30	0.154	4.27	0.023	199	37.8
Turkey, all classes, meat only, roasted	170	0.0	29.3	5.0	76	70	298	25	26	1.78	0.094	3.10	0.021	213	36.8
Turkey, all classes, wing, meat & skin, roasted	229	0.0	27.4	12.4	81	61	266	24	25	1.46	0.056	2.10	0.020	197	29.9
Turnip greens, boiled, w/o salt	20	4.4	1.1	0.2	0	29	203	137	22	0.80	0.253	0.14	0.337	29	0.9
Turnips, boiled, w/o salt	22	5.1	0.7	0.1	0	16	177	33	9	0.18	0.002	0.12	0.071	26	0.2
Turnips, raw	28	6.4	0.9	0.1	0	67	191	30	11	0.30	0.085	0.27	0.134	27	0.7
Veal, composite of retail cuts, fat, cooked	642	0.0	9.4	66.7	73	57	173	4	10	1.00	0.044	0.87	0.012	116	5.5
Veal, composite of retail cuts, lean & fat, cooked	231	0.0	30.1	11.4	114	87	325	22	26	1.15	0.114	4.76	0.036	239	12.3
Veal, composite of retail cuts, lean, cooked	196	0.0	31.9	6.6	118	89	338	24	28	1.16	0.120	5.10	0.038	250	13.0
Veal, ground, broiled	172	0.0	24.4	7.6	103	83	337	17	24	0.99	0.103	3.87	0.035	217	13.7
Veal, leg (top round), lean & fat, roasted	160	0.0	27.7	4.7	103	68	389	6	28	0.91	0.129	3.04	0.030	234	11.2
Veal, loin, lean & fat, roasted	217	0.0	24.8	12.3	103	93	325	19	25	0.87	0.110	3.03	0.029	212	11.0
Veal, rib, lean & fat, roasted	228	0.0	24.0	14.0	110	92	295	11	22	0.97	0.099	4.09	0.030	197	10.5
Veal, shoulder, arm, lean & fat, roasted	183	0.0	25.5	8.3	108	90	348	26	26	1.15	0.141	4.18	0.030	221	11.1
Veal, sirloin, lean & fat, roasted	202	0.0	25.1	10.5	102	83	351	13	26	0.92	0.129	3.35	0.029	223	11.1
Veal, sirloin, lean, roasted	168	0.0	26.3	6.2	104	85	365	14	27	0.91	0.136	3.54	0.030	231	11.5
Vinegar, distilled	18	0.0	0.0	0.0	0	2	2	6	1	0.03	0.006	0.01	0.055	4	0.5
Watercress, raw	11	1.3	2.3	0.1	0	41	330	120	21	0.20	0.077	0.11	0.244	60	0.9
Watermelon, raw	30	7.6	0.6	0.2	0	1	112	7	10	0.24	0.042	0.10	0.038	11	0.4
Wheat flour, whole-grain	339	72.6	13.7	1.9	0	5	405	34	138	3.88	0.382	2.93	3.799	346	70.7
Wheat flours, bread, unenriched	361	72.5	12.0	1.7	0	2	100	15	25	0.90	0.182	0.85	0.792	97	39.7
Wolffish, Atlantic, cooked, dry heat	123	0.0	22.4	3.1	59	109	385	8	38	0.12	0.037	1.00	0.019	256	46.8
Wolffish, Atlantic, raw	96	0.0	17.5	2.4	46	85	300	6	30	0.09	0.029	0.78	0.015	200	36.5
Yam, boiled, or baked, w/o salt	116	27.6	1.5	0.1	0	8	670	14	18	0.52	0.152	0.20	0.371	49	0.7
Yellowtail, mixed species, cooked, dry heat	187	0.0	29.7	6.7	71	50	538	29	38	0.63	0.058	0.67	0.019	201	46.8
Yogurt, fruit varieties, non-fat	94	19.0	4.4	0.2	2	58	194	152	15	0.07	0.011	0.74	0.035	119	6.0
Yogurt, plain, low fat, 12 grams protein per 8 oz	63	7.0	5.3	1.6	6	70	234	183	17	0.08	0.013	0.89	0.004	144	3.3
Yogurt, plain, skim milk, 13 grams protein per 8 oz	56	7.7	5.7	0.2	2	77	255	199	19	0.09	0.015	0.97	0.005	157	3.6
Yogurt, plain, whole milk, 8 grams protein per 8 oz	61	4.7	3.5	3.3	13	46	155	121	12	0.05	0.009	0.59	0.004	95	2.2

Section 8
Analytical Chemistry

INTRODUCTION

Thomas J. Bruno

For the 93rd edition of the *Handbook*, we have undertaken a significant revision and reorganization of the information in the area of analytical chemistry. Some information that was previously in other sections (such as spectroscopy), but which is most often used in chemical analysis, has now been compiled into this chapter. Some older tables and charts have been replaced with updated revisions, and some of limited utility have been deleted. We recognize that very detailed information is available in other treatises to guide the design and interpretation of analytical results. Indeed, some of these works are specifically designed to assist the user at decision points in either analytical methods development (design) or interpretation of results. In this section of the *Handbook*, we have adopted a different philosophy, and have focused on providing general information for the interpretation of analytical data, especially from instrumental techniques. In keeping with the focus on interpretation, we have limited the information provided on the execution of analytical procedures, although we recognize that there will be overlap. We have deliberately chosen to exclude information that is merely interesting, but of little value for interpretation. Similarly, we have not included information that is highly specific, and focused on families of analytes rather than on individual analytes.

ABBREVIATIONS AND SYMBOLS USED IN ANALYTICAL CHEMISTRY

Abbreviations Used in Analytical Chemistry

AAS	atomic absorption spectroscopy
AC	alternating current
ACP	alternating current plasma
ADXPS	angular dependent X-ray photoelectron spectroscopy
AED	atomic emission detector
AEM	analytical electron microscope (microscopy)
AES	Auger electron spectroscopy, atomic emission spectroscopy
AFID	alkali flame ionization detector
AFM	atomic force microscopy
AFS	atomic force spectroscopy
AM	amplitude modulation
AMS	accelerator mass spectrometry
AOTF	acousto-optical tunable filter
APCI	atmospheric pressure chemical ionization
API	atmospheric pressure ionization
APSTM	analytical photon scanning tunneling microscope
ARM	atomic resolution microscopy
ARPES	angle resolved photoelectron spectroscopy
ARUPS	angle resolved ultraviolet photoelectron spectroscopy
ASE	accelerated solvent extraction
AsFlFFF	asymmetrical flow field flow fractionation
ATD	above-threshold dissociation
ATI	above-threshold ionization
ATR	attenuated total reflection
BB	band broadening
BET	Brunauer-Emmett-Teller (adsorption isotherm)
BIFL	burst integrated fluorescence lifetime
BIS	Bremsstrahlung isochromat spectroscopy
BL	bioluminescence
BLRF	bispectral luminescence radiance factor
CAR	continuous addition of reagent
CARS	coherent anti-Stokes Raman spectroscopy
CCC	counter-current chromatography
CCD	charge-coupled device
CCT	constant current topography
CD	circular dichroism
CE	capillary electrophoresis, counter electrode
CEC	capillary electrokinetic chromatography, capillary electrochromatography
CED	cohesive energy density
CFA	continuous flow analysis
CF-FAB	continuous flow-fast atom bombardment
CFM	chemical force microscopy
CGE	capillary gel electrophoresis
CHEMFET	chemical-sensing field effect transistor
CI	chemical ionization
CID	collision induced dissociation
CIEF	capillary isoelectric focusing
CITP	capillary isotachophoresis
CL	chemiluminescence
CLLE	continuous liquid-liquid extraction
CMA	cylindrical mirror analyzer
COSY	correlation spectroscopy
CPAA	charged particle activation analysis
CP/MAS	cross polarization/magic angle spinning
CRDS	cavity ring-down spectroscopy

CRM	certified reference material
CT	cryogenic trapping
CTD	charge transfer device
CV	cyclic voltammetry
CV-AAS	cold vapor atomic absorption spectrometry
CVD	chemical vapor deposition
CW	continuous wave
CZE	capillary zone electrophoresis
DA	diode array
DAD	diode array detector (UV-Vis)
DBE	double bond equivalent
DC	direct current
DCI	desorption chemical ionization
DCP	direct-current plasma
DEPT	distortionless enhancement by polarization transfer
DETA	dielectric thermal analysis
DIN	direct injection nebulizer
DLI	direct liquid introduction
DMA	dynamic mechanical analysis
DME	dropping mercury electrode
DNMR	dynamic nuclear magnetic resonance
DPP	differential pulse polarography
DRIFT	diffuse-reflectance infrared Fourier transform
DSC	differential scanning calorimetry
DTA	differential thermal analysis
DTC	differential thermal calorimetry
EC	electrochemical
ECD	electron capture detector
ECMS	electron capture mass spectrometry
ECNIMS	electron capture negative ionization mass spectrometry
EDL	electrodeless discharge lamp
EDS	energy-dispersive spectrometer
EDXRF	energy-dispersive X-ray fluorescence
EELS	electron energy-loss spectroscopy
EFFF	electric field flow fractionation
EGA	evolved gas analysis
EIA	enzyme-linked immunoassay
EI(I)	electron impact (ionization)
EIMS	electron impact mass spectrometry
ELCD	electrolytic conductivity detector
ELISA	enzyme-linked immunosorbent assay
ELSD	evaporative light scattering detector
EM	electron microscopy
EMIRS	electrochemically modulated IR spectroscopy
EOF	electro-osmotic flow
EPL	enhanced photoactivated luminescence
EPMA	electron-probe microanalysis
EPR	electron paramagnetic resonance
EPXMA	electron probe x-ray microanalysis
EQL	estimated quantitation limit
ERD	elastic recoil detection
ESA	electrostatic analyzer
ESCA	electron spectroscopy for chemical analysis
ESEM	environmental scanning electron microscope
ESI	electro-spray ionization
ESP	electrospray
ESR	electron spin resonance
ETA	electrothermal analyzer, emanation thermal analysis
EXAFS	extended X-ray absorption fine structure

FAA	flame atomic absorption
FAAS	flame atomic absorption spectroscopy
FABMS	fast-atom bombardment mass spectrometry
FAES	flame atomic emission spectroscopy
FAFS	flame atomic fluoroescence spectroscopy
FAM	field analytical method
FAS	flame absorption spectroscopy
FD	field desorption
FES	flame emission spectroscopy
FFEM	freeze-fracture electron microscopy
FFF	field flow fractionation
FFFF	flow field flow fractionation
FFM	friction force microscopy
FFS	flame fluorescence spectroscopy
FFT	fast Fourier transform
FGC	fast gas chromatography
FI	flow injection
FIA	flow injection analysis
FIB	focused ion beam
FID	flame ionization detector, free-induction decay
FIM	field ion microscopy
FlFFF	flow field flow fractionation
FNAA	fast neutron activation analysis
FOCS	fiber optic chemical sensor
FPD	flame photometric detector
FT	Fourier transform
FT-ICR	Fourier transform ion cyclotron resonance
FT-IR	Fourier transform infrared (often "FT/IR," "FTIR," "FT IR")
FT-IRRAS	FT-IR reflection-absorption spectroscopy
FT-MS	Fourier transform mass spectrometry
FWHM	full-width half-maximum
GC	gas chromatography
GC-IR	gas chromatography–infrared
GCMS	gas chromatography mass spectrometry
GDL	glow discharge lamp
GDMS	glow discharge mass spectrometry
GE	gel electrophoresis
GFAAS	graphite furnace atomic absorption spectroscopy
GLC	gas-liquid chromatography
GPC	gel permeation chromatography
GS	Gram-Schmidt (algorithm)
GSC	gas-solid chromatography
GSED	gaseous secondary electron detector
HCL	hollow cathode lamp
HDC	hydrodynamic chromatography
HETP	height equivalent of (a) theoretical plate(s)
HG	hydride generation
HIC	hydrophoric interaction chromatography
HPAC	high-performance affinity chromatography
HPIAC	high-performance immunoaffinity chromatography
HPLC	high-performance liquid chromatography, high-pressure liquid chromatography
HPTLC	high-performance thin-layer chromatography
HRCGC	high-resolution capillary gas chromatography
HRGC	high-resolution gas(-liquid) chromatography
HS	head space
HSA	hemispherical analyzer
IAC	immunoaffinity chromatography
IC	ion chromatography
ICP	inductively coupled plasma
ICP-OES	ICP optical emission spectrometry
ICR	ion cyclotron resonance
IDMS	isotope dilution mass spectrometry

IEC	ion-exchange chromatography
IEF	isoelectric focusing
IF	intermediate frequency
IGC	inverse gas chromatography
ILDA	intensified linear diode array
IMAC	immobilized metal-ion affinity chromatography
INAA	instrumental neutron activation analysis
IP	ion pairing
IPC	ion-pair chromatography
IPG	immobilized pH gradient
IR	infrared (spectrophotometry)
IRN	indicator radionuclide(s)
IRS	internal reflection spectroscopy
ISCA	ionization spectroscopy for chemical analysis
ISE	ion selective electrode
ISP	ion spray
ISS	ion scattering spectrometry
LAMMS	laser micro mass spectrometry
LARIMS	laser atomization resonance ionization mass spectrometry
LARIS	laser atomization resonance ionization spectroscopy
LASER	light amplification by stimulated emission of radiation
LBB	Lambert-Beer-Bouguer law
LC	liquid chromatography
LC-LS	multidimensional liquid chromatography
LDMS	laser desorption mass spectrometry
LDR	linear dynamic range
LEAFS	laser-excited atomic fluorescence spectrometry
LED	light-emitting diode
LEED	low energy electron diffraction
LEEM	low energy electron microscopy
LEI	laser-enhanced ionization
LESS	laser-excited Shpol'skii spectroscopy
LFM	lateral force microscopy
LIDAR	light detection and ranging
LIFD	laser-induced fluorescence detection
LIMS	laboratory information management system
LLC	liquid-liquid chromatography
LLD	lower-limit detection
LLE	liquid-liquid extraction
LNRI	laser non-resonant ionization
LO	local oscillator
LOC	lab on a chip
LOD	limit of detection
LPDA	linear photodiode array
LPSIRS	linear potential-sweep IR reflectance spectroscopy
LRI	laser resonance ionization
LRMA	laser Raman microanalysis
LSC	liquid-solid chromatography
LSE	liquid-solid extraction
LTP	low-temperature phosphorescence
MAE	microwave assisted extraction
MALDI	matrix-assisted laser desorption source
MCD	magnetic circular dichroism
MCP	microchannel plate
MDGC	multidimensional gas chromatography
MDL	method detection limit
MDM	minimum detectable mass
MDQ	minimum detectable quantity
MEIS	medium energy ion scattering
MEKC	micellar electrokinetic chromatography

MFM	magnetic force microscopy
MID	multiple ion detection
MIP	microwave induced plasma, mercury intrusion porosimetry
MIRS	multiple internal reflection spectroscopy
MLC	micellar liquid chromatography
MLLSQ	multiple linear least squares
MMF	minimum mass fraction
MMLLE	microporous membrane liquid-liquid extraction
MPI	multiphoton ionization
MRI	magnetic resonance imaging
MS	mass spectrometry
MS-MS	tandem mass spectrometry
MSPD	matric solid-phase dispersion
MSRTP	micelle-stabilized room-temperature phosphorescence
MWD	microwave (assisted) digestion
NAA	neutron activation analysis
NCIMS	negative chemical ionization mass spectrometry
NDP	neutron depth profiling
NEXAFS	near edge x-ray absorption fine structure
NHE	normal-hydrogen electrode
NICI	negative ion chemical ionization
NIR	near-infrared, near-IR
NIRA	near-infrared reflectance analysis
nm	nanometer
NMR	nuclear magnetic resonance
NPD	nitrogen-phosphorus detector
NPLC	normal phase liquid chromatography
ODS	octadecylsilane
OES	optical emission spectrometry, optical emission spectroscopy
OID	optoelectronic imaging device
OMA	optical multichannel analyzer
OPO	optical parametric oscillator
OPTLC	over-pressured thin-layer chromatography
ORD	optical rotary dispersion
OTE	optically transparent electrodes
PA	proton affinity
PAA	photon activation analysis
PAGE	polyacrylamide gel electrophoresis
PAH	polycyclic aromatic hydrocarbon
PAS	photoacoustic spectroscopy
PB	particle beam
PC	paper chromatography
PCA	principal component analysis
PCR	polymerase chain reaction
PCS	photon correlation spectroscopy
PCSE	partially coherent solvent evaporation
PD	plasma desorption
PDA	photodiode array
PDHID	pulsed discharge helium ionization detector
PDMS	plasma desorption mass spectrometry, polydimethyl siloxane
PED	pulsed electrochemical detection, plasma emission detector
PES	photoelectron spectroscopy
PET	positron emission tomography
PFIA	process flow injection analysis
PGC	packed-column gas chromatography
pH	negative logarithm of hydrogen ion concentration
PICI	positive ion chemical ionization

PID	photoionization detector
PIXE	particle-induced X-ray emission
pK	negative logarithm of an equilibrium constant
PLE	pressurized liquid extraction
PLOT	porous-layer open tubular
PMT	photomultiplier tube
ppb	parts per billion
ppm	parts per million
ppt	parts per thousand, parts per trillion
PSD	position sensitive detector
PTFE	polytetrafluoroethylene
PTV	programmable temperature vaporizer
PVD	pulsed voltammetric detection, physical vapor deposition
QCL	quantum cascade laser
QCM	quartz-crystal microbalance
QFAA	quartz furnace atomic adsorption
QIT	quadrupole ion trap
QTH	quartz tungsten halogen
RBS	Rutherford backscattering spectrometry
REELS	reflection electron energy loss spectrometry
RES	reflection electron spectrometry
RF	radio frequency
RHEED	reflection high energy electron diffraction
RIC	reconstructed ion chromatogram
RI	refractive index
RID	refractive-index detector
RIMS	resonance ionization mass spectrometry
RIS	resonance ionization spectroscopy
RM	reference material
RNAA	radiochemical neutron activation analysis
ROA	Raman optical activity
RPLC	reversed-phase liquid chromatography
RRDE	rotating ring-disk electrode
RS	Raman spectroscopy
RSF	relative sensitivity factor
RTP	room-temperature phosphorescence
S/N	signal-to-noise ratio
SAE	sonication assisted extraction
SAM	scanning Auger microscopy, self-assembly monolayers
SANS	small-angle neutron scattering
SAW	surface acoustic wave
SAXS	small-angle X-ray scattering
SBSE	stirbar sorptive extraction
SCE	standard calomel electrode, saturated calomel electrode
SCF	supercritical fluid
SCOT	support-coated open tubular
SDD	silicon drift detector
SdFFF	sedimentation field flow fractionation
SEC	size-exclusion chromatography
SEM	scanning electron microscope
SERS	surface-enhanced Raman spectroscopy
SFC	supercritical-fluid chromatography
SFE	supercritical-fluid extraction
SFFF	sedimentation field flow fractionation
SF-MS	sector field mass spectrometry
SFS	synchronous fluorescence spectroscopy
SHE	standard hydrogen electrode
SIA	sequential injection analysis
SIMS	secondary ion mass spectrometry
SIRIS	sputter initiated resonance ionization spectroscopy
SMDE	static mercury drop electrode

SNIFTIRS	subtractively normalized interfacial FT-IR spectroscopy
SNMS	sputtered neutral mass spectrometry
SPE	solid-phase extraction
SPME	solid-phase microextraction
SPR	surface plasmon resonance
SRE	stray radiant energy
SRM	standard reference material
SSMS	spark source mass spectrometry
SSRTF	solid-surface room-temperature fluorescence
SSRTP	solid-surface room-temperature phosphorescence
STEM	scanning transmission electron microscope
STM	scanning tunneling microscope
SVE	solvent vapor exit
SWE	supercritical water extraction
TCD	thermal-conductivity detector
TD	thermodilatometry
TDL	tunable diode laser
TED	thermionic emission detector
TEELS	transmission electron energy loss spectrometry
TEM	transmission electron microscope
TET	thermometric enthalpimetric titration
TFFF	thermal field flow fractionation
TGA	thermogravimetric analysis
TGA-IR	thermogravimetric analysis– infrared
THEED	transmission high energy electron diffraction
ThFFF	thermal field flow fractionation
TIC	total ion current chromatogram, tentatively identified compound
TIMS	thermoionization mass spectrometry
TLC	thin-layer chromatography
TLE	thin-layer electrode
TLM	thermal lens microscopy
TLV	threshold limit value
TMA	thermomechanical analysis
TMS	tetramethylsilane
TOF	time-of-flight
TOF-MS	time-of-flight mass spectrometry
TSP	thermospray
UHV	ultrahigh vacuum
USE	ultrasonic extraction
UV	ultraviolet
UVPES, UPS	ultraviolet photoelectron spectroscopy
UV-VIS, UV-Vis	ultraviolet-visible
VAR	variable angle reflectance
Vis	visible (radiation)
VOC	volatile organic compound(s)
VOX	volatile organic halogens
VUV	vacuum ultraviolet
WCOT	wall-coated open tubular
WDS	wavelength dispersive spectrometer

XANES	X-ray absorption near-edge spectroscopy
XPS	X-ray photoelectron spectroscopy
XRD	X-ray diffraction
XRF	X-ray fluorescence
XRFS	X-ray fluorescence spectroscopy
XRS	X-ray spectroscopy
ZAF	Z (element number) absorption fluorescence

Symbols Used in Analytical Chemistry

a	Auger yield
A	absorbance
A	peak asymmetry factor
B	magnetic field strength
$[c]$	concentration of component c
d_p	particle diameter (HPLC stationary phase)
D_{ab}	diffusion coefficient
e	electron elementary charge
ε	extinction coefficient
E	energy
E_b	binding energy
E	electrode potential
v	frequency
γ	gyromagnetic ratio
n	refractive index
H	enthalpy, plate height
I_0	incident intensity
J	coupling constant
k	coverage factor
k'	capacity factor
m/z	mass-to-elementary-charge ratio (mass spectrometry)
Q_{crit}	Q value (outlier test)
q	quadrupole parameter (mass spectrometry)
r	correlation coefficient
R	resolution
ρ	density
s	standard deviation
s^2	variance
δ	chemical shift
δ^*	solubility parameter
τ_{crit}	Chauvanet's criterion (outlier test)
$t_{1/2}$	half life
t_M	mobile-phase hold up
t_R^0	retention time
t_R^0	specific retention time
T	transmittance
$T1$	spin-lattice relaxation time
$T2$	spin-spin relaxation time
$\bar{u}$	carrier phase velocity
V_M	carrier hold-up volume
V_R	retention volume
V_R^0	specific retention volume
λ	wavelength, thermal conductivity

BASIC INSTRUMENTAL TECHNIQUES OF ANALYTICAL CHEMISTRY

Thomas J. Bruno

The following section provides brief descriptions of the major instrumental methods of chemical analysis. Please note that these paragraphs are general descriptions and are not meant to convey a comprehensive knowledge on these topics. The reader is referred to one of many excellent texts on instrumental methods of chemical analysis for additional details.

Suggested Reading

Skoog, D. A., Holler, F. J., and Crouch, S. R., *Principles of Instrumental Analysis, 6th Edition,* Thomson Brooks/Cole Publishing, Belmont, CA, 2007.

Robinson, J. W., Skelly Frame, E. M., and Frame II, G. M., *Undergraduate Instrumental Analysis, 6th Edition,* CRC Press, Boca Raton, FL, 2004.

Pungor, E., *A Practical Guide to Instrumental Analysis*, CRC Press, Boca Raton, FL, 1994.

Gas Chromatography (GC): A separation method in which the sample or solute is vaporized (usually in a solvent, but sometimes neat or free of solvent) and passed through a medium under the influence of a carrier gas. The medium is called the stationary phase, in contrast to the carrier gas, which is mobile. The most common modern stationary phases are based on open tubular or capillary columns, in which the separation medium coats the inside periphery of a tube (typically tenths of millimeters in inside diameter) that is between 25 m and 60 m long. Older media are packed columns, consisting of packed beds, which are still used for gas analysis. In these applications, a solid sorbent is very common, and this is called gas-solid chromatography. Some open tubular columns are available with solid sorbents as well. Interactions of the solute with the separation medium affect the separation of the components of the mixture. A wide variety of detectors are available for general or specific applications. One of the most useful combinations is gas chromatography coupled with mass spectrometry (GC-MS). Solutes amenable to analysis by GC are usually of moderate volatility and a relative molecular mass usually not exceeding 400. The most common stationary phases are cross-linked polymers based on dimethyl polysiloxane, the backbone of which can be derivatized with ligands to provide specific interactions. It is also possible to incorporate stereogenic (chiral) stationary phases as well.

Liquid Chromatography (LC, HPLC): A separation method in which a sample or solute (usually in a solvent, but sometimes neat or free of solvent) is passed through a medium under the influence of a carrier liquid. The medium is called the stationary phase, in contrast to the carrier, which is the mobile phase. Unlike gas chromatography, where the carrier gas plays little role other than mass transfer, the mobile phase in liquid chromatography is a controllable variable whose polarity and other properties are varied, in addition to the interactions with the stationary phase, to affect separation. The stationary phase in liquid chromatography is usually a micrometer-size particle packed bed that requires a high-pressure solvent system to cause mass flow. Liquid chromatographic systems have therefore been called high-pressure liquid chromatography (HPLC),

although the acronym is usually taken to mean high-performance liquid chromatography. Many variations of the method have been developed for specific analysis. For example, gel permeation chromatography is an adaptation used for the separation of polymers. Affinity chromatography is similar in concept to gel permeation chromatography, but uses the specific interactions between an antibody and an antigen. The use of stereogenic (or chiral) stationary phases is also an important development, especially in the analysis of pharmaceuticals.

Thin-Layer Chromatography (TLC): A separation method in which a stationary phase (typically a polar adsorbent such as alumina or silica gel) coated on a sheet of plastic, aluminum, or glass is used with a mobile phase usually consisting of a solvent or mixture of solvents in a beaker. The solute is applied as a blotted spot just above the end of the adsorbent-coated plate, and then the end is immersed into the solvent (but not so far as to immerse the solute spot). Commercial plates are robust, plastic sheets that can be cut to the desired size. In earlier applications, filter paper has been used in TLC, giving rise to the term "paper chromatography." This is rarely used today. Solvent is then drawn up through the adsorbent coating by capillary action. Separation results from a combination of interactions with the adsorbent and solvent. Commonly, the separated components are rendered visible by spraying stains or reactants on the plate after separation has been completed. It is also common to view the "developed" plate under an ultraviolet lamp, to visualize spots that may be fluorescent.

Supercritical Fluid Chromatography (SFC) and Supercritical Fluid Extraction (SFE): A separation technology similar to other extraction and chromatographic methods, but in which the mobile phase is actually a fluid in its supercritical fluid state. A supercritical fluid is a fluid that is held above its critical temperature and pressure, and for which no application of additional pressure can result in the development of a liquid phase. Supercritical fluids are unique in that while they possess liquid-like densities, the mass transfer behavior is superior to that of liquids. Supercritical fluid chromatography remains a niche method that is applicable to pharmaceuticals and other high relative molecular mass solutes. Supercritical extraction, on the other hand, is more widely used as a sample preparation method, especially in pharmaceutical analyses, polymers, and environmental analyses.

Electrophoresis: A family of separation methods based on the motion induced in particles by an applied, uniform electric field. The most common application of electrophoresis is gel electrophoresis, in which sample (typically proteins, amino acids, nucleic acids, etc.) is applied to a channel that is formed in a cross-linked polymer, usually polyacrylamide or agarose (the gel). The speed at which the individual species move through the gel under the influence of the field is determined largely by the size of the species, as expressed by the mass-to-charge ratio. After separation, the individual species usually appear as discrete bands that may be better visualized by staining with ethidium bromide, silver, or Coomassie Brilliant Blue dye. Other related and more specific techniques include isoelectric focusing, pulsed-field gel electrophoresis, immunoelectrophoresis, and isotachophoresis.

Mass Spectrometry (MS): An analytical technique in which charged particles or radical ions are produced from a sample by either electron impact (bombardment with a stream of electrons) or chemical ionization (interaction with a small charged ion). Analysis on the basis of the mass-to-charge ratio is performed on fragments of the molecule that develop after the initial ionization. The method is very useful for mass determination, and structure determination on the basis of the induced fragmentation pattern. The charged fragments can be separated or analyzed by a magnetic sector, a quadrupole, an ion trap, by time of flight, or by cyclotroning. When coupled to separation techniques such as gas or liquid chromatography, a nearly universal qualitative detection capability is provided. Structure determination can be performed by comparison to well know fragmentation patterns or characteristics. When a mass spectrometer is capable of high mass resolution (0.0001 RMM units), a nearly unequivocal identification of a compound is possible. While direct interfaces and gas chromatographs are the most common sample introduction techniques, many others are available for specific applications. Matrix-assisted laser desorption/ionization (MALDI) is often used in time-of-flight instruments, and is especially useful for analysis of biopolymers. Inductively coupled plasma ionization is capable of producing mass spectra of high sensitivity for many metals and some non-metals. When coupled to liquid chromatography instrumentation, thermospray and electrospray methods have been used with HPLC for analytes that are not amenable to gas chromatography separation.

Ultraviolet Spectophotometry (UV, UV-Vis): A spectroscopic technique that focuses on electronic transitions in the visible and ultraviolet regions of the electromagnetic spectrum used for excitation and detection. The practical ultraviolet region extends from 190 nm to 400 nm in wavelength. The UV region can be divided into subranges: near-UV: 300 nm to 400 nm, mid-UV: 300 nm to 200 nm, far-UV: 200 nm to 122 nm, and vacuum-UV: 200 nm to 100 nm. Other divisions are possible, but these are less important for analytical chemistry. The visible region, so called because of the response of human vision, extends from about 390 nm to 750 nm in wavelength. Although it is possible to use ultraviolet visible spectrophotometry for structure determination, it is most often used as a quantitative tool. The wavelength-structure correlations are not as detailed, nor the spectra as sharp, as with other spectroscopic methods such as infrared spectrophotometry. The utility of UV-Vis absorptions for many organic compounds has led to this instrument being adapted as a detector for liquid chromatography. Related to ultraviolet spectrophotometry are fluorescence spectroscopic methods. In these methods, the energy emitted is at a different wavelength (usually longer, of lower energy) than the incident radiation, and this is typically detected perpendicular to the incident excitation beam. This makes fluorescence spectroscopy more sensitive than UV-Vis spectroscopy. This type of instrument is also incorporated as a detector for liquid chromatography.

Infrared Spectrophotometry (IR, FTIR): A spectroscopic technique that focuses on molecular vibrations (with a concurrent change in dipole moment) in the infrared (IR) region of the electromagnetic spectrum for excitation and detection. This region is further divided into three separate but overlapping ranges. The near-infrared (high energy IR, approximately 12 500 cm^{-1} to 4 000 cm^{-1}, 0.8 μm to 2.5 μm wavelength) is used to study overtone or harmonic vibrations. The mid-infrared (mid-range energy IR, approximately 4 000 cm^{-1} to 400 cm^{-1}, 2.5 μm to 25 μm wavelength) is used to study the fundamental vibrations and associated rotational-vibrational combinations. The far-infrared (low energy IR, approximately 400 cm^{-1} to 10 cm^{-1}, 25 μm to 1000 μm wavelength) is adjacent to the microwave region and is used to study rotational transitions. Most modern instruments use the Fourier transform (FT) technique to record the spectrum over all wavelengths, rather than by scanning through the wavelengths. The absorbances of the IR radiation are associated with specific chemical moieties, and a study of the spectra can be used to aid in structure determination. One often uses structure correlation charts to aid in assignment of absorbance bands. An analysis of the intensity of the absorptions can also be used for quantitative analysis.

Raman Spectroscopy: A vibrational spectroscopic method that arises from the inelastic scattering of monochromatic radiation by molecules that undergo a change in polarizability during the vibration. This is in contrast to infrared spectrophotometry, in which a change in the dipole moment occurs during the vibration. When radiation (typically light from a laser in the visible, near-infrared, or near-ultraviolet range) is scattered, a small fraction of the scattered radiation is observed to have a different frequency (the Raman effect). The variations of Raman spectroscopy are used to locate functional groups or chemical bonds in molecules. There are several variations in the approach to Raman spectroscopy. In resonance Raman spectroscopy, the excitation wavelength is matched to an electronic transition of the molecule, enhancing the vibrational modes. In coherent anti-Stokes Raman spectroscopy (CARS), two laser beams are used to generate a coherent anti-Stokes frequency beam. In surface-enhanced Raman spectroscopy (SERS), surface plasmons (a quantum of plasma oscillation) on a silver or gold colloid on a surface (such as a mirror) are excited by the laser, resulting in an increase in the electric fields surrounding the metal.

Nuclear Magnetic Resonance Spectrometry (NMR): A spectroscopic method that takes advantage of the fact that magnetic nuclei (nuclei with an odd number of protons and/or neutrons, having an intrinsic magnetic moment and angular momentum), when placed in a magnetic field, will absorb pulses of electromagnetic radiation and then radiate this energy back out for detection. For these nuclei, the energy and signal intensity are proportional to the applied magnetic field. The power of NMR results from the ability to probe the molecular environment around a particular nucleus, thus making it a favorite tool of the organic chemist. This is done by measurement of the chemical shift (or frequency) of an absorption, and by the analysis of splitting patterns, which are caused by the influence of adjacent nuclei. New high-field, high-sensitivity instruments have given this technique more applications in analytical chemistry. The most commonly studied nuclei are ^1H (proton, the most NMR-sensitive isotope after the radioactive ^3H) and ^{13}C. With high-field instruments, additional nuclei are accessible: ^2H, ^{10}B, ^{11}B, ^{14}N, ^{15}N, ^{17}O, ^{19}F, ^{23}Na, ^{29}Si, ^{31}P, ^{35}Cl, ^{113}Cd, ^{129}Xe, and ^{195}Pt.

Purge-and-Trap Sampling: Purge-and-trap sampling represents a family of methods that are used to capture the headspace above a condensed phase for subsequent analysis, most often for complex mixtures, environmental samples, etc. The headspace is the vapor space that develops above any condensed (solid or liquid) phase. Thermodynamics assures us that the concentration of a particular analyte found in the headspace will be different than that found in the condensed phase, but often the relationship is predictable. The value in the method comes from the simplicity; sample preparation is usually far simpler than the cleanup that is typically required for many complex mixtures. Purge-and-trap methods fall into two general categories: static and dynamic. The dynamic method typically uses a sweep gas to continuously purge vapor analytes into a cold (cryogenic) trap or an adsorbent. Modern dynamic purge-and-trap methods include porous-layer open tubular (PLOT) column cryoadsorption, which uses a combined adsorbent and cryotrapping on a high efficiency platform. Static methods typically employ a syringe to pressurize the headspace above a condensed phase, followed by uptake of the pressurized headspace into a trap. A modern static method (that usually is done without pressurization) is solid-phase microextraction (SPME). This method utilizes a fiber coated with a stationary phase (similar to stationary phases used for gas and liquid chromatography) at the end of a wire mounted in a syringe needle.

Atomic Absorption Spectroscopy (AAS) and Atomic Emission Spectroscopy (AES): Two related spectroscopic methods applied primarily to the analysis of inorganic compounds. Atomic absorption procedures use the absorption of optical radiation (light) by free atoms in the gaseous state. The light can be produced by a hollow cathode lamp, an electrodeless discharge lamp, or a deuterium lamp. The light is absorbed by the analyte during an electronic transition, the wavelength of which corresponds to only one element in the analyte, and the width of an absorption line is of the order of only a few picometers. This method can be used for the quantitative determination (on the basis of a calibration curve) of approximately 70 different elements in solution or directly in solid samples. Atomic emission spectroscopy (AES) uses the light emitted by a vaporized sample in a flame, plasma, arc, spark, or laser, at a particular wavelength, to determine the atomic spectrum (for determination of the elemental composition) and to determine the quantity of an element in a sample. The wavelength of the atomic spectral line gives the identity of the element while the intensity of the emitted light is proportional to the number of atoms of the element. No single source, as described above, is optimal for a given sample, and it is the choice of source that distinguishes the various techniques.

Thermal Analysis (TA) Methods: A family of analytical techniques in which various properties of a sample are examined as a function of changing temperature at a particular rate of change. For chemical analysis, the most common thermal analysis techniques include differential thermal analysis (DTA) and thermogravimetric analysis (TGA). In chemical analysis, TGA is used to determine a mass change as a function of temperature. This is used to determine decomposition or degradation temperatures, moisture content of materials (although Karl Fisher coulombic titrimetry is also used for this), the level of inorganic and organic components in materials, decomposition points of explosives, and solvent residues in materials. DTA, on the other hand, monitors a temperature change rather than a mass change and is used for determination of phase transitions. In this respect, it is useful as a complementary technique to probe the energetics of decomposition, moisture loss, etc. Other thermal analysis methods, such as differential scanning calorimetry (DSC), are not primarily analytical tools but rather thermophysical property measurement tools.

X-Ray Methods: A family of techniques that utilize radiation in the x-ray region, with wavelengths between 0.01 nm to 10 nm (corresponding to frequencies in the 3×10^{16} Hz to 3×10^{19} Hz range) to analyze for the presence of elements in a sample. In x-ray fluorescence spectroscopy (XRFS), short wavelength x-rays are used to excite secondary or fluorescent x-rays from a sample. The wavelength of x-rays used to excite the sample must be shorter than the expected fluorescence wavelength. The sample is typically presented as a powdered solid on a glass plate. Spectra are generated by changing the incident angle between the source and the sample with a device called a goniometer. The fluorescent signals obtained are very precise and specific, and can rival wet chemical methods for the identification of elements. Auger electron spectroscopy (AES) is a surface analysis method that measures the electrons emitted from a surface by electron bombardment of the surface. This method is considered to be in the same family as x-ray methods because during electron bombardment, the surface can lose energy either by electron emission (the Auger effect, via Auger electron emission) or x-ray emission. These methods are in contrast to x-ray diffraction methods, in which a crystal structure measurement is desired.

Flow Injection Analysis (FIA): An analytical protocol that seeks to replace the manual "test tube and beaker" aspects of wet chemical analysis by injecting an analyte in a flowing stream of a carrier reactant. As the analyte flows with the reactant stream, it diffuses into the reactant and product forms. Ultimately, the product zone, under the influence of the moving reactant, is passed into a detector section. The detection devices can consist of the same wide variety as is used in HPLC. The major advantage of flow injection analysis is the automation and decreased uncertainty associated with sampling and reagent addition. Strict control of reagent concentration, flow rate, and analyte volume is possible. Modern applications of FIA include the sequential addition of analyte and reactant in a stream so that the two are "stacked" in an inert carrier. They then mix by the parabolic flow profile of a laminar flowing stream in a tube. This arrangement can be miniaturized within a sampling valve, forming the so-called lab-in-a-valve approach.

ANALYTICAL STANDARDIZATION AND CALIBRATION

Thomas J. Bruno

Overview

Most modern instrumental techniques used in analytical chemistry produce an output or signal that is not absolute; the signal or peak is not a direct quantitative measure of concentration or target analyte quantity. Thus, to perform quantitative analysis, one must convert the raw output from an instrument (information) into a quantity (knowledge). This is done by standardizing or calibrating the raw response from an instrument (Refs. 1-4). Here, we briefly summarize the most common methods applied in analytical chemistry, recognizing that this is a very large field. We note that the common use of the term "standardization" is not to be confused with the application of standard methods as specified by regulatory or consensus standard organizations.

Samples

In all of the discussion to follow, we assume that the sample has been properly drawn from the parent population material, and properly prepared. Clearly, the most precise analytical methods and the most painstaking calibration methods are useless if applied to a sample that does not represent reality. Nevertheless, the term "sampling," which describes the process of obtaining the sample (from the population material), implies the existence of a sampling uncertainty (arising mainly from population material heterogeneity) (Refs. 5 and 6). Thus, the analytical result is an estimate of what would be obtained from the parent population material. The theory, concepts, and nomenclature regarding samples and sampling constitute a complex, statistically based, subspecialty of analytical chemistry well beyond the scope presented here. We begin with some simplified definitions (Ref. 7):

Amount of Substance: The amount of substance is the fundamental quantity of material measured in the number of moles.

Analyte: The analyte is the target component or compound in the sample for which one desires a measurement.

Aliquot: An aliquot is a known fraction of a homogeneous mass or volume.

Bias: As applied to sampling, bias refers to a systematic displacement, error, uncertainty. or mistake caused by a flaw in the sampling procedure.

Determination: The determination is the entire analytical procedure or method performed on a test portion.

Matrix: The matrix is the background or carrier material of the sample that includes all components except the analyte(s) of interest.

Phase: The phase describes the physical state of a substance: primarily solid, liquid, gas, but this term might include more detailed descriptions to include supercritical fluid, plasma, etc. Note that the term "vapor" typically refers to a gas phase above and in equilibrium with a condensed phase (often called the headspace in chemical analysis).

Population Material: The population material is the entirety of the bulk material from which the sample is drawn. This might be a plot of earth, a warehouse full of sugar, or a tank of jet fuel.

Quantity: The quantity refers to the mass or volume of a substance.

Test Portion: The test portion is the actual material removed from a sample for analysis.

Unknown: The unknown is a term that describes the target measurement or unknown quantity that is desired for the analyte, or the analyte itself.

Sampling uncertainty is that part of the total uncertainty in an analytical procedure or determination that results from using only a fraction of the population material. In this respect, sampling by any method is an extrapolation process. Since the sampling uncertainty is usually ignored for an individual analysis on an individual test portion, the sampling uncertainty is considered as being due entirely to the variability of the test portion. It is therefore assessed, when necessary, by replication of the sampling from the parent population material, and statistically isolating the uncertainty thus introduced by analysis of the variance. Typically, the problems associated with liquid population material are less complex but must not be ignored. Sample stratification, concentration and thermal gradients, poor mixing, and gradients associated with flow are all real effects that must be considered. Sampling uncertainty is often minimized by field and laboratory processing, with procedures that can include mixing, reduction, coning and quartering, riffling, milling, and grinding.

Another aspect that must be considered subsequent to sampling is sample preservation and handling. The integrity of the sample must be preserved during the inevitable delay between sampling and analysis. Sample preservation may include the addition of preservatives or buffer solutions, pH adjustment, use of an inert gas "blanket," and cold storage or freezing.

Calibration and Standardization

External Standard Methods

The external standard method can be applied to nearly all instrumental techniques, within the general limits discussed here, and the specific limitations that may be applicable with individual techniques. This method is conceptually simple; the user constructs a calibration curve of instrumental response with prepared mixtures containing the analyte(s) over a range of concentrations, an example of which is shown in Fig. 1a. Thus, the curve represents the raw instrumental response of the analyte as a function of analyte concentration or amount. Each point on this plot must be measured several times so that the repeatability can be assessed. Only random uncertainty should be observed in the replicates; trends of increasing or decreasing response (hysteresis) must be remedied by identifying the source and adjusting the method accordingly. The calibration solutions should be randomized (that is, measured in random order). Although called a calibration "curve," ideally the signal versus concentration plot is linear, or substantially linear (that is, areas of nonlinearity are unimportant, otherwise they are localized, minor, and properly treated by the measurement technique). In some cases, the response may be linearizable (for example, by calculating the logarithm of the raw response). If a curve shows nonlinearity in an area that is important for the analysis, one must measure more concentrations (data points) in the region of curvature.

In practice, the line that results from the calibration is fit with an appropriate model, and the desired value for the unknown concentration is calculated. The curve can be used graphically if

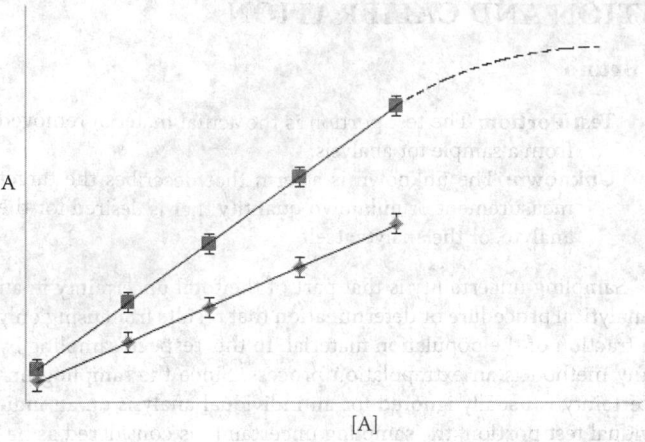

A

[A]

Fig. 1a: An example calibration curve prepared by use of the external standard method. The instrument response is represented by A, and the concentration resulting in that response is [A]. While curves for two analytes are shown, in principle, one can plot as many analytes as desired. While five points per analyte have been shown, one can measure as many as required. Note that a region of nonlinearity is shown in the latter part of the curve for one of the components. One would require a larger number of points to adequately represent and fit any nonlinear areas.

approximation suffices. Mixtures prepared for external standard calibration can contain one or many analytes. Once a calibration curve is prepared, it can often be used for some time period, provided such a procedure has been previously validated (that is, the stability of the standards and the instrument over the time of use has been assessed). Otherwise, it is best to measure the unknown and the standards within a short period of time. Moreover, if any major change is made to the instrumentation (changing a detector or detector parameters, changing a chromatographic column, etc.), the standards must be re-measured.

To successfully use the method, the standard mixtures must be in a concentration range that is comparable to that of the unknown analyte, and ideally should bracket the unknown. Multiple measurements of each standard mixture should be made to establish repeatability of points on the curve. Many instrumental methods have operation ranges (frequency, temperature, etc.) in which the uncertainty is minimized, so components and concentrations for standard mixtures must respect this. The standard mixtures should be in the same matrix as the unknown, and the matrix must not interfere with the unknown or other standard mixture components. Any pretreatment of the unknown must also be reflected in the standard mixtures. As with any calibration method, components in the standard mixtures must be available at a high (or at least known) purity, they must be stable during preparation, and must be soluble in the required matrix. Unless the physical phenomenon of a measurement is well understood, extrapolation beyond the curve is not recommended (and indeed is usually strongly discouraged); nevertheless, extrapolation is occasionally done in practice. In those cases, one must be cautious, report exactly how the extrapolation was done, and assess any increase in uncertainty that may result. Note that the curve might not extrapolate through the origin. This is usually the result of adsorption (of components on container walls), carryover hysteresis, absorption (of components in seals or septa), or component degradation or evaporation.

A major consideration with external standardization is that typically, the sample size (for example, the injection volume in chromatography) must be maintained constant for standard mixtures and the unknowns. If the sample size varies slightly, it is often possible to apply a correction to the raw signal. One should not attempt to generate a calibration curve by varying the sample size (that is, for example, injecting increasing volumes into a gas chromatograph). This caution does not preclude serial dilution methods (see below), in which multiple solutions are generated for separate measurement. Other issues that can hinder successful application of the external standards method include instrumental aspects that might not be readily apparent. In chromatographic methods, for example, one can overload the column or detector. In older instruments, settings of signal attenuation were typically made manually, while in newer instruments, this may occur through software, sometimes without operator interaction or knowledge.

Note, *inter alia*, that in Fig. 1a (and indeed all the examples presented here), the uncertainty is only indicated for the variable on the y-axis. In reality, we must recognize that there is uncertainty for the values plotted on the x-axis as well, but we often only treat the largest uncertainty, or the uncertainty that is most important for our application. Note, also, that it is critical to maintain the integrity of standards; decomposition, degradation, moisture uptake, etc., will adversely affect the validity of the calibration.

Abbreviated External Standard Methods

In many situations in chemical analysis, a full calibration curve is not prepared because of the complexity, time, or cost. In such situations, abbreviated external standard methods are often used. Under no circumstances can an abbreviated method be used if the raw signal response is nonlinear. Moreover, these methods are not generally appropriate for analyses in regulatory, forensic, or health care environments where the consequences can be far-reaching.

Single Standard

This method uses a simple proportion approach to standardize an instrument response. It can be used only when the system has no constant, determinate error or bias[*], and when the reagents used give a zero blank response (that is, the instrument response from the matrix and measurement system only, without the analyte). A standard should be prepared such that the concentration is close to that of the unknown. One then calculates the concentration of the unknown, [X], as:

$$[X] = (A_x/A_s) \, [S] \qquad (1)$$

where A_x is the instrument response of the unknown, A_s is the instrument response of the standard, and [S] is the concentration of the standard.

Single Standard Plus Assumed Zero

This method, illustrated schematically in Fig. 1b, assumes that the blank reading will be zero. One uses a two-point calibration in which the origin is included as the first point. It is important to ensure, by experiment or experience, that such a method is adequate to the task.

Single Standard Plus Blank

If the analytical method has no determinate error or bias, but does produce a finite blank value, then one must also perform a blank measurement, which is subtracted from the instrument response of the standard and the unknown. Then the same procedure (Eq. 1) is used as for the single standard. If multiple samples are to be measured, it is important to measure the blank between each measurement.

---

[*] Determinate error and bias are related terms that describe uncertainty that arises from a fixed cause, and that can, in principle, be eliminated if recognized. Determinate error (or systematic error) is most often associated with a measurement, while bias can be associated with either a measurement or with the sampling procedure.

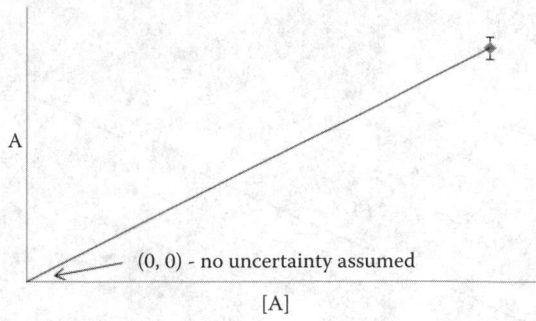

Fig. 1b: An example of a single-point calibration curve. The instrument response is represented by A, and the concentration resulting in that response is [A]. The origin (0,0) is assigned as part of the curve, and is assumed to have no uncertainty.

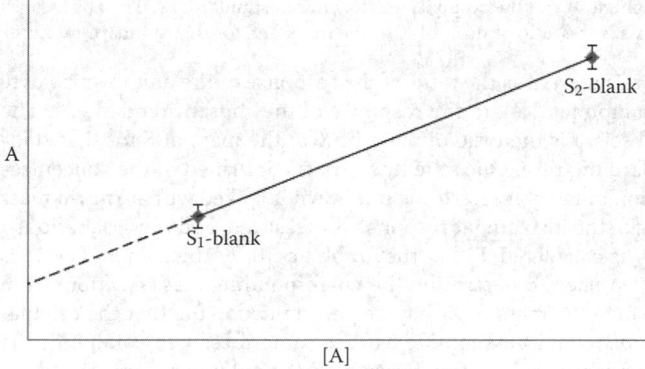

Fig. 1c: An example of two standards plus a blank calibration curve. The blank is subtracted from each of the standards. The instrument response is represented by A, and the concentration resulting in that response is [A].

Two Standards Plus Blank

When the analytical method has both a determinate error (or bias), and a finite blank value, at least three calibrations must be made: two standards and one blank. The standard concentrations are typically prepared widely spaced in concentration, and the higher concentration should be chosen to represent the limit of linearity of the instrument or method. If this is not practical, the higher concentration should simply be the highest expected concentration of the analyte (unknown). This method is illustrated schematically in Fig. 1c. If multiple samples are to be measured, it is important to measure the blank between each measurement.

Internal Normalization Method

As mentioned above, the raw signal from an analytical instrument is typically not an absolute measure of concentration of the analyte(s), because the instrument may respond differently to each component. In some cases, such as with chromatographic methods, it is possible to apply response factors, determined from a standard mixture containing all constituents of the unknown sample, for standardization (Ref. 8). The standard mixture is gravimetrically prepared (with known mass percents for each component), and the instrument response is measured, for example as chromatographic areas. The total mass percent and the total area percent each sum to 100. One calculates the ratio of each mass percent to each area, choosing one component as the reference, which is assigned a response factor of unity. To obtain the response factors of all the other components, one divides its (mass%-to-area ratio) with that of the reference. This is done for all components,

producing a response factor for all components, except of course for the reference, defined as unity. When the unknown sample is measured, the response factor is multiplied by each raw area, and the resulting area percent provides the normalized mass percent of each component in the unknown.

This method corrects for minor variations in sample size (earlier defined as the test portion), although large differences in sample size must be avoided so that one is assured of consistent instrument performance. Although the method corrects for the different responses of samples, large differences must be avoided. This also means that the detector must respond linearly to the concentrations of each component, even if the concentrations are very different. This may require dilution or concentration of the sample in some situations. In chromatographic applications, all components of a mixture must be analyzed and standardized, since normalization must be performed on the entire sample.

Some techniques, such as gas chromatography with flame ionization detection and thermal conductivity detection, have well defined physical phenomena associated with output signals. With these techniques, there are some limited, published response factor data that can be used in an approximate way to standardize the response from these devices.

In Situ Standardization

While it is rare that an analytical method can be calibrated by use of a single solution, some instances of spectrophotometry and electroanalytical methods can qualify. To use this method one sequentially and incrementally adds known masses of standard analytes to a solution, with an instrument response being measured after each addition. This procedure can only be used if the analytical method itself does not change the analyte concentration (nondestructive) and does not lead to a loss of solution volume. A solid crystalline analyte is an example. One must also minimize changes in solution volume over the course of the standardization.

Standard Addition Methods

Samples presented for analysis often are contained in complex matrices with many impurities that may interact with the analyte, potentially enhancing or diminishing a signal from an instrumental technique. In such cases, the preparation of an external standard calibration curve will be impossible, because it might be very difficult to reproduce the matrix. In these cases, the standard addition method may be used. A standard solution containing the target analyte is prepared and added to the sample, thus accounting for the unknown impurities and their effects. While the quantity of target analyte in the target sample is unknown, the added quantity is known, and its incremental additive effect on the instrument signal can be measured. Then, the quantity of the unknown analyte is determined by what is effectively an extrapolation. In practice, the volume of the standard solution added is kept small to avoid dilution of the unknown impurities by no more than 1% of the total signal. This method can only be used if there is a verified linear relationship between the signal and quantity of analyte. If a determinate error is present, then the slope of the line must be known. Moreover, the sample cannot contain any components that can respond as the analyte (that is, masquerade).

Single Standard Addition

In the simplest case, one addition of analyte is made after first measuring the response of the analyte in the unknown sample. Thus, two measurements are required:

$$A_{xo} = m[X_0] \tag{2}$$

$$A_{xi} = m([X_o] + [S]), \qquad (3)$$

where A_{xo} is the instrument response of the analyte in the unknown sample, $[X_0]$ is the concentration in the unknown sample, and A_{xi} is the instrument response upon the addition of the standard, $[S]$ (additive in equation because X and S are the same compound). The assumed slope is the proportionality constant, m. The two equations are solved simultaneously for $[X_0]$. This technique is very rapid and economical, but there are serious drawbacks. There is no built in check for mistakes on the part of the analyst, there is no means to average random uncertainties, and there is no way to detect interference (mentioned above as masquerade).

Multiple Standard Addition

This standard addition method alleviates some of the problems inherent in single standard addition. Here, the unknown sample is first measured in the instrument. Then that sample is "spiked" with incrementally increasing concentrations of the analyte, generating a curve such as that shown in Fig. 1d. The curve should extrapolate to zero signal at zero concentration. The concentration of the analyte in the unknown is read or calculated from the abscissa (*x*-axis).

Internal Standard Methods

An internal standard is a compound added to a sample at a known concentration, the purpose of which is to exhibit a similar signal when measured in an instrument, but be distinguishable from the signal of the desired analyte. It provides the highest level of reliability in quantitation by chromatographic methods, and is not affected by large differences in sample size (Ref. 8). Unlike the internal normalization method, it is not necessary to elute or measure all the components of the sample, one need focus only on the component(s) of interest. In atomic spectrometry, this method is not affected by changes in gas flow rates, sample aspiration rates, and flame suppression or enhancement. Another situation in which this method is valuable is when the sample matrix is either unknown or very complex, precluding the preparation of external standards.

Multiple Internal Standards

A set of calibration solutions is prepared by mass, containing the target analyte, X, and a standard that is not present in the unknown sample, A. The instrument response (for example, a chromatographic area) is measured for each calibration solution, and a plot is made to establish linearity as in Fig. 1a. The

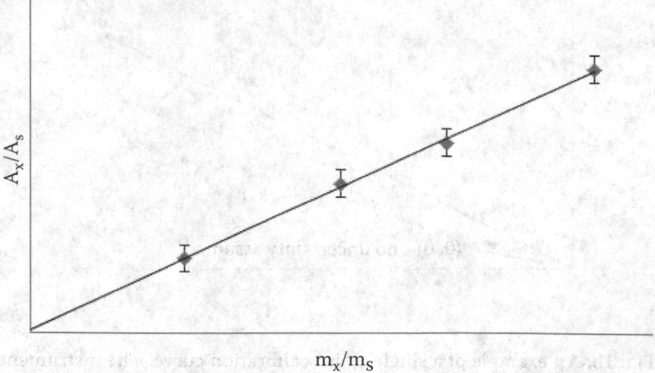

An example of the multiple internal standard method. The ordinate (*y*) axis is the ratio of the response of the unknown analyte component, A_x, to the response of the chosen standard, A_s. The abscissa (*x*) axis is the ratio of mass of X to the mass of S for that standard mixture.

ordinate axis is the ratio of the response of the unknown analyte component, A_x, to the response of the chosen standard, A_s. The abscissa is the ratio of mass of X to the mass of S for that standard mixture. Once the linearity is confirmed in the concentration range of interest, the unknown is spiked with a known mass of S, the instrument response is measured, and the area ratio A_x/A_s is calculated. Either the graph or a fit of the data on Fig. 1e is then used to determine the corresponding mass fraction, from which the mass of X may be determined. Note that the calculations could be simplified if the same mass per volume of the internal standard is added to both the unknown samples and the calibration standards.

Single Internal Standard

In practice, once the linearity is established for a given mixture, it is no longer necessary to use multiple standards, although this is the most precise method. Subsequent to the verification of linearity, one standard solution can be used to fix the slope, provided it is close in concentration to that of the target analyte. In this case, the mass of the unknown can be found from:

$$X/S = (A_x/A_s)(1/R), \qquad (4)$$

where X is the mass of the unknown analyte in the sample, S is the mass of the added internal standard in the sample, A_x and A_s are the instrument responses (areas) of the unknown and internal standard, respectively. R is a ratio determined from the standard solution prepared with both X and S: (mass, unknown analyte/mass, internal standard)/(signal, unknown analyte/signal, internal standard) = R.

$$\left(\frac{\left(\dfrac{\text{mass, unknown analyte}}{\text{mass, internal standard}} \right)}{\left(\dfrac{\text{signal, unknown analyte}}{\text{signal, internal standard}} \right)} \right) = R \qquad (5)$$

Since R is the slope of the calibration curve discussed above, once linearity is established, one solution suffices. There are many conditions that must be fulfilled in order to use the internal standard method, and it is rare that all of them can actually be met. Indeed, in practice, one tries to meet as many as possible, but those that are mandatory are italicized. The compound chosen *must not be present* already in the unknown. The compound chosen *must be separable from the analyte* present in the unknown.

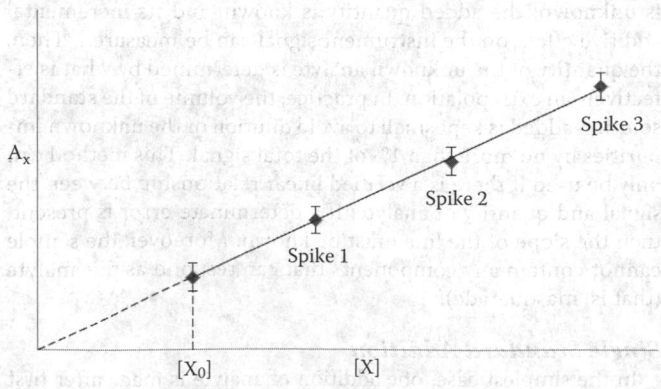

A_x

Spike 3

Spike 2

Spike 1

$[X_0]$ $[X]$

An example of calibration by multiple standard addition. Three additions (spikes) of the analyte X are shown, as is the extrapolation to the unknown concentration, X_0.

An exception occurs when an isotopically labeled standard is used, in conjunction with mass discrimination or radioactive counting detection. In a chromatographic measurement, this is typically at least baseline resolution, although this would be a minimally acceptable degree of separation. On the other hand, the unknown analyte peak and the internal standard peak should be close to each other (temporally) on the chromatogram. The compound chosen *must be miscible* with the solvent at the temperature of reagent preparation and measurement. The compound chosen must not react chemically with the sample or solvent, or interfere in any way with the analysis. It is critical to maintain the integrity of standards; decomposition, degradation, moisture uptake, etc., will adversely affect the validity of the calibration. In the case of a chromatographic measurement, the same applies to interactions with the stationary phase. The compound chosen must be chemically similar (for example, in functionality, thermophysical properties) to the analyte. If such a compound is not available (for example, in a chromatographic measurement), an appropriate hydrocarbon should be chosen as a surrogate. The standard solution should be prepared at a similar concentration as in the unknown matrix; ratio correction of large differences is no substitute for an appropriate concentration. In a chromatographic measurement, the compound chosen must elute as closely as possible to the analyte, and should not be the last peak to elute (the final peak often shows different geometry such as tailing). The compound chosen must be sufficiently nonvolatile to allow for storage as needed. When there is the potential for the unknown analyte to be lost by adsorption, absorption, or some other interaction with the matrix or container, a compound called a carrier is sometimes added in large excess. The carrier is similar, chemically and physically, to the unknown analyte, but easily separated from it. Its purpose is to saturate or season the matrix and prevent analyte loss.

Serial Dilution

Serial dilution is less a standardization method as it is a method of generating solutions to be used for standardizations. Nevertheless, its importance and utility, as well as the popularity of its application, warrants mention in this section. A serial dilution is the stepwise dilution of a substance, observant of a specified, constant progression, usually geometric (or logarithmic). One first prepares a known volume of stock solution of a known concentration, followed by withdrawing some small fraction of it to another container or vial. This subsequent container is then filled to the same volume as the stock solution with the same solvent or buffer. The process is then repeated for as many standard solutions as are desired. A ten-fold serial dilution could be 1 M, 0.1 M, 0.01 M, 0.001 M, etc. A ten-fold dilution for each step is called a logarithmic dilution or log-dilution, a 3.16 fold ($10^{0.5}$ fold) dilution is called a half-logarithmic dilution or half-log dilution, and a 1.78 fold ($10^{0.25}$ fold) dilution is called a quarter-logarithmic dilution or quarter-log dilution. In practice, the ten-fold dilution is the most common. The serial dilution procedure is not only used in chemical analysis but also in serological preparations in which cellular materials such as bacteria are diluted. A critical aspect of serial dilution is that the initial solution concentration must be prepared and determined with great care, since any mistake here will be propagated into all resulting solutions.

References

1. Chalmers, R. A., *Chapter 2: Standards and Standardization in Chemical Analysis*, Vol. 3, Elsevier, Amsterdam, 1975.
2. Danzer, K., and Currie, L. A., *Pure Appl. Chem.* 70, 993, 1998.
3. Danzer, K., Otto, M., and Currie, L. A., *Pure Appl. Chem.* 76, 1215, 2004.
4. Woodget, B. W., and Cooper, D., *Samples and Standards, Analytical Chemistry by Open Learning*, John Wiley and Sons, Chichester, 1987.
5. Gy, P., *Sampling for Analytical Purposes*, John Wiley and Sons, Chichester, 1998.
6. Vitt, J. E., and Engstrom, R. C., *J. Chem. Educ.* 76, 99, 1999.
7. Horowitz, W., *Pure Appl. Chem.* 62, 1193, 1990.
8. Grob, R. L., *Modern Practice of Gas Chromatography*, Wiley Interscience, New York, 1995.

MASS- AND VOLUME-BASED CONCENTRATION UNITS

Thomas J. Bruno and Paris D. N. Svoronos

A variety of concentration units are used in analytical chemistry, and the most common are provided in this table (Ref. 1). The reference below provides additional details.

Reference

1. Bruno, T. J., and Svoronos, P. D. N., *CRC Handbook of Basic Tables for Chemical Analysis*, 3rd Edition, CRC Press, Boca Raton, FL, 2011.

Parts per Million

Parts per Million		Percent
1 ppm	=	0.0001 %
10 ppm	=	0.001 %
100 ppm	=	0.01 %
1 000 ppm	=	0.1 %
10 000 ppm	=	1.0 %
100 000 ppm	=	10.0 %
1 000 000 ppm	=	100.0 %

Parts per Billion

Parts per Billion		Percent
10	=	0.000 001 %
100	=	0.000 01 %
1 000	=	0.000 1 %
10 000	=	0.001 %
100 000	=	0.01 %
1 000 000	=	0.1 %

Parts per Trillion

Parts per Trillion		Percent
100	=	1×10^{-8} %
10 000	=	0.000 001 %
1 000 000	=	0.000 1 %
100 000 000	=	0.01 %

Because the mass of one liter of water is approximately one kg, mg/L units of dilute aqueous solution are nearly equal to ppm units. The precise equivalence is obtained by dividing by the density, ρ:

$$ppm = (mg/L)/\rho$$

where the solution density, ρ, is in g cm^{-3}. Some sources will substitute specific gravity for density in the above equation. The specific gravity is the ratio of the solution density to that of the density of pure water at 4 °C. Since the density of pure water at 4 °C is very nearly 1 g cm^{-3}, the specific gravity is numerically equal (within an uncertainty of 25 ppm) to the solution density when the latter is expressed in units of g cm^{-3}.

Concentration Units Nomenclature

The following table provides guidance in the use of base-10 concentration units (presented in the three preceding tables), since there are differences in usage worldwide.

Number	Number of Zeros	Name (Scientific Community)	Name (United Kingdom, France, Germany)
1000.	3	thousand	thousand
1 000 000.	6	million	million
1 000 000 000.	9	billion	milliard, or thousand million
1 000 000 000 000.	12	trillion	billion
1 000 000 000 000 000.	15	quadrillion	thousand billion

Molar-Based Concentration Units

Molarity*, M: (moles of solute)/(liters of solution)
Molality, *m*: (moles of solute)/(kilograms of solvent)
Normality, N: (equivalents* of solute)/(liters of solution)
Formality, F: (moles of solute)/(kilograms of solution)

To convert from ppm to formality units:

$F = ppm/(1000\ RMM)$, where RMM is the relative molecular mass of the solute i

To convert from ppm to molality units:

$$m = [ppm/(1000\ RMM)]\ [1/(1 - tds/1\ 000\ 000)]$$

where tds is the total dissolved solids in ppm in the solution

To convert from ppm to molarity units:

$$M = [ppm/(1000\ RMM)]\ \rho$$

where ρ is the solution density

---

* This unit is temperature dependent.

DETECTION OF OUTLIERS IN MEASUREMENTS

Thomas J. Bruno and Paris D. N. Svoronos

The field of outlier detection and treatment is considerable, and a rigorous mathematical discussion is well beyond any treatment that is possible here. Moreover, the practice in the treatment of analytical results is usually simplified, since the number of observations is often not very large. The two most common methods used by analysts to detect outliers in measured data are versions of the Q-test (Refs. 1–3, 6) and Chauvanet's criterion (Refs. 4–6), both of which assume that the data are sampled from a population that is normally distributed.

References

1. Dean, R. B., and Dixon, W. J., *Anal. Chem.* 23, 636, 1951.
2. Day, R. A., and Underwood, A.L., *Quantitative Analysis*, 6th Edition, Prentice Hall, Englewood Cliffs, NJ, 1991.
3. Efstathiou, C. E., *Dixon's Q-test: Detection of a single outlier, http://www.chem.uoa.gr/applets/AppletQtest/Text_Qtest2.htm*, Laboratory of Analytical Chemistry, Department of Chemistry, National and Kapodistrian University of Athens, 2008.
4. Taylor, J. R., *An Introduction to Error Analysis*, 2nd edition, University Science Books, Sausalito, CA, 1997.
5. Benziger, J. B., and Aksay, I.A., *Notes on Data Analysis, http://www.princeton.edu/~che346/Notes/Analysis.pdf*, Department of Chemical Engineering, Princeton University, 1999.
6. Bruno, T. J., and Svoronos, P. D. N., *CRC Handbook of Basic Tables for Chemical Analysis*, 3rd Edition, CRC Press, Boca Raton, FL, 2011.

Q-Test

To perform the Q-test, one calculates the Q value given by:

$$Q = Q_{gap}/R$$

where Q_{gap} is the difference between the suspected outlier and the measured value closest to it, and R is the range of all the measured values in the data set. One then compares the calculated Q value with the critical Q values in the following table:

Number of Observations	Q_{crit}, 90% Confidence Level	Q_{crit}, 95% Confidence Level	Q_{crit}, 99% Confidence Level
3	0.941	0.970	0.994
4	0.765	0.829	0.926
5	0.642	0.710	0.821
6	0.560	0.625	0.740
7	0.507	0.568	0.680
8	0.468	0.526	0.634
9	0.437	0.493	0.598
10	0.412	0.466	0.568

If the calculated value of Q is greater than the appropriate value of Qcrit, then the value is a suspected outlier.

Chauvenet's Criterion

To perform Chauvenet's test on a set of measurements, one first must calculate the mean and standard deviation of the data. Then one calculates:

$$\tau = (x_i - x_{ave})/\sigma$$

where x_i is the suspected outlier, x_{ave} is the mean of all the measurements, and σ is the standard deviation. One then compares the calculated value of τ with τ_{crit} in the following table:

Number of Observations, N	τ_{crit}
5	1.65
6	1.73
7	1.81
8	1.86
9	1.91
10	1.96
15	2.12
20	2.24
25	2.33
50	2.57
100	2.81
150	2.93
200	3.02
500	3.29
1000	3.48

If the calculated value of τ is greater than the value of τ_{crit}, then the value is a suspected outlier.

For numbers of observations between those given in the table, especially for a large number of observations, one may use the following plot to estimate the value of Chauvenet's τ_{crit}:

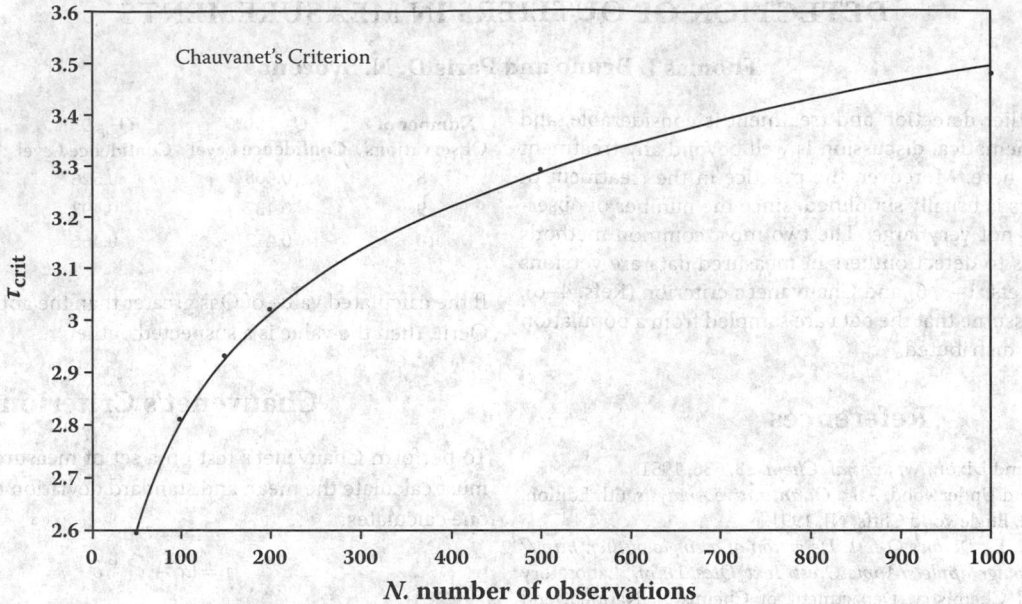

PROPERTIES OF CARRIER GASES FOR GAS CHROMATOGRAPHY

The following is a list of carrier gases sometimes used in gas chromatography, with properties relevant to the design of chromatographic systems. All data refer to normal atmospheric pressure (101.325 kPa). Additional properties related to the carrier gas, including split ratio, pressure drop in capillary columns, the Martin-James compressibility factor, and the Giddings plate height correction factor can be found in Ref. 2.

M_r: Molecular weight (relative molar mass)
ρ_{25}: Density at 25 °C in g L^{-1}

λ: Thermal conductivity in mW m^{-1} °C^{-1}
η: Viscosity in µPa s (equal to 10^{-3} cP)
c_p: Specific heat at 25 °C in J g^{-1} °C^{-1}

References

1. Lide, D. R., and Kehiaian, H. V., *CRC Handbook of Thermophysical and Thermochemical Data*, CRC Press, Boca Raton, FL, 1994.
2. Bruno, T. J., and Svoronos, P. D. N., *CRC Handbook of Basic Tables for Chemical Analysis*, 3rd Edition, CRC Press, Boca Raton, FL, 2011.

| Gas | M_r | ρ_{25} g L^{-1} | At 25 °C | | At 250 °C | | c_p (25 °C) J g^{-1} °C^{-1} |
			λ mW m^{-1} °C^{-1}	η µPa s	λ mW m^{-1} °C^{-1}	η µPa s	
Hydrogen	2.016	0.0824	185.9	8.9	280	13.1	14.3
Helium	4.003	0.1636	154.6	19.9	230	29.5	5.20
Argon	39.95	1.6329	17.8	22.7	27.7	35.3	0.521
Nitrogen	28.01	1.1449	25.9	17.9	39.6	26.8	1.039
Oxygen	32.00	1.3080	26.2	20.7	42.6	31.8	0.919
Carbon monoxide	28.01	1.1449	24.8	17.8	40.7	26.5	1.039
Carbon dioxide	44.01	1.7989	16.7	14.9	35.5	24.9	0.843
Sulfur hexafluoride	146.05	5.9696	13.1	28.1	15.3	24.8	0.664
Methane	16.04	0.6556	34.5	11.1	75.0	17.6	2.23
Ethane	30.07	1.2291	20.9	9.4	57.7	15.5	1.75
Ethylene	28.05	1.1465	20.5	10.3	53.8	17.2	1.53
Propane	44.10	1.8025	17.9	8.3	49.2	14.0	1.67

PROPERTIES OF COMMON CROSS-LINKED SILICONE STATIONARY PHASES

Thomas J. Bruno and Paris D. N. Svoronos

Modern gas chromatography is most often performed with high efficiency capillary (open tubular) columns coated with a cross-linked polymeric stationary phase. We provide chromatographic data on the two most common cross-linked phases (Ref. 1). These data are useful for the interpretation of chromatograms, and for quality control and assurance. The retention indices presented in these tables were measured at 120 °C isothermally. Retention indices are temperature dependent; the temperature dependence of the Kovats indices have been studied for many compounds (Ref. 2). For more extensive information on other cross-linked phases, other silicone phases, mesogenic phases, and solid sorbents, the reader is advised to consult Ref. 1.

References

1. Bruno, T. J., and Svoronos, P. D. N., *CRC Handbook of Basic Tables for Chemical Analysis*, 3rd Edition, CRC Press, Boca Raton, FL, 2011.
2. Bruno, T. J., Wertz, K. H., and Caciari, M., *Anal. Chem.* 68, 1347, 1996.

Phase: 5% Phenyl Dimethylpolysiloxane

Temperature Ranges:

–60 °C to 325 °C isothermally, –60 °C to 350 °C programmed for < 0.32 mm I.D. columns
–60 °C to 300 °C isothermally, –60 °C to 320 °C programmed for 0.53 mm I.D. columns
–60 °C to 260/280 °C for >2.0 µm films

Similar Phases:

DB-5, Ultra-2, SPB-5, CP-Sil 8 CB, Rtx-5, BP-5, OV-5, 007-2 (MPS-5), SE-52, SE-54, XTI-5, PTE-5, HP-5MS, ZB-5, AT-5, MDN-5

Notes: This phase is probably the most commonly used stationary phase in gas chromatography, because it combines boiling point separation with a minor contribution of a specific interaction; typically used as the first phase in any method development; versatile for hydrocarbons and more polar compounds.

Probe Compound	McReynolds Constant	McReynolds Code	Kovats Retention Index
Hexane			600
1-Butanol	66	y'	656
Benzene	31	x'	684
2-Pentanone	61	z'	688
Heptane			700
1,4-Dioxane	64	L	718
2-Methyl-2-pentanol	41	H	731
1-Nitropropane	93	u'	745
Pyridine	62	s'	761
Octane			800
Iodobutane	22	J	840
2-Octyne	35	K	876
Nonane			900
Decane			1000

Phase: Dimethylpolysiloxane

Temperature Range:

–60 °C to 325 °C for normal operations, periodic operation to 350 °C can be used to facilitate column clean-up
–60 °C to 260/280 °C for >2.0 µm films

Similar Phases:

DB-1, OV-1, HP-1, DB-1ms, HP-1ms, Rtx-1, Rtx-1ms, CP-Sil 5 CB Low Bleed/MS, MDN-1, AT-1

Notes: Useful for the separation of hydrocarbons, pesticides, PCBs, phenols, sulfur compounds, flavors and fragrances, and some amines; columns are typically stable and low bleed; a good all-purpose column used to begin method development protocols.

Probe Compound	McReynolds Constant	McReynolds Code	Kovats Retention Index
Hexane			600
1-Butanol	54	y'	644
Benzene	16	x'	669
2-Pentanone	44	z'	671
Heptane			700
1,4-Dioxane	46	L	700
2-Methyl-2-pentanol	31	H	721
1-Nitropropane	62	u'	714
Pyridine	44	s'	743
Octane			800
Iodobutane	3	J	821
2-Octyne	23	K	864
Nonane			900
Decane			1000

DETECTORS FOR GAS CHROMATOGRAPHY

Thomas J. Bruno and Paris D. N. Svoronos

The following table provides some comparative data to aid in interpreting results from the more common detectors applied to capillary and packed-column gas chromatography (Refs. 1–8). For more detailed information regarding operation and interpretation of results, see Ref. 8.

References

1. Hill, H.H., and McMinn, D., eds., *Detectors for Capillary Chromatography*, John Wiley & Sons, New York, 1992.
2. Buffington, R., and Wilson, M. K., *Detectors for Gas Chromatography — a Practical Primer*, Hewlett Packard Corp., Avondale, PA, 1987.
3. Buffington, R., *GC-Atomic Emission Spectroscopy using Microwave Plasmas*, Hewlett Packard Corp., Avondale, PA, 1988.
4. Liebrand, R. J., Ed., *Basics of GC/IRD and GC/IRD/MS*, Hewlett Packard Corp., Avondale, PA, 1993.
5. Bruno, T. J., *Sep. Purif. Method* 29, 63, 2000.
6. Bruno, T. J., *Sep. Purif. Method* 29, 27, 2000.
7. Sevcik, J., Detectors in Gas Chromatography, *Journal of Chromatography Library*, Vol. 4, Elsevier, Amsterdam, 1976.
8. Bruno, T. J., and Svoronos, P. D. N., *CRC Handbook of Basic Tables for Chemical Analysis*, 3rd Edition, CRC Press, Boca Raton, FL, 2011.

Detector	Limit of Detection	Linearity	Selectivity	Comments
Thermal conductivity detector (TCD, katharometer)	1×10^{-10} g propane (in helium carrier gas)	1×10^{6}	Universal response, concentration detector	Ultimate sensitivity depends on analyte thermal conductivity difference with carrier gas Since thermal conductivity is temperature dependent, response depends on cell temperature Wire selection depends on chemical nature of analyte Helium is recommended as carrier and make-up gas. When analyzing mixtures containing hydrogen, one can use a mixture of 8.5% (mass/mass) hydrogen in helium
Gas density balance detector (GADE)	1×10^{-9} g, H_2 with SF_6 as carrier gas	1×10^{6}	Universal response, concentration detector	Response and sensitivity are based on difference in relative molecular mass of analyte with that of the carrier gas; approximate calibration can be done on the basis of relative density The sensing elements (hot wires) never touch sample, thus making GADE suitable for the analysis of corrosive analytes such as acid gases; gold sheathed tungsten wires are most common Best used with SF_6 as a carrier gas, switched between nitrogen when analyses are required Detector can be sensitive to vibrations, and should be isolated on a cushioned base
Flame ionization detector (FID)	1×10^{-11} g to 1×10^{-10} g	1×10^{7}	Organic compounds with C–H bonds	Ultimate sensitivity depends on the number of C–H bonds on analyte Nitrogen is recommended as carrier gas and make-up gas to enhance sensitivity Sensitivity depends on carrier, make-up and jet gas flow-rates Column must be positioned 1 mm to 2 mm below the base of the flame tip Jet gases must be of high purity
Nitrogen-phosphorous detector (NPD, thermionic detector, alkali flame ionization detector)	4×10^{-13} g to 1×10^{-11} g of nitrogen compounds 1×10^{-13} g to 1×10^{-12} g of phosphorous compounds	1×10^{4}	10^5 to 10^6 by mass selectivity of N or P over carbon	Does not respond to inorganic nitrogen such as N_2 or NH_3 Jet gas flow rates are critical to optimization Response is temperature dependent Used for trace analysis only, and is very sensitive to contamination Avoid use of phosphate detergents or leak detectors Avoid tobacco use nearby Solvent-quenching is often a problem
Electron capture detector (ECD)	5×10^{-14} g to 1×10^{-12} g	1×10^{4}	Selective for compounds with high electron affinity, such as chlorinated organics; concentration detector	Sensitivity depends on number of halogen atoms on analyte Used with nitrogen or argon/methane (95/5, mass/mass) carrier and make-up gases Carrier and make-up gases must be pure and dry The radioactive ^{63}Ni source is subject to regulation and periodic inspection

Detector	Limit of Detection	Linearity	Selectivity	Comments
Flame photometric detector (FPD)	2×10^{-11}g of sulfur compounds, 9×10^{-13}g of phosphorous compounds	1×10^3 for sulfur compounds 1×10^4 for phosphorous compounds	10^5 to 1 by mass selectivity of S or P over carbon	Hydrocarbon quenching can result from high levels of CO_2 in the flame Self-quenching of S and P analytes can occur with large samples Gas flows are critical to optimization Response is temperature dependent Condensed water can be a source of window fogging and corrosion
Photoionization detector (PID)	1×10^{-12} g to 1×10^{-11}g	1×10^7	Depends on ionization potentials of analytes	Used with lamps with energies of 10.0 eV to 10.2 eV Detector will have response to ionizable compounds such as aromatics and unsaturated organics, some carboxylic acids, aldehydes, esters, ketones, silanes, iodo- and bromoalkanes, alkylamines and amides, and some thiocyanates
Sulfur chemilumenescence detector (SCD)	1×10^{-12}g of sulfur in sulfur compounds	1×10^4	10^7 by mass selectivity of S over carbon	Equimolar response to all sulfur compounds to within 10% Requires pure hydrogen and oxygen combustion gases Instrument generates ozone *in situ*, which must be catalytically destroyed at detector outlet Catalyst operates at 950 °C to 975 °C Detector operated at reduced pressure (10^{-3} Pa)
Electrolytic conductivity detector (ECD, Hall detector)	10×10^{-13} g to 1×10^{-12} g of chlorinated compounds, 2×10^{-12} g of sulfur compounds, 4×10^{-12} g of nitrogen compounds	1×10^6 for chlorinated compounds, 10^4 for sulfur and nitrogen compounds	10^6 by mass selectivity of Cl over carbon, 10^5 to 10^6 by mass selectivity of S and N over carbon	Only high purity solvents should be used Carbon particles in conductivity chamber can be problematic Frequent cleaning and maintenance is required Often used in conjunction with a photoionization detector For chlorine, use hydrogen as the reactant gas and n-propanol as the electrolyte For nitrogen or sulfur, hydrogen or oxygen can be used as reactant gas, and water or methanol as the electrolyte Ultrahigh purity reactant gases are required
Ion mobility detector (IMD)	1×10^{-12} g	1×10^3 to 1×10^4	10^3	Amenable to use in handheld instruments Linear dynamic range of 10^3 for radioactive sources and 10^5 for photo-ionization sources Selectivity depends on mobility differences of ions Has been used for a wide variety of compounds including amino acids, halogenated organics, explosives The radioactive ^{63}Ni source is subject to regulation and periodic inspection
Mass selective detector (MSD, mass spectrometer, MS)	1×10^{-11} g (single ion monitoring) 1×10^{-8} g (scan mode)	1×10^5	Universal	Quadrupole, dual quadrupole, and magnetic sector instruments available Must operate under moderate vacuum (1×10^{-4} Pa) Requires a molecular jet separator to operate with packed columns Amenable to library searching for qualitative identification Requires tuning of electronic optics over the entire m/e range of interest
Infrared detector (IRD)	1×10^{-9} g of a strong infrared absorber	1×10^3	Universal for compounds with mid-infrared active functionality	A costly and temperamental instrument that requires high purity carrier gas, a nitrogen purge of optical components (purified air will, in general, not be adequate) Must be isolated from vibrations Presence of carbon dioxide is a typical impurity band at 2200 cm^{-1} to 2300 cm^{-1} Requires frequent cleaning and optics maintenance Amenable to library searching for qualitative identification
Atomic emission detector (AED)	1×10^{-13} g to 2×10^{-11} g of each element	1×10^3 to 1×10^4	10^3 to 10^5, element to element	Requires the use of ultra high purity carrier and plasma gases Plasma produced in a microwave cavity operated at 2450 MHz Scavenger gases (H_2, O_2) are used as dopants Photodiode array is used to detect emitted radiation

SOLID-PHASE MICROEXTRACTION SORBENTS

Thomas J. Bruno and Paris D. N. Svoronos

While trapping sorbents have been used for many years in head-space analysis (most commonly with gas chromatography), the modern techniques of solid-phase microextraction (SPME) are particularly applicable to survey analyses (Ref. 1). In the following tables, we provide information for the selection and application of the various fibers, and data on salting out reagents (Refs. 1–5). For information on other trapping sorbents, chelating agents, resins, and polymeric phases used for headspace analysis, see Ref. 1.

References

1. Bruno, T. J., and Svoronos, P. D. N., *CRC Handbook of Basic Tables for Chemical Analysis*, 3rd Edition, CRC Press, Boca Raton, FL, 2011.
2. Haynes, W. M., Ed., *CRC Handbook for Chemistry and Physics*, 92nd Edition, CRC Press, Boca Raton, FL, 2011.
3. *NIST Chemistry Web Book*, www.webbook.nist.gov/chemistry/, 2009.
4. Machata, G., *Clin. Chem. Newsletter* 4, 29, 1972.
5. Ioffe, B. V., and Vitenberg, A. G., *Head Space Analysis and Related Methods in Gas Chromatography*, Wiley Interscience, New York, 1983.

Fiber Selection Criteria

The main fiber selection parameters are polarity and relative molecular mass. The table below provides general guidelines on the applicability of available fibers relative to these two parameters. The fibers are characterized by the extraction mechanism, either adsorption or absorption. Adsorbent fibers contain particles suspended in polydimethyl siloxane (PDMS) or polyethylene glycol (PEG, Carbowax).

Fiber	Type of Fiber	Polarity	RMM Range
7 μm PDMS	Absorbent	Nonpolar	150–700
30 μm PDMS	Absorbent	Nonpolar	80–600
85 μm Polyacrylate	Absorbent	Moderately polar	60–450
100 μm PDMS	Absorbent	Nonpolar	55–400
50 μm Carbowax (PEG)	Adsorbent	Polar	50–400
PDMS-DVB	Adsorbent	Bipolar	50–350
Carbowax-DVB	Adsorbent	Polar	50–350
PDMS-DVB-Carboxen	Adsorbent	Bipolar	40–270
PDMS-Carboxen	Adsorbent	Bipolar	35–180
Carbopak Z-PDMS	Adsorbent	Nonpolar	50–500

PDMS – Polydimethylsiloxane
DVB – Divinylbenzene (3 μm to 5 μm particles)
PEG – Polyethylene glycol
Carboxen – Carboxen 1006 (contains micro-, meso-, and macro-tapered pores) (3 μm to 5 μm particles)

RMM **Range** – Relative molecular mass range that is the ideal range for optimum extraction. Ranges can be extended by varying extraction times, but results will not be optimized

Phase Material Characteristics

Polydimethylsiloxane (PDMS):
Similar in properties to the OV-1 or SE-30 silicone phases (Ref. 1); non-polar fluid suitable for non-polar or slightly polar analytes; thicker coatings extract more analyte, but require longer extraction times.

Polyacrylate:
Rigid solid material; moderate polarity; diffusion of analytes through bulk is relatively slow because of rigidity of material; relatively higher desorption temperatures required because of rigidity of material; can be oxidized easily at higher temperatures; must use oxygen-free carrier gas and ensure gas chromatographic system is leak-free; fibers are very solvent resistant; darkens to a brown color upon exposure to temperatures in excess of 280 °C, but fiber is generally still usable until color becomes black.

Carbowax (polyethylene glycol, PEG):
Similar in properties to the PEG coatings used extensively in chromatography; moderately polar; highly crosslinked to counteract water solubility; sensitive to attack by oxygen at temperatures in excess of 220 °C, at which point the fiber will darken and become powdery; requires use of high purity carrier gas (typically He at 99.999 % mass/mass) treated for oxygen contamination.

Divinlybenzene (DVB):
Similar to the properties of divinylbenzene porous polymer phases; higher polarity than Carbowax, and when combined with Carbowax results in a more polar phase; like polyacrylate, it is a solid particle that must be carried in a liquid to coat on a fiber.

Carboxen:
Similar to the material used in Carboxen porous-layer open tubular (PLOT) columns; structure has an approximately even distribution of macro-, meso-, and micro-pores, making it valuable for smaller analytes; larger analytes can show hysteresis that must be addressed by desorption at 280 °C.

Extraction Capability of Solid-Phase Microextraction Sorbents

This table shows the extraction capability of the fibers for acetone, a small, moderately polar analyte, for 4-nitrophenol, a medium size polar analyte, and benzo(GHI)perylene, a large nonpolar analyte. This provides a general guideline for fiber selection.

Fiber	Approx. Linear Conc. Range Acetone 10 min Ext[a] (FID)	Approx. Linear Conc. Range 4-Nitrophenol 20 min Ext[b] (GC/MS)	Approx. Linear Conc. Range Benzo(GHI) Perylene 20 min Ext
7 μm PDMS	100 ppm and up	Not extracted	100 ppt to 500 ppb
30 μm PDMS	10 ppm and up	10 ppm and up	100 ppt to 10 ppm
85 μm Polyacrylate	1 ppm to 1000 ppm	5 ppb to 100 ppm	500 ppt to 10 ppm
100 μm PDMS	500 ppb to 1000 ppm	500 ppb to 500 ppm	500 ppt to 10 ppm
50 μm Carbowax (PEG)	1 ppm to 1000 ppm	5 ppb to 50 ppm	25 ppb to 10 ppm
PDMS-DVB	50 ppb to 100 ppm	25 ppb to 10 ppm	10 ppb to 1 ppm
Carbowax-DVB	100 ppb to 100 ppm	5 ppb to 10 ppm	50 ppb to 5 ppm
PDMS-DVB-Carboxen	25 ppb to 10 ppm	50 ppb to 10 ppm	100 ppb to 1 ppm poorly desorbed
PDMS-Carboxen	5 ppb to 5 ppm	100 ppb to 10 ppm	Not desorbed
Carbopak Z-PDMS	10 ppm to 500 ppm	5 ppm to 100 ppm	500 ppt to 100 ppb

Note: In each case, the concentration is expressed on a mass basis (e.g., ppm mass/mass).
[a] Water sample contains 25% NaCl (mass/mass)
[b] Water sample contains 2% NaCl (mass/mass) acidified to pH = 2 with 0.05 M phosphoric acid
   1 ppm = 1 part in 1×10^6
   1 ppb = 1 part in 1×10^9
   1 ppt = 1 part in 1×10^{12}

Typical Phase Volumes of SPME Fiber Coatings

Fiber Coating Thickness/Type	Type of Fiber Core	Fiber Core Diameter/mm	Phase Volume/mm³ or μL
PDMS	Fused silica	0.110	0.612
100 μm PDMS	Metal	0.130	0.598
30 μm PDMS	Fused silica	0.110	0.132
30 μm PDMS	Metal	0.130	0.136
7 μm PDMS	Fused silica	0.110	0.028
7 μm PDMS	Metal	0.130	0.030
85 μm PA	Fused silica	0.110	0.543
60 μm PEG	Metal	0.130	0.358
15 μm Carbopack Z/PDMS	Metal	0.130	0.068
65 μm PDMS/DVB	Fused silica	0.120	0.418
65 μm PDMS/DVB	Proprietary	0.130	0.440
65 μm PDMS/DVB	Metal	0.130	0.440
75 μm Carboxen-PDMS	Fused silica	0.120	0.502
85 μm Carboxen-PDMS	Proprietary	0.130	0.528
85 μm Carboxen-PDMS	Metal	0.130	0.528
50/30 μm DVB/Carboxen	Metal		
Carboxen layer		0.130	0.151
DVB layer		0.190	0.377
50/30 μm DVB/Carboxen	Metal		
Carboxen layer		0.130	0.151
DVB layer		0.190	0.377
60 μm PDMS-DVB HPLC	Proprietary	0.160	0.459

Salting-Out Reagents for Headspace Analysis

The following table provides data on the common salts used for salting out in chromatographic headspace analysis, as applied to direct injection methods and to solid-phase microextraction. Data are provided for the most commonly available salts, although others are possible. Sodium citrate, for example, occurs as the dihydrate and the pentahydrate. The pentahydrate is not as stable as the dihydrate, however, and dries out on exposure to air, forming cakes. Potassium carbonate occurs as the dihydrate, trihydrate, and sesquihydrate; however, data are provided only for the anhydrous material. The solubility is provided as the number of grams that can dissolve in 100 mL of water at the indicated temperature. The vapor enhancement cited is the degree of increase of the concentration of vapor over the solution of a 2 % (mass/mass) ethanol solution in water at 60 °C.

Salt	Formula	Rel. Mol. Mass	Density/g cm⁻³	Solubility/g mL⁻¹ H₂O Cold Water	Solubility/g mL⁻¹ H₂O Hot Water	Vapor Enhancement
Potassium carbonate	K_2CO_3	138.21	2.428 at 14 °C	112[a]	156[b]	8
Ammonium sulfate	$(NH_4)_2SO_4$	132.13	1.769 at 50 °C	70.6[c]	103.8[b]	5
Sodium citrate (dihydrate)	$Na_3C_6H_5O_7 \times 2H_2O$	294.10		72[d]	167[b]	5
Sodium chloride	NaCl	58.44	2.165[e]	37.5[a]	39.12[b]	3
Ammonium chloride	NH_4Cl	53.49	1.527	29.7[c]	75.8[b]	2

[a] 20 °C
[b] 100 °C
[c] 0 °C
[d] 25 °C
[e] Specific gravity, 25 °C/4 °C

ELUOTROPIC VALUES OF SOLVENTS ON OCTADECYLSILANE AND OCTYLSILANE

Thomas J. Bruno and Paris D. N. Svoronos

The following table provides, for comparative and interpretive purposes, eluotropic values on bonded octadecylsilane (ODS) and octylsilane (OS) for common solvents used in HPLC (Refs. 1–3). For additional information on common, specific and chiral stationary phases for HPLC, and for solvents, derivatizing reagents, and detectors, see Ref. 3.

References

1. Krieger, P. A., *High Purity Solvent Guide*, Burdick and Jackson Laboratories, McGaw Park, IL, 1984.
2. Ahuja, S., *Trace and Ultratrace Analysis by HPLC*, John Wiley and Sons, New York, 1992.
3. Bruno, T. J., and Svoronos, P. D. N., *CRC Handbook of Basic Tables for Chemical Analysis*, 3rd Edition, CRC Press, Boca Raton, FL, 2011.

Solvent	Eluotropic Value, ODS	Eluotropic Value, OS
Acetic acid	—	2.7
Acetone	8.8	9.3
Acetonitrile	3.1	3.3
1,4-dioxane	11.7	13.5
Dimethyl-formamide	7.6	9.4
Methanol	1.0	1.0
Ethanol	3.1	3.2
n-Propanol	10.1	10.8
2-Propanol	8.3	8.4
Tetrahydrofuran	3.7	—

SOLVENTS FOR ULTRAVIOLET SPECTROPHOTOMETRY

This table lists some solvents commonly used for sample preparation for ultraviolet spectrophotometry. The properties given are:

λ_c: cutoff wavelength, below which the solvent absorption becomes excessive.

ε: dielectric constant (relative permittivity); the temperature in °C is given as a superscript.

t_b: normal boiling point.

References

1. Bruno, T. J., and Svoronos, P. D. N., *CRC Handbook of Basic Tables for Chemical Analysis*, 3rd Edition, CRC Press, Boca Raton, FL, 2011.
2. *Landolt-Börnstein, Numerical Data and Functional Relationships in Science and Technology, New Series*, IV/6, *Static Dielectric Constants of Pure Liquids and Binary Liquid Mixtures*, Springer–Verlag, Heidelberg, 1991.

Name	λ_c/nm	ε	t_b/°C
Acetic acid	260	6.20^{20}	117.9
Acetone	330	21.01^{20}	56.0
Acetonitrile	190	36.64^{20}	81.6
Benzene	280	2.28^{20}	80.0
2-Butanol	260	17.26^{20}	99.5
Butyl acetate	254	5.07^{20}	126.1
Carbon disulfide	380	2.63^{20}	46
Carbon tetrachloride	265	2.24^{20}	76.8
1-Chlorobutane	220	7.28^{20}	78.6
Chloroform	245	4.81^{20}	61.1
Cyclohexane	210	2.02^{20}	80.7
1,2-Dichloroethane	226	10.42^{20}	83.5
Dichloromethane	235	8.93^{25}	40
Diethyl ether	218	4.27^{20}	34.5
N,N-Dimethylacetamide	268	38.85^{21}	165
N,N-Dimethylformamide	270	38.25^{20}	153
Dimethyl sulfoxide	265	47.24^{20}	189
1,4-Dioxane	215	2.22^{20}	101.5
Ethanol	210	25.3^{20}	78.2
Ethyl acetate	255	6.08^{20}	77.1
Ethylene glycol dimethyl ether	240	7.30^{24}	85
Ethylene glycol monoethyl ether	210	13.38^{25}	135
Ethylene glycol monomethyl ether	210	17.2^{25}	124.1
Glycerol	207	46.53^{20}	290
Heptane	197	1.92^{20}	98.5
Hexadecane	200	2.05^{20}	286.8
Hexane	210	1.89^{20}	68.7
Methanol	210	33.0^{20}	64.6
Methylcyclohexane	210	2.02^{20}	100.9
Methyl ethyl ketone	330	18.56^{20}	79.5
Methyl isobutyl ketone	335	13.11^{20}	116.5
2-Methyl-1-propanol	230	17.93^{20}	107.8
N-Methyl-2-pyrrolidone	285	32.55^{20}	202
Nitromethane	380	37.27^{20}	101.1
Pentane	210	1.84^{20}	36.0
Pentyl acetate	212	4.79^{20}	149.2
1-Propanol	210	20.8^{20}	97.2
2-Propanol	210	20.18^{20}	82.3
Pyridine	330	13.26^{20}	115.2
Tetrachloroethylene	290	2.27^{30}	121.3
Tetrahydrofuran	220	7.52^{22}	65
Toluene	286	2.38^{23}	110.6
1,1,2-Trichloro-1,2,2-trifluoroethane	231	2.41^{25}	47.7
2,2,4-Trimethylpentane	215	1.94^{20}	99.2
Water	191	80.10^{20}	100.0
o-Xylene	290	2.56^{20}	144.5
m-Xylene	290	2.36^{20}	139.1
p-Xylene	290	2.27^{20}	138.3

CORRELATION TABLE FOR ULTRAVIOLET ACTIVE FUNCTIONALITIES

Thomas J. Bruno and Paris D. N. Svoronos

The following table presents a correlation between common chromophoric functional groups and the expected absorptions from ultraviolet (UV) spectrophotometry. While not as informative as infrared correlations, UV can often provide valuable qualitative information. In these tables, λ_{max} is the wavelength in nm at which the maximum absorption occurs, and ε_{max} is the extinction coefficient.

References

1. Willard, H. H., Merritt, Jr., L. L., Dean, J. A., and Settle, F. A., *Instrumental Methods of Analysis*, 7th Edition, Wadsworth Publishing Co., Belmont, CA, 1988.

2. Silverstein, R. M., and Webster, F. X., *Spectrometric Identification of Organic Compounds*, 6th Edition, Wiley, New York, 1998.

3. Lambert, J. B., Shurvell, H. F., Lightner D. A., Verbit, L. and Cooks, R. G., *Organic Structural Spectroscopy*, Prentice Hall, Upper Saddle River, NJ, 1998.

4. Bruno, T. J., and Svoronos, P. D. N., *CRC Handbook of Basic Tables for Chemical Analysis*, 3rd Edition, CRC Press, Boca Raton, FL, 2011.

5. Woodward, R. B., *J. Am. Chem. Soc.* 63, 1123, 1941.

6. Woodward, R. B., *J. Am. Chem. Soc.* 64, 72, 1942.

7. Woodward, R. B., *J. Am. Chem. Soc.* 64, 76, 1942.

8. Fieser, L. F., and Fieser, M., *Natural Products Related to Phenanthrene*, Third Edition, Reinhold, New York, 1949.

Chromophore	Functional Group	λ_{max}/nm	ε_{max}	λ_{max}/nm	ε_{max}	λ_{max}/nm	ε_{max}
Ether	$-O-$	185	1000				
Thioether	$-S-$	194	4600	215	1600		
Amine	$-NH_2-$	195	2800				
Amide	$-CONH_2$	<210	—				
Thiol	$-SH$	195	1400				
Disulfide	$-S-S-$	194	5500	255	400		
Bromide	$-Br$	208	300				
Iodide	$-I$	260	400				
Nitrile	$-C\equiv N$	160	—				
Acetylide (alkyne)	$-C\equiv C-$	175–180	6000				
Sulfone	$-SO_2-$	180	—				
Oxime	$-NOH$	190	5000				
Azido	$>C=N-$	190	5000				
Alkene	$-C=C-$	190	8000				
Ketone	$>C=O$	195	1000	270–285	18–30		
Thioketone	$>C=S$	205	strong				
Esters	$-COOR$	205	50				
Aldehyde	$-CHO$	210	strong	280–300	11–18		
Carboxyl	$-COOH$	200–210	50–70				
Sulfoxide	$>S{\rightarrow}O$	210	1500				
Nitro	$-NO_2$	210	strong				
Nitrite	$-ONO$	220–230	1000–2000	300–4000	10		
Azo	$-N=N-$	285–400	3–25				
Nitroso	$-N=O$	302	100				
Nitrate	$-ONO_2$	270 (shoulder)	12				
Conjugated hydrocarbon	$-(C=C)_2-$ (acyclic)	210–230	21 000				
Conjugated hydrocarbon	$-(C=C)_3-$	260	35 000				
Conjugated hydrocarbon	$-(C=C)_4-$	300	52 000				
Conjugated hydrocarbon	$-(C=C)_5-$	330	118 000				
Conjugated hydrocarbon	$-(C=C)_2-$ (alicyclic)	230–260	3000–8000				
Conjugated hydrocarbon	$C=C-C\equiv C$	219	6500				
Conjugated system	$C=C-C=N$	220	23 000				
Conjugated system	$C=C-C=O$	210–250	10 000–20 000			300–350	weak
Conjugated system	$C=C-NO_2$	229	9500				
Phenyl		184	46 700	202	6900	255	170

Chromophore	Functional Group	λ_{max}/nm	ε_{max}	λ_{max}/nm	ε_{max}	λ_{max}/nm	ε_{max}
Diphenyl				246	20 000		
Naphthalene		220	112 000	275	5600	312	175
Anthracene		252	199 000	375	7900		
Pyridine		174	80 000	195	6000	251	1700
Quinoline		227	37 000	270	3600	314	2750
Isoquinoline		218	80 000	266	4000	317	3500

Note: φ denotes a phenyl group.

Wavenumber Adjustments for Bathochromic Shifts (Woodward's Rules)

Conjugated systems show bathochromic shifts in their π→π* transition bands. Empirical methods for predicting those shifts were originally formulated by Woodward (References 5–7) and Fieser and Fieser (Reference 8). This section includes the most important conjugated system rules. The reader should consult References 6 and 8 for more details on how to apply the wavelength increment data.

(a) Rules for Diene Absorption

Base value for diene: 214 nm
Increments for each (in nm):

Heteroannular diene	+ 0
Homoannular diene	+39
Extra double bond	+30
Alkyl substituent or ring residue	+5
Exocyclic double bond	+5
Polar groups:	
−OOCR	+0
−OR	+6
−S−R	+30
halogen	+5
−NR$_2$	+60
λ Calculated	= Total

(b) Rules for Enone Absorption*

$$\overset{\delta}{-}C=\overset{\gamma}{C}-\overset{\beta}{C}=\overset{\alpha}{C}-C-$$
with O double-bonded to the final C

Base value for acyclic (or six-membered) α,β-unsaturated ketone: 215 nm
Base value for five-membered α,β-unsaturated ketone: 202 nm
Base value for α,β-unsaturated aldehydes: 210 nm

Base value for α,β-unsaturated esters or carboxylic acids: 195 nm
Increments for each (in nm):

Heteroannular diene	+0
Homoannular diene	+39
Double bond	+30
Alkyl group:	
α−	+10
β−	+12
γ− and higher	+18
Polar groups:	
−OH	
α−	+35
β−	+30
δ−	+50
−OOCR	
α, β, γ, δ	+6
−OR	
α−	+35
β−	+30
γ−	+17
δ−	+31
−SR	
β−	+85
−Cl	
α−	+15
β−	+12
−Br	
α−	+25
β−	+30
−NR$_2$	
β−	+95
Exocyclic double bond	+5
λ Calculated	= Total

---

*Solvent corrections should be included. These are: water (-8), chloroform (+1), dioxane (+5), ether (+7), hexane (+11), and cyclohexane (+11). No correction is included for methanol or ethanol.

(c) Rules for monosubstituted benzene derivatives
   Parent Chromophore (benzene): 250 nm

Substituent	Increment (in nm)
–R	–4
–COR	–4
–CHO	0
–OH	–16
–OR	–16
–COOR	–16

where R is an alkyl group, and the substitution is on $C_6H_5–$.

(d) Rules for disubstituted benzene derivatives
   Parent Chromophore (benzene): 250 nm

	Increment (in nm)		
Substituent	*o–*	*m–*	*p–*
–R	+3	+3	+10
–COR	+3	+3	+10
–OH	+7	+7	+25
–OR	+7	+7	+25
$–O^-$	+11	+20	+78 (variable)
–Cl	+0	+0	+10
–Br	+2	+2	+15
$–NH_2$	+13	+13	+58
$–NHCOCH_3$	+20	+20	+45
$–NHCH_3$	—	—	+73
$–N(CH_3)_2$	+20	+20	+85

R indicates an alkyl group.

WAVELENGTH-WAVENUMBER CONVERSION TABLE

Thomas J. Bruno and Paris D. N Svoronos

The following table provides a conversion between wavelength and wavenumber units, for use in infrared spectrophotometry (Refs. 1 and 2). Because spectra are presented in different formats, this table is an aid in interpretation.

References

1. Bruno, T. J., and Svoronos, P. D. N., *CRC Handbook of Basic Tables for Chemical Analysis*, 3rd Edition, CRC Press, Boca Raton, FL, 2011.
2. Bruno, T. J., and Svoronos, P. D. N., *CRC Handbook of Fundamental Spectroscopic Correlation Charts*, CRC Press, Boca Raton, FL, 2006.

Wavelength/μm	Wavenumber/cm⁻¹									
	0	1	2	3	4	5	6	7	8	9
2.0	5000	4975	4950	4926	4902	4878	4854	4831	4808	4785
2.1	4762	4739	4717	4695	4673	4651	4630	4608	4587	4566
2.2	4545	4525	4505	4484	4464	4444	4425	4405	4386	4367
2.3	4348	4329	4310	4292	4274	4255	4237	4219	4202	4184
2.4	4167	4149	4232	4115	4098	4082	4065	4049	4032	4016
2.5	4000	3984	3968	4953	3937	3922	3006	3891	3876	3861
2.6	3846	3831	3817	3802	3788	3774	3759	3745	3731	3717
2.7	3704	3690	3676	3663	3650	3636	3623	3610	3597	3584
2.8	3571	3559	3546	3534	3521	3509	3497	3484	3472	3460
2.9	3448	3436	3425	3413	3401	3390	3378	3367	3356	3344
3.0	3333	3322	3311	3300	3289	3279	3268	3257	3247	3236
3.1	3226	3215	3205	3195	3185	3175	3165	3155	3145	3135
3.2	3125	3115	3106	3096	3086	3077	3067	3058	3049	3040
3.3	3030	3021	3012	3003	2994	2985	2976	2967	2959	2950
3.4	2941	2933	2924	2915	2907	2899	2890	2882	2874	2865
3.5	2857	2849	2841	2833	2825	2817	2809	2801	2793	2786
3.6	2778	2770	2762	2755	2747	2740	2732	2725	2717	2710
3.7	2703	2695	2688	2681	2674	2667	2660	2653	2646	2639
3.8	2632	2625	2618	2611	2604	2597	2591	2584	2577	2571
3.9	2654	2558	2551	2545	2538	2532	2525	2519	2513	2506
4.0	2500	2494	2488	2481	2475	2469	2463	2457	2451	2445
4.1	2439	2433	2427	2421	2415	2410	2404	2398	2387	2387
4.2	2381	2375	2370	2364	2358	2353	2347	2342	2336	2331
4.3	2326	2320	2315	2309	2304	2299	2294	2288	2283	2278
4.4	2273	2268	2262	2257	2252	2247	2242	2237	2232	2227
4.5	2222	2217	2212	2208	2203	2198	2193	2188	2183	2179
4.6	2174	2169	2165	2160	2155	2151	2146	2141	2137	2132
4.7	2128	2123	2119	2114	2110	2105	2101	2096	2092	2088
4.8	2083	2079	2075	2070	2066	2062	2058	2053	2049	2045
4.9	2041	2037	2033	2028	2024	2020	2016	2012	2008	2004
5.0	2000	1996	1992	1988	1984	1980	1976	1972	1969	1965
5.1	1961	1957	1953	1949	1946	1942	1938	1934	1931	1927
5.2	1923	1919	1916	1912	1908	1905	1901	1898	1894	1890
5.3	1887	1883	1880	1876	1873	1869	1866	1862	1859	1855
5.4	1852	1848	1845	1842	1838	1835	1832	1828	1825	1821
5.5	1818	1815	1812	1808	1805	1802	1799	1795	1792	1788
5.6	1786	1783	1779	1776	1773	1770	1767	1764	1761	1757
5.7	1754	1751	1748	1745	1742	1739	1736	1733	1730	1727
5.8	1724	1721	1718	1715	1712	1709	1706	1704	1701	1698
5.9	1695	1692	1689	1686	1684	1681	1678	1675	1672	1669
6.0	1667	1664	1661	1668	1656	1653	1650	1647	1645	1642
6.1	1639	1637	1634	1631	1629	1626	1623	1621	1618	1616
6.2	1613	1610	1608	1605	1603	1600	1597	1595	1592	1590
6.3	1587	1585	1582	1580	1577	1575	1572	1570	1567	1565
6.4	1563	1560	1558	1555	1553	1550	1548	1546	1543	1541
6.5	1538	1536	1534	1531	1529	1527	1524	1522	1520	1517
6.6	1515	1513	1511	1508	1506	1504	1502	1499	1497	1495
6.7	1493	1490	1488	1486	1484	1481	1479	1477	1475	1473
6.8	1471	1468	1466	1464	1462	1460	1458	1456	1453	1451
6.9	1449	1447	1445	1443	1441	1439	1437	1435	1433	1431

Wavelength/μm	Wavenumber/cm⁻¹									
	0	1	2	3	4	5	6	7	8	9
7.0	1429	1427	1425	1422	1420	1418	1416	1414	1412	1410
7.1	1408	1406	1404	1403	1401	1399	1397	1395	1393	1391
7.2	1389	1387	1385	1383	1381	1379	1377	1376	1374	1372
7.3	1370	1368	1366	1364	1362	1361	1359	1357	1355	1353
7.4	1351	1350	1348	1346	1344	1342	1340	1339	1337	1335
7.5	1333	1332	1330	1328	1326	1325	1323	1321	1319	1318
7.6	1316	1314	1312	1311	1309	1307	1305	1304	1302	1300
7.7	1299	1297	1295	1294	1292	1290	1289	1287	1285	1284
7.8	1282	1280	1279	1277	1276	1274	1272	1271	1269	1267
7.9	1266	1264	1263	1261	1259	1258	1256	1255	1253	1252
8.0	1250	1248	1247	1245	1244	1242	1241	1239	1238	1236
8.1	1235	1233	1232	1230	1229	1227	1225	1224	1222	1221
8.2	1220	1218	1217	1215	1214	1212	1211	1209	1208	1206
8.3	1205	1203	1202	1200	1199	1198	1196	1195	1193	1192
8.4	1190	1189	1188	1186	1185	1183	1182	1181	1179	1178
8.5	1176	1175	1174	1172	1171	1170	1168	1167	1166	1164
8.6	1163	1161	1160	1159	1157	1156	1155	1153	1152	1151
8.7	1149	1148	1147	1145	1144	1143	1142	1140	1139	1138
8.8	1136	1135	1134	1133	1131	1130	1129	1127	1126	1125
8.9	1124	1122	1121	1120	1119	1117	1116	1115	1114	1112
9.0	1111	1110	1109	1107	1106	1105	1104	1103	1101	1100
9.1	1099	1098	1096	1095	1094	1093	1092	1091	1089	1088
9.2	1087	1086	1085	1083	1082	1081	1080	1079	1078	1076
9.3	1075	1074	1073	1072	1071	1070	1068	1067	1066	1065
9.4	1064	1063	1062	1060	1059	1058	1057	1056	1055	1054
9.5	1053	1052	1050	1049	1048	1047	1046	1045	1044	1043
9.6	1042	1041	1040	1038	1037	1036	1035	1034	1033	1032
9.7	1031	1030	1029	1028	1027	1026	1025	1024	1022	1021
9.8	1020	1019	1018	1017	1016	1015	1014	1013	1012	1011
9.9	1010	1009	1008	1007	1006	1005	1004	1003	1002	1001
10.0	1000	999	998	997	996	995	994	993	992	991
10.1	990	989	988	987	986	985	984	983	982	981
10.2	980	979	978	978	977	976	975	974	973	972
10.3	971	970	969	968	967	966	965	964	963	962
10.4	962	961	960	959	958	957	956	955	954	953
10.5	952	951	951	950	949	948	947	946	945	944
10.6	943	943	942	941	940	939	938	937	936	935
10.7	935	934	933	932	931	930	929	929	928	927
10.8	926	925	924	923	923	922	921	920	919	918
10.9	917	917	916	915	914	913	912	912	911	910
11.0	909	908	907	907	906	905	904	903	903	902
11.1	901	900	899	898	898	897	896	895	894	894
11.2	893	892	891	890	890	889	888	887	887	886
11.3	885	884	883	883	882	881	880	880	879	878
11.4	877	876	876	875	874	873	873	872	871	870
11.5	870	869	868	867	867	866	865	864	864	863
11.6	862	861	861	860	859	858	858	857	856	855
11.7	855	854	853	853	852	851	850	850	849	848
11.8	847	847	846	845	845	844	843	842	842	841
11.9	840	840	839	838	838	837	836	835	835	834
12.0	833	833	832	831	831	830	829	829	828	827
12.1	826	826	825	824	824	823	822	822	821	820
12.2	820	819	818	818	817	816	816	815	814	814
12.3	813	812	812	811	810	810	809	808	808	807
12.4	806	806	805	805	804	803	803	802	801	801
12.5	800	799	799	798	797	797	796	796	795	794
12.6	794	793	792	792	791	791	790	789	789	788
12.7	787	787	786	786	785	784	784	783	782	782
12.8	781	781	780	779	779	778	778	777	776	776
12.9	775	775	774	773	773	772	772	771	770	770

Wavelength/μm	Wavenumber/cm⁻¹									
	0	1	2	3	4	5	6	7	8	9
13.0	769	769	768	767	767	766	766	765	765	764
13.1	763	763	762	762	761	760	760	759	759	758
13.2	758	757	756	756	755	755	754	754	753	752
13.3	752	751	751	750	750	749	749	748	747	747
13.4	746	746	745	745	744	743	743	742	742	741
13.5	741	740	740	739	739	738	737	737	736	736
13.6	735	735	734	734	733	733	732	732	731	730
13.7	730	729	729	728	728	727	727	726	726	725
13.8	725	724	724	723	723	722	722	721	720	720
13.9	719	719	718	718	717	717	716	716	715	715
14.0	714	714	713	713	712	712	711	711	710	710
14.1	709	709	708	708	707	707	706	706	705	705
14.2	704	704	703	703	702	702	702	701	701	700
14.3	699	699	698	698	697	697	696	696	695	695
14.4	694	694	693	693	693	692	692	691	691	690
14.5	690	689	689	688	688	687	687	686	686	685
14.6	685	684	684	684	683	683	682	682	681	681
14.7	680	680	679	679	678	678	678	677	677	676
14.8	676	675	675	674	674	673	673	672	672	672
14.9	671	671	670	670	669	669	668	668	668	667

MIDDLE-RANGE INFRARED ABSORPTION CORRELATION CHARTS

Thomas J. Bruno and Paris D.N. Svoronos

The following charts provide characteristic middle-range infrared absorptions obtained from particular functional groups on molecules (Refs. 1 and 2). These include a general mid-range correlation chart, a chart for aromatic absorptions, and a chart for carbonyl moieties. Charts for near infrared absorptions and for inorganic moieties can be found in the cited references.

References

1. Bruno, T. J. and Svoronos, P. D. N., *CRC Handbook of Basic Tables for Chemical Analysis*, 3rd Edition, CRC Press, Boca Raton, FL, 2011.
2. Bruno, T. J., and Svoronos, P. D. N., *CRC Handbook of Fundamental Spectroscopic Correlation Charts*, CRC Press, Boca Raton, FL, 2006.

Notes:

AR = aromatic
b = broad
sd = solid
sn = solution
sp = sharp
? = unreliable

Strong

Medium

Weak

Variable

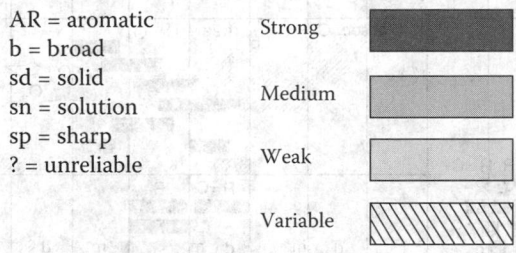

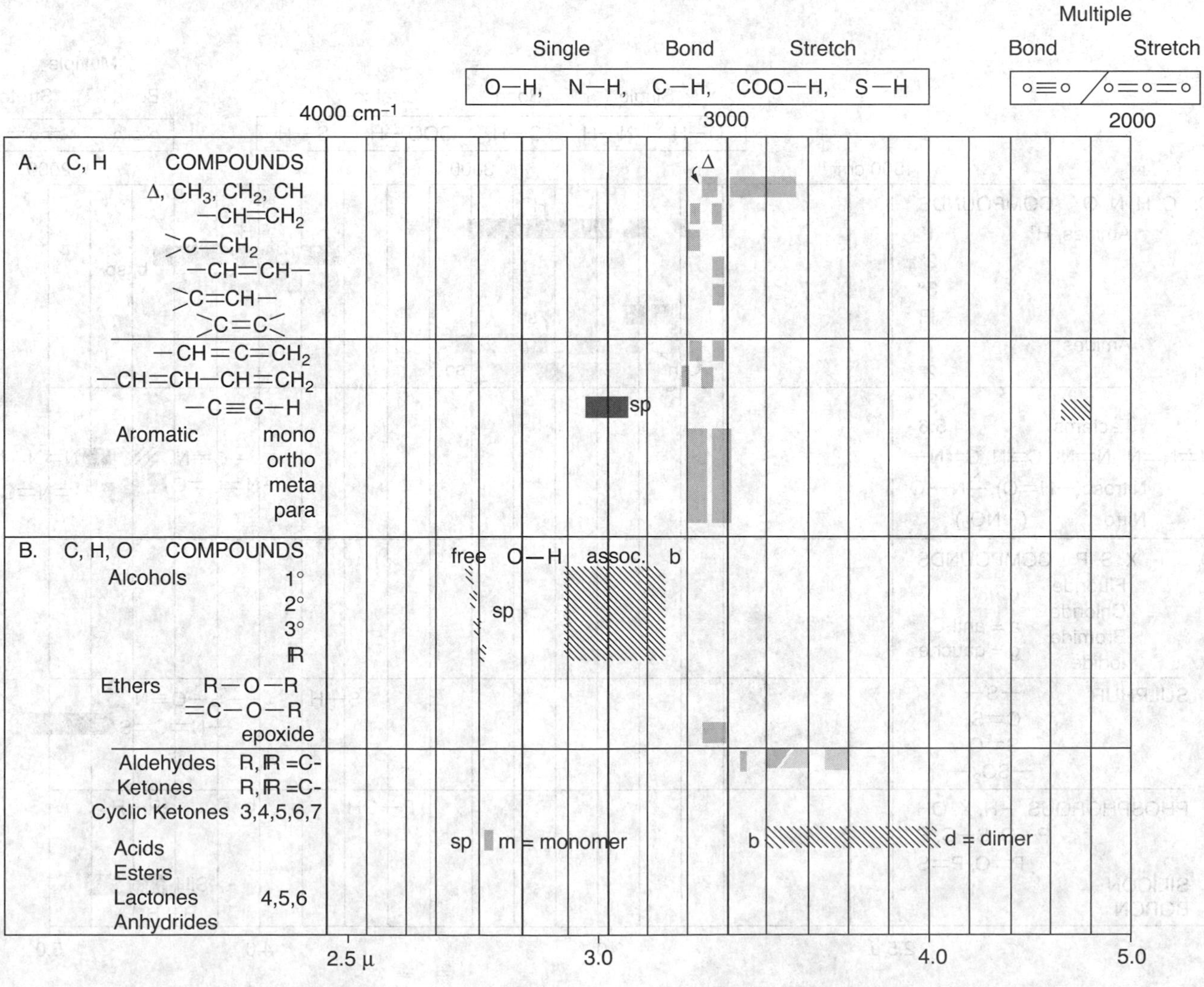

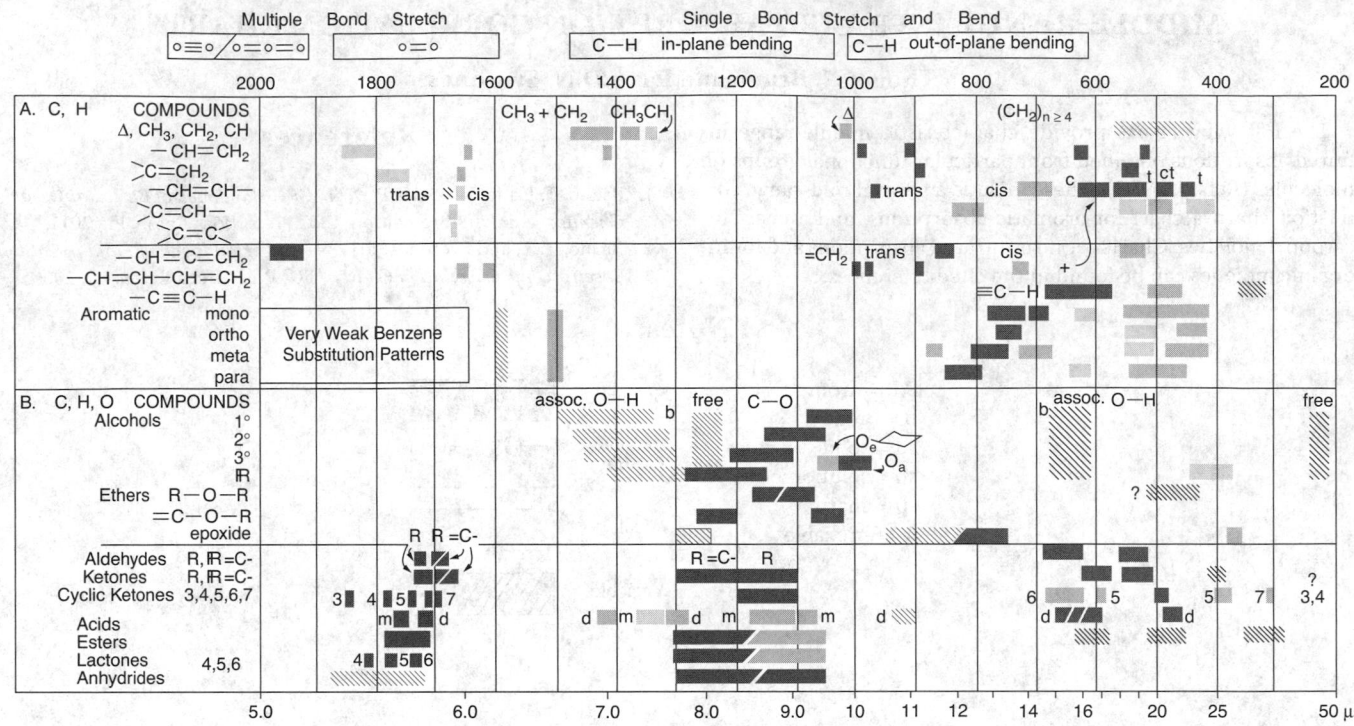

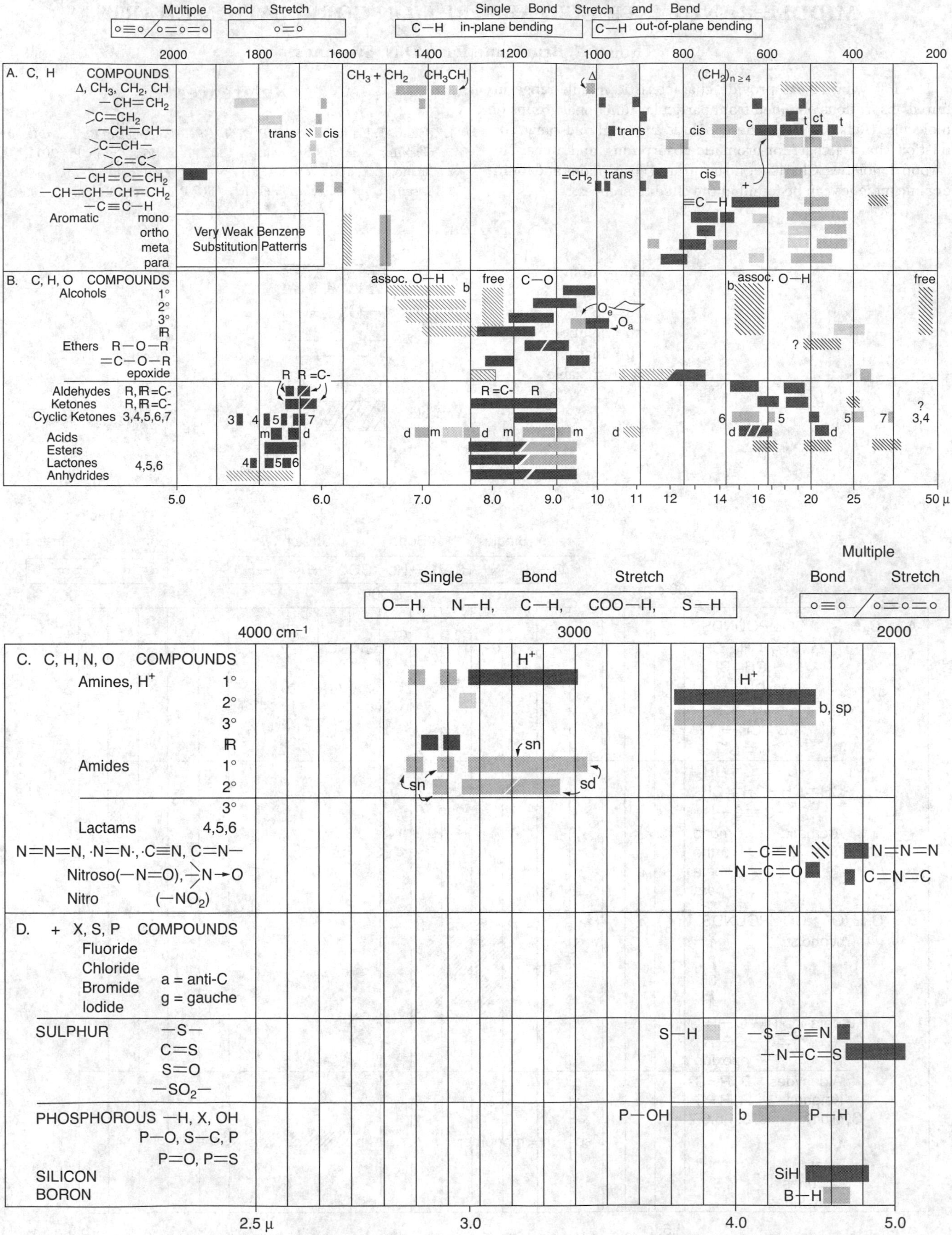

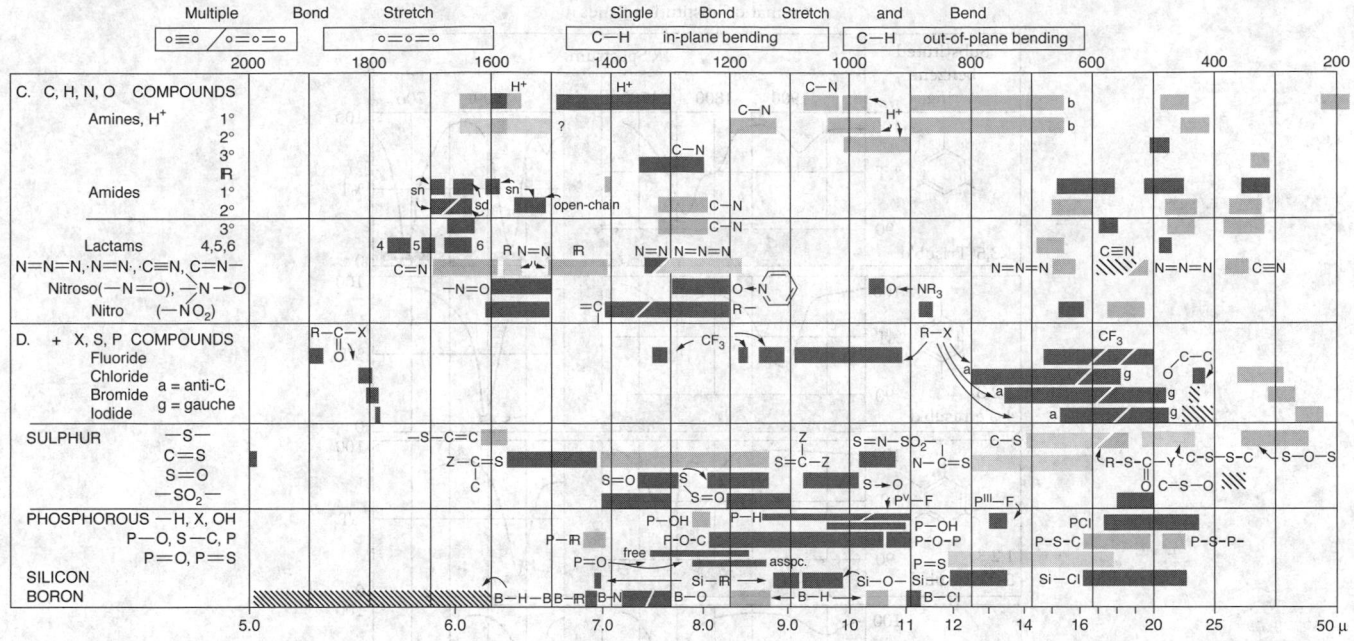

Aromatic Substitution Bands

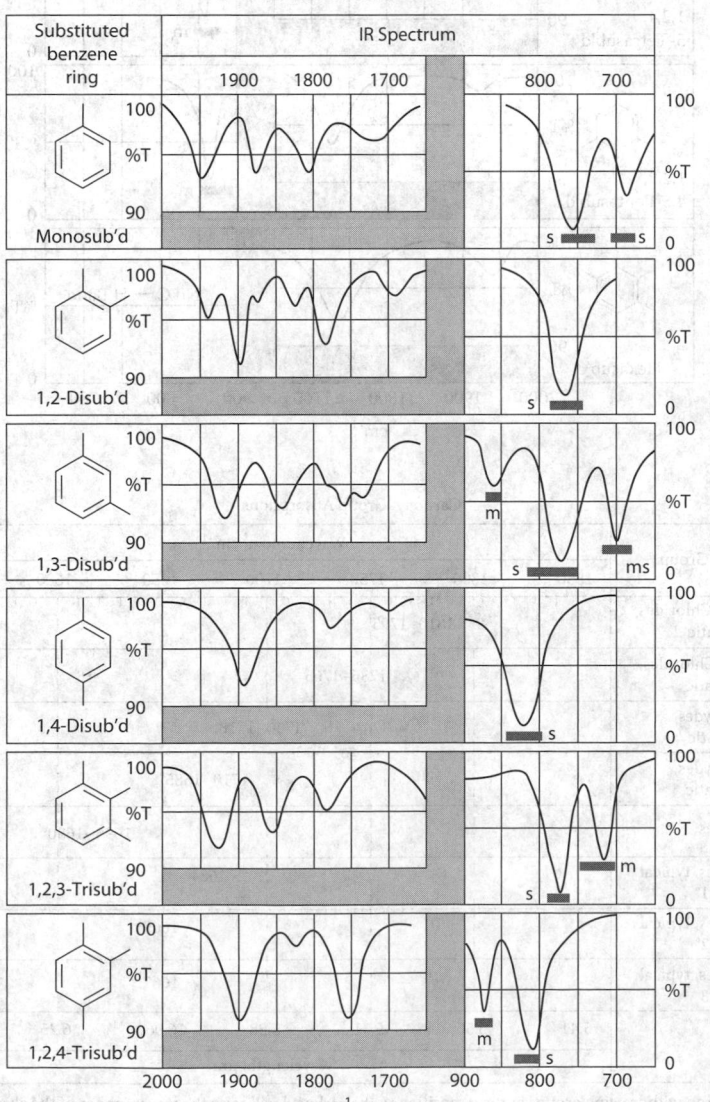

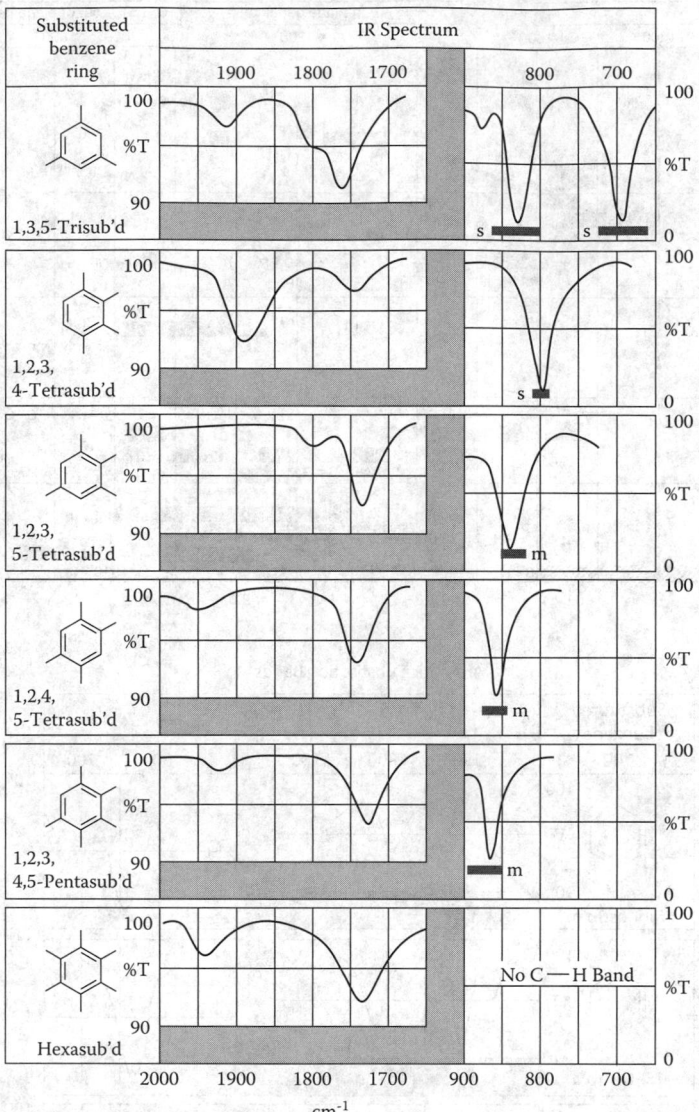

Aromatic Substitution Bands

Carbonyl Group Absorptions

Group	Wavenumber, cm^{-1}						
	1850	1800	1750	1700	1650	1600	1550
Acid, Chlorides, Aliphatic		1810–1795					
Acid Chlorides, Aromatic			1785–1765				
Aldehydes, Aliphatic				1740–1718			
Aldehydes, Aromatic				1710–1685			
Amides					1695–1630*		
Amides, typical value, 1°				1684			
Amides, typical value, 2°					1669		
Amides, typical value, 3°					1667		
	5.41	5.56	5.71	5.88	6.06	6.25	6.45
	Wavelength, µm						

* Electron withdrawing groups at the α-position to the carbonyl will raise the wavenumber of the absorption.

Carbonyl Group Absorptions (continued)

Group	Wavenumber, cm^{-1}
Anhydrides, acyclic, non-conjugated	1825–1815***; 1755–1745**
Anhydrides, acyclic, conjugated	1780–1770***; 1725–1715**
Anhydrides, ayclic non-conjugated	1870–1845; 1800–1775**
Anhydrides, cyclic conjugated	1860–1850; 1780–1760**
Carbamates	1740–1683
Carbonates, acyclic	1780–1740
Carbonates, five-membered ring	1850–1790
Carbonates, vinyl, typical value	1761

Wavenumber scale: 1850, 1800, 1750, 1700, 1650, 1600, 1550

Wavelength, µm scale: 5.41, 5.56, 5.71, 5.88, 6.06, 6.25, 6.45

** This band is the more intense of the two.
*** Intensity weakens as colinearity is approached.

Carbonyl Group Absorptions (continued)

Group	Wavenumber, cm^{-1}
Carboxylic acid, monomer	1800–1740
Carboxylic acid, dimer	1720–1680
Carboxylic acid, salts	1650–1540; 1450–1360
Carboxylic acid, conjugated	1695–1680
Carboxylic acid, non-conjugated	1720–1700
Esters, formate	1725–1720
Esters, saturated	1750–1735
Esters, conjugated	1735–1715*

Wavenumber scale: 1800, 1750, 1700, 1650, 1600, 1550, 1450, 1400, 1350

Wavelength, µm scale: 5.56, 5.71, 5.88, 6.06, 6.25, 6.45, 6.90, 7.14, 7.41

* Electron withdrawing groups in the α-position to the carbonyl will raise the wavenumber adsorption.

Carbonyl Group Absorptions (continued)

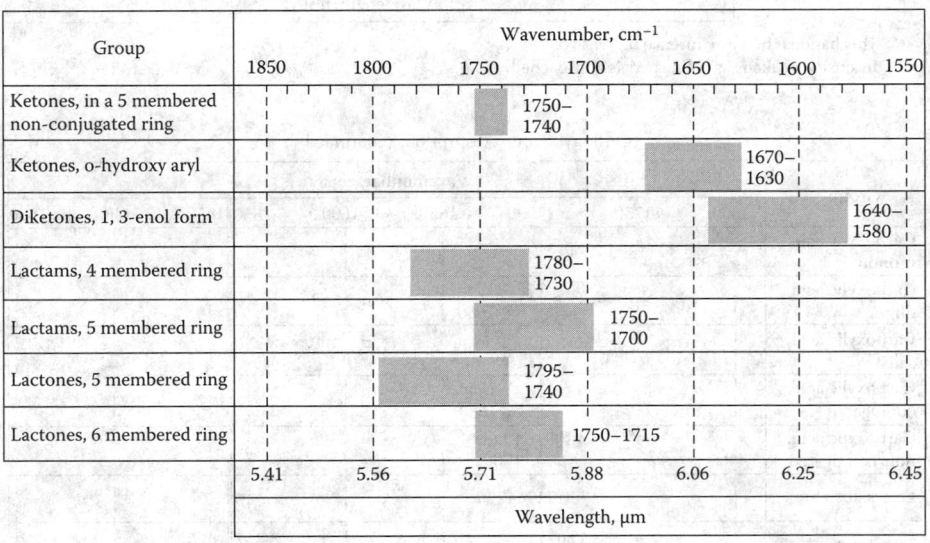

Group	Wavenumber, cm⁻¹							
	1800	1750	1700	1650	1600	1550	1450	1400	1350
Esters, phenyl, typical value	1770							
Esters, thiol, non-conjugated			1710–1680					
Esters, thiol, conjugated				1700–1640				
Esters, vinyl, typical value	1770							
Esters, vinylidene, typical value	1764							
Ketones, dialkyl			1725–1705					
Ketones, α, β- unsaturated			1700–1670					
Ketones, α, β, and α', β' conjugated				1680 1640				
Wavelength, μm	5.56	5.71	5.88	6.06	6.25	6.45 6.90	7.14	7.41

Carbonyl Group Absorptions (continued)

Group	Wavenumber, cm⁻¹						
	1850	1800	1750	1700	1650	1600	1550
Ketones, in a 5 membered non-conjugated ring			1750–1740				
Ketones, o-hydroxy aryl					1670–1630		
Diketones, 1, 3-enol form						1640–1580	
Lactams, 4 membered ring			1780–1730				
Lactams, 5 membered ring			1750–1700				
Lactones, 5 membered ring			1795–1740				
Lactones, 6 membered ring			1750–1715				
Wavelength, μm	5.41	5.56	5.71	5.88	6.06	6.25	6.45

COMMON SPURIOUS INFRARED ABSORPTION BANDS

Thomas J. Bruno and Paris D. N. Svoronos

The following table provides some of the common potential sources of spurious infrared absorptions that might appear on a spectrum (Refs. 1 and 2). Occasionally, the spectral lines of some impurities can be used as diagnostics; the reader is referred to the references for more details.

References

1. Bruno, T. J., and Svoronos, P. D. N., *CRC Handbook of Basic Tables for Chemical Analysis*, 3rd Edition, CRC Press, Boca Raton, FL, 2011.
2. Bruno, T. J., and Svoronos, P. D. N., *CRC Handbook of Fundamental Spectroscopic Correlation Charts*, CRC Press, Boca Raton, FL, 2006.

Approximate Wavenumber/cm^{-1}	Wavelength/μm	Compound or Group	Origin
3700	2.70	H_2O	Water in solvent (thick layers)
3650	2.74	H_2O	Water in some quartz windows
3450	2.9	H_2O	Hydrogen-bonded water, usually in KBr disks
2900	3.44	$-CH_3$, $>CH_2$	Paraffin oil, residual from previous mulls
2350	4.26	CO_2	Atmospheric absorption, or dissolved gas from a dry ice bath
2330	4.30	CO_2	
2300 and 2150	4.35 and 4.65	CS_2	Leaky cells, previous analysis of samples dissolved in carbon disulfide
1996	5.01	BO_2^-	Metaborate in the halide window
1400–2000	5–7	H_2O	Atmospheric absorption
1820	5.52	$COCl_2$	Phosgene, decomposition product in purified $CHCl_3$
1755	5.7	Phthalic anhydride	Decomposition product of phthalate esters or resins; paint off-gas product
1700–1760	5.7–5.9	$>C=O$	Bottle-cap liners leached by sample
1720	5.8	Phthalates	Phthalate polymer plastic tubing
1640	6.1	H_2O	Water of crystallization entrenched in sample
1520	6.6	CO_2	Leaky cells, previous analysis
1430	7.0	CO_3^{-2}	Contaminant in halide window
1360	7.38	NO_3^-	Contaminant in halide window
1270	7.9	$>SiO-$	Silicone oil or grease
1000–1110	9–10	$->Si-O-Si<-$	Glass; silicones
980	10.2	SO_4^{-2}	From decomposition of sulfates in KBr pellets
935	10.7	$(CH_2O)_x$	Deposit from gaseous formaldehyde
907	11.02	$->C-Cl$	Dissolved R-12 (Freon-12, CCl_2F_2)
837	11.95	NO_3^-	Contaminant in halide window
823	12.15	KNO_3	From decomposition of nitrates in KBr pellets
794	12.6	CCl_4 vapor	Leaky cells, from CCl_4 used as a solvent
788	12.7	CCl_4 liquid	Incomplete drying of cell or contamination, from CCl_4 used as a solvent
720 and 730	13.7 and 13.9	Polyethylene	Various experimental sources
728	13.75	$->Si-F$	SiF_4, found in NaCl windows
667	14.98	CO_3^{-2}	Atmospheric carbon dioxide
Any	Any	Fringes	If refractive index of windows is too high, or if the cell is partially empty, or the solid sample is not fully pulverized

NUCLEAR SPINS, MOMENTS, AND OTHER DATA RELATED TO NMR SPECTROSCOPY

David R. Lide

This table presents the following data relevant to nuclear magnetic resonance spectroscopy:

Z: Atomic number

Isotope: Element symbol and mass number

Abundance: Natural abundance of the isotope in percent. An * indicates a radioactive nuclide; if no value is given, the nuclide is not present in nature or its abundance is highly variable.

I: Nuclear spin

v: Resonant frequency in megahertz for an applied field H_0 of 1 tesla (in cgs units, 10 kilogauss). The resonant frequency scales with H_0.

Relative sensitivity: Sensitivity relative to ^{1}H (=1) assuming an equal number of nuclei and constant temperature. Values were calculated from the expressions:

For constant H_0: $0.0076508(\mu/\mu_N)^3(I + 1)/I^2$
For constant v: $0.23871(\mu/\mu_N)(I + 1)$

μ/μ_N: Nuclear magnetic moment in units of the nuclear magneton μ_N

Q: Nuclear quadrupole moment in units of femtometers squared (1 fm^2 = 10^{-2} barn). Because the determination of quadrupole moments requires knowledge of the electron configuration near the nucleus, values of Q in the literature tend to scatter considerably. The values quoted here come mainly from the review of Pyykkö (Ref. 3), otherwise from Ref. 1.

The table includes all stable nuclides of non-zero spin for which spin and magnetic moment values have been measured, as well as selected radioactive nuclides of current or potential interest. At least one isotope is included for each element through $Z = 95$ for which data are available. See Reference 1 for a complete listing of spins and moments.

The assistance of P. Pyykkö in providing data on nuclear quadrupole moments is gratefully acknowledged.

References

1. Holden, N. E., "Table of the Isotopes", in Haynes, W. M., Ed., *CRC Handbook of Chemistry and Physics*, 93rd Ed., CRC Press, Boca Raton, FL, 2012.
2. Raghavan, P., *At. Data Nucl. Data Tables* 42, 189, 1989.
3. Pyykkö, P., *Mol. Phys.* 106, 1965, 2008.
4. Stone, N. J., *At. Data Nucl. Data Tables* 90, 75, 2005.
5. IUPAC Commission on Physiochemical Symbols, Terminology and Units, *Quantities, Units, and Symbols in Physical Chemistry*, Third Edition, Royal Society of Chemistry, Cambridge, 2007.

Z	Isotope	Abundance %	I	v/MHz for H_0 = 1 T	Relative Sensitivity Const. H_0	Const. v	μ/μ_N	Q/fm^2
1	^{1}n		1/2	29.1647	0.32139	0.6850	−1.91304272	
1	^{1}H	99.9885	1/2	42.5775	1.00000	1.0000	+2.792847337	
1	^{2}H	0.0115	1	6.5359	0.00965	0.4094	+0.857438228	+0.2860
1	^{3}H	*	1/2	45.4148	1.21354	1.0667	+2.9789625	
2	^{3}He	0.000134	1/2	32.4380	0.44220	0.7619	−2.127750	
3	^{6}Li	7.59	1	6.2661	0.00850	0.3925	+0.8220467	−0.0808
3	^{7}Li	92.41	3/2	16.5483	0.29356	1.9434	+3.25644	−4.01
4	^{9}Be	100	3/2	5.9842	0.01388	0.7028	−1.1776	+5.288
5	^{10}B	19.9	3	4.5752	0.01985	1.7193	+1.800645	+8.459
5	^{11}B	80.1	3/2	13.6630	0.16522	1.6045	+2.688649	+4.059
6	^{13}C	1.07	1/2	10.7084	0.01591	0.2515	+0.7024118	
7	^{14}N	99.636	1	3.0777	0.00101	0.1928	+0.4037610	+2.044
7	^{15}N	0.364	1/2	4.3173	0.00104	0.1014	−0.2831888	
8	^{17}O	0.038	5/2	5.7742	0.02910	1.5822	−1.89379	−2.558
9	^{19}F	100	1/2	40.0776	0.83400	0.9413	+2.628868	
10	^{21}Ne	0.27	3/2	3.3631	0.00246	0.3949	−0.661797	+10.155
11	^{23}Na	100	3/2	11.2688	0.09270	1.3234	+2.217522	+10.4
12	^{25}Mg	10.00	5/2	2.6083	0.00268	0.7147	−0.85545	+19.94
13	^{27}Al	100	5/2	11.1031	0.20689	3.0424	+3.641507	+14.66
14	^{29}Si	4.685	1/2	8.4655	0.00786	0.1988	−0.55529	
15	^{31}P	100	1/2	17.2515	0.06652	0.4052	+1.13160	
16	^{33}S	0.75	3/2	3.2717	0.00227	0.3842	+0.6438212	−6.78
17	^{35}Cl	75.76	3/2	4.1765	0.00472	0.4905	+0.8218743	−8.165
17	^{37}Cl	24.24	3/2	3.4765	0.00272	0.4083	+0.6841236	−6.435
18	^{37}Ar	*	3/2	5.819	0.01276	0.6833	+1.145	+7.6
18	^{39}Ar	*	7/2	3.46	0.01130	1.7080	−1.59	−12
19	^{39}K	93.2581	3/2	1.9893	0.00051	0.2336	+0.3914662	+5.85
19	^{40}K	0.0117	4	2.4737	0.00523	1.5493	−1.298100	−7.3
19	^{41}K	6.7302	3/2	1.0919	0.00008	0.1282	+0.2148701	+7.11

Z	Isotope	Abundance %	I	ν/MHz for $H_0 = 1$ T	Relative Sensitivity		μ/μ_N	Q/fm^2
					Const. H_0	Const. ν		
20	^{43}Ca	0.135	7/2	2.8697	0.00643	1.4154	−1.317643	−4.08
21	^{45}Sc	100	7/2	10.3591	0.30244	5.1094	+4.756487	−22.0
22	^{47}Ti	7.44	5/2	2.4041	0.00210	0.6588	−0.78848	+30.2
22	^{49}Ti	5.41	7/2	2.4048	0.00378	1.1861	−1.10417	+24.7
23	^{50}V	0.250	6	4.2505	0.05571	5.5905	+3.345689	+21
23	^{51}V	99.750	7/2	11.2133	0.38360	5.5307	+5.1487057	−5.2
24	^{53}Cr	9.501	3/2	2.4115	0.00091	0.2832	−0.47454	−15
25	^{55}Mn	100	5/2	10.5763	0.17881	2.8981	+3.46872	+33
26	^{57}Fe	2.119	1/2	1.3816	0.00003	0.0324	+0.0906230	+16
27	^{59}Co	100	7/2	10.077	0.27841	4.9703	+4.627	+42
28	^{61}Ni	1.1399	3/2	3.8114	0.00359	0.4476	−0.75002	+16.2
29	^{63}Cu	69.15	3/2	11.3188	0.09393	1.3292	+2.2273456	−22.0
29	^{65}Cu	30.85	3/2	12.1027	0.11484	1.4213	+2.38161	−20.4
30	^{67}Zn	4.102	5/2	2.6685	0.00287	0.7312	+0.875205	+15.0
31	^{69}Ga	60.108	3/2	10.2478	0.06971	1.2035	+2.01659	+17.1
31	^{71}Ga	39.892	3/2	13.0208	0.14300	1.5291	+2.56227	+10.7
32	^{73}Ge	7.76	9/2	1.4897	0.00141	1.1547	−0.8794677	−19.6
33	^{75}As	100	3/2	7.3150	0.02536	0.8590	+1.439475	+31.4
34	^{77}Se	7.63	1/2	8.1568	0.00703	0.1916	+0.5350422	
35	^{79}Br	50.69	3/2	10.7042	0.07945	1.2570	+2.106400	+31.3
35	^{81}Br	49.31	3/2	11.5384	0.09951	1.3550	+2.270562	+26.2
36	^{83}Kr	11.500	9/2	1.6442	0.00190	1.2744	−0.970669	+25.9
37	^{85}Rb	72.17	5/2	4.1253	0.01061	1.1304	+1.35298	+27.6
37	^{87}Rb	27.83	3/2	13.9814	0.17704	1.6419	+2.75131	+13.35
38	^{87}Sr	7.00	9/2	1.8525	0.00272	1.4358	−1.093603	+30.5
39	^{89}Y	100	1/2	2.0949	0.00012	0.0492	−0.1374154	
40	^{91}Zr	11.22	5/2	3.9748	0.00949	1.0892	−1.30362	−17.6
41	^{93}Nb	100	9/2	10.4523	0.48821	8.1013	+6.1705	−32
42	^{95}Mo	15.90	5/2	2.7874	0.00327	0.7638	−0.9142	−2.2
42	^{97}Mo	9.56	5/2	2.8463	0.00349	0.7799	−0.9335	+25.5
43	^{99}Tc	*	9/2	9.6294	0.38174	7.4635	+5.6847	−12.9
44	^{99}Ru	12.76	5/2	1.9553	0.00113	0.5358	−0.6413	+7.9
44	^{101}Ru	17.06	5/2	2.1916	0.00159	0.6005	−0.7188	+45.7
45	^{103}Rh	100	1/2	1.3477	0.00003	0.0317	−0.08840	
46	^{105}Pd	22.33	5/2	1.957	0.00113	0.5364	−0.642	+66.0
47	^{107}Ag	51.839	1/2	1.7331	0.00007	0.0407	−0.1136796	
47	^{109}Ag	48.161	1/2	1.9924	0.00010	0.0468	−0.1306906	
48	^{111}Cd	12.80	1/2	9.0692	0.00966	0.2130	−0.5948861	
48	^{113}Cd	12.22	1/2	9.4871	0.01106	0.2228	−0.6223009	
49	^{113}In	4.29	9/2	9.3655	0.35121	7.2589	+5.5289	+75.9
49	^{115}In	95.71	9/2	9.3856	0.35348	7.2745	+5.5408	+77.0
50	^{115}Sn	0.34	1/2	14.0077	0.03561	0.3290	−0.91883	
50	^{117}Sn	7.68	1/2	15.2610	0.04605	0.3584	−1.00104	
50	^{119}Sn	8.59	1/2	15.9660	0.05273	0.3750	−1.04728	
51	^{121}Sb	57.21	5/2	10.2551	0.16302	2.8101	+3.3634	−54.3
51	^{123}Sb	42.79	7/2	5.5532	0.04659	2.7390	+2.5498	−69.2
52	^{123}Te	0.89	1/2	11.2349	0.01837	0.2639	−0.7369478	
52	^{125}Te	7.07	1/2	13.5446	0.03219	0.3181	−0.8884509	
53	^{127}I	100	5/2	8.5778	0.09540	2.3504	+2.813273	−69.6
54	^{129}Xe	26.4006	1/2	11.8604	0.02162	0.2786	−0.7779763	
54	^{131}Xe	21.2324	3/2	3.5159	0.00282	0.4129	+0.6918619	−11.4
55	^{133}Cs	100	7/2	5.6234	0.04838	2.7736	+2.582025	−0.343
56	^{135}Ba	6.592	3/2	4.2617	0.00501	0.5005	+0.838627	+16.0
56	^{137}Ba	11.232	3/2	4.7634	0.00700	0.5594	+0.937365	+24.5
57	^{138}La	0.090	5	5.6615	0.09404	5.3189	+3.713646	+45
57	^{139}La	99.910	7/2	6.0612	0.06058	2.9895	+2.7830455	+20.0
58	^{137}Ce	*	3/2	4.88	0.00752	0.5729	0.96	
58	^{139}Ce	*	3/2	5.39	0.01012	0.6326	1.06	
58	^{141}Ce	*	7/2	2.37	0.00364	1.1709	1.09	
59	^{141}Pr	100	5/2	13.0359	0.33483	3.5720	+4.2754	−5.9

Z	Isotope	Abundance %	I	v/MHz for $H_0 = 1$ T	Relative Sensitivity Const. H_0	Const. v	μ/μ_N	Q/fm^2
60	^{143}Nd	12.2	7/2	2.319	0.00339	1.1440	−1.065	−63
60	^{145}Nd	8.3	7/2	1.429	0.00079	0.7047	−0.656	−33
61	^{143}Pm	*	5/2	11.59	0.23510	3.1748	+3.80	
61	^{147}Pm	*	7/2	5.62	0.04827	2.7714	+2.58	+74
62	^{147}Sm	14.99	7/2	1.7748	0.00152	0.8754	−0.8149	−26
62	^{149}Sm	13.82	7/2	1.4631	0.00085	0.7216	−0.6718	+7.4
63	^{151}Eu	47.81	5/2	10.5856	0.17929	2.9006	+3.4718	+90.3
63	^{153}Eu	52.19	5/2	4.6745	0.01544	1.2809	+1.5331	+241
64	^{155}Gd	14.80	3/2	1.312	0.00015	0.1541	−0.2582	+127
64	^{157}Gd	15.65	3/2	1.720	0.00033	0.2020	−0.3385	+135
65	^{159}Tb	100	3/2	10.23	0.06945	1.2019	+2.014	+143.2
66	^{161}Dy	18.889	5/2	1.4654	0.00048	0.4015	−0.4806	+250.7
66	^{163}Dy	24.896	5/2	2.0508	0.00130	0.5619	+0.6726	+265
67	^{165}Ho	100	7/2	9.0883	0.20423	4.4826	+4.173	+358
68	^{167}Er	22.869	7/2	1.2281	0.00050	0.6057	−0.5639	+356.5
69	^{169}Tm	100	1/2	3.531	0.00057	0.0829	−0.2316	−120
70	^{171}Yb	14.28	1/2	7.5261	0.00552	0.1768	+0.49367	
70	^{173}Yb	16.13	5/2	2.0730	0.00135	0.5680	−0.67989	+280
71	^{175}Lu	97.41	7/2	4.8626	0.03128	2.3984	+2.2327	+349
71	^{176}Lu	2.59	7	3.451	0.03975	6.0518	+3.169	+497
72	^{177}Hf	18.60	7/2	1.7282	0.00140	0.8524	+0.7935	+336.5
72	^{179}Hf	13.62	9/2	1.0856	0.00055	0.8414	−0.6409	+379.3
73	^{181}Ta	99.988	7/2	5.1627	0.03744	2.5464	+2.3705	+317
74	^{183}W	14.31	1/2	1.7957	0.00008	0.0422	+0.1177848	
75	^{185}Re	37.40	5/2	9.7176	0.13870	2.6628	+3.1871	+218
75	^{187}Re	62.60	5/2	9.8170	0.14300	2.6900	+3.2197	+207
76	^{187}Os	1.96	1/2	0.9856	0.00001	0.0231	+0.06465189	
76	^{189}Os	16.15	3/2	3.3536	0.00244	0.3938	+0.659933	+85.6
77	^{191}Ir	37.3	3/2	0.7658	0.00003	0.0899	+0.1507	+81.6
77	^{193}Ir	62.7	3/2	0.8319	0.00004	0.0977	+0.1637	+75.1
78	^{195}Pt	33.832	1/2	9.2922	0.01039	0.2182	+0.60952	
79	^{197}Au	100	3/2	0.7406	0.00003	0.0870	+0.145746	+54.7
80	^{199}Hg	16.87	1/2	7.7123	0.00594	0.1811	+0.5058855	
80	^{201}Hg	13.18	3/2	2.8469	0.00149	0.3343	−0.5602257	+38.7
81	^{203}Tl	29.52	1/2	24.7316	0.19598	0.5809	+1.6222579	
81	^{205}Tl	70.48	1/2	24.9749	0.20182	0.5866	+1.6382146	
82	^{207}Pb	22.1	1/2	9.0340	0.00955	0.2122	+0.59258	
83	^{209}Bi	100	9/2	6.9630	0.14433	5.3968	+4.1106	−51.6
84	^{209}Po	*	1/2	11.7	0.02096	0.2757	+0.77	
86	^{211}Rn	*	1/2	9.16	0.00997	0.2152	+0.601	
87	^{223}Fr	*	3/2	5.95	0.01362	0.6982	+1.17	+117
88	^{223}Ra	*	3/2	1.3746	0.00017	0.1614	+0.2705	+121
88	^{225}Ra	*	1/2	11.187	0.01814	0.2627	−0.7338	
89	^{227}Ac	*	3/2	5.6	0.01131	0.6565	+1.1	+170
90	^{229}Th	*	5/2	1.40	0.00042	0.3843	+0.46	+430
91	^{231}Pa	100	3/2	10.2	0.06903	1.1995	2.01	−172
92	^{235}U	0.7204	7/2	0.83	0.00015	0.4082	−0.38	+493.6
93	^{237}Np	*	5/2	9.57	0.13264	2.6234	+3.14	+388.6
94	^{239}Pu	*	1/2	3.09	0.00038	0.0727	+0.203	
95	^{243}Am	*	5/2	4.6	0.01446	1.2532	+1.5	+421

PROPERTIES OF IMPORTANT NMR NUCLEI

Thomas J. Bruno and Paris D. N. Svoronos

The following table lists the magnetic properties at higher field strengths required for choosing the nuclei to be used in NMR experiments (Refs. 1–15). The reader is referred to several excellent texts and the literature for guidelines in nucleus selection. For more detailed information on these and other less common nuclei at 10 kG, the reader should consult the table entitled *Nuclear Spins, Moments and Other Data Related to NMR Spectroscopy* in this section.

References

1. Silverstein, R. M., Bassler, G. C., and Morrill, T. C., *Spectrometric Identification of Organic Compounds*, 5th Edition, John Wiley and Sons, New York, 1991.
2. Yoder, C. H., and Shaeffer, C. D., *Introduction to Multinuclear NMR*, Benjamin/Cummings, Menlo Park, CA, 1987.
3. Gordon, A. J., and Ford, R. A., *The Chemist's Companion*, Wiley Interscience, New York, 1971.
4. Silverstein, R. M., and Webster F. X., *Spectrometric Identification of Organic Compounds*, 6th Edition, John Wiley and Sons, New York, 1998.
5. Becker, E. D., *High Resolution NMR, Theory and Chemical Applications*, 2nd Edition, Academic Press, New York, 1980.
6. Gunther, H., *NMR Spectroscopy: Basic Principles, Concepts and Applications in Chemistry*, John Wiley and Sons, New York, 2003.
7. Rahman, A.-u., *Nuclear Magnetic Resonance*, Springer-Verlag, New York, 1986.
8. Harris, R. K., *Chem. Soc. Rev.* 5, 1, 1976.
9. Kitamaru, R., *Nuclear Magnetic Resonance: Principles and Theory*, Elsevier Science, 1990.
10. Lambert, J. B., Holland, L. N., and Mazzola, E. P., *Nuclear Magnetic Resonance Spectroscopy: Introduction to Principles, Applications and Experimental Methods*, Prentice Hall, Englewood Cliffs, NJ, 2003.
11. Bovey, F. A., and Mirau, P. A., *Nuclear Magnetic Resonance Spectroscopy*, 2nd Edition, Academic Press, New York, 1988.
12. Harris, R. K., and Mann, B. E., *NMR and the Periodic Table*, Academic Press, London, 1978.
13. Hore, P. J., *Nuclear Magnetic Resonance*, Oxford University Press, Oxford, 1995.
14. Nelson, J. H., *Nuclear Magnetic Resonance Spectroscopy*, 2nd Edition, John Wiley and Sons, New York, 2003.
15. Bruno, T. J., and Svoronos, P. D. N., *CRC Handbook of Basic Tables for Chemical Analysis*, 3rd Edition, CRC Press, Boca Raton, FL, 2011.

Isotope	Natural Abundance	Spin Number I	NMR Frequency[a] at Indicated Field Strength in kG							
			10.000	14.092	21.139	23.487	51.567	93.950	140.925	223.131
$_1H^1$	99.985	1/2	42.5759	60.0000	90.0000	100.0000	220.0000	400.0000	600.0000	950.0000
$_1H^2$	0.015	1	6.53566	9.21037	13.81555	15.35061	33.77134	61.40262	92.10380	145.9830
$_1H^{3*}$	—	1/2	45.4129	63.9980	95.9971	106.6634	234.6595	426.6542	639.9813	1013.3024
$_6C^{13}$	1.108	1/2	10.7054	15.0866	22.6298	25.1443	55.3174	100.5735	150.8659	2388.5150
$_7N^{14}$	99.635	1	3.0756	4.3343	6.5014	7.2238	15.924	28.9104	43.3615	68.6557
$_7N^{15}$	0.365	1/2	4.3142	6.0798	9.1197	10.1330	22.2925	40.5306	60.7960	96.2601
$_8O^{17}$	0.037	5/2	5.772	8.134	12.201	13.557	29.825	54.1811	81.3186	128.5801
$_9F^{19}$	100	1/2	40.0541	42.3537	63.5305	94.0769	206.9692	376.2515	564.3781	893.5963
$_{14}Si^{29}$	4.70	1/2	8.4578	11.9191	17.8787	19.8652	43.7035	79.4638	119.1956	188.72
$_{15}P^{31}$	100	1/2	17.235	24.288	36.433	40.481	89.057	161.9828	242.9741	384.7086
$_{16}S^{33}$	0.76	3/2	3.2654	4.6018	6.9026	7.6696	16.8731	30.6826	46.0238	72.8710
$_{16}S^{35*}$	—	3/2	5.08	7.16	10.74	11.932	26.250	47.7267	71.5875	113.3508
$_{17}Cl^{35}$	75.53	3/2	4.1717	5.8790	8.8184	9.7983	21.5562	39.1948	58.7902	93.0876
$_{17}Cl^{36*}$	—	2	4.8931	6.8956	10.3434	11.4927	25.2838	45.9638	68.9432	109.1639
$_{35}Br^{76*}$	—	1	4.18	5.89	8.84	9.82	21.60	39.2768	58.9130	93.2822
$_{35}Br^{79}$	50.54	3/2	10.667	15.032	22.549	25.054	55.119	100.2133	150.3202	238.0064
$_{35}Br^{81}$	49.46	3/2	11.498	16.204	24.305	27.006	59.413	108.0258	162.0386	256.5608
$_{74}W^{183}$	14.40	1/2	1.7716	2.4966	3.7449	4.1610	9.1543	16.6430	24.9646	39.5272

* Nucleus is radioactive

[a] 1 kG = 10^{-1} T, the corresponding SI unit
1 b = 10^{-28} m^2

PROTON NMR ABSORPTION OF MAJOR CHEMICAL FAMILIES

Thomas J. Bruno and Paris D. N. Svoronos

The following table gives the region of the expected nuclear magnetic resonance absorptions of major chemical families (Refs. 1–12). These absorptions are reported in the dimensionless units of parts per million (ppm) versus the standard compound tetramethylsilane (TMS, structure provided), which is recorded as 0.0 ppm.

$$CH_3-Si-CH_3 \text{ (with } CH_3 \text{ above and below)}$$

The use of this unit of measure makes the chemical shifts independent of the applied magnetic field strength or the radio frequency. For most proton NMR spectra, the protons in TMS are more shielded than almost all other protons. The chemical shift in this dimensionless unit system is then defined by:

$$\delta = \frac{\nu_s - \nu_r}{\nu_r} \times 10^6$$

where ν_s and ν_r are the absorption frequencies of the sample proton and the reference (TMS) protons (twelve, magnetically equivalent), respectively. In these tables, the proton(s) whose proton NMR shifts are cited are indicated by underscore. Reference 1 provides additional details on the absorptions of other moieties, as well as correlation charts.

References

1. Bruno, T. J., and Svoronos, P. D. N., *CRC Handbook of Basic Tables for Chemical Analysis*, 3rd Edition, CRC Press, Boca Raton, FL, 2011.
2. Silverstein, R. M., and Webster, F. X., *Spectrometric Identification of Organic Compounds*, 6th Edition, Wiley, New York, 1998.
3. Rahman, A.-u., *Nuclear Magnetic Resonance*, Springer Verlag, New York, 1986.
4. Gordon, A. J., and Ford, R. A., *The Chemist's Companion*, Wiley Interscience, New York, 1971.
5. Becker, E. D., *High Resolution NMR, Theory and Chemical Applications*, 2nd Edition, Academic Press, New York, 1980.
6. Gunther, H., *NMR Spectroscopy: Basic Principles, Concepts and Applications in Chemistry*, Wiley, New York, 2003.
7. Kitamaru, R., *Nuclear Magnetic Resonance: Principles and Theory*, Elsevier Science, 1990.
8. Lambert, J. B., Holland, L. N., and Mazzola, E. P., *Nuclear Magnetic Resonance Spectroscopy: Introduction to Principles, Applications and Experimental Methods*, Prentice Hall, Englewood Cliffs, NJ , 2003.
9. Bovey, F. A., and Mirau, P. A., *Nuclear Magnetic Resonance Spectroscopy*, 2nd Edition, Academic Press, New York, 1988.
10. Hore, P. J., *Nuclear Magnetic Resonance*, Oxford University Press, Oxford, 1995.
11. Nelson, J. H., *Nuclear Magnetic Resonance Spectroscopy*, 2nd Edition, Wiley, New York, 2003.
12. Abraham, R. J., Fisher, J., and Loftus, P., *Introduction to NMR Spectroscopy*, Wiley, New York, 1988.

Family			δ of Protons Underlined		
Alkanes	$\underline{CH_3}$–R	~0.8 ppm			
	–$\underline{CH_2}$–R	~1.1 ppm			
	>$\underline{CH}$–R	~1.4 ppm			
	(Cyclopropane 0.2 ppm)				
Alkenes	$\underline{CH_3}$–C=C<	~1.6 ppm	$\underline{CH_3}$–C–C=C<	~1.0 ppm	
	–$\underline{CH_2}$–C=C<	~2.1 ppm	–$\underline{CH_2}$–C–C=C<	~1.4 ppm	
	>$\underline{CH}$–C=C<	~2.5 ppm	>$\underline{CH}$–C–C=C<	~1.8 ppm	
	>C=C–$\underline{H}$	4.2 ppm to 6.2 ppm			
Alkynes	$\underline{CH_3}$–C≡C–	~1.7 ppm	$\underline{CH_3}$–C–C≡C–	~1.2 ppm	
	–$\underline{CH_2}$–C≡C–	~2.2 ppm	>$\underline{CH_2}$–C–C≡C–	~1.5 ppm	
	>$\underline{CH}$–C≡C–	~2.7 ppm	>$\underline{CH}$–C–C≡C–	~1.8 ppm	
	R–C≡C–$\underline{H}$	~2.4 ppm			
Aromatics	C_6H_5–G		Range: 8.5 ppm to 6.9 ppm		

G above benzene ring with positions o-, m-, p- labeled.

When <u>G</u>=Electron withdrawing (e.g., >C=O, $-NO_2$, –C≡N) o- and p-hydrogens relative to –G are closer to 8.5 ppm (more downfield)

When <u>G</u>=Electron donating (e.g., $-NH_2$, –OH, –OR, –R) o- and p-hydrogens relative to –G are closer to 6.9 ppm (more upfield)

Organic Oxygen Compounds

Family	Approximate δ of Protons Underlined					

Alcohols

$\underline{CH}_3$–OH — 3.2 ppm   R$\underline{CH}_2$–OH — 3.4 ppm   R$_2\underline{CH}$–OH — 3.6 ppm

$\underline{CH}_3$–C–OH — 1.2 ppm   R$\underline{CH}_2$–C–OH — 1.5 ppm   R$_2\underline{CH}$–C–OH — 1.8 ppm

R–$\underline{O}$–$\underline{H}$ — (1 ppm to 5 ppm — depending on concentration)

Aldehydes

$\underline{CH}_3$–CHO — 2.2 ppm   R$\underline{CH}_2$–CHO — 2.4 ppm   R$_2\underline{CH}$–CHO — 2.5 ppm

$\underline{CH}_3$–C–CHO — 1.1 ppm   R$\underline{CH}_2$–C–CHO — 1.6 ppm

---

Amides   See organic nitrogen compounds

Anhydrides, acyclic

$\underline{CH}_3$–C(=O)O– — 1.8 ppm   R$\underline{CH}_2$–C(=O)O– — 2.1 ppm   R$_2\underline{CH}$–C(=O)O– — 2.3 ppm

$\underline{CH}_3$–C–C(=O)O– — 1.2 ppm   R$\underline{CH}_2$–C–C(=O)O– — 1.8 ppm   R$_2\underline{CH}$–C–C(=O)O– — 2.0 ppm

---

Anhydrides, cyclic   3.0 ppm   7.1 ppm

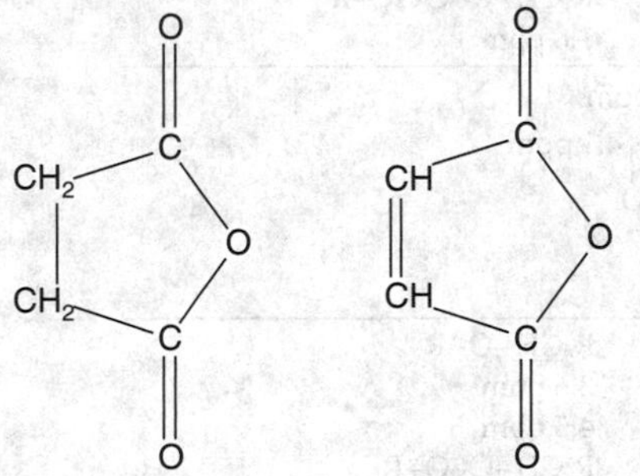

Carboxylic acids

$\underline{CH}_3$–COOH — 2.1 ppm   R$\underline{CH}_2$–COOH — 2.3 ppm   R$_2\underline{CH}$–COOH — 2.5 ppm

$\underline{CH}_3$–C–COOH — 1.1 ppm   R–$\underline{CH}_2$–C–COOH — 1.6 ppm   R$_2\underline{CH}$–C–COOH — 2.0 ppm

R–COO–$\underline{H}$ — 11 ppm to 12 ppm

Cyclic ethers

Oxacyclopropane (oxirane) — 2.5 ppm

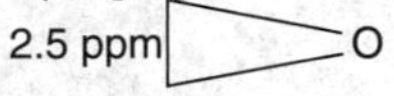

Oxacyclobutane (oxetane) — 2.7 ppm — 4.7 ppm

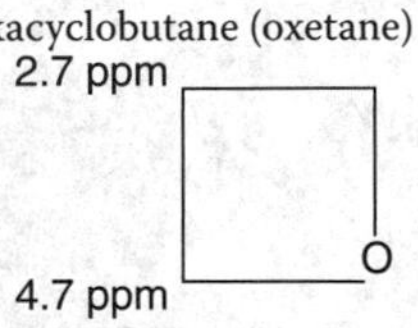

Oxacyclopentane (tetrahydrofuran) — 1.9 ppm — 3.8 ppm

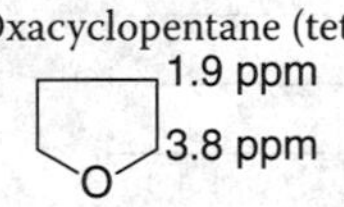

Oxacyclohexane (tetrahydropyran) — 1.6 ppm — 1.6 ppm — 3.6 ppm

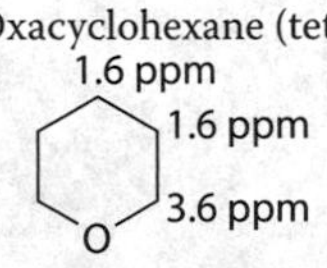

1,4-dioxane — 3.6 ppm

1,3-dioxane — 1.7 ppm — 3.8 ppm — 4.7 ppm

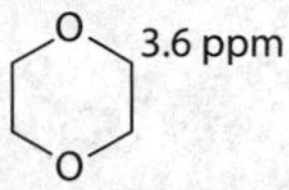

Furan — 6.3 ppm — 7.4 ppm

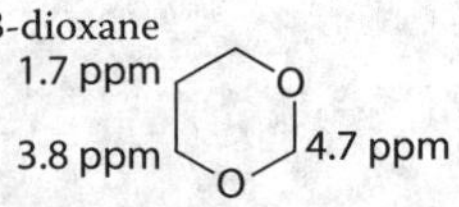

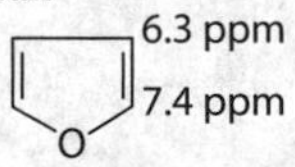

Family	Approximate δ of Protons Underlined

Cyclic ethers (continued)   Dihydropyran
1.9 ppm
4.5 ppm
6.2 ppm

Epoxides	See cyclic ethers

Esters

	CH_3–COOR	RCH_2–COOR	R_2CH–COOR
R = alkyl	1.9 ppm	2.1 ppm	2.3 ppm
R = aryl	2.0 ppm	2.2 ppm	2.4 ppm
	CH_3–C–COOR	RCH_2–C–COOR	R_2CH–C–COOR
	1.1 ppm	1.7 ppm	1.9 ppm
	CH_3–OOC–R	RCH_2–OOC–R	R_2CH–OOC–R
	3.6 ppm	4.1 ppm	4.8 ppm
	CH_3–C–OOC–R	RCH_2–C–OOC–R	R_2CH–C–OOC–R
	1.3 ppm	1.6 ppm	1.8 ppm

Cyclic esters

2.1 ppm   4.4 ppm
2.3 ppm
O

1.6 ppm
1.6 ppm   4.1 ppm
2.3 ppm
O

Ethers

	CH_3–O–R	RCH_2–O–R	R_2CH–O–R
R = alkyl	3.2 ppm	3.4 ppm	3.6 ppm
R = aryl	3.9 ppm	4.1 ppm	4.5 ppm
	CH_3–C–O–R	RCH_2–C–O–R	R_2CH–C–O–R
R = alkyl	1.2 ppm	1.5 ppm	1.8 ppm
R = aryl	1.3 ppm	1.6 ppm	2.0 ppm

Isocyanates	See nitrogen compounds

Ketones

	CH_3–C(=O)–		RCH_2–C(=O)–	R_2CH–C(=O)–
	1.9 ppm	R = alkyl	2.1 ppm	2.3 ppm
	2.4 ppm	R = aryl	2.7 ppm	3.4 ppm
	CH_3–C(=O)–		RCH_2–C(=O)–	R_2CH–C(=O)–
	1.1 ppm	R = alkyl	1.6 ppm	2.0 ppm
	1.2 ppm	R = aryl	1.6 ppm	2.1 ppm

Cyclic ketones (n = number of ring carbons)

$(CH_2)_n$ =O

α–hydrogens	2.0 ppm to 2.3 ppm (n > 5)
	3.0 ppm (n = 4)
	1.7 ppm (n = 3)
β–hydrogens	1.9 ppm to 1.5 ppm

Lactones	See esters, cyclic
Nitro-compounds	See organic nitrogen compounds
Phenols	Ar–O–H 9 ppm to 10 ppm (Ar = aryl group)

Organic Nitrogen Compounds

Amides:

δ of Proton(S) (Underlined)	Primary R–C(=O)NH_2 δ/ppm	Secondary R–C(=O)NHR_1 δ/ppm	Tertiary R–C(=O)NR_1R_2 δ/ppm
(i) N-substitution			—
R–C(=O)N–H	5–12	5–12	—
(a) alpha			
–C(=O)N–CH_3	—	~2.9	~2.9
–C(=O)N–CH_2–	—	~3.4	~3.4
–C(=O)N–CH–	—	~3.8	~3.8
(b) Beta			
–C(=O)N–C–CH_3	~1.1	~1.1	~1.1
–C(=O)N–C–CH_2–	~1.5	~1.5	~1.5
–C(=O)N–C–CH–	~1.9	~1.9	~1.9

δ of Proton(S) (Underlined)	Primary R–C(=O)NH$_2$ δ/ppm	Secondary R–C(=O)NHR$_1$ δ/ppm	Tertiary R–C(=O)NR$_1$R$_2$ δ/ppm
(ii) C-substitution			
(a) alpha			
CH$_3$–C(=O)N	~1.9	~2.0	~2.1
RCH$_2$–C(=O)N	~2.1	~2.1	~2.1
R$_2$CH–C(=O)N	~2.2	~2.2	~2.2
(b) Beta			
CH$_3$–C–C(=O)N	~1.1	~1.1	~1.1
CH$_2$–C–C(=O)N	~1.5	~1.5	~1.5
–CH–C–C(=O)N	~1.8	~1.8	~1.8

Amines:

δ of Proton(s) (Underlined)	Primary R–NH$_2$ δ/ppm	Secondary RN–HR δ/ppm	Tertiary RRRN δ/ppm
(i) Apha protons			
>N–CH$_2$	~2.5	2.3–3.0	~2.2
>N–CH$_2$–	~2.7	2.6–3.4	~2.4
>N–CH<	~3.1	2.9–3.6	~2.8
(ii) Beta protons			
>N–C–CH$_3$			~1.1
>N–C–CH$_2$–			~1.4
>N–C–CH<			~1.7

Cyanocompounds (nitriles):

(i) Alpha hydrogens δ/ppm		(ii) Beta hydrogens δ/ppm	
CH$_3$–C≡N	~2.1	CH$_3$–C–C≡N	~1.2
–CH$_2$–C≡N	~2.5	–CH$_2$–C–C≡N	~1.6
–CH–C≡N	~2.9	CH–C–C≡N	~2.0

Imides:

(i) Alpha hydrogens δ/ppm		(ii) Beta hydrogens δ/ppm	
CH$_3$–C(=O)NHC(=O)–	~2.0	CH$_3$–C(=O)C–NH–C(=O)–	~1.2
CH$_2$–C(=O)NHC(=O)–	~2.1	CH$_2$–C(=O)C–NH–C(=O)–	~1.3
CH–C(=O)NHC(=O)–	~2.2	–CH–C(=O)C–NH–C(=O)–	~1.4

Isocyanates:

Alpha hydrogens δ/ppm

CH$_3$–N=C=O ~3.0

–CH$_2$–N=C=O ~3.3

–CH–N=C=O ~3.6

Isocyanides (isonitriles):

Alpha hydrogens/ppm

CH$_3$–N=C<	~2.9
CH$_2$–N=C<	~3.3
CH–N=C<	~4.9

Isothiocyanates:

Alpha hydrogens/ppm

CH$_3$–N=C=S	~3.4
CH$_2$–N=C=S	~3.7
>CH–N=C=S	~4.0

Nitriles δ/ppm

–CH$_2$–O–N=O ~4.8

Nitrocompounds δ/ppm

CH₃–NO₂	~ 4.1	–CH₂–NO₂	~4.2	–CH–NO₂	~4.4
CH₃–C–NO₂	~1.6	–CH₂–C–NO₂	~2.1	–CH–C–NO₂	~2.5

Organic Sulfur Compounds

Family		d of Proton(S) Underlined			
Benzothiopyrans					
2H–1–	sp³ C–H	~3.3 ppm	sp² C–H	5.8–6.4	aromatic ~6.8
4H–1–	sp³ C–H	~3.2 ppm	sp² C–H	5.9–6.3	aromatic ~6.9
2,3,4H–1–	sp³ C–H	1.9 ppm to 2.8 ppm			aromatic ~7.1
Disulfides	CH₃–S–S–R	~2.4 ppm	CH₃–C–S–S–R	~1.2 ppm	
	CH₂–S–S–R	~2.7 ppm	CH₂–C–S–S–R	~1.6 ppm	
	CH–S–S–R	~3.0 ppm	CH–C–S–S–R	~2.0 ppm	
Isothiocyanates	CH₃–N=C=S	~2.4 ppm			
	–CH₂–N=C=S	~2.7 ppm			
	–CH–N=C=S	~3.0 ppm			
Mercaptans (thiols)	CH₃–S–H	~2.1 ppm	CH₃–C–S–H	~1.3 ppm	
	–CH₂–S–H	~2.6 ppm	–CH₂–C–S–H	~1.6 ppm	
	–CH–S–H	~3.1 ppm	–CH–C–S–H	~1.7 ppm	
S-methyl salts	>S⁺–CH₃	~3.2 ppm			
Sulfates	(CH₃–O)₂S(=O)₂	~3.4 ppm			

Family		δ/ppm of Proton(S) Underlined		
Sulfides	CH₃–S–	1.8–2.1	CH₂–CH₂–S–	1.1–1.2
	R–CH₂–S–	1.9–2.4	CH₃–CHR–S–	0.8–1.2
	R–CHR–S–	2.8–3.4	CH₃–CHAr–S–	1.3–1.4
	Ar–CH₂–S–	4.1–4.2	CH₃–CR₂–S–	1.0
	Ar–CHR–S–	3.6–4.2	Ar–CH₂–CHR–S–	3.0–3.2
	Ar₂–CH–S–	5.1–5.2	>C=C–CH₂–CHAr–S–	2.4–2.6
			>C=C–CH₂–CAr₂–S–	2.5
			R₂CH–CH₂–S–	2.6–3.0
			Ar₂ CH–CH₂–S–	4.0–4.2
			>C=C–CHR–CHAr–S–	2.3–2.4
			>C=C–CHR–CAr₂–S–	2.8–3.2
Sulfilimines	CH₂(R)S=N–R²	~2.5 ppm		
Sulfonamides	CH₂–SO₂NH₂	~3.0 ppm		
Sulfonates	CH₃–SO₂–OR	~3.0 ppm		
Sulfones	CH₃–SO₂–R²	~2.6 ppm		
Sulfonic acids	CH₃–SO₃H	~3.0 ppm		
Sulfoxides	CH₃–S(=O)R	~2.5 ppm		
	–CH₂–S(=O)R	~3.1 ppm		
Thiocyanates	CH₃–S–C≡N	~2.7 ppm		
	–CH₂–S–C≡N	~3.0 ppm		
	–CH–S–C≡N	~3.3 ppm		
Thiols	See mercaptans			

Note: Ar represents aryl.

Some Useful ¹H Coupling Constants

The following chart gives the values of some useful proton NMR coupling constants (in Hz). The single numbers indicate a typical or average value, while in some cases, the range is provided.

1. Freely rotating chains.

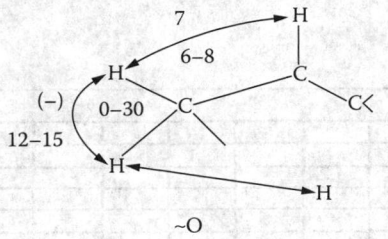

2. Alcohols with no exchange as in DMSO.
   1° = triplet
   2° = doublet (broad)
   3° = singlet
   Upon addition of TFA, a sharp singlet results.

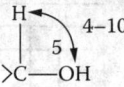

3. Alkenes

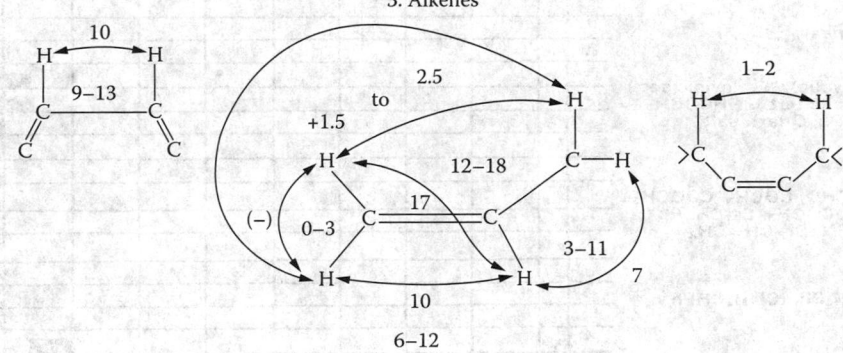

4. Alkynes

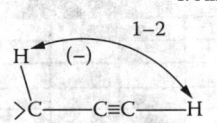

5. Aldehydes

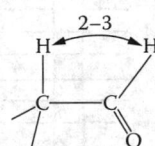
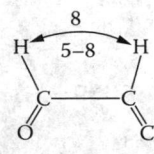

6. Aromatic

PROTON NMR CORRELATION CHART FOR MAJOR ORGANIC FUNCTIONAL GROUPS

The chart below summarizes the range of chemical shifts for protons in several classes or organic compounds and substituent groups. The chemical shifts δ are given in parts per million relative to tetramethylsilane.

Reference

Mohacsi, E., *J. Chem. Edu.* 41, 38, 1964 (with permission).

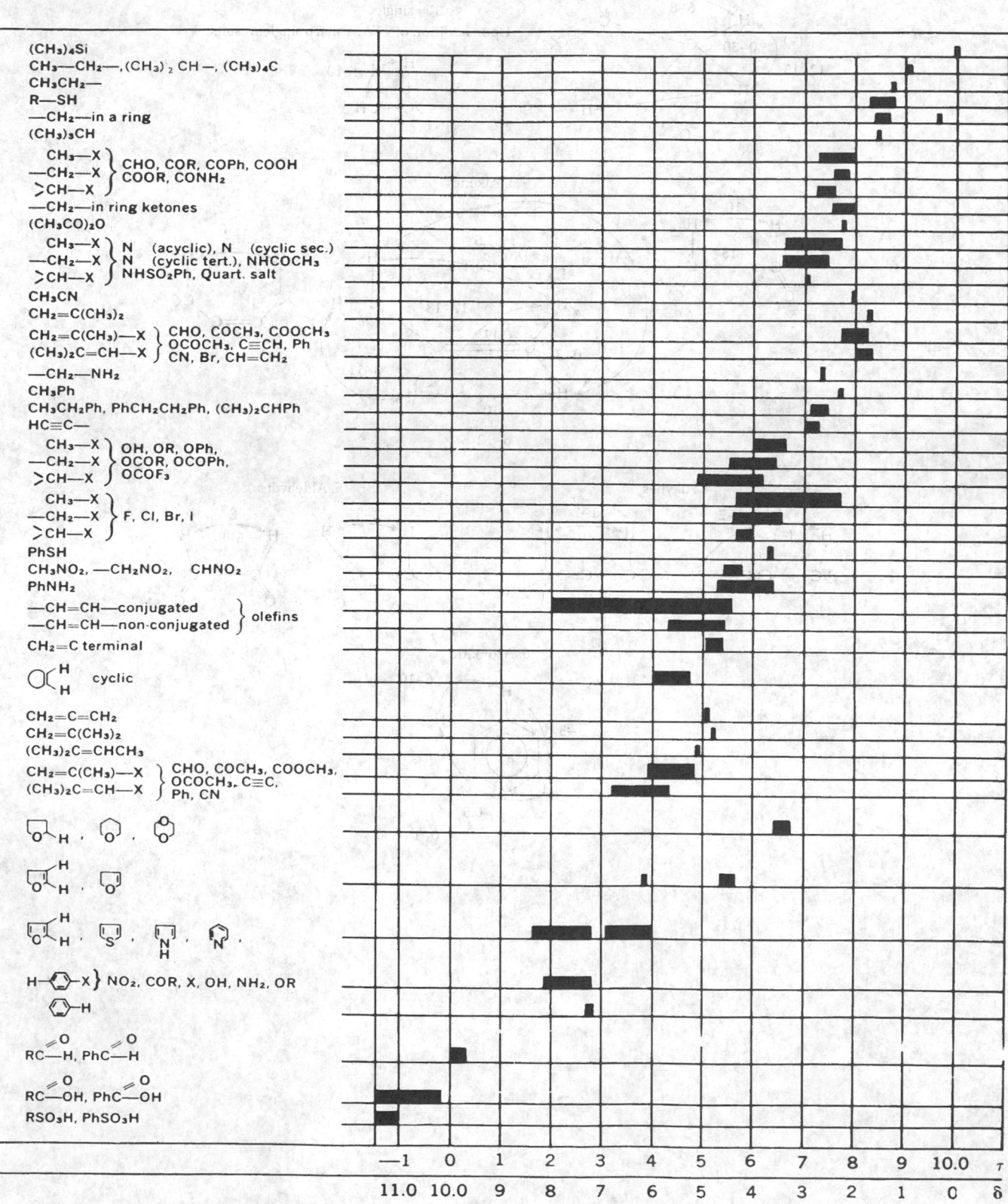

PROTON NMR SHIFTS OF COMMON ORGANIC SOLVENTS

The table below lists the ^{1}H chemical shifts for over 300 organic solvents and liquid reagents. The solvent in which the shift was measured is given in the second column. Shifts are given in parts per million relative to tetramethylsilane (TMS) and are listed in order of smallest to largest shift. In many cases the peaks show additional small splittings. Compounds are listed by the name used in this *Handbook*, with other common names given in parentheses.

References

1. Lide, D. R., Editor, *Properties of Organic Compounds*, <www.chem-netbase.com/scripts/pocweb.exe>; Lide, D. R., and Milne, G. W. A., Editors, *Handbook of Data on Organic Compounds, Third Edition*, CRC Press, Boca Raton, FL, 1993.
2. Spectral Database for Organic Compounds, SDBS, National Institute of Advanced Industrial Science and Technology (AIST), Japan, <riodb01.ibase.aist.go.jp>.

Compound	Solvent	^{1}H NMR shifts (ppm relative to TMS)					Ref.
Acetic acid	CDCl₃	2.10	11.4				2
Acetic anhydride	CCl₄	2.2					1
Acetone	CDCl₃	2.1					1
Acetonitrile	CCl₄	1.9					1
Acrolein (2-Propenal)	CDCl₃	6.4	9.5				1
Acrylonitrile	CDCl₃	6.3					1
Allyl alcohol	CDCl₃	3.6	4.1	5.1	5.3	6.0	1
Allylamine	CDCl₃	1.5	3.3	5.0	5.1	5.9	1
2-Amino-2-methyl-1-propanol (2-Aminoisobutanol)	CCl₄	1.1	2.8	3.2			1
Aniline (Benzenamine)	CCl₄	3.3	6.4	6.6	7.0		1
Anisole (Methoxybenzene)	CDCl₃	3.8	7.1				1
Benzaldehyde	CDCl₃	7.7	10.0				1
Benzene	CDCl₃	7.34					2
Benzeneacetonitrile (Benzyl cyanide)	CCl₄	0.8	0.9	1.6	2.3		1
Benzenethiol (Phenyl mercaptan)	CCl₄	3.2	6.9				1
Benzonitrile	CCl₄	7.5					1
Benzyl acetate	CDCl₃	2.1	5.1	7.3			1
Benzyl alcohol	CDCl₃	2.4	4.6	7.3			1
Bis(2-aminoethyl)amine (Diethylenetriamine)	CDCl₃	1.23	2.69	2.79			2
Bis(2-chloroethyl) ether	CDCl₃	3.66	3.77				2
Bis(2-ethylhexyl) phthalate	CCl₄	0.9	1.5	4.2	7.4	7.7	1
Bis(2-hydroxyethyl) sulfide	CDCl₃	2.8	3.8	4.2			1
Bromobenzene	CCl₄	7.1	7.4				1
1-Bromobutane (Butyl bromide)	CCl₄	1.0	1.4	1.8	3.4		1
2-Bromobutane (*sec*-Butyl bromide)	CDCl₃	1.1	1.7	1.8	4.1		1
1-Bromo-2-chloroethane	CDCl₃	3.3	4.0				1
Bromochloromethane	CCl₄	5.2					1
1-Bromodecane (Decyl bromide)	CCl₄	0.9	1.8	3.3			1
Bromoethane (Ethyl bromide)	CDCl₃	1.7	3.4				1
2-Bromo-2-methylpropane (*tert*-Butyl bromide)	CCl₄	1.8					1
1-Bromonaphthalene	CCl₄	7.4	8.1				1
1-Bromopentane (Pentyl bromide)	CCl₄	0.9	1.4	1.9	3.3		1
1-Bromopropane (Propyl bromide)	CCl₄	1.0	1.9	3.4			1
2-Bromopropane (Isopropyl bromide)	CDCl₃	1.7	4.3				1
2-Bromopropene	CDCl₃	2.3	5.3	5.5			1
Butanal	CDCl₃	1.0	1.7	2.4	9.7		1
Butanenitrile	CCl₄	1.1	1.7	2.3			1
1-Butanethiol (Butyl mercaptan)	CDCl₃	0.9	1.2	1.5	2.5		1
Butanoic acid	CCl₄	0.9	1.7	2.3	12.0		1
Butanoic anhydride	CCl₄	1.0	1.7	2.4			1
1-Butanol (Butyl alcohol)	CDCl₃	0.94	1.39	1.53	2.24	3.63	2

Compound	Solvent	¹H NMR shifts (ppm relative to TMS)						Ref.
2-Butanol (*sec*-Butyl alcohol)	CDCl₃	0.93	1.17	1.46	2.37	3.71		2
2-Butanone (Methyl ethyl ketone)	CDCl₃	1.06	2.14	2.45				2
trans-2-Butenal (*trans*-Crotonaldehyde)	CDCl₃	2.0	6.1	6.9	9.5			1
2-Butoxyethanol (Ethylene glycol monobutyl ether)	CCl₄	0.9	1.3	3.3	3.7			1
Butyl acetate	CDCl₃	0.9	1.4	2.0	4.1			1
Butylamine	CDCl₃	0.92	1.33	1.43	1.77	2.68		1
tert-Butylamine	CDCl₃	1.1	1.2					1
Butylbenzene	CCl₄	0.9	1.4	2.6	7.1			1
sec-Butylbenzene	CCl₄	0.8	1.2	1.6	2.5	7.1		1
tert-Butylbenzene	CCl₄	1.3	7.2					1
Butyl formate	CDCl₃	0.9	1.5	4.2	8.1			1
1-*tert*-Butyl-4-methylbenzene	CDCl₃	1.30	2.31	7.11	7.26			2
Butyl vinyl ether	CCl₄	0.9	1.4	3.6	3.8	4.0	6.3	1
γ-Butyrolactone	CDCl₃	4.4						1
Caprolactam	CDCl₃	1.7	2.4	3.2	7.8			1
2-Chloroaniline	CCl₄	3.8	6.8					1
Chlorobenzene	CDCl₃	7.3						2
2-Chlorobutane (*sec*-Butyl chloride)	CCl₄	1.1	1.5	1.7	3.9			1
Chloroethane (Ethyl chloride)	CDCl₃	1.5	3.6					1
2-Chloroethanol (Ethylene chlorohydrin)	CDCl₃	2.8	3.7	3.8				1
(Chloromethyl)benzene (Benzyl chloride)	CCl₄	4.5	7.3					1
1-Chloro-3-methylbutane (Isopentyl chloride)	CDCl₃	0.9	1.7	3.6				1
1-Chloro-2-methylpropane (Isobutyl chloride)	CCl₄	1.0	1.9	3.3				1
2-Chloro-2-methylpropane (*tert*-Butyl chloride)	CCl₄	1.6						1
1-Chloronaphthalene	CCl₄	7.1	7.5	8.2				1
1-Chlorooctane (Octyl chloride)	CCl₄	0.9	1.3	1.8	3.5			1
1-Chloropentane (Pentyl chloride)	CCl₄	0.9	1.6	3.4				1
1-Chloropropane (Propyl chloride)	CCl₄	1.0	1.8	3.4				1
3-Chloropropene (Allyl chloride)	CCl₄	4.0	5.2	5.3	5.9			1
2-Chlorotoluene	CDCl₃	2.4	7.2					1
3-Chlorotoluene	CCl₄	2.3	7.1					1
Cyclohexane	CDCl₃	1.43						2
Cyclohexanol	CCl₄	1.6	3.5	4.2				1
Cyclohexanone	CCl₄	1.8	2.3					1
Cyclohexene	CCl₄	1.6	2.0	5.6				1
Cyclohexylamine	CCl₄	1.4	1.5	2.6				1
Cyclopentane	CCl₄	1.5						1
Cyclopentanone	CCl₄	2.0						1
cis-Decahydronaphthalene (*cis*-Decalin)	CDCl₃	1.42	1.62					2
trans-Decahydronaphthalene (*trans*-Decalin)	CDCl₃	0.87	0.93	1.23	1.54	1.67		2
Decane	CCl₄	0.9	1.3					1
Diacetone alcohol	CDCl₃	1.3	2.2	2.6	3.7			1
1,2-Dibromoethane	CDCl₃	3.65						2
Dibromomethane	CCl₄	4.9						1
1,2-Dibromopropane	CCl₄	1.8	3.5	3.8	4.2			1
Dibutylamine	CCl₄	0.5	0.9	1.4	2.5			1
Dibutyl ether	CCl₄	0.9	1.4	3.3				1
Dibutyl sebacate	CCl₄	1.0	1.5	2.2	4.0			1
o-Dichlorobenzene	CCl₄	7.2						1
m-Dichlorobenzene	CCl₄	7.2	7.4					1
1,1-Dichloroethane	CDCl₃	2.06	5.90					2
1,2-Dichloroethane	CCl₄	3.7						1

Compound	Solvent	¹H NMR shifts (ppm relative to TMS)					Ref.
1,1-Dichloroethene	CCl₄	5.5					1
cis-1,2-Dichloroethene	CDCl₃	6.28					2
trans-1,2-Dichloroethene	(CH₃)₄Si	6.24					2
Dichloromethane (Methylene chloride)	CCl₄	5.3					1
(Dichloromethyl)benzene (Benzal chloride)	CCl₄	6.6	7.4				1
1,2-Dichloropropane	CDCl₃	1.61	3.59	3.74	4.14		2
2,4-Dichlorotoluene	CCl₄	2.3	7.0	7.3			1
3,4-Dichlorotoluene	CCl₄	2.3	7.0				1
Diethanolamine	D₂O	2.7	3.7				1
1,1-Diethoxyethane (Acetal)	CDCl₃	1.2	1.3	3.5	3.7	4.7	2
1,2-Diethoxyethane (Ethylene glycol diethyl ether)	CDCl₃	1.22	3.54	3.58			1
Diethylamine	CCl₄	0.9	1.0	2.6			1
Diethyl carbonate	CDCl₃	1.3	4.2				1
Diethylene glycol	CDCl₃	3.7	4.2				1
Diethylene glycol dimethyl ether (Diglyme)	CDCl₃	3.3	3.5				1
Diethylene glycol monoethyl ether (Carbitol)	CCl₄	1.2	3.1	3.5	3.6		1
Diethylene glycol monoethyl ether acetate	CDCl₃	1.22	2.08	3.54	3.71	4.23	2
Diethylene glycol monomethyl ether	CDCl₃	3.3	3.4	3.6			1
Diethyl ether	CDCl₃	1.21	3.47				2
Diethyl sulfide	CCl₄	1.2	2.5				1
Diisopropylamine	CCl₄	0.7	1.0	2.9			1
Diisopropyl ether	CCl₄	1.0	3.5				1
1,2-Dimethoxybenzene (Veratrole)	CCl₄	3.7	6.8				1
1,2-Dimethoxyethane (Ethylene glycol dimethyl ether)	CCl₄	3.3	3.4				1
Dimethoxymethane (Methylal)	CCl₄	3.2	4.4				1
N,N-Dimethylacetamide	CDCl₃	2.1	2.9	3.0			1
2,4-Dimethylaniline (2,4-Xylidine)	CCl₄	2.0	2.2	3.4	6.4	6.7	1
2,2-Dimethylbutane (Neohexane)	CCl₄	0.9	1.1	1.3	1.1	1.3	1
2,3-Dimethylbutane	CCl₄	0.9	1.5				1
N,N-Dimethylformamide	CDCl₃	2.9	3.0	8.0			1
Dimethyl glutarate	CDCl₃	2.0	2.4	3.7			1
2,6-Dimethyl-4-heptanone (Isovalerone)	CCl₄	0.9	2.1				1
2,5-Dimethylhexane	CCl₄	0.9	1.4				1
Dimethyl maleate	CCl₄	3.7	6.2				1
2,2-Dimethylpentane	CDCl₃	0.9	0.9	1.2			1
2,4-Dimethylpentane	CCl₄	0.9	1.1	1.6			1
2,4-Dimethyl-3-pentanone (Diisopropyl ketone)	CCl₄	1.0	2.6				1
2,4-Dimethylpyridine (2,4-Lutidine)	CDCl₃	2.3	2.5	7.0	7.0	8.4	1
2,6-Dimethylpyridine (2,6-Lutidine)	CDCl₃	2.51	6.93	7.42			2
Dimethyl sulfoxide	CDCl₃	2.62					2
1,4-Dioxane	CDCl₃	3.69					2
1,3-Dioxolane	CDCl₃	3.88	4.90				2
Dipentyl ether (Amyl ether)	CDCl₃	0.9	1.4	3.4			1
Dipropylamine	CDCl₃	1.5	2.6				1
Dodecane	CCl₄	0.9	1.3				1
1-Dodecene	CCl₄	0.9	1.3	2.0	5.4		1
Epichlorohydrin	CCl₄	2.6	2.8	3.2	3.5	3.6	1
1,2-Epoxybutane (Ethyloxirane)	CCl₄	1.0	1.5	2.3	2.6	2.7	1
1,2-Ethanediamine	CCl₄	1.2	2.6				1
1,2-Ethanediol (Ethylene glycol)	D₂O	3.7					1
1,2-Ethanediol, diacetate (Ethylene glycol diacetate)	CCl₄	2.0	4.2				1
Ethanol	CDCl₃	1.23	2.61	3.69			2

Proton NMR Shifts of Common Organic Solvents

Compound	Solvent	¹H NMR shifts (ppm relative to TMS)						Ref.
Ethanolamine	CDCl₃	2.7	2.8	3.5				1
Ethoxybenzene (Phenetole)	CCl₄	1.3	3.9	6.9				1
2-Ethoxyethanol (Ethylene glycol monoethyl ether (Cellosolve)	CDCl₃	1.22	2.70	3.55	3.72			2
2-Ethoxyethyl acetate (Ethylene glycol monoethyl ether acetate)	CCl₄	1.2	2.0	3.4	3.5	4.1		1
Ethyl acetate	CDCl₃	1.26	2.04	4.12				2
Ethyl acetoacetate	CDCl₃	1.3	1.9	2.2	3.3	4.1	4.9	1
Ethyl acrylate (Ethyl propenoate)	CCl₄	1.3	4.1	5.7	6.1	6.3		1
Ethylamine	D₂O	1.1	2.6					1
Ethylbenzene	CDCl₃	1.3	2.7	7.2				1
Ethyl benzoate	CCl₄	1.3	4.3	7.4	8.0			1
Ethyl butanoate	CCl₄	0.9	1.2	1.7	2.2	4.1		1
Ethyl cyanoacetate	CCl₄	1.3	3.4	4.3				1
Ethylcyclohexane	CDCl₃	0.9	1.9	1.4				1
Ethylene carbonate	CDCl₃	4.5						1
Ethyl formate	CCl₄	1.3	4.2	7.9				1
2-Ethyl-1,3-hexanediol	CDCl₃	1.0	1.4	3.8				1
2-Ethyl-1-hexanol	CDCl₃	0.9	1.3	1.8	3.5			1
Ethyl 3-methylbutanoate	CDCl₃	1.0	1.3	1.9	4.1			1
3-Ethyl-2-methylpentane	CCl₄	0.9	0.9	1.5				1
Fluorobenzene	CCl₄	7.0						1
2-Fluorotoluene	CCl₄	2.2	6.9					1
3-Fluorotoluene	CCl₄	2.3	6.9					1
4-Fluorotoluene	CCl₄	2.2	6.8	7.0				1
Furan	CDCl₃	6.38	7.44					2
Furfural	CDCl₃	6.6	7.3	7.7	9.7			1
Furfuryl alcohol	CDCl₃	2.8	4.6	6.3	7.4			1
Glycerol	D₂O	3.6						1
Glycerol triacetate (Triacetin)	CDCl₃	2.1	4.2	4.3	5.2			1
Heptane	CDCl₃	0.88	1.27	1.30				2
1-Heptanol	CCl₄	0.9	1.4	3.4	3.5			1
3-Heptanol	CCl₄	0.9	1.4	2.3	3.4			1
2-Heptanone (Methyl pentyl ketone)	CCl₄	0.9	1.3	2.0	2.3			1
3-Heptanone (Ethyl butyl ketone)	CCl₄	1.0	1.4	2.3				1
1-Heptene	CCl₄	0.9	1.4	2.0	4.9	5.7		1
Hexane	CDCl₃	0.89	1.27	1.29				2
Hexanedinitrile (Adiponitrile)	CDCl₃	1.8	2.5					1
Hexanenitrile	CDCl₃	0.9	1.5	2.3				1
Hexanoic acid (Caproic acid)	CDCl₃	0.9	1.4	2.4	11.4			1
1-Hexanol	CDCl₃	0.90	1.32	1.56	1.79	3.62		2
Hexyl acetate	CCl₄	0.9	1.4	2.0	4.0			1
3-Hydroxypropanenitrile (Hydracrylonitrile)	CDCl₃	2.6	3.4	3.9				1
Iodobenzene	CCl₄	6.8	7.5	7.7				1
1-Iodobutane (Butyl iodide)	CDCl₃	1.0	1.7	1.9	4.2			1
2-Iodobutane (sec-Butyl iodide)	CDCl₃	1.0	1.7	1.9	4.2			1
Iodoethane (Ethyl iodide)	CDCl₃	1.2	2.6	3.7				1
Iodomethane (Methyl iodide)	CDCl₃	2.2						1
1-Iodopropane (Propyl iodide)	CCl₄	1.0	1.8	3.2				1
2-Iodopropane (Isopropyl iodide)	CDCl₃	1.9	4.3					1
Isobutanal (2-Methyl-1-propanal)	CCl₄	1.1	2.4	9.6				1
Isobutyl acetate	CCl₄	0.9	1.9	2.0	3.8			1
Isobutylbenzene	CCl₄	0.9	1.9	2.4	7.1			1
Isobutyl formate	CDCl₃	1.0	2.0	3.9	8.0			1

Compound	Solvent	¹H NMR shifts (ppm relative to TMS)								Ref.
Isobutyl isobutanoate	CCl$_4$	0.9	1.2	1.9	2.5	3.8				1
Isopentyl acetate	CDCl$_3$	0.9	1.5	2.0	4.0					1
Isophorone	CDCl$_3$	1.04	1.95	2.19	5.88					2
Isopropyl acetate	CDCl$_3$	0.9	1.4	1.6	2.4					1
Isopropylbenzene (Cumene)	CDCl$_3$	1.3	2.4	2.9	7.3					1
1-Isopropyl-4-methylbenzene (*p*-Cymene)	CDCl$_3$	1.2	2.3	2.9	7.1					1
Isoquinoline	CDCl$_3$	8.5	9.3							1
d-Limonene (Citrene)	CCl$_4$	1.4	1.7	1.9	4.6	5.3				1
Mesityl oxide	CDCl$_3$	1.89	2.14	2.16	6.09					2
Methanol	CDCl$_3$	3.43	3.66							2
2-Methoxyethanol (Ethylene glycol monomethyl ether)	CDCl$_3$	2.5	3.4	3.5	3.7	3.5	3.7			1
2-Methoxyethyl acetate (Ethylene glycol monomethyl ether acetate)	CDCl$_3$	2.09	3.39	3.59	4.22					2
Methyl acetate	CCl$_4$	2.0	3.7							1
2-Methylacrylonitrile	CDCl$_3$	2.0	5.7	5.8						1
2-Methylaniline (*o*-Toluidine)	CCl$_4$	2.0	3.2	6.7						1
3-Methylaniline (*m*-Toluidine)	CCl$_4$	2.2	3.3	6.4	6.9					1
N-Methylaniline	CCl$_4$	2.7	3.3	6.4	6.6	7.1				1
Methyl benzoate	CCl$_4$	3.8	7.4	8.0						1
3-Methylbutanoic acid (Isovaleric acid)	CCl$_4$	1.0	2.2	11.0						1
3-Methyl-1-butanol (Isopentyl alcohol)	CCl$_4$	0.9	1.5	3.5	4.1					1
Methyl cyanoacetate	CDCl$_3$	3.5	3.8							1
Methylcyclohexane	CCl$_4$	0.9	1.4							1
cis-4-Methylcyclohexanol	CDCl$_3$	0.9	1.5	2.9	3.9					1
N-Methylformamide	CDCl$_3$	2.82	7.4	8.16						2
Methyl formate	CDCl$_3$	3.76	8.07							2
2-Methylheptane	CCl$_4$	0.9	1.3							1
4-Methylheptane	CDCl$_3$	0.8	0.9	1.4						1
2-Methylhexane	CDCl$_3$	0.9	0.9	1.4						1
5-Methyl-2-hexanone (Methyl isopentyl ketone)	CCl$_4$	0.9	1.4	2.0	2.3					1
Methyl methacrylate (Methyl 2-methyl-2-propenoate)	CDCl$_3$	2.0	3.8	5.6	6.1					1
2-Methyloctane	CCl$_4$	0.9	1.0	1.3						1
2-Methylpentane	CCl$_4$	0.8	0.9	1.5						1
3-Methylpentane	CCl$_4$	0.8	1.5							1
2-Methyl-2,4-pentanediol (Hexylene glycol)	CCl$_4$	1.1	1.2	1.3	1.4	1.6	4.2	4.7	4.9	1
4-Methylpentanenitrile	CCl$_4$	1.0	1.6	2.3						1
4-Methyl-2-pentanol	CCl$_4$	0.9	1.1	1.3	1.8	3.5	3.7			1
3-Methyl-3-pentanol	CCl$_4$	0.9	1.1	1.4	1.8					1
4-Methyl-2-pentanone (Methyl isobutyl ketone)	CDCl$_3$	0.9	2.1	2.3						1
2-Methylpropanenitrile (Isobutyronitrile)	CDCl$_3$	1.3	2.7							1
2-Methylpropanoic acid (Isobutyric acid)	CCl$_4$	1.2	2.6							1
2-Methyl-1-propanol (Isobutyl alcohol)	CCl$_4$	0.9	1.7	3.3	4.0					1
2-Methyl-2-propanol (*tert*-Butyl alcohol)	CDCl$_3$	1.3	1.4							1
2-Methylpyridine (2-Picoline)	CCl$_4$	2.5	6.9	7.4	8.4					1
3-Methylpyridine (3-Picoline)	CCl$_4$	2.3	7.0	7.4	8.4					1
N-Methyl-2-pyrrolidinone	CDCl$_3$	2.1	2.4	2.8	3.4					1
Methyl salicylate	CCl$_4$	3.9	6.7	6.9	7.3	7.7	10.6			1
2-Methylthiophene	CDCl$_3$	2.5	6.7	6.9	7.0					1
Morpholine	CDCl$_3$	2.59	2.86	3.67						2
Nitrobenzene	CCl$_4$	7.52	7.65	8.19						2
Nitroethane	CDCl$_3$	1.6	4.3							1
Nitromethane	CCl$_4$	4.2								1
1-Nitropropane	CDCl$_3$	1.0	2.1	4.4						1

Compound	Solvent	^{1}H NMR shifts (ppm relative to TMS)								Ref.
Nonane	CDCl$_3$	0.9	1.3							1
Octane	CDCl$_3$	0.88	1.26							1
1-Octanol	CDCl$_3$	0.88	1.29	1.5	2.40	3.60				2
2-Octanone (Hexyl methyl ketone)	CCl$_4$	0.9	1.4	2.1	2.4					1
1-Octene	CCl$_4$	0.9	1.3	2.1	4.8	4.9	5.7			1
Pentane	CDCl$_3$	0.88	1.26	1.30						2
1,5-Pentanediol (Pentamethylene glycol)	D$_2$O	1.5	3.6							1
2,4-Pentanedione (Acetylacetone)	CCl$_4$	2.0	2.2	3.5	5.4	14.7				1
Pentanenitrile (Valeronitrile)	CCl$_4$	1.0	1.6	2.3						1
Pentanoic acid (Valeric acid)	CCl$_4$	0.9	1.5	2.3	11.7					1
1-Pentanol (Amyl alcohol)	CCl$_4$	0.9	1.4	3.5	4.4					1
2-Pentanol (sec-Amyl alcohol)	CDCl$_3$	0.9	1.2	1.3	2.2	3.7				1
3-Pentanol (Diethyl carbinol)	CCl$_4$	0.9	1.4	3.3	3.4					1
2-Pentanone (Methyl propyl ketone)	CCl$_4$	0.9	1.6	2.0	2.3					1
3-Pentanone (Diethyl ketone)	CCl$_4$	1.0	2.4							1
Pentyl acetate (Amyl acetate)	CCl$_4$	0.9	1.4	2.0	4.0					1
Pentylamine (Amylamine)	CDCl$_3$	0.9	1.4	2.7						1
α-Pinene	CDCl$_3$	0.84	1.16	1.27	1.66	1.93	2.19	2.33	5.19	2
Piperidine	CDCl$_3$	1.53	2.18	2.79						2
Propanal	CDCl$_3$	1.1	2.4	9.8						1
1,2-Propanediol (1,2-Propylene glycol)	CDCl$_3$	1.1	3.4	3.9	4.3					1
1,3-Propanediol (Trimethylene glycol)	D$_2$O	1.8	3.7							1
Propanenitrile	CCl$_4$	1.3	2.3							1
Propanoic acid	CDCl$_3$	1.1	2.4	10.5						1
Propanoic anhydride	CCl$_4$	1.2	2.4							1
1-Propanol (Propyl alcohol)	CDCl$_3$	0.9	1.6	2.3	3.6					1
2-Propanol (Isopropyl alcohol)	CDCl$_3$	1.2	1.6	4.0						1
Propargyl alcohol (3-Hydroxy-1-propyne)	CDCl$_3$	2.5	2.8	4.3						1
Propyl acetate	CCl$_4$	0.9	1.6	2.0	4.0					1
Propylamine	CCl$_4$	0.9	1.5	2.1	2.7					1
Propylbenzene	CDCl$_3$	0.9	1.6	2.6						1
Propyl formate	CCl$_4$	1.0	1.7	4.1	7.9					1
Pyridine	CDCl$_3$	7.23	7.62	8.59						2
Pyrrole	CDCl$_3$	6.2	6.7	8.0						1
Pyrrolidine	C$_6$H$_6$	1.5	2.4	2.7						1
2-Pyrrolidone	CDCl$_3$	2.2	3.4	7.7						1
Quinoline	CCl$_4$	7.1	7.8	8.8						1
Salicylaldehyde (2-Hydroxybenzaldehyde)	CDCl$_3$	7.0	7.4	9.8	11.0					1
Styrene	CCl$_4$	5.1	5.6	6.6	7.2					1
Sulfolane	CDCl$_3$	2.2	3.0							1
α-Terpinene	CDCl$_3$	1.02	1.77	2.09	2.28	5.59	5.62			2
1,1,1,2-Tetrachloroethane	CDCl$_3$	4.29								2
1,1,2,2-Tetrachloroethane	CDCl$_3$	5.91								2
Tetraethylene glycol	CCl$_4$	3.5	3.6							1
Tetrahydrofuran	CDCl$_3$	1.84	3.73							2
1,2,3,4-Tetrahydronaphthalene	CDCl$_3$	1.8	2.8	7.1						1
Tetrahydropyran	CCl$_4$	1.6	3.6							1
Tetrahydrothiophene	CDCl$_3$	1.9	2.8							1
Tetramethylsilane	CCl$_4$	0.0								1
Tetramethylurea	CCl$_4$	2.8								1
Thiophene	CDCl$_3$	7.1	7.3							1
Toluene	CDCl$_3$	2.34	7.18							2

Compound	Solvent	¹H NMR shifts (ppm relative to TMS)						Ref.
Tribromomethane (Bromoform)	CCl$_4$	6.8						1
Tributylamine	CCl$_4$	0.9	1.3	2.3				1
1,1,1-Trichloroethane	CCl$_4$	2.7						1
1,1,2-Trichloroethane	CDCl$_3$	4.0	5.8					1
Trichloroethene	CCl$_4$	6.5						1
Trichloroethylsilane	CCl$_4$	1.3						1
Trichloromethane (Chloroform)	CCl$_4$	7.2						1
(Trichloromethyl)benzene (Benzotrichloride)	CCl$_4$	7.3	7.8					1
Tridecane	CCl$_4$	0.9	1.3					1
Triethanolamine	D$_2$O	2.7	3.6					1
Triethylamine	CCl$_4$	1.0	2.4					1
Triethylene glycol	CDCl$_3$	3.5	3.7					1
Triethyl phosphate	CDCl$_3$	1.4	4.1					1
2,2,2-Trifluoroethanol	CDCl$_3$	3.4	3.9					1
(Trifluoromethyl)benzene (Benzotrifluoride)	CCl$_4$	7.5						1
Trimethylamine	CCl$_4$	2.12						2
1,2,3-Trimethylbenzene (Hemimellitene)	CDCl$_3$	2.2	2.3	7.0				1
1,2,4-Trimethylbenzene (Pseudocumene)	CCl$_4$	2.2	6.8					1
1,3,5-Trimethylbenzene (Mesitylene)	CDCl$_3$	2.3	6.8					1
2,2,3-Trimethylbutane (Triptane)	CCl$_4$	0.8	1.3					1
2,2,5-Trimethylhexane	CCl$_4$	0.9	1.2					1
2,3,3-Trimethylpentane	CCl$_4$	0.8	0.8	1.4				1
2,3,4-Trimethylpentane	CCl$_4$	0.8	1.9					1
Trimethyl phosphate	CDCl$_3$	3.78						2
2,4,6-Trimethylpyridine (2,4,6-Collidine)	CCl$_4$	2.2	2.4	6.6				1
1-Undecene	CCl$_4$	0.9	1.3	2.0	4.8	4.9	5.6	1
Vinyl acetate	CDCl$_3$	2.1	4.6	4.9	7.3			1
o-Xylene	CDCl$_3$	2.22	7.07					2
m-Xylene	CDCl$_3$	2.28	6.95	7.11				2
p-Xylene	CDCl$_3$	2.30	7.05					2

^{13}C-NMR ABSORPTIONS OF MAJOR FUNCTIONAL GROUPS

Thomas J. Bruno and Paris D. N. Svoronos

The table below lists the ^{13}C chemical shift ranges (in ppm) with the corresponding functional groups in descending order. Some typical simple compounds for every family are given to illustrate the corresponding range. The shifts for the carbons of interest are given in parentheses, either for each carbon as it appears from left to right in the formula, or by the underscore (Refs. 1–17). In Ref. 1, the reader can find additional details, correlation charts, additivity rules and the expected ^{13}C peaks attributed to common solvents.

References

1. Bruno, T. J., and Svoronos, P. D. N., *CRC Handbook of Basic Tables for Chemical Analysis*, 3rd Edition, CRC Press, Boca Raton, FL, 2011.
2. Yoder, C. H., and Schaeffer, C. D., Jr., *Introduction to Multinuclear NMR: Theory and Application*, The Benjamin/Cummings Publishing Co., Menlo Park, CA, 1987.
3. Brown, D. W., *J. Chem. Educ.* 62, 209, 1985.
4. Silverstein, R. M., and Webster F. X., *Spectrometric Identification of Organic Compounds*, 6th Edition, John Wiley and Sons, New York, 1998.
5. Becker, E. D., *High Resolution NMR, Theory and Chemical Applications*, 2nd Edition, Academic Press, New York, 1980.
6. Gunther, H., *NMR Spectroscopy: Basic Principles, Concepts and Applications in Chemistry*, Wiley, New York, 2003.
7. Kitamaru, R., *Nuclear Magnetic Resonance: Principles and Theory*, Elsevier Science, 1990.
8. Lambert, J. B., Holland, L. N., and Mazzola, E. P., *Nuclear Magnetic Resonance Spectroscopy: Introduction to Principles, Applications and Experimental Methods*, Prentice Hall, Englewood Cliffs, NJ, 2003.
9. Bovey, F. A., and Mirau, P. A., *Nuclear Magnetic Resonance Spectroscopy*, 2nd Edition, Academic Press, New York, 1988.
10. Harris, R. K., and Mann, B. E., *NMR and the Periodic Table*, Academic Press, London, 1978.
11. Hore, P. J., *Nuclear Magnetic Resonance*, Oxford University Press, Oxford, 1995.
12. Nelson, J. H., *Nuclear Magnetic Resonance Spectroscopy*, 2nd Edition, Wiley, New York, 2003.
13. Levy, G. C., Lichter, R. L., and Nelson, G. L., *Carbon-13 Nuclear Magnetic Resonance Spectroscopy*, 2nd Edition, Wiley, New York, 1980.
14. Pihlaja, K., and Kleinpeter, E., *Carbon-13 NMR Chemical Shifts in Structural and Stereochemical Analysis*, VCH, New York, 1994.
15. *Aldrich Library of ^{1}H and ^{13}C FT-NMR Spectra*, Aldrich Chemical Company, Milwaukee, WI, 1996.
16. Balci, M., *Basic 1H- and 13C-NMR Spectroscopy*, Elsevier, London, 2005.
17. http://www.chem.wisc.edu/areas/reich/handouts/nmr-c13/cdata.htm, 2009.

δ/ppm	Group	Family		Example (δ of Underlined Carbon)
220–165	>C=O	Ketones	$(CH_3)_2\underline{C}O$	(206.0)
			$(CH_3)_2CH\underline{C}OCH_3$	(212.1)
		Aldehydes	$CH_3\underline{C}HO$	(199.7)
		α,β-unsaturated carbonyls	$CH_3CH=CH\underline{C}HO$	(192.4)
			$CH_2=CH\underline{C}OCH_3$	(169.9)
		Carboxylic acids	$H\underline{C}O_2H$	(166.0)
			$CH_3\underline{C}O_2H$	(178.1)
		Amides	$H\underline{C}ONH_2$	(165.0)
			$CH_3\underline{C}ONH_2$	(172.7)
		Esters	$CH_3\underline{C}O_2CH_2CH_3$	(170.3)
			$CH_2=CH\underline{C}O_2CH_3$	(165.5)
140–120	>C=C<	Aromatics, alkenes	C_6H_6	(128.5)
			$CH_2=CH_2$	(123.2)
			$CH_2=\underline{C}HCH_3$	(115.9, 136.2)
			$CH_2=\underline{C}HCH_2Cl$	(117.5, 133.7)
			$CH_3CH=\underline{C}HCH_2CH_3$	(132.7)
125–115	–C≡N	Nitriles	$CH_3-\underline{C}≡N$	(117.7)
80–70	–C≡C–	Alkynes	$H\underline{C}≡CH$	(71.9)
			$CH_3\underline{C}≡CCH_3$	(73.9)
70–45	>C–O⁻	Esters, alcohols	$\underline{C}H_3OOCCH_2CH_3$	(57.6, 67.9)
			$HO\underline{C}H_3$	(49.0)
			$HO\underline{C}H_2CH_3$	(57.0)
	>C–OH			
40–20	>C–NH$_2$	Amines	$\underline{C}H_3NH_2$	(26.9)
			$CH_3\underline{C}H_2NH_2$	(35.9)
30–15	–S–CH$_3$	Sulfides (thioethers)	$C_6H_5-S-\underline{C}H_3$	15.6
30–(–2.3)	>CH–	Alkanes, cycloalkanes	$\underline{C}H_4$	(–2.3)
			$\underline{C}H_3CH_3$	(5.7)
			$\underline{C}H_3\underline{C}H_2CH_3$	(15.8, 16.3)
			$\underline{C}H_3\underline{C}H_2CH_2CH_3$	(13.4, 25.2)
			$\underline{C}H_3\underline{C}H_2\underline{C}H_2CH_2CH_3$	(13.9, 22.8, 34.7)
			Cyclohexane	(26.9)

^{13}C NMR CHEMICAL SHIFTS OF COMMON ORGANIC SOLVENTS

The following table gives the expected carbon-13 chemical shifts, relative to tetramethylsilane, for various useful NMR solvents. In some solvents, slight changes can occur with change of concentration.[2,3]

References

1. Bruno, T. J., and Svoronos, P. D. N., *CRC Handbook of Basic Tables for Chemical Analysis*, 3rd Edition, CRC Press, Boca Raton, FL, 2011.

2. Silverstein, R. M., Bassler, G. C., and Morrill, T. C., *Spectrometric Identification of Organic Compounds*, John Wiley & Sons, Now York, 1981.
3. Rahman, A.-U., *Nuclear Magnetic Resonance. Basic Principles*, Springer-Verlag, New York, 1986.
4. Pretsch, E., Clerc, T., Seibl, J., and Simon, W., *Spectral Data for Structure Determination of Organic Compounds, Second Edition*, Springer-Verlag, Heidelberg, 1989.

Solvent	Formula	Chemical shift (ppm)
Acetic acid-d_4	CD_3COOD	20.0 (CD_3) 205.8 (C=O)
Acetone	$(CH_3)_2C=O$	30.7 (CH_3) 206.7 (C=O)
Acetone-d_6	$(CD_3)_2C=O$	29.2 (CD_3) 204.1 (C=O)
Acetonitrile-d_3	$CD_3C\equiv N$	1.3 (CD_3) 117.1 (C≡N)
Benzene	C_6H_6	128.5
Benzene-d_6	C_6D_6	128.4
Carbon disulfide	CS_2	192.3
Carbon tetrachloride	CCl_4	96.0
Chloroform	$CHCl_3$	77.2
Chloroform-d_3	$CDCl_3$	77.05
Cyclohexane-d_{12}	C_6D_{12}	27.5
Dichloromethane-d_2	CD_2Cl_2	53.6
Dimethylformamide-d_7	$(CD_3)_2NCDO$	31 (CD_3) 36 (CD_3) 162.4 (C=O)
Dimethylsulfoxide-d_6	$(CD_3)_2S=O$	39.6
Dioxane-d_8	$C_4D_8O_2$	67.4
Formic acid-d_2	$DCOOD$	165.5
Methanol-d_4	CD_3OD	49.3
Nitromethane-d_3	CD_3NO_2	57.3
Pyridine	C_5H_5N	123.6 (C_3) 135.7 (C_4) 149.8 (C_2)
Pyridine-d_5	C_5D_5N	123.9 (C_3) 135.9 (C_4) 150.2 (C_2)
1,1,2,2-Tetrachloroethane-d_2	$CDCl_2CDCl_2$	75.5
Tetrahydrofuran-d_8	C_4D_8O	25.8 (C_2) 67.9 (C_1)
Trichlorofluoromethane	$CFCl_3$	117.6

^{15}N-NMR CHEMICAL SHIFTS OF MAJOR CHEMICAL FAMILIES

Thomas J. Bruno and Paris D. N. Svoronos

The following table contains ^{15}N NMR chemical shifts of various organic nitrogen compounds. Chemical shifts are expressed relative to different standards (NH_3, NH_4Cl, CH_3NO_2, NH_4NO_3, HNO_3, etc.) and are interconvertible. Chemical shifts are sensitive to hydrogen bonding and are solvent dependent as seen in the case of pyridine (see footnote a below). Consequently, the reference as well as the solvent should always accompany chemical shift data. All shifts are relative to ammonia unless otherwise specified (Refs. 1–16). A section of "miscellaneous" data gives the chemical shift of special compounds relative to unusual standards. In Ref. 16, the reader can find additional details, correlation charts, and spin-spin coupling ranges.

References

1. Levy, G. C., and Lichter, R. L., *Nitrogen-15 Nuclear Magnetic Resonance Spectroscopy*, John Wiley and Sons, New York, 1979.
2. Yoder, C. H., and Schaeffer, C. D., Jr., *Introduction to Multinuclear NMR*, Benjamin/Cummings, Menlo Park, CA, 1987.
3. Duthaler, R. O., and Roberts, J. D., *J. Am. Chem. Soc.* 100, 4969, 1978.
4. Duthaler, R. O., and Roberts, J. D., *J. Am. Chem. Soc.* 100, 3889, 1978.
5. Kozerski, L., and von Philipsborn, W., *Org. Magn. Res.* 17, 306, 1981.
6. Duthaler, R. O., and Roberts, J. D., *J. Am. Chem. Soc.* 100, 3882, 1978.
7. Duthaler, R. O., and Roberts, J. D., *J. Magn. Res.* 34, 129, 1979.
8. Psota, L., Franzen-Sieveking, M., Turnier, J., and Lichter, R. L., *Org. Magn. Res.* 11, 401, 1978.
9. Subramanian, P. K., Chandra Sekara, N., and Ramalingam, K., *J. Org. Chem.* 47, 1933, 1982.
10. Schuster, I. I., and Roberts, J. D., *J. Org. Chem.* 45, 284, 1980.
11. Kupce, E., Liepins, E., Pudova, O., and Lukevics, E., *J. Chem. Soc., Chem. Commun.*, 581, 1984.
12. Allen, M., and Roberts, J. D., *J. Org. Chem.* 45, 130, 1980.
13. Brownlee, R. T. C., and Sadek, M., *Magn. Res. Chem.* 24, 821, 1986.
14. Dega-Szafran, Z., Szafran, M., Stefaniak, L., Brevard, C., and Bourdonneau, M., *Magn. Res. Chem.* 24, 424, 1986.
15. Lambert, J. B., Shurvell, H. F., Verbit, L., Cooks, R. G., and Stout, G. H., *Organic Structural Analysis*, MacMillan, New York, 1976.
16. Bruno, T. J., and Svoronos, P. D. N., *CRC Handbook of Basic Tables for Chemical Analysis*, 3rd Edition, CRC Press, Boca Raton, FL, 2011.

Chemical Shift Range/ppm	Family	Example (δ)
<930	Nitroso compounds	C_6H_5–NO (913, 930)
608	Sodium nitrite	$NaNO_2$
~500	Azo compounds	C_6H_5–N=N–C_6H_5 (510)
380–350	Nitro compounds	$C_6H_5NO_2$ (370.3); CH_3NO_2 (380.2); 4-F–C_6H_4–NO_2 (368.5); 1,3-$(NO_2)_2C_6H_4$ (365.4)
367	Nitric acid (8.57 M)	HNO_3
360–325	Nitramines	CH_3NHNO_2 (355.6); $CH_3O_2CNHNO_2$ (334.9)
350–300	Pyridines	C_5H_5N (317)[a] (gas); 4-CH_3–C_5H_4N (309.3); 4-NH_2–C_5H_4N (271.5); 4-NC–C_5H_4N (327.9)
~310	Imines (aromatic)	$(C_6H_5)_2$C=NH (308); C_6H_5CH=NCH_3 (318); C_6H_5CH=NC_6H_5 (326)
310.1	Nitrogen (gas)	N_2
250–200	Pyridinium salts	$C_5H_5NH^+$ (215)
260–175	Cyanides (nitriles)	CH_3CN (239.5, 245); C_6H_5CN (258.7); KCN 177.8
~160	Pyrroles, isonitriles	C_4H_4NH (158)CH_3NC (162)
~150	Thioamides	CH_3C(=S)NH_2 (150.2)
120–110	Lactams	$HN(CH_2)_3$C=O (5-membered ring; 114.7) $HN(CH_2)_6$C=O (8-membered ring; 117.7)
110–100	Amides	$C_6H_5CONH_2$ (100); CH_3CONH_2 (103.4); $CH_3CONHCH_3$ (105.8); $CH_3CON(CH_3)_2$ (103.8); $HCONH_2$ (108.5)
125–90	Sulfonamides	$CH_3SO_2NH_2$ (95); $C_6H_5SO_2NH_2$ (94.3)
~100	Hydrazines	$C_6H_5NHNHC_6H_5$ (96)
110–60	Ureas	$[H_2N]_2CO$ (75, 82); $[(CH_3)_2N]_2CO$ (63.5); $[C_6H_5NH]_2CO$ (107.7)
100–70	Aminophospines, aminophosphine oxides	$C_6H_5NHP(CH_3)_2$ (71.1) $C_6H_5NHPO(CH_3)_2$ (86.6)
70–50	Aromatic amines	$C_6H_5NH_2$ (55, 59), (–322.3)[b]; $C_6H_5NH_3^+$ (48), (–326.4)[b], 26.1[c]; p–O_2N–C_6H_4–NH_2 (70)
40–0	Aliphatic amines	CH_3NH_2 (1.3)[d], (–371)[b]; $(CH_3)_2NH$ (–363.3)[b], (–364.9)[e], 6.7[d]; $(CH_3)_3N$ (–356.9)[b], (–360.7)[e], 13.0[d]
50–10	Isonitriles	CH_3NCO (14.1); C_6H_5NCO (46.5)
65–20	Ammonium salts	NH_4Cl (26.1)[d]; CH_3NH_3Cl (24.5); $(CH_3)_2NH_2Cl$ (26.6); $(CH_3)_3NHCl$ (33.8); $(CH_3)_4NCl$ (44.7)
~15	Isocyanates	CH_3NCO (14.1)

Miscellaneous:

(–130)–(–110) and ~(–212)	Imidazoles	*N*-methylimdazole (–111.4, pyridine N and -215.7, pyrrole N)[b]
(–345)–(–310)[b]	Piperidine, hydrochloride salts	piperidinium hydorchloride (–344.8); 2-methyl piperdinium hydrochloride (–322.1)[e]
	Decahydroquinolines, hydrochloride salts	*trans*-decahydroquinolinium hydrochloride (-322.5); *cis*-decahydroquinolinium hydrochloride (–328.5)

Chemical Shift Range/ppm	Family	Example (δ)
$(-293)-(-280)^f$	Enaminones	CH$_3$C(=O)CH=CHNHCH$_3$
		[(E)-(-294.2); (Z)-(-285.9)]
35–15^g	4-Aminotetrahydropyrans,	2,6-diphenyl 4-aminotetrahydropyran (34.5)
	4-aminotetrahydro-thiopyrans	2,6-diphenyl 4-aminotetrahydrothiopyran (33.6)
$(-325)-(-310)^c$	1-Naphthylamines	8-nitro-1-naphthylamine (313.9)
$(-350)-(-300)^h$	Silylamines	HN[Si(CH$_3$)$_3$]$_2$ (-354.2)

[a] Varies with solvent. For instance: cyclohexane (315.5), benzene (312.1), chloroform (304.5), methanol (292.1), water (289), 2,2,2-trifluoroethanol (277.1). All chemical shifts relative to ammonia.

[b] Upfield from external HNO$_3$ 1M (CH$_3$OH).

[c] In ppm upfield from external 1M D^{15}NO$_3$ in D$_2$O (DMSO) (Ref. 10).

[d] Downfield from anhydrous liquid ammonia, ±0.2 ppm unless otherwise specified.

[e] Upfield from external HNO$_3$ 1M (cyclohexane).

[f] Relative to external CH$_3^{15}$NO$_2$.

[g] With respect to an external standard of 5M ^{15}NH$_4$NO$_3$ in 2M HNO$_3$ (^{15}NH$_4$NO$_3$ = 21.6 ppm relative to anhydrous ammonia) (Ref. 9).

[h] Relative to N(SiH$_3$) (50 % in CDCl$_3$) (Ref. 11).

NATURAL ABUNDANCE OF IMPORTANT ISOTOPES

Thomas J. Bruno and Paris D. N. Svoronos

The following table lists the atomic masses and relative percent concentrations of naturally occurring isotopes of importance in mass spectroscopy (Refs. 1–6).

References

1. de Hoffmann, E., and Stroobant, V., *Mass Spectrometry: Principles and* Applications, 3rd Edition, Wiley Interscience, Winchester, United Kingdom, 2007.

2. Johnstone, R. A. W., and Rose, M. E., *Mass Spectrometry for Chemists and Biochemists*, Cambridge University Press, 1996.
3. Lide, D. R., Ed., *CRC Handbook for Chemistry and Physics*, 90th Edition, CRC Press, Boca Raton, FL, 2010.
4. McLafferty, F. W., and Turecek, F., *Interpretation of Mass Spectra*, 4th Edition, University Science Books, Mill Valley, CA, 1993.
5. Watson, J. T., *Introduction to Mass Spectrometry*, 3rd Edition, Lippincott-Raven, Philadelphia, PA, 1997.
6. Bruno, T. J., and Svoronos, P. D. N., *CRC Handbook of Basic Tables for Chemical Analysis*, 3rd Edition, CRC Press, Boca Raton, FL, 2011.

Element	More Prominent Isotopes (Mass in u, Percent Abundance)		
Hydrogen	^{1}H (1.00783, 99.985)	^{2}H (2.01410, 0.015)	
Boron	^{10}B (10.01294, 19.8)	^{11}B (11.00931, 80.2)	
Carbon	^{12}C (12.00000, 98.9)	^{13}C (13.00335, 1.1)	
Nitrogen	^{14}N (14.00307, 99.6)	^{15}N (15.00011, 0.4)	
Oxygen	^{16}O (15.99491, 99.8)		^{18}O (17.9992, 0.2)
Fluorine	^{19}F (18.99840, ≈100.0)		
Silicon	^{28}Si (27.97693, 92.2)	^{29}Si (28.97649, 4.7)	^{30}Si (29.97376, 3.1)
Phosphorus	^{31}P (30.97376, ≈100.0)		
Sulfur	^{32}S (31.972017, 95.0)	^{33}S (32.97146, 0.7)	^{34}S (33.96786, 4.2)
Chlorine	^{35}Cl (34.96885, 75.5)		^{37}Cl (36.96590, 24.5)
Bromine	^{79}Br (78.9183, 50.5)		^{81}Br (80.91642, 49.5)
Iodine	^{127}I (126.90466, ≈100.0)		

COMMON MASS SPECTRAL FRAGMENTATION PATTERNS OF ORGANIC COMPOUND FAMILIES

Thomas J. Bruno and Paris D. N. Svoronos

The following table provides a guide to the identification and interpretation of commonly observed mass spectral fragmentation patterns for common organic functional groups (Refs. 1–10). It is of course highly desirable to augment mass spectroscopic data with as much other structural information as possible. Especially useful in this regard will be the confirmatory information of infrared and ultraviolet spectrophotometry as well as nuclear magnetic resonance spectrometry.

The reader is referred to other tables in this book for additional information that can be used in assessing or predicting fragmentation patterns, particularly the table of Proton Affinities (p. 10-167) and ionization energies of Gas-Phase Molecules (p. 10-199).

References

1. Bowie, J. H., Williams, D. H., Lawesson, S. O., Madsen, J. O., Nolde, C., and Schroll, G., *Tetrahedron* 22, 3515, 1966.

2. Johnstone, R. A. W., and Rose, M. E., *Mass Spectrometry for Chemical and Biochemists*, Cambridge University Press, New York, 1996.

3. Lee, T. A., *A Beginner's Guide to Mass Spectral Interpretation*, Wiley, New York, 1998

4. McLafferty, F. W., and Turecek, F., *Interpretation of Mass Spectra*, 4th Edition, University Science Books, Mill Valley, CA, 1993.

5. Pasto, D. J., and Johnson, C. R., *Organic Structure Determination*, Prentice-Hall, Englewood Cliffs, NJ, 1969.

6. Silverstein, R. M., Bassler, G. C., and Morrill, T. C., *Spectroscopic Identification of Organic Compounds*, 6th Edition, John Wiley and Sons, New York, 1998.

7. Smakman, R., and deBoer, T. J., *Org. Mass Spec.* 3, 1561, 1970.

8. Smith, R. M., *Understanding Mass Spectra: A Basic Approach*, John Wiley and Sons, New York, 1999.

9. Watson, T. J., and Watson, J. T., *Introduction to Mass Spectrometry*, Lippincott, Williams and Wilkins, Philadelphia, PA, 1997.

10. Bruno, T. J., and Svoronos, P. D. N., *CRC Handbook of Basic Tables for Chemical Analysis*, 3rd Edition, CRC Press, Boca Raton, FL, 2011.

Family	Molecular Ion Peak	Common Fragments; Characteristic Peaks
Acetals		Cleavage of all C–O, C–H and C–C bonds around the original aldehydic carbon
Alcohols	Weak for 1° and 2°; not detectable for 3°; strong for benzyl alcohols	Loss of 18 (H_2O — usually by cyclic mechanism); loss of H_2O and olefin simultaneously with four (or more) carbon-chain alcohols; prominent peak at $m/z = 31(CH_2\ddot{O}H)^+$ for 1° alcohols; prominent peak at $m/z = (RCH\ddot{O}H)^+$ for 2° alcohols and $m/z = (R_2\ddot{C}OH)^+$ for 3° alcohols
Aldehydes	Low intensity	Loss of aldehydic hydrogen (strong M-1 peak, especially with aromatic aldehydes); strong peak at $m/z = 29(HC{\equiv}O^+)$; loss of chain attached to the alpha carbon (beta cleavage); McLafferty rearrangement via beta cleavage if gamma hydrogen is present
Alkanes		
(a) Chain	Low intensity	Loss of 14 mass units (CH_2)
(b) Branched	Low intensity	Cleavage at the point of branch; low intensity ions from random rearrangements
(c) Alicyclic	Intense	Loss of 28 mass units ($CH_2{=}CH_2$) and side chains
Alkenes (olefins)	Rather high intensity (loss of π-electron) especially in case of cyclic olefins	Loss of units of general formula C_nH_{2n-1}; formation of fragments of the composition C_nH_{2n} (via McLafferty rearrangement); retro Diels-Alder fragmentation
Alkyl halides	Abundance of molecular ion F<Cl<Br<I; intensity decreases with increase in size and branching	Loss of fragments equal to the mass of the halogen until all halogens are cleaved off
(a) Fluorides	Very low intensity	Loss of 20 (HF); loss of 26 (C_2H_2) in case of fluorobenzenes
(b) Chlorides	Low intensity; characteristic isotope cluster	Loss of 35 (Cl) or 36 (HCl); loss of chain attached to the gamma carbon to the carbon carrying the Cl
(c) Bromides	Low intensity; characteristic isotope cluster	Loss of 79 (Br); loss of chain attached to the gamma carbon to the carbon carrying the Br
(d) Iodides	Higher than other halides	Loss of 127 (I)
Alkynes	Rather high intensity (loss of π-electron)	Fragmentation similar to that of alkenes
Amides	High intensity	Strong peak at $m/z = 44$ indicative of a 1° amide ($O{=}C{=}NH^+_2$); base peak at $m/z = 59$ ($CH_2{=}C(OH)N^+H_2$); possibility of McLafferty rearrangement; loss of 42 (C_2H_2O) for amides of the form $RNHCOCH_3$ when R is aromatic ring
Amines	Hardly detectable in case of acyclic aliphatic amines; high intensity for aromatic and cyclic amines	Beta cleavage yielding >C=N+<; base peak for all 1° amines at $m/z = 30$ ($CH_2{=}N^+H_2$); moderate M-1 peak for aromatic amines; loss of 27 (HCN) in aromatic amines; fragmentation at alpha carbons in cyclic amines
Aromatic hydrocarbons (arenes)	Intense	Loss of side chain; formation of RCH=CHR′ (via McLafferty rearrangement); cleavage at the bonds beta to the aromatic ring; peaks at $m/z = 77$ (benzene ring; especially mono-substituted), 91 (tropyllium); the ring position of alkyl substitution has very little effect on the spectrum
Carboxylic acids	Weak for straight-chain monocarboxylic acids; large if aromatic acids	Base peak at $m/z = 60$ ($CH_2{=}C(OH)_2$) if alpha hydrogen is present; peak at $m/z = 45$ (COOH); loss of 17 (–OH) in case of aromatic acids or short-chain acids
Disulfides	Low intensity	Loss of olefins (m/z equal to R–S–S–H+); strong peak at $m/z = 66$ (HSSH+)
Phenols	Highly intense peak (base peak* generally)	Loss of 28 (C=O) and 29 (CHO); strong peak at $m/z = 65$ ($C_5H_5^+$)

Family	Molecular Ion Peak	Common Fragments; Characteristic Peaks
Sulfides (thioethers)	Rather low intensity peak but higher than that of corresponding ether	Similar to those of ethers ($-O-$ substituted by $-S-$); aromatic sulfides show strong peaks at m/z = 109 ($C_6H_5S^+$.); 65 ($C_5H_5^+$.); 91 (tropyllium ion)
Sulfonamides	Intense	Loss of m/z = 64 ($SONH_2$) and m/z = 27 (HCN) in case of benzenesulfonamide
Esters	Weak intensity	Base peak at m/z equal to the mass of $R-C\equiv O^+$; peaks at m/z equal to the mass of $^+O\equiv C-OR'$, the mass of OR' and R'; McLafferty rearrangement possible in case of: (a) presence of a beta hydrogen in R' (peak at m/z equal to the mass of $R-C(^+OH)OH$, and (b) presence of a gamma hydrogen in R (peak at m/z equal to the mass of $(CH_2=C(^+OH)OR)$; loss of 42 ($CH_2=C=O$) in case of benzyl esters; loss of ROH via the ortho effect in case of o–substituted benzoates
Ketones	High intensity	Loss of R–groups attached to the >C=O (alpha cleavage); peak at m/z = 43 for all methyl ketones (CH_3CO^+); McLafferty rearrangement via beta cleavage if gamma hydrogen is present; loss of m/z = 28 (C=O) for cyclic ketones after initial alpha cleavage and McLafferty rearrangement
Mercaptan (thiols)	Rather low intensity but higher than that of corresponding alcohol	Similar to those of alcohols (–OH substituted by –SH); loss of m/z = 45 (CHS) and m/z = 44 (CS) for aromatic thiols
Nitriles	Unlikely to be detected except in case of acetonitrile (CH_3CN) and propionitrile (C_2H_5CN)	M+1 ion may appear (especially at higher pressures); M-1 peak is weak but detectable ($R-CH=C=N^+$); base peak at m/z = 41 ($CH_2=C=N^+H$); McLafferty rearrangement possible; loss of HCN is case of cyanobenzenes
Nitrites	Absent (or very weak at best)	Base peak at m/z = 30 (NO^+); large peak at m/z = 60 ($CH_2=O^+NO$) in all unbranched nitrites at the alpha carbon; absence of m/z = 46 permits differentiation from nitro compounds
Nitro compounds	Seldom observed	Loss of 30 (NO); subsequent loss of CO (in case of aromatic nitro- compounds); loss of NO_2, m/z = 46, from molecular ion peak
Sulfones	High intensity	Similar to sulfoxides; loss of mass equal to RSO_2; aromatic heterocycles show peaks at M-32 (sulfur), M-48 (SO), M-64 (SO_2)
Sulfoxides	High intensity	Loss of 17 (OH); loss of alkene (m/z equal to $RSOH^+$.); peak at m/z = 63 ($CH_2=SOH$)$^+$; aromatic sulfoxides show peak at m/z = 125 ($^+S-CH=CHCH=CHCHC=O$), 97($C_5H_5S^+$), 93(C_6H_5O); aromatic heterocycles show peaks at M-16 (oxygen), M-29(COH); M-48(SO)

[*] The base peak is the most intense peak in the mass spectrum, and is often the molecular ion peak, M^+.

COMMON MASS SPECTRAL FRAGMENTS LOST

Thomas J. Bruno and Paris D. N. Svoronos

The following table gives a list of neutral species that are most commonly lost when measuring the mass spectra of organic compounds. The list is suggestive rather than comprehensive, and should be used in conjunction with other sources [1–5]. The listed fragments include only combinations of carbon, hydrogen, oxygen, nitrogen, sulfur, and the halogens.

References

1. Hamming, M., and Foster, N., *Interpretation of Mass Spectra of Organic Compounds*, Academic Press, New York, 1972.
2. McLafferty, F. W., Turecek, F., *Interpretation of Mass Spectra*, 4th Edition, University Science Books, Mill Valley, CA, 1993.
3. Silverstein, R. M., Bassler, G. C., and Morrill, T. C., *Spectroscopic Identification of Organic Compounds*, 6th Edition, John Wiley and Sons, New York, 1996.
4. Bruno, T. J., *CRC Handbook for the Analysis and Identification of Alternative Refrigerants*, CRC Press, Boca Raton, FL, 1995.
5. Bruno, T. J., and Svoronos, P. D. N., *CRC Handbook of Basic Tables for Chemical Analysis*, 3rd Edition, CRC Press, Boca Raton, FL, 2011.

Mass Lost	Fragment Lost
1	$H^{\bullet}$
15	$CH_3^{\bullet}$
17	$OH^{\bullet}$
18	H_2O
19	$F^{\bullet}$
20	HF
26	$HC{\equiv}CH$; $^{\bullet}C{\equiv}N$
27	$CH_2{-}CH^{\bullet}$; $HC{\equiv}N$
28	$CH_2{=}CH_2$; $C{=}O$; $(HCN$ and $H^{\bullet})$
29	$CH_3CH_2^{\bullet}$; $H{-}^{\bullet}C{=}O$
30	$^{\bullet}CH_2NH_2$; $HCHO$; NO
31	$CH_3O^{\bullet}$; $^{\bullet}CH_2OH$; CH_3NH_2
32	CH_3OH; S
33	$HS^{\bullet}$
34	H_2S
35	$Cl^{\bullet}$
36	HCl, H_2O
37	H_2Cl
38	$C_3H_2^{\bullet}$; C_2N; F_2
39	C_3H_3; HC_2N
40	$CH_3C{\circ}CH$
41	$CH_2{=}CHCH_2^{\bullet}$
42	$CH_2{=}CHCH_3$; $CH_2{=}C{=}O$; $(CH_2)_3$; NCO; $NCNH_2$
43	$C_3H_7^{\bullet}$; $CH_3C{=}O^{\bullet}$; $CH_2{=}CH{-}O^{\bullet}$; $HCNO$
44	$CH_2{=}CHOH$; CO_2; N_2O; $CONH_2$; $NHCH_2CH_3$
45	CH_3CHOH; $CH_3CH_2O^{\bullet}$; CO_2H; $CH_3CH_2NH_2$
46	CH_3CH_2OH; $^{\bullet}NO_2$
47	$CH_3S^{\bullet}$
48	CH_3SH; SO; O_3
49	$^{\bullet}CH_2Cl$

Mass Lost	Fragment Lost
51	$^{\bullet}CHF_2$
52	$C_4H_4^{\bullet}$, C_2N_2
54	$CH_2{=}CHCH{=}CH_2$
55	$CH_2{=}CH{-}CH^{\bullet}CH_3$
56	$CH_2{=}CH{-}CH_2CH_3$; $CH_3CH{=}CHCH_3$; CO (2 moles)
57	$C_4H_9^{\bullet}$
58	$^{\bullet}NCS$; $(CH_3)_2C{=}O$; $(NO$ and $CO)$
59	$CH_3OC{=}O^{\bullet}$; CH_3CONH_2; $C_2H_3S^{\bullet}$
60	C_3H_7OH
61	$CH_3CH_2S^{\bullet}$; $(CH_3)_2S^{\bullet}H$
62	$[H_2S$ and $CH_2{=}CH_2]$
63	$^{\bullet}CH_2CH_2Cl$
64	$S_2^{\bullet}$; $SO_2^{\bullet}$; $C_5H_4^{\bullet}$
68	$CH_2{=}CHC(CH_3){=}CH_2$
69	$CF_3^{\bullet}$; $C_5H_9^{\bullet}$
71	$C_5H_{11}^{\bullet}$
73	$CH_3CH_2OC^{\bullet}{=}O$
74	C_4H_9OH
75	C_6H_3
76	C_6H_4; CS_2
77	C_6H_5; HCS_2
78	$C_6H_6^{\bullet}$; $H_2CS_2^{\bullet}$, C_5H_4N
79	$Br^{\bullet}$; C_5H_5N
80	HBr
85	$^{\bullet}CClF_2$
100	$CF_2{=}CF_2$
119	$CF_3CF_2^{\bullet}$
122	$C_6H_5CO_2H$
127	$I^{\bullet}$
128	HI

MAJOR REFERENCE MASSES IN THE SPECTRUM OF HEPTACOSAFLUOROTRIBUTYLAMINE (PERFLUOROTRIBUTYLAMINE)

Thomas J. Bruno and Paris D. N. Svoronos

The following list tabulates the major reference masses (with their relative intensities and formulas) of the mass spectrum of heptacosafluorotributylamine (CAS No. 311-89-7) (Ref. 1). This is one of the most widely used reference compounds in mass spectrometry. For guidance in the selection of additional reference compounds, the reader is advised to consult Ref. 1.

Reference

1. Bruno, T. J., and Svoronos, P. D. N., *CRC Handbook of Basic Tables for Chemical Analysis*, 3rd Edition, CRC Press, Boca Raton, FL, 2011.

Mass	Relative Intensity	Formula	Mass	Relative Intensity	Formula
613.9647	2.6	$C_{12}F_{24}N$	180.9888	1.9	C_4F_7
575.9679	1.7	$C_{12}F_{22}N$	175.9935	1.0	C_4F_6N
537.9711	0.4	$C_{12}F_{20}N$	168.9888	3.6	C_3F_7
501.9711	8.6	$C_9F_{20}N$	163.9935	0.7	C_3F_6N
463.9743	3.8	$C_9F_{18}N$	161.9904	0.3	C_4F_6
425.9775	2.5	$C_9F_{16}N$	149.9904	2.1	C_3F_6
413.9775	5.1	$C_8F_{16}N$	130.9920	31	C_3F_5
375.9807	0.9	$C_8F_{14}N$	118.9920	8.3	C_2F_5
325.9839	0.4	$C_7F_{12}N$	113.9967	3.7	C_2F_4N
313.9839	0.4	$C_6F_{12}N$	111.9936	0.7	C_3F_4
263.9871	10	$C_5F_{10}N$	99.9936	12	C_2F_4
230.9856	0.9	C_5F_9	92.9952	1.1	C_3F_3
225.9903	0.6	C_5F_8N	68.9952	100	CF_3
218.9856	62	C_4F_9	49.9968	1.0	CF_2
213.9903	0.6	C_4F_8N	30.9984	2.3	CF

MASS SPECTRAL PEAKS OF COMMON ORGANIC SOLVENTS

David R. Lide

The strongest peaks in the mass spectra of 375 important organic solvents and other liquid reagents are listed in this table. The *m/z* value for each peak is followed by the relative intensity in parentheses, with the strongest peak assigned an intensity of 100. The peaks for each compound are listed in order of decreasing intensity. Compounds are listed by the name used in this *Handbook*, with other common names given in parentheses.

Data on the physical properties of the same compounds may be found in Section 15 in the table *Laboratory Solvents and Other Liquid Reagents*.

References

1. NIST/EPA/NIH (NIST 08) Mass Spectral Database, NIST Standard Reference Database 1, National Institute of Standards and Technology, Gaithersburg, MD, 20899.
2. Lide, D. R., and Milne, G. W. A., Editors, *Handbook of Data on Organic Compounds, Third Edition*, CRC Press, Boca Raton, FL, 1994. (Also available as a CD-ROM database.)
3. Lide, D. R., Editor, *Properties of Organic Compounds*, <www.chemnetbase.com/scripts/pocweb.exe>.

Compound	*m/z* (intensity)									
Acetic acid	43(100)	45(87)	60(57)	15(42)	42(14)	29(13)	14(13)	28(7)	18(6)	16(6)
Acetic anhydride	43(100)	42(35)	45(29)	60(18)	29(9)	41(8)	40(2)	26(2)	87(1)	61(1)
Acetone	43(100)	15(34)	58(23)	27(9)	14(9)	42(8)	26(7)	29(5)	28(5)	39(4)
Acetonitrile	41(100)	40(46)	39(13)	14(9)	38(6)	28(4)	26(4)	25(3)	42(2)	27(2)
Acrolein (2-Propenal)	27(100)	56(74)	28(65)	26(54)	55(52)	29(37)	25(8)	53(5)	38(5)	57(4)
Acrylonitrile	53(100)	26(85)	52(79)	51(34)	27(13)	50(8)	25(7)	38(5)	54(3)	37(3)
Allyl alcohol	57(100)	31(34)	29(32)	28(31)	58(25)	39(22)	27(20)	30(16)	32(14)	26(11)
Allylamine	30(100)	56(80)	28(76)	57(33)	39(21)	29(20)	27(18)	26(13)	41(8)	18(8)
2-Amino-2-methyl-1-propanol (2-Aminoisobutanol)	58(100)	41(18)	18(17)	42(13)	28(11)	56(10)	30(10)	29(8)	43(6)	59(5)
Aniline (Benzenamine)	93(100)	66(32)	65(16)	39(13)	92(10)	94(7)	41(5)	40(5)	67(4)	64(3)
Anisole (Methoxybenzene)	108(100)	65(76)	78(60)	39(44)	51(21)	77(20)	93(16)	79(14)	50(13)	63(12)
Benzaldehyde	51(100)	77(81)	50(55)	106(44)	105(43)	52(26)	78(16)	39(13)	27(10)	74(8)
Benzene	78(100)	77(20)	52(19)	51(17)	50(15)	39(12)	79(6)	76(5)	74(4)	38(4)
Benzeneacetonitrile (Benzyl cyanide)	117(100)	90(43)	116(35)	89(22)	51(13)	39(11)	63(10)	118(9)	91(8)	50(8)
Benzenethiol (Phenyl mercaptan)	110(100)	66(28)	109(23)	39(15)	77(14)	51(14)	84(13)	69(11)	50(11)	45(11)
Benzonitrile	103(100)	76(34)	50(13)	104(9)	75(7)	51(7)	77(5)	52(4)	39(4)	74(3)
Benzyl acetate	108(100)	43(76)	91(60)	90(48)	79(27)	107(17)	77(17)	65(15)	51(15)	89(14)
Benzyl alcohol	79(100)	108(83)	77(74)	107(66)	51(35)	105(23)	106(22)	50(17)	78(16)	39(16)
Bis(2-aminoethyl)amine (Diethylenetriamine)	44(100)	73(59)	30(35)	19(18)	56(16)	28(16)	27(16)	42(11)	99(8)	43(8)
Bis(2-chloroethyl) ether	93(100)	63(74)	27(38)	95(32)	65(24)	31(9)	49(4)	28(4)	94(3)	62(3)
Bis(2-ethylhexyl) phthalate	149(100)	57(32)	167(29)	71(21)	43(21)	70(18)	150(11)	113(10)	55(10)	41(9)
Bis(2-hydroxyethyl) sulfide	61(100)	45(68)	31(38)	104(36)	91(34)	47(26)	44(26)	27(24)	60(18)	43(17)
Bromobenzene	77(100)	158(64)	156(64)	51(39)	50(17)	78(8)	76(6)	75(6)	28(5)	159(4)
1-Bromobutane (Butyl bromide)	57(100)	41(56)	29(45)	27(29)	56(13)	39(12)	28(12)	55(7)	43(7)	138(6)
2-Bromobutane (*sec*-Butyl bromide)	57(100)	41(60)	29(57)	27(34)	39(20)	28(9)	26(9)	136(1)		
1-Bromo-2-chloroethane	63(100)	27(85)	65(31)	26(23)	144(8)	81(8)	79(8)	28(7)	142(6)	93(6)
Bromochloromethane	49(100)	130(67)	128(52)	51(31)	93(23)	81(20)	79(20)	95(17)	132(16)	47(8)
1-Bromodecane (Decyl bromide)	43(100)	135(94)	137(91)	57(81)	41(58)	55(56)	71(38)	69(36)	85(33)	29(27)
Bromoethane (Ethyl bromide)	108(100)	110(97)	29(62)	27(51)	28(35)	26(14)	93(6)	32(6)	95(5)	81(5)
2-Bromo-2-methylpropane (*tert*-Butyl bromide)	57(100)	41(67)	29(45)	39(30)	27(18)	28(8)	40(5)	38(5)	58(4)	55(4)
1-Bromonaphthalene	44(100)	206(39)	127(39)	208(37)	36(31)	69(29)	131(13)	29(13)	126(12)	63(12)
1-Bromopentane (Pentyl bromide)	43(100)	71(80)	41(56)	27(51)	42(37)	29(34)	55(33)	39(30)	28(12)	26(9)
1-Bromopropane (Propyl bromide)	43(100)	41(77)	28(69)	27(60)	39(49)	124(42)	122(41)	42(34)	32(20)	29(15)
2-Bromopropane (Isopropyl bromide)	43(100)	27(47)	41(43)	39(22)	124(8)	122(8)	26(7)	81(6)	79(6)	38(6)
2-Bromopropene	41(100)	39(58)	122(37)	120(36)	38(12)	37(8)	40(6)	81(5)	79(5)	42(4)
Butanal	44(100)	43(74)	72(57)	41(56)	27(55)	29(48)	57(23)	39(22)	28(15)	42(11)
Butanenitrile	41(100)	29(62)	27(37)	28(10)	39(9)	26(7)	40(5)	42(4)	38(4)	15(4)
1-Butanethiol (Butyl mercaptan)	56(100)	41(74)	90(66)	47(43)	27(43)	28(36)	29(33)	57(17)	39(16)	61(15)
Butanoic acid	60(100)	27(50)	73(27)	42(25)	41(24)	43(22)	29(21)	45(19)	39(15)	28(11)
Butanoic anhydride	71(100)	43(59)	27(26)	41(19)	42(10)	39(10)	73(9)	28(7)	72(5)	55(5)
1-Butanol (Butyl alcohol)	31(100)	56(81)	41(62)	43(60)	27(50)	42(31)	29(31)	28(17)	39(16)	55(12)
2-Butanol (*sec*-Butyl alcohol)	45(100)	31(22)	27(22)	59(20)	29(18)	43(13)	41(12)	44(8)	18(8)	28(5)
2-Butanone (Methyl ethyl ketone)	43(100)	72(24)	29(19)	27(12)	57(7)	42(5)	26(4)	28(3)	44(2)	39(2)
trans-2-Butenal (*trans*-Crotonaldehyde)	41(100)	39(97)	70(82)	69(65)	27(49)	29(39)	42(30)	38(29)	40(27)	37(18)
2-Butoxyethanol (Ethylene glycol monobutyl ether)	57(100)	45(38)	29(35)	41(31)	87(16)	27(12)	56(11)	31(9)	75(7)	28(7)

Compound	m/z (intensity)									
2-Butoxyethyl acetate (Ethylene glycol monobutyl ether acetate)	57(100)	43(86)	56(50)	87(33)	41(26)	29(22)	85(18)	88(11)	44(11)	27(7)
Butyl acetate	43(100)	56(34)	41(17)	27(16)	29(15)	73(11)	61(10)	28(7)	55(6)	39(6)
sec-Butyl acetate	43(100)	56(21)	87(15)	41(14)	29(8)	57(6)	73(4)	61(4)	55(4)	27(4)
Butylamine	30(100)	73(10)	28(5)	41(3)	27(3)	18(3)	44(2)	42(2)	31(2)	29(2)
tert-Butylamine	58(100)	41(21)	42(15)	18(9)	30(8)	15(8)	39(7)	57(6)	28(6)	59(4)
Butylbenzene	91(100)	92(55)	134(20)	65(13)	27(10)	39(9)	105(8)	51(7)	78(6)	41(6)
sec-Butylbenzene	105(100)	134(18)	91(14)	77(10)	27(9)	106(9)	51(7)	79(7)		
tert-Butylbenzene	119(100)	91(65)	41(40)	134(24)	39(15)	79(14)	77(13)	51(13)	120(11)	65(7)
Butyl butanoate	43(100)	71(90)	56(80)	89(69)	41(67)	27(52)	29(47)	57(29)	39(22)	60(19)
Butyl formate	56(100)	41(60)	31(58)	29(53)	27(45)	43(34)	28(21)	39(19)	55(11)	42(11)
1-tert-Butyl-4-methylbenzene	133(100)	105(38)	41(23)	148(18)	93(16)	91(14)	115(13)	134(11)	39(11)	116(10)
Butyl vinyl ether	29(100)	41(74)	56(56)	57(43)	27(42)	44(26)	15(16)	85(14)	39(14)	43(13)
γ-Butyrolactone	28(100)	42(74)	29(48)	27(33)	41(27)	56(25)	86(24)	26(18)	85(10)	39(10)
Caprolactam	55(100)	113(87)	30(81)	56(66)	84(60)	85(57)	42(51)	41(33)	28(26)	43(17)
Carbon disulfide	76(100)	32(22)	44(17)	78(9)	38(6)	28(5)	77(3)	64(1)	46(1)	39(1)
2-Chloroaniline	127(100)	129(32)	92(17)	65(16)	128(10)	91(9)	64(9)	39(8)	63(7)	99(6)
Chlorobenzene	112(100)	77(63)	114(33)	51(29)	50(14)	75(8)	113(7)	78(5)	76(5)	28(4)
1-Chlorobutane (Butyl chloride)	56(100)	41(65)	27(50)	43(35)	29(24)	39(18)	28(16)	26(11)	55(8)	15(8)
2-Chlorobutane (sec-Butyl chloride)	56(100)	57(100)	41(90)	27(77)	29(57)	63(46)	39(34)	92(1)		
1-Chloro-1,1-difluoroethane	65(100)	45(31)	85(14)	31(10)	64(8)	44(7)	35(6)	26(6)	87(5)	81(4)
Chloroethane (Ethyl chloride)	64(100)	28(91)	29(84)	27(75)	66(32)	26(28)	49(25)	51(8)	63(6)	65(4)
2-Chloroethanol (Ethylene chlorohydrin)	31(100)	15(13)	29(10)	28(10)	27(9)	43(8)	44(7)	26(5)	18(5)	14(5)
(Chloromethyl)benzene (Benzyl chloride)	91(100)	126(20)	65(14)	92(9)	39(9)	63(8)	128(6)	45(6)	89(5)	125(3)
1-Chloro-3-methylbutane (Isopentyl chloride)	43(100)	55(56)	41(55)	27(51)	70(49)	42(37)	29(34)	57(30)	39(30)	56(15)
1-Chloro-2-methylpropane (Isobutyl chloride)	43(100)	41(67)	42(50)	27(33)	39(26)	15(11)	29(10)	56(7)	49(6)	38(5)
2-Chloro-2-methylpropane (tert-Butyl chloride)	57(100)	41(80)	77(44)	29(26)	39(24)	79(14)	27(14)	59(7)	56(7)	38(6)
1-Chloronaphthalene	162(100)	127(36)	164(30)	126(20)	163(10)	77(8)	101(6)	75(6)	128(5)	28(4)
1-Chlorooctane (Octyl chloride)	91(100)	41(83)	43(76)	27(62)	29(56)	55(55)	39(34)	93(32)	57(32)	69(29)
Chloropentafluoroethane	85(100)	69(61)	31(38)	87(32)	50(17)	35(8)	119(6)	66(4)	100(3)	47(3)
1-Chloropentane (Pentyl chloride)	42(100)	41(90)	70(89)	55(87)	27(73)	29(55)	43(40)	39(40)	28(19)	57(18)
1-Chloropropane (Propyl chloride)	42(100)	29(46)	27(37)	41(23)	28(15)	43(14)	39(12)	78(6)	63(6)	49(5)
3-Chloropropene (Allyl chloride)	41(100)	39(73)	76(28)	38(16)	37(13)	40(12)	27(12)	26(11)	78(9)	49(5)
2-Chlorotoluene	91(100)	126(68)	89(23)	128(22)	90(18)	65(17)	125(15)	92(13)	127(10)	63(6)
3-Chlorotoluene	91(100)	126(27)	63(15)	65(12)	89(11)	39(11)	128(9)	125(8)	92(8)	62(6)
Cyclohexane	56(100)	84(71)	41(70)	27(37)	55(36)	39(35)	42(30)	69(23)	28(18)	43(14)
Cyclohexanol	57(100)	44(68)	41(68)	39(51)	32(40)	43(38)	31(32)	42(22)	67(18)	82(16)
Cyclohexanone	55(100)	42(85)	41(34)	27(33)	98(31)	39(27)	69(26)	70(20)	43(14)	28(14)
Cyclohexene	67(100)	54(72)	82(37)	41(35)	39(33)	27(15)	53(12)	81(9)	51(8)	79(6)
Cyclohexylamine	56(100)	43(23)	28(17)	99(10)	70(8)	57(6)	30(6)	93(5)	54(4)	41(4)
Cyclopentane	42(100)	70(30)	55(29)	41(29)	39(22)	27(15)	40(7)	29(5)	28(4)	43(3)
Cyclopentanone	55(100)	28(50)	84(42)	41(38)	56(29)	27(24)	39(19)	42(15)	26(9)	29(7)
cis-Decahydronaphthalene (cis-Decalin)	67(100)	81(87)	41(81)	138(67)	96(62)	82(62)	39(50)	55(45)	27(44)	95(42)
trans-Decahydronaphthalene (trans-Decalin)	41(100)	68(91)	67(88)	82(67)	27(65)	96(61)	95(55)	138(51)	81(51)	29(51)
Decane	43(100)	57(90)	41(41)	71(33)	29(30)	85(24)	27(20)	56(17)	55(14)	42(14)
Diacetone alcohol	43(100)	59(41)	58(17)	101(10)	41(9)	31(9)	83(6)	56(6)	55(6)	29(6)
1,2-Dibromoethane	27(100)	107(77)	109(72)	26(24)	28(10)	81(5)	79(5)	25(5)	95(4)	93(4)
Dibromofluoromethane	111(100)	113(98)	192(29)	43(16)	41(16)	190(15)	194(14)	81(9)	79(9)	122(7)
Dibromomethane	174(100)	93(96)	95(84)	172(53)	176(50)	91(11)	81(9)	79(9)	94(5)	65(5)
1,2-Dibromopropane	41(100)	121(66)	123(65)	39(48)	27(28)	107(11)	38(10)	26(10)	109(9)	42(9)
1,2-Dibromotetrafluoroethane	179(100)	181(97)	129(34)	131(33)	100(17)	31(13)	260(12)	50(8)	69(7)	262(6)
Dibutylamine	86(100)	72(52)	30(48)	44(40)	29(31)	57(24)	41(21)	73(15)	28(15)	43(13)
Dibutyl ether	57(100)	41(34)	29(30)	56(25)	87(21)	27(9)	58(8)	55(6)	39(5)	28(5)
Dibutyl oxalate	57(100)	41(61)	56(24)	44(9)	55(8)	43(7)	58(5)	42(4)	103(2)	73(2)
Dibutyl phthalate	149(100)	86(18)	57(18)	223(17)	205(17)	150(17)	104(17)	56(17)	41(17)	65(16)
Dibutyl sebacate	241(100)	185(71)	41(37)	56(35)	55(32)	57(31)	242(23)	143(21)	98(21)	125(20)
o-Dichlorobenzene	146(100)	148(64)	111(38)	75(23)	113(12)	74(12)	50(11)	150(10)	73(9)	147(7)
m-Dichlorobenzene	146(100)	148(65)	111(36)	75(25)	50(19)	74(16)	150(11)	113(11)	73(11)	147(8)
1,1-Dichloroethane	63(100)	27(71)	65(31)	26(19)	83(11)	85(7)	61(7)	35(6)	98(5)	62(5)
1,2-Dichloroethane	62(100)	27(91)	49(40)	64(32)	26(31)	63(19)	98(14)	51(13)	61(12)	100(9)
1,1-Dichloroethene	61(100)	96(61)	98(38)	63(32)	26(16)	60(15)	62(7)	25(7)	100(6)	35(6)
cis-1,2-Dichloroethene	61(100)	96(73)	98(47)	63(32)	26(30)	60(21)	25(13)	35(12)	62(9)	100(8)

Compound	m/z (intensity)									
trans-1,2-Dichloroethene	61(100)	96(67)	98(43)	26(34)	63(32)	60(24)	25(15)	62(10)	100(7)	47(7)
Dichlorofluoromethane	67(100)	69(32)	47(13)	35(13)	31(9)	32(8)	48(7)	83(5)	102(4)	49(4)
Dichloromethane (Methylene chloride)	49(100)	84(64)	86(39)	51(31)	47(14)	48(8)	88(6)	50(3)	85(2)	83(2)
(Dichloromethyl)benzene (Benzal chloride)	125(100)	127(32)	160(14)	89(13)	162(9)	63(9)	126(8)	62(7)	105(5)	39(5)
1,2-Dichloropropane	63(100)	62(71)	27(57)	41(49)	39(32)	65(31)	76(27)	64(25)	49(13)	77(12)
1,2-Dichloro-1,1,2,2-tetrafluoroethane	85(100)	135(52)	87(33)	137(17)	101(9)	31(9)	103(6)	100(6)	50(5)	69(4)
2,4-Dichlorotoluene	125(100)	160(61)	162(40)	127(32)	89(23)	159(16)	161(14)	63(13)	62(11)	126(10)
3,4-Dichlorotoluene	125(100)	160(47)	127(32)	162(31)	89(15)	159(11)	161(10)	63(10)	126(8)	62(8)
Diethanolamine	30(100)	74(82)	28(77)	56(69)	18(50)	42(46)	29(36)	27(34)	45(30)	43(19)
1,1-Diethoxyethane (Acetal)	44(100)	43(92)	29(77)	31(76)	45(74)	27(52)	72(48)	73(23)	28(17)	46(15)
1,2-Diethoxyethane (Ethylene glycol diethyl ether)	31(100)	59(71)	29(58)	45(43)	27(33)	74(27)	43(15)	15(14)	28(12)	44(10)
Diethylamine	30(100)	58(81)	44(28)	73(18)	29(18)	28(17)	72(12)	42(11)	27(11)	59(4)
Diethyl carbonate	29(100)	45(70)	31(53)	27(39)	91(24)	28(15)	63(11)	26(10)	30(6)	43(5)
Diethylene glycol	45(100)	75(23)	31(20)	44(16)	27(14)	76(12)	29(12)	43(11)	42(9)	41(4)
Diethylene glycol dimethyl ether (Diglyme)	59(100)	58(43)	31(34)	29(32)	45(28)	28(19)	89(15)	43(9)	27(5)	60(4)
Diethylene glycol monobutyl ether acetate	43(100)	87(93)	57(82)	41(28)	45(25)	56(18)	29(18)	72(14)	85(12)	101(10)
Diethylene glycol monoethyl ether (Carbitol)	45(100)	59(56)	72(37)	73(22)	60(14)	31(13)	75(11)	44(9)	104(8)	103(7)
Diethylene glycol monoethyl ether acetate	43(100)	29(51)	31(42)	45(40)	59(24)	72(18)	44(10)	73(9)	42(9)	30(6)
Diethylene glycol monomethyl ether	45(100)	31(42)	59(41)	29(38)	28(32)	58(21)	43(14)	27(13)	44(11)	32(10)
Diethyl ether	31(100)	29(63)	59(40)	27(35)	45(33)	74(23)	15(17)	43(9)	28(9)	26(9)
Diethyl oxalate	29(100)	31(16)	45(14)	27(14)	74(11)	28(9)	43(4)	30(4)	73(3)	75(2)
Diethyl sulfide	75(100)	47(81)	90(73)	62(60)	29(55)	61(54)	27(48)	28(23)	46(17)	45(17)
Diisopropylamine	44(100)	86(30)	58(14)	42(13)	28(13)	41(12)	43(11)	27(11)	15(11)	39(6)
Diisopropyl ether	45(100)	43(39)	87(15)	41(12)	59(10)	27(8)	39(4)	69(3)	42(3)	31(3)
1,2-Dimethoxybenzene (Veratrole)	138(100)	95(65)	77(48)	123(44)	52(42)	41(33)	65(30)	51(29)	39(19)	63(17)
1,2-Dimethoxyethane (Ethylene glycol dimethyl ether)	45(100)	60(13)	29(13)	90(7)	58(6)	31(5)	28(5)	43(4)	59(3)	46(2)
Dimethoxymethane (Methylal)	45(100)	75(61)	29(59)	31(13)	30(6)	15(6)	47(5)	76(2)	46(2)	44(2)
N,N-Dimethylacetamide	44(100)	87(69)	43(46)	45(23)	42(19)	72(15)	15(11)	30(8)	28(5)	88(4)
Dimethylamine	44(100)	45(81)	18(32)	28(30)	43(19)	42(15)	15(9)	46(5)	41(5)	27(5)
2,4-Dimethylaniline (2,4-Xylidine)	121(100)	120(78)	106(57)	77(15)	28(15)	91(12)	122(9)	18(8)	93(6)	118(5)
2,2-Dimethylbutane (Neohexane)	43(100)	57(98)	71(73)	41(61)	29(51)	27(36)	56(29)	39(25)	55(15)	15(12)
2,3-Dimethylbutane	43(100)	42(97)	41(27)	71(25)	27(11)	86(10)	39(9)	29(6)	55(5)	44(4)
Dimethyl disulfide	94(100)	45(63)	79(59)	46(38)	47(26)	15(18)	48(14)	61(12)	64(11)	96(9)
N,N-Dimethylformamide	73(100)	44(86)	42(36)	30(22)	28(20)	29(8)	43(7)	72(6)	58(5)	74(4)
Dimethyl glutarate	59(100)	100(51)	55(49)	42(33)	129(32)	101(32)	41(26)	43(22)	128(19)	87(15)
2,6-Dimethyl-4-heptanone (Isovalerone)	57(100)	85(82)	41(46)	43(39)	58(33)	28(30)	26(30)	39(22)	42(12)	142(11)
2,5-Dimethylhexane	57(100)	43(93)	42(31)	41(26)	99(19)	71(19)	29(11)	70(10)	55(9)	27(9)
Dimethyl maleate	113(100)	59(83)	26(72)	29(40)	85(37)	54(31)	114(25)	53(24)	55(13)	82(9)
2,2-Dimethylpentane	57(100)	43(73)	41(46)	56(40)	85(34)	29(31)	27(23)	39(15)	15(7)	55(6)
2,4-Dimethylpentane	43(100)	57(73)	56(41)	41(35)	42(24)	85(17)	29(16)	27(12)	39(9)	58(3)
2,4-Dimethyl-3-pentanone (Diisopropyl ketone)	43(100)	71(31)	41(13)	27(9)	70(6)	39(6)	114(5)	42(5)	44(3)	72(2)
2,4-Dimethylpyridine (2,4-Lutidine)	107(100)	106(63)	79(44)	92(22)	65(22)	51(19)	77(17)	80(12)	52(11)	50(11)
2,6-Dimethylpyridine (2,6-Lutidine)	107(100)	39(39)	106(29)	66(22)	92(18)	65(18)	38(12)	27(11)	79(9)	63(9)
Dimethyl sulfoxide	63(100)	78(70)	15(40)	45(35)	29(16)	61(13)	46(12)	31(11)	48(10)	47(10)
1,4-Dioxane	28(100)	29(37)	88(31)	58(24)	31(17)	15(17)	27(15)	30(13)	43(11)	26(9)
1,3-Dioxolane	73(100)	29(56)	44(53)	45(28)	28(21)	43(20)	27(13)	31(7)	74(5)	42(3)
Dipentene	68(100)	93(50)	67(44)	94(22)	39(22)	107(18)	92(18)	53(18)	136(16)	79(16)
Dipentyl ether (Amyl ether)	71(100)	43(92)	29(43)	70(40)	41(36)	27(34)	42(23)	69(17)	55(16)	39(16)
Dipropylamine	30(100)	72(79)	44(40)	43(32)	27(25)	28(24)	41(22)	86(11)	58(10)	42(10)
Dodecane	57(100)	43(91)	71(53)	41(45)	85(31)	29(27)	55(19)	56(18)	28(16)	42(14)
1-Dodecene	43(100)	56(83)	55(83)	41(75)	69(62)	70(58)	57(57)	83(47)	29(38)	84(33)
Epichlorohydrin	57(100)	27(39)	29(32)	49(25)	31(22)	62(18)	28(16)	92(1)		
1,2-Epoxybutane (Ethyloxirane)	43(100)	44(58)	27(52)	29(50)	45(46)	26(27)	28(27)	72(5)		
1,2-Ethanediamine	30(100)	18(13)	42(6)	43(5)	27(5)	44(4)	29(4)	17(4)	15(4)	41(3)
1,2-Ethanediol (Ethylene glycol)	31(100)	33(35)	29(13)	32(11)	43(6)	27(5)	28(4)	62(3)	30(3)	44(2)
1,2-Ethanediol, diacetate (Ethylene glycol diacetate)	43(100)	86(11)	42(7)	15(7)	116(4)	73(4)	44(3)	29(3)	103(2)	45(2)
Ethanol	31(100)	45(44)	46(18)	27(18)	29(15)	43(14)	30(6)	42(3)	19(3)	14(3)
Ethanolamine	30(100)	18(30)	28(15)	42(7)	31(6)	17(6)	61(5)	15(5)	43(3)	29(3)
Ethoxybenzene (Phenetole)	94(100)	122(39)	28(12)	66(11)	39(9)	77(8)	95(7)	65(7)	51(7)	29(6)
2-Ethoxyethanol (Ethylene glycol monoethyl ether; Cellosolve)	31(100)	29(52)	59(50)	27(27)	45(26)	72(14)	43(14)	15(14)	28(8)	26(6)
2-Ethoxyethyl acetate (Ethylene glycol monoethyl ether acetate)	43(100)	31(34)	59(31)	72(28)	44(25)	29(24)	45(12)	27(11)	15(11)	87(7)

Compound	m/z (intensity)									
Ethyl acetate	43(100)	29(46)	27(33)	45(32)	61(28)	28(25)	42(18)	73(11)	88(10)	70(10)
Ethyl acetoacetate	43(100)	29(24)	88(18)	28(16)	85(14)	27(12)	42(11)	60(9)	130(6)	45(6)
Ethyl acrylate (Ethyl propenoate)	55(100)	27(32)	29(15)	56(12)	45(9)	73(8)	28(8)	26(6)	99(5)	85(5)
Ethylamine	30(100)	28(32)	44(20)	45(19)	27(13)	15(10)	42(9)	29(8)	41(5)	40(5)
Ethylbenzene	91(100)	106(31)	51(14)	39(10)	77(8)	65(8)	105(7)	92(7)	78(7)	27(6)
Ethyl benzoate	105(100)	77(65)	122(34)	51(34)	27(17)	150(16)	29(14)	50(13)	106(12)	78(6)
Ethyl butanoate	43(100)	71(88)	29(83)	27(43)	88(40)	41(28)	60(22)	45(20)	73(17)	42(17)
Ethyl cyanoacetate	29(100)	68(59)	27(34)	40(21)	28(16)	41(14)	15(13)	45(10)	43(9)	26(9)
Ethylcyclohexane	83(100)	55(65)	82(42)	41(36)	112(23)	56(11)	67(10)	39(10)	42(9)	84(8)
Ethylene carbonate	29(100)	44(62)	43(54)	88(40)	30(16)	28(11)	45(7)	58(6)	42(6)	73(4)
Ethyl formate	31(100)	28(73)	27(51)	29(38)	45(34)	26(17)	74(11)	43(9)	47(8)	56(4)
2-Ethyl-1,3-hexanediol	56(100)	55(71)	41(60)	43(55)	29(46)	27(44)	31(34)	57(33)	73(32)	39(20)
2-Ethyl-1-hexanol	57(100)	43(41)	41(40)	29(29)	55(28)	83(27)	56(23)	70(20)	27(17)	31(13)
N-Ethyl-N-isopropyl-2-propanamine	72(100)	114(77)	44(41)	129(22)	42(21)	30(20)	43(18)	41(17)	27(15)	70(11)
Ethyl 3-methylbutanoate	29(100)	41(52)	27(51)	57(43)	43(42)	60(39)	88(38)	85(38)	61(26)	45(24)
3-Ethyl-2-methylpentane	43(100)	70(50)	41(27)	71(25)	29(19)	27(19)	85(18)	55(18)	57(15)	42(14)
Ethyl propanoate	57(100)	29(84)	102(17)	27(17)	75(15)	28(14)	45(13)	74(12)	73(7)	43(6)
Fluorobenzene	96(100)	70(17)	97(7)	95(6)	75(6)	50(6)	51(4)	39(4)	69(3)	57(3)
2-Fluorotoluene	109(100)	110(55)	83(18)	57(11)	63(10)	39(9)	51(6)	50(6)	107(5)	62(5)
3-Fluorotoluene	109(100)	110(54)	83(11)	57(5)	111(4)	107(4)	63(3)	39(3)	108(2)	89(2)
4-Fluorotoluene	109(100)	110(60)	83(13)	57(10)	108(8)	39(8)	63(7)	107(5)	51(5)	50(5)
Furan	68(100)	39(64)	40(9)	38(9)	42(6)	29(6)	37(5)	69(4)	34(2)	67(1)
Furfural	39(100)	96(55)	95(52)	38(38)	29(35)	37(29)	40(11)	97(9)	50(7)	42(7)
Furfuryl alcohol	98(100)	41(65)	39(59)	81(55)	53(53)	97(51)	42(49)	69(39)	70(36)	29(28)
Glycerol	61(100)	43(90)	31(57)	44(54)	29(38)	18(32)	27(12)	42(11)	60(10)	45(10)
Glycerol triacetate (Triacetin)	43(100)	103(44)	145(34)	116(17)	115(13)	44(10)	86(9)	28(8)	73(7)	42(7)
Heptane	43(100)	41(56)	29(49)	57(47)	27(46)	71(45)	56(27)	42(26)	39(23)	70(18)
1-Heptanol	41(100)	70(87)	56(86)	31(78)	43(72)	29(70)	55(67)	27(65)	42(54)	69(41)
3-Heptanol	59(100)	69(73)	87(37)	41(37)	29(33)	55(30)	31(24)	116(0)		
2-Heptanone (Methyl pentyl ketone)	43(100)	58(60)	71(14)	41(11)	27(11)	59(9)	39(8)	29(8)	42(5)	114(4)
3-Heptanone (Ethyl butyl ketone)	57(100)	29(76)	41(32)	85(29)	72(22)	43(15)	39(11)	114(10)	27(7)	55(5)
1-Heptene	41(100)	56(88)	29(70)	55(59)	42(52)	27(45)	39(41)	70(37)	69(27)	57(27)
Hexane	57(100)	43(78)	41(77)	29(61)	27(57)	56(45)	42(39)	39(27)	28(16)	86(14)
Hexanedinitrile (Adiponitrile)	41(100)	68(50)	54(42)	40(21)	55(20)	27(17)	39(16)	28(13)	52(7)	42(6)
Hexanenitrile	41(100)	54(68)	27(59)	55(55)	29(44)	39(35)	57(30)	43(30)	68(24)	28(22)
Hexanoic acid (Caproic acid)	60(100)	73(42)	27(36)	41(33)	43(27)	29(26)	45(20)	39(16)	42(15)	55(14)
1-Hexanol	56(100)	43(78)	31(74)	41(71)	27(64)	29(59)	55(58)	42(53)	39(37)	69(27)
2-Hexanone (Butyl methyl ketone)	43(100)	58(60)	57(17)	100(16)	29(15)	41(13)	85(8)	27(8)	71(7)	59(5)
Hexyl acetate	43(100)	56(66)	41(38)	55(37)	61(35)	42(33)	84(31)	69(19)	73(17)	57(7)
3-Hydroxypropanenitrile (Hydracrylonitrile)	41(100)	31(97)	29(20)	42(17)	52(15)	40(13)	53(10)	51(9)	39(7)	26(7)
Iodobenzene	204(100)	77(82)	51(32)	50(20)	127(8)	205(6)	78(6)	74(6)	102(5)	76(5)
1-Iodobutane (Butyl iodide)	57(100)	29(78)	41(56)	27(39)	184(36)	39(17)	28(16)	26(7)	127(6)	55(6)
2-Iodobutane (sec-Butyl iodide)	57(100)	29(36)	184(36)	41(36)	27(10)	39(9)	127(6)	58(4)		
Iodoethane (Ethyl iodide)	156(100)	29(75)	27(63)	127(31)	26(14)	28(9)	128(8)	141(2)	25(2)	140(1)
Iodomethane (Methyl iodide)	142(100)	127(38)	141(14)	15(13)	139(5)	140(4)	128(3)	14(1)	13(1)	71(0)
1-Iodopropane (Propyl iodide)	43(100)	170(68)	41(35)	27(26)	127(14)	39(11)	44(4)	141(3)	128(3)	42(3)
2-Iodopropane (Isopropyl iodide)	43(100)	170(46)	41(45)	27(44)	39(19)	127(17)	42(4)	38(4)	128(3)	44(3)
Isobutanal (2-Methyl-1-propanal)	43(100)	41(84)	27(47)	72(46)	39(30)	29(25)	42(10)	70(5)	38(5)	28(5)
Isobutyl acetate	43(100)	56(26)	73(15)	41(10)	29(5)	71(3)	57(3)	39(3)	27(3)	86(2)
Isobutylamine	30(100)	28(9)	41(6)	73(5)	27(5)	39(4)	29(3)	15(3)	58(2)	56(2)
Isobutylbenzene	91(100)	92(58)	43(22)	134(20)	65(12)	41(12)	39(11)	27(8)	51(7)	93(5)
Isobutyl formate	43(100)	56(82)	41(78)	31(65)	29(64)	27(53)	60(39)	39(36)	42(26)	15(14)
Isobutyl isobutanoate	43(100)	41(94)	56(76)	71(70)	57(68)	27(62)	29(59)	89(37)	39(32)	42(17)
Isopentyl acetate	43(100)	70(49)	55(38)	61(15)	42(15)	41(14)	27(12)	87(11)	29(10)	73(9)
Isophorone	82(100)	39(20)	138(17)	54(13)	27(12)	41(10)	53(8)	83(7)	29(7)	55(6)
Isopropyl acetate	43(100)	61(17)	41(14)	87(9)	59(8)	27(8)	42(7)	39(4)	45(3)	44(2)
Isopropylbenzene (Cumene)	105(100)	120(25)	77(13)	51(12)	79(10)	106(9)	39(9)	27(8)	103(6)	91(5)
1-Isopropyl-4-methylbenzene (p-Cymene)	119(100)	91(42)	134(33)	39(27)	41(20)	117(18)	65(18)	77(17)	27(16)	120(15)
Isoquinoline	129(100)	102(26)	51(20)	128(18)	50(11)	130(10)	75(10)	76(9)	103(8)	74(7)
d-Limonene (Citrene)	68(100)	93(50)	67(49)	41(22)	94(21)	79(21)	39(21)	136(20)	53(19)	121(16)
Mesityl oxide	55(100)	83(89)	43(73)	29(42)	98(36)	39(32)	27(28)	53(11)	41(10)	56(5)

Compound	m/z (intensity)									
Methanol	31(100)	29(72)	32(67)	15(42)	28(12)	14(10)	30(9)	13(6)	12(3)	16(2)
2-Methoxyethanol (Ethylene glycol monomethyl ether)	45(100)	31(15)	29(14)	28(11)	47(9)	76(6)	43(6)	58(4)	46(4)	27(4)
2-Methoxyethyl acetate (Ethylene glycol monomethyl ether acetate)	43(100)	45(48)	58(42)	29(10)	42(4)	31(4)	73(3)	27(3)	59(2)	26(2)
Methyl acetate	43(100)	74(52)	28(38)	42(19)	59(17)	44(8)	32(8)	29(6)	31(4)	75(2)
2-Methylacrylonitrile	41(100)	39(54)	67(44)	40(26)	38(24)	52(23)	27(23)	37(20)	66(17)	51(12)
Methylamine	30(100)	31(87)	28(56)	29(19)	32(15)	15(12)	27(9)			
2-Methylaniline (o-Toluidine)	106(100)	107(83)	77(17)	79(13)	39(12)	53(10)	52(10)	54(9)	51(9)	28(9)
3-Methylaniline (m-Toluidine)	106(100)	107(84)	79(17)	77(17)	108(7)	78(6)	80(5)	89(4)	65(4)	53(4)
N-Methylaniline	106(100)	107(79)	77(23)	51(12)	79(11)	65(9)	39(9)	78(8)	108(7)	50(6)
Methyl benzoate	105(100)	77(81)	51(45)	136(24)	50(18)	106(8)	78(6)	28(6)	39(5)	27(5)
Methyl butanoate	43(100)	74(90)	71(66)	41(32)	27(31)	59(28)	87(19)	42(15)	28(15)	39(14)
3-Methylbutanoic acid (Isovaleric acid)	60(100)	43(61)	41(54)	27(33)	45(31)	29(27)	74(24)	39(24)	87(21)	57(20)
3-Methyl-1-butanol (Isopentyl alcohol)	55(100)	42(90)	43(82)	41(81)	70(71)	31(61)	29(59)	27(59)	39(44)	57(31)
Methyl tert-butyl ether	41(100)	73(78)	57(71)	29(60)	43(48)	39(34)	28(28)	56(23)	55(18)	45(16)
Methyl cyanoacetate	59(100)	15(65)	68(60)	40(38)	28(17)	29(16)	55(11)	54(10)	39(10)	67(8)
Methylcyclohexane	83(100)	55(82)	41(60)	98(44)	42(35)	56(30)	27(29)	39(27)	69(23)	70(22)
cis-4-Methylcyclohexanol	57(100)	58(54)	70(38)	81(36)	96(33)	55(25)	41(23)	114(17)	71(15)	56(12)
N-Methylformamide	59(100)	30(54)	28(34)	29(13)	58(8)	15(7)	60(3)	41(3)	27(3)	31(2)
Methyl formate	31(100)	29(63)	32(34)	60(28)	30(7)	28(7)	44(2)	18(2)	61(1)	59(1)
2-Methylheptane	43(100)	57(91)	42(39)	41(27)	29(16)	70(15)	27(13)	71(12)	99(10)	55(9)
4-Methylheptane	43(100)	71(53)	70(46)	41(27)	29(23)	27(23)	55(15)	57(14)	42(14)	39(10)
6-Methyl-1-heptanol (Isooctyl alcohol)	41(100)	55(94)	43(93)	69(84)	57(81)	56(73)	29(50)	84(48)	70(47)	27(46)
2-Methylhexane	43(100)	42(38)	41(35)	85(32)	57(26)	29(22)	27(22)	56(20)	39(11)	55(5)
5-Methyl-2-hexanone (Methyl isopentyl ketone)	43(100)	58(34)	27(14)	41(13)	15(13)	57(11)	39(9)	71(8)	59(8)	29(8)
Methyl methacrylate (Methyl 2-methyl-2-propenoate)	41(100)	69(66)	39(40)	100(34)	15(20)	40(10)	59(8)	99(6)	38(6)	55(5)
2-Methyloctane	43(100)	57(51)	41(35)	71(28)	42(28)	29(23)	27(21)	85(17)	56(15)	84(13)
2-Methylpentane	43(100)	42(53)	41(35)	27(31)	71(29)	39(20)	29(18)	57(11)	15(10)	70(7)
3-Methylpentane	57(100)	56(76)	41(68)	29(60)	27(40)	43(29)	39(22)	55(9)	15(9)	28(8)
2-Methyl-2,4-pentanediol (Hexylene glycol)	59(100)	43(61)	56(25)	45(17)	41(16)	57(13)	42(13)	85(11)	61(10)	31(10)
4-Methylpentanenitrile	55(100)	41(52)	43(46)	27(39)	39(29)	57(27)	54(26)	29(22)	82(13)	28(12)
4-Methyl-2-pentanol	45(100)	43(47)	69(30)	41(27)	27(19)	39(13)	29(12)	87(11)	84(10)	57(10)
3-Methyl-3-pentanol	73(100)	55(38)	43(35)	45(28)	27(25)	29(21)	41(12)	87(11)	31(11)	15(9)
4-Methyl-2-pentanone (Methyl isobutyl ketone)	43(100)	58(84)	29(65)	41(56)	57(44)	27(42)	39(31)	85(19)	100(14)	42(14)
2-Methylpropanenitrile (Isobutyronitrile)	42(100)	68(45)	28(45)	54(26)	41(26)	27(26)	29(25)	26(15)	39(13)	15(13)
Methyl propanoate	57(100)	29(72)	59(31)	88(26)	27(18)	28(9)	31(5)	44(4)	26(4)	58(3)
2-Methylpropanoic acid (Isobutyric acid)	43(100)	41(42)	27(40)	73(22)	39(15)	45(14)	42(11)	29(9)	88(7)	28(6)
2-Methyl-1-propanol (Isobutyl alcohol)	43(100)	33(73)	31(72)	41(66)	42(60)	27(43)	29(18)	39(17)	28(8)	74(6)
2-Methyl-2-propanol (tert-Butyl alcohol)	59(100)	31(33)	41(22)	43(18)	29(13)	27(11)	57(10)	42(4)	60(3)	28(3)
2-Methylpyridine (2-Picoline)	93(100)	66(41)	39(31)	92(20)	78(19)	51(19)	65(16)	38(13)	50(12)	52(11)
3-Methylpyridine (3-Picoline)	93(100)	39(51)	66(46)	92(31)	65(29)	40(19)	38(18)	67(13)	63(11)	51(11)
N-Methyl-2-pyrrolidinone	99(100)	44(89)	98(80)	42(60)	41(38)	43(17)	28(17)	71(13)	39(11)	70(10)
Methyl salicylate	120(100)	92(59)	152(47)	121(32)	65(22)	39(22)	93(15)	64(14)	18(14)	63(13)
2-Methylthiophene	97(100)	98(57)	45(22)	39(14)	53(9)	99(8)	27(8)	69(6)	58(6)	59(5)
Morpholine	57(100)	29(100)	87(69)	28(69)	30(38)	56(33)	86(28)	31(28)	27(12)	15(7)
Nitrobenzene	77(100)	51(59)	123(42)	50(25)	30(15)	65(14)	39(10)	93(9)	74(7)	78(6)
Nitroethane	29(100)	30(12)	28(11)	26(9)	27(8)	43(5)	41(5)	14(5)	15(3)	46(2)
Nitromethane	30(100)	61(64)	46(39)	28(30)	45(8)	27(8)	44(7)	29(7)	60(5)	43(4)
1-Nitropropane	43(100)	27(93)	41(90)	39(34)	30(25)	44(20)	42(20)	26(20)	28(13)	54(12)
2-Nitropropane	43(100)	41(73)	27(71)	39(30)	30(18)	15(11)	42(9)	28(8)	26(8)	38(6)
Nonane	43(100)	57(75)	41(29)	84(26)	85(22)	29(22)	71(18)	56(16)	27(13)	42(12)
Octane	43(100)	57(30)	85(25)	41(25)	71(19)	29(17)	56(14)	70(10)	42(10)	27(10)
1-Octanol	41(100)	56(85)	43(82)	55(81)	31(69)	27(69)	29(68)	42(62)	70(53)	69(48)
2-Octanone (Hexyl methyl ketone)	43(100)	58(79)	41(56)	59(52)	71(49)	27(46)	29(36)	39(27)	57(18)	55(17)
1-Octene	43(100)	41(82)	55(80)	56(67)	42(67)	70(54)	29(44)	27(31)	69(30)	39(29)
Pentachloroethane	167(100)	165(91)	117(90)	119(89)	83(58)	169(54)	130(43)	132(42)	60(40)	85(37)
Pentane	43(100)	42(55)	41(45)	27(42)	29(26)	39(19)	57(13)	28(9)	15(9)	72(8)
1,5-Pentanediol (Pentamethylene glycol)	31(100)	56(85)	41(67)	57(59)	55(51)	44(45)	29(37)	43(31)	68(29)	27(26)
2,4-Pentanedione (Acetylacetone)	43(100)	85(31)	100(20)	27(12)	42(10)	29(10)	41(7)	39(7)	31(5)	26(5)
Pentanenitrile (Valeronitrile)	41(100)	43(97)	54(54)	27(34)	55(21)	28(19)	29(16)	39(15)	42(5)	26(5)
Pentanoic acid (Valeric acid)	60(100)	73(34)	27(33)	29(28)	41(21)	43(17)	45(16)	28(14)	42(12)	39(12)
1-Pentanol (Amyl alcohol)	42(100)	70(72)	55(65)	41(56)	31(47)	29(41)	27(26)	57(22)	28(22)	43(21)

Compound	m/z (intensity)									
2-Pentanol (sec-Amyl alcohol)	45(100)	43(20)	55(18)	27(17)	29(11)	41(9)	31(9)	15(9)	44(8)	39(8)
3-Pentanol (Diethyl carbinol)	59(100)	31(83)	41(42)	27(35)	29(34)	58(15)	43(15)	57(14)	39(12)	15(9)
2-Pentanone (Methyl propyl ketone)	43(100)	41(17)	86(12)	42(12)	27(11)	39(8)	71(7)	58(7)	45(7)	44(3)
3-Pentanone (Diethyl ketone)	57(100)	29(50)	86(26)	27(13)	58(4)	56(4)	28(4)	26(3)	43(2)	42(2)
Pentyl acetate (Amyl acetate)	43(100)	70(90)	42(52)	28(51)	61(50)	55(41)	73(21)	41(20)	29(14)	69(11)
Pentylamine (Amylamine)	30(100)	87(8)	41(4)	28(4)	45(3)	42(3)	27(3)	56(2)	44(2)	43(2)
Pentyl lactate	43(100)	70(44)	71(32)	55(28)	41(22)	29(20)	27(15)	45(12)	42(9)	57(7)
α-Pinene	93(100)	92(30)	39(24)	41(23)	77(22)	91(21)	27(21)	79(18)	121(13)	53(10)
β-Pinene	93(100)	41(64)	69(47)	39(33)	27(31)	79(20)	77(18)	53(14)	94(13)	91(13)
Piperidine	84(100)	85(53)	56(46)	57(43)	28(41)	29(37)	44(34)	42(30)	30(30)	43(25)
Propanal	29(100)	58(59)	28(58)	27(39)	57(20)	31(4)	30(4)	42(3)	39(3)	59(2)
1,2-Propanediol (1,2-Propylene glycol)	45(100)	18(46)	29(21)	43(19)	31(18)	27(17)	28(11)	19(8)	44(6)	61(5)
1,3-Propanediol (Trimethylene glycol)	28(100)	58(93)	31(76)	57(70)	29(40)	27(26)	45(24)	43(23)	19(18)	30(17)
Propanenitrile	28(100)	54(63)	26(20)	27(17)	52(11)	55(10)	51(9)	15(9)	53(7)	25(7)
Propanoic acid	28(100)	29(84)	74(79)	27(62)	45(56)	73(48)	57(30)	26(21)	55(17)	56(16)
Propanoic anhydride	57(100)	29(55)	28(20)	27(20)	74(19)	73(12)	45(11)	26(5)	30(4)	58(3)
1-Propanol (Propyl alcohol)	31(100)	27(19)	29(18)	59(11)	42(9)	60(7)	41(7)	28(7)	43(3)	32(3)
2-Propanol (Isopropyl alcohol)	45(100)	43(19)	27(17)	29(12)	41(7)	31(6)	19(6)	42(5)	44(4)	59(3)
Propargyl alcohol (3-Hydroxy-1-propyne)	55(100)	39(25)	28(20)	27(19)	29(16)	38(14)	26(11)	37(8)	53(6)	56(4)
Propyl acetate	43(100)	61(19)	31(18)	27(15)	42(11)	73(9)	41(9)	29(9)	59(5)	39(5)
Propylamine	30(100)	28(13)	59(8)	27(7)	41(5)	42(3)	39(3)	29(3)	26(3)	18(3)
Propylbenzene	91(100)	120(21)	92(10)	38(10)	65(9)	78(6)	51(6)	27(5)	63(4)	105(3)
Propylene carbonate	28(100)	57(69)	43(66)	29(51)	27(45)	30(42)	26(19)	42(17)	31(16)	58(13)
Propyl formate	27(100)	29(92)	31(81)	42(60)	41(53)	43(29)	39(29)	26(28)	47(22)	30(16)
Pyridine	79(100)	52(62)	51(31)	50(19)	78(11)	53(7)	39(7)	80(6)	27(3)	77(2)
Pyrrole	67(100)	41(58)	39(58)	40(51)	28(42)	38(20)	37(12)	66(7)	68(5)	27(3)
Pyrrolidine	43(100)	28(52)	70(33)	71(26)	42(22)	41(20)	27(16)	39(15)	29(10)	30(9)
2-Pyrrolidone	85(100)	42(43)	41(36)	28(33)	30(29)	56(16)	84(14)	40(12)	27(12)	29(9)
Quinoline	129(100)	51(28)	76(25)	128(24)	44(24)	50(20)	32(19)	75(18)	74(12)	103(11)
Salicylaldehyde (2-Hydroxybenzaldehyde)	28(100)	122(98)	121(92)	39(79)	65(52)	76(31)	32(31)	44(30)	93(26)	38(23)
Styrene	104(100)	103(41)	78(32)	51(28)	77(23)	105(12)	50(12)	52(11)	39(11)	102(10)
Sulfolane	41(100)	28(94)	56(82)	55(72)	120(37)	27(32)	39(19)	29(17)	26(11)	48(5)
α-Terpinene	121(100)	93(85)	136(43)	91(40)	77(34)	39(33)	27(33)	79(27)	41(26)	43(18)
1,1,1,2-Tetrachloro-2,2-difluoroethane	167(100)	169(96)	117(85)	119(82)	171(31)	85(29)	121(26)	82(14)	47(14)	101(13)
1,1,2,2-Tetrachloro-1,2-difluoroethane	101(100)	103(64)	167(54)	169(52)	117(19)	119(18)	171(17)	105(11)	31(11)	132(9)
1,1,1,2-Tetrachloroethane	131(100)	133(96)	117(76)	119(73)	95(34)	135(31)	121(23)	97(23)	61(19)	60(18)
1,1,2,2-Tetrachloroethane	83(100)	85(63)	95(11)	87(10)	168(8)	133(8)	131(8)	96(8)	61(8)	60(8)
Tetrachloroethene	166(100)	164(82)	131(71)	129(71)	168(45)	94(38)	47(31)	96(24)	133(20)	59(17)
Tetrachloromethane (Carbon tetrachloride)	117(100)	119(98)	121(31)	82(24)	47(23)	84(16)	35(14)	49(8)	28(8)	36(6)
Tetraethylene glycol	45(100)	89(10)	44(8)	43(6)	31(6)	29(6)	27(6)	101(5)	75(5)	28(5)
Tetrahydrofuran	42(100)	41(52)	27(33)	72(29)	71(27)	39(24)	43(22)	29(22)	40(13)	15(10)
1,2,3,4-Tetrahydronaphthalene	104(100)	132(53)	91(43)	51(17)	39(17)	131(15)	117(15)	115(14)	78(13)	77(13)
Tetrahydropyran	41(100)	28(64)	56(57)	45(57)	29(51)	27(49)	85(47)	86(42)	39(28)	55(23)
Tetrahydrothiophene	60(100)	88(54)	45(37)	46(32)	47(26)	27(24)	59(18)	87(16)	39(14)	54(13)
Tetramethylsilane	73(100)	43(14)	45(12)	74(8)	29(7)	15(5)	75(4)	44(4)	42(4)	31(4)
Tetramethylurea	72(100)	44(27)	116(24)	42(13)	15(13)	17(7)	28(5)	73(4)	56(4)	18(4)
1H-Tetrazole	42(100)	28(60)	27(25)	29(24)	41(13)	43(11)	70(8)	26(7)	40(2)	38(2)
Thiophene	84(100)	58(65)	45(58)	39(29)	57(13)	38(8)	69(7)	37(7)	83(6)	50(6)
Toluene	91(100)	92(73)	39(20)	65(14)	63(11)	51(11)	50(7)	27(6)	93(5)	90(5)
Tribromomethane (Bromoform)	173(100)	171(50)	175(49)	93(22)	91(22)	79(18)	81(17)	94(13)	92(13)	254(11)
Tributylamine	142(100)	100(19)	143(11)	29(8)	185(7)	57(6)	44(6)	41(6)	30(5)	86(4)
1,2,4-Trichlorobenzene	180(100)	182(96)	184(30)	145(30)	109(26)	147(19)	75(11)	74(9)	111(8)	181(7)
1,1,1-Trichloroethane	97(100)	99(64)	61(58)	26(31)	27(24)	117(19)	63(19)	119(18)	35(17)	62(11)
1,1,2-Trichloroethane	97(100)	83(95)	99(62)	85(60)	61(58)	26(23)	96(21)	63(19)	27(17)	98(15)
Trichloroethene	95(100)	130(90)	132(85)	60(65)	97(64)	35(40)	134(27)	47(26)	62(21)	59(13)
Trichloroethylsilane	135(100)	133(100)	126(67)	128(46)	137(37)	98(22)	63(21)	35(16)	100(14)	127(11)
Trichlorofluoromethane	101(100)	103(66)	66(13)	105(11)	35(11)	47(9)	31(8)	82(4)	68(4)	37(4)
Trichloromethane (Chloroform)	83(100)	85(64)	47(35)	35(19)	48(16)	49(12)	87(10)	37(6)	50(5)	84(4)
(Trichloromethyl)benzene (Benzotrichloride)	159(100)	161(64)	89(14)	163(11)	28(10)	63(9)	160(8)	123(8)	62(8)	124(6)
1,2,3-Trichloropropane	75(100)	39(58)	49(42)	110(38)	61(34)	77(33)	112(22)	27(16)	97(15)	38(15)
1,1,2-Trichloro-1,2,2-trifluoroethane	101(100)	151(68)	103(64)	85(45)	31(45)	153(44)	35(20)	66(19)	47(18)	87(14)

Compound	m/z (intensity)								
Tridecane	57(100)	43(91)	71(51)	41(34)	85(24)	29(23)	56(14)	55(12)	27(11)	42(10)
Triethanolamine	118(100)	56(69)	45(60)	42(56)	44(27)	43(25)	41(14)	116(8)	57(8)	86(7)
Triethylamine	86(100)	30(68)	58(37)	28(24)	29(23)	27(19)	44(18)	101(17)	42(16)	56(8)
Triethylene glycol	45(100)	58(11)	89(9)	31(8)	29(8)	75(7)	44(7)	43(7)	27(7)	28(5)
Triethyl phosphate	99(100)	81(71)	155(56)	82(45)	45(45)	109(44)	127(41)	43(24)	125(16)	111(14)
Trifluoroacetic acid	45(100)	69(70)	51(36)	28(28)	50(15)	44(11)	43(7)	97(5)	31(5)	29(5)
2,2,2-Trifluoroethanol	31(100)	33(24)	61(19)	29(19)	51(16)	69(9)	32(6)	49(5)	83(4)	81(4)
(Trifluoromethyl)benzene (Benzotrifluoride)	146(100)	145(40)	127(34)	96(28)	77(10)	51(10)	147(8)	75(6)	50(6)	128(3)
Trimethylamine	58(100)	59(47)	30(29)	42(26)	44(17)	15(14)	28(10)	18(10)	43(8)	57(7)
1,2,3-Trimethylbenzene (Hemimellitene)	105(100)	120(47)	39(22)	77(17)	91(14)	51(14)	27(14)	79(12)	119(11)	106(9)
1,2,4-Trimethylbenzene (Pseudocumene)	105(100)	120(56)	119(17)	77(15)	39(15)	51(11)	91(10)	27(10)	106(9)	79(7)
1,3,5-Trimethylbenzene (Mesitylene)	105(100)	120(64)	119(15)	77(13)	39(11)	106(9)	91(9)	51(8)	27(7)	121(6)
2,2,3-Trimethylbutane (Triptane)	57(100)	43(71)	56(63)	41(53)	85(30)	29(26)	27(18)	39(14)	15(6)	55(5)
2,2,5-Trimethylhexane	57(100)	56(35)	71(18)	41(17)	43(14)	29(8)	70(4)	58(4)	113(3)	55(3)
2,2,4-Trimethylpentane (Isooctane)	57(100)	41(31)	56(28)	43(24)	29(16)	27(9)	39(7)	58(4)	55(4)	99(2)
2,3,3-Trimethylpentane	43(100)	71(45)	70(36)	57(36)	41(29)	85(25)	27(18)	55(16)	29(16)	39(11)
2,3,4-Trimethylpentane	43(100)	71(62)	70(41)	41(25)	27(18)	55(17)	57(16)	29(14)	39(10)	42(7)
Trimethyl phosphate	110(100)	109(35)	79(34)	95(25)	80(23)	15(20)	140(18)	47(10)	31(7)	139(5)
2,4,6-Trimethylpyridine (2,4,6-Collidine)	121(100)	39(27)	79(26)	120(24)	106(17)	27(16)	77(13)	51(11)	42(11)	122(10)
1-Undecene	41(100)	43(87)	55(80)	70(67)	56(67)	69(55)	29(55)	83(51)	57(50)	27(46)
Vinyl acetate	43(100)	28(45)	42(26)	44(24)	86(20)	31(10)	32(7)	29(7)	45(2)	41(2)
o-Xylene	91(100)	106(40)	39(21)	105(17)	51(17)	77(15)	27(12)	65(10)	92(8)	79(8)
m-Xylene	91(100)	106(65)	105(29)	39(18)	51(15)	77(14)	27(10)	92(8)	79(8)	78(8)
p-Xylene	91(100)	106(62)	105(30)	51(16)	39(16)	77(13)	27(11)	92(7)	78(7)	65(7)

COMMON SPURIOUS SIGNALS OBSERVED IN MASS SPECTROMETERS

Thomas J. Bruno and Paris D. N. Svoronos

The following table provides guidance in the recognition and interpretation of potentially spurious signals (m/z peaks) that will sometimes be observed in measured mass spectra (Ref. 1). Often, the occurrence of these signals can be predicted by the recent history of the instrument or the method being used. This is especially true if the mass spectrometer is interfaced to a gas chromatograph.

Reference

1. Bruno, T. J., and Svoronos, P. D. N., *CRC Handbook of Basic Tables for Chemical Analysis*, 3rd Edition, CRC Press, Boca Raton, FL, 2011.

Ions Observed, m/z	Possible Compound	Possible Source
13, 14, 15, 16	Methane*	Chlorine reagent gas
18	Water*	Residual impurity, outgassing of ferrules, septa and seals
14, 28	Nitrogen*	Residual impurity, outgassing of ferrules, septa and seals; leaking seal
16, 32	Oxygen*	Residual impurity, outgassing of ferrules, septa and seals; leaking seal
44	Carbon dioxide*	Residual impurity, outgassing of ferrules, septa and seals; leaking seal; note it may be mistaken for propane in a sample
31, 51, 69, 100, 119, 131, 169, 181, 214, 219, 264, 376, 414, 426, 464, 502, 576, 614	Perfluorotributyl amine (PFTBA), and related ions**	This is a common tuning compound; may indicate a leaking valve
31	Methanol	Solvent; can be used as a leak detector
43, 58	Acetone	Solvent; can be used as a leak detector
78	Benzene	Solvent; can be used as a leak detector
91, 92	Toluene	Solvent; can be used as a leak detector
105, 106	Xylenes	Solvent; can be used as a leak detector
151, 153	Trichloroethane	Solvent; can be used as a leak detector
69	Rough (fore) pump fluid, PFTBA	Back diffusion of fore pump fluid, possible leaking valve of tuning compound vial
73, 147, 207, 221, 281, 295, 355, 429	Dimethylpolysiloxane	Bleed from a column or septum, often during high temperature program methods in GC-MS
77, 94, 115, 141, 168, 170, 262, 354, 446	Diffusion pump fluid	Back diffusion from diffusion pump, if present
149	Phthalates	Plasticizer in vacuum seals, gloves
X – 14 peaks	Hydrocarbons	Loss of a methylene group indicates a hydrocarbon sample

* It is possible to operate the analyzer to ignore these common background impurities. They will be present to contribute to poor vacuum if these impurities result from a significant leak.

** See the table "Major Reference Masses in the Spectrum of Heptacosafluorotributylamine (Perfluorotributylamine) in this section for additional details.

REDUCTION OF WEIGHINGS IN AIR TO VACUO

When the mass M of a body is determined in air, a correction is necessary for the buoyancy of the air. The corrected mass is given by $M + kM/1000$, where k is a function of the material used for the weights, given by

$$k = 1000\rho_{air}(1/\rho_{body} - 1/\rho_{weight})$$

and ρ is density. The table below is computed for an air density of 0.0012 g/cm³ and for densities of three common weights:

platinum-iridium (21.6 g/cm³), brass (8.5 g/cm³), and aluminum or quartz (2.65 g/cm³).

References

1. Kaye, G. W. C., and Laby, T. H., *Tables of Physical and Chemical Constants, 16th Edition,* pp. 25–28, Longman, London, 1995.
2. Giacomo, P., *Metrologia,* 18, 33, 1982.
3. Davis, R. S., *Metrologia,* 29, 67, 1992.

Density of body (g cm⁻³)	Value of k for weights of:			Density of body (g cm⁻³)	Value of k for weights of:		
	Pt-Ir	Brass	Quartz or Al		Pt-Ir	Brass	Quartz or Al
0.5	2.34	2.26	1.95	1.8	0.61	0.53	0.21
0.6	1.94	1.86	1.55	1.9	0.58	0.49	0.18
0.7	1.66	1.57	1.26	2.0	0.54	0.46	0.15
0.8	1.44	1.36	1.05	2.5	0.42	0.34	0.03
0.9	1.28	1.19	0.88	3.0	0.34	0.26	-0.05
1.0	1.14	1.06	0.75	4.0	0.24	0.16	-0.15
1.1	1.04	0.95	0.64	6.0	0.14	0.06	-0.25
1.2	0.94	0.86	0.55	8.0	0.09	0.01	-0.30
1.3	0.87	0.78	0.47	10.0	0.06	-0.02	-0.33
1.4	0.80	0.72	0.40	15.0	0.02	-0.06	-0.37
1.5	0.74	0.66	0.35	20.0	0.00	-0.08	-0.39
1.6	0.69	0.61	0.30	22.0	0.00	-0.09	-0.40
1.7	0.65	0.56	0.25				

For a more accurate calculation, use the following values of the density of air (assuming 50% relative humidity and 0.04% CO_2).

	Air temperature		
P/kPa	10 °C	20 °C	30 °C
85	0.001043	0.001005	0.000968
90	0.001105	0.001065	0.001025
95	0.001166	0.001124	0.001083
100	0.001228	0.001184	0.001140
105	0.001290	0.001243	0.001198

Formulas for calculating the density of air over more extended ranges of temperature, pressure, and humidity may be found in the references.

STANDARDS FOR LABORATORY WEIGHTS

Thomas J. Bruno and Paris D. N. Svoronos

The following table provides a summary of the requirements for metric weights and mass standards commonly used in chemical analysis (Refs. 1–3). The actual specifications are under the jurisdiction of ASTM Committee E-41 on General Laboratory Apparatus and are the direct responsibility of subcommittee E-41.06 that deals with weighing devices. These standards do not generally refer to instruments used in commerce. Weights are classified according to type (either Type I or Type II), grade (S, O, P, or Q), and class (1, 2, 3, 4, 5, or 6). Information on these mass standards is presented to allow the user to make appropriate choices when using analytical weights for the calibration of electronic analytical balances, for making large-scale mass measurements (such as those involving gas cylinders), and in the use of dead-weight pressure balances. Some historical context can be found in Ref. 4.

References

1. *Annual Book of ASTM Standards, ANSI/ASTM E617-97 Standard Specification for Laboratory Weights and Precision Mass Standards*, Book of Standards Vol. 14.04, 2008.
2. Battino, R., and Williamson, A. G., *J. Chem. Educ.* 61, 51, 1984.
3. Bruno, T. J., and Svoronos, P. D. N., *CRC Handbook of Basic Tables for Chemical Analysis*, 3rd Edition, CRC Press, Boca Raton, FL, 2011.
4. NBS Circular 547, *Precision Laboratory Standards of Mass and Laboratory Weights*, Washington, D.C., 1954 (reprinted as NIST-IR 78-1476, Washington, D.C., 1978)

Type — Classification by Design

Type I — One-piece construction; contains no added adjusting material; used for highest accuracy work.

Type II — Can be of any appropriate and convenient design, incorporating plugs, knobs, rings, etc.; adjusting material can be added if it is contained so that it cannot become separated from the weight.

Grade — Classification by Physical Property

Grade S:	Density:	7.7 g cm^{-3} to 8.1 g cm^{-3} (for 50 mg and larger)
	Surface area:	not to exceed that of a cylinder of equal height and diameter
	Surface finish:	highly polished
	Surface protection:	none permitted
	Magnetic properties:	no more magnetic than 300 series stainless steels
	Corrosion resistance:	same as 303 stainless steel
	Hardness:	at least as hard as brass
Grade O:	Density:	7.7 g cm^{-3} to 9.1 g cm^{-3} (for 1 g and larger)
	Surface area:	same as grade S
	Surface finish:	same as grade S
	Surface protection:	may be plated with suitable material such as platinum or rhodium
	Magnetic properties:	same as grade S
	Corrosion resistance:	same as grade S
	Hardness:	at least as hard as brass when coated; smaller weights at least as hard as aluminum
Grade P:	Density:	7.2 g cm^{-3} to 10 g cm^{-3} (for 1 g or larger)
	Surface area:	no restriction
	Surface finish:	smooth, no irregularities
	Surface protection:	may be plated or lacquered
	Magnetic properties:	same as grades S and O
	Corrosion resistance:	surface must resist corrosion and oxidation
	Hardness:	same as grade O
Grade Q:	Density:	7.2 g cm^{-3} to 10 g cm^{-3} (for 1 g or larger)
	Surface area:	same as grade P
	Surface finish:	same as grade P
	Surface protection:	may be plated, lacquered, or painted
	Magnetic properties:	no more magnetic than unhardened unmagnetized steel
	Corrosion resistance:	same as grade P
	Hardness:	same as grades O and P

Tolerance — Classification by Deviation[1]

	CLASS 1			CLASS 2	
Grams	Individual Tolerance/mg	Group Tolerance/mg	Grams	Individual Tolerance/mg	Group Tolerance/mg
500	1.2		500	2.5	
300	0.75		300	1.5	
200	0.50		200	1.0	
100	0.25	1.35	100	0.5	2.7
50	0.12		50	0.25	
30	0.074		30	0.15	
20	0.074		20	0.10	
10	0.050	0.16	10	0.074	0.29
5	0.034		5	0.054	
3	0.034		3	0.054	
2	0.034		2	0.054	
1	0.034	0.065	1	0.054	0.105

CLASS 3		CLASS 4		CLASS 5		CLASS 6	
Grams	Tolerance/mg	Grams	Tolerance/mg	Grams	Tolerance/mg	Grams	Tolerance/mg
500	5.0	500	10	500	30	500	50
300	3.0	300	6.0	300	20	300	30
200	2.0	200	4.0	200	15	200	20
100	1.0	100	2.0	100	9	100	10
50	0.6	50	1.2	50	5.6	50	7
30	0.45	30	0.9	30	4.0	30	5
20	0.35	20	0.7	20	3.0	20	3
10	0.25	10	0.5	10	2.0	10	2
				5	1.3	5	2
				3	0.95	3	2
				2	0.75	2	2
				1	0.50	1	2

[1] In simple terms, the permitted deviation between the assigned nominal mass value of the weight and the actual mass of the weight. Verification of tolerance should be possible on reasonably precise equipment, without using a buoyancy correction, within the political jurisdiction or organizational bounds of a given weight specification.

Applications for Weights and Mass Standards[1]

Application	Type	Grade	Class
Reference standards used for calibrating other weights	I	S	1,2,3, or 4[1]
High-precision standards for calibration of weights and precision balances	I or II[2]	S or O[2]	1 or 2[3]
Working standards for calibration and precision analytical work, dead-weight pressure balances	I or II[2]	S or O	2
Laboratory weights for routine analytical work	II	O	2 or 3
Built-in weights, high-quality analytical balances	I or II	S	2
Moderate precision laboratory balances	II	P	3 or 4
Dial scales and trip balances	II	Q	4 or 5
Platform scales	II	Q	5 or 6

[1] Primary standards are for reference use only and should be calibrated. Since the actual values for each weight are stated, close tolerances are neither required nor desirable.

[2] Type I and Grade S will have a higher constancy but will probably be higher priced.

[3] Since working standards are used for the calibration of measuring instruments, the choice of tolerance depends upon the requirements of the instrument. The weights are usually used at the assumed nominal values and appropriate tolerances should be chosen.

Source: Reprinted (with modification) with permission of the ASTM International (formerly American Society for Testing and Materials), 100 Barr Harbor Drive, West Conshohocken, Pennsylvania, USA.

INDICATORS FOR ACIDS AND BASES

Thomas J. Bruno and Paris D. N Svoronos

The following table lists the most common indicators together with their pH range and colors in acidic and basic media. Since the color change is not instantaneous at the pK_a value, a pH range is given where a combination of colors is present. This pH range, which varies between indicators, generally falls between the pK_a with a spread or uncertainty of 1 pH unit. All solutions are either aqueous or ethanol/aqueous (% ethanol, vol/vol) (Refs. 1–4). Reference 4 provides additional solution properties that are related to buffer and solvent properties, and Ref. 5 lists the exact quantities needed for the indicator solutions.

References

1. Lange, N. A., *Lange's Handbook of Chemistry*, 8th Edition, Handbook Publishers, New York, 1952.
2. Kolthoff, I. M., and V. A. Stenger (translated in English by N. H. Furman), *Volumetric Analysis*, 2nd Edition, Interscience Publishers, New York, 1942.
3. Sabnis, R.W., *Handbook of Acid-Base Indicators*, CRC Press, Taylor and Francis, Boca Raton, FL, 2008.
4. Bruno, T. J., and Svoronos, P. D. N., *CRC Handbook of Basic Tables for Chemical Analysis*, 3rd Edition, CRC Press, Boca Raton, FL, 2011.
5. http://www.csudh.edu/oliver/chemdata/ind-prep.htm, accessed June 2011.

Indicator	pH Range	Solvent	Acid	Base
Gentian violet (crystal violet)	0.0–2.0	Aqueous	Yellow	Blue-violet
Thymol blue	1.2–2.8	Aqueous	Red	Yellow
Pentamethoxy red	1.2–2.3	70% ethanol	Red-violet	Colorless
Tropeolin OO	1.3–3.2	Aqueous	Red	Yellow
2,4-dinitrophenol	2.4–4.0	50% ethanol	Colorless	Yellow
Methyl yellow	2.9–4.0	90% ethanol	Red	Yellow
Methyl orange	3.1–4.4	Aqueous	Red	Orange
Bromophenol blue	3.0–4.6	Aqueous	Yellow	Blue-violet
Tetrabromphenol blue, sodium salt	3.0–4.6	Aqueous	Yellow	Blue
Congo red	3.0–5.0	Aqueous	Blue-violet	Red
Alizarin sodium sulfonate	3.7–5.2	Aqueous	Yellow	Violet
α-Naphthyl red	3.7–5.0	70% ethanol	Red	Yellow
p-Ethoxychrysoidine	3.5–5.5	Aqueous	Red	Yellow
Bromocresol green, sodium salt	4.0–5.6	Aqueous	Yellow	Blue
Methyl red, sodium salt	4.4–6.2	Aqueous	Red	Yellow
Bromocresol purple	5.2–6.8	Aqueous	Yellow	Purple
Chlorphenol red	5.4–6.8	Aqueous	Yellow	Red
Bromophenol blue, sodium salt	6.2–7.6	Aqueous	Yellow	Blue
p-nitrophenol	5.0–7.0	Aqueous	Colorless	Yellow
Azolitmin	5.0–8.0	Aqueous	Red	Blue
Bromothymol blue, sodium salt	6.0–7.6	70% ethanol	Yellow	Blue
Phenol red, sodium salt	6.4–8.0	Aqueous	Yellow	Red
Neutral red	6.8–8.0	70% ethanol	Red	Yellow
Rosolic acid	6.8–8.0	90% ethanol	Yellow	Red
Cresol red, sodium salt	7.2–8.8	Aqueous	Yellow	Red
α-naphtholphthalein	7.3–8.7	70% ethanol	Rose	Green
Tropeolin OOO	7.6–8.9	Aqueous	Yellow	Rose-red
Thymol blue, sodium salt	8.0–9.6	Aqueous	Yellow	Blue
Phenolphthalein	8.0–10.0	70% ethanol	Colorless	Red
α-naphtholbenzein	9.0–11.0	70% ethanol	Yellow	Blue
Thymolphthalein	9.4–10.6	90% ethanol	Colorless	Blue
Nile blue	10.1–11.1	Aqueous	Blue	Red
Alizarin yellow R	10.0–12.0	Aqueous	Yellow	Lilac
Salicyl yellow	10.0–12.0	90% ethanol	Yellow	Orange-brown
Diazo violet	10.1–12.0	Aqueous	Yellow	Violet
Tropeolin O	11.0–13.0	Aqueous	Yellow	Orange-brown
Nitramine	11.0–13.0	70% ethanol	Colorless	Orange-brown
Poirrier's blue	11.0–13.0	Aqueous	Blue	Violet-pink
Trinitrobenzoic acid	12.0–13.4	Aqueous	Colorless	Orange-red

PREPARATION OF SPECIAL ANALYTICAL REAGENTS

Paris D. N. Svoronos and Thomas J. Bruno

This listing of analytical reagents has been updated to include formulations based on more recent research, and to also include safety notes (Refs. 1–3). In the absence of specific cautions, the user must observe sound laboratory practice and housekeeping to avoid exposure and environmental contamination. When less familiar reagents are described, we include a chemical formula and a CAS registry number to avoid ambiguity. When a reagent calls for 95% ethanol, the azeotrope with water is specified unless noted. Additional details can be found in the references listed below.

References

1. Lide, D. R., Ed., *CRC Handbook for Chemistry and Physics*, 90th Edition, CRC Press, Boca Raton, FL, 2010.
2. Bruno, T. J., and Svoronos, P. D. N., *CRC Handbook of Basic Tables for Chemical Analysis*, 3rd Edition, CRC Press, Boca Raton, FL, 2011.
3. Svoronos, P. D. N., Sarlo, E., and Kulawiec, R., *Organic Chemistry Laboratory Manual*, WCB McGraw Hill, New York, 1996.

Aluminon (qualitative test for aluminum). Aluminon is the name for the triammonium salt of aurintricarboxylic acid (5-[(3-carboxy-4-hydroxyphenyl)(3-carboxy-4-oxo-2,5-cyclohexadienylidene)methyl]-2-hydroxybenzoic acid triammonium salt, $C_{22}H_{23}N_3O_9$, CAS No. 569-58-4). Dissolve 1 g of the salt in 1 L of distilled water. Shake the solution well to ensure thorough mixing.

Bang's reagent (for glucose estimation). Dissolve (in the exact order) 100 g of potassium carbonate, 66 g of potassium chloride and 160 g of anhydrous potassium bicarbonate in approximately 700 mL of distilled water at 30 °C. Add (while stirring) 4.4 g of copper (II) sulfate and dilute to 1 L with distilled water after all CO_2 has been released. This solution should be shaken and saved in an air-tight flask. After 24 hours, 300 mL of it are diluted to 1 L with a saturated aqueous potassium chloride solution, shaken gently and used after 24 hours. 50 mL of this solution is equivalent to 10 mg glucose.

Barfoed's reagent (test for glucose). See Cupric acetate.

Baudisch's reagent. See Cupferron.

Benedict's solution (qualitative reagent for glucose). Dissolve 173 g of sodium citrate and 100 g of sodium carbonate in 800 mL of distilled water. Filter, if necessary, and dilute to 850 mL with distilled water. Dissolve 17.3 g of copper (II) sulfate pentahydrate in 100 mL of distilled water. Pour the latter solution, with constant stirring, into the carbonate-citrate solution and dilute to 1 L with distilled water.

Benzidine hydrochloride solution (for sulfate determination). Prepare a paste of 8 g of benzidine hydrochloride ($C_{12}H_8(NH_2)_2 \cdot 2HCl$) and 20 mL of distilled water, add 20 mL of 20% (mass/mass) HCl and dilute to 1 L with distilled water. Each mL of this solution is equivalent to 0.00357 g of H_2SO_4. Note that the reagent is often called benzidine dihydrochloride.

Bertrand's reagent (glucose estimation). Consists of the following solutions:

1. Dissolve 200 g of Rochelle salt (potassium sodium tartrate, $NaKC_4H_4O_6$) and 150 g of NaOH in sufficient distilled water to make 1.0 L of solution.
2. Dissolve 40 g of copper (II) sulfate pentahydrate in sufficient distilled water to a total of 1.0 L of solution.

3. Dissolve 50 g of iron (III) sulfate and 200 g of H_2SO_4 (sp. gr. 1.84) in sufficient distilled water to a total of 1.0 L of solution.
4. Dissolve 5 g of potassium permanganate in sufficient distilled water to a total of 1.0 L solution.

Bial's reagent (for pentose). Dissolve 1 g of orcinol (5-methyl-1,3-benzenediol, $C_7H_8O_2$, CAS No. 504-15-4) in 500 mL of 30% (mass/mass) HCl to which 30 drops of a 10% aqueous solution of iron (III) chloride have been added.

Boutron — Boudet soap solution: Consists of the following solutions:

1. Dissolve 100 g of pure castile soap (olive oil based) in about 2.5 L of 56% (vol/vol) aqueous ethanol.
2. Dissolve 0.59 g of barium nitrate in 1.0 L of water. Adjust the castile soap solution (1) so that 2.4 mL of it will produce a permanent lather with 40 mL of solution (2). When adjusted, 2.4 mL of this soap solution is equivalent to 220 parts per million of hardness (as calcium carbonate) for a 40 mL sample. See also Soap solution.

Brucke's reagent (protein precipitation). See Potassium iodide mercuric iodide.

Cobaltic cyanide paper (Rinnmann's test for zinc detection). Dissolve 4 g of potassium cobalt (III) hexacyanide ($K_3Co(CN)_6$) and 1 g of potassium chlorate in 100 mL of water. Soak filter paper in solution and dry at 100 °C. Apply a drop of the zinc solution and heat in an evaporating dish. A green disk on the filter paper is obtained if zinc is present.

Congo red. Dissolve 0.5 g of Congo red (3,3'-([1,1'-biphenyl]-4,4'-diyl)bis(4-amino naphthalene-1-sulfonic acid, $C_{32}H_{22}N_6Na_2O_6S_2$, CAS No. 573-58-0) in 90 mL of distilled water and 10 mL of ethanol.

Cupferron (Baudisch's reagent for iron analysis). Dissolve 6 g of Cupferron, the ammonium salt of N-hydroxy-N-nitrosoaniline ($NH_4[C_6H_5N(O)NO]$) in 100 mL of distilled water. The reagent is good only for one week and must be kept in the dark.

Cupric acetate (Barfoed's reagent for reducing monosaccharides). Dissolve 66 g of copper (II) acetate and 10 mL of glacial acetic acid in water and dilute to 1.0 L.

Cupric oxide, ammoniacal; Schweitzer's reagent (dissolves cotton, linen, and silk, but not wool). Dissolve 5 g of cupric sulfate in 100 mL of boiling water, and add sodium hydroxide until precipitation is complete. Wash the precipitate well, and dissolve it in a minimal quantity of ammonium hydroxide. Bubble a slow stream of air through 300 mL of concentrated ammonium hydroxide solution containing 50 g of fine copper turnings. Continue stirring for 1 hour.

Cupric sulfate in glycerin-potassium hydroxide (reagent for silk). Dissolve 10 g of copper (II) sulfate pentahydrate in 100 mL of water and add 5 g of glycerol. Add 6N KOH solution slowly until a deep blue solution is obtained.

Cupron (precipitates copper). Dissolve 5 g of benzoinoxime in 100 mL 95% ethanol.

Cuprous chloride, acidic (reagent for CO in gas analysis).

1. Cover the bottom of a 2 L flask with a layer of copper (II) oxide about 1.5 cm deep, suspend a coil of copper wire so as to reach from the bottom to the top of the solution, and fill the flask with hydrochloric acid 20% (vol/vol) HCl

(aq) with continuous stirring. When the solution becomes nearly colorless, transfer to reagent bottles, which should also contain copper wire. The stock bottle may be refilled with dilute (6N) HCl until either the cupric oxide or the copper wire is used up. Copper (II) sulfate may be substituted for copper oxide in the above procedure.

2. Dissolve 340 g of copper (II) chloride dihydrate in 600 mL of concentrated HCl and reduce the cupric chloride by adding 190 mL of a saturated solution of stannous chloride or until the solution is colorless. The stannous chloride is prepared by treating 300 g of metallic tin in a 500 mL flask with concentrated HCl until no more tin is dissolved.

3. (Winkler method). Add a mixture of 86 g of copper (II) chloride and 17 g of finely divided metallic copper, prepared by the reduction of 8 g to12 g of cupper (II) oxide with hydrogen gas, to a solution of HCl, made by diluting 650 mL of concentrated HCl with 325 mL of distilled water. After the mixture has been added slowly and with frequent stirring, a spiral of copper wire is suspended in the bottle, reaching all the way to the bottom. The solution is ready to use when the solution becomes colorless.

Cuprous chloride, ammoniacal (reagent for CO in gas analysis).

1. The acid solution of copper (II) chloride as prepared above is neutralized with 6N ammonium hydroxide until the ammonia odor persists. An excess of metallic copper must be kept in the solution.

2. Pour 800 mL of acidic copper (II) chloride, prepared by the Winkler method, into approximately 4 L of water. Transfer the precipitate to a 250 mL graduated cylinder. After several hours, siphon off the liquid above the 50 mL mark and refill with 7.5% ammonium hydroxide solution which may be prepared by diluting 50 mL of concentrated ammonium hydroxide with 150 mL of distilled water. The solution, which should have a faint odor of ammonia should be well stirred and allowed to stand for several hours.

Note: Proper precautions must be taken for the safe handling of gaseous samples that contain CO.

Dichlorofluorescein indicator. Dissolve 1 g dichlorofluorescein ($C_{20}H_{10}Cl_2O_5$, CAS No. 76-54-0) in 1 L 70% (vol/vol) ethanol or 1 g of the sodium salt in 1 L of distilled water.

Dimethyglyoxime, 0.01 N. Dissolve 0.6 g of dimethylglyoxime (2,3-butanedione oxime) in 500 mL of 95% ethanol. This is an especially sensitive test for nickel (II) and produces a very characteristic crimson color.

Diphenylamine (reagent for rayon). Dissolve 0.2 g diphenylamine in 100 mL of concentrated sulfuric acid.

Diphenylamine sulfonate (for titration of iron with $K_2Cr_2O_7$). Dissolve 0.32 g of the barium salt of diphenylamine sulfonic acid in 100 mL of distilled water, add 0.5 g of sodium sulfate and filter off the precipitated barium sulfate.

Diphenylcarbazide. Dissolve 0.2 g of diphenylcarbazide[(C_6H_5N HNH)$_2$CO] in 10 mL of glacial acetic acid and dilute to 100 mL with 95% ethanol.

2,4-Dinitrophenylhydrazine (2,4-DNP) Reagent. Dissolve 3 g of 2,4-dinitrophenylhydrazine in 15 mL to 20 mL of concentrated sulfuric acid. Add this solution with stirring to a solution of 10 mL distilled water in 70 mL of 95% hydrous ethanol. Use the filtrate to test for the presence of aldehydes and ketones as follows: To 2 mL of the solution add 10 drops of the unknown. A precipitate (which may be recrystallized from 95% hydrous ethanol) is indicative of a positive test and its melting point can be matched to the corresponding melting points of various aldehydes and ketones.

Esbach's reagent (estimation of protein). Dissolve 10 g of picric acid and 20 g of citric acid in sufficient distilled water to make a 1.0 L of solution.

Eschka's compound. Two parts of calcinated ("light") magnesia are thoroughly mixed with 1 part of anhydrous sodium carbonate.

Fehling's solution (reagent for reducing sugars).

1. Copper (II) sulfate solution: Dissolve 34.66 g of copper (II) sulfate pentahydrate in distilled water and dilute to 500 mL.

2. Alkaline tartrate solution: Dissolve 173 g of potassium sodium tartrate (Rochelle salt, $KNaC_4H_4O_6 \cdot 4H_2O$) and 50 g of NaOH in distilled water, cool and dilute to 500 mL. Mix equal volumes of solutions (1) and (2) just before using.

Ferric-alum indicator. Dissolve 140 g of iron (III) ammonium sulfate crystals in 400 mL of hot distilled water. Cool, filter, and add enough 6N nitric acid until the 500 mL mark.

Folin's mixture (for uric acid). To 650 mL of distilled water add 500 g of ammonium sulfate, 5 g of uranium (VI) acetate, and 6 g of glacial acetic acid and then dilute to 1.0 L with distilled water.

Formaldehyde — sulfuric acid (Marquis' reagent for alkaloids). Add 100 mL of concentrated sulfuric acid to 5 mL of 37% (mass/mass) aqueous formaldehyde solution.

Froehde's reagent. See Sulfomolybdic acid.

Fuchsin (reagent for linen). Dissolve 1 g of fuchsin (an aniline dye) in 100 mL of ethanol.

Fuchsin - sulfurous acid (Schiff's reagent for aldehydes). Dissolve 0.5 g of fuchsin and 9 g of sodium bisulfite in 500 mL of distilled water, and add 10 mL of concentrated HCl. Keep in well-stoppered bottles and protect from light. The fuchsin used to make the Schiff reagent should incorporate a high content of pararosanilin.

Gunzberg's reagent (detection of HCl in gastric juice). Dissolve 4 g of phloroglucinol (1,3,5-benzenetriol, $C_6H_3(OH)_3$) and 2 g of vanillin ($C_8H_8O_3$) in 100 mL of absolute ethanol.

Hager's reagent. See Picric acid.

Hanus solution (for iodine number). Dissolve 13.2 g of resublimed iodine in 1 L of glacial acetic acid, which should pass the dichromate test for reducible matter. Add sufficient bromine to double the halogen content, as determined by titration (approximately 3 mL). The iodine may be dissolved by heating, but the solution should be cold when the bromine is added.

Hopkins-Cole Reagent (test for presence of the indole ring in the tryptophan moiety of a protein or peptide). Add 5 mL cold water to 10 g of magnesium in a conical flask and add 250 mL of a cold saturated aqueous oxalic acid solution with vigorous stirring. Filter, add 25 mL of glacial acetic acid and dilute with distilled water to the 1.0 L mark. Place 10 drops of the saturated protein or peptide solution in a test tube, add 10 drops of concentrated nitric acid and heat in a water bath for three minutes. Allow the mixture to cool and add 4 mL 6N NaOH. A positive test is confirmed by the formation of a yellow color.

Iodine, tincture of. To 50 mL of distilled water add 70 g of iodine and 50 g of potassium iodide. Dilute to 1 L with absolute ethanol.

Iodo-potassium iodide (Wagner's reagent for alkaloids). Dissolve 2 g of iodine and 6 g of KI in 100 mL of distilled water.

Jones Reagent (test is positive for 1° and 2° alcohols, aldehydes, mercaptans, sulfides and sulfoxides). To 1 mL of 5% (mass/mass) sodium (or potassium) dichromate add 10 drops of the unknown compound. The formation of a green color (sometimes appearing as brown due to the mixture of the green chromium (III) and the unreacted orange chromium (VI) colors) indicates a positive test.

Litmus (indicator). Extract litmus powder three times with boiling alcohol (ethanol or isopropanol), each treatment lasting for an hour. Discard the alcoholic extract. Treat the residue with an equal mass of cold distilled water and filter; then add five times its weight of boiling distilled water. Cool, filter, and combine all aqueous extracts.

Magnesia mixture (reagent for phosphates and arsenates). Dissolve 50 g of magnesium chloride (MgCl$_2$·6H2O) and 100 g of ammonium chloride (NH$_4$Cl) in 500 mL of distilled water, add a slight excess of NH$_4$OH, and allow to stand overnight. Filter, make the solution just acidic with HCl, and dilute with distilled water to 1 L. Let stand for several days and separate the precipitated solid by decanting. The decantate should be used just prior to testing; otherwise, if stored for any period of time it becomes turbid.

Magnesium uranyl acetate. Dissolve 100 g of uranyl (VI) acetate dehydrate (UO$_2$(C$_2$H$_3$O$_2$)$_2$·2H$_2$O) in 60 mL of glacial acetic acid and dilute to 500 mL with distilled water. Dissolve 330 g magnesium acetate tetrahydrate (Mg(C$_2$H$_3$O$_2$)$_2$·4H$_2$O) in 60 mL of glacial acetic acid and dilute to 200 mL. Heat both solutions to boiling until clear; pour the magnesium acetate solution into the uranyl acetate solution, cool and dilute with distilled water to 1 L. Let stand overnight and filter if necessary.

Marme's reagent. See Potassium-cadmium iodide.

Marquis' reagent. See Formaldehyde-sulfuric acid.

Mayer's reagent (white precipitate with most alkaloids in slightly acid solutions). Dissolve 1.358 g of mercury (II) chloride in 60 mL of distilled water and pour into a solution of 5 g of potassium iodide in 10 mL of distilled water. Add sufficient distilled water to a final volume of 100 mL.

Methyl orange indicator. Dissolve 1 g of methyl orange in 1 L of distilled water. Filter, if necessary.

Methyl orange, modified. Dissolve 2 g of methyl orange and 2.8 g of xylene cyanole FF (C$_{25}$H$_{27}$N$_2$NaO$_6$S$_2$, CAS No: 2650-17-1) in 1 L of 50% (vol/vol) ethanol.

Methyl red indicator. Dissolve 1 g of methyl red in 600 mL of ethanol and dilute with 400 mL of distilled water.

Methyl red, modified. Dissolve 0.50 g of methyl red and 1.25 g of xylene cyanole FF (C$_{25}$H$_{27}$N$_2$NaO$_6$S$_2$) in 1 L of 90% (vol/vol) ethanol. Alternatively dissolve 1.25 g of methyl red and 0.825 g of methylene blue in 1 L of 90% (vol/vol) ethanol.

Millon's reagent (for albumins and phenols). Dissolve 1 part of mercury in 1 part of cold fuming nitric acid. Dilute with twice the volume of distilled water and decant the clear solution after several hours. All appropriate precautions must be observed when handling mercury.

Molisch's reagent. See 1-Naphthol.

1-Naphthol (Molisch's reagent for wool). Dissolve 15 g of 1-naphthol in 100 mL of 95% (vol/vol) ethanol or chloroform.

Nessler's reagent (for ammonia). Dissolve 50 g of potassium iodide in the smallest possible quantity of cold distilled water (50 mL). Add a saturated solution of mercury (II) chloride (about 22 g in 350 mL of distilled water will be needed) until an excess is indicated by the formation of a precipitate. Then add 200 mL of 5 N NaOH and dilute to 1 L with distilled water. Allow to settle and draw off the clear liquid.

Nickel oxide, ammoniacal (reagent for silk). Dissolve 5 g of nickel (II) sulfate in 100 mL of distilled water, and add 6N NaOH solution until nickel hydroxide is completely precipitated. Wash the precipitate well and dissolve in 25 mL of concentrated ammonium hydroxide and 25 mL of distilled water.

Nitron (detection of the nitrate radical). Dissolve 10 g of nitron (1,4-diphenyl-3-(phenylamino)-1,2,4-triazolium hydroxide, C$_{20}$H$_{18}$N$_4$O) in 5 mL of glacial acetic acid and 95 mL of distilled water. The solution may be filtered with slight suction through an Alundum thimble crucible and kept in a dark bottle.

1-Nitroso-2-naphthol. Make a saturated solution 1-nitroso-2-naphthol in 50% (vol/vol) aqueous acetic acid. The reagent should be used as soon as it is prepared.

Nylander's solution (carbohydrates). Dissolve 20 g of bismuth subnitrate (Bi$_5$O(OH)$_9$(NO$_3$)$_4$) and 40 g of Rochelle salt (potassium sodium tartrate, KNaC$_4$H$_4$O$_6$·4H$_2$O)in 1 L of 8% (mass/mass) aqueous NaOH solution. Cool and filter.

Obermayer's reagent (for indoxyl in urine). Dissolve 4 g of iron (III) chloride in 1 L of aqueous (vol/vol) 40% HCl.

Oxine. Dissolve 14 g of 8-hydroxyquinoline (C$_9$H$_7$NO) in 30 mL of glacial acetic acid. Warm slightly, if necessary to dissolve. Dilute to 1 L with distilled water.

Oxygen absorbent. Dissolve 300 g of ammonium chloride in 1 L distilled water and add 1 L of concentrated ammonium hydroxide solution. Shake the solution thoroughly. For use as an oxygen absorbent, the gas to be tested is passed through a bottle that is half full of copper turnings filled nearly to the top with this ammonium chloride - ammonium hydroxide solution.

Pasteur's salt solution. To 1 L of distilled water add 2.5 g of potassium phosphate, 0.25 g of calcium phosphate, 0.25 g of magnesium sulfate, and 12.00 g of ammonium tartrate.

Pavy's solution (glucose reagent). To 120 mL of Fehling's solution, add 300 mL of 6N ammonium hydroxide (sp. gr. 0.88) and dilute to 1 L with distilled water.

Phenanthroline ferrous ion indicator. Dissolve 1.485 g of 1,10-phenanthroline monohydrate (C$_{12}$H$_8$N$_2$·H$_2$O) in 100 mL of 0.025 M aqueous iron (II) sulfate solution.

Phenolphthalein. Dissolve 1 g of phenolphthalein in 50 mL of ethanol and add 50 mL of distilled water.

Phenolsulfonic acid (determination of nitrogen as nitrate). Dissolve 25 g of phenol in 150 mL of concentrated H$_2$SO$_4$, add 75 mL of fuming sulfuric acid (approximately 15% SO$_3$), stir well and heat for 2 hours at 100 °C.

Phloroglucinol solution (pentosans). Make a 3% phloroglucinol (1,3,5-benzenetriol) solution in alcohol. Store in a dark bottle.

Phosphomolybdic acid (Sonnenschein's reagent for alkaloids).

1. Prepare ammonium phosphomolybdate and after washing with distilled water, boil with nitric acid to expel all ammonia. Evaporate to dryness and dissolve in 2 M HNO$_3$.

2. Dissolve ammonium molybdate in HNO$_3$ and treat with phosphoric acid. Filter, wash the precipitate, and boil with aqua regia (a mixture of concentrated nitric acid and concentrated hydrochloric acid, 1:3 vol/vol) until the ammonium salt is decomposed. Evaporate to dryness. The residue is dissolved in 10% (vol/vol) HNO$_3$. The solution constitutes Sonnenschein's reagent.

Phosphoric acid — sulfuric acid mixture. Dilute 150 mL of concentrated sulfuric acid and 100 mL of concentrated phosphoric acid with distilled water to a final volume of 1 L.

Phosphotungstic acid (Scheibler's reagent for alkaloids).

1. Dissolve 20 g of sodium tungstate and 15 g of sodium phosphate in 100 mL of distilled water slightly acidified with dilute nitric acid.

2. The reagent is a 10% (mass/mass) solution of phosphotungstic acid in distilled water. The phosphotungstic acid is prepared by evaporating a mixture of 10 g of sodium tungstate dissolved in 5 g of phosphoric acid (sp. gr. 1.13) and enough boiling water to a complete solution. Crystals of phosphotungstic acid will separate.

Picric acid (Hager's reagent for alkaloids, wool and silk). Dissolve 1 g of picric acid in 100 mL of distilled water.

Potassium antimonate (reagent for sodium). Boil 22 g of potassium antimonate with 1 L of distilled water until nearly all of the salt has dissolved, cool quickly, and add 35 mL of 10% (mass/mass) potassium hydroxide. Filter after standing overnight.

Potassium-cadmium iodide (Marme's reagent for alkaloids). Add 2 g of cadmium (II) chloride to a boiling solution of 4 g of potassium iodide in 12 mL of distilled water, and then mix with 12 mL of saturated aqueous potassium iodide solution.

Potassium hydroxide (for CO_2 absorption). Dissolve 360 g of potassium hydroxide in distilled water and dilute to 1 L.

Potassium iodide — mercuric iodide (Brucke's reagent for proteins). Dissolve 50 g of potassium iodide in 500 mL of distilled water, and saturate with mercury (II) iodide (about 120 g). Dilute to 1 L with distilled water.

Potassium pyrogallate (for oxygen absorption). For mixtures of gases containing less than 28% (mass/mass) oxygen, add 100 mL of potassium hydroxide solution (50 g of KOH to 100 mL of distilled water) to 5 g of pyrogallol. For mixtures containing more than 28% (mass/mass) oxygen, the KOH solution should contain instead 120 g of KOH in 100 mL of distilled water.

Pyrogallol, alkaline. The reagent is a mixture of two solutions

1. Dissolve 75 g of pyrogallic acid in 75 mL of distilled water.

2. Dissolve 500 g of KOH in 250 mL distilled water. When cooled, adjust the solution with distilled water until the concentration is 50% (vol/vol).

For use, add 270 mL of solution (2) to 30 mL of solution (1).

Rosolic acid (indicator). Dissolve 1 g of rosolic acid ($C_{19}H_{14}O_3$) in 10 mL of ethanol and add 100 mL of distilled water.

Sakaguchi Reagent (for the presence of arginine in proteins or peptides). To 10 drops of the saturated protein or peptide solution, add 5 drops of 6 N NaOH followed by 5 drops of 0.05% (mass/mass) ethanolic 1-naphthol solution and 10 drops of 0.5% aqueous sodium hypochlorite. A positive test is indicated by the formation of a red color that fades away upon standing.

Scheibler's reagent. See Phosphotungstic acid.

Schiff's reagent. See Fuchsin-sulfurous acid.

Schweitzer's reagent. See Cupric oxide, ammoniacal.

Soap solution (reagent for hardness in water). Dissolve 100 g of dry castile soap in 1 L of 80% (vol/vol) ethanol (4 parts ethanol to 1 part distilled water). Allow to stand for several days and dilute with 70% ethanol until 6.4 mL of this solution produces a permanent lather with 20 mL of standardized calcium solution. The latter solution is made by dissolving 0.2 g of calcium carbonate in a small amount of dilute HCl, evaporating to dryness and then dissolving the precipitate with distilled water to a final volume of 1 L.

Sodium bismuthate (for the oxidation of manganese). Heat 20 parts of sodium hydroxide nearly to redness in an iron or nickel crucible and add slowly 10 parts of basic bismuth (III) nitrate which has been previously dried. Add 2 parts of sodium peroxide, and pour the brownish-yellow fused mass onto an iron plate to cool. When cooled, break up in a mortar, extract with distilled water, and collect on an asbestos filter.

Sodium hydroxide (for CO_2 absorption). Dissolve 330 g of sodium hydroxide in distilled water and dilute to 1 L.

Sodium nitroprusside (reagent for hydrogen sulfide and wool). Use a freshly prepared solution of 1 g of sodium nitroferricyanide in 10 mL of distilled water.

Sodium oxalate (primary standard). Dissolve 30 g of the commercial salt of sodium oxalate in 1 L of distilled water, make slightly alkaline with sodium hydroxide, and let stand until clear. Filter and evaporate the filtrate to 100 mL. Cool and filter. Pulverize the residue and wash it several times with small volumes of distilled water. The procedure is repeated until the mother liquor is sulfate-free and is neutral to phenolphthalein.

Sodium plumbite (reagent for wool). Dissolve 5 g of sodium hydroxide in 100 mL distilled water. Add 5 g of litharge (lead (II) oxide) and boil until dissolved.

Sodium polysulfide. Dissolve 480 g of sodium sulfide nonahydrate in 500 mL of distilled water, add 40 g of NaOH and 18 g of sulfur. Stir thoroughly and dilute to 1 L with distilled water.

Sonnenschein's reagent. See Phosphomolybdic acid.

Starch solution.

1. Make a paste with 2 g of soluble starch and 0.01 g of mercury (II) iodide with a small amount of distilled water. Add the mixture slowly to 1 L of boiling distilled water and boil further for a few minutes. Keep in a glass stoppered bottle. If other than soluble starch is used, the solution will not be clear on boiling; it should then be allowed to stand and the clear liquid decanted.

2. A solution of starch that keeps stable indefinitely is made as follows: Mix 500 mL of aqueous saturated NaCl solution (filtered), 80 mL of glacial acetic acid, 20 mL of distilled water and 3 g of starch. Bring slowly to a boil and further heat for 2 minutes.

3. Make a paste with 1 g of soluble starch and 5 mg of mercury (II) iodide using as little cold distilled water as possible. Then pour about 200 mL of boiling distilled water on the paste and stir immediately. This will give a clear solution if the paste is prepared correctly and the water is actually boiling. Cool and add 4 g of potassium iodide. Starch solution decomposes on standing due to bacterial action, but this solution will be stable if stored under a layer of toluene.

Stoke's reagent. Dissolve 30 g of iron (II) sulfate and 20 g of tartaric acid in distilled water and dilute to 1 L. Just before using, add concentrated ammonium hydroxide until the precipitate that is initially formed is redissolved.

Sulfanilic acid (reagent for nitrites). Dissolve 0.5 g of sulfanilic acid in a mixture of 15 mL of glacial acetic acid and 135 mL of recently boiled distilled water.

Sulfomolybdic acid (Froehde's reagent for alkaloids and glucosides). Dissolve 10 g of molybdic acid or sodium molybdate in 100 mL of concentrated sulfuric acid.

Tannic acid (reagent for albumin, alkaloids, and gelatin). Dissolve 10 g of tannic acid in 10 mL of ethanol and dilute with distilled water to 100 mL.

Titration mixture (residual chlorine in water analysis). Prepare 1 L of dilute HCl (100 mL of HCl (sp. gr. 1.19) in sufficient distilled water to make 1 L). Dissolve 1 g of o-tolidine (3,3'-dimethylbenzidine , CAS No. 119-93-7) in 100 mL dilute hydrochloride, stir well and dilute to 1 L using dilute HCl solution.

Tollen's Reagent (confirming the presence of aldehydes). 5 drops of 5% NaOH is added to 2 mL 10% aqueous silver nitrate in a test tube. The insoluble silver (I) oxide dissolved by the drop wise addition of 10% aqueous ammonia (ammonium hydroxide) yields a clear solution. Excess ammonium hydroxide should be avoided as it may give false positive result. Approximately 10 drops of the aldehyde will yield a silver mirror coating on the test tube inner wall, especially if the mixture is warmed up to 50 °C.

Trinitrophenol solution. See Picric acid.

Turmeric tincture (reagent for borates). Digest ground turmeric root with several quantities of distilled water which are discarded. Dry the residue and digest it several days with six times its weight of ethanol. Filter.

Uffelmann's reagent (turns yellow in presence of lactic acid). To a 2% solution of pure phenol in distilled water, add an aqueous solution of iron (III) chloride until the phenol solution becomes violet in color.

Wagner's reagent. See Iodo-potassium iodide.

Wagner's solution (used in phosphate rock analysis to prevent precipitation of iron and aluminum). Dissolve 25 g of citric acid and 1 g of salicylic acid in distilled water and dilute to 1 L. Use 50 mL of the reagent.

Wij's iodine monochloride solution (for iodine number). Dissolve 13 g of resublimed iodine in 1 L of glacial acetic acid that will pass the dichromate test for reducible matter. Set aside 25 mL of this solution. Bubble into the remainder of the solution dry chlorine gas (dried and washed by passing through concentrated sulfuric acid) until the characteristic color of free iodine has been dissipated. Add the 25 mL of the iodine solution that was set aside, until all free chlorine has been dissipated. A slight excess of iodine does little or no harm, but an excess of chlorine must be avoided. Preserve in well stoppered, amber colored bottles. Avoid the use of solutions that have been stored for more than 30 days.

Wij's special solution (for iodine number). To 200 mL of glacial acetic acid that will pass the dichromate test for reducible matter, add 12 g of dichloramine T (N,N-dichloro-4-methylbenzenesulfonamide, CAS No. 473-34-7), and 16.6 g of dry potassium crystals (in small quantities with continual shaking until all the potassium iodide has dissolved). Dilute to 1 L with the same quality of acetic acid used above and preserve in a dark colored bottle.

Zimmermann-Reinhardt reagent (determination of iron). Dissolve 70 g of manganese (II) sulfate tetrahydrate in 500 mL of water, add slowly 125 mL of concentrated sulfuric acid and 125 mL of 85% phosphoric acid, and dilute to 1 L with water.

Zinc chloride solution, basic (reagent for silk). Dissolve 1000 g of zinc chloride in 850 mL of distilled water, and add 40 g of zinc oxide. Heat until complete dissolving is complete.

Zinc uranyl acetate (reagent for sodium). Dissolve 10 g of $UO_2(C_2H_3O_2)2 \cdot 2H_2O$ in 6 g of 30% acetic acid, heat, if necessary, and dilute to 50 mL. Dissolve 30 g of zinc acetate dehydrate in 3 g of 30% acetic acid and dilute to 50 mL. Mix the two solutions, add 50 mg of sodium chloride, allow to stand overnight and filter. Use distilled water in all dilutions.

ORGANIC ANALYTICAL REAGENTS FOR THE DETERMINATION OF INORGANIC CATIONS

Paris D. N. Svoronos and Thomas J. Bruno

The table entitled Organic Analytical Reagents for the Determination of Inorganic Substances, by G. Ackermann, L. Sommer, and D. Thorburn Burns, has been revamped and considerably expanded for this edition. The many recent advancements in this area have been surveyed, and in this table, specifically for inorganic cations, we present the major tests, reagents, and some guidance as to the expected result or method of observation. The abbreviations used here are defined in the abbreviations table in this section. In addition, for brevity, when a determination calls for a spectrophotometric measurement at a particular wavelength, for example at 500 nm, we denote this as: "spec λ = 500 nm." No wavelength is specified if it is variable, for example, with pH; here we indicate: "spec determination." When a determination calls for spectrofluorimetric determination, the excitation and emission wavelengths are provided, thus: specf λ_{ex} = 303.5 nm, λ_{em} = 353 nm. Note that a common surfactant used in many of these tests is cetyltrimethyl ammonium bromide, abbreviated (CTAB). Some of the procedures listed here require the use of hazardous chemicals (carcinogens such as benzene, strong acids such as HF). Appropriate precautions must be observed.

While a great deal of the information presented here is from the recent literature, the reader is referred to several excellent reviews and monographs for additional information (Refs. 1–9).

References

1. Marczenko, Z., *Separation and Spectrophotometric Determination of Elements*, Ellis Horwood, Chichester, 1986.
2. Sandell, E. B., and Onishi, H., *Photometric Determination of Traces of Metals. General Aspects, Part I*, 4th Edition, John Wiley and Sons, New York, 1986.
3. Onishi, H., *Photometric Determination of Traces of Metals. Part IIa: Individual Metals, Aluminium to Lithium*, 4th Edition, John Wiley and Sons, New York, 1986.
4. Onishi, H., *Photometric Determination of Traces of Metals. Part IIb: Individual Metals, Magnesium to Zinc*, 4th Edition, John Wiley and Sons, New York, 1986.
5. Townshend, A., Burns, D. T., Guilbault, G. G., Lobinski, R., Marczenko, Z., Newman, E., and Onishi, H., *Dictionary of Analytical Reagents*, Chapman and Hall, London, 1993.
6. West, T. S., and Nürnberg, H. W., *The Determination of Trace Metals in Natural Waters*, Blackwell, Oxford, 1988.
7. Savvin, S. B., Shtykov, S. N., and Mikhailova, A. V., Russ. Chem. Rev. 75, 341, 2006.
8. Ueno, K., Imamura, T., and Cheng, K. L., *Handbook of Organic Analytical Reagents*, CRC Press, Boca Raton, FL, 1992.
9. American Chemical Society, *Reagent chemicals: specifications and procedures: American Chemical Society specifications, official from January 1, 2006*, American Chemican Society, Washington, D.C., 2006.

Determination	Reagents	Results
Aluminum	Alizarin red S	Red color develops; spectrophotometric determination preferred
	Aluminon	Lake pigment stabilized with CTAB
	Chrome azurol S	Spec λ = 500 nm, stabilized with CTAB
	Chromazol KS	Spec λ = 625 nm, stabilized with cetylpyridinium bromide
	Eriochrome cyanine R (also known as mordant blue 3) + CTAB	Red dye lake (pH = 6) stabilized with CTAB
	Eriochrome cyanine R (also known as mordant blue 3) + N,N-dodecyltrimethylammonium bromide (DTAB)	Spec determination by use of cationic surfactants
	8-Hydroxyquinoline	Produces tris(8-hydroxyquinolinato) aluminum (Alq3), found in organic light emitting diodes (OLEDs)
	Bromopyrogallol red + cetyltrimethyl ammonium bromide	Spec λ = 627 nm
	Bromopyrogallol red + nonylphenol tetradecaethylene glycol ether	Spec λ = 612 nm
	Pyrocatechol violet	Spec λ = 578 nm after separation of aluminum from the matrix materials by chloroform extraction of its acetylacetone complex (pH = 6.5), from an ammonium acetate-hydrogen peroxide medium
	2,2′,3,4-Tetrahydroxy-3′,5′-disulfoazobenzene	Spec determination of the binary system (pH = 5)
Antimony	Brilliant green	Isolated as the hexachloroantimonate (V) salt extracted by either toluene or benzene
	Bromopyrogallol red	Used for the determination of antimony (III) with EDTA, cyanide or fluoride ions as masking agents
	Catechol violet	Ternary complex stabilized with CTAB
	Malachite green (basic green 4)	Isolated as the hexachloantimonate (V) complex after benzene extraction
	Phenyl fluorene	Sensitive color reaction with antimony (III) by use of cationic surfactants
	Potassium iodide	Antimony (III)-iodide complex formation in the presence of ascorbic acid
	Rhodamine B	Ion pair or ion association extraction by use of toluene or benzene as solvents
	Silver diethyldithiocarbamate	Spec λ = 504 nm to avoid arsenic interference
	Thiourea	Determination by hydride generation inductively coupled plasma atomic emission spectrometry after reduction of antimony (V) to antimony (III) by thiourea
Arsenic	Michler's ketone	Spec λ = 640 nm trophotometric determination of trace arsenic (V) in water

Determination	Reagents	Results
	Silver diethyldithiocarbamate	Spec λ = 600 nm to avoid antimony interference
	Thiourea	Determination by hydride generation inductively coupled plasma atomic emission spectrometry after reduction of arsenic (V) to arsenic (III) by thiourea
Barium	Dimethylsulfonazo-III (DMSA-III)	Spec λ = 662 nm of the chelate complex
	Alizarin S red	Red color used spectrophotometrically as a stain to determine the amount of barium in bone
Beryllium	Aluminon	Lake pigment derivative
	Ammonium bifluoride	Derivative detected by fluorescence
	Beryllon II	A resin phase spectrophotometric method that detects a change in absorbance of the resin phase immobilized with beryllon II
	Beryllon III	Determination achieved via third –derivative spectrophotometry and decolorization of excess reagent
	Chrome azurol S	Spec determination with chrome azurol S in the presence of EDTA or CTAB
	Eriochrome cyanine R + (CTAB)	Spec λ = 590 nm of ternary complex
	Sulfon black F	Derivative has a long color-development time
Bismuth	Amberlite XAD-7	Determination system implemented with (HG-ICP-AES) associated with flow injection (FI)
	Sodium azide	Azidodimethylbismuthine precipitate formed by the reaction of the corresponding bismuthine with sodium azide
	Diethyldithiocarbamate	Heterometric micro-determination of lead with sodium diethyldithiocarbamate by use of a mixture of EDTA, cyanide and ammonium hydroxide (λ = 400 nm)
	Dithizone	Orange-red derivative that is extracted in carbon tetrachloride
	Pyrocatechol violet	Spec determination of bismuth (III) with pyrocatechol violet in the presence of septonex CTAB
	Quinolin-8-ol	Determination system implemented with HG-ICP-AES associated with flow injection (FI)
	Thiourea	Determination by HG-ICP-AES
	Xylenol orange	Derivative used for sol-gel thin films that serve as bismuth (III) sensors
Boron	Azomethine H	Spec determination
	Carminic acid	Spec determination
	Curcumin	Spec determination of rosocyanine and rubrocurcumin formed by the reaction between borates and curcumin
	Methylene blue	Spec determination of complex formed between fluoroborate ions and methylene blue after treatment with hydrofluoric and sulfuric acids and extraction with ethylene chloride
Cadmium	2-(5-Bromo-2-pyridylazo)-5-diethylaminophenol (PAR)	Spec determination in the presence of cationic surfactant cetylpyridinium chloride
	Cadion	Determination by β-correction spectrophotometry with cadion and surfactant Triton-X
	Dithizone	Determination of dithizonate derivative by the extraction – spectrophotometric method
	1-(2-Pyridylazo)-2-naphthol (PAN)	Two-dimensional absorption spec determination of complex in aqueous micellar solutions
	4–(2-Pyridylazo)resorcinol	Preconcentration by cloud point extraction of the complex; determination by ICP optic emission spectrometry. Simultaneous spectrophotometric determination of cadmium and mercury
Calcium	Alizarin S	Red dye used in staining bones for calcium determination
	Chlorophosphonazo III	Spec λ = 667.5 nm (pH = 2.2)
	Eriochrome black T + EDTA	Complexometric titration where the initially formed red calcium-eriochrome black T color is replaced with the blue calcium-EDTA color at the end point
	Glyoxal-bis(2-hydroxyanil)	Spectrophotometric titration of the complex without preliminary extraction
	Murexide	Spec λ = 506 nm (pH = 11.3)
	Phthalein purple	High-performance chelation ion chromatography involving dye-coated resins
Cerium	Butaperazine dimaleate propericiazine	Spec determination of the colored complex in a phosphoric acid medium
	Persulfate oxidation to cerium (IV)	UV spec λ = 320 nm
	Propericiazine	Spec determination of the colored complex in a phosphoric acid medium
	Propionyl promazine phosphate (PPP)	Spec λ = 513 nm of the red-colored radical cation formed upon the reaction of PPP with cerium(IV) in a phosphoric acid medium
	N-Benzoyl-N-phenylhydroxylamine	Spec titration of the cerium (IV) complex (pH = 8 to 10)
	Sodium triphosphate	Specf λ_{ex} = 303.5 nm, λ_{em} = 353 nm of the cerium (III) complex
	8-Hydroxyquinoline	Spec determination of the metal-ligand complex
Chromium	Alizarin S	Lake pigment complex formation
	1,5-Diphenylcarbazide	Spec determination during sonication in carbonated aqueous solutions saturated with CCl_4 that produces chlorine radicals

Determination	Reagents	Results
	3-(2-Pyridyl)-5,6 bis(5-(2 furyl disulfonic acid))-1,2,4-triazine disodium salt (ferene-TM)	Indirect spec λ = 593 nm in aqueous samples with a chromogen ferene-TM
	4-(2-Pyridylazo)resorcinol (PAR)	Spec determination of the ternary chromium-peroxo-par ternary complex
	4-(2-Pyridylazo)resorcinol (PAR) + hydrogen peroxide	Spec determination of the ternary chromo-peroxo-PAR mixture after ethyl acetate extraction in 0.1 M sulfuric acid
	4-(2-Pyridylazo)resorcinol (PAR) + xylometazolonium (XMH) chloride	Spec determination of the orange-red anionic complex formed in a heated acetate buffer medium (pH = 4.0 to 5.5) and extracted with the xylometazolonium (XMH) chloride
	Sulfanilic acid	Spec λ = 360 nm on the catalytic effect of chromium (VI) in the oxidation of sulfanilic acid by hydrogen peroxide with p-aminobenzoic acid as an activator
Cobalt	8-Hydroxyquinoline	Spec determination of the metal-ligand complex
	p-Nitroso-N,N-dimethylaniline	Spec determination of the binary complex
	Nitroso-R salt	AAS by use of a continuous on-line precipitation-dissolution procedure
	1-Nitroso-2-naphthol	Spec determination by use of non-ionic surfactant Triton X-100
	1-Nitroso-2-naphthol	Spec Tween 80 micelar determination
	1-Nitroso-2-naphthol	AAS by use of a continuous on-line precipitation-dissolution procedure based on 1-nitroso-2-naphthol
	2-Nitroso-1-naphthol	Spec λ = 530 nm after isoamyl acetate extraction
	1-(2-Pyridylazo)-2-naphthol + surfactants	Spec λ = 620 nm of the cobalt complex in the presence of surfactants (Triton X-100 combined with sodium dodecylbenzene sulfonate (DBS)) and trace of ammonium persulfate (pH = 5.0)
	4-(2-Pyridylazo)resorcinol (PAR)	Spec determination of complex at both pH = 7.2 to 7.9 and in 1 M H_2SO_4
	4-(2-Pyridylazo)resorcinol (PAR) + triethanolamine	Ion-pair reversed-phase high-performance liquid chromatography of the complex
Copper	Bathocuproine disulfonic acid	Spec λ = 470 nm to 550 nm of the Bathocuproine-disulfonic acid complex after extraction with chloroform and methanol by use of (PLS2)
	Dithizone	Spec determination of the dithizone complex at pH = 2.3
	Neocuproine	Spec determination of the deep orange-red Cu(neocuproine)$_2$$^+$ complex color
	Cuprizone	Spec determination of the highly chromogenic copper (III) cuprizone complex
	p-Nitroso-N,N-dimethylaniline	Spec determination of the binary complex
	1-Nitroso-2-naphthol	Spec determination by use of non-ionic surfactant Triton X-100
	4-(2-Pyridylazo)resorcinol	Preconcentration of copper by cloud point extraction of the complex and determination by ICP optic emission spectrometry
Europium	1-Nitroso-2-naphthol	Spec Tween 80 micelar determination
	ChromAsurol S	Spec determination of the binary complex
	4-(2-Pyridylazo)resorcinol (PAR) + tetradecyl-dimethylbenzylammonium chloride (TDBA)	Spec λ = 510 nm of the ion-associate complex extracted with chloroform at pH = 9.7
Gallium	Chrome azurol S + (CTAB)	Spec λ = 640 nm of the ternary complex
	Haematoxylin or its oxidized form + (CTAB)	Spec determination of the ternary complex of indium with haematoxylin or its oxidized form in the presence of cationic, anionic and non-ionic surfactants such as (CTAB)
	Pyrocatechol violet + diphenylguanidine	Spec determination of the ternary complex
	8-hydroxyquinoline	Spec determination of the gallium complex
	1-(2-Pyridylazo)-2-naphthol	Spectrofluorimetric determination of Ga(III) with 1-(2-pyridylazo)-2-naphthol in sodium dodecyl sulfate micellar medium
	4-(2-Pyridylazo)resorcinol (PAR)	Extraction and spec λ = 510 nm of gallium with 4-(2-pyridylazo)resorcinol
	Rhodamine B	Comparison of the determination of gallium by a rhodamine B spectrophotometric method and by an AA method based on preliminary solvent extraction
	Xylenol orange + 8-hydroxyquinoline	Spectrophotometric determination of the ternary complex
Germanium	Brilliant Green + Molybdate	Spec λ = 430 nm of the yellow germanomolybdic acid
	Phenylfluorone	Spec determination of the complex previously extracted with carbon tetrachloride at pH = 3.1
Gold	5-(4-Diethylaminobenzylidene) rhodanine	Immobilized 5-(4-dimethylamino-benzylidene) rhodanine serves as a stable solid sorbent for trace amounts of gold (III) ions (pH = 2 to 4)
	Di(methylheptyl)methyl phosphonate (DMHMP)	Trace gold determination by on-line preconcentration with flow injection atomic absorption spectrometry, by use of di(methylheptyl)methyl phosphonate (DMHMP) as the immobilized phase loaded onto a macroporous resin
	2-Mercaptobenzothiazole	Radiochemical separation and determination of gold complex matrices employing substoichiometric thermal neutron activation analysis
	Molybdate + nile blue (NB)	A spectrophotometric method based on the reaction of gold(III) with molybdate and nile blue (NB) to form an ion-association complex in the presence of poly(vinyl alcohol)
	Rhodamine B	Aqueous spec determination of gold with rhodamine B and surfactant

Determination	Reagents	Results
Hafnium	Arsenazo III	Spec determination of the arsenazo III complex in 10 M HCl or H_2SO_4
Indium	Bromopyrogallol red	Spec determination of indium after an ether extraction from hydrobromic acid, and benzyl alcohol extraction of its complex with bromopyrogallol red (pH 9.0)
	5-Bromine-salicylaldehyde salicyloylhydrazone (5-Br-SASH)	Specf λ_{ex} = 395 nm, λ_{em} = 461 nm of the indium: 5-Br-SASH chelate in a water–ethanol (63%) medium (pH = 4.6)
	Chrome azurol S	Spec determination of the binary complex
	Chrome azurol S + benzyldodecyldimethyl-ammonium bromide (BDDMAB)	Spec determination of the mixed complex with chrome azurol S and (BDDMAB)
	Chrome azurol S + (CTAB)	Spec λ = 630 nm of the ternary complex after n-butyl acetate extraction from hydrobromic acid
	Chrome azurol S + cationic surfactants	Spec determination by use of chrome azurol S and surfactants such as CTA, CP, or zephiramine
	Dithizone	Spec determination of the binary complex
	Haematoxylin or its oxidized form + (CTAB)	Spec determination of the ternary complex of indium with haematoxylin or its oxidized form in the presence of cationic, anionic and non-ionic surfactants such as (CTAB)
	8-Hydroxyquinoline	Spec determination of the binary complex
	Methylthymol blue + zepheramine	Spec determination of the ternary complex
	1-(2-Pyridylazo)-2-naphthol (PAN)	Spec determination of the chelate complex after chloroform extraction (pH = 6)
	4-(2-Pyridylazo)resorcinol	Spec λ = 520 nm after indium extraction from the aqueous phase (pH 5.0 to 5.5) into chloroform with N-p-chlorophenyl-2-furohydroxamic acid and formation of the 4-(2-pyridylazo)resorcinol red chelate
	Pyrocatechol violet	Spec determination of the binary complex
	Pyrocatechol violet + tridodecylammonium bromide	Spec determination of the ternary complex
	Quinalizarin (1,2,5,8-tetrahydroxyanthraquinone)	Spec λ = 565 nm of the binary 3:1 Quinalizarin:In(III) colored complex in dimethylformamide-water solution
	2,2',3,4-Tetrahydroxy-3',5'-disulfoazobenzene	Spec determination of the binary system (pH = 5)
	Xylenol orange	Spec determination of the ternary complex
Iridium	1,5-Diphenylcarbazide	Spec determination of the complex (pH = 5.0)
	1-(2-Pyridylazo)-2-naphthol	Spec λ = 550 nm of the red complex (pH = 5.1) after chloroform extraction
Iron	Bathophenanthroline	Determination of iron(II) in the presence of thousand-to-one ratio of iron(III) by use of bathophenanthroline
	Bathophenantroline-disulfonic acid	Spec λ = 470 nm to 550 nm of the bathophenantroline-disulfonic acid complex after extraction with chloroform and methanol
	2,2'-Bipyridine	Spec determination of the iron (II)-2,2'-bipyridine dark red complex
	FerroZine	Spec λ = 562 nm of iron (II)-ferrozine complex after all iron(III) has been reduced by ascorbic acid
	Hematoxylin + (CTAB)	Spec of the ternary complex. Addition of CTAB shifts λ_{max} from 630 nm to 640 nm
	1-Nitroso-2-naphthol	Spec determination by use of non-ionic surfactant Triton X-100
	1-Nitroso-2-naphthol	Spec determination Tween 80 micelar determination
	1,10-Phenanthroline (o-Phen)	Spec λ = 508 nm of the Fe(o-Phen)$_3$$^{+2}$ complex
	1,10-Phenanthroline + bromothymol blue	Spec determination of the Fe(o-Phen)$_3$$^{+2}$ complex in the presence of bromothymol blue
	Phenylfluorone	Spec λ = 530 nm of the binary complex (pH = 9.0)
	Phenylfluorone + Triton X	Spec λ = 555 nm of the binary complex sensitized with Triton X-100 (pH = 9.0)
Lanthanum	Ammonium purpurate	Spec determination of lanthanum with ammonium purpurate as a chromogenic reagent
	Arsenazo III	Spec determination of the lanthanum complexation with reagents of the arsenazo III group on the solid phase of fibrous ion exchangers
	Eriochrome cyanine R (ECR)	Spec λ = 540 nm of the binary complex
	N-Phenylbenzohydroxamic acid + xylenol orange	Solvent extraction followed by the spec λ = 600 nm of the ternary complex (pH = 8.8 to 9.5)
Lead	Dithizone	Orange-red derivative that is extracted in carbon tetrachloride
	Sodium diethyldithiocarbamate	Electrothermal AAS determination of the diethyldithiocarbamate derivative extracted in carbon tetrachloride
	Sodium diethyldithiocarbamate	Heterometric micro-determination of lead diethyldithiocarbamate derivative
	4-(2-Pyridylazo)resorcinol	Spec λ = 520 nm of 4-(2-pyridylazo)-resorcinol:lead (1:1) complex in an ammonia-ammonium chloride medium at pH = 10 after extracting the lead in isobutyl methyl ketone
Lithium	1-(o-Arsenophenylazo)-2-naphthol-3, 6-disulfonate (Thoron)	Spec λ = 480 nm of Thoron-lithium complex in an alkaline acetone medium is measured against the reagent as reference
Magnesium	Chlorophosphonazo III	Spec λ = 669 nm (pH = 7.0)

Determination	Reagents	Results
	Eriochrome black T + EDTA	Complexometric titration where the initially formed red magnesium-eriochrome black T color is replaced with the blue magnesium-EDTA color at the end point
	8-Hydroxyquinoline	Volumetric, titrimetric, and colorimetric determination
	8-Hydroxyquinoline + butylamine	Spec λ = 380 nm of the complex after chloroform extraction
	Titan yellow	Spec determination of magnesium by titan yellow in biological fluids
	Xylidyl blue	Spec determination of magnesium in biological fluids
Manganese	Formaldoxime	Spec determination of the formaldoxime/ammonia complex (pH = 8.8 to 8.9)
	1-(2-Pyridylazo)-2-naphthol (PAN)	Preconcentration determination by use of (PAN) anchored SiO_2 nanoparticles
Mercury	Dithizone	Spec λ = 500 nm of dithizone complex after chloroform extraction at pH = 0.3
	Michler's thioketone	Spec λ = 560 nm of the binary complex in acetate buffer
	4–(2-Pyridylazo)resorcinol	Simultaneous spec determination of cadmium and mercury with 4-(2-pyridylazo)resorcinol
	Rhodamine 6G	Photoelectrochemical determination of mercury (II) in aqueous solutions by use of a rhodamine 6G derivative (RS) and polyaniline (PANI) coated optical probe in a photoelectrochemical cell
	Xylenol orange + amine buffer	Spec λ = 590 nm of Hg(II)/xylenol orange complex display a sharp hyperchromic effect in the presence of amine buffers (pH= 7.5)
	Xylenol orange + citric acid-phosphate buffer	Spec λ = 580 nm of Hg(II)/xylenol orange complex display a sharp hypochromic effect upon substituting amine buffers with a citric acid-phosphate (pH = 7.5)
Molybdenum	Bromopyrogallol red + cetylpyridium chloride	Sequential injection analysis to the determination of CPC based on the sensitized molybdenum-bromopyrogallol red reaction
	Phenylfluorone	Spec λ = 560 nm of molybdenum with phenylfluorone (pH of 1.5 to 3)
	8-Hydroxyquinoline-5-sulfonic acid and phenylfluorone	Diffuse reflection spectrometry with 8-hydroxyquinoline-5-sulfonic acid and phenylfluorone after sorbing on a disk of an anion exchange fibrous material
	Pyrocatechol violet	Preconcentration (by use of basic anion exchanger AV-17-10P) and determination via diffuse reflection spectroscopy. The colored surface compound to be determined involves Mo(VI) sorption on the resin and subsequent treatment of the concentrate obtained with pyrocatechol violet
Neodymium	Semi-xylenol orange + cetylpyridinium chloride	Fourth-order derivative spectrophotometric determination of the ternary complex
Neptunium	Arsenazo III	Spec after separation by use of thenoyltrifluoroacetone extraction method and determination in 5M HNO_3
Nickel	2-(5-Bromo-2-pyridylazo)-5-diethylaminophenol	Spec λ = 520 nm and 560 nm of the red-violet complex in water-ethanol (pH = 5.5)
	Dimethylglyoxime + ammonia	Spec λ = 543 nm of the complex
	Dimethylglyoxime, voltammetry	Nickel voltammetric determination at a chemically modified electrode based on dimethylglyoxime-containing carbon paste
	2,2′-Furildioxime	Spec λ = 438 nm after separation by adsorption of its α-furildioxime complex on naphthalene
	Hematoxylin	Spec λ = 595 nm of the binary system (pH = 7.8 to 8.3)
	Hematoxylin + (CTAB)	Spec λ = 608 nm of the ternary system (pH = 7.4 to 8.1)
	1-Nitroso-2-naphthol	Spec Tween 80 micelar determination
	p-Nitroso-N,N-dimethylaniline	Spec determination of the binary complex
	2-(2-Pyridylazo)-2-naphthol	Derivative spectrophotometry λ= 569 nm determination of nickel complex in Tween 80 micellar solutions
	4-(2-Pyridylazo)resorcinol	Preconcentration of nickel by cloud point extraction of the complex and determination by ICP optic emission spectrometry
	1-(2-Thiazolylazo)-2-naphthol (TAN)	Flow-injection solid phase spectrophotometry by use of TAN immobilized on C_{18}-bonded silica (λ = 595 nm)
	Xylenol orange	Spec determination by mean centering of ratio kinetic profile (pH = 5.3)
Niobium	N-Benzoyl-N-phenylhydroxylamine	Separation and determination of the complex from a tartrate solution at pH > 2
	Bromopyrogallol red	Spec λ = 610 nm of complex extracted into isopentyl acetate containing di-n-octylmethylamine
	O-Hydroxyhydroquinonephthalein (Qnph) + hexadecyltrimethylammonium chloride (HTAC)	Spec λ = 520 nm of the complex in strong acidic media
	4-(2-Pyridylazo)resorcinol + citrate	Determination of the ternary complex by ion-interaction reversed-phase HPLC on a C18 column with a 5 mM citrate buffer (pH = 6.5, λ = 540 nm)
	Pyrocatechol violet	Extraction and spec determination of niobium complex in the presence of pyridine and trichloroacetic acid
	Sulfochlorophenol S	Spec determination of the complex after extraction into amyl alcohol
	Xylenol orange	Spec λ = 530 nm of chelate (pH = 5.0)
Osmium	1,5-Diphenylcarbazide	Spec λ = 560 nm of the bluish-violet complex after extraction with isobutyl methyl ketone
	Thiourea	Spec λ = 540 nm of the red-rose complex in acid medium
Palladium	Ammonia + iodide	Thermogravimetric determination of palladium as $Pd(NH_3)_2I_2$

Determination	Reagents	Results
	2,2-*bis*-[3-(2-Thiazolylazo)-4-4-hydroxyphenyl-propane], (TAPHP)	Application of TAPHP immobilized on silica beads to determine the palladium concentration in a trans-luminance configuration (pH = 2) by use of flow-through spectrophotometric sensing phase
	2-(5-Bromo-2-pyridylazo)-5-diethylaminophenol	Spec determination of the complex in a sulfuric acid medium in the presence of ethanol
	Dithizone	Volumetric determination of palladium with dithizone in acid medium after extraction of its dimethylglyoxime complex with chloroform; PAS determination of the dithizone extraction solution into a thermally thin solid film
	Dithiazone + iodide	Spec determination of palladium with dithizone, by use of an iodide medium in the presence of sulfite at pH = 3 to 5, to separate platinum
	Dithizone + stannous chloride	Extractive separation and spec determination of palladium in the presence of stannous chloride to separate platinum
	Isonitrosobenzoylacetone	Radiochemical separation and determination of palladium in complex matrices employing substoichiometric thermal NAA
	2-Nitroso-1-naphthol	Spec λ = 370 nm of the violet complex
	4-(2-Pyridylazo)resorcinol	Spec determination of complex at both pH = 7.2 to 7.9 and in 1 M H$_2$SO$_4$
	4-(2-Pyridylazo)resorcinol + diphenylguanidine	Spec determination after extraction of the red Pd(II) chelate with 4-(2-pyridylazo) resorcinol in the presence of *N,N*′-diphenylguanidine into *n*-butanol
Platinum	*N*-Phenylbenzimidoylthiourea (PBITU)	Flow injection analysis spec λ = 345 nm of the Pd:PBITU complex in (0.2 to 2.0) M HCl in10% (vol/vol) ethanol solution
	Dithizone + iodide	Spec determination of platinum with Dithizone, by use of an iodide medium in the presence of sulfite at pH = 3 to 5, to separate palladium
	Dithizone + stannous chloride	Extractive separation and spectrophotometric determination of platinum in the presence of stannous chloride to separate palladium
	2-Mercaptobenzothiazole	Radiochemical separation and determination of platinum complex matrices employing substoichiometric thermal NAA
Protactinium	Arsenazo III	Spec λ = 680 nm of the complex after extraction with 7 N H$_2$SO$_4$ and isoamyl alcohol
	EDTA + tannic acid	Gravimetric analysis of the tannic acid precipitate (pH = 5.0)
Rhenium	2,2′-Furildioxime	Spec λ = 532 nm of the complex
	N,N-Diethyl-*N*′-benzoylthiourea	Spec λ = 383 nm of the green complex in hydrochloric acid medium in the presence of tin(II) chloride
	Tin(II)	Spectrophotometric titration for the determination of rhenium by use of tin(II) as the titrant
Rhodium	1,5-Diphenylcarbazide	Spec determination of the complex after isobutyl alcohol extraction (pH = 5.0)
	p-Nitrosodimethylaniline	Spec λ = 510 nm of the cherry-red binary complex (pH = 4.4)
	1-(2-Pyridylazo)-2-naphthol	Spec λ = 598 nm of the green complex (pH = 5.1) after chloroform extraction
Ruthenium	4-Benzylideneamino-3-mercapto-6-methyl-1,2,4-triazine (4H)-5-one	Spec λ = 620 nm
	1,10-Phenanthroline	Chemiluminescence determination of chlorpheniramine by use of tris(1,10-phenanthroline)-ruthenium(II) peroxydisulfate system and sequential injection analysis; Spec determination of the complex after reducing Ru (IV) to Ru(II)
	Thiourea	Spec λ = 640 nm of ruthenium complex in carbon supported Pt-Ru-Ge catalyst in 5 M HCl
	1,4-Diphenylthiosemicarbazide	Spec determination of the bright red complex
Scandium	Alizarin red S	Cathodic adsorptive stripping of the scandium-alizarin red S complex onto a carbon paste electrode
	Ammonium purpurate	Spec determination of scandium with ammonium purpurate as a chromogenic reagent
	Chrome azurol S	Spec determination of the binary complex
	Xylenol orange	Spec λ = 553 nm of the binary mixture
Selenium	3,3′-Diaminobenzidine	Spec λ = 350 nm of the yellow complex (pH = 3.0)
	2,3-Diaminonaphthaline	Fluorometric determination of the binary complex in water
	Thiourea	Determination of selenium by hydride generation inductively coupled plasma atomic emission spectrometry
Silver	Dithizone	Graphite-furnace atomic-absorption spectrophotometric determination of silver on suspended dithizone particles from acidic sample solutions with ultrasonics to facilitate the separation
	Eosin + 1,10-phenanthroline	Spec λ = 540 nm to 555 nm of the 1,10-phenanthroline (PHEN) and eosin (2,4,5,7-tetrabromofluorescein) association complexes
Strontium	Phthalein purple	High-performance chelation ion chromatography involving dye-coated resins
Tantalum	*N*-Benzoyl-*N*-phenylhydroxylamine	Separation and determination of the complex from a tartrate solution at pH < 1.5
	Crystal violet (CV) + *N,N*′-diphenylbenzamidine (DPBA)	Spec λ = 600 nm of Ta(V)-F-CV+ cation complex with a benzene solution of DPBA from sulfuric acid solution

Determination	Reagents	Results
	O-Hydroxyhydroquinonephthalein (Qnph) + hexadecyltrimethylammonium chloride (HTAC)	Spec λ = 510 nm of the complex in strong acidic media
	Malachite green	Spec λ = 623 nm of the complex in HF after extraction with benzene or toluene
	Methyl violet	Spec determination of the complex
	4-(2-Pyridylazo)resorcinol + citrate	Determination of the ternary complex by ion-interaction reversed-phase HPLC on a C18 column in 5 mM citrate buffer (pH = 6.5, λ = 540 nm)
	Phenylfluorone	Spectrophotometric determination from HF-HCl solution with methyl isobutyl ketone
	Victoria blue	Spectrophotometric determination of the complex after benzene extraction
Tellurium	Ammonium pyrrolidinedithiocarbamate	Differential determination of tellurium(IV) and tellurium(VI) by AAS with a carbon-tube atomizer
	Bismuthiol II	Spec λ = 330 mm of the yellow complex in acidic medium (pH = 3.5) after chloroform extraction
	Dithizone	Differential determination of tellurium(IV) and tellurium(VI) by AAS with a carbon-tube atomizer
	Sodium diethyldithiocarbamate	Differential determination of tellurium(IV) and tellurium(VI) by AAS with a carbon-tube atomizer
	Thiourea	Determination of selenium by hydride generation inductively coupled plasma atomic emission spectrometry
Thallium	Brilliant green	Spec determination of the binary complex after toluene extraction
	Dithizone	Spec λ = 505 mm of the complex after chloroform extraction from a citrate-sulfite-cyanide medium at pH = 10.6
	8-Hydroxyquinoline	Spectrophotometric determination of the metal-ligand complex
	Rhodamine B	Fluorimetric determination of thallium (in silicate rocks) with rhodamine B after separation by adsorption on a crown ether polymer
	Rhodamine B hydroxide	Spec λ = 565 nm via oxidation of rhodamine B hydrazide by thallium(I) in acidic medium to give a pinkish violet radical cation
Thorium	Arsenazo III	Spec determination of the arsenazo III complex in 10 M HCl or H_2SO_4
	Eriochrome cyanine R (ECR)	Spec λ = 540 nm of the binary complex
	Thoron	Spectrophotometric determination of binary complex in tartaric acid
	Xylenol orange	Spec λ = 570 nm of the binary complex (pH = 4.0)
	Xylenol orange + CTAB bromide	Spec λ = 600 nm of the complex sensitized by CTMAB (pH = 2.5)
Tin	Catechol violet + CTAB bromide	Spectrophotometric determination of the green complex
	Pyrocatechol violet (and + CTAB bromide)	Spec λ = 660 nm of the complex (pH = 2.0)
	Gallein	Spectrophotometric determination of complex in acid medium
	Phenylfluorone	Spec determination of the complex (pH = 1) in 36% aqueous ethanol after a preliminary solvent extraction of the tin as the iodide
	Toluene-3,4-dithiol + dispersant	Spec determination of the complex
Titanium	Chromotropic acid	Spec λ = 443 nm of the complex by use of a flow injection manifold
	Diantipyrinylmethane	Spec λ = 390 nm of the yellow complex
	3-Hydroxy-2-methyl-1-(4-tolyl)-4-pyridone (HY)	Spec λ = 355 nm of the ternary complex formed in perchloric acid and is extracted by chloroform
	Methylene blue-ascorbic acid redox reaction	Spec λ = 665 nm of the titanium-methylene blue-ascorbic acid redox reaction
	Tiron	Spec λ = 420 nm of the yellow Tiron derivative (pH = 5.2 to 5.6)
Tungsten	Cyanate	Spectrophotometric determination of tungsten with thiocyanate after both tungsten(VI) and molybdenum(VI) are extracted into chloroform as benzoin α-oxime complexes
	Pyrocatechol violet	Determination of the 2:1 green complex in acid medium which is fixed on a dextran-type anion-exchange resin (Sephadex QAEA-25) by first-derivative solid-phase spectrophotometry (λ = 674 nm)
	Tetraphenylarsonium chloride + thiocyanate	Gravimetric determination of tungsten with tetraphenylarsonium chloride after its extraction as thiocyanate (pH = 2 to 4)
	Toluene-3,5-dithiol	Spec λ = 630 nm of the toluene-3,4-dithiol derivative after extraction with isoamyl acetate
Uranium	Arsenazo III (1,8-dihydroxynaphthalene-3,6-disulfonic acid-2,7-bis[(azo-2)-phenylarsonic acid])	Spec determination of the binary complex
	2-(5-Bromo-2-pyridylazo)diethylaminophenol	Spec λ = 578 nm of the complex (pH = 7.6)
	Chlorophosphonazo III	Spec λ = 673 nm of the complex after extraction into 3-methyl-1-butanol from (1.5 to 3.0) M HCl
	8-Hydroxyquinoline	Spectrophotometric determination of the metal (UO_2 (II))-ligand complex
	2-(2-Pyridylazo)-5-diethylaminophenol (PADAP)	Spec λ = 564 nm of the complex after extraction into methyl isobutyl ketone (pH = 8.2)

Determination	Reagents	Results
	1-(2-Pyridylazo)-2-naphthol (PAN)	Spec λ = 560 nm of the deep red precipitate in ammoniacal solutions extracted with chloroform
	4-(2-Pyridylazo)-resorsinol (PAR)	Spec λ = 530 nm of the intensely deep red complex after extraction into methyl isobutyl ketone (pH = 8.0)
	2-(2-Thiazolylazo)-p-cresol + surfactants	Spec λ = 588 nm of the complex (pH = 6.5) with surfactants such as Triton X-100 or N-cetyl-N,N,N-trimethyl ammonium bromide (CTAB)
Vanadium	Chrome azurol S	Spectrophotometric determination of the binary complex
	N-Benzoyl-N-phenylhydroxylamine	Spec determination of the violet complex in acidic medium
	3,5-Dinitrocatechol (DNC) + brilliant green	Extraction-spectrophotometric determination of the system V(V)-3,5-dinitrocatechol (DNC)-brilliant green chelate complex
	8-Hydroxyquinoline (oxine)	Extraction with n-butanol and spec λ = 390 nm of a ternary complex (vanadium:oxine:n-butanol = 1:2:2)
	8-Hydroxyquinoline-5-sulfonic acid and phenylfluorone	Diffuse reflection spectrometry with 8-hydroxyquinoline-5-sulfonic acid and phenylfluorone after sorbing on a disk of an anion exchange fibrous material
	4-(2-Pyridylazo)resorcinol (PAR)	Spec determination of the binary complex based on the extraction tetraphenylphosphonium or tetraphenylarsonium chloride
	Xylenol orange	Spec λ = 490 nm of chelate (pH = 5.0)
Yttrium	Alizarin red S	Spec λ = 550 nm of the complex
	Ammonium purpurate	Spec determination of yttrium with ammonium purpurate as a chromogenic reagent
	Arsenazo III	Spec λ = 660 nm of the blue colored complex after preliminary purification with hydroxide and subsequent acidification
	Eriochrome cyanine R (ECR)	Spec λ = 540 nm of the binary complex
	Pyrocatechol violet	Spec λ = 665 nm of complex
Zinc	2-(5-Bromo-2-pyridylazo)-5-diethylaminophenol	Spec determination by use of 2-(5-bromo-2-pyridylazo)-5-diethyl aminophenol in the presence of cationic surfactant cetylpyridinium chloride
	Carbonic anhydrase	Enzymatic determination by use of carbonic anhydrase after removing zinc by dialysis against dipicolinic acid
	Dithizone	Spec λ = 530 nm of binary complex after chloroform extraction
	Eriochrome black T + EDTA	Complexometric titration where the initially formed red zinc-eriochrome black T color is replaced with the blue zinc-EDTA color at the end point (pH = 10)
	7-(4-Nitrophenylazo)-8-hydroxyquinoline-5-sulfonic acid (p-NIAZOXS)	Spec λ = 520 nm of the zinc derivative (pH = 9.2, borax buffer)
	Phenylglyoxal mono(2-pyridyl) hydrazone (PGMPH)	Spec λ = 464 nm to 470 nm of the yellow-orange complex (pH = 7.2 to 8.5) in 40% (v/v) ethanol
	1-(2-Pyridylazo)-2-naphthol (PAN)	Two-dimensional absorption spec determination of complex in aqueous micellar solutions
	1-(2-Pyridylazo)-2-naphthol (PAN)	Preconcentration determination by use of 1-(2-pyridylazo)-2-naphthol (PAN) anchored. SiO_2 nanoparticles
	4-(2-Pyridylazo)resorcinol	Spec determination at pH = 7.0; preconcentration of zinc by cloud point extraction of the complex and determination by ICP optic emission spectrometry
	1-(2-Thiazolylazo)-2-naphthol (TAN)	Flow-injection solid phase spec λ = 595 nm with TAN immobilized on C18-bonded silica
	Xylenol orange	Spec determination of 1:1 zinc-xylenol orange red-violet complex (pH= 5.8 to 6.2)
	Xylenol orange	Sequential injection analysis (SIA) based on the spec λ = 568 nm of zinc by use of xylenol orange as a color reagent
	Xylenol orange	Spec determination by mean centering of ratio kinetic profile (pH = 5.3)
	Xylenol orange + cetylpyridinium chloride	Spec λ = 580 nm of the 1:2:4 ratio for the metal:ligand: surfactant ternary complex (pH = 5.0 to 6.0)
Zirconium	Alizarin red S	Spec determination of the binary complex after phosphate extraction
	Arsenazo III (1,8-dihydroxynaphthalene-3,6-disulfonic acid-2,7-bis[(azo-2)-phenylarsonic acid])	Spec determination of the binary complex
	N-p-Chlorophenylbenzohydroxamic acid (N-p-Cl-BHA) + morin	Spec λ = 420 nm and fluorimetric determination of the N-p-Cl-BHA greenish yellow complex after extraction with isoamyl alcohol from (0.2 to 0.5) N H_2SO_4 followed by morin addition; it is critical that the morin be very pure
	Pyrocatechol violet	Spec λ = 650 nm of the blue complex in sulfuric acid solution
	Pyrocatechol violet + tri-n-octylphosphine oxide (TOPO)	Spec λ = 655 nm of the blue complex after extraction with tri-n-octylphosphine oxide (TOPO) in cyclohexane
	Morin	Fluorimetric determination of the complex after EDTA addition
	Xylenol orange	Flow injection spectrophotometric determination in sulfuric acid medium

Section 9
Molecular Structure and Spectroscopy

BOND LENGTHS IN CRYSTALLINE ORGANIC COMPOUNDS

The following table gives average interatomic distances for bonds between the elements H, B, C, N, O, F, Si, P, S, Cl, As, Se, Br, Te, and I as determined from X-ray and neutron diffraction measurements on organic crystals. The table has been derived from an analysis of high-precision structure data on about 10,000 crystals contained in the 1985 version of the Cambridge Structural Database, which is maintained by the Cambridge Crystallographic Data Center. The explanation of the columns is:

Column 1: Specification of elements in the bond, with coordination number given in parentheses, and bond type (single, double, etc.). For carbon, the hybridization state is given.

Column 2: Substructure in which the bond is found. The target bond is set in boldface. Where X is not specified, it denotes any element type. C# indicates any sp^3 carbon atom, and C* denotes an sp^3 carbon whose bonds, in addition to those specified in the linear formulation, are to C and H atoms only.

Column 3: d is the unweighted mean in Å units of all the values for that bond length found in the sample.

Column 4: m is the median in Å units of all values.
Column 5: σ is the standard deviation in the sample.
Column 6: q_1 is the lower quartile for the sample (i.e., 25% of values are less than q_1 and 75% exceed it).
Column 7: q_u is the upper quartile for the sample.
Column 8: n is number of observations in the sample.
Column 9: Notes refer to the footnotes in Appendix 1.

References to special cases are given in a shorthand form and listed in Appendix 2. Further information on the method of analysis of the data may be found in the reference cited below.

The table is reprinted with permission of the authors, the Royal Society of Chemistry, and the International Union of Crystallography.

Reference

Frank H. Allen, Olga Kennard, David G. Watson, Lee Brammer, A. Guy Orpen, and Robin Taylor, *J. Chem. Soc. Perkin Trans.* II, S1–S19, 1987.

Bond	Substructure	d	m	σ	q_1	q_u	n	Note
As(3)–As(3)	X₂–**As–As**–X₂	2.459	2.457	0.011	2.456	2.466	8	
As–B	see CUDLOC (2.065), CUDLUI (2.041)							
As–BR	see CODDEE, CODDII (2.346–3.203)							
As(4)–C	X₃–**As–CH₃**	1.903	1.907	0.016	1.893	1.916	12	
	(X)₂(C,O,S=)**As–C**sp^3	1.927	1.929	0.017	1.921	1.937	16	
	As–Car in Ph₄As⁺	1.905	1.909	0.012	1.897	1.912	108	
	(X)₂(C,O,S=)**As–C**ar	1.922	1.927	0.016	1.908	1.934	36	
As(3)–C	X₂–**As–C**sp^3	1.963	1.965	0.017	1.948	1.978	6	
	X₂–**As–C**ar	1.956	1.956	0.015	1.944	1.964	41	
As(3)–Cl	X₂–**As–Cl**	2.268	2.256	0.039	2.247	2.281	10	
As(6)–F	in **AsF₆⁻**	1.678	1.676	0.020	1.659	1.695	36	
As(3)–I	see OPIMAS (2.579, 2.590)							
As(3)–N(3)	X₂–**As–N**–X₂	1.858	1.858	0.029	1.839	1.873	19	
As(4)–N(2)	see TPASSN (1.837)							
As(4)–O	(X)₂(O=)**As–OH**	1.710	1.712	0.017	1.695	1.726	6	
As(3)–O	see ASAZOC, PHASOC01 (1.787–1.845)							
As(4)=O	X₃–**As=O**	1.661	1.661	0.016	1.652	1.667	9	
As(3)–P(3)	see BELNIP (2.350, 2.362)							†
As(3)–P(3)	see BUTHAZ10 (2.124)							†
As(3)–S	X₂–**As–S**	2.275	2.266	0.032	2.247	2.298	14	
As(4)=S	X₃–**As=S**	2.083	2.082	0.004	2.080	2.086	9	
As(3)–Se(2)	see COSDIX, ESEARS (2.355–2.401)							†
As(3)–Si(4)	see BICGEZ, MESIAD (2.351–2.365)							†
As(3)–Te(2)	see ETEARS (2.571, 2.576)							†
B(n)–B(n)	n = 5–7 in boron cages	1.775	1.773	0.031	1.763	1.786	688	
B(4)–B(4)	see CETTAW (2.041)							
B(4)–B(3)	see COFVOI (1.698)							
B(3)–B(3)	X₂–**B–B**–X₂	1.701	1.700	0.014	1.691	1.712	8	
B(6)–BR		1.967	1.971	0.014	1.954	1.979	7	†
B(4)–BR		2.017	2.008	0.031	1.990	2.044	15	†
B(n)–C	n = 5–7: **B–C** in cages	1.716	1.717	0.020	1.707	1.728	96	
	n = 3–4: **B–C**sp^3 not cages	1.597	1.599	0.022	1.585	1.611	29	1
	n = 4: **B–C**ar	1.606	1.607	0.012	1.596	1.615	41	
	n = 4: **B–C**ar in Ph₄B⁻	1.643	1.643	0.006	1.641	1.645	16	
B(n)–C	n = 3: **B–C**ar	1.556	1.552	0.015	1.546	1.566	24	
B(n)–Cl	**B(5)–Cl** and **B(3)–Cl**	1.751	1.751	0.011	1.743	1.761	14	

Bond	Substructure	d	m	σ	q_1	q_u	n	Note
	B(4)−**Cl**	1.833	1.833	0.013	1.821	1.843	22	
B(4)−F	**B−F** (B neutral)	1.366	1.368	0.017	1.356	1.375	25	
	B⁻**−F** in BF₄⁻	1.365	1.372	0.029	1.352	1.390	84	
B(4)−I	see TMPBTI (2.220, 2.253)							
B(4)−N(3)	X₃−**B−N**(=C)(X)	1.611	1.617	0.013	1.601	1.625	8	
	in pyrazaboles	1.549	1.552	0.015	1.536	1.560	10	
B(3)−N(3)	X₂−**B−N**−C₂: all coplanar	1.404	1.404	0.014	1.389	1.408	40	2
	for τ(BN) > 30° see BOGSUL, BUSHAY, CILRUK							
	(1.434–1.530)							
	S₂−**B−N**−X₂	1.447	1.443	0.013	1.435	1.470	14	
B(4)−O	**B**⁻**−O** in BO₄⁻	1.468	1.468	0.022	1.453	1.479	24	
	for neutral B−O see Note 3							3
B(3)−O(2)	X₂−**B−O**−X	1.367	1.367	0.024	1.349	1.382	35	
B(n)−P	n = 4: **B−P**	1.922	1.927	0.027	1.900	1.954	10	
	n = 3: see BUPSIB10 (1.892, 1.893)							
B(4)−S	**B**(4)−**S**(3)	1.930	1.927	0.009	1.925	1.934	10	
	B(4)−**S**(2)	1.896	1.896	0.004	1.893	1.899	6	
B(3)−S	N−**B−S₂**	1.806	1.806	0.010	1.799	1.816	28	
	(=X−)(N−)**B−S**	1.851	1.854	0.013	1.842	1.859	10	
Br−Br	see BEPZEB, TPASTB	2.542	2.548	0.015	2.526	2.551	4	
Br−C	**Br−C**˙	1.966	1.967	0.029	1.951	1.983	100	4
	Br−Csp^3 (cyclopropane)	1.910	1.910	0.010	1.900	1.914	8	
	Br−Csp^2	1.883	1.881	0.015	1.874	1.894	31	4
	Br−Car (mono-Br + $m.p$-Br₂)	1.899	1.899	0.012	1.892	1.906	119	4
	Br−Car (o-Br₂)	1.875	1.872	0.011	1.864	1.884	8	4
⁻Br(2)−Cl	see TEACBR (2.362−2.402)							†
Br−I	see DTHIBR10 (2.646), TPHOSI (2.695)							
Br−N	see NBBZAM (1.843)							
Br−O	see CIYFOF	1.581	1.581	0.007	1.574	1.587	4	
Br−P	see CISTED (2.366)							
Br−S(2)	see BEMLIO (2.206)							†
Br−S(3)	see CIWYIQ (2.435, 2.453)							†
Br−S(3)⁺	see THINBR (2.321)							†
Br−SE	see CIFZUM (2.508, 2.619)							
Br−Si	see BIZJAV (2.284)							
Br−Te	In **Br₆Te**²⁻ see CUGBAH (2.692−2.716)							
	Br−Te(4) see BETUTE10 (3.079, 3.015)							
	Br−Te(3) see BTUPTE (2.835)							
Csp^3−**C**sp^3	C#−**CH₂−CH₃**	1.513	1.514	0.014	1.507	1.523	192	
	(C#)₂−**CH−CH₃**	1.524	1.526	0.015	1.518	1.534	226	
	(C#)₃−**C−CH₃**	1.534	1.534	0.011	1.527	1.541	825	
	C#−**CH₂−CH₂**−C#	1.524	1.524	0.014	1.516	1.532	2459	
	(C#)₂−**CH−CH₂**−C#	1.531	1.531	0.012	1.524	1.538	1217	
	(C#)₃−**C−CH₂**−C#	1.538	1.539	0.010	1.533	1.544	330	
	(C#)₂−**CH−CH**−(C#)₂	1.542	1.542	0.011	1.536	1.549	321	
	(C#)₃−**C−CH**−(C#)₂	1.556	1.556	0.011	1.549	1.562	215	
	(C#)₃−**C−C**−(C#)₃	1.588	1.580	0.025	1.566	1.610	21	
	C˙−**C**˙ (overall)	1.530	1.530	0.015	1.521	1.539	5777	5,6
	in cyclopropane (any subst.)	1.510	1.509	0.026	1.497	1.523	888	7
	in cyclobutane (any subst.)	1.554	1.553	0.021	1.540	1.567	679	8
	in cyclopentane (C,H-subst.)	1.543	1.543	0.018	1.532	1.554	1641	
	in cyclohexane (C,H-subst.)	1.535	1.535	0.016	1.525	1.545	2814	
	cyclopropyl-**C**˙ (exocyclic)	1.518	1.518	0.019	1.505	1.531	366	7
	cyclobutyl-**C**˙ (exocyclic)	1.529	1.529	0.016	1.519	1.539	376	8
	cyclopentyl-**C**˙ (exocyclic)	1.540	1.541	0.017	1.527	1.549	956	
	cyclohexyl-**C**˙ (exocyclic)	1.539	1.538	0.016	1.529	1.549	2682	
	in cyclobutene (any subst.)	1.573	1.574	0.017	1.566	1.586	25	8
	in cyclopentene (C,H-subst.)	1.541	1.539	0.015	1.532	1.549	208	
	in cyclohexene (C,H-subst.)	1.541	1.541	0.020	1.528	1.554	586	
	in oxirane (epoxide)	1.466	1.466	0.015	1.458	1.474	249	9
	in aziridine	1.480	1.481	0.021	1.465	1.496	67	9

Bond	Substructure	d	m	σ	q_1	q_u	n	Note
	in oxetane	1.541	1.541	0.019	1.527	1.557	16	
	in azetidine	1.548	1.543	0.018	1.536	1.558	22	
	oxiranyl-C* (exocyclic)	1.509	1.507	0.018	1.497	1.519	333	9
	aziridinyl-C* (exocyclic)	1.512	1.512	0.018	1.496	1.526	13	9
Csp^3-Csp^2	**CH₃–C**=C	1.503	1.504	0.011	1.497	1.509	215	
	C#–**CH₂–C**=C	1.502	1.502	0.013	1.494	1.510	483	
	(C#)₂–**CH–C**=C	1.510	1.510	0.014	1.501	1.518	564	
	(C#)₃–**C–C**=C	1.522	1.522	0.016	1.511	1.533	193	
Csp^3-Csp^2	**C*–C**=C (overall)	1.507	1.507	0.015	1.499	1.517	1456	5
	C*–C=C (endocyclic)							
	in cyclopropene	1.509	1.508	0.016	1.500	1.516	20	10
	in cyclobutene	1.513	1.512	0.018	1.500	1.525	50	8
	in cyclopentene	1.512	1.512	0.014	1.502	1.521	208	
	in cyclohexene	1.506	1.505	0.016	1.495	1.516	391	
	in cyclopentadiene	1.502	1.503	0.019	1.490	1.515	18	
	in cyclohexa-1,3-diene	1.504	1.504	0.017	1.491	1.517	56	
	C*–C=C (exocyclic):							
	cyclopropenyl-C*	1.478	1.475	0.012	1.470	1.485	7	10
	cyclobutenyl-C*	1.489	1.483	0.015	1.479	1.496	11	8
	cyclopentenyl-C*	1.504	1.506	0.012	1.495	1.512	115	
	cyclohexenyl-C*	1.511	1.511	0.013	1.502	1.519	292	
	C*CH=O in aldehydes	1.510	1.510	0.008	1.501	1.518	7	
	(**C***)₂–**C**=O							
	in ketones	1.511	1.511	0.015	1.501	1.521	952	11
	in cyclobutanone	1.529	1.530	0.016	1.514	1.545	18	
	in cyclopentanone	1.514	1.514	0.016	1.505	1.523	312	
	acyclic and 6 + rings	1.509	1.509	0.016	1.499	1.519	626	
	C*–COOH in carboxylic acids	1.502	1.502	0.014	1.495	1.510	176	
	C*–COO⁻ in carboxylate anions	1.520	1.521	0.011	1.516	1.528	57	
	C*–C(=O)(–OC*)							
	in acyclic esters	1.497	1.496	0.018	1.484	1.509	553	12
	in β-lactones	1.519	1.519	0.020	1.500	1.538	4	13
	in γ-lactones	1.512	1.512	0.015	1.501	1.521	110	12
	in δ-lactones	1.504	1.502	0.013	1.495	1.517	27	12
	cyclopropyl (**C**)–**C**=O in ketones, acids and esters	1.486	1.485	0.018	1.474	1.497	105	7
	C*–C(=O)(–NH₂) in acyclic amides	1.514	1.512	0.016	1.506	1.526	32	14
	C*–C(=O)(–NHC*) in acyclic amides	1.506	1.505	0.012	1.498	1.515	78	14
	C*–C(=O)[–N(C*)₂] in acyclic amides	1.505	1.505	0.011	1.496	1.517	15	14
Csp^3-Car	**CH₃–C**ar	1.506	1.507	0.011	1.501	1.513	454	
	C#–**CH₂–C**ar	1.510	1.510	0.009	1.505	1.516	674	
	(C#)₂–**CH–C**ar	1.515	1.515	0.011	1.508	1.522	363	
	(C#)₃–**C–C**ar	1.527	1.530	0.016	1.517	1.539	308	
	C*–Car (overall)	1.513	1.513	0.014	1.505	1.521	1813	
	cyclopropyl (**C**)–**C**ar	1.490	1.490	0.015	1.479	1.503	90	7
Csp^3-Csp^1	**C*–C**≡C	1.466	1.465	0.010	1.460	1.469	21	15
	C#–C≡C	1.472	1.472	0.012	1.464	1.481	88	15
	C*–C≡N	1.470	1.469	0.013	1.463	1.479	106	7b
	cyclopropyl (**C**)–**C**≡N	1.444	1.447	0.010	1.436	1.451	38	7
Csp^2-Csp^2	**C**=C–**C**=C							
	(conjugated)	1.455	1.455	0.011	1.447	1.463	30	16,18
	(unconjugated)	1.478	1.476	0.012	1.470	1.479	8	17,18
	(overall)	1.460	1.460	0.015	1.450	1.470	38	
	C=**C–C**=C–C=C	1.443	1.445	0.013	1.431	1.454	29	18
	C=**C–C**=C (endocyclic in TCNQ)	1.432	1.433	0.012	1.424	1.441	280	19
	C=**C–C**(=O)(–C*)							
	(conjugated)	1.464	1.462	0.018	1.453	1.476	211	16,18
	(unconjugated)	1.484	1.486	0.017	1.475	1.497	14	17,18
	(overall)	1.465	1.462	0.018	1.453	1.478	226	
	C=**C–C**(=O)–C=C							
	in benzoquinone (C,H-subst. only)	1.478	1.476	0.011	1.469	1.488	28	
	in benzoquinone (any subst.)	1.478	1.478	0.031	1.464	1.498	172	

Bond	Substructure	d	m	σ	q_1	q_u	n	Note
	non-quinonoid	1.456	1.455	0.012	1.447	1.464	28	
	C=**C**–COOH	1.475	1.476	0.015	1.461	1.488	22	
	C=**C**–COOC*	1.488	1.489	0.014	1.478	1.497	113	
	C=**C**–COO⁻	1.502	1.499	0.017	1.488	1.510	11	
	HOO**C**–**C**OOH	1.538	1.537	0.007	1.535	1.541	9	
	HOO**C**–**C**OO⁻	1.549	1.552	0.009	1.546	1.553	13	
	⁻OO**C**–**C**OO⁻	1.564	1.559	0.022	1.554	1.568	9	
	formal Csp^2–Csp^2 single bond in selected non-fused heterocycles:							
	in 1*H*-pyrrole (C3–C4)	1.412	1.410	0.016	1.401	1.427	29	
	in furan (C3–C4)	1.423	1.423	0.016	1.412	1.433	62	
	in thiophene (C3–C4)	1.424	1.425	0.015	1.415	1.433	40	
	in pyrazole (C3–C4)	1.410	1.412	0.016	1.400	1.418	20	
	in isoxazole (C3–C4)	1.425	1.425	0.016	1.413	1.438	9	
	in furazan (C3–C4)	1.428	1.427	0.007	1.422	1.435	6	
	in furoxan (C3–C4)	1.417	1.417	0.006	1.412	1.422	14	
Csp^2–C*ar*	C=**C**–C*ar*							
	(conjugated)	1.470	1.470	0.015	1.463	1.480	37	16,18
Csp^2–C*ar*		1.488	1.490	0.012	1.480	1.496	87	17,18
	(overall)	1.483	1.483	0.015	1.472	1.494	124	
	cyclopropenyl (C=**C**)–C*ar*	1.447	1.448	0.006	1.441	1.452	8	10
	C*ar*–**C**(=O)–C*	1.488	1.489	0.016	1.478	1.500	84	
	C*ar*–**C**(=O)–C*ar*	1.480	1.481	0.017	1.468	1.494	58	
	C*ar*–**C**OOH	1.484	1.485	0.014	1.474	1.491	75	
	C*ar*–**C**(=O)(–OC*)	1.487	1.487	0.012	1.480	1.494	218	
	C*ar*–**C**OO⁻	1.504	1.509	0.014	1.495	1.512	26	
	C*ar*–**C**(–O)–NH$_2$	1.500	1.503	0.020	1.498	1.510	19	
	C*ar*–**C**=N–C#							
	(conjugated)	1.476	1.478	0.014	1.466	1.486	27	16
	(unconjugated)	1.491	1.490	0.008	1.485	1.496	48	17
	(overall)	1.485	1.487	0.013	1.481	1.493	75	
	in indole (C3–C3a)	1.434	1.434	0.011	1.428	1.439	40	
Csp^2–Csp^1	C=**C**–**C**≡C	1.431	1.427	0.014	1.425	1.441	11	7b
	C=**C**–**C**≡N in TCNQ	1.427	1.427	0.010	1.420	1.433	280	19
C*ar*–C*ar*	in biphenyls (*ortho* subst. all **H**)	1.487	1.488	0.007	1.484	1.493	30	
	(≥1 non-**H** *ortho*-subst.)	1.490	1.491	0.010	1.486	1.495	212	
C*ar*–Csp^1	C*ar*–**C**≡C	1.434	1.436	0.006	1.430	1.437	37	
	C*ar*–**C**≡N	1.443	1.444	0.008	1.436	1.448	31	
Csp^1–Csp^1	C≡**C**–**C**=C	1.377	1.378	0.012	1.374	1.384	21	
Csp^2=Csp^2	C*–**C**H=**C**H$_2$	1.299	1.300	0.027	1.280	1.311	42	
	(**C***)$_2$–**C**=**C**H$_2$	1.321	1.321	0.013	1.313	1.328	77	
	C*–**C**H=**C**H–C*							
	(*cis*)	1.317	1.318	0.013	1.310	1.323	106	
	(*trans*)	1.312	1.311	0.011	1.304	1.320	19	
	(overall)	1.316	1.317	0.015	1.309	1.323	127	
	(**C***)$_2$–**C**=**C**H–C*	1.326	1.328	0.011	1.319	1.334	168	
	(**C***$_2$–**C**=**C**–(**C***)$_2$	1.331	1.330	0.009	1.326	1.334	89	
	(**C***,**H**)$_2$–**C**=**C**–(**C***,**H**)$_2$ (overall)	1.322	1.323	0.014	1.315	1.331	493	5
	in cyclopropene (any subst.)	1.294	1.288	0.017	1.284	1.302	10	10
	in cyclobutene (any subst.)	1.335	1.335	0.019	1.324	1.347	25	8
	in cyclopentene (C,H-subst.)	1.323	1.324	0.013	1.314	1.331	104	
	in cyclohexene (C,H-subst.)	1.326	1.325	0.012	1.318	1.334	196	
	C=**C**=**C** (allenes, any subst.)	1.307	1.307	0.005	1.303	1.310	18	
	C=**C**–**C**=**C** (C,H subst., conjugated)	1.330	1.330	0.014	1.322	1.338	76	16
	C=**C**–**C**=**C**–**C**=**C** (C,H subst., conjugated)	1.345	1.345	0.012	1.337	1.350	58	16
	C=**C**–C*ar* (C,H subst., conjugated)	1.339	1.340	0.011	1.334	1.346	124	16
	C=**C** in cyclopenta-1,3-diene (any subst.)	1.341	1.341	0.017	1.328	1.356	18	
	C=**C** in cyclohexa-1,3-diene (any subst.)	1.332	1.332	0.013	1.323	1.341	56	
	in **C**=**C**–C=O							
	(C,**H** subst., conjugated)	1.340	1.340	0.013	1.332	1.348	211	16,18
	(C,**H** subst., unconjugated)	1.331	1.330	0.008	1.326	1.339	14	17,18

Bond	Substructure	d	m	σ	q_1	q_u	n	Note
	(C,**H** subst., overall)	1.340	1.339	0.013	1.332	1.348	226	
	in cyclohexa-2,5-dien-1-ones	1.329	1.327	0.011	1.321	1.335	28	
	in p-benzoquinones							
	(C*,**H** subst.)	1.333	1.337	0.011	1.325	1.338	14	
	(any subst.)	1.349	1.339	0.030	1.330	1.364	86	
	in TCNQ							
	(endocyclic)	1.352	1.353	0.010	1.345	1.358	142	19
	(exocyclic)	1.392	1.391	0.017	1.379	1.405	139	19
	C=C–OH in enol tautomers	1.362	1.360	0.020	1.349	1.370	54	
	in heterocycles (any subst.):							
	1H-pyrrole (C2–C3, C4–C5)	1.375	1.377	0.018	1.361	1.388	58	
	furan (C2–C3, C4–C5)	1.341	1.342	0.021	1.329	1.351	125	
	thiophene (C2–C3, C4–C5)	1.362	1.359	0.025	1.346	1.377	60	
	pyrazole (C4–C5)	1.369	1.372	0.019	1.362	1.383	20	
	imidazole (C4–C5)	1.360	1.361	0.014	1.352	1.367	44	
	isoxazole (C4–C5)	1.341	1.336	0.012	1.331	1.355	9	
	indole (C2–C3)	1.364	1.363	0.012	1.355	1.371	40	
$Car \simeq Car$	in phenyl rings with C*, **H** subst. only							
	H–**C** $\simeq$ **C**–H	1.380	1.381	0.013	1.372	1.388	2191	
	C*–**C** $\simeq$ **C**–H	1.387	1.388	0.010	1.382	1.393	891	
	C*–**C** $\simeq$ **C**–C*	1.397	1.397	0.009	1.392	1.403	182	
	C $\simeq$ **C** (overall)	1.384	1.384	0.013	1.375	1.391	3264	
	F–**C** $\simeq$ **C**–F	1.372	1.374	0.011	1.366	1.380	84	4
	Cl–**C** $\simeq$ **C**–Cl	1.388	1.389	0.014	1.380	1.398	152	4
	in naphthalene (D_{2h}, any subst.)							
	C1–C2	1.364	1.364	0.014	1.356	1.373	440	
	C2–C3	1.406	1.406	0.014	1.397	1.415	218	
	C1–C8a	1.420	1.419	0.012	1.412	1.426	440	
	C4a–C8a	1.422	1.424	0.011	1.417	1.429	109	
$Car \simeq Car$	in anthracene ($D_{2h,}$ any subst.)							
	C1–C2	1.356	1.356	0.009	1.350	1.360	56	
	C2–C3	1.410	1.410	0.010	1.401	1.416	34	
	C1–C9a	1.430	1.430	0.006	1.426	1.434	56	
	C4a–C9a	1.435	1.436	0.007	1.429	1.440	34	
	C9–C9a	1.400	1.402	0.009	1.395	1.406	68	
	in pyridine (C,H subst.)	1.379	1.381	0.012	1.371	1.387	276	20
	(any subst.)	1.380	1.380	0.015	1.371	1.389	537	20
	in pyridinium cation							
	(N$^+$ –H; C,H subst. on C)							
	C2–C3	1.373	1.375	0.012	1.368	1.380	30	
	C3–C4	1.379	1.380	0.011	1.371	1.388	30	
	(N$^+$ –X; C,H subst. on C)							
	C2–C3	1.373	1.372	0.019	1.362	1.382	151	
	C3–C4	1.383	1.385	0.019	1.372	1.394	151	
	in pyrazine (H subst. on C)	1.379	1.377	0.010	1.370	1.388	10	
	(any subst. on C)	1.405	1.405	0.024	1.388	1.420	60	
	in pyrimidine (C,H subst. on C)	1.387	1.389	0.018	1.379	1.400	28	
$Csp^1{\equiv}Csp^1$	X–**C≡C**–X	1.183	1.183	0.014	1.174	1.193	119	15
	C,H–**C≡C**–C,H	1.181	1.181	0.014	1.173	1.192	104	15
	in **C≡C**–C(sp^2,ar)	1.189	1.193	0.010	1.181	1.195	38	15
	in **C≡C**–**C≡C**	1.192	1.192	0.010	1.187	1.197	42	15
	in **CH≡C**–C#	1.174	1.174	0.011	1.167	1.180	42	15
Csp^3–Cl	Omitting 1,2-dichlorides:							
	C–**CH$_2$–Cl**	1.790	1.790	0.007	1.783	1.795	13	4
	C$_2$–**CH–Cl**	1.803	1.802	0.003	1.800	1.807	8	4
	C$_3$–**C–Cl**	1.849	1.856	0.011	1.837	1.858	5	4
	X–**CH$_2$–Cl** (X = C,H,N,O)	1.790	1.791	0.011	1.783	1.797	37	4
	X$_2$–**CH–Cl** (X = C,H,N,O)	1.805	1.803	0.014	1.800	1.812	26	4
	X$_3$–**C–Cl** (X = C,H,N,O)	1.843	1.838	0.014	1.835	1.858	7	4
	X$_2$–**C–Cl$_2$** (X = C,H,N,O)	1.779	1.776	0.015	1.769	1.790	18	4

Bond	Substructure	d	m	σ	q_1	q_u	n	Note
	X–**C**–**Cl**$_3$ (X = C,H,N,O)	1.768	1.765	0.011	1.761	1.776	33	4
	Cl–**CH**(–C)–**CH**(–C)–**Cl**	1.793	1.793	0.013	1.786	1.800	66	4
	Cl–**C**(–C$_2$)–**C**(–C$_2$)–**Cl**	1.762	1.760	0.010	1.757	1.765	54	4
	cyclopropyl–Cl	1.755	1.756	0.011	1.749	1.763	64	
Csp^2–Cl	C=**C**–**Cl** (C,H,N,O subst. on C)	1.734	1.729	0.019	1.719	1.748	63	4
	C=**C**–**Cl**$_2$ (C,H,N,O subst. on C)	1.720	1.716	0.013	1.708	1.729	20	4
	Cl–**C**=**C**–**Cl**	1.713	1.711	0.011	1.705	1.720	80	4
Car–Cl	**Car**–**Cl** (mono-Cl + m,p-Cl$_2$)	1.739	1.741	0.010	1.734	1.745	340	4
	Car–**Cl** (o-Cl$_2$)	1.720	1.720	0.010	1.713	1.717	364	4
Csp^1Cl	see HCLENE10 (1.634, 1.646)							
Csp^3–F	Omitting 1,2-difluorides							
	C–**CH**$_2$–**F** and C$_2$–**CH**–**F**	1.399	1.399	0.017	1.389	1.408	25	4
	C$_3$–**C**–**F**	1.428	1.431	0.009	1.421	1.435	11	4
	(C*,H)$_2$–**C**–**F**$_2$	1.349	1.347	0.012	1.342	1.356	58	4
	C*–**C**–**F**$_3$	1.336	1.334	0.007	1.330	1.344	12	4
	F–**C***–**C***–**F**	1.371	1.374	0.007	1.362	1.375	26	4
	X$_3$–**C**–**F** (X = C,H,N,O)	1.386	1.389	0.033	1.373	1.408	70	4
	X$_2$–**C**–**F**$_2$ (X = C,H,N,O)	1.351	1.349	0.013	1.342	1.356	58	4
	X–**C**–**F**$_3$ (X = C,H,N,O)	1.322	1.323	0.015	1.314	1.332	309	4
	F–**C**(–X)$_2$–**C**(–X)$_2$–**F** (X = C,H,N,O)	1.373	1.374	0.009	1.362	1.377	30	4
	F–**C**(–X)$_2$–**NO**$_2$ (X = any subst.)	1.320	1.319	0.009	1.312	1.327	18	
Csp^2–F	C=**C**–**F** (C,H,N,O subst. on C)	1.340	1.340	0.013	1.334	1.346	34	4
Car–F	**Car**–**F** (mono-F + m,p-F$_2$)	1.363	1.362	0.008	1.357	1.368	38	4
	Car–**F** (o-F$_2$)	1.340	1.340	0.009	1.336	1.344	167	4
Csp^3–H	C–**C**–**H**$_3$ (methyl)	1.059	1.061	0.030	1.039	1.083	83	21
	C$_2$–**C**–**H**$_2$ (primary)	1.092	1.095	0.013	1.088	1.099	100	21
	C$_3$–**C**–**H** (secondary)	1.099	1.097	0.004	1.095	1.103	14	21
	C$_{2,3}$–**C**–**H** (primary and secondary)	1.093	1.095	0.012	1.089	1.100	118	21
	X–**C**–**H**$_3$ (methyl)	1.066	1.074	0.028	1.049	1.087	160	21
	X$_2$–**C**–**H**$_2$ (primary)	1.092	1.095	0.012	1.088	1.099	230	21
	X$_3$–**C**–**H** (secondary)	1.099	1.099	0.007	1.095	1.103	117	21
	X$_{2,3}$–**C**–**H** (primary and secondary)	1.094	1.096	0.011	1.091	1.100	348	21
Csp^2–H	C–C=**C**–**H**	1.077	1.079	0.012	1.074	1.085	14	21
Car–H	**Car**–**H**	1.083	1.083	0.011	1.080	1.087	218	21
Csp^3–I	**C***–**I**	2.162	2.159	0.015	2.149	2.179	15	4
Car–I	**Car**–**I**	2.095	2.095	0.015	2.089	2.104	51	4
Csp^3–N(4)	**C***–**NH**$_3^+$	1.488	1.488	0.013	1.482	1.495	298	
	(**C***)$_2$–**NH**$_2^+$	1.494	1.493	0.016	1.484	1.503	249	
	(**C***)$_3$–**NH**$^+$	1.502	1.502	0.015	1.491	1.512	509	
	(**C***)$_4$–**N**$^+$	1.510	1.509	0.020	1.496	1.523	319	
	C*–**N**$^+$ (overall)	1.499	1.498	0.018	1.488	1.510	1370	
Csp^3–N(3)	**C***–**N**$^+$ in N-subst. pyridinium	1.485	1.484	0.009	1.477	1.490	32	
	C*–**NH**$_2$ (Nsp^3: pyramidal)	1.469	1.470	0.010	1.462	1.474	19	22
	(**C***)$_2$–**NH** (Nsp^3: pyramidal)	1.469	1.467	0.012	1.461	1.477	152	5,22
	(**C***)$_3$–**N** (Nsp^3: pyramidal)	1.469	1.468	0.014	1.460	1.476	1042	5,22
	C*–**N**sp^3 (overall)	1.469	1.468	0.014	1.460	1.476	1201	
	Csp^3–**N**sp^3							
	in aziridine	1.472	1.471	0.016	1.464	1.482	134	
	in azetidine	1.484	1.481	0.018	1.472	1.495	21	
	in tetrahydropyrrole	1.475	1.473	0.016	1.464	1.483	66	
	in piperidine	1.473	1.473	0.013	1.460	1.479	240	
	Csp^3–**N**sp^2 (N planar) in:							23
	acyclic amides **C***–**NH**–C=O	1.454	1.451	0.011	1.446	1.461	78	14
	β-lactams **C***–**N**(–X)–C=O (endo)	1.464	1.465	0.012	1.458	1.475	23	13
	γ-lactams							
	C*–**NH**–C=O (endo)	1.457	1.458	0.011	1.449	1.465	20	13
	C*–**N**(–C*)–C=O (endo)	1.462	1.461	0.010	1.453	1.466	15	13
	C*–**N**(–C*)–C=O (exo)	1.458	1.456	0.014	1.448	1.465	15	13
	δ-lactams							
	C*–**NH**–C=O (endo)	1.478	1.472	0.016	1.467	1.491	6	14
	C*–**N**(–C*)–C=O (endo)	1.479	1.476	0.007	1.475	1.482	15	14

Bond	Substructure	d	m	σ	q_1	q_u	n	Note
	C*–**N**(–**C***)–C=O (exo)	1.468	1.471	0.009	1.462	1.477	15	14
	nitro compounds (1,2-dinitro omitted):							
	C–**CH₂**–**NO₂**	1.485	1.483	0.020	1.478	1.502	8	
	C₂–**CH**–**NO₂**	1.509	1.509	0.011	1.502	1.511	12	
	C₃–**C**–**NO₂**	1.533	1.533	0.013	1.530	1.539	17	
	C₂–**C**–(**NO₂**)₂	1.537	1.536	0.016	1.525	1.550	19	
	1,2-dinitro: **NO₂**–**C***–**C***–**NO₂**	1.552	1.550	0.023	1.536	1.572	32	
Csp^3–N(2)	**C#**–**N**=N	1.493	1.493	0.020	1.477	1.506	54	
	C*–**N**=C–Car	1.465	1.468	0.011	1.461	1.472	75	
Csp^2–N(3)	C=**C**–**NH₂** Nsp^2 planar	1.336	1.344	0.017	1.317	1.348	10	23
	C=**C**–**NH**–C# Nsp^2 planar	1.339	1.340	0.016	1.327	1.351	17	23
	C=**C**–**N**–(C#)₂							
	Nsp^2 planar	1.355	1.358	0.014	1.341	1.363	22	23
	Nsp^3 pyramidal	1.416	1.418	0.018	1.397	1.432	18	22
	Csp^2–Nsp^2 (N planar) in:							23
	acyclic amides							
	NH₂–**C**=O	1.325	1.323	0.009	1.318	1.331	32	14
	C*–**NH**–**C**=O	1.334	1.333	0.011	1.326	1.343	78	14
	(C*)₂–**N**–**C**=O	1.346	1.342	0.011	1.339	1.356	5	14
	β-lactams C*–**NH**–**C**=O	1.385	1.388	0.019	1.374	1.396	23	13
	γ-lactams							
	C*–**NH**–**C**=O	1.331	1.331	0.011	1.326	1.337	20	13
	C*–**N**(–**C***)–**C**=O	1.347	1.344	0.014	1.335	1.359	15	13
	δ-lactams							
	C*–**NH**–**C**=O	1.334	1.334	0.006	1.330	1.339	6	14
	(C*)–**N**(–**C***)–**C**=O	1.352	1.353	0.010	1.344	1.356	15	14
	peptides **C#**–**N**(–X)–**C**(–C#)(=O)	1.333	1.334	0.013	1.326	1.340	380	24
	ureas							
	(**NH₂**)₂–**C**=O	1.334	1.334	0.008	1.329	1.339	48	25,26
	(C#–**NH**)₂–**C**=O	1.347	1.345	0.010	1.341	1.354	26	25
	[(C#)ₙ–**N**]₂–**C**=O	1.363	1.359	0.014	1.354	1.370	40	25,27
	thioureas	1.346	1.343	0.023	1.328	1.361	192	
	(X₂**N**)₂–**C**=S							
	imides							
	[C#–**C**(=O)]₂–**NH**	1.376	1.377	0.012	1.369	1.383	64	
	[C#–**C**(=O)]₂–**N**–C#	1.389	1.383	0.017	1.376	1.404	38	
	[Csp^2–**C**(=O)]₂–**N**–C#	1.396	1.396	0.010	1.389	1.403	46	
	[Csp^2–**C**(=O)]₂–**N**–Csp^2	1.409	1.406	0.020	1.391	1.419	28	
	guanidinium [**C**–(**NH₂**)₃]⁺ (unsubst.)	1.321	1.320	0.008	1.314	1.327	39	
	(any subst.)	1.328	1.325	0.015	1.317	1.333	140	
	in heterocyclic systems (any subst.)							
	1*H*-pyrrole (N1–C2, N1–C5)	1.372	1.374	0.016	1.363	1.384	58	
	indole (N1–C2)	1.370	1.370	0.012	1.364	1.377	40	
	pyrazole (N1–C5)	1.357	1.359	0.012	1.347	1.365	20	
	imidazole (N1–C2)	1.349	1.349	0.018	1.338	1.358	44	
	imidazole (N1–C5)	1.370	1.370	0.010	1.365	1.377	44	
Csp^2–N(2)	in imidazole (N3–C4)	1.376	1.377	0.011	1.369	1.384	44	
Car–N(4)	**Car**–**N**⁺–(C,H)₃	1.465	1.466	0.007	1.461	1.470	23	
Car–N(3)	**Car**–**NH₂**							
	(Nsp^2: planar)	1.355	1.360	0.020	1.340	1.372	33	23
	(Nsp^3: pyramidal)	1.394	1.396	0.011	1.385	1.403	25	22
	(overall)	1.375	1.377	0.025	1.363	1.394	98	28
Car–N(3)	**Car**–**NH**–C#							
	(Nsp^2: planar)	1.353	1.353	0.007	1.347	1.359	16	23
	(Nsp^3: pyramidal)	1.419	1.423	0.017	1.412	1.432	8	22
	(overall)	1.380	1.364	0.032	1.353	1.412	31	28
	Car–**N**–(C#)₂							
	(Nsp^2: planar)	1.371	1.370	0.016	1.363	1.382	41	23
	(Nsp^3: pyramidal)	1.426	1.425	0.011	1.421	1.431	22	22
	(overall)	1.390	1.385	0.030	1.366	1.420	69	28
	in indole (N1–C7a)	1.372	1.372	0.007	1.367	1.376	40	

Bond	Substructure	d	m	σ	q_1	q_u	n	Note
	Car–NO_2	1.468	1.469	0.014	1.460	1.476	556	
Car–N(2)	Car–N=N	1.431	1.435	0.020	1.422	1.442	26	
Csp^2=N(3)	in furoxan ($^+$N2=C3)	1.316	1.316	0.009	1.311	1.324	14	
Csp^2=N(2)	Car–C=N–C#	1.279	1.279	0.008	1.275	1.285	75	
	(C,H)$_2$–C=N–OH in oximes	1.281	1.280	0.013	1.273	1.288	67	
	S–C=N–X	1.302	1.302	0.021	1.285	1.319	36	
	in pyrazole (N2=C3)	1.329	1.331	0.014	1.315	1.339	20	
	in imidazole (C2=N3)	1.313	1.314	0.011	1.307	1.319	44	
	in isoxazole (N2=C3)	1.314	1.315	0.009	1.305	1.320	9	
	in furazan (N2=C3, C4=N5)	1.298	1.299	0.006	1.294	1.303	12	
	in furoxan (C4=N5)	1.304	1.306	0.008	1.300	1.308	14	
$Car \simeq$ N(3)	C $\simeq$ N$^+$–H (pyrimidinium)	1.335	1.334	0.015	1.325	1.342	30	
	C $\simeq$ N$^+$–C* (pyrimidinium)	1.346	1.346	0.010	1.340	1.352	64	
	C $\simeq$ N$^+$–O$^-$ (pyrimidinium)	1.362	1.359	0.013	1.353	1.369	56	
$Car \simeq$ N(2)	C $\simeq$ N (pyridine)	1.337	1.338	0.012	1.330	1.344	269	
	C $\simeq$ N (pyrazine)	1.336	1.335	0.022	1.319	1.347	120	
	C $\simeq$ N $\simeq$ C (pyrimidine)	1.339	1.338	0.015	1.333	1.342	28	
	N $\simeq$ C $\simeq$ N (pyrimidine)	1.333	1.335	0.013	1.326	1.337	28	
	C $\simeq$ N (pyrimidine) (overall) in any 6-membered	1.336	1.337	0.014	1.331	1.339	56	
	N-containing aromatic ring:							
	H–C $\simeq$ N $\simeq$ C–H	1.334	1.334	0.014	1.327	1.341	146	
	H–C $\simeq$ N $\simeq$ C–C*	1.339	1.341	0.013	1.336	1.345	38	
	C*–C $\simeq$ N $\simeq$ C–C*	1.345	1.345	0.008	1.342	1.348	24	
	C $\simeq$ N $\simeq$ C (overall)	1.336	1.337	0.014	1.329	1.344	204	
Csp^1≡N(2)	X–S–N≡C$^-$ (isothiocyanide)	1.144	1.147	0.006	1.140	1.148	6	
Csp^1≡N(1)	C*–C≡N	1.136	1.137	0.010	1.131	1.142	140	
	C=C–C≡N in TCNQ	1.144	1.144	0.008	1.139	1.149	284	19
	Car–C≡N	1.138	1.138	0.007	1.133	1.143	31	
	X–C≡N	1.144	1.141	0.012	1.138	1.151	10	
	(S–C≡N)$^-$	1.155	1.156	0.012	1.147	1.165	14	
Csp^3–O(2)	in alcohols							
	CH$_3$–OH	1.413	1.414	0.018	1.395	1.425	17	
	C–CH$_2$–OH	1.426	1.426	0.011	1.420	1.431	75	
	C$_2$–CH–OH	1.432	1.431	0.011	1.425	1.439	266	
	C$_3$–C–OH	1.440	1.440	0.012	1.432	1.449	106	
	C*–OH (overall)	1.432	1.431	0.013	1.424	1.441	464	
	in dialkyl ethers							29
	CH$_3$–O–C*	1.416	1.418	0.016	1.405	1.426	110	
	C–CH$_2$–O–C*	1.426	1.424	0.011	1.418	1.435	34	
	C$_2$–CH–O–C*	1.429	1.430	0.010	1.420	1.437	53	
	C$_3$–C–O–C*	1.452	1.450	0.011	1.445	1.458	39	
	C*–O–C* (overall)	1.426	1.425	0.019	1.414	1.437	236	5
	in aryl alkyl ethers							29
	CH$_3$–O–Car	1.424	1.424	0.012	1.417	1.431	616	
	C–CH$_2$–O–Car	1.431	1.430	0.013	1.422	1.438	188	
	C$_2$–CH–O–Car	1.447	1.446	0.020	1.435	1.466	58	
	C$_3$–C–O–Car	1.470	1.469	0.018	1.456	1.483	55	
	C*–O–Car (overall)	1.429	1.427	0.018	1.419	1.436	917	
	in alkyl esters of carboxylic acids							12,29
	CH$_3$–O–C(=O)–C*	1.448	1.449	0.010	1.442	1.455	200	
	C–CH$_2$–O–C(=O)–C*	1.452	1.453	0.009	1.445	1.458	32	
	C$_2$–CH–O–C(=O)–C*	1.460	1.460	0.010	1.454	1.465	78	
	C$_3$–C–O–C(=O)–C*	1.477	1.475	0.008	1.472	1.484	6	
	C*–O–C(=O)–C* (overall)	1.450	1.451	0.014	1.442	1.459	314	
	in alkyl esters of α,β-unsaturated acids:							
	C*–O–C(=O)–C=C (overall)	1.453	1.452	0.013	1.444	1.459	112	
	in alkyl esters of benzoic acid							
	C*–O–C(=O)–C(phenyl) (overall)	1.454	1.454	0.012	1.446	1.463	219	
	in ring systems							
	oxirane (epoxides) (any subst.)	1.446	1.446	0.014	1.438	1.456	498	9

Bond	Substructure	d	m	σ	q_l	q_u	n	Note
	oxetane (any subst.)	1.463	1.460	0.015	1.451	1.474	16	
	tetrahydrofuran (C,H subst.)	1.442	1.441	0.017	1.430	1.451	154	
Csp^3–O(2)	tetrahydropyran (C,H subst.)	1.441	1.442	0.015	1.431	1.451	22	
	β-lactones: **C*–O**–C(=O)	1.492	1.494	0.010	1.481	1.501	4	16
	γ-lactones: **C*–O**–C(=O)	1.464	1.464	0.012	1.455	1.473	110	12
	δ-lactones: **C*–O**–C(=O)	1.461	1.464	0.017	1.452	1.473	27	12
	O–C–O system in *gem*-diols, and pyranose and							30,31
	furanose sugars:							
	HO–C*–OH	1.397	1.401	0.012	1.388	1.405	18	
	C$_5$–O$_5$–C$_1$–O$_1$H in pyranoses							
	O$_1$ axial (α):							
	C$_5$–O$_5$	1.439	1.440	0.008	1.432	1.445	29	
	O$_5$–C$_1$	1.427	1.426	0.012	1.421	1.432	29	
	C$_1$–O$_1$	1.403	1.400	0.012	1.391	1.412	29	
	O$_1$ equatorial (β):							
	C$_5$–O$_5$	1.435	1.436	0.008	1.429	1.440	17	
	O$_5$–C$_1$	1.430	1.431	0.010	1.424	1.436	17	
	C$_1$–O$_1$	1.393	1.393	0.007	1.386	1.399	17	
	α + β (overall):							
	C$_5$–O$_5$	1.439	1.440	0.008	1.432	1.446	60	
	O$_5$–C$_1$	1.430	1.429	0.012	1.421	1.436	60	
	C$_1$–O$_1$	1.401	1.399	0.011	1.392	1.407	60	
	C$_4$–O$_4$–C$_1$–O$_1$H in furanoses (overall values)							
	C$_4$–O$_4$	1.442	1.446	0.012	1.436	1.449	18	
	O$_4$–C$_1$	1.432	1.432	0.012	1.421	1.443	18	
	C$_1$–O$_1$	1.404	1.405	0.013	1.397	1.409	18	
	C$_5$–O$_5$–C$_1$–O$_1$–C* in pyranoses							
	O$_1$ axial (α):							
	C$_5$–O$_5$	1.439	1.438	0.010	1.433	1.446	67	
	O$_5$–C$_1$	1.417	1.417	0.009	1.410	1.424	67	
	C$_1$–O$_1$	1.409	1.409	0.014	1.401	1.417	67	
	O$_1$–C*	1.435	1.435	0.013	1.427	1.443	67	
	O$_1$ equatorial (β):							
	C$_5$–O$_5$	1.434	1.435	0.006	1.429	1.439	39	
	O$_5$–C$_1$	1.424	1.424	0.008	1.418	1.431	39	
	C$_1$–O$_1$	1.390	1.390	0.011	1.381	1.400	39	
	O$_1$–C*	1.437	1.438	0.013	1.428	1.445	39	
	α + β (overall):							
	C$_5$–O$_5$	1.436	1.436	0.009	1.431	1.442	126	
	O$_5$–C$_1$	1.419	1.419	0.011	1.412	1.426	126	
	C$_1$–O$_1$	1.402	1.403	0.016	1.391	1.413	126	
	O$_1$–C*	1.436	1.436	0.013	1.428	1.445	126	
	C$_4$–O$_4$–C$_1$–O$_1$–C* in furanoses (overall values)							
	C$_4$–O$_4$	1.443	1.445	0.013	1.429	1.453	23	
	O$_4$–C$_1$	1.421	1.418	0.012	1.413	1.431	23	
	C$_1$–O$_1$	1.410	1.409	0.014	1.401	1.420	23	
	O$_1$–C*	1.439	1.437	0.014	1.429	1.449	23	
	Miscellaneous:							
	C#–O–SiX$_3$	1.416	1.416	0.017	1.405	1.428	29	
	C*–O–SO$_2$–C	1.465	1.461	0.014	1.454	1.475	33	
Csp^2–O(2)	in enols: C=**C–OH**	1.333	1.331	0.017	1.324	1.342	53	
	in enol esters: C=**C–O**–C*	1.354	1.353	0.016	1.341	1.363	40	
	in acids:							
	C*–**C**(=O)–**OH**	1.308	1.311	0.019	1.298	1.320	174	
	C=C–**C**(=O)–**OH**	1.293	1.295	0.019	1.279	1.307	22	
	Car–**C**(=O)–**OH**	1.305	1.311	0.020	1.291	1.317	75	
	in esters:							
	C*–**C**(=O)–**O**–C*	1.336	1.337	0.014	1.328	1.346	551	12,29
	C=C–**C**(=O)–**O**–C*	1.332	1.331	0.011	1.324	1.339	112	
	Car–**C**(=O)–**O**–C*	1.337	1.335	0.013	1.329	1.344	219	12
	C*–**C**(=O)–**O**–C=C	1.362	1.359	0.018	1.351	1.374	26	

Bond	Substructure	d	m	σ	q_1	q_u	n	Note
	C*–**C**(=O)–**O**–C=C	1.407	1.405	0.017	1.394	1.420	26	
	C*–**C**(=O)–**O**–Car	1.360	1.359	0.011	1.355	1.367	40	12
	in anhydrides: O=**C**–**O**–C=O	1.386	1.386	0.011	1.379	1.393	70	
	in ring systems:							
	furan (O1–C2, O1–C5)	1.368	1.369	0.015	1.359	1.377	125	
	isoxazole (O1–C5)	1.354	1.354	0.010	1.345	1.360	9	
	β-lactones: C*–**C**(=O)–**O**–C*	1.359	1.359	0.013	1.348	1.371	4	13
	γ-lactones: C*–**C**(=O)–**O**–C*	1.350	1.349	0.012	1.342	1.359	110	12
	δ-lactones: C*–**C**(=O)–**O**–C*	1.339	1.339	0.016	1.332	1.347	27	12
Car–O(2)	in phenols: **Car**–**OH**	1.362	1.364	0.015	1.353	1.373	551	
	in aryl alkyl ethers: **Car**–**O**–C*	1.370	1.370	0.011	1.363	1.377	920	29,32
Car–O(2)	in diaryl ethers: **Car**–**O**–Car	1.384	1.381	0.014	1.375	1.391	132	
	in esters: **Car**–**O**–C(=O)–C*	1.401	1.401	0.010	1.394	1.408	40	12
Csp^2=O(1)	in aldehydes and ketones:							
	C*–**CH**=**O**	1.192	1.192	0.005	1.188	1.197	7	
	(C*)$_2$–**C**=**O**	1.210	1.210	0.008	1.206	1.215	474	5
	(C#)$_2$–**C**=**O**							
	in cyclobutanones	1.198	1.198	0.007	1.194	1.204	12	
	in cyclopentanones	1.208	1.208	0.007	1.203	1.212	155	
	in cyclohexanones	1.211	1.211	0.009	1.207	1.216	312	
	C=C–**C**=**O**	1.222	1.222	0.010	1.216	1.229	225	
	(C=C)$_2$–**C**=**O**	1.233	1.229	0.010	1.226	1.242	28	
	Car–**C**=**O**	1.221	1.218	0.014	1.212	1.229	85	
	(Car)$_2$–**C**=**O**	1.230	1.226	0.015	1.220	1.238	66	
	C=**O** in benzoquinones	1.222	1.220	0.013	1.211	1.231	86	
	delocalized double bonds in carboxylate anions:							
	H–**C** ≃ **O**$_2^-$ (formate)	1.242	1.243	0.012	1.234	1.252	24	
	C*–**C** ≃ **O**$_2^-$	1.254	1.253	0.010	1.247	1.261	114	
	C=C–**C** ≃ **O**$_2^-$	1.250	1.248	0.017	1.238	1.261	52	
	Car–**C** ≃ **O**$_2^-$	1.255	1.253	0.010	1.249	1.262	22	
	HOOC–**C** ≃ **O**$_2^-$ (hydrogen oxalate)	1.243	1.247	0.015	1.232	1.256	26	
	$^-$**O**$_2$ ≃ **C**–**C** ≃ **O**$_2^-$ (oxalate)	1.251	1.251	0.007	1.248	1.254	18	
	in carboxylic acids (X–COOH)							
	C*–**C**(=**O**)–OH	1.214	1.214	0.019	1.203	1.224	175	
	C=C–**C**(=**O**)–OH	1.229	1.226	0.017	1.218	1.237	22	
	Car–**C**(=**O**)–OH	1.226	1.223	0.020	1.211	1.241	75	
	in esters:							
	C*–**C**(=**O**)–O–C*	1.196	1.196	0.010	1.190	1.202	551	12
	C=C–**C**(=**O**)–O–C*	1.199	1.198	0.009	1.193	1.203	113	
	Car–**C**(=**O**)–O–C*	1.202	1.201	0.009	1.196	1.207	218	12
	C*–**C**(=**O**)–O–C=C	1.190	1.190	0.014	1.184	1.198	26	
	C*–**C**(=**O**)–O–Car	1.187	1.188	0.011	1.181	1.195	40	12
	in anhydrides: **O**=**C**–O–C=O	1.187	1.187	0.010	1.184	1.193	70	
	in β-lactones: C*–**C**(=**O**)–O–C*	1.193	1.193	0.006	1.187	1.198	4	13
	γ-lactones: C*–**C**(=**O**)–O–C*	1.201	1.202	0.009	1.196	1.206	109	12
	δ-lactones: C*–**C**(=**O**)–O–C*	1.205	1.207	0.008	1.201	1.209	27	12
	in amides:							
	NH$_2$–**C**(–C*)=**O**	1.234	1.233	0.012	1.225	1.243	32	14
	(C*–)(C*,H–)N–**C**(–C*)=**O**	1.231	1.231	0.012	1.224	1.238	378	14
	β-lactams: C*–NH–**C**=**O**	1.198	1.200	0.012	1.193	1.204	23	13
	γ-lactams:							
	C*–NH–**C**=**O**	1.235	1.235	0.008	1.232	1.240	20	13
	C*–N(–C*)–**C**=**O**	1.225	1.226	0.011	1.217	1.233	15	13
	δ-lactams:							
	C*–NH–**C**=**O**	1.240	1.241	0.003	1.237	1.243	6	14
	C*–N(–C*)–**C**=**O**	1.233	1.233	0.007	1.229	1.239	15	14
	in ureas:							
	(NH)$_2$)$_2$–**C**=**O**	1.256	1.256	0.007	1.249	1.261	24	25,26
	(C#–NH)$_2$–**C**=**O**	1.241	1.237	0.011	1.235	1.245	13	25
	[(C#)$_n$–N]$_2$–**C**=**O**	1.230	1.230	0.007	1.224	1.234	20	25,27
Csp^3–P(4)	C$_3$–**P**$^+$–**C***	1.800	1.802	0.015	1.790	1.812	35	33

Bond	Substructure	d	m	σ	q_l	q_u	n	Note
	$C_2-P(=O)-CH_3$	1.791	1.790	0.006	1.786	1.795	10	
	$C_2-P(=O)-CH_2-C$	1.806	1.806	0.009	1.801	1.813	45	
	$C_2-P(=O)-CH-C_2$	1.821	1.821	0.009	1.815	1.828	15	
	$C_2-P(=O)-C-C_3$	1.841	1.842	0.008	1.835	1.847	14	
	$C_2-P(=O)-C^*$ (overall)	1.813	1.811	0.017	1.800	1.822	84	
$Csp^3-P(3)$	C_2-P-C^*	1.855	1.857	0.019	1.840	1.870	23	
$Car-P(4)$	C_3-P^+-Car	1.793	1.792	0.011	1.786	1.800	276	
	$C_2-P(=O)-Car$	1.801	1.802	0.011	1.796	1.807	98	
	$Ph_3-P=N^+=P-Ph_3$	1.795	1.795	0.008	1.789	1.800	197	
$Car-P(3)$	$C_2-P-Car$	1.836	1.837	0.010	1.830	1.844	102	
	$(N\approx)_2P-Car$ (P ≈ N aromatic)	1.795	1.793	0.011	1.788	1.803	43	
$Csp^3-S(4)$	C^*-SO_2-C ($C^* = CH_3$ excluded)	1.786	1.782	0.018	1.774	1.797	75	
	C^*-SO_2-C (overall)	1.779	1.778	0.020	1.764	1.790	94	
	C^*-SO_2-O-X	1.745	1.744	0.009	1.738	1.754	7	34
	$C^*-SO_2-N-X_2$	1.758	1.756	0.018	1.746	1.773	17	34
$Csp^3-S(3)$	$C^*-S(=O)-C$ ($C^* = CH_3$ excluded)	1.818	1.814	0.024	1.802	1.829	69	
	$C^*-S(=O)-C$ (overall)	1.809	1.806	0.025	1.793	1.820	88	
	$CH_3-S^+-X_2$	1.786	1.787	0.007	1.779	1.792	21	
	$C^*-S^+-X_2$ ($C^* = CH_3$ excluded)	1.823	1.820	0.016	1.812	1.834	18	
	$C^*-S^+-X_2$ (overall)	1.804	1.794	0.025	1.788	1.820	41	
$Csp^3-S(2)$	C^*-SH	1.808	1.805	0.010	1.800	1.819	6	
	CH_3-S-C^*	1.789	1.787	0.008	1.784	1.794	9	
$Csp^3-S(2)$	$C-CH_2-S-C^*$	1.817	1.816	0.013	1.808	1.824	92	
	$C_2-CH-S-C^*$	1.819	1.819	0.011	1.811	1.825	32	
	$C_3-C-S-C^*$	1.856	1.860	0.011	1.854	1.863	26	
	C^*-S-C^* (overall)	1.819	1.817	0.019	1.809	1.827	242	
	in thiirane	1.834	1.835	0.025	1.810	1.858	4	9
	in thiirane: see ZCMXSP (1.817, 1.844)							
	in tetrahydrothiophene	1.827	1.826	0.018	1.811	1.837	20	
	in tetrahydrothiopyran	1.823	1.821	0.014	1.812	1.832	24	
	$C-CH_2-S-S-X$	1.823	1.820	0.014	1.813	1.832	41	
	$C_3-C-S-S-X$	1.863	1.865	0.015	1.848	1.878	11	
	$C^*-S-S-X$ (overall)	1.833	1.828	0.022	1.818	1.848	59	
$Csp^2-S(2)$	$C=C-S-C^*$	1.751	1.755	0.017	1.740	1.764	61	
	$C=C-S-C=C$ (in tetrathiafulvalene)	1.741	1.741	0.011	1.733	1.750	88	
	$C=C-S-C=C$ (in thiophene)	1.712	1.712	0.013	1.703	1.722	60	
	$O=C-S-C\#$	1.762	1.759	0.018	1.747	1.778	20	
$Car-S(4)$	$Car-SO_2-C$	1.763	1.764	0.009	1.756	1.769	96	
	$Car-SO_2-O-X$	1.752	1.750	0.008	1.749	1.756	27	
	$Car-SO_2-N-X_2$	1.758	1.759	0.013	1.749	1.765	106	35
$Car-S(3)$	$Car-S(=O)-C$	1.790	1.790	0.010	1.783	1.798	41	
	$Car-S^+-X_2$	1.778	1.779	0.010	1.771	1.787	10	
$Car-S(2)$	$Car-S-C^*$	1.773	1.774	0.009	1.765	1.779	44	
	$Car-S-Car$	1.768	1.767	0.010	1.762	1.774	158	
	$Car-S-Car$ (in phenothiazine)	1.764	1.764	0.008	1.760	1.769	48	
	$Car-S-S-X$	1.777	1.777	0.012	1.767	1.785	47	
$Csp^1-S(2)$	$N\equiv C-S-X$	1.679	1.683	0.026	1.645	1.698	10	
$Csp^1-S(1)$	$(N\equiv C-S)^-$	1.630	1.630	0.014	1.619	1.641	14	
$Csp^2=S(1)$	$(C^*)_2-C=S$: see IPMUDS (1.599)							
	$(Car)_2-C=S$: see CELDOM (1.611)							
	$(X)_2-C=S$ (X = C,N,O,S)	1.671	1.675	0.024	1.656	1.689	245	
	$X_2N-C(=S)-S-X$	1.660	1.660	0.016	1.648	1.674	38	
	$(X_2N)_2-C=S$ (thioureas)	1.681	1.684	0.020	1.669	1.693	96	
	$N-C(\approx S)_2$	1.720	1.721	0.012	1.709	1.731	20	
Csp^3-Se	$C\#-Se$	1.970	1.967	0.032	1.948	1.998	21	
$Csp^2-Se(2)$	$C=C-Se-C=C$ (in tetraselenafulvalene)	1.893	1.895	0.013	1.882	1.902	32	
$Car-Se(3)$	Ph_3-Se^+	1.930	1.929	0.006	1.924	1.936	13	
$Csp^3-Si(5)$	$C\#-Si^--X_4$	1.874	1.876	0.015	1.859	1.884	9	
$Csp^3-Si(4)$	CH_3-Si-X_3	1.857	1.857	0.018	1.848	1.869	552	
	C^*-Si-X_3 ($C^* = CH_3$ excluded)	1.888	1.887	0.023	1.872	1.905	124	
	C^*-Si-X_3 (overall)	1.863	1.861	0.024	1.850	1.875	681	

Bond	Substructure	d	m	σ	q_1	q_u	n	Note
Car–Si(4)	**Car–Si–X$_3$**	1.868	1.868	0.014	1.857	1.878	178	
Csp^1–Si(4)	C≡C–**Si**–X$_3$	1.837	1.840	0.012	1.824	1.849	8	
Csp^3–Te	**C#–Te**	2.158	2.159	0.030	2.128	2.177	13	
Car–Te	Car–**Te**	2.116	2.115	0.020	2.104	2.130	72	
Csp^2=Te	see CEDCUJ (2.044)							
Cl–Cl	see PHASCL (2.306, 2.227)							
Cl–I	see CMBIDZ (2.563), HXPASC (2.541, 2.513), METAMM (2.552), BQUINI (2.416, 2.718)							
Cl–N	see BECTAE (1.743–1.757), BOGPOC (1.705)							
Cl–O(1)	in ClO$_4^-$	1.414	1.419	0.026	1.403	1.431	252	
Cl–P	(N≃)$_2$**P–Cl** (N ≃ P aromatic)	1.997	1.994	0.015	1.989	2.004	46	
	Cl–P (overall)	2.008	2.001	0.035	1.986	2.028	111	
Cl–S	**Cl–S** (overall)	2.072	2.079	0.023	2.047	2.091	6	
	see also longer bonds in CILSAR (2.283), BIHXIZ (2.357), CANLUY (2.749)							
Cl–Se	see BIRGUE10, BIRHAL10, CTCNSE (2.234–2.851)							
Cl–Si(4)	**Cl–Si**–X$_3$ (monochloro)	2.072	2.075	0.009	2.066	2.078	5	
	Cl$_2$–Si–X$_2$ and **Cl$_3$–Si**–X	2.020	2.012	0.015	2.007	2.036	5	
Cl–Te	Cl–Te in range 2.34–2.60	2.520	2.515	0.034	2.493	2.537	22	36
	see also longer bonds in BARRIV, BOJPUL, CETUTE, EPHTEA, OPNTEC10 (2.73–2.94)							
F–N(3)	**F–N**–C$_2$ and **F$_2$–N**–C	1.406	1.404	0.016	1.395	1.416	9	
F–P(6)	in hexafluorophosphate, PF$_6^-$	1.579	1.587	0.025	1.563	1.598	72	
F–P(3)	(N≃)$_2$**P–F** (N ≃ P aromatic)	1.495	1.497	0.016	1.481	1.510	10	
F–S	43 observations in range 1.409–1.770 in a wide variety of environments; F–S(6) in F$_2$–SO$_2$–C$_2$ (see FPSULF10, BETJOZ)	1.640	1.646	0.011	1.626	1.649	6	
	F–S(4) in **F$_2$–S**(=O)–N (see BUDTEZ)	1.527	1.528	0.004	1.524	1.530	24	37
F–Si(6)	in SiF$_6^{2-}$	1.694	1.701	0.013	1.677	1.703	6	
F–Si(5)	**F–Si$^-$**–X$_4$	1.636	1.639	0.035	1.602	1.657	10	
F–Si(4)	**F–Si**–X$_3$	1.588	1.587	0.014	1.581	1.599	24	
F–Te	see CUCPlZ (F–Te(6) = 1.942, 1.937), FPHTEL(F–Te(4) = 2.006)							
H–N(4)	X$_3$–**N$^+$–H**	1.033	1.036	0.022	1.026	1.045	87	21
H–N(3)	X$_3$–**N–H**	1.009	1.010	0.019	0.997	1.023	95	21
H–O(2)	in alcohols C*–**O–H**	0.967	0.969	0.010	0.959	0.974	63	21
	C#–**O–H**	0.967	0.970	0.010	0.959	0.974	73	21
	in acids O=C–**O–H**	1.015	1.017	0.017	1.001	1.031	16	21,38
I–I	in I$_3^-$	2.917	2.918	0.011	2.907	2.927	6	
I–N	see BZPRIB, CMBIDZ, HMTITI, HMTNTI, IFORAM, IODMAM (2.042–2.475)							
I–O	X–**I–O**(see BZPRIB, CAJMAB, IBZDAC11)	2.144	2.144	0.028	2.127	2.164	6	
	for IO$_6^-$ see BOVMEE (1.829–1.912)							
I–P(3)	see CEHKAB (2.490–2.493)							†
I–S	sec DTHIBR10 (2.687), ISUREA10 (2.629), BZTPPI (3.251)							
I–Te(4)	**I–Te**–X$_3$	2.926	2.928	0.026	2.902	2.944	8	
N(4)–N(3)	X$_3$–**N$^+$–N^0**–X$_2$ (N^0 planar)	1.414	1.414	0.005	1.412	1.418	13	
N(3)–N(3)	(C)(C,H)–**N$_a$–N$_b$**(C)(C,H)							5,39
	N$_a$, N$_b$ pyramidal	1.454	1.452	0.021	1.444	1.457	44	40
	N$_a$ pyramidal, **N$_b$** planar	1.420	1.420	0.015	1.407	1.433	68	40
	N$_a$, N$_b$ planar	1.401	1.401	0.018	1.384	1.418	40	40
	overall	1.425	1.425	0.027	1.407	1.443	139	
N(3)–N(2)	in pyrazole (N1–N2)	1.366	1.366	0.019	1.350	1.375	20	
	in pyridaznium (Nl$^+$=N2)	1.350	1.349	0.010	1.345	1.361	7	
N(2) ≃ N(2)	**N ≃ N** (aromatic) in pyridazine							
	with C,H as *ortho* substituents	1.304	1.300	0.019	1.287	1.326	6	
	with N,Cl as *ortho* substituents	1.368	1.373	0.011	1.362	1.375	9	
N(2)=N(2)	C#–**N=N**–C#							
	cis	1.245	1.244	0.009	1.239	1.252	21	
	trans	1.222	1.222	0.006	1.218	1.227	6	
	(overall)	1.240	1.241	0.012	1.230	1.251	27	

Bond	Substructure	*d*	*m*	σ	q_l	q_u	*n*	Note
	Car–N=N–*Car*	1.255	1.253	0.016	1.247	1.262	13	
	X–N=N=N (azides)	1.216	1.226	0.028	1.202	1.237	19	
N(2)=N(1)	X–N=N=N (azides)	1.124	1.128	0.015	1.114	1.137	19	
N(3)–O(2)	(C,H)$_2$–**N–OH** (Nsp^2: planar)	1.396	1.394	0.012	1.390	1.401	28	
	C$_2$–**N–O**–C							
	(Nsp^3: pyramidal)	1.463	1.465	0.012	1.457	1.468	22	
	(Nsp^2: planar)	1.397	1.394	0.011	1.388	1.409	12	
	in furoxan (N2–O1)	1.438	1.436	0.009	1.430	1.447	14	
N(3)–O(1)	(C≈)$_2$N$^+$–O$^-$ in pyridine *N*-oxides	1.304	1.299	0.015	1.291	1.316	11	
	in furoxan ($^+$N2–O6$^-$)	1.234	1.234	0.008	1.228	1.240	14	
N(2)–O(2)	in oximes							
	(C#)$_2$–C=**N–OH**	1.416	1.418	0.006	1.416	1.420	7	
	(H)(Csp^2)–C=**N–OH**	1.390	1.390	0.011	1.380	1.401	20	
	(C#)(Csp^2)–C=**N–OH**	1.402	1.403	0.010	1.393	1.410	18	
	(Csp^2)$_2$–C=**N–OH**	1.378	1.377	0.017	1.365	1.393	16	
	(C,H)$_2$–C=**N–OH** (overall)	1.394	1.395	0.018	1.379	1.408	67	
	in furazan (O1–N2, O1–N5)	1.385	1.383	0.013	1.378	1.392	12	
	in furoxan (O1–N5)	1.380	1.380	0.011	1.370	1.388	14	
	in isoxazole (O1–N2)	1.425	1.425	0.010	1.417	1.434	9	
N(3)=O(1)	in nitrate ions **NO$_3^-$**	1.239	1.240	0.020	1.227	1.251	105	
	in nitro groups							
	C*–**NO$_2$**	1.212	1.214	0.012	1.206	1.221	84	
	C#–**NO$_2$**	1.210	1.210	0.011	1.203	1.218	251	
	Car–**NO$_2$**	1.217	1.218	0.011	1.211	1.215	1116	
	C–NO$_2$ (overall)	1.218	1.219	0.013	1.210	1.226	1733	
N(3)–P(4)	X$_2$–**P**(=X)–NX$_2$							
	Nsp^2: planar	1.652	1.651	0.024	1.634	1.670	205	
	Nsp^3: pyramidal	1.683	1.683	0.005	1.680	1.686	6	
	(overall)	1.662	1.662	0.029	1.639	1.682	358	
	subsets of this group are:							
	O$_2$–**P**(=S)–NX$_2$	1.628	1.624	0.015	1.615	1.634	9	
	C–**P**(=S)–(NX$_2$)$_2$	1.691	1.694	0.018	1.678	1.703	28	
	O–**P**(=S)–(NX$_2$)$_2$	1.652	1.654	0.014	1.642	1.664	28	
	P(=O)–(NX$_2$)$_3$	1.663	1.668	0.026	1.640	1.679	78	
N(3)–P(3)	–NX–**P**(–X)–NX–**P**(–X)–(P$_2$N$_2$ ring)	1.730	1.721	0.017	1.716	1.748	20	
	–NX–**P**(=S)–NX–**P**(=S)–(P$_2$N$_2$ ring)	1.697	1.697	0.015	1.690	1.703	44	
	in P-substituted phosphazenes:							
	(N≈)$_2$**P–N** (amino) (aziridinyl)	1.637	1.638	0.014	1.625	1.651	16	
		1.672	1.674	0.010	1.665	1.676	15	
N(2)=P(4)	Ph$_3$–**P**=N$^+$=P–Ph$_3$	1.571	1.573	0.013	1.563	1.580	66	
N(2)=P(3)	Ph$_3$–**P**=N–C,S	1.599	1.597	0.018	1.580	1.615	7	
N(2) ≈ P(3)	**N ≈ P** aromatic							
	in phosphazenes	1.582	1.582	0.019	1.571	1.594	126	
	in P ≈ N ≈ S	1.604	1.606	0.009	1.594	1.612	36	
N(3)–S(4)	C–SO$_2$–**NH$_2$**	1.600	1.601	0.012	1.591	1.610	14	35
	C–SO$_2$–**NH**–C#	1.633	1.633	0.019	1.615	1.652	47	35
	C–SO$_2$–**N**–C(#)$_2$	1.642	1.641	0.024	1.623	1.659	38	35
N(3)–S(2)	C–S–**NX$_2$** Nsp^2: planar	1.710	1.707	0.019	1.698	1.722	22	23
	(for Nsp^3 pyramidal see MODIAZ: 1.765)							
	X–S–**NX$_2$** Nsp^2: planar	1.707	1.705	0.012	1.699	1.715	30	23
N(2)–S(2)	C=**N–S**–X	1.656	1.663	0.027	1.632	1.677	36	
N(2) ≈ S(2)	**N ≈ S** aromatic in **P ≈ N ≈ S**	1.560	1.558	0.011	1.554	1.563	37	
N(2)=S(2)	**N=S** in **N=S=N** and **N=S=S**	1.541	1.546	0.022	1.521	1.558	37	
N(3)–SE	see COJCUZ (1.830), DSEMOR10 (1.846, 1.852), MORTRS10 (1.841)							
N(2)–Se	see SEBZQI (1.805), NAPSEZ10 (1.809, 1.820)							
N(2)=Se	see CISMUM (1.790, 1.791)							
N(3)–Si(5)	see DMESIP01, BOJLER, CASSAQ, CASYOK, CECXEN, CINTEY, CIPBUY, FMESIB, MNPSIL, PNPOSI (1.973–2.344)							
N(3)–Si(4)	X$_3$–**Si**–NX$_2$ (overall)	1.748	1.746	0.022	1.735	1.757	170	
	subsets of this group are:							

Bond	Substructure	d	m	σ	q_l	q_u	n	Note
	X$_3$–**Si**–NHX	1.714	1.719	0.014	1.702	1.727	16	
	X$_3$–**Si**–NX–Si–X$_3$ acyclic	1.743	1.744	0.016	1.731	1.755	45	
	N–**Si**–N in 4-membered rings	1.742	1.742	0.009	1.735	1.748	53	
	N–**Si**–N in 5-membered rings	1.741	1.742	0.019	1.726	1.749	33	
N(2)–Si(4)	X$_3$–**Si**–N$^-$–Si–X$_3$	1.711	1.712	0.019	1.693	1.729	15	
N–Te	see ACLTEP (2.402), BIBLAZ (1.980), CESSAU (2.023)							
O(2)–O(2)	C*–**O**–**O**–C*,H							
	τ(OO) = 70–85°	1.464	1.464	0.009	1.458	1.472	12	
	τ(OO) *ca.* 180°	1.482	1.480	0.005	1.478	1.486	5	
	overall	1.469	1.471	0.012	1.461	1.478	17	
	O=C–**O**–**O**–C=O see ACBZPO01 (1.446), CEYLUN (1.452), CIMHIP (1.454)							
	Si–**O**–**O**–Si	1.496	1.499	0.005	1.490	1.499	10	
O(2)–P(5)	X–**P**–(O**X**)$_4$							41
	trigonal bipyramidal:							
	axial	1.689	1.685	0.024	1.675	1.712	20	
	equatorial	1.619	1.622	0.024	1.604	1.628	20	
	square pyramidal	1.662	1.661	0.020	1.649	1.673	28	
O(2)–P(4)	C–**O**–**P**(≈O)$_3$$^{2-}$	1.621	1.622	0.007	1.615	1.628	12	
	(H–**O**)$_2$–**P**(≈O)$_2$$^-$	1.560	1.561	0.009	1.555	1.566	16	
	(C–**O**)$_2$–**P**(≈O)$_2$$^-$	1.608	1.607	0.013	1.599	1.615	16	
	(C#–**O**)$_3$–**P**=O	1.558	1.554	0.011	1.550	1.564	30	
	(Car–**O**)$_3$–**P**=O	1.587	1.588	0.014	1.572	1.599	19	
	X–**O**–**P**(=O)–(C,N)$_2$	1.590	1.585	0.016	1.577	1.601	33	
	(X–**O**)$_2$–**P**(=O)–(C,N)	1.571	1.572	0.013	1.563	1.579	70	
O(2)–P(3)	(N≈)$_2$**P**–**O**–C (N ≈ P aromatic)	1.573	1.573	0.011	1.563	1.584	16	
O(1)=P(4)	C–O–**P**(≈**O**)$_3$$^{2-}$ (delocalized)	1.513	1.512	0.008	1.508	1.518	42	
	(H–O)$_2$–**P**(≈**O**)$_2$$^-$ (delocalized)	1.503	1.503	0.005	1.499	1.508	16	
	(C–O)$_2$–**P**(≈**O**)$_2$$^-$ (delocalized)	1.483	1.485	0.008	1.474	1.490	16	
	(C–O)$_3$–**P**=**O**	1.449	1.448	0.007	1.446	1.452	18	
	C$_3$–**P**=**O**	1.489	1.486	0.010	1.481	1.496	72	
	N$_3$–**P**=**O**	1.461	1.462	0.014	1.449	1.470	26	
	(C)$_2$(N)–**P**=**O**	1.487	1.489	0.007	1.479	1.493	5	
	(C,N)$_2$(O)–**P**=**O**	1.467	1.462	0.007	1.462	1.472	33	
	(C,N)(O)$_2$–**P**=**O**	1.457	1.458	0.009	1.454	1.462	35	
O(2)–S(4)	C–**O**–**S**O$_2$–C	1.577	1.576	0.015	1.566	1.584	41	
	C–**O**–**S**O$_2$–CH$_3$	1.569	1.569	0.013	1.556	1.582	7	
	C–**O**–**S**O$_2$–Car	1.580	1.578	0.015	1.571	1.588	27	
O(1)=S(4)	C–**S**O$_2$–C	1.436	1.437	0.010	1.431	1.442	316	42
	X–**S**O$_2$–NX$_2$	1.428	1.428	0.010	1.422	1.434	326	
	C–**S**O$_2$–N–(C,H)$_2$	1.430	1.430	0.009	1.425	1.435	206	
	C–**S**O$_2$–O–C	1.423	1.423	0.008	1.418	1.428	82	
	in **S**O$_4$$^{2-}$	1.472	1.473	0.013	1.463	1.481	104	
O(1)=S(3)	C–**S**(=**O**)–C	1.497	1.498	0.013	1.489	1.505	90	5
O–Se	see BAPPAJ, BIRGUE10, BIRHAL10, CXMSEO, DGLYSE, SPSEBU (1.597 for **O**=**Se** to 1.974 for **O**–**Se**)							
O(2)–Si(5)	(X–**O**)$_3$–**Si**–(N)(C)	1.663	1.658	0.023	1.650	1.665	21	
O(2)–Si(4)	X$_3$–**Si**–**O**–X (overall)	1.631	1.630	0.022	1.617	1.646	191	
O(2)–Si(4)	subsets of this group are:							
	X$_3$–**Si**–**O**–C#	1.645	1.647	0.012	1.634	1.652	29	
	X$_3$–**Si**–**O**–Si–X$_3$	1.622	1.625	0.014	1.614	1.631	70	
	X$_3$–**Si**–**O**–O–Si–X$_3$	1.680	1.676	0.008	1.673	1.688	10	
O(2)–Te(6)	(X–**O**)$_6$–**Te**	1.927	1.927	0.020	1.908	1.942	16	
O(2)–Te(4)	(X–**O**)$_2$–**Te**–X$_2$	2.133	2.136	0.054	2.078	2.177	12	
P(4)–P(4)	X$_3$–**P**–**P**–X$_3$	2.256	2.259	0.025	2.243	2.277	6	
P(4)–P(3)	see CECHEX (2.197), COZPIQ (2.249)							
P(3)–P(3)	X$_2$–**P**–**P**–X$_2$	2.214	2.210	0.022	2.200	2.224	41	
P(4)=P(4)	see BUTSUE (2.054)							
P(3)=P(3)	see BALXOB (2.034)							

Bond	Substructure	d	m	σ	q_1	q_u	n	Note
P(4)=S(1)	C_3-**P=S**	1.954	1.952	0.005	1.950	1.957	13	
	$(N,O)_2(C)-$**P=S**	1.922	1.924	0.014	1.913	1.927	26	
	$(N,O)_3-$**P=S**	1.913	1.914	0.014	1.906	1.921	50	
P(4)=Se(1)	X_3-**P=Se**	2.093	2.099	0.019	2.075	2.108	12	
P(3)−Si(4)	X_2-**P−Si**$-X_3$: 3- and 4-rings	2.264	2.260	0.019	2.249	2.283	22	
	excluded (see BOPFER, BOPFIV, CASTOF10, COZVIW: 2.201–2.317)							
P(4)=Te(1)	see MOPHTE (2.356), TTEBPZ (2.327)							
S(2)−S(2)	C−**S−S**−C							
	τ(SS) = 75–105°	2.031	2.029	0.015	2.021	2.038	46	
	τ(SS) = 0–20°	2.070	2.068	0.022	2.057	2.077	28	
	(overall)	2.048	2.045	0.026	2.028	2.068	99	
	in polysulphide chain−**S−S**−S−	2.051	2.050	0.022	2.037	2.065	126	
S(2)−S(1)	X−N=**S−S**	1.897	1.896	0.012	1.887	1.908	5	
S−Se(4)	see BUWZUO (2.264, 2.269)							
S−Se(2)	X−**Se−S** (any)	2.193	2.195	0.015	2.174	2.207	9	
S(2)−Si(4)	X_3-**Si−S**$-X$	2.145	2.138	0.020	2.130	2.158	19	
S(2)−Te	X−**S−Te** (any)	2.405	2.406	0.022	2.383	2.424	10	
	X=**S−Te** (any)	2.682	2.686	0.035	2.673	2.694	28	
Se(2)−Se(2)	X−**Se−Se**−X	2.340	2.340	0.024	2.315	2.361	15	
Se(2)−Te(2)	see BAWFUA, BAWGAH (2.524–2.561)							†
Si(4)−Si(4)	X_3-**Si−Si**$-X_3$ 3-membered rings excluded: see CIHRAM (2.511)	2.359	2.359	0.012	2.349	2.366	42	
Te−Te	see CAHJOK (2.751, 2.704)							

† The standard deviation in the sample for the bond type is greater than for the other entries.

Appendix 1. (Footnotes to Table)

1. Sample dominated by B−CH$_3$. For longer bonds in B$^-$−CH$_3$ see LITMEB10 [B(4)−CH$_3$ = 1.621–1.644Å].
2. p(π)–p(π) Bonding with Bsp^2 and Nsp^2 coplanar (τBN = 0 ± 15°) predominates. See G. Schmidt, R. Boese, and D. Bläser, *Z. Naturforsch.*, 1982, **37b**, 1230.
3. 84 observations range from 1.38 to 1.61 Å and individual values depend on substituents on B and O. For a discussion of borinic acid adducts see S. J. Rettig and J. Trotter, *Can. J. Chem.*, 1982, **60**, 2957.
4. See M. Kaftory in *The Chemistry of Functional Groups. Supplement D: The Chemistry of Halides, Pseudohalides, and Azides*, S. Patai and Z. Rappoport, Eds., Wiley: New York, 1983, Part 2, ch. 24.
5. Bonds which are endocyclic or exocyclic to any 3- or 4-membered rings have been omitted from all averages in this section.
6. The overall average given here is for Csp^3–Csp^3 bonds which carry only C or H substituents. The value cited reflects the relative abundance of each 'substitution' group. The 'mean of means' for the 9 subgroups is 1.538 (σ = 0.022) Å.
7. See F. H. Allen, (a) *Acta Crystallogr.*, 1980, **B36**, 81; (b) 1981, **B37**, 890.
8. See F. H. Allen, *Acta Crystallogr.*, 1984, **B40**, 64.
9. See F. H. Allen, *Tetrahedron*, 1982, **38**, 2843.
10. See F. H. Allen, *Tetrahedron*, 1982, **38**, 645.
11. Cyclopropanones and cyclobutanones excluded.
12. See W. B. Schweizer and J. D. Dunitz, *Helv. Chim. Acta*, 1982, **65**, 1547.
13. See L. Norskov-Lauritsen, H.-B. Bürgi, P. Hoffmann, and H. R. Schmidt, *Helv. Chim. Acta*, 1985, **68**, 76.
14. See P. Chakrabarti and J. D. Dunitz, *Helv. Chim. Acta*, 1982, **65**, 1555.
15. See J. L. Hencher in *The Chemistry of the C≡C Triple Bond*, S. Patai, Ed., Wiley, New York, 1978, ch. 2.
16. Conjugated: torsion angle about central C−C single bond is 0 ± 20° (*cis*) or 180 ± 20° (*trans*).
17. Unconjugated: torsion angle about central C−C single bond is 20–160°.
18. Other conjugative substituents excluded.
19. TCNQ is tetracyanoquinodimethane.
20. No difference detected between C2 ≃ C3 and C3 ≃ C4 bonds.
21. Derived from neutron diffraction results only.
22. Nsp^3: pyramidal; mean valence angle at N is in range 108–114°.
23. Nsp^2: planar; mean valence angle at N is ≥ 117.5°.
24. Cyclic and acyclic peptides.
25. See R. H. Blessing, *J. Am. Chem. Soc.*, 1983, **105**, 2776.
26. See L. Lebioda, *Acta Crystallogr.*, 1980, **B36**, 271.
27. n = 3 or 4, i.e. tri- or tetra-substituted ureas.
28. Overall value also includes structures with mean valence angle at N in the range 115–118°.
29. See F. H. Allen and A. J. Kirby, *J. Am. Chem. Soc.*, 1984, **106**, 6197.
30. See A. J. Kirby, 'The Anomeric Effect and Related Stereoelectronic Effects at Oxygen,' Springer, Berlin, 1983.
31. See B. Fuchs, L. Schleifer, and E. Tartakovsky, *Nouv. J. Chim.*, 1984, **8**, 275.
32. See S. C. Nyburg and C. H. Faerman, *J. Mol. Struct.*, 1986, **140**, 347.
33. Sample dominated by P−CH$_3$ and P−CH$_2$−C.
34. Sample dominated by C* = methyl.
35. See A. Kalman, M. Czugler, and G. Argay, *Acta Crystallogr.*, 1981, **B37**, 868.
36. Bimodal distribution resolved into 22 'short' bonds and 5 longer outliers.
37. All 24 observations come from BUDTEZ.
38. 'Long' O−H bonds in centrosymmetric O---H---O H−bonded dimers are excluded.
39. N−N bond length also dependent on torsion angle about N−N bond and on nature of substituent C atoms; these effects are ignored here.
40. N pyramidal has average angle at N in range 100–113.5°; N planar has average angle of ≥ 117.5°.
41. See R. R. Holmes and J. A. Deiters, *J. Amer. Chem. Soc.*, 1977, **99**, 3318.
42. No detectable variation in S=O bond length with type of C-substituent.

Appendix 2

Short-form references to individual CSD entries cited by reference code in the Table. A full list of CSD bibliographic entries is given in SUP 56701.

ACBZPO01	*J. Am. Chem. Soc.*, 1975, **97**, 6729.	CIWYIQ	*Inorg. Chem.*, 1984, **23**, 1946.
ACLTEP	*J. Organomet. Chem.*, 1980, **184**, 417.	CIYFOF	*Inorg. Chem.*, 1984, **23**, 1790.
ASAZOC	*Dokl. Akad. Nauk SSSR*, 1979, **249**, 120.	CMBIDZ	*J. Org. Chem.*, 1979, **44**, 1447.
BALXOB	*J. Am. Chem. Soc.*, 1981, **103**, 4587.	CODDEE	*Z. Naturforsch., Teil B*, 1984, **39**, 1257.
BAPPAJ	*Inorg. Chem.*, 1981, **20**, 3071.	CODDII	*Z. Naturforsch., Teil B*, 1984, **39**, 1257.
BARRIV	*Acta Chem. Scand., Ser. A*, 1981, **35**, 443.	COFVOI	*Z. Naturforsch., Teil B*, 1984, **39**, 1027.
BAWFUA	*Cryst. Struct. Commun.*, 1981, **10**, 1345.	COJCUZ	*Chem. Ber.*, 1984, **117**, 2686.
BAWGAH	*Cryst. Struct. Commun.*, 1981, **10**, 1353.	COSDIX	*Z. Naturforsch., Teil B*, 1984, **39**, 1344.
BECTAE	*J. Org. Chem.*, 1981, **46**, 5048, 1981.	COZPIQ	*Chem. Ber.*, 1984, **117**, 2063.
BELNIP	*Z. Naturforsch., Teil B*, 1982, **37**, 299.	COZVIW	*Z. Anorg. Allg. Chem.*, 1984, **515**, 7.
BEMLIO	*Chem. Ber.*, 1982, **115**, 1126.	CTCNSE	*J. Am. Chem. Soc.*, 1980, **102**, 5430.
BEPZEB	*Cryst. Struct. Commun.*, 1982, **11**, 175.	CUCPIZ	*J. Am. Chem. Soc.*, 1984, **106**, 7529.
BETJOZ	*J. Am. Chem. Soc.*, 1982, **104**, 1683.	CUDLOC	*J. Cryst. Spectrosc.*, 1985, **15**, 53.
BETUTE10	*Acta Chem. Scand., Ser. A*, 1976, **30**, 719.	CUDLUI	*J. Cryst. Spectrosc.*, 1985, **15**, 53.
BIBLAZ	*Zh. Strukt. Khim.*, 1981, **22**, 118.	CUGBAH	*Acta Crystallogr., Sect. C*, 1985, **41**, 476.
BICGEZ	*Z. Anorg. Allg. Chem.*, 1982, **486**, 90.	CXMSEO	*Acta Crystallogr., Sect. B*, 1973, **29**, 595.
BIHXIZ	*J. Chem. Soc., Chem. Commun.*, 1982, 982.	DGLYSE	*Acta Crystallogr., Sect. B*, 1975, **31**, 1785.
BIRGUE10	*Z. Naturforsch., Teil B*, 1983, **38**, 20.	DMESIP01	*Acta Crystallogr., Sect. C*, 1984, **40**, 895.
BIRHAL10	*Z. Naturforsch., Teil B*, 1982, **37**, 1410.	DSEMOR10	*J. Chem. Soc., Dalton Trans.*, 1980, 628.
BIZJAV	*J. Organomet. Chem.*, 1982, **238**, C1.	DTHIBR10	*Inorg. Chem.*, 1971, **10**, 697.
BOGPOC	*Z. Naturforsch., Teil B*, 1982, **37**, 1402.	EPHTEA	*Inorg. Chem.*, 1980, **19**, 2487.
BOGSUL	*Z. Naturforsch., Teil B*, 1982, **37**, 1230.	ESEARS	*J. Chem. Soc. C*, 1971, 1511.
BOJLER	*Z. Anorg. Allg. Chem.*, 1982, **493**, 53.	ETEARS	*J. Chem. Soc. C*, 1971, 1511.
BOJPUL	*Acta Chem. Scand., Ser. A*, 1982, **36**, 829.	FMESIB	*J. Organomet. Chem.*, 1980, **197**, 275.
BOPFER	*Chem. Ber.*, 1983, **116**, 146.	FPHTEL	*J. Chem. Soc., Dalton Trans.*, 1980, 2306.
BOPFIV	*Chem. Ber.*, 1983, **116**, 146.	FPSULF10	*J. Am. Chem. Soc.*, 1982, **104**, 1683.
BOVMEE	*Acta Crystallogr., Sect. B*, 1982, **38**, 1048.	HCLENE10	*Acta Crystallogr., Sect. B*, 1982, **38**, 3139.
BQUINI	*Acta Crystallogr., Sect. B*, 1979, **35**, 1930.	HMTITI	*Acta Crystallogr., Sect. B*, 1975, **31**, 1505.
BTUPTE	*Acta Chem. Scand., Ser. A*, 1975, **29**, 738.	HMTNTI	*Z. Anorg. Allg. Chem.*, 1974, **409**, 237.
BUDTEZ	*Z. Naturforsch., Teil B*, 1983, **38**, 454.	HXPASC	*J. Chem. Soc., Dalton Trans.*, 1975, 1381.
BUPSIB10	*Z. Anorg. Allg. Chem.*, 1981, **474**, 31.	IBZDAC11	*J. Chem. Soc., Dalton Trans.*, 1979, 854.
BUSHAY	*Z. Naturforsch., Teil. B*, 1983, **38**, 692.	IFORAM	*Monatsh. Chem.*, 1974, **105**, 621.
BUTHAZ10	*Inorg. Chem.*, 1984, **23**, 2582.	IODMAM	*Acta Crystallogr., Sect. B*, 1977, **33**, 3209.
BUTSUE	*J. Chem. Soc., Chem. Commun.*, 1983, 862.	IPMUDS	*Acta Crystallogr., Sect. B*, 1973, **29**, 2128.
BUWZUO	*Acta Chem. Scand., Ser A*, 1983, **37**, 219.	ISUREA10	*Acta Crystallogr., Sect. B*, 1972, **28**, 643.
BZPRIB	*Z. Naturforsch., Teil B*, 1981, **36**, 922.	LITMEB10	*J. Am. Chem. Soc.*, 1975, **97**, 6401.
BZTPPI	*Inorg. Chem.*, 1978, **17**, 894.	MESIAD	*Z. Naturforsch., Teil B*, 1980, **35**, 789.
CAHJOK	*Inorg. Chem.*, 1983, **22**, 1809.	METAMM	*Acta Crystallogr.*, 1964, **17**, 1336.
CAJMAB	*Chem. Z*, 1983, **107**, 169.	MNPSIL	*J. Am. Chem. Soc.*, 1969, **91**, 4134.
CANLUY	*Tetrahedron Lett.*, 1983, **24**, 4337.	MODIAZ	*J. Heterocycl. Chem.*, 1980, **17**, 1217.
CASSAQ	*J. Struct. Chem.*, 1983, **2**, 101.	MOPHTE	*Acta Chem. Scand., Ser. A*, 1980, **34**, 333.
CASTOF10	*Acta Crystallogr., Sect. C*, 1984, **40**, 1879.	MORTRS10	*J. Chem. Soc., Dalton Trans.*, 1980, 628.
CASYOK	*J. Struct. Chem.*, 1983, **2**, 107.	NAPSEZ10	*J. Am. Chem. Soc.*, 1980, **102**, 5070.
CECHEX	*Z. Anorg. Allg. Chem.*, 1984, **508**, 61.	NBBZAM	*Z. Naturforsch., Teil B*, 1977, **32**, 1416.
CECXEN	*J. Struct. Chem.*, 1983, **2**, 207.	OPIMAS	*Aust. J. Chem.*, 1977, **30**, 2417.
CEDCUJ	*J. Org. Chem.*, 1983, **48**, 5149.	OPNTEC10	*J. Chem. Soc., Dalton Trans.*, 1982, 251.
CEHKAB	*Z. Naturforsch., Teil B*, 1984, **39**, 139.	PHASCL	*Acta Crystallogr., Sect. B*, 1981, **37**, 1357.
CELDOM	*Acta Crystallogr., Sect. C*, 1984, **40**, 556.	PHASOC01	*Aust. J. Chem.*, 1975, **28**, 15.
CESSAU	*Acta Crystallogr., Sect. C*, 1984, **40**, 653.	PNPOSI	*J. Am. Chem. Soc.*, 1968, **90**, 5102.
CETTAW	*Chem. Ber.*, 1984, **117**, 1089.	SEBZQI	*J. Chem. Soc., Chem. Commun.*, 1977, 325.
CETUTE	*Acta Chem. Scand., Ser A*, 1975, **29**, 763.	SPSEBU	*Acta Chem. Scand., Ser. A*, 1979, **33**, 403.
CEYLUN	*Izv. Akad. Nauk SSSR, Ser. Khim.*, 1983, 2744.	TEACBR	*Cryst. Struct. Commun.*, 1974, **3**, 753.
CIFZUM	*Acta Chem. Scand., Ser A*, 1984, **38**, 289.	THINBR	*J. Am. Chem. Soc.*, 1970, **92**, 4002.
CIHRAM	*Angew. Chem., Int. Ed. Engl.*, 1984, **23**, 302.	TMPBTI	*Acta Crystallogr., Sect. B*, 1975, **31**, 1116.
CILRUK	*J. Chem. Soc., Chem. Commun.*, 1984, 1023.	TPASSN	*J. Chem. Soc., Dalton Trans.*, 1977, 514.
CILSAR	*J. Chem. Soc., Chem. Commun.*, 1984, 1021.	TPASTB	*Cryst. Struct. Commun.*, 1976, **5**, 39.
CIMHIP	*Acta Crystallogr., C*, 1984, **40**, 1458.	TPHOSI	*Z. Naturforsch., Teil B*, 1979, **34**, 1064.
CINTEY	*Dokl. Akad. Nauk SSSR*, 1984, **274**, 615.	TTEBPZ	*Z. Naturforsch., Teil B*, 1979, **34**, 256.
CIPBUY	*J. Struct. Chem.*, 1983, **2**, 281.	ZCMXSP	*Cryst. Struct. Commun.*, 1977, **6**, 93.
CISMUM	*Z. Naturforsch., Teil B*, 1984, **39**, 485.		
CISTED	*Z. Anorg. Allg. Chem.*, 1984, **511**, 95.		

BOND LENGTHS IN ORGANOMETALLIC COMPOUNDS

This table summarizes the average values of interatomic distances of representative metal–ligand bonds. Sigma bonds between *d*- and *f*-block metals and the elements C, N, O, P, S, and As are included. The values are extracted from a much larger list in Reference 1. The tabulated values are the unweighted means of reported measurements on compounds in each category. If four or more measurements are available, the standard deviation is given in parentheses. All values are in Ångstrom units (10^{-10} m).

The first part of the table covers metal–carbon bonds in different ligand categories, while the second part covers metal bonds to other elements. R stands for any alkyl group; Me for a CH_3 group; C_6R_5 indicates an aryl group; and C(=O)R an acyl group. Metals are listed in atomic number order.

Reference

1. Orpen, A. G., Brammer, L., Allen, F. H., Kennard, O., Watson, D. G., and Taylor, R., *J. Chem. Soc. Dalton Trans.*, 1989, S1-S83.

M	M–CH$_3$	M–CH$_2$R	M–CR=CR$_2$	M–C$_6$R$_5$	M–C(=O)R
Ti		2.167	2.215(0.042)	2.148	
V				2.114(0.012)	
Cr	2.168		2.035(0.009)	2.075(0.019)	
Mn	2.095(0.030)	2.176(0.024)	2.007	2.064(0.021)	2.044
Fe	2.074	2.091(0.030)	1.991(0.039)	2.031(0.062)	1.997(0.033)
Co	2.014(0.023)	2.039(0.032)	1.934(0.019)	1.974	1.990
Ni	2.029	1.964	1.892(0.017)	1.917(0.038)	1.850(0.059)
Cu				2.020	
Zn		1.964			
Zr	2.292(0.049)		2.257		
Nb	2.336	1.319			
Mo	2.254(0.065)	2.250(0.061)	2.204(0.049)	2.193(0.054)	2.109
Ru	2.179(0.045)	2.036(0.010)	2.063	2.092(0.057)	2.091
Rh	2.092(0.027)	2.100	2.040(0.054)	2.011(0.026)	1.995(0.031)
Pd		2.028	2.000(0.024)	1.981(0.032)	1.982(0.029)
Hf	2.275(0.049)		2.205		
Ta	2.217(0.035)	2.225(0.056)		2.199(0.073)	
W	2.189(0.039)	2.175	2.224		
Re	2.173(0.051)	2.290		2.027	2.190(0.027)
Os		2.221	2.052	2.090(0.032)	2.161
Ir	2.175		2.071(0.044)	2.070(0.038)	2.019
Pt	2.083(0.045)	2.062(0.031)	2.024(0.037)	2.049(0.046)	1.991(0.025)
Au	2.066(0.045)		2.042	2.059(0.024)	
Hg	2.072(0.026)	2.125		2.086(0.040)	
Th	2.567				

M	M–NH₃	M–OH₂	M–PMe₃	M–SR	M–AsR₃
Ti		2.066(0.052)		2.369	2.686
V		2.129(0.131)	2.510(0.010)	2.378(0.007)	
Cr	2.069(0.008)	1.997(0.070)	2.389(0.069)	2.362	2.460(0.040)
Mn		2.189(0.040)	2.455(0.164)	2.366(0.054)	2.400(0.013)
Fe		2.085(0.066)	2.246(0.042)	2.271(0.028)	2.352(0.043)
Co	1.965(0.021)	2.085(0.064)	2.217(0.043)	2.254(0.025)	2.323(0.021)
Ni	2.074(0.093)	2.079(0.038)	2.204(0.031)	2.187(0.007)	2.333(0.035)
Cu	1.987(0.017)	2.186(0.215)			2.367(0.016)
Zn	2.044	2.090(0.061)		2.295	
Y		2.398(0.068)			
Zr			2.692		
Nb		2.248(0.137)			2.741(0.008)
Mo	2.217	2.201(0.094)	2.462(0.046)	2.401(0.050)	2.582(0.036)
Ru	2.126(0.024)	2.074(0.051)	2.307(0.050)		2.446(0.031)
Rh	2.114(0.018)	2.190(0.096)	2.266(0.036)		2.416(0.039)
Pd	2.032	2.200	2.287(0.018)		2.386(0.052)
Ag		2.350			
Cd		2.318(0.065)		2.444	
La		2.556(0.062)			
Ce		2.565(0.063)			
Pr		2.518(0.038)			
Nd		2.533(0.058)			
Sm		2.459(0.050)			
Eu		2.441(0.055)			
Gd		2.443(0.074)			
Tb		2.455			
Dy		2.409(0.074)			
Ho		2.407(0.069)			
Er		2.404(0.083)			
Yb		2.353(0.066)			
Lu		2.404(0.116)			
Ta			2.589(0.044)		
W		2.115(0.065)	2.485(0.039)		
Re	2.253	2.199(0.091)	2.369(0.065)		2.575(0.006)
Os	2.136	2.166	2.328(0.029)		
Ir	2.050(0.021)		2.323(0.028)	2.461	
Pt			2.295(0.036)	2.320(0.015)	2.366(0.058)
Au		2.157		2.293	
Hg		2.690(0.083)		2.402(0.065)	
Th		2.483(0.032)			
U		2.455(0.047)			

STRUCTURE OF FREE MOLECULES IN THE GAS PHASE

David R. Lide

This table gives information on the geometric structure of selected molecules in the gas phase, including the overall geometry, interatomic distances, and bond angles. The molecules have been chosen to provide data on a wide variety of chemical bonds and to illustrate the influence of molecular environment on bond distances and angles. The table is restricted to molecules with conventional covalent or ionic bonds, but it should be pointed out that structure data on many loosely bonded complexes of the van der Waals type have recently become available. The references below contain data on many molecules that are not included here and give additional information such as uncertainties and isotopic variations.

The two techniques for gas phase structure determination are spectroscopy and electron diffraction. The following codes are used to indicate the method used for each set of data:

ED – Gas phase electron diffraction
MW – Microwave spectroscopy, including both measurements in bulk gases and molecular beams
IR – Infrared spectroscopy
R – Raman spectroscopy
UV – Electronic spectroscopy in the ultraviolet and visible regions, including fluorescence measurements
ESR – Electron spin resonance.

In some cases data from two sources have been combined to derive the structure; these are labeled by "ED, MW," for example.

Because of the internal vibrations that are present in all molecules, even in their lowest energy state, the definition of interatomic distance is not a simple matter. The ideal measure is the equilibrium distance in the hypothetical non-vibrating state, designated by r_e. This is the value of the separation of the atoms at the minimum of the potential function that describes the forces between the two atoms. All other measures represent some form of average, generally complex, over the vibrational motions. Since the potential function is asymmetric and less steep at distances beyond the potential minimum, the average distance is normally greater than r_e. Distances determined by electron diffraction (ED) represent an average over all vibrational states that are populated at the temperature of the measurement; the most common measure is designated r_g. Distances determined by spectroscopy (MW, IR, R, or UV) through measurements on the ground vibrational state of the molecule, designated by r_0, describe some form of average, not easily defined, over the zero-point vibrations. Another measure that is frequently used in microwave spectroscopy is the "substitution" distance r_s, which is operationally defined through a series of measurements on different isotopic species. In simple cases, r_s often lies between r_0 and r_e and is therefore a closer approximation to r_e. Several other types of averages have been used; good discussions can be found in Volumes II/25 and II/28 of the *Landolt-Börnstein* series (Reference 1) and in References 4 and 5.

Unless otherwise specified, distances and angles given in this table are r_0 values if the method is spectroscopic and r_g values if the method is electron diffraction. When given, equilibrium and substitution distances are designated by r_e and r_s, respectively.

Many interatomic distances and angles calculated by *ab initio* techniques have been reported in the recent literature. However, it should be emphasized that all data in this table are obtained from direct experimental measurements. In a few cases, *ab initio* calculations of vibration-rotation interaction constants have been combined with the primary experimental measurements to derive r_e values in the table.

The number of significant figures in the values is an indication of the precision of the measurement; thus a distance quoted to three decimal places is probably reliable to about 0.005 Å or better. However, discrepancies between r_e, r_0, and r_g values for the same bond are often the order of 0.01 Å because of vibrational averaging considerations, so care must be taken in comparing bond distances in different molecules. Some distances in simple molecules are given here to four or five decimal places, but little chemical significance can be attached to differences beyond the third decimal place.

The table is presented in two parts: Part A covers molecules that do not contain carbon while Part B lists carbon-containing molecules. Because many of the entries in Part A are free radicals or other transient species whose systematic chemical names are unfamiliar, the listing in Part A is in order of chemical formula. Part B is ordered by name. In both parts the second column gives information on the overall configuration of the molecule, often indicated by the point group of the equilibrium geometry. Columns 3 through 8 give the values of the bond distances and angles, and the last column indicates the experimental method. Distances are given in Å units, where 1 Å = 10^{-10} m or 0.1 nm. Angles are given in degrees.

The efforts of Kozo Kuchitsu in preparing an earlier version of this table and in giving advice on the new version are gratefully acknowledged.

References

1. *Landolt-Börnstein Numerical Data and Functional Relationships in Science and Technology*, Springer-Verlag, Berlin. The following volumes are in the series *Structure Data of Free Polyatomic Molecules*:
   II/7, 1976
   II/15, 1987
   II/21, *Supplement to II/7 and II/15*, 1992
   II/23, *Supplement to II/7, II/15, and II/21*, 1995
   II/25A, *Inorganic Molecules*, 1998
   II/25B, *Molecules Containing One or Two Carbon Atoms*, 1999
   II/25C, *Molecules Containing Three or Four Carbon Atoms*, 2000
   II/25D, *Molecules Containing Five or More Carbon Atoms*, 2003
   II/28A, *Inorganic Molecules*, 2006
   II/28B, *Molecules Containing One or Two Carbon Atoms*, 2006
   II/28C, *Molecules Containing Three or Four Carbon Atoms*, 2007
   II/28D, *Molecules Containing Five or More Carbon Atoms*, 2007.
2. Harmony, M. D., Laurie, V. W., Kuczkowski, R. L., Schwendeman, R. H., Ramsay, D. A., Lovas, F. J., Lafferty, W. J., and Maki, A. G., "Molecular Structure of Gas-Phase Polyatomic Molecules Determined by Spectroscopic Methods", *J. Phys. Chem. Ref. Data*, 8, 619, 1979.
3. Huber, K. P., and Herzberg, G., *Molecular Spectra and Molecular Structure IV. Constants of Diatomic Molecules*, Van Nostrand Reinhold, London, 1979.
4. Hargittai, M., "Molecular Structure of Metal Halides," *Chem. Rev*. 100, 2233-2301, 2000.
5. Harmony, M. D., and Berry, R. J., *Struct. Chem*. 1, 49, 1989.

Part 1 Molecules Not Containing Carbon

Formula	Structure	Bond distances in Å and angles in degrees						Method
AgBr		Ag—Br (r_e)	2.3931					MW
AgCl		Ag—Cl (r_e)	2.2808					MW
AgF		Ag—F (r_e)	1.9832					MW
AgH		Ag—H (r_e)	1.617					UV
AgI		Ag—I (r_e)	2.5446					MW
AgLi		Ag—Li	2.41					UV
AgO		Ag—O (r_e)	2.0030					UV
AgOH	bent	Ag—O	2.016	O—H	0.952	∠HOAg	108.3 (ass.)	MW
AlBr		Al—Br (r_e)	2.295					UV
AlBr₃	D_{3h}	Al—Br	2.221					ED
AlCa		Al—Ca	3.148					UV
AlCl		Al—Cl (r_e)	2.1301					MW
AlCl₃	D_{3h}	Al—Cl	2.063					ED
AlCo		Al—Co	2.283					UV
AlCu		Al—Cu	2.339					UV
AlF		Al—F (r_e)	1.6544					MW
AlF₃	D_{3h}	Al—F	1.633					ED
AlH		Al—H (r_e)	1.6482					UV
AlI		Al—I (r_e)	2.5371					MW
AlI₃	D_{3h}	Al—I	2.461					ED
AlK		Al—K	3.88					UV
AlMn		Al—Mn	2.638					UV
AlNi		Al—Ni	2.321					UV
AlO		Al—O (r_e)	1.6176					UV
AlS		Al—S (r_e)	2.029					UV
AlV		Al—V	2.620					UV
AlZn		Al—Zn	2.696					UV
Al₂		Al—Al (r_e)	2.701					UV
Al₂Br₆	Br$_a$ Br$_b$ Br$_a$ Al Al Br$_a$ Br$_b$ Br$_a$ D_{2h}	Al—Br$_a$ ∠Br$_b$AlBr$_b$	2.234 91.6	Al—Br$_b$ ∠Br$_a$AlBr$_a$	2.433 122			ED
Al₂Cl₆	See Al₂Br₆ D_{2h}	Al—Cl$_a$ ∠Cl$_a$AlCl$_b$	2.061 90.0	Al—Cl$_b$ ∠Cl$_a$AlCl$_a$	2.250 122			ED
AsBr₃	C_{3v}	As—Br	2.324	∠BrAsBr	99.6			ED
AsCl₃	C_{3v}	As—Cl	2.165	∠ClAsCl	98.6			ED, MW
AsF₃	C_{3v}	As—F	1.710	∠FAsF	95.9			ED, MW
AsF₅	F$_a$ F$_b$ F$_b$—As F$_b$ F$_a$ F$_b$ D_{3h}	As—F$_a$	1.711	As—F$_b$	1.656			ED
AsH		As—H (r_e)	1.5232					UV
AsH₃	C_{3v}	As—H (r_e)	1.511	∠HAsH (θ_e)	92.1			MW, IR
AsI₃	C_{3v}	As—I	2.557	∠IAsI	100.2			ED
AsN		As—N (r_e)	1.6184					UV
AsO		As—O (r_e)	1.6236					UV
AsP		As—P (r_e)	1.99954					MW
As₂		As—As (r_e)	2.1026					UV
AuH		Au—H (r_e)	1.5237					UV
Au₂		Au—Au (r_e)	2.4719					UV
BBr		B—Br (r_e)	1.888					UV
BBr₃	D_{3h}	B—Br	1.893					ED
BCl		B—Cl (r_e)	1.7153					UV
BClF₂	C_{2v}	B—Cl (r_s)	1.728	B—F	1.315	∠FBF	118.1	MW

Formula	Structure	Bond distances in Å and angles in degrees						Method
BCl_3	D_{3h}	B—Cl	1.742					ED
BF		B—F (r_e)	1.2626					UV
BF_2H		B—H	1.189	B—F	1.311	∠FBF	118.3	MW
BF_2OH	F_aF_bBOH	B—F_a (r_e)	1.3229	B—F_b (r_e)	1.3129	B—O (r_e)	1.3448	MW
	planar	∠FBF (θ_e)	118.36	∠F_aBO (θ_e)	122.25	∠BOH (θ_e)	113.14	
	F_a *cis* to OH	O—H (r_e)	0.9574					
BF_3	D_{3h}	B—F	1.313					ED, IR
BH		B—H (r_e)	1.2325					UV
BH_2NH_2	planar	B—N	1.391	B—H	1.195	N—H	1.004	MW
		∠HBH	122.2	∠HNH	114.2			
BH_3	planar	B—H	1.1900					IR
BH_3PH_3	staggered form	B—P	1.937	B—H	1.212	P—H	1.399	MW
		∠PBH	103.6	∠BPH	116.9	∠HBH	114.6	
		∠HPH	101.3					
BI_3	D_{3h}	B—I	2.118					ED
BN		B—N (r_e)	1.281					UV
BO		B—O (r_e)	1.2045					EPR
BO_2	linear	B—O	1.265					UV
BS		B—S	1.6091					UV
B_2		B—B (r_e)	1.590					UV
B_2H_6	(diborane structure)	B—H_a	1.19	B—H_b	1.33	B⋯B	1.77	IR, ED
		∠H_aBH_a	122	∠H_bBH_b	97			
$B_3H_3O_3$		B—O	1.376	∠BOB	120	∠OBO	120	ED
$B_3H_6N_3$	C_2	B—N	1.435	B—H	1.26	N—H	1.05	ED
		∠BNB	121	∠NBN	118			
BaBr		Ba—Br (r_e)	2.8445					UV
$BaBr_2$		Ba—Br	2.912	∠BrBaBr	137.0			ED
BaCl		Ba—Cl (r_e)	2.6828					UV
BaF		Ba—F (r_e)	2.163					UV
BaH		Ba—H (r_e)	2.2318					UV
BaI		Ba—I (r_e)	3.0848					UV
BaI_2		Ba—I	3.150	∠IBaI	137.6			ED
BaO		Ba—O (r_e)	1.9397					MW
BaOH	linear	Ba—O	2.200	O—H	0.927			UV
BaS		Ba—S (r_e)	2.5074					MBE
$BeCl_2$	linear	Be—Cl (r_e)	1.791					ED, IR
BeF		Be—F (r_e)	1.3609					UV
BeF_2	linear	Be—F (r_e)	1.3730					IR
BeH		Be—H (r_e)	1.3431					UV
BeH_2	linear	Be—H (r_e)	1.3264					IR
BeO		Be—O (r_e)	1.3308					UV
BeS		Be—S (r_e)	1.7415					UV
BiBr		Bi—Br (r_e)	2.6095					MW
$BiBr_3$	C_{3v}	Bi—Br	2.577	∠BrBiBr	98.6			ED
BiCl		Bi—Cl (r_e)	2.4716					MW
$BiCl_3$	C_{3v}	Bi—Cl	2.424	∠ClBiCl	97.5			ED
BiF		Bi—F (r_e)	2.0516					MW
BiF_3	C_{3v}	Bi—F	1.987	∠FBiF	96.1			ED
BiH		Bi—H (r_e)	1.805					UV
BiI		Bi—I (r_e)	2.8005					MW
BiI_3	C_{3v}	Bi—I	2.807	∠IBiI	99.5			ED
BiO		Bi—O (r_e)	1.934					UV
BiP		Bi—P (r_e)	2.29345					IR
Bi_2		Bi—Bi (r_e)	2.6596					UV
BrCl		Br—Cl (r_e)	2.1361					MW
BrF		Br—F (r_e)	1.7590					MW
BrF_3	F_a—Br—F_a	Br—F_a	1.810	∠F_{ax}BrF_{eq}	85.1	∠F_aBrF_b	86.2	MW
		Br—F_b	1.721					
	F_b							
	C_{2v}							

Formula	Structure	Bond distances in Å and angles in degrees						Method
BrF_5	C_{4v}	Br—F (av.)	1.753	(Br—F_{eq}) − (Br—F_{ax})	0.069	$\angle F_{ax}BrF_{eq}$	85.1	ED, MW
BrN_3	$BrN_aN_bN_c$ planar	N_a—N_b	1.113 (ass.)	N_b—N_c	1.247	N_a—Br	1.899	ED
		$\angle NNN$	170.7	$\angle BrNN$	109.7			
BrO		Br—O (r_e)	1.7172					MW
BrO_2	C_{2v}	Br—O (r_e)	1.644	$\angle OBrO$ (θ_e)	114.3			MW
Br_2		Br—Br (r_e)	2.2811					R
$CaBr_2$	linear	Ca—Br	2.62					ED
CaCl		Ca—Cl (r_e)	2.43676					UV
$CaCl_2$	linear	Ca—Cl	2.483					ED
CaF		Ca—F (r_e)	1.967					UV
CaH		Ca—H (r_e)	2.002					UV
CaI		Ca—I (r_e)	2.8286					UV
CaI_2	linear	Ca—I	2.840					ED
CaO		Ca—O (r_e)	1.8221					UV
CaOH	linear	Ca—O	1.985	O—H	0.921			UV
CaS		Ca—S (r_e)	2.3178					UV
CdH		Cd—H (r_e)	1.781					EPR
CdH_2	linear	Cd—H	1.6792					IR
$CdBr_2$	linear	Cd—Br	2.394					ED
$CdCl_2$	linear	Cd—Cl	2.284					ED
CdI_2	linear	Cd—I	2.582					ED
CeF_4	T_d	Ce—F	2.036					ED
CeI_3	quasiplanar	Ce—I	2.948					ED
ClBS	linear	B—Cl	1.681	B—S	1.606			MW
ClF		Cl—F (r_e)	1.6283					MW
ClF_3	F_a—Cl—F_a | F_b	Cl—F_a	1.698	Cl—F_b	1.598	$\angle F_aClF_b$	87.5	MW
ClN_3	$ClN_aN_bN_c$ planar	N_a—N_b	1.253	N_b—N_c	1.113	N_a—Cl	1.746	MW
		$\angle NNN$	171.0	$\angle ClNN$	108.7			
ClO		Cl—O (r_e)	1.5696					MW, UV
ClO_2	C_{2v}	Cl—O	1.470	$\angle OClO$	117.38			MW
Cl_2		Cl—Cl (r_e)	1.9878					UV
Cl_2O	C_{2v}	Cl—O	1.6959	$\angle ClOCl$	110.89			MW
$CoBr_2$	linear	Co—Br	2.241					ED
$CoCl_2$	linear	Co—Cl	2.113					ED
CoF_2	linear	Co—F	1.754	[Co—F (r_e)]	1.738			ED
CoF_3	D_{3h}	Co—F	1.732					ED
CoH		Co—H (r_e)	1.542					UV
CrF_2	linear	Cr—F	1.795					ED
CrF_3	D_{3h}	Cr—F	1.732					ED
CrF_4	T_d	Cr—F	1.706					ED
CrH		Cr—H (r_e)	1.656					UV
CrO		Cr—O (r_e)	1.615					UV
CsBr		Cs—Br (r_e)	3.0723					MW
CsCl		Cs—Cl (r_e)	2.9063					MW
CsF		Cs—F (r_e)	2.3454					MW
CsH		Cs—H (r_e)	2.4938					UV
CsI		Cs—I (r_e)	3.3152					MW
CsO		Cs—O (r_e)	2.3007					MW
CsOH	linear; large amplitude bending mode	Cs—O (r_e)	2.395	O—H (r_e)	0.97			MW
Cs_2		Cs—Cs (r_e)	4.47					UV
CuBr		Cu—Br (r_e)	2.1734					MW
CuCl		Cu—Cl (r_e)	2.0512					MW
CuF		Cu—F (r_e)	1.7449					MW
CuF_2	linear	Cu—F	1.713					ED
CuH		Cu—H (r_e)	1.4626					UV
CuI		Cu—I (r_e)	2.3383					MW
CuLi		Cu—Li	2.26					UV
CuO		Cu—O (r_e)	1.7244					UV

Formula	Structure	Bond distances in Å and angles in degrees						Method
CuOH	bent	Cu—O (r_s)	1.769	O—H	0.952	∠HOCu	110.24 (θ_s)	MW
CuS		Cu—S	2.051					UV
Cu_2		Cu—Cu (r_e)	2.2197					UV
$DyBr_3$	quasiplanar	Dy—Br	2.609					ED
$DyCl_3$	quasiplanar	Dy—Cl	2.461					ED
FN_3	$FN_aN_bN_c$	N_a—N_b	1.253	N_b—N_c	1.132	N_a—F	1.439	MW
	planar	∠NNN	170.3	∠FNN	103.8			
F_2		F—F (r_e)	1.4119					R
$FeBr_2$	linear	Fe—Br	2.294					ED
$FeCl_2$	linear	Fe—Cl	2.132					UV,ED
FeF_2	linear	Fe—F	1.769	[Fe—F (r_e)]	1.755			ED
FeF_3	D_{3h}	Fe—F	1.763					ED
FeH		Fe—H	1.620					IR
FeO		Fe—O	1.444					UV
FeS		Fe—S	2.017					MW
GaBr		Ga—Br (r_e)	2.3525					MW
$GaBr_3$	D_{3h}	Ga—Br	2.249					ED
GaCl		Ga—Cl (r_e)	2.2017					MW
$GaCl_3$	D_{3h}	Ga—Cl	2.110					ED
GaF		Ga—F (r_e)	1.7744					MW
GaF_3	D_{3h}	Ga—F	1.725					ED
GaH		Ga—H (r_e)	1.663					UV
GaI		Ga—I (r_e)	2.5747					MW
GaI_3	D_{3h}	Ga—I	2.458					ED
GaO		Ga—O	1.744					UV
Ga_2Br_6	See Al_2Br_6	Ga—Br_a	2.250	Ga—Br_b	2.453			ED
	D_{2h}	∠Br_aGaBr_a	92.7	∠Br_bGaBr_b	123			
Ga_2Cl_6	See Al_2Br_6	Ga—Cl_a	2.116	Ga—Cl_b	2.305			ED
	D_{2h}	∠Cl_aGaCl_a	90	∠Cl_bGaCl_b	124.5			
$GdBr_3$	C_{3v}	Gd—Br	2.641					ED
$GdCl_3$	C_{3v}	Gd—Cl	2.488					ED
GdF_3	C_{3v}	Gd—F	2.053					ED
GdI_3	C_{3v}	Gd—I	2.840	∠IGdI	108			ED
$GeBrH_3$	C_{3v}	Ge—H	1.526	Ge—Br	2.299	∠HGeH	106.2	MW, IR
$GeBr_2$		Ge—Br (r_e)	2.359	∠BrGeBr	101.0			ED
$GeBr_4$	T_d	Ge—Br	2.272					ED
$GeClH_3$	C_{3v}	Ge—H	1.537	Ge—Cl	2.150	∠HGeH	111.0	IR, MW
$GeCl_2$		Ge—Cl (r_e)	2.186	∠ClGeCl	100.3			ED
$GeCl_4$	T_d	Ge—Cl	2.113					ED
$GeFH_3$	C_{3v}	Ge—H	1.522	Ge—F	1.732	∠HGeH	113.0	MW, IR
GeF_2		Ge—F (r_e)	1.7321	∠FGeF (θ_e)	97.15			MW
GeH		Ge—H (r_e)	1.5880					UV
GeHI		Ge—I	2.525	Ge—H	1.593	∠HGeI	93.5	UV
GeH_4	T_d	Ge—H	1.5251					IR, R
GeI_2		Ge—I	2.540	∠IGeI	102.1			ED
GeI_4	T_d	Ge—I	2.515					ED
GeO		Ge—O (r_e)	1.6246					MW
GeS		Ge—S (r_e)	2.0121					MW
GeSe		Ge—Se (r_e)	2.1346					MW
GeTe		Ge—Te (r_e)	2.3402					MW
Ge_2H_6		Ge—Ge	2.403	Ge—H	1.541			ED
		∠HGeH	106.4	∠GeGeH	112.5			
HBr		H—Br (r_e)	1.4145					MW
HCl		H—Cl (r_e)	1.2746					MW
HClO	ClOH (bent)	Cl—O	1.690	O—H	0.975	∠HOCl	102.5	MW, IR
$HClO_4$		Cl—O_a	1.407	Cl—O_b	1.639			ED
		∠O_aClO_a	114.3	∠O_aClO_b	104.1			

Formula	Structure	Bond distances in Å and angles in degrees						Method
HF		H—F (r_e)	0.9169					MW
HFO	FOH (bent)	F—O	1.442	O—H	0.96	∠HOF	97.2	MW
HI		H—I (r_e)	1.6090					MW
HIO	IOH (bent)	I—O	1.9941	O—H	0.967	∠HOI	103.9	MW
HNO	bent	N—O	1.212	N—H	1.063	∠HNO	108.6	UV
HNO$_2$	O$_a$ N—O$_b$H	*s-trans* conformer		*s-cis* conformer				MW
		O$_b$—H	0.958	O$_b$—H	0.98			
		N—O$_b$	1.432	N—O$_b$	1.39			
		N—O$_a$	1.170	N—O$_a$	1.19			
		∠O$_a$NO$_b$	110.7	∠O$_a$NO$_b$	114			
		∠NO$_b$H	102.1	∠NO$_b$H	104			
HNO$_3$	planar	N—O$_a$	1.20	N—O$_b$	1.21	N—O$_c$	1.41	MW
		O$_c$—H	0.96	∠O$_c$NO$_b$	115.9	∠HO$_c$N	102.2	
		∠O$_c$NO$_a$	113.9					
HNSO	planar *cis*	N—S	1.512	S—O	1.451	N—H	1.029	MW
		∠NSO	120.4	∠HNS	115.8			
HN$_3$	HN$_a$N$_b$N$_c$ planar	N$_a$—N$_b$	1.245	N$_b$—N$_c$	1.134	N$_a$—H	1.015	MW
		∠NNN	171.8	∠HNN	109.2			
HPO		P—O	1.4843	P—H	1.473	∠HPO	104.57	MW
H$_2$		H—H (r_e)	0.74144					UV
H$_2$O	C$_{2v}$	O—H (r_e)	0.9575	∠HOH (θ_e)	104.51			MW, IR
H$_2$O$_2$	C$_2$	O—O	1.475	∠OOH	94.8	dihedral angle	119.8	IR
H$_2$S	C$_{2v}$	H—S (r_e)	1.3356	∠HSH (θ_e)	92.12			MW, IR
H$_2$SO$_4$	C$_2$	O—H	0.97	S—O$_a$	1.574	S—O$_c$	1.422	MW
		∠O$_a$SO$_b$	101.3	∠O$_c$SO$_d$	123.3	∠O$_a$SO$_c$	108.6	
		∠O$_a$H$_a$S	106.4	∠H$_a$O$_a$S	108.5	dihedral angle	20.8	
		dihedral angle	90.9	dihedral angle	88.4	between the H$_a$O$_a$S and O$_a$SO$_c$ planes		
		between the H$_a$O$_a$S and O$_a$SO$_b$ planes		between the H$_a$SO$_b$ and O$_c$SO$_d$ planes				
H$_2$S$_2$	C$_2$	S—S	2.055	S—H	1.327	∠SSH	91.3	ED, MW
		dihedral angle	90.6					
HfBr$_4$	T$_d$	Hf—Br	2.450					ED
HfCl$_4$	T$_d$	Hf—Cl	2.316					ED
HfF		Hf—F	1.8596					UV
HfF$_4$	T$_d$	Hf—F	1.909					ED
HfI$_4$	T$_d$	Hf—I	2.662					ED
HgBr$_2$	linear	Hg—Br	2.384					ED
HgCl$_2$	linear	Hg—Cl	2.252					ED
HgH		Hg—H (r_e)	1.7404					UV
HgI$_2$	linear	Hg—I	2.568					ED
HoCl$_3$		Ho—Cl	2.462					ED
HoF$_3$		Ho—F	2.007					ED
HoO		Ho—O	1.797					UV
IBr		I—Br (r_e)	2.4691					MW
ICl		I—Cl (r_e)	2.3210					MW
IF		I—F (r_e)	1.9098					UV
IF$_5$	C$_{4v}$	I—F (av.)	1.860	(I—F$_{eq}$) – (I—F$_{ax}$)	0.03	∠F$_{ax}$IF$_{eq}$	82.1	ED, MW
IO		I—O (r_e)	1.8676					MW
I$_2$		I—I (r_e)	2.6663					R
InBr		In—Br (r_e)	2.5432					MW
InCl		In—Cl (r_e)	2.4012					MW
InCl$_3$		In—Cl	2.291					ED
InF		In—F (r_e)	1.9854					MW
InH		In—H (r_e)	1.8376					UV

Formula	Structure	Bond distances in Å and angles in degrees						Method
InI		In—I (r_e)	2.7537					MW
IrF$_6$	O$_h$	Ir—F	1.831					ED
KBH$_4$	H$_a$(BH$_3$)K (C$_{3v}$)	B—H (BH$_3$)	1.272	B—H$_a$	1.233	K—B	2.656	MW
KBr		K—Br (r_e)	2.8208					MW
KCl		K—Cl (r_e)	2.6667					MW
KF		K—F (r_e)	2.1716					MW
KH		K—H (r_e)	2.244					UV
KI		K—I (r_e)	3.0478					MW
KOH	linear; large amplitude bending mode	K—O	2.212	O—H	0.91			MW
K$_2$		K—K (r_e)	3.9051					UV
KrF$_2$	linear	Kr—F	1.89					ED
LaBr		La—Br (r_e)	2.65208					MW
LaBr$_3$	C$_{3v}$	La—Br	2.742					ED
LaCl		La—Cl (r_e)	2.49804					MW
LaCl$_3$	C$_{3v}$	La—Cl	2.589					ED
LaF		La—F (r_e)	2.02338					MW
LaI		La—I (r_e)	2.87885					MW
LaO		La—O (r_e)	1.82591					UV
LiBH$_4$	H$_a$(BH$_3$)Li (C$_{3v}$)	B—H (H$_3$)	1.257	B—H$_a$	1.218	Li—B	1.939	MW
LiBr		Li—Br (r_e)	2.1704					MW
LiCl		Li—Cl (r_e)	2.0207					MW
LiF		Li—F (r_e)	1.5639					MW
LiH		Li—H (r_e)	1.5949					MW
LiI		Li—I (r_e)	2.3919					MW
LiO		Li—O (r_e)	1.68822					UV
LiOH	linear	Li—O (r_e)	1.5776	O—H (r_e)	0.949			MW
Li$_2$		Li—Li (r_e)	2.6729					UV
Li$_2$Cl$_2$	Li / Cl Cl / Li	Li—Cl	2.23	Cl—Cl	3.61	∠ClLiCl	108	ED
Li$_2$O	linear	Li—O	1.606					UV
LuBr$_3$	C$_{3v}$	Lu—Br	2.557					ED
LuCl$_3$	C$_{3v}$	Lu—Cl	2.417	∠ClLuCl	112			ED
LuI$_3$	C$_{3v}$	Lu—I	2.768					ED
MgBr		Mg—Br (r_e)	2.34742					MW
MgCl		Mg—Cl (r_e)	2.1964					UV
MgCl$_2$	linear	Mg—Cl	2.179					ED
MgF		Mg—F (r_e)	1.7500					UV
MgF$_2$	linear	Mg—F	1.771					ED
MgH		Mg—H (r_e)	1.7297					UV
MgO		Mg—O (r_e)	1.749					UV
MgOH	linear	Mg—O	1.770	O—H	0.912			UV
Mg$_2$		Mg—Mg (r_e)	3.891					UV
MnBr$_2$	linear	Mn—Br	2.344					ED
MnCl$_2$	linear	Mn—Cl	2.202					ED
MnF$_2$	linear	Mn—F	1.811	[Mn—F (r_e)]	1.797			ED
MnH		Mn—H (r_e)	1.7308					UV
MnI$_2$	linear	Mn—I	2.538					ED
MoCl$_4$O	C$_{4v}$	Mo—Cl	2.279	Mo—O	1.658			ED
		∠ClMoCl	87.2					
MoF$_4$		Mo—F	1.851					ED
MoF$_6$	O$_h$	Mo—F	1.821					ED
NBr		N—Br (r_e)	1.79					UV
NCl		N—Cl (r_e)	1.6107					UV
NClH$_2$		N—H	1.017	N—Cl	1.748			MW, IR
		∠HNCl	103.7	∠HNH	107			
NCl$_3$		N—Cl	1.759	∠ClNCl	107.1			ED

Formula	Structure	Bond distances in Å and angles in degrees						Method
NF		N—F (r_e)	1.3170					UV
NF$_2$		N—F	1.3528	∠FNF	103.18			MW
NH$_2$		N—H	1.024	∠HNH	103.3			UV
NH$_2$NO$_2$		N—N	1.427	N—H	1.005			MW
		dihedral angle between NH$_2$ and NNO$_2$ planes	128.2	∠HNH	115.2	∠ONO	130.1	
NH$_3$	C$_{3v}$	N—H (r_e)	1.012	∠HNH (θ_e)	106.7			IR
NH$_4$Cl	H$_3$N····HCl (C$_{3v}$)	N—Cl	3.136					MW
NH		N—H (r_e)	1.0362					LMR
NH$_2$OH	bisector of HNH angle is *trans* to OH bond	N—O	1.453	N—H	1.02	O—H	0.962	MW
		∠HNO	103.3	∠HNH	107	∠NOH	101.4	
NO		N—O (r_e)	1.1506					IR
NOCl		N—O	1.14	N—Cl	1.975	∠ONCl	113	MW
NOF		N—O	1.136	N—F	1.512	∠FNO	110.1	MW
NO$_2$		N—O	1.193	∠ONO	134.1			MW
NO$_2$Cl	C$_{2v}$	N—O	1.202	N—Cl	1.840	∠ONO	130.6	MW
NO$_2$F	C$_{2v}$	N—O	1.1798	N—F	1.467	∠ONO	136	MW
NS		N—S (r_e)	1.4940					IR
N$_2$		N—N (r_e)	1.0977					UV
N$_2$H$_4$	H$_a$ atom is closer to the C$_2$ axis, H$_b$ is farther from the C$_2$ axis	N—N	1.449	N—H	1.021	∠NNH$_b$	106	ED, MW
		∠HNH	106.6 (ass.)	∠NNH$_a$	112			
		dihedral angle of internal rotation	91					
N$_2$O		N—N (r_e)	1.1284	N—O (r_e)	1.1841			MW, IR
N$_2$O$_3$	O$_a$ O$_b$ N$_a$—N$_b$ O$_c$	N$_a$—N$_b$	1.864	N$_a$—O$_a$	1.142			MW
		N$_b$—O$_b$	1.202	N$_b$—O$_c$	1.217			
		∠O$_a$N$_a$N$_b$	105.05	∠N$_a$N$_b$O$_b$	112.72	∠N$_a$N$_b$O$_c$	117.47	
N$_2$O$_4$	O O N—N O O D$_{2h}$	N—N	1.782	N—O	1.190	∠ONO	135.4	ED
NaBH$_4$	H$_a$(BH$_3$)Na (C$_{3v}$)	B—H (BH$_3$)	1.278	B—H$_a$	1.238	Na—B	2.308	MW
NaBr		Na—Br (r_e)	2.5020					MW
NaCl		Na—Cl (r_e)	2.3609					MW
NaF		Na—F (r_e)	1.9260					MW
NaH		Na—H (r_e)	1.8873					UV
NaI		Na—I (r_e)	2.7115					MW
NaO		Na—O (r_e)	2.05155					UV
Na$_2$		Na—Na (r_e)	3.0789					UV
NbCl$_4$	T$_d$	Nb—Cl	2.279					ED
NbCl$_5$	D$_{3h}$	Nb—Cl$_{ax}$	2.307	Nb—Cl$_{eq}$	2.276			ED
NbO		Nb—O (r_e)	1.691					UV
NdI$_3$	C$_{3v}$	Nd—I	2.879					ED
NiBr		Ni—Br	2.1963					UV
NiBr$_2$	linear	Ni—Br	2.201					ED
NiCl$_2$	linear	Ni—Cl	2.076					ED
NiF$_2$	linear	Ni—F	1.729	[Ni—F (r_e)]	1.715			ED
NiH		Ni—H (r_e)	1.476					UV
NiI		Ni—I	2.348					UV
NpF$_6$	O$_h$	Np—F	1.982					ED
OF		O—F (r_e)	1.3579					LMR
OF$_2$	C$_{2v}$	O—F (r_e)	1.4053	∠FOF (θ_e)	103.07			MW
OH		O—H (r_e)	0.96966					UV
O(SiH$_3$)$_2$		Si—H	1.486	Si—O	1.634	∠SiOSi	144.1	ED
O$_2$		O—O (r_e)	1.2074					MW

Formula	Structure	Bond distances in Å and angles in degrees						Method
O_2F_2	C_2	O—O	1.217	F—O	1.575	∠OOF	109.5	MW
		dihedral angle of internal rotation	87.5					
O_3	C_{2v}	O—O (r_e)	1.2716	∠OOO (θ_e)	117.47			MW
OsF_6	O_h	Os—F	1.832					ED
OsO_4	T_d	Os—O	1.712					ED
PBr_3	C_{3v}	P—Br	2.220	∠BrPBr	101.0			ED
PCl		P—Cl (r_e)	2.01461					UV
PCl_3	C_{3v}	P—Cl	2.039	∠ClPCl	100.27			ED
PCl_5	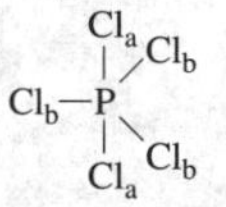 D_{3h}	P—Cl$_a$	2.124	P—Cl$_b$	2.020			ED
PF		P—F (r_e)	1.5896					UV
PF_3	C_{3v}	P—F	1.570	∠FPF	97.8			ED, MW
PF_5	D_{3h}	P—F$_{eq}$	1.534	P—F$_{ax}$	1.577			ED
PH		P—H (r_e)	1.4223					LMR
PH_2		P—H	1.418	∠HPH	91.70			UV
PH_3	c_{3v}	P—H	1.4200	∠HPH	93.345			MW
PN		N—P (r_e)	1.49087					MW
PO		O—P (r_e)	1.4759					UV
$POCl_3$	C_{3v}	P—O	1.449	P—Cl	1.993	∠ClPCl	103.3	ED
POF_3	C_{3v}	P—O	1.436	P—F	1.524	∠FPF	101.3	ED, MW
P_2		P—P (r_e)	1.8931					UV
P_2F_4	trans conformer	P—F	1.587	P—P	2.281	∠FPF	99.1	P_2F_4
		∠PPF	95.4					
P_4	T_d	P—P	2.21					ED
P_4O_6	T_d	P—O	1.638	∠POP	126.4			ED
$PbBr_2$	bent	Pb—Br (r_e)	2.598					ED
$PbCl_2$	bent	Pb—Cl (r_e)	2.444					ED
$PbCl_4$	T_d	Pb—Cl	2.369					ED
PbF		Pb—F (r_e)	2.0575					UV
PbF_2	bent	Pb—F (r_e)	2.041					ED
PbH		Pb—H (r_e)	1.839					UV
PbI_2	bent	Pb—I (r_e)	2.807					ED
PbO		Pb—O (r_e)	1.9218					MW
PbS		Pb—S (r_e)	2.2869					MW
PbSe		Pb—Se (r_e)	2.4022					MW
PbTe		Pb—Te (r_e)	2.5950					MW
$PrCl_3$	C_{3v}	Pr—Cl	2.554					ED
PrF_3	C_{3v}	Pr—F	2.091					ED
PrI_3	C_{3v}	Pr—I	2.901	∠IPrI	113			ED
PtC		Pt—C (r_e)	1.6767					UV
PtH		Pt—H (r_e)	1.52852					UV
PtN		Pt—N (r_e)	1.682					MW
PtO		Pt—O (r_e)	1.7273					UV
PtS		Pt—S (r_e)	2.03983					MW
PtSi		Pt—Si (r_e)	2.0612					MW
PuF_6	O_h	Pu—F	1.972					ED
RbBr		Rb—Br (r_e)	2.9447					MW
RbCl		Rb—Cl (r_e)	2.7869					MW
RbF		Rb—F (r_e)	2.2703					MW
RbH		Rb—H (r_e)	2.367					UV
RbI		Rb—I (r_e)	3.1768					MW
RbO		Rb—O (r_e)	2.25420					UV
RbOH	linear; large amplitude bending mode	Rb—O	2.301	O—H	0.957			MW
$ReClO_3$	C_{3v}	Re—O	1.702	Re—Cl	2.229	∠ClReO	109.4	MW

For PCl_5 structure:

Cl$_a$—Cl$_b$
Cl$_b$—P—Cl$_b$
Cl$_a$—Cl$_b$
D_{3h}

Formula	Structure	Bond distances in Å and angles in degrees						Method
ReClO$_4$	C$_{4v}$	Re—O	1.663	Re—Cl	2.270	∠ClReO	105.5	ED
ReCl$_5$	D$_{3h}$	Re—Cl$_{eq}$	2.238	Re—Cl$_{ax}$	2.263			ED
ReF$_6$	O$_h$	Re—F	1.832					ED
ReF$_7$	pseudorotation	Re—F	1.835					ED
RhB		Rh—B	1.691					UV
RhC		Rh—C	1.614					UV
RhS		Rh—S	2.059					UV
RuO$_4$	T$_d$	Ru—O	1.706					ED
SCl$_2$	C$_{2v}$	S—Cl	2.006	∠ClSCl	103.0			ED
SF		S—F (r_e)	1.6006					MW
SF$_2$		S—F	1.5921	∠FSF	98.20			MW
SF$_6$	O$_h$	S—F	1.561					ED
SH		S—H (r_e)	1.34066					UV
SO		S—O (r_e)	1.4811					MW
SOCl$_2$		S—O	1.44	S—Cl	2.072			MW
		∠ClSCl	97.2	∠OSCl	108.0			
SOF$_2$		S—O	1.420	S—F	1.583			ED
		∠FSF	92.2	∠OSF	106.2			
SOF$_4$		S—O	1.403	S—F$_a$	1.575	S—F$_b$	1.552	ED
	F$_b$—S—F$_b$; F$_a$—S—F$_a$; O	∠OSF$_a$	90.7	∠OSF$_b$	124.9			
		∠F$_a$SF$_b$	89.6	∠F$_b$SF$_b$	110.2			
	C$_{2v}$							
SO$_2$		S—O (r_e)	1.4308	∠OSO (θ_e)	119.329			MW
SO$_2$Cl$_2$	C$_{2v}$	S—Cl	2.011	S—O	1.404			ED
		∠ClSCl	100.0	∠OSO	123.5			
SO$_2$F$_2$	C$_{2v}$	S—F	1.530	S—O	1.397			ED
		∠FSF	97	∠OSO	123			
SO$_3$	D$_{3h}$	S—O	1.4198					IR
S(SiH$_3$)$_2$		Si—S	2.136	Si—H	1.494	∠SiSSi	97.4	ED
S$_2$		S—S (r_e)	1.8892					R
S$_2$Br$_2$	C$_2$	S—Br	2.24	S—S	1.98	∠SSBr	105	ED
		dihedral angle of internal rotation	83.5					
S$_2$Cl$_2$	C$_2$	S—Cl	2.057	S—S	1.931	∠SSCl	108.2	ED
		dihedral angle of internal rotation	84.1					
S$_2$O$_2$	planar *cis* form	S—S	2.025	S—O	1.458	∠OSS	112.8	MW
S$_8$		S—S	2.07	∠SSS	105	(D$_{4d}$)		ED
SbBr$_3$	C$_{3v}$	Sb—Br	2.490	∠BrSbBr	98.2			ED
SbCl$_3$	C$_{3v}$	Sb—Cl	2.334	∠ClSbCl	97.1			ED
SbCl$_5$	D$_{3h}$	Sb—Cl$_{eq}$	2.277	Sb—Cl$_{ax}$	2.338			ED
SbF		Sb—F (r_e)	1.918					UV
SbF$_3$	C$_{3v}$	Sb—F	1.880	∠FSbF	94.9			ED
SbH		Sb—H	1.723					UV
SbH$_3$	C$_{3v}$	Sb—H	1.704	∠HSbH	91.6			MW
SbI$_3$	C$_{3v}$	Sb—I	2.721	∠ISbI	99.0			ED
SbO		Sb—O (r_e)	1.826					UV
SbP		Sb—P (r_e)	2.20544					MW
ScCl$_3$	D$_{3h}$	Sc—Cl	2.291					ED
ScF		Sc—F (r_e)	1.788					UV
ScF$_3$	D$_{3h}$	Sc—F	1.847					ED
SeF		Se—F	1.742					MW
SeF$_6$	O$_h$	Se—F	1.69					ED
SeH		Se—H (r_e)	1.48					UV
SeO		Se—O (r_e)	1.6393					MW

Formula	Structure	Bond distances in Å and angles in degrees						Method
SeOF$_2$		Se—O	1.576	Se—F	1.730			MW
		∠OSeF	104.82	∠FSeF	92.22			
SeO$_2$		Se—O (r_e)	1.6076	∠OSeO (θ_e)	113.83			MW
SeO$_3$	D$_{3h}$	Se—O	1.69					ED
Se$_2$		Se—Se (r_e)	2.1660					UV
Se$_6$	six-membered ring with chair conformation	Se—Se	2.34	∠SeSeSe	102			ED
SiBrF$_3$	C$_{3v}$	Si—F	1.559	Si—Br	2.156	∠FSiBr	108.5	MW
SiBrH$_3$	C$_{3v}$	Si—Br	2.210	Si—H	1.486	∠HSiBr	107.8	MW
SiCl		Si—Cl (r_e)	2.058					UV
SiClH$_3$	C$_{3v}$	Si—Cl	2.049	Si—H	1.486	∠HSiCl	107.9	MW
SiCl$_4$	T$_d$	Si—Cl	2.019					ED
SiF		Si—F	1.6008					UV
SiFH$_3$	C$_{3v}$	Si—F	1.593	Si—H	1.486	∠HSiH	110.63	MW, IR
SiF$_2$		Si—F (r_e)	1.590	∠FSiF (θ_e)	100.8			MW
SiF$_3$H	C$_{3v}$	Si—H ((r_e)	1.4468	Si—F (r_e)	1.5624	∠HSiF (θ_e)	110.64	MW
SiF$_4$	T$_d$	Si—F	1.553					ED
SiH		Si—H (r_e)	1.5201					UV
SiH$_3$I	C$_{3v}$	Si—I	2.437	Si—H	1.486	∠HSH	107.8	MW
SiH$_4$	T$_d$	Si—H	1.4798					IR
SiN		Si—N (r_e)	1.572					UV
SiO		Si—O (r_e)	1.5097					MW
SiS		Si—S (r_e)	1.9293					MW
SiSe		Si—Se (r_e)	2.0583					MW
Si$_2$		Si—Si (r_e)	2.246					UV
Si$_2$Cl$_6$		Si—Si	2.32	Si—Cl	2.009	∠ClSiCl	109.7	ED
Si$_2$F$_6$		Si—Si	2.317	Si—F	1.564	∠FSiF	108.6	ED
Si$_2$H$_6$		Si—Si	2.331	Si—H	1.492			ED
		∠SiSiH	110.3	∠HSiH	108.6			
SnBr$_2$		Sn—Br (r_e)	2.501	∠BrSnBr	100.0			ED
SnCl		Sn—Cl (r_e)	2.361					UV
SnCl$_2$		Sn—Cl (r_e)	2.335	∠ClSnCl	99.1			ED
SnCl$_4$	T$_d$	Sn—Cl	2.281					ED
SnF		Sn—F (r_e)	1.944					UV
SnH		Sn—H (r_e)	1.7815					UV
SnH$_4$	T$_d$	Sn—H	1.711					R, IR
SnI$_2$		Sn—I (r_e)	2.688					ED
SnO		Sn—O (r_e)	1.8325					MW,UV
SnS		Sn—S (r_e)	2.2090					MW
SnSe		Sn—Se (r_e)	2.3256					MW
SnTe		Sn—Te (r_e)	2.5228					MW
SrBr		Sr—Br (r_e)	2.7352					UV
SrBr$_2$	quasilinear	Sr—Br	2.783					ED
SrCl$_2$		Sr—Cl	2.630	∠ClSrCl	155			ED
SrF		Sr—F (r_e)	2.0754					UV
SrH		Sr—H (r_e)	2.1456					UV
SrI		Sr—I (r_e)	2.9436					UV
SrI$_2$	linear	Sr—I	3.01					ED
SrO		Sr—O (r_e)	1.9198					MW
SrOH		Sr—O	2.111	O—H	0.922			UV
SrS		Sr—S (r_e)	2.4405					UV
TaBr$_5$	D$_{3h}$	Ta—Br$_{eq}$	2.412	Ta—Br$_{ax}$	2.473			ED
TaCl$_5$	D$_{3h}$	Ta—Cl$_{eq}$	2.268	Ta—Cl$_{ax}$	2.315			ED
TaO		Ta—O (r_e)	1.6875					UV
TbCl$_3$	C$_{3v}$	Tb—Cl	2.476					ED
TeF$_6$	O$_h$	Te—F	1.815					ED
TeH		Te—H	1.74					UV
TeO		Te—O (r_e)	1.825					UV
Te$_2$		Te—Te (r_e)	2.5574					UV
ThCl$_4$	T$_d$	Th—Cl	2.567					ED

Formula	Structure	Bond distances in Å and angles in degrees						Method
ThF_4	T_d	Th—F	2.124					ED
ThO		Th—O (r_e)	1.84032					UV
$TiBr_4$	T_d	Ti—Br	2.339					ED
$TiCl_3$	D_{3h}	Ti—Cl	2.208					ED
$TiCl_4$	T_d	Ti—Cl	2.170					ED
TiF		Ti—F	1.8342					MW
TiF_4	T_d	Ti—F	1.756					ED
TiI_3	D_{3h}	Ti—I	2.568					ED
TiI_4	T_d	Ti—I	2.546					ED
TiO		Ti—O (r_e)	1.620					UV
TiS		Ti—S (r_e)	2.0825					UV
TlBr		Tl—Br (r_e)	2.6182					MW
TlCl		Tl—Cl (r_e)	2.4848					MW
TlF		Tl—F (r_e)	2.0844					MW
TlH		Tl—H (r_e)	1.870					UV
TlI		Tl—I (r_e)	2.8137					MW
UCl_4	T_d	U—Cl	2.506					ED
UCl_6	O_h	U—F	2.46					ED
UF_4	T_d	U—F	2.059					ED
UF_6	O_h	U—F	2.000					ED
UI_3	C_{3v}	U—I	2.88					ED
VCl_3O	C_{3v}	V—O	1.570	V—Cl	2.142	∠ClVCl	111.3	ED, MW
VBr_4	T_d (Jahn-Teller effect)	V—Br	2.276					ED
VCl_4	T_d (Jahn-Teller effect)	V—Cl	2.138					ED
VF_3	D_{3h}	V—F	1.751					ED
VF_5		$V-F_{eq}$	1.709	$V-F_{ax}$	1.736			ED
VMo		V—Mo	1.876					UV
VO		V—O (r_e)	1.5893					UV
$WClF_5$	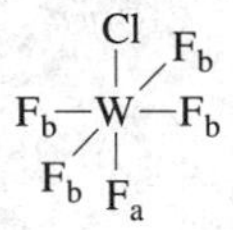 (see structure below)	W—F (av.)	1.836	W—Cl	2.251	$∠F_a WF_b$	88.7	MW

```
      Cl   F_b
       |  /
 F_b—W—F_b
      /|
   F_b F_a
```

Formula	Structure							Method
WCl_5	D_{3h}	$W-Cl_{eq}$	2.243	$W-Cl_{ax}$	2.293			ED
WCl_6	O_h	W—Cl	2.290					ED
WF_4O	C_{4v}	W—O	1.666	W—F	1.847	∠FWF	86.2	ED
WF_6	O_h	W—F	1.833					ED
XeF_2	linear	Xe—F	1.977					IR
XeF_4	D_{4h}	Xe—F	1.94					ED
XeF_6	O_h	Xe—F	1.890					ED
XeO_4	T_d	Xe—O	1.736					ED
YCl		Y—Cl	2.385					UV
YCl_3		Y—Cl	2.437					ED
YF		Y—F (r_e)	1.9257					UV
YI_3		Y—I	2.817					ED
YO		Y—O (r_e)	1.790					UV
YbBr		Yb—Br (r_e)	2.6454					UV
YbH		Yb—H (r_e)	2.0526					UV
$ZnBr_2$	linear	Zn—Br	2.204					ED
$ZnCl_2$	linear	Zn—Cl	2.072					ED
ZnF		Zn—F (r_e)	1.7677					MW
ZnF_2	linear	Zn—F	1.742	$[Zn-F\ (r_e)]$	1.729			ED
ZnH		Zn—H (r_e)	1.5949					UV
ZnI_2	linear	Zn—I	2.401					ED
$ZrBr_4$	T_d	Zr—Br	2.465					ED
$ZrCl_4$	T_d	Zr—Cl	2.328					ED
ZrF_4	T_d	Zr—F	1.902					ED
ZrI_4	T_d	Zr—I	2.660					ED
ZrO		Zr—O (r_e)	1.7116					UV

Part 2. Molecules containing carbon

Compound	Structure	Bond distances in Å and angles in degrees						Method
Acetaldehyde	$C_bH_3 - C_a$ (with =O and H)	$C_a - O$	1.210	$C_a - C_b$	1.515			ED, MW
		$C_a - H$	1.128	$C_b - H$	1.107			
		$\angle C_bC_aO$	124.1	$\angle C_bC_aH$	115.3	$\angle HC_bH$	109.8	
Acetamide	CH_3CONH_2	$C - O$	1.220	$C - N$	1.380			ED
		$C - C$	1.519	$N - H$	1.022	$C - H$	1.124	
		$\angle CCN$	115.1	$\angle NCO$	122.0			
Acetic acid	$CH_3 - C$ (with O_a and $O_b - H$)	$C - C$	1.520	$C - O_a$	1.214	$C - O_b$	1.364	ED
		$C - H$	1.10	$\angle CCO_a$	126.6	$\angle CCO_b$	110.6	
Acetone	$(CH_3)_2CO$	$C - C$	1.520	$C - O$	1.213	$C - H$	1.103	ED, MW
	Symmetry axis of each CH_3 is tilted 2° from the C—C bond	$\angle CCC$	116.0	$\angle HCH$	108.5			
Acetonitrile	CH_3CN (C_{3v})	$C - N$	1.159	$C - C$	1.468	$C - H$	1.107	ED, MW
		$\angle CCH$	109.7					
Acetonitrile-N-oxide	CH_3CNO (C_{3v})	$C - C$	1.442	$C - N$	1.169	$N - O$	1.217	MW
Acetyl chloride	CH_3COCl	$C - C$	1.506	$C - O$	1.187	$C - H$	1.105	ED, MW
		$C - Cl$	1.798	$\angle HCH$	108.6	$\angle OCCl$	121.2	
		$\angle CCCl$	111.6					
Acetylene	$HC \equiv CH$	$C - C$ (r_e)	1.203	$C - H$ (r_e)	1.060			IR
Acrolein	(planar s-trans form)	$C_a - C_b$	1.345	$C_b - C_c$	1.484	$C_c - O$	1.217	ED, MW
		$C_a - H$	1.10	$C_c - H$	1.13	$\angle HC_cC_b$	114	
		$\angle C_aC_bC_c$	120.3	$\angle C_bC_cO$	123.3	Other CCH (av.)	122	
Acrylonitrile		$C_a - C_b$	1.343	$C_b - C_c$	1.438	$C_c - N$	1.167	ED, MW
		$C_a - H$	1.114	$\angle C_bC_cN$	178	$\angle C_aC_bC_c$	121.7	
		$\angle HCC$	120					
Allene	$CH_2 = C = CH_2$	$C - C$	1.3084	$C - H$	1.087	$\angle HCH$	118.2	IR
Aniline	$C_6H_5NH_2$	$C - C$	1.392	$C - N$	1.431	$N - H$	0.998	MW
		$\angle HNH$	113.9	dihedral angle between NH_2 plane and N—C bond	140.6			
Azetidine	$CH_2 - CH_2$ / $CH_2 - NH$	$C - N$	1.482	$C - C$	1.553			ED
		$C - H$	1.107	$N - H$	1.03			
		$\angle CCC$	86.9	$\angle CCN$	85.8	$\angle CNC$	92.2	
		dihedral angle between CCC and CNC planes	147					
Benzamide	$C_6H_5 - C_aONH_2$	$C - C$ (ring)	1.401	C (ring)—C_a	1.511	$C_a - O$	1.225	ED
		$C - H$	1.112	$C - N$	1.380			
		$\angle CCN$	117.8	$\angle CCC$ (ring)	120(ass.)	$\angle CCO$	121.2	
Benzene	C_6H_6	$C - C$	1.399	$C - H$	1.101			ED, IR
p-Benzoquinone		$C_a - O$	1.225	$C_a - C_b$	1.481	$C_b - C_b$	1.344	ED
		$\angle C_bC_aC_b$	118.1					

Compound	Structure	Bond distances in Å and angles in degrees							Method
Bicyclo[1.1.0]butane		C_a—C_a	1.497	C_a—C_b	1.498	C_a—H_a	1.071		MW
		C_b—H_b	1.093	C_b—H_c	1.093	∠$H_b C_b H_c$	115.6		
		∠$C_b C_a H_a$	130.4	∠$C_a C_a H_a$	128.4	∠$C_a C_b C_a$	60.0		
		dihedral angle between the two $C_a C_a C_b$ planes	121.7						
Bicyclo[2.2.1]heptane	See preceeding structure C_7H_{12}	C_a—C_b	1.54	C_b—C_b	1.56	C_a—C_c	1.56		ED
		C—C (av.)	1.549	∠$C_a C_c C_a$	93.1				
		dihedral angle between the two $C_a C_b C_b C_a$ planes	113.1						
Bicyclo[2.2.0]hexa-2,5-diene		C_b—C_b	1.345	C_a—C_a	1.574	C_a—C_b	1.524		ED
		dihedral angle between the two $C_a C_b C_b C_a$ planes	117.3						
Bicyclo[2.2.2]octane	$HC_a(C_b H_2 C_b H_2)_3 C_a H$ large-amplitude torsional motion about D_{3h} symmetry axis	C_a—C_b	1.54	C_b—C_b	1.55	C—C (av.)	1.542		ED
		∠$C_a C_b C_b$	109.7						
Bicyclo[1.1.1]pentane	C_5H_8	C—C	1.557	∠CCC	74.2				ED
Bicyclo[2.1.0]pentane		C_a—C_a	1.536	C_b—C_b	1.565	C_a—C_c	1.507		MW
		C_a—C_b	1.528	dihedral angle between the $C_a C_a C_b C_b$ and $C_a C_a C_c$ planes	112.7				
Biphenyl		C—C (intra-ring)	1.396	C—C (inter-ring)	1.49				ED
		torsional dihedral angle between the two rings	≈40						
4,4′-Bipyridyl		C—C (inter-ring)	1.465	C—C (intra-ring)	1.375	C—N (intra-ring)	1.375		ED
		torsional dihedral angle between the two rings	≈37						
Bis(cyclopentadienyl) beryllium	$(C_5H_5)_2Be$ (C_{5v})	Be—(cyclopentadienyl plane)	1.470, 1.92	C—C	1.423				ED
Bis(cyclopentadienyl) iron	$(C_5H_5)_2Fe$ (D_{5h})	Fe—C	2.064	C—C	1.440	C—H	1.104		ED
Bis(cyclopentadienyl) lead	$(C_5H_5)_2Pb$ (D_{5h})	Pb—C	2.79	C—C	1.430				ED
		dihedral angle between the two C_5H_5 planes	40~50 (The two rings are not parallel)						
Bis(cyclopentadienyl) manganese	$(C_5H_5)_2Mn$ (D_{5h})	Mn—C	2.383	C—C	1.429				ED
Bis(cyclopentadienyl) nickel	$(C_5H_5)_2Ni$ (D_{5h})	Ni—C	2.196	C—C	1.430				ED
Bis(cyclopentadienyl) ruthenium	$(C_5H_5)_2Ru$ (D_{5h})	Ru—C	2.196	C—C	1.439				ED
Bis(cyclopentadienyl) tin	$(C_5H_5)_2Sn$ (D_{5h})	Sn—C	2.71	C—C	1.431	C—H	1.14		ED
Borane carbonyl	BH_3CO (C_{3v})	C—O	1.131	B—C	1.540	B—H	1.194		MW
		∠BCO	180	∠HBH	113.9				

Compound	Structure	Bond distances in Å and angles in degrees						Method
Bromobenzene	Br–C$_a$ (ring) HC_b C_bH / HC_c C_cH / C_d H	C_a—C_b	1.42	C_b—C_c	1.375	C_c—C_d	1.401	MW
		C—Br	1.85	C—H	1.072	∠$C_b$$C_a$$C_b$	117.4	
Bromochloroacetylene	ClC≡CBr	C—Cl	1.636	C—Br	1.784	C—C	1.206	ED
Bromoiodoacetylene	IC≡CBr	C—I	1.972	C—Br	1.795	C—C	1.206	ED
Bromomethane	CH$_3$Br	C—Br (r_e)	1.933	C—H (r_e)	1.086	∠HCH (θ_e)	111.2	MW, IR
Bromomethyl	CH$_2$Br (planar)	C—Br	1.848	C—H	1.084	∠HCH (ass.)	124.5	MW
Bromomethylene	CHBr (bent)	C—Br	1.857	C—H	1.110	∠HCH	101.0	UV
Bromomethylmercury	CH$_3$HgBr (C$_{3v}$)	C—Hg	2.07	Hg—Br	2.406			MW
1,3-Butadiene	C$_a$H$_2$ C_bH—C_bH C_aH_2 (C$_{2h}$)	C_a—C_b	1.349	C_b—C_b	1.467	C—H (av.)	1.108	ED
		∠CCC	124.4	∠$C_b$$C_a$H	120.9			
1,3-Butadiyne	HC$_a$≡C$_b$C$_b$≡C$_a$H (linear)	C_a—C_b	1.218	C_b—C_b	1.384	C—H	1.09	ED
Butane	CH$_3$CH$_2$CH$_2$CH$_3$	C—C	1.531	C—H	1.117	∠CCC	113.8	ED
		∠CCH	111.0	dihedral angle for the *gauche* conformer	65			
2,3-Butanedione	CH$_3$COCOCH$_3$ *trans* conformer	C—O	1.215	C—C (av.)	1.524	C—H	1.108	ED
		∠CCC	116.2	∠CCO	119.5			
2-Butanone	C$_a$H$_3$ C_c (=O) C_bH$_2$ C_dH$_3$	C—C (av.)	1.518	C_c—O	1.219	C—H (av.)	1.102	ED
	trans conformer	∠$C_a$$C_b$$C_c$	113.5	∠$C_b$$C_c$O	121.9	∠$C_d$$C_c$O	121.9	
1,2,3-Butatriene	H$_2$C$_a$=C$_b$=C$_b$=C$_a$H$_2$ (D$_{2h}$)	C_a—C_b	1.32	C_b—C_b	1.28	C—H	1.08	ED
cis-2-Butene	C$_a$H$_3$C$_b$H=C$_b$HC$_a$H$_3$	C_a—C_b	1.506	C_b—C_b	1.346	∠$C_a$$C_b$$C_b$	125.4	ED
trans-2-Butene	C$_a$H$_3$C$_b$H=C$_b$HC$_a$H$_3$	C_a—C_b	1.508	C_b—C_b	1.347	∠$C_a$$C_b$$C_b$	123.8	ED
1-Buten-3-yne	H$_a$ C$_a$=C$_b$ H$_c$ / H$_b$ C$_c$≡C$_d$ H$_d$	C_a—C_b	1.344	C_b—C_c	1.434	C_c—C_d	1.215	ED, MW
		C_a—H$_a$	1.11	C_d—H$_d$	1.09			
		∠$C_a$$C_b$$C_c$	123.1	∠$C_b$$C_c$$C_d$	178	∠H$_a$$C_a$$C_b$	119	
		∠H$_b$$C_a$$C_b$	122	∠H$_c$$C_b$$C_a$	122	∠$C_c$$C_dH_d$	182	
tert-Butyl chloride	(CH$_3$)$_3$CCl	C—C	1.528	C—Cl	1.828	C—H	1.102	ED, MW
		∠CCCl	107.3	∠CCH	110.8	∠CCC	111.6	
2-Butyne	C$_a$H$_3$—C$_b$≡C$_b$—C$_a$H$_3$	C_b—C_b	1.214	C_a—C_b	1.468	C—H	1.116	ED
		∠$C_b$$C_a$H	110.7					
Carbon dimer	C$_2$	C—C (r_e)	1.2425					UV
Carbon trimer	C$_3$ (linear)	C—C	1.277					UV
Carbon dioxide	CO$_2$ (linear)	C—O (r_e)	1.1600					IR
Carbon disulfide	CS$_2$ (linear)	C—S (r_e)	1.5526					IR
Carbon monobromide	CBr	C—Br	1.8209					UV
Carbon monoselenide	CSe	C—Se (r_e)	1.67609					UV
Carbon monosulfide	CS	C—S (r_e)	1.5349					MW
Carbon monoxide	CO	C—O (r_e)	1.1283					MW
Carbon oxyselenide	OCSe (linear)	C—O	1.159	C—Se	1.709			MW
Carbon oxysulfide	OCS (linear)	C—O (r_e)	1.1578	C—S (r_e)	1.5601			MW
Carbon phosphide	CP	C—P (r_e)	1.562					UV
Carbon sulfide selenide	SCSe (linear)	C—S	1.553	C—Se	1.693			MW
Carbon sulfide telluride	SCTe (linear)	C—S	1.557	C—Te	1.904			MW
Carbon suboxide	OCCCO (linear)	C—C	1.289	C—O	1.163			ED

Compound	Structure	Bond distances in Å and angles in degrees						Method
Carbonyl bromide	$COBr_2$	C—O	1.178	C—Br	1.923	∠BrCBr	112.3	ED, MW
Carbonyl chloride	$COCl_2$	C—O	1.179	C—Cl	1.742	∠ClCCl	111.8	ED, MW
Carbonyl chloride fluoride	COClF	C—O	1.173	C—F	1.334	C—Cl	1.725	ED, MW
		∠ClCO	127.5	∠FCCl	108.8			
Carbonyl dicyanide	$CO(CN)_2$	C—O	1.209	C—C	1.466	C—N	1.153	ED, MW
		∠CCC	115	∠CCN	180			
Carbonyl fluoride	COF_2	C—O	1.172	C—F	1.3157	∠FCF	107.71	ED, MW
Chloroacetylene	HC≡CCl	C—Cl	1.6368	C—C	1.2033	C—H	1.0550	MW
Chlorobenzene	C_6H_5Cl	C—C	1.400	C—Cl	1.737	C—H	1.083	ED
Chlorocyanoacetylene	ClC≡C—CN	C—Cl	1.624	C—N	1.160	C—C	1.205	ED
		C—CN	1.362					
Chloroethane	(structure diagram)	C—C	1.528	C—Cl	1.802	C—H	1.103	ED, MW
		∠CCCl	110.7	∠$H_bC_bH_b$	109.8	∠$H_aC_aH_a$	109.2	
		∠$C_bC_aH_a$	110.6	C_a—H_a = C_b—H_b (ass.)				
2-Chloroethanol	$ClCH_2CH_2OH$	C—O	1.413	C—C	1.519	C—Cl	1.801	ED
	(*gauche*)	O—H	1.033	C—H	1.093			
		∠CCCl	110.7	∠CCO	113.8	dihedral angle of internal rotation	62.4	
Chloroiodoacetylene	ClC≡CI	C—Cl	1.63	C—I	1.99	C—C	1.209 (ass)	MW
Chloromethane	CH_3Cl	C—Cl	1.785	C—H	1.090	∠HCH	110.8	MW, IR
Chloromethylidyne	CCl	C—Cl	1.6512					UV
Chloromethylmercury	CH_3HgCl (C_{3v})	C—Hg	2.06	Hg—Cl	2.282			MW
trans-1-Chloropropene	$CH_3CH=CHCl$	C—Cl	1.728	∠CCCl	121.9			MW
3-Chloropropene	$CH_2ClCH=CH_2$	C—Cl	1.811	∠CCCl	115.2			MW
	cis conformer							
	skew conformer	C—Cl	1.809	∠CCCl	109.6	dihedral angle of internal rotation	122.4	
Chlorotrifluoromethane	$CClF_3$ (C_{3v})	C—Cl	1.752	C—F	1.325	∠FCF	108.6	ED, MW
Chromium carbonyl	$Cr(CO)_6$	Cr—C	1.92	C—O	1.16	∠CrCO	180	ED
Cobalt cyanide	CoC≡N	Co—C	1.883	C—N	1.131			MW
Copper cyanide	CuC≡N	Cu—C	1.832	C—N	1.158			MW
Cyanamide	$H_2N_aCN_b$	N_a—C	1.346	C—N_b	1.160	N—H	1.00	MW
		∠HNH	114	dihedral angle between NH_2 plane and N—C bond	142			
Cyanide	CN	C—N (r_e)	1.1718					MW
Cyanoacetylene	$HC_a≡C_b—C_cN$	C_a—C_b	1.205	C_b—C_c	1.378	C—H	1.058	MW
		C_c—N	1.159					
Cyanocyclopropane	$C_3H_5C_aN$	C—C (ring)	1.513	C—C_a	1.472	C_a—N	1.157	MW
		C—H	1.107	∠C_aCH	119.6	∠HCH	114.6	
Cyanogen	N≡C—C≡N (linear)	C—N	1.163	C—C	1.393			ED
Cyanogen azide	N≡C—N=N≡N	C—N	1.312	N=N	1.252	N≡N	1.133	MW
	(planar)	C≡N	1.164	∠CNN	120.2	∠NCN	176.0	
Cyanogen bromide	BrCN (linear)	C—N (r_e)	1.157	C—Br (r_e)	1.790			MW
Cyanogen chloride	ClCN (linear)	C—Cl (r_e)	1.629	C—N (r_e)	1.160			MW
Cyanogen fluoride	FCN (linear)	C—F	1.262	C—N	1.159			MW
Cyanogen iodide	ICN (linear)	C—I	1.995	C—N	1.159			MW
1-Cyano-2-propyne	$HC_a≡C_bC_cH_2C_d≡N$	C_a—C_b	1.207 (ass.)	C_b—C_c(ass.)	1.465	C_c—C_d	1.454	MW
		C_d—N	1.159 (ass.)	C_a—H(ass.)	1.057	C_c—H(ass.)	1.090	
		∠$C_bC_cC_d$	113.4	∠HC_cH	109.4 (ass.)	∠C_bC_cH	111.3	

Compound	Structure	Bond distances in Å and angles in degrees							Method
Cyclobutane	$(CH_2)_4$	C—C	1.555	C—H	1.113				ED
		dihedral angle between the two CCC planes	145						
Cyclobutanone	C_bH_2 C_cH_2 C_a=O C_bH_2	C_a—C_b	1.527	C_b—C_c	1.556				MW
		$\angle C_bC_aC_b$	93.1	$\angle C_aC_bC_c$	88.0				
Cyclobutene	H_2C_a—C_aH_2 HC_b=C_bH	C_a—C_a	1.566	C_b—C_b	1.342	C_a—C_b	1.517		MW
		C_a—H	1.094	C_b—H	1.083				
		$\angle C_aC_bC_b$	94.2	$\angle C_bC_bH$	133.5	$\angle HC_aH$	109.2		
		$\angle C_aC_aH$	114.5	$\angle C_aC_aC_b$	85.8	dihedral angle between CH_2 plane and C_a—C_a bond	135.8		
2,4,6-Cycloheptatrien-1-one	(structure: O=C_a; HC_b=HC_c...C_bH=C_cH; C_d=C_d with H H) (C_{2v})	C_a—C_b	1.45	C_b—C_c	1.36	C_c—C_d	1.46		ED
		C_d—C_d	1.34	C_a—O	1.23	$\angle C_bC_aC_b$	122		
		$\angle C_aC_bC_c$	133	$\angle C_bC_cC_d$	126	$\angle C_cC_dC_d$	130		
Cyclohexane	C_6H_{12} (chair form)	C—C	1.536	C—H	1.119	$\angle$CCC	111.3		ED
Cyclohexene	HC_a=C_aH H_2C_b C_bH_2 C_cH_2—C_cH_2 half-chair form (C_2)	C_a—C_a	1.334	C_a—C_b	1.50	C_b—C_c	1.52		ED
		C_c—C_c	1.54	$\angle C_aC_aC_b$	123.4	$\angle C_aC_bC_c$	112.0		
		$\angle C_bC_cC_c$	110.9						
Cyclooctatetraene	(tub form (D_{2d}), labeled a and b)	C_a—C_b	1.476	C_a—C_a	1.340	C_b—C_b	1.340		ED
		C—H	1.100	$\angle C_bC_aC_a$	126.1	$\angle C_aC_bC_b$	126.1		
		dihedral angle between $C_aC_aC_aC_a$ and $C_aC_bC_bC_a$ planes	136.9						
1,3-Cyclopentadiene	C_aH_2 HC_b C_bH HC_c—C_cH	C_a—C_b	1.509	C_b—C_c	1.342	C_c—C_c	1.469		MW
		$\angle C_aC_bC_c$	109.3	$\angle C_bC_cC_c$	109.4	$\angle C_bC_aC_b$	102.8		
Cyclopentadienylindium	In HC—CH HC—CH CH	C—In	2.621	C—C	1.426	(C_{5v})			ED
Cyclopentane	$(CH_2)_5$	C—C	1.546	C—H	1.114	$\angle$CCH	111.7		ED
Cyclopentene	C_aH_2 H_2C_b C_bH_2 C_cH=C_cH	C_a—C_b	1.546	C_b—C_c	1.519	C_c—C_c	1.342		ED
		$\angle C_aC_bC_c$	103.0	$\angle C_bC_cC_c$	110.0	$\angle C_bC_aC_b$	104.0		
		dihedral angle between $C_bC_aC_b$ and $C_bC_cC_cC_b$ planes	151.2						
Cyclopropane	$(CH_2)_3$	C—C	1.512	C—H	1.083	$\angle$HCH	114.0		R

Compound	Structure	Bond distances in Å and angles in degrees						Method
Cyclopropanone		C_a—C_b	1.475	C_b—C_b	1.575	C_a—O	1.191	MW
		C—H	1.086	$\angle C_a C_b C_b$	57.7	$\angle HC_b H$	114	
		dihedral angle between CH_2 plane and C_b—C_b bond	151					
Cyclopropene		C_a—C_b	1.505	C_b—C_b	1.293	C_a—H	1.085	MW
		C_b—H	1.072	$\angle C_b C_b H$	150	$\angle HC_a H$	114.3	
Cyclopropenone		C_a—C_b (r_s)	1.423	C_b—C_c (r_s)	1.349	C_a—O (r_s)	1.212	MW
		C—H (r_s)	1.079	$\angle HC_b C_c$ (θ_s)	144.3	$C_b C_a C_c$ (θ_s)	56.6	
Decalin	$C_{10}H_{18}$	C—C (av.)	1.530	C—H (av.)	1.113	$\angle$CCC (av.)	111.4	ED
Diazirine		C—N	1.482	N—N	1.228	C—H	1.09	MW
		$\angle$HCH	117					
Diazoacetonitrile		C_a—C_b	1.424	C_a—N_a	1.165	C_b—N_b	1.280	MW
		N_b—N_c	1.132	C—H	1.082			
		$\angle C_a C_b H$	117	$\angle C_a C_b N_b$	119.5			
Diazomethane	CH_2N_2	C—N	1.32	N—N	1.12	C—H	1.075	MW, IR
		$\angle$HCH	126.0					
1,2-Dibromoethane	CH_2BrCH_2Br	C—C	1.506	C—Br	1.950	C—H	1.108	ED
		$\angle$CCBr	109.5	$\angle$CCH	110			
Dibromomethane	CH_2Br_2	C—Br	1.924	C—H	1.08	$\angle$HCBr	109	ED
		$\angle$BrCBr	113.2					
2,2'-Dichlorobiphenyl	C_6H_4Cl—C_6H_4Cl	C—C (rings)	1.398	C—C (inter-ring)	1.495	C—H	1.10	ED
		C—Cl	1.732	$\angle$CCCl	121.4	$\angle$CCH	126	
		dihedral angle between the two rings (defined as 0 for *cis* conformer)	74					
trans-1,4-Dichlorocyclohexane	$C_6H_{10}Cl_2$	C—C	1.530	C—Cl	1.810	C—H	1.102	ED
		$\angle$CCC	111.5					
	equatorial:	$\angle$CCCl	108.6	$\angle$HCCl	111.5			
	axial:	$\angle$CCCl	110.6	$\angle$HCCl	107.6			
1,1-Dichloroethane	$CHCl_2CH_3$	C—C	1.540	C—Cl	1.766			MW
		$\angle$ClCCl	112.0	$\angle$CCCl	111.0			
1,2-Dichloroethane	CH_2ClCH_2Cl	C—C	1.531	C—Cl	1.790	C—H	1.11	ED
		$\angle$CCCl	109.0	$\angle$CCH	113			
1,1-Dichloroethene	CH_2=CCl_2 (C_{2v})	C—C	1.32 (ass.)	C—Cl	1.73			MW
		$\angle$ClCC	123					
cis-1,2-Dichloroethene	CHCl=CHCl	C—C	1.354	C—Cl	1.718			ED
		$\angle$ClCC	123.8					
Dichloromethane	CH_2Cl_2	C—Cl (r_e)	1.765	C—H (r_e)	1.087			MW, IR
		$\angle$ClCCl (θ_e)	112.0	$\angle$HCH (θ_e)	111.5			

Compound	Structure	Bond distances in Å and angles in degrees						Method
1,2-Dicyanocyclobutene	$C_c \equiv N$ $H_2C_b - C_a$ $\|\quad\quad\|$ $H_2C_b - C_{a'}$ $C_{c'} \equiv N$ C_{2v}	$C_a - C_{a'}$ $C_a - C_c$ $\angle C_{a'}C_aC_b$ $\angle C_bC_aC_c$	1.361 1.420 93.9 133.3	$C_a - C_b$ $C_c - N$ $\angle C_aC_bC_{b'}$ $\angle C_aC_bH$	1.515 1.157 86.1 114.7	$C_b - C_{b'}$ $C_b - H$ $\angle C_aC_cN$ $\angle C_aC_cC_bH$	1.567 1.088 178.2 115.8	MW
Difluorocyanamide	$F_2N_b - C \equiv N_a$	$C - N_a$ $\angle N_aCN_b$	1.158 174	$C - N_b$ $\angle CN_bF$	1.386 105.4	$N_b - F$ $\angle FN_bF$	1.399 102.8	MW
Difluorocyclopropenone	F$\quad$F $C_b = C_c$ C_a $\|\|$ O C_{2v}	$C_a - C_b$ $C - F$	1.453 1.314	$C_b - C_c$ $\angle FC_bC_c$	1.324 145.7	$C_a - O$	1.192	MW
Difluorodimethylsilane	$(CH_3)_2SiF_2$	$C - Si$ $\angle CSiC$	1.844 115.2	$Si - F$ $\angle FSiF$	1.585 106.1	$C - H$ (ass.) $\angle SiCH$ (ass.)	1.093 110.8	MW
1,1-Difluoroethane	CH_3CHF_2	$C - C$ $\angle CCF$	1.498 110.7	$C - F$ $\angle CCH$ (av.)	1.364 111.0	$C - H$ (av.) dihedral angle between CCF planes	1.081 118.9	ED
1,2-Difluoroethane	CH_2FCH_2F	$C - C$ $\angle CCF$	1.503 110.3	$C - F$ $\angle CCH$	1.389 111	$C - H$ dihedral angle of internal rotation	1.103 109	ED
1,1-Difluoroethene	$CH_2 = CF_2$	$C - C$ $\angle CCF$	1.340 124.7	$C - F$ $\angle CCH$	1.315 119.0	$C - H$	1.091	ED, MW
cis-1,2-Difluoroethene	$CHF = CHF$	$C - C$ $\angle CCF$	1.33 122.0	$C - F$ $\angle CCH$	1.342 124.1	$C - H$	1.099	ED, MW
Difluoromethane	CH_2F_2	$C - F$ $\angle FCF$	1.357 108.3	$C - H$ $\angle HCH$	1.093 113.7			MW
Dimethoxymethane	H$\quad$O$\quad$O$\quad$H H$-C_a\quad C_b\quad C_a-$H H$\quad$H$\quad$H H$\quad$H	$C_a - O$ $\angle COC$	1.432 114.6	$C_b - O$ $\angle OCO$	1.382 114.3	$C - H$ (av.) $\angle OCH$	1.108 110.3	ED
Dimethylamine	$(CH)_2NH$	$C - N$ $\angle CNC$ $\angle HCH$	1.455 111.8 107	$N - H$ $\angle CNH$	1.00 107	$C - H$ $\angle NCH$	1.106 112	ED
Dimethylberyllium	$(CH_3)_2Be$ (CBeC linear)	$C - Be$	1.698	$C - H$	1.127	$\angle BeCH$	113.9	ED
Dimethyl cadmium	$(CH_3)_2Cd$	$C - Cd$	2.112	$\angle HCH$	108.4			R
Dimethyl carbonate	$(C_aH_3O)_2C_b = O_b$	$C_b - O_b$ $\angle O_aC_bO_a$	1.209 107	$C_b - O_a$ $\angle C_bO_aC_a$	1.34 114.5	$C_a - O_a$	1.42	ED
Dimethylcyanamide	$(C_aH_3)_2N_a - C_b \equiv N_b$	$C_b - N_b$	1.161	$C_a - N_a$	1.463	$C_b - N_a$	1.338	ED
trans-Dimethyldiazene	$CH_3N = NCH_3$	$C - N$ $\angle C_aNC_a$	1.482 115.5	$N - N$ $\angle C_aNC_b$	1.247 116.0	$\angle CNN$	112.3	ED
1,2-Dimethyldiborane	$CH_3\quad H_b\quad CH_3$ B$\quad$B $H_t\quad H_b\quad H_t$	$B - B$ $B - H_b$ (cis) $\angle BBC$ (cis)	1.799 1.358 122.6	$B - C$ $B - H_b$ (trans) $\angle BBC$ (trans)	1.580 1.365 121.8	$B - H_t$	1.24	ED
Dimethyl diselenide	$(CH_3)_2Se_2$	$C - Se$ $\angle CSeSe$	1.95 98.9	$Se - Se$ $\angle HCSe$	2.326 108	$C - H$ CSeSeC dihedral angle	1.13 88	ED
Dimethyl disulfide	$(CH_3)_2S_2$	$C - S$ $\angle SSC$	1.816 103.2	$S - S$ $\angle SCH$	2.029 111.3	$C - H$ CSSC dihedral angle	1.105 85	ED

Compound	Structure	Bond distances in Å and angles in degrees							Method
S,S'-Dimethyl dithiocarbonate	$C_aH_3SC_bSC_aH_3$ $\|$ O *syn-syn* conformer	C_a—S ∠OCS	1.802 124.9	C_b—S ∠CSC	1.777 99.3	C_b—O	1.206		ED
Dimethyl ether	$(CH_3)_2O$	C—O ∠COC	1.416 112	C—H ∠HCH	1.121 108				ED
N,N'-Dimethylhydrazine	CH_3NH—$NHCH_3$	C—N C—H	1.46 1.12	N—N ∠NNC	1.42 112	N—H CNNC dihedral angle	1.03 90		ED
Dimethyl mercury	$(CH_3)_2Hg$	C—Hg	2.083	C—H	1.160 (ass.)	Hg⋯H	2.71		ED
Dimethylphosphine	$(CH_3)_2PH$	C—P ∠CPC	1.848 99.7	P—H ∠CPH	1.419 97.0				MW
2,2-Dimethylpropanenitrile	$(C_cH_3)_3C_b$—$C_a\equiv N$	C_a—C_b ∠$C_cC_bC_c$	1.495 110.5	C_b—C_c	1.536	C_a—N	1.159		MW
Dimethyl selenide	$(CH_3)_2Se$	C—Se ∠CSeC	1.943 96.2	C—H ∠SeCH	1.093 108.7	∠HCH	110.3		MW
Dimethyl silane	$(CH_3)_2SiH_2$	C—Si ∠CSiC ∠HSiH	1.868 110.9 107.8	C—H ∠CSiH	1.089 109.5	Si—H ∠SiCH	1.482 110.9		MW
Dimethyl sulfide	$(CH_3)_2S$	C—S ∠CSC	1.802 98.80	C—H ∠HCH	1.090 109.3				ED, MW
Dimethyl sulfone	$(CH_3)_2SO_2$	C—S ∠CSC	1.771 102	S—O ∠OSO	1.435 121	C—H	1.114		ED
Dimethyl sulfoxide	$(CH_3)_2SO$	C—S ∠CSC dihedral angle between SCC plane and S—O bond	1.799 96.6 115.5	S—O ∠CSO	1.485 106.7	C—H ∠HCH	1.081 110.3		MW
Dimethyl zinc	$(CH_3)_2Zn$	C—Zn	1.929	∠HCH	107.7				R
1,4-Dioxane	CH_2CH_2 O〈 〉O CH_2CH_2 chair form	C—C ∠CCO	1.523 109.2	C—O ∠COC	1.423 112.45	C—H	1.112		ED
Ethane	C_2H_6 staggered conformation	C—C C—C (r_e)	1.5351 1.522	C—H	1.0940	∠CCH	111.17		MW
1,2-Ethanediamine	$H_2NCH_2CH_2NH_2$ *gauche* conformer	C—C ∠CCN	1.545 110.2	C—N dihedral angle between NCC and CCN planes	1.469 64	C—H	1.11		ED
Ethanethiol	C_bH_3—C_aH_2—SH	C_a—C_b C_a—H ∠C_bC_aS	1.530 1.090 108.3	C_a—S C_b—H ∠C_bC_aH	1.829 1.093 109.6	S—H ∠C_aSH ∠C_aC_bH	1.350 96.4 109.7		MW
Ethanol	$C_bH_3C_aH_2OH$ staggered conformation	C—C C_a—H ∠CCO	1.512 1.10 107.8	C—O C_b—H ∠C_bC_aH	1.431 1.09 111	O—H ∠COH ∠C_aC_bH	0.971 105 110		MW
Ethylene	CH_2=CH_2	C—C (r_s)	1.329	C—H (r_s)	1.082	∠HCH (θ_s)	117.2		MW, IR
Ethyleneimine	H_a N H_b〈 〉H_b C—C H_c H_c	C—C C—H ∠CNC ∠H_bCC ∠H_cCN	1.481 1.084 60.3 117.8 114.3	N—C N—H ∠H_aNC ∠H_bCN	1.475 1.016 109.3 118.3	∠H_bCH$_c$ ∠H_cCC	115.7 119.3		MW
Ethyl methyl ether	$C_2H_5OCH_3$	C—C ∠COC	1.520 111.9	C—O (av.) ∠OCC	1.418 109.4	C—H (av.) ∠HCH	1.118 109.0		ED

Compound	Structure	Bond distances in Å and angles in degrees						Method
Ethyl methyl sulfide	$C_2H_5SCH_3$	C—C	1.536	C—S (av.)	1.813	C—H	1.111	ED
	gauche conformer	∠CSC	97	∠SCC	114.0	∠HCH	110	
Fluoroketene	HFC=C=O	C—C	1.317	C—O	1.167	C—F	1.360	MW
		C—H	1.102	∠CCO	178.0	∠CCF	119.5	
		∠CCH	122.3					
Fluoromethane	CH_3F	C—F (r_e)	1.382	C—H (r_e)	1.095	∠HCH (θ_e)	110.45	MW, IR
Fluoromethylidyne	CF	C—F (r_e)	1.2718					UV
(Fluoromethylidyne) phosphine	FC≡P	C—F	1.285	C—P	1.541			MW
2-Fluoropropane	CH_3CHFCH_3	C—C	1.522	C—F	1.398			MW
		∠CCC	113.4	∠CCF	108.2			
Formaldehyde	H_2CO	C—O	1.208	C—H	1.116	∠HCH	116.5	MW
Formaldehyde azine	$H_2C=N—N=CH_2$	C—N	1.277	N—N	1.418	C—H	1.094	ED
	trans conformer	∠CNN	111.4	∠HCN	120.7			
Formaldehyde oxime	H_a ... OH_c / $C=N$ / H_b	C—N	1.276	N—O	1.408	O—H_c	0.956	MW
		C—H_a	1.085	C—H_b	1.086	∠CNO	110.2	
		∠H_aCN	121.8	∠H_bCN	115.6	∠NOH_c	102.7	
Formamide	H_c—N(H_b)—C(=O)H_a	C—N	1.368	C—O	1.212	C—H_a	1.125	ED, MW
		N—H	1.027	∠CNH (av.)	119.2	∠NCO	125.0	
Formic acid	H—C(=O_a)(O_b—H)	C—O_a	1.202	C—O_b	1.343	O_b—H	0.972	MW
		C—H	1.097					
	(planar)	∠O_aCO_b	124.9	∠HCO_a	124.1	∠CO_bH	106.3	
Formic acid dimer	O_a ··· H—O_b HC / CH O_b—H ··· O_a	C—O_a	1.220	C—O_b	1.323	O_a···O_b	2.703	ED
		∠O_aCO_b	126.2	∠C$O_a$$O_b$	108.5			
Formyl radical	HC=O	C—O	1.1712	C—H	1.110	∠HCO	127.43	MW
Fulvene	C_dH_2 ... C_a HC$_b$ C$_b$H HC$_c$—C$_c$H	C_a—C_d	1.349	C_a—C_b	1.470	C_b—C_c	1.355	MW
		C_c—C_c	1.476	C_b—H	1.078	C_c—H	1.080	
		C_d—H	1.13	∠$C_b$$C_a$$C_b$	106.6	∠$C_b$$C_c$$C_c$	109	
		∠$C_a$$C_b$$C_c$	107.7	∠$C_a$$C_b$H	124.7	∠$C_b$$C_c$H	126.4	
		∠HC$_d$H	117					
Furan	H_a O H_a / C_a C_a / C_b—C_b / H_b H_b	C_a—C_b	1.361	C_b—C_b	1.431	C_a—O	1.362	MW
		C_a—H_a	1.075	C_b—H_b	1.077			
		∠$C_a$$C_b$$C_b$	106.1	∠$C_b$$C_a$O	110.7	∠C_aOC_a	106.6	
		∠$C_b$$C_b$$H_b$	128.0	∠O$C_a$$H_a$	115.9			
Furfural	H H / C_c—C_b O_b / H—C_d C_a—C_e H / O_a	C_a—C_e	1.458	C_e—O_b	1.250	C_e—H	1.088	MW
		∠$C_a$$C_e$O	121.6	∠$C_e$$C_a$$C_b$	133.9	∠$C_a$$C_e$H	116.9	
		trans conformer (with respect to O_a and O_b atoms)						
Glycolaldehyde	H_c O_b / C_b / H_b C_a H_a / H_b O_a	C_a—C_b	1.499	C_a—O_a	1.437	C_b—O_b	1.209	MW
		C_a—H_b	1.093	C_b—H_c	1.102	O_a—H_a	1.051	
		∠$C_a$$C_b$$O_b$	122.7	∠$C_b$$C_a$$O_a$	111.5			
		∠$C_a$$C_b$$H_c$	115.3	∠$C_b$$C_a$$H_b$	109.2	∠$H_b$$C_a$$H_b$	107.6	
		∠$C_a$$O_a$$H_a$	101.6	∠$H_b$$C_a$$O_a$	109.7			
Glyoxal	CHOCHO	C—C	1.526	C—O	1.212	C—H	1.132	ED, UV
	trans conformer	∠CCO	121.2	∠HCO	112			
Hexachloroethane	Cl_3CCCl_3	C—C	1.56	C—Cl	1.769	∠CCCl	110.0	ED
2,4-Hexadiyne	$C_aH_3C_b≡C_cC_c≡C_bC_aH_3$	C_a—C_b	1.450	C_b—C_c	1.208	C_c—C_c	1.377	ED
		C_a—H	1.09					
Hexafluoroethane	F_3CCF_3	C—C	1.545	CF	1.326	∠CCF	109.8	ED

Compound	Structure	Bond distances in Å and angles in degrees						Method
Hexafluoropropene	$CF_2{=}CFCF_3$	C—C	1.513	C=C	1.329 (ass.)	C—F	1.329 (ass.)	ED
		∠CCC	127.8	∠FCC (CF)	120	∠FCC(CF_2)	124	
		∠FCC(CF_3)	110					
trans-1,3,5-Hexatriene	$H_2C_a{=}C_bHC_c H{=}C_c HC_b H{=}C_a H_2$	C_a—C_b	1.337	C_b—C_c	1.458	C_c—C_c	1.368	ED
		∠$C_a C_b C_c$	121.7	∠$C_b C_c C_c$	124.4			
Hydrogen cyanide	HCN (linear)	C—H (r_e)	1.0655	C—N (r_e)	1.1532			MW, IR
Iminocyanide radical	HNCN	N—H	1.034	N···N	2.470			UV
		∠HNC	116.5	∠NCN	~180			
Iodoacetylene	IC≡CH	C—C	1.218	C—I	1.980	C—H	1.059	IR
Iodocyanoacetylene	$IC_a{\equiv}C_b C_c{\equiv}N$ (linear)	C_a—C_b	1.207	C_b—C_c	1.370	C_c—N	1.160	MW
		C_a—I	1.985					
Iodomethane	CH_3I	C—I (r_e)	2.132	C—H (r_e)	1.084	∠HCH (θ_e)	111.2	MW, IR
Iron pentacarbonyl	$Fe(CO)_5$ (D_{3h})	Fe—C (av.)	1.821	$(Fe{-}C)_{eq} - (Fe{-}C)_{ax}$	0.020	C—O (av.)	1.153	ED
Isobutane	$(C_b H_3)_3 C_a H$	C_a—C_b	1.535	C_a—H	1.122	C_b—H	1.113	ED, MW
		∠$C_b C_a C_b$	110.8	∠$C_a C_b H$	111.4			
Isobutene	(see structure)	C_a—C_b	1.508	C_b—C_c	1.342	C_a—H	1.119	ED, MW
		C_c—H_c	1.10					
		∠$C_a C_b C_a$	115.6	∠$C_a C_b C_c$	122.2	∠$C_b C_c H$	121	
		∠$HC_a C_b$ (av.)	111.4	∠$HC_a H$	107.9	∠$H_c C_c H_c$	118.5	
Isocyanic acid	HNCO (bent)	N—C	1.209	C—O	1.166	N—H	0.986	MW
		∠NCO	180	∠HNC	128.0			
Isocyanomethane	$C_a H_3{-}N{\equiv}C_b$	C_a—N	1.424	N—C_b	1.166	C_a—H	1.102	MW
		∠$NC_a H$	109.12					
		∠HCH	123.0					
Isofulminic acid	HCNO (linear)	C—N	1.161	N—O	1.207	H—C	1.027	MW
Isothiocyanic acid	HNCS	N—C	1.216	C—S	1.561	N—H	0.989	MW
		∠NCS	180	∠HNC	135.0			
Ketene	$H_2C{=}C{=}O$	C—C	1.315	C—O	1.163			MW
		C—H	1.090	∠HCH	123.5			
Malononitrile	$CH_2(CN)_2$	C—C	1.480	C—N	1.147	C—H	1.091	MW
		∠CCC	110.4	∠CCN	176.6	∠HCH	108.4	
Methane	CH_4	C—H (r_e)	1.0870					IR
Methanethioamide	(see structure)	C—S	1.626	C—N	1.358	C—H_c	1.10	MW
		N—H_a	1.002	N—H_b	1.007			
		∠NCS	125.3	∠$H_a NC$	117.9	∠$H_b NC$	120.4	
		∠SCH_c	127	∠$H_a NH_b$	121.7	∠NCH_c	108	
Methanethiol	CH_3SH	C—S	1.819	S—H	1.34	C—H	1.09	MW
		∠HSC	96.5	∠HCH	109.8	angle between CH_3 symmetry axis and C—S bond	2.2	
Methanol	CH_3OH	C—O	1.4246	C—H	1.0936	O—H	0.9451	MW
		∠COH	108.53	∠HCH	108.63	angle between CH_3 symmetry axis and C—O bond	3.27	
Methyl	·CH_3 planar (D_{3h})	C—H	1.076					R
N-Methylacetamide	(see structure)	C_a—C_b	1.520	C_b—N	1.386	C_c—N	1.469	ED
		C_b—O	1.225	C—H	1.107			
		∠$C_b NC_c$	119.7	∠$NC_b O$	121.8	∠$C_a C_b N$	114.1	

Compound	Structure	Bond distances in Å and angles in degrees						Method
Methylamine	CH_3NH_2	C—N	1.471	N—H	1.019	C—H	1.095	MW
		∠HNC	110.3	∠HNH	106.6	∠HCH	108.1	
		angle between CH$_3$ symmetry axis and C—N bond	4.3					
Methyl azide	CH$_3$ N$_a$—N$_b$—N$_c$ NNN linear	C—N$_a$	1.468	N$_a$—N$_b$	1.216	N$_b$—N$_c$	1.113	ED
		C—H	1.09	∠CN$_a$N$_b$	116.8			
3-Methyl-3H-diazirine	CH$_3$CH ⟨N ‖ N⟩	C—C	1.501	C—N	1.481	N—N	1.235	MW
		∠NCN	49.3	dihedral angle between CNN plane and C—C bond	122.3			
Methylene	:CH$_2$	C—H (r_e)	1.0748	∠HCH (θ_e)	133.84			IR,MW
Methylenecyclopropane	C$_c$H$_2$, C$_b$=C$_a$H$_2$, C$_c$H$_2$	C$_a$—C$_b$	1.332	C$_b$—C$_c$	1.457	C$_c$—C$_c$	1.542	MW
		C$_c$—H	1.09	∠C$_c$C$_b$C$_c$	63.9	∠HC$_a$H	114.3	
		∠HC$_c$H	113.5	dihedral angle between C$_c$H$_2$ plane and C$_c$—C$_c$ bond	150.8			
3-Methyleneoxetane	O ⟨C$_c$H$_2$ C$_b$=C$_a$H$_2$ C$_c$H$_2$⟩	C$_a$—C$_b$	1.33	C$_b$—C$_c$	1.52	C$_c$—O	1.45	MW
		C—H	1.09 (ass)	∠HC$_c$H	114 (ass)	∠HC$_a$H	120 (ass)	
		∠C$_c$C$_b$C$_c$	87					
Methylenephosphine	CH$_c$H$_t$=PH planar	C—P	1.673	C—H$_c$	1.09	C—H$_t$	1.09	MW
		P—H	1.420	∠CPH	97.4			
		∠HCH	117.2	∠PCH$_c$	124.4	∠PCH$_t$	118.4	
Methyl formate	C$_a$H$_3$ O$_a$—C$_b$ O$_b$ H$_b$	C$_b$—O$_b$	1.206	C—O (av.)	1.393	C$_a$—H	1.08	ED
		C$_b$—H	1.101 (ass.)					
		∠COC	114	∠O$_a$C$_b$O$_b$	127	∠O$_a$C$_a$H	110	
Methylgermane	CH_3GeH_3	C—Ge	1.945	Ge—H	1.529	C—H	1.083	MW
		∠HGeH	109.3	∠HCH	108.4			
Methyl hypochlorite	CH_3OCl	C—O	1.389	O—Cl	1.674	C—H	1.103	MW
		∠COCl	112.8	∠HCH	109.6			
Methylidyne	:CH	C—H (r_e)	1.1198					UV
Methylidynephosphine	HCP	C—P (r_e)	1.5398	C—H (r_e)	1.0692			MW
Methylketene	C$_c$H$_3$ C$_b$=C$_a$=O H	C$_a$—C$_b$	1.306	C$_b$—C$_c$	1.518	C$_a$—O	1.171	MW
		C$_b$—H	1.083	C$_c$—H	1.10			
		∠OC$_a$C$_b$	180.5	∠C$_a$C$_b$C$_c$	122.6	∠C$_a$C$_b$H	113.7	
		∠C$_c$C$_b$H	123.7	∠HCH	109.2			
Methyl nitrate	H$_a$ H$_a$ O$_a$ C N O$_b$ H$_b$ O O$_b$	C—O	1.437	C—H$_a$	1.10	C—H$_b$	1.09	MW
		O—N	1.402	N—O$_a$	1.205	N—O$_b$	1.208	
		∠CON	112.7	∠ONO$_a$	118.1	∠ONO$_b$	112.4	
		∠OCH$_a$	110	∠OCH$_b$	103			
Methyloxirane	C$_a$H$_3$C$_b$H—C$_c$H$_2$ ⟨O⟩	C$_a$—C$_b$	1.51	∠C$_a$C$_b$C$_c$	121.0	dihedral angle between C$_b$C$_c$O plane and C$_a$C$_b$ bond	123.8	MW
Methylphosphine	CH_3PH_2	C—P	1.858	C—H	1.094			ED
Methylphosphonic difluoride	CH_3POF_2	C—P	1.770	P—O	1.444	P—F	1.545	ED,MW
		∠OPC	117.8	∠FPC	103.7	∠FPF	99.2	
Methylsilane	CH_3SiH_3	C—Si	1.867	Si—H	1.485	C—H	1.093	MW
		∠HCH	107.7	∠HSiH	108.3			
Methylstannane	CH_3SnH_3	C—Sn	2.143	Sn—H	1.700			MW

Compound	Structure	Bond distances in Å and angles in degrees						Method
Methyl thiocyanate	C_aH_3 $S-C_b-N$	$S-C_a$	1.824	$S-C_b$	1.684	C_b-N	1.170	MW
		$C-H$	1.081					
		$\angle C_aSC_b$	99.0	$\angle HCH$	110.6	$\angle HCS$	108.3	
Methyltrioxorhenium	CH_3ReO_3	$Re-C$	2.074	$Re-O$	1.703	$C-H$	1.088	MW
		$\angle ReCH$	108.9	$\angle CReO$	106.4			
Molybdenum carbide	MoC	$Mo-C$	1.676					UV
Molybdenum carbonyl	$Mo(CO)_6$ (O_h)	$Mo-C$	2.063	$C-O$	1.145			ED
Naphthalene		C_a-C_b	1.37	C_b-C_b	1.41	C_a-C_c	1.42	ED
		C_c-C_c	1.42	$C-C$ (av.)	1.40	$\angle C_aC_cC_c$	119.4	
Neopentane	$C(CH_3)_4$	$C-C$	1.537	$C-H$	1.114	$\angle CCH$	112	ED
Nickel carbonyl	$Ni(CO)_4$ (T_d)	$Ni-C$	1.839	$C-O$	1.121			IR
Nickel monocarbonyl	NiCO (linear)	$Ni-C$	1.64	$C-O$	1.19			IR
Nickel cyanide	$NiC\equiv N$ (linear)	$Ni-C$	1.828	$C-N$	1.158			MW
Nitromethane	CH_3NO_2	$C-N$	1.489	$N-O$	1.224	$C-H$	1.088 (ass.)	MW
		$\angle ONO$	125.3	$\angle NCH$	107			
N-Nitrosodimethylamine	$(CH_3)_2NNO$	$C-N$	1.461	$N-O$	1.235	$N-N$	1.344	ED
		$\angle CNC$	123.2	$\angle CNN$	116.4	$\angle ONN$	113.6	
Nitrosomethane	CH_3NO	$C-N$	1.49	$N-O$	1.22	$C-H$	1.084	MW
		$\angle CNO$	112.6	$\angle NCH$	109.0			
2,5-Norbornadiene	(C_{2v})	C_a-C_b	1.535	C_b-C_b	1.343	C_a-C_c	1.573	ED
		$C-H$	1.12	$\angle C_aC_cC_a$	94			
		dihedral angle between the two $C_aC_bC_bC_a$ planes	115.6					
1,2,5-Oxadiazole	(planar)	$C-C$	1.421	$C-N$	1.300	$O-N$	1.380	MW
		$C-H$	1.076	$\angle CCH$	130.2	$\angle NCH$	120.9	
		$\angle CCN$	109.0	$\angle NON$	110.4	$\angle ONC$	105.8	
1,3,4-Oxadiazole	(planar)	$C-O$	1.348	$C-N$	1.297	$N-N$	1.399	MW
		$C-H$	1.075	$\angle OCH$	118.1	$\angle NCH$	128.5	
		$\angle CNN$	105.6	$\angle COC$	102.0	$\angle OCN$	113.4	
Oxalic acid		$C-C$	1.544	$C-O_a$	1.205	$C-O_b$	1.336	ED
		O_b-H	1.05					
		$\angle CCO_a$	123.1	$\angle O_aCO_b$	125.0	$\angle CO_bH$	104	
Oxalyl chloride		$C-C$	1.534	$C-O$	1.182	$C-Cl$	1.744	ED
		$\angle CCO$	124.2	$\angle CCCl$	111.7	68% trans, 32% gauche at 0 °C		
Oxetane		$C-C$	1.546	$C-O$	1.448	$C-H$ (av.)	1.090	MW
		$\angle CCC$	85	$\angle COC$	92	$\angle OCC$	92	
		$\angle HCH$ (av.)	109.9					

Compound	Structure	Bond distances in Å and angles in degrees						Method
Oxirane	CH₂–O–CH₂ (ring)	C—C	1.466	C—O	1.431	C—H	1.085	MW
		∠HCH	116.6	dihedral angle between NH₂ plane and N—C bond	158.0			
Phenol	(phenol structure: OH on Cₐ ring with C_b, C_c, C_d)	C—C (av.)	1.397	Cₐ—O	1.364	O—H	0.956	MW
		C_b—H	1.084	C_c—H	1.076	C_d—H	1.082	
		∠COH	109.0					
Phosphirane	CH₂–PH–CH₂ (ring)	C—C	1.502	C—P	1.867	P—H	1.43	MW
		C—H	1.09	∠CPC	47.4	∠HPC	95.2	
		∠HCH	114.4	∠CCH	118	dihedral angle between PCC plane and PH bond	95.7	
Piperazine	CH₂—CH₂ / NH ... NH / CH₂—CH₂ (C₂ₕ)	C—C	1.540	C—N	1.467	C—H	1.110	ED
		∠CNC	109.0	∠CCN	110.4			
Palladium carbide	PdC	Pd—C	1.712					UV
Platinum carbide	PtC	Pt—C (r_e)	1.6767					UV
Potassium carbide	KC	K—C	2.528					MW
Propane	C₃H₈	C—C	1.532	C—H	1.107			ED
		∠CCC	112	∠HCH	107			
Propene	(propene structure with Cₐ, C_b, C_c and H_a–H_d)	Cₐ—C_b	1.341	C_b—C_c	1.506			ED, MW
		Cₐ—Hₐ	1.104	C_c—H_d	1.117			
		∠CₐC_bC_c	124.3	∠C_bCₐH_{a,b,c}	121.3	∠C_bC_cH_d	110.7	
2-Propenoyl chloride	(structure with Cₐ=C_b–C_c with Cl and O)	Cₐ—C_b	1.35	C_b—C_c	1.48	C_c—Cl	1.82	MW
		C_c—O	1.19	C—H	1.086 (ass.)			
		∠CₐC_bC_c	123	∠C_bC_cCl	116	∠C_bC_cO	127	
		∠CₐC_bH	120 (ass.)	∠C_bCₐH	121.5 (ass.)			
2-Propynal	H_aCₐ≡C_b–C_cH_cO (planar)	Cₐ—C_b	1.211	C_b—C_c	1.453	C_c—O	1.214	ED, MW
		Cₐ—Hₐ	1.085	C_c—H_c	1.130			
		∠CₐC_bC_c	178.6	∠C_bC_cO	124.2	∠C_bC_cH_c	113.7	
Propyne	H₃C_c–C_b≡CₐH	C_c—C_b	1.459	C_b—Cₐ	1.206			MW
		Cₐ—H	1.056	C_c—H	1.105	∠HC_cC_b	110.2	
Propynal isocyanide	H₃C_c–C_b≡Cₐ–N≡C	C_c—C_b (r_s)	1.456	C_b—Cₐ (r_s)	1.206	Cₐ—N (r_s)	1.316	MW
		N—C (r_s)	1.175	C_c—H (r_s)	1.090	∠HC_cC_b (θ_s)	110.7	
Pyrazine	(pyrazine ring, N at 1,4)	C—C	1.339	C—N	1.403	C—H	1.115	ED
		∠CCH	123.9	∠CCN	115.6			
Pyridazine	(pyridazine ring with C_b=C_b, HCₐ, N—N)	Cₐ—C_b	1.393	C_b—C_b	1.375	Cₐ—N	1.341	ED, MW
		N—N	1.330	∠NCC	123.7	∠NNC	119.3	

Compound	Structure	Bond distances in Å and angles in degrees							Method
Pyridine		$C_a—C_b$	1.395	$C_b—C_c$	1.394	$C_a—N$	1.340		MW
		$C_a—H_a$	1.084	$C_b—H_b$	1.081	$C_c—H_c$	1.077		
		$\angle C_aC_bC_c$	118.5	$\angle C_bC_cC_b$	118.3	$\angle C_cC_bH_b$	121.3		
		$\angle C_aNC_a$	116.8	$\angle NC_aC_b$	123.9	$\angle NC_aH_a$	115.9		
Pyrimidine	(C$_{2v}$ assumed)	C—C	1.393	C—N	1.340				ED
		$\angle$NCN	127.6	$\angle$CNC	115.5				
Pyrrole		$C_a—C_b$	1.382	$C_b—C_b$	1.417	$C_a—N$	1.370		MW
		$C_a—H_a$	1.076	$C_b—H_b$	1.077	N—H	0.996		
		$\angle C_aC_bC_b$	107.4	$\angle C_aNC_a$	109.8	$\angle NC_aC_b$	107.7		
		$\angle C_bC_bH$	127.1	$\angle NC_aH_a$	121.5				
Pyruvonitrile	$C_aH_3—C_b{\overset{O}{\|}}—C_c\equiv N$	$C_a—C_b$	1.518	$C_b—C_c$	1.477	C—H	1.12		ED, MW
		C—N	1.17	C—O	1.208	$\angle$HCH	109.2		
		$\angle C_aC_bC_c$	114.2	$\angle C_aC_bO$	124.5	$\angle$CCN	179		
Ruthenium carbide	RuC	Ru—C	1.607						UV
Silacyclobutane	$\begin{matrix} CH_2—CH_2 \\ \| \quad\quad \| \\ CH_2—SiH_2 \end{matrix}$	C—C	1.571	C—Si	1.885	C—H	1.100		ED
		Si—H	1.47	$\angle$CCC	99.8	$\angle$CSiC	77.2		
		$\angle$SiCC	84.8	dihedral angle between CCC and CSiC planes	146				
Silaethene	$H_2Si{=}CH_2$	Si—C (r_e)	1.704	Si—H (r_e)	1.467	C—H (r_e)	1.082		MW
		$\angle$HCSi	122.0	$\angle$HSiC	122.4				
Silicon dicarbide	CSiC (ring)	C—C(r_s)	1.269	Si—C (r_s)	1.832	$\angle$CSiC (θ_s)	40.5		MW
Silylchloroacetylene	$SiH_3C{\equiv}CCl$	C—C	1.234	Si—C	1.812	C—Cl	1.620		ED
		Si—H	1.488	$\angle$HSiC	109.4				
Silyl cyanide	$SiH_3C{\equiv}N$	Si—C	1.850	C—N	1.156	Si—H	1.487		ED,MW
		$\angle$HSiC	107.25						
Sodium carbide	NaC	Na—C	2.232						MW
Spiro[2.2]pentane	(D$_{2d}$)	$C_b—C_b$	1.52	$C_a—C_b$	1.47	C—H	1.09		ED
		$\angle C_bC_aC_b$	62	$\angle$HCH	118				
Strontium methyl	$SrCH_3$	Sr—C	2.487	C—H (ass.)	1.104	$\angle$HCH	105.8		UV
Succinonitrile	$\begin{matrix} CH_2CN \\ \| \\ CH_2CN \end{matrix}$	C—C	1.561	C—C(N)	1.465	C—N	1.161		ED
		C—H	1.09	$\angle$CCC	110.4	dihedral angle of CCCC for *gauche* conformer	75		
Tetrabromomethane	CBr_4 (T$_d$)	C—Br	1.935						ED
Tetrachloroethene	$CCl_2{=}CCl_2$	C—C	1.354	C—Cl	1.718	$\angle$ClCCl	115.7		ED
Tetrachloromethane	CCl_4 (T$_d$)	C—Cl	1.767						ED
Tetracyanoethene	$(CN)_2C{=}C(CN)_2$	C—C	1.435	C=C	1.357	C—N	1.162		ED
		$\angle$CC=C	121.1						
2,2,4,4-Tetrafluoro-1,3-dithietane	(D$_{2h}$ assumed)	C—S	1.785	C—F	1.314	$\angle$CSC	83.2		ED
		$\angle$FCS	113.7						
Tetrafluoroethene	$CF_2{=}CF_2$	C—C	1.31	C—F	1.319	$\angle$CCF	123.8		ED
Tetrafluoromethane	CF_4 (T$_d$)	C—F	1.323						ED

Compound	Structure	Bond distances in Å and angles in degrees						Method
Tetrahydrofuran	CH$_2$CH$_2$ / O / CH$_2$CH$_2$	C—C	1.536	C—O	1.428	C—H	1.115	ED
Tetrahydropyran	H$_2$C / H$_2$C CH$_2$ / H$_2$C CH$_2$ / O / chair form	C—C	1.531	C—O	1.420	C—H	1.116	ED
		∠COC	111.5	∠OCC	111.8	∠CCC (C)	108	
		∠CCC (O)	111					
Tetrahydrothiophene	CH$_2$CH$_2$ / S / CH$_2$CH$_2$	C—C	1.536	C—S	1.839	C—H	1.120	ED
		∠CCC	105.0	∠CSC	93.4	∠SCC	106.1	
Tetraiodomethane	CI$_4$ (T$_d$)	C—I	2.15					ED
Tetramethylgermane	(CH$_3$)$_4$Ge	C—Ge	1.945	C—H	1.12	∠GeCH	108	ED
Tetramethyl lead	(CH$_3$)$_4$Pb	C—Pb	2.238					ED
Tetramethylsilane	(CH$_3$)$_4$Si	C—Si	1.875	C—H	1.115	∠HCH	109.8	ED
Tetramethylstannane	(CH$_3$)$_4$Sn	C—Sn	2.144	C—H	1.12			ED
1,2,5-Thiadiazole	S / N N / HC—CH / (planar)	C—C	1.420	C—N	1.328	S—N	1.631	MW
		C—H	1.079					
		∠CCN	113.8	∠NSN	99.6	∠CCH	126.2	
1,3,4-Thiadiazole	S / HC CH / N—N / (planar)	C—S	1.721	C—N	1.302	N—N	1.371	MW
		C—H	1.08	∠CSC	86.4	∠SCN	114.6	
		∠CCN	112.2	∠NCH	123.5	∠SCH	121.9	
Thietane	CH$_2$—CH$_2$ / CH$_2$—S	C—C	1.549	C—S	1.847	C—H (av.)	1.100	ED, MW
		∠CSC	76.8	∠HCH (av.)	112	dihedral angle between CCC and CSC planes	154	
Thiirane	H$_2$C / S / H$_2$C	C—C	1.484	C—S	1.815	C—H	1.083	MW
		∠CSC	48.3	∠CCS	65.9	∠HCH	116	
		dihedral angle between CH$_2$ plane and C—C bond	152					
Thioacetaldehyde	H$_3$C$_b$ — C$_a$ (S, H)	C$_a$—S (r_s)	1.610	C$_a$—C$_b$ (r_s)	1.506			MW
		C$_a$—H (r_s)	1.089	C$_b$—H (r_s)	1.094 (av.)			
		∠C$_b$C$_a$S (θ_s)	125.3	∠C$_b$C$_a$H (θ_s)	119.4	∠HC$_b$C$_a$ (θ_s)	110.6 (av.)	
Thiocarbonyl fluoride	F$_2$CS	C—S	1.589	C—F	1.315	∠FCF	107.1	MW
Thioformaldehyde	CH$_2$S	C—S	1.611	C—H	1.093	∠HCH	116.9	MW
Thioketene	H$_2$C=C=S / C$_{2v}$	C—C (r_s)	1.314	C—S (r_s)	1.554	C—H (r_s)	1.080	IR
		∠HCH (θ_s)	119.8					
Thiophene	H$_b$ H$_b$ / C$_b$ C$_b$ / H$_a$C$_a$ C$_a$H$_a$ / S	C$_a$—C$_b$	1.370	C$_b$—C$_b$	1.423	C$_a$—S	1.714	MW
		C$_a$—H$_a$	1.078	C$_b$—H$_b$	1.081			
		∠C$_a$C$_b$C$_b$	112.5	∠C$_a$SC$_a$	92.2	∠SC$_a$C$_b$	115.5	
		∠SC$_a$H$_a$	119.9	∠C$_b$C$_b$H$_b$	124.3			
Toluene	C$_6$H$_5$—CH$_3$	C—C (ring)	1.399	C—CH$_3$	1.524	C—H (av.)	1.11	ED
1,1,1-Tribromoethane	CH$_3$CBr$_3$	C—C	1.51 (ass.)	C—Br	1.93	C—H	1.095 (ass.)	MW
		∠BrCBr	111	∠CCBr	108	∠CCH	109.0 (ass.)	

Compound	Structure	Bond distances in Å and angles in degrees							Method
Tribromomethane	CHBr$_3$ (C$_{3v}$)	C—Br	1.924	C—H	1.11	∠BrCBr	111.7		ED, MW
Tri-*tert*-butyl methane	HC$_a$[C$_b$(C$_c$H$_3$)$_3$]$_3$	C$_a$—C$_b$	1.611	C$_b$—C$_c$	1.548	C—H	1.111		ED
		∠C$_a$C$_b$C$_c$	113.0						
Trichloroacetonitrile	CCl$_3$CN	C—C	1.460	C—N	1.165	C—Cl	1.763		ED
		∠ClCCl	110.0						
1,1,1-Trichloroethane	CH$_3$CCl$_3$	C—C	1.541	C—Cl	1.771	C—H	1.090		MW
		∠CCCl	109.6	∠ClCCl	109.4	∠HCH	110.0		
		∠CCH	108.9						
Trichlorofluoromethane	CCl$_3$F	C—Cl	1.754	C—F	1.362	∠ClCCl	111		MW
Trichloromethane	CHCl$_3$	C—Cl	1.758	C—H	1.100	∠ClCCl	111.3		MW
Trichloromethylgermane	CH$_3$GeCl$_3$	C—Ge	1.89	Ge—Cl	2.132	C—H	1.103 (ass.)		ED, MW
		∠ClGeCl	106.4	∠GeCH	110.5 (ass.)				
Trichloromethylsilane	CH$_3$SiCl$_3$	C—Si	1.876	Si—Cl	2.021				MW
Trichloromethylstannane	CH$_3$SnCl$_3$	C—Sn	2.10	Sn—Cl	2.304	C—H	1.100		ED
1,1,1-Trichloro-2,2,2-trifluoroethane	CF$_3$CCl$_3$ (staggered configuration)	C—C	1.54	C—F	1.33	C—Cl	1.77		MW
		∠CCF	110	∠CCCl	109.6				
		∠CSnCl	113.9	∠ClSnCl	104.7	∠SnCH	108		
Triethylenediamine	CH$_2$CH$_2$ N—CH$_2$CH$_2$—N CH$_2$CH$_2$ (D$_{3h}$)	C—C	1.562	C—N	1.472	∠CNC	108.7		ED
		∠NCC	110.2						
Trifluoroacetic acid	O$_a$ CF$_3$C O$_b$H	C—C	1.546	C—O$_a$	1.192	C—O$_b$	1.35		ED
		C—F	1.325	O—H	0.96 (ass.)				
		∠CCO$_a$	126.8	∠CCO$_b$	111.1	∠CCF	109.5		
1,1,1-Trifluoroethane	CH$_3$CF$_3$	C—C	1.494	C—F	1.340	C—H	1.081		ED
Trifluoroiodomethane	CF$_3$I (C$_{3v}$)	C—F	1.330	C—I	2.138	∠FCF	108.1		ED, MW
Trifluoromethane	CHF$_3$ (C$_{3v}$)	C—F	1.332	C—H	1.098	∠FCF	108.8		MW
Trifluoromethanesulfonyl fluoride	CF$_3$SO$_2$F$_a$	C—S	1.835	C—F (av.)	1.325	S—O	1.410		ED
		S—F$_a$	1.543	∠CSF$_a$	95.4	∠CSO	108.5		
		∠OSO	124.1	∠FCF	109.8				
Trifluoromethylimino-sulfurdifluoride	CF$_3$N=SF$_2$	C—N	1.409	S—N	1.477	S—F	1.594		ED,MW
		C—F	1.331	∠CNS	127.2	∠NSF	112.7		
		∠FSF	92.8	∠FCF	108.1				
Trifluoromethyl peroxide	CF$_3$OOCF$_3$	O—O	1.42	C—O	1.399	C—F	1.320		ED
		∠COO	107	∠FCF	109.0	COOC dihedral angle of internal rotation	123		
		∠CCF	119.2	∠CCH	112				
Trimethyl aluminium	(CH$_3$)$_3$Al	C—Al	1.957	C—H	1.113				ED
		∠CAlC	120	∠AlCH	111.7				
Trimethylamine	(CH$_3$)$_3$N	C—N	1.458	C—H	1.100				ED
		∠CNC	110.9	∠HCH	110				
Trimethylarsine	(CH$_3$)$_3$As	C—As	1.979	∠CAsC	98.8	∠AsCH	111.4		ED
Trimethyl bismuth	(CH$_3$)$_3$Bi	C—Bi	2.263	C—H	1.07	∠CBiC	97.1		ED
Trimethylborane	(CH$_3$)$_3$B	C—B	1.578	C—H	1.114				ED
		∠CBC	120	∠BCH	112.5				
Trimethylphosphine	(CH$_3$)$_3$P	C—P	1.847	C—H	1.091				ED
		∠CPC	98.6	∠PCH	110.7				
1,3,5-Trioxane	O H$_2$C CH$_2$ O O C H$_2$	C—O	1.422	∠OCO	112.2	∠COC	110.3		MW

Compound	Structure	Bond distances in Å and angles in degrees						Method
Triphenylamine	$(C_6H_5)_3N$ (C_3)	C—C	1.392	C—N	1.42	∠CNC	116	ED
		torsional dihedral angle of phenyl rings	47					
Tungsten carbide	WC	W—C	1.7135					UV
Tungsten carbonyl	$W(CO)_6$ (O_h)	W—C	2.059	C—O	1.149			ED
Vanadium carbonyl	$V(CO)_6$ (O_h, involving dynamic Jahn-Teller effect)	V—C	2.015	C—O	1.138			ED
Vinyl bromide	See Vinyl chloride	C—C	1.3256	C—Br	1.8835	C—H_a	1.0780	MW
		C—H_b	1.0804	C—H_c	1.0794	∠CCBr	122.62	
		∠CCH_a	124.34	∠CCH_b	119.28	∠CCH_c	122.03	
Vinyl chloride		C—C	1.3262	C—Cl	1.7263	C—H_a	1.0783	MW
		C—H_b	1.0796	C—H_c	1.0796	∠CCCl	122.75	
		∠CCH_a	123.91	∠CCH_b	119.28	∠CCH_c	121.77	
Vinyl fluoride	See Vinyl chloride	C—C	1.3210	C—F	1.3428	C—H_a	1.0796	MW
		C—H_b	1.0774	C—H_c	1.0789	∠CCF	121.70	
		∠CCH_a	125.95	∠CCH_b	118.97	∠CCH_c	121.34	
Vinyl iodide	See Vinyl chloride	C—C	1.3276	C—I	2.0830	C—H_a	1.0787	MW
		C—H_b	1.0823	C—H_c	1.0799	∠CCI	122.97	
		∠CCH_a	123.54	∠CCH_b	119.36	∠CCH_c	122.30	
Zinc cyanide	ZnC≡N (linear)	Zn—C	1.955	C—N	1.146			MW

CHARACTERISTIC BOND LENGTHS IN FREE MOLECULES

David R. Lide

This is a summary of typical bond lengths in gas-phase molecules. The value given for each bond is near the mid-range of values found in simple molecules. Bond lengths usually vary by 1% or 2%, and often by more, depending on the nature of the other bonds attached to the two atoms in question. References 1 and 2 give measured bond lengths in individual gas-phase molecules, as determined by spectroscopic and electron diffraction methods.

All bond lengths are given in Å (1 Å = 10^{-10} m).

References

1. "Structure of Free Molecules in the Gas-Phase," *CRC Handbook of Chemistry and Physics*, 92nd Edition, 2011.
2. Harmony, M. D., Laurie, V. W., Kuczkowski, R. L., Schwendeman, R. H., Ramsay, D. A., Lovas, F. J., Lafferty, W. J., and Maki, A. G., *J. Phys. Chem. Ref. Data* 8, 619, 1979.
3. Lide, D. R., *Tetrahedron* 17, 125, 1962.

A. Characteristic lengths of single bonds in Å.

	As	B	Br	C	Cl	F	Ge	H	I	N	O	P	S	Sb	Se	Si
As	2.10															
Br	2.32	1.89	2.28													
C	1.98	1.58	1.93	1.53												
Cl	2.17	1.74	2.14	1.77	1.99											
F	1.71	1.31	1.76	1.35	1.63	1.41										
Ge		2.30		1.95	2.15	1.73	2.40									
H	1.51	1.19	1.41	1.09	1.28	0.92	1.53	0.74								
I		2.12	2.47	2.13	2.32	1.91	2.51	1.61	2.67							
N		1.39		1.46	1.90	1.37		1.02		1.45						
O		1.35		1.42	1.70	1.42		0.96		1.43	1.48					
P		1.94	2.22	1.85	2.04	1.57		1.42	1.65			2.25				
S		2.24		1.82	2.05	1.56		1.34					2.00			
Sb		2.49		2.33	1.88			1.70	2.27							
Se				1.95	1.71			1.47							2.33	
Si			2.21	1.87	2.05	1.58		1.48	2.44		1.63		2.14			2.33
Sn			2.50	2.14	2.30	1.94		1.71	2.67							
Te						1.82		1.66								

B. Lengths of multiple bonds (non-ring molecules) in Å.

C=C	1.34
C≡C	1.20
C=N	1.21
C≡N	1.16
C=O	1.21
C=S	1.61
N=N	1.24
N≡N	1.13
N=O	1.18
O=O	1.21

C. Effect of environment on carbon-carbon single bonds in Å (other single bonds not shown). From Reference 3.

Configuration	C—C Length	Examples of Molecules
C—C	1.526	$H_3C—CH_3$
C—C=	1.501	$H_3C—CH=CH_2$
C—C≡	1.459	$H_3C—C≡CH$
=C—C=	1.467	$H_2C=CH—CH=CH_2$
≡C—C=	1.445	$HC≡C—CH=CH_2$
≡C—C≡	1.378	$HC≡C—C≡CH$

D. Some metal-carbon bond lengths in gas-phase molecules in Å.

Al—C	1.96
Be—C	1.70
Bi—C	2.26
Cd—C	2.11
Co—C	1.88
Cr—C	1.92*
Cu—C	1.83
Fe—C	1.82*
Hg—C	2.08
Mo—C	2.06*
Ni—C	1.83
Pb—C	2.24
Sn—C	2.14
V—C	2.02*
W—C	2.06*
Zn—C	1.93

* In carbonyl molecules

ATOMIC RADII OF THE ELEMENTS

Manjeera Mantina, Rosendo Valero, Christopher J. Cramer, and Donald G. Truhlar

Atomic radii are not precisely defined, but are nevertheless very widely used parameters in modeling and understanding molecular structure and interactions. Three main classes of radii may be identified: van der Waals radii (which include radii used to characterize steric interactions), covalent radii, and ionic radii. This section is concerned with the first two; ionic radii are covered in another section of the *Handbook* called "Ionic Radii in Crystals."

There are many scales of van der Waals radii, but they are not fully consistent with one another. The van der Waals radii determined by Bondi [1] from x-ray diffraction data, crystal densities, gas kinetic collision cross sections, critical densities, and liquid-state properties are the most widely used values, but Bondi recommended radius values for only 28 of the 44 main-group elements in the periodic table plus 9 transition metals and one actinide. Rowland and Taylor [2] redetermined nine of the main-group radii from crystal structure data and recommended that Bondi's values be accepted except for H, for which they recommended a new value; we accepted their recommendation for H and adopted Bondi's 27 other values for main-group elements. Radii for the 16 remaining main-group elements were determined from electronic structure calculations on selected van der Waals molecules by analyzing the results in a way designed to yield radii compatible with Bondi's scale for main-group elements [3]. Bondi's values for the transition metals and actinide are smaller than expected based on ionization potentials and covalent radii, so we do not adopt them, but defer to later recommendations based on analysis of a more extensive set of crystal data. Van der Waals radii for the remaining elements through atomic number $Z = 93$ were taken from Hu et al. [4], who determined them from bond valence parameters (this gives radii usually within 0.1 Å to 0.15 Å of the radii from average atomic volumes in crystals as obtained by statistical analysis of the Cambridge Structural Database [5], but the results show a smoother variation with atomic number). Van der Waals radii for the elements with $Z = 94$ to 103 were based on the work of Guzei and Wendt [6], who modeled the zero-potential distance for the interaction potentials involved in nonbonded steric interactions; we increased their values by 6%, which is the value they found is needed to make their steric radii scale most consistent with Bondi's van der Waals radii.

Covalent radii are more straightforward, especially for elements that tend to form only single bonds, although some researchers distinguish metallic bonds from covalent bonds. The covalent radii tabulated here are recommendations for single covalent bonds, and they are based on a comprehensive evaluation of experimental data by Cordero et al. [7], who recommended covalent radii for all elements up to $Z = 96$, and on an analysis combining experimental data and theoretical calculations by Pyykkö and Atsumi [8], who recommended single-bond covalent radii for all elements up to $Z = 118$. If one is interested in a specific coordination number, oxidation state, or type of ligand, one might find a more appropriate radius in the specialized literature or in the Characteristic Bond Lengths in Free Molecules table in this *Handbook* since the values in the table below are generic average values. In particular, we give the Pyykkö-Atsumi values for carbon and for $Z = 97$ to 118 and an average of the values from the two sources for all other elements. For Mn, Fe, and Co, Cordero et al. give two values, and we used the lower of these on the average because it corresponds to a smoother periodic trend.

Please note that the van der Waals radii of bromine and lithium have typographical errors in Ref. 3; the correct values (1.85 for Br and 1.82 for Li) in the table below are from Ref. 1.

All values are rounded to the nearest 0.01 Å, but in most cases the uncertainty in the value is of the order of 0.1 Å.

References

1. Bondi, A., *J. Phys. Chem.* 68, 441, 1964.
2. Rowland, R. S., and Taylor, R., *J. Phys. Chem.* 100, 7384, 1996.
3. Mantina, M., Chamberlin, A. C., Valero, R., Cramer, C. J., and Truhlar, D. G., *J. Phys. Chem., A* 113, 5806, 2009.
4. Hu, S.-Z., Zhou, Z.-H., and Robertson, B. E., *Z. Kristallogr.* 224, 375, 2009.
5. Hu, S.-Z., Zhou, Z.-H., and Tsai, K.-R., *Acta Physico-Chimica Sinica* 19, 1073, 2003.
6. Guzei, I. A. and Wendt, M., *Dalton Trans.*, 2006, 3991, 2006.
7. Cordero, B., Gómez, V., Platero-Prats, A. E., Revés, M., Echeverría, J., Cremades, E., Barragán, F., and Alvarez, S., *Dalton Trans.* 2008, 2832, 2008.
8. Pyykkö, P., and Atsumi, M., *Chem. Eur. J.* 15, 186, 2009.

Element	Symbol	R_{vdW}/Å	R_{cov}/Å
Actinium	Ac	2.47	2.01
Aluminum	Al	1.84	1.24
Americium	Am	2.44	1.73
Antimony	Sb	2.06	1.40
Argon	Ar	1.88	1.01
Arsenic	As	1.85	1.20
Astatine	At	2.02	1.48
Barium	Ba	2.68	2.06
Berkelium	Bk	2.44	1.68
Beryllium	Be	1.53	0.99
Bismuth	Bi	2.07	1.50
Bohrium	Bh		1.41
Boron	B	1.92	0.84
Bromine	Br	1.85	1.17
Cadmium	Cd	2.18	1.40

Element	Symbol	R_{vdW}/Å	R_{cov}/Å
Calcium	Ca	2.31	1.74
Californium	Cf	2.45	1.68
Carbon	C	1.70	0.75
Cerium	Ce	2.42	1.84
Cesium	Cs	3.43	2.38
Chlorine	Cl	1.75	1.00
Chromium	Cr	2.06	1.30
Cobalt	Co	2.00	1.18
Copernicium	Cp		1.22
Copper	Cu	1.96	1.22
Curium	Cm	2.45	1.68
Darmstadtium	Ds		1.28
Dubnium	Db		1.49
Dysprosium	Dy	2.31	1.80
Einsteinium	Es	2.45	1.65

Element	Symbol	R_{vdW}/Å	R_{cov}/Å
Erbium	Er	2.29	1.77
Europium	Eu	2.35	1.83
Fermium	Fm	2.45	1.67
Fluorine	F	1.47	0.60
Francium	Fr	3.48	2.42
Gadolinium	Gd	2.34	1.82
Gallium	Ga	1.87	1.23
Germanium	Ge	2.11	1.20
Gold	Au	2.14	1.30
Hafnium	Hf	2.23	1.64
Hassium	Hs		1.34
Helium	He	1.40	0.37
Holmium	Ho	2.30	1.79
Hydrogen	H	1.10	0.32
Indium	In	1.93	1.42
Iodine	I	1.98	1.36
Iridium	Ir	2.13	1.32
Iron	Fe	2.04	1.24
Krypton	Kr	2.02	1.16
Lanthanum	La	2.43	1.94
Lawrencium	Lr	2.46	1.61
Lead	Pb	2.02	1.45
Lithium	Li	1.82	1.30
Lutetium	Lu	2.24	1.74
Magnesium	Mg	1.73	1.40
Manganese	Mn	2.05	1.29
Meitnerium	Mt		1.29
Mendelevium	Md	2.46	1.73
Mercury	Hg	2.23	1.32
Molybdenum	Mo	2.17	1.46
Neodymium	Nd	2.39	1.88
Neon	Ne	1.54	0.62
Neptunium	Np	2.39	1.80
Nickel	Ni	1.97	1.17
Niobium	Nb	2.18	1.56
Nitrogen	N	1.55	0.71
Nobelium	No	2.46	1.76
Osmium	Os	2.16	1.36
Oxygen	O	1.52	0.64
Palladium	Pd	2.10	1.30
Phosphorus	P	1.80	1.09
Platinum	Pt	2.13	1.30
Plutonium	Pu	2.43	1.80
Polonium	Po	1.97	1.42

Element	Symbol	R_{vdW}/Å	R_{cov}/Å
Potassium	K	2.75	2.00
Praseodymium	Pr	2.40	1.90
Promethium	Pm	2.38	1.86
Protactinium	Pa	2.43	1.84
Radium	Ra	2.83	2.11
Radon	Rn	2.20	1.46
Rhenium	Re	2.16	1.41
Rhodium	Rh	2.10	1.34
Roentgenium	Rg		1.21
Rubidium	Rb	3.03	2.15
Ruthenium	Ru	2.13	1.36
Rutherfordium	Rf		1.57
Samarium	Sm	2.36	1.85
Scandium	Sc	2.15	1.59
Seaborgium	Sg		1.43
Selenium	Se	1.90	1.18
Silicon	Si	2.10	1.14
Silver	Ag	2.11	1.36
Sodium	Na	2.27	1.60
Strontium	Sr	2.49	1.90
Sulfur	S	1.80	1.04
Tantalum	Ta	2.22	1.58
Technetium	Tc	2.16	1.38
Tellurium	Te	2.06	1.37
Terbium	Tb	2.33	1.81
Thallium	Tl	1.96	1.44
Thorium	Th	2.45	1.90
Thulium	Tm	2.27	1.77
Tin	Sn	2.17	1.40
Titanium	Ti	2.11	1.48
Tungsten	W	2.18	1.50
Ununhexium	Uuh		1.75
Ununoctium	Uuo		1.57
Ununpentium	Uup		1.62
Ununquadium	Uuq		1.43
Ununseptium	Uus		1.65
Ununtrium	Uut		1.36
Uranium	U	2.41	1.83
Vanadium	V	2.07	1.44
Xenon	Xe	2.16	1.36
Ytterbium	Yb	2.26	1.78
Yttrium	Y	2.32	1.76
Zinc	Zn	2.01	1.20
Zirconium	Zr	2.23	1.64

DIPOLE MOMENTS

David R. Lide

This table gives selected values of the electric dipole moment for over 800 molecules. When available, values determined by microwave spectroscopy, molecular beam electric resonance, and other high-resolution spectroscopic techniques were selected. Otherwise, the values come from measurements of the dielectric constant in the gas phase or, if these do not exist, in the liquid phase. Entries are listed alphabetically; compounds not containing carbon are listed first, followed by compounds containing carbon.

The dipole moment is given in debye units (D). The conversion factor to SI units is $1\ D = 3.33564 \times 10^{-30}$ C m.

Dipole moments of individual conformers (rotational isomers) are given when they have been measured. The conformers are designated as *gauche*, *trans*, *axial*, etc. The meaning of these terms can be found in the references. In some cases an average value, obtained from measurements on the bulk gas, is also given. Other information on molecules that have been studied by spectroscopy, such as the components of the dipole moment in the molecular framework and the variation with vibrational state and isotopic species, is given in the references.

When the accuracy of a value is explicitly stated (i.e., 1.234 ± 0.005), the stated uncertainty generally indicates two or three standard deviations. When no uncertainty is given, the value may be assumed to be precise to a few units in the last decimal place. However, if more than three decimal places are given, the exact interpretation of the final digits may require analysis of the vibrational averaging.

Values measured in the gas phase that are questionable because of undetermined error sources are indicated as approximate (≈). Values obtained by liquid phase measurements, which sometimes have large errors because of association effects, are enclosed in brackets, e.g., [1.8].

References

1. Nelson, R. D., Lide, D. R., and Maryott, A. A., *Selected Values of Electric Dipole Moments for Molecules in the Gas Phase*, Natl. Stand. Ref. Data Ser. — Nat. Bur. Stnds. 10, 1967.
2. *Landolt-Börnstein, Numerical Data and Functional Relationships in Science and Technology, New Series*, II/6 (1974), Springer-Verlag, Heidelberg.
3. *Landolt-Börnstein, Numerical Data and Functional Relationships in Science and Technology, New Series*, II/14a (1982), Springer-Verlag, Heidelberg.
4. *Landolt-Börnstein, Numerical Data and Functional Relationships in Science and Technology, New Series*, II/14b (1983), Springer-Verlag, Heidelberg.
5. *Landolt-Börnstein, Numerical Data and Functional Relationships in Science and Technology, New Series*, II/19c (1992), Springer-Verlag, Heidelberg.
6. *Landolt-Börnstein, Numerical Data and Functional Relationships in Science and Technology, New Series*, II/24c (2002), Springer-Verlag, Heidelberg.
7. Riddick, J. A., Bunger, W. B., and Sakano, T. K., *Organic Solvents, Fourth Edition*, John Wiley & Sons, New York, 1986.
8. Kasuya, T., Lafferty, W. J., and Lide, D. R., *J. Chem. Phys.* 48, 1, 1968.
9. Kirchhoff, W. H., and Lide, D. R., *J. Chem. Phys.* 51, 467, 1969.
10. Durig, J. R., Li, Y. S., and Rizzolo, J. J., *J. Chem. Phys.* 77, 5885, 1982.
11. Ogata, T., Mochizuki, A. and Yamashita, E., *J. Chem. Phys.* 87, 2531, 1987.
12. Rego, A., and Cox, A. P., *J. Chem. Phys.* 89, 124, 1988.
13. Tyblewski, M., et al., *J. Chem. Phys.* 97, 6168, 1992.
14. Kawashima, Y., et al., *J. Chem. Phys.* 99, 820, 1993.
15. Caminati, W., Melandri, S., and Favero, L., *J. Chem. Phys.* 100, 8569, 1994.
16. Cederberg, J., et al., *J. Chem. Phys.* 105, 3361, 1996.
17. Bauder, A., et al., *J. Chem. Phys.* 106, 7558, 1997.
18. Muller, H. S. P., Miller, C. E., and Cohen, E. A., *J. Chem. Phys.* 107, 8292, 1997.
19. Burgh, D. J., Suenram, R. D., and Stevens, W. J., *J. Chem. Phys.* 111, 3526, 1999.
20. Blake, T. A., et al., *J. Chem. Phys.* 98, 6031, 1993.
21. Ruoff, R. S., et al., *J. Chem. Phys.* 89, 138, 1988.
22. Muenter, J. S., *J. Chem. Phys.* 90, 4048, 1989.
23. Peterson, J. I., Suenram, R. D., and Lovas, F. J., *J. Chem. Phys.* 90, 5964, 1989.
24. Suenram, R. D., Lovas, F. J., and Matsumura, K., *Astrophys. J. Lett.* 342, 103, 1989.
25. Groner, P., et al., *J. Chem. Phys.* 91, 1434, 1989.
26. Suenram, R. D., Lovas, F. J., Fraser, G. T., and Matsumura, K., *J. Chem. Phys.* 92, 4724, 1990.
27. Andrews, A. M., et al., *J. Chem. Phys.* 93, 7030, 1990.
28. Peterson, K. I., Suenram, R. D., and Lovas, F. J., *J. Chem. Phys.* 94, 106, 1991.
29. Iida, M., Ohshima, Y., and Endo, Y., *J. Chem. Phys.* 95, 4772, 1991.
30. Andrews, A. M., Hillig, K. W., and Kuczkowski, R. L., *J. Chem. Phys.* 96, 1784, 1992.
31. Ruoff, R. S., et al., *J. Chem. Phys.* 96, 3441, 1992.
32. Germann, T. C., Tschopp, S. L., and Gutowsky, H. S., *J. Chem. Phys.* 97, 1619, 1992.
33. Taleb-Bendiab, A., Hillig, K. W., and Kuczkowski, R. L., *J. Chem. Phys.* 97, 2996, 1992.
34. Taleb-Bendiab, A., Hillig, K. W., and Kuczkowski, R. L., *J. Chem. Phys.* 98, 3627, 1993.
35. Xu, L-W., and Kuczkowski, R. L., *J. Chem. Phys.* 100, 15, 1994.
36. Peterson, K. I., Suenram, R. D., and Lovas, F. J., *J. Chem. Phys.* 102, 7807, 1995.
37. Tatamitani, Y., and Ogata, T., *J. Chem. Phys.* 121, 9885, 2004.
38. Medvedev, I., et al., *Astrophys. J. Suppl.* 148, 593, 2003.
39. Lesarri, A., Suenram, R. D., and Brugh, D., *J. Chem. Phys.* 117, 9651, 2002.
40. Arunan, E., et al., *J. Chem. Phys.* 117, 9766, 2002.
41. Smith, T. C., Clouthier, D. J., and Steimle, T. C., *J. Chem. Phys.* 115, 817, 2001.
42. Peebles, S. A., Sun, L., and Kuczkowski, R. L., *J. Chem. Phys.* 110, 6804, 1999.
43. Namiki, K. C., Robinson, J. S., and Steimle, T. C., *J. Chem. Phys.* 109, 5283, 1998.
44. Peebles, S. A., and Kuczkowski, R. L., *J. Chem. Phys.* 109, 5276, 1998.
45. Sauer, B. E., Wang, J., and Hinds, E. A., *J. Chem. Phys.* 105, 7412, 1996.
46. Fry, J. L., Drouin, B. J., and Miller, C. E., *J. Chem. Phys.* 124, 084304, 2006.
47. Christiansen, J. J., *J. Mol. Spectrosc.* 231, 131, 2005.
48. Kisiel, Z., et al., *Chem. Phys. Lett.* 325, 523, 2000.
49. Lovas, F. J., et al., *Astrophys. J. Lett.* 455, 201, 1995.
50. Suenram, R. D., and Lovas, F. J., *Astrophys. J. Lett.* 429, 89, 1994.
51. Biermann, S., et al., *J. Chem. Phys.* 105, 9754, 1996.
52. McGlone, S., and Bauder, A., *J. Chem. Phys.* 109, 5383, 1998.
53. Peebles, S. A., and Kuczkowski, R. L., *J. Chem. Phys.* 111, 10511, 1999.
54. Plusquellic, D. F., et al., *J. Chem. Phys.* 115, 3057, 2001.
55. Muller, H. S. P., and Cohen, E. A., *J. Chem. Phys.* 116, 2407, 2002.
56. Andrews, A. M., and Kuczkowski, R. L., *J. Chem. Phys.* 98, 791, 1993.
57. Lovas, F. J., et al., *J. Chem. Phys.* 92, 891, 1990.
58. Klots, T. D., Emilsson, T., and Gutowsky, H. S., *J. Chem. Phys.* 97, 5335, 1992.
59. Careless, A. J., Kroto, H. W., and Landsberg, B. M., *Chem. Phys.* 1, 371, 1973.
60. Costain, C. C., and Kroto, H. W., *Can. J. Phys.* 50, 1453, 1972.
61. Kroto, H. W., Nixon, J. F., and Ohno, K., *J. Mol. Spectrosc.* 90, 367, 1981.
62. Kroto, H. W., Nixon, J. F., and Simmons, N. P. C., *J. Mol. Spectrosc.* 82, 185, 1980.
63. Kroto, H. W., Nixon, J. F., and Ohno, K., *J. Mol. Spectrosc.* 77, 270, 1979.
64. Cox, A. P., Ewart, I. C., and Gayton, T. R., *J. Mol. Spectrosc.* 125, 76, 1987.
65. Cohen, E. A., and Pickett, H. M., *J. Mol. Spectrosc.* 87, 582, 1981.
66. Peebles, S. A., and Peebles, R. A., *J. Mol. Struct.* 607, 19, 2002.
67. Suenram, R. D., Lovas, F. J., and Pickett, H. M., *J. Mol. Spectrosc.* 116, 406, 1986.
68. Kroto, H. W., and Landsberg, B. M., *J. Mol. Spectrosc.* 62, 346, 1976.

Name	Mol. Form.	μ/D	Ref.
Compounds not containing carbon			
Aluminum monofluoride	AlF	1.53 ± 0.15	1
Ammonia	H_3N	1.4718 ± 0.0002	5
Arsenic(III) chloride	$AsCl_3$	1.59 ± 0.08	1
Arsenic(III) fluoride	AsF_3	2.59 ± 0.05	1
Arsine	AsH_3	0.217 ± 0.003	5
Barium oxide	BaO	7.954 ± 0.003	5
Barium sulfide	BaS	10.86 ± 0.02	3
Bromine chloride	BrCl	0.519 ± 0.004	3
Bromine dioxide	BrO_2	2.8 ± 0.1	18
Bromine fluoride	BrF	1.422 ± 0.016	3
Bromine oxide	BrO	1.76 ± 0.04	2
Bromine pentafluoride	BrF_5	1.51 ± 0.15	1
Bromosilane	BrH_3Si	1.319	3
Bromotrifluorosilane	BrF_3Si	0.835 ± 0.007	64
Calcium monochloride	CaCl	≈3.6	4
Cesium chloride	ClCs	10.387 ± 0.004	2
Cesium fluoride	CsF	7.884 ± 0.001	2
Cesium sodium	CsNa	4.75 ± 0.20	2
Chlorine fluoride	ClF	0.888061	5
Chlorine oxide	ClO	1.297 ± 0.001	5
Chlorine trifluoride	ClF_3	0.6 ± 0.1	1
Chloroborane	$BClH_2$	0.75 ± 0.05	14
Chlorogermane	$ClGeH_3$	2.13 ± 0.02	1
Chlorosilane	ClH_3Si	1.31 ± 0.01	1
Chlorosyl fluoride	ClFO	1.93 ± 0.02	55
Chlorotrifluorosilane	ClF_3Si	0.636 ± 0.004	5
Chromium monoxide	CrO	3.88 ± 0.13	5
Copper(I) fluoride	CuF	5.77 ± 0.29	2
Copper(II) oxide	CuO	4.5 ± 0.5	5
Dichlorosilane	Cl_2H_2Si	1.17 ± 0.02	1
Difluoramine	F_2HN	1.92 ± 0.02	1
Difluorine dioxide	F_2O_2	1.44 ± 0.07	1
Difluoroborane	BF_2H	0.971 ± 0.010	8
cis-Difluorodiazine	F_2N_2	0.16 ± 0.01	1
Difluorosilane	F_2H_2Si	1.55 ± 0.02	1
Difluorosilylene	F_2Si	1.23 ± 0.02	2
Disiloxane	H_6OSi_2	0.24 ± 0.02	1
Fluoramine	FH_2N	2.27 ± 0.18	5
Fluorine azide	FN_3	≈1.3	5
Fluorine monoxide	F_2O	0.308180	5
Fluorine oxide	FO	0.0043 ± 0.0004	5
Fluoroborane	BF	≈0.5	2
Fluorogermane	$FGeH_3$	2.33 ± 0.12	2
Fluorosilane	FH_3Si	1.2969 ± 0.0006	5
Gallium monofluoride	FGa	2.45 ± 0.05	2
Germanium(II) fluoride	F_2Ge	2.61 ± 0.02	2
Germanium(II) oxide	GeO	3.2823 ± 0.0001	2
Germanium(II) selenide	GeSe	1.65 ± 0.05	2
Germanium(II) sulfide	GeS	2.00 ± 0.06	2
Germanium(II) telluride	GeTe	1.06 ± 0.07	2
Germylazide	GeH_3N_3	2.58 ± 0.02	25
Hafnium monoxide	HfO	3.431 ± 0.005	26
Hafnium(IV) oxide	HfO_2	7.92 ± 0.01	39
Hexaborane(10)	B_6H_{10}	2.50 ± 0.05	3

Name	Mol. Form.	μ/D	Ref.
Hydrazine	H_4N_2	1.75 ± 0.09	1
Hydrazoic acid	HN_3	1.70 ± 0.09	3
Hydrogen bromide	BrH	0.8272 ± 0.0003	3
Hydrogen chloride	ClH	1.1086 ± 0.0003	3
Hydrogen fluoride	FH	1.826178	2
Hydrogen iodide	HI	0.448 ± 0.001	2
Hydrogen peroxide	H_2O_2	1.573 ± 0.001	65
Hydrogen sulfide	H_2S	0.97833	5
Hydroxyl	HO	1.655 ± 0.001	5
Hydroxylamine	H_3NO	0.59 ± 0.05	2
Hypochlorous acid	ClHO	≈1.3	2
Hypofluorous acid	FHO	2.23 ± 0.11	3
Imidogen	HN	1.39 ± 0.07	3
Indium(I) chloride	ClIn	3.79 ± 0.19	2
Indium(I) fluoride	FIn	3.40 ± 0.07	2
Iodine bromide	BrI	0.726 ± 0.003	5
Iodine chloride	ClI	1.24 ± 0.02	5
Iodine fluoride	FI	1.948 ± 0.020	3
Iodine monoxide	IO	2.45 ± 0.05	2
Iodine pentafluoride	F_5I	2.18 ± 0.11	1
Lanthanum monoxide	LaO	3.207 ± 0.011	26
Lead(II) oxide	OPb	4.64 ± 0.50	2
Lead(II) sulfide	PbS	3.59 ± 0.18	2
Lithium bromide	BrLi	7.268 ± 0.001	2
Lithium chloride	ClLi	7.12887	2
Lithium fluoride	FLi	6.3274 ± 0.0002	3
Lithium fluoride–sodium fluoride complex	FLi•FNa	2.62 ± 0.02	51
Lithium hydride	HLi	5.884 ± 0.001	2
Lithium hydroxide	HLiO	4.754 ± 0.002	3
Lithium iodide	ILi	7.428 ± 0.001	2
Lithium monoxide	LiO	6.84 ± 0.03	2
Lithium potassium	KLi	3.45 ± 0.20	2
Lithium rubidium	LiRb	4.0 ± 0.1	2
Lithium sodium	LiNa	0.463 ± 0.002	2
Magnesium oxide	MgO	6.2 ± 0.6	5
Mercapto	HS	0.7580 ± 0.0001	3
Nitric acid	HNO_3	2.17 ± 0.02	1
Nitric oxide	NO	0.15872	2
Nitrogen dioxide	NO_2	0.316 ± 0.010	1
Nitrogen sulfide	NS	1.81 ± 0.02	2
Nitrogen trichloride	Cl_3N	0.39 ± 0.01	3
Nitrogen trifluoride	F_3N	0.235 ± 0.004	1
Nitrogen trioxide	N_2O_3	2.122 ± 0.010	2
Nitrosyl bromide	BrNO	≈1.8	1
Nitrosyl fluoride	FNO	1.730 ± 0.003	3
Nitrosyl hydride	HNO	1.62 ± 0.03	3
Nitrous acid (cis)	HNO_2	1.423 ± 0.005	2
Nitrous acid (trans)	HNO_2	1.855 ± 0.016	2
Nitrous oxide	N_2O	0.16083	3
Nitryl chloride	$ClNO_2$	0.53	1
Nitryl fluoride	FNO_2	0.466 ± 0.005	2
Ozone	O_3	0.53373	3
Pentaborane(9)	B_5H_9	2.13 ± 0.04	1
Perchloryl fluoride	$ClFO_3$	0.023 ± 0.001	3
Peroxynitrous acid	HNO_3	1.07 ± 0.002	46

Name	Mol. Form.	μ/D	Ref.	Name	Mol. Form.	μ/D	Ref.
Peroxynitric acid	HNO$_4$	1.99 ± 0.02	67	Tin(II) oxide	OSn	4.32 ± 0.22	2
Phosphine	H$_3$P	0.5740 ± 0.0003	3	Tin(II) sulfide	SSn	3.18 ± 0.16	2
Phosphorothioc trifluoride	F$_3$PS	0.64 ± 0.02	1	Titanium(II) oxide	OTi	2.96 ± 0.05	5
Phosphorus(III) chloride	Cl$_3$P	0.56 ± 0.02	2	Trichlorofluorosilane	Cl$_3$FSi	0.49 ± 0.01	2
Phosphorus(III) fluoride	F$_3$P	1.03 ± 0.01	1	Trichlorosilane	Cl$_3$HSi	0.86 ± 0.01	2
Phosphorus monoxide	OP	1.88 ± 0.07	5	Trifluoramine oxide	F$_3$NO	0.0390 ± 0.0004	9
Phosphorus nitride	NP	2.7470 ± 0.0001	2	1,1,1-Trifluorodisilane	F$_3$H$_3$Si$_2$	2.03 ± 0.10	3
Phosphoryl chloride	Cl$_3$OP	2.54 ± 0.05	2	Trifluoroiodosilane	F$_3$ISi	1.11 ± 0.03	5
Phosphoryl fluoride	F$_3$OP	1.8685 ± 0.0001	3	Trifluorosilane	F$_3$HSi	1.27 ± 0.03	1
Potassium bromide	BrK	10.628 ± 0.001	2	Water	H$_2$O	1.8546 ± 0.0040	3
Potassium chloride	ClK	10.269 ± 0.001	2	Water dimer–hydrogen bromide complex	H$_4$O$_2$•BrH	2.281 ± 0.003	48
Potassium fluoride	FK	8.585 ± 0.003	2				
Potassium hydroxide	HKO	7.415 ± 0.002	16	Water dimer–hydrogen chloride complex	H$_4$O$_2$•ClH	2.328 ± 0.003	48
Potassium iodide	IK	≈10.8	2				
Potassium sodium	KNa	2.693 ± 0.014	3	Ytterbium monofluoride	FYb	3.91 ± 0.04	45
Rubidium bromide	BrRb	≈10.9	2	Yttrium monoxide	OY	4.524 ± 0.007	26
Rubidium chloride	ClRb	10.510 ± 0.005	2	Zirconium(II) oxide	OZr	2.55 ± 0.01	26
Rubidium fluoride	FRb	8.5465 ± 0.0005	2	Zirconium(IV) oxide	O$_2$Zr	7.80 ± 0.02	19
Rubidium iodide	IRb	≈11.5	2				
Rubidium sodium	NaRb	3.1 ± 0.3	2	*Compounds containing carbon*			
Selenium dioxide	O$_2$Se	2.62 ± 0.05	2	Acenaphthene	C$_{12}$H$_{10}$	≈0.85	1
Selenium tetrafluoride	F$_4$Se	1.78 ± 0.09	2	Acetaldehyde	C$_2$H$_4$O	2.750 ± 0.006	3
Silicon monosulfide	SSi	1.73 ± 0.09	2	Acetamide	C$_2$H$_5$NO	3.68 ± 0.03	5
Silicon monoxide	OSi	3.0982	2	Acetic acid	C$_2$H$_4$O$_2$	1.70 ± 0.03	2
Silver(I) bromide	AgBr	5.62 ± 0.03	5	Acetic anhydride	C$_4$H$_6$O$_3$	≈2.8	1
Silver(I) chloride	AgCl	6.08 ± 0.06	5	Acetone	C$_3$H$_6$O	2.88 ± 0.03	1
Silver(I) fluoride	AgF	6.22 ± 0.30	2	Acetonitrile	C$_2$H$_3$N	3.92519	5
Silver(I) iodide	AgI	4.55 ± 0.05	5	Acetophenone	C$_8$H$_8$O	3.02 ± 0.06	1
Sodium bromide	BrNa	9.1183 ± 0.0006	2	Acetyl chloride	C$_2$H$_3$ClO	2.72 ± 0.14	1
Sodium chloride	ClNa	9.00117	2	Acetylene–carbon dioxide complex	C$_2$H$_2$•CO$_2$	0.161 ± 0.001	22
Sodium fluoride	FNa	8.156 ± 0.001	2				
Sodium iodide	INa	9.236 ± 0.003	2	Acetylene–carbon monoxide complex	C$_2$H$_2$•CO	0.311 ± 0.001	32
Stibine	H$_3$Sb	0.12 ± 0.05	1				
Strontium oxide	OSr	8.900 ± 0.003	2	Acetylene–carbon oxysulfide trimer complex	C$_2$H$_2$•C$_3$O$_3$S$_3$	1.23 ± 0.02	53
Sulfur dichloride	Cl$_2$S	0.36 ± 0.01	3				
Sulfur difluoride	F$_2$S	1.05 ± 0.05	2	Acetylene–hydrogen cyanide complex	C$_2$H$_2$•CHN	3.29 ± 0.03	32
Sulfur dioxide	O$_2$S	1.63305	3				
Sulfur monofluoride	FS	0.794 ± 0.02	3	Acetyl fluoride	C$_2$H$_3$FO	2.96 ± 0.03	1
Sulfur monoxide	OS	1.55 ± 0.02	1	Acrolein (*cis*)	C$_3$H$_4$O	2.552 ± 0.003	5
Sulfur oxide (SSO)	OS$_2$	1.47 ± 0.03	1	Acrolein (*trans*)	C$_3$H$_4$O	3.117 ± 0.004	5
Sulfur tetrafluoride	F$_4$S	0.632 ± 0.003	1	Acrylonitrile	C$_3$H$_3$N	3.92 ± 0.07	5
Sulfuryl chloride	Cl$_2$O$_2$S	1.81 ± 0.04	1	Allyl alcohol (*gauche*)	C$_3$H$_6$O	1.55 ± 0.08	3
Sulfuryl fluoride	F$_2$O$_2$S	1.12 ± 0.02	1	Allyl alcohol (*average*)	C$_3$H$_6$O	1.60 ± 0.08	1
Tetraborane(10)	B$_4$H$_{10}$	0.486 ± 0.002	3	Allylamine	C$_3$H$_7$N	≈1.2	1
Tetrafluorohydrazine (*gauche*)	F$_4$N$_2$	0.257 ± 0.002	5	Aniline	C$_6$H$_7$N	1.13 ± 0.02	3
				Anisole	C$_7$H$_8$O	1.38 ± 0.07	1
Tetrafluorosilane–ammonia complex	F$_4$Si•H$_3$N	5.61 ± 0.02	31	Azulene	C$_{10}$H$_8$	0.80 ± 0.02	1
				Benzaldehyde	C$_7$H$_6$O	[3.0]	7
Thallium(I) bromide	BrTl	4.49 ± 0.05	2	Benzeneacetonitrile	C$_8$H$_7$N	[3.5]	7
Thallium(I) chloride	ClTl	4.54299	2	Benzene–hydrogen sulfide complex	C$_6$H$_6$•H$_2$S	1.14 ± 0.02	40
Thallium(I) fluoride	FTl	4.2282 ± 0.0008	2				
Thallium(I) iodide	ITl	4.61 ± 0.07	2	Benzene–krypton complex	C$_6$H$_6$•Kr	0.136 ± 0.002	58
Thionitrosyl chloride (NSCl)	ClNS	1.87 ± 0.02	2	Benzene–sulfur dioxide complex	C$_6$H$_6$•O$_2$S	2.061 ± 0.002	33
Thionitrosyl fluoride (NSF)	FNS	1.902 ± 0.012	2				
Thionyl chloride	Cl$_2$OS	1.45 ± 0.03	1	Benzenethiol	C$_6$H$_6$S	[1.23]	7
Thionyl fluoride	F$_2$OS	1.63 ± 0.01	1	Benzonitrile	C$_7$H$_5$N	4.18 ± 0.08	1
				Benzyl acetate	C$_9$H$_{10}$O$_2$	[1.22]	7
				Benzyl alcohol	C$_7$H$_8$O	1.71 ± 0.09	1

Name	Mol. Form.	μ/D	Ref.
Benzyl benzoate	$C_{14}H_{12}O_2$	[2.06]	7
Bis(2-aminoethyl)amine	$C_4H_{13}N_3$	[1.89]	7
Bis(2-chloroethyl) ether	$C_4H_8Cl_2O$	[2.58]	7
Bis(2-ethylhexyl) phthalate	$C_{24}H_{38}O_4$	[2.84]	7
Borane carbonyl	CH_3BO	1.698 ± 0.020	3
Bromoacetylene	C_2HBr	0.22962	5
Bromobenzene	C_6H_5Br	1.70 ± 0.03	1
1-Bromobutane	C_4H_9Br	2.08 ± 0.10	1
2-Bromobutane	C_4H_9Br	2.23 ± 0.11	1
1-Bromo-2-chloroethane	C_2H_4BrCl	[1.2]	7
Bromochlorofluoromethane	$CHBrClF$	1.5 ± 0.3	17
Bromochloromethane	CH_2BrCl	[1.66]	7
1-Bromodecane	$C_{10}H_{21}Br$	[1.93]	7
Bromoethane	C_2H_5Br	2.04 ± 0.02	5
Bromoethene	C_2H_3Br	1.42 ± 0.03	1
Bromofluoroacetylene	C_2BrF	0.448 ± 0.002	5
1-Bromoheptane	$C_7H_{15}Br$	2.16 ± 0.11	1
Bromomethane	CH_3Br	1.8203 ± 0.0004	5
2-Bromo-2-methylpropane	C_4H_9Br	[2.17]	7
1-Bromonaphthalene	$C_{10}H_7Br$	[1.55]	7
1-Bromopentane	$C_5H_{11}Br$	2.20 ± 0.11	1
1-Bromopropane	C_3H_7Br	2.18 ± 0.11	1
2-Bromopropane	C_3H_7Br	2.21 ± 0.11	1
2-Bromopropene	C_3H_5Br	[1.51]	7
3-Bromopropene	C_3H_5Br	≈1.9	1
Bromotrifluoromethane	$CBrF_3$	0.65 ± 0.05	1
1,2-Butadiene	C_4H_6	0.403 ± 0.002	1
Butanal	C_4H_8O	2.72 ± 0.05	1
1,4-Butanediol	$C_4H_{10}O_2$	[2.58]	7
Butanenitrile (gauche)	C_4H_7N	3.91 ± 0.04	5
Butanenitrile (anti)	C_4H_7N	3.73 ± 0.06	5
1-Butanethiol	$C_4H_{10}S$	[1.53]	7
Butanoic acid	$C_4H_8O_2$	[1.65]	7
1-Butanol	$C_4H_{10}O$	1.66 ± 0.03	1
2-Butanone	C_4H_8O	2.779 ± 0.015	2
trans-2-Butenal	C_4H_6O	3.67 ± 0.07	1
1-Butene (cis)	C_4H_8	0.438 ± 0.007	2
1-Butene (skew)	C_4H_8	0.359 ± 0.011	2
cis-2-Butene	C_4H_8	0.253 ± 0.005	2
cis-2-Butene-1,4-diol	$C_4H_8O_2$	[2.48]	7
trans-2-Butene-1,4-diol	$C_4H_8O_2$	[2.45]	7
trans-2-Butenoic acid	$C_4H_6O_2$	[2.13]	7
cis-2-Buten-1-ol	C_4H_8O	1.96 ± 0.03	5
trans-2-Buten-1-ol	C_4H_8O	1.90 ± 0.02	5
1-Buten-3-yne	C_4H_4	0.22 ± 0.02	3
2-Butoxyethanol	$C_6H_{14}O_2$	[2.08]	7
Butyl acetate	$C_6H_{12}O_2$	[1.87]	7
sec-Butyl acetate	$C_6H_{12}O_2$	[1.87]	7
Butylamine	$C_4H_{11}N$	≈1.0	1
sec-Butylamine	$C_4H_{11}N$	[1.28]	7
tert-Butylamine	$C_4H_{11}N$	[1.29]	7
tert-Butylbenzene	$C_{10}H_{14}$	≈0.83	1
Butyl ethyl ether	$C_6H_{14}O$	[1.24]	7
Butyl formate	$C_5H_{10}O_2$	[2.03]	7
Butyl stearate	$C_{22}H_{44}O_2$	[1.88]	7

Name	Mol. Form.	μ/D	Ref.
Butyl vinyl ether	$C_6H_{12}O$	[1.25]	7
1-Butyne	C_4H_6	0.782 ± 0.004	5
γ-Butyrolactone	$C_4H_6O_2$	4.27 ± 0.03	3
Calcium methoxide	CH_3CaO	1.58 ± 0.08	43
Camphor, (+)	$C_{10}H_{16}O$	[3.1]	7
Caprolactam	$C_6H_{11}NO$	[3.9]	7
Carboimidic difluoride	CHF_2N	1.393 ± 0.001	5
Carbon dioxide dimer–water complex	$C_2O_4·H_2O$	1.989 ± 0.002	23
Carbon dioxide–mercury complex	$CO_2·Hg$	0.107 ± 0.003	29
Carbon dioxide–water dimer complex	$CO_2·H_4O_2$	1.746 ± 0.010	28
Carbon disulfide–sulfur dioxide complex	$CO_2·O_2S$	1.096 ± 0.001	42
Carbon monoselenide	CSe	1.99 ± 0.04	3
Carbon monosulfide	CS	1.958 ± 0.005	2
Carbon monoxide	CO	0.10980	3
Carbon monoxide dimer–water complex	$C_2O_2·H_2O$	1.57 ± 0.05	36
Carbon oxyselenide	$COSe$	0.73 ± 0.02	1
Carbon oxysulfide	COS	0.715189	5
Carbon oxysulfide–carbon dioxide dimer complex	$COS·C_2O_4$	0.69 ± 0.05	44
Carbon oxysulfide–water complex	$COS·H_2O$	2.668 ± 0.003	37
Carbonyl chloride	CCl_2O	1.17 ± 0.01	1
Carbonyl fluoride	CF_2O	0.95 ± 0.01	1
Chloroacetyl chloride	$C_2H_2Cl_2O$	2.23 ± 0.11	1
Chloroacetylene	C_2HCl	0.44408	5
2-Chloroaniline	C_6H_6ClN	[1.77]	7
Chlorobenzene	C_6H_5Cl	1.69 ± 0.03	1
1-Chlorobutane	C_4H_9Cl	2.05 ± 0.04	1
2-Chlorobutane	C_4H_9Cl	2.04 ± 0.10	1
Chlorocyclohexane (axial)	$C_6H_{11}Cl$	1.91 ± 0.02	5
Chlorocyclohexane (equitorial)	$C_6H_{11}Cl$	2.44 ± 0.07	5
1-Chloro-1,1-difluoroethane	$C_2H_3ClF_2$	2.14 ± 0.04	1
Chlorodifluoromethane	$CHClF_2$	1.42 ± 0.03	1
Chloroethane	C_2H_5Cl	2.05 ± 0.02	1
2-Chloroethanol	C_2H_5ClO	1.78 ± 0.09	1
Chloroethene	C_2H_3Cl	1.45 ± 0.03	1
1-Chloro-4-fluorobenzene	C_6H_4ClF	0.12 ± 0.01	66
1-Chloro-1-fluoroethane	C_2H_4ClF	2.068 ± 0.014	3
Chlorofluoromethane	CH_2ClF	1.82 ± 0.04	1
Chloromethane	CH_3Cl	1.8963 ± 0.0002	5
(Chloromethyl)benzene	C_7H_7Cl	[1.82]	7
1-Chloro-3-methylbutane	$C_5H_{11}Cl$	[1.92]	7
1-Chloro-2-methylpropane	C_4H_9Cl	2.00 ± 0.10	1
2-Chloro-2-methylpropane	C_4H_9Cl	2.13 ± 0.04	1
1-Chloronaphthalene	$C_{10}H_7Cl$	[1.57]	7
1-Chloro-2-nitrobenzene	$C_6H_4ClNO_2$	4.64 ± 0.09	1
1-Chloro-3-nitrobenzene	$C_6H_4ClNO_2$	3.73 ± 0.07	1
1-Chloro-4-nitrobenzene	$C_6H_4ClNO_2$	2.83 ± 0.06	1
1-Chlorooctane	$C_8H_{17}Cl$	[2.00]	7
Chloropentafluoroethane	C_2ClF_5	0.52 ± 0.05	1
1-Chloropentane	$C_5H_{11}Cl$	2.16 ± 0.11	1
4-Chlorophenol	C_6H_5ClO	2.11 ± 0.11	1

Name	Mol. Form.	μ/D	Ref.	Name	Mol. Form.	μ/D	Ref.
1-Chloropropane (*gauche*)	C_3H_7Cl	2.02 ± 0.03	5	Dibutyl sulfide	$C_8H_{18}S$	[1.61]	7
1-Chloropropane (*trans*)	C_3H_7Cl	1.95 ± 0.02	5	*o*-Dichlorobenzene	$C_6H_4Cl_2$	2.50 ± 0.05	1
1-Chloropropane (*average*)	C_3H_7Cl	2.05 ± 0.04	1	*m*-Dichlorobenzene	$C_6H_4Cl_2$	1.72 ± 0.09	1
2-Chloropropane	C_3H_7Cl	2.17 ± 0.11	1	1,4-Dichlorobutane	$C_4H_8Cl_2$	2.22 ± 0.11	1
cis-1-Chloropropene	C_3H_5Cl	1.67 ± 0.08	1	1,1-Dichloro-2,2-difluoroethene	$C_2Cl_2F_2$	0.50	7
trans-1-Chloropropene	C_3H_5Cl	1.97 ± 0.10	1	Dichlorodifluoromethane	CCl_2F_2	0.51 ± 0.05	1
2-Chloropropene	C_3H_5Cl	1.647 ± 0.010	3	1,1-Dichloroethane	$C_2H_4Cl_2$	2.06 ± 0.04	1
3-Chloropropene	C_3H_5Cl	1.94 ± 0.10	1	1,2-Dichloroethane	$C_2H_4Cl_2$	[1.83]	7
4-Chloropyridine	C_5H_4ClN	0.756 ± 0.005	3	1,1-Dichloroethene	$C_2H_2Cl_2$	1.34 ± 0.01	1
2-Chlorotoluene	C_7H_7Cl	1.56 ± 0.08	1	*cis*-1,2-Dichloroethene	$C_2H_2Cl_2$	1.90 ± 0.04	1
3-Chlorotoluene	C_7H_7Cl	[1.82]	7	Dichlorofluoromethane	$CHCl_2F$	1.29 ± 0.03	1
4-Chlorotoluene	C_7H_7Cl	2.21 ± 0.04	1	1,1-Dichloro-2-fluoropropene	$C_3H_3Cl_2F$	2.43 ± 0.02	3
Chlorotrifluoroethene	C_2ClF_3	0.40 ± 0.10	1	Dichloromethane	CH_2Cl_2	1.60 ± 0.03	1
Chlorotrifluoromethane	$CClF_3$	0.50 ± 0.01	1	(Dichloromethyl)benzene	$C_7H_6Cl_2$	[2.07]	7
o-Cresol	C_7H_8O	[1.45]	7	Dichloromethylborane	CH_3BCl_2	1.419 ± 0.013	5
m-Cresol	C_7H_8O	[1.48]	7	1,2-Dichloropropane	$C_3H_6Cl_2$	[1.85]	7
p-Cresol	C_7H_8O	[1.48]	7	1,3-Dichloropropane	$C_3H_6Cl_2$	2.08 ± 0.04	1
Cyanamide	CH_2N_2	4.28 ± 0.10	5	1,2-Dichloro-1,1,2,2-tetrafluoroethane	$C_2Cl_2F_4$	≈0.5	1
Cyanoacetylene	C_3HN	3.73172	5	2,4-Dichlorotoluene	$C_7H_6Cl_2$	[1.70]	7
Cyanoformamide	$C_2H_2N_2O$	4.10 ± 0.12	47	3,4-Dichlorotoluene	$C_7H_6Cl_2$	[2.95]	7
Cyanogen azide (NCN$_3$)	CN_4	2.96 ± 0.07	60	Diethanolamine	$C_4H_{11}NO_2$	[2.8]	7
Cyanogen chloride	$CClN$	2.8331 ± 0.0002	3	1,1-Diethoxyethane	$C_6H_{14}O_2$	[1.38]	7
Cyanogen fluoride	CFN	2.120 ± 0.001	3	Diethylamine	$C_4H_{11}N$	0.92 ± 0.05	1
Cyanogen iodide	CIN	3.67 ± 0.02	5	Diethyl carbonate	$C_5H_{10}O_3$	1.10 ± 0.06	1
Cyanomethylmercury	C_2H_3HgN	4.7 ± 0.1	12	Diethylene glycol	$C_4H_{10}O_3$	[2.31]	7
Cyclobutanecarbonitrile	C_5H_7N	4.04 ± 0.04	5	Diethylene glycol dimethyl ether	$C_6H_{14}O_3$	[1.97]	7
Cyclobutanone	C_4H_6O	2.89 ± 0.03	2	Diethylene glycol monoethyl ether	$C_6H_{14}O_3$	[1.6]	7
Cyclobutene	C_4H_6	0.132 ± 0.001	1	Diethylene glycol monoethyl ether acetate	$C_8H_{16}O_4$	[1.8]	7
1,3-Cycloheptadiene	C_7H_{10}	0.740	3	Diethylene glycol monomethyl ether	$C_5H_{12}O_3$	[1.6]	7
2,4,6-Cycloheptatrien-1-one	C_7H_6O	4.1 ± 0.3	3				
3,5-Cyclohexadiene-1,2-dione	$C_6H_4O_2$	4.23 ± 0.02	3	Diethyl ether	$C_4H_{10}O$	1.098 ± 0.001	38
Cyclohexanone	$C_6H_{10}O$	3.246 ± 0.006	5	Diethyl malonate	$C_7H_{12}O_4$	[2.54]	7
Cyclohexene (*half-chair*)	C_6H_{10}	0.332 ± 0.012	2	Diethyl oxalate	$C_6H_{10}O_4$	[2.49]	7
Cyclohexylamine	$C_6H_{13}N$	[1.26]	7	Diethyl sulfide (*trans-trans*)	$C_4H_{10}S$	1.556 ± 0.004	54
1,3-Cyclopentadiene	C_5H_6	0.419 ± 0.004	1	Diethyl sulfide (*trans-gauche*)	$C_4H_{10}S$	1.591 ± 0.009	54
2,4-Cyclopentadien-1-one	C_5H_4O	3.132 ± 0.007	3	Diethyl sulfide (*gauche-gauche*)	$C_4H_{10}S$	1.645 ± 0.001	54
Cyclopentanone	C_5H_8O	≈3.3	1				
Cyclopentene	C_5H_8	0.20 ± 0.02	1				
3-Cyclopenten-1-one	C_5H_6O	2.79 ± 0.03	3	*o*-Difluorobenzene	$C_6H_4F_2$	2.46 ± 0.05	2
Cyclopropane-sulfur dioxide complex	$C_3H_6 \cdot O_2S$	1.681 ± 0.001	30	*m*-Difluorobenzene	$C_6H_4F_2$	1.51 ± 0.02	2
Cyclopropanone	C_3H_4O	2.67 ± 0.13	2	1,1-Difluorocyclohexane	$C_6H_{10}F_2$	2.556 ± 0.010	3
Cyclopropene	C_3H_4	0.454 ± 0.010	1	3,3-Difluorocyclopropene	$C_3H_2F_2$	2.98 ± 0.02	3
Cyclopropylamine	C_3H_7N	1.19 ± 0.01	2	1,1-Difluoroethane	$C_2H_4F_2$	2.27 ± 0.05	1
Cyclopropyl methyl ketone	C_5H_8O	2.62 ± 0.25	2	1,2-Difluoroethane (*gauche*)	$C_2H_4F_2$	2.67 ± 0.13	2
Diacetone alcohol	$C_6H_{12}O_2$	[3.24]	7	1,1-Difluoroethene	$C_2H_2F_2$	1.3893 ± 0.0002	5
Diazomethane	CH_2N_2	1.50 ± 0.01	1	*cis*-1,2-Difluoroethene	$C_2H_2F_2$	2.42 ± 0.02	1
Dibromodifluoromethane	CBr_2F_2	0.66 ± 0.05	1	Difluoromethane	CH_2F_2	1.9785 ± 0.02	3
1,2-Dibromoethane	$C_2H_4Br_2$	[1.19]	7	Difluoromethylborane	CH_3BF_2	1.668 ± 0.003	3
Dibromomethane	CH_2Br_2	1.43 ± 0.03	1	Difluoromethylene	CF_2	0.47 ± 0.02	3
1,2-Dibromopropane	$C_3H_6Br_2$	[1.2]	7	1,1-Difluoro-1-propene	$C_3H_4F_2$	0.889 ± 0.007	2
Dibutylamine	$C_8H_{19}N$	[0.98]	7	2,3-Dihydro-1,4-dioxin	$C_4H_6O_2$	0.939 ± 0.008	3
Dibutyl ether	$C_8H_{18}O$	1.17 ± 0.06	1	3,6-Dihydro-1,2-dioxin	$C_4H_6O_2$	2.329 ± 0.001	3
Dibutyl phthalate	$C_{16}H_{22}O_4$	[2.82]	7				
Dibutyl sebacate	$C_{18}H_{34}O_4$	[2.48]	7				

Name	Mol. Form.	μ/D	Ref.
2,3-Dihydrofuran	C_4H_6O	1.32 ± 0.03	2
2,5-Dihydrofuran	C_4H_6O	1.63 ± 0.01	5
Dihydro-3-methyl-2(3H)-furanone	$C_5H_8O_2$	4.56 ± 0.02	5
Dihydro-5-methyl-2(3H)-furanone	$C_5H_8O_2$	4.71 ± 0.05	5
3,4-Dihydro-2H-pyran	C_5H_8O	1.400 ± 0.008	5
3,6-Dihydro-2H-pyran	C_5H_8O	1.283 ± 0.005	3
2,3-Dihydrothiophene	C_4H_6S	1.61 ± 0.20	5
2,5-Dihydrothiophene	C_4H_6S	1.75 ± 0.01	3
Diiodomethane	CH_2I_2	[1.08]	7
Diisopentyl ether	$C_{10}H_{22}O$	[1.23]	7
Diisopropylamine	$C_6H_{15}N$	[1.15]	7
Diisopropyl ether	$C_6H_{14}O$	1.13 ± 0.10	1
Diketene	$C_4H_4O_2$	3.53 ± 0.07	1
1,2-Dimethoxybenzene	$C_8H_{10}O_2$	[1.29]	7
Dimethoxymethane	$C_3H_8O_2$	[0.74]	7
N,N-Dimethylacetamide	C_4H_9NO	[3.7]	7
Dimethylamine	C_2H_7N	1.01 ± 0.02	2
N,N-Dimethylaniline	$C_8H_{11}N$	1.68 ± 0.17	1
2,4-Dimethylaniline	$C_8H_{11}N$	[1.40]	7
2,6-Dimethylaniline	$C_8H_{11}N$	[1.63]	7
3,3-Dimethyl-1-butyne	C_6H_{10}	0.661 ± 0.004	1
1,1-Dimethylcyclopropane	C_5H_{10}	0.142 ± 0.001	3
3,3-Dimethylcyclopropene	C_5H_8	0.287 ± 0.003	3
Dimethyl disulfide	$C_2H_6S_2$	[1.85]	7
Dimethyl ether	C_2H_6O	1.30 ± 0.01	1
N,N-Dimethylformamide	C_3H_7NO	3.82 ± 0.08	1
2,6-Dimethyl-4-heptanone	$C_9H_{18}O$	[2.66]	7
Dimethyl maleate	$C_6H_8O_4$	[2.48]	7
2,4-Dimethyl-3-pentanone	$C_7H_{14}O$	[2.74]	7
2,2-Dimethylpropanal	$C_5H_{10}O$	2.66 ± 0.05	2
2,2-Dimethylpropanenitrile	C_5H_9N	3.95 ± 0.04	1
2,4-Dimethylpyridine	C_7H_9N	[2.30]	7
2,6-Dimethylpyridine	C_7H_9N	[1.66]	7
Dimethyl sulfide	C_2H_6S	1.554 ± 0.004	3
Dimethyl sulfoxide	C_2H_6OS	3.96 ± 0.04	1
1,3-Dioxane	$C_4H_8O_2$	2.06 ± 0.04	2
1,3-Dioxolane	$C_3H_6O_2$	1.19 ± 0.06	3
Dipentyl ether	$C_{10}H_{22}O$	[1.20]	7
Diphenyl ether	$C_{12}H_{10}O$	≈1.3	1
Dipropylamine	$C_6H_{15}N$	[1.03]	7
Dipropyl ether	$C_6H_{14}O$	1.21 ± 0.06	1
1,3-Dithiane	$C_4H_8S_2$	2.14 ± 0.04	5
Divinyl ether	C_4H_6O	0.78 ± 0.05	2
Epichlorohydrin	C_3H_5ClO	[1.8]	7
1,2-Epoxybutane	C_4H_8O	1.891 ± 0.011	3
1,2-Ethanediamine	$C_2H_8N_2$	1.99 ± 0.10	1
1,2-Ethanediol, diacetate	$C_6H_{10}O_4$	[2.34]	7
1,2-Ethanedithiol	$C_2H_6S_2$	2.03 ± 0.08	5
Ethanethiol (gauche)	C_2H_6S	1.61 ± 0.08	3
Ethanethiol (trans)	C_2H_6S	1.58 ± 0.08	3
Ethanol (gauche)	C_2H_6O	1.68 ± 0.03	3
Ethanol (trans)	C_2H_6O	1.44 ± 0.03	2
Ethanol (average)	C_2H_6O	1.69 ± 0.03	1
Ethanolamine	C_2H_7NO	[2.27]	7

Name	Mol. Form.	μ/D	Ref.
Ethoxybenzene	$C_8H_{10}O$	1.45 ± 0.15	1
2-Ethoxyethanol	$C_4H_{10}O_2$	[2.08]	7
2-Ethoxyethyl acetate	$C_6H_{12}O_3$	[2.25]	7
Ethyl acetate	$C_4H_8O_2$	1.78 ± 0.09	1
Ethyl acrylate	$C_5H_8O_2$	[1.96]	7
Ethylamine (gauche)	C_2H_7N	1.210 ± 0.015	5
Ethylamine (trans)	C_2H_7N	1.304 ± 0.011	5
Ethylamine (average)	C_2H_7N	1.22 ± 0.10	1
Ethylbenzene	C_8H_{10}	0.59 ± 0.05	1
Ethyl benzoate	$C_9H_{10}O_2$	2.00 ± 0.10	1
Ethyl butanoate	$C_6H_{12}O_2$	[1.74]	7
Ethyl trans-cinnamate	$C_{11}H_{12}O_2$	[1.84]	7
Ethyl cyanate	C_3H_5NO	4.72 ± 0.09	5
Ethyl cyanoacetate	$C_5H_7NO_2$	[2.17]	7
Ethylene carbonate	$C_3H_4O_3$	[4.9]	7
Ethylene glycol (average)	$C_2H_6O_2$	2.36 ± 0.10	5
Ethyleneimine	C_2H_5N	1.90 ± 0.01	1
Ethylene–sulfur dioxide complex	$C_2H_4 \cdot O_2S$	1.650 ± 0.003	27
Ethylene–water complex	$C_2H_4 \cdot H_2O$	1.10 ± 0.01	56
Ethyl formate (gauche)	$C_3H_6O_2$	1.81 ± 0.02	2
Ethyl formate (trans)	$C_3H_6O_2$	1.98 ± 0.02	2
Ethyl formate (average)	$C_3H_6O_2$	1.93	1
2-Ethyl-1-hexanol	$C_8H_{18}O$	[1.74]	7
2-Ethylhexyl acetate	$C_{10}H_{20}O_2$	[1.8]	7
Ethyl lactate	$C_5H_{10}O_3$	[2.4]	7
Ethyl methyl ether (trans)	C_3H_8O	1.17 ± 0.02	3
Ethyl methyl sulfide (gauche)	C_3H_8S	1.593 ± 0.004	5
Ethyl methyl sulfide (trans)	C_3H_8S	1.56 ± 0.03	3
Ethyl propanoate	$C_5H_{10}O_2$	[1.74]	7
Ethyl vinyl ether	C_4H_8O	[1.26]	7
Fluoroacetylene	C_2HF	0.7207 ± 0.0003	3
Fluorobenzene	C_6H_5F	1.60 ± 0.08	1
Fluorocyclohexane (equitorial)	$C_6H_{11}F$	2.11 ± 0.04	2
Fluorocyclohexane (axial)	$C_6H_{11}F$	1.81 ± 0.04	2
1-Fluorocyclohexene	C_6H_9F	1.942 ± 0.010	5
Fluoroethane	C_2H_5F	1.937 ± 0.007	5
Fluoroethene	C_2H_3F	1.468 ± 0.003	5
Fluoromethane	CH_3F	1.858 ± 0.002	3
Fluoromethylidyne	CF	0.645 ± 0.005	3
(Fluoromethylidyne) phosphine (FCP)	CFP	0.279 ± 0.001	62
Fluoromethylsilane	CH_5FSi	1.700 ± 0.008	5
1-Fluoro-4-nitrobenzene	$C_6H_4FNO_2$	2.87 ± 0.06	1
1-Fluoropropane (gauche)	C_3H_7F	1.90 ± 0.10	1
1-Fluoropropane (trans)	C_3H_7F	2.05 ± 0.04	1
2-Fluoropropane	C_3H_7F	1.958 ± 0.001	5
cis-1-Fluoropropene	C_3H_5F	1.46 ± 0.03	1
trans-1-Fluoropropene	C_3H_5F	≈1.9	1
2-Fluoropropene	C_3H_5F	1.61 ± 0.03	1
3-Fluoropropene (gauche)	C_3H_5F	1.939 ± 0.015	1
3-Fluoropropene (cis)	C_3H_5F	1.765 ± 0.014	1
3-Fluoropropyne	C_3H_3F	1.73 ± 0.02	5
3-Fluoropyridine	C_5H_4FN	2.09 ± 0.26	3
2-Fluorotoluene	C_7H_7F	1.37 ± 0.07	1

Name	Mol. Form.	μ/D	Ref.	Name	Mol. Form.	μ/D	Ref.
3-Fluorotoluene	C_7H_7F	1.82 ± 0.04	2	2-Isocyanopropane	C_4H_7N	4.055 ± 0.001	5
4-Fluorotoluene	C_7H_7F	2.00 ± 0.10	1	Isopentane	C_5H_{12}	0.13 ± 0.05	1
Formaldehyde	CH_2O	2.332 ± 0.002	3	Isopentyl acetate	$C_7H_{14}O_2$	[1.86]	7
Formaldehyde dimer	$C_2H_4O_2$	0.858 ± 0.005	57	Isopropylamine	C_3H_9N	1.19 ± 0.06	3
Formamide	CH_3NO	3.73 ± 0.07	1	Isopropylbenzene	C_9H_{12}	≈0.79	1
Formic acid	CH_2O_2	1.425 ± 0.002	5	Isopropyl methyl ether	$C_4H_{10}O$	1.247 ± 0.003	5
Formyl fluoride	CHFO	2.081 ± 0.001	5	Isoquinoline	C_9H_7N	2.73 ± 0.14	1
Fulminic acid	CHNO	3.09934	5	Isoxazole	C_3H_3NO	2.95 ± 0.04	3
Fulvene	C_6H_6	0.4236 ± 0.013	2	Isoxazole–carbon monoxide complex	$C_3H_3NO•CO$	2.873 ± 0.004	52
Furan	C_4H_4O	0.66 ± 0.01	1	Ketene	C_2H_2O	1.42215	3
Furfural	$C_5H_4O_2$	[3.54]	7	Mesityl oxide	$C_6H_{10}O$	[2.79]	7
Furfuryl alcohol	$C_5H_6O_2$	[1.92]	7	Methacrylic acid	$C_4H_6O_2$	[1.65]	7
Glycerol	$C_3H_8O_3$	[2.56]	7	Methanethiol	CH_4S	1.52 ± 0.08	1
Glycine (Conformer I)	$C_2H_5NO_2$	1.147 ± 0.005	49	Methanol	CH_4O	1.70 ± 0.02	1
Glycine (Conformer II)	$C_2H_5NO_2$	5.45 ± 0.05	49	2-Methoxyethanol (*gauche*)	$C_3H_8O_2$	2.36 ± 0.05	2
Glycolaldehyde	$C_2H_4O_2$	2.73 ± 0.05	2	2-Methoxyethyl acetate	$C_5H_{10}O_3$	[2.13]	7
Glyoxal (*cis*)	$C_2H_2O_2$	4.8 ± 0.2	2	1-Methoxy-1,2-propadiene	C_4H_6O	0.963 ± 0.020	5
2-Heptanol	$C_7H_{16}O$	[1.71]	7	*N*-Methylacetamide	C_3H_7NO	[4.3]	7
3-Heptanol	$C_7H_{16}O$	[1.71]	7	Methyl acetate	$C_3H_6O_2$	1.72 ± 0.09	1
2-Heptanone	$C_7H_{14}O$	[2.59]	7	Methyl acrylate	$C_4H_6O_2$	[1.77]	7
3-Heptanone	$C_7H_{14}O$	[2.78]	7	2-Methylacrylonitrile	C_4H_5N	3.69 ± 0.18	1
Hexamethylphosphoric triamide	$C_6H_{18}N_3OP$	[5.5]	7	Methylamine	CH_5N	1.31 ± 0.03	1
Hexanoic acid	$C_6H_{12}O_2$	[1.13]	7	2-Methylaniline	C_7H_9N	[1.60]	7
2-Hexanone	$C_6H_{12}O$	[2.66]	7	3-Methylaniline	C_7H_9N	[1.45]	7
sec-Hexyl acetate	$C_8H_{16}O_2$	[1.9]	7	4-Methylaniline	C_7H_9N	[1.52]	7
1-Hexyne	C_6H_{10}	0.83 ± 0.05	1	Methyl azide	CH_3N_3	2.17 ± 0.04	2
Hydrogen cyanide	CHN	2.985188	5	Methyl benzoate	$C_8H_8O_2$	[1.94]	7
Hydrogen cyanide trimer	$C_3H_3N_3$	10.6	21	2-Methyl-1,3-butadiene	C_5H_8	0.25 ± 0.01	1
Hydrogen isocyanide	CHN	3.05 ± 0.15	3	3-Methylbutanoic acid	$C_5H_{10}O_2$	[0.63]	7
p-Hydroquinone	$C_6H_6O_2$	2.38 ± 0.05	15	2-Methyl-1-butanol	$C_5H_{12}O$	[1.88]	7
3-Hydroxypropanenitrile (*gauche*)	C_3H_5NO	3.17 ± 0.02	5	2-Methyl-2-butanol	$C_5H_{12}O$	[1.82]	7
Imidazole	$C_3H_4N_2$	3.8 ± 0.4	2	3-Methyl-1-butene (*gauche*)	C_5H_{10}	0.398 ± 0.004	3
Iodoacetylene	C_2HI	0.02525	5	3-Methyl-1-butene (*trans*)	C_5H_{10}	0.320 ± 0.010	3
Iodobenzene	C_6H_5I	1.70 ± 0.09	1	3-Methyl-2-butenenitrile	C_5H_7N	4.61 ± 0.13	10
1-Iodobutane	C_4H_9I	[1.93]	7	2-Methyl-1-buten-3-yne	C_5H_6	0.513 ± 0.02	2
2-Iodobutane	C_4H_9I	2.12 ± 0.11	1	Methyl cyanate	C_2H_3NO	4.26 ± 0.18	5
Iodoethane	C_2H_5I	1.976 ± 0.002	5	*cis*-3-Methylcyclohexanol	$C_7H_{14}O$	[1.91]	7
Iodoethene	C_2H_3I	1.311 ± 0.005	5	*trans*-3-Methylcyclohexanol	$C_7H_{14}O$	[1.75]	7
Iodomethane	CH_3I	1.6406 ± 0.0004	5	3-Methylcyclopentanone	$C_6H_{10}O$	3.14 ± 0.03	5
1-Iodo-2-methylpropane	C_4H_9I	[1.87]	7	3-Methyl-2-cyclopenten-1-one	C_6H_8O	4.33 ± 0.002	5
Iodomethylsilane	CH_5ISi	1.862 ± 0.005	5	Methylcyclopropane	C_4H_8	0.139 ± 0.004	2
1-Iodopropane	C_3H_7I	2.04 ± 0.10	1	Methyldiborane(6)	CH_8B_2	0.566 ± 0.006	3
2-Iodopropane	C_3H_7I	[1.95]	7	Methyldifluorophosphine	CH_3F_2P	2.056 ± 0.006	3
Isobutanal (*gauche*)	C_4H_8O	2.69 ± 0.01	5	Methylenecyclohexane	C_7H_{12}	0.62 ± 0.01	5
Isobutanal (*trans*)	C_4H_8O	2.86 ± 0.01	5	Methylenecyclopropene	C_4H_4	1.90 ± 0.01	5
Isobutane	C_4H_{10}	0.132 ± 0.002	1	Methylenephosphine ($CH_2 = PH$)	CH_3P	0.869 ± 0.003	61
Isobutene	C_4H_8	0.503 ± 0.010	1	*N*-Methylformamide	C_2H_5NO	3.83 ± 0.08	1
Isobutyl acetate	$C_6H_{12}O_2$	[1.86]	7	Methyl formate	$C_2H_4O_2$	1.77 ± 0.04	1
Isobutylamine	$C_4H_{11}N$	[1.27]	7	2-Methylfuran	C_5H_6O	0.65 ± 0.05	2
Isobutyl formate	$C_5H_{10}O_2$	[1.88]	7	3-Methylfuran	C_5H_6O	1.03 ± 0.02	2
Isobutyl isobutanoate	$C_8H_{16}O_2$	[1.9]	7	5-Methyl-2(3*H*)-furanone	$C_5H_6O_2$	4.08 ± 0.02	5
Isocyanic acid (HNCO)	CHNO	≈1.6	2	Methyl hydroperoxide	CH_4O_2	≈0.65	13
Isocyanobenzene	C_7H_5N	4.018 ± 0.003	5	Methylidyne	CH	≈1.46	2
Isocyanocyclopropane	C_4H_5N	4.03 ± 0.10	3				

Name	Mol. Form.	μ/D	Ref.
Methyl isocyanate	C_2H_3NO	≈2.8	1
Methyl isothiocyanate	C_2H_3NS	3.453 ± 0.003	5
4-Methylisoxazole	C_4H_5NO	3.583 ± 0.005	5
Methyl methacrylate	$C_5H_8O_2$	[1.67]	7
2-Methyloxazole	C_4H_5NO	1.37 ± 0.07	5
4-Methyloxazole	C_4H_5NO	1.08 ± 0.05	5
5-Methyloxazole	C_4H_5NO	2.16 ± 0.04	5
Methyloxirane	C_3H_6O	2.01 ± 0.02	1
2-Methyl-2,4-pentanediol	$C_6H_{14}O_2$	[2.9]	7
4-Methylpentanenitrile	$C_6H_{11}N$	[3.5]	7
Methylphosphonic difluoride	CH_3F_2OP	3.69 ± 0.26	3
N-Methylpropanamide	C_4H_9NO	3.61	7
2-Methylpropanenitrile	C_4H_7N	4.29 ± 0.09	3
2-Methyl-2-propanethiol	$C_4H_{10}S$	1.66 ± 0.03	3
2-Methylpropanoic acid	$C_4H_8O_2$	[1.08]	7
2-Methyl-1-propanol	$C_4H_{10}O$	1.64 ± 0.08	1
2-Methyl-2-propanol	$C_4H_{10}O$	[1.66]	7
2-Methylpropenal	C_4H_6O	2.68 ± 0.13	1
2-Methyl-2-propenol (skew)	C_4H_8O	1.295 ± 0.022	5
Methyl propyl ether (trans-trans)	$C_4H_{10}O$	1.107 ± 0.013	3
2-Methylpyridine	C_6H_7N	1.85 ± 0.04	2
3-Methylpyridine	C_6H_7N	[2.40]	7
4-Methylpyridine	C_6H_7N	2.70 ± 0.02	2
2-Methylpyrimidine	$C_5H_6N_2$	1.676 ± 0.010	3
5-Methylpyrimidine	$C_5H_6N_2$	2.881 ± 0.006	3
N-Methylpyrrolidine	$C_5H_{11}N$	0.572 ± 0.003	5
N-Methyl-2-pyrrolidinone	C_5H_9NO	[4.1]	7
Methyl salicylate	$C_8H_8O_3$	[2.47]	7
Methylsilane	CH_6Si	0.73456	5
Methyl silyl ether	CH_6OSi	1.15 ± 0.02	2
3-Methylthietane	C_4H_8S	2.046 ± 0.009	5
2-Methylthiophene	C_5H_6S	0.674 ± 0.005	2
3-Methylthiophene	C_5H_6S	0.914 ± 0.015	3
Methyl vinyl ether	C_3H_6O	0.965 ± 0.002	5
Morpholine	C_4H_9NO	1.55 ± 0.03	3
2-Nitroanisole	$C_7H_7NO_3$	[5.0]	7
Nitrobenzene	$C_6H_5NO_2$	4.22 ± 0.08	1
Nitroethane	$C_2H_5NO_2$	3.23 ± 0.03	2
Nitromethane	CH_3NO_2	3.46 ± 0.02	1
1-Nitropropane	$C_3H_7NO_2$	3.66 ± 0.07	1
2-Nitropropane	$C_3H_7NO_2$	3.73 ± 0.07	1
Nonanoic acid	$C_9H_{18}O_2$	[0.79]	7
2,5-Norbornadiene	C_7H_8	0.0587 ± 0.0001	5
cis-9-Octadecenoic acid	$C_{18}H_{34}O_2$	[1.18]	7
Octanoic acid	$C_8H_{16}O_2$	[1.15]	7
1-Octanol	$C_8H_{18}O$	[1.76]	7
2-Octanol	$C_8H_{18}O$	[1.71]	7
2-Octanone	$C_8H_{16}O$	[2.70]	7
1,4-Oxathiane	C_4H_8OS	0.295 ± 0.003	3
Oxazole	C_3H_3NO	1.503 ± 0.030	3
Oxetane	C_3H_6O	1.94 ± 0.01	1
2-Oxetanone	$C_3H_4O_2$	4.18 ± 0.03	1
3-Oxetanone	$C_3H_4O_2$	0.887 ± 0.005	2
Oxirane	C_2H_4O	1.89 ± 0.01	1
Paraldehyde	$C_6H_{12}O_3$	1.43 ± 0.07	1

Name	Mol. Form.	μ/D	Ref.
Pentachloroethane	C_2HCl_5	0.92 ± 0.05	1
cis-1,3-Pentadiene	C_5H_8	0.500 ± 0.015	2
trans-1,3-Pentadiene	C_5H_8	0.585 ± 0.010	2
1,3-Pentadiyne	C_5H_4	1.207 ± 0.001	5
1,5-Pentanediol	$C_5H_{12}O_2$	[2.5]	7
2,4-Pentanedione	$C_5H_8O_2$	[2.78]	7
Pentanenitrile	C_5H_9N	4.12 ± 0.08	1
Pentanoic acid	$C_5H_{10}O_2$	[1.61]	7
1-Pentanol	$C_5H_{12}O$	[1.7]	7
2-Pentanol	$C_5H_{12}O$	[1.66]	7
3-Pentanol	$C_5H_{12}O$	[1.64]	7
2-Pentanone	$C_5H_{10}O$	[2.70]	7
3-Pentanone	$C_5H_{10}O$	[2.82]	7
1,2,3-Pentatriene	C_5H_6	0.51 ± 0.05	11
1-Pentene	C_5H_{10}	≈0.5	1
1-Penten-3-yne	C_5H_6	0.66 ± 0.02	2
cis-3-Penten-1-yne	C_5H_6	0.78 ± 0.02	2
trans-3-Penten-1-yne	C_5H_6	1.06 ± 0.05	2
Pentyl acetate	$C_7H_{14}O_2$	1.75 ± 0.10	1
Pentyl formate	$C_6H_{12}O_2$	1.90 ± 0.10	1
1-Pentyne (gauche)	C_5H_8	0.769 ± 0.028	2
1-Pentyne (trans)	C_5H_8	0.842 ± 0.010	2
Perfluoropyridine	C_5F_5N	0.98 ± 0.08	3
Phenol	C_6H_6O	1.224 ± 0.008	3
Phenylacetylene	C_8H_6	0.656 ± 0.005	3
Phenylsilane	C_6H_8Si	0.845 ± 0.012	3
1-Phosphapropyne (CH₃CP)	C_2H_3P	1.499 ± 0.001	63
Piperidine (equitorial)	$C_5H_{11}N$	0.82 ± 0.02	3
Piperidine (axial)	$C_5H_{11}N$	1.19 ± 0.02	3
Piperidine (average)	$C_5H_{11}N$	[1.19]	3
Propanal (gauche)	C_3H_6O	2.86 ± 0.01	5
Propanal (cis)	C_3H_6O	2.52 ± 0.05	1
Propanal (average)	C_3H_6O	2.72	1
Propane	C_3H_8	0.084 ± 0.001	1
1,2-Propanediol	$C_3H_8O_2$	[2.25]	7
1,3-Propanediol	$C_3H_8O_2$	[2.55]	7
Propanenitrile	C_3H_5N	4.05 ± 0.03	3
1-Propanethiol (gauche)	C_3H_8S	1.683 ± 0.010	3
1-Propanethiol (trans)	C_3H_8S	1.60 ± 0.08	3
2-Propanethiol (gauche)	C_3H_8S	1.53 ± 0.03	3
2-Propanethiol (trans)	C_3H_8S	1.61 ± 0.03	3
Propanoic acid (cis)	$C_3H_6O_2$	1.46 ± 0.07	2
Propanoic acid (average)	$C_3H_6O_2$	1.75 ± 0.09	1
1-Propanol (gauche)	C_3H_8O	1.58 ± 0.03	2
1-Propanol (trans)	C_3H_8O	1.55 ± 0.03	2
2-Propanol (trans)	C_3H_8O	1.58 ± 0.03	2
Propargyl alcohol	C_3H_4O	1.13 ± 0.06	2
Propene	C_3H_6	0.366 ± 0.001	1
Propene–sulfur dioxide complex	$C_3H_6 \cdot O_2S$	1.34 ± 0.003	35
Propyl acetate	$C_5H_{10}O_2$	[1.78]	7
Propylamine	C_3H_9N	1.17 ± 0.06	1
Propylene carbonate	$C_4H_6O_3$	[4.9]	7
Propyleneimine (cis)	C_3H_7N	1.77 ± 0.09	2
Propyleneimine (trans)	C_3H_7N	1.57 ± 0.03	2
Propyl formate	$C_4H_8O_2$	[1.89]	7

Name	Mol. Form.	μ/D	Ref.
2-Propynal	C$_3$H$_2$O	2.78 ± 0.02	5
Propyne	C$_3$H$_4$	0.784 ± 0.001	3
Propyne-argon complex	C$_3$H$_4$•Ar	0.730 ± 0.005	20
4H-Pyran-4-one	C$_5$H$_4$O$_2$	3.79 ± 0.02	5
4H-Pyran-4-thione	C$_5$H$_4$OS	3.95 ± 0.05	5
1H-Pyrazole	C$_3$H$_4$N$_2$	2.20 ± 0.01	3
Pyridazine	C$_4$H$_4$N$_2$	4.22 ± 0.02	2
Pyridine	C$_5$H$_5$N	2.215 ± 0.010	3
2-Pyridinecarbonitrile	C$_6$H$_4$N$_2$	5.78 ± 0.11	3
3-Pyridinecarbonitrile	C$_6$H$_4$N$_2$	3.66 ± 0.11	3
4-Pyridinecarbonitrile	C$_6$H$_4$N$_2$	1.96 ± 0.03	3
3-Pyridinecarboxaldehyde	C$_6$H$_5$NO	1.44	3
4-Pyridinecarboxaldehyde	C$_6$H$_5$NO	1.66	3
2-Pyridinecarboxaldehyde	C$_6$H$_5$NO	3.56 ± 0.07	3
Pyrimidine	C$_4$H$_4$N$_2$	2.334 ± 0.010	2
Pyrrole	C$_4$H$_5$N	1.767 ± 0.001	5
Pyrrolidine	C$_4$H$_9$N	[1.57]	7
2-Pyrrolidone	C$_4$H$_7$NO	[3.5]	7
Quinoline	C$_9$H$_7$N	2.29 ± 0.11	1
Salicylaldehyde	C$_7$H$_6$O$_2$	[2.86]	7
Selenoformaldehyde	CH$_2$Se	1.41 ± 0.01	5
Silicon dicarbide	C$_2$Si	2.393 ± 0.006	24
Silicon methylidyne	CHSi	0.066 ± 0.002	41
Styrene	C$_8$H$_8$	0.123 ± 0.003	5
Succinonitrile	C$_4$H$_4$N$_2$	[3.7]	7
Sulfolane	C$_4$H$_8$O$_2$S	[4.8]	7
1,1,2,2-Tetrabromoethane	C$_2$H$_2$Br$_4$	[1.38]	7
1,1,2,2-Tetrachloroethane	C$_2$H$_2$Cl$_4$	1.32 ± 0.07	1
1,2,3,4-Tetrafluorobenzene	C$_6$H$_2$F$_4$	2.42 ± 0.05	3
1,2,3,5-Tetrafluorobenzene	C$_6$H$_2$F$_4$	1.46 ± 0.06	3
1,1,1,2-Tetrafluoroethane	C$_2$H$_2$F$_4$	1.80 ± 0.22	5
Tetrahydrofuran	C$_4$H$_8$O	1.75 ± 0.04	2
Tetrahydrofurfuryl alcohol	C$_5$H$_{10}$O$_2$	[2.1]	7
Tetrahydropyran (chair)	C$_5$H$_{10}$O	1.58 ± 0.03	3
Tetrahydro-4H-pyran-4-one	C$_5$H$_8$O$_2$	1.720 ± 0.003	3
1,2,5,6-Tetrahydropyridine	C$_5$H$_9$N	1.007 ± 0.003	3
Tetrahydrothiophene	C$_4$H$_8$S	[1.90]	7
Tetramethylurea	C$_5$H$_{12}$N$_2$O	[3.5]	7
1H-Tetrazole	CH$_2$N$_4$	2.19 ± 0.05	3
Thiacyclohexane	C$_5$H$_{10}$S	1.781 ± 0.010	3
1,2,5-Thiadiazole	C$_2$H$_2$N$_2$S	1.579 ± 0.007	3
Thietane	C$_3$H$_6$S	1.85 ± 0.09	1
Thietane 1,1-dioxide	C$_3$H$_6$O$_2$S	4.8 ± 0.1	5
Thioacetaldehyde	C$_2$H$_4$S	2.33 ± 0.02	68
Thiocarbonyl fluoride	CF$_2$S	0.080	59
Thioformaldehyde	CH$_2$S	1.6491 ± 0.0004	3
Thiophene	C$_4$H$_4$S	0.55 ± 0.01	2
2-Thiophenecarbonitrile	C$_5$H$_3$NS	4.59 ± 0.02	3
3-Thiophenecarbonitrile	C$_5$H$_3$NS	4.13 ± 0.02	3
4H-Thiopyran-4-thione	C$_5$H$_4$S$_2$	3.9 ± 0.2	5

Name	Mol. Form.	μ/D	Ref.
Toluene	C$_7$H$_8$	0.375 ± 0.010	3
Toluene-sulfur dioxide complex	C$_7$H$_8$•O$_2$S	1.87 ± 0.03	34
1H-1,2,4-Triazole	C$_2$H$_3$N$_3$	2.7 ± 0.1	3
Tribromomethane	CHBr$_3$	0.99 ± 0.02	1
Tributylamine	C$_{12}$H$_{27}$N	[0.78]	7
Tributyl borate	C$_{12}$H$_{27}$BO$_3$	[0.77]	7
Tributyl phosphate	C$_{12}$H$_{27}$O$_4$P	[3.07]	7
Tricarbon monosulfide	C$_3$S	3.704 ± 0.009	50
1,1,1-Trichloroethane	C$_2$H$_3$Cl$_3$	1.755 ± 0.015	2
1,1,2-Trichloroethane	C$_2$H$_3$Cl$_3$	[1.4]	7
Trichloroethene	C$_2$HCl$_3$	[0.8]	7
Trichloroethylsilane	C$_2$H$_5$Cl$_3$Si	[2.04]	7
Trichlorofluoromethane	CCl$_3$F	0.46 ± 0.02	3
Trichloromethane	CHCl$_3$	1.04 ± 0.02	2
(Trichloromethyl)benzene	C$_7$H$_5$Cl$_3$	[2.03]	7
Trichloromethylsilane	CH$_3$Cl$_3$Si	1.91 ± 0.01	2
Tri-o-cresyl phosphate	C$_{21}$H$_{21}$O$_4$P	[2.87]	7
Tri-m-cresyl phosphate	C$_{21}$H$_{21}$O$_4$P	[3.05]	7
Tri-p-cresyl phosphate	C$_{21}$H$_{21}$O$_4$P	[3.18]	7
Triethanolamine	C$_6$H$_{15}$NO$_3$	[3.57]	7
Triethylamine	C$_6$H$_{15}$N	0.66 ± 0.05	1
Triethyl phosphate	C$_6$H$_{15}$O$_4$P	[3.12]	7
Trifluoroacetic acid	C$_2$HF$_3$O$_2$	2.28 ± 0.25	1
Trifluoroacetonitrile	C$_2$F$_3$N	1.262 ± 0.010	3
1,2,4-Trifluorobenzene	C$_6$H$_3$F$_3$	1.402 ± 0.009	5
1,1,1-Trifluoroethane	C$_2$H$_3$F$_3$	2.347 ± 0.005	3
Trifluoroethene	C$_2$HF$_3$	1.32 ± 0.03	2
Trifluoroiodomethane	CF$_3$I	1.048 ± 0.003	3
Trifluoroisocyanomethane	C$_2$F$_3$N	1.153 ± 0.010	5
Trifluoromethane	CHF$_3$	1.65150	3
(Trifluoromethyl)benzene	C$_7$H$_5$F$_3$	2.86 ± 0.06	1
Trifluoromethylsilane	CH$_3$F$_3$Si	2.3394 ± 0.0002	5
(Trifluoromethyl)silane	CH$_3$F$_3$Si	2.32 ± 0.02	5
3,3,3-Trifluoropropene	C$_3$H$_3$F$_3$	2.45 ± 0.05	1
3,3,3-Trifluoro-1-propyne	C$_3$HF$_3$	2.317 ± 0.013	5
Trimethylamine	C$_3$H$_9$N	0.612 ± 0.003	1
Trimethyl phosphate	C$_3$H$_9$O$_4$P	[3.18]	7
2,4,6-Trimethylpyridine	C$_8$H$_{11}$N	[2.05]	7
1,3,5-Trioxane	C$_3$H$_6$O$_3$	2.08 ± 0.02	1
Vinyl acetate	C$_4$H$_6$O$_2$	[1.79]	7
Vinyl formate	C$_3$H$_4$O$_2$	1.49 ± 0.01	1
2-Vinylfuran	C$_6$H$_6$O	0.69 ± 0.07	5
Vinylsilane	C$_2$H$_6$Si	0.657 ± 0.002	5
o-Xylene	C$_8$H$_{10}$	0.640 ± 0.005	2
2,4-Xylenol	C$_8$H$_{10}$O	[1.4]	7
2,5-Xylenol	C$_8$H$_{10}$O	[1.45]	7
2,6-Xylenol	C$_8$H$_{10}$O	[1.40]	7
3,4-Xylenol	C$_8$H$_{10}$O	[1.56]	7
3,5-Xylenol	C$_8$H$_{10}$O	[1.55]	7

HINDERED INTERNAL ROTATION

I. Ozier and N. Moazzen-Ahmadi

In asymmetric tops like methyl alcohol, CH_3OH, and symmetric rotors like CH_3SiH_3, the methyl group can undergo internal rotation relative to the rest of the molecule, traditionally called the frame (Refs. 1 and 2). Although various different tops are considered here, all have three-fold symmetry. In such cases, the potential V hindering the internal rotation can be written:

$$V(\alpha) = V_3(\tfrac{1}{2})(1 - \cos 3\alpha) + V_6(\tfrac{1}{2})(1 - \cos 6\alpha) + V_9(\tfrac{1}{2})(1 - \cos 9\alpha) + \ldots,$$

where α is the deviation from equilibrium of the angle between the top and frame that measures the torsional motion. If only the first two terms are retained, then V_3 is the height of the hindering potential and V_6 is the shape parameter. For symmetric tops like CH_3CH_3 where the top and frame are identical, α is replaced by 2γ and the origin for γ is often taken as the eclipsed configuration. In the expansion, $-\cos 6n\gamma$ is then replaced by $(-1)^{n+1}\cos 6n\gamma$, where $n=1,2,\ldots$ In cases where different forms of the expansion have been used in the original works, the values of the parameters published there have been converted to the conventions defined here.

In Tables 1 and 2, values are given for V_3 for a selection of asymmetric and symmetric tops, respectively. In cases where the higher order parameters have been determined, these are given in the Comments column. Where appropriate, this column also indicates the specific top, isomer, state, and/or isotopomer that has been studied. For ethane, three symmetric top isotopomer are listed to illustrate the isotopic dependence of V_3 and V_6. In all other cases, only one isotopomer is listed, even if several have been studied. In all but one of these cases, the isotopomer reported is the one with the highest natural abundance. However, CH_3OCDO is listed because the results obtained are more precise than for CH_3OCHO. The molecules are listed alphabetically in Hill order according to the molecular formula.

The determinations listed for the potential parameters are effective values that incorporate to varying degrees effects from other molecular parameters. For example, the apparent value of V_3 can be changed significantly if the reduced rotational constant F is calculated from the structure, rather than being determined independently (Ref. 1). Other examples include such mechanisms as coupling to excited skeletal vibrations (Ref. 2) and redundancies connecting some of the torsional parameters (Refs. 3 and 4). The experimental uncertainties quoted are taken from the original works; no attempt has been made to standardize the definitions. All the potential parameters are given in cm^{-1}. Where the original work has reported these values in other units, the conversion to cm^{-1} has been carried out using standard factors: 1 calorie = 4.1868 joules and 1 calorie/mole = 0.34998915 cm^{-1} (Ref. 5).

A variety of different methods have been used to measure V_3, V_6, and V_9 (Refs. 1 and 2); only a few of the more important will be discussed here. *For asymmetric rotors*, both the pure rotational spectrum and its torsion-rotation counterpart are electric dipole allowed and are affected in lowest order by the leading terms in the torsional Hamiltonian. Both types of spectra have been used extensively to determine V_3 (Ref. 1). *For symmetric tops* with a single torsional degree of freedom, either the permanent electric dipole moment vanishes, as in CH_3CH_3, or the normal rotational spectrum is independent of V_3 in lowest order, as in CH_3SiH_3. In the latter case, the molecular beam avoided crossing method can often be used (Ref. 2). The torsion-rotation spectrum is forbidden

in lowest order, but becomes weakly allowed through interactions with the infrared active skeletal vibrations (Ref. 2). By employing long absorption path lengths, this spectrum has been used to determine V_3 in a number of molecules. For both asymmetric and symmetric tops, the most precise determinations of the molecular parameters have been made in cases where both rotational and torsion-rotation spectra have been investigated.

References

1. Lin, C. C., and Swalen, J. D., *Rev. Mod. Phys.* 31, 841 1959.
2. Ozier, I., and Moazzen-Ahmadi, N., Internal rotation in symmetric tops, in: Arimondo, E., Berman, P. R., and Lin, C. C., (Eds.), *Advances in Atomic, Molecular and Optical Physics*, vol. 54, p. 423, Elsevier, Amsterdam, 2007.
3. Lees, R. M., and Baker, J. G., *J. Chem. Phys.* 48, 5299, 1968.
4. Moazzen-Ahmadi, N., and Ozier, I., *J. Mol. Spectrosc.* 126, 99, 1987.
5. Demaison, J., and Wlodarczak, G., Hindered rotation-Asymmetric top molecules, in: Hüttner, W. (Ed), *Landolt-Börnstein Numerical Data and Functional Relationships in Science and Technology, New Series: Group II: Molecules and Radicals, volume 24, subvolume C, Molecular Constants mostly from Microwave, Molecular Beam, and Sub-Doppler Laser Spectroscopy*, Springer-Verlag, Heidelberg, 2002.
6. Suenram, R. D., Lovas, F. J., Plusquellic, D. F., Ellzy, M. W., Lochner, J. M., Jensen, J. O., and Samuels, A. C., *J. Mol. Spectrosc.* 235, 18, 2006.
7. Sastry, K. V. L. N., Herbst, E., Booker, R. A., and De Lucia, F. C., *J. Mol. Spectrosc.* 116, 120, 1986.
8. Tyblewksi, M., Ha, T.-K., and Bauder, A., *J. Mol. Spectrosc.* 115, 353, 1986.
9. Durig, J. R., Guirgis, G. A., and Van Der Veken, B. J., *J. Raman Spectrosc.* 18, 549, 1987.
10. Eltayeb, S., Guirgis, G. A., Fanning, A. R., and Durig, J. R., *J. Raman Spectrosc.* 27, 111, 1996.
11. Krisher, L. C., *J. Chem. Phys.* 33, 1237, 1960.
12. Alonso, J. L., López, J. C., Blanco, S., and Guarnieri, A., *J. Mol. Spectrosc.* 182, 148, 1997.
13. Pierce, L., and Krisher, L. C., *J. Chem. Phys.* 31, 875, 1959.
14. Moloney, M. J., and Krisher, L. C., *J. Chem. Phys.* 45, 3277, 1966.
15. Kleiner, I., Hougen, J. T., Grabow, J.-U., Belov, S. P., Tretyakov, M. Yu., and Cosléou, J., *J. Mol. Spectrosc.* 179, 41, 1996.
16. Ilyushin, V. V., Alekseev, E. A., Dyubko, S. F., and Kleiner, I., *J. Mol. Spectrosc.* 220, 170, 2003.
17. Fliege, E., Dreizler, H., Demaison, J., Boucher, D., Burie, J., and Dubrulle, A., *J. Chem. Phys.* 78, 3541, 1983
18. Schnell, M., Grabow, J.-U., Hartwig, H., Heineking, N., Meyer, M., Stahl, W., and Caminati, W., *J. Mol. Spectrosc.* 229, 1, 2005.
19. Niide, Y., and Hayashi, M., *J. Mol. Spectrosc.* 223, 152, 2004.
20. Shiki, Y., Hasegawa, A., and Hayashi, M., *J. Mol. Structure* 78, 185, 1982.
21. Groner, P., Gillies, C. W., Gillies, J. Z., Zhang, Y., and Block, E., *J. Mol. Spectrosc.* 226, 169, 2004.
22. Alonso, J. L., Lesarri, A., López, J. C., Blanco, S., Kleiner, I., and Demaison, J., *Molec. Phys.* 91, 731, 1997.
23. Demaison, J., Maes, H., van Eijck, B. P., Wlodarczak, G. and Lasne, M. C., *J. Mol. Spectrosc.* 125, 214, 1987.
24. Antolínez, S., López, J. C., and Alonso, J. L., *J. Chem. Soc., Faraday Trans.* 93, 1291, 1997.
25. Butcher, S. S., and Wilson, E. B., *J. Chem. Phys.* 40, 1671, 1964.
26. Groner, P., *J. Mol. Structure* 550–551, 473, 2000.
27. Marstokk, K.-M., Mollendal, H., Samdal, S., and Steinborn, D., *J. Mol. Structure* 567, 41, 2001.
28. Stiefvater, O. L., *J. Chem. Phys.* 62, 233, 1975.
29. de Luis, A., Eugenia Sanz, M., Lorenzo, F. J., López, J. C., and Alonso, J. L., *J. Mol. Spectrosc.* 184, 60, 1997.

30. Kasten, W., and Dreizler, H., *Z. Naturforsch.* A 41, 944, 1986.
31. Vormann, K., and Dreizler, H., *Z. Naturforsch.* A 43, 338, 1988.
32. Marstokk, K.-M., Møllendal, H., and Samdal, S., *J. Mol. Structure* 376, 11, 1996.
33. Bestmann, G., Lalowski, W., and Dreizler, H., *Z. Naturforschung* A 40, 271, 1985.
34. Dreizler, H., and Scappini, F., *Z. Naturforsch.* A 36, 1187, 1981.
35. Typke, V., Botskor, I., and Wiedenmann, K.-H., *J. Mol. Spectrosc.* 120, 435, 1986.
36. Suenram, R. D., Lovas, F. J., Plusquellic, D. F., Lesarri, A., Kawashima, Y., Jensen, J. O., and Samuels, A. C., *J. Mol. Spectrosc.* 211, 110, 2002.
37. Ohashi, N., and Hougen, J. T., *J. Mol. Spectrosc.* 243, 162, 2007.
38. Grabow, J.-U., Hartwig, H., Heineking, N., Jäger, W., Mäder, H., Nicolaisen, H. W., and Stahl, W., *J. Mol. Structure* 612, 349, 2002.
39. Charro, M. E., and Alonso, J. L., *J. Mol. Spectrosc.* 176, 251, 1996.
40. Nair, K. P. R., Demaison, J., Wlodarczak, G., and Merke, I., *J. Mol. Spectrosc.* 237, 137, 2006.
41. Suenram, R. D., DaBell, R. S., Hight Walker, A. R., Lavrich, R. J., Plusquellic, D. F., Ellzy, M. W., Lochner, J. M., Cash, L., Jensen, J. O., and Samuels, A. C., *J. Mol. Spectrosc.* 224, 176, 2004.
42. Groner, P., Attia, G. M., Mohamad, A. B., Sullivan, J. F., Li, Y. S., and Durig, J. R., *J. Chem. Phys.* 91, 1434, 1989.
43. Varma, R., Ramaprasad, K. R., and Nelson, J. F., *J. Chem. Phys.* 63, 915, 1975.
44. Odom, J. D., Kalasinsky, V. F., and Durig, J. R., *Inorg. Chem.* 14, 2837, 1975.
45. Kuczkowski, R. L., and Lide, D. R., *J. Chem. Phys.* 46, 357, 1967.
46. Durig, J. R., Li, Y. S., Carreira, L. A., and Odom, J. D., *J. Amer. Chem. Soc.* 95, 2491, 1973.
47. Lide, D. R., Johnson, D. R., Sharp, K. G., and Coyle, T. D., *J. Chem. Phys.* 57, 3699, 1972.
48. Krisher, L. C., Watson, W. A., and Morrison, J. A., *J. Chem. Phys.* 61, 3429, 1974.
49. Styger, C., Ozier, I., Wang, S.-X., and Bauder, A., *J. Mol. Spectrosc.* 239, 115, 2006.
50. Laurie, V. W., *J. Chem. Phys.* 30, 1210, 1959.
51. Cahill, P., and Butcher, S., *J. Chem. Phys.* 35, 2255, 1961.
52. Wang, S.-X., Schroderus, J., Ozier, I., Moazzen-Ahmadi, N., Horneman, V.-M., Ilyushyn, V. V., Alekseev, E. A., Katrich, A. A., and Dyubko, S. F., *J. Mol. Spectrosc.* 214, 69, 2002.
53. Nakagawa, J., Yamada, K., Bester, M., and Winnewisser, G., *J. Mol. Spectrosc.* 110, 74, 1985.
54. Merke, I., Stahl, W., Kassi, S., Petitprez, D., and Wlodarczak, G., *J. Mol. Spectrosc.* 216, 437, 2002.
55. di Lauro, C., Bunker, P. R., Johns, J. W. C., and McKellar, A. R. W., *J. Mol. Spectrosc.* 184, 177, 1997.
56. Voges, K., Gripp, J., Hartwig, H., and Dreizler, H., *Z. Naturforsch.* A 51, 299, 1996.
57. Borvayeh, L., Moazzen-Ahmadi, N., and Horneman, V.-M., *J. Mol. Spectrosc.* 242, 77, 2007.

TABLE 1. Asymmetric Top Potential Parameters

	Name	Molecular Formula	Line Formula	Ref.	V_3/cm^{-1}	Comments
1	Trifluoromethanethiol	CHF_3S	CF_3SH	5	500.83 ± 0.03	
2	Methylphosphonic difluoride	CH_3F_2OP	$CH_3P(=O)F_2$	6	676 ± 25	
3	Methanol	CH_4O	CH_3OH	5	373.594 ± 0.007	$V_6 = -1.597 \pm 0.051$ $V_9 = 1.04 \pm 0.20$
4	Methanethiol	CH_4S	CH_3SH	7	443.029 ± 0.070	$V_6 = -1.6451 \pm 0.0144$
5	Methyldisulfane	CH_4S_2	CH_3SSH	8	609.0 ± 14.0	
6	Trifluoromethyl isocyanate	C_2F_3NO	$CF_3N=C=O$	5	47.8769 ± 0.0051	
7	Trifluoroacetaldehyde	C_2HF_3O	$CF_3C(H)=O$	9	298 ± 10	
8	Pentafluoroethane	C_2HF_5	CF_3CHF_2	10	1190 ± 4	
9	Acetyl bromide	C_2H_3BrO	$CH_3C(Br)=O$	11	456.7 ± 10.5	
10	1-Chloro-1,1-difluoroethane	$C_2H_3ClF_2$	CH_3CClF_2	12	1311.8 ± 1.4	
11	Acetyl chloride	C_2H_3ClO	$CH_3C(Cl)=O$	5	442.74 ± 1.05	^{35}Cl
12	Acetyl fluoride	C_2H_3FO	$CH_3C(F)=O$	13	364.3 ± 2.1	
13	Methyl fluoroformate	$C_2H_3FO_2$	$CH_3OC(F)=O$	5	374.1 ± 0.2	
14	Methyl trifluoromethyl ether	$C_2H_3F_3O$	CH_3OCF_3	5	382 ± 10	CH_3
15	Acetyl iodide	C_2H_3IO	$CH_3C(=O)I$	14	455.3 ± 10.5	
16	Methyl cyanate	C_2H_3NO	$CH_3OC\equiv N$	5	399.0 ± 17.5	
17	1-Chloro-1-fluoroethane	C_2H_4ClF	CH_3CHClF	5	1334.9 ± 3.8	
18	1,1-Difluoroethane	$C_2H_4F_2$	CH_3CHF_2	5	1163.0 ± 2.5	
19	Acetaldehyde	C_2H_4O	$CH_3C(H)=O$	15	407.716 ± 0.010	$V_6 = -12.068 \pm 0.037$
20	Thioacetaldehyde S-oxide	C_2H_4OS	$CH_3C(H)=S=O$	5	285.6 ± 0.3	Z isomer
21	Acetic acid	$C_2H_4O_2$	CH_3COOH	16	170.1742 ± 0.0002	$V_6 = -6.4725 \pm 0.0001$
22	Methyl formate	$C_2H_3DO_2$	$CH_3OC(D)=O$	5	400.60 ± 0.03	deuterated
23	Fluoroethane	C_2H_5F	CH_3CH_2F	17	1172.1 ± 1.4	
24	Nitrosoethane	C_2H_5NO	$CH_3CH_2N=O$	5	903 ± 25	*gauche* conformer
		C_2H_5NO	$CH_3CH_2N=O$	5	911 ± 25	*cis* conformer
25	Acetamide	C_2H_5NO	$CH_3C(NH_2)=O$	5	24.949 ± 0.008	
26	Difluorodimethylsilane	$C_2H_6F_2Si$	$(CH_3)_2SiF_2$	18	439.4 ± 2.5	

	Name	Molecular Formula	Line Formula	Ref.	V_3/cm^{-1}	Comments
27	N-Nitrosodimethylamine	$C_2H_6N_2O$	$(CH_3)_2NN=O$	5	145.8 ± 0.25	cis CH_3
		$C_2H_6N_2O$	$(CH_3)_2NN=O$	5	737.4 ± 13.3	trans CH_3
28	Ethanol	C_2H_6O	CH_3CH_2OH	5	1173.76 ± 2.20	trans isomer
29	Dimethyl ether	C_2H_6O	$(CH_3)_2O$	19	926.0 ± 3.5	
30	Dimethyl sulfide	C_2H_6S	$(CH_3)_2S$	19	751.1 ± 4.8	
31	Vinylsilane	C_2H_6Si	$SiH_3C(H)=CH_2$	20	520.1 ± 1.8	
32	Dimethyl disulfide	$C_2H_6S_2$	CH_3SSCH_3	5	535.1 ± 1.8	
33	Dimethyl diselenide	$C_2H_6Se_2$	$CH_3SeSeCH_3$	21	395 ± 2	
34	Dimethylsilane	C_2H_8Si	$(CH_3)_2SiH_2$	19	578.0 ± 3.5	
35	3,3,3-Trifluoropropene	$C_3H_3F_3$	$CF_3C(H)=CH_2$	22	653.06 ± 0.83	
36	Methyl cyanoformate	$C_3H_3NO_2$	$CH_3OC(C\equiv N)=O$	5	406.6 ± 1.1	s-trans conformer
37	(Methylthio)acetylene	C_3H_4S	$CH_3SC\equiv CH$	23	592.0 ± 3.3	
38	1,1,1-Trifluoropropane	$C_3H_5F_3$	$CH_3CH_2CF_3$	24	922.2 ± 1.4	
39	2-Iodopropene	C_3H_5I	$CH_3C(I)=CH_2$	5	905.8 ± 4.2	
40	Ethyl isocyanide	C_3H_5N	$CH_3CH_2N\equiv C$	5	1167.6 ± 18.2	
41	Propene	C_3H_6	$CH_3C(H)=CH_2$	5	697.499 ± 0.048	$V_6 = -13.0$ (fixed)
42	Propanal	C_3H_6O	$CH_3CH_2C(H)=O$	25	798 ± 39	cis conformer
43	Acetone	C_3H_6O	$(CH_3)_2C=O$	26	251.4 ± 2.6	$V_6 = -6.92 \pm 0.65$
44	(Methylthio)ethene	C_3H_6S	$CH_3SC(H)=CH_2$	27	1138 ± 13	
45	Propanoic acid	$C_3H_6O_2$	CH_3CH_2COOH	28	819.0 ± 10.5	cis conformer
46	Methyl mercaptoacetate	$C_3H_6O_2S$	$CH_3OC(=O)C(H_2)SH$	5	411 ± 8	state 0^+
		$C_3H_6O_2S$	$CH_3OC(=O)C(H_2)SH$	5	412 ± 9	state 0^-
47	2-Bromopropane	C_3H_7Br	$(CH_3)_2CHBr$	5	1437.0 ± 2.5	^{79}Br
48	1-Chloropropane	C_3H_7Cl	$CH_3C(H_2)C(H_2)Cl$	29	1017.8 ± 1.4	gauche conformer
		C_3H_7Cl	$CH_3C(H_2)C(H_2)Cl$	29	966.0 ± 7.0	trans conformer
49	2-Chloropropane	C_3H_7Cl	$(CH_3)_2CHCl$	5	1374.03 ± 1.00	^{35}Cl
50	1-Fluoropropane	C_3H_7F	$CH_3C(H_2)C(H_2)F$	30	965.3 ± 12.2	gauche conformer
		C_3H_7F	$CH_3C(H_2)C(H_2)F$	30	948.5 ± 2.8	trans conformer
51	2-Fluoropropane	C_3H_7F	$(CH_3)_2CHF$	5	1162.79 ± 0.84	
52	Butanenitrile	C_4H_7N	$CH_3C(H_2)C(H_2)C\equiv N$	31	1087.4 ± 8.4	gauche conformer
		C_4H_7N	$CH_3C(H_2)C(H_2)C\equiv N$	31	1088.5 ± 13.3	trans conformer
53	Propanamide	C_3H_7NO	$CH_3CH_2C(=O)NH_2$	32	761 ± 42	syn conformer
54	N,N-Dimethylformamide	C_3H_7NO	$(CH_3)_2NC(H)=O$	5	366.04 ± 0.26	cis CH_3
		C_3H_7NO	$(CH_3)_2NC(H)=O$	5	772.4 ± 7.4	trans CH_3
55	Propane	C_3H_8	$(CH_3)_2CH_2$	33	1108.1 ± 9.5	
56	Cyclopropylgermane	C_3H_8Ge	$\underset{\rule{2em}{0.4pt}}{C(H_2)C(H_2)C(H)(GeH_3)}$	5	466.6 ± 16.7	GeH_3
57	N-Nitrosoethylmethylamine	$C_3H_8N_2O$	$CH_3CH_2N(CH_3)N=O$	5	310 ± 30	N-methyl top, OGM conformer
58	1-Propanol	C_3H_8O	$CH_3C(H_2)C(H_2)OH$	34	956 ± 21	trans conformer
59	Cyclopropylsilane	C_3H_8Si	$\underset{\rule{2em}{0.4pt}}{C(H_2)C(H_2)C(H)(SiH_3)}$	35	670.9 ± 1.5	
60	Dimethyl(methylene)silane	C_3H_8Si	$(CH_3)_2Si=CH_2$	5	351.4 ± 5.9	
61	Dimethyl methylphosphonate	$C_3H_9O_3P$	$(OCH_3)_2P(=O)CH_3$	36	662 ± 6	P-methyl top
		$C_3H_9O_3P$	$(OCH_3)_2P(=O)CH_3$	37	278.82 ± 0.06	O-methyl top #1
		$C_3H_9O_3P$	$(OCH_3)_2P(=O)CH_3$	37	181.82 ± 0.01	O-methyl top #2
62	But-2-ynoyl fluoride	C_4H_3FO	$CH_3C\equiv CC(F)=O$	5	2.20 ± 0.12	
63	cis-2-Butenenitrile	C_4H_5N	$CH_3C(H)=C(H)C\equiv N$	5	485.50 ± 0.25	
64	2-Methylacrylonitrile	C_4H_5N	$CH_2=C(CH_3)C\equiv N$	5	695.2 ± 2.1	
65	2-Methyloxazole	C_4H_5NO	$\underset{\rule{2em}{0.4pt}}{N=C(CH_3)OC(H)=C(H)}$	5	251.70 ± 1.17	
66	4-Methyloxazole	C_4H_5NO	$\underset{\rule{2em}{0.4pt}}{N=C(H)OC(H)=C(CH_3)}$	5	429.44 ± 0.33	

	Name	Molecular Formula	Line Formula	Ref.	V_3/cm^{-1}	Comments
67	5-Methyloxazole	C_4H_5NO	N=C(H)OC(CH$_3$)=C(H)	5	477.90 ± 1.34	
68	5-Methylisoxazole	C_4H_5NO	C(H)=NOC(CH$_3$)=C(H)	5	272.05 ± 1.00	
69	2-Methylthiazole	C_4H_5NS	N=C(CH$_3$)SC(H)=C(H)	38	34.938 ± 0.020	
70	4-Methylisothiazole	C_4H_5NS	N=C(H)C(CH$_3$)=C(H)S	5	105.767 ± 0.043	
71	4-Methyl-2-oxetanone	$C_4H_6O_2$	OC(=O)C(H$_2$)C(H)(CH$_3$)	5	1256.5 ± 10.5	
72	*trans*-1-Fluoro-2-butene	C_4H_7F	CH$_3$C(H)=C(H)CH$_2$F	5	596 ± 7	anticlinal conformer
73	1-Isocyanopropane	C_4H_7N	CH$_3$C(H$_2$)C(H$_2$)N≡C	5	1012.3 ± 8.4	*gauche* conformer
		C_4H_7N	CH$_3$C(H$_2$)C(H$_2$)N≡C	5	1033.8 ± 7.7	*trans* conformer
74	Isobutene	C_4H_8	(CH$_3$)$_2$C=CH$_2$	5	761.58 ± 1.05	
75	*cis*-2-Butene	C_4H_8	CH$_3$CH=CHCH$_3$	5	259.89 ± 0.42	
76	3-Methoxy-1-propene	C_4H_8O	CH$_3$OC(H$_2$)C(H)=CH$_2$	5	728.0 ± 10.5	*skew-gauche* conformer
		C_4H_8O	CH$_3$OC(H$_2$)C(H)=CH$_2$	5	829.5 ± 10.5	*syn-trans* conformer
77	2,2-Dimethyloxirane	C_4H_8O	OC(CH$_3$)(CH$_3$)C(H$_2$)	5	945.61 ± 0.75	
78	*cis*-2,3-Dimethyloxirane	C_4H_8O	OC(H)(CH$_3$)C(H)(CH$_3$)	5	577.80 ± 1.84	*cis* conformer
		C_4H_8O	OC(H)(CH$_3$)C(H)(CH$_3$)	5	862.52 ± 1.84	*trans* conformer
79	2-Methyloxetane	C_4H_8O	OC(H$_2$)C(H$_2$)C(H)(CH$_3$)	5	1166.5 ± 4.9	
80	3-Methyloxetane	C_4H_8O	OC(H$_2$)C(H)(CH$_3$)C(H$_2$)	5	1149.4 ± 4.2	
81	3-Methoxythietane	C_4H_8OS	SC(H$_2$)C(H)(OCH$_3$)C(H$_2$)	5	1071.0 ± 10.5	
82	3-(Methylthio)-1-propene	C_4H_8S	CH$_3$SC(H$_2$)C(H)=CH$_2$	5	619 ± 28	
83	2,2-Dimethylthiirane	C_4H_8S	SC(CH$_3$)(CH$_3$)C(H$_2$)	5	1268.3 ± 3.0	
84	Butane	C_4H_{10}	CH$_3$C(H$_2$)C(H$_2$)CH$_3$	5	948 ± 24	
85	*N*-Methyl-*N*-nitrosopropylamine	$C_4H_{10}N_2O$	CH$_3$C(H$_2$)C(H$_2$)N(CH$_3$)N=O	5	320 ± 30	*N*-methyl top, conformer OMGA
86	Dihydro-3-methyl-2(3*H*)-furanone	$C_5H_8O_2$	OC(=O)C(H)(CH$_3$)C(H$_2$)C(H$_2$)	5	913.8 ± 2.5	
87	Dihydro-4-methyl-2(3*H*)-furanone	$C_5H_8O_2$	OC(=O)C(H$_2$)C(H)(CH$_3$)C(H$_2$)	39	1437.8 ± 8.4	
88	Dihydro-5-methyl-2(3*H*)-furanone	$C_5H_8O_2$	OC(=O)C(H$_2$)C(H$_2$)C(H)(CH$_3$)	39	1233.0 ± 4.2	
89	*tert*-Butyl isocyanate	C_5H_9NO	(CH$_3$)$_3$C≡N=C=O	5	41.510 ± 0.015	(CH$_3$)$_3$C group
90	Methyl *tert*-butyl ether	$C_5H_{12}O$	(CH$_3$)$_3$COCH$_3$	5	498.6 ± 1.5	*O*-methyl top
91	2-Methylcyclopentanone	$C_6H_{10}O$	C(=O)C(H)(CH$_3$)C(H$_2$)C(H$_2$)C(H$_2$)	5	844.2 ± 2.4	
92	3-Methylcyclopentanone	$C_6H_{10}O$	C(=O)C(H$_2$)C(H)(CH$_3$)C(H$_2$)C(H$_2$)	5	1233.8 ± 1.7	
93	*tert*-Butyl ethyl ether	$C_6H_{14}O$	(CH$_3$)$_3$COC(H$_2$)CH$_3$	5	1025 ± 3	ethyl CH$_3$
94	2,4-Difluorotoluene	$C_7H_6F_2$	C(H)=C(CH$_3$)C(F)=C(H)C(F)=C(H)	5	204.04 ± 0.23	
95	2-Chlorotoluene	C_7H_7Cl	C(H)=C(H)C(Cl)=C(CH$_3$)C(H)=C(H)	40	513.8 ± 2.7	^{35}Cl
96	2,6-Dimethylpyridine	C_7H_9N	C(H)=C(H)C(CH$_3$)=NC(CH$_3$)=C(H)	5	98.24 ± 0.27	
97	1,2,2-Trimethylpropyl methylphosphonofluoridate	$C_7H_{16}FO_2P$	(CH$_3$)$_3$CC(H)(CH$_3$)OP(O)(F)CH$_3$	41	821 ± 5	*P*-methyl top, conformer GD-I
		$C_7H_{16}FO_2P$	(CH$_3$)$_3$CC(H)(CH$_3$)OP(O)(F)CH$_3$	41	738 ± 5	*P*-methyl top, conformer GD-II
98	Germyl azide	GeH_3N_3	GeH$_3$-N=N≡N	42	86.598 ± 0.062	
99	Silylphosphine	H_5PSi	SiH$_3$PH$_2$	43	537.2 ± 14.0	

TABLE 2. Symmetric Top Potential Parameters

	Name	Molecular Formula	Line Formula	Ref.	V_3/cm^{-1}	Comments
1	Phosphine-trifluoroborane	BF_3H_3P	H_3PBF_3	44	1169 ± 123	
2	Trihydro(phosphorus trifluoride)boron	BF_3H_3P	F_3PBH_3	45	1134 ± 53	
3	Trihydro(phosphine)boron	BH_6P	H_3PBH_3	46	864.5 ± 17.5	
4	Trifluoro(trifluoromethyl)silane	CF_6Si	CF_3SiF_3	47	489 ± 50	
5	Trifluoromethylgermane	CH_3F_3Ge	CF_3GeH_3	48	448 ± 53	
6	Trifluoromethylsilane	CH_3F_3Si	CH_3SiF_3	49	414.147 ± 0.030	
7	Methylgermane	CH_6Ge	CH_3GeH_3	50	433.6 ± 8.8	
8	Methylsilane	CH_6Si	CH_3SiH_3	2	603.3878 ± 0.0037	
9	Methylstannane	CH_6Sn	CH_3SnH_3	51	227 ± 10	
10	1,1,1-Trifluoroethane	$C_2H_3F_3$	CH_3CF_3	52	1112.24 ± 0.16	
11	Ethane	C_2H_6	CH_3CH_3	2	1013.28 ± 0.10	$V_6 = 8.798 \pm 0.041$
12	Ethane-1,1,1-d_3	$C_2H_3D_3$	CH_3CD_3	2	1001.876 ± 0.023	$V_6 = 9.328 \pm 0.018$
13	Ethane-d_6	C_2D_6	CD_3CD_3	2	989.946 ± 0.090	$V_6 = 9.51 \pm 0.10$
14	1-Silylpropyne	C_3H_6Si	$CH_3C{\equiv}CSiH_3$	53	3.77 ± 0.70	
15	Trimethylchlorosilane	C_3H_9ClSi	$(CH_3)_3SiCl$	54	576.9 ± 0.9	
16	2-Butyne	C_4H_6	$CH_3C{\equiv}CCH_3$	55	6.067 ± 0.040	$V_6 = 0.1240 \pm 0.0144$ $V_9 = -0.0916 \pm 0.0180$
17	Ethynyltrimethylgermane	$C_5H_{10}Ge$	$(CH_3)_3GeC{\equiv}CH$	56	376.2 ± 16.7	
18	Disilane	H_6Si_2	SiH_3SiH_3	57	412.033 ± 0.010	

BOND DISSOCIATION ENERGIES

Yu-Ran Luo

The bond dissociation energy (enthalpy) is also referred to as bond disruption energy, bond energy, bond strength, or binding energy (abbreviation: BDE, BE, or D). It is defined as the standard enthalpy change of the following fission: R–X → R + X. The BDE, denoted by $D°(R–X)$, is usually derived by the thermochemical equation, $D°(R–X) = \Delta_f H°(R) + \Delta_f H°(X) – \Delta_f H°(RX)$. The enthalpy of formation $\Delta_f H°$ of a large number of atoms, free radicals, ions, clusters and compounds is available from the websites of NIST, NASA, CODATA, and IUPAC. Most authors prefer to use the BDE values at 298.15 K.

The following seven tables provide essential information of experimental BDE values of R–X and R+–X bonds.

(1) Table 1: Bond Dissociation Energies in Diatomic Molecules
(2) Table 2: Enthalpy of Formation of Gaseous Atoms
(3) Table 3: Bond Dissociation Energies in Polyatomic Molecules
(4) Table 4: Enthalpies of Formation of Free Radicals and Other Transient Species
(5) Table 5: Bond Dissociation Energies of Common Organic Molecules
(6) Table 6: Bond Dissociation Energies in Diatomic Cations
(7) Table 7: Bond Dissociation Energies in Polyatomic Cations

The data in these tables have been revised through September 2011.

TABLE 1. Bond Dissociation Energies in Diatomic Molecules

The BDEs in diatomic species have usually been measured by spectroscopy or mass spectrometry. In the absence of data on the enthalpy function, the values at 0 K, $D°(A–B)$, are converted to $D°_{298}$ by the approximate equation:

$$D°_{298}(A–B) \approx D°(A–B) + (3/2)RT = D°(A–B) + 3.7181 \text{ kJ mol}^{-1}$$

This table has been arranged in an alphabetical order of the atoms A in the diatomics A–B.

A–B	$D°_{298}$/kJ mol^{-1}	Ref.	A–B	$D°_{298}$/kJ mol^{-1}	Ref.	A–B	$D°_{298}$/kJ mol^{-1}	Ref.	A–B	$D°_{298}$/kJ mol^{-1}	Ref.
Ac–O	794	1	Al–Al	264.3 ± 0.5	1	Al–U	326 ± 29	1	As–Cl	448	1
Ac–S	505 ± 68	22	Al–Ar	5.69	1	Al–V	147.4 ± 1.0	1	As–D	270.3	1
Ag–Ag	162.9 ± 2.9	1	Al–As	202.7 ± 7.1	1	Al–Xe	7.39	1	As–F	410	1
Ag–Al	183.7 ± 9.2	1	Al–Au	325.9 ± 6.3	1	Am–O	582 ± 34	18	As–Ga	202.5 ± 4.8	1
Ag–Au	202.5 ± 9.6	1	Al–Br	429.2 ± 5.8	1	Am–S	375 ± 33	22	As–H	274.0 ± 2.9	1
Ag–Bi	192 ± 42	1	Al–C	267.7	1	Ar–Ar	4.91	1	As–I	296.6 ± 24	1
Ag–Br	280.3 ± 1.3	1	Al–Ca	52.7	1	Ar–Au	5.50 ± 0.16	19	As–In	201 ± 10	1
Ag–Cl	279.1 ± 8.4	1	Al–Cl	502	1	Ar–B	4.62	1	As–N	489 ± 2.1	1
Ag–Cu	171.5 ± 9.6	1	Al–Co	181.6 ± 0.2	1	Ar–Br	~5.0	1	As–O	484 ± 8	1
Ag–D	226.8	1	Al–Cr	222.9 ± 0.9	1	Ar–C	5.158	1	As–P	433.5 ± 12.6	1
Ag–Dy	130 ± 19	1	Al–Cu	227.1 ± 1.2	1	Ar–Ca	4.44 ± 0.60	1	As–S	379.5 ± 6.3	1
Ag–Eu	127 ± 13	1	Al–D	290.4	1	Ar–Cd	5.57 ± 0.05	1	As–Sb	330.5 ± 5.4	1
Ag–F	356.9 ± 5.8	1	Al–F	675	1	Ar–Ga	3.96	1	As–Se	96	1
Ag–Ga	159 ± 17	1	Al–H	288 ± 13	1	Ar–Ge	<5.4	1	As–Tl	198.3 ± 14.6	1
Ag–Ge	174.5 ± 21	1	Al–I	369.9 ± 2.1	1	Ar–He	3.96	1	Au–Au	226.2 ± 0.5	1
Ag–H	202.4 ± 9.1	1	Al–Kr	6.05	1	Ar–Hg	5.32	1	Au–B	367.8 ± 10.5	1
Ag–Ho	124 ± 19	1	Al–Li	76.1	1	Ar–I	~5.3	1	Au–Ba	254.8 ± 10.0	1
Ag–I	234 ± 29	1	Al–N	≤368 ± 15	1	Ar–In	4.18	1	Au–Be	237.7 ± 4.0	1
Ag–In	166.5 ± 4.9	1	Al–Ne	3.9	1	Ar–Kr	5.11	1	Au–Bi	293 ± 8.4	1
Ag–Li	186.1	1	Al–Ni	224.7 ± 4.8	1	Ar–Li	~7.82	1	Au–Br	213 ± 21	1
Ag–Mn	99.2 ± 21	1	Al–O	501.9 ± 10.6	1	Ar–Mg	~3.7	1	Au–Ca	250.4 ± 4.0	1
Ag–Na	133.1 ± 12.6	1	Al–P	216.7 ± 12.6	1	Ar–Na	~4.2	1	Au–Ce	322 ± 18	1
Ag–Nd	<213	1	Al–Pd	254.4 ± 12.1	1	Ar–Ne	4.27	1	Au–Cl	280 ± 13	1
Ag–O	221 ± 21	1	Al–S	332 ± 10	1	Ar–Si	5.86	1	Au–Co	218.0 ± 16.4	1
Ag–S	216.7 ± 14.6	1	Al–Sb	216.3 ± 6	1	Ar–Sn	<5.1	1	Au–Cr	223.7 ± 28.9	1
Ag–Se	210.0 ± 14.6	1	Al–Se	318 ± 13	1	Ar–Tl	4.09	1	Au–Cs	253 ± 3.5	1
Ag–Si	185.1 ± 9.6	1	Al–Si	246.9 ± 12.6	1	Ar–Xe	5.28	1	Au–Cu	227.1 ± 1.2	1
Ag–Sn	136 ± 21	1	Al–Te	268 ± 13	1	Ar–Zn	5.0	1	Au–D	322.2	1
Ag–Te	195.8 ± 14.6	1	Al–Ti	263.4	1	As–As	385.8 ± 10.5	1	Au–Dy	259 ± 24	1

A–B	D°_{298}/kJ mol^{-1}	Ref.	A–B	D°_{298}/kJ mol^{-1}	Ref.	A–B	D°_{298}/kJ mol^{-1}	Ref.	A–B	D°_{298}/kJ mol^{-1}	Ref.
Au–Eu	245 ± 12	1	B–Ru	446.9 ± 21	1	Br–Cl	219.32 ± 0.05	1	Br–Yb	297.7	23
Au–F	294.1	1	B–S	577 ± 9.2	1	Br–Co	326 ± 42	1	Br–Zn	138 ± 29	1
Au–Fe	187.0 ± 19.3	1	B–Sc	272 ± 63	1	Br–Cr	328.0 ± 24.3	1	Br–Zr	420	1
Au–Ga	290 ± 15	1	B–Se	462 ± 14.6	1	Br–Cs	389.1 ± 4.2	1	C–C	618.3 ± 15.4	1
Au–Ge	273.2 ± 14.6	1	B–Si	317 ± 12	1	Br–Cu	331 ± 25	1	C–Ce	443 ± 30	1
Au–H	300.5 ± 2.6	4	B–Te	354 ± 20	1	Br–D	370.74	1	C–Cl	394.9 ± 13.4	1
Au–Ho	267 ± 35	1	B–Th	297 ± 33	1	Br–Dy	315.7	23	C–D	341.4	1
Au–I	276	1	B–Ti	272 ± 63	1	Br–Er	363.2	23	C–F	513.8 ± 10.0	1
Au–In	286.0 ± 5.7	1	B–U	322 ± 33	1	Br–Eu	328.8	23	C–Fe	367.7 ± 4.2	20
Au–Kr	6.59 ± 0.23	19	B–Y	289 ± 63	1	Br–F	280 ± 12	1	C–Ge	455.7 ± 11	1
Au–La	457 ± 28	1	Ba–Br	359.9	23	Br–Fe	243 ± 84	1	C–H	338.4 ± 1.2	1
Au–Li	284.5 ± 6.7	1	Ba–Cl	439.3	23	Br–Ga	402 ± 13	1	C–Hf	540 ± 25	1
Au–Lu	332 ± 19	1	Ba–D	≤193.7	1	Br–Gd	374.5	23	C–I	253.1 ± 35.6	1
Au–Mg	179.1 ± 2.7	1	Ba–F	580.0	23	Br–Ge	347 ± 8	1	C–Ir	631 ± 5	1
Au–Mn	197.7 ± 21	1	Ba–H	192.0	1	Br–H	366.16 ± 0.20	1	C–La	463 ± 20	1
Au–Na	215.1 ± 12.6	1	Ba–I	321.0	23	Br–Hg	74.9	1	C–Mo	482 ± 16	1
Au–Nd	294 ± 29	1	Ba–O	562 ± 13.4	1	Br–Ho	323.9	23	C–N	750.0 ± 2.9	1
Au–Ni	247 ± 16.4	1	Ba–Pd	221.8 ± 5.0	1	Br–I	179.1 ± 0.4	1	C–Nb	523.8 ± 14.5	1
Au–O	223 ± 21	1	Ba–Rh	259.4 ± 25	1	Br–In	409 ± 10	1	C–Ni	337.0	1
Au–Pb	133 ± 42	1	Ba–S	418 ± 21	1	Br–K	379.1 ± 4.2	1	C–O	1076.38 ± 0.67	1
Au–Pd	142.7 ± 21	1	Be–Be	59	1	Br–La	448.6	23	C–Os	608 ± 25	1
Au–Pr	311 ± 25	1	Be–Br	316	1	Br–Li	418.8 ± 4.2	1	C–P	507.5 ± 8.8	1
Au–Rb	243 ± 3.5	1	Be–Cl	384	1	Br–Lu	303.3	23	C–Pd	436 ± 20	1
Au–Rh	232.6 ± 29	1	Be–D	203.1	1	Br–Mg	317.96	1	C–Pt	577.8 ± 6.8	13
Au–S	253.6 ± 14.6	1	Be–F	573	1	Br–Mn	314.2 ± 9.6	1	C–Rh	580 ± 4	1
Au–Sc	280 ± 40	1	Be–H	221	1	Br–Mo	313.4	1	C–Ru	648 ± 13	1
Au–Se	251.0 ± 14.6	1	Be–I	261	1	Br–N	280.8 ± 21	1	C–S	713.3 ± 1.2	1
Au–Si	304.6 ± 6.0	1	Be–O	437	1	Br–Na	363.1 ± 4.2	1	C–Sc	444 ± 21	1
Au–Sn	256.5 ± 7.2	1	Be–S	372 ± 59	1	Br–Nd	341.8	23	C–Se	590.4 ± 5.9	1
Au–Sr	264 ± 42	1	Be–T	204.4	1	Br–Ni	360 ± 13	1	C–Si	447	1
Au–Tb	285 ± 33	1	Bi–Bi	204.4	1	Br–O	237.6 ± 0.4	1	C–Tc	564 ± 29	1
Au–Te	237.2 ± 14.6	1	Bi–Br	240.2	1	Br–P	≤329	1	C–Th	453 ± 17	1
Au–U	318 ± 29	1	Bi–Cl	300.4 ± 4.2	1	Br–Pb	248.5 ± 14.6	1	C–Ti	423 ± 30	1
Au–V	246.0 ± 8.7	1	Bi–D	283.7	1	Br–Pm	337.6	23	C–U	455 ± 15	1
Au–Xe	11.33 ± 0.23	19	Bi–F	366.5 ± 12.5	1	Br–Pr	346.3	23	C–V	423 ± 24	1
Au–Y	310 ± 12	1	Bi–Ga	158.6 ± 16.7	1	Br–Rb	380.7 ± 4.2	1	C–Y	418 ± 14	1
B–B	290	1	Bi–H	≤283.3	1	Br–S	218 ± 17	1	C–Zr	495.8 ± 38.6	1
B–Br	390.9 ± 0.5	1	Bi–I	186.1 ± 5.8	1	Br–Sb	314 ± 59	1	Ca–Ca	16.52 ± 0.11	1
B–C	448 ± 29	1	Bi–In	153.6 ± 1.7	1	Br–Sc	444 ± 63	1	Ca–Cl	409 ± 8.7	1
B–Cd	301.0	1	Bi–Li	149.4	1	Br–Se	297 ± 84	1	Ca–D	≤169.9	1
B–Ce	305 ± 21	1	Bi–O	337.2 ± 12.6	1	Br–Si	358.2 ± 8.4	1	Ca–F	529	1
B–Cl	427	1	Bi–P	281.7 ± 13	1	Br–Sm	334.3	23	Ca–H	223.8	1
B–D	341.0 ± 6.3	1	Bi–Pb	142.4 ± 3.0	1	Br–Sn	337 ± 13	1	Ca–I	284.7 ± 8.4	1
B–F	732	1	Bi–S	315.5 ± 4.6	1	Br–Sr	365	1	Ca–Kr	5.15 ± 0.72	1
B–H	345.2 ± 2.5	1	Bi–Sb	252.7 ± 3.9	1	Br–T	372.77	1	Ca–Li	84.9 ± 8.4	1
B–I	361	1	Bi–Se	280.3 ± 5.9	1	Br–Tb	386.4	23	Ca–O	383.3 ± 5.0	1
B–Ir	512.2 ± 17	1	Bi–Sn	193 ± 13	1	Br–Th	364	1	Ca–Pd	347 - 360	1
B–La	335 ± 63	1	Bi–Te	232.2 ± 11.3	1	Br–Ti	373	1	Ca–S	335 ± 21	1
B–N	377.9 ± 8.7	1	Bi–Tl	120.9 ± 12.6	1	Br–Tl	331 ± 21	1	Ca–Xe	7.31 ± 0.96	1
B–Ne	3.97		Bk–O	598	1	Br–Tm	300.9	23	Cd–Cd	7.36	1
B–O	809	1	Br–Br	193.859 ± 0.120	1	Br–U	377 ± 15	1	Cd–Cl	208.4	1
B–P	347 ± 16.7	1	Br–C	318.0 ± 8.4	1	Br–V	439 ± 42	1	Cd–F	305 ± 21	1
B–Pd	351.5 ± 16.7	1	Br–Ca	339	1	Br–W	329.3	1	Cd–H	69.0 ± 0.4	1
B–Pt	477.8 ± 16.7	1	Br–Cd	159 ± 96	1	Br–Xe	5.94 ± 0.02	1	Cd–I	97.2 ± 2.1	1
B–Rh	475.8 ± 21	1	Br–Ce	375.2	23	Br–Y	481 ± 84	1	Cd–In	134	1

A–B	D°_{298}/kJ mol^{-1}	Ref.	A–B	D°_{298}/kJ mol^{-1}	Ref.	A–B	D°_{298}/kJ mol^{-1}	Ref.	A–B	D°_{298}/kJ mol^{-1}	Ref.
Cd–K	7.3	1	Cl–P	≤376	1	Cr–Nb	295.72 ± 0.06	1	D–T	444.91	1
Cd–Kr	5.17	1	Cl–Pb	301 ± 50	1	Cr–O	461 ± 8.7	1	D–Tl	193.0	1
Cd–Na	10.2	1	Cl–Pm	417.0	23	Cr–Pb	105 ± 2	1	D–Zn	88.7	1
Cd–Ne	3.97	1	Cl–Pr	425.7	23	Cr–S	331	1	Dy–Dy	70.3	1
Cd–O	236 ± 84	1	Cl–Ra	343 ± 75	1	Cr–Sn	141 ± 3	1	Dy–F	531.1	23
Cd–S	208.5 ± 20.9	1	Cl–Rb	427.6 ± 8.4	1	Cs–Cs	43.919 ± 0.010	1	Dy–I	277.2	23
Cd–Se	127.6 ± 25.1	1	Cl–S	241.8	1	Cs–F	517.1 ± 7.7	1	Dy–O	615	1
Cd–Te	100.0 ± 15.1	1	Cl–Sb	360 ± 50	1	Cs–H	175.364	1	Dy–S	414 ± 42	1
Cd–Xe	6.54	1	Cl–Sc	331	1	Cs–Hg	8	1	Dy–Se	322 ± 20	1
Ce–Ce	251.7	1	Cl–Se	322	1	Cs–I	338.5 ± 2.1	1	Dy–Te	234 ± 20	1
Ce–Cl	460.0	23	Cl–Si	416.7 ± 6.3	1	Cs–Li	72.9 ± 1.2	5	Er–Er	75 ± 29	1
Ce–F	621.6	23	Cl–Sm	422.1	23	Cs–Na	63.2 ± 1.3	1	Er–F	572.6	23
Ce–I	335.5	23	Cl–Sn	350 ± 8	1	Cs–O	293 ± 25	1	Er–I	317.6	23
Ce–Ir	575 ± 9	1	Cl–Sr	409	1	Cs–Rb	49.57 ± 0.01	1	Er–O	606	1
Ce–N	519 ± 21	1	Cl–T	438.64	1	Cu–Cu	201	1	Er–S	418 ± 21	1
Ce–O	790	1	Cl–Ta	544	1	Cu–D	270.3	1	Er–Se	326 ± 20	1
Ce–Os	524 ± 20	1	Cl–Tb	474.2	23	Cu–Dy	144 ± 18	1	Er–Te	238 ± 20	1
Ce–Pd	319 ± 21	1	Cl–Th	489	1	Cu–F	414	1	Es–O	460	1
Ce–Pt	550 ± 5	1	Cl–Ti	405.4 ± 10.5	1	Cu–Ga	215.9 ± 15	1	Eu–Eu	45.2	1
Ce–Rh	545 ± 7	1	Cl–Tl	372.8 ± 2.1	1	Cu–Ge	208.8 ± 21	1	Eu–F	543.0	23
Ce–Ru	494 ± 12	1	Cl–Tm	380.3	23	Cu–H	254.8 ± 6	1	Eu–I	290.4	23
Ce–S	569	1	Cl–U	439	1	Cu–Ho	144 ± 19	1	Eu–Li	268.1 ± 12.6	1
Ce–Se	494.5 ± 14.6	1	Cl–V	477 ± 63	1	Cu–I	289 ± 63	1	Eu–O	473	1
Ce–Te	189.4 ± 12.6	1	Cl–W	419	1	Cu–In	187.4 ± 7.9	1	Eu–Rh	238 ± 34	1
Cf–O	498	1	Cl–Xe	7.08	1	Cu–Li	191.9	1	Eu–S	365.7 ± 13.4	1
Cl–Cl	242.851 ± 0.096	8	Cl–Y	523 ± 84	1	Cu–Na	176.1 ± 16.7	1	Eu–Se	302.9 ± 14.6	1
Cl–Co	343.9	11	Cl–Yb	377.1	23	Cu–Ni	201.7 ± 9.6	1	Eu–Te	251.0 ± 14.6	1
Cl–Cr	380.3	11	Cl–Zn	229 ± 8	1	Cu–O	287.4 ± 11.6	1	F–F	158.670 ± 0.096	1
Cl–Cs	445.7 ± 7.7	1	Cl–Zr	530	1	Cu–S	274.5 ± 14.6	1	F–Fe	447	1
Cl–Cu	377.8 ± 7.5	1	Cm–O	709 ± 43	18	Cu–Se	255.2 ± 14.6	1	F–Ga	584 ± 13	1
Cl–D	436.303 ± 0.011	1	Cm–S	504 ± 25	22	Cu–Si	221.3 ± 6.3	1	F–Gd	594.6	23
Cl–Dy	395.1	23	Co–Co	<127	1	Cu–Sn	170 ± 10	1	F–Ge	523 ± 13	1
Cl–Er	451.0	23	Co–Cu	161.1 ± 16.4	1	Cu–Tb	191 ± 18	1	F–H	569.680 ± 0.011	1
Cl–Eu	408.4	23	Co–D	270.2 ± 5.8	1	Cu–Te	230.5 ± 14.6	1	F–Hf	650 ± 15	1
Cl–F	260.83	1	Co–F	431 ± 63	1	D–D	443.3197 ± 0.0003	1	F–Hg	~180	1
Cl–Fe	335.5	11	Co–Ge	230 ± 21	1	D–F	576.236 ± 0.011	1	F–Ho	517.7	23
Cl–Ga	463 ± 13	1	Co–H	244.9 ± 4.8	1	D–Ga	<276.5	1	F–I	≤271.5	1
Cl–Gd	453.9	23	Co–I	280 ± 21	1	D–Ge	≤322	1	F–In	516 ± 13	1
Cl–Ge	390.8 ± 9.6	1	Co–Mn	50 ± 8	1	D–H	439.2223 ± 0.0002	1	F–K	489.2	1
Cl–H	431.361 ± 0.013	1	Co–Nb	267.02 ± 0.10	1	D–Hg	42.05	1	F–Kr	6.6	1
Cl–Hg	92.0 ± 9.2	1	Co–O	397.4 ± 8.7	1	D–I	302.33	1	F–La	665.1	23
Cl–Ho	411.6	23	Co–S	331	1	D–In	246	1	F–Li	577 ± 21	1
Cl–I	211.3 ± 0.4	1	Co–Sc	240.1	7	D–K	182.4	1	F–Lu	523.4	23
Cl–In	436 ± 8	1	Co–Si	274.4 ± 17	1	D–Li	240.24	1	F–Mg	445.6	1
Cl–K	433.0 ± 8.4	1	Co–Ti	235.37 ± 0.10	1	D–Lu	302	1	F–Mn	445.2 ± 7.5	1
Cl–La	524.4	23	Co–Y	253.71 ± 0.10	1	D–Mg	161.33 ± 0.32	1	F–Mo	464	1
Cl–Li	469 ± 13	1	Co–Zr	306.39 ± 0.10	1	D–Mn	312 ± 6	1	F–N	≤349	1
Cl–Lu	383.3	23	Cr–Cr	152.0 ± 6	1	D–N	341.6	1	F–Na	477.3	1
Cl–Mg	312	1	Cr–Cu	154.4 ± 14.5	1	D–Ni	≤302.9	1	F–Nd	548.7	23
Cl–Mn	337.6	11	Cr–F	523 ± 19	1	D–O	429.64	1	F–Ni	439.7 ± 5.9	2
Cl–N	333.9 ± 9.6	1	Cr–Fe	~75	1	D–P	299.0	1	F–Np	430 ± 50	1
Cl–Na	412.1 ± 8.4	1	Cr–Ge	154 ± 7	1	D–Pt	≤350.2	1	F–O	220	1
Cl–Nd	421.1	23	Cr–H	189.9 ± 6.7	1	D–S	350.62 ± 1.20	1	F–P	≤405	1
Cl–Ni	372.3	11	Cr–I	287.0 ± 24.3	1	D–Si	302.5	1	F–Pb	355 ± 13	1
Cl–O	267.47 ± 0.08	1	Cr–N	377.8 ± 18.8	1	D–Sr	167.7	1	F–Pm	561.3	23

A–B	D^o_{298}/kJ mol⁻¹	Ref.	A–B	D^o_{298}/kJ mol⁻¹	Ref.	A–B	D^o_{298}/kJ mol⁻¹	Ref.	A–B	D^o_{298}/kJ mol⁻¹	Ref.
F–Pr	582.0	23	Ge–Ni	290.3 ± 10.9	1	Hg–Kr	5.75	1	In–Kr	4.85	1
F–Pu	538 ± 29	1	Ge–O	657.5 ± 4.6	4	Hg–Li	13.16 ± 0.38	1	In–Li	92.5 ± 14.6	1
F–Rb	494 ± 21	1	Ge–Pb	145.3 ± 6.9	6	Hg–Na	10.8	1	In–O	346 ± 30	1
F–Ru	402	1	Ge–Pd	254.7 ± 10.5	1	Hg–Ne	4.14	1	In–P	197.9 ± 8.4	1
F–S	343.5 ± 6.7	1	Ge–S	534 ± 3	1	Hg–O	269	1	In–S	287.9 ± 14.6	1
F–Sb	439 ± 96	1	Ge–Sc	270 ± 11	1	Hg–Rb	8.4	1	In–Sb	151.9 ± 10.5	1
F–Sc	599.1 ± 13.4	1	Ge–Se	484.7 ± 1.7	1	Hg–S	217.3 ± 22.2	1	In–Se	245.2 ± 14.6	1
F–Se	339 ± 42	1	Ge–Si	297	1	Hg–Se	144.3 ± 30.1	1	In–Te	215.5 ± 14.6	1
F–Si	576.4 ± 17	1	Ge–Sn	230.1 ± 13	1	Hg–T	43.14	1	In–Xe	6.48	1
F–Sm	565.2	23	Ge–Te	396.7 ± 3.3	1	Hg–Te	<142	1	In–Zn	32.2	1
F–Sn	476 ± 8	1	Ge–Y	279 ± 11	1	Hg–Tl	2.9	1	Ir–Ir	361 ± 68	1
F–Sr	538	1	H–H	435.7799 ± 0.0001	1	Hg–Xe	6.65	1	Ir–La	577 ± 12	1
F–T	579.009 ± 0.108	1	H–Hg	39.844	1	Hg–Zn	7.3	1	Ir–Nb	465 ± 25	1
F–Ta	573 ± 13	1	H–I	298.26 ± 0.10	1	Ho–Ho	70.3	1	Ir–O	414 ± 42	1
F–Tb	647.3	23	H–In	243.1	1	Ho–I	277.0	23	Ir–Si	462.8 ± 21	1
F–Th	652	1	H–K	174.576	1	Ho–O	606	1	Ir–Th	574 ± 42	1
F–Ti	569 ± 33	1	H–Li	238.039 ± 0.006	1	Ho–S	428.4 ± 14.6	1	Ir–Ti	422 ± 13	1
F–Tl	439 ± 21	1	H–Mg	127.18 ± 0.006	10	Ho–Se	333 ± 15	1	Ir–Y	457 ± 15	1
F–Tm	509.1	23	H–Mn	251 ± 5	1	Ho–Te	≤259 ± 15	1	K–K	56.96	1
F–U	648	1	H–Mo	202.5 ± 18.3	9	I–I	152.25 ± 0.57	1	K–Kr	4.6	1
F–V	590 ± 63	1	H–N	≤338.9	1	I–In	306.9 ± 1.1	1	K–Li	82.0 ± 4.2	1
F–W	≤544	1	H–Na	182.24 ± 0.06	21	I–K	322.5 ± 2.1	1	K–Na	65.994 ± 0.008	1
F–Xe	14.18	1	H–Nb	>221.9 ± 9.6	1	I–Kr	5.67	1	K–Zn	6.5	1
F–Y	685.3 ± 13.4	1	H–Ni	240 ± 8	1	I–La	414.8	23	K–O	271.5 ± 12.6	1
F–Yb	525.1	23	H–O	429.91 ± 0.29	1	I–Li	345.2 ± 4.2	1	K–Rb	53.723 ± 0.005	1
F–Zn	364 ± 63	1	H–P	297.0 ± 2.1	1	I–Lu	264.8	23	K–Xe	5.0	1
F–Zr	627.2 ± 10.5	1	H–Pb	≤157	1	I–Mg	229	1	Kr–Kr	5.39	1
Fe–Fe	118	1	H–Pd	234 ± 25	1	I–Mn	282.8 ± 9.6	1	Kr–Li	~12.1	1
Fe–Ge	210.9 ± 29	1	H–Pt	330	1	I–Mo	266.9	1	Kr–Mg	6.71 ± 0.96	1
Fe–H	148 ± 3	1	H–Rb	172.6	1	I–N	159 ± 17	1	Kr–Na	~4.53	1
Fe–I	123	1	H–Rh	241.0 ± 5.9	1	I–Na	304.2 ± 2.1	1	Kr–Ne	4.31	1
Fe–O	407.0 ± 1.0	1	H–Ru	223 ± 15	1	I–Nd	303.3	23	Kr–O	<8	1
Fe–S	328.9 ± 14.6	1	H–S	353.57 ± 0.30	1	I–Ni	293 ± 21	1	Kr–Tl	4.14	1
Fe–Si	297 ± 25	1	H–Sb	239.7 ± 4.2	1	I–O	233.4 ± 1.3	12	Kr–Xe	5.66	1
Fm–O	443	1	H–Sc	205 ± 17	1	I–Pb	194 ± 38	1	Kr–Zn	5.0	1
Ga–Ga	<106.4	1	H–Se	312.5	1	I–Pm	299.1	23	La–La	244.9	1
Ga–H	265.9 ± 5.9	4	H–Si	293.3 ± 1.9	1	I–Pr	307.8	23	La–N	519 ± 42	1
Ga–I	334 ± 13	1	H–Sn	264 ± 17	1	I–Rb	318.8 ± 2.1	1	La–O	798	1
Ga–In	94.0 ± 3	1	H–Sr	164 ± 8	1	I–Si	243.1 ± 8.4	1	La–Pt	505 ± 12	1
Ga–Kr	4.08	1	H–T	440.49	1	I–Sm	295.8	23	La–Rh	550 ± 12	1
Ga–Li	133.1 ± 14.6	1	H–Te	270.7 ± 1.7	1	I–Sn	235 ± 3	1	La–S	573.4 ± 1.7	1
Ga–O	374 ± 21	1	H–Ti	204.6 ± 8.8	1	I–Sr	301	1	La–Se	485.7 ± 14.6	1
Ga–P	229.7 ± 12.6	1	H–Tl	195.4 ± 4	1	I–Tb	339.6	23	La–Te	385.6 ± 15	1
Ga–Sb	192.0 ± 12.6	1	H–V	209.3 ± 6.8	1	I–Te	192 ± 42	1	La–Y	197 ± 21	1
Ga–Te	265 ± 21	1	H–Yb	183.1 ± 2.0	1	I–Th	361 ± 25	1	Li–Li	105.0	1
Ga–Xe	5.27	1	H–Zn	85.8 ± 2	1	I–Ti	306	1	Li–Mg	67.4 ± 6.3	1
Gd–Gd	206.3 ± 67.5	1	He–He	3.809	1	I–Tl	285 ± 21	1	Li–Na	87.181 ± 0.001	1
Gd–I	336.0	23	He–Hg	3.8	1	I–Tm	262.4	23	Li–O	340.5 ± 6.3	1
Gd–O	715	1	He–Xe	3.8	1	I–U	299 ± 27	1	Li–Pb	78.7 ± 8	1
Gd–S	526.8 ± 10.5	1	Hf–Hf	328 ± 58	1	I–Xe	~6.9	1	Li–S	312.5 ± 7.5	1
Gd–Se	430 ± 15	1	Hf–N	535 ± 30	1	I–Y	422.6 ± 12.5	1	Li–Sb	169.0 ± 10.0	1
Gd–Te	341 ± 15	1	Hf–O	801 ± 13	1	I–Yb	259.3	23	Li–Si	149	1
Ge–Ge	264.4 ± 6.8	1	Hg–Hg	8.10 ± 0.18	1	I–Zn	153.1 ± 6.3	1	Li–Sm	193.3 ± 18.8	1
Ge–H	263.2 ± 4.8	1	Hg–I	34.69 ± 0.96	1	I–Zr	127	1	Li–Tm	276.1 ± 14.6	1
Ge–I	268 ± 25	1	Hg–K	8.8	1	In–In	82.0 ± 5.7	1	Li–Xe	~12.1	1

A–B	D°_{298}/kJ mol^{-1}	Ref.	A–B	D°_{298}/kJ mol^{-1}	Ref.	A–B	D°_{298}/kJ mol^{-1}	Ref.	A–B	D°_{298}/kJ mol^{-1}	Ref.
Li–Yb	143.5 ± 12.6	1	Ne–Zn	3.92	1	P–Sb	356.9 ± 4.2	1	S–Th	608 ± 77	22
Lr–O	665	1	Ni–Ni	204		P–Se	363.7 ± 10.0	1	S–Ti	418 ± 3	1
Lu–Lu	142 ± 33	1	Ni–O	366 ± 30	1	P–Si	363.6		S–Tm	368 ± 21	1
Lu–O	669	1	Ni–Pd	140.9	1	P–Te	297.9 ± 10.0	1	S–U	510.4 ± 63	22
Lu–Pt	402 ± 34	1	Ni–Pt	273.7 ± 0.3	1	P–Th	372 ± 29	1	S–V	449.4 ± 14.6	1
Lu–S	508.4 ± 14.4	1	Ni–S	356 ± 21	1	P–Tl	209 ± 13	1	S–Y	528.4 ± 10.5	1
Lu–Se	418 ± 15	1	Ni–Si	318 ± 17	1	P–U	293 ± 21	1	S–Yb	167	1
Lu–Te	325 ± 15	1	Ni–V	206.3 ± 0.2	1	P–W	305 ± 4	1	S–Zn	224.8 ± 12.6	1
Md–O	418	1	Ni–Y	283.92 ± 0.10	1	Pa–S	545 ± 91	22	S–Zr	572.0 ± 11.6	1
Mg–Mg	11.3	1	Ni–Zr	279.8 ± 0.1	1	Pb–Pb	86.6 ± 0.8	1	Sb–Sb	301.7 ± 6.3	1
Mg–Ne	~4.1	1	No–O	268	1	Pb–S	398		Sb–Te	277.4 ± 3.8	1
Mg–O	358.2 ± 7.2	1	Np–O	744 ± 21	18	Pb–Sb	161.5 ± 10.5	1	Sb–Tl	126.7 ± 10.5	1
Mg–S	234	1	Np–S	495 ± 55	22	Pb–Se	302.9 ± 4.2	1	Sc–Sc	163 ± 21	1
Mg–Xe	9.70 ± 1.79	1	O–O	498.36 ± 0.17	1	Pb–Si	168.8 ± 7.3	6	Sc–Se	385 ± 17	1
Mn–Mn	61.6 ± 9.6	1	O–Os	575	1	Pb–Sn	126.3 ± 4.0	1	Sc–Si	227.2 ± 14	1
Mn–O	362 ± 25	1	O–P	589	1	Pb–Te	249.8 ± 10.5	1	Sc–Te	289 ± 17	1
Mn–S	301 ± 17	1	O–Pa	801 ± 59	18	Pd–Pd	>136		Se–Se	330.5	
Mn–Se	239.3 ± 9.2	1	O–Pb	382.4 ± 3.3	4	Pd–Pt	191.0		Se–Si	538 ± 13	1
Mo–Mo	435.5 ± 1.0	1	O–Pd	238.1 ± 12.6	1	Pd–Si	261 ± 12	1	Se–Sm	331.0 ± 14.6	1
Mo–Nb	452 ± 25	1	O–Pr	740	1	Pd–Y	241 ± 15	1	Se–Sn	401.2 ± 5.9	1
Mo–O	502		O–Pt	418.6 ± 11.6	13	Po–Po	187	1	Se–Sr	251.0 ± 12.6	1
N–N	944.84 ± 0.10	1	O–Pu	656.1	1	Pr–Pr	129.1	1	Se–Tb	423 ± 20	1
N–O	631.62 ± 0.18	1	O–Rb	276 ± 12.6	1	Pr–S	492.5 ± 4.6	1	Se–Te	293.3	
N–P	617.1 ± 20.9	1	O–Re	627 ± 84	1	Pr–Se	446.4 ± 23.0	1	Se–Ti	381 ± 42	1
N–Pt	374.2 ± 9.6	1	O–Rh	405 ± 42	1	Pr–Te	326 ± 20	1	Se–Tm	274 ± 40	1
N–Pu	469 ± 63	1	O–Ru	528 ± 42	1	Pt–Pt	306.7 ± 1.9	1	Se–V	347 ± 21	1
N–S	467 ± 24	1	O–S	517.90 ± 0.05	1	Pt–Si	501 ± 18	1	Se–Y	435 ± 13	1
N–Sb	460 ± 84	1	O–Sb	434 ± 42	1	Pt–Th	551 ± 42	1	Se–Zn	170.7 ± 25.9	1
N–Sc	464 ± 84	1	O–Sc	671.4 ± 1.0	1	Pt–Ti	397.5 ± 10.6	1	Si–Si	310	1
N–Si	437.1 ± 9.9	1	O–Se	429.7 ± 6.3	1	Pt–Y	474 ± 12	1	Si–Sn	242.1 ± 11.8	16
N–Ta	607 ± 84	1	O–Si	799.6 ± 13.4	1	Pu–S	446 ± 30	22	Si–Te	429.2	3
N–Th	577 ± 33	1	O–Sm	573		Rb–Rb	48.898 ± 0.005	1	Si–Y	258 ± 17	1
N–Ti	476 ± 33	1	O–Sn	528		Re–Re	432 ± 30	1	Sm–Sm	54 ± 21	1
N–U	531 ± 21	1	O–Sr	426.3 ± 6.3	1	Rh–Rh	235.85 ± 0.05	1	Sm–Te	272.4 ± 14.6	1
N–V	523 ± 38	1	O–Ta	839	1	Rh–Sc	444 ± 11	1	Sn–Sn	187.1 ± 0.3	1
N–Xe	26.9	1	O–Tb	694	1	Rh–Si	395.0 ± 18.0	1	Sn–Te	338.1 ± 6.3	1
N–Y	477 ± 63	1	O–Tc	548	1	Rh–Th	513 ± 21	1	Sr–Sr	16.64 ± 1.12	1
N–Zr	565 ± 25	1	O–Te	377 ± 21	1	Rh–Ti	390.8 ± 14.6	1	T–T	446.67	
Na–Na	74.805 ± 0.586	1	O–Th	871 ± 25	18	Rh–U	519 ± 17	1	Ta–Ta	390 ± 96	1
Na–Ne	~3.8	1	O–Ti	666.5 ± 5.6	1	Rh–V	364 ± 29	1	Tb–Tb	138.8	
Na–O	270 ± 4	1	O–Tl	213 ± 84	1	Rh–Y	446 ± 11	1	Tb–Te	339 ± 42	1
Na–Rb	63.887 ± 0.024	1	O–Tm	514	1	Ru–Ru	193.0 ± 19.3	1	Tc–Tc	330	
Na–Xe	~5.12	1	O–U	758 ± 13	18	Ru–Si	397.1 ± 21	1	Te–Te	257.6 ± 4.1	1
Nb–Nb	513		O–V	637	1	Ru–Th	592 ± 42	1	Te–Ti	289 ± 17	1
Nb–Ni	271.9 ± 0.1	1	O–W	720 ± 71	1	Ru–V	414 ± 29	1	Te–Tm	182 ± 40	1
Nb–O	726.5 ± 10.6	1	O–Xe	36.4	1				Te–Y	339 ± 13	1
Nb–Ti	302.0 ± 0.1	1	O–Y	714.1 ± 10.2	1	S–S	430.03 ± 0.03	17	Te–Zn	117.6 ± 18.0	1
Nb–V	369.3 ± 0.1	1	O–Yb	387.7 ± 10	1	S–Sb	378.7		Th–Th	≤289 ± 33	1
Nd–Nd	82.8	1	O–Zn	≤250	1	S–Sc	478.2 ± 12.6	1	Ti–Ti	117.6	
Nd–O	703	1	O–Zr	766.1 ± 10.6	1	S–Se	371.1 ± 6.7	1	Ti–V	203.2 ± 0.1	1
Nd–S	471.5 ± 14.6	1	Os–Os	415 ± 77	1	S–Si	617 ± 5	1	Ti–Zr	214.3 ± 0.1	1
Nd–Se	393.9		P–P	489.1		S–Sm	389		Tl–Tl	59.4	
Nd–Te	305 ± 15	1	P–Pt	≤416.7 ± 16.7	1	S–Sn	467		Tl–Xe	4.18	1
Ne–Ne	4.070	1	P–Rh	353.1 ± 16.7	1	S–Sr	338.5 ± 16.7	1	Tm–Tm	54 ± 17	1
Ne–Xe	4.31	1	P–S	442 ± 10	1	S–Ta	669.5 ± 13.5	1	U–U	222 ± 21	1
						S–Tb	515 ± 42	1	V–V	269.3 ± 0.1	1
						S–Te	335 ± 42	1	V–Zr	260.6 ± 0.3	1
									W–W	666	
									Xe–Xe	6.023	1
									Y–Y	~270 ± 39	1
									Yb–Yb	16.3	1
									Zn–Zn	22.2 ± 6.3	1
									Zr–Zr	298.2 ± 0.1	1

References

1. Luo, Y. R. *Comprehensive Handbook of Chemical Bond Energies*, CRC Press, Boca Raton, FL, 2007.
2. Hildenbrand, D. L., and Lau, K.H., *J. Phys. Chem. A* 110, 11886, 2006.
3. Chattopadhyaya, S., Pramanik, A., Banerjee, A., and Das, K. K., *J. Phys. Chem. A* 110, 12303, 2006.
4. Brutti, S., Balducci, G., and Gigli, G., *Rapid Commun. Mass Spectrom.* 21, 89, 2007.
5. Staanum, P., Pashov, A., Knöckel, H., and Tiemann, E., *Phys. Rev. A* 75, 042513, 2007.
6. Ciccioli, A., Gigli, G., Meloni, G., and Testani, E., *J. Chem. Phys.* 127, 054303/1, 2007.
7. Nagarajan, R., and Morse, M. D. *J. Chem. Phys.* 127, 074304/1, 2007.
8. Li, J., Hao, Y., Yang, J., Zhou, C., and Mo, Y., *J. Chem. Phys.* 127, 104307/1, 2007.
9. Armentrout, P. B., *Organometallics* 26, 5473, 2007.
10. Shayesteh, A., Henderson, R. D. E., Le Roy, R. J., and Bernath, P. F., *J. Phys. Chem. A* 111, 12495, 2007.
11. Hildenbrand, D. H., *J. Phys. Chem. A* 112, 3813, 2008.
12. Dooley, K. S., Geidosch, J. N., and North, S. W., *Chem. Phys. Lett.* 457, 303, 2008.
13. Citir, M., Metz, R. B., Belau, L., and Ahmed, M., *J. Phys. Chem. A* 112, 9584, 2008.
14. Hildenbrand, D. L., Lau, K. H., Perez-Mariano, J., and Sanjurjo, A., *J. Phys. Chem. A* 112, 9978, 2008.
15. Gibson, J. K., Haire, R. G., Santos, M., Pires de Matos, A., and Marçalo, J., *J. Phys. Chem. A* 112, 11373, 2008.
16. Ciccioli, A., Gigli, G., and Meloni, G., *Chem. Eur. J.* 15, 9543, 2009.
17. Frederix, P. W. J. M., Yang, C.-H., Groenenboom, G. C., Parker, D. H., Alnama, K., Western, C. M., and Orr-Ewing, A. J., *J. Phys. Chem. A* 113, 14995, 2009.
18. Marçalo, J., and Gibson, J. K., *J. Phys. Chem. A* 113, 12599, 2009.
19. Hopkins, W.S., Woodham, A. P., Plowright, R. J., Wright, T. G., and Mackenzie, S. R., *J. Chem. Phys.*, 132, 21403, 2010.
20. Tzeli, D., and Mavridis, A., *J. Chem. Phys.*, 132, 194312, 2010.
21. Huang, H.-Y., Lu, T.-L., Whang, T.-J., Chang, Y.-Y., and Tsai, C.-C., *J. Chem. Phys.* 133, 044301, 2010.
22. Pereira, C. C. L., Marsden, C. J., Marçalo, J., and Gibson, J. K., *Phys. Chem. Chem. Phys.* 13, 12940, 2011.
23. Mucklejohn, S. A., *J. Phys. D: Appl. Phys.* 44, 224010, 2011.

TABLE 2. Enthalpy of Formation of Gaseous Atoms

Atom	$\Delta_f H°_{298}$/kJ mol^{-1}	Ref.	Atom	$\Delta_f H°_{298}$/kJ mol^{-1}	Ref.	Atom	$\Delta_f H°_{298}$/kJ mol^{-1}	Ref.	Atom	$\Delta_f H°_{298}$/kJ mol^{-1}	Ref.
Ac	406	5	Dy	290.4 ± 2.1	4	N	472.68 ± 0.40	2	Se	227.2 ± 4	1
Ag	284.9 ± 0.8	2	Er	316.4 ± 2.1	4	Na	107.5 ± 0.7	3	Si	450.0 ± 8	2
Al	330.9 ± 4.0	2	Es	133	6	Nb	733.0 ± 8	3	Sm	206.7 ± 2.1	4
Am	284	6	Eu	177.4 ± 2.1	4	Nd	326.9 ± 2.1	4	Sn	301.2 ± 1.5	2
As	302.5 ± 13	1	F	79.38 ± 0.30	2	Ni	430.1 ± 8.4	3	Sr	164.0 ± 1.7	3
Au	368.2 ± 2.1	1	Fe	415.5 ± 1.3	3	Np	464.8	6	Ta	782.0 ± 2.5	1
B	565 ± 5	2	Ga	271.96 ± 2.1	3	O	249.229 ± 0.002	7	Tb	388.7 ± 2.1	4
Ba	179.1 ± 5.0	3	Gd	397.5 ± 2.1	4	Os	787 ± 6.3	1	Tc	678	5
Be	324 ± 5	2	Ge	372 ± 3	2	P	316.5 ± 1.0	2	Te	196.6 ± 2.1	1
Bi	209.6 ± 2.1	1	H	217.998 ± 0.006	2	Pa	563	5	Th	602 ± 6	2
Bk	310	6	Hf	618.4 ± 6.3	3	Pb	195.2 ± 0.8	2	Ti	473 ± 3	2
Br	111.87 ± 0.12	3	Hg	61.38 ± 0.04	2	Pd	376.6 ± 2.1	1	Tl	182.2 ± 0.4	1
C	716.68 ± 0.45	2	Ho	300.6 ± 2.1	4	Pr	356.9 ± 2.1	4	Tm	232.2 ± 2.1	4
Ca	177.8 ± 0.8	2	I	106.76 ± 0.04	2	Pt	565.7 ± 1.3	1			
Cd	111.80 ± 0.20	2	In	243 ± 4	1	Pu	345	6	U	533 ± 8	2
Ce	420.1 ± 2.1	4	Ir	669 ± 4	1	Ra	159	5	V	515.5 ± 8	3
Cf	196	6	K	89.0 ± 0.8	2	Rb	80.9 ± 0.8	2			
Cl	121.301 ± 0.008	2	La	431.0 ± 2.1	4	Re	774 ± 6.3	1	W	851.0 ± 6.3	3
Cm	386	6	Li	159.3 ± 1.0	2	Rh	556 ± 4	1	Y	424.7 ± 2.1	4
Co	426.7	3	Lu	427.6 ± 2.1	4	Ru	650.6 ± 6.3	1			
Cr	397.48 ± 4.2	3	Mg	147.1 ± 0.8	2	S	277.17 ± 0.15	2	Yb	155.6 ± 2.1	4
Cs	76.5 ± 1.0	2	Mn	283.3 ± 4.2	3	Sb	264.4 ± 2.5	1	Zn	130.40 ± 0.40	2
Cu	337.4 ± 1.2	2	Mo	658.98 ± 3.8	3	Sc	377.8 ± 4	1	Zr	610.0 ± 8.4	3

References

1. Brewer, L., and Rosenblatt, G. M., *Adv. High Temp. Chem.* 2, 1, 1969.
2. Cox, J. D., Wagman, D. D., and Medvedev, V. A., Eds., *CODATA Key Values for Thermodynamics*, Hemisphere Publishing Corporation, New York, 1989; updated e-version: http://www.codata.org/codata.
3. NIST Chemistry WebBook, http://webbook.nist.gov, *NIST-JANAF Thermochemical Table*, 4th Edn., Chase, Jr., M. W., Ed., ACS, AIP, New York, 1998.
4. Chandrasekharaiah, M.S., and Gingerich, K.A., Thermodynamic properties of gaseous species, in *Handbook on the Chemistry and Physics of Rare Earths*, Gschneidner, Jr., K.A., and Ering, L., Eds., Elsevier, Amsterdam, 1989, Vol. 12, Chap. 86, pp. 409–431.
5. Lias, S.G., Bartmess, J. E., Liebman, J. F., Holmes, J. L., Levin, R. D., and Mallard, W. G., *J. Phys. Chem. Ref. Data* 17, Suppl. 1, 1988.
6. Kleinschmidt, P.D., Ward, J. W., Matlack, G. M., and Haire, R. G., *High Temp. Sci.* 19, 267, 1985.
7. Ruscic, B., Pinzon, R.E., Morton, M. E., Srinivasan, N. K., Su, M.-C., Sutherland, J. W., and Michael, J. V., *J. Phys. Chem. A* 110, 6592, 2006.

TABLE 3. Bond Dissociation Energies in Polyatomic Molecules

The D°_{298} values in polyatomic molecules are notoriously difficult to measure accurately since the mechanism of the kinetic systems involved in many of the measurements are seldom straightforward. Thus, much lively controversy has taken place in the literature and is likely to continue for some time to come. We will continue updating and presenting our assessment of the most reliable BDE data every year.

The references relating to each of the D°_{298} values listed in Table 3 are contained in the *Comprehensive Handbook of Chemical Bond Energies*, by Yu-Ran Luo, CRC Press, 2007. Many D°_{298} in Table 3 are derived from the equation

$$D^\circ_{298}(\text{R-X}) = \Delta_f H^\circ(\text{R}) + \Delta_f H^\circ(\text{X}) - \Delta_f H^\circ(\text{RX})$$

Here, the enthalpies of formation of the atoms and radicals are taken from Tables 2 and 4, respectively, and the enthalpies of formation of the molecules are from reference sources listed in the above *Comprehensive Handbook of Chemical Bond Energies*.

Table 3 presents **H**-C, **C**-C, C-**halogen**, O-, N-, S-, Si-, Ge-, Sn-, Pb-, P-, As-, Sb-, Bi-, Se-, Te-, and **metal**-X BDEs. The **boldface** in the species indicates the dissociated fragment. The **metal**-X BDEs are arranged on the basis of the Periodic Table with the new IUPAC notation for Groups 1 to 18, see inside front cover of this *Handbook*.

Bond	D°_{298}/kJ mol^{-1}	Ref.	Bond	D°_{298}/kJ mol^{-1}	Ref.	Bond	D°_{298}/kJ mol^{-1}	Ref.
(1) C–H BDEs			H-*cyclo*-C$_3$H$_5$	444.8 ± 1.0	1	CHF$_2$CH$_2$–H	433.0 ± 14.6	1
CH$_3$–H	439.3 ± 0.4	1	H-CH$_2$-*cyclo*-C$_3$H$_5$	407.5 ± 6.7	1	CH$_2$FCH$_2$–H	433.5 ± 8.4	1
CH$_3$CH$_2$–H	420.5 ± 1.3	1	H-*cyclo*-C$_4$H$_7$	409.2 ± 1.3	1	CH$_3$CHF–H	410.9 ± 8.4	1
CH$_3$CH$_2$CH$_2$–H	422.2 ± 2.1	1	H-*cyclo*-C$_5$H$_9$	400.0 ± 4.2	1	CF$_3$CHCl–H	425.9 ± 6.3	1
CH$_3$CH$_2$CH$_3$	410.5 ± 2.9	1	H-*cyclo*-C$_6$H$_{11}$	416.3	1	CF$_3$CClBr–H	404.2 ± 6.3	1
CH$_3$CH$_2$CH$_2$CH$_2$–H	421.3	1	H–C$_6$H$_5$	472.2 ± 2.2	1	CClF$_2$CHF–H	412.1 ± 2.1	1
CH$_3$CH$_2$CH$_2$CH$_3$	411.1 ± 2.2	1	H–CH$_2$C$_6$H$_5$	375.5 ± 5.0	1	CCl$_3$CCl$_2$–H	397.5 ± 8.4	1
(CH$_3$)$_2$CHCH$_2$–H	419.2 ± 4.2	1	H–CH(CH$_3$)C$_6$H$_5$	357.3 ± 6.3	1	CHCl$_2$CCl$_2$–H	393.3 ± 8.4	1
(CH$_3$)$_3$C–H	400.4 ± 2.9	1	H–CH(C$_6$H$_5$)$_2$	353.5 ± 2.1	1	CH$_3$CCl$_2$–H	397.9 ± 5.0	1
(CH$_3$)$_3$CCH$_2$–H	419.7 ± 4.2	1	H–CH(C$_6$H$_4$-p-OH)$_2$	375.8 ± 4.7	1	CH$_3$CHCl–H	406.6 ± 1.5	1
(CH$_3$CH$_2$)CH(CH$_3$)$_2$	400.8	1	H–C(CH$_3$)$_2$C$_6$H$_5$	348.1 ± 4.2	1	CH$_2$ClCH$_2$–H	423.1 ± 2.4	1
CH$_3$CH$_2$(CH$_2$)$_2$CH$_3$	415.1	1	H–C(C$_6$H$_5$)$_3$	338.9 ± 8.4	1	CH$_3$CBr$_2$–H	397.1 ± 5.0	1
(C$_3$H$_7$)CH(CH$_3$)$_2$	396.2 ± 8.4	1	1-H-C$_{10}$H$_7$	469.4 ± 5.4	1	CH$_2$BrCH$_2$–H	415.1 ± 8.4	1
CH$_3$CH(CH$_3$)CH(CH$_3$)$_2$	399.2 ± 13.0	1	2-H-C$_{10}$H$_7$	468.2 ± 5.9	1	CH$_3$CHBr–H	415.0 ± 2.7	3
CH$_3$CH$_2$(CH$_2$)$_3$CH$_3$	410	1	H–CF$_3$	445.2 ± 2.9	1	CF$_2$=CF–H	464.4 ± 8.4	1
CH$_3$CH$_2$(CH$_2$)$_4$CH$_3$	410	1	H–CHF$_2$	431.8 ± 4.2	1	CF$_3$CF$_2$CF$_2$–H	432.2	1
HCC–H	557.81 ± 0.30	1	H–CH$_2$F	423.8 ± 4.2	1	CH$_3$CH$_2$CHCl–H	407.0 ± 3.5	1
HCCCC–H	539 ± 12	1	H–CClF$_2$	421.3 ± 8.4	1	CH$_2$=CH-CHF–H	370.7 ± 4.6	1
CHCCH$_2$–H	384.1 ± 4.2	1	H–CCl$_2$F	410.9 ± 8.4	1	CH$_2$=CHCHCl–H	370.7 ± 4.6	1
CH$_3$CCCH$_2$–H	379.5	1	H–CBrF$_2$	415.5 ± 12.6	1	CH$_2$=CHCHBr–H	374.0 ± 4.6	1
HCCCH$_2$CH$_3$	373.0	1	H–CHClF	421.7 ± 10.0	1	H–C$_6$F$_5$	487.4	1
CH$_2$=CHCCCH$_2$–H	363.3	1	H–CCl$_3$	392.5 ± 2.5	1	H–CH$_2$OH	401.92 ± 0.63	1
CH$_3$CCCH$_2$CH$_3$	365.3 ± 9.6	1	H–CHCl$_2$	400.6 ± 2.0	1	CH$_2$CHOH	467 ± 11	1
HCCCH$_2$CH$_2$CH$_3$	349.8 ± 8.4	1	H–CH$_2$Cl	419.0 ± 2.3	1	CH$_3$CH$_2$OH	401.2 ± 4.2	1
HCCCH(CH$_3$)$_2$	345.2 ± 8.4	1	H–CFClBr	413 ± 21	1	CH$_3$CH$_2$OH	421.7 ± 8	1
CH$_3$CCCH(CH$_3$)$_2$	344.3 ± 11.3	1	H–CHClBr	406.0 ± 2.4	1	CH$_3$CH$_2$CH$_2$OH	392	1
HCCCCCC–H	~543 ± 13	1	H–CCl$_2$Br	387 ± 21	1	CH$_3$CH$_2$CH$_2$OH	394.6 ± 8.4	1
H$_2$C=CH–H	464.2 ± 2.5	1	H–CClBr$_2$	371 ± 21	1	CH$_3$CH$_2$CH$_2$OH	406.3 ± 8.4	1
CH$_2$=C=CH–H	371.1 ± 12.6	1	H–CBr$_3$	399.2 ± 8.4	1	(CH$_3$)$_2$CHOH	383.7 ± 8.4	1
CH$_3$CH=CH–H	464.8	1	H–CHBr$_2$	412.6 ± 2.7	3	(CH$_3$)$_2$CHOH	394.6 ± 8.4	1
CH$_2$=CHCH$_2$–H	369 ± 3	1	H–CH$_2$Br	427.2 ± 2.4	1	CH$_2$=CHCH$_2$OH	341.4 ± 7.5	1
CH$_2$=CH-CH$_2$CH$_2$–H	410.5	1	H–CI$_3$	423 ± 29	1	(CH$_3$)$_3$COH	418.4 ± 8.4	1
CH$_2$=CHCH$_2$CH$_3$	350.6	1	H–CHI$_2$	431.0 ± 8.4	1	(CH$_2$=CH)$_2$CHOH	288.7	1
CH$_2$=C(CH$_3$)CH$_2$–H	372.8	1	H–CH$_2$I	431.6 ± 2.8	1	Ph$_2$CHOH	326	1
CH$_2$=CHCH=CHCH$_2$–H	347.3 ± 12.6	1	CF$_3$CF$_2$–H	429.7 ± 2.1	1	CH$_3$CH(OH)$_2$	~385	1
(CH$_2$=CH)$_2$CH–H	320.5 ± 4.2	1	CHF$_2$CF$_2$–H	431.0 ± 18.8	1	(CH$_2$OH)$_2$	385.3	1
CH$_2$=CHCH$_2$CH$_2$CH$_3$	348.8	1	CH$_2$FCF$_2$–H	433.0 ± 14.6	1	HOCH$_2$(CH$_2$)$_2$(OH)CH–H	399.2	1
CH$_2$=CHCH(CH$_3$)$_2$	332.6 ± 7.1	1	CHF$_2$CFH–H	426.8 ± 14.6	1	CH$_3$OCH$_3$	402.1	1
CH$_2$=C(CH$_3$CH$_2$)CH$_2$–H	356.1 ± 8.4	1	CF$_3$CH$_2$–H	446.4 ± 4.5	1	CHF$_2$OCF$_3$	443.5 ± 4.2	1
(CH$_2$=CH)$_2$C(CH$_3$) –H	322.2	1	CH$_3$CF$_2$–H	416.3 ± 4.2	1	CHF$_2$OCHF$_2$	435.1 ± 4.2	1
			CH$_2$FCHF–H	413.4 ± 12.6	1			

Bond	D°_{298}/kJ mol^{-1}	Ref.
CH_3OCF_3	426.8 ± 4.2	1
$CH_3OCH_2CH_3$	389.1	1
$(CH_3)_3COC(CH_3)_3$	402.1	1
$CH_3CH_2OCH_2CH_3$	389.1	1
$CH_3CH_2Ot\text{-}C(CH_3)_3$	405.4	1
CH_3OPh	385.0	1
H-2-oxiran-2-yl	420.5 ± 6.5	1
H-tetrahydrofuran-2-yl	385.3 ± 6.7	1
$HC(O)–H$	368.40 ± 0.67	1
$FC(O)–H$	423.0	1
$CH_3C(O)–H$	374.0 ± 1.3	1
$CF_3C(O)–H$	390.4	1
$C_2H_5C(O)–H$	374.5	1
$CH_2{=}CHC(O)–H$	372.8	1
$C_3H_7C(O)–H$	371.2	1
$iso\text{-}C_3H_7C(O)–H$	364.5	1
$C_4H_9C(O)–H$	372.0	1
$(CH_3)_2CHCH_2C(O)–H$	362.5	1
$C_2H_5CH(CH_3)C(O)–H$	360.8	1
$tert\text{-}BuC(O)–H$	375.1	1
$Et_2CHC(O)–H$	367.2	1
$CH_3(CH_2)_8C(O)–H$	373.3	1
$C_6H_5C(O)–H$	371.1 ± 10.9	1
$PhCH_2C(O)–H$	362.0	1
$PhC(CH_3)_2C(O)–H$	362.9	1
$H–CH{=}C{=}O$	448.1	1
$CH_3C(O)H$	394.5 ± 9.2	1
$CH_3C(O)Cl$	≤423.4	1
$CH_3CH_2C(O)H$	383.7	1
CH_3COCH_3	401.2 ± 2.9	1
$CF_3C(O)CH_3$	465.6	1
$CH_3COCH_2CH_3$	403.8	1
$MeCOCH_2Me$	386.2 ± 7.1	1
$EtCOCH_2Me$	396.5 ± 2.8	1
$CH_3CH_2COC_6H_5$	402.8 ± 3.6	1
$MeCH_2COPh$	388.7	1
$H–C(O)OH$	404.2	1
$CH_3C(O)OH$	398.7 ± 12.1	1
$ClCH_2C(O)OH$	398.9	1
$H–C(O)OCH_3$	399.2 ± 8.4	1
$CH_3C(O)OCH_3$	406.3 ± 10.5	1
$CH_3C(O)OCH_3$	404.6	1
$CH_3C(O)OCH_2CH_3$	401.7	1
$CH_3C(O)OPh$	419.2 ± 5.4	1
$CH_3CH_2C(O)OEt$	400	1
$PhCH_2C(O)OEt$	370.7	1
$Me_2CHC(O)OEt$	387.4	1
$PhCHMe(C(O)OEt)$	358.2	1
H-furaylmethyl	361.9 ± 8.4	1
CH_3NH_2	392.9 ± 8.4	1
$CH_3N{=}CH_2$	407.9 ± 14.6	1
$CH_3CH_2NH_2$	377.0 ± 8.4	1
$C_2H_5CH_2NH_2$	380.7 ± 8.4	1
$C_3H_7CH_2NH_2$	393.3 ± 8.4	1
$C_4H_9CH_2NH_2$	387.7 ± 8.4	1
$HOCH_2CH_2NH_2$	379.5 ± 8.4	1
$(CH_3CH_2)_2NH$	370.7 ± 8.4	1
$(C_3H_7CH_2)_2NH$	379.9 ± 8.4	1
$(C_4H_9CH_2)_2NH$	384.5 ± 8.4	1
$(C_2H_5)_2NCH_2CH_3$	379.5 ± 1.7	1
$(C_2H_5CH_2)_3N$	376.6 ± 8.4	1
$((CH_3)_3CCH_2)_3N$	388.3 ± 8.4	1
$(Bu)_2NCH_2(nPr)$	381 ± 10.0	1
$((CH_3)_2CH)_3N$	387.0 ± 8.4	1
$(CH_3)_2CHNH_2$	372.0 ± 8.4	1
CH_3NHCH_3	364.0 ± 8.4	1
$(CH_3)_3N$	380.7 ± 8.4	1
$tert\text{-}BuN(CH_3)_2$	376.6 ± 8.4	1
$((HOCH_2CH_2)_2(CH_3))N$	364.4 ± 8.4	1
$(HOCH_2CH_2)_3N$	379.9 ± 8.4	1
$((HOCH_2)CH(CH_3))_3N$	379.9 ± 8.4	1
$PhCH_2NH_2$	368.2	1
$PhN(CH_2CH_3)_2$	383.3 ± 4.2	1
Ph_2NCH_3	379.5 ± 1.7	1
$PhN(CH_2Ph)_2$	357.3 ± 8.8	1
$N(CH_2Ph)_3$	372.8 ± 2.5	1
$PhN(CH_2CH{=}CH_2)_2$	339.3 ± 2.9	1
$N(CH_2CH{=}CH_2)_3$	345.6 ± 3.3	1
$H_2NNH(CH_3)$	410	1
$HNN(CH_3)_2$	410	1
$(CH_3)_2NC_6H_5$	383.7 ± 5.4	1
$H–CN$	528.5 ± 0.8	1
CH_3CN	405.8 ± 4.2	1
CH_3CH_2CN	393.3 ± 12.6	1
$PhCH_2CN$	344.3	1
$C_6F_5CH_2CN$	350.6	1
$CH_2(CN)_2$	366.5	1
$CH_2(CN)(NH_2)$	355.2	1
$(CH_3)_2CHCN$	384.5	1
CH_3NC	389.1 ± 12.6	1
$H–HCNN$	405.8 ± 8.4	1
$H–CNN$	331 ± 17	1
CH_3NO_2	415.4	1
$CH_3CH_2NO_2$	410.5	1
$C_2H_5CH_2NO_2$	410.5	1
Me_2CHNO_2	394.9	1
$C_6H_5C(NO_2)CHCH_3$	357.3	1
$H–C(S)H$	399.6 ± 5.0	1
CH_3SH	392.9 ± 8.4	1
CH_3SCH_3	392.0 ± 5.9	1
$PhSCH_3$	389.1	1
$PhCH_2SPh$	352.3	1
$(PhS)_2CHPh$	341.0	1
$PhSCHPh_2$	344.8	1
CH_3SOCH_3	393.3	1
$CH_3SO_2CH_3$	414.2	1
$CH_3SO_2CF_3$	431.0	1
CH_3SO_2Ph	414.2	1
$PhCH_2SO_2Me$	380.7	1
$PhCH_2SO_2CF_3$	372.4	1
$PhCH_2SO_2tBu$	376.6	1
Ph_2CHSO_2Ph	365.3	1
$CH_2(SPh)_2$	372.4	1
$H–CH_2SiMe_3$	418 ± 6.3	1
$H–CH_2C(CH_3)_2SiMe_3$	409 ± 5	1
$H–CH_2SiMe_2Ph$	410.1	1
$H–CH((CH_3)_3Si)_2$	397 ± 13	1
$H–CH_2B(RO)_2$	412.5	1
$H–CH((CH_3)_2P)_2$	385 ± 13	1

(2) C–C BDEs

Bond	D°_{298}/kJ mol^{-1}	Ref.
$CH_3–CH_3$	377.4 ± 0.8	1
$CH_3–C_2H_5$	370.3 ± 2.1	1
$CH_3–C_3H_7$	372.0 ± 2.9	1
$CH_3–iso\text{-}C_3H_7$	369.0 ± 3.8	1
$CH_3–C_4H_9$	371.5 ± 2.9	1
$CH_3–iso\text{-}C_4H_9$	370.3 ± 4.6	1
$CH_3–sec\text{-}C_4H_9$	368.2 ± 2.9	1
$CH_3–tert\text{-}C_4H_9$	363.6 ± 2.9	1
$CH_3–C_5H_{11}$	368.4 ± 6.3	1
$CH_3–CH(C_2H_5)_2$	365.7 ± 4.2	1
$CH_3–C(CH_3)_2(CH_2CH_3)$	360.9 ± 6.3	1
$CH_3–C_6H_{13}$	368.2 ± 6.3	1
$C_2H_5–C_2H_5$	363.2 ± 2.5	1
$C_3H_7–C_3H_7$	366.1 ± 3.3	1
$iso\text{-}C_3H_7–iso\text{-}C_3H_7$	353.5 ± 4.6	1
$C_4H_9–C_4H_9$	364.0 ± 3.8	1
$iso\text{-}C_4H_9–iso\text{-}C_4H_9$	362.3 ± 6.3	1
$sec\text{-}C_4H_9–sec\text{-}C_4H_9$	348.5 ± 3.3	1
$tert\text{-}C_4H_9–tert\text{-}C_4H_9$	322.6 ± 4.2	1
$CH_3–cyclo\text{-}C_5H_9$	358.2 ± 5.0	1
$CH_3–cyclo\text{-}C_6H_{11}$	377.0 ± 7.5	1
$cyclo\text{-}C_6H_{11}–cyclo\text{-}C_6H_{11}$	369.0 ± 8.4	1
$CH_3–CH_2C{\equiv}CH$	320.5 ± 5.0	1
$CH_3–CH_2C{\equiv}CCH_3$	308.4 ± 6.3	1
$CH_3–CH(CH_3)C{\equiv}CH$	305.4 ± 8.4	1
$CH_3–CH(CH_3)C{\equiv}CCH_3$	320.9 ± 6.3	1
$CH_3–C(CH_3)_2C{\equiv}CH$	295.8 ± 6.3	1
$CH_3–C(CH_3)_2C{\equiv}CCH_3$	303.3 ± 6.3	1
$CH_3–CHCH_2$	426.3 ± 6.3	1
$CH_3–CH{=}CCH_2$	359.8 ± 5.9	1
$CH_3–cyclopro\text{-}en\text{-}1\text{-}yl$	340.6 ± 20.9	1
$CH_3–CH_2CH{=}CH_2$	317.6 ± 3.8	1
$CH_3–CH_2C(CH_3){=}CH_2$	310.0 ± 4.2	1
$CH_3–CH(CH_3)CH{=}CH_2$	302.5 ± 6.3	1
$CH_3–C(CH_3)_2CH{=}CH_2$	282.4 ± 6.3	1
$CH_3–cyclo\text{-}C_5H_7$	299.2 ± 8.4	1
$CH_3–C_6H_5$	426.8 ± 4.2	1
$HCC–C_6H_5$	590.8 ± 5.9	1
$C_2H_3–C_6H_5$	482.0 ± 5.4	1
$CH_3\text{-}CH_2C_6H_5$	325.1 ± 4.2	1
$CH_3–CH(CH_3)C_6H_5$	318.8 ± 8.4	1
$CH_3–C(CH_3)_2C_6H_5$	303.3 ± 8.4	1
$CH_3–CH_2CHCHPh$	295.4	1
$CH_3–CH(C_6H_5)_2$	315.9 ± 6.3	1
$CH_3–C(CH_3)(C_6H_5)_2$	290.8 ± 8.4	1

Bond	D^o_{298}/kJ mol^{-1}	Ref.
C_6H_5–C_6H_5	478.6 ± 6.3	1
C_6H_5–$CH_2C_6H_5$	383.7 ± 8.4	1
$C_6H_5CH_2$–$CH_2C_6H_5$	272.8 ± 9.2	1
C_6H_5–$CH(C_6H_5)_2$	361.1 ± 8.4	1
C_6H_5–$C(C_6H_5)_3$	324.3 ± 12.6	1
Ph_2CH–$CHPh_2$	247.3 ± 8.4	1
$PhCH_2$–CPh_3	234.7 ± 14.6	1
R-R, π-dimer, R = phenalenyl	42	1
R-R, σ-dimer, R = phenalenyl	42.7	1
R-R, R = 9-phenylfluorenyl	63.6	1
CF_3–CF_3	413.0 ± 5.0	1
CF_3–CHF_2	399.6 ± 8.4	1
CF_3–$CClF_2$	373.6 ± 12.5	1
CF_3–CH_2F	397.5 ± 8.4	1
CF_3–CCl_3	332.2 ± 5.4	1
CF_3–$CHBrCl$	377.0 ± 10.5	1
CF_3–CH_2Br	399.6 ± 8.4	1
CF_3–CH_2I	408.4 ± 10.5	1
CF_3–CH_3	429.3 ± 5.0	1
CHF_2–CHF_2	382.4 ± 15.5	1
$CClF_2$–$CClF_2$	378.7 ± 12.6	1
CF_2Cl–$CFCl_2$	358.6 ± 12.6	1
CHF_2–CH_2F	394.1 ± 16.7	1
CH_2F–CH_2F	368.2 ± 8.4	1
CHF_2–CH_3	405.0 ± 8.4	1
CH_2F–CH_3	388.3 ± 8.4	1
$CHClF$–CH_3	399.6 ± 12.6	1
CF_2Br–$CHClF$	369.4	1
CF_2Br–CH_3	396.6 ± 15.1	1
CCl_3–CCl_3	285.8 ± 6.3	1
CCl_3–$CClF_2$	282.0 ± 12.6	1
CCl_3–$CHCl_2$	303.3 ± 6.3	1
CCl_3–CH_2Cl	323.8 ± 8.4	1
CCl_3–CH_3	362.3 ± 6.3	1
$CHCl_2$–$CHCl_2$	326.9 ± 4.1	1
$CHCl_2$–CH_2Cl	352.2 ± 5.9	1
$CHCl_2$–CH_3	361.3 ± 2.5	1
$CHBrCl$–CH_3	384.5	1
$CHClBr$–$CHClBr$	317.1 ± 12.6	1
CH_2Cl–CH_2Cl	360.7 ± 8.4	1
CH_2Cl–CH_3	375.7 ± 9.2	1
Br_3C–CH_3	356.9 ± 12.6	1
Br_3C–CBr_3	278.7 ± 16.7	1
$CHBr_2$–CH_3	372.8	1
CH_2Br–CH_2Cl	378.2	1
CH_2Br–CH_2Br	379.9 ± 8.4	1
CH_2I–CH_2I	387.0 ± 10.5	1
CH_3–CH_2Br	381.6 ± 8.4	1
CH_3–CH_2I	384.5 ± 8.4	1
CF_3–CF_2CF_3	424.3 ± 13.6	1
CF_3–$CF=CF_2$	420.5	1
CH_3–CH_2CH_2Cl	371.4 ± 2.8	1
CH_3–$CHClCH_3$	367.5 ± 2.0	1

Bond	D^o_{298}/kJ mol^{-1}	Ref.
CH_2Cl–$CHClCH_3$	356.5 ± 8.4	1
CH_2Cl–CH_2CClH_2	369.0 ± 8.4	1
CH_3–CCl_2CH_3	362.8 ± 8.4	1
CH_2Br–$CHBrCH_3$	369.4 ± 8.4	1
CH_2ClCH_2–$CHClCH_3$	364.4 ± 8.4	1
CH_2ClCH_2–CH_2CClH_2	369.0 ± 8.4	1
CH_3CHBr–$CHBrCH_3$	355.6 ± 8.4	1
CF_3–C_6H_5	463.2 ± 12.6	1
CCl_3–C_6H_5	388.7 ± 8.4	1
CH_3–C_6F_5	439.3	1
CF_3–C_6F_5	435.1	1
CF_3–$CH_2C_6H_5$	365.7 ± 12.6	1
C_6F_5–C_6F_5	488.3	1
CF_3–$CHPh_2$	352.3 ± 16.7	1
CF_3–CPh_3	290.8 ± 16.7	1
CF_2CF–$CFCF_2$	558.1 ± 12.6	1
CH_2FCH_2–CPh_3	274.9 ± 16.7	1
CHF_2CH_2–CPh_3	264.0 ± 16.7	1
CH_3–CH_2OH	364.8 ± 4.2	1
CF_3–CH_2OH	405.4 ± 6.3	1
C_2H_5–CH_2OH	356.9 ± 5.0	1
C_3H_7–CH_2OH	357.3 ± 3.3	1
iso-C_3H_7–CH_2OH	354.8 ± 4.2	1
C_4H_9–CH_2OH	355.6 ± 4.2	1
sec-C_4H_9–CH_2OH	352.7 ± 4.2	1
iso-C_4H_9–CH_2OH	354.0 ± 5.4	1
C_6H_5–CH_2OH	413.4 ± 5.4	1
HOH_2C–CH_2OH	358.2 ± 6.3	1
NH_2CH_2–CH_2OH	335.6 ± 10.5	1
CH_3–CH_2OCH_3	363.2 ± 5.0	1
CH_3OCH_2–CH_2OCH_3	338.9 ± 10.5	1
CH_3–$C(O)H$	354.8 ± 1.7	1
CCl_3–$C(O)H$	309.2 ± 5.0	1
CH_3–$C(O)F$	417.6 ± 6.3	1
CH_3–$C(O)Cl$	367.8 ± 6.3	1
CCl_3–$C(O)Cl$	289.1 ± 6.3	1
$CHCl_2$–$C(O)Cl$	312.5 ± 8.4	1
$CClH_2$–$C(O)Cl$	340.2 ± 8.4	1
C_6H_5–$C(O)H$	408.4 ± 4.2	1
C_6H_5–$C(O)Cl$	417.6 ± 6.3	1
CH_3–$C(O)CH_3$	351.9 ± 2.1	1
C_2H_5–$C(O)CH_3$	347.3 ± 2.9	1
C_3H_7–$C(O)CH_3$	348.5 ± 2.9	1
iso-C_3H_7–$C(O)CH_3$	340.2 ± 3.8	1
C_4H_7–$C(O)CH_3$	346.9 ± 5.4	1
tert-C_4H_9–$C(O)CH_3$	329.3 ± 4.2	1
C_6H_5–$C(O)CH_3$	406.7 ± 4.6	1
$C_6H_5CH_2$–$C(O)CH_3$	299.7 ± 8.4	1
$HC(O)$–$C(O)H$	295.8 ± 6.3	1
$ClC(O)$–$C(O)Cl$	292.5 ± 8.4	1
$CH_3C(O)$–$C(O)H$	302.5 ± 8.4	1
$CH_3C(O)$–$C(O)CH_3$	307.1 ± 4.2	1
$C_6H_5C(O)$–$C(O)C_6H_5$	288.3 ± 16.7	1
CH_3–$C(O)OH$	384.9 ± 8.4	1
CF_3–$C(O)OH$	370.7 ± 8.4	1

Bond	D^o_{298}/kJ mol^{-1}	Ref.
CCl_3–$C(O)OH$	310.5 ± 12.6	1
$CClH_2$–$C(O)OH$	357.7 ± 8.4	1
CH_2Br–$C(O)OH$	358.2 ± 8.4	1
NH_2CH_2–$C(O)OH$	349.4 ± 8.4	1
CH_3NHCH_2–$C(O)OH$	300.4 ± 8.4	1
C_6H_5–$C(O)OH$	429.7 ± 8.4	1
C_6F_5–$C(O)OH$	470.0 ± 10.5	1
$HOCH_2$–$C(O)OH$	371.5 ± 5.4	1
$HOC(O)$–$C(O)OH$	334.7 ± 6.3	1
CH_3NHCH_2–$C(O)OH$	301.2 ± 16.7	1
$CH_3CH(NH_2)$–$C(O)OH$	331.4 ± 16.7	1
NH_2CH_2–$CH_2C(O)OH$	325.5 ± 16.7	1
CN–CN	571.9 ± 6.7	1
$HC(O)$–CN	455.2 ± 8.4	1
$HC(S)$–CN	530.1 ± 8.4	1
CF_3–CN	469.0 ± 4.2	1
CH_3–CN	521.7 ± 9.2	1
NCC–CN	462.3	1
C_2H_5–CN	506.7 ± 7.5	1
CH_3–CH_2CN	348.1 ± 12.6	1
C_6H_5–CH_2CN	386.6 ± 8.4	1
CH_3–$CH(CH_3)CN$	332.6 ± 8.4	1
CH_3–$C(CH_3)_2CN$	340.6 ± 16.7	1
CH_3–$C(CH_3)(CN)C_6H_5$	250.6	1
$(Ph)_2(CN)C$–$C(CN)(Ph)_2$	109.6	1
$(NO_2)_3C$–$C(NO_2)_3$	308.8	1
C_{58}–C_2	955.2 ± 14.5	1
(3) C–halogen BDEs		
F–CN	482.8	1
F–CF_3	546.8 ± 2.1	1
F–CHF_2	533.9 ± 5.9	1
F–CH_2F	496.2 ± 8.8	1
F–CF_2Cl	511.7	1
F–$CFCl_2$	482.0 ± 10.5	1
F–$CHFCl$	462.3 ± 10.0	1
F–CCl_3	439.3 ± 4	1
F–CH_2Cl	465.3 ± 9.6	1
F–CH_3	460.2 ± 8.4	1
F–$C≡CH$	521.3	1
F–$C≡CF$	519 ± 21	1
F–$CF=CF_2$	546.4 ± 12.6	1
F–CF_2CF_3	532.2 ± 6.3	1
F–CH_2CF_3	457.7	1
F–CF_2CH_3	522.2 ± 8.4	1
F–C_2H_3	517.6 ± 12.6	1
F–C_2H_5	467.4 ± 8.4	1
F–C_3H_7	474.9 ± 8.4	1
F–iso-C_3H_7	483.8 ± 8.4	1
F–tert-C_4H_9	495.8 ± 8.4	1
F–C_6H_5	525.5 ± 8.4	1
F–C_6F_5	485 ± 25	1
F–$CH_2C_6H_5$	412.8 ± 4.2	1
F–COH	497.9 ± 10.5	1
F–COF	510.3	1
F–$COCl$	484.5	1

Bond	D^o_{298}/kJ mol^{-1}	Ref.
F–C(O)CH$_3$	511.7 ± 12.6	1
Cl–CN	422.6 ± 8.4	1
Cl–CF$_3$	365.3 ± 3.8	1
Cl–CHF$_2$	364 ± 8	1
Cl–CH$_2$F	354.4 ± 11.7	1
Cl–CF$_2$Cl	333.9 ± 10.5	1
Cl–CFCl$_2$	320.9 ± 8.4	1
Cl–CHFCl	346.0 ± 13.4	1
Cl–CCl$_3$	296.6	1
Cl–CHCl$_2$	311.1 ± 2.0	1
Cl–CH$_2$Cl	338.0 ± 3.3	1
Cl–CBrCl$_2$	287 ± 10.5	1
Cl–CH$_2$Br	332.8 ± 4.6	1
Cl–CH$_2$I	328.2 ± 6.9	1
Cl–CH$_3$	350.2 ± 1.7	1
Cl–C≡CCl	443 ± 50	1
Cl–C≡CH	435.6 ± 8.4	1
Cl–CH$_2$CN	267.4	1
Cl–CCl=CCl$_2$	383.7	1
Cl–CH=CH$_2$	394.1 ± 3.1	2
Cl–CF=CF$_2$	434.7 ± 8.4	1
Cl–CF$_2$CF$_3$	346.0 ± 7.1	1
Cl–CF$_2$CF$_2$Cl	331.4 ± 20.9	1
Cl–CCl$_2$CF$_3$	307.9	1
Cl–CCl$_2$CCl$_3$	303.8	1
Cl–CHClCCl$_3$	330.5 ± 4.2	1
Cl–CCl$_2$CHCl$_2$	311.7	1
Cl–CHClCH$_3$	327.9 ± 1.8	1
Cl–CH$_2$CH$_2$Cl	345.1 ± 5.0	1
Cl–CHBrCH$_3$	331.8 ± 8.4	1
Cl–CH$_2$CH$_3$	352.3 ± 3.3	1
Cl–CH$_2$CH=CH$_2$	298.3 ± 5.0	1
Cl–C$_3$H$_7$	352.7 ± 4.2	1
Cl–CH$_2$CH$_2$CH$_2$Cl	348.9	1
Cl–iso-C$_3$H$_7$	354.8 ± 3.3	10
Cl–CH$_2$CHCH=CH$_2$	342.7	1
Cl–C$_4$H$_9$	350.6 ± 6.3	1
Cl–sec-C$_4$H$_9$	350.2 ± 6.3	1
Cl–tert-C$_4$H$_9$	351.9 ± 6.3	1
CH$_2$CHCHCl(CH$_3$)	300.0 ± 6.3	1
Cl–C$_5$H$_{11}$	350.6 ± 6.3	1
Cl–C(CH$_3$)$_2$(C$_2$H$_5$)	352.7 ± 6.3	1
Cl–cyclo-C$_6$H$_{11}$	360.2 ± 6.5	1
Cl–C$_6$H$_5$	399.6 ± 6.3	1
Cl–C$_6$F$_5$	383.3 ± 8.4	1
Cl–CH$_2$C$_6$H$_5$	299.9 ± 4.3	1
Cl–C(O)Cl	318.8 ± 8.4	1
Cl–COF	376.6	1
Cl–C(O)CH$_3$	354.0 ± 8.4	1
Cl–C(O)CH$_2$CH$_3$	353.3 ± 6.3	1
Cl–C(O)C$_6$H$_5$	341.0 ± 8.4	1
Cl–CH$_2$C(O)C$_6$H$_5$	309	1
Cl–CH$_2$C(O)OH	310.9 ± 2.2	1
Cl–C(O)OC$_6$H$_5$	364	1
Cl–C(NO$_2$)$_3$	302.1	1

Bond	D^o_{298}/kJ mol^{-1}	Ref.
Br–CN	364.8 ± 4.2	1
Br–CF$_3$	296.2 ± 1.3	1
Br–CHF$_2$	288.7 ± 8.4	1
Br–CF$_2$Cl	269.9 ± 6.3	1
Br–CCl$_3$	231.4 ± 4.2	1
Br–CH$_2$Cl	277.3 ± 3.6	1
Br–CBr$_3$	242.3 ± 8.4	1
Br–CHBr$_2$	274.9 ± 13.0	1
Br–CH$_2$Br	276.1 ± 5.3	1
Br–CH$_2$I	274.5 ± 7.5	1
Br–CH$_3$	294.1 ± 2.1	1
Br–C≡CH	410.5	1
Br–CH=CH$_2$	338.3 ± 3.1	1
Br–CF$_2$CF$_3$	283.3 ± 6.3	1
Br–CClBrCF$_3$	251.0 ± 6.3	1
Br–CF$_2$CF$_2$Br	282.8 ± 6.7	1
Br–CHClCF$_3$	274.9 ± 6.3	1
Br–CF$_2$CH$_3$	287.0 ± 5.4	1
Br–CH$_2$CH$_2$Cl	292.5 ± 8.4	1
Br–CHClCH$_3$	272.0 ± 8.4	1
Br–C$_2$H$_5$	292.9 ± 4.2	1
Br–CH$_2$CH=CH$_2$	237.2 ± 5.0	1
Br–C$_3$H$_7$	298.3 ± 4.2	1
Br–iso-C$_3$H$_7$	295.1 ± 3.3	10
Br–CH$_2$CH$_2$CH$_2$Br	324.7	1
Br–CF$_2$CF$_2$CF$_3$	278.2 ± 10.5	1
CF$_3$CFBrCF$_3$	274.2 ± 4.6	1
Br–C$_4$H$_9$	296.6 ± 4.2	1
Br–sec-C$_4$H$_9$	300.0 ± 4.2	1
Br–tert-C$_4$H$_9$	292.9 ± 6.3	1
Br–C$_6$H$_5$	336.4 ± 6.3	1
Br–C$_6$F$_5$	~328	1
Br–CH$_2$C$_6$H$_5$	239.3 ± 6.3	1
Br–CH$_2$C$_6$F$_5$	225.1 ± 6.3	1
Br–1–C$_{10}$H$_7$	339.7	1
Br–2–C$_{10}$H$_7$	341.8	1
Br–anthracenyl	322.6	1
Br–C(O)CH$_3$	292.0 ± 8.4	1
Br–C(O)C$_6$H$_5$	276.6 ± 8.4	1
Br–CH$_2$C(O)CH$_3$	257.9 ± 10.5	1
Br–CH$_2$C(O)C$_6$H$_5$	271	1
Br–CH$_2$C(O)OH	257.4 ± 3.7	1
Br–C(NO$_2$)$_3$	218.4	1
I–CN	320.1	1
I–CF$_3$	227.2 ± 1.3	1
I–CCl$_3$	168 ± 42	1
I–CH$_2$Cl	221.8 ± 4.2	1
I–CH$_2$Br	219.2 ± 5.4	1
I–CH$_2$I	216.9 ± 7.9	1
I–CH$_3$	238.9 ± 2.1	1
I–CH$_2$CN	187.0 ± 6.3	1
I–CF$_2$CF$_3$	219.2 ± 2.1	1
I–CF$_2$CF$_2$I	217.6 ± 6.7	1
I–CH$_2$CF$_3$	235.6 ± 4.2	1
I–CHFCClF$_2$	202 ± 2	1

Bond	D^o_{298}/kJ mol^{-1}	Ref.
I–CF$_2$CH$_3$	217.6 ± 4.2	1
I-CFICH3	218.0 ± 4.2	1
CF$_3$CFICF$_3$	215.1	1
I–CH=CH$_2$	259.0 ± 4.2	1
I–C$_2$H$_5$	233.5 ± 6.3	1
I–CH$_2$CH=CH$_2$	185.8 ± 6.3	1
I–C$_3$H$_7$	236.8 ± 4.2	1
I–iso-C$_3$H$_7$	233.1 ± 3.3	10
I–C$_4$F$_9$	205.8	1
I–tert-C$_4$H$_9$	227.2 ± 6.3	1
I–C$_6$H$_5$	272.0 ± 4.2	1
I–C$_6$F$_5$	<301.7	1
I–CH$_2$C$_6$H$_5$	187.8 ± 4.8	1
I–1-naphthyl	274.5 ± 10.5	1
I–2-naphthyl	272.0 ± 10.5	1
I–CH$_2$CN	187.0 ± 8.4	1
I–CH$_2$OCH$_3$	229.4 ± 8.4	1
I–CH$_2$SCH$_3$	216.8 ± 6.3	1
I–C(O)CH$_3$	223.0 ± 8.4	1
I–C(O)C$_6$H$_5$	212.1 ± 8.4	1
I–CH$_2$C(O)OH	197.5 ± 2.7	1
I–C(NO$_2$)$_3$	144.8	1

(4) O–X BDEs

Bond	D^o_{298}/kJ mol^{-1}	Ref.
HO–H	497.10 ± 0.29	1
FO–H	425.1	1
ClO–H	393.7	1
BrO–H	405	1
IO–H	403.3	1
CH$_3$O–H	440.2 ± 3	1
CF$_3$O–H	497.1	1
HC≡CO–H	443.1	1
C$_2$H$_5$O–H	441.0 ± 5.9	1
CH$_2$=CHO–H	355.6	1
CF$_3$CH$_2$O–H	447.7 ± 10.5	1
C$_3$H$_7$O–H	≤433 ± 2	1
iso-C$_3$H$_7$O–H	442.3 ± 2.8	1
C$_4$H$_9$O–H	432.3	1
sec-C$_4$H$_9$O–H	441.4 ± 4.2	1
tert-C$_4$H$_9$O–H	444.9 ± 2.8	1
tert-BuCH$_2$O–H	436.1	1
C$_6$H$_5$CH$_2$O–H	442.7 ± 8.8	1
CH$_3$C(OH)O–H	446.9 ± 6.3	1
(CH$_3$)$_2$C(OH)O–H	450.6 ± 6.3	1
HC(O)O–H	468.6 ± 12.6	1
CH$_3$C(O)O–H	468.6 ± 12.6	1
C$_2$H$_5$C(O)O–H	472.8	1
iso-C$_3$H$_7$C(O)O–H	472.8	1
C$_6$H$_5$C(O)O–H	464.4 ± 16.7	1
HOO–H	366.06 ± 0.29	1
CH$_3$OO–H	370.3 ± 2.1	1
CF$_3$OO–H	383	1
CH$_2$FOO–H	379	1
CCl$_3$OO–H	386	1
CHCl$_2$OO–H	383	1
CH$_2$ClOO–H	379	1

Bond	D^o_{298}/kJ mol^{-1}	Ref.	Bond	D^o_{298}/kJ mol^{-1}	Ref.	Bond	D^o_{298}/kJ mol^{-1}	Ref.
CBr$_3$OO–H	383	1	CF$_3$O–OCF$_3$	198.7 ± 2.1	1	CH$_3$O–C$_2$H$_5$	355.2 ± 5.4	1
CH$_2$BrOO–H	379	1	C$_2$H$_5$O–OC$_2$H$_5$	166.1	1	CH$_3$O–CHClCH$_3$	370.3 ± 8.4	1
C$_2$H$_5$OO–H	354.8 ± 9.2	1	C$_3$H$_7$O–OC$_3$H$_7$	155.2 ± 4.2	1	CH$_3$O–C$_3$H$_7$	358.6 ± 6.3	1
CH$_3$CHClOO–H	377	1	iso-C$_3$H$_7$O–O-iso-C$_3$H$_7$	157.7	1	CH$_3$O–iso-C$_3$H$_7$	360.7 ± 4.2	1
CH$_3$CCl$_2$OO–H	383	1	sec-C$_4$H$_9$O–O-sec-C$_4$H$_9$	152.3 ± 4.2	1	CH$_3$O–C$_4$H$_9$	346.0 ± 6.3	1
CF$_3$CHClOO–H	384	1	tert-BuO–O-tert-Bu	162.8 ± 2.1	1	CH$_3$O–tert-C$_4$H$_9$	353.1 ± 6.3	1
C$_2$Cl$_5$OO–H	383	1	tert-BuCH$_2$O–OCH$_2$-tert-Bu	152.3	1	C$_6$H$_5$–OCH$_3$	418.8 ± 5.9	1
iso-C$_3$H$_7$OO–H	356	1	EtC(Me)$_2$O–OC(Me)$_2$Et	164.4 ± 4.2	1	C$_6$H$_5$CH(CH$_3$)–OCH$_3$	313.4 ± 9.6	1
CH$_2$=CHCH$_2$OO–H	372.4	1	(CF$_3$)$_3$CO–OC(CF$_3$)$_3$	148.5 ± 4.6	1	C$_6$H$_5$–OC$_6$H$_5$	326.8 ± 4.2	1
tert-C$_4$H$_9$OO–H	352.3 ± 8.8	1	Ph$_3$CO–OCPh$_3$	131.4	1	CH$_3$–OC(O)H	383.7 ± 12.6	1
C$_6$H$_5$OO–H	384	1	SF$_5$O–OSF$_5$	155.6	1	HC(O)–OH	457.7 ± 2.1	1
C$_6$H$_5$CH$_2$OO–H	363	1	SF$_5$O–OOSF$_5$	126.8	1	CH$_3$C(O)–OH	459.4 ± 4.2	1
(C$_6$H$_5$)$_2$CHOO–H	370	1	(CH$_3$)$_3$CO–OSi(CH$_3$)$_3$	196.6	1	C$_6$H$_5$C(O)–OH	447.7 ± 10.5	1
trans-HC(O)OO–H	393 ± 13.8	6	tert-BuO–OGeEt$_3$	192.5	1	HO–CH$_2$C(O)OH	368.2 ± 10.5	1
cis-HC(O)OO–H	406.3 ± 13.8	6	tert-BuO–OSnEt$_3$	192.5	1	CH$_3$-OC(O)CH$_3$	380.3 ± 12.6	1
trans-CH$_3$C(O)OO–H	381 ± 4	8	CF$_3$OO–OCF$_3$	126.8 ± 8.4	1	HC(O)–OCH$_3$	423.8 ± 4.2	1
cis-CH$_3$C(O)OO–H	403 ± 14	8	HC(O)O–OH	199.2 ± 8.4	1	CH$_3$C(O)–OCH$_3$	424.3 ± 6.3	1
CCl$_2$(CN)OO–H	384	1	FC(O)O–OC(O)F	96.2	1	C$_6$H$_5$C(O)–OCH$_3$	421.3 ± 12.6	1
OHCH$_2$OO–H	368	1	CH$_3$C(O)O–ONO$_2$	131.4 ± 8.4	1	C$_6$H$_5$C(O)–OC$_6$H$_5$	307.5 ± 8.4	1
H–ONO	330.7	1	CH$_3$C(O)O–OC(O)CH$_3$	140.2 ± 21	1	CH$_3$OCH$_2$–OCH$_3$	367.5 ± 8.4	1
H–OONO	299.2	1	CF$_3$C(O)O–OC(O)CF$_3$	125.5	1	CH$_3$C(O)–OC(O)CH$_3$	382.4 ± 12.6	1
H–ONH$_2$	318	1	CF$_3$OC(O)O–OC(O)F	121.3 ± 4.2	1	C$_6$H$_5$C(O)–OC(O)C$_6$H$_5$	384.9 ± 16.7	1
H–ONO$_2$	426.8	1	CF$_3$OC(O)O–OCF$_3$	142.3 ± 2.9	1	CH$_3$–OOH	300.4 ± 12.6	1
H–ONNOH	189	1	CF$_3$OC(O)O–OC(O)OCF$_3$	119.2	1	C$_2$H$_5$–OOH	332.2 ± 20.9	1
H–OPO$_2$	465.7 ± 12.6	1	C$_2$H$_5$C(O)O–OC(O)C$_2$H$_5$	150.6	1	C$_3$H$_7$–OOH	364.4	1
H–OSO$_2$OH	441.4 ± 14.6	1	C$_3$H$_7$C(O)O–OC(O)C$_3$H$_7$	150.6	1	iso-C$_3$H$_7$–OOH	298.3	1
H–OSiMe3	495	1	FS(O)$_2$O–OS(O)$_2$F	92–100	1	tert-C$_4$H$_9$–OOH	309.2 ± 4.2	1
(CH$_3$)CHNO-H	354.4	1	HO–CF$_3$	≤482.0 ± 1.3	1	CH$_3$–OOCH$_3$	292.5 ± 8.4	1
(CH$_3$)$_2$CNO-H	354.0	1	FO–CF$_3$	408 ± 17	1	CF$_3$–OOCF$_3$	361.5 ± 8.4	1
(C$_6$H$_5$)CHNO-H	368.6	1	HO–CH$_3$	384.93 ± 0.71	1	CH$_3$–OO	137.0 ± 3.8	1
PhO–H	362.8 ± 2.9	1	HO–C$_2$H$_5$	391.2 ± 2.9	1	CF$_3$–OO	169.0	1
α-tocopherol RO-H	323.4	1	HO–CH$_2$CF$_3$	408.4 ± 8.4	1	CClF$_2$–OO	127.6	1
β-tocopherol RO-H	335.6	1	HO–CH$_2$CH=CH$_2$	332.6 ± 4.2	1	CCl$_2$F-OO	124.7	1
γ-tocopherol RO-H	335.1	1	HO–C$_3$H$_7$	392.0 ± 2.9	1	CH$_2$Cl–OO	122.4 ± 10.5	1
δ-tocopherol RO-H	342.8	1	HO–iso-C$_3$H$_7$	397.9 ± 4.2	1	CHCl$_2$–OO	108.2 ± 8.2	1
p-C$_6$H$_5$CH$_2$-C$_6$H$_4$O-H	356.2	1	HO–C$_4$H$_9$	389.9 ± 4.2	1	CCl$_3$–OO	92.0 ± 6.4	1
O–O$_2$	106.6	1	HO–sec-C$_4$H$_9$	396.1 ± 4.2	1	HC(O)–OOH	290.0	1
HO–OH	210.66 ± 0.42	1	HO–iso-C$_4$H$_9$	394.1 ± 4.2	1	CH$_3$C(O)–OOC(O)CH$_3$	315.1	1
HO–OF	199.7 ± 8.4	1	HO–tert-C$_4$H$_9$	398.3 ± 4.2	1	ClO–CF$_3$	≤369.9 ± 1.3	1
HO–OCl	~146	1	HO–CH(CH$_3$)(nC$_3$H$_7$)	398.3 ± 4.2	1	CH$_3$–ONO	245.2	1
HO–OBr	138.5 ± 8.4	1	HO–CH(C$_2$H$_5$)$_2$	399.2 ± 4.2	1	C$_2$H$_5$–ONO	260.2	1
FO–OF	199.6	1	HO–C(CH$_3$)$_2$(C$_2$H$_5$)	395.8 ± 6.3	1	C$_3$H$_7$–ONO	249.4 ± 6.3	1
ClO–OCl	72.4 ± 2.8	1	HO–C$_6$H$_5$	463.6 ± 4.2	1	iso-C$_3$H$_7$–ONO	254.4 ± 6.3	1
IO–OI	74.9 ± 17	1	HO–C$_6$F$_5$	446.9 ± 9.2	1	C$_4$H$_9$–ONO	256.5 ± 6.3	1
trans-perp-HO–ONO	≤67.8 ± 0.4	1	HO–CH$_2$C$_6$H$_5$	334.1 ± 2.6	1	iso-C$_4$H$_9$–ONO	254.0 ± 6.3	1
cis-cis-HO–ONO	83.3 ± 2.1	1	HO–C(CH$_3$)$_2$C$_6$H$_5$	339.3 ± 6.3	1	sec-C$_4$H$_9$–ONO	253.6 ± 6.3	1
HO–ONO$_2$	163.2 ± 8.4	1	cyclo-C$_5$H$_9$–OH	385.8 ± 6.3	1	tert-C$_4$H$_9$–ONO	252.7 ± 6.3	1
HO–OCH$_3$	189.1 ± 4.2	1	1-C$_{10}$H$_7$–OH	468.6 ± 6.3	1	(C$_2$H$_5$)(CH$_3$)$_2$C–ONO	254.0 ± 8.4	1
HO–OCF$_3$	201.3 ± 20.9	1	2-C$_{10}$H$_7$–OH	467.8 ± 6.3	1	CH$_3$–ONO$_2$	340.2	1
HO–OC$_2$H$_5$	178.7 ± 6.3	1	(CH$_3$)$_2$(NH$_2$)C–OH	310.4 ± 6.3	1	C$_2$H$_5$–ONO$_2$	344.8	1
HO–O-iso-C$_3$H$_7$	185.8 ± 6.3	1	CH$_3$C(O)-OH	459.4 ± 4.2	1	CH$_3$O–CH$_2$CN	393.3	1
HO–O-tert-C$_4$H$_9$	186.2 ± 4.2	1	HOCH$_2$–OH	411.3	1	O–N$_2$	167.4 ± 0.4	1
HO–OC(O)CH$_3$	169.9 ± 2.1	1	CH$_3$–OCH$_3$	351.9 ± 4.2	1	O–NO	306.21 ± 0.13	1
HO–OC(O)C$_2$H$_5$	169.9 ± 2.1	1	ICH$_2$–OCH$_3$	373.2 ± 12.6	1	O–NO$_2$	206.3	1
CH$_3$O–OCH$_3$	167.4 ± 6.3	1				NO–NO	40.6 ± 2.1	1

Bond	D^o_{298}/kJ mol^{-1}	Ref.
O$_2$N–ONO$_2$	95.4 ± 1.5	1
cis-HO–NO	207.0	1
trans-HO–NO	200.64 ± 0.19	1
FO–NO	132.5 ± 17	1
cis-ClO–NO	127.6 ± 8.4	1
trans-ClO–NO	116.6 ± 8.4	1
cis-BrO–NO	138.1 ± 8.4	1
trans-BrO–NO	121.6 ± 8.4	1
trans-perp-HOO–NO	114.2 ± 4	1
CH$_3$O–NO	176.6 ± 3.3	1
C$_2$H$_5$O–NO	185.4 ± 4.2	1
C$_3$H$_7$O–NO	179.1 ± 6.3	1
iso-C$_3$H$_7$O–NO	175.3 ± 4.2	1
C$_4$H$_9$O–NO	177.8 ± 6.5	1
iso-C$_4$H$_9$O–NO	175.7 ± 6.5	1
sec-C$_4$H$_9$O–NO	173.6 ± 3.3	1
tert-C$_4$H$_9$O–NO	176.1 ± 5.9	1
tert-AmO–NO	171.1 ± 0.4	1
C$_6$H$_5$O–NO	87.0	1
HO–NO$_2$	205.4	1
FO–NO$_2$	131.8 ± 12.6	1
ClO–NO$_2$	110.9	4
BrO–NO$_2$	118.0 ± 6.3	1
IO–NO$_2$	~100	1
CH$_3$O–NO$_2$	176.1 ± 4.2	1
C$_2$H$_5$O–NO$_2$	174.5 ± 4.2	1
C$_3$H$_7$O–NO$_2$	177.0 ± 4.2	1
iso-C$_3$H$_7$O–NO$_2$	175.7 ± 4.2	1
HOO–NO$_2$	99.2 ± 4.6	1
CH$_3$OO–NO$_2$	86.6 ± 8.4	1
CF$_3$OO–NO$_2$	105	1
CF$_2$ClOO–NO$_2$	106.7	1
CFCl$_2$OO–NO$_2$	106.7	1
CCl$_3$OO–NO$_2$	95.8	1
CH$_3$N(O)-O	305.3 ± 4.4	1
C$_6$H$_5$N(O)-O	392 ± 8	1
C$_5$H$_5$N-O	264.9 ± 2.0	1
C$_6$H$_5$N=N(O)(C$_6$H$_5$)-O	309.4 ± 3.5	1
C$_6$H$_5$(O)N=N(O) (C$_6$H$_5$)-O	309.4 ± 3.6	1
O–SO	551.1	1
O–SOF$_2$	513.3	1
O–SOCl$_2$	398.5	1
O–S(OH)$_2$	493.7 ± 25	1
HO–SH	293.3 ± 16.7	1
HO–SOH	313.4 ± 12.6	1
HO–S(OH)O$_2$	384.9 ± 8.4	1
HO–SCH$_3$	303.8 ± 12.6	1
HO–SO$_2$CH$_3$	360.2 ± 12.6	1
F–OH	215.1	1
F–OF	164.1	1
F–OCF$_3$	200.8 ± 4.2	1
F–OCH$_3$	>196.6	1
F–ONO$_2$	143.1	1
Cl–OH	233.5	1
Cl–OCl	142	1

Bond	D^o_{298}/kJ mol^{-1}	Ref.
Cl–OCF$_3$	≤220.9 ± 8.4	1
Cl–OCH$_3$	200.8	1
Cl–O-tert-C$_4$H$_9$	198.3	1
Cl–OOCl	91.2	1
Cl–ONO$_2$	172.0	1
Br–OH	209.6 ± 4.2	1
Br–OBr	125	1
Br–O-tert-C$_4$H$_9$	183.3	1
Br–ONO$_2$	143.1 ± 6.3	1
I–OH	213.4	1
I–OI	130.1	1
I–ONO$_2$	>140.6	1
(5) N–X BDEs		
H–NH$_2$	450.08 ± 0.24	1
H–NF$_2$	316.7 ± 10.5	1
H–NNH	254.4	1
H–N$_3$	≤389	1
H–N=CH$_2$	364 ± 25	1
H–NO	199.5	1
H–NHOH	341	1
H–NCO	460.7 ± 2.1	1
H–NCS	≤396.6 ± 4.6	1
H–NCS	347.3 ± 8.4	1
CH$_3$NH$_2$	425.1 ± 8.4	1
tert-BuNH$_2$	397.5 ± 8.4	1
C$_6$H$_5$CH$_2$NH$_2$	418.4	1
(CH$_3$)$_2$NH	395.8 ± 8.4	1
H–NHNH(CH$_3$)	276 ± 21	1
H–NHN(CH$_3$)$_2$	356 ± 21	1
NH$_2$CN	414.2	1
(NH$_2$)$_2$C=O	464.4	1
(NH$_2$)$_2$C=S	389.1	1
CH$_3$CSNH$_2$	380.7	1
PhCSNH$_2$	380.7	1
(PhNH)$_2$C=S	364.0	1
(NH$_2$)$_2$C=NH	435.1	1
Ph$_2$C=NH	489.5	1
H–N(SiMe$_3$)$_2$	464	1
H–NHPh	375.3	1
C$_6$H$_5$NHOH	292	1
C$_6$H$_5$NH(CONMe2)	387.9	1
H–NPh$_2$	364.8	1
HN–N$_2$	63	1
ON–N	480.7 ± 0.4	1
ON–NO	8.49 ± 0.12	1
ON–NO$_2$	42.5	1
O$_2$N–NO$_2$	57.3 ± 1	1
H$_2$N–NH$_2$	277.0 ± 1.3	1
F$_2$N–NF$_2$	92.9 ± 12.6	1
H$_2$N–NHCH$_3$	275.8 ± 8.4	1
H$_2$N–N(CH$_3$)$_2$	259.8 ± 8.4	1
H$_2$N–NHC$_6$H$_5$	227.6 ± 8.4	1
H$_2$N–NO$_2$	230	1
H$_2$NN(CH$_3$)–NO	179.6	1
(C$_6$H$_5$)$_2$N-NO	94.6	1

Bond	D^o_{298}/kJ mol^{-1}	Ref.
N$_3$–CH$_3$	335.1 ± 20.5	1
N$_3$–C$_6$H$_5$	375.7 ± 20.9	1
N$_3$–CH$_2$C$_6$H$_5$	211.3 ± 14.2	1
CH$_3$–NC	413.0 ± 3.3	1
C$_2$H$_5$–NC	413.4 ± 8.4	1
iso-C$_3$H$_7$–NC	423.0 ± 8.4	1
tert-C$_4$H$_9$–NC	399.6 ± 5.4	1
NC–NO	204.4	1
CH$_3$–NO	172	1
CF$_3$–NO	167	1
CCl$_3$–NO	125	1
C$_2$H$_5$–NO	171.5	1
CH$_2$CHCH$_2$–NO	110	1
iso-C$_3$H$_7$–NO	152.7 ± 12.6	1
tert-C$_4$H$_9$–NO	167	1
C$_6$H$_5$–NO	226.8 ± 2.1	1
C$_6$F$_5$–NO	211.3 ± 4.2	1
C$_6$H$_5$CH$_2$–NO	123	1
CH$_3$–NO$_2$	260.7 ± 2.1	1
C$_2$H$_5$–NO$_2$	254.4	1
C$_3$H$_7$–NO$_2$	256.5	1
iso-C$_3$H$_7$–NO$_2$	259.8	1
C$_4$H$_9$–NO$_2$	254.8	1
sec-C$_4$H$_9$–NO$_2$	263.2	1
tert-C$_4$H$_9$–NO$_2$	258.6	1
C$_6$H$_5$–NO$_2$	295.8 ± 4.2	1
C$_6$H$_5$CH$_2$–NO$_2$	210.3 ± 6.3	1
(NO$_2$)CH$_2$–NO$_2$	207.1	1
(NO$_2$)$_3$C–NO$_2$	176.1	1
CF$_3$–NF$_2$	280.7	1
C$_6$H$_5$CH$_2$–NF$_2$	237.2 ± 14.6	1
CH$_3$–NH$_2$	356.1 ± 2.1	1
C$_2$H$_5$–NH$_2$	352.3 ± 6.3	1
C$_3$H$_7$–NH$_2$	356.1 ± 2.9	1
iso-C$_3$H$_7$–NH$_2$	357.7 ± 3.8	1
C$_4$H$_9$–NH$_2$	356.1 ± 2.9	1
sec-C$_4$H$_9$–NH$_2$	359.0 ± 2.9	1
iso-C$_4$H$_9$–NH$_2$	254.8 ± 5.0	1
tert-C$_4$H$_9$–NH$_2$	355.6 ± 6.3	1
pyridin-2-yl–NH$_2$	431	1
C$_6$H$_5$–NH$_2$	429.3 ± 4.2	1
C$_6$H$_5$CH$_2$–NH$_2$	306.7 ± 6.3	1
C$_6$H$_5$CH(CH)$_3$–NH$_2$	307.5 ± 9.6	1
HC(O)–NH$_2$	421.7 ± 8.4	1
CH$_3$C(O)–NH$_2$	414.6 ± 8.4	1
HS–NO	138.9	1
CH$_3$S–NO	104.6 ± 4.2	1
tert-BuS–NO	115.1	1
PhCH$_2$S–NO	120.5	1
C$_6$H$_5$S–NO	81.2 ± 5.4	1
SCN–SCN	255.6	1
FSO$_2$–NF$_2$	163	1
F–NO	235.26	1
F–NO$_2$	221.3	1
F–NF$_2$	254.0	1

Bond	D^o_{298}/kJ mol^{-1}	Ref.
F–NH$_2$	286.6	1
Cl–NO	158.8 ± 0.8	1
Cl–NO$_2$	141.8 ± 1.3	1
Cl–NF$_2$	~134	1
Cl–NH$_2$	253.1	1
Br–NO	120.1 ± 0.8	1
Br–NO$_2$	82.0 ± 7.1	1
Br–NF$_2$	<227.2	1
I–NO	75.6 ± 4	1
I–NO$_2$	79.6 ± 4	1

(6) S–X BDEs

Bond	D^o_{298}/kJ mol^{-1}	Ref.
H–SH	381.18 ± 0.05	1
H–SCH$_3$	365.7 ± 2.1	1
H–SCHCH$_2$	351.5 ± 8.4	1
H–SC$_2$H$_5$	365.3	1
H–SC$_3$H$_7$	365.7	1
H–S-iso-C$_3$H$_7$	369.9 ± 8.4	1
H–S-tert-C$_4$H$_9$	362.3 ± 9.2	1
H–SOH	330.5 ± 14.6	1
H–SCOCH$_3$	370.7	1
H–SCOPh	364	1
H–SO$_2$CH$_3$	≤397	1
H–SSCH$_3$	330.5 ± 14.6	1
H–SPh	349.4 ± 4.5	1
H–SSH	318.0 ± 14.6	1
H–SSSH	292.9 ± 6.5	1
HS–SH	270.7 ± 8.4	1
FS–SF	362.3	1
ClS–SCl	329.7	1
HS–SCH$_3$	272.0	1
HS–SPh	255.2 ± 6.3	1
CH$_3$S–SCH$_3$	272.8 ± 3.8	1
C$_2$H$_5$S–SC$_2$H$_5$	276.6	1
MeS–SPh	272.0 ± 6.3	1
C$_6$H$_5$S–SC$_6$H$_5$	214.2 ± 12.6	1
F$_5$S–SF$_5$	305 ± 21	1
HS–CH$_3$	312.5 ± 4.2	1
HS–C$_2$H$_5$	307.9 ± 2.1	1
HS–C$_3$H$_7$	310.5 ± 2.9	1
HS–iso-C$_3$H$_7$	307.1 ± 3.8	1
HS–C$_4$H$_9$	309.2 ± 2.9	1
HS–sec-C$_4$H$_9$	307.5 ± 2.9	1
HS–iso-C$_4$H$_9$	310.0 ± 4.6	1
HS–tert-C$_4$H$_9$	301.2 ± 3.8	1
HS–C$_6$H$_5$	360.7 ± 6.3	1
HS–CH$_2$C$_6$H$_5$	258.2 ± 6.3	1
HS–C(O)H	309.6 ± 8.4	1
HS–C(O)CH$_3$	307.9 ± 6.3	1
CH$_3$S–CH$_3$	307.9 ± 3.3	1
HOS–CH$_3$	284.9 ± 12.6	1
CH$_3$SO–CH$_3$	221.8 ± 8.4	1
HOSO$_2$–CH$_3$	324.3 ± 12.6	1
CH$_3$SO$_2$–CH$_3$	279.5	1
F$_5$S–CF$_3$	392 ± 43	1
F–SF$_5$	391.6	1

Bond	D^o_{298}/kJ mol^{-1}	Ref.
F–SO$_2$(F)	379	1
Cl–SF$_5$	<272	1
Cl–SO$_2$CH$_3$	293	1
Cl–SO$_2$Ph	297	1
Br–SBr	259 ± 17	1
Br–SF$_5$	<230	1
I–SH	206.7 ± 8.4	1
I–SCH$_3$	206.3 ± 7.1	1

(7) Si-, Ge-, Sn-, and Pb–X BDEs

Bond	D^o_{298}/kJ mol^{-1}	Ref.
SiH$_3$–H	383.7 ± 2.1	1
Me$_3$Si–H	396 ± 7	1
H$_3$Si$_2$–H	373 ± 8	1
(C$_2$H$_5$)$_3$Si–H	396 ± 4	1
C$_6$H$_5$SiH$_2$–H	382 ± 5	1
(CH$_3$S)$_3$Si–H	364.0	1
(iPrS)$_3$Si–H	376.6	1
PhMe$_2$Si–H	377 ± 7	1
Ph$_2$SiH–H	379 ± 7	1
Ph$_2$MeSi–H	361 ± 10	1
SiF$_3$–H	432 ± 5	1
SiCl$_3$–H	391	5
SiBr$_3$–H	334 ± 8	1
SiH$_3$–SiH$_3$	321 ± 4	1
SiH$_3$–Si$_2$H$_5$	313 ± 8	1
Ph$_3$Si–SiPh$_3$	368.2	1
F$_3$Si–SiF$_3$	453.1 ± 25	1
SiH$_3$–CH$_3$	375 ± 5	1
SiF$_3$–CH$_3$	355.6	1
H$_3$Si–NO	158.2 ± 5.7	1
H$_3$Si–PH$_2$	331.4	1
SiH$_3$–F	638 ± 5	1
SiH$_3$–Cl	458 ± 7	1
SiH$_3$–Br	376 ± 9	1
SiH$_3$–I	299 ± 8	1
GeH$_3$–H	348.9 ± 8.4	1
Me$_3$Ge–H	364.0	1
Ph$_3$Ge–H	359.8	1
(CH$_3$)$_3$Ge-Ge(CH$_3$)$_3$	280.3	1
(CH$_3$)$_3$Ge–CH$_3$	288.7	1
Me$_3$Sn–H	307.5 ± 10.7	11
Ph$_3$Sn–H	294.6	1
(CH$_3$)$_3$Sn-Sn(CH$_3$)$_3$	252.6 ± 14.8	11
(CH$_3$)$_3$Sn–Cl	425 ± 17	1
(CH$_3$)$_3$Sn–CH$_3$	284.1 ± 9.9	11
(CH$_3$)$_3$Pb-Pb(CH$_3$)$_3$	228.4	1
Cl$_3$Pb-Cl	271 ± 84	1
(CH$_3$)$_3$Pb–CH$_3$	238 ± 21	1

(8) P-, As-, Sb-, Bi–X BDEs

Bond	D^o_{298}/kJ mol^{-1}	Ref.
H$_2$P–H	351.0 ± 2.1	1
CH$_3$PH–H	322.2 ± 12.6	1
H$_2$P–PH$_2$	256.1	1
(C$_2$H$_5$)$_2$P–P(C$_2$H$_5$)$_2$	359.8	1
F$_2$P–F	549	1
Cl$_2$P–Cl	356 ± 8	1

Bond	D^o_{298}/kJ mol^{-1}	Ref.
Br$_2$P–Br	<259	1
I$_2$P–I	217	1
H$_2$P–SiH$_3$	331.4	1
H$_2$As–H	319.2 ± 0.8	1
H$_2$Sb–H	288.3 ± 2.1	1
F$_2$Bi–F	435 ± 19	1
Br$_2$Bi–Br	>297.1	1

(9) Se- and Te–X BDEs

Bond	D^o_{298}/kJ mol^{-1}	Ref.
H–SeH	334.93 ± 0.75	1
H–SeC$_6$H$_5$	326.4 ± 16.7	1
PhSe–SePh	280 ± 19	1
H–TeH	277.0 ± 5.0	1
H–TeC$_6$H$_5$	≤264	1
PhTe–TePh	138.1 ± 12.6	1

(10) Metal-Centered BDEs

Arranged by the Periodic Table

(10.1) Group 1

Bond	D^o_{298}/kJ mol^{-1}	Ref.
Li–OH	431.0	1
Li–C$_2$H$_5$	214.6 ± 8.4	1
Li–nC$_4$H$_9$	197.9 ± 16.3	1
Na–OH	342.3	1
Na–O$_2$	<200	1
K–OH	359	1
Rb–OH	356.2 ± 4.2	1
Cs–OH	373	1

(10.2) Group 2

Bond	D^o_{298}/kJ mol^{-1}	Ref.
BeO–H	469	1
Be(OH)–OH	476	1
MgO–H	441	1
Mg(OH)–OH	349	1
BrMg–CH$_3$	253	1
BrMg–CH$_2$CH$_3$	205	1
BrMg–i-C$_3$H$_7$	184	1
BrMg–t-C$_4$H$_9$	174	1
BrMg–C$_6$H$_5$	289	1
BrMg–CH$_2$C$_6$H$_5$	201	1
BrMg–C(C$_6$H$_5$)$_3$	180	1
Ca(OH)–OH	409	1
Sr(OH)–OH	407	1
Ba(OH)–OH	443	1

(10.3) Group 3

Bond	D^o_{298}/kJ mol^{-1}	Ref.
Sc–CH$_3$	116 ± 29	1
Sc–C$_6$H$_6$	60.8	1
La(η^5-C$_5$Me$_5$)$_2$–CH(SiMe$_3$)$_2$	278.7 ± 10.5	1
Nd(η^5-C$_5$Me$_5$)$_2$–CH(SiMe$_3$)$_2$	236.8 ± 10.5	1
(η^5-C$_5$Me$_5$)$_2$Sm–H	226.8 ± 12.6	1
(η^5-C$_5$Me$_5$)$_2$Sm–OCH$_3$	343.1	1
(η^5-C$_5$Me$_5$)$_2$Sm–(η^3-C$_3$H$_5$)	188.3 ± 6.3	1
(η^5-C$_5$Me$_5$)$_2$Sm–S-nC$_3$H$_7$	295.4 ± 10.0	1
(η^5-C$_5$Me$_5$)$_2$Sm–N(CH$_3$)$_2$	201.7 ± 7.5	1

Bond	D^o_{298}/kJ mol^{-1}	Ref.
$(\eta^5\text{-}C_5Me_5)_2Sm\text{-}SiH$ $(SiMe_3)_2$	179.9 ± 21	1
$(\eta^5\text{-}C_5Me_5)_2Sm\text{-}P(Et)_2$	136.4 ± 8.4	1
$(\eta^5\text{-}C_5Me_5)_2Eu\text{-}I$	238.9 ± 8.4	1
$(\eta^5\text{-}C_5Me_5)_2Yb\text{-}I$	256.1 ± 6.3	1
$Lu(\eta^5\text{-}C_5Me_5)_2\text{-}$ $CH(SiMe_3)_2$	279.1 ± 10.5	1
$(\eta^5\text{-}C_5H_4SiMe_3)_3Th\text{-}H$	277 ± 6	1
$(\eta^5\text{-}C_5H_4SiMe_3)_3Th\text{-}O$	371 ± 24	1
$(\eta^5\text{-}C_5H_5)_3Th\text{-}CH_3$	375 ± 9	1
$(\eta^5\text{-}C_5H_5)_3Th\text{-}$ $CH_2Si(CH_3)_3$	369 ± 12	1
$(C_9H_7)_3Th\text{-}CH_2C_6H_5$	342 ± 9	1
$(\eta^5\text{-}C_5H_4tBu)_3U\text{-}H$	249.7 ± 5.7	1
$(\eta^5\text{-}C_5H_4SiMe_3)_3U\text{-}H$	253.7 ± 5.1	1
$[HB(3,5\text{-}Me_2Pz)_3]$ $U(Cl)_2\text{-}Cl$	422.6	1
$(\eta^5\text{-}C_5H_4SiMe_3)_3U\text{-}I$	265.6 ± 4.3	1
$(\eta^5\text{-}C_5H_4tBu)_3U\text{-}O$	307 ± 9	1
$(\eta^5\text{-}C_5H_4SiMe_3)_3U\text{-}CO$	43.1 ± 0.8	1
$(C_9H_7)_3U\text{-}CH_3$	196.3 ± 6.6	1
$(\eta^5\text{-}C_5Me_5)_2U(Cl)\text{-}C_6H_5$	358 ± 11	1
$(\eta^5\text{-}C_5H_4SiMe_3)_3U\text{-}THF$	41.0 ± 0.8	1

(10.4) Group 4

Bond	D^o_{298}/kJ mol^{-1}	Ref.
$Ti(\eta^5\text{-}C_5H_5)_2\text{-}Cl$	471	1
$Ti(Cl)(\eta^5\text{-}C_5H_5)_2\text{-}Cl$	390	1
$Ti(\eta^5\text{-}C_5Me_5)_2\text{-}I$	219	1
$Ti(\eta^5\text{-}C_5H_5)_2\text{-}CO$	174	1
$Ti(CO)(\eta^5\text{-}C_5H_5)_2\text{-}CO$	170	1
$Ti\text{-}CH_3$	174 ± 29	1
$Ti(Cl)(\eta^5\text{-}C_5H_5)_2\text{-}CH_3$	276	1
$Ti(Cl)((\eta^5\text{-}C_5H_5)_2\text{-}C_6H_5$	292	1
$Ti(C_6H_6)\text{-}C_6H_5$	308.7	1
$Zr(\eta^5\text{-}C_5Me_5)_2\text{-}H$	351.0 ± 7.5	1
$Zr(H)(\eta^5\text{-}C_5Me_5)_2\text{-}H$	326.4 ± 4	1
$Zr(\eta^5\text{-}C_5Me_5)_2\text{-}Cl$	481.2	1
$Zr(\eta^5\text{-}C_5Me_5)_2\text{-}Br$	410.0	1
$Zr(I)(\eta^5\text{-}C_5Me_5)_2\text{-}I$	336.4 ± 2.1	1
$Zr(\eta^5\text{-}C_5Me_5)_2(Ph)\text{-}OH$	482.4 ± 6.3	1
$Zr(\eta^5\text{-}C_5Me_5)_2(Ph)(OH)$ $\text{-}OH$	482.8 ± 10.5	1
$Zr(\eta^5\text{-}C_5Me_5)_2(NH_2)$ $H\text{-}NH_2$	421.3 ± 15.1	1
$Zr(\eta^5\text{-}C_5Me_5)_3\text{-}CH_3$	276 ± 10	1
$Zr(\eta^5\text{-}C_5H_5)_2(C_6H_5)\text{-}$ C_6H_5	300 ± 10	1
$Zr(\eta^5\text{-}C_5H_5)_2(Si(SiMe_3)_3)$ $\text{-}SiMe_3$	188 ± 30	1
$Hf(H)(\eta^5\text{-}C_5Me_5)_2\text{-}H$	346.0 ± 7.9	1
$Hf(\eta^5\text{-}C_5Me_5)(C_4H_9)\text{-}$ C_4H_9	274 ± 10	1

(10.5) Group 5

Bond	D^o_{298}/kJ mol^{-1}	Ref.
$(\eta^5\text{-}C_5H_5)(CO)_3V\text{-}\eta^2H_2$	90 ± 20	1
$(\eta^5\text{-}C_5H_5)(CO)_3V\text{-}CO$	146 ± 21	1
$V\text{-}CH_3$	169 ± 18	1
$V\text{-}C_6H_6$	76.2	1

Bond	D^o_{298}/kJ mol^{-1}	Ref.
$V(C_6H_6)\text{-}C_6H_6$	307.8	1
$Nb(\eta^5\text{-}C_5H_5)_2H_3\text{-}TFE$	18.8 ± 1.3	1
$Ta(CH_3)_5\text{-}CH_3$	261 ± 5	1
$(Me_3SiCH_2)_4Ta\text{-}$ (CH_2SiMe_3)	184.1 ± 8.4	1

(10.6) Group 6

Bond	D^o_{298}/kJ mol^{-1}	Ref.
$[Cr(CO)_3(\eta^5\text{-}C_5Me_5)]_2\text{-}$ Hg	61.5	1
$[Cr(CO)_3(\eta^5\text{-}C_5Me_5)]\text{-}$ Hg	111.3	1
$Cr(CO)_5\text{-}Xe$	37.7 ± 3.8	1
$(CO)_2(PPh_3)(\eta^5\text{-}C_5H_5)$ $Cr\text{-}H$	250.2 ± 4.2	1
$(\eta^5\text{-}C_5H_5)Cr(CO)_3\text{-}H,$	257	1
$Cr(CO)_5\text{-}H_2$	78 ± 4	1
$(P(C_6H_{11})_3)_2(CO)_3Cr\text{-}H_2$	30.5 ± 0.4	1
$(\eta^6\text{-}C_6H_6)(CO)_3Cr\text{-}H_2$	251 ± 17	1
$Cr(CO)_5\text{-}N_2$	81 ± 4	1
$(P(C_6H_{11})_3)_2(CO)_3Cr\text{-}N_2$	38.9 ± 0.8	1
$(\eta^5\text{-}C_5Me_5)(CO)_3Cr\text{-}SH$	193	1
$Cr(CO)_5\text{-}CO$	154.0 ± 8.4	1
$Cr(CO)_5\text{-}CH_4$	~33.5 ± 8	1
$Cr\text{-}C_6H_6$	9.6 ± 5.8	1
$Cr(C_6H_6)\text{-}C_6H_6$	268.2 ± 15.4	1
$Cr(CO)_5\text{-}C_6H_6$	57.3 ± 3.3	1
$(P(C_6H_{11})_3)_2(CO)_3Cr\text{-}$ $P(OMe_3)_3$	68.6 ± 2.5	1
$(\eta^5\text{-}C_5H_5))Mo(CO)_3\text{-}H$	290	1
$Mo(\eta^5\text{-}C_5H_5)_2\text{-}H$	246	1
$Mo(H)(\eta^5\text{-}C_5H_5)_2\text{-}H$	256.9 ± 8.4	1
$Mo(CO)_3(\eta^5\text{-}C_5H_5)\text{-}I$	216.7 ± 4.2	1
$(\eta^5\text{-}C_5Me_5)_2Mo\text{-}O$	272	1
$(P(C_6H_{11})_3)_2(CO)_3Mo\text{-}H_2$	27.2 ± 0.8	1
$(P(C_6H_{11})_3)_2(CO)_3Mo\text{-}N_2$	37.7 ± 2.5	1
$Mo(CO)_5\text{-}CO$	169.5 ± 8.4	1
$Mo(CO)_3(\eta^5\text{-}C_5H_5)\text{-}CH_3$	203 ± 8	1
$W(CO)_5\text{-}Xe$	35.1 ± 0.8	1
$W(CO)_3(\eta^5\text{-}C_5H_5)\text{-}H$	303	1
$W(H)(\eta^5\text{-}C_5H_5)_2\text{-}H$	310.9 ± 4.2	1
$W(I)(\eta^5\text{-}C_5H_5)_2\text{-}H$	273 ± 14	1
$(CO)_5W\text{-}H_2$	≥67	1
$(P(C_6H_{11})_3)(CO)_3W\text{-}$ $(\eta^2\text{-}H_2)$	28.5 ± 2.1	1
$W(CO)_5\text{-}CO$	192.5 ± 8.48.4	1
$W(CH_3)(\eta^5\text{-}C_5H_5)_2\text{-}CH_3$	220.9 ± 4	1

(10.7) Group 7

Bond	D^o_{298}/kJ mol^{-1}	Ref.
$F_3Mn\text{-}MnF_3$	210.9 ± 2.5	1
$(CO)_5Mn\text{-}Mn(CO)_5$	185 ± 8	1
$(CO)_5Mn\text{-}H$	284.5	1
$(PPh_3)Mn(CO)_4\text{-}H$	286.2	1
$MnBr(CO)_4\text{-}CO$	184	1
$(\eta^5\text{-}C_5H_5)(CO)_2Mn\text{-}CO$	195.8 ± 9.2	1
$Mn\text{-}CH_3$	>35 ± 12	1
$Mn(CO)_5\text{-}CH_3$	187.0 ± 3.8	1
$Mn(CO)_5\text{-}C_6H_5$	207 ± 11	1

Bond	D^o_{298}/kJ mol^{-1}	Ref.
$(CO)_5Mn\text{-}Re(CO)_5$	149 ± 11	1
$(\eta^5\text{-}C_5H_5)Mn(CO)_2\text{-}$ $PhMe$	59.4 ± 3.3	1
$(CO)_5Tc\text{-}Tc(CO)_5$	177.5 ± 1.9	1
$(CO)_5Re\text{-}Re(CO)_5$	187 ± 4.8	1
$(CO)_5Re\text{-}H$	313	1
$(CO)_5Re\text{-}CH_3$	220 ± 8	1

(10.8) Group 8

Bond	D^o_{298}/kJ mol^{-1}	Ref.
$(CO)_4Fe\text{-}Fe(CO)_5$	171.5	1
$(CO)_4Fe(H)_x\text{-}H$	259.4 ± 8.4	1
$(\eta^5\text{-}C_5H_5)(CO)_2Fe\text{-}H$	239	1
$Fe(CO)_3(N_2)\text{-}N_2$	37.7 ± 19.2	1
$Fe(C_2H_2)(CO)_4\text{-}CO$	88 ± 2.3	1
$Fe(CO)_2(PMe_3)\text{-}CO$	>125	1
$Fe(CO)_3(PPh_3)\text{-}CO$	<177.8 ± 5	1
$Fe\text{-}NH_3$	31.4 ± 4.2	1
$Fe\text{-}CH_2$	364 ± 29	1
$Fe\text{-}CH_3$	135 ± 29	1
$Fe(C_2H_4)(CO)_3\text{-}C_2H_4$	89.1 ± 8	1
$Fe\text{-}C_3H_5$	218	1
$Fe\text{-}C_3H_6$	79	1
$Fe(CO)_5\text{-}Ni(CO)_4$	37.7	1
$Fe(CO)_5\text{-}(\eta^3\text{-}C_3H_5)$	176	1
$Fe(C_3H_6)(CO)_3\text{-}C_3H_6$	~79.5	1
$(CO_2)(\eta^5\text{-}C_5H_5)Ru\text{-}H$	272	1
$(PMe_3)_2(\eta^5\text{-}C_5Me_5)Ru\text{-}H$	167.4	1
$(CO)_2(\eta^5\text{-}C_5Me_5)Ru\text{-}Cl$	337.6	1
$(\eta^5\text{-}C_5Me_5)(PMe_3)_2Ru\text{-}$ Cl	<138	1
$(\eta^5\text{-}C_5Me_5)(PMe_3)_2Ru\text{-}$ OH	204.6	1
$(CO)_4Ru\text{-}CO$	115 ± 1.7	1
$(\eta^5\text{-}C_5Me_5)(PMe_3)_2Ru\text{-}$ CH_3	142.3	1
$Os(H)(CO)_4\text{-}H$	326.4	1
$(CO)_4Os\text{-}CO$	133 ± 2.6	1
$Os(C_2H_2)(CO)_4\text{-}CO$	99.5 ± 0.8	1

(10.9) Group 9

Bond	D^o_{298}/kJ mol^{-1}	Ref.
$(CO)_4Co\text{-}Co(CO)_4$	83 ± 29	1
$(CO)_4Co\text{-}Mn(CO)_5$	96 ± 12	1
$(CO)_4Co\text{-}Re(CO)_5$	113 ± 15	1
$Co(CO)_4\text{-}H$	278	1
$Co(CO)_3(PPh_3)\text{-}H$	272	1
$(CO)_3HCo\text{-}CO$	~54	1
$(\eta^5\text{-}C_5H_5)Co(CO)\text{-}CO$	184.3 ± 4.8	1
$Co\text{-}CH_2$	331 ± 38	1
$Co\text{-}CH_3$	178 ± 8	1
$cobalamin\text{-}CH_3$	150.6	1
$cobinamide\text{-}iC_4H_9$	104	1
$Co\text{-}C$ bonds in B_{12}	123.8 ± 6.3	1
$Cl(CO)_2Rh\text{-}Rh(CO)_2Cl$	94.6	1
$HRh(m\text{-}xylyl)Rh\text{-}H$	255.6 ± 1.7	1
$(PiPr_3)_2(Cl)Rh\text{-}H_2$	136.0	1
$(PiPr_3)_2(Cl)Rh\text{-}N_2$	69.0	1
$(PiPr_3)_2(Cl)Rh\text{-}CO$	201.7	1

Bond	$D^o{}_{298}$/kJ mol⁻¹	Ref.	Bond	$D^o{}_{298}$/kJ mol⁻¹	Ref.	Bond	$D^o{}_{298}$/kJ mol⁻¹	Ref.
HRh(m-xylyl)Rh–CH₂OH	195.4 ± 7.5	1	Pt(η⁵-C₅H₅)(CH₃)₂–CH₃	163 ± 21	1	Hg–CH₃	22.6 ± 12.6	1
Ir(Cl)(CO)(PMe₃)₂–H	251	1	cis-Pt(PEt₃)₂(CH₃)–CH₃	269 ± 13	1	Hg(CH₃)–CH₃	239.3 ± 6.3	1
Ir(H)(η⁵-C₅Me₅)(PMe₃)–H	310.5 ± 21	1	**(10.11) Group 11**			ClHg–CH₃	280.0 ± 12.6	1
Ir(Cl)(H)(CO)(PEt₃)₂–H	243.1	1	Cu–OH	>406	1	BrHg–CH₃	270 ± 38	1
Ir(Cl)(H)(CO)(PPh₃)₂–H	246.9	1	Cu–CO	25 ± 5	1	IHg–CH₃	258.6 ± 12.6	1
(Cl)(CO)(PPh₃)₂Ir–H₂	62.8	1	Cu–CH₃	223 ± 5	1			
(Cl)(CO)(PPh₃)₂Ir–CO	45.2	1	Cu–NH₃	47 ± 15	1	**(10.13) Group 13**		
Ir(H)(η⁵-C₅Me₅)(PMe₃)–C₆H₅	321	1	Cu(NH₃)–NH₃	83.7 ± 4.2	1	H₃B–BH₃	172	1
			Cu–C₆H₆	16.4 ± 12.5	1	H₃B–NH₃	130.1 ± 4.2	1
(10.10) Group 10			Cu(C₆H₆)–C₆H₆	27.0 ± 19.3	1	(CH₃)₃B–NH₃	57.7 ± 1.3	1
Ni–H₂O	~29	1	Ag–CH₃	134.1 ± 6.8	1	F₃B–N(CH₃)₃	130 ± 4.6	1
Ni(CO)₃–N₂	~42	1	Ag–NH₃	8 ± 13	1	Cl₃B–N(CH₃)₃	127.6	1
Ni(CO)₃–CO	104.6 ± 8.4	1	Ag(NH₃)–NH₃	62.8 ± 4.2	1	F₂B–CH₃	397 - 418	1
Ni–CH₃	208 ± 8	1	Au–OH	>262	1	Al–OH	547 ± 13	1
Ni–C₂H₂	193 ± 25	1	Au–NH₃	76 ± 6	1	Al–C₂H₂	>54	1
Ni–C₂H₄	147.3 ± 17.6	1	Au–CH₃	≥191.6	1	Cl₃Al–N(CH₃)₃	198.7 ± 8.4	1
Ni–propyne	155 ± 21	1	Au–C₆H₆	8.4	1	(CH₃)₃Al–N(CH₃)₃	130	1
Ni–2-butyne	121 ± 21	1				(CH₃)₃Al–O(CH₃)₂	92	1
Pd–OH	213	1	**(10.12) Group 12**			CH₃Ga–GaH₃	59 ± 16	7
trans-Pt(PPh₃)₂(Cl)–H	307 ± 37	1	Zn–CH₃	70 ± 10	1	(CH₃)₃Ga–O(C₂H₅)₂	50.6 ± 0.8	1
[Ph₂PCH₂]₂MePt–H	104.6	1	Zn(CH₃)–CH₃	266.5 ± 6.3	1	Cl₃Ga–S(C₂H₅)₂	235.1	1
[Ph₂PCH₂]₂MePt–OH	167.4	1	Zn–C₂H₅	92.0 ± 17.6	1	In–CH₃	216.3	1
[Ph₂PCH₂]₂MePt–SH	90.0	1	Zn(C₂H₅)–C₂H₅	219.2 ± 8.4	1	In(CH₃)₁–CH₃	318.8	1
			Cd–CH₃	63.6 ± 10.0	1	In(CH₃)₂–CH₃	587.4	1
			Cd(CH₃)–CH₃	234.3 ± 6.3	1	(CH₃)₃In–N(CH₃)₃	83.3 ± 2.1	1
						Tl–OH	330 ± 30	1

References

1. Luo, Y. R. *Comprehensive Handbook of Chemical Bond Energies*, CRC Press, Boca Raton, FL, 2007.
2. Shuman, N. S., Ochieng, M. A., Sztáray, B., and Baer, T., *J. Phys. Chem. A* 112, 5647, 2008.
3. Seetula, J. A., and Eskola, A. J., *Chem. Phys.* 351, 141, 2008.
4. Golden, D. M., *Int. J. Chem. Kinet.* 41, 573, 2009.
5. Shuman, N. S., Spencer, A. P., and Baer, T., *J. Phys. Chem. A* 113, 9458, 2009.
6. Villano, S. M., Eyet, N., Wren, S. W., Ellison, G. B., Bierbaum, V. M., and Lineberger, W. C., *J. Phys. Chem. A* 114, 191, 2010.
7. Downs, A.J., Greene, T. M., Johnsen, E., Pulham, C. R., Robertson, H. E., and Wann, D. A., *Dalton Trans.* 39, 5637, 2010.
8. Villano, S. M., Eyet, N., Wren, S. W., Ellison, G. B., Bierbaum, V. M., and Lineberger, W. C., *Eur. J. Mass Spectrom.* 16, 255, 2010.
9. Bodi, A., Kercher, J. P., Bond, C., Meteesatien, P., Sztáray, B., and Baer, T., *J. Phys. Chem. A* 110, 13425, 2006. Derived from heats of formation of compounds.
10. Stevens, W. R., Bodi, A., and Baer, T., *J. Phys. Chem. A* 114, 11285, 2010. Derived from heats of formation of compounds.
11. Da'valos, J. Z., Herrero, R., Shuman, N. S., and Baer, T., *J. Phys. Chem. A* 115, 402, 2011.

TABLE 4. Enthalpies of Formation of Free Radicals and Other Transient Species

References: Yu-Ran Luo, *Comprehensive Handbook of Chemical Bond Energies*, CRC Press, 2007.

Radical	$\Delta_f H^o{}_{298}$/kJ mol⁻¹	Ref.	Radical	$\Delta_f H^o{}_{298}$/kJ mol⁻¹	Ref.
(1) Carbon-Centered Species			•C₃H₃, CH₂=C=CH• ↔ CH≡CC•H₂	351.9	2
CH	595.8 ± 0.6	1	•C₃H₃, cyclopro-2-en-1-yl	439.7 ± 17.2	1
CH₂ (triplet)	391.2 ± 1.6	1	•C₃H₅, allyl, CH₂=CHC•H₂	171.0 ± 3.0	1
CH₂ (singlet)	428.8 ± 1.6	1	•C₃H₅, CH₃CH=C•H	267 ± 6	1
•CH₃, methyl	146.7 ± 0.3	1	•C₃H₅, CH₃C•=CH₂	231.4	1
•C₂H, acetenyl, CH≡C•	567.4 ± 2.1	1	•C₃H₅, cyclopropyl	279.9 ± 10.5	1
•C₂H₂, vinylidene CH₂=C••	419.7 ± 16.7	1	n-C₃H₇•, n-propyl, CH₃CH₂C•H₂	100 ± 2	1
•C₂H₃, vinyl, CH₂=C•H	299.6 ± 3.3	1	i-C₃H₇•, i-propyl, CH₃C•HCH₃	88 ± 3	1
•C₂H₅, ethyl, CH₃C•H₂	118.8 ± 1.3	1	•n-C₄H₃, CH≡CCH=C•H	547.3	1
•C₃H₃, propargyl, CH≡CC•H₂	351.9	2	•i-C₄H₃, CH₂=C•C≡CH	499.2	1
•C₃H₃, CH₃C≡C•	515 ± 13	1	•C₄H₅, CH₃C≡CC•H₂	304.5	1
			•C₄H₅, CH≡CC•HCH₃	316.5	1

Radical	$\Delta_f H^o_{298}$/kJ mol^{-1}	Ref.
$^\bullet$C$_4$H$_5$, $^\bullet$CH=CHCHCH$_2$	364.4	1
$^\bullet$C$_4$H$_5$, CH$_2$=CHC$^\bullet$H$_2$	313.3	1
$^\bullet$C$_4$H$_7$, CH$_3$CH=CHC$^\bullet$H$_2$	146 ± 8	1
$^\bullet$C$_4$H$_7$, CH$_2$=CHCH$_2$C$^\bullet$H$_2$	192.5	1
$^\bullet$C$_4$H$_7$, CH$_2$=C(CH$_3$)C$^\bullet$H$_2$	137.9	1
$^\bullet$C$_4$H$_7$, CH$_2$=CHC$^\bullet$HCH$_3$	136.2	1
$^\bullet$C$_4$H$_7$, cyclopropylmethyl	213.8 ± 6.7	1
$^\bullet$C$_4$H$_7$, cyclobutyl	219.2 ± 4.2	1
n-C$_4$H$_9$$^\bullet$, n-butyl, CH$_3CH_2CH_2C^\bullet$H$_2$	77.8 ± 2.1	1
i-C$_4$H$_9$$^\bullet$, i-butyl, (CH$_3$)$_2CHC^\bullet$H$_2$	70 ± 4	1
s-C$_4$H$_9$$^\bullet$, s-butyl, CH$_3C^\bullet$HCH$_2CH_3$	67.8 ± 2.1	1
t-C$_4$H$_9$$^\bullet$, t-butyl, (CH$_3$)$_3C^\bullet$	48 ± 3	1
$^\bullet$C$_5$H$_3$, CH≡C-C≡CC$^\bullet$H$_2$	579.1	1
$^\bullet$C$_5$H$_3$, (CH≡C)$_2$C$^\bullet$H	573.2	1
$^\bullet$C$_5$H$_5$, CH$_2$=CHC≡CC$^\bullet$H$_2$	351.5	1
$^\bullet$C$_5$H$_5$, CH$_2$=CH-C$^\bullet$H-C≡CH	372.4	1
$^\bullet$C$_5$H$_5$, cyclopenta-1,3-dien-5-yl	274.1 ± 7.3	1
$^\bullet$C$_5$H$_7$, CH$_3$C≡CC$^\bullet$HCH$_3$	272.8 ± 9.2	1
$^\bullet$C$_5$H$_7$, CH≡CC$^\bullet$HC$_2$H$_5$	277.0 ± 8.4	1
$^\bullet$C$_5$H$_7$, CH≡CC$^\bullet$(CH$_3$)$_2$	257.3 ± 9.2	1
$^\bullet$C$_5$H$_7$, CH$_2$=CHCH=CHC$^\bullet$H$_2$	205.0 ± 12.6	1
$^\bullet$C$_5$H$_7$, (CH$_2$=CH)$_2$C$^\bullet$H	208.0 ± 4.2	1
$^\bullet$C$_5$H$_7$, CH$_3$CH=C=CHC$^\bullet$H$_2$	278.0	1
$^\bullet$C$_5$H$_7$, spiropentyl	380.7 ± 4.2	1
$^\bullet$C$_5$H$_7$, cyclopent-1-en-3-yl	160.7 ± 4.2	1
$^\bullet$C$_5$H$_9$, cyclopentyl	105.9 ± 4.2	1
$^\bullet$C$_5$H$_9$, CH$_2$=CHC$^\bullet$HCH$_2$CH$_3$	109.6 ± 8.4	1
$^\bullet$C$_5$H$_9$, CH$_3$CH=CHC$^\bullet$H(CH$_3$)	92	1
$^\bullet$C$_5$H$_9$, CH$_3$CH=C(CH$_3$)C$^\bullet$H$_2$	92.0	1
$^\bullet$C$_5$H$_9$, CH$_2$=CHC$^\bullet$(CH$_3$)$_2$	87.0 ± 8.4	1
$^\bullet$C$_5$H$_9$, CH$_2$=C(CH$_3$)C$^\bullet$H(CH$_3$)	93.7	1
$^\bullet$C$_5$H$_9$, CH$_2$=C(C$^\bullet$H$_2$)CH$_2$CH$_3$	114.2	1
$^\bullet$C$_5$H$_9$, CH$_2$=CH(CH$_2$)$_2$C$^\bullet$H$_2$	179.5	1
nC$_5$H$_{11}$$^\bullet$, CH$_3CH_2CH_2CH_2C^\bullet$H$_2$	54.4	1
$^\bullet$C$_5$H$_{11}$, (C$_2$H$_5$)$_2$C$^\bullet$H	47.0	1
$^\bullet$C$_5$H$_{11}$, (nC$_3$H$_7$)(CH$_3$)C$^\bullet$H	50.2	1
$^\bullet$C$_5$H$_{11}$, (CH$_3$)$_3$C$^\bullet$CH$_2$	36.4 ± 8.4	1
$^\bullet$C$_5$H$_{11}$, (C$_2$H$_5$)(CH$_3$)$_2$C$^\bullet$	29	1
$^\bullet$C$_6$H$_5$, phenyl	330.1 ± 3.3	1
$^\bullet$C$_6$H$_7$, cyclohexa-1,3-dien-5-yl	199.2	1
$^\bullet$C$_6$H$_7$, cyclohexa-1,4-dien-3-yl	208.0 ± 3.9	5
$^\bullet$C$_6$H$_9$, CH$_3$C≡CC$^\bullet$(CH$_3$)$_2$	221.8 ± 9.2	1
$^\bullet$C$_6$H$_9$, (CH$_2$=CH)$_2$C$^\bullet$(CH$_3$)	193.7	1
$^\bullet$C$_6$H$_9$, cyclohexa-1-en-3-yl	119.7	1
$^\bullet$C$_6$H$_{11}$, CH$_2$=CH(CH$_2$)$_3$C$^\bullet$H$_2$	158.6	1
$^\bullet$C$_6$H$_{11}$, CH$_2$=CHC$^\bullet$H(CH$_2$)$_2$CH$_3$	89.0	1
$^\bullet$C$_6$H$_{11}$, CH$_2$=C(CH$_3$)C$^\bullet$(CH$_3$)$_2$	37.7 ± 6.3	1
$^\bullet$C$_6$H$_{11}$, (CH$_3$)$_2$C=C(CH$_3$)C$^\bullet$H$_2$	39.7 ± 6.3	1
$^\bullet$C$_6$H$_{11}$, (CH$_3$)$_2$C=CHC$^\bullet$H(CH$_3$)	47.3	1
$^\bullet$C$_6$H$_{11}$, (Z)-CH$_3$CH=CHC$^\bullet$(CH$_3$)$_2$	54.4	1
$^\bullet$C$_6$H$_{11}$, cyclohexyl	75.3 ± 6.3	1
nC$_6$H$_{13}$$^\bullet$, CH$_3CH_2CH_2CH_2CH_2C^\bullet$H$_2$	33.5	1
$^\bullet$C$_6$H$_{13}$, (nC$_4$H$_9$)(CH$_3$)C$^\bullet$H	29.3	1
$^\bullet$C$_6$H$_{13}$, 2-methyl-2-pentyl	3.3 ± 8.4	1
$^\bullet$C$_6$H$_{13}$, 3-methyl-3-pentyl	14.2	1

Radical	$\Delta_f H^o_{298}$/kJ mol^{-1}	Ref.
$^\bullet$C$_6$H$_{13}$, 2,3-dimethyl-2-butyl	3.1 ± 10	1
$^\bullet$C$_7$H$_3$, (CH≡C)$_3$C$^\bullet$	784.5	1
$^\bullet$C$_7$H$_7$, benzyl, C$_6$H$_5$C$^\bullet$H$_2$	208.0 ± 1.7	1
$^\bullet$C$_7$H$_7$, quadricyclolan-5-yl	578.6 ± 5.4	1
$^\bullet$C$_7$H$_7$, quadricyclolan-4-yl	587.4 ± 5.4	1
$^\bullet$C$_7$H$_7$, norborna-2,5-dien-7-yl	511.7 ± 7.9	1
$^\bullet$C$_7$H$_7$, cyclohepta-1,3,5-trien-7-yl	285.3 ± 12.6	1
$^\bullet$C$_7$H$_9$, CH$_2$=CH(CH=CH)$_2$CC$^\bullet$H$_2$	251.0	1
$^\bullet$C$_7$H$_9$, (CH$_2$=CH)$_3$C$^\bullet$	274.0	1
$^\bullet$C$_7$H$_{11}$, norborn-1-yl	136.4 ± 10.5	1
$^\bullet$C$_7$H$_{11}$, cycloheptenyl	119.2	1
$^\bullet$C$_7$H$_{13}$, cycloheptyl	50.6 ± 4.2	1
$^\bullet$C$_7$H$_{13}$, cyclo-[C$^\bullet$(CH$_3$)(CH$_2$)$_5$]	22.6	1
$^\bullet$C$_7$H$_{13}$, cyclo-[C$^\bullet$(CH$_2$CH$_3$)(CH$_2$)$_4$]	47.0	1
$^\bullet$C$_7$H$_{15}$, (nC$_5$H$_{11}$)(CH$_3$)CH$^\bullet$	8.4	1
$^\bullet$C$_7$H$_{15}$, (CH$_3$)$_2$CHCHC$^\bullet$(CH$_3$)$_2$	−21.8 ± 5.2	1
$^\bullet$C$_8$H$_7$, cubyl	831.0 ± 16.7	1
$^\bullet$C$_8$H$_7$, C$_6$H$_5$C$^\bullet$=CH$_2$	309.6	1
$^\bullet$C$_8$H$_7$, C$_6$H$_5$CH=CH$^\bullet$	387.0	1
$^\bullet$C$_8$H$_9$, C$_6$H$_5$C$^\bullet$H(CH$_3$)	175.7 ± 7.5	1
$^\bullet$C$_8$H$_9$, C$_6$H$_5$CH$_2$C$^\bullet$H$_2$	236.0 ± 7.5	1
$^\bullet$C$_8$H$_9$, p-CH$_3$C$_6$H$_4$C$^\bullet$H$_2$	167.4	1
$^\bullet$C$_8$H$_9$, m-CH$_3$C$_6$H$_4$C$^\bullet$H$_2$	167.4	1
$^\bullet$C$_8$H$_9$, o-CH$_3$C$_6$H$_4$C$^\bullet$H$_2$	167.4	1
$^\bullet$C$_8$H$_9$, 1-vinyl-cyclohexa-2,4-dienyl	247.7 ± 14.2	1
$^\bullet$C$_8$H$_9$, 2-vinyl-cyclohexa-2,4-dienyl	249.8 ± 14.2	1
$^\bullet$C$_8$H$_9$, 3-vinyl-cyclohexa-2,4-dienyl	269.4 ± 14.2	1
$^\bullet$C$_8$H$_9$, 6-vinyl-cyclohexa-2,4-dienyl	284.5 ± 14.2	1
$^\bullet$C$_8$H$_{13}$, CH$_2$=CHCH=CHC$^\bullet$H(CH$_2$)$_2$CH$_3$	130.5	1
$^\bullet$C$_8$H$_{13}$, CH$_2$=CHC$^\bullet$H(CH$_2$)$_3$CH=CH$_2$	130.5	1
$^\bullet$C$_8$H$_{13}$, bicyclooct-1-yl	92.0	1
$^\bullet$C$_8$H$_{15}$, CH$_2$=CHC$^\bullet$H(CH$_2$)$_4$CH$_3$	49.8	1
$^\bullet$C$_8$H$_{15}$, (E)-CH$_3$CH=C$^\bullet$(CH$_2$)$_4$CH$_3$	29.7	1
$^\bullet$C$_8$H$_{15}$, (Z)-(CH$_3$)$_2$C$^\bullet$CH=CHCH(CH$_3$)$_2$	9.2	1
$^\bullet$C$_8$H$_{15}$, cyclooctanyl	59.4	1
$^\bullet$C$_8$H$_{15}$, cyclo-[C$^\bullet$(CH$_2$CH$_3$)(CH$_2$)$_5$]	10.0	1
$^\bullet$C$_9$H$_7$, indenyl	297.1	1
$^\bullet$C$_9$H$_9$, indanyl-1	204.2 ± 8.4	1
$^\bullet$C$_9$H$_{11}$, 2,6-dimethylbenzyl	124.7	1
$^\bullet$C$_9$H$_{11}$, 3,6-dimethylbenzyl	124.7	1
$^\bullet$C$_9$H$_{11}$, 3,5-dimethylbenzyl	124.7	1
$^\bullet$C$_9$H$_{11}$, C$_6$H$_5$C$^\bullet$(CH$_3$)$_2$	133.9 ± 4.2	1
$^\bullet$C$_9$H$_{11}$, o-$^\bullet$C$_6$H$_4$C$_2$H$_5$	279.5 ± 7.5	1
$^\bullet$C$_9$H$_{17}$, cyclononanyl	52.3	1
$^\bullet$C$_{10}$H$_7$, naphth-1-yl	401.7 ± 5.4	1
$^\bullet$C$_{10}$H$_7$, naphth-2-yl	400.4 ± 5.9	1
$^\bullet$C$_{10}$H$_{11}$, tetralin-1-yl	154.8 ± 5.0	1
$^\bullet$C$_{10}$H$_{13}$, 1-phenyl-but-4-yl	192.0	1
$^\bullet$C$_{10}$H$_{13}$, (C$_6$H$_5$CH$_2$)(C$_2$H$_5$)C$^\bullet$H	184.5	1
$^\bullet$C$_{10}$H$_{13}$, (C$_6$H$_5$CH$_2$CH$_2$)(CH$_3$)C$^\bullet$H	184.5	1
$^\bullet$C$_{10}$H$_{13}$, (C$_6$H$_5$C$^\bullet$HCH$_2$CH$_2$CH$_3$	134.7	1
$^\bullet$C$_{10}$H$_{15}$, 1-adamantyl	51.5	1
$^\bullet$C$_{10}$H$_{15}$, 2-adamantyl	61.9	1
$^\bullet$C$_{10}$H$_{19}$, cyclodecanyl	32.2	1
$^\bullet$C$_{11}$H$_9$, 1-naphthylmethyl	252.7	1

Radical	$\Delta_f H^o_{298}$/kJ mol^{-1}	Ref.	Radical	$\Delta_f H^o_{298}$/kJ mol^{-1}	Ref.
${}^{\bullet}C_{11}H_{21}$, cycloundecanyl	7.5	1	${}^{\bullet}C_2HF_4$, $CHF_2C^{\bullet}F_2$	−664.8	1
${}^{\bullet}C_{12}H_{23}$, cyclododecanyl	−38.5	1	${}^{\bullet}C_2H_2F_3$, $CF_3C^{\bullet}H_2$	−517.1 ± 8.4	1
${}^{\bullet}C_{13}H_9$, 9-fluorenyl	297.5	1	${}^{\bullet}C_2H_2F_3$, $CHF_2C^{\bullet}HF$	−456.0	1
${}^{\bullet}C_{13}H_{11}$, $(C_6H_5)_2C^{\bullet}H$	302.1 ± 4.2	1	${}^{\bullet}C_2H_2F_3$, $CH_2FC^{\bullet}F_2$	−449.8	1
${}^{\bullet}C_{13}H_{11}$, 9-methyl-9-fluorenyl	268.2	1	${}^{\bullet}C_2H_2F_2Cl$, $CF_2ClC^{\bullet}H_2$	−310.9 ± 7.0	1
${}^{\bullet}C_{14}H_{11}$, 9,10-dihydroanthracen-9-yl	261.0	1	${}^{\bullet}C_2H_3F_2$, $CH_3C^{\bullet}F_2$	−302.5 ± 8.4	1
${}^{\bullet}C_{15}H_{11}$, 9-anthracenylmethyl	337.6	1	${}^{\bullet}C_2H_3F_2$, $CHF_2C^{\bullet}H_2$	−285.8	1
${}^{\bullet}C_{15}H_{11}$, 9-phenanthrenylmethyl	311.3	1	${}^{\bullet}C_2H_3F_2$, $CH_2FC^{\bullet}HF$	−238.5	1
${}^{\bullet}C_{16}H_{31}$, $CH_2{=}CHC^{\bullet}H(CH_2)_{12}CH_3$	−118.8	1	${}^{\bullet}C_2H_4F$, $CH_3C^{\bullet}HF$	−70.3 ± 8.4	1
${}^{\bullet}C_{19}H_{15}$, trityl, $(C_6H_5)_3C^{\bullet}$	392.0 ± 8.4	1	${}^{\bullet}C_2H_4F$, $CH_2FC^{\bullet}H_2$	−59.4 ± 8.4	1
${}^{\bullet}C_{35}H_{25}$, pentamethylcyclopentadienyl	67.4	1	${}^{\bullet}C_2H_2F_2Cl$, $CF_2ClC^{\bullet}H_2$	−315.2 ± 6	1
CF	255.2 ± 8	1	${}^{\bullet}C_2F_4Cl$, $CF_2ClC^{\bullet}F_2$	−686.0	1
CF_2	−182.0 ± 6.3	1	${}^{\bullet}C_2HF_3Cl$, $CClF_2C^{\bullet}HF$	−450.6 ± 12.6	1
$FC^{\bullet}(O)$	−161.2 ± 8.4	1	${}^{\bullet}C_2F_4Cl$, $CF_3C^{\bullet}FCl$	−728.0	1
CHF	143.0 ± 12.6	1	${}^{\bullet}C_2F_3Cl_2$, $CF_3C^{\bullet}Cl_2$	−564.0	1
CClF	31.0 ± 13.4	1	${}^{\bullet}C_2F_3ClBr$, $CF_3C^{\bullet}ClBr$	−504.2 ± 8.4	1
CCl	443.1 ± 13.0	1	${}^{\bullet}C_2Cl$, $ClC{\equiv}C^{\bullet}$	534 ± 50	1
CCl_2	226	1	${}^{\bullet}C_2Cl_3$, $CCl_2{=}C^{\bullet}Cl$	190 ± 50	1
$ClC^{\bullet}(O)$	−21.8 ± 2.5	1	${}^{\bullet}C_2Cl_5$, $CCl_3C^{\bullet}Cl_2$	35.1 ± 5.4	1
CHCl	326.4 ± 8.4	1	${}^{\bullet}C_2HCl_4$, $CHCl_2C^{\bullet}Cl_2$	23.4 ± 8.4	1
CClBr	267	1	${}^{\bullet}C_2HCl_4$, $CCl_3C^{\bullet}HCl$	51.0	1
CBr	510 ± 63	1	${}^{\bullet}C_2H_2Cl_3$, $CH_2ClC^{\bullet}Cl_2$	26.4	1
CHBr	373 ± 18	1	${}^{\bullet}C_2H_2Cl_3$, $CHCl_2C^{\bullet}HCl$	46.4	1
CBr_2	343.5	1	${}^{\bullet}C_2H_2Cl_3$, $CCl_3C^{\bullet}H_2$	71.5 ± 8	1
CI	570 ± 35	1	${}^{\bullet}C_2H_3Cl_2$, $CH_3C^{\bullet}Cl_2$	42.5 ± 1.7	1
CI_2	468 ± 60	1	${}^{\bullet}C_2H_3Cl_2$, $CH_2ClC^{\bullet}ClH$	65.3	1
${}^{\bullet}CF_3$	−465.7 ± 2.1	1	${}^{\bullet}C_2H_3Cl_2$, $CHCl_2C^{\bullet}H_2$	90.1 ± 0.8	1
${}^{\bullet}CHF_2$	−238.9 ± 4.2	1	${}^{\bullet}C_2H_4Cl$, $CH_3C^{\bullet}HCl$	76.5 ± 1.6	1
${}^{\bullet}CH_2F$	−31.8 ± 4.2	1	${}^{\bullet}C_2H_4Cl$, $CH_2ClC^{\bullet}H_2$	93.0 ± 2.4	1
${}^{\bullet}CClF_2$	−279.0 ± 8.4	1	${}^{\bullet}C_2H_3Br_2$, $CH_3C^{\bullet}Br_2$	140.2 ± 5.4	1
${}^{\bullet}CCl_2F$	−89.0 ± 8.4	1	${}^{\bullet}C_2H_4Br$, $BrCH_2C^{\bullet}H_2$	135.1	1
${}^{\bullet}CBrClF$	−35.5 ± 6.3	1	${}^{\bullet}C_2H_4Br$, $CH_3C^{\bullet}HBr$	133.4 ± 3.4	3
${}^{\bullet}CHClF$	−60.7 ± 10.0	1	${}^{\bullet}C_2Br$, $CBrC^{\bullet}$	623.8	1
${}^{\bullet}CBrF_2$	−224.7 ± 12.6	1	${}^{\bullet}C_2Br_3$, $CBr_2C^{\bullet}Br$	385.3	1
${}^{\bullet}CCl_3$	71.1 ± 2.5	1	${}^{\bullet}C_2Br_5$, $CBr_3C^{\bullet}Br_2$	283.3	1
${}^{\bullet}CHCl_2$	87.1 ± 1.6	1	${}^{\bullet}C_3H_6Cl$, $CH_3CH_2C^{\bullet}HCl$	56.6	1
${}^{\bullet}CH_2Cl$	117.2 ± 2.9	1	${}^{\bullet}C_3H_6Cl$, $CH_3C^{\bullet}ClCH_3$	29.9 ± 0.6	1
${}^{\bullet}CHBrCl$	140 ± 4	1	${}^{\bullet}C_3H_6Br$, $C^{\bullet}H_2CH_2CH_2Br$	120.1 ± 1.3	1
${}^{\bullet}CHBr_2$	199.1 ± 2.7	3	${}^{\bullet}C_3H_6Br$, $CH_3C^{\bullet}HCH_2Br$	96.7 ± 5.9	1
${}^{\bullet}CBr_2Cl$	163 ± 8	1	${}^{\bullet}C_3H_6Br$, $CH_3CH_2C^{\bullet}HBr$	107.5 ± 2.5	1
${}^{\bullet}CBrCl_2$	124 ± 8	1	${}^{\bullet}C_6F_5$	−547.7 ± 8.4	1
${}^{\bullet}CBr_3$	214.8	1	${}^{\bullet}CH_3O$, $HOC^{\bullet}H_2$	−17.0 ± 0.7	1
${}^{\bullet}CH_2Br$	171.1 ± 2.7	1	${}^{\bullet}CH_2ClO$, $HOC^{\bullet}ClH$	−60.7 ± 7.5	1
${}^{\bullet}CI_3$	424.9 ± 2.8	1	${}^{\bullet}CHCl_2O$, $HOC^{\bullet}Cl_2$	−94.1 ± 7.5	1
${}^{\bullet}CHI_2$	314.4 ± 3.3	1	${}^{\bullet}CH_2ClO$, $ClOC^{\bullet}H_2$	135.6 ± 9.2	1
${}^{\bullet}CH_2I$	229.7 ± 8.4	1	${}^{\bullet}CH_2BrO$, $BrOC^{\bullet}H_2$	151 ± 16	1
${}^{\bullet}C_2F$, $FC{\equiv}C^{\bullet}$	460.0 ± 21.0	1	${}^{\bullet}C_2H_3O$, $C^{\bullet}H{=}CHOH$	121 ± 11	1
${}^{\bullet}C_2Cl$, $ClC{\equiv}C^{\bullet}$	568 ± 26	1	${}^{\bullet}C_2H_3O$, $C^{\bullet}H_2CHO$	13.0 ± 2	1
${}^{\bullet}C_2F_3$, $CF_2{=}C^{\bullet}F$	−192.0 ± 8.4	1	${}^{\bullet}C_2H_5O$, $CH_3C^{\bullet}HOH$	−54.0	1
${}^{\bullet}C_2F_2H$, $CF_2{=}C^{\bullet}H$	−92.9 ± 8.4	1	${}^{\bullet}C_2H_4ClO$, $CH_3C^{\bullet}ClOH$	−108.4 ± 8.8	1
${}^{\bullet}C_2F_2H$, $CHF{=}C^{\bullet}F$	−50.6 ± 8.4	1	${}^{\bullet}C_2H_4ClO$, $C^{\bullet}H_2CHClOH$	−73.2 ± 8.8	1
${}^{\bullet}CCl_2H$, $CHCl{=}C^{\bullet}Cl$	234.7 ± 8.4	1	${}^{\bullet}C_2H_3Cl_2O$, $C^{\bullet}H_2CCl_2OH$	−99.6 ± 8.8	1
${}^{\bullet}CClH_2$, $CH_2{=}C^{\bullet}Cl$	>251	1	${}^{\bullet}C_2H_5O$, $C^{\bullet}H_2CH_2OH$	−31 ± 7	1
${}^{\bullet}C_2F_5$, $CF_3C^{\bullet}F_2$	−892.9 ± 4.2	1	${}^{\bullet}C_2H_3O$, oxiran-2-yl	149.8 ± 6.3	1
${}^{\bullet}C_2HF_4$, $CF_3C^{\bullet}HF$	−680.8 ± 9.6	1	${}^{\bullet}C_3H_5O$, $CH_2{=}CHC^{\bullet}HOH$	0 ± 8.4	1

Radical	$\Delta_f H^o_{298}$/kJ mol^{-1}	Ref.	Radical	$\Delta_f H^o_{298}$/kJ mol^{-1}	Ref.
$^\bullet$C$_3$H$_7$O, CH$_3$CH$_2$C$^\bullet$HOH	-81 ± 4	1	$^\bullet$C(O)OCH$_3$	-161.5	1
$^\bullet$C$_3$H$_7$O, (CH$_3$)C$^\bullet$HCH$_2$OH	-78.7 ± 8.4	1	C$^\bullet$H$_2$C(O)OH	-248.9 ± 12.0	1
$^\bullet$C$_3$H$_7$O, HOCH$_2$CH$_2$C$^\bullet$H$_2$	-66.9 ± 8.4	1	C$^\bullet$H(CH$_3$)C(O)OH	-293 ± 3	1
$^\bullet$C$_3$H$_7$O, (CH$_3$)$_2$C$^\bullet$OH	-96.4	1	C$^\bullet$H$_2$C(O)OCH$_3$	-236.8 ± 8.4	1
$^\bullet$C$_3$H$_7$O, $^\bullet$CH$_2$CH(OH)CH$_3$	-62.8 ± 11.7	1	C$^\bullet$H$_2$C(O)OCH$_2$CH$_3$	-260.2 ± 12.6	1
$^\bullet$C$_4$H$_9$O, $^\bullet$CH$_2$C(OH)(CH$_3$)$_2$	-147.3 ± 8.4	1	C$^\bullet$H$_2$C(O)OPh	-28.0	1
$^\bullet$C$_2$H$_5$O$_3$, C$^\bullet$H$_2$OCH$_2$OOH	109.6 ± 4.2	1	$^\bullet$C$_4$H$_7$O, tetrahydrofuran-2-yl	-18.0 ± 6.3	1
PhCH$^\bullet$OH	29.3 ± 8.4	1	$^\bullet$C$_4$H$_8$O, cyclopentanon-2-yl	-41.8 ± 12.6	1
Ph$_2$C$^\bullet$OH	152.3 ± 6.3	1	$^\bullet$C$_4$H$_7$O$_2$, 1,4-dioxan-2-yl	-131.8 ± 12.6	1
$^\bullet$C$_2$H$_5$O, CH$_3$OC$^\bullet$H$_2$	0 ± 4.2	1	$^\bullet$C$_7$H$_5$O$_2$, 2-C(O)OH-$^\bullet$C$_6$H$_4$	-33.0	1
$^\bullet$C$_3$H$_7$O, CH$_3$OC$^\bullet$HCH$_3$	-57.7 ± 8.4	1	$^\bullet$C$_7$H$_5$O$_2$, 3-C(O)OH-$^\bullet$C$_6$H$_4$	-35.0	1
$^\bullet$C$_3$H$_7$O, CH$_3$CH$_2$OC$^\bullet$H$_2$	-45.2 ± 8.4	1	$^\bullet$C$_7$H$_5$O$_2$, 4-C(O)OH-$^\bullet$C$_6$H$_4$	-36.0	1
$^\bullet$C$_3$H$_7$O, C$^\bullet$H$_2$CH$_2$OCH$_3$	-7.1 ± 4.2	1	$^\bullet$CH$_3$O$_2$, C$^\bullet$H$_2$OOH	66.1	1
$^\bullet$C$_4$H$_9$O, (CH$_3$)$_2$CHOC$^\bullet$H$_2$	-70.3 ± 7.1	1	$^\bullet$C$_2$H$_5$O$_2$, C$^\bullet$H$_2$CH$_2$OOH	46.0 ± 4.6	1
$^\bullet$C$_4$H$_9$O, CH$_3$CH$_2$OC$^\bullet$HCH$_3$	-81.2 ± 4.2	1	$^\bullet$C$_2$H$_5$O$_2$, CH$_3$CH$^\bullet$OOH	26.9	1
$^\bullet$C$_4$H$_9$O, C$^\bullet$H$_2$CH(CH$_3$)OCH$_3$	-42.3 ± 3.8	1	$^\bullet$C$_3$H$_7$O$_2$, CH$_3$CH$^\bullet$CH$_2$OOH	10.9 ± 5.4	1
$^\bullet$C$_4$H$_9$O, (CH$_3$)$_2$C$^\bullet$OCH$_3$	-72.4 ± 10	1	$^\bullet$C$_3$H$_7$O$_2$, C$^\bullet$H$_2$CH(OOH)CH$_3$	2.9 ± 6.3	1
$^\bullet$C$_5$H$_{11}$O, (CH$_3$)$_3$COC$^\bullet$H$_2$	-102.5 ± 8.4	1	$^\bullet$C$_4$H$_9$O$_2$, (CH$_3$)$_2$C$^\bullet$CH$_2$OOH	-30.1 ± 5.4	1
$^\bullet$C$_2$H$_5$O$_2$, HOCH$_2$C$^\bullet$HOH	-220.1 ± 8.4	1	$^\bullet$C$_4$H$_9$O$_2$, C$^\bullet$H$_2$C(CH$_3$)$_2$OOH	-26.8 ± 5.4	1
C$^\bullet$H=C=O, ketenyl	177.5 ± 8.8	1	$^\bullet$C$_2$H$_3$O$_3$, C$^\bullet$H$_2$C(O)OOH	-137.9	1
HC$^\bullet$(O)	42.5 ± 0.5	1	$^\bullet$CHN$_2$	494.5	1
C$^\bullet$CO	381.2 ± 2.1	1	$^\bullet$CH$_2$N=CH$_2$	263.6 ± 12.6	1
CH$_3$C$^\bullet$(O)	-10.3 ± 1.8	1	$^\bullet$CH$_2$NH$_2$	151.9 ± 8.4	1
CF$_3$C$^\bullet$(O)	-608.7	1	CH$_3$C$^\bullet$HNH$_2$	111.7 ± 8.4	1
CH$_2$ClC$^\bullet$(O)	-21 ± 12.6	1	(CH$_3$)$_2$C$^\bullet$NH$_2$	69.9 ± 8.4	1
CHCl$_2$C$^\bullet$(O)	-17.6 ± 23	1	$^\bullet$CH$_2$NHCH$_3$	156.6	1
CCl$_3$C$^\bullet$(O)	-19.7	1	$^\bullet$CH$_2$N(CH$_3$)$_2$	148.0	1
CH$_3$CH$_2$C$^\bullet$(O)	-31.7 ± 3.4	1	(C$_2$H$_5$)$_2$NC$^\bullet$HCH$_3$	68.6 ± 2.1	1
CH$_2$CHC$^\bullet$(O)	88.5	1	$^\bullet$CH$_2$N(CH$_3$)Ph	266.0 ± 12.6	1
CH$_2$C(CH$_3$)C$^\bullet$(O)	58.6 ± 16.7	1	$^\bullet$CN	439.3 ± 2.9	1
CH$_3$CH$_2$CH$_2$C$^\bullet$(O)	54.4 ± 4.2	1	$^\bullet$CH$_2$CN	252.6 ± 4	1
(CH$_3$)$_2$CHC$^\bullet$(O)	-64.0 ± 3.8	1	CH$_3$C$^\bullet$HCN	226.7 ± 12.6	1
(CH$_3$)$_3$CC$^\bullet$(O)	-102.9 ± 6.3	1	$^\bullet$CH$_2$CH$_2$CN	245.4 ± 12.6	1
C$_6$H$_5$C$^\bullet$(O)	116.3 ± 10.9	1	(CH$_3$)$_2$C$^\bullet$CN	190.4 ± 12.6	1
HC(O)CH$_2$$^\bullet$	10.5 ± 9.2	1	Ph(CH$_3$)C$^\bullet$CN	248.5 ± 8.4	1
ClC(O)CH$_2$$^\bullet$	-52.7 ± 13	1	NCC$^\bullet$HCH$_2$CN	381.8 ± 12.6	1
E-C$^\bullet$HClC(O)H	-27.2 ± 10.5	1	$^\bullet$CH$_2$NC	334.7 ± 16.7	1
Z-C$^\bullet$HClC(O)H	-23.4 ± 10.5	1	$^\bullet$C(O)NC	210.0 ± 10	1
C$^\bullet$Cl$_2$C(O)H	-55.6 ± 14.2	1	$^\bullet$C(O)NH$_2$	-15.1 ± 4	1
E-C$^\bullet$HClC(O)Cl	-88.7 ± 15.1	1	C$^\bullet$NN	569 ± 21	1
C$^\bullet$H$_2$C(O)F	-273.0 ± 5.8	1	HC$^\bullet$NN	460 ± 8	1
Z-C$^\bullet$HClC(O)Cl	-84.9 ± 13.8	1	H$_2$C$^\bullet$NN	292.5 ± 2.1	1
C$^\bullet$Cl$_2$C(O)Cl	-101.7 ± 15.5	1	$^\bullet$CH$_2$NO	157 ± 4	1
CH$_3$C(O)CH$_2$$^\bullet$	-34 ± 3	1	$^\bullet$CH$_2$NO$_2$	115.1 ± 12.6	1
CH$_3$C(O)C$^\bullet$HCH$_3$	-70.3 ± 7.1	1	CH$_3$C$^\bullet$HNO$_2$	61.9 ± 12.6	1
CH$_3$C(O)C$^\bullet$=CH$_2$	113.4	1	(CH$_3$)$_2$C$^\bullet$NO$_2$	6.3 ± 12.6	1
C$_2$H$_5$C(O)C$^\bullet$HCH$_3$	-107.5 ± 20.9	1	PhC$^\bullet$HNO$_2$	169.0 ± 12.6	1
iPrC(O)C$^\bullet$(CH$_3$)$_2$	-173.6 ± 20.9	1	$^\bullet$C$_6$H$_6$N, 3-NH$_2$-C$_6$H$_4$	320.1	1
tC$_4$H$_9$C(O)C$^\bullet$H$_2$	-115.5 ± 12.6	1	$^\bullet$C$_6$H$_6$N, 4-NH$_2$-C$_6$H$_4$	327.8	1
PhC(O)C$^\bullet$H$_2$	84.5 ± 12.6	1	$^\bullet$C$_6$H$_4$NO$_2$, 3-NO$_2$-C$_6$H$_4$	340.6 ± 10.0	1
PhC(O)C$^\bullet$HCH$_3$	41.4 ± 20.9	1	$^\bullet$C$_6$H$_4$NO$_2$, 4-NO$_2$-C$_6$H$_4$	302.7	1
PhC$^\bullet$HC(O)CH$_2$Ph	134.3 ± 20.9	1	$^\bullet$C$_6$H$_4$CH$_3$, 2-Me-C$_6$H$_4$	315.1 ± 10.5	1
PhC(O)OC$^\bullet$H$_2$	-69.9	1	$^\bullet$C$_6$H$_4$CH$_3$, 4-Me-C$_6$H$_4$	296.6 ± 9.6	1
$^\bullet$C(O)OH-trans	$\geq -194.6 \pm 2.9$	1	$^\bullet$C$_6$H$_3$N$_2$O$_4$, 3,5-(NO$_2$)$_2$-C$_6$H$_3$	305.4	1
$^\bullet$C(O)OH-cis	-219.7	1	$^\bullet$C$_7$H$_6$NO$_2$, 2-Me-4-NO$_2$-C$_6$H$_3$	295.4 ± 8.4	1

Radical	$\Delta_f H^o_{298}$/kJ mol^{-1}	Ref.	Radical	$\Delta_f H^o_{298}$/kJ mol^{-1}	Ref.
•C$_4$H$_3$N, pyrrol-2-yl	385.8	1	ClOO•	98.0 ± 4	1
•C$_4$H$_3$N, pyrrol-3-yl	385.8	1	BrOO•	108 ± 40	1
•C$_4$H$_8$N, pyrrolidin-2-yl	142.7 ± 12.6	1	IOO•	96.6 ± 15	1
•C$_5$H$_4$N, pyrid-2-yl	362.0	1	OFO•	378.6 ± 20	1
•C$_5$H$_4$N, pyrid-3-yl	391.0	1	OClO•	95.4	1
•C$_5$H$_4$N, pyrid-4-yl	391.0	1	ClOOClO•	142 ± 12	1
•C$_4$H$_7$N$_2$, piperad-2-yl	119.7	1	ClClO•	90 ± 30	1
•C$_4$H$_3$N$_2$, pyrazin-2-yl	409.2 ± 12.6	1	NCO•	184.1	1
•C$_4$H$_3$N$_2$, pyrimid-2-yl	388.0 ± 12.6	1	CNO•	386.6	1
•C$_4$H$_3$N$_2$, pyrimid-4-yl	409.0 ± 12.6	1	HONNO•	172	1
•C$_4$H$_3$N$_2$, pyrimid-5-yl	446.4 ± 12.6	1	sym-ClO$_3$	217.2 ± 21	1
•CH(NO$_2$)$_2$	139.1	1	HSO•	−21.8 ± 2.1	1
•C(NO$_2$)$_3$	201.2	1	HSOO•	112	1
•CH$_2$C(NO$_2$)$_3$	150.6	1	CH$_3$SOO•	76	1
•CH$_2$CH(NO$_2$)$_2$	103.3	1	CF$_3$SO$_2$O•	−912	1
•CH$_2$CH$_2$C(NO$_2$)$_3$	133.9	1	NCO•	184.0	1
•CH$_2$N(NO$_2$)CH$_2$C(NO$_2$)$_3$	173.6	1	O$_2$NO•	73.7 ± 1.4	1
•CH$_2$N(NO$_2$)CH$_2$CH(NO$_2$)$_2$	126.4	1	ONOO•	82.8	1
•CH$_2$CH$_2$N(NO$_2$)CH$_2$C(NO$_2$)$_3$	168.6	1	HOS(O)$_2$O•	−511.7	1
•CH$_2$CH$_2$ONO$_2$	37.7	1	CH$_3$O•	21.0 ± 2.1	1
•CH$_2$(ONO$_2$)CHCH$_2$ONO$_2$	−25.5	1	CF$_3$O•	−635.1 ± 7.1	1
•CH(CH$_2$ONO$_2$)$_2$	−57.3	1	CCl$_3$O•	−38.1 ± 9.2	1
•CH$_2$C(CH$_2$ONO$_2$)$_3$	−158.2	1	CH$_2$ClO•	−21.3 ± 9.2	1
•CH$_2$NHNO$_2$	164.8	1	CHCl$_2$O•	−32.2 ± 9.2	1
•CH$_2$N(NO$_2$)CH$_3$	149.4	1	CH$_2$=CH-O•	18.4 ± 1.3	1
•CH$_2$N(NO$_2$)$_2$	210.5	1	CF$_3$CHFO•	−851.0	1
•CH$_2$CH$_2$N(NO$_2$)CH$_3$	144.3	1	C$_2$H$_5$O•	−13.6 ± 3.3	1
•CH$_2$N(NO$_2$)CH$_2$N(NO$_2$)CH$_3$	202.1	1	CH$_3$CHClO•	−61.9 ± 12.1	1
•CH$_2$N(NO$_2$)(CH$_2$)N(NO$_2$)CH$_3$	173.2	1	CH$_3$CCl$_2$O•	−91.6 ± 11.7	1
C•(S)H	300.4 ± 8.4	1	nC$_3$H$_7$O•	−30.1 ± 8.4	1
•CH$_2$SH	151.9 ± 8.4	1	iC$_3$H$_7$O•	−48.5 ± 3.3	1
•CH$_2$SCH$_3$	136.8 ± 5.9	1	(CH$_3$)$_2$CClO•	−108.4 ± 8.4	1
•CH$_2$SPh	268.6 ± 12.6	1	nC$_4$H$_9$O•	−62.8	1
•CH$_2$SOCH$_3$	23.8 ± 12.6	1	sC$_4$H$_9$O•	−69.5	1
HOC•(S)S	110.5	1	tC$_4$H$_9$O•	−85.8 ± 3.8	1
•CH$_2$SO$_2$CH$_3$	−177.0 ± 12.6	1	CH$_2$=CHCH$_2$O•	87.0	1
•CH$_2$SO$_2$Ph	−57.3 ± 12.6	1	C$_6$H$_5$O•	48.5 ± 2.9	1
PhC•HSO$_2$CH$_3$	−109.2 ± 12.6	1	o-Cl-C$_6$H$_4$O•	30.6	1
PhC•HSO$_2$Ph	7 ± 12.6	1	C$_6$Cl$_5$O•	~63	1
Ph$_2$C•SO$_2$Ph	102 ± 12.6	1	p-Cl-C$_6$H$_4$O•	~9	1
Ph$_2$C•SPh	435.6 ± 12.6	1	o-OH-C$_6$H$_4$O•	−186.3	1
NC•(O)	127.2	1	p-OH-C$_6$H$_4$O•	−143.6	1
•CNH	207.9 ± 12.1	1	o-CH$_3$O-C$_6$H$_4$O•	−125.5	1
•CNO	323 ± 30	1	p-CH$_3$O-C$_6$H$_4$O•	−81.1	1
•CH$_2$SiMe$_3$	−32 ± 6	1	C$_6$H$_5$CH$_2$O•	136.0 ± 12.6	1
•CH$_2$C(CH$_3$)$_2$SiMe$_3$	−125	1	C$_{10}$H$_7$O•, naphthoxy-1	165.3	1
•CP	450 ± 9	1	C$_{10}$H$_7$O•, naphthoxy-2	174.1	1
			HC(O)O•	−129.7 ± 12.6	1
(2) Oxygen-Centered Species			FC(O)O•	368.0	1
HO•	37.36 ± 0.13	1	CH$_3$C(O)O•	−179.9 ± 12.6	1
FO•	109 ± 10	1	CF$_3$C(O)O•	−797.0	1
ClO•	101.63 ± 0.1	1	CF$_3$OC(O)O•	−958.1 ± 16.7	1
BrO•	126.2 ± 1.7	1	C$_6$H$_5$C(O)O•	−50.2 ± 16.7	1
IO•	115.9 ± 5.0	1	CH$_3$OO•	20.1 ± 5.1	1
HOO•	12.30 ± 0.25	1	C$_2$H$_3$OO•, CH$_2$=CHOO•	101.7 ± 1.7	1
FOO•	25.4 ± 2	1			

Radical	$\Delta_f H^\circ_{298}$/kJ mol^{-1}	Ref.
$C_2H_5OO^\bullet$	-28.5 ± 9.6	1
$C_3H_5OO^\bullet$, $CH_2=CHCH_2OO^\bullet$	88.7	1
$iC_3H_7OO^\bullet$	-65.4 ± 11.3	1
$C_4H_7OO^\bullet$, $CH_3CH=CHCH_2OO^\bullet$	82.6 ± 5.3	1
$tC_4H_9OO^\bullet$	-109.7 ± 3.9	8
neo-$C_5H_{11}OO^\bullet$	-115.5	1
$HOCH_2OO^\bullet$	-162.1	1
$HOOCH_2CH_2OO^\bullet$	100	1
$C_6H_5CH_2OO^\bullet$	114.6 ± 4.2	1
c-$C_6H_{11}OO^\bullet$	-25.0 ± 10.5	1
$(C_2H_5)N(CH3)CHOO^\bullet$	-36.0 ± 12.6	1
$CF_3OO^\bullet$	-635.0	1
$CF_2ClOO^\bullet$	-406.7 ± 14.6	1
$CFCl_2OO^\bullet$	-213.7	1
$CH_2ClOO^\bullet$	-5.1 ± 13.6	1
$CHCl_2OO^\bullet$	-19.2 ± 11.2	1
$CCl_3OO^\bullet$	-20.9 ± 8.9	1
$CH_3CHClOO^\bullet$	-54.7 ± 3.4	1
$CH_3CCl_2OO^\bullet$	-63.8 ± 9.8	1
$CH_3OCH_2OO^\bullet$	-142.2 ± 4.2	1
$CH_3C(O)CH_2OO^\bullet$	-142.1 ± 4	1
cis-$HC(O)OO^\bullet$	85.8 ± 14.6	7
$trans$-$HC(O)OO^\bullet$	95.4 ± 14.6	7
$CH_3C(O)OO^\bullet$	-154.4 ± 5.8	1
$HOOO^\bullet$	19.3 ± 0.5	9
$CH_3OOO^\bullet$	33.4 ± 12.6	1
$C_2H_5OOO^\bullet$	5.4 ± 12.6	1

(3) Nitrogen-Centered Species

Radical	$\Delta_f H^\circ_{298}$/kJ mol^{-1}	Ref.
ON	91.04 ± 0.08	1
NO_2	33.97 ± 0.08	1
N_2O	82.05 ± 0.4	1
NH	357 ± 1	1
$^\bullet NH_2$	186.2 ± 1.0	1
$^\bullet NNH$	249.5	1
$^\bullet NCO$	131.8	1
$^\bullet N_3$	414.2 ± 20.9	1
$^\bullet N_2H_3$	243.5	1
(Z)-N_2H_2	213.0 ± 10.9	1
NF	209.2	1
$^\bullet NF_2$	42.3 ± 8	1
$^\bullet NHF$	112 ± 15	1
NBr	301 ± 21	1
HNO	107.1 ± 2.5	1
FNO	-65.7 ± 1.7	1
ClNO	51.71 ± 0.42	1
BrNO	82.13 ± 0.8	1
INO	112.1 ± 20.9	1
NCO	120.9	1
NCN	464.8 ± 2.9	1
NSi	372 ± 63	1
$NH_2C(O)N^\bullet H$	0.8 ± 12.6	1
$CH_3C(O)N^\bullet H$	-6.7 ± 12.6	1
$NH_2C(S)N^\bullet H$	194 ± 12.6	1
$CH_3C(S)N^\bullet H$	173 ± 12.6	1
$PhC(S)N^\bullet H$	307 ± 12.6	1

Radical	$\Delta_f H^\circ_{298}$/kJ mol^{-1}	Ref.
$HCON^\bullet H$	49.8 ± 12.6	1
$NH_2C(NH)N^\bullet H$	250.6 ± 12.6	1
$^\bullet NHCN$	319.2 ± 2.9	1
$CH_2N^\bullet H$	104.6 ± 12.6	1
$CH_3N^\bullet H$	184.1 ± 8.4	1
$tBuN^\bullet H$	95.4 ± 12.6	1
$C_6H_5CH_2N^\bullet H$	288.3 ± 12.6	1
$C_6H_5N^\bullet H$	244.3 ± 4.2	1
$(CH_3)_2N^\bullet$	158.2 ± 4.2	1
$(C_6H_5)(CH_3)N^\bullet$	241.0 ± 6.3	1
$(C_6H_5)_2N^\bullet$	366.0 ± 6.3	1
1-pyrrolyl	269.2 ± 12.6	1
1-pyrazolyl	413.0 ± 2.1	1
carbazol-9-yl	383.3 ± 8.4	1
$CH_3N_2^\bullet$	215.5 ± 7.5	1
$C_2H_5N_2^\bullet$	187.4 ± 10.5	1
$iC_3H_7N_2^\bullet$	146.0 ± 8.4	1
$nC_4H_9N_2^\bullet$	140.6 ± 8.4	1
$tC_4H_9N_2^\bullet$	97.5 ± 4.2	1
$(NO_2)HN^\bullet$	162.3	1
$(CH_3)(NO_2)N^\bullet$	139.0	1
$(NO_2)_2N^\bullet$	200.0	1
$CH_3N^\bullet CH_2N(NO_2)CH_3$	185.4	1

(4) Sulfur-Centered Species

Radical	$\Delta_f H^\circ_{298}$/kJ mol^{-1}	Ref.
$HOS^\bullet$	-6.7 ± 2.1	1
$HC(O)S^\bullet$	56.5	1
$HS^\bullet O_2$	-221.8	1
$HOS^\bullet O_2$	-384.9	1
$NCS^\bullet$	300 ± 8	1
$HS^\bullet$	143.0 ± 0.8	1
$CH_3S^\bullet$	124.7 ± 1.7	1
$C_2H_5S^\bullet$	101	1
$nC_3H_7S^\bullet$	80	1
$iC_3H_7S^\bullet$	74.9 ± 8.4	1
$tC_4H_9S^\bullet$	43.9 ± 8.4	1
$C_6H_5S^\bullet$	242.7 ± 4.6	1
$C_6Cl_5S^\bullet$	~184	1
$C_6H_5CH_2S^\bullet$	246	1
$CH_3S^\bullet O$	-67 ± 10	1
$CH_3S^\bullet O_2$	-239.3	1
$HSS^\bullet$	115.5 ± 14.6	1
$CH_3SS^\bullet$	68.6 ± 8.4	1
$C_2H_5SS^\bullet$	43.5 ± 8.4	1
$iC_3H_7SS^\bullet$	13.8 ± 8.4	1
$tC_4H_9SS^\bullet$	-19.2 ± 8.4	1
$HOC(S)S^\bullet$	110.5 ± 4.6	1
$HC(O)S^\bullet$	56.5	1
SF	13.0 ± 6.3	1
SF_2	-296.7 ± 16.7	1
SF_3	-503.0 ± 33.5	1
SF_4	-763.2 ± 20.9	1
SF_5	-879.9 ± 15.1	1
$ClS^\bullet$	156.5 ± 16.7	1
SN	263.6 ± 105	1
SCl	156.5 ± 16.7	1

(5) Si-, Ge-, Sn-, Pb-Centered Species

Radical	$\Delta_f H^\circ_{298}$/kJ mol^{-1}	Ref.
SiF	-20.1 ± 12.6	1

Radical	$\Delta_f H^o_{298}$/kJ mol^{-1}	Ref.	Radical	$\Delta_f H^o_{298}$/kJ mol^{-1}	Ref.
SiF$_2$	-638 ± 6	1	GeBr	137 ± 5	1
•SiF$_3$	-987 ± 20	1	GeBr$_2$	-61 ± 5	1
SiCl	198.3 ± 6.7	1	•GeBr$_3$	-119 ± 50	1
SiCl$_2$	-169 ± 3	1	GeI	211 ± 25	1
•SiCl$_3$	321 ± 8	6	GeI$_2$	50.2 ± 4	1
SiBr	235 ± 46	1	•GeI$_3$	42 ± 50	1
SiBr$_2$	46 ± 8	1	SnF	-95 ± 7.2	1
•SiBr$_3$	-201.7 ± 63	1	SnF$_2$	-511 ± 9.2	1
SiI	313.8 ± 42	1	•SnF$_3$	-647 ± 50	1
SiI$_2$	92.5 ± 8.4	1	SnCl	35 ± 12	1
•SiI$_3$	35.3 ± 63	1	SnCl$_2$	-202.6 ± 7.1	1
SiH	376.6 ± 8.4	1	•SnCl$_3$	-292 ± 50	1
SiH$_2$(^{1}A$_1$)	273 ± 2	1	SnBr	76 ± 12	1
SiH$_2$(^{3}B$_1$)	360.7	1	SnBr$_2$	-119 ± 2.8	1
•SiH$_3$	200.4 ± 2.5	1	•SnBr$_3$	-159 ± 50	1
MeSi•H$_2$	141 ± 6	1	SnI	173 ± 12	1
Me$_2$Si•H	78 ± 6	1	SnI$_2$	-8.1 ± 4.2	1
Me$_3$Si•	15 ± 7	1	•SnI$_3$	-8 ± 50	1
•Si$_2$H$_3$	~402	1	•Sn(CH$_3$)$_3$	116.6 ± 9.7	10
H$_3$SiSi•H$_2$	234 ± 6	1	•Sn(C$_6$H$_5$)$_3$	518.8 ± 21	1
C$_6$H$_5$Si•H$_2$	274	1	PbH	236.2 ± 19.2	1
H$_3$SiSi•H	312 ± 8	1	PbF	-80.3 ± 10.5	1
MeSi•	302.2	1	PbF$_2$	-435.1 ± 8.4	1
MeSi•H	202 ± 6	1	•PbF$_3$	-490 ± 60	1
Me$_2$Si••	135 ± 8	1	PbCl	15.1 ± 50	1
SiN	313.8 ± 42	1	PbCl$_2$	-174.1 ± 1.3	1
•GeH$_3$	221.8 ± 8.4	1	•PbCl$_3$	-178 ± 80	1
GeF	-71 ± 10	1	PbBr	70.9 ± 42	1
GeF$_2$	-574 ± 20	1	PbBr$_2$	-104.4 ± 6.3	1
•GeF$_3$	-807 ± 50	1	•PbBr$_3$	-104 ± 80	1
GeCl	69 ± 18	1	PbI	107.4 ± 37.7	1
GeCl$_2$	-171 ± 5	1	PbI$_2$	-3.2 ± 4.2	1
•GeCl$_3$	-268 ± 50	1	•PbI$_3$	22 ± 80	1

References

1. Luo, Y. R. *Comprehensive Handbook of Chemical Bond Energies*, CRC Press, Boca, Raton, FL, 2007.
2. Wheeler, S. E., Robertson, K. A., Allen, W. D., Schaefer III, H. F., Bomble, Y. J., and Stanton, J. F., *J. Phys. Chem. A* 111, 3819, 2007.
3. Seetula, J. A., and Eskola, A. J., *Chem. Phys.* 351, 141, 2008.
4. Denis, P. A., and Ornellas, F. R., *J. Phys. Chem. A* 113, 499, 2009.
5. Gao, Y., DeYonker, N. J., Garrett III, E. C., Wilson, A. K., Cundari, T. R., and Marshall, P., *J. Phys. Chem. A* 113, 6955, 2009.
6. Shuman, N. S., Spencer, A. P., and Baer, T., *J. Phys. Chem. A* 113, 9458, 2009.
7. Villano, S. M., Eyet, N., Wren, S. W., Ellison, G. B., Bierbaum, V. M., and Lineberger, W. C., *J. Phys. Chem. A* 114, 191, 2010.
8. Shuman, N. S., Bodi, A., and Baeer, T., *J. Phys. Chem. A* 114, 232, 2010.
9. Le Picard, S. D., Tizniti, M., Canosa, A., Sims, I. R., and Smith, I. W. M., *Science* 328, 1258, 2020.
10. Da'valos, J. Z., Herrero, R., Shuman, N. S., and Baer, T., *J. Phys. Chem. A* 115, 402, 2011.

TABLE 5. Bond Dissociation Energies of Some Organic Molecules

D^0_{298}(R-X)/ kJ mol^{-1} of some organic compounds are listed below. All data are from Tables 1 and 3.

	X=H	F	Cl	Br	I	OH	OCH$_3$	NH$_2$	NO	CH$_3$	COCH$_3$	CF$_3$	CCl$_3$
R=H	435.7799	569.658	431.361	366.16	298.26	497.10	440.2	450.08	199.5	439.3	374.0	445.2	392.5
CH$_3$	439.3	460.2	350.2	294.1	238.9	384.93	351.9	356.1	172.0	377.4	351.9	429.3	362.3
C$_2$H$_5$	420.5	447.4	352.3	292.9	233.5	391.2	355.2	355.2	171.5	370.3	347.3	—	—
i-C$_3$H$_7$	410.5	483.8	354.8	295.1	233.1	397.9	360.7	357.7	152.7	369.0	340.2	—	—
t-C$_4$H$_9$	400.4	495.8	351.9	292.9	227.2	398.3	353.1	355.6	167	363.6	329.3	—	—
C$_6$H$_5$	472.2	525.5	399.6	336.4	272.0	463.6	418.8	429.3	226.8	426.8	406.7	463.2	388.7
C$_6$H$_5$CH$_2$	375.5	412.8	299.9	239.3	187.8	334.1	—	306.7	123	325.1	299.7	365.7	—
CCl$_3$	392.5	439.3	296.6	231.4	168	—	—	—	125	362.3	—	332.2	285.8

	X=H	F	Cl	Br	I	OH	OCH$_3$	NH$_2$	NO	CH$_3$	COCH$_3$	CF$_3$	CCl$_3$
CF$_3$	445.2	546.8	365.3	296.2	227.2	≤482.0	—	—	167	429.3	—	413.0	332.2
C$_2$F$_5$	429.7	532.2	346.0	283.3	219.2	—	—	—	—	—	—	424.3	—
CH$_3$CO	374.0	511.7	354.0	292.0	223.0	459.4	424.3	414.6	—	351.9	307.1	—	—
CN	528.5	482.8	422.6	364.8	320.1	—	—	—	204.4	521.7	—	469.0	—
C$_6$F$_5$	487.4	485	383.3	~328	<301.7	446.9	—	—	211.3	439.3	—	435.1	—

TABLE 6. Bond Dissociation Energies in Diatomic Cations

From thermochemistry, we have

$$D^\circ_{298}(A^+ - B) \equiv \Delta_f H^\circ(A^+) + \Delta_f H^\circ(B) - \Delta_f H^\circ(AB^+) = D^\circ_{298}(A-B) + IP(A) - IP(AB)$$

Thus, $D^\circ_{298}(A^+ - B)$ may be derived using the Table 1 and the ionization potentials of species A and AB. The following Table has been arranged in an alphabetical order of the atoms. The **boldface** in the species indicates the dissociated fragment.

A$^+$–B	D°_{298}/kJ mol^{-1}	Ref.	A$^+$–B	D°_{298}/kJ mol^{-1}	Ref.	A$^+$–B	D°_{298}/kJ mol^{-1}	Ref.
Ac$^+$–S	465 ± 48	20	**B**$^+$–Pt	314 ± 98	1	**Ca**$^+$–Br	417.6 ± 10	1
Ag$^+$–Ag	167.9 ± 8.7	1	**B**$^+$–Se	298 ± 98	1	**Ca**$^+$–Ca	104.1	1
Ag$^+$–Cl	32 ± 30	1	**B**$^+$–Si	365 ± 15	1	**Ca**$^+$–Cl	433.4 ± 12	1
Ag$^+$–F	24 ± 27	1	**Ba**$^+$–Ar	11.85	1	**Ca**$^+$–F	556.5 ± 8.4	1
Ag$^+$–H	43.5 ± 5.9	1	**Ba**$^+$–Br	418 ± 10	1	**Ca**$^+$–H	284.2 ± 10	1
Ag$^+$–O	123 ± 5	1	**Ba**$^+$–Cl	468.2 ± 10	1	**Ca**$^+$–I	293.7 ± 10.8	1
Ag$^+$–S	123 ± 13	1	**Ba**$^+$–D	245.2 ± 9.6	1	**Ca**$^+$–Kr	18.60 ± 0.72	1
Al$^+$–Al	121	1	**Ba**$^+$–F	640 ± 29	1	**Ca**$^+$–Ne	4.95 ± 0.06	1
Al$^+$–Ar	15.47	1	**Ba**$^+$–I	335 ± 10	1	**Ca**$^+$–O	348 ± 5	1
Al$^+$–Ca	148.5	1	**Ba**$^+$–O	441.4 ± 15	1	**Ca**$^+$–Xe	25.38 ± 0.96	1
Al$^+$–Cl	173 ± 42	1	**Be**$^+$–Ar	49.0 ± 2.4	1	**Cd**$^+$–Cd	122.5 ± 10	1
Al$^+$–F	314 ± 21	1	**Be**$^+$–Au	410 ± 29	1	**Cd**$^+$–H	179.5	1
Al$^+$–Kr	5.54	1	**Be**$^+$–Be	196 ± 0.5	8	**Ce**$^+$–Au	278 ± 34	1
Al$^+$–O	166.7 ± 12.0	1	**Be**$^+$–Cl	417 ± 50	1	**Ce**$^+$–Br	341.0	1
Al$^+$–Se	114 ± 49	1	**Be**$^+$–F	575 ± 98	1	**Ce**$^+$–C	254 ± 96	1
Am$^+$–O	560 ± 28	14	**Be**$^+$–H	307.3 ± 5.0	1	**Ce**$^+$–Ce	207 ± 42	1
Am$^+$–S	334 ± 27	20	**Be**$^+$–O	362.0 ± 6.2	1	**Ce**$^+$–Cl	429.5	1
Ar$^+$–Ar	130.323 ± 0.087	1	**Bi**$^+$–Bi	199 ± 10	1	**Ce**$^+$–F	586 ± 63	1
Ar$^+$–He	2.9 ± 0.8	1	**Bi**$^+$–O	174	1	**Ce**$^+$–I	295.5	1
Ar$^+$–Ne	7.5 ± 0.8	1	**Bi**$^+$–S	179 ± 50	1	**Ce**$^+$–Ir	530 ± 96	1
As$^+$–As	364 ± 22	1	**Bi**$^+$–Se	184 ± 29	1	**Ce**$^+$–N	494 ± 63	1
As$^+$–H	290.8 ± 3.0	1	**Bi**$^+$–Te	125 ± 50	1	**Ce**$^+$–O	852 ± 15	1
As$^+$–O	495	1	**Bi**$^+$–Tl	100 ± 42	1	**Ce**$^+$–Pd	255 ± 53	1
As$^+$–P	367 ± 59	1	**Bk**$^+$–O	610	1	**Ce**$^+$–Pt	467 ± 96	1
As$^+$–S	433.2 ± 12.5	1	**Br**$^+$–Br	318.858 ± 0.024	1	**Ce**$^+$–Rh	423 ± 96	1
Au$^+$–Al	170 ± 30	1	**Br**$^+$–C	451.5 ± 8.6	1	**Ce**$^+$–S	524 ± 59	1
Au$^+$–Au	234.5	1	**Br**$^+$–Cl	303.000 ± 0.048	1	**Cl**$^+$–Ar	169	1
Au$^+$–B	329 ± 50	1	**Br**$^+$–F	251.5 ± 12.6	1	**Cl**$^+$–Cl	385.746 ± 0.096	6
Au$^+$–Be	401 ± 29	1	**Br**$^+$–H	379.26 ± 2.89	1	**Cl**$^+$–D	457.284 ± 0.017	1
Au$^+$–C	311.5 ± 7.7	4	**Br**$^+$–O	365.7 ± 3.1	1	**Cl**$^+$–F	291 ± 10	1
Au$^+$–F	79	1	**C**$^+$–Ar	72.3	1	**Cl**$^+$–H	452.714 ± 0.018	1
Au$^+$–Ge	292 ± 24	1	**C**$^+$–Br	398 ± 8.6	1	**Cl**$^+$–N	650 ± 10	1
Au$^+$–H	209.2 ± 10.6	18	**C**$^+$–C	601.9 ± 19.3	1	**Cl**$^+$–O	468.0 ± 2.1	1
Au$^+$–I	230~280	1	**C**$^+$–Cl	614	1	**Cm**$^+$–O	670 ± 38	14
Au$^+$–O	111.8 ± 7.7	19	**C**$^+$–F	721 ± 40	1	**Cm**$^+$–S	455 ± 16	20
Au$^+$–Xe	130 ± 13	1	**C**$^+$–H	397.848 ± 0.013	1	**Co**$^+$–Ar	52.89 ± 0.06	1
B$^+$–Ar	32.7	1	**C**$^+$–N	524.5 ± 4.2	1	**Co**$^+$–Br	>289	1
B$^+$–B	187	1	**C**$^+$–O	810.7 ± 0.8	1	**Co**$^+$–C	351 ± 29	1
B$^+$–Br	164 ± 21	1	**C**$^+$–P	587 ± 50	1	**Co**$^+$–Cl	285 ± 12	1
B$^+$–C	284 ± 58	1	**C**$^+$–S	706.6 ± 2.1	1	**Co**$^+$–Co	269	1
B$^+$–Cl	308 ± 21	1	**C**$^+$–Se	587 ± 50	1	**Co**$^+$–D	199.6 ± 5.8	1
B$^+$–F	460 ± 10	1	**Ca**$^+$–Al	144.7	1	**Co**$^+$–H	195 ± 6	1
B$^+$–H	198 ± 5	1	**Ca**$^+$–Ar	12.99 ± 0.60	1	**Co**$^+$–He	16.4 ± 0.4	1
B$^+$–O	326 ± 48	1	**Ca**$^+$–Au	306 ± 29	1	**Co**$^+$–I	211.7 ± 8.4	1

A^+-B	D^o_{298}/kJ mol^{-1}	Ref.	A^+-B	D^o_{298}/kJ mol^{-1}	Ref.	A^+-B	D^o_{298}/kJ mol^{-1}	Ref.
Co$^+$–Kr	68.37 ± 0.18	1	Eu$^+$–I	290.7	1	Hg$^+$–Ar	22.2 ± 1.2	1
Co$^+$–Ne	12.8 ± 0.4	1	Eu$^+$–O	393 ± 15	1	Hg$^+$–H	207	1
Co$^+$–O	317.3 ± 4.8	1	Eu$^+$–S	257 ± 32	1	Hg$^+$–Hg	134	1
Co$^+$–S	288.3 ± 8.7	1	F$^+$–Ar	161.1	1	Hg$^+$–Kr	37.9 ± 1.3	1
Co$^+$–Si	317.1 ± 6.7	1	F$^+$–F	325.393 ± 0.096	1	Hg$^+$–Xe	72.2 ± 1.3	1
Co$^+$–Xe	85.7 ± 6.8	1	F$^+$–He	181.62 ± 0.08	1	Ho$^+$–Ag	155 ± 61	1
Cr$^+$–Ar	31.7 ± 3.9	1	F$^+$–Kr	152.4	1	Ho$^+$–Au	250 ± 60	1
Cr$^+$–C	277 ± 24	1	F$^+$–Xe	188	1	Ho$^+$–Br	320.6	1
Cr$^+$–Cl	>211	1	Fe$^+$–Ar	14.2 ± 7.7	1	Ho$^+$–Cl	410.3	1
Cr$^+$–Cr	129	1	Fe$^+$–Br	>293	1	Ho$^+$–Cu	214 ± 35	1
Cr$^+$–D	135 ± 9	1	Fe$^+$–C	402.0 ± 4.2	16	Ho$^+$–F	542 ± 50	1
Cr$^+$–F	279 ± 42	1	Fe$^+$–Cl	>343	1	Ho$^+$–Ho	88 ± 96	1
Cr$^+$–H	136 ± 9	1	Fe$^+$–Co	259 ± 21	1	Ho$^+$–I	270.4	1
Cr$^+$–He	7.8 ± 0.4	1	Fe$^+$–Cr	209 ± 29	1	Ho$^+$–O	551 ± 25	1
Cr$^+$–Ne	9.5 ± 0.4	1	Fe$^+$–Cu	222 ± 29	1	I$^+$–Br	184.90 ± 0.02	1
Cr$^+$–O	359	1	Fe$^+$–D	227	1	I$^+$–Cl	247.5 ± 0.4	1
Cr$^+$–S	258.6 ± 16.4	1	Fe$^+$–F	360 – 423	1	I$^+$–F	262.9 ± 2.1	1
Cr$^+$–Si	203 ± 15	1	Fe$^+$–Fe	272	1	I$^+$–H	304.70 ± 0.10	1
Cr$^+$–Xe	71.9 ± 10.0	1	Fe$^+$–H	211.2 ± 9.6	1	I$^+$–I	262.90 ± 0.04	1
Cs$^+$–Ar	8.2	1	Fe$^+$–I	>239	1	I$^+$–O	316.3 ± 10.5	1
Cs$^+$–Br	60.5 ± 10	1	Fe$^+$–Kr	33.5 ± 6.7	1	In$^+$–Br	65.2 ± 12.6	1
Cs$^+$–Cl	107.4 ± 10	1	Fe$^+$–N	485	1	In$^+$–Cl	193 ± 21	1
Cs$^+$–Cs	62.6 ± 9.6	1	Fe$^+$–Nb	285 ± 21	1	In$^+$–F	148 ± 50	1
Cs$^+$–F	43.7 ± 10	1	Fe$^+$–Ni	268 ± 21	1	In$^+$–I	51.5 ± 21	1
Cs$^+$–He	5.1	1	Fe$^+$–O	343.3 ± 1.9	13	In$^+$–In	81 ± 30	1
Cs$^+$–I	29.3 ± 10	1	Fe$^+$–S	295.2 ± 5.8	1	In$^+$–S	171 ± 50	1
Cs$^+$–Kr	15.1	1	Fe$^+$–Sc	200 ± 21	1	In$^+$–Sb	73 ± 50	1
Cs$^+$–Na	48.1 ± 4.2	1	Fe$^+$–Si	277 ± 9	1	In$^+$–Se	118 ± 50	1
Cs$^+$–Ne	6.11	1	Fe$^+$–Ta	301 ± 21	1	In$^+$–Te	41 ± 50	1
Cs$^+$–O	59	1	Fe$^+$–Ti	251 ± 25	1	Ir$^+$–C	635.8 ± 4.8	3
Cs$^+$–Rb	68.3 ± 10	1	Fe$^+$–V	314 ± 21	1	Ir$^+$–D	302.8 ± 5.8	1
Cs$^+$–Xe	14.7	1	Fe$^+$–Xe	46.0 ± 5.8	1	Ir$^+$–H	305.7 ± 5.8	1
Cu$^+$–Ar	51.9 ± 6.8	1	Ga$^+$–Bi	62 ± 98	1	Ir$^+$–O	247	1
Cu$^+$–Cl	91 ± 10	1	Ga$^+$–Br	56.5 ± 16	1	K$^+$–Ar	14 ± 7	1
Cu$^+$–Cu	155.2 ± 7.7	1	Ga$^+$–Cl	86 ± 21	1	K$^+$–Br	35.7 ± 10.5	1
Cu$^+$–F	117 ± 21	1	Ga$^+$–F	136 ± 15	1	K$^+$–Cl	51 ± 19	1
Cu$^+$–Ge	231 ± 23	1	Ga$^+$–Ga	126.3	1	K$^+$–He	6.00	1
Cu$^+$–H	93 ± 13	1	Ga$^+$–I	41.6 ± 15	1	K$^+$–I	18 ± 45	1
Cu$^+$–Kr	24.3 ± 0.8	1	Ga$^+$–O	46 ± 50	1	K$^+$–K	83.86 ± 0.15	1
Cu$^+$–O	133.9 ± 11.6	1	Ga$^+$–Sb	38 ± 96	1	K$^+$–Kr	15.8	1
Cu$^+$–S	203.3 ± 14.5	1	Ga$^+$–Te	19 ± 29	1	K$^+$–Li	59.9 ± 5.9	1
Cu$^+$–Si	260 ± 8	1	Gd$^+$–Cd	122.5 ± 10	1	K$^+$–Na	58.69 ± 0.08	1
Cu$^+$–Xe	102.1 ± 5.8	1	Gd$^+$–H	179.5	1	K$^+$–Ne	7.79	1
D$^+$–D	263.4405 ± 0.0003	1	Ge$^+$–Br	398 ± 42	1	K$^+$–O	13	1
Dy$^+$–Br	324.2	1	Ge$^+$–C	223 ± 31	1	K$^+$–Xe	19.5	1
Dy$^+$–Cl	407.9	1	Ge$^+$–Cl	473 ± 50	1	Kr$^+$–Ar	55.31 ± 0.14	1
Dy$^+$–Cu	196 ± 42	1	Ge$^+$–F	565 ± 21	1	Kr$^+$–H	464	1
Dy$^+$–F	535 ± 24	1	Ge$^+$–Ge	274 ± 10	1	Kr$^+$–He	2.1 ± 0.8	1
Dy$^+$–I	279.9	1	Ge$^+$–H	377 ± 84	1	Kr$^+$–Kr	110.967 ± 0.033	1
Dy$^+$–O	597 ± 15	1	Ge$^+$–O	344 ± 21	1	Kr$^+$–N	136.9 ± 13	1
Er$^+$–Br	315.8	1	Ge$^+$–S	283 ± 21	1	Kr$^+$–Ne	3.8 ± 0.8	1
Er$^+$–Cl	406.7	1	Ge$^+$–Se	234 ± 10	1	La$^+$–Au	436 ± 97	1
Er$^+$–F	546 ± 34	1	Ge$^+$–Si	268 ± 21	1	La$^+$–Br	425.9	1
Er$^+$–I	271.6	1	Ge$^+$–Te	233 ± 19	1	La$^+$–C	427 ± 33	1
Er$^+$–O	583 ± 15	1	H$^+$–D	261.1021 ± 0.0002	1	La$^+$–Cl	503.6	1
Es$^+$–O	470 ± 60	1	H$^+$–H	259.4659 ± 0.0002	1	La$^+$–F	589 ± 34	1
Eu$^+$–Ag	85 ± 50	1	He$^+$–H	123.9	1	La$^+$–H	243 ± 9	1
Eu$^+$–Au	252 ± 97	1	He$^+$–He	229.687 ± 0.019	1	La$^+$–I	392.4	1
Eu$^+$–Br	333.8	1	Hf$^+$–C	311.5 ± 2.9	10	La$^+$–Ir	356 ± 97	1
Eu$^+$–Cl	430.7	1	Hf$^+$–H	207.3 ± 7.7	17	La$^+$–O	875 ± 25	1
Eu$^+$–F	543 ± 29	1	Hf$^+$–O	670.4 ± 10.6	10	La$^+$–Pt	522 ± 78	1

A⁺–B	D^o_{298}/kJ mol⁻¹	Ref.
La⁺–Rh	345 ± 97	1
La⁺–S	629 ± 96	1
La⁺–Si	277.0 ± 9.6	1
Li⁺–Ar	33 ± 14	1
Li⁺–Bi	91 ± 50	1
Li⁺–Br	41.8 ± 10.6	1
Li⁺–Cl	66 ± 15	1
Li⁺–F	7 ± 21	1
Li⁺–He	10.66	1
Li⁺–I	51.1 ± 6.3	1
Li⁺–Kr	48.1	1
Li⁺–Li	137.3 ± 6.3	1
Li⁺–Ne	15.32	1
Li⁺–O	38.9 ± 9.6	1
Li⁺–Sb	129.6 ± 13.9	1
Li⁺–Xe	56.4	1
Lu⁺–Br	86.1	1
Lu⁺–Cl	180.6	1
Lu⁺–F	376.8	1
Lu⁺–H	204 ± 15	1
Lu⁺–I	40.7	1
Lu⁺–O	524 ± 15	1
Lu⁺–Si	107 ± 13	1
Mg⁺–Ar	19.20	1
Mg⁺–Au	267 ± 29	1
Mg⁺–Cl	327 ± 6.5	1
Mg⁺–D	203.6 ± 0.8	1
Mg⁺–F	477 ± 50	1
Mg⁺–H	190.8 ± 5.8	1
Mg⁺–Kr	25.39	1
Mg⁺–Mg	125	1
Mg⁺–Ne	4.9 ± 0.6	1
Mg⁺–O	245.2 ± 10	1
Mg⁺–Xe	53.74	1
Mn⁺–Cl	>211	1
Mn⁺–F	321 ± 24	1
Mn⁺–H	202.5 ± 5.9	1
Mn⁺–I	>211	1
Mn⁺–Mn	129	1
Mn⁺–O	285 ± 13	1
Mn⁺–S	247 ± 23	1
Mn⁺–Se	165 ± 50	1
Mo⁺–C	442.7 ± 13.5	1
Mo⁺–F	376 ± 29	1
Mo⁺–H	170 ± 6	1
Mo⁺–Mo	449.4 ± 1.0	1
Mo⁺–O	488.2 ± 1.9	1
Mo⁺–S	355.1 ± 5.8	1
Mo⁺–Xe	>53.1 ± 6.8	1
N⁺–Ar	208.4 ± 9.6	1
N⁺–F	584 ± 42	1
N⁺–H	≥435.67 ± 0.77	1
N⁺–N	843.85 ± 0.10	1
N⁺–O	115	1
Na⁺–Ar	19 ± 8	1
Na⁺–Br	58.2 ± 10.6	1
Na⁺–Cl	20.3 ± 10	1
Na⁺–He	7.55	1
Na⁺–I	64.9 ± 3.0	1
Na⁺–Kr	~24.9	1
Na⁺–Li	95.8 ± 3.9	1

A⁺–B	D^o_{298}/kJ mol⁻¹	Ref.
Na⁺–Na	98.64 ± 0.29	1
Na⁺–Na	6.4	1
Na⁺–Ne	~9.04	1
Na⁺–O	37 ± 19	1
Na⁺–Xe	~28.6	1
Nb⁺–Ar	40.87 ± 0.13	1
Nb⁺–C	509 ± 15	1
Nb⁺–Fe	>251	1
Nb⁺–H	220 ± 7	1
Nb⁺–Nb	576.8 ± 9.6	1
Nb⁺–O	688 ± 11	1
Nb⁺–S	501.7 ± 20.3	1
Nb⁺–V	404.7 ± 0.2	1
Nb⁺–Xe	73.28 ± 0.12	1
Nd⁺–Au	267 ± 84	1
Nd⁺–Br	352.9	1
Nd⁺–Cl	441.4	1
Nd⁺–F	309.6	1
Nd⁺–I	596 ± 32	1
Nd⁺–O	753 ± 15	1
Ne⁺–H	1239	1
Ne⁺–He	13.0 ± 0.8	1
Ne⁺–Ne	125.29 ± 1.93	1
Ni⁺–Ar	53.9	1
Ni⁺–Br	>289	1
Ni⁺–C	418	1
Ni⁺–Cl	192 ± 4	1
Ni⁺–D	166.0 ± 7.7	1
Ni⁺–F	≥456	1
Ni⁺–H	158.1 ± 7.7	1
Ni⁺–He	12.4 ± 0.4	1
Ni⁺–I	>297	1
Ni⁺–Ne	9.9 ± 0.4	1
Ni⁺–Ni	208	1
Ni⁺–O	275.9 ± 7.7	1
Ni⁺–S	241.0 ± 3.9	1
Ni⁺–Si	326 ± 6.7	1
Np⁺–F	730 ± 100	1
Np⁺–O	760 ± 10	14
Np⁺–S	491 ± 52	20
O⁺–Ar	33.8	1
O⁺–F	301.8 ± 8.4	1
O⁺–H	487.9 ± 0.34	1
O⁺–N	1050.64 ± 0.13	1
O⁺–O	647.75 ± 0.17	1
Os⁺–H	238.9	1
Os⁺–O	418 ± 50	1
P⁺–C	512 ± 42	1
P⁺–Cl	289	1
P⁺–F	490.6 ± 8.4	1
P⁺–H	329.6 ± 2.1	1
P⁺–N	483 ± 21	1
P⁺–O	791.3 ± 8.4	1
P⁺–P	481 ± 50	1
P⁺–S	606 ± 34	1
Pa⁺–O	800 ± 50	14
Pa⁺–S	525 ± 86	20
Pb⁺–Br	260 ± 63	1
Pb⁺–Cl	285 ± 63	1
Pb⁺–F	347 ± 32	1
Pb⁺–O	247 ± 8.4	1

A⁺–B	D^o_{298}/kJ mol⁻¹	Ref.
Pb⁺–Pb	214 ± 29	1
Pb⁺–S	227.7 ± 10.6	12
Pb⁺–Se	169.4 ± 6.3	1
Pb⁺–Te	163 ± 63	1
Pd⁺–C	528 ± 5	1
Pd⁺–H	208.4 ± 8.7	1
Pd⁺–O	145 ± 11	1
Pd⁺–Pd	197 ± 29	1
Pd⁺–S	197 ± 6	1
Pd⁺–Si	289 ± 50	1
Pr⁺–Au	317 ± 81	1
Pr⁺–Br	357.7	1
Pr⁺–Cl	445.0	1
Pr⁺–F	557 ± 63	1
Pr⁺–I	317.0	1
Pr⁺–O	796 ± 15	1
Pt⁺–Ar	36.4 ± 8.7	1
Pt⁺–B	398 ± 105	1
Pt⁺–C	530.5 ± 4.8	1
Pt⁺–Cl	249.8 ± 14.5	1
Pt⁺–H	275 ± 5	1
Pt⁺–N	326.9 ± 9.6	1
Pt⁺–O	318.4 ± 6.7	1
Pt⁺–Pt	318 ± 23	1
Pt⁺–Si	515 ± 50	1
Pt⁺–Xe	86.6 ± 28.9	1
Pu⁺–F	562 ± 50	1
Pu⁺–O	651 ± 19	14
Pu⁺–S	420 ± 23	20
Rb⁺–Ar	12.0	1
Rb⁺–Br	17.6v5.1	1
Rb⁺–Cl	10.5 ± 10.5	1
Rb⁺–I	27 ± 42	1
Rb⁺–Kr	14.9	1
Rb⁺–Na	50.1 ± 3.9	1
Rb⁺–Ne	6.95	1
Rb⁺–O	29	1
Rb⁺–Rb	75.6 ± 9.6	1
Rb⁺–Xe	21.5	1
Re⁺–C	497.7 ± 3.9	1
Re⁺–H	224.7 ± 6.7	1
Re⁺–O	435 ± 59	1
Rh⁺–C	414 ± 17	1
Rh⁺–H	164.8 ± 3.8	1
Rh⁺–O	295.0 ± 5.8	1
Rh⁺–S	251.8 ± 11.6	11
Ru⁺–C	594.3 ± 6.8	15
Ru⁺–H	160.2 ± 5.0	1
Ru⁺–O	372 ± 5	1
Ru⁺–S	293.3 ± 9.6	15
S⁺–C	620.8 ± 1.3	1
S⁺–F	343.5 ± 4.8	1
S⁺–H	348.2 ± 1.7	1
S⁺–N	516 ± 34	1
S⁺–O	524.3 ± 0.4	1
S⁺–P	573 ± 21	1
S⁺–S	522.4 ± 0.5	1
Sc⁺–C	326 ± 6	1
Sc⁺–Cl	410 ± 42	1
Sc⁺–F	605 ± 32	1
Sc⁺–Fe	201 ± 21	1

A$^+$–B	D°_{298}/kJ mol^{-1}	Ref.	A$^+$–B	D°_{298}/kJ mol^{-1}	Ref.	A$^+$–B	D°_{298}/kJ mol^{-1}	Ref.
Sc$^+$–H	235 ± 8	1	Ta$^+$–Ta	666	1	V$^+$–Fe	314 ± 21	1
Sc$^+$–O	689 ± 5	1	Tb$^+$–Cu	245 ± 34	1	V$^+$–H	202 ± 6	1
Sc$^+$–S	529.7 ± 17.4	1	Tb$^+$–O	722 ± 15	1	V$^+$–Kr	49.46 ± 0.18	1
Sc$^+$–Se	475.8 ± 8.4	1	Tc$^+$–H	197.5	1	V$^+$–N	448.6 ± 5.8	1
Sc$^+$–Si	242.3 ± 10.5	1	Tc$^+$–O	>167	1	V$^+$–Nb	403.5 ± 0.2	1
Se$^+$–F	364 ± 42	1	Te$^+$–H	305 ± 12	1	V$^+$–O	581.6 ± 9.6	1
Se$^+$–H	304	1	Te$^+$–O	339 ± 50	1	V$^+$–S	358.9 ± 8.7	1
Se$^+$–P	514 ± 25	1	Te$^+$–P	415 ± 97	1	V$^+$–Si	229 ± 15	1
Se$^+$–S	392 ± 19	1	Te$^+$–Se	342 ± 19	1	V$^+$–V	302	1
Se$^+$–Se	413 ± 19	1	Te$^+$–Si	339.6	5	V$^+$–Xe	66.4 ± 0.6	1
Si$^+$–Au	175 ± 50	1	Te$^+$–Te	278 ± 29	1	W$^+$–C	463.0 ± 8.7	10
Si$^+$–B	351 ± 15	1	Th$^+$–Cl	499 ± 29	1	W$^+$–F	444 ± 96	1
Si$^+$–Br	276 ± 96	1	Th$^+$–F	682 ± 29	1	W$^+$–H	222.5 ± 5	1
Si$^+$–C	365 ± 50	1	Th$^+$–O	843 ± 25	14	W$^+$–O	656.9 ± 6.8	10
Si$^+$–Cl	591.0 ± 0.6	1	Th$^+$–Pt	388 ± 193	1	Xe$^+$–Ar	13.4	1
Si$^+$–F	684.1 ± 5.4	1	Th$^+$–Rh	504 ± 67	1	Xe$^+$–F	188	1
Si$^+$–H	316.6 ± 2.1	1	Th$^+$–S	570 ± 75	20	Xe$^+$–H	355	1
Si$^+$–O	478 ± 13.4	1	Ti$^+$–C	395 ± 23	1	Xe$^+$–Kr	41.65 ± 0.08	1
Si$^+$–P	272 ± 50	1	Ti$^+$–Cl	426.8	1	Xe$^+$–N	66.4 ± 9.6	1
Si$^+$–Pd	237 ± 50	1	Ti$^+$–F	≥456	1	Xe$^+$–Ne	2.1 ± 0.8	1
Si$^+$–Pt	525 ± 50	1	Ti$^+$–H	226.6 ± 10.6	1	Xe$^+$–Xe	99.6	1
Si$^+$–S	387.5 ± 6.0	1	Ti$^+$–N	501 ± 13	1	Y$^+$–C	281 ± 12	1
Si$^+$–Si	334 ± 19	1	Ti$^+$–O	667 ± 7	1	Y$^+$–F	677 ± 21	1
Si$^+$–Te	347 ± 50	1	Ti$^+$–Pt	82 ± 96	1	Y$^+$–H	260.5 ± 5.8	1
Sm$^+$–Br	343.3	1	Ti$^+$–S	461.1 ± 6.8	1	Y$^+$–O	718 ± 25	1
Sm$^+$–Cl	435.4	1	Ti$^+$–Si	249 ± 16	1	Y$^+$–Pt	466 ± 192	1
Sm$^+$–F	620.9	1	Ti$^+$–Ti	229	1	Y$^+$–S	533.9 ± 8	1
Sm$^+$–I	299.1	1	Tl$^+$–Br	52 ± 50	1	Y$^+$–Si	243 ± 13	1
Sm$^+$–O	569 ± 15	1	Tl$^+$–Cl	26 ± 4	1	Y$^+$–Te	360 ± 96	1
Sn$^+$–Br	335 ± 50	1	Tl$^+$–F	13 ± 21	1	Y$^+$–Y	281 ± 21	1
Sn$^+$–Cu	184 ± 96	1	Tl$^+$–I	133 ± 21	1	Yb$^+$–Br	307.4	1
Sn$^+$–F	364 ± 29	1	Tl$^+$–Tl	22 ± 50	1	Yb$^+$–Cl	399.6	1
Sn$^+$–O	281 ± 10	1	Tm$^+$–Br	312.2	1	Yb$^+$–F	557.5 ± 14.4	1
Sn$^+$–S	240 ± 19	1	Tm$^+$–Cl	407.9	1	Yb$^+$–I	262.0	1
Sn$^+$–Se	174 ± 6.3	1	Tm$^+$–F	537 ± 16	1	Yb$^+$–O	376 ± 15	1
Sn$^+$–Sn	193	1	Tm$^+$–I	266.8	1	Yb$^+$–Yb	238 ± 96	1
Sn$^+$–Te	168.7 ± 8.4	1	Tm$^+$–O	482 ± 15	1	Zn$^+$–Ar	28.7 ± 1.2	1
Sr$^+$–Ar	13.32 ± 2.92	1	U$^+$–Br	345 ± 29	1	Zn$^+$–H	216 ± 15	1
Sr$^+$–Br	378.1 ± 8.4	1	U$^+$–C	300 ± 96	1	Zn$^+$–O	161.1 ± 4.8	1
Sr$^+$–Cl	427 ± 8.4	1	U$^+$–Cl	431 ± 34	1	Zn$^+$–S	198 ± 12	1
Sr$^+$–F	615 ± 50	1	U$^+$–D	283.4 ± 9.6	1	Zn$^+$–Si	274.1 ± 9.6	1
Sr$^+$–H	209 ± 5	1	U$^+$–F	668 ± 29	1	Zn$^+$–Zn	60 ± 19	1
Sr$^+$–I	308.2	1	U$^+$–H	284 ± 8	1	Zr$^+$–Ar	36.09 ± 0.24	1
Sr$^+$–Kr	18.13 ± 6.94	1	U$^+$–N	~485	1	Zr$^+$–C	445.8 ± 15.4	1
Sr$^+$–Ne	4.52 ± 9.6	1	U$^+$–O	774 ± 13	14	Zr$^+$–H	218.8 ± 9.6	1
Sr$^+$–O	298.7	1	U$^+$–P	186	1	Zr$^+$–N	443 ± 46	1
Sr$^+$–Sr	108.5 ± 1.6	1	U$^+$–S	500 ± 60	20	Zr$^+$–O	753 ± 11	1
Ta$^+$–C	369.4 ± 3.9	10	V$^+$–Ar	39.39 ± 0.12	1	Zr$^+$–S	549.0 ± 9.6	1
Ta$^+$–H	230 ± 6	1	V$^+$–C	373 ± 13.5	1	Zr$^+$–Zr	407.0 ± 9.6	1
Ta$^+$–O	688.7 ± 11.6	10	V$^+$–D	202 ± 6	1			

References

1. Luo, Y. R., *Comprehensive Handbook of Chemical Bond Energies*, CRC Press, Boca, Raton, 2007.

2. Parke, L. G., Hinton, C. S., and Armentrout, P. B., *Int. J. Mass Spectrom.* 254, 168, 2006.

3. Li, F.-X., Zhang, X.-G., and Armentrout, P. B., *Int. J. Mass Spectrom.* 255/256, 279, 2006.

4. Li, F.-X., and Armentrout, P. B., *J. Chem. Phys.* 125, 133114/1, 2006.

5. Chattopadhyaya, S., Pramanik, A., Banerjee, A., and Das, K. K., *J. Phys. Chem. A* 110, 12303, 2006.

6. Li, J., Hao, Y., Yang, J., Zhou, C., and Mo, Y., *J. Chem. Phys.* 127, 104307/1, 2007.

7. Gibson, J. K., Haire, R. G., Santos, M., Pires de Matos, A., and Marçalo, J., *J. Phys. Chem. A* 112, 11373, 2008.

8. Merritt, J. M., Kaledin, A. L., Bondybey, V. E., and Heaven, M. C., *Phys. Chem. Chem. Phys.* 10, 4006, 2008.

9. Schröder, D., *J. Phys. Chem. A* 112, 13215, 2008.

10. Hinton, C. S., Li, F.-X., and Armentrout, P. B., *Int. J. Mass Spectrom.* 280, 226, 2009.

11. Armentrout, P. B., and Kretzschmar, I., *J. Phys. Chem. A* 113, 10955, 2009.
12. Armentrout, P. B., and Kretzschmar, I., *Inorg.. Chem.* 48, 10371, 2009.
13. Li, M., Liu, S.-R., and Armentrout, P. B., *J. Chem. Phys.* 131, 144310, 2009.
14. Marçalo, J., and Gibson, J. K., *J. Phys. Chem. A* 113, 12599, 2009.
15. Armentrout, P. B., and Kretzschmar, I., *Phys. Chem. Chem. Phys.* 12, 4078, 2010.
16. Tzeli, D., and Mavridis, A., *J. Chem. Phys.* 132, 194312, 2010.
17. Hinton, C. S., Armentrout, P. B., *J. Chem. Phys.* 133, 124307, 2010.
18. Li, F.-X., Hinton, C. S., Citir, M., Liu, F., and Armentrout, P. B., *J. Chem. Phys.* 134, 024310, 2011.
19. Li, F.-X., Gorham, K., and Armentrout, P. B., *J. Phys. Chem. A* 114, 11043, 2010.
20. Pereira, C. C. L., Marsden, C. J., Marçalo, J., and Gibson, J. K., *Phys. Chem. Chem. Phys.* 13, 12940, 2011.

TABLE 7. Bond Dissociation Energies in Polyatomic Cations

This Table has been arranged on the basis of the Periodic Table with the IUPAC notation for Groups 1 to 18, see inside front cover of this *Handbook*. The **boldface** in the species indicates the dissociated fragment.

Bond	D^o_{298}/kJ mol⁻¹	Ref.	Bond	D^o_{298}/kJ mol⁻¹	Ref.
(1) Group 1			$K^+(\mathbf{H_2O})_6$–H_2O	41.8	1
Li⁺–H₂	27.2	1	K^+–NH_3	79 ± 7	1
Li⁺–CO	57 ± 13	1	K^+–C_6H_6	80.3	1
Li⁺–H₂O	139 ± 8	1	K^+–adenine	95.1 ± 3.2	1
Li⁺–NH₃	156 ± 8	1	K^+–indole	104.6 ± 12.6	1
Li⁺–CH₄	130	1	K^+–Phe (phenylalanine)	150.5 ± 5.8	1
Li⁺–CH₃OH	156 ± 8	1	K^+–Tyr (tyrosine)	165.0 ± 5.8	1
Li⁺–CH₃OCH₃	167 ± 10	1	Rb^+–H_2O	66.9 ± 12.6	1
Li⁺–pyridine	183.0 ± 14.5	1	Rb^+–NH_3	78.2	1
Li⁺–Gly (glycine)	220 ± 9	1	Rb^+–CH_3CN	86.6 ± 1.3	1
Na⁺–H₂	10.4 ± 0.8	1	Rb^+–C_6H_5OH	70.2 ± 3.7	1
Na⁺–N₂	33.5	1	Cs^+–H_2O	57.3	1
Na⁺–CO	31 ± 8	1	Cs^+–$C_6H_5NH_2$	70.8 ± 4.5	1
Na⁺–CO₂	66.5	1	**(2) Group 2**		
Na⁺–SO₂	79.1	1	**CH₃Be⁺**–CH₃	192.9 ± 13.4	1
Na⁺–O₃	52.3	1	*tert*-**C(CH₃)₃Be⁺**–*tert*-C(CH₃)₃	121.8 ± 13.4	1
Na⁺–H₂O	91.2 ± 6.3	1	Mg^+–OH	314 ± 33	1
Na⁺(H₂O)–H₂O	82.0 ± 5.8	1	Mg^+–CO	43.1 ± 5.8	1
Na⁺(H₂O)₂–H₂O	66.1	1	Mg^+–CO_2	58.4 ± 5.8	1
Na⁺(H₂O)₃–H₂O	52.7 ± 0.8	1	Mg^+–H_2O	122.5 ± 12.5	1
Na⁺(glycine)–H₂O	75.1 ± 5.3	1	Mg^+–NH_3	158.9 ± 11.6	1
Na⁺(glutamine)–H₂O	52 ± 1	1	Mg^+–CH_4	29.8 ± 6.8	1
Na⁺–NH₃	106.2 ± 5.4	1	Mg^+–MeOH	147.6 ± 6.8	1
Na⁺–HNO₃	86.2	1	Mg^+–C_6H_6	155.2	1
Na⁺–CH₄	30.1	1	Mg^+–pyridine	200.0 ± 6.4	1
Na⁺–CH₃OH	98.8 ± 5.7	1	Mg^+–imidazole	243.9 ± 10.4	1
Na⁺–CH₃CN	125.5 ± 9.6	1	$Mg^{2+}(\mathbf{H_2O})_5$–H_2O	101.3	1
Na⁺–C₂H₄	44.6 ± 4.4	1	$Mg^{2+}(\mathbf{Me_2CO})_5$–$Me_2CO$	93.3	1
Na⁺–CH₃OCH₃	101.4 ± 5.7	1	Ca^+–OH	435.1 ± 14.5	1
Na⁺–CH₃C(O)H	114.4 ± 3.4	1	Ca^+–H_2O	117.2	1
Na⁺– MeCOMe	131.3 ± 4.1	1	Ca^+–C_6H_6	134	1
Na⁺–C₆H₆	97.0 ± 5.9	1	Ca^+–imidazole	186.3 ± 3.9	1
Na⁺–pyrrole	103.7 ± 4.8	1	$Ca^{2+}(\mathbf{H_2O})_4$–H_2O	110.0 ± 5.9	1
Na⁺–Gly (glycine)	166.7 ± 5.1	1	$Ca^{2+}(\mathbf{Me_2CO})_5$–$Me_2CO$	101.3	1
Na⁺–Ala (alanine)	167 ± 4	1	Sr^+–CO	20.3	1
Na⁺–GlyGly (glycylglycine)	203 ± 8	1	Sr^+–CO_2	41.9	1
K⁺–H₂	6.1 ± 0.8	1	Sr^+–H_2O	144.3	1
K⁺–CO₂	35.6	1	Sr^+–C_6H_6	117	1
K⁺–H₂O	74.9	1	$Sr^{2+}(\mathbf{H_2O})_5$–H_2O	87.4	1
K⁺(H₂O)₂–H₂O	67.4	1	Ba^+–OH	530.7 ± 19.3	1
K⁺(H₂O)₃–H₂O	55.2	1	$Ba^{2+}(\mathbf{H_2O})_4$–H_2O	90.8	1
K⁺(H₂O)₄–H₂O	11.8	1	**(3) Group 3**		
K⁺(H₂O)₅–H₂O	44.8	1	Sc^+–H₂	23.0 ± 1.3	1

Bond	D^o_{298}/kJ mol^{-1}	Ref.
Sc$^+$–CH$_2$	412 ± 22	1
Sc$^+$–CH$_3$	233 ± 10	1
Sc$^+$–C$_2$H$_2$	240 ± 20	1
Sc$^+$–C$_2$H$_4$	≥131	1
Sc$^+$–C$_6$H$_6$	222 ± 21	1
Sc$^+$–H$_2$O	131	1
Sc$^+$–NH	483 ± 10	1
Sc$^+$–NH$_2$	347 ± 5	1
Sc$^+$–pyridine	231.5 ± 10.3	1
Y$^+$–CH$_2$	398 ± 13	1
Y$^+$–CH$_3$	249 ± 5.0	1
Y$^+$–C$_2$H$_2$	218 ± 13	1
Y$^+$–C$_2$H$_4$	>138	1
Y$^+$–CO	29.9 ± 10.6	1
Y$^+$–CS	137.0 ± 7.7	1
Y$^+$(O)–CO$_2$	86 ± 5	1
La$^+$–CH	523 ± 33	1
La$^+$–CH$_2$	401 ± 7	1
La$^+$–CH$_3$	217 ± 15	1
La$^+$–C$_2$H$_2$	262 ± 30	1
La$^+$–C$_2$H$_4$	192.5	1
Lu$^+$–CH$_2$	>230 ± 6	1
Lu$^+$–CH$_3$	176 ± 20	1
U$^+$(F)–F	552 ± 44	1
U$^+$(F)$_2$–F	523 ± 38	1
U$^+$(F)$_3$–F	381 ± 19	1
U$^+$(F)$_4$–F	243 ± 17	1
U$^+$(F)$_5$–F	26 ± 11	1
(4) Group 4		
Ti$^+$–CH	478 ± 5	1
Ti$^+$–CH$_2$	391 ± 15	1
Ti$^+$–CH$_3$	213.8 ± 3	1
Ti$^+$–CH$_4$	70.3 ± 2.5	1
Ti$^+$–C$_2$H$_2$	213 ± 13	1
Ti$^+$–C$_2$H$_4$	146 ± 11	1
Ti$^+$–C$_6$H$_6$	259 ± 9	1
Ti$^+$–CO	117.7 ± 5.8	1
Ti$^+$–H$_2$O	157.7 ± 5.9	1
Ti$^+$–NH	466 ± 12	1
Ti$^+$–NH$_2$	356 ± 13	1
Ti$^+$–NH$_3$	197 ± 7	1
Ti$^+$–pyridine	217.2 ± 9.3	1
Ti$^+$–imidazole	≤232.4 ± 8.2	1
Zr$^+$–CH	568 ± 13	1
Zr$^+$–CH$_2$	444.8 ± 5	1
Zr$^+$–CH$_3$	227.7 ± 9.6	1
Zr$^+$–C$_2$H$_2$	273 ± 14	1
Zr$^+$–CO	77 ± 10	1
Zr$^+$–CS	257.6 ± 10.6	1
Hf$^+$–CH	492.1 ± 14.5	2
Hf$^+$–CH$_2$	421.6 ± 6.8	2
Hf$^+$–CH$_2$	204.5 ± 25.1	2
Hf$^+$–C$_2$H$_2$	150.6	1
(5) Group 5		
(CO)$_6$V$^+$–H	220 ± 14	1

Bond	D^o_{298}/kJ mol^{-1}	Ref.
V$^+$–H$_2$	42.7 ± 2.1	1
V$^+$–CH	470 ± 5	1
V$^+$–CH$_2$	326 ± 6	1
V$^+$–CH$_3$	193 ± 7	1
V$^+$–C$_2$H$_2$	172 ± 8	1
V$^+$–C$_2$H$_4$	124 ± 8	1
V$^+$–(η^5-C$_5$H$_5$)	530.7	
V$^+$–C$_6$H$_6$	234 ± 10	1
V$^+$–CO	114.8 ± 2.9	1
V$^+$–CO$_2$	72.4 ± 3.8	1
V$^+$–H$_2$O	149.8 ± 5.0	1
V$^+$–NH	423 ± 29	1
V$^+$–NH$_2$	293 ± 6	1
V$^+$–NH$_3$	192 ± 11	1
V$^+$–pyridine	218.7 ± 13.5	1
V$^+$–imidazole	≤243.4 ± 8.0	1
Nb$^+$–H$_2$	61.9	1
Nb$^+$–CH	581 ± 19	1
Nb$^+$–CH$_2$	428.4 ± 8.7	1
Nb$^+$–CH$_3$	198.8 ± 10.6	1
Nb$^+$–CH$_3$NH$_2$	134	1
Nb$^+$–C$_3$H$_6$	117.7	1
(NbFe)$^+$–C$_3$H$_4$	>163	1
Nb$^+$–CO	95.5 ± 4.8	1
Nb$^+$–CS	242.2 ± 10.6	1
Nb$_7$$^+$–N$_2$	<215	1
Ta$^+$–CH	561.5 ± 15.4	6
Ta$^+$–CH$_2$	464.1 ± 2.9	6
Ta$^+$–CH$_3$	259.5 ± 13.5	6
Ta$^+$–C$_6$H$_6$	251~301	1
(6) Group 6		
(CO)$_6$Cr$^+$–H	230 ± 10	1
(η^5-C$_5$H$_5$)(NO)(CO)$_2$Cr$^+$–H	207.1 ± 14	1
Cr$^+$–H$_2$	31.8 ± 2.1	1
Cr$^+$–CH	294 ± 29	1
Cr$^+$–CH$_2$	216 ± 4	1
Cr$^+$–CH$_3$	110 ± 4	1
Cr$^+$–C$_6$H$_6$	170 ± 10	1
Cr$^+$–indole	196.6 ± 16.7	1
Cr$^+$–CO	89.7 ± 5.8	1
Cr$^+$–OH	298 ± 14	1
Cr$^+$–H$_2$O	132.6 ± 8.8	1
Cr$^+$–N$_2$	59 ± 4	1
Cr$^+$–NH$_3$	183 ± 10	1
(CO)$_6$Mo$^+$–H	260 ± 9	1
Mo$^+$–CH	513.3 ± 13.5	1
Mo$^+$–CH$_2$	344.4 ± 10	1
Mo$^+$–CH$_3$	151.5 ± 8.7	1
Mo$^+$–CO	193.9 ± 9.6	1
Mo$^+$–CO$_2$	49.2 ± 7	1
Mo$^+$–CS	162 ± 18	1
Mo$^+$–CS$_2$	67.5 ± 12.5	1
Mo$^+$–NH	<385	1
Mo$^+$–pyrrole	>289	1
(CO)$_6$W$^+$–H	257 ± 9	1

Bond	D^{o}_{298}/kJ mol^{-1}	Ref.
$W^{+}-CH$	580 ± 27	1
$W^{+}-CH_2$	456.4 ± 5.8	1
$W^{+}-CH_3$	$\sim 222.9 \pm 9.6$	1
$(PMe_3)_3(CO)_3W^{+}-H$	259.4	1
W^{+}-pyrrole	>209	1
(7) Group 7		
$(CO)_5Mn^{+}-H$	172 ± 10	1
$Mn^{+}-H_2$	7.9 ± 1.7	1
$Mn^{+}-CH_2$	295 ± 13	1
$Mn^{+}-CH_3$	215 ± 10	1
$Mn^{+}(CO)_5-CH_3$	132 ± 15	1
$Mn^{+}(CO)_5-CH_4$	>30	1
$Mn^{+}-(\eta^5-C_5H_5)$	326.1 ± 9.6	1
$Mn^{+}-C_6H_6$	145 ± 10	1
$Mn^{+}-OH$	332 ± 24	1
$Mn^{+}-CO$	25 ± 10	1
$Mn^{+}-H_2O$	121.8 ± 5.9	1
$Mn^{+}-CH_3OH$	134 ± 29	1
$Mn^{+}-OC(CH_3)_2$	159 ± 14	1
$Mn^{+}-CS$	80.0 ± 21	1
$Mn^{+}-NH_2$	254 ± 20	1
$Mn^{+}-NH_3$	147 ± 8	1
$Tc^{+}-CH_2$	<464	1
$Tc^{+}-C_2H_2$	<320	1
$Re^{+}(CH_3)(CO)_5-H$	294 ± 13	1
$(PMe_3)(CO)_2Re^{+}-H$	300.4	1
(8) Group 8		
$Fe^{+}(O)-H$	444 ± 17	1
$Fe^{+}(CO)-H$	120 ± 23	1
$Fe^{+}(H_2O)-H$	215 ± 14	1
$Fe^{+}(\eta^5-C_5H_5)-H$	193 ± 21	1
$(CO_2)_2Fe^{+}-H$	299 ± 15	1
$Fe^{+}-H_2$	45.2 ± 2.5	1
$Fe^{+}-CH$	423 ± 29	1
$Fe^{+}-CH_2$	$\leq 342 \pm 2$	1
$Fe^{+}-CH_3$	229 ± 5	1
$Fe^{+}-CH_4$	73.2	1
$Fe^{+}-C_2H_2$	159.0 ± 2.1	1
$Fe^{+}-C_2H_3$	238 ± 10	1
$Fe^{+}-C_2H_4$	145 ± 11	1
$Fe^{+}-C_2H_5$	233 ± 9	1
$Fe^{+}-C_2H_6$	64 ± 6	1
$Fe^{+}-OH$	366 ± 12	1
$Fe^{+}-CO$	129.3 ± 3.9	1
$Fe^{+}D-CO$	53 ± 13	1
$Fe^{+}-CO_2$	74.3 ± 7.7	1
$Fe^{+}-H_2O$	128.9 ± 0.8	1
$Fe^{+}-N_2$	53 ± 4	1
$Fe^{+}-NH_3$	184 ± 12	1
$Fe^{+}-CS_2$	166.1 ± 4.6	1
Fe^{+}-imidazole	246.1 ± 13.8	1
$Fe^{+}-SiH$	254 ± 13	1
$Fe^{+}-SiH_2$	181 ± 9	1
$Fe^{+}-SiH_3$	183 ± 9	1

Bond	D^{o}_{298}/kJ mol^{-1}	Ref.
$Ru^{+}(\eta^5-C_5H_5)_2-H$	292 ± 16	1
$(\eta^5-C_5Me_5)_2Ru^{+}-H$	284.5	1
$Ru^{+}-CH$	501.7 ± 11.6	1
$Ru^{+}-CH_2$	344.4 ± 4.8	1
$Ru^{+}-CH_3$	160.2 ± 5.8	1
$Ru^{+}-CS$	244.9 ± 17.4	9
$OsO_4^{+}-H$	552 ± 13	1
(9) Group 9		
$(\eta^5-C_5H_5)(CO)_2Co^{+}-H$	245 ± 12	1
$(CH_3OD)Co^{+}-H$	147.6 ± 7.7	1
$Co^{+}-H_2$	76.1 ± 4.2	1
$(\eta^5-C_5H_5)Co^{+}-H_2$	67.8	1
$Co^{+}-CH$	420 ± 37	1
$Co^{+}-CH_2$	317 ± 5	1
$Co^{+}-CH_3$	203 ± 4	1
$Co^{+}-CH_4$	96.7	1
$Co^{+}-C_{60}$	243 ± 67	1
$Co^{+}-CO$	173.7 ± 6.7	1
$Co^{+}-H_2O$	164.4 ± 5.9	1
$Co^{+}-CS$	259 ± 33	1
$Co^{+}-N_2$	96.2 ± 7.1	1
$Co^{+}-NH_2$	247 ± 7	1
$Co^{+}-NH_3$	219 ± 16	1
$Co^{+}-CH_3CN$	$>255 \pm 17$	1
$Co^{+}-P(CH_3)_3$	278 ± 11	1
$Co^{+}-P(C_2H_5)_3$	339 ± 16	1
$(CH)Rh^{+}-H$	372 ± 21	1
$(\eta^5-C_5H_5)(CO)_2Rh^{+}-H$	287 ± 12	1
$Rh^{+}-CH$	444 ± 12	1
$Rh^{+}-CH_2$	356 ± 8	1
$Rh^{+}-CH_3$	142 ± 6	1
$Rh^{+}-NO$	167 ± 21	1
$Rh^{+}-CS$	256.6 ± 18.3	8
$(CO)(\eta^5-C_5H_5)(PPh_3)Ir^{+}-H$	313.4	1
$(CO)_2(\eta^5-C_5Me_5)Ir^{+}-H$	298.3	1
$Ir^{+}-CH$	666.7 ± 22.2	3
$Ir^{+}-CH_2$	474.7 ± 2.9	3
$Ir^{+}-CH_3$	313.6 ± 17.4	3
$Ir^{+}-C_2H_4$	234.3	1
(10) Group 10		
$(CO)_4Ni^{+}-H$	248 ± 9	1
$(\eta^5-C_5H_5)(NO)Ni^{+}-H$	315 ± 14	1
$(\eta^5-C_5H_5)(\eta^5-C_5H_5)Ni^{+}-H$	215 ± 13	1
$Ni^{+}-H_2$	72.4 ± 1.3	1
$Ni^{+}-CH$	301.0 ± 11.6	1
$Ni^{+}-CH_2$	306 ± 4	1
$Ni^{+}-CH_3$	169.8 ± 6.8	1
$Ni^{+}-CH_4$	96.5 ± 4	1
$Ni^{+}-OH$	235 ± 19	1
$Ni^{+}-CO$	175 ± 11	1
$Ni^{+}-CO_2$	104 ± 1	1
$Ni^{+}-H_2O$	183.7 ± 3.3	1
$Ni^{+}-CS$	234.5 ± 9.6	1
$Ni^{+}-N_2$	110.9 ± 10.5	1

Bond	D^o_{298}/kJ mol^{-1}	Ref.
Ni$^+$–NO	227.6 ± 7.5	1
Ni$^+$–NH$_2$	232.5 ± 7.7	1
Ni$^+$–NH$_3$	238 ± 19	1
Pd$^+$–CH	536 ± 10	1
Pd$^+$–CH$_2$	463 ± 3	1
Pd$^+$–CH$_3$	258 ± 8	1
Pd$^+$–CH$_4$	170.8 ± 7.7	1
Pd$^+$–CS	200 ± 14	1
Pd$^+$–C$_2$H$_2$	>28.9 ± 4.8	1
Pt$^+$–H$_2$	146.7 ± 11.6	1
Pt$^+$–CH	536.4 ± 9.6	1
Pt$^+$–CH$_2$	471	1
Pt$^+$–CH$_3$	257.6 ± 7.7	1
Pt$^+$–CH$_4$	170.8 ± 7.7	1
Pt$^+$–O$_2$	64.6 ± 4.8	1
Pt$^+$–CO	218.1 ± 8.7	1
Pt$^+$–CO$_2$	59.8 ± 4.8	1
Pt$^+$–NH$_3$	274 ± 12	1
Pt$^+$–C$_2$H$_4$	229.7	1

(11) Group 11

Bond	D^o_{298}/kJ mol^{-1}	Ref.
Cu$^+$–H$_2$	51.9 ± 0.4	1
Cu$^+$–CH$_2$	267.3 ± 6.8	1
Cu$^+$–CH$_3$	111 ± 7	1
Cu$^+$–C$_2$H$_2$	>21.2 ± 9.6	1
Cu$^+$–C$_2$H$_4$	176 ± 14	1
Cu$^+$–C$_6$H$_6$	218.0 ± 9.6	1
Cu$^+$–CO	149 ± 7	1
Cu$^+$–N$_2$	89 ± 30	1
Cu$^+$–NO	109.0 ± 4.8	1
Cu$^+$–H$_2$O	160.7 ± 7.5	1
Cu$^+$–NH$_2$	192 ± 13	1
Cu$^+$–NH$_3$	237 ± 15	1
Cu$^+$–CS	238.3 ± 11.6	1
Cu$^+$–SiH	246 ± 27	1
Cu$^+$–SiH$_2$	≥231 ± 7	1
Cu$^+$–SiH$_3$	97 ± 25	1
Ag$^+$–CH$_2$	≥107 ± 4	1
Ag$^+$–CH$_3$	66.6 ± 4.8	1
Ag$^+$–C$_2$H$_5$	65.7 ± 7.5	1
Ag$^+$–C$_6$H$_6$	167 ± 19	1
Ag$^+$–O$_2$	29.7 ± 0.8	1
Ag$^+$–CO	89 ± 5	1
Ag$^+$–H$_2$O	134 ± 8	1
Ag$^+$–CS	152 ± 20	1
Ag$^+$–NH$_3$	170 ± 13	1
Au$^+$–CH$_2$	357.0 ± 6.8	5
Au$^+$–CH$_3$	209.4 ± 23.2	5
Au$^+$–C$_2$H$_4$	344.5	1
Au$^+$–C$_6$H$_6$	289 ± 29	1
Au$^+$–CO	201 ± 8	1
Au$^+$–H$_2$O	164.0 ± 9.6	1
Au$^+$–H$_2$S	230 ± 25	1
Au$^+$–NH$_3$	297 ± 29	1
Au$^+$–PH$_3$	402 ± 33	1

Bond	D^o_{298}/kJ mol^{-1}	Ref.
(12) Group 12		
Zn$^+$–H$_2$	15.7 ± 1.7	1
Zn$^+$–CH$_3$	280 ± 7	1
Zn$^+$–OH	127.2	1
Zn$^+$–H$_2$O	163	1
Zn$^+$–NO	76.2 ± 9.6	1
Zn$^+$–pyrimidine	209.6 ± 7.7	1
Zn$^+$–CS	149 ± 23	1
Cd$^+$–CH$_3$	228 ± 3	1
Cd$^+$(CH$_3$)–CH$_3$	109 ± 3	1
Cd$^+$–C$_6$H$_6$	136 ± 19	1
Hg$^+$–CH$_3$	285 ± 3	1
Hg$^+$(CH$_3$)–CH$_3$	96 ± 3	1
(13) Group 13		
B$^+$–H$_2$	15.9 ± 0.8	1
HB$^+$–H$_2$	61.5 ± 2.1	1
(CH$_3$)$_2$B$^+$–CH$_3$	32.6 ± 4.2	1
Al$^+$–H$_2$	5.6 ± 0.6	1
Al$^+$–N$_2$	5.6	1
Al$^+$–CO$_2$	≥29.3	1
Al$^+$–H$_2$O	104 ± 15	1
Al$^+$–MeOH	139.7	1
Al$^+$–EtC(O)Et	191.2	1
Al$^+$–C$_6$H$_6$	147.3 ± 8.4	1
Al$^+$–pyridine	190.3 ± 10.3	1
Al$^+$–phenol	154.8 ± 16.7	1
Al$^+$–imidazole	232.4 ± 8.2	1
Ga$^+$–NH$_3$	122.5	1
In$^+$–NH$_3$	111.0	1
(14) Group 14		
C$_{58}$$^+$–C$_2$	955 ± 15	1
C$_{60}$$^+$–C$_2$	822.0 ± 12.5	1
C$_{62}$$^+$–C$_2$	846.2 ± 10.6	1
C$_{78}$$^+$–C$_2$	938.8 ± 10.6	1
HC$_2$$^+$–H	574.749	1
C$_6$H$_5$$^+$–H	376.3 ± 4.8	1
C$_2$H$_3$$^+$–Cl	249 ± 1.0	7
C$_2$H$_5$$^+$–Br	206.3 ± 1.0	7
C$_6$H$_5$$^+$–Br	266.3	1
C$_2$H$_3$$^+$–I	196.2 ± 1.4	7
CH$_3$$^+$–H$_2$	186	1
CH$_5$$^+$–H$_2$	7.9 ± 0.4	1
C$_2$H$_5$$^+$–H$_2$	17	1
CH$_3$$^+$–O$_2$	80 ± 7	4
CO$^+$–N$_2$	67.5 ± 19.3	1
H$_2$CH$^+$–N$_2$	31.8	1
CO$^+$–CO	173.7 ± 14.6	1
CO$^+$(CO)–CO	52.3	1
CO$^+$(CO)$_2$–CO	30.2	1
CO$^+$(CO)$_3$–CO	18.4	1
(CO$_2$)$^+$–CO$_2$	70.3	1
(CO$_2$)$^+$(CO$_2$)–CO$_2$	34.7	1
(CO$_2$)$^+$(CO$_2$)$_2$–CO$_2$	21.3	1
(CO$_2$)$^+$(CO$_2$)$_3$–CO$_2$	20.1 ± 1.3	1

Bond	D^o_{298}/kJ mol^{-1}	Ref.
$CH_3^+-N_2O$	221.3	1
$CH_3^+-SO_2$	253.6	1
CH_3^+-OCS	239.3	1
$CH_3^+-CS_2$	251.9	1
$CH_3^+-H_2O$	279	1
$CH_3^+(H_2O)-H_2O$	106.3	1
$CH_3^+(H_2O)_2-H_2O$	87.9	1
$CH_3^+(H_2O)_3-H_2O$	61.9	1
$CH_3^+(H_2O)_4-H_2O$	48.5	1
$CH_3^+-H_2S$	344.8	1
$CH_2^+-CH_2O$	303.0 ± 2.9	1
$CH_3^+-NH_3$	431.4	1
$(CH_3)^+-CH_3$	209.2 ± 4.2	1
$CH_3^+-CH_4$	166.5	1
$CF_3^+-CH_4$	19.0	1
$(CH_5)^+-CH_4$	28.7 ± 1.3	1
$C_6H_6^+-CH_4$	12.0	1
$CH_3^+-CH_3F$	230	1
$CH_3^+-CF_3Cl$	221	1
$CH_3^+-CH_3Cl$	259	1
$tert\text{-}C_4H_9^+-CH_3OH$	63	1
$tert\text{-}C_4H_9^+-CH_3CN$	85	1
$tert\text{-}C_4H_9^+-SO_2F_2$	43.5	1
$CH_3^+-C_2H_3O$	338.7 ± 2.9	1
$CH_3^+-CF_3ClOCl$	252	1
$tert\text{-}C_4H_9^+-(CH_3)_2S$	185	1
$tert\text{-}C_4H_9^+-C_2H_5OH$	85	1
$tert\text{-}C_4H_9^+-C_3H_8$	27.6	1
$tert\text{-}C_4H_9^+-t\text{-}C_4H_9Cl$	339	1
$tert\text{-}C_4H_9^+-(CH_3)_3CH$	30.1	1
$tert\text{-}C_4H_9^+-C_6H_6$	92	1
$(C_6H_6)^+-C_6H_6$	73.6	1
$(C_6H_6)^+-indole$	54.8	1
$C_6F_6^+-C_6F_6$	30.1 ± 4	1
$C_{60}^+-C_{60}$	35.89 ± 7.72	1
$PhSiH_2^+-H$	159	1
$Si^+(CH_3)_3-Cl$	178.5 ± 1.9	1
SiH_3^+-CO	≥151	1
SiF_3^+-CO	174.1 ± 1.3	1
$(CH_3)_3Si^+-H_2O$	125.9 ± 7.9	1
$(CH_3)_3Si^+-NH_3$	194.6	1
$Si^+(CH_3)(Cl)_2-CH_3$	60.8 ± 2.9	1
$Si^+(CH_3)_2(Cl)-CH_3$	41.5 ± 1.9	1
Si^+-CH_3	413.9 ± 5.8	1
$Si^+(CH_3)-CH_3$	123 ± 48	1
$Si^+(CH_3)_2-CH_3$	513 ± 27	1
$Si^+(CH_3)_3-CH_3$	66.6 ± 5.8	1
$(CH_3)_3Si^+-CH_3OH$	164.0	1
$(CH_3)_3Si^+-(C_2H_5)_2O$	184.9	1
$(CH_3)_3Si^+-C_6H_6$	100.0	1
$(CH_3)_3Si^+-CH_3NH_2$	231.8	1
$(CH_3)_3Ge^+-H_2O$	119.7 ± 2.1	1
$(C_2H_5)_3Ge^+-H_2O$	104.2 ± 2.1	1
$(CH_3)_3Sn^+-NH_3$	154	1
$(CH_3)_3Sn^+-H_2O$	108	1

Bond	D^o_{298}/kJ mol^{-1}	Ref.
$(CH_3)_3Sn^+-(CH_3)_2CO$	157	1
$(CH_3)_3Sn^+-C_3H_7SH$	143	1
Pb^+-H_2O	93.7	1
Pb^+-NH_3	118.4 ± 0.8	1
Pb^+-CH_3OH	97.5 ± 0.8	1
$Pb^+-CH_3NH_2$	148.1 ± 1.3	1
$Pb^+-C_6H_6$	110 ± 2	1
(15) Group 15		
H_2N^+-H	544.43 ± 0.10	1
H_3N^+-H	515.1	1
Me_3N^+-H	376	1
Et_3N^+-H	362	1
$(imidazole)^+-Zn$	216.1 ± 3.9	1
$N_2H^+-H_2$	24.7 ± 0.8	1
ON^+-O_2	14.2	1
N^+-N_2	303.8	1
ON^+-N_2	21.3	1
$N_2^+-N_2$	102.3 ± 14.6	1
$HN_2^+-N_2$	60.7	1
$N_3^+-N_2$	18.8 ± 1.3	1
$O_2N^+-N_2$	19.2 ± 1.3	1
$H_4N^+-N_2$	54 ± 21	1
ON^+-NO	59.4 ± 0.8	1
ON^+-CO	27.2 ± 1.3	1
ON^+-O_3	<58	1
ON^+-CO_2	32.2	1
$N_2O^+-ON_2$	72.8 ± 6.3	1
NO^+-ON_2	36.4 ± 0.8	1
$(HON_2)^+-ON_2$	69.9 ± 4	1
ON^+-H_2O	95	1
$ON^+(H_2O)-H_2O$	67.4	1
$ON^+(H_2O)_2-H_2O$	56.5	1
$H_4N^+-H_2O$	86.2 ± 4.2	1
$H_4N^+(H_2O)-H_2O$	72.8 ± 4.2	1
$H_4N^+(H_2O)_2-H_2O$	57.3 ± 4.2	1
$H_4N^+(H_2O)_3-H_2O$	51.0	1
$H_4N^+(H_2O)_4-H_2O$	44.4	1
$(glycine)H^+-H_2O$	77.2 ± 11.0	1
$(tryptophan)H^+-H_2O$	31.2 ± 2.5	1
$(tryptophanylglycine)H^+-H_2O$	56.0 ± 5.3	1
$H_4N^+-H_2S$	47.7	1
$H^+(NH_3)-NH_3$	108.8	1
$H^+(NH_3)_2-NH_3$	69.5	1
$H^+(NH_3)_3-NH_3$	57.3	1
$H^+(NH_3)_4-NH_3$	49.0	1
$H^+(NH_3)_5-NH_3$	29.3	1
$H^+(NH_3)_6-NH_3$	27.2	1
$NH_4^+-CH_4$	15.0	1
ON^+-CH_3OH	97.6	1
$O_2N^+-CH_3OH$	80.3 ± 9.6	1
$(CH_3CNH)^+-CH_3CN$	130.1 ± 9.6	1
$(pyridineH)^+-pyridine$	105.4 ± 4	1
$(valine H)^+-valine$	86.6 ± 8.4	1
$(betainH)^+-betaine$	139.9 ± 4.8	1
$H_4P^+-H_2O$	54.4	1

Bond	D^o_{298}/kJ mol^{-1}	Ref.
$(H_4P)^+-PH_3$	48.1	1
AsH_2^+-H	257	1
I_2As^+-acetone	106 ± 17	1
I_2As^+-benzene	77 ± 17	1
Bi^+-H_2O	95.4	1
Bi^+-NH_3	149	1
$Bi^+-C_6H_6$	≤149	1
(16) Group 16		
$(H_3O)^+-H_2$	14.6 ± 2.1	1
O^+-O_2	179.5	1
$O^+(O_2)_1-O_2$	28.9	1
$O^+(O_2)_2-O_2$	3.9	1
$O_2^+-O_2$	38.3 ± 2.1	1
$O_2^+(O_2)-O_2$	24.6 ± 1.3	1
$O_2^+(O_2)_2-O_2$	10.4 ± 0.8	1
$O_2^+(O_2)_3-O_2$	9.0 ± 0.8	1
$O_2^+(O_2)_4-O_2$	8.0 ± 0.8	1
$O_2^+(O_2)_5-O_2$	7.9 ± 1.3	1
O^+-N_2	231.4	1
$O_2^+-N_2$	22.6	1
$(H_3O)^+-N_2$	22.2 ± 2.1	1
$O_4^+-N_2$	12.3	1
O_2^+-CO	31.8	1
$O_2^+-CO_2$	41.0 ± 2.1	1
$CO_2^+-CO_2$	65.3 ± 4	1
$(H_3O)^+-CO_2$	64.0	1
$(H_3O)^+(CO_2)-CO_2$	51.9	1
$(H_3O)^+(CO_2)_2-CO_2$	43.9	1
$(H_3O)^+(CO_2)_3-CO_2$	18.0	1
$O_2^+-ON_2$	56.1 ± 4	1
$(H_3O)^+-ON_2$	70.7 ± 6.5	1
$(H_3O)^+(H_2O)-ON_2$	50.6 ± 2.1	1
$(H_3O)^+(H_2O)_2-ON_2$	42.7 ± 2.1	1
$O_3^+-O_3$	67.5 ± 39	1
$OClO^+-OClO$	246 ± 48	1
$O_2^+-H_2O$	>67	1
$(OH)^+(H_2O)_2-H_2O$	87.4	1
$(OH)^+(H_2SO_4)(H_2O)_4-H_2O$	56.9	1
$(OH)^+(H_2SO_4)(H_2O)_5-H_2O$	49.8	1
$(OH)^+(H_2SO_4)(H_2O)_6-H_2O$	44.8	1
$(H_2O)^+-H_2O$	164.0	1
$(H_3O)^+-H_2O$	140.2	1
$(H_3O)^+(H_2O)-H_2O$	93.3	1
$(H_3O)^+(H_2O)_2-H_2O$	71.1	1
$(H_3O)^+(H_2O)_3-H_2O$	64.0	1
$(H_3O)^+(H_2O)_4-H_2O$	54.4	1
$(H_3O)^+(H_2O)_5-H_2O$	49.0	1
$(H_3O)^+(H_2O)_6-H_2O$	43.1	1
$(HCOOH)H^+-H_2O$	100.8	1
$CH_3OH_2^+-H_2O$	115.6	1
$CH_3CHOH^+-H_2O$	104.6	1
$(CH_3)_2OH^+-H_2O$	100.4	1
$(tetrahydrofuranH)^+-H_2O$	82.8	1
$(furanH)^+-H_2O$	43.5	1
furane$^+-H_2O$	41.0	1

Bond	D^o_{298}/kJ mol^{-1}	Ref.
$(phenol)^+-H_2O$	78.0	1
$(1-naphthol)^+-H_2O$	66.4	1
$H_3O^+-HC(O)H$	137.7	1
$H_3O^+-NH_3$	229.3	1
$H_3O^+(NH_3)-NH_3$	77.0	1
$H_3O^+(NH_3)_2-NH_3$	71.5	1
$H_3O^+(NH_3)_3-NH_3$	62.8	1
$H_3O^+-PH_3$	144	1
$H_3O^+-SO_3$	74	1
$(HCOOH)^+-HCOOH$	96.5 ± 9.6	1
$H_3O^+-CH_4$	33.5	1
$(CH_3OH)^+-CH_3OH$	115.8 ± 19.3	1
$CH_3OH_2^+-CH_3OH$	136.4	1
$H_3O^+-CH_3CN$	195.4	1
furan$^+-$furan	94.1	1
BH^+-B, B = tetrahydrofuran	125.1	1
S^+-CS_2	166	1
CS^+-CS_2	150.6	1
$CS_2^+-CS_2$	104.2	1
$HCS_2^+-CS_2$	46.4	1
OS^+-SO_2	57.7	1
$O_2S^+-SO_2$	63.6	1
OCS^+-OCS	100.0	1
OCS^+-CO_2	72.0	1
$SO_2^+-CO_2$	42.7	1
$H_3S^+-H_2O$	91.6	1
thiopheneH$^+-H_2O$	42.7	1
$H_3S^+-H_2S$	53.6 ± 6.3	1
$H_3S^+-CH_4$	16.3	1
$(CH_3)_2Se^{\bullet+}-Se(CH_3)_2$	~95 ± 3	1
$(CH_3)_2Te^{\bullet+}-Te(CH_3)_2$	97 ± 2	1
(17) Group 17		
HF^+-HF	≥138	1
$(H_2Cl)^+-Cl$	39.6	1
HCl^+-HCl	83.9	1
Cl^+-CCl_3	446.7 ± 9.6	1
$Cl^+-C_2H_3$	685.0 ± 4.8	1
HBr^+-HBr	96	1
I^+-CH_3	330.0	1
$I^+(CH_3I)-CH_3$	51.1	1
$I^+(CH_3I)_2-CH_3$	112.9	1
(18) Group 18		
$He^+(He)_1-He$	17.6	1
$He^+(He)_2-He$	2.7 ± 0.6	1
$Ne^+(Ne)-Ne$	10.3 ± 0.6	1
$Ne^+(Ne)_2-Ne$	3.3 ± 0.6	1
$Ar^+(Ar)-Ar$	20.4 ± 0.6	1
$Ar^+(Ar)_2-Ar$	7.0 ± 0.6	1
$Ar^+(N_2)-Ar$	25.1	1
$Ar^+(N_2)(Ar)-Ar$	7.1	1
$Ar^+(N_2)(Ar)_2-Ar$	7.1	1
$Kr^+(Kr)-Kr$	23.3 ± 0.6	1
$Kr^+(Kr)_2-Kr$	9.0 ± 0.6	1
$Xe^+(Xe)-Xe$	25.2 ± 0.6	1

Bond	D^o_{298}/kJ mol^{-1}	Ref.
Xe$^+$(Xe)$_2$–Xe	11.0 ± 0.6	1
Ar$^+$–H$_2$	93.7	1
Ar$^+$–N$_2$	127.6	1
Ar$^+$(N$_2$)–N$_2$	31.0	1
Ar$^+$(N$_2$)$_2$–N$_2$	10.9	1

Bond	D^o_{298}/kJ mol^{-1}	Ref.
Ar$^+$–CO	75 ± 17	1
Ar$^+$(CO)–CO	13	1
Kr$^+$–CO	103.3 ± 7.5	1
Kr$^+$–CO$_2$	79.1 ± 2.9	1

References

1. Luo, Y. R., *Comprehensive Handbook of Chemical Bond Energies*, CRC Press, Boca Raton, FL, 2007.
2. Parke, L. G., Hinton, C. S., and Armentrout, P. B., *Int. J. Mass Spectrom.* 254, 168, 2006.
3. Li, F.-X., Zhang, X.-G., and Armentrout, P. B., *Int. J. Mass Spectrom.* 255/256, 279, 2006.
4. Meloni, G., Zou, P., Klippenstein, S. J., Ahmed, M., Leone, S. R., Taatjes, C. A., and Osborn, D. L., *J. Am. Chem. Soc.* 128, 13559, 2006.
5. Li, F.-X., and Armentrout, P. B., *J. Chem. Phys.* 125, 133114/1, 2006.
6. Parke, L. G., Hinton, C. S., and Armentrout, P. B., *J. Phys. Chem. C* 111, 17773, 2007.
7. Shuman, N. S., Ochieng, M. A., Sztáray, B., and Baer, T., *J. Phys. Chem. A* 112, 5647, 2008.
8. Armentrout, P. B., and Kretzschmar, I., *J. Phys. Chem. A* 113, 10955, 2009.
9. Armentrout, P. B., and Kretzschmar, I., *Phys. Chem. Chem. Phys.* 12, 4078, 2010.

ELECTRONEGATIVITY

Electronegativity is a parameter originally introduced by Pauling which describes, on a relative basis, the tendency of an atom in a molecule to attract bonding electrons. While electronegativity is not a precisely defined molecular property, the electronegativity difference between two atoms provides a useful measure of the polarity and ionic character of the bond between them. This table gives the electronegativity X, on the Pauling scale, for the most common oxidation state. Other scales are described in the references.

References

1. Pauling, L., *The Nature of the Chemical Bond, Third Edition*, Cornell University Press, Ithaca, NY, 1960.
2. Allen, L. C., *J. Am. Chem. Soc.*, 111, 9003, 1989.
3. Allred, A. L., *J. Inorg. Nucl. Chem.*, 17, 215, 1961.

Z	Symbol	X	Z	Symbol	X	Z	Symbol	X
1	H	2.20	33	As	2.18	65	Tb	—
2	He	—	34	Se	2.55	66	Dy	1.22
3	Li	0.98	35	Br	2.96	67	Ho	1.23
4	Be	1.57	36	Kr	—	68	Er	1.24
5	B	2.04	37	Rb	0.82	69	Tm	1.25
6	C	2.55	38	Sr	0.95	70	Yb	—
7	N	3.04	39	Y	1.22	71	Lu	1.0
8	O	3.44	40	Zr	1.33	72	Hf	1.3
9	F	3.98	41	Nb	1.6	73	Ta	1.5
10	Ne	—	42	Mo	2.16	74	W	1.7
11	Na	0.93	43	Tc	2.10	75	Re	1.9
12	Mg	1.31	44	Ru	2.2	76	Os	2.2
13	Al	1.61	45	Rh	2.28	77	Ir	2.2
14	Si	1.90	46	Pd	2.20	78	Pt	2.2
15	P	2.19	47	Ag	1.93	79	Au	2.4
16	S	2.58	48	Cd	1.69	80	Hg	1.9
17	Cl	3.16	49	In	1.78	81	Tl	1.8
18	Ar	—	50	Sn	1.96	82	Pb	1.8
19	K	0.82	51	Sb	2.05	83	Bi	1.9
20	Ca	1.00	52	Te	2.1	84	Po	2.0
21	Sc	1.36	53	I	2.66	85	At	2.2
22	Ti	1.54	54	Xe	2.60	86	Rn	—
23	V	1.63	55	Cs	0.79	87	Fr	0.7
24	Cr	1.66	56	Ba	0.89	88	Ra	0.9
25	Mn	1.55	57	La	1.10	89	Ac	1.1
26	Fe	1.83	58	Ce	1.12	90	Th	1.3
27	Co	1.88	59	Pr	1.13	91	Pa	1.5
28	Ni	1.91	60	Nd	1.14	92	U	1.7
29	Cu	1.90	61	Pm	—	93	Np	1.3
30	Zn	1.65	62	Sm	1.17	94	Pu	1.3
31	Ga	1.81	63	Eu	—			
32	Ge	2.01	64	Gd	1.20			

FORCE CONSTANTS FOR BOND STRETCHING

David R. Lide

Representative force constants (f) for stretching of chemical bonds are listed in this table. Except where noted, all force constants are derived from values of the harmonic vibrational frequencies ω_e. Values derived from the observed vibrational fundamentals v, which are noted by a, are lower than the harmonic force constants, typically by 2 to 3% in the case of heavy atoms (often by 5 to 10% if one of the atoms is hydrogen). Values are given in the SI unit newton per centimeter (N/cm), which is identical to the commonly used cgs unit mdyn/Å.

References

1. Huber, K. P., and Herzberg, G., *Molecular Spectra and Molecular Structure. IV. Constants of Diatomic Molecules*, Van Nostrand Reinhold, New York, 1979.
2. Shimanouchi, T., The Molecular Force Field, in Eyring, H., Henderson, D., and Yost, W., Eds., *Physical Chemistry: An Advanced Treatise*, Vol. IV, Academic Press, New York, 1970.
3. Tasumi, M., and Nakata, M., *Pure and Appl. Chem.*, 57, 121–147, 1985.

Bond	Molecule	f/(N/cm)	Note
H-H	H_2	5.75	
Be-H	BeH	2.27	
B-H	BH	3.05	
C-H	CH	4.48	
	CH_4	5.44	b
	C_2H_6	4.83	a,b,c
	CH_3CN	5.33	b
	CH_3Cl	5.02	a,b,c
	$CCl_2=CH_2$	5.57	b
	HCN	6.22	
N-H	NH	5.97	
O-H	OH	7.80	
	H_2O	8.45	
P-H	PH	3.22	
S-H	SH	4.23	
	H_2S	4.28	
F-H	HF	9.66	
Cl-H	HCl	5.16	
Br-H	HBr	4.12	
I-H	HI	3.14	
Li-H	LiH	1.03	
Na-H	NaH	0.78	
K-H	KH	0.56	
Rb-H	RbH	0.52	
Cs-H	CsH	0.47	
C-C	C_2	12.16	
	$CCl_2=CH_2$	8.43	
	C_2H_6	4.50	a,c
	CH_3CN	5.16	
C-F	CF	7.42	
	CH_3F	5.71	a,c
C-Cl	CCl	3.95	
	CH_3Cl	3.44	a,c
	$CCl_2=CH_2$	4.02	b
C-Br	CH_3Br	2.89	a,c
C-I	CH_3I	2.34	a,c
C-O	CO	19.02	
	CO_2	16.00	
	OCS	16.14	
	CH_3OH	5.42	a,c
C-S	CS	8.49	
	CS_2	7.88	

Bond	Molecule	f/(N/cm)	Note
	OCS	7.44	
C-N	CN	16.29	
	HCN	18.78	
	CH_3CN	18.33	
	CH_3NH_2	5.12	a,c
C-P	CP	7.83	
Si-Si	Si_2	2.15	
Si-O	SiO	9.24	
Si-F	SiF	4.90	
Si-Cl	SiCl	2.63	
N-N	N_2	22.95	
	N_2O	18.72	
N-O	NO	15.95	
	N_2O	11.70	
P-P	P_2	5.56	
P-O	PO	9.45	
O-O	O_2	11.77	
	O_3	5.74	a
S-O	SO	8.30	
	SO_2	10.33	a
S-S	S_2	4.96	
F-F	F_2	4.70	
Cl-F	ClF	4.48	
Br-F	BrF	4.06	
Cl-Cl	Cl_2	3.23	
Br-Cl	BrCl	2.82	
Br-Br	Br_2	2.46	
I-I	I_2	1.72	
Li-Li	Li_2	0.26	
Li-Na	LiNa	0.21	
Na-Na	Na_2	0.17	
Li-F	LiF	2.50	
Li-Cl	LiCl	1.43	
Li-Br	LiBr	1.20	
Li-I	LiI	0.97	
Na-F	NaF	1.76	
Na-Cl	NaCl	1.09	
Na-Br	NaBr	0.94	
Na-I	NaI	0.76	
Be-O	BeO	7.51	
Mg-O	MgO	3.48	
Ca-O	CaO	3.61	

[a] Derived from fundamental frequency, without anharmonicity correction.
[b] Average of symmetric and antisymmetric (or degenerate) modes.
[c] Calculated from Local Symmetry Force Field (see Reference 2).

FUNDAMENTAL VIBRATIONAL FREQUENCIES OF SMALL MOLECULES

David R. Lide

This table lists the fundamental vibrational frequencies of selected three-, four-, and five-atom molecules. Both stable molecules and transient free radicals are included. The data have been taken from evaluated sources. In general, the selected values are based on gas-phase infrared, Raman, or ultraviolet spectra; when these were not available, liquid-phase or matrix-isolation spectra were used.

Molecules are grouped by structural type. Within each group, related molecules appear together for convenient comparison.

The vibrational modes are described by their approximate character in terms of stretching, bending, deformation, etc. However, it should be emphasized that most such descriptions are only approximate, and that the true normal mode usually involves a mixture of motions. Abbreviations are:

sym.	symmetric
antisym.	antisymmetric
str.	stretch
deform.	deformation
scis.	scissors
rock.	rocking
deg.	degenerate

In the case of free radicals, strong interactions may exist between the electronic and bending vibrational motions. Details can be found in References 3 and 4. The references should be consulted for information on the accuracy of the data and for data on other molecules not listed here.

All fundamental frequencies (more precisely, wavenumbers) are given in units of cm^{-1}.

XY_2 Molecules

Point groups $D_{\infty h}$ (linear) and C_{2v} (bent)

Molecule	Structure	Sym. str.	Bend	Antisym. str.
CO_2	Linear	1333	667	2349
CS_2	Linear	658	397	1535
C_3	Linear	1224	63	2040
CNC	Linear		321	1453
NCN	Linear	1197	423	1476
BO_2	Linear	1056	447	1278
BS_2	Linear	510	120	1015
KrF_2	Linear	449	233	590
XeF_2	Linear	515	213	555
$XeCl_2$	Linear	316		481
H_2O	Bent	3657	1595	3756
D_2O	Bent	2671	1178	2788
F_2O	Bent	928	461	831
Cl_2O	Bent	639	296	686
O_3	Bent	1103	701	1042
H_2S	Bent	2615	1183	2626
D_2S	Bent	1896	855	1999
SF_2	Bent	838	357	813
SCl_2	Bent	525	208	535
SO_2	Bent	1151	518	1362
H_2Se	Bent	2345	1034	2358
D_2Se	Bent	1630	745	1696

XY_2 Molecules

Point groups $D_{\infty h}$ (linear) and C_{2v} (bent)

Molecule	Structure	Sym. str.	Bend	Antisym. str.
NH_2	Bent	3219	1497	3301
NO_2	Bent	1318	750	1618
NF_2	Bent	1075	573	942
ClO_2	Bent	945	445	1111
CH_2	Bent		963	
CD_2	Bent		752	
CF_2	Bent	1225	667	1114
CCl_2	Bent	721	333	748
CBr_2	Bent	595	196	641
SiH_2	Bent	2032	990	2022
SiD_2	Bent	1472	729	1468
SiF_2	Bent	855	345	870
$SiCl_2$	Bent	515		505
$SiBr_2$	Bent	403		400
GeH_2	Bent	1887	920	1864
$GeCl_2$	Bent	399	159	374
SnF_2	Bent	593	197	571
$SnCl_2$	Bent	352	120	334
$SnBr_2$	Bent	244	80	231
PbF_2	Bent	531	165	507
$PbCl_2$	Bent	314	99	299
ClF_2	Bent	500		576

XYZ Molecules

Point Groups $C_{\infty v}$ (linear) and C_s (bent)

Molecule	Structure	XY str.	Bend	YZ str.
HCN	Linear	3311	712	2097
DCN	Linear	2630	569	1925
FCN	Linear	1077	451	2323
ClCN	Linear	744	378	2216
BrCN	Linear	575	342	2198
ICN	Linear	486	305	2188
CCN	Linear	1060	230	1917
CCO	Linear	1063	379	1967
HCO	Bent	2485	1081	1868
HCC	Linear	3612		1848

XYZ Molecules

Point Groups $C_{\infty v}$ (linear) and C_s (bent)

Molecule	Structure	XY str.	Bend	YZ str.
OCS	Linear	2062	520	859
NCO	Linear	1270	535	1921
NNO	Linear	2224	589	1285
HNB	Linear	3675		2035
HNC	Linear	3653		2032
HNSi	Linear	3583	523	1198
HBO	Linear		754	1817
FBO	Linear		500	2075
ClBO	Linear	676	404	1958
BrBO	Linear	535	374	1937

XYZ Molecules

Point Groups C∞v (linear) and Cs (bent)

Molecule	Structure	XY str.	Bend	YZ str.
FNO	Bent	766	520	1844
ClNO	Bent	596	332	1800
BrNO	Bent	542	266	1799
HNF	Bent		1419	1000
HNO	Bent	2684	1501	1565
HPO	Bent	2095	983	1179
HOF	Bent	3537	886	1393
HOCl	Bent	3609	1242	725
HOO	Bent	3436	1392	1098
FOO	Bent	579	376	1490

XYZ Molecules

Point Groups C∞v (linear) and Cs (bent)

Molecule	Structure	XY str.	Bend	YZ str.
ClOO	Bent	407	373	1443
BrOO	Bent			1487
HSO	Bent		1063	1009
NSF	Bent	1372	366	640
NSCl	Bent	1325	273	414
HCF	Bent		1407	1181
HCCl	Bent		1201	815
HSiF	Bent	1913	860	834
HSiCl	Bent		808	522
HSiBr	Bent	1548	774	408

Symmetric XY₃ Molecules

Point Groups D₃h (planar) and C₃v (pyramidal)

Molecule	Structure	Sym. str.	Sym. deform.	Deg. str.	Deg. deform.
NH₃	Pyram.	3337	950	3444	1627
ND₃	Pyram.	2420	748	2564	1191
PH₃	Pyram.	2323	992	2328	1118
AsH₃	Pyram.	2116	906	2123	1003
SbH₃	Pyram.	1891	782	1894	831
NF₃	Pyram.	1032	647	907	492
PF₃	Pyram.	892	487	860	344
AsF₃	Pyram.	741	337	702	262
PCl₃	Pyram.	504	252	482	198
PI₃	Pyram.	303	111	325	79
AsI₃	Pyram.	219	94	224	71
AlCl₃	Pyram.	375	183	595	150
SO₃	Planar	1065	498	1391	530
BF₃	Planar	888	691	1449	480
BH₃	Planar	1125		2808	1640
CH₃	Planar	606		3161	1396
CD₃	Planar	453		2369	1029
CF₃	Pyram.	1090	701	1260	510
SiF₃	Pyram.	830	427	937	290

Linear XYYX Molecules

Point Group D∞h

Molecule	Sym. XY str.	Antisym. XY str.	YY str.	Bend	Bend
C₂H₂	3374	3289	1974	612	730
C₂D₂	2701	2439	1762	505	537
C₂N₂	2330	2158	851	507	233

Planar X₂YZ Molecules

Point Group C₂v

Molecule	Sym. XY str.	YZ str.	YX₂ scis.	Antisym. XY str.	YX₂ rock	YX₂ wag
H₂CO	2783	1746	1500	2843	1249	1167
D₂CO	2056	1700	1106	2160	990	938
F₂CO	965	1928	584	1249	626	774
Cl₂CO	567	1827	285	849	440	580
O₂NF	1310	822	568	1792	560	742
O₂NCl	1286	793	370	1685	408	652

Tetrahedral XY₄ Molecules

Point Group T_d

Molecule	Sym. str.	Deg. deform.(e)	Deg. str.(f)	Deg. deform.(f)
CH_4	2917	1534	3019	1306
CD_4	2109	1092	2259	996
CF_4	909	435	1281	632
CCl_4	459	217	776	314
CBr_4	267	122	672	182
CI_4	178	90	555	125
SiH_4	2187	975	2191	914
SiD_4	1558	700	1597	681
SiF_4	800	268	1032	389
$SiCl_4$	424	150	621	221
GeH_4	2106	931	2114	819
GeD_4	1504	665	1522	596
$GeCl_4$	396	134	453	172
$SnCl_4$	366	104	403	134
$TiCl_4$	389	114	498	136
$ZrCl_4$	377	98	418	113
$HfCl_4$	382	102	390	112
RuO_4	885	322	921	336
OsO_4	965	333	960	329

References

1. T. Shimanouchi, Tables of Molecular Vibrational Frequencies, Consolidated Volume I, Natl. Stand. Ref. Data Ser., Natl. Bur. Stand. (U.S.), 39, 1972.
2. T. Shimanouchi, Tables of Molecular Vibrational Frequencies, Consolidated Volume II, *J. Phys. Chem. Ref. Data*, 6, 993, 1977.
3. G. Herzberg, *Electronic Spectra and Electronic Structure of Polyatomic Molecules*, D. Van Nostrand Co., Princeton, NJ, 1966.
4. M. E. Jacox, Ground state vibrational energy levels of polyatomic transient molecules, *J. Phys. Chem. Ref. Data*, 13, 945, 1984.

SPECTROSCOPIC CONSTANTS OF DIATOMIC MOLECULES

David R. Lide

This table lists the leading spectroscopic constants and equilibrium internuclear distance r_e in the ground electronic state for selected diatomic molecules. The constants are those describing the vibrational and rotational energy through the expressions:

$$E_{vib}/hc = \omega_e(v + \tfrac{1}{2}) - \omega_e x_e(v + \tfrac{1}{2})^2 + \cdots$$

$$E_{rot}/hc = B_v J(J + 1) - D_v[J(J + 1)]^2 + \cdots$$

where

$$B_v = B_e - \alpha_e(v + \tfrac{1}{2}) + \cdots$$

$$D_v = D_e + \cdots$$

Here v and J are the vibrational and rotational quantum numbers, respectively, h is Planck's constant, and c is the speed of light. In this customary formulation the constants ω_e, B_e, etc. have dimensions of inverse length; in this table they are given in units of cm^{-1}.

Users should note that higher order terms in the above energy expressions are required for very precise calculations. The references contain constants for many of these higher terms, as well as more precise values of the lower constants. Also, if the ground electronic state is not $^1\Sigma$, additional terms are needed to account for the interaction between electronic and pure rotational angular momentum. For some molecules in the table the data have been analyzed in terms of the Dunham series expansion:

$$E/hc = \Sigma_{lm}\, Y_{lm}(v + \tfrac{1}{2})^l J^m(J + 1)^m$$

In such cases it has been assumed that $Y_{10} = \omega_e$, $Y_{01} = B_e$, etc., although in the highest approximations these identities are not precisely correct. Some of the values of r_e in the table have been corrected for breakdown of the Born-Oppenheimer approximation, which can affect the last decimal place. Because of differences in the method of data analysis and limitations in the model, care should be taken in comparing r_e values for different molecules to a precision beyond 0.001 Å.

Entries in the ω_e column that are marked by * give the interval between $v = 0$ and $v = 1$ states instead of a value of ω_e.

Molecules are listed in alphabetical order by formula as most commonly written. In most cases this form places the more electropositive element first, but there are exceptions such as OH, NH, CH, etc. References 1–5 are evaluated compilations covering many molecules and giving references to the original literature.

References

1. Huber, K. P., and Herzberg, G., *Molecular Spectra and Molecular Structure IV. Constants of Diatomic Molecules*, Van Nostrand Reinhold, New York, 1979.

2. *Landolt-Börnstein, Numerical Data and Functional Relationships in Science and Technology, New Series*, II/6 (1974), II/14a (1982), II/14b (1983), II/19a (1992), II/19d-1 (1995), II/24a (1998), *Molecular Constants*, Springer-Verlag, Heidelberg.

3. Lovas, F. J., and Tiemann, E., *J. Phys. Chem. Ref. Data* 3, 609, 1974.

4. Irikura, K. K., *J. Phys. Chem. Ref. Data* 35, 389, 2007.

5. Lovas, F. J., Tiemann, E., Coursey, J. S., Kotochigova, S. A., Chang, J., Olsen, K., and Dragoset, R. A., *Diatomic Spectral Database* (version 2.0). Available: http://physics.nist.gov/Diatomic, National Institute of Standards and Technology, Gaithersburg, MD, November 2009.

6. Wormsbecher, R. F., Hessel, M. M., and Lovas, F. J., *J. Chem. Phys.* 74, 6893, 1981.

7. Hedderich, H. G., Dulick, M., and Bernath, P. F., *J. Chem. Phys.* 99, 8363, 1993.

8. Ram, R. S., and Bernath, P. F., *J. Mol. Spectrosc.* 176, 320, 1996.

9. Miller, C. E., and Drouin, B. J., *J. Mol. Spectrosc.* 205, 312, 2001.

10. Tellinghuisen, P. C., Tellinghuisen, J., Coxon, J. A., Velazco, J. E., and Setser, D. W., *J. Chem. Phys.* 68, 5187, 1978.

11. Muntianu, A., Guo, B., and Bernath, P. F., *J. Mol. Spectrosc.* 176, 274, 1996.

12. Yamada, C., Chang, M. C., and Hirota, E., *J. Chem. Phys.* 86, 3804, 1987.

13. Wang, X., Magnes, J., Marjatta Lyyra, A., Ross, A. J., Martin, F., Dove, P. M., and Le Roy, R. J., *J. Chem. Phys.* 117, 9339, 2002.

14. Yamada, C., and Hirota, E., *J. Chem. Phys.* 99, 8489, 1993.

15. James, A. M., Kowalczyk, P., Fournier, R., and Simard, B., *J. Chem. Phys.* 99, 8504, 1993.

16. Campbell, J. M., Dulick, M., Klapstein, D., White, J. B., and Bernath, P. F., *J. Chem. Phys.* 99, 8379, 1993.

17. White, J. B., Dulick, M., and Bernath, P. F., *J. Chem. Phys.* 99, 8371, 1993.

18. Shayesteh, A., Appadoo, D. R. T., Gordon, I., Le Roy, R. J., and Bernath, P. F., *J. Chem. Phys.* 120, 10002, 2004.

19. Sanz, M. E., McCarthy, M. C., and Thaddeus, P., *J. Chem. Phys.* 119, 11715, 2003.

20. Müller, W., and Meyer, W., *J. Chem. Phys.* 80, 3311, 1984.

21. Staanum, P., Pashov, A., Knöckel, H., and Tiemann, E., *Phys. Rev. A* 75, 042513, 2007.

22. Lovas, F. J., Maki, A. G., and Olson, W. B., *J. Mol. Spectrosc.* 87, 449, 1981.

23. Bogey, M., Demuynck, C., and Destombes, J. L., *Chem. Phys.* 66, 99, 1982.

24. Babou, Y., Rivière, Ph., Perrin, M. Y., and Soufiani, A., *Int. J. Thermophys.* 30, 416, 2009.

25. Skatrud, D. D., DeLucia, F. C., Blake, G. A., and Sastry, K. V. L. N., *J. Mol. Spectrosc.* 99, 35, 1983.

26. Le Floch, A., *Mol. Phys.* 72, 133, 1991.

27. George, T., Urban, W., and Le Floch, A., *J. Mol. Spectrosc.* 165, 500, 1994.

28. Mürtz, P., Thümmel, H., Pfelzer, C., and Urban, W., *Mol. Phys.* 86, 1362, 1995.

29. Maki, A. G., Lovas, F. J., and Suenram, R. D., *J. Mol. Spectrosc.* 91, 424, 1982.

30. Engelke, F., Ennen, G., and Meiwes, K. H., *Chem. Phys.* 66, 391, 1982.

31. Tanaka, T., Tamura, M., and Tanaka, K., *J. Mol. Struct.* 413, 153, 1997.

Molecule	State	ω_e cm^{-1}	$\omega_e x_e$ cm^{-1}	B_e cm^{-1}	α_e cm^{-1}	D_e 10^{-6} cm^{-1}	r_e Å	Ref.
^{107}Ag^{79}Br	$^1\Sigma^+$	249.57	0.63	0.064833	0.0002361	0.0175	2.39311	1,2,3
^{107}Ag^{35}Cl	$^1\Sigma^+$	343.49	1.17	0.12298388	0.00059541	0.06305	2.28079	1,2,3
^{107}Ag^{19}F	$^1\Sigma^+$	513.45	2.59	0.2657020	0.0019206	0.284	1.98318	1,2,3
^{107}Ag^{1}H	$^1\Sigma^+$	1759.9	34.06	6.449	0.201	344	1.618	1,2
^{107}Ag^{2}H	$^1\Sigma^+$	1250.70	17.17	3.2572	0.0722	85.9	1.6180	1,2

Molecule	State	ω_e cm^{-1}	$\omega_e x_e$ cm^{-1}	B_e cm^{-1}	α_e cm^{-1}	D_e 10^{-6} cm^{-1}	r_e Å	Ref.
^{107}Ag^{127}I	$^1\Sigma^+$	206.50	0.46	0.04486821	0.0001414	0.00847	2.54463	1,2,3
^{107}Ag^{16}O	$^2\Pi_{1/2}$	490.2	3.1	0.3020	0.0025	0.45	2.003	1,2
^{27}Al$_2$	$^3\Pi_u$	285.8	0.9	0.17127	0.0008		2.701	1,2
^{27}Al^{79}Br	$^1\Sigma^+$	378.0	1.28	0.15919713	0.00086045	0.11285	2.29481	1,2,3
^{27}Al^{35}Cl	$^1\Sigma^+$	481.77	2.10	0.24393007	0.00161108	0.25017	2.13014	7
^{27}Al^{19}F	$^1\Sigma^+$	802.32	4.85	0.55248021	0.00498426	1.0464	1.65437	4,5
^{27}Al^{1}H	$^1\Sigma^+$	1682.37	29.05	6.3937842	0.1870527	368.53	1.64736	17
^{27}Al^{2}H	$^1\Sigma^+$	1211.77	15.06	3.3183929	0.0698773	99.42	1.64637	17
^{27}Al^{127}I	$^1\Sigma^+$	316.1	1.0	0.11769985	0.00055859		2.53710	1,2,3
^{27}Al^{16}O	$^2\Sigma^+$	979.49	7.01	0.6413856	0.0057796	1.08	1.61782	4
^{27}Al^{32}S	$^2\Sigma^+$	617.11	3.33	0.2800368	0.0017823	0.22	2.02828	4
^{75}As$_2$	$^1\Sigma_g^+$	429.55	1.12	0.10179	0.000333		2.1026	1,2
^{75}As^{1}H	$^3\Sigma^-$	2130*		7.3067	0.2117	327	1.52315	1,2
^{75}As^{2}H	$^3\Sigma^-$	1484*		3.6688		90	1.5306	1,2
^{75}As^{14}N	$^1\Sigma^+$	1068.54	5.41	0.54551	0.003366	0.53	1.6184	1,2
^{75}As^{16}O	$^2\Pi_{1/2}$	967.08	4.85	0.48482	0.003299	0.49	1.6236	1,2
^{197}Au$_2$	$^1\Sigma_g^+$	190.9	0.42	0.028013	0.0000723	0.00250	2.4719	1,2
^{197}Au^{1}H	$^1\Sigma^+$	2305.01	43.12	7.2401	0.2136	279	1.5239	1,2
^{197}Au^{2}H	$^1\Sigma^+$	1634.98	21.65	3.6415	0.07614	70.9	1.5238	1,2
^{11}B$_2$	$^3\Sigma_g^-$	1051.3	9.35	1.212	0.014		1.590	1,2
^{11}B^{79}Br	$^1\Sigma^+$	684.31	3.52	0.4894	0.0035	1.00	1.888	1,2
^{11}B^{35}Cl	$^1\Sigma^+$	840.29	5.49	0.684282	0.006812	1.80	1.71528	1,29
^{11}B^{19}F	$^1\Sigma^+$	1402.16	11.82	1.51674399	0.01904848	7.11	1.26267	4
^{11}B^{1}H	$^1\Sigma^+$	2366.73	49.34	12.025755	0.421565	1242	1.23217	4
^{11}B^{2}H	$^1\Sigma^+$	1703.3	28	6.54	0.17	400	1.2324	1,2
^{11}B^{14}N	$^3\Pi$	1514.6	12.3	1.666	0.025	8.1	1.281	1,2
^{11}B^{16}O	$^2\Sigma^+$	1885.29	11.69	1.781110	0.016516	6.32	1.20475	4
^{11}B^{32}S	$^2\Sigma^+$	1179.91	6.25	0.79478	0.00578	1.4	1.60935	4
^{138}Ba^{79}Br	$^2\Sigma^+$	193.77	0.41	0.0415082	0.0001219	0.00762	2.84449	1,2
^{138}Ba^{35}Cl	$^2\Sigma^+$	279.92	0.82	0.08396717	0.00033429	0.03022	2.68276	1,2
^{138}Ba^{19}F	$^2\Sigma^+$	468.9	1.79	0.2159	0.0012	0.175	2.163	1,2,3
^{138}Ba^{1}H	$^2\Sigma^+$	1168.31	14.50	3.38285	0.06599	112.67	2.23175	1,2
^{138}Ba^{2}H	$^2\Sigma^+$	829.77	7.32	1.7071	0.02363	28.77	2.2304	1,2
^{138}Ba^{127}I	$^2\Sigma^+$	152.14	0.27	0.02680587	0.00006634	0.00333	3.08476	1,2
^{138}Ba^{16}O	$^1\Sigma^+$	669.76	2.03	0.3126140	0.0013921	0.2724	1.93969	1,2,3
^{138}Ba^{32}S	$^1\Sigma^+$	379.42	0.88	0.10331	0.0003188	0.0306	2.5074	1,2
^{9}Be^{19}F	$^2\Sigma^+$	1247.36	9.12	1.4889	0.0176	8.28	1.3610	1,2
^{9}Be^{1}H	$^2\Sigma^+$	2061.24	37.33	10.31992	0.3084	1022	1.34241	4
^{9}Be^{2}H	$^2\Sigma^+$	1530.32	20.71	5.6872	0.1225	313.8	1.3419	1,2
^{9}Be^{16}O	$^1\Sigma^+$	1487.32	11.83	1.6510	0.0190	8.20	1.3309	1,2
^{9}Be^{32}S	$^1\Sigma^+$	997.94	6.14	0.79059	0.00664	2.00	1.7415	1
^{209}Bi$_2$	$^1\Sigma_g^+$	172.71	0.34	0.022781	0.000055	0.00150	2.6596	1,2
^{209}Bi^{79}Br	O$^+$	209.62	0.52	0.04321526	0.00013269	0.007347	2.60950	5
^{209}Bi^{35}Cl	O$^+$	308.18	1.09	0.9212553	0.0004020	0.0329	2.47152	5
^{209}Bi^{19}F	O$^+$	513.0	2.35	0.22998897	0.00150262	0.185	2.05154	5
^{209}Bi^{1}H	$^3\Sigma^-$	1635.73	31.6	5.137	0.148	183	1.805	1,2
^{209}Bi^{2}H	$^3\Sigma^-$	1173.32	16.1	2.592	0.054	50.6	1.804	1,2
^{209}Bi^{127}I	O$^+$	164.12	0.32	0.02722281	0.00006979	0.00300	2.80050	5
^{79}Br$_2$	$^1\Sigma_g^+$	325.32	1.08	0.082107	0.0003187	0.02092	2.2811	1,2
^{79}Br^{35}Cl	$^1\Sigma^+$	444.28	1.84	0.152470	0.000770	0.07183	2.13607	1,2,3
^{79}Br^{19}F	$^1\Sigma^+$	670.75	4.05	0.35584	0.00261	0.401	1.75894	1,2,3
^{79}Br^{16}O	$^2\Pi_{3/2}$	779	6.8	0.429598	0.003639	0.523	1.717	1,2,3
^{12}C$_2$	$^1\Sigma_g^+$	1855.01	13.56	1.82010	0.01801	6.96	1.24244	24
^{12}C^{35}Cl	$^2\Pi_{1/2}$	876.90	5.45	0.697137	0.006853	1.9	1.64518	4
^{12}C^{19}F	$^2\Pi_{1/2}$	1307.93	11.08	1.41626	0.01844	6.6	1.27218	4
^{12}C^{1}H	$^2\Pi_{1/2}$	2860.75	64.44	14.45988	0.53654	1450	1.1199	4
^{12}C^{2}H	$^2\Pi_{1/2}$	2101.05	34.73	7.8079823	0.212240	420	1.11887	4
^{12}C^{14}N	$^2\Sigma^+$	2068.65	13.10	1.8997830	0.0173717	6.4034	1.17181	24,25
^{12}C^{16}O	$^1\Sigma^+$	2169.81	13.29	1.931280985	0.01750439	6.1216	1.12832	24,26,27
^{12}C^{31}P	$^2\Sigma^+$	1239.79	6.83	0.79886775	0.00596933	1.33	1.56198	4

Molecule	State	ω_e cm^{-1}	$\omega_e x_e$ cm^{-1}	B_e cm^{-1}	α_e cm^{-1}	D_e 10^{-6} cm^{-1}	r_e Å	Ref.
^{12}C^{32}S	$^1\Sigma^+$	1285.15	6.50	0.82004356	0.00591835	1.336	1.53482	1,4
^{12}C^{80}Se	$^1\Sigma^+$	1035.36	4.86	0.5750	0.00379	0.71	1.67609	1,2,3
^{40}Ca^{79}Br	$^2\Sigma^+$	285.3	0.86	0.09446622	0.00040360	0.0413	2.59358	1,5
^{40}Ca^{35}Cl	$^2\Sigma^+$	367.53	1.31	0.1522302	0.0007990	0.1029	2.43676	1,2
^{40}Ca^{19}F	$^2\Sigma^+$	581.1	2.74	0.339	0.0026	0.45	1.967	1,2
^{40}Ca^{1}H	$^2\Sigma^+$	1298.34	19.10	4.2766	0.0970	183.7	2.0025	1,2
^{40}Ca^{2}H	$^2\Sigma^+$	910*		2.1769	0.035	47.9	2.002	1,2
^{40}Ca^{127}I	$^2\Sigma^+$	238.70	0.63	0.0693263	0.0002634	0.0234	2.82859	1,2
^{40}Ca^{16}O	$^1\Sigma^+$	732.03	4.83	0.444441	0.003282	0.6541	1.8221	1,2
^{40}Ca^{32}S	$^1\Sigma^+$	462.23	1.78	0.1766757	0.0008270	0.1032	2.31775	1,2
^{114}Cd^{1}H	$^2\Sigma^+$	1337.1*		5.323		314	1.781	1,2
^{114}Cd^{2}H	$^2\Sigma^+$			2.704		76	1.775	1,2
^{35}Cl$_2$	$^1\Sigma_g^+$	559.75	2.69	0.24415	0.00152	0.186	1.9872	4
^{35}Cl^{19}F	$^1\Sigma^+$	783.45	4.95	0.5164805	0.0043585	0.88	1.62831	4
^{35}Cl^{16}O	$^2\Pi_{3/2}$	853.64	5.52	0.62345797	0.0059357	1.33	1.56962	4
^{52}Cr^{1}H	$^6\Sigma^+$	1581*		6.220	0.179	347	1.656	1,2
^{52}Cr^{2}H	$^6\Sigma^+$	1182*		3.14		88.8	1.664	1,2
^{52}Cr^{16}O	$^5\Pi$	898.4	6.8	0.5231	0.0070		1.615	1,2
^{133}Cs$_2$	$^1\Sigma_g^+$	42.02	0.08	0.0127	0.0000264	0.00464	4.47	1,2
^{133}Cs^{79}Br	$^1\Sigma^+$	149.66	0.37	0.03606925	0.00012401	0.00838	3.07225	1,2,3
^{133}Cs^{35}Cl	$^1\Sigma^+$	214.17	0.73	0.07209149	0.00033756	0.03268	2.90627	1,2,3
^{133}Cs^{19}F	$^1\Sigma^+$	352.56	1.62	0.18436969	0.0011756	0.20168	2.34535	1,2,3
^{133}Cs^{1}H	$^1\Sigma^+$	891.0	12.9	2.7099	0.0579	113	2.4938	1,2
^{133}Cs^{2}H	$^1\Sigma^+$	619.1*		1.354		20	2.505	1,2
^{133}Cs^{127}I	$^1\Sigma^+$	119.18	0.25	0.02362736	0.00006826	0.00371	3.31519	1,2,3
^{133}Cs^{16}O	$^2\Sigma^+$	357.5*		0.223073	0.001303	0.348	2.3007	1,2
^{63}Cu$_2$	$^1\Sigma_g^+$	264.55	1.02	0.10874	0.000614	0.0716	2.2197	1,2
^{63}Cu^{79}Br	$^1\Sigma^+$	314.8	0.96	0.10192625	0.00045214	0.04274	2.17344	1,2
^{65}Cu^{35}Cl	$^1\Sigma^+$	415.29	1.58	0.17628802	0.00099647	0.12706	2.05118	1,2
^{63}Cu^{19}F	$^1\Sigma^+$	622.7	3.95	0.3794029	0.0032298	0.563	1.74493	1,2,3
^{63}Cu^{1}H	$^1\Sigma^+$	1941.26	37.51	7.9441	0.2563	520	1.46263	1,2
^{63}Cu^{2}H	$^1\Sigma^+$	1384.14	18.97	4.0381	0.0917	136.2	1.4626	1,2
^{63}Cu^{127}I	$^1\Sigma^+$	264.5	0.60	0.07328742	0.00028390	0.02244	2.33832	1,2
^{63}Cu^{16}O	$^2\Pi_{3/2}$	640.17	4.43	0.44454	0.00456	0.85	1.7244	1,2
^{63}Cu^{32}S	$^2\Pi_{3/2}$	415.0	1.75	0.1891		0.18	2.051	1,2
^{19}F$_2$	$^1\Sigma_g^+$	916.93	11.32	0.889294	0.0125952	3.3	1.41264	4
^{19}F^{16}O	$^2\Pi_{3/2}$	1053.01	9.92	1.0587076	0.013295	4.2823	1.35411	9
^{56}Fe^{16}O	$^5\Delta$	965*		0.650		0.72	1.444	1,2
^{69}Ga^{81}Br	$^1\Sigma^+$	263.0	0.81	0.081839	0.0003207	0.032	2.35248	1,2,3
^{69}Ga^{35}Cl	$^1\Sigma^+$	365.67	1.25	0.1499046	0.0007936	0.1008	2.20169	1,2,3
^{69}Ga^{19}F	$^1\Sigma^+$	622.2	3.2	0.3595161	0.0028642	0.50	1.77437	1,2,3
^{69}Ga^{1}H	$^1\Sigma^+$	1603.94	28.41	6.1434095	0.1906376	359.70	1.66208	16
^{69}Ga^{2}H	$^1\Sigma^+$	1143.23	14.43	3.1218854	0.0689978	93.021	1.66113	16
^{69}Ga^{127}I	$^1\Sigma^+$	216.38	0.47	0.0569359	0.0001897	0.015770	2.57464	1,2,3
^{69}Ga^{16}O	$^2\Sigma$	767.5	6.24	0.4271		0.37	1.744	1,2
^{74}Ge^{79}Br	$^2\Pi_{1/2}$	295	0.7					1,2
^{74}Ge^{35}Cl	$^2\Pi_{1/2}$	407.6	1.36					1,2
^{72}Ge^{1}H	$^2\Pi_{1/2}$	1833.77	37	6.726	0.192	326	1.5880	1,2
^{72}Ge^{2}H	$^2\Pi_{1/2}$	1320.09	19	3.415	0.070	83.2	1.5874	1,2
^{74}Ge^{16}O	$^1\Sigma^+$	986.49	4.47	0.4856981	0.0030787	0.4709	1.62464	1,2
^{74}Ge^{32}S	$^1\Sigma^+$	575.8	1.80	0.18656576	0.00074910	0.07883	2.01209	1,2
^{74}Ge^{80}Se	$^1\Sigma^+$	408.7	1.36	0.09634051	0.00028904	0.02207	2.13463	1,2
^{74}Ge^{130}Te	$^1\Sigma^+$	323.9	0.75	0.06533821	0.00017246	0.012	2.34017	1,2
^{1}H$_2$	$^1\Sigma_g^+$	4401.21	121.34	60.853	3.062	47100	0.74144	1,2
^{2}H$_2$	$^1\Sigma_g^+$	3115.50	61.82	30.444	1.0786	11410	0.74152	1,2
^{3}H$_2$	$^1\Sigma_g^+$	2546.5	41.23	20.335	0.5887		0.74142	1,2
^{1}H^{81}Br	$^1\Sigma^+$	2648.97	45.22	8.46488	0.23328	345.8	1.41444	1,2,3
^{2}H^{81}Br	$^1\Sigma^+$	1884.75	22.72	4.245596	0.084	88.32	1.4145	1,2
^{1}H^{35}Cl	$^1\Sigma^+$	2990.92	52.80	10.5933002	0.3069985	531.94	1.27456	4
^{2}H^{35}Cl	$^1\Sigma^+$	2145.16	27.18	5.448796	0.113292	140	1.27458	1,2

Molecule	State	ω_e cm^{-1}	$\omega_e x_e$ cm^{-1}	B_e cm^{-1}	α_e cm^{-1}	D_e 10^{-6} cm^{-1}	r_e Å	Ref.
^{1}H^{19}F	$^1\Sigma^+$	4138.39	89.94	20.953712	0.7933704	2150	0.91685	4
^{2}H^{19}F	$^1\Sigma^+$	2998.19	45.76	11.0102	0.3017	594	0.91694	1,2
^{1}H^{127}I	$^1\Sigma^+$	2309.01	39.64	6.4263650	0.1689	206.9	1.60916	1,2,3
^{202}Hg^{1}H	$^2\Sigma^+$	1203.24*		5.3888		395.3	1.7662	1,2
^{202}Hg^{2}H	$^2\Sigma^+$	896.12*		2.739		91	1.757	1,2
^{127}I$_2$	$^1\Sigma_g^+$	214.50	0.61	0.03737	0.000114	0.0043	2.666	1,2
^{127}I^{79}Br	$^1\Sigma^+$	268.64	0.81	0.0568325	0.0001969	0.0102	2.46899	1,2,3
^{127}I^{35}Cl	$^1\Sigma^+$	384.29	1.50	0.1141587	0.0005354	0.0403	2.32088	1,2,3
^{127}I^{19}F	$^1\Sigma^+$	610.24	3.12	0.2797111	0.0018738	0.2356	1.90976	1,2,3
^{127}I^{16}O	$^2\Pi_{3/2}$	681.5	4.3	0.34026	0.00270	0.36	1.8676	1,2
^{115}In^{81}Br	$^1\Sigma^+$	221.0	0.65	0.05489468	0.00018672	0.01350	2.54315	1,2,3
^{115}In^{35}Cl	$^1\Sigma^+$	317.39	1.03	0.1090583	0.0005177	0.0515	2.40117	1,2,3
^{115}In^{19}F	$^1\Sigma^+$	535.4	2.6	0.2623241	0.0018798	0.252	1.98540	1,2,3
^{115}In^{1}H	$^1\Sigma^+$	1476.0	25.61	4.995	0.143	223	1.8380	1,2
^{115}In^{2}H	$^1\Sigma^+$	1048.2	12.4	2.523	0.051	58	1.837	1,2
^{115}In^{127}I	$^1\Sigma^+$	177.08	0.34	0.03686702	0.00010411	0.00639	2.75364	1,2,3
^{39}K$_2$	$^1\Sigma_g^+$	92.02	0.28	0.056743	0.000165	0.0863	3.9051	1,2
^{39}K^{79}Br	$^1\Sigma^+$	213	0.80	0.08122109	0.00040481	0.04462	2.82078	1,2,3
^{39}K^{35}Cl	$^1\Sigma^+$	281	1.30	0.1286348	0.0007899	0.1087	2.66665	1,2,3
^{39}K^{19}F	$^1\Sigma^+$	426.26	2.45	0.27993741	0.00233492	0.4829	2.17146	1,2,3
^{39}K^{1}H	$^1\Sigma^+$	983.6	14.3	3.416400	0.085313	163.55	2.243	1,2
^{39}K^{2}H	$^1\Sigma^+$	707	7.7	1.754	0.0318	50	2.240	1,2
^{39}K^{127}I	$^1\Sigma^+$	186.53	0.57	0.06087473	0.00026776	0.02593	3.04784	1,2,3
^{39}K^{23}Na	$^1\Sigma^+$	124.03	0.50	0.09519989	0.00044966	0.2206	3.49958	6
^{139}La^{16}O	$^2\Sigma^+$	812.8	2.22	0.35252001	0.00142365	0.2626	1.82591	1,2
^{7}Li$_2$	$^1\Sigma_g^+$	351.41	2.58	0.672530	0.007046	9.79	2.6733	13
^{7}Li^{79}Br	$^1\Sigma^+$	563.2	3.5	0.555399	0.005644	2.159	2.17043	1,2,3
^{7}Li^{35}Cl	$^1\Sigma^+$	642.95	4.47	0.7065225	0.0080102	3.409	2.02067	1,5
^{7}Li^{133}Cs	$^1\Sigma^+$	184.70	1.00	0.188003	0.001248	0.7784	3.6681	21
^{7}Li^{19}F	$^1\Sigma^+$	910.57	8.21	1.34525715	0.02028749	11.75	1.56386	4,5
^{7}Li^{1}H	$^1\Sigma^+$	1405.50	21.17	7.5137315	0.2163911	859	1.59490	4,5
^{7}Li^{2}H	$^1\Sigma^+$	1054.94	13.06	4.23308131	0.09149428	272	1.59526	4,5
^{7}Li^{127}I	$^1\Sigma^+$	496.85	2.85	0.4431766	0.0040862	1.4104	2.39192	1,2,3
^{7}Li^{39}K	$^1\Sigma^+$	207		0.265			3.27	1,20
^{7}Li^{23}Na	$^1\Sigma^+$	256.99	1.66	0.376833	0.003810	3.340	2.88851	4,30
^{7}Li^{16}O	$^2\Pi$	814.62	7.78	1.212830	0.017899	0.1079	1.68822	3,14
^{24}Mg$_2$	$^1\Sigma_g^+$	51.12	1.64	0.09287	0.00378	1.22	3.891	1
^{24}Mg^{35}Cl	$^2\Sigma^+$	462.12*		0.2456154	0.0016204	0.2723	2.19639	1,2
^{24}Mg^{19}F	$^2\Sigma^+$	711.69*		0.51922	0.00470	1.080	1.7500	1,2
^{24}Mg^{1}H	$^2\Sigma^+$	1492.78	29.85	5.825523	0.177298	354.56	1.72972	18
^{24}Mg^{2}H	$^2\Sigma^+$	1077.30	15.52	3.034344	0.066607	96.25	1.72916	18
^{24}Mg^{16}O	$^1\Sigma^+$	785.21	5.13	0.5748414	0.0053223	1.233	1.74817	28,4
^{24}Mg^{32}S	$^1\Sigma^+$	528.74	2.70	0.26797	0.00176	0.276	2.1425	1
^{55}Mn^{1}H	$^7\Sigma$	1548.0	28.8	5.6841	0.1570	303.9	1.7311	1,2
^{55}Mn^{2}H	$^7\Sigma$	1103	13.9	2.8957	0.051	79.5	1.7310	1,2
^{14}N$_2$	$^1\Sigma_g^+$	2358.56	14.32	1.998236	0.017310	5.737	1.09769	24
^{14}N^{79}Br	$^3\Sigma^-$	691.75	4.72	0.444	0.0040		1.79	1,2
^{14}N^{35}Cl	$^3\Sigma^-$	827.96	5.30	0.64976739	0.00641432	1.596	1.61071	1,5
^{14}N^{19}F	$^3\Sigma^-$	1141.37	8.99	1.205679	0.014889	5.4	1.31698	4
^{14}N^{1}H	$^3\Sigma^-$	3282.72	79.04	16.66792	0.65038	1710	1.03719	4
^{14}N^{2}H	$^3\Sigma^-$	2399.13	42.11	8.9087	0.2546	491.7	1.03665	8
^{14}N^{16}O	$^2\Pi_{1/2}$	1904.20	14.07	1.67195	0.0171	0.5	1.15077	1,24
^{14}N^{32}S	$^2\Pi_{1/2}$	1218.7	7.28	0.769602	0.0064	1.2	1.4940	1,2,3
^{23}Na$_2$	$^1\Sigma_g^+$	159.09	0.71	0.15473537	0.0086375	0.58	3.07858	1,4,20
^{23}Na^{79}Br	$^1\Sigma^+$	302	1.5	0.1512533	0.0009410	0.1554	2.50204	1,2,3
^{23}Na^{35}Cl	$^1\Sigma^+$	364.68	1.78	0.21806302	0.00162479	0.31202	2.36080	4,5
^{23}Na^{19}F	$^1\Sigma^+$	535.66	3.58	0.43690153	0.0045592	1.16296	1.92595	11
^{23}Na^{1}H	$^1\Sigma^+$	1171.97	19.70	4.90327	0.1370	343.8	1.8870	4,5
^{23}Na^{2}H	$^1\Sigma^+$	826.1*		2.557089	0.051600	93.46	1.88654	1,2
^{23}Na^{127}I	$^1\Sigma^+$	258	1.1	0.1178056	0.0006478	0.0973	2.71145	1,2,3

Molecule	State	ω_e cm^{-1}	$\omega_e x_e$ cm^{-1}	B_e cm^{-1}	α_e cm^{-1}	D_e 10^{-6} cm^{-1}	r_e Å	Ref.
^{23}Na^{16}O	$^2\Pi$	492.3		0.424630	0.004506	1.2638	2.05155	1,2
^{93}Nb$_2$	$^3\Sigma_g^-$	424.89	0.94	0.084054	0.000242	0.016	2.0778	15
^{93}Nb^{16}O	$^4\Sigma^-$	989.0	3.8	0.4321	0.0021	0.22	1.691	1,2
^{58}Ni^{1}H	$^2\Delta_{5/2}$	1926.6	38	7.700	0.23	481	1.476	1,2
^{58}Ni^{2}H	$^2\Delta_{5/2}$	1390.1	19	3.992	0.092	130	1.465	1,2
^{16}O$_2$	$^3\Sigma_g^-$	1580.19	11.98	1.445622	0.015933	4.839	1.20752	24
^{16}O^{1}H	$^2\Pi_{3/2}$	3737.76	84.88	18.911	0.7242	1938	0.96966	1,2,3
^{16}O^{2}H	$^2\Pi_{3/2}$	2720.24	44.05	10.021	0.276	537.4	0.9698	1,2
^{31}P$_2$	$^1\Sigma_g^+$	780.77	2.84	0.30362	0.00149	0.188	1.8934	1
^{31}P^{35}Cl	$^3\Sigma^-$	551.38	2.23	0.2528748	0.0015119	0.2124	2.01461	1,2
^{31}P^{19}F	$^3\Sigma^-$	846.73	4.49	0.5667427	0.004639	1.0156	1.58933	12
^{31}P^{1}H	$^3\Sigma^-$	2363.77	43.91	8.53904	0.2534	4.462	1.42218	8
^{31}P^{2}H	$^3\Sigma^-$	1699.2	23.0	4.4081	0.0928	116	1.4220	1,2
^{31}P^{14}N	$^1\Sigma^+$	1336.95	6.90	0.7864844	0.0055337	1.091	1.49087	4
^{31}P^{16}O	$^2\Pi_{1/2}$	1233.34	6.56	0.733223657	0.005466162	1.3	1.47637	1,4
^{208}Pb$_2$		110.5	0.35					1,2
^{208}Pb^{79}Br	$^2\Pi_{1/2}$	207.5	0.50					1,2
^{208}Pb^{35}Cl	$^2\Pi_{1/2}$	303.9	0.88					1,2
^{208}Pb^{19}F	$^2\Pi_{1/2}$	502.73	2.28	0.22875	0.001473	0.183	2.0575	1,2
^{208}Pb^{1}H	$^2\Pi_{1/2}$	1564.1	29.75	4.971	0.144	201	1.839	1,2
^{208}Pb^{16}O	$^1\Sigma^+$	720.96	3.52	0.30730373	0.00190977	0.2138	1.92181	1,2,3
^{208}Pb^{32}S	$^1\Sigma^+$	429.17	1.26	0.11632307	0.00043510	0.03418	2.28678	1,2,3
^{208}Pb^{80}Se	$^1\Sigma^+$	277.6	0.51	0.05059953	0.00012993	0.0070	2.40218	1,2,3
^{208}Pb^{130}Te	$^1\Sigma^+$	212.0	0.43	0.03130774	0.00006743	0.0027	2.59492	1,2,3
^{195}Pt^{12}C	$^1\Sigma^+$	1051.13	4.86	0.53044	0.003273	0.546	1.6767	1,2
^{195}Pt^{1}H	$^2\Delta_{5/2}$	2294.68*		7.1963	0.1996	261	1.52852	1,2
^{195}Pt^{2}H	$^2\Delta_{5/2}$	1644.3*		3.640	0.071	66	1.524	1,2
^{85}Rb^{79}Br	$^1\Sigma^+$	169.46	0.46	0.04752798	0.00018596	0.01496	2.94474	1,2,3
^{85}Rb^{35}Cl	$^1\Sigma^+$	228	0.92	0.0876404	0.0004537	0.04947	2.78673	1,2,3
^{85}Rb^{19}F	$^1\Sigma^+$	376	1.9	0.2106640	0.0015228	0.2684	2.27033	1,2,3
^{85}Rb^{1}H	$^1\Sigma^+$	936.9	14.21	3.020	0.072	123	2.367	1,2
^{85}Rb^{127}I	$^1\Sigma^+$	138.51	0.33	0.03283293	0.00010946	0.00738	3.17688	1,2,3
^{85}Rb^{16}O	$^2\Sigma^+$	388.4*		0.246481	0.002174	0.397	2.25420	1,2
^{32}S$_2$	$^3\Sigma_g^-$	725.71	2.86	0.29539516	0.00159754	0.19	1.88941	4
^{32}S^{19}F	$^2\Pi_{3/2}$	837.64	4.47	0.555173	0.004459	0.975	1.59624	4,5
^{32}S^{1}H	$^2\Pi_{3/2}$	2696.25	48.74	9.60025	0.27990	480	1.34061	4
^{32}S^{2}H	$^2\Pi_{3/2}$	1885	31	4.95130	0.10308	130	1.34049	1,2
^{32}S^{16}O	$^3\Sigma^-$	1149.2	5.6	0.7208171	0.005737	1.134	1.48109	1,19,23
^{121}Sb^{35}Cl	$^3\Sigma^-$	374.7	0.6					1,2
^{121}Sb^{19}F	$^3\Sigma^-$	605.0	2.6	0.2792	0.0020	0.23	1.918	1,2
^{121}Sb^{1}H	$^3\Sigma^-$			5.684		240	1.723	1,2
^{121}Sb^{2}H	$^3\Sigma^-$			2.8782		45	1.7194	1,2
^{121}Sb^{14}N	$^1\Sigma^+$	942.0	5.6					1,2
^{121}Sb^{16}O	$^2\Pi_{1/2}$	816	4.2	0.3580	0.0022	0.270	1.826	1,2
^{45}Sc^{19}F	$^1\Sigma^+$	735.6	3.8	0.3950	0.00266		1.788	1,2
^{80}Se$_2$	$^3\Sigma_g^-$	385.30	0.96	0.08992	0.000288	0.024	2.166	1,2
^{80}Se^{1}H	$^2\Pi_{3/2}$	2400*		8.02	0.23	330	1.48	1,2
^{80}Se^{2}H	$^2\Pi_{3/2}$	1708*		3.94			1.48	1,2
^{80}Se^{16}O	$^3\Sigma^-$	914.69	4.52	0.4655	0.00323	0.5	1.648	1,2
^{28}Si$_2$	$^3\Sigma_g^-$	510.98	2.02	0.2390	0.0014	0.21	2.246	1
^{28}Si^{35}Cl	$^2\Pi_{1/2}$	535.59	2.18	0.256103	0.001582	0.25	2.05794	4
^{28}Si^{19}F	$^2\Pi_{1/2}$	857.33	4.83	0.58125735	0.00503859	1.065	1.60100	31
^{28}Si^{1}H	$^2\Pi_{1/2}$	2042.52	36.06	7.503898	0.21814	400	1.51966	4
^{28}Si^{2}H	$^2\Pi_{1/2}$	1469.32	18.23	3.8840	0.0781	105.4	1.5199	1,2
^{28}Si^{14}N	$^2\Sigma^+$	1151.28	6.46	0.730927	0.005685	1.2	1.57207	4
^{28}Si^{16}O	$^1\Sigma^+$	1241.54	5.97	0.7267521	0.0050379	0.9923	1.50975	1,19,22
^{28}Si^{32}S	$^1\Sigma^+$	749.64	2.58	0.30352788	0.00147308	0.201	1.92926	1,19
^{28}Si^{80}Se	$^1\Sigma^+$	580.0	1.78	0.1920117	0.0007767	0.0842	2.05832	1,2,3
^{120}Sn^{79}Br	$^2\Pi_{1/2}$	247.2	0.6					1,2
^{120}Sn^{35}Cl	$^2\Pi_{1/2}$	351.1	1.06	0.1117	0.0004		2.361	1,2

Molecule	State	ω_e cm^{-1}	$\omega_e x_e$ cm^{-1}	B_e cm^{-1}	α_e cm^{-1}	D_e 10^{-6} cm^{-1}	r_e Å	Ref.
^{118}Sn^{19}F	$^2\Pi_{1/2}$	577.6	2.69	0.2727	0.0014	0.26	1.944	1,2
^{120}Sn^{1}H	$^2\Pi_{1/2}$			5.31488		207.5	1.78146	1,2
^{120}Sn^{2}H	$^2\Pi_{1/2}$	1188.0*		2.6950	0.049	53.4	1.7770	1,2
^{120}Sn^{127}I	$^2\Pi_{1/2}$	199.0	0.6					1,2
^{120}Sn^{16}O	$^1\Sigma^+$	822.13	3.72	0.35571998	0.00214432	0.26638	1.83251	1,2,3
^{120}Sn^{32}S	$^1\Sigma^+$	487.26	1.36	0.13686139	0.00050563	0.0424	2.20898	1,2,3
^{120}Sn^{80}Se	$^1\Sigma^+$	331.2	0.74	0.0649978	0.0001705	0.011	2.32557	1,2,3
^{120}Sn^{130}Te	$^1\Sigma^+$	259.5	0.50	0.04247917	0.00009543	0.0055	2.52280	1,2,3
^{88}Sr^{79}Br	$^2\Sigma^+$	216.60	0.52	0.0541847	0.0001827	0.01356	2.73522	1,2
^{88}Sr^{35}Cl	$^2\Sigma^+$	302.3	0.95					1,2
^{88}Sr^{19}F	$^2\Sigma^+$	502.4	2.3	0.2505346	0.0015513	0.2498	2.07537	1,2
^{88}Sr^{1}H	$^2\Sigma^+$	1206.2	17.0	3.6751	0.0814	135	2.1456	1,2
^{88}Sr^{2}H	$^2\Sigma^+$	841	8.6	1.8609	0.0292	34.7	2.1449	1,2
^{88}Sr^{127}I	$^2\Sigma^+$	173.77	0.35	0.0367097	0.0001060	0.00655	2.94364	1,2
^{88}Sr^{16}O	$^1\Sigma^+$	653.5	3.96	0.33798	0.00219	0.36	1.91983	1,2
^{181}Ta^{16}O	$^2\Delta_{3/2}$	1028.69	3.51	0.40284	0.00182	0.2450	1.68746	1,2
^{130}Te$_2$	$^3\Sigma_g^-$	247.07	0.51	0.039681	0.000106	0.0044	2.5574	1,2
$^{(130)}$Te^{1}H	$^2\Pi_{3/2}$			5.56			1.74	1,2
^{130}Te^{16}O	0^+	797.11	4.00	0.3554	0.00237	0.27	1.825	1,2
^{232}Th^{16}O	$^1\Sigma^+$	895.77	2.39	0.332644	0.001302	0.1833	1.84032	1,2
^{48}Ti^{16}O	$^3\Delta_1$	1009.02	4.50	0.53541	0.00301	0.603	1.6202	1,2
^{205}Tl^{81}Br	$^1\Sigma^+$	192.10	0.39	0.0423899	0.0001276	0.0083	1.61817	1,2,3
^{205}Tl^{35}Cl	$^1\Sigma^+$	284.71	0.86	0.09139702	0.00039784	0.0377	2.48483	1,2,3
^{205}Tl^{19}F	$^1\Sigma^+$	476.86	2.24	0.22315014	0.00150380	0.1955	2.08439	1,2,3
^{205}Tl^{1}H	$^1\Sigma^+$	1390.7	22.7	4.806	0.154	254	1.870	1,2
^{205}Tl^{2}H	$^1\Sigma^+$	987.7	12.04	2.419	0.057	60	1.869	1,2
^{205}Tl^{127}I	$^1\Sigma^+$	150*		0.0271676	0.0000664	0.0036	2.81361	1,2,3
^{51}V^{16}O	$^4\Sigma^-$	1011.3	4.86	0.54825	0.00352	0.6	1.5893	1,2
$^{(132)}$Xe^{19}F	$^2\Sigma$	225.4	10.9	0.19326	0.00699	0.536	2.293	10
^{89}Y^{35}Cl	$^1\Sigma$	380.7	1.3	0.1160	0.0003	0.09	2.41	1,2
^{89}Y^{19}F	$^1\Sigma^+$	631.29	2.50	0.29042	0.00163	0.237	1.9257	1,2
^{89}Y^{16}O	$^2\Sigma^+$	861.0	2.9	0.3881	0.0018	0.32	1.790	1,2
^{174}Yb^{1}H	$^2\Sigma^+$	1249.54	21.06	3.9931	0.0957	161.8	2.0526	1,2
^{174}Yb^{2}H	$^2\Sigma^+$	886.6	10.57	2.01162	0.03425	41.60	2.0516	1,2
^{64}Zn^{35}Cl	$^2\Sigma$	390.5	1.6					1,2
^{64}Zn^{19}F	$^2\Sigma$	628	3.5					1,2
^{64}Zn^{1}H	$^2\Sigma^+$	1607.6	55.14	6.6794	0.2500	466	1.5949	1,2
^{64}Zn^{2}H	$^2\Sigma^+$	1072	28	3.350		124	1.6054	1,2
^{64}Zn^{127}I	$^2\Sigma$	223.4	0.6					1,2
^{90}Zr^{16}O	$^1\Sigma^+$	969.8	4.9	0.42263	0.0023	0.319	1.7116	1,2

* Indicates a value for the interval between $v = 0$ and $v = 1$ states instead of a value of ω_e.

Section 10
Atomic, Molecular, and Optical Physics

LINE SPECTRA OF THE ELEMENTS

Joseph Reader and Charles H. Corliss

The original tables from which this table was derived were prepared under the auspices of the Committee on Line Spectra of the Elements of the National Academy of Sciences National Research Council. The table contains the outstanding spectral lines of neutral (I) and singly ionized (II) atoms of the elements from hydrogen through plutonium (Z = 1–94); selected strong lines from doubly ionized (III), triply ionized (IV), and quadruply ionized (V) atoms are also included. Listed are lines that appear in emission from the vacuum ultraviolet to the far infrared. These lines were selected from much larger lists in such a way as to include the stronger observed lines in each spectral region. A more extensive list may be found in Reference 1.

The data were compiled by the following contributors.

J. G. Conway — Lawrence Berkeley Laboratory
C. H. Corliss — National Bureau of Standards
R. D. Cowan — Los Alamos Scientific Laboratory
C. R. Cowley — University of Michigan
Henry M. and Hannah Crosswhite — Argonne National Laboratory
S. P. Davis — University of California, Berkeley
V. Kaufman — National Bureau of Standards
R. L. Kelly — Naval Postgraduate School
J. F. Kielkopf — University of Louisville
W. C. Martin — National Bureau of Standards
T. K. McCubbin — Pennsylvania State University
L. J. Radziemski — Los Alamos Scientific Laboratory
J. Reader — National Bureau of Standards
C. J. Sansonetti — National Bureau of Standards
G. V. Shalimoff — Lawrence Berkeley Laboratory
R. W. Stanley — Purdue University
J. O. Stoner, Jr. — University of Arizona
H. H. Stroke — New York University
D. R. Wood — Wright State University
E. F. Worden — Lawrence Livermore Laboratory
J. J. Wynne — International Business Machines Corporation
R. Zalubas — National Bureau of Standards

All wavelengths are given in Ångstrom units (10^{-10} m). Below 2000 Å the wavelengths are in vacuum (except for the Cu II line at 1999.698 Å, which is in air); above 2000 Å the wavelengths are in air. Wavelengths given to three decimal places have an uncertainty of less than 0.001 Å and are therefore suitable for calibration purposes. In the air region, the elements used most commonly for calibration are Ne, Ar, Kr, Fe, Th, and Hg; in the vacuum region, the most common are C, N, O, Si, Cu.

All data refer to natural isotopic abundance of the elements except that Kr I and Kr II lines below 11,000 Å given to three decimal places are for ^{86}Kr. A separate table for ^{198}Hg contains accurately known wavelengths that are frequently used for calibration.

A large number of the lines for neutral and singly ionized atoms were extracted from the National Bureau of Standards (NBS) *Tables of Spectral Line Intensities* (Reference 2). The intensities of these lines represent quantitative estimates of relative line strengths that take into account varying detection sensitivity at different wavelengths. They are on a linear scale. For nearly all of the other lines the intensities represent qualitative estimates of the relative strengths of lines not greatly separated in wavelength. Because different observers frequently use different scales for their intensity estimates, these intensities are useful only as a rough indication of the appearance of a spectrum. In some cases the intensity scale is not intended to be linear. In the first and second spectra the intensities of the lines of the singly ionized atom (II) relative to those of the neutral atom (I) should be used with caution, inasmuch as the concentration of ions in a light source depends greatly on the excitation conditions.

Descriptive symbols that follow the wavelength have the following meanings:

c — complex
d — line consists of two unresolved lines
h — hazy
l — shaded to longer wavelengths
s — shaded to shorter wavelengths
p — perturbed by a close line
r — easily reversed
w — wide

The table is arranged alphabetically by element name (not symbol); for each element the lines are listed by wavelength. References to the sources of data for each element are given at the end of the table, starting on page 10-89.

General References

1. Reader, J., Corliss, C. H., Wiese, W. L., and Martin, G. A., *Tables of Line Spectra of the Elements, Part 1. Wavelengths and Intensities*, Nat. Stand. Ref. Data Sys.- Nat. Bur. Standards (U.S.), No. 68, 1980.
2. Meggers, W. F., Corliss, C. H., and Scribner, B. F., *Tables of Spectral Line Intensities, Part 1. Arranged by Elements*, Nat. Bur. Stand. (U.S.), Monograph 145, 1975.
3. Fuhr, J. R., Martin, W. C., Musgrove, A., Sugar, J., and Wiese, W. L., "NIST Atomic Spectroscopic Database" ver. 1.1, January 1996. *NIST Physical Reference Data*, National Institute of Standards and Technology, Gaithersburg, MD. Available at the WWW address: http://physics.nist.gov/PhysRefData/contents.html

Intensity	Wavelength/Å	
Actinium Ac Z = 89		
2000 h	2952.55	III
2000 h	3392.78	III
3000	3487.59	III
2000 s	3863.12	II
3000 s	4088.44	II
3000 s	4168.40	II
100	4179.98	I
20	4183.12	I
20	4194.40	I
20 l	4384.53	I
20	4396.71	I
2000 h	4413.09	III
20	4462.73	I
3000 h	4569.87	III
1000	5910.85	II
20	6359.86	I
20 l	6691.27	I
Aluminum Al Z = 13		
900	125.53	V
800	126.07	V
800	130.41	V
1000	130.85	V
900	131.00	V
900	131.44	V
800	160.07	IV
1000	278.69	V
900	281.39	V
70	486.884	III
30	486.912	III
250	511.138	III
150	511.191	III
500	560.317	III
200	560.433	III
100	670.068	III
200	671.118	III
500	695.829	III
400	696.217	III
200	725.683	III
300	726.915	III
400	855.034	III
500	856.746	III
400	892.024	III
50	893.887	III
450	893.897	III
800	1042.17	IV
50	1191.812	II
900	1237.19	IV
900	1257.62	IV
800	1264.18	IV
1000	1272.76	IV
150	1350.18	II
800	1384.13	III
800	1447.51	IV
800	1494.79	IV
1000	1526.14	V
800	1537.54	IV
800	1539.830	II
1000	1557.25	IV
100	1569.385	II

Intensity	Wavelength/Å	
900	1582.04	IV
800	1584.46	IV
125	1596.059	II
700	1605.766	III
100	1611.814	III
800	1611.874	III
150	1625.627	II
800	1639.06	IV
100	1644.235	II
100	1644.809	II
1000	1670.787	II
100	1686.250	II
800	1719.440	II
500	1721.244	II
900	1721.271	II
500	1724.952	II
900	1724.984	II
350	1760.104	II
300	1761.975	II
290	1763.00	I
500	1763.869	II
700	1763.952	II
450	1765.64	I
300	1765.815	II
450	1766.38	I
400	1767.731	II
450	1769.14	I
1000	1818.56	IV
600	1828.588	II
400	1832.837	II
250	1834.808	II
1000	1854.716	III
300	1855.929	II
700	1858.026	II
120	1859.980	II
1000	1862.311	II
600	1862.790	III
200	1929.978	II
150	1931.048	II
200	1932.377	II
400	1934.503	II
150	1934.713	II
300	1935.840	III
200	1935.949	III
150	1936.907	II
220	1939.261	II
700	1990.531	II
150	2016.052	II
150	2016.234	II
100	2016.368	II
200	2074.008	II
700	2094.264	II
150	2094.744	II
300	2094.791	II
100	2095.104	II
200	2095.141	II
400	2269.10	I
120	2269.22	I
140	2321.56	I
460	2367.05	I
110	2367.61	I

Intensity	Wavelength/Å	
110	2368.11	I
180	2369.30	I
140	2370.22	I
160	2372.07	I
850	2373.12	I
170	2373.35	I
110	2373.57	I
240	2567.98	I
480	2575.10	I
110	2637.70	II
150	2652.48	I
200	2660.39	I
160	2669.17	II
650	2816.19	II
150	3041.28	II
360	3050.07	I
450	3057.14	I
150	3074.64	II
4500 r	3082.153	II
7200 r	3092.710	I
1800 r	3092.839	I
150	3428.92	II
150	3443.64	I
900	3492.23	IV
800	3508.46	IV
450	3586.56	II
360	3587.07	II
290	3587.45	II
870	3601.63	III
220	3651.06	II
110	3651.10	II
150	3654.98	II
290	3655.00	II
450	3900.68	II
4500 r	3944.006	I
9000 r	3961.520	I
110	3995.86	II
290	4226.81	II
870	4529.19	III
150	4585.82	II
110	4588.19	II
550	4666.80	II
110	4898.76	II
110	4902.77	II
150	5280.21	II
290	5283.77	II
150	5285.85	II
110	5312.32	II
220	5316.07	II
150	5371.84	II
180	5557.06	I
110	5557.95	I
450	5593.23	II
1200	5696.60	III
1000	5722.73	III
110	5853.62	II
220	5971.94	II
290	6001.76	II
220	6001.88	II
450	6006.42	II
150	6061.11	II

Intensity	Wavelength/Å	
290	6068.43	II
110	6068.53	II
450	6073.23	II
110	6181.57	II
150	6181.68	II
290	6182.28	II
220	6182.45	II
450 h	6183.42	II
450	6201.52	II
360	6201.70	II
290	6226.18	II
360	6231.78	II
450	6243.36	II
450	6335.74	II
360	6696.02	I
230	6698.67	I
110	7361.57	I
140	7362.30	I
230	7835.31	I
290	7836.13	I
110	8075.35	I
290	8640.70	II
360	8772.87	I
450	8773.90	I
110	8828.91	I
180	8841.28	I
140	8923.56	I
150	9290.65	II
110	9290.75	II
150	10076.29	II
110	10768.36	I
140	10782.04	I
110	10872.98	I
230	10891.73	I
450	11253.19	I
570	11254.88	I
570	13123.41	I
450	13150.76	I
230	16718.96	I
300	16750.56	I
140	16763.36	I
300	21093.04	I
360	21163.75	I
Antimony Sb Z = 51		
15	722.86	III
15	732.33	III
	861.5	IV
4	876.84	II
4	921.07	II
6	983.57	II
15	999.62	III
6	1001.13	II
6	1009.43	II
40	1011.94	III
6	1052.21	II
8	1056.27	II
8	1057.32	II
40	1065.90	III
6	1073.81	II
30	1075.82	III
	1087.6	IV

Intensity	Wavelength/Å	
8	1104.32	V
30	1151.49	III
40	1157.74	III
	1199.1	IV
50	1205.20	III
50	1210.64	III
12	1226.00	V
6	1230.30	II
8	1274.98	II
20	1306.69	III
8	1327.40	II
6	1358.04	II
8	1384.70	II
20	1404.18	III
6	1407.83	II
8	1436.49	II
20 r	1486.57	I
40 h	1491.36	I
	1499.2	IV
12	1505.70	V
50 r	1512.57	I
12	1524.47	V
120 r	1532.74	I
80 r	1535.06	I
6	1565.51	II
8	1576.11	II
7	1581.36	II
80 r	1599.96	I
10	1606.98	II
200 w	1612.8	I
100 w	1623.3	I
20	1657.04	II
100 w	1662.6	I
15	1673.89	III
15	1711.84	III
80 r	1716.93	I
150 r	1717.45	I
150 r	1723.43	I
15	1725.33	III
100 r	1736.19	I
100 h	1765.76	I
100 r	1780.87	I
100 r	1788.24	I
150	1800.18	I
50 r	1810.50	I
80 r	1814.20	I
100	1829.50	I
50 r	1868.17	I
300 r	1871.15	I
150 r	1882.56	I
100	1927.08	I
200 r	1950.39	I
60 r	2029.49	I
70 r	2039.77	I
150 r	2049.57	I
1000 r	2068.33	I
100	2079.56	I
50 r	2098.41	I
80 r	2118.48	I
100 r	2127.39	I
50 r	2137.05	I
100 r	2139.69	I
10	2141.80	II
50 r	2141.83	I
100 r	2144.86	I
1500 r	2175.81	I
250 r	2179.19	I
200 r	2201.32	I
300 r	2208.45	I
150 r	2220.73	I
100	2221.98	I
120 r	2224.93	I
300 r	2262.51	I
120	2288.98	I
150 r	2293.44	I
300 r	2306.46	I
2500 r	2311.47	I
150	2315.89	I
400 h	2373.67	I
300 h	2383.64	I
100	2395.22	I
150	2422.13	I
250	2426.35	I
400 r	2445.51	I
400	2478.32	I
150	2480.44	I
100	2510.54	I
2000 r	2528.52	I
15	2528.54	II
10	2567.75	II
150	2574.06	I
15	2590.13	III
1500 r	2598.05	I
500 r	2598.09	I
300 r	2612.31	I
12	2617.17	III
200 r	2652.60	I
20	2669.39	III
300 r	2670.64	I
200 r	2682.76	I
120	2692.25	I
150 r	2718.90	I
400 r	2769.95	I
1000 r	2877.92	I
15	2980.96	II
500 r	3029.83	I
600 r	3232.52	I
20	3241.28	II
700 r	3267.51	I
15	3498.46	II
25	3637.80	II
250	3637.83	I
20	3722.78	II
200 r	3722.79	I
20	3850.22	II
200	4033.55	I
20	4033.56	II
20	4133.63	II
15	4140.54	II
15	4195.17	II
20	4219.07	II
20	4314.32	II
15	4514.50	II
30	4596.90	II
20	4599.09	II
15	4604.77	II
30	4647.32	II
20	4675.74	II
40	4711.26	II
20	4757.81	II
20	4765.36	II
30	4784.03	II
20	4802.01	II
20	4832.82	II
20	4877.24	II
15	4947.40	II
15	5044.56	II
20	5238.94	II
20	5354.24	II
40 h	5556.10	I
100 l	5632.02	I
30	5639.75	II
60 h	5830.34	I
100	6005.21	II
20	6053.41	II
30	6079.80	II
50	6130.04	II
20	6154.94	II
20	6611.49	I
30	6647.44	II
30 h	7648.28	I
80	7844.44	I
200	7924.65	I
60	8411.69	I
150	8572.64	I
100	8619.55	I
400	9518.68	I
400	9949.14	I
200	10078.49	I
300	10261.01	I
200	10585.60	I
1000	10677.41	I
800	10741.94	I
80	10794.11	I
600	10839.73	I
200	10868.58	I
400	10879.55	I
300	11012.79	I
150	11266.23	I
5	12116.06	I

Argon Ar Z = 18

Intensity	Wavelength/Å	
3	336.56	V
3	337.56	V
6	338.00	V
2	338.43	V
2	339.01	V
3	339.89	V
3	350.88	V
4	396.87	IV
4	398.55	IV
2	436.67	V
5	446.00	V
8	446.95	V
4	447.53	V
18	449.06	V
4	449.49	V
3	458.12	V
2	458.98	V
6 p	461.23	V
3	462.42	V
7	463.94	V
30	487.227	II
50	490.650	II
30	490.701	II
30	519.327	II
3	522.09	V
5	524.19	V
6	527.69	V
30	542.912	II
200	543.203	II
70	547.461	II
2	554.50	V
70	556.817	II
5	558.48	V
70	573.362	II
30	576.736	II
70	580.263	II
30	583.437	II
70	597.700	II
30	602.858	II
30	612.372	II
6	623.77	IV
3	635.12	V
500	661.867	II
30	664.562	II
200	666.011	II
1000	670.946	II
3000	671.851	II
70	676.242	II
30	677.952	II
30	679.218	II
200	679.401	II
10	683.28	IV
7	688.39	IV
12 p	689.01	IV
6	699.41	IV
8	700.28	IV
3	705.35	V
5	709.20	V
4	715.60	V
3	715.65	V
200	718.090	II
3000	723.361	II
2	725.11	V
500	725.548	II
70	730.930	II
200	740.269	II
200	744.925	II
70	745.322	II
4	754.20	IV
5	761.47	IV
12	769.15	III
5	800.57	IV
10	801.09	IV

Intensity	Wavelength/Å		Intensity	Wavelength/Å		Intensity	Wavelength/Å		Intensity	Wavelength/Å	
10	801.41	IV	12	2293.03	III	12	2913.00	IV	35	3718.206	II
5	801.91	IV	4	2299.72	IV	11	2926.33	IV	70	3729.309	II
20	802.859	I	10	2300.85	III	200	2942.893	II	50	3737.889	II
100	806.471	I	15	2302.17	III	100	2979.050	II	150	3765.270	II
60	806.869	I	9	2317.00	III	10	3010.02	III	50	3766.119	II
30	807.218	I	15	2317.47	III	12	3024.05	III	20	3770.369	I
40	807.653	I	12	2318.04	III	50	3033.508	II	20	3770.520	II
50	809.927	I	10	2319.13	III	6	3037.98	IV	25	3780.840	II
120	816.232	I	10	2319.37	III	12	3054.82	III	20	3795.37	III
70	816.464	I	9	2345.17	III	10	3064.77	III	25	3803.172	II
80	820.124	I	7	2351.67	III	8	3077.40	IV	50	3809.456	II
4	822.16	V	9	2360.26	III	10	3078.15	III	7	3834.679	I
120	825.346	I	10	2395.63	III	50	3093.402	II	70	3850.581	II
120	826.365	I	12	2399.15	III	7	3110.41	III	10	3858.32	III
5	827.05	V	10	2413.20	III	7	3127.90	III	35	3868.528	II
3	827.35	V	7	2415.61	III	8	3200.37	I	7	3907.84	III
150	834.392	I	10	2418.82	III	20	3243.689	II	35	3925.719	II
4 p	834.88	V	5	2420.456	II	25	3285.85	III	50	3928.623	II
100	835.002	I	12	2423.52	III	25	3293.640	II	25	3932.547	II
2	836.13	V	12	2423.93	III	20	3301.88	III	70	3946.097	II
15	840.03	IV	7	2443.69	III	20	3307.228	II	7	3947.505	I
100	842.805	I	8	2447.71	IV	15	3311.25	III	35	3948.979	I
20	843.77	IV	8	2472.95	III	7	3319.34	I	8	3960.53	III
25	850.60	IV	7	2476.10	III	7	3323.59	III	20	3979.356	II
180	866.800	I	12	2488.86	III	25	3336.13	III	35	3994.792	II
150	869.754	I	12	2513.28	IV	20	3344.72	III	50	4013.857	II
10	871.10	III	10	2516.789	II	25	3350.924	II	6	4023.60	III
9	875.53	III	6	2518.40	IV	15	3358.49	III	50	4033.809	II
180 r	876.058	I	9	2525.69	IV	7	3361.28	III	20	4035.460	II
12	878.73	III	10	2534.709	II	7	3373.47	I	150	4042.894	II
8	879.62	III	15	2562.087	II	25	3376.436	II	50	4044.418	I
180 r	879.947	I	12	2562.17	IV	25	3388.531	II	100	4052.921	II
9	883.18	III	10	2568.07	IV	15	3391.85	III	200	4072.005	II
10	887.40	III	7	2569.53	IV	7	3393.73	I	70	4072.385	II
150	894.310	I	12	2599.47	IV	7	3417.49	III	25	4076.628	II
5	900.36	IV	10	2608.06	IV	9	3424.25	III	35	4079.574	II
9	901.17	IV	7	2608.44	IV	8	3438.04	III	25	4082.387	II
1000	919.781	II	12	2615.68	IV	7	3461.07	I	150	4103.912	II
1000	932.054	II	6	2619.98	IV	9	3471.32	III	300	4131.724	II
1000 r	1048.220	I	12	2621.36	IV	70	3476.747	II	5	4146.70	III
500 r	1066.660	I	12	2624.92	IV	20	3478.232	II	35	4156.086	II
7	1669.67	III	7	2631.90	III	20	3480.55	III	400	4158.590	I
7	1673.42	III	15	2640.34	IV	50	3491.244	II	50	4164.180	I
7	1675.48	III	10	2654.63	III	100	3491.536	II	35	4179.297	II
9	1914.40	III	8	2674.02	III	12	3499.67	III	50	4181.884	I
7	1915.56	III	9	2678.38	III	15	3503.58	III	100	4190.713	I
10	2125.16	III	9	2682.63	IV	70	3509.778	II	50	4191.029	I
15	2133.87	III	10	2724.84	III	8	3511.12	III	200	4198.317	I
10	2138.59	III	14	2757.92	IV	70	3514.388	II	400	4200.674	I
10	2148.73	III	7	2762.23	III	70	3545.596	II	25	4218.665	II
15	2166.19	III	10	2776.26	IV	70	3545.845	II	25	4222.637	II
10	2168.26	III	12	2784.47	IV	7	3554.306	I	25	4226.988	II
20	2170.23	III	14	2788.96	IV	100	3559.508	II	100	4228.158	II
25	2177.22	III	7	2797.11	IV	100	3561.030	II	100	4237.220	II
8	2184.06	III	16	2809.44	IV	70	3576.616	II	25	4251.185	I
10	2188.22	III	10	2830.25	IV	25	3581.608	II	200	4259.362	I
15	2192.06	III	7	2842.88	III	50	3582.355	II	100	4266.286	I
7	2248.73	III	8	2855.29	III	70	3588.441	II	70	4266.527	II
10	2279.10	III	6	2874.40	IV	7	3606.522	I	150	4272.169	I
7	2281.22	III	9	2884.12	III	25	3622.138	II	550	4277.528	II
7	2282.21	III	25	2891.612	II	20	3639.833	II	20	4282.898	II

Intensity	Wavelength/Å		Intensity	Wavelength/Å		Intensity	Wavelength/Å		Intensity	Wavelength/Å	
100	4300.101	I	25	5165.773	II	7	6951.478	I	200	10673.565	I
25	4300.650	II	20	5187.746	I	7	6960.250	I	11	10681.773	I
70	4309.239	II	20	5216.814	II	10000	6965.431	I	7	10683.034	II
200	4331.200	II	7	5221.271	I	150	7030.251	I	30	10733.87	I
50	4332.030	II	5	5421.352	I	10000	7067.218	I	30	10759.16	I
100	4333.561	I	10	5451.652	I	100	7068.736	I	7	10812.896	II
50	4335.338	I	25	5495.874	I	25	7107.478	I	11	11078.869	I
25	4345.168	I	5	5506.113	I	25	7125.820	I	30	11106.46	I
800	4348.064	II	25	5558.702	I	1000	7147.042	I	12	11441.832	I
50	4352.205	II	10	5572.541	I	15	7158.839	I	400	11488.109	I
25	4362.066	II	35	5606.733	I	70	7206.980	I	200	11668.710	I
50	4367.832	II	20	5650.704	I	15	7265.172	I	12	11719.488	I
200	4370.753	II	10	5739.520	I	7	7270.664	I	200	12112.326	I
70	4371.329	II	5	5834.263	I	2000	7272.936	I	50	12139.738	I
50	4375.954	II	10	5860.310	I	35	7311.716	I	50	12343.393	I
150	4379.667	II	15	5882.624	I	25	7316.005	I	200	12402.827	I
50	4385.057	II	25	5888.584	I	5	7350.814	I	200	12439.321	I
70	4400.097	II	50	5912.085	I	70	7353.293	I	100	12456.12	I
200	4400.986	II	15	5928.813	I	200	7372.118	I	200	12487.663	I
400	4426.001	II	5	5942.669	I	20	7380.426	II	150	12702.281	I
150	4430.189	II	7	5987.302	I	10000	7383.980	I	30	12733.418	I
50	4430.996	II	5	5998.999	I	20	7392.980	I	12	12746.232	I
50	4433.838	II	5	6025.150	I	15	7412.337	I	200	12802.739	I
20	4439.461	II	70	6032.127	I	10	7425.294	I	50	12933.195	I
35	4448.879	II	35	6043.223	I	25	7435.368	I	500	12956.659	I
100	4474.759	II	10	6052.723	I	10	7436.297	I	200	13008.264	I
200	4481.811	II	20	6059.372	I	20000	7503.869	I	200	13213.99	I
100	4510.733	I	7	6098.803	I	15000	7514.652	I	200	13228.107	I
20	4522.323	I	10	6105.635	I	25000	7635.106	I	100	13230.90	I
20	4530.552	II	100	6114.923	II	15000	7723.761	I	500	13272.64	I
400	4545.052	II	10	6145.441	I	10000	7724.207	I	1000	13313.210	I
20	4564.405	II	7	6170.174	I	10	7891.075	I	1000	13367.111	I
400	4579.350	II	150	6172.278	II	20000	7948.176	I	30	13499.41	I
400	4589.898	II	10	6173.096	I	20000	8006.157	I	1000	13504.191	I
15	4596.097	I	10	6212.503	I	25000	8014.786	I	11	13573.617	I
550	4609.567	II	5	6215.938	I	7	8053.308	I	30	13599.333	I
7	4628.441	I	25	6243.120	II	20000	8103.693	I	400	13622.659	I
35	4637.233	II	7	6296.872	I	35000	8115.311	I	200	13678.550	I
400	4657.901	II	15	6307.657	I	10000	8264.522	I	1000	13718.577	I
15	4702.316	I	7	6369.575	I	20	8392.27	I	10	13825.715	I
20	4721.591	II	20	6384.717	I	15000	8408.210	I	10	13907.478	I
550	4726.868	II	70	6416.307	I	20000	8424.648	I	200	14093.640	I
50	4732.053	II	25	6483.082	II	15000	8521.442	I	100	15046.50	I
300	4735.906	II	15	6538.112	I	7	8605.776	I	25	15172.69	I
800	4764.865	II	15	6604.853	I	4500	8667.944	I	10	15329.34	I
550	4806.020	II	25	6638.221	II	20	8771.860	II	30	15989.49	I
150	4847.810	II	20	6639.740	II	180	8849.91	I	30	16519.86	I
50	4865.910	II	50	6643.698	II	20	9075.394	I	500	16940.58	I
800	4879.864	II	5	6660.676	I	35000	9122.967	I	12	18427.76	I
70	4889.042	II	5	6664.051	I	550	9194.638	I	50	20616.23	I
20	4904.752	II	25	6666.359	II	15000	9224.499	I	30	20986.11	I
35	4933.209	II	100	6677.282	I	400	9291.531	I	20	23133.20	I
200	4965.080	II	35	6684.293	II	1600	9354.220	I	20	23966.52	I
50	5009.334	II	150	6752.834	I	25000	9657.786	I			
70	5017.163	II	5	6756.163	I	4500	9784.503	I	*Arsenic As Z = 33*		
70	5062.037	II	15	6766.612	I	180	10052.06	I	510	871.7	III
20	5090.495	II	20	6861.269	II	30	10332.72	I	325	889.0	III
100	5141.783	II	150	6871.289	I	100	10467.177	II	325	927.5	III
70	5145.308	II	5	6879.582	I	1600	10470.054	I	325	937.2	III
5	5151.391	I	10	6888.174	I	13	10478.034	I	325	953.6	III
15	5162.285	I	50	6937.664	I	180	10506.50	I	325	963.8	III
									250	987.7	V

Intensity	Wavelength/Å		Intensity	Wavelength/Å		Intensity	Wavelength/Å		Intensity	Wavelength/Å	
340	1021.96	II	300	2831.164	II	14	587.57	III	10	3193.91	I
250	1029.5	V	100 r	2860.44	I	18	647.27	III	25 h	3203.70	I
340	1082.35	II	300	2884.406	II	300	719.86	V	30	3221.63	I
500	1139.40	II	80	2926.3	III	150	721.85	V	40	3222.19	I
615	1149.31	II	615	2959.572	II	1000	766.87	V	50	3261.96	I
555	1181.51	II	300	3003.819	II	40000	794.89	IV	60 r	3262.34	I
555	1189.87	II	300	3116.516	II	300	877.41	V	40	3281.50	I
615	1196.38	II	340	3842.60	II	50000	923.74	IV	15	3281.77	I
615	1196.56	II	325	3922.6	III	200	946.26	V	50	3322.80	I
340	1207.44	II	715	4190.082	II	200	1486.72	II	80 h	3356.80	I
800	1211.17	II	615	4197.40	II	400	1504.01	II	50	3368.18	III
800	1218.10	II	615	4242.982	II	300	1554.38	II	60 r	3377.08	I
340	1223.15	II	500	4315.657	II	200	1572.73	II	20	3377.39	I
760	1241.31	II	500	4323.867	II		1573.92	II	70 r	3420.32	I
965	1243.08	II	500	4336.64	II		1630.40	II	25	3421.01	I
870	1245.67	II	500	4352.145	II	100	1674.51	II	30 h	3421.48	I
800	1258.58	II	425	4352.864	II	400	1694.37	II	40	3463.74	I
965	1263.77	II	375	4371.17	II		1697.16	II	200 r	3501.11	I
800	1266.34	II	615	4427.106	II		1761.75	II	80 h	3524.97	I
800	1267.59	II	615	4431.562	II		1771.03	II	30 h	3531.35	I
715	1280.99	II	715	4458.469	II		1786.93	II	80 h	3544.66	I
715	1287.54	II	340	4461.075	II	100	1904.15	II	20 h	3547.68	I
715	1305.70	II	715	4466.348	II	500	1924.70	II	100	3552.45	II
340	1307.74	II	500	4474.46	II		1985.60	II	200	3567.73	II
760	1333.15	II	800	4494.230	II	300	1999.54	II	100	3576.28	II
965	1341.55	II	850	4507.659	II	10	2001.30	III	30	3577.62	I
760	1355.93	II	615	4539.74	II		2009.20	II	80 h	3579.67	I
965	1369.77	II	715	4543.483	II	400	2023.95	II	200	3596.57	II
800	1373.65	II	615	4602.427	II		2052.68	II	40	3630.64	I
1000	1375.07	II	340	4629.787	II		2054.57	II	40 h	3636.83	I
760	1375.78	II	340	4707.586	II	500	2214.7	II	20 h	3688.47	I
800	1394.64	II	340	4730.67	II	800	2245.61	II	400	3735.75	II
800	1400.31	II	340	4888.557	II	1000	2254.73	II	200	3816.69	II
500	1448.59	II	340	5105.58	II	1400	2304.24	II	200	3842.80	II
500	1558.88	II	500	5107.55	II	60	2331.10	III	100	3854.76	II
500	1570.99	II	425	5231.38	II	2000	2335.27	II	20	3889.33	I
100 r	1593.60	I	500	5331.23	II	190	2347.58	II	1400 l	3891.78	II
500	1660.55	II	340	5497.727	II	40	2512.28	III	20	3892.65	I
340	1860.34	II	425	5558.09	II	40	2523.83	III	40	3909.91	I
1000 r	1890.42	I	425	5651.32	II	60	2528.51	III	500	3914.73	II
500	1912.94	II	425	6110.07	II	50	2559.54	III	25	3926.85	III
800 r	1937.59	I	500	6170.27	II	8 h	2596.64	I	50	3935.72	I
585 r	1972.62	I	300	6511.74	II	100	2634.78	II	20	3937.87	I
170 r	1990.35	I	300	7092.27	II	40	2681.89	III	200	3939.67	II
100 r	1991.13	I	300	7102.72	II	8	2702.63	I	500	3949.51	II
100 r	1995.43	I	340	7990.53	II	18	2771.36	II	25	3993.06	III
230 r	2003.34	I	300	8174.51	II	15	2785.28	I	80	3993.40	I
100 r	2009.19	I	200	9300.61	I	100 r	3071.58	I	30	3995.66	I
200	2263.2	IV	230	9597.95	I	40	3079.14	III	300	4036.26	II
350 r	2288.12	I	290	9626.70	I	10 h	3108.21	I	200	4083.77	II
200	2301.0	IV	230	9833.76	I	8	3132.60	I	30 h	4084.86	I
350 r	2349.84	I	170	9915.71	I	8 h	3135.72	I	1500 h	4130.66	II
100 r	2370.77	I	290	9923.05	I	10	3137.70	I	20	4132.43	I
135 r	2381.18	I	290	10024.04	I	10	3155.34	I	200	4166.00	II
250	2417.5	IV	170	10614.07	I	10	3155.67	I	500	4216.04	II
250	2454.0	IV				12	3158.05	I	800	4267.95	II
170 r	2456.53	I	*Astatine At* Z = 85			12 h	3158.54	I	100	4283.10	I
200	2461.4	IV	8	2162.25	I	25	3165.60	I	300	4287.80	II
340	2602.00	II	10	2244.01	I	15 h	3173.69	I	200	4297.60	II
170 r	2780.22	I				30	3183.16	I	800	4309.32	II
300	2830.359	II	*Barium Ba* Z = 56			15	3183.96	I	20 h	4323.00	I
			14	555.48	III						

Intensity	Wavelength/Å		Intensity	Wavelength/Å		Intensity	Wavelength/Å		Intensity	Wavelength/Å	
600	4325.73	II	300	6110.78	I	120 h	11697.45	I		107.38	I
200	4326.74	II	400	6135.83	II	120	13207.30	I	3	509.99	III
300	4329.62	II	20000	6141.72	II	120	13810.50	I	2	549.31	III
80	4350.33	I	150	6341.68	I	120	14077.90	I	6	582.08	III
60	4402.54	I	500	6378.91	II	120	15000.40	I	4	661.32	III
400	4405.23	II	10	6383.76	III	120	20712.00	I	8	675.59	III
40	4431.89	I	90	6450.85	I	150	25515.70	I		714.0	II
60 h	4488.98	I	150	6482.91	I	150	29223.90	I	4	725.59	III
50 h	4493.64	I	12000	6496.90	II				5	725.71	II
40	4505.92	I	300	6498.76	I	**Beryllium Be** *Z = 4*			5	743.58	II
200	4509.63	II	150	6527.31	I		58.13	IV	7	746.23	III
60 h	4523.17	I	3000	6595.33	I		58.57	IV	2	767.75	III
130	4524.93	II	150	6654.10	I		59.32	IV	8	775.37	II
65000	4554.03	II	1500	6675.27	I		60.74	IV	20	842.06	II
40	4573.85	I	1800	6693.84	I		64.06	IV		865.3	II
80	4579.64	I	1000	6769.62	II		75.93	IV	2	925.25	II
30	4599.75	I	600	6865.69	I	1 h	76.10	III	10	943.56	II
20 h	4619.92	I	300 h	6867.85	I	2	76.48	III	10	973.27	II
25 h	4628.33	I	1000	6874.09	II	3	78.53	III		981.4	II
300	4644.10	II	6000	7059.94	I	4	78.66	III		1020.1	II
30	4673.62	I	2400 hs	7120.33	I	1 h	78.92	III	8	1026.93	II
35	4691.62	I	600	7195.24	I	5	81.89	III	5	1036.32	II
20	4700.43	I	600 hl	7228.84	I	10	82.38	III	15	1048.23	II
800	4708.94	II	3000	7280.30	I		82.58	II	1	1114.69	III
40	4726.44	I	1200	7392.41	I	20	83.20	III	20	1143.03	II
800	4843.46	II	300	7417.53	I		83.66	II		1155.9	II
300	4847.14	II	900 hl	7459.78	I	30	84.76	III	60	1197.19	II
200	4850.84	II	600	7488.08	I	50	88.31	III	2	1213.12	III
30 h	4877.65	I	450 hl	7636.90	I		89.16	I	1	1214.32	III
400	4899.97	II	600 hl	7642.91	I		89.80	II	2	1362.25	III
15	4902.90	I	1800	7672.09	I		90.04	II	1	1401.52	III
20000	4934.09	II	1200	7780.48	I		90.21	I	10	1421.26	III
8	4947.35	I	180 h	7839.57	I		90.67	I	5	1422.86	III
1000	4957.15	II	1500	7905.75	I		91.06	II		1426.12	I
300	4997.81	II	600	7911.34	I		91.36	II	1	1435.17	III
1000	5013.00	II	900 h	8210.24	I		91.74	II	2	1440.77	III
20 h	5159.94	I	8	8308.69	III		92.19	I		1491.76	I
20	5267.03	I	1800 h	8559.97	I		92.61	II	20	1512.30	II
800	5361.35	I	100	8710.74	II		93.14	II	60	1512.43	II
1000	5391.60	II	100	8737.71	II		93.42	II	100	1661.49	I
200	5421.05	II	300 h	8799.76	I		93.93	II	2 h	1754.69	III
100	5424.55	I	300	8860.98	I		94.78	II	15	1776.12	II
200	5428.79	II	450	8914.99	I		95.76	II	20	1776.34	II
300	5480.30	II	300	9219.69	I		96.29	I		1907.	I
200	5519.05	I	300	9308.08	I		97.24	I		1909.0	II
1000 r	5535.48	I	300 h	9324.58	I		97.44	I		1912.	I
20 h	5620.40	I	1500	9370.06	I		97.86	I	3	1917.03	III
10	5680.18	I	300	9455.92	I		97.97	I		1919.	I
400	5777.62	I	8	9521.76	III		98.12	I	5	1929.67	I
800	5784.18	II	450	9589.37	I		98.37	I	10	1943.68	I
100	5800.23	I	900	9608.88	I		98.66	I	60 h	1954.97	III
20	5805.69	I	300 h	9645.72	I		98.94	I		1956.	I
150	5826.28	I	1500 hl	9830.37	I		99.19	I	50	1964.59	I
2800	5853.68	II	900	10001.08	I	100	100.25	III	5	1985.13	I
15	5907.64	I	600	10032.10	I		100.86	I		1997.95	I
100	5971.70	I	1200 h	10233.23	I		101.20	I		1997.98	I
800	5981.25	II	300	10471.26	I		102.13	I	60	1998.01	I
100	5997.09	I	120 hl	10791.25	I		102.49	II		2033.25	I
300	5999.85	II	180 hl	11012.69	I		104.40	II		2033.28	I
100	6019.47	I	150 h	11114.42	I		104.67	I		2033.38	I
200	6063.12	I	240	11303.04	I		105.80	I	50	2055.90	I
							107.26	I			

Intensity	Wavelength/Å		Intensity	Wavelength/Å		Intensity	Wavelength/Å		Intensity	Wavelength/Å	
100	2056.01	I		3163.	I	64	5270.28	II	60	14644.75	I
75 h	2076.94	III		3168.	I	500	5270.81	II	200	16157.72	I
60 h	2080.38	III		3180.7	II	20	5403.04	II	80	17855.38	I
25	2118.56	III		3187.	I	20	5410.21	II	120	17856.63	I
15 h	2122.27	III	20	3193.81	I		5558.	I	100	18143.54	I
10	2125.57	I	20	3197.10	II	140 h	6142.01	III	160	31775.05	I
20	2125.68	I	30	3197.15	II	10	6229.11	I	200	31778.70	I
15 h	2127.20	III	20	3208.60	I	16	6279.43	II			
5	2137.25	III		3220.	I	30	6279.73	II	*Bismuth Bi* Z = 83		
25	2145.	I	60	3229.63	I	30	6473.54	I	6	420.7	IV
55	2174.99	I	2	3233.52	II	60	6547.89	II	6	431.2	IV
55	2175.10	I	10	3241.62	II	60	6558.36	II	2	488.39	V
5	2191.57	III	30	3241.83	II	30	6564.52	I	3	563.62	V
	2273.5	II	15	3269.02	I	2 h	6636.44	II	5	670.76	III
	2324.6	II	100	3274.58	II	1	6756.72	II	6	686.88	V
	2337.0	I	30	3274.67	II	2	6757.13	II	5	730.71	V
950	2348.61	I	30	3282.91	I	30	6786.56	I	10	738.17	V
20	2350.66	I	30	3321.01	I	1 h	6884.22	I	4	775.16	III
60	2350.71	I	30	3321.09	I	6 h	6884.44	I	6	790.5	IV
200	2350.83	I	220	3321.34	I	100	6982.75	I	6	790.6	IV
2	2413.34	II	20	3345.43	I	6 h	7154.40	I	8	792.5	IV
16	2413.46	II	60	3367.63	I	40 h	7154.65	I	10	820.3	IV
20	2453.84	II		3405.6	II	100	7209.13	I	9	822.9	IV
	2480.6	I	5	3451.37	I	3	7401.20	II	12	824.9	IV
35	2494.54	I	300	3455.18	I	2	7401.43	II	15 d	864.45	V
35	2494.58	I	20	3476.56	I	10	7551.90	I	15	872.6	IV
100	2494.73	I	300	3515.54	I	10 h	7618.68	I	12	923.9	IV
16	2507.43	II	10	3555.	I	20 h	7618.88	I	15	943.3	IV
5	2617.99	II	100	3720.36	III	60	8090.06	I	25	1039.99	III
20	2618.13	II		3720.92	III	5 h	8158.99	I	50 h	1045.76	III
100	2650.45	I		3722.98	III	10 h	8159.24	I	30	1051.81	III
60	2650.55	I	100	3736.30	I	4	8254.07	I	15	1058.88	II
200	2650.62	I	700	3813.45	I	10 h	8287.07	I	20	1085.47	II
60	2650.69	I	40	3865.13	I	30	8547.36	I	10	1099.20	II
100	2650.76	I	80	3865.42	I	60	8547.67	I	24	1103.4	IV
5	2697.46	II	1	3865.51	I	300	8801.37	I	20	1139.01	III
20	2697.58	II	6	3865.72	I	6	8882.18	I	50	1224.64	III
20	2728.88	II	100	3866.03	I	40	9190.45	I	10	1225.43	II
30	2738.05	I	90 h	4249.14	III	20 h	9243.92	I	15	1232.78	II
	2764.2	II	100	4253.05	I	1 h	9343.89	II	10	1241.05	II
20	2898.13	I	60	4253.76	I	40	9392.74	I	10	1265.35	II
10	2898.19	I	300	4360.66	II	2	9476.43	II	15	1283.73	II
20	2898.25	I	500	4360.99	II	16	9477.03	II	10	1306.18	II
30	2986.06	I	400	4407.94	I	20	9847.32	I	60	1317.0	IV
10	2986.42	I	2	4485.52	III	10 h	9895.63	I	20	1325.46	II
60	3019.33	I	100 h	4487.30	III	20 h	9895.96	I	40	1326.84	III
30	3019.49	I	1	4495.09	III	80	9939.78	I	20	1329.47	II
30	3019.53	I	140 h	4497.8	III	16	10095.52	II	60	1346.12	III
20	3019.60	I		4526.6	I	20	10095.73	II	20	1350.07	II
10	3046.52	II		4548.	I	60	10119.92	II	25	1372.61	II
30	3046.69	II	12	4572.66	I	80	10331.03	I	15	1376.02	II
	3090.3	I	700	4673.33	II	30	11066.46	I	20	1393.92	II
10	3110.81	I	1000	4673.42	II		11173.	II	35	1423.33	III
10	3110.92	I	6	4709.37	I	1	11173.73	II	35	1423.52	III
20	3110.99	I	200	4828.16	II	120	11496.39	I	45	1436.83	II
	3120.	I	40	4849.16	I	2 h	11625.16	II	25	1447.94	II
480	3130.42	II	2 h	4858.22	II		11659.	II	50	1455.11	II
320	3131.07	II	80	5087.75	I	2	11660.25	II	60 h	1461.00	III
	3136.	I	8	5218.12	II	100	12095.36	II	25	1462.14	II
	3150.	I	20	5218.33	II	30	12098.18	II	35	1486.93	II
	3160.6	I	3	5255.86	II	100	14643.92	I	20	1502.50	II
									40	1520.57	II

Intensity	Wavelength/Å		Intensity	Wavelength/Å		Intensity	Wavelength/Å		Intensity	Wavelength/Å	
40	1533.17	II	100	2924.	IV	60 h	5144.3	II			
30	1536.77	II	100	2933.	IV	20	5201.5	II	**Boron B Z = 5**		
35	1538.06	II	100	2936.	IV	75 h	5209.2	II		41.00	V
20	1563.67	II	15	2936.7	II	40 h	5270.3	II	30	48.59	V
40	1573.70	II	3200	2938.30	I	10	5397.8	II	10	52.68	IV
60	1591.79	II	20	2950.4	II	10 c	5552.35	I	30	60.31	IV
25	1601.58	II	12	2963.4	II	3	5599.41	I		194.37	V
60 h	1606.40	III	2800	2989.03	I	20	5655.2	II		262.37	V
40	1609.70	II	700	2993.34	I	40 h	5719.2	II	160	344.0	IV
40	1611.38	II	100	3012.	IV	6	5742.55	I	450	385.0	IV
20	1652.81	II	2400	3024.64	I	12	5818.3	II	40	411.80	III
20	1749.29	II	60	3034.87	I	20	5860.2	II	285	418.7	IV
80	1777.11	II	100	3042.	IV	20	5973.0	II	20	510.77	III
60	1787.47	II	9000 c	3067.72	I	15	6059.1	II	40	510.85	III
70	1791.93	II	140	3076.66	I	15	6128.0	II		512.53	V
70	1823.80	II	35	3115.0	III	6	6134.82	I	150	518.24	III
100	1902.41	II	100	3239.	IV	3	6475.73	I	75	518.27	III
9000	1954.53	I	550 c	3397.21	I	3	6476.24	I	110	677.00	III
7000	1960.13	I	10	3430.83	II	15	6497.7	II	160	677.14	III
25	1989.35	II	12	3431.23	II	10	6577.2	II	40	693.95	II
7000	2021.21	I	40 h	3451.0	III	40 h	6600.2	II	40	731.36	II
9000	2061.70	I	40	3473.8	III	50 h	6808.6	II	40	731.44	II
45 h	2068.9	II	35	3485.5	III	4 h	6991.12	I		749.74	V
4600	2110.26	I	500 c	3510.85	I	12	7033.	II	40	758.48	III
2500	2133.63	I	380 c	3596.11	I	2	7036.15	I	70	758.67	III
15	2143.40	II	45	3613.4	III	10 h	7381.	II	110	882.54	II
15	2143.46	II	100	3643.	IV	2	7502.33	I	110	882.68	II
60	2186.9	II	12	3654.2	II	10 h	7637.	II	40	984.67	II
40 h	2214.0	II	100	3682.	IV	10	7750.	II	110	1081.88	II
360	2228.25	I	50	3695.32	III	3	7838.70	I	110	1082.07	II
1700	2230.61	I	50	3695.68	III	2	7840.33	I	70	1112.2	IV
340	2276.58	I	100	3734.	IV	20	7965.	II	450	1168.9	IV
100	2311.	IV	70 h	3792.5	II	40	8008.	III	70	1170.9	IV
100	2326.	IV	12	3811.1	II	12 h	8050.	II	110	1230.16	II
16	2368.12	II	20	3815.8	II	50	8070.	III	220	1362.46	II
12	2368.25	II	10	3845.8	II	15	8328.	II	70	1600.46	I
100	2376.	IV	30	3863.9	II	15	8388.	II	120	1600.73	I
190	2400.88	I	100	3868.	IV	30	8532.	II	160	1623.58	II
75 h	2414.6	III	40 h	4079.1	II	2	8544.54	I	110	1623.77	II
10	2501.0	II	10	4097.2	II	1	8579.74	I	220	1624.02	II
25	2515.69	I	140	4121.53	I	25	8653.	II	70	1624.16	II
70	2524.49	I	140	4121.86	I	2	8754.88	I	160	1624.34	II
20 h	2544.5	II	75 h	4259.4	II	3	8761.54	I	100	1663.04	I
700	2627.91	I	25	4272.0	II	25	8863.	II	150	1666.87	I
100	2629.	IV	70 h	4301.7	II	2	8907.81	I	200	1667.29	I
100	2677.	IV	12 h	4339.8	II	2000 d	9657.04	I	150	1817.86	I
12	2693.0	II	25 h	4340.5	II	40	9827.78	I	200	1818.37	I
280 c	2696.76	I	12 h	4379.4	II	20	10104.5	I	300	1825.91	I
20	2713.3	II	25 h	4476.8	II	15	10138.8	I	300	1826.41	I
140 d	2730.50	I	60 h	4705.3	II	20	10300.6	I	110	1842.81	II
100	2767.	IV	600 c	4722.52	I	20	10536.19	I	20	1953.83	III
100	2772.	IV	30	4730.3	II	50	11072.44	I	550	2065.78	III
360	2780.52	I	20	4749.7	II	1500 d	11710.37	I	250	2066.38	I
100	2786.	IV	40 h	4797.4	III	40	11999.49	I	250	2066.65	I
15	2803.42	II	12	4908.2	II	200	12165.08	I	100	2066.93	I
11	2803.70	II	10	4916.6	II	200	12690.04	I	300	2067.19	I
12	2805.3	II	12	4969.7	II	100	12817.8	I	450	2067.23	III
140 c	2809.62	I	20	4993.6	II	200	14330.5	I	160	2077.09	III
100	2842.	IV	45 h	5079.3	III	50	16001.5	I	500	2088.91	I
80 h	2855.6	III	10	5091.6	II	60	22551.6	I	500	2089.57	I
4000	2897.98	I	50 h	5124.3	II				70	2220.30	II
									40	2234.09	III

Intensity	Wavelength/Å		Intensity	Wavelength/Å		Intensity	Wavelength/Å		Intensity	Wavelength/Å	
70	2234.59	III	1000	619.87	IV	500	3041.18	IV	1200	7827.23	I
40	2323.03	II	1000	630.14	IV	500	3074.42	III	2500 s	7881.45	I
40	2328.67	II	1000	642.23	IV	500	3349.64	III	2500	7881.57	I
40	2393.20	II	1000	661.53	IV	500	3380.56	IV	2500	7925.81	I
220	2395.05	II	1000	683.51	IV	500	3540.16	III	30000 c	7938.68	I
40	2459.69	II	1000	697.72	IV	500	3562.43	III	3000	7947.94	I
40	2459.90	II	1000	715.39	IV	1200	3815.65	I	3000	7950.18	I
1000	2496.77	I	1000	731.00	IV	1500	3992.36	I	8000	7978.44	I
1000	2497.73	I	1000	800.12	IV	1000	4223.89	II	10000	7978.57	I
70	2524.7	IV	700	812.95	V	2000	4365.14	I	30000	7989.94	I
160	2530.3	IV	1000	813.66	IV	1000	4365.60	II	2000	8026.35	I
450	2821.68	IV	1000	850.81	V	1500	4425.14	I	2500	8026.54	I
70	2824.57	IV	1000	889.23	II	10000	4441.74	I	30000	8131.52	I
285	2825.85	IV	1000	948.97	II	10000	4472.61	I	1000 c	8152.65	I
160	2918.08	II	1000	1015.54	II	20000	4477.72	I	10000	8153.75	I
110	3032.26	II	1000	1049.00	II	1000	4490.42	I	25000	8154.00	I
70	3179.33	II	1000	1069.15	V	3000	4513.44	I	5000	8246.86	I
110	3323.18	II	900	1112.13	V	15000	4525.59	I	15000	8264.96	I
110	3323.60	II	1000	1143.56	V	3000	4575.74	I	75000 c	8272.44	I
450	3451.29	II	1000	1189.28	I	2500	4614.58	I	20000	8334.70	I
285	4121.93	II	1000	1189.50	I	2500	4752.28	I	10000	8343.70	I
110	4194.79	II	1000	1210.73	I	4000	4780.31	I	1200	8384.04	I
40	4242.98	III	1000	1221.13	I	1600	4785.19	I	40000	8446.55	I
70	4243.61	III	1000	1223.24	I	4000	4979.76	I	4000	8477.45	I
110	4472.10	II	1200	1224.41	I	1200	5395.48	I	1500	8513.38	I
110	4472.85	II	1200	1226.90	I	1200	5466.22	I	1000	8557.73	I
220	4487.05	III	7500	1232.43	I	1800	5852.08	I	1000	8566.28	I
360	4497.73	III	1200	1243.90	I	1600	5940.48	I	20000	8638.66	I
70	4784.21	II	1500	1251.66	I	2400	6122.14	I	4000	8698.53	I
110	4940.38	II	1000	1255.80	I	40000	6148.60	I	10000 c	8793.47	I
110	6080.44	II	1500	1259.20	I	2000	6177.39	I	15000	8819.96	I
70	6285.47	II	1200	1261.66	I	1500	6335.48	I	25000	8825.22	I
70	7030.20	II	1200	1266.20	I	60000	6350.73	I	4000	8888.98	I
40	7031.90	II	1000	1279.48	I	2500	6410.32	I	30000	8897.62	I
110	7835.25	III	1000	1286.26	I	1800	6483.56	I	6000	8932.40	I
70	7841.41	III	3000	1309.91	I	1000	6514.62	I	1800	8949.39	I
20	8667.22	I	3000	1316.74	I	20000	6544.57	I	9000	8964.00	I
70	8668.57	I	1000	1317.37	I	1500	6548.09	I	30000	9166.06	I
800	11660.04	I	2000	1317.70	I	50000 c	6559.80	I	15000	9173.63	I
570	11662.47	I	12000	1384.60	I	1000	6571.31	I	20000	9178.16	I
125	15629.08	I	3000	1449.90	I	1800	6579.14	I	40000	9265.42	I
200	16240.38	I	50000	1488.45	I	20000	6582.17	I	15000	9320.86	I
250	16244.67	I	30000	1531.74	I	1500	6620.47	I	6000	9793.48	I
235	18994.33	I	25000	1540.65	I	50000 c	6631.62	I	10000	9896.40	I
			30000	1574.84	I	20000	6682.28	I	3000	10140.08	I
Bromine Br *Z = 35*			20000	1576.39	I	10000	6692.13	I	6000	10237.74	I
700	379.73	IV	25000	1582.31	I	8000	6728.28	I	1000	10299.62	I
700	400.37	IV	75000	1633.40	I	2000	6760.06	I	1500	10377.65	I
800	482.11	V	1000	2133.79	IV	2000	6779.48	I	30000	10457.96	I
900	531.97	V	1000	2145.02	IV	2200	6786.74	I	1000	10742.14	I
1000	545.43	IV	1000	2257.21	IV	6500	6790.04	I	3000	10755.92	I
1000	547.90	V	1000	2272.73	IV	1600 c	6791.48	I	1700	13217.17	I
1000	559.76	IV	1000	2307.40	IV	1800	6861.15	I	1800	14354.57	I
1000	569.19	IV	1000	2408.16	IV	10000	7005.19	I	1250	14888.70	I
1000	576.59	IV	1000	2411.58	IV	2000	7260.45	I	1800	16731.19	I
1000	585.10	IV	700	2491.14	IV	10000	7348.51	I	1200	18568.31	I
1000	586.71	IV	1000	2581.19	IV	40000	7512.96	I	3500	19733.62	I
1000	597.51	IV	600	2661.40	IV	1600	7591.61	I	1000	20281.73	I
1000	600.09	IV	1000	2842.88	IV	1800	7595.07	I	1000	20624.67	I
1000	601.27	IV	1100 h	2907.71	IV	2000	7616.41	I	1200	21787.24	I
1000	607.03	IV	500 h	2972.26	II	30000	7803.02	I	4000	22865.65	I
1000	617.85	IV									

Intensity	Wavelength/Å		Intensity	Wavelength/Å		Intensity	Wavelength/Å		Intensity	Wavelength/Å	
1000	23513.15	I	75	1793.40	III	30	2823.19	II	1000	5378.13	II
500	28346.50	I	40	1823.41	III	200	2836.900	I	200	5381.89	II
500	30380.85	I	100	1827.70	II	25	2856.46	II	40	5843.30	II
600	31630.13	I	50	1844.66	III	100	2868.180	I	50	5880.22	II
150	38345.75	I	40	1851.13	III	200 r	2880.767	I	300	6099.142	I
120	39964.36	I	40	1855.85	III	50 r	2881.224	I	100	6111.49	I
			200	1856.67	III	200	2914.67	II	100	6325.166	I
Cadmium Cd Z = 48			150	1874.08	III	50	2927.87	II	30	6330.013	I
50	427.01	IV	300	1922.23	II	200	2929.27	II	400	6354.72	II
50	447.85	IV	100	1943.54	II	1000 r	2980.620	I	500	6359.98	II
60	480.90	IV	40	1965.54	II	200 r	2981.362	I	2000	6438.470	I
70	493.00	IV	30	1986.89	II	50	2981.845	I	400	6464.94	II
70	495.13	IV	200	1995.43	II	50	3030.60	II	25	6567.65	II
70	498.14	IV	100	2007.49	II	150	3080.822	I	500	6725.78	II
70	498.53	IV	50	2032.45	II	25	3081.48	II	100	6759.19	II
80	504.09	IV	75	2036.23	II	30	3082.593	I	30	6778.116	I
70	504.20	IV	40	2039.83	III	100	3092.34	II	50	7237.01	II
70	504.50	IV	50	2045.61	III	200	3133.167	I	100	7284.38	II
80	506.31	IV	75	2087.91	III	50	3146.79	II	1000	7345.670	I
60	508.01	IV	150	2096.00	II	150	3250.33	II	50	8066.99	II
50	508.95	IV	50	2111.60	III	300	3252.524	I	5	8200.309	I
70	509.55	IV	1000 r	2144.41	II	300	3261.055	I	20	9289.	I
70	511.40	IV	50	2155.06	II	50	3343.21	II	15	11652.	I
80	513.00	IV	100	2187.79	II	50	3385.49	II	35	14487.	I
70	514.50	IV	1000	2194.56	II	30	3388.88	II	80	15708.	I
60	519.42	IV	1000	2265.02	II	800	3403.652	I	55 d	19120.	I
80	524.41	IV	1500 r	2288.022	I	50	3417.49	II	25	24371.	I
70	524.47	IV	1000	2312.77	II	50	3442.42	II	35	25448.	I
70	525.10	IV	200	2321.07	II	100	3464.43	II			
60	525.19	IV	40	2376.82	II	1000	3466.200	I	**Calcium Ca Z = 20**		
70	527.07	IV	50	2418.69	II	800	3467.655	I	250	190.46	V
80	531.09	IV	50	2469.73	II	25	3483.08	II	250	196.97	V
80	531.51	IV	40	2487.93	II	150	3495.44	II	300	199.55	V
70	534.29	IV	40	2495.58	II	25	3499.952	I	250	200.51	V
70	536.77	IV	50	2509.11	II	100	3524.11	II	265	257.98	V
60	540.90	IV	30	2516.22	II	100	3535.69	II	400	267.77	V
70	541.74	IV	25 h	2525.196	I	1000	3610.508	I	300	270.31	V
80	542.60	IV	50	2544.613	I	800	3612.873	I	400	280.99	V
80	546.55	IV	50	2551.98	II	60	3614.453	I	300	284.98	V
60	553.06	IV	25	2553.465	I	20	3649.558	I	450 c	286.96	V
80	554.05	IV	3	2565.789	I	10	3981.926	I	500	322.17	V
60	567.01	IV	500	2572.93	II	100	4029.12	II	300	323.22	V
150	1118.16	IV	50	2580.106	I	200	4134.77	II	300	330.94	V
100	1164.65	IV	30	2592.026	I	50	4141.49	II	300	334.55	V
100	1183.40	IV	25 h	2602.048	I	100	4285.08	II	250 c	342.45	IV
100	1256.00	II	50	2628.979	I	8	4306.672	I	250	343.93	IV
150	1296.43	II	40	2632.190	I	100	4412.41	II	450	352.92	V
100	1326.50	II	75	2639.420	I	3	4412.989	I	250	377.18	V
60	1370.48	IV	40	2659.23	II	1000	4415.63	II	200	387.08	V
150	1370.91	II	50 h	2660.325	I	30	4440.45	II	750	425.00	V
60	1418.89	IV	25	2668.20	II	8	4662.352	I	600	434.57	IV
200	1514.26	II	50	2672.62	II	200	4678.149	I	250	437.77	IV
50	1545.17	III	100	2677.540	I	30	4744.69	II	750	443.82	IV
200	1571.58	II	25	2677.748	I	300	4799.912	I	500	450.57	IV
100	1668.60	II	50	2707.00	II	50	4881.72	II	500	558.60	V
50	1702.47	II	75	2712.505	I	50	5025.50	II	400	637.93	V
40	1707.16	III	50	2733.820	I	1000 h	5085.822	I	300	643.12	V
40	1722.95	III	1000	2748.54	II	6	5154.660	I	400	646.57	V
50	1724.41	II	100 h	2763.894	I	100	5268.01	II	750	656.00	IV
40	1747.67	III	50 h	2764.230	I	100	5271.60	II	300	656.76	V
40	1773.06	III	50	2774.958	I	1000	5337.48	II	500	669.70	IV
100	1785.84	II							24	1341.89	II

Intensity	Wavelength/Å		Intensity	Wavelength/Å		Intensity	Wavelength/Å		Intensity	Wavelength/Å	
12	1342.54	II	20	4489.18	II	40	7820.78	II	250	371.75	III
20	1433.75	II	19	4499.88	III	60	7843.38	II	150	371.78	III
20	1545.29	III	23	4526.94	I	20	8017.50	II	650	384.03	IV
60	1649.86	II	22	4578.55	I	20	8020.50	II	700	384.18	IV
20	1807.34	II	23	4581.40	I	70	8133.05	II	500	386.203	III
40	1814.50	II	23	4581.47	I	100	8201.72	II	400	419.52	IV
40	1838.01	II	24	4585.87	I	110	8248.80	II	500	419.71	IV
60	1840.06	II	24	4585.96	I	70	8254.73	II	200	450.734	III
20	1843.09	II	20	4685.27	I	130	8498.02	II	400	459.46	III
40	1850.69	II	30	4716.74	II	170	8542.09	II	500	459.52	III
17	2123.03	III	40	4721.03	II	160	8662.14	II	570	459.63	III
16	2152.43	III	40	4799.97	II	100	8912.07	II	250	511.522	III
16	2687.76	III	25	4878.13	I	110	8927.36	II	250	535.288	III
19	2881.78	III	70	5001.48	II	110	9213.90	II	300	538.080	III
21	2899.79	III	80	5019.97	II	90	9312.00	II	350	538.149	III
19	2924.33	III	40	5021.14	II	100	9319.56	II	400	538.312	III
20	2988.63	III	23	5041.62	I	110	9320.65	II	350	574.281	III
10	3006.86	I	25	5188.85	I	25	9416.97	I	9	595.022	II
15	3028.59	III	22	5261.71	I	100	9567.97	II	30	687.053	II
3	3055.32	I	23	5262.24	I	110	9599.24	II	50	687.345	II
19	3119.67	III	22	5264.24	I	80	9601.82	II	10	858.092	II
170	3158.87	II	24	5265.56	I	80	9854.74	II	20	858.559	II
180	3179.33	II	25	5270.27	I	110	9890.63	II	30	903.624	II
150	3181.28	II	60	5285.27	II	90	9931.39	II	60	903.962	II
20	3316.51	II	70	5307.22	II	100	10223.04	II	150	904.142	II
12	3361.92	I	50	5339.19	II	20	10343.81	I	30	904.480	II
19	3372.67	III	27	5349.47	I	20	11838.99	II	800	977.03	III
20	3461.87	II	23	5512.98	I	25	12816.04	I	9	1009.86	II
13	3487.60	I	25	5581.97	I	24	12823.86	I	10	1010.08	II
18	3537.77	III	27	5588.76	I	25	12909.10	I	10	1010.37	II
20	3644.41	I	24	5590.12	I	30	13033.57	I	80	1036.337	II
30	3683.70	II	26	5594.47	I	21	13086.44	I	150	1037.018	II
40	3694.11	II	25	5598.49	I	24	13134.95	I	150	1157.910	I
170	3706.03	II	24	5601.29	I	20	16150.77	I	150	1158.019	I
180	3736.90	II	24	5602.85	I	22	16157.36	I	150	1158.035	I
20	3755.67	II	30	5857.45	I	21	16197.04	I	370	1174.93	III
30	3758.39	II	27	6102.72	I	20	18925.47	I	350	1175.26	III
230	3933.66	II	29	6122.22	I	24	18970.14	I	330	1175.59	III
220	3968.47	II	22	6161.29	I	30	19046.14	I	500	1175.71	III
50	4097.10	II	30	6162.17	I	48	19309.20	I	350	1175.99	III
60	4109.82	II	22	6163.76	I	49	19452.99	I	370	1176.37	III
30	4110.28	II	24	6166.44	I	47	19505.72	I	150	1188.992	I
40	4206.18	II	26	6169.06	I	50	19776.79	I	150	1189.447	I
50	4220.07	II	28	6169.56	I	35	19853.10	I	200	1189.631	I
50	4226.73	I	35	6439.07	I	34	19862.22	I	300	1193.009	I
24	4283.01	I	30	6449.81	I	23	19917.19	I	300	1193.031	I
22	4289.36	I	22	6455.60	I	24	19933.70	I	300	1193.240	I
22	4298.99	I	80	6456.87	II	25	22624.93	I	300	1193.264	I
25	4302.53	I	34	6462.57	I	30	22651.23	I	300	1193.393	I
20	4302.81	III	29	6471.66	I				150	1193.649	I
23	4307.74	I	32	6493.78	I	*Carbon C Z = 6*			150	1193.679	I
22	4318.65	I	28	6499.65	I	110	34.973	V	100	1194.064	I
20	4355.08	I	23	6572.78	I	450	40.268	V	100	1194.488	I
19	4399.59	III	30	6717.69	I	110	227.19	V	100	1261.552	I
25	4425.44	I	33	7148.15	I	250	244.91	IV	250	1277.245	I
26	4434.96	I	31	7202.19	I	160	248.66	V	250	1277.282	I
25	4435.69	I	33	7326.15	I	160	248.74	V	300	1277.513	I
30	4454.78	I	30	7575.81	II	200	289.14	IV	300	1277.550	I
28	4455.89	I	60	7581.11	II	250	289.23	IV	200	1280.333	I
20	4456.61	I	80	7601.30	II	570	312.42	IV	100	1311.363	I
20	4472.04	II	20	7602.32	II	500	312.46	IV	9	1323.951	II
						250	371.69	III			

Intensity	Wavelength/Å	
120	1329.578	I
120	1329.600	I
150	1334.532	II
300	1335.708	II
100	1354.288	I
150	1355.84	I
120	1364.164	I
100	1459.032	I
200	1463.336	I
120	1467.402	I
150	1481.764	I
1000	1548.202	IV
900	1550.774	IV
150	1560.310	I
400	1560.683	I
400	1560.708	I
100	1561.341	I
400	1561.438	I
150	1656.266	I
120	1656.928	I
300	1657.008	I
120	1657.380	I
120	1657.907	I
150	1658.122	I
500	1751.823	I
1000	1930.905	I
250	2162.94	III
40	2270.91	V
5	2277.25	V
20	2277.92	V
800	2296.87	III
800	2478.56	I
250	2509.12	II
350	2512.06	II
200 l	2524.41	IV
300 s	2529.98	IV
250 h	2574.83	II
150	2697.75	III
110 l	2724.85	III
150 l	2725.30	III
150 l	2725.90	III
350 l	2741.28	II
250	2746.49	II
1000	2836.71	II
800	2837.60	II
200	2982.11	III
800 h	2992.62	II
350	3876.19	II
350	3876.41	II
350	3876.66	II
570	3918.98	II
800	3920.69	II
150	4056.06	III
200	4067.94	III
250	4068.91	III
250	4070.26	III
250	4074.52	II
350 l	4075.85	II
150	4162.86	III
250 h	4186.90	III
800	4267.00	II

Intensity	Wavelength/Å	
1000	4267.26	II
200	4325.56	III
600	4647.42	III
520	4650.25	III
375	4651.47	III
200 w	4658.30	IV
200	4665.86	III
200	4771.75	I
200	4932.05	I
5	4943.88	V
5	4944.56	V
200	5052.17	I
350	5132.94	II
350	5133.28	II
350	5143.49	II
570	5145.16	II
400	5151.09	II
300	5380.34	I
250	5648.07	II
350	5662.47	II
450	5695.92	III
250	5801.33	IV
200	5811.98	IV
150	5826.42	III
570	5889.77	II
350	5891.59	II
200	6001.13	I
250	6006.03	I
110	6007.18	I
150	6010.68	I
300	6013.22	I
250	6014.84	I
800	6578.05	II
570	6582.88	II
200	6587.61	I
150	6744.38	III
250	6783.90	II
150 h	7037.25	III
250	7113.18	I
250	7115.19	I
250	7115.63	II
200	7116.99	I
350	7119.90	II
800	7231.32	II
1000	7236.42	II
150	7612.65	III
90 w	7726.2	IV
200	7860.89	I
200	8058.62	I
300 h	8196.48	III
150	8332.99	III
520	8335.15	I
300	8500.32	III
250	9061.43	I
200	9062.47	I
200	9078.28	I
250	9088.51	I
450	9094.83	I
300	9111.80	I
800	9405.73	I
150	9603.03	I

Intensity	Wavelength/Å	
250	9620.80	I
300	9658.44	I
200	10683.08	I
300	10691.25	I
12	11619.29	I
23	11628.83	I
13	11658.85	I
47	11659.68	I
24	11669.63	I
85	11748.22	I
142	11753.32	I
114	11754.76	I
11	11777.54	I
17	11892.91	I
30	11895.75	I
26	12614.10	I
20	13502.27	I
38	14399.65	I
16	14403.25	I
61	14420.12	I
12	14429.03	I
13	14442.24	I
12	16559.66	I
50	16890.38	I
10	17338.56	I
11	17448.60	I
13	18139.80	I
23	19721.99	I

Cerium Ce Z = 58

Intensity	Wavelength/Å	
300	399.36	V
200	482.96	V
40	741.79	IV
30	754.60	IV
75	1332.16	IV
75	1372.72	IV
100	2000.42	IV
100	2009.94	IV
10000	2318.64	III
10000	2372.34	III
10000	2380.12	III
10000	2431.45	III
15000	2439.80	III
10000	2454.32	III
10000	2469.95	III
10000	2483.82	III
10000	2497.50	III
20000	2531.99	III
10000	2603.59	III
340	2651.01	II
270	2830.90	II
250	2874.14	II
10000	2923.81	III
10000	2931.54	III
400	2976.91	II
10000	3022.75	III
50000	3031.58	III
95000	3055.59	III
20000	3056.56	III
40000	3057.23	III
20000	3057.58	III
680	3063.01	II

Intensity	Wavelength/Å	
40000	3085.10	III
20000	3106.98	III
30000	3110.53	III
30000	3121.56	III
20000	3141.29	III
20000	3143.96	III
20000	3147.06	III
710	3194.83	II
990	3201.71	II
710	3218.94	II
880	3221.17	II
710	3227.11	II
20000	3228.57	III
710	3234.16	II
990	3272.25	II
20000	3353.29	III
10000	3395.77	III
30000	3427.36	III
40000	3443.63	III
30000	3454.39	III
40000	3459.39	III
60000	3470.92	III
710	3485.05	II
50000	3497.81	III
60000	3504.64	III
770	3539.08	II
50000	3544.07	III
1200	3560.80	II
1000	3577.45	II
1800	3655.85	II
880	3660.64	II
880	3667.98	II
1000	3709.29	II
1000	3709.93	II
1400	3716.37	II
800	3728.42	II
860	3786.63	II
2500	3801.52	II
800	3803.09	II
1000	3808.11	II
1100	3838.54	II
860	3848.59	II
860	3853.15	II
1200	3854.18	II
1200	3854.31	II
1100	3878.36	II
1500	3882.45	II
1000	3889.98	II
770	3907.29	II
980	3912.44	II
770	3918.28	II
770	3931.09	II
770	3940.34	II
2000	3942.15	II
2700	3942.75	II
770	3943.89	II
3100	3952.54	II
980	3956.28	II
770	3960.91	II
770	3967.05	II
770	3978.65	II

Intensity	Wavelength/Å	
770	3984.68	II
700	3992.39	II
910	3993.82	II
2800	3999.24	II
910	4003.77	II
2700	4012.39	II
910	4014.90	II
840	4024.49	II
840	4028.41	II
840	4031.34	II
2100	4040.76	II
910	4042.58	II
700	4053.51	II
1100	4071.81	II
1800	4073.48	II
1500	4075.71	II
1500	4075.85	II
910	4083.23	II
770	4118.14	II
980	4123.87	II
980	4127.37	II
2700	4133.80	II
2000	4137.65	II
770	4142.40	II
980	4149.94	II
1400	4151.97	II
1300	4165.61	II
3500	4186.60	II
840	4198.72	II
910	4202.94	II
1500	4222.60	II
770	4227.75	II
980	4239.92	II
1100	4248.68	II
2000	4289.94	II
1500	4296.67	II
770	4300.33	II
770	4306.72	II
980	4337.77	II
700	4349.79	II
910	4364.66	II
910	4382.17	II
700	4386.84	II
1700	4391.66	II
980	4418.78	II
770	4449.34	II
2400	4460.21	II
1400	4471.24	II
700	4479.36	II
700	4483.90	II
840	4486.91	II
770	4523.08	II
840	4527.35	II
840	4528.47	II
840	4539.75	II
2100	4562.36	II
1100	4572.28	II
840	4593.93	II
1700	4628.16	II
310	4737.28	II
470	5079.68	II

Intensity	Wavelength/Å	
280	5159.69	I
280	5161.48	I
370	5187.46	II
260	5223.46	I
260	5245.92	I
340	5274.23	II
450	5353.53	II
300	5393.40	II
280	5409.23	II
260	5512.08	II
300	5696.99	I
370	5699.23	I
240	5719.03	I
230	5940.86	I
55	6001.90	I
55	6005.86	I
55	6006.82	I
75	6013.42	I
110	6024.20	I
10000	6032.54	III
110	6043.39	II
55	6047.40	I
10000	6060.91	III
45	6098.34	II
45	6123.67	I
35	6143.36	II
35	6186.17	I
35	6208.98	I
35	6228.94	I
23	6232.45	II
28	6237.45	I
45	6272.05	II
35	6295.58	I
28	6299.51	II
23	6300.21	I
35	6310.01	I
35	6343.95	II
35	6371.11	II
28	6386.84	I
23	6393.02	II
35	6430.07	I
23	6436.40	I
35	6458.03	I
28	6467.39	I
35	6473.72	I
23	6513.59	II
45	6555.65	I
23	6579.10	I
22	6612.06	I
30	6628.93	I
22	6652.72	II
26	6700.66	I
35	6704.27	I
30	6774.28	II
35	6775.59	I
30	6924.81	I
30	6986.02	I
35	7061.75	II
35	7086.35	II
22	7238.36	II
25	7252.75	I

Intensity	Wavelength/Å	
25	7329.91	I
25	7397.77	I
25	7616.11	II
25	7689.17	II
22	7844.94	II
22	7857.54	II
30	8025.56	II
25	8772.14	II
30	8891.20	II

Cesium Ce Z = 55

Intensity	Wavelength/Å	
10000	614.01	III
2000	638.17	III
2500	666.25	III
5000	691.60	III
3500	703.89	III
15000	718.14	II
20000	721.79	III
20000	722.20	III
5000	731.56	III
12000	740.29	III
15000	808.76	II
15000	813.84	II
7500	830.39	III
35000	901.27	II
15000	920.35	III
40000	926.66	II
25000 c	1054.79	III
17 c	1673.99	III
12	1705.25	III
10	1801.83	III
20 c	1822.40	III
11	1823.93	III
12	1824.70	III
12	1841.80	III
25	1915.50	III
25 c	1923.29	III
12	1961.33	III
17	1996.56	III
710	2035.11	III
120	2056.43	III
330	2076.43	III
540	2077.30	III
410	2088.68	III
210	2101.63	III
200	2141.47	III
1000	2316.88	III
230	2325.95	III
390	2340.49	III
1600	2455.81	III
1600	2477.57	III
890	2485.45	III
410	2495.07	III
1400	2525.67	III
430	2573.05	III
16000	2596.86	III
390	2610.12	III
6200	2630.51	III
370	2700.32	III
710	2701.20	III
390	2776.44	III
270	2810.87	III

Intensity	Wavelength/Å	
630	2845.70	III
3100	2859.32	III
200	2893.85	III
180	2921.13	III
3200	2976.86	III
210	3001.28	III
1700	3066.59	III
1100 c	3149.36	III
1400	3152.36	III
8400	3268.32	III
1300	3315.51	III
550	3340.60	III
430	3344.02	III
1200	3349.46	III
400	3463.45	III
580	3476.83	III
480	3559.82	III
7200	3597.45	III
1300	3608.31	III
2300	3618.19	III
300 c	3641.34	III
520	3651.08	III
4800	3661.40	III
640	3699.50	III
430	3837.46	III
2100 c	3876.15	I
2900	3888.37	III
600 c	3888.61	I
2700	3925.60	III
680 c	4001.70	III
3100	4006.55	III
420	4006.78	III
520	4043.42	III
14000	4264.70	II
18000 w	4277.13	II
370	4403.86	III
1200	4410.22	III
940	4425.68	III
530	4471.48	III
12000	4501.55	II
1200	4506.72	III
590	4522.86	III
20000	4526.74	II
1000 c	4555.28	I
460 c	4593.17	I
99900	4603.79	II
420 h	4620.61	III
210	4665.52	III
25000	4830.19	II
140	4851.59	III
19000	4870.04	II
37000	4952.85	II
370	5035.72	III
27000	5043.80	II
75000	5227.04	II
29000	5249.38	II
11000	5274.05	II
10000 c	5349.13	II
22000	5370.99	II
230	5380.79	III
60 c	5465.94	I

Intensity	Wavelength/Å		Intensity	Wavelength/Å		Intensity	Wavelength/Å		Intensity	Wavelength/Å	
37	5502.88	I	1500	17012.32	I	2500	728.951	II	12000	1396.527	I
39000	5563.02	II	760	20138.47	I	2000	777.562	II	500	1441.470	II
100	5635.21	I	880	22811.86	I	5000	787.580	II	500	1528.569	II
210 c	5664.02	I	1100	23037.98	I	5000	788.740	II	500	1542.942	II
27	5745.72	I	3900	23344.47	I	5000	793.342	II	500	1558.144	II
24000	5831.14	II	4400	24251.21	I	500	834.84	IV	500	1565.050	II
59 c	5838.83	I	850	24374.96	I	500	834.97	IV	600	1822.50	III
300	5845.14	I	890 d	25763.51	I	6000	839.297	II	500	1828.40	III
51000	5925.63	II	500	25764.73	I	8000	839.599	II	500	1857.488	II
140	5950.14	III	680 c	29310.06	I	600	840.93	IV	500	1901.61	III
110	5979.97	III	2800	30103.27	I	5000	851.691	II	500	1983.61	III
640 c	6010.49	I	610 c	30953.06	I	2000	888.026	II	450 h	1997.370	II
86	6034.09	I	1100	34900.13	I	2000	893.549	II	450	2032.116	II
150	6043.99	III	190	36131.00	I	2000	961.499	II	350 h	2088.583	II
870	6079.86	III	2 c	39177.28	I	500	973.21	IV	350 h	2091.458	II
9800	6128.61	II	2 d	39421.25	I	600	977.56	IV	700	2253.07	III
330	6150.42	III	1	39424.11	I	40	978.284	I	500	2268.95	III
1000	6213.10	I				700	984.95	IV	500	2278.34	III
170	6217.60	I	*Chlorine Cl* Z = 17			25	998.372	I	700	2283.93	III
450	6242.96	III	500	392.43	V	25	998.432	I	600	2323.50	III
320 c	6354.55	I	800	486.17	IV	75	1002.346	I	500	2336.45	III
510	6456.33	III	800	534.73	IV	500	1005.28	III	600	2340.64	III
8300	6495.53	II	700	535.67	IV	600	1008.78	III	600	2359.67	III
10000 w	6536.44	II	600	536.15	IV	150	1013.664	I	600	2370.37	III
490	6586.51	I	900	537.61	IV	700	1015.02	III	700	2416.42	III
97	6628.66	I	500	538.03	V	90	1025.553	I	600	2447.14	III
8800	6646.57	II	600	538.12	IV	6000	1063.831	II	600	2448.58	III
3300 c	6723.28	I	800	542.23	V	3000	1067.945	II	500	2486.91	III
9600	6724.47	II	600	542.30	V	9000	1071.036	II	500	2532.48	III
400	6753.12	III	1000	545.11	V	6000	1071.767	II	600	2580.67	III
200	6824.65	I	600	546.33	V	5000	1075.230	II	500	2603.59	III
300	6870.45	I	1000	547.63	V	5000	1079.080	II	500	2632.67	III
37000	6955.50	II	500	549.22	IV	200	1084.667	I	500	2633.18	III
4800	6973.30	I	700	552.02	IV	200	1085.171	I	600	2665.54	III
16000	6979.67	II	600	553.30	IV	250	1085.304	I	700	2710.37	III
980	6983.49	I	700	554.62	IV	400	1088.06	I	500	2724.03	IV
13000 w	7149.54	II	600	556.23	III	350	1090.271	I	500	2751.23	IV
1900 c	7219.60	III	700	556.61	III	250	1090.982	I	700	2782.47	IV
790	7228.53	I	700	557.12	III	250	1092.437	I	600	2965.56	III
130	7279.90	I	350	559.305	II	400	1094.769	I	500	3063.13	IV
1100	7279.96	I	700	561.53	III	350	1095.148	I	600	3076.68	IV
2600 c	7608.90	I	700	561.68	III	350	1095.662	I	600	3104.46	III
3300	7943.88	I	700	561.74	III	400	1095.797	I	800	3139.34	III
22000	7997.44	II	400	571.904	II	250	1096.810	I	900	3191.45	III
3500	8015.73	I	800	574.406	II	300	1097.369	I	700	3289.80	III
510	8078.94	I	500	601.50	IV	200	1098.068	I	700	3320.57	III
4500	8079.04	I	500	604.59	IV	200	1099.523	I	800	3329.06	III
59000 c	8521.13	I	500	606.35	III	500	1107.528	I	900	3340.42	III
15000 c	8761.41	I	700	618.057	II	800	1139.214	II	800	3392.89	III
61000 c	8943.47	I	600	619.982	II	800	1167.148	I	800	3393.45	III
18000	9172.32	I	800	620.298	II	3000	1179.293	I	900	3530.03	III
5200	9208.53	I	700	626.735	II	1200	1188.774	I	800	3560.68	III
19000	10024.36	I	800	635.881	II	900	1201.353	I	900	3602.10	III
4800	10123.41	I	1000	636.626	II	3000	1335.726	I	800	3612.85	III
26000	10123.60	I	1000	650.894	II	10000	1347.240	I	700	3622.69	III
2900	13424.31	I	1000	659.811	II	5000	1351.657	I	700	3656.95	III
38000 c	13588.29	I	1300	661.841	II	12000	1363.447	I	700	3670.28	III
8400	13602.56	I	2000	663.074	II	2500	1373.116	I	700	3682.05	III
5700	13758.81	I	1500	682.053	II	20000	1379.528	I	600	3705.45	III
55000 c	14694.91	I	1500	687.656	II	25000	1389.693	I	600	3707.34	III
820	16535.63	I	1500	693.594	II	20000	1389.957	I	800	3720.45	III
			2000	725.271	II						

Intensity	Wavelength/Å		Intensity	Wavelength/Å		Intensity	Wavelength/Å		Intensity	Wavelength/Å	
800	3748.81	III	2300	7899.31	I	310	13296.0	I	130	2383.33	I
500	3779.35	III	1800	7915.08	I	550	13346.8	I	140	2408.62	I
10000	3850.99	II	3000	7924.645	I	525	13821.7	I	170	2496.31	I
25000	3860.83	II	2100	7933.89	I	294	14931.7	I	110	2502.53	I
500	3925.87	III	1700	7935.012	I	269	15108.0	I	190	2504.31	I
700	3991.50	III	650	7952.52	I	381	15465.1	I	110	2516.92	I
600	4018.50	III	1500	7974.72	I	1094	15520.3	I	390	2519.52	I
600	4059.07	III	1300	7976.97	I	1487	15730.1	I	190	2527.12	I
500	4104.23	III	600	7980.60	I	2780	15869.7	I	160	2549.54	I
500	4106.83	III	2900	7997.85	I	277	15883.3	I	130	2560.69	I
10000 h	4132.50	II	2200	8015.61	I	342	15928.9	I	150	2571.74	I
500	4608.21	III	1100	8023.33	I	735	15960.0	I	100	2577.65	I
40	4623.938	I	400	8051.07	I	283	15970.5	I	380	2591.85	I
50	4654.040	I	1700	8084.51	I	259	16198.5	I	250	2653.59	II
80	4661.208	I	2200	8085.56	I	717	19755.3	I	250	2658.59	II
45	4691.523	I	3000	8086.67	I	100	24470.0	I	320	2663.42	II
40	4721.255	I	1300	8087.73	I		39716.0	I	440	2666.02	II
45	4740.729	I	2500	8194.42	I		40085.5	I	280	2668.71	II
13000	4781.32	II	2200	8199.13	I		40089.5	I	350	2671.81	II
99000	4794.55	II	2200	8200.21	I		40532.2	I	280	2672.83	II
29000	4810.06	II	800	8203.78	I				1800	2677.16	II
16000	4819.47	II	18000	8212.04	I	*Chromium Cr Z = 24*			320	2678.79	II
81000	4896.77	II	3000	8220.45	I	100	438.62	V	230	2687.09	II
47000	4904.78	II	20000	8221.74	I	100	464.02	V	280	2691.04	II
26000	4917.73	II	18000	8333.31	I	100	620.66	IV	180	2698.41	II
10000	4995.48	II	99900	8375.94	I	100	629.26	IV	180	2698.69	II
26000	5078.26	II	400	8406.199	I	80	630.30	IV	110	2701.99	I
30	5099.789	I	15000	8428.25	I	100	666.55	IV	140	2712.31	II
56000	5217.94	II	2200	8467.34	I	100	693.92	IV	170	2722.75	II
23000	5221.36	II	2200	8550.44	I	60	1030.47	III	420 h	2726.51	I
15000	5392.12	II	20000	8575.24	I	100	1033.69	III	280 h	2731.91	I
99000	5423.23	II	750	8578.02	I	100	1036.03	III	170 h	2736.47	I
10000	5423.51	II	75000	8585.97	I	80	1055.89	IV	250	2743.64	II
19000	5443.37	II	450	8628.54	I	80	1068.41	III	110 h	2748.29	I
10000	5444.21	II	300	8641.71	I	100	1116.48	V	330	2748.98	II
40	5532.162	I	3500	8686.26	I	150	1121.07	V	390	2750.73	II
50 d	5796.305	I	2200	8912.92	I	150	1127.63	V	280	2751.87	II
45	5799.914	I	3000	8948.06	I	100	1263.50	V	110 h	2752.88	I
30	5856.742	I	2000	9038.982	I	100	1417.42	IV	150	2757.10	I
50	6019.812	I	2500	9045.43	I	150	1465.86	V	350	2757.72	II
200	6140.245	I	1000	9069.656	I	150	1497.97	V	750	2762.59	II
160	6194.757	I	2000	9073.17	I	170	1519.03	V	750	2766.54	II
150	6434.833	I	7500	9121.15	I	220	1579.70	V	250 h	2769.92	I
300	6932.903	I	3000	9191.731	I	170	1591.72	V	610	2780.70	I
300	6981.886	I	500	9197.596	I	150	1603.19	V	180	2822.37	II
600	7086.814	I	4000	9288.86	I	120	1672.66	IV	180	2830.47	II
7500	7256.62	I	1500	9393.862	I	120	1758.51	IV	2500	2835.63	II
5000	7414.11	I	3500	9452.10	I	140	1802.72	IV	110	2840.02	II
550	7462.370	I	500	9486.964	I	130	1812.41	IV	1700	2843.25	II
550	7489.47	I	1000	9584.801	I	200	1837.44	V	1200	2849.84	II
700	7492.118	I	3500	9592.22	I	140	1873.89	IV	120	2851.36	II
11000	7547.072	I	250	9632.509	I	140	1967.18	IV	880	2855.68	II
2300	7672.42	I	1000	9702.439	I	120	1972.07	IV	610	2858.91	II
450	7702.828	I	250	9744.426	I	19000	2055.52	II	440	2860.93	II
7000	7717.581	I	200	9807.057	I	14000	2061.49	II	790	2862.57	II
10000	7744.97	I	400	9875.970	I	8900	2065.42	II	750	2865.11	II
2200	7769.16	I	331	10392.549	I	200	2226.72	III	610	2866.74	II
650	7771.09	I	300	11123.05	I	200	2235.91	III	480	2867.65	II
2200	7821.36	I	269	11409.69	I	150	2237.59	III	210	2870.44	II
1700	7830.75	I	1000	11436.33	I	150	2244.10	III	110	2871.63	I
3000	7878.22	I	350	13243.8	I	150	2284.44	III	160	2873.48	II
						150	2324.88	III			

Intensity	Wavelength/Å		Intensity	Wavelength/Å		Intensity	Wavelength/Å		Intensity	Wavelength/Å	
320	2875.99	II	220	3197.08	II	110	3807.93	I	110	4374.16	I
230	2876.24	II	170	3209.18	II	180	3815.43	I	530	4384.98	I
180	2877.98	II	140	3217.40	II	180	3819.56	I	110	4458.54	I
120	2879.27	I	120	3245.54	I	130	3826.42	I	660	4496.86	I
170	2887.00	I	130	3251.84	I	130	3830.03	I	380	4526.47	I
700	2889.29	I	130	3257.82	I	380	3841.28	I	380	4530.74	I
370	2893.25	I	130	3339.80	II	190	3848.98	I	240	4535.72	I
190	2894.17	I	110	3342.59	II	140	3849.36	I	240	4540.50	I
210	2896.75	I	170	3358.50	II	290	3850.04	I	240	4540.72	I
180	2905.49	I	160	3360.30	II	140	3852.22	I	140	4544.62	I
260	2909.05	I	430	3368.05	II	190	3854.22	I	600	4545.96	I
260	2910.90	I	140	3382.68	II	110	3855.29	I	120	4565.51	I
250	2911.14	I	170	3403.32	II	140	3855.57	I	120	4571.68	I
480	2967.64	I	360	3408.76	II	260	3857.63	I	360	4580.06	I
480	2971.11	I	210	3421.21	II	660	3883.29	I	360	4591.39	I
210	2971.91	II	270	3422.74	II	570	3885.22	I	480	4600.75	I
480	2975.48	I	140	3433.31	II	380	3886.79	I	240	4613.37	I
190	2979.74	II	270	3433.60	I	260	3894.04	I	600	4616.14	I
350	2980.79	I	160	3436.19	I	360	3902.92	I	550	4626.19	I
110	2985.32	II	140	3441.44	I	960	3908.76	I	1600	4646.17	I
480	2985.85	I	170	3445.62	I	120 hd	3911.82	I	570	4651.28	I
1500	2986.00	I	170	3447.43	I	120	3915.84	I	840	4652.16	I
2100	2986.47	I	190	3453.33	I	190	3916.24	I	240 d	4698.46	I
660	2988.65	I	130	3455.60	I	1900	3919.16	I	190	4708.04	I
160	2989.19	II	100	3460.43	I	600	3921.02	I	240	4718.43	I
480	2991.89	I	120	3550.64	I	600	3928.64	I	120	4730.71	I
230	2994.07	I	130	3566.16	I	410	3941.49	I	140	4737.35	I
300	2995.10	I	130	3573.64	I	1900	3963.69	I	340	4756.11	I
700	2996.58	I	330 h	3574.80	I	120	3969.06	I	190	4789.32	I
210	2998.79	I	19000	3578.69	I	1600	3969.75	I	120	4801.03	I
1100	3000.89	I	160 h	3584.33	I	1600	3976.66	I	110	4829.38	I
750	3005.06	I	130	3585.30	II	960	3983.91	I	140	4870.80	I
140	3013.03	I	17000	3593.49	I	190	3984.34	I	130	4887.01	I
710	3013.71	I	350	3601.67	I	160	3989.99	I	260	4922.27	I
710	3014.76	I	13000	3605.33	I	960	3991.12	I	110	4936.33	I
1400	3014.92	I	130	3632.84	I	160	3991.67	I	70	4942.50	I
710	3015.19	I	350	3636.59	I	190	3992.84	I	110	4954.81	I
2800	3017.57	I	630	3639.80	I	160	4001.44	I	60	5013.32	I
430	3018.50	I	220	3641.83	I	120	4012.47	II	70	5166.23	I
240	3018.82	I	220	3649.00	I	120	4026.17	I	70	5184.59	I
430	3020.67	I	170	3653.91	I	190	4039.10	I	70	5192.00	I
2800	3021.56	I	220	3656.26	I	160	4048.78	I	85	5196.44	I
1100	3024.35	I	130	3663.21	I	120	4058.77	I	5300	5204.52	I
170	3029.16	I	120	3685.55	I	140	4126.52	I	8400	5206.04	I
710	3030.24	I	130	3686.80	I	120	4153.82	I	11000	5208.44	I
140	3031.35	I	130	3687.25	I	140	4163.62	I	85	5224.94	I
390	3034.19	I	130	3730.81	I	170	4174.80	I	290	5247.56	I
550	3037.04	I	150	3732.03	I	170	4179.26	II	530	5264.15	I
550	3040.85	I	480	3743.58	I	110	4209.37	I	180	5265.72	I
110	3050.14	II	570	3743.88	I	20000	4254.35	I	95 h	5275.17	I
710	3053.88	I	340	3749.00	I	110	4263.14	I	70 h	5276.03	I
240	3118.65	II	230	3757.66	I	16000	4274.80	I	340	5296.69	I
430	3120.37	II	260	3768.24	I	10000	4289.72	I	70 h	5297.36	I
470	3124.94	II	130	3791.38	I	780	4337.57	I	660	5298.27	I
120	3128.70	II	130	3792.14	I	1100	4339.45	I	85	5300.75	I
590	3132.06	II	120	3793.29	I	380	4339.72	I	340 h	5328.34	I
140	3136.68	II	130	3793.88	I	1900	4344.51	I	70 h	5329.17	I
140	3147.23	II	140	3797.13	I	380	4351.05	I	780	5345.81	I
100	3155.15	I	200	3797.72	I	2300	4351.77	I	380	5348.32	I
100	3163.76	I	530	3804.80	I	570	4359.63	I	40	5400.61	I
240	3180.70	II	110	3806.83	I	530	4371.28	I	1400	5409.79	I

Intensity	Wavelength/Å		Intensity	Wavelength/Å		Intensity	Wavelength/Å		Intensity	Wavelength/Å	
24	5628.64	I	1800	1955.17	I	200	2347.39	II	800	2580.32	II
7	5642.36	I	1500	1958.55	I	1600	2352.85	I	300 d	2582.22	II
24	5664.04	I	1500	1961.59	I	200 d	2353.41	II	500	2587.22	II
24	5694.73	I	1500 h	1968.69	I	2000	2353.42	I	500	2587.52	II
40	5698.33	I	1500 h	1968.93	I	500	2363.80	II	200	2588.91	II
24	5702.31	I	3000	1970.71	I	400	2378.62	II	100 p	2605.71	II
24	5712.78	I	1800 h	1971.16	I	1400	2380.48	I	100	2612.50	II
24 h	5783.11	I	1800 h	1972.52	I	200	2381.76	II	100	2614.36	II
30 h	5783.93	I	1500	1973.85	I	300 p	2383.45	II	100 p	2628.77	II
24 h	5785.00	I	1800	1976.97	I	1400	2384.86	I	100	2632.26	II
19 h	5785.82	I	2400 h	1980.89	I	200	2386.36	II	100	2636.07	II
60 h	5787.99	I	1500	1989.80	I	500	2388.92	II	310	2646.42	I
180 h	5791.00	I	1800	1990.34	I	200	2397.38	II	770	2648.64	I
35	6330.10	I	1500 l	1998.49	I	1100 d	2402.06	I	100	2653.72	II
22	6362.87	I	1500	2002.32	I	200 p	2404.16	II	100	2663.53	II
19	6661.08	I	900	2008.04	I	5300	2407.25	I	200	2666.73	II
21 h	6883.03	I	50	2011.51	II	5300	2411.62	I	100	2675.85	II
27 h	6924.13	I	1200 h	2014.58	I	1600	2412.76	I	100	2684.42	II
30 h	6978.48	I	900	2016.17	I	4800	2414.46	I	100	2702.02	II
85	7355.90	I	50	2022.35	II	4800	2415.30	I	200	2706.62	II
130	7400.21	I	50	2027.04	II	300	2417.65	II	200	2707.35	II
150	7462.31	I	900	2031.96	I	4100	2424.93	I	190	2715.99	I
40	8947.15	I	1500	2039.95	I	3300	2432.21	I	100	2727.78	II
19	8976.83	I	1200	2041.11	I	2900	2436.66	I	80	2734.54	II
			50	2065.54	II	2400	2439.05	I	190	2745.10	I
Cobalt Co Z = 27			1500 h	2077.76	I	200	2442.63	II	100	2753.22	II
20	355.52	V	900	2085.67	I	200 d	2446.03	II	190	2764.19	I
18	355.88	V	900	2087.55	I	200 p	2447.69	II	100	2766.70	II
12	356.06	V	900	2089.35	I	200	2450.00	II	100	2774.97	II
66	609.16	IV	900	2093.40	I	200	2464.20	II	100	2791.00	II
70	609.21	IV	900	2094.86	I	200	2486.44	II	100	2793.73	II
64	609.28	IV	900	2095.77	I	200	2498.82	II	150	2815.56	I
10	1018.36	V	1200	2097.51	I	570	2504.52	I	80	2835.63	II
10	1021.14	V	1500	2104.73	I	500	2506.46	II	80	2847.35	II
15	1231.73	V	1500	2106.80	I	360	2506.88	I	80	2871.22	II
50	1277.01	V	900	2108.98	I	200	2511.16	II	190	2886.44	I
80	1299.58	II	900 s	2117.68	I	860	2517.87	I	100	2918.38	II
80	1306.95	II	900	2137.78	I	500	2519.82	II	100	2930.24	II
50	1345.67	V	900	2138.97	I	4300	2521.36	I	100	2954.73	II
1000	1696.01	III	900	2163.03	I	200 h	2524.65	II	690	2987.16	I
800	1697.99	III	1100	2174.60	I	300	2524.97	II	690	2989.59	I
1000	1707.35	III	200	2193.60	II	500	2528.62	II	60	3022.59	II
5000	1760.35	III	200	2256.73	II	2900	2528.97	I	3100	3044.00	I
5000	1773.57	III	150	2260.00	II	200 p	2530.09	II	1700	3061.82	I
2000	1780.05	III	200	2283.52	II	720	2530.13	I	80	3387.70	II
3000	1782.97	III	1000	2286.15	II	860	2532.18	I	1100	3388.17	I
1000	1787.08	III	200	2291.98	II	200 d	2533.82	II	2200	3395.38	I
1000	1789.07	III	300 d	2293.38	II	2900	2535.96	I	11000	3405.12	I
1000	1823.08	III	300	2301.40	II	860	2536.49	I	4500	3409.18	I
2000	1830.09	III	800 d	2307.85	II	300	2541.94	II	6700	3412.34	I
2000	1831.44	III	2600	2309.02	I	1700	2544.25	I	2200	3412.63	I
5000	1835.00	III	500	2311.60	II	200	2546.74	II	2700	3417.16	I
1500	1842.34	I	500	2314.05	II	340	2548.34	I	50	3423.84	II
1800	1847.89	I	300	2314.96	II	310	2553.37	I	2500	3431.58	I
1800	1852.71	I	200 p	2317.06	II	310	2555.07	I	4500	3433.04	I
2400	1855.05	I	2400	2323.14	I	300	2559.41	II	1600	3442.93	I
2000	1863.83	III	300 p	2324.31	II	200	2560.03	II	8800	3443.64	I
1500	1878.28	I	200 d	2326.11	II	960	2562.15	I	50	3446.39	II
1800	1936.58	I	500	2326.47	II	500	2564.04	II	4100	3449.17	I
1500	1946.79	I	1400	2335.99	I	1100	2567.35	I	2100	3449.44	I
1500	1951.90	I	1600	2338.67	I	960	2574.35	I	21000	3453.50	I
1800	1954.22	I									

Intensity	Wavelength/Å		Intensity	Wavelength/Å		Intensity	Wavelength/Å		Intensity	Wavelength/Å	
1000	3455.23	I	90	4549.66	I	200	974.759	II	250	1418.426	II
5100	3462.80	I	140	4565.59	I	250	977.567	II	250	1421.759	II
5100	3465.80	I	190	4581.60	I	100	987.657	II	200	1427.829	II
8000	3474.02	I	120	4629.38	I	250	992.953	II	400	1430.243	II
1900	3483.41	I	85	4663.41	I	300	1004.055	II	250	1434.904	II
4800	3489.40	I	110	4792.86	I	300	1008.569	II	150	1436.236	II
2400	3495.69	I	100	4840.27	I	300	1008.728	II	150	1442.139	II
50	3501.72	II	150	4867.88	I	300	1010.269	II	200	1445.984	II
9600	3502.28	I	80 h	4964.18	II	250	1012.597	II	200	1449.058	II
7000	3506.32	I	50	5212.71	I	500	1018.707	II	250	1450.304	II
50	3507.77	II	50	5230.22	I	500	1027.831	II	200	1452.294	II
2900	3509.84	I	50	5247.93	I	250	1028.328	II	300	1458.002	II
1400	3510.43	I	50	5342.71	I	200	1030.263	II	250	1459.412	II
4800	3512.64	I	50	5352.05	I	600	1036.470	II	200	1463.752	II
3800	3513.48	I				600	1039.348	II	400	1463.838	II
4800	3518.35	I	*Copper Cu Z = 29*			600	1039.582	II	200	1466.070	II
1300	3520.08	I	80	685.141	II	800	1044.519	II	400	1470.697	II
2700	3521.57	I	100	709.313	II	800	1044.744	II	200	1472.395	II
3800	3523.43	I	100	718.179	II	500	1049.755	II	250	1473.978	II
60	3523.51	II	150	724.489	II	600	1054.690	II	200	1474.935	II
6400	3526.85	I	200	735.520	II	400	1055.797	II	150	1476.059	II
2700	3529.03	I	250	736.032	II	600	1056.955	II	300 r	1481.23	III
7300	3529.81	I	80	779.295	II	400	1058.799	II	200	1481.544	II
1900	3533.36	I	100	797.455	II	600	1059.096	II	200	1485.328	II
50	3545.03	II	150	810.998	II	600	1060.634	II	750	1488.831	II
1100	3560.89	I	200	813.883	II	600	1063.005	II	300	1492.834	II
80	3561.07	II	300	826.996	II	200	1065.782	II	250	1493.366	II
8800	3569.38	I	150	848.808	II	200	1066.134	II	250	1495.430	II
50	3574.95	II	250	851.303	II	500	1069.195	II	350	1496.687	II
1600	3574.96	I	250	858.487	II	300	1073.745	II	150	1503.368	II
60	3575.32	II	400	861.994	II	200	1088.395	II	250	1504.757	II
2500	3575.36	I	400	865.390	II	300	1094.402	II	200	1505.388	II
60	3577.96	II	250	869.336	II	250	1097.053	II	300	1508.632	II
1000	3585.16	I	150	873.263	II	150	1119.947	II	350	1510.506	II
6700	3587.19	I	200	876.723	II	200	1142.640	II	200	1512.465	II
1900	3594.87	I	250	877.012	II	300	1144.856	II	200	1513.366	II
1600	3602.08	I	200	877.555	II	100	1250.048	II	500	1514.492	II
100	3621.21	II	500	878.699	II	150	1265.506	II	200	1517.631	II
1000	3627.81	I	100	884.133	II	300	1275.572	II	500	1519.492	II
80	3643.61	II	250	885.847	II	150	1282.455	II	600	1519.837	II
60	3681.35	II	600	886.943	II	150	1287.468	II	200	1520.540	II
1100	3745.50	I	600	890.567	II	150	1298.395	II	200	1524.860	II
1400	3842.05	I	500	892.414	II	300	1308.297	II	150	1525.764	II
6900	3845.47	I	800	893.678	II	300	1314.337	II	500	1531.856	II
5500	3873.12	I	400	894.227	II	100	1320.686	II	300	1532.131	II
2800	3873.96	I	600	896.759	II	100	1326.395	II	250	1533.986	II
7900	3894.08	I	400	896.976	II	150	1350.594	II	250	1535.002	II
1500	3935.97	I	600	901.073	II	250	1351.837	II	500	1537.559	II
80 h	3963.10	II	400	906.113	II	150	1355.305	II	200	1540.239	II
6000	3995.31	I	800	914.213	II	300	1358.773	II	300	1540.389	II
970	3997.91	I	600	922.019	II	200	1359.009	II	300	1540.588	II
350	4020.90	I	500	924.239	II	200	1362.600	II	750	1541.703	II
370	4045.39	I	400	935.232	II	250	1367.951	II	400	1544.677	II
350	4066.37	I	600	935.898	II	200	1371.840	II	100	1547.958	II
830	4092.39	I	600	943.335	II	300 r	1376.79	III	300	1550.653	II
550	4110.54	I	600	945.525	II	200 r	1377.49	III	300	1551.389	II
2800	4118.77	I	500	945.965	II	100	1393.128	II	500	1552.646	II
4400	4121.32	I	200	954.383	II	100	1398.642	II	250	1553.896	II
90	4190.71	I	250	956.290	II	150	1402.777	II	400	1555.134	II
90	4469.56	I	400	958.154	II	150	1407.169	II	500	1555.703	II
690	4530.96	I	200	960.414	II	100	1414.898	II	300	1558.345	II
			250	968.042	II						

Intensity	Wavelength/Å		Intensity	Wavelength/Å		Intensity	Wavelength/Å		Intensity	Wavelength/Å	
400	1565.924	II	900	2135.981	II	800	2769.669	II	500	4043.751	II
400	1566.415	II	400	2148.984	II	200	2791.795	II	2000	4062.64	I
100	1569.416	II	150	2161.320	II	170	2799.528	II	120	4068.106	II
300	1579.492	II	1300 r	2165.09	I	100	2810.804	II	500	4131.363	II
300	1580.626	II	250	2174.982	II	1250 r	2824.37	I	200	4143.017	II
400	1581.995	II	1600 r	2178.94	I	350	2837.368	II	300	4153.623	II
500	1583.682	II	700	2179.410	II	100	2857.748	II	500	4161.140	II
400	1590.165	II	1700 r	2181.72	I	600	2877.100	II	370	4164.284	II
600	1593.556	II	700	2189.630	II	270	2884.196	II	400	4171.851	II
500 r	1593.75	III	900	2192.268	II	2500 r	2961.16	I	500	4179.512	II
400	1598.402	II	400	2195.683	II	100	2986.335	II	500	4211.866	II
400	1602.388	II	1700 r	2199.58	I	2000	2997.36	I	320	4230.449	II
200	1604.848	II	1300 r	2199.75	I	2000	3010.84	I	200	4255.635	II
300	1605.281	II	100	2200.509	II	2500	3036.10	I	950	4275.11	I
400	1606.834	II	200	2209.806	II	2500	3063.41	I	300	4279.962	II
250	1608.639	II	750	2210.268	II	1400	3073.80	I	500	4292.470	II
150	1610.296	II	1600 r	2214.58	I	1500	3093.99	I	400	4365.370	II
200	1617.915	II	250	2215.106	II	1250	3099.93	I	100	4444.831	II
600	1621.426	II	1000 r	2215.65	I	2000	3108.60	I	400	4506.002	II
400	1622.428	II	750	2218.108	II	1400 h	3126.11	I	150	4516.049	II
250	1630.268	II	2100 r	2225.70	I	1500	3194.10	I	150	4541.032	II
100	1636.605	II	150	2226.780	II	1400	3208.23	I	500	4555.920	II
1000 r	1642.21	III	1600 r	2227.78	I	1500 h	3243.16	I	100	4596.906	II
250	1649.458	II	350	2228.868	II	10000 r	3247.54	I	120	4649.271	II
30 r	1655.32	I	2500 r	2230.08	I	10000 r	3273.96	I	2000	4651.12	I
200	1656.322	II	1100 r	2238.45	I	1400 h	3282.72	I	120	4661.363	II
200	1660.001	II	900	2242.618	II	400	3290.418	II	320	4671.702	II
300	1663.002	II	2300 r	2244.26	I	1500 h	3290.54	I	300	4673.577	II
100	1672.776	II	1000	2247.002	II	110	3300.881	II	450	4681.994	II
30	1688.09	I	1300 r	2260.53	I	250	3301.229	II	100	4758.433	II
30	1691.08	I	2200 r	2263.08	I	2500 h	3307.95	I	400	4812.948	II
30 r	1703.84	I	150	2263.786	II	200	3316.276	II	120	4851.262	II
50 r	1713.36	I	200	2276.258	II	1500	3337.84	I	300	4854.988	II
150	1717.721	II	100	2286.645	II	150	3338.648	II	100	4873.304	II
50 r	1725.66	I	2500 r	2293.84	I	200	3365.648	II	150	4901.427	II
100	1736.551	II	170	2294.368	II	450	3370.454	II	1000	4909.734	II
50 r	1741.57	I	1000	2303.12	I	300	3374.952	II	500	4918.376	II
150	1753.281	II	150	2369.890	II	200	3380.712	II	200	4926.424	II
200 r	1774.82	I	2500 r	2392.63	I	100	3384.945	II	900	4931.698	II
100 r	1825.35	I	120	2403.337	II	1250 h	3483.76	I	120	4943.026	II
250	1929.751	II	1500	2406.66	I	1250	3524.23	I	700	4953.724	II
250	1944.597	II	1000 r	2441.64	I	2000	3530.38	I	500	4985.506	II
100	1946.493	II	100	2485.792	II	1400	3599.13	I	400	5006.801	II
200	1957.518	II	2000 r	2492.15	I	1400	3602.03	I	350	5009.851	II
150	1970.495	II	150	2506.273	II	1000	3686.555	II	400	5012.620	II
150	1977.027	II	120	2526.593	II	150	3786.270	II	350	5021.279	II
500	1979.956	II	300	2544.805	II	170	3797.849	II	200	5039.016	II
300	1989.855	II	100	2571.756	II	100	3818.879	II	300	5047.348	II
250	1999.698	II	150	2590.529	II	140	3826.921	II	900	5051.793	II
270	2035.854	II	200	2600.270	II	160	3864.137	II	400	5058.910	II
250	2037.127	II	2500 r	2618.37	I	280	3884.131	II	500	5065.459	II
350	2043.802	II	200	2666.291	II	150	3892.924	II	450	5067.094	II
300	2054.980	II	750	2689.300	II	170	3903.177	II	350	5072.302	II
100	2078.663	II	700	2700.962	II	140	3920.654	II	450	5088.277	II
110	2098.398	II	650	2703.184	II	120	3933.268	II	420	5093.816	II
320	2104.797	II	700	2713.508	II	120	3987.024	II	350	5100.067	II
300	2112.100	II	650	2718.778	II	150	3993.302	II	1500	5105.54	I
320	2117.310	II	300	2721.677	II	140	4003.476	II	250	5124.476	II
350	2122.980	II	120	2737.342	II	1250	4022.63	I	2000	5153.24	I
350	2126.044	II	270	2745.271	II	100	4032.647	II	100	5158.093	II
420	2134.341	II	2500 r	2766.37	I	600	4043.484	II	100	5183.367	II

Intensity	Wavelength/Å		Intensity	Wavelength/Å		Intensity	Wavelength/Å		Intensity	Wavelength/Å	
2500	5218.20	I	270	6879.404	II	1200	3156.52	II	830	3577.98	II
100	5269.991	II	220	6937.553	II	670	3162.83	II	440	3580.04	II
100	5276.525	II	150	6952.871	II	1000	3169.99	II	3300	3585.06	II
1650	5292.52	I	150	6977.572	II	470	3215.19	II	1400	3585.78	II
100	5368.383	II	200	7022.860	II	830	3216.63	II	560	3586.11	II
1500	5700.24	I	300	7194.896	II	490	3235.89	II	1100	3591.41	II
1500	5782.13	I	400	7326.008	II	490	3245.12	II	560	3591.81	II
150	5805.989	II	300	7331.694	II	1200	3251.27	II	560	3592.11	II
100	5833.515	II	250	7382.277	II	890	3280.09	II	1800	3595.04	II
200	5897.971	II	1000	7404.354	II	490	3282.77	II	560	3600.38	II
120	5937.577	II	270	7434.156	II	1100	3308.88	II	1800	3606.12	II
400	5941.196	II	500	7562.015	II	780	3316.32	II	440	3618.51	II
100	5993.260	II	700	7652.333	II	1000	3319.88	II	560	3620.16	II
650	6000.120	II	1000	7664.648	II	780	3341.00	II	470	3624.27	II
100	6023.264	II	150	7681.788	II	510	3353.58	II	1100	3629.42	II
250	6072.218	II	450	7744.097	II	510	3368.11	II	4000	3630.24	II
150	6080.343	II	800	7778.738	II	5300	3385.02	II	440	3632.78	II
150	6099.990	II	750	7805.184	II	610	3388.85	II	1100	3640.25	II
160	6107.412	II	1500	7807.659	II	3800	3393.57	II	11000	3645.40	II
300	6114.493	II	1000	7825.654	II	1300	3396.16	II	1000	3648.78	II
600	6150.384	II	350	7860.577	II	5300	3407.80	II	700	3664.62	II
750	6154.222	II	300	7890.567	II	1300	3413.78	II	990	3672.30	II
500	6172.037	II	700	7902.553	II	530	3414.82	II	420	3672.70	II
550	6186.884	II	1500	7933.13	I	780	3419.63	II	1400	3674.08	II
400	6188.676	II	400	7944.438	II	530	3425.06	II	2200	3676.59	II
300	6198.092	II	400	7972.033	II	1900	3434.37	II	640	3678.51	I
470	6204.261	II	1200	7988.163	II	1300	3441.45	II	820	3684.85	I
450	6208.457	II	2000	8092.63	I	3800	3445.57	II	1300	3685.78	I
750	6216.939	II	500	8277.560	II	830	3446.99	II	4700	3694.81	II
700	6219.844	II	800	8283.160	II	2700	3454.32	II	990	3698.21	II
500	6261.848	II	250	8503.396	II	1300	3456.56	II	540	3701.63	II
1000	6273.349	II	750	8511.061	II	4400	3460.97	II	440	3707.57	II
350	6288.696	II	200	8609.134	II	720	3468.43	II	440	3708.22	II
900	6301.009	II	500	9813.213	II	560	3471.14	II	420	3710.07	II
550	6305.972	II	250	9827.978	II	560 d	3471.53	II	1600	3724.45	II
400	6312.492	II	200	9830.798	II	1300	3477.07	II	930	3739.34	I
120	6326.466	II	600	9861.280	II	4400	3494.49	II	1200	3747.82	II
400	6373.268	II	600	9864.137	II	560	3496.34	II	1400	3753.51	II
750	6377.840	II	200	9883.969	II	830	3498.71	II	1400	3753.75	II
400	6403.384	I	550	9916.419	II	830	3504.53	II	1200	3757.05	I
850	6423.884	II	500	9917.954	II	830	3505.45	II	4700	3757.37	II
200	6442.965	II	550	9925.594	II	1300	3506.81	II	640	3767.63	I
750	6448.559	II	450	9938.998	II	560	3517.26	II	640	3773.05	I
170	6466.246	II	500	9960.354	II	4400	3523.98	II	420	3781.47	I
950	6470.168	II	450	10006.588	II	22000	3531.70	II	3300	3786.18	II
750	6481.437	II	550	10022.969	II	4400	3534.96	II	1600	3788.44	II
400	6484.421	II	550	10038.093	II	5500	3536.02	II	700	3791.87	II
220	6517.317	II	650	10054.938	II	4400	3538.52	II	510	3804.14	II
400	6530.083	II	450	10080.354	II	1700	3542.33	II	580	3806.27	II
120	6551.286	II				1400	3546.83	II	470	3812.27	I
200	6577.080	II	*Dysprosium Dy Z = 66*			4400	3550.22	II	470	3813.67	II
750	6624.292	II	260	2356.91	II	2200	3551.62	II	1400	3816.76	II
800	6641.396	II	240	2410.01	II	440 h	3558.23	II	700	3825.68	II
450	6660.962	II	260	2439.84	II	440	3559.30	II	2300	3836.50	II
100	6770.362	II	220	2585.30	I	2200	3563.15	II	1400	3841.31	II
300	6806.216	II	440	2634.80	II	560	3563.69	II	420	3846.34	II
400	6809.647	II	220	2755.75	II	780	3573.83	II	420	3847.02	I
320	6823.202	II	300	2816.39	II	1400	3574.15	II	1200	3853.03	II
250	6844.157	II	390	2913.95	II	4400	3576.24	II	420	3858.40	I
320	6868.791	II	610	3038.28	II	1700	3576.87	II	560	3868.45	II
270	6872.231	II	830	3135.38	II				1600	3868.81	I
			500	3141.14	II						

Intensity	Wavelength/Å		Intensity	Wavelength/Å		Intensity	Wavelength/Å		Intensity	Wavelength/Å	
820	3869.86	II	990	4612.26	I	1000	3070.40	III	3200	3944.42	I
7000	3872.11	II	170	4731.84	II	610	3073.34	II	2700	3973.04	I
1200	3873.99	II	120 h	4775.79	I	720	3082.08	II	3200	3973.58	I
470	3879.11	II	480	4957.34	II	610	3084.02	II	1400	3974.72	II
5800	3898.53	II	70	5022.12	I	770	3122.72	II	810	3977.02	I
540	3914.87	II	160	5042.63	I	1500	3166.25	III	1100	3982.33	I
540	3915.59	II	95	5070.68	I	870	3181.92	II	810	3987.66	I
540 d	3917.29	I	120	5077.67	I	870	3220.73	II	14000	4007.96	I
420	3927.86	I	80	5090.38	II	610	3223.31	II	1100	4012.58	I
540	3930.14	I	80	5110.32	I	2300	3230.58	II	3000	4020.51	I
2100	3931.52	II	130 h	5120.04	I	2700	3264.78	II	1000	4046.96	I
10000	3944.68	II	190	5139.60	II	720	3279.33	II	940	4055.47	II
800	3957.79	II	110	5169.69	II	720	3280.22	II	690	4059.78	II
14000	3968.39	II	80	5185.30	I	2000	3301.23	III	3500	4087.63	I
2700	3978.57	II	290	5192.86	II	2300	3312.42	II	1100	4098.10	I
1400	3981.92	II	95	5197.66	II	770	3323.19	II	6900	4151.11	I
1600	3983.65	II	70	5259.88	I	770	3332.70	II	1000	4190.70	I
800	3984.21	II	130	5260.56	I	1300	3346.04	II	1400	4218.43	I
540	3991.32	II	65	5267.11	I	1400	3364.08	II	690	4286.56	I
1600	3996.69	II	55	5282.07	I	1400 d	3368.02	II	40000	4290.06	III
8000	4000.45	II	160	5301.58	I	7700	3372.71	II	20000	4386.86	III
420	4005.84	I	65	5340.30	I	970	3374.17	II	810	4409.34	I
540	4011.29	II	85	5389.58	II	1700	3385.08	II	1000	4606.61	I
540	4013.82	I	80	5419.13	I	2300	3392.00	II	570	4675.62	II
540	4014.70	II	70	5423.32	I	770	3441.13	II	15000	4735.56	III
420	4027.78	II	95	5451.11	I	970	3471.71	II	2000	4783.12	III
520 d	4028.32	II	65	5547.27	I	610	3479.41	II	250	5007.25	I
520	4032.47	II	100	5639.50	I	970	3485.85	II	200	5035.94	I
420	4033.65	II	55 h	5645.99	I	6700	3499.10	II	210	5042.05	II
420	4036.32	II	80	5652.01	I	610	3502.78	I	120	5124.56	I
12000	4045.97	I	70 h	5718.46	I	610	3524.91	II	130	5127.41	II
1600	4050.56	II	55	5745.53	I	820	3549.84	II	120	5131.53	I
520	4055.14	II	55 h	5868.11	II	1500	3558.02	I	130	5133.83	II
2500	4073.12	II	70	5945.80	I	1000	3559.90	II	170	5164.77	II
7400	4077.96	II	120	5974.49	I	920	3570.75	II	130	5172.78	I
3900	4103.30	II	140	5988.56	I	1000	3580.52	II	160	5188.90	II
860	4103.87	I	140	6088.26	I	610	3590.76	I	150	5206.52	I
1500	4111.34	II	100	6168.43	I	610	3599.50	II	140	5255.93	II
490	4124.63	II	270	6259.09	I	1000	3599.83	II	80	5272.91	I
990	4129.42	II	160	6579.37	I	3100	3616.56	II	90	5348.06	I
1200	4143.10	II	75	6667.86	I	720	3628.04	I	60	5414.63	II
990	4146.06	I	180	6835.42	I	1000	3633.54	II	180	5456.62	I
5700	4167.97	I	80	6852.96	I	1600	3638.68	I	90	5468.32	I
930	4183.72	I	65	6899.32	II	900	3645.94	II	80	5485.97	II
12000	4186.82	I	55	7426.86	II	7900	3692.65	II	80	5593.46	I
2200	4191.64	I	55	7543.73	I	1300	3729.52	II	60	5611.82	I
6800	4194.84	I	80	7662.36	I	900	3742.64	II	70	5622.01	I
800	4198.02	I	100	8201.57	II	900	3747.43	I	80	5626.53	II
680	4201.30	I	45	8791.39	II	1800	3786.84	II	90	5640.36	I
680	4202.24	I				1600	3810.33	I	70	5664.95	I
16000	4211.72	I	*Erbium Er Z = 68*			4000	3816.78	III	70	5719.55	I
1800	4213.18	I	600	2277.65	III	3600	3830.48	II	100	5739.19	I
3700	4215.16	I	290	2586.73	II	680	3855.90	I	290	5762.80	I
4400	4218.09	I	490	2670.26	II	7500	3862.85	I	70	5784.66	I
4400	4221.11	I	500	2739.27	III	1500	3880.61	II	70	5800.79	I
2700	4225.16	I	610	2755.63	II	1200	3882.89	II	430	5826.79	I
1000	4308.63	II	1000	2904.47	II	4200	3892.68	I	100	5850.07	I
540	4409.38	II	1500	2910.36	II	5200	3896.23	I	120	5855.31	I
740	4449.70	II	1500	2964.52	II	11000	3906.31	II	140	5872.35	I
420	4577.78	I	1200	3002.41	II	3200	3937.01	I	120	5881.14	I
2100	4589.36	I	1000	3055.10	III	2100	3938.63	II	8000	5903.30	III

Intensity	Wavelength/Å		Intensity	Wavelength/Å		Intensity	Wavelength/Å		Intensity	Wavelength/Å	
70	6022.56	I	140	2893.03	I	9800	4627.22	I	230	6303.41	II
70	6061.25	I	360	2893.83	I	8300	4661.88	I	120 cw	6350.04	I
60	6076.45	II	3200	2906.68	II	110	4867.62	I	120 cw	6400.93	I
360	6221.02	I	160	2908.99	I	150	4907.18	I	180	6410.04	I
55	6262.56	I	850	2925.04	II	180	4911.40	I	140	6411.32	I
60	6268.87	I	200 cw	2952.68	II	180	5013.17	I	830	6437.64	II
130	6308.77	I	260	2960.21	II	170	5022.91	I	120	6457.96	I
55	6326.13	I	300	2991.33	II	110	5029.54	I	1400	6645.11	II
55	6492.35	I	100 c	3023.93	III	170	5114.37	I	50	6666.35	III
60	6583.48	I	200 c	3026.79	III	170	5129.10	I	140	6802.72	I
70	6601.11	I	320 cw	3054.94	II	210	5133.52	I	360	6864.54	I
70	6759.87	I	120	3058.98	I	270	5160.07	I	120	7040.20	I
35	6790.92	I	220	3077.36	II	210	5166.70	I	330	7077.10	II
70	6848.10	I	120	3097.45	II	200	5199.85	I	570	7194.81	II
55	6865.13	I	320	3106.18	I	110	5200.96	I	570	7217.55	II
55	7459.55	I	950	3111.43	I	120	5206.44	I	540	7301.17	II
120	7469.51	I	120	3130.73	II	750	5215.10	I	720	7370.22	II
35	7680.01	I	50 c	3171.00	III	300	5223.49	I	300	7426.57	II
35	7797.47	I	50 c	3183.78	III	120	5239.24	I	160	7583.91	I
35	7921.85	I	420	3210.57	I	200	5266.40	I	60 cw	7742.57	I
30	7937.84	I	1000	3212.81	I	390	5271.96	I	70	7746.19	I
35	8312.82	I	420	3213.75	I	110	5272.48	I	35	7887.99	I
55	8409.90	I	150	3272.77	II	150	5282.82	I	24 cw	8209.80	I
9	8866.84	II	210	3277.78	II	120	5291.26	I	21 cw	8642.67	I
			150	3301.95	II	120	5294.64	I	18	8870.30	I
Europium Eu Z = 63			140	3308.02	II	540	5357.61	I			
30	2124.69	III	140	3313.33	II	120	5361.61	I	*Fluorine F Z = 9*		
200	2350.51	III	950	3334.33	I	110	5376.94	I	50	148.00	V
4000	2375.46	III	110	3350.40	I	120	5392.94	I	50	163.56	V
100 d	2435.14	III	140	3369.06	II	450	5402.77	I	90	165.98	V
1000	2444.38	III	190	3391.99	II	380	5451.51	I	100	166.18	V
4000	2445.99	III	280	3396.58	II	260	5452.94	I	50	186.84	V
2000	2513.76	III	150	3425.02	II	120	5488.65	I	60	190.57	V
200	2522.14	III	150	3441.00	II	120	5510.52	I	70	190.84	V
160	2564.17	II	130	3461.38	II	200	5547.44	I	50	196.39	IV
110	2568.17	II	470 cw	3521.09	II	150	5570.33	I	60	196.45	IV
230	2577.14	II	150	3542.15	II	200	5577.14	I	70	200.09	IV
1000	2638.77	II	180	3552.52	II	120	5580.03	I	80	201.16	IV
380	2641.27	II	150	3603.20	II	210	5645.80	I	90	208.25	IV
640	2668.34	II	6400	3688.42	II	330	5765.20	I	90	240.08	IV
110	2673.42	II	20000 cw	3724.94	II	180	5783.69	I	100	251.03	IV
250	2678.29	II	350	3741.31	II	170	5818.74	II	140	419.65	IV
250	2685.66	II	260	3761.12	II	600 cw	5830.98	I	150	420.05	IV
550	2692.03	II	39000 cw	3819.67	II	330	5966.07	II	160	420.73	IV
700	2701.14	II	140	3844.23	II	480 cw	5967.10	I	100	429.51	III
800	2701.90	II	190	3865.57	I	170	5972.75	I	110	430.15	III
240	2705.28	II	150	3884.75	I	240	5992.83	I	150	430.76	IV
180	2709.99	I	28000 cw	3907.10	II	110	6012.56	I	90	464.29	III
700	2716.98	II	32000 cw	3930.48	II	420	6018.15	I	120	465.98	IV
4200	2727.78	II	30000 cw	3971.96	II	170	6029.00	I	130	490.57	IV
160	2740.62	II	180	4011.69	II	420	6049.51	II	160	491.00	IV
120	2744.26	II	150	4017.58	II	140	6057.36	I	50	497.38	IV
480	2781.89	II	120	4039.19	I	240	6083.84	I	60	497.83	IV
1900	2802.84	II	120	4085.38	II	240	6099.35	I	70	498.80	IV
220	2811.75	II	33000 cw	4129.70	II	120	6118.78	I	90	506.16	V
3400	2813.94	II	60000 cw	4205.05	II	330	6173.05	II	100	508.08	V
550	2816.18	II	150	4298.73	I	110	6178.76	I	120	508.39	III
2000	2820.78	II	240	4355.09	II	260 cw	6188.13	I	60	514.08	V
400 cw	2828.72	II	14000 cw	4435.56	II	140	6195.07	I	90	525.29	V
260	2859.67	II	3000	4522.57	II	240	6262.25	I	100	526.30	V
280	2862.57	II	11000	4594.03	I	170	6299.77	I	120	567.69	III
200	2892.54	I							110	567.75	III

Intensity	Wavelength/Å		Intensity	Wavelength/Å		Intensity	Wavelength/Å		Intensity	Wavelength/Å	
140	570.64	IV	110	1839.30	III	270	3847.09	II	8000	6870.22	I
140	571.30	IV	120	1839.97	III	260	3849.99	II	15000	6902.48	I
150	571.39	IV	110	1840.14	III	250	3851.67	II	6000	6909.82	I
160	572.66	IV	100	2027.44	III	5	3898.48	I	4000	6966.35	I
90	605.67	II	120	2030.32	III	8	3930.69	I	45000	7037.47	I
100	606.80	II	120	2217.17	III	5	3934.26	I	30000	7127.89	I
90	630.20	III	50	2298.29	IV	5	3948.56	I	15000	7202.36	I
100	647.77	V	40	2451.58	IV	240	4024.73	II	1000	7309.03	I
110	647.87	V	120	2452.07	III	220	4025.01	II	15000	7311.02	I
130	654.03	V	50	2456.92	IV	230	4025.49	II	700	7314.30	I
120	656.12	III	130	2464.85	III	200	4103.51	II	5000	7331.96	I
130	656.87	III	130	2470.29	III	200	4246.23	II	120	7336.77	III
110	657.23	V	120	2478.73	III	200	4299.17	II	130	7354.94	III
140	657.33	V	150	2484.37	III	140 h	4420.30	III	10000	7398.69	I
140	658.33	III	120	2542.77	III	120 h	4427.35	III	4000	7425.65	I
140	676.12	IV	120	2580.04	III	120 h	4432.32	III	2200	7482.72	I
130	677.15	IV	130	2583.81	III	140 h	4479.99	III	2500	7489.16	I
150	677.22	IV	120	2593.23	III	6	4960.65	I	900	7514.92	I
130	678.99	IV	130	2595.53	III	150	5012.54	III	5000	7552.24	I
160	679.21	IV	140	2599.28	III	160	5110.99	III	5000	7573.38	I
60	757.04	V	130	2625.01	III	15	5230.41	I	7000	7607.17	I
150	806.96	I	140	2629.70	III	12	5279.01	I	18000	7754.70	I
125	809.60	I	120	2656.44	III	18	5540.52	I	15000	7800.21	I
500	951.87	I	130	2755.55	III	12	5552.43	I	300	7879.18	I
1000	954.83	I	160	2759.63	III	10	5577.33	I	500	7898.59	I
750	955.55	I	120	2788.15	III	20	5624.06	I	350	7936.31	I
500	958.52	I	160	2811.45	III	12	5626.93	I	300	7956.32	I
20	972.40	I	40	2820.74	IV	15	5659.15	I	80	8016.01	II
350	973.90	I	50	2826.13	IV	40	5667.53	I	1000	8040.93	I
100	976.22	I	140	2833.99	III	90	5671.67	I	900	8075.52	I
40	976.51	I	150	2835.63	III	18	5689.14	I	350	8077.52	I
100	977.75	I	150	2860.33	III	25	5700.82	I	350	8126.56	I
60	1082.31	V	120	2862.86	III	25	5707.31	I	600	8129.26	I
70	1088.39	V	140	2887.58	III	140	5753.17	III	300	8159.51	I
80	1219.03	III	150	2889.45	III	120	5761.20	III	600	8179.34	I
80	1266.87	III	120	2905.30	III	12	5950.15	I	300	8191.24	I
90	1267.71	III	140	2913.29	III	25	5959.19	I	350	8208.63	I
70	1297.54	III	160	2916.34	III	70	5965.28	I	2500	8214.73	I
70	1359.92	III	140	2932.49	III	50	5994.43	I	3000	8230.77	I
110	1498.93	III	140	2994.28	III	150	6015.83	I	500	8232.19	I
120	1502.01	III	120 h	2997.21	III	80	6038.04	I	1500	8274.62	I
110	1504.18	III	130	2997.53	III	900	6047.54	I	2000	8298.58	I
140	1504.79	III	120	2999.47	III	100	6080.11	I	600	8302.40	I
130	1506.30	III	130	3039.25	III	150	6091.82	III	900	8807.58	I
110	1506.77	III	120	3039.75	III	140	6125.50	III	1000	8900.92	I
100	1553.02	III	160	3042.80	III	800	6149.76	I	300	8912.78	I
110	1557.59	III	150	3049.14	III	400	6210.87	I	350	9025.49	I
100	1563.73	III	140	3113.62	III	130	6233.57	III	400	9042.10	I
100	1565.54	III	160	3115.70	III	13000	6239.65	I	350	9178.68	I
100	1623.40	III	180	3121.54	III	10000	6348.51	I	200	9433.67	I
100	1650.76	III	140	3124.79	III	140	6363.05	III	25	9505.30	I
130	1670.39	III	140	3134.23	III	8000	6413.65	I	12	9662.04	I
140	1677.40	III	140	3146.99	III	450	6569.69	I	25	9734.34	I
100	1716.99	III	180	3174.17	III	300	6580.39	I	15	9822.11	I
120	1770.09	III	170	3174.76	III	400	6650.41	I	12	9902.65	I
150	1770.67	III	120	3214.00	III	1800	6690.48	I	80 h	10047.98	II
110	1772.93	III	200	3501.45	II	400	6708.28	I	15	10285.45	I
140	1773.36	III	200	3501.57	II	7000	6773.98	I	20	10862.31	I
160	1791.65	III	200	3502.96	II	1500	6795.53	I			
110	1803.03	III	6	3594.10	I	9000	6834.26	I			
170	1805.90	III	12	3668.17	I	50000	6856.03	I			

Francium Fr Z = 87

	7177.	I

Intensity	Wavelength/Å		Intensity	Wavelength/Å		Intensity	Wavelength/Å		Intensity	Wavelength/Å	
Gadolinium Gd Z = 64			2000	3684.13	I	700	4694.33	I	50	6752.67	II
1200	1007.24	IV	3100	3687.74	II	410	4743.65	I	26	6786.33	II
1200	1063.84	IV	2000	3697.73	II	470	4767.24	I	100	6828.25	I
1600	1476.98	IV	1300	3699.73	II	300	4784.62	I	100	6916.57	I
1500	1705.03	IV	2700	3712.70	II	320	4821.69	I	50	6985.89	II
1600	1706.01	IV	2000	3713.57	I	280	4934.12	I	75	6991.92	I
2000	1736.24	IV	1400	3716.36	II	750	5015.04	I	60	6996.76	II
1500	1815.32	IV	2000	3717.48	I	75	5039.09	I	45	7006.16	II
2200	1975.24	III	1800 d	3719.45	II	5000	5091.70	III	35	7122.57	I
3400	2018.07	III	1500	3730.84	II	130	5098.38	II	170	7168.37	I
2800	2359.31	III	4500	3743.47	II	910	5103.45	I	28	7189.57	II
1400	2397.87	IV	1400	3758.31	II	180	5108.91	II	28	7262.66	I
2800	2697.39	III	8700	3768.39	II	120	5125.56	II	35	7441.85	I
2800	2703.28	III	1400	3770.69	II	860	5155.84	I	40	7464.36	I
2700	2727.89	III	2900	3783.05	I	190	5176.28	II	55	7562.97	I
9000	2904.73	III	5100	3796.37	II	410	5197.77	I	80	7733.50	I
9500	2955.53	III	3700	3813.97	II	280	5219.40	I	35	7846.35	II
1200	2999.04	II	3300	3850.69	II	130	5233.93	I	35	7856.93	I
2100	3010.13	II	5100	3850.97	II	320	5251.18	I	25	7930.25	II
1900	3027.60	II	4300	3852.45	II	120	5252.14	II	18	8146.15	I
2100	3032.84	II	1600	3866.99	I	140	5255.80	I	21	8668.63	I
1600	3034.05	II	1500	3894.70	II	280	5283.08	I	21 h	8832.06	II
2100	3081.99	II	2200	3916.51	II	280	5301.67	I	14 h	8849.14	I
3500	3100.50	II	1200	3934.79	I	220	5302.76	I	18 h	8867.31	I
930	3145.00	II	1400	3945.54	I	280	5307.30	I	5000	14332.88	III
980	3156.53	II	1200	3957.67	II	130	5321.50	I			
980	3161.37	II	1100	4023.14	I	280	5321.78	I	*Gallium Ga Z = 31*		
4000	3176.66	III	1100	4028.15	I	110	5327.32	I	14	294.53	IV
1400	3331.38	II	1400	4037.33	II	170	5333.30	I	61	295.67	IV
1100	3336.18	II	1600	4045.01	I	300	5343.00	I	30	298.44	V
5400	3350.47	II	1300	4049.43	II	200	5348.67	I	30	300.01	V
4300	3358.62	II	2200	4049.86	II	300	5350.38	I	30	302.86	V
5400	3362.23	II	2600	4053.64	I	240	5353.26	I	41	304.99	IV
1100	3392.53	II	2600	4058.22	I	3000	5365.96	III	30	307.03	V
1100 d	3407.56	II	1900	4063.39	II	150	5370.63	I	30	308.26	V
6900	3422.47	II	1300	4078.44	II	4000	5553.30	III	30	311.79	V
1700	3439.21	II	2800	4078.70	I	3000	5587.88	III	30	313.68	V
2700	3439.99	II	1500	4085.56	II	190	5617.91	I	40	319.41	V
1400	3450.38	II	1100	4092.71	I	110	5632.25	I	40	322.31	V
1100	3451.23	II	2600	4098.61	II	260	5643.24	I	50	322.99	V
2700	3463.98	II	2200	4130.37	II	3000	5658.98	III	30	323.10	V
1700	3467.27	II	1100	4132.28	II	390	5696.22	I	40	324.25	V
1700	3468.99	II	2400	4175.54	I	120	5733.86	II	40	324.95	V
1400	3473.22	II	2400	4184.25	II	240	5791.38	I	40	326.14	V
2200	3481.28	II	2200	4190.78	I	220	5851.63	I	30	326.77	V
1700	3481.80	II	1300	4212.00	II	280	5856.22	I	30	328.65	V
1700	3494.40	II	4800	4225.85	I	110	5904.56	I	25	423.18	IV
1400	3505.51	II	1700	4251.73	II	170	5911.45	II	16	439.92	IV
1100	3512.50	II	1600	4262.09	I	85	5930.29	I	50	620.00	III
4300	3545.80	II	1100	4306.34	I	85	5936.84	I	40	622.01	III
3900	3549.36	II	1800	4313.84	I	65	5937.71	I	90	806.51	III
1400	3557.05	II	2600 d	4325.57	II	110 h	5988.02	I	90	817.30	III
5400	3584.96	II	1900	4327.12	I	430	6114.07	I	50	828.70	III
1100	3592.71	II	1000	4344.30	II	75	6305.15	II	20	878.17	V
1100	3604.87	I	2200	4346.46	I	40	6331.35	I	40	973.21	V
6100	3646.19	II	1400	4401.86	I	40	6380.95	II	40	989.75	V
3900	3654.62	II	1400	4422.41	I	40 h	6538.15	I	90	1014.47	V
3100	3656.15	II	1100	4430.63	I	55	6564.78	I	90	1019.71	V
1400	3662.26	II	1100	4519.66	I	50	6634.36	II	120	1050.48	V
2700	3664.60	II	910	4537.81	I	35	6681.23	II	80	1054.56	V
2000	3671.20	II	520	4614.50	I	85	6730.73	I	90	1058.12	V
									80	1066.69	V

Intensity	Wavelength/Å		Intensity	Wavelength/Å		Intensity	Wavelength/Å		Intensity	Wavelength/Å	
60	1069.60	V	85	1338.09	IV				100	1742.195	I
80	1073.77	V	77	1347.03	IV	*Germanium Ge* Z = 32			50	1746.065	I
90	1078.83	V	76	1351.06	IV	700	294.51	V	200	1750.043	I
110	1079.60	V	70	1353.92	III	1000	295.64	V	100	1758.279	I
60	1080.99	V	74	1364.63	IV	200	304.98	V	100 h	1764.185	I
80	1085.00	III	60	1395.54	IV	20	621.52	V	100 h	1765.284	I
250	1085.01	V	77	1402.55	IV	50	724.21	V	50 h	1766.433	I
80	1087.37	V	70	1405.32	IV	60	746.88	V	200	1774.176	I
90	1091.71	V	73	1465.87	IV	60	760.05	V	200	1785.046	I
100	1094.36	V	90	1495.07	III	10	862.234	II	100 h	1793.071	I
80	1095.10	V	50	1534.46	III	15	875.493	II	75 h	1801.432	I
160	1102.83	V	10	1813.98	II	15	905.977	II	200 h	1841.328	I
140	1103.03	V	15	1845.30	II	20	920.554	II	200 h	1842.410	I
60	1105.61	III	20	2091.34	II	300	971.35	V	100 h	1844.410	I
75	1105.62	V	90	2417.70	III	300	990.66	V	100 h	1845.872	I
70	1106.17	V	90	2423.98	III	50	999.101	II	100 h	1846.958	I
80	1118.34	V	15	2424.36	III	300	1004.38	V	200	1853.134	I
120	1126.40	V	10	2632.66	I	100	1016.638	II	500 r	1860.086	I
130	1128.10	V	10	2665.05	I	900	1045.71	V	100	1865.052	I
120	1128.53	V	20	2700.47	II	700	1072.66	V	300 r	1874.256	I
100	1129.94	V	15	2780.15	II	100	1075.072	II	100	1895.197	I
130	1136.07	V	50	3521.77	III	300	1085.51	II	500 r	1904.702	I
67	1137.06	IV	80	3581.19	III	40	1088.45	III	50 h	1908.434	I
130	1150.23	V	100	3589.34	III	800	1089.49	II	30	1912.409	I
90	1150.27	III	10	3731.10	III	200	1098.71	II	300 r	1917.592	I
70	1156.10	IV	10	3806.60	III	500	1106.74	II	100 h	1923.467	I
120	1156.51	V	10	4032.99	I	1000	1116.94	V	500 r	1929.826	I
35	1157.74	V	10	4172.04	I	500	1120.46	II	10 h	1934.048	I
70	1163.60	IV	15	4251.16	II	700	1163.39	V	100 r	1937.483	I
25	1169.40	V	10 h	4254.04	II	200	1164.27	II	500	1938.008	II
75	1170.58	IV	10	4255.77	II	500	1181.19	II	100 r	1938.300	I
48	1171.71	IV	40	4262.00	II	500	1181.65	II	500	1938.891	II
40	1178.95	V	100	4380.69	III	200	1188.73	II	30 s	1944.116	I
68	1185.23	IV	150	4381.76	III	20	1188.99	IV	200	1944.731	I
40	1186.06	IV	100	4863.00	III	300	1191.26	II	200	1955.115	I
73	1190.89	IV	150	4993.78	III	700	1222.30	V	500	1962.013	I
73	1193.02	IV	10	5808.28	III	20	1229.81	IV	30 h	1963.373	I
75	1195.02	IV	20	5848.25	III	500	1237.059	II	30	1965.383	I
69	1201.54	IV	15	5993.51	III	500	1261.905	II	200	1970.880	I
72	1206.89	IV	10	6334.2	II	100	1264.710	II	200	1979.274	II
80	1213.17	V	2000	6396.56	I	200	1401.24	II	300 h	1987.849	I
63	1216.15	IV	1000	6413.44	I	200	1538.091	II	300	1988.267	I
50	1228.03	IV	10 h	7251.4	I	500	1576.855	II	500 r	1998.887	I
60	1236.38	IV	20 h	7403.0	I	75	1581.070	II	200	2011.29	I
60	1238.59	IV	30 h	7464.0	I	100	1602.486	II	1700	2019.068	I
75	1245.53	IV	10 h	7620.5	I	3 r	1615.57	I	2400 r	2041.712	I
83	1258.77	IV	50 h	7734.77	I	2 r	1624.130	I	1600 r	2043.770	I
81	1264.66	IV	100 h	7800.01	I	2 r	1630.173	I	420	2054.461	I
82	1267.15	IV	15 h	8002.55	I	3 r	1636.31	I	220 h	2057.238	I
90	1267.16	III	20 h	8074.25	I	4 r	1639.730	I	750 r	2065.215	I
81	1279.24	IV	100 h	8311.86	I	2	1647.531	I	2600 r	2068.656	I
15	1283.64	V	200 h	8386.49	I	200	1649.194	II	420	2086.021	I
80	1285.33	IV	200 h	9492.92	I	2	1651.528	I	2000 r	2094.258	I
80	1293.46	III	200 h	9493.12	I	4 r	1651.955	I	25	2104.45	III
60	1295.36	III	300 h	9589.36	I	3	1661.345	I	240	2105.824	I
82	1295.86	IV	100 h	10905.95	I	4 r	1663.539	I	95 h	2124.744	I
83	1299.46	IV	400	11949.12	I	10 h	1665.275	I	50 h	2186.451	I
82	1303.53	IV	200	12109.78	I	4	1667.802	I	340 r	2198.714	I
80	1309.68	IV	60	14996.64	I	3 r	1670.608	I	15	2220.375	I
80	1314.82	IV	60	22016.81	I	100 r	1691.090	I	18	2256.001	I
60	1323.15	III	70	22568.71	I	200 r	1716.784	I	18	2314.201	I
						100 h	1739.102	I			

Intensity	Wavelength/Å		Intensity	Wavelength/Å		Intensity	Wavelength/Å		Intensity	Wavelength/Å	
24	2327.918	I	1000	5893.389	II	60	1368.62	I	100	1624.34	I
15	2359.233	I	500	6021.041	II	70	1374.82	I	300 d	1629.13	III
20	2379.144	I	75	6283.452	II	50	1375.76	I	250	1638.88	III
10	2389.472	I	100	6336.377	II	180	1377.73	III	50	1639.90	I
15	2397.885	I	100	6484.181	II	150	1378.69	III	100	1644.17	III
130	2417.367	I	6	6557.488	I	150	1379.98	III	150	1646.67	I
30	2436.412	I	50	7049.369	II	125	1380.53	III	250	1652.74	III
30	2488.25	IV	30	7145.390	II	200	1381.36	III	250	1664.77	III
90	2497.962	I	5	7353.334	I	80	1382.75	I	100	1665.76	I
500	2500.54	II	7	7384.208	I	50	1385.33	I	100	1668.11	III
70	2533.230	I	10	7833.575	I	300	1385.79	III	125	1673.93	III
20	2542.44	IV	10	8031.039	I	100	1389.41	III	1000	1693.94	III
3	2556.298	I	6	8044.165	I	180	1391.46	III	150	1697.09	III
28	2589.188	I	10	8256.013	I	60	1392.27	I	200	1698.98	III
500	2592.534	I	10	8482.21	I	180	1396.00	III	200	1699.34	I
8	2644.184	I	9	8700.60	I	50	1402.12	I	200	1700.00	III
1200	2651.172	I	5	9068.785	I	100	1402.91	III	200	1702.25	III
500	2651.568	I	5	9095.957	I	70	1407.38	I	100	1707.53	III
500	2691.341	I	6	9398.868	I	100	1408.45	I	250	1710.16	III
850	2709.624	I	20	9474.993	II	225	1409.50	III	200	1715.69	III
400	2729.78	II	20	9475.645	II	250	1413.80	III	100	1716.71	III
40	2740.426	I	4	9492.559	I	100	1414.27	III	300	1717.83	III
650	2754.588	I	7	9625.664	I	100	1417.09	III	500	1727.31	III
70	2793.925	I	5	10039.436	I	125	1417.39	III	100 d	1733.17	III
80	2829.008	I	4	10200.952	I	150	1427.42	III	300	1738.48	III
1000	2831.843	II	10	10382.427	I	300	1428.93	III	150	1744.39	III
1000	2845.527	II	10	10404.913	I	80	1429.19	I	500	1746.10	III
750	3039.067	I	8	10734.068	I	250	1430.06	III	500	1756.92	III
600	3067.021	I	8	10947.416	I	275	1433.37	III	500	1761.95	III
20	3124.816	I	10	11125.130	I	50	1435.79	I	300	1767.42	III
35	3211.86	III	230	11252.83	I	250	1435.81	III	100	1774.42	III
40	3255.05	III	600	11714.76	I	300	1439.12	III	800	1775.17	III
110	3269.489	I	1300	12069.20	I	200	1441.21	III	200	1776.40	III
40	3434.03	III	1050	12391.58	I	150	1446.37	III	100	1780.57	III
300	3499.21	II	235	13107.61	I	250	1448.42	III	60	1783.22	II
60	3554.19	IV	470	14822.38	I	250	1454.95	III	300	1786.11	III
50	3676.65	IV	150	16759.79	I	100	1464.72	III	150	1792.65	III
200	4178.96	III	135	17214.34	I	150	1471.28	III	500	1793.76	III
70	4226.562	I	70	18811.86	I	100	1474.73	III	200	1801.98	III
200	4260.85	III	62	19279.24	I	150	1481.10	III	400	1805.24	III
150	4291.71	III	28	20673.64	I	100	1481.76	I	100	1809.81	III
10	4685.829	I				300	1487.15	III	400	1821.17	III
1000	4741.806	II	*Gold Au Z = 79*			250	1487.91	III	400	1844.89	III
1000	4814.608	II	100	843.44	III	200	1489.47	III	150	1848.83	III
50	4824.097	II	100	845.14	III	250	1500.37	III	500	1861.80	III
100	5131.752	II	200	945.10	III	200	1502.47	III	150	1871.92	III
200	5178.648	II	100	1040.63	III	200	1503.74	III	100	1879.83	I
3	5194.583	I	80	1044.49	III	100	1542.00	III	150	1918.28	III
6	5265.892	I	80	1046.81	III	100	1548.50	III	100	1932.04	III
6	5513.263	I	100 h	1239.96	III	200	1567.54	III	100	1935.42	III
8	5564.741	I	100	1278.51	III	200	1574.85	III	200	1948.79	III
8	5607.010	I	100	1314.84	III	200	1579.44	III	100	1958.47	III
6	5616.135	I	100 h	1328.37	I	150	1584.10	III	400	1989.63	III
7	5621.426	I	200	1336.72	III	200	1587.16	I	150	1996.85	III
6	5664.226	I	180	1341.68	III	200	1589.56	III	11000	2012.00	I
5	5664.842	I	100	1348.89	III	150	1593.41	III	2600	2021.38	I
9	5691.954	I	150	1350.32	III	70	1598.24	I	150	2082.09	II
6	5701.776	I	150	1355.61	III	200	1600.51	III	300	2083.09	III
5	5717.877	I	150	1356.13	III	250	1617.16	III	60	2110.68	II
6	5801.029	I	50	1363.98	I	100	1617.78	III	100	2159.08	III
9	5802.093	I	500	1365.40	III	500	1621.93	III	80	2167.33	III
			200	1367.17	III						

Intensity	Wavelength/Å		Intensity	Wavelength/Å		Intensity	Wavelength/Å		Intensity	Wavelength/Å	
200	2172.20	III	100	3927.69	I	370	1437.27	V	200	2537.33	II
100	2184.11	III	400	4040.93	I	500	1437.73	V	320	2551.40	II
500	2188.97	III	700	4065.07	I	370	1445.40	V	400 h	2560.74	III
70	2248.56	II	100	4084.10	I	270	1457.91	V	250	2563.61	II
80	2263.62	II	100	4241.80	I	100	1717.21	IV	300 h	2567.46	III
300	2322.27	III	200	4315.11	I	270	1719.32	V	890	2571.67	II
180	2352.65	I	120 h	4437.27	I	550	1729.08	V	320	2573.90	II
100	2382.40	III	250	4488.25	I	750	1731.83	V	320	2576.82	II
120	2387.75	I	900 h	4607.51	I	750	1733.96	V	300	2578.14	II
150	2402.71	III	100 h	4620.56	I	440	1741.74	V	320	2582.54	II
150	2405.12	III	500	4792.58	I	1000	1749.11	V	390	2606.37	II
2600	2427.95	I	100	4811.60	I	1000	1750.19	V	450	2607.03	II
60	2533.52	II	100	5147.44	I	500	1760.89	V	230	2613.60	II
250	2641.48	I	300	5230.26	I	370	1765.62	V	450	2622.74	II
3400	2675.95	I	100 h	5261.76	I	270	1774.02	V	160	2637.00	I
1100	2748.25	I	100	5655.77	I	6200	2012.78	II	1100	2638.71	II
100	2780.82	I	100 h	5721.36	I	8500	2028.18	II	1100	2641.41	II
1000	2802.04	II	300	5837.37	I	300	2070.94	III	160	2642.75	I
300	2819.79	II	100 h	5862.93	I	1200	2096.18	II	670	2647.29	II
100	2822.55	II	300 h	5956.96	I	200 h	2099.30	III	160	2657.84	II
100 h	2825.44	II	600	6278.17	I	200 h	2110.31	III	210	2661.88	II
300	2837.85	II	100	6562.68	I	200	2155.66	III	290	2683.35	II
100	2846.92	II	600	7510.73	I	200	2183.50	III	200	2687.22	III
100	2856.74	II	10	8145.06	I	200	2195.44	III	670	2705.61	I
300	2883.45	I	10	9254.28	I	540	2210.82	II	210	2712.42	II
300	2891.96	I				200	2234.59	III	250	2718.59	I
100	2893.25	II	*Hafnium Hf* Z = 72			320	2254.01	II	710	2738.76	II
100	2907.04	II	220	545.41	V	160	2255.15	II	200	2743.64	I
300	2913.52	II	180	600.00	V	250	2266.83	II	360	2751.81	II
300	2918.24	II	200	618.27	IV	620	2277.16	II	500	2753.60	III
100	2954.22	II	200	644.54	IV	200 h	2313.44	III	450	2761.63	I
100 h	2990.27	II	400	647.39	IV	230	2321.14	II	160	2766.96	I
300	2994.80	II	600	665.65	IV	580	2322.47	II	170	2773.02	I
320	3029.20	I	200	673.49	IV	300	2323.25	II	980	2773.36	II
300	3065.42	I	270	867.25	V	300	2324.89	II	180	2774.02	II
100	3122.50	II	180	875.88	V	200	2332.97	II	390	2779.37	I
1600	3122.78	I	180	885.58	V	300	2336.47	III	230	2808.00	II
100	3194.72	I	180	896.14	V	200	2337.33	II	230	2813.86	II
100	3227.99	III	180	901.54	V	230	2343.32	II	170	2814.48	II
300	3230.63	I	180	919.10	V	320	2347.44	II	230	2817.68	I
300	3308.30	I	270	921.67	V	540	2351.22	II	200	2819.74	I
300	3309.64	I	245	951.62	V	250	2380.30	II	1200	2820.22	II
100	3309.86	III	180	960.12	V	250	2383.540	III	490	2822.68	II
100	3320.12	I	180	964.74	V	170	2393.18	II	180	2833.28	I
100	3355.15	I	160	971.51	V	450	2393.36	II	410	2845.83	I
100 h	3391.31	I	160	1092.76	V	670	2393.83	II	270	2849.21	II
100	3395.40	I	160	1201.76	V	540	2405.42	II	270	2850.96	I
100	3467.21	I	270	1232.03	V	370	2410.14	II	180	2851.21	II
300	3557.36	I	200	1233.59	V	320	2417.69	II	180	2860.56	I
300	3586.73	I	160	1237.42	V	390	2447.25	II	760	2861.01	II
100	3611.57	I	160	1239.53	V	450	2460.49	II	760	2861.70	II
100 h	3631.31	I	160	1244.46	V	400	2461.74	III	2100	2866.37	I
300	3637.90	I	440	1396.66	V	430	2464.19	II	210	2887.14	I
100 h	3645.02	I	270	1400.09	V	210	2469.18	II	800	2889.62	I
100	3650.74	I	160	1401.70	V	2000	2495.16	III	1800	2898.26	I
100	3709.62	I	370	1407.17	V	290	2496.99	II	1200	2904.41	I
100	3796.01	I	370	1408.38	V	580	2512.69	II	890	2904.75	I
100	3874.73	I	270	1412.28	V	580	2513.03	II	2000	2916.48	I
100 h	3892.26	I	270	1413.51	V	1000	2515.16	III	580	2918.58	I
400	3897.86	I	160	1421.96	V	890	2516.88	II	320	2919.59	II
300	3909.38	I	220	1422.53	V	340	2531.19	II	180	2924.62	I
			370	1433.43	V						

Intensity	Wavelength/Å	
490	2929.63	II
450	2929.90	I
710	2937.80	II
2000	2940.77	I
160	2944.71	I
1200	2950.68	I
1100	2954.20	I
540	2958.02	I
1400	2964.88	I
620	2966.93	I
710	2968.81	II
890	2975.88	II
1100	2980.81	I
210	2982.72	I
170	3000.10	II
800	3005.56	I
1100	3012.90	II
540	3016.78	I
1100	3016.94	II
980	3018.31	I
1200	3020.53	I
410	3031.16	II
710	3050.76	I
1100	3057.02	I
850	3067.41	I
2100	3072.88	I
170	3074.10	I
250	3074.79	I
430	3080.84	I
200	3096.76	I
340	3101.40	II
710	3109.12	II
710	3131.81	I
850	3134.72	II
170	3139.65	II
220	3145.32	II
220	3148.41	I
450	3156.63	I
270	3159.82	I
710	3162.61	II
450	3168.39	I
890	3172.94	I
450	3176.86	II
220	3181.01	I
360	3193.53	II
670	3194.19	II
200	3196.93	I
310	3206.11	I
180	3210.98	I
180	3217.30	II
180	3220.61	II
360	3247.66	I
220	3249.53	I
890	3253.70	II
270	3255.28	II
180	3273.66	II
200 h	3279.67	III
270	3279.98	II
160	3291.05	I
210	3306.12	I
340	3310.27	I

Intensity	Wavelength/Å	
670	3312.86	I
180	3317.99	II
890	3332.73	I
370	3352.06	II
230	3358.91	I
180	3360.06	I
180	3378.93	I
230	3384.70	II
170	3386.21	I
800	3389.83	II
230	3392.81	I
230	3394.59	II
230	3397.26	I
230	3397.60	I
2300	3399.80	II
170	3400.21	I
180	3402.51	I
230	3410.17	II
230	3417.34	I
410	3419.18	I
200	3428.37	II
250	3438.24	II
710	3472.40	I
200	3478.99	II
480	3479.28	II
250	3495.75	II
250	3497.16	I
980	3497.49	I
1200	3505.23	II
980	3523.02	I
980	3535.54	II
760	3536.62	I
180	3548.81	I
540	3552.70	II
1300	3561.66	II
270	3567.36	I
1100	3569.04	II
210	3597.42	II
540	3599.87	I
800	3616.89	I
320	3630.87	II
800	3644.36	II
320	3649.10	I
200	3651.84	I
220	3665.35	II
200	3672.27	I
480	3675.74	I
2200	3682.24	I
280	3696.51	I
240	3699.72	II
340	3701.15	II
1000	3717.80	I
650	3719.28	II
160	3729.10	I
460	3733.79	I
160	3737.88	II
400	3746.80	I
170	3766.92	II
200	3768.25	I
1400	3777.64	I
1400	3785.46	I

Intensity	Wavelength/Å	
650	3793.37	II
850 d	3800.38	I
320	3811.78	I
1300	3820.73	I
280	3830.02	I
800	3849.18	I
600	3858.31	I
230	3860.91	I
200	3872.55	II
160	3877.10	II
380	3880.82	II
200	3882.52	I
200	3889.23	I
200	3889.33	I
620	3899.94	I
620	3918.09	II
200	3923.90	II
320	3931.38	I
410	3951.83	I
160	3968.01	I
200	3973.48	I
180	4032.27	I
230	4062.84	I
180	4083.35	I
540	4093.16	II
1100	4174.34	I
160	4206.58	II
190	4209.70	I
170	4228.08	I
170	4232.44	II
170	4260.98	I
200	4263.39	I
170	4272.85	II
320	4294.79	I
160	4330.27	I
180	4336.66	II
250	4356.33	I
180	4370.97	II
160	4417.91	I
200	4438.04	I
250	4565.94	I
500 d	4598.80	I
230	4620.86	I
210	4655.19	I
120	4699.01	I
160	4782.74	I
310	4800.50	I
130	4859.24	I
120	4975.25	I
95	5018.20	I
95	5047.45	I
75	5170.18	I
230	5181.86	I
110	5243.99	I
120	5294.87	I
110	5354.73	I
110	5373.86	I
75	5452.92	I
230	5550.60	I
230	5552.12	I
95	5613.27	I

Intensity	Wavelength/Å	
160	5719.18	I
195	6098.67	I
95	6185.13	I
85	6789.27	I
160	6818.94	I
160	7063.83	I
570	7131.81	I
650	7237.10	I
410	7240.87	I
360	7624.40	I
110	7740.17	I
310	7845.35	I
130	7920.71	I
250	7994.73	I
130	8204.58	I
150	8546.48	I
160	8640.06	I
40	8711.24	I
65	9004.73	I
Helium He Z = 2		
15	231.454	II
20	232.584	II
30	234.347	II
50	237.331	II
100	243.027	II
300	256.317	II
1000	303.780	II
500	303.786	II
10	320.293	I
2	505.500	I
3	505.684	I
4	505.912	I
5	506.200	I
7	506.570	I
10	507.058	I
15	507.718	I
20	508.643	I
25	509.998	I
35	512.098	I
50	515.616	I
100	522.213	I
400	537.030	I
1000	584.334	I
50	591.412	I
5	958.70	II
6	972.11	II
8	992.36	II
15	1025.27	II
30	1084.94	II
35	1215.09	II
50	1215.17	II
120	1640.34	II
180	1640.47	II
7	2385.40	II
9	2511.20	II
50	2577.6	I
1	2723.19	I
12	2733.30	II
2	2763.80	I
10	2818.2	I
4	2829.08	I

Intensity	Wavelength/Å		Intensity	Wavelength/Å		Intensity	Wavelength/Å		Intensity	Wavelength/Å	
10	2945.11	I	2000	10830.34	I	1600	3461.97	II	1300	4125.65	I
40	3013.7	I	9	10913.05	I	810 c	3473.91	II	4300	4127.16	I
20	3187.74	I	3	10917.10	I	5400 c	3474.26	II	1500	4136.22	I
3	3202.96	II	4	11626.4	II	6300	3484.84	II	980 cw	4152.61	II
15	3203.10	II	30	11969.12	I	2500 c	3494.76	II	8100	4163.03	I
1	3354.55	I	20	12527.52	I	810 c	3498.88	II	2500	4173.23	I
2	3447.59	I	50	12784.99	I	810	3510.73	I	540	4194.35	I
1	3587.27	I	20	12790.57	I	4100 c	3515.59	II	2000	4227.04	I
3	3613.64	I	7	12845.96	I	410 c	3519.94	II	1300 cw	4254.43	I
2	3634.23	I	10	12968.45	I	630	3540.76	II	490	4264.05	I
3	3705.00	I	2	12984.89	I	1600	3546.05	II	1300	4350.73	I
1	3732.86	I	12	15083.64	I	1100 c	3556.78	II	300	4477.64	II
10	3819.607	I	200	17002.47	I	410	3560.15	II	290	4629.10	II
1	3819.76	I	1	18555.55	I	410 c	3573.24	II	80	4701.69	II
500	3888.65	I	6	18636.8	II	630 c	3574.80	II	130	4709.84	II
20	3964.729	I	500	18685.34	I	810	3579.12	I	130 c	4717.52	I
1	4009.27	I	200	18697.23	I	410	3580.75	II	290	4742.04	II
50	4026.191	I	100	19089.38	I	410	3581.83	II	100 c	4757.01	I
5	4026.36	I	20	19543.08	I	630 c	3592.23	II	65	4782.92	I
12	4120.82	I	1000	20581.30	I	1100 cw	3598.77	II	290	4939.01	I
2	4120.99	I	80	21120.07	I	540 c	3600.95	II	250 c	4967.21	II
3	4143.76	I	10	21121.43	I	410	3618.43	I	220	4979.97	I
10	4387.929	I	20	21132.03	I	430 c	3626.69	II	90	4995.05	I
3	4437.55	I	3	30908.5	II	490	3627.25	II	130	5042.37	I
200	4471.479	I	4	40478.90	I	430 c	3631.76	II	80	5093.07	I
25	4471.68	I				430 c	3638.30	II	140	5127.81	I
6	4685.4	II	**Holmium Ho Z = 67**			1600 c	3662.29	I	130	5142.59	II
30	4685.7	II	170	2502.91	II	1400	3667.97	I	110	5143.22	II
30	4713.146	I	170	2533.80	I	720	3679.19	I	160	5149.59	II
4	4713.38	I	190	2605.86	II	670	3679.70	I	90 c	5167.88	I
20	4921.931	I	270	2750.35	II	720	3682.65	I	130 c	5182.11	I
100	5015.678	I	270	2769.89	II	580	3690.65	I	90	5190.11	II
10	5047.74	I	300	2824.20	II	410	3700.04	I	65	5251.82	I
5	5411.52	II	270 c	2831.69	II	490 c	3702.35	II	90	5301.25	I
500	5875.62	I	270	2849.10	II	430	3712.88	I	80	5330.11	I
100	5875.97	I	360	2880.26	II	450	3720.72	I	90	5359.99	I
8	6560.10	II	460	2880.98	II	1100	3731.40	I	100	5407.08	I
100	6678.15	I	570 c	2909.41	II	810	3736.35	I	70	5566.52	I
3	6867.48	I	410 c	2979.63	II	3200 cw	3748.17	II	65	5627.60	I
200	7065.19	I	410	2987.64	II	8900 c	3796.75	II	140	5659.58	I
30	7065.71	I	480 c	3049.38	II	8900 c	3810.73	II	140 c	5691.47	I
50	7281.35	I	410 c	3054.00	II	410 c	3835.35	II	140 c	5696.57	I
1	7816.15	I	500 c	3057.45	II	1300 cw	3837.51	II	140 c	5860.28	I
2	8361.69	I	500 c	3082.34	II	410 c	3842.05	II	70 c	5882.99	I
2	9063.27	I	910	3084.36	II	1100	3843.86	II	70	5921.76	I
2	9210.34	I	430 c	3086.54	II	490 c	3846.73	II	70 cw	5948.03	I
10	9463.61	I	760	3118.50	II	1800 c	3854.07	II	70	5972.76	I
4	9516.60	I	580 c	3166.62	II	2700 c	3861.68	II	90	5973.52	I
3	9526.17	I	810	3173.78	II	3000 c	3888.96	II	230 c	5982.90	I
1	9529.27	I	810 c	3181.50	II	13000 c	3891.02	II	120	6081.79	I
1	9603.42	I	980 c	3281.97	II	1300 cw	3905.68	II	70	6208.65	I
3	9702.60	I	630 c	3337.23	II	580	3955.73	I	70 c	6305.36	I
6	10027.73	I	980 c	3343.58	II	490	3959.68	I	70	6550.97	I
2	10031.16	I	8100 c	3398.98	II	2700	4040.81	I	260	6604.94	I
15	10123.6	II	810 c	3410.26	II	5400 c	4045.44	II	120	6628.99	I
1	10138.50	I	1400 c	3414.90	II	8100	4053.93	I	55 cw	6694.32	I
10	10311.23	I	5400	3416.46	II	1700	4065.09	II	55 c	6785.43	I
2	10311.54	I	1200	3421.63	II	720	4068.05	I	40 cw	6939.49	I
3	10667.65	I	2000 c	3425.34	II	8900	4103.84	I	45 cw	6950.39	I
300	10829.09	I	2000 c	3428.13	II	2900	4108.62	I	140	7555.09	I
1000	10830.25	I	3200	3453.14	II	1500	4120.20	I	40 c	7628.42	I
			16000 c	3456.00	II						

Intensity	Wavelength/Å		Intensity	Wavelength/Å		Intensity	Wavelength/Å		Intensity	Wavelength/Å	
50 c	7693.15	I	82	1096.81	IV	80	2865.68	II	380 c	4638.16	II
60 cw	7815.48	I	84	1097.18	IV	120 d	2890.18	II	220 c	4644.58	II
60	7894.64	I	85	1116.10	IV	1100	2932.63	I	360 c	4655.62	II
50	8512.94	I	80	1124.06	IV	100	2941.05	II	320 w	4656.74	II
40	8670.19	I	90	1131.46	IV	100	2982.80	III	190 c	4681.11	II
90	8915.98	II	85	1144.43	IV	110 c	2999.40	II	450 w	4684.8	II
			80	1145.41	IV	100	3008.08	III	90 d	4907.06	II
Hydrogen H Z = 1			89	1146.62	IV	8000	3039.36	I	150 c	4973.77	II
15	926.226	I	83	1154.11	IV	110 d	3099.80	II	80 h	5109.36	II
20	930.748	I	84	1154.60	IV	180 c	3101.8	II	100 w	5115.14	II
30	937.803	I	90	1157.71	IV	130 c	3138.60	II	140 c	5117.40	II
50	949.743	I	90	1157.82	IV	80 c	3142.75	II	270 c	5120.80	II
100	972.537	I	85	1159.78	IV	130 d	3146.70	II	200 w	5121.75	II
300	1025.722	I	88	1176.50	IV	150	3155.77	II	80 d	5129.85	II
1000	1215.668	I	85	1191.58	IV	100 c	3158.40	II	240 c	5175.42	II
500	1215.674	I	83	1204.87	IV	90 c	3176.30	II	140 c	5184.44	II
5	3835.384	I	90	1206.55	IV	90 d	3198.11	II	200	5248.77	III
6	3889.049	I	88	1221.50	IV	13000	3256.09	I	150 c	5309.45	II
8	3970.072	I	85	1221.90	IV	3000	3258.56	I	80	5411.41	II
15	4101.74	I	85	1233.58	IV	90 c	3338.50	II	140 c	5418.45	II
30	4340.47	I	87	1235.84	IV	100 c	3404.28	II	220 w	5436.70	II
80	4861.33	I	90	1373.20	IV	110 d	3438.40	II	130 c	5497.50	II
120	6562.72	I	88	1398.77	IV	180 c	3693.91	II	140 c	5507.08	II
180	6562.852	I	81	1412.09	IV	95 c	3708.13	II	320 c	5513.00	II
5	9545.97	I	100	1625.42	III	380 w	3716.14	II	250 w	5523.28	II
7	10049.4	I	100	1748.83	III	120 c	3718.30	II	130 c	5536.50	II
12	10938.1	I	30	1842.41	III	160 c	3718.72	II	190 w	5555.45	II
20	12818.1	I	40	1850.30	III	160 c	3723.40	II	240 c	5576.90	II
40	18751.0	I	30	2154.08	III	170 w	3795.21	II	200 w	5636.70	II
5	21655.3	I	50	2205.28	II	230 c	3799.21	II	100	5645.15	III
8	26251.5	I	40	2281.64	II	250 c	3834.65	II	160 c	5708.50	II
15	40511.6	I	100 c	2306.05	II	200 c	3842.18	II	100 c	5721.80	II
4	46525.1	I	90 d	2313.21	II	100	3852.82	III	100	5819.50	III
6	74578.	I	70 d	2327.95	II	100	3889.78	II	210 c	5853.15	II
			80 h	2334.57	II	100 c	3902.07	II	490 w	5903.4	II
Indium In Z = 49			110 d	2382.63	II	250 w	3962.35	II	260 w	5915.4	II
17	378.61	V	70 h	2427.20	II	120 c	4004.66	II	120 c	5918.78	II
17	386.21	V	100	2447.90	II	140 d	4013.92	II	130 c	6062.9	II
14	388.91	V	110 d	2488.62	II	100	4023.77	III	250 c	6095.95	II
25	393.89	V	90	2488.95	II	150	4032.32	III	210 c	6108.66	II
25	400.57	V	80	2498.59	II	410 w	4056.94	II	180 w	6115.9	II
25	402.39	V	100	2499.60	II	100	4071.57	III	230 w	6128.7	II
622	472.71	IV	90 d	2500.99	II	100	4072.93	III	240 w	6129.4	II
689	479.39	IV	110 d	2512.31	II	17000	4101.76	I	320 w	6132.1	II
709	498.62	IV	100	2521.37	I	140 c	4205.14	II	150 c	6140.0	II
10	882.24	III	100	2527.41	III	100 d	4213.04	II	90	6143.23	II
10	890.84	III	160 d	2554.44	II	110 d	4219.66	II	140 c	6148.10	II
10	915.87	III	1100	2560.15	I	100	4252.68	III	190 w	6149.5	II
85	954.67	IV	200	2601.76	I	150 d	4372.87	II	80	6161.15	II
87	973.50	IV	90 d	2654.70	II	150 c	4500.78	II	180 w	6162.45	II
86	991.60	IV	100 d	2662.63	II	18000	4511.31	I	200	6197.72	III
89	1024.68	IV	140 d	2668.65	II	110 c	4549.01	II	100 c	6224.28	II
85	1024.79	IV	140 d	2674.56	II	140 c	4570.85	II	280 w	6228.3	II
88	1031.45	IV	80	2683.12	II	180 w	4578.02	II	140 w	6231.1	II
82	1031.98	IV	1600	2710.26	I	180 w	4578.40	II	270 w	6304.8	II
80	1054.43	IV	300	2713.94	I	140 c	4616.08	II	290 w	6362.3	II
84	1063.03	IV	80	2726.15	III	170 c	4617.17	II	300 w	6469.0	II
83	1068.25	IV	130 d	2749.75	II	250 c	4620.14	II	210 c	6541.20	II
82	1069.82	IV	700	2753.88	I	150 w	4620.70	II	190 c	6751.88	II
86	1077.64	IV	90 d	2818.97	II	170 c	4627.30	II	180 c	6765.9	II
90	1082.10	IV	180 c	2836.92	I	140 c	4637.04	II	100 c	6783.72	II
83	1086.33	IV									

Intensity	Wavelength/Å	
320 w	6891.5	II
380 w	7182.9	II
180 c	7255.0	II
210 c	7276.5	II
180 c	7303.4	II
320 c	7350.6	II
100 c	7632.7	II
100 c	7682.9	II
210 c	7740.7	II
100 c	7776.96	II
180 c	7789.0	II
90 c	7840.9	II
240 c	8227.0	II
100 w	8813.5	II
80 c	8832.6	II
120 c	9197.7	II
120 c	9202.0	II
220 w	9213.0	II
160 d	9241.1	II
100	9977.86	I
200	10257.03	I
100 h	10744.31	I
20	11334.72	I
20	11731.48	I
10	12912.59	I
9	13429.96	I
7	14719.08	I
6	22291.06	I
7	23879.13	I

Iodine I Z = 53

Intensity	Wavelength/Å	
30	363.78	V
36	380.74	V
45	565.53	V
50	607.57	V
6	612.46	IV
6	666.81	III
8	705.11	III
7	784.64	III
7	784.80	III
8	795.52	III
7	919.28	IV
10000	1034.66	II
8	1094.20	III
10000	1139.80	II
10000	1160.56	II
20000	1166.48	II
10000	1178.65	II
15000	1187.34	II
10000	1190.85	II
200	1218.41	I
20000	1220.89	II
600	1224.05	I
600	1224.08	I
500	1228.89	I
20000	1234.06	II
600	1251.34	I
8	1252.35	III
2500	1259.15	I
3000	1259.51	I
800	1261.27	I
600	1267.57	I
600	1267.60	I
1500	1275.26	I
3000	1289.40	I
10000	1300.34	I
3000	1302.98	I
3000	1313.95	I
3000	1317.54	I
2000	1330.19	I
20000	1336.52	II
5000	1355.10	I
3000	1357.97	I
5000	1360.97	I
3000	1361.11	I
2500	1367.71	I
2500	1368.22	I
4000	1383.23	I
3000	1390.75	I
2000	1392.90	I
2000	1400.01	I
8000	1425.49	I
5000	1446.26	I
5000	1453.18	I
5000	1457.39	I
5000	1457.47	I
10000	1457.98	I
2500	1458.79	I
4000	1459.15	I
2500	1465.83	I
1000	1485.92	I
5000	1492.89	I
5000	1507.04	I
5000	1514.68	I
15000	1518.05	I
2500	1526.45	I
5000	1593.58	I
5000	1617.60	I
2500	1640.78	I
15000	1702.07	I
12000	1782.76	I
5000	1799.09	I
75000	1830.38	I
15000	1844.45	I
2000	2061.63	I
7	2361.13	IV
6	2372.45	IV
7	2376.46	IV
8	2387.11	IV
9	2426.10	IV
8	2475.35	IV
8	2519.74	IV
8	2545.67	IV
7	2545.71	III
8	2652.23	IV
5000	3078.75	II
200	4102.23	I
200	4129.21	I
100 d	4134.15	I
500 d	4321.84	I
250	4763.31	I
1000	4862.32	I
200	4916.94	I
10000	5119.29	I
3000 c	5161.20	II
1000	5234.57	I
3000 c	5245.71	II
10000	5338.22	II
5000 c	5345.15	II
600 c	5427.06	I
3000	5435.83	II
2000 c	5464.62	II
10000	5625.69	II
2000 c	5690.91	II
4000 c	5710.53	II
1000 d	5764.33	I
2000	5894.03	I
5000	5950.25	II
300	5984.86	I
2000 d	6024.08	I
2000 c	6074.98	II
1000	6082.43	I
2000 c	6127.49	II
800	6191.88	I
500	6213.10	I
800	6244.48	I
1000	6293.98	I
500	6313.13	I
800	6330.37	I
400	6333.50	I
2000	6337.85	I
1000	6339.44	I
500	6359.16	I
1000	6566.49	I
2000	6583.75	I
1000	6585.27	I
5000	6619.66	I
500	6661.11	I
500 c	6697.29	I
400	6732.03	I
4000	6812.57	II
500	6989.78	I
500	7120.05	I
1200	7122.05	I
2000	7142.06	I
1000	7164.79	I
400 d	7191.66	I
700	7227.30	I
1000	7236.78	I
500	7237.84	I
5000	7402.06	I
1000	7410.50	I
500	7416.48	I
5000	7468.99	I
500 c	7490.52	I
2000	7554.18	I
500 d	7556.65	I
2000 c	7700.20	I
600	7897.98	I
500	7969.48	I
1000	8003.63	I
99000	8043.74	I
300 d	8065.70	I
1000	8090.76	I
800 c	8169.38	I
500 d	8222.57	I
4000	8240.05	I
10000 c	8393.30	I
1000	8486.11	I
1500 c	8664.95	I
500 c	8700.80	I
250 d	8748.22	I
1000	8853.24	I
2000	8853.80	I
3000	8857.50	I
1000 d	8898.50	I
400	8964.69	I
400	8993.13	I
5000	9022.40	I
15000	9058.33	I
1000	9098.86	I
12000	9113.91	I
600	9128.03	I
600	9227.74	I
1000	9335.05	I
4000	9426.71	I
3000	9427.15	I
2000	9598.22	I
2000	9649.61	I
3000 d	9653.06	I
5000	9731.73	I
500	10003.05	I
750	10131.16	I
1000	10238.82	I
400	10375.20	I
400	10391.74	I
5000	10466.54	I
400	11236.56	I
350	11558.46	I
320	11778.34	I
450	11996.86	I
300	12033.69	I
150	12304.58	I
60	13149.16	I
140	13958.27	I
200	14287.02	I
100	14460.00	I
225	15032.57	I
105	15528.65	I
150	16037.33	I
15	18275.71	I
20	18348.52	I
15	18982.41	I
35	19070.17	I
110	19105.12	I
50	19370.02	I
10	20648.69	I
220	22183.03	I
150	22226.53	I
30	22309.21	I
32	24420.82	I
12	27365.42	I
9	27573.05	I
10	30361.93	I
8	30383.88	I

Intensity	Wavelength/Å		Intensity	Wavelength/Å		Intensity	Wavelength/Å		Intensity	Wavelength/Å	
10	34295.73	I	540	2333.30	I	230	2572.70	I	330	3039.26	I
9	34513.11	I	740	2333.84	I	740	2577.26	I	300	3047.16	I
3	40228.54	I	580	2334.50	I	740	2592.06	I	300	3049.44	I
2	41633.80	I	1600	2343.18	I	740	2599.04	I	300	3057.28	I
			740	2343.61	I	700	2608.25	I	1600	3068.89	I
Iridium Ir Z = 77			580	2355.00	I	1800	2611.30	I	320	3083.22	I
9900	2010.65	I	230	2357.53	II	210	2614.98	I	240	3086.44	I
8700	2022.35	I	410	2358.16	I	330	2617.78	I	390	3088.04	I
15000	2033.57	I	500	2360.73	I	210	2619.88	I	510	3100.29	I
6200	2052.22	I	2500	2363.04	I	250	2625.32	I	510	3100.45	I
5000	2060.64	I	370	2368.04	II	700	2634.17	I	340	3120.76	I
3700	2083.22	I	3500	2372.77	I	250	2639.42	I	200	3121.78	I
3100	2085.74	I	290	2375.09	II	3500	2639.71	I	3400	3133.32	I
17000	2088.82	I	250	2377.28	I	210	2644.19	I	490	3168.88	I
14000	2092.63	I	250	2377.98	I	1800	2661.98	I	370	3177.58	I
2700	2112.68	I	500	2379.38	I	350	2662.63	I	370	3198.92	I
1800	2119.54	I	540	2381.62	I	2700	2664.79	I	610	3212.12	I
2000	2125.44	I	210	2383.17	I	520	2669.91	I	370	3219.51	I
4500	2126.81	II	1300	2386.89	I	520	2671.84	I	5100	3220.78	I
2000	2127.52	I	2500	2390.62	I	330	2673.61	I	300	3229.28	I
4500	2127.94	I	2700	2391.18	I	270	2692.34	I	470	3241.52	I
3700	2148.22	I	230	2407.59	I	3000	2694.23	I	200	3262.01	I
2500	2150.54	I	290	2409.37	I	330	2772.46	I	390	3266.44	I
3500	2152.68	II	290	2410.17	I	250	2775.55	I	200	3322.60	I
2900	2155.81	I	290	2410.73	I	520	2781.29	I	560	3368.48	I
7900	2158.05	I	540	2413.31	I	330	2785.22	I	660	3437.02	I
2100	2162.88	I	370	2415.86	I	540	2797.35	I	410	3448.97	I
5800	2169.42	II	620	2418.11	I	1600	2797.70	I	3200	3513.64	I
4500	2175.24	I	210	2424.89	I	380	2798.18	I	220	3515.95	I
2700	2178.17	I	370	2424.99	I	410	2800.82	I	410	3522.03	I
1600	2187.43	II	290	2425.66	I	680	2823.18	I	320	3558.99	I
1100	2190.38	II	540	2427.61	I	1200	2824.45	I	1200	3573.72	I
740	2191.64	I	540	2431.24	I	820	2836.40	I	320	3594.39	I
910	2208.09	II	1300	2431.94	I	1100	2839.16	I	220	3609.77	I
1300	2220.37	I	270	2435.14	I	820	2840.22	I	660	3628.67	I
790	2221.07	II	250	2445.34	I	3800	2849.72	I	220	3636.20	I
2500	2242.68	II	250	2447.76	I	380	2875.60	I	300	3661.71	I
620	2245.76	II	910	2452.81	I	380	2875.98	I	300	3664.62	I
2100	2253.38	I	1300	2455.61	I	270	2877.68	I	320	3674.98	I
2100	2255.10	I	230	2455.87	I	820	2882.64	I	200	3687.08	I
1400	2255.81	I	210	2457.03	I	650	2897.15	I	200	3731.36	II
350	2258.51	I	210	2457.23	I	260	2901.95	I	530	3747.20	I
1400	2258.86	I	870	2467.30	I	260	2904.80	I	3100	3800.12	I
830	2264.61	I	3300	2475.12	I	200	2907.24	I	230	3817.24	I
1100	2266.33	I	210	2478.11	I	440	2916.36	I	480	3902.51	I
1000	2268.90	I	2100	2481.18	I	230	2918.57	I	480	3915.38	I
660	2280.00	I	620	2493.08	I	4400	2924.79	I	400	3934.84	I
950	2281.02	II	210	2496.27	I	1200	2934.64	I	590	3976.31	I
660	2281.91	I	250	2502.63	I	880	2936.68	I	460	3992.12	I
330	2284.60	I	4100	2502.98	I	250	2938.47	I	350	4033.76	I
330	2295.08	I	210	2513.71	I	2700	2943.15	I	370	4069.92	I
790	2298.05	I	990	2533.13	I	230	2946.97	I	150	4070.68	I
460	2299.53	I	1100	2534.46	I	200	2949.76	I	100	4092.61	I
910	2300.50	I	580	2537.22	I	1200	2951.22	I	140	4115.78	I
2700	2304.22	I	580	2542.02	I	200	2974.95	I	90	4172.56	I
410	2305.47	I	7900	2543.97	I	440	2980.65	I	260	4268.10	I
210	2307.27	I	790	2546.03	I	300	2996.08	I	220	4311.50	I
910	2308.93	I	210	2551.40	I	220	3002.25	I	160	4399.47	I
460	2315.38	I	210	2555.35	I	600	3003.63	I	65	4403.78	I
410	2321.45	I	910	2564.18	I	270	3017.31	I	110	4426.27	I
410	2321.58	I	210	2569.88	I	380	3029.36	I	55	4478.48	I
210	2327.98	I									

Intensity	Wavelength/Å		Intensity	Wavelength/Å		Intensity	Wavelength/Å		Intensity	Wavelength/Å	
55	4545.68	I	10	844.28	III	500	1479.47	V	15	1651.58	IV
30	4548.48	I	10 p	861.83	III	10 h	1505.17	III	15	1652.90	IV
35	4568.09	I	10	891.17	III	13	1526.60	IV	13	1653.41	IV
75	4616.39	I	10	950.33	III	13	1530.26	IV	13	1656.11	IV
26	4656.18	I	10	981.37	III	14	1532.63	IV	15	1656.65	IV
50	4728.86	I	10 w	983.88	III	13	1532.91	IV	14	1660.10	IV
26	4756.46	I	12	1055.27	II	15	1533.86	IV	13	1662.32	IV
65	4778.16	I	15	1068.36	II	13	1533.95	IV	13	1662.52	IV
30	4795.67	I	15	1071.60	II	14	1536.58	IV	13	1663.54	IV
50	4938.09	I	15	1096.89	II	10 h	1538.63	III	13	1668.09	IV
26	4970.48	I	12	1099.12	II	13	1542.16	IV	12	1670.74	II
25	4999.74	I	18	1112.09	II	14	1542.70	IV	14	1671.04	IV
25	5002.74	I	12	1121.99	II	12 h	1550.20	III	13	1673.68	IV
30	5014.98	I	12	1122.86	II	13	1566.26	IV	14	1675.66	IV
30	5123.66	I	12	1128.07	II	14	1568.27	IV	13	1681.36	IV
35	5364.32	I	12	1130.43	II	13	1591.51	IV	15	1687.69	IV
75	5449.50	I	15	1133.41	II	13	1592.05	IV	15	1698.88	IV
45	5625.55	I	12	1133.68	II	13	1598.01	IV	12	1702.04	II
35	5894.06	I	12	1138.64	II	13	1600.58	IV	13	1709.81	IV
20	6110.67	I	12	1142.33	II	10 h	1601.21	III	15	1711.41	IV
12	6288.28	I	12	1143.23	II	13	1601.67	IV	14	1712.76	IV
10	6686.08	I	18	1144.95	II	13	1603.18	IV	14	1717.90	IV
6	7834.32	I	12	1147.41	II	13	1603.73	IV	14	1718.16	IV
			15	1148.29	II	13	1604.88	IV	14	1719.46	IV
Iron Fe Z = 26			12	1151.16	II	13	1605.68	IV	14	1722.71	IV
350	386.16	V	12	1267.44	II	15	1605.97	IV	14	1724.06	IV
350	386.88	V	12	1272.00	II	13	1606.98	IV	16	1725.63	IV
400	387.20	V	400	1317.86	V	17	1609.10	IV	13	1761.08	IV
400	387.50	V	400	1323.27	V	14	1609.83	IV	12	1761.38	II
400	387.76	V	400	1330.40	V	13	1610.47	IV	20	1785.26	II
400	387.78	V	400	1359.01	V	13	1611.20	IV	20	1786.74	II
400	395.90	V	600	1361.82	V	13	1613.64	IV	18	1788.07	II
400	404.62	V	700	1373.59	V	15	1614.02	IV	13	1792.10	IV
400	405.50	V	600	1373.67	V	13	1614.64	IV	13	1796.93	IV
800	407.42	V	500	1376.34	V	13	1615.00	IV	13	1827.98	IV
600	407.44	V	500	1378.56	V	16	1616.68	IV	10	1869.83	III
400	407.49	V	800	1387.94	V	14	1617.68	IV	12	1877.99	III
500	407.75	V	400	1397.97	V	12	1618.47	II	10	1882.05	III
400	409.71	V	600	1400.24	V	14	1619.02	IV	12	1886.76	III
400	410.20	V	800	1402.39	V	13	1621.16	IV	13	1890.67	III
600	411.55	V	400	1406.67	V	14	1621.57	IV	11	1893.98	III
700	417.39	V	500	1406.82	V	13	1623.38	IV	20	1895.46	III
700	418.04	V	400	1407.25	V	13	1623.53	IV	10 s	1907.58	III
500	418.47	V	600	1409.45	V	15	1626.47	IV	19	1914.06	III
700	421.06	V	400	1415.20	V	14	1626.90	IV	15	1915.08	III
500	421.78	V	600	1420.46	V	13	1628.54	IV	15	1922.79	III
500	422.31	V	800	1430.57	V	13	1630.18	IV	10 p	1926.01	III
500	426.06	V	13	1431.43	IV	17	1631.08	IV	18	1926.30	III
500	426.11	V	800	1440.53	V	15	1631.12	II	15	1930.39	III
350	426.97	V	400	1442.22	V	14	1632.40	IV	14	1931.51	III
17	525.69	IV	800	1446.62	V	13	1634.01	IV	30	1934.538	I
15	526.29	IV	700	1448.85	V	18	1635.40	II	25	1937.269	I
13	526.63	IV	400	1449.93	V	15	1636.32	II	14	1937.34	III
14	536.61	IV	700	1456.16	V	15	1639.40	II	10 l	1938.90	III
15	537.10	IV	500	1459.83	V	14	1639.40	IV	14 s	1943.48	III
13	537.26	IV	400	1460.73	V	16	1640.04	IV	12	1945.34	III
14	537.79	IV	500	1462.63	V	14	1640.16	IV	50	1946.988	I
13	537.94	IV	700	1464.68	V	12	1641.76	II	10	1950.33	III
13	552.14	IV	500	1465.38	V	15	1641.87	IV	12	1951.01	III
14	607.53	IV	400	1466.65	V	15	1647.09	IV	25	1951.571	I
13	608.80	IV	500	1469.00	V	12	1647.16	II	30	1952.59	I
10	813.38	III									

Intensity	Wavelength/Å		Intensity	Wavelength/Å		Intensity	Wavelength/Å		Intensity	Wavelength/Å	
11	1952.65	III	60	2264.389	I	80	2406.97	II	500	2493.26	II
30	1953.005	I	80	2267.085	I	300	2410.52	II	60	2494.000	I
13	1953.32	III	10	2267.42	III	200	2411.07	II	50	2494.251	I
10	1953.49	III	80	2267.469	I	150	2413.31	II	100	2495.87	I
10 w	1954.22	III	50	2270.862	I	80	2417.87	II	600	2496.533	I
60	1957.823	I	150	2272.070	I	60	2420.396	I	150	2498.90	I
11	1958.58	III	150	2276.026	I	60	2423.089	I	1000	2501.132	I
60	1960.144	I	80	2279.937	I	150	2424.14	II	50	2501.693	I
13	1960.32	III	150	2284.086	I	120	2428.36	II	80	2506.09	II
30	1961.25	I	150	2287.250	I	120	2430.08	II	500	2507.900	I
50	1962.111	I	300	2292.524	I	80	2432.26	II	50	2508.753	I
12	1963.11	II	10	2293.06	III	60	2438.182	I	1000	2510.835	I
15	1987.50	III	15	2295.86	III	150	2439.30	II	120	2511.76	II
14	1991.61	III	200	2297.787	I	150	2439.74	I	80	2512.275	I
13	1994.07	III	600	2298.169	I	100	2442.57	I	400	2512.365	I
12	1995.56	III	80	2299.220	I	250	2443.872	I	80	2516.570	I
12	1996.42	III	300	2300.142	I	100	2444.51	II	300	2517.661	I
10	2061.55	III	50	2301.684	I	50	2445.212	I	800	2518.102	I
12	2068.24	III	100	2303.424	I	100	2445.57	II	150	2519.629	I
14	2078.99	III	150	2303.581	I	60	2447.709	I	50	2522.480	I
100	2084.122	I	120	2308.999	I	100	2453.476	I	4000	2522.849	I
10	2084.35	III	150	2313.104	I	1500	2457.598	I	200	2523.66	I
12	2090.14	III	10 p	2317.70	III	150	2458.78	II	500	2524.293	I
15	2097.48	III	10	2319.22	III	80	2461.28	II	100	2525.02	I
12	2097.69	III	200	2320.358	I	100	2461.86	II	200	2525.39	II
12	2103.80	III	10 p	2321.71	III	100	2462.181	I	300	2526.29	II
10	2107.32	III	10	2326.95	III	1500	2462.647	I	2000	2527.435	I
15	2151.78	III	100	2327.40	II	50	2463.730	I	800	2529.135	I
12	2157.71	III	100	2331.31	II	800	2465.149	I	250	2529.55	II
50	2157.794	I	300	2332.80	II	60	2467.732	I	150	2529.836	I
12	2158.47	III	10 p	2336.77	III	600	2468.879	I	200	2530.687	I
10	2161.27	III	200	2338.01	II	80	2470.67	II	120	2533.63	II
40	2166.773	I	10	2338.96	III	80	2470.965	I	100	2534.42	II
12	2166.95	III	600	2343.49	II	800	2472.336	I	120	2535.49	II
12	2171.04	III	80	2343.96	II	1000	2472.895	I	400	2535.607	I
15	2174.66	III	150	2344.28	II	200	2473.16	I	200	2536.792	I
300	2178.118	I	200	2348.11	II	600	2474.814	I	200	2536.80	II
12	2180.41	III	250	2348.30	II	60	2476.657	I	100	2538.80	II
250	2186.486	I	200	2359.12	II	120	2479.480	I	100	2538.91	II
60	2186.892	I	150	2360.00	II	1200	2479.776	I	150	2538.99	II
120	2187.195	I	120	2360.29	II	100	2480.16	II	50	2539.357	I
250	2191.839	I	200	2364.83	III	80	2482.12	II	200	2540.66	II
150	2196.043	I	80	2365.76	II	100	2482.66	II	600	2540.972	I
80	2200.390	I	80	2368.59	II	10000	2483.271	I	80	2541.10	II
80	2200.724	I	80	2369.456	I	300	2483.533	I	300	2542.10	I
15	2208.41	II	80	2369.95	II	1000	2484.185	I	250	2543.92	I
10 p	2208.85	III	120	2371.430	I	50	2485.990	I	150	2544.70	I
20	2213.65	II	300	2373.624	I	800	2486.373	I	800	2545.978	I
12	2218.26	II	150	2373.74	II	100	2486.691	I	80	2546.67	II
20	2220.38	II	120	2374.518	I	100	2487.066	I	100	2548.74	II
10	2221.83	III	120	2376.43	II	120	2487.370	I	80	2549.08	II
10	2229.27	III	80	2379.27	II	4000	2488.143	I	80	2549.39	II
10	2232.43	III	120	2380.76	II	100	2488.945	I	600	2549.613	I
10	2232.69	III	150	2381.835	I	80	2489.48	II	400	2562.53	II
10	2235.91	III	1000	2382.04	II	1000	2489.750	I	200	2563.48	II
10	2238.16	III	300	2388.63	II	50	2489.913	I	150	2574.36	II
12 p	2241.54	III	200	2389.973	I	3000	2490.644	I	300	2576.691	I
50	2250.790	I	1000	2395.62	II	100	2490.71	II	100	2582.58	II
60	2251.874	I	300	2399.24	II	2000	2491.155	I	1500	2584.54	I
300	2259.511	I	800	2404.88	II	100	2491.40	II	650	2585.88	II
12	2261.59	III	250	2406.66	II	100	2493.18	II	90	2591.54	II

Intensity	Wavelength/Å		Intensity	Wavelength/Å		Intensity	Wavelength/Å		Intensity	Wavelength/Å	
90	2593.73	II	80	2744.527	I	500	2999.512	I	80	3246.005	I
650	2598.37	II	300	2746.48	II	120	3000.451	I	80	3265.046	I
2000	2599.40	II	100	2749.32	II	800	3000.948	I	50	3265.617	I
300	2599.57	I	500	2749.48	II	12	3001.62	III	13	3266.88	III
60	2605.657	I	1200	2750.140	I	60	3001.655	I	50	3271.000	I
300	2606.51	II	80	2753.29	II	12 h	3007.28	III	11	3276.08	III
800	2606.827	I	150	2754.032	I	200	3007.282	I	150	3286.75	I
650	2607.09	II	100	2754.426	I	500	3008.14	I	10	3288.81	III
600	2611.87	II	800	2755.73	II	120	3009.569	I	120	3305.97	I
320	2613.82	II	250	2756.328	I	15	3013.17	III	200	3306.343	I
320	2617.62	II	100	2757.316	I	60	3017.627	I	400	3355.227	I
250	2618.018	I	120	2761.780	I	60	3018.983	I	80	3355.517	I
90	2620.41	II	150	2761.81	II	500	3020.491	I	60	3369.546	I
400	2623.53	I	150	2762.026	I	1500	3020.639	I	120	3370.783	I
200	2625.67	II	120	2762.772	I	600	3021.073	I	50	3378.678	I
150	2628.29	II	120	2763.109	I	500	3024.032	I	50	3380.110	I
250	2631.05	II	80	2766.910	I	150	3025.638	I	60	3383.978	I
250	2631.32	II	250	2767.522	I	500	3025.842	I	50	3392.304	I
100	2632.237	I	300	2772.07	I	80	3030.148	I	150	3392.651	I
300	2635.809	I	600	2778.220	I	60	3031.214	I	150	3399.333	I
50	2641.646	I	3000	2788.10	I	60	3034.484	I	80	3404.353	I
200	2643.998	I	200	2797.78	I	800	3037.389	I	500	3407.458	I
300	2666.812	I	400	2804.521	I	80	3041.637	I	250	3413.131	I
60	2666.965	I	1500	2806.98	I	800	3047.604	I	60	3424.284	I
600	2679.062	I	10 p	2813.24	III	600	3057.446	I	500	3427.119	I
500	2684.75	II	2500	2813.287	I	1000	3059.086	I	60	3428.748	I
400	2689.212	I	300	2823.276	I	250	3067.244	I	6000	3440.606	I
10 h	2695.13	III	600	2825.56	I	120	3075.719	I	2500	3440.989	I
200	2699.106	I	50	2825.687	I	120	3091.577	I	1000	3443.876	I
80	2706.012	I	120	2828.808	I	80	3098.189	I	200	3445.149	I
400	2706.582	I	1500	2832.436	I	100	3099.895	I	1200	3465.860	I
60	2708.571	I	120	2835.950	I	100	3099.968	I	2000	3475.450	I
200	2711.655	I	200	2838.119	I	60	3100.303	I	500	3476.702	I
80	2714.41	II	200	2843.631	I	100	3100.665	I	2500	3490.574	I
50	2716.257	I	1000	2843.977	I	10 p	3136.43	III	500	3497.840	I
50	2717.786	I	100	2845.594	I	10	3174.09	III	10	3501.76	III
250	2718.436	I	800	2851.797	I	80	3175.445	I	250	3513.817	I
4000	2719.027	I	50	2869.307	I	10	3175.99	III	300	3521.261	I
100	2719.420	I	50	2872.334	I	10	3178.01	III	400	3526.040	I
50	2720.197	I	80	2874.172	I	150	3184.895	I	100	3526.166	I
1500	2720.903	I	50	2894.504	I	250	3191.659	I	60	3526.237	I
400	2723.578	I	12	2904.43	III	500	3193.226	I	60	3526.381	I
150	2724.953	I	10	2907.50	III	800	3193.299	I	60	3526.467	I
50	2726.235	I	12	2907.70	III	200	3196.928	I	100	3533.199	I
80	2727.54	II	120	2912.157	I	80	3199.500	I	200	3536.556	I
200	2728.020	I	120	2929.007	I	50	3205.398	I	300	3541.083	I
50	2728.820	I	1200	2936.903	I	100	3211.88	I	250	3542.075	I
80	2728.90	II	60	2941.343	I	200	3214.011	I	80	3553.739	I
1000	2733.581	I	1000	2947.876	I	200	3214.396	I	400	3554.925	I
60	2734.005	I	600	2953.940	I	60	3215.938	I	200	3556.878	I
50	2734.268	I	250	2957.364	I	50	3217.377	I	400	3558.515	I
500	2735.475	I	150	2965.254	I	80	3219.583	I	1000	3565.379	I
50	2735.612	I	1500	2966.898	I	60	3219.766	I	1200	3570.097	I
500	2737.310	I	120	2969.36	I	300	3222.045	I	800	3570.25	I
120	2737.83	I	800	2970.099	I	600	3225.78	I	120	3571.996	I
400	2739.55	II	1200	2973.132	I	80	3227.796	I	100	3573.393	I
250	2742.254	I	500	2973.235	I	50	3233.967	I	60	3573.829	I
800	2742.405	I	600	2981.445	I	120	3234.613	I	60	3573.888	I
200	2743.20	II	1000	2983.570	I	300	3236.222	I	4000	3581.19	I
150	2743.565	I	1000	2994.427	I	100	3239.433	I	150	3582.199	I
200	2744.068	I	250	2994.502	I	80	3244.187	I	150	3584.660	I

Intensity	Wavelength/Å		Intensity	Wavelength/Å		Intensity	Wavelength/Å		Intensity	Wavelength/Å	
120	3584.929	I	6000	3737.131	I	1200	3927.920	I	40	4195.329	I
300	3585.319	I	100	3738.306	I	2000	3930.296	I	150	4198.304	I
150	3585.705	I	400	3743.362	I	60	3948.774	I	40	4199.095	I
10	3586.04	III	6000	3745.561	I	60	3949.953	I	300	4202.029	I
200	3586.103	I	1200	3745.899	I	50	3951.164	I	40	4203.984	I
400	3586.984	I	3000	3748.262	I	50	3952.601	I	80	4206.696	I
100	3594.633	I	80	3748.964	I	16	3954.33	III	80	4210.343	I
11	3600.94	III	3000	3749.485	I	60	3956.454	I	400	4216.183	I
150	3603.204	I	1500	3758.232	I	250	3956.68	I	100	4219.360	I
11	3603.88	III	400	3760.05	I	60	3966.614	I	50	4222.212	I
200	3605.454	I	1500	3763.788	I	11	3968.72	III	11	4222.27	III
500	3606.680	I	400	3765.54	I	100	3969.257	I	50	4225.956	I
1500	3608.859	I	600	3767.191	I	80	3977.741	I	200	4227.423	I
250	3610.16	I	60	3776.452	I	10 w	3979.42	III	100	4233.602	I
60	3612.068	I	250	3785.95	I	40	3981.771	I	13	4235.56	III
150	3617.788	I	100	3786.68	I	50	3983.956	I	250	4235.936	I
1500	3618.768	I	250	3787.880	I	60	3994.114	I	50	4238.809	I
200	3621.462	I	250	3790.092	I	200	3997.392	I	12	4243.75	III
150	3622.004	I	150	3794.34	I	40	3998.053	I	50	4247.425	I
150	3623.19	I	400	3795.002	I	400	4005.241	I	200	4250.118	I
100	3631.096	I	120	3797.518	I	60	4009.713	I	300	4250.787	I
1200	3631.463	I	250	3798.511	I	100	4021.867	I	40	4258.315	I
60	3632.041	I	400	3799.547	I	10	4035.42	III	800	4260.473	I
100	3638.298	I	200	3805.345	I	50	4040.638	I	250	4271.153	I
200	3640.389	I	80	3806.696	I	4000	4045.813	I	1200	4271.759	I
80	3643.717	I	600	3812.964	I	11	4053.11	III	12 h	4273.40	III
1500	3647.842	I	60	3813.059	I	1500	4063.594	I	12	4279.72	III
250	3649.506	I	1500	3815.840	I	50	4066.975	I	1200	4282.402	I
80	3650.279	I	2500	3820.425	I	50	4067.977	I	14 h	4286.16	III
200	3651.467	I	150	3821.179	I	1200	4071.737	I	80	4291.462	I
120	3670.024	I	80	3824.306	I	40	4076.629	I	16 h	4296.85	III
150	3670.089	I	2500	3824.444	I	12	4081.00	III	250	4299.234	I
100	3676.311	I	1500	3825.880	I	40	4100.737	I	18 h	4304.78	III
150	3677.629	I	1200	3827.823	I	40	4107.489	I	1200	4307.901	I
1500	3679.913	I	1000	3834.222	I	150	4118.544	I	20 h	4310.36	III
200	3682.242	I	120	3839.257	I	10	4120.90	III	150	4315.084	I
120	3683.054	I	500	3840.437	I	11	4122.02	III	1500	4325.761	I
150	3684.107	I	800	3841.047	I	11	4122.78	III	80	4352.734	I
120	3685.998	I	120	3843.256	I	40	4127.608	I	80	4369.771	I
500	3687.456	I	80	3846.800	I	400	4132.058	I	11 h	4372.31	III
120	3689.477	I	200	3849.96	I	80	4134.676	I	14 h	4372.53	III
150	3694.008	I	120	3850.817	I	40	4136.997	I	18 h	4372.81	III
120	3695.051	I	2500	3856.372	I	15	4137.76	III	800	4375.929	I
150	3701.086	I	150	3859.212	I	13	4139.35	III	3000	4383.544	I
80	3704.462	I	10000	3859.911	I	200	4143.415	I	1200	4404.750	I
1200	3705.566	I	150	3865.523	I	800	4143.869	I	300	4415.122	I
60	3707.041	I	60	3867.215	I	40	4153.898	I	12	4419.60	III
150	3707.821	I	250	3872.501	I	50	4154.500	I	600	4427.299	I
300	3707.919	I	150	3873.761	I	60	4156.799	I	400	4461.652	I
600	3709.246	I	250	3878.018	I	18	4164.73	III	120	4466.551	I
120	3716.442	I	2000	3878.573	I	13	4166.84	III	80	4476.017	I
8000	3719.935	I	4000	3886.282	I	50	4172.744	I	80	4482.169	I
1500	3722.563	I	200	3887.048	I	13	4174.26	III	200	4482.252	I
120	3724.377	I	300	3888.513	I	60	4174.912	I	50	4489.739	I
60	3725.491	I	800	3895.656	I	50	4175.635	I	50	4528.613	I
60	3727.093	I	1200	3899.707	I	50	4177.593	I	30	4647.433	I
500	3727.619	I	400	3902.945	I	120	4181.754	I	30	4736.771	I
150	3732.396	I	250	3906.479	I	50	4184.891	I	50	4859.741	I
1200	3733.317	I	80	3916.731	I	120	4187.038	I	120	4871.317	I
5000	3734.864	I	600	3920.258	I	120	4187.795	I	60	4872.136	I
120	3735.324	I	1200	3922.911	I	80	4191.430	I	30	4878.208	I

Intensity	Wavelength/Å		Intensity	Wavelength/Å		Intensity	Wavelength/Å		Intensity	Wavelength/Å	
100	4890.754	I	10	5368.06	III	80	7187.313	I	50	625.02	III
250	4891.492	I	400	5371.489	I	30	7207.381	I	30	625.76	III
30	4903.309	I	11 l	5375.47	III	30	7445.746	I	45	628.59	III
150	4918.992	I	40	5393.167	I	40	7495.059	I	50	630.04	III
500	4920.502	I	300	5397.127	I	60	7511.045	I	35	633.09	III
1500	4957.597	I	250	5405.774	I	80	7937.131	I	120	637.87	V
80	5001.862	I	250	5429.695	I	60	7945.984	I	50	639.98	III
30	5005.711	I	100	5434.523	I	80	7998.939	I	60	646.41	III
100	5006.117	I	200	5446.871	I	60	8046.047	I	50	651.20	III
60	5012.067	I	120	5455.609	I	50	8085.176	I	50	659.72	III
30	5014.941	I	25	5497.516	I	150	8220.41	I	30	664.86	III
150	5041.755	I	20	5501.464	I	120	8327.053	I	40	672.34	III
30	5049.819	I	30	5506.778	I	20	8331.908	I	35	672.85	III
30	5051.634	I	30	5569.618	I	120	8387.770	I	35	676.57	III
25	5074.748	I	60	5572.841	I	30	8468.404	I	35	680.13	III
150	5110.357	I	120	5586.755	I	15	8514.069	I	35	683.68	III
40	5139.251	I	200	5615.644	I	60	8661.898	I	35	686.25	III
100	5139.462	I	20	5624.541	I	150	8688.621	I	45	687.98	III
25	5151.910	I	50	5662.515	I	52	11422.32	I		690.86	V
12	5156.12	III	11	5719.88	III	87	11439.12	I		691.75	V
80	5166.281	I	10	5756.38	III	91	11593.59	I	45	691.93	III
2500	5167.487	I	20	5762.990	I	255	11607.57	I	50	695.61	III
80	5168.897	I	18	5833.93	III	160	11638.26	I	30	698.05	III
500	5171.595	I	10	5854.62	III	230	11689.98	I	50	708.36	III
50	5191.454	I	30	5862.353	I	160	11783.26	I	600	708.85	V
80	5192.343	I	15	5891.91	III	580	11882.84	I	50	714.00	III
200	5194.941	I	30	5914.114	I	225	11884.08	I	100 p	722.04	III
10	5199.08	III	10 p	5920.13	III	1030	11973.05	I	60	729.40	II
30	5204.582	I	18 p	5929.69	III	96	14400.56	I	30	746.70	III
25	5215.179	I	10	5952.31	III	72	14512.23	I	200	761.18	III
150	5216.274	I	14	5953.62	III	50	14555.06	I	100	763.98	II
60	5226.862	I	12	5979.32	III	40	14826.43	I	60	766.20	II
1000	5227.150	I	30	5986.956	I	94	15294.58	I	200	771.03	II
250	5232.939	I	12 h	5989.08	III	41	15769.42	I	60 p	773.69	II
10	5235.66	III	18	5999.54	III	105	18856.65	I	200	782.10	II
18	5243.31	III	16	6032.59	III				100	783.72	II
13 l	5260.34	III	13	6036.56	III	***Krypton Kr*** *Z = 36*			60	785.97	III
100	5266.555	I	11	6048.72	III	30	467.35	III		793.44	IV
1200	5269.537	I	11	6054.18	III	150	472.16	V		794.11	IV
800	5270.357	I	40	6065.482	I	100	484.39	V	7	805.76	IV
14	5272.98	III	30	6102.159	I	250	496.25	V		810.70	V
15	5276.48	III	40	6136.614	I	120	500.77	V	18	816.82	IV
30	5281.789	I	40	6137.694	I	200	507.20	V	60	818.15	II
16	5282.30	III	40	6191.558	I	30	540.86	III	60	830.38	II
60	5283.621	I	30	6213.429	I	60	548.04	V	50	837.66	III
12	5284.83	III	30	6219.279	I	30	565.64	III	22	842.04	IV
11	5298.12	III	40	6230.726	I	30	569.16	III	100	844.06	II
12	5299.93	III	20	6246.317	I	30	571.98	III	50	854.73	III
25	5302.299	I	80	6247.56	II	30	579.83	III	60	862.58	III
14 w	5302.60	III	30	6252.554	I	30	585.14	III	60	864.82	II
10	5306.76	III	20	6393.602	I	30	585.96	III	60	868.87	II
10	5322.74	III	30	6399.999	I	30	593.70	III	40	870.84	III
150	5324.178	I	20	6411.647	I	30	594.10	III	50	876.08	III
800	5328.038	I	20	6421.349	I	30	596.41	III	200	884.14	II
300	5328.531	I	30	6430.844	I	40	600.17	III	1000	886.30	II
100	5332.899	I	200	6456.38	II	30	603.67	III	400	891.01	II
80	5339.928	I	60	6494.981	I	50	605.86	III	75	897.81	III
500	5341.023	I	20	6546.239	I	35	606.47	III	200	911.39	II
11	5346.88	III	20	6592.913	I	50	611.12	III	2000	917.43	II
12	5353.77	III	40	6677.989	I	35	616.72	III	50	945.44	I
12	5363.76	III	25	7164.443	I	40	621.45	III	50	946.54	I
						45	622.80	III			

Intensity	Wavelength/Å		Intensity	Wavelength/Å		Intensity	Wavelength/Å		Intensity	Wavelength/Å	
20	951.06	I	60	2992.22	III	300	4057.037	II	200	5468.17	II
50	953.40	I	50	3022.30	III	300	4065.128	II	10	5501.43	III
50	963.37	I	80	3024.45	III	50	4067.37	III	500	5562.224	I
2000	964.97	II	50	3046.93	III	500	4088.337	II	2000	5570.288	I
50	987.29	III	30	3056.72	III	250	4098.729	II	80	5580.386	I
100	1001.06	I	60	3063.13	III	100	4109.248	II	100	5649.561	I
100	1003.55	I	40	3097.16	III	40	4131.33	III	400	5681.89	II
100	1030.02	I	60	3112.25	III	250	4145.122	II	200 h	5690.35	II
30	1158.74	III	30	3120.61	III	40	4154.46	III	100	5832.855	I
200	1164.87	I	100	3124.39	III	150	4250.580	II	3000	5870.914	I
650	1235.84	I	60	3141.35	III	1000	4273.969	I	200	5992.22	II
6	1638.82	III	3	3142.01	IV	100	4282.967	I	60	5993.849	I
6	1914.09	III	100	3189.11	III	600	4292.923	II	10 h	6037.17	III
3	2237.34	IV	80	3191.21	III	200	4300.49	II	60	6056.125	I
6	2291.26	IV	6	3224.99	IV	500 h	4317.81	II	10 h	6078.38	III
3	2329.3	IV	40	3239.52	III	400	4318.551	I	10	6310.22	I
4	2336.75	IV	40	3240.44	III	1000	4319.579	I	300	6420.18	I
4	2348.27	IV	300	3245.69	III	150 h	4322.98	II	100	6421.026	I
3	2358.5	IV	3	3261.70	IV	100	4351.359	I	200	6456.288	I
3	2388.05	IV	150	3264.81	III	3000	4355.477	II	150	6570.07	II
40	2393.94	III	100	3268.48	III	500	4362.641	I	60	6699.228	I
4	2416.9	IV	30	3271.65	III	200	4369.69	II	100	6904.678	I
3	2428.04	IV	30	3285.89	III	800	4376.121	I	250	7213.13	II
5	2442.68	IV	30	3304.75	III	300 h	4386.54	II	100	7224.104	I
4	2451.7	IV	50	3311.47	III	200	4399.965	I	80	7287.258	I
6	2459.74	IV	200	3325.75	III	100	4425.189	I	400	7289.78	II
100 h	2464.77	II	60	3330.76	III	500	4431.685	II	400	7407.02	II
5	2474.06	IV	50	3342.48	III	600	4436.812	II	60	7425.541	I
60	2492.48	II	100	3351.93	III	600	4453.917	I	200	7435.78	II
40	2494.01	III	40	3374.96	III	800	4463.689	I	100	7486.862	I
4	2517.0	IV	100	3439.46	III	800	4475.014	II	300	7524.46	II
5	2518.02	IV	70	3474.65	III	400 h	4489.88	II	1000	7587.411	I
6	2519.38	IV	100	3488.59	III	600	4502.353	I	2000	7601.544	I
5	2524.5	IV	200	3507.42	III	400 h	4523.14	II	150	7641.16	II
5	2546.0	IV	100	3564.23	III	200 h	4556.61	II	1000	7685.244	I
6	2547.0	IV	100 h	3607.88	II	800	4577.209	II	1200	7694.538	I
4	2558.08	IV	200	3631.889	II	300	4582.978	II	250	7735.69	II
30	2563.25	III	30	3641.34	III	150 h	4592.80	II	150	7746.827	I
3	2586.9	IV	250	3653.928	II	500	4615.292	II	800	7854.821	I
5	2606.17	IV	80	3665.324	I	1000	4619.166	II	200	7913.423	I
10	2609.5	IV	150	3669.01	II	800	4633.885	II	180	7928.597	I
8	2615.3	IV	100	3679.559	I	2000	4658.876	III	200	7933.22	II
7	2621.11	IV	80	3686.182	II	500	4680.406	II	120	7973.62	II
60	2639.76	III	30	3690.65	III	100	4691.301	II	100	7982.401	I
30	2680.32	III	300 h	3718.02	II	200	4694.360	II	1500	8059.503	I
40	2681.19	III	200	3718.595	II	3000	4739.002	II	4000	8104.364	I
80 h	2712.40	II	150	3721.350	II	300	4762.435	II	6000	8112.899	I
3	2730.55	IV	200	3741.638	II	1000	4765.744	II	60	8132.967	I
8	2748.18	IV	150	3744.80	II	300	4811.76	II	3000	8190.054	I
6	2774.70	IV	80	3754.245	II	300	4825.18	II	200	8202.72	II
3	2829.60	IV	500	3778.089	II	800	4832.077	II	80	8218.365	I
100	2833.00	II	500	3783.095	II	700	4846.612	II	3000	8263.240	I
3	2836.08	IV	3	3809.30	IV	150	4857.20	II	100	8272.353	I
30	2841.00	III	5	3860.58	IV	300	4945.59	II	1500	8281.050	I
30	2851.16	III	40 h	3868.70	III	20 h	5016.45	III	5000	8298.107	I
5	2853.0	IV	150 h	3875.44	II	200	5022.40	II	100	8412.430	I
3	2859.3	IV	150	3906.177	II	250	5086.52	II	3000	8508.870	I
50	2870.61	III	200	3920.081	II	400 h	5125.73	II	150	8764.110	I
100	2892.18	III	5	3934.29	IV	500	5208.32	II	6000	8776.748	I
30	2909.17	III	100	3994.840	II	200	5308.66	II	2000	8928.692	I
50	2952.56	III	100 h	3997.793	II	500	5333.41	II	500	9238.48	II

Intensity	Wavelength/Å		Intensity	Wavelength/Å		Intensity	Wavelength/Å		Intensity	Wavelength/Å	
500 hl	9293.82	II	2600	18167.315	I	10000	1523.79	III	410	4619.88	II
200 h	9320.99	II	100	18399.786	I	4000	1808.66	IV	540	4655.50	II
300	9361.95	II	150	18580.896	I	5000	1902.97	IV	360	4662.51	II
100	9362.082	I	300	18696.294	I	4000 c	2197.45	IV	230	4692.50	II
200 h	9402.82	II	170	18785.460	I	770	2256.76	II	230	4728.42	II
200 h	9470.93	II	200	18797.703	I	25000 w	2417.58	IV	500	4740.28	II
500	9577.52	II	140	20209.878	I	50000	2532.75	IV	390	4743.09	II
500 h	9605.80	II	300	20423.964	I	45000	2582.05	IV	320	4748.73	II
400 h	9619.61	II	140	20446.971	I	95000 w	2597.50	IV	320	4860.91	II
200	9663.34	II	600	21165.471	I	70000 w	2662.75	IV	850	4899.92	II
200 h	9711.60	II	1800	21902.513	I	420	2808.39	II	1000	4920.98	II
2000	9751.758	I	120	22485.775	I	50000 w	2848.30	IV	1000	4921.79	II
500	9803.14	II	180	23340.416	I	30000 c	2962.58	IV	370	4949.77	I
500	9856.314	I	120	24260.506	I	70000 w	3009.51	IV	340	4970.39	II
1000	10221.46	II	180	24292.221	I	90000 c	3056.68	IV	370	4986.83	II
100	11187.108	I	600	25233.820	I	1000	3171.63	III	720	4999.47	II
200	11257.711	I	180	28610.55	I	1500	3171.74	III	210	5050.57	I
150	11259.126	I	1000	28655.72	I	510	3245.13	II	470	5114.56	II
500	11457.481	I	150	28769.71	I	550	3265.67	II	470	5122.99	II
150	11792.425	I	140	28822.49	I	800	3303.11	II	450	5145.42	I
1500	11819.377	I	300	29236.69	I	1500	3337.49	II	290	5158.69	I
600	11997.105	I	300	30663.54	I	870	3344.56	II	580	5177.31	I
160	12077.224	I	300	30979.16	I	1500	3380.91	II	850	5183.42	II
100	12861.892	I	500	39300.6	I	320	3628.83	II	260	5188.22	II
1100 hl	13177.412	I	1100	39486.52	I	1000	3645.42	II	720	5211.86	I
1000	13622.415	I	220	39557.25	I	550	3713.54	II	520	5234.27	I
2400	13634.220	I	100	39572.60	I	2400	3759.08	II	340	5253.46	I
800	13658.394	I	1400	39588.4	I	3700	3790.83	II	370	5271.19	I
200	13711.036	I	1100	39589.6	I	3900	3794.78	II	370	5301.98	II
600	13738.851	I	500	39954.8	I	600	3840.72	II	180	5303.55	II
150	13974.027	I	300	39966.6	I	1600	3849.02	II	500	5455.15	I
550	14045.657	I	1300	40306.1	I	3400	3871.64	II	470	5501.34	I
140	14104.298	I	250	40685.16	I	1700	3886.37	II	240	5648.25	I
180	14402.22	I				1300	3916.05	II	180	5740.66	I
2000	14426.793	I	*Lanthanum La Z = 57*			1100	3921.54	II	370	5769.34	I
100	14517.84	I	100	344.12	IV	2200	3929.22	II	320	5789.24	I
1600	14734.436	I	400	390.72	V	9000	3949.10	II	450	5791.34	I
550	14762.672	I	1000	432.11	V	4400	3988.52	II	140	5821.99	I
450	14765.472	I	2500	435.28	V	3600	3995.75	II	320	5930.62	I
400	14961.894	I	10000	463.14	IV	2800	4031.69	III	720	6249.93	I
120	15005.307	I	5000	482.16	V	3000	4042.91	II	260 d	6262.30	II
140	15209.526	I	7000	498.08	V	850	4067.39	II	450	6394.23	I
1700	15239.615	I	15000	499.54	IV	2800	4077.35	II	250	6455.99	I
130	15326.480	I	10000	503.58	V	5500	4086.72	II	180	6709.50	I
1500	15334.958	I	12000	526.76	V	4400	4123.23	II	110	7045.96	I
700	15372.037	I	10000	531.07	V	550	4141.74	II	160	7066.23	II
200	15474.026	I	15000	533.23	V	1100	4151.97	II	50	7161.25	I
180	15681.02	I	8000	547.44	V	1500	4196.55	II	110 w	7282.34	II
120	15820.09	I	40000	552.02	IV	1600	4238.38	II	110 w	7334.18	I
200	16726.513	I	5000	600.24	V	480	4269.50	II	75 cw	7483.50	II
2000	16785.128	I	30000	631.26	IV	600	4286.97	II	50	7498.83	I
1000	16853.488	I	400	796.99	III	600	4296.05	II	85	7539.23	I
2400	16890.441	I	2000	870.40	III	440	4322.51	II	40	7964.83	I
1600	16896.753	I	1000	882.34	III	4600	4333.74	II	75	8086.05	I
1800	16935.806	I	400	942.86	III	550	4354.40	III	85	8324.69	I
600	17098.771	I	50000	1081.61	III	2000	4429.90	II	95	8346.53	I
700	17367.606	I	95000	1099.73	III	850	4522.37	II	65	8545.44	I
120	17404.443	I	2000	1255.63	III	420	4526.12	II	300	8583.45	III
150	17616.854	I	10000	1349.18	III	400	4558.46	II	40	8674.43	I
650	17842.737	I	25000	1368.04	IV	400	4574.88	II	35	8825.82	I
700	18002.229	I	20000	1463.47	IV	410	4613.39	II	120	9184.38	III
			15000	1507.87	IV						

Line Spectra of the Elements

Intensity	Wavelength/Å		Intensity	Wavelength/Å		Intensity	Wavelength/Å		Intensity	Wavelength/Å	
100	9212.63	III	15	1056.53	IV	20	2259.01	V	100	5006.572	I
140	10284.79	III	10	1060.66	II	150	2332.418	I	50	5089.484	I
			12	1072.09	IV	16	2359.53	IV	10	5107.242	I
Lead Pb Z = 82			18	1080.81	IV	180	2388.797	I	2000	5201.437	I
10	496.38	IV	20	1084.17	IV	550 r	2393.792	I	10	5372.099	II
12	499.94	IV	10	1088.86	V	140	2399.597	I	40	5692.346	I
14	529.78	IV	10	1103.94	II	320 r	2401.940	I	200	5895.624	I
20	570.16	IV	10	1108.43	II	320 r	2411.734	I	2000	6001.862	I
10	648.50	IV	10	1109.84	II	16	2417.61	IV	500	6011.667	I
20	703.73	V	20	1116.08	IV	15	2424.81	V	500	6059.356	I
12	749.46	V	10	1119.57	II	150 r	2443.829	I	40	6081.409	II
10	752.52	V	10	1121.36	II	160 r	2446.181	I	50	6110.520	I
10	761.09	IV	10	1133.14	II	130 r	2476.378	I	100	6235.266	I
18	767.45	V	18	1137.84	IV	80 r	2577.260	I	50 c	6660.20	II
18	769.49	V	14	1144.93	IV	500 r	2613.655	I	20000	7228.965	I
14	771.42	V	12	1157.88	V	900 r	2614.175	I	10	7346.676	I
14	782.79	V	14	1185.43	V	160	2628.262	I	20	7809.259	I
15	797.02	V	20	1189.95	IV	4	2634.256	II	5	7896.737	I
18	802.07	IV	10	1203.63	II	10	2657.094	I	10	8168.001	I
12	802.82	IV	10	1231.20	II	700	2663.154	I	6	8191.886	I
18	809.63	V	11	1233.50	V	10	2697.541	I	5	8217.711	I
10	812.59	IV	10	1291.10	IV	25000 r	2801.995	I	40	8272.690	I
10	827.41	IV	20	1313.05	IV	100	2822.58	I	20	8409.384	I
12	832.60	IV	10	1331.65	II	14000 r	2823.189	I	10	8478.492	I
12	845.94	IV	10	1335.20	II	35000 r	2833.053	I	5	8722.810	I
18	857.64	IV	12	1343.06	IV	6	2840.557	II	10	8857.457	I
16	862.33	IV	10	1348.37	II	14000 r	2873.311	I	8	9293.476	I
20	863.97	V	16	1388.94	IV	3	2914.442	II	15	9438.05	I
14	870.44	IV	18	1400.26	IV	15	2966.460	I	15	9604.297	I
6	873.71	II	10	1404.34	IV	15	2972.991	I	15	9674.351	I
12	879.96	IV	10	1433.96	II	15	2980.157	I	200	10290.458	I
18	883.90	IV	10	1512.42	II	4	2986.876	II	100	10498.965	I
14	884.96	IV	14	1535.71	IV	10	3043.85	III	50	10649.249	I
14	884.99	IV	20	1553.1	III	150	3118.894	I	15	10886.688	I
14	888.37	V	10	1671.53	II	10	3137.81	III	40	10969.53	I
8	889.68	II	10	1682.15	II	10	3176.50	III		13512.6	I
16	890.72	IV	20	1726.75	II	600	3220.528	I		14743.0	I
14	894.40	V	10	1796.670	II	100	3229.613	I		15349.6	I
12	896.08	V	10	1822.050	II	400	3240.186	I		39039.4	I
12	908.51	IV	10	1904.77	I	200	3262.355	I			
14	915.71	V	7	1921.471	II	35000	3572.729	I	*Lithium Li Z = 3*		
12	917.90	IV	12	1959.34	IV	50000 r	3639.568	I		102.9	III
12	918.09	V	16	1973.16	IV	20000	3671.491	I		103.4	III
12	920.28	V	10	1998.83	V	70000 r	3683.462	I		104.1	III
12	920.66	V	5 r	2022.02	I	10	3713.982	II		105.5	III
10	922.12	IV	10	2042.58	IV	25000	3739.935	I		108.0	III
12	922.49	IV	12	2049.34	IV	12	3854.08	III		113.9	III
10	927.64	IV	8 r	2053.28	I	15000	4019.632	I		125.5	II
14	932.20	IV	12	2079.22	IV	95000	4057.807	I		135.0	III
12	954.35	V	6	2111.758	I	14000	4062.136	I		136.5	II
10	967.23	II	10	2115.066	I	10	4157.814	I		140.5	II
10	986.71	II	15	2154.01	IV	10000	4168.033	I		167.21	II
10	995.89	II	500 r	2170.00	I	8	4272.66	III		168.74	II
10	1016.61	II	7	2175.580	I	200	4340.413	I		171.58	II
14	1028.61	IV	12	2177.46	IV	10	4496.15	IV		178.02	II
20	1032.05	IV	7	2187.888	I	6	4499.34	III		199.28	II
16	1041.24	IV	8	2189.603	I	16	4534.60	IV		207.5	II
18	1044.14	IV	10	2203.534	II	7	4571.21	III		456.	II
12	1048.9	III	20	2237.425	I	10	4579.051	II		483.	II
10	1049.82	II	20	2246.86	I	6	4761.12	III		540.	II
10	1050.77	II	25	2246.89	I	1000	5005.416	I		540.0	III
10	1051.26	V								729.	II

Intensity	Wavelength/Å		Intensity	Wavelength/Å		Intensity	Wavelength/Å		Intensity	Wavelength/Å	
	729.1	III		2313.49	I	5	3199.33	II	48	8126.23	I
	800.	II		2315.08	I	2	3199.43	II	48	8126.45	I
	820.	II		2316.95	I	17	3232.66	I		8517.37	II
	861.	II		2319.18	I		3249.87	II		9581.42	II
	905.5	II		2321.88	I		3306.28	II		10120.	II
	917.5	II		2325.11	I		3488.	I		12232.	I
	936.	II		2329.02	I		3579.8	I		12782.	I
	945.	II		2329.84	II		3618.	I		13566.	I
	965.	II		2333.94	I		3662.	I		17552.	I
	972.	II	3	2336.88	II		3684.32	II		18697.	I
	988.	II	5	2336.91	II	1	3714.00	II		19290.	I
	1018.	II	2	2337.00	II	5	3714.16	II		24467.	I
	1032.	II		2340.15	I	6 d	3714.27	II		40475.	I
	1036.	II		2348.22	I	8	3714.29	II			
	1093.	II		2358.93	I	7 d	3714.40	II	*Lutetium Lu Z = 71*		
	1103.	II		2373.54	I	10	3714.41	II	100	563.72	V
	1109.	II		2381.54	II	1	3714.51	II	500	810.73	III
	1116.	II		2383.20	II		3714.58	II	2000	832.28	III
	1132.1	II	1	2394.39	I	3	3718.7	I	100	861.92	V
	1141.	II		2402.33	II	6	3794.72	I	400	876.80	IV
	1166.4	II		2410.84	II	20	3915.30	I	100	880.32	V
	1198.09	II	3	2425.43	I	20	3915.35	I	100	891.81	V
	1215.	II		2429.81	II	10	3985.48	I	100	914.72	V
	1238.	II		2460.2	I	10	3985.54	I	400	1001.18	III
	1253.8	II	10	2475.06	I	40	4132.56	I	100	1272.42	IV
	1420.89	II		2506.94	II	40	4132.62	I	800	1333.79	IV
	1424.	II		2508.78	II		4196.	I	400	1429.38	IV
3	1492.93	II		2518.	I	20	4273.07	I	200	1441.76	V
5	1492.97	II		2539.49	II	20	4273.13	I	200	1453.35	V
1	1493.04	II	24	2551.7	II	5	4325.42	II	200	1468.99	V
	1555.	II		2559.	II	5	4325.47	II	400	1472.12	V
3	1653.08	II	15	2562.31	I	1	4325.54	II	200	1473.71	V
5	1653.13	II		2605.08	II		4516.45	II	200	1485.58	V
1	1653.21	II	2	2657.29	II	13	4602.83	I	400	1511.26	IV
	1681.66	II	3	2657.30	II	13	4602.89	I	600	1772.57	IV
	1755.33	II		2674.46	II		4607.34	II	100 c	1786.25	V
	2009.	II		2728.24	II		4671.51	II	1000	1854.57	III
	2039.	I	5	2728.29	II	6	4671.65	II	1500	2065.35	III
	2068.	II	2	2728.32	II	2	4671.70	II	1500 c	2070.56	III
	2131.	II	3	2730.47	II	3	4678.06	II	600 c	2086.47	IV
	2164.	II	1	2730.55	II	1	4678.29	II	1000 c	2104.41	IV
	2173.4	I	5	2741.20	I		4760.	I	1000 c	2108.31	IV
	2183.	II		2766.99	II		4763.	II	1700 h	2195.54	II
	2214.	II		2790.31	II		4788.36	II	1000	2236.14	III
	2222.	II		2801.	I		4843.0	II	2000	2236.22	III
	2237.	II		2846.	I	4	4881.32	II	95	2276.94	II
h	2249.21	II		2868.	I	4	4881.39	II	190	2297.41	II
	2286.82	II		2895.	I	1	4881.49	II	1300	2392.19	II
	2302.57	II	2	2934.02	II	8	4971.66	I	120	2399.14	II
	2303.33	II	2	2934.07	II	8	4971.75	I	80	2419.21	II
	2304.59	I	5	2934.12	II		5037.92	II	130	2459.64	II
	2304.92	I	1	2934.25	II		5271.	I	370	2536.95	II
	2305.36	I		2968.	I		5315.	I	930	2571.23	II
	2305.83	I	3	3029.12	II		5395.	I	1700	2578.79	II
	2306.29	I	3	3029.14	II		5440.	I	4500 c	2603.35	III
	2306.82	I		3144.	I	600 c	5483.55	II	1800	2613.40	II
	2307.44	I	3	3155.31	II	600 c	5485.65	II	18000	2615.42	II
	2308.97	I	4	3155.33	II	320	6103.54	I	1800	2619.26	II
	2309.88	I	1	3196.26	II	320	6103.65	I	2700	2657.80	II
	2310.94	I	9	3196.33	II	3600	6707.76	I	570 h	2685.08	I
	2312.11	I	4	3196.36	II	3600	6707.91	I	4200	2701.71	II
									180 d	2719.09	I

Intensity	Wavelength/Å		Intensity	Wavelength/Å		Intensity	Wavelength/Å		Intensity	Wavelength/Å	
480 h	2728.95	I	250	4281.03	I	3000	323.31	IV	1	2613.36	I
3600	2754.17	II	330 d	4295.97	I	30	353.09	V	2	2614.73	I
750 h	2765.74	I	150	4309.57	I	150	857.29	IV	3	2617.51	I
2000	2772.55	III	190 c	4430.48	I	50	919.03	IV	3	2628.66	I
2700	2796.63	II	190	4450.81	I	250	1037.41	IV	6	2630.05	I
270 c	2834.35	II	3300	4518.57	I	300	1210.99	IV	8	2632.87	I
330 h	2845.13	I	100 h	4648.21	I	300	1342.19	IV	2	2644.80	I
3000	2847.51	II	1000	4658.02	I	800	1346.57	IV	3	2646.21	I
570 h	2885.14	I	85 h	4659.03	I	300	1346.68	IV	4	2649.06	I
6300	2894.84	II	150	4785.42	II	600	1352.05	IV	8	2660.76	II
4500	2900.30	II	85	4815.05	I	900	1384.46	IV	8	2660.82	II
300	2903.05	I	460	4904.88	I	500	1385.77	IV	6	2668.12	I
9000	2911.39	II	180	4942.34	I	800	1387.53	IV	8	2669.55	I
270 h	2949.73	I	800	4994.13	II	300	1404.68	IV	10	2672.46	I
1200	2951.69	II	800	5001.14	I	1000	1409.36	IV	3	2693.72	I
4200	2963.32	II	140	5134.05	I	500	1437.53	IV	5	2695.18	I
2400	2969.82	II	2700	5135.09	I	1000	1437.64	IV	6	2698.14	I
1800	2989.27	I	170	5196.61	I	300	1447.42	IV	8	2731.99	I
3000	3020.54	II	500	5402.57	I	300	1459.54	IV	10	2733.49	I
2100	3056.72	II	140 c	5421.90	I	400	1459.62	IV	12	2736.53	I
1000	3057.86	III	100	5437.88	I	400	1481.51	IV	5	2765.22	I
7500	3077.60	II	2100	5476.69	II	350	1490.45	IV	7	2768.34	I
390	3080.11	I	550	5736.55	I	300	1495.50	IV	38	2776.69	I
5100 h	3081.47	I	80	5800.59	I	300	1607.11	IV	32	2778.27	I
3000	3118.43	I	690 cw	5983.9	II	5	1668.43	I	90	2779.83	I
2400	3171.36	I	140	5997.13	I	500	1683.02	IV	8	2781.29	I
260	3191.80	II	1400	6004.52	I	10	1683.41	I	32	2781.42	I
1400	3198.12	II	440	6055.03	I	400	1698.81	IV	36	2782.97	I
4800	3254.31	II	150	6159.94	II	15	1707.06	I	1000	2795.53	II
3800	3278.97	I	600	6198.13	III	40	1734.84	II	600	2802.70	II
7600	3281.74	I	160	6199.66	II	50	1737.62	II	3	2809.76	I
6200	3312.11	I	2100	6221.87	II	20	1747.80	I	2	2811.11	I
7600	3359.56	I	80	6235.36	II	40	1750.65	II	1	2811.78	I
6200	3376.50	I	160	6242.34	II	50	1753.46	II	12	2846.72	I
950	3385.50	I	70 h	6345.35	I	30	1827.93	I	12	2846.75	I
160 h	3391.55	I	1100	6463.12	II	300	1844.17	IV	14	2848.34	I
1400	3396.82	I	29	6477.67	I	9	2025.82	I	14	2848.42	I
4100	3397.07	II	55 c	6523.18	I	25	2064.90	III	16	2851.65	I
4800	3472.48	II	35 cw	6611.28	II	20	2091.96	III	16	2851.66	I
8300 c	3507.39	II	23 c	6677.14	I	20	2177.70	III	6000	2852.13	I
1600	3508.42	I	30 c	6793.77	I	3	2329.58	II	2	2902.92	I
4800	3554.43	II	45	6917.31	I	20	2395.15	III	4	2906.36	I
4800	3567.84	I	23	7031.24	I	6	2449.57	II	3	2915.45	I
340	3596.34	I	45	7125.84	II	1	2557.23	I	10	2936.74	I
800	3623.99	II	14 ch	7237.98	I	1	2560.94	I	12	2938.47	I
680	3636.25	I	11 c	7441.52	I	1	2562.26	I	2	2942.00	I
2600	3647.77	I	9 c	8178.16	I	1	2564.94	I	13	2942.00	I
110	3756.70	I	17	8382.08	I	1	2570.91	I	20	3091.08	I
110	3756.79	I	35	8459.19	II	1	2572.25	I	22	3092.99	I
150	3800.67	I	10 d	8478.50	I	2	2574.94	I	14	3096.90	I
2700	3841.18	I	29 c	8508.08	I	1	2577.89	I	9	3104.71	II
530	3876.65	II	35 c	8610.98	I	1	2580.59	I	8	3104.81	II
50	3918.86	I				1	2584.22	I	6	3168.98	II
480	3968.46	I	*Magnesium Mg Z = 12*			2	2585.56	I	6	3172.71	II
670	4054.45	I	400	146.95	IV	3	2588.28	I	7	3175.78	II
310	4122.49	I	20	186.51	III	1	2591.89	I	2	3197.62	I
3100	4124.73	I	20	187.20	III	1	2593.23	I	17	3329.93	I
150 c	4131.79	I	10	188.53	III	2	2595.97	I	6	3332.15	I
460	4154.08	I	100	231.73	III	2	2602.50	I	9	3336.68	I
1600	4184.25	II	80	234.26	III	4	2603.85	I	7	3535.04	II
150	4277.50	I	35	276.58	V	5	2606.62	I	8	3538.86	II
			4000	320.99	IV						

Intensity	Wavelength/Å		Intensity	Wavelength/Å		Intensity	Wavelength/Å		Intensity	Wavelength/Å	
7	3549.52	II	12	8712.69	I	90	1264.41	IV	500	2066.38	III
8	3553.37	II	13	8717.83	I	500	1283.58	III	1000	2069.02	III
140	3829.30	I	10	8734.99	II	400	1287.59	III	30	2076.21	II
300	3832.30	I	17	8736.02	I	300	1291.62	III	900	2077.38	III
500	3838.29	I	11	8745.66	II	1000	1360.72	III	800	2084.23	III
8	3848.24	II	14	8806.76	I	800	1365.20	III	600	2090.05	III
7	3850.40	II	10	8824.32	II	500 h	1609.17	III	1500	2092.16	I
3	3878.31	I	11	8835.08	II	1000	1614.14	III	500	2094.78	III
3	3895.57	I	20	8923.57	I	2000	1620.60	III	20	2097.46	II
4	3903.86	I	10	8997.16	I	500	1633.80	III	500	2097.93	III
6	3938.40	I	14	9218.25	II	80	1667.00	IV	500	2099.97	III
8	3986.75	I	13	9244.27	II	80	1698.30	IV	20	2102.50	II
10	4057.50	I	12	9246.50	I	20	1726.47	II	1700	2109.58	I
15	4167.27	I	30	9255.78	I	30	1732.70	II	30	2113.96	II
20	4351.91	I	10	9327.54	II	50	1733.55	II	1000	2169.78	III
9	4384.64	II	10	9340.54	II	40	1734.49	II	700	2174.15	III
10	4390.59	II	25	9414.96	I	30	1737.93	II	900	2176.87	III
8	4428.00	II	17	9429.81	I	20	1740.16	II	800	2181.86	III
9	4433.99	II	19	9432.76	II	20	1742.00	II	800	2184.87	III
14	4481.16	II	20	9438.78	I	85	1742.10	IV	290	2208.81	I
13	4481.33	II	12	9631.89	II	85	1766.27	IV	540	2213.85	I
28	4571.10	I	11	9632.43	II	80	1795.65	IV	900	2220.55	III
10	4730.03	I	15	9953.20	I	80	1795.79	IV	770	2221.84	I
7	4851.10	II	15	9983.20	I	30	1853.27	II	1000	2227.42	III
75	5167.33	I	17	9986.47	I	20	1857.92	II	20	2373.36	II
220	5172.68	I	18	9993.21	I	50	1902.95	II	20	2427.38	II
400	5183.61	I	14	10092.16	II	20	1907.84	II	50	2427.72	II
8	5264.21	II	35	10811.08	I	75	1910.25	IV	30	2427.94	II
7	5264.37	II	11	10914.23	II	30	1911.41	II	30	2437.37	II
9	5401.54	II	10	10951.78	II	20 d	1914.68	II	20	2437.84	II
6	5528.41	I	25	10953.32	I	100	1915.10	II	30	2452.49	II
30	5711.09	I	27	10957.30	I	20	1918.64	II	50	2499.00	II
10	6318.72	I	28	10965.45	I	30	1919.64	II	30	2507.60	II
9	6319.24	I	15	11032.10	I	80	1921.25	II	20	2516.60	II
7	6319.49	I	14	11033.66	I	20	1923.07	II	30	2516.74	II
10	6346.74	II	45	11828.18	I	20	1923.34	II	20	2521.66	II
9	6346.96	II	30	12083.66	I	30	1925.52	II	20	2530.72	II
11	6545.97	II	28	14877.62	I	50	1926.59	II	20	2531.80	II
7	6781.45	II	35	15024.99	I	30	1931.40	II	50	2532.78	II
8	6787.85	II	30	15040.24	I	500	1941.28	III	50	2533.33	II
7	6812.86	II	25	15047.70	I	800	1943.21	III	30	2534.10	II
8	6819.27	II	10	15765.84	I	20	1945.15	II	80	2534.22	II
10	7193.17	I	30	17108.66	I	20	1947.93	II	100	2535.66	II
10	7291.06	I	5	26392.90	I	20	1950.14	II	30	2535.98	II
12	7387.69	I				500	1952.36	III	100	2537.92	II
20	7657.60	I	*Manganese Mn Z = 25*			1000	1952.52	III	50	2541.11	II
19	7659.15	I	600	410.30	V	30	1953.23	II	80	2542.92	II
17	7659.90	I	600	410.60	V	20 d	1954.81	II	50	2543.45	II
15	7691.55	I	600	415.62	V	30	1959.25	II	100	2548.75	II
12	7877.05	II	650	415.98	V	20	1969.24	II	50	2551.85	II
13	7896.37	II	600	428.59	V	500	1978.95	III	30	2553.27	II
10	8098.72	I	600	435.67	V	30	1994.23	II	75	2556.57	II
9	8115.22	II	1000	441.72	V	9700	1996.06	I	30	2556.89	II
8	8120.43	II	850	442.49	V	14000	1999.51	I	95	2558.59	II
10	8209.84	I	60	579.79	IV	18000	2003.85	I	30	2559.41	II
20	8213.03	I	60	581.44	IV	1000 w	2027.83	III	150	2563.65	II
10	8213.99	II	60	581.65	IV	500 w	2028.14	III	30	2565.22	II
11	8234.64	II	60	585.21	IV	50	2037.31	II	580	2572.76	I
10	8310.26	I	90	1242.25	IV	40	2037.64	II	480	2575.51	I
15	8346.12	I	90	1244.50	IV	40	2039.97	II	12000	2576.10	II
10	8710.18	I	95	1251.93	IV	500	2049.68	III	550	2584.31	I
			95	1257.28	IV						

Intensity	Wavelength/Å		Intensity	Wavelength/Å		Intensity	Wavelength/Å		Intensity	Wavelength/Å	
30	2588.97	II	70	2886.68	II	290	3619.28	I	800	4451.59	I
45	2589.71	II	160	2889.58	II	220	3623.79	I	160	4453.00	I
250	2592.94	I	55	2892.39	II	140	3629.74	I	130	4455.01	I
6200	2593.73	II	50	2898.70	II	100	3660.40	I	160	4455.32	I
250	2595.76	I	80	2900.16	II	280	3693.67	I	110	4455.82	I
95	2598.90	II	140 h	2914.60	I	180	3696.57	I	210	4457.55	I
30	2602.72	II	190 h	2925.57	I	210	3706.08	I	270	4458.26	I
45	2603.72	II	1100	2933.06	II	130	3718.93	I	150	4461.08	I
4300	2605.69	II	1500	2939.30	II	130	3731.93	I	510	4462.02	I
190	2610.20	II	250 h	2940.39	I	260	3790.22	I	290	4464.68	I
500	2618.14	II	1900	2949.20	II	110	3800.55	I	200	4470.14	I
140	2622.90	I	30	3019.92	II	3200	3806.72	I	130	4472.79	I
150	2624.04	I	55	3031.06	II	700	3809.59	I	170	4490.08	I
40	2624.80	II	30	3035.35	II	2100	3823.51	I	240	4498.90	I
200	2625.58	II	330	3044.57	I	390	3823.89	I	240	4502.22	I
190	2632.35	II	120	3045.59	I	200	3829.68	I	160	4709.72	I
130	2638.17	II	200	3047.04	I	480	3833.86	I	180	4727.48	I
80	2639.84	II	30	3050.65	II	1300	3834.36	I	130	4739.11	I
27	2650.99	II	250	3054.36	I	350	3839.78	I	1000	4754.04	I
60	2655.91	II	140	3062.12	I	670	3841.08	I	180	4761.53	I
30	2666.77	II	170	3066.02	I	350	3843.98	I	750	4762.38	I
30	2667.03	II	170	3070.27	I	120	3926.47	I	300	4765.86	I
110	2672.59	II	160	3073.13	I	130	3982.58	I	500	4766.43	I
55	2673.37	II	140 h	3178.50	I	150	3985.24	I	940	4783.42	I
55	2674.43	II	220	3212.88	I	190	3986.83	I	1000	4823.52	I
45	2680.34	II	1000	3228.09	I	150	3987.10	I	19	5004.91	I
30	2680.68	II	300	3230.72	I	1500	4018.10	I	30	5074.79	I
30	2681.25	II	850	3236.78	I	150	4026.44	I	200	5079.20	III
55	2684.55	II	330	3243.78	I	27000	4030.76	I	150	5100.03	III
55	2685.94	II	650	3248.52	I	19000	4033.07	I	60	5117.94	I
110	2688.25	II	100	3251.14	I	11000	4034.49	I	50	5150.89	I
27	2693.19	II	310	3252.95	I	1500	4035.73	I	50	5196.59	I
55	2695.36	II	310	3256.14	I	5600	4041.36	I	85	5255.32	I
27	2698.97	II	220	3258.41	I	210 d	4045.13	I	160	5341.06	I
85	2701.00	II	180	3260.23	I	1100	4048.76	I	19	5349.88	I
50	2701.17	II	180	3264.71	I	150	4055.21	I	95	5377.63	I
160	2701.70	II	200	3330.78	II	1900	4055.54	I	95	5394.67	I
100	2703.98	II	720	3441.99	II	210	4057.95	I	50	5399.49	I
130	2705.74	II	50	3460.03	II	1100	4058.93	I	95	5407.42	I
80	2707.53	II	360	3460.33	II	150	4059.39	I	35	5413.69	I
110	2708.45	II	360 h	3474.04	II	730	4061.74	I	85	5420.36	I
45	2709.96	II		3474.13	II	730	4063.53	I	35	5432.55	I
80	2710.33	II	290	3482.91	II	290	4070.28	I	150	5454.07	III
110	2711.58	II	180	3488.68	II	730	4079.24	I	12	5457.47	I
30	2716.80	II	140	3495.84	II	730	4079.42	I	60	5470.64	I
30	2717.53	II	50	3496.81	II	1100	4082.94	I	200	5474.68	III
30	2719.01	II	100	3497.54	II	1100	4083.63	I	40	5481.40	I
50	2719.74	II	360	3531.85	I	200	4110.90	I	30	5505.87	I
30	2722.10	II	1100	3532.12	I	150	4131.12	I	50	5516.77	I
30	2724.46	II	1300	3547.80	I	120	4135.04	I	40	5537.76	I
55	2728.61	II	1100	3548.03	I	150	4176.60	I	21	5551.98	I
6200	2794.82	I	390	3548.20	I	120	4189.99	I	200	5946.65	III
5100	2798.27	I	2200	3569.49	I	370	4235.14	I	140	6013.50	I
220	2799.84	I	720	3569.80	I	510	4235.29	I	200	6016.64	I
3700	2801.06	I	1400	3577.88	I	190	4239.72	I	290	6021.80	I
110	2809.11	I	720	3586.54	I	290	4257.66	I	200	6231.21	III
60	2815.02	II	290	3595.12	I	290	4265.92	I	17	6440.97	I
30	2816.33	II	150	3601.72	III	270	4281.10	I	24	6491.71	I
60	2870.08	II	420	3607.54	I	50	4323.63	II	14 h	6942.52	I
30	2872.94	II	420	3608.49	I	350	4414.88	I	12	6989.96	I
80	2879.49	II	360	3610.30	I	210	4436.35	I	14	7069.84	I

Intensity	Wavelength/Å		Intensity	Wavelength/Å		Intensity	Wavelength/Å		Intensity	Wavelength/Å	
12	7184.25	I	30	3801.660	I	40	1783.70	II	50	2698.83	I
24 h	7283.82	I	20	3901.867	I	30	1796.22	II	50	2699.38	I
35 h	7302.89	I	60	3906.372	I	200	1796.90	II	80	2705.36	II
50	7326.51	I	200	3983.839	II	60	1798.74	II	70	2724.43	III
12	7680.20	I	1800	4046.572	I	30	1803.89	II	80	2752.78	I
12 h	8672.06	I	150	4077.838	I	40	1808.29	II	20	2759.71	I
12 h	8701.05	I	40	4108.057	I	400	1820.34	II	6	2769.22	III
17 h	8703.76	I	250	4339.224	I	5	1832.74	I	40	2803.46	I
30 h	8740.93	I	400	4347.496	I	1000	1849.50	I	30	2804.43	I
			4000	4358.337	I	160	1869.23	II	2	2805.34	I
Mercury 198 Hg Z = 80			80	4916.068	I	300	1870.55	II	2	2806.77	I
80	1250.564	I	1100	5460.753	I	200	1875.54	II	150	2814.93	II
8	1259.242	I	160	5675.922	I	1	1894.77	III	3	2844.76	III
100	1268.825	I	240	5769.598	I	20	1900.28	II	750	2847.68	II
5	1307.751	I	280	5790.663	I	30	1927.60	II	50	2856.94	I
20	1402.619	I	20	6072.713	I	300	1942.27	II	150	2893.60	I
10	1435.503	I	30	6234.402	I	100	1972.94	II	150	2916.27	II
1000	1849.492	I	160	6716.429	I	200	1973.89	II	60	2925.41	I
60	2262.210	II	250	6907.461	I	150	1987.98	II	150	2935.94	II
20	2302.065	I	240	11287.407	I	90	2026.97	II	400	2947.08	II
20	2345.440	I				90	2052.93	II	1200	2967.28	I
100	2378.325	I	*Mercury Hg Z = 80*			70	2148.00	II	300	3021.50	I
20	2380.004	I	3	621.44	III	5	2247.55	I	120	3023.47	I
40	2399.349	I	2	679.68	III	60	2262.23	II	30	3025.61	I
20	2399.729	I	2	878.59	III	20	2302.06	I	50	3027.49	I
20	2446.900	I	1	886.48	III	7	2314.15	III	15	3090.05	III
15	2464.064	I	400	893.08	II	15	2323.20	I	400	3125.67	I
40	2481.999	I	300	915.83	II	5	2340.57	I	320	3131.55	I
30	2482.713	I	150	923.39	II	20	2345.43	I	320	3131.84	I
40	2483.821	I	200	940.80	II	20	2352.48	I	400	3208.20	II
90	2534.769	I	100	962.74	II	100	2378.32	I	400	3264.06	II
15000	2536.506	I	50	969.13	II	20	2380.00	I	5	3283.02	III
25	2563.861	I	1	988.89	III	4	2380.55	III	12	3312.28	III
25	2576.290	I	2	1009.29	III	40	2399.38	I	80	3341.48	I
250	2652.043	I	5	1068.03	III	20	2399.73	I	100	3385.25	II
400	2653.683	I	800	1099.26	II	10	2400.49	I	8	3389.01	III
100	2655.130	I	2	1161.95	III	60	2407.35	II	5	3450.77	III
50	2698.831	I	80	1250.58	I	50	2414.13	II	400	3451.69	II
80	2752.783	I	8	1259.24	I	8	2431.65	III	3	3500.35	III
20	2759.710	I	100	1268.82	I	5	2441.06	I	4	3538.88	III
40	2803.471	I	5	1307.75	I	20	2446.90	I	200	3549.42	II
30	2804.438	I	300	1307.93	II	15	2464.06	I	5	3557.24	III
750	2847.675	II	400	1321.71	II	5	2480.56	III	2800	3650.15	I
50	2856.939	I	400	1331.74	II	40	2482.00	I	300	3654.84	I
150	2893.598	I	80	1350.07	II	30	2482.72	I	80	3662.88	I
150	2916.227	II	200	1361.27	II	40	2483.82	I	240	3663.28	I
60	2925.413	I	20	1402.62	I	7	2484.50	III	30	3701.44	I
1200	2967.283	I	200	1414.43	II	90	2534.77	I	35	3704.17	I
300	3021.500	I	10	1435.51	I	15000	2536.52	I	30	3801.66	I
120	3023.476	I	15	1619.46	II	25	2563.86	I	15	3803.51	III
30	3025.608	I	120	1623.95	II	25	2576.29	I	100	3806.38	II
50	3027.490	I	20	1628.25	II	5	2578.91	I	20	3901.87	I
400	3125.670	I	150	1649.94	II	2	2612.92	III	60	3906.37	I
320	3131.551	I	50	1653.64	II	4	2617.97	III	100	3918.92	II
320	3131.842	I	200	1672.41	II	15	2625.19	I	200	3983.96	II
80	3341.481	I	9	1681.40	III	5	2639.78	I	1800	4046.56	I
2800	3650.157	I	100	1702.73	II	250	2652.04	I	150	4077.83	I
300	3654.839	I	100	1707.40	II	400	2653.69	I	40	4108.05	I
80	3662.883	I	120	1727.18	II	100	2655.13	I	70	4122.07	III
240	3663.281	I	250	1732.14	II	3	2670.49	III	10	4140.34	III
30	3701.432	I	15	1759.75	III	5	2674.91	I	100	4216.74	III
35	3704.170	I	20	1775.68	I						

Intensity	Wavelength/Å		Intensity	Wavelength/Å		Intensity	Wavelength/Å		Intensity	Wavelength/Å	
250	4339.22	I	120	13209.95	I	150	2269.71	III	80	2834.39	II
400	4347.49	I	140	13426.57	I	200	2294.97	III	80	2835.33	II
4000	4358.33	I	60	13468.38	I	160	2304.25	II	160	2842.15	II
100	4398.62	II	80	13505.58	I	160	2306.97	II	1700	2848.23	II
15	4470.58	III	500	13570.21	I	150	2330.93	III	370	2853.23	II
12	4552.84	III	450	13673.51	I	110	2332.12	II	370	2863.81	II
90	4660.28	II	200	13950.55	I	190	2341.59	II	220	2866.69	II
50	4797.01	III	500	15295.82	I	100	2359.76	III	1700	2871.51	II
80	4855.72	II	100	16881.48	I	110	2389.20	II	85	2872.88	II
10	4869.85	III	400	16920.16	I	140	2403.61	II	220	2879.05	II
5	4883.00	I	300	16942.00	I	120	2413.01	II	65	2888.15	II
5	4889.91	I	500	17072.79	I	85	2498.28	II	1300	2890.99	II
80	4916.07	I	400	17109.93	I	200	2506.19	III	95	2891.28	II
5	4970.37	I	20	17116.75	I	440	2538.46	II	190	2892.81	II
80	4973.57	III	20	17198.67	I	330	2542.67	II	950	2894.45	II
5	4980.64	I	20	17213.20	I	80	2558.88	II	140	2897.63	II
20	5102.70	I	70	17329.41	I	85	2564.34	II	70	2900.80	II
40	5120.64	I	30	17436.18	I	250	2593.70	II	290	2903.07	II
100	5128.45	II	50	18130.38	I	250	2602.80	II	80	2907.12	II
20	5137.94	I	40	19700.17	I	400	2616.78	I	600	2909.12	II
30	5210.82	III		22493.28	I	440	2629.85	I	1100	2911.92	II
20	5290.74	I	250	23253.07	I	330	2636.67	II	120	2918.83	II
5	5316.78	I		32148.06	I	720	2638.76	II	1300	2923.39	II
60	5354.05	I		36303.03	I	410	2640.99	I	140	2924.32	II
30	5384.63	I				600	2644.35	II	1100	2930.50	II
1100	5460.74	I	*Molybdenum Mo Z = 42*			370	2646.49	II	800	2934.30	II
30	5549.63	I	50	867.92	IV	640	2649.46	I	95	2940.10	II
160	5675.86	I	100	884.19	IV	480	2653.35	II	110	2941.22	II
6	5695.71	III	60	886.05	IV	560 h	2655.03	I	150	2944.82	II
240	5769.60	I	50	891.74	IV	640	2660.58	II	140	2946.69	II
100	5789.66	I	100	1169.33	III	720	2672.84	II	95	2947.28	II
280	5790.66	I	100	1254.93	III	250	2673.27	II	125	2947.32	III
140	5803.78	I	100	1258.52	III	1000	2679.85	I	95	2955.84	II
60	5859.25	I	100	1262.21	III	95	2681.36	II	240	2956.06	II
60	5871.73	II	100	1263.74	III	640	2683.23	II	70	2956.90	II
20	5871.98	I	100	1274.37	III	880	2684.14	II	95	2960.24	II
20	6072.72	I	100	1276.40	III	560	2687.99	II	250	2963.79	II
1000	6149.50	II	200	1277.40	III	480	2701.42	II	210	2965.27	II
25	6220.35	III	200	1277.58	III	190	2713.51	II	70	2971.91	II
30	6234.40	I	200	1278.40	III	290	2717.35	II	250	2972.61	II
35	6418.98	III	150	1281.90	III	85	2726.97	II	80	2975.40	II
40	6501.38	III	150	1283.60	III	140	2729.68	II	95	2992.84	II
80	6521.13	II	100	1854.73	III	80	2730.20	II	95	3027.77	II
10	6584.26	III	80	1926.26	IV	330	2732.88	II	100	3060.78	II
6	6610.12	III	100	1929.24	IV	160	2736.96	II	800	3064.28	I
30	6709.29	III	80	1971.06	IV	80 h	2737.88	II	250	3065.04	II
160	6716.43	I	70	2010.92	IV	290	2746.30	II	800	3074.37	I
250	6907.52	I	19000	2015.11	II	110	2756.07	II	85	3077.66	II
250	7081.90	I	40000	2020.30	II	220	2763.62	II	800	3085.62	I
200	7091.86	I	21000	2038.44	II	240	2769.76	II	270	3087.62	II
40	7346.37	II	17000	2045.98	II	160	2773.78	II	190	3092.07	II
100	7485.87	II	50	2060.38	IV	190	2774.39	II	560	3094.66	I
12	7517.46	III	4800	2081.68	II	1700	2775.40	II	560	3101.34	I
20	7728.82	I	2400	2089.52	II	65	2777.86	II	1400	3112.12	I
7	7808.10	III	2200	2092.50	II	880	2780.04	II	290	3122.00	II
100	7944.66	II	4000	2093.11	II	400	2784.99	II	14000	3132.59	I
25	7946.75	III	2700	2100.84	II	100	2807.74	III	110	3138.72	II
50	7984.51	III	1500	2104.29	II	400	2807.76	II	220	3152.82	II
5	8151.64	III	1400	2108.02	II	1700	2816.15	II	55	3155.64	II
2000	10139.75	I	100	2184.37	III	220	2817.44	II	6000	3158.16	I
240	11287.40	I	100	2211.02	III	80	2827.74	II	8700	3170.35	I
			400	2269.69	II						

Intensity	Wavelength/Å		Intensity	Wavelength/Å		Intensity	Wavelength/Å		Intensity	Wavelength/Å	
95	3172.03	II	580	3886.82	I	65	5279.65	I	95	3007.97	II
160	3172.74	II	19000	3902.96	I	210	5280.86	I	95	3014.19	II
120 d	3187.59	II	65	3941.48	II	55	5292.08	I	95	3018.35	II
7600	3193.97	I	1400	4062.08	I	55	5295.47	I	140	3056.71	II
880	3205.88	I	2300	4069.88	I	55	5313.89	I	130	3069.73	II
3000	3208.83	I	1300	4081.44	I	80	5354.88	I	160	3075.38	II
560	3215.07	I	940	4084.38	I	65	5356.48	I	240	3092.92	II
880	3228.22	I	730	4107.47	I	560 hl	5360.56	I	260	3115.18	II
600	3229.79	I	630	4120.10	I	110 hl	5364.28	I	290	3133.60	II
1100	3233.14	I	2900	4143.55	I	65	5394.52	I	220	3134.90	II
950	3237.08	I	480	4185.82	I	50	5400.47	I	170	3141.46	II
65	3240.71	II	2500	4188.32	I	55	5435.68	I	170	3142.44	II
950	3256.21	I	1500	4232.59	I	65	5437.75	I	150	3203.47	II
480	3264.40	I	890	4276.91	I	50	5501.54	I	220	3259.24	II
800	3270.90	I	1200	4277.24	I	7800	5506.49	I	220	3265.12	II
200	3271.69	III	1400	4288.64	I	5200	5533.05	I	320	3275.22	II
1100	3289.02	I	680	4292.13	I	50	5543.12	I	290	3285.10	II
950	3290.82	I	890	4293.21	I	55	5556.28	I	410	3328.28	II
190	3292.31	II	840	4326.14	I	2500	5570.45	I	320	3353.59	II
100	3313.62	II	1900	4381.64	I	100	5610.93	I	410	3560.75	II
190	3320.90	II	2500	4411.57	I	330	5632.47	I	470	3587.51	II
640	3323.95	I	990	4434.95	I	50	5634.86	I	370	3615.82	II
1300	3344.75	I	480	4457.36	I	230	5650.13	I	410	3653.15	II
95	3346.40	II	630	4474.56	I	55	5674.47	I	470	3662.26	II
1600	3358.12	I	400	4536.80	I	460	5689.14	I	540	3665.18	II
950	3363.78	I	460	4626.47	I	80	5705.72	I	540	3672.36	II
950	3379.97	I	640	4707.26	I	210	5722.74	I	580	3673.54	II
1900	3384.62	I	700	4731.44	I	620	5751.40	I	1200	3685.80	II
130	3395.36	II	770	4760.19	I	520	5791.85	I	440	3687.30	II
640	3404.34	I	410	4819.25	I	55 h	5849.73	I	410	3689.69	II
1300	3405.94	I	410	4830.51	I	50 h	5851.52	I	410	3697.56	II
640	3437.22	I	180	5014.60	I	520	5858.27	I	470	3713.70	II
130	3446.08	II	80	5029.00	I	50	5869.33	I	640 d	3714.73	II
3200	3447.12	I	65	5030.78	I	820	5888.33	I	470	3715.68	II
640	3449.07	I	100	5047.71	I	50 h	5893.38	I	410	3718.54	II
950	3456.39	I	50	5055.00	I	160 h	5928.88	I	410	3721.35	II
640	3460.78	I	200	5059.88	I	35	6025.49	I	780	3723.50	II
800	3504.41	I	100	5080.02	I	1300	6030.66	I	410	3724.87	II
560	3508.12	I	100	5096.65	I	40	6101.87	I	710	3728.13	II
480	3521.41	I	130	5097.52	I	40	6357.22	I	470	3730.58	II
640	3537.28	I	130	5109.71	I	35	6401.07	I	1000 d	3735.54	II
520	3558.10	I	80	5114.97	I	100	6424.37	I	440	3737.10	II
400	3563.14	I	150	5145.38	I	230	6619.13	I	1000	3738.06	II
1400	3581.89	I	110	5147.39	I	50	6650.38	I	580	3752.49	II
1400	3624.46	I	80	5163.19	I	110	6733.98	I	510	3757.82	II
1000	3635.43	I	100	5167.76	I	50	6746.27	I	930	3758.95	II
400	3657.35	I	160 d	5171.08	I	35	6753.97	I	930	3763.47	II
540	3664.81	I	230 h	5172.94	I	40	6838.88	I	510	3769.65	II
590	3672.82	I	160 h	5174.18	I	35	6914.01	I	1400	3775.50	II
1300	3680.60	I	110	5200.17	I	110	7109.87	I	710	3779.47	II
65	3688.31	II	50	5200.74	I	150	7242.50	I	580	3780.40	II
180	3692.64	II	50	5211.86	I	40	7245.85	I	510	3781.32	II
1400	3694.94	I	80	5219.40	I	40	7391.36	I	2400	3784.25	II
500	3727.69	I	65	5231.06	I	140	7485.74	I	370	3801.12	II
80	3744.37	II	100	5234.26	I	27	7720.77	I	1200	3803.47	II
29000	3798.25	I	460 h	5238.20	I	40 h	8328.44	I	2500	3805.36	II
520	3826.70	I	230 h	5240.88	I	45 h	8389.32	I	470	3807.23	II
940	3828.87	I	110 h	5242.81	I	45 h	8483.39	I	540	3808.77	II
1700	3833.75	I	100	5245.51	I				440	3809.06	II
29000	3864.11	I	150	5259.04	I	**Neodymium Nd Z = 60**			580	3810.49	II
580	3869.08	I	65	5261.14	I	75	2764.98	I	710	3814.73	II
						80	2993.20	II			

Intensity	Wavelength/Å		Intensity	Wavelength/Å		Intensity	Wavelength/Å		Intensity	Wavelength/Å	
410	3822.47	II	1000	4021.78	II	240	4719.02	I	20	7192.01	II
1200	3826.42	II	1200	4023.00	II	240	4811.34	II	15	7236.54	II
540	3828.85	II	410	4030.47	II	350	4825.48	II	12	7316.81	II
440	3829.16	II	1200	4031.82	II	280	4859.02	II	10	7406.62	II
510	3830.47	II	3000	4040.80	II	350	4883.81	I	10	7418.18	II
740	3836.54	II	410	4043.59	II	220	4890.70	II	12	7511.16	II
1700	3838.98	II	410	4048.81	II	240	4891.07	I	17	7513.73	II
410 d	3841.82	II	850	4051.15	II	280	4896.93	I	12	7528.99	II
1700 d	3848.24	II	850	4059.96	II	210	4901.84	I	10	7538.26	II
1500	3848.52	II	4700	4061.09	II	330	4920.68	II	12	7696.56	II
470	3850.22	II	1100	4069.28	II	470	4924.53	I	10	7750.95	II
2400 d	3851.66	II	710	4075.12	II	260	4944.83	I	10	7808.47	II
3700 d	3863.33	II	470	4075.28	II	290	4954.78	I	12	7863.04	II
850	3869.07	II	470	4080.23	II	290	4959.13	II	12	7917.01	II
470	3875.87	II	1400	4109.08	II	250	4989.94	II	12	7958.95	I
1100	3878.58	II	2500	4109.46	II	360	5076.59	II	12	7965.73	II
1000	3879.55	II	510	4110.48	II	360	5092.80	II	15	7982.09	II
780	3880.38	II	410	4123.88	II	360	5107.59	II	12	7982.68	II
1200	3880.78	II	470	4133.36	II	340	5123.79	II	12	8000.76	II
540	3887.87	II	510	4135.33	II	680	5130.60	II	10	8120.93	II
1300	3889.93	II	3000	4156.08	II	500	5191.45	II	12	8122.07	II
1300	3890.58	II	510	4156.26	II	630	5192.62	II	12	8141.75	II
1300	3890.94	II	410	4168.00	II	330	5200.12	II	12	8143.27	II
580	3891.51	II	810	4175.61	II	310	5212.37	II	10	8231.52	II
470	3892.06	II	2400	4177.32	II	450	5234.20	II	10	8307.72	II
810	3894.63	II	640	4179.59	II	250	5239.79	II	12	8346.36	II
440	3897.63	II	470	4205.60	II	720	5249.59	II	17	8839.10	II
2000	3900.21	II	470	4211.29	II	360	5255.51	II			
1300	3901.84	II	440	4227.73	II	590	5273.43	II	**Neon Ne Z = 10**		
1700	3905.89	II	1300	4232.38	II	680	5293.17	II	66	119.01	V
510	3907.84	II	2000	4247.38	II	220	5311.46	II	200	122.52	V
2000	3911.16	II	850	4252.44	II	500	5319.82	II	66	125.12	V
850	3912.23	II	410	4261.84	II	290	5361.47	II	45	131.99	V
440	3915.13	II	470	4282.44	II	160	5431.53	II	50	132.04	V
610	3915.95	II	710	4284.52	II	240	5594.43	II	150	140.76	V
1100	3920.96	II	5400	4303.58	II	220	5620.54	I	150	140.79	V
510	3927.10	II	470	4314.52	II	140 d	5675.97	I	100	142.44	V
610	3934.82	II	1100	4325.76	II	220	5688.53	II	100	142.50	V
410	3936.11	II	510	4327.93	II	130	5702.24	II	150	142.72	V
510	3938.86	II	540	4338.70	II	160	5708.28	II	100	143.27	V
2000	3941.51	II	680	4351.29	II	100	5729.29	I	150	143.34	V
2000	3951.16	II	850	4358.17	II	160	5804.02	II	150	147.13	V
810	3952.20	II	470 d	4374.93	II	80	5811.57	II	66	151.23	V
590	3958.00	II	710	4385.66	II	70	5825.87	II	120	151.42	V
510	3962.21	II	540	4400.83	II	80	5842.39	II	15	151.82	IV
1400	3963.12	II	510	4411.06	II	55	5858.91	I	15	152.23	IV
1100	3973.30	II	580	4446.39	II	45	6007.67	I	45	154.50	V
740	3973.69	II	1400	4451.57	II	45	6034.24	II	15	158.65	IV
740	3976.85	II	740	4462.99	II	55	6066.03	I	15	158.82	IV
740	3979.49	II	410	4501.82	II	45	6178.59	I	100	164.02	V
470	3986.25	II	250	4516.36	II	45	6223.39	I	100	164.14	V
1400	3990.10	II	340	4541.27	II	55	6310.49	I	80	172.62	IV
1000	3991.74	II	340	4542.61	II	65	6385.20	II	500	173.93	V
1100	3994.68	II	340	4563.22	II	45	6630.14	I	80	177.16	IV
410	4000.50	II	300	4621.94	I	45	6650.57	II	150	186.58	IV
540	4004.02	II	510	4634.24	I	40	6740.11	II	100	194.28	IV
410	4007.43	II	340	4641.10	I	40	6900.43	II	100	208.48	IV
3700	4012.25	II	250	4645.77	II	35	7037.30	II	100	208.73	IV
540	4012.70	II	300	4649.67	I	40	7066.89	II	80	208.90	IV
1000	4020.87	II	310	4683.45	I	29	7129.35	II	150	212.56	IV
1000	4021.34	II	470	4706.54	II	24	7189.42	II	140	223.24	IV
									120	223.60	IV

Intensity	Wavelength/Å		Intensity	Wavelength/Å		Intensity	Wavelength/Å		Intensity	Wavelength/Å	
140	234.32	IV	100	542.07	IV	300	2216.07	III	150	2963.24	II
120	234.70	IV	150	543.89	IV	10	2220.81	IV	150	2967.18	II
20	251.14	III	400	568.42	V	75	2227.42	V	100	2973.10	II
20	251.56	III	250	569.76	V	110	2232.41	V	15	2974.72	I
20	251.73	III	500	569.83	V	65	2245.48	V	100	2979.46	II
40	267.06	III	250	572.11	V	250	2258.02	IV	12	2982.67	I
40	267.52	III	800	572.34	V	65	2259.57	V	150	3001.67	II
20	267.71	III	35	587.213	I	175	2262.08	IV	120 p	3017.31	II
40	283.18	III	35	589.179	I	240	2263.21	III	300	3027.02	II
160	283.21	III	35	589.911	I	65	2263.39	V	300	3028.86	II
110	283.69	III	70	591.830	I	110	2264.54	IV	100	3030.79	II
40	283.89	III	100	595.920	I	200	2264.91	III	120	3034.46	II
220	301.12	III	75	598.706	I	250	2265.71	V	100	3035.92	II
220	313.05	III	35	598.891	I	550	2285.79	IV	100	3037.72	II
220	313.68	III	70	600.036	I	30	2293.14	IV	100	3039.59	II
40	313.95	III	170	602.726	I	250	2293.49	IV	100	3044.09	II
90	352.956	I	170	615.628	I	250	2350.84	IV	100	3045.56	II
60	354.962	I	170	618.672	I	450	2352.52	IV	120	3047.56	II
50	357.83	IV	120	619.102	I	700	2357.96	IV	100	3054.34	II
400	357.96	V	200	626.823	I	250	2362.68	IV	100	3054.68	II
500	358.47	V	200	629.739	I	250	2363.28	IV	100	3059.11	II
200	358.72	IV	1000	735.896	I	110	2365.49	IV	100	3062.49	II
500	359.38	V	400	743.720	I	350	2372.16	IV	100	3063.30	II
90	361.433	II	60	993.88	I	65	2384.20	IV	100	3070.89	II
60	362.455	II	70	1068.65	I	350	2384.95	IV	100	3071.53	II
1000	365.59	V	90	1131.72	I	300	2412.73	III	100	3075.73	II
220	379.31	III	100	1131.85	II	240	2412.94	III	120	3088.17	II
125	387.14	IV	90	1229.83	I	200	2413.78	III	100	3092.09	II
100	388.22	IV	20	1255.03	III	200	2473.40	III	120	3092.90	II
150	405.854	II	110	1255.68	III	80 p	2562.12	II	100	3094.01	II
120	407.138	II	160	1257.19	III	90 w	2567.12	II	100	3095.10	II
800	416.20	V	90	1418.38	I	800	2590.04	III	100	3097.13	II
150	421.61	IV	90	1428.58	I	600	2593.60	III	100	3117.98	II
200	445.040	II	90	1436.09	I	400	2595.68	III	120	3118.16	II
300	446.256	II	120	1681.68	II	300	2610.03	III	10	3126.199	I
250	446.590	II	180	1688.36	II	240	2613.41	III	300	3141.33	II
180	447.815	II	100	1888.11	III	200	2615.87	III	100	3143.72	II
150	454.654	II	100	1889.71	III	80	2623.11	II	100 p	3148.68	II
200	455.274	II	200	1907.49	II	80	2629.89	II	100	3164.43	II
10	456.275	II	500	1916.08	II	90 w	2636.07	II	100	3165.65	II
120	456.348	II	300	1930.03	II	80	2638.29	II	100	3188.74	II
90	456.896	III	200	1938.83	II	200	2638.70	III	120	3194.58	II
1000	460.728	II	100 c	1945.46	II	200	2641.07	III	500	3198.59	II
500	462.391	II	80	2007.01	II	80	2644.10	II	60	3208.96	II
140	469.77	IV	65	2018.44	IV	600	2677.90	III	120	3209.36	II
200	469.82	IV	110	2022.19	IV	500	2678.64	III	120	3213.74	II
180	469.87	IV	80	2025.56	II	80	2762.92	II	150	3214.33	II
140	469.92	IV	150	2085.47	II	90	2792.02	II	150	3218.19	II
250	480.41	V	200	2086.96	III	80	2794.22	II	120	3224.82	II
150	481.28	V	300	2089.43	III	100	2809.48	II	120	3229.57	II
250	481.36	V	240	2092.44	III	80	2906.59	II	200	3230.07	II
500	482.99	V	400	2095.54	III	80	2906.82	II	120	3230.42	II
285	488.10	III	180	2096.11	II	90	2910.06	II	120	3232.02	II
220	488.87	III	120	2096.25	II	90	2910.41	II	150	3232.37	II
450	489.50	III	200	2161.22	III	80	2911.14	II	100	3243.40	II
70	489.64	III	300	2163.77	III	80	2915.12	II	100	3244.10	II
220	490.31	III	200	2180.89	III	80	2925.62	II	100	3248.34	II
360	491.05	III	30	2203.88	IV	80 w	2932.10	II	100	3250.36	II
120	521.74	IV	200	2209.35	III	80	2940.65	II	150	3297.73	II
140	521.82	IV	200	2211.85	III	90	2946.04	II	150	3309.74	II
80	541.13	IV	240	2213.76	III	150	2955.72	II	300	3319.72	II

Intensity	Wavelength/Å		Intensity	Wavelength/Å		Intensity	Wavelength/Å		Intensity	Wavelength/Å	
1000	3323.74	II	250	3727.11	II	100	6074.338	I	1000	8571.352	I
150	3327.15	II	800	3766.26	II	80	6096.163	I	4000	8591.259	I
100	3329.16	II	1000	3777.13	II	60	6128.450	I	6000	8634.647	I
200	3334.84	II	100	3818.43	II	100	6143.063	I	3000	8647.041	I
150	3344.40	II	120	3829.75	II	120	6163.594	I	15000	8654.383	I
300	3345.45	II	150	4219.74	II	250	6182.146	I	4000	8655.522	I
150	3345.83	II	100	4233.85	II	150	6217.281	I	100	8668.26	II
200	3355.02	II	120	4250.65	II	150	6266.495	I	5000	8679.492	I
120	3357.82	II	120	4369.86	II	60	6304.789	I	5000	8681.921	I
200	3360.60	II	70	4379.40	II	100	6334.428	I	2000	8704.112	I
120	3362.16	II	150	4379.55	II	120	6382.992	I	4000	8771.656	I
100	3362.71	II	100	4385.06	II	200	6402.246	I	12000	8780.621	I
120	3367.22	II	200	4391.99	II	150	6506.528	I	10000	8783.753	I
12	3369.808	I	150	4397.99	II	60	6532.882	I	500	8830.907	I
40	3369.908	I	150	4409.30	II	150	6598.953	I	7000	8853.867	I
100	3371.80	II	100	4413.22	II	70	6652.093	I	1000	8865.306	I
500	3378.22	II	100	4421.39	II	90	6678.276	I	1000	8865.755	I
150	3388.42	II	100 p	4428.52	II	20	6717.043	I	3000	8919.501	I
120	3388.94	II	100 p	4428.63	II	100	6929.467	I	2000	8988.57	I
300	3392.80	II	150 p	4430.90	II	90	7024.050	I	100	9079.46	II
100	3404.82	II	150 p	4430.94	II	100	7032.413	I	6000	9148.67	I
120	3406.95	II	120	4457.05	II	50	7051.292	I	6000	9201.76	I
100	3413.15	II	100	4522.72	II	80	7059.107	I	4000	9220.06	I
120	3416.91	II	10	4537.754	I	100	7173.938	I	2000	9221.58	I
120	3417.69	II	10	4540.380	I	150	7213.20	II	2000	9226.69	I
50	3417.904	I	100	4569.06	II	150	7235.19	II	1000	9275.52	I
15	3418.006	I	15	4704.395	I	100	7245.167	I	200	9287.56	II
120	3428.69	II	12	4708.862	I	150	7343.94	II	6000	9300.85	I
60	3447.703	I	10	4710.067	I	40	7472.439	I	1500	9310.58	I
50	3454.195	I	10	4712.066	I	90	7488.871	I	3000	9313.97	I
100	3456.61	II	15	4715.347	I	100	7492.10	II	6000	9326.51	I
100	3459.32	II	10	4752.732	I	150	7522.82	II	2000	9373.31	I
25	3460.524	I	12	4788.927	I	80	7535.774	I	5000	9425.38	I
30	3464.339	I	10	4790.22	I	60	7544.044	I	3000	9459.21	I
30	3466.579	I	10	4827.344	I	100	7724.628	I	5000	9486.68	I
60	3472.571	I	10	4884.917	I	120	7740.74	II	5000	9534.16	I
150	3479.52	II	4	5005.159	I	300	7839.055	I	3000	9547.40	I
200	3480.72	II	10	5037.751	I	120	7926.20	II	120	9577.01	II
200	3481.93	II	10	5144.938	I	400	7927.118	I	1000	9665.42	I
25	3498.064	I	25	5330.778	I	700	7936.996	I	100	9808.86	II
30	3501.216	I	20	5341.094	I	2000	7943.181	I	800	10295.42	I
25	3515.191	I	8	5343.283	I	2000	8082.458	I	2000	10562.41	I
150	3520.472	I	60	5400.562	I	100	8084.34	II	1500	10798.07	I
120	3542.85	II	5	5562.766	I	1000	8118.549	I	2000	10844.48	I
120	3557.80	II	10	5656.659	I	600	8128.911	I	3000	11143.020	I
100	3561.20	II	5	5719.225	I	3000	8136.406	I	3500	11177.528	I
250	3568.50	II	12	5748.298	I	2500	8259.379	I	1600	11390.434	I
100	3574.18	II	80	5764.419	I	100	8264.81	II	1100	11409.134	I
200	3574.61	II	12	5804.450	I	2500	8266.077	I	3000	11522.746	I
50	3593.526	I	40	5820.156	I	800	8267.117	I	1500	11525.020	I
30	3593.640	I	500	5852.488	I	6000	8300.326	I	950	11536.344	I
15	3600.169	I	100	5872.828	I	100	8315.00	II	500	11601.537	I
20	3633.665	I	100	5881.895	I	1500	8365.749	I	1200	11614.081	I
150	3643.93	II	60	5902.462	I	100	8372.11	II	300	11688.002	I
200	3664.07	II	60	5906.429	I	8000	8377.606	I	2000	11766.792	I
20	3682.243	I	100	5944.834	I	1000	8417.159	I	1500	11789.044	I
12	3685.736	I	100	5965.471	I	4000	8418.427	I	500	11789.889	I
200	3694.21	II	100	5974.627	I	1500	8463.358	I	1000	11984.912	I
10	3701.225	I	120	5975.534	I	800	8484.444	I	3000	12066.334	I
150	3709.62	II	80	5987.907	I	5000	8495.360	I	800	12459.389	I
250	3713.08	II	100	6029.997	I	600	8544.696	I	1000	12689.201	I

Intensity	Wavelength/Å		Intensity	Wavelength/Å		Intensity	Wavelength/Å		Intensity	Wavelength/Å	
1100	12912.014	I	300 l	3986.89	I	300	8155.11	I	300	863.22	III
700	13219.241	I	300 s	5044.66	I	300 l	8167.42	I	300	867.51	III
800	15230.714	I	300 l	5601.70	I	300 l	8183.06	I	300	973.79	III
400	17161.930	I	300 l	5652.75	I	300 l	8188.61	I	400	979.59	III
400	18035.80	I	300 l	5784.39	I	300 l	8247.82	I	500	1317.22	II
1000	18083.21	I	300 l	5878.04	I	300 l	8287.11	I	76	1398.19	IV
350	18221.11	I	300 s	6011.22	I	300 s	8287.75	I	74	1411.45	IV
250	18227.02	I	300	6056.09	I	300 l	8306.22	I	70	1438.82	IV
2500	18276.68	I	300 s	6073.90	I	300 s	8313.66	I	73	1449.01	IV
2000	18282.62	I	300 s	6080.05	I	1000 l	8339.12	I	76	1452.22	IV
1200	18303.97	I	300 l	6120.49	I	300	8356.79	I	73	1482.25	IV
250	18359.12	I	300	6188.59	I	300 l	8367.11	I	72	1489.83	IV
1200	18384.85	I	300 l	6200.00	I	3000	8372.88	I	75	1525.31	IV
2000	18389.95	I	300 s	6215.90	I	3000	8529.96	I	74	1527.68	IV
1000	18402.84	I	300 s	6317.84	I	1000 s	8696.23	I	74	1527.80	IV
1200	18422.39	I	300 l	6341.38	I	1000 s	8906.02	I	76	1534.71	IV
300	18458.65	I	300 l	6566.11	I	1000	8942.70	I	73	1537.25	IV
400	18475.79	I	300 l	6720.68	I	1000 s	9004.75	I	75	1543.41	IV
900	18591.55	I	300 s	6751.32	I	1000 l	9006.31	I	74	1546.23	IV
1600	18597.70	I	300 s	6795.21	I	10000 l	9016.18	I	300	1604.54	III
350	18618.96	I	300 l	6802.62	I	3000 l	9141.30	I	300	1652.87	III
550	18625.16	I	300 l	6805.81	I	3000 s	9379.33	I	400	1687.90	III
1200	21041.295	I	300 s	6816.44	I	3000 l	9468.66	I	1000	1692.51	III
750	21708.145	I	300 l	6865.45	I	3000 s	9679.13	I	800	1709.90	III
300	22247.35	I	300 s	6907.13	I	3000 l	9930.55	I	650	1715.30	III
350	22428.13	I	300 h	6912.91	I	10000 l	10091.99	I	500	1719.46	III
2250	22530.40	I	1000 s	6930.31	I	10000 s	10817.45	I	400	1722.28	III
400	22661.81	I	300 l	6963.63	I	10000 l	11695.15	I	500	1738.25	III
600	23100.51	I	3000 s	6972.09	I	10000 l	11776.64	I	300	1739.78	III
1000	23260.30	I	300	7014.02	I	10000 s	12148.18	I	1000	1741.55	II
1050	23373.00	I	300 l	7018.91	I	10000 s	12377.42	I	300	1741.96	III
850	23565.36	I	300 s	7039.14	I	10000 l	12407.99	I	550	1747.01	III
3500	23636.52	I	300 s	7080.01	I	10000 l	13834.33	I	300	1752.43	III
300	23701.64	I	300 l	7174.83	I				400	1753.01	III
1100	23709.2	I	300 l	7184.93	I	*Nickel Ni Z = 28*			800	1764.69	III
1800	23951.42	I	300 l	7284.28	I	55	315.24	V	500	1767.94	III
600	23956.46	I	300 l	7292.29	I	56	315.71	V	2000	1769.64	III
1000	23978.12	I	300 l	7332.52	I	72	354.18	V	400	1776.07	III
200	24098.54	I	300 s	7370.60	I	76	354.42	V	300	1807.24	III
500	24161.42	I	300 l	7381.03	I	68	354.49	V	300	1819.28	III
600	24249.64	I	300 l	7381.65	I	500	630.71	III	800	1823.06	III
1500	24365.05	I	300 l	7402.70	I	500	676.94	III	400	1830.01	III
800	24371.60	I	300 s	7512.22	I	300	713.33	III	650	1847.28	III
400	24447.85	I	300 l	7515.15	I	300	713.38	III	800	1854.15	III
700	24459.4	I	300 l	7546.05	I	500	718.48	III	300	1858.75	III
300	24776.46	I	300 l	7624.83	I	300	722.09	III	1000	2165.55	II
550	24928.88	I	300	7626.85	I	500	729.82	III	2000	2169.10	II
250	25161.69	I	300 s	7681.01	I	400	731.70	III	2000	2174.67	II
650	25524.37	I	300 s	7685.25	I	300	732.16	III	1500	2175.15	II
125	28386.21	I	1000 l	7735.14	I	300	747.99	III	2500	2185.50	II
150	30200.	I	300 l	7761.61	I	300	750.05	III	3000	2192.09	II
250	33173.09	I	1000 l	7765.75	I	300	757.80	III	5000	2205.55	II
450	33352.35	I	300 s	7776.07	I	400	770.22	III	4000	2206.72	II
1300	33901.	I	300	7787.46	I	500	778.81	III	6000	2216.48	II
2200	33912.10	I	1000 l	7791.38	I	300	788.04	III	1000	2264.46	II
600	34131.31	I	300 l	7851.44	I	500	811.57	III	2000	2270.21	II
100	34471.44	I	300 l	7887.88	I	500	826.14	III	1600	2289.98	I
120	35834.78	I	300 l	7901.71	I	500	842.14	III	630	2300.78	I
			300 l	7975.98	I	400	845.24	III	1000	2303.00	II
Neptunium Np Z = 93			300 h	8080.32	I	300	847.43	III	2000	2310.96	I
300	3481.93	I	300 s	8124.59	I	300	860.64	III	1700	2312.34	I
300 h	3501.50	I				300	862.88	III			

Intensity	Wavelength/Å		Intensity	Wavelength/Å		Intensity	Wavelength/Å		Intensity	Wavelength/Å	
1400	2313.66	I	530	3612.74	I	500 w	9900.92	II	100	2362.06	III
1400	2313.98	I	6600	3619.39	I				80	2362.50	III
1000	2316.04	II	200	3664.10	I	*Niobium Nb* *Z = 41*			80	2365.70	III
1400	2317.16	I	130	3669.24	I	80	464.55	V	100	2372.73	III
2600	2320.03	I	180	3670.43	I	80	468.32	V	170	2376.40	II
1900	2321.38	I	260	3674.15	I	80	763.77	V	110	2387.09	II
1400	2325.79	I	160	3688.42	I	80	774.02	V	100	2387.41	III
940	2329.96	I	80	3693.93	I	60	993.54	IV	140	2387.52	II
1200	2345.54	I	120	3722.48	I	400	1005.72	IV	80	2388.23	III
400	2347.52	I	150	3736.81	I	500	1007.05	IV	45	2388.27	II
1000	2375.42	II	60	3739.23	I	500	1010.19	IV	160	2398.48	II
240	2386.58	I	600	3775.57	I	100	1116.08	IV	80	2404.89	III
1000	2394.52	II	700	3783.53	I	150	1120.02	IV	55	2405.34	II
2000	2416.13	II	700	3807.14	I	100	1258.87	V	55	2405.85	II
240	2419.31	I	110	3831.69	I	60	1314.56	III	140	2412.46	II
160	2472.06	I	1200	3858.30	I	80	1445.43	III	100	2413.94	III
150	2798.65	I	110	3973.56	I	80	1445.98	III	160	2416.99	II
250	2821.29	I	110	4401.55	I	80	1447.09	III	140	2418.69	II
500	2943.91	I	85	4459.04	I	100	1456.68	III	100	2421.91	III
570	2981.65	I	55	4470.48	I	80	1484.73	III	75	2433.80	II
500	2992.60	I	65	4605.00	I	100	1495.94	III	40	2435.95	II
1000	2994.46	I	75	4648.66	I	80	1498.02	III	45	2437.42	II
4000	3002.49	I	110	4714.42	I	80	1499.45	III	40	2442.14	II
2200	3003.63	I	45	4786.54	I	100	1501.99	III	28	2442.68	II
3700	3012.00	I	45	4855.41	I	60	1502.30	IV	65	2451.87	II
1700	3037.94	I	40	4904.41	I	80	1513.81	III	65	2453.95	II
3500	3050.82	I	45	4980.16	I	60	1524.36	IV	100	2456.99	III
1500	3054.32	I	45	4984.13	I	100	1524.91	III	55	2458.09	II
1900	3057.64	I	50	5017.59	I	100	1590.21	III	65	2462.89	I
500	3064.62	I	100	5035.37	I	80	1598.86	III	80	2468.72	III
2600	3101.55	I	100	5080.52	I	80	1604.72	III	80	2475.87	III
1300	3101.88	I	65	5081.11	I	80	1639.51	III	110	2477.38	II
2900	3134.11	I	40 h	5146.48	I	100	1682.77	III	65	2478.29	II
1100	3232.96	I	40 h	5155.76	I	100	1705.44	III	65	2479.94	II
600	3243.06	I	180	5476.91	I	100	1707.14	III	35	2483.88	II
660	3315.66	I	23	5709.56	I	100	1758.33	V	100	2499.73	III
2000	3331.88	II	16	5754.68	I	100	1877.34	V	110	2511.00	II
2900	3369.57	I	10	5857.76	I	100	1892.92	III	110	2521.40	II
3300	3380.57	I	10	5892.88	I	60	1922.41	IV	390	2544.80	II
1300	3391.05	I	10	6108.12	I	100	1938.84	III	100	2545.64	III
3300	3392.99	I	10	6176.81	I	60	1978.22	IV	110	2551.38	II
8200	3414.76	I	10	6191.18	I	3300	2029.32	II	130	2556.94	II
1600	3423.71	I	13	6256.36	I	65	2032.53	IV	80	2557.94	III
2600	3433.56	I	16	6643.64	I	3000	2032.99	II	130	2562.41	II
990	3437.28	I	22	6767.77	I	2000	2109.42	II	110	2571.33	II
4800	3446.26	I	10	6914.56	I	1700	2125.21	II	390	2583.99	II
1300	3452.89	I	26	7122.20	I	1100	2126.54	II	390	2590.94	II
5000	3458.47	I	16	7393.60	I	80 h	2130.24	III	80	2598.86	III
5000	3461.65	I	16	7409.35	I	1500	2131.18	II	80	2633.17	III
1600	3472.54	I	23	7422.28	I	80	2273.92	III	200	2642.24	II
550	3483.77	I	13	7522.76	I	100	2275.23	III	320	2646.26	II
5500	3492.96	I	19	7555.60	I	80	2279.36	III	330	2647.50	I
660	3500.85	I	23	7617.00	I	100	2281.51	III	330	2654.45	I
2600	3510.34	I	16	7714.32	I	80	2284.40	III	310	2656.08	II
6600	3515.05	I	19	7727.61	I	100	2290.36	III	80	2657.99	III
660	3519.77	I	19	7748.89	I	370	2295.68	II	110	2665.25	II
8200	3524.54	I	10	7788.94	I	280	2302.08	II	110	2666.59	II
5000	3566.37	I	13	7797.59	I	100	2313.30	III	110	2667.30	II
990	3571.87	I	1000	8096.75	II	100	2338.09	III	400	2671.93	II
1300	3597.70	I	700	8121.48	II	80	2344.12	III	200	2673.57	II
1300	3610.46	I	9	8862.55	I	90	2349.21	III	200	2675.94	II
						80	2355.54	III			

Intensity	Wavelength/Å		Intensity	Wavelength/Å		Intensity	Wavelength/Å		Intensity	Wavelength/Å	
160	2691.77	II	80	3142.26	III	500	3584.97	I	350	4143.21	I
1000	2697.06	II	390	3145.40	II	750	3589.11	I	870	4150.12	I
320	2698.86	II	1200	3163.40	II	500	3589.36	I	4400	4152.58	I
320	2702.20	II	150	3175.78	II	500	3593.97	I	870	4163.47	I
150	2702.52	II	390	3180.29	II	500	3602.56	I	4400	4163.66	I
470	2716.62	II	300	3191.10	II	300	3619.51	II	4000	4164.66	I
470	2721.98	II	150	3191.43	II	420	3649.85	I	3500	4168.13	I
310	2733.26	II	1000	3194.98	II	400	3651.19	II	310	4184.44	I
110	2737.09	II	120	3203.35	II	200	3659.61	II	1200	4190.88	I
240	2768.13	II	300	3206.34	II	630	3660.37	I	870	4192.07	I
310	2773.20	I	390	3215.60	II	900	3664.70	I	870	4195.09	I
270	2780.24	II	800	3225.48	II	1500	3697.85	I	1300	4195.66	I
110	2793.05	II	140	3229.56	II	330	3711.34	I	310	4198.51	I
190	2827.08	II	400	3236.40	II	3300	3713.01	I	350	4201.52	I
250	2841.15	II	200	3247.47	II	480	3716.99	I	870	4205.31	I
280	2842.65	II	120	3248.94	II	2700	3726.24	I	350	4214.73	I
160	2846.28	II	320	3254.07	II	2700	3739.80	I	420	4217.94	I
240	2861.09	II	230	3260.56	II	670	3740.73	II	420	4229.15	I
100	2865.61	II	160	3263.37	II	1700	3742.39	I	770	4262.05	I
500	2868.52	II	200	3283.46	II	530	3763.49	I	420	4266.02	I
800	2875.39	II	160	3292.02	II	350	3765.08	I	400	4286.99	I
270	2876.95	II	320	3296.01	I	530	3771.85	I	580	4299.60	I
530	2877.03	II	400	3312.60	I	870	3781.01	I	580	4300.99	I
100	2880.72	II	120	3319.58	II	1700	3787.06	I	390	4311.27	I
570	2883.18	II	130	3341.60	II	1300	3790.15	I	350	4326.33	I
280	2888.83	II	1300	3341.97	I	3500	3791.21	I	390	4331.37	I
470	2897.81	II	1300	3343.71	I	2700	3798.12	I	330	4410.21	I
400	2899.24	II	1700	3349.06	I	2700	3802.92	I	150	4503.04	I
470	2908.24	II	420	3349.52	I	670	3803.88	I	530	4523.41	I
670	2910.59	II	340	3354.74	I	530	3804.74	I	480	4546.82	I
470	2911.74	II	1700	3358.42	I	670	3810.49	I	370	4564.53	I
1100	2927.81	II	130	3365.58	II	530	3811.03	I	720	4573.08	I
110	2931.47	II	340	3366.96	I	530	3815.51	I	480	4581.62	I
870	2941.54	II	130	3369.16	II	210	3818.86	II	1200	4606.77	I
110 h	2945.88	II	350	3374.92	I	670	3824.88	I	170	4616.17	I
110	2946.12	II	170	3386.24	II	350	3835.18	I	450	4630.11	I
110	2946.90	II	350	3392.34	I	350	3863.38	I	450	4648.95	I
1100	2950.88	II	230	3408.68	II	530	3877.56	I	450	4663.83	I
400	2972.57	II	180	3409.19	II	870	3878.82	I	340	4666.24	I
320	2974.10	II	230	3412.94	III	670	3883.14	I	240	4667.22	I
210	2977.68	II	230	3425.42	II	1100	3885.44	I	580	4672.09	I
200	2982.11	II	230	3426.57	II	670	3885.68	I	530	4675.37	I
330	2990.26	II	180	3432.70	II	580	3891.30	I	320	4685.14	I
470	2994.73	II	180	3440.59	II	670	3914.70	I	130 c	4706.14	I
80	3001.84	III	200	3479.56	II	530	3920.20	I	260	4708.29	I
140	3024.74	II	100	3484.05	II	670	3937.44	I	150	4713.50	I
350	3028.44	II	500	3498.63	I	520	3943.67	I	220 c	4749.70	I
300	3032.77	II	460	3507.96	I	910 d	3966.09	I	130 c	4967.78	I
100	3044.76	II	200	3510.26	II	1100	4032.52	I	190	4988.97	I
100	3055.52	II	200	3515.42	II	16000 c	4058.94	I	230	5017.75	I
220	3064.53	II	200	3517.67	II	350	4060.79	I	150	5026.36	I
110	3069.68	II	2000	3535.30	I	12000	4079.73	I	210	5039.04	I
100	3070.90	II	1300	3537.48	I	440	4100.40	I	170	5058.01	I
110	3071.56	II	250	3540.96	II	6700	4100.92	I	130	5065.25	I
100	3073.24	II	500	3544.02	I	310	4116.90	I	750	5078.96	I
400	3076.87	II	300	3550.45	I	5300	4123.81	I	420	5095.30	I
110	3080.35	II	1000	3554.66	I	670	4129.43	I	170	5100.16	I
1800	3094.18	II	630	3563.50	I	770	4129.93	I	170	5120.30	I
140	3099.19	II	630	3563.62	I	2300	4137.10	I	210	5134.75	I
270	3127.53	II	1500	3575.85	I	440	4139.44	I	250	5160.33	I
1500	3130.79	II	5000	3580.27	I	2700	4139.71	I	250	5164.38	I

Intensity	Wavelength/Å		Intensity	Wavelength/Å		Intensity	Wavelength/Å		Intensity	Wavelength/Å	
230	5180.31	I	600 w	234.25	IV	360	644.837	II	115	1098.095	I
190	5189.20	I	550	236.07	IV	450	645.178	II	115	1098.260	I
170	5193.08	I	500	237.99	IV	140	647.50	I	105	1100.360	I
150	5195.84	I	500 w	238.7	IV	360	660.286	II	40	1100.465	I
150	5232.81	I	600	238.80	IV	170	671.016	II	90	1101.291	I
150 d	5251.62	I	500 w	239.62	IV	285	671.386	II	360	1134.165	I
270	5271.53	I	900	247.20	IV	150	671.630	II	385	1134.415	I
130 c	5276.20	I	90	247.561	V	160	671.773	II	410	1134.980	I
250	5318.60	I	120	247.706	V	170	672.001	II	105	1143.65	I
460	5344.17	I	500 w	248.43	IV	500	684.996	III	130	1163.884	I
340	5350.74	I	500 w	248.46	IV	570	685.513	III	60	1164.206	I
110	5437.27	I	500 w	248.48	IV	650	685.816	III	105	1164.325	I
85	5551.35	I	500	257.95	III	500	686.335	III	270	1167.448	I
170	5642.11	I	650	258.50	III	350	692.70	I	105	1168.334	I
130	5664.71	I	700	259.19	III	90	713.518	V	60	1168.417	I
170	5665.63	I	800	260.09	III	150	713.860	V	195	1168.536	I
130	5729.19	I	600	260.45	IV	285	746.984	II	230	1176.510	I
110	5760.34	I	800	261.28	III	150	748.195	V	105	1176.630	I
110	5819.43	I	500	262.91	III	200	748.291	V	195	1177.695	I
130 d	5838.64	I	500	265.23	III	500	763.336	III	500	1183.031	III
190 cw	5900.62	I	500	265.27	III	570	764.359	III	570	1184.550	III
150	5983.22	I	150	266.196	V	570	765.148	IV	90	1188.01	IV
75	6221.96	I	200	266.379	V	250	771.544	III	410	1199.550	I
85 c	6430.46	I	500	268.70	III	300	771.901	III	385	1200.223	I
65	6544.61	I	650	270.99	IV	350	772.385	III	360	1200.710	I
210 cw	6660.84	I	250	283.42	IV	200	772.891	III	175	1225.026	I
150 cw	6677.33	I	300	283.48	IV	150	772.975	III	160	1225.37	I
130 c	6723.62	I	350	283.58	IV	650	775.965	II	130	1228.41	I
85	6828.11	I	600	285.56	IV	90	885.67	I	160	1228.79	I
85	6990.32	I	600 w	297.7	IV	90	909.697	I	1000	1238.821	V
190 c	7046.81	I	700	297.82	IV	80	910.278	I	900	1242.804	V
130	7159.43	I	650	300.32	IV	40	910.645	I	360	1243.179	I
190 cw	7372.50	I	90	303.123	IV	450	915.612	II	315	1243.306	I
65	7515.93	I	500	303.28	IV	450	915.962	II	290	1310.540	I
170 c	7574.58	I	150	314.715	III	550	916.012	II	250	1310.95	I
75 c	7726.68	I	200	314.850	III	650	916.701	II	230	1319.00	I
35	7885.31	I	90	314.877	III	520	921.992	IV	315	1319.68	I
40	8135.20	I	150	315.053	IV	500	922.519	IV	115	1326.57	I
29 cw	8320.93	I	120	322.503	IV	480	923.057	IV	115	1327.92	I
29	8346.08	I	150	322.570	IV	520	924.283	IV	150	1387.371	III
35	8905.78	I	200	322.724	IV	90	953.415	I	360	1411.94	I
			120	323.175	IV	100	953.655	I	700	1492.625	I
Nitrogen N Z = 7			600	323.26	III	130	953.970	I	490	1492.820	I
400	181.75	IV	300	335.050	IV	1000	955.335	IV	640	1494.675	I
52	186.069	V	500	338.35	III	130	963.990	I	90	1549.336	V
62	186.153	V	500	340.20	III	115	964.626	I	200 l	1616.33	V
400	191.7	IV	500 w	351.93	IV	70	965.041	I	350 l	1619.69	V
400	192.9	IV	500	351.98	III	650	979.842	III	1000	1718.55	IV
500	196.87	IV	700	353.06	IV	700	979.919	III	250	1729.945	III
500	197.23	IV	120	362.833	III	900	989.790	III	775	1742.729	I
500	202.60	IV	150	362.881	III	700	991.514	III	700	1745.252	I
500	205.94	IV	150	362.946	III	1000	991.579	III	570	1747.848	III
500	205.97	IV	90	362.985	III	150 w	1036.16	IV	350	1751.218	III
500	206.03	IV	300	374.204	III	90	1067.614	I	650	1751.657	III
90	209.303	V	350	374.441	III	60	1068.612	I	150	1804.486	III
500	217.20	IV	500	387.48	III	90	1078.71	IV	200	1805.669	III
500 d	217.90	IV	500	420.77	IV	450	1083.990	II	150	1846.42	III
500 d	223.4	IV	250	451.869	III	600	1084.580	II	90 w	1860.37	V
800 w	225.12	IV	300	452.226	III	430	1085.546	II	350	1885.06	III
800	225.21	IV	650	463.74	IV	650	1085.701	II	400	1885.22	III
600 w	234.12	IV	285	644.634	II	175	1097.237	I	200	1907.99	III
600 w	234.20	IV									

Intensity	Wavelength/Å		Intensity	Wavelength/Å		Intensity	Wavelength/Å		Intensity	Wavelength/Å	
150	1919.55	III	360	3919.00	II	185	5281.20	I	450	7762.24	II
150	1919.77	III	90	3938.52	III	140	5292.68	I	400	8184.87	I
300	1920.65	III	450	3955.85	II	90	5314.35	III	400	8188.02	I
150	1920.84	III	1000	3995.00	II	200	5320.82	III	250	8200.36	I
200	1921.30	III	150	3998.63	III	150	5327.18	III	300	8210.72	I
200	2064.01	III	200	4003.58	III	450	5495.67	II	570	8216.34	I
250	2064.42	III	360	4035.08	II	285	5535.36	II	400	8223.14	I
120	2068.68	III	550	4041.31	II	650	5666.63	II	400	8242.39	I
90	2071.09	III	360	4043.53	II	550	5676.02	II	550	8438.74	II
90	2080.34	IV	150	4057.76	IV	870	5679.56	II	500	8567.74	I
160	2095.53	II	250	4097.33	III	450	5686.21	II	570	8594.00	I
70	2096.20	II	140	4099.94	I	450	5710.77	II	650	8629.24	I
110	2096.86	II	200	4103.43	III	285	5747.30	II	500	8655.89	I
90	2117.59	III	185	4109.95	I	700	5752.50	I	220	8676.08	II
90	2121.50	III	285	4176.16	II	240	5764.75	I	700	8680.28	I
110	2130.18	II	120	4195.76	III	265	5829.54	I	650	8683.40	I
160	2142.78	II	150	4200.10	III	235	5854.04	I	500	8686.15	I
90	2147.31	III	285	4227.74	II	360	5927.81	II	110	8687.43	II
200	2188.20	III	285	4236.91	II	550	5931.78	II	110 h	8699.00	II
150	2188.38	III	220	4237.05	II	285	5940.24	II	500	8703.25	I
160	2206.09	II	450	4241.78	II	650	5941.65	II	160 h	8710.54	II
160	2286.69	II	90	4332.91	III	285	5952.39	II	570	8711.70	I
110	2288.44	II	120	4345.68	III	160	5999.43	I	500	8718.83	I
220	2316.49	II	300	4379.11	III	210	6008.47	I	250	8728.89	I
160	2316.69	II	285	4432.74	II	285	6167.76	II	200	8747.36	I
285	2317.05	II	650	4447.03	II	360	6379.62	II	500	9386.80	I
90 w	2318.09	IV	90	4510.91	III	150	6380.77	IV	570	9392.79	I
160	2461.27	II	120	4514.86	III	185	6411.65	I	250	9460.68	I
150	2477.69	IV	360	4530.41	II	210	6420.64	I	200	9863.33	I
110	2496.83	II	550	4601.48	II	210	6423.02	I	160 h	9865.41	II
70	2496.97	II	350	4603.73	V	210	6428.32	I	110 h	9868.21	II
110	2520.22	II	90	4606.33	IV	185	6437.68	I	160 h	9887.39	II
160	2520.79	II	450	4607.16	II	235	6440.94	I	220 h	9891.09	II
220	2522.23	II	360	4613.87	II	90	6454.11	III	160 h	9961.86	II
110	2590.94	II	250	4619.98	V	185	6457.90	I	220 h	9969.34	II
250	2645.65	IV	450	4621.39	II	120	6467.02	III	285 h	10023.27	II
300	2646.18	IV	870	4630.54	II	300	6468.44	I	220 h	10035.45	II
350	2646.96	IV	90	4634.14	III	265	6481.71	I	220 h	10065.15	II
250 w	2682.18	III	120	4640.64	III	750	6482.05	II	160 h	10070.12	II
90	2689.20	III	550	4643.08	II	360	6482.70	I	250	10105.13	I
160	2709.84	II	285	4788.13	II	300	6483.75	I	300	10108.89	I
110	2799.22	II	450	4803.29	II	325	6484.80	I	350	10112.48	I
110	2823.64	II	180	4847.38	I	160	6491.22	I	400	10114.64	I
60 l	2859.16	V	90	4858.82	III	210	6499.54	I	110 h	10126.27	II
160	2885.27	II	150	4867.15	III	185	6506.31	I	250	10539.57	I
90 l	2974.52	V	285	4895.11	II	750	6610.56	II	200	12074.51	I
150 w	2980.78	V	160	4914.94	I	185	6622.54	I	380	12186.82	I
250 w	2981.31	V	210	4935.12	I	185	6636.94	I	225	12288.97	I
60 w	2998.43	V	200 w	4944.56	V	235	6644.96	I	290	12328.76	I
220	3006.83	II	160	4950.23	I	185	6646.50	I	310	12381.65	I
90	3078.25	IV	350	4963.98	I	235	6653.46	I	180	12438.40	I
120	3367.34	III	285	4987.37	II	210	6656.51	I	510	12461.25	I
360	3437.15	II	450	4994.36	II	185	6722.62	I	920	12469.62	I
90	3463.37	IV	650	5001.48	II	210	7398.64	I	500	13429.61	I
570	3478.71	IV	360	5002.70	II	160	7406.12	I	840	13581.33	I
500	3482.99	IV	870	5005.15	II	265	7406.24	I	180	13587.73	I
400	3484.96	IV	550	5007.32	II	685	7423.64	I	180	13602.27	I
90	3747.54	IV	450	5010.62	II	785	7442.29	I	290	13624.18	I
90	3754.67	III	360	5016.39	II	900	7468.31	I	250	14757.07	I
120	3771.05	III	360	5025.66	II	185	7608.80	I	100	14868.87	I
285	3838.37	II	550	5045.10	II	60 w	7618.46	V	160	14966.60	I

Intensity	Wavelength/Å	
180	15582.27	I
120 s	17516.58	I
100 l	17584.86	I
100	17878.26	I

Osmium Os Z = 76

Intensity	Wavelength/Å	
9600	2001.45	I
13000	2003.73	I
17000	2010.15	I
29000	2018.14	I
14000	2022.76	I
14000	2028.23	I
18000	2034.44	I
26000	2045.36	I
8600	2058.69	I
13000	2061.69	I
7800	2067.21	II
4200	2070.67	II
7200	2076.95	I
14000	2079.97	I
2900	2082.54	I
2900	2089.03	I
2900	2089.21	I
6000	2097.60	I
5300	2100.63	I
2100	2117.66	I
4800	2117.96	I
5300	2137.11	I
2600	2154.59	I
1300	2157.84	I
1200	2158.53	I
3100	2166.90	I
1100	2167.75	I
2100	2171.65	I
1100	2234.61	I
1300	2252.15	I
2000	2255.85	II
1400	2264.60	I
1400	2282.26	II
500	2367.35	II
2600	2377.03	I
1700	2387.29	I
1100	2395.88	I
200	2423.07	II
1400	2424.97	I
110	2454.91	II
1800	2461.42	I
110	2468.90	II
530	2486.24	II
4500	2488.55	I
2600	2498.41	I
2400	2513.25	I
780	2538.00	II
1000	2542.51	I
1000	2590.76	I
1800	2613.06	I
3800	2637.13	I
1900	2644.11	I
1900	2658.60	I
2100	2689.82	I
3000	2714.64	I
1300	2720.04	I

Intensity	Wavelength/Å	
960	2770.71	I
2800	2806.91	I
5100	2838.63	I
2300	2844.40	I
1500	2850.76	I
1500	2860.96	I
9600	2909.06	I
2100	2912.33	I
2100	2919.79	I
1100 h	2948.23	I
1400	2949.53	I
4400	3018.04	I
1100	3030.70	I
2900	3040.90	I
120	3042.74	II
8600	3058.66	I
1100	3077.72	I
3100	3156.25	I
180	3173.93	II
150	3213.31	II
1900	3232.06	I
3100	3262.29	I
3100	3267.94	I
1200	3290.26	I
7600	3301.56	I
960	3336.15	I
960	3370.59	I
620	3387.84	I
620	3401.86	I
620	3504.66	I
1200	3528.60	I
1200	3560.86	I
620	3598.11	I
95	3604.48	II
480	3670.89	I
3700	3752.52	I
2100	3782.20	I
730	3876.77	I
1000	3963.63	I
730	3977.23	I
960	4066.69	I
1200	4112.02	I
2500	4135.78	I
1200	4173.23	I
1200	4211.86	I
4900	4260.85	I
560	4293.95	I
560	4311.40	I
4900	4420.47	I
540	4550.41	I
670	4793.99	I
55	5031.83	I
45	5039.12	I
35	5072.88	I
35	5074.77	I
35	5079.09	I
90	5103.50	I
55	5110.81	I
140	5149.74	I
40	5193.52	I
270	5202.63	I

Intensity	Wavelength/Å	
35	5203.23	I
45	5255.82	I
55	5265.15	I
40	5298.78	I
110	5376.79	I
120	5416.34	I
45	5416.69	I
28	5417.51	I
55	5443.31	I
22	5446.93	I
22	5457.30	I
28	5470.00	I
22	5509.33	I
270	5523.53	I
22	5546.82	I
80	5584.44	I
35	5620.08	I
22	5642.56	I
28	5645.25	I
28	5680.88	I
170	5721.93	I
22	5765.05	I
170	5780.82	I
40	5800.60	I
110	5857.76	I
28	5860.64	I
65	5996.00	I
35	6227.70	I
22	6269.41	I
22	6403.15	I
27	6729.56	I
22	7145.54	I
26	7602.95	I
7	8041.29	I

Oxygen O Z = 8

Intensity	Wavelength/Å	
80	124.616	V
110	135.523	V
80	138.109	V
110	139.029	V
80	151.447	V
110	151.477	V
150	151.546	V
80	164.574	V
110	164.657	V
80	164.709	V
80	166.235	V
150	167.99	V
110	170.219	V
450	172.169	V
250	185.745	V
375	192.751	V
450	192.799	V
520	192.906	V
80	193.003	V
200	194.593	V
150	195.86	IV
200	196.01	IV
80	202.161	V
80	202.224	V
80	202.283	V
80	202.334	V

Intensity	Wavelength/Å	
150	202.393	V
110	203.78	V
150	203.82	V
100	203.85	V
200	203.89	V
100	203.94	V
110	207.18	IV
150	207.24	IV
300	207.794	V
150	215.040	V
200	215.103	V
250	215.245	V
250	216.018	V
520	220.352	V
80	227.372	V
80	227.469	V
150	227.511	V
80	227.549	V
80	227.634	V
80	227.689	V
150	231.823	V
140	233.46	IV
150	233.50	IV
110	233.52	IV
200	233.56	IV
110	233.60	IV
90	238.36	IV
180	238.57	IV
110	248.459	V
110	252.56	IV
110	252.95	IV
150	253.08	IV
300	260.39	IV
250	260.56	IV
80 d	264.34	III
110	264.48	III
110	266.97	III
150	266.98	III
150	267.03	III
150	277.38	III
300	279.63	IV
375	279.94	IV
110	285.71	IV
150	285.84	IV
110	286.448	V
80	295.62	III
110	295.66	III
120	295.72	III
150	303.41	III
150	303.46	III
140	303.52	III
160	303.62	III
160	303.69	III
250	303.80	III
200	305.60	III
250	305.66	III
190	305.70	III
300	305.77	III
190	305.84	III
200	306.62	IV
150	306.88	IV

Intensity	Wavelength/Å		Intensity	Wavelength/Å		Intensity	Wavelength/Å		Intensity	Wavelength/Å	
450	320.979	III	775	760.445	V	220	1763.22	III	775	2789.85	V
300	328.45	III	640	761.128	V	220	1764.48	III	160	2836.26	IV
250	328.74	III	700	762.003	V	750	1767.78	III	160	2921.45	IV
300	345.31	III	70	770.793	I	550	1768.24	III	200	2941.33	V
110	355.14	III	90	771.056	I	360	1771.67	III	210	2941.65	V
90	355.33	III	520	774.518	V	110	1773.00	III	80	2959.68	III
80	355.47	III	70	775.321	I	110	1773.85	III	265	2972.29	I
200	359.02	III	200	779.734	IV	220	1779.16	III	250	2983.78	III
190	359.22	III	315	779.821	IV	160	1781.03	III	80	3017.63	III
150	359.38	III	360	779.912	IV	160	1784.85	III	80	3023.45	III
210	373.80	III	200	779.997	IV	220	1789.66	III	80	3043.02	III
200	374.00	III	640	787.711	IV	110	1848.26	III	200	3047.13	III
300	374.08	III	520	790.109	IV	110	1856.62	III	110	3059.30	III
190	374.16	III	700	790.199	IV	285	1872.78	III	460	3063.42	IV
200	374.33	III	70	791.973	I	285	1872.87	III	410	3071.61	IV
210	374.44	III	300	796.66	II	285	1874.94	III	80	3121.71	III
450	395.558	III	200	802.200	IV	160	1920.04	III	160	3122.62	II
300	434.98	III	160	802.255	IV	110	1920.75	III	220	3129.44	II
800	507.391	III	90	804.267	I	110	1921.52	III	110	3132.86	III
900	507.683	III	70	804.848	I	220	1923.49	III	450	3134.82	II
1000	508.182	III	70	805.295	I	110	1923.82	III	285	3138.44	II
1000	525.795	III	80	805.810	I	110	1926.94	III	160	3144.66	V
250	537.83	II	240	832.762	II	360	2013.27	III	160	3209.66	IV
300	538.26	II	600	832.927	III	160	2026.96	III	80	3238.57	III
220	539.09	II	450	833.332	II	220	2045.67	III	200	3260.98	III
200	539.55	II	780	833.742	III	160	2052.74	III	300	3265.46	III
150	539.85	II	600	834.467	II	30 d	2283.42	II	80	3267.31	III
700	553.330	IV	600	835.096	III	30 d	2284.89	II	220	3270.98	II
775	554.075	IV	800	835.292	III	110	2293.32	II	220	3273.52	II
850	554.514	IV	40	877.879	I	200	2300.35	II	220	3277.69	II
700	555.261	IV	130	921.296	IV	30 d	2313.05	II	360	3287.59	II
700	597.818	III	160	921.366	IV	30 d	2316.12	II	160	3305.15	II
1000	599.598	III	80	922.008	I	30 d	2316.79	II	160	3306.60	II
580	608.398	IV	200	923.367	IV	50 d	2319.68	II	80	3312.30	III
110	609.70	III	130	923.433	IV	30 d	2322.15	II	110	3340.74	III
640	609.829	IV	90	935.193	I	30 d	2339.31	II	230	3348.08	IV
160	610.04	III	40	948.686	I	200 d	2390.44	III	270	3349.11	IV
200	610.75	III	90	971.738	I	80	2394.33	III	160	3354.27	IV
100	610.85	III	40	976.448	I	110	2411.60	II	200	3375.40	IV
270	616.952	IV	160	988.773	I	80	2422.84	III	220	3377.20	II
150	617.005	IV	40	990.204	I	80	2425.55	II	130	3378.06	IV
200	617.036	IV	250	1025.762	I	250	2433.56	II	360	3381.20	IV
520	624.617	IV	90	1027.431	I	80 d	2436.06	II	360	3385.52	IV
580	625.130	IV	160	1039.230	I	80 d	2438.83	III	285	3390.25	II
640	625.852	IV	60	1040.942	I	80	2444.26	II	270	3396.79	IV
1000	629.730	V	40	1152.152	I	300	2445.55	II	360	3403.52	IV
150	644.148	II	900	1302.168	I	200	2449.372	IV	220	3407.38	II
200	672.95	II	600	1304.858	I	200	2450.040	IV	230	3409.66	IV
150	673.77	II	300	1306.029	I	200	2454.99	III	160	3409.84	II
230	681.272	V	200	1338.612	IV	200	2493.44	IV	410	3411.69	IV
70	685.544	I	130	1342.992	IV	200	2493.77	IV	230	3413.64	IV
800	702.332	III	230	1343.512	IV	200	2507.73	IV	80	3444.10	III
800	702.822	III	640	1371.292	V	230	2509.19	IV	80	3455.12	III
900	702.899	III	160	1476.89	III	200	2517.2	IV	285	3470.81	II
1000	703.850	III	160 w	1506.72	V	200	2558.06	III	200	3489.83	IV
900	718.484	II	285	1590.01	III	80	2687.53	III	160	3492.24	IV
600	718.562	II	160	1591.33	III	110	2695.49	III	230	3560.39	IV
70	744.794	I	315 w	1643.68	V	300	2733.34	II	270	3563.33	IV
700	758.678	V	160	1707.996	V	110	2747.46	II	80	3698.70	III
640	759.441	V	220	1760.12	III	1000	2781.01	V	80	3702.75	III
580	760.228	V	110	1760.42	III	920	2786.99	V	80	3703.37	III

Intensity	Wavelength/Å		Intensity	Wavelength/Å		Intensity	Wavelength/Å		Intensity	Wavelength/Å	
110	3707.24	III	50	4469.41	II	235	7947.55	I	540	11297.68	I
220	3712.75	II	360	4590.97	II	210	7950.80	I	590	11302.38	I
110	3715.08	III	285	4596.17	II	185	7952.16	I	265	11358.69	I
315 w	3725.93	IV	80 d	4609.39	II	110	7981.94	I	490	12464.02	I
285	3727.33	II	160	4638.85	II	135	7982.40	I	450	12570.04	I
360	3729.03	IV	360	4641.81	II	190	7986.98	I	120	12990.77	I
410	3736.85	IV	450	4649.14	II	135	7987.33	I	160	13076.91	I
160	3739.92	II	160	4650.84	II	250	7995.07	I	700	13163.89	I
110	3744.00	III	360	4661.64	II	400	8221.82	I	750	13164.85	I
230	3744.89	IV	285	4676.23	II	265	8227.65	I	640	13165.11	I
360	3749.49	II	220	4699.21	II	265	8230.02	I	160	16212.06	I
150	3754.67	III	285	4705.36	II	325	8233.00	I	120	17966.70	I
80	3757.21	III	160	4924.60	II	120	8235.35	I	590	18021.21	I
250	3759.87	III	230 w	4930.27	V	120	8426.16	I	120	18041.48	I
110	3791.26	III	220	4943.06	II	810	8446.25	I	120	18042.19	I
160	3803.14	II	135	5329.10	I	1000	8446.36	I	120	18046.23	I
120	3823.41	I	160	5329.68	I	935	8446.76	I	140	18229.23	I
450	3911.96	II	190	5330.74	I	325	8820.43	I	540	18243.63	I
160	3919.29	II	90	5435.18	I	160 d	9057.01	I	140	26173.56	I
185	3947.29	I	110	5435.78	I	120	9118.29	I			
160	3947.48	I	135	5436.86	I	80	9134.71	I	*Palladium Pd Z = 46*		
140	3947.59	I	120	5577.34	I	80	9150.14	I	200	705.49	III
220	3954.37	II	110	5592.37	III	80	9151.48	I	200	727.72	III
100	3954.61	I	130	5597.91	V	235	9156.01	I	500	763.06	III
200	3961.59	III	160	5958.39	I	450	9260.81	I	500	766.42	III
450	3973.26	II	190	5958.58	I	490	9260.84	I	2000	781.02	III
220	3982.20	II	80	5995.28	I	450	9260.94	I	500	794.08	III
160	4069.90	II	160	6046.23	I	400	9262.58	I	500	797.52	III
285	4072.16	II	190	6046.44	I	540	9262.67	I	500	800.03	III
450	4075.87	II	110	6046.49	I	590	9262.77	I	500	800.10	III
80 d	4083.91	II	100	6106.27	I	490	9265.94	I	500	803.67	III
50 d	4087.14	II	400	6155.98	I	640	9266.01	I	500	825.35	III
150 d	4089.27	II	450	6156.77	I	185	9399.19	I	500	840.58	III
110	4097.24	II	490	6158.18	I	120	9481.16	I	500	856.47	III
220	4105.00	II	80	6256.83	I	120 d	9482.88	I	500	864.04	III
285	4119.22	II	100	6261.55	I	235	9487.43	I	500	880.59	III
100	4123.99	V	100	6366.34	I	140	9492.71	I	500	888.84	III
160	4132.81	II	100	6374.32	I	265	9497.97	I	1000	889.29	III
50	4146.06	II	320	6453.60	I	160	9499.30	I	300	1596.89	III
220	4153.30	II	360	6454.44	I	235	9505.59	I	500	1741.62	III
285	4185.46	II	400	6455.98	I	210	9521.96	I	4000	1782.55	III
450	4189.79	II	130	6500.24	V	120	9523.36	I	400	1843.49	III
80	4233.27	I	80	6604.91	I	120	9523.96	I	1500	1851.59	III
50 d	4253.74	II	100	6653.83	I	100	9528.28	I	2000	1852.27	III
50 d	4253.98	II	360	7001.92	I	100	9622.13	I	1000	1859.21	III
50 d	4275.47	II	450	7002.23	I	120	9625.29	I	1500	1874.63	III
50 d	4303.78	II	210	7156.70	I	160	9677.38	I	2000	1885.83	III
285	4317.14	II	400	7254.15	I	80	9694.66	I	1000	1887.40	III
160	4336.86	II	450	7254.45	I	65	9694.91	I	1500	1891.34	III
220	4345.56	II	320	7254.53	I	235	9741.50	I	4000	1914.62	III
285	4349.43	II	210	7476.44	I	235	9760.65	I	1000	1930.33	III
220	4366.90	II	100	7477.24	I	120	9909.05	I	2000	1941.64	III
100	4368.25	I	120	7479.08	I	140	9936.98	I	800	2002.16	III
220	4395.95	II	120	7480.67	I	120	9940.41	I	1000	2004.47	III
450	4414.91	II	100	7706.75	I	160	9995.31	I	500	2055.11	III
285	4416.98	II	870	7771.94	I	120 d	10421.18	I	500	2149.82	III
160	4448.21	II	810	7774.17	I	590	11286.34	I	500	2177.55	III
160	4452.38	II	750	7775.39	I	640	11286.91	I	500	2177.63	III
50	4465.45	II	80	7886.27	I	490	11287.02	I	100 r	2231.59	II
50 d	4466.28	II	100	7943.15	I	490	11287.32	I	200 r	2296.53	II
50	4467.83	II	100	7947.17	I	490	11295.10	I	100	2426.87	II
									100	2430.94	II

Intensity	Wavelength/Å	
100	2433.11	II
100	2435.32	II
150	2446.17	II
1100	2447.91	I
100	2457.29	II
150	2469.29	II
100	2471.18	II
1700	2476.42	I
250	2486.52	II
300	2488.92	II
200	2498.81	II
150	2505.73	II
150	2551.84	II
150	2565.51	II
100	2569.56	II
150	2658.75	II
1900	2763.09	I
150 h	2776.85	II
100 h	2787.92	II
200	2854.59	II
100 h	2871.37	II
100 h	2878.01	II
520	2922.49	I
650	3002.65	I
1500	3027.91	I
1100	3065.31	I
2600	3114.04	I
11000	3242.70	I
2700	3251.64	I
3500	3258.78	I
3600	3302.13	I
5000	3373.00	I
24000	3404.58	I
13000	3421.24	I
5000	3433.45	I
6400	3441.40	I
7700	3460.77	I
10000	3481.15	I
2000	3489.77	I
12000	3516.94	I
12000	3553.08	I
4500	3571.16	I
20000	3609.55	I
20000	3634.70	I
5500	3690.34	I
1400	3718.91	I
1500	3799.19	I
1500	3832.29	I
2200	3894.20	I
1500	3958.64	I
290	4087.34	I
2500	4212.95	I
180	4473.59	I
160	5163.84	I
120	5295.63	I
55	5542.80	I
75	5670.07	I
55 h	5695.09	I
65	6784.52	I
75	7368.12	I
120	7764.03	I

Intensity	Wavelength/Å	
45	7915.80	I
55	8132.82	I
45	8300.83	I
65	8761.35	I

Phosphorus P Z = 15

Intensity	Wavelength/Å	
250	328.78	V
150	359.899	IV
500	388.318	IV
250	389.50	V
300	390.70	V
300	445.158	IV
375	475.60	V
120	498.180	III
520	542.57	V
600	544.92	V
200	569.853	III
200	581.831	III
350	629.008	IV
400	629.914	IV
500	631.779	IV
450	673.90	V
10	810.24	II
650	823.179	IV
700	824.730	IV
800	827.932	IV
300	847.669	III
350	855.624	III
500	859.652	III
10	865.44	II
450	865.45	V
600	871.39	V
700	877.476	IV
300	913.971	III
300	917.120	III
350	918.665	III
1000	950.655	IV
250	1003.598	III
570	1025.563	IV
500	1028.096	IV
570	1030.517	IV
500	1033.111	IV
500	1035.517	IV
900	1117.98	V
570	1118.551	IV
700	1128.01	V
20	1249.82	II
20	1301.87	II
20	1304.47	II
15	1304.68	II
35	1305.48	II
60	1310.70	II
500	1334.808	III
650	1344.327	III
300	1344.845	III
500	1366.695	IV
15	1372.033	I
400	1372.674	IV
15	1373.500	I
10	1374.732	I
15	1377.080	I
15	1377.937	I

Intensity	Wavelength/Å	
25	1379.429	I
25	1381.469	I
15	1381.637	I
500	1484.507	IV
400	1487.788	IV
350	1502.228	III
80	1532.51	II
120	1535.90	II
450	1610.50	V
150	1618.632	III
200	1618.907	III
140	1671.070	I
100	1671.510	I
180	1671.680	I
140	1672.035	I
140	1672.474	I
600	1674.591	I
600	1679.695	I
140	1685.976	I
100	1694.028	I
100	1694.486	I
100	1706.376	I
100	1707.553	I
600	1774.951	I
500	1782.838	I
400	1787.656	I
140	1834.801	I
140	1847.165	IV
100	1849.820	I
140	1851.194	I
100	1852.069	I
500	1858.886	I
400	1859.393	I
140	1864.348	I
650	1888.523	IV
180	1905.481	I
140	1906.403	I
280	1907.665	I
280	2023.489	I
180	2024.516	I
400	2032.432	I
400	2033.477	I
400	2135.465	I
400	2136.182	I
400	2149.145	I
280	2152.940	I
500	2154.080	I
180	2235.732	I
450	2440.93	V
250	2478.256	IV
750	2533.976	I
950	2535.603	I
750	2553.262	I
500	2554.915	I
250	2605.506	IV
300	2632.713	III
400	2644.295	IV
400	2728.770	IV
500	2739.309	IV
250	2739.872	IV
450	2978.55	V

Intensity	Wavelength/Å	
700	3175.09	V
520	3204.04	V
300	3219.307	III
400	3233.602	III
650	3347.736	IV
570	3364.467	IV
400	3371.122	IV
300	3957.641	III
350	3978.307	III
400	4059.312	III
300	4080.084	III
500	4222.195	III
350	4246.720	III
400	4420.71	II
250	4479.776	III
250	4540.288	IV
250	4541.112	IV
500	4588.04	II
500	4589.86	II
600	4602.08	II
300	4626.70	II
300	4658.31	II
500	4943.53	II
300	4954.39	II
300	4969.71	II
100	5079.381	I
100	5098.221	I
100	5100.974	I
140	5109.628	I
140	5154.844	I
180	5162.290	I
300	5253.52	II
140	5293.539	I
400	5296.13	II
250	5316.07	II
300	5344.75	II
180	5345.851	I
100	5364.631	I
250	5378.20	II
300	5386.88	II
400	5425.91	II
100	5428.094	I
400	5450.74	II
140	5458.305	I
180	5477.672	I
140	5477.860	I
140	5478.267	I
100	5514.774	I
100	5516.997	I
250	5588.34	II
500	6024.18	II
400	6034.04	II
500	6043.12	II
250	6055.50	II
150	6083.409	III
350	6087.82	II
180	6097.690	I
350	6165.59	II
500	6199.024	I
180	6210.499	I
140	6375.681	I

Intensity	Wavelength/Å		Intensity	Wavelength/Å		Intensity	Wavelength/Å		Intensity	Wavelength/Å	
100	6388.579	I	228	15962.53	I	240	2308.04	I	70	2738.48	I
250	6435.32	II	296	16254.77	I	50	2310.96	II	70	2747.61	I
600	6459.99	II	203	16292.97	I	90	2315.50	I	80	2753.86	I
600	6503.46	II	1627	16482.92	I	220	2318.29	I	200	2754.92	I
600	6507.97	II	588	16590.07	I	100	2326.10	I	30	2769.84	I
100	6717.411	I	225	16613.05	I	170	2340.18	I	500	2771.67	I
150	6992.690	III	221	16738.68	I	280	2357.10	I	40	2773.24	I
100	7102.200	I	419	16803.39	I	180	2368.28	I	20	2774.00	I
100	7158.367	I	471	17112.48	I	50	2377.28	II	50	2774.77	II
180	7165.465	I	289	17286.91	I	130	2383.64	I	50	2793.27	I
180	7175.102	I	299	17423.67	I	40	2386.81	I	100	2794.21	II
180	7176.660	I	287	23844.97	I	120	2389.53	I	40 h	2799.98	II
200	7443.657	IV	311	29097.16	I	35	2396.17	I	140	2803.24	I
250	7845.63	II				70	2401.87	I	10	2808.51	I
100	8046.801	I	*Platinum Pt* *Z = 78*			200	2403.09	I	50	2818.25	I
150	8113.528	III	30	1621.66	II	100	2418.06	I	30 h	2822.27	III
140	8278.058	I	30	1723.13	II	50	2424.87	II	1400	2830.30	I
100	8367.856	I	30	1751.70	II	80	2428.04	I	70	2834.71	I
140	8531.475	I	50 r	1777.09	II	50	2428.20	I	16	2853.11	I
140	8613.835	I	30	1781.86	II	25	2429.10	I	80 h	2860.68	II
180	8637.578	I	30	1879.09	II	180	2436.69	I	40 h	2865.05	II
400	8741.529	I	40	1883.05	II	650	2440.06	I	40 h	2875.85	II
100	8872.174	I	50	1889.52	II	60	2450.97	I	100 h	2877.52	II
180	9175.819	I	50	1911.70	II	440	2467.44	I	25	2888.20	I
950	9193.85	I	30	1929.25	II	35	2471.01	I	25	2893.22	I
600	9278.88	I	30	1929.68	II	1000	2487.17	I	600	2893.86	I
1250	9304.94	I	30	1939.80	II	25	2488.74	II	300	2897.87	I
500	9323.50	I	30	1949.90	II	200	2490.12	I	60	2905.90	I
950	9435.069	I	30	1983.74	II	160	2495.82	I	120	2912.26	I
950	9441.86	I	40	2014.93	II	240	2498.50	I	120	2913.54	I
600	9452.83	I	3200	2030.63	I	50	2505.93	I	70	2919.34	I
1250	9493.56	I	4400	2032.41	I	120	2508.50	I	30	2921.38	I
1700	9525.73	I	100	2036.46	II	50	2514.07	I	1700	2929.79	I
1500	9545.18	I	40	2041.57	II	60	2515.03	I	30	2942.76	I
280	9556.81	I	5500	2049.37	I	240	2515.58	I	30	2944.75	I
1700	9563.439	I	1500	2067.50	I	140	2524.30	I	25	2959.10	I
280	9593.50	I	3000	2084.59	I	40	2529.41	I	60	2960.75	I
750	9609.04	I	1000	2103.33	I	50	2536.49	I	1800	2997.97	I
400	9638.939	I	30	2115.57	II	160	2539.20	I	35	3001.17	II
500	9676.24	I	950	2128.61	I	18	2549.46	I	220	3002.27	I
180	9706.533	I	30	2130.69	II	50	2552.25	I	30	3017.88	I
1500	9734.750	I	1900	2144.23	I	50	2596.00	I	30 h	3031.22	II
280	9736.680	I	100	2144.24	II	70	2603.14	I	130	3036.45	I
1500	9750.77	I	600	2165.17	I	30	2616.76	II	800	3042.64	I
600	9790.21	I	1500	2174.67	I	50	2619.57	I	3200	3064.71	I
1700	9796.85	I	30	2190.32	II	30	2625.34	II	30	3071.94	I
280	9834.80	I	400	2202.22	I	1100	2628.03	I	130	3100.04	I
400	9903.68	I	50 h	2202.58	II	130	2639.35	I	320	3139.39	I
280	9976.67	I	320	2222.61	I	1000	2646.89	I	140	3156.56	I
229	10084.27	I	50 h	2233.11	II	500	2650.86	I	120	3200.71	I
458	10511.58	I	30 h	2240.99	II	20	2658.17	I	320	3204.04	I
962	10529.52	I	100	2245.52	II	2800	2659.45	I	30	3230.29	I
1235	10581.57	I	150	2249.30	I	40	2674.57	I	20	3233.42	I
415	10596.90	I	30	2251.52	II	440	2677.15	I	20	3250.36	I
435	10681.40	I	30 h	2251.92	II	200	2698.43	I	40	3251.98	I
265	10813.13	I	190	2268.84	I	2000	2702.40	I	160	3255.92	I
764	11183.23	I	30 h	2271.72	II	1600	2705.89	I	25	3268.42	I
402	11186.75	I	280	2274.38	I	60	2713.13	I	25	3281.97	I
479	14241.64	I	50 h	2287.50	II	1300	2719.04	I	120	3290.22	I
256	14307.83	I	30	2288.20	II	130	2729.92	I	500	3301.86	I
714	15711.52	I	150	2289.27	I	1800	2733.96	I	60	3315.05	I
			150	2292.40	I						

Intensity	Wavelength/Å	
35	3323.80	I
340	3408.13	I
35	3427.93	I
60	3483.43	I
160	3485.27	I
120	3628.11	I
70	3638.79	I
70	3643.17	I
50	3663.10	I
80	3671.99	I
80	3674.04	I
35	3699.91	I
18	3706.53	I
80	3818.69	I
40	3900.73	I
110	3922.96	I
35	3948.40	I
100	3966.36	I
20	3996.57	I
110	4118.69	I
80	4164.56	I
40	4192.43	I
18	4327.06	I
18	4391.83	I
80	4442.55	I
14	4445.55	I
25	4498.76	I
12	4520.90	I
35	4552.42	I
12	4879.53	I
14	5044.04	I
30	5059.48	I
35	5227.66	I
40	5301.02	I
12	5368.99	I
12	5390.79	I
14	5475.77	I
14	5478.50	I
6	5763.57	I
20	5840.12	I
8	5844.84	I
6	6026.04	I
7	6318.37	I
8	6326.58	I
9	6523.45	I
10	6710.42	I
20	6760.02	I
60	6842.60	I
20	7113.73	I
10	8224.74	I

Plutonium Pu Z = 94

Intensity	Wavelength/Å	
10000	2806.11	II
10000	2950.06	II
10000	3000.31	II
10000	3200.23	II
10000	3418.88	II
10000	3805.93	I
10000	4097.12	I
10000	4170.95	I
10000	4367.41	I
10000	5590.54	I

Intensity	Wavelength/Å	
10000	7068.90	I
10000	8691.94	I
3000	9533.07	I
3000	12144.46	I
3000	16897.38	I

Polonium Po Z = 84

Intensity	Wavelength/Å	
1500 w	2450.08	I
1500 w	2558.01	I
2500 w	3003.21	I
1200	4170.52	I
800	4493.21	I
500	8618.26	I

Potassium K Z = 19

Intensity	Wavelength/Å	
100	214.35	V
150	271.82	IV
100	273.06	IV
150	282.35	V
150	293.33	V
300	294.84	V
200	296.17	V
200	297.06	V
200	300.25	V
200	300.50	V
200	311.24	V
250	312.77	V
200	315.18	V
250	327.38	V
25	330.68	III
300	340.46	IV
150	340.74	IV
30	341.92	III
15	348.00	III
200	349.50	V
300	354.93	IV
150	356.26	IV
300	359.73	IV
200	359.91	IV
250	362.08	IV
150	362.15	IV
150	363.02	IV
500	372.15	V
200	372.46	V
200	372.77	V
300	375.96	IV
300	375.96	V
250	377.76	V
30	379.12	III
300	379.12	V
300	379.88	IV
25	380.48	III
250	380.48	IV
200	381.70	IV
30	382.23	III
300	382.23	IV
150	382.49	IV
200	382.65	IV
300	382.91	IV
250	384.10	IV
200	386.61	IV
300	387.80	V

Intensity	Wavelength/Å	
250	388.92	IV
250	389.07	IV
250	389.07	V
250	390.11	V
250	390.42	IV
300	390.57	IV
200	391.46	IV
200	392.47	IV
500	393.14	IV
250	395.40	V
200	398.36	V
15	398.63	III
200	398.88	V
200	399.75	V
400	400.21	IV
20	402.10	III
300	402.91	IV
250	403.97	IV
150	404.41	IV
30	406.48	III
250	408.08	IV
40	408.96	III
50	413.79	III
30	414.87	III
250	415.05	V
200	415.79	V
30	416.00	III
150	417.28	IV
30	417.54	III
30	418.62	III
400	422.18	V
300	425.16	V
500	425.59	V
75	434.72	III
50	435.68	III
250	438.02	V
25	441.81	II
200	442.30	IV
300	443.57	IV
75	444.34	III
200	445.61	IV
250	446.83	IV
75	448.60	III
750	448.60	IV
200	449.71	V
200	452.90	V
250	455.67	V
400	456.33	IV
400	456.33	V
75	466.79	III
100	470.09	III
75	471.57	III
45	474.92	III
10	476.03	II
40	479.18	III
10	482.11	III
10	482.41	III
200	482.71	V
200	483.75	V
30	495.14	II
75	497.10	III

Intensity	Wavelength/Å	
10	514.94	III
50	520.61	III
250	523.00	IV
25	523.79	III
200	526.45	IV
150	527.62	IV
40	529.80	III
15	539.71	III
15	546.12	III
750	580.32	V
250	585.51	V
500	586.32	V
30	600.77	II
250	602.27	V
400	603.43	V
25	607.93	II
30	612.62	II
250	638.67	V
750	646.19	IV
300	687.50	V
20	708.84	III
300	720.43	V
400	724.42	V
600	731.86	V
500	737.14	IV
500	741.95	IV
500	745.26	IV
400	746.35	IV
300	749.99	IV
150	754.19	IV
400	754.67	IV
20	765.31	III
30	765.64	III
150	770.29	V
150	771.46	V
35	778.53	III
20	872.31	III
10	873.86	III
15	874.04	III
6	2550.02	III
5	2635.11	III
5	2689.90	III
5	2938.45	III
5	2986.20	III
6	2992.42	III
6	3052.07	III
5	3056.84	III
5	3062.18	II
4	3101.79	I
3	3102.04	I
7	3217.16	I
6	3217.62	I
11	3446.37	I
10	3447.38	I
3	3648.84	I
4	3648.98	I
18	4044.14	I
17	4047.21	I
10	4641.88	I
11	4642.37	I

Intensity	Wavelength/Å		Intensity	Wavelength/Å		Intensity	Wavelength/Å		Intensity	Wavelength/Å	
4	4740.91	I	8	10487.11	I	680	3880.47	II	620	4171.82	II
6	4744.35	I	17	11019.87	I	440 c	3885.19	II	730	4172.25	II
5	4753.93	I	16	11022.67	I	440 c	3889.34	II	5200	4179.39	II
7	4757.39	I	17	11690.21	I	770 c	3908.05	II	2500	4189.48	II
5	4786.49	I	16	11769.62	I	630	3912.90	II	560 c	4191.60	II
7	4791.05	I	17	11772.83	I	310	3913.55	II	2500 c	4206.72	II
6	4799.75	I		12522.11	I	1300 c	3918.85	II	500	4208.32	II
8	4804.35	I		13377.86	I	420	3919.63	II	320	4211.86	II
7	4849.86	I		13397.09	I	960	3925.47	II	320	4217.81	II
8	4856.09	I		15163.08	I	480	3927.46	II	3800	4222.93	II
8	4863.48	I		15168.40	I	370	3929.29	II	3800	4225.35	II
9	4869.76	I		40158.37	I	370	3935.82	II	320	4233.11	II
8	4942.02	I				730 c	3947.63	II	320 c	4236.15	II
9	4950.82	I	*Praseodymium Pr Z = 59*			900 c	3949.43	II	960	4241.01	II
9	4956.15	I	7000	865.90	V	900 c	3953.51	II	340	4243.51	II
10	4965.03	I	5000	869.17	V	380	3956.75	II	840 c	4247.63	II
10	5084.23	I	2000	1228.59	IV	470	3962.45	II	500	4254.40	II
11	5097.17	I	5000	1293.22	IV	560	3964.26	II	320	4269.09	II
11	5099.20	I	5000	1295.28	IV	1600 c	3964.81	II	790 c	4272.27	II
12	5112.25	I	5000	1321.36	IV	560 c	3966.57	II	470 c	4280.07	II
12	5323.28	I	5000	1333.57	IV	500	3971.16	II	790 c	4282.42	II
13	5339.69	I	5000	1354.66	IV	320	3971.67	II	450 c	4298.98	II
12	5342.97	I	2000	1360.64	IV	620 c	3972.14	II	1500	4305.76	II
14	5359.57	I	2000	1365.77	IV	320	3974.85	II	1300	4333.97	II
16	5782.38	I	5000	1374.41	IV	1300 c	3989.68	II	360	4338.70	II
17	5801.75	I	5000	1435.56	IV	340	3992.16	II	620 cw	4344.30	II
15	5812.15	I	2000	1520.98	IV	1600	3994.79	II	470 c	4347.49	II
17	5831.89	I	5000	1574.55	IV	560 c	3997.04	II	340	4350.40	II
8	6120.27	II	5000	1575.10	IV	320	3999.12	II	450	4354.91	II
7	6307.29	II	3000	1578.38	IV	620 c	4000.17	II	410 c	4359.79	II
19	6911.08	I	2000	1622.30	IV	730	4004.70	II	1200	4368.33	II
12	6936.28	I	10000	1884.87	IV	1900	4008.69	II	320	4371.62	II
20	6938.77	I	2000	2083.23	IV	620	4010.60	II	430	4405.83	II
7	6964.18	I	3300	2246.20	V	730	4015.39	II	1700	4408.82	II
12	6964.67	I	2000 c	2378.98	IV	620	4020.96	II	410	4413.77	II
25	7664.90	I	40 h	2598.04	II	470	4022.71	II	1200 c	4429.13	II
24	7698.96	I	100 h	2707.37	II	360	4025.54	II	730	4449.83	II
5	7955.37	I	60	2760.35	II	360 c	4029.72	II	960	4468.66	II
4	7956.83	I	270	3168.24	II	730 c	4031.75	II	1100	4496.46	II
7	8078.11	I	200 d	3195.99	II	960	4033.83	II	790	4510.15	II
6	8079.62	I	190	3219.48	II	730	4038.45	II	340 c	4534.15	II
9	8250.18	I	200	3584.21	II	470	4039.34	II	340	4535.92	II
8	8251.74	I	250	3645.66	II	1300	4044.81	II	270 c	4628.74	II
3	8390.22	I	250	3646.30	II	340	4047.08	II	270 c	4672.09	II
11	8503.45	I	370	3668.83	II	450	4051.13	II	290	4695.77	I
10	8505.11	I	290	3714.05	II	2200	4054.88	II	250	4736.69	I
4	8763.96	I	410	3739.18	II	2200	4056.54	II	200	4924.60	I
3	8767.05	I	680	3761.87	II	450	4058.80	II	320	4939.74	I
13	8902.19	I	680	3800.30	II	3400	4062.81	II	380	4951.37	I
12	8904.02	I	390	3811.84	II	500 c	4079.77	II	270	5034.41	II
5	8923.31	I	1300 h	3816.02	II	500 c	4080.98	II	320	5045.52	I
4	8925.44	I	680	3818.28	II	790	4081.85	II	360	5110.38	II
7	9347.24	I	310	3821.80	II	500	4083.34	II	560	5110.76	II
3	9349.25	I	960	3830.72	II	560	4096.82	II	410	5129.52	II
6	9351.59	I	480	3840.99	II	380	4098.40	II	620	5173.90	II
15	9595.70	I	580	3846.59	II	2900 c	4100.72	II	360	5206.55	II
14	9597.83	I	1200	3850.79	II	1700 c	4118.46	II	360	5219.05	II
6	9949.67	I	720 c	3851.55	II	340	4130.77	II	560	5220.11	II
5	9954.14	I	960	3852.80	II	1500 c	4141.22	II	680	5259.73	II
9	10479.63	I	480 c	3865.45	II	2700	4143.11	II	340 c	5292.02	II
5	10482.15	I	480	3876.19	II	1700 c	4164.16	II	340	5292.62	II
			1700 c	3877.18	II						

Intensity	Wavelength/Å		Intensity	Wavelength/Å		Intensity	Wavelength/Å		Intensity	Wavelength/Å	
430	5322.76	II	1000	3957.74	II	3000	7076.27	I	4900	2167.94	I
65	5509.15	II	1000 r	3998.96	II	3000 h	7100.94	I	3400	2176.21	I
150	5535.17	II	1000	4417.96	II	10000 s	7114.89	I	4200 c	2214.26	II
110	5623.05	II	900 r	4728.36	I	3000 h	7171.55	I	5200 c	2275.25	II
90	5624.45	II	900	6100.21	I	3000	7227.13	I	2900	2287.51	I
90	5756.17	II	1000 d	6520.45	I	3000	7318.79	I	2700	2294.49	I
90	5779.28	I				10000 l	7368.25	I	390	2298.09	II
160 d	5815.17	II	**Protactinium Pa Z = 91**			3000 h	7471.89	I	610	2302.99	I
90	5823.72	II	3000	2599.16	II	10000 h	7493.15	I	680	2306.54	I
90	5859.68	II	3000	2699.22	II	3000 h	7558.26	I	800	2322.49	I
160	5939.90	II	3000	2822.79	II	10000 h	7608.20	I	300	2328.66	I
7000 w	5956.05	III	3000 h	2871.42	II	10000	7626.79	I	860	2344.78	I
90	5956.60	II	3000 h	2891.14	II	10000 s	7635.18	I	230	2349.39	I
110	5967.82	II	3000 l	3011.10	II	10000	7669.34	I	680	2352.07	I
90	6006.33	II	3000 s	3033.59	II	3000	7679.20	I	250	2356.50	I
150	6017.80	II	3000 l	3071.24	II	10000 h	7749.19	I	1200	2365.90	I
150	6025.72	II	3000 l	3093.23	II	3000	7872.95	I	570	2367.68	I
140	6055.13	I	3000 l	3126.23	II	3000 l	7945.56	I	520	2369.27	I
65	6087.52	II	3000 l	3146.28	II	10000	8039.34	I	220	2370.76	II
9000 w	6090.02	III	3000 l	3170.89	II	10000 h	8099.84	I	320	2375.07	I
65	6114.38	II	3000 l	3171.54	II	10000	8199.04	I	370	2379.77	I
65	6148.23	I	3000 l	3240.58	II	10000	8271.87	I	340	2388.57	I
5000	6160.24	III	3000	3274.46	II	3000 s	8358.98	I	230	2393.65	I
190	6161.18	II	3000 l	3332.69	II	3000 s	8369.60	I	320	2394.37	I
270	6165.94	II	3000 s	3346.66	II	3000 h	8441.04	I	320	2396.79	I
45	6244.35	II	3000 l	3452.82	II	10000 h	8532.66	I	210 d	2400.72	I
110	6281.28	II	3000	3504.97	I	10000 s	8572.96	I	210	2401.68	I
55 c	6359.03	I	3000 s	3530.65	II	3000 h	8639.91	I	1500	2405.06	I
55	6411.23	I	3000	3570.56	I	3000 h	8653.51	I	740	2405.60	I
45	6429.63	II	3000	3571.82	I	10000	8735.27	I	320	2406.70	I
45	6431.84	II	3000	3618.07	I	3000	10923.32	I	270	2410.37	I
45	6486.55	I	10000	3636.52	I	10000	11791.73	I	1200	2419.81	I
45	6566.77	II	3000	3702.74	I	10000	14344.76	I	300	2421.73	I
55	6616.67	I	3000	3752.67	I	3000	18478.61	I	300	2421.88	I
75	6656.83	II	3000	3873.35	I				2500	2428.58	I
55	6673.41	II	3000	3931.83	I	**Radium Ra Z = 88**			490	2431.54	I
75	6673.78	II	3000 s	3952.62	II	100	3649.55	II	420	2432.18	I
35 c	6747.09	I	10000 l	3957.85	II	200	3814.42	II	340 c	2441.47	I
55 cw	6798.60	I	3000 s	3970.07	II	100	4340.64	II	230	2442.51	I
35 cw	6827.60	II	3000	3981.82	I	100	4682.28	II	250	2444.94	I
7000	6910.14	III	10000	3982.23	I	100	4825.91	I	610	2446.98	I
40	7021.51	II	3000 l	4012.96	II	50	5660.81	I	610	2449.71	II
5000	7030.39	III	3000 s	4018.21	II	50	7141.21	I	390	2461.20	I
4500	7076.62	III	3000	4030.16	II	50	8019.70	II	800 c	2461.84	II
20	7114.55	I	3000 s	4046.93	II				1200	2483.92	I
24	7227.70	II	10000 s	4056.20	II	**Radon Rn Z = 86**			390	2485.81	I
16	7407.56	II	10000 s	4070.40	II	100	4349.60	I	980	2487.33	I
20 c	7451.74	II	3000 l	4176.18	II	200	7055.42	I	370	2496.04	I
14	7541.02	II	10000 l	4217.23	II	100	7268.11	I	370	2501.72	I
20	7645.66	II	10000 s	4248.08	II	300	7450.00	I	570	2502.35	II
16	7721.84	I	3000 s	4291.34	II	100	7809.82	I	230	2504.60	II
14	7871.67	I	3000 s	4601.43	II	100	8099.51	I	270	2505.94	I
14	8067.44	I	3000 l	6035.78	I	100	8270.96	I	1800 c	2508.99	I
10 cw	8122.78	II	3000	6162.56	I	100	8600.07	I	570	2520.01	I
11	8141.10	I	3000 l	6358.61	I				540	2521.50	I
5000 w	8602.74	III	3000	6379.25	I	**Rhenium Re Z = 75**			370	2534.80	I
10	8714.59	II	3000 l	6438.97	I	25000	2003.53	I	570	2540.51	I
			3000 h	6792.75	I	16000	2017.87	I	740 d	2544.74	I
Promethium Pm Z = 61			10000	6945.72	I	27000	2049.08	I	370	2545.48	I
1000	3892.15	II	3000	6960.09	I	10000	2085.59	I	300	2552.02	I
1000	3910.26	II	3000 h	6961.78	I	9800	2097.12	I	370	2554.63	II
1000	3919.10	II	3000 s	6992.73	I	3400	2139.04	II			
						3700	2156.67	I			

Intensity	Wavelength/Å		Intensity	Wavelength/Å		Intensity	Wavelength/Å		Intensity	Wavelength/Å	
1000	2556.51	I	320	3069.94	I	240	3453.50	I	110 c	5752.93	I
250	2559.08	I	260	3071.16	I	55000 c	3460.46	I	110 cw	5776.83	I
340	2564.19	I	550	3082.43	I	40000 c	3464.73	I	550	5834.31	I
540	2568.64	II	340	3088.76	I	400	3467.96	I	200	6307.70	I
370	2571.81	II	700	3100.67	I	240	3476.44	I	200	6321.90	I
380	2586.79	I	700	3108.81	I	400	3480.38	I	100 cw	6605.19	I
290	2599.86	I	340	3110.86	I	320	3480.85	I	180 c	6813.41	I
290	2603.89	I	340 c	3118.19	I	240	3482.23	I	260	6829.90	I
660	2608.50	II	340	3121.36	I	560	3503.06	I	50 cw	7640.94	I
610 d	2611.54	I	420	3128.94	I	320	3516.65	I	65 cw	7912.94	I
310	2635.83	II	260	3134.02	I	320	3517.33	I			
550	2636.64	I	250	3141.38	I	320	3537.46	I	*Rhodium Rh Z = 45*		
270	2642.75	I	440	3151.64	I	240	3549.89	I	50	813.44	III
270	2649.05	I	330	3153.79	I	240	3570.26	I	80	882.51	III
660	2651.90	I	360 c	3158.31	I	360	3579.12	I	100	925.75	III
400	2654.12	I	220	3164.52	I	810 c	3580.15	II	150	937.28	III
220	2663.63	I	700	3168.37	I	650	3580.97	I	500	991.62	III
940	2674.34	I	220	3174.61	I	810	3583.02	I	400	992.48	III
220	2688.53	I	440	3177.71	I	320	3617.08	I	500 d	1009.60	III
1300	2715.47	I	260	3178.61	I	810	3637.84	I	200	1012.22	III
220	2732.21	I	600	3182.87	I	440	3651.97	I	200	1015.17	III
610	2733.04	II	1100	3184.76	I	320	3670.53	I	200	1073.87	III
220	2758.00	I	1100	3185.57	I	860 c	3689.50	I	150	1784.24	III
210	2763.79	I	260	3190.78	I	1500 c	3691.48	I	200	1784.94	III
310	2767.74	I	260	3192.36	I	520	3703.24	I	150	1796.50	III
220	2768.85	I	220	3198.58	I	240	3709.93	I	200	1816.03	III
220	2769.32	I	1100 c	3204.25	I	360 c	3717.28	I	1000	1832.05	III
350	2770.42	I	380	3235.94	I	4000	3725.76	I	500	1859.85	III
550	2783.57	I	600	3258.85	I	240 c	3735.01	I	800	1880.66	III
220	2791.29	I	600	3259.55	I	810	3735.31	I	500	1884.91	III
220	2814.68	I	300	3268.89	I	910	3740.10	I	500	1887.36	III
880	2819.95	I	280	3296.70	I	300 cw	3745.44	I	700	1888.62	III
310	2834.08	I	280	3296.99	I	700	3787.52	I	800	1901.32	III
220	2843.00	I	280	3301.60	I	240	3869.94	I	500	1910.16	III
270	2850.98	I	240	3302.23	I	240	3875.26	I	600	1919.37	III
240	2867.19	I	320	3303.21	II	240	3876.86	I	500	1927.07	III
2900	2887.68	I	280	3303.75	I	380 c	3917.27	I	700	1931.79	III
490	2896.01	I	240	3313.95	I	550	3929.85	I	500	1954.25	III
830 c	2902.48	I	600	3322.48	I	280	3961.04	I	500	1994.26	III
210	2905.58	I	2000	3338.18	I	350 c	3962.48	I	800	2013.71	III
550	2909.82	I	1600	3342.24	I	220	4033.31	I	500	2017.47	III
830 c	2927.42	I	810	3344.32	I	240	4081.43	I	500	2028.53	III
270	2930.61	I	320	3346.20	I	240 c	4110.89	I	800	2036.72	III
440	2943.14	I	240 d	3356.33	I	240 cw	4133.42	I	600	2037.61	III
270	2962.27	I	240	3377.74	I	1800	4136.45	I	1000	2040.18	III
720	2965.11	I	320	3379.06	II	700	4144.36	I	3000	2048.67	III
1500	2965.76	I	320	3379.70	I	220	4182.90	I	2000	2064.11	III
310	2976.29	I	240	3389.43	I	220	4183.06	I	800	2076.84	III
210	2978.15	I	4000	3399.30	I	650	4221.08	I	1000	2118.53	III
220	2980.82	I	650	3404.72	I	3600 c	4227.46	I	1000	2118.63	III
220	2982.19	I	650	3405.89	I	260 c	4257.60	I	1000	2139.44	III
220	2988.47	I	240	3408.67	I	380	4358.69	I	1000	2152.23	III
1800	2992.36	I	320	3409.83	I	360 cw	4394.38	I	3000	2158.17	III
5500	2999.60	I	320	3417.77	I	2600	4513.31	I	3000	2163.19	III
350	3001.14	I	810	3419.41	I	260	4516.64	I	3000	2167.33	III
220	3004.14	I	8000	3424.62	I	500	4522.73	I	150	2276.21	II
500	3016.02	I	400	3426.19	I	2200 cw	4889.14	I	140	2288.57	I
300	3016.49	I	300	3427.61	I	220	4923.90	I	110	2309.82	I
380	3030.45	I	320	3437.71	I	1300	5270.95	I	350	2322.58	I
240	3047.25	I	400	3449.37	I	1600 cw	5275.56	I	140	2326.47	I
1600	3067.40	I	16000 c	3451.88	I	100	5667.88	I	190	2334.77	II
									300	2361.92	I

Intensity	Wavelength/Å		Intensity	Wavelength/Å		Intensity	Wavelength/Å		Intensity	Wavelength/Å	
110	2368.34	I	2300	3280.55	I	240	3754.12	I	16	6253.72	I
270	2382.89	I	2300	3283.57	I	380	3754.27	I	29	6319.53	I
230	2383.40	I	280	3289.14	I	490	3755.58	I	40	6752.35	I
270	2386.14	II	210	3294.28	I	1000	3760.40	I	13	6827.33	I
80	2415.84	II	260	3300.46	I	2300	3765.08	I	11	6857.68	I
130	2427.68	I	50	3310.69	III	490	3769.97	I	20	6879.94	I
230	2429.52	I	4200	3323.09	I	380	3778.13	I	65	6965.67	I
110	2437.90	I	330	3338.54	I	1000	3788.47	I	16	6979.15	I
330	2440.34	I	280	3360.80	I	1300	3792.18	I	16	7001.58	I
90	2461.04	II	420	3368.38	I	3800	3793.22	I	18	7101.64	I
130	2473.09	I	1100	3372.25	I	4900	3799.31	I	15	7104.45	I
150	2487.47	I	110	3377.14	I	760	3805.92	I	18	7268.18	I
100	2490.77	II	110	3385.78	I	1300	3806.76	I	35	7270.82	I
130	2502.46	I	5600	3396.82	I	470	3815.01	I	18 h	7442.39	I
300	2504.29	II	820	3399.70	I	760	3816.47	I	12	7475.74	I
150	2505.67	I	160	3406.55	I	1300	3818.19	I	12	7495.24	I
350	2509.70	I	820	3412.27	I	3800	3822.26	I	11	7557.67	I
300	2511.03	II	330	3421.22	I	2300	3828.48	I	29	7791.61	I
200	2515.75	I	120 d	3424.38	I	2000	3833.89	I	55	7824.91	I
130	2520.53	II	8200	3434.89	I	5900	3856.52	I	21	8029.91	I
110	2537.04	II	1400	3440.53	I	490	3870.01	I	29	8045.36	I
350	2545.70	I	120	3447.74	I	380	3877.34	I	15	8136.20	I
550	2555.36	I	120	3450.29	I	120	3913.51	I	8	8425.59	I
150	2622.58	I	400	3455.22	I	240	3922.19	I			
230	2625.88	I	180	3457.07	I	2000	3934.23	I	*Rubidium Rb* Z = 37		
100	2630.42	I	220	3457.93	I	590	3942.72	I	30	465.85	III
110	2647.28	I	5900	3462.04	I	3800	3958.86	I	40	481.118	II
400	2652.66	I	180	3469.62	I	45	3964.54	II	500	482.83	III
100	2680.63	I	4700	3470.66	I	380	3975.31	I	500	489.66	III
400	2703.73	I	120	3472.25	I	240	3984.40	I	600	493.48	III
100	2715.31	II	4700	3474.78	I	240	3995.61	I	90	497.430	II
180	2718.54	I	2100	3478.91	I	380	3996.15	I	20	508.434	II
160	2728.94	I	110	3494.44	I	120	4023.14	I	150	513.266	II
100	2771.51	I	1200	3498.73	I	560	4082.78	I	300	530.173	II
130	2783.03	I	5900	3502.52	I	140	4097.52	I	75	533.801	II
150	2826.43	I	2800	3507.32	I	120	4119.68	I	1200	535.86	III
180	2826.68	I	8800	3528.02	I	1100	4121.68	I	40	542.887	II
280	2862.94	I	880 d	3538.14	I	1500	4128.87	I	200	555.036	II
110	2878.66	I	280	3541.91	I	2100	4135.27	I	1200	556.19	III
140	2882.37	I	1200	3543.95	I	240	4154.37	I	1500	566.71	III
160	2907.21	I	1800	3549.54	I	330	4196.50	I	1000	572.82	III
65	2910.17	II	1200	3570.18	I	3300	4211.14	I	1500	576.65	III
180	2924.02	I	4700	3583.10	I	820	4288.71	I	2500	579.63	III
130	2929.11	I	4700	3596.19	I	4200	4374.80	I	1500	581.26	III
130	2931.94	I	5900	3597.15	I	130	4569.00	I	2500	589.419	II
230	2968.66	I	3100	3612.47	I	150	4675.03	I	1000	594.94	III
160	2977.68	I	1800	3626.59	I	70	4745.11	I	1300	595.88	III
450	2986.20	I	8200	3657.99	I	70	5090.63	I	1200	598.49	III
110	3004.46	I	1300	3666.22	I	60	5155.54	I	1500	643.878	II
50	3006.43	III	560	3681.04	I	60	5175.97	I	25	663.76	IV
130	3023.91	I	1900	3690.70	I	95	5193.14	I	3000	697.049	II
50	3052.44	III	9400	3692.36	I	130	5354.40	I	6000	711.187	II
180	3083.96	I	940	3695.52	I	95	5390.44	I	25	716.24	IV
140	3121.76	I	280	3698.26	I	160	5599.42	I	50	740.85	IV
240	3123.70	I	380	3698.60	I	40	5686.38	I	10000	741.456	II
130	3155.78	I	7600	3700.91	I	29	5792.66	I	5000	769.04	III
140	3189.05	I	940	3713.02	I	40	5806.91	I	25	776.89	IV
470	3191.19	I	650	3735.28	I	35	5831.58	I	2500	815.28	III
190	3197.13	I	420	3737.27	I	130	5983.60	I	15	850.18	IV
520	3263.14	I	420	3744.17	I	35	6102.72	I	1000	1604.12	II
520	3271.61	I	1200	3748.22	I	14	6199.99	I	5000	1760.50	II
									2000	2068.92	II

Intensity	Wavelength/Å		Intensity	Wavelength/Å		Intensity	Wavelength/Å		Intensity	Wavelength/Å	
10000	2075.95	II	40 l	8271.41	I	110	2517.32	II	650	3669.49	I
30000	2143.83	II	30	8271.71	I	150	2535.59	II	550	3726.10	I
10000	2217.08	II	40 l	8868.512	I	550	2549.58	I	8700	3726.93	I
5000	2291.71	II	30	8868.852	I	370	2609.06	I	11000	3728.03	I
50000	2472.20	II	30 l	9522.65	I	830	2612.07	I	7100	3730.43	I
1000	2631.75	III	20 l	9540.18	I	460	2642.96	I	3500	3742.28	I
2000	2956.07	III	2000 c	9689.05	II	330	2651.84	I	870	3742.78	I
500	3086.84	III	35 l	10075.282	I	400	2659.62	I	2800	3745.59	I
500	3111.36	III	30 l	10075.708	I	330	2661.61	II	760	3753.54	I
5000 c	3148.90	II	100	13235.17	I	690	2678.76	II	870	3755.93	I
25	3157.54	I	20	13442.81	I	330	2692.06	II	1200	3759.84	I
50	3227.98	I	30	13443.57	I	200	2712.41	II	600	3761.51	I
500	3286.41	III	75	13665.01	I	690	2719.52	I	600	3767.35	I
60	3348.72	I	1000	14752.41	I	140	2725.47	II	1500	3777.59	I
75	3350.82	I	800	15288.43	I	310	2734.35	II	600	3782.74	I
100	3587.05	I	150	15289.48	I	1800	2735.72	I	3900	3786.06	I
40	3591.57	I	20	22529.65	I	100	2778.38	II	6000	3790.51	I
5000	3600.60	II	4	27314.31	I	110	2787.83	II	760	3798.05	I
10000	3600.64	II				350	2810.03	I	7600	3798.90	I
25000	3940.51	II	*Ruthenium Ru Z = 44*			1700	2810.55	I	7600	3799.35	I
1000	4201.80	I	250	850.09	III	350	2818.36	I	600	3812.72	I
500	4215.53	I	200	850.30	III	400	2829.16	I	760	3817.27	I
90000	4244.40	II	250	919.74	III	640	2854.07	I	760	3819.03	I
15000	4273.14	II	500	940.09	III	420	2861.41	I	650	3822.09	I
20000	4571.77	II	500	966.54	III	550	2866.64	I	550	3824.93	I
10000	4648.57	II	750	974.14	III	1800	2874.98	I	760	3831.80	I
30000	4775.95	II	900	979.43	III	740	2886.54	I	930	3839.70	I
2	5087.987	I	500	981.35	III	370	2908.88	I	760	3850.43	I
2	5132.471	I	900	986.84	III	1100	2916.26	I	1300	3857.55	I
10	5150.134	I	900	994.56	III	180	2945.67	II	650	3862.69	I
10000	5152.08	I	300	1001.65	III	370	2949.50	I	1300	3867.84	I
1	5165.023	I	500	1009.13	III	550	2965.16	I	650	3892.21	I
2	5165.142	I	900	1009.87	III	170	2965.55	II	760	3909.08	I
15	5195.278	I	500	1014.68	III	140	2976.59	II	1500	3923.47	I
2	5233.968	I	800	1190.51	III	550	2976.92	I	3300	3925.92	I
20	5260.034	I	500	1200.07	III	1400	2988.95	I	600	3931.76	I
1	5260.228	I	500	1207.17	III	460	2994.96	I	760	3945.57	I
3	5322.380	I	500	1209.77	III	440	3006.59	I	600	3978.44	I
40	5362.601	I	300	1211.31	III	330	3017.24	I	600	3979.42	I
4	5390.568	I	500	1941.35	III	310	3020.88	I	870	3984.86	I
75	5431.532	I	500	2009.28	III	390	3064.84	I	1500	4022.16	I
3	5431.830	I	2400	2076.43	I	330	3096.57	I	600	4023.83	I
6	5578.788	I	2600	2083.77	I	830	3099.28	I	1400	4051.40	I
40	5647.774	I	2400	2090.89	I	740	3100.84	I	710	4054.05	I
20	5653.750	I	690	2255.52	I	490	3294.11	I	760	4068.37	I
60	5724.121	I	780	2272.09	I	370	3301.59	I	980	4076.73	I
3	5724.614	I	780	2279.57	I	930	3339.55	I	6000	4080.60	I
75	6070.755	I	480	2317.80	I	3100	3417.35	I	930	4097.79	I
30 c	6159.626	I	120	2334.96	II	4900	3428.31	I	1900	4112.74	I
75 c	6206.309	I	190 h	2342.85	II	6400	3436.74	I	2000	4144.16	I
120 c	6298.325	I	310	2351.33	I	8300	3498.94	I	650	4145.74	I
5	6299.224	I	170	2357.91	II	640	3514.49	I	870	4167.51	I
10000	6458.33	II	780	2402.72	II	790	3539.37	I	550	4197.58	I
5000	6560.81	II	150	2407.92	II	690	3570.59	I	550	4198.88	I
100 l	7279.997	I	180	2455.53	II	6400	3589.22	I	7600	4199.90	I
150	7408.173	I	150	2456.44	II	6900	3593.02	I	1500	4206.02	I
200 l	7618.933	I	370	2456.57	II	6400	3596.18	I	5400	4212.06	I
300	7757.651	I	280	2478.93	II	1300	3599.76	I	760	4214.44	I
60	7759.436	I	140	2498.42	II	3100	3634.93	I	930	4217.27	I
90000 c	7800.27	I	140	2498.57	II	6200	3661.35	I	550	4230.31	I
45000 c	7947.60	I	260	2507.01	II	830	3663.37	I	760	4241.05	I
			110	2513.32	II						

Intensity	Wavelength/Å		Intensity	Wavelength/Å		Intensity	Wavelength/Å		Intensity	Wavelength/Å	
760	4243.06	I	26	6982.01	I	480	3854.56	I	730	4716.10	I
760	4284.33	I	26	7027.98	I	400	3858.74	I	770	4728.42	I
550	4295.93	I	35	7238.92	I	3700	3885.29	II	470	4745.68	II
3700	4297.71	I	16	7393.93	I	1600	3896.98	II	730	4760.27	I
930	4307.60	I	18	7468.91	I	1300	3903.42	II	580	4783.10	I
550	4319.87	I	26	7485.79	I	2500	3922.40	II	350	4785.86	I
550	4342.07	I	70	7499.75	I	1900	3928.28	II	970	4841.70	I
710	4354.13	I	26	7559.61	I	1300	3941.87	II	730	4883.97	I
870	4361.21	I	18	7621.50	I	470	3951.89	I	630	4910.40	I
2400	4372.21	I	18	7722.87	I	1500	3963.00	II	350	4913.25	II
870	4385.39	I	22	7791.86	I	1500	3971.40	II	430	4918.99	I
1300	4385.65	I	30	7847.80	I	620	3974.66	I	400	5044.28	I
1700	4390.44	I	80	7881.49	I	1000	3976.43	II	540	5071.20	I
1600	4410.03	I	18	8264.96	I	1500	3990.00	II	510	5117.16	I
1100	4460.04	I	22	8710.84	I	1400	4064.58	II	350	5122.14	I
5400	4554.51	I				1000	4092.27	II	360	5155.03	II
1700	4584.44	I	*Samarium Sm Z = 62*			1900	4118.55	II	470	5175.42	I
720	4647.61	I	150	2789.38	II	1200	4152.21	II	250	5200.59	I
1400	4709.48	I	410	3152.52	II	1000	4188.13	II	260	5251.92	I
500	4757.84	I	720	3183.92	II	1100	4203.05	II	400	5271.40	I
550	4869.15	I	600	3211.73	II	1000	4225.33	II	250	5282.91	I
160	5011.23	I	530	3216.85	II	1200	4236.74	II	220	5453.00	I
450	5057.33	I	600	3218.61	II	2100	4256.39	II	230	5493.72	I
120	5076.32	I	720	3230.56	II	1300	4262.68	II	230	5516.09	I
200	5093.83	I	720	3236.64	II	1200	4279.68	II	140	5550.40	I
530	5136.55	I	720	3239.66	II	2200	4280.79	II	140	5659.86	I
170	5142.76	I	720	3250.37	II	710	4282.21	I	120	5696.73	I
250	5147.24	I	850	3254.38	II	470	4282.83	I	85	5706.20	I
110	5151.07	I	1700	3306.39	II	1600	4296.74	I	70	5773.77	I
500	5155.14	I	1200	3321.18	II	1900	4318.94	II	60	5778.33	I
920	5171.03	I	1200	3365.86	II	470	4319.53	I	70 d	5786.98	II
180	5195.02	I	1200	3382.40	II	1800	4329.02	II	60	5788.38	I
130	5284.08	I	4200	3568.27	II	440	4330.02	I	60	5800.52	I
260	5309.27	I	4200	3592.60	II	1300	4334.15	II	65	5802.84	I
110	5335.93	I	1700	3604.28	II	880	4336.14	I	65	5867.79	I
130	5361.77	I	3400	3609.49	II	1100	4347.80	II	50	5874.21	I
110 h	5401.04	I	1700	3621.23	II	440	4362.91	I	50	5898.96	I
80	5484.32	I	3400	3634.29	II	530	4380.42	I	65	5965.71	II
130	5510.71	I	2200	3661.36	II	1600	4390.86	II	50	6045.00	I
90	5559.75	I	2200	3670.84	II	410	4401.17	I	50	6070.06	I
290	5636.24	I	1100	3693.99	II	470	4419.33	I	45	6084.12	I
180	5699.05	I	1600	3728.47	II	1500	4420.53	II	45 h	6159.56	I
65	5814.98	I	2100	3731.26	II	2900	4424.34	II	45	6256.54	I
55	5919.34	I	1600	3735.98	II	470	4429.66	I	100	6267.28	II
80	5921.45	I	2900	3739.12	II	1600	4433.88	II	140	6569.31	II
21 h	5973.38	I	1200	3743.87	II	1800	4434.32	II	110	6589.72	II
16	5988.67	I	930	3745.46	I	530	4441.81	I	50	6671.51	I
35	5993.65	I	800	3756.41	I	440	4442.28	I	120 d	6731.84	II
18	6116.77	I	1200	3757.53	II	710	4445.15	I	95	6794.20	II
26	6199.42	I	1900	3760.69	II	1300	4452.73	II	120	6860.93	I
26	6225.20	I	1100	3764.37	II	1200	4454.63	II	120	6955.29	II
18	6295.22	I	370 d	3773.33	I	1000	4458.52	II	90	7020.44	II
16	6390.23	I	1100	3778.14	II	2200	4467.34	II	90	7039.22	II
26 h	6444.84	I	1500	3788.12	II	810	4470.89	I	90	7042.24	II
21	6663.14	I	1600	3793.97	II	370	4499.11	I	90	7051.52	II
55	6690.00	I	1600	3797.73	II	440	4581.73	I	90	7082.37	II
21	6766.95	I	1600	3826.20	II	380	4649.49	I	26	7088.30	I
30	6775.02	I	1100	3831.50	II	470 d	4670.75	I	30	7095.50	I
21	6824.17	I	560	3834.48	I	1100	4674.60	II	30	7104.54	I
26	6911.48	I	1600	3843.50	II	370	4688.73	I	26	7115.96	I
110	6923.23	I	530	3853.30	I	530	4704.40	II	85 d	7149.60	II
			2700	3854.21	II						

Intensity	Wavelength/Å		Intensity	Wavelength/Å		Intensity	Wavelength/Å		Intensity	Wavelength/Å	
23	7213.82	I	1200	2974.01	I	2700	4047.79	I	270	5089.89	I
60	7240.90	II	1400	2980.75	I	120	4049.95	I	390	5096.73	I
26	7347.30	I	340	2988.95	I	5500	4054.55	I	620	5099.23	I
30	7444.56	I	2200	3015.36	I	220	4056.59	I	370	5101.12	I
26	7445.41	I	2700	3019.34	I	100	4068.66	III	180	5109.06	I
13	7470.76	I	360	3030.76	I	160 h	4074.97	I	150	5112.86	I
45	7645.09	II	120 h	3056.31	I	160	4078.57	I	320	5116.69	I
12	7645.82	I	130	3065.11	II	6100	4082.40	I	390	5210.52	I
40 w	7835.08	II	990	3251.32	II	200	4086.67	I	280	5219.67	I
16	7895.96	I	1500	3255.69	I	400	4087.16	I	350	5239.82	II
90	7928.14	II	4400	3269.91	I	440 h	4133.00	I	280	5258.33	I
40	8048.70	II	5500	3273.63	I	530 h	4140.30	I	210	5285.76	I
16	8065.16	I	110 d	3343.28	II	720	4152.36	I	120	5341.05	I
45	8068.46	II	270	3352.05	II	1100 h	4165.19	I	350	5349.30	I
40 w	8305.79	II	9900	3353.73	II	110 h	4218.26	I	120	5349.71	I
19	8383.71	I	2000	3359.68	II	110 h	4219.73	I	210	5355.75	I
45 w	8485.99	II	1700	3361.27	II	180	4231.93	I	530	5356.10	I
45 w	8708.43	II	1700	3361.94	II	200	4233.61	I	270	5375.35	I
95	8913.66	II	4000	3368.95	II	400	4238.05	I	370	5392.08	I
			6600	3372.15	II	15000	4246.83	II	270	5446.20	I
Scandium Sc Z = 21			130	3418.51	I	290	4294.77	II	120	5451.34	I
350	180.14	V	200	3429.21	I	350	4305.71	II	750	5481.99	I
500	243.87	V	200	3429.48	I	4200	4314.09	II	530	5484.62	I
500	252.85	V	270	3431.36	I	3300	4320.74	II	570	5514.22	I
500	253.73	V	530	3435.56	I	2400	4325.01	II	660	5520.50	I
900	283.91	V	270	3457.45	I	180	4354.61	II	660	5526.82	II
800	284.45	V	180	3462.19	I	110	4358.64	I	70	5564.86	I
600	288.29	V	130 d	3469.65	I	2000	4374.46	II	110	5591.33	I
900	289.59	V	110	3471.13	I	130	4384.81	II	80	5640.98	II
15	289.85	IV	200	3498.91	I	1100	4400.37	II	250	5657.88	II
1000 d	291.93	V	2700	3535.73	II	880	4415.56	II	1500	5671.81	I
800	293.25	V	6600	3558.55	II	120 h	4557.24	I	1200	5686.84	I
15	296.31	IV	6100	3567.70	II	160 h	4573.99	I	1100	5700.21	I
15	299.04	IV	13000	3572.53	II	350	4670.40	II	10	5706.82	IV
700	300.00	V	9900	3576.35	II	120	4706.97	I	190	5708.61	I
1000	573.36	V	7700	3580.94	II	120	4709.34	I	880	5711.75	I
600	587.94	V	4000	3589.64	II	200	4728.77	I	230	5717.28	I
10	785.12	IV	4000	3590.48	II	490	4729.23	I	180	5724.08	I
25	1168.61	III	28000	3613.84	II	590	4734.10	I	14	5771.63	IV
15	1550.80	IV	110	3617.43	I	690	4737.65	I	620	6210.68	I
180	1603.06	III	20000	3630.75	II	790	4741.02	I	320	6239.78	I
150	1610.19	III	13000	3642.79	II	1200	4743.81	I	120	6245.63	II
160	2010.42	III	6600	3645.31	II	200	4753.16	I	110	6249.96	I
12	2118.97	IV	110	3646.90	I	220	4779.35	I	80	6256.01	III
11	2185.43	IV	5300	3651.80	II	170	4839.44	I	250	6258.96	I
11	2205.46	IV	110	3664.25	II	90	4909.76	I	750	6305.67	I
14	2222.22	IV	290	3666.54	II	90	4922.84	I	60	6378.82	I
11	2271.33	IV	75 h	3717.10	I	90	4934.25	I	90	6413.35	I
110	2438.62	I	270	3833.07	II	170	4954.06	I	60	6604.60	II
560	2545.22	II	610	3843.03	II	120	4973.66	I	65	6737.87	I
2900	2552.37	II	20000	3907.49	I	150	4980.37	I	50	6819.52	I
560	2555.82	II	23000	3911.81	I	140	4991.92	I	50	6835.03	I
2300	2560.25	II	4400	3933.38	I	530	5031.02	II	90	7449.16	III
1100	2563.21	II	5500	3996.61	I	250	5064.32	I	55 h	7741.17	I
11	2586.93	IV	530	4014.49	II	530	5070.23	I	30	7800.44	I
120	2692.78	I	20000	4020.40	I	250	5075.81	I	19 h	8761.40	I
350	2699.07	III	20000	4023.69	I	2100	5081.56	I	50	8829.78	III
360	2706.77	I	220	4030.67	I	1200	5083.72	I	30 h	8834.35	I
210	2707.95	I	140	4031.39	I	1100	5085.55	I	400	22051.86	I
580	2711.35	I	220	4043.80	I	750	5086.95	I	150	22065.05	I
230	2734.05	III	200	4046.48	I	390	5087.14	I			
340	2965.86	I									

Selenium Se Z = 34

Intensity	Wavelength/Å	
360	613.0	V
360	652.7	IV
450	670.1	IV
360	724.3	III
450	746.4	IV
450	759.1	V
360	808.7	V
360	830.3	V
360	832.7	II
450	839.5	V
360	843.0	III
360	845.8	V
360	912.9	II
360	959.6	IV
360	974.8	III
450	996.7	IV
360	1013.4	II
360	1014.0	II
450	1033.6	II
450	1049.6	II
360	1057.4	II
360	1094.7	V
360	1099.1	III
450	1119.2	III
360	1141.9	II
450	1192.3	II
450	1227.6	V
285	1291.0	II
285	1308.9	II
285	1314.4	IV
120	1435.3	I
120	1435.8	I
150	1449.2	I
150	1500.9	I
250	1530.4	I
150	1531.3	I
200	1531.8	I
150	1575.3	I
150	1577.6	I
150	1577.9	I
150	1579.5	I
200	1580.0	I
150	1587.5	I
150	1593.2	I
250	1606.5	I
200	1617.4	I
150	1621.2	I
150	1643.4	I
250	1671.2	I
250	1675.3	I
250	1690.7	I
250	1793.3	I
300	1795.3	I
300	1855.2	I
250	1858.8	I
400	1898.6	I
350	1913.8	I
300	1919.2	I
500	1960.9	I
500	2039.8	I

Intensity	Wavelength/Å	
285	2057.5	III
500	2074.8	I
285	2136.6	IV
500	2164.2	I
600	2413.5	I
300	2548.0	I
360	2665.5	IV
285	2724.3	IV
285	2767.2	III
220	2773.8	III
160	2951.6	IV
450	3387.2	III
450	3413.9	III
450	3457.8	III
450	3637.6	III
450	3738.7	III
450	3800.9	III
450	4169.1	III
360	4175.3	II
450	4180.9	II
285	4382.9	II
285	4446.0	II
220	4449.2	II
285	4467.6	II
500	4730.8	I
400	4739.0	I
300	4742.2	I
285	4840.6	II
360	4845.0	II
450	5227.5	II
360	5305.4	II
100	5365.5	I
120	5369.9	I
110	5374.1	I
285	5522.4	II
285	5566.9	II
285	5866.3	II
450	6056.0	II
285	6303.8	III
200	6325.6	I
360	6444.2	II
285	6490.5	II
285	6535.0	II
150	6831.3	I
120	6990.690	I
100	6991.792	I
200	7010.809	I
150	7013.875	I
300	7062.065	I
200	7575.1	I
250	7583.4	I
150	7592.2	I
300	8001.0	I
200	8036.4	I
150	8093.2	I
150	8094.7	I
180	8149.3	I
150	8152.0	I
200	8157.7	I
180	8163.1	I
150	8182.9	I

Intensity	Wavelength/Å	
150	8440.47	I
150	8450.38	I
150	8742.33	I
300	8918.86	I
200	9001.97	I
200	9038.61	I
100	9432.50	I
200	10217.25	I
377	10307.45	I
900	10327.26	I
640	10386.36	I
275	11946.87	I
170	11952.64	I
205	11972.93	I
315	14817.93	I
410	14917.47	I
500	15151.44	I
320	15471.00	I
265	15520.97	I
395	15618.40	I
360	16659.44	I
505	16813.78	I
205	16866.54	I
235	21374.24	I
680	21442.56	I
415	21473.48	I
270	21716.36	I
240	21730.60	I
150	23388.85	I
265	24148.18	I
375	24385.99	I
255	25017.51	I
510	25127.43	I

Silicon Si Z = 14

Intensity	Wavelength/Å	
10	85.18	V
15	96.44	V
10	97.14	V
20	117.86	V
20	118.97	V
4	457.82	IV
8	566.61	III
8	653.33	III
7	815.05	IV
8	818.13	IV
9	823.41	III
40 h	845.77	II
100	889.72	II
200	892.00	II
9	967.95	III
100	989.87	II
200	992.68	II
10	993.52	III
13	994.79	III
16	997.39	III
50	1023.69	II
8	1066.63	IV
14	1108.37	III
16	1109.97	III
18	1113.23	III
8	1122.49	IV
10	1128.34	IV

Intensity	Wavelength/Å	
100	1190.42	II
200	1193.28	II
250	1194.50	II
100	1197.39	II
30	1206.51	III
30	1206.53	III
9	1207.52	III
10	1210.46	III
50	1226.81	II
100	1227.60	II
150	1228.75	II
200	1229.39	II
100	1246.74	II
150	1248.43	II
100	1250.09	II
150	1250.43	II
200	1251.16	II
40	1256.49	I
50	1258.80	I
1000	1260.42	II
2000	1264.73	II
200	1265.02	II
17	1294.54	III
14	1296.73	III
15	1298.89	III
18	1298.96	III
14	1301.15	III
16	1303.32	III
100	1304.37	II
50 h	1305.59	II
200	1309.27	II
13	1312.59	III
100	1346.87	II
100	1348.54	II
150	1350.06	II
100	1352.64	II
100	1353.72	II
15	1393.76	IV
12	1402.77	IV
13	1417.24	III
90 h	1485.02	II
100 h	1485.51	II
12	1500.24	III
10	1501.19	III
9	1501.87	III
100 h	1509.10	II
50 h	1512.07	II
60 p	1516.91	II
500	1526.72	II
1000	1533.45	II
150	1594.55	I
100	1622.87	I
300	1629.43	I
200	1629.92	I
100	1667.62	I
100	1668.52	I
100	1672.59	I
200	1675.20	I
200	1696.20	I
200	1697.94	I
100 h	1770.92	I

Intensity	Wavelength/Å		Intensity	Wavelength/Å		Intensity	Wavelength/Å		Intensity	Wavelength/Å	
100 h	1776.83	I	300	2904.28	II	16	4716.65	III	90	6131.850	I
100 h	1799.12	I	500	2905.69	II	50	4782.991	I	100	6142.487	I
150	1808.00	II	55	2970.355	I	35	4792.212	I	100	6145.015	I
500 h	1814.07	I	150	2987.645	I	80	4792.324	I	160	6155.134	I
200	1816.92	II	50	3006.739	I	15	4813.33	III	160	6237.320	I
200	1836.51	I	75	3020.004	I	16	4819.72	III	40	6238.287	I
200	1841.44	I	100 h	3030.00	II	18	4828.97	III	125	6243.813	I
9	1842.55	III	9	3040.93	III	30	4947.607	I	125	6244.468	I
200	1843.77	I	100 h	3043.69	II	40	5006.061	I	180	6254.188	I
300	1845.51	I	50 h	3048.30	II	1000	5041.03	II	45	6331.954	I
400	1847.47	I	150 h	3053.18	II	1000	5055.98	II	1000	6347.10	II
200	1848.14	I	25	3086.24	III	10 h	5091.42	III	1000	6371.36	II
500	1850.67	I	20	3093.42	III	100	5181.90	II	45	6526.609	I
200	1852.46	I	16	3096.83	III	100 h	5185.25	II	45	6527.199	I
500 h	1874.84	I	9	3165.71	IV	200 h	5192.86	II	45	6555.462	I
200	1881.85	I	16	3185.13	III	500 h	5202.41	II	50 h	6660.52	II
200	1887.70	I	13	3186.02	III	100 h	5405.34	II	100	6671.88	II
200 h	1893.25	I	150	3188.97	II	100 h	5438.62	II	7	6701.21	IV
1000 h	1901.33	I	150	3193.09	II	100 h	5456.45	II	50 h	6717.04	II
100 h	1902.46	II	100	3195.41	II	500 h	5466.43	II	100	6721.853	I
50 h	1910.62	II	14	3196.50	III	500 h	5466.87	II	50	6829.82	II
50	1941.67	II	200	3199.51	II	100 h	5469.21	II	30	6848.568	I
100	1949.56	II	100 h	3203.87	II	200 h	5496.45	II	80	6976.523	I
100	1954.97	I	200 h	3210.03	II	35	5517.535	I	180	7003.567	I
50	2058.65	II	15	3210.55	III	100 h	5540.74	II	180	7005.883	I
50	2059.01	II	75	3214.66	II	150 h	5576.66	II	90	7017.646	I
200	2072.02	II	12	3230.50	III	30	5622.221	I	250	7034.903	I
200	2072.70	II	14	3233.95	III	100 h	5632.97	II	6 h	7047.94	IV
100	2124.12	I	15	3241.62	III	200 h	5639.48	II	200	7165.545	I
50 h	2136.56	II	12	3258.66	III	90	5645.611	I	100	7226.206	I
110	2207.98	I	10	3276.26	III	150 h	5660.66	II	100	7235.326	I
115	2210.89	I	300	3333.14	II	80	5665.554	I	180	7250.625	I
110	2211.74	I	500	3339.82	II	1000 h	5669.56	II	160	7275.294	I
120	2216.67	I	15	3486.91	III	120	5684.484	I	400	7289.173	I
120	2218.06	I	9	3525.94	III	300 h	5688.81	II	375	7405.774	I
10	2296.87	III	20	3590.47	III	100	5690.425	I	200	7409.082	I
10	2308.19	III	8	3762.44	IV	90	5701.105	I	275	7415.946	I
100 h	2356.30	II	20 c	3791.41	III	200 h	5701.37	II	425	7423.497	I
30 h	2357.18	II	25	3796.11	III	100 h	5706.37	II	9 h	7466.32	III
50 h	2357.97	II	30	3806.54	III	160	5708.397	I	12 h	7612.36	III
300	2435.15	I	100 h	3853.66	II	20	5739.73	III	100	7680.267	I
11	2449.48	III	500 h	3856.02	II	45	5747.667	I	6 h	7723.82	IV
425	2506.90	I	200 h	3862.60	II	45	5753.625	I	30	7800.008	I
375	2514.32	I	300	3905.523	I	45	5754.220	I	400	7848.80	II
500	2516.113	I	20	3924.47	III	45	5762.977	I	500	7849.72	II
7	2517.51	IV	10	4088.85	IV	70	5772.145	I	30	7849.967	I
350	2519.202	I	70	4102.936	I	70	5780.384	I	90	7918.386	I
425	2524.108	I	9	4116.10	IV	90	5793.071	I	120	7932.349	I
450	2528.509	I	300 h	4128.07	II	100	5797.859	I	140	7944.001	I
110	2532.381	I	500 h	4130.89	II	150 h	5800.47	II	35	7970.306	I
25	2541.82	III	100 h	4190.72	II	200	5806.74	II	35	8035.619	I
10	2546.09	III	50	4198.13	II	50	5846.13	II	70	8093.241	I
14	2559.21	III	9	4338.50	III	300 h	5868.40	II	9 h	8102.86	III
30	2563.679	I	30	4552.62	III	40	5873.764	I	11 h	8103.45	III
85	2568.641	I	25	4567.82	III	10 h	5898.79	III	35	8230.642	I
45	2577.151	I	20	4574.76	III	150	5915.22	II	9 h	8262.57	III
190	2631.282	I	100	4621.42	II	200	5948.545	I	40	8443.982	I
11	2640.79	III	150	4621.72	II	500	5957.56	II	40	8501.547	I
14	2655.51	III	9 h	4631.24	IV	500	5978.93	II	60	8502.221	I
9	2817.11	III	10 h	4654.32	IV	90	6125.021	I	40	8536.165	I
1000	2881.579	I	9	4683.02	III	85	6131.574	I	120	8556.780	I

Intensity	Wavelength/Å	
50	8648.462	I
40	8728.011	I
75	8742.451	I
100	8752.009	I
35	8790.389	I
100	9412.72	II
100	9413.506	I
30	10371.269	I
120	10585.141	I
120	10603.431	I
120	10660.975	I
30	10694.251	I
30	10727.408	I
60	10749.384	I
30	10784.550	I
80	10786.856	I
140	10827.091	I
60	10843.854	I
130	10869.541	I
30	10882.802	I
30	10885.336	I
80	10979.308	I
30	10982.061	I
80	11017.965	I
370	11984.19	I
220	11991.57	I
440	12031.51	I
190	15888.39	I
95	16060.03	I
110	19722.50	I

Silver Ag Z = 47

Intensity	Wavelength/Å	
25	730.83	II
30	752.80	II
400	799.41	III
15	1005.32	II
10	1065.49	II
12	1072.23	II
250	1074.22	II
150	1107.03	II
150	1112.46	II
60	1195.83	II
50	1223.33	II
50	1240.80	II
50	1246.87	II
55	1256.81	II
55	1257.55	II
50	1266.63	II
70	1273.67	II
65	1297.51	II
85	1311.20	II
55	1313.81	II
50	1314.61	II
60	1323.84	II
60	1342.09	II
50	1342.57	II
70	1346.62	II
50	1353.54	II
150	1364.50	II
100	1396.00	II
100	1410.93	II
90	1419.72	II

Intensity	Wavelength/Å	
95	1432.60	II
100	1464.72	II
50	1466.23	II
50 r	1507.37	I
100 r	1515.63	I
50 r	1548.58	I
100	1555.16	II
100	1644.50	II
60	1651.52	I
50	1652.10	I
700	1656.18	III
120	1682.82	II
500	1693.51	III
10	1708.11	I
50	1709.27	I
125	1736.44	II
750	1751.03	III
10 h	1766.14	I
75	1790.37	II
600	1917.08	III
700	1957.62	III
100	1967.38	II
600	1975.92	III
500	1977.03	III
600	2000.24	III
150	2015.96	II
150	2033.98	II
200	2061.17	I
100	2069.85	I
80 r	2113.82	II
60	2145.60	II
600	2161.89	III
50	2186.76	II
60	2229.53	II
100 r	2246.43	II
500	2246.51	III
75 r	2248.74	II
75	2280.03	II
30 h	2309.56	I
700	2310.04	III
70 r	2317.05	II
80 r	2320.29	II
70 r	2324.68	II
80 r	2331.40	II
70	2357.92	II
50 h	2375.02	I
75	2411.41	II
90 r	2413.23	II
100 r	2437.81	II
80	2447.93	II
80	2473.84	II
60	2506.63	II
50 h	2575.63	I
60	2660.49	II
60	2721.77	I
75	2767.54	II
100 h	2824.39	I
30 h	3130.02	I
90	3180.70	II
100	3267.35	II
55000 r	3280.68	I

Intensity	Wavelength/Å	
28000 r	3382.89	I
30	3469.16	I
70	3475.82	II
80	3495.28	II
50	3542.61	I
50 h	3624.68	I
75	3682.46	II
30	3682.50	I
80	3683.34	II
50 h	3709.20	I
200	3810.94	I
50	3811.78	I
100 h	3840.74	I
50 h	3907.41	I
50	3909.31	II
50 h	3914.40	I
70	3920.10	II
60	3949.43	II
100 h	3981.58	I
70	3985.19	II
100 h	4055.48	I
80	4085.91	II
100	4185.48	II
90 h	4210.96	I
100	4212.82	I
50	4311.07	I
50 h	4476.04	I
30 h	4615.69	I
80	4620.04	II
50	4620.46	II
60 h	4668.48	I
30 h	4677.60	I
100	4788.40	II
30 h	4847.82	I
100	4874.10	I
80	5027.35	II
1000	5209.08	I
1000	5465.50	I
100	5471.55	I
100	5667.34	I
10 h	6268.50	I
320	7687.78	I
25	8005.4	II
500	8273.52	I
25	8403.8	II
30 h	8645.70	I
10 h	8704.85	I
12	8747.6	II
15	9000.9	II
10	12551.0	I
60	16819.5	I
20	17416.7	I
15	18307.9	I
15	18382.3	I

Sodium Na Z = 11

Intensity	Wavelength/Å	
7	142.232	IV
8	146.064	IV
9	150.298	IV
8	150.687	IV
8	155.510	IV
8	156.537	IV

Intensity	Wavelength/Å	
12	162.448	IV
10	163.190	IV
12	168.411	IV
10	168.546	IV
5	183.95	III
10	190.445	IV
10	199.772	IV
8	202.49	III
8	202.76	III
8 p	203.06	III
8	203.28	III
8	203.33	III
15	229.87	III
50 c	250.52	III
30	251.37	III
25	266.90	III
70	267.65	III
50	267.87	III
50	268.63	III
20 p	272.08	III
20	272.45	III
10	319.644	IV
300	372.08	II
350	376.38	II
100	378.14	III
70	380.10	III
12	408.684	IV
10	409.614	IV
15	410.372	IV
10	411.334	IV
13	412.242	IV
11	1582.18	IV
11 d	1583.98	IV
12	1584.14	IV
12 d	1587.05	IV
11	1615.92	IV
12	1618.57	IV
11	1655.47	IV
15 c	1701.97	IV
20 d	1887.47	III
12	1960.76	IV
11	1965.08	IV
12 d	2106.33	IV
30	2230.33	III
16	2232.19	III
20 h	2246.70	III
300	2315.65	II
18	2386.99	III
17	2394.03	III
300	2420.99	II
300	2424.73	II
25	2459.31	III
18	2468.85	III
20	2474.73	III
1000	2493.15	II
25	2497.03	III
17	2510.26	III
20	2543.84	I
10	2543.87	I
70	2593.87	I
35	2593.92	I

Intensity	Wavelength/Å		Intensity	Wavelength/Å		Intensity	Wavelength/Å		Intensity	Wavelength/Å	
850	2611.81	II	20	4747.941	I	1875	491.79	III	40	5257.71	III
850	2661.00	II	30	4751.822	I	1250	507.04	III	40	5443.48	III
1000	2671.83	II	200	4978.541	I	3750	514.38	III	1500	5450.84	I
200	2680.34	I	400	4982.813	I	10	517.28	V	7000	5480.84	I
100	2680.43	I	40	5148.838	I	2500	562.75	III	3500	5504.17	I
1000	2841.72	II	80	5153.402	I	25	578.01	V	2600	5521.83	I
400	2852.81	I	280	5682.633	I	30	624.93	V	2000	5534.81	I
200	2853.01	I	70	5688.193	I	25	642.23	V	2000	5540.05	I
2	2893.62	I	560	5688.205	I	50	649.21	V	1000	6380.75	I
1100	2904.92	II	80000	5889.950	I	25	660.94	V	900 h	6386.50	I
1100	2917.52	II	40000	5895.924	I	200	664.43	IV	600 h	6388.24	I
1100	2919.05	II	120	6154.225	I	35	686.23	V	9000	6408.47	I
1200	2919.85	II	240	6160.747	I	100	710.35	IV	5500	6504.00	I
1300	2920.95	II	130	6530.70	II	12	747.82	V	1000	6546.79	I
1000	2923.49	II	130	6544.04	II	50	1025.23	III	1700	6550.26	I
1200	2951.24	II	130	6545.75	II	35	1125.49	III	3000	6617.26	I
1100	2952.40	II	20	7373.23	I	50	1236.23	III	1800	6791.05	I
1000	2977.13	II	10	7373.49	I	1400	2152.84	II	4800	6878.38	I
1100	2979.66	II	50	7809.78	I	1400	2165.96	II	1200	6892.59	I
1100	2980.63	II	25	7810.24	I	100	2273.71	III	5500	7070.10	I
1300	2984.19	II	4400	8183.256	I	100	2340.13	III	2500	7309.41	I
1700	3124.42	II	800	8194.790	I	50	2346.97	IV	500	7621.50	I
2500	3135.48	II	8800	8194.824	I	160	2428.10	I	400 h	7673.06	I
1700	3137.86	II	100	8649.92	I	100	2486.52	III	200 h	8422.80	I
2000	3149.28	II	60	8650.89	I	40	2555.60	IV	120	8505.69	II
2000	3163.74	II	25	8942.96	I	40	2571.04	IV	200	8688.91	II
1000	3179.06	II	40	9153.88	I	100	3002.61	III	100	9294.10	I
1700	3189.79	II	60	9465.94	I	200	3012.32	III	400 h	9448.95	I
1600	3212.19	II	80	9961.28	I	10	3019.29	IV	600	9596.00	I
1500	3257.96	II	20	10566.00	I	100	3021.73	III	300	9624.70	I
1700	3285.60	II	60	10572.28	I	50	3061.43	III	100	9638.10	I
1700	3301.35	II	200	10746.44	I	50	3182.61	III	100 h	9647.70	II
1200	3302.37	I	80	10749.29	I	100	3235.39	III	300	10036.66	II
600	3302.98	I	120	10834.87	I	400	3351.25	I	1000	10327.31	II
1500	3304.96	II	35	11190.19	I	650	3380.71	II	200	10914.88	II
1000	3318.04	II	50	11197.21	I	50	3430.76	III	700	11241.25	I
50	3426.86	I	400	11381.45	I	950	3464.46	II	100	12014.76	II
1500	3533.05	II	1000	11403.78	I	600	3969.26	I	60	12445.90	II
1200	3631.27	II	400	12679.17	I	1300	4030.38	I	40	12495.00	I
6	4238.99	I	60	14767.48	I	46000	4077.71	II	75	12974.70	II
10	4242.08	I	100	14779.73	I	32000	4215.52	II	100	13123.80	II
1	4249.41	I	60	16373.85	I	9	4298.57	IV	50	17447.40	I
2	4252.52	I	100	16388.85	I	340	4305.45	II	230	20261.40	I
15	4273.64	I	400	18465.25	I	65000	4607.33	I	120	20700.70	I
20	4276.79	I	50	22056.44	I	9	4685.08	IV	30	26023.60	I
2	4287.84	I	25	22083.67	I	3200	4722.28	I			
3	4291.01	I	60	23348.41	I	2200	4741.92	I	*Sulfur S Z = 16*		
30	4321.40	I	100	23379.13	I	1400	4784.32	I	5	437.4	V
40	4324.62	I	*Strontium Sr Z = 38*			4800	4811.88	I	5	438.2	V
3	4341.49	I	15	298.12	IV	3600	4832.08	I	5	439.6	V
5	4344.74	I	15	300.12	IV	3000	4872.49	I	20	519.3	IV
40	4390.03	I	125	330.67	III	2000	4876.32	I	20	520.1	IV
60	4393.34	I	500	351.62	III	1000	4891.98	I	40	520.8	IV
5	4419.88	I	75	358.80	III	8000	4962.26	I	20	522.0	IV
8	4423.25	I	250	363.49	III	1300	4967.94	I	20	522.5	IV
60	4494.18	I	150	371.21	III	800 h	5156.07	I	20	551.2	IV
100	4497.66	I	20	378.53	IV	1400	5222.20	I	40	652.5	IV
10	4541.63	I	75	392.44	IV	2000	5225.11	I	40	653.0	IV
15	4545.19	I	50	393.00	IV	2000	5229.27	I	70	653.6	IV
120	4664.811	I	50	396.22	IV	2800	5238.55	I	40	654.0	IV
200	4668.560	I	1000	437.24	III	4800	5256.90	I	70	655.6	IV
									20	655.9	IV

Intensity	Wavelength/Å		Intensity	Wavelength/Å		Intensity	Wavelength/Å		Intensity	Wavelength/Å	
110	657.3	IV	355	1316.542	I	285	3902.0	I	160	9035.9	I
40	658.3	V	290	1316.618	I	160	3928.6	III	450	9212.9	I
70	659.8	V	375	1323.515	I	360	3933.3	II	450	9228.1	I
40	660.9	IV	355	1326.643	I	450	4120.8	I	450	9237.5	I
160	661.4	IV	775	1381.552	I	280	4142.3	II	285	9413.5	I
110	663.2	V	710	1385.510	I	360	4145.1	II	285	9421.9	I
40	663.7	IV	960	1388.435	I	450	4153.1	II	285	9437.1	I
40	664.8	IV	640	1389.154	I	450	4162.7	II	650	9649.9	I
70	666.1	IV	775	1392.588	I	360	4253.6	III	450	9672.3	I
20	678.1	V	1000	1396.112	I	450	4694.1	I	450	9680.8	I
40	680.3	V	300	1409.337	I	285	4695.4	I	450	9693.7	I
110	680.9	V	510	1425.030	I	160	4696.2	I	285	9697.3	I
40	681.6	V	425	1433.280	I	280	4716.2	II	285	9739.7	I
20	693.5	V	300	1436.968	I	450	4815.5	II	285	9932.3	I
70	729.5	III	300	1448.229	I	360	4924.1	II	285	9949.8	I
110	732.42	III	425	1472.972	I	450	4925.3	II	285	9958.9	I
70	735.2	III	550	1473.995	I	285	4993.5	I	285	10455.5	I
70	738.5	III	300	1474.380	I	360	5428.6	II	285	10459.5	I
110	744.9	IV	355	1481.665	I	650	5432.8	II			
110	748.4	IV	485	1483.039	I	1000	5453.8	II	*Tantalum Ta Z = 73*		
110	750.2	IV	300	1483.233	I	1000	5473.6	II	60	493.07	V
110	753.8	IV	330	1485.622	I	1000	5509.7	II	1000	890.87	V
285	786.5	V	390	1487.150	I	280	5564.9	II	500	947.30	V
70	789.0	III	20	1624.0	IV	1000	5606.1	II	67	999.34	IV
70	796.7	III	20	1629.2	IV	450	5640.0	II	79	1116.10	IV
70	800.5	IV	680	1666.688	I	450	5640.3	II	78	1136.17	IV
70	804.0	IV	640	1687.530	I	280	5647.0	II	85	1175.51	IV
70	809.7	IV	710	1807.311	I	650	5659.9	II	80	1189.28	IV
110	816.0	IV	680	1820.343	I	450	5664.7	II	80	1192.67	IV
70	824.9	III	640	1826.245	I	160	5706.1	I	85	1213.09	IV
70	836.3	III	710	1900.286	I	450	5819.2	II	500	1213.42	V
160	849.2	V	550	1914.698	I	450	6052.7	I	85	1215.53	IV
110	852.2	V	20	2387.0	IV	280	6286.4	II	90	1223.73	IV
220	854.8	V	40	2398.9	IV	450	6287.1	II	88	1238.12	IV
110	857.9	V	110	2460.5	III	450	6305.5	II	95	1240.06	IV
110	860.5	V	110	2489.6	III	450	6312.7	II	87	1258.34	IV
40	906.9	II	160	2496.2	III	280	6384.9	II	94	1264.91	IV
40	910.5	II	160	2499.1	III	280	6397.3	II	98	1272.42	IV
40	912.7	II	220	2508.2	III	280	6398.0	II	94	1275.48	IV
40	937.4	II	70	2636.9	III	360	6413.7	II	86	1275.94	IV
40	937.7	II	220	2665.4	III	160	6743.6	I	92	1308.51	IV
160	1062.7	IV	110	2691.8	III	285	6748.8	I	87	1315.58	IV
160	1073.0	IV	110	2702.8	III	450	6757.2	I	92	1332.38	IV
70	1073.5	IV	220	2718.9	III	450	7579.0	I	86	1343.30	IV
285	1077.1	III	110	2721.4	III	450	7629.8	I	92	1365.88	IV
40	1102.3	II	220	2726.8	III	285	7686.1	I	5000	1392.56	V
70	1194.0	III	220	2731.1	III	450	7696.7	I	91	1398.78	IV
70	1201.0	III	110	2741.0	III	1000	7924.0	I	93	1413.40	IV
40	1234.1	II	285	2756.9	III	160	7928.8	I	91	1454.32	IV
40	1250.5	II	110	2775.2	III	285	7930.3	I	92	1464.41	IV
110	1253.8	II	160	2785.5	III	450	7931.7	I	93	1469.82	IV
110	1259.5	II	110	2863.5	III	450	7967.4	I	90	1495.25	IV
275	1270.782	I	160	2904.3	III	450	7967.4	II	95	1514.19	IV
250	1277.216	I	160	2986.0	III	450	8314.7	I	85	1607.70	IV
280	1295.653	I	110	3097.5	IV	450	8314.7	II	7000	1709.10	V
275	1302.337	I	110	3497.3	III	450	8585.6	I	85	1712.16	IV
235	1302.863	I	160	3632.0	III	285	8680.5	I	85	1716.13	IV
235	1303.110	I	110	3709.4	III	450	8694.7	I	85	2055.75	IV
245	1303.430	I	160	3717.8	III	360	8874.5	I	1100	2140.13	II
260	1305.883	I	160	3838.3	III	110	8882.5	I	1500	2146.87	II
265	1310.194	I	285	3867.6	I	220	8884.2	I	1200	2182.71	II
									1100	2193.88	II

Intensity	Wavelength/Å		Intensity	Wavelength/Å		Intensity	Wavelength/Å		Intensity	Wavelength/Å	
1500	2196.03	II	770	2775.88	I	200	5037.37	I	75	7006.96	I
90	2199.58	IV	680	2796.34	I	100	5067.87	I	150	7148.63	I
1500	2199.67	II	680	2797.76	II	110	5115.84	I	110	7172.90	I
90	2207.64	IV	510	2806.58	I	100	5141.62	I	140	7301.74	I
1400 d	2210.03	II	640	2844.25	I	100	5143.69	I	160	7346.41	I
1400	2239.48	II	560	2848.52	I	330	5156.56	I	140 c	7352.86	I
1200	2250.76	II	1500	2850.49	I	110	5212.74	I	100	7356.96	I
840	2261.42	II	1900	2850.98	I	110 d	5218.45	I	90 cw	7369.09	I
990	2262.30	II	360	2861.98	I	140	5341.05	I	160	7407.89	I
990	2272.59	II	470	2871.42	I	200	5402.51	I	100	7882.37	I
790	2285.25	II	380	2880.02	I	130	5419.19	I	75	8026.50	I
600	2286.59	II	770	2891.84	I	90	5518.91	I	75	8281.62	I
990	2289.16	II	560	2902.05	I	150	5645.91	I			
440	2302.24	II	310	2915.49	I	130	5664.90	I	*Technetium Tc Z = 43*		
440	2302.93	II	410	2925.19	I	130	5776.77	I	10000 c	3636.07	I
440	2312.60	II	310	2932.70	I	90	5780.71	I	20000 c	4031.63	I
420	2315.46	II	1700	2933.55	I	130	5811.10	I	15000	4095.67	I
690	2331.98	II	470	2940.06	I	240	5877.36	I	20000	4262.27	I
550	2332.19	II	1200	2940.22	I	130	5882.30	I	30000	4297.06	I
250	2357.30	I	510	2951.92	I	90	5901.91	I	20000	4853.59	I
260	2361.09	I	340	2953.56	I	90	5918.95	I			
600	2364.24	II	1500	2963.32	I	130	5939.76	I	*Tellurium Te Z = 52*		
320	2371.58	I	770	2965.13	II	240	5944.02	I	8	802.28	II
1400	2387.06	II	770	2965.54	I	190 c	5997.23	I	8	1059.51	II
2400	2400.63	II	340	2969.47	I	100	6020.72	I	8	1077.66	II
320	2416.89	II	430	2975.56	I	250	6045.39	I	10	1161.42	II
360	2427.64	I	1800	3012.54	II	100	6047.25	I	10	1174.34	II
360	2429.71	II	290 d	3027.48	I	100	6101.58	I	12	1175.79	II
480	2432.70	II	530	3049.56	I	65	6144.56	I	9	1208.54	II
380	2470.90	II	530	3069.24	I	130	6154.50	I	9	1220.98	II
600	2474.62	I	360	3077.24	I	150	6256.68	I	9	1253.62	II
500	2484.95	I	560	3103.25	I	150	6268.70	I	9	1270.52	II
600	2488.70	II	380	3124.97	I	150	6309.58	I	10	1324.92	II
500	2490.46	I	380	3130.58	I	75	6325.08	I	9	1363.24	II
600	2504.45	I	270	3132.64	I	65	6341.17	I	8	1366.73	II
600	2507.45	I	320	3170.29	I	75	6356.16	I	10	1374.80	II
1200 d	2526.35	I	270	3173.59	I	65	6360.84	I	10	1608.41	II
600	2532.12	II	600	3180.95	I	90	6389.45	I	10	1613.15	II
1200	2559.43	I	300	3223.83	I	65	6428.60	I	5	1655.4	I
460	2562.10	I	1100	3311.16	I	250	6430.79	I	5	1688.5	I
600	2577.37	II	680	3318.84	I	200	6450.36	I	6	1700.0	I
600	2603.49	II	330 d	3330.99	II	380	6485.37	I	5	1708.0	I
1400	2608.63	I	640	3371.54	I	65	6505.52	I	10	1822.4	I
1200	2635.58	II	360	3385.05	I	100	6514.39	I	26000	2002.02	I
860	2636.90	I	450	3406.94	I	100	6516.10	I	6500	2081.16	I
2400	2647.47	I	490	3480.52	I	100	6574.84	I	18000	2142.81	I
2600	2653.27	I	380	3497.85	I	110	6611.95	I	3200	2147.25	I
1900	2656.61	I	490	3511.04	I	75	6621.30	I	500	2259.02	I
1500	2661.34	I	750	3607.41	I	100	6673.73	I	1200	2383.26	I
770	2675.90	II	980	3626.62	I	180	6675.53	I	1500	2385.78	I
1500	2685.17	II	500	3642.06	I	75 c	6740.73	I	50	2438.69	II
470	2694.52	II	210	3918.51	I	75	6771.74	I	120	2530.72	I
1000	2698.30	I	210	3970.10	I	160 c	6813.25	I	100	2649.66	II
1200	2710.13	I	210	3996.17	I	210	6866.23	I	80	2661.10	II
2600	2714.67	I	410	4061.40	I	180	6875.27	I	110	2677.13	I
470	2727.44	II	310	4067.91	I	150	6902.10	I	100	2858.29	II
1200	2748.78	I	300	4205.88	I	140	6927.38	I	150	2895.41	II
860	2749.83	I	360 c	4510.98	I	140	6928.54	I	70	2967.29	II
410	2752.49	II	340	4574.31	I	65	6951.26	I	70	3047.00	II
1000	2758.31	I	260	4619.51	I	180	6966.13	I	100	3175.14	I
430	2761.68	II	450	4681.88	I	110 d	6995.22	I	60	3256.80	II
									60	3329.22	II

Intensity	Wavelength/Å		Intensity	Wavelength/Å		Intensity	Wavelength/Å		Intensity	Wavelength/Å	
150	3406.79	II	689	9956.30	I	810 d	3472.79	II	390	4196.74	I
50	3442.25	II	325	9977.13	I	810	3500.84	II	650	4203.74	I
50	3521.11	II	5950	10051.41	I	5700	3509.17	II	600	4206.49	I
50	3552.19	II	4097	10091.01	I	1300	3523.66	II	480	4215.09	I
100	3611.78	II	381	10118.08	I	1100	3540.24	II	480	4232.82	I
50	3617.57	II	397	10300.56	I	810	3543.89	II	650	4266.34	I
50	4006.52	II	745	10493.57	I	3200	3561.74	II	760 cw	4278.52	II
70	4127.32	II	1880	10918.34	I	810	3567.35	II	450	4310.42	I
100	4169.77	II	10200	11089.56	I	4200	3568.52	II	2200	4318.83	I
80	4225.73	II	508	11163.74	I	1600	3568.98	II	600	4322.23	I
100	4261.11	II	6620	11487.23	I	1100	3579.20	II	600	4325.83	II
60	4273.43	II	1580	13247.75	I	710	3585.03	II	3000	4326.43	I
80	4285.85	II	1050	14513.51	I	810	3596.38	II	600	4332.12	I
150	4364.00	II	1480	15452.45	I	1600	3600.44	II	870	4336.43	I
75	4385.10	II	2430	15546.23	I	810	3625.54	II	600	4337.64	I
170	4478.63	II	3760	16403.90	I	2300	3650.40	II	1700	4338.41	I
80	4537.07	II	1960	17303.54	I	810	3654.88	II	700	4340.62	I
100	4557.78	II	2780	18291.59	I	2000	3658.88	II	870	4356.81	I
70	4630.62	II	1020	21043.73	I	3800	3676.35	II	330	4382.45	I
100	4641.12	II	464	21602.50	I	810	3682.26	II	300	4388.23	I
180	4654.37	II	74	22555.29	I	450	3693.58	I	260	4390.91	I
200	4686.91	II	38	26539.17	I	450	3700.12	I	350	4423.10	I
100	4696.38	II				4700	3702.86	II	240	4436.12	I
100	4706.53	II	*Terbium Tb Z = 65*			2400	3703.92	II	240	4448.04	I
100	4766.05	II	1000	1259.40	IV	1000 d	3711.76	II	430	4493.07	I
100	4784.87	II	1000	1327.67	IV	650	3745.04	I	75	4514.31	II
100	4827.14	II	1000	1373.86	IV	870	3747.17	II	110	4549.07	I
150	4831.28	II	5000	1595.39	IV	870	3747.34	II	110	4550.45	I
150	4842.90	II	2000	1633.19	IV	1100	3755.24	II	110	4556.46	I
130	4865.12	II	2000	2027.79	IV	650	3759.35	I	110	4563.69	II
200	4866.24	II	1000	2089.98	IV	1700	3765.14	I	210	4578.69	II
8	5083.0	I	1000	2332.54	IV	2100	3776.49	II	65	4584.84	II
50	5449.84	II	110	2584.61	II	600	3783.53	I	65	4591.56	II
50	5487.95	II	110	2608.57	II	410	3789.92	I	75 d	4626.32	II
150	5576.35	II	130	2628.69	II	760 d	3806.85	II	95	4626.94	II
150	5649.26	II	140	2669.29	II	1500	3830.26	I	65	4632.07	I
100	5666.20	II	190	2704.07	II	540	3833.42	I	65 h	4636.59	I
200	5708.12	II	270	2769.53	II	920 d	3842.50	II	85	4641.00	II
150	5755.85	II	320	2897.44	II	3700	3848.73	II	210	4641.98	II
100	5974.68	II	250	2956.21	II	3500 w	3874.17	II	260 cw	4645.31	II
50	6367.13	II	230	3010.59	II	480	3888.22	I	80	4647.23	I
10 h	6790.0	I	230	3016.18	II	490	3894.64	I	80	4662.79	I
20 h	6837.6	I	460	3053.55	II	2400	3899.20	II	80	4676.90	I
20 h	6854.7	I	460	3070.05	II	1600	3901.33	I	70 c	4681.87	I
15 h	7191.1	I	670	3078.86	II	480	3908.06	I	80	4688.63	II
20 h	7263.5	I	480	3082.36	II	650	3915.43	I	80	4693.11	II
12	7460.98	II	480	3089.58	II	760	3925.45	II	200	4702.41	II
15	7468.75	II	480	3102.96	II	810 d	3939.52	II	110	4707.94	II
15	7921.69	II	440	3139.64	II	2200 d	3976.84	II	80	4739.93	I
15	7943.14	II	480	3187.26	II	1800	3981.87	II	70	4747.80	I
10	7950.34	II	480	3199.56	II	970	4002.59	II	410 cw	4752.53	II
30 h	8061.4	I	1100	3218.93	II	1900	4005.47	II	180	4786.78	I
10	8122.44	II	1200	3219.98	II	760	4012.75	II	100	4813.77	I
20	8186.44	II	480	3252.32	II	870	4032.28	I	80	4875.57	II
15	8273.53	II	760	3280.31	II	2100	4033.03	II	80	4881.15	II
15	8672.95	II	760	3281.40	II	430	4054.12	I	95	4915.90	I
10	8733.81	II	1000	3285.04	II	410	4060.37	I	65	4931.79	I
205	8758.18	I	1500	3293.07	II	1300	4061.58	I	85	4993.82	II
81	9004.37	I	3800	3324.40	II	650	4105.37	I	110	5078.25	I
5660	9722.74	I	760	3349.42	II	1100	4144.41	II	75	5089.12	II
532	9868.92	I	760	3364.93	II	350	4158.53	I	85	5186.13	I
			810	3454.06	II						

Intensity	Wavelength/Å		Intensity	Wavelength/Å		Intensity	Wavelength/Å		Intensity	Wavelength/Å	
120	5228.12	I	20	1034.73	IV	15	4981.35	II	910	3256.274	II
75	5248.71	I	20	1036.61	IV	25	5078.54	II	180	3257.366	I
75 w	5262.11	II	10 r	1049.73	II	25	5152.14	II	910	3262.668	II
75	5281.05	I	8 r	1050.30	II	18000	5350.46	I	620	3287.789	II
65	5304.72	I	5 r	1074.97	II	15 d	5384.85	II	910	3291.739	II
110	5319.23	I	30	1079.68	IV	10	5410.97	II	620	3292.520	II
65 w	5337.90	I	10 r	1130.17	II	25	5949.48	II	240	3301.650	I
160	5354.88	I	15 r	1162.55	II	10	6179.98	II	480	3304.238	I
75	5369.72	I	10 r	1167.43	II	10	6378.32	II	510	3321.450	II
75	5375.98	I	10 r	1183.41	II	16 h	6549.84	I	840	3325.120	II
50	5424.10	II	12 r	1194.84	II	10	6966.5	II	250	3330.476	I
55	5459.81	I	5 r	1246.00	II	10	7815.80	I	620	3334.604	II
55	5509.61	I	10	1266.33	III	20	8373.6	I	620	3337.870	II
50	5514.54	I	15 r	1307.50	II	10	8474.27	I	310	3348.768	I
65	5524.12	I	8 r	1310.20	II	10	8664.1	II	980	3351.228	II
85 c	5747.58	I	25 r	1321.71	II	20	9130.	II	620	3358.602	II
75	5795.64	I	8 r	1330.40	II	20	9130.5	I	250	3374.974	I
75	5803.13	II	10 r	1373.52	II	40	9509.4	I	1300	3392.035	II
65	5815.36	I	10	1477.14	III	20	9930.4	I	200	3396.727	I
65	5851.07	I	8 r	1489.65	I	30	10011.9	I	250	3398.544	I
65	5870.62	I	10 r	1499.30	II	40	10488.80	I	200	3405.558	I
65 c	5920.78	I	10 r	1507.82	II	1000	11512.82	I	250	3413.012	I
75	5967.34	II	15 r	1561.58	II	150	12736.4	I	390	3421.210	I
35	6331.68	II	10 r	1568.57	II	700	13013.2	I	270	3423.989	I
35 cw	6518.68	I	7 r	1593.26	II				980	3433.998	II
35	6581.82	I	5 h	1616.	I	**Thorium Th Z = 90**			770	3435.976	I
90	6677.94	II	5	1685.40	I	150	1707.37	IV	1300	3469.920	II
40 cw	6702.61	I	10 r	1792.76	II	200	1959.02	IV	170	3471.218	I
130	6794.58	II	12 r	1814.85	II	200	2002.34	IV	200	3486.552	I
55	6896.37	II	25 r	1908.64	II	200	2413.50	III	670	3539.587	II
45 h	6899.95	I	100 r	2007.56	I	200	2427.94	III	180	3544.018	I
40	6901.98	I	100 r	2210.71	I	200	2431.68	III	170	3549.595	I
65	7204.28	I	30	2298.04	II	200	2441.24	III	200	3555.013	I
40	7257.73	I	140	2315.98	I	500	2565.593	II	530	3559.451	II
45	7348.88	II	900 h	2379.69	I	480	2692.415	II	200	3576.557	I
45	7496.12	I	20	2530.86	II	520	2747.156	II	270	3592.780	I
27 h	7582.03	II	700	2580.14	I	410	2752.166	II	270	3598.120	II
45	7590.24	I	420	2709.23	I	800	2832.315	II	980	3609.445	II
65	7596.44	I	4400 d	2767.87	I	1200	2837.295	II	200	3612.427	I
30	7627.81	I	10	2849.80	II	100	2848.084	I	480	3615.133	II
30	7737.63	I	2800	2918.32	I	550	2870.406	II	270	3635.943	I
30	7855.79	II	20	3091.56	II	100	2936.086	I	210	3642.248	I
27	7927.90	II	15	3185.51	II	100	2943.729	I	170	3649.735	I
30	8025.42	II	15	3186.56	II	420	3049.092	II	220	3663.202	I
30	8085.06	II	15	3187.74	II	450	3067.729	II	280	3669.968	I
65	8194.82	II	1200	3229.75	I	670	3078.828	II	700	3675.567	II
95	8212.57	I	15	3291.01	II	480	3080.217	II	150	3682.486	I
40	8450.06	II	15	3369.15	II	510	3108.296	II	170	3692.566	I
30 h	8511.80	I	9	3456.34	III	100	3116.263	I	180	3698.105	I
45	8583.45	II	20000	3519.24	I	510	3119.526	II	340	3706.767	I
30	8603.40	II	5000	3529.43	I	510	3122.963	II	590	3719.435	I
65	8765.74	II	8	3540.08	II	480	3125.507	II	770	3721.825	II
			9	3560.68	I	100	3136.216	I	1300	3741.183	I
Thallium Tl Z = 81			12000 w	3775.72	I	420	3139.306	II	310	3747.539	I
10	570.49	IV	10	3832.30	II	420	3142.835	II	650	3752.569	I
5 r	670.87	II	10	3887.15	II	420	3175.726	II	180	3770.056	I
15 r	696.30	II	7	4109.85	III	1100	3180.193	II	590	3803.075	I
5 r	709.23	II	6	4269.81	III	770	3188.233	II	450	3828.384	I
10 r	817.18	II	20	4274.98	II	560	3221.292	II	840	3839.746	II
5 r	836.34	II	40	4306.80	II	560	3229.009	II	450	3863.405	II
8 r	1018.85	II	20	4737.05	II	480	3235.84	II	210	3875.374	I
30	1028.69	IV				590	3238.116	II			

Intensity	Wavelength/Å		Intensity	Wavelength/Å		Intensity	Wavelength/Å		Intensity	Wavelength/Å	
340	3895.419	I	40	7191.132	II	1900	3151.04	II	6800	3883.13	I
590	3929.669	II	35	7208.006	I	1500	3157.34	II	1800	3883.44	II
200	3932.911	I	50	7525.508	II	450	3172.65	I	5400	3887.35	I
390	3967.392	I	30	7647.380	I	2300	3172.83	II	440	3896.62	I
200	3972.155	I	30	8330.451	I	1200	3236.81	II	3500	3916.48	I
150	3980.089	I	40	8967.641	I	1600	3240.23	II	1500	3949.27	I
530	3994.549	II	20	9833.42	I	2300	3241.54	II	1500	3958.10	II
220	4008.210	I	20	10726.93	I	320	3246.96	I	1800	3996.52	II
220	4009.056	I	20	10942.24	II	1900	3258.05	II	220	4024.23	I
280	4012.495	I	30	11230.259	I	1600	3266.64	II	380	4044.47	I
4200	4019.129	II	20	11984.67	II	1200	3267.40	II	10000	4094.19	I
250	4030.842	I	20	17208.22	II	1100	3276.81	II	9500	4105.84	I
250	4036.047	I	15	18811.88	I	1200	3283.40	II	1100	4138.33	I
250	4063.407	I	10	22264.35	II	1200	3285.61	II	8800	4187.62	I
700	4086.520	II				2300	3291.00	II	6000	4203.73	I
700	4094.747	II	*Thulium Tm Z = 69*			2000	3302.46	II	380	4222.67	I
150	4100.341	I	5000	2185.94	III	1200	3309.80	II	3000	4242.15	II
840	4108.421	II	360	2284.79	II	230	3349.99	I	270	4271.71	I
240	4112.754	I	20000	2296.21	III	4000	3362.61	II	150	4298.36	I
280	4115.758	I	5000	2305.03	III	1700	3397.50	II	2700	4359.93	I
1100	4116.713	II	20000	2311.16	III	850	3410.05	I	1400	4386.43	I
200	4127.411	I	5000	2312.72	III	340	3412.59	I	200	4394.42	I
200	4134.067	I	5000	2326.19	III	340	3416.59	I	140	4396.50	I
450	4149.986	II	6000	2328.50	III	6400	3425.08	II	120	4454.03	I
620	4178.060	II	6000	2329.29	III	340	3429.33	I	540	4481.26	II
620	4208.890	II	3000	2331.80	III	4900	3441.50	II	150	4519.60	I
110	4253.538	I	3000	2357.05	III	4900	3453.66	II	260	4522.57	II
110	4260.333	I	4000	2406.63	III	8500	3462.20	II	110	4548.60	I
480	4277.313	II	450	2409.02	II	210	3467.51	I	270	4599.02	I
700	4282.042	II	450	2426.17	II	340	3476.69	I	300	4615.94	II
130	4337.277	I	770	2480.13	II	340	3480.98	I	80	4626.33	II
1300	4381.860	II	30000	2489.44	III	420	3487.38	I	95	4626.56	I
1100	4391.110	II	2000	2504.71	III	340	3499.95	I	110	4634.26	II
110	4498.940	I	1300	2509.08	II	250	3517.60	I	120	4655.09	I
280	4510.527	II	3000	2519.78	III	1700	3535.52	II	160	4681.92	I
90	4723.438	I	130	2527.02	I	420	3537.91	I	120	4691.11	I
50	4840.843	I	10000	2552.46	III	210	3555.82	I	110	4724.26	I
280	4863.163	II	360	2552.76	I	340	3560.92	I	680	4733.34	I
260	5017.255	II	540	2561.65	II	420	3563.88	I	70	4759.90	I
110	5067.974	I	430	2588.27	II	1300	3566.47	II	80	4831.20	II
120	5148.211	II	170 h	2596.49	I	420	3567.36	I	140	4957.18	I
95	5216.596	II	810	2607.06	II	280	3586.07	I	160	5009.77	II
110	5231.160	I	730	2624.33	II	2100	3608.77	II	160	5034.22	II
95	5247.654	II	5000	2682.32	III	1000	3629.09	III	150	5060.90	I
60	5343.581	I	2000	2707.03	III	380	3638.41	I	95	5113.97	I
60	5587.026	I	3000	2719.47	III	1100	3668.09	II	80	5213.38	I
95	5707.103	II	540	2721.19	II	4800	3700.26	II	650	5307.12	I
70	5760.551	I	3000	2724.44	III	3800	3701.36	II	80	5346.49	II
85	5989.044	II	4000	2727.56	III	7700	3717.91	I	270	5631.41	I
60	6169.822	I	680	2794.60	II	2400	3734.12	II	520	5675.84	I
50	6182.622	I	730	2797.27	II	5000	3744.06	I	40	5684.76	II
50	6274.116	II	2000	2806.77	III	1700	3751.81	I	35	5709.97	II
50	6274.117	II	580	2827.92	II	6000	3761.33	II	190	5764.29	I
50	6355.911	II	200	2854.17	I	4800	3761.91	II	35	5838.76	II
60	6457.283	I	1600	2869.23	II	7100	3795.75	II	240	5895.63	I
50	6462.614	I	1000	2947.72	III	770	3798.54	I	140	5971.26	I
50 h	6531.342	I	490	2973.22	I	600	3807.72	I	200	6460.26	I
55	6989.656	I	1000	2998.28	III	290	3826.39	I	95	6604.96	I
30	7045.795	II	1500	3015.30	II	1300	3838.20	II	110	6779.77	I
30	7084.171	I	360	3081.12	I	290	3840.87	I	120	6844.26	I
30	7168.896	I	7400	3131.26	II	8900	3848.02	II	80	6845.76	I
			2300	3133.89	II						

Intensity	Wavelength/Å		Intensity	Wavelength/Å		Intensity	Wavelength/Å		Intensity	Wavelength/Å	
10	6937.37	I	120 r	1823.00	I	110	3330.62	I	15	498.26	V
10	7017.90	I	9	1831.89	II	60	3351.97	II	14	502.08	V
12	7034.34	I	50 r	1848.75	I	10	3472.46	II	13	526.57	V
10	7106.14	I	200 r	1860.32	I	11	3575.45	II	18	779.07	IV
17	7272.62	I	80	1886.05	I	280 r	3801.02	I	20	1298.66	III
14	7310.51	I	100	1891.40	I	10	5332.36	II	20	1298.97	III
14	7432.18	I	12	1899.91	II	20	5561.95	II	23	1455.19	III
75	7481.08	I	50	1909.30	I	25	5588.92	II	20	1467.34	IV
75	7490.20	I	80	1925.31	I	500	5631.71	I	11	1717.40	V
140	7558.33	I	500	1941.86	III	15	5799.18	II	10	1841.49	V
80	7731.53	I	150	1952.15	I	50	5925.44	I	20	2067.56	IV
40	7856.08	I	50 h	1977.6	I	100	5970.30	I	18	2103.16	IV
55	7927.51	I	80	1984.20	I	150	6037.70	I	180	2273.28	I
110	7930.84	I	50	2040.66	I	250	6069.00	I	190	2279.96	I
95	8017.90	I	50	2054.03	I	100	6073.46	I	190	2305.67	I
27	8472.01	II	70	2058.31	I	400	6149.71	I	22	2413.99	III
			80	2068.58	I	200	6154.60	I	25	2516.05	III
Tin Sn Z = 50			100	2072.89	I	150	6171.50	I	360	2525.60	II
7	169.47	II	100	2073.08	I	100	6310.78	I	24	2527.84	III
150	361.01	V	200	2096.39	I	70	6453.50	II	210	2529.85	I
100	753.01	III	100	2100.93	I	25	6844.05	II	190	2531.25	II
200	910.92	III	100 r	2113.93	I	20	7191.40	II	190	2534.62	II
500	956.25	IV	50	2121.26	I	10	7387.79	II	130	2535.87	II
7	985.13	II	40 r	2148.73	I	13	7741.80	II	23	2540.06	III
500	1019.72	IV	20 r	2151.43	I	100	7754.97	I	24	2563.44	III
1000	1044.49	IV	30	2151.54	II	100 h	8030.5	I	23	2565.42	III
1000	1073.41	IV	80	2171.32	I	200	8114.09	I	22	2567.56	III
200	1089.35	V	150 r	2194.49	I	80	8357.04	I	270	2599.92	I
8	1108.19	II	300 r	2199.34	I	300	8422.72	I	340	2605.15	I
1000	1119.34	IV	400 r	2209.65	I	400	8552.60	I	510	2611.28	I
1000	1139.29	III	80 r	2231.72	I	50 h	8681.7	I	300	2619.94	I
1000	1158.33	III	400 r	2246.05	I	50 h	9410.86	I	640	2641.10	I
200	1160.74	V	60	2251.17	I	80 h	9415.37	I	800	2644.26	I
10	1161.43	II	400 r	2268.91	I	150	9616.40	I	950	2646.64	I
1000	1184.25	III	200 r	2286.68	I	50	9741.1	I	250	2742.32	I
2000	1210.52	III	600 r	2317.23	I	100 h	9742.8	I	250	2802.50	I
9	1219.07	II	300 r	2334.80	I	300 h	9805.38	I	190	2841.94	II
13	1223.70	II	1000 r	2354.84	I	500	9850.52	I	180	2877.44	II
11	1243.00	II	22	2368.33	II	54	10894.00	I	280	2884.11	II
2000	1251.38	V	100	2408.15	I	70	11191.85	I	450	2912.08	I
1000	1259.92	III	800 r	2421.70	I	56	11277.66	I	340	2928.34	I
20	1290.86	II	1000 r	2429.49	I	200	11454.59	I	1100	2942.00	I
200	1294.36	V	15	2448.98	II	200	11616.26	I	1300	2948.26	I
1000	1305.97	III	300	2483.39	I	258	11739.78	I	1600	2956.13	I
1000	1314.55	IV	13	2483.48	II	96	11825.18	I	22	2984.75	III
20	1316.59	II	10	2486.99	II	106	11835.82	I	1300 d	3066.22	II
1000	1327.34	III	200	2495.70	I	254	11932.99	I	1100	3072.97	II
1000	1347.65	III	400	2546.55	I	48	12009.50	I	1600	3075.22	II
1000	1386.74	III	500 r	2571.58	I	111	12313.24	I	2300	3078.64	II
25	1400.52	II	200	2594.42	I	42	12530.87	I	3600	3088.02	II
1000	1437.52	IV	200 r	2661.24	I	42	12536.5	I	720	3119.72	I
20	1475.15	II	700 r	2706.51	I	89	12888.5	I	500	3161.20	II
9	1489.22	II	150	2779.81	I	187	12981.7	I	780	3161.77	II
1000	1570.36	III	1400 r	2839.99	I	187	13018.5	I	1000	3162.57	II
10 r	1737.21	I	1000 r	2863.32	I	68	13081.5	I	1600	3168.52	II
15 r	1751.46	I	700 r	3009.14	I	378	13460.2	I	2400	3186.45	I
20 r	1764.98	I	850 r	3034.12	I	144	13608.2	I	1000	3190.87	II
30 r	1790.75	I	12	3047.50	II	40	20861.7	I	3100	3191.99	I
80 r	1804.60	I	550 r	3175.05	I	4	24738.2	I	3800	3199.92	I
15	1811.34	II	550 r	3262.34	I				780	3202.54	II
500	1811.71	III	50	3283.21	II	*Titanium Ti Z = 22*			1100	3217.06	II
40 r	1815.74	I				17	252.96	V			

Intensity	Wavelength/Å		Intensity	Wavelength/Å		Intensity	Wavelength/Å		Intensity	Wavelength/Å	
1300	3222.84	II	950	3962.85	I	840	5020.03	I	20	8466.87	III
6600	3234.52	II	950	3964.27	I	840	5022.87	I	90	8675.39	I
5200	3236.57	II	4800	3981.76	I	1200	5035.91	I			
4100	3239.04	II	570	3982.48	I	840	5036.47	I	*Tungsten W Z = 74*		
2600	3241.99	II	5700	3989.76	I	740	5038.40	I	5800	2001.71	II
1200	3248.60	II	7800	3998.64	I	1200	5039.95	I	13000	2008.07	II
1200	3252.91	II	950	4008.93	I	1400	5064.66	I	5100	2009.98	II
1200	3254.25	II	1200	4024.57	I	1100	5173.75	I	4100	2010.23	II
1200	3261.60	II	840	4078.47	I	1300	5192.98	I	4100	2014.23	II
840	3314.42	I	890	4286.01	I	1400	5210.39	I	7300	2026.08	II
2900	3322.94	II	840	4287.40	I	17	5278.12	III	15000	2029.98	II
2100	3329.46	II	950	4289.07	I	20	5398.93	IV	5300	2049.63	II
1800	3335.20	II	840	4290.94	I	340	5512.53	I	9700	2079.11	II
1100	3340.34	II	840	4295.76	I	270	5514.35	I	6100	2094.75	II
5700	3341.88	I	2000	4298.66	I	320	5514.54	I	2100	2118.87	II
4300	3349.04	II	200	4300.05	II	250	5644.14	I	2400	2121.59	II
12000	3349.41	II	2900	4300.56	I	130	5675.44	I	1500	2166.32	II
4100	3354.64	I	4100	4301.09	I	95	5689.47	I	1300	2204.48	II
7200	3361.21	II	6000	4305.92	I	95	5715.13	I	460	2249.80	I
1100	3370.44	I	1200	4314.80	I	85	5739.51	I	510	2277.58	I
4300	3371.45	I	330	4395.04	II	400	5866.46	I	530 d	2294.49	I
5700	3372.80	II	890	4427.10	I	230	5899.32	I	340	2309.02	I
2900 d	3377.48	I	230	4443.80	II	120	5918.55	I	440	2313.17	I
1400	3380.28	II	840	4449.15	I	150	5922.12	I	460	2321.63	I
5700	3383.76	II	550	4450.90	I	120	5941.76	I	390 d	2326.56	I
1400	3385.95	I	840	4453.32	I	300	5953.17	I	320	2354.61	I
1400	3387.84	II	950	4455.33	I	200	5965.84	I	580	2360.44	I
1100	3394.58	II	1100	4457.43	I	270	5978.56	I	850	2363.07	I
890	3444.31	II	240	4468.50	II	340	5999.04	I	510	2374.47	I
600	3461.50	II	530	4481.26	I	110	6064.63	I	670	2384.82	I
600	3477.18	II	780	4512.74	I	120	6085.23	I	730	2397.09	II
480	3491.05	II	1000	4518.03	I	120	6091.17	I	560	2397.73	I
890	3504.89	II	1000	4522.80	I	120	6126.22	I	560	2397.98	I
600	3510.84	II	780	4527.31	I	17	6246.65	IV	1700 d	2405.58	I
17	3576.44	IV	6000	4533.24	I	380	6258.10	I	610	2415.68	I
600	3610.16	I	240	4533.97	II	380	6258.70	I	870	2424.21	I
4800	3635.46	I	3600	4534.78	I	300	6261.10	I	1800	2435.96	I
6600	3642.68	I	2400	4535.58	I	55	6546.28	I	580	2444.06	I
7200	3653.50	I	1200	4535.92	I	65	6554.23	I	780	2451.48	II
600	3671.67	I	1200	4536.05	I	75	6556.07	I	870	2452.00	I
3100	3685.20	II	720	4544.69	I	18	6621.58	III	630	2454.98	I
600	3689.91	I	950	4548.77	I	18	6667.99	III	780	2455.51	I
2900	3729.82	I	240	4549.63	II	80	6743.12	I	780	2456.53	I
3300	3741.06	I	15	4549.84	III	20	7072.64	III	1100	2459.30	I
330	3741.64	II	950	4552.46	I	18	7084.57	III	1400	2466.85	I
5200	3752.86	I	720	4555.49	I	260	7209.44	I	480	2472.51	I
3300	3759.30	II	240	4571.98	II	130	7244.86	I	1200	2474.15	I
2900	3761.32	II	15 d	4572.20	III	130	7251.72	I	870	2480.13	I
840	3786.04	I	950	4617.27	I	120	7344.72	I	1500	2481.44	I
500	3882.89	I	480	4623.09	I	90	7357.74	I	480 d	2482.10	I
530	3900.54	II	720	4656.47	I	60	7364.11	I	580	2484.74	I
2600	3904.78	I	840	4667.59	I	60	7978.88	I	390	2487.50	I
500	3913.46	II	950	4681.92	I	55	8024.84	I	390	2489.23	II
500	3914.34	I	470	4840.87	I	75	8364.24	I	630	2495.26	I
15	3915.47	III	400	4885.08	I	100	8377.85	I	680	2504.70	I
1100	3924.53	I	380	4899.91	I	100	8382.54	I	75	2510.47	II
890	3929.88	I	5800	4981.73	I	75	8396.87	I	310	2520.46	I
1100	3947.78	I	4600	4991.07	I	120	8412.36	I	780	2521.32	I
4500	3948.67	I	4000	4999.51	I	170	8426.52	I	270	2522.04	II
4500	3956.34	I	3600	5007.21	I	490	8434.94	I	780	2523.41	I
5200	3958.21	I	3200 d	5014.19	I	240	8435.70	I	430	2527.76	I
									780	2533.64	I

Intensity	Wavelength/Å		Intensity	Wavelength/Å		Intensity	Wavelength/Å		Intensity	Wavelength/Å	
1200	2547.14	I	440	3046.44	I	2200	4302.11	I	580	2931.41	II
780	2550.38	I	810	3049.69	I	200	4378.48	I	530 p	2940.37	II
2700	2551.35	I	180	3073.28	I	180	4384.85	I	1300	2941.92	II
730	2561.97	I	180 d	3084.83	I	200	4408.28	I	830	2943.90	II
870	2580.49	I	370	3093.50	I	640	4484.19	I	580	2956.06	II
390	2584.39	I	240	3107.23	I	170	4588.73	I	580	2967.94	II
390	2589.17	II	240	3108.02	I	640	4659.87	I	580	2971.06	II
370	2601.96	I	230	3117.57	I	640	4680.51	I	530	2984.61	II
680	2606.39	I	260	3120.18	I	790	4843.81	I	630	3022.21	II
370	2608.32	I	290	3163.42	I	380	4886.90	I	630	3031.99	II
970	2613.08	I	320	3176.60	I	220	4982.59	I	580	3050.20	II
480	2613.82	I	190	3181.82	I	820	5053.28	I	630	3057.91	II
400	2620.25	I	390	3191.57	I	770	5224.66	I	630	3062.54	II
400	2622.21	I	390	3198.84	I	220	5514.68	I	580	3072.78	II
400	2625.22	I	520	3207.25	I	65	5648.37		580	3093.01	II
400	2632.48	I	1000	3215.56	I	55	5735.09		580	3102.39	II
400	2632.70	I	190	3232.49	I	45	5804.85	I	970	3111.62	II
810	2633.13	I	210	3254.36	I	40	5902.64	I	530	3119.35	II
400 d	2638.62	I	210	3259.66	I	55	5947.57	I	680	3124.95	II
650	2646.18	I	210 d	3266.62	I	55	5965.86	I	530	3139.61	II
400	2646.73	I	730	3300.82	I	55	6012.78	I	680	3149.24	II
1600	2656.54	I	440	3311.38	I	40	6021.52	I	530	3153.11	II
810	2662.84	I	440	3326.20	I	45	6292.02	I	730	3229.50	II
810	2671.47	I	440	3331.69	I	35	6404.21	I	680	3232.16	II
650	2677.28	I	390	3373.75	I	40	6445.12	I	730	3291.33	II
2100	2681.42	I	230	3429.59	I	17	6611.62	I	1100	3305.89	II
650	2695.67	I	240	3443.00	I	13	6678.42	I	730	3390.38	I
650	2699.59	I	400	3495.24	I	15	6693.08	I	580	3424.56	II
400	2700.01	I	650	3545.22	I	13	6984.27	I	580	3435.49	I
400	2706.58	I	240	3570.65	I	15	7140.52	I	630	3466.30	I
400	2708.59	I	1900	3617.52	I	9	7162.64	I	680	3482.49	II
400 d	2708.80	I	650	3682.08	I	11	7200.16	I	1600	3489.37	I
400	2715.50	I	400	3683.30	I	10	7278.24	I	530	3496.41	II
2100	2718.91	I	570	3688.06	I	15	7285.81	I	630	3500.08	I
2600	2724.35	I	810	3707.92	I	15	7296.55	I	780	3507.34	I
400	2725.03	I	510	3757.92	I	10	7509.00	I	1600	3514.61	I
650	2748.84	I	680	3760.13	I	17	7569.92	I	630	3533.57	II
400	2762.34	I	1000	3768.45	I	17	7614.15	I	530	3540.47	II
400	2764.27	II	340	3773.71	I	13	7688.97	I	1200	3550.82	II
400	2769.74	I	1000	3780.77	I	11	7784.15	I	680	3555.32	I
810	2770.88	I	290	3809.22	I	22	8017.19	I	1200	3561.80	I
810	2774.00	I	190	3810.38	I	22	8055.64	I	2300	3566.59	I
810	2774.48	I	260	3810.79	I	13	8123.82	I	530	3569.08	I
810	2792.70	I	1400	3817.48	I	10	8338.08	I	630	3578.72	II
400	2799.93	I	1100	3835.06	I	27	8585.11	I	3200	3584.88	I
810	2818.06	I	730	3846.22	I	10	8594.42	I	840	3638.20	I
1600	2831.38	I	1800	3867.99	I	13	8865.53	I	2800	3670.07	II
810	2833.63	I	730	3881.41	I				1100	3701.52	II
810	2848.02	I	8600	4008.75	I	*Uranium U Z = 92*			600	3738.04	II
1500	2896.44	I	540	4015.22	I	440	2565.41	II	680	3746.42	II
690	2935.00	I	910	4045.59	I	610	2635.53	II	950	3748.68	II
2400	2944.40	I	730	4069.95	I	830	2793.94	II	600	3751.17	I
2400	2946.99	I	5000	4074.36	I	870	2802.56	II	1900	3782.84	II
730 d	2979.71	I	1000	4102.70	I	630	2807.05	II	570	3793.10	II
360	3013.79	I	540	4137.46	I	630	2817.96	II	1900	3811.99	I
520	3016.47	I	450	4171.17	I	870	2821.12	II	750	3826.51	II
770	3017.44	I	220	4207.05	I	680	2828.90	II	2000	3831.46	II
210	3024.93	I	250	4219.37	I	920	2832.06	II	1200	3839.63	I
310 d	3026.67	I	540	4244.36	I	970	2865.68	II	2400	3854.64	II
440 d	3041.73	I	1400	4269.38	I	1200	2889.62	II	4900	3859.57	II
270	3043.80	I	4100	4294.61	I	780	2906.80	II	1900	3865.92	II
						780	2908.28	II			

Intensity	Wavelength/Å		Intensity	Wavelength/Å		Intensity	Wavelength/Å		Intensity	Wavelength/Å	
1500	3871.03	I	500	1006.46	III	750	3190.68	II	510	4332.82	I
1000	3881.45	II	500	1149.94	III	1100	3267.70	II	760	4341.01	I
2200	3890.36	II	100	1426.65	IV	900	3271.12	II	1000	4352.87	I
2000	3932.02	II	1000	1643.03	III	750	3276.12	II	12000	4379.24	I
1200	3943.82	I	1000	1650.14	III	80 h	3514.25	IV	7000	4384.72	I
1200	3985.79	II	100	1680.20	V	560	3517.30	II	4800	4389.97	I
1000	4042.75	I	1000	1694.78	III	560	3533.68	I	3600	4395.23	I
1600	4050.04	II	1000	1760.07	III	560	3545.20	II	1400	4400.58	I
880	4062.54	II	1000	1788.26	III	560	3556.80	II	2300	4406.64	I
2200	4090.13	II	1000	1794.60	III	560	3589.76	II	2800	4407.64	I
810	4116.10	II	1000	1812.19	III	490	3592.02	II	3600	4408.20	I
880	4153.97	I	300	1861.56	IV	560	3592.53	I	4600	4408.51	I
1400	4171.59	II	500	1939.06	IV	100	3679.86	III	640	4416.47	I
1000	4241.67	II	400	1951.43	IV	1300	3688.07	I	640	4421.57	I
600	4472.33	II	500	1997.72	IV	1000	3690.28	I	640	4437.84	I
620	4543.63	II	2100	2092.44	I	1500	3692.22	I	830	4441.68	I
170	4689.07	II	500	2268.30	IV	1000	3695.86	I	640	4444.21	I
150	4756.81	I	1000	2292.86	III	3800	3703.58	I	610	4452.01	I
110	5008.21	II	2500	2330.42	III	1800	3704.70	I	1000	4459.76	I
170	5027.38	I	2500	2371.06	III	320	3715.47	II	2000	4460.29	I
80	5160.32	II	1000	2382.46	III	250	3727.34	II	610	4462.36	I
70	5280.38	I	240	2507.78	I	280	3732.76	II	510	4577.17	I
80	5475.70	II	410	2526.22	I	520	3790.32	I	640	4580.40	I
70	5480.26	II	210	2527.90	II	1100	3794.96	I	830	4586.36	I
70	5481.20	II	80 h	2570.72	IV	570	3799.91	I	1300	4594.11	I
160	5492.95	II	230	2574.02	I	570	3803.47	I	230	4619.77	I
70	5780.59	I	250	2593.05	III	1000	3813.49	I	100	4635.18	I
70	5798.53	II	250	2595.10	III	1300	3818.24	I	130	4646.40	I
230	5915.39	I	80 h	2645.54	IV	1700	3828.56	I	160	4670.49	I
100	5976.32	I	180	2661.42	I	2600	3840.75	I	130	4776.36	I
90	6077.29	I	1100	2687.96	II	1200	3855.37	I	110	4786.51	I
55	6372.46	I	680	2700.94	II	3000	3855.84	I	130	4796.92	I
90	6395.42	I	530	2706.17	II	1300	3864.86	I	130	4807.53	I
110	6449.16	I	640	2715.69	II	1500	3875.08	I	130	4827.45	I
90	6826.92	I	180	2731.35	I	700	3890.18	I	150	4831.64	I
45	7533.93	I	240	2864.36	I	2400	3902.25	I	120	4832.43	I
50	7881.94	I	900	2891.64	II	700	3909.89	I	320	4851.48	I
35	8445.39	I	900	2892.66	II	540	3990.57	I	480	4864.74	I
75	8607.95	I	1400	2893.32	II	430	3998.73	I	620	4875.48	I
30	8757.76	I	900	2906.46	II	170	4005.71	II	740	4881.56	I
100	10554.93	I	2400	2908.82	II	1100	4090.58	I	110	5128.53	I
75	11167.84	I	710	2923.62	I	1800	4092.69	I	110	5138.42	I
100	11384.13	I	2400	2924.02	II	890	4095.49	I	110	5192.99	I
100	11859.42	I	1700	2924.64	II	2800	4099.80	I	110	5194.83	I
100	11908.83	I	900	2941.37	II	590	4102.16	I	110	5234.07	I
100	12250.46	I	1100	2944.57	II	2800	4105.17	I	110	5240.87	I
100	13185.16	I	410	2962.77	I	2300	4109.79	I	100	5401.93	I
75	13306.23	I	600	2968.38	II	8900	4111.78	I	140	5415.26	I
100	13961.58	I	1200	3056.33	I	4300	4115.18	I	140	5584.50	I
75	18634.43	I	1400	3060.46	I	1800	4116.47	I	100	5592.42	I
75	21910.22	I	2400	3066.38	I	2000	4123.57	I	200	5624.60	I
			3800	3093.11	II	3100	4128.07	I	400	5627.64	I
Vanadium V Z = 23			3000	3102.30	II	3100	4132.02	I	110	5657.44	I
20	225.46	V	2600	3110.71	II	2300	4134.49	I	110	5668.36	I
20	251.66	V	2000	3118.38	II	20	4200.32	V	310	5670.85	I
20	286.84	V	1500	3125.28	II	360	4232.46	I	1200	5698.52	I
35	483.01	V	3200	3183.41	I	560	4268.64	I	920	5703.56	I
50	633.94	III	5300	3183.98	I	460	4271.55	I	570	5706.98	I
200	677.34	IV	3800	3185.40	I	460	4276.96	I	850	5727.03	I
500	684.37	IV	410	3187.71	II	430	4284.06	I	230	5731.25	I
400	737.85	IV	530	3188.51	II	460	4330.02	I	230	5737.06	I
100	864.27	III									

Intensity	Wavelength/Å		Intensity	Wavelength/Å		Intensity	Wavelength/Å		Intensity	Wavelength/Å	
450	6039.73	I	500	1032.44	II	2	3400.07	I	500 h	4310.51	II
480	6081.44	I	700	1037.68	II	2	3418.37	I	1000 l	4330.52	II
1300	6090.22	I	1100	1041.31	II	2	3420.00	I	200 h	4369.20	II
600	6119.52	I	10	1047.8	III	3	3442.66	I	100 l	4373.78	II
450	6199.19	I	1000	1048.27	II	60	3444.2	III	500 h	4393.20	II
450	6216.37	I	1200	1051.92	II	70	3454.2	III	500 l	4395.77	II
430	6230.74	I	12	1066.4	III	100 w	3458.7	III	200 l	4406.88	II
710	6243.10	I	2000	1074.48	II	100 h	3461.26	II	150 l	4416.07	II
280	6251.82	I	600	1083.86	II	40	3468.22	III	50	4434.2	III
130	6268.82	I	1200	1100.43	II	4	3469.81	I	500 h	4448.13	II
170	6274.65	I	30	1130.3	III	4	3472.36	I	100 w	4462.1	III
200	6285.16	I	600	1158.47	II	5	3506.74	I	1000 h	4462.19	II
200	6292.83	I	250	1169.63	II	80	3522.83	III	500 l	4480.86	II
170	6296.49	I	800 p	1183.05	II	50	3542.3	III	100 l	4521.86	II
110	6531.43	I	250	1192.04	I	10	3549.86	I	100 w	4569.1	III
65 c	6753.00	I	25	1232.1	III	50	3552.1	III	100 w	4570.1	III
50 c	6766.49	I	600	1244.76	II	10	3554.04	I	100 w	4641.4	III
40	6784.98	I	250	1250.20	I	40	3561.4	III	30	4673.7	III
40	7338.92	I	1000	1295.59	I	100	3579.7	III	60	4683.57	III
35	7356.54	I	600	1469.61	I	80	3583.6	III	30	4723.60	III
29 c	8027.39	I	80	2668.98	III	100 w	3595.4	III	600	4734.152	I
120 w	8116.80	I	100	2717.33	III	100	3606.06	III	100 w	4757.3	III
70 c	8161.07	I	30	2814.45	III	40	3607.0	III	150	4792.619	I
60 c	8919.85	I	40	2815.91	III	15	3610.32	I	500	4807.02	I
			30	2827.45	III	8	3613.06	I	400	4829.71	I
Xenon Xe Z = 54			40	2847.65	III	100 w	3615.9	III	300	4843.29	I
8	657.8	III	30	2862.40	III	40	3623.1	III	40	4869.5	III
8	660.1	III	200	2864.73	II	600	3624.08	III	500	4916.51	I
9	673.8	III	80 w	2871.10	III	6	3633.06	I	500	4923.152	I
9	674.0	III	60 w	2871.24	III	10	3669.91	I	200 l	4971.71	II
9	676.6	III	30	2871.7	III	50	3676.67	III	400	4972.71	II
10	694.0	III	150 h	2895.22	II	40	3685.90	I	300	4988.77	II
20	698.5	III	30	2896.62	III	40	3693.49	I	100 l	4991.17	II
12	705.1	III	50	2906.6	III	40	3776.3	III	200	5028.280	I
10	721.2	III	40	2911.89	III	300	3781.02	III	200	5044.92	II
15	731.0	III	80 w	2912.36	III	100	3841.5	III	1000	5080.62	II
10	733.3	III	40	2940.2	III	200	3877.8	III	300	5122.42	II
350	740.41	II	60	2945.2	III	60	3880.5	III	100	5125.70	II
15	742.6	III	40	2947.5	III	100 l	3907.91	II	100	5178.82	II
10	756.0	III	40	2948.1	III	500	3922.55	III	300	5188.04	II
10	761.5	III	80 w	2970.47	III	300	3950.59	III	400	5191.37	II
10	769.1	III	400	2979.32	II	100	4037.59	II	100	5192.10	II
25	779.1	III	40	2992.87	III	200	4050.07	III	60	5239.0	III
15	792.9	III	30	3004.25	III	200 l	4057.46	II	500	5260.44	II
12	796.1	III	100 h	3017.43	II	60	4060.4	III	500	5261.95	II
15	802.0	III	100	3023.81	III	100 h	4098.89	II	2000	5292.22	II
350	803.07	II	40	3083.5	III	100	4109.1	III	300	5309.27	II
25	823.2	III	50	3091.1	III	100	4145.7	III	1000	5313.87	II
30	824.9	III	30	3106.46	III	200 l	4158.04	II	2000	5339.33	II
25	853.0	III	300	3128.87	II	1000 h	4180.10	II	200	5363.20	II
600	880.80	II	100 w	3138.3	III	500 h	4193.15	II	30	5367.1	III
350	885.54	II	80 c	3150.82	III	300 h	4208.48	II	200	5368.07	II
15	889.3	III	40	3185.2	III	100 h	4209.47	II	500	5372.39	II
20	894.0	III	100	3242.86	III	300 h	4213.72	II	100	5392.80	I
20	896.0	III	80	3268.98	III	100	4215.60	II	50	5401.0	III
600	925.87	II	30	3287.82	III	300 h	4223.00	II	3000	5419.15	II
250	935.40	II	80 w	3301.55	III	400 h	4238.25	II	800	5438.96	II
10	965.5	III	40	3331.6	III	500 h	4245.38	II	300	5445.45	II
800	972.77	II	30	3358.0	III	100 l	4251.57	II	200	5450.45	II
700	976.68	II	200 h	3366.72	II	30	4285.9	III	400	5460.39	II
35	1003.4	III	80	3384.12	III	500 h	4296.40	II	1000	5472.61	II
35	1017.7	III									

Intensity	Wavelength/Å		Intensity	Wavelength/Å		Intensity	Wavelength/Å		Intensity	Wavelength/Å	
100 l	5494.86	II	200	6528.65	II	300	8739.39	I	500	30794.18	I
40	5524.4	III	100	6533.16	I	100	8758.20	I	6000	31069.23	I
200	5525.53	II	1000	6595.01	II	5000	8819.41	I	125	31336.01	I
600	5531.07	II	100	6595.56	I	300	8862.32	I	550	31607.91	I
100	5566.62	I	400	6597.25	II	200	8908.73	I	100	32293.08	I
300	5616.67	II	100	6598.84	II	200	8930.83	I	1800	32739.26	I
300	5659.38	II	150	6668.92	I	1000	8952.25	I	3500	33666.69	I
600	5667.56	II	300	6694.32	II	100	8981.05	I	150	34014.67	I
150	5670.91	II	200	6728.01	I	200	8987.57	I	450	34335.27	I
100	5695.75	I	150	6788.71	II	400	9045.45	I	170	34744.00	I
200	5699.61	II	100	6790.37	II	500	9162.65	I	5000	35070.25	I
200	5716.10	II	1000	6805.74	II	100	9167.52	I	110	35246.92	I
500	5726.91	II	200	6827.32	I	100	9374.76	I	250	36209.21	I
500	5751.03	II	100	6872.11	I	200	9513.38	I	150	36231.74	I
300	5758.65	II	300	6882.16	I	50 h	9591.35	II	450	36508.36	I
300	5776.39	II	80	6910.22	II	150	9685.32	I	850	36788.83	I
100	5815.96	II	100	6925.53	I	50 l	9698.68	II	140	38685.98	I
300	5823.89	I	800 h	6942.11	II	100	9718.16	I	175	38737.82	I
150	5824.80	I	100	6976.18	I	2000	9799.70	I	270	38939.60	I
100	5875.02	I	2000	6990.88	II	3000	9923.19	I	120	39955.14	I
300	5893.29	II	150	7082.15	II	100	10838.37	I			
100	5894.99	I	500	7119.60	I	90	11742.01	I	*Ytterbium Yb* Z = 70		
200	5905.13	II	50 s	7147.50	II	375	12235.24	I	1000	1050.24	IV
100	5934.17	I	200	7149.03	II	100	12257.76	I	1000	1054.46	IV
500	5945.53	II	500	7164.83	II	300	12590.20	I	5000	1134.43	IV
300	5971.13	II	100	7284.34	II	2500	12623.391	I	900	1316.04	IV
2000	5976.46	II	200	7301.80	II	250	13544.15	I	800	1326.36	IV
200	6008.92	II	200	7339.30	II	2000	13657.055	I	900	1350.26	IV
1000	6036.20	II	100	7386.00	I	1250	14142.444	I	80	1561.42	III
2000	6051.15	II	150	7393.79	I	800	14240.96	I	80 h	1765.21	III
600	6093.50	II	300	7548.45	II	375	14364.99	I	800	1791.06	IV
1500	6097.59	II	200	7584.68	I	140	14660.81	I	100	1863.32	III
400	6101.43	II	80	7618.57	II	3000	14732.806	I	800	1873.91	III
100	6115.08	II	500	7642.02	I	100	15099.72	I	500	1898.25	III
100	6146.45	II	100	7643.91	I	2500	15418.394	I	500	1998.82	III
150	6178.30	I	200	7670.66	II	150	15557.13	I	900	2116.65	IV
120	6179.66	I	60	7787.04	II	250	15979.54	I	2500	2116.67	II
300	6182.42	I	100	7802.65	I	100	16039.90	I	800	2123.32	IV
500	6194.07	II	100	7881.32	I	1000	16053.28	I	3000	2126.74	II
100	6198.26	I	300	7887.40	I	125	16554.49	I	800	2139.99	IV
60	6205.97	III	500	7967.34	I	1500	16728.15	I	20000	2144.77	IV
100	6220.02	II	100	8029.67	I	1500	17325.77	I	15000	2154.18	IV
25	6221.7	III	200	8057.26	I	350	18788.13	I	370	2161.60	II
60	6238.2	III	150	8061.34	I	150	20187.19	I	850	2185.71	II
60	6259.05	III	100	8101.98	I	3000	20262.242	I	640	2224.46	II
500	6270.82	II	150 h	8151.80	II	250	21470.09	I	300	2240.11	III
400	6277.54	II	100	8171.02	I	1250	23193.33	I	300	2305.32	III
100	6284.41	II	700	8206.34	I	110	23279.54	I	140	2320.81	II
100	6286.01	I	10000	8231.635	I	1800	24824.71	I	170	2390.74	II
250	6300.86	II	500	8266.52	I	175	25145.84	I	460	2464.50	I
500	6318.06	I	7000	8280.116	I	2000	26269.08	I	140	2512.06	II
400	6343.96	II	2000	8346.82	I	2500	26510.86	I	270	2538.67	II
600	6356.35	II	100	8347.24	II	250	28381.54	I	2000	2567.61	III
200	6375.28	II	2000	8409.19	I	750	28582.25	I	1000	2579.57	III
100	6397.99	II	50 h	8515.19	II	300	29384.41	I	800	2599.14	III
300	6469.70	I	200	8576.01	I	150	29448.06	I	600	2621.11	III
150	6472.84	I	50 h	8604.23	II	100	29649.58	I	1000	2642.56	III
120	6487.76	I	250	8648.54	I	100	29813.62	I	1000	2651.74	III
100	6498.72	I	100	8692.20	I	600	30253.14	I	700	2652.25	III
200 h	6504.18	I	200	8696.86	I	1500	30475.46	I	990	2653.75	II
300	6512.83	II	50 h	8716.19	II	100	30504.12	I	200	2665.04	II
									2000	2666.13	III

Intensity	Wavelength/Å		Intensity	Wavelength/Å		Intensity	Wavelength/Å		Intensity	Wavelength/Å	
2000	2666.99	III	140	3418.39	I	170	4786.61	II	500	344.59	V
390	2671.96	I	360	3426.04	I	35	4816.43	I	900	355.86	IV
390	2672.66	II	240	3431.11	I	40	4837.46	I	300	370.42	IV
170	2718.35	II	85	3452.40	I	40 h	4894.60	I	300	372.05	V
230	2748.66	II	500	3454.08	II	27	4912.36	I	400	379.96	V
1300	2750.48	II	190 d	3458.29	II	710	4935.50	I	500	386.82	IV
170	2776.28	II	360	3460.27	I	140	4966.90	I	300	403.45	V
600	2795.60	III	2400	3464.37	I	30	5067.80	I	300	420.74	V
1000	2803.43	III	500	3476.30	II	70	5069.14	I	600	425.03	IV
600	2816.92	III	500	3478.84	II	220	5074.34	I	300	473.10	IV
1000	2818.72	III	50	3517.00	I	50	5076.74	I	4000	584.98	V
140	2821.15	II	230	3520.29	II	60	5196.08	I	2000	630.97	V
190	2830.99	II	35	3559.03	I	85	5211.60	I	5000	805.20	III
230 h	2847.18	II	200	3560.33	II	100	5244.11	I	7000	809.92	III
360	2851.13	II	170	3560.70	II	150 h	5277.04	I	15000	989.21	III
430	2859.80	II	360	3585.47	II	170	5335.15	II	25000	996.37	III
140	2861.21	II	200	3619.80	II	30 h	5351.29	I	5000	1314.51	III
200	2867.06	II	240	3637.76	II	150	5352.95	II	4000	1334.04	III
45	2873.49	I	70	3648.15	I	30	5363.66	I	4000	2068.98	III
200	2888.04	II	90	3655.73	I	40	5449.27	II	10000	2127.98	III
3600	2891.38	II	240	3669.69	II	60	5481.92	I	16000	2191.16	III
600	2898.30	III	140	3675.08	II	40	5505.49	I	350	2243.06	II
1000	2906.31	III	32000	3694.19	II	17	5524.54	I	10000	2284.34	III
170	2914.21	II	70	3700.58	I	85 h	5539.05	I	10000	2327.31	III
140	2915.28	II	400	3711.91	III	2400	5556.47	I	50	2354.20	I
280	2919.35	II	180	3734.69	I	60	5651.98	II	50000	2367.23	III
35	2934.36	I	550	3770.10	I	220	5719.99	I	40000	2414.64	III
140	2945.91	II	80	3774.32	I	27	5771.66	II	560	2422.20	II
2000	2970.56	II	60 h	3791.74	I	35	5833.99	II	60	2694.21	I
200	2983.99	II	170	3839.91	I	35	5837.14	II	95	2723.00	I
170	2994.80	II	340	3872.85	I	27	5854.51	I	70	2742.53	I
800	2998.00	III	340	3900.85	I	17	5989.33	I	140	2760.10	I
310	3005.77	II	140	3911.27	I	40	5991.51	II	90000	2817.04	III
160	3017.56	II	500	3931.23	III	60	6152.57	II	45	2822.56	I
160	3026.67	II	32000	3987.99	I	60	6274.78	II	70	2854.43	II
2000	3029.49	III	930	3990.88	I	200	6328.52	III	95	2886.48	I
920	3031.11	II	50	4007.36	I	35 h	6400.35	I	160	2919.05	I
3000	3092.50	III	2000	4028.14	III	35 h	6417.91	I	99000	2946.01	III
28	3100.74	I	70	4052.28	I	340	6489.06	I	390	2948.40	I
170	3107.90	II	440	4089.68	I	180	6667.82	I	350	2964.96	I
190	3117.81	II	470	4149.07	I	25	6727.61	II	480	2974.59	I
4000	3126.01	III	120	4174.56	I	690	6799.60	I	750	2984.26	I
1000	3138.58	III	340	4180.81	II	9 h	7244.41	I	140	2996.94	I
230	3140.94	II	300	4213.64	III	8 h	7305.22	I	130	3021.73	I
28	3162.29	I	150 d	4218.56	II	10 h	7313.05	I	190	3045.37	I
800	3191.35	III	120	4231.97	I	16 h	7350.04	I	95	3095.88	II
390	3192.88	II	70	4277.74	I	25	7448.28	I	110	3173.06	II
240	3201.16	II	120	4305.97	I	30 h	7527.46	I	220	3179.41	II
2000	3228.58	III	60 h	4393.69	I	750	7699.48	I	70	3191.31	I
35	3239.58	I	60 h	4430.21	I	100	7971.46	III	2300	3195.62	II
18000	3289.37	II	440	4439.19	I	70 h	8922.56	II	2200	3200.27	II
130	3305.25	I	85 h	4482.42	I	200	10110.60	III	2200	3203.32	II
140	3305.73	II	100	4517.58	III	100	10830.36	III	3900	3216.69	II
80	3319.41	I	85 h	4563.95	I				6200	3242.28	II
2000	3325.51	III	640	4576.21	I	*Yttrium Y Z = 39*			4700	3327.89	II
240	3337.17	II	200	4582.36	I	150	264.64	IV	85	3388.59	I
280 d	3342.93	II	70	4589.21	I	150	273.03	IV	85	3412.47	I
240	3375.48	II	140	4590.83	I	900	333.09	V	170	3485.73	I
2000	3384.01	III	40	4684.27	I	500	333.80	V	1700	3496.09	II
140	3387.50	I	190	4726.08	II	400	335.14	V	3900	3549.01	II
50	3412.45	I	170 h	4781.87	I	500	336.62	V	130	3551.80	I
						500	339.02	V			

Intensity	Wavelength/Å		Intensity	Wavelength/Å		Intensity	Wavelength/Å		Intensity	Wavelength/Å	
540	3552.69	I	300	4487.47	I	160	5706.73	I	100	1767.69	III
170	3558.76	I	500	4505.95	I	90	5743.85	I	100	1797.64	II
190	3571.43	I	890	4527.25	I	75	5765.64	I	100 d	1811.05	II
260	3576.05	I	440	4527.80	I	100	5781.69	II	100 d	1833.57	II
3300	3584.52	II	100	4544.32	I	120	6009.19	I	100	1864.12	II
300	3587.75	I	100	4559.37	I	120	6023.41	I	100	1866.08	II
100	3589.69	I	130	4596.55	I	120	6135.04	I	100	1872.13	II
2800	3592.92	I	95	4604.80	I	150	6138.43	I	100 d	1918.96	II
10000	3600.73	II	2000	4643.70	I	1200	6191.73	I	100 d	1929.67	II
6200	3601.92	II	200 h	4658.32	I	300	6222.59	I	100	1969.40	III
7800	3611.05	II	2000	4674.84	I	1000	6435.00	I	100	1982.11	II
4300	3620.94	I	180	4696.81	I	90	6538.60	I	100	1986.99	II
1900	3628.71	II	170	4728.53	I	70	6557.39	I	500	2025.48	II
7800	3633.12	II	160	4752.79	I	95	6613.75	II	500	2062.00	II
3000	3664.61	II	410	4760.98	I	40	6650.61	I	200	2064.23	II
170	3692.53	I	120	4781.04	I	150	6687.58	I	120	2079.08	I
13000	3710.30	II	170	4786.89	I	70	6700.71	I	300	2099.94	II
1200	3747.55	II	180	4799.30	I	190	6793.71	I	200	2102.18	II
10000	3774.33	II	140	4819.64	I	21	6815.16	I	800 r	2138.56	I
1400	3776.56	II	120	4822.13	I	45	6845.24	I	1000	2501.99	II
7400	3788.70	II	770	4839.87	I	29	6887.22	I	150	2515.81	I
1300	3818.35	II	550	4845.68	I	24 h	6950.31	I	1000	2557.95	II
4000	3832.88	II	410	4852.69	I	24	6979.88	I	300	2582.49	I
80	3876.82	I	120	4854.25	I	29	7052.94	I	200	2608.56	I
480	3878.28	II	890	4854.87	II	35	7191.66	I	300	2608.64	I
4400	3950.36	II	330	4859.84	I	35	7264.17	II	200	2670.53	I
3600	3982.60	II	1900	4883.69	II	50	7346.46	I	300	2684.16	I
940	4039.83	I	95	4893.44	I	29	7450.30	II	300	2712.49	I
2400	4047.64	I	1100	4900.12	II	9000	7558.71	III	200	2756.45	I
9400	4077.38	I	100	4906.11	I	35	7563.13	I	300	2770.86	I
2000	4083.71	I	150	4921.87	I	29	7855.52	I	300	2770.98	I
9900	4102.38	I	120	4974.30	I	110	7881.90	II	400	2800.87	I
8900	4128.31	I	100	5006.97	I	10000	7991.43	III	100	2801.06	I
7500	4142.85	I	75	5070.21	I	24	8344.43	I	200	3035.78	I
100 h	4157.63	I	75	5072.19	I	10000	8796.21	III	200	3072.06	I
2400	4167.52	I	1100	5087.42	II	95	8800.62	I	300	3196.31	II
2000	4174.14	I	180	5135.20	I	19 h	8835.85	II	500 r	3282.33	I
8000	4177.54	II	960	5200.41	II				800	3302.58	I
160	4217.80	I	1500	5205.72	II	*Zinc Zn Z = 30*			700 r	3302.94	I
280 h	4220.63	I	10000	5238.10	III	200	425.90	IV	800	3345.02	I
600	4235.73	II	180	5240.81	I	200	428.54	IV	500	3345.57	I
2200	4235.94	I	75	5380.62	I	200	430.59	IV	50	3883.34	I
300	4251.20	I	220	5402.78	II	1000	677.63	III	300	4680.14	I
360 h	4302.30	I	90	5424.37	I	750	677.96	III	400	4722.15	I
2800	4309.63	II	190	5438.24	I	200	713.90	III	400	4810.53	I
110	4330.78	I	710	5466.46	I	60	1193.23	II	800	4911.62	II
440 h	4348.79	I	100	5468.47	I	50	1239.12	IV	500	4924.03	II
120	4357.73	I	240	5497.41	II	50	1249.69	IV	200	5181.98	I
800	4358.73	II	300	5503.45	I	500	1265.74	IV	500	5894.33	II
120	4366.03	I	250	5509.90	II	500	1306.66	IV	500	6021.18	II
12000	4374.94	II	120	5521.63	I	200	1456.72	III	500	6102.49	II
150 h	4375.61	I	740	5527.54	I	200	1459.98	IV	500	6214.61	II
100	4387.74	I	120	5544.50	I	300	1499.42	III	1000 h	6362.34	I
1800	4398.02	II	180	5577.42	I	300	1500.42	III	300	7588.5	II
890	4422.59	II	620	5581.87	I	300	1505.92	III	300	7732.5	II
100	4443.66	I	120	5606.33	I	300	1515.85	III	100	11054.25	I
130	4446.63	I	560	5630.13	I	300	1552.30	III	100	13053.63	I
170	4475.72	I	120	5644.69	I	90	1572.99	II	100	13150.59	I
180	4476.96	I	120	5648.47	I	200	1629.19	III	100	14038.70	I
160	4477.45	I	740	5662.94	II	200	1639.33	III	20	16483.45	I
110	4487.28	I	90	5675.27	I	200	1673.05	III	20	16491.98	I
						80 d	1735.61	II			

Intensity	Wavelength/Å		Intensity	Wavelength/Å		Intensity	Wavelength/Å		Intensity	Wavelength/Å	
20	16505.23	I	620	2814.90	I	1000	3430.53	II	770	3998.97	II
10	24375.02	I	390	2818.74	II	4700	3438.23	II	400	4023.98	I
	Zirconium Zr Z = 40		530	2825.56	II	600	3447.36	I	770	4024.92	I
500	304.01	V	710	2837.23	I	410	3457.56	II	990	4027.20	I
60	480.66	IV	660	2844.58	II	820	3463.02	II	400	4029.68	II
60	497.23	IV	350	2848.52	I	600	3471.19	I	490	4030.04	I
60	500.22	IV	350	2851.97	II	1200	3479.39	II	400	4035.89	I
600	628.66	IV	340	2869.81	II	1300	3481.15	II	610	4043.58	I
500	633.56	IV	490	2875.98	I	4100	3496.21	II	490	4044.56	I
50	690.39	III	300	2915.99	II	820	3505.67	II	400	4045.61	II
2000	740.61	V	270	2918.24	II	1000	3509.32	I	610	4048.67	II
10000	800.00	V	320	2926.99	II	2000	3519.60	I	770	4055.03	I
10000	806.89	V	320	2948.94	II	440	3525.81	II	600	4055.71	I
10000	812.05	V	320	2955.78	II	440	3533.22	I	1500	4064.16	I
3000	841.40	V	320	2960.87	I	630	3542.62	II	2000	4072.70	I
300	863.65	IV	320	2962.68	II	1800	3547.68	I	240	4078.31	I
500	864.59	IV	320	2968.96	II	630	3550.46	I	2000	4081.22	I
9000	1183.97	IV	320	2978.05	II	1800	3551.95	II	400	4121.46	I
9000	1201.77	IV	820	2985.39	I	2100	3556.60	II	1200	4149.20	II
10000	1219.86	IV	320	3003.74	II	1100	3566.10	I	400	4161.21	II
500	1303.93	V	820	3011.75	I	2100	3572.47	II	400	4166.36	I
500 p	1323.81	V	350	3020.47	II	1100	3575.79	I	660	4187.56	I
1000	1469.47	IV	500	3028.04	II	1300	3576.85	II	400	4194.76	I
10000	1546.17	IV	880	3029.52	I	880	3586.29	I	610	4199.09	I
10000	1598.95	IV	350 d	3036.39	II	3500	3601.19	I	610	4201.46	I
5000	1607.95	IV	690	3054.84	II	690	3611.89	II	610	4208.98	II
100	1612.38	III	690	3106.58	II	1100	3613.10	II	400	4213.86	I
700	1725.02	V	350	3120.74	I	1100	3614.77	II	2000	4227.76	I
200	1790.19	III	500	3129.18	II	1100	3623.86	I	2000	4239.31	I
150	1793.56	III	500	3129.76	II	1100	3663.65	I	770	4240.34	I
125	1798.13	III	350	3132.07	I	390	3671.27	II	770	4241.20	I
600	1860.86	V	690	3138.68	II	800	3674.72	II	1200	4241.69	I
200	1940.25	III	540	3164.31	II	390	3697.46	II	550	4282.20	I
600	2028.54	V	880	3165.97	II	960	3698.17	II	550	4294.79	I
125	2070.43	III	880	3182.86	II	720	3709.26	II	550	4341.13	I
200	2086.78	III	540	3191.21	I	560	3745.98	II	1000	4347.89	I
10000	2091.49	IV	540	3212.01	I	880	3751.60	II	290	4359.74	II
10000	2092.36	IV	760	3214.19	II	480	3764.39	I	310	4360.81	I
600	2132.42	V	630	3231.69	II	480	3766.72	I	350	4366.45	I
10000	2163.68	IV	630	3234.12	I	340	3766.82	II	550	4507.12	I
100	2175.80	III	760	3241.05	II	720	3780.54	I	610	4535.75	I
100	2191.15	III	1000	3273.05	II	560	3791.40	I	490	4542.22	I
10000	2286.67	IV	1300	3279.26	II	560	3822.41	I	490	4575.52	I
100	2301.60	III	880	3284.71	II	2200	3835.96	I	350	4602.57	I
90	2539.65	I	540	3305.15	II	1300	3836.76	II	700	4633.98	I
570	2567.64	II	880	3306.28	II	550	3843.02	II	2300	4687.80	I
1600	2568.87	II	380	3322.99	II	550	3847.01	I	510	4688.45	I
2100	2571.39	II	380	3326.80	II	550	3849.25	I	1900	4710.08	I
250	2620.56	III	380	3334.25	II	2900	3863.87	I	1400	4739.48	I
200	2643.79	III	760	3340.56	II	770	3864.34	I	870	4772.31	I
150	2664.26	III	380	3344.79	II	990	3877.60	I	700	4815.63	I
1800	2678.63	II	760	3356.09	II	1500	3885.42	I	250	5046.58	I
90	2687.75	I	540	3357.26	II	2900	3890.32	I	360	5064.91	I
750	2700.13	II	380	3374.73	II	2000	3891.38	I	470	5078.25	I
1300	2722.61	II	570	3387.87	II	610	3921.79	I	300	5155.45	I
800	2726.49	II	760	3388.30	II	1200	3929.53	I	200	5158.00	I
1400	2734.86	II	5700	3391.98	II	940	3958.22	II	100	5191.60	II
1100	2742.56	II	570	3393.12	II	490	3966.66	I	270	5385.14	I
660	2745.86	II	570	3404.83	II	990	3968.26	I	160	5664.51	I
660	2752.21	II	760	3410.25	II	660	3973.50	I	160	5797.74	I
530	2758.81	II	380	3414.66	I	770	3991.13	II	340	5879.80	I

Intensity	Wavelength/Å		Intensity	Wavelength/Å		Intensity	Wavelength/Å		Intensity	Wavelength/Å	
170	6045.85	I	300	6313.02	I	170	7103.72	I	790	8070.08	I
170	6121.91	I	150	6953.84	I	590	7169.09	I	390	8132.99	I
680	6127.44	I	150	6990.84	I	160	7944.61	I	280	8212.53	I
340	6134.55	I	540	7097.70	I	160	8005.27	I			
440	6143.20	I	280	7102.91	I	150	8063.09	I			

Sources of Data for Each Element

Numbers following the element name refer to the references on the following pages.

Actinium: 193
Aluminum: 6,8,81,89,127,144,146,227,228,282
Americium: 92
Antimony: 164,167,194,386,406
Argon: 190,203,204,219,367,368,372,373,374,375,414,421
Arsenic: 163,168,197,244,280
Astatine: 188
Barium: 1,78,111,252,259,277,279
Berkelium: 53,339
Beryllium: 15,44,73,102,115,134,135,171,175,198,335
Bismuth: 1,357,358,359,360,361
Boron: 66,69,74,94,104,171,221,222
Bromine: 42,122,124,139,142,240,243,246,248,249,250,316
Cadmium: 44,285,296,353,399
Calcium: 16,25,70,150,270
Californium: 52,331
Carbon: 22,66,211
Cerium: 1,136,166,261,305
Cesium: 78,82,154,155,200,201,259,263,325
Chlorine: 11,28,30,31,85,233,238,239
Chromium: 1,379,380,412
Cobalt: 1,100,125,159,236,276,291
Copper: 199,273,290,295,324
Curium: 51,332
Dysprosium: 1
Einsteinium: 333
Erbium: 1,301
Europium: 1,312
Fluorine: 68,169,224,225,226
Francium: 408
Gadolinium: 1,46,137,151,152
Gallium: 2,19,62,132,140,141,143,195,281
Germanium: 5,119,293,340,341,342
Gold: 38,72,234,393,395
Hafnium: 1,369,404,410,425
Helium: 16,94,173,183,317
Holmium: 1
Hydrogen: 214
Indium: 1,132,348,349,350,351,352,353,435,436
Iodine: 20,21,58,84,124,153,161,176,184
Iridium: 1
Iron: 56,63,71,101,105,138,174,278,381,382
Krypton: 61,121,123,147,208,232,366,390,409,417,421
Lanthanum: 1,78,79,220,309
Lead: 54,64,106,256,274,297,283,329,330
Lithium: 3,15,17,18,37,44,112,284,321,335
Lutetium: 1,148,310,401
Magnesium: 4,7,49,83,103,128,129,177,217,269,315,335
Manganese: 1,126,385,405,433

Mercury (198): 43,50,69,145,229,242
Mercury (Natural): 34,45,90,117,133,189,235,304,327,328,343
Molybdenum: 1,383,420
Neodymium: 1
Neon: 56,58,69,118,150,230,364,365,371,388,389,400,402,413,430
Neptunium: 93
Nickel: 1,294,415,416,422
Niobium: 1,392,407,431
Nitrogen: 66,107,108,212,213,318
Osmium: 1
Oxygen: 23,24,36,66,69,209,210,215
Palladium: 1,287,424
Phosphorus: 179,180,182,336
Platinum: 1,288
Plutonium: 91
Polonium: 47,48
Potassium: 32,59,60,75,76,86,150,160,172,268,314,322
Praseodymium: 1,149,306,308,337,338
Promethium: 196,260
Protactinium: 96
Radium: 253,254
Radon: 251
Rhenium: 1
Rhodium: 1,396
Rubidium: 12,109,130,241,257,258,262,264
Ruthenium: 1,423
Samarium: 1
Scandium: 1,88,150,298,323
Selenium: 9,80,181,216,245,247,275
Silicon: 87,170,237,292,319,320
Silver: 13,99,255,286,289,363,387,398
Sodium: 178,205,206,207,268,299,334
Strontium: 1,109,110,218,231,265,279,313
Sulfur: 29,144,202,209,210,266
Tantalum: 1,411,426
Technetium: 35
Tellurium: 1,344,345,346,347
Terbium: 1,302
Thallium: 1,195,348,354,355,356
Thorium: 1,97,98,156,157,165,434
Thulium: 1,307
Tin: 187,191,399,423
Titanium: 1,378,427,428
Tungsten: 1
Uranium: 1,303
Vanadium: 1,394,397,432
Xenon: 33,116,118,120,232,384,391,429
Ytterbium: 1,40,192,311
Yttrium: 1,77,265,419
Zinc: 39,55,113,131,185,186,370,376,377
Zirconium: 1,362,403,418

References

1. Meggers, W. F., Corliss, C. H., and Scribner, B. F., *Natl. Bur. Stand. (U.S.) Monogr.*, 145, Washington, D.C., 1975.

2. Aksenov, V. P. and Ryabtsev, A. N., *Opt. Spectrosc.*, 37, 860, 1970.

3. Andersen, N., Bickel, W. S., Carriveau, G. W., Jensen, K., and Veje, E., *Phys. Scr.*, 4, 113, 1971.

4. Andersson, E. and Johannesson, G. A., *Phys. Scr.*, 3, 203, 1971.

5. Andrew, K. L. and Meissner, K. W., *J. Opt. Soc. Am.*, 49, 146, 1959.

6. Artru, M. C. and Brillet, W. U. L., *J. Opt. Soc. Am.*, 64, 1063, 1974.

7. Artru, M. C. and Kaufman, V., *J. Opt. Soc. Am.*, 62, 949, 1972.

8. Artru, M. C. and Kaufman, V., *J. Opt. Soc. Am.*, 65, 594, 1975.

9. Badami, J. S. and Rao, K. R., *Proc. R. Soc. London*, 140(A), 387, 1933.

10. Baird, K. M. and Smith, D. S., *J. Opt. Soc. Am.*, 48, 300, 1958.

11. Bashkin, S. and Martinson, I., *J. Opt. Soc. Am.*, 61, 1686, 1971.

12. Beacham, J. R., Ph.D. thesis, Purdue University, 1970.

13. Benschop, H., Joshi, Y. N., and van Kleef, T. A. M., *Can. J. Phys.*, 53, 700, 1975.

14. Berry, H. G., Bromander, J., and Buchta, R., *Phys. Scr.*, 1, 181, 1970.

15. Berry, H. G., Bromander, J., Martinson, I., and Buchta, R., *Phys. Scr.*, 3, 63, 1971.

16. Berry, H. G., Desesquelles, J., and Dufay, M., *Phys. Rev. Sect A.*, 6, 600, 1972.

17. Berry, H. G., Desesquelles, J., and Dufay, M., *Nucl. Instrum. Methods*, 110, 43, 1973.

18. Berry, H. G., Pinnington, E. H., and Subtil, J. L., *J. Opt. Soc. Am.*, 62, 767, 1972.

19. Bidelman, W. P. and Corliss, C. H., *Astrophys. J.*, 135, 968, 1962.

20. Bloch, L. and Bloch, E., *Ann. Phys.* (Paris), 10(11), 141, 1929.

21. Bloch, L., Bloch, E., and Felici, N., *J. Phys. Radium*, 8, 355, 1937.

22. Bockasten, K., *Ark. Fys.*, 9, 457, 1955.

23. Bockasten, K., Hallin, R., Johansson, K. B., and Tsui, P., *Phys. Lett.* (Netherlands), 8, 181, 1964.

24. Bockasten, K. and Johansson, K. B., *Ark. Fys.*, 38, 563, 1969.

25. Borgstrom, A., *Ark. Fys.*, 38, 243, 1968.

26. Borgstrom, A., *Phys. Scr.*, 3, 157, 1971.

27. Bowen, I. S., *Phys. Rev.*, 29, 231, 1927.

28. Bowen, I. S., *Phys. Rev.*, 31, 34, 1928.

29. Bowen, I. S., *Phys. Rev.*, 39, 8, 1932.

30. Bowen, I. S., *Phys. Rev.*, 45, 401, 1934.

31. Bowen, I. S., *Phys. Rev.*, 46, 377, 1934.

32. Bowen, I. S., *Phys. Rev.*, 46, 791, 1934.

33. Boyce, J. C., *Phys. Rev.*, 49, 730, 1936.

34. Boyce, J. C. and Robinson, H. A., *J. Opt. Soc. Am.*, 26, 133, 1936.

35. Bozman, W. R., Meggers, W. F., and Corliss, C. H., *J. Res. Natl. Bur. Stand. Sect. A*, 71, 547, 1967.

36. Bromander, J., *Ark. Fys.*, 40, 257, 1969.

37. Bromander, J. and Buchta, R., *Phys. Scr.*, 1, 184, 1970.

38. Brown, C. M. and Ginter, M. L., *J. Opt. Soc. Am.*, 68, 243, 1978.

39. Brown, C. M., Tilford, S. G., and Ginter, M. L., *J. Opt. Soc. Am.*, 65, 1404, 1975.

40. Bryant, B. W., *Johns Hopkins Spectroscopic Report* No. 21, 1961.

41. Buchet, J. P., Buchet-Poulizac, M. C., Berry, H. G., and Drake, G. W. F., *Phys. Rev. Sect. A*, 7, 922, 1973.

42. Budhiraja, C. J. and Joshi, Y. N., *Can. J. Phys.*, 49, 391, 1971.

43. Burns, K. and Adams, K. B., *J. Opt. Soc. Am.*, 42, 56, 1952.

44. Burns, K. and Adams, K. B., *J. Opt. Soc. Am.*, 46, 94, 1956.

45. Burns, K., Adams, K. B., and Longwell, J., *J. Opt. Soc. Am.*, 40, 339, 1950.

46. Callahan, W. R., Ph.D. thesis, Johns Hopkins University, 1962.

47. Charles, G. W., *J. Opt. Soc. Am.*, 56, 1292, 1966.

48. Charles, G. W., Hunt, D. J., Pish, G., and Timma, D. L., *J. Opt. Soc. Am.*, 45, 869, 1955.

49. Codling, K., *Proc. Phys. Soc.*, 77, 797, 1961.

50. Comite Consultatif Pour La Definition du Metre, *J. Phys. Chem. Ref. Data*, 3, 852, 1974.

51. Conway, J. G., Blaise, J., and Verges, J., *Spectrochim. Acta Part B*, 31, 31, 1976.

52. Conway, J. G., Worden, E. F., Blaise, J., and Verges, J., *Spectrochim. Acta Part B*, 32, 97, 1977.

53. Conway, J. G., Worden, E. F., Blaise, J., Camus, P., and Verges, J., *Spectrochim. Acta Part B*, 32, 101, 1977.

54. Crooker, A. M., *Can. J. Res. Sect.* A, 14, 115, 1936.

55. Crooker, A. M. and Dick, K. A., *Can. J. Phys.*, 46, 1241, 1968.

56. Crosswhite, H. M., *J. Res. Natl. Bur. Stand. Sect. A*, 79, 17, 1975.

58. Crosswhite, H. M. and Dieke, G. H., *American Institute of Physics Handbook*, Section 7, 1972.

59. de Bruin, T. L., *Z. Phys.*, 38, 94, 1926.

60. de Bruin, T. L., *Z. Phys.*, 53, 658, 1929.

61. de Bruin, T. L., Humphreys, C. J., and Meggers, W. F., *J. Res. Natl. Bur. Stand.*, 11, 409, 1933.

62. Dick, K. A., *J. Opt. Soc. Am.*, 64, 702, 1973.

63. Dobbie, J. C., *Ann. Solar Phys. Observ.* (Cambridge), 5, 1, 1938.

64. Earls, L. T. and Sawyer, R. A., *Phys. Rev.*, 47, 115, 1935.

65. Edlen, B., *Z. Phys.*, 85, 85, 1933.

66. Edlen, B., *Nova Acta Reglae Soc. Sci. Ups.*, (IV) 9, No. 6, 1934.

67. Edlen, B., *Z. Phys.*, 93, 726, 1935.

68. Edlen, B., *Z. Phys.*, 94, 47, 1935.

69. Edlen, B., *Rep. Prog. Phys.*, 26, 181, 1963.

70. Edlen, B. and Risberg, P., *Ark. Fys.*, 10, 553, 1956.

71. Edlen, B. and Swings, P., *Astrophys. J.*, 95, 532, 1942.

72. Ehrhardt, J. C. and Davis, S. P., *J. Opt. Soc. Am.*, 61, 1342, 1971.

73. Eidelsberg, M., *J. Phys. B*, 5, 1031, 1972.

74. Eidelsberg, M., *J. Phys. B*, 7, 1476, 1974.

75. Ekberg, J. O. and Svensson, L. A., *Phys. Scr.*, 2, 283, 1970.

76. Ekefors, E., *Z. Phys.*, 71, 53, 1931.

77. Epstein, G. L. and Reader, J., *J. Opt. Soc. Am.*, 65, 310, 1975.

78. Epstein, G. L. and Reader, J., *J. Opt. Soc. Am.*, 66, 590, 1976.

79. Epstein, G. L. and Reader, J., unpublished.

80. Eriksson, K. B. S., *Phys. Lett. A.*, 41, 97, 1972.

81. Eriksson, K. B. S. and Isberg, H. B. S., *Ark. Fys.*, 23, 527, 1963.

82. Eriksson, K. B. S. and Wenaker, I., *Phys. Scr.*, 1, 21, 1970.

83. Esteva, J. M. and Mehlman, G., *Astrophys. J.*, 193, 747, 1974.

84. Even-Zohar, M. and Fraenkel, B. S., *J. Phys. B*, 5, 1596, 1972.

85. Fawcett, B. C., *J. Phys. B*, 3, 1732, 1970.

86. Fawcett, B. C., Culham Laboratory Report ARU-R4, 1971.

87. Ferner, E., *Ark. Mat. Astron. Fys.*, 28(A), 4, 1941.

88. Fischer, R. A., Knopf, W. C., and Kinney, F. E., *Astrophys. J.*, 130, 683, 1959.

89. Fowler, A., *Report on Series in Line Spectra*, Fleetway Press, London, 1922.

90. Fowles, G. R., *J. Opt. Soc. Am.*, 44, 760, 1954.

91. Fred, M., *Argonne Natl. Lab.*, unpublished, 1977.

92. Fred, M. and Tomkins, F. S., *J. Opt. Soc. Am.*, 47, 1076, 1957.

93. Fred, M., Tomkins, F. S., Blaise, J. E., Camus, P., and Verges, J., Argonne National Laboratory Report No. 76–68, 1976.

94. Garcia, J. D. and Mack, J. E., *J. Opt. Soc. Am.*, 55, 654, 1965.

96. Giacchetti, A., *Argonne Natl. Lab.*, unpublished, 1975.

97. Giacchetti, A., Blaise, J., Corliss, C. H., and Zalubas, R., *J. Res. Natl. Bur. Stand. Sect. A*, 78, 247, 1974.

98. Giacchetti, A., Stanley, R. W., and Zalubas, R., *J. Opt. Soc. Am.*, 69, 474, 1970.

99. Gilbert, W. P., *Phys. Rev.*, 47, 847, 1935.

100. Gilroy, H. T., *Phys. Rev.*, 38, 2217, 1931.

101. Glad, S., *Ark. Fys.*, 10, 291, 1956.

102. Goldsmith, S., *J. Phys. B*, 2, 1075, 1969.

103. Goorvitch, D., Mehlmam-Balloffet, G., and Valero, F. P. J., *J. Opt. Soc. Am.*, 60, 1458, 1970.

104. Goorvitch, D. and Valero, F. P. J., *Astrophys. J.*, 171, 643, 1972.

105. Green, L. C., *Phys. Rev.*, 55, 1209, 1939.

106. Gutman, F., *Diss. Abstr. Int.* B, 31, 363, 1970.

107. Hallin, R., *Ark. Fys.*, 31, 511, 1966.

108. Hallin, R., *Ark. Fys.*, 32, 201, 1966.

109. Hansen, J. E. and Persson, W., *J. Opt. Soc. Am.*, 64, 696, 1974.

110. Hansen, J. E. and Persson, W., *Phys. Scr.*, 13, 166, 1976.

111. Hellintin, P., *Phys. Scr.*, 13, 155, 1976.

112. Herzberg, G. and Moore, H. R., *Can. J. Phys.*, 37, 1293, 1959.

113. Hetzler, C. W., Boreman, R. W., and Burns, K., *Phys. Rev.*, 48, 656, 1935.

114. Holmstrom, J. E. and Johansson, L., *Ark. Fys.*, 40, 133, 1969.

115. Hontzeas, S., Martinson, I., Erman, P., and Buchta, R., *Nucl. Instrum. Methods*, 110, 51, 1973.
116. Humphreys, C. J., *J. Res. Natl. Bur. Stand.*, 22, 19, 1939.
117. Humphreys, C. J., *J. Opt. Soc. Am.*, 43, 1027, 1953.
118. Humphreys, C. J., *J. Phys. Chem. Ref. Data*, 2, 519, 1973.
119. Humphreys, C. J. and Andrew, K. L., *J. Opt. Soc. Am.*, 54, 1134, 1964.
120. Humphreys, C. J. and Meggers, W. F., *J. Res. Natl. Bur. Stand.*, 10, 139, 1933.
121. Humphreys, C. J. and Paul, E., Jr., *J. Opt. Soc. Am.*, 60, 200, 1970.
122. Humphreys, C. J. and Paul, E., Jr., *J. Opt. Soc. Am.*, 62, 432, 1972.
123. Humphreys, C. J., Paul, E., Jr., Cowan, R. D., and Andrew, K. L., *J. Opt. Soc. Am.*, 57, 855, 1967.
124. Humphreys, C. J., Paul, E., Jr., and Minnhagen, L., *J. Opt. Soc. Am.*, 61, 110, 1971.
125. Iglesias, L., Inst. of Optics, Madrid, unpublished, 1977.
126. Iglesias, L. and Velasco, R., *Publ. Inst. Opt. Madrid*, No. 23, 1964.
127. Isberg, B., *Ark. Fys.*, 35, 551, 1967.
128. Johannesson, G. A., Lundstrom, T., and Minnhagen, L., *Phys. Scr.*, 6, 129, 1972.
129. Johannesson, G. A. and Lundstrom, T., *Phys. Scr.*, 8, 53, 1973.
130. Johansson, I. *Ark. Fys.*, 20, 135, 1961.
131. Johansson, I. and Contreras, R., *Ark. Fys.*, 37, 513, 1968.
132. Johansson, I. and Litzen, U., *Ark. Fys.*, 34, 573, 1967.
133. Johansson, I. and Svensson, K. F., *Ark. Fys.*, 16, 353, 1960.
134. Johansson, L., *Ark. Fys.*, 20, 489, 1961.
135. Johansson, L., *Ark. Fys.*, 23, 119, 1963.
136. Johansson, S. and Litzen, U., *Phys. Scr.*, 6, 139, 1972.
137. Johansson, S. and Litzen, U., *Phys. Scr.*, 8, 43, 1973.
138. Johansson, S. and Litzen, U., *Phys. Scr.*, 10, 121, 1974.
139. Joshi, Y. N., St. Francis Xavier Univ., Nova Scotia, unpublished.
140. Joshi, Y. N., Bhatia, K. S., and Jones, W. E., *Sci. Light Tokyo*, 21, 113, 1972.
141. Joshi, Y. N., Bhatia, K. S., and Jones, W. E., *Spectrochim. Acta Part B*, 28, 149, 1973.
142. Joshi, Y. N. and Budhiraja, C. J., *Can. J. Phys.*, 49, 670, 1971.
143. Joshi, Y. N. and van Kleef, T. A. M., *Can. J. Phys.*, 52, 1891, 1974.
144. Kaufman, V., *Natl. Bur. Stand.*, unpublished.
145. Kaufman, V., *J. Opt. Soc. Am.*, 52, 866, 1962.
146. Kaufman, V., Artru, M. C., and Brillet, W. U. L., *J. Opt. Soc. Am.*, 64, 197, 1974.
147. Kaufman, V. and Humphreys, C. J., *J. Opt. Soc. Am.*, 59, 1614, 1969.
148. Kaufman, V. and Sugar, J., *J. Opt. Soc. Am.*, 61, 1693, 1971.
149. Kaufman, V. and Sugar, J., *J. Res. Natl. Bur. Stand. Sect. A*, 71, 583, 1967.
150. Kelly, R. L. and Palumbo, L. J., *Naval Research Laboratory Report 7599*, Washington, DC., 1973.
151. Kielkopf, J. F., *Univ. of Louisville*, unpublished. 1975.
152. Kielkopf, J. F., *Univ. of Louisville*, unpublished, 1976.
153. Kiess, C. C. and Corliss, C. H., *J. Res. Natl. Bur. Stand. Sect. A*, 63, 1, 1959.
154. Kleiman, H., *J. Opt. Soc. Am.*, 52, 441, 1962.
155. Eriksson, K. B., Johansson, I., and Norlen, G., *Ark. Fys.*, 28, 233, 1964.
156. Klinkenberg, P. F. A., *Physica*, 15, 774, 1949.
157. Klinkenberg, P. F. A., *Physica*, 16, 618, 1950.
158. Krishnamurty, S. G., *Proc. Phys. Soc. London*, 48, 277, 1936.
159. Kruger, P. G. and Gilroy, H. T., *Phys. Rev.*, 48, 720, 1935.
160. Kruger, P. G. and Pattin, H. S., *Phys. Rev.*, 52, 621, 1937.
161. Lacroute, P., *Ann. Phys. (Paris)*, 3, 5, 1935.
162. Lang, R. J., *Phys. Rev.*, 30, 762, 1927.
163. Lang, R. J., *Phys. Rev.*, 32, 737, 1928.
164. Lang, R. J., *Phys. Rev.*, 35, 445, 1930.
165. Lang, R. J., *Can. J. Res. Sect. A*, 14, 43, 1936.
166. Lang, R. J., *Can. J. Res. Sect. A*, 14, 127, 1936.
167. Lang, R. J. and Vestine, E. H., *Phys. Rev.*, 42, 233, 1932.
168. Li, H. and Andrew, K. L., *J. Opt. Soc. Am.*, 61, 96, 1971.
169. Liden, K., *Ark. Fys.*, 1, 229, 1949.
170. Litzen, U., *Ark. Fys.*, 28, 239, 1965.
171. Litzen, U., *Phys. Scr.*, 1, 251, 1970.
172. Litzen, U., *Phys. Scr.*, 1, 253, 1970.
173. Litzen, U., *Phys. Scr.*, 2, 103, 1970.
174. Litzen, U. and Verges, I., *Phys. Scr.*, 13, 240, 1976.

175. Lofstrand, B., *Phys. Scr.*, 8, 57, 1973.
176. Luc-Koenig, E., Morillon, C., and Verges, J., *Phys. Scr.*, 12, 199, 1975.
177. Lundstrom, T., *Phys. Scr.*, 7, 62, 1973.
178. Lundstrom, T. and Minnhagen, L., *Phys. Scr.*, 5, 243, 1972.
179. Magnusson, C. E. and Zetterberg, P. O., *Phys. Scr.*, 10, 177, 1974.
180. Magnusson, C. E., and Zetterberg, P. O., *Phys. Scr.*, 15, 237, 1977.
181. Martin, D. C., *Phys. Rev.*, 48. 938, 1935.
182. Svendenius, N., *Phys. Scr.*, 22, 240, 1980.
183. Martin, W. C., *J. Res. Natl. Bur. Stand. Sect. A*, 64, 19, 1960.
184. Martin, W. C. and Corliss, C. H., *J. Res. Natl. Bur. Stand. Sect. A*, 64, 443, 1960.
185. Martin, W. C. and Kaufman, V., *J. Res. Natl. Bur. Stand. Sect. A*, 74, 11, 1970.
186. Martin, W. C. and Kaufman, V., *J. Opt. Soc. Am.*, 60, 1096, 1970.
187. McCormick, W. W. and Sawyer, R. A., *Phys. Rev.*, 54, 71, 1938.
188. McLaughlin, R., *J. Opt. Soc. Am.*, 54, 965, 1964.
189. McLennan, J. C., McLay, A. B., and Crawford, M. F., *Proc. R. Soc. London Ser. A*, 134, 41, 1931.
190. Meissner, K. W., *Z. Phys.*, 39, 172, 1926.
191. Meggers, W. F., *J. Res. Natl. Bur. Stand.*, 24, 153, 1940.
192. Meggers, W. F. and Corliss, C. H., *J. Res. Natl. Bur. Stand. Sect. A*, 70, 63, 1966.
193. Meggers, W. F., Fred, M., and Tomkins, F. S., *J. Res. Natl. Bur. Stand.*, 58, 297, 1957.
194. Meggers, W. F. and Humphreys, C. J., *J. Res. Natl. Bur. Stand.*, 28, 463, 1942.
195. Meggers, W. F. and Murphy, R. J., *J. Res. Natl. Bur. Stand.*, 48, 334, 1952.
196. Meggers, W. F., Scribner, B. F., and Bozman, W. R., *J. Res. Natl. Bur. Stand.*, 46, 85, 1951.
197. Meggers, W. F., Shenstone, A. G., and Moore, C. E., *J. Res. Natl. Bur. Stand.*, 45, 346, 1950.
198. Mehlman, G. and Esteva, J. M., *Astrophys. J.*, 188, 191, 1974.
199. Meinders, E., *Physica*, 84(C), 117, 1976.
200. Sansonetti, C. J., Dissertation, Purdue University, 1981.
201. Sansonetti, C. J., *Natl. Bur. Stand. (U.S.)*, unpublished.
202. Millikan, R. A. and Bowen, I. S., *Phys. Rev.*, 25, 600, 1925.
203. Minnhagen, L., *J. Opt. Soc. Am.*, 61, 1257, 1925.
204. Minnhagen, L., *J. Opt. Soc. Am.*, 63, 1185, 1973.
205. Minnhagen, L., *Phys. Scr.*, 11, 38, 1975.
206. Minnhagen, L., *J. Opt. Soc. Am.*, 66, 659, 1976.
207. Minnhagen, L. and Nietsche, H., *Phys. Scr.*, 5, 237, 1972.
208. Minnhagen. L., Strihed, H., and Petersson, B., *Ark. Fys.*, 39, 471, 1969.
209. Moore, C. E., *Natl. Bur. Stand. (U.S.) Circ.*, 488, 1950.
210. Moore, C. E., *Revised Multiplet Table*, Princeton University Observatory No. 20, 1945.
211. Moore, C. E., National Standard Reference Data Series - National Bureau of Standards 3, Sect. 3, 1970.
212. Moore, C. E., National Standard Reference Data Series - National Bureau of Standards 3, Sect. 4, 1971.
213. Moore, C. E., National Standard Reference Data Series - National Bureau of Standards 3, Sect. 5, 1975.
214. Moore, C. E., National Standard Reference Data Series - National Bureau of Standards 3, Sect. 6, 1972.
215. Moore, C. E., National Standard Reference Data Series - National Bureau of Standards 3, Sect. 7, 1975.
216. Morillon, C. and Verges, J., *Phys. Scr.*, 10, 227, 1974.
217. Newsom, G. H., *Astrophys. J.*, 166, 243, 1971.
218. Newsom, G. H., O'Connor, S., and Learner, R. C. M., *J. Phys. B*, 6, 2162, 1973.
219. Norlen, G., *Phys. Scr.*, 8, 249, 1973.
220. Odabasi, H., *J. Opt. Soc. Am.*, 57, 1459, 1967.
221. Olme, A., *Ark. Fys.*, 40, 35, 1969.
222. Olme, A., *Phys. Scr.*, 1, 256, 1970.
223. Johansson, S., and Litzen, U., *J. Opt. Soc. Am.*, 61, 1427, 1971.
224. Palenius, H. P., *Ark. Fys.*, 39, 15, 1969.
225. Palenius, H. P., *Phys. Scr.*, 1, 113, 1970.
226. Palenius, H. P., *Univ. of Lund, Sweden*, unpublished.
227. Paschen, F., *Ann. Phys.*, Series 5, 12, 509, 1932.
228. Paschen, F. and Ritschl, R., *Ann. Phys.*, Series 5, 18, 867, 1933.

229. Peck, E. R., Khanna, B. N., and Anderholm, N. C., *J. Opt. Soc. Am.*, 52, 53, 1962.

230. Persson, W., *Phys. Scr.*, 3, 133, 1971.

231. Persson, W. and Valind, S., *Phys. Scr.*, 5, 187, 1972.

232. Petersson, B., *Ark. Fys.*, 27, 317, 1964.

233. Phillips, L. W. and Parker, W. L., *Phys. Rev.*, 60, 301, 1941.

234. Platt, J. R. and Sawyer, R. A., *Phys. Rev.*, 60, 866, 1941.

235. Plyer, E. K., Blaine, L. R., and Tidwell, E., *J. Res. Natl. Bur. Stand.*, 55, 279, 1955.

236. Poppe, R., van Kleef, T. A. M., and Raassen, A. J. J., *Physica*, 77, 165, 1974.

237. Radziemski, L. J., Jr. and Andrew, K. L., *J. Opt. Soc. Am.*, 55, 474, 1965.

238. Radziemski, L. J., Jr. and Kaufman, V., *J. Opt. Soc. Am.*, 59, 424, 1969.

239. Radziemski, L. J., Jr. and Kaufman, V., *J. Opt. Soc. Am.*, 64, 366, 1974.

240. Ramanadham, R. and Rao, K. R., *Indian J. Phys.*, 18, 317, 1944.

241. Ramb, R., *Ann. Phys.*, 10, 311, 1931.

242. Rank, D. H., Bennett, J. M., and Bennett, H. E., *J. Opt. Soc. Am.*, 40, 477, 1950.

243. Rao, A. S. and Krishnamurty, S. G., *Proc. Phys. Soc. London*, 46, 531, 1943.

244. Rao, K. R., *Proc. R. Soc. London, Ser. A*, 134, 604, 1932.

245. Rao, K. R. and Badami, J. S., *Proc. R. Soc. London Ser. A*, 131, 154, 1931.

246. Rao, K. R. and Krishnamurty, S. G., *Proc. R. Soc. London Ser. A*, 161, 38, 1937.

247. Rao, K. R. and Murti, S. G. K., *Proc. R. Soc. London* Ser. A, 145, 681, 1934.

248. Rao, Y. B., *Indian J. Phys.*, 32, 497, 1958.

249. Rao, Y. B., *Indian J. Phys.*, 33, 546, 1959.

250. Rao, Y. B., *Indian J. Phys.*, 35, 386, 1961.

251. Rasmussen, E., *Z. Phys.*, 80, 726, 1933.

252. Rasmussen, E., *Z. Phys.*, 83, 404, 1933.

253. Rasmussen, E., *Z. Phys.*, 86, 24, 1934.

254. Rasmussen, E., *Z. Phys.*, 87, 607, 1934.

255. Rasmussen, E., *Phys. Rev.*, 57, 840, 1940.

256. Rau, A. S. and Narayan, A. L., *Z. Phys.*, 59, 687, 1930.

257. Reader, J., *J. Opt. Soc. Am.*, 65, 286, 1975.

258. Reader, J., *J. Opt. Soc. Am.*, 65, 988, 1975.

259. Reader, J., *J. Opt Soc. Am.*, 73, 349, 1983.

260. Reader, J. and Davis, S., *J. Res. Natl. Bur. Stand. Sect. A*, 71, 587, 1967, and unpublished.

261. Reader, J. and Ekberg, J. O., *J. Opt. Soc. Am.*, 62, 464, 1972.

262. Reader, J. and Epstein, G. L., *J. Opt. Soc. Am.*, 62, 1467, 1972.

263. Reader, J. and Epstein, G. L., *J. Opt. Soc. Am.*, 65, 638, 1975.

264. Reader, J. and Epstein, G. L., *Natl. Bur. Stand.*, unpublished.

265. Reader, J., Epstein, G. L., and Ekberg, J. O., *J. Opt. Soc. Am.*, 62, 273, 1972.

266. Kaufman, V., *Phys. Scr.*, 26, 439, 1982.

267. Ricard, R., Givord, M., and George, F., *C. R. Acad. Sci. Paris*, 205, 1229, 1937.

268. Risberg, P., *Ark. Fys.*, 10, 583, 1956.

269. Risberg, G., *Ark. Fys.*, 28, 381, 1965.

270. Risberg, G., *Ark. Fys.*, 37, 231, 1968.

271. Robinson, H. A., *Phys. Rev.*, 49, 297, 1936.

272. Robinson, H. A., *Phys. Rev.*, 50, 99, 1936.

273. Ross, C. B., Jr., Doctoral dissertation, Purdue University, 1969.

274. Ross, C. B., Wood, D. R., and Scholl, P. S., *J. Opt. Soc. Am.*, 66, 36, 1976.

275. Ruedy, J. E. and Gibbs, R. C., *Phys. Rev.*, 46, 880, 1934.

276. Russell, H. N., King, R. B., and Moore, C. E., *Phys. Rev.*, 58, 407, 1940.

277. Russell, H. N. and Moore, C. E., *J. Res. Natl. Bur. Stand.*, 55, 299, 1955.

278. Russell, H. N., Moore, C. E., and Weeks, D. W., *Trans. Am. Philos. Soc.*, 34(2), 111, 1944.

279. Saunders, F., Schneider, E., and Buckingham, E., *Proc. Natl. Acad. Sci.*, 20, 291, 1934.

280. Sawyer, R. A. and Humphreys, C. J., *Phys. Rev.*, 32, 583, 1928.

281. Sawyer, R. A. and Lang, R. J., *Phys. Rev.*, 34, 712, 1929.

282. Sawyer, R. A. and Paschen, F., *Ann. Phys.*, 84(4),1, 1927.

283. Scholl, P. S., M.S. thesis, Wright State Univ., 1975.

284. Schurmann, D., *Z. Phys.*, 17, 4, 1975.

285. Seguier, J., *C. R. Acad. Sci. Paris*, 256, 1703, 1963.

286. Shenstone, A. G., *Phys. Rev.*, 31, 317, 1928.

287. Shenstone, A. G., *Phys. Rev.*, 32, 30, 1928.

288. Shenstone, A. G., *Trans. R. Soc. London*, 237(A), 57, 1938.

289. Shenstone, A. G., *Phys. Rev.*, 57, 894, 1940.

290. Shenstone, A. G. *Philos. Trans. R. Soc. London Ser. A*, 241, 297, 1948.

291. Shenstone, A. G., *Can. J. Phys.*, 38, 677, 1960.

292. Shenstone, A. G., *Proc. R. Soc. London*, 261(A), 153, 1961.

293. Shenstone, A. G., *Proc. R. Soc. London*, 276(A), 293, 1963.

294. Shenstone, A. G., *J. Res. Natl. Bur. Stand. Sect. A*, 74, 801, 1970.

295. Shenstone, A. G., *J. Res. Natl. Bur. Stand. Sect. A*, 79, 497, 1975.

296. Shenstone, A. G. and Pittenger, J. T., *J. Opt. Soc. Am.*, 39, 219, 1949.

297. Smith, S., *Phys. Rev.*, 36, 1, 1930.

298. Smitt, R., *Phys. Scr.*, 8, 292, 1973.

299. Soderqvist, J., *Ark. Mat. Astronom. Fys.*, 32(A), 1, 1946.

300. Sommer, L. A., *Ann. Phys.*, 75, 163, 1924.

301. Spector, N., *J. Opt. Soc. Am.*, 63, 358, 1973.

302. Spector, N. and Sugar, J., *J. Opt. Soc. Am.*, 66, 436, 1976.

303. Steinhaus, D. W., Radziemski, L. J., Jr., and Blaise, J., *Los Alamos Sci. Lab.*, unpublished, 1975.

304. Subbaraya, T. S., *Z. Phys.*, 78, 541, 1932.

305. Sugar, J., *J. Opt. Soc. Am.*, 55, 33, 1965.

306. Sugar, J., *J. Res. Natl. Bur. Stand. Sect. A*, 73, 333, 1969.

307. Sugar, J., *J. Opt. Soc. Am.*, 60, 454, 1970.

308. Sugar, J., *J. Res. Natl. Bur. Stand. Sect. A*, 78, 555, 1974.

309. Sugar, J. and Kaufman, V., *J. Opt. Soc. Am.*, 55, 1283, 1965.

310. Sugar, J. and Kaufman, V., *J. Opt. Soc. Am.*, 62, 562, 1972.

311. Sugar, J., Kaufman, V., and Spector, N., *J. Res. Natl. Bur. Stand., Sect. A*, 83, 233, 1978.

312. Sugar, J. and Spector, N., *J. Opt. Soc. Am.*, 64, 1484, 1974.

313. Sullivan, F. J. *Univ. Pittsburgh Bull.*, 35, 1, 1938.

314. Svensson, L. A. and Ekberg, J. O., *Ark. Fys.*, 37, 65, 1968.

315. Swensson, J. W. and Risberg, G., *Ark. Fys.*, 31, 237, 1966.

316. Tech, J. L., *J. Res. Natl. Bur. Stand. Sect. A*, 67, 505, 1963.

317. Tech, J. L. and Ward, J. F., *Phys. Rev. Lett.*, 27, 367, 1971.

318. Tilford, S. G., *J. Opt. Soc. Am.*, 53, 1051, 1963.

319. Toresson, Y. G., *Ark. Fys.*, 17, 179, 1960.

320. Toresson, Y. G., *Ark. Fys.*, 18, 389, 1960.

321. Toresson, Y. G. and Edlen, B., *Ark. Fys.*, 23, 117, 1963.

322. Tsien, W. Z., *Chin. J. Phys.*, Peiping, 3, 117, 1939.

323. van Deurzen, C. H. H., Conway, J., and Davis, S. P., *J. Opt. Soc. Am.*, 63, 158, 1973.

324. van Kleef, T. A. M., Raassen, A. J. J., and Joshi, Y. N., *Physica*, 84(C), 401, 1976.

325. Sansonetti, C. J., Andrew, K. L., and Verges, J., *J. Opt. Soc. Am.*, 71, 423, 1981.

326. Wheatley, M. A. and Sawyer, R. A., *Phys. Rev.*, 61, 591, 1942.

327. Wilkinson, P. G., *J. Opt. Soc. Am.*, 45, 862, 1955.

328. Wilkinson, P. G. and Andrew, K. L., *J. Opt. Soc. Am.*, 53, 710, 1963.

329. Wood, D. and Andrew, K. L., *J. Opt. Soc. Am.*, 58, 818, 1968.

330. Wood, D. R., Ron, C. B., Scholl, P. S., and Hoke, M., *J. Opt. Soc. Am.*, 64, 1159, 1974.

331. Worden, E. F. and Conway, J. G., *Lawrence Livermore Lab.*, unpublished, 1977.

332. Worden, E. F., Hulet, E. K., Gutmacher, R. G., Conway, J. G., *At. Data Nucl. Data Tables*, 18, 459, 1976.

333. Worden, E. F., Lougheed, R. W., Gutmacher, R. G., and Conway, J. G., *J. Opt. Soc. Am.*, 64, 77, 1974.

334. Wu, C. M., Ph.D. thesis, University of British Columbia, 1971.

335. Zaidel, A. N., Prokofev, V. K., Raiskii, S. M., Slavnyi, V. A., and Schreider, E. Y., *Tables of Spectral Lines*, 3rd ed., Plenum, New York, 1970.

336. Zetterberg, P. O. and Magnusson, C. E., *Phys. Scr.*, 15, 189, 1977.

337. Sugar, J., *J. Opt. Soc. Am.*, 55, 1058, 1965.

338. Sugar, J., *J. Opt. Soc. Am.*, 61, 727, 1971.

339. Worden, E. F., and Conway, J. G., *At. Data Nucl. Data Tables*, 22, 329, 1978.

340. Kaufman, V. and Edlen, B., *J. Phys. Chem. Ref. Data*, 3, 825, 1974.

341. Lang, R. J., *Phys. Rev.*, 34, 697, 1929.

342. Ryabtsev, A. N., *Opt. Spectros.*, 39, 455, 1975.

343. Foster, E. W., *Proc. R. Soc. London*, 200(A), 429, 1950.
344. Morillon, C. and Verges, J., *Phys. Scr.*, 12, 129, 1975.
345. Ruedy, J. E., *Phys. Rev.*, 41, 588, 1932.
346. McLennan, J. C., McLay, A. B., and McLeod, J. H., *Philos. Mag.*, 4, 486, 1927.
347. Handrup, M. B. and Mack, J. E., *Physica*, 30, 1245, 1964.
348. Clearman, H. E., *J. Opt. Soc. Am.*, 42, 373, 1952.
349. Paschen, F., *Ann. Physik*, 424, 148, 1938.
350. Paschen, F. and Campbell, J. S., *Ann. Phys.*, 31(5), 29, 1938.
351. Nodwell, R., *Univ. of British Columbia, Vancouver*, unpublished, 1955.
352. Gibbs, R. C. and White, H. E., *Phys. Rev.*, 31, 776, 1928.
353. Green, M., *Phys. Rev.*, 60, 117, 1941.
354. Ellis, C. B. and Sawyer, R. A., *Phys. Rev.*, 49, 145, 1936.
355. McLennan, J. C., McLay, A. B., and Crawford, M. F., *Proc. R. Soc. London Ser. A*, 125, 50, 1929.
356. Mack, J. E. and Fromer, M., *Phys. Rev.*, 48, 346, 1935.
357. Humphreys, C. J. and Paul, E., U.S. Nav. Ord. Lab., Navord Rep. 4589, 25, 1956.
358. Walters, F. M., *Sci. Pap. Bur. Stand.*, 17, 161, 1921.
359. Crawford, M. F. and McLay, A. B., *Proc. R. Soc. London Ser. A*, 143, 540, 1934.
360. McLay, A. D. and Crawford, M. F., *Phys. Rev.*, 44, 986, 1933.
361. Schoepfle, G. K., *Phys. Rev.*, 47, 232, 1935.
362. Acquista, N., and Reader, J., *J. Opt. Soc. Am.*, 70, 789, 1980.
363. Benschop, H., Joshi, Y. N., and van Kleef, T. A. M., *Can. J. Phys.*, 53, 498, 1975.
364. Bockasten, K., Hallin, R., and Hughes, T. P., *Proc. Phys. Soc.*, 81, 522, 1963.
365. Boyce, J. C., *Phys. Rev.*, 46, 378, 1934.
366. Boyce, J. C., *Phys. Rev.*, 47, 718, 1935.
367. Boyce, J. C., *Phys. Rev.*, 48, 396, 1935.
368. Boyce, J. C., *Phys. Rev.*, 49, 351, 1936.
369. Corliss, C. H. and Meggers, W. F., *J. Res. Natl. Bur. Stand.*, 61, 269, 1958.
370. Crooker, A. M. and Dick, K. A., *Can. J. Phys.*, 42, 766, 1964.
371. De Bruin, T. L., *Z. Physik*, 77, 505, 1932.
372. De Bruin, T. L., *Proc. Roy. Acad. Amsterdam*, 36, 727, 1933.
373. De Bruin, T. L., *Zeeman Verhandelingen*, (The Hague), 1935, p. 415.
374. De Bruin, T. L., *Physica*, 3, 809, 1936.
375. De Bruin, T. L., Proc. Roy. Acad. Amsterdam, 40, 339, 1937.
376. Dick, K. A., *Can. J. Phys.*, 46, 1291, 1968.
377. Dick, K. A., unpublished, 1978.
378. Edlen, B. and Swensson, J. W., *Phys. Scr.*, 12, 21, 1975.
379. Ekberg, J. O., *Phys. Scr.*, 7, 55, 1973.
380. Ekberg, J. O., *Phys. Scr.*, 7, 59, 1973.
381. Ekberg, J. O., *Phys. Scr.*, 12, 42, 1975.
382. Ekberg, J. O. and Edlen, B., *Phys. Scr.*, 18, 107, 1978.
383. Eliason, A. Y., *Phys. Rev.*, 43, 745, 1933.
384. Gallardo, M., Massone, C. A., Tagliaferri, A. A., Garavaglia, M., and Persson, W., *Phys. Scr.*, 19, 538, 1979.
385. Garcia-Riquelme, O., *Optica Pura Y Aplicada*, 1, 53, 1968.
386. Gibbs, R. C., Vieweg, A. M., and Gartlein, C. W., *Phys. Rev.*, 34, 406, 1929.
387. Gilbert, W. P., *Phys. Rev.*, 48, 338, 1935.
388. Goldsmith, S. and Kaufman, A. S., *Proc. Phys. Soc.*, 81, 544, 1963.

389. Hermansdorfer, H., *J. Opt. Soc. Am.*, 62, 1149, 1972.
390. Humphreys, C. J., *Phys. Rev.*, 47, 712, 1935.
391. Humphreys, C. J., *J. Res. Natl. Bur. Stand.*, 16, 639, 1936.
392. Iglesias, L., *J. Opt. Soc. Am.*, 45, 856, 1955.
393. Iglesias, L., *J. Res. Natl. Bur. Stand.*, 64A, 481, 1960.
394. Iglesias, L., Anales Fisica Y Quimica, 58A, 191, 1962.
395. Iglesias, L., *J. Res. Natl. Bur. Stand.*, 70A, 465, 1966.
396. Iglesias, L., *Can. J. Phys.*, 44, 895, 1966.
397. Iglesias, L., *J. Res. Natl. Bur. Stand.*, 72A, 295, 1968.
398. Joshi, Y. N., *Can. Spectrosc.*, 15, 96, 1970.
399. Joshi, Y. N. and van Kleef, T. A. M., *Can. J. Phys.*, 55, 714, 1977.
400. Kaufman, A. S., Hughes, T. P., and Williams, R. V., *Proc. Phys. Soc.*, 76, 17, 1960.
401. Kaufman, V. and Sugar, J., *J. Opt. Soc. Am.*, 68, 1529, 1978.
402. Keussler, V., *Z. Physik*, 85, 1, 1933.
403. Kiess, C. C., *J. Res. Natl. Bur. Stand.*, 56, 167, 1956.
404. Klinkenberg, P. F. A., van Kleef, T. A. M., and Noorman, P. E., *Physica*, 27, 1177, 1961.
405. Kovalev, V. I., Romanos, A. A., and Ryabtsev, A. N., *Opt. Spectrosc.*, 43, 10, 1977.
406. Lang, R. J., *Proc. Natl. Acad. Sci.*, 13, 341, 1927.
407. Lang, R. J., *Zeeman Verhandelingen*, (The Hague), 44, 1935.
408. Liberman, S., et al., *C. R. Acad. Sci.* (Paris), 286, 253, 1978.
409. Livingston, A. E., *J. Phys.*, B9, L215, 1976.
410. Meijer, F. G., *Physica*, 72, 431, 1974.
411. Meijer, F. G. and Metsch, B. C., *Physica*, 94C, 259, 1978.
412. Moore, F. L., thesis, Princeton, 1949.
413. Paul, F. W. and Polster, H. D., *Phys. Rev.*, 59, 424, 1941.
414. Phillips, L. W. and Parker, W. L., *Phys. Rev.*, 60, 301, 1941.
415. Poppe, R., *Physica*, 81C, 351, 1976.
416. Raassen, A. J. J., van Kleef, T. A. M., and Metsch, B. C., *Physica*, 84C, 133, 1976.
417. Rao, A. B. and Krishnamurty, S. G., *Proc. Phys. Soc. (London)*, 51, 772, 1939.
418. Reader, J. and Acquista, N., *J. Opt. Soc. Am.*, 69, 239, 1979.
419. Reader, J. and Epstein, G. L., *J. Opt. Soc. Am.*, 62, 619, 1972.
420. Rico, F. R., *Anales, Real Soc. Esp. Fis. Quim.*, 61, 103, 1965.
421. Schonheit, E., *Optik*, 23, 409, 1966.
422. Shenstone, A. G., *J. Opt. Soc. Am.*, 44, 749, 1954.
423. Shenstone, A. G., unpublished, 1958.
424. Shenstone, A. G., *J. Res. Natl. Bur. Stand.*, 67A, 87, 1963.
425. Sugar, J. and Kaufman, V., *J. Opt. Soc. Am.*, 64, 1656, 1974.
426. Sugar, J. and Kaufman, V., *Phys. Rev.*, C12, 1336, 1975.
427. Svensson, L. A., *Phys. Scr.*, 13, 235, 1976.
428. Swensson, J. W. and Edlen, B., *Phys. Scr.*, 9, 335, 1974.
429. Tagliaferri, A. A., Gallego Lluesma, E., Garavaglia, M., Gallardo, M., and Massone, C. A., *Optica Pura Y Aplica*, 7, 89.
430. Tilford, S. G. and Giddings, L. E., *Astrophys.* J., 141, 1222, 1965.
431. Trawick, M. W., *Phys. Rev.*, 46, 63, 1934.
432. Van Deurzen, C. H. H., *J. Opt. Soc. Am.*, 67, 476, 1977.
433. Yarosewick, S. L. and Moore, F. L., *J. Opt. Soc. Am.*, 57, 1381, 1967.
434. Zalubas, R., unpublished, 1979.
435. Bhatia, K. S., Jones, W. E., and Crooker, A. M., *Can. J. Phys.*, 50, 2421, 1972.
436. van Kleef, T. A. M. and Joshi, Y. N., *Phys. Scr.*, 24, 557, 1981.

ATOMIC TRANSITION PROBABILITIES

J. R. Fuhr, W. L. Wiese, L. I. Podobedova, and D. E. Kelleher

For the 91st edition of this *Handbook*, we include new, more accurate data for H, He, Li, Be, B, C I and C II, N I and N II, Na I – IV, Mg I – IV, Al I – III, and Si I – V. The new printed tables contain critically evaluated atomic transition probabilities for about 9000 selected lines of all elements for which reliable data are available on an absolute scale. The material is largely for neutral and singly ionized spectra, but also includes some prominent lines of doubly and more highly charged ions of important elements. A more extensive database can be found in the Internet and CD-ROM editions of the *CRC Handbook of Chemistry and Physics*.

Most of the data are obtained from comprehensive compilations of the Data Center on Atomic Transition Probabilities at the National Institute of Standards and Technology. Specifically, data have been taken from recent critical compilations on H, He, and Li (Ref. 1); on Be and B (Ref. 2); on neutral and singly-ionized C and N (Ref. 3); and on Na (Ref. 4), Mg (Ref. 4), Al (Ref. 5), and Si (Ref. 6). Material from earlier compilations for the elements H through Ne (Refs. 7 and 8) and Na through Ca (Ref. 9) was supplemented by some more recent material taken directly from the original literature. Most of the original literature is cited in the above tables and in recent bibliographies (Refs. 10 and 11); for lack of space, individual literature references are not cited here.

The wavelength range for the neutral species is normally the visible spectrum or shorter wavelengths; only the very prominent near infrared lines are included. For the higher ions, most of the strong lines are located in the far UV. The tabulation is limited to electric dipole - including intercombination - lines and comprises essentially the fairly strong transitions with estimated uncertainties in the 10 % to 50 % range. With the exception of hydrogen, helium, and the alkali metals, most transitions are between states with low principal quantum numbers.

The transition probability, A, is given in units of 10^8 s^{-1} and is listed with as many digits as is consistent with the indicated accuracy. Generally, the estimated uncertainties of the A-values are in the range from 25 % to 50 % for two-digit numbers, 10 % to 25 % for three-digit numbers and 1 % or better for four- and five-digit numbers.

Each transition is identified by the wavelength λ in angstroms and the statistical weights, g_i and g_k, of the lower (i) and upper (k) states [the product $g_k A$ (or $g_i f$) is needed for many applications]. Whenever the wavelengths of individual lines within a multiplet are extremely close, only an average wavelength for the multiplet as well as the multiplet A-value are given, and this is indicated by an asterisk (*) to the left of the wavelength. This also has been done when the transition probability for an entire multiplet has been taken from the literature and values for individual lines cannot be determined because of insufficient knowledge of the coupling of electrons. The wavelength data have been taken either from recent compilations or from the original literature cited in bibliographies published by the Atomic Energy Levels Data Center (Refs. 12 and 13) at the National Institute of Standards and Technology. Wavelength values are consistent with those given in the table "Line Spectra of the Elements," which appears elsewhere in this *Handbook*.

In addition to the transition probability A, the atomic oscillator strength f and the line strength S are often used in the literature. The conversion factors between these quantities are (for electric-dipole transitions):

$$gf = 1.499 \cdot 10^{-8} \, \lambda^2 g_k A = 303.8 \, \lambda^{-1} S$$

where λ is in Å, A is in 10^8 s^{-1}, and S is in atomic units, which are $a_0^2 e^2 = 7.188 \cdot 10^{-59}$ $m^2 C^2$.

The table for hydrogen is presented first, followed by the tables for other elements in alphabetic sequence by element name (not symbol). Within each element, the tables are ordered by increasing ionization stage (e.g., Al I, Al II, etc.).

The transition probabilities for hydrogen and hydrogen-like ions are known precisely. Because of the hydrogen degeneracy, a "transition" is actually the sum of all fine-structure transitions between the principal quantum numbers; therefore, the hydrogen table gives weighted average A-values. For hydrogen-like ions of nuclear charge Z, the following scaling laws hold:

$$A_Z = Z^4 A_{Hydrogen} \quad f_Z = f_{Hydrogen} \quad S_Z = Z^{-2} S_{Hydrogen} \quad \lambda_Z = Z^{-2} \lambda_{Hydrogen}$$

For very highly-charged hydrogen-like ions, starting at about $Z > 25$, relativistic values must be used.

References

1. Wiese, W. L., and Fuhr, J. R., *J. Phys. Chem. Ref. Data* 38, 565, 2009.
2. Fuhr, J. R., and Wiese, W. L., *J. Phys. Chem. Ref. Data* 39, 013101, 2010.
3. Wiese, W. L., and Fuhr, J. R., *J. Phys. Chem. Ref. Data* 36, 1287, 2007.
4. Kelleher, D. E., and Podobedova, L. I., *J. Phys. Chem. Ref. Data* 37, 267, 2008.
5. Kelleher, D. E., and Podobedova, L. I., *J. Phys. Chem. Ref. Data* 37, 709, 2008.
6. Kelleher, D. E., and Podobedova, L. I., *J. Phys. Chem. Ref. Data* 37, 1285, 2008.
7. Wiese, W. L., Smith, M. W., and Glennon, B. M., *Atomic Transition Probabilities* (*H through Ne - A Critical Data Compilation*), National Standard Reference Data Series, National Bureau of Standards 4, Vol. I, U.S. Government Printing Office, Washington, D.C., 1966.
8. Wiese, W. L., Fuhr, J. R., and Deters, T. M., *Atomic Transition Probabilities of Carbon, Nitrogen, and Oxygen, J. Phys. Chem. Ref. Data, Monograph 7*, 1996.
9. Wiese, W. L., Smith, M. W., and Miles, B. M., *Atomic Transition Probabilities* (*Na through Ca - A Critical Data Compilation*), National Standard Reference Data Series, National Bureau of Standards 22, Vol. II, U. S. Government Printing Office, Washington, D.C., 1969.
10. Fuhr, J. R., Miller, B. J., and Martin, G. A., *Bibliography on Atomic Transition Probabilities* (*1914 through October 1997*), National Bureau of Standards Special Publication 505, 1978; Miller, B. J., Fuhr, J. R., and Martin, G. A., *Bibliography on Atomic Transition Probabilities* (*November 1977 through February 1980*), National Bureau of Standards Special Publication 505, Supplement 1, 1980.
11. Wiese, W. L., Reports on Astronomy, *Trans. Int. Astron. Union* 18A, 116, 1982; 19A, 122, 1985; 20A, 117, 1988, Reidel, D., Ed., Kluwer, Dordrecht, The Netherlands.
12. Moore, C. E., *Bibliography on the Analyses of Optical Atomic Spectra*, National Bureau of Standards Special Publication 306 - Section 1, 1968; Sections 2-4, 1969.
13. Hagan, L., and Martin, W. C., *Bibliography on Atomic Energy Levels and Spectra* (*July 1968 through June 1971*), National Bureau of Standards Special Publication 363, 1972; Hagan, L., *Bibliography on Atomic Energy Levels and Spectra* (*July 1971 through June 1975*), National Bureau of Standards Special Publication 363, Supplement 1, 1977; Zalubas, R., and Albright, A., *Bibliography on Atomic Energy Levels and Spectra* (*July 1975 through June 1979*), National Bureau of Standards Special Publication 363, Supplement 2, 1980; Musgrove, A., and Zalubas, R., *Bibliography on Atomic Energy Levels and Spectra* (*July 1979 through December 1983*), National Bureau of Standards Special Publication 363, Supplement 3, 1985.

λ Å	Weights g_i	g_k	A 10^8 s^{-1}
Hydrogen			
H I			
912.765	2	1800	5.1673·10⁻⁶
912.837	2	1682	6.1221·10⁻⁶
912.916	2	1568	7.2967·10⁻⁶
913.004	2	1458	8.7524·10⁻⁶
913.102	2	1352	1.0571·10⁻⁵
913.212	2	1250	1.2862·10⁻⁵
913.337	2	1152	1.5776·10⁻⁵
913.478	2	1058	1.9519·10⁻⁵
913.639	2	968	2.4380·10⁻⁵
913.823	2	882	3.0769·10⁻⁵
914.036	2	800	3.9276·10⁻⁵
914.284	2	722	5.0767·10⁻⁵
914.574	2	648	6.6540·10⁻⁵
914.917	2	578	8.8574·10⁻⁵
915.327	2	512	1.1997·10⁻⁴
915.821	2	450	1.6572·10⁻⁴
916.427	2	392	2.3409·10⁻⁴
917.178	2	338	3.3927·10⁻⁴
918.127	2	288	5.0659·10⁻⁴
919.349	2	242	7.8340·10⁻⁴
920.961	2	200	1.2631·10⁻³
923.148	2	162	2.1425·10⁻³
926.223	2	128	3.8694·10⁻³
930.748	2	98	7.5684·10⁻³
937.803	2	72	1.6440·10⁻²
949.743	2	50	4.1250·10⁻²
972.537	2	32	1.2785·10⁻¹
1025.72	2	18	5.5751·10⁻¹
1215.67	2	8	4.6986
3662.23	8	1800	2.8474·10⁻⁶
3663.37	8	1682	3.3742·10⁻⁶
3664.65	8	1568	4.0224·10⁻⁶
3666.07	8	1458	4.8261·10⁻⁶
3667.65	8	1352	5.8304·10⁻⁶
3669.43	8	1250	7.0963·10⁻⁶
3671.45	8	1152	8.7069·10⁻⁶
3673.73	8	1058	1.0777·10⁻⁵
3676.33	8	968	1.3467·10⁻⁵
3679.32	8	882	1.7005·10⁻⁵
3682.78	8	800	2.1719·10⁻⁵
3686.80	8	722	2.8093·10⁻⁵
3691.52	8	648	3.6851·10⁻⁵
3697.12	8	578	4.9101·10⁻⁵
3703.82	8	512	6.6583·10⁻⁵
3711.94	8	450	9.2102·10⁻⁵
3721.91	8	392	1.3032·10⁻⁴
3734.34	8	338	1.8927·10⁻⁴
3750.12	8	288	2.8337·10⁻⁴
3770.60	8	242	4.3972·10⁻⁴
3797.87	8	200	7.1225·10⁻⁴
3835.35	8	162	1.2156·10⁻³
3889.02	8	128	2.2148·10⁻³
3970.08	8	98	4.3889·10⁻³
4101.74	8	72	9.7320·10⁻³
4340.47	8	50	2.5304·10⁻²
4861.34	8	32	8.4193·10⁻²

λ Å	Weights g_i	g_k	A 10^8 s^{-1}
6562.83	8	18	4.4101·10⁻¹
8392.19	18	800	1.5167·10⁻⁵
8413.11	18	722	1.9643·10⁻⁵
8437.75	18	648	2.5804·10⁻⁵
8467.04	18	578	3.4442·10⁻⁵
8502.27	18	512	4.6801·10⁻⁵
8545.17	18	450	6.4901·10⁻⁵
8598.18	18	392	9.2117·10⁻⁵
8664.80	18	338	1.3431·10⁻⁴
8750.25	18	288	2.0207·10⁻⁴
8862.55	18	242	3.1558·10⁻⁴
9014.67	18	200	5.1558·10⁻⁴
9228.77	18	162	8.9050·10⁻⁴
9545.70	18	128	1.6506·10⁻³
10049.4	18	98	3.3585·10⁻³
10938.1	18	72	7.7829·10⁻³
12818.1	18	50	2.2008·10⁻²
16406.4	32	288	1.6205·10⁻⁴
16805.7	32	242	2.5565·10⁻⁴
17361.2	32	200	4.2347·10⁻⁴
18173.2	32	162	7.4593·10⁻⁴
18751.0	18	32	8.9860·10⁻²
19444.5	32	128	1.4242·10⁻³
21655.2	32	98	3.0415·10⁻³
26251.4	32	72	7.7110·10⁻³
27573.0	50	288	1.4024·10⁻⁴
28719.8	50	242	2.2460·10⁻⁴
30381.1	50	200	3.7999·10⁻⁴
32957.8	50	162	6.9078·10⁻⁴
37391.4	50	128	1.3877·10⁻³
40511.4	32	50	2.6993·10⁻²
43747.2	72	288	1.2884·10⁻⁴
46524.9	50	98	3.2528·10⁻³
46706.2	72	242	2.1096·10⁻⁴
51279.2	72	200	3.6881·10⁻⁴
59072.4	72	162	7.0652·10⁻⁴
74598.3	50	72	1.0254·10⁻²
75009.1	72	128	1.5609·10⁻³
123718.6	72	98	4.5608·10⁻³
Aluminum			
Al I			
2118.312	2	4	1.03·10⁻¹
2123.362	4	6	1.22·10⁻¹
2129.663	2	4	1.52·10⁻¹
2134.733	4	6	1.81·10⁻¹
2145.555	2	4	2.06·10⁻¹
2150.699	4	6	2.46·10⁻¹
2168.805	2	4	2.96·10⁻¹
2174.028	4	6	3.53·10⁻¹
2199.150	2	2	1.75·10⁻²
2204.590	4	2	3.49·10⁻²
2204.660	2	4	4.37·10⁻¹
2210.046	4	6	5.20·10⁻¹
2257.999	2	2	3.77·10⁻²
2263.462	2	4	6.83·10⁻¹
2263.731	4	2	7.50·10⁻²
2269.096	4	6	7.58·10⁻¹
2269.220	4	4	1.26·10⁻¹

λ Å	Weights g_i	g_k	A 10^8 s^{-1}
2367.052	2	4	7.61·10⁻¹
2372.070	2	2	5.76·10⁻²
2373.124	4	6	9.07·10⁻¹
2373.349	4	4	1.51·10⁻¹
2378.368	4	2	1.14·10⁻¹
2567.984	2	4	1.92·10⁻¹
2575.094	4	6	3.60·10⁻¹
2575.393	4	4	5.99·10⁻²
2652.484	2	2	1.42·10⁻¹
2660.393	4	2	2.84·10⁻¹
3082.1529	2	4	5.87·10⁻¹
3092.7099	4	6	7.29·10⁻¹
3092.8386	4	4	1.16·10⁻¹
3944.0058	2	2	4.99·10⁻¹
3961.5200	4	2	9.85·10⁻¹
5557.063	2	4	2.30·10⁻³
5557.948	2	2	2.29·10⁻³
6696.015	2	4	1.00·10⁻²
6698.673	2	2	1.00·10⁻²
7835.309	4	6	3.71·10⁻²
7836.134	6	8	3.97·10⁻²
8772.866	4	6	6.47·10⁻²
8773.896	6	8	6.95·10⁻²
8828.909	2	2	6.72·10⁻³
8841.277	4	2	1.34·10⁻²
8912.900	2	4	2.28·10⁻³
8923.555	4	6	2.73·10⁻³
8925.504	4	4	4.54·10⁻⁴
Al II			
1047.8893	1	3	2.33·10⁻¹
1048.5588	3	5	3.14·10⁻¹
1049.9233	5	7	4.17·10⁻¹
1189.1854	1	3	9.30·10⁻¹
1190.0518	3	5	1.12
1191.8111	5	7	1.48
1350.1782	3	5	4.80
1539.8303	3	5	6.70
1670.7867	1	3	1.41·10¹
1719.4400	1	3	6.55
1721.2435	3	3	4.93
1721.2714	3	5	8.82
1724.9519	5	5	2.97
1724.9838	5	7	1.18·10¹
1760.1044	3	5	3.13
1761.9751	1	3	4.12
1763.8692	3	3	3.04
1763.9521	5	5	9.21
1765.8150	3	1	1.23·10¹
1767.7308	5	3	5.13
1772.802	1	3	9.41
1774.002	3	5	1.26·10¹
1774.770	3	3	7.01
1776.975	5	7	1.67·10¹
1777.825	5	5	4.18
1818.352	7	7	5.45
1820.124	3	3	4.57
1855.9286	1	3	8.38·10⁻¹
1858.0262	3	3	2.49
1862.3111	5	3	4.08

λ Å	g_i	g_k	A 10^8 s^{-1}
1904.326	1	3	2.70
1906.4082	3	5	2.00
1906.596	3	3	2.00
1906.674	3	1	8.20
1910.8252	5	5	5.80
1911.013	5	3	3.40
1931.0481	3	1	$1.04 \cdot 10^1$
1958.77	7	5	5.70
1990.5310	3	5	$1.38 \cdot 10^1$
2192.604	7	9	2.47
2194.189	7	7	$2.74 \cdot 10^{-1}$
2195.502	3	5	2.07
2816.185	3	1	3.57
2994.277	1	3	$4.04 \cdot 10^{-2}$
2995.525	3	5	$5.44 \cdot 10^{-2}$
2998.150	5	7	$7.26 \cdot 10^{-2}$
3088.516	3	5	$1.08 \cdot 10^{-1}$
3649.184	1	3	$1.26 \cdot 10^{-1}$
3649.232	1	3	$1.26 \cdot 10^{-1}$
3651.065	3	5	$1.70 \cdot 10^{-1}$
3651.096	3	5	$1.70 \cdot 10^{-1}$
3654.981	5	7	$2.26 \cdot 10^{-1}$
3654.998	5	7	$2.26 \cdot 10^{-1}$
3703.219	3	5	$3.80 \cdot 10^{-1}$
3731.952	1	3	$4.30 \cdot 10^{-2}$
3733.908	3	3	$1.30 \cdot 10^{-1}$
3738.015	5	3	$2.10 \cdot 10^{-1}$
3866.160	3	1	$3.70 \cdot 10^{-1}$
3900.675	3	5	$4.80 \cdot 10^{-3}$
4663.056	5	3	$5.81 \cdot 10^{-1}$
5388.48	1	3	$1.20 \cdot 10^{-2}$
5593.302	3	5	$9.26 \cdot 10^{-1}$
5613.291	5	7	$3.43 \cdot 10^{-2}$
5853.62	7	9	$1.28 \cdot 10^{-1}$
5861.53	5	7	$1.14 \cdot 10^{-1}$
5867.81	3	5	$2.00 \cdot 10^{-1}$
5971.980	3	5	$4.90 \cdot 10^{-2}$
5999.70	1	3	$2.07 \cdot 10^{-2}$
5999.83	1	3	$2.07 \cdot 10^{-2}$
6001.76	3	5	$2.79 \cdot 10^{-2}$
6001.88	3	3	$1.55 \cdot 10^{-2}$
6006.410	5	7	$3.40 \cdot 10^{-2}$
6061.124	3	1	$7.60 \cdot 10^{-2}$
6066.32	1	3	$8.88 \cdot 10^{-3}$
6066.44	1	3	$8.88 \cdot 10^{-3}$
6068.43	3	3	$2.66 \cdot 10^{-2}$
6068.53	3	3	$2.66 \cdot 10^{-2}$
6073.198	5	3	$4.20 \cdot 10^{-2}$
6226.18	1	3	$6.20 \cdot 10^{-1}$
6231.745	3	5	$8.40 \cdot 10^{-1}$
6243.36	5	7	1.11
6335.701	5	3	$1.40 \cdot 10^{-1}$
6816.69	1	3	$1.10 \cdot 10^{-1}$
6823.48	3	3	$3.40 \cdot 10^{-1}$
6837.14	5	3	$5.70 \cdot 10^{-1}$
6917.93	5	7	$1.60 \cdot 10^{-1}$
6919.96	3	1	$9.60 \cdot 10^{-1}$
7042.06	3	5	$5.78 \cdot 10^{-1}$
7056.60	3	3	$5.74 \cdot 10^{-1}$
7063.64	3	1	$5.73 \cdot 10^{-1}$

λ Å	g_i	g_k	A 10^8 s^{-1}
7449.42	3	5	$1.20 \cdot 10^{-1}$
7471.41	5	7	$5.57 \cdot 10^{-1}$
7624.48	1	3	$4.59 \cdot 10^{-2}$
7627.85	3	5	$6.20 \cdot 10^{-2}$
7635.33	5	7	$9.00 \cdot 10^{-2}$
8354.318	7	9	$4.27 \cdot 10^{-1}$
8359.23	7	7	$4.75 \cdot 10^{-2}$
8359.492	5	7	$3.79 \cdot 10^{-1}$
8363.251	5	5	$6.64 \cdot 10^{-2}$
8363.469	3	5	$4.20 \cdot 10^{-1}$
8640.705	1	3	$3.00 \cdot 10^{-1}$

Al III

λ Å	g_i	g_k	A 10^8 s^{-1}
560.390	2	4	$6.10 \cdot 10^{-1}$
560.390	2	2	$6.31 \cdot 10^{-1}$
695.817	2	4	$6.83 \cdot 10^{-1}$
696.212	2	2	$7.17 \cdot 10^{-1}$
1162.66	6	6	$1.39 \cdot 10^{-1}$
1162.66	6	8	2.09
1162.67	4	6	1.95
1352.857	6	8	4.26
1352.857	6	6	$2.84 \cdot 10^{-1}$
1352.857	4	6	3.98
1379.670	2	2	4.61
1384.140	4	2	9.22
1605.7661	2	4	$1.16 \cdot 10^1$
1611.8141	4	4	2.30
1611.8735	4	6	$1.38 \cdot 10^1$
1854.7164	2	4	5.44
1862.7895	2	2	5.36
1935.8404	6	8	$1.19 \cdot 10^1$
1935.8404	6	6	$7.92 \cdot 10^{-1}$
1935.9489	4	6	$1.11 \cdot 10^1$
3283.316	2	4	$1.99 \cdot 10^{-2}$
3287.302	4	6	$2.29 \cdot 10^{-2}$
3601.628	6	4	1.31
3601.926	4	4	$1.46 \cdot 10^{-1}$
3612.356	4	2	1.45
3702.106	2	2	1.13
3713.123	4	2	2.27
3980.14	6	8	$2.33 \cdot 10^{-1}$
3980.14	4	6	$2.17 \cdot 10^{-1}$
4149.915	6	8	2.05
4150.173	4	6	1.91
4357.562	2	4	$7.40 \cdot 10^{-2}$
4364.642	4	6	$8.67 \cdot 10^{-2}$
4512.564	2	4	2.09
4528.942	4	4	$4.15 \cdot 10^{-1}$
4529.194	4	6	2.49
4701.148	6	4	$7.67 \cdot 10^{-2}$
4701.412	8	6	$7.31 \cdot 10^{-2}$
4904.10	6	8	$3.51 \cdot 10^{-1}$
4904.10	4	6	$3.27 \cdot 10^{-1}$
5696.603	2	4	$8.77 \cdot 10^{-1}$
5722.728	2	2	$8.65 \cdot 10^{-1}$

Argon

Ar I

λ Å	g_i	g_k	A 10^8 s^{-1}
1048.22	1	3	5.36

λ Å	g_i	g_k	A 10^8 s^{-1}
1066.66	1	3	1.29
3948.98	5	3	$4.55 \cdot 10^{-3}$
4044.42	3	5	$3.33 \cdot 10^{-3}$
4158.59	5	5	$1.40 \cdot 10^{-2}$
4181.88	1	3	$5.61 \cdot 10^{-3}$
4190.71	5	5	$2.80 \cdot 10^{-3}$
4191.03	1	3	$5.39 \cdot 10^{-3}$
4198.32	3	1	$2.57 \cdot 10^{-2}$
4200.67	5	7	$9.67 \cdot 10^{-3}$
4259.36	3	1	$3.98 \cdot 10^{-2}$
4266.29	3	5	$3.12 \cdot 10^{-3}$
4272.17	3	3	$7.97 \cdot 10^{-3}$
4300.10	3	5	$3.77 \cdot 10^{-3}$
4333.56	3	5	$5.68 \cdot 10^{-3}$
4335.34	3	3	$3.87 \cdot 10^{-3}$
4510.73	3	1	$1.18 \cdot 10^{-2}$
4752.94	3	3	$4.5 \cdot 10^{-3}$
4768.68	3	5	$8.6 \cdot 10^{-3}$
4798.74	7	9	$8.8 \cdot 10^{-4}$
4835.97	7	9	$9.3 \cdot 10^{-4}$
4876.26	3	5	$7.8 \cdot 10^{-3}$
4886.29	7	9	$1.2 \cdot 10^{-3}$
4887.95	3	3	$1.3 \cdot 10^{-2}$
4894.69	3	1	$1.8 \cdot 10^{-2}$
4956.75	7	9	$1.8 \cdot 10^{-3}$
4989.95	5	7	$1.1 \cdot 10^{-3}$
5048.81	3	5	$4.6 \cdot 10^{-3}$
5054.18	3	3	$4.5 \cdot 10^{-3}$
5060.08	7	9	$3.7 \cdot 10^{-3}$
5070.99	5	3	$2.6 \cdot 10^{-3}$
5087.09	5	7	$1.6 \cdot 10^{-3}$
5118.21	5	7	$2.7 \cdot 10^{-3}$
5151.39	3	1	$2.39 \cdot 10^{-2}$
5162.29	3	3	$1.90 \cdot 10^{-2}$
5177.54	7	5	$2.4 \cdot 10^{-3}$
5194.02	3	1	$7.8 \cdot 10^{-3}$
5210.49	7	7	$1.1 \cdot 10^{-3}$
5214.77	5	3	$2.1 \cdot 10^{-3}$
5221.27	7	9	$8.8 \cdot 10^{-3}$
5241.09	5	5	$1.3 \cdot 10^{-3}$
5246.24	5	7	$1.2 \cdot 10^{-3}$
5252.79	5	7	$5.4 \cdot 10^{-3}$
5254.47	3	5	$3.6 \cdot 10^{-3}$
5286.07	5	7	$9.6 \cdot 10^{-4}$
5309.52	5	5	$1.2 \cdot 10^{-3}$
5317.73	5	7	$2.6 \cdot 10^{-3}$
5373.50	3	5	$2.7 \cdot 10^{-3}$
5410.48	5	7	$2.0 \cdot 10^{-3}$
5421.35	7	5	$6.0 \cdot 10^{-3}$
5439.99	3	3	$1.9 \cdot 10^{-3}$
5442.24	7	7	$9.3 \cdot 10^{-4}$
5451.65	3	5	$4.7 \cdot 10^{-3}$
5457.42	5	3	$3.6 \cdot 10^{-3}$
5473.46	5	3	$2.0 \cdot 10^{-3}$
5492.09	3	1	$5.6 \cdot 10^{-3}$
5495.87	7	9	$1.69 \cdot 10^{-2}$
5506.11	5	7	$3.6 \cdot 10^{-3}$
5524.96	7	7	$1.7 \cdot 10^{-3}$
5534.49	5	3	$2.7 \cdot 10^{-3}$
5558.70	3	5	$1.42 \cdot 10^{-2}$

λ Å	Weights g_i	g_k	A 10^8 s^{-1}	λ Å	Weights g_i	g_k	A 10^8 s^{-1}	λ Å	Weights g_i	g_k	A 10^8 s^{-1}
5559.66	3	5	$2.2 \cdot 10^{-3}$	6754.37	3	3	$2.1 \cdot 10^{-3}$	8037.23	1	3	$3.59 \cdot 10^{-3}$
5572.54	5	7	$6.6 \cdot 10^{-3}$	6756.10	5	5	$3.6 \cdot 10^{-3}$	8046.13	3	1	$1.12 \cdot 10^{-2}$
5588.72	5	5	$1.5 \cdot 10^{-3}$	6766.61	5	3	$4.0 \cdot 10^{-3}$	8053.31	5	3	$8.6 \cdot 10^{-3}$
5597.48	5	7	$4.2 \cdot 10^{-3}$	6779.93	1	3	$1.21 \cdot 10^{-3}$	8066.60	5	5	$1.4 \cdot 10^{-3}$
5606.73	3	3	$2.20 \cdot 10^{-2}$	6827.25	5	3	$2.4 \cdot 10^{-3}$	8103.69	3	3	$2.5 \cdot 10^{-1}$
5618.01	3	3	$2.1 \cdot 10^{-3}$	6851.88	3	5	$6.7 \cdot 10^{-4}$	8115.31	5	7	$3.31 \cdot 10^{-1}$
5623.78	5	5	$1.4 \cdot 10^{-3}$	6871.29	3	3	$2.78 \cdot 10^{-2}$	8264.52	3	3	$1.53 \cdot 10^{-1}$
5635.58	3	5	$9.6 \cdot 10^{-4}$	6879.59	3	5	$1.8 \cdot 10^{-3}$	8384.73	5	7	$2.4 \cdot 10^{-3}$
5639.12	1	3	$2.1 \cdot 10^{-3}$	6887.10	5	7	$1.3 \cdot 10^{-3}$	8408.21	3	5	$2.23 \cdot 10^{-1}$
5650.70	3	1	$3.20 \cdot 10^{-2}$	6888.17	3	5	$2.5 \cdot 10^{-3}$	8424.65	3	5	$2.15 \cdot 10^{-1}$
5659.13	5	5	$2.6 \cdot 10^{-3}$	6925.01	3	3	$1.2 \cdot 10^{-3}$	8490.30	3	5	$9.6 \cdot 10^{-4}$
5681.90	5	7	$2.0 \cdot 10^{-3}$	6937.67	3	1	$3.08 \cdot 10^{-2}$	8521.44	3	3	$1.39 \cdot 10^{-1}$
5683.73	5	5	$2.0 \cdot 10^{-3}$	6951.46	5	5	$2.2 \cdot 10^{-3}$	8605.78	5	5	$1.04 \cdot 10^{-2}$
5700.87	5	7	$5.9 \cdot 10^{-3}$	6960.23	5	5	$2.4 \cdot 10^{-3}$	8620.46	1	3	$9.2 \cdot 10^{-3}$
5739.52	3	5	$8.7 \cdot 10^{-3}$	6965.43	5	3	$6.39 \cdot 10^{-2}$	8667.94	1	3	$2.43 \cdot 10^{-2}$
5772.11	5	7	$2.0 \cdot 10^{-3}$	6992.17	3	1	$7.5 \cdot 10^{-3}$	8761.69	3	5	$9.5 \cdot 10^{-3}$
5773.99	5	5	$1.1 \cdot 10^{-3}$	7030.25	7	5	$2.67 \cdot 10^{-2}$	8784.61	3	1	$2.4 \cdot 10^{-3}$
5802.08	5	3	$4.2 \cdot 10^{-3}$	7067.22	5	5	$3.80 \cdot 10^{-2}$	8799.08	5	3	$4.6 \cdot 10^{-3}$
5882.62	3	1	$1.23 \cdot 10^{-2}$	7068.73	5	3	$2.0 \cdot 10^{-2}$	8962.19	3	3	$1.6 \cdot 10^{-3}$
5888.58	7	5	$1.29 \cdot 10^{-2}$	7086.70	1	3	$1.5 \cdot 10^{-3}$	9075.42	3	1	$1.2 \cdot 10^{-2}$
5928.81	5	3	$1.1 \cdot 10^{-2}$	7107.48	5	5	$4.5 \cdot 10^{-3}$	9122.97	5	3	$1.89 \cdot 10^{-1}$
5942.67	5	5	$1.8 \cdot 10^{-3}$	7125.83	3	3	$6.0 \cdot 10^{-3}$	9194.64	3	3	$1.76 \cdot 10^{-2}$
5949.26	3	3	$1.5 \cdot 10^{-3}$	7147.04	5	3	$6.25 \cdot 10^{-3}$	9224.50	3	5	$5.03 \cdot 10^{-2}$
5968.32	3	3	$1.8 \cdot 10^{-3}$	7158.83	3	1	$2.1 \cdot 10^{-2}$	9291.53	3	1	$3.26 \cdot 10^{-2}$
5971.60	3	1	$1.1 \cdot 10^{-2}$	7206.98	5	3	$2.48 \cdot 10^{-2}$	9354.22	3	3	$1.06 \cdot 10^{-2}$
5987.30	7	7	$1.2 \cdot 10^{-3}$	7229.93	5	5	$6.6 \cdot 10^{-4}$	9657.78	3	5	$5.43 \cdot 10^{-2}$
5999.00	5	5	$1.4 \cdot 10^{-3}$	7265.17	3	3	$1.7 \cdot 10^{-3}$	9784.50	3	5	$1.47 \cdot 10^{-2}$
6005.73	5	3	$1.4 \cdot 10^{-3}$	7270.66	7	7	$1.1 \cdot 10^{-3}$	10470.05	1	3	$9.8 \cdot 10^{-3}$
6013.68	7	5	$1.4 \cdot 10^{-3}$	7272.93	3	3	$1.83 \cdot 10^{-2}$	10478.0	3	3	$2.44 \cdot 10^{-2}$
6025.15	5	3	$9.0 \cdot 10^{-3}$	7285.44	5	3	$1.2 \cdot 10^{-3}$	10950.7	5	3	$3.96 \cdot 10^{-3}$
6043.22	5	7	$1.47 \cdot 10^{-2}$	7311.72	3	3	$1.7 \cdot 10^{-2}$	11078.9	5	5	$8.3 \cdot 10^{-3}$
6052.73	3	5	$1.9 \cdot 10^{-3}$	7316.01	3	3	$9.6 \cdot 10^{-3}$	11393.7	3	1	$2.22 \cdot 10^{-2}$
6064.76	5	7	$5.8 \cdot 10^{-4}$	7350.78	3	1	$1.2 \cdot 10^{-2}$	11441.8	5	3	$1.39 \cdot 10^{-2}$
6090.79	1	3	$3.0 \cdot 10^{-3}$	7353.32	5	7	$9.6 \cdot 10^{-3}$	11467.5	3	5	$3.69 \cdot 10^{-3}$
6098.81	3	3	$5.2 \cdot 10^{-3}$	7372.12	7	9	$1.9 \cdot 10^{-2}$	11488.11	3	3	$1.9 \cdot 10^{-3}$
6101.16	3	3	$3.3 \cdot 10^{-3}$	7383.98	3	5	$8.47 \cdot 10^{-2}$	11668.7	5	5	$3.76 \cdot 10^{-2}$
6105.64	3	5	$1.21 \cdot 10^{-2}$	7392.97	5	3	$7.2 \cdot 10^{-3}$	11719.5	5	3	$9.52 \cdot 10^{-3}$
6128.73	3	5	$8.6 \cdot 10^{-4}$	7412.33	3	5	$3.9 \cdot 10^{-3}$	12026.6	1	3	$4.2 \cdot 10^{-3}$
6145.44	5	7	$7.6 \cdot 10^{-3}$	7422.26	3	5	$6.6 \cdot 10^{-4}$	12112.2	7	7	$3.1 \cdot 10^{-2}$
6155.24	5	3	$5.1 \cdot 10^{-3}$	7425.29	5	7	$3.1 \cdot 10^{-3}$	12139.8	3	3	$4.5 \cdot 10^{-2}$
6165.12	5	5	$9.89 \cdot 10^{-4}$	7435.33	5	5	$9.0 \cdot 10^{-3}$	12343.7	5	7	$2.0 \cdot 10^{-2}$
6170.17	5	5	$5.0 \cdot 10^{-3}$	7436.25	7	5	$2.7 \cdot 10^{-3}$	12402.9	3	3	$1.1 \cdot 10^{-1}$
6173.10	3	5	$6.7 \cdot 10^{-3}$	7484.24	3	5	$3.4 \cdot 10^{-3}$	12439.2	3	5	$4.9 \cdot 10^{-2}$
6212.50	5	7	$3.9 \cdot 10^{-3}$	7503.84	3	1	$4.45 \cdot 10^{-1}$	12456.1	5	3	$8.9 \cdot 10^{-2}$
6215.94	5	5	$5.7 \cdot 10^{-3}$	7510.42	5	5	$4.5 \cdot 10^{-3}$	12487.6	7	5	$1.1 \cdot 10^{-1}$
6248.41	3	5	$6.8 \cdot 10^{-4}$	7514.65	3	1	$4.02 \cdot 10^{-1}$	12554.4	7	5	$1.2 \cdot 10^{-3}$
6296.87	3	5	$9.0 \cdot 10^{-3}$	7618.33	3	5	$2.9 \cdot 10^{-3}$	12702.4	3	3	$7.1 \cdot 10^{-2}$
6307.66	5	5	$6.0 \cdot 10^{-3}$	7628.86	3	5	$2.9 \cdot 10^{-3}$	12733.6	5	5	$1.1 \cdot 10^{-2}$
6364.89	3	1	$5.6 \cdot 10^{-3}$	7635.11	5	5	$2.45 \cdot 10^{-1}$	12746.3	3	3	$2.0 \cdot 10^{-2}$
6369.58	5	3	$4.2 \cdot 10^{-3}$	7670.04	5	3	$2.8 \cdot 10^{-3}$	12802.7	5	5	$5.7 \cdot 10^{-2}$
6384.72	3	3	$4.21 \cdot 10^{-3}$	7704.81	5	7	$6.3 \cdot 10^{-4}$	12933.3	3	1	$1.0 \cdot 10^{-1}$
6416.31	3	5	$1.16 \cdot 10^{-2}$	7723.76	5	3	$5.18 \cdot 10^{-2}$	12956.6	3	3	$7.4 \cdot 10^{-2}$
6466.55	1	3	$1.5 \cdot 10^{-3}$	7724.21	1	3	$1.17 \cdot 10^{-1}$	13008.5	5	3	$8.9 \cdot 10^{-2}$
6538.11	7	7	$1.1 \cdot 10^{-3}$	7798.55	3	5	$8.7 \cdot 10^{-4}$	13214.7	3	1	$8.1 \cdot 10^{-2}$
6604.02	7	5	$2.8 \cdot 10^{-3}$	7868.20	1	3	$3.50 \cdot 10^{-3}$	13273.1	5	7	$1.5 \cdot 10^{-1}$
6660.68	3	1	$7.8 \cdot 10^{-3}$	7891.08	5	5	$9.5 \cdot 10^{-3}$	13313.4	3	5	$1.3 \cdot 10^{-1}$
6664.05	5	5	$1.5 \cdot 10^{-3}$	7916.45	3	3	$1.2 \cdot 10^{-3}$	13504.0	5	7	$1.1 \cdot 10^{-1}$
6698.88	5	3	$1.6 \cdot 10^{-3}$	7948.18	1	3	$1.86 \cdot 10^{-1}$	13599.2	5	5	$2.2 \cdot 10^{-2}$
6719.22	1	3	$2.4 \cdot 10^{-3}$	8006.16	3	5	$4.90 \cdot 10^{-2}$	13622.4	3	5	$7.3 \cdot 10^{-2}$
6752.84	3	5	$1.93 \cdot 10^{-2}$	8014.79	5	5	$9.28 \cdot 10^{-2}$	13678.5	3	5	$6.2 \cdot 10^{-2}$

λ Å	g_i	g_k	A 10^8 s⁻¹	λ Å	g_i	g_k	A 10^8 s⁻¹	λ Å	g_i	g_k	A 10^8 s⁻¹
14093.6	1	3	$4.3 \cdot 10^{-2}$	3844.7	6	8	$4.8 \cdot 10^{-2}$	5062.0	2	4	$2.23 \cdot 10^{-1}$
14739.1	5	7	$8.8 \cdot 10^{-4}$	3850.6	4	4	$3.87 \cdot 10^{-1}$	5141.8	6	8	$8.1 \cdot 10^{-2}$
15046.4	1	3	$5.2 \cdot 10^{-2}$	3868.5	4	6	1.4	5145.3	4	6	$1.06 \cdot 10^{-1}$
15172.3	1	3	$1.3 \cdot 10^{-2}$	3872.1	4	4	$1.5 \cdot 10^{-1}$	6114.9	10	8	$2.00 \cdot 10^{-1}$
15329.6	5	5	$1.2 \cdot 10^{-3}$	3880.3	2	2	$2.32 \cdot 10^{-1}$	6172.3	8	6	$2.00 \cdot 10^{-1}$
15555.5	5	7	$9.8 \cdot 10^{-5}$	3900.6	4	6	$7.2 \cdot 10^{-2}$	6243.1	8	6	$3.0 \cdot 10^{-2}$
15734.9	5	3	$2.9 \cdot 10^{-4}$	3928.6	2	4	$2.44 \cdot 10^{-1}$	6483.1	4	2	$1.06 \cdot 10^{-1}$
15816.8	5	3	$8.7 \cdot 10^{-4}$	3932.5	4	4	$9.3 \cdot 10^{-1}$	6638.2	6	4	$1.37 \cdot 10^{-1}$
15989.3	1	3	$1.9 \cdot 10^{-2}$	3952.7	4	4	$2.08 \cdot 10^{-1}$	6639.7	4	2	$1.69 \cdot 10^{-1}$
16122.7	5	3	$3.9 \cdot 10^{-4}$	3979.4	4	2	$9.8 \cdot 10^{-1}$	6643.7	10	8	$1.47 \cdot 10^{-1}$
16180.0	5	5	$1.2 \cdot 10^{-3}$	4013.9	8	8	$1.05 \cdot 10^{-1}$	6666.4	2	2	$8.8 \cdot 10^{-2}$
16264.1	3	3	$3.0 \cdot 10^{-4}$	4042.9	4	4	$4.06 \cdot 10^{-1}$	6684.3	8	6	$1.07 \cdot 10^{-1}$
16520.1	3	5	$2.6 \cdot 10^{-3}$	4052.9	2	4	$6.7 \cdot 10^{-1}$	6756.6	4	4	$2.0 \cdot 10^{-2}$
16739.8	3	5	$3.1 \cdot 10^{-3}$	4072.0	6	6	$5.8 \cdot 10^{-1}$	6863.5	6	6	$2.5 \cdot 10^{-2}$
16940.4	5	5	$2.5 \cdot 10^{-2}$	4079.6	6	4	$1.19 \cdot 10^{-1}$	7233.5	2	4	$3.7 \cdot 10^{-2}$
20317.0	1	3	$1.6 \cdot 10^{-3}$	4131.7	4	2	$8.5 \cdot 10^{-1}$	7380.4	4	4	$5.6 \cdot 10^{-2}$
20616.5	5	5	$3.9 \cdot 10^{-3}$	4228.2	4	6	$1.31 \cdot 10^{-1}$	7589.3	6	4	$1.07 \cdot 10^{-1}$
20812.0	5	7	$7.6 \cdot 10^{-4}$	4237.2	4	4	$1.12 \cdot 10^{-1}$				
21332.2	3	3	$3.2 \cdot 10^{-4}$	4266.5	6	6	$1.64 \cdot 10^{-1}$	*Ar III*			
21534.9	3	5	$1.1 \cdot 10^{-3}$	4277.5	6	4	$8.0 \cdot 10^{-1}$	769.15	5	3	6.0
22039.2	3	1	$1.2 \cdot 10^{-3}$	4282.9	4	2	$1.32 \cdot 10^{-1}$	871.10	5	3	1.59
22077.4	5	3	$1.4 \cdot 10^{-3}$	4300.6	6	6	$5.7 \cdot 10^{-2}$	875.53	3	1	3.74
23133.4	3	3	$1.7 \cdot 10^{-3}$	4331.2	4	4	$5.74 \cdot 10^{-1}$	878.73	5	5	2.79
23844.8	9	7	$1.1 \cdot 10^{-2}$	4332.0	4	2	$1.92 \cdot 10^{-1}$	879.62	3	3	$9.2 \cdot 10^{-1}$
23967.5	3	1	$3.6 \cdot 10^{-3}$	4348.1	6	8	1.17	883.18	1	3	1.22
				4352.2	2	2	$2.12 \cdot 10^{-1}$	887.40	3	5	$9.0 \cdot 10^{-1}$
Ar II				4362.1	4	6	$5.5 \cdot 10^{-2}$	3024.1	5	7	2.6
2942.9	4	4	$5.3 \cdot 10^{-1}$	4370.8	4	4	$6.6 \cdot 10^{-1}$	3027.2	5	5	$6.4 \cdot 10^{-1}$
2979.1	2	2	$4.16 \cdot 10^{-1}$	4371.3	6	4	$2.21 \cdot 10^{-1}$	3054.8	3	5	1.9
3139.0	6	6	$5.2 \cdot 10^{-1}$	4376.0	4	2	$2.05 \cdot 10^{-1}$	3064.8	3	3	1.0
3169.7	4	6	$4.9 \cdot 10^{-1}$	4379.7	2	2	1.00	3078.2	1	3	1.4
3181.0	6	4	$3.7 \cdot 10^{-1}$	4400.1	4	4	$1.60 \cdot 10^{-1}$	3285.9	5	7	2.0
3243.7	4	2	1.06	4401.0	8	6	$3.04 \cdot 10^{-1}$	3301.9	5	5	2.0
3249.8	2	4	$6.3 \cdot 10^{-1}$	4412.9	6	8	$6.1 \cdot 10^{-2}$	3311.3	5	3	2.0
3263.6	2	4	$1.55 \cdot 10^{-1}$	4426.0	4	6	$8.17 \cdot 10^{-1}$	3336.1	7	9	2.0
3281.7	2	2	$4.2 \cdot 10^{-1}$	4430.2	2	4	$5.69 \cdot 10^{-1}$	3344.7	5	7	1.8
3454.1	6	4	$3.14 \cdot 10^{-1}$	4431.0	6	6	$1.09 \cdot 10^{-1}$	3352.1	7	7	$2.2 \cdot 10^{-1}$
3476.7	6	6	1.25	4474.8	4	2	$2.90 \cdot 10^{-1}$	3358.5	3	5	1.6
3491.2	4	4	1.79	4481.8	6	6	$4.55 \cdot 10^{-1}$	3361.3	5	5	$3.0 \cdot 10^{-1}$
3491.5	6	8	2.31	4545.1	4	4	$4.71 \cdot 10^{-1}$	3472.6	5	7	$2.0 \cdot 10^{-1}$
3509.8	2	2	2.55	4579.4	2	2	$8.0 \cdot 10^{-1}$	3480.6	7	7	1.6
3514.4	4	6	1.36	4589.9	4	6	$6.64 \cdot 10^{-1}$	3499.7	3	3	1.3
3520.0	6	6	$5.2 \cdot 10^{-1}$	4598.8	4	4	$6.7 \cdot 10^{-2}$	3500.6	3	5	$2.6 \cdot 10^{-1}$
3521.3	8	8	$2.27 \cdot 10^{-1}$	4609.6	6	8	$7.89 \cdot 10^{-1}$	3502.7	5	3	$4.3 \cdot 10^{-1}$
3535.3	2	4	$5.7 \cdot 10^{-1}$	4637.2	6	6	$7.1 \cdot 10^{-2}$	3503.6	5	5	1.2
3548.5	4	4	$8.7 \cdot 10^{-1}$	4657.9	4	2	$8.92 \cdot 10^{-1}$	3511.7	7	5	$2.6 \cdot 10^{-1}$
3559.5	6	8	2.88	4726.9	4	4	$5.88 \cdot 10^{-1}$				
3565.0	2	4	$5.5 \cdot 10^{-1}$	4732.1	6	4	$6.7 \cdot 10^{-2}$	*Ar IV*			
3576.6	6	8	2.75	4735.9	6	4	$5.80 \cdot 10^{-1}$	840.03	4	2	2.73
3581.6	2	4	1.76	4764.9	2	4	$6.4 \cdot 10^{-1}$	843.77	4	4	2.70
3582.4	4	6	2.53	4806.0	6	6	$7.80 \cdot 10^{-1}$	850.60	4	6	2.63
3588.4	8	10	3.03	4847.8	4	2	$8.49 \cdot 10^{-1}$				
3656.0	6	6	$7.6 \cdot 10^{-2}$	4879.9	4	6	$8.23 \cdot 10^{-1}$	*Ar VI*			
3717.2	6	8	$5.2 \cdot 10^{-2}$	4889.0	2	2	$1.9 \cdot 10^{-1}$	292.15	2	2	$6.9 \cdot 10^{1}$
3729.3	6	4	$4.80 \cdot 10^{-1}$	4904.8	6	8	$3.7 \cdot 10^{-2}$	294.05	4	2	$1.36 \cdot 10^{2}$
3763.5	8	6	$1.78 \cdot 10^{-1}$	4933.2	4	4	$1.44 \cdot 10^{-1}$				
3780.8	8	8	$7.7 \cdot 10^{-1}$	4965.1	2	4	$3.94 \cdot 10^{-1}$	*Ar VII*			
3799.4	6	4	$1.7 \cdot 10^{-1}$	4972.2	2	2	$9.7 \cdot 10^{-2}$	*250.41	9	3	$2.78 \cdot 10^{2}$
3826.8	6	6	$2.81 \cdot 10^{-1}$	5009.3	4	6	$1.51 \cdot 10^{-1}$	*477.54	9	15	$9.92 \cdot 10^{1}$
3841.5	4	2	$2.69 \cdot 10^{-1}$	5017.2	4	6	$2.07 \cdot 10^{-1}$	585.75	1	3	$7.83 \cdot 10^{1}$

λ Å	g_i	g_k	A 10^8 s^{-1}
*637.30	9	9	$6.7 \cdot 10^1$

Ar VIII

λ Å	g_i	g_k	A 10^8 s^{-1}
158.92	2	4	$1.1 \cdot 10^2$
159.18	2	2	$1.11 \cdot 10^2$
229.44	2	2	$1.12 \cdot 10^2$
230.88	4	2	$2.21 \cdot 10^2$
337.09	4	4	$1.2 \cdot 10^1$
337.26	6	4	$1.0 \cdot 10^2$
338.22	4	2	$1.1 \cdot 10^2$
519.43	2	4	$6.3 \cdot 10^1$
526.46	4	6	$7.2 \cdot 10^1$
526.87	4	4	$1.2 \cdot 10^1$
700.24	2	4	$2.55 \cdot 10^1$
713.81	2	2	$2.4 \cdot 10^1$

Ar IX

λ Å	g_i	g_k	A 10^8 s^{-1}
48.739	1	3	$1.69 \cdot 10^3$

Ar XIII

λ Å	g_i	g_k	A 10^8 s^{-1}
162.96	5	3	$3.4 \cdot 10^2$
*163.08	9	3	$5.3 \cdot 10^2$
184.90	5	5	$1.66 \cdot 10^2$
186.38	1	3	$8.8 \cdot 10^1$
*207.89	9	9	$9.5 \cdot 10^1$
*245.10	9	15	$3.7 \cdot 10^1$

Ar XIV

λ Å	g_i	g_k	A 10^8 s^{-1}
180.29	2	4	$4.5 \cdot 10^1$
183.41	2	2	$1.69 \cdot 10^2$
187.95	4	4	$1.97 \cdot 10^2$
191.35	4	2	$7.5 \cdot 10^1$
194.39	2	2	$4.6 \cdot 10^1$
203.35	4	2	$7.8 \cdot 10^1$

Ar XV

λ Å	g_i	g_k	A 10^8 s^{-1}
25.05	1	3	$1.7 \cdot 10^4$
221.10	1	3	$9.55 \cdot 10^1$
*265.3	9	9	$8.1 \cdot 10^1$

Ar XVI

λ Å	g_i	g_k	A 10^8 s^{-1}
*23.52	2	6	$1.43 \cdot 10^4$
*24.96	6	10	$4.4 \cdot 10^4$
353.88	2	4	$1.5 \cdot 10^1$
389.11	2	2	$1.1 \cdot 10^1$
1268	2	4	1.9
1401	2	2	1.4
2975	2	4	$9.0 \cdot 10^{-2}$
3514	4	6	$6.5 \cdot 10^{-2}$

Arsenic

As I

λ Å	g_i	g_k	A 10^8 s^{-1}
1890.4	4	6	2.0
1937.6	4	4	2.0
1972.6	4	2	2.0
2288.1	6	4	2.8
2344.0	2	4	$3.5 \cdot 10^{-1}$
2349.8	4	2	3.1
2369.7	4	4	$6.0 \cdot 10^{-1}$
2370.8	4	6	$4.2 \cdot 10^{-1}$
2456.5	6	4	$7.2 \cdot 10^{-2}$
2492.9	4	2	$1.2 \cdot 10^{-1}$
2745.0	2	4	$2.6 \cdot 10^{-1}$
2780.2	4	4	$7.8 \cdot 10^{-1}$
2860.4	2	2	$5.5 \cdot 10^{-1}$
2898.7	4	2	$9.9 \cdot 10^{-2}$

Barium

Ba I

λ Å	g_i	g_k	A 10^8 s^{-1}
2427.41	1	3	$5.60 \cdot 10^{-3}$
2472.74	1	3	$4.60 \cdot 10^{-3}$
3071.58	1	3	$4.20 \cdot 10^{-1}$
3501.11	1	3	$3.50 \cdot 10^{-1}$
3889.33	1	3	$1.10 \cdot 10^{-2}$
4132.43	1	3	$1.50 \cdot 10^{-2}$
4181.09	3	5	$4.99 \cdot 10^{-4}$
4181.66	3	5	$5.42 \cdot 10^{-4}$
4182.27	3	5	$6.11 \cdot 10^{-4}$
4182.94	3	5	$6.65 \cdot 10^{-4}$
4183.64	3	5	$6.70 \cdot 10^{-4}$
4184.40	3	5	$7.93 \cdot 10^{-4}$
4185.25	3	5	$8.43 \cdot 10^{-4}$
4186.16	3	5	$9.24 \cdot 10^{-4}$
4187.15	3	5	$9.90 \cdot 10^{-4}$
4188.25	3	5	$1.03 \cdot 10^{-3}$
4189.44	3	5	$1.13 \cdot 10^{-3}$
4190.76	3	5	$1.28 \cdot 10^{-3}$
4192.20	3	5	$1.36 \cdot 10^{-3}$
4193.81	3	5	$1.58 \cdot 10^{-3}$
4195.59	3	5	$1.78 \cdot 10^{-3}$
4323.00	3	5	$8.80 \cdot 10^{-2}$
4402.54	3	5	$2.70 \cdot 10^{-1}$
4488.98	5	7	$2.80 \cdot 10^{-1}$
4493.64	5	5	$1.95 \cdot 10^{-1}$
4573.85	3	1	1.21
4579.64	5	5	$7.00 \cdot 10^{-1}$
4599.72	3	1	$4.07 \cdot 10^{-1}$
4619.92	1	3	$2.70 \cdot 10^{-2}$
4700.42	3	5	$6.10 \cdot 10^{-2}$
4726.43	5	3	$3.30 \cdot 10^{-1}$
4801.30	9	3	$1.39 \cdot 10^{-1}$
4902.85	5	3	$5.40 \cdot 10^{-2}$
5169.53	5	3	$9.00 \cdot 10^{-4}$
5519.04	3	5	$5.70 \cdot 10^{-1}$
5535.48	1	3	1.19
5777.62	5	7	$8.00 \cdot 10^{-1}$
5784.04	3	5	$2.10 \cdot 10^{-1}$
5800.23	5	5	$2.39 \cdot 10^{-1}$
5826.27	5	3	$4.50 \cdot 10^{-1}$
5971.70	5	5	$1.62 \cdot 10^{-1}$
5997.09	3	3	$2.80 \cdot 10^{-1}$
6019.47	3	1	$8.10 \cdot 10^{-1}$
6063.11	5	3	$5.60 \cdot 10^{-1}$
6083.39	3	1	$1.10 \cdot 10^{-1}$
6129.23	3	1	$6.00 \cdot 10^{-2}$
6341.68	5	7	$1.16 \cdot 10^{-1}$
6450.85	3	5	$1.10 \cdot 10^{-1}$
6498.76	7	7	$5.40 \cdot 10^{-1}$
6527.31	5	5	$3.30 \cdot 10^{-1}$
6527.40	15	15	$6.15 \cdot 10^{-1}$
6595.33	3	3	$3.80 \cdot 10^{-1}$
6675.27	5	3	$1.89 \cdot 10^{-1}$
6693.84	7	5	$1.46 \cdot 10^{-1}$
6986.80	5	3	$5.20 \cdot 10^{-3}$
7059.94	7	9	$5.00 \cdot 10^{-1}$
7120.33	3	5	$1.10 \cdot 10^{-1}$
7195.23	1	3	$5.60 \cdot 10^{-2}$
7213.60	5	5	$6.50 \cdot 10^{-4}$
7280.30	5	7	$3.20 \cdot 10^{-1}$
7392.41	3	3	$1.81 \cdot 10^{-1}$
7417.54	7	5	$7.70 \cdot 10^{-3}$
7488.08	7	7	$7.30 \cdot 10^{-2}$
7528.18	5	5	$2.70 \cdot 10^{-2}$
7610.48	5	5	$1.10 \cdot 10^{-2}$
7644.90	9	3	$5.03 \cdot 10^{-1}$
7672.09	3	5	$1.52 \cdot 10^{-1}$
7780.48	5	5	$7.60 \cdot 10^{-2}$
7877.80	3	5	$1.60 \cdot 10^{-2}$
7905.75	5	3	$2.65 \cdot 10^{-1}$
8147.70	5	5	$6.30 \cdot 10^{-2}$
8560.00	5	5	$2.00 \cdot 10^{-1}$
8654.08	5	7	$3.10 \cdot 10^{-3}$
9370.12	5	5	$7.60 \cdot 10^{-2}$
9645.60	7	5	$1.10 \cdot 10^{-1}$
9704.31	3	1	$1.60 \cdot 10^{-1}$
9821.48	3	1	$5.50 \cdot 10^{-2}$
10370.30	3	5	$1.30 \cdot 10^{-2}$
10540.10	5	3	$1.80 \cdot 10^{-2}$
10649.10	5	5	$2.70 \cdot 10^{-2}$
11303.00	5	3	$1.10 \cdot 10^{-3}$
11373.70	3	1	$1.30 \cdot 10^{-1}$
12342.30	3	3	$9.00 \cdot 10^{-4}$
14723.10	3	5	$8.60 \cdot 10^{-3}$
14999.90	5	3	$2.50 \cdot 10^{-3}$
17186.90	3	1	$2.70 \cdot 10^{-2}$
18202.80	5	3	$1.20 \cdot 10^{-2}$
21567.70	5	3	$2.60 \cdot 10^{-3}$
30685.30	5	3	$6.50 \cdot 10^{-3}$

Ba II

λ Å	g_i	g_k	A 10^8 s^{-1}
2528.41	2	4	$6.91 \cdot 10^{-1}$
2634.78	4	6	$7.33 \cdot 10^{-1}$
3891.78	2	4	2.17
4130.65	4	6	2.18
4166.00	4	4	$3.54 \cdot 10^{-1}$
4267.92	6	8	$3.10 \cdot 10^{-1}$
4309.26	8	10	$3.10 \cdot 10^{-1}$
4325.75	4	6	$5.65 \cdot 10^{-2}$
4524.93	2	2	$6.63 \cdot 10^{-1}$
4554.03	2	4	1.11
4708.90	2	4	$8.47 \cdot 10^{-2}$
4843.48	4	6	$9.34 \cdot 10^{-2}$
4899.93	4	2	1.04
4934.08	2	2	$9.53 \cdot 10^{-1}$
4957.09	6	8	$5.13 \cdot 10^{-1}$
5012.95	8	10	$5.15 \cdot 10^{-1}$
5361.35	4	6	$4.01 \cdot 10^{-2}$
5391.59	6	8	$4.22 \cdot 10^{-2}$

λ Å	g_i	g_k	A 10^8 s^{-1}	λ Å	g_i	g_k	A 10^8 s^{-1}	λ Å	g_i	g_k	A 10^8 s^{-1}
5784.15	2	4	$1.59\cdot10^{-1}$	1997.98	3	5	$1.76\cdot10^{-1}$	6725.96	5	3	$3.1\cdot10^{-3}$
5853.67	4	4	$6.00\cdot10^{-2}$	1997.98	3	3	$9.80\cdot10^{-2}$	6786.56	3	5	$1.7\cdot10^{-3}$
5981.26	4	6	$1.73\cdot10^{-1}$	1998.07	5	5	$5.88\cdot10^{-2}$	6786.56	3	3	$1.7\cdot10^{-3}$
6135.60	2	2	$6.64\cdot10^{-2}$	1998.07	5	7	$2.35\cdot10^{-1}$	6884.26	3	3	$3.81\cdot10^{-3}$
6141.71	6	4	$4.12\cdot10^{-1}$	2032.72	5	3	$5.3\cdot10^{-2}$	6884.26	3	5	$6.86\cdot10^{-3}$
6378.92	4	2	$1.18\cdot10^{-1}$	2055.88	1	3	$2.28\cdot10^{-1}$	6884.26	1	3	$5.08\cdot10^{-3}$
6496.90	4	2	$3.10\cdot10^{-1}$	2055.90	3	3	$1.71\cdot10^{-1}$	6884.44	5	7	$9.16\cdot10^{-3}$
6769.48	6	8	$9.35\cdot10^{-1}$	2055.90	3	5	$3.08\cdot10^{-1}$	6884.44	5	5	$2.29\cdot10^{-3}$
6874.08	8	10	$9.26\cdot10^{-1}$	2056.00	5	5	$1.03\cdot10^{-1}$	7049.72	1	3	$9.66\cdot10^{-4}$
7115.03	8	10	$8.80\cdot10^{-3}$	2056.00	5	7	$4.11\cdot10^{-1}$	7049.72	3	3	$2.90\cdot10^{-3}$
8496.80	2	4	$3.31\cdot10^{-2}$	2125.57	3	3	$6.1\cdot10^{-2}$	7049.91	5	3	$4.83\cdot10^{-3}$
8661.90	6	4	$1.27\cdot10^{-2}$	2125.68	5	3	$1.01\cdot10^{-1}$	7154.46	3	5	$5.55\cdot10^{-3}$
8703.69	4	6	$3.69\cdot10^{-2}$	2174.96	1	3	$4.51\cdot10^{-1}$	7154.46	1	3	$7.39\cdot10^{-3}$
8710.77	6	8	$7.88\cdot10^{-1}$	2174.99	3	3	$3.39\cdot10^{-1}$	7154.46	3	3	$9.99\cdot10^{-3}$
8737.75	4	6	$7.29\cdot10^{-1}$	2174.99	3	5	$6.09\cdot10^{-1}$	7154.65	5	7	$1.33\cdot10^{-2}$
8760.61	8	6	$1.17\cdot10^{-2}$	2175.10	5	5	$2.03\cdot10^{-1}$	7154.65	5	5	$3.33\cdot10^{-3}$
8897.46	6	6	$4.93\cdot10^{-2}$	2175.10	5	7	$8.13\cdot10^{-1}$	7209.13	5	7	$5.71\cdot10^{-2}$
9603.12	2	4	$4.16\cdot10^{-1}$	2348.61	1	3	5.54	7308.29	3	5	$3.33\cdot10^{-3}$
10115.00	4	6	$4.27\cdot10^{-1}$	2350.66	1	3	$4.68\cdot10^{-2}$	7434.42	1	3	$1.53\cdot10^{-3}$
10212.80	4	4	$6.92\cdot10^{-2}$	2350.70	3	3	$1.41\cdot10^{-1}$	7434.42	3	3	$4.59\cdot10^{-3}$
10768.00	2	4	$5.56\cdot10^{-2}$	2350.83	5	3	$2.35\cdot10^{-1}$	7434.63	5	3	$7.64\cdot10^{-3}$
11088.50	4	6	$6.11\cdot10^{-2}$	2494.54	1	3	1.07	7448.87	3	1	$5.63\cdot10^{-3}$
11127.50	4	4	$1.01\cdot10^{-2}$	2494.58	3	5	1.44	7498.42	3	5	$4.46\cdot10^{-3}$
11519.50	2	2	$2.47\cdot10^{-2}$	2494.58	3	3	$8.02\cdot10^{-1}$	7551.90	5	3	$4.3\cdot10^{-3}$
11577.10	2	2	$1.75\cdot10^{-1}$	2494.73	5	3	$5.35\cdot10^{-2}$	7618.66	3	5	$1.51\cdot10^{-2}$
11931.90	4	2	$4.44\cdot10^{-2}$	2494.73	5	7	1.93	7618.66	1	3	$1.12\cdot10^{-2}$
12475.00	4	2	$2.80\cdot10^{-1}$	2494.73	5	5	$4.81\cdot10^{-1}$	7618.66	3	3	$8.41\cdot10^{-3}$
13057.80	2	4	$2.14\cdot10^{-1}$	2650.45	3	5	1.06	7618.88	5	7	$2.02\cdot10^{-2}$
14211.50	2	2	$1.66\cdot10^{-1}$	2650.55	1	3	1.41	7618.88	5	5	$5.05\cdot10^{-3}$
17738.90	6	8	$2.16\cdot10^{-1}$	2650.60	3	3	1.06	7714.38	3	1	$8.21\cdot10^{-3}$
18530.70	8	10	$1.96\cdot10^{-1}$	2650.62	5	5	3.17	7792.05	3	5	$6.02\cdot10^{-3}$
18729.70	2	4	$1.23\cdot10^{-1}$	2650.70	3	1	4.23	8090.07	1	3	$3.29\cdot10^{-3}$
19642.60	4	6	$1.28\cdot10^{-1}$	2650.76	5	3	1.76	8153.74	3	1	$1.27\cdot10^{-2}$
19845.10	4	4	$2.07\cdot10^{-2}$	3193.83	3	5	$5.08\cdot10^{-2}$	8158.98	1	3	$2.65\cdot10^{-3}$
22994.70	2	2	$6.18\cdot10^{-2}$	3229.62	3	5	$7.17\cdot10^{-2}$	8158.98	3	3	$7.94\cdot10^{-3}$
24612.50	4	4	$4.75\cdot10^{-3}$	3282.91	3	5	$1.06\cdot10^{-1}$	8159.23	5	3	$1.32\cdot10^{-2}$
24699.00	4	2	$9.98\cdot10^{-2}$	3321.01	1	3	$1.70\cdot10^{-1}$	8254.07	3	1	$3.38\cdot10^{-1}$
25923.20	6	4	$3.66\cdot10^{-2}$	3321.08	3	3	$5.10\cdot10^{-1}$	8286.90	3	5	$8.00\cdot10^{-3}$
27687.20	2	4	$6.10\cdot10^{-2}$	3321.34	5	3	$8.50\cdot10^{-1}$	8547.36	3	3	$1.31\cdot10^{-2}$
29058.90	4	2	$2.89\cdot10^{-2}$	3367.63	3	5	$1.64\cdot10^{-1}$	8547.36	1	3	$1.74\cdot10^{-2}$
30196.00	2	2	$4.70\cdot10^{-2}$	3476.56	3	1	$3.11\cdot10^{-2}$	8547.36	3	5	$2.36\cdot10^{-2}$
42934.70	6	8	$4.82\cdot10^{-3}$	3515.54	3	5	$2.73\cdot10^{-1}$	8547.63	5	7	$3.14\cdot10^{-2}$
43294.30	4	6	$4.39\cdot10^{-3}$	3736.30	3	1	$5.09\cdot10^{-2}$	8547.63	5	5	$7.86\cdot10^{-3}$
47520.80	6	6	$2.37\cdot10^{-4}$	3813.45	3	5	$4.87\cdot10^{-1}$	8547.63	5	3	$8.73\cdot10^{-4}$
				4407.94	3	1	$1.01\cdot10^{-1}$	8801.37	5	7	$9.58\cdot10^{-2}$
Beryllium				4572.66	3	5	$7.61\cdot10^{-1}$	8882.16	7	5	$1.34\cdot10^{-3}$
				5252.81	1	3	$3.12\cdot10^{-3}$	8882.16	5	3	$1.06\cdot10^{-3}$
Be I				5365.49	1	3	$3.65\cdot10^{-3}$	8882.16	3	1	$1.59\cdot10^{-3}$
1661.48	1	3	$7.23\cdot10^{-2}$	5546.45	1	3	$5.4\cdot10^{-3}$	8979.19	3	1	$2.13\cdot10^{-2}$
1929.62	3	5	$5.12\cdot10^{-2}$	5857.01	1	3	$7.0\cdot10^{-3}$	9190.45	5	7	$1.43\cdot10^{-2}$
1929.71	5	7	$6.82\cdot10^{-2}$	6085.75	5	3	$1.4\cdot10^{-3}$	9190.45	3	5	$1.35\cdot10^{-2}$
1943.59	1	3	$5.43\cdot10^{-2}$	6229.11	5	7	$2.26\cdot10^{-2}$	9190.45	7	7	$1.79\cdot10^{-3}$
1943.62	3	5	$7.32\cdot10^{-2}$	6319.62	5	3	$2.2\cdot10^{-3}$	9190.45	5	5	$2.50\cdot10^{-3}$
1943.71	5	7	$9.77\cdot10^{-2}$	6473.54	1	3	$7.7\cdot10^{-3}$	9243.88	3	5	$9.47\cdot10^{-3}$
1964.54	1	3	$8.15\cdot10^{-2}$	6564.52	5	7	$3.49\cdot10^{-2}$	9392.74	5	3	$1.86\cdot10^{-3}$
1964.56	3	3	$6.12\cdot10^{-2}$	6711.74	1	3	$3.62\cdot10^{-3}$	9392.74	3	3	$6.19\cdot10^{-4}$
1964.56	3	5	$1.10\cdot10^{-1}$	6711.74	3	5	$4.89\cdot10^{-3}$	9392.74	5	5	$3.71\cdot10^{-4}$
1964.65	5	7	$1.47\cdot10^{-1}$	6711.74	3	3	$2.71\cdot10^{-3}$	9392.74	7	5	$2.08\cdot10^{-3}$
1964.65	5	5	$3.67\cdot10^{-2}$	6711.91	5	7	$6.52\cdot10^{-3}$	9392.74	3	1	$2.47\cdot10^{-3}$
1997.95	1	3	$1.31\cdot10^{-1}$	6711.91	5	5	$1.63\cdot10^{-3}$	9847.31	5	3	$5.2\cdot10^{-3}$

λ Å	Weights g_i	g_k	A 10^8 s^{-1}
9895.58	3	3	1.59·10⁻²
9895.58	1	3	5.30·10⁻³
9895.95	5	3	2.65·10⁻²
9939.78	5	7	2.51·10⁻²
9939.78	5	5	4.40·10⁻³
9939.78	7	7	3.14·10⁻³
9939.78	3	5	2.38·10⁻²

Be II

λ Å	g_i	g_k	A 10^8 s^{-1}
973.276	4	6	9.69·10⁻¹
1026.89	2	4	1.46
1026.96	4	6	1.76
1036.30	2	4	1.720
1142.96	2	4	3.14
1143.04	4	6	3.76
1512.27	2	4	9.217
1512.41	4	6	1.106·10¹
1512.42	4	4	1.843
1776.10	2	2	1.361
1776.31	4	2	2.722
2296.81	2	4	1.09·10⁻¹
2296.91	4	6	1.30·10⁻¹
2302.96	4	6	7.28·10⁻²
2302.98	6	8	7.80·10⁻²
2381.95	4	6	1.13·10⁻¹
2381.98	6	8	1.22·10⁻¹
2413.34	2	4	1.64·10⁻¹
2413.45	4	6	1.97·10⁻¹
2453.84	2	4	1.06·10⁻¹
2507.41	4	6	1.82·10⁻¹
2507.45	6	8	1.94·10⁻¹
2617.99	2	4	2.65·10⁻¹
2618.13	4	6	3.18·10⁻¹
2697.59	4	2	1.48·10⁻¹
2728.85	4	6	3.23·10⁻¹
2728.89	6	8	3.46·10⁻¹
3046.52	2	4	4.66·10⁻¹
3046.69	4	6	5.60·10⁻¹
3046.70	4	4	9.33·10⁻²
3130.42	2	4	1.1292
3131.07	2	2	1.1285
3197.10	4	6	6.82·10⁻¹
3197.15	6	8	7.31·10⁻¹
3197.16	6	6	4.87·10⁻²
3233.54	6	4	3.37·10⁻²
3241.63	2	2	1.39·10⁻¹
3241.83	4	2	2.78·10⁻¹
3274.59	2	4	1.41·10⁻¹
3274.67	2	2	1.41·10⁻¹
4360.66	2	4	9.12·10⁻¹
4360.99	4	6	1.09
4361.03	4	4	1.82·10⁻¹
4673.33	4	6	2.06
4673.42	6	8	2.21
4673.45	6	6	1.47·10⁻¹
4702.34	2	4	4.73·10⁻²
4702.52	4	6	5.68·10⁻²
4828.12	6	4	7.89·10⁻²
4828.18	4	2	8.77·10⁻²
5218.12	2	4	7.08·10⁻²

λ Å	g_i	g_k	A 10^8 s^{-1}
5218.34	4	6	8.49·10⁻²
5218.34	4	4	1.41·10⁻²
5270.27	2	2	3.234·10⁻¹
5270.81	4	2	6.466·10⁻¹
5416.12	2	2	2.84·10⁻²
5416.36	4	2	5.68·10⁻²
6279.42	2	4	1.12·10⁻¹
6279.74	4	4	2.23·10⁻²
6279.74	4	6	1.34·10⁻¹
6756.75	2	4	4.90·10⁻²
6757.12	4	2	9.80·10⁻²
7401.20	2	4	2.54·10⁻²
7401.43	2	2	2.54·10⁻²
9048.14	2	4	2.47·10⁻²
9048.49	4	6	2.96·10⁻²
9048.49	4	4	4.94·10⁻³
9476.41	2	4	1.81·10⁻¹
9477.03	4	6	2.17·10⁻¹
9477.14	4	4	3.62·10⁻²

Bismuth

Bi I

λ Å	g_i	g_k	A 10^8 s^{-1}
1954.5	4	6	1.2
2021.2	4	4	6.0·10⁻²
2061.7	4	6	9.9·10⁻¹
2110.3	4	2	9.1·10⁻¹
2177.3	4	2	2.6·10⁻²
2228.3	4	4	8.9·10⁻¹
2230.6	4	6	2.6
2276.6	4	4	2.5·10⁻¹
2515.7	4	6	4.3·10⁻²
2627.9	4	4	4.7·10⁻¹
2696.8	4	6	6.4·10⁻²
2780.5	4	2	3.09·10⁻¹
2798.7	6	6	3.6·10⁻²
2898.0	4	2	1.53
2938.3	6	4	1.23
2989.0	4	4	5.5·10⁻¹
2993.3	4	6	1.6·10⁻¹
3024.6	6	6	8.8·10⁻¹
3067.7	4	2	2.07
3076.7	4	4	3.5·10⁻²
3397.2	6	4	1.81·10⁻¹
3402.9	6	6	1.6·10⁻²
3510.9	6	4	6.8·10⁻²
3596.1	2	4	1.98·10⁻¹
3888.2	2	2	6.9·10⁻²
4121.5	2	2	1.64·10⁻¹
4308.5	2	4	1.6·10⁻²
4493.0	2	4	1.5·10⁻²
4722.5	4	2	1.17·10⁻¹
6134.8	4	4	1.8·10⁻²

Boron

B I

λ Å	g_i	g_k	A 10^8 s^{-1}
1151.42	4	4	2.44
1378.65	2	4	3.42
1378.87	2	2	1.37·10¹
1378.94	4	4	1.71·10¹

λ Å	g_i	g_k	A 10^8 s^{-1}
1379.17	4	2	6.84
1465.56	2	4	3.3
1465.66	4	4	6.6
1465.79	6	4	9.9
1566.66	4	6	3.36·10⁻¹
1587.38	2	4	6.9·10⁻¹
1587.45	4	2	1.4
1587.59	4	6	5.0·10⁻¹
1587.66	6	4	7.5·10⁻¹
1587.75	6	6	1.2
1600.37	2	4	4.67·10⁻¹
1600.76	4	6	5.60·10⁻¹
1666.85	2	4	8.66·10⁻¹
1667.27	4	6	1.04
1818.35	4	2	6.22·10⁻¹
1825.89	2	4	1.70
1826.40	4	6	2.04
1826.40	4	4	3.39·10⁻¹
2066.38	4	6	6.8·10⁻¹
2066.65	6	6	1.6
2066.73	2	4	9.4·10⁻¹
2066.93	4	4	3.0·10⁻¹
2067.20	6	4	1.0
2067.20	4	2	1.9
2088.89	2	4	3.61·10⁻¹
2089.57	4	6	4.32·10⁻¹
2496.77	2	2	8.37·10⁻¹
2497.72	4	2	1.67
5761.90	6	8	5.7·10⁻³
5942.62	6	8	8.4·10⁻³
5942.73	4	6	7.8·10⁻³
6244.56	6	8	1.3·10⁻²
6244.68	4	6	1.2·10⁻²
6819.52	6	8	2.2·10⁻²
6819.66	4	6	2.0·10⁻²
7208.59	4	2	1.72·10⁻²
8211.79	6	8	3.42·10⁻²
8211.79	6	6	2.28·10⁻³
8212.00	4	6	3.19·10⁻²
8667.23	2	2	1.83·10⁻²
8668.57	4	2	3.67·10⁻²
9576.21	4	6	5.01·10⁻³
9576.34	6	8	5.37·10⁻³

B II

λ Å	g_i	g_k	A 10^8 s^{-1}
882.676	5	7	2.44·10¹
984.698	5	7	2.62·10¹
1230.17	3	5	1.36·10¹
1624.02	5	5	6.48
2005.87	7	7	2.22
2123.86	5	7	1.58
2220.30	3	3	3.49
2918.08	7	9	6.57·10⁻¹
3323.21	3	5	8.62·10⁻¹
3323.60	5	7	1.15
3451.30	3	5	5.41·10⁻¹
4121.93	5	5	3.68·10⁻¹
4121.93	5	7	2.10
4121.93	7	9	2.36
4121.93	7	7	2.63·10⁻¹

λ Å	Weights g_i	g_k	A $10^8\ s^{-1}$
4121.93	3	5	1.98
4472.15	3	3	$5.44 \cdot 10^{-1}$
4472.86	5	3	$9.07 \cdot 10^{-1}$
4784.20	7	5	$1.94 \cdot 10^{-1}$
4940.37	5	7	1.88
6148.91	5	7	$1.96 \cdot 10^{-1}$
6148.91	3	5	$1.85 \cdot 10^{-1}$
6148.91	7	9	$2.20 \cdot 10^{-1}$
6285.51	5	3	$3.30 \cdot 10^{-1}$
6571.12	3	5	$1.14 \cdot 10^{-1}$
6717.65	5	7	$1.71 \cdot 10^{-1}$
7030.27	3	5	$3.97 \cdot 10^{-1}$
7032.03	3	3	$3.97 \cdot 10^{-1}$
7032.33	3	1	$3.97 \cdot 10^{-1}$
7159.55	3	5	$1.19 \cdot 10^{-1}$
7160.16	1	3	$1.59 \cdot 10^{-1}$
7165.11	3	3	$1.19 \cdot 10^{-1}$
7168.46	3	1	$4.75 \cdot 10^{-1}$
7170.45	5	5	$3.56 \cdot 10^{-1}$
7176.02	5	3	$1.97 \cdot 10^{-1}$
7228.46	1	3	$1.39 \cdot 10^{-1}$
7638.62	5	7	$1.71 \cdot 10^{-1}$
7638.62	3	5	$1.28 \cdot 10^{-1}$
7638.62	1	3	$9.5 \cdot 10^{-2}$
8655.83	5	3	$6.4 \cdot 10^{-2}$
9121.00	3	1	$3.23 \cdot 10^{-1}$
9226.43	7	9	$4.37 \cdot 10^{-1}$
9226.43	5	7	$3.88 \cdot 10^{-1}$
9226.43	3	5	$3.67 \cdot 10^{-1}$
9226.43	7	7	$4.87 \cdot 10^{-2}$
9226.43	5	5	$6.8 \cdot 10^{-2}$
9446.36	3	5	$2.08 \cdot 10^{-1}$
9659.49	3	5	$6.3 \cdot 10^{-2}$

Bromine

Br I

λ Å	Weights g_i	g_k	A $10^8\ s^{-1}$
1488.5	4	4	1.2
1540.7	4	4	1.4
1574.8	2	4	$2.0 \cdot 10^{-1}$
1576.4	4	6	$2.1 \cdot 10^{-2}$
1633.4	2	4	$8.1 \cdot 10^{-2}$
4365.1	2	4	$7.5 \cdot 10^{-3}$
4425.1	4	2	$4.2 \cdot 10^{-3}$
4441.7	6	4	$7.5 \cdot 10^{-3}$
4472.6	4	4	$9.3 \cdot 10^{-3}$
4477.7	6	8	$1.3 \cdot 10^{-2}$
4513.4	6	4	$2.8 \cdot 10^{-3}$
4525.6	6	6	$7.2 \cdot 10^{-3}$
4575.7	4	4	$1.6 \cdot 10^{-2}$
4614.6	4	6	$5.4 \cdot 10^{-3}$
4979.8	4	4	$2.6 \cdot 10^{-3}$
5245.1	2	4	$3.1 \cdot 10^{-3}$
5345.4	2	4	$7.6 \cdot 10^{-3}$
7348.5	4	6	$1.2 \cdot 10^{-1}$
7513.0	6	4	$1.2 \cdot 10^{-1}$
7803.0	2	4	$5.3 \cdot 10^{-2}$
7938.7	6	6	$1.9 \cdot 10^{-1}$
8131.5	2	4	$3.8 \cdot 10^{-2}$
8343.7	2	2	$2.2 \cdot 10^{-1}$

λ Å	Weights g_i	g_k	A $10^8\ s^{-1}$
8446.6	4	4	$1.2 \cdot 10^{-1}$
8638.7	6	4	$9.7 \cdot 10^{-2}$

Br II

λ Å	Weights g_i	g_k	A $10^8\ s^{-1}$
4704.9	5	7	1.1
4785.5	5	5	$9.4 \cdot 10^{-1}$
4816.7	5	3	1.1

Cadmium

Cd I

λ Å	Weights g_i	g_k	A $10^8\ s^{-1}$
2288.0	1	3	5.3
2836.9	1	3	$2.8 \cdot 10^{-1}$
2880.8	3	5	$4.2 \cdot 10^{-1}$
2881.2	3	3	$2.4 \cdot 10^{-1}$
2980.6	5	7	$5.9 \cdot 10^{-1}$
2981.4	5	5	$1.5 \cdot 10^{-1}$
3261.1	1	3	$4.06 \cdot 10^{-3}$
3403.7	1	3	$7.7 \cdot 10^{-1}$
3466.2	3	5	1.2
3467.7	3	3	$6.7 \cdot 10^{-1}$
3610.5	5	7	1.3
3612.9	5	5	$3.5 \cdot 10^{-1}$
4140.5	3	5	$4.7 \cdot 10^{-2}$
4662.4	3	5	$5.5 \cdot 10^{-2}$
4678.1	1	3	$1.3 \cdot 10^{-1}$
4799.9	3	3	$4.1 \cdot 10^{-1}$
5085.8	5	3	$5.6 \cdot 10^{-1}$
6438.5	3	5	$5.9 \cdot 10^{-1}$

Cd II

λ Å	Weights g_i	g_k	A $10^8\ s^{-1}$
2144.4	2	4	2.8
2265.0	2	2	3.0
2572.9	2	2	1.7
2748.5	4	2	2.8
4415.6	4	6	$1.4 \cdot 10^{-2}$

Calcium

Ca I

λ Å	Weights g_i	g_k	A $10^8\ s^{-1}$
3006.9	5	5	$7.5 \cdot 10^{-1}$
3361.9	5	7	$2.23 \cdot 10^{-1}$
3630.8	3	5	$2.97 \cdot 10^{-1}$
3644.4	5	7	$3.55 \cdot 10^{-1}$
4098.5	7	9	$1.3 \cdot 10^{-1}$
4108.5	5	7	$9.0 \cdot 10^{-1}$
4226.7	1	3	2.18
4283.0	3	5	$4.34 \cdot 10^{-1}$
4289.4	1	3	$6.0 \cdot 10^{-1}$
4299.0	3	3	$4.66 \cdot 10^{-1}$
4302.5	5	5	1.36
4307.7	3	1	1.99
4318.7	5	3	$7.4 \cdot 10^{-1}$
4355.1	5	7	$1.9 \cdot 10^{-1}$
4425.4	1	3	$4.98 \cdot 10^{-1}$
4435.0	3	5	$6.7 \cdot 10^{-1}$
4435.7	3	3	$3.42 \cdot 10^{-1}$
4454.8	5	7	$8.7 \cdot 10^{-1}$
4455.9	5	5	$2.0 \cdot 10^{-1}$
4526.9	5	3	$4.1 \cdot 10^{-1}$
4578.6	3	5	$1.76 \cdot 10^{-1}$

λ Å	Weights g_i	g_k	A $10^8\ s^{-1}$
4581.4	5	7	$2.09 \cdot 10^{-1}$
4585.9	7	9	$2.29 \cdot 10^{-1}$
4878.1	5	7	$1.88 \cdot 10^{-1}$
5041.6	5	3	$3.3 \cdot 10^{-1}$
5188.9	3	5	$4.0 \cdot 10^{-1}$
5261.7	3	3	$1.5 \cdot 10^{-1}$
5262.2	3	1	$6.0 \cdot 10^{-1}$
5264.2	5	5	$9.1 \cdot 10^{-2}$
5265.6	5	3	$4.4 \cdot 10^{-1}$
5270.3	7	5	$5.0 \cdot 10^{-1}$
5582.0	5	7	$6.0 \cdot 10^{-2}$
5588.8	7	7	$4.9 \cdot 10^{-1}$
5590.1	3	5	$8.3 \cdot 10^{-2}$
5594.5	5	5	$3.8 \cdot 10^{-1}$
5598.5	3	3	$4.3 \cdot 10^{-1}$
5601.3	7	5	$8.6 \cdot 10^{-2}$
5602.9	5	3	$1.4 \cdot 10^{-1}$
5857.5	3	5	$6.6 \cdot 10^{-1}$
6102.7	1	3	$9.6 \cdot 10^{-2}$
6122.2	3	3	$2.87 \cdot 10^{-1}$
6162.2	5	3	$3.54 \cdot 10^{-1}$
6169.1	5	3	$1.7 \cdot 10^{-1}$
6169.6	7	5	$1.9 \cdot 10^{-1}$
6439.1	7	9	$5.3 \cdot 10^{-1}$
6449.8	3	5	$9.0 \cdot 10^{-2}$
6462.6	5	7	$4.7 \cdot 10^{-1}$
6471.7	7	7	$5.9 \cdot 10^{-2}$
6493.8	3	5	$4.4 \cdot 10^{-1}$
6499.7	5	5	$8.1 \cdot 10^{-2}$

Ca II

λ Å	Weights g_i	g_k	A $10^8\ s^{-1}$
1341.9	2	4	$1.5 \cdot 10^{-2}$
1342.5	2	2	$1.5 \cdot 10^{-2}$
1649.9	2	4	$3.2 \cdot 10^{-3}$
1652.0	2	2	$3.1 \cdot 10^{-3}$
1673.9	2	4	$2.24 \cdot 10^{-1}$
1680.1	4	6	$2.65 \cdot 10^{-1}$
1680.1	4	4	$4.41 \cdot 10^{-2}$
1807.3	2	4	$3.54 \cdot 10^{-1}$
1814.5	4	6	$4.2 \cdot 10^{-1}$
1814.7	4	4	$7.0 \cdot 10^{-2}$
1843.1	2	2	$1.6 \cdot 10^{-1}$
1850.7	4	2	$3.08 \cdot 10^{-1}$
2103.2	2	4	$8.2 \cdot 10^{-1}$
2112.8	4	6	$9.7 \cdot 10^{-1}$
2113.2	4	4	$1.6 \cdot 10^{-1}$
2197.8	2	2	$3.1 \cdot 10^{-1}$
2208.6	4	2	$6.2 \cdot 10^{-1}$
3158.9	2	4	3.1
3179.3	4	6	3.6
3181.3	4	4	$5.8 \cdot 10^{-1}$
3706.0	2	2	$8.8 \cdot 10^{-1}$
3736.9	4	2	1.7
3933.7	2	4	1.47
3968.5	2	2	1.4

Ca III

λ Å	Weights g_i	g_k	A $10^8\ s^{-1}$
357.97	1	3	$8.8 \cdot 10^2$
439.69	1	3	$1.9 \cdot 10^{-1}$
490.55	1	3	$1.6 \cdot 10^{-2}$

λ Å	Weights g_i	g_k	A 10^8 s^{-1}		λ Å	Weights g_i	g_k	A 10^8 s^{-1}		λ Å	Weights g_i	g_k	A 10^8 s^{-1}
Ca V					*Ca XVII*					1139.43	5	7	4.31·10^{-2}
558.60	5	3	2.2·10^1		19.558	1	3	3.8·10^4		1139.51	3	5	2.72·10^{-1}
637.93	5	3	3.9		21.198	3	5	4.9·10^4		1139.65	5	3	2.26·10^{-2}
643.12	3	1	9.1		192.82	1	3	1.21·10^2		1139.77	3	5	4.63·10^{-2}
646.57	5	5	6.9		218.82	3	5	2.76·10^1		1139.79	1	3	2.11·10^{-1}
647.88	3	3	2.3		223.02	1	3	3.44·10^1		1139.81	5	7	2.96·10^{-2}
651.55	1	3	2.9		228.72	3	3	2.37·10^1		1139.86	5	5	1.30·10^{-2}
656.76	3	5	2.1		232.83	5	5	6.5·10^1		1140.01	3	3	9.45·10^{-2}
					244.06	**5**	3	3.28·10^1		1140.01	1	3	6.10·10^{-2}
Ca VII										1140.12	5	5	1.84·10^{-1}
550.20	5	5	1.8·10^1		*Ca XVIII*					1140.32	3	1	1.15·10^{-1}
624.39	1	3	3.3		*18.71	2	6	2.31·10^4		1140.36	3	5	6.39·10^{-2}
630.54	3	5	4.5		*19.74	6	10	7.0·10^4		1140.57	5	3	4.17·10^{-2}
630.79	3	3	2.2		302.19	2	4	2.0·10^1		1140.64	5	7	6.29·10^{-2}
639.15	5	7	5.7		344.76	2	2	1.3·10^1		1155.81	1	3	8.21·10^{-2}
640.41	5	5	1.3							1155.98	3	1	3.74·10^{-1}
					Carbon					1156.03	3	3	1.14·10^{-1}
Ca VIII										1156.20	3	5	2.39·10^{-2}
182.71	2	2	1.6·10^2		*C I*					1156.39	5	3	1.51·10^{-1}
184.16	4	2	3.2·10^2		1121.52	1	3	1.74·10^{-2}		1156.56	5	5	2.69·10^{-1}
					1121.64	3	1	1.10·10^{-1}		1157.33	5	7	3.99·10^{-2}
Ca IX					1121.66	3	3	3.50·10^{-2}		1157.41	3	3	7.50·10^{-2}
163.23	5	3	3.76·10^2		1121.92	3	3	3.60·10^{-2}		1157.77	5	3	2.50·10^{-2}
371.89	1	3	8.8·10^1		1122.00	5	3	3.96·10^{-2}		1157.77	3	5	2.62·10^{-1}
373.81	3	5	1.16·10^2		1122.00	3	5	8.49·10^{-2}		1157.91	1	3	3.52·10^{-1}
378.08	5	7	1.5·10^2		1122.10	5	5	6.71·10^{-2}		1158.02	5	7	5.57·10^{-1}
395.03	3	5	2.2·10^2		1122.18	5	7	3.38·10^{-2}		1158.03	3	5	3.10·10^{-1}
466.24	1	3	1.12·10^2		1122.33	5	7	1.06·10^{-1}		1158.13	3	3	1.98·10^{-1}
498.01	3	5	2.49·10^1		1122.45	1	3	9.04·10^{-2}		1158.13	5	5	2.58·10^{-1}
506.18	5	5	7.2·10^1		1122.52	1	3	2.03·10^{-2}		1158.32	1	3	1.09·10^{-1}
515.57	5	3	3.75·10^1		1122.65	3	3	2.73·10^{-2}		1158.40	5	5	8.20·10^{-2}
					1122.79	3	5	7.25·10^{-2}		1158.54	3	3	3.36·10^{-1}
Ca X					1123.11	5	7	3.58·10^{-2}		1158.67	3	1	2.19·10^{-1}
110.96	2	4	2.9·10^2		1128.07	1	3	2.37·10^{-2}		1158.73	3	5	5.11·10^{-2}
111.20	2	2	2.92·10^2		1128.17	1	3	1.83·10^{-2}		1158.91	5	3	8.37·10^{-2}
151.84	2	2	2.3·10^2		1128.25	3	1	1.55·10^{-1}		1158.97	5	7	7.22·10^{-2}
153.02	4	2	4.5·10^2		1128.28	3	3	4.96·10^{-2}		1188.83	1	3	1.95·10^{-1}
206.57	4	4	2.9·10^1		1128.63	5	3	5.80·10^{-2}		1188.99	3	1	7.18·10^{-1}
206.75	6	4	2.6·10^2		1128.69	3	3	4.79·10^{-2}		1189.07	3	3	2.01·10^{-1}
207.39	4	2	2.8·10^2		1128.75	5	5	1.02·10^{-1}		1189.25	3	5	1.09·10^{-1}
411.70	2	4	8.3·10^1		1128.82	3	5	1.61·10^{-1}		1189.45	5	3	2.96·10^{-1}
419.75	4	6	9.5·10^1		1128.90	5	7	3.98·10^{-2}		1189.63	5	5	5.28·10^{-1}
420.47	4	4	1.6·10^1		1129.03	5	3	1.82·10^{-2}		1191.84	5	7	2.91·10^{-2}
557.76	2	4	3.50·10^1		1129.13	5	7	1.70·10^{-1}		1192.22	1	3	2.15·10^{-2}
574.01	2	2	3.2·10^1		1129.20	1	3	1.35·10^{-1}		1192.83	5	3	2.19·10^{-2}
					1129.32	1	3	3.46·10^{-2}		1193.01	3	5	7.91·10^{-1}
Ca XI					1129.40	3	3	4.86·10^{-2}		1193.03	1	3	6.39·10^{-1}
30.448	1	3	6.2·10^3		1129.42	5	5	1.19·10^{-1}		1193.24	5	7	1.11
30.867	1	3	4.9·10^4		1129.59	3	1	6.41·10^{-2}		1193.26	3	3	4.29·10^{-1}
35.212	1	3	2.0·10^3		1129.62	3	5	7.29·10^{-2}		1193.39	5	5	3.46·10^{-1}
					1129.87	5	3	2.22·10^{-2}		1193.65	5	3	2.25·10^{-2}
Ca XII					1129.92	5	7	4.84·10^{-2}		1193.68	3	5	2.54·10^{-1}
140.05	4	2	3.7·10^2		1138.38	1	3	4.13·10^{-2}		1194.00	1	3	1.94·10^{-1}
147.27	2	2	1.6·10^2		1138.56	3	1	2.31·10^{-1}		1194.06	5	5	2.98·10^{-1}
					1138.56	1	3	1.70·10^{-2}		1194.23	3	3	8.34·10^{-2}
Ca XV					1138.60	3	3	7.31·10^{-2}		1194.30	3	5	4.10·10^{-2}
141.69	5	3	4.08·10^2		1138.95	5	3	8.98·10^{-2}		1194.41	3	1	4.40·10^{-1}
*142.23	9	3	6.3·10^2		1139.09	5	5	1.60·10^{-1}		1194.49	5	7	6.77·10^{-2}
161.00	5	5	1.9·10^2		1139.30	3	3	6.24·10^{-2}		1194.61	5	3	1.77·10^{-1}
										1260.74	1	3	5.70·10^{-1}

λ Å	g_i	g_k	A 10^8 s^{-1}	λ Å	g_i	g_k	A 10^8 s^{-1}	λ Å	g_i	g_k	A 10^8 s^{-1}
1260.93	3	1	1.81	806.861	6	4	3.79	6582.88	2	2	$3.66 \cdot 10^{-1}$
1261.00	3	3	$4.68 \cdot 10^{-1}$	858.092	2	2	1.49	7231.33	2	4	$3.49 \cdot 10^{-1}$
1261.12	3	5	$4.00 \cdot 10^{-1}$	858.559	4	2	2.93	7236.42	4	6	$4.18 \cdot 10^{-1}$
1261.43	5	3	$7.51 \cdot 10^{-1}$	903.623	2	4	6.78	7237.17	4	4	$6.96 \cdot 10^{-2}$
1261.55	5	5	1.34	903.962	2	2	$2.70 \cdot 10^{1}$				
1274.11	5	7	$1.20 \cdot 10^{-2}$	904.142	4	4	$3.39 \cdot 10^{1}$	**C III**			
1276.48	1	3	$5.04 \cdot 10^{-2}$	904.480	4	2	$1.36 \cdot 10^{1}$	310.170	1	3	6.56
1276.75	3	3	$9.38 \cdot 10^{-2}$	1009.86	2	4	5.65	386.203	1	3	$3.46 \cdot 10^{1}$
1277.19	5	3	$1.81 \cdot 10^{-2}$	1010.08	4	4	$1.13 \cdot 10^{1}$	459.466	1	3	$5.91 \cdot 10^{1}$
1277.25	1	3	1.26	1010.37	6	4	$1.69 \cdot 10^{1}$	459.514	3	5	$7.97 \cdot 10^{1}$
1277.28	3	5	1.70	1036.34	2	2	7.38	459.627	5	7	$1.06 \cdot 10^{2}$
1277.51	3	3	$9.24 \cdot 10^{-1}$	1037.02	4	2	$1.46 \cdot 10^{1}$	574.281	3	5	$6.24 \cdot 10^{1}$
1277.55	5	7	2.28	1065.89	6	4	$1.30 \cdot 10^{1}$	977.020	1	3	$1.767 \cdot 10^{1}$
1277.72	5	5	$6.30 \cdot 10^{-1}$	1065.92	4	4	1.44	1174.93	3	5	3.293
1277.95	5	3	$5.94 \cdot 10^{-2}$	1066.13	4	2	$1.45 \cdot 10^{1}$	1175.26	1	3	4.385
1279.89	3	5	$3.43 \cdot 10^{-1}$	1126.99	2	4	$1.01 \cdot 10^{-6}$	1175.59	3	3	3.287
1280.14	1	3	$3.55 \cdot 10^{-1}$	1127.13	2	2	$5.55 \cdot 10^{-7}$	1175.71	5	5	9.856
1280.33	5	5	$6.38 \cdot 10^{-1}$	1127.27	4	4	$1.91 \cdot 10^{-7}$	1175.99	3	1	$1.313 \cdot 10^{1}$
1280.40	3	3	$1.87 \cdot 10^{-1}$	1127.41	4	2	$9.06 \cdot 10^{-8}$	1176.37	5	3	5.468
1280.60	3	1	$8.82 \cdot 10^{-1}$	1127.63	6	4	$1.78 \cdot 10^{-6}$	1247.38	3	1	$2.082 \cdot 10^{1}$
1280.85	5	3	$3.62 \cdot 10^{-1}$	1323.86	6	4	$4.94 \cdot 10^{-1}$	2296.87	3	5	1.376
1328.83	1	3	$7.31 \cdot 10^{-1}$	1323.91	4	4	4.38	2849.05	3	1	$1.95 \cdot 10^{-1}$
1329.09	3	1	2.22	1323.95	6	6	4.55	3703.70	3	3	$5.90 \cdot 10^{-1}$
1329.10	3	5	$5.39 \cdot 10^{-1}$	1324.00	4	6	$3.27 \cdot 10^{-1}$	4325.56	3	5	$1.24 \cdot 10^{-1}$
1329.12	3	3	$5.55 \cdot 10^{-1}$	1334.53	2	4	2.41	4647.42	3	5	$7.26 \cdot 10^{-1}$
1329.58	5	5	1.64	1335.66	4	4	$4.76 \cdot 10^{-1}$	4650.25	3	3	$7.25 \cdot 10^{-1}$
1329.60	5	3	$9.19 \cdot 10^{-1}$	1335.71	4	6	2.88	4651.02	3	5	$2.28 \cdot 10^{-1}$
1459.03	5	3	$5.45 \cdot 10^{-1}$	1384.00	2	4	$4.07 \cdot 10^{-1}$	4651.47	3	1	$7.24 \cdot 10^{-1}$
1463.34	5	7	1.79	1384.36	2	2	$3.88 \cdot 10^{-1}$	4652.05	1	3	$3.04 \cdot 10^{-1}$
1467.40	5	3	$5.49 \cdot 10^{-1}$	1490.38	6	4	$2.02 \cdot 10^{-6}$	4659.06	3	3	$2.27 \cdot 10^{-1}$
1470.09	5	7	$1.41 \cdot 10^{-2}$	1720.46	2	4	$8.65 \cdot 10^{-1}$	4663.64	3	1	$9.05 \cdot 10^{-1}$
1472.23	5	3	$7.79 \cdot 10^{-3}$	1721.01	2	2	3.46	4665.86	5	5	$6.78 \cdot 10^{-1}$
1481.76	5	5	$3.45 \cdot 10^{-1}$	1721.68	4	4	4.33	4673.95	5	3	$3.75 \cdot 10^{-1}$
1560.31	1	3	$6.54 \cdot 10^{-1}$	1722.24	4	2	1.72	5244.66	1	3	$5.30 \cdot 10^{-2}$
1560.68	3	5	$8.82 \cdot 10^{-1}$	1760.40	6	4	$3.68 \cdot 10^{-1}$	5253.58	3	3	$1.58 \cdot 10^{-1}$
1560.71	3	3	$4.89 \cdot 10^{-1}$	1760.47	4	4	$4.08 \cdot 10^{-2}$	5272.52	5	3	$2.61 \cdot 10^{-1}$
1561.34	5	5	$2.93 \cdot 10^{-1}$	1760.82	4	2	$4.09 \cdot 10^{-1}$	5695.92	3	5	$4.27 \cdot 10^{-1}$
1561.37	5	3	$3.25 \cdot 10^{-2}$	1915.32	2	4	$9.36 \cdot 10^{-2}$	5858.34	3	1	$1.34 \cdot 10^{-1}$
1561.44	5	7	1.17	1916.01	2	2	$9.50 \cdot 10^{-2}$	5863.25	3	3	$3.35 \cdot 10^{-2}$
1656.27	3	5	$8.72 \cdot 10^{-1}$	2323.50	2	4	$1.40 \cdot 10^{-8}$	5871.68	5	3	$1.00 \cdot 10^{-1}$
1656.93	1	3	1.16	2324.69	2	2	$5.99 \cdot 10^{-7}$	5880.56	5	5	$1.99 \cdot 10^{-2}$
1657.01	5	5	2.61	2325.40	4	6	$4.43 \cdot 10^{-7}$	5894.07	7	5	$1.11 \cdot 10^{-1}$
1657.38	3	3	$8.66 \cdot 10^{-1}$	2326.93	4	4	$8.49 \cdot 10^{-8}$	6727.48	1	3	$1.12 \cdot 10^{-1}$
1657.91	3	1	3.47	2328.12	4	2	$6.78 \cdot 10^{-7}$	6731.04	3	5	$1.50 \cdot 10^{-1}$
1658.12	5	3	1.44	2509.13	2	4	$4.71 \cdot 10^{-1}$	6742.15	3	3	$8.32 \cdot 10^{-2}$
1751.83	1	3	$8.38 \cdot 10^{-1}$	2511.74	4	4	$9.27 \cdot 10^{-2}$	6744.39	5	7	$1.99 \cdot 10^{-1}$
1763.91	1	3	$2.44 \cdot 10^{-2}$	2512.06	4	6	$5.61 \cdot 10^{-1}$	6762.17	5	5	$4.95 \cdot 10^{-2}$
1930.90	5	3	3.39	2836.71	2	4	$3.30 \cdot 10^{-1}$	6773.39	5	3	$5.47 \cdot 10^{-3}$
2478.56	1	3	$2.80 \cdot 10^{-1}$	2837.60	2	2	$3.29 \cdot 10^{-1}$	6851.18	3	5	$7.60 \cdot 10^{-3}$
				3183.50	2	4	$2.40 \cdot 10^{-7}$	6853.68	5	7	$5.64 \cdot 10^{-3}$
C II				3187.70	4	4	$8.09 \cdot 10^{-7}$	6857.24	3	3	$3.79 \cdot 10^{-1}$
687.053	2	4	$2.35 \cdot 10^{1}$	3922.08	2	4	$1.44 \cdot 10^{-7}$	6862.69	5	5	$3.51 \cdot 10^{-2}$
687.345	4	6	$2.82 \cdot 10^{1}$	4309.31	4	4	$3.92 \cdot 10^{-3}$	6868.78	5	3	$1.26 \cdot 10^{-2}$
687.352	4	4	4.70	4309.58	6	4	$3.52 \cdot 10^{-2}$	6872.04	7	7	$4.46 \cdot 10^{-2}$
806.384	4	6	2.53	4312.80	4	2	$3.92 \cdot 10^{-2}$	6881.10	7	5	$7.80 \cdot 10^{-3}$
806.533	2	4	3.51	4735.46	2	4	$9.05 \cdot 10^{-5}$	7353.88	5	3	$3.09 \cdot 10^{-2}$
806.568	6	6	5.89	4737.97	2	2	$5.47 \cdot 10^{-4}$	7707.43	3	5	$1.30 \cdot 10^{-1}$
806.677	4	4	1.12	4744.77	4	4	$5.98 \cdot 10^{-4}$	7771.76	3	1	$1.77 \cdot 10^{-1}$
806.687	2	2	1.40	4747.28	4	2	$2.33 \cdot 10^{-4}$	7780.41	3	3	$1.76 \cdot 10^{-1}$
806.830	4	2	7.01	6578.05	2	4	$3.67 \cdot 10^{-1}$	7796.00	3	5	$1.75 \cdot 10^{-1}$

λ Å	Weights g_i	g_k	A 10^8 s^{-1}
8500.32	1	3	$1.01 \cdot 10^{-1}$
9593.32	3	3	$5.32 \cdot 10^{-3}$
9651.47	5	5	$1.57 \cdot 10^{-2}$
9696.48	5	7	$7.53 \cdot 10^{-3}$
9696.54	3	5	$7.12 \cdot 10^{-3}$
9699.57	7	9	$8.47 \cdot 10^{-3}$
9701.10	1	3	$4.40 \cdot 10^{-2}$
9705.41	3	5	$5.93 \cdot 10^{-2}$
9706.44	3	3	$3.29 \cdot 10^{-2}$
9715.09	5	7	$7.88 \cdot 10^{-2}$
9717.75	5	5	$1.97 \cdot 10^{-2}$
9718.79	5	3	$2.19 \cdot 10^{-3}$

C IV

λ Å	Weights g_i	g_k	A 10^8 s^{-1}
*312.43	2	6	$4.63 \cdot 10^{1}$
*384.13	6	10	$1.76 \cdot 10^{2}$
1548.19	2	4	2.65
1550.77	2	2	2.64
5801.31	2	4	$3.17 \cdot 10^{-1}$
5811.97	2	2	$3.16 \cdot 10^{-1}$

C V

λ Å	Weights g_i	g_k	A 10^8 s^{-1}
34.9728	1	3	$2.554 \cdot 10^{3}$
40.2678	1	3	$8.873 \cdot 10^{3}$
*227.19	3	9	$1.363 \cdot 10^{2}$
247.315	1	3	$1.278 \cdot 10^{2}$
*248.71	9	15	$4.247 \cdot 10^{2}$
*260.19	9	3	$6.680 \cdot 10^{1}$
267.267	3	5	$3.947 \cdot 10^{2}$
*2273.9	3	9	$5.646 \cdot 10^{-1}$
3526.66	1	3	$1.663 \cdot 10^{-1}$
8420.72	3	5	$6.898 \cdot 10^{-2}$
*8433.2	3	9	$6.868 \cdot 10^{-2}$
8448.12	3	1	$6.832 \cdot 10^{-2}$
8449.19	3	3	$6.829 \cdot 10^{-2}$

Cesium

Cs I

λ Å	Weights g_i	g_k	A 10^8 s^{-1}
3203.5	2	4	$7.6 \cdot 10^{-6}$
3205.3	2	4	$7.9 \cdot 10^{-6}$
3207.5	2	4	$8.5 \cdot 10^{-6}$
3210.0	2	4	$9.4 \cdot 10^{-6}$
3212.8	2	4	$1.19 \cdot 10^{-5}$
3216.2	2	4	$1.49 \cdot 10^{-5}$
3220.1	2	4	$1.7 \cdot 10^{-5}$
3220.2	2	2	$1.07 \cdot 10^{-7}$
3224.8	2	4	$2.0 \cdot 10^{-5}$
3225.0	2	2	$1.43 \cdot 10^{-7}$
3230.5	2	4	$2.5 \cdot 10^{-5}$
3230.7	2	2	$1.97 \cdot 10^{-7}$
3237.4	2	4	$2.8 \cdot 10^{-5}$
3237.6	2	2	$2.63 \cdot 10^{-7}$
3245.9	2	4	$3.45 \cdot 10^{-5}$
3246.2	2	2	$3.7 \cdot 10^{-7}$
3256.7	2	4	$4.25 \cdot 10^{-5}$
3257.1	2	2	$7.0 \cdot 10^{-7}$
3270.5	2	4	$5.6 \cdot 10^{-5}$
3271.0	2	2	$9.8 \cdot 10^{-7}$
3288.6	2	4	$1.0 \cdot 10^{-4}$
3289.3	2	2	$2.7 \cdot 10^{-6}$
3313.1	2	4	$1.6 \cdot 10^{-4}$
3314.0	2	2	$5.2 \cdot 10^{-6}$
3347.5	2	4	$2.2 \cdot 10^{-4}$
3348.8	2	2	$1.1 \cdot 10^{-5}$
3397.9	2	4	$4.0 \cdot 10^{-4}$
3400.0	2	2	$2.4 \cdot 10^{-5}$
3476.8	2	4	$6.6 \cdot 10^{-4}$
3480.0	2	2	$6.6 \cdot 10^{-5}$
3611.4	2	4	$1.5 \cdot 10^{-3}$
3617.3	2	2	$2.5 \cdot 10^{-4}$
3876.1	2	4	$3.8 \cdot 10^{-3}$
3888.6	2	2	$9.7 \cdot 10^{-4}$
4555.3	2	4	$1.88 \cdot 10^{-2}$
4593.2	2	2	$8.0 \cdot 10^{-3}$
8521.1	2	4	$3.276 \cdot 10^{-1}$
8943.5	2	2	$2.87 \cdot 10^{-1}$

Chlorine

Cl I

λ Å	Weights g_i	g_k	A 10^8 s^{-1}
1188.8	4	6	2.33
1188.8	4	4	$2.71 \cdot 10^{-1}$
1201.4	2	4	2.39
1335.7	4	2	1.74
1347.2	4	4	4.19
1351.7	2	2	3.23
1363.4	2	4	$7.5 \cdot 10^{-1}$
4323.3	4	4	$1.1 \cdot 10^{-2}$
4363.3	4	6	$6.8 \cdot 10^{-3}$
4379.9	4	4	$1.4 \cdot 10^{-2}$
4389.8	6	8	$1.4 \cdot 10^{-2}$
4526.2	4	4	$5.1 \cdot 10^{-2}$
4601.0	2	2	$4.2 \cdot 10^{-2}$
4661.2	2	4	$1.2 \cdot 10^{-2}$
7256.6	6	4	$1.5 \cdot 10^{-1}$
7414.1	6	4	$4.7 \cdot 10^{-2}$
7547.1	4	4	$1.2 \cdot 10^{-1}$
7717.6	4	4	$3.0 \cdot 10^{-2}$
7745.0	2	4	$6.3 \cdot 10^{-2}$
7769.2	6	6	$6.0 \cdot 10^{-2}$
7821.4	6	8	$9.8 \cdot 10^{-2}$
7830.8	4	4	$9.7 \cdot 10^{-2}$
7878.2	6	6	$1.8 \cdot 10^{-2}$
7899.3	4	6	$5.1 \cdot 10^{-2}$
7924.6	2	4	$2.1 \cdot 10^{-2}$
7935.0	6	8	$3.9 \cdot 10^{-2}$
7997.9	4	4	$2.1 \cdot 10^{-2}$

Cl II

λ Å	Weights g_i	g_k	A 10^8 s^{-1}
3329.1	5	7	1.5
3522.1	7	7	1.4
3798.8	5	7	1.6
3805.2	7	9	1.8
3809.5	3	5	1.5
3851.0	5	7	1.8
3851.4	5	5	1.6
3854.7	3	5	2.2
3861.9	5	7	2.4
3868.6	7	9	2.7
3913.9	9	9	$8.2 \cdot 10^{-1}$
3990.2	5	7	$8.4 \cdot 10^{-1}$
4132.5	5	5	1.6
4276.5	9	7	$7.6 \cdot 10^{-1}$
4768.7	3	5	$7.7 \cdot 10^{-1}$
4781.3	5	7	1.0
4794.6	5	7	1.04
4810.1	5	5	$9.9 \cdot 10^{-1}$
4819.5	5	3	1.00
4904.8	5	7	$8.1 \cdot 10^{-1}$
4917.7	3	5	$7.5 \cdot 10^{-1}$
5078.3	7	7	$7.7 \cdot 10^{-1}$
5219.1	3	9	$8.6 \cdot 10^{-1}$
5392.1	5	7	1.0

Cl III

λ Å	Weights g_i	g_k	A 10^8 s^{-1}
2298.5	4	4	4.2
2340.6	6	6	4.2
2370.4	8	6	2.8
2531.8	2	4	4.4
2532.5	4	6	5.3
2577.1	4	6	4.3
2580.7	6	8	4.7
2601.2	2	4	4.6
2603.6	4	6	5.0
2609.5	6	8	5.7
2617.0	8	10	6.6
2661.6	4	6	3.4
2665.5	6	8	4.8
2691.5	4	4	3.5
2710.4	4	6	3.5
3340.4	6	6	1.5
3392.9	4	4	1.9
3393.5	6	6	1.9
3530.0	6	8	1.8
3560.7	4	6	1.7
3602.1	6	8	1.7
3612.9	4	6	1.2
3720.5	4	6	1.7

Chromium

Cr I

λ Å	Weights g_i	g_k	A 10^8 s^{-1}
1999.95	9	9	1.4
2726.50	5	7	$7.5 \cdot 10^{-1}$
2769.90	7	5	1.1
2780.70	9	7	1.4
2889.22	9	9	$6.6 \cdot 10^{-1}$
2893.25	7	7	$5.2 \cdot 10^{-1}$
2967.64	7	9	$3.9 \cdot 10^{-1}$
2971.10	5	7	$7.1 \cdot 10^{-1}$
2975.48	3	5	$8.9 \cdot 10^{-1}$
2988.64	5	7	$5.2 \cdot 10^{-1}$
2996.57	5	3	2.0
3000.88	7	5	1.6
3005.06	9	7	$9.2 \cdot 10^{-1}$
3013.72	3	5	$8.3 \cdot 10^{-1}$
3015.20	1	3	1.63
3020.67	3	3	1.5
3021.58	9	11	2.91

λ Å	Weights g_i	g_k	A 10^8 s^{-1}
3024.36	5	5	1.27
3030.25	7	7	1.1
3037.05	9	9	$5.4 \cdot 10^{-1}$
3040.84	7	5	$7.4 \cdot 10^{-1}$
3053.87	9	7	$7.97 \cdot 10^{-1}$
3148.44	9	11	$5.6 \cdot 10^{-1}$
3155.16	11	13	$5.7 \cdot 10^{-1}$
3163.76	13	15	$6.0 \cdot 10^{-1}$
3237.73	9	9	1.3
3578.68	7	9	1.48
3593.48	7	7	1.50
3605.32	7	5	1.62
3639.80	13	11	1.8
3743.89	13	13	$7.61 \cdot 10^{-1}$
3757.66	7	7	$4.13 \cdot 10^{-1}$
3768.24	5	5	$5.10 \cdot 10^{-1}$
3804.80	9	9	$6.9 \cdot 10^{-1}$
3963.69	13	15	1.3
3969.75	11	13	1.2
3983.90	7	9	1.05
3991.12	5	7	1.07
4001.44	9	11	$6.8 \cdot 10^{-1}$
4039.10	15	15	$6.7 \cdot 10^{-1}$
4048.78	13	13	$6.4 \cdot 10^{-1}$
4058.78	11	11	$6.7 \cdot 10^{-1}$
4065.71	9	11	$3.5 \cdot 10^{-1}$
4165.52	11	13	$7.5 \cdot 10^{-1}$
4204.48	13	11	$3.1 \cdot 10^{-1}$
4254.33	7	9	$3.15 \cdot 10^{-1}$
4263.15	15	17	$6.4 \cdot 10^{-1}$
4274.81	7	7	$3.07 \cdot 10^{-1}$
4275.98	11	11	$2.2 \cdot 10^{-1}$
4280.42	13	15	$4.7 \cdot 10^{-1}$
4289.73	7	5	$3.16 \cdot 10^{-1}$
4291.97	7	5	$2.4 \cdot 10^{-1}$
4297.75	11	13	$4.9 \cdot 10^{-1}$
4298.05	9	9	$2.6 \cdot 10^{-1}$
4300.52	9	7	$1.9 \cdot 10^{-1}$
4301.19	11	9	$2.6 \cdot 10^{-1}$
4302.78	11	11	$2.5 \cdot 10^{-1}$
4337.25	5	7	$2.0 \cdot 10^{-1}$
4373.65	9	9	$2.8 \cdot 10^{-1}$
4376.80	13	13	$3.2 \cdot 10^{-1}$
4413.86	7	5	$2.7 \cdot 10^{-1}$
4422.70	5	5	$2.7 \cdot 10^{-1}$
4424.29	9	7	$2.1 \cdot 10^{-1}$
4432.77	15	15	$4.9 \cdot 10^{-1}$
4490.55	9	7	$3.9 \cdot 10^{-1}$
4492.31	5	3	$4.47 \cdot 10^{-1}$
4495.28	9	7	$2.0 \cdot 10^{-1}$
4500.29	7	7	$2.1 \cdot 10^{-1}$
4506.84	13	11	$2.7 \cdot 10^{-1}$
4540.72	11	11	$3.14 \cdot 10^{-1}$
4564.17	11	13	$5.1 \cdot 10^{-1}$
4595.60	13	13	$4.7 \cdot 10^{-1}$
4622.47	7	7	$4.1 \cdot 10^{-1}$
4665.90	3	3	$3.0 \cdot 10^{-1}$
4689.38	7	5	$2.3 \cdot 10^{-1}$
4698.46	9	7	$2.2 \cdot 10^{-1}$
4708.02	11	9	$4.31 \cdot 10^{-1}$

λ Å	Weights g_i	g_k	A 10^8 s^{-1}
4718.43	13	11	$3.4 \cdot 10^{-1}$
4730.69	7	5	$3.83 \cdot 10^{-1}$
4737.33	9	7	$3.38 \cdot 10^{-1}$
4741.09	3	5	$2.2 \cdot 10^{-1}$
4752.07	13	13	$6.2 \cdot 10^{-1}$
4756.09	11	9	$4.0 \cdot 10^{-1}$
4792.49	7	5	$2.6 \cdot 10^{-1}$
4801.02	9	7	$3.06 \cdot 10^{-1}$
4816.13	9	9	$1.8 \cdot 10^{-1}$
4870.79	7	9	$3.5 \cdot 10^{-1}$
4887.01	9	11	$3.2 \cdot 10^{-1}$
4922.28	11	13	$4.0 \cdot 10^{-1}$
5204.51	5	3	$5.09 \cdot 10^{-1}$
5206.02	5	5	$5.14 \cdot 10^{-1}$
5208.42	5	7	$5.06 \cdot 10^{-1}$
5243.38	5	3	$2.19 \cdot 10^{-1}$
5297.37	7	9	$3.88 \cdot 10^{-1}$
5297.99	7	7	$3.0 \cdot 10^{-1}$
5328.36	9	11	$6.2 \cdot 10^{-1}$
5329.17	9	9	$2.25 \cdot 10^{-1}$
5783.11	3	3	$2.1 \cdot 10^{-1}$
5783.89	5	5	$2.02 \cdot 10^{-1}$
5787.97	5	7	$2.35 \cdot 10^{-1}$

Cr II

λ Å	Weights g_i	g_k	A 10^8 s^{-1}
2653.57	4	6	$3.5 \cdot 10^{-1}$
2658.59	2	4	$5.8 \cdot 10^{-1}$
2666.02	6	8	$5.9 \cdot 10^{-1}$
2668.71	4	2	1.4
2671.80	6	4	1.0
2672.83	8	6	$5.5 \cdot 10^{-1}$
2744.97	4	6	$8.5 \cdot 10^{-1}$
2787.61	6	6	1.5
2822.38	14	16	2.3
2835.63	10	12	2.0
2840.01	10	12	2.7
2843.24	8	10	$6.4 \cdot 10^{-1}$
2849.83	6	8	$9.2 \cdot 10^{-1}$
2851.35	8	10	2.2
2856.77	4	6	$4.3 \cdot 10^{-1}$
2857.40	6	8	$2.8 \cdot 10^{-1}$
2860.92	2	4	$6.9 \cdot 10^{-1}$
2862.57	8	8	$6.3 \cdot 10^{-1}$
2866.72	4	4	1.2
2867.09	4	4	1.1
2867.65	2	2	1.1
2870.43	6	6	1.3
2873.81	4	2	$8.8 \cdot 10^{-1}$
2880.86	6	4	$7.9 \cdot 10^{-1}$
2898.53	10	12	1.2
2921.81	8	10	$9.0 \cdot 10^{-1}$
2930.83	2	4	1.1
2935.12	6	8	1.8
2953.34	2	2	1.8
2966.03	10	8	$5.4 \cdot 10^{-1}$
2971.90	14	14	2.0
2979.73	12	12	1.8
2985.32	10	10	2.2
2989.18	8	8	2.2
3118.64	2	4	1.7

λ Å	Weights g_i	g_k	A 10^8 s^{-1}
3120.36	4	6	1.5
3122.59	12	12	$4.4 \cdot 10^{-1}$
3128.69	4	4	$8.1 \cdot 10^{-1}$
3136.68	6	6	$6.4 \cdot 10^{-1}$
4588.22	8	6	$1.2 \cdot 10^{-1}$

Cobalt

Co I

λ Å	Weights g_i	g_k	A 10^8 s^{-1}
2407.25	10	12	3.6
2414.46	6	8	3.4
2415.29	4	6	3.6
2424.93	10	10	3.2
2432.21	8	8	2.6
2436.66	6	6	2.6
2439.04	4	4	2.7
2511.02	10	10	$9.2 \cdot 10^{-1}$
2521.36	10	8	3.0
2528.97	8	6	2.8
2535.96	6	4	1.9
3405.12	10	10	1.0
3409.17	8	8	$4.2 \cdot 10^{-1}$
3412.34	8	10	$6.1 \cdot 10^{-1}$
3433.05	4	4	1.0
3443.64	8	8	$6.9 \cdot 10^{-1}$
3449.17	6	6	$7.6 \cdot 10^{-1}$
3453.51	10	12	1.1
3462.80	4	6	$7.9 \cdot 10^{-1}$
3474.02	6	8	$5.6 \cdot 10^{-1}$
3489.40	8	6	1.3
3495.68	4	6	$4.9 \cdot 10^{-1}$
3502.28	10	8	$8.0 \cdot 10^{-1}$
3506.32	8	6	$8.2 \cdot 10^{-1}$
3512.64	6	4	1.0
3518.34	6	4	1.6
3529.82	8	10	$4.6 \cdot 10^{-1}$
3569.37	8	8	1.6
3587.19	6	6	1.4
3845.47	8	10	$4.6 \cdot 10^{-1}$
3894.07	6	8	$6.9 \cdot 10^{-1}$
3995.31	8	10	$2.5 \cdot 10^{-1}$
4121.32	8	10	$1.9 \cdot 10^{-1}$
5146.75	8	8	$1.5 \cdot 10^{-1}$
5212.70	10	10	$1.9 \cdot 10^{-1}$
5280.63	10	8	$2.8 \cdot 10^{-1}$
5352.05	12	10	$2.7 \cdot 10^{-1}$
6082.43	10	10	$5.4 \cdot 10^{-2}$
6455.00	8	10	$9.0 \cdot 10^{-2}$
7838.12	8	10	$5.4 \cdot 10^{-2}$
8093.93	12	10	$2.0 \cdot 10^{-1}$
8372.79	10	10	$8.7 \cdot 10^{-2}$

Co II

λ Å	Weights g_i	g_k	A 10^8 s^{-1}
2286.15	11	13	3.3
2307.85	9	11	2.6
2311.61	7	9	2.8
2314.05	5	7	2.8
2314.97	3	5	2.7
2330.36	5	3	1.32
2344.28	3	3	1.5

λ Å	Weights g_i	g_k	A $10^8\,s^{-1}$
2353.41	7	7	1.9
2363.80	9	9	2.1
2378.62	11	9	1.9
2383.45	9	7	1.8
2388.92	11	11	2.8
2389.54	5	3	1.5
2404.17	3	3	1.5
2417.66	9	9	$8.5\cdot10^{-1}$

Copper

Cu I

λ Å	g_i	g_k	A $10^8\,s^{-1}$
*2024.3	2	6	$9.8\cdot10^{-2}$
2165.1	2	4	$5.1\cdot10^{-1}$
2178.9	2	4	$9.13\cdot10^{-1}$
2181.7	2	2	1.0
2225.7	2	2	$4.6\cdot10^{-1}$
2244.3	2	4	$1.19\cdot10^{-2}$
2441.6	2	2	$2.0\cdot10^{-2}$
2492.2	2	4	$3.11\cdot10^{-2}$
2618.4	6	4	$3.07\cdot10^{-1}$
2766.4	4	4	$9.6\cdot10^{-2}$
2824.4	6	6	$7.8\cdot10^{-2}$
2961.2	6	8	$3.76\cdot10^{-2}$
3063.4	4	4	$1.55\cdot10^{-2}$
3194.1	4	4	$1.55\cdot10^{-2}$
3247.5	2	4	1.39
3274.0	2	2	1.37
3337.8	6	8	$3.8\cdot10^{-3}$
4022.6	2	4	$1.90\cdot10^{-1}$
4062.6	4	6	$2.10\cdot10^{-1}$
4249.0	2	2	$1.95\cdot10^{-1}$
4275.1	6	8	$3.45\cdot10^{-1}$
4480.4	2	2	$3.0\cdot10^{-2}$
4509.4	4	2	$2.75\cdot10^{-1}$
4530.8	4	2	$8.4\cdot10^{-2}$
4539.7	6	4	$2.12\cdot10^{-1}$
4587.0	8	6	$3.20\cdot10^{-1}$
4651.1	10	8	$3.80\cdot10^{-1}$
4704.6	8	8	$5.5\cdot10^{-2}$
5105.5	6	4	$2.0\cdot10^{-2}$
5153.2	2	4	$6.0\cdot10^{-1}$
5218.2	4	6	$7.5\cdot10^{-1}$
5220.1	4	4	$1.50\cdot10^{-1}$
5292.5	8	8	$1.09\cdot10^{-1}$
5700.2	4	4	$2.4\cdot10^{-3}$
5782.1	4	2	$1.65\cdot10^{-2}$

Cu II

λ Å	g_i	g_k	A $10^8\,s^{-1}$
2489.7	5	5	$1.5\cdot10^{-2}$
2544.8	9	7	1.1
2689.3	7	7	$4.1\cdot10^{-1}$
2701.0	5	5	$6.7\cdot10^{-1}$
2703.2	3	3	1.2
2713.5	5	5	$6.8\cdot10^{-1}$
2769.7	7	7	$6.1\cdot10^{-1}$

Dysprosium

Dy I

λ Å	g_i	g_k	A $10^8\,s^{-1}$
2862.7	17	15	$6.5\cdot10^{-2}$

λ Å	g_i	g_k	A $10^8\,s^{-1}$
2964.6	17	17	$6.5\cdot10^{-2}$
3147.7	15	17	$1.1\cdot10^{-1}$
3263.2	15	13	$1.4\cdot10^{-1}$
3511.0	15	13	$3.1\cdot10^{-1}$
3571.4	15	13	$2.0\cdot10^{-1}$
3757.1	17	19	3.0
3868.8	17	17	3.1
3967.5	17	19	$8.7\cdot10^{-1}$
4046.0	17	15	1.5
4103.9	13	11	1.7
4186.8	17	17	1.32
4194.8	17	17	$7.2\cdot10^{-1}$
4211.7	17	19	2.08
4218.1	15	15	1.85
4221.1	15	17	1.52
4225.2	13	15	4.5
4268.3	15	15	$3.6\cdot10^{-2}$
4276.7	13	13	$7.3\cdot10^{-1}$
4292.0	15	15	$5.8\cdot10^{-2}$
4577.8	17	19	$2.2\cdot10^{-2}$
4589.4	17	15	$1.3\cdot10^{-1}$
4612.3	17	15	$8.2\cdot10^{-2}$
5077.7	17	17	$5.7\cdot10^{-3}$
5301.6	17	15	$1.1\cdot10^{-2}$
5547.3	17	17	$2.7\cdot10^{-3}$
5639.5	17	19	$4.7\cdot10^{-3}$
5974.5	17	17	$4.0\cdot10^{-3}$
5988.6	17	15	$5.3\cdot10^{-3}$
6010.8	15	15	$2.6\cdot10^{-2}$
6088.3	15	13	$3.5\cdot10^{-2}$
6168.4	15	17	$2.5\cdot10^{-2}$
6259.1	17	19	$8.5\cdot10^{-3}$
6579.4	17	15	$7.5\cdot10^{-3}$

Erbium

Er I

λ Å	g_i	g_k	A $10^8\,s^{-1}$
3862.9	13	13	2.5
4008.0	13	15	2.6
4151.1	13	11	1.8

Europium

Eu I

λ Å	g_i	g_k	A $10^8\,s^{-1}$
2372.9	8	6	$1.9\cdot10^{-1}$
2375.3	8	8	$2.0\cdot10^{-1}$
2379.7	8	10	$2.0\cdot10^{-1}$
2710.0	8	10	$1.4\cdot10^{-1}$
2724.0	8	8	$1.2\cdot10^{-1}$
2892.5	8	8	$1.0\cdot10^{-1}$
2893.0	8	6	$1.0\cdot10^{-1}$
2909.0	8	10	$6.9\cdot10^{-2}$
3106.2	8	10	$5.5\cdot10^{-2}$
3111.4	8	10	$3.0\cdot10^{-1}$
3168.3	8	10	$6.9\cdot10^{-2}$
3210.6	8	8	$1.1\cdot10^{-1}$
3212.8	8	8	$2.9\cdot10^{-1}$
3213.8	8	6	$1.8\cdot10^{-1}$
3334.3	8	6	$3.4\cdot10^{-1}$
4594.0	8	10	1.4
4627.2	8	8	1.3

λ Å	g_i	g_k	A $10^8\,s^{-1}$
4661.9	8	6	1.3
5765.2	8	8	$1.1\cdot10^{-2}$
6018.2	8	10	$8.5\cdot10^{-3}$
6864.5	8	10	$5.8\cdot10^{-3}$

Fluorine

F I

λ Å	g_i	g_k	A $10^8\,s^{-1}$
806.96	4	6	3.3
809.60	2	4	2.8
951.87	4	2	2.6
954.83	4	4	5.77
955.55	2	2	5.1
958.52	2	4	1.3
6239.7	6	4	$2.5\cdot10^{-1}$
6348.5	4	4	$1.8\cdot10^{-1}$
6413.7	2	4	$1.1\cdot10^{-1}$
6708.3	6	4	$1.4\cdot10^{-2}$
6774.0	6	6	$1.0\cdot10^{-1}$
6795.5	4	2	$5.2\cdot10^{-2}$
6834.3	4	4	$2.1\cdot10^{-1}$
6856.0	6	8	$4.94\cdot10^{-1}$
6870.2	2	2	$3.8\cdot10^{-1}$
6902.5	4	6	$3.2\cdot10^{-1}$
6909.8	2	4	$2.2\cdot10^{-1}$
6966.4	4	2	$1.1\cdot10^{-1}$
7037.5	4	4	$3.0\cdot10^{-1}$
7127.9	2	2	$3.8\cdot10^{-1}$
7309.0	6	8	$4.7\cdot10^{-1}$
7311.0	4	2	$3.9\cdot10^{-1}$
7314.3	4	6	$4.8\cdot10^{-1}$
7332.0	6	4	$3.1\cdot10^{-1}$
7398.7	6	6	$2.85\cdot10^{-1}$
7425.7	4	2	$3.4\cdot10^{-1}$
7482.7	4	4	$5.6\cdot10^{-2}$
7489.2	2	2	$1.1\cdot10^{-1}$
7514.9	2	2	$5.2\cdot10^{-2}$
7552.2	4	6	$7.8\cdot10^{-2}$
7573.4	2	4	$1.0\cdot10^{-1}$
7607.2	4	4	$7.0\cdot10^{-2}$
7754.7	4	6	$3.82\cdot10^{-1}$
7800.2	2	4	$2.1\cdot10^{-1}$

Gallium

Ga I

λ Å	g_i	g_k	A $10^8\,s^{-1}$
2195.4	2	2	$1.9\cdot10^{-2}$
2199.7	4	2	$3.3\cdot10^{-2}$
2214.4	4	6	$1.2\cdot10^{-2}$
2235.9	4	2	$4.3\cdot10^{-2}$
2255.0	2	2	$3.1\cdot10^{-2}$
2259.2	4	6	$3.1\cdot10^{-2}$
2294.2	2	4	$7.0\cdot10^{-2}$
2297.9	4	2	$5.8\cdot10^{-2}$
2338.2	4	4	$9.8\cdot10^{-2}$
2371.3	2	2	$5.7\cdot10^{-2}$
2418.7	4	2	$1.0\cdot10^{-1}$
2450.1	2	4	$2.8\cdot10^{-1}$
2500.2	4	6	$3.4\cdot10^{-1}$
2659.9	2	2	$1.2\cdot10^{-1}$
2719.7	4	2	$2.3\cdot10^{-1}$

λ Å	Weights g_i	g_k	A 10^8 s^{-1}
2874.2	2	4	1.2
2943.6	4	6	1.4
2944.2	4	4	$2.7 \cdot 10^{-1}$
4033.0	2	2	$4.9 \cdot 10^{-1}$
4172.0	4	2	$9.2 \cdot 10^{-1}$

Ga II

λ Å	g_i	g_k	A 10^8 s^{-1}
829.60	1	3	$2.2 \cdot 10^{-1}$
1414.4	1	3	$1.88 \cdot 10^1$

Germanium

Ge I

λ Å	g_i	g_k	A 10^8 s^{-1}
1944.7	3	1	$7.0 \cdot 10^{-1}$
1955.1	3	3	$2.8 \cdot 10^{-1}$
1988.3	5	3	$2.5 \cdot 10^{-1}$
1998.9	5	5	$5.5 \cdot 10^{-1}$
2041.7	1	3	1.1
2065.2	3	3	$8.5 \cdot 10^{-1}$
2068.7	3	5	1.2
2086.0	3	5	$4.0 \cdot 10^{-1}$
2094.3	5	7	$9.7 \cdot 10^{-1}$
2105.8	5	5	$1.7 \cdot 10^{-1}$
2256.0	5	5	$3.2 \cdot 10^{-2}$
2417.4	5	5	$9.6 \cdot 10^{-1}$
2498.0	1	3	$1.3 \cdot 10^{-1}$
2533.2	3	3	$1.0 \cdot 10^{-1}$
2589.2	5	3	$5.1 \cdot 10^{-2}$
2592.5	3	5	$7.1 \cdot 10^{-1}$
2651.2	5	5	2.0
2651.6	1	3	$8.5 \cdot 10^{-1}$
2691.3	3	3	$6.1 \cdot 10^{-1}$
2709.6	3	1	2.8
2754.6	5	3	1.1
3039.1	5	3	2.8
3124.8	5	3	$3.1 \cdot 10^{-2}$
3269.5	5	3	$2.9 \cdot 10^{-1}$
4226.6	1	3	$2.1 \cdot 10^{-1}$
4685.8	1	3	$9.5 \cdot 10^{-2}$

Ge II

λ Å	g_i	g_k	A 10^8 s^{-1}
999.10	2	4	1.9
1016.6	4	6	2.1
1017.1	4	4	$3.5 \cdot 10^{-1}$
1055.0	2	2	$6.9 \cdot 10^{-1}$
1075.1	4	2	1.3
1237.1	2	4	$1.9 \cdot 10^1$
1261.9	4	6	$2.2 \cdot 10^1$
1264.7	4	4	3.5
1602.5	2	2	3.4
1649.2	4	2	6.5
4741.8	2	4	$4.6 \cdot 10^{-1}$
4814.6	4	6	$5.1 \cdot 10^{-1}$
4824.1	4	4	$8.6 \cdot 10^{-2}$
5131.8	4	6	1.9
5178.5	6	6	$1.3 \cdot 10^{-1}$
5178.6	6	8	2.0
5893.4	2	4	$9.2 \cdot 10^{-1}$
6021.0	2	2	$8.4 \cdot 10^{-1}$
6336.4	2	2	$4.4 \cdot 10^{-1}$

λ Å	g_i	g_k	A 10^8 s^{-1}
6484.2	4	2	$8.5 \cdot 10^{-1}$

Gold

Au I

λ Å	g_i	g_k	A 10^8 s^{-1}
2427.95	2	4	1.99
2675.95	2	2	1.64
3122.78	6	4	$1.90 \cdot 10^{-1}$
6278.30	4	2	$3.4 \cdot 10^{-2}$

Helium

He I

λ Å	g_i	g_k	A 10^8 s^{-1}
507.058	1	3	$1.5929 \cdot 10^{-1}$
507.718	1	3	$2.1826 \cdot 10^{-1}$
508.643	1	3	$3.1031 \cdot 10^{-1}$
509.998	1	3	$4.6224 \cdot 10^{-1}$
512.099	1	3	$7.3174 \cdot 10^{-1}$
515.617	1	3	1.2582
522.213	1	3	2.4356
537.030	1	3	5.6634
584.334	1	3	$1.7989 \cdot 10^1$
*2677.13	3	9	$4.4174 \cdot 10^{-3}$
*2696.12	3	9	$6.0234 \cdot 10^{-3}$
*2723.19	3	9	$8.4996 \cdot 10^{-3}$
*2763.80	3	9	$1.2508 \cdot 10^{-2}$
*2829.08	3	9	$1.9389 \cdot 10^{-2}$
*2945.10	3	9	$3.2006 \cdot 10^{-2}$
*3187.74	3	9	$5.6361 \cdot 10^{-2}$
3231.270	1	3	$5.1015 \cdot 10^{-3}$
3258.273	1	3	$6.9627 \cdot 10^{-3}$
3296.773	1	3	$9.8432 \cdot 10^{-3}$
3354.555	1	3	$1.4537 \cdot 10^{-2}$
3447.589	1	3	$2.2691 \cdot 10^{-2}$
*3554.42	9	15	$7.5971 \cdot 10^{-3}$
*3562.99	9	3	$4.8363 \cdot 10^{-3}$
*3587.28	9	15	$1.8107 \cdot 10^{-2}$
*3599.32	9	3	$6.7245 \cdot 10^{-3}$
3613.642	1	3	$3.8022 \cdot 10^{-2}$
*3634.25	9	15	$2.6062 \cdot 10^{-2}$
*3652.00	9	3	$9.7444 \cdot 10^{-3}$
*3705.02	9	15	$3.9528 \cdot 10^{-2}$
*3732.88	9	3	$1.4895 \cdot 10^{-2}$
*3819.62	9	15	$6.4351 \cdot 10^{-2}$
3833.549	3	5	$9.6470 \cdot 10^{-3}$
3838.100	3	1	$3.7425 \cdot 10^{-3}$
*3867.49	9	3	$2.4466 \cdot 10^{-2}$
3871.786	3	5	$1.3386 \cdot 10^{-2}$
3878.177	3	1	$5.1753 \cdot 10^{-3}$
*3888.64	3	9	$9.4746 \cdot 10^{-2}$
3926.544	3	5	$1.9371 \cdot 10^{-2}$
3935.945	3	1	$7.4475 \cdot 10^{-3}$
3964.729	1	3	$6.9507 \cdot 10^{-2}$
4009.256	3	5	$2.9612 \cdot 10^{-2}$
4023.980	3	1	$1.1281 \cdot 10^{-2}$
*4026.21	9	15	$1.1600 \cdot 10^{-1}$
*4120.84	9	3	$4.4529 \cdot 10^{-2}$
4143.759	3	5	$4.8812 \cdot 10^{-2}$
4168.971	3	1	$1.8298 \cdot 10^{-2}$
4387.929	3	5	$8.9889 \cdot 10^{-2}$
4437.553	3	1	$3.2689 \cdot 10^{-2}$

λ Å	g_i	g_k	A 10^8 s^{-1}
*4471.50	9	15	$2.4578 \cdot 10^{-1}$
*4713.17	9	3	$9.5209 \cdot 10^{-2}$
4921.931	3	5	$1.9863 \cdot 10^{-1}$
5015.678	1	3	$1.3372 \cdot 10^{-1}$
5047.738	3	1	$6.7712 \cdot 10^{-2}$
*5875.66	9	15	$7.0703 \cdot 10^{-1}$
6678.152	3	5	$6.3705 \cdot 10^{-1}$
*7065.25	9	3	$2.7853 \cdot 10^{-1}$
7281.350	3	1	$1.8299 \cdot 10^{-1}$
*7298.04	3	9	$1.2913 \cdot 10^{-3}$
*7499.85	3	9	$1.7942 \cdot 10^{-3}$
*7816.14	3	9	$2.5748 \cdot 10^{-3}$
8094.115	1	3	$1.3791 \cdot 10^{-3}$
8265.701	1	3	$1.8722 \cdot 10^{-3}$
*8361.73	3	9	$3.8126 \cdot 10^{-3}$
8518.036	1	3	$2.6252 \cdot 10^{-3}$
*8582.64	9	15	$4.1927 \cdot 10^{-3}$
*8632.74	9	3	$2.7471 \cdot 10^{-3}$
*8776.74	9	15	$5.7758 \cdot 10^{-3}$
*8849.18	9	3	$3.8377 \cdot 10^{-3}$
8914.772	1	3	$3.8260 \cdot 10^{-3}$
8999.736	5	7	$2.8406 \cdot 10^{-3}$
*9063.32	9	15	$8.2702 \cdot 10^{-3}$
9085.421	3	5	$3.6807 \cdot 10^{-3}$
9111.026	3	1	$2.2000 \cdot 10^{-3}$
*9174.52	9	3	$5.5996 \cdot 10^{-3}$
*9210.34	15	21	$4.7381 \cdot 10^{-3}$
9303.163	3	5	$5.1030 \cdot 10^{-3}$
9340.143	3	1	$3.0562 \cdot 10^{-3}$
*9463.58	3	9	$5.6868 \cdot 10^{-3}$
*9516.60	9	15	$1.2439 \cdot 10^{-2}$
9603.441	1	3	$5.8286 \cdot 10^{-3}$
9625.697	3	5	$7.3744 \cdot 10^{-3}$
9682.388	3	1	$4.4271 \cdot 10^{-3}$
*9702.65	9	3	$8.6511 \cdot 10^{-3}$
10138.424	3	5	$1.1248 \cdot 10^{-2}$
10233.102	3	1	$6.7731 \cdot 10^{-3}$
*10311.27	9	15	$1.9945 \cdot 10^{-2}$
*10667.71	9	3	$1.4471 \cdot 10^{-2}$
*10830.17	3	9	$1.0216 \cdot 10^{-1}$
*10913.00	15	21	$1.9801 \cdot 10^{-2}$
*10996.65	15	9	$1.4253 \cdot 10^{-3}$
11013.072	1	3	$9.2496 \cdot 10^{-3}$
11044.983	3	5	$1.8457 \cdot 10^{-2}$
11225.937	3	1	$1.1168 \cdot 10^{-2}$
*10969.11	9	15	$3.4781 \cdot 10^{-2}$
*12527.48	3	9	$7.0932 \cdot 10^{-3}$
12755.688	5	3	$1.2754 \cdot 10^{-3}$
*12846.01	9	3	$2.7317 \cdot 10^{-2}$
12968.430	3	5	$3.3615 \cdot 10^{-2}$
*12984.88	15	9	$2.7292 \cdot 10^{-3}$
13411.683	3	1	$2.0572 \cdot 10^{-2}$

Indium

In I

λ Å	g_i	g_k	A 10^8 s^{-1}
2560.2	2	4	$4.0 \cdot 10^{-1}$
2710.3	4	6	$4.0 \cdot 10^{-1}$
3039.4	2	4	1.3
3256.1	4	6	1.3

λ Å	Weights g_i	g_k	A 10^8 s^{-1}
4101.8	2	2	5.6·10^{-1}
4511.3	4	2	1.02

In II

λ Å	g_i	g_k	A
2941.1	3	1	1.4

Iodine

I I

λ Å	g_i	g_k	A
1782.8	4	4	2.71
1830.4	4	6	1.6·10^{-1}

Iridium

Ir I

λ Å	g_i	g_k	A
2475.12	10	10	2.1·10^{-1}
2502.98	10	12	3.2·10^{-1}
2639.71	10	10	4.7·10^{-1}
2661.98	10	10	2.5·10^{-1}
2664.79	10	8	4.0·10^{-1}
2694.23	10	12	4.8·10^{-1}
2849.72	10	10	2.2·10^{-1}
2924.79	10	12	1.42·10^{-1}
2934.64	8	10	2.0·10^{-1}
3220.78	10	8	2.4·10^{-1}
3573.72	8	10	5.4·10^{-2}
3661.71	8	10	4.0·10^{-2}
4033.76	8	10	2.7·10^{-2}
4069.92	6	8	3.6·10^{-2}
4913.35	12	12	3.3·10^{-2}

Iron

Fe I

λ Å	g_i	g_k	A
2166.77	9	7	2.7
2191.84	5	5	1.2
2196.04	3	3	1.2
2200.39	1	3	8.9·10^{-1}
2298.17	9	9	3.09·10^{-1}
2439.74	13	13	3.46
2442.57	11	11	3.12
2457.60	11	11	4.81·10^{-1}
2462.65	9	9	5.85·10^{-1}
2465.15	9	9	4.35·10^{-1}
2468.88	11	11	2.40·10^{-1}
2472.89	7	7	1.30
2474.81	7	7	6.13·10^{-1}
2479.78	5	5	1.74
2483.27	9	11	4.80
2484.19	3	3	2.26
2487.07	3	3	6.40·10^{-1}
2488.14	7	9	4.20
2489.75	1	3	2.31
2490.64	5	7	3.44
2491.16	3	5	2.91
2491.99	9	9	3.25·10^{-1}
2496.53	9	11	2.15·10^{-1}
2501.13	9	7	6.75·10^{-1}
2505.01	9	11	2.56·10^{-1}
2506.57	7	9	2.04·10^{-1}
2507.90	7	9	1.93·10^{-1}

λ Å	g_i	g_k	A
2510.83	7	5	1.29
2518.10	5	3	1.93
2522.85	9	9	2.13
2524.29	3	1	3.23
2527.27	13	13	3.46·10^{-1}
2527.44	7	7	1.93
2529.14	5	5	9.91·10^{-1}
2529.31	5	7	4.86
2533.14	11	11	2.07·10^{-1}
2535.61	1	3	9.59·10^{-1}
2537.17	13	15	3.70
2540.97	3	5	9.59·10^{-1}
2542.10	11	13	4.47
2543.92	9	11	4.70
2545.98	5	7	7.16·10^{-1}
2549.61	7	9	2.31·10^{-1}
2584.54	11	13	3.15·10^{-1}
2606.83	9	11	2.43·10^{-1}
2609.22	7	7	4.60·10^{-1}
2623.53	7	9	2.13·10^{-1}
2635.81	5	7	2.11·10^{-1}
2656.15	13	15	1.63·10^{-1}
2669.49	11	13	1.34·10^{-1}
2679.06	11	11	1.50·10^{-1}
2706.01	13	13	2.28·10^{-1}
2706.58	7	5	2.69·10^{-1}
2708.57	9	9	6.49·10^{-1}
2719.03	9	7	1.42
2719.06	7	7	7.40·10^{-1}
2719.42	11	11	3.20·10^{-1}
2720.90	7	5	1.04
2723.58	5	3	5.69·10^{-1}
2728.82	9	9	2.98·10^{-1}
2733.58	11	9	7.10·10^{-1}
2735.48	9	7	5.03·10^{-1}
2737.31	3	3	7.25·10^{-1}
2737.64	13	11	1.14·10^{-1}
2742.25	7	5	3.41·10^{-1}
2742.41	5	5	4.70·10^{-1}
2750.14	7	7	2.74·10^{-1}
2762.03	7	7	1.76·10^{-1}
2767.52	9	9	1.48·10^{-1}
2769.30	13	13	1.80·10^{-1}
2788.10	11	13	5.92·10^{-1}
2789.80	11	9	2.36·10^{-1}
2804.86	9	7	2.40·10^{-1}
2806.98	9	11	1.15·10^{-1}
2813.29	9	11	3.42·10^{-1}
2825.56	7	9	1.32·10^{-1}
2832.44	7	9	2.38·10^{-1}
2843.98	5	7	3.17·10^{-1}
2851.80	3	5	3.37·10^{-1}
2894.50	5	5	4.83·10^{-1}
2899.41	5	3	4.68·10^{-1}
2901.91	11	11	1.78·10^{-1}
2907.52	9	11	1.61·10^{-1}
2918.02	13	13	1.18
2923.29	11	11	1.39
2923.85	11	11	2.97·10^{-1}
2925.36	7	9	1.69·10^{-1}

λ Å	g_i	g_k	A
2929.12	9	9	1.53
2936.90	9	9	1.40·10^{-1}
2947.88	7	7	1.83·10^{-1}
2948.43	9	9	3.32·10^{-1}
2953.49	7	7	3.64·10^{-1}
2959.99	11	13	5.02·10^{-1}
2966.90	9	11	2.72·10^{-1}
2973.24	7	9	1.83·10^{-1}
2980.53	7	7	1.66·10^{-1}
2981.85	7	9	1.86·10^{-1}
2983.57	9	7	2.79·10^{-1}
2990.39	9	11	3.5·10^{-1}
2994.43	7	5	4.39·10^{-1}
2999.51	11	11	1.70·10^{-1}
3000.95	5	3	6.42·10^{-1}
3008.14	3	1	1.07
3009.57	9	9	1.43·10^{-1}
3011.48	7	9	3.79·10^{-1}
3020.49	5	5	1.94·10^{-1}
3020.64	9	9	7.59·10^{-1}
3021.07	7	7	4.55·10^{-1}
3025.64	13	13	5.86·10^{-1}
3025.84	1	3	3.48·10^{-1}
3030.15	11	11	5.04·10^{-1}
3037.39	3	5	2.91·10^{-1}
3047.60	5	7	2.84·10^{-1}
3057.45	11	9	3.13·10^{-1}
3059.09	7	9	1.63·10^{-1}
3067.00	11	13	1.71·10^{-1}
3067.24	9	7	3.12·10^{-1}
3075.72	7	5	3.14·10^{-1}
3079.99	9	11	8.35·10^{-2}
3083.74	5	3	3.08·10^{-1}
3100.30	5	5	1.87·10^{-1}
3100.67	7	7	1.35·10^{-1}
3125.68	13	11	8.46·10^{-2}
3132.52	9	7	3.39·10^{-1}
3143.99	9	9	6.10·10^{-1}
3156.27	7	7	6.36·10^{-1}
3157.04	9	11	1.26·10^{-1}
3157.89	5	7	1.61·10^{-1}
3160.66	9	9	1.93·10^{-1}
3166.44	9	7	1.14·10^{-1}
3171.35	9	7	1.85·10^{-1}
3175.44	11	11	1.44·10^{-1}
3178.01	11	9	1.28·10^{-1}
3180.22	7	9	4.42·10^{-1}
3181.52	7	5	1.84·10^{-1}
3182.97	5	7	1.42·10^{-1}
3188.82	3	5	2.53·10^{-1}
3192.80	3	5	5.01·10^{-1}
3193.30	5	7	3.07·10^{-1}
3196.12	11	9	1.40·10^{-1}
3196.93	9	11	5.97·10^{-1}
3199.53	9	9	2.23·10^{-1}
3205.40	3	3	9.77·10^{-1}
3210.23	9	11	1.15·10^{-1}
3210.83	5	3	9.24·10^{-1}
3211.99	11	9	4.64·10^{-1}
3214.01	7	7	8.38·10^{-1}

λ (Å)	g_i	g_k	A (10^8 s^{-1})
3214.06	7	5	1.18
3215.94	5	5	$6.19 \cdot 10^{-1}$
3217.38	11	9	$1.50 \cdot 10^{-1}$
3219.58	7	9	$4.64 \cdot 10^{-1}$
3219.80	9	7	$3.61 \cdot 10^{-1}$
3222.07	11	11	$8.65 \cdot 10^{-1}$
3225.79	11	13	1.18
3227.80	9	7	$4.96 \cdot 10^{-1}$
3228.25	5	3	$3.72 \cdot 10^{-1}$
3229.99	9	11	$1.06 \cdot 10^{-1}$
3230.21	5	5	$2.06 \cdot 10^{-1}$
3230.96	7	5	$3.7 \cdot 10^{-1}$
3233.05	13	15	$4.19 \cdot 10^{-1}$
3233.97	9	9	$2.08 \cdot 10^{-1}$
3239.43	9	9	$2.95 \cdot 10^{-1}$
3244.19	9	11	$3.06 \cdot 10^{-1}$
3248.20	7	7	$1.92 \cdot 10^{-1}$
3253.60	7	9	$1.62 \cdot 10^{-1}$
3254.36	11	13	$4.24 \cdot 10^{-1}$
3265.62	7	5	$3.06 \cdot 10^{-1}$
3271.00	5	3	$6.4 \cdot 10^{-1}$
3280.26	9	11	$4.21 \cdot 10^{-1}$
3282.89	3	5	$3.42 \cdot 10^{-1}$
3286.75	7	7	$5.99 \cdot 10^{-1}$
3292.02	7	9	$5.77 \cdot 10^{-1}$
3292.59	3	3	$3.0 \cdot 10^{-1}$
3305.97	5	7	$4.05 \cdot 10^{-1}$
3306.34	9	9	$5.74 \cdot 10^{-1}$
3306.35	3	5	$4.84 \cdot 10^{-1}$
3307.23	13	13	$1.97 \cdot 10^{-1}$
3314.74	5	7	$7.25 \cdot 10^{-1}$
3322.47	9	11	$8.21 \cdot 10^{-2}$
3323.74	5	5	$2.8 \cdot 10^{-1}$
3328.87	11	11	$2.21 \cdot 10^{-1}$
3355.23	9	9	$2.59 \cdot 10^{-1}$
3369.55	9	9	$2.15 \cdot 10^{-1}$
3370.78	11	11	$2.89 \cdot 10^{-1}$
3380.11	7	7	$1.66 \cdot 10^{-1}$
3392.65	7	7	$1.88 \cdot 10^{-1}$
3399.33	5	5	$2.76 \cdot 10^{-1}$
3402.26	13	13	$2.19 \cdot 10^{-1}$
3404.35	5	7	$1.09 \cdot 10^{-1}$
3406.44	3	5	$2.7 \cdot 10^{-1}$
3407.46	7	9	$6.09 \cdot 10^{-1}$
3410.17	3	5	$5.07 \cdot 10^{-1}$
3413.13	5	7	$3.23 \cdot 10^{-1}$
3417.84	3	3	$4.01 \cdot 10^{-1}$
3418.51	3	1	$9.88 \cdot 10^{-1}$
3422.66	3	5	$1.38 \cdot 10^{-1}$
3424.28	7	7	$1.61 \cdot 10^{-1}$
3425.01	9	7	$2.57 \cdot 10^{-1}$
3426.67	11	11	$1.07 \cdot 10^{-1}$
3427.12	7	9	$5.04 \cdot 10^{-1}$
3428.19	5	5	$1.71 \cdot 10^{-1}$
3440.61	9	7	$1.71 \cdot 10^{-1}$
3440.99	7	5	$1.24 \cdot 10^{-1}$
3445.15	5	7	$2.34 \cdot 10^{-1}$
3450.33	3	3	$2.34 \cdot 10^{-1}$
3459.91	5	3	$2.17 \cdot 10^{-1}$
3469.01	9	9	$8.58 \cdot 10^{-2}$
3476.34	7	7	$2.70 \cdot 10^{-1}$
3489.67	11	13	$7.47 \cdot 10^{-2}$
3495.29	9	7	$9.46 \cdot 10^{-2}$
3497.10	7	7	$9.02 \cdot 10^{-2}$
3508.47	9	11	$6.46 \cdot 10^{-2}$
3526.24	7	9	$1.70 \cdot 10^{-1}$
3526.38	7	7	$4.13 \cdot 10^{-1}$
3526.47	5	5	$1.29 \cdot 10^{-1}$
3526.67	5	5	$5.26 \cdot 10^{-1}$
3527.79	9	9	$2.17 \cdot 10^{-1}$
3529.82	3	3	$7.75 \cdot 10^{-1}$
3530.39	13	13	$4.65 \cdot 10^{-2}$
3533.01	1	3	$8.52 \cdot 10^{-1}$
3533.20	3	5	$8.25 \cdot 10^{-1}$
3536.56	5	7	$9.95 \cdot 10^{-1}$
3537.89	11	11	$8.0 \cdot 10^{-2}$
3540.12	7	9	$9.48 \cdot 10^{-2}$
3541.08	9	11	$8.65 \cdot 10^{-1}$
3542.08	7	9	$9.51 \cdot 10^{-1}$
3543.67	3	5	$1.6 \cdot 10^{-1}$
3545.64	9	9	$2.05 \cdot 10^{-1}$
3547.19	9	9	$7.13 \cdot 10^{-2}$
3552.83	5	5	$1.74 \cdot 10^{-1}$
3553.74	11	9	1.09
3554.92	11	13	1.40
3556.88	9	11	$4.1 \cdot 10^{-1}$
3558.52	5	7	$1.77 \cdot 10^{-1}$
3559.50	3	3	$2.2 \cdot 10^{-1}$
3560.70	7	9	$7.4 \cdot 10^{-2}$
3565.38	7	9	$4.29 \cdot 10^{-1}$
3567.03	5	7	$8.34 \cdot 10^{-2}$
3568.82	7	9	$6.72 \cdot 10^{-2}$
3570.10	9	11	$6.76 \cdot 10^{-1}$
3572.00	11	11	$2.89 \cdot 10^{-1}$
3573.39	5	7	$1.05 \cdot 10^{-1}$
3573.89	9	7	$5.73 \cdot 10^{-1}$
3575.25	11	9	$7.43 \cdot 10^{-2}$
3575.37	5	5	$3.06 \cdot 10^{-1}$
3576.76	11	9	$8.8 \cdot 10^{-2}$
3581.19	11	13	1.02
3582.20	13	11	$2.35 \cdot 10^{-1}$
3584.66	11	11	$3.29 \cdot 10^{-1}$
3584.79	7	5	$1.56 \cdot 10^{-1}$
3584.96	11	9	$6.74 \cdot 10^{-1}$
3585.32	7	7	$1.17 \cdot 10^{-1}$
3586.11	13	11	$7.02 \cdot 10^{-1}$
3586.98	5	5	$1.66 \cdot 10^{-1}$
3587.24	7	9	$7.73 \cdot 10^{-2}$
3588.61	11	11	$1.19 \cdot 10^{-1}$
3588.92	5	3	$2.15 \cdot 10^{-1}$
3589.45	9	7	$1.05 \cdot 10^{-1}$
3594.63	9	9	$3.14 \cdot 10^{-1}$
3597.02	5	3	$1.8 \cdot 10^{-1}$
3599.63	11	9	$2.33 \cdot 10^{-1}$
3602.46	7	7	$1.02 \cdot 10^{-1}$
3602.53	7	5	$2.12 \cdot 10^{-1}$
3603.20	11	11	$2.59 \cdot 10^{-1}$
3605.45	9	9	$4.66 \cdot 10^{-1}$
3605.50	13	11	$2.12 \cdot 10^{-1}$
3606.68	11	13	$8.29 \cdot 10^{-1}$
3608.14	9	11	$6.22 \cdot 10^{-2}$
3608.86	3	5	$8.13 \cdot 10^{-1}$
3610.16	13	13	$5.90 \cdot 10^{-1}$
3612.07	11	13	$1.11 \cdot 10^{-1}$
3617.79	5	7	$7.09 \cdot 10^{-1}$
3618.39	9	9	$8.88 \cdot 10^{-2}$
3618.77	5	7	$7.22 \cdot 10^{-1}$
3621.46	9	11	$4.45 \cdot 10^{-1}$
3621.72	11	9	$1.07 \cdot 10^{-1}$
3622.00	7	7	$5.14 \cdot 10^{-1}$
3623.19	13	13	$6.68 \cdot 10^{-2}$
3625.14	11	9	$8.15 \cdot 10^{-2}$
3630.35	9	7	$1.04 \cdot 10^{-1}$
3631.10	11	11	$2.15 \cdot 10^{-1}$
3631.46	7	9	$5.17 \cdot 10^{-1}$
3632.04	3	5	$6.74 \cdot 10^{-1}$
3634.33	9	7	$1.05 \cdot 10^{-1}$
3636.22	5	7	$2.20 \cdot 10^{-1}$
3637.87	9	9	$5.9 \cdot 10^{-2}$
3638.30	7	9	$2.36 \cdot 10^{-1}$
3640.39	9	11	$3.57 \cdot 10^{-1}$
3645.82	1	3	$4.87 \cdot 10^{-1}$
3647.42	3	3	$3.38 \cdot 10^{-1}$
3647.84	9	11	$2.91 \cdot 10^{-1}$
3649.51	11	9	$3.94 \cdot 10^{-1}$
3650.03	7	7	$2.26 \cdot 10^{-1}$
3650.28	11	11	$6.15 \cdot 10^{-2}$
3651.47	7	9	$5.83 \cdot 10^{-1}$
3655.46	5	5	$1.18 \cdot 10^{-1}$
3659.52	9	9	$6.31 \cdot 10^{-2}$
3667.25	9	7	$1.3 \cdot 10^{-1}$
3669.15	9	7	$8.03 \cdot 10^{-2}$
3669.52	9	7	$2.34 \cdot 10^{-1}$
3670.09	11	13	$7.20 \cdot 10^{-2}$
3676.31	9	11	$4.63 \cdot 10^{-2}$
3677.31	5	7	$2.28 \cdot 10^{-1}$
3677.63	7	5	$6.08 \cdot 10^{-1}$
3682.17	7	5	$1.04 \cdot 10^{-1}$
3682.24	5	5	1.5
3684.11	9	7	$2.97 \cdot 10^{-1}$
3684.14	9	7	$9.29 \cdot 10^{-2}$
3686.00	9	11	$3.34 \cdot 10^{-1}$
3687.46	11	9	$8.00 \cdot 10^{-2}$
3687.66	9	9	$7.38 \cdot 10^{-2}$
3688.46	7	9	$7.3 \cdot 10^{-2}$
3689.46	9	9	$3.70 \cdot 10^{-1}$
3690.73	11	11	$2.99 \cdot 10^{-1}$
3694.01	5	7	$8.35 \cdot 10^{-1}$
3695.05	7	9	$2.01 \cdot 10^{-1}$
3697.43	7	7	$1.94 \cdot 10^{-1}$
3701.09	7	9	$6.35 \cdot 10^{-1}$
3703.69	9	11	$6.31 \cdot 10^{-2}$
3704.46	11	9	$1.42 \cdot 10^{-1}$
3707.92	7	5	$3.32 \cdot 10^{-1}$
3709.25	9	7	$1.56 \cdot 10^{-1}$
3711.41	3	5	$1.28 \cdot 10^{-1}$
3716.44	9	7	$3.49 \cdot 10^{-1}$
3719.93	9	11	$1.62 \cdot 10^{-1}$
3721.50	5	5	$1.94 \cdot 10^{-1}$
3724.38	5	7	$1.04 \cdot 10^{-1}$

λ Å	g_i	g_k	A 10^8 s^{-1}	λ Å	g_i	g_k	A 10^8 s^{-1}	λ Å	g_i	g_k	A 10^8 s^{-1}
3726.93	5	5	4.57·10⁻¹	3865.52	3	3	1.55·10⁻¹	4070.77	7	5	1.1·10⁻¹
3727.09	9	7	1.71·10⁻¹	3867.22	5	5	3.16·10⁻¹	4071.74	5	5	7.64·10⁻¹
3727.62	7	5	2.24·10⁻¹	3871.75	11	11	5.83·10⁻²	4073.76	5	3	1.68·10⁻¹
3727.81	7	5	1.91·10⁻¹	3872.50	5	5	1.05·10⁻¹	4076.63	9	9	1.32·10⁻¹
3730.39	9	11	9.73·10⁻²	3873.76	11	9	6.57·10⁻²	4084.49	11	9	8.66·10⁻²
3732.40	5	5	2.69·10⁻¹	3878.02	7	7	7.72·10⁻²	4085.30	7	7	8.92·10⁻²
3734.86	11	11	9.01·10⁻¹	3878.67	9	7	7.02·10⁻²	4098.18	7	7	7.49·10⁻²
3735.32	9	9	2.70·10⁻¹	3878.73	3	3	5.34·10⁻¹	4107.49	5	3	1.74·10⁻¹
3737.13	7	9	1.41·10⁻¹	3883.28	7	7	1.28·10⁻¹	4109.80	3	3	1.51·10⁻¹
3738.31	11	13	3.44·10⁻¹	3888.51	5	5	2.50·10⁻¹	4112.96	11	13	1.1·10⁻¹
3740.24	7	9	1.3·10⁻¹	3888.82	5	3	1.95·10⁻¹	4118.55	11	13	4.96·10⁻¹
3742.62	9	9	6.75·10⁻²	3891.93	3	3	2.71·10⁻¹	4125.62	9	11	9.9·10⁻²
3743.36	5	3	2.60·10⁻¹	3893.39	11	11	1.00·10⁻¹	4126.18	11	11	4.2·10⁻²
3743.47	11	11	6.05·10⁻¹	3894.01	5	5	1.03·10⁻¹	4127.61	1	3	1.43·10⁻¹
3744.10	5	3	3.17·10⁻¹	3897.89	11	13	6.20·10⁻²	4132.06	5	7	1.18·10⁻¹
3745.56	5	7	1.15·10⁻¹	3900.52	7	7	7.9·10⁻²	4132.90	3	5	7.70·10⁻²
3746.93	7	7	2.33·10⁻¹	3902.95	7	7	2.14·10⁻¹	4134.68	5	7	1.25·10⁻¹
3748.96	9	11	1.48·10⁻¹	3903.90	9	9	7.61·10⁻²	4137.00	3	5	2.75·10⁻¹
3749.49	9	9	7.63·10⁻¹	3906.75	5	7	7.05·10⁻²	4142.59	3	5	7.5·10⁻²
3753.61	7	5	1.22·10⁻¹	3916.73	13	11	9.83·10⁻²	4143.41	9	9	2.70·10⁻¹
3756.94	11	11	2.2·10⁻¹	3918.42	3	1	4.22·10⁻¹	4143.87	7	9	1.33·10⁻¹
3758.23	7	7	6.34·10⁻¹	3918.64	7	7	1.17·10⁻¹	4149.37	11	13	4.23·10⁻²
3759.15	13	11	4.55·10⁻²	3925.94	1	3	1.67·10⁻¹	4153.90	7	9	2.05·10⁻¹
3760.05	13	15	4.47·10⁻²	3926.01	7	7	7.26·10⁻²	4154.50	5	3	2.64·10⁻¹
3763.79	5	5	5.44·10⁻¹	3928.08	9	9	5.64·10⁻²	4154.81	9	11	1.40·10⁻¹
3765.54	13	15	9.51·10⁻¹	3935.81	5	5	1.14·10⁻¹	4156.80	5	5	1.20·10⁻¹
3767.19	3	3	6.39·10⁻¹	3941.28	5	5	9.1·10⁻²	4157.78	5	7	2.18·10⁻¹
3778.51	7	5	1.17·10⁻¹	3942.44	3	5	9.62·10⁻²	4158.79	3	5	1.6·10⁻¹
3785.95	11	13	4.14·10⁻²	3946.99	9	11	3.91·10⁻²	4172.12	7	5	9.80·10⁻²
3786.19	5	5	1.3·10⁻¹	3948.10	7	9	1.31·10⁻¹	4175.64	3	5	1.14·10⁻¹
3787.16	5	5	9.9·10⁻²	3948.77	11	9	2.08·10⁻¹	4181.75	5	7	2.32·10⁻¹
3787.88	3	5	1.29·10⁻¹	3951.16	3	5	4.29·10⁻¹	4184.89	5	5	1.03·10⁻¹
3793.48	7	7	7.92·10⁻²	3955.34	3	3	1.5·10⁻¹	4187.04	7	5	2.15·10⁻¹
3795.00	5	7	1.15·10⁻¹	3956.46	13	11	1.76·10⁻¹	4187.80	9	7	1.52·10⁻¹
3797.51	13	13	4.57·10⁻¹	3956.68	11	13	1.22·10⁻¹	4191.43	5	3	2.73·10⁻¹
3799.55	7	9	7.31·10⁻²	3957.02	5	7	1.67·10⁻¹	4195.33	11	11	1.11·10⁻¹
3801.98	11	13	3.7·10⁻²	3963.10	3	5	1.5·10⁻¹	4196.21	7	7	1.09·10⁻¹
3805.34	9	11	8.60·10⁻¹	3967.42	9	7	1.52·10⁻¹	4198.25	9	9	1.47·10⁻¹
3806.22	3	3	2.5·10⁻¹	3967.96	7	9	6.09·10⁻²	4198.30	11	9	8.03·10⁻²
3806.70	11	11	4.35·10⁻¹	3969.26	9	7	2.26·10⁻¹	4198.63	5	5	1.25·10⁻¹
3807.54	3	5	9.37·10⁻²	3971.32	11	9	4.97·10⁻²	4199.10	9	11	4.92·10⁻¹
3810.76	5	3	1.94·10⁻¹	3973.65	5	7	5.81·10⁻²	4200.92	7	9	6.25·10⁻²
3813.88	13	11	6.62·10⁻²	3976.61	3	5	1.20·10⁻¹	4202.03	9	9	8.22·10⁻²
3815.84	9	7	1.12	3983.96	9	7	5.72·10⁻²	4203.94	13	13	2.97·10⁻²
3817.64	11	11	7.7·10⁻²	3985.39	5	5	8.53·10⁻²	4203.98	3	5	7.37·10⁻²
3820.43	11	9	6.67·10⁻¹	3996.96	9	9	7.95·10⁻²	4210.34	3	3	1.48·10⁻¹
3821.18	11	13	5.54·10⁻¹	3997.39	9	11	1.26·10⁻¹	4217.55	3	5	2.46·10⁻¹
3825.88	9	7	5.97·10⁻¹	3998.05	11	9	5.70·10⁻²	4219.36	11	13	2.88·10⁻¹
3827.82	7	5	1.05	4005.24	7	5	2.04·10⁻¹	4222.21	7	7	5.76·10⁻²
3834.22	7	5	4.52·10⁻¹	4006.31	11	9	5.1·10⁻²	4224.17	9	11	1.06·10⁻¹
3836.33	5	5	3.29·10⁻¹	4014.53	9	7	1.53·10⁻¹	4224.51	3	5	6.81·10⁻²
3839.26	9	9	2.35·10⁻¹	4021.87	7	9	8.55·10⁻²	4225.45	5	7	1.65·10⁻¹
3840.44	5	3	4.70·10⁻¹	4024.73	7	9	8.09·10⁻²	4227.43	11	13	5.29·10⁻¹
3841.05	5	3	1.36	4030.49	9	11	1.04·10⁻¹	4233.60	3	5	1.85·10⁻¹
3843.26	9	7	3.70·10⁻¹	4043.90	7	7	8.69·10⁻²	4235.94	9	9	1.88·10⁻¹
3846.41	11	9	1.68·10⁻¹	4045.59	9	7	7.39·10⁻²	4238.81	7	9	2.41·10⁻¹
3846.80	7	7	6.20·10⁻¹	4045.81	9	9	8.62·10⁻¹	4247.43	9	11	1.94·10⁻¹
3849.97	3	1	6.05·10⁻¹	4062.44	3	3	1.85·10⁻¹	4250.12	5	7	2.07·10⁻¹
3859.21	13	11	7.25·10⁻²	4063.59	7	7	6.65·10⁻¹	4250.79	7	7	1.02·10⁻¹
3859.91	9	9	9.69·10⁻²	4067.98	9	9	1.51·10⁻¹	4260.47	11	11	3.99·10⁻¹

λ Å	g_i	g_k	A 10^8 s^{-1}	λ Å	g_i	g_k	A 10^8 s^{-1}	λ Å	g_i	g_k	A 10^8 s^{-1}
4271.15	7	9	1.82·10⁻¹	5139.25	7	5	9.16·10⁻²	6496.47	5	5	7.8·10⁻²
4271.76	9	11	2.28·10⁻¹	5139.46	9	9	8.69·10⁻²	6569.22	7	9	6.0·10⁻²
4282.40	7	5	1.21·10⁻¹	5184.27	5	7	3.8·10⁻²	6633.75	7	7	3.44·10⁻²
4299.23	9	11	1.29·10⁻¹	5191.46	5	3	2.32·10⁻¹	6841.34	5	7	3.4·10⁻²
4307.90	7	9	3.38·10⁻¹	5192.34	7	7	1.34·10⁻¹	6855.16	7	9	2.86·10⁻²
4315.08	5	5	7.76·10⁻²	5208.59	7	5	6.23·10⁻²	7187.32	9	11	8.36·10⁻²
4325.76	5	7	5.16·10⁻¹	5215.18	5	3	1.10·10⁻¹	7511.02	11	11	1.35·10⁻¹
4327.10	5	5	1.12·10⁻¹	5226.86	5	5	1.36·10⁻¹	8220.38	13	11	1.69·10⁻¹
4369.77	9	9	6.09·10⁻²	5232.94	9	11	1.94·10⁻¹	8699.45	7	9	4.08·10⁻²
4383.55	9	11	5.00·10⁻¹	5235.39	9	7	3.75·10⁻²	9012.07	11	9	4.46·10⁻²
4388.41	7	7	1.03·10⁻¹	5242.49	13	11	2.38·10⁻²	9401.11	9	11	2.64·10⁻²
4401.29	7	7	6.4·10⁻²	5263.31	5	5	6.36·10⁻²	9414.04	7	9	3.98·10⁻²
4404.75	7	9	2.75·10⁻¹	5266.56	7	9	1.10·10⁻¹	9443.80	5	7	6.39·10⁻²
4415.12	5	7	1.19·10⁻¹	5273.16	1	3	8.12·10⁻²	9569.91	11	11	2.50·10⁻²
4433.22	5	3	2.1·10⁻¹	5281.79	5	7	5.00·10⁻²	9626.50	9	9	4.51·10⁻²
4443.19	1	3	1.02·10⁻¹	5283.62	7	7	1.02·10⁻¹	9738.57	11	13	7.64·10⁻²
4455.03	9	7	4.1·10⁻²	5302.30	3	5	9.04·10⁻²	9763.38	3	5	5.42·10⁻²
4466.55	5	7	1.20·10⁻¹	5324.18	9	9	2.06·10⁻¹	9861.74	7	9	5.49·10⁻²
4469.38	5	7	1.59·10⁻¹	5339.93	5	7	6.36·10⁻²	9889.04	9	11	2.22·10⁻²
4476.02	3	5	1.01·10⁻¹	5364.87	5	7	5.59·10⁻¹	22473.28	11	11	3.32·10⁻²
4484.22	7	9	5.04·10⁻²	5367.47	7	9	7.13·10⁻¹	23566.67	9	11	2.21·10⁻²
4485.68	3	3	1.1·10⁻¹	5369.96	9	11	7.22·10⁻¹	24547.95	11	9	3.72·10⁻²
4528.61	7	9	5.44·10⁻²	5373.71	7	9	3.7·10⁻²	24729.10	13	11	5.08·10⁻²
4547.85	5	7	4.48·10⁻²	5383.37	11	13	7.81·10⁻¹				
4556.13	7	5	1.05·10⁻¹	5393.17	7	9	4.91·10⁻¹	**Fe II**			
4619.29	7	5	5.2·10⁻²	5398.28	5	5	9.0·10⁻²	1055.26	10	8	4.6·10⁻¹
4654.61	7	7	3.68·10⁻²	5404.15	9	11	6.92·10⁻¹	1063.97	10	8	3.5·10⁻¹
4667.45	7	9	6.03·10⁻²	5410.91	7	9	6.33·10⁻¹	1068.35	8	8	1.59
4673.16	5	7	3.81·10⁻²	5415.20	11	13	7.67·10⁻¹	1071.58	6	8	1.14
4678.85	7	9	4.97·10⁻²	5463.28	9	9	2.9·10⁻¹	1096.88	10	8	2.26
4736.77	9	11	4.78·10⁻²	5473.90	7	7	5.2·10⁻²	1112.05	10	12	2.0·10⁻¹
4789.65	5	7	4.57·10⁻²	5476.29	7	9	2.87·10⁻²	1121.97	10	8	1.92
4800.65	7	9	3.01·10⁻²	5476.56	9	9	8.70·10⁻²	1122.84	8	6	1.81
4859.74	5	3	1.62·10⁻¹	5563.60	5	7	3.4·10⁻²	1125.45	10	8	1.03
4871.32	7	5	2.44·10⁻¹	5569.62	5	3	2.34·10⁻¹	1128.05	2	4	1.40
4872.14	3	3	2.54·10⁻¹	5572.84	7	5	2.28·10⁻¹	1130.44	6	8	3.1·10⁻¹
4878.21	1	3	1.21·10⁻¹	5576.09	3	1	2.5·10⁻¹	1133.40	8	10	2.6·10⁻¹
4890.76	5	5	2.25·10⁻¹	5586.76	9	7	2.19·10⁻¹	1133.67	10	8	3.1·10⁻¹
4891.49	9	7	3.08·10⁻¹	5594.66	9	9	5.20·10⁻²	1138.63	8	8	5.5·10⁻¹
4903.31	3	5	6.58·10⁻²	5602.95	3	3	1.00·10⁻¹	1142.37	10	8	2.6·10⁻¹
4918.99	7	7	1.79·10⁻¹	5615.64	11	9	2.64·10⁻¹	1143.23	10	10	9.8·10⁻¹
4920.50	11	9	3.58·10⁻¹	5624.54	5	5	7.41·10⁻²	1144.94	10	12	3.52
4957.30	9	9	1.18·10⁻¹	5633.95	11	13	7.7·10⁻²	1147.41	8	8	1.24
4957.60	13	11	4.22·10⁻¹	5638.26	9	7	4.4·10⁻²	1148.28	8	10	3.35
4966.09	11	11	3.31·10⁻²	5649.99	3	5	5.1·10⁻²	1151.15	6	8	2.23
4973.10	3	3	1.1·10⁻¹	5655.18	7	9	4.7·10⁻²	1267.42	8	6	9.3·10⁻¹
4978.60	5	3	1.19·10⁻¹	5658.82	7	7	4.34·10⁻²	1272.61	6	4	3.3·10⁻¹
4985.25	5	5	1.48·10⁻¹	5662.52	11	9	6.18·10⁻²	1371.02	14	12	1.74
4988.95	7	7	5.2·10⁻²	5679.02	5	7	3.7·10⁻²	1563.79	8	8	1.33
5001.86	9	7	3.7·10⁻¹	5686.53	9	11	6.71·10⁻²	1580.63	8	10	5.8·10⁻¹
5006.12	11	11	5.87·10⁻²	5705.99	7	9	6.1·10⁻²	1608.45	10	8	1.91
5014.94	7	5	2.64·10⁻¹	5753.12	3	5	8.26·10⁻²	1610.92	10	10	1.94·10⁻¹
5021.59	7	9	6.18·10⁻²	5762.99	5	7	9.6·10⁻²	1618.47	8	8	5.53·10⁻¹
5022.24	5	3	2.4·10⁻¹	5816.37	9	11	4.49·10⁻²	1621.69	8	6	1.32
5048.44	3	5	4.88·10⁻²	5905.67	5	3	1.1·10⁻¹	1623.09	8	8	1.99·10⁻¹
5068.77	9	7	3.37·10⁻²	6301.50	5	5	6.43·10⁻²	1625.52	8	10	4.04·10⁻¹
5074.75	9	11	1.4·10⁻¹	6400.00	7	9	9.27·10⁻²	1625.91	6	8	1.02·10⁻¹
5090.77	7	5	1.9·10⁻¹	6411.65	5	7	4.43·10⁻²	1629.16	6	6	8.66·10⁻¹
5121.64	5	7	7.9·10⁻²	6419.95	7	7	1.2·10⁻¹	1631.13	6	4	6.93·10⁻¹
5137.38	11	9	1.0·10⁻¹	6469.19	3	3	8.3·10⁻²	1633.91	6	8	3.85·10⁻¹

λ Å	Weights g_i	g_k	A 10^8 s⁻¹	λ Å	Weights g_i	g_k	A 10^8 s⁻¹	λ Å	Weights g_i	g_k	A 10^8 s⁻¹
1634.35	4	6	$3.21 \cdot 10^{-1}$	2364.83	8	8	$5.90 \cdot 10^{-1}$	2429.86	8	8	1.51
1635.40	8	6	2.28	2365.76	6	6	2.16	2430.08	8	10	1.91
1636.33	4	4	$9.63 \cdot 10^{-1}$	2366.59	6	6	$1.01 \cdot 10^{-1}$	2432.26	6	8	1.57
1637.40	10	8	$3.57 \cdot 10^{-1}$	2366.88	8	10	$3.51 \cdot 10^{-2}$	2432.87	14	14	2.86
1639.40	2	4	$6.85 \cdot 10^{-1}$	2368.60	6	4	$6.06 \cdot 10^{-1}$	2433.50	10	12	$1.30 \cdot 10^{-1}$
1641.76	6	4	1.76	2369.95	10	12	5.9	2434.06	8	6	$7.2 \cdot 10^{-1}$
1647.16	6	6	$4.98 \cdot 10^{-1}$	2370.50	4	4	$1.73 \cdot 10^{-1}$	2434.24	8	10	2.01
1670.75	10	8	1.06	2373.74	10	10	$4.25 \cdot 10^{-1}$	2434.73	12	12	2.79
1676.86	8	8	$6.75 \cdot 10^{-2}$	2375.19	4	2	$9.81 \cdot 10^{-1}$	2434.95	4	6	1.39
1702.04	10	12	1.02	2376.43	12	14	6.4	2435.00	8	8	2.02
1761.37	8	8	1.42	2378.55	8	8	$1.70 \cdot 10^{-1}$	2436.62	6	8	2.70
1785.27	6	8	$1.2 \cdot 10^{1}$	2378.70	8	8	$1.49 \cdot 10^{-1}$	2439.30	12	14	2.25
1786.75	6	6	$1.2 \cdot 10^{1}$	2379.28	8	8	$2.73 \cdot 10^{-1}$	2440.42	6	8	1.18
1788.08	6	4	4.6	2379.42	10	10	$3.68 \cdot 10^{-1}$	2441.13	10	10	$8.95 \cdot 10^{-1}$
1818.52	8	8	$5.70 \cdot 10^{-2}$	2380.76	6	8	$3.10 \cdot 10^{-1}$	2442.38	10	12	2.75
2020.75	6	8	$1.83 \cdot 10^{-1}$	2382.04	10	12	3.13	2443.71	8	10	1.44
2078.16	10	10	$2.84 \cdot 10^{-2}$	2382.36	4	6	$3.19 \cdot 10^{-2}$	2444.52	6	8	2.78
2162.02	10	10	$2.54 \cdot 10^{-1}$	2382.90	12	14	$1.62 \cdot 10^{-1}$	2445.11	12	12	2.03
2182.36	10	8	$8.6 \cdot 10^{-2}$	2383.06	8	6	$1.0 \cdot 10^{-1}$	2445.57	4	6	2.07
2191.98	8	8	$7.54 \cdot 10^{-1}$	2383.24	6	6	$3.59 \cdot 10^{-1}$	2445.80	4	6	1.23
2201.59	6	8	$7.77 \cdot 10^{-1}$	2384.39	4	4	$3.22 \cdot 10^{-1}$	2446.11	8	8	1.06
2208.41	10	10	1.59	2385.01	6	8	$3.60 \cdot 10^{-2}$	2446.47	12	14	$2.99 \cdot 10^{-1}$
2209.03	10	8	1.27	2388.39	10	12	$2.02 \cdot 10^{-1}$	2447.21	6	6	1.15
2213.66	14	14	$3.26 \cdot 10^{-1}$	2388.63	8	8	1.05	2447.33	4	2	2.56
2218.26	8	10	1.57	2390.10	14	16	5.5	2447.76	12	10	1.97
2220.38	12	12	$4.19 \cdot 10^{-1}$	2390.76	6	6	1.17	2449.96	4	4	1.24
2228.73	6	8	1.59	2391.48	8	10	$3.77 \cdot 10^{-2}$	2450.21	2	4	1.26
2249.18	10	8	$3.00 \cdot 10^{-2}$	2394.00	12	10	$9.4 \cdot 10^{-2}$	2453.98	8	10	1.31
2253.13	8	8	$4.41 \cdot 10^{-2}$	2395.42	6	4	$2.67 \cdot 10^{-1}$	2454.58	14	12	1.16
2255.77	6	4	$4.75 \cdot 10^{-1}$	2395.63	8	10	2.59	2455.71	8	8	1.01
2260.08	10	10	$3.18 \cdot 10^{-2}$	2396.72	10	12	$2.15 \cdot 10^{-1}$	2455.90	4	6	1.73
2267.59	6	8	$3.69 \cdot 10^{-2}$	2399.24	6	6	1.39	2457.10	6	4	$4.71 \cdot 10^{-1}$
2279.92	8	10	$4.49 \cdot 10^{-2}$	2400.05	12	14	4.57	2458.78	10	12	2.31
2327.40	6	4	$6.55 \cdot 10^{-1}$	2401.29	6	8	1.89	2458.97	6	4	2.51
2327.88	10	12	1.08	2402.45	10	10	$5.8 \cdot 10^{-1}$	2460.44	10	12	5.39
2331.31	10	8	$3.17 \cdot 10^{-1}$	2402.63	8	8	$8.19 \cdot 10^{-1}$	2461.28	6	8	2.34
2332.80	8	6	1.31	2404.43	4	2	$6.44 \cdot 10^{-1}$	2461.86	8	10	2.43
2338.01	4	4	1.13	2404.89	6	8	1.96	2463.28	12	10	$7.1 \cdot 10^{-1}$
2338.54	10	12	$5.6 \cdot 10^{-2}$	2406.66	4	4	1.61	2464.01	10	8	1.32
2343.50	10	8	1.73	2410.27	8	8	$7.65 \cdot 10^{-1}$	2464.91	6	4	2.22
2343.96	8	6	$3.13 \cdot 10^{-1}$	2410.52	4	6	1.55	2465.91	8	6	1.62
2344.28	2	4	$9.27 \cdot 10^{-1}$	2411.07	2	2	2.37	2466.50	2	4	2.40
2345.34	14	12	$7.3 \cdot 10^{-1}$	2411.81	10	12	4.33	2466.67	4	2	2.64
2348.12	10	8	$6.50 \cdot 10^{-1}$	2412.01	6	8	$1.66 \cdot 10^{-1}$	2466.82	6	4	1.77
2348.30	6	6	1.15	2413.31	2	4	1.02	2468.30	10	10	$9.8 \cdot 10^{-2}$
2351.20	12	10	$7.19 \cdot 10^{-1}$	2416.45	8	10	2.38	2469.37	10	8	$2.23 \cdot 10^{-2}$
2351.67	6	6	1.80	2417.87	12	12	$9.5 \cdot 10^{-1}$	2469.52	8	6	2.58
2352.31	2	4	4.38	2418.44	6	8	2.28	2470.41	6	6	$6.0 \cdot 10^{-1}$
2353.47	12	14	4.98	2419.89	10	10	$2.2 \cdot 10^{-2}$	2470.67	8	6	1.54
2353.68	8	8	1.30	2422.69	6	8	1.46	2471.28	10	8	$4.15 \cdot 10^{-1}$
2354.48	10	8	$8.13 \cdot 10^{-1}$	2422.93	10	8	$2.94 \cdot 10^{-2}$	2472.61	8	10	3.22
2354.89	6	4	$2.67 \cdot 10^{-1}$	2423.21	4	6	1.40	2473.32	2	2	2.74
2359.11	4	6	$5.0 \cdot 10^{-1}$	2424.15	10	12	2.21	2475.12	4	6	3.72
2359.60	10	10	$2.25 \cdot 10^{-1}$	2424.39	6	8	$1.61 \cdot 10^{-1}$	2475.54	6	8	3.18
2360.00	10	10	$3.59 \cdot 10^{-1}$	2424.59	6	6	1.24	2476.27	8	10	$9.7 \cdot 10^{-2}$
2360.29	8	6	$6.23 \cdot 10^{-1}$	2424.65	8	8	$6.55 \cdot 10^{-2}$	2477.35	8	8	$1.70 \cdot 10^{-1}$
2360.53	6	8	$2.22 \cdot 10^{-1}$	2428.36	8	10	2.68	2478.57	6	6	$9.1 \cdot 10^{-1}$
2361.73	8	8	$2.40 \cdot 10^{-1}$	2428.80	4	4	1.38	2480.16	10	8	1.55
2362.02	8	8	$1.41 \cdot 10^{-1}$	2429.04	2	4	1.23	2481.05	12	12	$1.46 \cdot 10^{-1}$
2363.86	8	10	5.3	2429.39	4	4	$6.9 \cdot 10^{-1}$	2482.12	14	14	$6.5 \cdot 10^{-1}$

λ Å	Weights g_i	g_k	A 10^8 s⁻¹	λ Å	Weights g_i	g_k	A 10^8 s⁻¹	λ Å	Weights g_i	g_k	A 10^8 s⁻¹
2482.33	4	4	2.23	2540.52	2	2	1.26	2588.19	2	2	$1.5 \cdot 10^{-1}$
2482.66	12	10	1.25	2540.66	6	8	1.70	2588.80	8	8	$8.4 \cdot 10^{-2}$
2482.87	6	4	1.69	2541.10	8	6	$9.6 \cdot 10^{-1}$	2590.55	4	6	$7.9 \cdot 10^{-2}$
2483.72	8	10	$5.4 \cdot 10^{-1}$	2541.84	8	6	$8.2 \cdot 10^{-1}$	2591.54	6	6	$5.72 \cdot 10^{-1}$
2484.24	4	6	$8.3 \cdot 10^{-2}$	2542.74	2	2	1.61	2592.79	14	16	2.74
2484.44	8	8	2.16	2543.38	10	12	$6.7 \cdot 10^{-1}$	2593.73	2	4	$1.63 \cdot 10^{-1}$
2489.48	12	12	$5.1 \cdot 10^{-1}$	2543.43	6	4	$8.3 \cdot 10^{-1}$	2594.96	8	8	$1.0 \cdot 10^{-1}$
2489.83	12	12	1.94	2544.97	4	6	$3.93 \cdot 10^{-1}$	2598.37	8	6	1.43
2490.71	10	12	1.44	2545.22	8	10	$5.3 \cdot 10^{-1}$	2599.40	10	10	2.35
2490.86	8	8	$8.8 \cdot 10^{-1}$	2545.44	8	10	$1.52 \cdot 10^{-1}$	2604.05	8	8	$1.49 \cdot 10^{-1}$
2491.40	10	8	1.01	2546.67	8	8	$7.98 \cdot 10^{-1}$	2605.04	6	8	2.34
2492.34	10	12	$2.30 \cdot 10^{-1}$	2547.34	8	8	$2.28 \cdot 10^{-1}$	2605.31	4	4	1.99
2493.26	14	16	3.04	2548.32	4	6	$2.69 \cdot 10^{-1}$	2605.43	6	6	$3.40 \cdot 10^{-1}$
2493.88	6	6	1.74	2548.59	10	10	$2.67 \cdot 10^{-1}$	2605.90	4	2	1.27
2494.12	12	10	$2.97 \cdot 10^{-2}$	2548.74	4	2	2.43	2606.52	6	6	2.31
2497.82	6	6	1.68	2548.92	12	10	$6.0 \cdot 10^{-1}$	2607.09	6	4	1.73
2500.92	6	8	2.41	2549.08	10	8	1.89	2608.85	10	8	$5.0 \cdot 10^{-2}$
2501.35	2	2	1.48	2549.40	4	4	1.65	2609.13	8	10	$2.77 \cdot 10^{-1}$
2502.39	8	8	1.43	2549.46	6	6	1.12	2609.44	6	8	$6.0 \cdot 10^{-2}$
2503.33	12	12	$7.3 \cdot 10^{-1}$	2549.77	8	6	$2.35 \cdot 10^{-1}$	2609.87	8	8	$1.34 \cdot 10^{-1}$
2503.54	8	8	$3.32 \cdot 10^{-1}$	2550.03	10	10	1.74	2611.07	4	6	$7.28 \cdot 10^{-2}$
2503.57	10	10	$2.53 \cdot 10^{-1}$	2550.15	8	10	$3.91 \cdot 10^{-1}$	2611.87	8	8	1.20
2503.88	10	10	2.23	2550.57	12	12	$1.6 \cdot 10^{-2}$	2613.57	10	12	$2.0 \cdot 10^{-2}$
2506.09	10	10	$9.9 \cdot 10^{-1}$	2550.68	12	12	1.07	2613.82	4	2	2.12
2506.80	8	10	1.98	2551.20	10	8	$2.48 \cdot 10^{-1}$	2614.19	8	10	$3.3 \cdot 10^{-2}$
2508.34	8	10	$3.79 \cdot 10^{-1}$	2554.94	8	8	$2.6 \cdot 10^{-2}$	2614.59	10	8	$3.37 \cdot 10^{-2}$
2510.57	8	8	$1.54 \cdot 10^{-1}$	2555.07	6	8	$1.96 \cdot 10^{-1}$	2614.87	8	6	$3.5 \cdot 10^{-2}$
2511.76	8	10	2.30	2555.45	4	6	$2.49 \cdot 10^{-1}$	2617.62	6	6	$4.88 \cdot 10^{-1}$
2513.15	10	8	$2.49 \cdot 10^{-1}$	2557.08	8	10	$2.8 \cdot 10^{-2}$	2619.08	10	10	$2.48 \cdot 10^{-1}$
2514.38	8	8	2.11	2557.51	10	8	$1.53 \cdot 10^{-1}$	2620.17	6	6	$1.1 \cdot 10^{-1}$
2517.14	2	4	$9.2 \cdot 10^{-1}$	2559.24	8	8	$6.4 \cdot 10^{-2}$	2620.41	4	4	$4.30 \cdot 10^{-2}$
2519.05	8	6	2.10	2559.77	6	8	$2.42 \cdot 10^{-1}$	2620.70	8	8	$3.43 \cdot 10^{-1}$
2521.09	6	4	2.05	2559.93	6	8	$2.47 \cdot 10^{-1}$	2621.67	2	2	$5.60 \cdot 10^{-1}$
2521.82	8	8	2.36	2560.28	4	4	1.77	2623.13	14	14	$8.8 \cdot 10^{-2}$
2525.39	14	14	1.91	2562.09	4	2	1.62	2623.72	6	6	$1.92 \cdot 10^{-1}$
2525.92	8	8	$7.4 \cdot 10^{-1}$	2562.54	8	6	1.79	2625.49	12	14	2.55
2526.08	10	8	$3.52 \cdot 10^{-1}$	2563.48	6	4	1.51	2625.67	8	10	$3.52 \cdot 10^{-1}$
2526.29	6	6	2.47	2566.22	8	10	2.61	2626.50	4	6	$3.48 \cdot 10^{-1}$
2527.10	12	10	$3.67 \cdot 10^{-1}$	2566.40	8	6	2.29	2626.70	8	8	$1.94 \cdot 10^{-2}$
2527.71	10	8	$9.1 \cdot 10^{-1}$	2566.62	10	12	$7.1 \cdot 10^{-2}$	2628.29	2	4	$8.74 \cdot 10^{-1}$
2528.68	10	8	$2.3 \cdot 10^{-2}$	2566.91	4	2	1.15	2628.58	6	6	$3.4 \cdot 10^{-2}$
2529.08	4	6	1.80	2568.41	2	4	$4.77 \cdot 10^{-1}$	2629.59	6	8	$7.2 \cdot 10^{-1}$
2529.23	12	10	$3.27 \cdot 10^{-1}$	2568.89	8	8	$2.8 \cdot 10^{-2}$	2630.07	4	6	$5.1 \cdot 10^{-1}$
2529.55	10	10	2.20	2569.78	2	4	1.11	2631.05	4	6	$8.16 \cdot 10^{-1}$
2530.10	4	6	$6.6 \cdot 10^{-1}$	2570.85	8	6	1.84	2631.32	6	8	$6.29 \cdot 10^{-1}$
2533.63	12	12	1.92	2571.55	10	10	$2.89 \cdot 10^{-2}$	2631.61	10	12	$6.6 \cdot 10^{-1}$
2534.42	8	8	1.83	2572.97	6	8	$7.89 \cdot 10^{-2}$	2633.20	6	4	1.21
2535.36	6	4	2.46	2573.21	8	10	$1.42 \cdot 10^{-1}$	2636.70	4	4	$8.8 \cdot 10^{-2}$
2535.49	10	8	$7.47 \cdot 10^{-1}$	2573.76	8	8	$2.3 \cdot 10^{-2}$	2637.50	6	6	$6.2 \cdot 10^{-1}$
2536.67	12	12	$5.7 \cdot 10^{-1}$	2574.37	6	4	2.43	2637.64	2	4	$6.6 \cdot 10^{-1}$
2536.81	10	10	1.69	2576.86	10	12	1.32	2639.57	2	2	$8.0 \cdot 10^{-1}$
2536.84	12	14	$6.8 \cdot 10^{-1}$	2577.92	2	2	1.24	2641.12	4	4	$3.7 \cdot 10^{-2}$
2537.14	10	10	1.44	2581.11	6	6	$7.61 \cdot 10^{-2}$	2642.01	6	6	$2.29 \cdot 10^{-1}$
2538.21	14	12	1.26	2582.41	6	8	$2.22 \cdot 10^{-1}$	2646.21	12	10	$1.44 \cdot 10^{-2}$
2538.40	6	8	$3.7 \cdot 10^{-2}$	2582.58	4	4	$8.80 \cdot 10^{-1}$	2649.47	6	8	1.98
2538.50	8	6	$5.9 \cdot 10^{-1}$	2583.05	8	10	$2.16 \cdot 10^{-2}$	2650.48	6	8	1.60
2538.68	6	8	$7.4 \cdot 10^{-1}$	2585.62	10	10	$3.09 \cdot 10^{-1}$	2652.57	10	8	$4.45 \cdot 10^{-2}$
2538.91	10	8	1.28	2585.88	10	8	$8.94 \cdot 10^{-1}$	2654.63	4	4	$8.1 \cdot 10^{-1}$
2538.99	14	12	1.93	2586.06	6	4	$5.8 \cdot 10^{-2}$	2657.92	10	10	$3.2 \cdot 10^{-2}$
2539.81	8	8	$5.6 \cdot 10^{-2}$	2587.95	8	10	1.69	2658.25	8	8	$2.12 \cdot 10^{-1}$

λ Å	g_i	g_k	A 10^8 s^{-1}	λ Å	g_i	g_k	A 10^8 s^{-1}	λ Å	g_i	g_k	A 10^8 s^{-1}
2662.56	2	2	1.33	2761.81	2	4	$1.38 \cdot 10^{-1}$	2849.61	10	12	$4.6 \cdot 10^{-2}$
2664.66	8	10	1.91	2762.33	6	6	$6.0 \cdot 10^{-1}$	2853.21	6	6	$2.3 \cdot 10^{-2}$
2666.64	6	8	1.87	2762.45	6	4	$3.3 \cdot 10^{-2}$	2855.67	8	10	$9.2 \cdot 10^{-2}$
2667.22	4	6	1.02	2763.66	14	12	1.34	2856.15	10	10	$5.0 \cdot 10^{-2}$
2669.93	2	4	$5.2 \cdot 10^{-1}$	2763.91	8	6	$2.9 \cdot 10^{-2}$	2856.38	6	8	$4.42 \cdot 10^{-1}$
2670.38	6	8	$6.0 \cdot 10^{-2}$	2764.79	12	12	$1.1 \cdot 10^{-2}$	2856.91	8	8	1.32
2671.39	2	4	$6.5 \cdot 10^{-1}$	2765.13	10	8	1.47	2857.17	6	8	$1.22 \cdot 10^{-1}$
2680.23	6	6	$1.10 \cdot 10^{-1}$	2767.50	12	14	1.58	2857.42	6	6	$2.0 \cdot 10^{-2}$
2682.51	8	10	$9.2 \cdot 10^{-1}$	2768.93	4	6	$4.75 \cdot 10^{-2}$	2858.34	10	12	$4.85 \cdot 10^{-1}$
2683.00	4	6	$7.3 \cdot 10^{-1}$	2769.15	8	10	$6.6 \cdot 10^{-2}$	2864.97	8	8	$4.3 \cdot 10^{-2}$
2684.75	8	10	1.57	2769.35	12	14	$2.07 \cdot 10^{-1}$	2869.16	8	10	$1.4 \cdot 10^{-2}$
2691.74	10	8	$5.04 \cdot 10^{-2}$	2770.50	12	10	$4.08 \cdot 10^{-2}$	2869.31	4	6	$4.04 \cdot 10^{-1}$
2692.60	10	12	1.40	2771.19	10	12	$4.3 \cdot 10^{-2}$	2871.06	10	12	$2.2 \cdot 10^{-2}$
2693.86	8	6	$4.2 \cdot 10^{-2}$	2774.69	2	4	$2.73 \cdot 10^{-1}$	2871.13	12	10	$3.0 \cdot 10^{-2}$
2697.33	4	4	$2.48 \cdot 10^{-1}$	2776.18	6	8	$2.66 \cdot 10^{-2}$	2872.38	10	8	$1.70 \cdot 10^{-1}$
2697.46	4	2	1.65	2776.91	8	8	$4.08 \cdot 10^{-1}$	2873.40	8	10	$4.56 \cdot 10^{-1}$
2697.73	10	8	$2.6 \cdot 10^{-2}$	2779.30	10	8	1.00	2875.35	8	10	$1.35 \cdot 10^{-1}$
2699.20	4	4	$6.2 \cdot 10^{-1}$	2779.91	2	4	$2.56 \cdot 10^{-1}$	2876.80	8	8	$9.56 \cdot 10^{-2}$
2703.99	8	8	1.38	2780.05	2	2	$3.3 \cdot 10^{-1}$	2879.25	10	8	$3.6 \cdot 10^{-2}$
2704.58	8	8	$1.66 \cdot 10^{-2}$	2783.69	12	10	1.06	2880.76	8	8	$2.21 \cdot 10^{-2}$
2707.13	4	6	$8.3 \cdot 10^{-1}$	2784.28	2	4	$3.4 \cdot 10^{-2}$	2883.71	12	14	$1.48 \cdot 10^{-1}$
2709.06	4	6	$3.88 \cdot 10^{-1}$	2785.19	12	10	1.53	2884.76	6	8	$2.46 \cdot 10^{-1}$
2711.84	12	14	$4.36 \cdot 10^{-1}$	2787.24	8	6	$1.83 \cdot 10^{-1}$	2885.93	14	12	$3.8 \cdot 10^{-2}$
2712.39	10	12	$1.29 \cdot 10^{-1}$	2790.56	8	10	$2.1 \cdot 10^{-2}$	2888.10	4	6	$6.1 \cdot 10^{-2}$
2714.41	8	6	$5.70 \cdot 10^{-1}$	2793.89	10	12	$1.26 \cdot 10^{-1}$	2894.78	10	12	$5.7 \cdot 10^{-2}$
2716.22	6	6	1.15	2796.63	10	10	$2.0 \cdot 10^{-1}$	2895.22	8	10	$1.09 \cdot 10^{-1}$
2716.44	6	6	$2.8 \cdot 10^{-2}$	2797.92	10	10	$3.2 \cdot 10^{-2}$	2897.27	6	4	$1.8 \cdot 10^{-1}$
2716.57	14	12	1.35	2799.29	10	8	$1.55 \cdot 10^{-1}$	2902.46	10	10	$3.2 \cdot 10^{-2}$
2717.88	16	14	1.51	2805.32	4	6	$2.5 \cdot 10^{-2}$	2906.12	2	4	$4.4 \cdot 10^{-1}$
2718.64	10	8	1.18	2805.79	8	8	$3.22 \cdot 10^{-1}$	2910.76	8	8	$1.5 \cdot 10^{-2}$
2719.30	6	8	$4.44 \cdot 10^{-1}$	2809.78	8	8	$3.10 \cdot 10^{-1}$	2917.08	6	8	$2.7 \cdot 10^{-2}$
2721.81	12	10	$5.1 \cdot 10^{-2}$	2811.27	12	10	$1.2 \cdot 10^{-2}$	2922.02	8	10	$3.8 \cdot 10^{-2}$
2722.06	8	8	$1.42 \cdot 10^{-1}$	2812.49	4	4	$2.9 \cdot 10^{-2}$	2926.59	8	10	$5.1 \cdot 10^{-2}$
2722.74	6	8	$8.2 \cdot 10^{-1}$	2813.61	8	10	$3.40 \cdot 10^{-2}$	2944.40	4	2	$3.5 \cdot 10^{-1}$
2724.88	6	6	$9.58 \cdot 10^{-2}$	2817.09	6	4	$3.37 \cdot 10^{-1}$	2947.65	6	4	$2.01 \cdot 10^{-1}$
2726.52	6	8	$5.0 \cdot 10^{-2}$	2819.34	12	12	$9.7 \cdot 10^{-3}$	2949.18	10	8	$2.45 \cdot 10^{-1}$
2727.38	12	10	$3.12 \cdot 10^{-1}$	2826.03	8	6	$4.5 \cdot 10^{-2}$	2953.77	6	8	$5.2 \cdot 10^{-2}$
2727.54	6	4	$9.38 \cdot 10^{-1}$	2827.43	12	14	$2.4 \cdot 10^{-2}$	2954.05	8	8	$1.2 \cdot 10^{-2}$
2728.91	8	10	$1.25 \cdot 10^{-1}$	2828.63	12	10	$6.9 \cdot 10^{-2}$	2959.60	8	6	$9.7 \cdot 10^{-2}$
2730.73	4	4	$2.79 \cdot 10^{-1}$	2831.56	4	6	$7.6 \cdot 10^{-1}$	2959.84	8	6	$1.36 \cdot 10^{-1}$
2732.01	10	8	$7.05 \cdot 10^{-2}$	2833.09	6	6	$4.55 \cdot 10^{-1}$	2964.13	8	6	$4.6 \cdot 10^{-2}$
2732.94	8	6	$9.5 \cdot 10^{-1}$	2835.71	4	6	$5.1 \cdot 10^{-1}$	2964.62	2	2	$6.5 \cdot 10^{-2}$
2736.97	4	2	1.22	2836.19	4	4	$5.4 \cdot 10^{-2}$	2965.03	4	4	$9.43 \cdot 10^{-2}$
2739.55	8	8	2.21	2836.51	2	4	$9.8 \cdot 10^{-2}$	2969.94	8	6	$2.28 \cdot 10^{-1}$
2741.39	6	6	$2.03 \cdot 10^{-1}$	2837.30	10	12	$1.9 \cdot 10^{-2}$	2970.52	4	6	$2.70 \cdot 10^{-2}$
2743.20	2	4	1.97	2838.22	4	2	$8.6 \cdot 10^{-1}$	2970.69	10	8	$4.15 \cdot 10^{-2}$
2744.90	6	8	$3.62 \cdot 10^{-2}$	2839.51	10	8	1.47	2982.06	4	6	$2.41 \cdot 10^{-1}$
2746.48	4	6	2.05	2839.80	8	10	$5.8 \cdot 10^{-1}$	2984.82	6	6	$4.29 \cdot 10^{-1}$
2746.98	6	6	1.69	2840.34	12	12	$7.7 \cdot 10^{-2}$	2985.54	2	4	$2.39 \cdot 10^{-1}$
2749.18	4	4	1.21	2840.65	2	4	$7.6 \cdot 10^{-1}$	2997.30	6	6	$8.6 \cdot 10^{-2}$
2749.32	6	8	2.16	2840.76	10	12	$1.49 \cdot 10^{-1}$	3000.06	8	6	$3.0 \cdot 10^{-2}$
2749.49	2	2	1.16	2842.08	8	8	$1.5 \cdot 10^{-2}$	3002.32	6	8	$2.0 \cdot 10^{-2}$
2750.01	10	10	$1.8 \cdot 10^{-2}$	2843.32	10	10	$1.40 \cdot 10^{-1}$	3002.65	4	6	$1.79 \cdot 10^{-1}$
2751.13	4	4	$2.92 \cdot 10^{-1}$	2843.48	4	6	$9.6 \cdot 10^{-2}$	3036.96	6	6	$2.22 \cdot 10^{-1}$
2752.15	4	4	$7.7 \cdot 10^{-1}$	2844.96	2	2	$5.5 \cdot 10^{-1}$	3044.84	8	10	$1.2 \cdot 10^{-2}$
2753.29	10	12	1.89	2845.60	8	6	1.57	3048.99	4	4	$3.84 \cdot 10^{-1}$
2754.89	8	6	1.21	2847.77	4	4	$5.1 \cdot 10^{-1}$	3056.80	14	12	$1.7 \cdot 10^{-2}$
2755.74	8	10	2.15	2848.11	6	6	$9.9 \cdot 10^{-1}$	3062.24	12	10	$1.36 \cdot 10^{-1}$
2756.51	6	8	$7.3 \cdot 10^{-2}$	2848.32	6	4	1.59	3065.32	6	6	$2.9 \cdot 10^{-2}$
2757.03	10	8	$8.07 \cdot 10^{-2}$	2848.91	12	10	$5.3 \cdot 10^{-2}$	3070.69	10	8	$1.28 \cdot 10^{-2}$

λ Å	g_i	g_k	A $10^8 \, s^{-1}$	λ Å	g_i	g_k	A $10^8 \, s^{-1}$	λ Å	g_i	g_k	A $10^8 \, s^{-1}$
3071.12	2	4	$2.59 \cdot 10^{-1}$	5018.44	6	6	$2.0 \cdot 10^{-2}$	1987.50	13	13	4.9
3076.44	4	6	$3.75 \cdot 10^{-1}$	5030.63	10	10	$7.1 \cdot 10^{-1}$				
3077.17	14	12	$1.35 \cdot 10^{-1}$	5035.71	10	12	$9.4 \cdot 10^{-1}$	*Fe VII*			
3078.68	6	8	$5.5 \cdot 10^{-1}$	5144.35	4	6	$8.5 \cdot 10^{-1}$	150.807	5	7	$1.3 \cdot 10^3$
3089.38	6	8	$2.2 \cdot 10^{-2}$	5149.47	8	10	$9.0 \cdot 10^{-1}$	150.852	7	9	$1.3 \cdot 10^3$
3096.29	8	8	$1.9 \cdot 10^{-2}$	5169.03	6	8	$4.22 \cdot 10^{-2}$	151.023	9	11	$1.6 \cdot 10^3$
3105.17	4	2	$7.5 \cdot 10^{-2}$	5197.58	6	4	$5.4 \cdot 10^{-3}$	151.046	7	7	$2.2 \cdot 10^2$
3105.55	2	2	$7.0 \cdot 10^{-2}$	5227.48	12	14	1.22	151.145	9	9	$2.1 \cdot 10^2$
3106.57	8	8	$1.88 \cdot 10^{-2}$	5247.95	4	6	1.43	151.432	5	7	$2.2 \cdot 10^2$
3114.30	4	4	$6.4 \cdot 10^{-2}$	5251.23	6	8	$8.0 \cdot 10^{-1}$	151.512	5	5	$5.3 \cdot 10^2$
3114.69	2	4	$2.5 \cdot 10^{-2}$	5264.18	8	10	$4.76 \cdot 10^{-1}$	151.675	7	7	$3.9 \cdot 10^2$
3116.58	6	4	$5.5 \cdot 10^{-2}$	5272.40	6	6	$3.9 \cdot 10^{-3}$	151.782	9	9	$2.4 \cdot 10^2$
3133.05	4	6	$1.5 \cdot 10^{-2}$	5276.00	10	8	$3.76 \cdot 10^{-3}$	154.307	3	1	$8.9 \cdot 10^2$
3135.36	6	6	$8.8 \cdot 10^{-2}$	5306.18	6	8	$3.28 \cdot 10^{-1}$	154.335	5	7	$1.2 \cdot 10^3$
3144.75	8	6	$2.7 \cdot 10^{-2}$	5316.22	14	14	$3.69 \cdot 10^{-1}$	154.363	3	3	$4.2 \cdot 10^2$
3154.20	10	10	$2.06 \cdot 10^{-1}$	5316.62	12	10	$3.89 \cdot 10^{-3}$	154.565	5	3	$3.5 \cdot 10^2$
3162.80	8	8	$5.5 \cdot 10^{-2}$	5387.06	12	14	$5.2 \cdot 10^{-1}$	154.650	5	5	$8.8 \cdot 10^2$
3167.86	8	8	$1.59 \cdot 10^{-1}$	5395.86	6	8	$5.5 \cdot 10^{-1}$	154.848	1	3	$7.7 \cdot 10^2$
3177.53	8	8	$1.74 \cdot 10^{-1}$	5402.06	10	12	$5.6 \cdot 10^{-1}$	154.921	3	5	$9.7 \cdot 10^2$
3179.50	6	8	$1.11 \cdot 10^{-1}$	5427.83	12	10	$5.9 \cdot 10^{-3}$	154.941	3	3	$2.4 \cdot 10^2$
3180.15	4	6	$7.7 \cdot 10^{-2}$	5429.99	8	10	$6.0 \cdot 10^{-1}$	154.949	5	7	$1.0 \cdot 10^3$
3186.74	4	4	$3.85 \cdot 10^{-2}$	5465.93	6	8	$6.2 \cdot 10^{-1}$	155.994	9	11	$1.8 \cdot 10^3$
3187.30	10	10	$5.0 \cdot 10^{-2}$	5482.31	10	12	$4.78 \cdot 10^{-1}$	158.481	9	9	$2.3 \cdot 10^2$
3192.91	6	6	$1.27 \cdot 10^{-2}$	5493.83	8	10	$4.01 \cdot 10^{-1}$	165.087	1	3	$6.9 \cdot 10^2$
3193.80	2	2	$5.4 \cdot 10^{-2}$	5506.19	12	14	1.14	165.919	7	5	$2.8 \cdot 10^3$
3193.86	8	8	$3.86 \cdot 10^{-2}$	5510.78	10	12	$2.28 \cdot 10^{-1}$	166.365	9	7	$2.9 \cdot 10^3$
3196.07	6	8	$1.61 \cdot 10^{-1}$	5529.05	6	6	$2.01 \cdot 10^{-1}$	173.441	9	9	$3.6 \cdot 10^3$
3210.44	2	4	$3.63 \cdot 10^{-2}$	5544.76	12	12	$2.49 \cdot 10^{-1}$	176.744	9	9	$2.7 \cdot 10^3$
3213.31	4	6	$6.12 \cdot 10^{-2}$	5783.63	8	10	$4.62 \cdot 10^{-1}$	176.928	7	7	$2.4 \cdot 10^3$
3227.74	6	8	$8.9 \cdot 10^{-2}$	5885.01	4	6	$6.4 \cdot 10^{-1}$	177.172	5	5	$1.5 \cdot 10^3$
3231.71	6	8	$1.4 \cdot 10^{-2}$	5902.83	8	10	$4.98 \cdot 10^{-1}$	235.221	5	3	$1.7 \cdot 10^2$
3232.78	8	6	$5.0 \cdot 10^{-2}$	5955.70	6	8	$4.19 \cdot 10^{-1}$	240.053	3	1	$1.3 \cdot 10^2$
3237.82	2	4	$6.8 \cdot 10^{-2}$	5961.71	10	12	$7.4 \cdot 10^{-1}$	243.379	9	7	$2.1 \cdot 10^2$
3243.72	10	8	$5.1 \cdot 10^{-2}$	5965.62	10	10	$2.19 \cdot 10^{-1}$				
3247.17	4	6	$7.1 \cdot 10^{-2}$	6175.15	8	8	$1.8 \cdot 10^{-3}$	*Fe VIII*			
3258.77	6	8	$9.39 \cdot 10^{-2}$	6305.30	10	10	$1.4 \cdot 10^{-3}$	112.472	4	4	$3.6 \cdot 10^2$
3259.05	8	10	$6.7 \cdot 10^{-2}$	6331.95	6	8	$1.8 \cdot 10^{-3}$	112.486	6	6	$4.3 \cdot 10^2$
3276.60	6	8	$1.0 \cdot 10^{-2}$	6446.41	8	10	$1.3 \cdot 10^{-3}$	116.196	4	6	$4.5 \cdot 10^2$
3289.35	8	8	$2.1 \cdot 10^{-2}$	6456.38	8	6	$1.7 \cdot 10^{-3}$	117.197	6	8	$3.8 \cdot 10^2$
3323.06	8	10	$1.4 \cdot 10^{-2}$					167.486	4	4	$3.0 \cdot 10^3$
3366.97	8	6	$2.2 \cdot 10^{-2}$	*Fe III*				168.172	6	6	$3.1 \cdot 10^3$
3381.01	6	4	$3.0 \cdot 10^{-2}$	1843.4	9	7	4.8	168.545	6	4	$2.0 \cdot 10^3$
3453.62	8	10	$8.5 \cdot 10^{-3}$	1844.3	7	5	4.9	168.929	4	2	$2.1 \cdot 10^3$
3468.68	8	8	$2.0 \cdot 10^{-2}$	1846.9	5	3	5.5	185.213	6	8	$1.0 \cdot 10^3$
3493.47	10	10	$3.2 \cdot 10^{-2}$	1854.38	3	1	5.7	186.601	4	6	$9.4 \cdot 10^2$
3621.27	2	4	$2.2 \cdot 10^{-2}$	1865.20	7	7	6.1				
3748.48	6	4	$3.4 \cdot 10^{-2}$	1893.98	11	9	5.5	*Fe X*			
3759.46	4	2	$3.2 \cdot 10^{-2}$	1896.80	13	11	5.0	76.822	2	2	$1.8 \cdot 10^3$
3906.04	6	8	$1.1 \cdot 10^{-2}$	1904.3	5	5	5.7	77.865	4	6	$1.6 \cdot 10^3$
3935.96	8	10	$8.3 \cdot 10^{-3}$	1907.58	15	13	5.3	100.026	8	10	$2.6 \cdot 10^3$
3938.97	4	6	$8.4 \cdot 10^{-3}$	1915.08	13	15	6.0	101.733	6	8	$1.8 \cdot 10^3$
4233.17	6	8	$7.22 \cdot 10^{-3}$	1922.79	11	13	5.5	101.846	4	6	$1.7 \cdot 10^3$
4522.63	6	4	$8.4 \cdot 10^{-3}$	1930.39	9	11	5.1	102.095	10	12	$2.9 \cdot 10^3$
4549.19	4	6	$9.2 \cdot 10^{-3}$	1931.51	9	11	5.3	102.192	10	12	$2.9 \cdot 10^3$
4549.47	8	6	$1.00 \cdot 10^{-2}$	1937.35	7	9	5.1	102.829	4	6	$2.1 \cdot 10^3$
4583.84	10	8	$7.22 \cdot 10^{-3}$	1943.48	5	7	5.0	103.319	6	8	$2.6 \cdot 10^3$
4635.32	6	8	$1.0 \cdot 10^{-2}$	1950.33	13	15	5.5	103.724	6	8	$1.7 \cdot 10^3$
4923.93	6	4	$4.28 \cdot 10^{-2}$	1951.01	11	11	5.3	104.638	8	10	$2.1 \cdot 10^3$
4990.51	6	8	$5.2 \cdot 10^{-1}$	1952.65	9	9	4.9	174.534	4	6	$1.8 \cdot 10^3$
5001.96	12	14	1.57	1953.32	7	7	5.1	175.266	2	4	$1.72 \cdot 10^3$

λ Å	Weights g_i	g_k	A 10^8 s^{-1}	λ Å	Weights g_i	g_k	A 10^8 s^{-1}	λ Å	Weights g_i	g_k	A 10^8 s^{-1}
				219	2	4	4.8·10²	96.348	6	8	9.3·10²
Fe XI				219	4	6	2.4·10²	117.2	2	4	3.93·10²
72.166	5	7	2.9·10³	219.123	4	6	3.9·10²	117.7	2	2	3.9·10²
72.310	5	5	1.5·10³	220	4	4	3.2·10²	123.4	2	4	5.9·10²
72.635	5	7	1.6·10³	221	4	6	5.9·10²	124.5	4	6	7.0·10²
91.394	5	7	2.6·10³	226	2	4	3.9·10²	144.06	4	6	1.6·10³
91.472	7	9	2.5·10³	264.787	4	4	3.38·10²	144.25	6	8	1.6·10³
91.63	3	5	2.3·10³	268	6	6	2.1·10²	148	4	2	6.5·10²
91.63	7	9	3.4·10³	280	4	6	2.8·10²	266.7	4	6	3.9·10²
91.63	5	7	2.8·10³	283	6	8	2.7·10²	267.0	6	8	4.3·10²
91.733	9	11	4.1·10³	288.45	6	4	1.6·10²				
92.81	9	11	3.7·10³								
92.87	11	13	3.9·10³	**Fe XV**				**Fe XVII**			
93.433	9	11	3.2·10³	38.95	1	3	1.69·10³	11.023	1	3	2.1·10⁴
179.762	5	7	1.67·10³	52.911	1	3	2.94·10³	12.123	1	3	8.0·10⁴
				59.404	3	5	3.4·10³	12.264	1	3	5.9·10⁴
Fe XII				63.959	5	7	1.6·10³	12.526	1	3	3.0·10³
65.905	4	4	2.0·10³	65.370	1	3	3.2·10²	12.681	1	3	3.5·10³
66.526	6	8	1.7·10³	65.612	3	3	9.8·10²	13.823	1	3	3.3·10⁴
66.960	4	6	1.6·10³	66.238	5	3	1.6·10³	13.891	1	3	3.4·10³
67.164	4	2	1.1·10³	68.860	9	11	9.2·10³	15.015	1	3	2.28·10⁵
67.821	4	6	1.4·10³	69.7	3	1	1.9·10³	15.262	1	3	6.0·10⁴
68.382	2	4	1.7·10³	69.942	3	5	7.4·10³	16.777	1	3	8.29·10³
80.541	6	6	8.7·10²	69.989	5	7	7.9·10³	17.054	1	3	9.33·10³
81.943	6	4	1.4·10³	70.052	7	9	8.8·10³	41.37	9	11	4.8·10³
82.226	4	2	1.9·10³	70.224	1	3	4.13·10³	49.427	3	3	4.0·10³
84.48	4	6	4.5·10³	70.53	7	5	2.6·10²	50.26	7	9	6.0·10³
84.48	8	10	4.9·10³	70.59	7	7	1.7·10³	58.76	9	11	1.2·10⁴
84.52	10	12	5.2·10³	73.199	7	9	8.8·10³				
84.52	6	8	4.0·10³	73.473	5	7	6.2·10³	**Fe XIX**			
84.85	6	8	2.3·10³	233.857	5	7	2.2·10²	13.413	5	3	1.3·10⁴
85.14	8	10	3.4·10³	235	1	3	2.5·10²	13.426	5	7	4.8·10⁴
85.477	10	12	4.6·10³	243	1	3	2.4·10²	13.47	3	1	1.5·10⁵
186.880	6	8	1.0·10³	243	5	7	2.3·10²	13.520	5	7	2.0·10⁵
192.394	4	2	9.0·10²	243.790	3	5	4.2·10²	13.56	3	5	1.0·10⁴
193.509	4	4	9.1·10²	248	3	1	5.4·10²	13.68	3	1	8.0·10⁴
195.119	4	6	8.6·10²	284.160	1	3	2.28·10²	13.69	5	7	2.3·10⁴
								13.700	1	3	2.7·10⁵
Fe XIII				**Fe XVI**				13.71	5	5	2.2·10⁴
62.353	1	3	2.0·10³	39.827	2	4	2.1·10³	13.738	5	7	1.0·10⁴
62.46	5	7	1.2·10³	40.153	4	6	2.5·10³	13.796	5	7	7.0·10⁴
62.699	3	5	2.3·10³	40.199	4	6	1.7·10³	13.83	5	5	1.4·10⁴
63.188	5	7	3.9·10³	40.245	6	8	1.8·10³	13.934	1	3	4.51·10⁴
64.139	1	3	2.1·10³	46.661	4	6	3.46·10³	13.961	3	3	2.0·10⁴
74.845	5	5	1.0·10³	46.718	6	8	3.7·10³	14.668	5	7	1.1·10⁴
75.892	5	3	7.7·10²	50.350	2	4	1.86·10³	14.671	5	3	1.1·10⁴
76.117	5	3	2.1·10³	54.142	2	4	3.41·10³	14.929	3	3	1.2·10⁴
78.452	9	11	6.3·10³	54.728	4	6	4.16·10³	14.966	5	3	2.5·10⁴
84.270	7	9	5.5·10³	62.879	2	2	1.05·10³	14.995	5	5	2.2·10⁴
107.384	7	5	1.8·10³	63.719	4	2	2.18·10³	15.015	1	3	1.4·10⁴
				66.263	4	6	9.39·10³	16.668	3	1	1.1·10⁴
Fe XIV				66.368	6	8	1.00·10⁴				
72.80	10	12	7.9·10³	66.392	6	6	6.69·10²	**Fe XX**			
76.022	4	6	6.6·10³	76.502	6	4	6.7·10²	12.77	4	4	2.1·10⁵
76.152	6	8	7.0·10³	76.796	4	2	7.72·10²	12.78	4	2	6.9·10⁴
190	6	8	2.8·10²	80.192	4	6	5.2·10²	12.78	2	4	1.4·10⁵
211.316	2	4	3.6·10²	80.270	6	8	5.4·10²	12.82	4	4	1.1·10⁵
216	6	8	1.7·10²	85.587	2	4	4.0·10²	12.88	6	4	2.7·10⁴
217	6	8	4.0·10²	86.133	4	6	4.8·10²	12.89	4	4	4.4·10⁴
217	6	6	2.6·10²	96.256	4	6	8.7·10²	12.90	4	6	1.4·10⁵
								12.93	4	6	1.6·10⁵

λ Å	Weights g_i	g_k	A 10^8 s^{-1}
12.98	2	2	6.7·10^4
12.99	6	6	5.1·10^4
13.01	2	4	3.0·10^4
13.03	4	2	8.6·10^4
13.13	2	4	8.9·10^4
13.79	6	6	1.2·10^4
Fe XXI			
8.56	5	7	2.0·10^4
8.64	5	7	1.5·10^4
8.65	5	7	3.9·10^4
9.42	3	3	3.3·10^4
9.44	3	5	1.7·10^4
9.45	1	3	5.2·10^4
9.47	5	7	4.9·10^4
9.67	1	3	5.7·10^4
12.02	1	3	1.3·10^4
12.13	3	3	1.8·10^4
12.18	5	7	2.2·10^4
12.21	3	1	1.5·10^5
12.21	3	3	1.2·10^5
12.25	1	3	2.1·10^5
12.28	5	3	5.2·10^4
12.30	5	7	2.1·10^5
12.36	3	3	3.6·10^4
12.37	5	7	3.1·10^5
12.47	5	7	5.8·10^4
12.47	5	3	1.3·10^4
12.49	5	7	1.3·10^4
12.53	5	5	1.5·10^4
12.57	1	3	7.2·10^4
12.73	5	5	8.2·10^3
12.95	3	5	6.2·10^3
13.03	5	5	1.3·10^4
Fe XXII			
9.002	4	6	5.5·10^4
9.006	6	8	5.7·10^4
9.006	6	6	5.3·10^4
9.163	4	6	6.9·10^4
9.183	6	8	8.3·10^4
9.241	4	6	5.1·10^4
11.748	4	4	1.2·10^5
11.748	4	6	1.6·10^5
11.748	4	2	1.8·10^5
11.763	2	4	1.6·10^5
11.789	2	2	2.6·10^5
11.789	6	8	1.2·10^5
11.797	2	4	1.7·10^5
11.823	6	4	7.9·10^4
11.837	6	8	2.3·10^5
11.837	6	6	1.7·10^5
11.886	4	6	1.3·10^5
11.898	2	4	8.2·10^4
11.922	4	6	1.8·10^5
11.976	6	8	5.9·10^4
12.027	2	4	6.9·10^4
12.045	6	8	2.4·10^5
12.045	4	4	9.7·10^4
12.053	4	6	6.1·10^4

λ Å	Weights g_i	g_k	A 10^8 s^{-1}
12.077	2	4	1.0·10^5
12.077	4	6	2.4·10^5
12.095	6	6	7.8·10^4
12.193	2	4	7.2·10^4
12.193	4	6	9.9·10^4
12.325	2	2	1.5·10^5
Fe XXIII			
8.614	5	7	7.7·10^4
8.752	5	7	1.2·10^5
10.927	5	7	6.0·10^4
10.934	3	5	5.4·10^4
11.165	3	5	6.7·10^4
11.298	1	3	1.3·10^5
11.325	3	5	1.7·10^5
11.338	3	3	9.3·10^4
11.433	3	3	1.2·10^5
11.441	5	7	2.2·10^5
11.445	5	5	5.6·10^4
11.485	3	5	1.40·10^5
11.519	5	5	1.16·10^5
11.520	1	3	2.16·10^5
11.524	5	7	2.3·10^5
11.593	5	7	3.58·10^5
11.613	3	5	1.0·10^5
11.691	5	7	7.7·10^4
11.698	5	5	7.3·10^4
11.737	3	5	1.8·10^5
11.898	1	3	2.03·10^5
Fe XXIV			
1.8523	2	2	1.0·10^5
1.8552	2	4	4.82·10^6
1.8563	4	2	2.43·10^6
1.8572	2	2	3.06·10^6
1.858	2	4	1.2·10^5
1.8614	4	4	6.24·10^6
1.8626	2	4	3.16·10^6
1.8627	2	2	5.47·10^6
1.8637	2	2	1.91·10^6
1.8655	4	6	2.14·10^6
1.8672	4	2	1.63·10^6
1.8678	4	4	3.5·10^5
1.8721	4	6	3.2·10^5
1.8721	2	2	2.0·10^5
1.8730	2	4	1.5·10^5
1.8739	4	4	8.3·10^4
1.891	2	2	9.7·10^4
1.897	4	2	9.8·10^4
8.231	2	4	6.10·10^4
8.316	4	6	7.07·10^4
10.619	2	4	7.28·10^4
10.663	2	2	7.51·10^4
11.030	2	4	1.84·10^5
11.171	4	6	2.18·10^5
Fe XXV			
1.4607	1	3	2.54·10^5
1.4945	1	3	5.05·10^5
1.5730	1	3	1.24·10^6

λ Å	Weights g_i	g_k	A 10^8 s^{-1}
1.5749	1	3	1.5·10^5
1.778	3	3	8.7·10^4
1.782	3	1	4.69·10^6
1.787	1	3	2.57·10^6
1.787	5	5	1.19·10^6
1.788	3	5	2.68·10^6
1.788	3	5	1.63·10^6
1.789	1	3	1.78·10^6
1.790	3	3	1.23·10^6
1.791	3	5	4.10·10^6
1.791	3	3	2.59·10^6
1.792	3	1	4.92·10^6
1.792	5	5	2.81·10^6
1.793	3	1	2.67·10^6
1.794	5	3	2.22·10^6
1.797	3	5	8.8·10^5
1.798	3	3	1.0·10^5
1.800	1	3	8.6·10^4
1.802	3	1	4.1·10^5
1.810	3	1	5.9·10^5
1.8502	1	3	4.57·10^6
1.8593	1	3	4.42·10^5
10.038	3	3	8.08·10^4
Krypton			
Kr I			
1164.9	1	3	3.16
1235.8	1	3	3.12
4274.0	5	5	2.6·10^{-2}
4351.4	3	1	3.2·10^{-2}
4362.6	5	3	8.4·10^{-3}
4376.1	3	1	5.6·10^{-2}
4400.0	3	5	2.0·10^{-2}
4410.4	3	3	4.4·10^{-3}
4425.2	3	3	9.7·10^{-3}
4453.9	3	5	7.8·10^{-3}
4463.7	3	3	2.3·10^{-2}
4502.4	3	5	9.2·10^{-3}
5562.2	5	5	2.8·10^{-3}
5570.3	5	3	2.1·10^{-2}
5649.6	1	3	3.7·10^{-3}
5870.9	3	5	1.8·10^{-2}
6904.7	3	5	1.3·10^{-2}
7224.1	3	5	1.4·10^{-2}
7587.4	3	1	5.1·10^{-1}
7601.5	5	5	3.1·10^{-1}
7685.2	3	1	4.9·10^{-1}
7694.5	5	3	5.6·10^{-2}
7854.8	1	3	2.3·10^{-1}
8059.5	1	3	1.9·10^{-1}
8104.4	5	5	1.3·10^{-1}
8112.9	5	7	3.6·10^{-1}
8190.1	3	5	1.1·10^{-1}
8263.2	3	5	3.5·10^{-1}
8281.1	3	5	1.9·10^{-1}
8298.1	3	3	3.2·10^{-1}
8508.9	3	3	2.4·10^{-1}
8776.7	3	5	2.7·10^{-1}
8928.7	5	3	3.7·10^{-1}

λ Å	Weights g_i	g_k	A $10^8\,\text{s}^{-1}$
Kr II			
4250.6	4	4	$1.2\cdot10^{-1}$
4292.9	4	4	$9.6\cdot10^{-1}$
4355.5	6	8	1.0
4431.7	2	2	1.8
4436.8	2	4	$6.6\cdot10^{-1}$
4577.2	6	8	$9.6\cdot10^{-1}$
4583.0	6	4	$7.6\cdot10^{-1}$
4615.3	4	4	$5.4\cdot10^{-1}$
4619.2	4	6	$8.1\cdot10^{-1}$
4633.9	4	6	$7.1\cdot10^{-1}$
4658.9	6	4	$6.5\cdot10^{-1}$
4739.0	6	6	$7.6\cdot10^{-1}$
4762.4	2	4	$4.2\cdot10^{-1}$
4765.7	4	6	$6.7\cdot10^{-1}$
4811.8	2	4	$1.7\cdot10^{-1}$
4825.2	2	4	$1.9\cdot10^{-1}$
4832.1	4	2	$7.3\cdot10^{-1}$
5208.3	4	4	$1.4\cdot10^{-1}$
5308.7	4	6	$2.4\cdot10^{-2}$
7407.0	6	6	$7.0\cdot10^{-2}$
Lead			
Pb I			
2022.0	1	3	$5.2\cdot10^{-2}$
2053.3	1	3	$1.2\cdot10^{-1}$
2170.0	1	3	1.5
2401.9	3	3	$1.9\cdot10^{-1}$
2446.2	3	3	$2.5\cdot10^{-1}$
2476.4	3	5	$2.8\cdot10^{-1}$
2577.3	5	3	$5.0\cdot10^{-1}$
2613.7	3	3	$2.7\cdot10^{-1}$
2614.2	3	5	1.9
2628.3	5	3	$3.1\cdot10^{-2}$
2657.1	3	5	$9.8\cdot10^{-4}$
2663.2	5	5	$7.1\cdot10^{-1}$
2802.0	5	7	1.6
2823.2	5	5	$2.6\cdot10^{-1}$
2833.1	1	3	$5.8\cdot10^{-1}$
2873.3	5	5	$3.7\cdot10^{-1}$
3572.7	5	3	$9.9\cdot10^{-1}$
3639.6	3	3	$3.4\cdot10^{-1}$
3671.5	5	3	$4.4\cdot10^{-1}$
3683.5	3	1	1.5
3739.9	5	5	$7.3\cdot10^{-1}$
4019.6	5	7	$3.5\cdot10^{-2}$
4057.8	5	3	$8.9\cdot10^{-1}$
4062.1	5	3	$9.2\cdot10^{-1}$
4168.0	5	5	$1.2\cdot10^{-2}$
5005.4	1	3	$2.7\cdot10^{-1}$
5201.4	1	3	$1.9\cdot10^{-1}$
7229.0	5	3	$8.9\cdot10^{-3}$
Lithium			
Li I			
*2394.4	2	6	$2.664\cdot10^{-3}$
*2425.4	2	6	$3.823\cdot10^{-3}$
*2475.1	2	6	$5.735\cdot10^{-3}$

λ Å	Weights g_i	g_k	A $10^8\,\text{s}^{-1}$
*2562.3	2	6	$8.865\cdot10^{-3}$
*2741.2	2	6	$1.248\cdot10^{-2}$
*3232.7	2	6	$1.002\cdot10^{-2}$
*3671.7	6	10	$1.678\cdot10^{-2}$
*3720.9	6	10	$2.413\cdot10^{-2}$
*3746.6	6	2	$1.01\cdot10^{-2}$
*3795.1	6	10	$3.649\cdot10^{-2}$
*3835.6	6	2	$1.56\cdot10^{-2}$
*3915.3	6	10	$5.957\cdot10^{-2}$
*3985.5	6	2	$2.59\cdot10^{-2}$
*4132.6	6	10	$1.08\cdot10^{-1}$
*4273.1	6	2	$4.76\cdot10^{-2}$
*4602.9	6	10	$2.322\cdot10^{-1}$
*4971.7	6	2	$1.038\cdot10^{-1}$
*6103.6	6	10	$6.8563\cdot10^{-1}$
*6707.8	2	6	$3.6891\cdot10^{-1}$
*8126.4	6	2	$3.3466\cdot10^{-1}$
*10510	6	10	$1.97\cdot10^{-2}$
*11032	6	2	$1.46\cdot10^{-2}$
*12237	6	10	$3.49\cdot10^{-2}$
*12782	10	14	$4.578\cdot10^{-2}$
*12929	10	6	$2.28\cdot10^{-3}$
*13557	6	2	$2.84\cdot10^{-2}$
*17545	6	10	$6.791\cdot10^{-2}$
*18697	10	14	$1.383\cdot10^{-1}$
*19276	10	6	$5.375\cdot10^{-3}$
*24463	6	2	$7.453\cdot10^{-2}$
*26880	2	6	$3.738\cdot10^{-2}$
*38079	6	10	$1.37\cdot10^{-2}$
*41792	10	6	$2.77\cdot10^{-3}$
Li II			
199.279	1	3	$2.5569\cdot10^2$
*935.88	9	15	$6.1345\cdot10^{-1}$
*944.72	3	9	1.4329
*965.13	9	15	1.0002
*1017.9	9	15	1.8076
1093.43	1	3	1.3533
1102.46	3	5	1.4070
*1131.9	9	15	3.8492
*1198.1	3	9	2.8969
1237.28	3	5	3.1179
1420.89	1	3	2.8309
*1493.0	9	15	$1.1215\cdot10^1$
*1653.1	9	3	2.8585
1681.66	3	5	$1.0069\cdot10^1$
1755.33	3	1	2.0499
*2329.8	3	9	$1.1758\cdot10^{-1}$
*2674.4	3	9	$1.9081\cdot10^{-1}$
*3684.7	3	9	$3.0580\cdot10^{-1}$
*5484.5	3	9	$2.2727\cdot10^{-1}$
9581.43	1	3	$5.1423\cdot10^{-2}$
*21061	3	9	$2.5664\cdot10^{-2}$
Lutetium			
Lu I			
3376.5	4	4	2.23
3567.8	4	6	$5.9\cdot10^{-1}$
3620.3	6	4	$1.1\cdot10^{-2}$

λ Å	Weights g_i	g_k	A $10^8\,\text{s}^{-1}$
3841.2	6	6	$2.5\cdot10^{-1}$
4518.6	4	4	$2.1\cdot10^{-1}$
Magnesium			
Mg I			
1683.412	1	3	$1.88\cdot10^{-2}$
1707.061	1	3	$3.28\cdot10^{-2}$
1747.794	1	3	$6.62\cdot10^{-2}$
1827.934	1	3	$1.60\cdot10^{-1}$
2025.824	1	3	$6.12\cdot10^{-1}$
2731.993	1	3	$6.97\cdot10^{-1}$
2733.493	3	5	$9.37\cdot10^{-2}$
2736.542	5	7	$1.25\cdot10^{-1}$
2776.690	3	5	1.32
2778.270	1	3	1.82
2779.831	3	3	1.36
2779.831	5	5	4.09
2781.416	3	1	5.43
2782.972	5	3	2.14
2809.761	7	7	2.50
2811.112	5	5	1.96
2811.781	3	3	2.11
2846.716	1	3	$1.31\cdot10^{-1}$
2848.342	3	5	$1.77\cdot10^{-1}$
2851.660	5	7	$2.35\cdot10^{-1}$
2852.127	1	3	4.91
2915.453	5	5	4.09
2936.739	1	3	$1.37\cdot10^{-2}$
2938.473	3	3	$4.12\cdot10^{-2}$
2941.995	5	3	$6.83\cdot10^{-2}$
3091.065	1	3	$3.09\cdot10^{-2}$
3092.984	3	5	$3.74\cdot10^{-1}$
3096.890	5	7	$4.96\cdot10^{-1}$
3329.919	1	3	$3.09\cdot10^{-2}$
3332.146	3	3	$1.02\cdot10^{-1}$
3336.674	5	3	$1.70\cdot10^{-1}$
3829.3549	1	3	$8.99\cdot10^{-1}$
3832.2996	3	3	$6.74\cdot10^{-1}$
3832.3037	3	5	1.21
3838.2918	5	7	1.61
3838.2943	5	5	$4.03\cdot10^{-1}$
3890.241	1	3	1.31
3891.906	3	5	1.77
3893.304	3	3	$9.81\cdot10^{-1}$
3895.572	5	7	2.35
3898.120	5	5	$5.88\cdot10^{-1}$
3938.400	3	5	$5.47\cdot10^{-2}$
3986.7533	3	5	$7.30\cdot10^{-2}$
4057.5052	3	5	$1.02\cdot10^{-1}$
4167.2712	3	5	$1.38\cdot10^{-1}$
4351.9056	3	5	$1.84\cdot10^{-1}$
4571.0956	1	3	$2.54\cdot10^{-6}$
4702.9909	3	5	$2.19\cdot10^{-1}$
4730.0285	3	1	$1.34\cdot10^{-2}$
5167.3216	1	3	$1.13\cdot10^{-1}$
5172.6843	3	3	$3.37\cdot10^{-1}$
5183.6042	5	3	$5.61\cdot10^{-1}$
5528.4047	3	5	$1.39\cdot10^{-1}$
5711.0880	3	1	$3.86\cdot10^{-2}$

λ (Å)	g_i	g_k	A ($10^8\,s^{-1}$)	λ (Å)	g_i	g_k	A ($10^8\,s^{-1}$)	λ (Å)	g_i	g_k	A ($10^8\,s^{-1}$)
6318.716	3	5	$2.63\cdot10^{-3}$	1271.943	2	2	$4.49\cdot10^{-2}$	4739.712	4	6	$7.55\cdot10^{-2}$
6319.236	3	3	$2.64\cdot10^{-3}$	1272.725	4	6	$1.44\cdot10^{-2}$	4868.845	6	4	$1.50\cdot10^{-3}$
6319.493	3	1	$2.63\cdot10^{-3}$	1273.427	4	2	$8.97\cdot10^{-2}$	4868.845	8	6	$1.41\cdot10^{-3}$
7291.060	1	3	$6.27\cdot10^{-4}$	1306.711	2	4	$2.19\cdot10^{-2}$	5068.937	6	4	$3.97\cdot10^{-3}$
7657.603	3	5	$1.23\cdot10^{-2}$	1307.877	2	2	$6.72\cdot10^{-2}$	5069.802	4	2	$4.35\cdot10^{-3}$
7659.152	3	3	$1.23\cdot10^{-2}$	1308.282	4	6	$2.58\cdot10^{-2}$	5264.215	6	8	$1.27\cdot10^{-1}$
7659.902	3	1	$1.23\cdot10^{-2}$	1309.439	4	2	$1.34\cdot10^{-1}$	5264.368	4	6	$1.19\cdot10^{-1}$
7881.667	5	7	$4.56\cdot10^{-3}$	1365.545	2	4	$4.53\cdot10^{-2}$	5434.039	6	4	$2.45\cdot10^{-3}$
7930.806	7	9	$5.38\cdot10^{-3}$	1367.260	4	6	$5.35\cdot10^{-2}$	5434.039	8	6	$2.33\cdot10^{-3}$
7930.806	3	5	$4.52\cdot10^{-3}$	1367.704	2	2	$1.08\cdot10^{-1}$	5451.259	2	4	$1.03\cdot10^{-2}$
7947.07	1	3	$2.91\cdot10^{-4}$	1369.425	4	2	$2.17\cdot10^{-1}$	5460.019	2	2	$1.11\cdot10^{-2}$
7953.39	5	3	$1.45\cdot10^{-3}$	1476.004	2	4	$1.10\cdot10^{-1}$	5464.136	4	2	$2.22\cdot10^{-2}$
8047.73	1	3	$3.53\cdot10^{-3}$	1478.013	4	6	$1.30\cdot10^{-1}$	5916.429	6	4	$6.42\cdot10^{-3}$
8049.854	3	5	$4.78\cdot10^{-3}$	1480.890	2	2	$1.93\cdot10^{-1}$	5918.158	4	2	$7.04\cdot10^{-3}$
8054.232	5	7	$6.36\cdot10^{-3}$	1482.902	4	2	$3.85\cdot10^{-1}$	5923.366	2	4	$1.40\cdot10^{-2}$
8098.724	7	9	$7.68\cdot10^{-3}$	1734.845	2	4	$4.29\cdot10^{-1}$	5928.233	4	6	$1.66\cdot10^{-2}$
8098.724	3	5	$6.46\cdot10^{-3}$	1737.618	4	6	$5.09\cdot10^{-1}$	5938.629	2	2	$1.36\cdot10^{-2}$
8209.839	5	3	$1.81\cdot10^{-3}$	1750.654	2	4	$4.00\cdot10^{-1}$	5943.499	4	2	$2.71\cdot10^{-2}$
8213.034	5	7	$4.38\cdot10^{-2}$	1753.456	4	2	$7.98\cdot10^{-1}$	6346.737	6	8	$2.20\cdot10^{-1}$
8303.313	1	3	$5.16\cdot10^{-3}$	2329.578	6	8	$1.36\cdot10^{-1}$	6346.962	4	6	$2.05\cdot10^{-1}$
8305.596	3	5	$9.28\cdot10^{-3}$	2329.578	4	6	$1.27\cdot10^{-1}$	6620.440	6	4	$4.61\cdot10^{-3}$
8310.264	5	7	$9.24\cdot10^{-3}$	2449.590	6	8	$2.16\cdot10^{-1}$	6620.569	8	6	$4.37\cdot10^{-3}$
8346.120	7	9	$1.15\cdot10^{-2}$	2449.590	4	6	$2.02\cdot10^{-1}$	6781.451	2	4	$2.49\cdot10^{-2}$
8346.120	3	5	$9.59\cdot10^{-3}$	2660.755	6	8	$3.81\cdot10^{-1}$	6787.851	4	6	$2.96\cdot10^{-2}$
8710.175	1	3	$7.97\cdot10^{-3}$	2660.817	4	6	$3.56\cdot10^{-1}$	6812.860	2	2	$2.11\cdot10^{-2}$
8712.689	3	5	$1.07\cdot10^{-2}$	2790.776	2	4	4.01	6819.270	4	2	$4.21\cdot10^{-2}$
8717.825	5	7	$1.43\cdot10^{-2}$	2795.528	2	4	2.60	7786.500	2	4	$1.97\cdot10^{-3}$
8736.021	7	9	$1.83\cdot10^{-2}$	2797.998	4	6	4.79	7790.978	2	2	$1.84\cdot10^{-3}$
8736.021	3	5	$1.54\cdot10^{-2}$	2802.704	2	2	2.57	7877.051	2	4	$6.58\cdot10^{-1}$
8806.757	3	5	$1.27\cdot10^{-1}$	2928.634	2	2	1.15	7896.368	4	6	$7.86\cdot10^{-1}$
8923.569	1	3	$5.86\cdot10^{-3}$	2936.509	4	2	2.30	8115.220	6	4	$1.22\cdot10^{-2}$
9246.499	5	3	$2.99\cdot10^{-3}$	2968.020	4	6	$1.98\cdot10^{-2}$	8120.434	4	2	$1.34\cdot10^{-2}$
9255.778	5	7	$7.95\cdot10^{-2}$	2969.145	2	2	$2.25\cdot10^{-2}$	8213.989	2	2	$2.65\cdot10^{-1}$
9414.964	5	7	$2.88\cdot10^{-2}$	2971.839	4	2	$4.49\cdot10^{-2}$	8234.639	4	2	$5.29\cdot10^{-1}$
9414.964	7	9	$3.24\cdot10^{-2}$	3104.722	6	8	$7.97\cdot10^{-1}$	8734.990	2	4	$5.34\cdot10^{-2}$
9414.964	3	5	$2.72\cdot10^{-2}$	3104.809	4	6	$7.44\cdot10^{-1}$	8745.657	4	6	$6.37\cdot10^{-2}$
9429.814	1	3	$1.34\cdot10^{-2}$	3165.878	2	4	$2.89\cdot10^{-2}$	8824.323	2	2	$3.69\cdot10^{-2}$
9432.764	3	5	$1.81\cdot10^{-2}$	3168.951	4	6	$3.43\cdot10^{-2}$	8835.082	4	2	$7.36\cdot10^{-2}$
9438.783	5	7	$2.41\cdot10^{-2}$	3172.706	2	2	$3.41\cdot10^{-2}$	9218.248	2	4	$3.64\cdot10^{-1}$
9665.54	3	5	$2.44\cdot10^{-3}$	3175.783	4	2	$6.81\cdot10^{-2}$	9244.266	2	2	$3.61\cdot10^{-1}$
9983.20	1	3	$1.57\cdot10^{-3}$	3534.972	2	4	$5.80\cdot10^{-2}$	9631.888	6	8	$4.21\cdot10^{-1}$
9986.475	3	3	$4.70\cdot10^{-3}$	3538.813	4	6	$6.90\cdot10^{-2}$	9632.435	4	6	$3.93\cdot10^{-1}$
9993.209	5	3	$7.81\cdot10^{-3}$	3549.516	2	2	$5.64\cdot10^{-2}$				
				3553.366	4	2	$1.12\cdot10^{-1}$	***Mg III***			
Mg II				3613.781	2	4	$1.79\cdot10^{-3}$	186.5149	1	3	$1.86\cdot10^{2}$
870.2	2	4	$1.04\cdot10^{-2}$	3615.583	2	2	$1.56\cdot10^{-3}$	187.1977	1	3	$1.26\cdot10^{2}$
870.2	2	2	$1.08\cdot10^{-2}$	3848.209	6	4	$2.96\cdot10^{-2}$	188.5296	1	3	2.5
884.7	2	4	$1.38\cdot10^{-2}$	3848.335	4	4	$3.29\cdot10^{-3}$	231.7333	1	3	$9.12\cdot10$
884.7	2	2	$1.44\cdot10^{-2}$	3850.385	4	2	$3.24\cdot10^{-2}$	234.2631	1	3	4.98
907.4	2	4	$1.94\cdot10^{-2}$	4384.637	2	4	$1.45\cdot10^{-1}$	1229.389	3	3	$5.72\cdot10^{-1}$
907.4	2	2	$2.02\cdot10^{-2}$	4390.564	4	6	$1.73\cdot10^{-1}$	1239.827	3	1	1.82
946.703	2	4	$2.69\cdot10^{-2}$	4427.994	2	2	$1.05\cdot10^{-1}$	1274.831	3	5	2.53
946.769	2	2	$2.81\cdot10^{-2}$	4433.990	4	2	$2.10\cdot10^{-1}$	1350.156	5	3	$3.50\cdot10^{-1}$
1025.962	2	4	$3.43\cdot10^{-2}$	4436.486	6	8	$6.38\cdot10^{-2}$	1378.700	3	1	5.32
1026.108	2	2	$3.63\cdot10^{-2}$	4436.598	4	6	$5.95\cdot10^{-2}$	1393.391	7	5	6.38
1239.936	2	4	$1.35\cdot10^{-2}$	4481.130	6	8	2.33	1405.170	5	5	1.54
1240.399	2	2	$1.52\cdot10^{-2}$	4481.327	4	6	2.17	1422.118	3	5	$1.80\cdot10^{-1}$
1248.511	2	2	$3.83\cdot10^{-2}$	4630.878	6	4	$2.75\cdot10^{-3}$	1431.136	5	3	4.30
1249.932	4	2	$7.64\cdot10^{-2}$	4631.405	4	2	$3.01\cdot10^{-3}$	1435.550	3	1	3.04
1271.243	2	4	$1.23\cdot10^{-2}$	4739.588	6	8	$8.09\cdot10^{-2}$	1439.770	1	3	$3.54\cdot10^{-1}$

λ Å	Weights g_i	g_k	A 10^8 s^{-1}	λ Å	Weights g_i	g_k	A 10^8 s^{-1}	λ Å	Weights g_i	g_k	A 10^8 s^{-1}
1443.738	3	3	1.81	1879.492	5	5	1.76	3044.57	10	8	5.7·10^{-1}
1446.254	3	1	4.15·10^{-1}	1887.308	5	3	7.56·10^{-1}	3045.59	10	10	6.7·10^{-1}
1447.260	5	3	2.22	1896.304	5	3	4.20·10^{-1}	3047.03	12	12	6.1·10^{-1}
1458.172	3	1	3.21	1901.572	3	5	3.78·10^{-1}	3082.71	14	14	2.9·10^{-1}
1462.305	5	5	1.21	1901.572	3	3	5.47·10^{-1}	3228.09	10	12	6.4·10^{-1}
1467.188	3	3	9.37·10^{-1}	1908.500	3	1	5.15	3230.72	8	8	3.5·10^{-1}
1482.67	3	5	1.55·10^{-1}	1918.777	5	5	1.98	3243.78	6	6	5.3·10^{-1}
1493.097	5	5	1.02	1921.374	3	3	2.26·10^{-1}	3256.14	4	6	5.0·10^{-1}
1506.826	3	5	3.91·10^{-1}	1923.896	3	5	1.37	3267.79	14	14	3.5·10^{-1}
1550.82	3	1	1.04·10	1930.374	3	1	8.30·10^{-1}	3268.72	6	8	3.3·10^{-1}
1572.712	3	5	6.92	1930.672	5	5	1.70	3270.35	12	12	2.6·10^{-1}
1586.237	3	3	8.02	1937.843	1	3	1.46	3273.02	10	10	2.7·10^{-1}
1592.360	3	1	8.53	1938.936	5	3	8.97·10^{-1}	3463.66	8	8	3.2·10^{-1}
1626.093	5	3	1.77·10^{-1}	1941.500	3	5	1.23·10^{-1}	3511.83	12	12	2.7·10^{-1}
1635.946	7	5	2.44·10^{-1}	1941.500	3	3	3.27·10^{-1}	3577.87	10	8	9.4·10^{-1}
1642.826	7	7	9.17·10^{-1}	1954.831	1	3	7.97·10^{-1}	3601.27	12	10	2.3·10^{-1}
1648.822	3	3	5.38·10^{-1}	1962.145	3	3	7.76·10^{-1}	3608.49	6	6	3.6·10^{-1}
1652.218	5	5	9.10·10^{-1}	1971.514	3	1	2.06	3660.40	12	14	9.1·10^{-1}
1659.244	5	7	8.48·10^{-1}	1977.554	3	5	4.94·10^{-1}	3675.67	6	8	2.2·10^{-1}
1663.287	5	5	1.68·10^{-1}	1979.327	1	3	1.22	3676.96	10	12	7.3·10^{-1}
1675.710	3	5	4.19·10^{-1}	1979.43	1	3	4.10·10^{-1}	3680.15	12	10	1.9·10^{-1}
1679.470	5	3	5.88·10^{-1}	2004.860	5	3	3.33·10^{-1}	3682.09	8	10	7.6·10^{-1}
1687.091	3	5	1.41	2039.553	5	5	1.54	3684.87	6	8	2.6·10^{-1}
1697.282	7	7	2.14	2055.491	3	3	2.35	3706.08	12	14	1.4
1703.108	5	3	1.71·10^{-1}	2064.902	5	7	4.21	3718.92	10	12	9.6·10^{-1}
1703.731	3	3	2.97	2085.891	3	3	1.61	3731.94	8	10	1.0
1704.368	7	5	1.79·10^{-1}	2091.963	3	5	2.57	3771.44	14	14	1.9·10^{-1}
1714.783	5	7	4.76·10^{-1}	2094.207	3	1	1.21·10^{-1}	3773.86	12	12	2.5·10^{-1}
1722.041	5	5	3.23	2097.936	1	3	1.50	3800.55	6	8	2.7·10^{-1}
1730.733	7	7	1.03	2112.773	3	5	1.69	3806.72	10	12	5.9·10^{-1}
1730.778	3	3	1.44	2134.054	3	3	2.36	3823.51	8	10	5.21·10^{-1}
1731.786	5	5	1.60	2177.694	3	5	2.12	3834.37	6	8	4.29·10^{-1}
1738.835	7	9	1.14·10	2273.414	3	3	1.27·10^{-2}	3841.07	4	6	3.3·10^{-1}
1739.475	5	7	3.26·10^{-4}	2318.125	3	5	4.44·10^{-2}	3889.46	12	14	3.1·10^{-1}
1743.947	5	5	3.76·10^{-1}	2395.149	5	3	1.67	3924.08	2	4	9.4·10^{-1}
1745.009	5	3	5.37·10^{-2}	2467.751	3	3	6.91·10^{-1}	3926.48	6	8	5.4·10^{-1}
1747.561	3	5	7.38	2490.534	1	3	1.61	3952.84	6	6	4.1·10^{-1}
1748.932	5	7	8.71	2529.190	1	3	1.89·10^{-1}	3982.90	6	4	5.5·10^{-1}
1757.176	7	5	2.04·10^{-1}	2618.011	1	3	8.72·10^{-1}	4011.91	8	8	2.3·10^{-1}
1757.888	1	3	2.10	2905.419	1	3	6.84·10^{-3}	4018.11	10	8	2.54·10^{-1}
1761.740	5	3	3.16·10^{-3}					4030.76	6	8	1.7·10^{-1}
1763.805	3	3	3.17	*Mg IV*				4041.36	10	10	7.87·10^{-1}
1772.982	3	5	7.67	129.857	4	6	5.61·10^{2}	4048.75	6	4	7.5·10^{-1}
1775.942	5	5	1.28·10^{-1}	320.9943	4	2	1.13·10^{2}	4052.48	6	8	3.8·10^{-1}
1783.253	5	7	9.39	323.3076	2	2	5.52·10^{1}	4055.55	8	8	4.31·10^{-1}
1787.927	5	5	1.46	840.366	6	4	1.24·10^{1}	4058.94	4	2	7.25·10^{-1}
1791.375	3	3	2.32	842.087	8	6	1.19·10^{1}	4065.08	12	14	2.5·10^{-1}
1793.207	5	3	3.36·10^{-1}	1346.543	6	8	1.07·10^{1}	4066.24	10	8	2.2·10^{-1}
1794.582	3	5	7.73	1352.020	4	6	6.59	4079.42	2	4	3.8·10^{-1}
1800.662	5	7	8.21	1384.425	8	10	1.45·10^{1}	4082.95	4	6	2.95·10^{-1}
1803.087	3	5	1.03·10^{-1}	1437.61	6	8	1.21·10^{1}	4083.63	6	8	2.8·10^{-1}
1820.421	1	3	2.98					4089.94	8	10	1.7·10^{-1}
1820.896	3	3	3.05·10^{-2}	*Manganese*				4105.37	10	8	1.7·10^{-1}
1826.750	3	3	1.32·10^{-1}					4135.03	12	12	3.0·10^{-1}
1828.974	3	1	4.97·10^{-1}	*Mn I*				4141.06	10	10	2.6·10^{-1}
1838.336	5	7	1.09	2794.82	6	8	3.7	4148.80	8	8	2.3·10^{-1}
1839.878	3	5	4.55·10^{-1}	2798.27	6	6	3.6	4176.61	14	12	2.4·10^{-1}
1847.561	5	7	3.74·10^{-2}	2801.08	6	4	3.7	4189.99	12	10	2.0·10^{-1}
1858.186	5	3	1.28	3016.45	10	12	2.9·10^{-1}	4201.78	10	8	2.3·10^{-1}
1868.225	5	5	1.94	3043.36	8	8	5.9·10^{-1}	4235.30	8	6	9.17·10^{-1}

λ Å	Weights g_i	g_k	A 10^8 s^{-1}	λ Å	Weights g_i	g_k	A 10^8 s^{-1}	λ Å	Weights g_i	g_k	A 10^8 s^{-1}
4265.93	4	4	4.92·10^{-1}	328.431	5	5	4.4·10^1	2906.06	3	3	8.04·10^{-1}
4281.10	6	6	2.3·10^{-1}	328.558	3	5	1.2·10^1	2915.38	5	3	7.31·10^{-1}
4411.87	12	10	2.6·10^{-1}	329.043	1	3	1.1·10^1	2936.50	11	11	2.33·10^{-1}
4414.89	8	6	2.93·10^{-1}	1236.23	5	3	1.3·10^1	2945.43	7	7	3.66·10^{-1}
4419.77	10	8	2.1·10^{-1}	1255.77	3	1	1.2·10^1	2959.48	9	11	1.75·10^{-1}
4436.36	6	4	4.37·10^{-1}	1285.10	5	7	1.1·10^1	2977.27	9	7	3.28·10^{-1}
4451.58	8	8	7.98·10^{-1}	1333.87	7	9	1.0·10^1	2987.92	3	5	8.43·10^{-1}
4453.01	4	2	5.44·10^{-1}					2988.23	5	7	4.28·10^{-1}
4455.82	4	6	1.7·10^{-1}	*Mercury*				2989.80	9	7	9.27·10^{-1}
4457.55	6	6	4.27·10^{-1}					3000.85	5	7	2.58·10^{-1}
4458.26	6	8	4.62·10^{-1}	*Hg I*				3013.39	7	5	6.06·10^{-1}
4461.09	8	8	1.7·10^{-1}	2536.52	1	3	8.00·10^{-2}	3016.78	9	9	2.75·10^{-1}
4462.03	8	10	7.00·10^{-1}	2652.04	3	5	3.88·10^{-1}	3025.00	5	5	8.49·10^{-1}
4464.68	6	6	4.39·10^{-1}	2655.13	3	5	1.1·10^{-1}	3036.31	3	5	5.81·10^{-1}
4470.14	4	4	3.00·10^{-1}	2752.78	1	3	6.10·10^{-2}	3041.70	13	11	5.94·10^{-1}
4479.40	8	10	3.4·10^{-1}	2856.94	3	1	1.1·10^{-2}	3046.80	13	11	1.63·10^{-1}
4490.08	2	4	2.49·10^{-1}	2893.60	3	3	1.6·10^{-1}	3047.31	11	9	5.01·10^{-1}
4498.90	4	6	2.49·10^{-1}	2925.4	5	3	7.7·10^{-2}	3055.32	9	7	4.29·10^{-1}
4502.22	6	8	1.86·10^{-1}	2967.3	1	3	4.5·10^{-1}	3061.59	7	5	4.41·10^{-1}
4605.37	10	12	3.6·10^{-1}	3021.50	5	7	5.09·10^{-1}	3064.27	13	13	8.46·10^{-1}
4626.54	12	14	3.6·10^{-1}	3023.48	5	5	9.4·10^{-2}	3065.04	13	13	3.08·10^{-1}
4709.71	8	8	1.72·10^{-1}	3027.49	5	5	2.0·10^{-2}	3069.96	11	11	2.72·10^{-1}
4727.46	6	6	1.7·10^{-1}	3125.66	3	5	6.56·10^{-1}	3070.89	9	11	1.87·10^{-1}
4739.11	4	4	2.40·10^{-1}	3341.48	5	3	1.68·10^{-1}	3074.37	11	11	1.42
4754.05	6	8	3.03·10^{-1}	3650.15	5	7	1.3	3079.88	9	11	9.55·10^{-1}
4761.53	2	4	5.35·10^{-1}	3654.83	5	5	1.8·10^{-1}	3080.40	7	9	3.61·10^{-1}
4762.38	8	10	7.83·10^{-1}	4046.56	1	3	2.1·10^{-1}	3085.62	9	9	1.63
4765.86	4	6	4.1·10^{-1}	4077.81	3	1	4.0·10^{-2}	3089.71	5	7	2.34·10^{-1}
4766.43	6	8	4.6·10^{-1}	4108.1	3	1	3.0·10^{-2}	3094.66	7	7	1.63
4783.43	8	8	4.01·10^{-1}	4339.22	3	5	2.88·10^{-2}	3100.88	7	9	1.20
4823.53	10	8	4.99·10^{-1}	4347.50	3	5	8.4·10^{-2}	3101.34	5	5	1.92
6013.48	4	6	1.72·10^{-1}	4358.34	3	5	5.57·10^{-1}	3117.54	13	13	1.89·10^{-1}
6021.79	8	6	3.32·10^{-1}	4916.07	3	1	5.8·10^{-2}	3132.59	7	9	1.79
				5025.64	3	3	2.7·10^{-4}	3135.90	9	11	3.68·10^{-1}
Mn II				5460.75	5	3	4.87·10^{-1}	3136.75	9	11	1.57·10^{-1}
2593.72	7	7	2.6	5769.59	3	5	2.36·10^{-1}	3142.75	3	5	4.10·10^{-1}
2605.68	7	5	2.7	6234.4	1	3	5.3·10^{-3}	3147.35	13	11	2.41·10^{-1}
2933.05	5	3	2.0	6716.4	1	3	4.3·10^{-3}	3155.19	7	7	2.75·10^{-1}
2939.31	5	5	1.9	6907.5	3	5	2.8·10^{-2}	3158.17	7	7	4.63·10^{-1}
2949.20	5	7	1.9	7728.8	1	3	9.7·10^{-3}	3170.34	7	7	1.37
3441.99	9	7	4.3·10^{-1}	10139.79	3	1	2.71·10^{-1}	3171.38	5	7	2.03·10^{-1}
3460.32	7	5	3.2·10^{-1}					3175.59	13	11	8.40·10^{-1}
3474.13	5	3	1.5·10^{-1}	*Molybdenum*				3179.78	11	13	2.33·10^{-1}
3482.90	5	5	2.0·10^{-1}					3183.03	11	9	3.98·10^{-1}
3488.68	3	3	2.5·10^{-1}	*Mo I*				3184.58	7	5	2.77·10^{-1}
				2616.79	3	5	7.34·10^{-1}	3185.10	7	7	2.54·10^{-1}
Mn VI				2629.85	5	7	7.75·10^{-1}	3185.71	5	3	6.10·10^{-1}
307.999	9	9	3.7·10^1	2638.30	5	5	7.57·10^{-1}	3188.10	7	9	3.45·10^{-1}
309.440	9	7	5.7·10^1	2640.98	7	5	1.20	3188.41	5	7	4.40·10^{-1}
309.579	7	5	4.4·10^1	2649.46	7	9	9.84·10^{-1}	3192.79	9	11	1.88·10^{-1}
310.058	7	7	3.4·10^1	2655.02	9	7	4.08·10^{-1}	3193.98	7	5	1.53
310.182	5	5	2.8·10^1	2658.11	7	7	6.43·10^{-1}	3194.88	9	11	1.75·10^{-1}
311.748	5	3	5.7·10^1	2679.85	9	11	1.31	3195.96	9	7	4.10·10^{-1}
320.598	3	5	1.5·10^1	2684.16	9	9	4.18·10^{-1}	3198.85	15	13	7.22·10^{-1}
320.681	1	3	2.2·10^1	2733.39	5	7	2.95·10^{-1}	3205.22	1	3	4.27·10^{-1}
320.874	3	1	7.8·10^1	2751.47	7	9	2.54·10^{-1}	3205.43	9	11	2.55·10^{-1}
320.979	3	3	2.2·10^1	2761.53	9	11	2.06·10^{-1}	3205.89	9	9	5.35·10^{-1}
321.176	5	5	6.0·10^1	2787.83	9	7	2.85·10^{-1}	3208.84	7	5	2.77·10^{-1}
321.541	5	3	2.7·10^1	2826.75	7	7	4.23·10^{-1}	3210.97	7	5	6.94·10^{-1}
325.146	9	7	1.3·10^2	2876.54	9	9	2.84·10^{-1}	3214.44	9	7	2.01·10^{-1}
				2886.60	11	11	4.74·10^{-1}				

λ (Å)	g_i	g_k	A (10^8 s^{-1})	λ (Å)	g_i	g_k	A (10^8 s^{-1})	λ (Å)	g_i	g_k	A (10^8 s^{-1})
3215.07	3	5	$4.20 \cdot 10^{-1}$	3475.03	3	3	$4.68 \cdot 10^{-1}$	3720.25	7	9	$2.86 \cdot 10^{-1}$
3216.78	15	13	$2.10 \cdot 10^{-1}$	3479.42	7	5	$2.26 \cdot 10^{-1}$	3725.55	7	7	$1.60 \cdot 10^{-1}$
3221.73	3	1	1.41	3489.43	7	7	$3.27 \cdot 10^{-1}$	3727.68	9	11	$1.51 \cdot 10^{-1}$
3228.21	5	7	$3.85 \cdot 10^{-1}$	3504.41	7	9	$8.06 \cdot 10^{-1}$	3728.50	7	9	$2.20 \cdot 10^{-1}$
3229.79	9	11	$1.44 \cdot 10^{-1}$	3505.31	7	9	$2.25 \cdot 10^{-1}$	3733.02	7	7	$1.45 \cdot 10^{-1}$
3233.14	13	13	$6.33 \cdot 10^{-1}$	3508.11	9	9	$1.59 \cdot 10^{-1}$	3733.41	13	13	$2.80 \cdot 10^{-1}$
3237.06	7	9	$2.95 \cdot 10^{-1}$	3510.77	13	13	$4.75 \cdot 10^{-1}$	3735.62	11	11	$1.66 \cdot 10^{-1}$
3251.65	3	5	$3.05 \cdot 10^{-1}$	3517.55	11	11	$5.41 \cdot 10^{-1}$	3742.28	7	7	$1.56 \cdot 10^{-1}$
3256.21	5	3	$6.89 \cdot 10^{-1}$	3518.21	3	3	$3.64 \cdot 10^{-1}$	3747.19	5	7	$3.07 \cdot 10^{-1}$
3259.16	11	13	$1.62 \cdot 10^{-1}$	3521.38	9	9	$1.39 \cdot 10^{-1}$	3748.48	9	11	$3.95 \cdot 10^{-1}$
3262.63	7	9	$3.62 \cdot 10^{-1}$	3521.41	9	11	$6.06 \cdot 10^{-1}$	3755.16	9	9	$2.48 \cdot 10^{-1}$
3264.40	11	9	$5.42 \cdot 10^{-1}$	3524.98	7	9	$2.25 \cdot 10^{-1}$	3758.52	9	9	$1.22 \cdot 10^{-1}$
3265.14	5	7	$2.60 \cdot 10^{-1}$	3538.92	11	11	$2.24 \cdot 10^{-1}$	3759.60	9	7	$1.82 \cdot 10^{-1}$
3266.16	9	11	$1.95 \cdot 10^{-1}$	3540.57	5	3	$4.46 \cdot 10^{-1}$	3760.88	9	9	$2.16 \cdot 10^{-1}$
3270.90	7	7	$3.59 \cdot 10^{-1}$	3542.17	7	5	$4.93 \cdot 10^{-1}$	3768.73	9	9	$2.88 \cdot 10^{-1}$
3285.35	9	7	$4.49 \cdot 10^{-1}$	3552.71	9	7	$3.64 \cdot 10^{-1}$	3769.99	7	9	$2.46 \cdot 10^{-1}$
3289.01	9	9	$5.08 \cdot 10^{-1}$	3555.64	3	3	$3.46 \cdot 10^{-1}$	3777.72	13	11	$1.66 \cdot 10^{-1}$
3290.82	7	5	$5.44 \cdot 10^{-1}$	3558.09	5	7	$5.43 \cdot 10^{-1}$	3788.25	7	9	$2.87 \cdot 10^{-1}$
3305.91	7	9	$3.06 \cdot 10^{-1}$	3566.05	9	9	$2.67 \cdot 10^{-1}$	3794.43	9	9	$1.22 \cdot 10^{-1}$
3323.95	9	7	$2.82 \cdot 10^{-1}$	3566.74	7	7	$1.43 \cdot 10^{-1}$	3798.25	7	9	$6.90 \cdot 10^{-1}$
3336.56	9	9	$1.64 \cdot 10^{-1}$	3570.64	15	15	$7.18 \cdot 10^{-1}$	3801.84	9	7	$3.16 \cdot 10^{-1}$
3344.73	3	5	$6.04 \cdot 10^{-1}$	3573.88	3	5	$3.58 \cdot 10^{-1}$	3805.99	5	5	$2.44 \cdot 10^{-1}$
3346.83	11	11	$1.13 \cdot 10^{-1}$	3580.54	13	11	$5.49 \cdot 10^{-1}$	3819.78	9	11	$1.47 \cdot 10^{-1}$
3358.12	5	7	$7.59 \cdot 10^{-1}$	3581.88	11	13	$3.81 \cdot 10^{-1}$	3824.78	5	7	$1.40 \cdot 10^{-1}$
3361.37	9	9	$1.38 \cdot 10^{-1}$	3585.57	7	5	$3.95 \cdot 10^{-1}$	3827.15	7	7	$1.94 \cdot 10^{-1}$
3363.78	5	7	$2.74 \cdot 10^{-1}$	3590.74	7	9	$2.23 \cdot 10^{-1}$	3828.88	7	7	$1.35 \cdot 10^{-1}$
3375.65	7	9	$1.56 \cdot 10^{-1}$	3595.55	5	5	$2.32 \cdot 10^{-1}$	3830.81	5	5	$1.83 \cdot 10^{-1}$
3378.46	13	13	$3.75 \cdot 10^{-1}$	3598.88	13	11	$5.67 \cdot 10^{-1}$	3831.07	7	9	$1.20 \cdot 10^{-1}$
3379.96	5	5	$4.11 \cdot 10^{-1}$	3600.73	9	9	$2.07 \cdot 10^{-1}$	3832.11	9	9	$3.05 \cdot 10^{-1}$
3384.61	7	9	$7.32 \cdot 10^{-1}$	3601.88	7	9	$1.15 \cdot 10^{-1}$	3833.75	9	9	$1.70 \cdot 10^{-1}$
3385.87	9	11	$3.30 \cdot 10^{-1}$	3602.94	5	7	$2.96 \cdot 10^{-1}$	3846.18	7	7	$1.26 \cdot 10^{-1}$
3389.79	5	7	$1.85 \cdot 10^{-1}$	3604.07	9	7	$3.25 \cdot 10^{-1}$	3848.30	9	9	$1.26 \cdot 10^{-1}$
3392.17	9	9	$1.97 \cdot 10^{-1}$	3615.16	7	9	$1.96 \cdot 10^{-1}$	3851.99	11	9	$1.78 \cdot 10^{-1}$
3393.65	11	11	$2.08 \cdot 10^{-1}$	3623.22	11	9	$5.58 \cdot 10^{-1}$	3864.10	7	7	$6.24 \cdot 10^{-1}$
3404.33	7	7	$2.10 \cdot 10^{-1}$	3624.46	9	11	$5.27 \cdot 10^{-1}$	3866.69	3	5	$1.74 \cdot 10^{-1}$
3413.37	11	11	$1.25 \cdot 10^{-1}$	3624.62	5	7	$1.37 \cdot 10^{-1}$	3874.15	7	5	$1.67 \cdot 10^{-1}$
3415.27	9	9	$1.83 \cdot 10^{-1}$	3638.20	5	3	$3.51 \cdot 10^{-1}$	3902.95	7	5	$6.17 \cdot 10^{-1}$
3415.61	7	9	$1.29 \cdot 10^{-1}$	3638.21	5	3	$3.33 \cdot 10^{-1}$	3909.54	9	7	$1.13 \cdot 10^{-1}$
3416.14	9	11	$2.45 \cdot 10^{-1}$	3640.62	7	5	$1.94 \cdot 10^{-1}$	3919.55	11	13	$2.24 \cdot 10^{-1}$
3420.04	5	5	$3.28 \cdot 10^{-1}$	3647.84	7	7	$2.11 \cdot 10^{-1}$	3955.48	13	11	$1.71 \cdot 10^{-1}$
3422.31	9	9	$2.52 \cdot 10^{-1}$	3657.36	5	7	$2.03 \cdot 10^{-1}$	3973.76	11	13	$4.39 \cdot 10^{-1}$
3425.13	11	11	$2.29 \cdot 10^{-1}$	3658.13	9	9	$1.86 \cdot 10^{-1}$	3977.90	9	7	$1.35 \cdot 10^{-1}$
3427.90	11	13	$4.09 \cdot 10^{-1}$	3659.36	7	9	$6.70 \cdot 10^{-1}$	3980.20	5	3	$2.70 \cdot 10^{-1}$
3434.79	7	7	$1.75 \cdot 10^{-1}$	3662.15	7	9	$1.45 \cdot 10^{-1}$	3991.85	11	9	$1.29 \cdot 10^{-1}$
3435.45	15	15	1.50	3662.99	11	11	$3.48 \cdot 10^{-1}$	4010.13	5	3	$4.38 \cdot 10^{-1}$
3437.21	11	9	$8.06 \cdot 10^{-1}$	3663.27	7	5	$2.30 \cdot 10^{-1}$	4021.01	9	11	$2.65 \cdot 10^{-1}$
3445.03	7	9	$1.53 \cdot 10^{-1}$	3664.81	11	13	$9.54 \cdot 10^{-1}$	4051.18	13	11	$1.36 \cdot 10^{-1}$
3445.26	7	5	$2.96 \cdot 10^{-1}$	3669.34	9	7	$2.16 \cdot 10^{-1}$	4062.08	11	9	$1.96 \cdot 10^{-1}$
3447.12	9	11	$8.75 \cdot 10^{-1}$	3672.81	9	11	$1.95 \cdot 10^{-1}$	4069.88	13	11	$3.25 \cdot 10^{-1}$
3449.07	7	9	$1.52 \cdot 10^{-1}$	3672.82	9	9	$1.13 \cdot 10^{-1}$	4076.19	9	9	$1.16 \cdot 10^{-1}$
3449.85	5	7	$1.65 \cdot 10^{-1}$	3680.68	11	11	$2.96 \cdot 10^{-1}$	4084.37	9	7	$1.94 \cdot 10^{-1}$
3452.60	7	7	$2.48 \cdot 10^{-1}$	3681.72	9	7	$1.68 \cdot 10^{-1}$	4107.46	7	5	$2.02 \cdot 10^{-1}$
3456.15	5	5	$3.60 \cdot 10^{-1}$	3687.96	5	7	$2.12 \cdot 10^{-1}$	4120.09	13	15	$6.05 \cdot 10^{-1}$
3460.78	9	7	$6.03 \cdot 10^{-1}$	3688.97	11	9	$3.26 \cdot 10^{-1}$	4131.92	9	11	$1.56 \cdot 10^{-1}$
3466.19	9	7	$2.11 \cdot 10^{-1}$	3690.59	11	9	$2.07 \cdot 10^{-1}$	4148.98	9	11	$1.56 \cdot 10^{-1}$
3466.96	7	7	$1.52 \cdot 10^{-1}$	3694.94	5	7	$6.36 \cdot 10^{-1}$	4157.40	13	11	$2.17 \cdot 10^{-1}$
3467.85	5	7	$2.63 \cdot 10^{-1}$	3696.04	11	11	$3.59 \cdot 10^{-1}$	4157.90	9	11	$1.60 \cdot 10^{-1}$
3469.22	5	3	$6.96 \cdot 10^{-1}$	3708.55	7	9	$1.28 \cdot 10^{-1}$	4185.82	11	13	$3.82 \cdot 10^{-1}$
3469.63	13	15	$1.51 \cdot 10^{-1}$	3715.75	9	7	$2.38 \cdot 10^{-1}$	4188.32	11	13	$3.32 \cdot 10^{-1}$
3470.92	3	5	$2.91 \cdot 10^{-1}$	3718.48	5	7	$1.34 \cdot 10^{-1}$	4194.56	11	11	$2.70 \cdot 10^{-1}$

λ Å	Weights g_i	g_k	A 10^8 s^{-1}
4232.59	9	11	3.17·10^{-1}
4240.83	5	5	1.68·10^{-1}
4246.02	11	13	2.00·10^{-1}
4251.88	13	11	1.76·10^{-1}
4254.95	7	9	2.01·10^{-1}
4269.28	11	11	1.36·10^{-1}
4276.91	7	9	2.85·10^{-1}
4277.24	9	11	1.35·10^{-1}
4317.92	15	15	1.28·10^{-1}
4325.80	3	3	1.84·10^{-1}
4326.14	5	7	2.56·10^{-1}
4340.74	5	7	1.23·10^{-1}
4381.63	13	13	2.93·10^{-1}
4382.41	11	13	3.83·10^{-1}
4409.94	13	13	1.38·10^{-1}
4411.69	11	11	2.63·10^{-1}
4434.95	9	9	2.51·10^{-1}
4446.42	11	11	1.90·10^{-1}
4457.35	7	7	1.28·10^{-1}
4474.57	5	5	2.10·10^{-1}
4491.65	11	11	2.09·10^{-1}
4536.80	13	15	5.03·10^{-1}
4598.23	1	3	1.47·10^{-1}
4624.23	9	9	1.32·10^{-1}
4633.08	3	5	2.35·10^{-1}
4652.24	5	7	1.55·10^{-1}
4686.08	3	3	1.72·10^{-1}
4688.21	13	15	1.54·10^{-1}
4707.25	7	9	3.63·10^{-1}
4718.86	5	5	2.17·10^{-1}
4723.05	9	9	1.23·10^{-1}
4731.44	9	11	4.49·10^{-1}
4758.50	11	9	3.01·10^{-1}
4760.18	11	13	4.67·10^{-1}
4764.11	9	7	2.16·10^{-1}
4811.05	13	11	4.36·10^{-1}
4819.25	11	9	2.71·10^{-1}
4830.51	9	7	4.07·10^{-1}
4858.39	13	11	1.24·10^{-1}
4868.02	7	5	3.11·10^{-1}
5037.18	9	7	1.14·10^{-1}
5044.36	7	5	1.31·10^{-1}
5163.18	9	11	2.03·10^{-1}
5171.06	5	7	1.84·10^{-1}
5172.94	5	5	4.11·10^{-1}
5174.18	5	3	5.83·10^{-1}
5191.45	7	9	1.62·10^{-1}
5238.21	7	9	3.74·10^{-1}
5240.87	7	7	3.89·10^{-1}
5242.80	7	5	2.01·10^{-1}
5261.53	5	7	1.13·10^{-1}
5280.85	5	5	1.28·10^{-1}
5355.52	9	9	1.21·10^{-1}
5356.46	11	11	2.11·10^{-1}
5360.51	9	11	6.19·10^{-1}
5364.28	9	9	2.26·10^{-1}
5460.50	5	3	3.46·10^{-1}
5493.76	7	5	2.13·10^{-1}
5506.49	5	7	3.61·10^{-1}
5533.03	5	5	3.72·10^{-1}

λ Å	Weights g_i	g_k	A 10^8 s^{-1}
5570.44	5	3	3.30·10^{-1}
5849.71	3	3	3.02·10^{-1}
5851.50	3	5	1.55·10^{-1}
5893.36	5	5	2.60·10^{-1}
5895.93	5	7	3.12·10^{-1}
5926.37	7	7	1.63·10^{-1}
5928.88	7	9	5.32·10^{-1}
7154.11	9	9	3.45·10^{-1}

Neodymium

Nd II

λ Å	Weights g_i	g_k	A 10^8 s^{-1}
3780.4	16	18	1.4·10^{-1}
3805.4	14	16	6.9·10^{-1}
3807.2	10	12	4.9·10^{-2}
3863.3	8	10	1.5·10^{-1}
3941.5	10	10	6.1·10^{-1}
3951.2	12	12	6.0·10^{-1}
3973.3	18	18	6.3·10^{-1}
3979.5	10	12	2.7·10^{-1}
3990.1	16	16	5.2·10^{-1}
4012.3	18	20	5.5·10^{-1}
4061.1	16	18	4.4·10^{-1}
4106.6	14	16	6.8·10^{-2}
4109.5	14	16	3.7·10^{-1}
4133.4	14	12	1.5·10^{-1}
4156.1	12	14	3.4·10^{-1}
4205.6	18	16	1.8·10^{-1}
4284.5	18	18	8.5·10^{-2}
4303.6	8	10	4.7·10^{-1}
4325.8	16	16	1.6·10^{-1}
4358.2	14	14	1.5·10^{-1}
4382.7	12	10	4.0·10^{-2}
4400.8	10	10	6.8·10^{-2}
4451.6	12	14	2.5·10^{-1}
4456.4	16	18	6.4·10^{-2}
4463.0	14	16	1.8·10^{-1}
4958.1	12	10	1.2·10^{-2}
5130.6	22	20	1.6·10^{-1}
5192.6	20	18	1.7·10^{-1}
5249.6	18	16	1.8·10^{-1}
5276.9	12	10	1.2·10^{-1}
5293.2	16	14	1.2·10^{-1}
5302.3	20	18	1.1·10^{-1}
5311.5	14	12	1.1·10^{-1}
5319.8	12	10	1.6·10^{-1}
5357.0	18	16	1.8·10^{-1}
5371.9	20	20	5.1·10^{-2}
5485.7	18	18	5.7·10^{-2}
5594.4	16	16	7.0·10^{-2}
5620.6	18	18	1.3·10^{-1}
5688.5	14	14	5.9·10^{-2}
5718.1	16	16	8.7·10^{-2}
5726.8	10	10	5.6·10^{-2}
5740.9	12	12	7.2·10^{-2}
5804.0	10	10	4.6·10^{-2}
5865.1	16	18	1.3·10^{-2}
6051.9	12	10	1.1·10^{-2}

Neon

Ne I

λ Å	Weights g_i	g_k	A 10^8 s^{-1}
615.63	1	3	3.8·10^{-1}
618.67	1	3	9.3·10^{-1}
619.10	1	3	3.3·10^{-1}
626.82	1	3	7.4·10^{-1}
629.74	1	3	4.8·10^{-1}
735.90	1	3	6.11
743.72	1	3	4.86·10^{-1}
3369.8	5	5	1.0·10^{-3}
3369.9	5	3	7.6·10^{-3}
3375.6	5	3	2.2·10^{-3}
3417.9	3	5	9.2·10^{-3}
3418.0	3	3	2.2·10^{-3}
3423.9	3	3	1.0·10^{-3}
3447.7	5	5	2.1·10^{-2}
3450.8	5	3	4.9·10^{-3}
3454.2	3	1	3.7·10^{-2}
3460.5	1	3	7.0·10^{-3}
3464.3	5	5	6.7·10^{-3}
3466.6	1	3	1.3·10^{-2}
3472.6	5	7	1.7·10^{-2}
3498.1	3	5	5.1·10^{-3}
3501.2	3	3	1.2·10^{-2}
3510.7	5	3	2.2·10^{-3}
3515.2	3	5	6.9·10^{-3}
3520.5	3	1	9.3·10^{-2}
3593.5	3	5	9.9·10^{-3}
3593.6	3	3	6.6·10^{-3}
3600.2	3	3	4.3·10^{-3}
3633.7	3	1	1.1·10^{-2}
3682.2	3	5	1.6·10^{-3}
3685.7	3	3	3.9·10^{-3}
3701.2	3	5	2.2·10^{-3}
4536.3	3	3	5.0·10^{-3}
4702.5	3	3	2.1·10^{-3}
4708.9	3	3	4.2·10^{-3}
4955.4	3	3	3.3·10^{-3}
5113.7	3	3	1.0·10^{-2}
5120.5	3	3	5.6·10^{-3}
5154.4	3	3	1.9·10^{-2}
5191.3	3	3	1.3·10^{-2}
5326.4	3	3	6.8·10^{-3}
5333.3	3	3	5.3·10^{-3}
5341.1	3	3	1.1·10^{-1}
5400.6	3	1	9.0·10^{-3}
5418.6	3	3	5.2·10^{-3}
5433.7	3	3	2.83·10^{-3}
5652.6	3	3	8.9·10^{-3}
5662.5	3	3	6.9·10^{-3}
5852.5	3	1	6.82·10^{-1}
5868.4	3	3	1.4·10^{-2}
5881.9	5	3	1.15·10^{-1}
5913.6	3	3	4.8·10^{-2}
5939.3	5	3	2.00·10^{-3}
5944.8	5	5	1.13·10^{-1}
5961.6	3	3	3.3·10^{-2}
5975.5	5	3	3.51·10^{-2}

λ Å	g_i	g_k	A $10^8\,s^{-1}$	λ Å	g_i	g_k	A $10^8\,s^{-1}$	λ Å	g_i	g_k	A $10^8\,s^{-1}$
6030.0	3	3	$5.61 \cdot 10^{-2}$	11767	3	3	$6.9 \cdot 10^{-2}$	3214.3	4	6	2.2
6046.1	3	3	$2.26 \cdot 10^{-3}$	12459	3	3	$1.5 \cdot 10^{-2}$	3218.2	8	10	3.6
6074.3	3	1	$6.03 \cdot 10^{-1}$					3224.8	6	8	3.5
6096.2	3	5	$1.81 \cdot 10^{-1}$	*Ne II*				3229.5	8	8	$1.3 \cdot 10^{-1}$
6118.0	5	3	$6.09 \cdot 10^{-3}$	*357.03	6	10	$3.8 \cdot 10^{1}$	3229.6	8	10	3.6
6128.5	3	3	$6.7 \cdot 10^{-3}$	*361.77	6	2	$1.6 \cdot 10^{1}$	3230.1	6	6	1.8
6143.1	5	5	$2.82 \cdot 10^{-1}$	*406.28	6	10	$1.8 \cdot 10^{1}$	3230.4	4	6	$1.4 \cdot 10^{-1}$
6150.3	3	3	$1.5 \cdot 10^{-2}$	*446.37	6	6	$4.07 \cdot 10^{1}$	3232.0	6	4	$2.7 \cdot 10^{-1}$
6163.6	1	3	$1.46 \cdot 10^{-1}$	460.73	4	2	$4.7 \cdot 10^{1}$	3232.4	4	4	1.6
6217.3	5	3	$6.37 \cdot 10^{-2}$	462.39	2	2	$2.3 \cdot 10^{1}$	3243.4	6	6	$2.3 \cdot 10^{-1}$
6266.5	1	3	$2.49 \cdot 10^{-1}$	1907.5	4	2	$2.8 \cdot 10^{-1}$	3244.1	6	8	1.5
6273.0	3	3	$9.7 \cdot 10^{-3}$	1916.1	4	4	$6.9 \cdot 10^{-1}$	3248.1	4	4	$2.4 \cdot 10^{-1}$
6293.7	3	3	$6.39 \cdot 10^{-3}$	1930.0	2	2	$5.7 \cdot 10^{-1}$	3255.4	6	4	$3.8 \cdot 10^{-2}$
6304.8	3	5	$4.16 \cdot 10^{-2}$	1938.8	2	4	$1.3 \cdot 10^{-1}$	3263.4	2	4	$3.9 \cdot 10^{-1}$
6328.2	5	3	$3.39 \cdot 10^{-2}$	2858.0	6	6	$7.9 \cdot 10^{-1}$	3269.9	4	6	$5.1 \cdot 10^{-1}$
6330.9	3	3	$2.3 \cdot 10^{-2}$	2870.0	6	6	$1.7 \cdot 10^{-1}$	3270.8	6	4	$5.7 \cdot 10^{-2}$
6334.4	5	5	$1.61 \cdot 10^{-1}$	2873.0	6	4	$3.8 \cdot 10^{-1}$	3297.7	6	6	$4.3 \cdot 10^{-1}$
6351.9	1	3	$3.45 \cdot 10^{-3}$	2876.3	4	6	$7.8 \cdot 10^{-1}$	3309.7	4	2	$3.1 \cdot 10^{-1}$
6383.0	3	3	$3.21 \cdot 10^{-1}$	2876.5	6	4	$3.3 \cdot 10^{-1}$	3310.5	4	6	$6.9 \cdot 10^{-2}$
6401.1	3	3	$1.39 \cdot 10^{-2}$	2878.1	2	6	$6.9 \cdot 10^{-1}$	3311.3	4	2	$2.6 \cdot 10^{-1}$
6402.2	5	7	$5.14 \cdot 10^{-1}$	2888.4	4	6	$7.0 \cdot 10^{-1}$	3314.7	6	6	$4.4 \cdot 10^{-2}$
6506.5	3	5	$3.00 \cdot 10^{-1}$	2891.5	4	4	$6.1 \cdot 10^{-1}$	3319.7	4	2	1.6
6532.9	1	3	$1.08 \cdot 10^{-1}$	2897.0	6	8	$5.2 \cdot 10^{-2}$	3320.2	8	6	$2.1 \cdot 10^{-1}$
6599.0	3	3	$2.32 \cdot 10^{-1}$	2906.8	2	4	$5.5 \cdot 10^{-1}$	3323.7	4	4	1.6
6602.9	3	5	$5.9 \cdot 10^{-3}$	2910.1	4	2	1.7	3327.2	4	4	$9.1 \cdot 10^{-1}$
6652.1	3	1	$2.9 \cdot 10^{-3}$	2910.4	2	4	$5.9 \cdot 10^{-1}$	3329.2	8	8	$8.8 \cdot 10^{-1}$
6678.3	3	5	$2.33 \cdot 10^{-1}$	2916.2	6	4	$9.6 \cdot 10^{-2}$	3330.7	6	6	$3.9 \cdot 10^{-2}$
6717.0	3	3	$2.17 \cdot 10^{-1}$	2925.6	2	2	$5.6 \cdot 10^{-1}$	3334.8	6	8	1.8
6721.1	3	3	$4.9 \cdot 10^{-4}$	2933.7	6	6	$6.9 \cdot 10^{-2}$	3336.1	4	6	1.1
6929.5	3	5	$1.74 \cdot 10^{-1}$	2955.7	6	4	1.2	3344.4	2	2	1.5
7024.1	3	3	$1.89 \cdot 10^{-2}$	3001.7	4	4	$8.7 \cdot 10^{-1}$	3345.5	6	4	1.4
7032.4	5	3	$2.53 \cdot 10^{-1}$	3017.3	6	4	$3.5 \cdot 10^{-1}$	3345.8	4	4	$2.2 \cdot 10^{-1}$
7051.3	3	3	$3.0 \cdot 10^{-2}$	3027.0	6	6	1.4	3353.6	4	2	$1.2 \cdot 10^{-1}$
7059.1	3	5	$6.8 \cdot 10^{-2}$	3028.7	4	2	$8.5 \cdot 10^{-1}$	3355.0	4	6	1.3
7173.9	3	5	$2.87 \cdot 10^{-2}$	3028.9	2	4	$4.7 \cdot 10^{-1}$	3356.3	6	6	$2.0 \cdot 10^{-1}$
7245.2	3	3	$9.35 \cdot 10^{-2}$	3034.5	6	8	3.1	3357.8	6	6	$5.0 \cdot 10^{-1}$
7304.8	1	3	$2.55 \cdot 10^{-3}$	3037.7	4	4	2.1	3360.3	2	4	$8.6 \cdot 10^{-1}$
7438.9	1	3	$2.31 \cdot 10^{-2}$	3045.6	2	2	2.5	3360.6	2	4	$8.2 \cdot 10^{-1}$
7472.4	3	3	$4.0 \cdot 10^{-2}$	3047.6	4	6	1.8	3362.9	4	2	$3.5 \cdot 10^{-1}$
7535.8	3	3	$4.3 \cdot 10^{-1}$	3054.7	2	4	$9.4 \cdot 10^{-1}$	3371.8	4	6	$2.2 \cdot 10^{-1}$
7937.0	5	5	$7.8 \cdot 10^{-3}$	3092.9	6	6	1.3	3374.1	4	4	$3.0 \cdot 10^{-1}$
8082.5	3	3	$1.2 \cdot 10^{-3}$	3097.1	8	8	1.3	3378.2	2	2	1.7
8118.5	3	3	$4.9 \cdot 10^{-2}$	3118.0	8	6	$4.2 \cdot 10^{-2}$	3379.3	2	2	$3.0 \cdot 10^{-1}$
8128.9	3	5	$7.2 \cdot 10^{-3}$	3134.1	6	4	$2.6 \cdot 10^{-1}$	3386.2	4	6	$5.5 \cdot 10^{-2}$
8259.4	5	5	$2.03 \cdot 10^{-2}$	3140.4	8	6	$2.4 \cdot 10^{-1}$	3388.4	4	6	2.2
8571.4	3	5	$5.5 \cdot 10^{-2}$	3151.1	6	4	$4.8 \cdot 10^{-1}$	3390.6	2	4	$7.7 \cdot 10^{-2}$
8582.9	3	5	$1.00 \cdot 10^{-2}$	3154.8	8	6	$1.8 \cdot 10^{-2}$	3392.8	2	4	$4.4 \cdot 10^{-1}$
8647.0	5	5	$3.91 \cdot 10^{-2}$	3164.4	8	8	$1.6 \cdot 10^{-1}$	3404.8	4	6	1.9
8681.9	3	3	$2.1 \cdot 10^{-1}$	3165.7	6	6	$1.2 \cdot 10^{-1}$	3407.0	6	8	2.3
8767.5	3	3	$1.1 \cdot 10^{-3}$	3173.6	6	4	$4.5 \cdot 10^{-1}$	3411.4	4	2	$6.1 \cdot 10^{-1}$
8771.7	3	3	$1.6 \cdot 10^{-1}$	3176.1	4	6	$6.0 \cdot 10^{-2}$	3413.2	4	4	1.8
8783.8	3	5	$3.13 \cdot 10^{-1}$	3187.6	4	6	$1.4 \cdot 10^{-2}$	3414.9	4	6	$1.8 \cdot 10^{-2}$
8865.3	3	3	$9.4 \cdot 10^{-3}$	3188.7	6	6	$3.9 \cdot 10^{-1}$	3416.9	6	6	$6.4 \cdot 10^{-1}$
9201.8	3	3	$9.1 \cdot 10^{-2}$	3190.9	4	6	$1.5 \cdot 10^{-1}$	3417.7	6	8	1.6
9433.0	3	3	$1.1 \cdot 10^{-3}$	3194.6	4	4	$5.2 \cdot 10^{-1}$	3438.9	2	2	1.4
9486.7	3	3	$2.5 \cdot 10^{-2}$	3198.6	6	8	1.7	3440.7	2	4	$3.5 \cdot 10^{-1}$
9534.2	3	3	$6.3 \cdot 10^{-2}$	3198.9	4	4	$2.3 \cdot 10^{-1}$	3453.1	4	4	$4.6 \cdot 10^{-1}$
10621	3	3	$2.4 \cdot 10^{-3}$	3209.0	8	8	$1.6 \cdot 10^{-1}$	3454.8	4	4	1.6
11409	3	3	$4.2 \cdot 10^{-2}$	3209.4	2	4	$6.0 \cdot 10^{-1}$	3456.6	2	4	$9.6 \cdot 10^{-1}$
11525	3	3	$8.4 \cdot 10^{-2}$	3213.7	2	4	1.7	3457.1	4	6	$9.9 \cdot 10^{-2}$

λ Å	Weights g_i	g_k	A 10^8 s^{-1}	λ Å	Weights g_i	g_k	A 10^8 s^{-1}	λ Å	Weights g_i	g_k	A 10^8 s^{-1}
3459.3	6	6	1.6	2265.7	5	7	2.4	3458.46	3	5	$6.1 \cdot 10^{-1}$
3475.2	4	4	$1.2 \cdot 10^{-2}$					3461.66	7	9	$2.7 \cdot 10^{-1}$
3477.6	4	6	$4.3 \cdot 10^{-1}$	*Ne VII*				3472.55	5	7	$1.2 \cdot 10^{-1}$
3481.9	4	2	1.4	97.502	1	3	$1.07 \cdot 10^{3}$	3492.96	5	3	$9.8 \cdot 10^{-1}$
3503.6	2	2	2.0	*115.46	9	3	$4.8 \cdot 10^{2}$	3510.33	3	1	1.2
3522.7	4	2	$2.3 \cdot 10^{-2}$	116.69	3	5	$1.6 \cdot 10^{3}$	3515.05	5	7	$4.2 \cdot 10^{-1}$
3538.0	4	2	$7.6 \cdot 10^{-1}$	127.66	3	1	$1.9 \cdot 10^{2}$	3524.54	7	5	1.0
3539.9	4	4	$3.6 \cdot 10^{-2}$	465.22	1	3	$4.09 \cdot 10^{1}$	3566.37	5	5	$5.6 \cdot 10^{-1}$
3542.2	6	4	$6.0 \cdot 10^{-1}$	558.61	3	5	8.11	3619.39	5	7	$6.6 \cdot 10^{-1}$
3542.9	4	6	1.2	559.95	1	3	$1.07 \cdot 10^{1}$	4027.67	5	7	$1.3 \cdot 10^{-1}$
3546.2	2	4	$6.3 \cdot 10^{-2}$	561.38	3	3	7.99	4295.88	9	7	$1.7 \cdot 10^{-1}$
3551.6	2	4	$3.7 \cdot 10^{-2}$	561.73	5	5	$2.39 \cdot 10^{1}$	4401.54	9	11	$3.8 \cdot 10^{-1}$
3557.8	2	2	$1.9 \cdot 10^{-1}$	562.99	3	1	$3.17 \cdot 10^{1}$	4462.46	3	5	$1.7 \cdot 10^{-1}$
3561.2	4	6	$2.1 \cdot 10^{-1}$	564.53	5	3	$1.31 \cdot 10^{1}$	4470.48	5	7	$1.9 \cdot 10^{-1}$
3565.8	4	4	$6.2 \cdot 10^{-1}$					4600.37	5	3	$2.6 \cdot 10^{-1}$
3568.5	6	8	1.4	*Ne VIII*				4604.99	9	7	$2.3 \cdot 10^{-1}$
3571.2	4	4	$6.3 \cdot 10^{-1}$	*88.09	2	6	$8.4 \cdot 10^{2}$	4606.23	5	3	$1.0 \cdot 10^{-1}$
3574.2	6	6	$1.0 \cdot 10^{-1}$	*98.208	6	10	$2.77 \cdot 10^{3}$	4648.66	11	9	$2.4 \cdot 10^{-1}$
3574.6	4	6	1.3	770.41	2	4	5.90	4686.22	5	5	$1.4 \cdot 10^{-1}$
3590.4	4	6	$3.6 \cdot 10^{-2}$	780.32	2	2	5.69	4701.54	9	9	$1.4 \cdot 10^{-1}$
3594.2	4	2	1.3	2820.7	2	4	$7.20 \cdot 10^{-1}$	4714.42	13	11	$4.6 \cdot 10^{-1}$
3612.3	2	4	$2.6 \cdot 10^{-1}$	2860.1	2	2	$6.88 \cdot 10^{-1}$	4715.78	7	7	$2.0 \cdot 10^{-1}$
3628.0	4	4	$6.0 \cdot 10^{-1}$					4732.47	7	9	$9.3 \cdot 10^{-2}$
3632.7	4	4	$1.3 \cdot 10^{-1}$	*Nickel*				4752.43	3	3	$2.0 \cdot 10^{-1}$
3643.9	4	4	$3.2 \cdot 10^{-1}$					4756.52	9	9	$1.5 \cdot 10^{-1}$
3644.9	2	4	$9.9 \cdot 10^{-1}$	*Ni I*				4786.54	11	11	$1.8 \cdot 10^{-1}$
3659.9	4	6	$6.7 \cdot 10^{-2}$	1976.87	7	9	1.1	4829.03	5	7	$1.9 \cdot 10^{-1}$
3664.1	6	4	$7.0 \cdot 10^{-1}$	1990.25	5	7	$8.3 \cdot 10^{-1}$	4831.18	9	7	$1.6 \cdot 10^{-1}$
3679.8	4	2	$3.2 \cdot 10^{-1}$	2014.25	3	5	$9.3 \cdot 10^{-1}$	4838.64	9	7	$2.2 \cdot 10^{-1}$
3694.2	6	6	1.0	2085.57	5	5	2.6	4855.41	5	5	$5.7 \cdot 10^{-1}$
3697.1	2	2	$2.8 \cdot 10^{-1}$	2158.31	7	5	$6.9 \cdot 10^{-1}$	4904.41	5	3	$6.2 \cdot 10^{-1}$
3701.8	4	6	$2.7 \cdot 10^{-1}$	2289.98	9	7	2.1	4912.03	3	3	$1.5 \cdot 10^{-1}$
3709.6	4	2	1.1	2300.77	7	7	$7.5 \cdot 10^{-1}$	4913.97	1	3	$2.2 \cdot 10^{-1}$
3713.1	4	6	1.3	2312.34	7	7	5.5	4918.36	9	7	$2.3 \cdot 10^{-1}$
3721.8	4	6	$2.0 \cdot 10^{-1}$	2313.98	5	5	5.0	4935.83	7	5	$2.4 \cdot 10^{-1}$
3726.9	4	4	$1.2 \cdot 10^{-1}$	2317.16	7	5	3.8	4937.34	9	9	$1.2 \cdot 10^{-1}$
3727.1	2	4	$9.8 \cdot 10^{-1}$	2320.03	9	11	6.9	4953.20	5	5	$1.2 \cdot 10^{-1}$
3734.9	4	4	$1.9 \cdot 10^{-1}$	2321.38	5	7	5.6	4980.17	9	11	$1.9 \cdot 10^{-1}$
3744.6	2	4	$2.6 \cdot 10^{-1}$	2325.79	7	9	3.5	5000.34	7	7	$1.4 \cdot 10^{-1}$
3751.2	2	2	$1.8 \cdot 10^{-1}$	2329.96	5	3	5.3	5012.46	7	7	$1.1 \cdot 10^{-1}$
3753.8	4	6	$4.5 \cdot 10^{-1}$	2345.54	9	7	2.2	5017.58	11	11	$2.0 \cdot 10^{-1}$
3766.3	4	6	$2.9 \cdot 10^{-1}$	2346.63	7	5	$5.5 \cdot 10^{-1}$	5035.37	7	9	$5.7 \cdot 10^{-1}$
3777.1	2	4	$4.2 \cdot 10^{-1}$	3002.48	7	7	$8.0 \cdot 10^{-1}$	5042.20	3	5	$1.4 \cdot 10^{-1}$
3800.0	4	4	$3.7 \cdot 10^{-1}$	3003.62	5	5	$6.9 \cdot 10^{-1}$	5048.85	7	7	$1.6 \cdot 10^{-1}$
3818.4	2	4	$6.1 \cdot 10^{-1}$	3012.00	5	5	1.3	5080.53	9	11	$3.2 \cdot 10^{-1}$
3829.8	4	6	$8.4 \cdot 10^{-1}$	3037.93	7	7	$2.8 \cdot 10^{-1}$	5081.11	7	9	$5.7 \cdot 10^{-1}$
3942.3	4	6	$1.0 \cdot 10^{-2}$	3050.82	7	9	$6.0 \cdot 10^{-1}$	5082.35	3	3	$2.5 \cdot 10^{-1}$
				3054.31	5	5	$4.0 \cdot 10^{-1}$	5084.08	7	9	$3.1 \cdot 10^{-1}$
Ne V				3057.64	3	3	1.0	5099.95	7	7	$2.9 \cdot 10^{-1}$
*142.61	9	9	$6.7 \cdot 10^{2}$	3101.56	5	7	$6.3 \cdot 10^{-1}$	5115.40	11	9	$2.2 \cdot 10^{-1}$
*143.32	9	15	$1.2 \cdot 10^{3}$	3101.88	5	7	$4.9 \cdot 10^{-1}$	5129.37	7	5	$1.2 \cdot 10^{-1}$
147.13	5	7	$1.5 \cdot 10^{3}$	3134.11	3	5	$7.3 \cdot 10^{-1}$	5155.14	5	5	$1.1 \cdot 10^{-1}$
151.23	5	5	$3.38 \cdot 10^{2}$	3369.56	9	7	$1.8 \cdot 10^{-1}$	5155.76	5	7	$2.9 \cdot 10^{-1}$
154.50	1	3	$7.0 \cdot 10^{2}$	3380.57	5	3	1.3	5176.57	5	5	$1.8 \cdot 10^{-1}$
*167.69	9	9	$1.5 \cdot 10^{2}$	3392.98	7	7	$2.4 \cdot 10^{-1}$	5371.33	7	7	$1.6 \cdot 10^{-1}$
*358.93	9	3	$2.1 \cdot 10^{2}$	3414.76	7	9	$5.5 \cdot 10^{-1}$	5476.91	1	3	$9.5 \cdot 10^{-2}$
365.59	5	3	$1.35 \cdot 10^{2}$	3423.71	3	3	$3.3 \cdot 10^{-1}$	5637.12	3	3	$1.1 \cdot 10^{-1}$
*482.15	9	9	$3.01 \cdot 10^{1}$	3433.56	7	7	$1.7 \cdot 10^{-1}$	5664.02	5	7	$1.1 \cdot 10^{-1}$
*571.04	9	15	$1.0 \cdot 10^{1}$	3446.26	5	5	$4.4 \cdot 10^{-1}$	5695.00	3	3	$1.7 \cdot 10^{-1}$
2259.6	3	5	1.9	3452.88	5	7	$9.8 \cdot 10^{-2}$	6086.29	3	5	$1.1 \cdot 10^{-1}$

λ (Å)	g_i	g_k	A (10^8 s^{-1})
6175.42	3	3	$1.7 \cdot 10^{-1}$
7122.24	5	7	$2.1 \cdot 10^{-1}$
7381.94	9	11	$9.7 \cdot 10^{-2}$
7422.30	7	5	$1.8 \cdot 10^{-1}$
7727.66	7	7	$1.1 \cdot 10^{-1}$

Ni II

λ (Å)	g_i	g_k	A (10^8 s^{-1})
2165.55	10	10	2.4
2169.10	8	8	1.58
2174.67	8	10	1.43
2175.15	6	6	1.77
2184.61	4	4	2.90
2201.41	4	6	1.3
2206.72	6	8	1.66
2216.48	10	12	3.4
2220.40	6	8	2.3
2222.96	10	10	$9.8 \cdot 10^{-1}$
2224.86	8	8	1.55
2226.33	6	6	1.3
2253.85	4	6	1.98
2264.46	6	8	1.43
2270.21	8	10	1.56
2278.77	8	6	2.8
2287.09	6	4	2.8
2296.55	8	8	1.98
2297.14	6	4	2.70
2297.49	4	2	3.0
2298.27	6	6	2.8
2303.00	8	6	2.9
2316.04	10	8	2.88
2334.58	8	8	$8.0 \cdot 10^{-1}$
2375.42	6	8	$6.6 \cdot 10^{-1}$
2394.52	8	10	1.70
2416.13	6	8	2.1
2437.89	8	10	$5.4 \cdot 10^{-1}$
2510.87	8	10	$5.8 \cdot 10^{-1}$

Ni III

λ (Å)	g_i	g_k	A (10^8 s^{-1})
1692.51	11	13	7.9
1709.90	9	11	6.3
1719.46	5	7	6.0
1722.28	3	5	5.9
1724.52	3	1	6.7
1741.96	9	7	5.7
1752.43	7	5	5.5
1760.56	5	3	6.5
1769.64	11	11	6.2
1823.06	9	9	5.6

Nitrogen

N I

λ (Å)	g_i	g_k	A (10^8 s^{-1})
951.079	4	6	$8.29 \cdot 10^{-3}$
951.295	4	4	$1.71 \cdot 10^{-3}$
952.303	4	6	$1.12 \cdot 10^{-1}$
952.415	4	4	$1.45 \cdot 10^{-1}$
952.523	4	2	$7.62 \cdot 10^{-2}$
953.415	4	2	1.90
953.655	4	4	1.81
953.970	4	6	1.62

λ (Å)	g_i	g_k	A (10^8 s^{-1})
954.104	4	6	$1.95 \cdot 10^{-1}$
955.264	4	6	$3.37 \cdot 10^{-3}$
955.437	4	4	$1.40 \cdot 10^{-4}$
955.529	4	2	$2.63 \cdot 10^{-3}$
955.882	4	4	$4.29 \cdot 10^{-3}$
959.494	4	4	$3.75 \cdot 10^{-3}$
960.201	4	2	$1.69 \cdot 10^{-3}$
963.990	4	6	$5.94 \cdot 10^{-1}$
964.626	4	4	$5.66 \cdot 10^{-1}$
965.041	4	2	$5.52 \cdot 10^{-1}$
1003.37	4	4	$1.86 \cdot 10^{-6}$
1003.38	4	6	$8.40 \cdot 10^{-6}$
1134.17	4	2	1.51
1134.41	4	4	1.49
1134.98	4	6	1.44
1159.82	4	4	$4.94 \cdot 10^{-4}$
1160.94	4	2	$2.72 \cdot 10^{-4}$
1163.88	6	6	$3.25 \cdot 10^{-1}$
1164.00	4	6	$2.53 \cdot 10^{-3}$
1164.21	6	4	$2.98 \cdot 10^{-2}$
1164.32	4	4	$2.82 \cdot 10^{-1}$
1165.59	6	8	$1.69 \cdot 10^{-2}$
1165.72	6	6	$8.31 \cdot 10^{-4}$
1165.84	6	6	$1.99 \cdot 10^{-3}$
1165.88	6	4	$2.43 \cdot 10^{-4}$
1166.00	4	4	$6.79 \cdot 10^{-4}$
1167.45	6	8	1.10
1167.74	6	4	$1.17 \cdot 10^{-4}$
1167.86	4	4	$7.03 \cdot 10^{-4}$
1168.22	6	6	$1.30 \cdot 10^{-2}$
1168.33	4	6	$1.24 \cdot 10^{-1}$
1168.42	6	6	$3.38 \cdot 10^{-2}$
1168.54	4	6	$9.32 \cdot 10^{-1}$
1169.69	6	8	$2.79 \cdot 10^{-2}$
1170.16	6	6	$3.03 \cdot 10^{-3}$
1170.28	4	6	$3.06 \cdot 10^{-2}$
1170.42	6	4	$4.91 \cdot 10^{-4}$
1170.54	4	4	$1.11 \cdot 10^{-3}$
1170.67	4	2	$5.86 \cdot 10^{-2}$
1171.08	6	4	$1.25 \cdot 10^{-1}$
1171.20	4	4	$3.93 \cdot 10^{-3}$
1176.51	6	4	$8.52 \cdot 10^{-1}$
1176.63	4	4	$9.91 \cdot 10^{-2}$
1177.69	4	2	1.03
1183.28	6	6	$1.88 \cdot 10^{-5}$
1183.40	4	6	$2.16 \cdot 10^{-6}$
1184.24	6	4	$2.45 \cdot 10^{-3}$
1184.36	4	4	$3.34 \cdot 10^{-4}$
1184.98	4	2	$1.78 \cdot 10^{-3}$
1199.55	4	6	4.07
1200.22	4	4	4.03
1200.71	4	2	4.00
1243.17	6	4	$3.33 \cdot 10^{-1}$
1243.18	6	6	3.22
1243.31	4	4	3.10
1243.31	4	6	$2.36 \cdot 10^{-1}$
1310.54	4	6	$7.68 \cdot 10^{-1}$
1310.94	2	4	$6.05 \cdot 10^{-1}$
1310.95	4	4	$1.75 \cdot 10^{-1}$
1312.87	4	6	$5.53 \cdot 10^{-4}$

λ (Å)	g_i	g_k	A (10^8 s^{-1})
1313.07	2	4	$2.79 \cdot 10^{-3}$
1313.08	4	4	$7.66 \cdot 10^{-5}$
1313.28	2	2	$2.37 \cdot 10^{-3}$
1313.28	4	2	$1.15 \cdot 10^{-3}$
1314.97	2	2	$1.14 \cdot 10^{-3}$
1314.98	4	2	$6.83 \cdot 10^{-4}$
1315.43	2	4	$3.65 \cdot 10^{-3}$
1315.44	4	4	$7.40 \cdot 10^{-4}$
1316.04	4	6	$1.06 \cdot 10^{-3}$
1316.29	4	6	$1.12 \cdot 10^{-2}$
1318.50	4	6	$1.36 \cdot 10^{-4}$
1318.82	2	4	$1.45 \cdot 10^{-5}$
1318.83	4	4	$1.00 \cdot 10^{-2}$
1319.00	2	2	$4.59 \cdot 10^{-1}$
1319.00	4	2	$2.28 \cdot 10^{-1}$
1319.67	2	4	$1.70 \cdot 10^{-1}$
1319.68	4	4	$5.76 \cdot 10^{-1}$
1326.56	2	4	$9.72 \cdot 10^{-3}$
1326.57	4	4	$6.71 \cdot 10^{-2}$
1327.92	2	2	$9.47 \cdot 10^{-2}$
1327.92	4	2	$4.43 \cdot 10^{-2}$
1335.18	4	6	$3.25 \cdot 10^{-6}$
1336.39	2	4	$1.02 \cdot 10^{-4}$
1336.40	4	4	$5.93 \cdot 10^{-4}$
1337.19	2	2	$3.16 \cdot 10^{-4}$
1337.20	4	2	$1.22 \cdot 10^{-4}$
1411.93	2	4	$4.46 \cdot 10^{-1}$
1411.94	4	4	$1.01 \cdot 10^{-1}$
1411.95	4	6	$5.34 \cdot 10^{-1}$
1492.63	6	4	3.11
1492.82	4	4	$3.26 \cdot 10^{-1}$
1494.68	4	2	3.46
1742.72	2	4	$2.12 \cdot 10^{-1}$
1742.73	4	4	1.05
1745.25	2	2	$8.35 \cdot 10^{-1}$
1745.26	4	2	$4.01 \cdot 10^{-1}$
6884.31	2	4	$2.90 \cdot 10^{-6}$
6900.36	4	4	$1.27 \cdot 10^{-5}$
7380.10	4	6	$1.94 \cdot 10^{-5}$
7403.39	2	4	$1.14 \cdot 10^{-4}$
7405.68	6	6	$4.39 \cdot 10^{-5}$
7421.94	4	4	$1.63 \cdot 10^{-4}$
7423.64	2	4	$5.64 \cdot 10^{-2}$
7442.30	4	4	$1.19 \cdot 10^{-1}$
7447.81	6	4	$1.49 \cdot 10^{-4}$
7468.31	6	4	$1.96 \cdot 10^{-1}$
8184.86	4	6	$8.21 \cdot 10^{-2}$
8188.01	2	4	$1.25 \cdot 10^{-1}$
8200.36	2	2	$4.68 \cdot 10^{-2}$
8210.72	4	4	$5.23 \cdot 10^{-2}$
8216.34	6	6	$2.26 \cdot 10^{-1}$
8223.13	4	2	$2.62 \cdot 10^{-1}$
8242.39	6	4	$1.31 \cdot 10^{-1}$
8567.74	2	4	$4.86 \cdot 10^{-2}$
8594.00	2	2	$2.09 \cdot 10^{-1}$
8629.24	4	4	$2.67 \cdot 10^{-1}$
8655.88	4	2	$1.07 \cdot 10^{-1}$
8664.39	2	4	$1.63 \cdot 10^{-3}$
8680.28	6	8	$2.53 \cdot 10^{-1}$
8683.40	4	6	$1.88 \cdot 10^{-1}$

λ Å	Weights g_i	g_k	A $10^8\,\mathrm{s}^{-1}$	λ Å	Weights g_i	g_k	A $10^8\,\mathrm{s}^{-1}$	λ Å	Weights g_i	g_k	A $10^8\,\mathrm{s}^{-1}$
8686.15	2	4	$1.15\cdot10^{-1}$	536.365	5	7	$1.52\cdot10^{-1}$	1275.25	5	5	$9.34\cdot10^{-2}$
8703.25	2	2	$2.16\cdot10^{-1}$	536.536	5	5	$2.35\cdot10^{-2}$	1275.28	3	5	$6.32\cdot10^{-3}$
8711.70	4	4	$1.29\cdot10^{-1}$	572.069	5	3	$4.24\cdot10^{-3}$	1276.20	5	3	$4.54\cdot10^{-1}$
8718.84	6	6	$6.54\cdot10^{-2}$	574.650	5	7	$3.83\cdot10^{1}$	1276.22	3	3	$1.54\cdot10^{-1}$
8728.90	4	2	$3.75\cdot10^{-2}$	576.060	5	3	$4.22\cdot10^{-4}$	1276.80	3	1	$6.08\cdot10^{-1}$
8747.37	6	4	$9.65\cdot10^{-3}$	576.232	5	5	$3.23\cdot10^{-2}$	1299.81	5	5	$7.33\cdot10^{-6}$
8758.21	2	4	$2.62\cdot10^{-4}$	580.802	5	7	$1.39\cdot10^{-2}$	1300.04	5	7	$4.14\cdot10^{-5}$
8767.35	2	2	$1.22\cdot10^{-3}$	580.904	5	5	$1.38\cdot10^{-1}$	1304.77	5	3	$6.74\cdot10^{-4}$
9020.69	2	4	$3.68\cdot10^{-3}$	582.156	5	5	$2.74\cdot10^{1}$	1304.79	3	3	$2.32\cdot10^{-4}$
9028.92	2	2	$3.20\cdot10^{-1}$	583.925	5	7	$2.32\cdot10^{-2}$	1306.71	1	3	$2.06\cdot10^{-5}$
9060.48	2	4	$3.21\cdot10^{-1}$	584.128	5	5	$2.35\cdot10^{-1}$	1343.34	7	7	$6.40\cdot10^{-2}$
9386.81	2	4	$2.13\cdot10^{-1}$	599.644	1	3	$1.02\cdot10^{-4}$	1343.57	5	7	$7.55\cdot10^{-3}$
9392.79	4	6	$2.51\cdot10^{-1}$	599.819	3	3	$1.69\cdot10^{-3}$	1345.08	7	5	$1.27\cdot10^{-2}$
9419.39	2	4	$2.93\cdot10^{-4}$	600.115	5	3	$3.15\cdot10^{-4}$	1345.31	5	5	$4.90\cdot10^{-2}$
9460.68	4	4	$3.73\cdot10^{-2}$	635.197	1	3	$2.58\cdot10^{1}$	1345.34	3	5	$1.03\cdot10^{-2}$
9464.17	2	2	$3.50\cdot10^{-3}$	640.121	1	3	$2.30\cdot10^{-2}$	1346.41	5	3	$1.94\cdot10^{-2}$
9493.77	4	4	$4.06\cdot10^{-6}$	644.634	1	3	$1.19\cdot10^{1}$	1346.44	3	5	$5.28\cdot10^{-2}$
9660.12	2	4	$2.49\cdot10^{-4}$	644.837	3	3	$3.56\cdot10^{1}$	1381.97	5	3	$1.30\cdot10^{-7}$
9708.45	2	2	$3.18\cdot10^{-5}$	645.178	5	3	$5.94\cdot10^{1}$	1382.00	3	3	$7.07\cdot10^{-5}$
9716.46	6	4	$1.34\cdot10^{-4}$	646.209	1	3	$1.43\cdot10^{-2}$	1538.57	3	5	$1.59\cdot10^{-5}$
9740.39	4	2	$9.03\cdot10^{-5}$	660.286	5	3	$3.17\cdot10^{1}$	1627.35	3	5	$7.03\cdot10^{-3}$
9742.12	8	6	$4.91\cdot10^{-7}$	670.296	1	3	$3.80\cdot10^{-1}$	1627.38	5	5	$2.39\cdot10^{-2}$
9776.90	2	4	$1.40\cdot10^{-2}$	670.515	3	3	$3.02\cdot10^{-1}$	1628.90	3	3	$5.17\cdot10^{-3}$
9786.78	4	6	$1.18\cdot10^{-2}$	670.884	5	3	$3.68\cdot10^{-1}$	1628.92	5	3	$1.98\cdot10^{-2}$
9788.29	2	2	$3.83\cdot10^{-2}$	671.016	3	5	2.84	1629.08	1	3	$7.93\cdot10^{-3}$
9798.56	4	4	$3.33\cdot10^{-2}$	671.386	5	5	8.52	1629.83	3	1	$3.21\cdot10^{-2}$
9810.01	4	2	$5.42\cdot10^{-2}$	671.411	1	3	3.40	1675.73	3	3	$2.70\cdot10^{-1}$
9814.02	6	8	$7.00\cdot10^{-3}$	671.630	3	3	2.53	1675.75	5	3	$4.39\cdot10^{-1}$
9822.75	6	6	$5.74\cdot10^{-2}$	671.773	3	1	$1.13\cdot10^{1}$	1675.92	1	3	$9.09\cdot10^{-2}$
9834.61	6	4	$4.50\cdot10^{-2}$	672.001	5	3	4.38	1678.89	1	3	$3.62\cdot10^{-6}$
9863.33	8	8	$1.03\cdot10^{-1}$	693.774	3	5	$5.66\cdot10^{-5}$	1740.31	5	7	$2.21\cdot10^{-1}$
9872.15	8	6	$2.97\cdot10^{-2}$	694.169	5	5	$4.14\cdot10^{-4}$	1743.20	3	5	$1.67\cdot10^{-1}$
9883.38	2	2	$1.83\cdot10^{-2}$	715.254	5	3	$1.06\cdot10^{-3}$	1743.23	5	5	$5.40\cdot10^{-2}$
9905.52	4	2	$1.83\cdot10^{-3}$	745.841	1	3	$1.35\cdot10^{1}$	1745.05	3	3	$9.07\cdot10^{-2}$
9909.22	2	4	$4.07\cdot10^{-3}$	746.984	5	3	$3.69\cdot10^{1}$	1745.08	5	3	$5.97\cdot10^{-3}$
9931.47	4	4	$2.02\cdot10^{-2}$	747.606	5	5	$7.17\cdot10^{-5}$	1745.26	1	3	$1.23\cdot10^{-1}$
9965.75	4	6	$3.48\cdot10^{-3}$	748.369	5	3	3.81	1805.24	3	3	$2.48\cdot10^{-4}$
9968.51	6	4	$2.23\cdot10^{-3}$	775.965	5	5	$3.14\cdot10^{1}$	1805.28	5	3	$3.05\cdot10^{-5}$
9980.42	4	6	$4.86\cdot10^{-3}$	816.740	1	3	$1.85\cdot10^{-4}$	1805.47	1	3	$4.06\cdot10^{-5}$
				834.070	5	7	$8.13\cdot10^{-7}$	2139.01	3	5	$5.15\cdot10^{-7}$
N II				834.740	5	5	$2.42\cdot10^{-7}$	2142.77	5	5	$1.27\cdot10^{-6}$
525.983	1	3	$3.94\cdot10^{-2}$	835.163	5	3	$3.04\cdot10^{-8}$	3329.70	5	5	$2.01\cdot10^{-2}$
526.118	3	3	$8.74\cdot10^{-4}$	858.376	1	3	$2.56\cdot10^{-1}$	3408.13	3	1	$1.91\cdot10^{-1}$
526.345	5	3	$5.91\cdot10^{-3}$	860.205	1	3	$2.65\cdot10^{-2}$	3437.14	3	1	1.91
528.529	5	7	$1.86\cdot10^{-2}$	915.612	1	3	4.23	3775.61	5	5	$6.61\cdot10^{-6}$
529.355	1	3	7.17	915.962	3	1	$1.27\cdot10^{1}$	3919.00	3	3	$7.56\cdot10^{-1}$
529.413	3	1	$2.44\cdot10^{1}$	916.012	3	5	3.14	3955.85	3	5	$1.21\cdot10^{-1}$
529.491	3	3	6.86	916.020	3	3	3.21	3977.31	5	5	$7.79\cdot10^{-5}$
529.637	3	5	4.83	916.701	5	5	9.55	3995.00	3	5	1.22
529.722	5	3	$1.04\cdot10^{1}$	916.710	5	3	5.27	4046.74	5	3	$1.44\cdot10^{-5}$
529.867	5	5	$1.96\cdot10^{1}$	1064.95	5	5	$2.30\cdot10^{-5}$	4109.59	3	1	$1.24\cdot10^{-6}$
533.511	1	3	$2.46\cdot10^{1}$	1064.96	5	3	$5.16\cdot10^{-5}$	4114.33	3	3	$1.65\cdot10^{-3}$
533.581	3	5	$3.29\cdot10^{1}$	1083.99	1	3	2.10	4123.12	3	5	$4.37\cdot10^{-4}$
533.650	3	3	$1.70\cdot10^{1}$	1084.56	3	3	1.54	4236.36	3	3	$1.93\cdot10^{-3}$
533.729	5	7	$4.24\cdot10^{1}$	1084.58	3	5	2.82	4319.06	3	1	$5.59\cdot10^{-5}$
533.815	5	5	9.35	1085.53	5	3	$9.96\cdot10^{-2}$	4374.99	3	5	$6.31\cdot10^{-3}$
533.884	5	3	$9.54\cdot10^{-1}$	1085.55	5	5	$9.10\cdot10^{-1}$	4379.58	3	3	$1.76\cdot10^{-3}$
534.637	3	5	$4.96\cdot10^{-2}$	1085.70	5	7	3.72	4393.85	5	7	$5.21\cdot10^{-4}$
534.872	5	5	$7.14\cdot10^{-4}$	1162.50	3	1	$8.42\cdot10^{-5}$	4412.50	7	7	$2.93\cdot10^{-5}$
536.300	3	5	$6.40\cdot10^{-2}$	1275.04	7	5	$5.10\cdot10^{-1}$	4445.03	5	7	$1.20\cdot10^{-6}$

λ Å	gi	gk	A 10^8 s^-1	λ Å	gi	gk	A 10^8 s^-1	λ Å	gi	gk	A 10^8 s^-1
4447.03	3	5	1.12	5676.02	1	3	$2.80 \cdot 10^{-1}$	1751.66	4	6	1.51
4459.94	3	1	$7.99 \cdot 10^{-2}$	5679.56	5	7	$4.96 \cdot 10^{-1}$	2972.55	2	2	$6.67 \cdot 10^{-1}$
4464.13	5	5	$5.85 \cdot 10^{-8}$	5686.21	3	3	$1.78 \cdot 10^{-1}$	2977.33	4	2	$3.32 \cdot 10^{-1}$
4465.53	3	3	$1.59 \cdot 10^{-2}$	5710.77	5	5	$1.17 \cdot 10^{-1}$	2978.84	2	4	$1.66 \cdot 10^{-1}$
4475.89	3	5	$3.96 \cdot 10^{-4}$	5730.66	5	3	$1.26 \cdot 10^{-2}$	2983.64	4	4	$8.24 \cdot 10^{-1}$
4476.27	5	3	$3.40 \cdot 10^{-5}$	5747.30	3	5	$3.27 \cdot 10^{-2}$	3342.76	2	2	$3.80 \cdot 10^{-1}$
4477.68	5	3	$6.44 \cdot 10^{-2}$	5767.45	3	3	$2.39 \cdot 10^{-2}$	3353.98	2	4	$7.66 \cdot 10^{-1}$
4488.09	5	5	$8.62 \cdot 10^{-3}$	5927.81	1	3	$3.19 \cdot 10^{-1}$	3354.32	4	6	$5.51 \cdot 10^{-1}$
4507.56	7	5	$7.39 \cdot 10^{-2}$	5931.78	3	5	$4.23 \cdot 10^{-1}$	3355.46	4	2	$7.51 \cdot 10^{-1}$
4564.76	3	5	$1.65 \cdot 10^{-2}$	5940.24	3	3	$2.22 \cdot 10^{-1}$	3358.78	2	2	$3.05 \cdot 10^{-1}$
4601.48	3	5	$2.22 \cdot 10^{-1}$	5941.65	5	7	$5.47 \cdot 10^{-1}$	3360.98	4	4	$2.44 \cdot 10^{-1}$
4607.15	1	3	$3.15 \cdot 10^{-1}$	5952.39	5	5	$1.24 \cdot 10^{-1}$	3365.80	4	2	1.52
4613.87	3	3	$2.12 \cdot 10^{-1}$	5960.91	5	3	$1.29 \cdot 10^{-2}$	3367.36	6	6	1.27
4621.39	3	1	$9.04 \cdot 10^{-1}$	6065.00	3	5	$2.57 \cdot 10^{-3}$	3374.07	6	4	$8.13 \cdot 10^{-1}$
4630.54	5	5	$7.48 \cdot 10^{-1}$	6086.54	5	5	$5.87 \cdot 10^{-5}$	3745.95	2	4	$1.90 \cdot 10^{-1}$
4643.09	5	3	$4.39 \cdot 10^{-1}$	6284.32	5	3	$4.63 \cdot 10^{-2}$	3752.63	2	2	$6.67 \cdot 10^{-2}$
4654.53	3	5	$1.92 \cdot 10^{-2}$	6285.69	5	7	$1.10 \cdot 10^{-3}$	3754.69	4	4	$3.78 \cdot 10^{-1}$
4667.21	3	3	$2.31 \cdot 10^{-2}$	6286.11	3	5	$4.50 \cdot 10^{-4}$	3762.60	4	4	$4.24 \cdot 10^{-2}$
4674.91	3	1	$8.54 \cdot 10^{-2}$	6309.25	5	5	$1.71 \cdot 10^{-4}$	3771.03	6	4	$5.59 \cdot 10^{-1}$
4709.44	3	3	$1.49 \cdot 10^{-3}$	6366.79	1	3	$8.14 \cdot 10^{-5}$	3771.36	6	4	$8.28 \cdot 10^{-2}$
4774.24	3	5	$3.07 \cdot 10^{-2}$	6379.62	3	3	$3.52 \cdot 10^{-2}$	3792.97	8	6	$1.03 \cdot 10^{-1}$
4779.72	3	3	$2.49 \cdot 10^{-1}$	6433.44	3	5	$1.22 \cdot 10^{-4}$	3934.50	2	4	$7.49 \cdot 10^{-1}$
4781.19	5	7	$1.92 \cdot 10^{-2}$	6457.68	3	3	$1.17 \cdot 10^{-4}$	3938.51	4	6	$8.96 \cdot 10^{-1}$
4788.14	5	5	$2.50 \cdot 10^{-1}$	6472.43	3	1	$1.13 \cdot 10^{-4}$	3942.88	4	4	$1.49 \cdot 10^{-1}$
4793.65	5	3	$7.73 \cdot 10^{-2}$	6482.05	3	3	$2.58 \cdot 10^{-1}$	4097.36	2	4	$8.70 \cdot 10^{-1}$
4803.29	7	7	$3.17 \cdot 10^{-1}$	6610.56	5	7	$6.01 \cdot 10^{-1}$	4103.39	2	2	$8.67 \cdot 10^{-1}$
4810.30	7	5	$4.75 \cdot 10^{-1}$	6802.17	5	3	$1.43 \cdot 10^{-5}$	4195.74	2	4	$9.37 \cdot 10^{-1}$
4860.17	3	5	$1.87 \cdot 10^{-2}$	6826.23	5	5	$2.49 \cdot 10^{-4}$	4200.07	4	6	1.12
4874.57	5	5	$2.17 \cdot 10^{-5}$	7262.55	3	3	$1.42 \cdot 10^{-7}$	4215.77	4	4	$1.85 \cdot 10^{-1}$
4895.12	5	3	$3.04 \cdot 10^{-2}$	7528.12	5	7	$6.61 \cdot 10^{-5}$	4318.78	2	4	$5.40 \cdot 10^{-2}$
4897.54	7	5	$5.61 \cdot 10^{-6}$	7545.36	5	5	$4.53 \cdot 10^{-4}$	4321.22	2	2	$1.08 \cdot 10^{-1}$
4987.38	3	1	$6.98 \cdot 10^{-1}$	7559.05	5	3	$1.90 \cdot 10^{-5}$	4321.39	4	6	$5.03 \cdot 10^{-2}$
4994.37	3	3	$7.11 \cdot 10^{-1}$	7762.24	5	5	$8.49 \cdot 10^{-2}$	4325.43	4	4	$8.60 \cdot 10^{-2}$
5001.13	3	5	$9.65 \cdot 10^{-1}$	8089.08	5	7	$1.01 \cdot 10^{-4}$	4327.69	6	8	$3.06 \cdot 10^{-2}$
5001.47	5	7	1.04	8128.14	5	5	$5.55 \cdot 10^{-4}$	4327.88	4	2	$1.07 \cdot 10^{-1}$
5002.70	1	3	$8.33 \cdot 10^{-2}$	8687.43	3	1	$1.16 \cdot 10^{-2}$	4332.95	6	6	$1.23 \cdot 10^{-1}$
5005.15	7	9	1.14	9399.64	1	3	$1.04 \cdot 10^{-4}$	4337.01	6	4	$7.47 \cdot 10^{-2}$
5007.33	3	5	$7.43 \cdot 10^{-1}$					4345.81	8	8	$1.82 \cdot 10^{-1}$
5010.62	3	3	$2.10 \cdot 10^{-1}$	*N III*				4351.11	8	6	$4.01 \cdot 10^{-2}$
5016.38	5	5	$1.59 \cdot 10^{-1}$	374.198	2	4	$9.89 \cdot 10^{1}$	4510.88	2	4	$2.84 \cdot 10^{-1}$
5025.66	7	7	$1.04 \cdot 10^{-1}$	451.871	2	2	$1.03 \cdot 10^{1}$	4510.96	4	6	$4.77 \cdot 10^{-1}$
5040.71	7	5	$3.65 \cdot 10^{-3}$	452.227	4	2	$2.05 \cdot 10^{1}$	4514.85	6	8	$6.80 \cdot 10^{-1}$
5045.10	5	3	$3.37 \cdot 10^{-1}$	684.998	2	4	9.63	4518.14	2	5	$5.65 \cdot 10^{-1}$
5073.59	3	3	$2.43 \cdot 10^{-2}$	685.515	2	2	$3.83 \cdot 10^{1}$	4523.56	4	4	$3.61 \cdot 10^{-1}$
5114.28	1	3	$3.88 \cdot 10^{-4}$	685.817	4	4	$4.54 \cdot 10^{1}$	4530.86	4	2	$1.12 \cdot 10^{-1}$
5123.53	3	3	$5.69 \cdot 10^{-5}$	686.336	4	2	$1.95 \cdot 10^{1}$	4534.58	6	4	$2.01 \cdot 10^{-1}$
5138.90	5	3	$3.34 \cdot 10^{-5}$	763.334	2	2	9.58	4547.30	6	4	$3.33 \cdot 10^{-2}$
5238.18	3	5	$3.16 \cdot 10^{-5}$	764.351	4	2	$1.85 \cdot 10^{1}$	4634.13	2	4	$6.36 \cdot 10^{-1}$
5355.00	5	7	$7.02 \cdot 10^{-5}$	771.545	2	4	8.19	4640.64	4	6	$7.60 \cdot 10^{-1}$
5383.72	3	5	$3.69 \cdot 10^{-3}$	771.901	4	4	$1.64 \cdot 10^{1}$	4641.85	4	4	$1.26 \cdot 10^{-1}$
5390.69	3	3	$2.04 \cdot 10^{-3}$	772.384	6	4	$2.45 \cdot 10^{1}$	4858.70	2	4	$4.35 \cdot 10^{-1}$
5452.07	1	3	$9.82 \cdot 10^{-2}$	772.889	6	4	$2.09 \cdot 10^{1}$	4858.98	4	6	$4.66 \cdot 10^{-1}$
5454.22	3	1	$3.70 \cdot 10^{-1}$	772.955	4	2	$2.34 \cdot 10^{1}$	4861.27	6	8	$5.32 \cdot 10^{-1}$
5462.58	3	3	$1.11 \cdot 10^{-1}$	979.832	4	4	8.84	4867.12	4	4	$1.73 \cdot 10^{-1}$
5478.09	3	5	$5.22 \cdot 10^{-2}$	979.905	6	6	9.21	4867.17	8	10	$6.18 \cdot 10^{-1}$
5480.05	5	3	$1.44 \cdot 10^{-1}$	989.799	2	4	4.18	4873.60	6	6	$1.50 \cdot 10^{-1}$
5493.23	3	5	$3.11 \cdot 10^{-4}$	991.511	4	4	$8.17 \cdot 10^{-1}$	4881.78	6	4	$1.22 \cdot 10^{-2}$
5495.65	5	5	$2.66 \cdot 10^{-1}$	991.577	4	6	4.97	4884.14	8	8	$8.71 \cdot 10^{-2}$
5666.63	3	5	$3.45 \cdot 10^{-1}$	1747.85	2	4	1.28	4896.58	8	6	$5.86 \cdot 10^{-3}$
5674.00	3	5	$4.82 \cdot 10^{-7}$	1751.22	4	4	$2.48 \cdot 10^{-1}$	5260.86	2	2	$2.80 \cdot 10^{-2}$

λ (Å)	g_i	g_k	A (10^8 s^{-1})
5270.57	2	4	$6.95 \cdot 10^{-2}$
5272.68	4	2	$1.39 \cdot 10^{-1}$
5282.43	4	4	$2.21 \cdot 10^{-2}$
5297.75	4	6	$4.93 \cdot 10^{-2}$
5298.95	6	4	$7.38 \cdot 10^{-2}$
5314.36	6	6	$1.14 \cdot 10^{-1}$
5320.87	6	8	$5.68 \cdot 10^{-1}$
5327.19	4	6	$5.29 \cdot 10^{-1}$
5352.46	6	6	$3.72 \cdot 10^{-2}$
6365.84	2	2	$2.18 \cdot 10^{-1}$
6394.75	2	4	$2.15 \cdot 10^{-1}$
6445.34	2	4	$8.89 \cdot 10^{-2}$
6450.79	2	2	$1.77 \cdot 10^{-1}$
6454.08	4	6	$1.49 \cdot 10^{-1}$
6463.09	4	4	$1.13 \cdot 10^{-1}$
6467.02	6	8	$2.11 \cdot 10^{-1}$
6468.57	4	2	$3.52 \cdot 10^{-2}$
6478.76	6	6	$6.31 \cdot 10^{-2}$
6487.84	6	4	$1.05 \cdot 10^{-2}$
7371.51	4	4	$3.53 \cdot 10^{-2}$
7404.54	6	6	$3.61 \cdot 10^{-2}$
8307.51	2	4	$1.65 \cdot 10^{-2}$
8344.95	2	2	$6.52 \cdot 10^{-2}$
8386.39	4	4	$8.03 \cdot 10^{-2}$
8424.56	4	2	$3.17 \cdot 10^{-2}$

N IV

λ (Å)	g_i	g_k	A (10^8 s^{-1})
247.205	1	3	$1.19 \cdot 10^{2}$
*283.52	9	15	$3.05 \cdot 10^{2}$
*322.64	9	3	$8.99 \cdot 10^{1}$
335.047	3	5	$1.845 \cdot 10^{2}$
387.356	3	1	$2.55 \cdot 10^{1}$
765.147	1	3	$2.320 \cdot 10^{1}$
*923.16	9	9	$1.759 \cdot 10^{1}$
955.334	3	1	$2.919 \cdot 10^{1}$
1718.55	3	5	2.321
2649.88	3	3	1.07
3052.20	1	3	$1.33 \cdot 10^{-1}$
3059.60	3	3	$3.95 \cdot 10^{-1}$
3075.19	5	3	$6.48 \cdot 10^{-1}$
3443.61	3	5	$3.46 \cdot 10^{-1}$
3445.22	1	3	$4.60 \cdot 10^{-1}$
3454.65	3	3	$3.42 \cdot 10^{-1}$
3461.36	3	1	1.36
3463.36	5	5	1.02
3474.53	5	3	$5.61 \cdot 10^{-1}$
3478.72	3	5	1.06
*3480.8	3	9	1.06
3483.00	3	3	1.06
3484.93	3	1	1.06
3689.94	3	1	$9.10 \cdot 10^{-2}$
3694.14	3	3	$2.27 \cdot 10^{-2}$
3707.39	5	3	$6.73 \cdot 10^{-2}$
3714.43	5	5	$1.34 \cdot 10^{-1}$
3735.43	7	5	$7.37 \cdot 10^{-2}$
3747.54	3	5	$9.92 \cdot 10^{-1}$
4057.76	3	5	$6.62 \cdot 10^{-1}$
4740.26	3	5	$1.53 \cdot 10^{-2}$
4747.96	3	3	$7.60 \cdot 10^{-2}$
4752.49	5	7	$1.13 \cdot 10^{-2}$
4762.09	5	5	$6.99 \cdot 10^{-2}$
4769.86	5	3	$2.50 \cdot 10^{-2}$
4786.92	7	7	$8.79 \cdot 10^{-2}$
4796.66	7	5	$1.53 \cdot 10^{-2}$
5200.41	3	5	$2.67 \cdot 10^{-1}$
5204.28	5	7	$3.55 \cdot 10^{-1}$
5205.15	1	3	$1.97 \cdot 10^{-1}$
5226.70	3	3	$1.46 \cdot 10^{-1}$
5245.60	5	5	$8.66 \cdot 10^{-2}$
5272.35	5	3	$9.48 \cdot 10^{-3}$
5288.25	5	3	$3.22 \cdot 10^{-2}$
5736.93	3	5	$1.84 \cdot 10^{-1}$
5776.31	1	3	$1.85 \cdot 10^{-2}$
5784.76	3	1	$5.51 \cdot 10^{-2}$
5795.09	3	3	$1.37 \cdot 10^{-2}$
5812.31	3	5	$1.36 \cdot 10^{-2}$
5826.43	5	3	$2.25 \cdot 10^{-2}$
5843.84	5	5	$4.01 \cdot 10^{-2}$
6380.75	1	3	$1.42 \cdot 10^{-1}$
7103.24	1	3	$6.28 \cdot 10^{-2}$
7109.35	3	5	$8.46 \cdot 10^{-2}$
7111.28	3	3	$4.70 \cdot 10^{-2}$
*7116.8	9	15	$1.12 \cdot 10^{-1}$
7122.98	5	7	$1.12 \cdot 10^{-1}$
7127.25	5	5	$2.80 \cdot 10^{-2}$
7127.25	5	3	$3.11 \cdot 10^{-3}$
9165.07	3	5	$4.23 \cdot 10^{-2}$
9182.16	5	7	$4.45 \cdot 10^{-2}$
9222.99	7	9	$4.95 \cdot 10^{-2}$
9247.04	5	5	$7.66 \cdot 10^{-3}$
9311.55	7	7	$5.36 \cdot 10^{-3}$

N V

λ (Å)	g_i	g_k	A (10^8 s^{-1})
*209.29	2	6	$1.21 \cdot 10^{2}$
*247.66	6	10	$4.26 \cdot 10^{2}$
1238.82	2	4	3.40
1242.80	2	2	3.37
4603.74	2	4	$4.14 \cdot 10^{-1}$
4619.97	2	2	$4.10 \cdot 10^{-1}$

N VI

λ (Å)	g_i	g_k	A (10^8 s^{-1})
24.8980	1	3	$5.158 \cdot 10^{3}$
28.7870	1	3	$1.809 \cdot 10^{4}$
*161.220	3	9	$2.859 \cdot 10^{2}$
173.275	1	3	$2.697 \cdot 10^{2}$
*173.93	9	15	$8.756 \cdot 10^{2}$
185.192	3	5	$8.205 \cdot 10^{2}$
*1901	3	9	$6.780 \cdot 10^{-1}$
2896.4	1	3	$2.079 \cdot 10^{-1}$
*6991.1	3	9	$8.384 \cdot 10^{-2}$
9622.0	1	3	$3.276 \cdot 10^{-2}$

Oxygen

O I

λ (Å)	g_i	g_k	A (10^8 s^{-1})
791.973	5	5	4.94
792.938	1	3	2.19
792.967	3	5	1.64
877.798	5	3	2.85
877.879	5	5	5.12
922.008	5	7	1.23
935.193	5	5	1.33
1028.16	1	3	$4.22 \cdot 10^{-1}$
1152.15	5	5	5.28
1217.65	1	3	2.06
1302.17	5	3	3.41
1304.86	3	3	2.03
1306.03	1	3	$6.76 \cdot 10^{-1}$
3823.41	7	7	$6.63 \cdot 10^{-3}$
3823.87	5	3	$1.87 \cdot 10^{-3}$
3824.35	5	5	$5.19 \cdot 10^{-3}$
3825.02	3	3	$5.59 \cdot 10^{-3}$
3825.19	5	7	$8.31 \cdot 10^{-4}$
3855.01	5	5	$1.63 \cdot 10^{-2}$
3947.29	5	7	$4.91 \cdot 10^{-3}$
3947.48	5	5	$4.88 \cdot 10^{-3}$
3947.59	5	3	$4.87 \cdot 10^{-3}$
3951.93	3	1	$3.10 \cdot 10^{-3}$
3952.98	5	3	$1.29 \cdot 10^{-3}$
3953.00	1	3	$1.03 \cdot 10^{-3}$
3954.52	3	5	$7.73 \cdot 10^{-4}$
3954.61	5	5	$2.32 \cdot 10^{-3}$
3997.95	5	3	$2.41 \cdot 10^{-2}$
4217.09	3	1	$5.44 \cdot 10^{-3}$
4222.77	5	3	$2.26 \cdot 10^{-3}$
4222.82	1	3	$1.81 \cdot 10^{-3}$
4233.27	5	5	$4.04 \cdot 10^{-3}$
4368.19	3	1	$7.56 \cdot 10^{-3}$
4368.24	3	5	$7.59 \cdot 10^{-3}$
4967.38	3	5	$4.43 \cdot 10^{-3}$
4967.88	5	7	$8.44 \cdot 10^{-3}$
4968.79	7	9	$1.27 \cdot 10^{-2}$
5019.29	5	5	$7.13 \cdot 10^{-3}$
5020.22	7	5	$9.98 \cdot 10^{-3}$
5329.11	3	5	$9.48 \cdot 10^{-3}$
5329.69	5	7	$1.81 \cdot 10^{-2}$
5330.74	7	9	$2.71 \cdot 10^{-2}$
5435.18	3	5	$7.74 \cdot 10^{-3}$
5435.77	5	5	$1.29 \cdot 10^{-2}$
5436.86	7	5	$1.80 \cdot 10^{-2}$
5512.60	3	5	$2.69 \cdot 10^{-3}$
5512.77	5	7	$3.58 \cdot 10^{-3}$
5554.83	3	3	$5.83 \cdot 10^{-3}$
5555.00	5	3	$9.71 \cdot 10^{-3}$
5958.39	3	5	$6.80 \cdot 10^{-3}$
5958.58	5	7	$9.06 \cdot 10^{-3}$
6046.23	3	3	$1.05 \cdot 10^{-2}$
6046.44	5	3	$1.75 \cdot 10^{-2}$
6046.49	1	3	$3.50 \cdot 10^{-3}$
6155.99	3	5	$2.67 \cdot 10^{-2}$
6156.78	5	7	$5.08 \cdot 10^{-2}$
6158.19	7	9	$7.62 \cdot 10^{-2}$
6324.84	7	5	$3.76 \cdot 10^{-5}$
6453.60	3	5	$1.65 \cdot 10^{-2}$
6454.44	5	5	$2.75 \cdot 10^{-2}$
6455.98	7	5	$3.85 \cdot 10^{-2}$
6726.28	5	5	$1.18 \cdot 10^{-5}$
6726.54	5	3	$6.44 \cdot 10^{-6}$
7001.92	3	5	$2.65 \cdot 10^{-2}$
7002.23	5	7	$3.53 \cdot 10^{-2}$

λ Å	g_i	g_k	A 10⁸ s⁻¹
7254.15	3	3	$2.24 \cdot 10^{-2}$
7254.45	5	3	$3.73 \cdot 10^{-2}$
7254.53	1	3	$7.45 \cdot 10^{-3}$
7771.94	5	7	$3.69 \cdot 10^{-1}$
7774.17	5	5	$3.69 \cdot 10^{-1}$
7775.39	5	5	$3.69 \cdot 10^{-1}$
7981.94	3	3	$2.33 \cdot 10^{-4}$
7982.40	1	3	$3.09 \cdot 10^{-4}$
7986.98	3	5	$4.19 \cdot 10^{-4}$
7987.33	5	5	$1.41 \cdot 10^{-4}$
7995.07	5	7	$5.63 \cdot 10^{-4}$
8221.82	7	7	$2.89 \cdot 10^{-1}$
8227.65	5	3	$8.13 \cdot 10^{-2}$
8230.00	5	5	$2.26 \cdot 10^{-1}$
8233.00	3	3	$2.43 \cdot 10^{-1}$
8235.35	3	5	$4.86 \cdot 10^{-2}$
8446.25	3	1	$3.22 \cdot 10^{-1}$
8446.36	3	5	$3.22 \cdot 10^{-1}$
8446.76	3	3	$3.22 \cdot 10^{-1}$
8820.42	5	7	$2.93 \cdot 10^{-1}$
9260.81	3	1	$4.46 \cdot 10^{-1}$
9260.85	3	3	$3.34 \cdot 10^{-1}$
9260.94	3	5	$1.56 \cdot 10^{-1}$
9262.58	5	3	$1.11 \cdot 10^{-1}$
9262.67	5	5	$2.60 \cdot 10^{-1}$
9262.78	5	7	$2.97 \cdot 10^{-1}$
9265.83	7	5	$2.97 \cdot 10^{-2}$
9265.93	7	7	$1.48 \cdot 10^{-1}$
9266.01	7	9	$4.45 \cdot 10^{-1}$
9482.89	5	3	$2.34 \cdot 10^{-1}$
9622.11	5	3	$5.22 \cdot 10^{-4}$
9622.16	3	3	$1.57 \cdot 10^{-3}$
9625.26	7	5	$3.25 \cdot 10^{-4}$
9625.30	7	7	$1.85 \cdot 10^{-3}$
9694.66	5	7	$4.54 \cdot 10^{-4}$
9694.91	5	5	$4.54 \cdot 10^{-4}$
9695.06	5	3	$4.54 \cdot 10^{-4}$

O II

λ Å	g_i	g_k	A 10⁸ s⁻¹
429.918	4	2	$4.25 \cdot 10^{1}$
430.041	4	4	$4.13 \cdot 10^{1}$
430.176	4	6	$4.36 \cdot 10^{1}$
483.760	4	2	$2.05 \cdot 10^{1}$
483.980	6	4	$1.80 \cdot 10^{1}$
484.027	4	4	3.22
485.087	6	8	$2.60 \cdot 10^{1}$
485.470	6	6	1.20
485.518	4	6	$1.93 \cdot 10^{1}$
2290.85	2	4	$7.41 \cdot 10^{-2}$
2293.30	2	2	$3.25 \cdot 10^{-1}$
2300.33	4	4	$4.17 \cdot 10^{-1}$
2302.81	4	2	$1.67 \cdot 10^{-1}$
2365.14	4	2	$1.52 \cdot 10^{-1}$
2375.72	6	4	$1.35 \cdot 10^{-1}$
2406.38	6	4	$1.85 \cdot 10^{-1}$
2407.48	4	4	$2.25 \cdot 10^{-1}$
2411.60	4	2	$2.05 \cdot 10^{-1}$
2411.64	2	2	$1.10 \cdot 10^{-1}$
2415.13	4	2	$2.20 \cdot 10^{-1}$
2418.46	6	6	$2.30 \cdot 10^{-1}$

λ Å	g_i	g_k	A 10⁸ s⁻¹
2425.57	6	6	$1.77 \cdot 10^{-1}$
2433.54	2	4	$4.21 \cdot 10^{-1}$
2436.06	4	4	$1.69 \cdot 10^{-1}$
2444.25	4	4	$7.56 \cdot 10^{-2}$
2445.53	4	6	$4.98 \cdot 10^{-1}$
2517.96	4	6	$7.72 \cdot 10^{-2}$
2523.21	2	2	$9.63 \cdot 10^{-2}$
2526.87	4	4	$1.20 \cdot 10^{-1}$
2530.28	6	8	$8.16 \cdot 10^{-2}$
2571.46	2	4	$1.15 \cdot 10^{-1}$
2575.28	4	6	$1.37 \cdot 10^{-1}$
3134.73	8	6	1.23
3273.43	8	6	$9.99 \cdot 10^{-1}$
3377.15	2	2	1.27
3390.21	2	4	1.22
3407.28	6	6	1.02
3712.74	2	4	$2.84 \cdot 10^{-1}$
3727.32	4	4	$5.81 \cdot 10^{-1}$
3749.48	6	4	$9.31 \cdot 10^{-1}$
3833.07	6	8	$1.02 \cdot 10^{-2}$
3842.81	2	4	$7.45 \cdot 10^{-2}$
3843.58	4	6	$3.55 \cdot 10^{-1}$
3847.89	2	2	$1.95 \cdot 10^{-1}$
3850.80	4	6	$6.00 \cdot 10^{-3}$
3851.03	4	4	$1.59 \cdot 10^{-1}$
3851.47	8	8	$2.72 \cdot 10^{-1}$
3856.13	4	2	$2.28 \cdot 10^{-1}$
3857.16	6	6	$6.59 \cdot 10^{-2}$
3863.50	6	8	$6.49 \cdot 10^{-2}$
3864.13	2	2	$9.12 \cdot 10^{-2}$
3864.43	6	6	$2.15 \cdot 10^{-1}$
3864.67	6	4	$1.80 \cdot 10^{-1}$
3874.09	2	4	$3.26 \cdot 10^{-2}$
3875.80	8	6	$3.38 \cdot 10^{-2}$
3882.19	8	8	$5.50 \cdot 10^{-1}$
3882.45	4	4	$8.94 \cdot 10^{-2}$
3883.14	8	6	$1.13 \cdot 10^{-1}$
3893.52	4	6	$1.89 \cdot 10^{-2}$
3907.45	6	6	$8.64 \cdot 10^{-2}$
3911.96	6	4	1.09
3912.12	4	4	$1.41 \cdot 10^{-1}$
3919.27	4	2	1.22
3945.04	2	4	$2.05 \cdot 10^{-1}$
3954.36	2	2	$8.57 \cdot 10^{-1}$
3973.26	4	4	1.04
3982.71	4	2	$4.27 \cdot 10^{-1}$
4069.62	2	4	1.52
4069.88	4	6	1.53
4072.15	6	8	1.98
4075.86	8	10	2.11
4078.84	4	4	$5.52 \cdot 10^{-1}$
4084.65	6	8	$7.28 \cdot 10^{-2}$
4085.11	6	6	$4.55 \cdot 10^{-1}$
4092.93	8	8	$2.65 \cdot 10^{-1}$
4094.14	6	4	$4.70 \cdot 10^{-2}$
4096.53	4	6	$1.73 \cdot 10^{-1}$
4097.22	2	4	$3.62 \cdot 10^{-1}$
4103.00	2	2	$5.09 \cdot 10^{-1}$
4104.72	4	6	$3.14 \cdot 10^{-1}$
4104.99	4	4	$9.14 \cdot 10^{-1}$

λ Å	g_i	g_k	A 10⁸ s⁻¹
4106.02	8	6	$1.70 \cdot 10^{-2}$
4109.84	6	6	$1.21 \cdot 10^{-2}$
4110.19	6	4	$2.54 \cdot 10^{-1}$
4110.79	4	2	$7.70 \cdot 10^{-1}$
4112.02	6	6	$1.81 \cdot 10^{-1}$
4113.83	8	6	$2.41 \cdot 10^{-1}$
4119.22	6	8	1.33
4120.28	6	6	$2.15 \cdot 10^{-1}$
4120.55	6	4	$2.60 \cdot 10^{-1}$
4121.46	2	2	$5.60 \cdot 10^{-1}$
4129.32	4	2	$1.79 \cdot 10^{-1}$
4132.80	2	4	$9.13 \cdot 10^{-1}$
4140.70	4	4	$4.09 \cdot 10^{-2}$
4153.30	4	6	$7.91 \cdot 10^{-1}$
4156.53	6	4	$2.11 \cdot 10^{-1}$
4169.22	6	6	$2.71 \cdot 10^{-1}$
4185.44	6	8	1.91
4189.58	8	8	$7.06 \cdot 10^{-2}$
4189.79	8	10	1.98
4192.51	6	4	$3.21 \cdot 10^{-1}$
4196.27	4	4	$3.56 \cdot 10^{-2}$
4196.70	4	2	$3.56 \cdot 10^{-1}$
4317.14	2	4	$3.70 \cdot 10^{-1}$
4319.63	4	6	$2.55 \cdot 10^{-1}$
4319.87	2	2	$5.62 \cdot 10^{-1}$
4325.76	2	2	$1.47 \cdot 10^{-1}$
4327.46	6	6	$6.76 \cdot 10^{-1}$
4327.85	6	4	$7.24 \cdot 10^{-2}$
4328.59	4	2	1.12
4331.47	4	6	$4.82 \cdot 10^{-2}$
4331.86	4	4	$6.50 \cdot 10^{-1}$
4336.86	4	4	$1.57 \cdot 10^{-1}$
4345.56	4	2	$8.31 \cdot 10^{-1}$
4347.22	6	4	$1.19 \cdot 10^{-1}$
4347.41	4	4	$9.32 \cdot 10^{-1}$
4349.43	6	6	$6.91 \cdot 10^{-1}$
4351.26	6	6	$9.89 \cdot 10^{-1}$
4351.46	4	6	$5.82 \cdot 10^{-2}$
4359.40	4	6	$1.44 \cdot 10^{-2}$
4366.89	6	4	$3.98 \cdot 10^{-1}$
4369.27	4	4	$3.57 \cdot 10^{-1}$
4395.93	6	6	$3.91 \cdot 10^{-1}$
4405.98	6	4	$4.30 \cdot 10^{-2}$
4414.90	4	6	$8.34 \cdot 10^{-1}$
4416.97	2	4	$7.13 \cdot 10^{-1}$
4443.01	6	6	$5.05 \cdot 10^{-1}$
4443.52	6	8	$1.89 \cdot 10^{-2}$
4447.68	8	6	$2.52 \cdot 10^{-2}$
4448.19	8	8	$5.10 \cdot 10^{-1}$
4452.38	4	4	$1.37 \cdot 10^{-1}$
4466.24	2	4	$9.00 \cdot 10^{-1}$
4467.46	2	2	$9.00 \cdot 10^{-1}$
4563.18	4	4	$7.18 \cdot 10^{-3}$
4590.97	6	8	$8.85 \cdot 10^{-1}$
4595.96	6	6	$4.87 \cdot 10^{-2}$
4596.18	4	6	$8.34 \cdot 10^{-1}$
4638.86	2	4	$3.71 \cdot 10^{-1}$
4641.81	4	6	$5.96 \cdot 10^{-1}$
4649.13	6	8	$7.81 \cdot 10^{-1}$
4650.84	2	2	$6.86 \cdot 10^{-1}$

λ (Å)	g_i	g_k	A (10^8 s^{-1})
4661.63	4	4	$4.10\cdot10^{-1}$
4673.73	4	2	$1.35\cdot10^{-1}$
4676.23	6	6	$2.05\cdot10^{-1}$
4690.89	2	4	$1.86\cdot10^{-1}$
4691.42	2	2	$7.43\cdot10^{-1}$
4696.35	6	4	$3.25\cdot10^{-2}$
4698.44	6	6	$6.59\cdot10^{-2}$
4699.01	6	8	$9.88\cdot10^{-1}$
4699.22	4	6	$9.36\cdot10^{-1}$
4701.18	4	4	$9.23\cdot10^{-1}$
4701.71	4	2	$3.69\cdot10^{-1}$
4703.16	4	6	$9.20\cdot10^{-1}$
4705.35	6	8	1.10
4710.01	4	6	$2.98\cdot10^{-1}$
4741.70	6	6	$4.71\cdot10^{-2}$
4751.28	6	8	$6.39\cdot10^{-2}$
4752.69	6	6	$1.45\cdot10^{-2}$
4844.92	4	6	$1.02\cdot10^{-2}$
4856.39	4	6	$5.58\cdot10^{-2}$
4856.76	4	4	$1.00\cdot10^{-1}$
4860.97	2	4	$4.70\cdot10^{-1}$
4864.88	4	2	$8.07\cdot10^{-2}$
4871.52	4	6	$5.60\cdot10^{-1}$
4872.02	4	4	$9.34\cdot10^{-2}$
4890.86	4	2	$4.80\cdot10^{-1}$
4906.83	4	4	$4.54\cdot10^{-1}$
4924.53	4	6	$5.43\cdot10^{-1}$
4941.07	2	4	$5.87\cdot10^{-1}$
4943.01	4	6	$7.78\cdot10^{-1}$
4955.71	4	4	$1.82\cdot10^{-1}$
5159.94	2	2	$3.29\cdot10^{-1}$
5175.90	4	2	$1.49\cdot10^{-1}$
5190.50	2	4	$1.26\cdot10^{-1}$
5206.65	4	4	$3.58\cdot10^{-1}$
5583.22	2	4	$2.17\cdot10^{-2}$
5611.07	2	2	$2.14\cdot10^{-2}$
6627.37	4	4	$1.73\cdot10^{-1}$
6641.03	2	2	$9.88\cdot10^{-2}$
6666.66	4	2	$6.78\cdot10^{-2}$
6677.87	2	4	$3.37\cdot10^{-2}$
6717.75	2	2	$1.33\cdot10^{-1}$
6721.39	4	2	$1.81\cdot10^{-1}$
6810.48	6	8	$1.64\cdot10^{-3}$
6844.10	4	6	$2.97\cdot10^{-3}$
6846.80	8	8	$3.17\cdot10^{-2}$
6869.48	6	6	$5.35\cdot10^{-2}$
6884.88	4	4	$6.12\cdot10^{-2}$
6895.10	10	8	$2.72\cdot10^{-1}$
6906.44	8	6	$2.48\cdot10^{-1}$
6907.87	4	2	$3.03\cdot10^{-1}$
6910.56	6	4	$2.43\cdot10^{-1}$

O III

λ (Å)	g_i	g_k	A (10^8 s^{-1})
305.656	3	5	$1.62\cdot10^{2}$
305.767	5	7	$2.16\cdot10^{2}$
320.978	5	7	$2.17\cdot10^{2}$
328.448	5	5	$1.04\cdot10^{2}$
345.312	1	3	$1.35\cdot10^{2}$
507.680	3	3	$4.82\cdot10^{1}$
508.178	5	3	$8.04\cdot10^{1}$
525.794	5	3	$9.60\cdot10^{1}$
599.590	5	5	$5.41\cdot10^{1}$
835.289	5	7	5.99
1760.41	3	5	$8.38\cdot10^{-1}$
1764.46	5	5	2.50
1766.63	1	3	1.11
1772.28	3	1	3.29
1772.97	5	3	1.37
2390.43	3	3	1.62
2454.97	3	1	3.43
2665.68	3	5	$6.75\cdot10^{-1}$
2674.58	5	5	1.11
2683.66	3	1	1.85
2686.15	7	5	1.54
2687.55	3	3	1.84
2695.48	3	5	1.82
2959.69	3	5	1.83
2983.78	3	5	2.15
2996.48	3	3	$4.64\cdot10^{-1}$
3004.34	5	5	$4.27\cdot10^{-1}$
3017.62	7	7	$5.38\cdot10^{-1}$
3023.43	3	5	$4.79\cdot10^{-1}$
3024.54	1	3	$6.16\cdot10^{-1}$
3035.41	3	3	$4.59\cdot10^{-1}$
3042.07	3	1	1.94
3047.10	5	5	1.49
3059.28	5	3	$8.72\cdot10^{-1}$
3064.98	1	3	$2.17\cdot10^{-1}$
3068.13	3	1	$6.49\cdot10^{-1}$
3068.67	3	5	$2.27\cdot10^{-1}$
3074.14	5	7	$1.84\cdot10^{-1}$
3074.72	5	3	$3.76\cdot10^{-1}$
3075.13	5	5	$1.61\cdot10^{-1}$
3075.95	7	9	$1.07\cdot10^{-1}$
3083.65	7	7	$3.20\cdot10^{-1}$
3084.64	7	5	$2.55\cdot10^{-1}$
3088.04	9	9	$5.30\cdot10^{-1}$
3095.79	9	7	$1.35\cdot10^{-1}$
3115.67	3	1	1.39
3121.63	3	3	1.38
3132.79	3	5	1.37
3201.14	3	3	$4.77\cdot10^{-1}$
3207.61	5	5	$4.40\cdot10^{-1}$
3216.07	7	7	$5.58\cdot10^{-1}$
3260.86	5	7	1.68
3265.33	7	9	1.88
3267.20	3	5	1.58
3281.83	5	5	$2.89\cdot10^{-1}$
3284.45	7	7	$2.06\cdot10^{-1}$
3299.39	1	3	$1.64\cdot10^{-1}$
3312.33	3	3	$4.60\cdot10^{-1}$
3326.06	3	3	$2.65\cdot10^{-1}$
3330.30	3	5	$6.81\cdot10^{-1}$
3330.32	3	5	$4.76\cdot10^{-1}$
3332.41	5	3	$7.92\cdot10^{-1}$
3332.93	5	7	$5.04\cdot10^{-1}$
3336.67	3	3	$3.76\cdot10^{-1}$
3340.76	5	3	$6.57\cdot10^{-1}$
3344.20	5	5	$1.25\cdot10^{-1}$
3344.51	5	7	$3.48\cdot10^{-1}$
3347.98	7	5	$4.86\cdot10^{-1}$
3350.62	5	3	1.12
3350.92	7	7	$9.91\cdot10^{-1}$
3355.86	7	7	$6.89\cdot10^{-1}$
3362.31	7	5	$6.87\cdot10^{-1}$
3376.61	3	1	1.49
3376.76	3	3	1.12
3377.26	3	5	$5.20\cdot10^{-1}$
3382.61	5	7	$9.86\cdot10^{-1}$
3383.31	5	3	$3.70\cdot10^{-1}$
3383.81	5	5	$8.62\cdot10^{-1}$
3384.90	7	9	1.48
3394.22	7	7	$4.88\cdot10^{-1}$
3395.43	7	5	$9.75\cdot10^{-2}$
3406.88	1	3	$1.93\cdot10^{-1}$
3408.13	3	1	$5.79\cdot10^{-1}$
3415.26	3	3	$1.44\cdot10^{-1}$
3428.63	3	5	$1.42\cdot10^{-1}$
3430.57	5	3	$2.37\cdot10^{-1}$
3444.05	5	5	$4.21\cdot10^{-1}$
3446.68	3	5	$9.71\cdot10^{-1}$
3447.15	1	3	$8.09\cdot10^{-1}$
3447.97	5	7	1.19
3450.91	7	9	1.44
3451.30	3	3	$8.06\cdot10^{-1}$
3454.84	5	5	$6.89\cdot10^{-1}$
3454.99	9	11	1.72
3459.94	7	7	$5.14\cdot10^{-1}$
3466.13	9	9	$2.84\cdot10^{-1}$
3520.94	1	3	$1.50\cdot10^{-1}$
3531.22	3	1	$4.45\cdot10^{-1}$
3534.90	3	5	$1.11\cdot10^{-1}$
3555.24	5	3	$1.82\cdot10^{-1}$
3556.78	5	5	$3.26\cdot10^{-1}$
3695.38	3	5	$4.01\cdot10^{-1}$
3698.72	5	7	$7.62\cdot10^{-1}$
3703.36	7	9	1.14
3704.75	3	3	$8.53\cdot10^{-1}$
3707.27	3	5	$7.34\cdot10^{-1}$
3709.54	3	1	1.13
3712.49	5	5	$6.59\cdot10^{-1}$
3714.03	3	3	$4.06\cdot10^{-1}$
3715.09	5	7	$9.73\cdot10^{-1}$
3720.89	7	7	$3.74\cdot10^{-1}$
3721.95	5	3	$2.80\cdot10^{-1}$
3725.31	5	5	$2.41\cdot10^{-1}$
3728.51	5	7	1.29
3728.84	7	9	1.45
3729.80	3	5	1.22
3734.83	7	5	$7.40\cdot10^{-2}$
3742.63	5	5	$2.24\cdot10^{-1}$
3746.90	7	7	$1.59\cdot10^{-1}$
3754.70	3	5	$7.53\cdot10^{-1}$
3757.23	1	3	$5.56\cdot10^{-1}$
3759.88	5	7	$9.79\cdot10^{-1}$
3774.03	3	3	$3.91\cdot10^{-1}$
3791.28	5	5	$2.24\cdot10^{-1}$
3961.57	5	7	1.25
4072.64	1	3	$3.37\cdot10^{-1}$
4073.98	3	5	$4.54\cdot10^{-1}$

λ Å	Weights g_i	g_k	A 10^8 s^{-1}	λ Å	Weights g_i	g_k	A 10^8 s^{-1}	λ Å	Weights g_i	g_k	A 10^8 s^{-1}
4081.02	5	7	6.02·10^{-1}	3736.68	4	4	2.23·10^{-1}	3746.64	7	7	1.18·10^{-1}
4089.30	3	3	2.49·10^{-1}	3736.85	8	10	7.95·10^{-1}	3761.58	7	5	1.61·10^{-2}
4103.07	5	5	1.48·10^{-1}	3744.89	6	6	1.92·10^{-1}	4119.37	3	5	3.66·10^{-1}
4440.09	5	3	4.42·10^{-1}	3758.39	8	8	1.11·10^{-1}	4120.49	3	1	3.33·10^{-1}
4447.69	5	5	4.40·10^{-1}	3974.58	4	6	6.62·10^{-2}	4123.96	5	7	4.81·10^{-1}
4461.61	5	7	4.36·10^{-1}	3977.09	6	4	9.91·10^{-2}	4125.49	1	3	2.70·10^{-1}
4524.22	3	1	3.38·10^{-1}	3995.08	6	6	1.52·10^{-1}	4134.11	3	3	3.34·10^{-1}
4532.78	5	3	1.40·10^{-1}	4687.03	2	4	2.79·10^{-1}	4153.27	3	3	1.92·10^{-1}
4535.29	3	3	8.40·10^{-2}	4772.60	2	4	1.23·10^{-1}	4158.86	3	5	3.39·10^{-1}
4555.39	5	5	2.49·10^{-1}	4779.10	2	2	2.45·10^{-1}	4178.46	5	5	1.12·10^{-1}
4557.91	3	5	8.27·10^{-2}	4783.42	4	6	2.06·10^{-1}	4213.35	5	3	1.19·10^{-2}
5268.30	1	3	3.50·10^{-1}	4794.18	4	4	1.56·10^{-1}	4522.66	5	3	1.02·10^{-2}
5508.24	5	5	1.06·10^{-1}	4798.27	6	8	2.91·10^{-1}	4554.53	3	5	2.41·10^{-1}
5592.25	3	3	3.27·10^{-1}	4813.15	6	6	8.65·10^{-2}	5114.06	1	3	1.80·10^{-1}
				5305.51	4	4	6.10·10^{-2}	5339.94	1	3	1.85·10^{-2}
O IV				5362.51	6	6	6.12·10^{-2}	5349.74	3	1	7.04·10^{-2}
238.570	4	6	3.54·10^2	6931.60	2	2	7.35·10^{-2}	5372.71	3	3	1.42·10^{-2}
554.513	4	4	6.06·10^1	7004.11	4	4	8.90·10^{-2}	5414.59	3	5	9.29·10^{-3}
625.853	6	4	3.19·10^1	7061.30	4	2	3.48·10^{-2}	5428.38	5	3	2.68·10^{-2}
779.820	4	4	1.31·10^1					5471.12	5	5	4.86·10^{-2}
779.912	6	6	1.36·10^1	**O V**				5571.81	1	3	8.33·10^{-2}
923.367	4	4	1.10·10^1	172.169	1	3	2.94·10^2	5580.12	3	5	1.11·10^{-1}
1343.51	4	6	2.57	*192.85	9	15	6.90·10^2	5583.23	3	3	6.20·10^{-2}
2132.64	4	4	1.29	*215.17	9	3	1.83·10^2	*5589.9	9	15	1.49·10^{-1}
2493.39	2	4	1.18	220.353	3	5	4.292·10^2	5597.89	5	7	1.48·10^{-1}
2493.75	4	6	8.48·10^{-1}	248.460	3	1	5.59·10^1	5604.27	5	5	3.68·10^{-2}
2507.73	4	2	2.32	629.732	1	3	2.872·10^1	5607.41	5	3	4.08·10^{-3}
2509.22	6	6	1.94	758.677	3	5	5.547	6330.05	5	7	1.21·10^{-1}
2510.58	4	2	1.19	759.442	1	3	7.373	6460.12	3	5	9.37·10^{-2}
2517.37	6	4	1.24	760.227	3	3	5.514	6466.14	5	7	1.01·10^{-1}
2805.87	2	4	2.90·10^{-1}	760.446	5	5	1.652·10^1	6500.24	7	9	1.11·10^{-1}
2816.53	4	4	5.74·10^{-1}	761.128	3	1	2.197·10^1	6543.77	5	5	1.64·10^{-2}
2836.27	6	4	8.43·10^{-1}	762.004	5	3	9.125	6601.28	7	7	1.14·10^{-2}
2916.31	2	4	1.06	774.518	3	1	3.804·10^1	6764.72	1	3	4.37·10^{-2}
2921.46	4	6	1.27	1371.30	3	5	3.336	6789.62	3	5	5.79·10^{-2}
3063.43	2	4	1.30	2729.31	3	5	4.52·10^{-1}	6817.40	3	3	3.00·10^{-2}
3071.60	2	2	1.29	2731.45	1	3	5.90·10^{-1}	6828.95	5	7	7.35·10^{-2}
3194.78	6	6	1.71·10^{-1}	2743.61	3	3	4.38·10^{-1}	6878.76	5	5	1.65·10^{-2}
3209.65	8	8	2.53·10^{-1}	2752.23	3	1	1.82				
3348.06	2	4	8.51·10^{-1}	2755.13	5	5	1.37	**O VI**			
3349.11	4	6	1.02	2769.69	5	3	7.88·10^{-1}	*150.10	2	6	2.62·10^2
3354.27	4	2	7.71·10^{-1}	2781.01	3	5	1.40	*173.03	6	10	8.78·10^2
3362.55	4	4	7.65·10^{-1}	*2784.0	3	9	1.40	1031.91	2	4	4.16
3375.40	4	6	7.56·10^{-1}	2786.99	3	3	1.39	1037.61	2	2	4.09
3378.02	4	4	1.66·10^{-1}	2789.85	3	1	1.38	3811.35	2	4	5.14·10^{-1}
3381.21	4	6	7.19·10^{-1}	3058.68	3	5	1.39	3834.24	2	2	5.05·10^{-1}
3381.30	2	4	4.28·10^{-1}	3144.66	3	5	8.86·10^{-1}				
3385.52	6	8	1.02	3219.24	3	1	1.54·10^{-1}	**O VII**			
3390.19	2	2	8.49·10^{-1}	3222.29	1	3	1.16·10^{-1}	18.6270	1	3	9.365·10^3
3396.80	4	4	5.40·10^{-1}	3227.54	3	3	3.38·10^{-2}	21.6020	1	3	3.309·10^4
3409.70	6	6	3.00·10^{-1}	3239.21	3	3	3.28·10^{-1}	*120.33	3	9	5.334·10^2
3411.30	4	4	1.69·10^{-1}	3248.28	5	3	1.18·10^{-1}	128.411	1	3	8.982·10^2
3411.69	4	6	1.02	3263.54	5	5	1.86·10^{-2}	*128.46	9	15	1.615·10^3
3489.89	4	6	7.29·10^{-1}	3275.64	5	3	4.76·10^{-1}	135.820	3	5	1.523·10^3
3492.21	2	4	6.06·10^{-1}	3297.62	7	5	1.30·10^{-1}	*1630.3	3	9	7.935·10^{-1}
3560.39	4	6	1.03	3690.17	3	5	1.97·10^{-2}	2448.98	1	3	2.514·10^{-1}
3563.33	6	8	1.10	3698.36	3	3	1.03·10^{-1}	*5933.1	3	9	1.002·10^{-1}
3725.89	2	4	5.61·10^{-1}	3702.72	5	7	1.41·10^{-2}	8241.76	1	3	3.864·10^{-2}
3725.94	4	6	6.01·10^{-1}	3717.31	5	5	9.63·10^{-2}				
3729.03	6	8	6.86·10^{-1}	3725.63	5	3	2.91·10^{-2}				

λ Å	Weights g_i	g_k	A $10^8 s^{-1}$
Phosphorus			
P I			
1671.7	4	2	$3.9 \cdot 10^{-1}$
1674.6	4	4	$4.0 \cdot 10^{-1}$
1679.7	4	6	$3.9 \cdot 10^{-1}$
1775.0	4	6	2.17
1782.9	4	4	2.14
1787.7	4	2	2.13
2135.5	4	4	$2.11 \cdot 10^{-1}$
2136.2	6	4	2.83
2149.1	4	2	3.18
2152.9	2	4	$4.85 \cdot 10^{-1}$
2154.1	4	4	$1.73 \cdot 10^{-1}$
2154.1	4	6	$5.8 \cdot 10^{-1}$
2534.0	2	4	$2.00 \cdot 10^{-1}$
2535.6	4	4	$9.5 \cdot 10^{-1}$
2553.3	2	2	$7.1 \cdot 10^{-1}$
2554.9	4	2	$3.00 \cdot 10^{-1}$
P II			
1301.9	1	3	$5.0 \cdot 10^{-1}$
1304.5	3	1	1.5
1304.7	3	3	$3.7 \cdot 10^{-1}$
1305.5	3	5	$3.8 \cdot 10^{-1}$
1309.9	5	3	$6.2 \cdot 10^{-1}$
1310.7	5	5	1.1
4475.3	5	7	1.3
4499.2	5	7	1.4
4530.8	3	5	1.0
4554.8	3	5	$9.6 \cdot 10^{-1}$
4588.0	5	7	1.7
4589.9	3	5	1.6
4602.1	7	9	1.9
4943.5	7	5	$6.3 \cdot 10^{-1}$
5253.5	3	5	1.0
5425.9	5	5	$6.9 \cdot 10^{-1}$
6024.2	3	5	$5.1 \cdot 10^{-1}$
6043.1	5	7	$6.8 \cdot 10^{-1}$
P III			
1334.8	2	4	$5.5 \cdot 10^{-1}$
1344.3	4	6	$6.4 \cdot 10^{-1}$
1344.8	4	4	$1.1 \cdot 10^{-1}$
4057.4	4	4	$1.0 \cdot 10^{-1}$
4059.3	6	4	$9.0 \cdot 10^{-1}$
4080.1	4	2	$9.9 \cdot 10^{-1}$
Potassium			
K I			
4044.1	2	4	$1.24 \cdot 10^{-2}$
4047.2	2	2	$1.24 \cdot 10^{-2}$
5084.2	2	2	$3.50 \cdot 10^{-3}$
5099.2	4	2	$7.0 \cdot 10^{-3}$
5323.3	2	2	$6.3 \cdot 10^{-3}$
5339.7	4	2	$1.26 \cdot 10^{-2}$
5343.0	2	4	$4.0 \cdot 10^{-3}$
5359.6	4	6	$4.6 \cdot 10^{-3}$
5782.4	2	2	$1.23 \cdot 10^{-2}$

λ Å	Weights g_i	g_k	A $10^8 s^{-1}$
5801.8	4	2	$2.46 \cdot 10^{-2}$
5812.2	2	4	$2.8 \cdot 10^{-3}$
5831.9	4	6	$3.2 \cdot 10^{-3}$
6911.1	2	2	$2.72 \cdot 10^{-2}$
6938.8	4	2	$5.4 \cdot 10^{-2}$
7664.9	2	4	$3.87 \cdot 10^{-1}$
7699.0	2	2	$3.82 \cdot 10^{-1}$
K II			
607.93	1	3	$1.3 \cdot 10^{-2}$
K III			
2550.0	6	4	2.0
2635.1	4	4	1.2
2992.4	6	8	2.5
3052.1	4	6	1.7
3202.0	4	4	1.8
3289.1	4	6	2.0
3322.4	6	6	1.3
3421.8	2	4	1.5
K XVI			
206.27	1	3	$9.4 \cdot 10^{1}$
K XVII			
22.020	2	4	$4.7 \cdot 10^{4}$
22.163	4	6	$5.6 \cdot 10^{4}$
22.18	4	4	$9.3 \cdot 10^{3}$
22.60	2	2	$2.5 \cdot 10^{3}$
22.76	4	2	$4.7 \cdot 10^{3}$
Praseodymium			
Pr II			
3997.0	15	15	$1.87 \cdot 10^{-1}$
4062.8	13	15	1.00
4100.7	17	19	$8.4 \cdot 10^{-1}$
4143.1	15	17	$5.8 \cdot 10^{-1}$
4179.4	13	15	$5.2 \cdot 10^{-1}$
4222.9	11	13	$3.91 \cdot 10^{-1}$
4241.0	17	15	$2.30 \cdot 10^{-1}$
4359.8	15	15	$1.1 \cdot 10^{-1}$
4405.8	17	17	$9.0 \cdot 10^{-2}$
4429.3	15	15	$2.28 \cdot 10^{-1}$
4449.8	13	13	$1.24 \cdot 10^{-1}$
4468.7	11	13	$1.54 \cdot 10^{-1}$
4510.2	13	15	$1.16 \cdot 10^{-1}$
4534.2	15	17	$4.9 \cdot 10^{-2}$
4734.2	15	13	$2.5 \cdot 10^{-2}$
4879.1	15	15	$1.8 \cdot 10^{-2}$
4886.0	15	15	$1.3 \cdot 10^{-2}$
4912.6	17	15	$5.7 \cdot 10^{-2}$
5034.4	19	19	$1.1 \cdot 10^{-1}$
5110.8	21	19	$2.78 \cdot 10^{-1}$
5135.1	17	17	$1.25 \cdot 10^{-1}$
5173.9	19	17	$3.18 \cdot 10^{-1}$
5219.1	15	15	$9.5 \cdot 10^{-2}$
5220.1	17	15	$2.35 \cdot 10^{-1}$
5251.7	15	13	$1.1 \cdot 10^{-2}$
5259.7	15	13	$2.24 \cdot 10^{-1}$

λ Å	Weights g_i	g_k	A $10^8 s^{-1}$
5292.6	13	13	$9.3 \cdot 10^{-2}$
5810.6	17	19	$2.3 \cdot 10^{-2}$
5879.3	15	15	$7.6 \cdot 10^{-2}$
6200.8	15	17	$1.8 \cdot 10^{-2}$
6278.7	13	15	$2.6 \cdot 10^{-2}$
6398.0	11	13	$1.9 \cdot 10^{-2}$
Rhodium			
Rh I			
3121.76	6	6	$1.1 \cdot 10^{-1}$
3189.05	6	6	$3.03 \cdot 10^{-1}$
3263.14	6	6	$1.3 \cdot 10^{-1}$
3271.61	6	4	$2.0 \cdot 10^{-1}$
3280.55	8	8	$2.36 \cdot 10^{-1}$
3283.57	6	8	$4.4 \cdot 10^{-1}$
3323.09	8	10	$6.3 \cdot 10^{-1}$
3396.82	10	10	$6.5 \cdot 10^{-1}$
3399.70	6	8	$1.2 \cdot 10^{-1}$
3462.04	6	6	$6.2 \cdot 10^{-1}$
3470.66	4	4	$8.5 \cdot 10^{-1}$
3478.91	6	6	$3.32 \cdot 10^{-1}$
3498.73	4	6	$2.12 \cdot 10^{-1}$
3502.52	10	10	$4.3 \cdot 10^{-1}$
3507.32	6	8	$3.4 \cdot 10^{-1}$
3528.02	8	8	$8.5 \cdot 10^{-1}$
3543.95	4	4	$4.65 \cdot 10^{-1}$
3549.54	6	6	$2.22 \cdot 10^{-1}$
3570.18	4	6	$1.82 \cdot 10^{-1}$
3583.10	8	10	$2.6 \cdot 10^{-1}$
3596.19	6	4	$5.5 \cdot 10^{-1}$
3597.15	6	8	$5.9 \cdot 10^{-1}$
3612.47	4	2	$8.90 \cdot 10^{-1}$
3654.87	8	8	$6.0 \cdot 10^{-2}$
3657.99	8	6	$8.8 \cdot 10^{-1}$
3666.22	6	8	$8.4 \cdot 10^{-2}$
3690.70	6	4	$3.23 \cdot 10^{-1}$
3692.36	10	8	$9.1 \cdot 10^{-1}$
3700.91	8	10	$3.9 \cdot 10^{-1}$
3788.47	4	6	$1.4 \cdot 10^{-1}$
3793.22	8	6	$4.2 \cdot 10^{-1}$
3799.31	8	8	$5.5 \cdot 10^{-1}$
3806.76	6	6	$6.2 \cdot 10^{-2}$
3818.19	6	4	$5.8 \cdot 10^{-1}$
3822.26	6	6	$8.5 \cdot 10^{-1}$
3828.48	6	6	$6.2 \cdot 10^{-1}$
3833.89	6	4	$5.8 \cdot 10^{-1}$
3856.52	8	10	$5.9 \cdot 10^{-1}$
3934.23	8	8	$1.58 \cdot 10^{-1}$
3942.72	4	2	$7.15 \cdot 10^{-1}$
3958.86	6	6	$5.5 \cdot 10^{-1}$
3984.40	4	4	$1.1 \cdot 10^{-1}$
4082.78	6	4	$1.4 \cdot 10^{-1}$
4121.68	6	6	$9.8 \cdot 10^{-2}$
4128.87	6	8	$1.73 \cdot 10^{-1}$
4135.27	8	8	$1.0 \cdot 10^{-1}$
4196.50	6	8	$3.9 \cdot 10^{-2}$
4211.14	8	10	$1.62 \cdot 10^{-1}$
4288.71	6	8	$6.1 \cdot 10^{-2}$
4374.80	8	10	$1.64 \cdot 10^{-1}$

λ Å	g_i	g_k	A 10^8 s^{-1}
5983.60	10	10	$2.1 \cdot 10^{-2}$

Rubidium

Rb I

λ Å	g_i	g_k	A 10^8 s^{-1}
3022.5	2	4	$4.13 \cdot 10^{-5}$
3032.0	2	4	$4.93 \cdot 10^{-5}$
3044.2	2	4	$8.2 \cdot 10^{-5}$
3060.2	2	4	$1.05 \cdot 10^{-4}$
3082.0	2	4	$1.49 \cdot 10^{-4}$
3112.6	2	4	$2.5 \cdot 10^{-4}$
3113.1	2	2	$1.3 \cdot 10^{-4}$
3157.5	2	4	$3.38 \cdot 10^{-4}$
3158.3	2	2	$2.0 \cdot 10^{-4}$
3228.0	2	4	$6.4 \cdot 10^{-4}$
3229.2	2	2	$3.8 \cdot 10^{-4}$
3348.7	2	4	$1.37 \cdot 10^{-3}$
3350.8	2	2	$8.9 \cdot 10^{-4}$
3587.1	2	4	$3.97 \cdot 10^{-3}$
3591.6	2	2	$2.9 \cdot 10^{-3}$
4201.8	2	4	$1.8 \cdot 10^{-2}$
4215.5	2	2	$1.5 \cdot 10^{-2}$
7800.3	2	4	$3.70 \cdot 10^{-1}$
7947.6	2	2	$3.40 \cdot 10^{-1}$

Scandium

Sc I

λ Å	g_i	g_k	A 10^8 s^{-1}
3015.37	4	6	$7.8 \cdot 10^{-1}$
3019.35	6	8	$8.7 \cdot 10^{-1}$
3269.90	4	2	3.13
3273.63	6	4	2.81
3907.48	4	6	1.66
3911.81	6	8	1.79
4020.39	4	4	1.63
4023.68	6	6	1.65
4031.38	6	6	$2.9 \cdot 10^{-1}$
4043.80	8	8	$3.11 \cdot 10^{-1}$
4067.00	6	8	$1.91 \cdot 10^{-1}$
4074.96	4	6	$3.7 \cdot 10^{-1}$
4078.56	2	4	$4.3 \cdot 10^{-1}$
4086.66	6	8	$3.7 \cdot 10^{-1}$
4132.98	4	6	1.19
4140.27	6	8	1.17
4161.85	8	8	$1.77 \cdot 10^{-1}$
4233.59	6	6	$4.0 \cdot 10^{-1}$
4238.05	8	8	$7.1 \cdot 10^{-1}$
4706.94	4	6	$2.81 \cdot 10^{-1}$
4709.31	6	8	$4.0 \cdot 10^{-1}$
4728.77	8	8	$1.16 \cdot 10^{-1}$
4729.24	6	6	$1.93 \cdot 10^{-1}$
4734.11	4	2	1.10
4737.65	6	4	$8.8 \cdot 10^{-1}$
4741.02	8	6	$9.1 \cdot 10^{-1}$
4743.82	10	8	$9.8 \cdot 10^{-1}$
4973.67	4	2	$8.4 \cdot 10^{-1}$
4980.36	6	4	$5.6 \cdot 10^{-1}$
4983.43	4	4	$2.58 \cdot 10^{-1}$
4991.91	6	6	$3.8 \cdot 10^{-1}$
5018.41	6	4	$2.09 \cdot 10^{-1}$
5021.52	4	4	$2.30 \cdot 10^{-1}$

λ Å	g_i	g_k	A 10^8 s^{-1}
5070.17	6	8	$1.16 \cdot 10^{-1}$
5081.56	10	10	$7.6 \cdot 10^{-1}$
5083.72	8	8	$6.2 \cdot 10^{-1}$
5085.55	6	6	$5.7 \cdot 10^{-1}$
5086.94	4	4	$6.6 \cdot 10^{-1}$
5099.27	4	6	$1.50 \cdot 10^{-1}$
5339.43	6	6	$1.06 \cdot 10^{-1}$
5341.07	4	2	$3.8 \cdot 10^{-1}$
5349.34	6	4	$5.9 \cdot 10^{-1}$
5355.79	6	4	$3.0 \cdot 10^{-1}$
5356.10	8	6	$5.7 \cdot 10^{-1}$
5375.37	8	6	$3.4 \cdot 10^{-1}$
5392.06	10	8	$4.2 \cdot 10^{-1}$
5446.20	8	8	$2.8 \cdot 10^{-1}$
5451.37	6	6	$1.50 \cdot 10^{-1}$
5472.19	8	6	$9.7 \cdot 10^{-2}$
5482.01	8	8	$5.2 \cdot 10^{-1}$
5484.63	6	6	$5.2 \cdot 10^{-1}$
5514.23	6	8	$4.1 \cdot 10^{-1}$
5520.52	8	10	$4.3 \cdot 10^{-1}$
5671.83	10	12	$5.4 \cdot 10^{-1}$
5686.86	8	10	$4.9 \cdot 10^{-1}$
5700.19	6	8	$4.6 \cdot 10^{-1}$
5711.79	4	6	$4.5 \cdot 10^{-1}$
5717.31	8	8	$7.5 \cdot 10^{-2}$
6262.22	4	6	$8.4 \cdot 10^{-2}$
7741.16	10	10	$3.8 \cdot 10^{-2}$
7800.42	8	8	$5.1 \cdot 10^{-2}$

Sc II

λ Å	g_i	g_k	A 10^8 s^{-1}
1880.6	5	3	5.0
2064.3	7	5	2.2
2068.0	5	3	2.0
2273.1	1	3	7.7
2545.20	5	5	$4.0 \cdot 10^{-1}$
2552.35	7	5	2.21
2555.79	3	3	$6.9 \cdot 10^{-1}$
2560.23	5	3	2.01
2563.19	3	1	2.70
2611.19	5	5	2.2
2667.70	3	5	1.5
2746.36	3	1	3.9
2782.31	5	5	1.3
2789.15	7	7	1.3
2801.31	9	9	1.3
2819.49	3	5	2.3
2822.12	5	7	2.5
2826.64	7	9	2.8
2870.85	5	3	1.1
2912.98	5	3	1.1
2979.68	3	5	1.2
2988.92	5	7	2.9
3039.92	7	9	3.5
3045.73	5	7	3.68
3052.92	7	9	3.92
3060.54	7	7	$3.0 \cdot 10^{-1}$
3065.12	9	11	4.00
3075.36	9	9	$2.5 \cdot 10^{-1}$
3128.27	3	3	1.9
3133.07	5	5	1.8

λ Å	g_i	g_k	A 10^8 s^{-1}
3139.72	7	7	2.1
3190.98	3	3	1.1
3199.33	5	3	1.9
3312.72	5	7	1.2
3320.40	5	3	1.2
3343.23	9	7	1.1
3353.72	5	7	1.51
3359.67	5	5	$2.16 \cdot 10^{-1}$
3361.26	3	3	$3.4 \cdot 10^{-1}$
3361.93	3	1	1.17
3368.94	5	3	$8.3 \cdot 10^{-1}$
3372.15	7	5	$9.9 \cdot 10^{-1}$
3379.16	3	3	2.5
3535.71	5	3	$6.1 \cdot 10^{-1}$
3558.53	5	7	$3.0 \cdot 10^{-1}$
3567.70	3	5	$3.5 \cdot 10^{-1}$
3572.53	7	7	1.38
3576.34	5	5	1.06
3580.93	3	3	1.23
3589.63	5	3	$4.6 \cdot 10^{-1}$
3590.47	7	5	$2.9 \cdot 10^{-1}$
3613.83	7	9	1.48
3630.74	5	7	1.20
3642.78	3	5	1.13
3645.31	7	7	$2.74 \cdot 10^{-1}$
3651.80	5	5	$3.0 \cdot 10^{-1}$
3859.59	7	5	1.1
4246.82	5	5	1.29
4314.08	9	7	$4.1 \cdot 10^{-1}$
4320.75	7	5	$4.0 \cdot 10^{-1}$
4325.00	5	3	$4.3 \cdot 10^{-1}$
4374.46	9	9	$1.48 \cdot 10^{-1}$
4400.39	7	7	$1.43 \cdot 10^{-1}$
4415.54	5	5	$1.47 \cdot 10^{-1}$
4670.41	5	7	$1.16 \cdot 10^{-1}$
5031.01	5	3	$3.5 \cdot 10^{-1}$
5239.81	1	3	$1.39 \cdot 10^{-1}$
5526.79	9	7	$3.3 \cdot 10^{-1}$
5657.91	5	5	$1.04 \cdot 10^{-1}$
5669.06	3	1	$1.31 \cdot 10^{-1}$

Silicon

Si I

λ Å	g_i	g_k	A 10^8 s^{-1}
2207.98	1	3	$2.62 \cdot 10^{-1}$
2210.89	3	5	$3.46 \cdot 10^{-1}$
2211.74	3	3	$1.81 \cdot 10^{-1}$
2216.67	5	7	$4.54 \cdot 10^{-1}$
2218.06	5	5	$1.09 \cdot 10^{-1}$
2218.91	5	5	$1.05 \cdot 10^{-2}$
2438.77	1	3	$7.91 \cdot 10^{-3}$
2443.36	3	3	$1.32 \cdot 10^{-2}$
2452.12	5	3	$1.73 \cdot 10^{-3}$
2506.90	3	5	$5.47 \cdot 10^{-1}$
2514.32	1	3	$7.39 \cdot 10^{-1}$
2516.113	5	5	1.68
2519.202	3	5	$5.49 \cdot 10^{-1}$
2524.108	3	1	2.22
2528.509	5	3	$9.04 \cdot 10^{-1}$
2881.579	5	3	2.17

λ Å	g_i	g_k	A 10^8 s^{-1}	λ Å	g_i	g_k	A 10^8 s^{-1}	λ Å	g_i	g_k	A 10^8 s^{-1}
2970.355	5	5	$6.00 \cdot 10^{-4}$	1228.44	4	2	5.53	5957.56	2	2	$5.60 \cdot 10^{-1}$
2987.645	5	3	$1.34 \cdot 10^{-2}$	1228.62	4	4	$1.77 \cdot 10^{1}$	5978.93	4	2	1.13
3006.739	3	5	$1.1 \cdot 10^{-5}$	1228.75	4	6	$2.32 \cdot 10^{1}$	6347.10	2	4	$5.84 \cdot 10^{-1}$
3020.004	5	5	$3.3 \cdot 10^{-5}$	1229.39	6	8	$2.25 \cdot 10^{1}$	6371.36	2	2	$6.80 \cdot 10^{-1}$
3905.523	1	3	$1.33 \cdot 10^{-1}$	1246.74	2	4	4.03	6660.52	4	6	$3.64 \cdot 10^{-1}$
4102.936	1	3	$6.09 \cdot 10^{-4}$	1248.43	4	4	8.30	6665.00	2	4	$2.16 \cdot 10^{-1}$
4782.991	5	3	$1.7 \cdot 10^{-2}$	1251.16	6	4	$1.30 \cdot 10^{1}$	6671.88	6	8	$4.80 \cdot 10^{-1}$
4792.212	3	1	$2.2 \cdot 10^{-2}$	1260.42	2	4	$2.57 \cdot 10^{1}$	6699.38	2	2	$4.20 \cdot 10^{-1}$
4792.324	5	5	$1.7 \cdot 10^{-2}$	1264.73	4	6	$3.04 \cdot 10^{1}$	6750.28	6	6	$1.49 \cdot 10^{-1}$
4947.607	3	1	$4.2 \cdot 10^{-2}$	1265.02	4	4	4.73	6818.45	2	4	$1.08 \cdot 10^{-1}$
5006.061	3	5	$2.8 \cdot 10^{-2}$	1304.37	2	2	3.64	6829.82	4	4	$2.16 \cdot 10^{-2}$
5622.221	3	3	$1.6 \cdot 10^{-2}$	1309.27	4	2	6.23	7848.80	4	6	$3.73 \cdot 10^{-1}$
5645.611	3	5	$9.7 \cdot 10^{-3}$	1348.54	2	4	3.36	7849.72	6	8	$3.99 \cdot 10^{-1}$
5665.554	1	3	$6.31 \cdot 10^{-3}$	1350.06	6	6	5.34	9412.72	8	8	$4.65 \cdot 10^{-2}$
5684.484	5	3	$2.6 \cdot 10^{-2}$	1350.52	4	4	1.61				
5690.425	3	3	$9.26 \cdot 10^{-3}$	1350.66	2	2	2.02	**Si III**			
5701.105	3	1	$1.83 \cdot 10^{-2}$	1352.64	4	2	6.12	566.61	1	3	1.21
5708.397	5	5	$1.4 \cdot 10^{-2}$	1353.72	6	4	3.22	652.22	3	3	2.67
5754.220	5	3	$1.23 \cdot 10^{-2}$	1410.22	6	4	3.47	653.33	5	3	4.40
5772.145	3	1	$3.6 \cdot 10^{-2}$	1509.10	6	4	2.85	673.48	5	5	$8.44 \cdot 10^{-2}$
5780.384	1	3	$9.8 \cdot 10^{-3}$	1512.07	4	2	3.15	673.48	5	3	$9.33 \cdot 10^{-3}$
5793.071	3	5	$1.3 \cdot 10^{-2}$	1526.72	2	2	3.81	800.07	3	1	4.13
5797.859	5	7	$2.53 \cdot 10^{-3}$	1533.45	4	2	7.52	823.41	3	5	5.85
5948.545	3	5	$2.2 \cdot 10^{-2}$	1808.00	2	4	$2.54 \cdot 10^{-2}$	883.40	5	7	$5.79 \cdot 10^{2}$
6331.954	3	3	$1.0 \cdot 10^{-4}$	1816.92	4	6	$2.65 \cdot 10^{-2}$	939.09	5	3	$1.11 \cdot 10^{1}$
6555.462	7	9	$6.9 \cdot 10^{-3}$	1817.45	4	4	$3.23 \cdot 10^{-3}$	967.95	5	7	4.68
6721.853	3	5	$3.4 \cdot 10^{-5}$	2072.02	4	6	$9.6 \cdot 10^{-1}$	993.52	1	3	2.68
6976.523	3	5	$3.26 \cdot 10^{-2}$	2072.70	6	8	1.0	994.79	3	3	8.05
7003.567	5	7	$3.42 \cdot 10^{-2}$	2334.40	2	2	$5.51 \cdot 10^{-5}$	997.39	5	3	$1.35 \cdot 10^{1}$
7005.883	7	9	$3.83 \cdot 10^{-2}$	2334.61	4	6	$2.44 \cdot 10^{-5}$	1005.37	5	5	$7.06 \cdot 10^{-1}$
7680.267	3	5	$4.62 \cdot 10^{-2}$	2344.20	4	4	$1.31 \cdot 10^{-5}$	1031.16	3	5	2.72
7918.386	3	5	$5.22 \cdot 10^{-2}$	2350.17	4	2	$4.70 \cdot 10^{-5}$	1033.92	5	5	8.11
7932.349	5	7	$5.13 \cdot 10^{-2}$	2682.21	4	4	$3.49 \cdot 10^{-2}$	1037.05	5	3	4.46
7944.001	7	9	$5.75 \cdot 10^{-2}$	2887.51	6	4	$6.39 \cdot 10^{-2}$	1083.22	5	3	3.65
7970.306	5	5	$7.11 \cdot 10^{-3}$	2904.28	4	6	$3.58 \cdot 10^{-1}$	1108.37	1	3	$1.54 \cdot 10^{1}$
8035.619	7	7	$8.11 \cdot 10^{-3}$	2905.69	6	8	$3.83 \cdot 10^{-1}$	1109.97	3	5	$2.07 \cdot 10^{1}$
8093.241	3	3	$1.5 \cdot 10^{-2}$	3203.87	2	4	$4.45 \cdot 10^{-1}$	1113.23	5	7	$2.74 \cdot 10^{1}$
9413.506	3	1	$2.26 \cdot 10^{-1}$	3210.03	4	6	$5.29 \cdot 10^{-1}$	1140.55	1	3	$1.96 \cdot 10^{1}$
				3333.14	2	2	$1.00 \cdot 10^{-1}$	1141.58	3	5	$2.67 \cdot 10^{1}$
Si II				3339.82	4	2	$2.00 \cdot 10^{-1}$	1142.28	3	3	$1.75 \cdot 10^{1}$
843.72	2	4	$7.05 \cdot 10^{-1}$	3853.66	4	4	$5.11 \cdot 10^{-2}$	1144.31	5	7	$3.86 \cdot 10^{1}$
845.77	4	4	$1.42 \cdot 10^{-1}$	3856.02	6	4	$4.40 \cdot 10^{-1}$	1144.96	5	5	$1.19 \cdot 10^{1}$
889.72	2	4	1.43	3862.60	4	2	$3.91 \cdot 10^{-1}$	1145.18	5	5	1.38
892.00	4	4	$2.88 \cdot 10^{-1}$	4075.45	6	4	$4.00 \cdot 10^{-2}$	1155.00	1	3	9.34
899.41	2	2	$4.63 \cdot 10^{-1}$	4076.78	4	2	$4.00 \cdot 10^{-2}$	1155.96	3	1	$2.22 \cdot 10^{1}$
901.74	4	2	$9.23 \cdot 10^{-1}$	4128.07	4	6	1.49	1158.10	3	5	7.88
989.87	2	4	6.81	4130.89	6	8	1.74	1160.26	5	3	8.65
992.68	4	6	7.11	4621.42	4	6	$1.28 \cdot 10^{-1}$	1161.58	5	5	$1.42 \cdot 10^{1}$
1020.70	2	2	$8.91 \cdot 10^{-1}$	4621.72	6	8	$1.37 \cdot 10^{-1}$	1174.37	3	3	1.00
1023.69	4	2	1.77	5041.03	2	4	$7.00 \cdot 10^{-1}$	1174.43	3	5	1.00
1190.42	2	4	6.53	5055.98	4	6	1.45	1206.51	1	3	$2.55 \cdot 10^{1}$
1193.28	2	2	$2.69 \cdot 10^{1}$	5466.43	4	6	$2.16 \cdot 10^{-1}$	1206.53	3	5	$4.17 \cdot 10^{1}$
1194.50	4	4	$3.45 \cdot 10^{1}$	5466.87	6	6	$1.54 \cdot 10^{-2}$	1207.52	5	5	$2.43 \cdot 10^{1}$
1197.39	4	2	$1.40 \cdot 10^{1}$	5632.97	8	8	$4.00 \cdot 10^{-2}$	1210.46	5	7	$1.60 \cdot 10^{1}$
1224.25	4	4	6.72	5660.66	6	6	$7.00 \cdot 10^{-2}$	1235.43	1	3	$2.77 \cdot 10^{1}$
1224.97	4	6	4.39	5669.56	10	8	$5.00 \cdot 10^{-1}$	1280.35	5	7	$1.04 \cdot 10^{1}$
1226.81	2	2	$2.77 \cdot 10^{1}$	5681.44	4	4	$1.00 \cdot 10^{-1}$	1294.54	3	5	5.35
1226.89	6	4	6.55	5688.81	8	6	$4.60 \cdot 10^{-1}$	1296.73	1	3	7.10
1226.99	2	4	$1.39 \cdot 10^{1}$	5701.37	6	4	$4.50 \cdot 10^{-1}$	1298.89	3	3	5.29
1227.60	6	6	$1.46 \cdot 10^{1}$	5706.37	4	2	$6.10 \cdot 10^{-1}$	1298.96	5	5	$1.59 \cdot 10^{1}$

λ Å	Weights g_i	g_k	A $10^8\,\text{s}^{-1}$	λ Å	Weights g_i	g_k	A $10^8\,\text{s}^{-1}$	λ Å	Weights g_i	g_k	A $10^8\,\text{s}^{-1}$
1301.15	3	1	$2.11\cdot10^1$	3216.25	5	7	$1.90\cdot10^{-1}$	6314.46	3	1	1.18
1303.32	5	3	8.71	3230.50	1	3	$4.75\cdot10^{-1}$	6524.36	7	9	$3.82\cdot10^{-1}$
1312.59	3	1	6.66	3233.95	3	3	1.42	6831.56	5	3	$5.83\cdot10^{-1}$
1341.47	7	7	8.96	3241.62	5	3	2.35	7461.89	3	1	$4.99\cdot10^{-1}$
1342.39	5	5	7.54	3253.40	5	5	$1.82\cdot10^{-1}$	7462.62	5	3	$3.75\cdot10^{-1}$
1343.39	3	3	8.16	3253.74	7	5	$1.75\cdot10^{-1}$	7466.32	7	5	$4.19\cdot10^{-1}$
1361.60	3	5	$1.06\cdot10^1$	3254.80	9	7	$1.81\cdot10^{-1}$	7612.36	3	5	$9.94\cdot10^{-1}$
1362.37	3	1	$1.06\cdot10^1$	3258.66	7	5	1.01	8190.43	5	7	$8.09\cdot10^{-1}$
1363.47	5	3	8.53	3270.46	3	3	$2.99\cdot10^{-1}$	8191.16	7	7	$7.05\cdot10^{-2}$
1365.26	7	5	9.40	3276.26	5	3	$8.91\cdot10^{-1}$	8191.68	9	11	$8.78\cdot10^{-1}$
1367.05	3	3	$1.05\cdot10^1$	3279.26	3	1	1.18	8262.57	5	7	$7.18\cdot10^{-1}$
1369.44	3	1	$1.04\cdot10^1$	3486.91	5	7	1.54	8265.64	5	5	$1.80\cdot10^{-1}$
1373.03	5	5	8.00	3525.94	3	3	$2.48\cdot10^{-1}$	8269.32	3	5	$5.36\cdot10^{-1}$
1387.99	3	3	$2.09\cdot10^{-1}$	3569.67	3	1	$5.78\cdot10^{-1}$	8271.38	3	3	$2.98\cdot10^{-1}$
1417.24	3	1	$2.17\cdot10^1$	3590.47	3	5	2.53	8271.94	1	3	$3.96\cdot10^{-1}$
1433.69	5	7	8.67	3622.54	7	7	$1.25\cdot10^{-1}$				
1435.77	5	7	4.22	3639.45	7	9	$2.24\cdot10^{-1}$	*Si IV*			
1436.17	1	3	7.79	3645.12	5	5	$9.58\cdot10^{-2}$	457.82	2	4	3.90
1457.25	5	3	$1.07\cdot10^1$	3681.40	5	3	$2.51\cdot10^{-1}$	458.16	2	2	4.05
1500.24	7	9	$2.10\cdot10^1$	3682.15	3	3	$1.50\cdot10^{-1}$	515.12	2	2	4.66
1501.19	5	7	$1.86\cdot10^1$	3791.41	1	3	1.76	516.35	4	2	9.32
1501.87	3	5	$1.76\cdot10^1$	3796.11	3	5	2.36	645.76	6	6	$4.61\cdot10^{-1}$
1506.06	3	3	$1.24\cdot10^1$	3806.54	5	7	3.14	645.76	6	8	6.92
1673.32	5	7	6.70	3842.46	3	5	$1.73\cdot10^{-1}$	645.76	4	6	6.46
1842.55	5	3	2.99	3924.47	7	9	3.47	749.94	6	6	$9.33\cdot10^{-1}$
2176.89	9	7	1.80	3947.49	5	7	$5.59\cdot10^{-2}$	749.94	6	8	$1.40\cdot10^1$
2295.48	5	7	$6.69\cdot10^{-1}$	3963.84	7	5	$3.63\cdot10^{-2}$	815.05	2	2	$1.18\cdot10^1$
2300.93	7	7	$5.80\cdot10^{-2}$	3981.24	3	3	$2.09\cdot10^{-3}$	818.13	4	2	$2.37\cdot10^1$
2308.19	9	11	$7.16\cdot10^{-1}$	4102.42	5	3	$2.46\cdot10^{-1}$	1066.63	6	6	2.54
2449.48	5	5	$1.81\cdot10^{-1}$	4115.50	7	9	$4.07\cdot10^{-1}$	1122.49	2	4	$2.14\cdot10^1$
2483.20	7	9	$1.33\cdot10^{-1}$	4338.50	1	3	$1.47\cdot10^{-1}$	1128.34	4	6	$2.53\cdot10^1$
2541.82	3	5	$3.22\cdot10^{-1}$	4341.40	3	1	$5.25\cdot10^{-3}$	1393.76	2	4	8.80
2546.09	5	5	$6.1\cdot10^{-1}$	4377.63	7	5	$4.14\cdot10^{-2}$	1402.77	2	2	8.63
2559.21	5	7	1.63	4405.90	5	5	$4.84\cdot10^{-2}$	1722.53	6	4	4.92
2640.79	5	3	1.83	4406.72	7	9	$3.10\cdot10^{-1}$	1727.38	4	2	5.47
2655.51	7	9	1.35	4494.05	3	3	$4.19\cdot10^{-1}$	2120.18	2	2	3.06
2817.11	9	7	$2.22\cdot10^{-1}$	4552.62	3	5	1.26	2127.47	4	2	6.14
2831.49	7	5	$2.12\cdot10^{-1}$	4554.00	5	3	$6.75\cdot10^{-1}$	2287.04	6	6	$4.27\cdot10^{-1}$
2839.62	5	3	$2.37\cdot10^{-1}$	4567.82	3	3	1.25	2287.04	6	8	6.41
2959.15	5	3	$1.48\cdot10^{-1}$	4574.76	3	1	1.24	2287.04	4	6	5.98
2980.52	5	3	$2.37\cdot10^{-3}$	4619.66	3	5	$2.89\cdot10^{-1}$	2366.76	2	2	$5.18\cdot10^{-1}$
3013.09	5	5	$5.67\cdot10^{-2}$	4638.28	1	3	$3.79\cdot10^{-1}$	2370.99	4	2	1.04
3034.73	5	7	$4.50\cdot10^{-1}$	4665.87	3	3	$2.79\cdot10^{-1}$	2482.82	2	4	$6.62\cdot10^{-2}$
3037.29	7	7	$3.93\cdot10^{-2}$	4683.02	5	5	$8.30\cdot10^{-1}$	2485.38	2	2	$7.06\cdot10^{-2}$
3040.93	9	11	$4.87\cdot10^{-1}$	4683.80	3	1	1.10	2672.19	2	4	$2.49\cdot10^{-2}$
3043.93	7	5	$2.51\cdot10^{-1}$	4716.65	5	7	1.32	2675.12	6	4	$2.74\cdot10^{-1}$
3045.08	5	3	$2.24\cdot10^{-1}$	4730.52	5	3	$4.47\cdot10^{-1}$	2675.25	8	6	$2.61\cdot10^{-1}$
3068.24	5	3	$5.15\cdot10^{-1}$	4800.43	1	3	$2.76\cdot10^{-1}$	2677.57	4	4	$4.51\cdot10^{-3}$
3086.24	7	5	1.46	4813.33	5	7	2.10	2723.81	6	8	1.10
3086.46	5	5	$2.61\cdot10^{-1}$	4819.72	7	9	2.12	2723.81	6	6	$7.34\cdot10^{-2}$
3093.42	5	3	1.30	4828.97	9	11	2.25	2971.52	6	6	$4.62\cdot10^{-3}$
3093.65	3	3	$4.33\cdot10^{-1}$	5113.76	5	7	$3.82\cdot10^{-1}$	2971.52	6	4	$9.70\cdot10^{-2}$
3096.83	3	1	1.73	5114.12	9	11	$4.17\cdot10^{-1}$	2971.52	8	6	$9.24\cdot10^{-2}$
3126.27	5	5	$4.67\cdot10^{-1}$	5197.26	3	5	$2.47\cdot10^{-1}$	3149.56	2	4	4.02
3147.37	5	3	$2.53\cdot10^{-1}$	5451.46	3	5	$4.60\cdot10^{-1}$	3165.71	4	6	4.77
3161.61	5	3	$4.91\cdot10^{-1}$	5473.05	5	7	$6.07\cdot10^{-1}$	3244.19	4	2	$4.21\cdot10^{-1}$
3185.13	3	1	4.04	5704.60	7	5	$1.86\cdot10^{-1}$	3762.44	4	4	$2.33\cdot10^{-1}$
3186.02	5	7	1.13	5716.29	9	7	$1.91\cdot10^{-1}$	3773.15	4	2	2.33
3196.50	7	9	1.14	5739.73	1	3	$5.41\cdot10^{-1}$	4031.39	2	2	$2.29\cdot10^{-1}$
3210.55	9	11	1.20	5898.79	7	9	$4.86\cdot10^{-1}$	4038.06	4	2	$4.59\cdot10^{-1}$

λ Å	g_i	g_k	A $10^8\,s^{-1}$	λ Å	g_i	g_k	A $10^8\,s^{-1}$	λ Å	g_i	g_k	A $10^8\,s^{-1}$
4088.85	2	4	1.56	2853.01	2	2	$5.31\cdot10^{-3}$	9961.28	6	8	$1.27\cdot10^{-2}$
4116.10	2	2	1.53	3302.37	2	4	$2.75\cdot10^{-2}$				
4212.41	6	8	1.63	3302.98	2	2	$2.73\cdot10^{-2}$	*Na II*			
4212.41	6	6	$1.09\cdot10^{-1}$	4238.99	2	4	$2.90\cdot10^{-3}$	300.15	1	3	$1.18\cdot10^{1}$
4314.10	2	2	1.06	4242.08	4	4	$5.8\cdot10^{-4}$	300.20	1	3	$1.17\cdot10^{1}$
4328.18	4	2	2.12	4242.08	4	6	$3.46\cdot10^{-3}$	301.44	1	3	$3.33\cdot10^{1}$
4403.73	6	8	$4.09\cdot10^{-1}$	4249.41	2	2	$8.7\cdot10^{-4}$	302.45	1	3	1.4
4403.73	6	6	$2.73\cdot10^{-2}$	4252.52	4	2	$1.73\cdot10^{-3}$	372.08	1	3	$3.13\cdot10^{1}$
4403.73	4	6	$3.82\cdot10^{-1}$	4273.64	2	4	$3.91\cdot10^{-3}$	376.38	1	3	1.70
4411.65	2	4	$1.74\cdot10^{-2}$	4276.79	4	4	$7.8\cdot10^{-4}$	2315.65	3	1	$1.05\cdot10^{-1}$
4611.27	4	6	$2.23\cdot10^{-2}$	4276.79	4	6	$4.69\cdot10^{-3}$	2493.15	3	1	4.15
4611.27	4	4	$3.71\cdot10^{-3}$	4287.84	2	2	$1.19\cdot10^{-3}$	2506.30	3	3	$2.73\cdot10^{-1}$
4950.11	6	6	$9.46\cdot10^{-3}$	4291.01	4	2	$2.38\cdot10^{-3}$	2515.46	3	5	$2.25\cdot10^{-1}$
4950.11	8	6	$1.89\cdot10^{-1}$	4321.40	2	4	$5.5\cdot10^{-3}$	2531.54	3	1	$8.44\cdot10^{-1}$
4950.11	6	4	$1.99\cdot10^{-1}$	4324.62	4	4	$1.09\cdot10^{-3}$	2594.96	3	3	$3.71\cdot10^{-1}$
5304.97	6	4	$1.90\cdot10^{-1}$	4324.62	4	6	$6.6\cdot10^{-3}$	2661.00	3	5	1.65
5304.97	4	4	$2.11\cdot10^{-2}$	4341.49	2	2	$3.26\cdot10^{-3}$	2671.83	3	3	2.64
5309.49	4	2	$2.10\cdot10^{-1}$	4344.74	4	2	$6.50\cdot10^{-3}$	2678.09	3	1	3.01
6667.56	2	4	1.15	4390.03	2	4	$9.83\cdot10^{-3}$	2808.71	7	7	$1.93\cdot10^{-2}$
6701.21	4	6	1.36	4393.34	4	4	$1.95\cdot10^{-3}$	2829.87	5	5	$3.36\cdot10^{-1}$
6998.36	6	6	$3.65\cdot10^{-2}$	4393.34	4	6	$1.17\cdot10^{-2}$	2839.56	5	7	1.11
6998.36	6	8	$5.48\cdot10^{-1}$	4419.88	2	2	$2.82\cdot10^{-3}$	2872.95	3	5	$2.63\cdot10^{-1}$
6998.36	4	6	$5.11\cdot10^{-1}$	4423.25	4	2	$5.61\cdot10^{-3}$	2881.15	3	1	2.50
7047.94	6	4	$9.05\cdot10^{-1}$	4494.18	2	4	$1.23\cdot10^{-2}$	2886.26	3	5	1.07
7068.41	4	2	1.00	4497.66	4	4	$2.44\cdot10^{-3}$	2893.95	3	1	1.48
7630.50	2	2	$4.40\cdot10^{-1}$	4497.66	4	6	$1.46\cdot10^{-2}$	2901.14	7	7	$2.89\cdot10^{-1}$
7654.56	4	2	$8.82\cdot10^{-1}$	4541.63	2	2	$3.76\cdot10^{-3}$	2904.72	7	5	$1.04\cdot10^{-1}$
8240.61	6	6	$5.85\cdot10^{-3}$	4545.19	4	2	$7.50\cdot10^{-3}$	2919.05	7	5	$7.94\cdot10^{-2}$
8240.61	6	4	$1.23\cdot10^{-1}$	4664.811	2	4	$2.08\cdot10^{-2}$	2920.95	1	3	$6.66\cdot10^{-1}$
8240.61	8	6	$1.17\cdot10^{-1}$	4668.560	4	4	$4.14\cdot10^{-3}$	2930.88	5	3	$9.83\cdot10^{-1}$
8957.25	2	4	$4.26\cdot10^{-1}$	4668.560	4	6	$2.49\cdot10^{-2}$	2934.08	5	7	3.17
9018.16	2	2	$4.17\cdot10^{-1}$	4747.941	2	2	$6.19\cdot10^{-3}$	2937.74	5	5	1.28
				4751.822	4	2	$1.23\cdot10^{-2}$	2945.70	7	7	$2.50\cdot10^{-2}$
Si V				4978.541	2	4	$4.09\cdot10^{-2}$	2951.24	7	9	4.33
96.44	1	3	$2.36\cdot10^{3}$	4982.813	4	6	$4.88\cdot10^{-2}$	2952.40	5	5	1.29
97.14	1	3	$2.42\cdot10^{2}$	5148.838	2	2	$1.14\cdot10^{-2}$	2960.12	5	5	$5.60\cdot10^{-1}$
98.21	1	3	6.88	5153.402	4	2	$2.27\cdot10^{-2}$	2970.73	5	7	$1.03\cdot10^{-1}$
117.86	1	3	$3.57\cdot10^{2}$	5682.633	2	4	$1.01\cdot10^{-1}$	2974.24	5	5	$2.02\cdot10^{-1}$
118.97	1	3	$3.84\cdot10^{1}$	5688.193	4	4	$2.02\cdot10^{-2}$	2974.99	1	3	$6.47\cdot10^{-1}$
				5688.205	4	6	$1.21\cdot10^{-1}$	2977.13	3	3	$1.05\cdot10^{-1}$
Silver				5889.950	2	4	$6.16\cdot10^{-1}$	2979.66	5	7	1.96
				5895.924	2	2	$6.14\cdot10^{-1}$	2980.63	7	5	$3.30\cdot10^{-3}$
Ag I				6154.225	2	2	$2.50\cdot10^{-2}$	3004.15	3	3	$9.44\cdot10^{-2}$
2061.2	2	4	$3.1\cdot10^{-2}$	6160.747	4	2	$4.98\cdot10^{-2}$	3009.14	3	3	$1.73\cdot10^{-1}$
2069.9	2	2	$1.5\cdot10^{-2}$	7373.23	2	4	$5.42\cdot10^{-4}$	3015.40	5	5	$1.51\cdot10^{-1}$
3280.7	2	4	1.4	7373.49	2	2	$5.30\cdot10^{-4}$	3053.67	5	7	2.99
3382.9	2	2	1.3	7809.78	2	4	$9.91\cdot10^{-4}$	3055.35	1	3	$8.80\cdot10^{-2}$
5209.1	2	4	$7.5\cdot10^{-1}$	7810.24	2	2	$9.72\cdot10^{-4}$	3057.38	5	5	$4.51\cdot10^{-1}$
5465.5	4	6	$8.6\cdot10^{-1}$	8183.256	2	4	$4.29\cdot10^{-1}$	3058.72	5	3	$2.18\cdot10^{-1}$
5471.6	4	4	$1.4\cdot10^{-1}$	8194.790	4	4	$8.57\cdot10^{-2}$	3064.38	3	5	$7.62\cdot10^{-2}$
				8194.824	4	6	$5.14\cdot10^{-1}$	3066.54	3	3	$3.74\cdot10^{-1}$
Sodium				8649.92	2	4	$2.25\cdot10^{-3}$	3080.25	3	5	2.81
				8650.89	2	2	$2.21\cdot10^{-3}$	3087.06	3	1	$1.66\cdot10^{-1}$
Na I				8942.96	6	6	$2.47\cdot10^{-4}$	3094.45	5	5	$1.91\cdot10^{-3}$
2543.84	2	4	$4.46\cdot10^{-4}$	8942.96	6	8	$3.71\cdot10^{-3}$	3095.55	3	5	$3.97\cdot10^{-3}$
2543.87	2	2	$4.35\cdot10^{-4}$	9153.88	6	6	$3.5\cdot10^{-4}$	3104.40	3	1	$5.63\cdot10^{-1}$
2593.87	2	4	$8.13\cdot10^{-4}$	9153.88	6	8	$5.3\cdot10^{-3}$	3124.42	5	7	2.56
2593.92	2	2	$7.96\cdot10^{-4}$	9465.94	6	6	$6.38\cdot10^{-4}$	3125.21	3	3	$9.72\cdot10^{-2}$
2680.34	2	4	$1.84\cdot10^{-3}$	9465.94	6	8	$9.57\cdot10^{-3}$	3135.48	1	3	$7.42\cdot10^{-1}$
2680.43	2	2	$1.81\cdot10^{-3}$	9961.28	6	6	$8.45\cdot10^{-4}$	3137.86	3	3	1.41
2852.81	2	4	$5.38\cdot10^{-3}$								

λ Å	Weights g_i	g_k	A $10^8\,\text{s}^{-1}$	λ Å	Weights g_i	g_k	A $10^8\,\text{s}^{-1}$	λ Å	Weights g_i	g_k	A $10^8\,\text{s}^{-1}$
3145.71	3	5	$7.23 \cdot 10^{-3}$	1946.43	6	8	8.28	168.411	3	3	$1.37 \cdot 10^2$
3163.74	5	5	1.05	1950.91	8	10	8.55	168.546	1	3	$1.81 \cdot 10^2$
3179.06	5	3	$3.94 \cdot 10^{-1}$	1951.24	6	4	2.38	319.644	5	3	$2.52 \cdot 10^2$
3234.93	3	5	$1.83 \cdot 10^{-1}$	1977.16	4	6	8.46	360.76	1	3	$1.38 \cdot 10^1$
3257.96	5	5	1.03	1985.57	4	4	1.78	408.684	5	3	$2.66 \cdot 10^1$
3260.21	3	1	$4.65 \cdot 10^{-1}$	1995.68	6	6	$8.12 \cdot 10^{-1}$	409.614	3	1	$6.34 \cdot 10^1$
3274.22	5	3	$4.29 \cdot 10^{-1}$	2004.21	2	4	2.07	410.372	5	5	$4.73 \cdot 10^1$
3301.35	3	5	$4.54 \cdot 10^{-2}$	2005.22	2	4	$9.67 \cdot 10^{-1}$	411.334	1	3	$2.09 \cdot 10^1$
3304.96	1	3	$3.60 \cdot 10^{-1}$	2008.47	4	6	1.45	412.242	3	5	$1.55 \cdot 10^1$
3318.04	3	3	$4.14 \cdot 10^{-1}$	2011.87	6	8	6.37	1580.50	7	9	6.27
3327.69	3	1	$9.45 \cdot 10^{-1}$	2014.17	4	6	4.88	1582.18	3	3	7.16
3711.07	1	3	$1.02 \cdot 10^{-1}$	2017.03	6	8	5.44	1582.33	3	5	3.34
4123.08	1	3	$3.71 \cdot 10^{-1}$	2028.56	8	8	1.93	1583.98	5	5	5.55
4344.11	1	3	$4.32 \cdot 10^{-1}$	2031.13	4	6	3.08	1584.14	5	7	6.34
4368.60	1	3	$3.59 \cdot 10^{-1}$	2035.90	2	2	6.23	1587.05	7	9	9.46
				2041.66	6	8	1.64	1613.95	5	7	8.74
Na III				2043.29	4	2	4.02	1615.92	7	9	8.89
202.15	4	4	$1.40 \cdot 10^1$	2044.82	4	4	4.00	1618.57	9	11	9.23
202.49	4	2	$1.59 \cdot 10^2$	2045.44	6	6	1.05	1655.47	7	9	8.72
202.71	2	4	$6.40 \cdot 10^1$	2051.48	6	6	4.36	1701.97	5	7	7.93
202.72	4	2	$6.72 \cdot 10^1$	2060.36	4	6	$8.12 \cdot 10^{-1}$	1702.41	3	5	5.95
202.76	4	4	$1.26 \cdot 10^2$	2066.60	4	4	1.53	1960.76	5	7	3.71
203.28	2	2	$9.24 \cdot 10^1$	2140.72	6	6	3.80	1965.08	5	5	3.68
215.86	2	4	$1.72 \cdot 10^1$	2144.54	4	4	3.88	1967.60	5	3	3.67
216.12	4	4	$1.57 \cdot 10^1$	2202.83	6	6	$8.53 \cdot 10^{-1}$	2018.39	3	5	3.44
229.87	4	2	$2.91 \cdot 10^1$	2225.93	4	4	1.82				
267.87	2	4	$7.56 \cdot 10^1$	2230.33	6	8	3.64	**Strontium**			
378.14	4	2	$8.42 \cdot 10^1$	2232.19	4	4	2.34				
380.10	2	4	$4.13 \cdot 10^1$	2246.70	4	6	2.72	**Sr I**			
1336.76	4	2	4.27	2251.47	2	4	1.66	2206.2	1	3	$6.6 \cdot 10^{-3}$
1337.36	4	4	1.61	2278.42	2	2	2.98	2211.3	1	3	$8.5 \cdot 10^{-3}$
1340.67	6	4	2.29	2285.66	2	4	$7.73 \cdot 10^{-1}$	2217.8	1	3	$1.2 \cdot 10^{-2}$
1342.39	2	2	$8.44 \cdot 10^{-1}$	2309.99	4	2	3.08	2226.3	1	3	$1.6 \cdot 10^{-2}$
1342.73	4	2	1.06	2386.99	6	8	3.05	2237.7	1	3	$2.3 \cdot 10^{-2}$
1355.28	6	6	3.44	2394.03	4	6	2.80	2253.3	1	3	$3.7 \cdot 10^{-2}$
1361.90	2	2	2.24	2406.59	4	4	$5.1 \cdot 10^{-1}$	2275.3	1	3	$6.7 \cdot 10^{-2}$
1372.34	2	4	$5.81 \cdot 10^{-1}$	2459.31	4	6	3.0	2307.3	1	3	$1.2 \cdot 10^{-1}$
1420.89	8	6	5.73	2468.85	2	4	2.4	2354.3	1	3	$1.8 \cdot 10^{-1}$
1444.19	6	4	4.07	2474.73	6	4	1.38	2428.1	1	3	$1.7 \cdot 10^{-1}$
1449.31	8	6	5.12	2497.03	6	6	1.99	2569.5	1	3	$5.3 \cdot 10^{-2}$
1562.87	4	4	2.98	2510.26	4	2	2.19	2931.8	1	3	$1.9 \cdot 10^{-2}$
1565.29	4	4	$3.95 \cdot 10^{-1}$	2530.25	4	4	$3.65 \cdot 10^{-1}$	4607.3	1	3	2.01
1598.18	4	6	$7.96 \cdot 10^{-1}$	2542.80	2	2	$3.39 \cdot 10^{-1}$				
1711.12	6	4	1.52					**Sr II**			
1728.27	4	2	4.10	**Na IV**				2018.7	2	2	$1.2 \cdot 10^{-1}$
1731.11	2	4	1.87	136.551	5	5	$2.08 \cdot 10^2$	2051.9	4	2	$2.4 \cdot 10^{-1}$
1835.22	6	6	3.31	136.854	5	7	$1.56 \cdot 10^2$	2282.0	2	4	$8.3 \cdot 10^{-1}$
1838.94	8	8	3.31	142.232	5	7	$3.06 \cdot 10^2$	2322.4	4	6	$9.1 \cdot 10^{-1}$
1844.36	6	6	3.00	142.359	5	5	$2.77 \cdot 10^2$	2324.5	4	4	$1.5 \cdot 10^{-1}$
1849.56	6	8	6.87	146.064	5	5	$6.69 \cdot 10$	2423.5	2	2	$2.4 \cdot 10^{-1}$
1850.38	4	4	4.37	146.302	3	5	$2.00 \cdot 10^2$	2471.6	4	2	$4.8 \cdot 10^{-1}$
1855.92	2	2	5.77	150.298	5	7	$2.38 \cdot 10^2$	3464.5	4	6	3.1
1856.71	4	6	4.22	150.543	3	5	$1.71 \cdot 10^2$	3474.9	4	4	$5.1 \cdot 10^{-1}$
1861.21	2	4	2.34	150.687	5	5	$9.87 \cdot 10$	4077.7	2	4	1.42
1887.47	10	12	$1.25 \cdot 10^1$	151.299	5	5	$6.70 \cdot 10$	4161.8	2	2	$6.5 \cdot 10^{-1}$
1918.45	4	4	2.29	155.083	5	3	$1.33 \cdot 10^2$	4215.5	2	2	1.27
1926.26	8	10	8.94	155.240	5	3	$1.88 \cdot 10^2$	4305.5	4	2	1.4
1927.24	2	4	5.66	156.537	5	7	$2.92 \cdot 10^2$	4414.8	4	6	$1.1 \cdot 10^{-1}$
1932.74	4	6	5.37	162.448	5	7	$6.04 \cdot 10^2$	4417.5	4	4	$1.8 \cdot 10^{-2}$
1933.89	6	8	6.33	163.190	5	5	$4.32 \cdot 10^2$	4585.9	4	2	$7.0 \cdot 10^{-2}$

λ Å	g_i	g_k	A 10⁸ s⁻¹
5303.1	2	4	$1.9 \cdot 10^{-1}$
5379.1	4	6	$2.2 \cdot 10^{-1}$
5385.5	4	4	$3.7 \cdot 10^{-2}$
5723.7	2	2	$7.1 \cdot 10^{-2}$
5819.0	4	2	$1.4 \cdot 10^{-1}$
8688.9	4	6	$5.5 \cdot 10^{-1}$
8719.6	4	4	$9.7 \cdot 10^{-2}$

Sulfur

S I

λ Å	g_i	g_k	A 10⁸ s⁻¹
1295.7	5	5	4.9
1296.2	5	3	2.7
1302.3	3	5	1.8
1302.9	3	3	1.6
1303.1	3	1	6.6
1303.4	5	3	1.9
1305.9	1	3	2.4
1401.5	5	3	$9.1 \cdot 10^{-1}$
1409.3	3	3	$5.0 \cdot 10^{-1}$
1412.9	1	3	$1.6 \cdot 10^{-1}$
1425.0	5	7	4.5
1425.2	5	5	1.2
1433.3	3	5	3.3
1433.3	3	3	1.9
1437.0	1	3	2.4
1448.2	5	3	7.3
1473.0	5	7	$4.2 \cdot 10^{-1}$
1474.0	5	7	1.6
1474.4	5	5	$5.0 \cdot 10^{-1}$
1474.6	5	3	$6.2 \cdot 10^{-2}$
1481.7	3	5	$1.7 \cdot 10^{-1}$
1483.0	3	5	1.2
1483.2	3	3	$7.5 \cdot 10^{-1}$
1487.2	1	3	$8.7 \cdot 10^{-1}$
1666.7	5	5	6.3
1687.5	1	3	$9.4 \cdot 10^{-1}$
1782.3	1	3	1.9
1807.3	5	3	3.8
1820.3	3	3	2.2
1826.2	1	3	$7.2 \cdot 10^{-1}$
4694.1	5	7	$6.7 \cdot 10^{-3}$
4695.4	5	5	$6.7 \cdot 10^{-3}$
4696.2	5	3	$6.5 \cdot 10^{-3}$
6403.6	3	5	$5.7 \cdot 10^{-3}$
6408.1	5	5	$9.5 \cdot 10^{-3}$
6415.5	7	5	$1.3 \cdot 10^{-2}$
*6751.2	15	25	$7.9 \cdot 10^{-2}$
7679.6	3	5	$1.2 \cdot 10^{-2}$
7686.1	5	5	$2.0 \cdot 10^{-2}$
7696.7	7	5	$2.8 \cdot 10^{-2}$

S II

λ Å	g_i	g_k	A 10⁸ s⁻¹
1124.4	2	4	1.0
1125.0	4	4	4.6
1131.0	2	2	3.5
1131.6	4	2	1.4
1250.5	4	2	$4.6 \cdot 10^{-1}$
1253.8	4	4	$4.2 \cdot 10^{-1}$
1259.5	4	6	$3.4 \cdot 10^{-1}$

λ Å	g_i	g_k	A 10⁸ s⁻¹
4463.6	8	6	$5.3 \cdot 10^{-1}$
4483.4	6	4	$3.1 \cdot 10^{-1}$
4486.7	4	2	$6.6 \cdot 10^{-1}$
4524.7	4	4	$9.3 \cdot 10^{-2}$
4525.0	6	4	1.2
4552.4	4	2	1.2
4656.7	2	4	$9.0 \cdot 10^{-2}$
4716.2	4	4	$2.9 \cdot 10^{-1}$
4815.5	6	4	$8.8 \cdot 10^{-1}$
4885.6	2	4	$1.7 \cdot 10^{-1}$
4917.2	2	2	$6.6 \cdot 10^{-1}$
4924.1	4	6	$2.2 \cdot 10^{-1}$
4925.3	2	4	$2.4 \cdot 10^{-1}$
4942.5	2	2	$1.5 \cdot 10^{-1}$
4991.9	4	4	$1.5 \cdot 10^{-1}$
5009.5	4	2	$7.0 \cdot 10^{-1}$
5014.0	4	4	$8.4 \cdot 10^{-1}$
5027.2	4	2	$2.6 \cdot 10^{-1}$
5032.4	6	6	$8.1 \cdot 10^{-1}$
5047.3	4	2	$3.6 \cdot 10^{-1}$
5103.3	6	4	$5.0 \cdot 10^{-1}$
5142.3	2	2	$1.9 \cdot 10^{-1}$
5201.0	4	4	$7.5 \cdot 10^{-1}$
5201.3	6	4	$6.5 \cdot 10^{-2}$
5212.6	4	6	$9.8 \cdot 10^{-2}$
5212.6	6	6	$8.5 \cdot 10^{-1}$
5320.7	6	8	$9.2 \cdot 10^{-1}$
5345.7	4	6	$8.8 \cdot 10^{-1}$
5345.7	6	6	$1.1 \cdot 10^{-1}$
5428.6	2	4	$4.2 \cdot 10^{-1}$
5432.8	4	6	$6.8 \cdot 10^{-1}$
5453.8	6	8	$8.5 \cdot 10^{-1}$
5473.6	2	2	$7.3 \cdot 10^{-1}$
5509.7	4	4	$4.0 \cdot 10^{-1}$
5526.2	8	8	$8.1 \cdot 10^{-2}$
5536.8	4	6	$6.6 \cdot 10^{-2}$
5556.0	4	2	$1.1 \cdot 10^{-1}$
5564.9	6	6	$1.7 \cdot 10^{-1}$
5578.8	6	6	$1.1 \cdot 10^{-1}$
5606.1	10	8	$5.4 \cdot 10^{-1}$
5616.6	4	4	$1.2 \cdot 10^{-1}$
5640.0	4	6	$6.6 \cdot 10^{-1}$
5645.6	6	4	$1.8 \cdot 10^{-2}$
5647.0	2	4	$5.7 \cdot 10^{-1}$
5659.9	6	4	$4.6 \cdot 10^{-1}$
5664.7	4	2	$5.8 \cdot 10^{-1}$
5819.2	4	4	$8.5 \cdot 10^{-2}$
6305.5	8	6	$1.8 \cdot 10^{-1}$
6312.7	6	4	$3.0 \cdot 10^{-1}$

S III

λ Å	g_i	g_k	A 10⁸ s⁻¹
2496.2	7	5	2.5
2508.2	5	3	2.3
2636.9	3	5	$4.5 \cdot 10^{-1}$
2665.4	5	5	1.4
2680.5	1	3	$6.2 \cdot 10^{-1}$
2691.8	3	3	$4.6 \cdot 10^{-1}$
2702.8	3	1	1.9
2718.9	3	3	1.2
2721.4	5	3	$7.7 \cdot 10^{-1}$

λ Å	g_i	g_k	A 10⁸ s⁻¹
2726.8	3	5	$6.0 \cdot 10^{-1}$
2731.1	5	5	1.1
2756.9	7	7	1.4
2785.5	3	3	$6.1 \cdot 10^{-1}$
2856.0	5	7	5.1
2863.5	7	9	5.7
2872.0	3	5	4.7
2950.2	3	5	3.0
2964.8	5	7	4.0
3662.0	3	3	$6.4 \cdot 10^{-1}$
3717.8	5	3	1.0
3778.9	3	5	$4.4 \cdot 10^{-1}$
3831.8	1	3	$5.6 \cdot 10^{-1}$
3837.8	3	3	$4.2 \cdot 10^{-1}$
3838.3	5	5	1.3
3860.6	3	1	1.6
3899.1	5	3	$6.7 \cdot 10^{-1}$
4253.6	5	7	1.2
4285.0	3	5	$9.0 \cdot 10^{-1}$

Tantalum

Ta I

λ Å	g_i	g_k	A 10⁸ s⁻¹
3170.3	8	10	$8.5 \cdot 10^{-2}$
3406.9	4	6	$6.8 \cdot 10^{-2}$
3419.7	8	8	$1.91 \cdot 10^{-2}$
3463.8	4	6	$2.62 \cdot 10^{-2}$
3497.9	6	8	$4.9 \cdot 10^{-2}$
3505.0	8	6	$2.72 \cdot 10^{-2}$
3607.4	6	8	$4.6 \cdot 10^{-2}$
3626.6	8	10	$7.1 \cdot 10^{-2}$
3642.1	10	12	$5.5 \cdot 10^{-2}$
3784.3	4	6	$4.3 \cdot 10^{-2}$
3848.1	10	8	$1.30 \cdot 10^{-2}$
3922.8	4	4	$3.98 \cdot 10^{-2}$
3996.2	2	4	$3.35 \cdot 10^{-2}$
4026.9	4	4	$3.60 \cdot 10^{-2}$
4029.9	10	10	$2.8 \cdot 10^{-2}$
4040.9	10	12	$7.3 \cdot 10^{-3}$
4061.4	2	4	$6.5 \cdot 10^{-2}$
4064.6	4	4	$3.83 \cdot 10^{-2}$
4136.2	8	6	$1.82 \cdot 10^{-2}$
4147.9	10	8	$1.79 \cdot 10^{-2}$
4175.2	6	8	$2.8 \cdot 10^{-2}$
4205.9	8	10	$8.9 \cdot 10^{-3}$
4303.0	6	6	$2.08 \cdot 10^{-2}$
4386.1	4	6	$1.0 \cdot 10^{-2}$
4402.5	6	6	$2.28 \cdot 10^{-2}$
4415.7	2	4	$2.53 \cdot 10^{-2}$
4441.7	10	8	$9.0 \cdot 10^{-3}$
4473.5	6	8	$1.36 \cdot 10^{-2}$
4511.0	10	12	$1.56 \cdot 10^{-2}$
4530.9	4	6	$2.42 \cdot 10^{-2}$
4553.7	6	8	$9.5 \cdot 10^{-3}$
4565.9	8	8	$2.5 \cdot 10^{-2}$
4619.5	6	4	$5.3 \cdot 10^{-2}$
4669.1	6	4	$2.85 \cdot 10^{-2}$
4681.9	6	6	$1.5 \cdot 10^{-2}$
4691.9	2	4	$4.08 \cdot 10^{-2}$
4706.1	6	6	$1.4 \cdot 10^{-2}$

λ Å	Weights g_i	g_k	A $10^8 s^{-1}$
4740.2	4	4	5.0·10⁻²
4758.0	4	6	7.5·10⁻³
4769.0	8	8	2.8·10⁻²
4780.9	10	8	2.16·10⁻²
4812.8	4	4	1.2·10⁻²
4825.4	6	6	2.63·10⁻²
4832.2	4	4	1.7·10⁻²
4852.2	4	4	1.7·10⁻²
4884.0	6	8	1.1·10⁻²
4904.6	12	10	1.95·10⁻²
4921.3	2	4	1.2·10⁻²
4926.0	4	4	1.5·10⁻²
4936.4	8	6	4.5·10⁻²
4969.7	4	4	1.0·10⁻²
5012.5	4	4	1.9·10⁻²
5037.4	10	8	4.4·10⁻²
5043.3	6	4	2.73·10⁻²
5067.9	8	6	2.92·10⁻²
5087.4	6	4	1.5·10⁻²
5090.7	8	6	9.5·10⁻³
5136.5	2	2	4.5·10⁻²
5143.7	6	4	1.7·10⁻²
5147.6	6	4	9.0·10⁻³
5161.8	4	6	6.3·10⁻³
5218.7	8	6	8.2·10⁻³
5295.0	6	6	7.5·10⁻³
5336.1	6	8	5.5·10⁻³
5349.6	6	4	2.2·10⁻²
5435.3	4	6	1.1·10⁻²
5499.4	10	10	6.1·10⁻³
5518.9	8	10	3.8·10⁻²
5620.7	8	10	6.0·10⁻³
5640.2	6	8	4.9·10⁻³
5645.9	6	8	1.43·10⁻²
5811.1	8	6	5.7·10⁻³
5877.4	10	12	2.3·10⁻²
5939.8	2	4	1.6·10⁻²
5944.0	4	6	2.13·10⁻²
5997.2	10	10	2.4·10⁻²
6020.7	2	4	1.0·10⁻²
6045.4	6	8	2.6·10⁻²
6047.3	8	10	9.0·10⁻³
6249.8	6	6	3.5·10⁻³
6258.7	6	8	3.3·10⁻³
6309.6	4	6	1.83·10⁻²
6360.8	6	8	4.6·10⁻³
6428.6	6	6	6.0·10⁻³
6430.8	8	6	2.9·10⁻²
6450.4	8	10	2.2·10⁻²
6485.4	10	10	5.8·10⁻³
6514.4	6	4	2.2·10⁻²
6516.1	6	8	1.25·10⁻²
6612.0	6	4	1.9·10⁻²
6673.7	2	4	9.0·10⁻³
6771.7	4	4	5.8·10⁻³
6866.2	8	6	2.58·10⁻²
6927.4	10	12	1.01·10⁻²
6928.5	10	8	1.69·10⁻²
6951.3	10	10	3.7·10⁻³
6953.9	6	8	8.3·10⁻³

λ Å	Weights g_i	g_k	A $10^8 s^{-1}$
6966.1	8	8	1.2·10⁻²
6969.5	10	10	2.9·10⁻³
7407.9	6	4	2.0·10⁻²

Thallium

Tl I

λ Å	Weights g_i	g_k	A $10^8 s^{-1}$
2104.6	2	4	4.0·10⁻²
2118.9	2	2	2.0·10⁻²
2129.3	2	4	5.8·10⁻²
2151.9	2	2	3.1·10⁻¹
2168.6	2	4	9.8·10⁻²
2237.8	2	4	1.9·10⁻¹
2316.0	2	2	7.8·10⁻²
2379.7	2	4	4.4·10⁻¹
2507.9	4	2	1.1·10⁻²
2538.2	4	2	1.6·10⁻²
2580.1	2	2	1.8·10⁻¹
2609.0	4	6	1.0·10⁻¹
2609.8	4	4	1.9·10⁻²
2665.6	4	2	5.7·10⁻²
2709.2	4	6	1.7·10⁻¹
2710.7	4	4	3.7·10⁻²
2767.9	2	4	1.26
2826.2	4	2	8.0·10⁻²
2918.3	4	6	4.2·10⁻¹
2921.5	4	4	7.6·10⁻²
3229.8	4	2	1.73·10⁻¹
3519.2	4	6	1.24
3529.4	4	4	2.20·10⁻¹
3775.7	2	2	6.25·10⁻¹
5350.5	4	2	7.05·10⁻¹

Thulium

Tm I

λ Å	Weights g_i	g_k	A $10^8 s^{-1}$
2513.8	8	10	6.9·10⁻²
2527.0	8	8	1.7·10⁻¹
2596.5	8	10	1.6·10⁻¹
2601.1	8	6	1.7·10⁻¹
2622.5	8	10	6.1·10⁻²
2841.1	6	6	2.0·10⁻¹
2854.2	8	6	2.7·10⁻¹
2914.8	8	8	7.7·10⁻²
2933.0	8	6	1.0·10⁻¹
2973.2	8	8	2.3·10⁻¹
3046.9	8	8	1.8·10⁻¹
3081.1	8	8	1.9·10⁻¹
3122.5	6	6	5.2·10⁻¹
3142.4	6	6	8.8·10⁻²
3172.7	8	8	1.8·10⁻¹
3233.7	8	10	5.1·10⁻²
3247.0	6	8	3.0·10⁻¹
3251.8	6	4	5.2·10⁻¹
3380.7	6	8	2.0·10⁻¹
3406.0	6	8	1.5·10⁻¹
3410.1	8	10	1.0·10⁻¹
3416.6	8	8	5.7·10⁻²
3418.6	6	6	1.1·10⁻¹
3563.9	8	6	9.8·10⁻²
3567.4	8	10	4.2·10⁻²

λ Å	Weights g_i	g_k	A $10^8 s^{-1}$
3744.1	8	8	9.5·10⁻¹
3751.8	8	10	1.9·10⁻¹
3798.5	6	4	1.2
3807.7	6	6	3.9·10⁻¹
3883.1	8	6	1.0
3887.4	8	8	3.8·10⁻¹
3916.5	6	8	1.5
3949.3	6	6	1.0
4022.6	6	8	4.0·10⁻²
4044.5	6	4	2.9·10⁻¹
4094.2	8	6	9.0·10⁻¹
4105.8	8	10	6.0·10⁻¹
4138.3	6	4	7.0·10⁻¹
4158.6	6	8	5.5·10⁻²
4187.6	8	8	6.1·10⁻¹
4203.7	8	10	2.5·10⁻¹
4222.7	6	8	1.5·10⁻¹
4271.7	6	6	1.1·10⁻¹
4359.9	8	6	1.3·10⁻¹
4386.4	8	8	4.2·10⁻²
4394.4	6	4	1.1·10⁻¹
4643.1	6	6	3.4·10⁻²
4681.9	6	8	3.9·10⁻¹
4691.1	6	6	3.9·10⁻¹
5307.1	8	10	2.3·10⁻¹
5658.3	6	8	1.0·10⁻¹
5675.8	8	10	1.3·10⁻²
5760.2	6	6	1.3·10⁻²

Tin

Sn I

λ Å	Weights g_i	g_k	A $10^8 s^{-1}$
2073.1	1	3	3.6·10⁻²
2199.3	3	5	2.9·10⁻¹
2209.7	5	5	5.6·10⁻¹
2246.1	1	3	1.6
2268.9	5	7	1.2
2286.7	5	5	3.1·10⁻¹
2317.2	5	7	2.0
2334.8	3	3	6.6·10⁻¹
2354.8	3	5	1.7
2380.7	3	5	3.1·10⁻²
2408.2	5	3	1.8·10⁻¹
2421.7	5	7	2.5
2429.5	5	7	1.5
2433.5	5	3	8.0·10⁻³
2455.2	5	5	1.1·10⁻²
2476.4	5	3	1.1·10⁻²
2483.4	5	5	2.1·10⁻¹
2491.8	1	3	1.7·10⁻¹
2495.7	5	5	6.2·10⁻¹
2523.9	5	3	7.4·10⁻²
2546.6	1	3	2.1·10⁻¹
2558.0	1	3	3.4·10⁻¹
2571.6	5	7	4.5·10⁻¹
2594.4	5	5	3.0·10⁻¹
2636.9	1	3	1.1·10⁻¹
2661.2	3	3	1.1·10⁻¹
2706.5	3	5	6.6·10⁻¹
2761.8	5	5	3.7·10⁻³

λ Å	Weights g_i	g_k	A 10^8 s^{-1}
2779.8	5	7	1.8·10^{-1}
2785.0	5	3	1.4·10^{-1}
2788.0	1	3	1.4·10^{-1}
2812.6	1	3	2.3·10^{-1}
2813.6	5	5	1.2·10^{-1}
2840.0	5	5	1.7
2850.6	5	5	3.3·10^{-1}
2863.3	1	3	5.4·10^{-1}
2913.5	1	3	8.3·10^{-1}
3009.1	3	3	3.8·10^{-1}
3032.8	1	3	6.2·10^{-1}
3034.1	3	1	2.0
3141.8	1	3	1.9·10^{-1}
3175.1	5	3	1.0
3218.7	1	3	4.7·10^{-2}
3223.6	5	5	1.2·10^{-3}
3262.3	5	3	2.7
3330.6	5	5	2.0·10^{-1}
3655.8	1	3	4.1·10^{-2}
3801.0	5	3	2.8·10^{-1}
4524.7	1	3	2.6·10^{-1}
5631.7	1	3	2.4·10^{-2}
5970.3	5	3	9.6·10^{-2}
6037.7	5	5	5.0·10^{-2}
6069.0	1	3	4.6·10^{-2}
6073.5	3	1	6.3·10^{-2}
6171.5	3	3	4.9·10^{-2}

Sn II

λ Å	g_i	g_k	A 10^8 s^{-1}
2368.3	4	2	4.4·10^{-3}
2449.0	4	6	3.7·10^{-1}
2487.0	6	8	5.5·10^{-1}
3283.2	4	6	1.0
3352.0	6	8	1.0
3472.5	2	4	1.6·10^{-1}
3575.5	4	6	1.3·10^{-1}
5332.4	2	4	8.6·10^{-1}
5562.0	4	6	1.2
5588.9	4	6	8.5·10^{-1}
5596.2	4	4	1.5·10^{-1}
5797.2	6	6	2.8·10^{-1}
5799.2	6	8	8.1·10^{-1}
6453.5	2	4	1.2
6761.5	2	2	3.2·10^{-1}
6844.1	2	2	6.6·10^{-1}

Titanium

Ti I

λ Å	g_i	g_k	A 10^8 s^{-1}
2644.28	7	5	1.4
2646.65	9	7	1.5
2733.27	5	5	1.9
2912.07	5	7	1.3
2942.00	5	5	1.0
2948.26	7	7	9.3·10^{-1}
2956.13	9	9	9.7·10^{-1}
3186.45	5	7	8.0·10^{-1}
3191.99	7	9	8.5·10^{-1}
3199.92	9	11	9.4·10^{-1}
3341.88	5	7	6.5·10^{-1}

λ Å	g_i	g_k	A 10^8 s^{-1}
3354.63	7	9	6.9·10^{-1}
3371.45	9	11	7.2·10^{-1}
3377.58	7	5	6.9·10^{-1}
3385.94	9	7	5.0·10^{-1}
3635.46	5	7	8.04·10^{-1}
3642.68	7	9	7.74·10^{-1}
3653.50	9	11	7.54·10^{-1}
3724.57	9	9	9.1·10^{-1}
3741.06	7	7	4.17·10^{-1}
3752.86	9	9	5.04·10^{-1}
3786.04	5	3	1.4
3958.21	9	7	4.05·10^{-1}
3981.76	5	5	3.76·10^{-1}
3989.76	7	7	3.79·10^{-1}
3998.64	9	9	4.08·10^{-1}
4186.12	9	9	2.10·10^{-1}
4266.23	5	5	3.1·10^{-1}
4284.99	5	5	3.2·10^{-1}
4289.07	5	5	3.0·10^{-1}
4393.93	9	11	3.3·10^{-1}
4417.27	11	9	3.6·10^{-1}
4449.14	11	11	9.7·10^{-1}
4450.90	9	9	9.6·10^{-1}
4453.31	5	5	5.98·10^{-1}
4453.71	7	7	4.7·10^{-1}
4455.32	7	7	4.8·10^{-1}
4457.43	9	9	5.6·10^{-1}
4465.81	5	7	3.28·10^{-1}
4481.26	7	7	5.7·10^{-1}
4496.15	7	5	4.4·10^{-1}
4518.02	7	9	1.72·10^{-1}
4522.80	5	7	1.9·10^{-1}
4533.24	11	11	8.83·10^{-1}
4534.78	9	9	6.87·10^{-1}
4548.76	7	5	2.85·10^{-1}
4552.45	9	7	2.1·10^{-1}
4563.43	9	11	2.1·10^{-1}
4617.27	7	9	8.51·10^{-1}
4623.10	5	7	5.74·10^{-1}
4639.94	3	3	6.64·10^{-1}
4742.79	9	9	5.3·10^{-1}
4758.12	11	11	7.13·10^{-1}
4759.27	13	13	7.40·10^{-1}
4778.26	9	9	2.0·10^{-1}
4805.42	5	7	5.8·10^{-1}
4856.01	13	15	5.2·10^{-1}
4885.08	11	13	4.90·10^{-1}
4913.62	7	9	4.44·10^{-1}
4928.34	3	5	6.2·10^{-1}
4981.73	11	13	6.60·10^{-1}
4989.14	7	5	3.25·10^{-1}
4991.07	9	11	5.84·10^{-1}
4999.50	7	9	5.27·10^{-1}
5000.99	9	7	3.52·10^{-1}
5007.21	5	7	4.92·10^{-1}
5014.28	3	5	6.8·10^{-1}
5036.47	7	9	3.94·10^{-1}
5038.40	5	7	3.87·10^{-1}
5224.30	11	11	3.6·10^{-1}
5259.98	5	7	2.3·10^{-1}

λ Å	g_i	g_k	A 10^8 s^{-1}
5351.07	7	7	3.4·10^{-1}
5503.90	11	9	2.6·10^{-1}
5774.04	9	11	5.5·10^{-1}
5785.98	11	13	6.1·10^{-1}
5804.27	13	15	6.8·10^{-1}
6098.66	9	7	2.5·10^{-1}
6220.46	9	7	1.8·10^{-1}

Ti II

λ Å	g_i	g_k	A 10^8 s^{-1}
2635.44	4	4	1.9
2638.56	6	6	1.7
2642.02	8	8	1.9
2645.86	10	10	2.7
2746.54	6	8	2.6
2751.59	8	10	3.7
2752.68	8	10	1.1
2757.62	6	8	7.2·10^{-1}
2758.35	4	6	9.9·10^{-1}
2804.82	6	8	4.6
2810.30	8	10	5.1
2817.83	10	12	3.8
2819.87	8	8	6.5·10^{-1}
2821.26	6	8	7.9·10^{-1}
2827.12	8	10	1.0
2828.06	12	14	4.4
2828.64	6	6	1.2
2828.83	10	10	9.1·10^{-1}
2834.02	10	12	7.9·10^{-1}
2836.47	8	8	1.2
2839.64	12	12	8.3·10^{-1}
2845.93	10	10	1.2
2856.10	12	12	1.5
2877.47	8	8	5.7·10^{-1}
2884.13	10	10	5.2·10^{-1}
2926.64	10	8	8.9·10^{-1}
2931.10	6	6	3.2
2936.02	4	6	2.7
2938.57	6	8	2.4
2941.90	8	10	1.8
2942.97	8	8	1.1
2945.30	10	12	2.7
2954.59	10	12	4.0
2958.80	8	10	4.0
2979.06	4	6	1.2
2990.06	6	8	5.6·10^{-1}
3017.17	12	12	3.6·10^{-1}
3022.64	10	10	1.2
3023.67	8	8	1.0
3075.23	6	4	1.13
3078.65	8	6	1.09
3081.52	10	8	1.1
3088.04	10	8	1.25
3089.44	8	6	1.3
3103.81	10	8	1.1
3106.26	6	6	7.8·10^{-1}
3127.86	6	6	1.6
3128.50	8	8	1.1
3168.55	10	8	4.1·10^{-1}
3181.73	6	8	4.6·10^{-1}
3189.49	4	4	9.2·10^{-1}

λ Å	Weights g_i	g_k	A $10^8\,s^{-1}$
3190.91	6	8	1.3
3202.56	4	6	1.1
3224.25	12	10	$7.0 \cdot 10^{-1}$
3228.62	4	2	2.0
3232.29	8	6	$6.0 \cdot 10^{-1}$
3234.51	10	10	1.38
3236.58	8	8	1.11
3239.04	6	6	$9.87 \cdot 10^{-1}$
3239.66	6	4	$9.4 \cdot 10^{-1}$
3241.99	4	4	1.16
3278.28	4	4	$9.6 \cdot 10^{-1}$
3278.91	6	4	1.0
3282.32	2	2	1.6
3287.66	8	10	1.4
3321.70	4	4	$7.2 \cdot 10^{-1}$
3322.94	10	10	$3.96 \cdot 10^{-1}$
3332.11	6	4	1.1
3361.23	8	10	1.1
3372.80	6	8	1.11
3383.77	4	6	1.09
3456.40	4	4	$8.2 \cdot 10^{-1}$
3483.63	10	8	$9.7 \cdot 10^{-1}$
3492.37	8	6	$9.8 \cdot 10^{-1}$
3504.90	10	10	$8.2 \cdot 10^{-1}$
3510.86	8	8	$9.3 \cdot 10^{-1}$
3535.41	4	6	$5.5 \cdot 10^{-1}$
3741.64	6	6	$6.2 \cdot 10^{-1}$
3759.30	8	8	$9.4 \cdot 10^{-1}$
3761.33	6	6	$9.9 \cdot 10^{-1}$
4911.18	6	4	$3.2 \cdot 10^{-1}$
Ti III			
865.79	5	3	$6.6 \cdot 10^{1}$
1002.37	5	5	7.6
1004.67	7	5	$4.3 \cdot 10^{1}$
1005.80	3	3	$1.3 \cdot 10^{1}$
1007.16	5	3	$3.8 \cdot 10^{1}$
1008.12	3	1	$5.1 \cdot 10^{1}$
1286.37	9	9	2.0
1289.30	7	7	2.2
1291.62	5	5	2.4
1293.23	9	7	1.0
1298.97	7	5	4.9
1327.59	5	3	3.2
1420.44	1	3	1.2
1421.63	3	1	4.0
1422.41	5	5	3.0
1424.14	5	3	1.6
1455.19	9	7	6.4
1498.70	5	5	2.8
2007.36	3	3	3.4
2007.60	1	3	1.2
2010.80	5	3	5.4
2097.30	5	7	3.3
2099.86	3	5	2.5
2104.86	3	3	1.1
2105.09	1	3	1.7
2199.22	3	3	5.7
2237.77	7	7	2.4
2331.35	3	1	4.3

λ Å	Weights g_i	g_k	A $10^8\,s^{-1}$
2331.66	3	3	1.2
2339.00	5	3	3.0
2346.79	7	5	3.3
2374.99	5	3	4.0
2413.99	5	7	3.8
2516.05	7	9	3.4
2567.56	3	3	2.3
2984.75	5	5	1.9
3066.51	3	3	2.5
3228.89	3	3	1.5
3278.31	7	9	3.4
3320.94	3	5	2.8
3340.20	7	9	3.7
3346.18	9	11	3.7
3354.71	11	13	4.4
3397.24	3	1	1.8
3404.46	3	3	1.8
3417.62	3	5	1.9
3915.47	9	11	2.1
4119.14	5	5	$9.9 \cdot 10^{-1}$
4213.26	9	11	2.2
4215.53	9	11	2.2
4247.62	11	13	1.1
4248.54	5	7	2.3
4250.09	3	5	$9.5 \cdot 10^{-1}$
4259.01	11	13	$9.4 \cdot 10^{-1}$
4269.84	9	11	1.7
4285.61	13	15	3.0
4288.66	11	13	1.1
4296.70	11	13	1.6
4319.56	9	11	1.1
4343.25	3	1	1.0
4378.94	3	5	1.6
4433.91	11	13	1.8
4440.66	1	3	1.2
4533.26	3	5	1.5
4576.53	9	7	1.3
4628.07	3	1	1.5
4652.86	7	9	2.6
4874.00	5	7	1.5
4914.32	3	3	1.1
4971.19	9	11	2.1
5083.80	5	3	$9.7 \cdot 10^{-1}$
5278.33	3	3	$9.4 \cdot 10^{-1}$
7506.87	11	13	1.1
Ti IV			
423.49	4	6	$4.9 \cdot 10^{1}$
424.16	6	8	$5.3 \cdot 10^{1}$
433.63	4	2	5.5
433.76	6	4	5.0
729.36	4	2	5.7
1183.64	2	2	6.9
1195.21	4	2	$1.4 \cdot 10^{1}$
1451.74	2	4	$1.8 \cdot 10^{1}$
1467.34	4	6	$2.1 \cdot 10^{1}$
2067.56	2	4	5.1
2103.16	2	2	5.0
2541.79	4	6	6.9
2546.88	6	8	7.4

λ Å	Weights g_i	g_k	A $10^8\,s^{-1}$
2862.60	4	2	4.1
3576.44	4	6	4.6
Tungsten			
W I			
2879.4	1	3	$2.4 \cdot 10^{-1}$
2911.0	1	3	$7.7 \cdot 10^{-2}$
2923.5	7	9	$1.54 \cdot 10^{-2}$
2935.0	3	5	$1.5 \cdot 10^{-1}$
3013.8	7	9	$6.4 \cdot 10^{-2}$
3016.5	9	11	$9.27 \cdot 10^{-2}$
3017.4	7	9	$1.21 \cdot 10^{-1}$
3024.9	3	3	$1.4 \cdot 10^{-1}$
3046.4	3	5	$5.8 \cdot 10^{-2}$
3049.7	7	5	$1.7 \cdot 10^{-1}$
3064.9	5	7	$1.1 \cdot 10^{-2}$
3084.9	5	5	$1.3 \cdot 10^{-2}$
3093.5	7	9	$4.4 \cdot 10^{-2}$
3107.2	5	7	$2.33 \cdot 10^{-2}$
3108.0	7	9	$1.58 \cdot 10^{-2}$
3145.5	9	9	$4.8 \cdot 10^{-3}$
3170.2	7	5	$6.0 \cdot 10^{-3}$
3176.6	3	5	$2.12 \cdot 10^{-2}$
3183.5	7	7	$2.64 \cdot 10^{-3}$
3184.4	5	3	$2.3 \cdot 10^{-2}$
3191.6	1	3	$3.2 \cdot 10^{-2}$
3198.8	7	9	$4.6 \cdot 10^{-2}$
3207.3	7	9	$3.0 \cdot 10^{-2}$
3208.3	5	5	$4.4 \cdot 10^{-2}$
3215.6	9	11	$2.1 \cdot 10^{-1}$
3221.9	5	7	$1.61 \cdot 10^{-2}$
3223.1	5	3	$3.53 \cdot 10^{-3}$
3232.5	9	9	$2.4 \cdot 10^{-2}$
3235.1	7	5	$2.68 \cdot 10^{-3}$
3259.7	7	7	$1.3 \cdot 10^{-2}$
3300.8	7	9	$8.1 \cdot 10^{-2}$
3311.4	7	5	$5.6 \cdot 10^{-2}$
3363.3	9	7	$6.6 \cdot 10^{-3}$
3371.0	7	5	$1.0 \cdot 10^{-2}$
3371.4	3	3	$6.7 \cdot 10^{-3}$
3386.1	7	7	$2.64 \cdot 10^{-3}$
3413.0	7	9	$9.7 \cdot 10^{-3}$
3459.5	9	9	$2.04 \cdot 10^{-3}$
3510.0	7	9	$5.2 \cdot 10^{-3}$
3545.2	1	3	$3.2 \cdot 10^{-2}$
3570.6	5	3	$6.7 \cdot 10^{-3}$
3606.1	3	5	$9.6 \cdot 10^{-3}$
3617.5	7	7	$1.1 \cdot 10^{-1}$
3631.9	3	5	$1.3 \cdot 10^{-2}$
3675.6	9	11	$1.20 \cdot 10^{-2}$
3682.1	9	11	$2.0 \cdot 10^{-2}$
3707.9	7	7	$2.9 \cdot 10^{-2}$
3757.9	7	9	$1.38 \cdot 10^{-2}$
3760.1	5	7	$1.99 \cdot 10^{-2}$
3768.5	3	3	$3.47 \cdot 10^{-2}$
3780.8	7	5	$4.2 \cdot 10^{-2}$
3809.2	7	5	$9.0 \cdot 10^{-3}$
3817.5	7	7	$3.1 \cdot 10^{-2}$
3829.1	3	3	$3.83 \cdot 10^{-3}$

λ Å	g_i	g_k	A 10^8 s^{-1}
3835.1	5	5	$5.2 \cdot 10^{-2}$
3846.3	3	5	$2.14 \cdot 10^{-2}$
3847.5	1	3	$8.3 \cdot 10^{-3}$
3864.3	5	5	$5.6 \cdot 10^{-3}$
3868.0	7	9	$4.6 \cdot 10^{-2}$
3881.4	7	7	$3.6 \cdot 10^{-2}$
3968.5	1	3	$5.07 \cdot 10^{-3}$
3975.5	9	11	$4.1 \cdot 10^{-3}$
4001.4	9	9	$5.6 \cdot 10^{-3}$
4008.8	7	9	$1.63 \cdot 10^{-1}$
4019.3	5	3	$6.7 \cdot 10^{-3}$
4028.8	1	3	$2.0 \cdot 10^{-2}$
4045.6	7	5	$2.88 \cdot 10^{-2}$
4055.2	7	9	$1.79 \cdot 10^{-3}$
4070.0	7	5	$3.60 \cdot 10^{-2}$
4070.6	3	5	$5.6 \cdot 10^{-3}$
4074.4	7	7	$1.0 \cdot 10^{-1}$
4088.3	5	3	$4.13 \cdot 10^{-3}$
4102.7	9	7	$4.9 \cdot 10^{-2}$
4115.6	11	11	$4.8 \cdot 10^{-3}$
4137.5	5	7	$8.4 \cdot 10^{-3}$
4171.2	7	9	$8.6 \cdot 10^{-3}$
4203.8	9	7	$4.9 \cdot 10^{-3}$
4219.4	9	7	$6.1 \cdot 10^{-3}$
4244.4	9	11	$1.38 \cdot 10^{-2}$
4269.4	7	5	$3.04 \cdot 10^{-2}$
4283.8	9	7	$1.69 \cdot 10^{-3}$
4294.6	7	5	$1.2 \cdot 10^{-1}$
4302.1	7	7	$3.6 \cdot 10^{-2}$
4355.2	9	9	$5.1 \cdot 10^{-3}$
4361.8	9	7	$1.64 \cdot 10^{-3}$
4378.5	7	5	$3.48 \cdot 10^{-3}$
4458.1	3	5	$4.2 \cdot 10^{-3}$
4466.3	7	5	$1.5 \cdot 10^{-2}$
4472.5	13	11	$1.55 \cdot 10^{-3}$
4484.2	3	5	$5.6 \cdot 10^{-3}$
4492.3	9	11	$3.6 \cdot 10^{-3}$
4495.3	11	11	$3.3 \cdot 10^{-3}$
4504.8	9	7	$7.0 \cdot 10^{-3}$
4552.5	9	9	$1.42 \cdot 10^{-3}$
4586.8	1	3	$4.20 \cdot 10^{-3}$
4592.6	7	9	$3.4 \cdot 10^{-3}$
4609.9	7	9	$1.42 \cdot 10^{-2}$
4613.3	9	9	$2.9 \cdot 10^{-3}$
4634.8	9	9	$8.8 \cdot 10^{-3}$
4659.9	1	3	$1.0 \cdot 10^{-2}$
4680.5	7	7	$1.4 \cdot 10^{-2}$
4720.4	3	5	$3.22 \cdot 10^{-3}$
4729.6	7	5	$7.8 \cdot 10^{-3}$
4752.6	3	3	$5.20 \cdot 10^{-3}$
4757.5	7	5	$2.72 \cdot 10^{-3}$
4757.8	11	9	$4.1 \cdot 10^{-3}$
4788.4	9	11	$2.6 \cdot 10^{-3}$
4843.8	5	5	$1.9 \cdot 10^{-2}$
4886.9	9	11	$8.1 \cdot 10^{-3}$
4924.6	13	11	$1.75 \cdot 10^{-3}$
4931.6	7	5	$1.0 \cdot 10^{-2}$
4948.6	9	11	$1.36 \cdot 10^{-3}$
4972.6	9	11	$3.9 \cdot 10^{-3}$
4982.6	1	3	$4.17 \cdot 10^{-3}$

λ Å	g_i	g_k	A 10^8 s^{-1}
4986.9	11	9	$6.3 \cdot 10^{-3}$
5006.2	9	7	$1.2 \cdot 10^{-2}$
5015.3	7	9	$5.4 \cdot 10^{-3}$
5040.4	3	5	$5.2 \cdot 10^{-3}$
5053.3	3	3	$1.9 \cdot 10^{-2}$
5071.5	13	11	$3.4 \cdot 10^{-3}$
5117.6	11	11	$1.61 \cdot 10^{-3}$
5124.2	5	5	$4.0 \cdot 10^{-3}$
5141.2	7	9	$1.12 \cdot 10^{-3}$
5224.7	7	5	$1.2 \cdot 10^{-2}$
5243.0	9	7	$1.1 \cdot 10^{-2}$
5254.5	7	5	$3.86 \cdot 10^{-3}$
5268.6	9	9	$1.4 \cdot 10^{-3}$
5500.5	11	9	$6.9 \cdot 10^{-3}$
5514.7	5	3	$7.3 \cdot 10^{-3}$
5537.7	9	11	$2.2 \cdot 10^{-3}$
5617.1	7	7	$1.47 \cdot 10^{-3}$
5631.9	9	7	$1.43 \cdot 10^{-3}$
5660.7	13	11	$6.8 \cdot 10^{-3}$
5675.4	5	5	$2.20 \cdot 10^{-3}$
5796.5	9	7	$2.21 \cdot 10^{-3}$
5891.6	7	7	$1.47 \cdot 10^{-3}$
5947.6	5	7	$2.40 \cdot 10^{-3}$
5965.9	7	5	$1.0 \cdot 10^{-2}$
6021.5	5	3	$8.7 \cdot 10^{-3}$
6081.4	5	3	$4.7 \cdot 10^{-3}$
6203.5	7	7	$3.0 \cdot 10^{-3}$
6285.9	7	5	$6.6 \cdot 10^{-3}$
6292.0	3	5	$2.26 \cdot 10^{-3}$
6303.2	9	9	$1.84 \cdot 10^{-3}$
6404.2	5	7	$1.50 \cdot 10^{-3}$
6439.7	9	9	$1.29 \cdot 10^{-3}$
6445.1	7	5	$6.4 \cdot 10^{-3}$
6532.4	3	5	$4.6 \cdot 10^{-3}$
6538.1	11	9	$2.7 \cdot 10^{-3}$
6563.2	5	5	$2.04 \cdot 10^{-3}$
6814.9	9	9	$1.46 \cdot 10^{-3}$
7285.8	13	11	$1.47 \cdot 10^{-3}$
7569.9	5	3	$3.73 \cdot 10^{-3}$
7664.9	5	3	$3.80 \cdot 10^{-3}$
8017.2	5	7	$1.6 \cdot 10^{-3}$
8358.7	5	7	$1.89 \cdot 10^{-3}$
9381.4	9	7	$1.53 \cdot 10^{-3}$

Uranium

U I

λ Å	g_i	g_k	A 10^8 s^{-1}
3553.0	13	13	$2.0 \cdot 10^{-2}$
3553.0	9	7	$1.4 \cdot 10^{-2}$
3553.4	15	13	$2.2 \cdot 10^{-2}$
3554.5	11	9	$8.4 \cdot 10^{-3}$
3554.9	15	17	$7.9 \cdot 10^{-3}$
3555.3	13	15	$2.7 \cdot 10^{-2}$
3555.8	13	11	$4.1 \cdot 10^{-3}$
3556.9	13	11	$7.5 \cdot 10^{-3}$
3557.8	13	13	$2.9 \cdot 10^{-2}$
3558.0	11	13	$1.6 \cdot 10^{-2}$
3558.6	9	7	$3.9 \cdot 10^{-2}$
3559.4	7	9	$1.5 \cdot 10^{-2}$
3560.3	9	7	$6.4 \cdot 10^{-2}$

λ Å	g_i	g_k	A 10^8 s^{-1}
3561.4	15	13	$5.5 \cdot 10^{-2}$
3561.5	9	9	$2.5 \cdot 10^{-2}$
3561.8	13	11	$5.7 \cdot 10^{-2}$
3563.7	13	13	$2.9 \cdot 10^{-2}$
3563.8	7	7	$1.1 \cdot 10^{-2}$
3565.0	13	11	$2.9 \cdot 10^{-2}$
3566.0	13	15	$1.7 \cdot 10^{-2}$
3566.6	11	11	$2.4 \cdot 10^{-1}$
3568.8	13	13	$3.8 \cdot 10^{-2}$
3569.1	17	15	$1.1 \cdot 10^{-1}$
3569.4	9	9	$1.5 \cdot 10^{-2}$
3570.1	13	11	$1.3 \cdot 10^{-2}$
3570.2	11	9	$5.3 \cdot 10^{-3}$
3570.6	13	15	$2.7 \cdot 10^{-2}$
3570.7	15	15	$1.2 \cdot 10^{-2}$
3571.2	11	11	$6.3 \cdot 10^{-3}$
3571.6	17	15	$1.3 \cdot 10^{-1}$
3572.9	13	15	$1.5 \cdot 10^{-2}$
3573.9	13	11	$4.0 \cdot 10^{-2}$
3574.1	13	15	$3.5 \cdot 10^{-2}$
3574.8	13	15	$1.9 \cdot 10^{-2}$
3577.1	17	15	$4.3 \cdot 10^{-1}$
3577.5	15	13	$7.8 \cdot 10^{-3}$
3577.8	11	11	$8.3 \cdot 10^{-3}$
3577.9	13	13	$2.3 \cdot 10^{-2}$
3578.3	13	11	$2.0 \cdot 10^{-2}$
3580.0	9	9	$1.2 \cdot 10^{-2}$
3580.2	11	9	$2.9 \cdot 10^{-2}$
3580.4	11	13	$7.5 \cdot 10^{-3}$
3580.9	13	13	$2.1 \cdot 10^{-2}$
3582.6	13	13	$2.9 \cdot 10^{-2}$
3584.6	7	5	$2.4 \cdot 10^{-2}$
3584.9	13	15	$1.8 \cdot 10^{-1}$
3585.4	11	11	$1.9 \cdot 10^{-2}$
3585.8	11	9	$2.8 \cdot 10^{-2}$
3587.8	9	11	$1.3 \cdot 10^{-2}$
3588.3	7	9	$1.8 \cdot 10^{-2}$
3589.7	11	13	$2.1 \cdot 10^{-2}$
3589.8	15	13	$5.9 \cdot 10^{-2}$
3590.7	9	7	$2.2 \cdot 10^{-2}$
3591.7	11	9	$5.3 \cdot 10^{-2}$
3593.0	11	11	$1.4 \cdot 10^{-2}$
3593.2	13	15	$4.2 \cdot 10^{-2}$
3593.7	11	11	$7.2 \cdot 10^{-2}$

Vanadium

V I

λ Å	g_i	g_k	A 10^8 s^{-1}
3053.65	4	4	1.3
3056.33	6	6	1.3
3060.46	8	8	1.4
3066.37	10	10	2.1
3183.41	6	8	2.4
3183.96	8	10	2.5
3183.98	4	6	2.4
3185.38	10	12	2.7
3205.58	8	10	1.3
3212.43	10	12	1.4
3377.62	6	6	$6.0 \cdot 10^{-1}$
3533.68	6	8	$5.2 \cdot 10^{-1}$

λ Å	g_i	g_k	A 10^8 s^{-1}
3663.60	4	6	3.1
3667.74	6	8	2.7
3672.41	12	12	9.2·10^{-1}
3673.41	8	10	2.7
3676.70	14	14	1.3
3680.12	10	12	2.2
3686.26	10	12	2.3·10^{-1}
3687.50	12	14	2.9
3688.07	8	8	3.5·10^{-1}
3692.22	6	6	5.4·10^{-1}
3695.34	14	16	2.8
3695.86	4	4	6.6·10^{-1}
3703.57	10	8	9.2·10^{-1}
3704.70	8	6	6.6·10^{-1}
3706.03	10	10	5.2·10^{-1}
3708.71	12	12	4.4·10^{-1}
3794.96	10	10	2.3·10^{-1}
3806.79	10	10	2.5·10^{-1}
3840.75	8	6	5.48·10^{-1}
3855.85	10	8	5.78·10^{-1}
3871.07	10	8	2.8·10^{-1}
3902.26	10	10	2.68·10^{-1}
3930.02	10	10	3.3·10^{-1}
3934.01	8	8	6.2·10^{-1}
3992.80	12	10	1.2
3998.73	14	12	1.0
4050.96	10	10	1.4
4051.35	12	12	1.3
4090.57	8	10	8.5·10^{-1}
4092.68	8	10	2.30·10^{-1}
4095.48	6	8	7.2·10^{-1}
4099.78	6	8	4.10·10^{-1}
4102.15	4	6	7.1·10^{-1}
4104.77	10	8	2.1
4105.16	4	6	4.9·10^{-1}
4109.78	2	4	5.00·10^{-1}
4111.78	10	10	1.01
4115.18	8	8	5.80·10^{-1}
4116.47	6	6	3.2·10^{-1}
4123.50	4	2	1.00
4128.06	6	4	7.70·10^{-1}
4131.99	8	6	5.5·10^{-1}
4134.49	10	8	2.90·10^{-1}
4232.46	10	10	9.8·10^{-1}
4232.95	8	8	7.7·10^{-1}
4268.64	14	14	1.2
4271.55	12	12	9.6·10^{-1}
4276.95	10	10	9.4·10^{-1}
4284.05	8	8	1.2
4291.82	12	14	8.8·10^{-1}
4296.10	10	12	7.7·10^{-1}
4297.67	8	10	7.0·10^{-1}
4298.03	6	8	7.8·10^{-1}
4379.23	10	12	1.1
4384.71	8	10	1.1
4389.98	6	8	6.9·10^{-1}
4395.22	4	6	5.5·10^{-1}
4406.64	10	10	2.2·10^{-1}
4407.63	8	8	4.4·10^{-1}
4408.20	6	6	6.0·10^{-1}

λ Å	g_i	g_k	A 10^8 s^{-1}
4452.01	14	16	9.2·10^{-1}
4457.75	10	12	2.7·10^{-1}
4460.33	10	8	3.0·10^{-1}
4462.36	12	14	7.6·10^{-1}
4468.00	8	10	2.3·10^{-1}
4469.71	10	12	6.2·10^{-1}
4474.04	10	8	4.7·10^{-1}
4496.06	8	6	4.0·10^{-1}
4524.21	12	10	3.0·10^{-1}
4529.58	10	8	2.4·10^{-1}
4545.40	10	12	7.6·10^{-1}
4560.72	8	10	7.0·10^{-1}
4571.79	6	8	6.0·10^{-1}
4578.73	4	6	6.8·10^{-1}
4757.47	4	2	7.6·10^{-1}
4766.62	6	4	5.6·10^{-1}
4776.36	8	6	5.1·10^{-1}
4786.50	10	8	4.7·10^{-1}
4796.92	12	10	4.8·10^{-1}
4807.52	14	12	5.8·10^{-1}
5193.00	12	12	4.0·10^{-1}
5195.39	8	8	2.3·10^{-1}
5234.08	10	10	4.9·10^{-1}
5240.87	12	12	4.3·10^{-1}
5415.25	12	14	3.1·10^{-1}
5487.91	12	10	2.9·10^{-1}
5507.75	10	8	3.5·10^{-1}
6090.21	8	6	2.60·10^{-1}

V II

λ Å	g_i	g_k	A 10^8 s^{-1}
2527.90	13	13	6.1·10^{-1}
2528.47	9	9	5.2·10^{-1}
2528.83	11	11	5.3·10^{-1}
2554.04	9	9	5.4·10^{-1}
2589.10	9	9	7.7·10^{-1}
2640.86	5	7	1.2
2677.80	3	5	3.4·10^{-1}
2679.33	7	7	3.4·10^{-1}
2683.09	1	3	3.4·10^{-1}
2687.96	9	9	7.6·10^{-1}
2689.88	3	1	9.2·10^{-1}
2690.25	7	5	3.4·10^{-1}
2690.79	5	3	5.2·10^{-1}
2700.94	9	11	3.5·10^{-1}
2706.17	7	9	3.4·10^{-1}
2734.22	9	7	6.2·10^{-1}
2753.41	13	11	4.2·10^{-1}
2784.20	9	9	1.3
2787.91	7	9	5.0·10^{-1}
2825.86	9	7	1.2
2843.82	7	5	9.9·10^{-1}
2847.57	9	7	4.6·10^{-1}
2854.34	11	9	5.0·10^{-1}
2862.31	11	11	3.6·10^{-1}
2868.11	5	3	2.1
2869.13	13	11	4.8·10^{-1}
2882.49	5	5	4.2·10^{-1}
2884.78	3	3	5.6·10^{-1}
2889.61	3	1	1.9
2891.64	5	3	1.4

λ Å	g_i	g_k	A 10^8 s^{-1}
2892.43	9	9	3.6·10^{-1}
2892.65	7	5	1.3
2893.31	9	7	1.2
2903.07	3	5	3.4·10^{-1}
2906.45	7	7	7.8·10^{-1}
2908.81	11	9	1.6
2910.01	5	5	1.1
2910.38	3	3	1.2
2911.05	7	9	3.7·10^{-1}
2912.46	11	9	5.0·10^{-1}
2915.88	9	7	4.9·10^{-1}
2924.02	11	11	1.7
2924.63	9	9	1.2
2930.80	7	7	5.8·10^{-1}
2941.37	11	9	3.5·10^{-1}
2944.57	9	7	7.6·10^{-1}
2948.08	9	11	4.0·10^{-1}
2952.07	7	5	7.2·10^{-1}
2955.58	7	9	3.3·10^{-1}
2968.37	7	9	7.0·10^{-1}
2972.26	5	7	5.2·10^{-1}
2973.98	9	11	3.5·10^{-1}
2985.18	7	9	4.4·10^{-1}
3001.20	7	7	7.5·10^{-1}
3014.82	5	3	8.9·10^{-1}
3016.78	7	5	5.0·10^{-1}
3020.21	9	7	5.0·10^{-1}
3048.21	11	13	7.0·10^{-1}
3063.25	9	11	1.0
3100.94	7	7	5.8·10^{-1}
3113.56	11	11	5.0·10^{-1}
3122.89	11	13	7.6·10^{-1}
3134.93	13	13	5.9·10^{-1}
3136.50	11	11	5.3·10^{-1}
3139.73	9	9	5.2·10^{-1}
3151.32	3	5	4.4·10^{-1}
3190.69	9	9	3.3·10^{-1}
3250.78	11	9	5.2·10^{-1}
3251.87	5	7	3.5·10^{-1}
3271.12	7	9	6.9·10^{-1}
3276.12	9	11	5.2·10^{-1}
3279.84	9	11	5.8·10^{-1}
3287.71	5	7	7.5·10^{-1}
3337.85	5	7	5.3·10^{-1}
3517.30	9	7	3.8·10^{-1}
3530.77	5	3	4.5·10^{-1}
3545.19	7	5	4.3·10^{-1}
3556.80	9	7	5.1·10^{-1}
3592.01	7	5	4.4·10^{-1}
3618.92	3	5	3.3·10^{-1}

V III

λ Å	g_i	g_k	A 10^8 s^{-1}
2318.06	8	10	4.6
2323.82	6	8	3.8
2330.42	10	10	3.2
2331.75	8	8	2.5
2334.21	6	6	2.2
2337.13	4	4	2.7
2343.10	6	8	3.6
2358.73	6	8	4.2

λ Å	g_i	g_k	A 10^8 s^-1
2366.31	8	10	4.2
2371.06	10	12	5.2
2373.06	4	6	2.9
2382.46	8	10	5.0
2393.58	6	8	4.3
2404.18	4	6	2.5
2516.14	10	10	3.7
2521.55	8	8	3.5
2548.21	6	4	2.0
2554.22	8	6	1.2
2593.05	6	6	2.8
2595.10	8	8	2.8

V IV

λ Å	g_i	g_k	A 10^8 s^-1
677.345	9	9	6.7
680.632	9	7	$1.2 \cdot 10^1$
681.145	7	5	$1.1 \cdot 10^1$
682.455	7	7	6.5
682.923	5	5	6.9
684.450	7	5	7.7
691.530	5	3	$1.1 \cdot 10^1$
723.537	3	1	$1.5 \cdot 10^1$
724.068	5	5	$1.1 \cdot 10^1$
724.809	5	3	5.6
737.854	9	7	$2.4 \cdot 10^1$
750.110	5	5	$1.0 \cdot 10^1$
884.146	1	3	4.7
1071.05	5	5	6.1
1110.72	3	3	5.0
1112.20	7	7	6.3
1112.44	5	5	5.0
1127.84	7	5	8.9
1131.26	9	7	9.4
1194.46	7	5	$1.0 \cdot 10^1$
1226.52	5	5	$1.5 \cdot 10^1$
1243.72	3	1	9.4
1247.07	5	3	4.7
1272.97	3	1	$2.7 \cdot 10^1$
1304.17	3	5	$1.5 \cdot 10^1$
1305.42	5	7	7.0
1308.06	7	9	7.9
1309.50	5	5	8.7
1312.72	7	7	8.6
1317.57	5	7	8.7
1321.92	7	9	9.9
1326.81	3	5	4.0
1329.29	5	5	$1.5 \cdot 10^1$
1329.97	3	3	4.8
1330.36	1	3	6.0
1331.67	3	1	$1.7 \cdot 10^1$
1332.46	5	3	7.5
1334.49	9	9	8.3
1355.13	7	9	$2.5 \cdot 10^1$
1356.53	5	3	4.9
1395.00	5	7	$1.4 \cdot 10^1$
1400.42	5	7	7.5
1403.62	7	9	8.4
1412.69	3	3	$1.1 \cdot 10^1$
1414.41	5	7	$1.2 \cdot 10^1$
1414.84	5	5	4.6
1418.53	7	7	5.2
1419.58	7	9	$1.3 \cdot 10^1$
1423.72	3	5	7.1
1426.65	9	11	$2.2 \cdot 10^1$
1429.11	5	5	5.0
1434.84	7	7	5.4
1451.04	3	3	7.0
1454.00	5	3	$1.1 \cdot 10^1$
1520.14	5	7	7.2
1522.49	3	5	5.5
1601.92	3	3	$1.2 \cdot 10^1$
1611.88	7	7	5.2
1806.18	5	3	7.3
1809.85	3	1	7.2
1817.68	5	3	4.8
1825.84	7	5	5.3
1861.56	5	7	6.6
1939.07	7	9	5.8
1951.43	5	7	5.0
1963.10	3	5	4.8
1997.72	7	7	4.7
2084.43	5	5	4.0
2120.05	7	9	8.1
2141.20	3	5	7.0
2146.83	7	9	6.6
2149.85	5	7	5.1
2151.09	7	9	4.3
2155.34	11	13	$1.2 \cdot 10^1$
2446.80	9	11	5.3
2570.72	9	11	7.6
3284.56	7	9	5.3
3496.42	7	9	4.4
3514.25	9	11	4.7

Xenon

Xe I

λ Å	g_i	g_k	A 10^8 s^-1
1043.8	1	3	$5.9 \cdot 10^{-1}$
1047.1	1	3	1.3
1050.1	1	3	$8.5 \cdot 10^{-2}$
1056.1	1	3	2.45
1061.2	1	3	$1.9 \cdot 10^{-1}$
1068.2	1	3	3.99
1085.4	1	3	$4.10 \cdot 10^{-1}$
1099.7	1	3	$4.34 \cdot 10^{-1}$
1110.7	1	3	1.5
1129.3	1	3	$4.4 \cdot 10^{-2}$
1170.4	1	3	1.6
1192.0	1	3	6.2
1250.2	1	3	$1.4 \cdot 10^{-1}$
1295.6	1	3	2.46
1469.6	1	3	2.81
4501.0	5	3	$6.2 \cdot 10^{-3}$
4524.7	5	5	$2.1 \cdot 10^{-3}$
4624.3	5	5	$7.2 \cdot 10^{-3}$
4671.2	5	7	$1.0 \cdot 10^{-2}$
4807.0	3	1	$2.4 \cdot 10^{-2}$
7119.6	7	9	$6.6 \cdot 10^{-2}$
7967.3	1	3	$3.0 \cdot 10^{-3}$
8409.2	5	3	$1.0 \cdot 10^{-2}$

Xe II

λ Å	g_i	g_k	A 10^8 s^-1
4180.1	4	4	2.2
4330.5	6	8	1.4
4414.8	6	6	1.0
4603.0	4	4	$8.2 \cdot 10^{-1}$
4844.3	6	8	1.1
4876.5	6	8	$6.3 \cdot 10^{-1}$
5260.4	2	4	$2.2 \cdot 10^{-1}$
5262.0	4	4	$8.5 \cdot 10^{-1}$
5292.2	6	6	$8.9 \cdot 10^{-1}$
5372.4	4	2	$7.1 \cdot 10^{-1}$
5419.2	4	6	$6.2 \cdot 10^{-1}$
5439.0	4	2	$7.4 \cdot 10^{-1}$
5472.6	8	8	$9.9 \cdot 10^{-2}$
5531.1	8	6	$8.8 \cdot 10^{-2}$
5719.6	4	6	$6.1 \cdot 10^{-2}$
5976.5	4	4	$2.8 \cdot 10^{-1}$
6036.2	6	6	$7.5 \cdot 10^{-2}$
6051.2	8	6	$1.7 \cdot 10^{-1}$
6097.6	6	4	$2.6 \cdot 10^{-1}$
6270.8	4	6	$1.8 \cdot 10^{-1}$
6277.5	4	6	$3.6 \cdot 10^{-2}$
6805.7	8	6	$6.1 \cdot 10^{-2}$
6990.9	10	8	$2.7 \cdot 10^{-1}$

Ytterbium

Yb I

λ Å	g_i	g_k	A 10^8 s^-1
2464.5	1	3	$9.1 \cdot 10^{-1}$
2672.0	1	3	$1.18 \cdot 10^{-1}$
3464.4	1	3	$6.2 \cdot 10^{-1}$
3988.0	1	3	1.76
5556.5	1	3	$1.14 \cdot 10^{-2}$

Yb II

λ Å	g_i	g_k	A 10^8 s^-1
3289.4	2	4	1.8
3694.2	2	2	1.4

Yttrium

Y I

λ Å	g_i	g_k	A 10^8 s^-1
2984.25	6	8	$4.8 \cdot 10^{-1}$
4077.36	4	6	1.1
4102.36	6	8	1.3
4128.30	6	6	1.6
4142.84	4	4	1.6
4167.51	6	6	$2.38 \cdot 10^{-1}$
4235.93	6	4	$3.0 \cdot 10^{-1}$
4379.33	6	4	$7.83 \cdot 10^{-1}$
4476.95	8	6	$2.8 \cdot 10^{-1}$
4514.01	4	6	$3.34 \cdot 10^{-1}$
4527.78	8	6	$8.33 \cdot 10^{-1}$
4544.31	6	6	$4.10 \cdot 10^{-1}$
4559.36	2	4	$4.0 \cdot 10^{-1}$
4643.70	4	6	$1.8 \cdot 10^{-1}$
4653.78	4	6	$1.6 \cdot 10^{-1}$
4674.85	6	8	$1.3 \cdot 10^{-1}$
4781.03	8	10	$1.0 \cdot 10^{-1}$
4799.30	6	8	$1.6 \cdot 10^{-1}$
4804.31	6	4	$2.6 \cdot 10^{-1}$

λ Å	Weights g_i	g_k	A $10^8 s^{-1}$	λ Å	Weights g_i	g_k	A $10^8 s^{-1}$	λ Å	Weights g_i	g_k	A $10^8 s^{-1}$
4804.80	4	4	$3.84 \cdot 10^{-1}$	3611.04	5	5	1.04	5320.78	9	7	$3.9 \cdot 10^{-3}$
4845.67	8	8	$6.8 \cdot 10^{-1}$	3628.70	5	3	$3.3 \cdot 10^{-1}$	5473.39	3	5	$4.3 \cdot 10^{-2}$
4852.68	6	6	$6.2 \cdot 10^{-1}$	3664.62	7	5	$3.7 \cdot 10^{-1}$	5480.73	1	3	$7.62 \cdot 10^{-2}$
4856.71	6	6	$2.0 \cdot 10^{-1}$	3710.29	7	9	1.5	5497.41	5	5	$1.2 \cdot 10^{-1}$
4859.84	4	4	$7.26 \cdot 10^{-1}$	3747.55	3	3	$1.9 \cdot 10^{-1}$	5509.90	5	5	$4.24 \cdot 10^{-2}$
4893.44	6	4	$2.2 \cdot 10^{-1}$	3774.34	5	7	1.1	5544.61	3	1	$1.8 \cdot 10^{-1}$
4900.08	8	6	$2.0 \cdot 10^{-1}$	3776.56	5	3	$2.42 \cdot 10^{-1}$	5546.01	5	3	$5.8 \cdot 10^{-2}$
4906.11	10	8	$1.2 \cdot 10^{-1}$	3788.70	3	5	$8.1 \cdot 10^{-1}$	5728.89	5	5	$3.0 \cdot 10^{-2}$
5380.63	6	4	$3.2 \cdot 10^{-1}$	3818.34	5	5	$9.70 \cdot 10^{-2}$	6613.74	5	7	$1.7 \cdot 10^{-2}$
5424.36	6	4	$3.47 \cdot 10^{-1}$	3832.90	7	7	$3.0 \cdot 10^{-1}$	6832.48	5	5	$3.3 \cdot 10^{-3}$
5466.47	10	12	$6.3 \cdot 10^{-1}$	3878.29	7	5	$2.9 \cdot 10^{-2}$	7264.16	5	3	$1.3 \cdot 10^{-2}$
5513.65	6	6	$2.39 \cdot 10^{-1}$	3930.66	5	5	$2.1 \cdot 10^{-2}$				
5527.56	8	10	$5.4 \cdot 10^{-1}$	3950.36	3	5	$2.80 \cdot 10^{-1}$	*Zinc*			
5606.34	10	10	$5.84 \cdot 10^{-2}$	3951.59	5	3	$1.5 \cdot 10^{-2}$				
5630.14	4	6	$4.9 \cdot 10^{-1}$	3982.60	5	5	$2.7 \cdot 10^{-1}$	*Zn I*			
5675.27	6	6	$9.3 \cdot 10^{-2}$	4124.91	5	7	$1.8 \cdot 10^{-2}$	748.29	1	3	$6.0 \cdot 10^{-2}$
5732.09	6	6	$7.5 \cdot 10^{-2}$	4177.54	5	5	$5.27 \cdot 10^{-1}$	765.60	1	3	$7.6 \cdot 10^{-2}$
6087.94	6	4	$1.1 \cdot 10^{-1}$	4199.27	3	5	$5.36 \cdot 10^{-3}$	792.05	1	3	$5.7 \cdot 10^{-2}$
6437.17	10	8	$4.8 \cdot 10^{-2}$	4204.69	1	3	$2.20 \cdot 10^{-2}$	793.85	1	3	$1.8 \cdot 10^{-1}$
6538.57	10	10	$1.5 \cdot 10^{-1}$	4235.73	5	5	$2.3 \cdot 10^{-2}$	809.92	1	3	$2.6 \cdot 10^{-1}$
6815.15	2	4	$7.18 \cdot 10^{-2}$	4309.62	7	5	$1.29 \cdot 10^{-1}$	1109.1	1	3	$3.05 \cdot 10^{-1}$
7035.15	4	4	$6.3 \cdot 10^{-2}$	4358.73	3	3	$5.55 \cdot 10^{-2}$	2138.6	1	3	7.09
				4374.95	5	5	$9.97 \cdot 10^{-1}$	3075.9	1	3	$3.29 \cdot 10^{-4}$
Y II				4398.01	5	3	$1.16 \cdot 10^{-1}$	3282.3	1	3	$9.0 \cdot 10^{-1}$
3112.03	1	3	$1.3 \cdot 10^{-2}$	4422.59	3	1	$1.83 \cdot 10^{-1}$	3302.6	3	5	1.2
3179.42	3	5	$3.8 \cdot 10^{-2}$	4682.33	5	5	$1.9 \cdot 10^{-2}$	3302.9	3	3	$6.7 \cdot 10^{-1}$
3195.62	3	3	$8.23 \cdot 10^{-1}$	4786.58	7	7	$2.1 \cdot 10^{-2}$	3345.0	5	7	1.7
3200.27	5	5	$4.8 \cdot 10^{-1}$	4823.31	5	5	$4.3 \cdot 10^{-2}$	3345.6	5	5	$4.0 \cdot 10^{-1}$
3203.32	3	1	2.77	4854.87	5	3	$3.9 \cdot 10^{-1}$	3345.9	5	3	$4.5 \cdot 10^{-2}$
3216.69	5	3	2.0	4881.44	5	3	$1.5 \cdot 10^{-3}$	6362.3	3	5	$4.74 \cdot 10^{-1}$
3242.28	7	5	2.0	4883.69	9	7	$4.7 \cdot 10^{-1}$	11054	3	1	$2.43 \cdot 10^{-1}$
3448.81	5	5	$4.1 \cdot 10^{-2}$	4900.11	7	5	$4.51 \cdot 10^{-1}$				
3467.88	5	3	$2.7 \cdot 10^{-2}$	4982.13	7	9	$1.5 \cdot 10^{-2}$	*Zn II*			
3496.08	1	3	$3.49 \cdot 10^{-1}$	5087.42	9	9	$2.0 \cdot 10^{-1}$	2025.5	2	4	3.3
3549.01	5	7	$3.97 \cdot 10^{-1}$	5119.11	5	7	$1.6 \cdot 10^{-2}$	2064.2	2	4	4.6
3584.51	3	5	$4.02 \cdot 10^{-1}$	5200.41	5	5	$1.3 \cdot 10^{-1}$	2099.9	4	6	5.6
3600.74	7	7	1.4	5205.73	7	7	$1.6 \cdot 10^{-1}$	2102.2	4	4	$9.3 \cdot 10^{-1}$
3601.91	3	3	1.13	5289.82	7	5	$6.7 \cdot 10^{-3}$	4911.6	4	6	1.6

ELECTRON AFFINITIES

Thomas M. Miller

Electron affinity is defined as the energy difference between the lowest (ground) state of the neutral and the lowest state of the corresponding negative ion. The accuracy of electron affinity measurements has been greatly improved since the advent of laser photodetachment experiments with negative ions. Electron affinities can be determined with optical precision, though a detailed understanding of atomic and molecular states and splittings is required to specify the photodetachment threshold corresponding to the electron affinity.

Atomic and molecular electron affinities are discussed in two excellent articles reviewing photodetachment studies which appear in *Gas Phase Ion Chemistry*, Vol. 3, Bowers, M. T., Ed., Academic Press, Orlando, 1984: Chapter 21 by Drzaic, P. S., Marks, J., and Brauman, J. I., "Electron Photodetachment from Gas Phase Negative Ions," p. 167, and Chapter 22 by Mead, R. D., Stevens, A. E., and Lineberger, W. C., "Photodetachment in Negative Ion Beams," p. 213. Persons interested in photodetachment details should consult these articles and the critical reviews of Andersen, T., Haugen, H. K., and Hotop, H., *J. Phys. Chem. Ref. Data* 28, 1511, 1999, Hotop, H., and Lineberger, W. C., J. *Phys. Chem. Ref. Data* 14, 731, 1985, and Andersen, T., Haugen, H. K., and Hotop, H. *J. Phys. Chem. Ref. Data* 28, 1511, 1999. For simplicity in the tables below, any electron affinity which was discussed in the articles by Drzaic et al. or Hotop and Lineberger is referenced to these sources, where original references are given. The development of cluster-ion photodetachment apparatuses has brought an

explosion of electron affinity estimates for atomic and molecular clusters. The policy in this tabulation is to list the electron affinities for the atoms, diatoms, and triatoms, if adiabatic electron affinities have been determined, but to refer the reader to original sources for higher-order clusters. Additional data on molecular electron affinities may be found in Lias, S. G., Bartmess, J. E., Liebman, J. F., Holmes, J. L., Levin, R. D., and Mallard, W. G., Gas Phase Ion and Neutral Thermochemistry, *J. Phys. Chem. Ref. Data* 17, (Supplement No. 1), 1988 and on the NIST WebBook at the Internet address http://webbook.nist.gov/.

For the present tabulation, the 2010 CODATA value $e/(hc)$ = 8065.54429 ± 0.00018 cm^{-1} eV^{-1} (http://physics.nist.gov/constants/) has been used to convert electron affinities from the units used in spectroscopic work, cm^{-1}, into eV for these tables. Experimental measurements have improved to the level that the 25 ppb uncertainty in $e/(hc)$ will make a difference in a few cases. For this reason, very accurate electron affinities will be given in cm^{-1} with the relevant references.

Abbreviations used in the tables: calc = calculated value; PT = photodetachment threshold using a lamp as a light source; LPT = laser photodetachment threshold; LPES = laser photoelectron spectroscopy; DA = dissociative attachment; attach = electron attachment/detachment equilibrium; e-scat = electron scattering; kinetic = dissociation kinetics; Knud = Knudsen cell; CT = charge transfer; CD = collisional detachment; and ZEKE = zero electron kinetic energy spectroscopy.

TABLE 1. Atomic Electron Affinities

Atomic number	Atom	Electron affinity in eV	Uncertainty in eV	Method	Ref.	
1	H	0.754195	0.000019	LPT	89	
		0.75420817	—	calc	205	
	D	0.754593	0.000074	LPT	89	deuterium
	D	0.75465629	—	calc	205	deuterium
	T	0.75480545	—	calc	205	tritium
2	He	not stable	—	calc	1	
3	Li	0.618049	0.000020	LPT	185	
4	Be	not stable	—	calc	1	
5	B	0.279723	0.000025	LPES	191	
6	C	1.262119	0.000020	LPT	28	
7	N	not stable	—	DA	1	
8	O	1.4611135	0.0000009	LPT	4	
9	F	3.4011897	0.0000024	LPT	227	
10	Ne	not stable	—	calc	1	
11	Na	0.547926	0.000025	LPT	1	
12	Mg	not stable	—	e-scat	1	
13	Al	0.43283	0.00005	LPES	208	
14	Si	1.3895211	0.0000013	LPES	4	
15	P	0.746607	0.000010	LPT	377	
16	S	2.07710403	0.00000051	LPT	334	^{32}S
	S	2.0771043	0.0000011	LPT	334	^{34}S

Atomic number	Atom	Electron affinity in eV	Uncertainty in eV	Method	Ref.
17	Cl	3.612725	0.000027	LPT	52
18	Ar	not stable	—	calc	1
19	K	0.50147	0.00010	LPT	1
20	Ca	0.02455	0.00010	LPT	44
21	Sc	0.188	0.020	LPES	1
22	Ti	0.079	0.014	LPES	1
23	V	0.525	0.012	LPES	1
24	Cr	0.666	0.012	LPES	1
25	Mn	not stable	—	calc	1
26	Fe	0.151	0.003	LPES	27
27	Co	0.662	0.003	LPES	27
28	Ni	1.156	0.010	LPES	1
29	Cu	1.235	0.005	LPES	37
30	Zn	not stable	—	e-scat	1
31	Ga	0.43	0.03	LPES	183
32	Ge	1.232712	0.000015	LPES	28
33	As	0.804	0.002	LPES	352
34	Se	2.020670	0.000025	LPT	1
35	Br	3.3635882	0.0000019	LPT	74
36	Kr	not stable	—	calc	1
37	Rb	0.48592	0.00002	LPT	1
38	Sr	0.048	0.006	LPT	122
39	Y	0.307	0.012	LPES	1
40	Zr	0.426	0.014	LPES	1
41	Nb	0.916	0.005	LPES	311
42	Mo	0.748	0.002	LPES	127
43	Tc	0.55	0.20	calc	1
44	Ru	1.05	0.15	calc	1
45	Rh	1.137	0.008	LPES	1
46	Pd	0.562	0.005	LPES	116
47	Ag	1.302	0.007	LPES	1
48	Cd	not stable	—	e-scat	1
49	In	0.3	0.2	PT	1
50	Sn	1.112067	0.000015	LPES	28
51	Sb	1.046	0.005	LPES	108
52	Te	1.970876	0.000007	LPT	261
53	I	3.0590368	0.0000010	LPES	92
54	Xe	not stable	—	calc	1
55	Cs	0.471626	0.000025	LPT	1
56	Ba	0.14462	0.00006	LPT	195
57	La	0.47	0.02	LPT	184
58	Ce	0.65	0.03	LPT	269
59	Pr	0.962	0.024	LPES	225
60	Nd	>1.916	—	LPES	342
63	Eu	0.864	0.024	LPES	268
65	Tb	>1.165	—	LPES	342
66	Dy	>0	—	LPES	342
69	Tm	1.029	0.022	LPES	264
70	Yb	−0.020	—	calc	196
71	Lu	0.34	0.01	LPT	223
72	Hf	0.014	—	calc	343
73	Ta	0.322	0.012	LPES	1
74	W	0.81626	0.00007	LPES	360
75	Re	0.15	0.15	calc	1
76	Os	1.1	0.2	calc	1
77	Ir	1.5638	0.0005	LPT	141
78	Pt	2.128	0.002	LPT	1
79	Au	2.30863	0.00003	LPT	1
80	Hg	not stable	—	e-scat	1

Atomic number	Atom	Electron affinity in eV	Uncertainty in eV	Method	Ref.
81	Tl	0.377	0.013	LPES	341
82	Pb	0.364	0.008	LPES	1
83	Bi	0.942362	0.000013	LPT	262
84	Po	1.9	0.3	calc	1
85	At	2.8	0.2	calc	1
86	Rn	not stable	—	calc	1
87	Fr	0.486	0.002	calc	82
88	Ra	0.10	—	calc	273
89	Ac	0.35	—	calc	207
118	ekaradon	0.056	0.01	calc	140
114	—	<0	—	calc	354
121	ekaactinium	0.57	—	calc	207
57–71	lanthanides	—	—	calc	355
89–103	actinides	—	—	calc	355

TABLE 2. Electron Affinities for Diatomic Molecules

Molecule	Electron affinity in eV	Uncertainty in eV	Method	Ref.	Molecule	Electron affinity in eV	Uncertainty in eV	Method	Ref.
Ag_2	1.023	0.007	LPES	37	Cu_2	0.836	0.006	LPES	37
AgO	1.654	0.002	LPES	233	CuD	0.439	0.006	LPES	308
Al_2	1.10	0.15	LPES	68	CuD_2	2.60	0.05	LPES	308
AlO	2.60	0.02	LPES	143	CuH	0.444	0.006	LPES	308
AlP	2.043	0.020	LPES	218	CuH_2	2.60	0.05	LPES	308
AlS	2.60	0.03	LPES	129	CuO	1.777	0.006	LPES	118
As_2	0.739	0.008	LPES	200	F_2	3.01	0.07	kinetic	331
AsH	1.0	0.1	PT	2	FO	2.272	0.006	LPES	88
AsO	1.286	0.008	LPES	198	Fe_2	0.902	0.008	LPES	27
Au_2	1.938	0.007	LPES	37	FeD	0.932	0.015	LPES	9
AuH	0.758	0.020	LPES	276	FeH	0.934	0.011	LPES	9
AuO	2.374	0.007	LPES	282	FeO	1.493	0.005	LPES	45
AuPd	1.88	—	LPES	220	GaAs	1.949	0.020	LPES	218
AuS	2.469	0.006	LPES	282	GaO	2.612	0.008	LPES	279
BN	3.160	0.005	LPES	189	GaP	1.988	0.020	LPES	218
BO	2.508	0.008	LPES	6	Ge_2	2.035	0.001	LPES	123
BeH	0.7	0.1	PT	2	I_2	2.524	0.015	LPES	305
Bi_2	1.271	0.008	LPES	119	IBr	2.512	0.003	LPES	350
BiIn	1.72	—	LPES	364	IO	2.378	0.006	LPES	88
Br_2	2.55	0.10	CT	2	InP	1.845	0.020	LPES	218
BrO	2.353	0.006	LPES	88	K_2	0.497	0.012	LPES	104
C_2	3.269	0.006	LPES	87	KBr	0.642	0.010	LPES	30
CH	1.238	0.008	LPES	2	KCl	0.582	0.010	LPES	30
CN	3.862	0.004	LPES	111	KCs	0.471	0.020	LPES	104
CRh	1.46	0.02	LPES	206	KI	0.728	0.010	LPES	30
CS	0.205	0.021	LPES	2	KRb	0.486	0.020	LPES	104
CaH	0.93	0.05	PT	2	LiCl	0.593	0.010	LPES	30
Cl_2	2.38	0.10	CT	2	LiD	0.337	0.012	LPES	102
ClO	2.2775	0.0013	LPES	339	LiH	0.342	0.012	LPES	102
Co_2	1.110	0.008	LPES	27	MgCl	1.589	0.011	LPES	31
CoD	0.680	0.010	LPES	29	MgH	1.05	0.06	PT	2
CoH	0.671	0.010	LPES	29	MgI	1.899	0.018	LPES	31
Cr_2	0.505	0.005	LPES	114	MgO	1.630	0.025	LPES	178
CrD	0.568	0.010	LPES	29	MnD	0.866	0.010	LPES	9
CrH	0.563	0.010	LPES	29	MnH	0.869	0.010	LPES	9
CrO	1.221	0.006	LPES	5	MnO	1.375	0.010	LPES	158
Cs_2	0.469	0.015	LPES	104	MoO	1.290	0.006	LPES	127
CsCl	0.455	0.010	LPES	30	NH	0.370	0.004	LPT	32
CsO	0.273	0.012	LPES	133	NO	0.026	0.005	LPES	73

Molecule	Electron affinity in eV	Uncertainty in eV	Method	Ref.
NRh	1.51	0.02	LPES	206
NS	1.194	0.011	LPES	2
Na$_2$	0.430	0.015	LPES	104
NaBr	0.788	0.010	LPES	30
NaCl	0.727	0.010	LPES	30
NaF	0.520	0.010	LPES	30
NaI	0.865	0.010	LPES	30
NaK	0.465	0.030	LPES	104
NbO	1.29	0.02	LPES	174
Ni$_2$	0.926	0.010	LPES	112
NiCu	0.889	0.010	LPES	128
NiAg	0.979	0.010	LPES	128
NiD	0.477	0.007	LPES	29
NiH	0.481	0.007	LPES	29
NiO	1.470	0.003	LPES	146
O$_2$	0.450	0.002	LPES	222
OD	1.825533	0.000037	LPT	142
OH	1.8276488	0.0000009	LPT	226
ORh	1.58	0.02	LPES	206
P$_2$	0.589	0.025	LPES	42
PH	1.027	0.006	LPES	281
PO	1.092	0.010	LPES	2
Pb$_2$	1.366	0.010	LPES	117
PbO	0.722	0.006	LPES	105
PbS	1.049	0.010	LPES	228
Pd$_2$	1.685	0.008	LPES	112
PdCO	0.604	0.010	LPES	160
PdO	1.570	0.006	LPES	290
Pt$_2$	1.898	0.008	LPES	112
PtN	1.240	0.010	LPES	46
Rb$_2$	0.498	0.015	LPES	104
RbCl	0.544	0.010	LPES	30
RbCs	0.478	0.020	LPES	104
Re$_2$	1.571	0.008	LPES	33
S$_2$	1.670	0.015	LPES	53
SD	2.315	0.002	LPES	10
SF	2.285	0.006	LPES	93
SH	2.3147282	0.0000015	LPT	47
SO	1.125	0.005	LPES	84
Sb$_2$	1.282	0.008	LPES	108
ScO	1.35	0.02	LPES	171
Se$_2$	1.94	0.07	LPES	38
SeH	2.212519	0.000025	LPT	48
SeO	1.456	0.020	LPES	41
Si$_2$	2.201	0.010	LPES	100
SiF	0.81	0.02	LPES	278
SiH	1.277	0.009	LPES	2
SiN	2.949	0.008	LPES	274
Sn$_2$	1.962	0.010	LPES	117
SnO	0.598	0.006	LPES	168
SnPb	1.569	0.008	LPES	117
TaO	1.07	0.06	LPES	360
Te$_2$	1.92	0.07	LPES	38
TeH	2.102	0.015	LPES	39
TeO	1.697	0.022	LPES	40
TiO	1.30	0.03	LPES	172
VO	1.229	0.008	LPES	170
YO	1.35	0.02	LPES	171
ZnF	1.974	0.008	LPES	179
ZnH	<0.95	—	PT	2
ZnO	2.087	0.008	LPES	179
ZrO	1.3	0.3	LPES	173

TABLE 3. Electron Affinities for Triatomic Molecules

Molecule	Electron affinity in eV	Uncertainty in eV	Method	Ref.	
Ag$_3$	2.32	0.05	LPES	37	
AgCN	1.588	0.010	LPES	163	
Al$_3$	1.916	0.004	LPES	332	
AlO$_2$	4.23	0.02	LPES	143	
AlP$_2$	1.933	0.007	LPES	217	
AlTiO	1.70	0.08	LPES	359	
Al$_2$N	2.571	0.008	LPES	297	
Al$_2$P	2.513	0.020	LPES	217	
Al$_2$S	0.80	0.12	LPES	129	
As$_3$	1.45	0.03	LPES	200	
AsH$_2$	1.27	0.03	PT	2	
Au$_3$	3.7	0.3	LPES	37	
AuBO	1.46	0.02	LPES	336	
AuBr$_2$	4.46	0.07	LPES	294	
AuCl$_2$	4.60	0.07	LPES	294	
AuI$_2$	4.226	0.010	LPES	372	
AuO$_2$	3.40	0.03	LPES	373	OAuO
AuOH	1.771	0.015	LPES	312	
AuS$_2$	3.42	0.03	LPES	373	SAuS
AuS$_2$	2.24	0.03	LPES	373	Au(S$_2$)
Au$_2$H	3.437	0.003	LPES	276	
Au$_2$Pd	3.80	—	LPES	220	
BO$_2$	4.46	0.03	LPES	338	
B$_2$N	3.098	0.005	LPES	193	
B$_3$	2.82	0.02	LPES	221	
Bi$_3$	1.60	0.03	LPES	119	
BiIn$_2$	2.13	—	LPES	364	
Bi$_2$In	2.11	—	LPES	364	
C$_3$	1.981	0.020	LPES	11	
CBr$_2$	1.78	0.10	LPES	249	
CCl$_2$	1.593	0.006	LPES	249	
CD$_2$	0.645	0.006	LPES	12	
CDF	0.535	0.005	LPES	95	
CF$_2$	0.180	0.020	LPES	235	
CH$_2$	0.652	0.006	LPES	12	
CHBr	1.454	0.005	LPES	95	
CHCl	1.210	0.005	LPES	95	
CHF	0.542	0.005	LPES	95	
CHI	1.42	0.17	LPES	95	
CI$_2$	2.09	0.07	LPES	235	
C$_2$Cr	2.30	1.617	0.015	271	
C$_2$H	2.969	0.006	LPES	87	
C$_2$N	2.7489	0.0010	LPES	346	
C$_2$Nb	1.380	0.025	LPES	243	
C$_2$O	2.3107	0.0006	LPES	323	CCO
C$_2$S	2.7475	0.0006	LPES	323	CCS

Molecule	Electron affinity in eV	Uncertainty in eV	Method	Ref.	
COS	−0.04	—	LPES	272	
CS$_2$	0.58	0.05	LPES	278	
C$_2$Ti	1.542	0.020	LPES	147	
ClO$_2$	1.6600	0.0002	LPES	339	ClOO
CoD$_2$	1.465	0.013	LPES	34	
CoH$_2$	1.450	0.014	LPES	34	
CrH$_2$	>2.5	—	LPES	34	
Cr$_2$D	1.464	0.005	LPES	107	
Cr$_2$H	1.474	0.005	LPES	107	
Cr$_2$O	0.9	0.1	LPES	306	
CrO$_2$	2.413	0.008	LPES	144	OCrO
CrO$_2$	1.5	0.06	LPES	241	Cr(O$_2$)
Cs$_3$	0.864	0.030	LPES	18	
CsI$_2$	4.52	0.02	LPES	372	
Cu$_3$	2.11	0.05	LPES	37	
CuBr$_2$	4.35	0.05	LPES	177	
CuCN	1.466	0.010	LPES	163	
CuCl$_2$	4.35	0.05	LPES	177	
CuI$_2$	4.256	0.010	LPES	372	
DCO	0.301	0.005	LPES	35	
DNO	0.330	0.015	LPES	14	
DO$_2$	1.077	0.005	LPES	15	
DS$_2$	1.918	0.015	LPES	347	
Fe$_3$	1.43	0.06	LPES	149	
FeC$_2$	1.9782	0.0006	LPES	254	
FeCO	1.157	0.005	LPES	103	
FeD$_2$	1.038	0.013	LPES	34	
FeH$_2$	1.049	0.014	LPES	34	
FeO$_2$	2.358	0.030	LPES	130	
Fe$_2$H	0.564	0.019	LPES	254	
Fe$_2$O	1.60	0.02	LPES	152	
GaAs$_2$	1.894	0.033	LPES	192	
GaP$_2$	1.666	0.041	LPES	192	
Ga$_2$As	2.428	0.020	LPES	192	
Ga$_2$N	2.506	0.008	LPES	302	
Ga$_2$P	2.481	0.015	LPES	192	
Ge$_3$	2.23	0.01	LPES	123	
GeH$_2$	1.097	0.015	LPES	28	
HCO	0.313	0.005	LPES	35	
HCl$_2$	4.896	0.005	LPES	69	
HNO	0.338	0.015	LPES	14	
HO$_2$	1.078	0.006	LPES	15	
HS$_2$	1.916	0.015	LPES	347	
HfO$_2$	2.14	0.03	LPES	319	
I$_3$	4.226	0.013	LPES	162	
I$_2$Au	4.226	0.010	LPES	372	
InP$_2$	1.61	0.05	LPES	137	
In$_2$P	2.36	0.05	LPES	137	
K$_3$	0.956	0.050	LPES	18	
MnD$_2$	0.465	0.014	LPES	34	
MnH$_2$	0.444	0.016	LPES	34	
MnO$_2$	2.06	0.03	LPES	158	
N$_3$	2.68	0.01	LPT	255	

Molecule	Electron affinity in eV	Uncertainty in eV	Method	Ref.	
NCN	2.484	0.006	LPES	154	
NCO	3.609	0.005	LPES	111	
NCS	3.537	0.005	LPES	111	
NH$_2$	0.771	0.005	LPES	58	
N$_2$O	−0.03	0.10	calc	59	
NO$_2$	2.273	0.005	LPES	63	
(NO)R	R=Ar,Kr,Xe	—	LPES	90	
Na$_3$	1.019	0.060	LPES	18	
Nb$_3$	1.032	0.010	LPES	175	
Ni$_3$	1.41	0.05	LPES	55	
NiCN	1.771	0.010	LPES	287	
NiCO	0.804	0.012	LPES	2	
NiD$_2$	1.926	0.007	LPES	34	
NiH$_2$	1.934	0.008	LPES	34	
NiO$_2$	3.05	0.01	LPES	214	ONiO
NiO$_2$	0.82	0.03	LPES	214	Ni(O$_2$)
O$_3$	2.1028	0.0025	LPT	2	
O$_2$Ar	0.52	0.02	LPES	75	
OClO	2.1451	0.0025	LPES	339	
OIO	2.577	0.008	LPES	88	
PH$_2$	1.263	0.006	LPES	281	
P$_2$H	1.514	0.010	LPES	281	
PO$_2$	3.42	0.01	LPES	124	
Pb$_3$	1.70	0.09	LPES	345	
Pd$_3$	<1.5	0.1	LPES	55	
PdCN	2.543	0.007	LPES	287	
PdCO	0.606	0.010	LPES	293	
Pt$_3$	1.87	0.02	LPES	55	
PtCN	3.191	0.003	LPES	287	
PtCO	1.212	0.010	LPES	293	
Rb$_3$	0.920	0.030	LPES	18	
ReO$_2$	2.5	0.1	LPES	216	
S$_3$	2.093	0.025	LPES	16	
SO$_2$	1.107	0.008	LPES	16	
S$_2$O	1.877	0.008	LPES	16	
Sb$_3$	1.85	0.03	LPES	108	
ScO$_2$	2.32	0.02	LPES	171	
SeO$_2$	1.823	0.050	LPES	38	
SiF$_2$	0.10	0.10	LPES	278	
Si$_2$F	1.99	0.28	LPES	17	
SiH$_2$	1.124	0.020	LPES	2	
Si$_2$H	2.31	0.01	LPES	182	
Si$_3$	2.29	0.02	LPES	110	
Sn$_3$	2.24	0.01	LPES	289	
SnCN	1.922	0.006	LPES	292	
Ta$_3$	1.36	0.03	LPES	169	
TaO$_2$	2.40	0.06	LPES	360	
TiO$_2$	1.59	0.03	LPES	172	
V$_3$	1.107	0.010	LPES	176	
VO$_2$	2.3	0.2	CT	101	
WO$_2$	1.998	0.010	LPES	299	
YO$_2$	2.00	0.03	LPES	171	
ZrO$_2$	1.64	0.03	LPES	319	

TABLE 4. Electron Affinities for Larger Polyatomic Molecules

Molecule	Electron Affinity in eV	Uncertainty in eV	Method	Ref.	Name
Ag_n	n=1–60	—	LPES	37	
Al_5	2.23	0.05	LPES	238	
Al_n	n=3–32	—	LPES	68	
$AlTiO_2$	1.70	0.08	LPES	359	
$AlTiO_3$	2.47	0.08	LPES	359	
Al_2C_2	0.64	0.05	LPES	239	acetylide
Al_2TiO_2	1.17	0.08	LPES	359	
Al_2TiO_3	2.2	0.1	LPES	359	
Al_3C	2.56	0.06	LPES	161	
Al_3C_2	2.19	0.03	LPES	244	
Al_3Ge_2	2.43	0.03	LPES	244	
Al_3O	1.00	0.15	LPES	68	
Al_3Si_2	2.36	0.03	LPES	244	
$Al_5H_2O_4$	3.10	0.10	LPES	283	
Al_5O_4	3.50	0.05	LPES	283	
Al_6N	2.58	0.04	LPES	337	
Al_8N	2.75	0.05	LPES	348	
Al_nO_m	n=1,2	m=1–5	LPES	143	
Al_nO_m	n=3–7	m=2–5	LPES	267	
Al_nP_m	n=1–4	m=1–4	LPES	217	
Al_nS_m	n=1–5	m=1–3	LPES	129	
$Ar(H_2O)_n$	n=2,6,7	—	LPES	77	
Ar_nBr	n=2–9	—	ZEKE	212	
Ar_nI	n=2–19	—	ZEKE	212	
As_4	<0.8	—	LPES	200	
As_5	≈1.7	—	LPES	200	
As_5	≈3.5	—	LPES	253	
Au_6	2.06	0.02	LPES	288	
Au_n	n=1–233	—	LPES	37	
$Au(BO_2)$	2.8	0.1	LPES	371	
$Au(BO_2)_2$	5.7	0.1	LPES	371	
AuF_6	7.5	estimate	CT	98	
$AuO(BO_2)$	4.0	0.1	LPES	371	
Au_2BO	4.32	0.02	LPES	336	
Au_3BO	3.08	0.02	LPES	336	
$Au_3(BO_2)$	3.1	0.1	LPES	371	
$Au_3O(BO_2)$	4.9	0.1	LPES	371	
Au_3Pd	2.51	—	LPES	220	
Au_4Pd	2.69	—	LPES	220	
$Au_6(CO)$	2.04	0.05	LPES	288	
$Au_6(CO)_2$	2.03	0.05	LPES	288	
$Au_6(CO)_3$	1.95	0.05	LPES	288	
$Au_{12}Nb$	3.70	0.03	LPES	275	
$Au_{12}Ta$	3.77	0.03	LPES	275	
$Au_{12}V$	3.76	0.03	LPES	275	
B_5	2.33	0.02	LPES	245	
BD_3	0.027	0.014	LPES	62	
BH_3	0.038	0.015	LPES	62	
BO_2Fe_n	n=1–5	—	LPES	358	
B_3N	2.098	0.035	LPES	193	
B_6Li	2.3	0.1	LPES	298	
Bi_4	1.05	0.010	LPES	119	
Bi_5	2.87	0.02	LPES	253	
Bi_n	n=2–9	—	LPES	213	
Bi_2In_2	1.82	—	LPES	364	
Bi_2In_3	2.36	—	LPES	364	
Bi_3Ga	1.87	0.06	LPES	363	
Bi_3Ga_2	2.39	0.05	LPES	363	

Molecule	Electron Affinity in eV	Uncertainty in eV	Method	Ref.	Name
Bi_3Ga_3	1.84	0.06	LPES	363	
Bi_3Ga_4	2.29	0.08	LPES	363	
Bi_3In_2	2.42	—	LPES	364	
$Br(CO_2)$	3.582	0.017	LPES	131	
$Br(H_2O)_n$	n=1–4	—	LPES	250	
Br_3Yb	4.0	0.2	Knud	379	
Br_7Au_2	3.52	0.02	LPES	301	
C_n	n=2–84	—	LPES	70	
CAl_3Ge	2.70	0.06	LPES	224	
CAl_3Si	2.77	0.06	LPES	224	
CCl_4	≤1.14	—	CT	266	
$CCoNO_3$	1.73	0.03	LPES	199	$Co(CO_2)NO$
$CDBr_2$	1.9	0.2	LPES	367	
$CDCl_2$	1.3	0.2	LPES	367	
CDO_2	3.510	0.015	LPES	109	
CD_3O	1.5546	0.0019	LPES	194	
CD_3O_2	1.154	0.004	LPES	188	d_3-methyl peroxyl radical
CD_3S	1.856	0.006	LPT	2	
CD_3S_2	1.748	0.022	LPES	53	
CFO_2	4.277	0.030	LPES	131	
CF_3	1.82	0.05	LPES	187	
CF_3Br	0.91	0.2	CD	2	
CF_3I	1.57	0.2	CD	2	
$CHBr_2$	1.9	0.2	LPES	367	
$CHCl_2$	1.3	0.2	LPES	367	
$CHCl_3$	≤0.78	—	CT	266	
CHI_2	1.9	0.2	LPES	367	
CHO_2	3.498	0.015	LPES	109	
CH_2O_4	2.1	0.2	PT	2	$CO_3(H_2O)$
CH_2S	0.465	0.023	LPES	53	
CH_3	0.08	0.03	LPES	2	
CH_3I	0.11	0.02	LPES	277	
CH_3NO_2	0.172	0.006	LPES	211	
CH_3O	1.5690	0.0019	LPES	194	
CH_3O_2	1.161	0.005	LPES	188	methyl peroxyl radical
CH_3S	1.867	0.004	LPES	166	
CH_3S_2	1.757	0.022	LPES	53	
CH_3Si	0.852	0.010	LPES	97	CH_3-Si
CH_3Si	2.010	0.010	LPES	97	CH_2=SiH
CH_4N	0.432	0.015	LPES	215	
CH_5Si	1.19	0.04	LPT	65	CH_3SiH_2
CO_3	2.69	0.14	LPES	2	
C_2DN	2.009	0.020	LPES	219	DCCN
C_2DN	1.877	0.010	LPES	219	DCNC
C_2DO	2.350	0.020	LPES	13	
C_2D_2	0.492	0.006	LPES	83	vinylidene-d_2
C_2D_2N	1.538	0.012	LPES	21	cyanomethyl-d_2 radical
C_2D_2N	1.070	0.024	LPES	21	isocyanomethyl-d_2 radical
C_2D_3	2.7300	0.0010	LPES	311	1-propynyl-d_3
C_2D_3O	1.8191	0.0012	LPT	22	vinoxy-d_3
C_2D_5O	1.699	0.004	LPES	194	ethoxide-d_3
C_2F_2	2.255	0.006	LPES	106	difluorovinylidene
C_2HD	0.489	0.006	LPES	83	vinylidene-d_1
C_2HF	1.718	0.006	LPES	106	monofluorovinylidene
C_2HFe	1.4512	0.0025	LPES	254	
C_2HN	2.003	0.014	LPES	219	HCCN
C_2HN	1.883	0.013	LPES	219	HCNC
C_2HNPd	2.17	0.03	LPES	291	
C_2HNi	1.063	0.019	LPES	254	
C_2HO	2.338	0.008	LPES	190	

Molecule	Electron Affinity in eV	Uncertainty in eV	Method	Ref.	Name
C_2HPd	1.98	0.03	LPES	287	
C_2HPt	2.650	0.010	LPES	287	
C_2H_2	0.490	0.006	LPES	83	vinylidene
$C_2H_2BrO_2$	3.97	0.03	LPES	310	
$C_2H_2ClO_2$	3.93	0.03	LPES	310	
C_2H_2FO	2.22	0.09	PT	2	acetyl fluoride enolate
$C_2H_2FO_2$	3.80	0.03	LPES	310	
C_2H_2Fe	1.328	0.019	LPES	254	(ended)
C_2H_2N	1.548	0.005	LPES	316	cyanomethyl radical
C_2H_2N	1.059	0.024	LPES	21	isocyanomethyl radical
$C_2H_2N_3$	3.447	0.004	LPES	316	1,2,3-triazolyl
C_2H_2Ni	2.531	0.005	LPES	287	$HNiC_2H$
C_2H_3	0.667	0.024	LPES	90	vinyl
C_2H_3	2.7355	0.0010	LPES	311	1-propynyl
C_2H_3Fe	1.587	0.019	LPES	254	
C_2H_3Ni	1.103	0.019	LPES	254	
C_2H_3O	1.8249	0.0012	LPT	22	vinoxy
C_2H_5N	0.56	0.01	PT	2	ethyl nitrine
$C_2H_5NO_2$	0.191	0.006	LPES	368	nitroethane
$(C_2H_5NO_2)_n$	n=0–4	—	LPES	322	$(nitroethane)_n$
C_2H_5O	1.712	0.004	LPES	194	ethoxide
$C_2H_5O_2$	1.186	0.004	LPES	188	ethyl peroxyl radical
C_2H_5S	1.953	0.006	LPT	2	ethyl sulfide
C_2H_5S	0.868	0.051	LPES	53	CH_3SCH_2
$C_2H_7O_2$	2.26	0.08	PT	50	MeOHOMe
C_2O	2.3107	0.0006	LPES	323	CCO
C_2S	2.7475	0.0006	LPES	323	CCO
C_3D	1.997	0.005	LPES	315	$c-C_3D$
C_3D_2H	0.907	0.023	LPES	24	$propargyl-d_2$ radical
C_3D_5	0.464	0.006	LPES	138	allyl-d5
C_3D_5O	1.603	0.002	LPES	363	$cis-n-methylvinoxy-d_5$
C_3D_5O	1.561	0.001	LPES	363	$trans-n-methylvinoxy-d_5$
C_3Fe	1.69	0.08	LPES	132	
C_3H	1.999	0.003	LPES	315	$c-C_3H$
C_3HFe	1.58	0.06	LPES	132	
C_3HN	1.84	0.01	LPES	335	CCHCN
C_3H_2	1.794	0.008	LPES	153	
C_3H_2D	0.88	0.15	LPES	24	$propargyl-d_1$ radical
$C_3H_2F_3O$	2.625	0.010	LPT	113	1,1,1-trifluoroacetone enolate
C_3H_3	0.918	0.008	LPES	153	propargyl
C_3H_3	2.718	0.008	LPES	153	1-propynyl
C_3H_3N	1.247	0.012	LPES	21	CH_3CH-CN
C_3H_4D	0.373	0.019	LPES	25	$allyl-d_1$
C_3H_5	0.481	0.008	LPES	138	allyl
C_3H_5	0.397	0.069	kinetic	155	cyclopropyl
C_3H_5O	1.758	0.019	LPT	113	acetone enolate
C_3H_5O	1.6106	0.0008	LPES	363	$cis-n-methylvinoxy$
C_3H_5O	1.570	0.002	LPES	363	$trans-n-methylvinoxy$
$C_3H_5O_2$	1.80	0.06	PT	2	methyl acetate enolate
C_3H_7O	1.789	0.033	LPES	23	$n-propyl$ oxide
C_3H_7O	1.847	0.004	LPES	194	isopropyl oxide
C_3H_7S	2.00	0.02	PT	2	$n-propyl$ sulfide
C_3H_7S	2.02	0.02	PT	2	isopropyl sulfide
C_3O	1.237	0.003	LPES	351	
C_3O_2	0.85	0.15	LPES	11	
C_3S	1.5957	0.0010	LPES	351	
C_3Ti	1.561	0.015	LPES	147	
C_4D	3.5308	0.0012	LPES	314	
C_4D_4	0.909	0.015	LPES	125	$vinylvinylidene-d_4$
$C_4F_4Cl_2$	0.87	0.08	attach	258	1,2-dichlorotetrafluorocyclobutene

Molecule	Electron Affinity in eV	Uncertainty in eV	Method	Ref.	Name
$C_4F_4O_3$	0.5	0.2	CD	2	tetrafluorosuccinic anhydride
C_4F_8	0.63	0.05	attach	256	octafluorocyclobutane
C_4Fe	<2.2	0.2	LPES	132	
C_4H	3.5332	0.0010	LPES	314	
C_4HFe	1.67	0.06	LPES	132	
C_4H_2Fe	1.633	0.019	LPES	254	
$C_4H_2O_3$	1.44	0.10	CT	61	maleic anhydride
C_4H_3Fe	1.182	0.019	LPES	254	
C_4H_3Ni	0.824	0.019	LPES	254	
C_4H_3O	1.853	0.004	LPES	361	α-furanyl
C_4H_4	0.914	0.015	LPES	125	vinylvinylidene
C_4H_4N	2.145	0.010	LPES	265	pyrrolyl
$C_4H_4N_3O$	0.75	—	LPES	285	NO·pyrimidine
$C_4H_4N_3O$	3.037	0.015	LPES	309	dehydrogenated cytosine
C_4H_5DO	1.67	0.05	PT	2	2-butanone-3-d_1 enolate
$C_4H_5D_2O$	1.75	0.06	PT	2	2-butanone-3,3-d_2 enolate
C_4H_5O	1.801	0.008	LPT	113	cyclobutanone enolate
C_4H_6	0.431	0.006	LPES	135	trimethylenemethane
C_4H_6D	0.493	0.008	LPES	138	2-methylallyl-d_7
$C_4H_6O_2$	0.69	0.10	CT	61	2,3-butanedione
C_4H_7	0.505	0.006	LPES	138	2-methylallyl
C_4H_7O	1.67	0.05	PT	2	butyraldehyde enolate
C_4H_9O	1.909	0.004	LPES	194	t-butoxyl
C_4H_9S	2.03	0.02	PT	2	n-butyl sulfide
C_4H_9S	2.07	0.02	PT	2	t-butyl sulfide
C_4N	3.1113	0.0010	LPES	346	
C_4O	2.05	0.15	LPES	11	
C_4O_2	2.0	0.2	LPES	11	
C_4Ti	1.494	0.020	LPES	147	
C_5	2.853	0.001	LPT	99	
C_5D_5	1.790	0.008	LPES	11	cyclopentadienyl-d5
C_5F_5N	0.70	0.05	attach	259	pentafluoropyridine
$C_5F_6O_3$	1.5	0.2	CD	2	hexafluoroglutaric anhydride
C_5H	2.421	0.019	LPES	317	linear
C_5H	2.857	0.028	LPES	317	cyclic
C_5HF_4N	0.40	0.08	attach	259	tetrafluoropyridine
C_5H_5	1.804	0.007	LPES	11	cyclopentadienyl
$(C_5H_5N)_nCo_m$	n=1–4	m=1–3	LPES	327	
$C_5H_5NO_2$	1.39	—	LPES	285	O_2·pyridine
$C_5H_5NO_2$	3.250	0.015	LPES	309	dehydrogenated thymine
$C_5H_5N_2O$	0.62	—	LPES	285	NO·pyridine
C_5H_7	0.91	0.03	PT	2	pentadienyl
$C_5H_7NO_3$	1.87	—	LPES	285	O_2·pyridine·H_2O
C_5H_7O	1.598	0.007	LPT	113	cyclopentanone enolate
C_5H_9O	1.69	0.05	PT	2	3-penanone enolate
$C_5H_{11}O$	1.93	0.05	LPT	2	neopentoxyl
$C_5H_{11}S$	2.09	0.02	PT	2	n-pentyl sulfide
C_5O_2	1.2	0.2	LPES	11	
C_5Ti	1.748	0.050	LPES	147	
C_6	4.180	0.001	LPT	8	
$C_6Br_4O_2$	2.44	0.20	CT	2	tetrabromo-BQ
$C_6Cl_4O_2$	2.78	0.10	CT	61	tetrachloro-BQ
C_6D_4	0.551	0.010	LPES	36	o-benzyne-d_4
C_6D_5	1.092	0.020	LPES	26	phenyl-d_5
C_6D_5N	1.44	0.02	LPES	96	phenylnitrene-d_5
$C_6F_4O_2$	2.70	0.10	CT	61	tetrafluoro-BQ
C_6F_5Br	1.15	0.11	CT	67	pentafluorobromobenzene
C_6F_5Cl	0.75	0.05	attach	260	pentafluorochlorobenzene
C_6F_5I	1.41	0.11	CT	67	pentafluoroiodobenzene
$C_6F_5NO_2$	1.52	0.11	CT	67	pentafluoro-NB

Molecule	Electron Affinity in eV	Uncertainty in eV	Method	Ref.	Name
C_6F_6	0.53	0.05	attach	257	hexafluorobenzene
C_6F_{10}	>1.4	0.3	CT	2	perfluorocyclohexene
$C_6H_2Cl_2O_2$	2.48	0.10	CT	61	2,6-dichloro-BQ
$C_6H_2O_2$	1.859	0.005	LPES	232	dehydrobenzoquinone
$C_6H_3F_2NO_2$	1.17	0.10	CT	61	2,4-difluoro-NB
$C_6H_3O_2$	<2.18	—	LPES	232	benzoquinonide
C_6H_4	0.560	0.010	LPES	36	o-benzyne
$C_6H_4BrNO_2$	1.16	0.10	CT	61	o-bromo-NB
$C_6H_4BrNO_2$	1.32	0.10	CT	61	m-bromo-NB
$C_6H_4BrNO_2$	1.29	0.10	CT	61	p-bromo-NB
$C_6H_4ClNO_2$	1.14	0.10	CT	61	o-chloro-NB
$C_6H_4ClNO_2$	1.28	0.10	CT	61	m-chloro-NB
$C_6H_4ClNO_2$	1.26	0.10	CT	61	p-chloro-NB
C_6H_4ClO	≤2.58	0.08	PT	2	o-chlorophenoxide
$C_6H_4FNO_2$	1.07	0.10	CT	61	o-fluoro-NB
$C_6H_4FNO_2$	1.23	0.10	CT	61	m-fluoro-NB
$C_6H_4FNO_2$	1.12	0.10	CT	61	p-fluoro-NB
$C_6H_4N_2O_4$	1.65	0.10	CT	61	o-diNB
$C_6H_4N_2O_4$	1.65	0.10	CT	61	m-diNB
$C_6H_4N_2O_4$	2.00	0.10	CT	61	p-diNB
$C_6H_4O_2$	1.860	0.005	LPES	284	1,4-benzoquinone (BQ)
C_6H_5	1.096	0.006	LPES	26	phenyl
C_6H_5N	1.429	0.011	LPT	115	phenylnitrene
C_6H_5NH	1.70	0.03	PT	2	anilide
$C_6H_5NO_2$	1.00	0.01	LPES	164	nitrobenzene (NB)
C_6H_5O	2.253	0.006	LPES	26	phenoxyl
$C_6H_5O_2$	2.315	0.010	LPES	375	o-HO(C_6H_4)O
$C_6H_5O_2$	2.330	0.010	LPES	375	m-HO(C_6H_4)O
$C_6H_5O_2$	1.990	0.010	LPES	375	p-HO(C_6H_4)O
C_6H_5S	<2.47	0.06	PT	2	thiophenoxide
C_6H_6NO	0.44	—	LPES	285	NO·(benzene)
C_6H_6Nb	0.893	0.006	LPES	311	
$C_6H_6O_2$	1.06	—	LPES	285	O₂·(benzene)
$(C_6H_6)_nCo_m$	n=1–4	m=1–5	LPES	326	
$(C_6H_6)_nFe_m$	n=1–4	m=1–7	LPES	329	
C_6H_7	<1.67	0.04	PT	2	methylchylopentadienyl
C_6H_8	0.855	0.010	LPES	203	$(CH_2)_2C-C(CH_2)_2$
C_6H_8Si	1.435	0.004	LPT	65	$C_6H_5SiH_3$
C_6H_9	0.654	0.010	LPES	203	$CH_2=C(CH_3)-C(CH_2)_2$
C_6H_9O	1.526	0.010	LPT	113	cyclohexanone enolate
C_6H_{10}	0.645	0.015	LPES	126	t-butyl vinylidene
$C_6H_{11}O$	1.755	+0.05/−0.005	LPT	113	pinacolone enolate
$C_6H_{11}O$	1.82	0.06	PT	2	3,3-dimethylbutananl enolate
C_6N	3.3715	0.0010	LPES	346	
C_6N_4	2.3	0.3	PT	2	TCNE
C_6O_6	2.54	0.05	LPES	333	
C_7F_5N	1.11	0.11	CT	67	pentafluorobenzonitrile
C_7F_8	0.86	0.11	CT	67	octafluorotoluene
C_7F_{14}	1.06	0.15	CT	56	perfluoromethylcyclohexane
C_7HF_5O	1.10	0.11	CT	67	pentafluorobenzaldehyde
$C_7H_3N_3O_4$	2.16	0.10	CT	61	3,5-(NO2)2-benzonitrile
$C_7H_4F_3NO_2$	1.41	0.10	CT	61	m-trifluoromethyl-NB
$C_7H_4N_2O_2$	1.61	0.10	CT	61	o-cyano-NB
$C_7H_4N_2O_2$	1.56	0.10	CT	61	m-cyno-NB
$C_7H_4N_2O_2$	1.72	0.10	CT	61	p-cyano-NB
C_7H_6Br	1.308	0.008	LPES	167	o-bromobenzyl
C_7H_6Br	1.307	0.008	LPES	167	m-bromobenzyl
C_7H_6Br	1.229	0.008	LPES	167	p-bromobenzyl
C_7H_6Cl	1.257	0.008	LPES	167	o-chlorobenzyl
C_7H_6Cl	1.272	0.008	LPES	167	m-chlorobenzyl

Molecule	Electron Affinity in eV	Uncertainty in eV	Method	Ref.	Name
C_7H_6Cl	1.174	0.008	LPES	167	p-chlorobenzyl
C_7H_6F	1.091	0.008	LPES	167	o-fluorobenzyl
C_7H_6F	1.173	0.008	LPES	167	m-fluorobenzyl
C_7H_6F	0.937	0.008	LPES	167	p-fluorobenzyl
C_7H_6FO	2.218	0.010	LPT	2	m-fluoroacetophenone enolate
C_7H_6FO	2.176	0.010	LPT	2	p-fluoroacetophenone enolate
$C_7H_6FeO_3$	0.990	0.10	CT	120	h_4-1,3-butadiene-Fe(CO)$_3$
$C_7H_6N_2O_4$	1.77	0.05	PT	60	3,4-dintrotoluene
$C_7H_6N_2O_4$	1.77	0.05	PT	60	2,3-dinitrotoluene
$C_7H_6N_2O_4$	1.60	0.05	PT	60	2,4-dinitrotoluene
$C_7H_6N_2O_4$	1.55	0.05	PT	60	2,6-dinitrotoluene
$C_7H_6O_2$	1.85	0.10	CT	61	o-CH_3-BQ
C_7H_7	0.912	0.006	LPES	26	benzyl
C_7H_7	0.868	0.006	LPES	136	1-quadricyclanide
C_7H_7	0.962	0.006	LPES	136	2-quadricyclanide
C_7H_7	1.286	0.006	LPES	136	norbornadienide
C_7H_7	0.39	0.04	LPES	136	cycloheptatrienide
C_7H_7	3.046	0.006	LPES	136	1-(1,6-heptadiynide)
C_7H_7	>1.140	0.006	LPES	136	3-(1,6-heptadiynide)
$C_7H_7NO_2$	0.92	0.10	CT	61	o-methyl-NB
$C_7H_7NO_2$	0.99	0.10	CT	61	m-methyl-NB
$C_7H_7NO_2$	0.95	0.10	CT	61	p-methyl-NB
$C_7H_7NO_3$	1.04	0.10	CT	61	m-OCH_3-NB
$C_7H_7NO_3$	0.91	0.10	CT	61	p-OCH_3-NB
C_7H_7O	<2.36	0.06	PT	2	o-methyl phenoxide
C_7H_7O	2.14	0.02	PT	50	benzyloxide
C_7H_8FO	<3.05	0.06	PT	50	$PhCH_2OHF$
C_7H_9	1.27	0.03	PT	2	heptatrienyl
C_7H_9O	1.61	0.05	PT	2	2-norbornanone enolate
C_7H_9Si	1.33	0.04	LPT	65	$C_6H_5(CH_3)SiH$
$C_7H_{11}O$	1.598	0.007	LPT	113	cycloheptanone enolate
$C_7H_{11}O$	1.49	0.04	PT	2	2,5-dimethylcyclopentanone enolate
$C_7H_{13}O$	1.72	0.06	PT	2	4-heptanone enolate
$C_7H_{13}O$	1.46	0.04	PT	2	di-isopropyl ketone enolate
$C_8F_{14}N_2$	1.89	0.10	CT	51	1,4-$(CN)_2C_6F_4$
$C_8H_3F_5O$	0.88	0.11	CT	67	pentafluoroacetophenone
$C_8H_3F_6NO_2$	1.79	0.10	CT	61	3,5-$(CF_3)_2$-NB
$C_8H_4F_3N$	0.70	0.05	attach	263	o-trifluoromethylbenzonitrile
$C_8H_4F_3N$	0.67	0.05	attach	263	m-trifluoromethylbenzonitrile
$C_8H_4F_3N$	0.83	0.05	attach	263	p-trifluoromethylbenzonitrile
$C_8H_4O_3$	1.21	0.10	CT	61	phthalic anhydride
C_8H_6	1.044	0.008	LPES	148	
C_8H_7	1.091	0.008	LPES	134	
C_8H_7O	2.057	0.010	PT	2	acetophenone enolate
C_8H_7O	2.10	0.08	LPT	2	phenylacetaldehyde enolate
C_8H_8	0.55	0.02	CT	134	cyclooctatetraene
C_8H_8	0.919	0.008	LPES	139	m-xylylene
$C_8H_9NO_2$	1.21	0.05	PT	60	3,5-dimethyl-NB
$C_8H_9NO_2$	2.61	0.05	PT	60	2,6-dimethyl-NB
$C_8H_9NO_2$	0.86	0.10	CT	61	2,3-dimethyl-NB
$C_8H_{13}O$	1.63	0.06	PT	2	cyclooctanone enolate
$C_8N_4NiS_4$	4.56	0.04	LPES	307	Ni-bis(dithiolene)
$C_8N_4PdS_4$	4.55	0.04	LPES	307	Ni-bis(dithiolene)
$C_8N_4PtS_4$	4.45	0.04	LPES	307	Ni-bis(dithiolene)
C_8S_2	0.049	0.005	LPES	230	bithiophene
$C_9H_8FeO_3$	0.76	0.10	CT	120	h_4-1,3-cyclohexadiene-Fe(CO)$_3$
C_9H_9O	2.030	0.010	LPT	2	m-methylacetophenone enolate
C_9H_9SiN	1.43	0.10	PT	2	trimethylsilylnitrene
$C_9H_{11}NO_2$	0.70	0.10	CT	61	2,4,6-trimethyl-NB
$C_9H_{15}O$	1.69	0.06	PT	2	cyclononanone enolate

Molecule	Electron Affinity in eV	Uncertainty in eV	Method	Ref.	Name
$C_{10}H_4C_{l2}O_2$	2.19	0.10	CT	61	2,3-dichloro-1,4-naphthoquinone
$C_{10}H_6N_2O_4$	1.78	0.10	CT	61	1,3-dinitronaphthalene
$C_{10}H_6N_2O_4$	1.77	0.10	CT	61	1,5-dinitronaphthalene
$C_{10}H_6O_2$	1.81	0.10	CT	61	1,4-naphthoquinone
$C_{10}H_7$	1.403	0.015	LPES	197	1-naphthyl radical
$C_{10}H_7NO_2$	1.23	0.10	CT	61	1-nitronaphthalene
$C_{10}H_7NO_2$	1.18	0.10	CT	61	2-nitronaphthalene
$C_{10}H_8$	0.790	0.008	LPES	230	azulene
$C_{10}H_8CrO_3$	0.93	0.10	CT	120	h_4-1,3,5-cycloheptatriene Cr(CO)$_3$
$C_{10}H_8FeO_3$	0.98	0.10	CT	120	h_4-1,3,5-cycloheptatriene-Fe(CO)$_3$
$C_{10}H_8NO$	0.66	—	LPES	285	NO·naphthlene
$C_{10}H_8O_2$	1.41	—	LPES	285	O$_2$·naphthlene
$C_{10}H_{10}O_3$	2.09	—	LPES	285	O$_2$·naphthalene·H$_2$O
$C_{10}H_{12}O_4$	2.72	—	LPES	285	O$_2$·naphthalene·(H$_2$O)$_2$
$C_{10}H_{17}O$	1.83	0.06	PT	2	cyclodecanone enolate
$C_{11}H_8FeO_3$	1.29	0.10	CT	120	h_4-1,3-butadiene-Fe(CO)$_3$
$C_{12}F_{10}$	0.82	0.11	CT	67	decafluorobiphenyl
$C_{12}H_4N_4$	2.8	0.3	CD	2	TCNQ
$C_{12}H_9$	1.07	0.10	PT	2	perinaphthenyl
$C_{12}H_{12}$	0.84	0.15	kinetic	378	p-bisallyl benzene
$C_{12}H_{12}$	0.90	0.15	kinetic	378	m-bisallyl benzene
$C_{12}H_{12}NO$	0.79	—	LPES	285	NO·(benzene)$_2$
$C_{12}H_{15}O$	2.032	0.010	LPT	2	t-butylacetophenone enolate
$C_{12}H_{21}O$	1.90	0.07	PT	2	cyclododecanone enolate
$C_{13}F_{10}O$	1.52	0.11	CT	67	decafluorobenzophenone
$C_{13}H_9FO$	0.64	0.10	CT	61	4-fluorobenzophenone
$C_{13}H_{10}O$	0.62	0.10	CT	61	benzophenone
$C_{14}H_9NO_2$	1.43	0.10	CT	61	9-nitroanthracene
$C_{14}H_{10}$	0.530	0.005	LPES	286	anthracene
$C_{14}H_{12}O$	0.770	0.005	LPES	286	anthracene·H$_2$O
$(C_{14}H_{10})_n$	n=1–16	—	LPES	231	anthracene clusters
$C_{16}H_{10}$	0.406	0.010	LPES	270	pyrene
$(C_{16}H_{10})_nCo_m$	n=1,2	m=1,2	LPES	330	
$C_{18}H_{12}$	1.058	0.005	LPES	313	tetracene
$C_{18}H_{12}$	0.32	0.01	LPES	303	chrysene
$C_{20}H_{12}$	0.79	0.10	CT	66	benz[a]pyrene
$C_{20}H_{12}$	0.973	0.005	LPES	236	perylene
$C_{20}H_{16}NO$	1.06	—	LPES	285	NO·(naphthalene)$_2$
$C_{22}H_{14}$	1.35	0.10	CT	66	pentacene
$C_{24}H_{12}$	0.47	0.09	LPES	328	coronene
$C_{24}H_{12}Co$	1.15	0.15	LPES	324	Co·(coronene)
$C_{24}H_{12}Co_2$	1.15	0.10	LPES	324	Co$_2$·(coronene)
$C_{24}H_{12}Fe$	1.06	—	LPES	324	Fe·(coronene)
$C_{24}H_{12}Fe_2$	1.59	—	LPES	324	Fe$_2$·(coronene)
$C_{44}Cl_8F_{20}FeN_4$	3.21	0.03	CT	186	FeTPPbCl$_8$
$C_{44}Cl_9F_{20}FeN_4$	3.35	0.03	CT	186	FeTPPF$_{20}$bCl$_8$Cl
$C_{44}Cl_{28}FeN_4$	2.59	0.11	CT	186	FeTPPCl$_{28}$
$C_{44}H_8ClF_{20}FeN_4$	3.14	0.03	CT	186	FeTPPF$_{20}$Cl
$C_{44}H_8Cl_{21}FeN_4$	2.93	0.23	CT	186	FeTPPoCl$_{20}$Cl
$C_{44}H_8F_{20}FeN_4$	2.15	0.15	CT	186	FeTPPF$_{20}$
$C_{44}H_{12}Cl_{17}FeN_4$	3.14	0.03	CT	186	FeTPPoCl$_8$bCl$_8$Cl
$C_{44}H_{20}Cl_8FeN_4$	1.86	0.03	CT	186	FeTPPoCl$_8$
$C_{44}H_{20}Cl_9FeN_4$	2.10	0.19	CT	186	FeTPPoCl$_8$Cl
$C_{44}H_{28}ClFeN_4$	2.15	0.15	CT	186	FeTPPCl
$C_{44}H_{28}FeN_4$	1.87	0.03	CT	186	iron tetraphenylporphyrin (FeTPP)
$C_{44}H_{28}NiN_4$	1.51	0.01	CT	186	nickel tetraphenylporphyrin (NiTPP)
$C_{44}H_{30}N_4$	1.69	0.01	CT	186	H$_2$ tetraphenylporphyrin
$C_{45}H_{29}NiN_4O$	1.74	0.01	CT	186	NiTPPCHO
$C_{48}H_{24}$	0.67	0.09	LPES	328	coronene dimer
$C_{48}H_{24}Fe$	1.50	—	LPES	324	Fe·(coronene)$_2$

Molecule	Electron Affinity in eV	Uncertainty in eV	Method	Ref.	Name
$C_{48}H_{24}Fe$	1.48	—	LPES	324	$Fe_2\cdot(coronene)_2$
$C_{52}H_{39}FeN_7O$	1.97	0.03	CT	186	FeTPP-val
C_{60}	2.683	0.008	LPES	201	
$C_{60}F_2$	2.74	0.07	Knud	202	
$C_{64}H_{64}FeN_8O_4$	2.07	0.03	CT	186	FeTPP-piv
$C_{70}F_2$	2.80	0.07	Knud	202	
$C_{70}F_{30}$	2.81	0.14	Knud	376	$C_{60}(CF_3)_{10}$
$C_{72}F_{36}$	2.57	0.17	Knud	376	$C_{60}(CF_3)_{12}$
C_nCr	n=2–8	—	LPES	271	
C_nNb	n=2–7	—	LPES	243	
$(CO_2)_n$	n=1,2	—	LPES	75	
$(CS)_n$	n=2	—	LPES	75	
$(CS_2)_n$	n=1,2	—	LPES	75	
$(benzene)_n$	n=53–124	—	LPES	248	
$(toluene)_n$	n=33–139	—	LPES	248	
CeF_4	3.8	0.4	CT	98	
$Cl(CO_2)$	3.907	0.010	LPES	131	
$Cl(H_2O)$	n=1–4	—	LPES	250	
ClO_3	4.25	0.10	LPES	340	
ClO_4	5.25	0.10	LPES	340	
Co_n	n=1–108	—	LPES	251	
$CoBr_3$	4.6	0.1	LPES	249	
$CoCl_3$	4.7	0.1	LPES	249	
CoF_4	6.4	0.3	CT	98	
$Cr(CO)_3$	1.349	0.006	LPES	94	
CrO_3	3.66	0.02	LPES	241	
CrO_4	4.98	0.09	LPES	241	
CrO_5	4.4	0.1	LPES	241	
Cr_2O_n	n=1–7		LPES	306	
CsO_4	2.5	0.2	LPES	252	
Cu_n	n=1–411	—	LPES	37	
$CuBO_2$	1.90	0.08	LPES	362	
$Cu(BO_2)_2$	5.07	0.08	LPES	362	
Cu_2BO_2	3.53	0.08	LPES	362	
$Cu_2(BO_2)_2$	2.74	0.08	LPES	362	
$Cu_n(CN)_m$	n=1–6	m=1–6	LPES	159	
$EuSi_n$	n=3–17	—	LPES	321	
$F(H_2O)_n$	n=1–4	—	LPES	242	
$F(H_2O)_n$	n=1–4	—	LPES	250	
Fe_n	n=3–34	—	LPES	149	
$FeBr_3$	4.26	0.06	LPES	249	
$FeBr_4$	5.50	0.08	LPES	249	
$Fe(CO)_2$	1.22	0.02	LPES	2	
$Fe(CO)_3$	1.8	0.2	LPES	2	
$Fe(CO)_4$	2.4	0.3	LPES	2	
$FeCl_3$	4.22	0.06	LPES	249	
$FeCl_4$	6.00	0.08	LPES	249	
FeF_3	3.6	0.1	CT	98	
FeF_4	6.0	estimate	CT	98	
Fe_2H_2	0.942	0.019	LPES	254	
Fe_nO_m	n=1–4	m=1–6	LPES	152	
Ga_2As_3	2.783	0.024	LPES	192	
Ga_2P_3	2.991	0.026	LPES	192	
Ga_xAs_y	n=2–50	n=x+y	LPES	229	
Ge_n	n=3–15	—	LPES	71	
GeH_3	<1.74	0.04	PT	2	
Ge_xAs_y	n=5–30	n=x+y	LPES	72	
HNO_3	0.57	0.15	CD	2	
$H(NH_3)_n$	n=1,2	—	LPES	76	
$(H_2O)_n$	n=2–19	—	LPES	77	

Molecule	Electron Affinity in eV	Uncertainty in eV	Method	Ref.	Name
$I(CO_2)$	3.225	0.001	LPES	131	
$I(CO_2)_n$	n=1–3	—	LPES	350	
$I(H_2O)_n$	n=1–4	—	LPES	250	
In_2Sb_3	2.45	0.08	LPES	365	
In_xP_y	n=2–8	n=x+y	LPES	137	
IrF_4	4.7	0.3	CT	98	
IrF_6	6.5	0.4	CT	98	
K_n	n=2–7	—	LPES	18	
KO_4	2.8	0.2	LPES	252	
$LaCl_4$	7.03	0.01	LPES	145	
LiO_4	3.3	0.2	LPES	252	
$MnBr_3$	5.03	0.06	LPES	249	
$MnCl_3$	5.07	0.06	LPES	249	
MnF_4	5.5	0.2	CT	98	
MnO_3	3.335	0.010	LPES	158	
$Mo(CO)_3$	1.337	0.006	LPES	94	
MoF_5	3.5	0.2	CT	98	
MoF_6	3.8	0.2	CT	98	
MoO_3	3.17	0.02	LPES	280	
MoO_4	5.20	0.07	LPES	86	
MoO_5	5.10	0.07	LPES	86	
MoW_2O_6	2.76	0.05	LPES	353	
Mo_2O_2	2.24	0.02	LPES	280	
Mo_2O_3	2.33	0.07	LPES	280	
Mo_2O_4	2.13	0.04	LPES	280	
Mo_2WO_6	2.85	0.05	LPES	353	
Mo_3O_3	1.91	0.03	LPES	369	
Mo_3O_4	1.77	0.05	LPES	369	
Mo_3O_5	2.72	0.03	LPES	369	
Mo_3O_6	2.75	0.02	LPES	353	
$NH_2(NH_3)_n$	n=1,2	—	LPES	78	
$(NH_3)_n$	n=41–1100	—	LPES	77	
$NO(H_2O)_n$	n=1,2	—	LPES	75	
$NO(N_2O)_n$	n=1–5	—	LPES	79	
NO_3	3.937	0.014	LPES	85	
$NO_3(H_2O)_n$	n=0–6	—	LPES	240	
N_2CD	2.622	0.005	LPES	154	NCND
N_2CH	2.622	0.005	LPES	154	NCNH
$(NO)_2$	>2.1	—	LPES	75	
$(N_2O)_n$	n=1,2	—	LPES	81	
Na_n	n=2–5	—	LPES	18	
$NaCS_2$	0.80	0.05	LPES	278	
NaO_4	3.1	0.2	LPES	252	
NaO_5	3.2	0.2	LPES	252	
$NaSO_3$	2.3	0.2	LPES	252	
$NaSn_n$	n=5–7	—	LPES	380	
Na_2CS_2	0.25	0.05	LPES	278	
$(NaF)_n$	n=1–7,12	—	LPES	64	
$Na(NaF)_n$	n=5,7–12	—	LPES	64	
Na_nSn	n=0–4	—	LPES	380	
Nb_8	1.513	0.008	LPES	157	
Nb_n	n=6–17	—	LPES	181	
Nb_2O_2	0.97	0.01	LPES	370	
Nb_2O_3	1.61	0.01	LPES	370	
Nb_2O_4	1.62	0.01	LPES	370	
Nb_2O_5	3.35	0.05	LPES	370	
Nb_3O	1.393	0.006	LPES	169	
Nb_3O_3	1.54	0.02	LPES	374	
Nb_3O_n	n=3–8	—	LPES	374	
Ni_n	n=1–100	—	LPES	247	

Molecule	Electron Affinity in eV	Uncertainty in eV	Method	Ref.	Name
NiBr$_3$	4.94	0.08	LPES	249	
NiCl$_3$	5.20	0.08	LPES	249	
Ni(CO)H	1.126	0.010	LPES	293	HNiCO
Ni(CO)$_2$	0.643	0.014	LPES	2	
Ni(CO)$_3$	1.077	0.013	LPES	2	
Ni$_n$(benzene)$_m$	n=1–3	m=1,2	LPES	295	
OH(H$_2$O)	<2.95	0.15	PT	2	
OH(NH$_3$)	2.35	0.07	LPES	234	
OH(N$_2$O)	2.14	0.02	LPES	209	
OH(N$_2$O)$_n$	n=1–5	—	LPES	209	
OsF$_4$	3.9	0.3	CT	98	
OsF$_6$	6.0	0.3	CT	98	
P$_5$	3.88	0.03	LPES	253	
PBr$_2$Cl	1.63	0.20	CD	2	
PBr$_3$	1.59	0.15	CD	2	
PCl$_2$Br	1.52	0.20	CD	2	
PCl$_3$	0.82	0.10	CD	2	
PF$_5$	0.75	0.15	CT	121	
PO$_3$	4.95	0.06	LPES	156	
POCl$_2$	3.83	0.25	CD	2	
POCl$_3$	1.41	0.20	CD	2	
P$_2$H$_2$	1.00	0.01	LPES	281	trans-P$_2$H$_2$
P$_2$H$_2$	1.03	0.01	LPES	281	cis-P$_2$H$_2$
Pb$_4$	1.55	0.09	LPES	345	
PtF$_4$	5.5	0.3	CT	98	
PtF$_6$	7.0	0.4	CT	98	
ReF$_6$	4.7	estimate	CT	98	
ReO$_3$	3.6	0.1	LPES	216	
RhF$_4$	5.4	0.3	CT	98	
RuF$_4$	4.8	0.3	CT	98	
RuF$_5$	5.2	0.4	CT	98	
RuF$_6$	7.5	0.3	CT	98	
SF$_4$	1.5	0.2	CT	91	
SF$_5$	4.23	0.12	e-scat	204	
SF$_6$	1.20	0.05	attach	318	
SO$_3$	1.97	0.10	LPES	165	
(SO$_2$)$_2$	0.6	0.2	LPES	80	
Sb$_5$	3.46	0.03	LPES	253	
Sb$_n$	n=2–9	—	LPES	213	
ScBr$_4$	6.13	0.08	LPES	249	
ScCl$_4$	6.84	0.01	LPES	145	
Sc$_2$Si$_n$	n=2–6	—	LPES	356	
SeF$_6$	2.9	0.2	CD	2	
Si$_4$	2.13	0.01	LPES	110	
Si$_5$	2.59	0.02	LPES	110	
Si$_7$	1.85	0.02	LPES	110	
Si$_n$	n=3–20	—	LPES	71	
SiD$_3$	1.386	0.022	LPES	43	
SiF$_3$	2.41	0.22	LPES	17	
SiF$_4$	≤0	—	LPES	17	
SiF$_5$	≥4.66	—	LPES	17	
SiH$_3$	1.406	0.014	LPES	43	
Si$_2$C$_3$	1.766	0.012	LPES	296	linear Si-C$_3$-Si
Si$_3$H	2.53	0.01	LPES	182	
Si$_4$H	2.68	0.01	LPES	182	
Si$_n$F	n=2–11	—	LPES	17	
Si$_n$Na$_m$	n=4–11	m=1–3	LPES	210	
Sn$_n$	n=1–12	—	LPES	289	
SnCH$_2$CN	1.57	0.02	LPES	292	
Sn(CN)(CH$_2$CN)	2.29	0.05	LPES	292	

Molecule	Electron Affinity in eV	Uncertainty in eV	Method	Ref.	Name
$Sn(CN)_2$	2.622	0.004	LPES	292	
Ta_3O	1.583	0.010	LPES	169	
TeF_6	3.34	0.17	CD	2	
Ti_n	n=1–130	—	LPES	151	
TiO_3	4.2	—	LPES	172	
UF_5	3.7	0.2	CT	98	
UF_6	5.1	0.2	CT	98	
UO_3	<2.1	—	CT	98	
V_n	n=3–65	—	LPES	150	
VF_4	3.5	0.2	CT	98	
VSi_n	n=3–6	—	LPES	357	
V_2O_n	n=3–7	—	LPES	246	
V_2Si_n	n=3–6	—			

REFERENCES

1. Hotop, H., and Lineberger, W. C., *J. Phys. Chem. Ref. Data* 14, 731, 1985.
2. Drzaic, P. S., Marks, J., and Brauman, J. I., in *Gas Phase Ion Chemistry*, Vol. 3, Bowers, M. T., Ed., Academic Press, Orlando, 1984, p. 167. The reference for C_6H_4ClO should read "Richardson et al., 1975c."
3. Schulz, P. A., Mead, R. D., Jones, P. L., and Lineberger, W. C., *J. Chem. Phys.* 77, 1153, 1982.
4. Chaibi, W., Pelàez, R. J., Blondel, C., Drag, C., and Delsart, C., *Eur. Phys. J. D* 58, 29, 2010. EA(^{16}O) = 1 178 467.6 ± 0.7 cm^{-1}, EA(^{32}S) = 16 752.975 3 ± 0.004 1 cm^{-1}, EA(^{34}S) = 16 752.977 6 ± 0.008 5 cm^{-1}, and EA(^{28}Si) = 11 207.294 ± 0.006 cm^{-1}. EAs given in the paper in eV were obtained using an older CODATA value for hc/e, which affects EA values in the Tables 1 in the final decimal place. See also Neumark, D. M., Lykke, K. R., Anderson, T., and Lineberger, W. C., *Phys. Rev. A* 32, 1890, 1985. EA(O) = 11 784.645 ± 0.008 cm^{-1}.
5. Wenthold, P. G., Gunion, R. F., and Lineberger, W. C., *Chem. Phys. Lett.* 258, 101, 1996.
6. Wenthold, P. G., Kim, J. B., Jonas, K. L., and Lineberger, W. C., *J. Phys. Chem. A* 101, 4472, 1997.
7. Klein, R., McGinnis, R. P., and Leone, S. R., *Chem. Phys. Lett.* 100, 475, 1983.
8. Arnold, D. W., Bradforth, S. E., Kitsopoulos, T. N., and Neumark, D. M., *J. Chem. Phys.* 95, 8753, 1991; linear C_n.
9. Stevens, A. E., Fiegerle, C. S., and Lineberger, W. C., *J. Chem. Phys.* 78, 5420, 1983.
10. Breyer, F., Frey, P., and Hotop, H., *Z. Phys. A* 300, 7, 1981.
11. Oakes, J. M., and Ellison, G. B., *Tetrahedron* 42, 6263, 1986.
12. Leopold, D. G., Murray, K. K., Miller, A. E. S., and Lineberger, W. C., *J. Chem. Phys.* 83, 4849, 1985.
13. Oakes, J. M., Jones, M.E., Bierbaum, V. M., and Ellison, G. B., *J. Phys. Chem.* 87, 4810, 1983.
14. Ellis, H. B., Jr., and Ellison, G. B., *J. Chem. Phys.* 78, 6541, 1983.
15. Ramond, T. M., Blanksby, S. J., Kato, S., Bierbaum, V. M., Davico, G. E., Schwartz, R. L., Lineberger, W. C., and Ellison, G. B., *J. Phys. Chem. A* 106, 9641, 2002.
16. Nimlos, M. E., and Ellison, G. B., *J. Chem. Phys.* 90, 2574, 1986.
17. Kawamata, H., Negishi, Y., Kishi, R., Iwata, S., Nakajima, A., and Kaya, K., *J. Chem. Phys.* 105, 5369, 1996.
18. McHugh, K. M., Eaton, J. G., Lee, G. H., Sarkas, H. W., Kidder, L. H., Snodgrass, J. T., Manaa, M. R., and Bowen, K. H., *J. Chem. Phys.* 91, 3792, 1989. See also Ref. 104.
19. George, P. M., and Beauchamp, J. L., *Chem. Phys.* 36, 345, 1979. The lower limit given in this paper (3.4 eV) may be increased to 3.5 eV, as rapid charge transfer from HCO_2^- to WF_6 has since been observed (Miller, T. M., and Viggiano, A. A., unpublished).
20. Burnett, S. M., Stevens, A. E., Fiegerle, C. S., and Lineberger, W. C., *Chem. Phys. Lett.* 100, 124, 1983.
21. Moran, S., Ellis, H. B., DeFrees, D. J., McLean, A. D., and Ellison, G. B., *J. Am. Chem. Soc.* 109, 5996, 1987; Moran, S., Ellis, H. B., DeFrees, D. J., McLean, A. D., Paulson, S. E., and Ellison, G. B., *J. Am. Chem. Soc.*

109, 6004, 1987; see also Lykke, K. R., Neumark, D. M., Andersen, T., Trapa, V. J., and Lineberger, W. C., *J. Chem. Phys.* 87, 6842, 1987.
22. Yacovitch, T. I., Garand, E., and Neumark, D. M., *J. Chem. Phys.* 130, 244309, 2009 for C_2H_3O (14,719 ± 9 cm^{-1}). EA(C_2D_3O) was deduced from this and the difference measured by Mead, R. D., Lykke, K. R., Lineberger, W. C., Marks, J., and Brauman, J. I., *J. Chem. Phys.* 81, 4883, 1984.
23. Ellison, G. B., Engelking, P. C., and Lineberger, W. C., *J. Chem. Phys.* 86, 4873, 1982.
24. Oakes, J. M., and Ellison, G. B., *J. Am. Chem. Soc.* 105, 2969, 1983.
25. Ellison, G. B., and Oakes, J. M., *J. Am. Chem. Soc.* 106, 7734, 1984. EA(allyl) and EA(allyl-d_5) are 0.119 and 0.083 eV too low, respectively, in this work, according to Ref. 138. Therefore, EA(allyl-d_1) is likely too low by a similar amount.
26. Gunion, R. F., Gilles, M. K., Polak, M. L., and Lineberger, W. C., *Int. J. Mass Spectrom. Ion Process.* 117, 601, 1992.
27. Leopold, D. G., and Lineberger, W. C., *J. Chem. Phys.* 85, 51, 1986.
28. Scheer, M., Bilodeau, R. C., Brodie, C. A., and Haugen, H. K., *Phys. Rev. A* 58, 2844, 1998.
29. Miller, A. E. S., Fiegerle, C. S., and Lineberger, W. C., *J. Chem. Phys.* 87, 1549, 1987.
30. Miller, T. M., Leopold, D. G., Murray, K. K., and Lineberger, W. C., *J. Chem. Phys.* 85, 2368, 1986.
31. Miller, T. M., and Lineberger, W. C., *Chem. Phys. Lett.* 146, 364, 1988.
32. Neumark, D. M., Lykke, K. R., Andersen, T., and Lineberger, W. C., *J. Chem. Phys.* 83, 4364, 1985.
33. Leopold, D. G., Miller, T. M., and Lineberger, W. C., *J. Am. Chem. Soc.* 108, 178, 1986.
34. Miller, A. E. S., Fiegerle, C. S., and Lineberger, W. C., *J. Chem. Phys.* 84, 4127, 1986.
35. Murray, K. K., Miller, T. M., Leopold, D. G., and Lineberger, W. C., *J. Chem. Phys.* 84, 2520, 1986.
36. Leopold, D. G., Miller, A. E. S., and Lineberger, W. C., *J. Am. Chem. Soc.* 108, 1379, 1986.
37. Li, J., Li, X., Zhai, H. J., and Wang, L.-S., *Science* 299, 864, 2003; Hakkinen, H., Yoon, B., Landman, U., Li, X., Zhai, H. J., and Wang, L.-S., *J. Phys. Chem.* 107, 6168, 2003; Taylor, K. J., Pettiette-Hall, C. L., Cheshnovsky, O., and Smalley, R. E., *J. Chem. Phys.* 96, 3319, 1992; Handschuh, H., Cha, C.-Y., Bechthold, P. S., Ganteför, G., and Eberhardt, W., J., *Chem. Phys.* 102, 6406, 1995; Cha, C.-Y., Ganteför, G., and Eberhardt, W., J., *Chem. Phys.* 99, 6308, 1993; Ho, J., Ervin, K. M., and Lineberger, W. C., J., *Chem. Phys.* 93, 6987, 1990; Leopold, D. G., Ho, J., and Lineberger, W. C., *J. Chem. Phys.* 86, 1715, 1987; Pettiette, C. L., Yang, S. H., Craycraft, M. J., Conceicao, J., Laaksonen, R. T., Cheshnovsky, O., and Smalley, R. E., *J. Chem. Phys.* 88, 5377, 1988.
38. Snodgrass, J. T., Coe, J. V., McHugh, K. M., Friedhoff, C. B., and Bowen, K. H., *J. Phys. Chem.* 93, 1249, 1989.
39. Friedhoff, C. B., Snodgrass, J. T., Coe, J. V., McHugh, K. M., and Bowen, K. H., *J. Chem. Phys.* 84, 1051, 1986.
40. Friedhoff, C. B., Coe, J. V., Snodgrass, J. T., McHugh, K. M., and Bowen, K. H., *Chem. Phys. Lett.* 124, 268, 1986.

41. Coe, J. V., Snodgrass, J. T., Friedhoff, C. B., McHugh, K. M., and Bowen, K. H., *J. Chem. Phys.* 84, 619, 1986.

42. Snodgrass, J. T., Coe, J. V., Friedhoff, C. B., McHugh, K. M., and Bowen, K. H., *Chem. Phys. Lett.* 122, 352, 1985.

43. Nimlos, M. R., and Ellison, G. B., *J. Am. Chem. Soc.* 108, 6522, 1986.

44. Petrunin, V., Andersen, H., Balling, P., and Andersen, T., *Phys. Rev. Lett.* 76, 744, 1996.

45. Andersen, T., Lykke, K. R., Neumark, D. M., and Lineberger, W. C., *J. Chem. Phys.* 86, 1858, 1987.

46. Murray, K. K., Lykke, K. R., and Lineberger, W. C., *Phys. Rev. A* 36, 699, 1987.

47. Chaibi, W., Delsart, C., Drag, C., and Blondel, C., *J. Mol. Spectrosc.* 239, 11, 2006. EA(^{32}SH) = 18669.543(12) cm^{-1}.

48. Stonemann, R. C., and Larson, D. J., *Phys. Rev. A* 35, 2928, 1987. EA(SeH) = 17,845.17 ± 0.20 cm^{-1}.

49. Nimlos, M. R., Harding, L. B., and Ellison, G. B., *J. Chem. Phys.* 87, 5116, 1987.

50. Moylan, C. R., Dodd, J. A., Han, C.-C., and Braumann, J. I., *J. Chem. Phys.* 86, 5350, 1987.

51. Chowdhury, S., Grimsrud, E. P., Heinis, T., and Kebarle, P., *J. Am. Chem. Soc.* 108, 3630, 1986.

52. Berzinsh, U., Gustafsson, M., Hanstorp, D., Klinkmueller, A. E., Ljungblad, and U., Maartensson-Pendrill, A.-M., *Phys. Rev. A* 51, 231, 1995. EA(Cl) = 29138.59 ± 0.22 cm^{-1}.

53. Moran, S., and Ellison, G. B., *J. Phys. Chem.* 92, 1794, 1988.

54. Murray, K. K., Leopold, D. G., Miller, T. M., and Lineberger, W. C., *J. Chem. Phys.* 89, 5442, 1988.

55. Ervin, K. M., Ho, J., and Lineberger, W. C., *J. Chem. Phys.* 89, 4514, 1988.

56. Grimsrud, E. P., Chowdhury, S., and Kebarle, P., *J. Chem. Phys.* 83, 1059, 1985.

57. Fischer, C. F., *Phys. Rev. A* 39, 963, 1989.

58. Wickham-Jones, C. T., Ervin, K. M., Ellision, G. B., and Lineberger, W. C., *J. Chem. Phys.* 91, 2762, 1989.

59. Kryachko, E. S., Vinckier, C., and Nguyen, M. T., *J. Chem. Phys.* 114, 7911, 2001.

60. Mock, R. S., and Grimsrud, E. P., *J. Am. Chem. Soc.* 111, 2861, 1989.

61. Chowdhury, S., Heinis, T., Grimsrud, E. P., and Kebarle, P., *J. Phys. Chem.* 90, 2747, 1986. The uncertainty and other results are quoted in Ref. 60.

62. Wickham-Jones, C. T., Moran, S., and Ellison, G. B., *J. Chem. Phys.* 90, 795, 1989.

63. Ervin, K. M., Ho, J., and Lineberger, W. C., *J. Phys. Chem.* 92, 5405, 1988.

64. Miller, T. M., and Lineberger, W. C., *Int. J. Mass Spectrom. Ion Process.* 102, 239, 1990.

65. Wetzel, D. M., Salomon, K. E., Berger, S., and Brauman, J. I., *J. Am. Chem. Soc.* 111, 3835, 1989.

66. Crocker, L., Wang, T., and Kebarle, P., *J. Am. Chem. Soc.* 115, 7818, 1993.

67. Dillow, G. W., and Kebarle, P., *J. Am. Chem. Soc.* 111, 5592, 1989.

68. Gantefor, G., Gausa, M., Meiwes-Broer, K. H., and Lutz, H. O., *Z. Phys. D* 9, 253, 1988; Taylor, K. J., Petteitte, C. L., Craycraft, M. J., Chesnovsky, O., and Smalley, R. E., *Chem. Phys. Lett.* 152, 347, 1988.

69. Metz, R. B., Kitsopoulos, T., Weaver, A., and Neumark, D. M., *J. Chem. Phys.* 88, 1463, 1988.

70. Yang, S., Pettiette, C. L., Conceicao, J., Cheshnovsky, O., and Smalley, R. E., *Chem. Phys. Lett.* 139, 233, 1987; Yang, S., Taylor, K. J., Craycraft, M. J., Conceicao, J., Pettiette, C. L., Cheshnovsky, O., and Smalley, R. E., *Chem. Phys. Lett.* 144, 431, 1988; Arnold, D. W., Bradforth, S. E., Kitsopoulos, T. N., and Neumark, D. M., *J. Chem. Phys.* 95, 5479, 1991.

71. Cheshnovsky, O., Yang, S., Pettiette, C. L., Craycraft, M. J., Liu, Y., and Smalley, R. E., *Chem. Phys. Lett.* 138, 119, 1987.

72. Liu, Y., Zhang, Q.-L., Tittel, F. K., Curl, R. F., and Smalley, R. E., *J. Chem. Phys.* 85, 7434, 1986.

73. Travers, M. J., Cowles, D. C., and Ellison, G. B., *Chem. Phys. Lett.* 164, 449, 1989.

74. Blondel, C., Cacciani, P., Delsart, C., and Trainham, R., *Phys. Rev. A* 40, 3698, 1989. EA(Br) = 27,129.170 ± 0.015 cm^{-1} and EA(F) = 27,432.440 ± 0.025 cm^{-1}.

75. Bowen, K. H., and Eaton, J. G., in *The Structure of Small Molecules and Ions*, Naaman, R., and Vager, Z., Eds., Plenum, New York, 1988, pp. 147-169; Arnold, S. T., Eaton, J. G., Patel-Mistra, D., Sarkas, H. W., and Bowen, K. H., in *Ion and Cluster Ion Spectroscopy and Structure*, Maier, J. P., Ed., Elsevier Science, New York, 1989, p. 417.

76. Snodgrass, J. T., Coe, J. V., Friedhoff, C. B., McHugh, K. M., and Bowen, K. H., *Faraday Disc. Chem. Soc.* 88, 1988.

77. Lee, G. H., Arnold, S. T., Eaton, J. G., Sarkas, H. W., Bowen, K. H., Ludewigt, C., and Haberland, H., *Z. Phys. D – At., Mol. Clusters* 20, 9, 1991; Coe, J. V., Lee, G. H., Eaton, J. G., Arnold, S. T., Sarkas, H. W., Bowen, K. H., Ludewigt, C., Haberland, H., and Worsnop, D. R., *J. Chem. Phys.* 92, 3980, 1990.

78. Snodgrass, J. T., Coe, J. V., Freidhoff, C. B., McHugh, K. M., Arnold, S. T., and Bowen, K. H., *J. Phys. Chem.* **99**, 9675, 1995.

79. Hendricks, J. H., de Clercq, H. L., Freidhoff, C. B., Arnold, S. T., Eaton, J. G., Fancher, C., Lyapustina, S. A., Snodgrass, J. T., and Bowen, K. H., *J. Chem. Phys.* **116**, 7926, 2002.

80. Snodgrass, J. T., Coe, J. V., Friedhoff, C. B., McHugh, K. M., and Bowen, K. H., *J. Chem. Phys.* 88, 8014, 1988.

81. Coe, J. V., Snodgrass, J. T., Friedhoff, C. B., McHugh, K. M., and Bowen, K. H., *Chem. Phys. Lett.* 124, 274, 1986.

82. Eliav, E., Vilkas, M. J., Ishikawa, Y., and Kaldor, U., *J. Chem. Phys.* 123, 224113(5), 2005.

83. Ervin, K. M., Ho, J., and Lineberger, W. C., *J. Chem. Phys.* 91, 5974, 1989.

84. Polak, M. L., Fiala, B. L., Ervin, K. M., and Lineberger, W. C., *J. Chem. Phys.* 94, 6924, 1991.

85. Weaver, A., Arnold, D. W., Bradforth, S. E., Neumark, D. M., *J. Chem. Phys.* 94, 1740, 1991.

86. Zhai, H.-J., Kirian, B., Cui, L.-F., Li, X., Dixon, D. A., and Wang, L.-S, *J. Am. Chem. Soc.* 126, 16134, 2004.

87. Ervin, K. M., and Lineberger, W. C., *J. Phys. Chem.* 95, 1167, 1991.

88. Gilles, M. K., Polak, M. L., and Lineberger, W. C., *J. Chem. Phys.* 96, 8012, 1992.

89. Lykke, K. R., Murray, K. K., and Lineberger, W. C., *Phys. Rev. A* 43, 6104, 1991. EA(H) = 6082.99 ± 0.15 cm^{-1} and EA(D) = 6086.2 ± 0.6 cm^{-1}.

90. Ervin, K. M., Gronert, S., Barlow, S. E., Gilles, M. K., Harrison, A. G., Bierbaum, V. M., DePuy, C. H., Lineberger, W. C., and Ellison, G. B., *J. Am. Chem. Soc.* 112, 5750, 1990.

91. Viggiano, A. A., Miller, T. M., Miller, A. E. S., Morris, R. A., Van Doren, J. M., and Paulson, J. F., *Int. J. Mass Spectrom. Ion Process.* 109, 327, 1991.

92. Peláez, R. J., Blondel, C., Delsart, C., and Drag, C., *J. Phys. B: At. Mol. Opt. Phys.* 42, 125001, 2009. EA(I) = 24,672.874 ± 0.029 cm^{-1}.

93. Polak, M. L., Gilles, M. K., and Lineberger, W. C., *J. Chem. Phys.* 96, 7191, 1992.

94. Bengali, A. A., Casey, S. M., Cheng, C.-L., Dick, J. P., Fenn, P. T., Villalta, P. W., and Leopold, D. G., *J. Am. Chem. Soc.* 114, 5257, 1992.

95. Gilles, M. K., Ervin, K. M., Ho, J., and Lineberger, W. C., *J. Phys. Chem.* 96, 1130, 1992.

96. Travers, M. J., Cowles, D. C., Clifford, E. P., and Ellison, G. B., *J. Am. Chem. Soc.* 114, 8699, 1992.

97. Bengali, A. A., and Leopold, D. G., *J. Am. Chem. Soc.* 114, 9192, 1992.

98. Rudnyi, E. B., Kaibicheva, E. A., and Sidorov, L. N., *Rapid Commun. Mass Spectrom.* 6, 356, 1992; Sidorov, L. N., *High Temp. Sci.* 29, 153, 1990. See also Srivastava, R. D., Uy, O. M., and Farber, M., *Trans. Faraday Soc.* 67, 2941, 1971.

99. Kitsopoulos, T. N., Chick, C. J., Zhao, Y., and Neumark, D. M., *J. Chem. Phys.* 95, 5479, 1991.

100. Arnold, C. C., Kitsopoulos, T. N., and Neumark, D. M., *J. Chem. Phys.* 99, 766, 1993.

101. Rudnyi, E. B., Kaibicheva, E. A., and Sidorov, L. N., *J. Chem. Thermodyn.* 25, 929, 1993.

102. Sarkas, H. W., Hendricks, J. H., Arnold, S. T., and Bowen, K. H., *J. Chem. Phys.* 100, 1884, 1994.

103. Villalta, P. W., and Leopold, D. G., *J. Chem. Phys.* 98, 7730, 1993.

104. Eaton, J. G., Sarkas, H. W., Arnold, S. T., McHugh, K. M., and Bowen, K. H., *Chem. Phys. Lett.* 193, 141, 1992. See also Ref. 18.

105. Polak, M. L., Gilles, M. K., Gunion, R. F., and Lineberger, W. C., *Chem. Phys. Lett.* 210, 55, 1993.

106. Gilles, M. K., Lineberger, W. C., and Ervin, K. M., *J. Am. Chem. Soc.* 115, 1031, 1993.

107. Casey, S. M., and Leopold, D. G., *Chem. Phys. Lett.* 201, 205, 1993.

108. Polak, M. L., Gerber, G., Ho, J., and Lineberger, W. C., *J. Chem. Phys.* 97, 8990, 1992.

109. Garand, E., Klein, K., Stanton, J. F., Zhou, J., Yacovitch, T. I., and Neumark, D. M., *J. Phys. Chem. A* 114, 1374, 2010.

110. Xu, C., Taylor, T. R., Burton, G. R., and Neumark, D. M., *J. Chem. Phys.* 108, 1395, 1998.

111. Bradforth, S. E., Kim, E. H., Arnold, D. W., and Neumark, D. M., *J. Chem. Phys.* 98, 800, 1993.

112. Ho, J., Polak, M. L., Ervin, K. M., and Lineberger, W. C., *J. Chem. Phys.* 99, 8542, 1993.

113. Brinkman, E. A., Berger, S., Marks, J., and Brauman, J. I., *J. Chem. Phys.* 99, 7586, 1993.

114. Casey, S. M., and Leopold, D. G., *J. Phys. Chem.* 97, 816, 1993.

115. McDonald, R. N., and Davidson, S. J., *J. Am. Chem. Soc.* 115, 10857, 1993.

116. Ho, J., Ervin, K. M., Polak, M. L., Gilles, M. K., and Lineberger, W. C., *J. Chem. Phys.* 95, 4845, 1991.

117. Ho, J., Polak, M. L., and Lineberger, W. C., *J. Chem. Phys.* 96, 144, 1992. See also Ref. 289.

118. Polak, M. L., Gilles, M. K., Ho, J., and Lineberger, W. C., *J. Phys. Chem.* 95, 3460, 1991.

119. Polak, M. L., Ho, J., Gerber, G., and Lineberger, W. C., *J. Chem. Phys.* 95, 3053, 1991. See also Ref. 363 confirmation of EA(Bi$_3$).

120. Sharpe, P., and Kebarle, P., *J. Am. Chem. Soc.* 115, 782, 1993.

121. Miller, T. M., Miller, A. E. S., Viggiano, A. A., Morris, R. A., and Paulson, J. F., *J. Chem. Phys.* 100, 7200, 1994. Accurate calculations have yielded a higher result (0.90 eV); see Lau, J. K.-C. and Li, W.-K., *J. Mol. Struct.* (*Theochem*) 578, 221, 2002.

122. Berkovits, D., Boaretto, E., Gehlberg, S., Heber, O., and Paul, M., *Phys. Rev. Lett.* 75, 414, 1995.

123. Arnold, C. C., Xu, C., Burton, G. R., and Neumark, D. M., *J. Chem. Phys.* 102, 6982, 1995; Burton, G. R., Xu, C., Arnold, C. C., and Neumark, D. M., *J. Chem. Phys.* 104, 2757 1996.

124. Xu, C., de Beer, E., and Neumark, D. M., *J. Chem. Phys.* 104, 2749, 1996.

125. Gunion, R. F., Koppel, H., Leach, G. W., and Lineberger, W. C., *J. Chem. Phys.* 103, 1250, 1995.

126. Gunion, R. F., and Lineberger, W. C., *J. Phys. Chem.* 100, 4395, 1996.

127. Gunion, R. F., Dixon-Warren, St. J., and Lineberger, W. C., *J. Chem. Phys.* 104, 1765, 1996.

128. Dixon-Warren, St. J., Gunion, R. F., and Lineberger, W. C., *J. Chem. Phys.* 104, 4902, 1996.

129. Nakajima, A., Zhang, N., Kawamata, H., Hayase, T., Nakao, K., and Kaya, K., *Chem. Phys. Lett.* 241, 295, 1995; Nakajima, A., Taguwa, T., Nakao, K., Hoshino, K., Iwata, S., and Kaya, K., *J. Chem. Phys.* 102, 660, 1995.

130. Fan, J., and Wang, L.-S., *J. Chem. Phys.* 102, 8714, 1995.

131. Arnold, D. W., Bradforth, S. E., Kim, E. H., and Neumark, D. M., *J. Chem. Phys.* 102, 3493, 1995; Zhao, Y., Arnold, C. C., and Neumark, D. M., *J. Chem. Soc. Faraday Trans.* 2 89, 1449, 1992.

132. Fan, J., Lou, L., and Wang, L.-S., *J. Chem. Phys.* 102, 2701, 1995.

133. Sarkas, H. W., Hendricks, J. H., Arnold, S. T., Slager, V. L., and Bowen, K. H., *J. Chem. Phys.* 100, 3358, 1994.

134. Kato, S., Lee, H. S., Gareyev, R., Wenthold, P. G., Lineberger, W. C., DePuy, C. H., and Bierbaum, V. M., *J. Am. Chem. Soc.* 119, 7863, 1997. See also Miller, T. M., Viggiano, A. A., and Miller, A. E. S., *J. Phys. Chem. A* 106, 10200, 2002.

135. Wenthold, P. G., Hu, J., Squires, R. R., and Lineberger, W. C., *J. Am. Chem. Soc.* 118, 475, 1996.

136. Gunion, R. F., Karney, W., Wenthold, P. G., Borden, W. T., and Lineberger, W. C., *J. Am. Chem. Soc.* 118, 5074, 1996. The numbers in the abstract for 1,6-heptadiyne were misprinted. EA(cycloheptatrienide) quoted here derives from the LPES data combined with other thermochemical data in Ref. 136.

137. Xu, C., de Beer, E., Arnold, D. W., Arnold, C. C., and Neumark, D. M., *J. Chem. Phys.* 101, 5406, 1996.

138. Wenthold, P. G., Polak, M. L., and Lineberger, W. C., *J. Phys. Chem.* 100, 6920, 1996.

139. Wenthold, P. G., Kim, J. B., and Lineberger, W. C., *J. Am. Chem. Soc.* 119, 1354, 1997.

140. Eliav, E., Kaldor, U., Ishikawa, Y., and Pyykko, P., *Phys. Rev. Lett.* 77, 5350, 1996.

141. Davies, B. J., Ingram, C. W., Larson, D. J., and Ljungblad, U., *J. Chem. Phys.* 106, 5783, 1997. EA(Ir) = 12,613 ± 4 cm^{-1}.

142. Smith, J. R., Kim, J. B., and Lineberger, W. C., *Phys. Rev. A* 55, 2036, 1997. EA(OH) = 14,741.02 ± 0.03 cm^{-1}. Schulz, P. A., Mead, R. D., Jones, P. L., and Lineberger, W. C., *J. Chem. Phys.* 77, 1153, 1982. EA(OD) = 14,723.92 ± 0.30 cm^{-1}. See also Rudmin, J. D., Ratliff, L. P., Yukich, J. N., and Larson, D. J., *J. Phys. B: At. Mol. Opt. Phys.* 29, L881, 1996.

143. Desai, S. R., Wu, H., Rohlfing, C. M., and Wang, L.-S., *Int. J. Chem. Phys.* 106, 1309, 1997.

144. Wenthold, P. G., Jonas, K.-L., and Lineberger, W. C., *J. Chem. Phys.* 106, 9961, 1997.

145. Yang, J., Wang, X.-B., Xing, X.-P., and Wang, L.-S., *J. Chem. Phys.* 128, 201102, 2008.

146. Moravec, V. D., and Jarrold, C. C., *J. Chem. Phys.* 108, 1804, 1998.

147. Wang, X.-B., Ding, C.-F., and Wang, L.-S., *J. Phys. Chem. A* 101, 7699, 1997.

148. Wenthold, P. G., and Lineberger, W. C., *J. Am. Chem. Soc.* 19, 7772, 1997.

149. Wang, L.-S., Li, X., and Zhang, H.-F., *Chem. Phys.* 262, 53, 2000; Wang, L.-S., Cheng, H.-S., and Fan, J., *J. Chem. Phys.* 102, 9480, 1995.

150. Wu, H., Desai, S. R., and Wang, L.-S., *Phys. Rev. Lett.* 77, 2436, 1996.

151. Liu, S.-R., Zhai, H.-J., Castro, M., and Wang, L.-S., *J. Chem. Phys.* 118, 2108, 2003.

152. Gutsev, G. L., Bauschlicher, C. W., Zhai, H.-J., and Wang, L.-S., *J. Chem. Phys.* 119, 11135, 2003; Wang, L.-S., Wu, H., and Desai, S. R., *Phys. Rev. Lett.* 76, 4853, 1996.

153. Robinson, M. S., Polak, M. L., Bierbaum, V. M., DePuy, C. H., and Lineberger, W. C., *J. Am. Chem. Soc.* 117, 6766, 1995.

154. Clifford, E. P., Wenthold, P. G., Lineberger, W. C., Petersson, G. A., and Ellison, G. B., *J. Phys. Chem.* 101, 4338, 1997.

155. Seburg, R. A., and Squires, R. R., *Int. J. Mass Spectrom. Ion Process.* 167/168, 541, 1997.

156. Wang, X.-B., and Wang, L.-S., *Chem. Phys. Lett.* 313, 179, 1999.

157. Marcy, T. P., and Leopold, D. G., *Int. J. Mass Spectrom.* 195/196, 653, 2000.

158. Gutsev, G. L., Rao, B. K., Jena, P., Li, X., and Wang, L.-S., *J. Chem. Phys.* 113, 1473, 2000.

159. Negishi, Y., Yasuike, T., Hayakawa, F., Kizawa, M., Yabushita, S., and Nakajima, A., *J. Chem. Phys.* 113, 1725, 2000.

160. Klopcic, S. A., Moravec, V. D., and Jarrold, C. C., *J. Chem. Phys.* 110, 8986, 1999.

161. Boldyrev, A. I., Simons, J., Li, X., Chen, W., and Wang, L.-S., *J. Chem. Phys.* 110, 8980, 1999.

162. Taylor, T. R., Asmis, K. R., Zanni, M. T., and Neumark, D. M., *J. Chem. Phys.* 110, 7607, 1999.

163. Boldyrev, A., Li, X., and Wang, L.-S., *J. Chem. Phys.* 112, 3627, 2000.

164. Defrançois, C., Périquet, V., Lyapustina, S. A., Lippa, T. P., Robinson, D. W., Bowen, K. H., Nonaka, H., and Compton, R. N., *J. Chem. Phys.* 111, 4569, 1999.

165. Dobrin, S., Boo, B. H., Alconcel, L. S., and Continetti, R. E., *J. Phys. Chem. A* 104, 10695, 2000.

166. Schwartz, R. L., Davico, G. E., and Lineberger, W. C., *J. Electron Spectrosc. Relat. Phenom.* 108, 163, 2000.

167. Kim, J. B., Wenthold, P. G., and Lineberger, W. C., *J. Phys. Chem.* 103, 10833, 1999.

168. Davico, G. E., Ramond, T. M., and Lineberger, W. C., *J. Chem. Phys.* 113, 8852, 2000.

169. Green, S. M. E., Alex, S., Fleischer, N. L., Millam, E. L., Marcy, T. P., and Leopold, D. G., *J. Chem. Phys.* 114, 2653, 2001.

170. Wu, H., and Wang, L.-S., *J. Chem. Phys.* 108, 5310, 1998.

171. Wu, H., and Wang, L.-S., *J. Phys. Chem. A* 102, 9129, 1998.

172. Wu, H., and Wang, L.-S., *J. Chem. Phys.* 107, 8221, 1997.

173. Thomas, O. C., Xu, S. J., Lippa, T. P., and Bowen, K. H., *J. Clust. Sci.* 10, 525, 1999.

174. Wang, L.-S., private communication quoted in Ref. 169.

175. Marcy, T. P., Ph.D. dissertation, quoted in Ref. 169.

176. Alex, S., Green, M. E., and Leopold, D. G., unpublished, quoted in Ref. 169.

177. Wang, X.-B., Wang, L.-S., Brown, R., Schwerdtfeger, P., Schröder, D., and Schwarz, H., *J. Chem. Phys.* 114, 7388, 2001.

178. Kim, J. H., Li, X., Wang, L.-S., de Clercq, H. L., Fancher, C. A., Thomas, O. C., and Bowen, K. H., *J. Phys. Chem. A* 105, 5709, 2001.

179. Moravec, V. D., Klopcic, S. A., Chatterjee, B., and Jarrold, C. C., *Chem. Phys. Lett.* 341, 313, 2001.

180. Zengin, V., Persson, B. J., Strong, K. M., and Continetti, R. E., *J. Chem. Phys.* 105, 9740, 1996.

181. Kietzmann, H., Morenzin, J., Bechthold, P. S., Ganteför, G., and Eberhardt, W., *J. Chem. Phys.* 109, 2275, 1998.

182. Xu, C., Taylor, T. R., Burton, G. R., and Neumark, D. M., *J. Chem. Phys.* 108, 7645, 1998.

183. Williams, W. W., Carpenter, D. L., Covington, A. M., Koepnick, M. C., Calabrese, D., and Thompson, J. S., *J. Phys. B: At. Mol. Opt. Phys.* 31, L341, 1998.

184. Covington, A. M., Calabrese, D., Thompson, J. S., and Kvale, T. J., *J. Phys. B: At. Mol. Opt. Phys.* 31, L855, 1998. High-level caculations imply EA(La) ≈ 0.545 eV; see O'Malley, S. M., and Beck, D. R., *Phys. Rev. A* 79, 012511, 2009.

185. Haeffler, G., Hanstrorp, D., Kiyan, I., Klinkmueller, A. E., Ljungblad, U., and Pegg, D. J., *Phys. Rev. A* 53, 4127, 1996. EA(Li) = 4984.90 ± 0.17 cm^{-1}.

186. Chen, H. L., Ellis, Jr., P. E., Wijesekera, T., Hagan, T. E., Groh, S. E., Lyons, J. E., and Ridge, D. P., *J. Am. Chem. Soc.* 116, 1086, 1994.

187. Deyerl, H.-J., Alconcel, L. S., and Continetti, R. E., *J. Phys. Chem. A* 105, 552, 2001.

188. Blanksby, S. J., Ramond, T. M., Davico, G. E., Nimlos, M. R., Kato, S., Bierbaum, V. M., Lineberger, W. C., Ellison, G. B., and Okumura, M., *J. Am. Chem. Soc.* 123, 9585, 2001.

189. Asmis, K. R., Taylor, T. R., Xu, C., and Neumark, D. M., *Chem. Phys. Lett.* 295, 75, 1998.

190. Schäfer-Bung, B., Engels, B., Taylor, T. R., Neumark, D. M., Botschwina, P., and Peric, M., *J. Chem. Phys.* 115, 1777, 2001.

191. Scheer, M., Bilodeau, R. C., and Haugen, H. K., *Phys. Rev. Lett.* 80, 2562, 1998.

192. Taylor, T. R., Gómez, H., Asmis, K. R., and Neumark, D. M., *J. Chem. Phys.* 115, 4620, 2001.

193. Asmis, K. R., Taylor, T. R., and Neumark, D. M., *J. Chem. Phys.* 111, 8838, 1999 and 111, 10491, 1999.

194. Ramond, T. M., Davico, G. E., Schwartz, R. L., and Lineberger, W. C., *J. Chem. Phys.* 112, 1158, 2000.

195. Petrunin, V. V., Voldstad, J. D., Balling, P., Kristensen, P., Andersen, T., and Haugen, H. K., *Phys. Rev. Lett.* 75, 1911, 1995.

196. Dzuba, V. A., and Gribakin, G. F., *J. Phys. B: At. Mol. Opt. Phys.* 31, L483, 1998.

197. Ervin, K. M., Ramond, T. M., Davico, G. E., G. E., Schwartz, R. L., Casey, S. M., and Lineberger, W. C., *J. Phys. Chem.* 105, 10822, 2001.

198. Lippa, T. P., Xu, S.-J., Lyapustina, S. A., and Bowen, K. H., *J. Chem. Phys.* 109, 9263, 1998.

199. Turner, N. J., Martel, A. A., and Waller, I. M., *J. Phys. Chem.* 98, 474, 1994.

200. Lippa, T. P., Xu, S.-J., Lyapustina, S. A., Nilles, J. M., and Bowen, K. H., *J. Chem. Phys.* 109, 10727, 1998.

201. Wang, X.-B., Woo, H.-K., Wang, L.-S., *J. Chem. Phys.* 123, 051106, 2005.

202. Boltalina, O. V., Sidorov, L. N., Sukhanova, E. V., and Sorokin, I. D., *Chem. Phys. Lett.* 230, 567, 1994.

203. Clifford, E. P., Wenthold, P. G., Lineberger, W. C., Ellison, G. B., Wang, C. X., Grabowski, J. J., Vila, F., and Jordan, K. D., *J. Chem. Soc. Perkin Trans.* 2, 1015, 1998.

204. Spanel, P., Matejcik, S., and Smith, D., *J. Phys. B: At. Mol. Phys.* 28, 2941, 1995. See Miller, A. E. S., Miller, T. M., Viggiano, A. A., Morris, R. A., Van Doren, J. M., Arnold, S. T., and Paulson, J. F., *J. Chem. Phys.* 102, 8865, 1995 for interpretation in terms of EA(SF$_5$).

205. Kinghom, D. B., and Adamowicz, L., J., *Chem. Phys.* 106, 4589, 1997. EA(H) = 6083.0994 cm^{-1}, EA(D) = 6086.7137 cm^{-1}, and EA(T) = 6087.9168 cm^{-1}.

206. Xi, L., and Wang, L.-S., *J. Chem. Phys.* 109, 5264, 1998.

207. Ephraim, E., Shmulyian, S., Kaldor, U., and Isikawa, Y., *J. Chem. Phys.* 109, 3954, 1998. Also EA(La) = 0.35 eV.

208. Scheer, M., Bilodeau, R. C., Thogersen, J., and Haugen, H. K., *Phys. Rev. A* 57, R1493, 1998.

209. Kim, J. B., Wenthold, P. G., and Lineberger, W. C., *J. Chem. Phys.* 108, 830, 1998.

210. Kishi, R., Kawamata, H., Negishi, Y., Iwata, S., Nakajima, A., and Kaya, K., *J. Chem. Phys.* 107, 10029, 1997.

211. Adams, C. L., Schneider, H., Ervin, K. M. and Weber, J. M., *J. Chem. Phys.* 130, 074307, 2009. Earlier work by Compton, R. N., Carman, Jr., H. S., Desfrançois, C., Abdoul-Carmine, J., Schermann, J. P., Hendricks, J. H., Lyapustina, S. A., and Bowen, K. H., *J. Chem. Phys.* 105, 3472, 1996, found EA(CD$_3$NO$_2$) to be 0.02 eV lower than EA(CH$_3$NO$_2$). High-level calculations yield 0.172 ± 0.006 eV; see Bull, J. N., Maclagan, R. G. A. R., and Harland, P. W., *J. Phys. Chem. A* 114, 3622, 2010.

212. Yourshaw, I., Zhao, Y., and Neumark, D. M., *J. Chem. Phys.* 105, 351, 1996.

213. Gausa, M., Kaschner, R., Seifert, G., Faehrmann, J. H., Lutz, H. O., and Meiwes-Broer, K., *J. Chem. Phys.* 104, 9719, 1996.

214. Wu, H., and Wang, L.-S., *J. Chem. Phys.* 107, 16, 1997.

215. Radisic, D., Xu, S., and Bowen, K. H., *Chem. Phys. Lett.* 354, 9, 2002.

216. Pramann, A., and Rademann, K., *Chem. Phys. Lett.* 343, 99, 2001.

217. Gómez, H., Taylor, T. R., and Neumark, D. M., *J. Phys. Chem. A* 105, 6886, 2001.

218. Gómez, H., Taylor, T. R., Zhao, Y., and Neumark, D. M., *J. Chem. Phys.* 117, 8644, 2002.

219. Nimlos, M. R., Davico, G., Geise, C. M., Wenthold, P. G., Lineberger, W. C., Blanksby, S. J., Hadad, C. M., Petersson, G. A., and Ellison, G. B., *J. Chem. Phys.* 117, 4323, 2002.

220. Koyasu, K., Mitsui, M., Nakajima, A., and Kaya, K., *Chem. Phys. Lett.* 358, 224, 2002.

221. Zhai, H.-J., Wang, L.-S., Alexandrova, A. N., Boldyrev, A. I., and Zakrzewski, V. G., *J. Phys. Chem. A* 107, 9319, 2003.

222. Schiedt, J., and Weinkauf, R., *Z. Naturforsch. A* 50, 1041, 1995. See also Ervin, K. M., Anusiewicz, I., Skurski, P., Simons, J., and Lineberger, W. C., *J. Phys. Chem. A* 107, 8521, 2003 [EA(O$_2$) = 0.448 ± 0.006 eV].

223. Davis, V. T., and Thompson, J. S., *J. Phys. B* 34, L433, 2001. Recent high-level calculations give EA(Lu) = 0.029 eV; see Ref. 343.

224. Li, X., Zhai, H.-J., and Wang, L.-S., *Chem. Phys. Lett.* 357, 415, 2002.

225. Davis, V. T., and Thompson, J. S., *J. Phys. B: At. Mol. Opt. Phys.* 35, L11, 2002. High-level caculations imply EA(Pr) ≈ 0.177 eV; see O'Malley, S. M., and Beck, D. R., *Phys. Rev. A* 78, 012510, 2008.

226. Goldfarb, F., Drag, C., Chaibi, W., Kröger, S., Blondel, C., and Delsart, C., *J. Chem. Phys.* 122, 014308, 2005. EA(OH) = 14740.982(7) cm^{-1}.

227. Blondel, C., Delsart, C., and Goldfarb, F., *J. Phys. B* 34, L281, 2001. EA(F) = 27432.446(19) cm^{-1}.

228. Fancher, C. A., de Clercq, H. L., and Bowen, K. H., *Chem. Phys. Lett.* 366, 197, 2002.

229. Jin, C., Taylor, K. J., Conceicao, J., and Smalley, R. E., *Chem. Phys. Lett.* 175, 17, 1990.

230. Schiedt, J., Knott, W. J., Le Barbu, K., Schlag, E. W., and Weinkauf, R., *J. Chem. Phys.* 113, 9470, 2000.

231. Song, J. K., Lee, N. K., Kim, J. H., Han, S. Y., and Kim, S. K., *J. Chem. Phys.* 119, 3071, 2003.

232. Davico, G. E., Schwartz, R. L., Ramond, T. M., and Lineberger, W. C., *J. Amer. Chem. Soc.* 121, 6047, 1999.

233. Andrews, D. H., Gianola, A. J., and Lineberger, W. C., *J. Chem. Phys.* 117, 4074, 2002.

234. Schwartz, R. L., Davico, G. E., Kim, J. B., and Lineberger, W. C., *J. Chem. Phys.* 112, 4966, 2000.

235. Schwartz, R. L., Davico, G. E., Ramond, T. M., and Lineberger, W. C., *J. Phys. Chem. A* 103, 8213, 1999.

236. Schiedt, J., and Weinkauf, R., *Chem. Phys. Lett.* 274, 18, 1997.

237. Goebbert, D. J., Pichugin, K., Khuseynov, D., Wenthold, P. G., and Sanov, A., *J. Chem. Phys.* 132, 224301, 2010.

238. Geske, G. D., Boldyrev, A. I., Li, X., and Wang, L.-S., *J. Chem. Phys.* 113, 5130, 2000.

239. Cannon, N. A., Boldyrev, A. I., Li, X., and Wang, L.-S., *J. Chem. Phys.* 113, 2671, 2000.

240. Wang, X.-B., Yang, X., and Wang, L.-S., *J. Chem. Phys.* 116, 561, 2002.

241. Gutsev, G. L., Jena, P., Zhai, H.-J., and Wang, L.-S., *J. Chem. Phys.* 115, 7935, 2001.

242. Yang, X., Wang, X.-B., and Wang, L.-S., *J. Chem. Phys.* 115, 2889, 2001.
243. Zhai, H.-J., Liu, S.-R., Li, X., and Wang, L.-S., *J. Chem. Phys.* 115, 5170, 2001.
244. Li, X., Wang, L.-S., Cannon, N. A., and Boldyrev, A. I., *J. Chem. Phys.* 116, 1330, 2002.
245. Zhai, H.-J., Wang, L.-S., Alexandrova, A. N., and Boldyrev, A. I., *J. Chem. Phys.* 117, 7917, 2002.
246. Zhai, H.-J., and Wang, L.-S., *J. Chem. Phys.* 117, 7882, 2002.
247. Liu, S.-R., Zhai, H.-J., and Wang, L.-S., *J. Chem. Phys.* 117, 9758, 2002.
248. Mitsui, M., Nakajima, A., and Kaya, K., *J. Chem. Phys.* 117, 9740, 2002.
249. Yang, X., Wang, X.-B., Wang, L.-S., Niu, S., and Ichiye, T., *J. Chem. Phys.* 119, 8311, 2003.
250. Kim, J., Lee, H. M., Suh, S. B., Majumdar, D., and Kim, K. S., *J. Chem. Phys.* 113, 5259, 2000.
251. Liu, S. R., Zhai, H. J., and Wang, L.-S., *Phys. Rev. B* 64, 153402, 2001.
252. Zhai, H. J., Yang, X., Wang, X. B., Wang, L.-S., Elliott, B., and Boldyrev, A. I., *J. Am. Chem. Soc.* 124, 6742, 2002.
253. Zhai, H. J., Wang, L.-S., Kuznetsov, A. E., and Boldyrev, A. I., *J. Phys. Chem. A* 106, 5600, 2002.
254. Drechsler, G., and Boesl, U., *Int. J. Mass Spectrom.* 228, 1067, 2003.
255. Illenberger, E, Comita, P. B., Brauman, J. I., Fenzlaff, H. P., Heni, M., Heinrich, N., Koch, W., and Fenking, G., *Ber. Bunsen-Ges. Phys. Chem.* 89, 1026, 1985; Jackson, R. L., Pellerite, M. J., and Brauman, J. I., *J. Am. Chem. Soc.* 103, 1802, 1981.
256. Miller, T. M., Friedman, J. F., and Viggiano, A. A., *J. Chem. Phys.* 120, 7024, 2004.
257. Miller, T. M., Van Doren, J. M., and Viggiano, A. A., *Int. J. Mass Spectrom.* 233, 67, 2004.
258. Van Doren, J. M., McSweeney, S. A., Hargus, M. D., Kerr, D. M., Miller, T. M., Arnold, S. T., and Viggiano, A. A., *Int. J. Mass Spectrom.* 228, 541, 2003.
259. Van Doren, J. M., Miller, T. M., and Viggiano, A. A., *J. Chem. Phys.* 123, 114303, 2005.
260. Miller, T. M., and Viggiano, A. A., *Phys. Rev. A* 71, 012702, 2005.
261. Haeffler, G., Klinkmueller, A. E., Rangell, J., Berzinsh, U., and Hanstorp, D., *Z. Phys. D* 38, 211, 1996.
262. Bilodeau, R. C., and Haugen, H. K., *Phys. Rev. A* 64, 024501, 2001.
263. Miller, T. M., Viggiano, A. A., Friedman, J. F., and Van Doren, J. M., *J. Chem. Phys.* 121, 9993, 2004.
264. Davis, V. T., and Thompson, J. S., *Phys. Rev A* 65, 010501, 2001. Theoretical work implies that the measured EA(Tm) was actually for a long-lived excited anion state, and that Tm does not form a stable anion. See O'Malley, S. M., and Beck, D. R., *Phys. Rev. A* 70, 022502, 2004.
265. Gianola, A. J., Ichino, T., Hoenigman, R. L., Kato, S., Bierbaum, V. M., and Lineberger, W. C., *J. Phys. Chem. A* 108, 10326, 2004.
266. Staneke, P. O., Groothuis, G., Ingemann, S., and Nibbering, N. M. M., *Int. J. Mass Spectrom. Ion Process.* 142, 83, 1995. A Gaussian-3 calculation yields EA(CCl₄) = 0.994 eV [Ed.].
267. Meloni, G., Ferguson, M. J., and Neumark, D. M., *Phys. Chem. Chem. Phys.* 5, 4073, 2003.
268. Davis, V. T., and Thompson, J. S., *J. Phys. B: At. Mol. Opt. Phys.* 37, 1961, 2004.
269. Walter, C. W., Gibson, N. D., Janczak, C. M., Starr, K. A., Snedden, A. P., Field III, R. L., and Andersson, P., *Phys. Rev. A* 76, 052702, 2007.
270. Ando, N., Kokubo, S., Mitsui, M., and Nakajima, A., *Chem. Phys. Lett.* 389, 279, 2004.
271. Zhai, H.-J., and Wang, L.-S., *J. Chem. Phys.* 120, 8996, 2004.
272. Surber, E., and Sanov, A., *J. Chem. Phys.* 116, 5921, 2002.
273. Dzuba, V. A., and Gribakin, G. F., *Phys. Rev. A* 55, 2443, 1997.
274. Meloni, G., Sheehan, S. M., Ferguson, M. J., and Neumark, D. M., *J. Phys. Chem. A* 108, 9750, 2004.
275. Zhai, H.-J., Li, J., and Wang, L.-S., *J. Chem. Phys.* 121, 8369, 2004.
276. Wu, X., Qin, Z., Xie, H., Cong, R., Wu, X., Tang, Z., and Fan, H., *J. Chem. Phys.* 133, 044303, 2010.
277. Kim, J., Kelley, J. A., Ayotte, P., Nielsen, S. B., Weddle, G. H., and Johnson, M. A., *J. Am. Soc. Mass Spectrom.* 10, 810, 1999.
278. Misaizu, F., Tsunoyama, H., Yasumura, Y., Ohshimo, K., and Ohno, K., *Chem. Phys. Lett.* 389, 241, 2004. See also Ref. 236 for comments on EA(CS₂).
279. Meloni, G., Sheehan, S. M., and Neumark, D. M., *J. Chem. Phys.* 122, 074317, 2005.
280. Yoder, B. L., Maze, J. T., Raghavachari, K., and Jarrold, C. C., *J. Chem. Phys.* 122, 094313, 2005.
281. Ervin, K. M., and Lineberger, W. C., *J. Chem. Phys.* 122, 194303, 2005.
282. Ichino, T., Gianola, A. J., Andrews, D. H., and Lineberger, W. C., *J. Phys. Chem. A* 108, 11307, 2004. See also Ref. 373.
283. Das, U., Raghavachari, K., and Jarrold, C. C., *J. Chem. Phys.* 122, 014313, 2005.
284. Schiedt, J., and Weinkauf, R., *J. Chem. Phys.* 110, 304, 1999.
285. Le Barbu, K., Schiedt, J., Weinkauf, R., Schlag, E. W., Nilles, J. M., Xu, S.-J., Thomas, O. C., and Bowen, J. H., *J. Chem. Phys.* 116, 9663, 2002. Uncertainties not stated.
286. Schiedt, J., and Weinkauf, R., *Chem. Phys. Lett.* 266, 201, 1997. The uncertainty for EA(anthracene) quoted as ±0.008 eV in a later paper (Ref. 230).
287. Chatterjee, B., Akin, F. A., Jarrold, C. C., and Raghavachari, K., *J. Phys. Chem. A* 109, 6880, 2005.
288. Zhai, H.-J., Kiran, B., Dai, B., Li, J., and Wang, L.-S., *J. Am. Chem. Soc.* 127, 12098, 2005.
289. Moravec, V. D., Klopcic, S. A., and Jarrold, C. C., *J. Chem. Phys.* 110, 5079, 1999.
290. Klopcic, S. A., Moravec, V. D., and Jarrold, C. C., *J. Chem. Phys.* 110, 10216, 1999.
291. Moravec, V. D., and Jarrold, C. C., *J. Chem. Phys.* 112, 792, 2000.
292. Moravec, V. D., and Jarrold, C. C., *J. Chem. Phys.* 113, 1035, 2000.
293. Chatterjee, B., Akin, F. A., Jarrold, C. C., and Raghavachari, K., *J. Chem. Phys.* 119, 10591, 2003.
294. Schröder, D., Brown, R., Schwerdtfeger, P., Wang, X. B., Yang, X., Wang, L. S., and Schwarz, H., *Angew. Chem. Int. Ed.* 42, 311, 2003.
295. Zheng, W., Nilles, J. M., Thomas, O. C., and Bowen, K. H., *J. Chem. Phys.* 122, 044306, 2005.
296. Duan, X., Burggraf, L. W., Weeks, D. E., Davico, G. E., Schwartz, R. L., and Lineberger, W. C., *J. Chem. Phys.* 116, 3601, 2002.
297. Meloni, G., Sheehan, S. M., Parsons, B. F., and Neumark, D. M., *J. Phys. Chem. A* 110, 3527, 2006.
298. Alexandrova, A. N., Boldyrev, A. I., Zhai, H.-J., and Wang, L.-S., *J. Chem. Phys.* 122, 054313, 2005.
299. Davico, G. E., Schwartz, R. L., Raymond, T. M., and Lineberger, W. C., *J. Phys. Chem. A* 103, 6167, 1999. EA(WO₂) may be lower than 1.998 eV by either 0.040 eV or 0.080 eV depending on the precise assignment of structures in the photoelectron spectrum.
300. Huang, X., Zhai, H.-J., Li, J., and Wang, L.-S., *J. Phys. Chem. A* 110, 85, 2006.
301. Zhai, H.-J., Wang, L.-S., Zubarev, D. Yu., and Boldyrev, A. I., *J. Phys. Chem. A* 110, 1689, 2006.
302. Sheehan, S. M., Meloni, G., Parsons, B. F., Wehres, N., and Neumark, D. M., *J. Chem. Phys.* 124, 064303, 2006.
303. Tschurl, M., and Boesl, U., *Int. J. Mass Spectrom.* 249/250, 364, 2006.
304. Nee, M. J., Osterwalder, A., Zhou, J., and Neumark, D. M., *J. Chem. Phys.* 125, 014306, 2006.
305. Zanni, M. T., Taylor, T. R., Greenblatt, B. J., Soep, B., and Neumark, D. M., *J. Chem. Phys.* 110, 3748, 1997.
306. Zhai, H.-J., and Wang, L.-S., *J. Chem. Phys.* 125, 164315, 2006.
307. Waters, T., Woo, H.-K., Wang, X.-B., and Wang, L.-S., *J. Am. Chem. Soc.* 128, 4282, 2006.
308. Calvi, R. M. D., Andrews, D. H., and Lineberger, W. C., *Chem. Phys. Lett.* 442, 12, 2007.
309. Parsons, B. F., Sheehan, S. M., Yen, T. A., Neumark, D. M., Wehres, N., and Weinkauf, R., *Phys. Chem. Chem. Phys.* 9, 3291, 2007.
310. Yu, W., Lin, Z., and Ding, C., *J. Chem. Phys.* 126, 114301, 2007.
311. Zhou, J., Garand, E., Eisfeld, W., and Neumark, D. M., *J. Chem. Phys.* 127, 034304, 2007.
312. Zheng, W., Li, X., Eustis, S., Grubisic, A., Thomas, O., de Clercq, H., and Bowen, K., *Chem. Phys. Lett.* 444, 232, 2007.
313. Mitsui, M., Ando, N., and Nakajima, A., *J. Phys. Chem. A* 111, 9644, 2007.
314. Zhou, J. Garand, E., and Neumark, D. M., *J. Chem. Phys.* 127, 154320, 2007. The measured values are EA(C₄D) = 28478(10) cm⁻¹ and EA(C₄H) = 28497(8) cm⁻¹.

315. Sheehan, S. M., Parsons, B. F., Zhou, J., Garand, E., Yen, T. A., Moore, D. T., and Neumark, D. M., J., *Chem. Phys.* 128, 034301, 2008.

316. Ichino, T., Andrews, D. H., Rathbone, G. J., Misaizu, F., Calvi, R. M. D., Wren, S. W., Kato, S., Bierbaum, V. M., and Lineberger, W. C., *J. Phys. Chem. B* 112, 545, 2008.

317. Sheehan, S. M., Parsons, B. F., Yen, T. A., Furlanetto, M. R., and Neumark, D. M., *J. Chem. Phys.* 128, 174301, 2008.

318. Viggiano, A. A., Miller, T. M., Friedman, J. F., and Troe, J., *J. Chem. Phys.* 127, 244305, 2007. This result utilizes theoretical anion frequencies for entropy and thermal corrections; a new, high-level calculation by Eisfeld, W., *J. Chem. Phys.* 134, 054303, 2011, would imply a lower value for $EA(SF_6)$, around 1.075 eV.

319. Zheng, W., Bowen, K. H., Li, J., Dąbkowska, I., Gutowski, M., *J. Phys. Chem. A* 109, 482, 2005. See also Mok, D. K. W., Chau, F.-t., Dyke, J. M., and Lee, E. P. F., *Chem. Phys. Lett.* 458, 11, 2008 and Mok, D. K. W., Lee, E. P. F., Chau, F.-t., and Dyke, J. M., *Phys. Chem. Chem. Phys.* 10, 7270, 2008.

320. Zheng, W., Li, X., Eustis, S., and Bowen, K., *Chem. Phys. Lett.* 460, 68, 2008.

321. Grubisic, A., Wang, H., Ko, Y. J., and Bowen, K. H., *J. Chem. Phys.* 129, 054302, 2008.

322. Stokes, S. T., Bowen, K. H., Sommerfeld, T., Ard, S., Mirsaleh-Kohan, N., Steill, J. D., and Compton, R. N., *J. Chem. Phys.* 129, 064308, 2008.

323. Garand, E., Yacovitch, T. I., and Neumark, D. M., *J. Chem. Phys.* 129, 074312, 2008.

324. Li, X., Eustis, S., Bowen, K. H., Kandalam, A. K., and Jena, P., *J. Chem. Phys.* 129, 074313, 2008. Uncertainties not stated.

325. Kandalam, A. K., Kiran, B., Jena, P., Li, X., Grubisic, A., and Bowen, K. H., *J. Chem. Phys.* 126, 084306, 2007.

326. Gerhards, M., Thomas, O. C., Nilles, J. M., Zhang, W.-J., and Bowen, K. H., *J. Chem. Phys.* 116, 10247, 2002.

327. Edmonds, B. D., Kandalam, A. K., Khanna, S. N., Li, X., Grubisic, A., Khanna, I., and Bowen, K. H., *J. Chem. Phys.* 124, 074316, 2006.

328. Duncan, M. A., Knight, A. M., Negishi, Y., Nagao, S., Nakamura, Y., Kato, A., Nakajima, A., and Kaya, K., *Chem. Phys. Lett.* 309, 49, 1999.

329. Zheng, W., Eustis, S. N., Li, X., Nilles, J. M., Thomas, O. C., Bowen, K. H., and Kandalam, A. K., *Chem. Phys. Lett.* 462, 35, 2008.

330. Kandalam, A. K., Jena, P., Li, X., Eustis, S. N., and Bowen, K. H., *J. Chem. Phys.* 129, 134308, 2008.

331. Wenthold, P. G., and Squires, R. R., *J. Phys. Chem.* 99, 2002, 1995. See also Artau, A., Nizzi, K. E., Hill, B. T., Sunderlin, L. S., and Wenthold, P. G., *J. Am. Chem. Soc.* 122, 10667, 2000.

332. Villalta, P. W., and Leopold, D. G., *J. Chem. Phys.* 130, 024303, 2009.

333. Wyrwas, R. B., and Jarrold, C. C., *J. Am. Chem. Soc.* 128, 13688, 2008.

334. Carette, T., Drag, C., Scharf, O., Blondel, C., Delsart, C., Fischer, C. F., and Godefroid, M., *Phys. Rev. A* 81, 042522, 2010. $EA(^{32}S) = 16$ 752.975 3(41) cm^{-1}, and $EA(^{34}S) = 16$ 752.977 6(85) cm^{-1}.

335. Goebbert, D. J., Khuseynov, D., and Sanov, A., *J. Phys. Chem. A* 114, 2259, 2010.

336. Zubarev, D. Yu., Boldyrev, A. I., Li, J., Zhai, H.-J., and Wang, L.-S., *J. Phys. Chem. A* 111, 1648, 2007.

337. Averkiev, B. B., Boldyrev, A. I., Li, Y., and Wang, L.-S., *J. Phys. Chem. A* 111, 34, 2007.

338. Zhai, H.-J., Wang, L.-M., Li, S.-D., and Wang, L.-S., *J. Phys. Chem. A* 111, 1030, 2007.

339. Distelrath, V., and Boesl, U., *Faraday Discuss.* 115, 161, 2000.

340. Wang, X.-B., and Wang, L.-S., *J. Chem. Phys.* 113, 10928, 2000.

341. Carpenter, D. L., Covington, A. M., and Thompson, J. S., *Phys. Rev. A* 61, 042501, 2000.

342. Davis, V. T., Thompson, J., and Covington, A., *Nucl. Instrum. Methods Phys. Res. B* 241, 118, 2005. High-level calculations give a lower value, EA(Nd) = 0.162 eV (Ref. 343) and 0.169 eV (O'Malley, S. M., and Beck, D. R., *Phys. Rev. A* 77, 012505, 2008).

343. Pan, L., and Beck, D. R., *J. Phys. B* 43, 025002, 2010.

344. Sobhy, M. A., Reveless, J. U., Gupta, U., Khanna, S. N., and Castleman, A. W., *J. Chem. Phys.* 130, 054304, 2009.

345. Ganteför, G., Gausa, M., Meiwes-Broer, K. H., and Lutz, H. O., *Z. Phys. D* 12, 405, 1989.

346. Garand, E., Yacovitch, T. I., and Neumark, D. M., *J. Chem. Phys.* 130, 064304, 2009.

347. Entfellner, M., and Boesl, U., *Phys. Chem. Chem. Phys.* 11, 2657, 2009.

348. Wang, L.-M., Huang, W., Wang, L.-S., Averkiev, B. B., and Boldyrev, A. I., *J. Chem. Phys.* 130, 134303, 2009.

349. Wren, S. W., Vogelhuber, K. M., Ervin, K. M., and Lineberger, W. C., *Phys. Chem. Chem. Phys.* 11, 4745, 2009.

350. Sheps, L., Miller, E. M., and Lineberger, W. C., *J. Chem. Phys.* 131, 064304, 2009.

351. Garand, E., Yacovitch, T. I., and Neumark, D. M., *J. Chem. Phys.* 131, 054312, 2009.

352. Walter, C. W., Gibson, N. D., Field III, R. L., Snedden, A. P., Shapiro, J. Z., Janczak, C. M., and Hanstrorp, D., *Phys. Rev. A* 80, 014501, 2009.

353. Rothgeb, D. W., Hossain, E., Kuo, A. T., Troyer, J. L., and Jarrold, C. C., *J. Chem. Phys.* 131, 044310, 2009. See also Ref. 369.

354. Borschevsky, A., Pershina, V., Bliav, E., and Kaldor, U., *Chem. Phys. Lett.* 480, 49, 2009.

355. O'Malley, S. M., and Beck, D. M., *Phys. Rev. A* 79, 012511, 2009 and 80, 032514, 2009.

356. Xu, H.-G., Zhang, Z.-F., Feng, Y., and Zheng, W., *Chem. Phys. Lett.* 498, 22, 2010.

357. Xu, H.-G., Zhang, Z.-F., Feng, Y., Yuan, J., Zhao, Y., and Zheng, W., *Chem. Phys. Lett.* 487, 204, 2010.

358. Feng, Y., Xu, H.-G., Zhang, Z.-G., Gao, Z., and Zheng, W., *J. Chem. Phys.* 132, 074308, 2010.

359. Zhang, Z.-G., Xu, H.-G., Zhao, Y., and Zheng, W., *J. Chem. Phys.* 133, 154314, 2010.

360. Lindahl, A. O., Andersson, P., Diehl, C., Forstner, O., Klason, P., and Hanstorp, D., *Eur. Phys. J. D* 60, 219, 2010. EA(W) = 6583.6 ± 0.6 cm^{-1}.

361. Vogelhuber, K. M., Wren, S. W., Sheps, L., and Lineberger, W. C., *J. Chem. Phys.* 134, 064302, 2011.

362. Feng, Y., Xu, H.-G., Zheng, W., Zhao, H., Kandalam, A. K., and Jena, P., *J. Chem. Phys.* 134, 094309, 2011.

363. Gupta, U., Reveles, J. U., Melko, J. J., Khanna, S. N., and Castleman, A. W., *Chem. Phys. Lett.* 467, 223, 2009.

364. Gupta, U., Reveles, J. U., Melko, J. J., Khanna, S. N., and Castleman, A. W., *J. Phys. Chem. C* 114, 15963, 2010. Uncertainties not given, but are typically ± 0.05 to ± 0.08 eV (See Refs. 360, 363, and 365).

365. Gupta, U., Reber, A. C., Melko, J. J., Khanna, S. N., and Castleman, A. W., *Chem. Phys. Lett.* 505, 92, 2011.

366. Yacovich, T. I., Kim, J. B., Garand, E., van der Poll, D. G., and Neumark, D. M., *J. Chem. Phys.* 134, 134307, 2011.

367. Vogelhuber, K. M., Wren, S. W., McCoy, A. B., Ervin, K. M., and Lineberger, W. C., *J. Chem. Phys.* 134, 184306, 2011.

368. Adams, C. L., and Weber, J. M., *J. Chem. Phys.* 134, 244301, 2011.

369. Rothgeb, D. W., Mann, J. E., Waller, S. E., and Jarrold, C. C., *J. Chem. Phys.* 135, 104312, 2011.

370. Mann, J. E., Waller, S. E., Rothgeb, D. W., and Jarrold, C. C., *J. Chem. Phys.* 135, 104317, 2011.

371. Willis, M., Götz, M., Kandalam, A. K., Ganteför, G. F., and Jena, P., *Angew. Chem. Int. Ed.* 49, 8966, 2010.

372. Wang, Y.-L., Wang, X.-B., Xing, X.-P., Wei, F., Li, J., and Wang, L.-S., *J. Phys. Chem. A* 114, 11244, 2010.

373. Zhai, H.-J., Bürgel, C., Bonacid-Koutecky, V., and Wang, L.-S., *J. Am. Chem. Soc.* 130, 9156 (2008).

374. Chen, W.-J., Zhai, H.-J., Zhang, Y.-F., Huang, X., and Wang, L.-S., *J. Phys. Chem. A* 114, 5958, 2010.

375. Wang, X.-B., Fu, Q., and Yang, J., *J. Phys. Chem. A* 114, 9083, 2010.

376. Gruzinskaya, N. I., Aleshina, V. E., Borshchevskii, A. Ya., and Sidorov, L. N., *J. Anal. Chem.* 65, 1328, 2010.

377. Peláez, R. J., Blondel, C., Vandevraye, M., Drag, C., and Delsart, C., *J. Phys. B* 44, 1950009, 2011. EA(P) = 6021.79 ± 0.08 cm^{-1}.

378. Lenington, M. J., and Wenthold, P. G., *J. Phys. Chem. A* 114, 1334, 2010.

379. Butman, M. F., Sergeev, D. N., Motalov, V. B., Kudin, L. S., Kryuchkov, A. S., and Krämer, K. W., *Russ. J. Phys. Chem. A* 85, 922, 2011.

380. Zheng, W.-J., Thomas, O. C., Nilles, J. M., Bowen, K. H., Reber, A. C., and Khanna, S. N., *J. Chem. Phys.* 134, 224307, 2011.

PROTON AFFINITIES

Proton affinity is a useful parameter for describing gas phase ion-molecule reactions in fields such as atmospheric chemistry, plasma chemistry, mass spectrometry, and astrophysics. The proton affinity E_{pa} (often designated in the literature as PA) of a molecular species M is defined as the negative of the enthalpy change for the gas phase reaction

$$M + H^+ \rightarrow MH^+.$$

A closely related quantity is the gas phase basicity $\Delta_{base}G°$ (often designated as GB), which is the negative of the Gibbs energy change for the same reaction. Thus the two are related by

$$\Delta_{base}G° = E_{pa} + T\Delta S,$$

where T is the temperature and ΔS is the entropy change in the reaction (which can be calculated if the molecular structure of M and M^+ is known).

Direct measurement of the proton affinity is possible for only a few molecules, mainly olefins and carbonyl compounds. However, these measurements have been used to establish a scale of E_{pa} values that permits proton affinities to be determined for many other molecules, including unstable species and reaction intermediates. The basis for this scale is described by Hunter and Lias in Reference 1.

The E_{pa} and $\Delta_{base}G°$ values at a temperature of 298 K are tabulated below for selected molecules. Many values are given to one decimal place, but the majority are not accurate to better than one or two kilojoules per mole. The methods of measurement are described in Reference 1, which contains a much more extensive and detailed tabulation.

Compounds are listed by molecular formula in the Hill order, but with all compounds that do not contain carbon appearing before those that do contain carbon.

References

1. Hunter, E. P. L., and Lias, S. G., *J. Phys. Chem. Ref. Data* 27, 413, 1998.
2. Hunter, E. P., and Lias, S. G., "Proton Affinity Evaluation", in *NIST Chemistry WebBook*, NIST Standard Reference Database No. 69, Linstrom, P. J., and Mallard, W. G., Eds., March 2003, National Institute of Standards and Technology, Gaithersburg, MD 20899, <http://webbook.nist.gov>.
3. Do, K., Klein, T. P., Pommerening, C. A., Bachrach, S. M., and Sunderlin, L. S., *J. Am. Chem. Soc.* 120, 6093, 1998.
4. Kim, H.-T., Green, R. J., Qian, J., and Anderson, S. L., *J. Chem. Phys.* 112, 5717, 2000.
5. Park, S. T., Kim, S. K., and Kim, M. S., *J. Chem. Phys.* 114, 5568, 2001.
6. Hiraoka, K., Mizuno, T., Eguchi, D., Takao, T., and Ino, S., *J. Chem. Phys.* 116, 7574, 2002.
7. Oresmaa, L. O., Haukka, M., Vainiotalo, P., and Pakkanen, T. A., *J. Org. Chem.* 67, 8216, 2002.
8. Wang, F., Ma, S., Zhang, D., and Cooks, R. G., *J. Phys. Chem. A* 102, 2988, 1998.
9. Bouchoux, G., Gal, J.-F., Szulejko, J. E., McMahon, T. B., Tortajada, J., Luna, A., Yanez, M., and Mo, O., *J. Phys. Chem. A* 102, 9183, 1998.
10. van Beelen, E., Koblenz, T. A., Ingemann, S. and Hammerum, S., *J. Phys. Chem. A* 108, 2728, 2004.

Molecular formula	Name	E_{pa} kJ/mol	$\Delta_{base}G°$ kJ/mol	Notes
Ar	Argon	369.2	346.3	
AsF$_3$	Arsenic(III) fluoride	636.7	604.2	
AsH$_3$	Arsine	747.9	712.0	
BHO$_2$	Metaboric acid	763.0	730.5	
BH$_3$O$_3$	Boric acid	728.1	698.4	
B$_2$H$_6$	Diborane	615	586.0	
B$_3$H$_6$N$_3$	Borazine	802.5	772.8	
B$_4$H$_{10}$	Tetraborane(10)	605	572.5	
B$_5$H$_9$	Pentaborane(9)	699.4	666.9	
BaO	Barium oxide	1215.4	1187.6	
Br	Bromine (atomic)	554.4	531.2	
BrH	Hydrogen bromide	584.2	557.7	
BrLi	Lithium bromide	819	792.5	
CaO	Calcium oxide	1190.6	1162.3	
Cl	Chlorine (atomic)	513.6	490.1	
ClH	Hydrogen chloride	556.9	530.1	
ClLi	Lithium chloride	827	800.5	
Co	Cobalt	742.7	719.8	
Cr	Chromium	791.3	768.4	
CsHO	Cesium hydroxide	1117.9	1092.2	
Cs$_2$O	Cesium oxide	1442.9	1412.2	
Cu	Copper	655.3	632.4	
F	Fluorine (atomic)	340.1	315.1	
FH	Hydrogen fluoride	484	456.7	
FO	Fluorine oxide	508.7	482.2	
F$_2$	Fluorine	332	305.5	
F$_2$O$_2$S	Sulfuryl fluoride	605.5	580.5	

Molecular formula	Name	E_{pa} kJ/mol	$\Delta_{base}G°$ kJ/mol	Notes
F_3N	Nitrogen trifluoride	568.4	538.6	
F_3OP	Phosphoryl fluoride	694.0	664.2	
F_3P	Phosphorus(III) fluoride	695.3	662.8	
F_4Si	Tetrafluorosilane	502.9	476.6	
F_6S	Sulfur hexafluoride	575.3	550.7	
Fe	Iron	754	731.1	
FeO	Iron(II) oxide	907	880.5	
GeH_4	Germane	713.4	687.1	
HI	Hydrogen iodide	627.5	601.3	
HKO	Potassium hydroxide	1101.8	1075.4	
HLi	Lithium hydride	1021.7	996.4	
HLiO	Lithium hydroxide	1000.1	972.1	
HNO_3	Nitric acid	751.4	731.5	
HN_3	Hydrazoic acid	756.0	723.5	
HNa	Sodium hydride	1095	1070.6	
HNaO	Sodium hydroxide	1071.8	1044.8	
HO	Hydroxyl	593.2	564.0	
HO_2	Hydroperoxy	660	627.5	
HP	Phosphorus monohydride	670.3	639.6	
H_2	Hydrogen	422.3	394.7	
$H_2N_2O_2$	Nitramide	757.4	725.0	
H_2O	Water	691	660.0	
H_2O_2	Hydrogen peroxide	674.5	643.8	
H_2O_4S	Sulfuric acid	717	681	Ref. 3
H_2P	Phosphino	709.2	675.7	
H_2S	Hydrogen sulfide	705	673.8	
H_2Se	Hydrogen selenide	707.8	676.4	
H_2Si	Silylene	839.2	804.1	
H_2Te	Hydrogen telluride	735.9	704.5	
H_3N	Ammonia	853.6	819.0	
H_3P	Phosphine	785	750.9	
H_4N_2	Hydrazine	853.2	822.4	
H_4Si	Silane	639.7	613.4	
H_6OSi_2	Disiloxane	749	718.3	
He	Helium	177.8	148.5	
I	Iodine (atomic)	608.2	583.5	
K_2O	Potassium oxide	1342.5	1311.8	
Kr	Krypton	424.6	402.4	
La	Lanthanum	1013	991.9	
Li_2	Dilithium	1162	1133.1	
Li_2O	Lithium oxide	1206	1175.3	
Lu	Lutetium	992	970.6	
Mg	Magnesium	819.6	797.3	
MgO	Magnesium oxide	988	959.4	
Mg_2	Dimagnesium	919	886.5	
Mn	Manganese	797.3	774.4	
N	Nitrogen (atomic)	342.2	318.7	
NO	Nitric oxide	531.8	505.3	
NO_2	Nitrogen dioxide	591.0	560.3	
NP	Phosphorus nitride	789.4	757.0	
N_2	Nitrogen	493.8	464.5	
N_2O	Nitrous oxide	549.8	523.3	Protonation at N
N_2O	Nitrous oxide	575.2	548.7	Protonation at O
Na_2	Disodium	1146.8	1118.2	
Na_2O	Sodium oxide	1375.9	1345.2	
Ne	Neon	198.8	174.4	
Ni	Nickel	737	714.1	
O	Oxygen (atomic)	485.2	459.6	
OP	Phosphorus monoxide	682	649.5	
OSi	Silicon monoxide	777.8	750.4	Protonation at O

Molecular formula	Name	E_{pa} kJ/mol	$\Delta_{base}G°$ kJ/mol	Notes
OSi	Silicon monoxide	533	500.5	Protonation at Si
OSr	Strontium oxide	1209	1180.7	
O_2	Oxygen	421	396.3	
O_2S	Sulfur dioxide	672.3	643.3	
O_3	Ozone	625.5	595.9	
O_3S	Sulfur trioxide	588.3	560.3	
O_4Os	Osmium(VIII) oxide	676.9	650.6	
P	Phosphorus	626.8	604.8	
Pd	Palladium	696	673.4	
Rh	Rhodium	768	745.4	
Ru	Ruthenium	774	751.4	
S	Sulfur	664.3	640.2	
SSi	Silicon monosulfide	627	596.6	Protonation at Si
SSi	Silicon monosulfide	683	660.2	Protonation at S
Sc	Scandium	914	892.0	
Si	Silicon	837	814.1	
Ti	Titanium	876	853.7	
U	Uranium	995.2	973.2	
V	Vanadium	859.4	836.8	
Xe	Xenon	499.6	478.1	
Y	Yttrium	967	945.9	
Zn	Zinc	608.6	586.0	
$CBrF_3$	Bromotrifluoromethane	580.0	550.3	
CBrN	Cyanogen bromide	749.8	719.2	
$CClF_3$	Chlorotrifluoromethane	571.3	541.5	
CClN	Cyanogen chloride	722.1	691.5	
CCl_2	Dichloromethylene	861	828.5	
CCl_2S	Carbonothioic dichloride	752.5	721.8	
CFN	Cyanogen fluoride	632	601.3	
CF_2	Difluoromethylene	765	732.5	
CF_2O	Carbonyl fluoride	666.7	637.0	
CF_3I	Trifluoroiodomethane	628.0	598.2	
CF_3NO	Trifluoronitrosomethane	703.3	670.8	
CF_4	Tetrafluoromethane	529.3	503.7	
CHCl	Chloromethylene	874.1	839.9	
CHF	Fluoromethylene	797.9	763.8	
CHF_3	Trifluoromethane	619.5	589.7	
CHF_3O_3S	Trifluoromethanesulfonic acid	699.4	666.9	
CHN	Hydrogen cyanide	712.9	681.6	
CHN	Hydrogen isocyanide	772.3	739.8	
CHNO	Isocyanic acid (HNCO)	753	718.8	
CHNO	Fulminic acid	758	725.5	
CHO	Oxomethyl (HCO)	636	601.8	
CHO_2	Formyloxyl	623.4	590.9	
CH_2F_2	Difluoromethane	620.5	589.7	
CH_2N_2	Diazomethane	858.9	826.7	
CH_2N_2	Cyanamide	805.6	774.9	
CH_2O	Formaldehyde	712.9	683.3	
CH_2O_2	Formic acid	742.0	710.3	
CH_2S	Thioformaldehyde	759.7	730.5	
CH_2Se	Selenoformaldehyde	764.0	734.9	
CH_3Br	Bromomethane	664.2	638.0	
CH_3Cl	Chloromethane	647.3	621.1	
CH_3F	Fluoromethane	598.9	571.5	
CH_3I	Iodomethane	691.7	665.5	
CH_3NO	Formamide	822.2	791.2	
CH_3NO_2	Nitromethane	754.6	721.6	
CH_3NO_2	Methyl nitrite	798.9	766.4	
CH_3NO_3	Methyl nitrate	733.6	714.8	
CH_3N_3	Methyl azide	833	800.5	

Molecular formula	Name	E_{pa} kJ/mol	$\Delta_{base}G°$ kJ/mol	Notes
CH_4	Methane	543.5	520.6	
CH_4N	Methylamidogen	832.8	801.6	
CH_4N_2O	Urea	873.5	841.6	Protonation at O; Ref. 8
CH_4N_2S	Thiourea	893.7	863.9	
CH_4O	Methanol	754.3	724.5	
CH_4O_3S	Methanesulfonic acid	761.3	728.9	
CH_4S	Methanethiol	773.4	742	
CH_5N	Methylamine	899.0	864.5	
CH_5NO	O-Methylhydroxylamine	844.8	812.3	
CH_5N_3	Guanidine	986.3	949.4	
CH_5P	Methylphosphine	851.5	817.6	
CH_6N_2	Methylhydrazine	898.8	866.4	
CN	Cyanide	>595	>564	Protonation at N
CNS	Thiocyanate	751	718.5	
CO	Carbon monoxide	594	562.8	Protonation at C
CO	Carbon monoxide	426.3	402.2	Protonation at O
COS	Carbon oxysulfide	628.5	602.6	Protonation at S
$COSe$	Carbon oxyselenide	670	644.1	Protonation at Se
CO_2	Carbon dioxide	540.5	515.8	
CS	Carbon monosulfide	791.5	760	
CS_2	Carbon disulfide	681.9	657.7	
CSe	Carbon monoselenide	831.8	800.2	Protonation at C
CSe_2	Carbon diselenide	725	700.9	
C_2ClF_3O	Trifluoroacetyl chloride	681.6	649.8	
C_2Cl_3N	Trichloroacetonitrile	723.2	692.6	
C_2F_3N	Trifluoroacetonitrile	688.4	657.7	
C_2H	Ethynyl	753	720.8	
C_2HCl_3O	Trichloroacetaldehyde	722.3	690.5	
$C_2HCl_3O_2$	Trichloroacetic acid	770.0	739.1	
C_2HF	Fluoroacetylene	686	661.3	
C_2HF_3	Trifluoroethene	699.4	666.9	
$C_2HF_3O_2$	Trifluoroacetic acid	711.7	680.7	
C_2H_2	Acetylene	641.4	616.7	
C_2H_2ClN	Chloroacetonitrile	745.7	715.1	
$C_2H_2F_2$	1,1-Difluoroethene	734	705.1	
$C_2H_2F_2$	trans-1,2-Difluoroethene	688.6	657.9	
C_2H_2O	Ketene	825.3	793.6	
$C_2H_3ClO_2$	Chloroacetic acid	765.4	734.5	
$C_2H_3Cl_3O$	2,2,2-Trichloroethanol	729.3	698.9	
C_2H_3F	Fluoroethene	729	700.1	
$C_2H_3FO_2$	Fluoroacetic acid	765.4	734.5	
$C_2H_3F_3O$	2,2,2-Trifluoroethanol	700.2	669.9	
$C_2H_3F_3O$	Methyl trifluoromethyl ether	719.2	690.0	
C_2H_3N	Acetonitrile	779.2	748	
C_2H_3N	Isocyanomethane	839.1	806.6	
C_2H_3NO	Methyl isocyanate	764.4	732.0	
C_2H_3NS	Methyl thiocyanate	796.7	766.1	
C_2H_3NS	Methyl isothiocyanate	799.2	766.7	
$C_2H_3N_3$	1H-1,2,3-Triazole	879.3	847.4	
$C_2H_3N_3$	1H-1,2,4-Triazole	886.0	855.9	
C_2H_4	Ethylene	680.5	651.5	
$C_2H_4F_2O$	2,2-Difluoroethanol	727.4	697.0	
$C_2H_4F_3N$	2,2,2-Trifluoroethylamine	846.8	812.9	
$C_2H_4N_2$	Aminoacetonitrile	824.9	791.0	
C_2H_4O	Acetaldehyde	768.5	736.5	
C_2H_4O	Oxirane	774.2	745.3	
$C_2H_4O_2$	Acetic acid	783.7	752.8	
$C_2H_4O_2$	Methyl formate	782.5	751.5	
C_2H_4S	Thiirane	807.4	777.6	
C_2H_5Br	Bromoethane	696.2	669.7	

Molecular formula	Name	E_{pa} kJ/mol	$\Delta_{base}G°$ kJ/mol	Notes
C_2H_5BrO	2-Bromoethanol	766.1	735.7	
C_2H_5Cl	Chloroethane	693.4	666.9	
C_2H_5ClO	2-Chloroethanol	766.1	735.7	
C_2H_5F	Fluoroethane	683.4	655.8	
C_2H_5FO	2-Fluoroethanol	715.6	685.2	
C_2H_5I	Iodoethane	724.8	698.3	
C_2H_5N	Ethenamine	898.9	866.5	
C_2H_5N	Ethyleneimine	905.5	872.5	
C_2H_5NO	Acetamide	863.6	832.6	
C_2H_5NO	N-Methylformamide	851.3	820.3	
$C_2H_5NO_2$	Nitroethane	765.7	733.2	
$C_2H_5NO_2$	Ethyl nitrite	818.9	786.4	
$C_2H_5NO_2$	Glycine	886.5	852.2	
$C_2H_5NO_2$	Acetohydroxamic acid	854.0	823.0	
C_2H_5NS	Thioacetamide	884.6	852.8	
C_2H_6	Ethane	596.3	569.9	
C_2H_6Hg	Dimethyl mercury	771.6	740.8	
$C_2H_6N_2$	Ethanimidamide	970.7	938.2	
$C_2H_6N_2$	trans-Dimethyldiazene	865.1	834.4	
$C_2H_6N_2O$	2-Aminoacetamide		882.3	
$C_2H_6N_2O_2$	N-Methyl-N-nitromethanamine	828.3	795.8	
C_2H_6O	Ethanol	776.4	746	
C_2H_6O	Dimethyl ether	792	764.5	
C_2H_6OS	Dimethyl sulfoxide	884.4	853.7	
$C_2H_6O_2$	1,2-Ethanediol	815.9	773.6	
C_2H_6S	Ethanethiol	789.6	758.4	
C_2H_6S	Dimethyl sulfide	830.9	801.2	
$C_2H_6S_2$	Dimethyl disulfide	815.3	782.8	
C_2H_7N	Ethylamine	912.0	878	
C_2H_7N	Dimethylamine	929.5	896.5	
C_2H_7NO	Ethanolamine	930.3	896.8	
$C_2H_7O_3P$	Dimethyl hydrogen phosphite	894.8	862.4	
C_2H_7P	Dimethylphosphine	912.0	877.9	
$C_2H_8N_2$	1,2-Ethanediamine	951.6	912.5	
$C_2H_8N_2$	1,1-Dimethylhydrazine	927.1	894.7	
C_2N_2	Cyanogen	674.7	645.8	
C_2O	Dicarbon monoxide	774.7	747.0	
C_3	Carbon trimer	767.0	736.3	
C_3F_6O	Perfluoroacetone	670.4	639.7	
C_3HN	Cyanoacetylene	751.2	720.5	
$C_3H_2F_6O$	1,1,1,3,3,3-Hexafluoro-2-propanol	686.6	656.2	
$C_3H_2N_2$	Malononitrile	723.0	694.1	
C_3H_3	2-Propynyl	741	708.5	
$C_3H_3Cl_3O$	1,1,1-Trichloro-2-propanone	768.3	736.3	
$C_3H_3F_3O$	1,1,1-Trifluoroacetone	723.9	692.0	
$C_3H_3F_3O_2$	Methyl trifluoroacetate	740.5	709.6	
C_3H_3N	Acrylonitrile	784.7	753.7	
C_3H_3NO	Oxazole	876.4	844.5	
C_3H_3NO	Isoxazole	848.6	816.8	
C_3H_3NO	2-Oxopropanenitrile	746.9	716.2	
C_3H_3NS	Thiazole	904	872.1	
$C_3H_3N_3$	1,3,5-Triazine	848.8	819.6	
C_3H_4	Allene	775.3	745.8	
C_3H_4	Propyne	748.2	723.0	
C_3H_4	Cyclopropene	818.5	787.8	
C_3H_4ClN	3-Chloropropanenitrile	773.1	742.4	
$C_3H_4N_2$	1H-Pyrazole	894.1	860.5	
$C_3H_4N_2$	Imidazole	942.8	909.2	
$C_3H_4N_2S$	2-Thiazolamine	930.6	898.7	
C_3H_4O	Acrolein	797.0	765.1	

Molecular formula	Name	E_{pa} kJ/mol	$\Delta_{base}G°$ kJ/mol	Notes
C_3H_4O	1-Propen-1-one	834.1	803.4	
$C_3H_4O_3$	Ethylene carbonate	814.2	784.4	
C_3H_5	Allyl	736	707.4	
C_3H_5	Cyclopropyl	738.9	702.0	
$C_3H_5ClO_2$	Ethyl chloroformate	764.8	733.8	
C_3H_5FO	1-Fluoro-2-propanone	795.4	763.5	
$C_3H_5F_3O$	2,2,2-Trifluoroethyl methyl ether	747.6	718.4	
C_3H_5N	Propanenitrile	794.1	763.0	
C_3H_5N	2-Propyn-1-amine	887.4	853.5	
C_3H_5N	Ethyl isocyanide	851.3	818.9	
C_3H_5NO	Acrylamide	870.7	839.8	
C_3H_5NO	Methoxyacetonitrile	758.1	727.4	
C_3H_5NO	2-Azetidinone	852.6	821.7	
C_3H_5NS	(Methylthio)acetonitrile	784.8	754.1	
$C_3H_5N_3$	1*H*-Pyrazol-3-amine	921.5	889.6	
$C_3H_5N_3$	1*H*-Pyrazol-4-amine	907.6	874.0	
C_3H_6	Propene	741.6		Ref. 5
C_3H_6	Cyclopropane	750.3	722.2	
$C_3H_6N_2$	3-Aminopropanenitrile	866.4	832.5	
$C_3H_6N_2$	Dimethylcyanamide	852.1	821.4	
$C_3H_6N_2S$	2-Imidazolidinethione	921.9	891.2	
C_3H_6O	Methyl vinyl ether	859.2	830.3	
C_3H_6O	Propanal	786.0	754.0	
C_3H_6O	Acetone	812	782.1	
C_3H_6O	Oxetane	801.3	773.9	
$C_3H_6O_2$	Propanoic acid	797.2	766.2	
$C_3H_6O_2$	Ethyl formate	799.4	768.4	
$C_3H_6O_2$	Methyl acetate	821.6	790.7	
$C_3H_6O_3$	Dimethyl carbonate	830.2	799.2	
C_3H_6S	(Methylthio)ethene	858.2	829.3	
C_3H_6S	Thietane	834.8	805.0	
C_3H_6S	Methylthiirane	833.3	801.5	
C_3H_7N	Allylamine	909.5	875.5	
C_3H_7N	Cyclopropylamine	904.7	869.9	
C_3H_7N	Azetidine	943.4	908.6	
C_3H_7N	1-Methylaziridine	934.8	904.1	
C_3H_7N	Propyleneimine	925.1	892.1	
C_3H_7NO	*N,N*-Dimethylformamide	887.5	856.6	
C_3H_7NO	*N*-Methylacetamide	888.5	857.6	
C_3H_7NO	Propanamide	876.2	845.3	
$C_3H_7NO_2$	Isopropyl nitrite	845.5	813.0	
$C_3H_7NO_2$	*L*-Alanine	901.6	867.7	
$C_3H_7NO_2$	Sarcosine	921.2	888.7	
$C_3H_7NO_2S$	*L*-Cysteine	903.2	869.3	
$C_3H_7NO_3$	*L*-Serine	914.6	880.7	
C_3H_8	Propane	625.7	607.8	
$C_3H_8N_2O$	*N,N'*-Dimethylurea	903.3	873.5	
$C_3H_8N_2S$	*N,N'*-Dimethylthiourea	926.0	895.1	
C_3H_8O	1-Propanol	786.5	756.1	
C_3H_8O	2-Propanol	793.0	762.6	
C_3H_8O	Ethyl methyl ether	808.6	781.2	
$C_3H_8O_2$	1,3-Propanediol	876.2	825.9	
$C_3H_8O_2$	2-Methoxyethanol	768.8	729.8	
$C_3H_8O_3$	Glycerol	874.8	820	
C_3H_8S	1-Propanethiol	794.9	763.6	
C_3H_8S	2-Propanethiol	803.6	772.3	
C_3H_8S	Ethyl methyl sulfide	846.5	815.3	
C_3H_9As	Trimethylarsine	897.3	864.9	
$C_3H_9BO_3$	Trimethyl borate	815.8	783.4	
C_3H_9N	Propylamine	917.8	883.9	

Molecular formula	Name	E_{pa} kJ/mol	$\Delta_{base}G°$ kJ/mol	Notes
C_3H_9N	Isopropylamine	923.8	889.0	
C_3H_9N	Ethylmethylamine	942.2	909.2	
C_3H_9N	Trimethylamine	948.9	918.1	
C_3H_9NO	2-Methoxyethylamine	928.6	894.6	
C_3H_9NO	Trimethylamine oxide	983.2	953.5	
C_3H_9NO	3-Amino-1-propanol	962.5	917.3	
$C_3H_9O_3P$	Trimethyl phosphite	929.7	899.9	
$C_3H_9O_4P$	Trimethyl phosphate	890.6	860.8	
C_3H_9P	Trimethylphosphine	958.8	926.3	
$C_3H_{10}N_2$	1,3-Propanediamine	987.0	940.0	
$C_3H_{10}OSi$	Trimethylsilanol	814.0	781.5	
C_4F_8	Perfluorocyclobutane	>544		Ref. 6
C_4H_2	1,3-Butadiyne	737.2	712.8	
$C_4H_4F_6O$	Bis(2,2,2-trifluoroethyl) ether	702.3	674.9	
$C_4H_4N_2$	Pyrazine	877.1	847.0	
$C_4H_4N_2$	Pyrimidine	885.8	855.7	
$C_4H_4N_2$	Pyridazine	907.2	877.1	
$C_4H_4N_2O_2$	Uracil	872.7	841.7	
$C_4H_4N_2S_2$	2,4(1H,3H)-Pyrimidinedithione	911.4	880.5	
C_4H_4O	Furan	812	781	Ref. 10
$C_4H_4O_3$	Succinic anhydride	797		Ref. 9
C_4H_4S	Thiophene	815.0	784.3	
$C_4H_5Cl_3O_2$	Ethyl trichloroacetate	790.4	759.4	
$C_4H_5F_3O_2$	Ethyl trifluoroacetate	758.8	727.9	
C_4H_5N	Pyrrole	875.4	843.8	
C_4H_5N	Cyclopropanecarbonitrile	808.2	777.5	
$C_4H_5NO_2$	Ethyl cyanoformate	745.7	714.7	
C_4H_5NS	2-Methylthiazole	930.6	898.7	
$C_4H_5N_3O$	Cytosine	949.9	918	
C_4H_6	1,2-Butadiene	778.9	749.8	
C_4H_6	1,3-Butadiene	783.4	757.6	
C_4H_6	2-Butyne	775.8	745.1	
C_4H_6	Cyclobutene	784.4	753.6	
$C_4H_6F_3NO$	2,2,2-Trifluoro-N,N-dimethylacetamide	849.0	818.0	
$C_4H_6N_2$	1-Methylimidazol	959.6	927.7	
$C_4H_6N_2$	2-Methyl-1H-imidazole	963.4	929.6	
$C_4H_6N_2$	4-Methyl-1H-imidazole	952.8	920.9	
$C_4H_6N_2$	1-Methyl-1H-pyrazole	912.0	880.1	
$C_4H_6N_2$	3-Methyl-1H-pyrazole	906.0	874.2	
$C_4H_6N_2$	4-Methyl-1H-pyrazole	906.8	873.4	
C_4H_6O	2-Methylpropenal	808.7	776.8	
C_4H_6O	3-Buten-2-one	834.7	802.8	
C_4H_6O	Cyclobutanone	802.5	772.7	
C_4H_6O	2,3-Dihydrofuran	866.9	834.4	
C_4H_6O	2,5-Dihydrofuran	823.4	796	
$C_4H_6O_2$	trans-2-Butenoic acid	824.0	793	
$C_4H_6O_2$	Methacrylic acid	816.7	785.7	
$C_4H_6O_2$	Cyclopropanecarboxylic acid	821.4	790.4	
$C_4H_6O_2$	Vinyl acetate	813.9	782.9	
$C_4H_6O_2$	Methyl acrylate	825.8	794.8	
$C_4H_6O_2$	2,3-Butanedione	801.9	770.1	
$C_4H_6O_2$	γ-Butyrolactone	840.0	808.1	
$C_4H_6O_2$	2,3-Dihydro-1,4-dioxin	823.5	792.8	
$C_4H_6O_3$	Acetic anhydride	844		Ref. 9
C_4H_7	2-Methylallyl	778	747.3	
C_4H_7N	Butanenitrile	798.4	767.7	
C_4H_7N	2-Methylpropanenitrile	803.6	772.8	
C_4H_7N	1-Isocyanopropane	856.8	824.3	
C_4H_7NO	2-Butenamide	887.1	856.1	
C_4H_7NO	2-Methyl-2-propenamide	880.4	849.4	

Molecular formula	Name	E_{pa} kJ/mol	$\Delta_{base}G°$ kJ/mol	Notes
$C_4H_7NO_4$	L-Aspartic acid	908.9	875	
C_4H_8	trans-2-Butene	747	719.9	
C_4H_8	Isobutene	802.1	775.6	
$C_4H_8N_2$	(Dimethylamino)acetonitrile	884.5	853.7	
$C_4H_8N_2O_3$	L-Asparagine	929	891.5	
$C_4H_8N_2O_3$	N-Glycylglycine		882	
C_4H_8O	Ethyl vinyl ether	870.1	840.4	
C_4H_8O	2-Methoxy-1-propene	894.9	866.1	
C_4H_8O	Butanal	792.7	760.8	
C_4H_8O	Isobutanal	797.3	765.5	
C_4H_8O	2-Butanone	827.3	795.5	
C_4H_8O	Tetrahydrofuran	822.1	794.7	
$C_4H_8O_2$	Propyl formate	804.9	773.9	
$C_4H_8O_2$	Isopropyl formate	811.3	780.3	
$C_4H_8O_2$	Ethyl acetate	835.7	804.7	
$C_4H_8O_2$	Methyl propanoate	830.2	799.2	
$C_4H_8O_2$	1,3-Dioxane	825.4	796.2	
$C_4H_8O_2$	1,4-Dioxane	797.4	770.0	
$C_4H_8O_3$	Ethyl methyl carbonate	842.7	810.8	
C_4H_8S	Tetrahydrothiophene	849.1	819.3	
C_4H_9N	Pyrrolidine	948.3	915.3	
C_4H_9NO	N-Methylpropanamide	920.4	889.4	
C_4H_9NO	2-Methylpropanamide	878.6	846.7	
C_4H_9NO	N-Ethylacetamide	898.0	867.0	
C_4H_9NO	N,N-Dimethylacetamide	908.0	877.0	
C_4H_9NO	Morpholine	924.3	891.2	
$C_4H_9NO_2$	tert-Butyl nitrite	863.9	831.4	
$C_4H_9NO_2$	Ethyl N-methylcarbamate	888.8	857.8	
$C_4H_9NO_3$	L-Threonine	922.5	888.5	
C_4H_9NS	N,N-Dimethylthioacetamide	925.3	894.4	
C_4H_{10}	Isobutane	677.8	671.3	
$C_4H_{10}N_2$	Piperazine	943.7	914.7	
$C_4H_{10}N_2$	3-Ethyl-3-methyldiaziridine	903.8	871.3	
$C_4H_{10}O$	1-Butanol	789.2	758.9	
$C_4H_{10}O$	2-Butanol	815.7	784.6	
$C_4H_{10}O$	2-Methyl-1-propanol	793.7	762.2	
$C_4H_{10}O$	2-Methyl-2-propanol	802.6	772.2	
$C_4H_{10}O$	Diethyl ether	828.4	801	
$C_4H_{10}O$	Methyl propyl ether	814.9	785.7	
$C_4H_{10}O$	Isopropyl methyl ether	826.3	797.1	
$C_4H_{10}O_2$	1,4-Butanediol	915.6	854.9	
$C_4H_{10}O_2$	1,2-Dimethoxyethane	858.0	820.2	
$C_4H_{10}O_3$	1,2,4-Butanetriol	905.9	841	
$C_4H_{10}S$	1-Butanethiol	801.7	770.5	
$C_4H_{10}S$	2-Methyl-1-propanethiol	802.6	771.4	
$C_4H_{10}S$	2-Methyl-2-propanethiol	816.4	785.1	
$C_4H_{10}S$	Diethyl sulfide	856.7	827.0	
$C_4H_{11}N$	Butylamine	921.5	886.6	
$C_4H_{11}N$	tert-Butylamine	934.1	899.9	
$C_4H_{11}N$	Isobutylamine	924.8	890.8	
$C_4H_{11}N$	Diethylamine	952.4	919.4	
$C_4H_{11}N$	Isopropylmethylamine	952.4	919.4	
$C_4H_{11}N$	Ethyldimethylamine	960.1	929.1	
$C_4H_{11}NO$	N-Ethyl-N-hydroxyethanamine	914.7	882.2	
$C_4H_{11}NO$	4-Amino-1-butanol	984.5	932.1	
$C_4H_{11}NO_2$	Diethanolamine	953	920	
$C_4H_{12}N_2$	1,4-Butanediamine	1005.6	954.3	
$C_4H_{12}N_2$	N,N'-Dimethyl-1,2-ethanediamine	989.2	946.9	
$C_4H_{12}Sn$	Tetramethylstannane	823.7	797.4	
$C_4H_{14}OSi_2$	1,1,3,3-Tetramethyldisiloxane	845.3	814.6	

Molecular formula	Name	E_{pa} kJ/mol	$\Delta_{base}G°$ kJ/mol	Notes
C_4NiO_4	Nickel carbonyl	742.3	716.0	
C_5F_5N	Perfluoropyridine	764.9	733.0	
C_5FeO_5	Iron pentacarbonyl	833.0	798.5	
$C_5H_3ClN_4$	6-Chloro-1H-purine	873.6	841.7	
C_5H_4BrN	2-Bromopyridine	904.8	873.0	
C_5H_4BrN	3-Bromopyridine	910.0	878.2	
C_5H_4BrN	4-Bromopyridine	917.8	886.0	
C_5H_4ClN	2-Chloropyridine	900.9	869	
C_5H_4ClN	3-Chloropyridine	903.4	871.5	
C_5H_4ClN	4-Chloropyridine	916.1	884.2	
C_5H_4FN	3-Fluoropyridine	902.0	870.1	
C_5H_4FN	2-Fluoropyridine	884.6	852.7	
$C_5H_4N_2O_2$	4-Nitropyridine	874.3	842.5	
$C_5H_4N_2O_3$	4-Nitropyridine 1-oxide	868.0	837.3	
$C_5H_4N_4$	1H-Purine	920.1	888.2	
$C_5H_4N_4O$	Hypoxanthine	912.3	880.5	
C_5H_5	Cyclopentadienyl	831.5	799.1	
C_5H_5N	Pyridine	930	898.1	
C_5H_5NO	3-Pyridinol	929.5	897.7	
C_5H_5NO	Pyridine-1-oxide	923.6	892.9	
$C_5H_5N_5$	Adenine	942.8	912.5	
$C_5H_5N_5O$	Guanine	959.5	927.6	
C_5H_6	1,3-Cyclopentadiene	821.6	798.4	
$C_5H_6N_2$	2-Pyridinamine	947.2	915.3	
$C_5H_6N_2$	3-Pyridinamine	954.4	922.6	
$C_5H_6N_2$	4-Pyridinamine	979.7	947.8	
$C_5H_6N_2O_2$	Thymine	880.9	850.0	
C_5H_6O	2-Methylfuran	865.9	833.5	
C_5H_6O	3-Methylfuran	854.0	821.5	
$C_5H_6O_3$	Glutaric anhydride	816		Ref. 9
$C_5H_6O_3$	3-Methylsuccinic anhydride	807		Ref. 9
C_5H_6S	2-Methylthiophene	859.0	826.5	
$C_5H_7F_3O_2$	Propyl trifluoroacetate	763.9	732.9	
C_5H_8	$trans$-1,3-Pentadiene	834.1	804.4	
C_5H_8	2-Methyl-1,3-butadiene	826.4	797.6	
C_5H_8	2-Pentyne	810.2	778.0	
C_5H_8	3-Methyl-1-butyne	814.9	787.8	
C_5H_8	Cyclopentene	766.3	733.8	
C_5H_8	1-Methylcyclobutene	841.5	807.3	
C_5H_8	Vinylcyclopropane	816.3	787.5	
C_5H_8	3,3-Dimethylcyclopropene	847.8	817.1	
$C_5H_8N_2$	1,3-Dimethyl-1H-pyrazole	933.9	902.3	
$C_5H_8N_2$	1,4-Dimethyl-1H-imidazole	976.7	944.9	
$C_5H_8N_2$	1,5-Dimethyl-1H-pyrazole	934.3	902.8	
$C_5H_8N_2$	3,4-Dimethyl-1H-pyrazole	927.3	895.4	
$C_5H_8N_2$	3,5-Dimethyl-1H-pyrazole	933.5	900.1	
$C_5H_8N_2$	1,2-Dimethyl-1H-imidazole	984.7	952.6	
$C_5H_8N_2$	1,5-Dimethyl-1H-imidazole	977.6	945.8	
C_5H_8O	$trans$-2-Pentenal	839.0	807.2	
C_5H_8O	3-Methyl-2-butenal	856.9	825.0	
C_5H_8O	3-Methyl-3-buten-2-one	843.1	811.3	
C_5H_8O	Cyclopropyl methyl ketone	854.9	823	
C_5H_8O	Cyclopentanone	823.7	794.0	
C_5H_8O	3,4-Dihydro-2H-pyran	865.8	833.4	
$C_5H_8O_2$	3-Methyl-2-butenoic acid	822.9	791.9	
$C_5H_8O_2$	cis-2-Methyl-2-butenoic acid	822.5	791.5	
$C_5H_8O_2$	Cyclobutanecarboxylic acid	817.4	786.4	
$C_5H_8O_2$	Methyl trans-2-butenoate	851.3	820.4	
$C_5H_8O_2$	Methyl methacrylate	831.4	800.5	
$C_5H_8O_2$	Methyl cyclopropanecarboxylate	842.1	811.2	

Molecular formula	Name	E_{pa} kJ/mol	$\Delta_{base}G°$ kJ/mol	Notes
$C_5H_8O_2$	2,4-Pentanedione	873.5	836.8	
C_5H_9N	2-Isocyano-2-methylpropane	870.7	838.3	
C_5H_9N	3-(Dimethylamino)-1-propyne	940.3	909.5	
C_5H_9N	Pentanenitrile	802.4	771.7	
C_5H_9N	2,2-Dimethylpropanenitrile	810.9	780.2	
C_5H_9NO	3-Ethoxypropanenitrile	807.2	776.5	
C_5H_9NO	N,N-Dimethyl-2-propenamide	904.3	873.4	
C_5H_9NO	N-Methyl-2-pyrrolidone	923.5	891.6	
$C_5H_9NO_2$	L-Proline	920.5	886.0	
$C_5H_9NO_4$	L-Glutamic acid	913.0	879.1	
$C_5H_9N_3$	Histamine	999.8	961.9	
C_5H_{10}	2-Methyl-2-butene	808.8	779.9	
$C_5H_{10}N_2O$	1,3-Dimethyl-2-imidazolidinone	918.4	886.0	
$C_5H_{10}N_2O_3$	L-Glutamine	937.8	900	
$C_5H_{10}O$	Allyl ethyl ether	833.7	804.5	
$C_5H_{10}O$	Pentanal	796.6	764.8	
$C_5H_{10}O$	2-Pentanone	832.7	800.9	
$C_5H_{10}O$	3-Pentanone	836.8	807	
$C_5H_{10}O$	3-Methyl-2-butanone	836.3	804.4	
$C_5H_{10}O$	Tetrahydropyran	822.8	795.4	
$C_5H_{10}O$	2-Methyltetrahydrofuran	840.8	811.6	
$C_5H_{10}O_2$	Butyl formate	806.0	775	
$C_5H_{10}O_2$	Propyl acetate	836.6	805.6	
$C_5H_{10}O_2$	Isopropyl acetate	836.6	805.6	
$C_5H_{10}O_2$	Methyl butanoate	836.4	805.4	
$C_5H_{10}O_2$	Methyl isobutanoate	836.6	805.7	
$C_5H_{10}O_2$	cis-1,2-Cyclopentanediol	885.6	853.1	
$C_5H_{10}S$	Thiacyclohexane	855.8	826.0	
$C_5H_{11}N$	Allyldimethylamine	957.8	926.8	
$C_5H_{11}N$	Piperidine	954.0	921	
$C_5H_{11}N$	N-Methylpyrrolidine	965.6	934.8	
$C_5H_{11}NO$	2,2-Dimethylpropanamide	889.0	857.2	
$C_5H_{11}NO_2$	L-Valine	910.6	876.7	
$C_5H_{11}NO_2S$	L-Methionine	935.4	901.5	
$C_5H_{12}N_2O$	Tetramethylurea	930.6	899.6	
$C_5H_{12}N_2S$	Tetramethylthiourea	947.6	916.6	
$C_5H_{12}O$	2,2-Dimethyl-1-propanol	795.5	765.2	
$C_5H_{12}O$	Butyl methyl ether	820.3	791.2	
$C_5H_{12}O$	Methyl tert-butyl ether	841.6	812.4	
$C_5H_{12}O$	Ethyl isopropyl ether	842.7	813.5	
$C_5H_{12}S$	2,2-Dimethyl-1-propanethiol	809.5	778.2	
$C_5H_{12}Si$	Vinyltrimethylsilane	833	804.1	
$C_5H_{13}N$	Pentylamine	923.5	889.5	
$C_5H_{13}N$	2-Methyl-2-butanamine	937.8	903.6	
$C_5H_{13}N$	2,2-Dimethylpropylamine	928.3	894.0	
$C_5H_{13}N$	Ethylisopropylamine	960.0	926.7	
$C_5H_{13}N$	N,N-Dimethyl-1-propanamine	962.8	931.9	
$C_5H_{13}N$	Diethylmethylamine	971.0	940.0	
$C_5H_{13}N_3$	1,1,3,3-Tetramethylguanidine	1031.6	997.4	
$C_5H_{14}N_2$	N,N,N',N'-Tetramethylmethanediamine	952.2	919.8	
$C_5H_{14}N_2$	N,N-Dimethyl-1,3-propanediamine	1025.0	975.3	
$C_5H_{14}N_2$	1,5-Pentanediamine	999.6	946.2	
C_6CrO_6	Chromium carbonyl	739.2	714.6	
C_6F_6	Hexafluorobenzene	648.0	624.4	
C_6HF_5	Pentafluorobenzene	690.4	662.7	
$C_6H_2F_4$	1,2,3,4-Tetrafluorobenzene	700.4	672.7	
$C_6H_2F_4$	1,2,3,5-Tetrafluorobenzene	747.3	719.6	
$C_6H_2F_4$	1,2,4,5-Tetrafluorobenzene	746.5	718.8	
$C_6H_3F_3$	1,2,3-Trifluorobenzene	724.3	696.6	
$C_6H_3F_3$	1,2,4-Trifluorobenzene	729.5	699.4	

Molecular formula	Name	E_{pa} kJ/mol	$\Delta_{base}G°$ kJ/mol	Notes
$C_6H_3F_3$	1,3,5-Trifluorobenzene	741.9	715.4	
C_6H_4	Benzyne	841	808.5	
$C_6H_4F_2$	o-Difluorobenzene	731.2	703.5	
$C_6H_4F_2$	m-Difluorobenzene	749.7	722	
$C_6H_4F_2$	p-Difluorobenzene	718.7	692.8	
$C_6H_4N_2$	2-Pyridinecarbonitrile	872.9	841	
$C_6H_4N_2$	3-Pyridinecarbonitrile	877.0	845.1	
$C_6H_4N_2$	4-Pyridinecarbonitrile	880.6	848.8	
$C_6H_4O_2$	p-Benzoquinone	799.1	769.3	
C_6H_5	Phenyl	884	851.5	
C_6H_5Br	Bromobenzene	754.1	725.8	
C_6H_5Cl	Chlorobenzene	753.1	724.6	
C_6H_5F	Fluorobenzene	755.9	726.6	
C_6H_5NO	Nitrosobenzene	854.3	823.6	
C_6H_5NO	4-Pyridinecarboxaldehyde	904.6	872.8	
$C_6H_5NO_2$	Nitrobenzene	800.3	769.5	
$C_6H_5N_3$	Azidobenzene	820	787.5	
C_6H_5O	Phenoxy	873.2		Ref. 4
C_6H_6	Benzene	750.4	725.4	
C_6H_6BrN	3-Bromoaniline	873.2	841.4	
C_6H_6ClN	3-Chloroaniline	868.1	836.3	
C_6H_6ClN	4-Chloroaniline	873.8	842.0	
C_6H_6ClN	2-Chloro-4-methylpyridine	921.2	889.4	
C_6H_6ClN	2-Chloro-6-methylpyridine	908.0	876.2	
C_6H_6ClNO	2-Chloro-6-methoxypyridine	909.9	878.0	
C_6H_6FN	3-Fluoroaniline	867.3	835.5	
C_6H_6FN	4-Fluoroaniline	871.5	839.7	
C_6H_6IN	3-Iodoaniline	878.7	846.8	
C_6H_6N	Anilino	949.8	917.4	
$C_6H_6N_2O$	3-Pyridinecarboxamide	918.3	886.4	
$C_6H_6N_2O_2$	4-Nitroaniline	866.0	834.2	
$C_6H_6N_4$	6-Methyl-1H-purine	939.2	907.3	
C_6H_6O	Bis(2-propynyl) ether	783.9	756.5	
C_6H_6O	Phenol	817.3	786.3	
C_6H_7N	Bis(2-propynyl)amine	910.0	876.9	
C_6H_7N	Aniline	882.5	850.6	
C_6H_7N	2-Methylpyridine	949.1	917.3	
C_6H_7N	3-Methylpyridine	943.4	911.6	
C_6H_7N	4-Methylpyridine	947.2	915.3	
C_6H_7NO	1-Methyl-2(1H)-pyridinone	925.8	894.8	
C_6H_7NO	2-Aminophenol	898.8	866.9	
C_6H_7NO	3-Aminophenol	898.8	866.9	
C_6H_7NO	2-Methoxypyridine	934.7	902.8	
C_6H_7NO	3-Methoxypyridine	942.7	910.9	
C_6H_7NO	4-Methoxypyridine	961.7	929.8	
C_6H_7NO	3-Methylpyridine-1-oxide	935.2	902.8	
C_6H_8	1,3-Cyclohexadiene	837	804.5	
C_6H_8	1,4-Cyclohexadiene	837	808.0	
$C_6H_8N_2$	1,2-Benzenediamine	896.5	865.8	
$C_6H_8N_2$	1,3-Benzenediamine	929.9	899.2	
$C_6H_8N_2$	1,4-Benzenediamine	905.9	874.0	
$C_6H_8N_2O$	Bis(2-cyanoethyl) ether	813.8	786.4	
C_6H_8O	2,4-Dimethylfuran	894.7	862.3	
C_6H_8O	2,5-Dimethylfuran	865.9	835.2	
C_6H_8O	3,4-Dimethylfuran	869.0	838.3	
$C_6H_8O_2$	1,3-Cyclohexanedione	881.2	849.4	
$C_6H_8O_2$	1,4-Cyclohexanedione	812.5	782.7	
$C_6H_8O_2$	1,2-Cyclohexanedione	849.6	818.9	
$C_6H_8O_3$	4-Methylglutaric anhydride	820		Ref. 9
$C_6H_9F_3O_2$	Butyl trifluoroacetate	764.8	733.8	

Molecular formula	Name	E_{pa} kJ/mol	$\Delta_{base}G°$ kJ/mol	Notes
C_6H_9N	2,5-Dimethylpyrrole	918.7	887.1	
$C_6H_9N_3O_2$	*L*-Histidine	988	950.2	
C_6H_{10}	Methylenecyclopentane	832.4	803.5	
C_6H_{10}	(1-Methylvinyl)cyclopropane	871.6	842.7	
C_6H_{10}	2-Methyl-1,3-pentadiene	864.9	836	
C_6H_{10}	3-Methyl-1,3-pentadiene	852.3	823.4	
C_6H_{10}	2,3-Dimethyl-1,3-butadiene	835.0	807.8	
C_6H_{10}	1-Hexyne	799.8	774.8	
C_6H_{10}	2-Hexyne	806.1	781.1	
C_6H_{10}	Cyclohexene	784.5	752.0	
C_6H_{10}	1-Methylcyclopentene	816.5	787.1	
$C_6H_{10}N_2$	1,3,5-Trimethyl-1*H*-pyrazole	949.3	917.4	
$C_6H_{10}N_2$	3,4,5-Trimethyl-1*H*-pyrazole	949.3	916.0	
$C_6H_{10}O$	7-Oxabicyclo[2.2.1]heptane	844.2	816.8	
$C_6H_{10}O$	7-Oxabicyclo[4.1.0]heptane	848.1	815.6	
$C_6H_{10}O$	*trans*-3-Hexen-2-one	865.6	833.8	
$C_6H_{10}O$	Diallyl ether	827.4	800.0	
$C_6H_{10}O$	Cyclohexanone	841.0	811.2	
$C_6H_{10}O$	Mesityl oxide	878.7	846.9	
$C_6H_{10}O_2$	Cyclopentanecarboxylic acid	817.4	786.4	
$C_6H_{10}O_2$	2,5-Hexanedione	892.0	851.8	
$C_6H_{11}N$	*N*-Allyl-2-propen-1-amine	949.3	916.3	
$C_6H_{11}NO$	1-Methyl-2-piperidinone	924.4	892.6	
$C_6H_{11}N_3O_4$	*N*-(*N*-Glycylglycyl)glycine	966.8	916.8	
C_6H_{12}	1-Hexene	805.2	776.3	
C_6H_{12}	2-Methyl-2-pentene	812	783.1	
C_6H_{12}	2,3-Dimethyl-2-butene	813.9	785.9	
C_6H_{12}	Cyclohexane	686.9	666.9	
$C_6H_{12}N_2$	Triethylenediamine	963.4	934.6	
$C_6H_{12}N_2O_3$	*N-L*-Alanyl-*L*-alanine		905.6	
$C_6H_{12}O$	Oxepane	834.2	806.8	
$C_6H_{12}O$	3-Hexanone	843.2	811.3	
$C_6H_{12}O$	3,3-Dimethyl-2-butanone	840.1	808.2	
$C_6H_{12}O_2$	*cis*-1,3-Cyclohexanediol	882.2	849.7	
$C_6H_{12}O_2$	*trans*-1,3-Cyclohexanediol	828.6	797.9	
$C_6H_{12}O_2$	Methyl 2,2-dimethylpropanoate	845.2	814.2	
$C_6H_{12}O_2$	Diacetone alcohol	822.9	791.1	
$C_6H_{12}O_6$	α-*D*-Glucose		778.9	
$C_6H_{12}O_6$	β-*D*-Glucose		778.9	
$C_6H_{13}N$	*N,N*,2-Trimethylpropenylamine	967.0	934.5	
$C_6H_{13}N$	Cyclohexylamine	934.4	899.6	
$C_6H_{13}N$	1-Methylpiperidine	971.1	940.1	
$C_6H_{13}N$	Hexahydro-1*H*-azepine	956.7	923.5	
$C_6H_{13}NO$	*N,N*-Dimethylbutanamide	921.7	890.8	
$C_6H_{13}NO$	*N,N*-Diethylacetamide	925.4	894.4	
$C_6H_{13}NO_2$	*L*-Leucine	914.6	880.6	
$C_6H_{13}NO_2$	*L*-Isoleucine	917.4	883.5	
$C_6H_{14}N_2O_2$	*L*-Lysine	996	951.0	
$C_6H_{14}N_4O_2$	*L*-Arginine	1051.0	1006.6	
$C_6H_{14}O$	Dipropyl ether	837.9	810.5	
$C_6H_{14}O$	Diisopropyl ether	855.5	828.1	
$C_6H_{14}O$	*tert*-Butyl ethyl ether	856.0	826.9	
$C_6H_{14}O_3$	Diethylene glycol dimethyl ether	918.8	870.9	
$C_6H_{14}S$	Dipropyl sulfide	864.7	834.9	
$C_6H_{14}S$	Diisopropyl sulfide	876.4	846.6	
$C_6H_{15}N$	Butyldimethylamine	969.2	938.2	
$C_6H_{15}N$	Isobutyldimethylamine	968.7	937.8	
$C_6H_{15}N$	Hexylamine	927.5	893.5	
$C_6H_{15}N$	Dipropylamine	962.3	929.3	
$C_6H_{15}N$	Diisopropylamine	971.9	938.6	

Molecular formula	Name	E_{pa} kJ/mol	$\Delta_{base}G°$ kJ/mol	Notes
$C_6H_{15}N$	Triethylamine	981.8	951	
$C_6H_{15}NO$	6-Amino-1-hexanol	969.0	915.7	
$C_6H_{15}OP$	Triethylphosphine oxide	936.6	906.8	
$C_6H_{15}O_4P$	Triethyl phosphate	909.3	879.6	
$C_6H_{15}P$	Triethylphosphine	984.5	952.0	
$C_6H_{16}N_2$	1,6-Hexanediamine	999.5	946.2	
$C_6H_{16}N_2$	N,N,N',N'-Tetramethyl-1,2-ethanediamine	1012.8	970.6	
$C_6H_{16}OSi$	Triethylsilanol	822.1	794.8	
$C_6H_{18}N_3OP$	Hexamethylphosphoric triamide	958.6	928.7	
$C_6H_{18}N_3P$	Hexamethylphosphorous triamide	930.1	897.7	
$C_6H_{18}OSi_2$	Hexamethyldisiloxane	846.4	816.2	
C_6MoO_6	Molybdenum hexacarbonyl	762.6	738.1	
C_6O_6W	Tungsten carbonyl	758.0	733.4	
$C_7H_4N_2O_2$	3-Nitrobenzonitrile	781.4	750.7	
$C_7H_4N_2O_2$	4-Nitrobenzonitrile	775.7	745.1	
C_7H_5ClO	3-Chlorobenzaldehyde	813.0	781.1	
C_7H_5ClO	4-Chlorobenzaldehyde	831.3	799.4	
C_7H_5FO	3-Fluorobenzaldehyde	814.3	782.5	
C_7H_5FO	4-Fluorobenzaldehyde	827.1	795.3	
C_7H_5N	Benzonitrile	811.5	780.9	
C_7H_5N	Isocyanobenzene	868.4	836.0	
C_7H_5NO	Benzoxazole	891.6	859.8	
$C_7H_5NO_3$	4-Nitrobenzaldehyde	795.1	763.2	
C_7H_6ClNO	3-Chlorobenzamide	877.2	846.3	
C_7H_6ClNO	4-Chlorobenzamide	877.2	846.3	
C_7H_6F	m-Fluorobenzyl	836.5	804	
C_7H_6FNO	3-Fluorobenzamide	877.2	846.3	
C_7H_6FNO	4-Fluorobenzamide	877.2	846.3	
$C_7H_6F_3N$	3-(Trifluoromethyl)aniline	856.9	825.1	
$C_7H_6N_2$	1H-Benzimidazole	953.8	920.5	
$C_7H_6N_2$	1H-Indazole	900.8	868.9	
$C_7H_6N_2$	3-Aminobenzonitrile	842.3	810.4	
$C_7H_6N_2$	1H-Pyrrolo[2,3-b]pyridine	940.2	908.3	
$C_7H_6N_2O_3$	4-Nitrobenzamide	845.3	814.4	
$C_7H_6N_2O_3$	3-Nitrobenzamide	854.2	823.2	
C_7H_6O	Benzaldehyde	834.0	802.1	
C_7H_6O	2,4,6-Cycloheptatrien-1-one	920.8	891.0	
$C_7H_6O_2$	Benzoic acid	821.1	790.1	
C_7H_7	Benzyl	831.4	800.7	
C_7H_7Br	2-Bromotoluene	775.3	745.8	
C_7H_7Br	3-Bromotoluene	782.0	752.5	
C_7H_7Br	4-Bromotoluene	775.3	745.8	
C_7H_7Cl	2-Chlorotoluene	790.5	761.1	
C_7H_7Cl	3-Chlorotoluene	783.9	754.5	
C_7H_7Cl	4-Chlorotoluene	762.9	735.2	
C_7H_7F	2-Fluorotoluene	773.3	743.8	
C_7H_7F	3-Fluorotoluene	785.4	756.0	
C_7H_7F	4-Fluorotoluene	763.8	736.1	
C_7H_7I	1-Iodo-2-methylbenzene	780.3	750.8	
C_7H_7N	4-Vinylpyridine	944.1	912.3	
C_7H_7NO	1-(3-Pyridinyl)ethanone	916.2	884.3	
C_7H_7NO	1-(4-Pyridinyl)ethanone	914.7	882.9	
C_7H_7NO	4-Aminobenzaldehyde	910.4	878.6	
C_7H_7NO	Benzamide	892.1	861.2	
$C_7H_7NO_2$	Methyl 3-pyridinecarboxylate	925.6	893.8	
$C_7H_7NO_2$	Methyl 4-pyridinecarboxylate	926.6	894.7	
$C_7H_7NO_2$	Aniline-2-carboxylic acid	901.5	869.0	
$C_7H_7NO_2$	Aniline-3-carboxylic acid	864.7	832.3	
$C_7H_7NO_2$	Aniline-4-carboxylic acid	864.7	832.3	
$C_7H_7NO_2$	4-Nitrotoluene	815.2	782.7	

Molecular formula	Name	E_{pa} kJ/mol	$\Delta_{base}G°$ kJ/mol	Notes
$C_7H_7NO_3$	4-Nitrobenzenemethanol	810.5	778.0	
$C_7H_7N_3$	1-Methyl-1H-benzotriazole	931.2	898.7	
C_7H_7O	2-Methylphenoxy	874.5	842	
C_7H_8	Toluene	784.0	756.3	
C_7H_8	2,5-Norbornadiene	849.3	820.3	
$C_7H_8N_2O$	4-Aminobenzamide	927.9	896.9	
$C_7H_8N_2O$	3-Aminobenzamide	900.9	869.9	
$C_7H_8N_2O_2$	N-Methyl-4-nitroaniline	891.6	865.1	
C_7H_8O	o-Cresol	832	800	Ref. 10
C_7H_8O	m-Cresol	841	809	Ref. 10
C_7H_8O	p-Cresol	814	782	Ref. 10
C_7H_8O	Benzyl alcohol	778.3	748.0	
C_7H_8O	Anisole	839.6	807.2	
$C_7H_8O_2$	2,6-Dimethyl-4H-pyran-4-one	941.5	907.3	
$C_7H_8O_2S$	Methyl phenyl sulfone	812.7	780.3	
C_7H_8S	(Methylthio)benzene	872.6	843.7	
C_7H_9N	Benzylamine	913.3	879.4	
C_7H_9N	2-Methylaniline	890.9	859.1	
C_7H_9N	3-Methylaniline	895.8	864.0	
C_7H_9N	4-Methylaniline	896.7	864.8	
C_7H_9N	N-Methylaniline	916.6	890.1	
C_7H_9N	2-Ethylpyridine	952.4	920.6	
C_7H_9N	3-Ethylpyridine	947.4	915.5	
C_7H_9N	4-Ethylpyridine	951.1	919.2	
C_7H_9N	2,3-Dimethylpyridine	958.9	927.0	
C_7H_9N	2,4-Dimethylpyridine	962.9	930.8	
C_7H_9N	2,5-Dimethylpyridine	958.8	926.9	
C_7H_9N	2,6-Dimethylpyridine	963.0	931.1	
C_7H_9N	3,4-Dimethylpyridine	957.3	925.5	
C_7H_9N	3,5-Dimethylpyridine	955.4	923.5	
C_7H_9NO	2-Methoxyaniline	905.2	873.3	
C_7H_9NO	3-Methoxyaniline	913.0	881.1	
C_7H_9NO	4-Methoxyaniline	900.3	868.5	
C_7H_{10}	Bicyclo[2.2.1]hept-2-ene	836.5	804.0	
$C_7H_{10}N_2$	N,N-Dimethyl-2-pyridinamine	968.2	941.6	
$C_7H_{10}N_2$	N,N-Dimethyl-4-pyridinamine	997.6	971.1	
$C_7H_{10}O$	Dicyclopropyl ketone	880.4	850.6	
$C_7H_{10}O$	Bicyclo[2.2.1]heptan-2-one	847.4	815.5	
$C_7H_{11}N$	Cyclohexanecarbonitrile	815.0	784.4	
C_7H_{12}	2,4-Dimethyl-1,3-pentadiene	886.5	857.6	
C_7H_{12}	1-Methylcyclohexene	825.1	792.6	
C_7H_{12}	1,2-Dimethylcyclopentene	822.6	791.9	
$C_7H_{12}N_2$	2,3,4,6,7,8-Hexahydropyrrolo[1,2-a]pyrimidine	1038.3	1005.9	
$C_7H_{12}O$	Cycloheptanone	845.6	815.9	
$C_7H_{12}O$	4-Methylcyclohexanone	844.9	813.0	
$C_7H_{12}O_2$	Cyclohexanecarboxylic acid	823.8	792.8	
$C_7H_{13}N$	1-Azabicyclo[2.2.2]octane	983.3	952.5	
C_7H_{14}	2,4-Dimethyl-2-pentene	812	783.1	
$C_7H_{14}O$	Methoxycyclohexane	840.5	811.3	
$C_7H_{14}O$	4-Heptanone	845.0	815.3	
$C_7H_{14}O$	2,4-Dimethyl-3-pentanone	850.3	820.5	
$C_7H_{14}O$	Cyclohexanemethanol	802.1	771.7	
$C_7H_{15}N$	Cyclohexanemethanamine	926.6	895.8	
$C_7H_{16}O$	tert-Butyl isopropyl ether	870.7	841.5	
$C_7H_{17}N$	Heptylamine	923.2	889.3	
$C_7H_{17}N$	Methyldipropylamine	983.5	950.9	
$C_7H_{17}N$	Diethylpropylamine	978.8	947.9	
$C_7H_{18}N_2$	1,7-Heptanediamine	998.5	944.9	
$C_7H_{18}N_2$	N,N,N',N'-Tetramethyl-1,3-propanediamine	1035.2	985.4	

Molecular formula	Name	E_{pa} kJ/mol	$\Delta_{base}G°$ kJ/mol	Notes
$C_8H_4F_3N$	3-(Trifluoromethyl)benzonitrile	791.4	760.8	
$C_8H_4F_3N$	4-(Trifluoromethyl)benzonitrile	787.2	758.3	
$C_8H_4N_2$	m-Dicyanobenzene	779.3	750.4	
$C_8H_4N_2$	p-Dicyanobenzene	779.0	751.8	
C_8H_5Cl	1-Chloro-4-ethynylbenzene	832.4	801.7	
$C_8H_5Cl_3O$	2,2,2-Trichloro-1-phenylethanone	818.9	787.0	
$C_8H_5F_3O$	2,2,2-Trifluoro-1-phenylethanone	799.2	767.4	
$C_8H_5F_3O$	4-(Trifluoromethyl)benzaldehyde	805.6	773.8	
C_8H_5NO	4-Formylbenzonitrile	796.9	766.3	
C_8H_6	Phenylacetylene	832.0	801.3	
C_8H_6ClN	4-(Chloromethyl)benzonitrile	812.8	782.1	
C_8H_6ClN	3-(Chloromethyl)benzonitrile	811.2	780.6	
$C_8H_6N_2$	Quinoxaline	903.8	873.7	
$C_8H_6N_2$	Cinnoline	936.3	904.4	
C_8H_7Br	1-Bromo-4-vinylbenzene	838.7	809.8	
C_8H_7Br	1-Bromo-3-vinylbenzene	822.4	793.5	
C_8H_7ClO	1-(3-Chlorophenyl)ethanone	846.9	815.1	
C_8H_7ClO	1-(4-Chlorophenyl)ethanone	856.6	824.8	
$C_8H_7ClO_2$	Methyl 4-chlorobenzoate	842.1	811.1	
$C_8H_7ClO_2$	Methyl 3-chlorobenzoate	835.4	804.4	
C_8H_7FO	1-(4-Fluorophenyl)ethanone	858.6	826.8	
C_8H_7N	Benzeneacetonitrile	805.5	774.8	
C_8H_7N	1H-Indole	933.4	901.9	
$C_8H_7NO_3$	1-(4-Nitrophenyl)ethanone	824.3	792.5	
$C_8H_7NO_3$	1-(3-Nitrophenyl)ethanone	826.0	794.1	
$C_8H_7NO_4$	Methyl 3-nitrobenzoate	815.7	784.7	
$C_8H_7NO_4$	Methyl 4-nitrobenzoate	813.2	782.3	
C_8H_8	Styrene	839.5	809.2	
$C_8H_8N_2$	1-Methyl-1H-benzimidazole	967.0	935.2	
$C_8H_8N_2$	2-Methyl-2H-indazole	941.4	909.6	
$C_8H_8N_2$	1-Methyl-1H-indazole	922.4	890.5	
C_8H_8O	3-Methylbenzaldehyde	840.0	808.1	
C_8H_8O	4-Methylbenzaldehyde	851.8	820.0	
C_8H_8O	Acetophenone	861.1	829.3	
$C_8H_8O_2$	o-Toluic acid	838.8	807.8	
$C_8H_8O_2$	m-Toluic acid	829.8	798.8	
$C_8H_8O_2$	p-Toluic acid	836.7	805.7	
$C_8H_8O_2$	Methyl benzoate	850.5	819.5	
$C_8H_8O_2$	3-Methoxybenzaldehyde	844.1	812.2	
$C_8H_8O_2$	4-Methoxybenzaldehyde	881.1	849.3	
$C_8H_8O_2$	1-(3-Hydroxyphenyl)ethanone	863.6	831.8	
$C_8H_8O_2$	1-(4-Hydroxyphenyl)ethanone	883.7	851.9	
$C_8H_8O_3$	Methyl 4-hydroxybenzoate	863.4	832.5	
$C_8H_8O_3$	Methyl 3-hydroxybenzoate	850.0	819.1	
C_8H_9N	2,3-Dihydro-1H-indole	957.1	926.3	
C_8H_9NO	3-Methylbenzamide	900.9	869.9	
C_8H_9NO	4-Methylbenzamide	900.9	869.9	
C_8H_9NO	1-(4-Aminophenyl)ethanone	908.8	877.0	
$C_8H_9NO_2$	3-Methoxybenzamide	900.9	869.9	
$C_8H_9NO_2$	2,4-Dimethyl-1-nitrobenzene	831.0	798.5	
$C_8H_9NO_2$	Methyl 4-aminobenzoate	883.9	853.0	
$C_8H_9NO_2$	4-Methoxybenzamide	900.3	869.4	
C_8H_{10}	Ethylbenzene	788.0	760.3	
C_8H_{10}	o-Xylene	796.0	768.3	
C_8H_{10}	m-Xylene	812.1	786.2	
C_8H_{10}	p-Xylene	794.4	766.8	
$C_8H_{10}ClN$	4-Chloro-N,N-dimethylaniline	922.9	896.4	
$C_8H_{10}N_2O_2$	N,N-Dimethyl-4-nitroaniline	896.7	870.2	
$C_8H_{10}N_2O_2$	N,N-Dimethyl-3-nitroaniline	894.1	867.6	
$C_8H_{10}O$	Benzyl methyl ether	816.7	787.5	

Molecular formula	Name	E_{pa} kJ/mol	$\Delta_{base}G°$ kJ/mol	Notes
$C_8H_{10}O$	2-Methylanisole	850	818	Ref. 10
$C_8H_{10}O$	3-Methylanisole	860	828	Ref. 10
$C_8H_{10}O$	4-Methylanisole	841	809	Ref. 10
$C_8H_{11}N$	4-Isopropylpyridine	955.7	923.8	
$C_8H_{11}N$	3-Ethylaniline	897.9	866.1	
$C_8H_{11}N$	N-Ethylaniline	924.8	892.9	
$C_8H_{11}N$	N,N-Dimethylaniline	941.1	909.2	
$C_8H_{11}N$	2,6-Dimethylaniline	901.7	869.8	
$C_8H_{11}N$	Benzeneethanamine	936.2	902.3	
$C_8H_{11}N$	2-Propylpyridine	955.7	923.8	
C_8H_{12}	2-Methyl-2-norbornene	845	812.5	
$C_8H_{12}N_2$	N,N-Dimethyl-1,4-benzenediamine	955.0	928.4	
$C_8H_{12}N_2O_2$	Ethyl 1,5-dimethyl-1H-pyrazole-3-carboxylate	933.4	901.5	
$C_8H_{14}O$	Cyclooctanone	849.4	819.6	
$C_8H_{14}O$	1-Cyclohexylethanone	841.4	809.5	
$C_8H_{14}O_2$	Methyl cyclohexanecarboxylate	846.2	815.3	
$C_8H_{16}O$	2,2,4-Trimethyl-3-pentanone	856.9	825.0	
$C_8H_{17}N$	Cyclohexyldimethylamine	983.6	952.6	
$C_8H_{18}O$	Dibutyl ether	845.7	818.3	
$C_8H_{18}O$	Di-sec-butyl ether	865.9	838.5	
$C_8H_{18}O$	Di-tert-butyl ether	887.4	860.0	
$C_8H_{18}O_4$	Triethylene glycol dimethyl ether	946.6	892.4	
$C_8H_{18}O_5$	Tetraethylene glycol		>910	
$C_8H_{18}S$	Dibutyl sulfide	871.8	842.1	
$C_8H_{18}S$	Di-tert-butyl sulfide	893.8	864.0	
$C_8H_{19}N$	N-Ethyl-N-isopropyl-2-propanamine	994.3	963.5	
$C_8H_{19}N$	Octylamine	928.9	895.0	
$C_8H_{19}N$	Dibutylamine	968.5	935.3	
$C_8H_{19}N$	Di-sec-butylamine	980.7	947.5	
$C_8H_{19}N$	Diisobutylamine	958.1	925.1	
$C_8H_{20}N_2$	N,N,N′,N′-Tetramethyl-1,4-butanediamine	1046.3	992.7	
$C_8H_{20}N_2$	Tetraethylhydrazine	964.3	935.3	
$C_9H_7MnO_3$	Manganese 2-methylcyclopentadienyl tricarbonyl	833.8	801.3	
C_9H_7N	Quinoline	953.2	921.4	
C_9H_7N	Isoquinoline	951.7	919.9	
C_9H_7NO	4-Acetylbenzonitrile	826.8	795.0	
C_9H_8	Indene	848.8	819.6	
C_9H_8O	2-Methylbenzofuran	859.6	827.2	
$C_9H_8O_3$	Methyl 4-formylbenzoate	832.9	801.9	
C_9H_9Cl	1-Chloro-4-isopropenylbenzene	854.3	825.4	
$C_9H_9ClO_2$	3-Chloro-4-methoxyacetophenone	883.7	851.9	
C_9H_{10}	2-Methylstyrene	855.2	826.3	
C_9H_{10}	3-Methylstyrene	849.4	820.5	
C_9H_{10}	4-Methylstyrene	861.7	832.8	
C_9H_{10}	cis-1-Propenylbenzene	836.4	807.5	
C_9H_{10}	trans-1-Propenylbenzene	834.2	805.3	
C_9H_{10}	Isopropenylbenzene	864.2	835.3	
C_9H_{10}	Cyclopropylbenzene	834.9	802.4	
$C_9H_{10}N_2$	4-(Dimethylamino)benzonitrile	889.1	862.6	
$C_9H_{10}O$	1-(3-Methylphenyl)ethanone	868.2	836.4	
$C_9H_{10}O$	1-Phenyl-1-propanone	867.4	835.6	
$C_9H_{10}O$	1-Phenyl-2-propanone	842.6	810.8	
$C_9H_{10}O$	4-Methylacetophenone	875.5	843.6	
$C_9H_{10}OS$	4-Acetylthioanisole	888.2	856.3	
$C_9H_{10}O_2$	Methyl 2-methylbenzoate	858.3	827.3	
$C_9H_{10}O_2$	Methyl 3-methylbenzoate	857.7	826.8	
$C_9H_{10}O_2$	1-(3-Methoxyphenyl)ethanone	871.2	839.3	
$C_9H_{10}O_2$	Methyl 4-methylbenzoate	861.5	830.6	

Molecular formula	Name	E_{pa} kJ/mol	$\Delta_{base}G°$ kJ/mol	Notes
$C_9H_{10}O_2$	4-Acetylanisole	895.6	863.7	
$C_9H_{10}O_3$	Methyl 4-methoxybenzoate	870.6	839.6	
$C_9H_{10}O_3$	Methyl 3-methoxybenzoate	856.7	825.8	
$C_9H_{11}N$	5,6,7,8-Tetrahydroquinoline	966.0	934.1	
$C_9H_{11}N$	5,6,7,8-Tetrahydroisoquinoline	966.6	934.7	
$C_9H_{11}NO$	4-(Dimethylamino)benzaldehyde	924.8	898.3	
$C_9H_{11}NO$	N,N-Dimethylbenzamide	932.7	901.8	
$C_9H_{11}NO_2$	1,3,5-Trimethyl-2-nitrobenzene	823.8	793.1	
$C_9H_{11}NO_2$	L-Phenylalanine	922.9	888.9	
$C_9H_{11}NO_3$	L-Tyrosine	926	892.1	
C_9H_{12}	Propylbenzene	790.1	762.4	
C_9H_{12}	Isopropylbenzene	791.6	763.9	
C_9H_{12}	1,3,5-Trimethylbenzene	836.2	808.6	
$C_9H_{12}N_2$	3-(2-Pyrrolidinyl)pyridine, (S)-	964.0	931.0	
$C_9H_{12}N_2O_6$	Uridine	947.6	916.6	
$C_9H_{12}O_3$	1,3,5-Trimethoxybenzene	926.7	898.2	
$C_9H_{13}N$	N-Ethyl-N-methylaniline	939.0	912.4	
$C_9H_{13}N$	2,6-Diethylpyridine	972.3	940.4	
$C_9H_{13}N$	4-tert-Butylpyridine	957.7	925.8	
$C_9H_{13}N$	2-tert-Butylpyridine	961.7	929.8	
$C_9H_{13}N$	2-Methyl-N,N-dimethylaniline	951.8	925.3	
$C_9H_{13}N$	3-Methyl-N,N-dimethylaniline	942.1	915.7	
$C_9H_{13}N$	4-Methyl-N,N-dimethylaniline	950.0	918.1	
$C_9H_{13}N$	N,N-Dimethylbenzylamine	968.4	937.4	
$C_9H_{13}NO$	4-Methoxy-N,N-dimethylaniline	949.1	922.4	
$C_9H_{13}N_3O_5$	Cytidine	982.5	950.0	
$C_9H_{14}O$	Isophorone	893.5	861.6	
$C_9H_{15}N$	N,N-Diallyl-2-propen-1-amine	972.3	941.3	
$C_9H_{15}N$	N-(1-Cyclopenten-1-yl)pyrrolidine	1019.2	988.4	
$C_9H_{16}O$	Cyclononanone	852.6	822.8	
$C_9H_{17}N_3O_4$	N-(N-L-Alanyl-L-alanyl)-L-alanine		924.1	
$C_9H_{18}O$	5-Nonanone	853.7	821.9	
$C_9H_{18}O$	Di-tert-butyl ketone	861.3	831.5	
$C_9H_{19}N$	1-Isobutylpiperidine	974.5	943.5	
$C_9H_{19}N$	2,2,6,6-Tetramethylpiperidine	987.0	953.9	
$C_9H_{21}N$	Tripropylamine	991.0	960.1	
$C_{10}H_8$	Naphthalene	802.9	779.4	
$C_{10}H_8$	Azulene	925.2	896	
$C_{10}H_8N_2$	2,2'-Bipyridine	965		Ref. 7
$C_{10}H_9N$	1-Naphthylamine	907.0	875.1	
$C_{10}H_{10}Fe$	Ferrocene	863.6	841.3	
$C_{10}H_{10}N_2$	1,8-Naphthalenediamine	944.5	912.1	
$C_{10}H_{10}N_2$	1-Methyl-3-phenyl-1H-pyrazole	932.6	900.8	
$C_{10}H_{10}N_2$	1-Methyl-5-phenyl-1H-pyrazole	932.4	900.5	
$C_{10}H_{10}Ni$	Nickelocene	935.7	907.3	
$C_{10}H_{10}O_2$	1,4-Diacetylbenzene	850.8	821.0	
$C_{10}H_{10}O_2$	1,3-Diacetylbenzene	852.0	822.3	
$C_{10}H_{10}O_3$	4-Acetylphenyl acetate	853.2	821.3	
$C_{10}H_{10}O_4$	Dimethyl isophthalate	843.5	814.3	
$C_{10}H_{10}O_4$	Dimethyl terephthalate	843.2	812.3	
$C_{10}H_{10}Ru$	Ruthenocene	899.1	876.8	
$C_{10}H_{12}$	1-Methyl-3-(1-methylvinyl)benzene	867.6	838.7	
$C_{10}H_{12}$	1-Methyl-4-(1-methylvinyl)benzene	881.8	852.9	
$C_{10}H_{12}$	1-Methyl-2-(1-methylvinyl)benzene	857.8	828.9	
$C_{10}H_{12}$	1,2,3,4-Tetrahydronaphthalene	809.7	782.1	
$C_{10}H_{12}O$	1-(2,4-Dimethylphenyl)ethanone	882.6	850.8	
$C_{10}H_{12}O$	1-(2,5-Dimethylphenyl)ethanone	873.5	841.6	
$C_{10}H_{12}O$	1-(3,4-Dimethylphenyl)ethanone	882.8	851.0	
$C_{10}H_{12}O_2$	Methyl 2,5-dimethylbenzoate	864.7	833.7	
$C_{10}H_{12}O_2$	Methyl 2,4-dimethylbenzoate	868.2	837.2	

Molecular formula	Name	E_{pa} kJ/mol	$\Delta_{base}G°$ kJ/mol	Notes
$C_{10}H_{12}O_2$	Methyl 3,5-dimethylbenzoate	864.3	833.4	
$C_{10}H_{13}N$	1-Phenylpyrrolidine	941.6	915.1	
$C_{10}H_{13}NO$	4'-(Dimethylamino)acetophenone	932.8	906.3	
$C_{10}H_{13}NO$	N,N,3-Trimethylbenzamide	927.0	896.0	
$C_{10}H_{13}NO$	N,N,4-Trimethylbenzamide	927.0	896.0	
$C_{10}H_{13}NO$	1-[3-(Dimethylamino)phenyl]ethanone	928.0	901.5	
$C_{10}H_{13}N_5O_3$	2'-Deoxyadenosine	991.5	959.1	
$C_{10}H_{13}N_5O_4$	Adenosine	989.3	956.8	
$C_{10}H_{13}N_5O_5$	Guanosine	993.4	960.9	
$C_{10}H_{14}$	Butylbenzene	791.9	764.2	
$C_{10}H_{14}$	1,2,3,5-Tetramethylbenzene	845.6	816.5	
$C_{10}H_{14}ClN$	4-Chloro-N,N-diethylaniline	931.0	899.2	
$C_{10}H_{14}N_2$	L-Nicotine	963.4	932.6	
$C_{10}H_{14}N_2O$	N,N-Diethyl-3-pyridinecarboxamide	940.9	909.0	
$C_{10}H_{14}N_2O_5$	Thymidine	948.3	915.9	
$C_{10}H_{15}N$	N,N,2,6-Tetramethylaniline	954.1	923.2	
$C_{10}H_{15}N$	N,N,3,5-Tetramethylaniline	956.1	924.3	
$C_{10}H_{15}N$	N,N-Diethylaniline	959.8	927.9	
$C_{10}H_{16}N_2$	N,N,N',N'-Tetramethyl-1,2-benzenediamine	982.6	950.2	
$C_{10}H_{17}N$	Tricyclo[3.3.1.1³,⁷]decan-1-amine	948.8	916.3	
$C_{10}H_{22}O$	Dipentyl ether	852.7	825.3	
$C_{10}H_{22}O_5$	Tetraethylene glycol dimethyl ether	953.8	897.8	
$C_{10}H_{23}N$	Decylamine	930.4	896.5	
$C_{10}H_{24}N_2$	N,N,N',N'-Tetramethyl-1,6-hexanediamine	1035.8	982.2	
$C_{11}H_9N$	4-Phenylpyridine	939.7	907.8	
$C_{11}H_{10}$	1-Methylnaphthalene	834.8	805.3	
$C_{11}H_{10}$	2-Methylnaphthalene	831.9	802.4	
$C_{11}H_{12}N_2O_2$	L-Tryptophan	948.9	915	
$C_{11}H_{14}O_2$	Methyl 2,4,6-trimethylbenzoate	866.3	835.3	
$C_{11}H_{15}N$	1-Phenylpiperidine	952.9	926.4	
$C_{11}H_{15}N$	Tricyclo[3.3.1.1³,⁷]decane-1-carbonitrile	834.4	803.8	
$C_{11}H_{16}$	Pentamethylbenzene	850.7	823.5	
$C_{11}H_{17}N$	N,N-Diethyl-4-methylaniline	962.8	931.0	
$C_{11}H_{17}N$	2-Hexylpyridine	963.6	931.7	
$C_{11}H_{18}O$	1,4,7,7-Tetramethylbicyclo[2.2.1]heptan-2-one	863.3	831.4	
$C_{11}H_{24}O_4$	2,6,10,14-Tetraoxapentadecane		895.1	
$C_{12}H_8N_2$	Phenazine	938.4	908.3	
$C_{12}H_9NO$	Phenyl-3-pyridinylmethanone	934.1	902.3	
$C_{12}H_{10}$	Acenaphthene	851.7	821.0	
$C_{12}H_{10}$	Biphenyl	813.6	782.9	
$C_{12}H_{16}O$	1-(4-tert-Butylphenyl)ethanone	882.5	850.6	
$C_{12}H_{18}$	Hexamethylbenzene	860.6	836.0	
$C_{12}H_{18}O$	1-Tricyclo[3.3.1.1³,⁷]dec-1-ylethanone	864.9	833.1	
$C_{12}H_{19}N$	N,N-Dipropylaniline	963.0	931.1	
$C_{12}H_{20}O$	2,5-Di-tert-butylfuran	894.7	863.9	
$C_{12}H_{27}N$	Tributylamine	998.5	967.6	
$C_{13}H_9N$	Acridine	972.6	940.7	
$C_{13}H_{10}$	9H-Fluorene	831.5	803.8	
$C_{13}H_{10}O$	Benzophenone	882.3	852.5	
$C_{13}H_{12}$	2-Methylbiphenyl	815.9	783.4	
$C_{13}H_{12}$	3-Methylbiphenyl	828.0	795.5	
$C_{13}H_{12}$	4-Methylbiphenyl	817.9	785.4	
$C_{13}H_{12}$	Diphenylmethane	802.0	769.5	
$C_{13}H_{13}P$	Methyldiphenylphosphine	972.1	939.7	
$C_{13}H_{21}N$	2,4-Di-tert-butylpyridine	983.8	952.0	
$C_{13}H_{21}N$	2,6-Di-tert-butylpyridine	982.9	951	
$C_{13}H_{21}NO$	N,N-Dimethyltricyclo[3.3.1.1³,⁷]decane-1-carboxamide	949.4	917.6	

Molecular formula	Name	E_{pa} kJ/mol	$\Delta_{base}G°$ kJ/mol	Notes
$C_{14}H_{10}$	Anthracene	877.3	846.6	
$C_{14}H_{10}$	Phenanthrene	825.7	795.0	
$C_{14}H_{12}$	1,1-Diphenylethene	885.7	856.9	
$C_{14}H_{14}$	1,2-Diphenylethane	801.8	774.1	
$C_{14}H_{18}$	1,2,3,4,5,6,7,8-Octahydroanthracene	845.4	814.7	
$C_{14}H_{18}$	1,2,3,4,5,6,7,8-Octahydrophenanthrene	846.2	815.5	
$C_{14}H_{23}N$	4-Octylaniline	894.5	862	
$C_{15}H_{12}$	2-Methylanthracene	887.5	855.1	
$C_{15}H_{12}$	9-Methylanthracene	896.5	865.8	
$C_{15}H_{12}N_2$	3,5-Diphenyl-1H-pyrazole	946.3	912.7	
$C_{15}H_{16}$	1,3-Diphenylpropane	820.1	787.6	
$C_{15}H_{18}$	1,4-Dimethyl-7-isopropylazulene	983.1	950.6	
$C_{15}H_{24}$	1,3-Di-tert-butyl-5-methylbenzene	853.7	826.0	
$C_{16}H_{10}$	Fluoranthene	828.6	800.9	
$C_{16}H_{10}$	Pyrene	869.2	840.1	
$C_{16}H_{18}$	1,4-Diphenylbutane	822.0	779.8	
$C_{17}H_{20}$	1,5-Diphenylpentane	824.7	782.4	
$C_{18}H_{12}$	Chrysene	840.9	810.1	
$C_{18}H_{12}$	Naphthacene	905.5	876.5	
$C_{18}H_{12}$	Triphenylene	819.2	791.2	
$C_{18}H_{15}As$	Triphenylarsine	908.9	876.4	
$C_{18}H_{15}AsO$	Triphenylarsine oxide	906.2	876.4	
$C_{18}H_{15}N$	Triphenylamine	908.9	876.4	
$C_{18}H_{15}OP$	Triphenylphosphine oxide	906.2	876.4	
$C_{18}H_{15}P$	Triphenylphosphine	972.8	940.4	
$C_{18}H_{15}PS$	Triphenylphosphine sulfide	906.2	876.4	
$C_{18}H_{15}Sb$	Triphenylstibine	845.5	813.1	
$C_{18}H_{22}$	1,6-Diphenylhexane	826.1	783.8	
$C_{18}H_{30}$	1,3,5-Tri-tert-butylbenzene	848.8	822.3	
$C_{20}H_{12}$	Perylene	888.6	859.6	
$C_{22}H_{12}$	Benzo[ghi]perylene	876.0	845.2	
$C_{22}H_{14}$	Picene	851.3	820.6	
$C_{24}H_{12}$	Coronene	861.3	835.0	
C_{60}	Carbon (fullerene-C_{60})		827.5	
C_{70}	Carbon (fullerene-C_{70})		827.5	

ATOMIC AND MOLECULAR POLARIZABILITIES

Thomas M. Miller

The *polarizability* of an atom or molecule describes the response of the electron cloud to an external field. The atomic or molecular energy shift ΔW due to an external electric field E is proportional to E^2 for external fields that are weak compared to the internal electric fields between the nucleus and electron cloud. The *electric dipole polarizability* α is the constant of proportionality defined by $\Delta W = -\alpha E^2/2$. The induced electric dipole moment is αE. *Hyperpolarizabilities*, coefficients of higher powers of E, are less often required. Technically, the polarizability is a tensor quantity but for spherically symmetric charge distributions reduces to a single number. In any case, an *average polarizability* is usually adequate in calculations. Frequency-dependent or *dynamic polarizabilities* are needed for electric fields that vary in time, except for frequencies that are much lower than electron orbital frequencies, where *static polarizabilities* suffice.

Polarizabilities for atoms and molecules in excited states are found to be larger than for ground states and may be positive or negative. Molecular polarizabilities are very slightly temperature dependent since the size of the molecule depends on its rovibrational state. Only in the case of dihydrogen has this effect been studied enough to warrant consideration in Table 3.

Polarizabilities are normally expressed in c.g.s. units of cm³. Ground state polarizabilities are in the range of 10^{-24} cm³ = 1 Å³ and hence are often given in Å³ units. Theorists tend to use atomic units of a_o^3 where a_o is the Bohr radius. The conversion is $\alpha(\text{cm}^3) = 0.148184 \times 10^{-24} \times \alpha(a_o^3)$. Polarizabilities are only recently encountered in SI units, C m²/V = J/(V/m)². The conversion from c.g.s. units to SI units is $\alpha(\text{C m}^2/\text{V}) = 4\pi\varepsilon_o \times 10^{-6} \alpha(\text{cm}^3)$, where ε_o is the permittivity of free space in SI units and the factor 10^{-6} simply converts cm³ into m³. Thus, $\alpha(\text{C m}^2/\text{V}) = 1.11265 \times 10^{-16} \times \alpha(\text{cm}^3)$. Persons measuring excited state polarizabilities by optical methods tend to use units of MHz/(V/cm)², where the energy shift, ΔW, is expressed in frequency units with a factor of h understood. The polarizability is $-2\,\Delta W/E^2$. The conversion into c.g.s. units is $\alpha(\text{cm}^3) = 5.955214 \times 10^{-16} \times \alpha[\text{MHz}/(\text{V/cm})^2]$.

The polarizability appears in many formulas for low-energy processes involving the valence electrons of atoms or molecules. These formulas are given below in c.g.s. units: the polarizability α is in cm³; masses m or μ are in grams; energies are in ergs; and electric charges are in esu, where $e = 4.8032 \times 10^{-10}$ esu. The symbol $\alpha(\nu)$ denotes a frequency (ν) dependent polarizability, where $\alpha(\nu)$ reduces to the static polarizability a for $\nu = 0$. For further information, see Bonin, K. D., and Kresin, V. V., *Electric Dipole Polarizabilities of Atoms, Molecules, and Clusters*, World Scientific, Singapore, 1997; Bonin, K. D., and Kadar-Kallen, *Int. J. Mod. Phys. B*, 24, 3313, 1994; and Miller, T. M., and Bederson, B., *Advances in Atomic and Molecular Physics*, 13, 1, 1977, and Gould, H., and Miller, T. M., *Advances in Atomic, Molecular, and Optical Physics*, 51, 243, 2005. Details on polarizability-related interactions, especially in regard to hyperpolarizabilities and nonlinear optical phenomena, are given by Bogaard, M. P., and Orr, B. J., in *Physical Chemistry, Series Two, Vol. 2, Molecular Structure and Properties*, Buckingham, A. D., Ed., Butterworths, London, 1975, pp. 149–194. A tabulation of tensor and hyperpolarizabilities is included. The gas number density, n, in Table 1 is usually taken to be that of 1 atm at 0 °C in reporting experimental data.

TABLE 1. Formulas Involving Polarizability

Description	Formula	Remarks
Lorentz-Lorenz relation	$\alpha(\nu) = \dfrac{3}{4\pi n}\left[\dfrac{\eta^2(\nu)-1}{\eta^2(\nu)+2}\right]$	For a gas of atoms or nonpolar molecules; the index of refraction is $\eta(\nu)$
Refraction by polar molecules	$\alpha(\nu) + \dfrac{d^2}{3kT} = \dfrac{3}{4\pi n}\left[\dfrac{\eta^2(\nu)-1}{\eta^2(\nu)+2}\right]$	The dipole moment is d, in esu·cm (= 10^{-18} D)
Dielectric constant (dimensionless)	$\kappa(\nu) = 1 + 4\pi n\ \alpha(\nu)$	From the Lorentz-Lorenz relation for the usual case of $\kappa(\nu) \approx 1$
Index of refraction (dimensionless)	$\eta(\nu) = 1 + 2\pi n\ \alpha(\nu)$	From $\eta^3(\nu) = \kappa(\nu)$
Diamagnetic susceptibility	$\chi_m = e^2 (a_o N\alpha)^{1/2} / 4m_e c^2$	From the approximation that the static polarizability is given by the variational formula $\alpha = (4/9a_o)\Sigma(N_i r_i^2)^2$; N is the number of electrons, m_e is the electron mass; a crude approximation is $\chi_m = (E_i/4m_e c^2)\alpha$, where E_i is the ionization energy
Long-range electron- or ion-molecule interaction energy	$V(r) = -e^2\alpha / 2r^4$	The target molecule polarizability is α
Ion mobility in a gas	$\kappa = -13.87 / (\alpha\mu)^{1/2}\ \text{cm}^2 / \text{V}\cdot\text{s}$	This one formula is not in c.g.s. units. Enter α in Å³ or 10^{-24} cm³ units and the reduced mass μ of the ion-molecule pair in amu. Classical limit; pure polarization potential
Langevin capture cross section	$\sigma(\nu_o) = (2\pi e / \nu_o)(\alpha / \mu)^{1/2}$	The relative velocity of approach for an ion-molecule pair is ν_o; the target molecular polarizability is α and the reduced mass of the ion-molecule pair is μ
Langevin reaction rate coefficient	$k = 2\pi e(\alpha / \mu)^{1/2}$	Collisional rate coefficient for an ion-molecule reaction
Rate coefficient for polar molecules	$k_d = 2\pi e\left[(\alpha / \mu)^{1/2} + cd(2 / \mu\pi kT)^{1/2}\right]$	The dipole moment of the neutral is d in esu cm; the number c is a "locking factor" that depends on α and d, and is between 0 and 1

Description	Formula	Remarks
Modified effective range cross section for electron-neutral scattering	$\sigma(k) = 4\pi A^2$ $+ 32\pi^4 \mu e^2 \alpha A k / 3h^2$ $+ \ldots$	Here, k is the electron momentum divided by $h/2\pi$, where h is Planck's constant; A is called the "scattering length"; the reduced mass is μ
van der Waals constant between two systems A, B	$C_6 = \dfrac{3}{2}\left[\dfrac{\alpha^A \alpha^B E^A E^B}{E^A + E^B}\right]$	For the interaction potential term $V_6(r) = -C_6 r^6$; $E^{A,B}$ represents average dipole transition energies and $\alpha^{A,B}$ the respective polarizabilities of A, B
Dipole-quadrupole constant between two systems A, B	$C_8 = \dfrac{15}{4}\left[\dfrac{\alpha^A \alpha_q^{\ B} E^A E_q^{\ B}}{E^A + E_q^{\ B}}\right]$ $+ \dfrac{15}{4}\left[\dfrac{\alpha_q^{\ A} \alpha^B E_q^{\ A} E^B}{E_q^{\ A} + E^B}\right]$	For the interaction potential term $V_8(r) = -C_8 r^8$; $E_q^{A,B}$ represents average quadrupole transition energies and $\alpha_q^{\ A,B}$ are the respective quadrupole polarizabilities of A, B
van der Waals constant between an atom and a surface	$C_3 = \dfrac{\alpha g E^A E^S}{8(E^A + E^S)}$	For an interaction potential $V_3(r) = -C_3 r^3$; $E^{A,S}$ are characteristic energies of the atom and surface; $g = 1$ for a free-electron metal and $g = (\varepsilon_\infty - 1)/(\varepsilon_\infty + 1)$ for an ionic crystal
Relationship between $\alpha(\nu)$ and oscillator strengths	$\alpha(\nu) = \dfrac{e^2 h^2}{4\pi^2 m_e} \sum \dfrac{f_k}{E_k^2 - (h\nu)^2}$	Here, f_k is the oscillator strength from the ground state to an excited state k, with excitation energy E_k. This formula is often used to estimate static polarizabilities ($\nu = 0$)
Dynamic polarizability	$\alpha(\nu) = \dfrac{\alpha E_r^2}{E_r^2 - (h\nu)^2}$	Approximate variation of the frequency-dependent polarizability $\alpha(\nu)$ from $\nu = 0$ up to the first dipole-allowed electronic transition, of energy E_r; the static dipole polarizability is $\alpha(0)$; infrared contributions ignored
Rayleigh scattering cross section	$\alpha(\nu) = \dfrac{8\pi}{9c^4}(2\pi\nu)^4$ $\times\left[3\alpha^2(\nu) + 2\gamma^2(\nu)/3\right]$	The photon frequency is ν; the polarizability anisotropy (the difference between polarizabilities parallel and perpendicular to the molecular axis) is $\gamma(\nu)$
Verdet constant	$V(\nu) = \dfrac{\nu n}{2m_e c^2}\left[\dfrac{d\alpha(\nu)}{d\nu}\right]$	Defined from $\theta = V(\nu)B$, where θ is the angle of rotation of linearly polarized light through a medium of number density n, per unit length, for a longitudinal magnetic field strength B (Faraday effect)

TABLE 2. Static Average Electric Dipole Polarizabilities for Ground State Atoms (in Units of 10^{-24} cm³)

Atomic number	Atom	Polariz- ability	Estimated accuracy (%)	Method	Ref.	Atomic number	Atom	Polariz- ability	Estimated accuracy (%)	Method	Ref.
1	H	0.666793	"exact"	calc	1	19	K	43.06	0.49	inter-ferom	8
2	He	0.2050522	"exact"	calc	2			43.4	2	beam	1
		0.2050519	0.0009	diel	3	20	Ca	22.8	2	calc	1
3	Li	24.33	0.16	beam	4			29.4	6	calc	10
4	Be	5.60	2	calc	1			25.0	8	beam	1
5	B	3.03	2	calc	1	21	Sc	17.8	25	calc	14
6	C	1.67	2	calc	5	22	Ti	14.6	25	calc	14
7	N	1.10	2	calc/index	1	23	V	12.4	25	calc	14
8	O	0.802	2	calc/index	1	24	Cr	11.6	25	calc	14
						25	Mn	9.4	25	calc	14
9	F	0.557	2	calc	1	26	Fe	8.4	25	calc	14
10	Ne	0.39432	0.003	diel	6	27	Co	7.5	25	calc	14
11	Na	24.11	0.12	inter-ferom	7	28	Ni	6.8	25	calc	14
						29	Cu	6.2	6	calc	10
		24.11	0.33	inter-ferom	8			6.1	25	calc	14
12	Mg	10.6	2	calc	1	30	Zn	5.75	2	index	15
		11.1	5	calc	9			6.1	6	calc	10
		10.6	5	calc	10			5.6	25	calc	14
13	Al	6.8	4.4	beam	11	31	Ga	8.12	2	calc	1
14	Si	5.53	2	calc	5	32	Ge	5.84	2	calc	5
15	P	3.63	2	calc	1	33	As	4.31	2	calc	1
16	S	2.90	2	calc	1	34	Se	3.77	2	calc	1
17	Cl	2.18	2	calc	1	35	Br	3.05	2	calc	1
18	Ar	1.6411	0.05	index/diel	12/13	36	Kr	2.4844	0.05	diel	13
						37	Rb	47.24	0.44	inter-ferom	8
								47.3	2	beam	1

Atomic number	Atom	Polariz- ability	Estimated accuracy (%)	Method	Ref.	Atomic number	Atom	Polariz- ability	Estimated accuracy (%)	Method	Ref.
38	Sr	27.6	8	beam	1	71	Lu	21.9	25	calc	14
		23.5	6	calc	10	72	Hf	16.2	25	calc	14
39	Y	22.7	25	calc	14	73	Ta	13.1	25	calc	14
40	Zr	17.9	25	calc	14	74	W	11.1	25	calc	14
41	Nb	15.7	25	calc	14	75	Re	9.7	25	calc	14
42	Mo	12.8	25	calc	14	76	Os	8.5	25	calc	14
43	Tc	11.4	25	calc	14	77	Ir	7.6	25	calc	14
44	Ru	9.6	25	calc	14	78	Pt	6.5	25	calc	14
45	Rh	8.6	25	calc	14	79	Au	5.8	25	calc	14
46	Pd	4.8	25	calc	14	80	Hg	5.02	1	index	22
47	Ag	6.78	18	index	16			5.7	25	calc	14
		7.2	25	calc	14	81	Tl	7.6	15	beam	23
48	Cd	7.36	3	index	17			7.5	25	calc	14
		7.4	6	calc	10	82	Pb	6.98	15	beam	5
		7.2	25	calc	14			7.01	2	calc	5
49	In	10.2	12	beam	18	83	Bi	7.4	25	calc	14
		9.1	25	calc	14	84	Po	6.8	25	calc	14
50	Sn	6.28	26	beam	5	85	At	6.0	25	calc	14
		7.84	2	calc	5	86	Rn	5.3	25	calc	14
51	Sb	6.6	25	calc	14	87	Fr	48.60	2	calc	24
52	Te	5.5	25	calc	14			47.1	5	calc	25
53	I	5.35	25	index	19	88	Ra	38.3	25	calc	26
		4.7	25	calc	14	89	Ac	32.1	25	calc	14
54	Xe	4.044	0.5	diel	1	90	Th	32.1	25	calc	14
55	Cs	59.42	0.13	beam	20	91	Pa	25.4	25	calc	14
56	Ba	39.7	8	beam	1	92	U	24.9	6	beam	27
57	La	31.1	25	calc	14	93	Np	24.8	25	calc	14
58	Ce	29.6	25	calc	14	94	Pu	24.5	25	calc	14
59	Pr	28.2	25	calc	14	95	Am	23.3	25	calc	14
60	Nd	31.4	25	calc	14	96	Cm	23.0	25	calc	14
61	Pm	30.1	25	calc	14	97	Bk	22.7	25	calc	14
62	Sm	28.8	25	calc	14	98	Cf	20.5	25	calc	14
63	Eu	27.7	25	calc	14	99	Es	19.7	25	calc	14
64	Gd	23.5	25	calc	14	100	Fm	23.8	25	calc	14
65	Tb	25.5	25	calc	14	101	Md	18.2	25	calc	14
66	Dy	24.5	25	calc	14	102	No	16.4	1	calc	21
67	Ho	23.6	25	calc	14	112	E112	4.06	2	calc	28
68	Er	22.7	25	calc	14	114	E114	4.59	2	calc	5
69	Tm	21.8	25	calc	14			4.37	2	calc	28
70	Yb	20.9	2	calc	21	119	ekafrancium	24.26	2	cal	24

[a] Methods: calc = calculated value; beam = atomic beam deflection technique; interferom = atomic beam interference; index = determination based on the measured index of refraction; diel = determination based on the measured dielectric constant.

References

1. Miller, T. M., and Bederson, B., *Adv. At. Mol. Phys.* 13, 1, 1977. For simplicity, any value in Table 2 which has not changed since this 1977 review is referenced as 1. Persons interested in original references and further details should consult 1.
2. Lim, I. S., and Schwerdfeger, P., *Phys. Rev. A* 70, 062501, 2004.
3. Schmidt, J. W., Glavioso, R. M., May, E. F., and Moldover, M. R., *Phys. Rev. Lett.* 98, 254504, 2007.
4. Miffre, A., Jacquey, M., Büchner, M., Trénec, G., and Vigué, J., *Phys. Rev. A* 73, 011603(R), 2006.
5. Thierfelder, C., Assadollahzadeh, B., Schwerdtfeger, P., Schäfer, S., and Schäfer, R., *Phys. Rev. A* 78, 052506, 2008. Relativistic calculations, Dirac-Coulomb.
6. Gaiser, C., and Fellmuth, B., *Eur. Phys. Lett.* 90, 63002, 2010.
7. Ekstrom, C. R., Schmiedmayer, J., Chapman, M. S., Hammond, T. D., and Pritchard, D. E., *Phys. Rev. A* 51, 3883, 1995. See theoretical work by Thakkar, A. J., and Lupinetti, C., *Chem. Phys. Lett.* 402, 270, 2005.
8. Holmgren, W. F., Revelle, M. C., Lonij, V. P. A., and Cronin, A. D., *Phys. Rev. A* 81, 053607, 2010.
9. Stwalley, W. C., *J. Chem. Phys.* 54, 4517, 1971.
10. Bromley, M. W. J., and Mitroy, J., *Phys. Rev. A* 65, 062505, 2002; 062506, 2002.
11. Milani, P., Moullet, I., and de Heer, W. A., *Phys. Rev. A* 42, 5150, 1990. See theoretical comments on this result, in Fuentealba, P., *Chem. Phys. Lett.* 397, 459, 2004, and in Lupinetti, C., and Thakkar, A. J., *J. Chem. Phys.* 122, 044301, 2005.
12. Newell, A. C., and Baird, R. D., *J. Appl. Phys.* 36, 3751, 1965.
13. Orcutt, R. H., and Cole, R. H., *J. Chem. Phys.* 46, 697, 1967; see also the later references from this group, given following the tables.
14. Doolen, G. D., Los Alamos National Laboratory, unpublished. A relativistic linear response method was used. The method is that described by Zangwill, A., and Soven, P., *Phys. Rev. A* 21, 1561, 1980. Adjustments of less than 10% across the periodic table have been made to these results to bring them into agreement with accurate experimental values where available, for the purpose of presenting "recommended" polarizabilities in Table 2.
15. Goebel, D., Holm, U., and Maroulis, G., *Phys. Rev. A* 54, 1973, 1996.
16. Hu, M., and Kusse, B. R., *Phys. Rev. A* 66, 062506, 2002. Measured at 532 nm and extrapolated to a static value. The uncertainty is ±1.2 Å³.
17. Goebel, D., and Holm, U., *Phys. Rev. A* 52, 3691, 1995.
18. Guella, T. P., Miller, T. M., Bederson, B., Stockdale, J. A. D., and Jaduszliwer, B., *Phys. Rev. A* 29, 2977, 1984.
19. Atoji, M., *J. Chem. Phys.* 25, 174, 1956. Semiempirical method based on molecular polarizabilities and atomic radii.
20. Amini, J. M., and Gould, H., *Phys. Rev. Lett.* 91, 153001, 2003.

21. Thierfelder, C., and Schwerdtfeger, P., *Phys. Rev. A* 79, 032512, 2009. Relativistic coupled-cluster calculations.
22. Goebel, D., and Holm, U., *J. Chem. Phys.* 100, 7710, 1996.
23. Preliminary value from the New York University group. See Ref. 18.
24. Lim, L. S., Schwerdtfeger, P., Metz, B., Stoll, H., *J. Chem. Phys.* 122, 104103, 2005.
25. Derevianko, A., Johnson, W. R., Safronova, M. S., and Babb, J. F., *Phys. Rev. Lett.* 82, 3589, 1999.
26. Łach, G., Jezionski, B., and Szalewicz, K., *Phys. Rev. Lett.* 92, 233001, 2004.
27. Kadar-Kallen, M. A., and Bonin, K. D., *Phys. Rev. Lett.* 72, 828, 1994.
28. Pershina, V., Borschevsky, A., Eliav, E., and Kaldor, U., *J. Chem. Phys.* 128, 024707, 2008. Relativistic coupled-cluster calculations.

TABLE 3. Average Electric Dipole Polarizabilities for Ground State Diatomic Molecules (in Units of 10^{-24} cm^3)

Molecule	Polarizability	Ref.	Molecule	Polarizability	Ref.
Al_2	19	23		5.35	2
BH	3.32*	1	HgCl	7.4*	9
Br_2	7.02	2	ICl	12.3	2
CO	1.95	3	K_2	77	22
Cl_2	4.61	3		72	21
Cs_2	104	22	Li_2	32.8	29
CsK	89	22		34	22
D_2 (v=0,J=0)	0.7921*	5	LiCl	3.46*	10
D_2 (293 K)	0.7954	6	LiF	10.8*	11
DCl	2.84	2	LiH	3.84*	12
F_2	1.38*	7		3.68*	13
H_2 (v=0,J=0)	0.8023*	5		3.88*	14
H_2 (293 K)	0.8045*	5	N_2	1.7403	6,8
H_2 (293 K)	0.8042	6	NO	1.70	2
H_2 (322 K)	0.8059	8	Na_2	40	22
HBr	3.61	3		38	21
HCl	2.63	3	NaK	51	22
	2.77	2	NaLi	40	4
HD (v=0,J=0)	0.7976*	5	O_2	1.5689	34
HF	0.80	27	Rb_2	79	22
HI	5.44	3			

TABLE 4. Average Electric Dipole Polarizabilities for Ground State Triatomic Molecules (in Units of 10^{-24} cm^3)

Molecule	Polarizability	Ref.	Molecule	Polarizability	Ref.
BeH_2	4.34*	14	HgI_2	19.1	2
CO_2	2.911	8	Li_3	34.5	29
CS_2	8.74	3	$LiNa_2$	61.2	30
	8.86	2	Li_2Na	35.4	30
D_2O	1.26	2	N_2O	3.03	8
H_2O	1.45	2	NO_2	3.02	2†
H_2S	3.782	3	Na_3	70	21
	3.95	2	O_3	3.21	2
HCN	2.59	3	OCS	5.71	2
	2.46	2		5.2	15
$HgBr_2$	14.5	2	SO_2	3.72	3
$HgCl_2$	11.6	2		4.28	2

TABLE 5. Average Electric Dipole Polarizabilities for Ground State Inorganic Polyatomic Molecules (Larger than Triatomic) (in Units of 10^{-24} cm^3)

Molecule	Polarizability	Ref.	Molecule	Polarizability	Ref.	Molecule	Polarizability	Ref.
$AsCl_3$	14.9	2	$(CsF)_2$	28.4	16	$(KBr)_2$	42.0	16
AsN_3	5.75	2	$(CsI)_2$	51.8	16	$(KCl)_2$	32.1	16
BCl_3	9.38	20	Ga_nAs_m	$n+m$=4–30	28	$(KF)_2$	21.0	16
BF_3	3.31	2	Ge_9	90	35	$(KI)_2$	36.3	16
$(BN_3)_2$	5.73	2	Ge_{10}	128	35	Li_n	n=2–22	29
$(BH_2N)_3$	8.0	2†	Ge_{15}	167	35	$(LiBr)_2$	18.9	16
ClF_3	6.32	2	$GeCl_4$	15.1	2	$(LiCl)_2$	13.1	16
Cs_nBr_{n-1}	n=3–32		GeH_3Cl	6.7	2†	$(LiF)_2$	6.9	16
$(CsBr)_2$	54.5	16	$(HgCl)_2$	14.7	9	$(LiI)_2$	23.4	16
$(CsCl)_2$	42.4	16	K_n	n=2,5,7–9,11,20	21	$LiNa_3$	75.6	30

Molecule	Polarizability	Ref.	Molecule	Polarizability	Ref.	Molecule	Polarizability	Ref.
Li_2Na_2	60.0	30	$(NaI)_2$	26.9	16	SeF_6	7.33	2
Li_3Na	54.8	30	OsO_4	8.17	2	SiF_4	5.45	2
ND_3	1.70	2	P_4	13.59	40	SiH_4	5.44	2
NF_3	3.62	2	PCl_3	12.8	2	$(SiH_3)_2$	11.1	2
NH_3	2.81	20	PF_5	6.10	2	$SiHCl_3$	10.7	2
	2.10	2	PH_3	4.84	2	SiH_2Cl_2	8.92	2
	2.26	3	$(RbBr)_2$	48.2	16	SiH_3Cl	7.02	2
	2.22*	33	$(RbCl)_2$	43.2	16	$SnBr_4$	22.0	2
$(NO_2)_2$	6.69	2	$(RbF)_2$	40.7	16	$SnCl_4$	18.0	2
Na_n	n=7–93	39	$(RbI)_2$	46.3	16		13.8	15
	n=1–40	21	SF_6	6.54	8	SnI_4	32.3	2
$(NaBr)_2$	26.8	16	$(SF_5)_2$	13.2	2	TeF_6	9.00	2
$(NaCl)_2$	23.4	16	SO_3	4.84	2	$TiCl_4$	16.4	2
$(NaF)_2$	20.7	16	SO_2Cl_2	10.5	2	UF_6	12.5	2

TABLE 6. Average Electric Dipole Polarizabilities for Ground State Hydrocarbon Molecules (in Units of 10^{-24} cm³)

Molecule	Name	Polarizability	Ref.	Molecule	Name	Polarizability	Ref.
CH_4	methane	2.593	8			12.3	2
C_2H_2	acetylene	3.33	3	C_7H_{12}	1-heptyne	12.8	2†
		3.93	2	C_7H_{14}	methylcyclohexane	13.1	2
C_2H_4	ethylene	4.252	8		1-heptene	13.51	27
C_2H_6	ethane	4.47	3	C_7H_{16}	heptane	13.61	2
		4.43	2	C_8H_8	styrene	15.0	2
C_3H_4	propyne	6.18	2			14.41	27
C_3H_6	propene	6.26	2	C_8H_{10}	ethylbenzene	14.2	2
	cyclopropane	5.66	2		o-xylene	14.9	2
C_3H_8	propane	6.29	3			14.1	15
		6.37	2		p-xylene	13.7	25
C_4H_6	1-butyne	7.41	2†			14.2	15
	1,3-butadiene	8.64	2			14.9	2
C_4H_8	1-butene	7.97	2		m-xylene	14.2	15
		8.52	2	C_8H_{16}	ethylcyclohexane	15.9	2
	trans-2-butene	8.49	2	C_8H_{18}	n-octane	15.9	2
	2-methylpropene	8.29	2		3-methylheptane	15.44	27
C_4H_{10}	butane	8.20	2		2,2,4-trimethylpentane	15.44	27
	isobutane	8.14	27	C_9H_{10}	α-methylstyrene	16.05	27
C_5H_6	1,3-cyclopentadiene	8.64	2	C_9H_{12}	isopropylbenzene	16.0	2†
C_5H_8	1-pentyne	9.12	2		1,3,5-trimethylbenzene	15.5	25
	trans-1,3-pentadiene	10.0	2			16.14	27
	isoprene	9.99	2	C_9H_{18}	isopropylcyclohexane	17.2	2
C_5H_{10}	cyclopentane	9.15	18	C_9H_{20}	nonane	17.36	27
	1-pentene	9.65	27	$C_{10}H_8$	naphthalene	16.5	17
	2-pentene	9.84	27			17.48	27
C_5H_{12}	pentane	9.99	2	$C_{10}H_{14}$	durene	17.3	25
	neopentane	10.20	18		tert-butylbenzene	17.2	25
C_6H_6	benzene	10.0	25			17.8	2†
		10.32	3	$C_{10}H_{20}$	tert-butylcyclohexane	19.8	2
		10.74	2	$C_{10}H_{22}$	decane	19.10	27
C_6H_{10}	1-hexyne	10.9	2†	$C_{11}H_{10}$	α-methylnaphthalene	19.35	27
	2-ethyl-1,3-butadiene	11.8	2†		β-methylnaphthalene	19.52	27
	3-methyl-1,3-pentadiene	11.8	2†	$C_{11}H_{14}$	α,β,β-trimethylstyrene	19.64	27
	2-methyl-1,3-pentadiene	12.1	2†	$C_{11}H_{16}$	pentamethylbenzene	19.1	25
	2,3-dimethyl-1,3-butadiene	11.8	2†	$C_{11}H_{24}$	undecane	21.03	27
	cyclohexene	10.7	2†	$C_{12}H_{10}$	acenaphthene	20.61	27
C_6H_{12}	cyclohexane	11.0	18	$C_{12}H_{12}$	α-ethylnaphthalene	21.19	27
		10.87	15		β-ethylnaphthalene	21.36	27
	1-hexene	11.65	27	$C_{12}H_{18}$	hexamethylbenzene	20.9	25
C_6H_{14}	hexane	11.9	2	$C_{12}H_{26}$	dodecane	22.75	27
C_7H_8	toluene	11.8	25	$C_{13}H_{10}$	fluorene	21.68	27
		12.26	15	$C_{14}H_{10}$	anthracene	25.4	17

Molecule	Name	Polarizability	Ref.	Molecule	Name	Polarizability	Ref.
		25.93	27	$C_{18}H_{12}$	naphthacene	32.27	27
	phenanthrene	36.8*	17		1,2-benzanthracene	32.86	27
		24.70	27		chrysene	33.06	27
$C_{14}H_{22}$	p-di-tert-butylbenzene	24.5	25		triphenylene	31.07	27
$C_{16}H_{10}$	pyrene	28.22	27	$C_{18}H_{30}$	1,3,5-tri-tert-butylbenzene	31.8	25
$C_{17}H_{12}$	2,3-benzfluorene	30.21	27	$C_{24}H_{12}$	coronene	42.50	27

TABLE 7. Average Electric Dipole Polarizabilities for Ground State Organic Halides (in Units of 10^{-24} cm^3)

Molecule	Name	Polarizability	Ref.	Molecule	Name	Polarizability	Ref.
CBr_2F_2	dibromodifluoromethane	9.0	2[†]	$C_2H_2Cl_4$	1,1,2,2-tetrachloroethane	12.1	2[†]
$CClF_3$	chlorotrifluoromethane	5.72	20	C_2H_2ClN	chloroacetonitrile	6.10	18
		5.59	8	$C_2H_2F_2$	1,1-difluoroethylene	5.01	20
CCl_2F_2	dichlorodifluoromethane	7.93	20	C_2H_3Br	bromoethylene	7.59	2
		7.81	2	C_2H_3Cl	chloroethylene	6.41	2
CCl_2O	phosgene	7.29	2	$C_2H_3ClF_2$	1-chloro-1,1-difluoroethane	8.05	2
CCl_2S	thiophosgene	10.2	2	C_2H_3ClO	acetyl chloride	6.62	2
CCl_3F	trichlorofluoromethane	9.47	2	$C_2H_3ClO_2$	methyl chloroformate	7.1	2[†]
CCl_3NO_2	trichloronitromethane	10.8	2[†]	$C_2H_3Cl_3$	1,1,1-trichloroethane	10.7	2
CCl_4	carbon tetrachloride	11.2	2	$C_2H_3F_3$	1,1,1-trifluoroethane	4.4	2[†]
		10.5	3	C_2H_3I	iodoethylene	9.3	2[†]
CF_4	carbon tetrafluoride	3.838	8	C_2H_4BrCl	1-bromo-2-chloroethane	9.5	2[†]
CF_2O	carbonylfluoride	1.88*	17	$C_2H_4Br_2$	1,2-dibromoethane	10.7	2[†]
$CHBr_3$	bromoform	11.8	27	C_2H_4ClF	1-chloro-2-fluoroethane	6.5	2[†]
$CHBrF_2$	bromodifluoromethane	5.7	2[†]	$C_2H_4ClNO_2$	1-chloro-1-nitroethane	10.9	2
$CHClF_2$	chlorodifluoromethane	6.38	20	$C_2H_4Cl_2$	1,1-dichloroethane	8.64	2
		5.91	2		1,2-dichloroethane	8.0	2[†]
$CHCl_2F$	dichlorofluoromethane	6.82	2	C_2H_5Br	bromoethane	8.05	2
$CHCl_3$	chloroform	9.5	8			7.28	27
		8.23	27	C_2H_5Cl	chloroethane	7.27	20
CHF_3	fluoroform	3.52	20			8.29	2
		3.57	8			6.4	15
$CHFO$	fluoroformaldehyde	1.76*	17	C_2H_5ClO	2-chloroethanol	7.1	2[†]
CHI_3	iodoform	18.0	27			6.88	27
CH_2Br_2	dibromomethane	9.32	2		chloromethyl methyl ether	7.1	2[†]
		8.68	27	C_2H_5F	fluoroethane	4.96	2
CH_2ClNO_2	chloronitromethane	6.9	2[†]	C_2H_5I	iodoethane	10.0	2
CH_2Cl_2	dichloromethane	6.48	3	$C_3H_4Cl_2$	dichloropropene	10.1	2[†]
		7.93	2	C_3H_5Cl	chloropropene	8.3	2
CH_2I_2	diiodomethane	12.90	27	C_3H_5ClO	chloroacetone	8.4	2[†]
CH_3Br	bromomethane	5.87	20	$C_3H_5ClO_2$	ethyl chloroformate	9.0	2[†]
		6.03	2	$C_3H_6ClNO_2$	1-chloro-1-nitropropane	10.4	2[†]
		5.55	15	$C_3H_6Cl_2$	dichloropropane	10.9	2[†]
CH_3Cl	chloromethane	5.35	20	C_3H_7Br	1-bromopropane	9.4	2[†]
		4.72	8			9.07	27
CH_3F	fluoromethane	2.97	8		2-bromopropane	9.6	2[†]
CH_3I	iodomethane	7.97	2	C_3H_7Cl	chloropropane	10.0	2
C_2ClF_5	chloropentafluoroethane	6.3	2[†]	C_3H_7ClO	β-chloroethyl methyl ether	8.71	27
$C_2Cl_2F_4$	1,2-dichlorotetrafluoroethane	8.5	2[†]		2-chloro-1-propanol	8.89	27
C_2Cl_3N	trichloroacetonitrile	10.42	18		3-chloro-1-propanol	8.84	27
C_2F_6	hexafluoroethane	6.82	2	C_3H_7I	1-iodopropane	11.5	2[†]
C_2HBr	bromoacetylene	7.39	2	C_4H_5Cl	4-chloro-1,2-butadiene	10.0	2[†]
C_2HCl	chloroacetylene	6.07	2	C_4H_7Cl	1-chloro-2-methylpropene	10.8	2
C_2HCl_3	trichloroethlyene	10.03	27	$C_4H_7ClO_2$	2-chlorobutyric acid	10.87	27
C_2HCl_5	pentachloroethane	14.0	2		3-chlorobutyric acid	10.80	27
$C_2H_2Cl_2$	1,1-dichloroethylene	7.83	27		4-chlorobutyric acid	10.69	27
	trans-dichloroethylene	8.15	27	$C_4H_8Cl_2$	1,4-dichlorobutane	12.0	2[†]
	cis-dichloroethylene	8.03	27	C_4H_9Br	bromobutane	13.9	2
$C_2H_2Cl_2F_2$	1,1-dichloro-2,2-difluoroethane	8.4	2[†]			10.86	27
$C_2H_2Cl_2O$	chloroacetyl chloride	8.92	2	C_4H_9Cl	1-chlorobutane	11.3	2
$C_2H_2Cl_3F$	1,2,2-trichloro-1-fluoroethane	10.2	2[†]		1-chloro-2-methylpropane	11.1	2

Molecule	Name	Polarizability	Ref.
	2-chloro-2-methylpropane	12.5	2[†]
	2-chlorobutane	12.4	2
C_4H_9ClO	β-chloroethyl ethyl ether	10.56	27
	2-chloro-1-butanol	10.70	27
	3-chloro-1-butanol	10.38	27
C_4H_9I	1-iodobutane	13.3	2[†]
		12.65	27
$C_5H_9ClO_2$	methyl 2-chlorobutanoate	12.33	27
	methyl 3-chlorobutanoate	12.31	27
	methyl 4-chlorobutanoate	12.27	27
	2-chloropentanoic acid	12.69	27
	3-chloropentanoic acid	12.57	27
	4-chloropentanoic acid	12.53	27
$C_5H_{11}Br$	1-bromopentane	13.1	2[†]
$C_5H_{11}Cl$	1-chloropentane	12.0	2[†]
$C_5H_{11}F$	fluoropentane	9.95	27
C_6F_6	hexafluorobenzene	9.58	27
C_6HF_5	pentafluorobenzene	9.63	27
$C_6H_2Cl_2O_2$	2,5-dichloro-1,4-benzoquinone	18.4	2
$C_6H_2F_4$	1,2,3,4-tetrafluorobenzene	9.69	27
	1,2,4,5-tetrafluorobenzene	9.69	27
$C_6H_3F_3$	1,3,5-trifluorobenzene	9.74	27
C_6H_4BrF	p-bromofluorobenzene	13.4	2[†]
$C_6H_4ClNO_2$	chloronitrobenzene	14.6	2[†]
$C_6H_4Cl_2$	o-dichlorobenzene	14.17	27
	m-dichlorobenzene	14.23	27
	p-dichlorobenzene	14.20	27
C_6H_4FI	p-fluoroiodobenzene	15.5	2[†]
$C_6H_4FNO_2$	p-fluoronitrobenzene	12.8	2[†]
$C_6H_4F_2$	o-difluorobenzene	9.80	27
	m-difluorobenzene	10.3	2[†]
	p-difluorobenzene	9.80	27
C_6H_5Br	bromobenzene	14.7	2
		13.62	27
C_6H_5Cl	chlorobenzene	14.1	2
		12.3	15
C_6H_5ClO	chlorophenol	13.0	2[†]
C_6H_5F	fluorobenzene	10.3	2

Molecule	Name	Polarizability	Ref.
C_6H_5I	iodobenzene	15.5	2[†]
$C_6H_{11}ClO_2$	ethyl 2-chlorobutanoate	14.16	27
	ethyl 3-chlorobutanoate	14.13	27
	ethyl 4-chlorobutanoate	14.11	27
$C_6H_{13}Br$	bromohexane	14.44	27
$C_6H_{13}F$	fluorohexane	11.80	27
C_7H_7Br	p-bromotoluene	14.80	27
C_7H_7Cl	p-chlorotoluene	13.70	27
C_7H_7F	p-fluorotoluene	11.70	27
C_7H_7I	p-iodotoluene	17.10	27
$C_7H_{15}Br$	1-bromoheptane	16.8	2[†]
		16.23	27
$C_7H_{15}F$	fluoroheptane	13.66	27
$C_8H_{17}Br$	bromooctane	18.02	27
$C_8H_{17}F$	fluorooctane	15.46	27
$C_9H_{19}Br$	bromononane	19.81	27
$C_9H_{19}F$	fluorononane	17.34	27
$C_{10}F_8$	octafluoronaphthalene	17.64	27
$C_{10}H_7Br$	α-bromonaphthalene	20.34	27
$C_{10}H_7Cl$	α-chloronaphthalene	19.30	27
	β-chloronaphthalene	19.58	27
$C_{10}H_7I$	α-iodonaphthalene	22.41	27
	β-iodonaphthalene	22.95	27
$C_{10}H_{21}Br$	bromodecane	21.60	27
$C_{10}H_{21}F$	fluorodecane	19.18	27
$C_{11}H_{23}F$	fluoroundecane	21.00	27
$C_{12}H_{25}Br$	bromododecane	25.18	27
$C_{12}H_{25}F$	fluorododecane	22.83	27
$C_{12}H_8Br_2O$	4,4'-dibromodiphenyl ether	27.8	2[†]
$C_{12}H_9BrO$	4-bromodiphenyl ether	24.2	2[†]
$C_{13}H_{11}BrO$	p-bromophenyl-p-tolyl ether	26.6	2[†]
$C_{14}H_9Br$	9-bromoanthracene	28.32	27
$C_{14}H_9Cl$	9-chloroanthracene	27.35	27
$C_{14}H_9F$	fluoranthracene	28.34	27
$C_{14}H_{29}F$	fluorotetradecane	26.57	27
$C_{16}H_{33}Br$	bromohexadecane	32.34	27
$C_{18}H_{37}Br$	bromooctadecane	35.92	27

TABLE 8. Static Average Electric Dipole Polarizabilities for Other Ground State Organic Molecules (in Units of 10^{-24} cm^3)

Molecule	Name	Polarizability	Ref.
CN_4O_8	tetranitromethane	15.3	2
CH_2O	formaldehyde	2.8	2[†]
		2.45	18
CH_2O_2	formic acid	3.4	2[†]
CH_3NO	formamide	4.2	2[†]
		4.08	18
CH_3NO_2	nitromethane	7.37	2
CH_4O	methanol	3.29	2
		3.23	15
		3.32	18
CH_5N	methyl amine	4.7	2
		4.01	19
		4.01*	33
C_2N_2	cyanogen	7.99	2
C_2H_2O	ketene	4.4	2[†]
C_2H_3N	acetonitrile	4.40	2[†]
		4.48	18
C_2H_4O	acetaldehyde	4.6	2[†]
		4.59	18

Molecule	Name	Polarizability	Ref.
	ethylene oxide	4.43	18
$C_2H_4O_2$	acetic acid	5.1	2[†]
	methyl formate	5.05	27
$C_2H_4O_4$	formic acid dimer	12.7	2
C_2H_5NO	acetamide	5.67	18
	N-methyl formamide	5.91	18
$C_2H_5NO_2$	nitroethane	9.63	2
	ethyl nitrite	7.0	15
C_2H_6O	ethanol	5.41	2
		5.11	18
	methyl ether	5.29	20
		5.84	2
		5.16	15
$C_2H_6O_2$	ethylene glycol	5.7	2[†]
		5.61	27
$C_2H_6O_2S$	dimethyl sulfone	7.3	2[†]
C_2H_6S	ethanethiol	7.41	2
C_2H_7N	ethyl amine	7.10	2
	dimethyl amine	6.37	2

Molecule	Name	Polarizability	Ref.
		5.90*	33
C₂H₈N₂	ethylene diamine	7.2	2†
C₃H₂N₂	malononitrile	5.79	18
C₃H₃N	acrylonitrile	8.05	2
C₃H₄N₂	pyrazole	7.23	27
C₃H₄O	propenal	6.38	2†
C₃H₅N	propionitrile	6.70	2
		6.24	18
		6.27*	32
C₃H₆O	acetone	6.33	15
		6.4	2†
		6.39	18
	allyl alcohol	7.65	2
	propionaldehyde	6.50	2
C₃H₆O₂	propionic acid	6.9	2†
	ethyl formate	8.01	2
		6.88	27
	methyl acetate	6.94	2
		6.81	27
C₃H₆O₃	dimethyl carbonate	7.7	2†
C₃H₇NO	N-methyl acetamide	7.82	18
	N,N-dimethyl formamide	7.81	18
C₃H₇NO₂	nitropropane	8.5	2†
C₃H₈O	2-propanol	7.61	2
		6.97	18
	1-propanol	6.74	2
	ethyl methyl ether	7.93	2
C₃H₈O₂	dimethoxymethane	7.7	2†
	ethylene glycol monomethyl ether	7.44	27
C₃H₉N	propylamine	7.70	27
		9.20	2
	isopropylamine	7.77	27
	trimethylamine	8.15	2
		7.78*	33
C₄H₂N₂	fumaronitrile	11.8	2
C₄H₄N₂	succinonitrile	8.1	2†
	pyrimidine	8.53*	17
	pyridazine	9.27*	17
C₄H₄O₂	diketene	8.0	2†
C₄H₄S	thiophene	9.67	2
C₄H₅N	methacrylonitrile	8.0	2†
	trans-crotononitrile	8.2	2†
C₄H₆N₂	N-methylpyrazole	8.99	27
C₄H₆O	crotonaldehyde	8.5	2†
	methacrylaldehyde	8.3	2†
C₄H₆O₂	biacetyl	8.2	2†
C₄H₆O₃	acetic anhydride	8.9	2†
C₄H₆S	divinyl sulfide	10.9	2†
C₄H₇N	butyronitrile	8.4	2†
	isobutyronitrile	8.05	18
		8.05*	32
C₄H₈O	butanal	8.2	2†
	methyl ethyl ketone	8.13	15
	trans-2,3-epoxy butane	8.22*	17
C₄H₈O₂	ethyl acetate	9.7	2
		8.62	27
	1,4-dioxane	10.0	2
	p-dioxane	8.60	18
	2-methyl-1,3-dioxolane	9.44	15
	butyric acid	8.58	27
	methyl propionate	8.97	27
C₄H₉NO₂	1-nitrobutane	10.4	2†

Molecule	Name	Polarizability	Ref.
	2-methyl-2-nitropropane	10.3	2†
C₄H₁₀O	ethyl ether	10.2	2
		8.73	15
	1-butanol	8.88	2
	2-methylpropanol	8.92	2
	methyl propyl ether	8.86	27
C₄H₁₀O₂	ethylene glycol monoethyl ether	9.28	27
C₄H₁₀S	ethyl sulfide	10.8	2
C₄H₁₁N	butylamine	13.5	2
	diethylamine	10.2	2
		9.61	27
C₅H₅N	pyridine	9.5	15
		9.18	27
	4-cyano-1,3-butadiene	10.5	2†
C₅H₈N₂	1,5-dimethylpyrazole	10.72	27
C₅H₈O₂	acetyl acetone	10.5	2†
C₅H₉N	valeronitrile	10.4	2
	22-DMPN	9.59	18
C₅H₁₀O	diethyl ketone	9.93	15
	methyl propyl ketone	9.93	15
C₅H₁₀O₂	ethyl propionate	10.41	27
	methyl butanoate	10.41	27
C₅H₁₀O₃	diethyl carbonate	11.3	2
C₅H₁₂O	ethyl propyl ether	10.68	27
C₅H₁₂O₄	tetramethyl orthocarbonate	13.0	2†
C₆H₄N₂O₄	p-dinitrobenzene	18.4	2
C₆H₄O₂	p-benzoquinone	14.5	2
C₆H₅NO₂	nitrobenzene	14.7	2
		12.92	15
C₆H₆O	phenol	11.1	2†
		9.94*	17
C₆H₇N	aniline	12.1	2†
C₆H₈N₂	phenylenediamine	13.8	2†
	phenylhydrazine	12.91	27
C₆H₁₀N₂	1-ethyl-5-methylpyrazole	12.50	27
C₆H₁₀O₃	ethyl acetoacetate	12.9	2†
C₆H₁₂N₂	dimethylketazine	15.6	2
C₆H₁₂O	cyclohexanol	11.56	18
C₆H₁₂O₂	amyl formate	14.2	2
C₆H₁₂O₃	paraldehyde	17.9	2
C₆H₁₄O	propyl ether	12.8	2
		12.5	15
C₆H₁₄O₂	1,1-diethoxyethane	13.2	2†
	1,2-diethoxyethane	11.3	2†
C₆H₁₅N	triethylamine	13.1	2
		13.38	27
	dipropylamine	13.29	27
C₇H₄N₂O₂	p-cyanonitrobenzene	19.0	2
C₇H₅N	benzonitrile	12.5	2†
C₇H₅NO₃	nitroanisole	15.7	2†
C₇H₈O	anisole	13.1	2†
C₇H₉NO	o-anisidine	14.2	2†
C₇H₁₀N₂	1,1-methylphenylhydrazine	14.81	27
C₇H₁₄O	cyclohexyl methyl ether	13.4	2†
	2,4-dimethyl-3-pentanone	13.5	15
C₇H₁₄O₂	pentyl acetate	14.9	2
C₈H₄N₂	p-dicyanobenzene	19.2	2
C₈H₆N₂	quinoxaline	15.13	27
C₈H₈O	acetophenone	15.0	2
C₈H₈O₂	2,5-dimethyl-1,4-benzoquinone	18.8	2
C₈H₁₀O	phenetole	14.9	2
C₈H₁₁N	N-dimethylaniline	16.2	2†
C₈H₁₂N₂	1,1-ethylphenylhydrazine	16.62	27

Molecule	Name	Polarizability	Ref.
$C_8H_{12}O_2$	ethyl sorbate	17.2	2†
	tetramethylcyclobutane-1,3-dione	18.6	2
$C_8H_{14}O_4$	diethyl succinate	16.8	2†
$C_8H_{18}O$	butyl ether	17.2	2
C_9H_7N	quinoline	15.70	27
	isoquinoline	16.43	27
$C_9H_{10}O_2$	ethyl benzoate	16.9	2†
$C_9H_{21}N$	tripropylamine	18.87	27
$C_{10}H_9N$	α-naphthylamine	19.50	27
	β-naphthylamine	19.73	27
	2-methylquinoline	18.65	27
	1-methylisoquinoline	18.28	27
$C_{10}H_{10}Fe$	ferrocene	17.1	26
$C_{10}H_{10}N_2$	2,3-dimethylquinoxaline	18.70	27
$C_{10}H_{14}BeO_4$	beryllium acetylacetonate	34.1	2
$C_{11}H_8O$	1-naphthaldehyde	19.75	27

Molecule	Name	Polarizability	Ref.
	2-naphthaldehyde	20.06	27
$C_{12}H_8N_2$	phenazine	23.43	27
$C_{12}H_9NO_3$	4-nitrodiphenyl ether	24.7	2†
$C_{14}H_8O_2$	anthraquinone	24.46	27
$C_{14}H_{14}O$	di-p-tolyl ether	24.9	2†
$C_{15}H_{15}Er$	$(C_5H_5)_3Er$	28.44	37
$C_{15}H_{15}Nd$	$(C_5H_5)_3Nd$	33.05	37
$C_{15}H_{15}Sm$	$(C_5H_5)_3Sm$	32.01	37
$C_{15}H_{21}AlO_6$	aluminum acetylacetonate	51.9	2
$C_{15}H_{21}CrO_6$	chromium acetylacetonate	53.7	2
$C_{15}H_{21}FeO_6$	ferric acetylacetonate	58.1	2
$C_{20}H_{28}O_8Th$	thorium acetylacetonate	79.0	2
C_{60}	buckminsterfullerene	88.9	36
		79	31
		76.5	24
C_{70}		108.5	36
		102	24

Note: All polarizabilities in the tables are experimental values except those values marked by an asterisk (*), which indicates a calculated result. The experimental polarizabilities are mostly determined by measurements of a dielectric constant or refractive index that are quite accurate (0.5% or better). However, one should treat many of the results with several percent of caution because of the age of the data and because some of the results refer to optical frequencies rather than static. Comments given with the references are intended to allow one to judge the degree of caution required. Interested persons should consult these references. In many cases, the reference given is to a theoretical paper in which the experimental results are quoted. These papers, noted in the References, contain valuable information on polarizability calculations and experimental data which often includes the tensor components of the polarizability.

An empirical additive formula for molecular polarizabilities at 589 nm frequency has been given in Bosque, R., and Sales, J., *J. Chem. Inf. Comput. Sci.* 42, 1154, 2002: a = 0.32 + 1.51#C + 0.17#H + 0.57#O + 1.05#N + 2.99#S + 2.48#P + 0.22#F + 2.16#Cl + 3.29#Br + 5.45#I, where #C denotes the number of carbon atoms in the molecule, etc. A helium-elimination additive method has been given by Kassimi, N. E.-B., and Thakkar, A. J., *Chem. Phys. Lett.* 472, 232, 2009.

References

1. McCullough, E. A., Jr., *J. Chem. Phys.*, 63, 5050, 1975. This calculation is for the parallel component, not the average polarizability.

2. Maryott, A. A., and Buckley, F., *U. S. National Bureau of Standards Circular No. 537*, 1953. A tabulation of dipole moments, dielectric constants, and molar refractions measured between 1910 and 1952, and used here to determine polarizabilities if no more recent result exists. The polarizability is $3/(4\pi N_A)$ times the molar polarization or molar refraction, where N_A is Avogadro's number. The value $3/(4\pi N_A)$ = 0.3964308 × 10^{-24} cm^3 was used for this conversion. A dagger (†) following the reference number in the tables indicates that the polarizability was derived from the molar refraction and hence may not include some low-frequency contributions to the static polarizability; these "static" polarizabilities are therefore low by 1 to 30%.

3. Hirschfelder, J. O., Curtis, C. F., and Bird, R. B., *Molecular Theory of Gases and Liquids*, Wiley, New York, 1954, p. 950. Fundamental information on molecular polarizabilities.

4. Miller, T. M., and Bederson, B., *Adv. At. Mol. Phys.*, 13, 1, 1977. Review emphasizing atomic polarizabilities and measurement techniques. The data quoted in Table 3 are accurate to 8 to 12%.

5. Kolos, W., and Wolniewicz, L., *J. Chem. Phys.*, 46, 1426, 1967. Highly accurate molecular hydrogen calculations. See also recent work by Machado, A.M., and Masilli, M., *J. Chem. Phys.* 120, 7505, 2004.

6. Newell, A. C., and Baird, R. C., *J. Appl. Phys.*, 36, 3751, 1965. Highly accurate refractive index measurements at 47.7 GHz (essentially static).

7. Jao, T. C., Beebe, N. H. F., Person, W. B., and Sabin, J. R., *Chem. Phys. Lett.*, 26, 474, 1974. Tensor polarizabilities, derivatives, and other results are reported.

8. Orcutt, R. H., and Cole, R. H., *J. Chem. Phys.*, 46, 697, 1967 (He, Ne, Ar, Kr, H_2, N_2); Sutter, H., and Cole, R. H., *J. Chem. Phys.*, 52, 132, 1970 (CF_3H, CFH_3, $CClF_3$, $CClH_3$); Bose, T. K., and Cole, R. H., *J. Chem. Phys.*, 52, 140, 1970 (CO_2), and 54, 3829, 1971 (C_2H_4); Nelson, R. D., and Cole, R. H., *J. Chem. Phys.*, 54, 4033, 1971 (SF_6, $CClF_3$); Bose, T. K., Sochanski, J. S., and Cole, R. H., *J. Chem. Phys.*, 57, 3592, 1972 (CH_4, CF_4); Kirouac, S., and Bose, T. K., *J. Chem. Phys.*, 59, 3043, 1973 (N_2O), and 64, 1580, 1976 (He). Highly accurate dielectric constant measurements. These modern data give the most accurate polarizabilities available. A criticism of the interpretation of these data in the case of polar molecules is given in Ref. 20, p. 2905.

9. Huestis, D. L., Technical Report #MP 78-25, SRI International (project PYU 6158), Menlo Park, CA 94025. Molar refractions for mercury-chlorine compounds are analyzed.

10. Bounds, D. G., Clarke, J. H. R., and Hinchliffe, A., *Chem. Phys. Lett.*, 45, 367, 1977. Theoretical tensor polarizability for LiCl.

11. Kolker, H. J., and Karplus, M., *J. Chem. Phys.*, 39, 2011, 1963. Theoretical.

12. Cutschick, V. P., and McKoy, V., *J. Chem. Phys.*, 58, 2397, 1973. Theoretical tensor polarizabilities.

13. Gready, J. E., Bacskay, G. B., and Hush, N. S., *Chem. Phys.*, 22, 141, 1977, and 23, 9, 1977. Theoretical.

14. Amos, A. T., and Yoffe, J. A., *J. Chem. Phys.*, 63, 4723, 1975. Theoretical.

15. Stuart, H. A., *Landolt-Börnstein Zahlenwerte and Funktionen*, Vol. 1, Part 3, Eucken, A., and Hellwege, K. H., Eds., Springer-Verlag, Berlin, 1951, p. 511. Tabulation of molecular polarizabilities. Two misprints in the chemical symbols have been corrected.

16. Guella, T., Miller, T. M., Stockdale, J. A. D., Bederson, B., and Vuskovic, L., *J. Chem. Phys.*, 94, 6857, 1991. Beam measurements with accuracies between 12 and 24%.

17. Marchese, F. T., and Jaff, H. H., *Theoret. Chim. Acta* (Berlin), 45, 241, 1977. Theoretical and experimental tensor polarizabilities are tabulated in this paper.

18. Applequist, J., Carl, J. R., and Fung, K.-K., *J. Am. Chem. Soc.*, 94, 2952, 1972. Excellent reference on the calculation of molecular polarizabilities, including extensive tables of tensor polarizabilities, both theoretical and experimental, at 589.3 nm wavelength.

19. Bridge, N. J., and Buckingham, A. D., *Proc. Roy. Soc.* (London), A295, 334, 1966. Measured tensor polarizabilities at 633 nm wavelength.

20. Barnes, A. N. M., Turner, D. J., and Sutton, L. E., *Trans. Faraday Soc.*, 67, 2902, 1971. Dielectric constants yielding polarizabilities accurate from 0.3–8%.

21. Rayane, D., Allouche, A. R., Benichou, E., Antoine, R., Aubert-Frecon, M., Dugourd, Ph., Broyer, M., Ristori, C., Chandezon, F., Huber, B. A., and Guet, C., *Eur. Phys. J.* D 9, 243, 1999. See also Knight, W. D., Clemenger, K., de Heer, W. A., and Saunders, W. A., *Phys. Rev. B* 31, 2539, 1985. These data probably correspond to a very low internal temperature.

22. Tarnovsky, V., Bunimovicz, M., Vuskovic, L., Stumpf, B., and Bederson, B., *J. Chem. Phys.*, 98, 3894, 1993. These data correspond to internal temperatures 480-948 K.

23. Milani, P., Moullet, I., and de Heer, W. A., *Phys. Rev. A*, 42, 5150, 1990. Beam measurements accurate to 11%.

24. Compagnon, I., Antoine, R., Broyer, M., Dugourd, P., Lermé, J., and Rayane, D., *Phys. Rev. A* 64, 025201, 2001. The uncertainties are ±8 Å³ for C_{60} and ±14 Å³ for C_{70}.

25. Aroney, M. J., and Pratten, S. J., *J. Chem. Soc., Faraday Trans.*, 1, 80, 1201, 1984. Uncertainties in the range 1–3%.

26. Le Fevre, R. J. W., Murthy, D. S. N., and Saxby, J. D., *Aust. J. Chem.*, 24, 1057, 1971. Kerr effect.

27. No, K. T., Cho, K. H., Jhon, M. S., and Scheraga, H. A., *J. Am. Chem. Soc.*, 115, 2005, 1993. Theoretical; these results are quoted in numerous valuable papers on calculated polarizabilities, e.g., Miller, K. J., and Savchik, J. A., *J. Am. Chem. Soc.*, 101, 7206, 1979.

28. Schlecht, S., Schäfer, R., Woenckhaus, J., Becker, J. A., *Chem. Phys. Lett.*, 246, 315, 1995.

29. Benichou, E., Antoine, R., Rayane, D., Vezin, B., Dalby, F. W., Dugourd, P., Ristori, C., Chandezon, F., Huber, B. A., Rocco, J. C., Blundell, S. A., and Guet, C., *Phys. Rev. A* 59, R1, 1999. See also Rayane, D., Allouche, A. R., Benichou, E., Antoine, R., Aubert-Frecon, M., Dugourd, Ph., Broyer, M., Ristori, C., Chandezon, F., Huber, B. A., and Guet, C., *Eur. Phys. J. D* 9, 243, 1999.

30. Antoine, R., Rayane, D., Allouche, A. R., Aubert-Frécon, M., Benichou, E., Dalby, F. W., Dugourd, P., Broyer, M., and Guet, C., *J. Chem. Phys.*, 110, 5568, 1999.

31. Ballard, A., Bonin, K., and Louderback, J., *J. Chem. Phys.* 113, 5732, 2000.

32. Ritchie, G.L.D., and Watson, J.N., *J. Phys. Chem. A* 108, 4515, 2004. These measurements are at 632.8 nm frequency, and are stated accurate to 0.4%.

33. Ritchie, G.L.D., and Blanch, E.W., *J. Phys. Chem. A* 107, 2093, 2003. These measurements are at 632.8 nm frequncy, and are stated accurate to better than 1%.

34. May, E. F., Moldover, M. R., and Schmidt, J. W., *Phys. Rev. A* 78, 032522 (2008). The uncertainty is 0.0003 Å³.

35. Heiles, S., Schäfer, S., and Schäfer, R., *J. Chem. Phys.* 135, 034303, 2011. The uncertainties are ±2.5 Å³ for Ge_9 and Ge_{10}, and ±6.1 Å³ for Ge_{15}.

36. Berninger, M., Stefanov, A., Deachapunya, S., and Arndt, M., *Phys. Rev. A* 76, 013607, 2007. The uncertainties (statistical, systematic) are ±(0.9, 5.1) Å³ for C_{60} and ±(2.0, 6.2) Å³ for C_{70}.

37. Hohm, U., and Loose, A., *Chem. Phys. Lett.* 348, 375, 2001. The uncertainties are ±0.25 Å³ for $Nd(C_5H_5)_3$, ±0.13 Å³ for $Sm(C_5H_5)_3$, and ±0.45 Å³ for $Er(C_5H_5)_3$.

38. Rayane, D., Antoine, R., Dugourd, P., and Broyer, M., *J. Chem. Phys.* 113, 4501, 2000. A 10% precision in the measurements is stated.

39. Tikhonov, G., Kasperovich, V., Wong, K., and Kresin, V. V., *Phys. Rev. A* 64, 063202, 2001. The uncertainties increase with cluster size, from ±2.8% (Na_7) to ±12.4% (Na_{93}).

40. Hohm, U., Loose, A., Maroulis, G., and Xenides, D., *Phys. Rev. A* 61, 053202, 2000. The uncertainty is ±0.14 Å³.

IONIZATION ENERGIES OF ATOMS AND ATOMIC IONS

The ionization energies (often called ionization potentials) of neutral and partially ionized atoms are listed in this table. Data were obtained from the compilations cited below, supplemented by results from the recent research literature. Values for the first and second ionization energies come from Reference 6. All values are given in electron volts (eV).

Following the traditional spectroscopic notation, columns are headed I, II, III, etc. up to XXX, where I indicates the neutral atom, II the singly ionized atom, III the doubly ionized atom, etc. The first section of the table includes spectra I to VIII of all the elements through rutherfordium; subsequent sections cover higher spectra (ionization stages) for those elements for which data are available.

References

1. Moore, C. E., *Ionization Potentials and Ionization Limits Derived from the Analysis of Optical Spectra*, Natl. Stand. Ref. Data Ser. — Natl. Bur. Stand. (U.S.) No. 34, 1970.
2. Martin, W. C., Zalubas, R., and Hagan, L., *Atomic Energy Levels — The Rare Earth Elements*, Natl. Stand. Ref. Data Ser. — Natl. Bur. Stand. (U.S.), No. 60, 1978.
3. Sugar, J. and Corliss, C., *Atomic Energy Levels of the Iron Period Elements: Potassium through Nickel, J. Phys. Chem. Ref. Data*, Vol.14, Suppl. 2, 1985.
4. References to papers in *J. Phys. Chem. Ref. Data*, in the period 1973–91 covering other elements may be found in the cumulative index to that journal.
5. Martin, W.C., and Wiese, W.L., in *Atomic, Molecular, and Optical Physics Handbook*, Drake, G.W.F., Ed., AIP Press, New York, 1996.
6. Sansonetti, J. E., Martin, W. C., and Young, S. L., *Handbook of Basic Atomic Spectroscopic Data* (version 1.1), NIST Physical Data web site <http://physics.nist.gov/PhysRefData/Handbook> (October 2004); *J. Phys. Chem. Ref. Data*, 34, 1559, 2005.

Neutral Atoms to +7 Ions

Z	Element	I	II	III	IV	V	VI	VII	VIII
1	H	13.598443							
2	He	24.587387	54.417760						
3	Li	5.391719	75.6400	122.45429					
4	Be	9.32270	18.21114	153.89661	217.71865				
5	B	8.29802	25.1548	37.93064	259.37521	340.22580			
6	C	11.26030	24.3833	47.8878	64.4939	392.087	489.99334		
7	N	14.5341	29.6013	47.44924	77.4735	97.8902	552.0718	667.046	
8	O	13.61805	35.1211	54.9355	77.41353	113.8990	138.1197	739.29	871.4101
9	F	17.4228	34.9708	62.7084	87.1398	114.2428	157.1651	185.186	953.9112
10	Ne	21.56454	40.96296	63.45	97.12	126.21	157.93	207.2759	239.0989
11	Na	5.139076	47.2864	71.6200	98.91	138.40	172.18	208.50	264.25
12	Mg	7.646235	15.03527	80.1437	109.2655	141.27	186.76	225.02	265.96
13	Al	5.985768	18.82855	28.44765	119.992	153.825	190.49	241.76	284.66
14	Si	8.15168	16.34584	33.49302	45.14181	166.767	205.27	246.5	303.54
15	P	10.48669	19.7695	30.2027	51.4439	65.0251	220.421	263.57	309.60
16	S	10.36001	23.33788	34.79	47.222	72.5945	88.0530	280.948	328.75
17	Cl	12.96763	23.8136	39.61	53.4652	67.8	97.03	114.1958	348.28
18	Ar	15.759610	27.62966	40.74	59.81	75.02	91.009	124.323	143.460
19	K	4.3406633	31.63	45.806	60.91	82.66	99.4	117.56	154.88
20	Ca	6.11316	11.87172	50.9131	67.27	84.50	108.78	127.2	147.24
21	Sc	6.56149	12.79977	24.75666	73.4894	91.65	110.68	138.0	158.1
22	Ti	6.82812	13.5755	27.4917	43.2672	99.30	119.53	140.8	170.4
23	V	6.74619	14.618	29.311	46.709	65.2817	128.13	150.6	173.4
24	Cr	6.76651	16.4857	30.96	49.16	69.46	90.6349	160.18	184.7
25	Mn	7.43402	15.6400	33.668	51.2	72.4	95.6	119.203	194.5
26	Fe	7.9024	16.1877	30.652	54.8	75.0	99.1	124.98	151.06
27	Co	7.88101	17.084	33.50	51.3	79.5	102.0	128.9	157.8
28	Ni	7.6398	18.16884	35.19	54.9	76.06	108	133	162
29	Cu	7.72638	20.2924	36.841	57.38	79.8	103	139	166
30	Zn	9.394199	17.96439	39.723	59.4	82.6	108	134	174
31	Ga	5.999301	20.51515	30.7258	63.241	86.01	112.7	140.9	169.9
32	Ge	7.89943	15.93461	34.2241	45.7131	93.5			
33	As	9.7886	18.5892	28.351	50.13	62.63	127.6		
34	Se	9.75239	21.19	30.8204	42.9450	68.3	81.7	155.4	
35	Br	11.8138	21.591	36	47.3	59.7	88.6	103.0	192.8
36	Kr	13.99961	24.35984	36.950	52.5	64.7	78.5	111.0	125.802
37	Rb	4.177128	27.2895	40	52.6	71.0	84.4	99.2	136
38	Sr	5.69485	11.0301	42.89	57	71.6	90.8	106	122.3
39	Y	6.2173	12.224	20.52	60.597	77.0	93.0	116	129
40	Zr	6.63390	13.1	22.99	34.34	80.348			
41	Nb	6.75885	14.0	25.04	38.3	50.55	102.057	125	

Neutral Atoms to +7 Ions

Z	Element	I	II	III	IV	V	VI	VII	VIII
42	Mo	7.09243	16.16	27.13	46.4	54.49	68.8276	125.664	143.6
43	Tc	7.28	15.26	29.54					
44	Ru	7.36050	16.76	28.47					
45	Rh	7.45890	18.08	31.06					
46	Pd	8.3369	19.43	32.93					
47	Ag	7.57623	21.47746	34.83					
48	Cd	8.99382	16.90831	37.48					
49	In	5.78636	18.8703	28.03	54				
50	Sn	7.34392	14.6322	30.50260	40.73502	72.28			
51	Sb	8.60839	16.63	25.3	44.2	56	108		
52	Te	9.0096	18.6	27.96	37.41	58.75	70.7	137	
53	I	10.45126	19.1313	33					
54	Xe	12.12984	20.9750	32.1230					
55	Cs	3.893905	23.15744						
56	Ba	5.211664	10.00383						
57	La	5.5769	11.059	19.1773	49.95	61.6			
58	Ce	5.5387	10.85	20.198	36.758	65.55	77.6		
59	Pr	5.473	10.55	21.624	38.98	57.53			
60	Nd	5.5250	10.72	22.1	40.4				
61	Pm	5.582	10.90	22.3	41.1				
62	Sm	5.6437	11.07	23.4	41.4				
63	Eu	5.67038	11.25	24.92	42.7				
64	Gd	6.14980	12.09	20.63	44.0				
65	Tb	5.8638	11.52	21.91	39.79				
66	Dy	5.9389	11.67	22.8	41.47				
67	Ho	6.0215	11.80	22.84	42.5				
68	Er	6.1077	11.93	22.74	42.7				
69	Tm	6.18431	12.05	23.68	42.7				
70	Yb	6.25416	12.176	25.05	43.56				
71	Lu	5.42586	13.9	20.9594	45.25	66.8			
72	Hf	6.82507	15	23.3	33.33				
73	Ta	7.54957							
74	W	7.86403	16.1						
75	Re	7.83352							
76	Os	8.43823							
77	Ir	8.96702							
78	Pt	8.9588	18.563						
79	Au	9.22553	20.20						
80	Hg	10.4375	18.7568	34.2					
81	Tl	6.108194	20.4283	29.83					
82	Pb	7.41663	15.03248	31.9373	42.32	68.8			
83	Bi	7.2855	16.703	25.56	45.3	56.0	88.3		
84	Po	8.414							
85	At								
86	Rn	10.7485							
87	Fr	4.072741							
88	Ra	5.278423	10.14715						
89	Ac	5.17	11.75						
90	Th	6.3067	11.9	20.0	28.8				
91	Pa	5.89							
92	U	6.1941	10.6						
93	Np	6.2657							
94	Pu	6.0260	11.2						
95	Am	5.9738							
96	Cm	5.9914							
97	Bk	6.1979							
98	Cf	6.2817	11.8						
99	Es	6.42	12.0						
100	Fm	6.50							
101	Md	6.58							
102	No	6.65							
103	Lr	4.9							
104	Rf	6.0							

+8 Ions to +15 Ions

Z	Element	IX	X	XI	XII	XIII	XIV	XV	XVI
9	F	1103.1176							
10	Ne	1195.8286	1362.1995						
11	Na	299.864	1465.121	1648.702					
12	Mg	328.06	367.50	1761.805	1962.6650				
13	Al	330.13	398.75	442.00	2085.98	2304.1410			
14	Si	351.12	401.37	476.36	523.42	2437.63	2673.182		
15	P	372.13	424.4	479.46	560.8	611.74	2816.91	3069.842	
16	S	379.55	447.5	504.8	564.44	652.2	707.01	3223.78	3494.1892
17	Cl	400.06	455.63	529.28	591.99	656.71	749.76	809.40	3658.521
18	Ar	422.45	478.69	538.96	618.26	686.10	755.74	854.77	918.03
19	K	175.8174	503.8	564.7	629.4	714.6	786.6	861.1	968
20	Ca	188.54	211.275	591.9	657.2	726.6	817.6	894.5	974
21	Sc	180.03	225.18	249.798	687.36	756.7	830.8	927.5	1009
22	Ti	192.1	215.92	265.07	291.500	787.84	863.1	941.9	1044
23	V	205.8	230.5	255.7	308.1	336.277	896.0	976	1060
24	Cr	209.3	244.4	270.8	298.0	354.8	384.168	1010.6	1097
25	Mn	221.8	248.3	286.0	314.4	343.6	403.0	435.163	1134.7
26	Fe	233.6	262.1	290.2	330.8	361.0	392.2	457	489.256
27	Co	186.13	275.4	305	336	379	411	444	511.96
28	Ni	193	224.6	321.0	352	384	430	464	499
29	Cu	199	232	265.3	369	401	435	484	520
30	Zn	203	238	274	310.8	419.7	454	490	542
31	Ga	210.8	244.0	280.7	319.2	357.2	471.2	508.8	548.3
36	Kr	230.85	268.2	308	350	391	447	492	541
37	Rb	150	277.1						
38	Sr	162	177	324.1					
39	Y	146.2	191	206	374.0				
42	Mo	164.12	186.4	209.3	230.28	279.1	302.60	544.0	570

+16 Ions to +23 Ions

Z	Element	XVII	XVIII	XIX	XX	XXI	XXII	XXIII	XXIV
17	Cl	3946.2960							
18	Ar	4120.8857	4426.2296						
19	K	1033.4	4610.8	4934.046					
20	Ca	1087	1157.8	5128.8	5469.864				
21	Sc	1094	1213	1287.97	5674.8	6033.712			
22	Ti	1131	1221	1346	1425.4	6249.0	6625.82		
23	V	1168	1260	1355	1486	1569.6	6851.3	7246.12	
24	Cr	1185	1299	1396	1496	1634	1721.4	7481.7	7894.81
25	Mn	1224	1317	1437	1539	1644	1788	1879.9	8140.6
26	Fe	1266	1358	1456	1582	1689	1799	1950	2023
27	Co	546.58	1397.2	1504.6	1603	1735	1846	1962	2119
28	Ni	571.08	607.06	1541	1648	1756	1894	2011	2131
29	Cu	557	633	670.588	1697	1804	1916	2060	2182
30	Zn	579	619	698	738	1856			
36	Kr	592	641	786	833	884	937	998	1051
42	Mo	636	702	767	833	902	968	1020	1082

+24 Ions to +29 Ions

Z	Element	XXV	XXVI	XXVII	XXVIII	XXIX	XXX
25	Mn	8571.94					
26	Fe	8828	9277.69				
27	Co	2219.0	9544.1	10012.12			
28	Ni	2295	2399.2	10288.8	10775.40		
29	Cu	2308	2478	2587.5	11062.38	11567.617	
36	Kr	1151	1205.3	2928	3070	3227	3381
42	Mo	1263	1323	1387	1449	1535	1601

IONIZATION ENERGIES OF GAS-PHASE MOLECULES

Sharon G. Lias

This table presents values for the first ionization energies (IP) of approximately 1000 molecules and atoms. Substances are listed by molecular formula in the modified Hill order (see Preface). Values enclosed in parentheses are considered not to be well established. Data appearing in the 1988 reference were updated in 1996 for inclusion in the database of ionization energies available at the Internet site of the Standard Reference Data program of the National Institute of Standards and Technology (http://webbook.nist.gov). The list appearing here includes these updates.

The list also includes values for enthalpies of formation of the ions at 298 K, $\Delta_f H_{ion}$, given according to the ion convention used by mass spectrometrists; to convert these values to the electron convention used by thermodynamicists, add 6 kJ/mol. Details on the calculation of $\Delta_f H_{ion}$, as well as data for a much larger number of molecules, may be found in the reference and on the Internet site.

Reference

Lias, S. G., Bartmess, J. E., Liebman, J. F., Holmes, J. L., Levin, R. D., and Mallard, W.G., *Gas-Phase Ion and Neutral Thermochemistry, J. Phys. Chem. Ref. Data*, Vol. 17, Suppl. No. 1, 1988.

Mol. form.	Name	IP/eV	$\Delta_f H_{ion}$ kJ/mol
Substances not containing carbon			
Ac	Actinium	5.17	905
Ag	Silver	7.57624	1016
AgCl	Silver(I) chloride	(≤ 10.08)	≤ 1065
AgF	Silver(I) fluoride	(11.0 ± 0.3)	1071
Al	Aluminum	5.98577	905
AlBr	Aluminum monobromide	(9.3)	913
AlBr$_3$	Aluminum tribromide	(10.4)	593
AlCl	Aluminum monochloride	9.4	855
AlCl$_3$	Aluminum trichloride	(12.01)	573
AlF	Aluminum monofluoride	9.73 ± 0.01	673
AlF$_3$	Aluminum trifluoride	≤ 15.45	≤ 282
AlI	Aluminum monoiodide	9.3 ± 0.3	965
AlI$_3$	Aluminum triiodide	(9.1)	673
Am	Americium	5.9738 ± 0.0002	860
Ar	Argon	15.75962	1521
As	Arsenic	9.8152	1250
AsCl$_3$	Arsenic(III) chloride	(10.55 ± 0.025)	754
AsF$_3$	Arsenic(III) fluoride	(12.84 ± 0.05)	452
AsH$_3$	Arsine	(9.89)	1021
Au	Gold	9.22567	1254
B	Boron	8.29803	1363
BBr$_3$	Boron tribromide	(10.51)	809
BCl$_3$	Boron trichloride	11.60 ± 0.02	718
BF	Fluoroborane	11.12 ± 0.01	957
	Difluoroborane	(9.4)	317
BF$_3$	Boron trifluoride	15.7 ± 0.3	365
BH	Boron monohydride	(9.77)	1385
BH$_3$	Borane	12.026 ± 0.024	1261
BI$_3$	Boron triiodide	(9.25 ± 0.03)	964
BO$_2$	Boron dioxide	(13.5 ± 0.3)	1001
B$_2$H$_6$	Diborane	11.38 ± 0.05	1134
B$_2$O$_3$	Boron oxide	13.5 ± 0.15	460
B$_4$H$_{10}$	Tetraborane	10.76 ± 0.04	1105
B$_5$H$_9$	Pentaborane(9)	9.90 ± 0.04	1028
B$_6$H$_{10}$	Hexaborane	(9.0)	965
Ba	Barium	5.21170	683
BaO	Barium oxide	6.91 ± 0.06	543
Be	Beryllium	9.32263	1224
BeO	Beryllium oxide	(10.1 ± 0.4)	1111
Bi	Bismuth	7.2855	908
BiCl$_3$	Bismuth trichloride	(10.4)	736

Mol. form.	Name	IP/eV	$\Delta_f H_{ion}$ kJ/mol
Bk	Berkelium	6.23	911
Br	Bromine (atomic)	11.81381	1252
BrCl	Bromine chloride	11.01	1079
BrF	Bromine fluoride	11.86	1086
BrF$_5$	Bromine pentafluoride	13.172 ± 0.002	840
BrH	Hydrogen bromide	11.66 ± 0.03	1087
BrH$_3$Si	Bromosilane	10.6	943
BrI	Iodine bromide	9.790 ± 0.004	986
BrK	Potassium bromide	7.85 ± 0.1	578
BrLi	Lithium bromide	(8.7)	685
BrNO	Nitrosyl bromide	10.17 ± 0.03	1065
BrNa	Sodium bromide	8.31 ± 0.1	660
BrO	Bromine monoxide	10.46 ± 0.02	1135
BrRb	Rubidium bromide	7.94 ± 0.03	583
BrTl	Thallium(I) bromide	9.14 ± 0.02	844
Br$_2$	Bromine	10.516 ± 0.005	1046
Br$_2$Hg	Mercury(II) bromide	10.560 ± 0.003	935
Br$_2$Sn	Tin(II) bromide	9.0	839
Br$_3$Ga	Gallium(III) bromide	10.40	711
Br$_3$P	Phosphorus(III) bromide	9.7	798
Br$_4$Hf	Hafnium(IV) bromide	(10.9)	366
Br$_4$Sn	Tin(IV) bromide	10.6	709
Br$_4$Ti	Titanium(IV) bromide	10.3	375
Br$_4$Zr	Zirconium(IV) bromide	(10.7)	388
Ca	Calcium	6.11316	768
CaCl	Calcium monochloride	5.86 ± 0.07	462
CaO	Calcium oxide	6.66 ± 0.18	668
Cd	Cadmium	8.99367	980
Ce	Cerium	5.5387	957
Cf	Californium	6.30	805
Cl	Chlorine (atomic)	12.96764	1373
ClCs	Cesium chloride	(7.84 ± 0.05)	510
ClF	Chlorine fluoride	12.66 ± 0.01	1171
ClFO$_3$	Perchloryl fluoride	(12.945 ± 0.005)	1224
ClF$_2$	Chlorine difluoride	(12.77 ± 0.05)	1128
ClF$_3$	Chlorine trifluoride	(12.65 ± 0.05)	1057
ClF$_5$S	Sulfur chloride pentafluoride	(12.335 ± 0.005)	144
ClH	Hydrogen chloride	12.749 ± 0.009	1137
ClHO	Hypochlorous acid	(11.12 ± 0.01)	993
ClH$_3$Si	Chlorosilane	11.4	899
ClI	Iodine chloride	10.088 ± 0.01	991
ClIn	Indium(I) chloride	(9.51)	842
ClK	Potassium chloride	(8.0 ± 0.4)	557
ClLi	Lithium chloride	9.57	727
ClNO	Nitrosyl chloride	10.87 ± 0.01	1099
ClNO$_2$	Nitryl chloride	(11.84)	1155
ClNa	Sodium chloride	8.92 ± 0.06	681
ClO	Chlorine monoxide	10.95	1159
ClO$_2$	Chlorine dioxide	10.33 ± 0.02	1093
ClRb	Rubidium chloride	(8.50 ± 0.03)	590
ClTl	Thallium(I) chloride	9.70 ± 0.03	869
Cl$_2$	Chlorine	11.480 ± 0.005	1108
Cl$_2$CrO$_2$	Chromyl chloride	11.6	580
Cl$_2$Ge	Germanium(II) chloride	(10.20 ± 0.05)	813
Cl$_2$H$_2$Si	Dichlorosilane	11.4	765
Cl$_2$Hg	Mercury(II) chloride	11.380 ± 0.003	952
Cl$_2$O	Chlorine oxide	10.94	1135
Cl$_2$OS	Thionyl chloride	10.96	844
Cl$_2$O$_2$S	Sulfuryl chloride	12.05	807
Cl$_2$Pb	Lead(II) chloride	(10.2)	791

Mol. form.	Name	IP/eV	$\Delta_f H_{ion}$ kJ/mol
Cl_2S	Sulfur dichloride	9.45 ± 0.03	895
Cl_2Si	Dichlorosilylene	(10.93 ± 0.10)	887
Cl_2Sn	Tin(II) chloride	(10.0)	760
Cl_3Ga	Gallium(III) chloride	11.52	664
Cl_3HSi	Trichlorosilane	(11.7)	648
Cl_3N	Nitrogen trichloride	(10.12 ± 0.1)	1244
Cl_3OP	Phosphorus(V) oxychloride	11.36 ± 0.02	540
Cl_3OV	Vanadyl trichloride	(11.6)	425
Cl_3P	Phosphorus(III) chloride	9.91	668
Cl_3PS	Phosphorus(V) sulfide trichloride	9.71 ± 0.03	573
Cl_3Sb	Antimony(III) chloride	(≤ 10.7)	s719
Cl_4Ge	Germanium(IV) chloride	11.68 ± 0.05	629
Cl_4Hf	Hafnium(IV) chloride	(11.7)	246
Cl_4Si	Tetrachlorosilane	11.79 ± 0.01	527
Cl_4Sn	Tin(IV) chloride	11.7 ± 0.2	656
Cl_4Ti	Titanium(IV) chloride	(11.5)	349
Cl_4V	Vanadium(IV) chloride	(9.2)	361
Cl_4Zr	Zirconium(IV) chloride	(11.2)	210
Cl_5Mo	Molybdenum(V) chloride	(8.7)	392
Cl_5Nb	Niobium(V) chloride	(10.97)	356
Cl_5P	Phosphorus(V) chloride	(10.2)	608
Cl_5Ta	Tantalum(V) chloride	(11.08)	303
Cl_6W	Tungsten(VI) chloride	(9.5)	348
Cm	Curium	6.02	966
Co	Cobalt	7.8810	1187
Cr	Chromium	6.76664	1050
Cs	Cesium	3.89390	452
CsF	Cesium fluoride	(8.80 ± 0.10)	489
$CsNa$	Cesium sodium	(4.05 ± 0.04)	535
Cu	Copper	7.72638	1084
CuF	Copper(I) fluoride	10.15 ± 0.02	984
Dy	Dysprosium	5.9389	862
Er	Erbium	6.1078	907
Es	Einsteinium	6.42	753
Eu	Europium	5.6704	723
F	Fluorine (atomic)	17.42282	1761
FGa	Gallium monofluoride	(9.6 ± 0.5)	700
FH	Hydrogen fluoride	16.044 ± 0.003	1276
FHO	Hypofluorous acid	12.71 ± 0.01	1130
FH_3Si	Fluorosilane	11.7	752
FI	Iodine fluoride	10.54 ± 0.01	922
FIn	Indium monofluoride	(9.6 ± 0.5)	740
FNO	Nitrosyl fluoride	12.63 ± 0.03	1152
FNO_2	Nitryl fluoride	(13.09)	1154
FNS	Thionitrosyl fluoride (NSF)	11.51 ± 0.04	1090
FO	Fluorine monoxide	12.78 ± 0.03	1342
FO_2	Fluorine superoxide (FOO)	(12.6 ± 0.2)	1228
FS	Sulfur fluoride	10.09	986
FTl	Thallium(I) fluoride	10.52	835
F_2	Fluorine	15.697 ± 0.003	1515
F_2Ge	Germanium(II) fluoride	(≤ 11.65)	551
F_2HN	Difluoramine	(11.53 ± 0.08)	1046
F_2H_2Si	Difluorosilane	(12.2)	386
F_2Mg	Magnesium fluoride	(13.6 ± 0.3)	588
F_2N	Difluoroamidogen	11.628 ± 0.01	1155
F_2N_2	trans-Difluorodiazine	(12.8)	1315
F_2O	Fluorine monoxide	13.11 ± 0.01	1290
F_2OS	Thionyl fluoride	12.25	688
F_2O_2S	Sulfuryl fluoride	13.04 ± 0.01	501
F_2Pb	Lead(II) fluoride	(11.5)	679

Mol. form.	Name	IP/eV	$\Delta_f H_{ion}$ kJ/mol
F_2S	Sulfur difluoride	(10.08)	676
F_2Si	Difluorosilylene	10.78 ± 0.05	450
F_2Sn	Tin(II) fluoride	(11.1)	586
F_2Xe	Xenon difluoride	12.35 ± 0.01	1083
F_3HSi	Trifluorosilane	(14.0)	150
F_3N	Nitrogen trifluoride	13.00 ± 0.02	1125
F_3NO	Trifluoramine oxide	13.31 ± 0.06	1121
F_3OP	Phosphorus(V) oxyfluoride	12.76 ± 0.01	−24
F_3P	Phosphorus(III) fluoride	11.60 ± 0.05	161
F_3PS	Phosphorus(V) sulfide trifluoride	≤ 11.05 ± 0.035	≤ 58
F_3Si	Trifluorosilyl	(9.99)	− 32
F_4Ge	Germanium(IV) fluoride	(15.5)	307
F_4N_2	Tetrafluorohydrazine	11.94 ± 0.03	1119
F_4S	Sulfur tetrafluoride	12.0 ± 0.3	399
F_4Si	Tetrafluorosilane	15.24 ± 0.14	−144
F_4Xe	Xenon tetrafluoride	12.65 ± 0.1	1016
F_5I	Iodine pentafluoride	12.943 ± 0.005	408
F_5P	Phosphorus(V) fluoride	(15.1)	−137
F_5S	Sulfur pentafluoride	9.60 ± 0.05	10
F_6Mo	Molybdenum(VI) fluoride	(14.5 ± 0.1)	−159
F_6S	Sulfur hexafluoride	15.32 ± 0.02	258
F_6U	Uranium(VI) fluoride	14.00 ± 0.10	−796
Fe	Iron	7.9024	1177
Fm	Fermium	6.50	627
Ga	Gallium	5.99930	851
GaI_3	Gallium(III) iodide	9.40	765
Gd	Gadolinium	6.1500	991
Ge	Germanium	7.900	1139
GeH_4	Germane	≤ 10.53	≤ 1108
GeI_4	Germanium(IV) iodide	(9.42)	850
GeO	Germanium(II) oxide	11.25 ± 0.01	1044
GeS	Germanium(II) sulfide	(9.98)	1055
H	Hydrogen (atomic)	13.59844	1530
HI	Hydrogen iodide	10.386 ± 0.001	1028
HLi	Lithium hydride	7.7	882
HN	Imidogen	≤ 13.49 ± 0.01	1678
HNO	Nitrosyl hydride	(10.1)	1075
HNO_2	Nitrous acid	≤ 11.3	≤ 1011
HNO_3	Nitric acid	11.95 ± 0.01	1019
HN_3	Hydrazoic acid	10.72 ± 0.025	1328
HO	Hydroxyl	13.0170 ± 0.0002	1294
HO_2	Hydroperoxy	11.35 ± 0.01	1106
HS	Mercapto	10.4219 ± 0.0004	1145
H_2	Hydrogen	15.42593 ± 0.00005	1488
H_2N	Amidogen	11.14 ± 0.01	1264
H_2O	Water	12.6206 ± 0.0020	976
H_2O_2	Hydrogen peroxide	10.58 ± 0.04	885
H_2S	Hydrogen sulfide	10.457 ± 0.012	989
H_2Se	Hydrogen selenide	9.892 ± 0.005	984
H_2Si	Silylene	8.244 ± 0.025	1084
H_3N	Ammonia	10.070 ± 0.020	925
H_3NO	Hydroxylamine	(10.00)	923
H_3P	Phosphine	9.869 ± 0.002	958
H_3Sb	Stibine	9.54 ± 0.03	1067
H_4N_2	Hydrazine	8.1 ± 0.15	877
H_4Si	Silane	11.00 ± 0.02	1095
H_4Sn	Stannane	(10.75)	1200
H_6Si_2	Disilane	9.74 ± 0.02	1019
H_8Si_3	Trisilane	(9.2)	1009
He	Helium	24.58741	2372

Mol. form.	Name	IP/eV	$\Delta_f H_{ion}$ kJ/mol
Hf	Hafnium	6.82507 ± 0.00004	1278
Hg	Mercury	10.43750	1069
HgI_2	Mercury(II) iodide	9.5088 ± 0.0022	900
Ho	Holmium	6.0216	882
I	Iodine (atomic)	10.45126	1115
IK	Potassium iodide	(7.21 ± 0.3)	570
ILi	Lithium iodide	(7.5)	633
INa	Sodium iodide	7.64 ± 0.02	659
ITl	Thallium(I) iodide	8.47 ± 0.02	826
I_2	Iodine	9.3074 ± 0.0002	960
I_4Ti	Titanium(IV) iodide	(9.1)	602
I_4Zr	Zirconium(IV) iodide	(9.3)	500
In	Indium	5.78636	802
Ir	Iridium	9.1	1543
K	Potassium	4.34066	508
KLi	Lithium potassium	4.57 ± 0.04	512
KNa	Potassium sodium	4.41636 ± 0.00017	561
K_2	Dipotassium	4.0637 ± 0.0002	519
Kr	Krypton	13.99961	1351
La	Lanthanum	5.5770	969
Li	Lithium	5.39172	680
LiNa	Lithium sodium	5.05 ± 0.04	571
LiO	Lithium monoxide	(8.44)	894
LiRb	Lithium rubidium	4.3 ± 0.1	486
Li_2	Dilithium	5.1127 ± 0.0003	709
Lu	Lutetium	5.42585	950
Md	Mendelevium	6.58	635
Mg	Magnesium	7.64624	885
MgO	Magnesium oxide	(8.76 ± 0.22)	901
Mn	Manganese	7.43402	998
Mo	Molybdenum	7.09243	1343
N	Nitrogen (atomic)	14.53414	1875
NO	Nitric oxide	9.26438 ± 0.00005	985
NO_2	Nitrogen dioxide	9.586 ± 0.002	958
NP	Phosphorus nitride	11.84 ± 0.04	1247
NS	Nitrogen sulfide	8.87 ± 0.01	1119
N_2	Nitrogen	15.5808	1503
N_2O	Nitrous oxide	12.886	1325
N_2O_4	Nitrogen tetroxide	(10.8)	1050
N_2O_5	Nitrogen pentoxide	(11.9)	1161
Na	Sodium	5.13908	603
NaRb	Rubidium sodium	4.32 ± 0.04	480
Na_2	Disodium	4.894 ± 0.003	614
Nb	Niobium	6.75885	1384
Nd	Neodymium	5.5250	859
Ne	Neon	21.56454	2081
Ni	Nickel	7.6398	1167
No	Nobelium	6.65	642
Np	Neptunium	6.2657 ± 0.0003	1069
O	Oxygen (atomic)	13.61806	1563
OPb	Lead(II) oxide	9.08 ± 0.10	939
OS	Sulfur monoxide	10.294 ± 0.004	998
OS_2	Sulfur oxide (SSO)	10.584 ± 0.005	971
OSi	Silicon monoxide	11.49 ± 0.20	1008
OSn	Tin(II) oxide	9.60 ± 0.02	944
OSr	Strontium oxide	6.6 ± 0.2	623
O_2	Oxygen	12.0697 ± 0.0002	1165
O_2S	Sulfur dioxide	12.349 ± 0.001	894
O_2Th	Thorium(IV) oxide	(8.7 ± 0.15)	342
O_2Ti	Titanium(IV) oxide	(9.54 ± 0.1)	623

Mol. form.	Name	IP/eV	$\Delta_f H_{ion}$ kJ/mol
O_2U	Uranium(IV) oxide	(5.4 ± 0.1)	57
O_3	Ozone	12.43	1342
O_3S	Sulfur trioxide	12.82 ± 0.03	841
O_3U	Uranium(VI) oxide	(10.5 ± 0.5)	214
O_4Os	Osmium(VIII) oxide	(12.32)	850
O_4Ru	Ruthenium(VIII) oxide	12.15 ± 0.03	988
O_7Re_2	Rhenium(VII) oxide	(12.7 ± 0.2)	125
Os	Osmium	8.7	1630
P	Phosphorus	10.48669	1328
P_2	Diphosphorus	10.53	1160
Pa	Protactinium	5.89	1133
Pb	Lead	7.41666	911
PbS	Lead(II) sulfide	(8.5 ± 0.5)	954
Pd	Palladium	8.3367	1181
Pm	Promethium	5.55	536
Pr	Praseodymium	5.464	883
Pt	Platinum	9.0	1433
Pu	Plutonium	6.025	926
Ra	Radium	5.27892	668
Rb	Rubidium	4.17713	484
Re	Rhenium	7.88	1530
Rh	Rhodium	7.45890	1276
Rn	Radon	10.74850	1037
Ru	Ruthenium	7.36050	1355
S	Sulfur	10.36001	1277
SSn	Tin(II) sulfide	(8.8)	966
S_2	Disulfur	9.356 ± 0.002	1031
Sb	Antimony	8.64	1096
Sc	Scandium	6.56144	1010
Se	Selenium	9.75238	1168
Si	Silicon	8.15169	1238
Sm	Samarium	5.6437	751
Sn	Tin	7.34381	1011
Sr	Strontium	5.69484	713
Ta	Tantalum	7.89	1544
Tb	Terbium	5.8639	955
Tc	Technetium	7.28	1380
Te	Tellurium	9.0096	1066
Th	Thorium	6.308 ± 0.003	1207
Ti	Titanium	6.8282	1127
Tl	Thallium	6.10829	771
Tm	Thulium	6.18431	827
U	Uranium	6.19405	1129
V	Vanadium	6.746 ± 0.002	1166
W	Tungsten	7.98	1621
Xe	Xenon	12.12987	1170
Y	Yttrium	6.217	1022
Yb	Ytterbium	6.25416	754
Zn	Zinc	9.39405	1037
Zr	Zirconium	6.63390	1251

Substances containing carbon

Mol. form.	Name	IP/eV	$\Delta_f H_{ion}$ kJ/mol
C	Carbon	11.26030	1803
$CBrClF_2$	Bromochlorodifluoromethane	(11.21)	642
$CBrCl_3$	Bromotrichloromethane	(10.6)	980
$CBrF_3$	Bromotrifluoromethane	(11.40)	451
CBr_2F_2	Dibromodifluoromethane	11.03 ± 0.04	683
CBr_4	Tetrabromomethane	(10.31 ± 0.02)	1079
CCl	Chloromethylidyne	(8.9 ± 0.2)	1244
$CClF_3$	Chlorotrifluoromethane	12.6 ± 0.2	505

Mol. form.	Name	IP/eV	$\Delta_f H_{ion}$ kJ/mol
CClN	Cyanogen chloride	12.34 ± 0.01	1329
CCl$_2$	Dichloromethylene	(9.27)	1058
CCl$_2$F$_2$	Dichlorodifluoromethane	12.05 ± 0.24	685
CCl$_2$O	Carbonyl chloride	(11.5)	888
CCl$_3$F	Trichlorofluoromethane	11.77 ± 0.02	868
CCl$_4$	Tetrachloromethane	11.47 ± 0.01	1010
CF	Fluoromethylidyne	9.11 ± 0.01	1134
CFN	Cyanogen fluoride	13.34 ± 0.02	1325
CF$_2$	Difluoromethylene	11.44 ± 0.03	899
CF$_2$O	Carbonyl fluoride	13.035 ± 0.030	617
CF$_3$	Trifluoromethyl	8.7 ± 0.2	379
CF$_3$I	Trifluoroiodomethane	10.23	397
CH	Methylidyne	10.64 ± 0.01	1622
CHBrCl$_2$	Bromodichloromethane	10.6	973
CHBr$_2$Cl	Chlorodibromomethane	10.59 ± 0.01	1030
CHBr$_3$	Tribromomethane	10.48 ± 0.02	1035
CHCl	Chloromethylene	9.84	1247
CHClF$_2$	Chlorodifluoromethane	(12.2)	693
CHCl$_2$F	Dichlorofluoromethane	(11.5)	829
CHCl$_3$	Trichloromethane	11.37 ± 0.02	992
CHF	Fluoromethylene	10.06 ± 0.05	1121
CHF$_3$	Trifluoromethane	(13.86)	643
CHI$_3$	Triiodomethane	9.25 ± 0.02	1010
CHN	Hydrogen cyanide	13.60 ± 0.01	1447
CHN	Hydrogen isocyanide	(12.5 ± 0.1)	1407
CHNO	Isocyanic acid	11.595 ± 0.005	1016
CHNO	Fulminic acid	(10.83)	1263
CHO	Oxomethyl (HCO)	(8.55)	826
CH$_2$	Methylene	10.396 ± 0.003	1392
CH$_2$BrCl	Bromochloromethane	10.77 ± 0.01	1085
CH$_2$Br$_2$	Dibromomethane	(10.50 ± 0.02)	1013
CH$_2$ClF	Chlorofluoromethane	11.71 ± 0.01	870
CH$_2$Cl$_2$	Dichloromethane	11.32 ± .01	996
CH$_2$F$_2$	Difluoromethane	12.71	774
CH$_2$I$_2$	Diiodomethane	9.46 ± 0.02	1030
CH$_2$N$_2$	Diazomethane	8.999 ± 0.001	1098
CH$_2$N$_2$	Cyanamide	(10.4)	1137
CH$_2$O	Formaldehyde	10.88 ± 0.01	941
CH$_2$O$_2$	Formic acid	11.33 ± 0.01	715
CH$_3$	Methyl	9.843 ± 0.002	1095
CH$_3$BO	Borane carbonyl	11.14 ± 0.02	962
CH$_3$Br	Bromomethane	10.541 ± 0.003	979
CH$_3$Cl	Chloromethane	11.22 ± 0.01	1001
CH$_3$Cl$_3$Si	Methyltrichlorosilane	(11.36 ± 0.03)	548
CH$_3$F	Fluoromethane	12.47 ± 0.02	956
CH$_3$I	Iodomethane	9.538	936
CH$_3$NO	Formamide	10.16 ± 0.06	796
CH$_3$NO$_2$	Nitromethane	11.08 ± 0.07	994
CH$_3$N$_3$	Methyl azide	9.81 ± 0.02	1227
CH$_3$O	Methoxy	(10.72)	1050
CH$_4$	Methane	12.61 ± 0.01	1143
CH$_4$N$_2$O	Urea	9.7	690
CH$_4$O	Methanol	10.85 ± 0.01	845
CH$_4$S	Methanethiol	9.44 ± 0.005	888
CH$_5$N	Methylamine	(8.80)	826
CH$_6$N$_2$	Methylhydrazine	7.7 ± 0.15	835
CH$_6$Si	Methylsilane	(10.7)	1003
CN	Cyanide	13.5984	1748
CNO	Cyanate	11.76 ± 0.01	1290
CO	Carbon monoxide	14.014 ± 0.0003	1242

Mol. form.	Name	IP/eV	$\Delta_f H_{ion}$ kJ/mol
COS	Carbon oxysulfide	11.18 ± 0.01	936
COSe	Carbon oxyselenide	10.36 ± 0.01	929
CO_2	Carbon dioxide	13.773 ± 0.002	935
CS	Carbon sulfide	11.33 ± 0.01	1361
CS_2	Carbon disulfide	10.0685 ± 0.0020	1089
C_2	Dicarbon	(11.4 ± 0.3)	2000
$C_2Br_2F_4$	1,2-Dibromotetrafluoroethane	(11.1)	280
C_2ClF_3	Chlorotrifluoroethylene	9.81 ± 0.03	373
C_2ClF_5	Chloropentafluoroethane	(12.6)	99
C_2Cl_2	Dichloroacetylene	9.9	1165
$C_2Cl_2F_4$	1,2-Dichlorotetrafluoroethane	12.2	252
$C_2Cl_3F_3$	1,1,1-Trichlorotrifluoroethane	11.5	386
$C_2Cl_3F_3$	1,1,2-Trichlorotrifluoroethane	11.99 ± 0.02	429
C_2Cl_4	Tetrachloroethylene	9.326 ± 0.001	887
$C_2Cl_4F_2$	1,1,2,2-Tetrachloro-1,2-difluoroethane	(11.3)	563
C_2Cl_4O	Trichloroacetyl chloride	(11.0)	827
C_2Cl_6	Hexachloroethane	(11.1)	920
C_2F_3N	Trifluoroacetonitrile	13.93 ± 0.07	845
C_2F_4	Tetrafluoroethylene	10.12 ± 0.02	315
C_2F_6	Hexafluoroethane	(13.6)	-30
C_2H	Ethynyl	(11.61 ± 0.07)	1685
C_2HBr	Bromoacetylene	10.31 ± 0.02	1242
$C_2HBrClF_3$	2-Bromo-2-chloro-1,1,1-trifluoroethane	(11.0)	363
C_2HCl	Chloroacetylene	10.58 ± 0.02	1276
C_2HClF_2	1-Chloro-2,2-difluoroethylene	9.80 ± 0.04	628
C_2HCl_3	Trichloroethylene	9.46 ± 0.02	894
C_2HCl_3O	Dichloroacetyl chloride	(10.9)	809
C_2HCl_5	Pentachloroethane	(11.0)	919
C_2HF	Fluoroacetylene	11.26	1195
C_2HF_3	Trifluoroethylene	10.14	489
$C_2HF_3O_2$	Trifluoroacetic acid	11.46	75
C_2H_2	Acetylene	11.400 ± 0.002	1328
$C_2H_2Cl_2$	1,1-Dichloroethylene	9.81 ± 0.04	949
$C_2H_2Cl_2$	cis-1,2-Dichloroethylene	9.66 ± 0.01	936
$C_2H_2Cl_2$	trans-1,2-Dichloroethylene	9.64 ± 0.02	934
$C_2H_2Cl_2O$	Chloroacetyl chloride	(≤ 10.3)	815
$C_2H_2Cl_4$	1,1,1,2-Tetrachloroethane	(11.1)	920
$C_2H_2Cl_4$	1,1,2,2-Tetrachloroethane	(≤ 11.62)	≤ 971
$C_2H_2F_2$	1,1-Difluoroethylene	10.29 ± 0.01	650
$C_2H_2F_2$	cis-1,2-Difluoroethylene	10.23 ± 0.02	690
C_2H_2O	Ketene	9.617 ± 0.003	880
$C_2H_2O_2$	Glyoxal	10.2	773
$C_2H_2S_2$	Thiirene	8.61	892
C_2H_3Br	Bromoethylene	9.83 ± 0.02	1028
C_2H_3Cl	Chloroethylene	9.99 ± 0.02	985
$C_2H_3ClF_2$	1-Chloro-1,1-difluoroethane	(11.98)	626
C_2H_3ClO	Acetyl chloride	10.82 ± 0.04	801
C_2H_3ClO	Chloroacetaldehyde	(10.48)	815
$C_2H_3ClO_2$	Chloroacetic acid	(10.7)	597
$C_2H_3Cl_3$	1,1,1-Trichloroethane	(11.0)	917
$C_2H_3Cl_3$	1,1,2-Trichloroethane	(11.0)	911
C_2H_3F	Fluoroethylene	10.36 ± 0.01	861
C_2H_3FO	Acetyl fluoride	(11.5)	667
$C_2H_3F_3$	1,1,1-Trifluoroethane	13.3 ± 0.5	536
C_2H_3N	Acetonitrile	12.20 ± 0.01	1253
C_2H_3NO	Methylisocyanate	(10.67)	900
C_2H_4	Ethylene	10.5138 ± 0.0006	1067
$C_2H_4Br_2$	1,2-Dibromoethane	10.35 ± 0.04	961
$C_2H_4Cl_2$	1,1-Dichloroethane	11.04 ± 0.02	935
$C_2H_4Cl_2$	1,2-Dichloroethane	11.04 ± 0.02	931

Mol. form.	Name	IP/eV	$\Delta_f H_{ion}$ kJ/mol
$C_2H_4F_2$	1,1-Difluoroethane	(11.87)	643
C_2H_4O	Acetaldehyde	10.229 ± 0.0007	821
C_2H_4O	Ethylene oxide	10.56 ± 0.01	966
$C_2H_4O_2$	Acetic acid	10.65 ± 0.02	595
$C_2H_4O_2$	Methyl formate	10.835 ± 0.005	690
C_2H_5Br	Bromoethane	10.29 ± 0.01	931
C_2H_5Cl	Chloroethane	10.98 ± 0.02	947
C_2H_5ClO	2-Chloroethanol	(10.5)	756
C_2H_5F	Fluoroethane	(11.78)	873
C_2H_5I	Iodoethane	9.3492 ± 0.0006	893
C_2H_5N	Ethyleneimine	(9.5 ± 0.3)	1044
C_2H_5NO	Acetamide	9.65 ± 0.03	693
C_2H_5NO	N-Methylformamide	9.83 ± 0.04	760
$C_2H_5NO_2$	Nitroethane	10.88 ± 0.05	948
C_2H_6	Ethane	11.56 ± 0.02	1031
$C_2H_6Cl_2Si$	Dichlorodimethylsilane	(10.7)	576
C_2H_6O	Ethanol	10.43 ± 0.05	772
C_2H_6O	Dimethyl ether	10.025 ± 0.025	783
C_2H_6OS	Dimethyl sulfoxide	9.10 ± 0.03	727
$C_2H_6O_2$	Ethylene glycol	10.16	593
C_2H_6S	Ethanethiol	9.31 ± 0.03	851
C_2H_6S	Dimethyl sulfide	8.69 ± 0.02	801
$C_2H_6S_2$	Dimethyl disulfide	(7.4 ± 0.3)	690
C_2H_7N	Ethylamine	8.86 ± 0.02	808
C_2H_7N	Dimethylamine	8.24 ± 0.08	777
C_2H_7NO	Ethanolamine	8.96	664
$C_2H_8N_2$	1,2-Ethanediamine	(8.6)	812
$C_2H_8N_2$	1,1-Dimethylhydrazine	7.29 ± 0.05	787
C_2N_2	Cyanogen	13.37 ± 0.01	1597
C_3F_6	Perfluoropropene	10.60 ± 0.03	−103
C_3F_6O	Perfluoroacetone	(11.57 ± 0.13)	−282
C_3F_8	Perfluoropropane	(13.38)	−491
C_3HN	Cyanoacetylene	11.64 ± 0.01	1475
C_3H_2O	2-Propynal	(10.7 ± 0.1)	1145
$C_3H_3F_3$	3,3,3-Trifluoropropene	(10.9)	437
C_3H_3N	2-Propenenitrile	10.91 ± 0.01	1237
C_3H_3NO	Oxazole	(9.9)	940
C_3H_3NO	Isoxazole	(9.93)	1038
C_3H_4	Allene	9.692 ± 0.004	1126
C_3H_4	Propyne	10.37 ± 0.01	1187
C_3H_4	Cyclopropene	9.67 ± 0.01	1209
$C_3H_4N_2$	Imidazole	(8.81)	997
C_3H_4O	Propargyl alcohol	10.49 ± 0.02	1060
C_3H_4O	Acrolein	10.103 ± 0.006	900
C_3H_4O	Cyclopropanone	(9.1 ± 0.1)	895
$C_3H_4O_2$	Propenoic acid	10.60	701
$C_3H_4O_2$	2-Oxetanone	(9.70 ± 0.01)	653
C_3H_5Br	3-Bromopropene	(9.96)	1008
C_3H_5Cl	3-Chloropropene	10.04 ± 0.01	965
C_3H_5ClO	Epichlorohydrin	(10.64)	919
$C_3H_5ClO_2$	Methyl chloroacetate	(10.3)	575
C_3H_5F	3-Fluoropropene	(10.11)	821
C_3H_5N	Propanenitrile	11.84 ± 0.02	1194
C_3H_5NO	Acrylamide	(9.5)	720
C_3H_6	Propene	9.73 ± 0.02	959
C_3H_6	Cyclopropane	9.86	1005
$C_3H_6Br_2$	1,2-Dibromopropane	(10.1)	903
$C_3H_6Br_2$	1,3-Dibromopropane	(≤ 10.2)	≤ 919
$C_3H_6Cl_2$	1,2-Dichloropropane	10.8 ± 0.1	886
$C_3H_6Cl_2$	1,3-Dichloropropane	10.89 ± 0.04	892

Mol. form.	Name	IP/eV	$\Delta_f H_{ion}$ kJ/mol
C_3H_6O	Allyl alcohol	9.67 ± 0.05	808
C_3H_6O	Methyl vinyl ether	8.95 ± 0.01	763
C_3H_6O	Propanal	9.96 ± 0.01	772
C_3H_6O	Acetone	9.703 ± 0.006	719
C_3H_6O	Methyloxirane	(10.22)	892
C_3H_6O	Oxetane	9.65 ± 0.01	851
$C_3H_6O_2$	Propanoic acid	10.525 ± 0.003	568
$C_3H_6O_2$	Ethyl formate	10.61 ± 0.01	639
$C_3H_6O_2$	Methyl acetate	10.25 ± 0.02	579
$C_3H_6O_2$	1,3-Dioxolane	(9.9)	658
$C_3H_6O_3$	1,3,5-Trioxane	(10.3)	528
C_3H_7Br	1-Bromopropane	10.18 ± 0.01	898
C_3H_7Br	2-Bromopropane	10.10 ± 0.03	877
C_3H_7Cl	1-Chloropropane	10.81 ± 0.01	911
C_3H_7Cl	2-Chloropropane	10.79 ± 0.02	896
C_3H_7F	1-Fluoropropane	(11.3)	806
C_3H_7F	2-Fluoropropane	(11.08)	776
C_3H_7I	1-Iodopropane	9.25 ± 0.01	860
C_3H_7I	2-Iodopropane	9.19 ± 0.02	845
C_3H_7N	Allylamine	(8.76)	891
C_3H_7N	Cyclopropylamine	(8.8)	926
C_3H_7N	Propyleneimine	(9.0)	960
C_3H_7NO	N,N-Dimethylformamide	(9.12)	688
$C_3H_7NO_2$	1-Nitropropane	(10.81)	919
$C_3H_7NO_2$	2-Nitropropane	(10.71)	894
C_3H_8	Propane	10.95 ± 0.05	952
C_3H_8O	1-Propanol	10.18 ± 0.06	727
C_3H_8O	2-Propanol	10.17 ± 0.02	709
C_3H_8O	Ethyl methyl ether	9.72 ± 0.07	722
$C_3H_8O_2$	Dimethoxymethane	9.7	588
C_3H_8S	1-Propanethiol	9.20 ± 0.01	819
C_3H_8S	2-Propanethiol	9.145 ± 0.005	806
C_3H_8S	Ethyl methyl sulfide	(8.55)	765
$C_3H_9BO_3$	Trimethyl borate	(10.0)	65
C_3H_9ClSi	Trimethylchlorosilane	(10.15)	624
C_3H_9N	Propylamine	(8.78)	777
C_3H_9N	Isopropylamine	(8.72)	758
C_3H_9N	Trimethylamine	7.82 ± 0.06	731
C_3H_9NO	3-Amino-1-propanol	(9.0)	651
$C_4H_2O_3$	Maleic anhydride	(10.8)	645
C_4H_4	1-Buten-3-yne	9.58 ± 0.02	1230
$C_4H_4N_2$	Succinonitrile	(12.1 ± 0.25)	1377
$C_4H_4N_2$	Pyrimidine	9.23	1087
$C_4H_4N_2$	Pyridazine	8.67 ± 0.03	1112
C_4H_4O	Furan	8.883 ± 0.003	822
$C_4H_4O_2$	Diketene	(9.6 ± 0.02)	736
$C_4H_4O_3$	Succinic anhydride	(10.6)	500
$C_4H_4O_4$	Fumaric acid	(10.7)	355
C_4H_4S	Thiophene	8.86 ± 0.02	970
C_4H_5N	Methylacrylonitrile	10.34	1127
C_4H_5N	Pyrrole	8.207 ± 0.005	900
C_4H_5N	Cyclopropanecarbonitrile	(10.25)	1173
C_4H_6	1,2-Butadiene	(9.03)	1034
C_4H_6	1,3-Butadiene	9.082 ± 0.004	986
C_4H_6	1-Butyne	10.19 ± 0.02	1148
C_4H_6	2-Butyne	9.59 ± 0.03	1071
C_4H_6	Cyclobutene	9.43 ± 0.02	1067
C_4H_6O	Divinyl ether	(8.7)	827
C_4H_6O	trans-2-Butenal	9.73 ± 0.01	835
C_4H_6O	2-Methylpropenal	(9.92)	834

Mol. form.	Name	IP/eV	$\Delta_f H_{ion}$ kJ/mol
C_4H_6O	Cyclobutanone	(9.35)	815
$C_4H_6O_2$	*cis*-Crotonic acid	(10.08)	625
$C_4H_6O_2$	*trans*-Crotonic acid	(9.9)	604
$C_4H_6O_2$	Methacrylic acid	(10.15)	611
$C_4H_6O_2$	Vinyl acetate	9.19 ± 0.05	572
$C_4H_6O_2$	Methyl acrylate	(9.9)	641
$C_4H_6O_3$	Acetic anhydride	(10.0)	398
$C_4H_6O_4$	Dimethyl oxalate	(10.0)	287
C_4H_6S	2,5-Dihydrothiophene	(8.4)	898
C_4H_7N	Butanenitrile	(11.2)	1110
C_4H_7N	2-Methylpropanenitrile	(11.3)	1115
C_4H_7NO	2-Pyrrolidone	(9.2)	674
C_4H_8	1-Butene	9.55 ± 0.06	921
C_4H_8	*cis*-2-Butene	9.11 ± 0.01	871
C_4H_8	*trans*-2-Butene	9.10 ± 0.01	866
C_4H_8	Isobutene	9.239 ± 0.003	875
C_4H_8	Cyclobutane	(9.82 ± 0.05)	976
C_4H_8	Methylcyclopropane	(9.46)	936
$C_4H_8Br_2$	1,4-Dibromobutane	(10.15)	879
C_4H_8O	Ethyl vinyl ether	(8.98)	709
C_4H_8O	1,2-Epoxybutane	(≤ 10.15)	862
C_4H_8O	Butanal	9.84 ± 0.02	742
C_4H_8O	Isobutanal	9.71 ± 0.01	721
C_4H_8O	2-Butanone	9.52 ± 0.04	678
C_4H_8O	Tetrahydrofuran	9.38 ± 0.05	721
$C_4H_8O_2$	Butanoic acid	10.17 ± 0.05	509
$C_4H_8O_2$	2-Methylpropanoic acid	10.33 ± 0.03	516
$C_4H_8O_2$	Propyl formate	10.52 ± 0.02	555
$C_4H_8O_2$	Ethyl acetate	10.01 ± 0.05	522
$C_4H_8O_2$	Methyl propanoate	10.15 ± 0.03	548
$C_4H_8O_2$	1,3-Dioxane	9.8	607
$C_4H_8O_2$	1,4-Dioxane	9.19 ± 0.01	571
$C_4H_8O_2S$	Sulfolane	(9.8)	577
C_4H_8S	Tetrahydrothiophene	8.38	774
C_4H_9Br	1-Bromobutane	(10.12)	869
C_4H_9Br	2-Bromobutane	10.01 ± 0.02	845
C_4H_9Br	1-Bromo-2-methylpropane	10.09 ± 0.02	861
C_4H_9Br	2-Bromo-2-methylpropane	9.92 ± 0.03	823
C_4H_9Cl	1-Chlorobutane	10.67 ± 0.03	875
C_4H_9Cl	2-Chlorobutane	10.53	857
C_4H_9Cl	1-Chloro-2-methylpropane	10.73 ± 0.07	877
C_4H_9Cl	2-Chloro-2-methylpropane	(10.61)	842
C_4H_9I	1-Iodobutane	9.23 ± 0.01	840
C_4H_9I	2-Iodobutane	9.10 ± 0.02	815
C_4H_9I	1-Iodo-2-methylpropane	9.19 ± 0.01	824
C_4H_9I	2-Iodo-2-methylpropane	(9.02)	798
C_4H_9N	Pyrrolidine	(8.0)	769
C_4H_9NO	*N,N*-Dimethylacetamide	8.81 ± 0.03	616
C_4H_9NO	Morpholine	(8.2)	841
C_4H_{10}	Butane	10.53 ± 0.10	890
C_4H_{10}	Isobutane	(10.57)	886
$C_4H_{10}O$	1-Butanol	9.99 ± 0.05	689
$C_4H_{10}O$	2-Butanol	9.88 ± 0.03	658
$C_4H_{10}O$	2-Methyl-1-propanol	10.02 ± 0.04	683
$C_4H_{10}O$	2-Methyl-2-propanol	9.90 ± 0.02	642
$C_4H_{10}O$	Diethyl ether	9.51 ± 0.03	666
$C_4H_{10}O$	Methyl propyl ether	9.41 ± 0.07	670
$C_4H_{10}O$	Isopropyl methyl ether	9.45 ± 0.04	661
$C_4H_{10}O_2$	Ethylene glycol monoethyl ether	(9.6)	529
$C_4H_{10}O_2$	Ethylene glycol dimethyl ether	(9.3)	558

Mol. form.	Name	IP/eV	$\Delta_f H_{ion}$ kJ/mol
$C_4H_{10}S$	1-Butanethiol	9.14 ± 0.01	794
$C_4H_{10}S$	2-Butanethiol	(9.10)	781
$C_4H_{10}S$	2-Methyl-1-propanethiol	(9.12)	783
$C_4H_{10}S$	2-Methyl-2-propanethiol	(9.03)	762
$C_4H_{10}S$	Diethyl sulfide	(8.43)	730
$C_4H_{10}S$	Methyl propyl sulfide	(8.8)	767
$C_4H_{10}S$	Isopropyl methyl sulfide	(8.7)	749
$C_4H_{10}S_2$	Diethyl disulfide	(8.27)	724
$C_4H_{11}N$	Butylamine	8.7 ± 0.1	748
$C_4H_{11}N$	sec-Butylamine	8.46 ± 0.1	711
$C_4H_{11}N$	tert-Butylamine	8.46 ± 0.1	695
$C_4H_{11}N$	Isobutylamine	8.50 ± 0.1	721
$C_4H_{11}N$	Diethylamine	7.85 ± 0.1	684
$C_4H_{12}Si$	Tetramethylsilane	9.80 ± 0.04	713
$C_4H_{12}Sn$	Tetramethylstannane	8.89 ± 0.05	837
C_4NiO_4	Nickel carbonyl	8.27 ± 0.04	200
$C_5H_4O_2$	Furfural	9.22 ± 0.01	739
C_5H_5N	Pyridine	9.25	1031
C_5H_6	1-Penten-3-yne	9.00 ± 0.01	1119
C_5H_6	cis-3-Penten-1-yne	9.14 ± 0.04	1137
C_5H_6	trans-3-Penten-1-yne	9.05 ± 0.01	1128
C_5H_6	2-Methyl-1-buten-3-yne	9.25 ± 0.02	1152
C_5H_6	1,3-Cyclopentadiene	8.55 ± 0.02	955
C_5H_6O	2-Methylfuran	8.38 ± 0.02	729
C_5H_6O	3-Methylfuran	(8.64)	763
C_5H_6S	2-Methylthiophene	(8.14)	867
C_5H_6S	3-Methylthiophene	(8.40)	893
C_5H_8	cis-1,3-Pentadiene	8.63 ± 0.03	914
C_5H_8	trans-1,3-Pentadiene	8.59 ± 0.02	905
C_5H_8	1,4-Pentadiene	9.60 ± 0.02	1032
C_5H_8	2-Methyl-1,3-butadiene	8.84 ± 0.01	928
C_5H_8	1-Pentyne	10.10 ± 0.01	1119
C_5H_8	Cyclopentene	9.01 ± 0.01	905
C_5H_8	Spiropentane	(9.26)	1078
C_5H_8O	Cyclopropyl methyl ketone	(≤ 9.46)	796
C_5H_8O	Cyclopentanone	9.26 ± 0.01	701
C_5H_8O	3,4-Dihydro-2H-pyran	8.35 ± 0.01	681
$C_5H_8O_2$	Ethyl acrylate	(≤ 10.3)	617
$C_5H_8O_2$	Methyl methacrylate	(9.7)	589
$C_5H_8O_2$	2,4-Pentanedione	8.85 ± 0.01	469
C_5H_9NO	N-Methyl-2-pyrrolidone	(≤ 9.17)	≤ 676
C_5H_{10}	1-Pentene	9.51 ± 0.01	896
C_5H_{10}	cis-2-Pentene	9.01 ± 0.03	843
C_5H_{10}	trans-2-Pentene	9.04 ± 0.01	841
C_5H_{10}	2-Methyl-1-butene	9.12 ± 0.01	844
C_5H_{10}	3-Methyl-1-butene	9.52 ± 0.01	891
C_5H_{10}	2-Methyl-2-butene	8.69 ± 0.01	796
C_5H_{10}	Cyclopentane	(10.33 ± 0.15)	918
$C_5H_{10}O$	2,2-Dimethylpropanal	9.51 ± 0.01	675
$C_5H_{10}O$	Cyclopentanol	(9.72)	695
$C_5H_{10}O$	Pentanal	9.74 ± 0.04	709
$C_5H_{10}O$	2-Pentanone	9.38 ± 0.01	646
$C_5H_{10}O$	3-Pentanone	9.31 ± 0.01	640
$C_5H_{10}O$	3-Methyl-2-butanone	9.30 ± 0.01	635
$C_5H_{10}O$	Tetrahydropyran	9.25 ± 0.01	670
$C_5H_{10}O_2$	Pentanoic acid	(≤ 10.53)	≤ 527
$C_5H_{10}O_2$	3-Methylbutanoic acid	(≤ 10.51)	≤ 499
$C_5H_{10}O_2$	Butyl formate	10.52 ± 0.02	584
$C_5H_{10}O_2$	Propyl acetate	(≤ 9.92)	501
$C_5H_{10}O_2$	Isopropyl acetate	9.99 ± 0.03	482

Mol. form.	Name	IP/eV	$\Delta_f H_{ion}$ kJ/mol
$C_5H_{10}O_2$	Ethyl propanoate	(10.00)	500
$C_5H_{10}O_2$	Methyl butanoate	(10.07)	520
$C_5H_{10}S$	Thiacyclohexane	(8.2)	728
$C_5H_{11}Br$	1-Bromopentane	10.10 ± 0.01	846
$C_5H_{11}I$	1-Iodopentane	9.20 ± 0.01	817
$C_5H_{11}N$	Piperidine	8.03 ± 0.11	726
$C_5H_{11}N$	N-Methylpyrrolidine	≤ 8.41 ± 0.02	≤ 809
C_5H_{12}	Pentane	10.28 ± 0.10	845
C_5H_{12}	Isopentane	10.32 ± 0.05	843
C_5H_{12}	Neopentane	(≤ 10.2)	≤ 818
$C_5H_{12}O$	1-Pentanol	(10.00)	668
$C_5H_{12}O$	2-Pentanol	(9.78)	630
$C_5H_{12}O$	3-Pentanol	9.78	628
$C_5H_{12}O$	2-Methyl-1-butanol	(9.86)	649
$C_5H_{12}O$	2-Methyl-2-butanol	(9.8)	615
$C_5H_{12}O$	3-Methyl-2-butanol	(9.88 ± 0.13)	637
$C_5H_{12}O$	Butyl methyl ether	(9.4 ± 0.1)	648
$C_5H_{12}O$	Methyl tert-butyl ether	(9.24)	608
$C_5H_{12}O$	Ethyl propyl ether	(9.45)	640
$C_5H_{12}S$	tert-Butyl methyl sulfide	(8.38)	687
$C_5H_{12}S$	Ethyl propyl sulfide	(8.50)	716
$C_5H_{12}S$	Ethyl isopropyl sulfide	(8.35)	689
C_6BrF_5	Bromopentafluorobenzene	(9.67)	222
C_6ClF_5	Chloropentafluorobenzene	(9.72)	126
C_6Cl_6	Hexachlorobenzene	(8.98)	822
C_6F_6	Hexafluorobenzene	9.89 ± 0.04	8
C_6F_{12}	Perfluorocyclohexane	(13.2)	−1095
C_6HF_5	Pentafluorobenzene	(9.63)	122
C_6HF_5O	Pentafluorophenol	(9.20)	−71
$C_6H_2F_4$	1,2,3,4-Tetrafluorobenzene	(9.53)	284
$C_6H_2F_4$	1,2,3,5-Tetrafluorobenzene	(9.53)	263
$C_6H_2F_4$	1,2,4,5-Tetrafluorobenzene	(9.35)	254
$C_6H_3Cl_3$	1,2,4-Trichlorobenzene	(9.04)	880
$C_6H_3Cl_3$	1,3,5-Trichlorobenzene	9.32 ± 0.02	899
$C_6H_4ClNO_2$	1-Chloro-3-nitrobenzene	(9.92 ± 0.1)	995
$C_6H_4ClNO_2$	1-Chloro-4-nitrobenzene	(9.96 ± 0.1)	999
$C_6H_4Cl_2$	o-Dichlorobenzene	9.06 ± 0.02	907
$C_6H_4Cl_2$	m-Dichlorobenzene	9.10 ± 0.02	906
$C_6H_4Cl_2$	p-Dichlorobenzene	8.92 ± 0.02	885
$C_6H_4FNO_2$	1-Fluoro-4-nitrobenzene	(9.90)	826
$C_6H_4F_2$	o-Difluorobenzene	9.29 ± 0.01	602
$C_6H_4F_2$	m-Difluorobenzene	9.33 ± 0.01	591
$C_6H_4F_2$	p-Difluorobenzene	9.1589 ± 0.0003	577
$C_6H_4O_2$	p-Benzoquinone	10.01 ± 0.06	844
C_6H_5Br	Bromobenzene	9.00 ± 0.02	971
C_6H_5Cl	Chlorobenzene	9.07 ± 0.02	930
C_6H_5ClO	m-Chlorophenol	8.655 ± 0.001	680
C_6H_5ClO	p-Chlorophenol	(≤ 8.69)	≤ 692
C_6H_5F	Fluorobenzene	9.20 ± 0.01	772
C_6H_5I	Iodobenzene	8.685	1003
$C_6H_5NO_2$	Nitrobenzene	9.86 ± 0.02	1019
$C_6H_5NO_3$	o-Nitrophenol	(9.1)	782
$C_6H_5NO_3$	m-Nitrophenol	(9.0)	755
$C_6H_5NO_3$	p-Nitrophenol	(9.1)	761
C_6H_6	Benzene	9.24378 ± 0.00007	975
C_6H_6	Fulvene	(8.36)	1031
C_6H_6ClN	o-Chloroaniline	(8.50)	883
C_6H_6ClN	m-Chloroaniline	(8.09)	835
C_6H_6ClN	p-Chloroaniline	(≤ 8.18)	≤ 844
$C_6H_6N_2O_2$	o-Nitroaniline	(8.27)	861

Mol. form.	Name	IP/eV	$\Delta_f H_{ion}$ kJ/mol
$C_6H_6N_2O_2$	m-Nitroaniline	(8.31)	865
$C_6H_6N_2O_2$	p-Nitroaniline	(8.34)	859
C_6H_6O	Phenol	8.49 ± 0.02	723
$C_6H_6O_2$	p-Hydroquinone	7.94 ± 0.01	503
C_6H_6S	Benzenethiol	(8.32)	915
C_6H_7N	Aniline	7.720 ± 0.002	832
C_6H_7N	2-Methylpyridine	(9.02)	970
C_6H_7N	3-Methylpyridine	(9.04)	979
C_6H_7N	4-Methylpyridine	(9.04)	976
$C_6H_8N_2$	o-Phenylenediamine	(7.2)	787
$C_6H_8N_2$	m-Phenylenediamine	(7.14)	777
$C_6H_8N_2$	p-Phenylenediamine	(6.87 ± 0.05)	759
C_6H_{10}	1,5-Hexadiene	9.27 ± 0.05	978
C_6H_{10}	1-Hexyne	10.03 ± 0.05	1089
C_6H_{10}	3,3-Dimethyl-1-butyne	9.90 ± 0.04	1060
C_6H_{10}	Cyclohexene	8.945 ± 0.01	859
$C_6H_{10}O$	Cyclohexanone	9.14 ± 0.01	656
$C_6H_{10}O$	Mesityl oxide	9.10 ± 0.01	694
$C_6H_{10}O_4$	Diethyl oxalate	(9.8)	205
$C_6H_{11}NO$	Caprolactam	(9.07 ± 0.02)	629
C_6H_{12}	1-Hexene	9.44 ± 0.04	869
C_6H_{12}	cis-2-Hexene	(8.97 ± 0.01)	818
C_6H_{12}	trans-2-Hexene	(8.97 ± 0.01)	814
C_6H_{12}	2-Methyl-1-pentene	(9.08 ± 0.01)	817
C_6H_{12}	4-Methyl-1-pentene	9.45 ± 0.01	862
C_6H_{12}	2-Methyl-2-pentene	(8.58)	761
C_6H_{12}	4-Methyl-cis-2-pentene	8.98 ± 0.01	809
C_6H_{12}	4-Methyl-trans-2-pentene	(8.97 ± 0.01)	804
C_6H_{12}	2-Ethyl-1-butene	(9.06 ± 0.02)	818
C_6H_{12}	2,3-Dimethyl-1-butene	(9.07 ± 0.01)	812
C_6H_{12}	2,3-Dimethyl-2-butene	8.27 ± 0.01	729
C_6H_{12}	Cyclohexane	9.86 ± 0.03	828
C_6H_{12}	Methylcyclopentane	(9.85)	845
$C_6H_{12}O$	Hexanal	9.72 ± 0.05	691
$C_6H_{12}O$	2-Hexanone	9.3 ± 0.1	626
$C_6H_{12}O$	3-Hexanone	9.12 ± 0.02	600
$C_6H_{12}O$	3-Methyl-2-pentanone	9.21 ± 0.01	600
$C_6H_{12}O$	4-Methyl-2-pentanone	9.30 ± 0.01	609
$C_6H_{12}O$	2-Methyl-3-pentanone	9.10 ± 0.01	592
$C_6H_{12}O$	3,3-Dimethyl-2-butanone	9.12 ± 0.02	589
$C_6H_{12}O$	Cyclohexanol	(9.75)	651
$C_6H_{12}O_2$	Hexanoic acid	≤ 10.12	≤ 463
$C_6H_{12}O_2$	Butyl acetate	(9.92 ± .05)	471
$C_6H_{12}O_2$	sec-Butyl acetate	9.90	453
$C_6H_{12}O_2$	Methyl 2,2-dimethylpropanoate	(9.90 ± 0.04)	466
$C_6H_{13}I$	1-Iodohexane	9.179	794
$C_6H_{13}N$	Cyclohexylamine	(8.86)	750
C_6H_{14}	Hexane	10.13	810
C_6H_{14}	2-Methylpentane	(10.12)	802
C_6H_{14}	3-Methylpentane	(10.08)	801
C_6H_{14}	2,2-Dimethylbutane	(10.06)	787
C_6H_{14}	2,3-Dimethylbutane	(10.02)	791
$C_6H_{14}O$	1-Hexanol	(9.89)	639
$C_6H_{14}O$	2-Hexanol	(9.80 ± 0.03)	611
$C_6H_{14}O$	3-Hexanol	(9.63 ± 0.03)	599
$C_6H_{14}O$	Dipropyl ether	(9.27)	602
$C_6H_{14}O$	Diisopropyl ether	9.20 ± 0.05	569
$C_6H_{14}O$	Butyl ethyl ether	(9.36)	610
$C_6H_{14}O$	Methyl pentyl ether	(≤ 9.67)	≤ 657
$C_6H_{14}O_2$	1,1-Diethoxyethane	(9.2)	434

Mol. form.	Name	IP/eV	$\Delta_f H_{ion}$ kJ/mol
$C_6H_{14}O_3$	Diethylene glycol dimethyl ether	≤ 9.8	≤ 448
$C_6H_{14}S$	Dipropyl sulfide	8.30 ± 0.02	676
$C_6H_{14}S$	Diisopropyl sulfide	(8.2 ± 0.2)	649
$C_6H_{15}N$	Hexylamine	(8.63 ± 0.05)	699
$C_6H_{15}N$	Dipropylamine	(7.84 ± 0.02)	641
$C_6H_{15}N$	Diisopropylamine	(7.73 ± 0.03)	602
$C_6H_{15}N$	Triethylamine	(7.50 ± 0.02)	631
$C_6H_{15}NO_3$	Triethanolamine	(7.9)	206
$C_7H_3F_5$	2,3,4,5,6-Pentafluorotoluene	(9.4)	64
C_7H_5ClO	Benzoyl chloride	(9.53)	815
$C_7H_5Cl_3$	(Trichloromethyl)benzene	(≤ 9.60)	≤ 914
$C_7H_5F_3$	(Trifluoromethyl)benzene	9.685 ± 0.005	335
C_7H_5N	Benzonitrile	9.70 ± 0.01	1154
C_7H_6O	Benzaldehyde	9.49 ± 0.02	878
$C_7H_6O_2$	Benzoic acid	(9.3)	604
C_7H_7Br	p-Bromotoluene	8.67 ± 0.02	908
C_7H_7Cl	o-Chlorotoluene	(8.7 ± 0.1)	856
C_7H_7Cl	m-Chlorotoluene	(8.83)	869
C_7H_7Cl	p-Chlorotoluene	(8.69)	855
C_7H_7Cl	(Chloromethyl)benzene	9.10 ± 0.02	897
C_7H_7F	o-Fluorotoluene	8.91 ± 0.01	709
C_7H_7F	m-Fluorotoluene	8.91 ± 0.01	709
C_7H_7F	p-Fluorotoluene	8.79 ± 0.01	701
C_7H_7NO	Benzamide	(9.25)	792
$C_7H_7NO_2$	o-Nitrotoluene	9.24	946
$C_7H_7NO_2$	m-Nitrotoluene	9.45 ± 0.1	941
$C_7H_7NO_2$	p-Nitrotoluene	9.46 ± 0.05	942
C_7H_8	Toluene	8.8276 ± 0.0006	901
C_7H_8O	o-Cresol	(8.24)	670
C_7H_8O	m-Cresol	8.29 ± 0.07	668
C_7H_8O	p-Cresol	(8.3)	675
C_7H_8O	Benzyl alcohol	(8.3)	701
C_7H_8O	Anisole	8.22 ± 0.03	725
C_7H_9N	Benzylamine	(8.64)	917
C_7H_9N	o-Methylaniline	(7.44 ± 0.02)	772
C_7H_9N	m-Methylaniline	(7.50 ± 0.02)	778
C_7H_9N	p-Methylaniline	(7.24 ± 0.02)	753
C_7H_9N	N-Methylaniline	7.34 ± 0.04	792
C_7H_9N	2,3-Dimethylpyridine	(8.85 ± 0.02)	922
C_7H_9N	2,4-Dimethylpyridine	(8.85 ± 0.03)	918
C_7H_9N	2,5-Dimethylpyridine	(≤ 8.80 ± 0.05)	≤ 916
C_7H_9N	2,6-Dimethylpyridine	8.86 ± 0.03	913
C_7H_9N	3,4-Dimethylpyridine	(≤ 9.15)	≤ 953
C_7H_9N	3,5-Dimethylpyridine	(≤ 9.25)	≤ 965
$C_7H_{10}O$	Dicyclopropyl ketone	(9.1)	1041
C_7H_{14}	1-Heptene	9.34 ± 0.10	839
C_7H_{14}	trans-3-Heptene	(8.92)	790
C_7H_{14}	Cycloheptane	9.97	844
C_7H_{14}	Methylcyclohexane	9.64	775
C_7H_{14}	cis-1,2-Dimethylcyclopentane	(9.92 ± 0.05)	828
C_7H_{14}	trans-1,2-Dimethylcyclopentane	9.7 ± 0.2	799
$C_7H_{14}O$	1-Heptanal	(9.65)	668
$C_7H_{14}O$	2-Heptanone	9.28 ± 0.10	594
$C_7H_{14}O$	3-Heptanone	9.18 ± 0.08	589
$C_7H_{14}O$	4-Heptanone	9.10 ± 0.06	577
$C_7H_{14}O$	5-Methyl-2-hexanone	(9.28)	586
$C_7H_{14}O$	2,4-Dimethyl-3-pentanone	8.95 ± 0.01	552
$C_7H_{14}O$	1-Methylcyclohexanol	(9.8 ± 0.2)	586
C_7H_{16}	Heptane	9.93 ± 0.10	771
$C_7H_{16}O$	1-Heptanol	(9.84)	614

Mol. form.	Name	IP/eV	$\Delta_f H_{ion}$ kJ/mol
$C_7H_{16}O$	2-Heptanol	(9.70)	580
$C_7H_{16}O$	3-Heptanol	(9.68)	578
$C_7H_{16}O$	4-Heptanol	(9.61)	572
$C_7H_{16}O$	Ethyl pentyl ether	(≤ 9.49)	≤ 602
$C_8H_4O_3$	Phthalic anhydride	(10.1)	603
$C_8H_6O_4$	Isophthalic acid	(9.98)	268
$C_8H_6O_4$	Terephthalic acid	(9.86)	232
C_8H_7N	2-Methylbenzonitrile	(≤ 9.38)	1085
C_8H_7N	3-Methylbenzonitrile	(≤ 9.34)	1085
C_8H_7N	4-Methylbenzonitrile	9.32 ± 0.02	1083
C_8H_7N	Indole	7.7602 ± 0.0006	908
C_8H_8	Styrene	8.464 ± 0.001	964
C_8H_8O	p-Tolualdehyde	(9.33)	825
C_8H_8O	Acetophenone	9.29 ± 0.03	810
$C_8H_8O_2$	o-Toluic acid	(9.1)	558
$C_8H_8O_2$	m-Toluic acid	(9.43)	579
$C_8H_8O_2$	p-Toluic acid	(9.23)	560
$C_8H_8O_2$	Benzeneacetic acid	(8.26)	479
$C_8H_8O_2$	Methyl benzoate	9.32 ± 0.03	611
C_8H_{10}	Ethylbenzene	8.77 ± 0.01	876
C_8H_{10}	o-Xylene	8.56 ± 0.01	844
C_8H_{10}	m-Xylene	8.56 ± 0.01	843
C_8H_{10}	p-Xylene	8.44 ± 0.01	832
$C_8H_{10}O$	p-Ethylphenol	(7.84)	613
$C_8H_{10}O$	2,3-Xylenol	(8.26)	640
$C_8H_{10}O$	2,4-Xylenol	(8.0)	609
$C_8H_{10}O$	2,6-Xylenol	(8.05)	615
$C_8H_{10}O$	3,4-Xylenol	(8.09)	624
$C_8H_{10}O$	Phenetole	(8.13)	683
$C_8H_{11}N$	2,4,6-Trimethylpyridine	(≤ 8.9)	≤ 880
$C_8H_{11}N$	N-Ethylaniline	(≤ 7.67)	≤ 794
$C_8H_{11}N$	N,N-Dimethylaniline	7.12 ± 0.02	787
C_8H_{14}	1-Octyne	(9.95 ± 0.02)	1040
C_8H_{14}	2-Octyne	9.31 ± 0.01	961
C_8H_{14}	3-Octyne	9.22 ± 0.01	952
C_8H_{14}	4-Octyne	9.20 ± 0.01	946
C_8H_{16}	1-Octene	9.43 ± 0.01	829
C_8H_{16}	Cyclooctane	9.75 ± 0.05	816
C_8H_{16}	Ethylcyclohexane	(9.54)	748
C_8H_{16}	1,1-Dimethylcyclohexane	(9.42)	728
C_8H_{16}	cis-1,2-Dimethylcyclohexane	(<9.78)	772
C_8H_{16}	trans-1,2-Dimethylcyclohexane	9.41	728
C_8H_{16}	cis-1,3-Dimethylcyclohexane	(<9.98)	778
C_8H_{16}	trans-1,3-Dimethylcyclohexane	9.53	743
C_8H_{16}	cis-1,4-Dimethylcyclohexane	(<9.93)	782
C_8H_{16}	trans-1,4-Dimethylcyclohexane	(9.56)	738
C_8H_{16}	Propylcyclopentane	(9.34)	753
$C_8H_{16}O$	2,2,4-Trimethyl-3-pentanone	(8.80)	511
C_8H_{18}	Octane	9.80 ± 0.10	737
C_8H_{18}	2-Methylheptane	(9.84)	734
C_8H_{18}	2,2,4-Trimethylpentane	(9.86)	713
C_8H_{18}	2,2,3,3-Tetramethylbutane	9.8	720
$C_8H_{18}O$	Dibutyl ether	(9.28)	s 560
$C_8H_{18}O$	Di-sec-butyl ether	(9.11)	511
$C_8H_{18}O$	Di-tert-butyl ether	8.88 ± 0.07	493
$C_8H_{18}S$	Dibutyl sulfide	(8.2)	624
$C_8H_{18}S$	Di-tert-butyl sulfide	(8.0)	583
$C_8H_{18}S$	Diisobutyl sulfide	(8.34)	625
$C_8H_{19}N$	Dibutylamine	(7.69)	586
$C_8H_{19}N$	Diisobutylamine	(7.8)	574

Mol. form.	Name	IP/eV	$\Delta_f H_{ion}$ kJ/mol
$C_8H_{20}Si$	Tetraethylsilane	(8.9)	595
C_9H_7N	Quinoline	8.62 ± 0.01	1041
C_9H_7N	Isoquinoline	8.53 ± 0.03	1032
C_9H_8	Indene	8.14 ± 0.01	949
C_9H_{10}	o-Methylstyrene	(8.20)	908
C_9H_{10}	m-Methylstyrene	(8.15)	899
C_9H_{10}	p-Methylstyrene	(8.1)	895
C_9H_{10}	Cyclopropylbenzene	(8.35)	956
C_9H_{10}	Indan	(8.3)	864
$C_9H_{10}O_2$	Ethyl benzoate	(8.9)	537
C_9H_{12}	Propylbenzene	8.713 ± 0.010	848
C_9H_{12}	Isopropylbenzene	8.73 ± 0.01	847
C_9H_{12}	1,2,3-Trimethylbenzene	8.42 ± 0.02	803
C_9H_{12}	1,2,4-Trimethylbenzene	8.27 ± 0.01	784
C_9H_{12}	1,3,5-Trimethylbenzene	8.41 ± 0.01	796
$C_9H_{13}N$	N,N-Dimethyl-o-toluidine	7.40 ± 0.02	814
$C_9H_{14}O$	Isophorone	(≤ 9.07)	≤ 670
C_9H_{18}	Butylcyclopentane	(9.95)	793
C_9H_{18}	Propylcyclohexane	(9.46)	720
C_9H_{18}	Isopropylcyclohexane	(9.33)	704
$C_9H_{18}O$	2-Nonanone	(9.16)	545
$C_9H_{18}O$	5-Nonanone	(9.07)	530
$C_9H_{18}O$	2,6-Dimethyl-4-heptanone	9.01 ± 0.06	512
C_9H_{20}	Nonane	9.71 ± 0.10	709
$C_{10}F_8$	Perfluoronaphthalene	8.85	−368
$C_{10}H_7Br$	1-Bromonaphthalene	8.08 ± 0.03	955
$C_{10}H_7Cl$	1-Chloronaphthalene	(8.13)	906
$C_{10}H_8$	Naphthalene	8.1442 ± 0.0009	936
$C_{10}H_8$	Azulene	7.38 ± 0.05	1001
$C_{10}H_8O$	1-Naphthol	7.76 ± 0.03	719
$C_{10}H_8O$	2-Naphthol	7.87 ± 0.06	729
$C_{10}H_{10}O_4$	Dimethyl phthalate	(9.64 ± 0.07)	277
$C_{10}H_{12}$	1,2,3,4-Tetrahydronaphthalene	8.46 ± 0.02	841
$C_{10}H_{14}$	Butylbenzene	8.69 ± 0.02	826
$C_{10}H_{14}$	sec-Butylbenzene	8.68 ± 0.02	820
$C_{10}H_{14}$	tert-Butylbenzene	8.68 ± 0.05	816
$C_{10}H_{14}$	Isobutylbenzene	8.69 ± 0.02	817
$C_{10}H_{14}$	p-Cymene	(8.29)	771
$C_{10}H_{14}$	o-Diethylbenzene	(≤ 8.51)	≤ 804
$C_{10}H_{14}$	m-Diethylbenzene	(8.49)	798
$C_{10}H_{14}$	p-Diethylbenzene	(8.40)	790
$C_{10}H_{14}$	1,2,4,5-Tetramethylbenzene	8.04 ± 0.02	730
$C_{10}H_{14}O$	p-tert-Butylphenol	(7.8)	552
$C_{10}H_{16}$	α-Pinene	(8.07)	808
$C_{10}H_{16}O$	Camphor	(8.76)	577
$C_{10}H_{18}$	cis-Decahydronaphthalene	9.36 ± 0.04	734
$C_{10}H_{18}$	trans-Decahydronaphthalene	9.34 ± 0.04	720
$C_{10}H_{20}$	1-Decene	9.42 ± 0.05	786
$C_{10}H_{20}$	Butylcyclohexane	(9.41)	695
$C_{10}H_{22}$	Decane	(9.65)	682
$C_{11}H_{10}$	1-Methylnaphthalene	7.97 ± 0.03	882
$C_{11}H_{10}$	2-Methylnaphthalene	7.91 ± 0.08	877
$C_{11}H_{16}$	p-tert-Butyltoluene	(8.12)	730
$C_{11}H_{24}$	Undecane	(9.56)	650
$C_{11}H_{24}$	2-Methyldecane	(9.7)	658
$C_{12}H_8$	Acenaphthylene	(8.22)	1053
$C_{12}H_9N$	Carbazole	(7.57)	961
$C_{12}H_{10}$	Acenaphthene	7.75 ± 0.07	903
$C_{12}H_{10}$	Biphenyl	8.23 ± 0.10	977
$C_{12}H_{10}N_2O$	trans-Azoxybenzene	(8.1)	1123

Mol. form.	Name	IP/eV	$\Delta_f H_{ion}$ kJ/mol
$C_{12}H_{10}O$	Diphenyl ether	(8.09)	766
$C_{12}H_{11}N$	Diphenylamine	7.16 ± 0.04	908
$C_{12}H_{18}$	5,7-Dodecadiyne	(8.67)	1079
$C_{12}H_{18}$	Hexamethylbenzene	7.85 ± 0.01	670
$C_{12}H_{22}$	Cyclohexylcyclohexane	(9.41)	690
$C_{12}H_{27}N$	Tributylamine	(7.4)	492
$C_{13}H_{10}$	9H-Fluorene	7.91 ± 0.02	952
$C_{13}H_{10}O$	Benzophenone	9.08 ± 0.05	926
$C_{13}H_{12}$	Diphenylmethane	(8.55)	963
$C_{14}H_{10}$	Anthracene	7.439 ± 0.006	948
$C_{14}H_{10}$	Phenanthrene	7.8914 ± 0.0006	966
$C_{14}H_{10}$	Diphenylacetylene	7.94 ± 0.03	1168
$C_{14}H_{12}$	cis-Stilbene	(7.80)	1005
$C_{14}H_{12}$	trans-Stilbene	7.656 ± 0.001	973
$C_{14}H_{14}$	1,2-Diphenylethane	8.9 ± 0.1	1002
$C_{16}H_{10}$	Fluoranthene	7.9 ± 0.1	1052
$C_{16}H_{10}$	Pyrene	7.4256 ± 0.0006	935
$C_{18}H_{12}$	Chrysene	7.60 ± 0.01	1017
$C_{18}H_{14}$	o-Terphenyl	(7.99)	1056
$C_{18}H_{14}$	m-Terphenyl	(8.01)	1057
$C_{18}H_{14}$	p-Terphenyl	7.80 ± 0.03	1037
$C_{20}H_{12}$	Perylene	6.960 ± 0.001	981
$C_{24}H_{12}$	Coronene	7.29 ± 0.01	1026

X-RAY ATOMIC ENERGY LEVELS

The energy levels in this table are the values recommended by Bearden and Burr on the basis of a thorough review of the literature on x-ray wavelengths and related data. All values are in electron volts (eV). Values in parentheses are interpolated, and an asterisk * indicates a level that is not resolved from the level above it. See Reference 1 for uncertainties in the levels and a complete description of how the recommended values were obtained.

References

1. Bearden, J. A., and Burr, A. F., *Rev. Mod. Phys.*, 39, 125, 1967; also published as *X-Ray Wavelengths and X-Ray Atomic Energy Levels*, Natl. Stand. Ref. Data Sys. — Natl. Bur. Standards (U.S.), No. 14, 1967.
2. Gray, D. E., Ed., *American Institute of Physics Handbook, Third Edition*, pp. 7-158 to 7-167, McGraw-Hill, New York, 1972.

Level	¹H	²He	³Li	⁴Be	⁵B	⁶C	⁷N	⁸O
K	13.59811	24.58678	54.75	111.0	188.0	283.8	401.6	532.0
L_I								23.7
$L_{II,III}$					4.7	6.4	9.2	7.1

Level	⁹F	¹⁰Ne	¹¹Na	¹²Mg	¹³Al	¹⁴Si	¹⁵P	¹⁶S
K	685.4	866.9	1072.1	1305.0	1559.6	1838.9	2145.5	2472.0
L_I	(31)	(45)	63.3	89.4	117.7	148.7	189.3	229.2
$L_{II,III}$	8.6	18.3	31.1	51.4	73.1	99.2	132.2	164.8

Level	¹⁷Cl	¹⁸Ar	¹⁹K	²⁰Ca	²¹Sc	²²Ti	²³V	²⁴Cr
K	2822.4	3202.9	3607.4	4038.1	4492.8	4966.4	5465.1	5989.2
L_I	270.2	320	377.1	437.8	500.4	563.7	628.2	694.6
L_{II}	201.6	247.3	296.3	350.0	406.7	461.5	520.5	583.7
L_{III}	200.0	245.2	293.6	346.4	402.2	455.5	512.9	574.5
M_I	17.5	25.3	33.9	43.7	53.8	60.3	66.5	74.1
$M_{II,III}$	6.8	12.4	17.8	25.4	32.3	34.6	37.8	42.5
$M_{IV,V}$					6.6	3.7	2.2	2.3

Level	²⁵Mn	²⁶Fe	²⁷Co	²⁸Ni	²⁹Cu	³⁰Zn	³¹Ga	³²Ge
K	6539.0	7112.0	7708.9	8332.8	8978.9	9658.6	10367.1	11103.1
L_I	769.0	846.1	925.6	1008.1	1096.1	1193.6	1297.7	1414.3
L_{II}	651.4	721.1	793.6	871.9	951.0	1042.8	1142.3	1247.8
L_{III}	640.3	708.1	778.6	854.7	931.1	1019.7	1115.4	1216.7
M_I	83.9	92.9	100.7	111.8	119.8	135.9	158.1	180.0
M_{II}	48.6	54.0	59.5	68.1	73.6	86.6	106.8	127.9
M_{III}	48.6*	54.0*	59.5*	68.1*	73.6*	86.6*	102.9	120.8
$M_{IV,V}$	3.3	3.6	2.9	3.6	1.6	8.1	17.4	28.7

Level	³³As	³⁴Se	³⁵Br	³⁶Kr	³⁷Rb	³⁸Sr	³⁹Y	⁴⁰Zr
K	11866.7	12657.8	13473.7	14325.6	15199.7	16104.6	17038.4	17997.6
L_I	1526.5	1653.9	1782.0	1921.0	2065.1	2216.3	2372.5	2531.6
L_{II}	1358.6	1476.2	1596.0	1727.2	1863.9	2006.8	2155.5	2306.7
L_{III}	1323.1	1435.8	1549.9	1674.9	1804.4	1939.6	2080.0	2222.3
M_I	203.5	231.5	256.5		322.1	357.5	393.6	430.3
M_{II}	146.4	168.2	189.3	222.7	247.4	279.8	312.4	344.2
M_{III}	140.5	161.9	181.5	213.8	238.5	269.1	300.3	330.5
M_{IV}	41.2	56.7	70.1	88.9	111.8	135.0	159.6	182.4
M_V	41.2*	56.7*	69.0	88.9*	110.3	133.1	157.4	180.0
N_I			27.3	24.0	29.3	37.7	45.4	51.3
N_{II}	2.5	5.6	5.2	10.6	14.8	19.9	25.6	28.7
N_{III}	2.5*	5.6*	4.6	10.6*	14.0	19.9*	25.6*	28.7*

Level	⁴¹Nb	⁴²Mo	⁴³Tc	⁴⁴Ru	⁴⁵Rh	⁴⁶Pd	⁴⁷Ag	⁴⁸Cd
K	18985.6	19999.5	21044.0	22117.2	23219.9	24350.3	25514.0	26711.2
L_I	2697.7	2865.5	3042.5	3224.0	3411.9	3604.3	3805.8	4018.0
L_{II}	2464.7	2625.1	2793.2	2966.9	3146.1	3330.3	3523.7	3727.0
L_{III}	2370.5	2520.2	2676.9	2837.9	3003.8	3173.3	3351.1	3537.5
M_I	468.4	504.6		585.0	627.1	669.9	717.5	770.2
M_{II}	378.4	409.7	444.9	482.8	521.0	559.1	602.4	650.7
M_{III}	363.0	392.3	425.0	460.6	496.2	531.5	571.4	616.5
M_{IV}	207.4	230.3	256.4	283.6	311.7	340.0	372.8	410.5

Level	^{41}Nb	^{42}Mo	^{43}Tc	^{44}Ru	^{45}Rh	^{46}Pd	^{47}Ag	^{48}Cd
M$_V$	204.6	227.0	252.9	279.4	307.0	334.7	366.7	403.7
N$_I$	58.1	61.8		74.9	81.0	86.4	95.2	107.6
N$_{II}$	33.9	34.8	38.9	43.1	47.9	51.1	62.6	66.9
N$_{III}$	33.9*	34.8*	38.9*	43.1*	47.9*	51.1*	55.9	66.9*
N$_{IV,V}$	3.2	1.8		2.0	2.5	1.5	3.3	9.3

Level	^{49}In	^{50}Sn	^{51}Sb	^{52}Te	^{53}I	^{54}Xe	^{55}Cs	^{56}Ba
K	27939.9	29200.1	30491.2	31813.8	33169.4	34561.4	35984.6	37440.6
L$_I$	4237.5	4464.7	4698.3	4939.2	5188.1	5452.8	5714.3	5988.8
L$_{II}$	3938.0	4156.1	4380.4	4612.0	4852.1	5103.7	5359.4	5623.6
L$_{III}$	3730.1	3928.8	4132.2	4341.4	4557.1	4782.2	5011.9	5247.0
M$_I$	825.6	883.8	943.7	1006.0	1072.1		1217.1	1292.8
M$_{II}$	702.2	756.4	811.9	869.7	930.5	999.0	1065.0	1136.7
M$_{III}$	664.3	714.4	765.6	818.7	874.6	937.0	997.6	1062.2
M$_{IV}$	450.8	493.3	536.9	582.5	631.3		739.5	796.1
M$_V$	443.1	484.8	527.5	572.1	619.4	672.3	725.5	780.7
N$_I$	121.9	136.5	152.0	168.3	186.4		230.8	253.0
N$_{II}$	77.4	88.6	98.4	110.2	122.7	146.7	172.3	191.8
N$_{III}$	77.4*	88.6*	98.4*	110.2*	122.7*	146.7*	161.6	179.7
N$_{IV}$	16.2	23.9	31.4	39.8	49.6		78.8	92.5
N$_V$	16.2*	23.9*	31.4*	39.8*	49.6*		76.5	89.9
O$_I$	0.1	0.9	6.7	11.6	13.6		22.7	39.1
O$_{II}$	0.8	1.1	2.1	2.3	3.3		13.1	16.6
O$_{III}$	0.8*	1.1*	2.1*	2.3*	3.3*		11.4	14.6

Level	^{57}La	^{58}Ce	^{59}Pr	^{60}Nd	^{61}Pm	^{62}Sm	^{63}Eu	^{64}Gd
K	38924.6	40443.0	41990.6	43568.9	45184.0	46834.2	48519.0	50239.1
L$_I$	6266.3	6548.8	6834.8	7126.0	7427.9	7736.8	8052.0	8375.6
L$_{II}$	5890.6	6164.2	6440.4	6721.5	7012.8	7311.8	7617.1	7930.3
L$_{III}$	5482.7	5723.4	5964.3	6207.9	6459.3	6716.2	6976.9	7242.8
M$_I$	1361.3	1434.6	1511.0	1575.3		1722.8	1800.0	1880.8
M$_{II}$	1204.4	1272.8	1337.4	1402.8	1471.4	1540.7	1613.9	1688.3
M$_{III}$	1123.4	1185.4	1242.2	1297.4	1356.9	1419.8	1480.6	1544.0
M$_{IV}$	848.5	901.3	951.1	999.9	1051.5	1106.0	1160.6	1217.2
M$_V$	831.7	883.3	931.0	977.7	1026.9	1080.2	1130.9	1185.2
N$_I$	270.4	289.6	304.5	315.2		345.7	360.2	375.8
N$_{II}$	205.8	223.3	236.3	243.3	242	265.6	283.9	288.5
N$_{III}$	191.4	207.2	217.6	224.6	242*	247.4	256.6	270.9
N$_{IV,V}$	98.9	110.0	113.2	117.5	120.4	129.0	133.2	140.5
N$_{VI,VII}$		0.1	2.0	1.5		5.5	0.0	0.1
O$_I$	32.3	37.8	37.4	37.5		37.4	31.8	36.1
O$_{II,III}$	14.4	19.8	22.3	21.1		21.3	22.0	20.3

Level	^{65}Tb	^{66}Dy	^{67}Ho	^{68}Er	^{69}Tm	^{70}Yb	^{71}Lu	^{72}Hf
K	51995.7	53788.5	55617.7	57485.5	59389.6	61332.3	63313.8	65350.8
L$_I$	8708.0	9045.8	9394.2	9751.3	10115.7	10486.4	10870.4	11270.7
L$_{II}$	8251.6	8580.6	8917.8	9264.3	9616.9	9978.2	10348.6	10739.4
L$_{III}$	7514.0	7790.1	8071.1	8357.9	8648.0	8943.6	9244.1	9560.7
M$_I$	1967.5	2046.8	2128.3	2206.5	2306.8	2398.1	2491.2	2600.9
M$_{II}$	1767.7	1841.8	1922.8	2005.8	2089.8	2173.0	2263.5	2365.4
M$_{III}$	1611.3	1675.6	1741.2	1811.8	1884.5	1949.8	2023.6	2107.6
M$_{IV}$	1275.0	1332.5	1391.5	1453.3	1514.6	1576.3	1639.4	1716.4
M$_V$	1241.2	1294.9	1351.4	1409.3	1467.7	1527.8	1588.5	1661.7
N$_I$	397.9	416.3	435.7	449.1	471.7	487.2	506.2	538.1
N$_{II}$	310.2	331.8	343.5	366.2	385.9	396.7	410.1	437.0
N$_{III}$	385.0	292.9	306.6	320.0	336.6	343.5	359.3	380.4
N$_{IV}$	147.0	154.2	161.0	176.7	179.6	198.1	204.8	223.8
N$_V$	147.0*	154.2*	161.0*	167.6	179.6*	184.9	195.0	213.7
N$_{VI,VII}$	2.6	4.2	3.7	4.3	5.3	6.3	6.9	17.1
O$_I$	39.0	62.9	51.2	59.8	53.2	54.1	56.8	64.9
O$_{II}$	25.4	26.3	20.3	29.4	32.3	23.4	28.0	38.1
O$_{III}$	25.4*	26.3*	20.3*	29.4*	32.3*	23.4*	28.0*	30.6

Level	^{73}Ta	^{74}W	^{75}Re	^{76}Os	^{77}Ir	^{78}Pt	^{79}Au	^{80}Hg
K	67416.4	69525.0	71676.4	73870.8	76111.0	78394.8	80724.9	83102.3
L$_I$	11681.5	12099.8	12526.7	12968.0	13418.5	13879.9	14352.8	14839.3
L$_{II}$	11136.1	11544.0	11958.7	12385.0	12824.1	13272.6	13733.6	14208.7
L$_{III}$	9881.1	10206.8	10535.3	10870.9	11215.2	11563.7	11918.7	12283.9
M$_I$	2708.0	2819.6	2931.7	3048.5	3173.7	3296.0	3424.9	3561.6
M$_{II}$	2468.7	2574.9	2681.6	2792.2	2908.7	3026.5	3147.8	3278.5
M$_{III}$	2194.0	2281.0	2367.3	2457.2	2550.7	2645.4	2743.0	2847.1
M$_{IV}$	1793.2	1871.6	1948.9	2030.8	2116.1	2201.9	2291.1	2384.9
M$_V$	1735.1	1809.2	1882.9	1960.1	2040.4	2121.6	2205.7	2294.9
N$_I$	565.5	595.0	625.0	654.3	690.1	722.0	758.8	800.3
N$_{II}$	464.8	491.6	517.9	546.5	577.1	609.2	643.7	676.9
N$_{III}$	404.5	425.3	444.4	468.2	494.3	519.0	545.4	571.0
N$_{IV}$	241.3	258.8	273.7	289.4	311.4	330.8	352.0	378.3
N$_V$	229.3	245.4	260.2	272.8	294.9	313.3	333.9	359.8
N$_{VI}$	25.0	36.5	40.6	46.3	63.4	74.3	86.4	102.2
N$_{VII}$	25.0*	33.6	40.6*	46.3*	60.5	71.1	82.8	98.5
O$_I$	71.1	77.1	82.8	83.7	95.2	101.7	107.8	120.3
O$_{II}$	44.9	46.8	45.6	58.0	63.0	65.3	71.7	80.5
O$_{III}$	36.4	35.6	34.6	45.4	50.5	51.7	53.7	57.6
O$_{IV,V}$	5.7	6.1	3.5		3.8	2.2	2.5	6.4

Level	^{81}Tl	^{82}Pb	^{83}Bi	^{84}Po	^{85}At	^{86}Rn	^{87}Fr	^{88}Ra
K	85530.4	88004.5	90525.9	93105.0	95729.9	98404	101137	103921.9
L$_I$	15346.7	15860.8	16387.5	16939.3	17493	18049	18639	19236.7
L$_{II}$	14697.9	15200.0	15711.1	16244.3	16784.7	17337.1	17906.5	18484.3
L$_{III}$	12657.5	13035.2	13418.6	13813.8	14213.5	14619.4	15031.2	15444.4
M$_I$	3704.1	3850.7	3999.1	4149.4	(4317)	(4482)	(4652)	4822.0
M$_{II}$	3415.7	3554.2	3696.3	3854.1	4008	4159	4327	4489.5
M$_{III}$	2956.6	3066.4	3176.9	3301.9	3426	3538	3663	3791.8
M$_{IV}$	2485.1	2585.6	2687.6	2798.0	2908.7	3021.5	3136.2	3248.4
M$_V$	2389.3	2484.0	2579.6	2683.0	2786.7	2892.4	2999.9	3104.9
N$_I$	845.5	893.6	938.2	995.3	(1042)	(1097)	(1153)	1208.4
N$_{II}$	721.3	763.9	805.3	851	886	929	980	1057.6
N$_{III}$	609.0	644.5	678.9	705	740	768	810	879.1
N$_{IV}$	406.6	435.2	463.6	500.2	533.2	566.6	603.3	635.9
N$_V$	386.2	412.9	440.0	473.4			577	602.7
N$_{VI}$	122.8	142.9	161.9					298.9
N$_{VII}$	118.5	138.1	157.4					298.9*
O$_I$	136.3	147.3	159.3					254.4
O$_{II}$	99.6	104.8	116.8					200.4
O$_{III}$	75.4	86.0	92.8					152.8
O$_{IV}$	15.3	21.8	26.5	31.4				67.2
O$_V$	13.1	19.2	24.4	31.4*				67.2*
P$_I$		3.1						43.5
P$_{II,III}$		0.7	2.7					18.8

Level	^{89}Ac	^{90}Th	^{91}Pa	^{92}U	^{93}Np	^{94}Pu	^{95}Am	^{96}Cm
K	106755.3	109650.9	112601.4	115606.1	118678	121818	125027	128220
L$_I$	19840	20472.1	21104.6	21757.4	22426.8	23097.2	23772.9	24460
L$_{II}$	19083.2	19693.2	20313.7	20947.6	21600.5	22266.2	22944.0	23779
L$_{III}$	15871.0	16300.3	16733.1	17166.3	17610.0	18056.8	18504.1	18930
M$_I$	(5002)	5182.3	5366.9	5548.0	5723.2	5932.9	6120.5	6288
M$_{II}$	4656	4830.4	5000.9	5182.2	5366.2	5541.2	5710.2	5895
M$_{III}$	3909	4046.1	4173.8	4303.4	4434.7	4556.6	4667.0	4797
M$_{IV}$	3370.2	3490.8	3611.2	3727.6	3850.3	3972.6	4092.1	4227
M$_V$	3219.0	3332.0	3441.8	3551.7	3665.8	3778.1	3886.9	3971
N$_I$	(1269)	1329.5	1387.1	1440.8	1500.7	1558.6	1617.1	1643
N$_{II}$	1080	1168.2	1224.3	1272.6	1327.7	1372.1	1411.8	1440
N$_{III}$	890	967.3	1006.7	1044.9	1086.8	1114.8	(1135.7)	1154
N$_{IV}$	674.9	714.1	743.4	780.4	815.9	848.9	878.7	

Level	^{89}Ac	^{90}Th	^{91}Pa	^{92}U	^{93}Np	^{94}Pu	^{95}Am	^{96}Cm
N$_V$		676.4	708.2	737.7	770.3	801.4	827.6	
N$_{VI}$		344.4	371.2	391.3	415.0	445.8		
N$_{VII}$		335.2	359.5	380.9	404.4	432.4		
O$_I$		290.2	309.6	323.7		351.9		385
O$_{II}$		229.4	222.9	259.3	283.4	274.1		
O$_{III}$		181.8	222.9*	195.1	206.1	206.5		
O$_{IV}$		94.3	94.1	105.0	109.3	116.0	115.8	
O$_V$		87.9	94.1*	96.3	101.3	105.4	103.3	
P$_I$		59.5		70.7				
P$_{II}$		49.0		42.3				
P$_{III}$		43.0		32.3				

Level	^{97}Bk	^{98}Cf	^{99}Es	^{100}Fm	^{101}Md	^{102}No	^{103}Lr
K	131590	135960	139490	143090	146780	150540	154380
L$_I$	25275	26110	26900	27700	28530	29380	30240
L$_{II}$	24385	25250	26020	26810	27610	28440	29280
L$_{III}$	19452	19930	20410	20900	21390	21880	22360
M$_I$	6556	6754	6977	7205	7441	7675	7900
M$_{II}$	6147	6359	6574	6793	7019	7245	7460
M$_{III}$	4977	5109	5252	5397	5546	5688	5710
M$_{IV}$	4366	4497	4630	4766	4903	5037	5150
M$_V$	4132	4253	4374	4498	4622	4741	4860
N$_I$	1755	1799	1868	1937	2010	2078	2140
N$_{II}$	1554	1616	1680	1747	1814	1876	1930
N$_{III}$	1235	1279	1321	1366	1410	1448	1480
O$_I$	398	419	435	454	472	484	490

ELECTRON BINDING ENERGIES OF THE ELEMENTS

Gwyn P. Williams

This table gives the binding energies in electron volts (eV) for selected electronic levels of the elements. For metallic elements the binding energy is referred to the Fermi level; for semiconductors, to the valence band maximum; and for gases and insulators, to the vacuum level. The atomic number is listed after the element name.

References

1. Fluggle, J. C., and Martensson, N., *J. Elect. Spect.*, 21, 275, 1980.
2. Cardona, M. and Ley, L., *Photoemission from Solids*, Springer-Verlag, Heidelberg, 1978.
3. Bearden, J. A. and Burr, A. F., *Rev. Mod. Phys.*, 39, 125, 1967.

Actinium (89)

Level	Orbital	Energy
K	1s	106755
L I	2s	19840
L II	$2p_{1/2}$	19083
L III	$2p_{3/2}$	15871
M I	3s	5002
M II	$3p_{1/2}$	4656
M III	$3p_{3/2}$	3909
M IV	$3d_{3/2}$	3370
M V	$3d_{5/2}$	3219
N I	4s	1269[a]
N II	$4p_{1/2}$	1080[a]
N III	$4p_{3/2}$	890[a]
N IV	$4d_{3/2}$	675[a]
N V	$4d_{5/2}$	639[a]
N VI	$4f_{5/2}$	319[a]
N VII	$4f_{7/2}$	319[a]
O I	5s	272[a]
O II	$5p_{1/2}$	215[a]
O III	$5p_{3/2}$	167[a]
O IV	$5d_{3/2}$	80[a]
O V	$5d_{5/2}$	80[a]
P I	6s	—
P II	$6p_{1/2}$	—
P III	$6p_{3/2}$	—

Aluminum (13)

Level	Orbital	Energy
K	1s	1559.0
L I	2s	117.8[a]
L II	$2p_{1/2}$	72.9[a]
L III	$2p_{3/2}$	72.5[a]

Antimony (51)

Level	Orbital	Energy
K	1s	30419
L I	2s	4698
L II	$2p_{1/2}$	4380
L III	$2p_{3/2}$	4132
M I	3s	946[b]
M II	$3p_{1/2}$	812.7[b]
M III	$3p_{3/2}$	766.4[b]
M IV	$3d_{3/2}$	537.5[b]
M V	$3d_{5/2}$	528.2[b]
N I	4s	153.2[b]
N II	$4p_{1/2}$	95.6[b,c]
N III	$4p_{3/2}$	95.6[b]
N IV	$4d_{3/2}$	33.3[b]
N V	$4d_{5/2}$	32.1[b]

Argon (18)

Level	Orbital	Energy
K	1s	3205.9[a]
L I	2s	326.3[a]
L II	$2p_{1/2}$	250.6[a]
L III	$2p_{3/2}$	248.4[a]
M I	3s	29.3[a]
M II	$3p_{1/2}$	15.9[a]
M III	$3p_{3/2}$	15.7[a]

Arsenic (33)

Level	Orbital	Energy
K	1s	11867
L I	2s	1527.0[a,d]
L II	$2p_{1/2}$	1359.1[a,d]
L III	$2p_{3/2}$	1323.6[a,d]
M I	3s	204.7[a]
M II	$3p_{1/2}$	146.2[a]
M III	$3p_{3/2}$	141.2[a]
M IV	$3d_{3/2}$	41.7[a]
M V	$3d_{5/2}$	41.7[a]

Astatine (85)

Level	Orbital	Energy
K	1s	95730
L I	2s	17493
L II	$2p_{1/2}$	16785
L III	$2p_{3/2}$	14214
M I	3s	4317
M II	$3p_{1/2}$	4008
M III	$3p_{3/2}$	3426
M IV	$3d_{3/2}$	2909
M V	$3d_{5/2}$	2787
N I	4s	1042[a]
N II	$4p_{1/2}$	886[a]
N III	$4p_{3/2}$	740[a]
N IV	$4d_{3/2}$	533[a]
N V	$4d_{5/2}$	507[a]
N VI	$4f_{5/2}$	210[a]
N VII	$4f_{7/2}$	210[a]
O I	5s	195[a]
O II	$5p_{1/2}$	148[a]
O III	$5p_{3/2}$	115[a]
O IV	$5d_{3/2}$	40[a]
O V	$5d_{5/2}$	40[a]

Barium (56)

Level	Orbital	Energy
K	1s	37441
L I	2s	5989
L II	$2p_{1/2}$	5624
L III	$2p_{3/2}$	5247
M I	3s	1293[a,d]
M II	$3p_{1/2}$	1137[a,d]
M III	$3p_{3/2}$	1063[a,d]
M IV	$3d_{3/2}$	795.7[a]
M V	$3d_{5/2}$	780.5[a]
N I	4s	253.5[b]
N II	$4p_{1/2}$	192
N III	$4p_{3/2}$	178.6[b]
N IV	$4d_{3/2}$	92.6[b]
N V	$4d_{5/2}$	89.9[b]
N VI	$4f_{5/2}$	—
N VII	$4f_{7/2}$	—
O I	5s	30.3[a]
O II	$5p_{1/2}$	17.0[a]
O III	$5p_{3/2}$	14.8[b]

Beryllium (4)

Level	Orbital	Energy
K	1s	111.5[a]

Bismuth (83)

Level	Orbital	Energy
K	1s	90526
L I	2s	16388
L II	$2p_{1/2}$	15711
L III	$2p_{3/2}$	13419
M I	3s	3999
M II	$3p_{1/2}$	3696
M III	$3p_{3/2}$	3177
M IV	$3d_{3/2}$	2688
M V	$3d_{5/2}$	2580
N I	4s	939[b]
N II	$4p_{1/2}$	805.2[b]
N III	$4p_{3/2}$	678.8[b]
N IV	$4d_{3/2}$	464.0[b]
N V	$4d_{5/2}$	440.1[b]
N VI	$4f_{5/2}$	162.3[b]
N VII	$4f_{7/2}$	157.0[b]
O I	5s	159.3[a,d]
O II	$5p_{1/2}$	119.0[b]
O III	$5p_{3/2}$	92.6[b]
O IV	$5d_{3/2}$	26.9[b]
O V	$5d_{5/2}$	23.8[b]

Boron (5)

Level	Orbital	Energy
K	1s	188[a]

Bromine (35)

Level	Orbital	Energy
K	1s	13474
L I	2s	1782[a]
L II	$2p_{1/2}$	1596[a]
L III	$2p_{3/2}$	1550[a]
M I	3s	257[a]
M II	$3p_{1/2}$	189[a]
M III	$3p_{3/2}$	182[a]
M IV	$3d_{3/2}$	70[a]
M V	$3d_{5/2}$	69[a]

Cadmium (48)

Level	Orbital	Energy
K	1s	26711
L I	2s	4018
L II	$2p_{1/2}$	3727
L III	$2p_{3/2}$	3538
M I	3s	772.0[b]
M II	$3p_{1/2}$	652.6[b]
M III	$3p_{3/2}$	618.4[b]
M IV	$3d_{3/2}$	411.9[b]
M V	$3d_{5/2}$	405.2[b]
N I	4s	109.8[b]
N II	$4p_{1/2}$	63.9[b,c]
N III	$4p_{3/2}$	63.9[b,c]
N IV	$4d_{3/2}$	11.7[b]
N V	$4d_{5/2}$	10.7[b]

Calcium (20)

Level	Orbital	Energy
K	1s	4038.5[a]
L I	2s	438.4[b]
L II	$2p_{1/2}$	349.7[b]
L III	$2p_{3/2}$	346.2[b]
M I	3s	44.3[b]
M II	$3p_{1/2}$	25.4[b]
M III	$3p_{3/2}$	25.4[b]

Carbon (6)

Level	Orbital	Energy
K	1s	284.2[a]

Cerium (58)

Level	Orbital	Energy
K	1s	40443
L I	2s	6548
L II	$2p_{1/2}$	6164
L III	$2p_{3/2}$	5723
M I	3s	1436[a,d]
M II	$3p_{1/2}$	1274[a,d]
M III	$3p_{3/2}$	1187[a,d]
M IV	$3d_{3/2}$	902.4[a]
M V	$3d_{5/2}$	883.8[a]
N I	4s	291.0[a]
N II	$4p_{1/2}$	223.3
N III	$4p_{3/2}$	206.5[a]
N IV	$4d_{3/2}$	109[a]
N V	$4d_{5/2}$	—
N VI	$4f_{5/2}$	0.1
N VII	$4f_{7/2}$	0.1
O I	5s	37.8
O II	$5p_{1/2}$	19.8[a]
O III	$5p_{3/2}$	17.0[a]

Cesium (55)

Level	Orbital	Energy
K	1s	35985
L I	2s	5714
L II	$2p_{1/2}$	5359
L III	$2p_{3/2}$	5012
M I	3s	1211[a,d]
M II	$3p_{1/2}$	1071[a]

M III	$3p_{3/2}$	1003[a]
M IV	$3d_{3/2}$	740.5[a]
M V	$3d_{5/2}$	726.6[a]
N I	4s	232.3[a]
N II	$4p_{1/2}$	172.4[a]
N III	$4p_{3/2}$	161.3[a]
N IV	$4d_{3/2}$	79.8[a]
N V	$4d_{5/2}$	77.5[a]
N VI	$4f_{5/2}$	—
N VII	$4f_{7/2}$	—
O I	5s	22.7
O II	$5p_{1/2}$	14.2[a]
O III	$5p_{3/2}$	12.1[a]

Chlorine (17)

K	1s	2822.0
L I	2s	270[a]
L II	$2p_{1/2}$	202[a]
L III	$2p_{3/2}$	200[a]

Chromium (24)

K	1s	5989
L I	2s	696.0[b]
L II	$2p_{1/2}$	583.8[b]
L III	$2p_{3/2}$	574.1[b]
M I	3s	74.1[b]
M II	$3p_{1/2}$	42.2[b]
M III	$3p_{3/2}$	42.2[b]

Cobalt (27)

K	1s	7709
L I	2s	925.1[b]
L II	$2p_{1/2}$	793.2[b]
L III	$2p_{3/2}$	778.1[b]
M I	3s	101.0[b]
M II	$3p_{1/2}$	58.9[b]
M III	$3p_{3/2}$	58.9[b]

Copper (29)

K	1s	8979
L I	2s	1096.7[b]
L II	$2p_{1/2}$	952.3[b]
L III	$2p_{3/2}$	932.5[b]
M I	3s	122.5[b]
M II	$3p_{1/2}$	77.3[b]
M III	$3p_{3/2}$	75.1[b]

Dysprosium (66)

K	1s	53789
L I	2s	9046
L II	$2p_{1/2}$	8581
L III	$2p_{3/2}$	7790
M I	3s	2047
M II	$3p_{1/2}$	1842
M III	$3p_{3/2}$	1676
M IV	$3d_{3/2}$	1333
M V	$3d_{5/2}$	1292[a]
N I	4s	414.2[a]
N II	$4p_{1/2}$	333.5[a]
N III	$4p_{3/2}$	293.2[a]
N IV	$4d_{3/2}$	153.6[a]
N V	$4d_{5/2}$	153.6[a]
N VI	$4f_{5/2}$	8.0[a]
N VII	$4f_{7/2}$	4.3[a]
O I	5s	49.9[a]

O II	$5p_{1/2}$	26.3
O III	$5p_{3/2}$	26.3

Erbium (68)

K	1s	57486
L I	2s	9751
L II	$2p_{1/2}$	9264
L III	$2p_{3/2}$	8358
M I	3s	2206
M II	$3p_{1/2}$	2006
M III	$3p_{3/2}$	1812
M IV	$3d_{3/2}$	1453
M V	$3d_{5/2}$	1409
N I	4s	449.8[a]
N II	$4p_{1/2}$	366.2
N III	$4p_{3/2}$	320.2[a]
N IV	$4d_{3/2}$	167.6[a]
N V	$4d_{5/2}$	167.6[a]
N VI	$4f_{5/2}$	—
N VII	$4f_{7/2}$	4.7[a]
O I	5s	50.6[a]
O II	$5p_{1/2}$	31.4[a]
O III	$5p_{3/2}$	24.7[a]

Europium (63)

K	1s	48519
L I	2s	8052
L II	$2p_{1/2}$	7617
L III	$2p_{3/2}$	6977
M I	3s	1800
M II	$3p_{1/2}$	1614
M III	$3p_{3/2}$	1481
M IV	$3d_{3/2}$	1158.6[a]
M V	$3d_{5/2}$	1127.5[a]
N I	4s	360
N II	$4p_{1/2}$	284
N III	$4p_{3/2}$	257
N IV	$4d_{3/2}$	133
N V	$4d_{5/2}$	1227[a]
N VI	$4f_{5/2}$	0
N VII	$4f_{7/2}$	0
O I	5s	32
O II	$5p_{1/2}$	22
O III	$5p_{3/2}$	22

Fluorine (9)

K	1s	696.7[a]

Francium (87)

K	1s	101137
L I	2s	18639
L II	$2p_{1/2}$	17907
L III	$2p_{3/2}$	15031
M I	3s	4652
M II	$3p_{1/2}$	4327
M III	$3p_{3/2}$	3663
M IV	$3d_{3/2}$	3136
M V	$3d_{5/2}$	3000
N I	4s	1153[a]
N II	$4p_{1/2}$	980[a]
N III	$4p_{3/2}$	810[a]
N IV	$4d_{3/2}$	603[a]
N V	$4d_{5/2}$	577[a]
N VI	$4f_{5/2}$	268[a]
N VII	$4f_{7/2}$	268[a]

O I	5s	234[a]
O II	$5p_{1/2}$	182[a]
O III	$5p_{3/2}$	140[a]
O IV	$5d_{3/2}$	58[a]
O V	$5d_{5/2}$	58[a]
P I	6s	34
P II	$6p_{1/2}$	15
P III	$6p_{3/2}$	15

Gadolinium (64)

K	1s	50239
L I	2s	8376
L II	$2p_{1/2}$	7930
L III	$2p_{3/2}$	7243
M I	3s	1881
M II	$3p_{1/2}$	1688
M III	$3p_{3/2}$	1544
M IV	$3d_{3/2}$	1221.9[a]
M V	$3d_{5/2}$	1189.6[a]
N I	4s	378.6[a]
N II	$4p_{1/2}$	286
N III	$4p_{3/2}$	271
N IV	$4d_{3/2}$	—
N V	$4d_{5/2}$	142.6[a]
N VI	$4f_{5/2}$	8.6[a]
N VII	$4f_{7/2}$	8.6[a]
O I	5s	36
O II	$5p_{1/2}$	20
O III	$5p_{3/2}$	20

Gallium (31)

K	1s	10367
L I	2s	1299.0[a,d]
L II	$2p_{1/2}$	1143.2[b]
L III	$2p_{3/2}$	1116.4[b]
M I	3s	159.5[b]
M II	$3p_{1/2}$	103.5[b]
M III	$3p_{3/2}$	100.0[b]
M IV	$3d_{3/2}$	18.7[b]
M V	$3d_{5/2}$	18.7[b]

Germanium (32)

K	1s	11103
L I	2s	1414.6[a,d]
L II	$2p_{1/2}$	1248.1[a,d]
L III	$2p_{3/2}$	1217.0[a,d]
M I	3s	180.1[a]
M II	$3p_{1/2}$	124.9[a]
M III	$3p_{3/2}$	120.8[a]
M IV	$3d_{3/2}$	29.8[a]
M V	$3d_{5/2}$	29.2[a]

Gold (79)

K	1s	80725
L I	2s	14353
L II	$2p_{1/2}$	13734
L III	$2p_{3/2}$	11919
M I	3s	3425
M II	$3p_{1/2}$	3148
M III	$3p_{3/2}$	2743
M IV	$3d_{3/2}$	2291
M V	$3d_{5/2}$	2206
N I	4s	762.1[b]
N II	$4p_{1/2}$	642.7[b]
N III	$4p_{3/2}$	546.3[b]

N IV	$4d_{3/2}$	353.2[b]
N V	$4d_{5/2}$	335.1[b]
N VI	$4f_{5/2}$	87.6[b]
N VII	$4f_{7/2}$	83.9[b]
O I	5s	107.2[a,d]
O II	$5p_{1/2}$	74.2[b]
O III	$5p_{3/2}$	57.2[b]

Hafnium (72)

K	1s	65351
L I	2s	11271
L II	$2p_{1/2}$	10739
L III	$2p_{3/2}$	9561
M I	3s	2601
M II	$3p_{1/2}$	2365
M III	$3p_{3/2}$	2107
M IV	$3d_{3/2}$	1176
M V	$3d_{5/2}$	1662
N I	4s	538[a]
N II	$4p_{1/2}$	438.2[b]
N III	$4p_{3/2}$	380.7[b]
N IV	$4d_{3/2}$	220.0[b]
N V	$4d_{5/2}$	211.5[b]
N VI	$4f_{5/2}$	15.9[b]
N VII	$4f_{7/2}$	14.2[b]
O I	5s	64.2[b]
O II	$5p_{1/2}$	38[a]
O III	$5p_{3/2}$	29.9[b]

Helium (2)

K	1s	24.6[a]

Holmium (67)

K	1s	55618
L I	2s	9394
L II	$2p_{1/2}$	8918
L III	$2p_{3/2}$	8071
M I	3s	2128
M II	$3p_{1/2}$	1923
M III	$3p_{3/2}$	1741
M IV	$3d_{3/2}$	1392
M V	$3d_{5/2}$	1351
N I	4s	432.4[a]
N II	$4p_{1/2}$	343.5
N III	$4p_{3/2}$	308.2[a]
N IV	$4d_{3/2}$	160[a]
N V	$4d_{5/2}$	160[a]
N VI	$4f_{5/2}$	8.6[a]
N VII	$4f_{7/2}$	5.2[a]
O I	5s	49.3[a]
O II	$5p_{1/2}$	30.8[a]
O III	$5p_{3/2}$	24.1[a]

Hydrogen (1)

K	1s	13.6

Indium (49)

K	1s	27940
L I	2s	4238
L II	$2p_{1/2}$	3938
L III	$2p_{3/2}$	3730
M I	3s	827.2[b]
M II	$3p_{1/2}$	703.2[b]
M III	$3p_{3/2}$	665.3[b]
M IV	$3d_{3/2}$	451.4[b]

M V	3d$_{5/2}$	443.9[b]
N I	4s	122.9[b]
N II	4p$_{1/2}$	73.5[b,c]
N III	4p$_{3/2}$	73.5[b,c]
N IV	4d$_{3/2}$	17.7[b]
N V	4d$_{5/2}$	16.9[b]

Iodine (53)

K	1s	33169
L I	2s	5188
L II	2p$_{1/2}$	4852
L III	2p$_{3/2}$	4557
M I	3s	1072[a]
M II	3p$_{1/2}$	931[a]
M III	3p$_{3/2}$	875[a]
M IV	3d$_{3/2}$	631[a]
M V	3d$_{5/2}$	620[a]
N I	4s	186[a]
N II	4p$_{1/2}$	123[a]
N III	4p$_{3/2}$	123[a]
N IV	4d$_{3/2}$	50[a]
N V	4d$_{5/2}$	50[a]

Iridium (77)

K	1s	76111
L I	2s	13419[b]
L II	2p$_{1/2}$	12824
L III	2p$_{3/2}$	11215
M I	3s	3174
M II	3p$_{1/2}$	2909
M III	3p$_{3/2}$	2551
M IV	3d$_{3/2}$	2116
M V	3d$_{5/2}$	2040
N I	4s	691.1[b]
N II	4p$_{1/2}$	577.8[b]
N III	4p$_{3/2}$	495.8[b]
N IV	4d$_{3/2}$	311.9[b]
N V	4d$_{5/2}$	296.3[b]
N VI	4f$_{5/2}$	63.8[b]
N VII	4f$_{7/2}$	60.8[b]
O I	52	95.2[a,d]
O II	5p$_{1/2}$	63.0[a,d]
O III	5p$_{3/2}$	48.0[b]

Iron (26)

K	1s	7112
L I	2s	844.6[b]
L II	2p$_{1/2}$	719.9[b]
L III	2p$_{3/2}$	706.8[b]
M I	3s	91.3[b]
M II	3p$_{1/2}$	52.7[b]
M III	3p$_{3/2}$	52.7[b]

Krypton (36)

K	1s	14326
L I	2s	1921
L II	2p$_{1/2}$	1730.9[a]
L III	2p$_{3/2}$	1678.4[a]
M I	3s	292.8[a]
M II	3p$_{1/2}$	222.2[a]
M III	3p$_{3/2}$	214.4[a]
M IV	3d$_{3/2}$	95.0[a]
M V	3d$_{5/2}$	93.8[a]
N I	4s	27.5[a]
N II	4p$_{1/2}$	14.1[a]

N III	4p$_{3/2}$	14.1[a]

Lanthanum (57)

K	1s	38925
L I	2s	6266
L II	2p$_{1/2}$	5891
L III	2p$_{3/2}$	5483
M I	3s	1362[a,d]
M II	3p$_{1/2}$	1209[a,d]
M III	3p$_{3/2}$	1128[a,d]
M IV	3d$_{3/2}$	853[a]
M V	3d$_{5/2}$	836[a]
N I	4s	247.7[a]
N II	4p$_{1/2}$	205.8
N III	4p$_{3/2}$	196.0[a]
N IV	4d$_{3/2}$	105.3[a]
N V	4d$_{5/2}$	102.5[a]
N VI	4f$_{5/2}$	—
N VII	4f$_{7/2}$	—
O I	5s	34.3[a]
O II	5p$_{1/2}$	19.3[a]
O III	5p$_{3/2}$	16.8[a]

Lead (82)

K	1s	88005
L I	2s	15861
L II	2p$_{1/2}$	15200
L III	2p$_{3/2}$	13055
M I	3s	3851
M II	3p$_{1/2}$	3554
M III	3p$_{3/2}$	3066
M IV	3d$_{3/2}$	2586
M V	3d$_{5/2}$	2484
N I	4s	891.8[b]
N II	4p$_{1/2}$	761.9[b]
N III	4p$_{3/2}$	643.5[b]
N IV	4d$_{3/2}$	434.3[b]
N V	4d$_{5/2}$	412.2[b]
N VI	4f$_{5/2}$	141.7[b]
N VII	4f$_{7/2}$	136.9[b]
O I	5s	147[a,d]
O II	5p$_{1/2}$	106.4[b]
O III	5p$_{3/2}$	83.3[b]
O IV	5d$_{3/2}$	20.7[b]
O V	5d$_{5/2}$	18.1[b]

Lithium (3)

K	1s	54.7[a]

Lutetium (71)

K	1s	63314
L I	2s	10870
L II	2p$_{1/2}$	10349
L III	2p$_{3/2}$	9244
M I	3s	2491
M II	3p$_{1/2}$	2264
M III	3p$_{3/2}$	2024
M IV	3d$_{3/2}$	1639
M V	3d$_{5/2}$	1589
N I	4s	506.8[a]
N II	4p$_{1/2}$	412.4[a]
N III	4p$_{3/2}$	359.2[a]
N IV	4d$_{3/2}$	206.1[a]
N V	4d$_{5/2}$	196.3[a]
N VI	4f$_{5/2}$	8.9[a]

N VII	4f$_{7/2}$	7.5[a]
O I	5s	57.3[a]
O II	5p$_{1/2}$	33.6[a]
O III	5p$_{3/2}$	26.7[a]

Magnesium (12)

K	1s	1303.0[b]
L I	2s	88.6[a]
L II	2p$_{1/2}$	49.6[b]
L III	2p$_{3/2}$	49.2[a]

Manganese (25)

K	1s	6539
L I	2s	769.1[b]
L II	2p$_{1/2}$	649.9[b]
L III	2p$_{3/2}$	638.7[b]
M I	3s	82.3[b]
M II	3p$_{1/2}$	47.2[b]
M III	3p$_{3/2}$	47.2[b]

Mercury (80)

K	1s	83102
L I	2s	14839
L II	2p$_{1/2}$	14209
L III	2p$_{3/2}$	12284
M I	3s	3562
M II	3p$_{1/2}$	3279
M III	3p$_{3/2}$	2847
M IV	3d$_{3/2}$	2385
M V	3d$_{5/2}$	2295
N I	4s	802.2[b]
N II	4p$_{1/2}$	680.2[b]
N III	4p$_{3/2}$	576.6[b]
N IV	4d$_{3/2}$	378.2[b]
N V	4d$_{5/2}$	358.8[b]
N VI	4f$_{5/2}$	104.0[b]
N VII	4f$_{7/2}$	99.9[b]
O I	5s	127[b]
O II	5p$_{3/2}$	83.1[b]
O III	5p$_{3/2}$	64.5[b]
O IV	5d$_{3/2}$	9.6[b]
O V	5d$_{5/2}$	7.8[b]

Molybdenum (42)

K	1s	20000
L I	2s	2866
L II	2p$_{1/2}$	2625
L III	2p$_{3/2}$	2520
M I	3s	506.3[b]
M II	3p$_{1/2}$	411.6[b]
M III	3p$_{3/2}$	394.0[b]
M IV	3d$_{3/2}$	231.1[b]
M V	3d$_{5/2}$	227.9[b]
N I	4s	63.2[b]
N II	4p$_{1/2}$	37.6[b]
N III	4p$_{3/2}$	35.5[b]

Neodymium (60)

K	1s	43569
L I	2s	7126
L II	2p$_{1/2}$	6722
L III	2p$_{3/2}$	6208
M I	3s	1575
M II	3p$_{1/2}$	1403
M III	3p$_{3/2}$	1297

M IV	3d$_{3/2}$	1003.3[a]
M V	3d$_{5/2}$	980.4[a]
N I	4s	319.2[a]
N II	4p$_{1/2}$	243.3
N III	4p$_{3/2}$	224.6
N IV	4d$_{3/2}$	120.5[a]
N V	4d$_{5/2}$	120.5[a]
N VI	4f$_{5/2}$	1.5
N VII	4f$_{7/2}$	1.5
O I	5s	37.5
O II	5p$_{1/2}$	21.1
O III	5p$_{3/2}$	21.1

Neon (10)

K	1s	870.2[a]
L I	2s	48.5[a]
L II	2p$_{1/2}$	21.7[a]
L III	2p$_{3/2}$	21.6[a]

Nickel (28)

K	1s	8333
L I	2s	1008.6[b]
L II	2p$_{1/2}$	870.0[b]
L III	2p$_{3/2}$	852.7[b]
M I	3s	110.8[b]
M II	3p$_{1/2}$	68.0[b]
M III	3p$_{3/2}$	66.2[b]

Niobium (41)

K	1s	18986
L I	2s	2698
L II	2p$_{1/2}$	2465
L III	2p$_{3/2}$	2371
M I	3s	466.6[b]
M II	3p$_{1/2}$	376.1[b]
M III	3p$_{3/2}$	360.6[b]
M IV	3d$_{3/2}$	205.0[b]
M V	3d$_{5/2}$	202.3[b]
N I	4s	56.4[b]
N II	4p$_{1/2}$	32.6[b]
N III	4p$_{3/2}$	30.8[b]

Nitrogen (7)

K	1s	409.9[a]
L I	2s	37.3[a]

Osmium (76)

K	1s	73871
L I	2s	12968
L II	2p$_{1/2}$	12385
L III	2p$_{3/2}$	10871
M I	3s	3049
M II	3p$_{1/2}$	2792
M III	3p$_{3/2}$	2457
M IV	3d$_{3/2}$	2031
M V	3d$_{5/2}$	1960
N I	4s	658.2[b]
N II	4p$_{1/2}$	549.1[b]
N III	4p$_{3/2}$	470.7[b]
N IV	4d$_{3/2}$	293.1[b]
N V	4d$_{5/2}$	278.5[b]
N VI	4f$_{5/2}$	53.4[b]
N VII	4f$_{7/2}$	50.7[b]
O I	5s	84[a]
O II	5p$_{1/2}$	58[a]

O III	$5p_{3/2}$	44.5[b]

Oxygen (8)

K	1s	543.1[a]
L I	2s	41.6[a]

Palladium (46)

K	1s	24350
L I	2s	3604
L II	$2p_{1/2}$	3330
L III	$2p_{3/2}$	3173
M I	3s	671.6[b]
M II	$3p_{1/2}$	559.9[b]
M III	$3p_{3/2}$	532.3[b]
M IV	$3d_{3/2}$	340.5[b]
M V	$3d_{5/2}$	335.2[b]
N I	4s	87.1[a,d]
N II	$4p_{1/2}$	55.7[b,c]
N III	$4p_{3/2}$	50.9[b,c]

Phosphorus (15)

K	1s	2145.5
L I	2s	189[a]
L II	$2p_{1/2}$	136[a]
L III	$2p_{3/2}$	135[a]

Platinum (78)

K	1s	78395
L I	2s	13880
L II	$2p_{1/2}$	13273
L III	$2p_{3/2}$	11564
M I	3s	3296
M II	$3p_{1/2}$	3027
M III	$3p_{3/2}$	2645
M IV	$3d_{3/2}$	2202
M V	$3d_{5/2}$	2122
N I	4s	725.4[b]
N II	$4p_{1/2}$	609.1[b]
N III	$4p_{3/2}$	519.4[b]
N IV	$4d_{3/2}$	331.6[b]
N V	$4d_{5/2}$	314.6[b]
N VI	$4f_{5/2}$	74.5[b]
N VII	$4f_{7/2}$	71.2[b]
O I	5s	101.7[a,d]
O II	$5p_{1/2}$	65.3[a,b]
O III	$5p_{3/2}$	51.7[b]

Polonium (84)

K	1s	93105
L I	2s	16939
L II	$2p_{1/2}$	16244
L III	$2p_{3/2}$	13814
M I	3s	4149
M II	$3p_{1/2}$	3854
M III	$3p_{3/2}$	3302
M IV	$3d_{3/2}$	2798
M V	$3d_{5/2}$	2683
N I	4s	995[a]
N II	$4p_{1/2}$	851[a]
N III	$4p_{3/2}$	705[a]
N IV	$4d_{3/2}$	500[a]
N V	$4d_{5/2}$	473[a]
N VI	$4f_{5/2}$	184[a]
N VII	$4f_{7/2}$	184[a]
O I	5s	177[a]

O II	$5p_{1/2}$	132[a]
O III	$5p_{3/2}$	104[a]
O IV	$5d_{3/2}$	31[a]
O V	$5d_{5/2}$	31[a]

Potassium (19)

K	1s	3608.4[a]
L I	2s	378.6[a]
L II	$2p_{1/2}$	297.3[a]
L III	$2p_{3/2}$	294.6[a]
M I	3s	34.8[a]
M II	$3p_{1/2}$	18.3[a]
M III	$3p_{3/2}$	18.3[a]

Praseodymium (59)

K	1s	41991
L I	2s	6835
L II	$2p_{1/2}$	6440
L III	$2p_{3/2}$	5964
M I	3s	1511
M II	$3p_{1/2}$	1337
M III	$3p_{3/2}$	1242
M IV	$3d_{3/2}$	948.3[a]
M V	$3d_{5/2}$	928.8[a]
N I	4s	304.5
N II	$4p_{1/2}$	236.3
N III	$4p_{3/2}$	217.6
N IV	$4d_{3/2}$	115.1[a]
N V	$4d_{5/2}$	115.1[a]
N VI	$4f_{5/2}$	2.0
N VII	$4f_{7/2}$	2.0
O I	5s	37.4
O II	$5p_{1/2}$	22.3
O III	$5p_{3/2}$	22.3

Promethium (61)

K	1s	45184
L I	2s	7428
L II	$2p_{1/2}$	7013
L III	$2p_{3/2}$	6459
M I	3s	—
M II	$3p_{1/2}$	1471.4
M III	$3p_{3/2}$	1357
M IV	$3d_{3/2}$	1052
M V	$3d_{5/2}$	1027
N I	4s	—
N II	$4p_{1/2}$	242
N III	$4p_{3/2}$	242
N IV	$4d_{3/2}$	120
N V	$4d_{5/2}$	120

Protactinium (91)

K	1s	112601
L I	2s	21105
L II	$2p_{1/2}$	20314
L III	$2p_{3/2}$	16733
M I	3s	5367
M II	$3p_{1/2}$	5001
M III	$3p_{3/2}$	4174
M IV	$3d_{3/2}$	3611
M V	$3d_{5/2}$	3442
N I	4s	1387[a]
N II	$4p_{1/2}$	1224[a]
N III	$4p_{3/2}$	1007[a]
N IV	$4d_{3/2}$	743[a]

N V	$4d_{5/2}$	708[a]
N VI	$4f_{5/2}$	371[a]
N VII	$4f_{7/2}$	360[a]
O I	5s	310[a]
O II	$5p_{1/2}$	232[a]
O III	$5p_{3/2}$	232[a]
O IV	$5d_{3/2}$	94[a]
O V	$5d_{5/2}$	94[a]
P I	6s	—
P II	$6p_{1/2}$	—
P III	$6p_{3/2}$	—

Radium (88)

K	1s	103922
L I	2s	19237
L II	$2p_{1/2}$	18484
L III	$2p_{3/2}$	15444
M I	3s	4822
M II	$3p_{1/2}$	4490
M III	$3p_{3/2}$	3792
M IV	$3d_{3/2}$	3248
M V	$3d_{5/2}$	3105
N I	4s	1208[a]
N II	$4p_{1/2}$	1058
N III	$4p_{3/2}$	879[a]
N IV	$4d_{3/2}$	636[a]
N V	$4d_{5/2}$	603[a]
N VI	$4f_{5/2}$	299[a]
N VII	$4f_{7/2}$	299[a]
O I	5s	254[a]
O II	$5p_{1/2}$	200[a]
O III	$5p_{3/2}$	153[a]
O IV	$5d_{3/2}$	68[a]
O V	$5d_{5/2}$	68[a]
P I	6s	44
P II	$6p_{1/2}$	19
P III	$6p_{3/2}$	19

Radon (86)

K	1s	98404
L I	2s	18049
L II	$2p_{1/2}$	17337
L III	$2p_{3/2}$	14619
M I	3s	4482
M II	$3p_{1/2}$	4159
M III	$3p_{3/2}$	3538
M IV	$3d_{3/2}$	3022
M V	$3d_{5/2}$	2892
N I	4s	1097[a]
N II	$4p_{1/2}$	929[a]
N III	$4p_{3/2}$	768[a]
N IV	$4d_{3/2}$	567[a]
N V	$4d_{5/2}$	541[a]
N VI	$4f_{5/2}$	238[a]
N VII	$4f_{7/2}$	238[a]
O I	5s	214[a]
O II	$5p_{1/2}$	164[a]
O III	$5p_{3/2}$	127[a]
O IV	$5d_{3/2}$	48[a]
O V	$5d_{5/2}$	48[a]
P I	6s	26

Rhenium (75)

K	1s	71676

L I	2s	12527
L II	$2p_{1/2}$	11959
L III	$2p_{3/2}$	10535
M I	3s	2932
M II	$3p_{1/2}$	2682
M III	$3p_{3/2}$	2367
M IV	$3d_{3/2}$	1949
M V	$3d_{5/2}$	1883
N I	4s	625.4[b]
N II	$4p_{1/2}$	518.7[b]
N III	$4p_{3/2}$	446.8[b]
N IV	$4d_{3/2}$	273.9[b]
N V	$4d_{5/2}$	260.5[b]
N VI	$4f_{5/2}$	42.9[a]
N VII	$4f_{7/2}$	40.5[a]
O I	5s	83[b]
O II	$5p_{1/2}$	45.6[b]
O III	$5p_{3/2}$	34.6[a,d]

Rhodium (45)

K	1s	23220
L I	2s	3412
L II	$2p_{1/2}$	3146
L III	$2p_{3/2}$	3004
M I	3s	628.1[b]
M II	$3p_{1/2}$	521.3[b]
M III	$3p_{3/2}$	496.5[b]
M IV	$3d_{3/2}$	311.9[b]
M V	$3d_{5/2}$	307.2[b]
N I	4s	81.4[a,d]
N II	$4p_{1/2}$	50.5[b]
N III	$4p_{3/2}$	47.3[b]

Rubidium (37)

K	1s	15200
L I	2s	2065
L II	$2p_{1/2}$	1864
L III	$2p_{3/2}$	1804
M I	3s	326.7[a]
M II	$3p_{1/2}$	248.7[a]
M III	$3p_{3/2}$	239.1[a]
M IV	$3d_{3/2}$	113.0[a]
M V	$3d_{5/2}$	112[a]
N I	4s	30.5[a]
N II	$4p_{1/2}$	16.3[a]
N III	$4p_{3/2}$	15.3[a]

Ruthenium (44)

K	1s	22117
L I	2s	3224
L II	$2p_{1/2}$	2967
L III	$2p_{3/2}$	2838
M I	3s	586.2[b]
M II	$3p_{1/2}$	483.3[b]
M III	$3p_{3/2}$	461.5[b]
M IV	$3d_{3/2}$	284.2[b]
M V	$3d_{5/2}$	280.0[b]
N I	4s	75.0[b]
N II	$4p_{1/2}$	46.5[b]
N III	$4p_{3/2}$	43.2[b]

Samarium (62)

K	1s	46834
L I	2s	7737
L II	$2p_{1/2}$	7312

L III	$2p_{3/2}$	6716
M I	3s	1723
M II	$3p_{1/2}$	1541
M III	$3p_{3/2}$	1419.8
M IV	$3d_{3/2}$	1110.9[a]
M V	$3d_{5/2}$	1083.4[a]
N I	4s	347.2[a]
N II	$4p_{1/2}$	265.6
N III	$4p_{3/2}$	247.4
N IV	$4d_{3/2}$	129.0
N V	$4d_{5/2}$	129.0
N VI	$4f_{5/2}$	5.2
N VII	$4f_{7/2}$	5.2
O I	5s	37.4
O II	$5p_{1/2}$	21.3
O III	$5p_{3/2}$	21.3

Scandium (21)

K	1s	4492
L I	2s	498.0[a]
L II	$2p_{1/2}$	403.6[a]
L III	$2p_{3/2}$	389.7[a]
M I	3s	51.1[a]
M II	$3p_{1/2}$	28.3[a]
M III	$3p_{3/2}$	28.3[a]

Selenium (34)

K	1s	12658
L I	2s	1652.0[a,d]
L II	$2p_{1/2}$	1474.3[a,d]
L III	$2p_{3/2}$	1433.9[a,d]
M I	3s	229.6[a]
M II	$3p_{1/2}$	166.5[a]
M III	$3p_{3/2}$	160.7[a]
M IV	$3d_{3/2}$	55.5[a]
M V	$3d_{5/2}$	54.6[a]

Silicon (14)

K	1s	1839
L I	2s	149.7[a,d]
L II	$2p_{1/2}$	99.8[a]
L III	$2p_{3/2}$	99.2[a]

Silver (47)

K	1s	25514
L I	2s	3806
L II	$2p_{1/2}$	3524
L III	$2p_{3/2}$	3351
M I	3s	719.0[b]
M II	$3p_{1/2}$	603.8[b]
M III	$3p_{3/2}$	573.0[b]
M IV	$3d_{3/2}$	374.0[b]
M V	$3d_{5/2}$	368.0[b]
N I	4s	97.0[b]
N II	$4p_{1/2}$	63.7[b]
N III	$4p_{3/2}$	58.3[b]

Sodium (11)

K	1s	1070.8[b]
L I	2s	63.5[b]
L II	$2p_{1/2}$	30.4[b]
L III	$2p_{3/2}$	30.5[a]

Strontium (38)

K	1s	16105
L I	2s	2216

L II	$2p_{1/2}$	2007
L III	$2p_{3/2}$	1940
M I	3s	358.7[b]
M II	$3p_{1/2}$	280.3[b]
M III	$3p_{3/2}$	270.0[b]
M IV	$3d_{3/2}$	136.0[b]
M V	$3d_{5/2}$	134.2[b]
N I	4s	38.9[b]
N II	$4p_{1/2}$	21.6[b]
N III	$4p_{3/2}$	20.1[b]

Sulfur (16)

K	1s	2472
L I	2s	230.9[a,d]
L II	$2p_{1/2}$	163.6[a]
L III	$2p_{3/2}$	162.5[a]

Tantalum (73)

K	1s	67416
L I	2s	11682
L II	$2p_{1/2}$	11136
L III	$2p_{3/2}$	9881
M I	3s	2708
M II	$3p_{1/2}$	2469
M III	$3p_{3/2}$	2194
M IV	$3d_{3/2}$	1793
M V	$3d_{5/2}$	1735
N I	4s	563.4[b]
N II	$4p_{1/2}$	463.4[b]
N III	$4p_{3/2}$	400.9[b]
N IV	$4d_{3/2}$	237.9[b]
N V	$4d_{5/2}$	226.4[b]
N VI	$4f_{5/2}$	23.5[b]
N VII	$4f_{7/2}$	21.6[b]
O I	5s	69.7[b]
O II	$5p_{1/2}$	42.2[a]
O III	$5p_{3/2}$	32.7[b]

Technetium (43)

K	1s	21044
L I	2s	3043
L II	$2p_{1/2}$	2793
L III	$2p_{3/2}$	2677
M I	3s	586.1[a]
M II	$3p_{1/2}$	447.6[a]
M III	$3p_{3/2}$	417.7[a]
M IV	$3d_{3/2}$	257.6[a]
M V	$3d_{5/2}$	253.9[a]
N I	4s	69.5[a]
N II	$4p_{1/2}$	42.3[a]
N III	$4p_{3/2}$	39.9[a]

Tellurium (52)

K	1s	31814
L I	2s	4939
L II	$2p_{1/2}$	4612
L III	$2p_{3/2}$	4341
M I	3s	1006[b]
M II	$3p_{1/2}$	870.8[b]
M III	$3p_{3/2}$	820.0[b]
M IV	$3d_{3/2}$	583.4[b]
M V	$3d_{5/2}$	573.0[b]
N I	4s	169.4[b]
N II	$4p_{1/2}$	103.3[b,c]
N III	$4p_{3/2}$	103.3[b,c]

N IV	$4d_{3/2}$	41.9[b]
N V	$4d_{5/2}$	40.4[b]

Terbium (65)

K	1s	51996
L I	2s	8708
L II	$2p_{1/2}$	8252
L III	$2p_{3/2}$	7514
M I	3s	1968
M II	$3p_{1/2}$	1768
M III	$3p_{3/2}$	1611
M IV	$3d_{3/2}$	1267.9[a]
M V	$3d_{5/2}$	1241.1[a]
N I	4s	396.0[a]
N II	$4p_{1/2}$	322.4[a]
N III	$4p_{3/2}$	284.1[a]
N IV	$4d_{3/2}$	150.5[a]
N V	$4d_{5/2}$	150.5[a]
N VI	$4f_{5/2}$	7.7[a]
N VII	$4f_{7/2}$	2.4[a]
O I	5s	45.6[a]
O II	$5p_{1/2}$	28.7[a]
O III	$5p_{3/2}$	22.6[a]

Thallium (81)

K	1s	85530
L I	2s	15347
L II	$2p_{1/2}$	14698
L III	$2p_{3/2}$	12658
M I	3s	3704
M II	$3p_{1/2}$	3416
M III	$3p_{3/2}$	2957
M IV	$3d_{3/2}$	2485
M V	$3d_{5/2}$	2389
N I	4s	846.2[b]
N II	$4p_{1/2}$	720.5[b]
N III	$4p_{3/2}$	609.5[b]
N IV	$4d_{3/2}$	405.7[b]
N V	$4d_{5/2}$	385.0[b]
N VI	$4f_{5/2}$	122.2[b]
N VII	$4f_{7/2}$	117.8[b]
O I	5s	136[a,d]
O II	$5p_{1/2}$	94.6[b]
O III	$5p_{3/2}$	73.5[b]
O IV	$5d_{3/2}$	14.7[b]
O V	$5d_{5/2}$	12.5[b]

Thorium (90)

K	1s	109651
L I	2s	20472
L II	$2p_{1/2}$	19693
L III	$2p_{3/2}$	16300
M I	3s	5182
M II	$3p_{1/2}$	4830
M III	$3p_{3/2}$	4046
M IV	$3d_{3/2}$	3491
M V	$3d_{5/2}$	3332
N I	4s	1330[a]
N II	$4p_{1/2}$	1168[a]
N III	$4p_{3/2}$	966.4[b]
N IV	$4d_{3/2}$	712.1[b]
N V	$4d_{5/2}$	675.2[b]
N VI	$4f_{5/2}$	342.4[b]
N VII	$4f_{7/2}$	333.1[b]

O I	5s	290[a,c]
O II	$5p_{1/2}$	229[a,c]
O III	$5p_{3/2}$	182[a,c]
O IV	$5d_{3/2}$	92.5[b]
O V	$5d_{5/2}$	85.4[b]
P I	6s	41.4[b]
P II	$6p_{1/2}$	24.5[b]
P III	$6p_{3/2}$	16.6[b]

Thulium (69)

K	1s	59390
L I	2s	10116
L II	$2p_{1/2}$	9617
L III	$2p_{3/2}$	8648
M I	3s	2307
M II	$3p_{1/2}$	2090
M III	$3p_{3/2}$	1885
M IV	$3d_{3/2}$	1515
M V	$3d_{5/2}$	1468
N I	4s	470.9[a]
N II	$4p_{1/2}$	385.9[a]
N III	$4p_{3/2}$	332.6[a]
N IV	$4d_{3/2}$	175.5[a]
N V	$4d_{5/2}$	175.5[a]
N VI	$4f_{5/2}$	—
N VII	$4f_{7/2}$	4.6
O I	5s	54.7[a]
O II	$5p_{1/2}$	31.8[a]
O III	$5p_{3/2}$	25.0[a]

Tin (50)

K	1s	29200
L I	2s	4465
L II	$2p_{1/2}$	4156
L III	$2p_{3/2}$	3929
M I	3s	884.7[b]
M II	$3p_{1/2}$	756.5[b]
M III	$3p_{3/2}$	714.6[b]
M IV	$3d_{3/2}$	493.2[b]
M V	$3d_{5/2}$	484.9[b]
N I	4s	137.1[b]
N II	$4p_{1/2}$	83.6[b,c]
N III	$4p_{3/2}$	83.6[b,c]
N IV	$4d_{3/2}$	24.9[b]
N V	$4d_{5/2}$	23.9[b]

Titanium (22)

K	1s	4966
L I	2s	560.9[b]
L II	$2p_{1/2}$	460.2[b]
L III	$2p_{3/2}$	453.8[b]
M I	3s	58.7[b]
M II	$3p_{1/2}$	32.6[b]
M III	$3p_{3/2}$	32.6[b]

Tungsten (74)

K	1s	69525
L I	2s	12100
L II	$2p_{1/2}$	11544
L III	$2p_{3/2}$	10207
M I	3s	2820
M II	$3p_{1/2}$	2575
M III	$3p_{3/2}$	2281
M IV	$3d_{3/2}$	1949

M V	$3d_{5/2}$	1809
N I	4s	594.1[b]
N II	$4p_{1/2}$	490.4[b]
N III	$4p_{3/2}$	423.6[b]
N IV	$4d_{3/2}$	255.9[b]
N V	$4d_{5/2}$	243.5[b]
N VI	$4f_{5/2}$	33.6[a]
N VII	$4f_{7/2}$	31.4[b]
O I	5s	75.6[b]
O II	$5p_{1/2}$	453[a,d]
O III	$5p_{3/2}$	36.8[b]

Uranium (92)

K	1s	115606
L I	2s	21757
L II	$2p_{1/2}$	20948
L III	$2p_{3/2}$	17166
M I	3s	5548
M II	$3p_{1/2}$	5182
M III	$3p_{3/2}$	4303
M IV	$3d_{3/2}$	3728
M V	$3d_{5/2}$	3552
N I	4s	1439[a,d]
N II	$4p_{1/2}$	1271[a,d]
N III	$4p_{3/2}$	1043[b]
N IV	$4d_{3/2}$	778.3[b]
N V	$4d_{5/2}$	736.2[b]
N VI	$4f_{5/2}$	388.2[a]
N VII	$4f_{7/2}$	377.4[b]
O I	5s	321[a,c,d]

O II	$5p_{1/2}$	257[a,c,d]
O III	$5p_{3/2}$	192[a,c,d]
O IV	$5d_{3/2}$	102.8[b]
O V	$5d_{5/2}$	94.2[b]
P I	6s	43.9[b]
P II	$6p_{1/2}$	26.8[b]
P III	$6p_{3/2}$	16.8[b]

Vanadium (23)

K	1s	5465
L I	2s	626.7[b]
L II	$2p_{1/2}$	519.8[b]
L III	$2p_{3/2}$	521.1[b]
M I	3s	66.3[b]
M II	$3p_{1/2}$	37.2[b]
M III	$3p_{3/2}$	37.2[b]

Xenon (54)

K	1s	34561
L I	2s	5453
L II	$2p_{1/2}$	5107
L III	$2p_{3/2}$	4786
M I	3s	1148.7[a]
M II	$3p_{1/2}$	1002.1[a]
M III	$3p_{3/2}$	940.6[a]
M IV	$3d_{3/2}$	689.0[a]
M V	$3d_{5/2}$	676.4[a]
N I	4s	213.2[a]
N II	$4p_{1/2}$	146.7
N III	$4p_{3/2}$	145.5[a]
N IV	$4d_{3/2}$	69.5[a]
N V	$4d_{5/2}$	67.5[a]
N VI	$4f_{5/2}$	—

N VII	$4f_{7/2}$	—
O I	5s	23.3[a]
O II	$5p_{1/2}$	13.4[a]
O III	$5p_{3/2}$	12.1[a]

Ytterbium (70)

K	1s	61332
L I	2s	10486
L II	$2p_{1/2}$	9978
L III	$2p_{3/2}$	8944
M I	3s	2398
M II	$3p_{1/2}$	2173
M III	$3p_{3/2}$	1950
M IV	$3d_{3/2}$	1576
M V	$3d_{5/2}$	1528
N I	4s	480.5[a]
N II	$4p_{1/2}$	388.7[a]
N III	$4p_{3/2}$	339.7[a]
N IV	$4d_{3/2}$	191.2[a]
N V	$4d_{5/2}$	182.4[a]
N VI	$4f_{5/2}$	2.5[a]
N VII	$4f_{7/2}$	1.3[a]
O I	5s	52.0[a]
O II	$5p_{1/2}$	30.3[a]
O III	$5p_{3/2}$	24.1[a]

Yttrium (39)

K	1s	17038
L I	2s	2373
L II	$2p_{1/2}$	2156
L III	$2p_{3/2}$	2080
M I	3s	392.0[a,d]
M II	$3p_{1/2}$	310.6[a]

M III	$3p_{3/2}$	298.8[a]
M IV	$3d_{3/2}$	157.7[b]
M V	$3d_{5/2}$	155.8[b]
N I	4s	43.8[a]
N II	$4p_{1/2}$	24.4[a]
N III	$4p_{3/2}$	23.1[a]

Zinc (30)

K	1s	9659
L I	2s	1196.2[a]
L II	$2p_{1/2}$	1044.9[a]
L III	$2p_{3/2}$	1021.8[a]
M I	3s	139.8[a]
M II	$3p_{1/2}$	91.4[a]
M III	$3p_{3/2}$	88.6[a]
M IV	$3d_{3/2}$	10.2[a]
M V	$3d_{5/2}$	10.1[a]

Zirconium (40)

K	1s	17998
L I	2s	2532
L II	$2p_{1/2}$	2307
L III	$2p_{3/2}$	2223
M I	3s	430.3[b]
M II	$3p_{1/2}$	343.5[b]
M III	$3p_{3/2}$	329.8[b]
M IV	$3d_{3/2}$	181.1[b]
M V	$3d_{5/2}$	178.8[b]
N I	4s	50.6[b]
N II	$4p_{1/2}$	28.5[b]
N III	$4p_{3/2}$	27.1[b]

[a] Reference 1.
[b] Reference 2 (remaining values from Reference 3).
[c] One-particle approximation not valid.
[d] Derived using energy differences from Reference 3.

NATURAL WIDTH OF X-RAY LINES

Natural widths of K X-ray lines in eV:

Element	Kα₁	Kα₂	Kβ₁	Kβ₃	Element	Kα₁	Kα₂	Kβ₁	Kβ₃
Ca	1.00	0.98			Ce	18.60	19.50	20.60	18.60
Ti	1.45	2.13			Nd	21.50	21.50	23.25	21.33
Cr	2.05	2.64			Sm	26.00	24.70	25.65	24.65
Fe	2.45	3.20			Gd	29.50	28.00	29.37	28.00
Ni	3.00	3.70			Dy	33.90	32.20	32.73	32.00
Zn	3.40	3.96			Er	35.00	35.50	36.20	35.70
Ge	3.75	4.18			Yb	38.80	40.60	41.43	41.15
Se	4.10	4.43			Hf	42.70	44.30	46.00	46.10
Kr	4.23	4.62			W	46.80	48.00	51.83	51.50
Sr	5.17	4.97			Os	49.00	49.40	55.90	55.95
Zr	5.70	5.25			Pt	54.10	54.30	59.98	62.13
Mo	6.82	6.80			Hg	64.75	68.20	65.75	68.95
Ru	7.41	7.96			Pb	67.10	72.30	72.20	73.80
Pd	8.80	9.20			Po	73.20	75.10	78.60	80.10
Cd	9.80	10.40			Rn	80.00	81.50	85.50	86.50
Sn	11.20	12.40	11.80	11.00	Ra	87.00	88.20	94.20	95.50
Te	12.80	14.20	13.30	13.10	Th	94.70	95.00	99.70	101.00
Xe	14.20	15.10	15.30	14.50	U	103.00	104.30	105.00	107.30
Ba	16.10	16.80	18.15	16.70					

From Salem, S. I. and Lee, P. L., *At. Data Nucl. Data Tables*, 18, 233, 1976.

Natural widths of L X-ray lines in eV:

Element	Lα₁	Lα₂	Lβ₁	Lβ₂	Lβ₃	Lβ₄	Lγ₁
Zr	1.68	1.52	1.87	5.13	5.50	5.60	3.34
Mo	1.86	1.80	2.03	5.30	5.90	5.78	3.76
Ru	2.03	1.98	2.18	5.45	6.35	5.96	4.15
Pd	2.21	2.16	2.36	5.63	6.80	6.18	4.50
Cd	2.43	2.40	2.54	5.82	7.23	6.28	4.83
Sn	2.62	2.62	2.75	6.10	7.70	6.60	5.23
Tc	2.88	2.88	2.96	6.25	8.22	6.82	5.60
Xe	3.15	3.15	3.20	6.43	8.70	7.15	5.95
Ba	3.39	3.45	3.45	6.70	9.20	7.42	6.35
Ce	3.70	3.78	3.73	6.86	9.70	7.82	6.75
Nd	3.93	4.08	4.00	7.18	10.30	8.15	7.16
Sm	4.13	4.50	4.33	7.42	10.80	8.60	7.50
Gd	4.46	4.90	4.63	7.70	11.20	9.08	7.83
Dy	4.81	5.35	5.03	7.90	11.50	9.60	8.30
Er	5.17	5.73	5.45	8.28	11.85	10.03	8.75
Yb	5.40	6.22	5.90	8.58	12.20	11.00	9.20
Hf	5.83	6.70	6.36	8.92	12.40	12.80	9.63
W	6.50	7.20	6.90	9.06	13.10	14.60	10.20
Os	7.04	7.70	7.42	9.60	14.60	16.50	10.65
Pt	7.60	8.28	8.00	9.95	16.10	18.00	11.20
Hg	8.10	8.80	8.70	10.40	17.40	19.70	11.80
Pb	8.82	9.35	9.35	10.75	18.65	21.30	12.30
Po	9.50	9.95	10.10	11.25	19.90	22.70	13.05
Rn	10.03	10.50	10.65	11.65	21.00	24.00	13.55
Ra	11.00	11.20	11.60	12.20	22.00	25.20	14.30
Th	11.90	11.80	12.40	12.80	22.85	26.35	15.00
U	12.40	12.40	13.50	13.30	23.70	27.50	15.70
Pu	13.20	13.00	14.10	13.90	24.10	28.30	16.40
Cm	14.80	13.60	15.70	14.60	25.00	29.40	17.10

PHOTON ATTENUATION COEFFICIENTS

Martin J. Berger and John H. Hubbell

This table gives mass attenuation coefficients for photons for all elements at energies between 1 keV (soft x-rays) and 1 GeV (hard gamma rays). The mass attenuation coefficient μ describes the attenuation of radiation as it passes through matter by the relation

$$I(x)/I_o = e^{-\mu\rho x}$$

where I_o is the initial intensity, $I(x)$ the intensity after path length x, and ρ is the mass density of the element in question. To a high approximation the mass attenuation coefficient is additive for the elements present, independent of the way in which they are bound in chemical compounds.

The power of ten is indicated beside each number in the table; i.e., 7.41 + 03 means 7.41×10^3. A vertical line between two columns indicates that an absorption edge lies between those energy values. The various edges are labeled at the bottom of the table.

The attenuation coefficients were calculated with the computer program XCOM (Reference 1), which uses a cross-section database compiled at the Photon and Charged Particle Data Center at the National Institute of Standards and Technology. Their accuracy has been confirmed at all energies by extensive comparisons with experimental attenuation coefficients. Such comparisons for X-ray energies up to 100 keV can be found in Reference 2.

References

1. Berger, M. J. and Hubbell, J. H., *National Bureau of Standards Report* NBSIR-87-3597, 1987.
2. Saloman, E. B., Hubbell, J. H., and Scofield, J. H., *Atomic Data and Nuclear Data Tables*, 38, 1, 1988.

Mass attenuation coefficient, cm²/g
Photon energy, MeV

	Atomic no.	0.001	0.002	0.005	0.01	0.02	0.05	0.1	0.2	0.5
H	1	7.21 + 00	1.06 + 00	4.19-01	3.85-01	3.69-01	3.36-01	2.94-01	2.43-01	1.73-01
He	2	6.08 + 01	6.86 + 00	5.77-01	2.48-01	1.96-01	1.70-01	1.49-01	1.22-01	8.71-02
Li	3	2.34 + 02	2.71 + 01	1.62 + 00	3.40-01	1.86-01	1.49-01	1.29-01	1.06-01	7.53-02
Be	4	6.04 + 02	7.47 + 01	4.37 + 00	6.47-01	2.25-01	1.55-01	1.33-01	1.09-01	7.74-02
B	5	1.23 + 03	1.60 + 02	9.68 + 00	1.25 + 00	3.01-01	1.66-01	1.39-01	1.14-01	8.07-02
C	6	2.21 + 03	3.03 + 02	1.91 + 01	2.37 + 00	4.42-01	1.87-01	1.51-01	1.23-01	8.72-02
N	7	3.31 + 03	4.77 + 02	3.14 + 01	3.88 + 00	6.18-01	1.98-01	1.53-01	1.23-01	8.72-02
O	8	4.59 + 03	6.95 + 02	4.79 + 01	5.95 + 00	8.65-01	2.13-01	1.55-01	1.24-01	8.73-02
F	9	5.65 + 03	9.05 + 02	6.51 + 01	8.21 + 00	1.13 + 00	2.21-01	1.50-01	1.18-01	8.27-02
Ne	10	7.41 + 03	1.24 + 03	9.34 + 01	1.20 + 01	1.61 + 00	2.58-01	1.60-01	1.24-01	8.66-02
Na	11	6.54 + 02	1.52 + 03	1.19 + 02	1.56 + 01	2.06 + 00	2.80-01	1.59-01	1.20-01	8.37-02
Mg	12	9.22 + 02	1.93 + 03	1.58 + 02	2.11 + 01	2.76 + 00	3.29-01	1.69-01	1.24-01	8.65-02
Al	13	1.19 + 03	2.26 + 03	1.93 + 02	2.62 + 01	3.44 + 00	3.68-01	1.70-01	1.22-01	8.44-02
Si	14	1.57 + 03	2.78 + 03	2.45 + 02	3.39 + 01	4.46 + 00	4.38-01	1.84-01	1.28-01	8.75-02
P	15	1.91 + 03	3.02 + 02	2.86 + 02	4.04 + 01	5.35 + 00	4.92-01	1.87-01	1.25-01	8.51-02
S	16	2.43 + 03	3.85 + 02	3.49 + 02	5.01 + 01	6.71 + 00	5.85-01	2.02-01	1.30-01	8.78-02
Cl	17	2.83 + 03	4.52 + 02	3.90 + 02	5.73 + 01	7.74 + 00	6.48-01	2.05-01	1.27-01	8.45-02
Ar	18	3.18 + 03	5.12 + 02	4.23 + 02	6.32 + 01	8.63 + 00	7.01-01	2.04-01	1.20-01	7.96-02
K	19	4.06 + 03	6.59 + 02	5.19 + 02	7.91 + 01	1.09 + 01	8.68-01	2.34-01	1.32-01	8.60-02
Ca	20	4.87 + 03	8.00 + 02	6.03 + 02	9.34 + 01	1.31 + 01	1.02 + 00	2.57-01	1.38-01	8.85-02
Sc	21	5.24 + 03	8.70 + 02	6.31 + 02	9.95 + 01	1.41 + 01	1.09 + 00	2.58-01	1.31-01	8.31-02
Ti	22	5.87 + 03	9.86 + 02	6.84 + 02	1.11 + 02	1.59 + 01	1.21 + 00	2.72-01	1.31-01	8.19-02
V	23	6.50 + 03	1.11 + 03	9.29 + 01	1.22 + 02	1.77 + 01	1.35 + 00	2.88-01	1.32-01	8.07-02
Cr	24	7.40 + 03	1.28 + 03	1.08 + 02	1.39 + 02	2.04 + 01	1.55 + 00	3.17-01	1.38-01	8.28-02
Mn	25	8.09 + 03	1.42 + 03	1.21 + 02	1.51 + 02	2.25 + 01	1.71 + 00	3.37-01	1.39-01	8.19-02
Fe	26	9.09 + 03	1.63 + 03	1.40 + 02	1.71 + 02	2.57 + 01	1.96 + 00	3.72-01	1.46-01	8.41-02
Co	27	9.80 + 03	1.78 + 03	1.54 + 02	1.84 + 02	2.80 + 01	2.14 + 00	3.95-01	1.48-01	8.32-02
Ni	28	9.86 + 03	2.05 + 03	1.79 + 02	2.09 + 02	3.22 + 01	2.47 + 00	4.44-01	1.58-01	8.70-02
Cu	29	1.06 + 04	2.15 + 03	1.90 + 02	2.16 + 02	3.38 + 01	2.61 + 00	4.58-01	1.56-01	8.36-02
Zn	30	1.55 + 03	2.37 + 03	2.12 + 02	2.33 + 02	3.72 + 01	2.89 + 00	4.97-01	1.62-01	8.45-02
Ga	31	1.70 + 03	2.52 + 03	2.27 + 02	3.42 + 01	3.93 + 01	3.08 + 00	5.20-01	1.62-01	8.24-02
Ge	32	1.89 + 03	2.71 + 03	2.47 + 02	3.74 + 01	4.22 + 01	3.34 + 00	5.55-01	1.66-01	8.21-02
As	33	2.12 + 03	2.93 + 03	2.71 + 02	4.12 + 01	4.56 + 01	3.63 + 00	5.97-01	1.72-01	8.26-02
Se	34	2.32 + 03	3.10 + 03	2.90 + 02	4.41 + 01	4.82 + 01	3.86 + 00	6.28-01	1.74-01	8.13-02
Br	35	2.62 + 03	3.41 + 03	3.21 + 02	4.91 + 01	5.27 + 01	4.26 + 00	6.86-01	1.84-01	8.33-02
Kr	36	2.85 + 03	3.60 + 03	3.43 + 02	5.26 + 01	5.55 + 01	4.52 + 00	7.22-01	1.87-01	8.23-02
Rb	37	3.17 + 03	3.41 + 03	3.74 + 02	5.77 + 01	5.98 + 01	4.92 + 00	7.80-01	1.96-01	8.36-02
Sr	38	3.49 + 03	2.59 + 03	4.06 + 02	6.27 + 01	6.39 + 01	5.31 + 00	8.37-01	2.04-01	8.44-02
Y	39	3.86 + 03	7.42 + 02	4.42 + 02	6.87 + 01	6.86 + 01	5.76 + 00	9.05-01	2.15-01	8.61-02
Zr	40	4.21 + 03	8.12 + 02	4.76 + 02	7.42 + 01	7.24 + 01	6.17 + 00	9.66-01	2.24-01	8.69-02
Nb	41	4.60 + 03	8.89 + 02	5.13 + 02	8.04 + 01	7.71 + 01	6.64 + 00	1.04 + 00	2.34-01	8.83-02
Mo	42	4.94 + 03	9.60 + 02	5.45 + 02	8.58 + 01	1.31 + 01	7.04 + 00	1.10 + 00	2.42-01	8.85-02
Tc	43	5.36 + 03	1.04 + 03	5.84 + 02	9.23 + 01	1.41 + 01	7.52 + 00	1.17 + 00	2.53-01	8.97-02
Ru	44	5.72 + 03	1.12 + 03	6.17 + 02	9.80 + 01	1.50 + 01	7.92 + 00	1.23 + 00	2.62-01	8.99-02
Rh	45	6.17 + 03	1.21 + 03	6.59 + 02	1.05 + 02	1.61 + 01	8.45 + 00	1.31 + 00	2.74-01	9.13-02
Pd	46	6.54 + 03	1.29 + 03	6.91 + 02	1.11 + 02	1.70 + 01	8.85 + 00	1.38 + 00	2.83-01	9.13-02
Ag	47	7.04 + 03	1.40 + 03	7.39 + 02	1.19 + 02	1.84 + 01	9.45 + 00	1.47 + 00	2.97-01	9.32-02
Cd	48	7.35 + 03	1.47 + 03	7.69 + 02	1.24 + 02	1.92 + 01	9.78 + 00	1.52 + 00	3.04-01	9.25-02
In	49	7.81 + 03	1.58 + 03	8.13 + 02	1.32 + 02	2.04 + 01	1.03 + 01	1.61 + 00	3.17-01	9.37-02
Sn	50	8.16 + 03	1.66 + 03	8.47 + 02	1.38 + 02	2.15 + 01	1.07 + 01	1.68 + 00	3.26-01	9.37-02

L₃ L₁
L₂

K EDGE

Mass attenuation coefficient, cm²/g
Photon energy, MeV

	Atomic no.	1.0	2.0	5.0	10.0	20.0	50.0	100.0	500.0	1000.0
H	1	1.26-01	8.77-02	5.05-02	3.25-02	2.15-02	1.42-02	1.19-02	1.14-02	1.16-02
He	2	6.36-02	4.42-02	2.58-02	1.70-02	1.18-02	8.61-03	7.78-03	7.79-03	7.95-03
Li	3	5.50-02	3.83-02	2.26-02	1.53-02	1.11-02	8.68-03	8.21-03	8.61-03	8.87-03
Be	4	5.65-02	3.94-02	2.35-02	1.63-02	1.23-02	1.02-02	9.94-03	1.08-02	1.12-02
B	5	5.89-02	4.11-02	2.48-02	1.76-02	1.37-02	1.19-02	1.19-02	1.32-02	1.37-02
C	6	6.36-02	4.44-02	2.71-02	1.96-02	1.58-02	1.43-02	1.46-02	1.64-02	1.70-02
N	7	6.36-02	4.45-02	2.74-02	2.02-02	1.67-02	1.57-02	1.63-02	1.85-02	1.92-02
O	8	6.37-02	4.46-02	2.78-02	2.09-02	1.77-02	1.71-02	1.79-02	2.06-02	2.13-02
F	9	6.04-02	4.23-02	2.66-02	2.04-02	1.77-02	1.75-02	1.86-02	2.14-02	2.21-02
Ne	10	6.32-02	4.43-02	2.82-02	2.20-02	1.95-02	1.96-02	2.11-02	2.43-02	2.51-02
Na	11	6.10-02	4.28-02	2.75-02	2.18-02	1.97-02	2.03-02	2.19-02	2.53-02	2.62-02
Mg	12	6.30-02	4.43-02	2.87-02	2.31-02	2.13-02	2.23-02	2.42-02	2.81-02	2.90-02
Al	13	6.15-02	4.32-02	2.84-02	2.32-02	2.17-02	2.31-02	2.52-02	2.93-02	3.03-02
Si	14	6.36-02	4.48-02	2.97-02	2.46-02	2.34-02	2.52-02	2.76-02	3.23-02	3.34-02
P	15	6.18-02	4.36-02	2.91-02	2.45-02	2.36-02	2.58-02	2.84-02	3.33-02	3.45-02
S	16	6.37-02	4.50-02	3.04-02	2.59-02	2.53-02	2.79-02	3.08-02	3.62-02	3.75-02
Cl	17	6.13-02	4.33-02	2.95-02	2.55-02	2.52-02	2.81-02	3.11-02	3.67-02	3.80-02
Ar	18	5.76-02	4.07-02	2.80-02	2.45-02	2.45-02	2.76-02	3.07-02	3.62-02	3.75-02
K	19	6.22-02	4.40-02	3.05-02	2.70-02	2.74-02	3.11-02	3.46-02	4.09-02	4.24-02
Ca	20	6.39-02	4.52-02	3.17-02	2.84-02	2.90-02	3.32-02	3.71-02	4.40-02	4.56-02
Sc	21	5.98-02	4.24-02	3.00-02	2.72-02	2.80-02	3.23-02	3.62-02	4.30-02	4.45-02
Ti	22	5.89-02	4.18-02	2.98-02	2.73-02	2.84-02	3.30-02	3.71-02	4.40-02	4.56-02
V	23	5.79-02	4.11-02	2.96-02	2.74-02	2.88-02	3.36-02	3.78-02	4.49-02	4.65-02
Cr	24	5.93-02	4.21-02	3.06-02	2.86-02	3.03-02	3.56-02	4.01-02	4.76-02	4.93-02
Mn	25	5.85-02	4.16-02	3.04-02	2.87-02	3.07-02	3.63-02	4.09-02	4.86-02	5.04-02
Fe	26	5.99-02	4.26-02	3.15-02	2.99-02	3.22-02	3.83-02	4.33-02	5.15-02	5.33-02
Co	27	5.91-02	4.20-02	3.13-02	3.00-02	3.26-02	3.88-02	4.40-02	5.23-02	5.41-02
Ni	28	6.16-02	4.39-02	3.29-02	3.18-02	3.48-02	4.17-02	4.73-02	5.61-02	5.81-02
Cu	29	5.90-02	4.20-02	3.18-02	3.10-02	3.41-02	4.10-02	4.66-02	5.53-02	5.72-02
Zn	30	5.94-02	4.24-02	3.22-02	3.18-02	3.51-02	4.24-02	4.82-02	5.72-02	5.91-02
Ga	31	5.77-02	4.11-02	3.16-02	3.13-02	3.48-02	4.22-02	4.80-02	5.70-02	5.89-02
Ge	32	5.73-02	4.09-02	3.16-02	3.16-02	3.53-02	4.30-02	4.89-02	5.80-02	6.00-02
As	33	5.73-02	4.09-02	3.19-02	3.21-02	3.60-02	4.40-02	5.01-02	5.95-02	6.15-02
Se	34	5.62-02	4.01-02	3.14-02	3.19-02	3.60-02	4.41-02	5.03-02	5.97-02	6.17-02
Br	35	5.73-02	4.09-02	3.23-02	3.29-02	3.74-02	4.60-02	5.24-02	6.22-02	6.43-02
Kr	36	5.63-02	4.02-02	3.20-02	3.28-02	3.74-02	4.61-02	5.26-02	6.25-02	6.46-02
Rb	37	5.69-02	4.06-02	3.25-02	3.36-02	3.85-02	4.75-02	5.43-02	6.45-02	6.67-02
Sr	38	5.71-02	4.08-02	3.29-02	3.41-02	3.93-02	4.87-02	5.56-02	6.61-02	6.83-02
Y	39	5.80-02	4.14-02	3.35-02	3.50-02	4.05-02	5.03-02	5.75-02	6.83-02	7.06-02
Zr	40	5.81-02	4.15-02	3.38-02	3.55-02	4.12-02	5.13-02	5.87-02	6.98-02	7.22-02
Nb	41	5.87-02	4.18-02	3.44-02	3.63-02	4.22-02	5.27-02	6.03-02	7.17-02	7.42-02
Mo	42	5.84-02	4.16-02	3.44-02	3.65-02	4.26-02	5.33-02	6.10-02	7.26-02	7.51-02
Tc	43	5.88-02	4.19-02	3.48-02	3.71-02	4.35-02	5.45-02	6.24-02	7.43-02	7.68-02
Ru	44	5.85-02	4.16-02	3.48-02	3.73-02	4.39-02	5.50-02	6.30-02	7.51-02	7.77-02
Rh	45	5.89-02	4.20-02	3.53-02	3.80-02	4.48-02	5.63-02	6.45-02	7.69-02	7.94-02
Pd	46	5.85-02	4.16-02	3.52-02	3.80-02	4.50-02	5.66-02	6.49-02	7.73-02	8.00-02
Ag	47	5.92-02	4.21-02	3.58-02	3.88-02	4.61-02	5.81-02	6.67-02	7.93-02	8.20-02
Cd	48	5.83-02	4.14-02	3.54-02	3.85-02	4.59-02	5.79-02	6.64-02	7.91-02	8.18-02
In	49	5.85-02	4.15-02	3.56-02	3.90-02	4.65-02	5.88-02	6.75-02	8.04-02	8.32-02
Sn	50	5.80-02	4.11-02	3.55-02	3.90-02	4.66-02	5.90-02	6.78-02	8.07-02	8.35-02

Mass attenuation coefficient, cm²/g
Photon energy, MeV

	Atomic no.	0.001	0.002	0.005	0.01	0.02	0.05	0.1	0.2	0.5
Sb	51	8.58 + 03	1.77 + 03	8.85 + 02	1.46 + 02	2.27 + 01	1.12 + 01	1.76 + 00	3.38-01	9.45-02
Te	52	8.43 + 03	1.83 + 03	9.01 + 02	1.50 + 02	2.34 + 01	1.14 + 01	1.80 + 00	3.43-01	9.33-02
I	53	9.10 + 03	2.00 + 03	8.43 + 02	1.63 + 02	2.54 + 01	1.23 + 01	1.94 + 00	3.66-01	9.70-02
Xe	54	9.41 + 03	2.09 + 03	6.39 + 02	1.69 + 02	2.65 + 01	1.27 + 01	2.01 + 00	3.76-01	9.70-02
Cs	55	9.37 + 03	2.23 + 03	2.30 + 02	1.79 + 02	2.82 + 01	1.34 + 01	2.12 + 00	3.94-01	9.91-02
ZB	56	8.54 + 03	2.32 + 03	2.41 + 02	1.86 + 02	2.94 + 01	1.38 + 01	2.20 + 00	4.05-01	9.92-02
La	57	9.09 + 03	2.46 + 03	2.58 + 02	1.97 + 02	3.12 + 01	1.45 + 01	2.32 + 00	4.24-01	1.01-01
Ce	58	9.71 + 03	2.61 + 03	2.74 + 02	2.08 + 02	3.31 + 01	1.52 + 01	2.45 + 00	4.45-01	1.04-01
Pr	59	1.06 + 04	2.77 + 03	2.92 + 02	2.21 + 02	3.53 + 01	1.60 + 01	2.59 + 00	4.69-01	1.07-01
Nd	60	6.63 + 03	2.88 + 03	3.06 + 02	2.30 + 02	3.68 + 01	1.65 + 01	2.69 + 00	4.84-01	1.08-01
Pm	61	2.06 + 03	3.05 + 03	3.26 + 02	2.44 + 02	3.92 + 01	1.73 + 01	2.84 + 00	5.10-01	1.12-01
Sm	62	2.11 + 03	3.12 + 03	3.36 + 02	2.50 + 02	4.03 + 01	1.77 + 01	2.90 + 00	5.19-01	1.11-01
Eu	63	2.22 + 03	3.28 + 03	3.54 + 02	2.63 + 02	4.24 + 01	1.85 + 01	3.04 + 00	5.43-01	1.14-01
Gd	64	2.29 + 03	3.36 + 03	3.65 + 02	2.69 + 02	4.36 + 01	3.86 + 00	3.11 + 00	5.54-01	1.14-01
Tb	65	2.40 + 03	3.51 + 03	3.84 + 02	2.82 + 02	4.59 + 01	4.06 + 00	3.25 + 00	5.77-01	1.17-01
Dy	66	2.49 + 03	3.47 + 03	3.99 + 02	2.90 + 02	4.76 + 01	4.23 + 00	3.36 + 00	5.95-01	1.18-01
Ho	67	2.62 + 03	3.59 + 03	4.17 + 02	3.01 + 02	4.98 + 01	4.43 + 00	3.49 + 00	6.18-01	1.20-01
Er	68	2.75 + 03	3.52 + 03	4.36 + 02	3.13 + 02	5.20 + 01	4.63 + 00	3.63 + 00	6.41-01	1.23-01
Tm	69	2.90 + 03	3.69 + 03	4.57 + 02	2.83 + 02	5.45 + 01	4.87 + 00	3.78 + 00	6.68-01	1.26-01
Yb	70	3.02 + 03	3.80 + 03	4.72 + 02	2.94 + 02	5.63 + 01	5.04 + 00	3.88 + 00	6.86-01	1.27-01
Lu	71	3.19 + 03	3.45 + 03	4.94 + 02	2.21 + 02	5.88 + 01	5.28 + 00	4.03 + 00	7.13-01	1.30-01
Hf	72	3.34 + 03	3.60 + 03	5.11 + 02	2.30 + 02	6.09 + 01	5.48 + 00	4.15 + 00	7.34-01	1.32-01
Ta	73	3.51 + 03	3.77 + 03	5.33 + 02	2.38 + 02	6.33 + 01	5.72 + 00	4.30 + 00	7.60-01	1.35-01
W	74	3.68 + 03	3.92 + 03	5.53 + 02	9.69 + 01	6.57 + 01	5.95 + 00	4.44 + 00	7.84-01	1.38-01
Re	75	3.87 + 03	3.77 + 03	5.76 + 02	1.01 + 02	6.84 + 01	6.21 + 00	4.59 + 00	8.12-01	1.41-01
Os	76	4.03 + 03	2.22 + 03	5.93 + 02	1.04 + 02	7.04 + 01	6.41 + 00	4.70 + 00	8.33-01	1.43-01
Ir	77	4.24 + 03	1.03 + 03	6.18 + 02	1.09 + 02	7.32 + 01	6.69 + 00	4.86 + 00	8.63-01	1.46-01
Pt	78	4.43 + 03	1.08 + 03	6.40 + 02	1.13 + 02	7.57 + 01	6.95 + 00	4.99 + 00	8.90-01	1.49-01
Au	79	4.65 + 03	1.14 + 03	6.66 + 02	1.18 + 02	7.88 + 01	7.26 + 00	5.16 + 00	9.22-01	1.53-01
Hg	80	4.83 + 03	1.18 + 03	6.87 + 02	1.22 + 02	8.12 + 01	7.50 + 00	5.28 + 00	9.46-01	1.56-01
Tl	81	5.01 + 03	1.23 + 03	7.07 + 02	1.26 + 02	8.36 + 01	7.75 + 00	5.40 + 00	9.69-01	1.58-01
Pb	82	5.21 + 03	1.29 + 03	7.30 + 02	1.31 + 02	8.64 + 01	8.04 + 00	5.55 + 00	9.99-01	1.61-01
Bi	83	5.44 + 03	1.35 + 03	7.58 + 02	1.36 + 02	8.95 + 01	8.38 + 00	5.74 + 00	1.03 + 00	1.66-01
Po	84	5.72 + 03	1.42 + 03	7.93 + 02	1.43 + 02	9.35 + 01	8.80 + 00	5.99 + 00	1.08 + 00	1.71-01
At	85	5.87 + 03	1.49 + 03	8.25 + 02	1.49 + 02	9.70 + 01	9.19 + 00	6.17 + 00	1.12 + 00	1.77-01
Rn	86	5.83 + 03	1.49 + 03	8.16 + 02	1.48 + 02	9.56 + 01	9.12 + 00	6.09 + 00	1.10 + 00	1.73-01
Fr	87	6.08 + 03	1.56 + 03	8.49 + 02	1.54 + 02	9.93 + 01	9.52 + 00	1.66 + 00	1.14 + 00	1.78-01
Ra	88	6.20 + 03	1.62 + 03	8.74 + 02	1.59 + 02	1.02 + 02	9.85 + 00	1.71 + 00	1.17 + 00	1.82-01
Ac	89	6.47 + 03	1.70 + 03	8.69 + 02	1.65 + 02	1.06 + 02	1.03 + 01	1.79 + 00	1.21 + 00	1.87-01
Th	90	6.61 + 03	1.74 + 03	8.88 + 02	1.69 + 02	9.37 + 01	1.05 + 01	1.83 + 00	1.23 + 00	1.90-01
Pa	91	6.53 + 03	1.83 + 03	8.76 + 02	1.77 + 02	7.03 + 01	1.10 + 01	1.92 + 00	1.29 + 00	1.97-01
U	92	6.63 + 03	1.86 + 03	8.89 + 02	1.79 + 02	7.11 + 01	1.12 + 01	1.95 + 00	1.30 + 00	1.98-01
Np	93	6.95 + 03	1.96 + 03	9.32 + 02	1.87 + 02	7.45 + 01	1.18 + 01	2.05 + 00	1.35 + 00	2.05-01
Pu	94	7.19 + 03	2.04 + 03	9.65 + 02	1.94 + 02	7.71 + 01	1.22 + 01	2.13 + 00	1.39 + 00	2.10-01
Am	95	7.37 + 03	2.10 + 03	9.90 + 02	1.98 + 02	7.93 + 01	1.25 + 01	2.19 + 00	1.42 + 00	2.14-01
Cm	96	7.54 + 03	2.15 + 03	1.02 + 03	2.03 + 02	8.14 + 01	1.28 + 01	2.25 + 00	1.44 + 00	2.18-01
Bk	97	7.84 + 03	2.25 + 03	1.06 + 03	2.10 + 02	8.39 + 01	1.34 + 01	2.35 + 00	1.50 + 00	2.25-01
Cf	98	7.89 + 03	2.31 + 03	9.27 + 02	2.15 + 02	8.58 + 01	1.37 + 01	2.41 + 00	1.52 + 00	2.29-01
Es	99	7.79 + 03	2.40 + 03	9.59 + 02	2.22 + 02	4.01 + 01	1.42 + 01	2.51 + 00	1.57 + 00	2.36-01
Fm	100	7.13 + 03	2.46 + 03	9.77 + 02	2.26 + 02	4.09 + 01	1.45 + 01	2.57 + 00	1.59 + 00	2.39-01

N₅
N₄
N₃
N₂
N₁

M₅
M₄

M₃
M₂
M₁

L₃
L₂
L₁

K EDGE

Mass attenuation coefficient, cm²/g

Photon energy, MeV

	Atomic no.	1.0	2.0	5.0	10.0	20.0	50.0	100.0	500.0	1000.0
Sb	51	5.80-02	4.10-02	3.56-02	3.92-02	4.70-02	5.96-02	6.85-02	8.16-02	8.44-02
Te	52	5.67-02	4.01-02	3.49-02	3.86-02	4.64-02	5.89-02	6.77-02	8.07-02	8.35-02
I	53	5.84-02	4.12-02	3.61-02	4.00-02	4.82-02	6.13-02	7.04-02	8.40-02	8.69-02
Xe	54	5.78-02	4.08-02	3.58-02	3.99-02	4.82-02	6.12-02	7.04-02	8.40-02	8.69-02
Cs	55	5.85-02	4.12-02	3.64-02	4.06-02	4.91-02	6.25-02	7.19-02	8.58-02	8.88-02
ZB	56	5.80-02	4.08-02	3.61-02	4.04-02	4.90-02	6.25-02	7.19-02	8.58-02	8.88-02
La	57	5.88-02	4.12-02	3.66-02	4.11-02	5.00-02	6.37-02	7.34-02	8.76-02	9.06-02
Ce	58	5.96-02	4.18-02	3.73-02	4.19-02	5.10-02	6.52-02	7.50-02	8.96-02	9.27-02
Pr	59	6.07-02	4.24-02	3.80-02	4.29-02	5.23-02	6.68-02	7.69-02	9.19-02	9.50-02
Nd	60	6.07-02	4.24-02	3.81-02	4.30-02	5.26-02	6.72-02	7.74-02	9.25-02	9.56-02
Pm	61	6.19-02	4.31-02	3.88-02	4.40-02	5.38-02	6.89-02	7.94-02	9.48-02	9.81-02
Sm	62	6.11-02	4.24-02	3.83-02	4.35-02	5.34-02	6.84-02	7.88-02	9.41-02	9.73-02
Eu	63	6.19-02	4.28-02	3.88-02	4.42-02	5.42-02	6.96-02	8.02-02	9.57-02	9.90-02
Gd	64	6.12-02	4.23-02	3.84-02	4.38-02	5.38-02	6.91-02	7.97-02	9.51-02	9.83-02
Tb	65	6.20-02	4.27-02	3.89-02	4.45-02	5.47-02	7.03-02	8.11-02	9.67-02	1.00-01
Dy	66	6.20-02	4.26-02	3.90-02	4.46-02	5.49-02	7.06-02	8.15-02	9.72-02	1.00-01
Ho	67	6.26-02	4.29-02	3.93-02	4.50-02	5.55-02	7.14-02	8.24-02	9.83-02	1.02-01
Er	68	6.32-02	4.32-02	3.96-02	4.55-02	5.61-02	7.23-02	8.34-02	9.95-02	1.03-01
Tm	69	6.40-02	4.36-02	4.01-02	4.61-02	5.70-02	7.35-02	8.48-02	1.01-01	1.04-01
Yb	70	6.40-02	4.35-02	4.00-02	4.61-02	5.70-02	7.35-02	8.49-02	1.01-01	1.04-01
Lu	71	6.48-02	4.39-02	4.05-02	4.66-02	5.77-02	7.45-02	8.60-02	1.02-01	1.06-01
Hf	72	6.50-02	4.39-02	4.05-02	4.68-02	5.80-02	7.48-02	8.64-02	1.03-01	1.06-01
Ta	73	6.57-02	4.41-02	4.08-02	4.72-02	5.85-02	7.56-02	8.73-02	1.04-01	1.07-01
W	74	6.62-02	4.43-02	4.10-02	4.75-02	5.89-02	7.62-02	8.80-02	1.05-01	1.08-01
Re	75	6.69-02	4.46-02	4.14-02	4.79-02	5.95-02	7.70-02	8.89-02	1.06-01	1.09-01
Os	76	6.71-02	4.46-02	4.13-02	4.79-02	5.96-02	7.71-02	8.90-02	1.06-01	1.10-01
Ir	77	6.79-02	4.50-02	4.17-02	4.84-02	6.02-02	7.80-02	9.01-02	1.07-01	1.11-01
Pt	78	6.86-02	4.52-02	4.20-02	4.87-02	6.06-02	7.86-02	9.08-02	1.08-01	1.12-01
Au	79	6.95-02	4.57-02	4.24-02	4.93-02	6.14-02	7.95-02	9.19-02	1.09-01	1.13-01
Hg	80	6.99-02	4.57-02	4.25-02	4.94-02	6.15-02	7.98-02	9.22-02	1.10-01	1.13-01
Tl	81	7.03-02	4.58-02	4.25-02	4.94-02	6.16-02	8.00-02	9.24-02	1.10-01	1.14-01
Pb	82	7.10-02	4.61-02	4.27-02	4.97-02	6.21-02	8.06-02	9.31-02	1.11-01	1.15-01
Bi	83	7.21-02	4.66-02	4.32-02	5.03-02	6.28-02	8.15-02	9.42-02	1.12-01	1.16-01
Po	84	7.39-02	4.75-02	4.40-02	5.12-02	6.40-02	8.32-02	9.61-02	1.15-01	1.18-01
At	85	7.54-02	4.82-02	4.46-02	5.20-02	6.49-02	8.44-02	9.76-02	1.16-01	1.20-01
Rn	86	7.30-02	4.65-02	4.30-02	5.01-02	6.26-02	8.14-02	9.42-02	1.12-01	1.16-01
Fr	87	7.45-02	4.72-02	4.36-02	5.08-02	6.35-02	8.26-02	9.56-02	1.14-01	1.18-01
Ra	88	7.53-02	4.75-02	4.38-02	5.10-02	6.38-02	8.31-02	9.61-02	1.15-01	1.19-01
Ac	89	7.69-02	4.82-02	4.44-02	5.17-02	6.47-02	8.43-02	9.75-02	1.16-01	1.20-01
Th	90	7.71-02	4.81-02	4.42-02	5.15-02	6.45-02	8.40-02	9.72-02	1.16-01	1.20-01
Pa	91	7.94-02	4.93-02	4.52-02	5.26-02	6.59-02	8.60-02	9.95-02	1.19-01	1.23-01
U	92	7.90-02	4.88-02	4.46-02	5.19-02	6.51-02	8.49-02	9.83-02	1.17-01	1.21-01
Np	93	8.13-02	4.99-02	4.56-02	5.30-02	6.65-02	8.68-02	1.01-01	1.20-01	1.24-01
Pu	94	8.26-02	5.05-02	4.60-02	5.34-02	6.71-02	8.76-02	1.01-01	1.21-01	1.25-01
Am	95	8.33-02	5.06-02	4.60-02	5.34-02	6.70-02	8.77-02	1.02-01	1.21-01	1.25-01
Cm	96	8.41-02	5.08-02	4.60-02	5.34-02	6.70-02	8.77-02	1.02-01	1.21-01	1.26-01
Bk	97	8.62-02	5.18-02	4.68-02	5.42-02	6.81-02	8.92-02	1.03-01	1.24-01	1.28-01
Cf	98	8.70-02	5.20-02	4.68-02	5.42-02	6.81-02	8.92-02	1.04-01	1.24-01	1.28-01
Es	99	8.89-02	5.28-02	4.74-02	5.48-02	6.89-02	9.04-02	1.05-01	1.25-01	1.29-01
Fm	100	8.94-02	5.28-02	4.72-02	5.45-02	6.86-02	9.00-02	1.05-01	1.25-01	1.29-01

CLASSIFICATION OF ELECTROMAGNETIC RADIATION

Hans Dolezalek

Basic Conversions:

$$c = \lambda\nu = \nu/k$$
$$\nu = c/\lambda = ck$$

$$\lambda = c/\nu = 1/k$$
$$k = \nu/c = 1/\lambda$$

c = speed of light = 2.99792458×10^8 m/s

Frequency (ν)	Wavelength (λ)	Wave number (k)	Names of bands	Approximate photon energies
$3 \times 10^0 - 3 \times 10^1$ Hz 3 – 30 Hz	$10^8 - 10^7$ m 100 – 10 Mm	$10^{-8} - 10^{-7}$ m^{-1} 10 – 100 Gm^{-1}	ELF-(ELF 1), ITU band no. 1	
$3 \times 10^1 - 3 \times 10^2$ Hz 30 – 300 Hz	$10^7 - 10^6$ m 10 – 1 Mm	$10^{-7} - 10^{-6}$ m^{-1} 100 Gm^{-1} – 1Mm^{-1}	SLF-(ELF 2), ITU band no. 2, megameter waves	
$3 \times 10^2 - 3 \times 10^3$ Hz 300 Hz – 3 kHz	$10^6 - 10^5$ m 1 Mm – 100 km	$10^{-6} - 10^{-5}$ m^{-1} 1 – 10 Mm^{-1}	ULF-(ELF 3), ITU band no. 3	
$3 \times 10^3 - 3 \times 10^4$ Hz 3 – 30 kHz	$10^5 - 10^4$ m 100 – 10 km	$10^{-5} - 10^{-4}$ m^{-1} 10 – 100 Mm^{-1}	VLF, ITU band no. 4, myriameter waves	
$3 \times 10^4 - 3 \times 10^5$ Hz 30 – 300 kHz	$10^4 - 10^3$ m 10 – 1 km	$10^{-4} - 10^{-3}$ m^{-1} 100 Mm^{-1} – 1 km^{-1}	LF, ITU band no. 5, kilometer waves	
$3 \times 10^5 - 3 \times 10^6$ Hz 300 kHz – 3 MHz	$10^3 - 10^2$ m 1 km – 100 m	$10^{-3} - 10^{-2}$ m^{-1} 1 – 10 km^{-1}	MF, ITU band no. 6, hectometer waves	
$3 \times 10^6 - 3 \times 10^7$ Hz 3 – 30 MHz	$10^2 - 10^1$ m 100 – 10 m	$10^{-2} - 10^{-1}$ m^{-1} 10 – 100 km^{-1}	HF, ITU band no. 7, decameter waves	
$3 \times 10^7 - 3 \times 10^8$ Hz 30 – 300 MHz	$10^1 - 10^0$ m 10 – 1 m	$10^{-1} - 10^0$ m^{-1} 100 km^{-1} – 1 mm^{-1}	VHF, ITU band no. 8, meter waves	
$3 \times 10^8 - 3 \times 10^9$ Hz 300 MHz – 3 GHz	$10^0 - 10^{-1}$ m 1 m – 100 mm	$10^0 - 10^1$ m^{-1} 1 – 10 m^{-1}	UHF, ITU band no. 9, decimeter waves[a]	
$3 \times 10^9 - 3 \times 10^{10}$ Hz 3 – 30 GHz	$10^{-1} - 10^{-2}$ m 100 –10 mm	$10^1 - 10^2$ m^{-1} 10 – 100 m^{-1}	SHF, ITU band no. 10, centimeter waves[a]	
$3 \times 10^{10} - 3 \times 10^{11}$ Hz 30 – 300 GHz	$10^{-2} - 10^{-3}$ m 10 –1 mm	$10^2 - 10^3$ m^{-1} 100 m^{-1} – 1 mm^{-1} (1 – 10 cm^{-1})	EHF, ITU band no. 11, millimeter waves	
$3 \times 10^{11} - 3 \times 10^{12}$ Hz 300 GHz – 3 THz	$10^{-3} - 10^{-4}$ m 1 mm – 100 μm	$10^3 - 10^4$ m^{-1} (1 – 10 mm^{-1}) (10 – 100 cm^{-1})	Part of micrometer waves, includes part of far or thermal infrared; ITU band no. 12	
$3 \times 10^{12} - 3 \times 10^{13}$ Hz 3 – 30 THz	$10^{-4} - 10^{-5}$ m 100 – 10 μm	$10^4 - 10^5$ m^{-1} 10 – 100 mm^{-1} (100 – 1000 cm^{-1})	Part of micrometer waves includes part of far (thermal) infrared	
$3 \times 10^{13} - 3 \times 10^{14}$ Hz 30 – 300 THz	$10^{-5} - 10^{-6}$ m 10 – 1 μm (100,000 – 10,000 Å)	$10^5 - 10^6$ m^{-1} 100 mm^{-1} – 1 μm^{-1}	Part of μm waves, part of infrared	$(1.6 - 16) \times 10^{-20}$ joule {0.1 – 1 eV}
$3 \times 10^{14} - 3 \times 10^{15}$ Hz 300 THz – 3 PHz	$10^{-6} - 10^{-7}$ m 1 μm – 100 nm (10,000 – 1000 Å)	$10^6 - 10^7$ m^{-1} 1 – 10 μm^{-1}	Near infrared, visible, near ultraviolet	$(1.6 - 16) \times 10^{-19}$ joule {1 – 10 eV}
$3 \times 10^{15} - 3 \times 10^{16}$ Hz 3 – 30 PHz	$10^{-7} - 10^{-8}$ m 100 – 10 nm (1000 – 100 Å)	$10^7 - 10^8$ m^{-1} 10 – 100 μm^{-1}	Part of vacuum ultraviolet	$(1.6 - 16) \times 10^{-18}$ joule {10 – 100 eV}
$3 \times 10^{16} - 3 \times 10^{17}$ Hz 30 – 300 PHz	$10^{-8} - 10^{-9}$ m 10 –1 nm (100 – 10 Å)	$10^8 - 10^9$ m^{-1} 100 μm^{-1} – 1 nm^{-1}	Part of soft X-rays	$(1.6 - 16) \times 10^{-17}$ joule {100 – 1000 eV}
$3 \times 10^{17} - 3 \times 10^{18}$ Hz 300 PHz – 3 EHz	$10^{-9} - 10^{-10}$ m 1 nm – 100 pm (10 – 1 Å)	$10^9 - 10^{10}$ m^{-1} 1 – 10 nm^{-1}	Part of soft X-rays	$(1.6 - 16) \times 10^{-16}$ joule {1 – 10 keV}
$3 \times 10^{18} - 3 \times 10^{19}$ Hz 3 – 30 EHz	$10^{-10} - 10^{-11}$ m 100 – 10 pm (1 – 0.1 Å)	$10^{10} - 10^{11}$ m^{-1} 10 – 100 nm^{-1}	Hard X-rays and part of soft γ-rays	$(1.6 - 16) \times 10^{-15}$ joule {10 – 100 keV}
$3 \times 10^{19} - 3 \times 10^{20}$ Hz 30 – 300 EHz	$10^{-11} - 10^{-12}$ m 10 – 1 pm (0.1 – 0.01 Å)	$10^{11} - 10^{12}$ m^{-1} 100 nm^{-1} – 1 pm^{-1}	Part of soft and part of hard γ-rays (limit at 510 keV)	$(1.6 - 16) \times 10^{-14}$ joule {100 keV – 1 MeV}
$3 \times 10^{20} - 3 \times 10^{21}$ Hz 300 – 3000 EHz	$10^{-12} - 10^{-13}$ m 1 pm – 100 fm (0.01 – 0.001 Å)	$10^{12} - 10^{13}$ m^{-1} 1 – 10 pm^{-1}	Part of hard γ-rays and part of "cosmic" γ-rays	$(1.6 - 16) \times 10^{-13}$ joule {1 – 10 MeV}
$3 \times 10^{21} - 3 \times 10^{22}$ Hz 3000 – 30,000 EHz	$10^{-13} - 10^{-14}$ m 100 – 10 fm (0.001 – 0.0001 Å)	$10^{13} - 10^{14}$ m^{-1} 10 – 100 pm^{-1}	γ-rays produced by cosmic rays	$(1.6 - 16) \times 10^{-12}$ joule {10 – 100 MeV}

Note: Abbreviations used in this table: Å— ångstrom (1 Å=10^{-10} m); EHz—exahertz (10^{18} hertz); EHF—extremely high frequency; ELF— extremely low frequency; eV—electron volt (1 eV = 1.60218×10^{-19} joule); fm—femtometer (10^{-15} m); GHz—gigahertz (10^{9} hertz); Gm—gigameter (10^{9} m); HF—high frequency; Hz—hertz (s^{-1}); ITU—International Telecommunications Union; keV— kiloelectron volt (10^{3} eV); km—kilometer (10^{3} m); LF—low frequency; m—meter; MeV— megaelectron volt (10^{6} eV); MF—medium frequency; MHz—megahertz (10^{6} hertz); Mm—megameter (10^{6} meter); mm—millimeter (10^{-3} meter); μm—micrometer (10^{-6} meter); nm—nanometer (10^{-9} meter); PHz—petahertz (10^{15} hertz); pm—pico-meter (10^{-12} meter); SHF—super high frequency; SLF—super low frequency; THz—terahertz; UHF— ultra high frequency; ULF—ultra low frequency; VHF—very high frequency; VLF—very low frequency.

[a] Also called "microwaves"; not to be confused with "micrometer waves".

Letter Designations of Microwave Bands

Frequency (GHz)	Wavelength (cm)	Wavenumber (cm^{-1})	Band
1—2	30—15	0.033—0.067	L-Band
2—4	15—7.5	0.067—0.133	S-Band
4—8	7.5—3.7	0.133—0.267	C-Band
8—12	3.7—2.5	0.267—0.4	X-Band
12—18	2.5—1.7	0.4—0.6	Ku-Band
18—27	1.7—1.1	0.6—0.9	K-Band
27—40	1.1—0.75	0.9—1.33	Ka-Band

SENSITIVITY OF THE HUMAN EYE TO LIGHT OF DIFFERENT WAVELENGTHS

The human eye responds to electromagnetic radiation in the wavelength range from about 360 nm (violet) to 820 nm (red), with a peak sensitivity near 555 nm (green). While the detailed shape of this response curve depends on the individual person, studies on representative samples of human subjects have led to adoption of a standard function relating the perceived brightness (luminous flux) to the actual power of the spectral radiation. This function is referred to as $V(\lambda)$, the photopic spectral luminous efficiency function, and it plays an important role in photometry.

The function $V(\lambda)$, as adopted by the International Commission on Illumination (CIE), is tabulated and plotted below.

References

1. *The Basis for Physical Photometry*, CIE Publication #18.2, 1983.
2. *CIE Standard Colorimetric Observers*, ISO/CIE #10527, 1991.
3. *Kaye and Laby Tables of Physical and Chemical Constants, Sixteenth Edition*, Longman Group Ltd., Harlow, Essex, 1995.

λ/nm	$V(\lambda)$	λ/nm	$V(\lambda)$	λ/nm	$V(\lambda)$
360	0.000004	520	0.710000	670	0.032000
370	0.000012	530	0.862000	680	0.017000
380	0.000039	540	0.954000	690	0.008210
390	0.000120	550	0.994950	700	0.004102
400	0.000396	555	1.000000	710	0.002091
410	0.001210	560	0.995000	720	0.001047
420	0.004000	570	0.952000	730	0.000520
430	0.011600	580	0.870000	740	0.000249
440	0.023000	590	0.757000	750	0.000120
450	0.038000	600	0.631000	760	0.000060
460	0.060000	610	0.503000	770	0.000030
470	0.090980	620	0.381000	780	0.000015
480	0.139020	630	0.265000	790	0.000007
490	0.208020	640	0.175000	800	0.000004
500	0.323000	650	0.107000	810	0.000002
510	0.503000	660	0.061000	820	0.000001

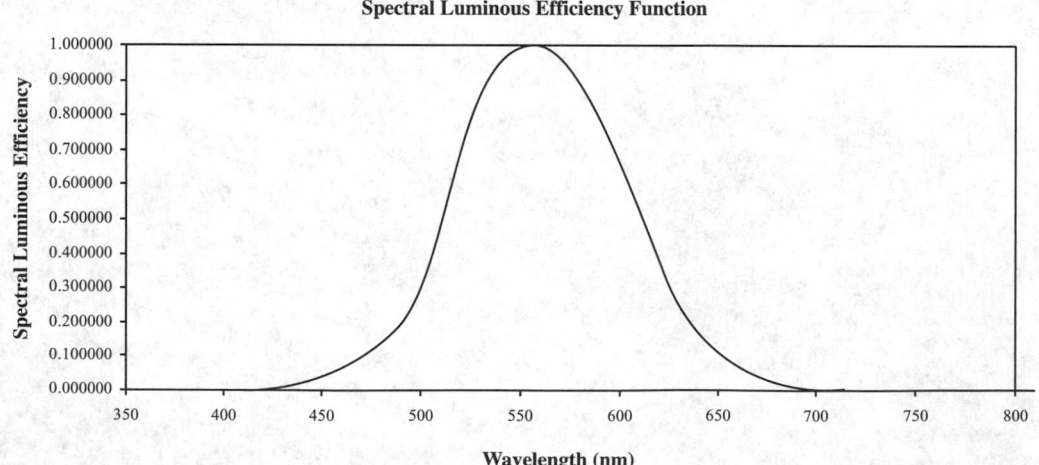

Spectral Luminous Efficiency Function

BLACKBODY RADIATION

The total power radiated from an ideal blackbody and the wavelength corresponding to maximum power are given here as a function of absolute temperature. Constants used in the calculation are taken from the table "Fundamental Physical Constants" in Section 1. The radiated power in a band $\Delta\lambda$ at λ_{max} may be calculated from:

$$P_{max} = 0.657548 \, (\Delta\lambda/\lambda_{max}) \, P_{tot}$$

T/K	P_{tot}	$\lambda_{max}/\mu m$	T/K	P_{tot}	$\lambda_{max}/\mu m$	T/K	P_{tot}	$\lambda_{max}/\mu m$
50	0.354 W/m²	57.955	740	17.004	3.916	1520	302.689	1.906
100	5.671	28.978	750	17.942	3.864	1540	318.937	1.882
150	28.707	19.318	760	18.918	3.813	1560	335.831	1.858
200	90.728	14.489	770	19.934	3.763	1580	353.387	1.834
250	221.504	11.591	780	20.989	3.715	1600	371.623	1.811
273	314.973	10.614	790	22.087	3.668	1620	390.555	1.789
280	348.541	10.349	800	23.226	3.622	1640	410.202	1.767
290	401.064	9.992	810	24.410	3.577	1660	430.581	1.746
300	459.311	9.659	820	25.638	3.534	1680	451.710	1.725
310	523.684	9.348	830	26.911	3.491	1700	473.607	1.705
320	594.596	9.055	840	28.232	3.450	1720	496.290	1.685
330	672.478	8.781	850	29.600	3.409	1740	519.779	1.665
340	757.771	8.523	860	31.018	3.369	1760	544.093	1.646
350	850.931	8.279	870	32.486	3.331	1780	569.249	1.628
360	952.428	8.049	880	34.006	3.293	1800	595.267	1.610
370	1.063 kW/m²	7.832	890	35.578	3.256	1820	622.168	1.592
380	1.182	7.626	900	37.204	3.220	1840	649.970	1.575
390	1.312	7.430	910	38.886	3.184	1860	678.694	1.558
400	1.452	7.244	920	40.623	3.150	1880	708.359	1.541
410	1.602	7.068	930	42.418	3.116	1900	738.987	1.525
420	1.764	6.899	940	44.272	3.083	1920	770.597	1.509
430	1.939	6.739	950	46.187	3.050	1940	803.210	1.494
440	2.125	6.586	960	48.162	3.018	1960	836.848	1.478
450	2.325	6.439	970	50.201	2.987	1980	871.531	1.464
460	2.539	6.299	980	52.303	2.957	2000	907.282	1.449
470	2.767	6.165	990	54.471	2.927	2020	944.121	1.435
480	3.010	6.037	1000	56.705	2.898	2040	982.071	1.420
490	3.269	5.914	1020	61.379	2.841	2060	1.021 MW/m²	1.407
500	3.544	5.796	1040	66.337	2.786	2080	1.061	1.393
510	3.836	5.682	1060	71.589	2.734	2100	1.103	1.380
520	4.146	5.573	1080	77.147	2.683	2120	1.145	1.367
530	4.474	5.467	1100	83.022	2.634	2140	1.189	1.354
540	4.822	5.366	1120	89.227	2.587	2160	1.234	1.342
550	5.189	5.269	1140	95.773	2.542	2180	1.281	1.329
560	5.577	5.175	1160	102.672	2.498	2200	1.328	1.317
570	5.986	5.084	1180	109.939	2.456	2220	1.377	1.305
580	6.417	4.996	1200	117.584	2.415	2240	1.428	1.294
590	6.871	4.911	1220	125.621	2.375	2260	1.479	1.282
600	7.349	4.830	1240	134.063	2.337	2280	1.532	1.271
610	7.851	4.750	1260	142.924	2.300	2300	1.587	1.260
620	8.379	4.674	1280	152.217	2.264	2320	1.643	1.249
630	8.933	4.600	1300	161.955	2.229	2340	1.700	1.238
640	9.514	4.528	1320	172.154	2.195	2360	1.759	1.228
650	10.122	4.458	1340	182.827	2.163	2380	1.819	1.218
660	10.760	4.391	1360	193.989	2.131	2400	1.881	1.207
670	11.427	4.325	1380	205.655	2.100	2420	1.945	1.197
680	12.124	4.261	1400	217.838	2.070	2440	2.010	1.188
690	12.853	4.200	1420	230.556	2.041	2460	2.077	1.178
700	13.615	4.140	1440	243.822	2.012	2480	2.145	1.168
710	14.410	4.081	1460	257.652	1.985	2500	2.215	1.159
720	15.239	4.025	1480	272.063	1.958	2550	2.398	1.136
730	16.103	3.970	1500	287.070	1.932	2600	2.591	1.115

T/K	P_{tot}	$\lambda_{max}/\mu m$	T/K	P_{tot}	$\lambda_{max}/\mu m$	T/K	P_{tot}	$\lambda_{max}/\mu m$
2650	2.796	1.093	3600	9.524	0.805	5100	38.362	0.568
2700	3.014	1.073	3650	10.065	0.794	5200	41.461	0.557
2750	3.243	1.054	3700	10.627	0.783	5300	44.743	0.547
2800	3.485	1.035	3750	11.214	0.773	5400	48.217	0.537
2850	3.741	1.017	3800	11.824	0.763	5500	51.889	0.527
2900	4.011	0.999	3850	12.458	0.753	5600	55.767	0.517
2950	4.294	0.982	3900	13.118	0.743	5700	59.858	0.508
3000	4.593	0.966	3950	13.804	0.734	5800	64.170	0.500
3050	4.907	0.950	4000	14.517	0.724	5900	68.712	0.491
3100	5.237	0.935	4100	16.024	0.707	6000	73.490	0.483
3150	5.583	0.920	4200	17.645	0.690	6500	101.222	0.446
3200	5.946	0.906	4300	19.386	0.674	7000	136.149	0.414
3250	6.326	0.892	4400	21.254	0.659	7500	179.418	0.386
3300	6.725	0.878	4500	23.253	0.644	8000	232.264	0.362
3350	7.142	0.865	4600	25.389	0.630	8500	296.004	0.341
3400	7.578	0.852	4700	27.670	0.617	9000	372.042	0.322
3450	8.033	0.840	4800	30.101	0.604	9500	461.867	0.305
3500	8.509	0.828	4900	32.689	0.591	10000	567.051	0.290
3550	9.006	0.816	5000	35.441	0.580			

The curves below show, for various temperatures, the fraction of radiant power as a function of wavelength. The function plotted is $P_\lambda/\Delta\lambda\, P_{tot}$, where P_λ is the power at wavelength λ in a small interval $\Delta\lambda$ (in μm), and P_{tot} is the total power.

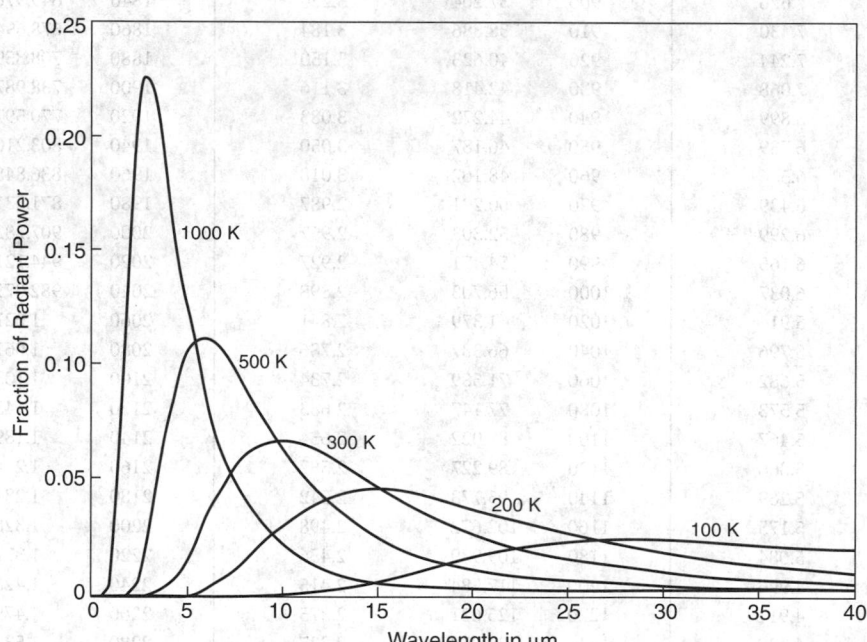

CHARACTERISTICS OF INFRARED DETECTORS

This graph summarizes the wavelength response of some semiconductors used as detectors for infrared radiation. The quantity $D^*(\lambda)$ is the signal-to-noise ratio for an incident radiant power density of 1 W/cm^2 and a bandwidth of 1 Hz (60° field of view). The Ge, InAs, and InSb detectors are photovoltaics, while the HgCdTe series are photoconductive devices. The cutoff wavelength of the latter can be varied by adjusting the relative amounts of Hg, Cd, and Te (three examples are shown at 77 K). The graph also shows the theoretical background limited sensitivity for ideal detectors which introduce no intrinsic noise.

Reference

Infrared Detectors 1995, EG&G Judson, Montgomeryville, PA.

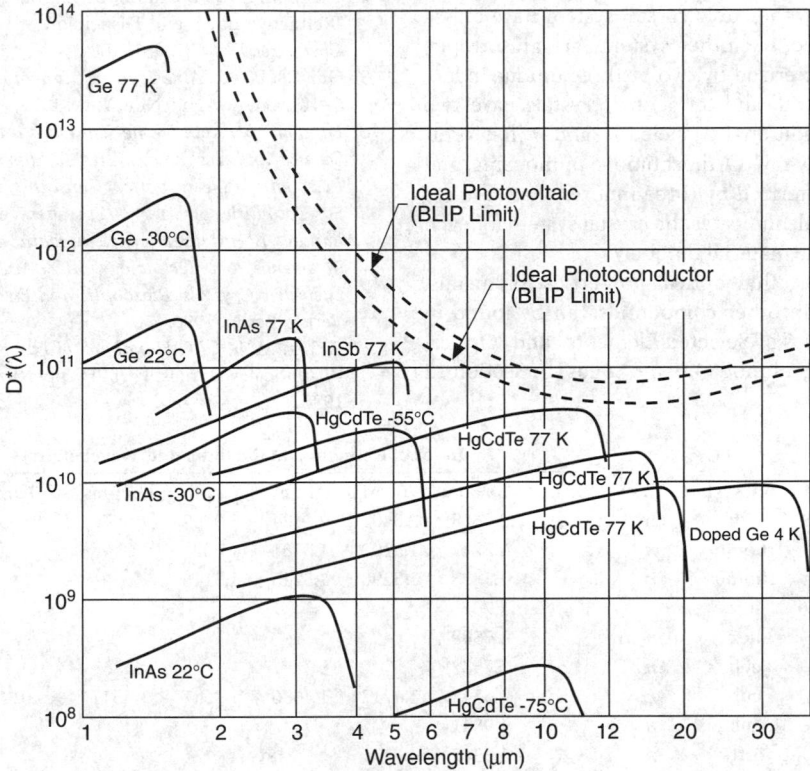

INDEX OF REFRACTION OF INORGANIC CRYSTALS

This table lists the index of refraction of selected crystalline inorganic compounds. When available, values are given as a function of wavelength in the range from the ultraviolet to the far infrared region. For each compound a value at 589 nm, the wavelength of the principal sodium line, is given. The data have been taken from the references indicated; in many cases, data from a reference have been refitted to generate the index of refraction at the wavelengths used in this table. All values refer to ambient temperature. Entries marked by * are based on extrapolation beyond the range of available experimental data.

Compounds belonging to the cubic crystal system have only a single refractive index value, but other systems are anisotropic, so that the crystal is characterized by two or three unique indexes. Hexagonal, rhombohedral, and tetragonal crystals have two unique indexes that are traditionally labeled n_o and n_e for "ordinary ray" and "extraordinary ray." Orthorhombic, monoclinic, and triclinic crystals are characterized by three indexes that here are called n_x, n_y, and n_z. The table indicates the crystal system for each entry in order to identify the material uniquely.

The refractive index and other optical properties for metals, semiconductors, and certain other compounds can be found in the tables "Optical Properties of Selected Elements" and "Optical Properties of Selected Inorganic and Organic Solids" in Section 12 of this *Handbook*.

References

1. Li, H. H., "Refractive Index of Alkali Halides and its Wavelength and Temperature Derivatives," *J. Phys. Chem. Ref. Data* 5, 329, 1976.
2. Li, H. H., "Refractive Index of Alkaline Earth Halides and its Wavelength and Temperature Derivatives," *J. Phys. Chem. Ref. Data* 9, 161, 1980.
3. Li, H. H., "Refractive Index of ZnS, ZnSe, and ZnTe and its Wavelength and Temperature Derivatives," *J. Phys. Chem. Ref. Data* 13, 103, 1984.
4. Shannon, R. D., Shannon, R. C., Medenbach, O., and Fischer, R. X., "Refractive Index and Dispersion of Fluorides and Oxides," *J. Phys. Chem. Ref. Data* 31, 931, 2002.
5. Gray, D. E., ed., *American Institute of Physics Handbook*, Sec. 6b, pp. 6–12, McGraw-Hill, New York, 1972.
6. *Landolt-Börnstein Numerical Data and Functional Relationships in Science and Technology, III/11, Elastic, Piezoelectric, Pyroelectric, Piezooptic, Electrooptic Constants, and Nonlinear Dielectric Susceptibilities of Crystals*, Springer-Verlag, Berlin, 1979.
7. *Landolt-Börnstein Numerical Data and Functional Relationships in Science and Technology, III/30A, High Frequency Properties of Dielectric Crystals. Piezooptic and Electrooptic Constants*, Springer-Verlag, Berlin, 1996.
8. Weber, M. J., *CRC Handbook of Laser Science and Technology*, Vol. IV. Optical Materials. Part 2: Properties, CRC Press, Boca Raton, FL, 1986.

| Compound | Crystal system | Ray | Index of Refraction at the Indicated Wavelength | | | | | | | | Ref. |
			300 nm	589 nm	750 nm	1 μm	2 μm	5 μm	10 μm	20 μm	
AgCl	cub	n		2.0668	2.0401	2.0224	2.0062	1.9975	1.9803	1.9069	5
AlPO$_4$	rhomb	n_o		1.5247	1.5203	1.5161	1.5034				6
	rhomb	n_e		1.5338	1.5290	1.5245	1.5116				6
Al$_2$O$_3$	hex	n_o		1.7673							4
	hex	n_e		1.7598							4
As$_2$O$_3$^{a}	cub	n		1.7537							4
BaF$_2$	cub	n	1.5010	1.4744	1.4712	1.4686	1.4647	1.4511	1.4014		2
BaO	cub	n		1.9841							4
BaSO$_4$	orth	n_x		1.6362							4
	orth	n_y		1.6374							4
	orth	n_z		1.6480							4
BaTiO$_3$	tetr	n_o		2.4405							4
	tetr	n_e		2.3831							4
BaWO$_4$	tetr	n_o		1.8426							4
	tetr	n_e		1.8405							4
BeO	hex	n_o		1.7184							4
	hex	n_e		1.7342							4
BeSO$_4$·4H$_2$O	tetr	n_o		1.4713							4
	tetr	n_e		1.4328							4
CaCO$_3$^{b}	hex	n_o	1.7216	1.6584	1.6503	1.6436	1.6249				5
	hex	n_e	1.5145	1.4864	1.4828	1.4801	1.4753				5
CaF$_2$	cub	n	1.4540	1.4338	1.4311	1.4289	1.4239	1.3990	1.299		2
CaO	cub	n		1.8396							4
CaSO$_4$	orth	n_x		1.5698							4
	orth	n_y		1.5755							4
	orth	n_z		1.6137							4
CaSO$_4$·2H$_2$O	monocl	n_x		1.5207							4
	monocl	n_y		1.5227							4
	monocl	n_z		1.5304							4
CaWO$_4$	tetr	n_o		1.9195							4
	tetr	n_e		1.9355							4
CdS	hex	n_o		2.507	2.390	2.334					5

Compound	Crystal system	Ray	300 nm	589 nm	750 nm	1 μm	2 μm	5 μm	10 μm	20 μm	Ref.
	hex	n_e		2.525	2.409	2.352					5
CdSe	hex	n_o			2.68*	2.5502	2.4682	2.4483	2.4331		7
	hex	n_e			2.69*	2.5696	2.4873	2.4676	2.4514		7
CdTe	cub	n							2.6724	2.6302	7
CeF_3	hex	n_o		1.6183							4
	hex	n_e		1.6113							4
CsBr	cub	n	1.8047	1.6974	1.6861	1.6784	1.6711	1.6678	1.6630	1.6439	1
CsCl	cub	n	1.712	1.640	1.631	1.626	1.620	1.616	1.606	1.563	1
$CsClO_4$	orth	n_x		1.4752							4
	orth	n_y		1.4788							4
	orth	n_z		1.4804							4
CsF	cub	n	1.506	1.477	1.474	1.472	1.469*	1.461*	1.436*	1.32*	1
CsI	cub	n	1.9790	1.7873	1.7694	1.7576	1.7465	1.7428	1.7396	1.7280	1
Cs_2SO_4	orth	n_x		1.5598							4
	orth	n_y		1.5644							4
	orth	n_z		1.5662							4
CuBr	cub	n		2.117							7
CuCl	cub	n		1.9727	1.9391				1.9245		7
$CuSO_4·5H_2O$	tricl	n_x		1.5140							4
	tricl	n_y		1.5367							4
	tricl	n_z		1.5436							4
Dy_2O_3	cub	n		1.9757							4
FeF_2	tetr	n_o		1.514							4
	tetr	n_e		1.524							4
Gd_2O_3	cub	n		1.96							4
HgS	rhomb	n_o		2.9413	2.7770	2.7120	2.6305		2.6018		6
	rhomb	n_e		3.3072	3.0896	3.0050	2.8776		2.8522		6
KBr	cub	n	1.6482	1.5598	1.5498	1.5444	1.5383	1.5345	1.5264	1.4924	1
KCl	cub	n	1.5455	1.4902	1.4840	1.4798	1.4753	1.4704	1.4564	1.3946	1
$KClO_4$	orth	n_x		1.4730							4
	orth	n_y		1.4736							4
	orth	n_z		1.4768							4
KF	cub	n	1.380	1.362	1.360	1.358	1.355	1.344	1.304*	1.09*	1
KH_2AsO_4	tetr	n_o		1.5674							7
	tetr	n_e		1.5179							7
KH_2PO_4	tetr	n_o	1.5450	1.5093	1.5030	1.4957					5
	tetr	n_e	1.4977	1.4682	1.4641	1.4606					5
KI	cub	n	1.834*	1.665	1.650	1.640	1.631	1.627	1.620	1.593	1
KIO_3	tricl	n_x		1.6959							7
	tricl	n_y		1.8317							7
	tricl	n_z		1.8343							7
KIO_4	tetr	n_o		1.6205							4
	tetr	n_e		1.6476							4
$KNbO_3$	orth	n_x		2.2480	2.3395	2.2612					7
	orth	n_y		2.3464	2.2959	2.2622					7
	orth	n_x		2.1803	2.1457	2.1288					7
K_2SO_4	orth	n_x		1.4934							4
	orth	n_y		1.4947							4
	orth	n_z		1.4973							4
LaF_3	hex	n_o		1.605							4
	hex	n_e		1.599							4
LiBr	cub	n	1.810	1.783	1.781	1.778	1.774*	1.756*	1.68*	1.33*	1
LiCl	cub	n	1.677	1.662	1.660	1.658	1.654*	1.62*	1.53*		1
$LiClO_4·3H_2O$	hex	n_o		1.4832							4
	hex	n_e		1.4384							4
LiF	cub	n	1.4087	1.3921	1.3895	1.3871	1.3786	1.3266	1.1005		1
LiI	cub	n	1.979	1.955	1.952	1.950	1.948*	1.940*	1.91*	1.77*	1
$LiIO_3$	hex	n_o		1.8875	1.8713	1.8589	1.8410				6
	hex	n_e		1.7400	1.7268	1.7179	1.7062				6

Compound	Crystal system	Ray	Index of Refraction at the Indicated Wavelength								Ref.
			300 nm	589 nm	750 nm	1 μm	2 μm	5 μm	10 μm	20 μm	
LiNbO$_3$	rhomb	n_o		2.3007	2.2632	2.2370					7
	rhomb	n_e		2.2116	2.1804	2.1567					7
LiTaO$_3$	rhomb	n_o		2.1864	2.1590	2.1391	2.1066				7
	rhomb	n_e		2.1908	2.1634	2.1432	2.1115				7
Li$_2$SO$_4$·H$_2$O	monocl	n_x		1.4615							4
	monocl	n_y		1.4765							4
	monocl	n_z		1.4863							4
Lu$_2$O$_3$	cub	n		1.9349							4
MgF$_2$	tetr	n_o	1.3930	1.3776	1.375	1.373	1.368	1.34	1.21		2
	tetr	n_e	1.4055	1.3894	1.387	1.385	1.379	1.34	1.21		2
MgO	cub	n		1.7355	1.7283	1.7228	1.7084	1.6361			5
MgSO$_4$·7H$_2$O	orth	n_x		1.4326							4
	orth	n_y		1.4555							4
	orth	n_z		1.4607							4
MnF$_2$	tetr	n_o		1.472							4
	tetr	n_e		1.501							4
NH$_4$H$_2$AsO$_4$	tetr	n_o	1.6401	1.5777	1.5704	1.5583					7
	tetr	n_e	1.5754	1.5232	1.5179	1.5101					7
NH$_4$H$_2$PO$_4$	tetr	n_o	1.5668	1.5247	1.5187	1.5084					7
	tetr	n_e	1.5137	1.4797	1.4754	1.4694					7
NaBr	cub	n	1.748	1.642	1.631	1.623	1.616	1.609	1.593*	1.520*	1
NaBrO$_3$	cub	n		1.6168							4
NaCl	cub	n	1.6066	1.5441	1.5369	1.5320	1.5265	1.5188	1.4947	1.382*	1
NaClO$_3$	cub	n		1.5151							7
NaF	cub	n	1.3424	1.3252	1.3231	1.3214	1.3179	1.3017	1.2400		1
NaH$_2$PO$_4$·2H$_2$O	orth	n_x		1.4400							7
	orth	n_y		1.4628							7
	orth	n_z		1.4814							7
NaI	cub	n	1.93*	1.774	1.758	1.74	1.73*	1.73*	1.71*	1.66*	1
NaNO$_2$	orth	n_x		1.6547							7
	orth	n_y		1.3455							7
	orth	n_z		1.4125							7
NaNO$_3$	rhomb	n_o		1.5840							5
	rhomb	n_e		1.3340							5
Na$_2$HPO$_4$·7H$_2$O	monocl	n_x		1.4411							4
	monocl	n_y		1.4423							4
	monocl	n_z		1.4525							4
Na$_2$SO$_4$	orth	n_x		1.4669							4
	orth	n_y		1.4730							4
	orth	n_z		1.4809							4
NdF$_3$	hex	n_o		1.6191							4
	hex	n_e		1.6132							4
Nd$_2$O$_3$	cub	n		1.92							4
NiF$_2$	tetr	n_o		1.526							4
	tetr	n_e		1.561							4
NiSO$_4$·6H$_2$O	tetr	n_o		1.5107							4
	tetr	n_e		1.4870							4
PbF$_2$	cub	n	1.94*	1.767	1.754	1.745	1.73	1.70	1.66	1.32	5
PbSO$_4$	orth	n_x		1.8780							4
	orth	n_y		1.8834							4
	orth	n_z		1.8945							4
PrF$_3$	hex	n_o		1.6207							4
	hex	n_e		1.6146							4
RbBr	cub	n	1.639	1.553	1.544	1.538	1.532	1.530	1.525	1.505*	1
RbCl	cub	n	1.549	1.493	1.487	1.483	1.479	1.475	1.465	1.424*	1
RbClO$_4$	orth	n_x		1.4691							4
	orth	n_y		1.4701							4
	orth	n_z		1.4732							4
RbF	cub	n	1.428*	1.397	1.394	1.391	1.388	1.379	1.346	1.19*	1

Compound	Crystal system	Ray	\multicolumn{8}{c}{Index of Refraction at the Indicated Wavelength}	Ref.
			300 nm	589 nm	750 nm	1 µm	2 µm	5 µm	10 µm	20 µm	
RbH_2AsO_4	tetr	n_o	1.6183	1.5603	1.5538	1.5432					7
	tetr	n_e	1.5718	1.5232	1.5184	1.5121					7
RbH_2PO_4	tetr	n_o	1.5434	1.5078	1.5021	1.4941					7
	tetr	n_e	1.5106	1.4791	1.4754	1.4704					7
RbI	cub	n	1.808	1.647	1.633	1.623	1.615	1.612	1.608	1.595	1
Rb_2SO_4	orth	n_x		1.5131							4
	orth	n_y		1.5133							4
	orth	n_z		1.5144							4
Sb_2O_3[c]	cub	n		2.8017							4
Sc_2O_3	cub	n		1.9943							4
SiO_2[d]	hex	n_o	1.5733	1.5442	1.5394	1.5350	1.5209				5
	hex	n_e	1.5882	1.5534	1.5484	1.5438	1.5291				5
SnO_2	tetr	n_o		1.993							4
	tetr	n_e		2.088							4
SrF_2	cub	n	1.459	1.4380	1.435	1.433	1.429	1.412	1.35		2
SrO	cub	n		1.8710							4
$SrSO_4$	orth	n_x		1.6214							4
	orth	n_y		1.6231							4
	orth	n_z		1.6303							4
$SrTiO_3$	cub	n		2.4082	2.3525	2.3160	2.2676	2.1205			5
$SrWO_4$	tetr	n_o		1.8618							4
	tetr	n_e		1.8719							4
TbF_3	hex	n_o		1.6034							4
	hex	n_e		1.5603							4
TeO_2	tetr	n_o		2.2738		2.2080					7
	tetr	n_e		2.4295		2.3520					7
ThO_2	cub	n		2.1113							4
TiO_2[e]	tetr	n_o		2.612	2.533	2.485	2.399	2.220			5
	tetr	n_e		2.910	2.805	2.748					5
	tetr	n_o		2.562							4
	tetr	n_e		2.489							4
TlBr	cub	n		2.418	2.350	2.289	2.103	1.984	2.339	2.322	5
TlCl	cub	n		2.247	2.198	2.145	1.986	1.891	2.193		5
$TlClO_4$	orth	n_x		1.6427							4
	orth	n_y		1.6446							4
	orth	n_z		1.6542							4
Tl_2SO_4	orth	n_x		1.8604							4
	orth	n_y		1.8676							4
	orth	n_z		1.8857							4
Y_2O_3	cub	n		1.930							4
Yb_2O_3	cub	n		1.9468							4
ZnF_2	tetr	n_o		1.495							4
	tetr	n_e		1.525							4
ZnO	hex	n_o		2.0036	1.9662	1.9435	1.9197				7
	hex	n_e		2.0199	1.9821	1.9589	1.9330				7
ZnS[f]	cub	n		2.3691	2.3232	2.2932	2.2633				7
ZnS[g]	hex	n_o		2.372	2.331	2.303	2.26	2.25	2.20		3, 5
	hex	n_e		2.368	2.327	2.301					5
ZnSe	cub	n		2.6222	2.5384	2.4888	2.4462	2.4296	2.4065		3
ZnTe	cub	n		3.060	2.880	2.789	2.719	2.698	2.684		3
$ZrSiO_4$[h]	tetr	n_o		1.9255							4
	tetr	n_e		1.9843							4

* Provisional value based on extrapolation beyond the range of experimental data.

[a] Arsenolite
[b] Calcite
[c] Senarmontite
[d] α-Quartz
[e] Rutile
[f] Sphalerite
[g] Wurtzite
[h] Zircon

REFRACTIVE INDEX AND TRANSMITTANCE OF REPRESENTATIVE GLASSES

Typical values of the index of refraction and internal transmittance (fraction of light transmitted through a one centimeter thickness) are tabulated here for selected types of glasses, as well as for synthetic fused (vitreous) silica. Nominal compositions are given in the first part of the table. The second part gives the index of refraction, relative to air, and the internal transmittance for representative samples of each glass at wavelengths in the infrared, visible, and near-ultraviolet regions. It should be emphasized that a wide variation of these parameters may be found among subtypes of each glass. More detailed data may be found in Reference 3.

Assuming that the Lambert-Beer Law is followed, the transmittance of a glass plate of thickness d (in centimeters) can be obtained by raising the transmittance value in the table to the power d.

References

1. Weber, M. J., *CRC Handbook of Laser Science and Technology*, Vol. IV, Part 2, CRC Press, Boca Raton, FL, 1988.
2. Gray, D. E., Ed., *American Institute of Physics Handbook, Third Edition*, McGraw Hill, New York, 1972.
3. *Schott Optical Glass*, Schott Glass Technologies, Inc., 400 York Ave., Duryea, PA.
4. Kaye, G. W. C., and Laby, T. H., *Tables of Physical and Chemical Constants, 15th Edition*, Longman, London, 1986.

Type	Name	Composition in percent by mass									
		SiO_2	B_2O_3	Al_2O_3	Na_2O	K_2O	CaO	BaO	ZnO	PbO	P_2O_5
PK	Phosphate crown		3	10		12	5				70
PSK	Dense phosphate crown		3	5			4	28			60
BK	Borosilicate crown	70	10		8	8	1	3			
K	Crown	74			9	11	6				
ZK	Zinc crown	71			17				12		
BaK	Barium crown	60	3		3	10		19	5		
SK	Dense crown	39	15	5				41			
KF	Crown flint	67		2	16				3	12	
BaLF	Barium light flint	51			6	5		20	14	4	
SSK	Extra dense crown	35	10	5				42	8		
LLF	Extra light flint	63			5	8				24	
BaF	Barium flint	46				8		16	8	22	
LF	Light flint	53			5	8				34	
F	Flint	47			2	7				44	
BaSF	Dense barium flint	43			1	7		11	5	33	
SF	Dense flint	33				5				62	
KzFS	Short flint										
SiO_2	Fused silica	100									

Type	Index of refraction				Transmittance of 1 cm plate			
	1.060 µm	546.1 nm	365.0 nm	312.6 nm	1.060 µm	546.1 nm	365.0 nm	310 nm
PK	1.51519	1.52736	1.54503	1.5574	0.997	0.998	0.987	0.46
PSK	1.54154	1.55440	1.57342	1.5868	0.996	0.998	0.984	0.46
BK	1.50669	1.51872	1.53627	1.5486	0.999	0.998	0.987	0.35
K	1.50091	1.51314	1.53189	1.5454	0.998	0.998	0.988	0.40
ZK	1.52220	1.53534	1.55588	1.5708	0.996	0.998	0.976	0.27
BaK	1.55695	1.57124	1.59407	1.6108	0.998	0.997	0.986	0.28
SK	1.59490	1.60994	1.63398		0.998	0.998	0.959	0.28
KF	1.50586	1.51978	1.54251	1.5600	0.998	0.996	0.989	0.49
BaLF	1.57579	1.59166	1.61804		0.996	0.998	0.933	0.010
SSK	1.60402	1.61993	1.64595		0.999	0.998	0.915	0.010
LaK	1.69710	1.71616	1.74573		0.999	0.998	0.882	0.17
LLF	1.52775	1.54344	1.57038		0.998	0.997	0.990	0.32
BaF	1.56873	1.58565	1.61524		0.999	0.997	0.992	0.004
LF	1.56594	1.58482	1.61926		0.999	0.998	0.981	0.008
F	1.58636	1.60718	1.64606		0.997	0.998	0.959	
BaSF	1.60889	1.62987	1.66926		0.999	0.998	0.857	
SF	1.71350	1.74620	1.8145		0.998	0.997	0.650	
KzFS	1.59680	1.61639	1.64849	1.6739		0.998	0.672	0.012
SiO_2	1.44968	1.46008	1.47435[a]	1.53430[b]				

[a] At 366.3 nm.
[b] At 213.9 nm.

INDEX OF REFRACTION OF WATER

This table gives the index of refraction of liquid water at atmospheric pressure, relative to a vacuum, at several temperatures and wavelengths. It is generated from the formulation in Reference 1, which covers a wide range of temperature, pressure, and wavelength. The wavelengths listed here correspond to prominent lines of cadmium (226.50 and 361.05 nm), potassium (404.41 nm), sodium (589.00 nm), Ne (632.80 nm, from a helium - neon laser), and mercury (1.01398 μm).

References

1. Schiebener, P., Straub, J., Levelt Sengers, J. M. H., and Gallagher, J. S., *J. Phys. Chem. Ref. Data*, 19, 677, 1990; 19, 1617, 1990.
2. Marsh, K. N., Editor, *Recommended Reference Materials for the Realization of Physicochemical Properties*, Blackwell Scientific Publications, Oxford, 1987.

$t/°C$	226.50 nm	361.05 nm	404.41 nm	589.00 nm	632.80 nm	1.01398 μm
0	1.39450	1.34896	1.34415	1.33432	1.33306	1.32612
10	1.39422	1.34870	1.34389	1.33408	1.33282	1.32591
20	1.39336	1.34795	1.34315	1.33336	1.33211	1.32524
30	1.39208	1.34682	1.34205	1.33230	1.33105	1.32424
40	1.39046	1.34540	1.34065	1.33095	1.32972	1.32296
50	1.38854	1.34373	1.33901	1.32937	1.32814	1.32145
60	1.38636	1.34184	1.33714	1.32757	1.32636	1.31974
70	1.38395	1.33974	1.33508	1.32559	1.32438	1.31784
80	1.38132	1.33746	1.33284	1.32342	1.32223	1.31576
90	1.37849	1.33501	1.33042	1.32109	1.31991	1.31353
100	1.37547	1.33239	1.32784	1.31861	1.31744	1.31114

INDEX OF REFRACTION OF LIQUIDS FOR CALIBRATION PURPOSES

This table gives the index of refraction of six liquids that are available in highly pure form and whose index of refraction has been accurately measured as a function of wavelength and temperature. They are therefore useful for calibration of refractometers. The estimated uncertainty in the values is:

2,2,4-Trimethylpentane	±0.00003
Hexadecane	±0.00008
trans-Bicyclo[4.0.0]decane	±0.00008
1-Methylnaphthalene	±0.00008
Toluene	±0.00003
Methylcyclohexane	±0.00003

Full details are given in the references. This table is reprinted from Reference 1 by permission of the International Union of Pure and Applied Chemistry.

References

1. Marsh, K. N., Editor, *Recommended Reference Materials for the Realization of Physicochemical Properties*, Blackwell Scientific Publications, Oxford, 1987.
2. Tilton, L. W., *J. Opt. Soc. Am.*, 32, 71, 1941.

λ	2,2,4-Trimethylpentane			Hexadecane		
nm	20 °C	25 °C	30 °C	20 °C	25 °C	30 °C
667.81	1.38916	1.38670	1.38424	1.43204	1.43001	1.42798
656.28	1.38945	1.38698	1.38452	1.43235	1.43032	1.42829
589.26	1.39145	1.38898	1.38650	1.43453	1.43250	1.43047
546.07	1.39316	1.39068	1.38820	1.43640	1.43436	1.43232
501.57	1.39544	1.39294	1.39044	1.43888	1.43684	1.43480
486.13	1.39639	1.39389	1.39138	1.43993	1.43788	1.43583
435.83	1.40029	1.39776	1.39523	1.44419	1.44213	1.44007

λ	trans-Bicyclo[4.4.0]decane			1-Methylnaphthalene		
nm	20 °C	25 °C	30 °C	20 °C	25 °C	30 °C
667.81	1.46654	1.46438	1.46222	1.60828	1.60592	1.60360
656.28	1.46688	1.46472	1.46256	1.60940	1.60703	1.60471
589.26	1.46932	1.46715	1.46498	1.61755	1.61512	1.61278
546.07	1.47141	1.46923	1.46705	1.62488	1.62240	1.62005
501.57	1.47420	1.47200	1.46980	1.63513	1.63259	1.63022
486.13	1.47535	1.47315	1.47095	1.63958	1.63701	1.63463
435.83	1.48011	1.47789	1.47567		1.65627	1.65386

λ	Toluene			Methylcyclohexane		
nm	20 °C	25 °C	30 °C	20 °C	25 °C	30 °C
667.81	1.49180	1.48903	1.48619	1.42064	1.41812	1.41560
656.28	1.49243	1.48966	1.48682	1.42094	1.41842	1.41591
589.26	1.49693	1.49413	1.49126	1.42312	1.42058	1.41806
546.07	1.50086	1.49803	1.49514	1.42497	1.42243	1.41989
501.57	1.50620	1.50334	1.50041	1.42744	1.42488	1.42233
486.13	1.50847	1.50559	1.50265	1.42847	1.42590	1.42334
435.83	1.51800	1.51506	1.51206	1.43269	1.43010	1.42752

INDEX OF REFRACTION OF AIR

This is a table of the index of refraction n of dry air at 15 °C and a pressure of 101.325 kPa and containing 0.045% by volume of carbon dioxide ("standard air"). The index of refraction is defined by $n = \lambda_{vac}/\lambda_{air}$ where λ is the wavelength of the radiation. The index is calculated from the expression

$$(n-1) \times 10^8 = 8342.54 + 2406147(130 - \sigma^2)^{-1} + 15998(38.9 - \sigma^2)^{-1}$$

where $\sigma = 1/\lambda_{vac}$ and λ_{vac} has units of μm. The equation is valid for λ_{vac} from 200 nm to 2 μm. The table also gives the correction $(n-1)\lambda_{air}$ that must be added to the wavelength in air to obtain λ_{vac}.

If the air is at a temperature t in °C (ITS-90) and a pressure p in pascals, a value of $(n-1)$ from this table should be multiplied by

$$p[1 + p(60.1 - 0.972t) \times 10^{-10}]/96095.43(1 + 0.003661t)$$

References

1. Birch, K. P., and Downs, M. J., *Metrologia*, 31, 315, 1994.
2. Edlen, B., *Metrologia* 2, 71, 1966.

λ_{vac}	$(n-1) \times 10^8$	$\lambda_{vac} - \lambda_{air}$	λ_{vac}	$(n-1) \times 10^8$	$\lambda_{vac} - \lambda_{air}$	λ_{vac}	$(n-1) \times 10^8$	$\lambda_{vac} - \lambda_{air}$
200 nm	32409	0.06480 nm	540	27804	0.15010	880	27462	0.24160
210	31748	0.06665	550	27784	0.15277	890	27458	0.24431
220	31226	0.06868	560	27765	0.15544	900	27454	0.24701
230	30801	0.07082	570	27747	0.15811	910	27449	0.24972
240	30447	0.07305	580	27730	0.16079	920	27445	0.25243
250	30148	0.07535	590	27714	0.16347	930	27441	0.25513
260	29892	0.07769	600	27698	0.16614	940	27437	0.25784
270	29670	0.08009	610	27684	0.16882	950	27434	0.26055
280	29477	0.08251	620	27670	0.17151	960	27430	0.26326
290	29307	0.08497	630	27657	0.17419	970	27427	0.26597
300	29157	0.08745	640	27644	0.17688	980	27423	0.26868
310	29023	0.08995	650	27632	0.17956	990	27420	0.27138
320	28904	0.09247	660	27621	0.18225			
330	28796	0.09500	670	27610	0.18494	1.00 μm	27417	0.0002741
340	28700	0.09755	680	27600	0.18763	1.05	27402	0.0002876
350	28612	0.10011	690	27590	0.19032	1.10	27390	0.0003012
360	28532	0.10269	700	27581	0.19301	1.15	27379	0.0003148
370	28460	0.10527	710	27572	0.19570	1.20	27370	0.0003283
380	28393	0.10786	720	27563	0.19840	1.25	27361	0.0003419
390	28332	0.11046	730	27555	0.20109	1.30	27354	0.0003555
400	28276	0.11307	740	27547	0.20379	1.35	27347	0.0003691
410	28224	0.11569	750	27539	0.20649	1.40	27341	0.0003827
420	28177	0.11831	760	27532	0.20918	1.45	27336	0.0003963
430	28132	0.12094	770	27525	0.21188	1.50	27331	0.0004099
440	28091	0.12357	780	27518	0.21458	1.55	27327	0.0004234
450	28053	0.12620	790	27511	0.21728	1.60	27323	0.0004370
460	28018	0.12885	800	27505	0.21998	1.65	27319	0.0004506
470	27985	0.13149	810	27499	0.22268	1.70	27316	0.0004642
480	27954	0.13414	820	27493	0.22538	1.75	27313	0.0004778
490	27925	0.13679	830	27488	0.22808	1.80	27310	0.0004914
500	27897	0.13945	840	27482	0.23079	1.85	27307	0.0005050
510	27872	0.14211	850	27477	0.23349	1.90	27305	0.0005187
520	27848	0.14477	860	27472	0.23619	1.95	27303	0.0005323
530	27825	0.14743	870	27467	0.23890	2.00	27301	0.0005459

INDEX OF REFRACTION OF GASES

This table gives the index of refraction of several gases at selected wavelengths ranging from the blue to the red region of the spectrum. The entries at 0.5893 μm correspond to the prominent sodium D line in the yellow region. All values refer to gas at a pressure of one atmosphere (101.325 kPa) and at a temperature of 0 °C.

References

1. Gray, D. E., ed., *American Institute of Physics Handbook, 3rd Edition*, p. 6-110, McGraw-Hill, New York, 1972.
2. Forsythe, W. E., *Smithsonian Physical Tables, Ninth Edition*, p. 533, Smithsonian Institution, Washington, 1954.
3. *Kaye and Laby Tables of Physical and Chemical Constants, Sixteenth Edition*, p. 131, Longman Group Ltd., Harlow, Essex, 1995.

Gas	Wavelength						
	0.4360 μm	0.4861 μm	0.5461 μm	0.5790 μm	0.5893 μm	0.6563 μm	0.6709 μm
Air	1.0002966	1.0002947	1.0002932	1.0002926	1.0002924	1.0002915	1.0002913
Ar					1.000281		
CH_4					1.000444		
Cl_2					1.000773		
CO_2	1.0004563		1.0004506		1.0004493		1.0004471
H_2	1.0001418	1.0001406	1.0001397	1.0001393	1.0001392	1.0001387	1.0001385
He					1.000036		
N_2		1.0003012	1.0002998		1.0002990	1.0002982	
N_2O					1.000516		
NO					1.000297		
O_2	1.0002743	1.0002734	1.0002717	1.0002710	1.0002709	1.0002698	1.0002683
SO_2					1.000686		

CHARACTERISTICS OF LASER SOURCES

William F. Krupke

Light amplification by stimulated emission of radiation was first demonstrated by Maiman in 1960, the result of a population inversion produced between energy levels of chromium ions in a ruby crystal when irradiated with a xenon flashlamp. Since then population inversions and coherent emission have been generated in literally thousands of substances (neutral and ionized gases, liquids, and solids) using a variety of incoherent excitation techniques (optical pumping, electrical discharges, gas-dynamic flow, electron-beams, chemical reactions, nuclear decay).

The extrema of laser output parameters that have been demonstrated to date and the laser media used are summarized in Table 1. Note that the extreme power and energy parameters listed in this table were attained with laser systems rather than with simple laser oscillators.

Laser sources are commonly classified in terms of the state-of-matter of the active medium: gas, liquid, and solid. Each of these classes is further subdivided into one or more types as shown in Table 2. A well-known representative example of each type of laser is also given in Table 2 together with its nominal operation wavelength and the methods by which it is pumped.

The various lasers together cover a wide spectral range from the far ultraviolet to the far infrared. The particular wavelength of emission (usually a narrow line) is presented for some six dozen lasers in Figures 1A and 1B.

By suitably designing the excitation source and/or by controlling the laser resonator structure, laser systems can provide continuous or pulsed radiation as shown in Table 3.

Besides the method of excitation and the temporal behavior of a laser, there are many other parameters that characterize its operation and efficiency, as shown in Tables 4 and 5.

Although many lasers only emit in one or more narrow spectral "lines," an increasing number of lasers can be tuned by changing the composition or the pressure of the medium, or by varying the wavelength of the pump bands. The spectral regions in which these tunable lasers operate are presented in Figure 2.

Reference

Krupke, W. F., in *Handbook of Laser Science and Technology*, Vol. I, Weber, M. J., Ed., CRC Press, Boca Raton, FL, 1986.

TABLE 1. Extrema of Output Parameters of Laser Devices or Systems

Parameter	Value	Laser medium
Peak power	1×10^{14} W (collimated)	Nd:glass
Peak power density	10^{18} W/cm^2 (focused)	Nd:glass
Pulse energy	$>10^5$ J	CO_2, Nd:glass
Average power	10^5 W	CO_2
Pulse duration	3×10^{-15} s continuous wave (cw)	Rh6G dye; various gases, liquids, solids
Wavelength	60 nm $\leftrightarrow$ 385 μm	Many required
Efficiency (nonlaser pumped)	70%	CO
Beam quality	Diffraction limited	Various gases, liquids, solids
Spectral linewidth	20 Hz (for 10^{-1} s)	Neon-helium
Spatial coherence	10 m	Ruby

TABLE 2. Classes, Types, and Representative Examples of Laser Sources

Class	Type (characteristic)	Representative example	Nominal operating wavelength (nm)	Method(s) of excitation
Gas	Atom, neutral (electronic transition)	Neon-Helium (Ne-He)	633	Glow discharge
	Atom, ionic (electronic transition)	Argon (Ar$^+$)	488	Arc discharge
	Molecule, neutral (electronic transition)	Krypton fluoride (KrF)	248	Glow discharge; e-beam
	Molecule, neutral (vibrational transition)	Carbon dioxide (CO_2)	10600	Glow discharge; gasdynamic flow
	Molecule, neutral (rotational transition)	Methyl fluoride (CH_3F)	496000	Laser pumping
	Molecule, ionic (electronic transition)	Nitrogen ion (N_2^+)	420	E-beam
Liquid	Organic solvent (dye-chromophore)	Rhodamine dye (Rh6G)	580–610	Flashlamp; laser pumping
	Organic solvent (rare earth chelate)	Europium:TTF	612	Flashlamp
	Inorganic solvent (trivalent rare earth ion)	Neodymium:POCl$_4$	1060	Flashlamp
Solid	Insulator, crystal (impurity)	Neodymium:YAG	1064	Flashlamp, arc lamp
	Insulator, crystal (stoichiometric)	Neodymium:UP(NdP$_5$O$_{14}$)	1052	Flashlamp
	Insulator, crystal (color center)	F$_2^-$:LiF	1120	Laser pumping
	Insulator, amorphous (impurity)	Neodymium:glass	1061	Flashlamp
	Semiconductor (p-n junction)	GaAs	820	Injection current
	Semiconductor (electron-hole plasma)	GaAs	890	E- beam, laser pumping

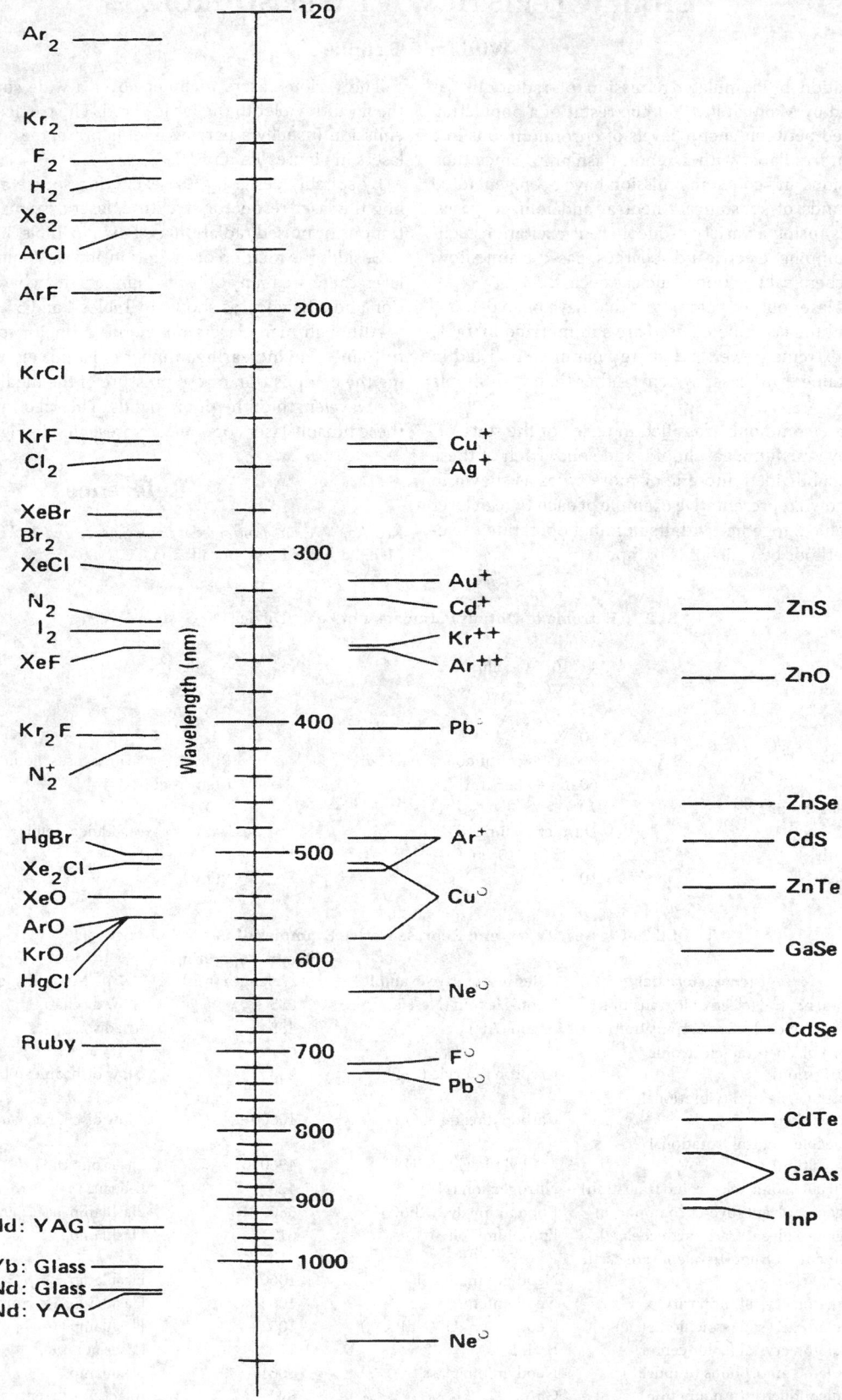

FIGURE 1A. Wavelengths of lasers operating in the 120 to 1200 nm spectral region.

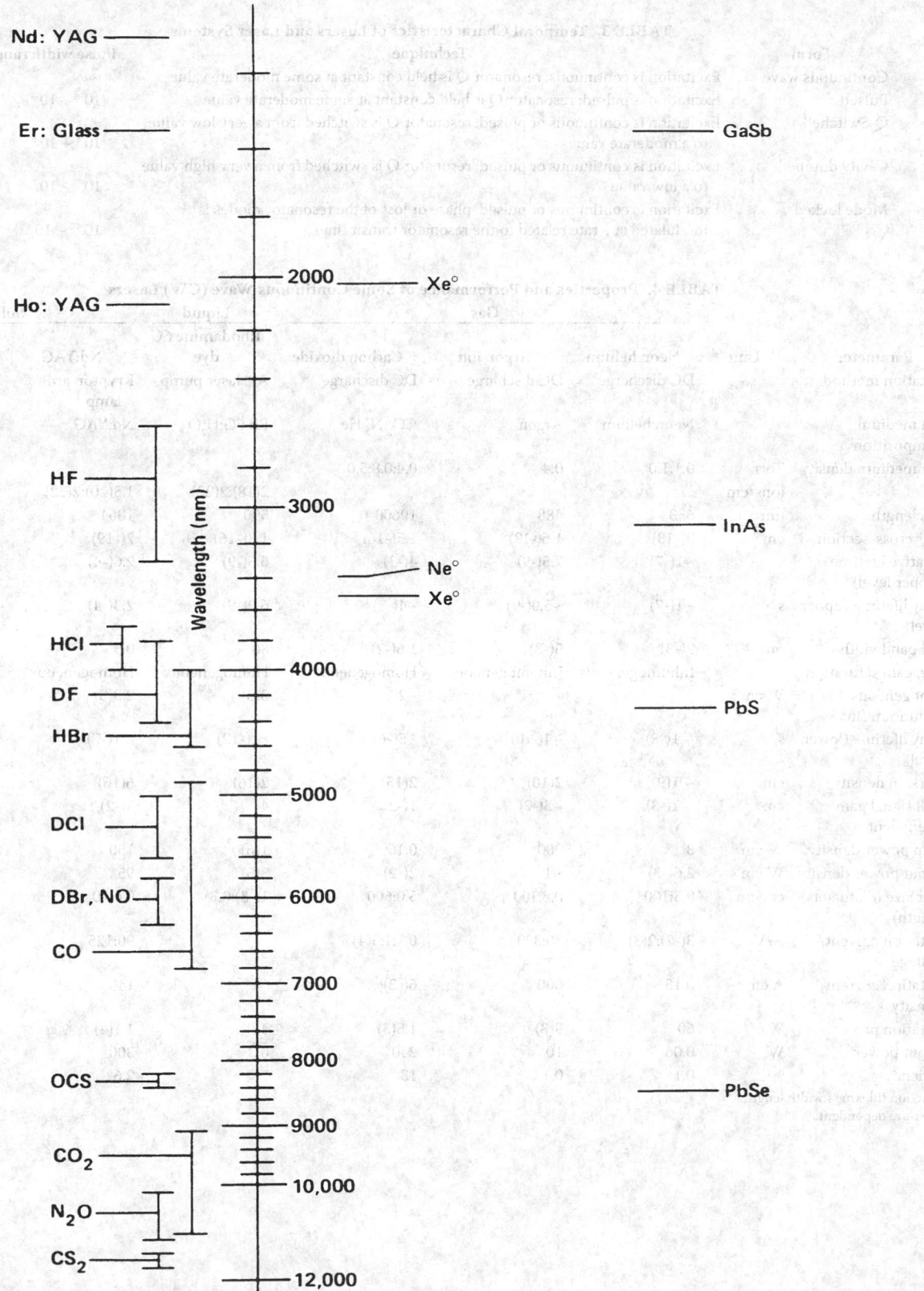

FIGURE 1B. Wavelength of lasers operating in the 1300 to 12,000 nm spectral region.

TABLE 3. Temporal Characteristics of Lasers and Laser Systems

Form	Technique	Pulse width range (s)
Continuous wave	Excitation is continuous; resonator Q is held constant at some moderate value	∞
Pulsed	Excitation is pulsed; resonator Q is held constant at some moderate value	$10^{-8} - 10^{-3}$
Q-Switched	Excitation is continuous or pulsed; resonator Q is switched from a very low value to a moderate value	$10^{-8} - 10^{-6}$
Cavity dumped	Excitation is continuous or pulsed; resonator Q is switched from a very high value to a low value	$10^{-7} - 10^{-5}$
Mode locked	Excitation is continuous or pulsed; phase or loss of the resonator modes is modulated at a rate related to the resonator transit time	$10^{-12} - 10^{-9}$

TABLE 4. Properties and Performance of Some Continuous Wave (CW) Lasers

Parameter	Unit	Gas			Liquid	Solid	
		Neon helium	Argon ion	Carbon dioxide	Rhodamine 6G dye	Nd:YAG	GaAs
Excitation method		DC discharge	DC discharge	DC discharge	Ar^+ laser pump	Krypton arc lamp	DC injection
Gain medium composition		Neon:helium	Argon	CO_2:N_2:He	Rh 6G:H_2O	Nd:YAG	p:n:GaAs
Gain medium density	Torr	0.1:1.0	0.4	0.4:0.8:5.0			
	ions/cm³				2(18):2(22)	1.5(20):2(22)	2(19):3(18):3(22)
Wavelength	nm	633	488	10600	590	1064	810
Laser cross-section	cm⁻²	3(-13)	1.6(-12)	1.5(-16)	1.8(-16)	7(-19)	~6(-15)
Radiative lifetime (upper level)	s	~1(-7)	7.5(-9)	4(-3)	6.5(-9)	2.6(-4)	~1(-9)
Decay lifetime (upper level)	s	~1(-7)	~5.0(-9)	~4(-3)	6.0(-9)	2.3(-4)	~1(-9)
Gain bandwidth	nm	2(-3)	5(-3)	1.6(-2)	80	0.5	10
Type, gain saturation		Inhomogeneous	Inhomogeneous	Homogeneous	Homogeneous	Homogeneous	Homogeneous
Homogeneous saturation flux	W cm⁻²			~20	3(5)	2.3(3)	~2(4)
Decay lifetime (lower level)	s	~ 1(-8)	~4(-10)	~5(-6)	<1(-12)	< 1(-7)	<1(-12)
Inversion density	cm⁻³	~ 1(9)	2(10)	2(15)	2(16)	6(16)	1(16)
Small signal gain coefficient	cm⁻¹	~ 1(-3)	~3(-2)	1(-2)	4	5(-2)	40
Pump power density	W cm⁻³	3	900	0.15	1(6)	150	7(7)
Output power density	W cm⁻³	2.6(-3)	~1	2(-2)	3(5)	95	5(6)
Laser size (diameter: length)	cm:cm	0.5:100	0.3:100	5.0:600	1(-3):0.3	0.6:10	5(-4):7(-3);2(-2)[a]
Excitation current/ voltage	A/V	3(-2):2(3)	30:300	0.1:1.5(4)		90:125	1.0/1.7
Excitation current density	A cm⁻²	0.15	600	6(-3)		140	4.5(3)
Excitation power	W	60	9(3)	1.5(3)	4	1.1(4)	1.7
Output power	W	0.06	10	240	0.3	300	0.12
Efficiency[b]	%	0.1	0.1	13	7	2.6	7

[a] Junction thickness:width:length.
[b] Pressure dependent.

TABLE 5. Properties and Performance of Some Pulsed Lasers

Parameter	Unit	Gas				Liquid	Solid	
		Carbon dixoxide		Krypton fluoride		Rhodamine 6G	Nd:YAG	Nd:glass
Excitation method		TEA-discharge	E- beam/sust.	Glow discharge	E-beam	Xenon flashlamp	Xenon flashlamp	Xenon flashlamp
Gain medium composition		$CO_2:N_2:He$	$CO_2:N_2:He$	$He:Kr:F_2$	$Ar:Kr:F_2$	Rh6G:alcohol	Nd:YAG	Nd:Glass
Gain medium density	torr	100:50:600	240:240:320	1070:70:3	1235:52:3			–
	ions/cm^3					1(18):1.5(22)	1.5(20):1(22)	3(20):2(22)
Wavelength	nm	10600	10600	249	249	590	1064	1061
Laser cross-section	cm^{-2}	2(-18)	2(-18)	2(-16)	2(-16)	1.8(-16)	7(-19)	2.8(-20)
Radiative lifetime (upper level)	s	4(-3)	4(-3)	7(-9)	7(-9)	6.5(-9)	2.6(-4)	4.1(-4)
Decay lifetime (upper level)	s	~1(-4)	5(-5)	2(-9)	3(-9)	6.0(-9)	2.3(-4)	3.7(-4)
Gain bandwidth	nm	1	1	2	2	80	0.5	26
Homogeneous saturation fluence	J/cm^2	0.2	0.2	4(-3)	4(-3)	2(-3)	0.6	~5
Decay lifetime (lower level)	s	5(-8)[a]	1(-8)[a]	<1(-12)	<1(-12)	<1(-12)	<1(-7)	<1(-8)
Inversion density	cm^{-3}	3(17)	6(17)	4(14)	2(14)	2(16)	4(17)	3(18)
Small signal gain coefficient	cm^{-1}	2(-2)	4(-2)	8–92)	4(-2)	4	0.3	8(-2)
Medium excitation energy density	J/cm^3	0.1	0.36	0.15	0.13	2.8	0.15	0.6
Output energy density	J/cm^3	2(-2)	1.8(-2)	1.5(-3)	1.2(-2)	0.85	5(-2)	2(-2)
Laser dimensions	cm:cm:cm	4.5:4.5:87	10:10:100	1.5:4.5:100	8.5:10:100	1.2:25	0.6:7.5	0.6:8.3
Excitation current/voltage	A/V	6(4)/3.3(3)	2.4(4)/4(4)	2.5(4)/1.5(5)	1.2(4)/2.5(5)	2(5)/2.5(4)		
Excitation current density	A cm^{-2}	8.5	22	170	11.5	2.6(3)		
Excitation peak power	W	2(8)	9(8)	4(9)	3(9)	5.4(9)	4(4)	9(4)
Output pulse energy	J	35	180	1	102	32	0.1	1.0
Output pulse length	s	1(-6)	4(-6)	2.5(-8)	6(-7)	3.2(-6)	2(-8)	1(-4)
Output pulse power	W	3.5(7)	4(7)	4(7)	2(8)	1(7)	5(6)	1(4)
Efficiency	%	17	5	1	10[b]	0.2	1.5	3.7

[a] Pressure dependent.
[b] Intrinsic efficiency ≡ energy output/energy deposited in gas.

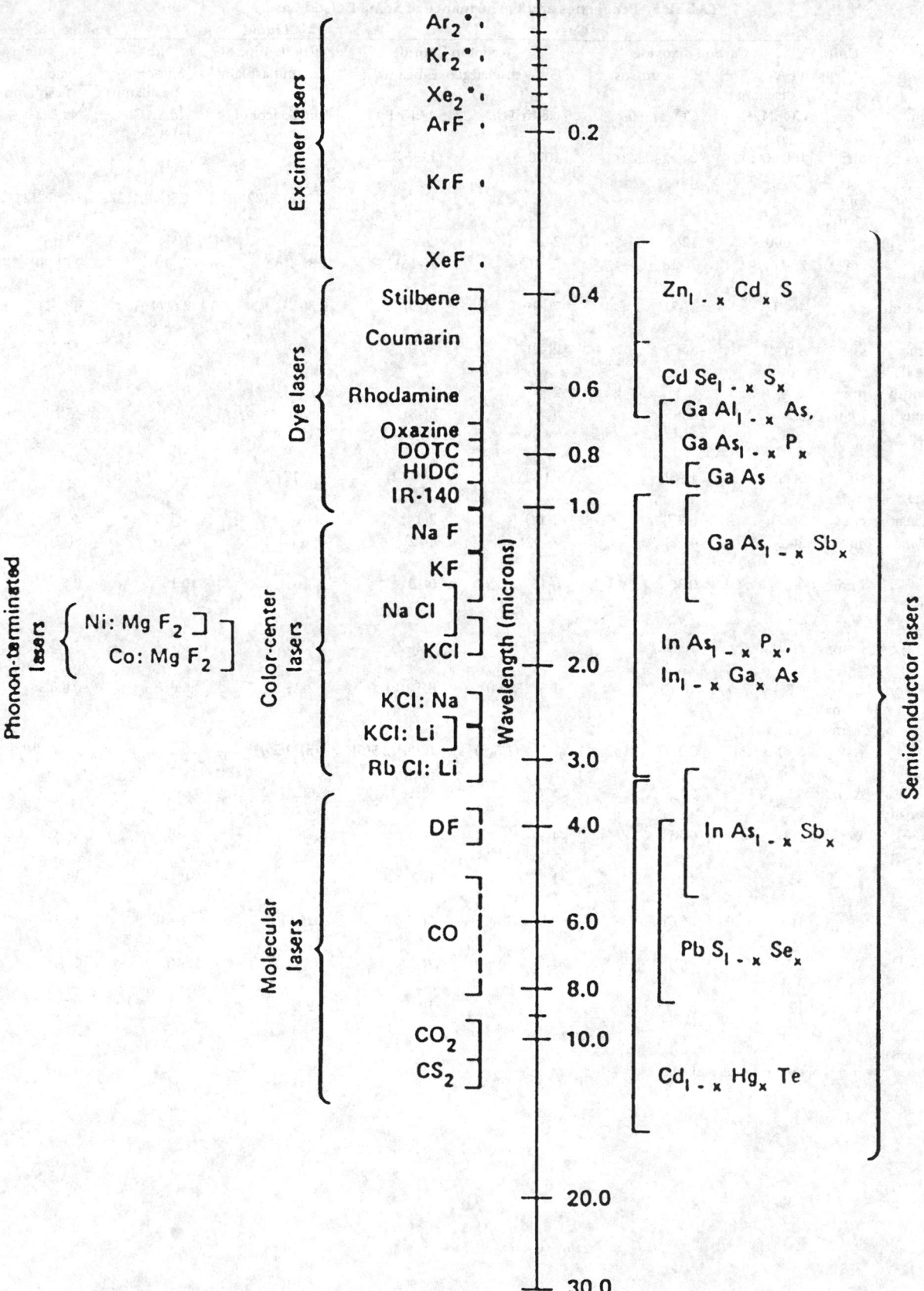

FIGURE 2. Spectral tuning ranges of various types of tunable lasers.

INFRARED LASER FREQUENCIES

Arthur Maki

The CO_2 laser has been the subject of a number of very accurate frequency measurements. Most of the earlier measurements are given by Bradley et al.[1] That analysis was based on a single absolute frequency measurement and many laser frequency differences. New measurements of the methane frequency[2-4] have made it necessary to slightly revise that single absolute frequency measurement. In addition, there have been several other absolute frequency measurements[5-7] that have been used here to improve the accuracy of the present tables. New frequency difference measurements have also been added to the database used for the present tables.[8]

References

1. Bradley, L. C., Soohoo, K. L., and Freed, C., *IEEE J. Quantum Electron.*, QE-22, 234–267, 1986.

2. Clairon, A., Dahmani, B., Filimon, A., and Rutman, J., *IEEE Trans. Inst. Meas.*, IM-34, 265–268, 1985.
3. Weiss, C. O., Kramer, G., Lipphardt, B., and Garcia, E., *IEEE J. Quantum Electron.*, QE-24, 1970–1972, 1988.
4. Bagayev, S. N., Baklanov, A. E., Chebotayev, V. P., and Dychkov, A. S., *Appl. Phys.*, B-48, 31–35, 1989.
5. Blaney, T. G., Bradley, C. C., Edwards, G. J., Jolliffe, B. W., Knight, D. J. E., Rowley, W. R. C., Shotten, K. C., and Woods, P. T., *Proc. R. Soc. Lond.*, A-355, 61–88, 1977.
6. Chardonnet, Ch., Van Lerberghe, A., and Bordé, Ch. J., *Opt. Comm.*, 58, 333–337, 1986.
7. Clairon, A., Acef, O., Chardonnet, Ch., and Bordé, Ch. J., *Frequency Standards and Metrology*, De Marchi, A., Ed., Springer-Verlag, Berlin, Heidelberg, 1989, p. 212.
8. Evenson, K., private communication.

Frequencies for the 00°1-(10°0,02°0)$_I$ and 00°1-(10°0,02°0)$_{II}$ Bands of $^{12}C^{16}O_2$ with the Estimated 2σ Uncertainties

Line	Band I frequency (MHz)	Uncertainty (MHz)	Line	Band II frequency (MHz)	Uncertainty (MHz)
P(70)	26721305.4647	0.1680	P(70)	29789856.3783	0.0308
P(68)	26794232.6712	0.1217	P(68)	29861850.7690	0.0192
P(66)	26866318.8073	0.0867	P(66)	29933216.1760	0.0122
P(64)	26937571.7234	0.0606	P(64)	30003944.2861	0.0086
P(62)	27007998.9216	0.0415	P(62)	30074026.9127	0.0072
P(60)	27077607.5643	0.0279	P(60)	30143456.0039	0.0066
P(58)	27146404.4834	0.0185	P(58)	30212223.6504	0.0061
P(56)	27214396.1873	0.0121	P(56)	30280322.0930	0.0055
P(54)	27281588.8696	0.0081	P(54)	30347743.7306	0.0049
P(52)	27347988.4161	0.0057	P(52)	30414481.1273	0.0044
P(50)	27413600.4119	0.0043	P(50)	30480527.0196	0.0041
P(48)	27478430.1487	0.0036	P(48)	30545874.3239	0.0039
P(46)	27542482.6310	0.0032	P(46)	30610516.1429	0.0039
P(44)	27605762.5826	0.0030	P(44)	30674445.7724	0.0039
P(42)	27668274.4525	0.0028	P(42)	30737656.7080	0.0039
P(40)	27730022.4206	0.0027	P(40)	30800142.6511	0.0039
P(38)	27791010.4036	0.0026	P(38)	30861897.5150	0.0038
P(36)	27851242.0594	0.0025	P(36)	30922915.4310	0.0037
P(34)	27910720.7927	0.0024	P(34)	30983190.7534	0.0037
P(32)	27969449.7593	0.0023	P(32)	31042718.0652	0.0037
P(30)	28027431.8708	0.0022	P(30)	31101492.1833	0.0036
P(28)	28084669.7981	0.0021	P(28)	31159508.1631	0.0037
P(26)	28141165.9762	0.0020	P(26)	31216761.3029	0.0037
P(24)	28196922.6067	0.0019	P(24)	31273247.1487	0.0037
P(22)	28251941.6622	0.0017	P(22)	31328961.4978	0.0037
P(20)	28306224.8888	0.0016	P(20)	31383900.4028	0.0037
P(18)	28359773.8090	0.0014	P(18)	31438060.1749	0.0037
P(16)	28412589.7245	0.0012	P(16)	31491437.3872	0.0036
P(14)	28464673.7184	0.0011	P(14)	31544028.8776	0.0036
P(12)	28516026.6574	0.0009	P(12)	31595831.7516	0.0036
P(10)	28566649.1935	0.0008	P(10)	31646843.3843	0.0035
P(8)	28616541.7661	0.0008	P(8)	31697061.4225	0.0035
P(6)	28665704.6027	0.0008	P(6)	31746483.7868	0.0035
P(4)	28714137.7205	0.0008	P(4)	31795108.6724	0.0035
P(2)	28761840.9272	0.0008	P(2)	31842934.5511	0.0035
R(0)	28832026.2198	0.0008	R(0)	31913172.5691	0.0035
R(2)	28877902.4382	0.0007	R(2)	31958996.0621	0.0034
R(4)	28923046.4303	0.0006	R(4)	32004017.3822	0.0034

Line	Band I frequency (MHz)	Uncertainty (MHz)	Line	Band II frequency (MHz)	Uncertainty (MHz)
R(6)	28967457.0657	0.0005	R(6)	32048236.2498	0.0034
R(8)	29011133.0054	0.0003	R(8)	32091652.6619	0.0034
R(10)	29054072.7010	0.0001	R(10)	32134266.8917	0.0034
R(12)	29096274.3935	0.0003	R(12)	32176079.4878	0.0034
R(14)	29137736.1129	0.0005	R(14)	32217091.2721	0.0035
R(16)	29178455.6759	0.0007	R(16)	32257303.3386	0.0036
R(18)	29218430.6852	0.0009	R(18)	32296717.0510	0.0037
R(20)	29257658.5269	0.0010	R(20)	32335334.0408	0.0038
R(22)	29296136.3689	0.0011	R(22)	32373156.2044	0.0039
R(24)	29333861.1583	0.0012	R(24)	32410185.7003	0.0041
R(26)	29370829.6191	0.0011	R(26)	32446424.9459	0.0042
R(28)	29407038.2491	0.0011	R(28)	32481876.6140	0.0042
R(30)	29442483.3168	0.0011	R(30)	32516543.6293	0.0042
R(32)	29477160.8582	0.0012	R(32)	32550429.1641	0.0042
R(34)	29511066.6733	0.0013	R(34)	32583536.6340	0.0042
R(36)	29544196.3221	0.0015	R(36)	32615869.6937	0.0041
R(38)	29576545.1205	0.0017	R(38)	32647432.2320	0.0040
R(40)	29608108.1360	0.0019	R(40)	32678228.3665	0.0039
R(42)	29638880.1831	0.0022	R(42)	32708262.4386	0.0038
R(44)	29668855.8183	0.0024	R(44)	32737539.0081	0.0039
R(46)	29698029.3350	0.0027	R(46)	32766062.8469	0.0041
R(48)	29726394.7582	0.0032	R(48)	32793838.9334	0.0045
R(50)	29753945.8385	0.0037	R(50)	32820872.4463	0.0055
R(52)	29780676.0464	0.0042	R(52)	32847168.7576	0.0071
R(54)	29806578.5659	0.0047	R(54)	32872733.4269	0.0099
R(56)	29831646.2878	0.0052	R(56)	32897572.1935	0.0141
R(58)	29855871.8032	0.0058	R(58)	32921690.9701	0.0202
R(60)	29879247.3960	0.0074	R(60)	32945095.8355	0.0288
R(62)	29901765.0357	0.0113	R(62)	32967793.0268	0.0407
R(64)	29923416.3695	0.0186	R(64)	32989788.9322	0.0567
R(66)	29944192.7145	0.0302	R(66)	33011090.0831	0.0780
R(68)	29964085.0488	0.0475	R(68)	33031703.1467	0.1060
R(70)	29983084.0036	0.0720	R(70)	33051634.9172	0.1423

Frequencies for the 00°1-(10°0,02°0)$_I$ and 00°1-(10°0,02°0)$_{II}$ Bands of $^{13}C^{16}O_2$ with the Estimated 2σ Uncertainties

P(66)	25523832.1808	0.7836	P(66)	28512082.5283	1.2894
P(64)	25590013.4703	0.5415	P(64)	28585121.9396	0.9194
P(62)	25655543.6502	0.3629	P(62)	28657449.4180	0.6420
P(60)	25720428.2487	0.2339	P(60)	28729056.6374	0.4375
P(58)	25784672.4840	0.1430	P(58)	28799935.4147	0.2897
P(56)	25848281.2771	0.0810	P(56)	28870077.7187	0.1853
P(54)	25911259.2627	0.0405	P(54)	28939475.6771	0.1135
P(52)	25973610.8005	0.0157	P(52)	29008121.5846	0.0659
P(50)	26035339.9857	0.0045	P(50)	29076007.9109	0.0357
P(48)	26096450.6582	0.0079	P(48)	29143127.3077	0.0180
P(46)	26156946.4123	0.0101	P(46)	29209472.6164	0.0090
P(44)	26216830.6053	0.0101	P(44)	29275036.8754	0.0058
P(42)	26276106.3655	0.0090	P(42)	29339813.3270	0.0050
P(40)	26334776.6003	0.0077	P(40)	29403795.4243	0.0044
P(38)	26392844.0030	0.0068	P(38)	29466976.8383	0.0037
P(36)	26450311.0599	0.0063	P(36)	29529351.4635	0.0032
P(34)	26507180.0565	0.0061	P(34)	29590913.4252	0.0029
P(32)	26563453.0836	0.0060	P(32)	29651657.0844	0.0028
P(30)	26619132.0428	0.0058	P(30)	29711577.0447	0.0028
P(28)	26674218.6515	0.0055	P(28)	29770668.1566	0.0031
P(26)	26728714.4479	0.0054	P(26)	29828925.5239	0.0035
P(24)	26782620.7952	0.0054	P(24)	29886344.5074	0.0041
P(22)	26835938.8858	0.0054	P(22)	29942920.7308	0.0046
P(20)	26888669.7451	0.0055	P(20)	29998650.0838	0.0051
P(18)	26940814.2347	0.0055	P(18)	30053528.7271	0.0054

Line	Band I frequency (MHz)	Uncertainty (MHz)	Line	Band II frequency (MHz)	Uncertainty (MHz)
P(16)	26992373.0555	0.0055	P(16)	30107553.0955	0.0055
P(14)	27043346.7508	0.0054	P(14)	30160719.9016	0.0055
P(12)	27093735.7083	0.0052	P(12)	30213026.1388	0.0054
P(10)	27143540.1624	0.0051	P(10)	30264469.0839	0.0054
P(8)	27192760.1962	0.0049	P(8)	30315046.2994	0.0054
P(6)	27241395.7431	0.0048	P(6)	30364755.6359	0.0055
P(4)	27289446.5880	0.0047	P(4)	30413595.2335	0.0056
P(2)	27336912.3682	0.0046	P(2)	30461563.5231	0.0057
R(0)	27407012.8882	0.0045	P(0)	30531879.5415	0.0057
R(2)	27453013.4589	0.0043	P(2)	30577664.6138	0.0056
R(4)	27498426.5430	0.0040	P(4)	30622575.1885	0.0054
R(6)	27543251.1200	0.0037	P(6)	30666611.0128	0.0051
R(8)	27587486.0225	0.0034	P(8)	30709772.1257	0.0047
R(10)	27631129.9356	0.0031	P(10)	30752058.8571	0.0045
R(12)	27674181.3963	0.0029	P(12)	30793471.8269	0.0044
R(14)	27716638.7917	0.0029	P(14)	30834011.9425	0.0043
R(16)	27758500.3577	0.0029	P(16)	30873680.3976	0.0044
R(18)	27799764.1770	0.0029	P(18)	30912478.6694	0.0044
R(20)	27840428.1773	0.0030	P(20)	30950408.5159	0.0044
R(22)	27880490.1283	0.0029	P(22)	30987471.9732	0.0043
R(24)	27919947.6395	0.0029	P(24)	31023671.3517	0.0042
R(26)	27958798.1567	0.0028	P(26)	31059009.2327	0.0042
R(28)	27997038.9591	0.0028	P(28)	31093488.4642	0.0042
R(30)	28034667.1551	0.0027	P(30)	31127112.1569	0.0043
R(32)	28071679.6785	0.0027	P(32)	31159883.6793	0.0045
R(34)	28108073.2842	0.0026	P(34)	31191806.6529	0.0046
R(36)	28143844.5432	0.0026	P(36)	31222884.9469	0.0048
R(38)	28178989.8377	0.0026	P(38)	31253122.6730	0.0053
R(40)	28213505.3554	0.0028	P(40)	31282524.1795	0.0061
R(42)	28247387.0838	0.0033	P(42)	31311094.0452	0.0077
R(44)	28280630.8035	0.0046	P(44)	31338837.0736	0.0108
R(46)	28313232.0818	0.0083	P(46)	31365758.2858	0.0173
R(48)	28345186.2652	0.0161	P(48)	31391862.9147	0.0295
R(50)	28376488.4720	0.0301	P(50)	31417156.3972	0.0505
R(52)	28407133.5839	0.0531	P(52)	31441644.3679	0.0845
R(54)	28437116.2372	0.0887	P(54)	31465332.6516	0.1366
R(56)	28466430.8141	0.1419	P(56)	31488227.2557	0.2138
R(58)	28495071.4324	0.2188	P(58)	31510334.3631	0.3247
R(60)	28523031.9357	0.3271	P(60)	31531660.3243	0.4800
R(62)	28550305.8819	0.4763	P(62)	31552211.6497	0.6932
R(64)	28576886.5323	0.6781	P(64)	31571995.0017	0.9805
R(66)	28602766.8393	0.9467	P(66)	31591017.1868	1.3619

Frequencies for the $00°1$-$(10°0,02°0)_I$ and $00°1$-$(10°0,02°0)_{II}$ Bands of $^{12}C^{18}O_2$ with the Estimated 2σ Uncertainties

Line	Band I frequency (MHz)	Uncertainty (MHz)	Line	Band II frequency (MHz)	Uncertainty (MHz)
P(70)	27045326.3119	0.4540	P(70)	30695237.5856	0.0858
P(68)	27114914.0922	0.3324	P(68)	30755520.2231	0.0570
P(66)	27183635.7945	0.2392	P(66)	30815311.4928	0.0364
P(64)	27251496.4118	0.1688	P(64)	30874607.2084	0.0223
P(62)	27318500.7361	0.1165	P(62)	30933403.2309	0.0131
P(60)	27384653.3618	0.0783	P(60)	30991695.4724	0.0075
P(58)	27449958.6881	0.0510	P(58)	31049479.9009	0.0049
P(56)	27514420.9224	0.0319	P(56)	31106752.5446	0.0041
P(54)	27578044.0828	0.0191	P(54)	31163509.4964	0.0040
P(52)	27640832.0010	0.0108	P(52)	31219746.9183	0.0040
P(50)	27702788.3248	0.0059	P(50)	31275461.0455	0.0039
P(48)	27763916.5206	0.0035	P(48)	31330648.1908	0.0039
P(46)	27824219.8762	0.0028	P(46)	31385304.7490	0.0039
P(44)	27883701.5029	0.0026	P(44)	31439427.2006	0.0039
P(42)	27942364.3379	0.0025	P(42)	31493012.1163	0.0038
P(40)	28000211.1464	0.0024	P(40)	31546056.1605	0.0038

Line	Band I frequency (MHz)	Uncertainty (MHz)	Line	Band II frequency (MHz)	Uncertainty (MHz)
P(38)	28057244.5242	0.0022	P(38)	31598556.0954	0.0037
P(36)	28113466.8992	0.0021	P(36)	31650508.7847	0.0037
P(34)	28168880.5335	0.0020	P(34)	31701911.1970	0.0037
P(32)	28223487.5256	0.0019	P(32)	31752760.4093	0.0037
P(30)	28277289.8118	0.0017	P(30)	31803053.6105	0.0037
P(28)	28330289.1679	0.0016	P(28)	31852788.1043	0.0038
P(26)	28382487.2111	0.0015	P(26)	31901961.3125	0.0038
P(24)	28433885.4012	0.0013	P(24)	31950570.7773	0.0038
P(22)	28484485.0420	0.0012	P(22)	31998614.1649	0.0038
P(20)	28534287.2828	0.0011	P(20)	32046089.2669	0.0037
P(18)	28583293.1193	0.0010	P(18)	32092994.0036	0.0037
P(16)	28631503.3952	0.0010	P(16)	32139326.4254	0.0036
P(14)	28678918.8025	0.0009	P(14)	32185084.7154	0.0036
P(12)	28725539.8830	0.0010	P(12)	32230267.1907	0.0036
P(10)	28771367.0288	0.0010	P(10)	32274872.3041	0.0037
P(8)	28816400.4829	0.0010	P(8)	32318898.6455	0.0038
P(6)	28860640.3403	0.0011	P(6)	32362344.9434	0.0039
P(4)	28904086.5477	0.0011	P(4)	32405210.0652	0.0041
P(2)	28946738.9048	0.0011	P(2)	32447493.0185	0.0041
R(0)	29009228.1702	0.0010	P(0)	32509824.0580	0.0042
R(2)	29049894.0586	0.0010	P(2)	32550648.1723	0.0042
R(4)	29089764.2368	0.0009	P(4)	32590887.7542	0.0042
R(6)	29128837.8426	0.0008	P(6)	32630542.4457	0.0041
R(8)	29167113.8668	0.0008	P(8)	32669612.0295	0.0041
R(10)	29204591.1529	0.0009	P(10)	32708096.4282	0.0040
R(12)	29241268.3964	0.0010	P(12)	32745995.7040	0.0040
R(14)	29277144.1444	0.0011	P(14)	32783310.0573	0.0040
R(16)	29312216.7955	0.0012	P(16)	32820039.8258	0.0040
R(18)	29346484.5984	0.0012	P(18)	32856185.4827	0.0040
R(20)	29379945.6517	0.0013	P(20)	32891747.6358	0.0040
R(22)	29412597.9024	0.0013	P(22)	32926727.0254	0.0040
R(24)	29444439.1458	0.0013	P(24)	32961124.5220	0.0040
R(26)	29475467.0236	0.0014	P(26)	32994941.1249	0.0040
R(28)	29505679.0230	0.0015	P(28)	33028177.9594	0.0040
R(30)	29535072.4755	0.0016	P(30)	33060836.2743	0.0040
R(32)	29563644.5557	0.0018	P(32)	33092917.4394	0.0041
R(34)	29591392.2794	0.0020	P(34)	33124422.9429	0.0043
R(36)	29618312.5023	0.0023	P(36)	33155354.3878	0.0046
R(38)	29644401.9182	0.0028	P(38)	33185713.4894	0.0049
R(40)	29669657.0575	0.0036	P(40)	33215502.0716	0.0056
R(42)	29694074.2853	0.0053	P(42)	33244722.0637	0.0068
R(44)	29717649.7992	0.0082	P(44)	33273375.4969	0.0092
R(46)	29740379.6276	0.0128	P(46)	33301464.5003	0.0134
R(48)	29762259.6274	0.0200	P(48)	33328991.2976	0.0199
R(50)	29783285.4820	0.0307	P(50)	33355958.2027	0.0294
R(52)	29803452.6988	0.0461	P(52)	33382367.6161	0.0427
R(54)	29822756.6072	0.0681	P(54)	33408222.0209	0.0607
R(56)	29841192.3558	0.0985	P(56)	33433523.9780	0.0848
R(58)	29858754.9100	0.1401	P(58)	33458276.1228	0.1165
R(60)	29875439.0495	0.1960	P(60)	33482481.1601	0.1576
R(62)	29891239.3658	0.2702	P(62)	33506141.8605	0.2104
R(64)	29906150.2589	0.3673	P(64)	33529261.0556	0.2775
R(66)	29920165.9352	0.4930	P(66)	33551841.6335	0.3621
R(68)	29933280.4042	0.6540	P(68)	33573886.5352	0.4679
R(70)	29945487.4756	0.8581	P(70)	33595398.7493	0.5992

Frequencies for the $00°1$-$(10°0,02°0)_I$ and $00°1$-$(10°0,02°0)_{II}$ Bands of $^{13}C^{18}O_2$ with the Estimated 2σ Uncertainties

Line			Line		
P(70)	25967863.7652	1.1146	P(70)	28960476.2278	0.4069
P(68)	26033448.2798	0.8152	P(68)	29022326.9578	0.2861
P(66)	26098273.9159	0.5860	P(66)	29083661.3546	0.1961

Line	Band I frequency (MHz)	Uncertainty (MHz)	Line	Band II frequency (MHz)	Uncertainty (MHz)
P(64)	26162346.4813	0.4129	P(64)	29144473.5795	0.1303
P(62)	26225671.5466	0.2844	P(62)	29204757.8761	0.0833
P(60)	26288254.4494	0.1906	P(60)	29264508.5768	0.0507
P(58)	26350100.2984	0.1237	P(58)	29323720.1086	0.0290
P(56)	26411213.9778	0.0772	P(56)	29382386.9988	0.0152
P(54)	26471600.1504	0.0459	P(54)	29440503.8809	0.0073
P(52)	26531263.2618	0.0258	P(52)	29498065.4997	0.0038
P(50)	26590207.5442	0.0138	P(50)	29555066.7172	0.0032
P(48)	26648437.0195	0.0077	P(48)	29611502.5178	0.0031
P(46)	26705955.5026	0.0057	P(46)	29667368.0132	0.0031
P(44)	26762766.6051	0.0055	P(44)	29722658.4475	0.0034
P(42)	26818873.7378	0.0055	P(42)	29777369.2022	0.0039
P(40)	26874280.1143	0.0056	P(40)	29831495.8006	0.0044
P(38)	26928988.7531	0.0056	P(38)	29885033.9125	0.0049
P(36)	26983002.4809	0.0056	P(36)	29937979.3584	0.0053
P(34)	27036323.9351	0.0055	P(34)	29990328.1139	0.0054
P(32)	27088955.5657	0.0054	P(32)	30042076.3132	0.0055
P(30)	27140899.6384	0.0051	P(30)	30093220.2534	0.0055
P(28)	27192158.2363	0.0049	P(28)	30143756.3978	0.0054
P(26)	27242733.2620	0.0047	P(26)	30193681.3793	0.0053
P(24)	27292626.4396	0.0044	P(24)	30242992.0038	0.0052
P(22)	27341839.3165	0.0042	P(22)	30291685.2529	0.0051
P(20)	27390373.2651	0.0040	P(20)	30339758.2870	0.0049
P(18)	27438229.4843	0.0037	P(18)	30387208.4477	0.0048
P(16)	27485409.0008	0.0035	P(16)	30434033.2603	0.0046
P(14)	27531912.6704	0.0033	P(14)	30480230.4356	0.0045
P(12)	27577741.1795	0.0031	P(12)	30525797.8725	0.0044
P(10)	27622895.0455	0.0031	P(10)	30570733.6593	0.0043
P(8)	27667374.6182	0.0031	P(8)	30615036.0750	0.0043
P(6)	27711180.0803	0.0033	P(6)	30658703.5912	0.0044
P(4)	27754311.4480	0.0034	P(4)	30701734.8727	0.0045
P(2)	27796768.5718	0.0036	P(2)	30744128.7785	0.0045
R(0)	27859189.3155	0.0036	P(0)	30806522.5414	0.0045
R(2)	27899959.0889	0.0035	P(2)	30847319.2956	0.0044
R(4)	27940052.7921	0.0033	P(4)	30887476.2168	0.0043
R(6)	27979469.5315	0.0031	P(6)	30926993.0424	0.0042
R(8)	28018208.2478	0.0028	P(8)	30965869.7046	0.0041
R(10)	28056267.7161	0.0026	P(10)	31004106.3298	0.0040
R(12)	28093646.5448	0.0025	P(12)	31041703.2379	0.0040
R(14)	28130343.1757	0.0025	P(14)	31078660.9408	0.0040
R(16)	28166355.8825	0.0025	P(16)	31114980.1420	0.0040
R(18)	28201682.7706	0.0025	P(18)	31150661.7340	0.0041
R(20)	28236321.7757	0.0025	P(20)	31185706.7976	0.0042
R(22)	28270270.6628	0.0024	P(22)	31220116.5992	0.0043
R(24)	28303527.0249	0.0024	P(24)	31253892.5891	0.0043
R(26)	28336088.2817	0.0023	P(26)	31287036.3991	0.0044
R(28)	28367951.6781	0.0024	P(28)	31319549.8396	0.0043
R(30)	28399114.2823	0.0025	P(30)	31351434.8973	0.0043
R(32)	28429572.9843	0.0026	P(32)	31382693.7318	0.0042
R(34)	28459324.4940	0.0028	P(34)	31413328.6728	0.0042
R(36)	28488365.3390	0.0029	P(36)	31443342.2165	0.0041
R(38)	28516691.8625	0.0029	P(38)	31472737.0219	0.0040
R(40)	28544300.2211	0.0031	P(40)	31501515.9074	0.0039
R(42)	28571186.3823	0.0032	P(42)	31529681.8467	0.0040
R(44)	28597346.1222	0.0032	P(44)	31557237.9646	0.0042
R(46)	28622775.0223	0.0038	P(46)	31584187.5329	0.0046
R(48)	28647468.4672	0.0071	P(48)	31610533.9656	0.0057
R(50)	28671421.6417	0.0148	P(50)	31636280.8146	0.0088
R(52)	28694629.5272	0.0286	P(52)	31661431.7650	0.0151
R(54)	28717086.8993	0.0510	P(54)	31685990.6298	0.0261
R(56)	28738788.3239	0.0852	P(56)	31709961.3449	0.0434

Line	Band I frequency (MHz)	Uncertainty (MHz)	Line	Band II frequency (MHz)	Uncertainty (MHz)
R(58)	28759728.1540	0.1355	P(58)	31733347.9642	0.0693
R(60)	28779900.5263	0.2075	P(60)	31756154.6537	0.1068
R(62)	28799299.3572	0.3078	P(62)	31778385.6867	0.1594
R(64)	28817918.3393	0.4447	P(64)	31800045.4375	0.2317
R(66)	28835750.9374	0.6283	P(66)	31821138.3761	0.3291
R(68)	28852790.3843	0.8707	P(68)	31841669.0622	0.4581
R(70)	28869029.6768	1.1863	P(70)	31861642.1394	0.6268

Frequencies for the $01^{1e}1$-$(11^{1e}0,03^{1e}0)_I$ and $01^{1e}1$-$(11^{1e}0,03^{1e}0)_{II}$ Bands of $^{12}C^{16}O_2$ with the Estimated 2σ Uncertainties

Line	Band I frequency (MHz)	Uncertainty (MHz)	Line	Band II frequency (MHz)	Uncertainty (MHz)
P(59)	26125213.2723	1.6633	P(59)	30427055.2899	0.1962
P(57)	26191576.6703	1.0880	P(57)	30494640.3229	0.1332
P(55)	26257240.7898	0.6844	P(55)	30561557.5929	0.0865
P(53)	26322208.2302	0.4094	P(53)	30627802.0344	0.0530
P(51)	26386481.4313	0.2286	P(51)	30693368.7014	0.0306
P(49)	26450062.6783	0.1155	P(49)	30758252.7710	0.0175
P(47)	26512954.1076	0.0498	P(47)	30822449.5469	0.0123
P(45)	26575157.7109	0.0191	P(45)	30885954.4624	0.0114
P(43)	26636675.3402	0.0160	P(43)	30948763.0834	0.0109
P(41)	26697508.7115	0.0182	P(41)	31010871.1119	0.0100
P(39)	26757659.4084	0.0177	P(39)	31072274.3882	0.0091
P(37)	26817128.8857	0.0160	P(37)	31132968.8940	0.0091
P(35)	26875918.4726	0.0144	P(35)	31192950.7549	0.0102
P(33)	26934029.3751	0.0131	P(33)	31252216.2430	0.0118
P(31)	26991462.6787	0.0119	P(31)	31310761.7788	0.0134
P(29)	27048219.3509	0.0106	P(29)	31368583.9339	0.0147
P(27)	27104300.2431	0.0096	P(27)	31425679.4328	0.0155
P(25)	27159706.0925	0.0093	P(25)	31482045.1550	0.0157
P(23)	27214437.5237	0.0097	P(23)	31537678.1367	0.0154
P(21)	27268495.0505	0.0104	P(21)	31592575.5725	0.0147
P(19)	27321879.0769	0.0108	P(19)	31646734.8172	0.0137
P(17)	27374589.8987	0.0108	P(17)	31700153.3868	0.0127
P(15)	27426627.7040	0.0104	P(15)	31752828.9602	0.0119
P(13)	27477992.5747	0.0098	P(13)	31804759.3803	0.0113
P(11)	27528684.4867	0.0096	P(11)	31855942.6551	0.0113
P(9)	27578703.3113	0.0101	P(9)	31906376.9582	0.0116
P(7)	27628048.8151	0.0113	P(7)	31956060.6304	0.0122
P(5)	27676720.6609	0.0127	P(5)	32004992.1796	0.0129
P(3)	27724718.4080	0.0141	P(3)	32053170.2819	0.0136
R(1)	27841759.7696	0.0152	P(1)	32170312.0391	0.0149
R(3)	27887393.2105	0.0146	P(3)	32215845.0845	0.0151
R(5)	27932349.2934	0.0135	P(5)	32260620.8121	0.0152
R(7)	27976627.0108	0.0124	P(7)	32304638.8261	0.0152
R(9)	28020225.2521	0.0115	P(9)	32347898.8990	0.0150
R(11)	28063142.8031	0.0110	P(11)	32390400.9714	0.0148
R(13)	28105378.3457	0.0109	P(13)	32432145.1513	0.0145
R(15)	28146930.4576	0.0109	P(15)	32473131.7137	0.0142
R(17)	28187797.6116	0.0107	P(17)	32513361.0997	0.0140
R(19)	28227978.1750	0.0103	P(19)	32552833.9153	0.0140
R(21)	28267470.4088	0.0099	P(21)	32591550.9309	0.0141
R(23)	28306272.4666	0.0099	P(23)	32629513.0796	0.0143
R(25)	28344382.3939	0.0107	P(25)	32666721.4564	0.0144
R(27)	28381798.1267	0.0122	P(27)	32703177.3164	0.0142
R(29)	28418517.4902	0.0141	P(29)	32738882.0732	0.0136
R(31)	28454538.1976	0.0165	P(31)	32773837.2976	0.0136
R(33)	28489857.8477	0.0213	P(33)	32808044.7156	0.0174
R(35)	28524473.9240	0.0312	P(35)	32841506.2063	0.0279
R(37)	28558383.7917	0.0486	P(37)	32874223.8000	0.0462
R(39)	28591584.6963	0.0754	P(39)	32906199.6761	0.0735
R(41)	28624073.7602	0.1131	P(41)	32937436.1606	0.1114
R(43)	28655847.9806	0.1644	P(43)	32967935.7238	0.1624
R(45)	28686904.2261	0.2328	P(45)	32997700.9775	0.2292

Line	Band I frequency (MHz)	Uncertainty (MHz)	Line	Band II frequency (MHz)	Uncertainty (MHz)
R(47)	28717239.2334	0.3239	P(47)	33026734.6728	0.3151
R(49)	28746849.6038	0.4465	P(49)	33055039.6965	0.4238
R(51)	28775731.7988	0.6142	P(51)	33082619.0689	0.5595
R(53)	28803882.1361	0.8465	P(53)	33109475.9403	0.7272

Frequencies for the $01^{1f}1$-$(11^{1f}0,03^{1f}0)_I$ and $01^{1f}1$-$(11^{1f}0,03^{1f}0)_{II}$ Bands of $^{12}C^{16}O_2$ with the Estimated 2σ Uncertainties

Line	Band I frequency (MHz)	Uncertainty (MHz)	Line	Band II frequency (MHz)	Uncertainty (MHz)
P(60)	26051570.0104	4.4521	P(60)	30355115.0204	0.2752
P(58)	26120964.4932	3.0629	P(58)	30425283.5969	0.1926
P(56)	26189552.8496	2.0516	P(56)	30494732.8293	0.1301
P(54)	26257339.6006	1.3305	P(54)	30563455.6325	0.0840
P(52)	26324329.0344	0.8289	P(52)	30631445.1076	0.0512
P(50)	26390525.2136	0.4901	P(50)	30698694.5456	0.0292
P(48)	26455931.9824	0.2698	P(48)	30765197.4310	0.0163
P(46)	26520552.9722	0.1334	P(46)	30830947.4444	0.0111
P(44)	26584391.6075	0.0551	P(44)	30895938.4662	0.0104
P(42)	26647451.1105	0.0181	P(42)	30960164.5794	0.0105
P(40)	26709734.5057	0.0151	P(40)	31023620.0723	0.0105
P(38)	26771244.6242	0.0174	P(38)	31086299.4415	0.0107
P(36)	26831984.1067	0.0157	P(36)	31148197.3941	0.0114
P(34)	26891955.4069	0.0126	P(34)	31209308.8510	0.0126
P(32)	26951160.7945	0.0105	P(32)	31269628.9481	0.0138
P(30)	27009602.3576	0.0096	P(30)	31329153.0395	0.0147
P(28)	27067282.0045	0.0092	P(28)	31387876.6994	0.0151
P(26)	27124201.4662	0.0090	P(26)	31445795.7236	0.0149
P(24)	27180362.2977	0.0089	P(24)	31502906.1318	0.0141
P(22)	27235765.8792	0.0090	P(22)	31559204.1695	0.0128
P(20)	27290413.4182	0.0093	P(20)	31614686.3091	0.0113
P(18)	27344305.9494	0.0096	P(18)	31669349.2515	0.0098
P(16)	27397444.3368	0.0097	P(16)	31723189.9280	0.0086
P(14)	27449829.2733	0.0096	P(14)	31776205.5007	0.0081
P(12)	27501461.2824	0.0096	P(12)	31828393.3642	0.0085
P(10)	27552340.7179	0.0101	P(10)	31879751.1463	0.0095
P(8)	27602467.7649	0.0111	P(8)	31930276.7092	0.0107
P(6)	27651842.4399	0.0125	P(6)	31979968.1497	0.0120
P(4)	27700464.5912	0.0139	P(4)	32028823.8002	0.0131
P(2)	27748333.8988	0.0148	P(2)	32076842.2290	0.0139
R(2)	27864709.8633	0.0146	P(2)	32193218.1935	0.0150
R(4)	27909939.2762	0.0135	P(4)	32238298.4853	0.0151
R(6)	27954412.3294	0.0122	P(6)	32282538.0393	0.0153
R(8)	27998127.7801	0.0112	P(8)	32325936.7244	0.0153
R(10)	28041084.2173	0.0107	P(10)	32368494.6458	0.0154
R(12)	28083280.0620	0.0108	P(12)	32410212.1438	0.0155
R(14)	28124713.5668	0.0110	P(14)	32451089.7941	0.0155
R(16)	28165382.8151	0.0112	P(16)	32491128.4063	0.0154
R(18)	28205285.7213	0.0111	P(18)	32530329.0234	0.0152
R(20)	28244420.0302	0.0110	P(20)	32568692.9211	0.0149
R(22)	28282783.3158	0.0114	P(22)	32606221.6061	0.0147
R(24)	28320372.9812	0.0129	P(24)	32642916.8154	0.0144
R(26)	28357186.2574	0.0149	P(26)	32678780.5147	0.0140
R(28)	28393220.2023	0.0168	P(28)	32713814.8971	0.0133
R(30)	28428471.6994	0.0175	P(30)	32748022.3813	0.0120
R(32)	28462937.4565	0.0165	P(32)	32781405.6101	0.0106
R(34)	28496614.0042	0.0142	P(34)	32813967.4482	0.0122
R(36)	28529497.6934	0.0163	P(36)	32845710.9809	0.0212
R(38)	28561584.6939	0.0309	P(38)	32876639.5111	0.0385
R(40)	28592870.9914	0.0593	P(40)	32906756.5580	0.0646
R(42)	28623352.3850	0.1054	P(42)	32936065.8540	0.1015
R(44)	28653024.4839	0.1779	P(44)	32964571.3426	0.1518
R(46)	28681882.7038	0.2907	P(46)	32992277.1760	0.2184
R(48)	28709922.2632	0.4635	P(48)	33019187.7118	0.3048
R(50)	28737138.1785	0.7231	P(50)	33045307.5105	0.4152

INFRARED AND FAR-INFRARED ABSORPTION FREQUENCY STANDARDS

Arthur Maki

Aside from the CO_2 laser transitions, the absorption spectrum of CO has been more accurately and thoroughly measured than any other spectrum. A bibliography of earlier measurements on CO is given by Maki and Wells,[1] and the present tables were calculated from the measurements referred to in that work. In addition, some new and very accurate frequency measurements[2,3] have been made and were incorporated in the present tables. The frequencies of the rotational transitions of HF and HCl were calculated from constants obtained from fitting the measurements of Evenson et al.[4,5] and Jennings and Wells.[6]

A new report on infrared wavenumber standards from the International Union of Pure and Applied Chemistry, Commission on Molecular Structure and Spectroscopy, may be found in Reference 7.

References

1. Maki, A. G. and Wells, J. S., *Wavenumber Calibration Tables from Heterodyne Frequency Measurements*, NIST Special Publication 821, U.S. Dept. of Commerce, Washington, D.C., 1991.
2. Evenson, K. and Stroh, F., private communication.
3. George, T., Urban, W., and co-workers, private communication.
4. Jennings, D. A., Evenson, K. M., Zink, L. R., Demuynck, C., Destombes, J. L., Lemoine, B., and Johns, J. W. C., *J. Mol. Spectrosc.*, 122, 477–480, 1987.
5. Nolt, I. G., Radostitz, J. V., DiLonardo, G., Evenson, K. M., Jennings, D. A., Leopold, K. R., Vanek, M. D., Zink, L. R., Hinz, A., and Chance, K. V., *J. Mol. Spectrosc.*, 125, 274–287, 1987.
6. Jennings, D. A. and Wells, J. S., *J. Mol. Spectrosc.*, 130, 267–268, 1988.
7. High Resolution Wavenumber Standards for the Infrared, *Pure Appl. Chemistry*, 68, 193, 1996.

Wavenumbers for the $v = 1 - 0$ Band of CO

Wavenumber (unc)*/cm⁻¹	Transition	Wavenumber (unc)/cm⁻¹	Transition
		2147.081132(01)	R(0)
2139.426071(01)	P(1)	2150.856006(01)	R(1)
2135.546178(01)	P(2)	2154.595581(01)	R(2)
2131.631574(01)	P(3)	2158.299710(01)	R(3)
2127.682404(01)	P(4)	2161.968245(01)	R(4)
2123.698816(01)	P(5)	2165.601041(01)	R(5)
2119.680957(01)	P(6)	2169.197949(01)	R(6)
2115.628973(01)	P(7)	2172.758824(01)	R(7)
2111.543012(01)	P(8)	2176.283519(01)	R(8)
2107.423221(01)	P(9)	2179.771887(01)	R(9)
2103.269746(01)	P(10)	2183.223782(01)	R(10)
2099.082734(01)	P(11)	2186.639057(01)	R(11)
2094.862333(01)	P(12)	2190.017565(01)	R(12)
2090.608688(01)	P(13)	2193.359161(01)	R(13)
2086.321947(01)	P(14)	2196.663698(01)	R(14)
2082.002256(01)	P(15)	2199.931030(01)	R(15)
2077.649762(01)	P(16)	2203.161010(01)	R(16)
2073.264612(01)	P(17)	2206.353492(01)	R(17)
2068.846952(01)	P(18)	2209.508331(02)	R(18)
2064.396929(01)	P(19)	2212.625379(02)	R(19)
2059.914688(02)	P(20)	2215.704492(02)	R(20)
2055.400377(02)	P(21)	2218.745522(02)	R(21)
2050.854140(02)	P(22)	2221.748326(03)	R(22)
2046.276126(03)	P(23)	2224.712755(03)	R(23)
2041.666479(03)	P(24)	2227.638666(03)	R(24)
2037.025345(03)	P(25)	2230.525912(04)	R(25)
2032.352870(04)	P(26)	2233.374349(04)	R(26)
2027.649200(04)	P(27)	2236.183829(04)	R(27)
2022.914480(04)	P(28)	2238.954210(05)	R(28)
2018.148857(05)	P(29)	2241.685344(05)	R(29)
2013.352474(05)	P(30)	2244.377088(06)	R(30)
2008.525477(06)	P(31)	2247.029296(07)	R(31)
2003.668012(06)	P(32)	2249.641824(08)	R(32)
1998.780224(07)	P(33)	2252.214527(10)	R(33)
1993.862257(09)	P(34)	2254.747262(14)	R(34)
1988.914257(11)	P(35)	2257.239883(18)	R(35)
1983.936367(14)	P(36)	2259.692248(24)	R(36)
1978.928733(18)	P(37)	2262.104213(33)	R(37)
1973.891500(25)	P(38)	2264.475634(45)	R(38)

Wavenumbers for the $v = 1 - 0$ Band of CO

Wavenumber (unc)*/cm⁻¹	Transition	Wavenumber (unc)/cm⁻¹	Transition
1968.824811(34)	P(39)	2266.806368(61)	R(39)
1963.728813(46)	P(40)	2269.096273(81)	R(40)
1958.603648(61)	P(41)	2271.345206(106)	R(41)
1953.449462(82)	P(42)	2273.553027(139)	R(42)

* The uncertainty in the last digits (twice the standard error) is given in parentheses.

Wavenumbers for the $v = 2 - 0$ Band of CO

Wavenumber (unc)*/cm⁻¹	Transition	Wavenumber (unc)/cm⁻¹	Transition
		4263.837198(02)	R(0)
4256.217140(02)	P(1)	4267.542066(02)	R(1)
4252.302244(02)	P(2)	4271.176630(02)	R(2)
4248.317633(02)	P(3)	4274.740746(02)	R(3)
4244.263453(02)	P(4)	4278.234264(02)	R(4)
4240.139852(02)	P(5)	4281.657039(02)	R(5)
4235.946975(02)	P(6)	4285.008924(02)	R(6)
4231.684972(02)	P(7)	4288.289772(02)	R(7)
4227.353987(02)	P(8)	4291.499437(02)	R(8)
4222.954169(02)	P(9)	4294.637773(02)	R(9)
4218.485665(02)	P(10)	4297.704631(02)	R(10)
4213.948620(02)	P(11)	4300.699868(02)	R(11)
4209.343182(02)	P(12)	4303.623334(02)	R(12)
4204.669499(02)	P(13)	4306.474886(02)	R(13)
4199.927716(02)	P(14)	4309.254375(02)	R(14)
4195.117980(02)	P(15)	4311.961657(02)	R(15)
4190.240439(02)	P(16)	4314.596584(02)	R(16)
4185.295239(02)	P(17)	4317.159011(02)	R(17)
4180.282526(02)	P(18)	4319.648791(02)	R(18)
4175.202447(02)	P(19)	4322.065779(03)	R(19)
4170.055149(03)	P(20)	4324.409829(03)	R(20)
4164.840777(03)	P(21)	4326.680794(03)	R(21)
4159.559478(03)	P(22)	4328.878530(03)	R(22)
4154.211398(03)	P(23)	4331.002889(04)	R(23)
4148.796683(04)	P(24)	4333.053728(04)	R(24)
4143.315479(04)	P(25)	4335.030899(05)	R(25)
4137.767932(04)	P(26)	4336.934259(06)	R(26)
4132.154187(05)	P(27)	4338.763661(07)	R(27)
4126.474391(06)	P(28)	4340.518961(09)	R(28)
4120.728689(07)	P(29)	4342.200014(11)	R(29)
4114.917226(09)	P(30)	4343.806675(16)	R(30)
4109.040148(12)	P(31)	4345.338799(21)	R(31)
4103.097600(16)	P(32)	4346.796243(29)	R(32)
4097.089728(21)	P(33)	4348.178862(40)	R(33)
4091.016676(29)	P(34)	4349.486513(54)	R(34)
4084.878591(40)	P(35)	4350.719052(73)	R(35)
4078.675618(54)	P(36)	4351.876336(96)	R(36)
4072.407901(73)	P(37)	4352.958224(127)	R(37)
4066.075588(97)	P(38)	4353.964572(166)	R(38)
4059.678822(127)	P(39)	4354.895240(214)	R(39)

* The uncertainty in the last digits (twice the standard error) is given in parentheses.

Wavenumbers for the $v = 3 - 0$ Band of CO

Wavenumber (unc)*/cm⁻¹	Transition	Wavenumber (unc)/cm⁻¹	Transition
		6354.179057(13)	R(0)
6346.594000(13)	P(1)	6357.813923(13)	R(1)
6342.644103(13)	P(2)	6361.343487(13)	R(2)
6338.589491(13)	P(3)	6364.767599(13)	R(3)
6334.430309(13)	P(4)	6368.086115(13)	R(4)

Wavenumbers for the $\nu = 3 - 0$ Band of CO

Wavenumber (unc)*/cm^{-1}	Transition	Wavenumber (unc)/cm^{-1}	Transition
6330.166705(13)	P(5)	6371.298887(13)	R(5)
6325.798826(13)	P(6)	6374.405768(12)	R(6)
6321.326819(13)	P(7)	6377.406611(12)	R(7)
6316.750831(12)	P(8)	6380.301271(12)	R(8)
6312.071008(12)	P(9)	6383.089600(12)	R(9)
6307.287498(12)	P(10)	6385.771452(12)	R(10)
6302.400447(12)	P(11)	6388.346680(13)	R(11)
6297.410003(12)	P(12)	6390.815139(13)	R(12)
6292.316311(13)	P(13)	6393.176681(13)	R(13)
6287.119520(13)	P(14)	6395.431160(13)	R(14)
6281.819775(13)	P(15)	6397.578430(13)	R(15)
6276.417224(13)	P(16)	6399.618344(13)	R(16)
6270.912012(13)	P(17)	6401.550757(13)	R(17)
6265.304287(13)	P(18)	6403.375523(13)	R(18)
6259.594194(13)	P(19)	6405.092495(14)	R(19)
6253.781880(13)	P(20)	6406.701527(14)	R(20)
6247.867492(14)	P(21)	6408.202474(14)	R(21)
6241.851176(14)	P(22)	6409.595189(15)	R(22)
6235.733077(14)	P(23)	6410.879527(15)	R(23)
6229.513342(15)	P(24)	6412.055343(16)	R(24)
6223.192117(15)	P(25)	6413.122491(17)	R(25)
6216.769547(16)	P(26)	6414.080825(19)	R(26)
6210.245778(17)	P(27)	6414.930201(23)	R(27)
6203.620957(19)	P(28)	6415.670474(28)	R(28)
6196.895229(23)	P(29)	6416.301500(37)	R(29)
6190.068739(28)	P(30)	6416.823133(50)	R(30)
6183.141633(37)	P(31)	6417.235231(67)	R(31)
6176.114058(50)	P(32)	6417.537649(90)	R(32)
6168.986159(67)	P(33)		
6161.758082(90)	P(34)		

* The uncertainty in the last digits (twice the standard error) is given in parentheses.

Frequencies and Wavenumbers for the Rotational Lines of CO

Frequency/MHz	Uncertainty*/MHz	J'	J''	Wavenumber/cm^{-1}	Uncertainty*/cm^{-1}
115271.2029	0.0004	1	0	3.84503345	0.00000001
230538.0016	0.0008	2	1	7.68991999	0.00000003
345795.9923	0.0012	3	2	11.53451273	0.00000004
461040.7712	0.0016	4	3	15.37866477	0.00000005
576267.9350	0.0019	5	4	19.22222923	0.00000006
691473.0809	0.0021	6	5	23.06505926	0.00000007
806651.8065	0.0023	7	6	26.90700800	0.00000008
921799.7104	0.0025	8	7	30.74792863	0.00000008
1036912.3919	0.0027	9	8	34.58767438	0.00000009
1151985.4515	0.0029	10	9	38.42609848	0.00000010
1267014.4906	0.0031	11	10	42.26305422	0.00000010
1381995.1119	0.0034	12	11	46.09839491	0.00000011
1496922.9195	0.0038	13	12	49.93197392	0.00000013
1611793.5189	0.0042	14	13	53.76364468	0.00000014
1726602.5173	0.0047	15	14	57.59326065	0.00000016
1841345.5237	0.0052	16	15	61.42067535	0.00000017
1956018.1486	0.0057	17	16	65.24574239	0.00000019
2070616.0050	0.0061	18	17	69.06831542	0.00000020
2185134.7075	0.0065	19	18	72.88824816	0.00000022
2299569.8733	0.0069	20	19	76.70539441	0.00000023
2413917.1217	0.0071	21	20	80.51960806	0.00000024
2528172.0747	0.0073	22	21	84.33074306	0.00000024
2642330.3567	0.0074	23	22	88.13865346	0.00000025
2756387.5949	0.0075	24	23	91.94319341	0.00000025

Frequencies and Wavenumbers for the Rotational Lines of CO

Frequency/MHz	Uncertainty*/MHz	J′	J″	Wavenumber/cm⁻¹	Uncertainty*/cm⁻¹
2870339.4194	0.0077	25	24	95.74421713	0.00000026
2984181.4631	0.0080	26	25	99.54157896	0.00000027
3097909.3621	0.0085	27	26	103.33513334	0.00000028
3211518.7558	0.0090	28	27	107.12473480	0.00000030
3325005.2869	0.0096	29	28	110.91023800	0.00000032
3438364.6013	0.0102	30	29	114.69149772	0.00000034
3551592.3489	0.0107	31	30	118.46836884	0.00000036
3664684.1829	0.0111	32	31	122.24070637	0.00000037
3777635.7608	0.0118	33	32	126.00836545	0.00000039
3890442.7435	0.0137	34	33	129.77120137	0.00000046
4003100.7965	0.0179	35	34	133.52906952	0.00000060
4115605.5892	0.0254	36	35	137.28182546	0.00000085
4227952.7954	0.0370	37	36	141.02932487	0.00000123
4340138.0932	0.0531	38	37	144.77142361	0.00000177
4452157.1657	0.0746	39	38	148.50797766	0.00000249
4564005.7001	0.1025	40	39	152.23884318	0.00000342

* The uncertainty given is twice the standard error.

Frequencies and Wavenumbers for the Rotational Lines of HF

Frequency/MHz	Uncertainty*/MHz	J′	J″	Wavenumber/cm⁻¹	Uncertainty*/cm⁻¹
1232476.21	0.12	1	0	41.110981	0.000004
2463428.09	0.19	2	1	82.171116	0.000006
3691334.81	0.25	3	2	123.129676	0.000008
4914682.58	0.51	4	3	163.936165	0.000017
6131968.11	1.10	5	4	204.540439	0.000037
7341702.00	2.00	6	5	244.892818	0.000067
8542412.1	3.21	7	6	284.944197	0.000107
9732646.8	4.72	8	7	324.646153	0.000157
10910978.2	6.51	9	8	363.951056	0.000217
12076004.8	8.55	10	9	402.81216	0.000285
13226355.2	10.81	11	10	441.18372	0.000361
14360689.8	13.25	12	11	479.02105	0.00044
15477704.4	15.86	13	12	516.28065	0.00053
16576131.8	18.61	14	13	552.92024	0.00062
17654744.4	21.48	15	14	588.89888	0.00072
18712356.5	24.44	16	15	624.17703	0.00082
19747825.6	27.43	17	16	658.71656	0.00092
20760054.3	30.32	18	17	692.4809	0.00101
21747991.7	32.91	19	18	725.4349	0.00110
22710634.7	34.94	20	19	757.5452	0.00117
23647028.7	36.08	21	20	788.7800	0.00120
24556268.8	35.93	22	21	819.1090	0.00120
25437499.9	34.12	23	22	848.5037	0.00114
26289917.4	30.32	24	23	876.9373	0.00101
27112767.2	24.41	25	24	904.38457	0.00081
27905345.6	16.88	26	25	930.82214	0.00056
28666999.3	10.80	27	26	956.22817	0.00036
29397124.8	14.65	28	27	980.58253	0.00049
30095168.2	24.62	29	28	1003.86676	0.00082
30760624.2	33.36	30	29	1026.0640	0.00111
31393035.7	36.17	31	30	1047.1590	0.00121

* The uncertainty given is twice the standard error.

Frequencies and Wavenumbers for the Rotational Lines of $H^{35}Cl$

Frequency/MHz	Uncertainty*/MHz	J'	J''	Wavenumber/cm^{-1}	Uncertainty*/cm^{-1}
1876226.517	0.065	3	2	62.584180	0.000002
2499864.439	0.066	4	3	83.386502	0.000002
3121986.563	0.064	5	4	104.138262	0.000002
3742216.601	0.076	6	5	124.826909	0.000003
4360180.042	0.098	7	6	145.439951	0.000003
4975504.51	0.11	8	7	165.964966	0.000004
5587820.10	0.12	9	8	186.389615	0.000004
6196759.76	0.22	10	9	206.701656	0.000007
6801959.63	0.50	11	10	226.888951	0.000017
7403059.41	1.02	12	11	246.939481	0.000034
7999702.7	1.8	13	12	266.841359	0.000062
8591537.3	3.1	14	13	286.582837	0.000103
9178215.8	4.8	15	14	306.152324	0.000161

* The uncertainty given is twice the standard error.

Frequencies and Wavenumbers for the Rotational Lines of $H^{37}Cl$

Frequency/MHz	Uncertainty*/MHz	J'	J''	Wavenumber/cm^{-1}	Uncertainty*/cm^{-1}
1873410.72	0.05	3	2	62.490255	0.000002
2496115.33	0.05	4	3	83.261445	0.000002
3117308.69	0.05	5	4	103.982225	0.000002
3736615.64	0.06	6	5	124.640082	0.000002
4353662.84	0.08	7	6	145.222561	0.000003
4968079.04	0.09	8	7	165.717279	0.000003
5579495.53	0.10	9	8	186.111938	0.000003
6187546.42	0.19	10	9	206.394332	0.000006
6791869.04	0.45	11	10	226.552365	0.000015
7392104.3	0.9	12	11	246.574057	0.000030
7987896.9	1.6	13	12	266.447561	0.000054
8578896.1	2.7	14	13	286.161170	0.000089

* The uncertainty given is twice the standard error.

Section 11
Nuclear and Particle Physics

SUMMARY TABLES OF PARTICLE PROPERTIES

Details of particle properties may be found in the following publication:

K. Nakamura *et al.* (Particle Data Group), *J. Phys. G* **37**, 075021 (2010).

Abstract. This biennial *Review* summarizes much of particle physics. Using data from previous editions, plus 2158 new measurements from 551 papers, we list, evaluate, and average measured properties of gauge bosons, leptons, quarks, mesons, and baryons. We also summarize searches for hypothetical particles such as Higgs bosons, heavy neutrinos, and supersymmetric particles. All the particle properties and search limits are listed in Summary Tables. We also give numerous tables, figures, formulae, and reviews of topics such as the Standard Model, particle detectors, probability, and statistics. Among the 108 reviews are many that are new or heavily revised, including those on CKM quark-mixing matrix, V_{ud} & V_{us}, V_{cb} & V_{ub}, fragmentation functions, particle detectors for accelerator and non-accelerator physics, magnetic monopoles, cosmological parameters, and big bang cosmology. A booklet is available containing the Summary Tables and abbreviated versions of some of the other sections of this full *Review*. An electronic version of this publication is available on the web site of the Particle Data Group: http://pdg.lbl.gov. The Summary Tables are included in the DVD and Internet versions of the 92nd Edition of the *CRC Handbook of Chemistry and Physics*.

TABLE OF THE ISOTOPES

Norman E. Holden

This table presents an evaluated set of values for the experimental quantities that characterize the decay of radioactive nuclides. A list of the major references used in this evaluation is given below. When uncertainties are not listed, they are assumed to be five or less in the last digit quoted. If the uncertainty in the value exceeds five in the last digit, the value is preceded by an approximate sign.

For quasi-stable nuclides, the measured width, Γ, of the resonance is given. To estimate the approximate half-life, the Heisenberg relationship may be used, the half-life = 4.56×10^{-22} seconds / Γ (MeV). The effective literature cutoff date for data in this edition of the table is December 2010.

Table Layout

Column number	Column title	Description
1	Isotope or Element	For elements, the atomic number and chemical symbol are listed. For nuclides, the mass number and chemical symbol are listed. Isomers are indicated by the addition of m, m1, or m2.
2	Isotopic Abundance	The abundance of an isotope in normal terrestrial samples of an element, listed in atom percent.
3	Atomic Mass or Atomic Weight	Atomic mass relative to $^{12}C = 12$. Atomic weight of elements is given on the same scale.
4	Half-life/Resonance Width	Half-life in decimal notation. μs = microseconds; ms = milliseconds; s = seconds; m = minutes; h = hours; d = days; and a = years. For quasi-stable nuclides, the measured width at half maximum of the energy resonance is given.
5	Decay Mode/Energy	Decay modes are α = alpha particle emission; β = negative beta emission; β+ = positron emission; EC = orbital electron capture; IT = isomeric transition from upper to lower isomeric state; n = neutron emission; p = proton emission; sf = spontaneous fission; ββ, ECEC, 2p, 2n, 3n = double beta, double EC, two proton and multiple neutron decay. Total disintegration energy in MeV units.
6	Particle Energy/Intensity	End point energies of beta transitions and discrete energies of alpha particles, neutrons and protons are given in MeV units and their intensities in percent.
7	Spin and Parity	Nuclear spin or angular momentum of the nuclides in units of $h/2\pi$; parity is positive or negative.
8	Magnetic Dipole Moment	Magnetic dipole moments in nuclear magneton units. An absolute value is indicated in the absence of a positive or a negative sign.
9	Electric Quadrupole Moment	Electric quadrupole moments in barn units (10^{-24} cm^2). An absolute value is indicated in the absence of a positive or a negative sign.
10	Gamma Ray Energy/Intensity	Gamma ray energies are given in MeV units and intensities in percent. Annihilation radiation (Ann.Rad.) refers to the 511.006 keV photons emitted in the annihilation of positrons in matter.

General Nuclear Data References

The following references represent the major sources of the nuclear data presented, along with subsequent published journal articles and reports:

1. G. Audi, O. Bersillon, J. Blachot, A.H. Wapstra, *The Nubase Evaluation of Nuclear and Decay Properties*, Nuclear Physics A729, 3 (2003).
2. G. Audi, A.H. Wapstra, C. Thibault, *The AME2003 Atomic Mass Evaluation (II)*, Nuclear Physics A729, 337 (2003).
3. M.E. Wieser, *Atomic Weights of the Elements - 2005*, Pure & Applied Chemistry 78, 2051 (2006).
4. E.M. Baum, H.D. Knox, T.R. Miller, *Chart of the Nuclides, 17th Edition*, Knolls Atomic Power Lab (2009).
5. N.E. Holden, *Total and Spontaneous Fission Half-lives for Uranium, Plutonium, Americium and Curium Nuclides*, Pure & Applied Chemistry 61, 1483 (1989).
6. N.E. Holden, *Half-lives of Selected Nuclides*, Pure & Applied Chemistry 62, 941 (1990).
7. M. Berglund, M. E. Wieser, *Isotopic Composition of the Elements – 2009, Pure & Applied Chemistry* 83, 397 (2011).
8. P. Raghavan, *Table of Nuclear Moments*, Atomic Data Nuclear Data Tables 42, 189 (1989).
9. E. Brown, R. Firestone, *Radioactivity Handbook*, Wiley Interscience Press (1986).
10. J.K. Tuli, *Nuclear Wallet Cards*, Brookhaven National Laboratory (April 2005).
11. N.E. Holden, D.C. Hoffman, *Spontaneous Fission Half-lives for Ground State Nuclides*, Pure & Applied Chemistry 72, 1525 (2000).
12. N.J. Stone, *Table of Nuclear Magnetic Dipole and Electrical Quadrupole Moments*, Atomic Data Nuclear Data Tables 90, 75 (2005).

This research was carried out under the auspices of the US Department of Energy Contract No. DE-AC02-98CH10886. The author is at Brookhaven National Laboratory, Upton, NY, and can be contacted at holden@bnl.gov.

Elem. or Isot.	Natural Abundance (Atom %)	Atomic Mass or Weight	Half–life/ Resonance Width (MeV)	Decay Mode/ Energy (/MeV)	Particle Energy/ Intensity (MeV/%)	Spin ($h/2\pi$)	Nuclear Magnetic Mom. (nm)	Elect. Quadr. Mom. (b)	γ-Energy/ Intensity (MeV/%)
$_0$n		1.008664916	610. s	β- /0.78235	0.782/100.	1/2+	−1.913043		
				β-, γ	/ 0.0031				
$_1$H		**1.00784– 1.00811**							
^{1}H	99.9885(70)	1.007825032	>2.8 × 10^{23} a			1/2+	+2.7928473		
^{2}H	0.0115(70)	2.014101778				1+	+0.8574382	+2.86 mb	
^{3}H		3.016049278	12.31 a	β-/0.01859	0.01860/100.	1/2+	+2.9789624		
^{4}H		4.0278	Γ = 1.2	n/	/100	2-			
^{5}H		5.0353	Γ < 0.5	n/	/100	(1/2+)			
^{6}H		6.0449	Γ ~ 6.	n/		(2-)			
^{7}H		7.053	Γ ~ 0.1						
$_2$He		**4.002602(2)**							
^{3}He	0.000134(3)	3.016029319				1/2+	−2.127750		
^{4}He	99.999866(3)	4.002603254				0+			
^{5}He		5.01222	Γ = 0.60(2)	n, α		3/2-			
^{6}He		6.018889	0.807 s	β-/3.508	3.510/100.	0+			
				β, d	0.35/0.00017				
^{7}He		7.02802	Γ ~ 0.19	n		(3/2)-			
^{8}He		8.03392	0.119 s	β-/10.65	/84.	0+			0.9807/84.
				β- n/	/16.				0.4776/5.
				β, t	/0.82				
^{9}He		9.04395	Γ = 0.10(6)	n	/100	(1/2-)			
^{10}He		10.0524	Γ = 0.1	2n	/100	0+			
$_3$Li		**6.938–6.997**							
^{4}Li		4.0272	Γ = 6.0	p/	/100	2-			
^{5}Li		5.01254	Γ = 1.2	p/α		3/2-			
^{6}Li	7.59(4)	6.01512280				1+	+0.822047	−0.82 mb	
^{7}Li	92.41(4)	7.0160046				3/2-	+3.25644	−0.0406	
^{8}Li		8.0224874	0.839 s	β-/16.004	12.5/100.	2+	+1.65334	+0.0314	
				α/	α(1.6)				
^{9}Li		9.026790	0.178 s	β-/13.606	13.5/75.	3/2-	3.4368	−0.0306	
				β-/	11/25.				
^{10}Li		10.03548	Γ = 0.11(5)	n	/7.	1+			
^{11}Li		11.04380	8.8 ms	β-/20.6	/8.3	3/2(-)	3.671	−0.033	3.368/33.
				β-, n	/85.7				0.320/7.
				β-, 2n	/4.1				2.590/8.
				β-, 3n	/1.9				5.958/3.
				β-, d	/0.013				2.895/1.5
				β-, t	/0.00009				2.811/1.1
^{12}Li		12.054	< 0.01 μs						
^{13}Li			Γ ~ 1.5						
$_4$Be		**9.012182(3)**							
^{5}Be		5.041		p, ^{3}He		(1/2+)			
^{6}Be		6.01973	Γ = 0.092(6)	2p,α		0+			
^{7}Be		7.0169298	53.28 d	EC/0.8618		3/2-	−1.40		0.4776/10.4
^{8}Be		8.00530510	Γ = 6.8(17) eV	2α/0.046		0+			
^{9}Be	100.	9.0121822				3/2-	−1.1776	+0.0529	
^{10}Be		10.0135338	1.39 × 10^6 a	β-/0.5559	0.555/100.	0+			
^{11}Be		11.02166	13.8 s	β-, β-α/11.51	11.48/61.	1/2+	−1.681		2.125/35.5
^{12}Be		12.02692	22.0 ms	β-, (n)/11.71	n//0.5	0+			(0.95–4.4)
^{13}Be		13.0357	Γ ~ 1.						
^{14}Be		14.0429	4.6 ms	β-/16.2		0+			3.5346/0.9
				β-, n	0.288/94.				3.6845/7.
				β-, 2n	/6.				

Elem. or Isot.	Natural Abundance (Atom %)	Atomic Mass or Weight	Half–life/ Resonance Width (MeV)	Decay Mode/ Energy (/MeV)	Particle Energy/ Intensity (MeV/%)	Spin ($h/2\pi$)	Nuclear Magnetic Mom. (nm)	Elect. Quadr. Mom. (b)	γ-Energy/ Intensity (MeV/%)
				β-, α	/<0.012				
				β-, t	/<0.04				
^{15}Be		15.053	< 0.2 μs	β-					
^{16}Be		16.062	< 0.2 μs	β-		0+			
$_5$**B**		**10.806–10.821**							
^{7}B		7.0299	Γ = 1.4(2)	p, α		(3/2-)			
^{8}B		8.024607	0.770 s	β+, 2α/17.979	13.7(β+)/93.	2+	1.0358	+0.065	ann.rad.
^{9}B		9.013329	Γ = 0.5(2) keV	p, 2α/		3/2-			
^{10}B	19.9(7)	10.0129370				3+	+1.8006448	+0.085	
^{11}B	80.1(7)	11.0093054				3/2-	+2.688649	+0.0407	
^{12}B		12.014352	0.0202 s	β-/13.369		1+	+1.0027	0.0132	4.438/1.3
				β- α/1.6/					3.215/0.00065
^{13}B		13.017780	0.0174 s	β- /13.437	13.4	3/2-	+3.1778	+0.037	3.68/7.6
				β- n/0.25/	2.43(n)/0.09				
					3.55(n)/0.16				
^{14}B		14.02540	14. ms	β-/20.64		2-	1.185	0.030	6.094/90.
^{15}B		15.03110	9.9 ms	β-, (n)/19.09	n//99.7	(3/2-)	2.66	0.038	
^{16}B		16.0398	Γ < 0.1	n					
^{17}B		17.0470	5.1 ms	β-, (n)/22.7			2.55	0.039	
^{18}B		18.056	< 0.026 μs			0-			
^{19}B		19.0637	2.9 ms	β-, (n)/26.5	1n//72.	(3/2-)			
					2n//16.				
					3n// < 9.				
$_6$**C**		**12.0096– 12.0116**							
^{8}C		8.03768	Γ = 0.25(4)	p		0+			
^{9}C		9.031037	127. ms	β+, p, 2α/16.498		(3/2-)	1.391		ann.rad.
^{10}C		10.0168532	19.3 s	β+/3.648	1.865	0+			ann.rad.
									0.71829/100.
^{11}C		11.011434	20.3 m	β+, EC/1.982	0.9608/99.	3/2-	−0.964	0.032	ann.rad.
^{12}C	98.93(8)	12.000000000				0+			
^{13}C	1.07(8)	13.003354838				½-	+0.702412		
^{14}C		14.003241989	5715. a	β-/0.15648	0.1565/100.	0+			
^{15}C		15.010599	2.45 s	β-/9.772	4.51/68.	½+	1.72		5.298/68.
					9.82/32.				(7.30–9.05)
^{16}C		16.014701	~ 0.750 s	β-/8.012	β/3.3, 4.3/84, 16	0+			
				β, n	n/0.8, 1.7/84, 16				
^{17}C		17.02259	0.19 s	β-/13.17		3/2+	0.758		1.375
				β-, n	n/1.6–3.7/11.				1.849
									1.906
^{18}C		18.02676	0.092 s	β-/11.81		0+			
				β-, n	n/0.88–4.59/21.				
^{19}C		19.0348	0.05 s	n		½+			
^{20}C		20.0403	0.02 s	β,n	1n// ~ 65.	0+			
					2n// < 19.				
^{21}C		21.049	< 0.03 μs						
^{22}C		22.057	6 ms	β-, n	1n// ~ 61.	0+			
					2n// < 37.				
$_7$**N**		**14.00643– 14.00728**							
^{10}N		10.0417	Γ = 2.3(16)						
^{11}N		11.02609	Γ ~ 1.			½+			
^{12}N		12.018613	11.00 ms	β+, β+α/17.338	16.38/95.	1+	+0.457	+10. mb	ann.rad.
									4.438/2.

Elem. or Isot.	Natural Abundance (Atom %)	Atomic Mass or Weight	Half–life/ Resonance Width (MeV)	Decay Mode/ Energy (/MeV)	Particle Energy/ Intensity (MeV/%)	Spin ($h/2\pi$)	Nuclear Magnetic Mom. (nm)	Elect. Quadr. Mom. (b)	γ-Energy/ Intensity (MeV/%)
^{13}N		13.0057386	9.97 m	β+ /2.2204	1.190/100.	½–	0.3222		
^{14}N	99.636(20)	14.003074005				1+	+0.403761	+0.02044	
^{15}N	0.364(20)	15.00010898				½–	−0.283189		
^{16}N		16.006102	7.13 s	β– /10.419	4.27/68.	2–			6.129/68.8
					10.44/26.		1.986	18 mb	7.115/4.7
				β-, α	1.85/.0012				(0.99–8.87)
					(0.6–2.3)				
^{17}N		17.00845	4.17 s	β-, β- n/8.68	3.7/100.	½–	0.3551		0.871/3.
				0.4–1.7n/95.					2.1842/0.3
				β- α/	8.0, 8.2				
^{18}N		18.01408	0.62 s	β- /13.90	9.4/100.	1–	0.3273	0.012	0.822/48.
				β-, α	(1.08–5.23)/12.				1.65/47.
				β-, n	/14				1.982/77.
									(0.535–7.13)
^{19}N		19.01703	0.336 s	β-/12.53			0.305		(0.096–3.14)
				β-, n	0.45–4.51/42.				
^{20}N		20.0234	0.136 s	β- /17.97					
				β-, n	1.10–3.68/43.				
^{21}N		21.0271	0.083 s	β-,n	4.98//90.5				1.222
				β-	//3.5				
^{22}N		22.0344	0.02 s	β-,n	1n// ~ 41.				
					2n// < 13.				
^{23}N		23.0412	15. ms	β-, n	n// ~ 42.				
					2n// ~ 8.				
					3n// < 3.4				
^{24}N		24.0510	< 0.052 µs						
^{25}N		25.061	< 0.26 µs						
$_{8}$O		**15.99903– 15.99977**							
^{12}O		12.03441	Γ = 0.51(16)	2p		0+			
^{13}O		13.02481	8.9 ms	β+, p/17.77	1.560 (p)	(3/2-)	1.389	0.011	ann.rad.
					p/(1.00–13.5)				4.438/0.56
^{14}O		14.0085963	70.63 s	β+ /5.1430	1.81/99.	0+			ann.rad.
									2.312/99.4
^{15}O		15.0030656	122.2 s	β+ /2.754	1.723/100.	1/2-	0.7195		ann.rad.
^{16}O	99.757(16)	15.9949146196				0+			
^{17}O	0.038(1)	16.9991317				5/2+	−1.8938	−0.026	
^{18}O	0.205(14)	17.999161				0+			
^{19}O		19.003580	26.9 s	β- /4.820	3.25/60.	5/2+	1.5320	3.7 mb	0.197/95.9
					4.60/40.				1.3569/50.4
									(0.11–4.18)
^{20}O		20.004077	13.5 s	β- /3.814		0+			1.057/100.
^{21}O		21.00866	3.4 s	β- /8.11					(0.28–4.6)
^{22}O		22.0100	2.2 s	β- /6.5		0+			0.072/100.
									0.638/98
									1.862/63
									(0.918-2.499)
^{23}O		23.0157	0.097 s	β-, n	/7.2%				2.234/51.5
									4.066/17.1
									(0.911-3.868)
^{24}O		24.0205	~ 65. ms	β-, n	n//18.	0+			1.83/28.
									0.52/14.
									1.31/12.
^{25}O		25.0295	< 0.05 µs						

Elem. or Isot.	Natural Abundance (Atom %)	Atomic Mass or Weight	Half–life/ Resonance Width (MeV)	Decay Mode/ Energy (/MeV)	Particle Energy/ Intensity (MeV/%)	Spin ($h/2\pi$)	Nuclear Magnetic Mom. (nm)	Elect. Quadr. Mom. (b)	γ-Energy/ Intensity (MeV/%)
^{26}O		26.0383	< 0.04 μs			0+			
^{27}O		27.048	< 0.026 μs						
^{28}O		28.058	< 0.10 μs			0+			
$_9$F		18.9984032(5)							
^{14}F		14.0351	Γ = 0.9						
^{15}F		15.0180	Γ = 0.8(3)	p		(1/2+)			
^{16}F		16.01147	Γ = 0.037(14)	p		0-			
^{17}F		17.0020952	64.5 s	β+ /2.761	1.75/	5/2+	+4.721	−0.08	ann.rad.
^{18}F		18.000938	1.8287 h	β+, EC/1.656	0.635/97.	1+			ann.rad.
^{19}F	100.	18.9984032				½+	+2.62887		
^{20}F		19.9999813	11.00 s	β- /7.0245	5.398/100.	2+	+2.0934	0.042	1.634/100.
									3.33/0.009
^{21}F		20.999949	4.16 s	β- /5.684	3.7/8.	5/2+	3.9		0.3507/90.
					5.0/63.				1.395/15.
					5.4/29.				(1.746–4.684)
^{22}F		22.00300	4.23 s	β- /10.82	3.48/15.	4+	2.694	0.003	1.2746/100.
					4.67/7.				2.0826/82.
					5.50/62.				(0.82–4.37)
^{23}F		23.0036	2.2 s	β- /8.5		5/2+			1.701/48.
									2.129/34.
									(0.493–3.83)
^{24}F		24.0081	0.3 s	β- /13.5					1.9816/
^{25}F		25.0121	~ 50. ms	β-, (n)	n//14.				1.70/39.
									(0.57–2.19)
^{26}F		26.0196	10. ms	β-, (n)	n//11.				2.02/67.
									1.67/19.
^{27}F		27.0268	5.0 ms	β-, (n)	n//90.				2.02/18.
^{28}F		28.036	< 0.04 μs						
^{29}F		29.043	2.5 ms	β-, (n)	n//100.				
^{30}F		30.053	< 0.26 μs						
^{31}F		31.060	> 0.26 μs						
$_{10}$Ne		20.1797(6)							
^{16}Ne		16.02576	Γ = 0.12(4)	2p		0+			
^{17}Ne		17.01767	109. ms	β+, p/14.53	1.4–10.6/6.9	1/2-	0.787		ann.rad./
				β+, α	/0.014				0.495
^{18}Ne		18.0057082	1.668 s	β+ /4.446	3.416/92.	0+			ann.rad./
									1.0413/7.8
									(0.658–1.70)
^{19}Ne		19.0018802	17.22 s	β+ /3.238	2.24/99.	1/2+	−1.885		ann.rad./
									(0.11–1.55)
^{20}Ne	90.48(3)	19.992440175				0+			
^{21}Ne	0.27(1)	20.99384668				3/2+	−0.66180	+0.103	
^{22}Ne	9.25(3)	21.99138511				0+			
^{23}Ne		22.9944669	37.2 s	β- /4.376	3.95/32.	5/2+	−1.08	+0.15	0.440/33.
					4.39/67.				(1.64–2.98)
^{24}Ne		23.9936108	3.38 m	β- /2.47	1.10/8.	0+			0.4723/100.
					1.98/92.				0.874/7.9
^{25}Ne		24.99774	0.61 s	β- /7.30	6.3/	1/2+	−1.006		0.0895/96.
					7.3/				(0.98–3.69)
^{26}Ne		26.00046	197 ms	β-, n/7.3	n //0.13	0+			0.082/100
									1.278/6
									0.233/5
									0.151/3
									1.211/1

Elem. or Isot.	Natural Abundance (Atom %)	Atomic Mass or Weight	Half–life/ Resonance Width (MeV)	Decay Mode/ Energy (/MeV)	Particle Energy/ Intensity (MeV/%)	Spin ($h/2\pi$)	Nuclear Magnetic Mom. (nm)	Elect. Quadr. Mom. (b)	γ-Energy/ Intensity (MeV/%)
									2.489/1
27Ne		27.0076	31. ms	β-, n/12.7	n //3.	(3/2+)			
28Ne		28.0121	19. ms	β-, n/12.3	n//12.	0+			2.06/19.
					2n//3.				0.86/3.
29Ne		29.0194	15. ms	β-, (n)/15.4	n//29.	(3/2+)			2.92/55.
				β-, 2n	2n//4.				(0.22–1.18)
30Ne		30.025	7. ms	β-, (n)	n//9.	0+			0.151/9.
31Ne		31.033	3. ms						
32Ne		32.040	~ 3.5 ms			0+			
33Ne		33.049	< 0.26 μs						
34Ne		34.057	> 1.5 μs			0+			
11 **Na**		**22.98976928(2)**							
18Na		18.02597	Γ = 0.34(9)						
19Na		19.01388	0.03 s	β+, p/11.18					
20Na		20.00735	0.446 s	β+ /13.89		2+	+0.3694	+0.10	ann.rad./
				α	2.15/16				1.634/79.
21Na		20.997655	22.48 s	β+ /3.547	2.50/95.	3/2+	+2.3863	+0.14	ann.rad./
									0.351/5.
22Na		21.9944364	2.605 a	β+ /90/2.842	0.545/90.	3+	+1.746	+0.19	ann.rad./
				EC/10/					1.2745/99.9
23Na	100.	22.989769281				3/2+	+2.21752	+0.106	
24mNa			20.2 ms	I.T., β-		1+	−1.93		0.4723/100.
24Na		23.9909628	14.96 h	β- /5.5158	1.389/>99.	4+	+1.690		1.3686/100.
									2.754/100.
									(0.997–4.238)
25Na		24.989954	59.3 s	β- /3.835	2.6/7.	5/2+	+3.683	−0.0015	0.3897/12.7
					3.15/25.				0.5850/13.
					4.0/65.				0.9747/14.9
									(0.836–2.80)
26Na		25.99263	1.071 s	β- /9.31		3+	+2.851	−5.3 mb	1.809/98.9
									(0.24-7.37)
27Na		26.994077	0.290 s	β- /9.01	7.95/	5/2+	+3.90	−7.2 mb	0.9847/87.4
				β-, n/					1.698/11.9
28Na		27.99894	31. ms	β- /14.0	12.3/	1+	+2.43	+0.040	1.473/37.
				β-, n/					2.389/18.6
29Na		29.00286	44. ms	β-, n/13.3	11.5/	3/2+	+2.45	+86. mb	2.560/36.
									(1.04–3.99)
30Na		30.00898	50. ms	β-, n/17.5	n//30.	2+	+2.08		1.483/46.
31Na		31.0136	17.2 ms	β-, n/15.9	n//37.	3/2-	+2.30		1.483/14.
									(0.05–3.54)
32Na		32.0205	13.5 ms	β- /19.1					0.240–3.935
				β-, n					0.171-0.894
				β-, 2n					1.483/4.2
33Na		33.027	8.0 ms	β- /20.	/ ~ 38				0.886/16
				β-, n	0.8,1.02/47(6)				0.546/6.4
				β-, 2n	/13(3)				0.050–2.55
34Na		34.035	5. ms	β- /24.					
35Na		35.042	1.5 ms	β- /24					
36Na		36.051	< 0.26 μs						
37Na		37.059	> 1.5 μs						
12 **Mg**		**24.3050(6)**							
19Mg		19.0355	4 x 10⁻¹² s	2p					
20Mg		20.01886	96. ms	β+ /10.73	/70	0+			
				β+, p	/30				

Elem. or Isot.	Natural Abundance (Atom %)	Atomic Mass or Weight	Half–life/ Resonance Width (MeV)	Decay Mode/ Energy (/MeV)	Particle Energy/ Intensity (MeV/%)	Spin ($h/2\pi$)	Nuclear Magnetic Mom. (nm)	Elect. Quadr. Mom. (b)	γ-Energy/ Intensity (MeV/%)
^{21}Mg		21.01171	122. ms	β+, p/13.10		5/2+	−0.98		0.332/51.
^{22}Mg		21.999574	3.876 s	β+ /4.786	3.05/	0+			0.0729/60.
									0.5820/100.
									(1.28−1.93)
^{23}Mg		22.994124	11.32 s	β+ /4.057	3.09/92.	3/2+	0.536	1.25	0.440/8.2
^{24}Mg	78.99(4)	23.98504170				0+			
^{25}Mg	10.00(1)	24.98583692				5/2+	−0.85545	+0.200	
^{26}Mg	11.01(3)	25.98259293				0+			
^{27}Mg		26.98434059	9.45 m	β- /2.6103	1.59/41.	1/2+	−0.411		0.17068/0.9
					1.75/58.				0.84376/72.
					2.65/0.3				1.01443/28.
^{28}Mg		27.983877	20.9 h	β- /1.832	0.459/95.	0+			0.0306/95.
									0.4006/36.
									0.9418/36.
									1.342/54.
^{29}Mg		28.98860	1.3 s	β-/7.55	5.4/	3/2+	+0.978		0.960/15.
									1.398/16.
									2.224/36.
^{30}Mg		29.99043	0.32 s	β- /7.0		0+			0.224/85.
^{31}Mg		30.99655	0.24 s	β- /11.7	8.4/29.9	1/2+	−0.8836		1.613/47.
									0.947/37.
									(0.666-4.640)
				β-, n	/1.7				
^{32}Mg		31.99898	0.12 s	β- /10.3		0+			2.765/25.
^{33}Mg		33.00525	91. ms	β- /13.7	/83.				1.848/
				β-, n	/17.				
^{34}Mg		34.0095	0.02 s	β- /11.3		0+			
^{35}Mg		35.0173	0.07 s			(7/2-)			
^{36}Mg		36.023	4. ms			0+			
^{37}Mg		37.031	> 0.26 μs			(7/2-)			
^{38}Mg		38.038	> 0.26 μs			0+			
^{39}Mg		39.047	< 0.26 μs						
^{40}Mg		40.054				0+			
$_{13}$Al		26.9815386(8)							
^{21}Al		21.0280	< 0.035 μs						
^{22}Al		22.0195	59. ms	β+ /18.6	p/1.3/18.	4+			ann.rad./
					p/0.48−7.52				0.584/47.
									(1.25-2.15)
				β+, p, 2p, α/	α/3.3/0.3				
^{23m}Al			~ 0.35 s	β+, p/0.17					0.554
									0.839
^{23}Al		23.00727	0.47 s	β+ /12.24		5/2+	3.9		ann.rad./
				β+, p/					
^{24m}Al			0.129 s	I.T./0.4259		1+	3.0		
				β+	13.3				1.3686/5.3
^{24}Al		23.999939	2.07 s	β+ /13.878,p	3.40/48.	4+			1.078(2)/16.
					4.42/41.				1.368(2)/96.
					6.80/3.				2.753(2)/43.
					8.74/8.				4.315(3)/15.
									5.392(3)/20.
									7.0662(2)/41.
^{25}Al		24.9904281	7.17 s	β+ /4.277	3.27/	5/2+	3.646	0.24	ann.rad./
									1.6115(2)/100.
									0.975(2)/5.

Elem. or Isot.	Natural Abundance (Atom %)	Atomic Mass or Weight	Half–life/ Resonance Width (MeV)	Decay Mode/ Energy (/MeV)	Particle Energy/ Intensity (MeV/%)	Spin ($h/2\pi$)	Nuclear Magnetic Mom. (nm)	Elect. Quadr. Mom. (b)	γ-Energy/ Intensity (MeV/%)
^{26m}Al			6.345 s	β+ /	3.2/	0+			ann.rad./
^{26}Al		25.9868917	7.1×10^5 a	β+ /82/4.0042	1.16/	5+	+2.804	+0.27	ann.rad./
				EC/18					1.8087/99.8
^{27}Al	100.	26.9815386				5/2+	+3.641507	+0.147	
^{28}Al		27.9819103	2.25 m	β- /4.6422	2.865/100.	3+	3.24	0.18	1.7778(6)/100.
^{29}Al		28.980445	6.5 m	β- /3.680	1.4/30.	5/2+			1.2732(8)/89.
					2.5/70.				2.0282(8)/4.
									2.4262(8)/7.
^{30}Al		29.98296	3.68 s	β- /8.56	5.05/	3+	3.01		1.26313(3)/35.
									2.23525(5)/65.
^{31}Al		30.98395	0.64 s	β- /8.00	6.25/	5/2+	3.8	0.134	0.75223(3)/18.
									1.69473(3)/59.
									2.31664(4)/73.
^{32}Al		31.9881	33. ms	β- /13.0		1+	1.96	0.024	
^{33}Al		32.9908	41.7 ms	β- /12.0	/91.5			~ 0.13	1.940/2.5
				β-, n	/8.5				(1.01–4.34)
^{34}Al		33.9969	56. ms	β- /17.1	4.255/44	4			0.929/57
				β-, n	/26.				(0.12–4.26)
^{35}Al		34.9999	38. ms	β-/14.3	0.974/48	5/2+			0.064/45.
				β-, n	/ 38.				(0.12–5.63)
^{36}Al		36.0062	0.09 s	β-/18.3					
				β-, n	/<31.				
^{37}Al		37.0107	11. ms	β-/16.					
^{38}Al		38.017	> 7.6 ms						
^{39}Al		39.023	> 8. ms						
^{40}Al		40.031	> 0.26 μs						
^{41}Al		41.038	> 0.26 μs						
^{42}Al									
^{43}Al									
$_{14}$**Si**		**28.084–28.086**							
^{22}Si		22.0345	29. ms	β+, p	1.99/20	0+			
^{23}Si		23.0255	40.7 ms	β+, p/5.9	1.32,(0.6–11.6)				
^{24}Si		24.01155	0.14 s	β+, p/10.81	1.44,3.92,1.09	0+			ann.rad./
					(1.66–4.47)				
^{25}Si		25.00411	221 ms	β+, p/12.74	p/4.25/9.5	5/2+			ann.rad./
					p/0.40/4.75				
					p/0.56–6.80				
^{26}Si		25.992330	2.24 s	β+ /5.066	3.282/	0+			ann.rad./
									0.8294(8)/22.
^{27}Si		26.9867049	4.14 s	β+ /4.8118	3.85/100.	5/2+	−0.865		ann.rad./
									2.211(5)/0.2
^{28}Si	92.223(19)	27.976926533				0+			
^{29}Si	4.685(8)	28.97649470				1/2+	−0.5553		
^{30}Si	3.092(11)	29.97377017				0+			
^{31}Si		30.97536323	2.62 h	β- /1.4920	1.471/99.9	3/2+			1.2662(5)/0.05
^{32}Si		31.97414808	1.5×10^2 a	β- /0.224	0.213/100.	0+			
^{33}Si		32.97800	6.1 s	β- /5.85	3.92	(3/2+)	1.21		1.4313(5)/13.
									1.8477/100.
									2.538(2)/10.
^{34}Si		33.97858	2.8 s	β- /4.60	3.09/	0+			0.42907(5)/60.
									1.17852(2)/64.
									1.60756(5)/36.
^{35}Si		34.98458	0.9 s	β- /10.50					
^{36}Si		35.9866	0.5 s	β- /7.9		0+			

Elem. or Isot.	Natural Abundance (Atom %)	Atomic Mass or Weight	Half–life/ Resonance Width (MeV)	Decay Mode/ Energy (/MeV)	Particle Energy/ Intensity (MeV/%)	Spin ($h/2\pi$)	Nuclear Magnetic Mom. (nm)	Elect. Quadr. Mom. (b)	γ-Energy/ Intensity (MeV/%)
				β-, n	/~ 12.				
^{37}Si		36.9929	~ 0.09 s	β- /12.5					
				β-, n	/~ 17.				
^{38}Si		37.9956	> 1 µs	β- /10.7		0$^+$			
				β-, n					
^{39}Si		39.0021	48. ms	β- /14.8					
^{40}Si		40.006	33. ms			0$^+$			
^{41}Si		41.015	20. ms						
^{42}Si		42.020	13. ms			0$^+$			
^{43}Si		43.029	> 0.26 µs						
^{44}Si									
$_{15}$P		30.973762(2)							
^{24}P		24.034							
^{25}P		25.0203	< 0.03 µs						
^{26}P		26.0118	44. ms	β+, p/18.1	p/0.41/18.0	3$^+$			
					p/1.98/2.4				
					p/0.78–7.49				
^{27}P		26.99923	0.3 s	β+, p/11.63	p/0.73, 0.61/0.07	1/2$^+$			
^{28}P		27.992315	270. ms	β+ /14.332	3.94/13.	3$^+$	0.31		ann.rad./
					5.25/13.				1.779(2)/98.
					6.96/16.				2.839(2)/2.8
					8.8/7.				3.040(2)/3.2
					11.49/52.				4.498(2)/12.
									7.537(2)/9.
^{29}P		28.981801	4.14 s	β+ /4.9431	3.945/98.	1/2$^+$	1.2349		ann.rad./
									1.273/1.32
									2.426/0.39
^{30}P		29.9783138	2.50 m	β+ /4.2323	3.245/99.9	1$^+$			ann.rad./
									2.230(3)/0.07
^{31}P	100.	30.9737616				1/2$^+$	+1.13160		
^{32}P		31.9739073	14.28 d	β- /1.7106	1.710/100.	1$^+$	−0.2524		
^{33}P		32.971726	25.3 d	β- /0.249	0.249/100.	1/2$^+$			
^{34}P		33.973636	12.4 s	β- /5.374	3.2/15.	1$^+$			1.78–4.1/
					5.1/85.				2.127(5)/15.
^{35}P		34.973314	47. s	β- /3.989	2.34/100.	1/2$^+$			1.572(1)/100.
^{36}P		35.97826	5.7 s	β- /10.41					0.902/77.
									3.291/100.
^{37}P		36.97961	2.3 s	β- /7.90					0.6462/
									1.5829/
^{38}P		37.9842	0.6 s	β- /12.4					1.2923/
				β-, n	/~ 12.				2.224/
^{39}P		38.9862	0.3 s	β- /10.5					
				β-, n	/26				
^{40}P		39.9913	0.15 s	β- /14.5					
				β-,n	/~ 30.				
^{41}P		40.9943	0.10 s	β-/~ 13.8					
				β-, n	/~ 30.				
^{42}P		42.0010	49. ms	β-/17.					
				β-, n	/ ~ 50.				
^{43}P		43.006	36. ms	β-/16.					
				β-, n	/100.				
^{44}P		44.013	19. ms						
^{45}P		45.019	> 0.2 µs						
^{46}P		46.027	> 0.2 µs						

Elem. or Isot.	Natural Abundance (Atom %)	Atomic Mass or Weight	Half–life/ Resonance Width (MeV)	Decay Mode/ Energy (/MeV)	Particle Energy/ Intensity (MeV/%)	Spin ($h/2\pi$)	Nuclear Magnetic Mom. (nm)	Elect. Quadr. Mom. (b)	γ-Energy/ Intensity (MeV/%)
$_{16}$S		32.059–32.076							
^{26}S		26.0279	~ 10 ms			0$^+$			
^{27}S		27.0188	16. ms	β+, 2p/18.3	p/2.26, 7.80				
^{28}S		28.0044	0.13 s			0+			
^{29}S		28.99661	0.188 s	β+ /13.79		5/2+			ann.rad./
				β+, p/					
^{30}S		29.984903	1.18 s	β+ /6.138	4.42/78.	0+			ann.rad./
					5.08/20.				0.678/79.
^{31}S		30.979555	2.56 s	β+ /5.396	4.39/99.	1/2+	0.48793		ann.rad./
									1.2662(5)/1.2
^{32}S	94.99(26)	31.9720710				0+			
^{33}S	0.75(2)	32.9714588				3/2+	+0.64382	−0.068	
^{34}S	4.25(24)	33.9678669				0+			
^{35}S		34.9690322	87.2 d	β- /0.1672	0.1674/100.	3/2+	+1.00	+0.047	
^{36}S	0.01(1)	35.9670808				0+			
^{37}S		36.9711256	5.05 m	β- /4.8653	1.64/94.	7/2-			0.9083(4)/0.06
					4.75/5.6				3.1033(2)/94.2
^{38}S		37.97116	2.84 h	β- /2.94	1.00/	0+			0.1962(4)/0.2
									1.9421(3)/84.
^{39}S		38.97513	11.5 s	β- /6.64					1.301/52.
									1.697/44.
^{40}S		39.9755	9. s	β- /4.7		0+			(0.2116-1.874)
^{41}S		40.9796	~ 2.6 s	β- /8.7					
				β-, n					
^{42}S		41.9810	1.03 s	β- /7.8		0+			(0.118-2.912)
				β-, n	/ < 4.				
^{43}S		42.9872	0.26 s	β-/12.					
				β-, n	/ ~ 40				
^{44}S		43.9902	0.10 s	β-/9.		0+			
				β-, n	/18.				
^{45}S		44.997	68. ms	β-/14.					
				β-, n	/54.				
^{46}S		46.001	0.05 s			0+			
^{47}S		47.009	> 0.2 μs						
^{48}S		48.014	> 0.2 μs			0+			
^{49}S		49.024	< 0.2 μs						
$_{17}$Cl		35.446–35.457							
^{28}Cl		28.029							
^{29}Cl		29.0141	< 0.02 μs						
^{30}Cl		30.0048	< 0.03 μs						
^{31}Cl		30.99241	0.15 s	β+, p/11.98	0.986, 1.52/0.7	3/2+			ann.rad./
									2.234
					0.762–2.75				1.249-4.045
^{32}Cl		31.98569	297. ms	β+ /12.69	9.47/50.	1+	1.11		ann.rad./
				β+, α	/0.05				2.2305/92
				β+, p	/0.026				(1.55–4.77)
^{33}Cl		32.9774519	2.511 s	β+ /5.583	4.51/98.	3/2+	+0.752		ann.rad./
									0.8409/0.52
									1.966/0.45
									2.866/0.44
^{34m}Cl			32.2 m	β+/	1.35/24.	3+			ann.rad./
					2.47/28.				
				I.T./					0.1457(8)/42.
									2.1276(5)/42.

Elem. or Isot.	Natural Abundance (Atom %)	Atomic Mass or Weight	Half–life/ Resonance Width (MeV)	Decay Mode/ Energy (/MeV)	Particle Energy/ Intensity (MeV/%)	Spin ($h/2\pi$)	Nuclear Magnetic Mom. (nm)	Elect. Quadr. Mom. (b)	γ-Energy/ Intensity (MeV/%)
^{34}Cl		33.9737628	1.528 s	β+ /5.4922	4.50/100.	0+			ann.rad./
^{35}Cl	75.76(10)	34.96885268				3/2+	+0.82187	−0.082	
^{36}Cl		35.9683070	3.01 × 10⁵ a	β- /0.7086	0.7093/98.	0+	+1.28547	−0.018	
				β+, EC/1.1421	0.115/0.002				ann.rad./
^{37}Cl	24.24(10)	36.96590259				3/2+	+0.68412	−0.0646	
^{38m}Cl			0.715 s	I.T./		5-			0.6714/100
^{38}Cl		37.9680104	37.2 m	β- /4.9168	1.11/31.	2-	2.05		1.64216/33.3
					2.77/11.				2.16760/44.8
					4.91/58.				
^{39}Cl		38.968008	55.6 m	β- /3.442	1.91/85.	3/2+			0.25026(1)/47.
					2.18/8.				1.26720(5)/54.
					3.45/7.				0.986−1.517
^{40}Cl		39.97042	1.38 m	β- /7.48		2-			0.6431(3)/6.
									1.4608(1)/77.
									2.8402(2)/17.
^{41}Cl		40.9707	34. s	β- /5.7	3.8/				(0.167−1.359)
^{42}Cl		41.9733	6.8 s	β- /9.4					
^{43}Cl		42.9741	3.3 s	β- /8.0					(0.352-4.247)
^{44}Cl		43.9783	~ 0.43 s	β-/12.3					
				β-, n	/ < 8.				
^{45}Cl		44.9803	0.40 s	β-/11.					
				β-, n	/24.				
^{46}Cl		45.984	0.23 s	β-/14.9					
				β-, n	/ ~ 60				
^{47}Cl		46.989	0.10 s	β-/15.					
				β-, n	/ < 3.				
^{48}Cl		47.995	> 0.2 μs						
^{49}Cl		48.000	> 0.17 s						
^{50}Cl		50.008							
^{51}Cl		51.014	> 0.2 μs						
$_{18}$Ar		39.948(1)							
^{30}Ar		30.0216	< 0.02 μs			0+			
^{31}Ar		31.0121	~ 14.1 ms	β+ /18.4	p/2.08/100.	5/2+			
				β+, p	/55.				
				β+, 2p	/2.5				
				β+, 3p	/0.11				
^{32}Ar		31.997638	98. ms	β+ /11.2		0+			ann.rad./
				β+, p	3.98−6.40/22.7				
^{33}Ar		32.9899257	174. ms	β+ /11.62	3.17,2.10	1/2+	−0.72		ann.rad./
				β+, p/	(1.32−5.72)				0.810(2)/48.
^{34}Ar		33.9802712	0.844 s	β+/6.061	5.0/95.	0+			ann.rad./
									0.6658(1)/2.5
									3.1290(1)/1.3
^{35}Ar		34.975258	1.77 s	β+/5.965	4.94/93.	3/2+	+0.6322	−0.08	ann.rad./
									1.2185(5)/1.22
									1.763(1)/0.25
									2.964(1)/0.2
^{36}Ar	0.3336(21)	35.96754511				0+			
^{37}Ar		36.9667763	35.0 d	EC/.813		3/2+	+1.15	+0.076	
^{38}Ar	0.0629(7)	37.9627324				0+			
^{39}Ar		38.964313	268. a	β-/0.565	0.565/100.	7/2-	−1.59	−0.12	
^{40}Ar	99.6035(25)	39.962383123				0+			
^{41}Ar		40.9645006	1.82 h	β-/2.492	1.198/	7/2-			1.29364(5)/99.
									1.6770(3)/0.05

Elem. or Isot.	Natural Abundance (Atom %)	Atomic Mass or Weight	Half–life/ Resonance Width (MeV)	Decay Mode/ Energy (/MeV)	Particle Energy/ Intensity (MeV/%)	Spin ($h/2\pi$)	Nuclear Magnetic Mom. (nm)	Elect. Quadr. Mom. (b)	γ-Energy/ Intensity (MeV/%)
^{42}Ar		41.96305	33. a	β-/0.60	0.60/100.	0+			
^{43}Ar		42.96564	5.4 m	β-/4.6					0.4791(2)/10.
									0.7380(1)/43.
									0.9752(1)/100.
									1.4400(3)/39.
^{44}Ar		43.964924	11.87 m	β-/3.55		0+			0.182–1.866
^{45}Ar		44.968040	21.5 s	β-/6.9		7/2-			0.0610/25.
									1.020/35.
									3.707/34.
^{46}Ar		45.96809	8.4 s	β-/5.70		0+			1.944/
^{47}Ar		46.9722	1.23 s	β-					0.36/100
									1.66/53
									1.74/41
									(2.02–4.01)
^{48}Ar		47.9745	0.48 s			0+			
^{49}Ar		48.981	0.17 s	β-,n	n// ~ 65.				
^{50}Ar		49.984	~ 0.085 s	β-,n	n// ~ 35.	0+			
^{51}Ar		50.992	> 0.2 μs	β-					
^{52}Ar		51.997	10 ms	β-		0+			
^{53}Ar		53.005		β-					
$_{19}$K		39.0983(1)							
^{32}K		32.022							
^{33}K		33.0073	< 0.025 μs						
^{34}K		33.9984	< 0.04 μs						
^{35}K		34.98801	0.19 s	β+ /11.88		3/2+	0.392		ann.rad./
				β+, p/	/0.37				1.751/14.
									2.5698/26.
									2.9827/51.
^{36}K		35.98129	0.342 s	β+ /12.81	5.3/42.	2+	+0.548		ann.rad./
					9.9/44.				1.97044(5)/82.
				β+, p	/0.048				2.20783(5)/30.
									2.43343(2)/32.
^{37}K		36.9733759	1.23 s	β+ /6.149	5.13/	3/2+	+0.2032	0.106	ann.rad./
									2.7944(8)/2.
									3.602(2)/0.05
^{38m}K			924.5 ms	β+ /6.742	5.02/100.	0+			ann.rad./
				IT	/0.033				
				β-	/<0.001				
^{38}K		37.9690812	7.63 m	β+ /5.913	2.60/99.8	3+	+1.37		ann.rad./
									2.1675(3)/99.8
									3.9356(5)/0.2
^{39}K	93.2581(44)	38.9637067				3/2+	+0.39146	+0.049	
^{40}K	0.0117(1)	39.9639985	1.248 × 10⁹ a	β- /1.3111	1.312/89.	4-	−1.29810	−0.074	ann.rad./
				β+, EC/1.505	1.50/10.7				1.4608/10.5
^{41}K	6.7302(44)	40.9618258				3/2+	+0.21487	+0.071	
^{42}K		41.9624028	12.36 h	β- /3.525	1.97/19.	2-	−1.1425		0.31260(2)/0.3
					3.523/81.				1.5246(3)/18.1
^{43}K		42.96072	22.3 h	β- /1.82	0.465/8.	3/2+	+0.163		0.2211(2)/4.
					0.825/87.				0.3729(2)/88.
					1.24/3.5				0.3971(2)/11.
					1.814/1.3				0.6178(2)/81.
^{44}K		43.96156	22.1 m	β- /5.66	5.66/34.	2-	−0.856		0.36821/2.2
									1.15700(1)/58.
									2.15079(2)/22.

Elem. or Isot.	Natural Abundance (Atom %)	Atomic Mass or Weight	Half–life/ Resonance Width (MeV)	Decay Mode/ Energy (/MeV)	Particle Energy/ Intensity (MeV/%)	Spin ($h/2\pi$)	Nuclear Magnetic Mom. (nm)	Elect. Quadr. Mom. (b)	γ-Energy/ Intensity (MeV/%)
⁴⁵K		44.96070	17.8 m	β- /4.20	1.1/23.	3/2+	+0.173		0.1743(5)/80.
					2.1/69.				1.2607(8)/7.
					4.0/8.				1.7056(6)/69.
									2.3542(5)/14.
⁴⁶K		45.96198	1.8 m	β- /7.72	6.3/	2-	−1.05		1.347(1)/91.
									3.700(5)/28.
⁴⁷K		46.96168	17.5 s	β- /6.64	4.1/99.	½+	+1.93		0.56474(3)/15.
					6.0/1.				0.58575(3)/85.
									2.0131/100
⁴⁸K		47.96551	6.8 s	β- /12.09	5.0/	(2-)			0.67122(1)/4.
									0.6723(5)/20.
									0.78016(1)/32.
									3.83153(7)/80.
⁴⁹K		48.9675	1.26 s	β- /11.0					2.025/
									2.252/
⁵⁰K		49.9728	0.472 s	β- /14.2					
⁵¹K		50.976	0.365 s	β- /					1.027/21.7
				β-, n	//68(10)				3.46/3.9
					/2.23/0.218				(1.976-4.035)
⁵²K		51.983	0.105 s	β-	//79(12)				2.563/25.2
				β-, n	/1.040/0.216				2.377/6.9
				β-, 2n					1.027/0.55
⁵³K		52.987	30. ms	β-		3/2+			2.22/15.3
				β-, n	//85(19)				2.56/51.5
⁵⁴K		53.994	10. ms	β-					
⁵⁵K									
⁵⁶K									
₂₀Ca		40.078(4)							
³⁴Ca		34.0141	< 0.035 µs			0+			
³⁵Ca		35.0049	25.7 ms	β+, p/15.6	p/1.43/49				
					1.9–8.8				
³⁶Ca		35.99309	0.100 s	β+, (p)/10.99	p/2.61/32	0+			ann.rad./
				β+, n/	p/1.71/7				
³⁷Ca		36.98587	0.182 s	β+ /11.64	p/3.19/40.7	3/2+			ann.rad./
				β+, n/	p/(0.899–2.00)				1.369
³⁸Ca		37.976318	444 ms	β+ /6.74		0+			ann.rad./
									1.5677(5)/25.
									3.210(2)/1.
³⁹Ca		38.970720	0.861 s	β+ /6.531	5.49/100.	3/2+	1.02168	0.04	ann.rad./
⁴⁰Ca	96.941(156)	39.9625910	5.92 × 10²¹ a	EC-EC		0+			
⁴¹Ca		40.9622781	1.02 × 10⁵ a	EC/0.4214		7/2-	−1.5948	−0.090	
⁴²Ca	0.647(23)	41.9586180				0+			
⁴³Ca	0.135(10)	42.9587666				7/2-	−1.31764	−0.055	
⁴⁴Ca	2.086(110)	43.9554818				0+			
⁴⁵Ca		44.9561866	162.7 d	β- /0.257	0.257/100.	7/2-	−1.327	+0.05	
⁴⁶Ca	0.004(3)	45.953693	>0.4 × 10¹⁶ a	β-β-		0+			
⁴⁷Ca		46.954546	4.536 d	β-/1.992	0.684/84.	7/2-	−1.38	+0.02	1.297/75
					1.98/16.				(0.041–1.88)
⁴⁸Ca	0.187(21)	47.952534	4.4 × 10¹⁹ a	β-β-		0+			
			>7.1 × 10¹⁹ a	β-					
⁴⁹Ca		48.955674	8.72 m	β- /5.262	0.89/7.	3/2-			3.0844(1)/90.7
					1.95/92.				4.0719(1)/8.12
									(0.143–4.738)
⁵⁰Ca		49.95752	14. s	β- /4.97	3.12/	0+			0.2569/98.

Elem. or Isot.	Natural Abundance (Atom %)	Atomic Mass or Weight	Half–life/ Resonance Width (MeV)	Decay Mode/ Energy (/MeV)	Particle Energy/ Intensity (MeV/%)	Spin ($h/2\pi$)	Nuclear Magnetic Mom. (nm)	Elect. Quadr. Mom. (b)	γ-Energy/ Intensity (MeV/%)
									(0.0715–1.59)
^{51}Ca		50.9615	10. s	β- /7.3		(3/2-)			
^{52}Ca		51.965	4.6 s	β- /8.0		0+			
^{53}Ca		52.9701	0.4 s	β- /10.9					2.11/56
^{54}Ca		53.974	0.1 s			0+			0.247/65
^{55}Ca		54.981	22 ms						
^{56}Ca		55.986	11 ms			0+			
^{57}Ca									
^{58}Ca									
$_{21}$Sc		44.955912(6)							
^{36}Sc		36.0149	0.102 s						
^{37}Sc		37.0031	0.181 s						
^{38}Sc		37.9947	< 0.3 µs						
^{39}Sc		38.98479	< 0.3 µs	p					
^{40}Sc		39.977967	0.182 s	β+ /14.320	5.73/50.	4-			ann.rad./
					7.53/15.				0.752/41.
					8.76/15.				3.732/99.5
					9.58/20.				(1.12–3.92)
^{41}Sc		40.9692511	0.596 s	β+ /6.4953	5.61/100.	7/2-	+5.431	−0.156	ann.rad./
^{42m}Sc			61.6 s	β+ /	2.82/	7+			ann.rad./
									0.4375(5)/100.
									1.2270(5)/100.
									1.5245(5)/100.
^{42}Sc		41.9655164	0.682 s	β+ /6.4259	5.32/100.	0+			ann.rad./
^{43}Sc		42.961151	3.89 h	β+, EC/2.221	0.82/22.	7/2-	+4.50	−0.21	ann.rad./
					1.22/78.				0.3729(1)/22.
^{44m}Sc			58.2 h	I.T./0.27		6+	+3.81	−0.20	0.27124(1)/87.
				EC/3.926					(1.00–1.16)
^{44}Sc		43.959403	3.93 h	β+, EC/3.653	1.47/	2+	+2.51	+0.18	ann.rad./
									1.157/100
^{45}Sc	100.	44.955912				7/2-	+4.75649	−0.220	
^{46m}Sc			18.7 s	I.T./0.14253		1-			0.14253(2)/62.
^{46}Sc		45.955172	83.81 d	β- /2.367	0.357/100.	4+	+3.04	+0.12	0.8893/100
									1.121/100
^{47}Sc		46.952408	3.349 d	β- /0.600	0.439/69.	7/2-	+5.34	−0.22	0.15938(1)/68.
					0.601/31.				
^{48}Sc		47.95223	43.7 h	β- /3.99	0.655/	6+	3.72		0.9835/100
									1.03750(1)/97.
									1.3121/100
^{49}Sc		48.950024	57.3 m	β- /2.006	2.00/99.9.	7/2-			1.7619(3)/0.05
^{50}Sc		49.95219	1.71 m	β- /6.89	3.05/76.	(5+)			0.5235(1)/88.
					3.60/24.				1.1210(1)/100.
									1.5537(2)/100.
^{51}Sc		50.95360	12.4 s	β- /6.51	4.4/	7/2-			1.4373(4)/52.
					5.0/				0.718–2.144
^{52}Sc		51.9567	8.2 s	β- /9.0		(3+)			
^{53}Sc		52.9596	> 3. ms	β- /8.1					
^{54m}Sc			2.8 µs	I.T.		(5+)			0.110/IT
^{54}Sc		53.9633	0.53 s	β- /11.6					0.100/50
									1.70/40
									0.50/40
^{55}Sc		54.968	0.103 s	β- /13					0.593(1)/40
^{56m}Sc			0.06 s						1.161/21
									0.690/19

Elem. or Isot.	Natural Abundance (Atom %)	Atomic Mass or Weight	Half–life/ Resonance Width (MeV)	Decay Mode/ Energy (/MeV)	Particle Energy/ Intensity (MeV/%)	Spin ($h/2\pi$)	Nuclear Magnetic Mom. (nm)	Elect. Quadr. Mom. (b)	γ-Energy/ Intensity (MeV/%)
^{56}Sc		55.973	35. ms	β-		(1+)			1.129/48
^{57}Sc		56.978	13. ms	β-					
^{58}Sc		57.984	12. ms	β-					
^{59}Sc									
^{60}Sc									
^{61}Sc									
$_{22}$Ti		47.867(1)							
^{38}Ti		38.0098	< 0.12 µs			0+			
^{39}Ti		39.0016	29. ms	β+ /15.4	p//94				
^{40}Ti		39.9905	52. ms	β+ /11.7	p/2.16/29	0+			2.467/8.5
				β+, p	3.73/23				
					1.70/22				
					0.242–5.74				
^{41}Ti		40.9832	83. ms	β+, p/12.93	p/4.73/107	3/2+			ann.rad./
					3.10/67				
					3.75/39				
					0.744–6.73				
^{42}Ti		41.97303	0.208 s	β+ /7.000	6.0/48	0+			ann.rad./
									0.6107(5)/56.
^{43}Ti		42.96852	0.50 s	β+ /6.87	5.80/	7/2-	0.85		ann.rad./
^{44}Ti		43.959690	59. a	EC/0.268		0+			0.06787/91
									0.07832/97
^{45}Ti		44.958126	3.078 h	β+/86/2.062	1.04	7/2-	0.095	~ 0.015	ann.rad./
				EC/14/					(0.36–1.66)
^{46}Ti	8.25(3)	45.952632				0+			
^{47}Ti	7.44(2)	46.951763				5/2-	−0.78848	+0.30	
^{48}Ti	73.72(3)	47.947946				0+			
^{49}Ti	5.41(2)	48.947870				7/2-	−1.10417	+0.25	
^{50}Ti	5.18(2)	49.944791				0+			
^{51}Ti		50.946615	5.76 m	β- /2.471	1.50/92.	3/2-			0.3197(2)/93.
					2.13/				0.6094–0.9291
^{52}Ti		51.94690	1.7 m	β- /1.97	1.8/100.	0+			0.0170(5)/100.
									0.1245/100
^{53}Ti		52.9497	33. s	β- /5.0	(2.2–3)/	3/2-			0.1008(1)/20.
									0.1276(1)/45.
									0.2284(1)/39.
									1.6755(5)/45.
									(1.72–2.8)/
^{54}Ti		53.9511	1.5 s	β- /4.3		0+			
^{55}Ti		54.9553	1.3 s	β- /7.4					0.672/44
									(0.32–1.83)
^{56}Ti		55.9582	0.20 s	β- /7.0		0+			
^{57}Ti		56.9640	98. ms	β- /11.					
^{58}Ti		57.967	53. ms	β-		0+			0.114
^{59}Ti		58.973	30. ms	β-					
^{60}Ti		59.978	22. ms	β-		0+			
^{61}Ti		60.983	> 0.3 µs						
^{62}Ti									
^{63}Ti									
$_{23}$V		50.9415(1)							
^{40}V		40.0111							
^{41}V		40.9998							
^{42}V		41.9912	< 0.055 µs						
^{43}V		42.9807	79. ms	β+ /11.3	p// < 2.5				

Elem. or Isot.	Natural Abundance (Atom %)	Atomic Mass or Weight	Half–life/ Resonance Width (MeV)	Decay Mode/ Energy (/MeV)	Particle Energy/ Intensity (MeV/%)	Spin ($h/2\pi$)	Nuclear Magnetic Mom. (nm)	Elect. Quadr. Mom. (b)	γ-Energy/ Intensity (MeV/%)
^{44}V		43.9741	0.09 s	β+, α/13.7					ann.rad./
^{45}V		44.96578	0.54 s	β+ /7.13		7/2-			
^{46}V		45.960201	0.4223 s	β+ /7.051	6.03/100.	0+			ann.rad./
^{47}V		46.954909	32.6 m	β+, EC/2.928	1.90/99.+	3/2-			ann.rad./
									1.7949(8)/0.19
									(0.2–2.16)
^{48}V		47.952254	15.98 d	β+ /4.012	0.698/50.	4+	2.01		ann.rad./
									0.9835/100
									(1.3–2.4)
^{49}V		48.948516	337. d	EC/0.602		7/2-	4.47		
^{50}V	0.250(4)	49.947159	1.4×10^{17} a	EC	/82.7	6+	+3.34569	+0.21	
				β-	/17.3				
^{51}V	99.750(4)	50.943960				7/2-	+5.148706	−0.04	
^{52}V		51.944776	3.76 m	β- /3.976	2.47/	3+			1.4341(1)/100.
^{53}V		52.944338	1.56 m	β- /3.436	2.52/	7/2-			1.0060(5)/90.
									1.2891(3)/10.
^{54m}V			0.9 μs	I.T.		(5+)			0.108/IT
^{54}V		53.94644	49.8 s	β- /7.04	1.00/5.	3+			0.8348/97.
					2.00/12.				0.9887/80.
					2.95/45.				2.259/46.
					5.20/11.				(0.56–3.38)
^{55}V		54.9472	6.5 s	β- /6.0	6.0/	(7/2-)			0.5177/73.
									(0.224–1.21)
^{56}V		55.9505	0.22 s	β- /9.1					1.01/30.
									0.688/26.
									(0.82 – 1.32)
^{57}V		56.9526	0.35 s	β- /8.1					0.268/52.
									0.692/20.
									(0.25 – 1.31)
^{58}V		57.9567	0.19 s	β- /11.6					0.880/62
									1.056/28
									2.217/13
									(1.04 – 1.57)
^{59}V		58.9602	97. ms	β- /9.9					0.90/80.
^{60m}V			0.12 s						
^{60}V		59.9650	0.07 s	β- /14.					0.102–0.208
^{61}V		60.9685	47. ms						(0.071-1.144)
^{62}V		61.9738	34. ms						
^{63}V		62.978	17. ms						
^{64}V		63.983	> 0.3 μs						
^{65}V									
^{66}V									
$_{24}$**Cr**		**51.9961(6)**							
^{42}Cr		42.0064	13. ms	β+, p	p/1.90/29	0+			1.623/35
					p/1.50–3.7				
^{43}Cr		42.9977	21. ms	β+, p	p/3.83/18				1.555/35
					p/4.29/15				0.838/6.2
					p/1.01–4.59				1.937/1.8
^{44}Cr		43.98555	43. ms	β+, (p)/10.3	p/0.95–3.1	0+			0.677/59.
^{45}Cr		44.9796	61. ms	β+, p/12.5	p/2.088/19.6	7/2-			ann.rad./
					p/(0.945–1.61)				(1.08-1.37)
^{46}Cr		45.96836	0.3 s	β+ /7.60		0+			ann.rad./
^{47}Cr		46.96290	0.51 s	β+ /7.45		3/2-			ann.rad./
^{48}Cr		47.95403	21.6 h	EC/1.66		0+			ann.rad./

Elem. or Isot.	Natural Abundance (Atom %)	Atomic Mass or Weight	Half–life/ Resonance Width (MeV)	Decay Mode/ Energy (/MeV)	Particle Energy/ Intensity (MeV/%)	Spin ($h/2\pi$)	Nuclear Magnetic Mom. (nm)	Elect. Quadr. Mom. (b)	γ-Energy/ Intensity (MeV/%)
									0.116(2)/95.
									0.305(10)/100.
^{49}Cr		48.951336	42.3 m	β+, EC/2.631	1.39/	5/2-	0.476		ann.rad./
					1.45/				0.09064(1)/51.
					1.54/				0.15293(1)/27.
									(0.062-1.6)
^{50}Cr	4.345(13)	49.946044	>1.3 × 10^{18} a	β+EC		0+			
^{51}Cr		50.944767	27.70 d	EC/0.7527		7/2-	−0.934		0.3201/9.8
									0.00543/2.6
									0.00495/0.02
^{52}Cr	83.789(18)	51.940508				0+			
^{53}Cr	9.501(17)	52.940649				3/2-	−0.47454	−0.22	
^{54}Cr	2.365(7)	53.938880				0+			
^{55}Cr		54.940840	3.497 m	β- /2.603	2.5/	3/2-			1.5282(2)/0.04
									(0.13–2.37)
^{56}Cr		55.940653	5.9 m	β- /1.62	1.50/100.	0+			0.026(2)/100.
									0.083(3)/100.
^{57}Cr		56.943613	21. s	β- /5.1	3.3/	3/2-	0.0834		0.850/8.
					3.5/				(0.083-2.62)
^{58}Cr		57.9444	7.0 s	β- /4.0		0+			(0.131–0.683)
^{59m}Cr			0.10 ms	I.T.		(9/2+)			0.208/IT
									0.193
									0.102
^{59}Cr		58.9486	1.0 s	β- /7.7					1.236
^{60}Cr		59.9500	0.6 s	β- /6.0		0+			
^{61}Cr		60.9547	0.23 s	β- /8.8					(0.157–2.378)
^{62}Cr		61.9566	0.19 s	β- /7.3		0+			(0.156-1.215)
^{63}Cr		62.9619	0.129 s	β-					(0.250-3.454)
^{64}Cr		63.9644	0.043 s	β-		0+			0.188
^{65}Cr		64.9702	0.027 s	β-					0.272, 1.368
^{66}Cr		65.973	0.01 s	β-		0+			
^{67}Cr		66.980	> 0.3 µs						
^{68}Cr									
$_{25}$Mn		54.938045(5)							
^{44}Mn		44.0069	< 0.105 µs						
^{45}Mn		44.9945	< 0.07 µs						
^{46}Mn		45.9867	36. ms	β+ /17.1	p/3.00/6.5				0.329/11.
				β+, p	// ~ 58				(0.0544-1.322)
^{47}Mn		46.9761	88. ms	β+ /12.3	p// <1.7				
^{48}Mn		47.9685	0.15 s	β+ /13.5	5.79/58.	4+			
					4.43/10.				
^{49}Mn		48.95962	0.38 s	β+ /7.72	6.69/	5/2-			ann.rad./
^{50m}Mn			1.74 m	β+ /7.887	3.54/	5+	+2.76	+0.8	ann.rad./
									1.0980/94.
									0.783/91.
									(0.66–3.11)
^{50}Mn		49.954238	0.283 s	β+ /7.6330	6.61/	0+			ann.rad./
^{51}Mn		50.948211	46.2 m	β+, EC/3.208	2.2/	5/2-	3.568	0.4	ann.rad./
									0.7491(1)/0.26
									(1.148–1.164)
^{52m}Mn			21.1 m	β+ /98/5.09	2.631/	2+	0.0077		ann.rad./
				I.T./2/0.378					0.3778 (I.T.)
									1.43406(1)/98.

Elem. or Isot.	Natural Abundance (Atom %)	Atomic Mass or Weight	Half–life/ Resonance Width (MeV)	Decay Mode/ Energy (/MeV)	Particle Energy/ Intensity (MeV/%)	Spin ($h/2\pi$)	Nuclear Magnetic Mom. (nm)	Elect. Quadr. Mom. (b)	γ-Energy/ Intensity (MeV/%)
									(0.7–4.8)
^{52}Mn		51.945566	5.591 d	β+ /4.712	0.575/	6+	+3.063	+0.5	ann.rad./
				EC/					0.74421(1)/90.
									1.4341/100
^{53}Mn		52.941290	3.7×10^6 a	EC/0.5970		7/2-	5.035	+0.16	
^{54}Mn		53.940359	312.2 d	EC/1.377		3+	+3.306	+0.37	0.8340/100
			6.7×10^8 a	β+	//1.3 × 10^{-7}				
^{55}Mn	100.	54.938045				5/2-	+3.4687	+0.32	
^{56}Mn		55.938905	2.579 h	β- /3.6954	0.718/18.	3+	+3.2266	+0.5	0.84675/98.9
					1.028/34.				1.81072(4)/26.3
									2.113/13.8
									(1.04 – 3.37)
^{57}Mn		56.938285	1.45 m	β- /2.691		5/2-			
^{58}Mn		57.93998	65 s	β- /6.25	3.8/	3+			0.45916(2)/20.
					5.1/				0.81076(1)/82.
									1.32309(5)/53.
^{59}Mn		58.94044	4.6 s	β- /5.19	4.5/	5/2-			0.726/
									0.473/
									0.287–2.35
^{60m}Mn			1.77 s	β- /IT	5.7/	3+			0.824/
^{60}Mn		59.9429	0.28 s	β- /8.6	8.2/88.	0+			0.8234/12.2
					6.2/5.0				1.150/5.0
					7.4/4.2				1.523/3.0
^{61}Mn		60.9447	0.67 s	β- /7.4		(5/2)-			
^{62}Mn		61.9484	0.67 s	β- /10.4		(3+)			0.877/
									0.942–1.299
^{63}Mn		62.9502	0.28 s	β- /8.8					0.356,0.450
^{64m}Mn			> 0.1 ms						0.135/IT
^{64}Mn		63.9543	87 ms	β- /11.8					0.746
^{65}Mn		64.9563	0.092 s	β- /10.					0.366
^{66}Mn		65.9611	64 ms						0.471
^{67}Mn		66.9641	45 ms						
^{68}Mn		67.969	~ 28 ms						
^{69}Mn		68.973	14 ms						
^{70}Mn									
^{71}Mn									
$_{26}$Fe		55.845(2)							
^{45}Fe		45.0146	2.6 ms	2p /1.14	p// ~ 59.				
^{46}Fe		46.0008	13. ms	β+ /13.1	p// 79.	0+			0.493/23.
^{47}Fe		46.9929	21.9 ms	β+ /15.6	p//87.				0.892/76.
^{48}Fe		47.9805	45. ms	β+ /11.2	p//16.	0+			0.313/63.
^{49}Fe		48.9736	65. ms	β+ /13.0	p/1.977/34.5	(7/2-)			0.797/23.7
					p/(1.161–1.550)				(0.261-1.279)
^{50}Fe		49.9630	0.15 s	β+ /8.2		0+			0.651
^{51}Fe		50.95682	0.31 s	β+ /8.02		(5/2-)			ann.rad./
^{52m}Fe			46. s	β+ /4.4		(12+)			ann.rad./
									(0.622–2.286)/
^{52}Fe		51.94811	8.28 h	β+ /57/2.37	0.804/	0+			ann.rad./
				EC/43/					0.16868(1)/99.
				I.T./					0.377 (I.T.)/
^{53m}Fe			2.6 m	I.T./3.0407		19/2-			0.7011(1)/99.
									1.0115(1)/87.
									1.3281(1)/87.
									2.3396(1)/13.

Elem. or Isot.	Natural Abundance (Atom %)	Atomic Mass or Weight	Half–life/ Resonance Width (MeV)	Decay Mode/ Energy (/MeV)	Particle Energy/ Intensity (MeV/%)	Spin ($h/2\pi$)	Nuclear Magnetic Mom. (nm)	Elect. Quadr. Mom. (b)	γ-Energy/ Intensity (MeV/%)
^{53}Fe		52.945308	8.51 m	β+ /3.743	2.40/42.	7/2-			ann.rad./
					2.80/57.				0.3779(1)/42.
									(1.2–3.2)
^{54}Fe	5.845(35)	53.939611	>3.1 × 10^{22} a	EC-EC		0+			
^{55}Fe		54.938293	2.73 a	EC/0.2314		3/2-			Mn x-ray
^{56}Fe	91.754(36)	55.934938				0+			
^{57}Fe	2.119(10)	56.935394				½-	+0.0906	0.16	
^{58}Fe	0.282(4)	57.933276				0+			
^{59}Fe		58.934876	44.51 d	β- /1.565	0.273/48.	3/2-	−0.336		1.099/57
					0.475/51.				1.292/43.
									(0.14–1.48)
^{60}Fe		59.934072	2.6 × 10^6 a	β- /0.237	0.184/100.	0+			0.0586/100
^{61m}Fe			0.25 µs	I.T.		(9/2+)			0.654/IT
									0.207
^{61}Fe		60.93675	6.0 m	β- /3.98	2.5/13.				1.205/44.
					2.63/54.				1.028/43.
					2.80/31.				(0.12–3.37)
^{62}Fe		61.93677	68. s	β- /2.53	2.5/100.	0+			0.5061(1)/100.
^{63}Fe		62.9404	6. s	β- /6.3		5/2-			0.995/
									(1.365–1.427)
^{64}Fe		63.9412	2.0 s	β- /4.9		0+			
^{65m}Fe			1.1 s						(0.413-2.996)
^{65}Fe		64.9454	0.8 s	β- /7.9					(0.128-1.997)
^{66}Fe		65.9468	0.44 s	β- /5.7		0+			0.471–1.425
^{67m}Fe			~ 0.04 ms	I.T.		(5/2-)			0.367/IT
^{67}Fe		66.9510	0.48 s	β- /8.8					0.189/85
									2.089/14
^{68}Fe		67.954	0.19 s	β- / ~ 7.6		0+			
^{69}Fe		68.959	0.11 s						
^{70}Fe		69.961	0.10 s			0+			
^{71}Fe		70.967	> 0.3 µs						
^{72}Fe		71.970	> 0.3 µs			0+			
^{73}Fe									
^{74}Fe									
$_{27}$Co		58.933195(5)							
^{47}Co		47.0115							
^{48}Co		48.0018							
^{49}Co		48.9897	< 0.035 µs						
^{50}Co		49.9815	39. ms	β+ /17.0	p/2.770/41.				0.2614/64.
					p/(1.874–2.296)				(0.482-1.308)
^{51}Co		50.9707	69. ms	β+ /12.8	p// < 3.8				
^{52}Co		51.9636	0.12 s	β+ /14.0					0.849–1.942
^{53m}Co			0.25 s	β+, p/		19/2-			ann.rad./
^{53}Co		52.95422	0.24 s	β+ /8.30		7/2-			ann.rad./
^{54m}Co			1.46 m	β+ /8.44	4.25/100.	7+			ann.rad./
									0.411(1)/99.
									1.130(1)/100.
									1.408(1)/100.
^{54}Co		53.948460	0.1932 s	β+ /8.2430	7.34/100.	0+			ann.rad./
^{55}Co		54.941999	17.53 h	β+ /3.4513	0.53/	7/2-	+4.822		ann.rad./
				EC/	1.03/				0.9312/75.
					1.50/				0.4772/20.
									(0.092–3.11)
^{56}Co		55.939839	77.3 d	β+ /4.566	1.459/18.	4+	3.85	~+0.25	ann.rad./

Elem. or Isot.	Natural Abundance (Atom %)	Atomic Mass or Weight	Half–life/ Resonance Width (MeV)	Decay Mode/ Energy (/MeV)	Particle Energy/ Intensity (MeV/%)	Spin ($h/2\pi$)	Nuclear Magnetic Mom. (nm)	Elect. Quadr. Mom. (b)	γ-Energy/ Intensity (MeV/%)
				EC/					0.8468/99.9
									1.2383/68.
									(0.26–3.61)
^{57}Co		56.936291	271.8 d	EC/0.8361		7/2-	+4.72	+0.5	0.12206/86
									(0.014–0.706)
^{58m}Co			9.1 h	I.T./		5+			0.02489/0.035
^{58}Co		57.935753	70.88 d	β+ /2.307		2+	+4.04	+0.22	ann.rad./
				EC/					0.81076/99
^{59}Co	100.	58.933195				7/2-	+4.63	+0.41	
^{60m}Co			10.47 m	I.T./99.8/0.059		2+	+4.4	~+0.3	0.0586/2.0
				β- /0.2/1.56					
^{60}Co		59.933817	5.271 a	β- /2.824	0.315/99.7	5+	+3.799	+0.44	1.1732/100
									1.3325/100
^{61}Co		60.932476	1.650 h	β- /1.322	1.22/95.	7/2-			0.0674/86.
									0.842–0.909
^{62m}Co			13.9 m	β- /	0.88/25.	5+			1.1635(3)/70.
					2.88/75.				1.1730(3)/98.
									2.0039(3)/19.
^{62}Co		61.93405	1.50 m	β- /5.32	1.03/10.	2+			1.1292(3)/13.
					1.76/5.				1.1730(3)/83.
					2.9/20.				1.9851(1)/3.
					4.05/60.				2.3020(1)/19.
^{63}Co		62.93361	27.5 s	β- /3.67	3.6/	7/2-			0.08713(1)/49.
									0.9817(3)/2.6
									0.156–2.17
^{64}Co		63.93581	0.30 s	β- /7.31	7.0/	1+			
^{65}Co		64.93648	1.14 s	β- /5.96		(7/2)-			(0.063-1.273)
^{66m2}Co			> 0.1 ms	I.T.		(8-)			0.252/IT
									0.214
									0.175
^{66m1}Co			1.2 μs	I.T.		(5+)			0.175/IT
^{66}Co		65.9398	0.25 s	β- /10.0					(1.245–1.425)
^{67}Co		66.9409	0.43 s	β- /8.4					0.694
									(0.189-2.769)
^{68}Co		67.9449	0.19 s	β- /11.7					
^{69}Co		68.9463	0.20 s	β- /9.3					
^{70}Co		69.951	0.12 s	β- 13.					1.26/102
									0.97/100
									(0.45 – 0.92)
^{71}Co		70.953	97. ms	β-					0.566/100
				β-,n	// > 3				(0.25 – 0.77)
^{72}Co		71.958	60. ms	β-					1.096/100
				β-,n	// > 6				0.845
									(0.455 – 1.197)
^{73}Co		72.960	41. ms	β-					0.524/100
				β-,n	// < 7.9				(0.24 – 0.76)
^{74}Co		73.965	30. ms	β-					0.739
				β-,n	// ~ 18				1.024
^{75}Co		74.968	0.03 s	β-,n	//< 16				
^{76}Co									
$_{28}$Ni		**58.6934(4)**							
^{48}Ni		48.020	~ 2.1 ms	2p	p // ~ 25	0+			
^{49}Ni		49.0097	12. ms		p //~83				0.965/82
^{50}Ni		49.9959	12. ms	β+, p	p //70.	0+			0.063

Elem. or Isot.	Natural Abundance (Atom %)	Atomic Mass or Weight	Half–life/ Resonance Width (MeV)	Decay Mode/ Energy (/MeV)	Particle Energy/ Intensity (MeV/%)	Spin ($h/2\pi$)	Nuclear Magnetic Mom. (nm)	Elect. Quadr. Mom. (b)	γ-Energy/ Intensity (MeV/%)
					p /1.97/14				0.090
^{51}Ni		50.9877	24. ms	β+ /16.0	p/4.66/8.7				0.765/73
					p/1.08−5.66				1.087/29
					p//87.				1.546-1.743
^{52}Ni		51.9757	38. ms	β+ /11.7	p//31.	0+			2.418/38
					p/1.35/9				0.142
^{53}Ni		52.9685	55. ms	β+, p/13.3	p//23.	7/2-			ann.rad./
					p/1.93/5.4				0.849/13
^{54}Ni		53.95791	0.10 s	β+ /8.80		0+			0.937
^{55}Ni		54.95133	0.20 s	β+ /8.70	7.66/	7/2-	0.98		ann.rad./
^{56}Ni		55.94213	6.08 d	EC/2.14		0+			0.15838/99
				β+ /<10^{-6}					0.81185(3)/87.
									0.2695−0.7500
^{57}Ni		56.939794	35.6 h	β+ /3.264	0.712/10.	3/2-	−0.798		ann.rad./
				EC/	0.849/76.				1.3776/78.
									(0.127−3.177)
^{58}Ni	68.077(19)	57.935343	>4 × 10^{19} a	EC-EC		0+			
^{59}Ni		58.934347	~ 7.6 × 10^4 a	EC/		3/2-			
^{60}Ni	26.223(15)	59.930786				0+			
^{61}Ni	1.1399(13)	60.931056				3/2-	−0.75002	+0.16	
^{62}Ni	3.6345(40)	61.928345				0+			
^{63}Ni		62.929669	101. a	β- /0.066945	0.065/	½-			
^{64}Ni	0.9255(19)	63.927966				0+			
^{65}Ni		64.930084	2.517 h	β- /2.137	0.65/30.	5/2-	0.69		0.36627(3)/5.
					1.020/11.				1.11553(4)/16.
					2.140/58.				1.48184(5)/23.
^{66}Ni		65.929139	54.6 h	β- /0.23		0+			
^{67m}Ni			13.3 μs	I.T.		9/2+	0.56		0.313/IT
									0.694
^{67}Ni		66.931569	21. s	β- /3.56	3.8/	½-	+0.601		1.0722/100.
									1.6539/100.
									(0.10−1.98)
^{68m2}Ni			0.34 μs						0.511
^{68m1}Ni			0.86 ms	I.T.		(5-)			0.814/IT
									2.033
^{68}Ni		67.931869	29. s	β- /2.06		0+			
^{69m2}Ni			0.44 μs	I.T.		(17/2)			0.148/IT
									0.593
									1.959
^{69m1}Ni			3.5 s						
^{69}Ni		68.935610	11. s	β- /5.4					0.6807(3)/100.
									(0.207−1.213)
^{70m}Ni			0.21 μs	I.T.		(8+)			0.183/IT
									0.448
									0.970
									1.259
^{70}Ni		69.9365	6.0 s	β- /3.5		0+			
^{71}Ni		70.9407	2.56 s	β- /6.9					
^{72}Ni		71.9421	1.6 s	β- /5.2		0+			
^{73}Ni		72.9465	0.84 s	β- /9.					
^{74}Ni		73.9481	0.9 s	β- /7.		0+			
^{75}Ni		74.9529	0.34 s	β⁻,n	/10				
^{76}Ni		75.955	0.24 s	β⁻,n	/14	0+			
^{77}Ni		76.961	0.13 s	β⁻,n	/ ~ 30				

Elem. or Isot.	Natural Abundance (Atom %)	Atomic Mass or Weight	Half–life/ Resonance Width (MeV)	Decay Mode/ Energy (/MeV)	Particle Energy/ Intensity (MeV/%)	Spin ($h/2\pi$)	Nuclear Magnetic Mom. (nm)	Elect. Quadr. Mom. (b)	γ-Energy/ Intensity (MeV/%)
78Ni		77.963	~ 0.11 s			0+			
79Ni									
29Cu		63.546(3)							
52Cu		51.9972							
53Cu		52.9856	< 0.3 μs						
54Cu		53.9767	< 0.075 μs						
55Cu		54.9661	~ 27. ms	β+ /13.2	p//15.				
56Cu		55.9586	93. ms	β+ /15.3					0.511/233
									2.700/100
									0.9507–3.287
57Cu		56.94921	196. ms	β+ /8.77		3/2-	+2.58		0.77–3.01
58Cu		57.944539	3.21 s	β+ /8.563	4.5/15.	1+	+0.48		ann.rad./
				EC/	7.439/83.				0.0403(4)/5.
									1.4483(2)/11.
									1.4546(2)/16.
59Cu		58.939498	1.36 m	β+ /4.800	1.9/	3/2-	+1.91		ann.rad./
					3.75/				0.3393(1)/8.
									0.8780(1)/12.
									1.3015(1)/15.
									(0.4–2.6)
60Cu		59.937365	23.7 m	β+ /6.127	2.00/69.	2+	+1.219		ann.rad./
				EC/	3.00/18.				1.3325/88.
					3.92/6.				1.7915/45.
									(0.12–5.048)
61Cu		60.933458	3.35 h	β+ /2.237	0.56/3.	3/2-	+2.14		ann.rad./
					0.94/5.				0.2830/13.
					1.15/2.				0.6560/11.
					1.220/51.				(0.067–2.123)
62Cu		61.932584	9.67 m	β+ /98/3.948	2.93/98.	1+	−0.380		ann.rad./
				EC/					1.17302(1)/0.6
									(0.87–3.37)
63Cu	69.15(15)	62.929598				3/2-	+2.2273	−0.211	
64Cu		63.929764	12.701 h	β- /38/0.579	0.578/	1+	−0.217		ann.rad./35.1
				β+ /19/1.6751	0.65/				1.3459(3)/0.47
				EC/41/					
65Cu	30.85(15)	64.927790				3/2-	+2.3817	−0.195	
66Cu		65.928869	5.09 m	β- /2.642	1.65/6.	1+	−0.282		0.8330(1)/0.22
					2.7/94.				1.0392(2)/9.2
67Cu		66.927730	2.580 d	β- /0.58	0.395/56.	3/2-	+2.54		0.09125(1)/7.
					0.484/23.				0.09325(1)/17.
					0.577/20.				0.18453(1)/47.
68mCu			3.79 m	I.T./86/		6-	+1.24		0.0843(5)/70.
				β- /14/1.8					0.1112(5)/18.
									0.5259(5)/74.
									(0.64–1.34)
68Cu		67.929611	31. s	β- /4.46	3.5/40.	1+	+2.48		1.0774(5)/58.
					4.6/31.				1.2613(5)/17.
									(0.15–2.34)
69mCu			0.36 μs	I.T.		(13/2+)	+1.5		0.075/IT
									0.190/IT
									0.680
									1.871
69Cu		68.929429	2.8 m	β- /2.68	2.48/80.	3/2-	+2.84		0.5307(3)/3.
									0.8340(5)/6.

Elem. or Isot.	Natural Abundance (Atom %)	Atomic Mass or Weight	Half–life/ Resonance Width (MeV)	Decay Mode/ Energy (/MeV)	Particle Energy/ Intensity (MeV/%)	Spin ($h/2\pi$)	Nuclear Magnetic Mom. (nm)	Elect. Quadr. Mom. (b)	γ-Energy/ Intensity (MeV/%)
									1.0065(8)/10.
70m2Cu			6.6 s	β/93		1+	+1.9		0.8849/100
									1.072/19
				IT/7					0.141/ IT
70m1Cu			33. s	β- /52	2.52/10.	3-	−3.5		0.8848(2)/100.
				IT/48					0.9017(2)/90.
									1.2517(5)/60.
									(0.39–3.06)
70Cu		69.932392	44.5 s	β- /6.60	5.42/54.	6-	+1.5		0.8848(2)/100.
					6.09/46.				0.9017/99.7
									(0.438–3.062)
71mCu			0.28 μs	I.T.		(19/2)			0.133/IT
									0.494
									0.939
									1.189
71Cu		70.932677	20. s	β- /4.56		3/2-	+2.275		0.490/
72mCu			1.76 μs	I.T.		(4-)			0.051/IT
									0.082
									0.138
72Cu		71.935820	6.6 s	β- /8.2		(1+)			0.652/
73Cu		72.936675	4.2 s	β- /6.3	5.8/43	(3/2)	+1.743		0.450/100
					6.25/42				0.307−1.559
74Cu		73.93988	1.59 s	β- /9.9					
75Cu		74.942	1.2 s	β- /7.9		(5/2)	+1.006		
76mCu			1.2 s						
76Cu		75.94528	0.64 s	β- /11.	/88.				
				β-,n	/7.				
77Cu		76.9479	0.47 s	β- / ~ 10.					0.5057
				β-,n	/30				(.1147-3827)
78Cu		77.9520	0.33 s	β- /12.	/35				
				β-,n	/65				
79Cu		78.9546	0.2 s	β- /11.					
				β-,n	/7				
80Cu		79.961	~ 0.17 s						
81Cu									
82Cu									
30Zn	65.38(2)								
54Zn		53.9930	~ 3.2 ms	2p	p//87	0+			
55Zn		54.9840	20. ms		p//91.				
56Zn		55.9724	30. ms		p//86.	0+			
57Zn		56.9648	47. ms	β+, p/14.6		(7/2-)			ann.rad./
58Zn		57.95459	0.09 s	β+		0+			
59Zn		58.94926	183. ms	β+, p/9.09	8.1/	3/2-			ann.rad./
									(0.491−0.914)
60Zn		59.94183	2.40 m	β+ /97/4.16		0+			ann.rad./
				EC/3/					0.669/47.
									(0.062−0.947)
61Zn		60.93951	1.485 m	β+ /5.64	4.38/68.	3/2-			ann.rad./
									0.4748/17.
									(0.15−3.52)
62Zn		61.93433	9.22 h	β+ /3/1.63	0.66/7.	0+			ann.rad./
				EC/93/					0.0408/25
									0.5967/26.
									(0.20−1.526)/

Elem. or Isot.	Natural Abundance (Atom %)	Atomic Mass or Weight	Half–life/ Resonance Width (MeV)	Decay Mode/ Energy (/MeV)	Particle Energy/ Intensity (MeV/%)	Spin ($h/2\pi$)	Nuclear Magnetic Mom. (nm)	Elect. Quadr. Mom. (b)	γ-Energy/ Intensity (MeV/%)
^{63}Zn		62.933212	38.5 m	β+ /93/3.367	1.02/	3/2-	–0.28164	+0.29	ann.rad./
				EC/7/	1.40/				0.66962(5)/8.4
					1.71/				0.96206(5)/6.6
					2.36/84.				(0.24–3.1)
^{64}Zn	49.17(75)	63.929142	>7. × 10^{20} a	EC-β+		0+			
^{65}Zn		64.929241	244. d	β+ /98/1.3514	0.325/	5/2-	+0.7690	–0.023	ann.rad./
				EC/1.5/					1.1155/50.2
^{66}Zn	27.73(98)	65.926033				0+			
^{67}Zn	4.04(16)	66.927127				5/2-	+0.8753	+0.15	
^{68}Zn	18.45(63)	67.924844				0+			
^{69m}Zn			13.76 h	I.T./99+/0.439		9/2+	1.157		0.4390(2)/95.
^{69}Zn		68.926550	56. m	β- /0.906	0.905/99.9	½-			0.318/
^{70}Zn	0.61(10)	69.925319	>2.3 × 10^{17} a	β-β-		0+			
^{71m}Zn			3.97 h	β- /	1.45/	9/2+	1.05		0.3864/93.
									0.4874/62.
									0.6203/57.
									(0.099–2.489)
^{71}Zn		70.92772	2.4 m	β- /2.81		½-			0.5116(1)/30.
									0.9103(1)/7.5
									(0.12–2.29)
^{72}Zn		71.92686	46.5 h	β- /0.46	0.25/14.	0+			0.0164(3)/8.
					0.30/86.				0.1447(1)/83.
									0.1915(2)/9.4
^{73m}Zn			6. s		I.T./0.196	(7/2+)			0.042
^{73}Zn		72.92978	24. s	β- /4.29	4.7/	(1/2-)			0.216(1)/100.
									0.496–0.911
^{74}Zn		73.92946	1.60 m	β- /2.3	2.1/	0+			0.0565/
									0.1401/
									(0.05–0.35)
^{75}Zn		74.9329	10.2 s	β- /6.0					0.229/
^{76}Zn		75.9333	5.7 s	β- /4.2	3.6/	0+			0.119/
^{77m}Zn			1.0 s	β- /		(1/2-)			0.772
^{77}Zn		76.9370	2.1 s	β- /7.3	4.8/				0.189/
^{78m}Zn			> 0.03 ms						1.070
^{78}Zn		77.9384	1.5 s	β- /6.4		0+			0.225/
^{79}Zn		78.9427	1.0 s	β- /8.6					0.702/
				β-,n	/2				
^{80}Zn		79.9443	0.56 s	β- /7.3		0+			0.713/
				β-,n	/ <18				0.2248/
^{81}Zn		80.9505	0.29 s	β- /11.9					
				β-,n	/ ~30				
^{82}Zn		81.9544	> 0.15 µs			0+			
^{83}Zn		82.9610	> 0.15 µs						
^{84}Zn									
^{85}Zn									
$_{31}$**Ga**		**69.723(1)**							
^{56}Ga		55.9949							
^{57}Ga		56.9829							
^{58}Ga		57.9743							
^{59}Ga		58.9634	< 0.043 µs						
^{60}Ga		59.9571	0.07 s	β+					1.004
				β+, p	// ~ 1.6				3.848
				β+, α	// ~ 0.02				1.555–2.559

Elem. or Isot.	Natural Abundance (Atom %)	Atomic Mass or Weight	Half–life/ Resonance Width (MeV)	Decay Mode/ Energy (/MeV)	Particle Energy/ Intensity (MeV/%)	Spin (h/2π)	Nuclear Magnetic Mom. (nm)	Elect. Quadr. Mom. (b)	γ-Energy/ Intensity (MeV/%)
^{61}Ga		60.9495	0.17 s	β+ /9.0		3/2$^-$			0.088–1.362
^{62}Ga		61.94418	116.1 ms	β+ /9.17	8.3/99.89	0+			ann.rad./
				EC/					0.954/0.085
									(0.851-5.92)
^{63}Ga		62.939294	32. s	β+ /5.5	4.5/				ann.rad./
				EC/					0.6271(2)/10.
									0.6370(2)/11.
									1.0652(4)/45.
^{64m}Ga			0.022 ms						0.0429
^{64}Ga		63.936839	2.63 m	β+ /7.165	2.79/	0+			ann.rad./
					6.05/				0.80785(1)/14.
									0.99152(1)/43.
									1.38727(1)/12.
									3.3659(1)/13.
^{65}Ga		64.932735	15.2 m	β+ /86/3.255	0.82/10.	3/2$^-$			ann.rad./
				EC/	1.39/19.				0.1151(2)/55.
					2.113/56.				0.1530(2)/96.
					2.237/15.				0.2069(2)/39.
									(0.06–2.4)
^{66}Ga		65.931589	9.3 h	β+ /56/5.175	0.74/1.	0+			ann.rad./
				EC/43/	1.84/54.				1.03935(8)/38.
					4.153/51.				2.7523(1)/23.
									(0.28–5.01)
^{67}Ga		66.928202	3.261 d	EC/1.001		3/2-	+1.8507	+0.20	0.09332/37.
									0.18459/20.
									0.30024/17.
									(0.091–0.89)
^{68}Ga		67.927980	1.130 h	β+ /90/2.921	1.83/	1+	0.01175	0.028	ann.rad./
				EC/10/					1.0774(1)/3.
									(0.57–2.33)/
^{69}Ga	60.108(9)	68.925574				3/2-	+2.01659	+0.165	
^{70}Ga		69.926022	21.1 m	EC/0.2/0.655		1+			0.1755(5)/0.15
				β- /99.8/1.656	1.65/99.				1.042(5)/0.48
^{71}Ga	39.892(9)	70.924701	>2.4 × 10^{26} a	β-		3/2-	+2.56227	+0.104	
^{72}Ga		71.926366	14.10 h	β- /4.001	0.64/40.	3-	−0.13224	+0.52	0.8340/95.53
					1.51/9.				2.202/26.9
					2.52/8.				0.630/26.2
					3.15/11.				(0.113–3.678)
^{73}Ga		72.925175	74.87 h	β- /1.59		3/2-	+0.209		0.05344(5)/10.
									0.29732(5)/47.
									(0.01–1.00)/
^{74m}Ga			10. s	I.T./		1+			0.0565(1)/75.
^{74}Ga		73.926946	8.1 m	β- /5.4	2.6/	3-			0.5959/92.
									2.354/45.
									(0.23–3.99)
^{75}Ga		74.926500	2.10 m	β- /3.39	3.3/	3/2-	+1.836	−0.29	0.2529/
									0.5746/
									(0.12–2.10)
^{76}Ga		75.928828	29. s	β- /7.0		3-			0.5629/66.
									0.5455/26.
									(0.34–4.25)
^{77}Ga		76.929154	13.0 s	β- /5.3	5.2/		+2.020	−0.21	0.469/
									0.459/
^{78}Ga		77.931608	5.09 s	β- /8.2		3+			0.619/77.

Elem. or Isot.	Natural Abundance (Atom %)	Atomic Mass or Weight	Half–life/ Resonance Width (MeV)	Decay Mode/ Energy (/MeV)	Particle Energy/ Intensity (MeV/%)	Spin ($h/2\pi$)	Nuclear Magnetic Mom. (nm)	Elect. Quadr. Mom. (b)	γ-Energy/ Intensity (MeV/%)
									1.187/20.
^{79}Ga		78.9329	2.85 s	β- /7.0	4.6/		+1.047	+0.16	0.465/
^{80}Ga		79.9365	1.68 s	β- /10.4	10./				0.659/
^{81}Ga		80.9378	1.22 s	β- /8.3	5.1/		+1.747	−0.05	0.217/
				β-,n	/ <21				
^{82}Ga		81.9430	0.599 s	β- /12.6					1.348/
				β-,n	/30				
^{83}Ga		82.9470	0.308 s	β- /~ 11.5					1.348
^{84}Ga		83.9527	~ 0.085 s	β- /14					0.624
				β-,n	/74				1.046
^{85}Ga		84.9570	> 0.3 μs						
^{86}Ga		85.963	> 0.3 μs						
^{87}Ga									
$_{32}$**Ge**		**72.63(1)**							
^{58}Ge		57.9910				0+			
^{59}Ge		58.9818							
^{60}Ge		59.9702	> 0.11 μs			0+			
^{61}Ge		60.9638	~ 44. ms	β+ /13.6					
^{62}Ge		61.9547	0.13 s			0+			
^{63}Ge		62.9496	0.15 s	β- /9.8					
^{64}Ge		63.94165	1.06 m	β+ /4.4	3.0/	0+			ann.rad./
				EC/					0.1282(2)/11.
				β+, p					0.4270(3)/37.
									0.6671(3)/17.
^{65}Ge		64.9394	31. s	β+ /6.2	0.82/10.				ann.rad./
				EC/	1.39/19.				0.0620/27.
				EC, p	2.113/56.				0.6497/33.
					2.237/15.				0.8091/21.
				β+, p	//0.011				(0.19–3.28)
^{66}Ge		65.93384	2.26 h	β+ /27/2.10		0+			ann.rad./
				EC/73/					0.0438/29.
									0.3819/28.
									(0.022–1.77)
^{67}Ge		66.932734	19.0 m	β+ /96/4.225	1.6/	½-			ann.rad./
				EC/4/	2.3/				0.1670/84.
					3.15/				(0.25–3.73)
^{68}Ge		67.92809	270.8 d	EC/0.11		0+			Ga k x-ray/39.
^{69}Ge		68.927965	1.63 d	β+ /36/2.2273	0.70/	5/2-	0.735	0.02	ann.rad./
				EC/64/	1.2/				0.574/13.
									1.1068/36.
									(0.2–2.04)
^{70}Ge	20.57(27)	69.924247				0+			
^{71m}Ge			20.4 ms	I.T./0.0234		9/2+	−1.041	~ 0.34	0.1749
^{71}Ge		70.924951	11.2 d	EC/0.229		½-	+0.547		
^{72}Ge	27.45(32)	71.922076				0+			
^{73}Ge	7.75(12)	72.923459	>1.8 × 10^{23} a	β-		9/2+	−0.879468	−0.17	
^{74}Ge	36.50(20)	73.921178				0+			
^{75m}Ge			48. s	I.T./		7/2+			0.13968(3)/39.
^{75}Ge		74.922859	1.380 h	β- /1.177	1.19/	½-	+0.510		0.26461(5)/11.
									0.41931(5)/0.2
^{76}Ge	7.73(12)	75.921403	1.6 × 10^{21} a	β-β-		0+			
^{77m}Ge			53. s	I.T./20/		½-			1.605/0.22
				β- /80/2.861	2.9/				1.676/0.16
									0.195–1.482

Elem. or Isot.	Natural Abundance (Atom %)	Atomic Mass or Weight	Half–life/ Resonance Width (MeV)	Decay Mode/ Energy (/MeV)	Particle Energy/ Intensity (MeV/%)	Spin (h/2π)	Nuclear Magnetic Mom. (nm)	Elect. Quadr. Mom. (b)	γ-Energy/ Intensity (MeV/%)
77Ge		76.923549	11.25 h	β- /2.702	0.71/23.	7/2+			0.2110/29.
					1.38/35.				0.2155/27.
					2.19/42.				0.2644/51.
									(0.15–2.35)
78Ge		77.922853	1.45 h	β- /0.95	0.70/	0+			0.2773(5)/96.
									0.2939(5)/4.
79mGe			39. s	β- /IT		7/2+			
79Ge		78.9254	19.1 s	β- /4.2	4.0/20.	½-			0.1096/21.
					4.3/80.				(0.10–2.59)
									0.5427(4)/15.
80Ge		79.92537	29.5 s	β- /2.67	2.4/	0+			0.1104(4)/6.
									0.2656(4)/25.
81mGe			~ 7.6 s	β- /	3.75/	½+			0.3362(4)/
									0.7935(4)/
81Ge		80.9288	~ 7.6 s	β- /6.2	3.44/	9/2+			0.1976(4)/21.
									0.3362(4)/100.
82Ge		81.9296	4.6 s	β- /4.7	1.093/80	0+			1.093/
83Ge		82.9346	1.9 s	β- /8.9					
				β-,n	/63				
84Ge		83.9375	0.98 s	β- /7.7		0+			
85Ge		84.9430	0.54 s	β- /10.					
86Ge		85.9465	> 0.3 µs			0+			
87Ge		86.9525	> 0.3 µs						
88Ge		87.957	> 0.3 µs			0+			
89Ge		88.964	> 0.3 µs						
90Ge									
33As		74.92160(2)							
60As		59.993							
61As		60.981							
62As		61.9732							
63As		62.9637	< 0.043 µs						
64As		63.9576	0.02 s						
65As		64.9496	0.13 s	β+ /9.4					
66m2As			8. µs						
66m1As			1.1 µs						
66As		65.945	95.8 ms	β+ /9.55					
67As		66.9392	42. s	β+ /6.0	5.0/	5/2-			0.121/
				EC/					0.123/
									0.244/
68As		67.93677	2.53 m	β+ /8.1		3+			ann.rad./
									0.652/32.
									0.762/33.
									1.016/77.
									(0.61–3.55)
69As		68.93227	15.2 m	β+ /98/4.01	2.95/	5/2-	+1.623		ann.rad./
				EC/2/					0.0868(5)/1.5
									0.1458(3)/2.4
70As		69.93092	52.6 m	β+ /84/6.22	1.44/	4+	+2.1061	+0.09	ann.rad./
				EC/16/2.14					1.0395(7)/82.
				/2.89					(0.17–4.4)/
71As		70.927112	2.72 d	β+ /32/2.013		5/2-	+1.6735	-0.02	ann.rad./
				EC/68/					0.1749(2)/84.
									1.0957(2)/4.2
72As		71.926752	26.0 h	β+ /77/4.356	0.669/5.	2-	-2.1566	-0.08	ann.rad./

Elem. or Isot.	Natural Abundance (Atom %)	Atomic Mass or Weight	Half–life/ Resonance Width (MeV)	Decay Mode/ Energy (/MeV)	Particle Energy/ Intensity (MeV/%)	Spin ($h/2\pi$)	Nuclear Magnetic Mom. (nm)	Elect. Quadr. Mom. (b)	γ-Energy/ Intensity (MeV/%)
					1.884/12.				0.83395(5)/80.
					2.498/62.				1.0507(1)/9.6
					3.339/19.				(0.1–4.0)
^{73}As		72.923825	80.3 d	EC/0.341		3/2-			0.0133/0.1
									0.0534/10.5
									Se k x-ray/90.
^{74}As		73.923829	17.78 d	β+ /31/2.562	0.94/26.	2-	−1.597		ann.rad./
				EC/37/	1.53/3.				0.59588(1)/60.
				β- /1.353	0.71/16.				0.6084(1)/0.6
					1.35/16.				0.6348(1)/15.
^{75m}As			0.017 s						
^{75}As	100.	74.921597				3/2-	+1.43947	+0.31	
^{76}As		75.922394	26.3 h	β- /2.962	0.54/3.	2-	−0.903		0.5591(1)/45.
					1.785/8.				0.65703(5)/6.2
					2.410/36.				1.21602(1)/3.4
					2.97/51.				(0.3–2.67)
^{77}As		76.920647	38.8 h	β- /0.683	0.70/98.	3/2-	+1.295		0.2391(2)/1.6
									0.2500(3)/0.4
									0.5208/0.43
^{78}As		77.92183	1.512 h	β- /4.21	3.00/12.	2-			0.6136(3)/54.
					3.70/17.				0.6954(3)/18.
					4.42/37.				1.3088(3)/10.
^{79m}As			1.21 μs	I.T.		9/2+			0.542/IT
									0.231
^{79}As		78.92095	9.0 m	β- /2.28	1.80/95.	3/2-			0.0955(5)/16.
									0.3645(5)/1.9
^{80}As		79.92253	16. s	β- /5.64	3.38/	1+			0.6662(2)/42.
									(2.5–3.0)
^{81}As		80.92213	33. s	β- /3.856		3/2-			0.4676(2)/20.
									0.4911(2)/8.
^{82m}As			13.7 s	β- /	3.6/	5-			0.6544(1)/77.
									0.344/65.
									(0.561 – 1.894)
^{82}As		81.9245	19. s	β- /7.4	7.2/80.	(2-)			0.6544(1)/54.
									(0.755 – 3.667)
^{83}As		82.9250	13.4 s	β- /5.5					0.7345/100.
									1.1131/34.
									2.0767/28.
^{84m}As			0.6 s	β-					
^{84}As		83.9291	4. s	β-, n/7.2		1-			0.6671(2)/21.
									1.4439(5)/49.
									(0.325–5.150)
^{85}As		84.9320	2.03 s	β-, n/8.9		3/2-			0.667(1)/42.
									1.4551(2)/100.
^{86}As		85.9365	0.95 s	β-, n/11.4					0.704/
^{87}As		86.9399	0.49 s	β-, n/10.					0.704/
^{88}As		87.9449	> 0.3 μs						
^{89}As		88.9494	> 0.3 μs						
^{90}As		89.956	> 0.3 μs						
^{91}As		90.960	> 0.3 μs						
^{92}As		91.967	> 0.3 μs						
$_{34}$Se		78.96(3)							
^{64}Se			> 0.18 μs			0+			

Elem. or Isot.	Natural Abundance (Atom %)	Atomic Mass or Weight	Half–life/ Resonance Width (MeV)	Decay Mode/ Energy (/MeV)	Particle Energy/ Intensity (MeV/%)	Spin ($h/2\pi$)	Nuclear Magnetic Mom. (nm)	Elect. Quadr. Mom. (b)	γ-Energy/ Intensity (MeV/%)
^{65}Se		64.965	0.011 s	β+ /60/14.					
				β+, p	3.55/				
^{66}Se		65.9552	0.03 s			0+			
^{67}Se		66.9501	0.13 s	β+ /10.2					ann.rad./
				β+, (p)/					0.352
^{68}Se		67.94180	36. s	β+ /4.7		0+			ann.rad./
									(0.050–0.426)
^{69}Se		68.93956	27.4 s	β+ /6.78	5.006/				ann.rad./
				EC/					0.0664(4)/27.
				β+, p	// ~ 0.045				0.0982(4)/63.
^{70}Se		69.9334	41.1 m	β+ /2.4		0+			ann.rad
									0.04951(5)/35.
									0.4262(2)/29.
^{71}Se		70.93224	4.7 m	β+ /4.4	3.4/36.	5/2-			ann.rad
				EC/					0.1472(3)/47.
									0.8309(3)/13.
									1.0960(3)/10.
^{72}Se		71.92711	8.5 d	EC/0.34		0+			0.0460(2)/57.
^{73m}Se			40. m	I.T./73/0.0257	0.85	3/2-			ann.rad.
				β+ /27/2.77	1.45/				0.0257(2)/27.
					1.70/				0.2538(1)/2.5
^{73}Se		72.92677	7.1 h	β+ /65/2.74	0.80/	9/2+	0.86		ann.rad
				EC/35/	1.32/95.				0.0670(1)/72.
					1.68/1.				0.3609(1)/97.
									(0.6–1.5)
^{74}Se	0.89(4)	73.922476	> 5.5×10^{18} a	EC-EC		0+			
^{75}Se		74.922523	119.78 d	EC/0.864		5/2+	0.68	1.0	0.13600/55
									0.26465/58
									(0.024–0.821)
^{76}Se	9.37(29)	75.919214				0+			
^{77m}Se			17.4 s	I.T./		7/2+			0.1619(2)/52.
^{77}Se	7.63(16)	76.919914				½-	+0.53506		
^{78}Se	23.77(28)	77.917309				0+			
^{79m}Se			3.92 m	I.T./					0.09573(3)/9.5
^{79}Se		78.918499	3.3 × 10^5 a	β- /0.151		7/2+	−1.02	+0.8	
^{80}Se	49.61(41)	79.916521				0+			
^{81m}Se			57.3 m	I.T./99/0.1031		7/2+			0.1031(3)/9.7
									0.2602(2)/0.06
									0.2760/0.06
^{81}Se		80.917993	18.5 m	β- /1.585	1.6/98.	½-			0.2759/0.85
									0.2901/0.75
									0.8283/0.32
^{82}Se	8.73(22)	81.916699	>9.5 × 10^{19} a	β-β-		0+			
^{83m}Se			1.17 m	β- /3.96	2.88/	½-			0.35666(6)/17.
					3.92/				0.9879(1)/15.
									1.0305(1)/21.
									2.0514(2)/11.
									(0.19–3.1)
^{83}Se		82.919118	22.3 m	β- /3.668	0.93/	9/2+			0.22516(6)/33.
					1.51/				0.35666(6)/69.
									0.51004(8)/45.
									(0.21–2.42)
^{84}Se		83.91846	3.3 m	β- /1.83	1.41/100.	0+			0.4088(5)/100.
^{85}Se		84.92225	32. s	β- /6.18	5.9/	5/2+			0.3450(1)/22.

Elem. or Isot.	Natural Abundance (Atom %)	Atomic Mass or Weight	Half–life/ Resonance Width (MeV)	Decay Mode/ Energy (/MeV)	Particle Energy/ Intensity (MeV/%)	Spin ($h/2\pi$)	Nuclear Magnetic Mom. (nm)	Elect. Quadr. Mom. (b)	γ-Energy/ Intensity (MeV/%)
									0.6094(1)/41.
^{86}Se		85.92427	15. s	β- /5.10		0+			2.0124(1)/24.
									2.4433(8)/100.
									2.6619(1)/49.
^{87}Se		86.92852	5.4 s	β- /7.28		5/2+			0.468(1)/100.
				n/					1.4979(1)/23.
^{88}Se		87.93142	1.5 s	β-, n/6.85		0+			0.5346/
^{89}Se		88.9365	0.41 s	β-, n/9.0					
^{90}Se		89.9400	> 0.3 µs			0+			
^{91}Se		90.9460	0.27 s	β-, n/8.					
^{92}Se		91.950	> 0.3 µs			0+			
^{93}Se		92.956	> 0.3 µs						
^{94}Se		93.960	> 0.3 µs			0+			
^{95}Se									
$_{35}$**Br**		**79.904(1)**							
^{67}Br		66.9648							
^{68}Br		67.9585	< 1.5 µs						
^{69}Br		68.9501	< 0.024 µs	β+ /9.6					
^{70m}Br			2.2 s			9+			
^{70}Br		69.9446	~ 0.08 s	β+ /10.0	/0.75				
^{71}Br		70.939	21. s	β+ /6.9					
^{72}Br		71.9366	1.31 m	β+ /8.7		3	0.6		0.4547–1.317
^{73}Br		72.93169	3.4 m	β+ /4.7	3.7/	3/2-			ann.rad
									0.065–0.700
^{74m}Br			46. m	β+ /	4.5/	4-	1.82		ann.rad
									0.6348
									0.7285
									(0.2–4.38)
^{74}Br		73.92989	25.4 m	β+ /6.91					ann.rad
									0.6341
									0.6348
									(0.2–4.7)
^{75}Br		74.92578	1.62 h	β+ /76/3.03		3/2-	+0.75		ann.rad
									0.28650
									(0.1–1.56)
^{76m}Br			1.4 s	I.T./5.05		4+			0.104548
									0.05711
^{76}Br		75.92454	16.0 h	β+ /57/4.96	1.9/	1-	0.54821	0.270	ann.rad
					3.68/				0.55911
									1.85368
									(0.4–4.6)
^{77m}Br			4.3 m	I.T./0.1059		9/2+			0.1059
^{77}Br		76.921379	2.376 d	EC/99/1.365		3/2-	0.973	+0.53	ann.rad.
									0.23898
									0.52069
									(0.08–1.2)
^{78}Br		77.921146	6.45 m	β+ /92/3.574	1.2/	1+	0.13		ann.rad.
				EC/8/	2.5/				0.61363
									(0.7–3.0)
^{79m}Br			4.86 s	I.T./0.207		9/2+			0.2072
^{79}Br	50.69(7)	78.918337				3/2-	+2.106400	+0.31	
^{80m}Br			4.42 h	I.T./0.04885		5-	+1.3177	+0.70	Br k x-ray
									0.03705/39.1
									0.04885/0.3

Elem. or Isot.	Natural Abundance (Atom %)	Atomic Mass or Weight	Half–life/ Resonance Width (MeV)	Decay Mode/ Energy (/MeV)	Particle Energy/ Intensity (MeV/%)	Spin ($h/2\pi$)	Nuclear Magnetic Mom. (nm)	Elect. Quadr. Mom. (b)	γ-Energy/ Intensity (MeV/%)
^{80}Br		79.918529	17.66 m	β- /92/2.004	1.38 β-/7.6	1+	0.5140	+0.18	ann.rad.
				EC/5.7/1.8706	1.99 β-/82				0.6169/6.7
				β+ /2.6/	0.85 β+ /2.8				(0.64–1.45)
^{81}Br	49.31(7)	80.916291				3/2-	+2.270562	+0.26	
^{82m}Br			6.1 m	I.T./98/0.046		2-			0.046/0.24
				β- /2 /3.139					(0.62–2.66)
^{82}Br		81.916804	1.471 d	β- /3.093	0.444/	5-	+1.6270	0.69	0.5544/71
									0.61905/43
									0.77649/84
									(0.013–1.96)
^{83}Br		82.915180	2.40 h	β- /0.972	0.395/1	3/2-			0.52964
					0.925/99				(0.12–0.68)
^{84m}Br			6.0 m	β- /4.97	2.2/100	(6-)			0.4240/100
									0.8817/98
									1.4637/101
^{84}Br		83.91648	31.8 m	β- /4.65	2.70/11	2-	2.		0.8816/41
					3.81/20				1.8976/13
					4.63/34				(0.23–4.12)
^{85}Br		84.91561	2.87 m	β- /2.87	2.57	3/2-			0.80241/2.56
									0.92463/1.6
									(0.09–2.4)
^{86}Br		85.91880	55.5 s	β- /7.63	3.3	(2-)			1.56460/64
					7.4				2.75106/21
									(0.5–6.8)
^{87}Br		86.92071	55.6 s	β- /6.85	6.1/	3/2-			1.41983/55.
				n/					1.578/45.
									(0.173–1.76)
^{88m}Br			5.1 μs						
^{88}Br		87.92407	16.3 s	β- /8.96		1-			0.7649
				n/					0.7753
									0.8021
									(0.1–6.99)
^{89}Br		88.92640	4.35 s	β- /8.16		3/2-			0.7753
				n/					1.0978
^{90}Br		89.9306	1.91 s	β- /10.4	8.3/	2-			0.6555
				n/	9.8/				0.7071
									1.3626
^{91}Br		90.9340	0.54 s	β- /90 /9.80					0.263
				β- n/10 /					0.803
^{92}Br		91.93926	0.31 s	β- /12.20					0.740
				β- n/					
^{93}Br		92.9431	0.10 s	β- /11					0.117
				β- n	//11				(0.237–3.606)
^{94}Br		93.9487	0.07 s	β- n/					
^{95}Br		94.9529	> 0.3 μs						
^{96}Br		95.959	> 0.3 μs						
^{97}Br		96.963	> 0.3 μs						
^{98}Br									
$_{36}$**Kr**		**83.798(2)**							
^{69}Kr		68.9652	0.03 s	β+, (p)	4.07/				
^{70}Kr		69.9553	0.06 s			0+			
^{71}Kr		70.950	100. ms	β+, EC/10.1					(0.198–0.207)
^{72}Kr		71.94209	17.1 s	β+ /5.0		0+			ann.rad
				EC/					0.3099/15.3

Elem. or Isot.	Natural Abundance (Atom %)	Atomic Mass or Weight	Half–life/ Resonance Width (MeV)	Decay Mode/ Energy (/MeV)	Particle Energy/ Intensity (MeV/%)	Spin (h/2π)	Nuclear Magnetic Mom. (nm)	Elect. Quadr. Mom. (b)	γ-Energy/ Intensity (MeV/%)
									0.4150/12.8
									(0.305 – 3.305)
73Kr		72.93929	28. s	β+ /6.7		5/2-			ann.rad.
				EC/					0.1781/66
				β+, p/	/0.25				(0.06–0.86)
74Kr		73.933084	11.5 m	β+ /3.1		0+			ann.rad.
				EC/					0.08970/31
									0.2030/20
									(0.010–1.06)
75Kr		74.93095	4.3 m	β+ /4.90	3.2/	5/2+	−0.531	+1.1	ann.rad.
				EC/					0.1325/68
									0.1547/21
									(0.02–1.7)
76Kr		75.925910	14.8 h	EC/1.31		0+			Br k x-ray
									0.270/21
									0.3158/39
									(0.03–1.07)
77Kr		76.924670	1.24 h	β+ /80/3.06		5/2+	−0.583	+0.9	ann.rad.
				EC/20/	1.55/				0.1297/80
					1.70/				0.1465/38
					1.87/				(0.02–2.3)
78Kr	0.355(3)	77.920365	>1.5 × 10²¹ a	EC-EC		0+			
79mKr			53. s	I.T./0.1299		7/2+	−0.786	+0.40	Kr x-ray
79Kr		78.920082	1.455 d	β+ /7 /1.626		½-	+0.536		ann.rad.
				EC/93 /					0.2613/13
									0.39756/19
									0.6061/8
									(0.04–1.3)
80Kr	2.286(10)	79.916379				0+			
81mKr			13.1 s	I.T./0.1904		½-	+0.586		0.1904
81Kr		80.916592	2.1 × 10⁵ a	EC/0.2807		7/2+	−0.908	+0.644	Br k x-ray
									0.2760
82Kr	11.593(31)	81.913484				0+			
83mKr			1.86 h	I.T./0.0416		½-	+0.591		Kr k x-ray
									0.00940
									0.03215/0.055
83Kr	11.500(19)	82.914136				9/2+	−0.970669	+0.259	
84Kr	56.987(15)	83.911507				0+			
85mKr			4.48 h	β- /79 /	0.83/79	½-	+0.633		0.30487
				I.T./21 /0.305					0.15118
85Kr		84.912527	10.73 a	β- /0.687	0.15/0.4	9/2+	1.005	+0.443	0.51399
86Kr	17.279(41)	85.9106107				0+			
87Kr		86.9133549	1.27 h	β- /3.887	1.33/8	5/2+	−1.023	−0.30	0.40258/49.6
					3.49/43				2.5548/9.2
					3.89/30				(0.13–3.31)
88Kr		87.91445	2.84 h	β- /2.91		0+			0.19632/26.
									2.392/34.6
									(0.03–2.8)
89Kr		88.9176	3.15 m	β- /4.99	3.8/	5/2+	−0.330	+0.16	0.19746
					4.6/				0.2209/19.9
					4.9/				0.5858/16.4
									1.4728/6.8
									(0.2–4.7)
90Kr		89.91952	32.3 s	β- /4.39	2.6/77	0+			0.12182/32.9

Elem. or Isot.	Natural Abundance (Atom %)	Atomic Mass or Weight	Half–life/ Resonance Width (MeV)	Decay Mode/ Energy (/MeV)	Particle Energy/ Intensity (MeV/%)	Spin ($h/2\pi$)	Nuclear Magnetic Mom. (nm)	Elect. Quadr. Mom. (b)	γ-Energy/ Intensity (MeV/%)
					2.8/6				0.5395/28.6
									1.1187/36.2
									(0.1–4.2)
^{91}Kr		90.9235	8.6 s	β- /6.4	4.33/	5/2+	−0.583	+0.30	0.10878/43.5
					4.59/				0.50658/19.
									(0.2–4.4)
^{92}Kr		91.92616	1.84 s	β- /5.99		0+			0.1424/66.
				n/					(0.14–3.7)
^{93}Kr		92.9313	1.29 s	β- /8.6	7.1/	½+	−0.413		0.1820
				n/					0.2534/42.
									0.32309/24.6
									(0.057–4.03)
^{94}Kr		93.9344	0.21 s	β- /7.3		0+			0.2196/67
				n	n//1.0				0.6293/100.
									(0.098–0.985)
^{95}Kr		94.9398	0.10 s	β- /9.7	n//2.9		−0.410		
^{96}Kr		95.9431	~ 80 ms	β-,n	n//3.7	0+			
^{97}Kr		96.9486	0.06 s	β-,n	n//7.				
^{98}Kr		97.952	0.05 s	β-,n	n//7.	0+			
^{99}Kr		98.958	0.04 s	β-,n	n// ~ 11.				
^{100}Kr		99.9611	> 0.34 μs			0+			
^{101}Kr									
$_{37}$Rb		85.4678(3)							
^{71}Rb		70.9653							
^{72}Rb		71.9591	< 1.5 μs						
^{73}Rb		72.9506	< 0.03 μs						
^{74}Rb		73.944265	64.8 ms	β+ /10.4					0.456/0.0025
									(0.053 – 4.244)
^{75}Rb		74.93857	19. s	β+ /7.02	2.31/				ann.rad
									0.179
^{76}Rb		75.935072	39. s	β+ /8.50	4.7/	1-	−0.372623	+0.4	ann.rad
									0.4240/92.
									(0.064–1.68)
^{77}Rb		76.93041	3.8 m	β+ /5.34	3.86/	3/2-	+0.654468	+0.70	ann.rad
									0.0665/59
									(0.04–2.82)
^{78m}Rb			5.7 m	I.T./0.1034		4-	+2.549	+0.81	ann.rad
				β+ /	3.4				0.4553/81.
				EC/					(0.103–4.01)
^{78}Rb		77.92814	17.7 m	β+ /7.22		0+			ann.rad
				EC/					0.4553/63.
									(0.42–5.57)
^{79}Rb		78.92399	23. m	β+ /84/3.65		5/2+	+0.3358	−0.10	ann.rad.
				EC/16 /					0.68812/23.
									(0.017–3.02)
^{80}Rb		79.92252	34. s	β+ /5.72	4.1/22	1+	−0.0836	+0.35	ann.rad.
					4.7/74				0.6167/25.
^{81m}Rb			30.5 m	I.T./0.85	1.4	9/2+	+5.598	−0.74	ann.rad.
				β+, EC/					(0.085–1.9)
^{81}Rb		80.91900	4.57 h	β+ /27/2.24	1.05/	3/2-	+2.060	+0.40	ann.rad./
				EC/73					0.19030/64.
									(0.05–1.9)
^{82m}Rb			6.47 h	β+/26/	0.80/	5-	+1.51001	+1.0	ann.rad./

Elem. or Isot.	Natural Abundance (Atom %)	Atomic Mass or Weight	Half–life/ Resonance Width (MeV)	Decay Mode/ Energy (/MeV)	Particle Energy/ Intensity (MeV/%)	Spin ($h/2\pi$)	Nuclear Magnetic Mom. (nm)	Elect. Quadr. Mom. (b)	γ-Energy/ Intensity (MeV/%)
				EC/74/					0.5544/63.
									0.7765/85.
									(0.092–2.3)
^{82}Rb		81.918209	1.258 m	β+/96/4.40	3.3/	1+	+0.554508	+0.19	ann.rad./
				EC/4/					0.7665/13.
									(0.47–3.96)
^{83}Rb		82.91511	86.2 d	EC/0.91		5/2-	+1.425	+0.20	Kr x-ray
									0.5205/46.
									(0.03–0.80)
^{84m}Rb			20.3 m	I.T./0.216		6-	+0.21293	+0.6	0.2163/34.
									0.2482/63.
									0.4645/32.
^{84}Rb		83.914385	32.9 d	β+/22/2.681	0.780/11	2-	−1.32412	−0.015	ann.rad./
				EC/75 /	1.658/11				0.8817/68.
				β-/3/0.894	0.893/				(1.02–1.9)
^{85}Rb	72.17(2)	84.91178974				5/2-	+1.353	+0.28	
^{86m}Rb			1.018 m	I.T./0.5560		6-	+1.815	+0.37	0.556/98.
^{86}Rb		85.9111674	18.65 d	β-/1.775	1.774/8.8	2-	−1.6920	+0.19	1.0768/8.8
^{87}Rb	27.83(2)	86.90918053	4.88 × 10^{10} a	β-/0.283	0.273/100	3/2-	+2.7512	+0.13	
^{88}Rb		87.9113156	17.7 m	β-/5.316	5.31	2-	0.508		0.8980/14.4
									1.8360/22.8
									(0.34–4.85)
^{89}Rb		88.91228	15.4 m	β-/4.50	1.26/38	3/2-	+2.38	+0.14	1.032/58.
					1.9/5				1.248/42.
					2.2/34				2.1960/13
					4.49/18				(0.12–4.09)
^{90m}Rb			4.3 m	β-/4.50	1.7/	4-	+1.616	+0.20	0.1069(IT)
					6.5/				0.8317/94
									(0.20–5.00)
^{90}Rb		89.91480	2.6 m	β-/6.59	6.6	1-			0.8317/28.
									(0.31–5.60)
^{91}Rb		90.91654	58.0 s	β-/5.861	5.9	3/2-	+2.182	+0.15	0.0936/34.
									(0.35–4.70)
^{92}Rb		91.91073	4.48 s	β-/8.11	8.1/94	1-			0.8148/8.
									(0.1–6.1)
^{93}Rb		92.92204	5.85 s	β-/7.46	7.4/	5/2-	+1.410	+0.18	0.2134/4.8
				n/1					0.4326/12.5
									0.9861/4.9
									(0.16–5.41)
^{94}Rb		93.92641	2.71 s	β-/10.31	9.5/	3	+1.498	+0.16	0.8369/87.
				n/10					1.5775/32.
									(0.12–6.35)
^{95}Rb		94.92930	0.377 s	β-/9.30	8.6/	5/2-	+1.334	+0.21	0.352/65.
				n/8					0.680/22.
									(0.20–2.27)
^{96m}Rb			1.7 μs						0.2999
									0.4612
									0.2400
									0.093–0.369
^{96}Rb		95.93427	0.199 s	β-/11.76	10.8/	2+	+1.466	+0.25	0.815/76.
				n/13/					(0.20–5.42)
^{97}Rb		96.93735	0.169 s	β-/10.42	10.0	3/2+	+1.841	+0.58	0.167/100.
				n/27/					0.585/79.
									0.599/56.

Elem. or Isot.	Natural Abundance (Atom %)	Atomic Mass or Weight	Half–life/ Resonance Width (MeV)	Decay Mode/ Energy (/MeV)	Particle Energy/ Intensity (MeV/%)	Spin ($h/2\pi$)	Nuclear Magnetic Mom. (nm)	Elect. Quadr. Mom. (b)	γ-Energy/ Intensity (MeV/%)
									1.258/52.
									(0.14–2.08)
^{98}Rb		97.94179	0.107 s	β-/12.34	0.144/				
				n/13					(0.07–3.68)
^{99}Rb		98.9454	59. ms	β-/11.3					
^{100}Rb		99.9499	53. ms	β- /13.5					0.129
									(0.058–4.483)
^{101}Rb		100.9532	0.03 s	β- /11.8					
^{102}Rb		101.9589	0.09 s	β-					
^{103}Rb									
$_{38}$**Sr**		**87.62(1)**							
^{73}Sr		72.966	> 25 ms						
^{74}Sr		73.9563	> 1.5 μs			0+			
^{75}Sr		74.9499	88. ms	β+ ,p	p//5.				0.144/4.5
^{76}Sr		75.94177	7.9 s	β+ /6.1		0+			
^{77}Sr		76.93795	9.0 s	β+ /6.9	5.6		−0.348	+1.4	0.147
				β+, p	//0.08				
^{78}Sr		77.93218	2.7 m	β+ /3.76		0+			(0.047–0.793)
^{79}Sr		78.92971	2.1 m	β+ /5.32	4.1	3/2-	−0.474	+0.71	ann.rad./
									0.039/28.
									0.105/22.
									(0.135–0.612)
^{80}Sr		79.92452	1.77 h	β+ /1.87		0+			ann.rad./
									0.174/10.
									0.589/39.
									(0.24–0.55)
^{81}Sr		80.92321	22.3 m	β+ /87/3.93	2.43/	1/2-	+0.544		ann.rad./
				EC/13/	2.68/				0.148/31.
									0.1534/35
									(0.06–1.7)
^{82}Sr		81.91840	25.36 d	EC/0.18		0+			Rb x-ray
^{83m}Sr			5.0 s	I.T./0.2591		½-	+0.582		0.2591/87.5
^{83}Sr		82.91756	1.350 d	β+/24/2.28	0.465/	7/2+	−0.8298	+0.76	ann.rad./
				EC/76/	0.803/				0.3816/12.
					1.227/				0.3816
									0.7627/30.
									(0.094–2.15)
^{84}Sr	0.56(1)	83.913425				0+			
^{85m}Sr			1.127 h	I.T./87/0.2387		½-	+0.600		0.2318/84.
				EC/13					(0.15–0.24)
^{85}Sr		84.912933	64.85 d	EC/1.065		9/2+	−1.001	+0.28	0.51399/99.3
^{86}Sr	9.86(1)	85.909260				0+			
^{87m}Sr			2.81 h	I.T./0.3884		½-	+0.63		0.3884(IT)
^{87}Sr	7.00(1)	86.908877				9/2+	−1.0936	+0.305	
^{88}Sr	82.58(1)	87.905612				0+			
^{89}Sr		88.907451	50.6 d	β-/1.497	1.492/100	5/2+	−1.148	−0.3	0.9092
^{90}Sr		89.907738	28.9 a	β-/0.546	0.546/100	0+			
^{91}Sr		90.910203	9.5 h	β-/2.70	0.61/7	5/2+	−0.885	+0.05	0.5556/61.
					1.09/33				0.7498/24.
					1.36/29				1.0243/33.
					2.66/26				(0.12–2.4)
^{92}Sr		91.911038	2.64 h	β-/1.91	0.55/96	0+			1.3831/90.
					1.5/3				(0.24–1.1)
^{93}Sr		92.91403	7.4 m	β-/4.08	2.2/10	5/2+	−0.793	+0.26	0.5903/

Elem. or Isot.	Natural Abundance (Atom %)	Atomic Mass or Weight	Half–life/ Resonance Width (MeV)	Decay Mode/ Energy (/MeV)	Particle Energy/ Intensity (MeV/%)	Spin ($h/2\pi$)	Nuclear Magnetic Mom. (nm)	Elect. Quadr. Mom. (b)	γ-Energy/ Intensity (MeV/%)
					2.6/25				0.7104
					3.2/65				0.87573
									0.8883/
									(0.17–3.97)
⁹⁴Sr		93.91536	1.25 m	β-/3.511	2.1/	0+			0.6219
					3.3/				0.7043
									0.7241
									0.8064
									1.4283
⁹⁵Sr		94.91936	25.1 s	β-/6.08		½+	−0.537		0.6859
					6.1/50				0.8269
									2.7173
									2.9332
⁹⁶Sr		95.92170	1.06 s	β-/5.37	4.2/	0+			0.1222
									0.5305
									0.8094
									0.9318
⁹⁷Sr		96.92615	0.42 s	β-/7.47	5.3	(1/2+)	−0.498		0.2164
									0.3071
									0.6522
									0.9538
									1.2580
									1.9050
⁹⁸Sr		97.92845	0.65 s	β-/5.83	5.1	0+			0.0365
									0.1190
									0.4286
									0.4447
									0.5636
⁹⁹Sr		98.9332	0.27 s	β-/8.0			−0.26	0.8	
¹⁰⁰Sr		99.9354	0.201 s	β-/7.1		0+			
¹⁰¹Sr		100.9405	0.115 s	β-/9.5					
¹⁰²Sr		101.9430	68. ms	β-/8.8		0+			
¹⁰³Sr		102.9490	> 0.3 µs						
¹⁰⁴Sr		103.952	> 0.3 µs			0+			
¹⁰⁵Sr		104.959	> 0.3 µs						
¹⁰⁶Sr									
¹⁰⁷Sr									
₃₉Y		88.90585(2)							
⁷⁶Y		75.9585	> 0.2 µs						
⁷⁷Y		76.9497	~ 57. ms						
⁷⁸ᵐY			5.8 s			(5+)			
⁷⁸Y		77.9436	53 ms	β+/10.5					0.279/100
									0.504/90
									0.713/40
⁷⁹Y		78.9374	15. s	β+/7.1					(0.152–1.106)
⁸⁰ᵐY			4.8 s						0.2285
⁸⁰Y		79.9343	30. s	β+/7.0	5.5	(4-)			ann.rad./
					5.0/				0.3858/100
									0.5951/42
									0.756–1.396
⁸¹Y		80.9291	1.21 m	β+/5.5	3.7/				ann.rad./
					4.2/				0.428
									0.469
⁸²Y		81.9268	9.5 s	β+/7.8	6.3/	1+			ann.rad./

Elem. or Isot.	Natural Abundance (Atom %)	Atomic Mass or Weight	Half–life/ Resonance Width (MeV)	Decay Mode/ Energy (/MeV)	Particle Energy/ Intensity (MeV/%)	Spin ($h/2\pi$)	Nuclear Magnetic Mom. (nm)	Elect. Quadr. Mom. (b)	γ-Energy/ Intensity (MeV/%)
									0.5736
									0.6017
									0.7375
83mY			2.85 m	β+/95/4.6	2.9	1/2-			ann.rad./
				EC/5 /					0.2591
									0.4218
									0.4945
83Y		82.92235	7.1 m	β+/4.47	3.3	9/2+			ann.rad./
				EC/					0.0355
									0.4899
									0.8821
									(0.03–3.4)
84mY			4.6 s	β+/		1+			ann.rad./
				EC/					0.7930
84Y		83.9204	40. m	β+/6.4	1.64/47	5-			ann.rad./
				EC/	2.24/25				0.4628
					2.64/21				0.6606
					3.15/7				0.7931
									0.9744
									1.0398
									(0.2–3.3)
85mY			4.9 h	β+/70/		9/2+	6.2		ann.rad./
				EC/30/					0.2317
									0.5356
									2.1238
									(0.1–3.1)
									0.7673
85Y		84.91643	2.6 h	β+/55/3.26	1.54/	1/2-			ann.rad./
				EC/45/					0.2317
									0.5045
									0.9140
									(0.07–1.4)
86mY			48. m	I.T./99/		8+	4.8		ann.rad./
				β+/					0.0102(IT)
				EC/					0.2080
									(0.09–1.1)
86Y		85.91489	14.74 h	β+/5.24		4-	<0.6		ann.rad./
				EC/					0.3070
									0.6277
									1.0766
									1.1531
									1.9207
									(0.1–3.8)
87mY			13. h	I.T./98/		9/2+	+6.24	-0.5	0.3807
				β+/0.7/	1.15/0.7				
				EC/					
87Y		86.910876	3.35 d	EC/99+/1.862	0.78/	1/2-	-0.19		0.3880
									0.4870
88Y		87.909501	106.6 d	EC/99+/3.623	0.76/	4-	-0.42	+0.16	ann.rad./
				β+/0.2/					0.89802
									1.83601
									2.73404
									3.2190
89mY			15.7 s	I.T./0.909		9/2+	+6.37	~-0.43	0.9092(IT)

Elem. or Isot.	Natural Abundance (Atom %)	Atomic Mass or Weight	Half–life/ Resonance Width (MeV)	Decay Mode/ Energy (/MeV)	Particle Energy/ Intensity (MeV/%)	Spin ($h/2\pi$)	Nuclear Magnetic Mom. (nm)	Elect. Quadr. Mom. (b)	γ-Energy/ Intensity (MeV/%)
[89]Y	100.	88.905848				1/2-	−0.13742		
[90m]Y			3.24 h	I.T./99+/	0.68204	7+	+5.28	~−0.65	0.2025
				β-/0.002/					0.4794
									0.6820
[90]Y		89.907152	2.669 d	β-/2.282	2.28/	2-	−1.63	−0.155	
[91m]Y			49.7 m	I.T./0.555		9/2+	5.96		0.5556(IT)
[91]Y		90.907305	58.5 d	β-/1.544	1.545/	1/2-	0.164		1.208
[92]Y		91.90895	3.54 h	β-/3.63	3.64/	2-	−0.67		0.4485
									0.5611
									0.9345
									1.4054
									(0.4–3.3)
[93m]Y			0.82 s	I.T./0.759		9/2+	+6.04	~−0.64	0.1686(IT)
									0.5902
[93]Y		92.90958	10.2 h	β-/2.87	2.88/90	1/2-	−0.12		0.2669
									0.9471
									1.9178
[94m]Y			1.4 µs						0.4322
									0.7699
									1.2024
[94]Y		93.91160	18.7 m	β-/4.919	4.92/	2-	−0.24	~ 0.03	0.3816
									0.9188
									1.1389
									(0.3–4.1)
[95]Y		94.91282	10.3 m	β-/4.42		1/2-	−0.16		0.4324
									0.9542
									2.1760
									3.5770
[96m]Y			9.6 s	β-/		(3+)			0.1467
									0.6174
									0.9150
									1.1071
									1.7507
[96]Y		95.91589	6.2 s	β- /7.09	7.12/	0-			1.594
[97m]Y			1.21 s	β- /7.4	4.8/	9/2+	+5.88	~ −0.76	0.1614
					6.0/				0.9700
									1.1030
[97]Y		96.91813	3.76 s	β- /6.69	6.7	1/2-	−0.12		0.2969
									1.9960
									3.2876
									3.4013
[98m]Y			2.1 s	β- /9.8	5.5/	(4-)	+2.98	+1.7	0.2415
									0.6205
									0.6473
									1.2228
									1.8016
[98]Y		97.92220	0.59 s	β- /8.83	8.7/	1+			0.2131
									1.2228
									1.5907
									2.9413
									4.4501
[99m]Y			0.011 ms						
[99]Y		98.92464	1.47 s	β- /7.57		1/2-			0.1218/43.8
				n	/2.5/				0.5362

Elem. or Isot.	Natural Abundance (Atom %)	Atomic Mass or Weight	Half–life/ Resonance Width (MeV)	Decay Mode/ Energy (/MeV)	Particle Energy/ Intensity (MeV/%)	Spin ($h/2\pi$)	Nuclear Magnetic Mom. (nm)	Elect. Quadr. Mom. (b)	γ-Energy/ Intensity (MeV/%)
									0.7242
									1.0130
^{100m}Y			0.94 s	β-, n /		3+			
^{100}Y		99.9278	0.73 s	β-, n/9.3	n/1.8/	1+			
^{101}Y		100.9303	0.43 s	β-, n/8.6	n/1.5/	(5/2+)	+3.22	−1.5	
^{102}Y		101.9336	0.36 s	β-, n/9.9	n/4.0/	0+			
^{103}Y		102.9367	0.23 s	β-, n	n/8.3/				
^{104}Y		103.9411	0.18 s						
^{105}Y		104.9449	> 0.15 μs						
^{106}Y		105.950	> 0.15 μs						
^{107}Y		106.9414	> 0.15 μs						
^{108}Y		107.959	> 0.15 μs						
^{109}Y									
$_{40}$Zr		91.224(2)							
^{78}Zr		77.9552	> 0.2 μs			0+			
^{79}Zr		78.9492	0.06 s						
^{80}Zr		79.940	~ 4.5 s	β+ /8.0		0+			0.290
									0.538
^{81}Zr		80.9372	5.3 s	β+ /7.2	6.1	(3/2-)			
^{82}Zr		81.9311	32. s	β+ /4.0	3.	0+			ann.rad./
^{83m}Zr			7. s	β+ /7.0		(7/2+)			ann.rad./
^{83}Zr		82.9287	44. s	β+ /5.9	4.8	(1/2-)			ann.rad./
				EC					0.0556
									0.1050
									0.2560
									0.474
									1.525
^{84}Zr		83.9233	26. m	β+ /2.7		0+			ann.rad./
				EC/					0.0449
									0.1125
									0.3729
									0.667
^{85m}Zr			10.9 s	I.T./0.2922		½-			ann.rad./
				β+, EC/					0.2922(IT)
									0.4165
^{85}Zr		84.9215	7.9 m	β+ /4.7	3.1	7/2+			ann.rad./
				EC/					0.2663
									0.4163
									0.4543
^{86}Zr		85.91647	16.5 h	EC/1.47		0+			0.0280
									0.243
									0.612
^{87m}Zr			14.0 s	I.T./0.3362		½-	+0.64		0.1352(IT)
									0.2010
^{87}Zr		86.91482	1.73 h	β+ /3.67	2.26	9/2+	−0.895	+0.42	ann.rad./
				EC/					0.3811
									1.228
^{88m}Zr			1.4 μs			(8+)	−1.81	+0.51	0.077
^{88}Zr		87.91023	83.4 d	EC/0.67		0+			0.3929
^{89m}Zr			4.18 m	I.T./94/0.5877		½-	+0.80		ann.rad./
				β+ /1.5/					0.5877(IT)
				EC/4.7/					1.507

Elem. or Isot.	Natural Abundance (Atom %)	Atomic Mass or Weight	Half–life/ Resonance Width (MeV)	Decay Mode/ Energy (/MeV)	Particle Energy/ Intensity (MeV/%)	Spin ($h/2\pi$)	Nuclear Magnetic Mom. (nm)	Elect. Quadr. Mom. (b)	γ-Energy/ Intensity (MeV/%)
^{89}Zr		88.908889	3.27 d	β+ /23/2.832	0.9/	9/2+	−1.05	+0.3	ann.rad./
				EC/77/					0.9092
^{90m}Zr			0.809 s	I.T./		5−	6.3		0.1326
									2.1862
									2.3189(IT)
^{90}Zr	51.45(40)	89.904704				0+			
^{91}Zr	11.22(5)	90.905646				5/2+	−1.30362	−0.176	
^{92}Zr	17.15(8)	91.905041				0+			
^{93}Zr		92.906476	1.5×10^6 a	β- /0.091		5/2+			0.0304
^{94}Zr	17.38(28)	93.906315	$>10^{17}$ a	β-β-		0+			
^{95}Zr		94.908043	64.02 d	β- /1.125	0.366/55	5/2+	1.13	+0.22	0.7242
					0.400/44				0.7567
^{96}Zr	2.80(9)	95.908273	2.3×10^{19} a	β-β-		0+			
			$>1.7 \times 10^{18}$ a	β-					
^{97}Zr		96.910953	16.75 h	β- /2.658	1.91/	½−	−0.937		0.7434
^{98}Zr		97.91274	30.7 s	β- /2.26	2.2/100	0+			
^{99}Zr		98.91651	2.2 s	β- /4.56	3.9/	½+	−0.930		0.4692/55.2
					3.5/				0.5459/48
									0.028–1.321
^{100}Zr		99.91776	7.1 s	β- /3.34		0+			0.4006/20
									0.5043/30
^{101}Zr		100.92114	2.1 s	β- /5.49	6.2/	3/2−	−0.27	+0.81	0.1194
									0.2057
									0.2089
^{102}Zr		101.92298	2.9 s	β- /4.61		0+			0.064/15
									0.599/14
									0.535/11
^{103}Zr		102.9266	1.3 s	β- /7.0					
^{104}Zr		103.9288	1.2 s	β- /5.9		0+			
^{105}Zr		104.9331	~ 1. s	β- /8.5					
^{106}Zr		105.9359	> 0.24 μs			0+			
^{107}Zr		106.9408	> 0.24 μs						
^{108}Zr		107.944	> 0.15 μs			0+			
^{109}Zr		108.9492	> 0.15 μs						
^{110}Zr		109.953	> 0.15 μs			0+			
^{111}Zr									
^{112}Zr									
$_{41}$**Nb**		**92.90638(2)**							
^{81}Nb		80.949	<0.08 μs						
^{82}Nb		81.9431	50 ms	β+ /11.					
^{83}Nb		82.9367	4.1 s	β+ /7.5					
^{84}Nb		83.9336	10. s	β+, EC/9.6		(3+)			0.540
									(0.456-1.427)
^{85m}Nb			3. s						0.069
^{85}Nb		84.9279	21. s	β+ /6.0					
^{86m}Nb			56. s	β+					
^{86}Nb		85.9250	1.46 m	β+ /8.0					ann.rad./
									0.751
									1.003
^{87m}Nb			3.7 m	β+ /		1/2−			ann.rad./
				EC/					0.1352
									0.2010
^{87}Nb		86.92036	2.6 m	β+ /5.2/		(9/2+)			ann.rad./
				EC/					0.2010

Elem. or Isot.	Natural Abundance (Atom %)	Atomic Mass or Weight	Half–life/ Resonance Width (MeV)	Decay Mode/ Energy (/MeV)	Particle Energy/ Intensity (MeV/%)	Spin ($h/2\pi$)	Nuclear Magnetic Mom. (nm)	Elect. Quadr. Mom. (b)	γ-Energy/ Intensity (MeV/%)
									0.4706
									0.6165
									1.0665
									1.8842
88mNb			7.7 m	β+ /		4-			ann.rad./
				EC/					0.2625
									0.3996
									1.0569
									1.0825
88Nb		87.9183	14.3 m	β+ /7.6	3.2/	8+			ann.rad./
				EC/					1.0570
									1.0828
									(0.07–2.5)
89mNb			2.0 h	β+ /	3.3/	9/2+			0.5880/10(D)
				EC/					(0.17–4.0)
89Nb		88.91342	1.10 h	β+ /74/4.29	2.8/	1/2-	+6.216		ann.rad./
				EC/26 /					0.5074
									0.5880
									0.7696
									1.2775
90mNb			18.8 s	I.T./0.1246		4-			0.002
									0.1225
90Nb		89.911265	14.6 h	β/53 /6.111	0.86/5	8+	4.961	+0.05	ann.rad./
				EC/47 /	1.5/92				0.1412
									1.1292
									2.1862
									2.3189
									(0.1–3.3)
91mNb			62. d	I.T./97 /		1/2-			0.1045(IT)
				EC/3 /					1.2050
91Nb		90.906996	7 × 10² a	EC/1.253		9/2+			Mo k x-ray
92mNb			10.13 d	EC/99+/		2+	6.14		0.9126
									0.9345
									1.8475
92Nb		91.907194	3.7 × 10⁷ a	EC/2.006		7+			0.5611
									0.9345
93mNb			16.1 a	I.T./0.0304		1/2-			Nb x-ray
									0.0304
93Nb	100.	92.906378				9/2+	+6.1705	−0.32	
94mNb			6.26 m	I.T./99+	/2.086	3+			Nb k x-ray
				β- /0.5/					0.0409
									0.87109
94Nb		93.907284	2.4 × 10⁴ a	β- /2.045	0.47/	6+			0.70263
									0.87109
95mNb			3.61 d	I.T./97.5/	0.2357	1/2-			0.2040
				β- /2.5 /					0.2356
95Nb		94.906836	34.97 d	β- /0.926	0.160/	9/2+	6.14		0.76578
96Nb		95.908101	23.4 h	β- /3.187	0.5/10	6+	4.976		0.7782
					0.75/90				0.2191–1.498
97mNb			58.1 s	I.T./0.7434	0.734/98	1/2-			0.7434
97Nb		96.908099	1.23 h	β- /1.934	1.27/98	9/2+	6.15		0.4809
									0.6579
98mNb			51. m	β- /4.67		5+			0.7874
									0.1726–1.89

Elem. or Isot.	Natural Abundance (Atom %)	Atomic Mass or Weight	Half–life/ Resonance Width (MeV)	Decay Mode/ Energy (/MeV)	Particle Energy/ Intensity (MeV/%)	Spin ($h/2\pi$)	Nuclear Magnetic Mom. (nm)	Elect. Quadr. Mom. (b)	γ-Energy/ Intensity (MeV/%)
^{98}Nb		97.91033	2.9 s	β- /4.59	4.6/	1+			0.6451
									0.7874
									1.0243
^{99m}Nb			2.6 m	β- /	3.2/	1/2-			0.0978/100
									(0.138–3.010)
^{99}Nb		98.91162	15.0 s	β- /3.64	3.5/100	9/2+			0.0977
									0.1378/3.1
^{100m2}Nb			0.013 ms						
^{100m1}Nb			3.0 s	β- /6.74	5.8				Nb k x-ray
									0.159
									0.6364
									1.0637
^{100}Nb		99.91418	1.5 s	β- /6.25	6.2/				0.5354
					5.3/				0.6001–1.566
^{101}Nb		100.91525	7.1 s	β- /4.57	4.3/				0.1105–0.810
^{102m}Nb			4.3 s	β- /					
^{102}Nb		101.91804	1.3 s	β- /7.21	7.2/				0.2960–2.184
^{103}Nb		102.9191	1.5 s	β- /5.53	5.3/	5/2+			
^{104m}Nb			0.9 s	β-, n/	n/0.06				
^{104}Nb		103.9225	4.8 s	β-, n/8.1	n/0.05				
^{105}Nb		104.9239	3.0 s	β-, n/6.5	n/1.7				
^{106}Nb		105.9280	1.0 s	β-, n/9.3	n/4.5				
^{107}Nb		106.9303	0.30 s	β-, n/7.9	n/6.0				
^{108}Nb		107.9348	0.19 s	β, n/	n/6.2				(0.193–0.590)
^{109}Nb		108.9376	0.19 s	β, n/	n/31				
^{110}Nb		109.9424	0.17 s	β, n/	n/40				
^{111}Nb		110.9457	> 0.15 µs						
^{112}Nb		111.951	> 0.15 µs						
^{113}Nb		112.955	> 0.15 µs						
^{114}Nb									
^{115}Nb									
$_{42}$Mo		95.96(2)							
^{83}Mo		82.9487	~ 6. ms						
^{84}Mo		83.9401	~ 2.2 s	β+ /6.		0+			
^{85}Mo		84.9366	3.2 s	β+/8.1		½+			
^{86}Mo		85.9307	19. s	β+ /4.8		0+			
^{87}Mo		86.9273	14. s	EC, β+/6.5					(0.752–1.004)
^{88}Mo		87.92195	8.0 m	β+ /3.4		0+	+0.5		ann.rad./
				EC					0.0800
									0.1399
									0.1707
^{89m}Mo			0.19 s	I.T./0.118		½-			0.118(IT)
									0.268
^{89}Mo		88.91948	2.2 m	β+ /5.58		9/2+			ann.rad./
				EC/					0.659
									0.803
									1.155
									1.272
^{90m}Mo			1.2 µs				-1.39	0.58	0.063
^{90}Mo		89.91394	5.7 h	β+ /25/2.489	1.085/	0+			ann.rad./
				EC/75 /					0.04274
									0.12237
									0.25734
^{91m}Mo			1.08 m	I.T./50/0.653		½-			ann.rad./

Elem. or Isot.	Natural Abundance (Atom %)	Atomic Mass or Weight	Half–life/ Resonance Width (MeV)	Decay Mode/ Energy (/MeV)	Particle Energy/ Intensity (MeV/%)	Spin ($h/2\pi$)	Nuclear Magnetic Mom. (nm)	Elect. Quadr. Mom. (b)	γ-Energy/ Intensity (MeV/%)
				β+, EC/50 /	2.5/				0.6529
					2.8/				1.2081
					4.0/				1.5080
									2.2407
^{91}Mo		90.91175	15.5 m	β+ /94/4.43	3.44/94	9/2-			ann.rad./
				EC/6/					1.6373
									2.6321
									3.0286
									(0.1–4.2)
^{92}Mo	14.53(30)	91.906811	> 3 × 10^{17} a	β+-EC		0+			
^{93m}Mo			6.9 h	I.T./99+ /2.425		21/2+	+9.9		0.26306(IT)
									0.68461
									1.47711
^{93}Mo		92.906813	3.5 × 10^3 a	EC/0.405		5/2+			0.0304
^{94}Mo	9.15(9)	93.905088				0+			
^{95}Mo	15.84(11)	94.905842				5/2+	−0.9142	−0.02	
^{96}Mo	16.67(15)	95.904680				0+			
^{97}Mo	9.60(14)	96.906022				5/2+	−0.9335	+0.26	
^{98}Mo	24.39(37)	97.905408				0+			
^{99}Mo		98.907712	2.7476 d	β- /1.357	0.45/14	½+	0.375		0.144048
					0.84/2				0.18109
					1.21/84				0.36644
									0.73947
^{100}Mo	9.82(31)	99.90748	6 × 10^{20} a	β-β-		0+			
^{101}Mo		100.91035	14.8 m	β- /2.82	2.23/	1/2+			0.0063
					0.7/				0.19193
									0.5909
									(0.0809–2.405)
^{102}Mo		101.91030	11.3 m	β- /1.01	1.2/	0+			0.1493/89.
									0.2116/100.
									0.2243/32.
^{103}Mo		102.9132	1.13 m	β- /3.8		3/2+			0.1028(2)/
									0.1440(2)
									0.2511(2)
^{104}Mo		103.9138	1.00 m	β- /2.16		0+			0.0686(1)/100.
									0.4239(4)/21.
^{105}Mo		104.9170	36. s	β- /4.95		3/2+			0.0642/
									0.0856/
									0.2495/
^{106}Mo		105.91814	8.4 s	β- /3.52		0+			0.1894(2)/22.
									0.3644(2)/6.
									0.3723(2)/12.
^{107}Mo		106.9217	3.5 s	β- /6.2					
^{108}Mo		107.9235	1.1 s	β- /5.1		0+			(0.028–0.636)
^{109}Mo		108.9278	0.5 s	β- /7.2					
^{110}Mo		109.9297	0.27 s	β- /5.7		0+			Tc k x-ray
									0.142
									(0.039–0.599)
^{111}Mo		110.9344	> 0.15 μs						
^{112}Mo		111.937	> 0.15 μs			0+			
^{113}Mo		112.942	> 0.15 μs						
^{114}Mo		113.945	> 0.15 μs			0+			
^{115}Mo		114.950	> 0.15 μs						
^{116}Mo			> 0.15 μs			0+			

Elem. or Isot.	Natural Abundance (Atom %)	Atomic Mass or Weight	Half–life/ Resonance Width (MeV)	Decay Mode/ Energy (/MeV)	Particle Energy/ Intensity (MeV/%)	Spin ($h/2\pi$)	Nuclear Magnetic Mom. (nm)	Elect. Quadr. Mom. (b)	γ-Energy/ Intensity (MeV/%)
^{117}Mo			> 0.15 µs						
$_{43}$Tc									
^{85}Tc		84.9488	< 0.1 ms						
^{86}Tc		85.9429	0.05 s	β+ /11.9					
^{87}Tc		86.9365	2.4 s	β+ /8.6					
^{88}Tc		87.9327	5.8 s	β+ /10.1					
^{89m}Tc			13. s						
^{89}Tc		88.9272	13. s	β+ /7.5					
^{90m}Tc			49.2 s	β+	5.3/	6+			ann.rad./
									0.9479/
									1.0542/
^{90}Tc		89.9236	8.3 s	β+ /8.9	7.0/15	1+			ann.rad./
					7.9/95.				0.9479/
^{91m}Tc			3.3 m	β+		½+			ann.rad./170.
				EC					0.8110(5)/5.
									1.6052(1)/7.8
									1.6339(1)/9.1
									1.9023(1)/6.
									2.4509(1)/13.5
^{91}Tc		90.9184	3.14 m	β+ /6.2	5.2	9/2+			ann.rad./200.
^{92}Tc		91.91526	4.4 m	β+ /7.87	4.1	8+			ann.rad./200.
				EC					0.0850/
									0.1475
									0.3293
									0.7731
									1.5096
^{93m}Tc			43. m	I.T./13		½-			0.3924(IT)
				EC/20					0.9437
									2.6445
^{93}Tc		92.910249	2.73 h	β+ /13/3.201	0.81	9/2+	6.3		ann.rad./
				EC/87/					1.3629
									1.4771
									1.5203
									(0.1–3.0)
^{94m}Tc			52. m	β+ /72/4.33		2+			ann.rad./
				EC/28/					0.8710
									1.8686
^{94}Tc		93.909657	4.88 h	β+ /11/4.256		7+	5.12		ann.rad./
				EC/89/					0.4491
									0.7026
									0.8496
									0.8710
^{95m}Tc			62.0 d	I.T./4/		1/2-			ann.rad./
				β+ /0.3	0.5/				0.0389(IT)
				EC/96	0.7/				0.2041
									0.5821
									0.5821
									0.8351
^{95}Tc		94.90766	20.0 h	EC/100/1.691		9/2+	5.9		0.7657
									1.0738
^{96m}Tc			52. m	I.T./90/		4+			0.0342(IT)
				β+, EC/2/					0.7782
									1.2002
^{96}Tc		95.90787	4.3 d	EC/2.973		7+	+5.1		Mo k x-ray

Elem. or Isot.	Natural Abundance (Atom %)	Atomic Mass or Weight	Half–life/ Resonance Width (MeV)	Decay Mode/ Energy (/MeV)	Particle Energy/ Intensity (MeV/%)	Spin ($h/2\pi$)	Nuclear Magnetic Mom. (nm)	Elect. Quadr. Mom. (b)	γ-Energy/ Intensity (MeV/%)
									0.7782
									0.8125
									0.8498
									1.12168
^{97m}Tc			91. d	I.T./0.0965		1/2-			Tc k x-ray
				EC	/3.9				0.0965
^{97}Tc		96.906365	4.2×10^6 a	EC/100/0.320		9/2+			Mo k x-ray
^{98}Tc		97.907216	$\sim 6.6 \times 10^6$ a	β- /1.80	0.40/100	6+			0.65241/100
				EC	//<0.036				0.74535/100
^{99m}Tc			6.01 h	I.T./100/0.142		1/2-			Tc k x-ray
									0.14049
									0.14261
^{99}Tc		98.906255	2.13×10^5 a	β- /0.294	0.293/100	9/2+	+5.6847	−0.129	
^{100}Tc		99.907658	15.8 s	β- /3.202	2.2/	1+			0.5396
				EC /2.6(10)$^{-3}$/	2.9/				0.5908
					3.3				(0.3 79–2.30)
^{101}Tc		100.90732	14. m	β- /1.61	1.32/	9/2+			0.1272
									0.1841
									0.3068
									0.5451
									(0.073–0.969)
^{102m}Tc			4.4 m	I.T./2/4.8	1.8/				0.4184
				β- /98/					0.4752
									0.6281
									0.6302
									1.0464
									1.1033
									1.6163
									2.2447
^{102}Tc		101.90922	5.3 s	β- /4.53	3.4/	1+			0.4686
					4.2				0.4751
					2.2/				1.1055
^{103}Tc		102.90918	54. s	β- /2.66	2.0/	5/2+			0.1361
					2.2/				0.1743
									0.2104
									0.3464
									0.5629
									(0.13–1.0)
^{104m}Tc			0.005 ms						
^{104}Tc		103.91145	18.2 m	β- /5.60	5.3/	(3+)			0.3483
									0.3580
									0.5305
									0.5351
									0.8844
									0.8931
									1.6768
									(0.3–3.7)
^{105}Tc		104.9117	7.6 m	β- /3.6	3.4/	5/2+			0.1079
									0.1432
									0.3215
^{106}Tc		105.91436	36. s	β- /6.55		2+			0.2703
									0.5222
									1.9694
									2.2393

Elem. or Isot.	Natural Abundance (Atom %)	Atomic Mass or Weight	Half–life/ Resonance Width (MeV)	Decay Mode/ Energy (/MeV)	Particle Energy/ Intensity (MeV/%)	Spin ($h/2\pi$)	Nuclear Magnetic Mom. (nm)	Elect. Quadr. Mom. (b)	γ-Energy/ Intensity (MeV/%)
									2.7893
^{107}Tc		106.9151	21.2 s	β- /4.8					0.1027
									0.1063
									0.1770
									0.4587
^{108}Tc		107.9185	5.1 s	β- /7.72		(3)			0.2422
									0.4656
									0.7078
									0.7326
									1.5835
^{109}Tc		108.9200	1.4 s	β- /6.3	p/0.08				
^{110}Tc		109.9238	0.83 s	β- /8.8	p/0.04				0.2407
^{111}Tc		110.9257	0.30 s	β- .n/7.0	n/0.85				0.150/92.7
									0.063–1.435
^{112}Tc		111.9292	0.26 s	β, n	n/2.6				
^{113}Tc		112.9316	0.15 s	β-, n/8.	/2.1				0.0985/100
									0.0658–1.520
^{114}Tc		113.936	0.15 s	β-, n	/1.3				
^{115}Tc		114.939	0.07 s						
^{116}Tc		115.943	> 0.15 µs						
^{117}Tc		116.946	> 0.15 µs						
^{118}Tc		117.951	> 0.15 µs						
^{119}Tc									
^{120}Tc									
$_{44}$**Ru**		**101.07(2)**							
^{87}Ru		86.949	> 1.5 µs						
^{88}Ru		87.9403	1.2 s			0+			
^{89}Ru		88.9361	1.4 s	β+ .p/8.					
^{90}Ru		89.9299	12. s	β+ /5.9		0+			ann.rad./
									0.155–1.551
^{91}Ru		90.9263	7.9 s	β+, EC/7.4		9/2+			ann.rad./
									(0.205-1.998)
^{92}Ru		91.9201	3.7 m	β+ /53/4.5		0+			ann.rad./
				EC/47/					0.1346
									0.2138
									0.2593
^{93m}Ru			10.8 s	I.T./21/		1/2-			ann.rad./
				β+, EC/79/	5.3/				0.7344
									1.1112
									1.3962
									2.0931
^{93}Ru		92.9171	1.0 m	β+ /6.3		9/2+			ann.rad./
				EC/					0.6807
									1.4349
									(0.5–4.2)weak
^{94}Ru		93.91136	52. m	EC/100/1.59		0+			0.3672
									0.5247
									0.8922
^{95}Ru		94.91041	1.64 h	EC/85/2.57	1.20/	5/2+	0.86		ann.rad./
				β+ /15/	0.91/				0.3364/71
									1.097/16.4
									0.6268/15.9
									0.036–2.424
^{96}Ru	5.54(14)	95.90760	>3.1 × 10^{16} a	β+β+		0+			

Elem. or Isot.	Natural Abundance (Atom %)	Atomic Mass or Weight	Half–life/ Resonance Width (MeV)	Decay Mode/ Energy (/MeV)	Particle Energy/ Intensity (MeV/%)	Spin ($h/2\pi$)	Nuclear Magnetic Mom. (nm)	Elect. Quadr. Mom. (b)	γ-Energy/ Intensity (MeV/%)
97Ru		96.90756	2.84 d	EC/1.12		5/2+	−0.79		Tc k x-ray
									0.2157
									0.3245
									0.4606
98Ru	1.87(3)	97.90529				0+			
99Ru	12.76(14)	98.905939				5/2+	−0.64	+0.079	
100Ru	12.60(7)	99.904220				0+			
101Ru	17.06(2)	100.905582				5/2+	−0.72	+0.46	
102Ru	31.55(14)	101.904349				0+			
103Ru		102.906324	39.26 d	β- /0.763	0.223	3/2+	0.206	+0.62	0.05329
									0.29498
									0.4438
									0.49708
									0.55704
									0.61033
									(0.04–1.6)
104Ru	18.62(27)	103.905433				0+			
105Ru		104.907753	4.44 h	β- /1.917	1.11/22	3/2+	−0.3		0.12968
					1.134/13				0.1491
					1.187/49				0.2629
									0.31664
									0.46943
									0.67634
									0.72420
									(0.1–1.8)
106Ru		105.90733	1.020 a	β- /0.0394	0.0394/100	0+			
107Ru		106.9099	3.8 m	β- /2.9	2.1/				0.1939
					3.2/				0.3741
									0.4625
									0.8488
108Ru		107.9102	4.5 m	β- /1.4	1.2/	0+			0.0923
									0.1651
									0.4339
									0.4975
									0.6189
109Ru		108.9132	34.5 s	β- /4.2					0.1164
									0.3584
110Ru		109.9141	15. s	β- /2.81		0+			0.1121
									0.3737
									0.4397
									0.7967
111Ru		110.9177	1.5 s	β- /5.5					
112Ru		111.9190	4.5 s	β- /4.5		0+			
113mRu			0.6 s						
113Ru		112.9225	0.80 s	β- /7.					0.2632
									0.048–2.418
114Ru		113.9243	0.57 s	β- /6.1		0+			0.127/24
									(0.053–0.180)
115mRu			0.08 s						0.0617
115Ru		114.9287	0.32 s	β- /8.					0.292
116Ru		115.931	0.20 s			0+			
117Ru		116.936	0.14 s						
118Ru		117.938	0.12 s			0+			
119Ru		118.943	> 0.15 µs						

Elem. or Isot.	Natural Abundance (Atom %)	Atomic Mass or Weight	Half–life/ Resonance Width (MeV)	Decay Mode/ Energy (/MeV)	Particle Energy/ Intensity (MeV/%)	Spin ($h/2\pi$)	Nuclear Magnetic Mom. (nm)	Elect. Quadr. Mom. (b)	γ-Energy/ Intensity (MeV/%)
120Ru		119.945	> 0.15 μs			0+			
121Ru									
122Ru									
123Ru									
124Ru									
45Rh		102.90550(2)							
89Rh		88.9488	> 0.15 μs						
90mRh			~ 12. ms						
90Rh		89.9429	1.0 s						
91mRh			1.5 s	IT					0.387
91Rh		90.9366	1.5 s						(0.438-0.973)
92mRh			0.5 s						0.866
92Rh		91.9320	4.7 s	β+ /11.1					(0.163-0.991)
93Rh		92.9257	12. s	β+ /8.1					(0.138–1.493)
94mRh			25.8 s	β+ /		8+			ann.rad./
									0.1264
									0.3117
									0.7562
									1.0752
									1.4307
94Rh		93.9217	1.18 m	β+ /9.6	6.4/	4+			ann.rad./
									0.1461
									0.3117
									0.7562
									1.4307
95mRh			1.96 m	I.T./88/		½+			ann.rad./
				β+, EC/12/					0.5433(IT)
									0.7837
95Rh		94.9159	5.0 m	β+ /5.1	3.2	9/2+			ann.rad./
									0.2293
									0.4103
									0.6610
									0.9416
									1.3520
									(0.2–3.8)
96mRh			1.51 m	I.T./60/0.052		2+			ann.rad./
				β+, EC/40/	4.70/				Tc,Ru x-rays
									0.8326
									1.0985
									1.6921
									(0.4–3.3)
96Rh		95.91446	9.6 m	β+/6.45	3.3/	5+			ann.rad./
				EC/					0.4299
									0.6315
									0.6853
									0.7418
									0.8326
									(0.2–3.4)
97mRh			46. m	I.T./5 /	2.6/	1/2-			ann.rad./
				β+, EC/95/					0.1886
									0.4215
									2.2452
97Rh		96.91134	31.0 m	β+ /3.52	2.1/	9/2+			ann.rad./
									0.1886

Elem. or Isot.	Natural Abundance (Atom %)	Atomic Mass or Weight	Half–life/ Resonance Width (MeV)	Decay Mode/ Energy (/MeV)	Particle Energy/ Intensity (MeV/%)	Spin ($h/2\pi$)	Nuclear Magnetic Mom. (nm)	Elect. Quadr. Mom. (b)	γ-Energy/ Intensity (MeV/%)
									0.3892
									0.4515
									0.8398
									0.8788
									(0.2–3.5)
^{98m}Rh			3.5 m	β+ /		5+			ann.rad./
									0.6154
									0.6524
									0.7452
^{98}Rh		97.91071	8.7 m	β+ /90/5.06	3.4/	2+			ann.rad./
									0.6524
									0.7623
^{99m}Rh			4.7 h	β+ /8/	.74/	9/2+	5.67		ann.rad./
				EC/92/					0.2766/
									0.3408
									0.6178
									1.2612
^{99}Rh		98.90813	16. d	β+/4/2.10	0.54/	1/2-			ann.rad./
				EC/97/	0.68/				0.0894/
									0.3530
									0.5277
									(0.1–2.0)
^{100m}Rh			4.7 m	I.T./99/		5+			ann.rad./
				β+ /0.4/					0.0748/
									0.2647(IT)
^{100}Rh		99.90812	20.8 h	β+ /3.63	2.62/	1-			0.4462
				EC/	2.07/				0.5396
									0.5882
									0.8225
									1.5534
									2.3761
^{101m}Rh			4.35 d	EC/92/		9/2+	+5.47		Rh k x-ray
				I.T./8/0.1573					0.1272/
									0.3069
									0.5451
^{101}Rh		100.90616	3.3 a	EC/0.54		1/2-			Ru k x-ray
									0.1272
									0.1980
									0.3252
^{102m}Rh			3.74 a	EC/2.323		6+	4.04		0.4751
				IT/0.0419					0.6313
			> 1.2×10^6 a	β+	/<0.00025				0.6975
									0.7668
									1.0466
									1.1032
^{102}Rh		101.906843	207. d	EC/62			~ 0.5		ann.rad./
				β- /19/					0.4686
				β+ /14/					0.4751
									0.5566
									0.6280
									1.1032
									(0.4–1.6)
^{103m}Rh			56.12 m	IT		7/2+	4.54		
^{103}Rh	100.	102.905504				1/2-	−0.0884		

Elem. or Isot.	Natural Abundance (Atom %)	Atomic Mass or Weight	Half–life/ Resonance Width (MeV)	Decay Mode/ Energy (/MeV)	Particle Energy/ Intensity (MeV/%)	Spin ($h/2\pi$)	Nuclear Magnetic Mom. (nm)	Elect. Quadr. Mom. (b)	γ-Energy/ Intensity (MeV/%)
104mRh			4.36 m	I.T./99+ /		5+			Rh k x-ray
				β-	1.3/				0.0514
									0.0971
									0.5558
104Rh		103.906656	42.3 s	β-/99+/2.441	1.88/2	1+			0.3581
				EC/0.4/1.141	2.44/98				0.5558
									1.2370
									(0.35−1.8)
105mRh			43. s	I.T./1.296		1/2-			Rh k x-ray
									0.1296
105Rh		104.905694	35.3 h	β- /0.567	0.247/30	7/2+	+4.45		0.2801
					0.567/70				0.3061/4.8
									0.3189/17.0
106mRh			2.18 h	β- /	0.92/	6+			0.2217
									0.4510
									0.5119
									0.6162
									0.7173
									0.7484
									1.0458
									1.5277
106Rh		105.90729	29.9 s	β- /3.54	2.4/2	1+	+2.58		0.51186/
					3.0/12				0.61612
					3.54/79				0.62187
									(0.05−3.04)
107Rh		106.90675	21.7 m	β- /1.51	1.20/65	7/2+			0.2776
					1.5/17				0.3028
									0.3925
108mRh			6.0 m	β- /	1.57/				0.4339
									0.4973
									0.6189
108Rh		107.9087	17. s	β- /4.5		1+			0.4046
									0.4339
									0.4973
									0.5811
									0.6146
									0.9014
									0.9471
109Rh		108.90874	1.34 m	β- /2.59	2.25/	7/2+			0.1134
									0.1780
									0.2914
									0.3254
									0.3268
									0.4261
									(0.1−1.6)
110mRh			29. s	β- /	6/				0.3737
									0.4397
									0.7967
110Rh		109.91114	3.1 s	β- /5.4	5.5/	1+			0.3737
									0.4400
									0.5463
									0.6877
									0.8381
									0.9045

Elem. or Isot.	Natural Abundance (Atom %)	Atomic Mass or Weight	Half–life/ Resonance Width (MeV)	Decay Mode/ Energy (/MeV)	Particle Energy/ Intensity (MeV/%)	Spin ($h/2\pi$)	Nuclear Magnetic Mom. (nm)	Elect. Quadr. Mom. (b)	γ-Energy/ Intensity (MeV/%)
[111]Rh		110.91159	11. s	β- /3.7					0.275
[112m]Rh			6.8 s	β- /					
[112]Rh		111.9144	3.5 s	β- /6.2		1+			0.3489
[113]Rh		112.91553	0.9 s	β- /4.9					0.1285
[114m]Rh			1.9 s	β- /					(0.103–1.923)
[114]Rh		113.9188	1.8 s	β- /6.5		1+			(0.276–0.783)
[115]Rh		114.9203	0.99 s	β- /6.0					
[116m]Rh			0.9 s	β- /					0.3405
[116]Rh		115.9241	0.7 s	β- /8.0		1+			0.340
									0.398–1.665
[117]Rh		116.9260	0.42 s	β- /7.					0.0346
									0.1317
[118]Rh		117.9301	0.27 s						0.379
									0.575
									0.370–1.037
[119]Rh		118.932	0.17 s						
[120]Rh		119.936	0.13 s						
[121]Rh		120.939	0.15 s						
[122]Rh		121.943							
[123]Rh									
[124]Rh									
[125]Rh									
[126]Rh									
46 Pd		106.42(1)							
[91]Pd		90.949	> 1.5 μs						
[92]Pd		91.9404	1.0 s			0+			
[93]Pd		92.9359	1.2 s	β+, p		9/2+			0.240/81
									0.382–0.864
[94]Pd		93.9288	9.6 s	EC, β+ /~ 6.6		0+			0.5582
									(0.055–0.798)
[95m]Pd		94.92684	13.4 s	EC, β+ /10.2		21/2+			
[95]Pd		94.9247							
[96]Pd		95.9182	2.03 m	EC, β+ /3.5	1.15/	0+			0.1248
									0.4995
[97]Pd		96.9165	3.1 m	β+, EC/4.8	3.5/	5/2+			ann.rad./
									0.2653
									0.4752
									0.7927
									(0.2–3.4)
[98]Pd		97.91272	17.7 m	β+ /1.87		0+			ann.rad./
				EC/					0.0677
									0.1125
									0.6630
									0.8379
[99]Pd		98.91177	21.4 m	β+ /49/3.37	2.18/	5/2+			ann.rad./
				EC/51/					0.1360
									0.2636
									0.6734
									(0.2–2.85)
[100]Pd		99.90851	3.7 d	EC/0.36		0+			0.03271
									0.0748
									0.0840
[101]Pd		100.90829	8.4 h	β+ /5/1.980	0.776/	5/2+	−0.66		ann.rad./
				EC/95/					0.0244

Elem. or Isot.	Natural Abundance (Atom %)	Atomic Mass or Weight	Half–life/ Resonance Width (MeV)	Decay Mode/ Energy (/MeV)	Particle Energy/ Intensity (MeV/%)	Spin ($h/2\pi$)	Nuclear Magnetic Mom. (nm)	Elect. Quadr. Mom. (b)	γ-Energy/ Intensity (MeV/%)
									0.2963
									0.5904
^{102}Pd	1.02(1)	101.905609				0+			
^{103}Pd		102.906087	16.99 d	EC/0.543		5/2+			Rh k x-ray
									0.03975
									0.3575
									0.4971
^{104}Pd	11.14(8)	103.904036				0+			
^{105}Pd	22.33(8)	104.905085				5/2+	−0.642	+0.66	
^{106}Pd	27.33(3)	105.903486				0+			
^{107m}Pd			20.9 s	I.T./0.2149		11/2-			Pd k x-ray
									0.2149(IT)
^{107}Pd		106.905133	6.5×10^6 a	β- /0.033	0.03/	5/2+			
^{108}Pd	26.46(9)	107.903893				0+			
^{109m}Pd			4.75 m	I.T./0.1889		11/2-			Pd x-ray
									0.1889(IT)
^{109}Pd		108.905950	13.5 h	β- /1.116	1.028	5/2+			0.0880
									(0.08–1.0)
^{110}Pd	11.72(9)	109.905153				0+			
^{111m}Pd			5.5 h	I.T./73/0.172		11/2-			0.0704
				β- /27/	0.35				0.1722
					0.77				0.3912
									(0.1–1.97)
^{111}Pd		110.90767	23.4 m	β- /2.19	2.2/95	5/2+			0.0598
									0.2454
									0.5800
									0.6504
									1.3885
									1.4590
^{112}Pd		111.90731	21.04 h	β- /0.29	0.28/	0+			0.018
^{113m}Pd			1.48 m	β- /		5/2+			0.0959
^{113}Pd		112.91015	1.64 m	β- /3.34					0.0958
									0.4824
									0.6436
									0.7394
^{114}Pd		113.91036	2.48 m	β- /1.45		0+			0.1266
									0.2320
									0.5582
									0.5760
^{115m}Pd			50. s			(9/2-)			0.089
^{115}Pd		114.9137	25. s	β- /4.58		(3/2+)			0.1255
									0.2554
									0.3428
^{116}Pd		115.9142	12.7 s	β- /2.61		0+			0.1015
									0.1147
									0.1778
^{117m}Pd			19. ms			(9/2-)			0.203
^{117}Pd		116.9178	4.4 s	β- /5.7		(3/2+)			0.2473
									0.077–0.403
^{118}Pd		117.9190	2.4 s	β- /4.1		0+			0.1254
									0.028–0.596
^{119}Pd		118.9231	0.9 s	β- /6.5					0.2566
									0.070–0.326
^{120}Pd		119.9247	0.49 s	β- /5.0		0+			0.1581

Elem. or Isot.	Natural Abundance (Atom %)	Atomic Mass or Weight	Half–life/ Resonance Width (MeV)	Decay Mode/ Energy (/MeV)	Particle Energy/ Intensity (MeV/%)	Spin ($h/2\pi$)	Nuclear Magnetic Mom. (nm)	Elect. Quadr. Mom. (b)	γ-Energy/ Intensity (MeV/%)
									0.053–0.595
^{121}Pd		120.9289	0.29 s						
^{122}Pd		121.9306	0.18 s			0+			
^{123}Pd		122.935	0.17 s						
^{124}Pd		123.9369	~ 0.04 s			0+			
^{125}Pd									
^{126}Pd									
^{127}Pd									
^{128}Pd									
$_{47}$**Ag**		**107.8682(2)**							
^{93}Ag		92.950							
^{94m2}Ag			0.40 s	β+	p/1.01/2.2	21+			(0.153-1.132)
				β+,p	p/0.79/1.9				
^{94m1}Ag			0.60 s	β+		7+			(0.659-0.905)
				β+,p	p//20.				
^{94}Ag		93.9428	0.03 s	β+		0+			
				β+,p					
^{95}Ag		94.9355	2.0 s	β+, p/					(0.089–2.940)
^{96m}Ag			4.4 s	β+		8+			
				β+, p	/8.				
^{96}Ag		95.9307	7. s	β+ /11.6		2+			ann.rad./
				EC/					0.1248
				β+, p	/18.				0.4995
									(0.1066–1.416)
^{97}Ag		96.9240	19. s	β+ /7.0					ann.rad./
				EC/					0.6862
									1.2941
									(0.352–3.294)
^{98}Ag		97.9216	47.6 s	β+ /8.4		5+			ann.rad./
				EC/	/36.				0.5711
				β+, p	/0.11				0.6786
									0.8631
									(0.153–1.185)
^{99m}Ag			11. s	I.T./100/		½-			Ag k x-ray
									0.1636(IT)
									0.3426
^{99}Ag		98.9176	2.07 m	β+ /87 5.4		9/2+			ann.rad./
				EC/13/					0.2199
									0.2645
									0.8056
									0.8323
									(0.2–3.5)
^{100m}Ag			2.3 m	β+ /		2+			ann.rad./
				EC/					0.6657
									1.6941
^{100}Ag		99.9161	2.0 m	β+/7.1	4.7/	5+			ann.rad./
				EC/					0.2807
									0.4503
									0.6657
									0.7508
									0.7732
^{101m}Ag			3.1 s	I.T./0.23		½-			Ag k x-ray
									0.0981
									0.176(IT)

Elem. or Isot.	Natural Abundance (Atom %)	Atomic Mass or Weight	Half–life/ Resonance Width (MeV)	Decay Mode/ Energy (/MeV)	Particle Energy/ Intensity (MeV/%)	Spin ($h/2\pi$)	Nuclear Magnetic Mom. (nm)	Elect. Quadr. Mom. (b)	γ-Energy/ Intensity (MeV/%)
^{101}Ag		100.9128	11.1 m	β+/69/4.2	2.7/	9/2+	5.7		ann.rad./
				EC/31/					0.2610
					2.18/				0.2747
					2.73/				0.3269
					3.38/				0.4392
									0.6673
									1.1739
									(0.2–3.1)
^{102m}Ag			7.8 m	β+/38/	3.4	2+	+4.1		ann.rad./
				EC/13/					0.5567
				I.T./49/					0.9777
									1.8347
									2.0545
									2.1594
									3.2386
^{102}Ag		101.91169	13.0 m	β+/78/5.92	2.26/	5+	~ 4.6		ann.rad./
				EC/22/					0.5564
									0.7193
									0.163–2.242
^{103m}Ag			5.7 s	I.T./0.134		1/2-			Ag k x-ray
									0.1344
^{103}Ag		102.90897	1.10 h	β+/28/2.69	1.7	7/2+	+4.47		ann.rad./
				EC/72/	1.3				0.1187
									0.1482
^{104m}Ag			33. m	β+/64/	2.71/	2+	+3.69		ann.rad./
				EC/36/					0.5558
				I.T./0.07/					0.7657
									(0.5–3.4)
^{104}Ag		103.90863	69. m	β+/16/4.28	0.99/	5+	3.92		ann.rad./
				EC/84/					0.5558
									0.9259
									0.9416
									(0.18–2.27)
^{105m}Ag			7.2 m	I.T./98/0.0255		7/2+	+4.41		Ag x-ray
				EC/2 /					0.3063
									0.3192
									(0.1–1.0)
^{105}Ag		104.90653	41.3 d	EC/1.35		1/2-	0.101		0.0640
									0.2804
									0.3445
									0.4434
^{106m}Ag			8.4 d	EC/		6+	3.71	+1.1	Pd k x-ray
									0.4510
									0.5118
									0.7173
									1.0458
^{106}Ag		105.90667	24.0 m	β+/59/2.965	/1.96	1+	+2.9		ann.rad./
				EC/41 /					0.5119
^{107m}Ag			44.2 s	I.T./0.093		7/2+	+4.40	1.0	Ag x-ray
									0.0931
^{107}Ag	51.839(8)	106.905097				1/2-	−0.113680		
^{108m}Ag			418. a	EC/92/		6+	3.58	+1.3	Ag k x-ray
				I.T./8 /0.079					Pd k x-ray
									0.43392

Elem. or Isot.	Natural Abundance (Atom %)	Atomic Mass or Weight	Half–life/ Resonance Width (MeV)	Decay Mode/ Energy (/MeV)	Particle Energy/ Intensity (MeV/%)	Spin (h/2π)	Nuclear Magnetic Mom. (nm)	Elect. Quadr. Mom. (b)	γ-Energy/ Intensity (MeV/%)
									0.61427
									0.72290
^{108}Ag		107.905956	2.39 m	β- /97/1.65	1.02/1.7	1+	+2.6884		ann.rad./
				EC/2/	1.65/96				0.43392
				β- /1/1.92	0.88/0.3				0.61885
									0.63298
^{109m}Ag			39.8 s	I.T./0.088		7/2+	+4.40	+1.0	Ag k x-ray
									0.0880
^{109}Ag	48.161(8)	108.904752				1/2-	−0.13069		
^{110m}Ag			249.8 d	β- /99/	0.087	6+	+3.61	+1.4	0.65774
				I.T./1 /0.1164	0.530				0.76393
									0.88467
									0.93748
									1.38427
									(0.447−1.56)
^{110}Ag		109.906107	24.6 s	β- /2.892	2.22/5	1+	+2.727	0.2	0.65774
					2.89/95				0.8154
									1.1257
^{111m}Ag			1.08 m	IT/99/0.0598		7/2+			Ag k x-ray
				β- /1/					0.0598
									0.2454
^{111}Ag		110.905294	7.47 d	β- /1.037	1.035/	1/2-	−0.146		0.2454
									0.3421
^{112}Ag		111.90701	3.13 h	β- /3.96	3.94/	2-	0.0547		0.6067
					3.4				0.6174
									1.3877
									(0.4−2.9)
^{113m}Ag			1.14 m	I.T./80 /0.043		7/2+			0.1422
				β- /20 /	1.5				0.2983
									0.3161
									0.3923
^{113}Ag		112.90657	5.3 h	β- /2.02	2.01/	1/2-	0.159		0.2588
									0.2986
^{114}Ag		113.90880	4.6 s	β- /5.08	4.9/	1+			0.5582
									0.5760
									1.9946
^{115m}Ag			18.7 s	β- /		7/2+			0.1134
									0.1315
									0.2288
									0.3887
^{115}Ag		114.90876	20. m	β- /3.10		1/2-			0.1316
									0.2128
									0.2291
									0.4727
									(0.13−2.49)
^{116m2}Ag			20. s	β-,IT/7	IT/0.0479				
^{116m}Ag			9.8 s	β-/92 /	3.2/	5+			0.5134
					2.9				0.7055
				I.T./8	IT/.0809				(0.255−2.838)
^{116}Ag		115.91136	2.68 m	β- /6.16	5.3	2-			0.5134
									0.6993
									2.4779
^{117m}Ag			5.3 s	β- /	3.2/	7/2+			0.1354
									0.2981

Elem. or Isot.	Natural Abundance (Atom %)	Atomic Mass or Weight	Half–life/ Resonance Width (MeV)	Decay Mode/ Energy (/MeV)	Particle Energy/ Intensity (MeV/%)	Spin (h/2π)	Nuclear Magnetic Mom. (nm)	Elect. Quadr. Mom. (b)	γ-Energy/ Intensity (MeV/%)
									0.3868
									0.1571
^{117}Ag		116.91168	1.22 m	β- /4.18	2.3	1/2-			0.1354
									0.3377
^{118m}Ag			2.8 s	β- /59/					0.1277
				I.T./41 /0.1277					0.4878
									0.6771
									0.7709
									(0.190-2.778)
^{118}Ag		117.9146	4.0 s	β- /7.1					0.4878
									0.6771
									3.2259
^{119}Ag		118.9157	2.1 s	β- /5.35		7/2+			0.0674
									0.3662
									0.3991
									0.6264.
^{120m}Ag			0.40 s	β- /63.					0.2030
				I.T./37.					0.5059
									0.6978
									0.8300
									(0.115-1.644)
^{120}Ag		119.9188	1.23 s	β- /8.2					0.5059
				β-,n	n//<0.0030%				0.6978
									0.8171
									(0.442-3.044)
^{121}Ag		120.9199	0.78 s	β- /6.4					0.1150
									0.3148
									0.3537
									0.3696
									0.5007
									1.5105
									(0.11–2.5)
^{122m}Ag			1. s	β- /					
^{122}Ag		121.9235	0.44 s	β- /9.2					0.7617
									(0.351-1.079)
^{123}Ag		122.9249	0.28 s	β- /7.4					
^{124}Ag		123.9286	0.17 s	β- /10.1					
^{125}Ag		124.9304	0.17 s	β-					0.686
									(0.672-0.731)
^{126}Ag		125.9345	0.11 s	β-					
^{127}Ag		126.9368	0.11 s	β-					
^{128}Ag		127.9412	58 ms	β-					
^{129m}Ag			0.16 s						
^{129}Ag		128.9437	~ 46. ms	β-, n					
^{130}Ag		129.9505	~ 35 ms						
$_{48}$**Cd**		**112.411(8)**							
^{95}Cd		94.950							
^{96}Cd		95.9398	1.0 s			0+			
^{97}Cd		96.9349	3. s	β+, (p)					
^{98}Cd		97.9274	9.2 s	β+ /5.4		0+			
				(p)	/0.025				
^{99}Cd		98.9250	16. s	β+, EC/6.9					ann.rad./
^{100}Cd		99.9203	1.1 m	β+, EC/3.9		0+			ann.rad./
									(0.090–1.043)

Elem. or Isot.	Natural Abundance (Atom %)	Atomic Mass or Weight	Half–life/ Resonance Width (MeV)	Decay Mode/ Energy (/MeV)	Particle Energy/ Intensity (MeV/%)	Spin ($h/2\pi$)	Nuclear Magnetic Mom. (nm)	Elect. Quadr. Mom. (b)	γ-Energy/ Intensity (MeV/%)
[101]Cd		100.9187	1.2 m	β+ /83/5.5	4.5	5/2+			In k x-ray
				EC/17/					0.0985
									1.7225
									0.31–2.84
[102]Cd		101.91446	5.8 m	β+ /27/2.59		0+			ann.rad./
				EC/73					0.0974
									0.4810
									1.0366
									1.3598
[103]Cd		102.91342	7.5 m	β+ /33/4.14		5/2+	−0.81	~ −0.8	ann.rad./
				EC/67/					Ag k x-ray
									1.0799
									1.4487
									1.4618
									(0.1–2.8)
[104]Cd		103.90985	58. m	EC/1.14		0+			Ag k x-ray
									0.0835
									0.7093
[105]Cd		104.90947	55.5 m	β+ /26/2.739	1.69/	5/2+	−0.7393	+0.43	Ag k x-ray
				EC/74/					0.3469
									0.6072
									0.9618
									1.3025
									(0.25–2.4)
[106]Cd	1.25(6)	105.90646	>1.9 × 10¹⁹ a	EC, EC		0+			
[107]Cd		106.90662	6.52 h	EC/99+/1.417		5/2+	−0.615055	~ +0.68	Ag k x-ray
				β+ /					0.0931
									0.8289
[108]Cd	0.89(3)	107.90418	>4.1 × 10¹⁷ a	EC EC		0+			
[109]Cd		108.904982	462.0 d	EC/0.214		5/2+	−0.827846	~ +0.69	Ag k x-ray
									0.08804/.0366
[110]Cd	12.49(18)	109.903002				0+			
[111m]Cd			48.5 m	I.T./		11/2-	−1.105	~ −0.85	Cd k x-ray
									0.1508(IT)
									0.2454
[111]Cd	12.80(12)	110.904178				1/2+	−0.594886		
[112]Cd	24.13(21)	111.902758				0+			
[113m]Cd			14.1 y	β- /99.9/0.59	0.59/99.9	11/2-	−1.087784	~ −0.71	0.2637
[113]Cd	12.22(12)	112.904402	8.04 × 10¹⁵ a	β-		1/2+	−0.622301		
[114]Cd	28.73(42)	113.903359	>1.3 × 10¹⁸ a	β-β-		0+			
[115m]Cd			44.6 d	β- /1.629	0.68/1.6	11/2-	−1.041034	~ −0.54	0.48450
					1.62/97				0.93381
									1.29064
[115]Cd		114.905431	2.228 d	β- /1.446	0.593/42	1/2+	−0.648426		0.23141
					1.11/58				0.26085
									0.33624
									0.49227
									0.52780
[116]Cd	7.49(18)	115.904756	3.8 × 10¹⁹ a	β-β-		0+			
[117m]Cd			3.4 h	β- /2.66	0.72/	11/2-			0.1586
									0.5529
									0.37–2.42

Elem. or Isot.	Natural Abundance (Atom %)	Atomic Mass or Weight	Half–life/ Resonance Width (MeV)	Decay Mode/ Energy (/MeV)	Particle Energy/ Intensity (MeV/%)	Spin ($h/2\pi$)	Nuclear Magnetic Mom. (nm)	Elect. Quadr. Mom. (b)	γ-Energy/ Intensity (MeV/%)
117Cd		116.907219	2.49 h	β- /2.52	0.67/51	1/2+			0.2209
					2.2/10				0.2733
									0.3445
									1.3033
118Cd		117.90692	50.3 m	β- /0.52		0+			
119mCd			2.20 m	β- /		11/2-			0.1056
									0.7208
									1.0250
									2.0213
119Cd		118.9099	2.69 m	β- /3.8	~ 3.5/	1/2+			0.1340
									0.2929
									0.3429
120Cd		119.90985	50.8 s	β- /1.76	1.5/	0+			
121mCd			8. s	β- /		11/2-			0.1008
									0.9878
									1.0209
									1.1815
									2.0594
121Cd		120.9130	13.5 s	β- /4.9		(3/2+)			0.2102
									0.3242
									0.3492
									1.0403
122Cd		121.91333	5.3 s	β- /3.0		0+			
123mCd			1.9 s	β- /					
123Cd		122.91700	2.09 s	β- /6.12		3+			
124Cd		123.9177	1.24 s	β- /4.17		0+			0.0365
									0.0628
									0.1799
125mCd			0.66 s	β- /					
125Cd		124.9213	0.68 s	β- /7.16		3/2+			
126Cd		125.9224	0.52 s	β- /5.49		0+			0.2601
127mCd			0.0175 ms						(0.110-0.849)
127Cd		126.9264	0.4 s	β- /8.5		3/2+			
128Cd		127.9278	0.28 s	β- /7.1		0+			0.247
129Cd		128.9322	0.24 s	β- /5.9					0.281
130Cd		129.9339	0.162 s	β- /		0+			
				β-, n	/~ 3.5				
131Cd		130.9407	68 ms						(0.844-6.039)
132Cd		131.9456	0.10 s	β-, n/	/60	0+			
133Cd			0.06 s						
49In		**114.818(3)**							
97In		96.950							
98mIn			0.7 s						
98In		97.9421	0.5 s						
99In		98.9342	~ 3.8 s	β+ /8.9					
100In		99.9311	5.9 s	β+, (p)/10.5					(0.297-1.365)
101In		100.9263	15. s	β+ /7.3					
102In		101.9241	23. s	EC/8.9		(5)			0.1566
									0.7767
									(0.397–0.923)
103mIn			34. s						
103In		102.91991	1.1 m	β+, EC/6.05	4.2	9/2+			ann.rad./
				EC	/45				0.1879
									(0.157–3.98)

Elem. or Isot.	Natural Abundance (Atom %)	Atomic Mass or Weight	Half–life/ Resonance Width (MeV)	Decay Mode/ Energy (/MeV)	Particle Energy/ Intensity (MeV/%)	Spin ($h/2\pi$)	Nuclear Magnetic Mom. (nm)	Elect. Quadr. Mom. (b)	γ-Energy/ Intensity (MeV/%)
[104m]In			16. s	IT/0.0935					
[104]In		103.9183	1.84 m	β+, EC/7.9	4.8	5+	+4.44	+0.7	ann.rad./
									0.6580
									0.8341
									0.8781
[105m]In			43. s	I.T.		½–			In k x-ray
									0.6740
[105]In		104.91467	5.1 m	β+, EC/4.85	3.7	9/2+	+5.675	+0.83	0.1310
									0.2600
									0.6038
[106m]In			5.3 m	β+ /85/	4.90	3+			ann.rad./
				EC/15/					0.6326
									0.8611
									1.7164
[106]In		105.91347	6.2 m	β+ /65/6.52	2.6	7+	+4.92	~ +0.97	ann.rad./
				EC/35/					0.2259
									0.6327
									0.8611
									0.9978
									1.0091
[107m]In			51. s	I.T./0.6786		½–			In k x-ray
									0.6785
[107]In		106.91030	32.4 m	β+ /35/3.43	2.20/	9/2+	+5.59	+0.81	ann.rad./
				E.C/65/					Cd k x-ray
									0.2050
									0.3209
									0.5055
									(0.2–2.99)
[108m]In			57. m	β+ /53/	1.3	6+	+4.94	+0.47	ann.rad./
				EC/47/					Cd k x-ray
									0.6329
									1.9863
									3.4522
[108]In		107.90970	40. m	β+ /33/5.15	3.49/	3+	+4.561	+1.01	ann.rad./
				EC/67/					Cd k x-ray
									0.2429
									0.6331
									0.8756
[109m]In			1.3 m	I.T./0.650		½–			In k x-ray
									0.6498
[109]In		108.90715	4.17 h	β+ /8/2.02	0.79/	9/2+	+5.54	+0.84	ann.rad./
				EC/92/					Cd k x-ray
									0.2035
									0.6235
[110m]In			4.9 h	EC/		7+	+4.71	+1.00	Cd k x-ray
									0.6577
									0.8847
									0.9375
									(0.1–1.98)
[110]In		109.90717	1.15 h	β+ /62/3.88	2.22/	2+	+4.37	+0.35	ann.rad./
				EC/38/					Cd k x-ray
									0.6577
									(0.6–3.6)
[111m]In			7.7 m	I.T./0.537		½–	+5.53		In k x-ray

Elem. or Isot.	Natural Abundance (Atom %)	Atomic Mass or Weight	Half–life/ Resonance Width (MeV)	Decay Mode/ Energy (/MeV)	Particle Energy/ Intensity (MeV/%)	Spin ($h/2\pi$)	Nuclear Magnetic Mom. (nm)	Elect. Quadr. Mom. (b)	γ-Energy/ Intensity (MeV/%)
									0.537
^{111}In		110.905103	2.8049 d	EC/0.866		9/2+	+5.50	+0.80	Cd k x-ray
									0.1712
									0.2453
^{112m}In			20.8 m	I.T./0.155		4+	+5.227	+0.71	In k x-ray
									0.1555
^{112}In		111.90553	14.4 m	β+ /22/2.586		1+	+2.82	~ +0.087	ann.rad./
				EC/34/					Cd k x-ray
				β- /0.663					0.6171
^{113m}In			1.658 h	I.T./0.3917		½-	−0.21074		In k x-ray
									0.3917
^{113}In	4.29(5)	112.904058				9/2+	+5.5289	+0.80	
^{114m}In			49.51 d	I.T./97/0.190		5+	+4.65	+0.74	In k x-ray
				EC/3 /					0.19027
^{114}In		113.904914	1.198 m	β- /97/1.989	1.984/	1+	+2.82		Cd k x-ray
				EC/3/1.453					0.5584
									0.5727
									1.2998
^{115m}In			4.486 h	I.T./95/0.336		½-	−0.2440		In k x-ray
				β- /5 /0.83					0.3362
									0.4974
^{115}In	95.71(5)	114.903878	4.4×10^{14} a	β- /0.495		9/2+	+5.5408	+0.81	
^{116m2}In			2.16 s	I.T./0.162		8-	+3.22	+0.31	In k x-ray
				EC	/0.023				0.1624
^{116m1}In			54.1 m	β- /	1.0	5+	+4.44	+0.80	0.13792
									0.41688/27
									1.09723/58.5
									1.29349/85
^{116}In		115.905260	14.1 s	β- /3.274	3.3/99	1+	2.788	0.11	0.46313
									1.2526
									1.29349
^{117m}In			1.94 h	β- /53/1.769	1.77/	½-	−0.25174		In k x-ray
				I.T./47 /					0.15855
									0.31531
									0.55294
^{117}In		116.90451	44. m	β- /1.455	0.74/	9/2+	+5.519	+0.83	0.15855/
									0.3966
									0.55294
^{118m2}In			8.5 s	I.T./98/		(8-)	+3.32	+0.44	In k x-ray
				β- /2/					0.1382
^{118m1}In			4.40 m	β- /	1.3	5+	+4.23	+0.80	0.2086
					2.0				0.6833
									1.2295
^{118}In		117.90635	5.0 s	β- /4.42	4.2/	1+			0.5282
									1.1734
									1.2295
									2.0432
^{119m}In			17.9 m	β- /97/	2.7/	½-	−0.32		0.3114
				I.T./3/0.311					0.7631
^{119}In		118.90585	2.3 m	β- /2.36	1.6/	9/2+	+5.52	+0.85	0.0239
									0.6495
									0.7631
									1.2149
^{120m2}In			47 s	β- /6.1		8-	+3.692	+0.53	1.171

Elem. or Isot.	Natural Abundance (Atom %)	Atomic Mass or Weight	Half–life/ Resonance Width (MeV)	Decay Mode/ Energy (/MeV)	Particle Energy/ Intensity (MeV/%)	Spin ($h/2\pi$)	Nuclear Magnetic Mom. (nm)	Elect. Quadr. Mom. (b)	γ-Energy/ Intensity (MeV/%)
[120m1]In			46. s	β- /5.8	2.2/	5+	+4.30	+0.81	1.023
									1.171
[120]In		119.90796	3.1 s	β- /5.37	5.6/	(1+)			1.023
					3.1/				0.4146
									0.5924
									0.8637
									1.0232
									1.1714
									(0.4–2.7%)
[121m]In			3.8 m	β- /99/	3.7/	1/2-	−0.355		0.0601
				I.T./1/0.313					0.3136
									0.9256
									1.0412
									1.1022
									1.1204
[121]In		120.90785	23. s	β- /3.36	2.5	9/2+	+5.50	+0.81	0.2620
									0.6573
									0.9256
[122m]In			10. s	β- /	4.4/	8-	+3.78	+0.59	1.0014
									1.1403
[122]In		121.91028	1.5 s	β- /6.37	5.3/	(1+)			0.2391
									1.0014
									1.1403
									1.164
									1.1903
[123m]In			47. s	β- /	4.6/	(1/2-)	−0.400		0.1258
									1.170
									3.234
[123]In		122.91044	6.0 s	β- /4.39	3.3/	(9/2+)	+5.49	+0.76	0.6188
									1.0197
									1.1305
[124m]In			3.4 s	β-		8-	+3.89	+0.66	0.1029
									0.9699
									1.0729
									1.1316
[124]In		123.91318	3.18 s	β- /7.36	5/	3+	+4.04	+0.61	0.7070
									0.9978
									1.1316
									3.2142
									(0.3–4.6)
[125m]In			12.2 s	β- /	5.5/	1/2-	−0.433		0.1876
[125]In		124.91360	2.33 s	β- /5.42	4.1/	9/2+	+5.50	+0.71	0.4260
									1.0318
									1.3350
[126m]In			1.53 s		4.9/	3+	+4.03	+0.49	0.9086
									0.9696
									1.1411
[126]In		125.91646	1.63 s	β- /8.21	4.2/	8-	+4.061		0.1118
									0.9086
									1.1411
[127m]In			3.73 s	β- /	6.4/	(1/2-)			0.2523
									3.074
[127]In		126.91735	1.14 s	β- /6.51	4.9/	(9/2+)	+5.52	+0.59	0.4680
									0.6461

Elem. or Isot.	Natural Abundance (Atom %)	Atomic Mass or Weight	Half–life/ Resonance Width (MeV)	Decay Mode/ Energy (/MeV)	Particle Energy/ Intensity (MeV/%)	Spin ($h/2\pi$)	Nuclear Magnetic Mom. (nm)	Elect. Quadr. Mom. (b)	γ-Energy/ Intensity (MeV/%)
									0.8051
									1.5977
^{128m}In			0.7 s	β- /	5.4/	(8-)			1.8670
									1.9739
									(0.1205–2.12)
^{128}In		127.92017	0.80 s	β- /8.98	5.0/	3+			0.9352
									1.1688
									3.5198
									4.2970
^{129m}In			1.23 s	β- /98/	~ 7.5/	1/2-			0.3153
				n/2/					0.9067
									1.2220
^{129}In		128.9217	0.63 s	β- /7.66	5.5/	9/2+			0.2853
									0.7693
									1.8650
									2.1180
^{130m2}In			0.53 s	β- /	8.8/	5+			0.0892
									0.7744
									1.2212
^{130m1}In			0.51 s	β- /	6.1/	10-			0.0892
									0.1298
									0.7744
									1.2212
									1.9052
^{130}In		129.92497	0.29 s	β- /10.25	10.0/	1-			
^{131m2}In			0.3 s	β- /		(21/2+)			
^{131m1}In			0.35 s	β- /		(1/2-)			
^{131}In		130.92685	0.28 s	β- /9.18	6.4/	(9/2+)			0.3328
									2.433
^{132}In		131.9330	~ 0.206 s	β- /13.6	6.0/	(7-)			0.1320
					8.8/				0.2992
									0.3747
									4.0406
^{133}In		132.9378	0.165 s	β-, (n)					
^{134}In		133.9442	0.14 s						(0.354–2.005)
^{135}In		134.9493	0.09 s						
$_{50}$**Sn**		**118.710(7)**							
^{99}Sn		98.949							
^{100}Sn		99.939	1.0 s	β+ /7.3	3.4/	0+			
^{101}Sn		100.9361	1.9 s	β+ /9.					0.352
									1.065
^{102}Sn		101.9303	3.8 s	β+ /5.8		0+			(0.069-1.425)
^{103}Sn		102.9281	7. s	β+ /7.7					1.3558
				β+,p	p//1.2				(0.351-2.813)
				EC	/ 20.				
^{104}Sn		103.9231	21. s	β+, EC/4.5		0+			(0.913-1.846)
^{105}Sn		104.9214	32.7 s	β+ /6.3	EC//42.				In-x-ray
				β$^+$, p	p//0.11				(0.288–3.819)
^{106}Sn		105.91688	2.0 m	β+ /20/3.18		0+			ann.rad./
				EC/80/					In k x-ray
									0.3865
									0.4772
^{107}Sn		106.9156	2.92 m	EC/5.0	1.2/				0.4218
				β+ /					0.6105

Elem. or Isot.	Natural Abundance (Atom %)	Atomic Mass or Weight	Half–life/ Resonance Width (MeV)	Decay Mode/ Energy (/MeV)	Particle Energy/ Intensity (MeV/%)	Spin ($h/2\pi$)	Nuclear Magnetic Mom. (nm)	Elect. Quadr. Mom. (b)	γ-Energy/ Intensity (MeV/%)
									0.6785
									1.0013
									1.1290
									1.542
^{108}Sn		107.91193	10.3 m	β+ /1/2.09	0.36/	0+			In k x-ray
				EC/99/					0.2724
									0.3965
									(0.105–1.68)
^{109}Sn		108.91128	18.0 m	β+ /9/3.85	1.52/	7/2+	−1.08	+0.3	ann.rad./
				EC/91/					In k x-ray
									0.6498
									1.0992
^{110}Sn		109.90784	4.17 h	EC/0.64		0+			In k x-ray
									0.283
^{111}Sn		110.90773	35. m	β+ /31/2.45	1.5/	7/2+	+0.61	+0.2	In k x-ray
				EC/69/					0.7620
									1.1530
									1.9147
^{112}Sn	0.97(1)	111.904818	> 1.8 × 10^{19} a	Ec/ec		0+			
^{113m}Sn			21.4 m	I.T./92/0.077		7/2+			Sn k x-ray
				EC/8/					In x-ray
									0.0774
^{113}Sn		112.905171	115.1 d	EC/1.036		½+	−0.879		In k x-ray
									0.25511
									0.39169
^{114}Sn	0.66(1)	113.902779				0+			
^{115}Sn	0.34(1)	114.903342				½+	−0.9188		
^{116}Sn	14.54(9)	115.901741				0+			
^{117m}Sn			14.0 d	I.T./0.3146		11/2-	−1.396	−0.4	Sn k x-ray
									0.15856
^{117}Sn	7.68(7)	116.902952				½+	−1.0010		
^{118}Sn	24.22(9)	117.901603				0+			
^{119m}Sn			293. d	I.T./0.0896		11/2-	−1.4	0.21	Sn k x-ray
									0.02387
^{119}Sn	8.59(4)	118.903308				½+	−1.0473		
^{120}Sn	32.58(9)	119.902195				0+			
^{121m}Sn			44. a	I.T./78/0.006		11/2-	−1.388	−0.14	Sn k x-ray
				β- /22/	0.354/				0.03715
^{121}Sn		120.904236	1.128 d	β- /0.388	0.383/100	3/2+	0.698	∼ −0.02	
^{122}Sn	4.63(3)	121.903439				0+			
^{123m}Sn			40.1 m	β- /1.428	1.26/99	3/2+			0.1603
									0.3814
^{123}Sn		122.905721	129.2 d	β- /1.404	1.42/99.4	11/2-	−1.370	∼ +0.03	0.1603
									1.0302
									1.0886
^{124}Sn	5.79(5)	123.905274	>2.2 × 10^{18} a	β-β-		0+			
^{125m}Sn			9.51 m	β- /2.387	2.03/98	3/2+	+0.764	+0.8	0.3321
									1.4040
^{125}Sn		124.907784	9.63 d	β- /2.364	2.35/82	11/2-	−1.348	∼ +0.1	1.0671
									(0.2–2.3)
^{126}Sn		125.90765	2.0 × 10^5 a	β- /0.38	0.25/100	0+			0.0643
									0.0876
									0.4148
									0.6663

Elem. or Isot.	Natural Abundance (Atom %)	Atomic Mass or Weight	Half–life/ Resonance Width (MeV)	Decay Mode/ Energy (/MeV)	Particle Energy/ Intensity (MeV/%)	Spin ($h/2\pi$)	Nuclear Magnetic Mom. (nm)	Elect. Quadr. Mom. (b)	γ-Energy/ Intensity (MeV/%)
									0.6950
^{127m}Sn			4.15 m	β- /3.21	2.72/	3/2+	+0.757	+0.60	0.4909
									1.3480
									1.5640
^{127}Sn		126.91036	2.12 h	β- /3.20	2.42/	11/2-	−1.33	+0.3	0.8231
					3.2/				1.0956
									(0.120–2.84)
^{128m}Sn			6.5 s	IT/0.091		(7-)			
^{128}Sn		127.91054	59.1 m	β- /1.27	0.48/	0+			0.4823
					0.63/				0.5573
									0.6805
^{129m}Sn			6.9 m	β- /		11/2-	−1.30	~ −0.2	1.1611
^{129}Sn		128.91348	2.4 m	β- /4.0		3/2+	+0.754	~ +0.05	0.6456
^{130m}Sn			1.7 m	β- /		(7-)	−0.381	−0.4	0.1449
									0.8992
^{130}Sn		129.91397	3.7 m	β- /2.15	1.10/	0+			0.0700
									0.1925
									0.7798
^{131m}Sn			1.02 m	β- /	3.4/	11/2-	−1.28		0.3043
									0.4500
									0.7985
									1.2260
									(0.08–3.21)
^{131}Sn		130.91700	39. s	β- /4.69	3.8/	3/2+	+0.747	~ −0.04	see ^{131m}Sn
^{132}Sn		131.91782	40. s	β- /3.12	1.8/	0+			0.0855
									0.2467
									0.3402
									0.8985
^{133}Sn		132.92383	1.44 s	β- /7.8	7.5/	7/2-			
^{134}Sn		133.9283	1.04 s	β- /6.8		0+			(0.053-2.417)
^{135}Sn		134.9347	0.53 s	β-					(0.053-0.830)
				β-,n	/21.				0.733–1.855
^{136}Sn		135.9393	0.25 s	β-, n	/30.	0+			
^{137}Sn		136.946	0.19 s	β-, n	/~ 58				
^{138}Sn			0.15 s			0+			
$_{51}$**Sb**		**121.760(1)**							
^{103}Sb		102.9397	> 1.5 µs						
^{104}Sb		103.9365	0.5 s						
^{105}Sb		104.9315	1.1 s	β+,p	p//<0.1				
^{106}Sb		105.9288	0.6 s	β+ /10.5					
^{107}Sb		106.9242	4.0 s	β+ /7.9					1.280
									0.1515
									0.6666
									0.553–2.046
^{108}Sb		107.9222	7.0 s	β+ /9.5					(0.151–1.280)
^{109}Sb		108.91813	17.3 s	β+ /6.38	4.42/	5/2+			0.925
				EC/	4.67/				1.062
					4.33/				0.261–2.127
^{110}Sb		109.9168	24. s	β+ /9.0	6.8/	3+			ann.rad./
				EC/					0.6365
									0.9847
									1.2117
									1.2433
^{111}Sb		110.91316	1.25 m	β+ /87/4.47	3.3/	5/2+			ann.rad./

Elem. or Isot.	Natural Abundance (Atom %)	Atomic Mass or Weight	Half–life/ Resonance Width (MeV)	Decay Mode/ Energy (/MeV)	Particle Energy/ Intensity (MeV/%)	Spin ($h/2\pi$)	Nuclear Magnetic Mom. (nm)	Elect. Quadr. Mom. (b)	γ-Energy/ Intensity (MeV/%)
				EC/13 /					0.1002
									0.1545
									0.4891
									1.0326
^{112}Sb		111.91240	51.4 s	β+ /90/7.06	4.75/	3+			ann. rad./
				EC/10/					0.6700
									0.9909
									1.2571
^{113}Sb		112.90937	6.7 m	β+ /65/3.91	2.42/	5/2+			ann. rad./
									(0.3–3.6)
				EC/35/					Sn k x-ray
									0.3324
									0.4980
^{114}Sb		113.90927	3.49 m	β+ /78/5.9	3.4/	3+	1.7		ann. rad./
				EC/22/					Sn k x-ray
									0.8876
									1.2999
^{115}Sb		114.90660	32.1 m	β+ /67/3.03	1.51/	5/2+	+3.46	−0.4	ann. rad./
				EC/33/					Sn k x-ray
									0.4973
^{116m}Sb			1.00 h	β+ /78/	1.16/	8-	2.6		ann. rad./
				EC/22/					Sn k x-ray
									0.4073
									0.5429
									0.9725
									1.2935
									(0.100–1.501)
^{116}Sb		115.90679	16. m	β+ /50/4.707	1.3/	3+	2.72		ann. rad./
				EC/50/	2.3/				Sn k x-ray
									0.93180
									1.29354
									(0.138–3.903)
^{117}Sb		116.90484	2.80 h	β+ /2/1.76	0.57/	5/2+	+3.4		Sn k x-ray
				EC/98/					0.1586
^{118m}Sb			5.00 h	EC/99/		8-	2.32		Sn k x-ray
									0.25368
									1.05069
									1.22964
^{118}Sb		117.905529	3.6 m	β+ /74/3.657	2.65/	1+	2.5		ann. rad./
				EC/26/					Sn k x-ray
									1.22964
^{119}Sb		118.90394	38.1 h	EC/0.59		5/2+	+3.45	−0.4	Sn k x-ray
									0.0239
^{120m}Sb			5.76 d	EC/		8-	2.34		Sn k x-ray
									0.0898
									0.19730
									1.02301
									1.17121
^{120}Sb		119.90507	15.89 m	β+ /41/2.68	1.72/	1+	+2.3		ann. rad./
				EC/59/					Sn k x-ray
									0.7038
									1.17121
^{121}Sb	57.21(5)	120.903816				5/2+	+3.3634	−0.4	
^{122m}Sb			4.19 m	I.T./0.162		8-			Sb x-ray

Elem. or Isot.	Natural Abundance (Atom %)	Atomic Mass or Weight	Half–life/ Resonance Width (MeV)	Decay Mode/ Energy (/MeV)	Particle Energy/ Intensity (MeV/%)	Spin (*h*/2π)	Nuclear Magnetic Mom. (nm)	Elect. Quadr. Mom. (b)	γ-Energy/ Intensity (MeV/%)
									0.0614
									0.0761
¹²²Sb		121.905174	2.72 d	β- /98/1.979	1.414/65	2-	−1.90	+0.9	0.56409
				β+ /2/1.620	1.980/26				0.69277
									1.14050
									1.2569
¹²³Sb	42.79(5)	122.904214				7/2+	+2.5498	−0.5	
¹²⁴m²Sb			20.3 m	I.T./0.035		8-			
¹²⁴m¹Sb			1.6 m	I.T./80/	1.2/	5+			0.4984
				β- /20/	1.7/				0.6027
									0.6458
									1.1010
¹²⁴Sb		123.905936	60.2 d	β- /2.905	0.61/52	3-	1.20	+1.9	0.60271/97.8
					2.301/23				0.64583/7.4
									0.72277/10.5
									1.69094/48.2
									(0.027−2.871)
¹²⁵Sb		124.905254	2.758 a	β- /0.767	0.13/30	7/2+	+2.63		0.0355
					0.302/45				0.17632
					0.62/13				0.38044
									0.42786
									0.46336
									0.60060
									0.63595
¹²⁶m²Sb			11. s	I.T./		3-			L x-ray
									0.0227
¹²⁶m¹Sb			19.0 m	β- /86 /	1.9	5+			0.4148
				I.T./14 /					0.6663
									0.6950
									(0.222-1.477)
¹²⁶Sb		125.90725	12.4 d	β- /3.67	1.9	8-	1.3		0.2786
									0.4148/83.3
									0.6663/99.7
									0.6950/99
									0.7205
¹²⁷Sb		126.90692	3.84 d	β- /1.581	0.89/	7/2+	2.70		0.2524
					1.10/				0.2908
					1.50/				0.4121
									0.4370
									0.6857
									0.7837
¹²⁸mSb			10.1 m	β- /96/	2.6/	5+			0.3140
				I.T./4/					0.5941
									0.7432
									0.7539
¹²⁸Sb		127.90917	9.1 h	β- /4.38	2.3/	8-	1.3		0.2148
									0.3141
									0.5265
									0.7433
									0.7540
¹²⁹mSb			17.7 m	β- /					0.4338
									0.6578
									0.7598
¹²⁹Sb		128.90915	4.40 h	β- /2.38	0.65/	7/2-	2.79		0.0278

Elem. or Isot.	Natural Abundance (Atom %)	Atomic Mass or Weight	Half–life/ Resonance Width (MeV)	Decay Mode/ Energy (/MeV)	Particle Energy/ Intensity (MeV/%)	Spin ($h/2\pi$)	Nuclear Magnetic Mom. (nm)	Elect. Quadr. Mom. (b)	γ-Energy/ Intensity (MeV/%)
									0.1808
									0.3594
									0.4596
									0.5447
									0.8128
									0.9146
									1.0301
130mSb			6.5 m	β- /2.6	2.12/		3.09		0.1023
									0.7934
									0.8394
130Sb		129.91166	38.4 m	β- /4.96	2.9/	8-			0.1823
									0.3309
									0.4680
									0.7394
									0.8394
131Sb		130.91198	23.0 m	β- /3.20	1.31/	7/2+	2.89		0.6423
					3.0/				0.6579
									0.9331
									0.9434
132mSb			2.8 m	β- /	3.9/	4+	3.18		0.1034
									0.3538
									0.6968
									0.9739
									0.9896
132Sb		131.91447	4.2 m	β- /5.49		8-			0.1034
									0.1506
									0.6968
									0.9739
133Sb		132.91525	2.5 m	β- /4.00	1.20/	7/2+	3.00		0.4235
									0.6318
									0.8165
									1.0764
134mSb			10.4 s	β- /	6.1	7-			
134Sb		133.92038	0.8 s	β- /8.4	8.4	0-			0.1152
									0.2970
									0.7063
									1.2791
135Sb		134.9252	1.71 s	β- /8.12		7/2+			1.127
									1.279
136Sb		135.9304	0.82 s	β- /9.3					
137Sb		136.9353	> 0.15 μs						
138Sb		137.9408	> 0.15 μs						
139Sb		138.9460	> 0.15 μs						
140Sb									
52Te		127.60(3)							
105Te		104.9436	~ 0.63 μs	α	A/4.70				
106Te		105.9375	0.07 ms	α/4.3	/100	0+			
107Te		106.9350	3.1 ms	α/ 70/	3.86(1)/				(0.090-0.721)
				β+, EC/10.1					
108Te		107.9294	2.1 s	α /68 /	3.314(4)/	0+			
				β+, EC/32 /6.8					
109Te		108.9274	4.6 s	β+ EC/96 /8.7					0.7523
				α/4 /	3.107(4)/				0.287–2.045
110Te		109.9224	19. s	β+, EC/4.5		0+			ann.rad./

Elem. or Isot.	Natural Abundance (Atom %)	Atomic Mass or Weight	Half–life/ Resonance Width (MeV)	Decay Mode/ Energy (/MeV)	Particle Energy/ Intensity (MeV/%)	Spin ($h/2\pi$)	Nuclear Magnetic Mom. (nm)	Elect. Quadr. Mom. (b)	γ-Energy/ Intensity (MeV/%)
									0.2191
									0.6059
^{111}Te		110.9211	19.3 s	β+, EC/8.0		(7/2+)			ann.rad./
									0.267
									0.322
									0.341
^{112}Te		111.9170	2.0 m	β+, EC/4.3		0+			ann.rad./
									0.2962
									0.3727
									0.4187
^{113}Te		112.9159	1.7 s	β+ /85/5.7	4.5/	(7/2+)			ann.rad./
				EC/15/					Sb k x-ray
									0.8144
									1.0181
									1.1812
^{114}Te		113.91209	15. m	β+ /40/3.2		0+			ann.rad./
				EC/60/					Sb k x-ray
									0.0838
									0.0903
^{115m}Te			6.7 m	β+ /45/		(1/2+)			ann.rad./
				EC/55/					Sb k x-ray
									0.7236
									0.7704
^{115}Te		114.91190	5.8 m	β+ /45/4.6	2.7/	7/2+			ann.rad./
				EC/55/					Sb k x-ray
									0.7236
									1.3268
									1.3806
									(0.22–2.7)
^{116}Te		115.90846	2.49 h	EC/1.5		0+			Sb k x-ray
									0.0937
^{117}Te		116.90865	1.03 h	EC/75/3.54	1.78/	½+			ann.rad./
				β+ /25/					Sb k x-ray
									0.9197
									1.7164
									2.3000
^{118}Te		117.90583	6.00 d	EC/0.28		0+			Sb k x-ray
^{119m}Te			4.69 d	EC/		11/2-	0.89		Sb k x-ray
									0.15360
									0.2705
									1.21271
^{119}Te		118.90640	16.0 h	β+ /2/2.293	0.627/	½+	0.25		ann.rad.
				EC/98/					Sb k x-ray
									0.6440
									0.6998
^{120}Te	0.09(1)	119.90402	1.9×10^{17} a	β$^+$ EC		0+			
^{121m}Te			~ 154. d	I.T. (89%)		11/2-	0.90		Te k x-ray
				EC (11%)					0.2122
^{121}Te		120.90494	16.8 d	EC/1.04		½+			Sb k x-ray
									0.5076
									0.5731
^{122}Te	2.55(12)	121.903044				0+			
^{123m}Te			119.7 d	I.T./0.247		11/2-	−0.93		Te k x-ray
									0.1590/84.1

Elem. or Isot.	Natural Abundance (Atom %)	Atomic Mass or Weight	Half–life/ Resonance Width (MeV)	Decay Mode/ Energy (/MeV)	Particle Energy/ Intensity (MeV/%)	Spin ($h/2\pi$)	Nuclear Magnetic Mom. (nm)	Elect. Quadr. Mom. (b)	γ-Energy/ Intensity (MeV/%)
123Te	0.89(3)	122.904270	>9.2 × 10^16 a	EC/0.051		½+	−0.736948		
124Te	4.74(14)	123.902818				0+			
125mTe			58. d	I.T./0.145		11/2-	−0.99	−0.06	Te k x-ray
									0.0355
125Te	7.07(15)	124.904431				½+	−0.8885		
126Te	18.84(25)	125.903312				0+			
127mTe			109. d	I.T./98/0.088		11/2-	−1.04	∼ 0.2	Te k x-ray
				β-/2/0.77					0.0883
127Te		126.905226	9.30 h	β-/0.698	0.696/	3/2+	0.635		0.3603
128Te	31.74(8)	127.904463	2.2 × 10^24 a	β-β-		0+			
129mTe			33.6 d	I.T./63/0.105		11/2-	−1.09	0.40	Te k x-ray
				β-/37/	1.60/				0.45984
									0.6959
129Te		128.906598	1.16 h	β-/1.498	0.99/9	3/2+	0.702	0.06	0.0278
					1.45/89				0.45984
									0.48728
130Te	34.08(62)	129.906224	8 × 10^20 a	β-β-		0+			
131mTe			1.35 d	β-/78/2.4	0.42/	11/2-	−1.04	∼ 0.25	0.0811
				I.T./22/0.18					0.1021
									0.14973
									0.77369
									0.79375
									0.85225
131Te		130.908524	25.0 m	β-/2.233	1.35/12	3/2+	0.70		0.14973
					1.69/22				0.45327
					2.14/60				0.49269
132Te		131.90855	3.26 d	β-/0.51	0.215	0+			0.049725
									0.11198
									0.22830
133mTe			55.4 m	β-/82/	2.4/30	11/2-	−1.13	0.3	Te k x-ray
				I.T./18/0.334					0.0949
									0.1689
									0.3121
									0.3341
133Te		132.91096	12.4 m	β-/2.94	2.25/25	3/2+	0.85	0.2	0.3121
					2.65				0.4079
									1.3334
134Te		133.91137	42. m	β-/1.51	0.6/	0+			0.7672/29
					0.7/				(0.079–0.926)
135Te		134.9165	19.0 s	β-/6.0	5.4/		0.7	0.3	0.267
					6.0				0.603
									0.870
136Te		135.92010	17.5 s	β-/5.1	2.5/	0+			2.0779/25
									(0.087–3.235)
137Te		136.9253	2.5 s	β-/98/6.9	6.8	7/2-			0.2436
				n/2 /					
138Te		137.9292	1.4 s	β-/6.4		0+			
139Te		138.9347	> 0.15 μs						
140Te		139.9389	> 0.15 μs			0+			
141Te		140.9447	> 0.15 μs						
142Te		141.949	> 0.15 μs			0+			
143Te									
53 I		126.90447(3)							
108I		107.9435	0.04 s	α/91/4.	3.95				

Elem. or Isot.	Natural Abundance (Atom %)	Atomic Mass or Weight	Half–life/ Resonance Width (MeV)	Decay Mode/ Energy (/MeV)	Particle Energy/ Intensity (MeV/%)	Spin ($h/2\pi$)	Nuclear Magnetic Mom. (nm)	Elect. Quadr. Mom. (b)	γ-Energy/ Intensity (MeV/%)
109I		108.9382	93.5 μs	p					0.593/100
				α	α//0.014				0.717/63
									(0.496–1.057)
110I		109.9352	0.65 s	β+, EC/83/11.4					ann.rad./
				α/17/~ 3.6	3.457(10)/				
				p/11/					
111I		110.9303	2.5 s	β+, E.C./8.5					ann.rad./
									0.2665
									0.3215
									0.3412
112I		111.9280	3.4 s	β+, EC/10.2					ann.rad./
									0.6889
									0.7869
113I		112.9236	5.9 s	β+, EC/7.6					ann.rad./
									0.4625/100
									0.6224/74
									0.0550–1.422
114I		113.9219	2.1 s	β+, EC/8.7					ann.rad./
									0.6826
									0.7088
115I		114.9181	1.3 m	β+, EC/6.7		5/2+			ann.rad./
									0.275
									0.284
									0.460
									0.709
116I		115.9168	2.9 s	β+ /97/7.8	6.7/	1+			ann.rad./
				EC/3/					0.5402
									0.6789
117I		116.91365	2.22 m	β+, EC/4.7	3.2/	(5/2+)	3.1		ann.rad./
									0.2744
									0.3259
118mI			8.5 m	β+, EC/	4.9/	7-	4.2		ann.rad./
				I.T.					0.104
									0.5998
									0.6052
									0.6138
118I		117.91307	14. m	β+, EC/7.0		2-	2.0		ann.rad./
									0.5448
									0.6052
									1.3384
119I		118.91007	19. m	β+ /54/3.5	2.4/	(5/2+)	+2.9		ann.rad./
				EC/46/					Te k x-ray
									0.2575
120mI			53. m	β+ /80/	3.8		4.2		ann.rad.
				EC/20/					Te k x-ray
									0.4257
									0.5604
									0.6147
									1.3459
120I		119.91005	1.36 h	β+ /56/5.62	4.03	2-	1.23		ann.rad./
				EC/	4.60				Te k x-ray
									0.5604
									0.6411
									1.5230

Elem. or Isot.	Natural Abundance (Atom %)	Atomic Mass or Weight	Half–life/ Resonance Width (MeV)	Decay Mode/ Energy (/MeV)	Particle Energy/ Intensity (MeV/%)	Spin ($h/2\pi$)	Nuclear Magnetic Mom. (nm)	Elect. Quadr. Mom. (b)	γ-Energy/ Intensity (MeV/%)
									(0.111–3.1)
121I		120.90737	2.12 h	β+ /13/2.27	1.2/	5/2+	2.3		ann.rad./
				EC/87/					Te k x-ray
									0.2122
									(0.14–1.1)
122I		121.90759	3.6 m	β+ /4.234	3.1/	1+	+0.94		ann.rad./
				EC/					Te k x-ray
									0.5641
123I		122.905589	13.2 h	EC/1.242		5/2+	2.82		Te k x-ray
									0.1590
124I		123.906210	4.18 d	β+ /23/3.160	1.54/	2-	1.446		ann.rad./
				EC/77/	2.14/				Te k x-ray
					0.75/				0.6027/62.9
									0.7228/10.3
									1.6910/11.2
									(0.31–1.73)
125I		124.904630	59.4 d	EC/0.1861		5/2+	2.82	−0.78	Te k x-ray
									0.0355
126I		125.905624	13.0 d	EC/		2-	1.438		ann.rad./
				β+ /2.155	1.13/				Te k x-ray
				β- /1.258/47	0.87/				0.3887
					1.25/				0.6622
127I	100.	126.904473				5/2+	+2.8133	−0.71	
128I		127.905809	25.00 m	β- /2.118	2.13/	1+			Te k x-ray
				EC/1.251					0.44287
									0.52658
129I		128.904988	1.7 × 10⁷ a	β- /0.194	0.15/	7/2+	+2.6210	−0.50	Xe k x-ray
									0.0396
130mI			9.0 m	I.T./83/0.048		2+			I k x-ray
				β- /17/					0.5361
130I		129.906674	12.36 h	β- /2.949	1.04/	5+	3.35		0.4180
					0.62				0.5361
									0.6685
									0.7395
131I		130.906125	8.021 d	β- /0.971	0.606/	7/2+	+2.742	−0.35	0.08017
									0.28431
									0.36446
									0.63699
132mI			1.39 h	IT		8-			
132I		131.90800	2.283 h	β- /14/3.58	0.80/	4+	3.09	0.08	I k x-ray
				I.T./86/	1.03/				0.0980
					1.2/				0.5059
					1.6/				0.52264
					2.16/				0.63019
									0.6506
									0.66768
									0.77260
									0.95457
133mI			9. s	I.T./1.63		19/2-			I kx-ray
									0.0730
									0.6474
									0.9126
133I		132.907797	20.8 h	β- /1.77	1.24/85	7/2+	+2.86	−0.24	0.51056
									0.52989

Elem. or Isot.	Natural Abundance (Atom %)	Atomic Mass or Weight	Half–life/ Resonance Width (MeV)	Decay Mode/ Energy (/MeV)	Particle Energy/ Intensity (MeV/%)	Spin ($h/2\pi$)	Nuclear Magnetic Mom. (nm)	Elect. Quadr. Mom. (b)	γ-Energy/ Intensity (MeV/%)
									0.87537
^{134m}I			3.7 m	I.T./98/0.316		8-			I k x-ray
				β- /2/					0.0444
									0.2719
^{134}I		133.90974	52.6 m	β- /4.05	1.2/	4+			0.1354
									0.84702
									0.88409
^{135}I		134.91005	6.57 h	β- /2.63	0.9/	7/2+	2.940		0.2884
					1.3/				0.41768
									0.52658
									1.13156
									1.26046
^{136m}I			47. s	β- /	4.7/	6-			0.1973
					5.2/				0.3468
									0.3701
									0.3814
									1.3130
									(0.16–2.36)
^{136}I		135.91465	1.39 m	β- /6.93	4.3/	2-			0.3447
					5.6/				1.3130
									1.3211
									2.2896
									(0.3–6.1)
^{137}I		136.91787	24.5 s	β- /5.88	5.0/	(7/2+)			0.6010
									1.2180
									1.2201
									1.3026
									1.5343
									(0.25–4.4)
^{138}I		137.9224	6.5 s	β- /7.8	6.9/	2-			0.4836
					7.4/				0.5888
									0.8752
									(0.4–5.3)
^{139}I		138.92610	2.30 s	β- /6.81					0.192
				n/					0.198
									0.273
									0.382
									0.386
									0.468
									0.683
									1.313
^{140}I		139.9310	0.86 s	β- /8.8		(3)			0.372
				n/					0.377
									0.457
^{141}I		140.9350	0.45 s	β- /7.8					
^{142}I		141.9402	~ 0.2 s	β-					
^{143}I		142.9446	> 0.15 μs						
^{144}I		143.9500	> 0.15 μs						
^{145}I									
$_{54}$Xe		131.293(6)							
^{109}Xe			13. ms	α	α/3.92				
					α/4.06				
^{110}Xe		109.9443	0.11 s	β+ /9.2		0+			
				α	/~ 64				

Elem. or Isot.	Natural Abundance (Atom %)	Atomic Mass or Weight	Half–life/ Resonance Width (MeV)	Decay Mode/ Energy (/MeV)	Particle Energy/ Intensity (MeV/%)	Spin ($h/2\pi$)	Nuclear Magnetic Mom. (nm)	Elect. Quadr. Mom. (b)	γ-Energy/ Intensity (MeV/%)
[111m]Xe			0.9 s	EC, β+					
[111]Xe		110.9416	0.7 s	EC, β+ /10.6					
				α/	3.58(1)/				
[112]Xe		111.9356	3. s	EC, β+ /7.2	α/0.8/	0+			
[113]Xe		112.9333	2.8 s	EC, β+ /9.1					
[114]Xe		113.92798	10.0 s	β+, EC/5.9		0+			ann.rad./
									0.1031
									0.1616
									0.3085
									0.6826
									0.7088
[115]Xe		114.92629	18. s	β+, EC/7.6		(5/2+)			ann.rad./
[116]Xe		115.92158	56. s	β+, EC/4.3	3.3/	0+			ann.rad./
									0.1042
									0.1916
									0.2477
									0.3107
									0.4127
[117]Xe		116.92036	1.02 m	β+, EC/6.5		(5/2+)	−0.594	+1.16	ann.rad./
									0.2214
									0.5190
									0.6389
									0.6613
[118]Xe		117.91618	~ 4. m	β+, EC/3.	2.7/	0+			ann.rad./
									0.0535
									0.0600
									0.1199
[119]Xe		118.91541	5.8 m	β+, EC/5.0	3.5/	7/2+	−0.654	+1.31	0.0873
									0.1000
									0.2318
									0.4615
[120]Xe		119.91178	46. m	β+, EC/97/1.96		0+			I k x-ray
				β+ /3/					0.0251
									0.0726
									0.1781
									(0.1–1.03)
[121]Xe		120.91146	39. m	β+ /44/3.73	2.8/	5/2+	−0.701	+1.33	ann.rad./
				EC/56/					I k x-ray
									0.1328
									0.2527
									0.4452
									(0.1–3.1)
[122]Xe		121.90837	20.1 h	EC/0.9		0+			I k x-ray
									0.3501
[123]Xe		122.90848	2.00 h	β+ /23/2.68	1.51/	½+	−0.150		ann.rad./
				EC/77/					I k x-ray
									0.1489
									0.1781
									(0.1–2.1)
[124]Xe	0.0952(3)	123.905893	> 10[17] a	β-β-		0+			
[125m]Xe			57. s	I.T./0.252		(9/2-)	−0.745	+0.42	xe k x-ray
									0.1111
									0.141
[125]Xe		124.906395	17.1 h	EC/1.653	0.47/	½+	−0.269		I k x-ray

Elem. or Isot.	Natural Abundance (Atom %)	Atomic Mass or Weight	Half–life/ Resonance Width (MeV)	Decay Mode/ Energy (/MeV)	Particle Energy/ Intensity (MeV/%)	Spin ($h/2\pi$)	Nuclear Magnetic Mom. (nm)	Elect. Quadr. Mom. (b)	γ-Energy/ Intensity (MeV/%)
									0.1884
									0.2434
^{126}Xe	0.0890(2)	125.90427				0+			
^{127m}Xe			1.15 m	I.T./0.297		(9/2-)	−0.884	+0.69	xe k x-ray
									0.1246
									0.1725
^{127}Xe		126.905184	36.34 d	EC/0.662		½+	−0.504		I k x-ray
									0.1721
									0.2029
									0.3750
^{128}Xe	1.9102(8)	127.903531				0+			
^{129m}Xe			8.89 d	I.T./0.236		11/2-	−0.89122	+0.64	xe k x-ray
									0.0396
									0.1966
^{129}Xe	26.4006(82)	128.904779				½+	−0.77798		
^{130}Xe	4.0710(13)	129.903508				0+			
^{131m}Xe			11.9 d	I.T./0.164		11/2-	−0.99405	+0.73	xe k x-ray
									0.16398
^{131}Xe	21.2324(30)	130.905082				3/2+	+0.69186	−0.12	
^{132}Xe	26.9086(33)	131.904153				0+			
^{133m}Xe			2.19 d	I.T./0.233		11/2-	−1.082	+0.77	xe k x-ray
									0.23325
^{133}Xe		132.905911	5.243 d	β- /0.427	0.346/99	3/2+	+0.813	+0.14	Cs k x-ray
									0.080998
									0.1606
^{134}Xe	10.4357(21)	133.905394	>1.1 × 10^{16} a	β- β-		0+			
^{135m}Xe			15.3 m	I.T./		11/2-	1.1030	+0.62	xe k x-ray
									0.52658
^{135}Xe		134.907227	9.10 h	β-/1.15	0.91/	3/2+	0.903	+0.21	0.24975
									0.60807
^{136}Xe	8.8573(44)	135.90722	>8.5 × 10^{21} a	β-β-		0+			
^{137}Xe		136.91156	3.82 m	β- /4.17	4.1/	7/2-	−0.97	−0.48	0.45549
					3.6/				0.8489
									0.9822
									1.2732
									1.7834
									2.8498
^{138}Xe		137.91395	14.1 m	β- /2.77	0.8/	0+			0.1538
					2.4/				0.2426
									0.2583
									0.4345
									1.76826
									2.0158
^{139}Xe		138.91879	39.7 s	β- /5.06	4.5/		−0.30	+0.40	0.1750
					5.0/				0.2186
									0.2965
									(0.1–3.37)
^{140}Xe		139.9216	13.6 s	β- /4.1	2.6	0+			0.0801
									0.6220
									0.8055
									1.4137
									(0.04–2.3)
^{141}Xe		140.9266	1.72 s	β- /6.2	6.2/	5/2+	+0.010	−0.58	0.1187
									0.9095

Elem. or Isot.	Natural Abundance (Atom %)	Atomic Mass or Weight	Half–life/ Resonance Width (MeV)	Decay Mode/ Energy (/MeV)	Particle Energy/ Intensity (MeV/%)	Spin ($h/2\pi$)	Nuclear Magnetic Mom. (nm)	Elect. Quadr. Mom. (b)	γ-Energy/ Intensity (MeV/%)
									(0.05–2.55)
^{142}Xe		141.9297	1.22 s	β- /5.0	3.7/	0+			0.0338
					4.2/				0.0729
									0.2038
									0.3091
									0.4145
									0.5382
									0.5718
									0.6181
									0.6448
^{143m}Xe			0.96 s	β-					
^{143}Xe		142.9351	0.30 s	β- /7.3			−0.460	+0.93	
^{144}Xe		143.9385	1.2 s	β- /6.1		0+			
^{145}Xe		144.9441	0.9 s	β-, (n)					
^{146}Xe		145.9478	> 0.15 μs			0+			
^{147}Xe		146.9536	> 0.15 μs						
^{148}Xe									
$_{55}$Cs		132.9054519(2)							
^{112}Cs		111.9503	0.5 ms	p	0.81				
^{113}Cs		112.9445	17. μs	p	0.96				
^{114}Cs		113.9414	0.58 s	β+, EC/11.8		1+			ann.rad./
									0.6826
									0.7088
^{115}Cs		114.9359	~ 1.4 s	β+, EC/8.4					ann.rad./
^{116m}Cs			0.7 s	β, EC/					ann.rad./
									0.3935
^{116}Cs		115.9334	3.8 s	β+, EC/10.8					ann.rad./
									0.3935
									0.5243
									0.6151
									0.6223
^{117m}Cs			6.5 s	β+, EC/					
^{117}Cs		116.9287	~ 8.4 s	β+, EC/7.5					ann.rad./
^{118m}Cs			17. s	β+, EC/			5.		
^{118}Cs		117.92656	14. s	β+, EC/9.		2	+3.88	+1.4	ann.rad./
									0.3372
									0.4727
									0.5865
									0.5906
^{119m}Cs			29. s			3/2	+0.84	+0.9	
^{119}Cs		118.92238	43. s	β+, EC/6.3		9/2+	+5.46	+2.8	ann.rad./
									0.169
									0.176
									0.224
									0.257
^{120m}Cs			60. s	β+, EC/					
^{120}Cs		119.92068	64. s	β+, EC/7.92		2+	+3.87	+1.45	ann.rad./
									0.3224
									0.4735
									0.5534
									(0.3–3.28)
^{121m}Cs			2.0 m	I.T./60/		(9/2+)	+5.41	+2.7	ann.rad./
				β+ /40/	4.4				0.1794
									0.1961

Elem. or Isot.	Natural Abundance (Atom %)	Atomic Mass or Weight	Half–life/ Resonance Width (MeV)	Decay Mode/ Energy (/MeV)	Particle Energy/ Intensity (MeV/%)	Spin ($h/2\pi$)	Nuclear Magnetic Mom. (nm)	Elect. Quadr. Mom. (b)	γ-Energy/ Intensity (MeV/%)
121Cs		120.91723	2.3 m	β+, EC/5.40	4.38/	3/2+	+0.77	+0.84	ann.rad./
									0.1537
									(0.08–0.56)
122m2Cs			4.4 m	β+, EC		8-	+4.77	+3.3	ann.rad./
122m1Cs			0.36 s	IT					0.3311
									0.4971
									0.6385
									(0.27–2.22)
122Cs		121.91611	21. s	β+, EC/7.1	5.8/	(1+)	−0.133	−0.19	ann.rad./
									0.3311
									0.5120
									0.8179
123mCs			1.6 s	I.T./		11/2-			Cs k x-ray
									0.0946
123Cs		122.91300	5.87 m	β+ /75/4.20	3.0/	1/2+	+1.38		ann.rad./
				EC/25/					Xe k x-ray
									0.0974
									0.5964
124mCs			6.3 s	IT		7+			
124Cs		123.91226	30. s	β+ /9 /5.92	~ 5.	1+	+0.673	−0.74	ann.rad./
				EC/8 /					Xe k x-ray
									0.3539
									0.4925
									0.9418
125Cs		124.90973	45. m	β+ /40/3.09	2.06/	1/2+	+1.41		ann.rad./
				EC/60/					Xe k x-ray
									0.112
									0.526
126Cs		125.90945	1.64 m.	β+ /81/4.83	3.4	1+	+0.78	−0.68	ann.rad./
				EC/19/	3.7/				Xe k x-ray
									0.3886
									0.4912
									0.9252
127Cs		126.90742	6.2 h	β+ /96/2.08	0.65/	1/2+	+1.46		Xe k x-ray
				EC/4/	1.06				0.1247
									0.4119
128Cs		127.90775	3.62 m	β+ /68/3.930	2.44/	1+	+0.97	−0.57	ann.rad./
				EC/32 /	2.88/				Xe k x-ray
									0.4429
129Cs		128.90606	1.336 d	EC/1.195		1/2+	+1.49		Xe k x-ray
									0.3719
									0.4115
130mCs			3.5 m	IT, β+, EC		5-	+0.629	+1.45	
130Cs		129.90671	29.21 m	β+ /55/2.98	1.98/	1+	+1.46	−0.06	ann.rad./
				EC/43/					Xe k x-ray
				β- /1.6/0.37	0.44/1.6				0.5361
131Cs		130.90546	9.69 d	EC/0.352		5/2+	+3.543	−0.58	Xe k x-ray
132Cs		131.906434	6.48 d	EC/98/		2-	+2.22	+0.51	Xe k x-ray
				β+ /0.3/2.120					0.4646
				β- / /1.280					0.6302
									0.66769
133Cs	100.	132.90545193				7/2+	+2.58291	−0.00355	
134mCs			2.91 h	I.T./0.139		8-	+1.0978	+1.0	Cs k x-ray
									0.12749

Elem. or Isot.	Natural Abundance (Atom %)	Atomic Mass or Weight	Half–life/ Resonance Width (MeV)	Decay Mode/ Energy (/MeV)	Particle Energy/ Intensity (MeV/%)	Spin ($h/2\pi$)	Nuclear Magnetic Mom. (nm)	Elect. Quadr. Mom. (b)	γ-Energy/ Intensity (MeV/%)
[134]Cs		133.90671848	2.065 a	β- /2.059	0.089/27	4+	+2.994	+0.389	0.56327
					0.658/70				0.56935
			EC/1.22						0.60473
									0.79584
[135m]Cs			53. m	I.T./1.627		19/2-	+2.18	+0.9	0.7869
									0.8402
[135]Cs		134.905977	2.3 × 10⁶ a	β- /0.269	0.205/100	7/2+	+2.7324	+0.050	
[136m]Cs			19. s	I.T./		8-	+1.32	+0.7	
[136]Cs		135.907312	13.16 d	β- /2.548	0.341/	5+	+3.71	+0.23	0.06691
									0.34057
									0.81850
									1.04807
[137]Cs		136.907089	30.2 a	β- /1.176	0.514/95	7/2+	+2.851	+0.051	Ba k x-ray
									0.66164
[138m]Cs			2.9 m	I.T./75	/0.080	6-	+1.71	−0.40	Cs k x-ray
				β- /25 /	3.3				0.0799
									0.1917
									0.4628
									1.43579
[138]Cs		137.91102	32.2 m	β- /5.37	2.9/	3-	+0.700	+0.12	0.1381
									0.46269
									1.00969
									1.43579
									2.21788
[139]Cs		138.913364	9.3 m	β- /4.213	4.21	7/2+	+2.70	−0.07	0.6272
									1.2832
									(0.4–3.66)
[140]Cs		139.91728	1.06 m	β- /6.22	5.7/	1-	+0.133895	−0.11	0.5283
					6.21/				0.6023
									0.9084
									(0.41–3.94)
[141]Cs		140.92005	24.9 s	β- /5.26	5.20/	7/2+	+2.44	−0.4	Ba k x-ray
									0.0485
									0.5616
									0.5887
									1.1940
									(0.05–3.33)
[142]Cs		141.92430	1.8 s	β- /7.31	6.9/				0.3596
					7.28				0.9668
									1.1759
									1.3265
[143]Cs		142.92735	1.78 s	β- /6.24	6.1	(3/2+)	+0.87	+0.47	0.1955
									0.2324
									0.3064
									(0.17–1.98)
[144]Cs		143.93208	1.01 s	β- /8.47	8.46/	1	−0.546	+0.30	0.1993
					7.9/				0.5598
									0.6392
									0.7587
[145]Cs		144.93553	0.59 s	β- /7.89	7.4/	3/2+	+0.784	+0.6	0.1126
					7.9/				0.1755
									0.1990
[146]Cs		145.9403	0.322 s	β-, (n)/9.38	~ 9.0	2-	−0.515	+0.22	
[147]Cs		146.9442	0.227 s	β-, (n)/9.3					(0.024-2.2798)

Elem. or Isot.	Natural Abundance (Atom %)	Atomic Mass or Weight	Half–life/ Resonance Width (MeV)	Decay Mode/ Energy (/MeV)	Particle Energy/ Intensity (MeV/%)	Spin ($h/2\pi$)	Nuclear Magnetic Mom. (nm)	Elect. Quadr. Mom. (b)	γ-Energy/ Intensity (MeV/%)
^{148}Cs		147.9492	0.15 s	β-, (n)/10.5					
^{149}Cs		148.9529	> 50 ms						
^{150}Cs		149.9582	> 50 ms						
^{151}Cs		150.9622	> 50 ms						
$_{56}$Ba		137.327(7)							
^{114}Ba		113.9507	0.43 s	β+, (p)	p/20	0+			
				α	/0.9				
^{115}Ba		114.947	0.45 s	β+, (p)	p/<15				
^{116}Ba		115.9414	1.3 s	β+, (p)	p/3	0+			
^{117}Ba		116.9385	1.8 s	β+, (p), EC/8.4	p/16	(3/2-)			(0.046–0.364)
^{118}Ba		117.9330	5.2 s	β+,		0+			(0.040–0.156)
^{119}Ba		118.9307	5.4 s	β+, EC/8.					
^{120}Ba		119.9260	24. s	β+, EC/5.0		0+			ann.rad./
									0.140
									(0.075–0.146)
^{121}Ba		120.9241	30. s	β+, EC/6.8		5/2+	+0.660	+1.8	ann.rad./
^{122}Ba		121.91990	2.0 m	β+, EC/3.8		0+			ann.rad./
^{123}Ba		122.91878	2.7 m	β+, EC/5.5			−0.680	+1.5	ann.rad./
									0.0306
									0.0927
									0.1161
									0.1235
^{124}Ba		123.91509	12. m	β+, EC/2.65		0+			ann.rad./
									0.1695
									0.1888
									1.2160
^{125m}Ba			8. m	β+, EC/	4.5		0.174		
^{125}Ba		124.9145	3.5 m	β+, EC/4.6	3.4	½+	+0.18		ann.rad./
									0.0550
									0.0776
									0.0854
									0.1409
^{126}Ba		125.91125	1.65 h	β+ /2/1.67		0+			Cs k x-ray
				EC/98 /					0.2179
									0.2336
									0.2576
^{127m}Ba			1.9 s	IT		7/2-	−0.723	+1.6	
^{127}Ba		126.91109	12.9 m	β+ /54/3.5		1/2+	+0.083		ann.rad./
				EC/46/					Cs k x-ray
									0.1148
									0.1808
									(0.07–2.5)
^{128}Ba		127.90832	2.43 d	EC/0.52		0+			Cs k x-ray
									0.27344
^{129m}Ba			2.17 h	EC/98/		7/2+	+0.93	+1.6	Cs k x-ray
				β+ /2/					0.1769
									0.1823
									0.2023
									1.4593
^{129}Ba		128.90868	2.2 h	β+ /20/2.43	1.42/	1/2+	−0.40		ann.rad./
				EC/80/					Cs k x-ray
									0.1291
									0.2143
									0.2208

Elem. or Isot.	Natural Abundance (Atom %)	Atomic Mass or Weight	Half–life/ Resonance Width (MeV)	Decay Mode/ Energy (/MeV)	Particle Energy/ Intensity (MeV/%)	Spin ($h/2\pi$)	Nuclear Magnetic Mom. (nm)	Elect. Quadr. Mom. (b)	γ-Energy/ Intensity (MeV/%)
[130m]Ba			9.5 ms	I.T./2.475	/100.	8-	−0.04	+2.8	0.080–0.802
[130]Ba	0.106(1)	129.906321	2.2×10^{21} a	β+β+		0+			
[131m]Ba			14.6 m	I.T./0.187		9/2-	−0.87	+1.5	Ba k x-ray
									0.1085
[131]Ba		130.906941	11.7 d	EC/1.37		1/2+	0.70811		Cs k x-ray
									0.12381/28.4
									0.21608/21.3
									0.49636/42.9
									(0.055–1.171)
[132]Ba	0.101(1)	131.905061	1.3×10^{21} a	EC EC		0+			
[133m]Ba			1.621 d	I.T./0.288		11/2-	−0.91	+0.9	Ba k x-ray
									0.2761
[133]Ba		132.906008	10.53 a	EC/0.517		1/2+	0.77167		Cs k x-ray
									0.08099
									0.35600
[134]Ba	2.417(18)	133.904508				0+			
[135m]Ba			1.20 d	I.T./0.2682		11/2-	−1.00	+1.0	Ba k x-ray
									0.2682
[135]Ba	6.592(12)	134.9056886				3/2+	+0.83863	+0.160	
[136m]Ba			0.308 s	I.T./2.0305		7-			Ba k x-ray
									0.8185
									1.0481
[136]Ba	7.854(24)	135.9045759				0+			
[137m]Ba			2.552 m	I.T./0.6617		11/2-	−0.99	+0.8	Ba k x-ray
									0.66164
[137]Ba	11.232(24)	136.9058274				3/2+	+0.93737	+0.245	
[138]Ba	71.698(42)	137.9052472				0+			
[139]Ba		138.9088412	1.396 h	β- /2.317	2.14/27	7/2-	−0.97	−0.57	0.16585
					2.27/72				1.2544
									1.42033
[140]Ba		139.91060	12.75 d	β- /1.05	0.48	0+			0.16268
					1.0/				0.30485
					1.02/				0.53727
[141]Ba		140.91441	18.3 m	β- /3.22	2.59/	3/2-	−0.34	+0.45	0.1903
					2.73/				0.2770
									0.3042
									(0.1–2.5)
[142]Ba		141.91645	10.7 m	β- /2.212	1.0/	0+			0.23152
					1.10/				0.25512
									0.3090
									1.2040
[143]Ba		142.92063	14.3 s	β- /4.24	4.2/	5/2+	+0.44	−0.88	0.1786
									0.21148
									0.7988
									(0.17–2.4)
[144]Ba		143.92295	11.4 s	β- /3.1	2.4/	0+			La k x-ray
					2.9/				0.10386
									0.1566
									0.1728
									0.3882
									0.43048
[145]Ba		144.9276	4.0 s	β- /4.9	4.9/	(5/2-)	−0.28	+1.22	La k x-ray
									0.0918
									0.09709

Elem. or Isot.	Natural Abundance (Atom %)	Atomic Mass or Weight	Half–life/ Resonance Width (MeV)	Decay Mode/ Energy (/MeV)	Particle Energy/ Intensity (MeV/%)	Spin (h/2π)	Nuclear Magnetic Mom. (nm)	Elect. Quadr. Mom. (b)	γ-Energy/ Intensity (MeV/%)
146Ba		145.9302	2.20 s	β- /4.12	3.9/	0+			0.0644
									0.2513
									0.3270
									0.3329
									0.3622
147Ba		146.9349	0.892 s	β- /5.75	5.5/				
148Ba		147.9377	0.64 s	β-, n/5.11		0+			
149Ba		148.9426	0.36 s	β-, (n)/7.3					
150Ba		149.9457	0.3 s			0+			
151Ba		150.9508	> 0.15 µs						
152Ba		151.9543				0+			
153Ba		152.960							
57La		138.90547(7)							
117La		116.9501	23 ms	p	0.806/	3/2+			
118La		117.9467							
119La		118.9410							
120La		119.9381	2.8 s	EC, β+ /11.					
121La		120.9330	5.3 s						
122La		121.9307	9. s	EC, β+/~ 9.7					
123La		122.9262	17. s	EC/7.					
124La		123.9246	30. s	EC/ ~ 8.8		(7+)			
125mLa			0.39 s						
125La		124.92082	1.2 m	β+, EC/5.6		11/2-			ann.rad./
									0.0436
									0.0676
126mLa			< 50. s						
126La		125.9195	54. s	β+, EC/7.6					ann.rad./
									0.256
									0.455
									0.117–3.853
127La		126.91638	3.8 m	β+, EC/4.7		3/2+			ann.rad./
									0.025
									0.0562
128La		127.9156	5.0 m	β+ /80/6.7		(5-)			ann.rad./
				EC/20/					Ba k x-ray
									0.2841/87
									0.4793/54
									(0.315–2.212)
129mLa			0.56 s	IT		(11/2-)			
129La		128.91269	11.6 m	β+ /58/3.72	2.42/	3/2+			ann.rad./
				EC/42/					Ba k x-ray
									0.1105
									0.2786
									(0.1–1.8)
130La		129.91237	8.7 m	β+ /78/5.6		3+			ann.rad./
				EC/22/					Ba k x-ray
									0.3573/81
									0.5506/27
									(0.1965–1.989)
131La		130.91007	59. m	β+ /76/3.0	1.42/	3/2+			ann.rad./
				EC/24/	1.94/				Ba k x-ray
									0.1085
									0.3658
									0.5263

Elem. or Isot.	Natural Abundance (Atom %)	Atomic Mass or Weight	Half–life/ Resonance Width (MeV)	Decay Mode/ Energy (/MeV)	Particle Energy/ Intensity (MeV/%)	Spin ($h/2\pi$)	Nuclear Magnetic Mom. (nm)	Elect. Quadr. Mom. (b)	γ-Energy/ Intensity (MeV/%)
[132m]La			24. m	I.T./76/		6-			La k x-ray
				β+, EC/24/					0.1352
									0.4645
[132]La		131.91010	4.8 h	β+ /40/4.71	2.6/	2-			ann.rad./
				EC/60/	3.2				Ba k x-ray
					3.7/				0.4645
									0.5671
[133]La		132.90822	3.91 h	β+ /4/2.2	1.2/	5/2+			Ba k x-ray
				EC/96/					0.2788
									0.2901
									0.3024
[134]La		133.90851	6.5 m	β+ /63/3.71	2.67/	1+			ann.rad./
				EC/37/					Ba k x-ray
									0.6047
									(0.5–1.9)
[135]La		134.90698	19.5 h	EC/1.20		5/2+			Ba k x-ray
									0.4805
[136]La		135.9076	9.87 m	β+ /36/2.9	1.8/	1+	+3.7	~ −0.4	ann.rad./
				EC/64/					Ba k x-ray
									0.8185
[137]La		136.90649	6 × 10⁴ a	EC/0.60		7/2+	+2.70	+0.21	0.2836
[138]La	0.08881(71)	137.907112	1.06 × 10¹¹ a			5+	+3.71365	+0.44	1.4358/65
									0.7887/35
[139]La	99.91119(71)	138.906353				7/2+	+2.783046	+0.20	
[140]La		139.909478	1.678 d	β- /3.762	1.35	3-	+0.73	+0.09	
					1.24/				
					1.67/				
[141]La		140.910962	3.90 h	β- /2.502	2.43/	7/2+			
[142]La		141.91408	1.54 h	β- /4.505	2.11/	2-			
					2.98/				
					4.52/				
[143]La		142.91606	14.1 m	β- /3.43	3.3/	7/2-			
[144]La		143.91960	40.7 s	β- /5.5	4.1/				
[145]La		144.9216	24. s	β- /4.1	4.1/	3/2+			
[146m]La			10.0 s	β- /6.7	5.5/	(6)			
[146]La		145.9258	6.3 s	β- /6.6	6.2/	(2-)			
[147]La		146.9282	4.02 s	β- /5.0	4.6/				
[148]La		147.9322	1.1 s	β- /7.26		2-			
[149]La		148.9347	1.10 s	β- /5.5					0.1335
									0.009–1.709
[150]La		149.9388	0.51 s						x-ray
									(0.097–0.209)
[151]La		150.9417	> 0.15 μs						
[152]La		151.9462	> 0.15 μs						
[153]La		152.950	> 0.15 μs						
[154]La		153.955							
[155]La		154.958							
₅₈Ce		140.116(1)							
[119]Ce		118.953							
[120]Ce		119.947				0+			
[121]Ce		120.943	1.1 s	β+, p					
[122]Ce		121.9379				0+			
[123]Ce		122.9354	3.8 s	β+, EC/~ 8.6					ann.rad./
[124]Ce		123.9304	6. s	EC/~ 5.6		0+			

Elem. or Isot.	Natural Abundance (Atom %)	Atomic Mass or Weight	Half–life/ Resonance Width (MeV)	Decay Mode/ Energy (/MeV)	Particle Energy/ Intensity (MeV/%)	Spin ($h/2\pi$)	Nuclear Magnetic Mom. (nm)	Elect. Quadr. Mom. (b)	γ-Energy/ Intensity (MeV/%)
^{125m}Ce			2. m						
^{125}Ce		124.9284	9.6 s	β+, EC/7.		7/2-			ann.rad./
									0.1346
									0.1666
									0.056–1.329
^{126}Ce		125.92397	50. s	EC/4.		0+			
^{127}Ce		126.9227	29. s	β+, EC/6.1					ann.rad./
									(0.058–1.961)
^{128}Ce		127.91891	4.1 m	β+, EC/3.2		0+			ann.rad./
									(0.023–0.880)
^{129}Ce		128.91810	3.5 m	β+, EC/5.6					ann.rad./
									(0.0675–1.015)
^{130}Ce		129.91474	26. m	β+, EC/2.2		0+			ann.rad./
									La k x-ray
									0.047–1.431
^{131m}Ce			5. m	β+ EC					ann.rad./
									0.2304
									0.3955
									0.4213
^{131}Ce		130.91442	10. m	β+, EC/4.0	2.8/				ann.rad.
									0.119
									0.169
									0.414
^{132m}Ce			9.4 ms	IT/2.340					0.3255
									0.10–0.955
^{132}Ce		131.91146	3.5 h	EC/1.3		0+			La k x-ray
									0.1554
									0.1821
^{133m}Ce			1.6 h	β+, EC/		½+			ann.rad.
									0.0769
									0.0973
									0.5577
^{133}Ce		132.91152	5.4 h	β+/8/2.9	1.3/	9/2-			ann.rad.
				EC/92/					0.0584
									0.1308
									0.4722
									0.5104
^{134}Ce		133.90892	3.16 d	EC/0.5		0+			La k x-ray
									0.1304
									0.1623
									0.6047
^{135m}Ce			20. s	I.T./0.446		11/2-			Ce k x-ray
									0.0826
									0.1497
									0.2134
^{135}Ce		134.90915	17.7 h	β+/1 /2.026	0.8/	1/2+			La k x-ray
				EC/99 /					0.0345
									0.2656
									0.3001
									0.6068
^{136}Ce	0.185(2)	135.90717	>0.7 × 10^{14} a	EC EC		0+			
			> 4.2 × 10^{15} a	β- β-					
^{137m}Ce			1.43 d	I.T./99 /0.254		11/2-	1.0		Ce k x-ray
				EC/0.8 /					0.1693

Elem. or Isot.	Natural Abundance (Atom %)	Atomic Mass or Weight	Half–life/ Resonance Width (MeV)	Decay Mode/ Energy (/MeV)	Particle Energy/ Intensity (MeV/%)	Spin ($h/2\pi$)	Nuclear Magnetic Mom. (nm)	Elect. Quadr. Mom. (b)	γ-Energy/ Intensity (MeV/%)
									0.2543
^{137}Ce		136.90781	9.0 h	β+/1.222		3/2+	0.96		La k x-ray
									0.4472
^{138}Ce	0.251(2)	137.90599	>3.7 × 10^{14} a	EC EC		0+			
^{139m}Ce			56.4 s	I.T./0.7542		11/2-			Ce k x-ray
									0.7542
^{139}Ce		138.90665	137.6 d	EC/0.28		3/2+	1.06		La k x-ray
									1.320/72.1
									0.255/59.6
									0.825/45.8
									(0.231-2.364)
^{140}Ce	88.450(51)	139.905439				0+			
^{141}Ce		140.908276	32.50 d	β-/0.581	0.436/69	7/2-	1.1		Pr k x-ray
					0.581/31				0.14544/48.0
^{142}Ce	11.114(51)	141.909244	>1.6 × 10^{17} a	β- β-		0+			
^{143}Ce		142.912386	1.38 d	β-/1.462	1.404/	3/2-	0.43		Pr k x-ray
					1.110/47				0.0574
									0.2933
^{144}Ce		143.913647	284.6 d	β-/0.319	0.185/20	0+			Pr k x-ray
					0.318/				0.0801
									0.1335
^{145}Ce		144.91723	3.00 m	β-/2.54	1.7/24	3/2-			Pr k x-ray
					1.3				0.0627
									0.7245
^{146}Ce		145.9188	13.5 m	β-/1.04	0.7/90	0+			Pr k x-ray
									0.0986
									0.2182
									0.3167
									0.0930
^{147}Ce		146.92267	56. s	β-/3.29	3.3/				0.2687
^{148}Ce		147.92443	56. s	β-/2.1	1.66/	0+			0.0904
									0.0985
									0.1212
									0.2918
^{149}Ce		148.9284	5.2 s	β-/4.2					0.0577
									0.0864
									0.3800
									0.1099
^{150}Ce		149.93041	4.4 s	β-/3.0		0+			0.0526
^{151}Ce		150.9340	1.8 s	β-/5.3					Pr k x-ray
									(0.035-0.637)
^{152}Ce		151.9365	1.4 s	β-/4.4		0+			0.098
									0.115
^{153}Ce		152.9406	> 0.15 μs						
^{154}Ce		153.9434	> 0.15 μs			0+			
^{155}Ce		154.948	> 0.15 μs						
^{156}Ce		155.951				0+			
^{157}Ce		156.956							
$_{59}$Pr		140.90765(2)							
^{121}Pr		120.955	0.01 s	p	p/0.882				
^{122}Pr		121.9518							
^{123}Pr		122.946							
^{124}Pr		123.943	1.2 s	β+, EC/12.					ann.rad./
^{125}Pr		124.9378	~ 3.3 s	β+					ann.rad./

Elem. or Isot.	Natural Abundance (Atom %)	Atomic Mass or Weight	Half–life/ Resonance Width (MeV)	Decay Mode/ Energy (/MeV)	Particle Energy/ Intensity (MeV/%)	Spin ($h/2\pi$)	Nuclear Magnetic Mom. (nm)	Elect. Quadr. Mom. (b)	γ-Energy/ Intensity (MeV/%)
									0.1358
^{126}Pr		125.9353	3.1 s	β+, EC/~ 10.4					ann.rad./
									(0.170–0.985)
^{127}Pr		126.9308	4.2 s	β+ /~ 7.5					ann.rad./
									(0.028–0.8949)
^{128}Pr		127.92879	3.0 s	β+, EC/~ 9.3					ann.rad./
									0.207/100
									0.400–1.373
^{129}Pr		128.92510	32 s	β+, EC/5.8					ann.rad./
									(0.0395–1.865)
^{130}Pr		129.9236	40. s	β+, EC/8.1					ann.rad./
^{131m}Pr			5.7 s						(0.06–0.16)
^{131}Pr		130.9203	1.7 m	β+, EC/5.3				~ 5.5	ann.rad./
									(0.059–0.980)
^{132}Pr		131.9193	1.6 m	β+, EC/7.1					ann.rad./
									0.325
									0.496
									0.533
^{133m}Pr			1.1 s	IT/0.192					0.1305
									0.0617
^{133}Pr		132.91633	6.5 m	β+, EC/4.3		5/2+			ann.rad./
									0.074
									0.1343
									0.2419
									0.3156
									0.3308
									0.4650
^{134m}Pr			~ 11. m	β+, EC/					ann.rad./
									0.294
									0.460
									0.495
									0.632
^{134}Pr		133.91571	17. m	β+, EC/6.2		2+			ann.rad./
									0.294
									0.495
^{135}Pr		134.91311	24. m	β+, EC/3.7	2.5/	3/2+			ann.rad./
									0.0826
									0.2135
									0.2961
									0.5832
^{136}Pr		135.91269	13.1 m	β+ /57 /5.13	2.98/	2+			ann.rad./
				EC/43					Ce k x-ray
									0.5398
									0.5522
^{137}Pr		136.91071	1.28 h	β+ /26 /2.70	1.68/	5/2+			ann.rad./
				EC/74 /					Ce k x-ray
									0.4339
									0.5140
									0.8367
									(0.16–1.8)
^{138m}Pr			2.1 h	β+ /24 /	1.65/	7-			ann.rad./
				EC/76 /					Ce k x-ray
									0.3027
									0.7887

Elem. or Isot.	Natural Abundance (Atom %)	Atomic Mass or Weight	Half–life/ Resonance Width (MeV)	Decay Mode/ Energy (/MeV)	Particle Energy/ Intensity (MeV/%)	Spin ($h/2\pi$)	Nuclear Magnetic Mom. (nm)	Elect. Quadr. Mom. (b)	γ-Energy/ Intensity (MeV/%)
									1.0378
									(0.07–2.0)
^{138}Pr		137.91075	1.45 m	β+ /75 /4.44	3.42/	1+			ann.rad./
				EC/25 /					Ce k x-ray
									0.7887
^{139}Pr		138.90894	4.41 h	β+ /8 /2.129	1.09/	5/2+			ann.rad./
				EC/92 /					Ce k x-ray
									0.2551
									1.3473
									1.6307
^{140}Pr		139.90908	3.39 m	β+ /51 /3.39	2.37/	1+			ann.rad./
				EC/49 /					Ce k x-ray
									0.3069
									1.5965
^{141}Pr	100.	140.907653				5/2+	+4.275	−0.08	
^{142m}Pr			14.6 m	I.T./0.004	c.e.	5-	2.2		
^{142}Pr		141.910045	19.12 h	β- /2.162	0.58/4	2-	+0.234	+0.03	0.5088
				EC/0.744	2.16/96				1.57580
^{143}Pr		142.910817	13.57 d	β- /0.934	0.933/	7/2+	+2.70	+0.8	0.7420
^{144m}Pr			7.2 m	IT/99+/0.059		3-			Pr k x-ray
				β- /					0.0590
									0.6965
									0.8142
^{144}Pr		143.913305	17.28 m	β- /2.998	0.807/1	0-			0.69649
					2.30/				1.48912
					2.996/98				2.18562
^{145}Pr		144.91451	5.98 h	β- /1.81	1.80/97	7/2+			0.0725
									0.6758
									0.7483
^{146}Pr		145.9176	24.2 m	β- /4.2	2.2/30	2-			0.4539/48
					3.7/10				1.5247
					4.2/40				
^{147}Pr		146.91900	13.4 m	β- /2.69	1.5/	3/2+			0.3146/24.
					2.1/				0.5779/16
									0.6413/19.
^{148m}Pr			2.0 m	β- /	4.0/	(4)			0.3016
					3.8/				0.4506
				IT	0.77/36				0.6975
^{148}Pr		147.92213	2.27 m	β- /4.9	4.8/	1⁻			0.3017
					4.5/				
^{149}Pr		148.9237	2.3 m	β- /3.40	3.0	(5/2+)			0.1085
									0.1385
									0.1651
^{150}Pr		149.92667	6.2 s	β- /5.7		1⁻			0.1302
					~ 5.5				0.8044
									0.8527
^{151}Pr		150.92832	22.4 s	β- /4.2					
^{152}Pr		151.9315	3.2 s	β- /6.7		4⁺			0.0726
									0.164
									0.285
^{153}Pr		152.9338	4.3 s	β- /5.5					
^{154}Pr		153.9375	2.3 s	β- /7.9					
^{155}Pr		154.9401	> 0.3 µs						
^{156}Pr		155.9443	> 0.3 µs						

Elem. or Isot.	Natural Abundance (Atom %)	Atomic Mass or Weight	Half–life/ Resonance Width (MeV)	Decay Mode/ Energy (/MeV)	Particle Energy/ Intensity (MeV/%)	Spin ($h/2\pi$)	Nuclear Magnetic Mom. (nm)	Elect. Quadr. Mom. (b)	γ-Energy/ Intensity (MeV/%)
^{157}Pr		156.9474							
^{158}Pr		157.952							
^{159}Pr		158.956							
$_{60}$Nd		144.242(3)							
^{124}Nd		123.952				0+			
^{125}Nd		124.9489	0.6 s	β+, p					
^{126}Nd		125.9432				0+			
^{127}Nd		126.9405	1.8 s	β+, EC/9.		(5/2)			ann.rad./
^{128}Nd		127.9354	4. s	β+, EC/6.					ann.rad./
^{129m}Nd			2.6 s						0.134,0.399
^{129}Nd		128.9332	4.9 s	β+, EC/8.		5/2(-)			ann.rad./
									(0.091–0.875)
^{130}Nd		129.92851	28. s	β+, EC/5.		0+			ann.rad./
^{131}Nd		130.92725	0.5 m	β+, EC/6.6					ann.rad./
									(0.09–0.36)
^{132}Nd		131.92332	1.5 m	β+, EC/3.7		0+			ann.rad./
									(0.099–0.567)
^{133}Nd		132.92235	1.2 m	β+, EC/5.6					ann.rad./
									(0.06–0.37)
^{134}Nd		133.91879	~ 8.5 m	β+ /17 /2.8		0+			ann.rad./
				EC/83 /					Pr k x-ray
									0.1631/58
									(0.09–1.00)
^{135m}Nd			5.5 m	β+ /					
^{135}Nd		134.91818	12. m	β+ /65 /4.8		9/2-	−0.78	+1.9	ann.rad./
				EC/35 /					Pr k x-ray
									0.0415/23.
									0.204/51.
									(0.11–1.8)
^{136}Nd		135.91498	50.6 m	EC/94 /2.21	1.04/	0+			Pr k x-ray
				β+ /6 /					0.0401/21.
									0.1091/35.
									(0.10–0.97)
^{137m}Nd			1.6 s	I.T./0.5196		11/2-			Nd k x-ray
									0.1084
									0.1775
									0.2337
^{137}Nd		136.91457	38. m	β+ /40 /3.69	1.7/20	1/2+	−0.63		ann.rad./
				EC/60 /	2.40/20				Pr k x-ray
									0.0755
									0.5806
^{138}Nd		137.91195	5.1 h	EC/1.1		0+			Pr k x-ray
									0.1995
									0.3258
^{139m}Nd			5.5 h	I.T./12 /0.231	1.17/	11/2-			Nd k x-ray
				β+ /88 /					Pr k x-ray
									0.1139/34.
									0.7382/30.
^{139}Nd		138.91198	30. m	β+ /25 /2.79	1.77/	3/2+	+0.91	+0.3	ann.rad./
				EC/75 /					Pr k x-ray
									0.4050
^{140}Nd		139.90955	3.37 d	EC /0.22		0+			Pr k x-ray
^{141m}Nd			1.04 m	IT/99+/0.756		11/2-			Nd k x-ray
									0.7565

Elem. or Isot.	Natural Abundance (Atom %)	Atomic Mass or Weight	Half–life/ Resonance Width (MeV)	Decay Mode/ Energy (/MeV)	Particle Energy/ Intensity (MeV/%)	Spin ($h/2\pi$)	Nuclear Magnetic Mom. (nm)	Elect. Quadr. Mom. (b)	γ-Energy/ Intensity (MeV/%)
141Nd		140.909610	2.49 h	EC/98 /1.823	0.802/	3/2+	+1.01	+0.3	Pr k x-ray
				β+ /2 /					(0.15–1.7)
142Nd	27.152(40)	141.907723				0+			
143Nd	12.174(26)	142.909814				7/2-	−1.07	−0.60	
144Nd	23.798(19)	143.910087	2.1×10^{15} a	α	1.83	0+			
145Nd	8.293(12)	144.912574				7/2-	−0.66	−0.31	
146Nd	17.189(32)	145.913117				0+			
147Nd		146.916100	10.98 d	β- /0.896	0.805/	5/2-	0.58	0.9	Pr k x-ray
									0.53102
									0.09111–0.686
148Nd	5.756(21)	147.916893	10^{20} a	β-β-		0+			
149Nd		148.920149	1.73 h	β- /1.691	1.03/25	5/2-	0.35	1.3	Pr k x-ray
					1.13/26				0.1143/19.
					1.42/				0.2113/27.
									(0.026–1.6)
150Nd	5.638(28)	149.920891	1.33×10^{20} a	β-β-		0+			
151Nd		150.923829	12.4 m	β- /2.442	1.2/	(3/2+)			Pm k x-ray
									0.1168
									0.2557
									1.1806
									(0.10–1.9)m
152Nd		151.92468	11.4 m	β- /1.1		0+			0.2785/29.
									0.2501/18.
									(0.016–0.66)
153Nd		152.92770	28.9 s	β- /3.6					0.418
154Nd		153.9295	25.9 s	β- /2.8		0+			0.1519
									0.7998
155Nd		154.9329	8.9 s	β- /5.0					0.1807
156Nd		155.9350	5.5 s	β- /4.1		0+			0.0848
157Nd		156.9390	> 0.3 µs						
158Nd		157.9416	> 0.3 µs			0+			
159Nd		158.946							
160Nd		159.949				0+			
161Nd		160.954							
61Pm									
128Pm		127.9484	1.0 s	β+, p					ann.rad.
129Pm		128.9432	~ 2.4 s						
130Pm		129.9405	2.5 s	β+, EC/11.					0.1589
									0.326–1.062
131Pm		130.9359	~ 6.3 s	β+					0.185
									0.220
									0.146
132Pm		131.9338	6. s	β+, EC/10.					ann.rad./
133Pm		132.92978	12. s	β+, EC/~ 7.0					ann.rad./
134Pm		133.9284	24. s	β+, EC/~ 8.9		(5+)			ann.rad./
									0.294
									0.495
135Pm		134.9249	0.8 m	β+, EC/6.0		11/2-			(0.13–0.47)
136Pm		135.9236	1.8 m	β+ /89 /7.9		(3+)			ann.rad./
				EC/11 /					Nd k x-ray
									0.3735
									0.6027
137Pm		136.92048	2.4 m	β+, EC/5.6		(11/2-)			ann.rad./
									0.1086

Elem. or Isot.	Natural Abundance (Atom %)	Atomic Mass or Weight	Half–life/ Resonance Width (MeV)	Decay Mode/ Energy (/MeV)	Particle Energy/ Intensity (MeV/%)	Spin ($h/2\pi$)	Nuclear Magnetic Mom. (nm)	Elect. Quadr. Mom. (b)	γ-Energy/ Intensity (MeV/%)
									0.1775
138mPm			3.2 m	β+ /50 /~ 7.0	3.9/	3+	3.		ann. rad./
				EC/50 /					Nd k x-ray
									0.5209
									0.7290
138Pm		137.91955	10. s	β+ /6.9	6.1/	1+			ann.rad./
139mPm			0.18 s	IT/		(11/2-)			0.1887
139Pm		138.91680	4.14 m	β+ /68 /4.52	3.52/	(5/2+)			ann.rad./
				EC/32 /					Nd k x-ray
									0.4028
140mPm			5.87 m	β+ /70 /	3.2	7/2-			(0.27–2.4)
				EC/30 /					ann.rad./
									Nd k x-ray
									0.4199
									0.7738
									1.0283
140Pm		139.91604	9.2 s	β+ /89 /6.09	5.07/74	1+			ann.rad./
				EC/11 /					Nd k x-ray
									0.7738
									1.4898
141Pm		140.91356	20.9 m	β+ /52 /3.72	2.71	5/2+			ann.rad./
				EC/48 /					Nd k x-ray
									0.8862
									1.2233
142mPm			67 μs						(0.208-0.882)
142Pm		141.91287	40.5 s	β+ /86 /4.87	3.8/	1+			ann.rad./
				EC/20 /					Nd k x-ray
									0.6414
									1.5758
143Pm		142.910933	265. d	EC/1.041		5/2+	3.8		Nd k x-ray
				β+ /< 6 × 10⁻⁶/					0.7420
144Pm		143.912591	360. d	EC/2.332		5-	1.7		Nd k x-ray
				β+ /7 × 10⁻⁶/					0.6180
									0.6965
145Pm		144.912749	17.7 a	EC/0.163		5/2+	+3.8	+0.2	Nd k x-ray
									0.0723
146Pm		145.914696	5.53 a	EC/63 /1.472		3-			Nd k x-ray
				β- /37 /1.542	0.795/				0.4538
									0.7362
									0.7474
147Pm		146.915139	2.623 a	β- /0.224	0.224/	7/2+	+2.6	+0.6	0.1213
									0.1974
148mPm			41.3 d	β- /95 /2.6	0.4/60	6-	1.8		0.5503/94.
				I.T./5 /0.137	0.5/17				0.6300/89.
					0.7/21				0.7257/33
148Pm		147.91748	5.37 d	β- /2.47	1.02/	1-	+2.0	~ +0.2	0.5503
					2.47/				0.9149
									1.4651
149Pm		148.918334	2.212 d	β- /1.071	0.78/9	7/2+	3.3		0.2859
					1.072/90				0.5909
									0.8594
150Pm		149.92098	2.68 h	β- /3.45	1.6/	(1-)			0.3339/69.
					2.3/				1.1658/16.
					1.8/				1.3245/17.

Elem. or Isot.	Natural Abundance (Atom %)	Atomic Mass or Weight	Half–life/ Resonance Width (MeV)	Decay Mode/ Energy (/MeV)	Particle Energy/ Intensity (MeV/%)	Spin ($h/2\pi$)	Nuclear Magnetic Mom. (nm)	Elect. Quadr. Mom. (b)	γ-Energy/ Intensity (MeV/%)
									(0.25–2.9)
[151]Pm		150.92121	1.183 d	β- /1.187	0.84/	5/2+	+1.8	1.9	0.1677/8
									0.2751/7
									0.3401/22
[152m2]Pm			15. m	β-, I.T./		(>6)			(0.14–1.4)
[152m1]Pm			7.5 m	β- /		(4-)			0.1218
									0.2447
									0.3404
									1.0971
									1.4375
[152]Pm		151.92350	4.1 m	β- /3.5	3.5/20	1+			0.1218
					3.50/60				(0.12–2.1)
[153]Pm		152.92412	5.4 m	β- /1.90	1.7/	(5/2-)			0.0910
									0.1198
									0.1273
[154m]Pm			2.7 m	β- /	2.0/				0.0820
									0.1848
									1.4403
[154]Pm		153.92646	1.7 m	β- /4.1	1.9/				0.0820
									0.8396
									1.3940
									2.0589
									(0.08–2.8)
[155]Pm		154.92810	48. s	β- /3.2		(5/2-)			(0.05–0.78)
[156]Pm		155.93106	26.7 s	β- /5.16					
[157]Pm		156.9330	10.9 s	β- /4.6					
[158]Pm		157.9366	5. s	β- /6.3					
[159]Pm		158.9390	1.5 s						(0.072-0.261)
[160]Pm		159.9430							
[161]Pm		160.9459							
[162]Pm		161.950							
[163]Pm		162.954							
[62]Sm		150.36(2)							
[129]Sm		128.954	~ 0.55 s	β+, p					
[130]Sm		129.9489				0+			
[131]Sm		130.9461	1.2 s	β+, EC/					ann.rad./
[132]Sm		131.9407	4.0 s	β+		0+			
[133]Sm		132.9387	2.9 s	β+, EC/~ 8.4		5/2+			ann.rad./
									0.3696
									0.0845
[134]Sm		133.9340	11. s	β+, EC/5.		0+			ann.rad./
[135]Sm		134.9325	10. s	β+, EC/7.		7/2+			ann.rad./
[136]Sm		135.92828	42. s	β+, EC/4.5		0+			ann.rad./
[137]Sm		136.92697	45. s	β+, EC/6.1					ann.rad./
[138]Sm		137.92324	3.0 m	β+, EC/3.9		0+			ann.rad./
									0.0536
									0.0747
[139m]Sm			10. s	I.T./94 /0.457		(11/2-)	1.1		Sm k x-ray
				β+ /6 /	4.7				0.1118
									0.1553
									0.1901
									0.2673
[139]Sm		138.92230	2.6 m	β+ /75 /5.5	4.1/	½+	−0.53		Pm k x-ray
				EC/25 /					0.3678

Elem. or Isot.	Natural Abundance (Atom %)	Atomic Mass or Weight	Half–life/ Resonance Width (MeV)	Decay Mode/ Energy (/MeV)	Particle Energy/ Intensity (MeV/%)	Spin ($h/2\pi$)	Nuclear Magnetic Mom. (nm)	Elect. Quadr. Mom. (b)	γ-Energy/ Intensity (MeV/%)
									0.4028
									(0.27–2.4)
140Sm		139.91900	14.8 m	β+, EC/3.4	1.9/	0+			ann.rad./
									Pm k x-ray
									0.1396
									0.2255
									(0.07–1.7)
141mSm			22.6 m	β+ /32 /	1.6/	11/2-	−0.84	+1.6	ann.rad./
				EC/68 /	2.19/				Pm k x-ray
				I.T./0.3 /0.1758					0.1966
									0.4318
									0.7774
141Sm		140.91848	10.2 m	β+ /52 /4.54	3.2/	½+	−0.74		ann.rad./
				EC/48 /					Pm k x-ray
									0.4382
142Sm		141.91520	1.208 h	β+ /6 /2.10	1.0/	0+			ann.rad./
				EC/94 /					Pm k x-ray
143mSm			1.10 m	IT/99/0.7540		11/2-			Sm k x-ray
									0.7540
143Sm		142.914628	8.83 m	β+ /46 /3.443	2.47/	3/2+	+1.01	+0.4	ann.rad./
				EC/54 /					Pm k x-ray
									1.0565
144Sm	3.07(7)	143.911999				0+			
145Sm		144.913410	340. d	EC/0.617		7/2-	−1.12	−0.6	Pm k x-ray
									0.0613
									0.4924
146Sm		145.913041	1.03×10^8 a	α/	2.50/	0+			
147Sm	14.99(18)	146.914898	1.06×10^{11} a	α/	2.23/	7/2-	−0.815	−0.26	
148Sm	11.24(10)	147.914823	7×10^{15} a	α/	1.96/	0+			
149Sm	13.82(7)	148.917185	10^{16} a	α/		7/2-	−0.672	+0.075	
150Sm	7.38(1)	149.917276				0+			
151Sm		150.919932	96. a	β- /0.0768	0.076/	5/2-	−0.363	+0.7	0.02154
152Sm	26.75(16)	151.919732				0+			
153Sm		152.922097	1.930 d	β- /0.808	0.64/	3/2+	−0.0216	+1.3	Eu k x-ray
					0.69/				0.0697/4.7
									0.10318/28
									0.075–0.714
154Sm	22.75(29)	153.922209				0+			
155Sm		154.924640	22.2 m	β- /1.627	1.52	3/2-		1.1	Eu k x-ray
									0.1043/75.
156Sm		155.92553	9.4 h	β- /0.72	0.43/	0+			0.0872
					0.71/				0.1657
									0.2038
157Sm		156.92836	8.0 m	β- /2.7	2.4/	3/2-			Eu k x-ray
									0.1964
									0.1978
									0.3942
158Sm		157.9300	5.5 m	β- /2.0		0+			0.1894/100.
									0.3636/82.
159Sm		158.9332	11.3 s	β- /3.8					0.1898
160Sm		159.9351	9.6 s	β- /3.6		0+			0.110
161Sm		160.9388	~ 4.8 s						0.264
162Sm		161.941	2.4 s			0+			(0.036–0.741)
163Sm		162.945							

Elem. or Isot.	Natural Abundance (Atom %)	Atomic Mass or Weight	Half–life/ Resonance Width (MeV)	Decay Mode/ Energy (/MeV)	Particle Energy/ Intensity (MeV/%)	Spin ($h/2\pi$)	Nuclear Magnetic Mom. (nm)	Elect. Quadr. Mom. (b)	γ-Energy/ Intensity (MeV/%)
^{164}Sm		163.948				0+			
^{165}Sm		164.953							
$_{63}$Eu		**151.964(1)**							
^{130}Eu		129.964	0.9 ms	p	1.027/				
^{131}Eu		130.9578	~ 26. ms	β+, p	p/0.95				
^{132}Eu		131.9544							
^{133}Eu		132.9492							
^{134}Eu		133.9465	0.5 s	EC, β+					ann.rad./
^{135}Eu		134.9418	1.5 s	EC, β+ /~ 8.7					ann.rad./
^{136m}Eu			~ 3.2 s			7+			0.255
^{136}Eu		135.9396	~ 3.9 s	EC, β+ /10.		1+			ann.rad./
^{137}Eu		136.9356	11. s	EC/~ 7.5		11/2-			ann.rad./
^{138}Eu		137.93371	12. s	EC, β+ /~ 9.2		7+	5		ann.rad./
^{139}Eu		138.92979	18. s	EC, β+ /6.7			6		ann.rad./
^{140m}Eu			0.125 s	EC, β+					ann.rad./
^{140}Eu		139.9281	1.51 s	EC, β+ /8.4		1-	+1.37	+0.31	ann.rad./
^{141m}Eu			3.0 s	β+ /58 /		11/2-			ann.rad./
				EC/9 /					Eu k x-ray
				I.T./33 /0.0964					(0.09–1.6)
^{141}Eu		140.92493	40. s	β+ /81 /5.6		5/2+	+3.49	+0.85	ann.rad./
				EC/15 /					Sm k x-ray
									0.3845
									0.3940
^{142m}Eu			1.22 m	β+ /83 /	4.8/	8-	+2.98	+1.4	ann.rad./
				EC/17 /					Sm k x-ray
									0.5566
									0.7680
									1.0233
^{142}Eu		141.92343	2.4 s	β- /94/7.4	7.0/	1+	+1.54	+0.12	ann.rad./
				EC/6 /					0.7680
^{143}Eu		142.92030	2.62 m	β+ /72/5.17	4.1/	5/2+	+3.67	+0.51	ann.rad./
				EC/28/	5.1/				Sm k x-ray
									0.1107/7
									1.5368/3.
									1.9127/2.
^{144}Eu		143.91882	10.2 s	β+ /86 /6.33	5.31/	1+	+1.89	+0.10	ann.rad./
				EC/13 /					Sm k x-ray
									1.6601
^{145}Eu		144.916265	5.93 d	β+ /2 /2.660	0.79/	5/2+	+4.00	+0.29	ann.rad./
				EC/98 /1.71					Sm k x-ray
									0.6535
									0.8937
									1.6587
^{146}Eu		145.91721	4.57 d	β+ /5 /3.88	1.47/	4-	+1.42	–0.18	ann.rad./
				EC/95 /					Sm k x-ray
									0.6336
									0.6341
									0.7470
									(0.27–2.64)
^{147}Eu		146.916746	24.4 d	EC/99. /1.722		5/2+	+3.73	+0.53	Sm k x-ray
				β+ /0.4 /					0.12113/20.6
									0.19725/24.0
									(0.601-1.077)
^{148}Eu		147.91809	54.5 d	EC/3.11	0.92	5-	+2.34	+0.35	Sm k x-ray

Elem. or Isot.	Natural Abundance (Atom %)	Atomic Mass or Weight	Half–life/ Resonance Width (MeV)	Decay Mode/ Energy (/MeV)	Particle Energy/ Intensity (MeV/%)	Spin (h/2π)	Nuclear Magnetic Mom. (nm)	Elect. Quadr. Mom. (b)	γ-Energy/ Intensity (MeV/%)
									0.5503/99.
									0.6299/71.
									(0.067–2.17)
^{149}Eu		148.917931	93.1 d	EC/0.692		5/2+	+3.57	+0.75	Sm k x-ray
									0.2770/4.1
									0.3275/4.8
^{150}Eu		149.91970	36. a	EC/2.26		5-	+2.71	+1.13	Sm k x-ray
									0.3340
									0.4394
									0.5843
									(0.25–1.8)
^{150m}Eu			12.8 h	β- /92 /	1.013/	0-			Sm k x-ray
				β+ /0.4 /	1.24/				0.3339
				EC/8 /					0.4065
^{151}Eu	47.81(6)	150.919850	>1.7x10^{18} a			5/2+	+3.472	+0.90	
^{152m2}Eu			1.60 h	I.T./0.1478		8-			Eu k x-ray
									0.0898
^{152m1}Eu			9.30 h	β- /72 /	1.85/	0-			Sm k x-ray
				EC/28 /	0.89/				0.12178
									0.84153
									0.96334
^{152}Eu		151.921745	13.5 a	EC/72 /1.874	0.69/	3-	−1.941	+2.71	Sm k x-ray
				β- /28 /1.818	1.47/				Gd k x-ray
									0.12178
									0.34427
									1.40802
									(0.252–1.528)
^{153}Eu	52.19(6)	152.921230				5/2+	+1.533	+2.41	
^{154m}Eu			46.1 m	I.T./~ 0.16		8-			Eu k x-ray
									0.0682
									0.1009
^{154}Eu		153.922979	8.59 a	β- /99.9/1.969	0.27/29	3-	−2.01	+2.8	Gd k x-ray
				EC/0.02/0.717	0.58/38				0.12299/40.
					0.84/17				0.72331/20.
					0.98/4				1.2745/36
					1.87/11				(0.059-1.90)
^{155}Eu		154.922893	4.76 a	β- /0.252	0.15/	5/2+	+1.520	+2.50	Gd k x-ray
									0.0865/30
									0.1053/20
^{156}Eu		155.92475	15.2 d	β- /2.451	0.30/11	1+	≈1.1		0.08899/9.
					0.49/30				0.64623/7.
					1.2/12				0.723441/6.
					2.45/31				0.8118/10.
^{157}Eu		156.92542	15.13 h	β- /1.36	0.98/	(5/2+)	+1.50	+2.6	Gd k x-ray
					1.30/41				0.0639/100.
									0.3705/48.
									0.4107/76.
^{158}Eu		157.9279	45.9 m	β- /3.5	2.5/	(1-)	+1.44	+0.7	0.0795
									0.8976
									0.9442
									0.9771
^{159}Eu		158.92909	18.1 m	β- /2.51	2.4/	(5/2+)	+1.38	+2.7	0.0678
					2.57/				0.0786
									0.0957

Elem. or Isot.	Natural Abundance (Atom %)	Atomic Mass or Weight	Half–life/ Resonance Width (MeV)	Decay Mode/ Energy (/MeV)	Particle Energy/ Intensity (MeV/%)	Spin ($h/2\pi$)	Nuclear Magnetic Mom. (nm)	Elect. Quadr. Mom. (b)	γ-Energy/ Intensity (MeV/%)
^{160}Eu		159.9320	38. s	β- /4.1	2.7/	(0-)			0.0753
					4.1/				0.1735
									0.4131
									0.5155
									0.8217
									0.9110
									0.9246
^{161}Eu		160.9337	27. s	β- /3.7					0.0719
^{162}Eu		161.9370	11. s	β- /5.6					
^{163}Eu		162.9392							
^{164}Eu		163.943							
^{165}Eu		164.946							
^{166}Eu		165.950							
^{167}Eu		166.953							
$_{64}$Gd		**157.25(3)**							
^{135}Gd		134.953	1.1 s	β+					(0.163−0.360)
^{136}Gd		135.9473				0+			
^{137}Gd		136.9450	7. s	EC, β+ /~ 8.8					ann.rad./
^{138}Gd		137.9401	~ 4.7 s	EC, β+		0+			0.0647
^{139m}Gd			~ 4.8 s						0.1216
^{139}Gd		138.9382	5. s	EC, β+ /~ 7.7					0.104−0.323
^{140}Gd		139.93367	16. s	EC/4.8		0+			0.1748
^{141m}Gd			25. s	EC, β+ /		11/2-			ann.rad./
^{141}Gd		140.93213	21. s	β+ /7.3		½+			ann.rad./
^{142}Gd		141.92812	1.17 m	EC, β+ /4.2		0+			ann.rad./
^{143m}Gd			1.84 m	β+ /67 /		11/2-			ann.rad./
				EC/33 /					Eu k x-ray
				I.T./					0.1176
									0.2719
									0.5880
									0.6681
									0.7999
^{143}Gd		142.9268	39. s	β+ /82 /6.0		1/2+			ann.rad./
				EC/18 /					Eu k x-ray
									0.2048
									0.2588
^{144}Gd		143.92296	4.5 m	β+ /45 /4.3	3.3/	0+			ann.rad./
				EC/55 /					Eu k x-ray
									0.3332
^{145m}Gd			1.44 m	I.T./95 /0.749		11/2-	−1.0		0.0273
				β+ /4 /5.7					0.3295
									0.3866
									0.7214
^{145}Gd		144.92171	23.4 m	β+ /33 /5.05	2.5/	1/2+	−0.74		ann.rad./
				EC/67 /					Eu k x-ray
									1.7579
									1.8806
									(0.32−3.69)
^{146}Gd		145.918311	48.3 d	EC/99.9 /1.03	0.35/	0+			Eu k x-ray
				β+ /0.2					0.1147
									0.1155
									0.1546
^{147}Gd		146.919094	1.588 d	EC/99.8 /2.188	0.93/	7/2-	1.0		Eu k x-ray
				EC/0.2 /					0.2293

Elem. or Isot.	Natural Abundance (Atom %)	Atomic Mass or Weight	Half–life/ Resonance Width (MeV)	Decay Mode/ Energy (/MeV)	Particle Energy/ Intensity (MeV/%)	Spin ($h/2\pi$)	Nuclear Magnetic Mom. (nm)	Elect. Quadr. Mom. (b)	γ-Energy/ Intensity (MeV/%)
									0.3699
									0.3960
									0.9289
									(0.1–1.8)
^{148}Gd		147.918115	71. a	α/3.27	3.1828/	0+			
^{149}Gd		148.919341	9.3 d	EC/1.32		7/2-	0.9		Eu k x-ray
									0.1496
									0.2985
									0.3465
^{150}Gd		149.91866	1.8×10^6 a	α/2.80	2.73/	0+			
^{151}Gd		150.920348	124. d	EC/0.464		7/2-	0.8		Eu k x-ray
									0.1536
									0.2432
^{152}Gd	0.20(1)	151.919791				0+			
^{153}Gd		152.921750	240. d	EC/0.485		3/2-	0.4		Eu k x-ray
									0.09743
									0.10318
^{154}Gd	2.18(3)	153.920867				0+			
^{155}Gd	14.80(12)	154.922622				3/2-	−0.258	+1.30	
^{156}Gd	20.47(9)	155.922123				0+			
^{157}Gd	15.65(2)	156.923960				3/2-	−0.340	+1.36	
^{158}Gd	24.84(7)	157.924104				0+			
^{159}Gd		158.926389	18.6 h	β- 0.971	0.971/58	3/2-	−0.44		Tb k x-ray
					0.913/29				0.36351
					0.607/12				0.058-0.855
^{160}Gd	21.86(19)	159.927054	$>1.9 \times 10^{19}$ a	β- β-		0+			
^{161}Gd		160.929669	3.66 m	β- /1.956	1.56/85	5/2-			Tb k x-ray
									0.1023
									0.3149
									0.3609
^{162}Gd		161.930985	8.4 m	β- /1.39	1.0/	0+			0.4030
									0.4421
^{163}Gd		162.9340	1.13 m	β- /3.1					0.2868
									0.214
									1.685
^{164}Gd		163.9359	45. s	β- /2.3		0+			
^{165}Gd		164.9394	10 s	β-					
^{166}Gd		165.942	∼ 4.8 s			0+			(0.040-1.015)
^{167}Gd		166.946							
^{168}Gd		167.948				0+			
^{169}Gd		168.953							
$_{65}$**Tb**		**158.92535(2)**							
^{135}Tb			0.9 ms	p	p/1.179				
^{138}Tb		137.9532							
^{139}Tb		138.9483	1.6 s						0.109
									0.120
^{140}Tb		139.946	2.4 s	β+, EC/11					0.329
									0.355-0.740
^{141}Tb		140.9415	3.5 s	β+, EC/∼ 8.3					
^{142m2}Tb			25 μs						
^{142m}Tb			0.30 s	β+, EC/		4-			
^{142}Tb		141.9387	0.60 s	β+, EC/10.		0+			
^{143}Tb		142.9351	12. s	β+, EC/7.4		11/2-			
^{144m}Tb			4.1 s	IT		5-			

Elem. or Isot.	Natural Abundance (Atom %)	Atomic Mass or Weight	Half–life/ Resonance Width (MeV)	Decay Mode/ Energy (/MeV)	Particle Energy/ Intensity (MeV/%)	Spin ($h/2\pi$)	Nuclear Magnetic Mom. (nm)	Elect. Quadr. Mom. (b)	γ-Energy/ Intensity (MeV/%)
144Tb		143.93305	< 1.5 s	β+, EC/8.4		1+			
145mTb			30. s	β+, EC/~ 6.6		11/2-			ann.rad./
									0.2577
									0.5370
									0.9876
145Tb		144.9293		β+, EC/6.5		½+			
146mTb			23. s	β+ /76 /		(5-)			ann.rad./
				EC/24 /					Gd k x-ray
									1.0789
									1.5795
146Tb		145.92725	~ 8. s	β+ /8.1		1+			
147mTb			1.8 m	β+ /35 /		11/2-			ann.rad./
				EC/65 /					Gd k x-ray
									1.3977
									1.7978
147Tb		146.92405	1.6 h	β+ /42 /4.61		5/2+	+1.70		ann.rad./
				EC/58 /					Gd k x-ray
									0.6944
									1.1522
									(0.120–3.318)
148mTb			2.3 m	β+ /25 /		9+			ann.rad./
				EC/75 /					Gd k x-ray
									0.3945
									0.6319
									0.7845
									0.8824
148Tb		147.92427	1.00 h	β+, EC/5.69		2-	–1.75	–0.3	ann.rad./
									Gd k x-ray
									0.4888
									0.7845
									(0.14–3.8)
149mTb			4.16 m	EC/88 /		11/2-			ann.rad./
				β+ /12 /					Gd k x-ray
									0.1650
									0.7960
149Tb		148.923246	4.13 h	β+ /4 /3.636	1.8/	½+	+1.35		Gd k x-ray
				α/16/	3.97/				0.1650
									0.3522
									0.3886
									(0.1–3.2)
150mTb			6.0 m	β+ /17 /					ann.rad./
				EC/83 /					Gd k x-ray
									0.4384
									0.6380
									0.6504
									0.8275
150Tb		149.92366	3.3 h	β+, EC/4.66		2-	–0.90		ann.rad./
									0.4963
									0.6380
									(0.3–4.29)
151mTb			25. s	I.T./95 /		11/2-			0.0229
				β+, EC/7 /					0.0495
									0.3797
									0.8305

Elem. or Isot.	Natural Abundance (Atom %)	Atomic Mass or Weight	Half–life/ Resonance Width (MeV)	Decay Mode/ Energy (/MeV)	Particle Energy/ Intensity (MeV/%)	Spin ($h/2\pi$)	Nuclear Magnetic Mom. (nm)	Elect. Quadr. Mom. (b)	γ-Energy/ Intensity (MeV/%)
^{151}Tb		150.923103	17.61 h	β+/1 /2.565	0.70/	1/2+	+0.92		Gd k x-ray
				EC/99 /					0.1083
									0.2517
									0.2870
									(0.1–1.8)
^{152m}Tb			4.3 m	I.T./79 /0.5018		(8+)			Tb k x-ray
				EC/21 /4.35					Gd k x-ray
									0.2833
									0.3443
									0.4111
^{152}Tb		151.92407	17.5 h	β+ /20 /3.99	2.5/	2-	−0.58	+0.3	ann.rad./
				EC/80 /	2.8/				Gd k x-ray
									0.3443
									(0.2–2.88)
^{153}Tb		152.923435	2.34 d	EC/1.570		5/2+	+3.44	+1.1	Gd k x-ray
									0.2119
									(0.05–1.1)
^{154m2}Tb			23.1 h	EC/98 /		(7-)	0.9		Gd k x-ray
				I.T./2 /					0.1231
									0.2479
									0.3467
									1.4199
^{154m1}Tb			10.0 h	β+ /78 /		(3-)	+1.7	+3.	Gd k x-ray
				I.T./22 /					0.1231
									0.2479
									0.5401
									(0.12–2.57)
^{154}Tb		153.92468	21.5 h	EC/99 /3.56	1.86/	0-			Gd k x-ray
				β+ /1 /	2.45				0.1231
									1.2744
									2.1872
									(0.12–3.14)
^{155}Tb		154.92351	5.3 d	EC/0.82		3/2+	+2.01	+1.41	Gd k x-ray
									0.08654
									0.10530
^{156m2}Tb			1.02 d	I.T./		(7-)			Tb k x-ray
									0.0496
^{156m1}Tb			5.3 h	I.T./0.0884		(0+)			Tb k x-ray
									0.0884
^{156}Tb		155.924747	5.3 d	EC/2.444		3-	~ 1.7	+2.	Gd k x-ray
									0.08896
									0.19921
									0.53435
									1.22245
^{157}Tb		156.924025	1.1 × 10² a	EC/0.0601		3/2+	+2.01	+1.4	Gd k x-ray
									0.0545
^{158m}Tb			10.5 s	I.T./0.11		0-			Gd k x-ray
									0.0110
^{158}Tb		157.925413	1.8 × 10² a	EC/80 /1.220		3-	+1.76	+2.7	Gd k x-ray
				β- /20 /0.937					0.0795
									0.9442
									0.9621
^{159}Tb	100.	158.925347				3/2+	+2.014	+1.43	
^{160}Tb		159.927168	72.3 d	β- /1.835	0.57/47	3-	+1.79	3.8	Dy k x-ray

Elem. or Isot.	Natural Abundance (Atom %)	Atomic Mass or Weight	Half–life/ Resonance Width (MeV)	Decay Mode/ Energy (/MeV)	Particle Energy/ Intensity (MeV/%)	Spin ($h/2\pi$)	Nuclear Magnetic Mom. (nm)	Elect. Quadr. Mom. (b)	γ-Energy/ Intensity (MeV/%)
					0.86/27				0.08678
									0.29857
									0.87936
									0.96615
^{161}Tb		160.927570	6.91 d	β- /0.593	0.46/23	3/2+	2.2	+1.2	Dy k x-ray
					0.52/66				0.02565
					0.6/10				0.04892
									0.07458
^{162}Tb		161.92949	7.6 m	β- /2.51	1.4	(1/2-)			Dy k x-ray
									0.2600
									0.8075
									0.8882
^{163}Tb		162.930648	19.5 m	β- /1.785	0.80/	3/2+			Dy k x-ray
									0.3511
									0.3897
									0.4945
^{164}Tb		163.9334	3.0 m	β- /3.9	1.7/	(5+)			Dy k x-ray
									0.1689
									0.2157
									0.6110
									0.6885
									0.7548
^{165}Tb		164.9349	2.1 m	β- /3.0		3/2+			0.5389
									1.1785
									1.2920
									1.6648
^{166}Tb		165.9380	26 s	β-/					
^{167}Tb		166.9401	19 s						0.057
									0.070
^{168}Tb		167.944	8 s						(0.075–0.227)
^{169}Tb		168.946							
^{170}Tb		169.950							
^{171}Tb		170.953							
$_{66}$Dy		**162.500(1)**							
^{139}Dy		138.960	0.6 s	β+, p					
^{140}Dy		139.954				0+			
^{141}Dy		140.9514	0.9 s	EC, β+ /9.					
^{142}Dy		141.9464	2.3 s	EC, β+ /7.1		0+			
^{143}Dy		142.9438	3.9 s	EC, β+ /~ 8.8					
^{144}Dy		143.93925	9.1 s	EC, β+ /~ 6.2		0+			
^{145m}Dy		144.9365	14. s	EC, β+		11/2-			
^{146m}Dy			0.15 s	I.T.		10+			
^{146}Dy		145.93285	30. s	EC, β+ /5.2		0+			
^{147m}Dy			56. s	I.T./40 /		(11/2-)	−0.66	+0.7	Dy k x-ray
				β+, EC/60 /					0.072
									0.6787
^{147}Dy		146.93109	75. s	EC, β+ /6.37		½+	−0.92		ann.rad./
									0.1007
									0.2534
									0.3653
^{148}Dy		147.92715	3.1 m	β+ /4 /2.68	1.2/	0+			ann.rad./
				EC/96 /					Tb k x-ray
									0.6202
^{149}Dy		148.92731	4.2 m	β+, EC/3.81		(7/2-)	−0.12	−0.62	ann.rad./

Elem. or Isot.	Natural Abundance (Atom %)	Atomic Mass or Weight	Half–life/ Resonance Width (MeV)	Decay Mode/ Energy (/MeV)	Particle Energy/ Intensity (MeV/%)	Spin ($h/2\pi$)	Nuclear Magnetic Mom. (nm)	Elect. Quadr. Mom. (b)	γ-Energy/ Intensity (MeV/%)
									0.1008
									0.1063
									0.2534
									0.6536
									0.7894
									1.7765
									1.8062
150Dy		149.925585	7.18 m	β+, EC/67 /1.79		0+			Tb k x-ray
				α/33 /	4.233/				0.3967
151Dy		150.926185	17. m	β+ /5 /2.871		7/2-	−0.95	−0.30	Tb k x-ray
				EC/89 /					0.1764
				α /6 /	4.067/				0.3030
									0.3861
									0.5463
									(0.16–2.09)
152Dy		151.92472	2.37 h	EC/0.60		0+			Tb k x-ray
				α /	3.63/				0.2569
153Dy		152.925765	6.3 h	β+ /1 /2.171	0.89/	(7/2-)	−0.78	~−0.15	Tb k x-ray
				EC/99 /					0.0807
				α /0.01 /	3.46/				0.0997
									0.2137
									(0.08–1.66)
154Dy		153.92442	3. × 10⁶ a	α/2.95	2.87/	0+			
155Dy		154.92575	9.9 h	β+ /2 /2.095	0.845/	3/2-	−0.385	+1.04	Tb k x-ray
				EC/98 /					0.0655
									0.2269
156Dy	0.056(3)	155.92428				0+			
157Dy		156.92547	8.1 h	EC/1.34		3/2-	−0.301	+1.30	Tb k x-ray
									(0.061–1.319)
158Dy	0.095(3)	157.924409				0+			
159Dy		158.925739	144. d	EC/0.366		3/2-	−0.354	+1.37	Tb k x-ray
									0.3262
160Dy	2.329(18)	159.925198				0+			
161Dy	18.889(42)	160.926933				5/2+	−0.480	+2.51	
162Dy	25.475(36)	161.926798				0+			
163Dy	24.896(42)	162.928731				5/2-	+0.673	+2.65	
164Dy	28.260(54)	163.929175				0+			
165mDy			1.26 m	I.T./98 /0.108		1/2-			Dy k x-ray
				β- /2 /					0.1082
									0.5155
165Dy		164.931703	2.33 h	β- /1.286	1.29/	7/2+	−0.52	−3.5	Ho k x-ray
									0.09468/3.8
166Dy		165.932807	3.400 d	β- /0.486	0.40/	0+			Ho k x-ray
									0.0282
									0.0825
167Dy		166.9357	6.2 m	β- /~ 2.35	1.78	(1/2-)			Ho k x-ray
									0.2593
									0.3103
									0.5697
									(0.06–1.4)
168Dy		167.9371	8.5 m	β- /1.6		0+			Ho k x-ray
									0.1925
									0.4867

Elem. or Isot.	Natural Abundance (Atom %)	Atomic Mass or Weight	Half–life/ Resonance Width (MeV)	Decay Mode/ Energy (/MeV)	Particle Energy/ Intensity (MeV/%)	Spin ($h/2\pi$)	Nuclear Magnetic Mom. (nm)	Elect. Quadr. Mom. (b)	γ-Energy/ Intensity (MeV/%)
^{169}Dy		168.9403	~ 39. s	β- /3.2					
^{170}Dy		169.9424				0+			
^{171}Dy		170.9462							
^{172}Dy		171.9488				0+			
^{173}Dy		172.953							
$_{67}$**Ho**		**164.93032(2)**							
^{140}Ho		139.969	6 ms	p/	p/1.09				
^{141m}Ho			8 μs	p/	p/1.23				
^{141}Ho		140.963	4.1 ms	β+, p	p/1.17/99.3				
					p/0.97/0.7				
^{142}Ho		141.960	0.4 s	EC/β+, p					0.307
^{143}Ho		142.9546	> 0.2 μs						
^{144}Ho		143.9515	0.7 s	β+, EC/12					
^{145}Ho		144.9472	2.4 s	β+					
^{146}Ho		145.9446	3.3 s	β+, EC/10.7		(10+)			ann.rad./
^{147}Ho		146.94006	5.8 s	β+, EC/8.2		11/2-			ann.rad./
^{148m}Ho			9.6 s	β+, EC/		4-			ann.rad./
^{148}Ho		147.9377	2. s	β+, EC/9.4		1+			ann.rad./
									0.6615
									1.6883
^{149m}Ho			21. s	β+, EC/		11/2-			ann.rad./
									1.0733
									1.0911
^{149}Ho		148.93378	> 30. s	β+, EC/6.01		1/2+			
^{150m}Ho			25. s	β+, EC/		(9+)			ann.rad./
									0.3939
									0.5511
									0.6534
									0.8034
^{150}Ho		149.93350	1.3 m	β+, EC/6.6					ann.rad./
									0.5913
									0.6534
									0.8034
^{151m}Ho			47. s	β+, EC/87 /					ann.rad./
				α/13	4.605/				0.2102
									0.4889
									0.6948
									0.7762
^{151}Ho		150.93169	35.2 s	β+, EC/80/5.13					ann.rad./
				α/20 /	4.519/				0.3522
									0.5274
									0.9676
									1.0471
^{152m}Ho			50. s	β+, EC/90/		(9+)	+5.9	−1.3	ann.rad./
				α/10/	4.453/				0.4929
									0.6138
									0.6474
									0.6835
^{152}Ho		151.93171	2.4 m	β+, EC/88/6.47		(3+)	−1.02	~ +0.1	ann.rad./
				α/12/	4.387/				0.6140
									0.6476
^{153m}Ho			9.3 m	β+, EC/99+/4.12		5/2	+1.19		ann.rad./
				α/	4.01/				0.0905

Elem. or Isot.	Natural Abundance (Atom %)	Atomic Mass or Weight	Half–life/ Resonance Width (MeV)	Decay Mode/ Energy (/MeV)	Particle Energy/ Intensity (MeV/%)	Spin ($h/2\pi$)	Nuclear Magnetic Mom. (nm)	Elect. Quadr. Mom. (b)	γ-Energy/ Intensity (MeV/%)
									0.1089
									0.1618
									0.2302
									0.2707
									0.3659
									0.4565
^{153}Ho		152.93020	2.0 m	β+, EC/99+/4.13		11/2-	+6.8	−1.1	ann.rad./
				α/	3.91/				0.2958
									0.3346
									0.4381
									0.6383
^{154m}Ho			3.3 m	β+, EC/		(8+)	5.7	−1.0	ann.rad./
									0.3346
									0.4124
									0.4771
^{154}Ho		153.93060	12. m	β+, EC/5.75		1-	−0.64	+0.2	ann.rad./
									Dy k x-ray
									0.3346
									0.5700
									0.8734
^{155}Ho		154.92910	48. m	β+/6/3.10		(5/2+)	+3.51	+1.5	ann.rad./
				EC/94 /					Dy k x-ray
									0.0474
									0.1363
									0.3254
									(0.06–2.24)
^{156m}Ho			5.8 m	I.T./0.0352			+2.99	+2.3	ann.rad./
				β+ /25 /	1.8/				Dy k x-ray
				EC/75 /	2.9/				0.1378
									0.2666
									(0.28–2.9)
^{156}Ho		155.92984	56. m	β+, EC/4.4		(5+)			ann.rad./
									0.1378
									0.2665
^{157}Ho		156.92826	12.6 m	β+/5/2.54	1.18/	7/2-	+4.35	+3.0	ann.rad./
				EC/95/					Dy k x-ray
									0.2800
									0.3411
^{158m2}Ho			28. m	I.T./44/		2-	+2.44	+1.6	ann.rad./
				EC/56/					Dy k x-ray
									0.0989
									0.2182
^{158m1}Ho			21. m	β+, EC/		(9+)			ann.rad./
									0.0981
									0.1664
									0.2182
									0.3205
									0.4062
									0.9774
									1.0532
									0.4846
^{158}Ho		157.92894	11.3 m	β+/8/4.24	1.30/	5+	+3.77	+4.1	ann.rad./
				EC/92/					Dy k x-ray

Elem. or Isot.	Natural Abundance (Atom %)	Atomic Mass or Weight	Half–life/ Resonance Width (MeV)	Decay Mode/ Energy (/MeV)	Particle Energy/ Intensity (MeV/%)	Spin ($h/2\pi$)	Nuclear Magnetic Mom. (nm)	Elect. Quadr. Mom. (b)	γ-Energy/ Intensity (MeV/%)
									0.0989
									0.2182
									0.9488
159mHo			8.3 s	IT/0.206		1/2+			Ho k x-ray
									0.1660
									0.2059
159Ho		158.927712	33.0 m	EC/1.838		7/2-	+4.28	+3.2	Dy k x-ray
									0.1210
									0.1320
									0.2529
									0.3096
									(0.06–1.2)
160m2Ho			3. s			1+			
160mHo			5.0 h	IT/67/0.060		2-	+2.52	+1.8	0.0868
				EC/33/3.35					0.1970
									0.6464
									0.7281
									0.8791
									0.9619
									0.9658
160Ho		159.92873	25.6 m	β+, EC/3.29	0.57/	5+	+3.71	+4.0	See Ho[166m]
									0.7282
									0.8794
161mHo			6.8 s	IT/0.211					Ho k x-ray
									0.2112
161Ho		160.927855	2.48 h	EC/0.859		7/2-	+4.25	+3.2	Dy k x-ray
									0.0256
									0.0592
									0.0774
									0.1031
162mHo			1.12 h	IT/61/		6-	+3.60	+4.	Dy k x-ray
				EC/39/					Ho k x-ray
									0.0807
									0.1850
									0.2828
									0.9372
									1.2200
162Ho		161.929096	15. m	EC/96 /0.295		1+			Dy k x-ray
				β+ /4 /					0.0807
									1.3196
									1.3728
163mHo			1.09 s	I.T./0.298		(1/2+)			Ho k x-ray
									0.2798
163Ho		162.928734	4.57×10^3 a	EC/0.00258		7/2-	+4.23	+3.6	Dy M x-ray
164mHo			36.4 m	I.T./0.140		(6-)			Ho k x-ray
									0.0373
									0.0566
									0.0940
164Ho		163.930234	29. m	EC/58 /0.987		1+			Dy k x-ray
				β- /42 /0.963					0.0734
									0.0914
165Ho	100.	164.930322				7/2-	+4.17	+3.6	
166mHo			1.2×10^3 a	β- /		7-	3.6	~ -3.	Er k x-ray
									0.18407

Elem. or Isot.	Natural Abundance (Atom %)	Atomic Mass or Weight	Half–life/ Resonance Width (MeV)	Decay Mode/ Energy (/MeV)	Particle Energy/ Intensity (MeV/%)	Spin ($h/2\pi$)	Nuclear Magnetic Mom. (nm)	Elect. Quadr. Mom. (b)	γ-Energy/ Intensity (MeV/%)
									0.71169
									0.81031
^{166}Ho		165.932284	1.117 d	β- /1.855	1.776/48	0-			Er k x-ray
					1.855/51				0.08057
									1.37943
^{167}Ho		166.93313	3.1 h	β- /1.007	0.31/43	(7/2-)			Er k x-ray
					0.61/21				0.0793
					0.96/15				0.0835
					0.97/15				0.2379
									0.3213
									0.3465
^{168m}Ho			2.2 m	I.T./					
^{168}Ho		167.93552	3.0 m	β- /2.91	2.0/	3+			Er k x-ray
									0.7413
									0.8159
									0.8211
									(0.08–2.34)
^{169}Ho		168.93687	4.7 m	β- /2.12	1.2/	(7/2-)			
					2.0/				0.1496
									0.7610
									0.7784
									0.7884
									0.8529
^{170m}Ho			43. s	β- /		1+			0.0787
									0.8123
									1.8940
									1.9726
^{170}Ho		169.93962	2.8 m	β- /3.87		6+			Er k x-ray
									0.1816
									0.2582
									0.8902
									0.9321
									0.9414
									1.1387
^{171}Ho		170.941	53 s	β- /					
^{172}Ho		171.9448	25. s	β- /					Er k x-ray
									(0.077–1.186)
^{173}Ho		172.9473							
^{174}Ho		173.951							
^{175}Ho		174.954							
$_{68}$**Er**		**167.259(3)**							
^{144}Er		143.9604	> 0.2 μs			0+			
^{145m}Er			1.0 s						0.067
^{145}Er		144.9574	0.9 s	β+					0.049
^{146}Er		145.9520	~ 1.7 s	β+		0+			
^{147m}Er			25. s						0.683
^{147}Er		146.9495	2.5 s	E.C, β+ /~ 9.1					
^{148}Er		147.9446	4.5 s	β+, EC/6.8		0+			
^{149m}Er			10. s	IT		11/2-			
^{149}Er		148.94231	10.7 s	ECβ+ /8.1		½+			
^{150}Er		149.93791	18. s	β+ /36 /4.11		0+			ann.rad./
				EC/64 /					Ho k x-ray
									0.4758
^{151}Er		150.93745	23. s	β+, EC/5.2		7/2-			ann.rad./

Elem. or Isot.	Natural Abundance (Atom %)	Atomic Mass or Weight	Half–life/ Resonance Width (MeV)	Decay Mode/ Energy (/MeV)	Particle Energy/ Intensity (MeV/%)	Spin ($h/2\pi$)	Nuclear Magnetic Mom. (nm)	Elect. Quadr. Mom. (b)	γ-Energy/ Intensity (MeV/%)
^{152}Er		151.93505	10.2 s	β+, EC/10/3.11		0+			ann.rad./
				α/90/	4.804/				
^{153}Er		152.935063	37.1 s	α/	4.674		−0.934	−0.42	0.351
				β+, EC/47/4.56	4.35/				(0.0945–1.70)
^{154}Er		153.93278	3.7 m	β+, EC/99+/2.03		0+			ann.rad./
				α/0.5/	4.166/				
^{155}Er		154.93321	5.3 m	β+, EC/47/3.84		(7/2-)	−0.669	−0.27	ann.rad./
				EC/53 /					Ho k x-ray
									0.1101
									0.2415
^{156}Er		155.93107	20. m	β+, EC/1.7		0+			ann.rad./
									0.0298
									0.0352
									0.0522
									0.1336
^{157}Er		156.93192	25. m	β+, EC/3.5		3/2-	−0.412	+0.92	ann.rad./
									0.117
									0.385
									1.320
									1.660
									1.820
									2.000
^{158}Er		157.92989	2.2 h	EC/99.5 /1.78	0.74/	0+			Ho k x-ray
				β+ /0.5 /					0.0719
									0.2486
									0.3868
^{159}Er		158.930684	36. m	β+ /7 /2.769		3/2-	−0.304	+1.17	ann.rad./
				EC/93 /					Ho k x-ray
									0.6245
									0.6493
									(0.07–2.5)
^{160}Er		159.92908	1.191 d	EC/0.33		0+			Ho k x-ray
									(0.05–0.96)
^{161}Er		160.93000	3.21 h	EC/2.00		3/2-	−0.37	+1.36	Ho k x-ray
									0.8265
									(0.07–1.74)
^{162}Er	0.139(5)	161.928778				0+			
^{163}Er		162.93003	1.25 h	EC/1.210		5/2-	+0.557	+2.55	Ho k x-ray
									0.4361
									0.4399
									1.1135
^{164}Er	1.601(3)	163.929200				0+			
^{165}Er		164.930726	10.36 h	EC/0.376		5/2-	+0.643	+2.71	Ho k x-ray
^{166}Er	33.503(36)	165.930293				0+			
^{167m}Er			2.27 s	I.T./0.208		½-			Er k x-ray
									0.2078
^{167}Er	22.869(9)	166.932048				7/2+	−0.5639	+3.57	
^{168}Er	26.978(18)	167.932370				0+			
^{169}Er		168.934590	9.40 d	β- /0.351	0.35/~ 100	½-	+0.485		Tm k x-ray
									0.1098
									0.1182
^{170}Er	14.910(36)	169.935464				0+			
^{171}Er		170.938030	7.52 h	β- /1.491		5/2-	0.66	2.9	Tm k x-ray

Elem. or Isot.	Natural Abundance (Atom %)	Atomic Mass or Weight	Half–life/ Resonance Width (MeV)	Decay Mode/ Energy (/MeV)	Particle Energy/ Intensity (MeV/%)	Spin ($h/2\pi$)	Nuclear Magnetic Mom. (nm)	Elect. Quadr. Mom. (b)	γ-Energy/ Intensity (MeV/%)
									0.11160
									0.29591
									0.30832
									(0.08–1.4)
^{172}Er		171.939356	2.05 d	β-/0.891	0.28/48	0+			Tm k x-ray
					0.36/46				0.0597
									0.4073
									0.6101
^{173}Er		172.9424	1.4 m	β- /2.6		(7/2-)			Tm k x-ray
									0.1928
									0.1992
									0.8952
^{174}Er		173.9442	3.1 m	β- /1.8		0+			Tm k x-ray
									(0.100–0.152)
^{175}Er		174.9478	1.2 m	β-					(0.0765–1.17)
^{176}Er		175.9501				0+			
^{177}Er		176.954							
$_{69}$Tm		168.93421(2)							
^{144}Tm			~ 1.9 μs	p	1.70, 1.43				
^{145}Tm		144.9701	3.1 μs	p// ~ 10	1.73/91				
					1.40/~9.6				
^{146m}Tm			0.198 s	β+, p	p/1.12/99				
					p/0.89/~1				
^{146}Tm		145.9664	73 ms	β+/14.	p/1.19/68				
				p	p/1.01/~18				
					p/0.94/~14				
^{147m}Tm			0.4 ms	β+, p	p/1.115				
^{147}Tm		146.9610	0.56 s	EC, β+/85	~ 10.7				
				p/15/	1.052/				
^{148m}Tm		147.9578	0.7 s	β+, EC/12.					ann.rad./
^{148}Tm									
^{149}Tm		148.9527	0.9 s	β+, EC/~ 9.2		11/2-			
^{150}Tm		149.9500	2.3 s	β+, EC/~ 11.5		6-			(0.101–2.177)
^{151}Tm		150.94548	4. s	β+, EC/7.5					ann.rad./
^{152m}Tm			8. s	β+, EC/		9+			
^{152}Tm		151.9444	5. s	β+, EC/8.8					ann.rad./
^{153}Tm		152.94201	1.6 s	β+,EC/10 /6.46		(11/2-)	6.9	~ +0.5	ann.rad./
				α/90 /	5.109/				
^{154m}Tm			3.3 s	β+, EC/15 /	α/5.031/100	(9+)	+5.9	~ -0.2	ann.rad./
				α/	4.84/0.24				0.4605–0.796
^{154}Tm		153.94157	8.1 s	β+, EC/56 /7.4	α/4.956/100	(2-)	-1.14	~ +0.4	ann.rad./
				α/44 /	4.83/0.45				
^{155}Tm		154.93920	30. s	β+, EC/5.58					0.0315
				α/	4.46/				0.0638
									0.0881
									0.2268
									0.5320
									0.6067
^{156m}Tm			19. s	α/	4.46/				
^{156}Tm		155.93898	1.40 m	β+, EC/7.6		2-	+0.40	-0.5	ann.rad./
				α/	4.23/				0.3446
									0.4529
									0.5860
^{157}Tm		156.93697	3.6 m	β+, EC/4.5	2.6	½+	+0.48		ann.rad./

Elem. or Isot.	Natural Abundance (Atom %)	Atomic Mass or Weight	Half–life/ Resonance Width (MeV)	Decay Mode/ Energy (/MeV)	Particle Energy/ Intensity (MeV/%)	Spin ($h/2\pi$)	Nuclear Magnetic Mom. (nm)	Elect. Quadr. Mom. (b)	γ-Energy/ Intensity (MeV/%)
				α/	3.97/				0.1104
									0.3484
									0.3855
									0.4550
									(0.1–1.58)
^{158}Tm		157.93698	4.0 m	β+, EC/74 /6.5		(2-)	+0.04	+0.7	ann.rad./
				EC/26 /					Er k x-ray
									0.1921
									0.3351
									0.6280
									1.1498
									(0.18–2.81)
^{159}Tm		158.93498	9.1 m	β+/23 /3.9		5/2+	+3.42	+1.9	ann.rad./
				EC/77 /					Er k x-ray
									0.0591
									0.0848
									0.2713
									(0.05–1.27)
^{160m}Tm			1.24 m	IT		(5)			
^{160}Tm		159.93526	9.4 m	β+/15 /5.9		1-	+0.16	+0.58	ann.rad./
				EC/85 /					Er k x-ray
									0.1264
									0.2642
									0.7285
									0.8544
									0.8614
									1.3685
^{161}Tm		160.93355	31. m	β+, EC/3.2		7/2+	+2.40	+2.9	ann.rad./
									Er k x-ray
									0.0595
									0.0844
									1.6481
									(0.04–2.15)
^{162m}Tm			24. s	I.T./90 /		5+			Tm k x-ray
				β+, EC/10 /					Er k x-ray
									0.0669
									0.8115
									0.9003
^{162}Tm		161.93400	21.7 m	β+ /8 /4.81		1-	+0.07	+0.69	ann.rad./
				EC/92 /					Er k x-ray
									0.1020
									0.7987
									(0.1–3.75)
^{163}Tm		162.93265	1.81 h	EC/98 /2.439		½+	−0.082		Er k x-ray
				β+ /1 /					0.0692
									0.1043
									0.2414
^{164m}Tm			5.1 m	I.T./80 /		6-			0.0914
				β+, EC/20 /					0.1394
									0.2081
									0.2405
									0.3149
^{164}Tm		163.93356	2.0 m	β+ /36 /3.96	2.94/	1+	+2.38	+0.71	ann.rad./
				EC/64 /					Er k x-ray

Elem. or Isot.	Natural Abundance (Atom %)	Atomic Mass or Weight	Half–life/ Resonance Width (MeV)	Decay Mode/ Energy (/MeV)	Particle Energy/ Intensity (MeV/%)	Spin ($h/2\pi$)	Nuclear Magnetic Mom. (nm)	Elect. Quadr. Mom. (b)	γ-Energy/ Intensity (MeV/%)
									0.0914
165Tm		164.932435	1.253 d	EC/1.593		½+	−0.139		Er k x-ray
									0.0472
									0.0544
									0.29728
									0.80636
166Tm		165.93355	7.70 h	EC/98 /3.04		2+	+0.092	+2.14	Er k x-ray
				β+ /2 /					0.0806
									0.1844
									0.7789
									1.2734
									2.0524
167Tm		166.932852	9.24 d	EC/0.748		½+	−0.197		Er k x-ray
									0.0571
									0.20778
168Tm		167.934173	93.1 d	EC/1.679		3+	+0.23	+3.2	Er k x-ray
									0.19825
									0.4475
									0.81595
169Tm	100	168.934213				½+	−0.231	−1.2	
170Tm		169.935801	128.6 d	β- /99.8/0.968	0.883/24	1-	+0.246	+0.74	Yb k x-ray
				EC/0.2 /0.314	0.968/76				0.08425
171Tm		170.936429	1.92 a	β- /0.096	0.03/2	½+	−0.228		0.06674
					0.096/98				
172Tm		171.93840	2.65 d	β- /1.88	1.79/36	2-			Yb k x-ray
					1.88/29				0.07879
									1.38722
									1.46601
									1.52982
									1.60861
173Tm		172.939604	8.2 h	β- /1.298	0.80/21	½+			Yb k x-ray
					0.86/71				0.3988
									0.4613
174mTm			2.29 s						
174Tm		173.94217	5.4 m	β- /3.08	0.70/14	(4-)			Yb k x-ray
					1.20/83				0.07664
									0.17669
									0.27332
									0.3666
									0.99205
									(0.08−1.6)
175Tm		174.94384	15.2 m	β- /2.39	0.9/36	(1/2+)			Yb k x-ray
					1.9/23				0.36396
									0.51487
									0.94125
									0.98247
176Tm		175.9470	1.9 m	β-/4.2	2.0/	(4+)			Yb k x-ray
					1.2/				0.1898
									0.3819
									1.0691
177Tm		176.9490	1.4 m	β-		(7/2-)			
178Tm		177.9526							
179Tm		178.955							
70Yb		173.054(5)							

Elem. or Isot.	Natural Abundance (Atom %)	Atomic Mass or Weight	Half–life/ Resonance Width (MeV)	Decay Mode/ Energy (/MeV)	Particle Energy/ Intensity (MeV/%)	Spin ($h/2\pi$)	Nuclear Magnetic Mom. (nm)	Elect. Quadr. Mom. (b)	γ-Energy/ Intensity (MeV/%)
[148]Yb		147.967				0+			
[149]Yb		148.964	0.7 s	β+, p	p/2.5–6.4/				0.647
[150]Yb		149.9584	> 0.2 μs			0+			
[151]Yb		150.9554	1.6 s	β+ /8.5		(1/2+)			
[152]Yb		151.9503	3.2 s	β+ EC/5.5		0+			
[153]Yb		152.9495	4. s	β+ EC/6.7					
[154]Yb		153.94639	0.40 s	β+ EC/7 /4.49		0+			ann.rad./
				α/93 /	5.32/				
[155]Yb		154.9458	1.7 s	β+, EC/16 /6.0		(7/2-)	−0.91	−0.5	ann.rad./
				α/84 /	5.19/				
[156]Yb		155.94282	26. s	β+, EC/21/3.57		0+			ann.rad./
				α/79 /	4.69/				
[157]Yb		156.94263	39. s	β+, EC/99+/5.5		7/2-	−0.64		ann.rad./
				α/0.5/	4.69/				0.231
									(0.035–0.670)
[158]Yb		157.93987	1.5 m	β+, EC/1.9		0+			ann.rad./
									0.0741
									0.2526
[159]Yb		158.94005	1.4 m	EC, β+/5.1		5/2-	−0.37	−0.22	Tm k x-ray
									0.1661
									0.1772
									0.3297
									0.3903
[160]Yb		159.93755	4.8 m	β+, EC/2.0		0+			ann.rad./
									0.1404
									0.1737
									0.2158
[161]Yb		160.93790	4.2 m	β+, EC/3.9		3/2-	−0.33	+1.03	ann.rad./
									Tm k x-ray
									0.0782
									0.5999
									0.6315
[162]Yb		161.93577	18.9 m	β+, EC/1.7		0+			ann.rad./
									Tm k x-ray
									0.1188
									0.1635
[163]Yb		162.93633	11.1 m	β+ /26 /3.4	1.4/	3/2-	−0.37	+1.24	ann.rad./
									Tm k x-ray
									0.0636
									0.8603
									(0.06–1.9)
[164]Yb		163.93449	1.26 h	EC/1.0		0+			Tm k x-ray
									0.0914
									0.6752
[165]Yb		164.93528	9.9 m	β+ /10 /2.76	1.58/	(5/2-)	+0.48	+2.48	ann.rad./
				EC/90 /					Tm k x-ray
									0.0801
									1.0903
[166]Yb		165.93388	2.363 d	EC/0.30		0+			Tm k x-ray
									0.0828
									0.1844
									0.7789
									1.2734
									2.0524

Elem. or Isot.	Natural Abundance (Atom %)	Atomic Mass or Weight	Half–life/ Resonance Width (MeV)	Decay Mode/ Energy (/MeV)	Particle Energy/ Intensity (MeV/%)	Spin ($h/2\pi$)	Nuclear Magnetic Mom. (nm)	Elect. Quadr. Mom. (b)	γ-Energy/ Intensity (MeV/%)
^{167}Yb		166.934950	17.5 m	β+ /0.5 /1.954	0.639/	5/2-	+0.62	+2.70	Tm k x-ray
				EC/99.5 /					0.06296
									0.10616
									0.11337
									0.17633
^{168}Yb	0.123(3)	167.933897				0+			
^{169m}Yb			46. s	I.T./0.0242		1/2-	+0.51		Yb L x-ray
									0.0242
^{169}Yb		168.935190	32.02 d	EC/0.909		7/2+	−0.63	+3.5	0.1979/35.9
									0.3078/10.05
									0.0207−0.261
^{170}Yb	2.982(39)	169.934762				0+			
^{171}Yb	14.09(14)	170.936326				1/2-	+0.49367		
^{172}Yb	21.68(13)	171.936382				0+			
^{173}Yb	16.103(63)	172.938211				5/2-	−0.67989	+2.80	
^{174}Yb	32.026(80)	173.938862				0+			
^{175}Yb		174.941277	4.19 d	β- /0.470	0.466/73	7/2-	0.77		Lu k x-ray
					0.071/21				0.3963/13
					0.353/6.2				(0.114−0.28)
^{176m}Yb			11.4 s	I.T./1.051		(8-)			Yb k x-ray
									0.0961
									0.1901
									0.2929
									0.3897
^{176}Yb	12.996(83)	175.942572	10^{26} a	β-β-		0+			
^{177m}Yb			6.41 s	I.T./0.3315		1/2-			Yb k x-ray
									0.1131
									0.2084
^{177}Yb		176.945261	1.9 h	β- /1.399	1.40	9/2+			Lu k x-ray
									0.1504
^{178}Yb		177.94665	1.23 h	β- /0.65	0.25/	0+			0.1415
									0.3246
									0.3516
									0.3815
									0.6125
^{179}Yb		178.9502	8. m	β- /2.4		(1/2-)			
^{180}Yb		179.9523	2. m	β-		0+			0.1028−0.442
^{181}Yb		180.9562							
^{182}Yb									
$_{71}$**Lu**		**174.9668(1)**							
^{150m}Lu			0.045 ms	p/1.29					
^{150}Lu		149.973	43. ms	p					
^{151m}Lu			16 μs	p/1.31		(3/2+)			
^{151}Lu		150.9676	0.08 s	p/1.231		(11/2-)			
^{152}Lu		151.9641	0.7 s						
^{153}Lu		152.9588	0.9 s			11/2-			
^{154}Lu		153.9575	1.0 s	β+, EC/10.8					
^{155m}Lu			2.6 ms	α/7.41		(25/2-)			
^{155}Lu		154.95432	0.07 s	EC/8.0		(11/2-)			
				α/	5.66/90				
^{156m}Lu			0.20 s	α/	5.57/	9+			
^{156}Lu		155.9530	~ 0.5 s	β+, EC/9.5		2-			ann.rad./
				α/	5.45/				
^{157m}Lu			~ 9.6 s	α	4.925/				

Elem. or Isot.	Natural Abundance (Atom %)	Atomic Mass or Weight	Half–life/ Resonance Width (MeV)	Decay Mode/ Energy (/MeV)	Particle Energy/ Intensity (MeV/%)	Spin ($h/2\pi$)	Nuclear Magnetic Mom. (nm)	Elect. Quadr. Mom. (b)	γ-Energy/ Intensity (MeV/%)
^{157}Lu		156.95010	4.8 s	β+, EC/94 /6.93					ann.rad./
				α/	5.00/				
^{158}Lu		157.94931	10.4 s	β+, EC/99 /8.0		2-			ann.rad./
				α/	4.67/				0.3682
									0.4770
^{159}Lu		158.94663	12.3 s	β+, EC/6.0					ann.rad./
									0.1505
									0.1875
									0.3693
^{160}Lu		159.9460	36.1 s	β+, EC/7.3					ann.rad./
									0.2434
									0.3957
									0.5773
^{161}Lu		160.94357	1.2 m	β+, EC/5.3		1/2+	+0.223		ann.rad./
									0.0437
									0.0671
									0.1003
									0.1108
									0.1562
									0.2562
^{162m}Lu			~ 1.5 m	EC/		4-			
^{162}Lu		161.9433	1.37 m	β+, EC/6.9		1-	+0.055	+0.52	ann.rad./
									0.1666
									0.6314
^{163}Lu		162.94118	4.1 m	β+, EC/4.6		1/2+	+0.077		ann.rad./
									0.0539
									0.0581
									0.1504
									0.1631
									0.3717
^{164}Lu		163.94134	3.14 m	β+, EC/6.3	1.6/	1-	+0.059	+0.61	0.1238
					3.8/				0.2621
									0.7404
									0.8639
									0.8804
^{165}Lu		164.93941	10.7 m	β+, EC/3.9	2.06/	1/2+	−0.0245		ann.rad./
									0.1206
									0.1324
									0.1742
									0.2036
									(0.04−2.0)
^{166m2}Lu			2.1 m	β+ /35 /		(0-)			ann.rad./
				EC/65 /					Yb k x-ray
									1.0673
									1.2566
									2.0986
^{166m1}Lu			1.4 m	β+, EC/58 /		(3-)	+0.189	+2.72	ann.rad./
				I.T./42 /0.0344					0.1024
									0.2281
									0.2861
									0.8119
									0.8301
^{166}Lu		165.93986	2.8 m	β+ /25 /5.5		(6-)	+2.91	+4.33	ann.rad./

Elem. or Isot.	Natural Abundance (Atom %)	Atomic Mass or Weight	Half–life/ Resonance Width (MeV)	Decay Mode/ Energy (/MeV)	Particle Energy/ Intensity (MeV/%)	Spin ($h/2\pi$)	Nuclear Magnetic Mom. (nm)	Elect. Quadr. Mom. (b)	γ-Energy/ Intensity (MeV/%)
				EC/75 /					Yb k x-ray
									0.1024
									0.2281
									0.3375
									0.3679
^{167}Lu		166.93827	52. m	β+ /2 /3.1	2.1/	7/2+	+2.325	+3.28	Yb k x-ray
				EC/98 /					0.0297
									0.2392
									(0.03–2.0)
^{168m}Lu			6.7 m	β+ /12 /		3+	+1.221	+2.43	ann.rad./
				EC/88 /					Yb k x-ray
				IT/<0.8					0.1988/190
									0.8960/100
									0.9792/128
									0.018–2.65
^{168}Lu		167.93874	5.5 m	β+ /6 /4.5	1.2/	(6-)	+3.02	+4.8	ann.rad./
				EC/94 /					Yb k x-ray
									0.1114
									0.1124
									0.2286
									0.3483
									1.4836
^{169m}Lu			2.7 m	I.T./0.0290		1/2-			Lu L x-ray
									0.0290
^{169}Lu		168.93765	1.419 d	EC/2.293	1.271/	7/2+	2.30	3.5	Yb k x-ray
									0.19121
									0.9606
									(0.08–2.1)
^{170m}Lu			0.7 s	I.T./0.0929		4-			Lu L x-ray
									0.04449
									0.0484
^{170}Lu		169.93848	2.01 d	EC/3.46	2.44/	0+			Yb k x-ray
									0.58711
									0.5908
									1.28029
									(0.1–3.38)
^{171m}Lu			1.31 m	I.T./0.0711		1/2-	+0.59		Lu k x-ray
									0.07119
^{171}Lu		170.937913	8.24 d	EC/1.479	0.362/	7/2+	+2.293	+3.48	Yb k x-ray
									0.01939
									0.66744
									(0.02–1.3)
^{172m}Lu			3.7 m	I.T./0.0419		1-	+1.98	+0.76	Lu L x-rays
									0.04186
^{172}Lu		171.939086	6.64 d	EC/2.519		4-	2.90	+3.80	Yb k x-ray
									0.18156
									1.09367
									(0.07–2.2)
^{173}Lu		172.938931	1.37 a	EC/0.671		7/2+	+2.281	+3.53	Yb k x-ray
									0.07860
									0.27198
^{174m}Lu			142. d	IT/99.3/	0.17086	6-	+1.49	+4.80	Lu k x-ray
				EC/0.7 /					0.067055
^{174}Lu		173.940338	3.3 a	EC/1.374		1-	+1.988	+0.773	Yb k x-ray

Elem. or Isot.	Natural Abundance (Atom %)	Atomic Mass or Weight	Half–life/ Resonance Width (MeV)	Decay Mode/ Energy (/MeV)	Particle Energy/ Intensity (MeV/%)	Spin ($h/2\pi$)	Nuclear Magnetic Mom. (nm)	Elect. Quadr. Mom. (b)	γ-Energy/ Intensity (MeV/%)
									0.07664
									1.2419
^{175}Lu	97.401(13)	174.940772				7/2+	+2.232	+3.49	
^{176m}Lu			3.66 h	β- /1.315	1.229/	1-	+0.318	−1.47	Hf k x-ray
					1.317/				0.088372
^{176}Lu	2.599(13)	175.942686	3.73 × 10^{10} a	β- /1.192		7-	+3.169	+4.92	Hf k x-ray
				β+/ < 0.9					0.20187
				EC/<0.36					0.30691
^{177m2}Lu			6. m	β-		39/2-			0.089
^{177m}Lu			160.7 d	IT/22/0.9702		23/2-	+2.32	+5.7	Lu k x-ray
				β- /78					Hf k x-ray
									0.11295
									0.20836
									0.37850
									0.41853
^{177}Lu		176.943758	6.65 d	β- /0.498	0.497/	7/2+	+2.238	+3.39	0.11295/0.062
									0.20836/0.104
^{178m}Lu			23.1 m	β- /		(9-)	+4.83	+5.39	0.2166
									0.3317
^{178}Lu		177.945955	28.5 m	β-/2.099	2.03/	1+	−1.38	+0.71	Hf k x-ray
									0.0932
									1.3099
									1.3408
									(0.09−1.7)
^{179}Lu		178.94733	4.6 h	β- /1.405	1.35/	7/2+	+2.38	+3.32	0.2143
									0.3377
^{180}Lu		179.9499	5.7 m	β- /3.1	1.49/				0.40795/50.
									(0.07−1.9)
^{181}Lu		180.9520	3.5 m	β- /2.5		(7/2+)			0.0458
									0.2059
									0.5749
^{182}Lu		181.9550	2.0 m	β- /~ 4.1					0.0978
									0.7208
									0.8182
^{183}Lu		182.9576	58. s	β- /		7/2+			
^{184}Lu		183.9609	20 s	β-					
^{185}Lu									
^{186}Lu									
$_{72}$Hf		178.49(2)							
^{153}Hf		152.971	> 0.2 μs						
^{154}Hf		153.965	2. s	EC, β+/~ 6.7		0+			
^{155}Hf		154.9634	0.9 s	EC, β+/8.					
^{156}Hf		155.9594	25. ms	α/		0+			
^{157}Hf		156.9584	0.11 s	α/					
^{158}Hf		157.95480	2.9 s	EC/54 /5.1		0+			
				α/46 /	5.27/				
^{159}Hf		158.95400	5.6 s	β+, EC/88 /6.9					ann.rad./
				α/12 /	5.09/				
^{160}Hf		159.95068	~ 12. s	β+, EC/97 /4.9		0+			ann.rad./
				α/4.78					
^{161}Hf		160.95028	17. s	α/	4.60/				
^{162}Hf		161.94721	38. s	β+, EC/3.7		0+			ann.rad./
									0.1739
									0.1963

Elem. or Isot.	Natural Abundance (Atom %)	Atomic Mass or Weight	Half–life/ Resonance Width (MeV)	Decay Mode/ Energy (/MeV)	Particle Energy/ Intensity (MeV/%)	Spin ($h/2\pi$)	Nuclear Magnetic Mom. (nm)	Elect. Quadr. Mom. (b)	γ-Energy/ Intensity (MeV/%)
									0.4101
^{163}Hf		162.94709	40. s	β+, EC/5.5					ann.rad./
									0.0454
									0.0621
									0.0710
									0.6882
^{164}Hf		163.94438	2.8 m	EC, β+/3.0		0+			
^{165}Hf		164.94457	1.32 m	EC/4.6		11/2−			
^{166}Hf		165.94218	6.8 m	EC/93 /2.3		0+			ann.rad./
				β+ /7 /					Lu k x-ray
									0.0788
^{167}Hf		166.94260	2.0 m	β+ /40 /4.0		(5/2−)			ann.rad./
				EC/60 /					Lu k x-ray
									0.1754
									0.3152
^{168}Hf		167.94057	25.9 m	β+, EC/1.8		0+			ann.rad./
									(0.014−1.311)
^{169}Hf		168.94126	3.25 m	EC/85 /3.3		(5/2−)			ann.rad./
				β+ /15 /					Lu k x-ray
									0.3695
									0.4929
^{170}Hf		169.93961	16.0 h	EC/1.1		0+			Lu k x-ray
									0.0985
									0.1202
									0.1647
									0.5729
									0.6207
^{171m}Hf			30. s			(1/2−)	+0.53		
^{171}Hf		170.94049	12.2 h	EC, β+ /2.4		7/2+	−0.67	+3.46	ann.rad./
									Lu k x-ray
									0.1221
									0.6620
									1.0714
^{172}Hf		171.93945	1.87 a	EC/0.35		0+			Lu k x-ray
									0.02399
									0.12582
									(0.082−0.123)
^{173}Hf		172.94051	23.6 h	EC/1.6		½−	+0.50		Lu k x-ray
									0.12367
									0.13963
									0.29697
									0.31124
									(0.1−2.1)
^{174}Hf	0.16(1)	173.940046	2.0×10^{15} a			0+			
^{175}Hf		174.941509	71. d	EC/0.686		5/2−	−0.68	+2.72	Lu k x-ray
									0.08936
									0.34340
^{176}Hf	5.26(7)	175.941409				0+			
^{177m2}Hf			51.4 m	I.T./2.740		37/2−			Hf k x-ray
									0.2140
									0.2951
									0.3115
									0.3267
^{177m1}Hf			1.1 s	I.T./		23/2+			Hf k x-ray

Elem. or Isot.	Natural Abundance (Atom %)	Atomic Mass or Weight	Half–life/ Resonance Width (MeV)	Decay Mode/ Energy (/MeV)	Particle Energy/ Intensity (MeV/%)	Spin ($h/2\pi$)	Nuclear Magnetic Mom. (nm)	Elect. Quadr. Mom. (b)	γ-Energy/ Intensity (MeV/%)
									0.20836
									0.22847
									0.37851
^{177}Hf	18.60(9)	176.943221				7/2-	+0.794	+0.337	
^{178m2}Hf			31. a	I.T./		16+	+8.16	+6.0	Hf k x-ray
									0.32555
									0.42635
									0.089–0.574
^{178m1}Hf			4.0 s	I.T./		8-	3		Hf k x-ray
									0.21342
									0.32555
									0.42635
^{178}Hf	27.28(7)	177.943699				0+			
^{179m2}Hf			25.1 d	I.T./1.1057		25/2-	7.4		Hf k x-ray
									0.1227
									0.1461
									0.3626
									0.4537
^{179m1}Hf			18.7 s	I.T./0.375		½-			Hf k x-ray
									0.1607
									0.2141
^{179}Hf	13.62(2)	178.945816				9/2+	−0.641	+3.86	
^{180m}Hf			5.54 h	I.T./1.1416		8-	+9.	+4.6	Hf k x-ray
									0.2152
									0.3323
									0.4432
^{180}Hf	35.08(16)	179.946550				0+			
^{181m}Hf			1.5 ms	/1.738		25/2-			
^{181}Hf		180.949101	42.4 d	β- /1.027	0.408/	1/2-			Ta k x-ray
									0.13294/36.6
									0.48200/67.7
									0.3459/13.5
^{182m}Hf			62. m	β- /54 /1.60	0.49/43	8-			Hf k x-ray
				IT/46 /1.173	0.95/10				0.0509
									0.2244
									0.3441
									0.4558
									0.5066
									0.9428
^{182}Hf		181.95055	8.9 × 10⁶ a	β- /0.37		0+			Ta k x-ray
									0.2704/79
									(0.098–0.270)
^{183}Hf		182.95353	1.07 h	β- /2.01	1.18/68	3/2-			Ta k x-ray
					1.54/25				0.0732
									0.4591
									0.7837
^{184}Hf		183.95545	4.1 h	β- /1.34	0.74/38	0+			Ta k x-ray
					0.85/16				0.0414
					1.10/46				0.1391
									0.3449
^{185}Hf		184.9588	~ 3.5 m	β- /					0.165
^{186}Hf		185.9609	~ 2.6 m			0+			0.738
^{187}Hf		186.9646	> 0.3 μs						
^{188}Hf		187.967	> 0.3 μs			0+			

Elem. or Isot.	Natural Abundance (Atom %)	Atomic Mass or Weight	Half–life/ Resonance Width (MeV)	Decay Mode/ Energy (/MeV)	Particle Energy/ Intensity (MeV/%)	Spin ($h/2\pi$)	Nuclear Magnetic Mom. (nm)	Elect. Quadr. Mom. (b)	γ-Energy/ Intensity (MeV/%)
¹⁸⁹Hf									
¹⁹⁰Hf									
₇₃Ta		180.94788(2)							
¹⁵⁵Ta		154.975	3. ms	p/1.44					
¹⁵⁶Ta		155.9723	0.11 s	β+ /~ 11.6					
				p/	1.02/~ 100				
¹⁵⁷Ta		156.9682	10 ms	α/	6.117				
				p/	0.927/3.4				
¹⁵⁸Ta		157.9667	37. ms	α/	6.05/100				
					5.97/100				
¹⁵⁹Ta		158.96302	0.6 s	β+, EC/20 /8.5	α/5.52/34				ann.rad./
				α/80 /	5.60/55				
¹⁶⁰Ta		159.9615	1.4 s	β+, EC/10.1					ann.rad./
				α	5.41/				
¹⁶¹Ta		160.9584	3.16 s	β+, EC/7.5					ann.rad./
				α/	5.15				
¹⁶²Ta		161.9573	4. s	EC/8.6					
¹⁶³Ta		162.95433	10.6 s	EC/6.8					
¹⁶⁴Ta		163.95353	14.2 s	β+ /8.5		3+			ann.rad./
				α/	4.62/				0.2110
									0.3768
¹⁶⁵Ta		164.95077	31. s	ECβ+/5.9					
¹⁶⁶Ta		165.95051	34. s	β+ /82 /7.7					ann.rad./
				EC/18 /					Hf k x-ray
									0.1587
									0.3117
									0.8101
¹⁶⁷Ta		166.94809	1.4 m	β+, EC/5.6					ann.rad./
¹⁶⁸Ta		167.94805	2.4 m	β+ /77 /6.7		3+			ann.rad./
				EC/23 /					Hf k x-ray
									0.1241
									0.2619
									0.7518
									(0.307−1.985)
¹⁶⁹Ta		168.94601	4.9 m	β+, EC/4.4					ann.rad./
									0.0288
									0.1535
									0.1924
¹⁷⁰Ta		169.94618	6.8 m	β +/70 /6.0		(3+)			ann.rad./
				EC/35 /					Hf k x-ray
									0.1008
									0.2212
¹⁷¹Ta		170.94448	23.3 m	β+, EC/3.7		(5/2-)			0.0496
									0.5018
									0.5064
									(0.05−1.02)
¹⁷²Ta		171.94490	36.8 m	β+ /25 /4.9		(3-)			ann.rad./
				EC/75 /					Hf k x-ray
									0.21396
									1.10923
									(0.09−3.8)
¹⁷³Ta		172.94375	3.6 h	β+ /24 /3.7		(5/2-)	1.70	−1.9	ann.rad./
				EC/76 /					Hf k x-ray
									0.06972

Elem. or Isot.	Natural Abundance (Atom %)	Atomic Mass or Weight	Half–life/ Resonance Width (MeV)	Decay Mode/ Energy (/MeV)	Particle Energy/ Intensity (MeV/%)	Spin ($h/2\pi$)	Nuclear Magnetic Mom. (nm)	Elect. Quadr. Mom. (b)	γ-Energy/ Intensity (MeV/%)
									0.17219
									(0.06–2.7)
174Ta		173.94445	1.12 h	β+ /27 /3.8		(3+)			ann.rad./
				EC/73 /					Hf k x-ray
									0.09089
									0.20638
									(0.09–3.64)
175Ta		174.94374	10.5 h	EC/2.0		7/2+	2.27	+3.6	Hf k x-ray
									0.2077
									0.2671
									0.3487
176Ta		175.94486	8.1 h	EC/3.1		1-			Hf k x-ray
									0.08837
									1.15735
177Ta		176.944472	2.356 d	EC/1.166		7/2+	2.25		Hf k x-ray
									0.11295
									(0.07–1.06)
178mTa			2.4 h	EC/		(7-)			Hf k x-ray
									0.08886
									0.21342
									0.32555
									0.42635
178Ta		177.94578	9.29 m	EC/99 /1.9		1+	+2.74	+0.65	ann.rad./
				β+ /1 /					Hf k x-ray
									0.09316
179Ta		178.945930	1.8 a	EC/0.110		7/2+	+2.29	3.37	Hf k x-ray
180mTa	0.01201(32)		3.65 × 10^16 a			(9-)	+4.82	+4.95	
			4.5 × 10^16 a	β-					
			> 2.0 × 10^16 a	EC					
180Ta		179.947465	8.15 h	EC/87 /0.854		1+			Hf k x-ray
				β- /13 /0.708	0.61/3				W k x-ray
					0.71/10				0.09333
									0.10340
181Ta	99.98799(32)	180.947996				7/2+	+2.370	+3.3	
182mTa			15.8 m	I.T./0.5198		10-			Ta k x-ray
									0.14678
									0.17157
182Ta		181.950152	114.43 d	β- /1.814	0.25/30	3-	+3.02	+2.6	W k x-ray
					0.44/20				1.12127/100
					0.52/40				1.22138/79
									0.085–1.289
183Ta		182.951373	5.1 d	β- /1.070	0.45/5	7/2+	+2.36		W k x-ray
					0.62/91				0.0847
									0.0991
									0.1079
									0.2461
									0.3540
184Ta		183.95401	8.7 h	β- /2.87	1.11/15	(5-)			W k x-ray
					1.17/81				0.2528/44.
									0.4140/74.
									(0.09–1.4)
185mTa			17 ms						0.280
185Ta		184.95556	49. m	β- /1.99	1.21/5	(7/2+)			W k x-ray
					1.77/81				0.0697

Elem. or Isot.	Natural Abundance (Atom %)	Atomic Mass or Weight	Half–life/ Resonance Width (MeV)	Decay Mode/ Energy (/MeV)	Particle Energy/ Intensity (MeV/%)	Spin ($h/2\pi$)	Nuclear Magnetic Mom. (nm)	Elect. Quadr. Mom. (b)	γ-Energy/ Intensity (MeV/%)
									0.1739
									0.1776
^{186}Ta		185.9586	10.5 m	β- /3.9	2.2/	(3-)			W k x-ray
									0.1979
									0.2149
									0.5106
									(0.09–1.5)
^{187}Ta		186.9605	> 0.3 µs						
^{188m}Ta			20 s						(0.143–0.434)
^{188}Ta		187.9637	~ 4.4 µs						0.292
^{189}Ta		188.9658	1.3 µs						
^{190}Ta			5. s						(0.207–0.357)
^{191}Ta									
^{192}Ta			2. s						0.219
$_{74}$**W**		**183.84(1)**							
^{157}**W**			0.28 s			(7/2-)			
^{158m}W			0.14 ms	α	8.28(3)/				
^{158}W		157.975	1.3 ms	α/	6.433/96	0+			
^{159}W		158.9729	7. ms	α/					
^{160}W		159.9685	0.08 s	α/	5.92/	0+			
^{161}W		160.9674	0.41 s	β+, EC/18 /8.1					
				α/82 /	5.78/				
^{162}W		161.9635	1.39 s	β+, EC/54 /5.8		0+			
				α/46 /	5.54/				
^{163}W		162.9625	2.7 s	β+, EC/59 /7.5					
				α/41 /	5.38/				
^{164}W		163.95895	6. s	β+, EC/97 /5.0		0+			ann.rad./
				α/3 /	5.15/				
^{165}W		164.95828	5.1 s	β+, EC/99 /7.0					ann.rad./
				α/1 /	4.91/				
^{166}W		165.95503	16. s	β+, EC/99 /4.2		0+			ann.rad./
				α/1 /	4.74/				
^{167}W		166.95482	20. s	EC/5.6					
^{168}W		167.95181	53. s	EC/3.8		0+			ann.rad./
				α/10^{-5}/	4.40(1)				Ta k x-ray
									0.1755
									(0.037–0.573)
^{169}W		168.95178	1.3 m	EC/5.4					ann.rad./
									Ta k x-ray
									0.123
									(0.097–0.699)
^{170}W		169.94923	2.4 m	EC/2.2		0+			ann.rad./
									Ta k x-ray
									0.3162
									(0.060–0.144)
^{171}W		170.94945	2.4 m	EC/4.6					ann.rad./
									Ta k x-ray
									0.1842
									(0.052–0.479)
^{172}W		171.94729	6.6 m	β+, EC/2.5		0+			ann.rad./
									Ta k x-ray
									0.0389
									(0.034–0.674)
^{173}W		172.94769	6.3 m	EC/4.0					ann.rad./

Elem. or Isot.	Natural Abundance (Atom %)	Atomic Mass or Weight	Half–life/ Resonance Width (MeV)	Decay Mode/ Energy (/MeV)	Particle Energy/ Intensity (MeV/%)	Spin (h/2π)	Nuclear Magnetic Mom. (nm)	Elect. Quadr. Mom. (b)	γ-Energy/ Intensity (MeV/%)
									Ta k x-ray
									0.4576
									(0.035–0.623)
^{174}W		173.94608	35. m	EC/1.9		0+			ann.rad./
									Ta k x-ray
									0.3287
									0.4288
									(0.056–0.429)
^{175}W		174.94672	35. m	EC/2.9		½-			(0.015–0.27)
^{176}W		175.94563	2.5 h	β+, EC/0.8		0+			0.03358
									0.06129
									0.09487
									0.10020
^{177}W		176.94664	2.21 h	EC/2.0		(1/2-)			Ta k x-ray
									0.15505
									0.18569
									0.42694
^{178}W		177.94588	21.6 d	EC/0.091		0+			Ta k x-ray
^{179m}W			6.4 m	IT/99.7/0.222		(1/2-)			W k x-ray
				EC/0.3/					0.2220
^{179}W		178.94707	38. m	EC/1.06		(7/2-)			Ta k x-ray
									0.0307
^{180}W	0.12(1)	179.946704	1.8×10^{18} a	α/		0+			
^{181}W		180.948197	121.1 d	EC/0.188		9/2+			Ta k x-ray
									0.13617
									0.15221
^{182}W	26.50(16)	181.948204	$>7.7 \times 10^{21}$ a	α/		0+			
^{183m}W			5.15 s	I.T./		(11/2+)			W k x-ray
									0.0465
									0.0526
									0.0991
									0.1605
^{183}W	14.31(4)	182.950223	$>4.1 \times 10^{21}$ a	α/		½-	+0.1177848		
^{184}W	30.64(2)	183.950931	$>8.9 \times 10^{21}$ a	α/		0+			
^{185m}W			1.6 m	I.T./0.1974		11/2+			W k x-ray
									0.0659
									0.1315
									0.1737
^{185}W		184.953419	74.8 d	β- /0.433	0.433/99.9	3/2-	+0.54		0.12536
^{186}W	28.43(19)	185.954364	$>8.2 \times 10^{21}$ a	α/		0+			
^{187m}W			1.6 µs	IT	0.411	11/2+			(0.014–0.287)
^{187}W		186.957161	23.9 h	β- /1.311	0.624/66	3/2-	0.62		Re k x-ray
					1.315/16				0.68572/33
					0.081–1.18				0.134–0.773
^{188}W		187.958489	69.78 d	β- /0.349	0.349/99	0+			0.0636
									0.2271
									0.2907
^{189}W		188.9619	9.7 m	β- /2.5	1.4/	(3/2-)			0.2604
					2.5/				(0.1262–1.466)
^{190m}W			~ 0.11 ms						(0.0585–0.694)
^{190}W		189.9632	30. m	β- /1.3	0.95/	0+			Re k x-ray
									0.1576
									0.1621
^{191}W		190.9666	> 0.3 µs						

Elem. or Isot.	Natural Abundance (Atom %)	Atomic Mass or Weight	Half–life/ Resonance Width (MeV)	Decay Mode/ Energy (/MeV)	Particle Energy/ Intensity (MeV/%)	Spin ($h/2\pi$)	Nuclear Magnetic Mom. (nm)	Elect. Quadr. Mom. (b)	γ-Energy/ Intensity (MeV/%)
^{192}W		191.968	> 0.3 μs			0+			
^{193}W									
^{194}W									
^{195}W									
$_{75}$Re		186.207(1)							
^{159}Re			~ 0.02 ms	α	α/6.78/7.5				
				p	p/1.80				
^{160}Re		159.9821	0.7 ms	p/	1.261(6)/91				
				α/	6.54/				
^{161}Re		160.9776	14 ms	α/	6.24				
				p	1.35				
^{162}Re		161.9760	0.10 s	α/	6.12/94				
					6.09/94				
^{163}Re		162.97208	0.26 s	β+, EC/9.0	α/5.87/32				
				α/	5.92/66				
^{164}Re		163.9703	~ 0.85 s	β+, EC/10.7					
				α/	5.78/				
^{165m}Re			~ 2.37 s	α/	5.502/				
^{165}Re		164.96709	2.6 s	β+, EC/87 /8.1					
				α/	5.49/ < 5.				
^{166}Re		165.9658	2.5 s	β+, EC/9.4					
				α/	5.50/				
^{167m}Re			6.2 s	α, EC/					
^{167}Re		166.9626	3.4 s	β+, EC/7.4					
				α/	5.015/				
^{168}Re		167.96157	4.4 s	β+, EC/9.1					
				α/	4.833/				0.1117
^{169m}Re			8.1 s	α	4.70/				
					4.87/				
^{169}Re		168.95879	16. s						
^{170}Re		169.95822	9.2 s	β+, EC/9.0					0.1560
									0.3055
									0.4125
^{171}Re		170.95572	15.2 s	EC/~ 5.7					
^{172m}Re			55. s	β+, EC/		(2)			ann.rad./
									0.1234
									0.2537
									0.3504
^{172}Re		171.9554	15. s	β+, EC/7.3					ann.rad./
									0.1234
									0.2537
^{173}Re		172.95324	2.0 m	EC/~ 3.9					ann.rad./
^{174}Re		173.95312	2.4 m	β+, EC/5.6					ann.rad./
									0.1119
									0.2430
^{175}Re		174.95138	5.8 m	β+, EC/4.3					ann.rad./
^{176}Re		175.95162	5.3 m	β+, EC/5.6		(3+)			ann.rad./
									0.1089
									0.2406
^{177}Re		176.95033	14. m	EC/78 /3.4		(5/2-)			ann.rad./
				β+ /22 /					W k x-ray
									0.0797
									0.0843
									0.1968

Elem. or Isot.	Natural Abundance (Atom %)	Atomic Mass or Weight	Half–life/ Resonance Width (MeV)	Decay Mode/ Energy (/MeV)	Particle Energy/ Intensity (MeV/%)	Spin ($h/2\pi$)	Nuclear Magnetic Mom. (nm)	Elect. Quadr. Mom. (b)	γ-Energy/ Intensity (MeV/%)
^{178}Re		177.95099	13.2 m	β+ /11 /4.7	3.3/	(3+)			ann.rad./
				EC/89 /					W k x-ray
									0.1059
									0.2373
									0.9391
^{179m}Re			0.47 ms						
^{179}Re		178.94999	19.7 m	EC/99 /2.71	0.95/	(5/2+)	2.8		W k x-ray
				β+ /1 /					0.1199
									0.2900
									0.4154
									0.4302
									1.6803
^{180}Re		179.95079	2.45 m	EC/92 /3.80	1.76/	1-	1.6		ann.rad./
				β+ /8 /					W k x-ray
									0.1036
									0.9028
									(0.07–2.2)
^{181}Re		180.95007	20. h	EC /1.74		5/2+	3.19		W k x-ray
									0.3607
									0.3655
									0.6390
^{182m}Re			12.7 h	EC/	0.55/	2+	3.3	+1.8	W k x-ray
					1.74/				0.0677
									1.1214
									1.2215
									(0.06–2.2)
^{182}Re		181.9512	2.67 d	EC/2.8		(7+)	2.8	+4.1	W k x-ray
									0.0678
									0.2293
									1.1213
									1.2214
^{183}Re		182.95082	70. d	EC/0.56		(5/2+)	+3.16	+2.2	W k x-ray
									0.16232
^{184m}Re			165. d	I.T./75 /0.188		8+	+2.9		Re k x-ray
				EC/25 /					0.1047
									0.2165
									0.92093
									(0.10–1.1)
^{184}Re		183.952521	35. d	EC/1.48		3-	+2.53	+3.0	W k x-ray
									0.79207
									08948
									0.90328
^{185}Re	37.40(2)	184.952955				5/2+	+3.1871	+2.19	
^{186m}Re			2.0 × 10^5 a	I.T./0.150		8+			Re k x-ray
									0.0590
^{186}Re		185.954986	3.718 d	β- /92 /1.070	0.973/21	1-	+1.739	+0.62	W k x-ray
				EC/8 /0.582	1.07/71				0.1227/0.6
									0.1372/9.5
									(0.63–0.77)
^{187}Re	62.60(2)	186.955753	4.16 × 10^{10} a	β- /0.00266	0.0025/	5/2+	+3.2197	+2.07	
^{188m}Re			18.6 m	I.T./0.172		(6-)			Re k x-ray
									0.0925
									0.1059
^{188}Re		187.958114	17.00 h	β- /2.120	1.962/20	1-	+1.788	+0.57	Os k x-ray

Elem. or Isot.	Natural Abundance (Atom %)	Atomic Mass or Weight	Half–life/ Resonance Width (MeV)	Decay Mode/ Energy (/MeV)	Particle Energy/ Intensity (MeV/%)	Spin ($h/2\pi$)	Nuclear Magnetic Mom. (nm)	Elect. Quadr. Mom. (b)	γ-Energy/ Intensity (MeV/%)
					2.118/79				0.15502
									0.309–2.022
^{189}Re		188.95923	24. h	β- /1.01	1.01/	(5/2+)			0.1471
									0.2167
									0.2194
									0.2451
^{190m}Re			3.0 h	β- /51 /		(6-)			Re k x-ray
				I.T./49 /					0.1191
									0.2238
									0.6731
									(0.1–1.79)
^{190}Re		189.9618	3.0 m	β- /3.2	1.8/	(2-)			Os k x-ray
									0.1867
									0.5580
									0.6051
^{191}Re		190.96313	9.7 m	β- /2.05	1.8/				
^{192m}Re			~ 0.12 ms	β⁻					(0.0606–0.146)
^{192}Re		191.9660	16. s	β- /4.2	~ 2.5/				(0.2–0.75)
^{193m}Re			~ 0.08 ms						(0.061–0.146)
^{193}Re		192.9675	0.07 ms	β⁻					
^{194}Re		193.9704	1.0 s	β⁻					
^{195}Re			6. s						
^{196}Re			3. s						
^{197}Re									
^{198}Re									
$_{76}$**Os**		**190.23(3)**							
^{161}**Os**			0.64 ms	α	/5.5	(7/2-)			
^{162}Os		161.984	1.8 ms	α/	6.60	0+			
^{163}Os		162.9827	5.5 ms	α/	6.51				
^{164}Os		163.9780	0.04 s	α	/96	0+			
^{165}Os		164.9768	0.07 s	α	/88				
^{166}Os		165.97269	0.18 s	β+, EC/28 /6.3	6.27/	0+			ann. rad./
				α/72 /	5.98/84				
^{167}Os		166.9716	0.84 s	β+, EC/76 /8.2					ann.rad./
				α/24 /	5.84/				
^{168}Os		167.96780	2.2 s	β+, EC/51 /5.7		0+			ann. rad./
				α/49 /					
^{169}Os		168.96702	3.3 s	β+, EC/89 /7.7	5.57/80				ann.rad./
				α/13 /	5.51/12				
					5.54/8				
^{170}Os		169.96358	7.1 s	β+, EC/5.0		0+			ann.rad./
				α/	5.40/				(0.162–0.216)
^{171}Os		170.96319	8.4 s	β+, EC/98 /7.1	α/5.24/93.5				ann.rad./
				α/19 /	5.17/6.5				0.190–0.705
^{172}Os		171.96002	19. s	β+, EC/99 /4.5		0+			ann.rad./
				α/1.1/	5.10/				(0.063–1.120)
^{173}Os		172.95981	16. s	β+, EC/6.3					ann.rad./
				α/0.4 /	4.94/				0.142–0.299
^{174}Os		173.95706	44. s	β+, EC/3.9		0+			0.118
				α/0.02 /	4.76/				0.138/0.001
									0.158
									0.325
^{175}Os		174.95695	1.4 m	β+, EC/5.3					0.125
									0.181

Elem. or Isot.	Natural Abundance (Atom %)	Atomic Mass or Weight	Half–life/ Resonance Width (MeV)	Decay Mode/ Energy (/MeV)	Particle Energy/ Intensity (MeV/%)	Spin ($h/2\pi$)	Nuclear Magnetic Mom. (nm)	Elect. Quadr. Mom. (b)	γ-Energy/ Intensity (MeV/%)
									0.248
^{176}Os		175.95481	3.6 m	β+, EC/3.2		0+			0.8155
									0.7758
									0.8573
									1.2093
									1.2909
^{177}Os		176.95497	2.8 m	β+, EC/4.5		(1/2-)			0.0848
									0.1958
									0.3002
									1.2686
^{178}Os		177.95325	5.0 m	β+, EC/2.3		0+			ann.rad./
									0.5946
									0.6850
									0.9687
									1.3311
^{179}Os		178.95382	7. m	β+, EC/3.7					ann.rad./
									0.0654
									0.2186
									0.5938
^{180}Os		179.95238	21.5 m	β+, EC/1.5		0+			Re k x-ray
									0.0202–0.7174
^{181m}Os			1.75 h	EC/		(1/2-)			ann.rad./
									0.0489
^{181}Os		180.95324	2.7 m	EC/2.9		(7/2-)			ann.rad./
									0.11794
									0.23868
									0.8267
									(0.07–2.64)
^{182}Os		181.95211	21.5 h	EC/0.9		0+			Re k x-ray
									0.1802
									0.5100
^{183m}Os			9.9 h	EC/84 /		½-			Os k x-ray
				I.T./16 /					Re k x-ray
									1.1020
									1.1080
^{183}Os		182.95313	13. h	EC/2.1		9/2+	−0.79	+3.1	Re k x-ray
									0.1144
									0.3818
^{184}Os	0.02(1)	183.952489				0+			
^{185}Os		184.954042	93.6 d	EC/1.013		½-			Re k x-ray
									0.6461
									0.8748
									0.8805
^{186}Os	1.59(3)	185.953838	2. × 10^{15} a	α/	~ 2.75/	0+			
^{187}Os	1.96(2)	186.955750				½-	+0.0646519		
^{188}Os	13.24(8)	187.955838				0+			
^{189m}Os			5.8 h	I.T./0.0308		9/2-			Os L x-ray
									0.0308
^{189}Os	16.15(5)	188.958148				3/2+	+0.659933	+0.9	
^{190m}Os			9.9 m	I.T./1.705		10-	−0.6		Os k x-ray
									0.1867
									0.3611
									0.5026
									0.6161

Elem. or Isot.	Natural Abundance (Atom %)	Atomic Mass or Weight	Half–life/ Resonance Width (MeV)	Decay Mode/ Energy (/MeV)	Particle Energy/ Intensity (MeV/%)	Spin ($h/2\pi$)	Nuclear Magnetic Mom. (nm)	Elect. Quadr. Mom. (b)	γ-Energy/ Intensity (MeV/%)
190Os	26.26(2)	189.958447				0+			
191mOs			13.1 h	I.T./0.0744		3/2-			Os k x-ray
									0.0744
191Os		190.960930	15.4 d	β- /0.314	0.140/100	9/2-	+0.96	+2.5	Ir k x-ray
									0.1294
192mOs			6.0 s	I.T./2.0154		(10-)			Os k x-ray
									0.2058/65.9
									0.5692/70
									(0.201–1.000)
192Os	40.78(19)	191.961481				0+			
193Os		192.964152	30.0 h	β- /1.141	1.04/20	3/2-	+0.730	+0.47	Ir k x-ray
									0.1389
									0.4605
194Os		193.965182	6.0 a	β- /0.097	0.054/33	0+			Ir L x-ray
					0.096/67				0.0429
195Os		194.968	6.5 m	β- /2.0	2.0/				
196Os		195.96964	34.9 m	β- /1.16	0.84/	0+			0.1262/5
									0.4079/5.9
197Os			2.8 m	β-					0.2239
									(0.0412–0.406)
198Os									
199Os			5. s						
200Os			6. s						
77 Ir		192.217(3)							
164Ir		163.9922	0.06 ms	p	1.78				
165Ir		164.9875	0.3 ms	p/87	1.71				
				α/13	6.72				
166mIr			14.3 ms	α/98.2	6.545				
				p/1.8	1.32				
166Ir		165.9858	0.010 s	α/93	6.56				
				p/6.9	1.15				
167mIr			26. ms	α/48, β+	6.39/90				
				p/32	1.25/0.42				
167Ir		166.98167	32. ms	α/80, β+	6.35/48				
				p/0.4	1.06/39.3				
168Ir		167.9799	0.17 s	α/82					
169mIr			280. ms	α/	6.12/59				
169Ir		168.97630	353. ms	α/	5.99/42				
170Ir		169.9750	0.81 s	α/	6.01/				0.175
					6.05/				
					5.95/				
					6.12/				
171Ir		170.97163	1.3 s	α/	5.91/				
172Ir		171.9705	2.1 s	α/	5.811/				0.228
									(0.379–0.475)
173Ir		172.96750	3.0 s	α/	5.665/				0.0493
									(0.092–0.296)
174Ir		173.96686	4. s	α/	5.478/				0.1587
									(0.276–1.33)
175Ir		174.96411	~ 4.5 s	α/	5.393/				0.1056
176Ir		175.96365	10.9 s	EC, β+/80					0.260
				α/3.2	5.118/				(0.135–0.415)
177Ir		176.96130	30. s	EC, β+/5.7					0.184
				α/0.06/	5.011/				(0.062–0.194)

Elem. or Isot.	Natural Abundance (Atom %)	Atomic Mass or Weight	Half–life/ Resonance Width (MeV)	Decay Mode/ Energy (/MeV)	Particle Energy/ Intensity (MeV/%)	Spin ($h/2\pi$)	Nuclear Magnetic Mom. (nm)	Elect. Quadr. Mom. (b)	γ-Energy/ Intensity (MeV/%)
^{178}Ir		177.96108	12. s	β+, EC/6.3					
									0.1320
									0.2667
									0.3633
^{179}Ir		178.95912	4. m	EC/4.9					0.0975
									(0.045–0.220)
^{180}Ir		179.95923	1.5 m	EC/6.4					0.2765
									(0.132–1.106)
^{181}Ir		180.95763	4.9 m	β+, EC/4.1		(7/2+)			ann.rad./
									0.1076
									(0.020–1.715)
^{182}Ir		181.95808	15. m	β+ /44 /5.6			+2.6	~ −1.7	ann.rad./
				EC/56 /					Os k x-ray
									0.1273
									0.2370
^{183}Ir		182.95685	57. m	β+, EC/3.5			+2.4	~ −1.8	ann.rad./
									0.0877
									0.2285
									0.2824
^{184}Ir		183.95748	3.0 h	β+ /12 /4.6	2.3/	5-	0.70	+2.41	ann.rad./
				EC/88 /	2.9/				Os k x-ray
									0.11968
									0.2640
									0.3904
^{185}Ir		184.95670	14. h	β+ /3 /2.4		(5/2-)	2.60	−2.1	ann.rad./
				EC/97 /					Os k x-ray
									0.2543
									1.8288
^{186m}Ir			1.7 h	EC /		(2-)	0.64	+1.46	Os k x-ray
									0.1371
									0.7675
^{186}Ir		185.95795	15.7 h	EC/98 /3.83		(5+)	3.9	−2.55	Os k x-ray
				β+ /2 /					0.1372
									0.2968
									0.4348
									(0.13–3.0)
^{187}Ir		186.95736	10.5 h	EC/1.50		3/2+	+0.17	+0.94	Os k x-ray
									0.0743
									0.4009
									0.4271
									0.6109
									0.9128
^{188}Ir		187.95885	1.72 d	β+ /2.81	1.13/	(2-)	0.30	+0.48	Os k x-ray
				EC/99+ /	1.64/				0.1550
									0.4780
									0.6330
									2.2146
^{189}Ir		188.95872	13.2 d	EC/0.53		3/2+	+0.14	+0.88	Os k x-ray
									0.2449
^{190m2}Ir			3.09 h	β+, EC/95 /		(11-)			0.376
				I.T./5 /					
^{190m1}Ir			1.12 h	I.T. /0.0263		7+			Ir L x-ray
^{190}Ir		189.960546	11.8 d	EC/2.0		(4+)	0.04	+2.8	Os k x-ray
									0.1867

Elem. or Isot.	Natural Abundance (Atom %)	Atomic Mass or Weight	Half–life/ Resonance Width (MeV)	Decay Mode/ Energy (/MeV)	Particle Energy/ Intensity (MeV/%)	Spin ($h/2\pi$)	Nuclear Magnetic Mom. (nm)	Elect. Quadr. Mom. (b)	γ-Energy/ Intensity (MeV/%)
									0.4072
									0.5186
									0.5580
									0.6051
									(0.2–1.4)
191mIr			4.93 s	I.T./0.1714		11/2-	+0.603		Ir k x-ray
									0.1294
191Ir	37.3(2)	190.960594				3/2+	+0.151	+0.82	
192m2Ir			241. a	I.T./0.161		(9+)			Ir k x-ray
192m1Ir			1.44 m	I.T./0.0580		(1+)			Ir L x-ray
									0.0580
									0.3165
192Ir		191.962605	73.83 d	β- /1.460		(4-)	+1.92	+2.15	Pt k x-ray
									0.31649/83.
									0.46806/48.
193mIr			10.53 d	I.T./0.0802		11/2-			Ir L x-ray
									0.0803
193Ir	62.7(2)	192.962926				3/2+	+0.164	+0.75	
194mIr			170. d	β- /		11			Pt k x-ray
									0.3284
									0.4829
									0.5624
194Ir		193.965078	19.3 h	β-/2.247	1.92/9	1-	+0.39	+0.34	0.2935
					2.25/86				0.3284
									0.6451
									(0.1–2.2)
195mIr			3.9 h	β- /	0.41/	(11/2-)			Pt k x-ray
					0.97/				0.3199/9.6
									0.3649/9.5
									0.4329/9.6
									0.6849/9.6
195Ir		194.965980	2.8 h	β- /1.120	1.0/80	(3/2+)			Pt k x-ray
					1.11/13				0.0989/9.7
196mIr			1.40 h	β-/	1.16/				Pt k x-ray
									0.3557
									0.3935
									0.4471
									0.5214
									0.6473
196Ir		195.96840	52. s	β- /3.21	2.1/15	0-			0.3329
					3.2/80				0.3557
									0.7796
197mIr			8.9 m	β- /		(11/2-)			0.3465
				I.T./					see Ir[197m]
197Ir		196.96965	5.8 m	β- /2.16	1.5/	(3/2+)			0.0531
					2.0/				0.1351
									0.4306
									0.4697
198Ir		197.9723	8. s	β- /4.1					0.4074
									0.5070
199Ir		198.97380	6. s	β- /					
200Ir									
201Ir									
202Ir			11 s						

Elem. or Isot.	Natural Abundance (Atom %)	Atomic Mass or Weight	Half–life/ Resonance Width (MeV)	Decay Mode/ Energy (/MeV)	Particle Energy/ Intensity (MeV/%)	Spin ($h/2\pi$)	Nuclear Magnetic Mom. (nm)	Elect. Quadr. Mom. (b)	γ-Energy/ Intensity (MeV/%)
[203]Ir									
[78]Pt		195.084(9)							
[166]Pt		165.995	0.3 ms	α/	7.11/	0+			
[167]Pt		166.930	0.9 ms	α/	6.98/				
[168]Pt		167.9882	2.1 ms	α	6.82	0+			0.582/69
									0.594/69
									0.725/62
[169]Pt		168.9867	7.0 ms	α	6.69				
[170]Pt		169.98250	14.0 ms	α	6.55	0+			0.509/100
									0.662/86
									0.214–0.726
[171]Pt		170.9812	0.048 s	α	6.45				0.4450
									(0.1564-1.208)
[172]Pt		171.97735	0.10 s	α/	6.31/94	0+			
[173]Pt		172.9764	0.36 s	β+, EC/8.2	6.23				
				α/	6.20/				
[174]Pt		173.97282	0.89 s	β+, EC/17 /5.6		0+			
				α/83 /	6.040/				
[175]Pt		174.97242	2.5 s	β+, EC/65 /7.6					0.0774
				α/35 /	5.831/5				0.1354
					5.96/54				0.2128
					6.038/				
[176]Pt		175.96895	6.3 s	β+, EC/60 /5.1		0+			ann.rad./
				α/40 /	5.528/0.6				0.2277
					5.750/41				
[177]Pt		176.96847	11. s	EC/91 /6.8	5.53/				0.0908
				α/9 /	5.485/3				
					5.525/6				
[178]Pt		177.96565	21. s	EC/93 /4.5		0+			
				α/7 /	5.286/0.2				
					5.442/7				
[179]Pt		178.96536	33. s	β+, EC/5.7			+0.43		
				α/	5.16/				
[180]Pt		179.96303	52. s	β+, EC/99.7 /3.7	0+				
				α/0.3 /	5.140/				
[181]Pt		180.96310	51. s	β+, EC/5.2			+0.48		
[182]Pt		181.96117	2.7 m	β+, EC/2.9		0+			ann.rad./
									0.1360
									0.1460
									0.2100
[183m]Pt			43. s	β+, EC/		(7/2-)	+0.78	+3.4	ann.rad./
				I.T./					0.3132/26
									0.3164/59
									0.6296/100
									0.058–1.75
[183]Pt		182.96160	7. m	β+, EC/4.6			+0.50		ann.rad./
									0.119/100
									0.307/93
									0.260/90
									0.058–1.377
[184]Pt		183.95992	17.3 m	β+, EC/2.3		0+			ann.rad./
									0.1549
									0.1919

Elem. or Isot.	Natural Abundance (Atom %)	Atomic Mass or Weight	Half–life/ Resonance Width (MeV)	Decay Mode/ Energy (/MeV)	Particle Energy/ Intensity (MeV/%)	Spin ($h/2\pi$)	Nuclear Magnetic Mom. (nm)	Elect. Quadr. Mom. (b)	γ-Energy/ Intensity (MeV/%)
									0.5484
^{185m}Pt			33. m	β+, EC/		½-	+0.50		
^{185}Pt		184.96062	1.18 h	β+, EC/3.8		(9/2+)	−0.75	+4.	ann.rad./
									0.1353
									0.1974
									0.2296
									0.2551
^{186}Pt		185.95935	2.0 h	β+, EC/1.38		0+			ann.rad./
									0.6115
									0.6892
^{187}Pt		186.96059	2.35 h	β+, EC/3.1		3/2-	−0.40	−1.0	ann.rad./
									Ir k x-ray
									0.1064
									0.1100
									0.2015
									0.2849
									0.7092
^{188}Pt		187.95940	10.2 d	EC/0.51		0+			Ir k x-ray
									0.1876
									0.1951
^{189}Pt		188.96083	10.9 h	β+, EC/1.97		3/2-	−0.43	−1.2	Ir k x-ray
									0.0943
									0.6076
									0.7214
									(0.09–1.47)
^{190}Pt	0.012(2)	189.95993	4.5×10^{11} a	α		0+			
^{191}Pt		190.961677	2.86 d	EC/1.02		(3/2-)	−0.50	−0.9	Ir k x-ray
									0.3599
									0.4094
									0.5389
^{192}Pt	0.782(24)	191.961038				0+			
^{193m}Pt			4.33 d	I.T./0.1498		13/2+	−0.75		Pt k x-ray
									0.1355
^{193}Pt		192.962988	60. a	EC/0.0566		(1/2-)	+0.60		Ir k x-rays
^{194}Pt	32.86(40)	193.962680				0+			
^{195m}Pt			4.01 d	I.T./0.2952		13/2+	−0.61	~ +1.4	Pt k x-ray
									0.0989
^{195}Pt	33.78(24)	194.964791				1/2-	+0.6095		
^{196}Pt	25.21(34)	195.964952				0+			
^{197m}Pt			1.590 h	I.T./97 /		13/2+			Pt k x-ray
				β- /3 /					0.0530
									0.3465
^{197}Pt		196.967340	19.9 h	β- /0.719		1/2-	0.51		Au k x-ray
									0.1914
									0.2688
^{198}Pt	7.356(130)	197.967893				0+			
^{199m}Pt			13.6 s	I.T./0.424		13/2+			Pt k x-ray
									0.3919
^{199}Pt		198.970593	30.8 m	β- /1.70	0.90/18	(5/2-)			0.3170/3.88
					1.14/14				0.49375/4.47
									0.5430/11.7
									(0.055–1.293)
^{200}Pt		199.971441	12.5 h	β- /~ 0.66		0+			Au k x-ray
									0.13590

Elem. or Isot.	Natural Abundance (Atom %)	Atomic Mass or Weight	Half–life/ Resonance Width (MeV)	Decay Mode/ Energy (/MeV)	Particle Energy/ Intensity (MeV/%)	Spin ($h/2\pi$)	Nuclear Magnetic Mom. (nm)	Elect. Quadr. Mom. (b)	γ-Energy/ Intensity (MeV/%)
									0.22747
									0.24371
²⁰¹Pt		200.97451	2.5 m	β- /2.66		(5/2-)			0.070
									0.152
									0.222
									1.760
²⁰²ᵐPt			0.3 ms						(0.535–0.719)
²⁰²Pt		201.9757	1.8 d			0+			0.440
²⁰³Pt									
²⁰⁴Pt			0.055 ms						
²⁰⁵Pt									
₇₉Au		196.966569(4)							
¹⁷⁰ᵐAu			0.62 ms	p/58	1.74/				
				α/42	7.11/				
¹⁷⁰Au		169.9961	0.30 ms	p/89	1.46/				
				α/11	7.00/				
¹⁷¹ᵐAu			1.09 ms	α/66	6.995				
				p/34	1.694				
¹⁷¹Au		170.99188	0.022 ms	p/100	1.437				
¹⁷²ᵐAu			8. ms		6.80				
¹⁷²Au		171.9900	22. ms	α/7.02	6.76				
¹⁷³ᵐAu			15 ms	α/92	6.732				
¹⁷³Au		172.98624	0.02 s	α/94	6.672				
¹⁷⁴Au		173.9848	0.14 s	α	6.54				
¹⁷⁵ᵐAu			0.14 s	α	6.43/90				
¹⁷⁵Au		174.98127	0.15 s	α					
¹⁷⁶Au		175.9801	0.9 s	β+, EC/10.5					
				α/	6.260/80				
					6.290/20				
¹⁷⁷ᵐAu			1.0 s	α	6.12/66				
¹⁷⁷Au		176.97687	1.5 s	α/	6.16/40				
					6.150/				
¹⁷⁸Au		177.9760	2.6 s	α/	5.920/				
¹⁷⁹Au		178.97321	7.5 s	α/	5.85/				
¹⁸⁰Au		179.97252	8.1 s	EC/8.6	5.65				0.1522
				α/	5.61				0.2564
					5.50				0.5242
									0.6765
									0.8084
									0.8597
¹⁸¹Au		180.97008	11.4 s	EC/97.5/6.3	5.482/				
				α/2.7/					
¹⁸²Au		181.96962	21. s	β+, EC/6.9		(2+)	1.3		ann.rad./
				α/0.13/					0.1549
									0.2649
									(0.13–1.4)
¹⁸³Au		182.96759	42. s	EC/5.5			+1.97		0.1630
				α/0.8/					0.2730
									0.3625
¹⁸⁴ᵐAu			48 s	I.T.		(2+)	+1.44	+1.9	0.069(IT)
¹⁸⁴Au		183.96745	21. s	EC, β+/7.1		(5+)	+2.07	+4.7	
				α/0.013/					
¹⁸⁵ᵐAu			6.8 m	β+, EC/					
				I.T./0.145					

Elem. or Isot.	Natural Abundance (Atom %)	Atomic Mass or Weight	Half–life/ Resonance Width (MeV)	Decay Mode/ Energy (/MeV)	Particle Energy/ Intensity (MeV/%)	Spin ($h/2\pi$)	Nuclear Magnetic Mom. (nm)	Elect. Quadr. Mom. (b)	γ-Energy/ Intensity (MeV/%)
^{185}Au		184.96579	4.3 m	β+, EC/4.71		(5/2-)	+2.17	−1.1	ann.rad./
				α/0.26/					
^{186m}Au			< 2. m	β+, EC/					0.1915
^{186}Au		185.96595	10.7 m	β+, EC/6.0		3-	−1.28	+3.1	ann.rad./
				α/8(10)$^{-4}$/					0.1915
									0.2988
^{187m}Au			2.3 s	IT		9/2-			
^{187}Au		186.96457	8.3 m	β+, EC/3.60		1/2+	+0.53		ann.rad./
									0.9152
									1.2668
									1.3321
									1.4081
^{188}Au		187.96532	8.8 m	β+, EC/5.3		(1-)	−0.07		ann.rad./
									0.2660
									0.3404
									0.6061
^{189m}Au			4.6 m	β+, EC/		11/2-	+6.19		0.1667
^{189}Au		188.96395	28.7 m	EC/96 /3.2		1/2+	+0.49		ann.rad./
				β+ /4 /					Pt k x-ray
									0.4478
									0.7133
									0.8128
^{190}Au		189.96470	43. m	β+ /2 /4.44		1-	−0.07		ann.rad./
				EC/98 /					Pt k x-ray
									0.2958
									0.3018
									0.5977
^{191m}Au			0.9 s	I.T./0.2663		(11/2-)	6.6		Au k x-ray
									0.2414
									0.2526
^{191}Au		190.96370	3.2 h	EC/1.83		3/2+	+0.137	+0.72	Pt k x-ray
									0.5864/16
									(0.088−1.30)
^{192}Au		191.96481	4.9 h	β+ /5 /3.52	2.19/	1-	−0.011	−0.23	ann.rad./
				EC/95 /	2.49/				Pt k x-ray
									0.2959
									0.3165
^{193m}Au			3.9 s	I.T./0.2901		11/2-	6.2	+1.98	Au k x-ray
									0.2580
^{193}Au		192.96415	17.6 h	EC/1.07		3/2+	+0.140	+0.66	Pt k x-ray
									0.1862
									0.2556
^{194}Au		193.96537	1.64 d	β+ /3 /2.49	1.49/	1-	+0.076	−0.24	ann.rad./
				EC/97 /					Pt k x-ray
									0.2935
									0.3284/61
^{195m}Au			30.5 s	I.T./0.3186		11/2-	6.2	+1.9	Au k x-ray
									0.2617
^{195}Au		194.965035	186.10 d	EC/0.227		3/2+	+0.149	+0.61	Pt k x-ray
^{196m2}Au			9.7 h	I.T./0.5954		12-	5.7		Au k x-ray
									0.1478
									0.1883
^{196m1}Au			8.1 s	I.T./0.0846		8+			0.0847
^{196}Au		195.966570	6.17 d	EC/92 /1.506		2-	+0.591	~ 0.81	Pt k x-ray

Elem. or Isot.	Natural Abundance (Atom %)	Atomic Mass or Weight	Half–life/ Resonance Width (MeV)	Decay Mode/ Energy (/MeV)	Particle Energy/ Intensity (MeV/%)	Spin ($h/2\pi$)	Nuclear Magnetic Mom. (nm)	Elect. Quadr. Mom. (b)	γ-Energy/ Intensity (MeV/%)
^{197m}Au			7.8 s	I.T./0.4094		11/2-	+6.0	+1.7	Au k x-ray
				β- /8 /0.686					0.1302
									0.2790
^{197}Au	100.	196.966569				3/2+	+0.14575	+0.55	
^{198m}Au			2.30 d	I.T./0.812		(12-)	+5.9		Au k x-ray
									0.0972
									0.1803
									0.2419
^{198}Au		197.968242	2.695 d	β- /1.372	0.290/1	2-	+0.5934	+0.64	Hg k x-ray
					0.961/99				0.411794/95.3
^{199}Au		198.968765	3.14 d	β- /0.453	0.25/22	3/2+	~ +0.2715	+0.51	Hg k x-ray
					0.292/72				0.15837
					0.462/6				0.20820
^{200m}Au			18.7 h	β- /84 /1.0	0.56/	12-	5.9		Au k x-ray
				I.T./16 /					0.2559/71
									0.3680/77
									0.4978/73
									0.5793/72
									0.084−0.904)
^{200}Au		199.97073	48.4 m	β- /2.24	0.7/15	1-			0.3679/19
					2.2/77				1.2254/10.6
									(0.077−1.570)
^{201}Au		200.971657	26. m	β- /1.28	1.27/82	3/2+			(0.027−0.732)
^{202}Au		201.9738	29. s	β- /3.0		(1-)			0.4396
^{203}Au		202.975155	1.0 m	β- /2.14	~ 1.9/	3/2+			(0.04−0.37)
^{204}Au		203.9777	40. s	β- /4.5		(2-)			0.4366
									1.5113
^{205}Au		204.9799	31. s	β- /					(0.38−1.33)
^{206}Au									
^{207}Au									
^{208}Au									
^{209}Au									
^{210}Au									
$_{80}$Hg		200.59(2)							
^{171}Hg		171.0038	0.06 ms	α	7.49				
^{172}Hg		171.9988	0.3 ms	α	7.36	0+			
^{173}Hg		172.9972	0.8 ms	α	7.20				
^{174}Hg		173.99286	1.9 ms	α	7.07	0+			
^{175}Hg		174.9914	0.02 s	α					
^{176}Hg		175.98736	21 ms	α	6.74/94	0+			
^{177m}Hg			1.5 μs	IT					0.246
^{177}Hg		176.9863	0.13 s	α	6.58				
^{178}Hg		177.98248	0.26 s	EC/50 /6.1		0+			
				α/50 /	6.43/				
^{179}Hg		178.98183	1.05 s	EC/8.0					
				α/	6.29/				
^{180}Hg		179.97827	2.6 s	EC/5.5		0+			0.1250
				α/	6.12/33				0.3005
					5.69/.03				0.3812
^{181m}Hg			0.48 ms			(13/2+)			
^{181}Hg		180.97782	3.6 s	β+ EC/76 /~ 7.3		(1/2-)	+0.507		0.0663
				α/24 /					0.0811
									0.0924

Elem. or Isot.	Natural Abundance (Atom %)	Atomic Mass or Weight	Half–life/ Resonance Width (MeV)	Decay Mode/ Energy (/MeV)	Particle Energy/ Intensity (MeV/%)	Spin ($h/2\pi$)	Nuclear Magnetic Mom. (nm)	Elect. Quadr. Mom. (b)	γ-Energy/ Intensity (MeV/%)
									0.1474
									0.1587
									0.2142
									0.2398
^{182}Hg		181.97469	10.8 s	β+, EC/85/5.0		0+			0.129/122
				α/15/	5.87/8.6				0.2176/66
					5.45/0.03				0.0256–0.543
^{183}Hg		182.97445	9. s	β+, EC/77/6.3		½-	+0.524		0.0714
				α/	5.83/				0.0874
					5.91/				0.1538
^{184}Hg		183.97171	30.9 s	β+, EC/99/4.1		0+			0.1565/102
				α/1/	5.54/1.3				0.2367/100
					5.07/0.002				0.2384/18
									(0.018–0.4227)
^{185m}Hg			21. s	β+, EC, IT, α/	5.37/	13/2+	−1.02	~ +0.2	0.211
									0.292
^{185}Hg		184.97190	51. s	β+, EC/95/5.8		½-	+0.509		0.02–0.55
^{186}Hg		185.96936	1.4 m	β+, EC/3.3		0+			0.1119
				α	5.09/0.02				0.2518
^{187m}Hg			1.9 m	β+, EC/		13/2+	−1.04	+0.5	see Hg[187m]
^{187}Hg		186.96981	2.4 m	β+, EC/4.9		3/2-	−0.594	−0.8	0.1034/32
									0.2334/100
									0.2403/33
									0.27151/31
									0.3763/38
									0.5254/30
									0.10–2.18
^{188}Hg		187.96758	3.2 m	β+, EC/2.3		0+			0.0988
				α	4.61				0.1148
									0.1424
									0.1900
^{189m}Hg			8.6 m	EC/		13/2+	−1.06	+0.7	0.0780
									0.3210
									0.4345
									0.5655
									(0.08–2.170)
^{189}Hg		188.96819	7.6 m	EC/4.2		3/2-	−0.609	−0.8	0.2005
									0.2038
									0.2386
									0.2485
^{190}Hg		189.96632	20.0 m	EC/1.5		0+			0.1296
									0.1426
^{191m}Hg			51. m	β+ /6 /		13/2+	−1.07	+0.6	ann.rad./
				EC/94 /					Au k x-ray
									0.2741
									0.4203
									0.5787
									(0.07–1.9)
^{191}Hg		190.96716	50. m	β+, EC/3.2		(3/2-)	−0.62	−0.8	0.1963
									0.2247
									0.2524
^{192}Hg		191.96563	5.0 h	EC/~ 0.5		0+			Au k x-ray
									0.1572
									0.2748

Elem. or Isot.	Natural Abundance (Atom %)	Atomic Mass or Weight	Half–life/ Resonance Width (MeV)	Decay Mode/ Energy (/MeV)	Particle Energy/ Intensity (MeV/%)	Spin ($h/2\pi$)	Nuclear Magnetic Mom. (nm)	Elect. Quadr. Mom. (b)	γ-Energy/ Intensity (MeV/%)
^{193m}Hg			11.8 h	β+, EC/91 /		13/2+	−1.058430	~ +0.92	0.3065
				I.T./9 /0.2901					Hg k x-ray
									0.1866
									0.2580
									0.4076
									0.5733
									0.9324
									(0.1–1.96)
^{193}Hg		192.96667	3.8 h	EC, B+/2.34		3/2-	−0.6276	−0.7	0.1866
									0.2580
									0.8611
^{194}Hg		193.96544	520. a	EC/0.04		0+			Au L x-ray
^{195m}Hg			1.67 d	I.T./(54)/0.3186		13/2+	−1.04465	+1.1	Hg k x-ray
				EC/(46)/					Au k x-ray
									0.2617
									0.5603
									0.7798
^{195}Hg		194.96672	10.5 h	EC/1.51		1/2-	+0.541475		Au k x-ray
									0.0614
									0.7798
^{196}Hg	0.15(1)	195.965833	>2.5 × 10^{18} a	α		0+			
^{197m}Hg			23.8 h	I.T./(93)/0.2989		13/2+	−1.027684	+1.2	Hg k x-ray
									Au k x-ray
									0.13398
^{197}Hg		196.967213	2.69 d	EC/0.600		1/2-	+0.527374		Au k x-ray
									0.07735
^{198}Hg	9.97(20)	197.9667690				0+			
^{199m}Hg			42.7 m	I.T./0.532		13/2+	−1.014703	+1.2	Hg k x-ray
									0.15841
^{199}Hg	16.87(22)	198.9682799				1/2-	+0.505885		
^{200}Hg	23.10(19)	199.9683260				0+			
^{201}Hg	13.18(9)	200.970302				3/2-	−0.560226	+0.37	
^{202}Hg	29.86(26)	201.970643				0+			
^{203}Hg		202.972873	46.61 d	β- /0.492	0.213/100	5/2-	+0.8489	+0.34	Tl k x-ray
									0.279188
^{204}Hg	6.87(15)	203.9734939				0+			
^{205}Hg		204.976073	5.2 m	β- /1.531	1.33/4	1/2-	+0.6009		0.20378
									(0.2–1.4)
^{206}Hg		205.97751	8.2 m	β- /1.31	0.935/34	0+			Tl k x-ray
					1.3/63				0.3052
									0.6502
^{207}Hg		206.9826	2.9 m	β- /4.8		(9/2+)			
^{208}Hg		207.9859	41. m	β-		0+			0.474
^{209}Hg		208.9910	36 s	β-					0.324
^{210}Hg		209.9945	> 0.3 μs	β-		0+			
^{211}Hg									
^{212}Hg									
^{213}Hg									
^{214}Hg									
^{215}Hg									
^{216}Hg									
$_{81}$**Tl**		**204.382– 204.385**							
^{176}Tl		176.0006	5 ms	p	1.26/~ 100				

Elem. or Isot.	Natural Abundance (Atom %)	Atomic Mass or Weight	Half–life/ Resonance Width (MeV)	Decay Mode/ Energy (/MeV)	Particle Energy/ Intensity (MeV/%)	Spin ($h/2\pi$)	Nuclear Magnetic Mom. (nm)	Elect. Quadr. Mom. (b)	γ-Energy/ Intensity (MeV/%)
[177m]Tl			0.23 ms	p/51	1.95				
				α/49	7.48				
[177]Tl		176.99643	0.017 s	α/73					
				p/27					
[178]Tl		177.9949	0.25 s	α/	6.704				
					6.785				
					6.62				
					6.859				
[179m]Tl			1.5 ms	α	/7.21/80				
				α	/7.10/20				
[179]Tl		178.99109	0.3 s	α	6.57/				
[180]Tl		179.9899	1.09 s	α//8	6.28/30				
					6.36/30				
					6.21/18				
					6.56/15				
					6.47/7				
[181m]Tl			1.4 ms	α	6.58/96				
					6.97/2.6				
[181]Tl		180.98626	3.2 ms	α/ < 10	6.19/100				
[182]Tl		181.9857	3. s	β+, EC/10.9					0.351
									(0.26–0.41)
[183m]Tl			53. ms	α//1.5	6.33/80	9/2-			0.0618
					6.38/16				(0.046–0.0894)
					6.46/4				
[183]Tl		182.98219	5. s	β+, EC/7.7		½+			0.208
[184]Tl		183.98187	11. s	β+, EC/(98)/9.2					0.2868
				α/(2)/	6.16/				0.3399
									0.3667
[185m]Tl			1.8 s	I.T./0.453		(9/2-)			0.1688
				α/5.97	6.01				0.2840
[185]Tl		184.9788	20. s	EC/β+/6.6					
[186m]Tl			4. s	I.T./0.374					0.3738
[186]Tl		185.9783	28. s	β+, EC/7.5					0.3567
									0.4026
									0.4053
[187m]Tl			15.6 s	I.T./~ 0.33		(9/2+)	+3.79	−2.4	0.2995
[187]Tl		186.97591	50. s	β+, EC/6.0		½+	1.6		
[188m]Tl			1.18 m	β+, EC/		(7+)			Hg k x-ray
									0.4129
									0.5043
									0.5921
[188]Tl		187.97601	1.2 m	β+, EC/7.8		(2-)	+0.48	+0.129	see Tl[188m]
									0.4129
[189m]Tl			1.4 m	β+, EC/		(9/2-)	+3.88	−2.29	0.2156
									0.2284
									0.3175
									0.4452
[189]Tl		188.97359	2.3 m	β+, EC/5.2		(1/2+)			0.3337
									0.4510
									0.5223
									0.9422
[190m]Tl			3.7 m	β+, EC/	4.2/	(7+)	+0.495	+0.29	0.1968
									0.4164
									0.7311

Elem. or Isot.	Natural Abundance (Atom %)	Atomic Mass or Weight	Half–life/ Resonance Width (MeV)	Decay Mode/ Energy (/MeV)	Particle Energy/ Intensity (MeV/%)	Spin ($h/2\pi$)	Nuclear Magnetic Mom. (nm)	Elect. Quadr. Mom. (b)	γ-Energy/ Intensity (MeV/%)
¹⁹⁰Tl		189.97388	2.6 m	β+, EC/7.0	5.7/	(2-)	+0.254	−0.33	0.4164
									0.6254
									0.6838
									1.0999
¹⁹¹ᵐTl			5.2 m	β+, EC/(98)/		(9/2+)	+3.903	−2.2	0.2157
									0.2647
									0.3256
									0.3359
¹⁹¹Tl		190.97179	2.2 m			(1/2)	+1.588		
¹⁹²ᵐTl			10.8 m	β+, EC/		(7+)	+0.518	+0.46	0.1740
									0.4228
									0.6348
									0.7863
									0.7455
¹⁹²Tl		191.97223	9.6 m	β+, EC/6.4		(2-)	+0.200	−0.33	0.3975
									0.4228
									0.6908
¹⁹³ᵐTl			2.1 m	I.T./(75)/		(9/2-)	+3.948	−2.20	0.3650
¹⁹³Tl		192.9707	22. m	β+, EC/3.6		(1/2+)	+1.591		0.2077
									0.3244
									0.3440
									0.6761
									1.0447
									1.5793
¹⁹⁴ᵐTl			32.8 m	β+ /(20)/∼ 0.30		(7+)	+0.540	+0.61	ann.rad./
				EC/(80)/					Hg k x-ray
									0.4282
									0.6363
									0.7490
¹⁹⁴Tl		193.9712	33.0 m	β+, EC/5.3		2-	+0.140	−0.28	0.4279/75.2
									0.6452/10.8
									(0.395–1.623)
¹⁹⁵ᵐTl			3.6 s	I.T./0.483		9/2-			Tl k x-ray
									0.0990
									0.3836
¹⁹⁵Tl		194.96977	1.16 h	EC/97/2.8		1/2+	+1.58		ann.rad./
				β+ /(3)/					Hg k x-ray
									0.2422
									0.5635
									0.8845
									1.3639
									(0.13–2.5)
¹⁹⁶ᵐTl			1.41 h	β+, EC/95/4.9		(7+)	+0.55	+0.76	0.0840
									0.4261
									0.6353
									0.6954
									(0.08–1.0)
¹⁹⁶Tl		195.97048	1.84 h	β+ /(15)/4.4		2-	+0.072	−0.18	ann.rad./
				EC/(85)/					Hg k x-ray
									0.4257
									0.6105
									(0.03–2.4)
¹⁹⁷ᵐTl			0.54 s	IT/53/0.608		9/2-			Tl k x-ray
				β+, EC/47/					0.2262

Elem. or Isot.	Natural Abundance (Atom %)	Atomic Mass or Weight	Half–life/ Resonance Width (MeV)	Decay Mode/ Energy (/MeV)	Particle Energy/ Intensity (MeV/%)	Spin ($h/2\pi$)	Nuclear Magnetic Mom. (nm)	Elect. Quadr. Mom. (b)	γ-Energy/ Intensity (MeV/%)
									0.4118
									0.5872
									0.6367
¹⁹⁷Tl		196.96958	2.83 h	β+ /(1)/2.18		1/2+	+1.58		Hg k x-ray
				EC/(99)/					0.1522/8.2
									0.4258
¹⁹⁸ᵐTl			1.87 h	β+, EC/(53)/		7+	+0.64		Hg k x-ray
				IT/47/0.5347					Tl k x-ray
									0.4118
									0.5872
									0.6367
¹⁹⁸Tl		197.9405	5.3 h	EC, β+ /(1)/3.5	1.4/	2-			Hg k x-ray
					2.1/				0.4118
					2.4/				0.6367
									0.6759
									(0.23–2.8)
¹⁹⁹Tl		198.96988	7.4 h	EC/1.4		1/2+	+1.60		Hg k x-ray
									0.2082
									0.2473
									0.4555
²⁰⁰Tl		199.97096	1.087 d	EC/2.46	1.07/	2-	0.04		Hg k x-ray
					1.44/				0.36799
									1.2057
									(0.11–2.3)
²⁰¹Tl		200.97082	3.038 d	EC/0.48		1/2+	+1.605		Hg k x-ray
									0.13528
									0.16740/10.0
²⁰²Tl		201.97211	12.47 d	EC/1.36		2-	0.06		Hg k x-ray
									0.43957
²⁰³Tl	29.52(1)	202.972344				1/2+	+1.622258		
²⁰⁴Tl		203.973864	3.78 a	β- /97/0.7637	0.763/97	2-	0.09		Hg k x-ray
				EC/(3)/0.347					
²⁰⁵Tl	70.48(1)	204.974428				1/2+	+1.638215		
²⁰⁶ᵐTl			3.76 m	I.T./2.644		12-			Tl k x-ray
									0.2166
									0.2661
									0.4534
									0.6866
									1.0219
²⁰⁶Tl		205.976110	4.20 m	β- /1.533	1.53/99.9	0-			Pb k x-ray
									0.80313
²⁰⁷ᵐTl			1.3 s	I.T./1.350		11/2-			Tl k x-ray
									0.3501
									1.0000
²⁰⁷Tl		206.97742	4.77 m	β- /1.423	1.43/99.8	1/2+	+1.88		0.89723
²⁰⁸Tl		207.982019	3.053 m	β- /5.001	1.28/23	(5+)	+0.29		Pb k x-ray
					1.52/22				0.27728
					1.796/51				0.51061
									0.58302
									2.61448
²⁰⁹Tl		208.98536	2.16 m	β-/3.98	1.8 /100	(1/2+)			Pb k x-ray
									1.5670/100
									0.4651/95
									(0.12–1.33)

Elem. or Isot.	Natural Abundance (Atom %)	Atomic Mass or Weight	Half–life/ Resonance Width (MeV)	Decay Mode/ Energy (/MeV)	Particle Energy/ Intensity (MeV/%)	Spin ($h/2\pi$)	Nuclear Magnetic Mom. (nm)	Elect. Quadr. Mom. (b)	γ-Energy/ Intensity (MeV/%)
^{210}Tl		209.99007	1.30 m	β- /5.48	1.3/25	(5+)			Pb k x-ray
					1.9/56				0.081
									0.2981
									0.79788
^{211}Tl		210.9935	> 0.3 μs	β-					
^{212}Tl		211.9982	> 0.3 μs	β-					
^{213}Tl									
^{214}Tl									
^{215}Tl									
^{216}Tl									
^{217}Tl									
$_{82}$Pb		207.2(1)							
^{178}Pb		178.00383	~ 0.2 ms			0+			
^{179}Pb		179.0022	~ 3.5 ms	α	7.35				
^{180}Pb		179.99792	4.2 ms	α/	7.25	0+			
^{181}Pb		180.9966	0.036 s	α/	7.02				
^{182}Pb		181.99267	55 ms	α	6.90	0+			
^{183m}Pb			0.42 s	α	6.70/82.7	13/2+	−1.25	~ 1.7	
					6.86/1.9				
^{183}Pb		182.99187	0.54 s	α/	6.57/4.3	(3/2-)	−1.16	~ 0.6	
					6.78/11.0				
^{184}Pb		183.98814	0.48 s	α/~ 80	6.63/	0+			
^{185m}Pb			4.3 s	α	6.41/100	13/2+	−1.19	~ 0.9	
^{185}Pb		184.98761	6.3 s	α/	6.29/56	3/2-	−1.10	~ 0.2	0.205
					6.49/44				0.269
					6.55/<1.4				
^{186}Pb		185.98424	5. s	β+, EC/95/5.5		0+			
				α/(5)/	6.32/				
					6.34/<100				
					6.01/<0.2				
^{187m}Pb			15.2 s	β+, EC/	5.99/	(1/2-)	−1.21	~ −0.5	0.0674
				α/12	6.19/				0.2080
									0.2755
									0.2995
									0.4487
									0.7477
^{187}Pb		186.98392	18.3 s	EC/7.2		13/2+	−1.13	~ −0.4	0.1930
				α/7	6.08/				0.3314
									0.3435
									0.3934
^{188}Pb		187.98087	23. s	EC/(78)/4.8		0+			0.1850
				α/(22)/	5.98/<10				0.7582
					5.61/<0.1				
^{189m}Pb			39 s			(13/2+)	−1.19	~ −1.3	
^{189}Pb		188.98081	51. s	EC/6.1		(3/2-)	−1.08	~ 0.5	
				α/	5.58/				
^{190}Pb		189.97808	1.2 m	β+ (13)/4.1		0+			ann.rad./
				EC/(86)/					Tl k x-ray
				α/(0.9)/	5.58/				0.1415
									0.1512
									0.9422
^{191m}Pb			2.2 m	β+, EC/		13/2+	−1.17	+0.085	ann.rad./
									0.3871
									0.6135

Elem. or Isot.	Natural Abundance (Atom %)	Atomic Mass or Weight	Half–life/ Resonance Width (MeV)	Decay Mode/ Energy (/MeV)	Particle Energy/ Intensity (MeV/%)	Spin ($h/2\pi$)	Nuclear Magnetic Mom. (nm)	Elect. Quadr. Mom. (b)	γ-Energy/ Intensity (MeV/%)
									0.7122
¹⁹¹Pb		190.97827	1.3 m	β+, EC/5.5					ann.rad./
									0.9368
¹⁹²Pb		191.97579	3.5 m	β+, EC/~ 3.4		0+			ann.rad./
				α/.006/	5.11				0.1675
									0.6082
									1.1954
¹⁹³ᵐPb			5.8 m	β+, EC/		13/2+	−1.15	+0.19	ann.rad./
									0.3650
									0.3922
¹⁹³Pb		192.97617	~ 2. m	EC/5.2		3/2 (-)			
¹⁹⁴Pb		193.97401	10. m	β+, EC/2.7		0+			ann.rad./
				α	4.64				0.2036
¹⁹⁵ᵐPb			15. m	β+ /(8)/		13/2+	−1.132	+0.31	ann.rad./
				EC/(92)/					Tl k x-ray
									0.3836
									0.3942
									0.8784
¹⁹⁵Pb		194.97454	~ 15. m	β+, EC/5.8					ann.rad./
									0.3836
									0.3937
									0.7776
¹⁹⁶Pb		195.97277	37. m	β+, EC/2.1		0+			Tl k x-ray
									0.2531
									0.5021
¹⁹⁷ᵐPb			43. m	EC/79/		13/2+	−1.105	+0.38	Tl k x-ray
				β+ /2/					0.3079
				IT/19/0.3193					0.3877
									0.7743
									(0.2–2.2)
¹⁹⁷Pb		196.97343	~ 8. m	EC/97/3.6		(3/2-)	−1.075	~ −0.08	Tl k x-ray
				β+ /3/					0.3755
									0.3858
									0.7611
¹⁹⁸Pb		197.97203	2.4 h	EC/1.4		0+			Tl k x-ray
									0.1734
									0.2903
									0.3654
¹⁹⁹ᵐPb			12.2 m	IT/93/0.4248		13/2+			Pb k x-ray
				β+, EC/(7)/					0.4255
¹⁹⁹Pb		198.97292	1.5 h	EC/(99)/2.9		5/2-	−1.074	~ +0.08	Tl k x-ray
				β+ /(1)/					0.3534
									0.7202
									1.1350
									(0.22–2.4)
²⁰⁰Pb		199.97183	21.5 h	EC/0.81		0+			Tl k x-ray
									0.14763
²⁰¹ᵐPb			1.02 m	I.T./0.6291		13/2+			Pb k x-ray
									0.6288
²⁰¹Pb		200.97289	9.33 h	EC/1.90		5/2-	+0.675	~ −0.009	Tl k x-ray
									0.33120
									0.36131
									(0.11–1.8)
²⁰²ᵐPb			3.53 h	IT/90/2.170		9-	−0.228	~ +0.58	Pb k x-ray

Elem. or Isot.	Natural Abundance (Atom %)	Atomic Mass or Weight	Half–life/ Resonance Width (MeV)	Decay Mode/ Energy (/MeV)	Particle Energy/ Intensity (MeV/%)	Spin ($h/2\pi$)	Nuclear Magnetic Mom. (nm)	Elect. Quadr. Mom. (b)	γ-Energy/ Intensity (MeV/%)
				β+ /10/					Tl k x-ray
									0.42219
									0.78700
									0.96271
^{202}Pb		201.97216	5.3×10^4 a	EC/0.05		0+			Tl L x-ray
^{203m}Pb			6.2 s	I.T./0.8252		13/2+			Pb k x-ray
									0.8203
									0.8252
^{203}Pb		202.97339	2.163 d	EC/0.98		5/2-	+0.686	+0.10	Tl k x-ray
									0.279188
^{204m}Pb			1.13 h	I.T./2.185		9-			Pb k x-ray
									0.37481
									0.89922
									0.91175
^{204}Pb	1.4(1)	203.973044				0+			
^{205}Pb		204.974482	1.51×10^7 a	EC/0.0512		5/2-	+0.712	+0.23	Tl L x-ray
^{206}Pb	24.1(1)	205.974465				0+			
^{207m}Pb			0.80 s	I.T./1.632		13/2+			Pb k x-ray
									0.56915
									1.06310
^{207}Pb	22.1(1)	206.975897				1/2-	+0.59258		
^{208}Pb	52.4(1)	207.976652	$> 2 \times 10^{19}$ a	sf		0+			
^{209}Pb		208.981090	3.25 h	β- /0.644	0.645/100	9/2+	−1.474	−0.3	
^{210}Pb		209.984189	22.6 a	β- /0.0635	0.017/81	0+			
					0.061/19				
				α	3.72				
^{211}Pb		210.988737	36.1 m	β- /1.37	0.57/5	(9/2+)	−1.404	+0.09	0.40486
					1.36/92				0.42700
									0.83186
									(0.09−1.27)
^{212}Pb		211.991898	10.64 h	β- /0.574	0.28/83	0+			Bi k x-ray
					0.57/12				0.23858
^{213}Pb		212.99658	10.2 m	β- /2.1					
^{214}Pb		213.999805	26.9 m	β- /1.0	0.67/48	0+			Bi k x-ray
					0.73/42				0.24192
									0.29509
									0.35187
^{215}Pb		215.0048	36 s						
^{216}Pb									
^{217}Pb									
^{218}Pb									
^{219}Pb									
^{220}Pb									
$_{83}$Bi		208.98040(1)							
^{184m}Bi			0.007 s	α	(7.22−7.85)				0.449
^{184}Bi		184.0011	13. ms	α					0.124
^{185}Bi		184.9976	60. µs	p/90	1.55				
				α/10	8.03				
^{186m}Bi			15. ms	α	7.07−7.23				(0.087−0.520)
				p/<0.5					
^{186}Bi		185.9966	9.8 ms	α	7.26				0.1085
					7.37				
^{187m}Bi			0.37 ms	α/12	7.72/100	1/2+			
^{187}Bi		186.99316	40. ms	α/7	7.00/88.	9/2-			

Elem. or Isot.	Natural Abundance (Atom %)	Atomic Mass or Weight	Half–life/ Resonance Width (MeV)	Decay Mode/ Energy (/MeV)	Particle Energy/ Intensity (MeV/%)	Spin ($h/2\pi$)	Nuclear Magnetic Mom. (nm)	Elect. Quadr. Mom. (b)	γ-Energy/ Intensity (MeV/%)
					7.61/9.0				
					7.34/3.0				
^{188}Bi		187.99227	0.271 s	α	6.81				(0.071–0.320)
^{189m}Bi			7.0 ms	α	7.30				
^{189}Bi		188.9892	0.68 s	α					
^{190m}Bi			5.7 s	α/90	6.43				(0.105–0.314)
					(6.23–6.72)				
^{190}Bi		189.9883	~ 5.9 s	β+, EC/(30)/8.7	α/6.45				(0.089–0.374)
				α/70	(6.39–6.82)				
^{191m}Bi			0.12 ms	α/	6.87/100				
^{191}Bi		190.98579	12.4 s	β+, EC/(60)/7.3					
				α/(40)/	6.31				
^{192}Bi		191.98546	40. s	β+, EC/(80)/9.0					
				α/(20)/	6.06/				
^{193m}Bi			3.2 s	β+, EC/		1/2+			
				α/	6.48/				
^{193}Bi		192.98296	1.11 m	β+, EC/40/7.1		9/2+			
				α/(60)/	5.91/				
^{194}Bi		193.98283	1.8 m	β+, EC/99.9/8.2		(10-)			0.1661
				α/0.1/					0.1740
									0.2802
									0.421
									0.5754
									0.9650
^{195m}Bi			1.45 m	β+, EC/(94)/					
				α/(6)/	6.11/				
^{195}Bi		194.98065	2.9 m	β+, EC/99.8/5.8		3/2-			
				α/(0.2)	5.45/				
^{196}Bi		195.98067	5. m	EC/~ 7.4					0.1376
									0.3720
									0.6880
									1.0486
^{197}Bi		196.97886	5. m	β+, EC/5.2		1/2+			
^{198m}Bi			7.7 s	I.T./0.2485		(10-)			0.2485
^{198}Bi		197.97921	11.8 m	β+, EC/6.6		(7+)			0.0900
									0.1976
									0.5624
									1.0635
^{199m}Bi			24.7 m	β+, EC/					ann.rad./
^{199}Bi		198.97767	27. m	β+, EC/4.3		9/2-	4.6		0.7203
									0.8374
									0.8417
									0.9460
									1.0528
									1.3056
									(0.12–3.2)
^{200m}Bi			31. m	β+, EC/		(2+)			0.2453
									0.4198
									0.4624
									1.0265
^{200}Bi		199.97813	36. m	EC/(90)/5.9		7+			ann.rad./
				β+ /(10)/					Pb k x-ray
									0.4198
									0.4623

Elem. or Isot.	Natural Abundance (Atom %)	Atomic Mass or Weight	Half–life/ Resonance Width (MeV)	Decay Mode/ Energy (/MeV)	Particle Energy/ Intensity (MeV/%)	Spin (h/2π)	Nuclear Magnetic Mom. (nm)	Elect. Quadr. Mom. (b)	γ-Energy/ Intensity (MeV/%)
									1.0265
^{201m}Bi			59.1 m	I.T./0.846		(1/2+)			Bi k x-ray
				β+, EC/					0.8464
^{201}Bi		200.97701	1.8 h	EC/3.84		9/2-	4.8		Pb k x-ray
									0.6288
									0.9357
									1.0138
									(0.13–2.4)
^{202}Bi		201.97774	1.72 h	β+ /(3)/5.16		5+	+4.26	−0.9	ann.rad./
				EC/(97)/					Pb k x-ray
									0.57860
									0.92734
									(0.08–3.5)
^{203}Bi		202.97688	11.8 h	EC/99.8/3.25		9/2-	+4.02	−0.8	Pb k x-ray
				β+ /(0.2)/	1.35/				0.1865
									0.8203
									0.8969
									1.8475
									(0.1–2.9)
^{204}Bi		203.97781	11.2 h	EC/4.44		6+	+4.32	−0.43	Pb k x-ray
									0.37481
									0.89922
									0.98409
^{205}Bi		204.97739	14.9 d	EC/2.71		9/2-	+4.07	−0.7	Pb k x-ray
									0.70347
									1.76435
									(0.550–1.862)
^{206}Bi		205.97850	6.243 d	EC/3.76		6+	+4.36	−0.4	Pb k x-ray
									0.51619
									0.80313
									0.88100
^{207}Bi		206.978471	31.55 a	EC/2.399		9/2-	4.092	−0.6	Pb k x-ray
									0.56915
									1.06310
^{208m}Bi			2.58 ms	IT		10-	2.67		0.921
^{208}Bi		207.979742	3.68×10^5 a	EC/2.880		5+	4.60	−0.64	Pb k x-ray
									2.61435
^{209}Bi	100.	208.980399	1.9×10^{19} a	α	3.13	9/2-	+4.111	−0.6	
^{210m}Bi			3.0×10^6 a	α/	4.420(3)/0.29	9-	+2.73	−0.6	Tl k x-ray
					4.569(3)/3.9				0.2661
					4.584(3)/1.4				0.3052
					4.908(4)/39				0.6502
					4.946(3)/55				
^{210}Bi		209.984120	5.01 d	β- /1.163	1.16/99	1-	−0.0445	+0.136	0.2661
									0.3.52
^{211}Bi		210.98727	2.14 m	α/(99.7)/	6.279/16	9/2-			Tl k x-ray
				β- /(0.3)/0.58	6.623/84				0.3501
^{212m2}Bi			7. m	β- /		(15-)			
^{212m1}Bi			25.0 m	α/(93)/	6.300/40	(9-)			0.120
				β- /(7)/	6.340/53				0.233
									0.275
									0.404
									0.727
^{212}Bi		211.991286	1.009 h	β- /(64)/2.254		(1-)	+0.32	~ +0.1	Tl k x-ray

Elem. or Isot.	Natural Abundance (Atom %)	Atomic Mass or Weight	Half–life/ Resonance Width (MeV)	Decay Mode/ Energy (/MeV)	Particle Energy/ Intensity (MeV/%)	Spin ($h/2\pi$)	Nuclear Magnetic Mom. (nm)	Elect. Quadr. Mom. (b)	γ-Energy/ Intensity (MeV/%)
				α/(36)/	6.051/25				Po k x-ray
					6.090/9.6				0.2881
									0.72725
									0.78551
									1.62066
^{213}Bi		212.994385	45.6 m	β- /(98)/1.43	1.02/31	9/2-	+3.72	−0.71	Po k x-ray
				α/(2)/	1.42/66				0.4404
					5.549/0.16				(0.15–1.328)
					5.869/2.0				
									1.10006
^{214}Bi		213.99871	19.7 m	β- /3.27					0.60931
									1.12027
									1.76449
									(0.19–3.2)
^{215m}Bi			37. s	β					(0.158–0.498)
^{215}Bi		215.00177	7.7 m	β- /2.3					0.2937/35.2
									(0.271–1.399)
^{216}Bi		216.00631	2.3 m	β-/4.0					0.5498
									0.4192
^{217}Bi		217.0095	98 s	β/					0.2646/100
									(0.254–1.017)
^{218}Bi		218.0143	33. s	β-					0.5097/134
									0.3857/100
									(0.174–0.703)
^{219}Bi									
^{220}Bi									
^{221}Bi									
^{222}Bi									
^{223}Bi									
^{224}Bi									
$_{84}$**Po**									
^{187}Po			1.4 ms	α	7.53/100				0.286
^{188}Po		187.99942	0.27 ms	α	7.91/80	0+			
					7.320				
^{189}Po		188.99848	5 ms	α	7.532/8				
					7.259/80				
					7.309/12				
^{190}Po		189.99510	2.4 ms	α/	7.53/96.4	0+			
					7.01/3.3				
^{191m}Po			93. ms	α	7.376/50				
					6.888/46				
^{191}Po		190.99457	22 ms	α/	7.334/77				
					6.97/8				
^{192}Po		191.99134	32. ms	α/8.5	7.17/98.6	0+			
					6.59/1.4				
^{193m}Po			~ 0.07 s	α/	7.00				
^{193}Po		192.99103	0.45 s	α/	6.95				
^{194}Po		193.98819	0.2 s	α/	6.84/93	0+			
					6.19/0.22				
^{195m}Po			~ 2.8 s	α/	6.70/				
^{195}Po		194.98811	~ 3.9 s	α/	6.62/				
^{196}Po		195.98554	5. s	α/(95)/	6.53/94	0+			
				β+, EC/(5)/~4.6	5.77/0.02				
^{197m}Po			25.8 s	α/(84)/	6.385(3)/55	13/2+			

Elem. or Isot.	Natural Abundance (Atom %)	Atomic Mass or Weight	Half–life/ Resonance Width (MeV)	Decay Mode/ Energy (/MeV)	Particle Energy/ Intensity (MeV/%)	Spin ($h/2\pi$)	Nuclear Magnetic Mom. (nm)	Elect. Quadr. Mom. (b)	γ-Energy/ Intensity (MeV/%)
				β+, EC/(16)/					
^{197}Po		196.98566	53. s	α/(44)/	6.282(4)/76	(3/2-)			
				β+, EC/(56)/6.2					
^{198}Po		197.98339	1.76 m	α/(70)/	6.18/57	0+			
				β+, EC/(30)/4.0	5.27/7.6 × 10^{-4}				
^{199m}Po			4.2 m	β+, EC/(51)/		13/2+	~ 0.99		ann.rad./
				α/(39)/	6.059/24				0.2745
									0.4998
									1.0020
^{199}Po		198.98367	5.2 m	β+, EC/(88)/7.		(3/2-)			Bi k x-ray
				α/(12)/	5.952/7.5				0.1877
									0.3616
									1.0214
									1.0344
^{200}Po		199.981780	11.5 m	β+, EC/85/3.4		0+			0.14748
				α/(15)/	5.863/11.1				0.32792
									0.6176
									0.6709
^{201m}Po			8.9 m	β+, EC/(57)/		13/2+	~ 1.00		Bi k x-ray
				IT/40/0.418					Po k x-ray
				α/(3)/	5.786/~ 3.				0.2726
									0.4123
									0.4179
									0.9670
^{201}Po		200.98226	15.3 m	β+, EC/98/4.9		3/2-	~ 0.94		Bi k x-ray
				α/(2)/	5.683(3)/1.1				0.2056
									0.2250
									0.8483
									0.9048
^{202}Po		201.98076	45. m	β+, EC/98/2.8		0+			0.0410
				α/(2)/	5.588/1.9				0.1656
									0.3158
									0.6884
^{203m}Po			1.2 m	IT/96/0.6414		13/2+			Bi k x-ray
				β-EC/(4)/					Po k x-ray
									0.6414
^{203}Po		202.98142	35. m	β+, EC/4.2		5/2-	+0.74		0.17516
									0.21477
									0.89350
									0.90863
									1.09095
^{204}Po		203.98032	3.53 h	EC/2.34		0+			Bi k x-ray
				α	5.377/0.66				0.2702
									0.8844
									1.0162
									(0.11–1.9)
^{205}Po		204.98120	1.7 h	β+, EC/3.53		5/2-	~ +0.76	+0.17	Bi k x-ray
									0.83681
									0.84983
									0.87241
									1.00124
									(0.12–2.7)
^{206}Po		205.98048	8.8 d	EC/(95)/1.85		0+			Bi k x-ray
				α/(5)/	5.223/5.5				0.28644

Elem. or Isot.	Natural Abundance (Atom %)	Atomic Mass or Weight	Half–life/ Resonance Width (MeV)	Decay Mode/ Energy (/MeV)	Particle Energy/ Intensity (MeV/%)	Spin ($h/2\pi$)	Nuclear Magnetic Mom. (nm)	Elect. Quadr. Mom. (b)	γ-Energy/ Intensity (MeV/%)
									0.31156
									0.51134
									0.80737
									1.03228
									(0.11–1.5)
^{207m}Po			2.8 s	I.T./1.383		19/2-			Po k x-ray
									0.2682
									0.30074
									0.81448
^{207}Po		206.98159	5.80 h	EC, β+/2.91		5/2-	~ +0.79	+0.28	Bi k x-ray
									0.74263
									0.91176
									0.99225
^{208}Po		207.981246	2.898 a	α/5.213	4.233/0.0002	0+			
					5.1158/100				
^{209}Po		208.982430	128. a	α/4.976	4.624/0.56	1/2-	+0.7		0.26049
					4.879/99.2				0.8964
^{210}Po		209.982874	138.4 d	α/5.407	4.516/0.001	0+			0.80313
					5.304/100				
^{211m}Po			25.2 s	α/	7.273/91	25/2+			Pb k x-ray
					7.994/1.7				0.32808
					8.316/0.25				0.56915
					8.875/7.0				0.89723
									1.06310
^{211}Po		210.986653	0.516 s	α/7.594	6.570/0.54	9/2+			0.56915
					6.892/0.55				0.89723
					7.450/98.9				
^{212m}Po			45. s	α/	8.514/2.0	16+			
					9.086/1.0				
					11.650/97				
^{212}Po		211.988868	0.298 μs	α/8.953	8.784/100	0+			
^{213}Po		212.992857	3.7 μs	α/8.537	7.614/0.003	9/2+			
					8.375/100				
^{214}Po		213.995201	163.7 μs	α/7.833	6.904/0.01	0+			0.7995
					7.686/99.99				0.298
^{215}Po		214.999420	1.780 ms	α/7.526	6.950/0.02	(9/2+)			
					6.957/0.03				
					7.386/100				
^{216}Po		216.001915	0.145 s	α/6.906	5.895/0.002	0+			
					6.778/99.99				
^{217}Po		217.00634	1.53 s	α/6.662	6.539/				
^{218}Po		218.008973	3.04 m	α/6.114	6.003/99.999	0+			
					5.181/0.11				
^{219}Po		219.0137	~ 2 m						
^{220}Po		220.0166	> 0.3 μs			0+			
^{221}Po									
^{222}Po									
^{223}Po									
^{224}Po									
^{225}Po									
^{226}Po									
^{227}Po									
$_{85}$At									
^{191m}At			2.1 ms	α	7.65/98				

Elem. or Isot.	Natural Abundance (Atom %)	Atomic Mass or Weight	Half–life/ Resonance Width (MeV)	Decay Mode/ Energy (/MeV)	Particle Energy/ Intensity (MeV/%)	Spin ($h/2\pi$)	Nuclear Magnetic Mom. (nm)	Elect. Quadr. Mom. (b)	γ-Energy/ Intensity (MeV/%)
					7.72/2				
^{191}At			~ 1.7 ms	α	7.55/100				
^{192m}At			0.012 s	α	7.435/56				0.036
					7.47/31				
^{192}At			0.09 s	α	7.22/82				0.165
					7.385/14				0.188
^{193m}At			21 ms	α	7.33/98				
					7.42/2				
^{193}At		192.9998	28 ms	α/	7.24/100				
^{194}At		193.9987	40 ms	α/					
^{195m}At			147 ms	α	7.07–7.22				
^{195}At		194.99627	0.33 s	α/	6.95				
^{196m}At			8 μs						0.158
^{196}At		195.9958	0.39 s	α/	7.05/				
^{197m}At			2.0 s	α	6.707	(1/2+)			
^{197}At		196.99319	0.39 s	β+, EC/7.8		(9/2-)			
				α/	6.960/				
^{198m}At			1.0 s	β+, EC/(75)/					
				α/(25)/	6.856/86				
^{198}At		197.99284	4.1 s	α/	6.755/94				
^{199}At		198.99053	6.9 s	β+, EC/8/5.6		9/2-			
				α/(92)/	6.643/				
^{200m2}At			47. s	α	6.411/				
^{200m}At			3.5 s	β+, EC/(80)		10-			
				α/(20)/	6.538/12				
^{200}At		199.99035	43. s	β+, EC/65/~ 8.0		5+			
				α/(35)/	6.412/44				
					6.465/57				
^{201}At		200.98842	1.48 m	β+, EC/29/5.9		9/2-			0.5918/100
				α/(71)/6.474	6.344/				(0.3585–0.761)
^{202m}At			0.46 s	I.T./0.391					
^{202}At		201.98863	3.02 m	β+, EC/88/7.2		5+			ann.rad./
				α/(12)/	6.135/7.7				0.4413
					6.225/4.3				0.5697
									0.6753
^{203}At		202.98694	7.4 m	β+, EC/69/5.1		9/2-			0.1458
				α/(31)/6.210	6.088/				0.2459
									0.6414
									1.0020
									1.0340
^{204}At		203.98725	9.1 m	β+, EC/95/6.5		(5+)			Po k x-ray
				α/(5)/	5.951/				0.3271
									0.4254
									0.5156
									0.6837
^{205}At		204.98607	26. m	β+, EC/90/4.54		(9/2-)			Po k x-ray
				α/(10)/6.020	5.902/				0.1543
									0.6696
									0.7194
^{206}At		205.98667	29.4 m	β+, EC/99/5.72		5+			Po k x-ray
				α/(1)/5.881	5.703/				0.20186
									0.39561
									0.47716
									0.70071

Table of the Isotopes

Elem. or Isot.	Natural Abundance (Atom %)	Atomic Mass or Weight	Half–life/ Resonance Width (MeV)	Decay Mode/ Energy (/MeV)	Particle Energy/ Intensity (MeV/%)	Spin ($h/2\pi$)	Nuclear Magnetic Mom. (nm)	Elect. Quadr. Mom. (b)	γ-Energy/ Intensity (MeV/%)
207At		206.98578	1.81 h	β+, EC/90/3.91		9/2-			Po k x-ray
				α/(10)/5.873	5.758/				0.16801
									0.58842
									0.81448
208At		207.98650	1.63 h	β+, EC/99/4.97		(6+)			Po k x-ray
				α/(1)/5.752	5.626/0.01				0.1770
					5.641/0.53				0.2060
									0.6601
									0.6852
									0.8450
									1.0281
209At		208.98617	5.4 h	β+, EC/96/3.49		(6+)			Po k x-ray
				α/(4)/5.757	5.647/4.1				0.10422
									0.54503
									0.78189
									0.79020
									(0.1–2.6)
210At		209.98715	8.1 h	EC/99.8/3.98		5+			Po k x-ray
				α/(0.2)/5.632	5.361/0.05				0.24535
					5.442/0.05				0.52758
									1.18143
									1.43678
									1.48335
									(0.04–2.4)
211At		210.987496	7.21 h	EC/(58)/0.787		9/2-			Po k x-ray
				α/(42)/5.980	5.211/0.004				0.66956
					5.868/42				0.6870
									0.74263
212mAt			0.119 s	α/	7.837/65	(9-)			
					7.897/33				
212At		211.99075	0.314 s	α/7.828	7.058/0.4	(1-)			
					7.088/0.6				
					7.618/15				
					7.681/84				
213At		212.992937	0.11 μs	α/9.254	9.080/	9/2-			
214mAt			0.76 μs	α/8.762		(9-)			
214At		213.996372	0.56 μs	α/8.987	8.819/100	(1-)			
215At		214.99865	0.10 ms	α/8.178	7.626/0.045	(9/2-)			0.40486
					8.023/99.9				
216At		216.002423	0.30 ms	α/7.947	7.595/0.2	(1-)			
					7.697/2.1				
					7.800/97				
217At		217.004719	32. ms	α/7.202	6.812/0.06	(9/2-)	3.8		0.2595
					7.067/99.9				0.3345
									0.5940
218At		218.00869	1.6 s	α/6.883	6.654/6				
					6.695/90				
					6.748/4				
219At		219.011162	50. s	α/6.390	6.275/				
220At		220.0154	3.71 m	β- /3.7					(0.24–0.70)
221At		221.0181	2.3 m	β					
222At		222.0223	0.9 m	β					
223At		223.0252	50. s	β					
224At									

Elem. or Isot.	Natural Abundance (Atom %)	Atomic Mass or Weight	Half–life/ Resonance Width (MeV)	Decay Mode/ Energy (/MeV)	Particle Energy/ Intensity (MeV/%)	Spin ($h/2\pi$)	Nuclear Magnetic Mom. (nm)	Elect. Quadr. Mom. (b)	γ-Energy/ Intensity (MeV/%)
^{225}At									
^{226}At									
^{227}At									
^{228}At									
^{229}At									
$_{86}$Rn									
^{193}Rn			1.2 ms	α	7.69/74				0.194
					7.88/26				
^{194}Rn			0.8 ms	α	7.70				
^{195m}Rn			5 ms	α	7.56				
^{195}Rn		195.00544	6 ms	α	7.54				
^{196}Rn		196.00212	4. ms	α/	7.46	0+			
^{197m}Rn			0.02 s	α	7.36				
^{197}Rn		197.0016	0.07 s	α/	7.26				
^{198}Rn		197.99868	64. ms	α	7.205	0+			
^{199m}Rn			0.32 s	α	7.060	(13/2+)			
^{199}Rn		198.9984	0.62 s	α/	6.989	3/2-			
^{200}Rn		199.99570	1.06 s	α/(98)/	6.901/	0+			0.4329
				EC/(2)/5.					0.5043
^{201m}Rn			3.8 s	EC/(10)/		13/2+			
				α/(90)/	6.773/				
^{201}Rn		200.9956	7.0 s	α/(80)/	6.725/	(3/2-)			
				EC/(20)/	α/6.778				
^{202}Rn		201.99326	9.9 s	α/(12)/	6.641/	0+			0.5695
				EC/(88)/					0.288–0.6255
^{203m}Rn			28. s	α/	6.551	13/2+	−0.96	+1.3	
^{203}Rn		202.99339	45. s	α/(66)/6.629	6.499/	3/2-			
				EC/(34)/~ 7.4					
^{204}Rn		203.99143	1.24 m	α/(68)/	6.420/	0+			
				EC/(32)/3.8					
^{205}Rn		204.99172	2.8 m	α/(23)/6.390	6.123(3)/0.02	(5/2-)	+0.80	+0.06	0.2652
				EC/(77)/5.2	6.262(3)/23				0.3553
									0.4648
									0.6205
									0.6753
									0.7300
^{206}Rn		205.99021	5.7 m	α/(68)/6.384	6.258(3)/	0+			0.06170
				EC/(32)/3.3					0.0968
									0.3245
									0.3862
									0.4822
									0.4973
									0.7728
^{207m}Rn			0.18 ms						
^{207}Rn		206.99073	9.3 m	β+, EC/77/4.6		5/2-	+0.82	+0.22	At k x-ray
				α/(23)/6.252	5.995(4)/0.02				0.32947
					6.068(3)/0.15				0.34455
					6.126(3)/22.8				0.36767
									0.40267
									0.74723
									(0.18–1.4)
^{208}Rn		207.98964	24.3 m	α/(60)/6.260	5.469(2)/0.003	0+			
				EC/(40)/2.85	6.140(2)/60				
^{209}Rn		208.99042	29. m	β+ /(83)/3.93	2.16/2.3	5/2-	+0.8388	+0.31	At k x-ray

Elem. or Isot.	Natural Abundance (Atom %)	Atomic Mass or Weight	Half–life/ Resonance Width (MeV)	Decay Mode/ Energy (/MeV)	Particle Energy/ Intensity (MeV/%)	Spin ($h/2\pi$)	Nuclear Magnetic Mom. (nm)	Elect. Quadr. Mom. (b)	γ-Energy/ Intensity (MeV/%)
				α/(17)/	5.887(3)/0.04				0.27933
					5.898(3)/0.02				0.33753
					6.039(2)/16.9				0.40841
									0.68942
									0.74594
									(0.18–3.2)
^{210}Rn		209.98970	2.4 h	α/(96)/6.157	5.351(2)/0.005	0+			At k x-ray
				EC/(4)/2.37	6.039(2)/96				0.19625
									0.45824
									0.57104
									0.64868
									(0.14–1.7)
^{211}Rn		210.99060	14.6 h	β+, EC/74/2.89		1/2-	+0.60		At k x-ray
				α/(26)/5.964	5.619(1)/0.7				0.16877
					5.784(1)/16.4				0.25022
					5.851(1)/8.8				0.37049
									0.67412
									0.67839
									1.36298
									(0.11–2.7)
^{212}Rn		211.990704	24. m	α/6.385	5.587(4)/0.05	0+			
					6.260(4)/99.95				
^{213}Rn		212.99388	19. ms	α/8.243	7.552(8)/1.0	9/2+			0.540
					8.087(8)/98.2				
					7.254/0.8				
^{214}Rn		213.99536	0.27 μs	α/9.209	9.037(9)/	0+			
^{215}Rn		214.99875	2.3 μs	α/8.840	8.674(8)/	(9/2+)			
^{216}Rn		216.00027	45. μs	α		0+			
^{217}Rn		217.003928	0.6 ms	α/7.885	7.500/0.1	9/2+			
					7.742(4)/100				
^{218}Rn		218.005601	35. ms	α/7.267	6.534(1)/0.16	0+			0.6093
					7.133(1)/99.8				0.6653
^{219}Rn		219.009480	3.96 s	α/6.946(1)	6.3130(5)/0.05	(5/2+)	−0.44	+1.0	Po k x-ray
					6.425(3)/7.5				0.13057
					6.5309(4)/0.12				0.27113
					6.5531(3)/12.2				0.40170
					6.8193(3)/81				(0.1–1.05)
^{220}Rn		220.011394	55.6 s	α/6.404	5.7486(5)/0.07	0+			
					6.2883(1)/99.9				
^{221}Rn		221.01554	25. m	α/(22)/6.148	5.778(3)/1.8	7/2+	−0.020	−0.4	Fr L x-ray
				β- /(78)/1.2	5.788(3)/2.2				0.07384
					6.037(3)/18				0.08323
									0.0610
									0.18639
^{222}Rn		222.017578	3.823 d	α/5.590	4.987(1)/0.08	0+			0.510
					5.4897(3)/99.9				
^{223}Rn		223.0218	23. m	β- /			−0.78	~ +0.80	
^{224}Rn		224.0241	1.8 h	β- /		0+			0.1085
									0.2601
									0.2655
^{225}Rn		225.0284	4.5 m	β- /		7/2-	−0.70	~ +0.84	
^{226}Rn		226.0309	7.4 m	β- /		0+			
^{227}Rn		227.0354	2. s	β- /					
^{228}Rn		228.0380	65. s	β- /		0+			

Elem. or Isot.	Natural Abundance (Atom %)	Atomic Mass or Weight	Half–life/ Resonance Width (MeV)	Decay Mode/ Energy (/MeV)	Particle Energy/ Intensity (MeV/%)	Spin ($\hbar/2\pi$)	Nuclear Magnetic Mom. (nm)	Elect. Quadr. Mom. (b)	γ-Energy/ Intensity (MeV/%)
^{229}Rn			12 s						
^{230}Rn									
^{231}Rn									
$_{87}$**Fr**									
^{199}Fr		199.00726	12 s	α	7.66				
^{200}Fr		200.0066	49 ms	α	7.47				
^{201m}Fr			~ 0.02 s	α/	7.454				
^{201}Fr		201.0039	~ 60 ms	α/	7.36/	(9/2-)			
^{202m}Fr			0.29 s	α	7.236/				
^{202}Fr		202.00337	0.30 s	α/7.590	7.24/100				
^{203}Fr		203.00093	0.54 s	α/7.280	7.132(5)/	(9/2-)			
^{204m2}Fr			0.8 s	α	7.01				
^{204m1}Fr			2. s	α	6.97				
^{204}Fr		204.00065	1.8 s	α/	7.03/96				
					6.97/90				
					7.01/74				
^{205}Fr		204.99859	3.9 s	α/7.050	6.914(5)/	(9/2-)			0.5647/100
									(0.356–0.657)
^{206m}Fr			0.7 s	α/	6.93				0.531(IT)
^{206}Fr		205.99867	16.0 s	α/7.416	6.792(5)/84				
^{207}Fr		206.99695	14.8 s	α/6.900	6.766(5)/	9/2-	+3.9	−0.16	
^{208}Fr		207.99714	59.1 s	α/(77)/6.770	6.636(5)/	7+	+4.8		
				EC/(23)/6.99					
^{209}Fr		208.99595	50.0 s	α/(89)/5.1	6.646(3)/	9/2-	+3.9	−0.24	0.7978
				EC/(11)/5.16					(0.110–1.384)
^{210}Fr		209.99641	3.2 m	α/6.670/71	6.543(5)/99.87	6+	+4.4	+0.19	0.2030
				EC/6.26	(5.90–6.42)				0.6438
									0.8175
									0.9008
^{211}Fr		210.99554	3.10 m	α/6.660/87	6.534(5)/99.94	9/2-	+4.0	−0.19	0.220
				EC/4.61	(5.87–6.20)				0.2799
									0.5389
									0.9169
^{212}Fr		211.99620	20. m	EC/(57)/5.12	6.261(1)/16	(5+)	+4.6	−0.10	Rn x-ray
				α/(43)/6.529	6.335(1)/4				0.08107
					6.335(1)/4				0.08378
					6.343(1)/1.3				0.2277
					6.383(1)/10				1.1856
					6.406(1)/9.5				1.2748
					6.08–6.18				0.014–1.178
^{213}Fr		212.99619	34.6 s	α/6.905	8.476(4)/51	9/2-	+4.0	−0.14	(0.408–0.577)
^{214m}Fr			3.4 ms	α/	8.547(4)/46	9-			
					6.775–8.046				
^{214}Fr		213.99897	5.0 ms	α/8.587	7.409(3)/0.3	(1-)			(0.073–0.966)
					7.605(8)/1.0				
					7.940(3)/1.0				
					8.355(3)/4.7				
					8.427(3)/93				
^{215}Fr		215.00034	0.12 μs	α/9.537	9.360(8)/	(9/2-)			
^{216}Fr		216.00320	0.70 μs	α/9.175	9.005(10)/95				(0.045–0.160)
^{217}Fr		217.00463	0.016 ms	α/8.471	8.315(8)/	(9/2-)			
^{218m}Fr			22. ms	α					
^{218}Fr		218.007578	1. ms	α/8.014	7.384(10)/0.5	(1-)			
					7.542(15)/1.0				

Elem. or Isot.	Natural Abundance (Atom %)	Atomic Mass or Weight	Half–life/ Resonance Width (MeV)	Decay Mode/ Energy (/MeV)	Particle Energy/ Intensity (MeV/%)	Spin ($h/2\pi$)	Nuclear Magnetic Mom. (nm)	Elect. Quadr. Mom. (b)	γ-Energy/ Intensity (MeV/%)
					7.572(10)/5				
					7.732(10)/0.5				
					7.867(2)/93				
^{219}Fr		219.00925	21. ms	α/8.132	6.802(2)/0.25	(9/2-)			
					6.967(2)/0.6				
					7.146(2)/0.25				
					7.313(2)/99				
^{220}Fr		220.012327	27.4 s	α/6.800	6.582(1)/10	1+	−0.67	+0.47	0.0450
					6.630(2)/6				0.061
					6.641(1)/12				0.1060
					6.686(1)/61				0.1539
					6.39–6.58				0.1617
^{221}Fr		221.014255	4.78 m	α/6.457	5.9393(7)/0.17	(5/2-)	+1.58	−1.0	At k x-ray
					5.9797(7)/0.49				0.0995
					6.0751(7)/0.15				0.21798
					6.1270(7)/				0.4091
					6.2433(3)/1.3				
					6.3410(7)/83.4				
^{222}Fr		222.01755	14.3 m	β- /2.03	1.78/	2-	+0.63	+0.51	
				α/5.850					
^{223}Fr		223.019736	22.0 m	β- /1.149	α/5.291	(3/2+)	+1.17	+1.17	0.1509
				α//0.006	5.314				0.0589
					5.403				0.1453
^{224}Fr		224.02325	3.0 m	β- /2.82		1-	+0.40	+0.517	0.13150
									0.21575
									0.8367
									(0.1–2.21)
^{225}Fr		225.02557	3.9 m	β- /1.87		3/2	+1.07	+1.3	
^{226}Fr		226.0294	49. s	β- /3.6		1	+0.071	−1.35	0.18606
									0.25373
^{227}Fr		227.0318	2.48 m	β- /2.5		1/2	+1.50		
^{228}Fr		228.0357	39. s	β- /~ 3.5		2-	−0.76	+2.4	
^{229}Fr		229.03845	50. s	β- /					
^{230}Fr		230.0425	19. s	β- /		(3)			
^{231}Fr		231.0454	17. s	β- /					
^{232}Fr		232.050	5. s	β- /					(0.0545–0.721)
^{233}Fr									
$_{88}$**Ra**									
^{201}Ra			~ 1.6 ms	α	7.91/				
^{202}Ra		202.0099	~ 0.02 ms	α	7.74	0+			
^{203m}Ra			24 ms	α	7.61				
^{203}Ra		203.0093	~ 31 ms	α	7.59				
^{204}Ra		204.0065	0.06 s	α	7.48	0+			
^{205m}Ra			~ 0.17 s						
^{205}Ra		205.0063	0.22 s	α	7.34				
^{206}Ra		206.00383	0.4 s	α/7.416	7.272(5)/	0+			
^{207}Ra		207.0038	1.3 s	α/7.270	7.133(5)/				
^{208}Ra		208.00184	1.1 s	α/7.273/87	7.133(5)/	0+			
^{209m}Ra			0.12 ms			(13/2+)			(0.2384–0.644)
^{209}Ra		209.00199	4.6 s	α/7.150	(6.50–7.14)	5/2-	+0.87	+0.40	(0.387–0.634)
^{210m}Ra			2.34 µs						(0.0967–0.775)
^{210}Ra		210.00050	3.7 s	α/7.610	7.020(5)/	0+			574.9
^{211m}Ra			5.1 µs						(0.396–0.802)
^{211}Ra		211.00090	13. s	α/7.046	6.907/99.	(5/2-)	+0.878	+0.48	

Elem. or Isot.	Natural Abundance (Atom %)	Atomic Mass or Weight	Half–life/ Resonance Width (MeV)	Decay Mode/ Energy (/MeV)	Particle Energy/ Intensity (MeV/%)	Spin ($h/2\pi$)	Nuclear Magnetic Mom. (nm)	Elect. Quadr. Mom. (b)	γ-Energy/ Intensity (MeV/%)
				EC/5.0	(6.26–6.79)				(0.120–0.665)
[212m]Ra			8.4 μs						(0.440–0.824)
[212]Ra		211.99979	13.0 s	α/7.033	6.901(2)/	0+			
[213m]Ra			2.20 ms	IT//99	α/8.47/63	17/2-	7.4		0.1612/41
				α//0.6	8.36/33				0.5462/100
					8.27/4				0.1062/97
[213]Ra		213.00038	2.7 m	EC/(20)/3.88	α/6.225/49	(1/2-)	+0.613		0.1024
				α/(80)/6.860	6.733/45				0.110
					6.522/5.4				0.215
					6.413/0.22				(0.105–0.511)
[214m]Ra			0.068 ms	α//0.09	8.95/91				(0.181–1.382)
[214]Ra		214.00011	2.46 s	α/7.272	7.14/99.8/	0+			0.642
					6.51/0.2				
[215m]Ra			7.6 μs						(0.196–1.048)
[215]Ra		215.00272	1.64 ms	α/8.864	7.883(6)/2.8	(9/2+)			0.773/100
					8.171(3)/1.4				0.852/74
					8.700(3)/95.9				0.055–1.048
[216]Ra		216.00353	0.18 μs	α/9.526	9.349(8)/	0+			
[217]Ra		217.00632	1.6 μs	α/9.161	8.992(8)/	9/2-			
[218]Ra		218.00714	26. μs	α/8.547	8.390(8)/	0+			
[219]Ra		219.01009	0.010 s	α/8.132	7.680(10)/65				
					7.982(9)/35				
[220]Ra		220.01103	18. ms	α/7.593	7.39/5	0+			0.465
					7.45/95				
[221]Ra		221.013917	29. s	α/6.879	6.254(10)/0.7	5/2+	−0.180	+2.0	
					6.578(5)/3				
					6.585(3)/8				
					6.608(3)/35				
					6.669(3)/21				
					6.758(3)/31				
[222]Ra		222.015375	36.2 s	α/5.590	6.237(2)/3.0	0+			0.324
					6.556(2)/97				0.145–0.8402
[223]Ra		223.018502	11.43 d	α/5.979	5.287(1)/0.15	(3/2+)	+0.271	+1.25	Rn k x-ray
					5.338(1)/0.13				0.12231
					5.365(1)/0.13				0.14418
					5.433(5)/2.3				0.15418
					5.502(1)/1.0				0.15859
					5.540(1)/9.2				0.26939
					5.607(3)/24				0.32388
					5.716(3)/52				0.33328
					5.747(1)/9				0.44494
					5.857(1)/0.32				(0.10–0.7)
					5.872(1)/0.85				
[224]Ra		224.020212	3.66 d	α/5.789	5.034(10)/0.003	0+			Rn k x-ray
					5.047(1)/0.007				0.2407
					5.164(5)/0.007				0.4093
					5.449(2)/4.9				0.6501
					5.685(2)/95				
[225]Ra		225.023612	14.9 d	β- /0.36	0.32/100	(3/2+)	−0.734		Ac k x-ray
				α	5.01 × 10⁻⁵				0.0434
					4.98 × 10⁻⁶				
[226]Ra		226.025410	1599. a	α/4.870	4.194(1)/0.001	0+			Rn k x-ray
			> 4 × 10¹⁸ a	sf/4 × 10⁻¹⁴	4.343(1)/0.006				0.1861/3.64
					4.601(1)/6.16				0.2624

Elem. or Isot.	Natural Abundance (Atom %)	Atomic Mass or Weight	Half–life/ Resonance Width (MeV)	Decay Mode/ Energy (/MeV)	Particle Energy/ Intensity (MeV/%)	Spin ($h/2\pi$)	Nuclear Magnetic Mom. (nm)	Elect. Quadr. Mom. (b)	γ-Energy/ Intensity (MeV/%)
					4.784(1)/93.8				0.053–2.448
^{227}Ra		227.029178	42. m	β- /1.325	1.03/	(3/2+)	−0.404	+1.5	Ac L x-ray
					1.30/				Ac k x-ray
									0.02739
^{228}Ra		228.031070	5.76 a	β- /0.046	0.039/50	0+			0.0135
			βf//5×10^{-12}	0.014/30					(0.006–0.031)
					0.026/20				
^{229}Ra		229.03496	4.0 m	β- /1.76	1.76/	(3/2+)	+0.503	+3.1	0.0145–0.172
^{230}Ra		230.03706	1.5 h	β- /1.0	0.7/	0+			0.0631
									0.0720
									0.2028
									0.4698
									0.4787
^{231}Ra		231.0412	1.7 m	β-					(0.018–1.155)
^{232}Ra		232.0436	4. m	β-		0+			
^{233}Ra		233.0481	30. s	β-					
^{234}Ra		234.051	~ 30. s	β-/		0+			
$_{89}$**Ac**									
^{206m}Ac			0.04 s	α	7.79				
^{206}Ac		206.0145	~ 26 ms	α	7.75				
^{207}Ac		207.0120	27 ms	α/	7.69				
^{208m}Ac			~ 25. ms	α/	7.72				
^{208}Ac		208.0116	~ 0.1 s	α/	7.62				
^{209}Ac		209.00949	~ 0.10 s	α/	7.58				
^{210}Ac		210.0094	0.34 s	α/7.610	7.462(8)/				
^{211}Ac		211.0077	0.20 s	α/7.620	7.480(8)/				
^{212}Ac		212.0078	0.9 s	α/7.520	7.379(8)/				
^{213}Ac		213.0066	0.73 s	α/7.500	7.364(8)/	(9/2-)			
^{214}Ac		214.00690	8.2 s	α/(86)/7.350	7.215/54	(5+)			(0.0626–0.754)
				EC/(14)/6.34	7.081/42				
					(6.48–7.15)				
^{215}Ac		215.00645	0.17 s	α/7.750	7.60/99.57	(9/2-)			0.399
					7.21/0.46				0.582
					7.03/0.20				0.654
					6.96/0.14				
^{216m}Ac			0.44 ms	α/	8.198(8)/1.7	(9-)			(0.0826–1.375)
					8.283(8)/2.5				
					9.028(5)/49				
					9.106(5)/46				
^{216}Ac		216.00872	44. ms	α/9.241	8.990(2)/10	(1-)			
					9.070(8)/90				
^{217m}Ac			0.7 μs	α/	10.540/100				
^{217}Ac		217.00935	0.07 μs	α/9.832	9.650(10)/100	9/2-			
^{218}Ac		218.01164	1.1 μs	α/9.380	9.205(15)/				
^{219}Ac		219.01242	0.012 ms	α/8.830	8.664(10)/	(9/2-)			
^{220}Ac		220.01476	26. ms	α/8.350	7.610(20)/23				
					4.680(20)/21				
					7.790(10)/13				
					7.850(10)/24				
					7.985(10)/4				
					8.005(10)/5				
					8.060(10)/6				
					8.195(10)/3				
^{221}Ac		221.01559	52. ms	α/7.790	7.170(10)/2				

Elem. or Isot.	Natural Abundance (Atom %)	Atomic Mass or Weight	Half–life/ Resonance Width (MeV)	Decay Mode/ Energy (/MeV)	Particle Energy/ Intensity (MeV/%)	Spin (h/2π)	Nuclear Magnetic Mom. (nm)	Elect. Quadr. Mom. (b)	γ-Energy/ Intensity (MeV/%)
					7.375(10)/10				
					7.440(15)/20				
					7.645(10)/70				
222mAc			63. s	α/(>89)/	6.710(20)/7				
				EC/(1)/	6.750(20)/13				
				I.T./(<10)/	6.810(20)/24				
					6.840(20)/9				
					6.890(20)/13				
					6.970(20)/7				
					7.000(20)/13				
222Ac		222.01784	5. s	α/7.141	6.967(10)/6	1-			
					7.013(2)/94				
223Ac		223.01914	2.1 m	α/(99)/6.783	6.131(2)/0.12	(5/2-)			0.0725
				EC/(1)/0.59	6.177(2)/0.94				0.0839
					6.293(1)/0.47				0.0927
					6.326(1)/0.3				0.0990
					6.332(2)/0.14				0.1917
					6.360(1)/0.22				0.2158
					6.397(1)/0.13				0.3588
					6.448(1)/0.2				0.4768
					6.473(1)/3.1				
					6.523(2)/0.6				
					6.528(1)/3.1				
					6.563(1)/13.6				
					6.582(3)/0.3				
					6.646(1)/44				
					6.661(1)/31				
224Ac		224.021723	2.7 h	EC/(90)/1.403	5.841(1)/0.5	0-			Ra L x-ray
				α/(10)/6.323	5.860(1)/0.75				Ra k x-ray
					5.875(1)/1.7				0.08426
					5.941(1)/4.4				0.13150
					6.000(1)/6.7				0.1571
					6.013(1)/1.4				0.21575
					6.056(1)/22				0.2619
					6.138(1)/26				(0.03–0.3)
					6.154(1)/1.0				
					6.204(1)/12				
					6.210(1)/20				
225Ac		225.023230	10.0 d	α/5.935	5.286(1)/0.2	3/2			Fr k x-ray
					5.444(3)/0.1				0.06296/0.48
					5.554(1)/0.1				0.09982/1.36
					5.608(1)/1.1				0.1084
					5.636(1)/4.5				0.1116
					5.681(1)/1.4				0.1451
					5.722(1)/2.9				0.150/0.691
					5.731(1)/10				0.15724
					5.791(1)/9				0.18795/0.54
					5.793(1)/18				0.0075–0.809
226Ac		226.026098	1.224 d	EC/(17)/0.640		(1-)			Ra k x-ray
				β- /(83)/1.116					Th k x-ray
				α/(0.006)/5.51	5.399(5)/0.006				0.07218
									0.15816
									0.23034
227Ac		227.027752	21.77 a	β- /98.6/0.045	0.045/54	(3/2-)	+1.1	+1.7	0.0838/23.

Elem. or Isot.	Natural Abundance (Atom %)	Atomic Mass or Weight	Half–life/ Resonance Width (MeV)	Decay Mode/ Energy (/MeV)	Particle Energy/ Intensity (MeV/%)	Spin ($h/2\pi$)	Nuclear Magnetic Mom. (nm)	Elect. Quadr. Mom. (b)	γ-Energy/ Intensity (MeV/%)
				α/(1.4)/5.043	4.869(1)/0.09				0.0811/14.
					4.938(1)/0.52				0.2696/13.
					4.951(1)/0.65				(0.044–1.27)
²²⁸Ac		228.031021	6.15 h	β- /2.127	1.11/32	(3+)			Th L x-ray
					1.85/12				Th k x-ray
					2.18/11				0.12903
									0.33842
									0.91116
									0.96897
									(0.2–1.96)
²²⁹Ac		229.03302	1.04 h	β- /1.10	1.1/	(3/2+)			0.09335/2.43
									0.16451/2.61
									0.56916/2.24
									0.0111–0.898
²³⁰Ac		230.0363	2.03 m	β- /2.7	1.4/	1+			Th k x-ray
				β-, sf	/0.000119				0.45497
									0.50820
									(0.12–2.5)
²³¹Ac		231.0386	7.5 m	β- /2.1	2.1/100	(1/2+)			0.14379
									0.18574
									0.22140
									0.28250
									0.3070
²³²Ac		232.0420	2.0 m	β- /3.7		(2-)			
²³³Ac		233.0446	2.4 m	β- /		(1/2+)			
²³⁴Ac		234.0484	40. s	β- /		(1+)			
²³⁵Ac									
²³⁶Ac			~ 1.2 m						
₉₀Th		**232.03806(2)**							
²⁰⁸Th			~ 1.7 ms	α	8.04				
²⁰⁹Th		209.0177	~ 2.5 ms	α	8.08				
²¹⁰Th		210.0158	16. ms	α	7.90	0+			
²¹¹Th		211.0149	0.04 s	α	7.79				
²¹²Th		212.01298	32. ms	α/	7.80/	0+			
²¹³Th		213.0130	0.14 s	α/7.840	7.692(10)/				
²¹⁴Th		214.01150	0.10 s	α/7.825	7.677(10)/	0+			
²¹⁵Th		215.01173	1.2 s	α/7.660	7.33(10)/8	(1/2-)			0.134
					7.395(8)/52				0.192
					7.524(8)/40				(0.069–0.295)
²¹⁶ᵐTh			0.14 ms	α	9.93/74				(0.0905–1.478)
					8.00, 9.31				
²¹⁶Th		216.01106	27. ms	α/8.071	7.92/99.46	0+			0.628
					7.30/0.54				
²¹⁷Th		217.01311	0.25 ms	α/9.424	9.27/94.6				(0.546–0.822)
					8.46/3.8				
					8.73/1.6				
²¹⁸Th		218.01328	0.11 μs	α/9.847	9.665(10)/	0+			
²¹⁹Th		219.01554	1.05 μs	α/9.510	9.340(20)/				
²²⁰Th		220.01575	10. μs	α/8.953	8.790(20)/	0+			
²²¹Th		221.01818	2. ms	α/8.628	7.732/7				
					8.142/72				
					8.469/21				
²²²Th		222.01847	2.24 ms	α/8.129	7.980/97.7	0+			
					7.599/2.3				

Elem. or Isot.	Natural Abundance (Atom %)	Atomic Mass or Weight	Half–life/ Resonance Width (MeV)	Decay Mode/ Energy (/MeV)	Particle Energy/ Intensity (MeV/%)	Spin ($h/2\pi$)	Nuclear Magnetic Mom. (nm)	Elect. Quadr. Mom. (b)	γ-Energy/ Intensity (MeV/%)
^{223}Th		223.02081	0.60 s	α/7.454	7.29(1)/41(5)				
					7.32(1)/29(5)				
					7.350(15)/20(5)				
					7.390(15)/10(4)				
^{224}Th		224.02147	1.05 s	α/7.305	6.768(5)/1.2	0+			
					6.997(5)/19				
					7.170(5)/7				
^{225}Th		225.023951	8.72 m	EC/(10)/0.68		(3/2+)			
				α/(90)/6.920	6.441(2)/15				
					6.479(2)/43				
					6.501(3)/14				
					6.627(3)/3				
					6.650(5)/3				
					6.700(5)/2				
					6.743(3)/7				
					6.796(2)/9				
^{226}Th		226.024903	30.83 m	α/6.454	6.026(1)/0.2	0+			Ra k x-ray
					6.041(1)/0.19				0.1112
					6.098(1)/1.3				0.2421
					6.2283(4)/23				0.1310
					6.3375(4)/75				0.1733–0.9295
^{227}Th		227.027704	18.72 d	α/6.146		(3/2+)			Ra L x-ray
									Ra k x-ray
									0.05014
									0.23597
									0.25624
									(0.02–1.0)
^{228}Th		228.028741	1.913 a	α/5.520	5.1770(2)/0.18	0+			
					5.2114(1)/0.4				
					5.3405(1)/26.7				
					5.4233(1)/73				
^{229m}Th			13.9 h	α	4.83–5.08				
^{229}Th		229.031762	7.9×10^3 a	α/5.168	4.814/9.3	5/2+	+0.46	+4.	0.1935/4.3
					4.845(5)/56				0.21089/277
					4.9008(5)/10.2				0.13697/1.21
					4.689–5.077				0.011–0.6036
^{230}Th		230.033134	7.56×10^4 a	α/4.771	4.4383(6)/0.03	0+			0.0677/0.46
					4.4798(6)/0.12				0.1439/0.078
				$> 2 \times 10^{18}$ a	SF// $< 4 \times 10^{-12}$	4.6211(6)/23.4				
					4.6876(6)/76.3				
^{231}Th		231.036304	1.063 d	β- /0.390	0.138/22	5/2+			Pa L x-ray
					0.218/20				Pa k x-ray
					0.305/52				0.02564
									0.084203/
									(0.02–0.3)
^{232}Th	100.	232.038055	1.40×10^{10} a	α/4.081	3.830(10)/0.2	0+			0.0590
			1.2×10^{21} a	sf/1.1×10^{-9}	3.952(5)/23				0.124
					4.010(5)/77				
^{233}Th		233.041582	22.3 m	β-/1.245	1.245/	½+			Pa L x-ray
									Pa k x-ray
									0.02938
									0.08653
									0.45930
									(0.02–1.2)

Elem. or Isot.	Natural Abundance (Atom %)	Atomic Mass or Weight	Half–life/ Resonance Width (MeV)	Decay Mode/ Energy (/MeV)	Particle Energy/ Intensity (MeV/%)	Spin ($h/2\pi$)	Nuclear Magnetic Mom. (nm)	Elect. Quadr. Mom. (b)	γ-Energy/ Intensity (MeV/%)
^{234}Th		234.043601	24.10 d	β- /0.273	0.102/20	0+			Pa L x-ray
					0.198/72				0.06329/4.1
									0.09235/2.4
									0.09278/2.4
^{235}Th		235.04751	7.2 m	β- /1.9					0.4162
									0.6594
									0.7272
									0.747
									0.9318
^{236}Th		236.0499	37.5 m	β- /~ 1.0		0+			Pa k x-ray
									0.1107
^{237}Th		237.0539	5.0 m	β-					
^{238}Th		238.0565	9.4 m			0+			0.0890
$_{91}$Pa		231.03588(2)							
^{212}Pa		212.0232	~ 5 ms	α	8.27				
^{213}Pa		213.0211	7 ms	α	8.24				
^{214}Pa		214.0209	17 ms	α	8.12				
^{215}Pa		215.0192	15. ms	α	8.08/100				
^{216}Pa		216.0191	0.19 s	α/	7.95/51				0.134
					7.82/45				
					7.79/4				
^{217m}Pa			1.08 ms	α/	10.16/72				0.4504–0.821
					8.306/11				
					9.55/6				
					9.69/2				
^{217}Pa		217.0183	3.8 ms	α/8.490	8.337/99				0.0466–0.634
					7.873/0.4				
					7.728/0.3				
					7.710/0.3				
^{218}Pa		218.02004	0.12 ms	α/	9.54/31				0.092
					9.61/69				
^{219}Pa		219.0199	0.05 μs	α					
^{220}Pa		220.0219	0.8 μs	α					
^{221}Pa		221.0219	6. μs	α	9.08(3)				
^{222}Pa		222.0237	~ 4.3 ms	α/8.700	8.180/50				
					8.330/20				
					8.540/30				
^{223}Pa		223.0240	~ 6.5 ms	α/8.340	8.006(10)/55				
					8.196(10)/45				
^{224}Pa		224.02563	0.84 s	α/7.630	7.555(10)/75(3)				0.1945
					7.46(1)/25(3)				(0.028–0.412)
^{225}Pa		225.0261	1.8 s	α/7.380	7.195(10)/30				
					7.245(10)/70				
^{226}Pa		226.02795	1.8 m	α/(74)/6.987	6.728(10)/0.7				
				EC/(26)/2.83	6.823(10)/35				
					6.863(10)/39				
^{227}Pa		227.02881	38.3 m	α/(85)/6.582	6.357(4)/7	(5/2-)			0.0649
				EC/(15)/1.02	6.376(10)/2.2				0.0669
					6.401(4)/8				0.1100
					6.416(4)/13				
					6.423(10)/10				
					6.465(4)/43				
^{228}Pa		228.031051	22. h	EC/(98)/2.111		(3+)	+3.5		Th k x-ray
				α/(2)	5.779/0.23				0.409/100

Elem. or Isot.	Natural Abundance (Atom %)	Atomic Mass or Weight	Half–life/ Resonance Width (MeV)	Decay Mode/ Energy (/MeV)	Particle Energy/ Intensity (MeV/%)	Spin (ℏ/2π)	Nuclear Magnetic Mom. (nm)	Elect. Quadr. Mom. (b)	γ-Energy/ Intensity (MeV/%)
					5.805/0.15				0.4631/222
					6.078/0.4				0.91116/242
					6.105/0.25				0.96464/120
					6.118/0.22				0.96897/149
									0.058−1.96
^{229}Pa		229.032097	1.5 d	EC/(99.8)/0.32		(5/2+)			0.04244
				α/(0.2)/5.836	5.536(2)/0.02				(0.024−0.18)
					5.579(2)/0.09				
					5.668(2)/0.05				
^{230}Pa		230.034541	17.4 d	EC/(90)/1.310	0.51/	(2-)	2.0		Th L x-ray
				β-/(10)/0.563					Th k x-ray
									0.4437
									0.45477
									0.89876
									0.91856
									0.95199
									(0.053−1.07)
^{231}Pa	100	231.035884	3.25 × 10^4 a	α/5.148	4.6781(5)/1.5	3/2-	2.01	−1.7	Ac L x-ray
					4.7102(5)/1.0				Ac k x-ray
				> 2 × 10^{17} a	sf/< 1.6 × 10^{-15} 4.7343(5)/8.4				0.01899
					4.8513(5)/1.4				0.027396
					4.9339(5)/3				0.03823
					4.9505(5)/22.8				0.04639
					4.9858(5)/1.4				0.25586
					5.0131(5)/25.4				0.26029
					5.0292(5)/20				0.28367
					5.0318(5)/2.5				0.30007
					5.0587(5)/11				0.30264
									0.33007
									(0.02−0.61)
^{232m}Pa			0.10 d						
^{232}Pa		232.03859	1.31 d	β- /1.34		(2-)			U k x-ray
									0.10900
									0.15009
									0.89439
									0.96934
									(0.10−1.17)
^{233}Pa		233.040247	26.97 d	β-/0.571	0.15/40	3/2-	+4.0	−3.0	U L x-ray
					0.256/60				U k x-ray
									0.30017
									0.31201/38.4
									(0.0286−0.456)
^{234m}Pa			1.17 m	β- /99.9/2.29		(0-)			U k x-ray
				IT/0.13/					0.25818/0.07
									0.76641/0.32
									1.0009/0.86
									(0.06−1.96)
^{234}Pa		234.043308	6.69 h	β- /2.197	0.51/	(4+)			U L x-ray
									U k x-ray
									0.1312/0.03
									0.5695/0.02
									0.9256/0.02
									(0.02−1.99)
^{235}Pa		235.04544	24.4 m	β- /1.41	1.4/97	(3/2-)			0.0308−0.659

Elem. or Isot.	Natural Abundance (Atom %)	Atomic Mass or Weight	Half–life/ Resonance Width (MeV)	Decay Mode/ Energy (/MeV)	Particle Energy/ Intensity (MeV/%)	Spin ($h/2\pi$)	Nuclear Magnetic Mom. (nm)	Elect. Quadr. Mom. (b)	γ-Energy/ Intensity (MeV/%)
^{236}Pa		236.0487	9.1 m	β- /2.9	1.1/40	(1-)			U k x-ray
					2.0/50				0.64235
					3.1/10				0.68759
									1.7630
									(0.04–2.18)
^{237}Pa		237.0512	8.7 m	β- /2.3	1.1/60	(1/2+)			0.4986
					1.6/30				0.5293
					2.3/10				0.5407
									0.8536
									0.8650
									(0.04–1.4)
^{238}Pa		238.0545	2.3 m	β- /3.5	1.2/	(3-)			0.10350
					1.7/				0.1785
									0.4484
									0.6350
									0.6800
									1.01446
									(0.04–2.5)
^{239}Pa		239.0573	1.8 h						
^{240}Pa									
$_{92}$U		**238.02891(3)**							
^{217}U		217.0244	~ 0.2 ms	α	8.02				
^{218m}U			~ 0.56 ms	α	10.68				
^{218}U		218.02354	0.5 ms	α	8.61	0+			
^{219}U		219.0249	~ 0.08 ms	α	9.77				
^{222}U		222.0261	~ 1.μs	α		0+			
^{223}U		223.0277	0.02 s	α/	8.78(4)/				
^{224}U		224.02761	~ 1. ms	α/	8.46/100	0+			
^{225}U		225.02939	84. ms	α/	7.87/83				
					7.82/15				
					7.63/2				
^{226}U		226.02934	0.26 s	α/7.560	7.56/86	0+			
					7.38/14				
^{227}U		227.03116	1.1 m	α/7.200	6.870/				
^{228}U		228.03137	9.1 m	α/6.803	6.404(6)/0.6	0+			0.095
					6.440(5)/0.7				0.152
					6.589(5)/29				0.187
					6.681(6)/70				0.246
^{229}U		229.03351	58. m	EC/(80)/1.31	6.223/3	(3/2+)			
				α/(20)/6.473	6.297(3)/11				
					6.332(3)/20				
					6.360(3)/64				
^{230}U		230.033940	20.8 d	α/5.992	5.5866(3)/0.01	0+			Th L x-ray
			> 4 × 10^{10} a	sf/< 10^{-10}	5.6624(3)/0.26				0.07218
					5.6663(3)/0.38				0.15421
					5.8178(3)/32				0.23034
					5.8887(3)/67				(0.081–0.8565)
^{231}U		231.036294	4.2 d	EC/0.36		(5/2-)			Pa L x-ray
				α/(10^{-3})	5.46/1.6 × 10^{-3}				Pa k x-ray
					5.47/1.4 × 10^{-3}				0.02564
					5.40/1. × 10^{-3}				0.08420
^{232}U		232.037156	70. a	α/5.414	4.9979(1)/0.003	0+			
			2.6 × 10^{15} a	sf/2.7 × 10^{-12}	5.1367(1)/0.3				
					5.2635(1)/31				

Elem. or Isot.	Natural Abundance (Atom %)	Atomic Mass or Weight	Half–life/ Resonance Width (MeV)	Decay Mode/ Energy (/MeV)	Particle Energy/ Intensity (MeV/%)	Spin ($h/2\pi$)	Nuclear Magnetic Mom. (nm)	Elect. Quadr. Mom. (b)	γ-Energy/ Intensity (MeV/%)
					5.3203(1)/69				
^{233}U		233.039635	1.590×10^5 a	α/4.909	4.7830(8)/13.2	5/2+	+0.59	3.66	Th L x-ray
			$>2.7 \times 10^{17}$ a	sf/6×10^{-11}	4.8247(8)/84.4				0.04244
					4.510–4.804				0.09714
									(0.0252–1.119)
^{234}U	0.0054(5)	234.040952	2.453×10^5 a	α/4.856	4.604(1)/0.24	0+			0.05323/0.156
			1.5×10^{16} a	sf/1.6×10^{-9}	4.7231(1)/27.5				0.12091
					4.776(1)/72.5				
^{235m}U			26. m	IT/0.0007		1/2+			
^{235}U	0.7204(6)	235.043930	7.03×10^8 a	α/4.6793	4.1525(9)/0.9	7/2-	−0.38	4.94	Th L x-ray
			1.0×10^{19} a	sf/7×10^{-9}	4.2157(9)/6.				Th k x-ray
					4.3237(9)/4.6				0.10917
					4.3641(9)/19.				0.14378/0.134
					4.370(4)/6				0.16338/0.067
					4.3952(9)/57.				0.18574/0.806
					4.4144(9)/2.1				0.1949/0.009
					4.5025(9)/1.7				0.20533/0.774
					4.5558(9)/4.2				0.2214/0.0014
					4.5970(9)/4.8				(0.03–0.79)
^{236}U		236.045568	2.342×10^7 a	α/4.569	4.332(8)/0.26	0+			Th L x-ray
			2.5×10^{16} a	SF// 9×10^{-8}	4.445(5)/26				0.04946/100
					4.494(3)/74				0.11279/24.1
									0.17115/0.080
^{237}U		237.048730	6.75 d	β- /0.519	0.24/	1/2+			Np L x-ray
					0.25/				Np k x-ray
									0.05953
									0.20801
^{238}U	99.2742(10)	238.050788	4.47×10^9 a	α	4.0395/0.23	0+			Th L x-ray
			8.2×10^{15} a	SF// 5×10^{-5}	4.147(5)/23				0.04955/.06
					4.196(5)/77				0.1135/.01
^{239}U		239.054293	23.5 m	β- /1.265	1.2/	5/2+			(0135–1.102)
					1.3/				
^{240}U		240.05659	14.1 h	β- /0.39	0.36/	0+			Np L x-ray
									0.04410
									0.05558
									0.06760
^{242}U		242.0629	16.8 m	β- /~ 1.2		0+			
$_{93}$**Np**									
^{225}Np		225.0339	> 2 µs						
^{226}Np		226.0352	0.03 s	α/	8.04(2)/				
^{227}Np		227.0350	0.51 s	α/	7.65(2)/				
					7.68(1)/				
^{228}Np		228.0362	61. s	EC/60(7)/					
				α/40(7)/, sf					
^{229}Np		229.0363	4.0 m	α/7.010	6.890(20)				
^{230}Np		230.0378	4.6 m	EC/97 /3.6					
				α/3	6.660(20)				
^{231}Np		231.03825	48.8 m	EC/98 /1.8		5/2			0.2629
				α/2 /6.368	6.280/2				0.3475
									0.3703
^{232}Np		232.0401	14.7 m	EC/99 /2.7		(4-)			U L x-ray
									U k x-ray
									0.3268
									0.81925

Elem. or Isot.	Natural Abundance (Atom %)	Atomic Mass or Weight	Half–life/ Resonance Width (MeV)	Decay Mode/ Energy (/MeV)	Particle Energy/ Intensity (MeV/%)	Spin ($h/2\pi$)	Nuclear Magnetic Mom. (nm)	Elect. Quadr. Mom. (b)	γ-Energy/ Intensity (MeV/%)
									0.86683
²³³Np		233.04074	36.2 m	EC/1.2		(5/2+)			U L x-ray
									U k x-ray
									0.29887
									0.31201
²³⁴Np		234.04290	4.4 d	β+, EC/1.81	0.79/	(0+)			U L x-ray
									U k x-ray
									1.5272
									1.5587
									1.6022
²³⁵Np		235.044063	1.085 a	EC/99.9 /0.124		5/2+			U k x-ray
				α/0.001/5.191					
²³⁶ᵐNp			22.5 h	EC/52 /		(1-)			U L x-ray
				β- /48 /					Pu L x-ray
									U k x-ray
									0.64235
									0.68759
²³⁶Np		236.04657	1.55 × 10⁵ a	EC/91 /0.94		(6-)			U L x-ray
				β- /9 /0.49					U k x-ray
									0.10423
									0.16031
²³⁷Np		237.048173	2.14 × 10⁶ a	α/4.957	4.6395(5)/6.5	5/2+	+3.14	+3.87	Pa L x-ray
			1 × 10¹⁸ a	sf/2.1 × 10⁻¹⁰	4.766(5)/9.7				Pa k x-ray
					4.7715(5)/22.7				0.029378/15
					4.7884(5)/47.8				0.08653/12
					4.558–4.873				(0.03–0.28)
²³⁸Np		238.050946	2.117 d	β- /1.292	1.2/	2+			Pu L x-ray
									Pu k x-ray
									0.98447/25.2
									1.02855/18.3
									(.044–1.026)
²³⁹Np		239.052939	2.355 d	β- /0.722	0.341/30	5/2+			Pu L x-ray
					0.438/48				Pu k x-ray
									0.10613
									0.228186/11
									0.27760/15
									(0.04–0.50)
²⁴⁰ᵐNp			7.22 m	β- /99.9 /	2.18/	(1+)			0.25143
				IT/0.1 /					0.26333
									0.55454
									0.59735
²⁴⁰Np		240.05616	1.032 h	β- /2.20	0.89/	5+			0.1471/
									0.5664
									0.6008
²⁴¹Np		241.0583	13.9 m	β- /1.3	1.3/	5/2+			0.1330/
									0.1740
									0.280
²⁴²ᵐNp			2.2 m	β- /		(1+)			0.15910
									0.2651/
									0.78570
									0.9448/
²⁴²Np		242.0616	5.5 m	β- /2.7	2.7/	6+			0.6209
									0.73620
									0.78074

Elem. or Isot.	Natural Abundance (Atom %)	Atomic Mass or Weight	Half–life/ Resonance Width (MeV)	Decay Mode/ Energy (/MeV)	Particle Energy/ Intensity (MeV/%)	Spin ($h/2\pi$)	Nuclear Magnetic Mom. (nm)	Elect. Quadr. Mom. (b)	γ-Energy/ Intensity (MeV/%)
									1.47340
									(0.04–2.37)
^{243}Np		243.06428	1.9 m						
^{244}Np		244.0679	2.3 m						
$_{94}$**Pu**									
^{228}Pu		228.03874	~ 1.1 s	α/	7.81(2)/	0+			
^{229}Pu		229.0402	1.1 m	α/50	7.46/				
				EC/50					
				SF < 7					
^{230}Pu		230.03965	1.7 m	α/	7.06/81	0+			
					7.00/19				
^{231}Pu		231.04110	8.6 m	EC/90					
				α/10	6.72				
^{232}Pu		232.04119	34. m	EC/>80/1.1		0+			
				α/<20/6.716	6.542(10)/38				
					6.600(10)/62				
^{233}Pu		233.04300	20.9 m	EC(99.9)/1.9					0.1503
				α/0.1 /6.416	6.300(20)/0.1				0.1804
									0.2353
									0.5002
									0.5346/
									1.0352/
^{234}Pu		234.04332	8.8 h	EC/94 /0.39		0+			
				α/6 /6.310	6.035(3)/0.024				
					6.149(3)/1.9				
					6.200(3)/4.				
^{235}Pu		235.04529	25.3 m	EC/99+ /1.2		(5/2+)			
				α/0.003/5.957	5.850(20)/0.003				
^{236m}Pu			1.2 μs						
^{236}Pu		236.046058	2.87 a	α/5.867	5.611/0.21	0+			0.0476/0.07
			1.5×10^9 a	sf/1.9×10^{-7}	5.7210/30.5				0.109/0.02
					5.7677(1)/69.3				(0.17–0.97)
^{237}Pu		237.048410	45.7 d	EC/99.9 /0.220		7/2-			Np L x-ray
				α/0.003 /5.747	5.334(4)/0.0015				Np k x-ray
					5.356(4)/0.0006				0.026344
					5.650(4)/0.0007				0.03319
									0.05954
									(0.03–0.5)
^{238}Pu		238.049560	87.7 a	α/5.593	5.3583(1)/0.10	0+			U k x-ray
			4.75×10^{10} a	sf/1.8×10^{-7}	5.465(1)/28.3				0.04347
					5.4992(1)/71.6				(0.04–1.1)
^{239}Pu		239.052163	2.410×10^4 a	α/5.244	5.055/0.047	1/2+	+0.203		U k x-ray
			$8. \times 10^{15}$ a	sf/3×10^{-10}	5.076/0.078				0.05162
					5.106/11.9				0.05682
					5.144/17.1				0.12928
					5.157/70.8				0.37502
					(4.74–5.03)				0.41369
^{240}Pu		240.053814	6.56×10^3 a	α/5.255	5.0212(1)/0.07	0+			U L x-ray
			1.14×10^{11} a	sf/5.7×10^{-6}	5.1237(1)/26.4				0.04524
					5.1681(1)/73.5				0.10423
					(4.492–4.863)				(0.04–0.97)
^{241}Pu		241.056852	14.33 a	β-/99+/0.0208	4.853/3 × 10^{-4}	5/2+	−0.68	+6.	0.14854
				α/0.002 /5.139	4.897/0.002				0.1600
				sf/> 2.4 × 10^{-14}					

Elem. or Isot.	Natural Abundance (Atom %)	Atomic Mass or Weight	Half–life/ Resonance Width (MeV)	Decay Mode/ Energy (/MeV)	Particle Energy/ Intensity (MeV/%)	Spin ($h/2\pi$)	Nuclear Magnetic Mom. (nm)	Elect. Quadr. Mom. (b)	γ-Energy/ Intensity (MeV/%)
^{242}Pu		242.058743	3.75×10^5 a	α/4.983	4.7546(7)/0.098	0+			U L x-ray
			6.77×10^{10} a	sf/5.5×10^{-4}	4.8564(7)/22.4				0.04491
					4.9006(7)/78				0.10350
^{243}Pu		243.062003	4.956 h	β- /0.582	0.49/21	7/2+			Am L x-ray
					0.58/60				0.0417
									0.0839
^{244}Pu		244.064204	8.12×10^7 a	α/99.9/4.665	4.546(1)/19.4	0+			U L x-ray
			6.6×10^{10} a	sf/0.12	4.589(1)/80.5				0.0439
^{245}Pu		245.06775	10.5 h	β- /1.21	0.93/57	(9/2-)			Am L x-ray
					1.21/11				Am k x-ray
									0.2804 /
									0.30832
									0.32752
									0.56014
									(0.03–1.2)
^{246}Pu		246.07021	10.85 d	β- /0.40	0.150/85	0+			Am L x-ray
					0.35/10				Am k x-ray
									0.04379
									0.22371
^{247}Pu		247.0741	2.3 d							
$_{95}$Am									
^{230}Am									
^{232}Am		232.0466	0.9 m	EC/~ 5.0					
^{233}Am		233.0464	~ 3.2 m	α	6.78				
^{234}Am		234.0478	2.3 m	EC/4.2					
^{235}Am		235.0480	10.3 m	EC					Pu K x-ray
				α	6.46/0.4				0.291/100
									(0.170–0.828)
^{236m}Am			2.9 m			(1-)			(0.583–0.713)
^{236}Am		236.0496	3.6 m	EC		(5-)			(0.158–1.038)
^{237}Am		237.0500	1.22 h	EC/99.98 /1.7		(5/2-)			Pu k x-ray
				α/0.02 /6.20	6.042(5)/0.02				0.14559
									0.28026
									0.43845
^{238}Am		238.05198	1.63 h	EC/2.26		1+			Pu L x-ray
				α/0.0001 /6.04	5.940/0.0001				Pu k x-ray
									0.91870
									0.96278
^{239}Am		239.053025	11.9 h	EC/99.99/0.803		5/2-			Pu L x-ray
				α/0.01/5.924	5.734(2)/0.001				Pu k x-ray
					5.776(2)/0.008				0.18172
									0.22818
									0.27760
^{240}Am		240.05530	2.12 d	EC/1.38		(3-)			Pu L x-ray
				α/5.592	5.378/16 × 10^{-4}				Pu k x-ray
									0.88878
									0.98764
									(0.1–1.3)
^{241}Am		241.056829	432.7 a	α/5.637	5.2443(1)/0.002	5/2-	+1.58	+3.1	Np L x-ray
			1.2×10^{14} a	sf/3.6×10^{-10}	5.3221(1)/0.015				0.02634 /.024
					5.3884(1)/1.4				0.0332/.00126
					5.4431(1)/12.8				0.05954/0.359
					5.4857(1)/85.2				(0.03–1.128)
					5.5116(1)/0.20				

Elem. or Isot.	Natural Abundance (Atom %)	Atomic Mass or Weight	Half–life/ Resonance Width (MeV)	Decay Mode/ Energy (/MeV)	Particle Energy/ Intensity (MeV/%)	Spin ($h/2\pi$)	Nuclear Magnetic Mom. (nm)	Elect. Quadr. Mom. (b)	γ-Energy/ Intensity (MeV/%)
					5.5442(1)/0.34				
^{242m}Am			141. a	IT/99.5/0.048		5−	+1.0	+7.	Am L x-ray
				α/0.5/5.62	5.141(4)/0.026				0.04863
			> 3 × 10^{12} a	sf/< 4.7 × 10^{-9}	5.2070(2)/0.4				0.08648
									0.10944
									0.16304
^{242}Am		242.059549	16.02 h	β− /83 /0.665	0.63/46	1−	+0.388	~ −2.4	Pu L x-ray
				EC/17 /0.750	0.67/37				Cm L x-ray
									Pu k x-ray
									0.0422
									0.04453
^{243}Am		243.061381	7.37 × 10^3 a	α/5.438	5.1798(5)/1.1	5/2−	+1.50	+2.86	0.04354
			2. × 10^{14} a	sf/3.7 × 10^{-9}	5.2343(5)/11				0.07467
					5.2766(5)/88				0.08657
					5.394(5)/0.12				0.11770
					5.3500(5)/0.16				0.14197
^{244m}Am			~ 26. m	β− /1.498		(1−)			0.0429
^{244}Am		244.064285	10.1 h	β− /1.428					Am L x-ray
									Cm k x-ray
									0.7460
									0.9000
^{245}Am		245.066452	2.05 h	β− /0.894	0.65/19	(5/2+)			Cm L x-ray
					0.90/77				Cm k x-ray
									0.25299
^{246m}Am			25.0 m	β− /	1.3/79.	2−			Cm L x-ray
					1.60/14				Cm k x-ray
					2.1/7				0.27002
									0.79881
									1.06201
									1.07885
									(0.04−2.29)
^{246}Am		246.06978	39. m	β− /2.38	1.2/	(7−)			Cm L x-ray
									Cm k x-ray
									0.1529
									0.2046
									0.6786
^{247}Am		247.0721	22. m	β− /1.7					Cm L x-ray
									Cm k x-ray
									0.2267 /
									0.2853 /
^{248}Am									
^{249}Am									
$_{96}$**Cm**									
^{232}Cm									
^{233}Cm		233.0508	~ 23 s	α/20	7.34/				
				EC/80					
^{234}Cm		234.05016	~ 51. s	α/27	7.24/	0+			
				EC/71					
				SF/~2					
^{235}Cm		235.0514							
^{236}Cm		236.0514	6.8 m	EC/1.7/82		0+			
				α/18	6.95				
				SF/< 0.1					
^{237}Cm		237.0529		EC/2.5					

Elem. or Isot.	Natural Abundance (Atom %)	Atomic Mass or Weight	Half–life/ Resonance Width (MeV)	Decay Mode/ Energy (/MeV)	Particle Energy/ Intensity (MeV/%)	Spin ($h/2\pi$)	Nuclear Magnetic Mom. (nm)	Elect. Quadr. Mom. (b)	γ-Energy/ Intensity (MeV/%)
				α/<1	6.66				
^{238}Cm		238.05303	2.4 h	EC/>90 /0.97	α/6.558/86	0+			
				α/<10 /6.632	6.520(50)/14				
^{239}Cm		239.0550	~ 3. h	EC/1.7					
				α/< 0.001					0.0407
									0.1466
									0.1874
^{240}Cm		240.055530	27. d	α/6.397	5.989/0.014	0+			
					6.147/0.05				
			1.9×10^6 a	sf/3.9×10^{-6}	6.2478(6) /28.8				
					6.2906(6) /70.6				
^{241}Cm		241.057653	32.8 d	EC/99 /0.768		1/2+			Am k x-ray
				α/1 /6.184	5.8842(4)/0.12				0.13241
					5.9291(4)/0.18				0.16505
					5.9389(4)/0.69				0.18028
									0.43063
									0.47181
^{242}Cm		242.058836	162.8 d	α/6.216	5.9694(1)/0.035	0+			Pu L x-ray
					6.069(1)/25				0.04408
			7.0×10^6 a	sf/6.4×10^{-6}	6.1129(1)/74				0.10189
									(0.04–1.2)
^{243}Cm		243.061389	29.1 a	α/6.167	5.6815(5) /0.2	5/2+	0.40		Pu L x-ray
					5.6856(5)/1.6				Pu k x-ray
			5.5×10^{11} a	sf/5.3×10^{-9}	5.7420(5)/10.6				0.10612
					5.7859(5)/73.5				0.20975
					5.9922(5)/6.5				0.22819
					6.0103(5)/1.0				0.27760
					6.0589(5)/5				0.28546
					6.0666(5)/1.5				0.33431
									(0.04–0.7)
^{244}Cm		244.062753	18.1 a	α/5.902	5.6656/0.02	0+			Pu L x-ray
					5.7528/23				0.04282
			1.32×10^7 a	sf/1.4×10^{-4}	5.8050/77				0.09885
					5.515/0.004				0.15262
^{245}Cm		245.065491	8.48×10^3 a	α/5.623	5.235(10)/0.3	7/2+	0.5		Pu L x-ray
					5.3038(10)/5.0				Pu k x-ray
			1.4×10^{12} a	sf/6.1×10^{-7}	5.3620(7)/93				0.04195
					5.4927(11)/0.8				0.13299
					5.5331(11)/0.6				0.13606
									0.17494
^{246}Cm		246.067224	4.75×10^3 a	α/5.476	5.343(3)/21	0+			Pu L x-ray
			1.8×10^7 a	sf/0.026	5.386(3)/79				0.04453
^{247}Cm		247.070354	1.56×10^7 a	α/5.352	4.818(4)/4.7	9/2-	0.36		Pu k x-ray
					4.8690(20)/71				0.2792
					4.941(4)/1.6				0.2886
					4.9820(20)/2.0				0.3471
					5.1436(20)/1.2				0.4035
					5.2104(20)/5.7				
					5.2659(20)/13.8				
^{248}Cm		248.072349	3.48×10^5 a	α/99.92 /5.162	4.931(5)/0.07	0+			
					5.0349(2)/16.5				
			4.15×10^6 a	sf/8.38	5.0784(2)/(75)/1				
^{249}Cm		249.075953	64.15 m	β- /0.900	0.9/	1/2+			Bk k x-ray
									0.56039/0.84

Elem. or Isot.	Natural Abundance (Atom %)	Atomic Mass or Weight	Half–life/ Resonance Width (MeV)	Decay Mode/ Energy (/MeV)	Particle Energy/ Intensity (MeV/%)	Spin ($h/2\pi$)	Nuclear Magnetic Mom. (nm)	Elect. Quadr. Mom. (b)	γ-Energy/ Intensity (MeV/%)
									0.63431/1.5
									(0.085–0.653)
^{250}Cm		250.07836	~ 9.7×10^3 a	sf/85.8		0+			
				α/5.27					
^{251}Cm		251.08229	16.8 m	β- /1.42	0.90/16	(1/2+)			0.3896 /
									0.5299
									0.5425
^{252}Cm		252.0849	< 2 d			0+			
$_{97}$Bk									
^{234}Bk									
^{236}Bk									
^{237}Bk									
^{238}Bk		238.0583	2.4 m	EC/5.0					
^{239}Bk		239.0583		EC//> 99					
				α//< 1					
				SF//< 1					
^{240}Bk		240.0598	~ 4.8 m						
^{241}Bk		241.0602	4.6 m	EC					(0.152–0.262)
^{242}Bk		242.0620	7.0 m	EC/3.0					
^{243}Bk		243.063008	4.5 h	EC/99.8 /1.508	6.542(4)/0.03	(3/2-)			0.1466
				α/0.15 /6.871	6.5738(2)/0.04				0.1874
					6.7180(22)/0.02				0.755
					6.7581(20)/0.02				0.840
									0.946
^{244}Bk		244.06518	4.4 h	EC/99.99 /2.26		(4-)			0.1445
				α/0.01 /6.778	6.625(4)/0.003				0.1876
					6.667(4)/0.003				0.2176
									0.9815
									0.9215/
^{245}Bk		245.066362	4.94 d	EC/99.9 /0.810		3/2-			Cm L x-ray
				α/0.1 /6.453	5.8851(5)/0.03				Cm k x-ray
					6.1176(9)/0.01				0.25299
					6.1467(5)/0.02				0.3809
					6.3087(5)/0.014				0.3851
					6.3492(5)/0.018				
^{246}Bk		246.0687	1.80 d	EC/1.35		(2-)			Cm L x-ray
									Cm k x-ray
									0.79881
									1.08142
^{247}Bk		247.07031	1.4×10^3 a	α/5.889	5.465(5)/1.5	(3/2-)			0.04175
					5.501(5)/7				0.0839
					5.532(5)/45				0.268
					5.6535(20)/5.5				
					5.678(2)/13				
					5.712(2)/17				
					5.753(2)/4.3				
					5.794(2)/5.5				
^{248}Bk		248.07310	23.7 h	β- /70 /0.87	0.86/	(1-)			Cm L x-ray
				EC/30 /0.72					Cf L x-ray
									Cm k x-ray
									Cf k x-ray
									0.5507
^{249}Bk		249.074987	320. d	β- /0.125	0.125/100	7/2+	2.0		0.327/10^{-5}
				α/0.001 /5.525	5.390(1)/0.0002				0.308/10^{-6}

Elem. or Isot.	Natural Abundance (Atom %)	Atomic Mass or Weight	Half–life/ Resonance Width (MeV)	Decay Mode/ Energy (/MeV)	Particle Energy/ Intensity (MeV/%)	Spin ($h/2\pi$)	Nuclear Magnetic Mom. (nm)	Elect. Quadr. Mom. (b)	γ-Energy/ Intensity (MeV/%)
			1.8×10^9 a	sf/4.9×10^{-8}	5.4174(6)/0.001				
^{250}Bk		250.078317	3.217 h	β- /1.780	0.74/	2-			Cf L x-ray
									Cf k x-ray
									0.98912
									1.03184
									(0.04–1.6)
^{251}Bk		251.08076	56. m	β- /1.09		(3/2-)			0.02481
									0.1528
									0.1776
^{252}Bk		252.0843	1.8 m						
^{253}Bk									
^{254}Bk									
$_{98}$**Cf**									
^{237}Cf		237.062	2.1 s	α//70	8.08				
				SF//30					
^{238}Cf		238.0614	21 ms	SF// > 95		0+			
				α// < 5					
^{239}Cf		239.0624	~ 0.7 m	α					
^{240}Cf		240.0623	1.1 m	α/7.719/98.5	7.590(10)/	0+			
				SF// 1.5					
^{241}Cf		241.0637	4. m	EC/3.3					
				α/7.60	7.335(5)/				
^{242}Cf		242.06370	3.5 m	α/7.509	7.351(6)/20	0+			
				sf/<0.014	7.385(4)/80				
^{243}Cf		243.0654	11. m	EC/86 /2.2	7.060(6)/20	(1/2+)			
				α/14 /7.40	7.170/4				
^{244}Cf		244.066001	20. m	α/7.328	7.168(5)/25	0+			
					7.210(5)/75				
^{245}Cf		245.068049	44. m	α/36 /7.255	7.14/91.7				Cm K x-ray
				EC/64 /1.569	6.983/0.31				0.5709
					7.09/7				0.6014
					7.065/0.68				0.6163
^{246}Cf		246.068805	1.49 d	α/6.869	6.6156(10)/0.18	0+			Cm L x-ray
					6.7086(7)/21.8				0.04221
			1.8×10^3 a	sf/2.3×10^{-4}	6.7501(7)/78.0				0.0945
									0.147
^{247}Cf		247.07100	3.11 h	EC/99.96 /0.65		7/2+			Bk k x-ray
				α/0.04 /6.55	6.301(5)/				0.2941
									0.4778
^{248}Cf		248.07219	334. d	α/6.369	6.220(5)/17	0+			
			3.2×10^4 a	sf/0.0029	6.262(5)/83				
^{249}Cf		249.074854	351. a	α/6.295	5.758/3.7	9/2-			Cm L x-ray
					5.812/85.7				Cm k x-ray
			$8. \times 10^{10}$ a	sf/4.4×10^{-7}	5.8488(2)/1.0				0.25299/2.5
					5.9029(2)/2.8				0.33351/13.6
					5.9451(2)/4.0				0.38832/63.6
					6.1401(2)/1.1				(0.0376–1.10)
					6.1940(2)/2.2				
^{250}Cf		250.076406	13.1 a	α/6.129	5.8913(4)/0.28	0+			Cm L x-ray
			1.7×10^4 a	sf/0.077	5.9889(4)/17.1				0.04285
					6.0310(4)/82.6				
^{251m}Cf			26.3 μs						
^{251}Cf		251.079587	9.0×10^2 a	α/6.172	5.56448(7)/1.5	1/2+			0.109/19.8
					5.632(1)/4.5				0.1775/17.3

Elem. or Isot.	Natural Abundance (Atom %)	Atomic Mass or Weight	Half–life/ Resonance Width (MeV)	Decay Mode/ Energy (/MeV)	Particle Energy/ Intensity (MeV/%)	Spin ($h/2\pi$)	Nuclear Magnetic Mom. (nm)	Elect. Quadr. Mom. (b)	γ-Energy/ Intensity (MeV/%)
					5.648(1)/3.5				(0.0385–0.354)
					5.6773(6)/35				
					5.762(3)/3.8				
					5.7937(7)/2.0				
					5.8124(8)/4.2				
					5.8514(6)/27				
					6.0140(7)/11.6				
					6.0744(7)/2.7				
^{252}Cf		252.081626	2.65 a	α/96.9 /6.217	5.7977(1)/0.23	0+			Cm L x-ray
			86. a	sf/3.1/	6.0756(4)/15.2				0.04339
					6.1184(4)/81.6				0.1002
^{253}Cf		253.08513	17.8 d	β- /99.7 /0.29	0.27/100	(7/2+)			
				α/0.3 /6.126	5.921(5)/0.02				
^{254}Cf		254.08732	60.5 d	sf/99.7/		0+			
				α/0.3/5.930	5.792(5)/0.05				
					5.834(5)/0.26				
^{255}Cf		255.0911	1.4 h	β- /0.7					
^{256}Cf		256.0934	12. m	sf		0+			
$_{99}$**Es**									
^{241}Es		241.0685	~ 8 s	α	8.11				
^{242}Es		242.0698	18 s	α//57	8.03				(0.087–0.122)
				EC//43					
^{243}Es		243.0696	22. s	α//61	7.89/79				
				EC/39 /4.0	7.85/16				
				SF// < 1	7.75/4.3				
^{244}Es		244.0709	37. s	EC/76 /4.6					
				α/4 /	7.57/4				
^{245}Es		245.0713	1. m	α/40 /7.858	7.74				
				EC/60 /3.1					
^{246}Es		246.0729	7.7 m	EC/90 /3.9					
				α/10 /	7.35				
^{247}Es		247.07366	4.8 m	EC/93 /2.48					
				α/7 /	7.32				
^{248}Es		248.0755	26. m	EC/99.7 /3.1					
				α/0.3 /	6.87				
^{249}Es		249.07641	1.70 h	EC/99.4 /1.45		(7/2+)			0.3795
				α/0.6 /	6.77				0.8132
^{250m}Es			2.2 h	EC/		(1-)			Cf L x-ray
				β+					Cf k x-ray
									0.9891
									1.0319
^{250}Es		250.0786	8.6 h	EC/2.1		(6+)			Cf L x-ray
									Cf k x-ray
									0.30339
									0.34948
									0.82883
^{251}Es		251.07999	1.38 d	EC/99.5 /0.38		(3/2-)			
				α/0.5 /	6.462/0.05				
					6.492/0.4				
^{252}Es		252.08298	1.29 a	α/76 /	6.632/61.0	(5-)			
				EC/24 /1.26	6.562/10.3				
^{253}Es		253.084825	20.47 d	α/	6.633/89.8	7/2+	~ +4.10	7.	0.04180/5.6
			6.3×10^5 a	sf/8.9 × 10^{-6}	6.5916/6.6				0.3892/2.7
									(0.0309–1.106)

Elem. or Isot.	Natural Abundance (Atom %)	Atomic Mass or Weight	Half–life/ Resonance Width (MeV)	Decay Mode/ Energy (/MeV)	Particle Energy/ Intensity (MeV/%)	Spin ($h/2\pi$)	Nuclear Magnetic Mom. (nm)	Elect. Quadr. Mom. (b)	γ-Energy/ Intensity (MeV/%)
^{254m}Es			1.64 d	β- /99.6 /	0.475	2+	2.9	3.7	Fm L x-ray
				α/0.3 /6.67	6.382	2+			Fm k x-ray
			> 10. a	sf/0.045					0.6488
									0.6938
^{254}Es		254.088022	276. d	α/	6.429	(7+)	4.35		0.064
			> 2.5 × 10^7 a	sf/< 3 × 10^{-6}					
^{255}Es		255.09027	40. d	β- /92 /0.29		(7/2+)			
				α/8 /	6.26				
			2.6 × 10^3 a	sf/0.0042	6.300				
^{256m}Es			7.6 h	β- /		(8+)			0.218
									0.232
									0.862
^{256}Es		256.0936	25. m	β-/1.7		(1+)			
^{257}Es		257.0960	7.7 d	β-					
^{258}Es									
$_{100}$**Fm**									
^{241}**Fm**									
^{242}Fm		242.0734	0.8 ms	SF// > 96		0+			
^{243}Fm		243.0744	0.22 s	α/	8.55				
				sf/< 0.4					
^{244}Fm		244.0741	3.2 ms	sf/> 97		0+			
^{245}Fm		245.0754	4. s	α/	8.15/				
				sf/<0.1					
^{246}Fm		246.07530	1.2 s	α/85/	8.24/	0+			
				sf/15/					
^{247m}Fm			5.1 s	α// 79	8.78/83				
				SF// 21	8.66/17				
^{247}Fm		247.0769	30. s	α/8.20	7.87/70				(0.082–0.167)
				EC/2.9	7.93/30				
^{248}Fm		248.07720	33. s	α/99.9 /8.001	7.83/20	0+			
				sf/0.1/	7.87/80				
^{249}Fm		249.0790	1.6 m	EC/2.4		(7/2+)			
				α/	7.57				
^{250m}Fm			1.8 s	IT/					
				sf/<8 × 10^{-5}					
^{250}Fm		250.07952	28. m	α/	7.43/	0+			
				EC/0.8					
				sf/0.007					
^{251}Fm		251.08158	5.3 h	EC/98 /1.47		(9/2-)			
				α/2 /	6.833				
^{252}Fm		252.08247	1.058 d	α/7.154	6.998/15	0+			
				sf/0.0023	7.039/85				
^{253}Fm		253.085185	3.0 d	EC/88/0.333	6.676/	½+			Es k x-ray
				α/12 /	6.943/				0.2719
^{254}Fm		254.086854	3.240 h	α/	7.150	0+			
				sf/0.059	7.192				
^{255}Fm		255.089962	20.1 h	α/	6.9635(5)/5.0	7/2+			0.08148/1.
			1.0 × 10^4 a	sf/2.3 × 10^{-5}	7.0225(5)/93.4				(0.041–0.900)
^{256}Fm		256.09177	2.63 h	sf/91		0+			
				α/19	6.92/				
^{257}Fm		257.09511	100.5 d	α/99.79	6.519	(9/2+)			0.1794
				sf/0.21					0.2410
^{258}Fm		258.0971	0.37 ms	sf/		0+			
^{259}Fm		259.1006	1.5 s	sf/					

Elem. or Isot.	Natural Abundance (Atom %)	Atomic Mass or Weight	Half–life/ Resonance Width (MeV)	Decay Mode/ Energy (/MeV)	Particle Energy/ Intensity (MeV/%)	Spin ($h/2\pi$)	Nuclear Magnetic Mom. (nm)	Elect. Quadr. Mom. (b)	γ-Energy/ Intensity (MeV/%)
^{260}Fm		260.103	~ 4 ms	sf/		0+			
$_{101}$**Md**									
^{245m}Md			~ 0.4 s	α	8.64, 8.68				
^{245}Md		245.0808	0.9 ms	sf					
^{246m}Md			~ 4.4 s	Sf					
				α	8.18				
^{246}Md		246.0819	1.3 s	α	8.74				(0.169−0.396)
					8.50−8.56				
^{247m}Md			0.26 s	α// 79	8.78/83				
				SF// 21	8.66/17				
^{247}Md		247.0816	1.3 s	α// > 99.9	8.42/92				0.2096/100
				SF// < 0.1	8.35/6				0.1575/11
					8.62/2				
^{248}Md		248.0828	7. s	EC/80 /5.3	8.32/15				
				α/20 /	8.36/5				
				SF// <0.05					
^{249}Md		249.0830	24. s	EC>/<80 /3.7					0.2532/100
				α/>20 /8.46	8.026(20)/				0.2004/20
									0.2232/10
^{250}Md		250.0844	50. s	EC/94 /4.6	7.75/4				0.1523
				α/6 /8.25	7.83/2				
^{251}Md		251.0848	4.3 m	EC/>94 /3.1	α/7.55/87	7/2-			0.293/100
				α/10	7.59/13				0.243/16
^{252}Md		252.0866	2. m	EC/>50 /3.9	7.73/				
				α/<50 /					
^{253}Md		253.0873	~ 6 m	EC/2.0	α/7.100				0.3532/100
^{254m}Md			30. m	EC/					
^{254}Md		254.0897	10. m	EC/2.7					
^{255}Md		255.09108	27. m	EC/92 /1.04	α/7.33/93	(7/2-)			0.121/100
				α/8 /	7.27/5				0.115/65
				SF// < 0.15	7.75/1				0.136/35
					7.71/1				0.141−0.453
^{256}Md		256.0941	1.30 h	EC/89 /2.13	7.21/71				Fm k x-ray
				α/11 /	7.14/22				0.121/409
				SF// < 2.6	7.68/2.5				0.115/266
					7.25/2.5				0.136/143
					7.64/2.1				0.634/119
									0.141−1.37
^{257}Md		257.095541	5.5 h	EC/85 /0.41	7.074	(7/2-)			Fm k x-ray
				α/15, SF// < 1	7.014				(0.181−0.389)
^{258m}Md			57. m	EC/		(1-)			Fm k x-ray
				SF// < 30					
^{258}Md		258.098431	51.5 d	α/7.40	6.718(2)/	(8-)			0.3678
				SF// < 0.003	6.763(4)/				0.057−0.448
^{259}Md		259.1005	1.64 h	SF// >98.7		7/2+			
				α// <1.3					
^{260}Md		260.1037	~ 27.8 d	SF// 73−100					
^{261}Md									
$_{102}$**No**									
^{248}No		248.0866	< 1.0 μs	SF		0+			
^{249}No		249.0878	0.05 ms	SF					
				α/ < 20					
^{250m}No			0.04 ms	SF					
^{250g}No		250.0875	~ 0.005 ms	SF		0+			

Elem. or Isot.	Natural Abundance (Atom %)	Atomic Mass or Weight	Half–life/ Resonance Width (MeV)	Decay Mode/ Energy (/MeV)	Particle Energy/ Intensity (MeV/%)	Spin ($h/2\pi$)	Nuclear Magnetic Mom. (nm)	Elect. Quadr. Mom. (b)	γ-Energy/ Intensity (MeV/%)
				α/ < 10					
^{251m}No			1.0 s	α	8.668/~98				
					8.625/~2				
^{251}No		251.0890	0.80 s	α// 91	8.612/~98				
				SF// 0.26	8.552/~1				
^{252m}No			0.11 s						0.710
^{252g}No		252.08898	2.44 s	α/75/8.551	8.42	0+			
				SF/24/	8.37				
				EC, β+/<1.6					
^{253m2}No			~ 0.97 ms						
^{253m1}No			0.03 ms	IT		5/2+			0.167
^{253}No		253.0907	1.56 m	α/	8.00	(9/2-)			0.222/100
				EC/3.2	α/8.28				(0.151–0.280)
^{254m2}No			0.17 ms	SF// <0.012					
^{254m1}No			0.27 s	I.T./					0.778
				SF// 0.02					0.856
				α// 0.01					
^{254}No		254.09096	49. s	α/	8.09	0+			0.102
				EC/1.1					0.152
				SF// 0.17					
^{255}No		255.09324	3.1 m	α/62 /	8.095/58	½+			0.187
				EC/38/2.01	7.742/19				(0.163–0.358)
					7.903/11				
^{256}No		256.09428	2.9 s	α/	8.43	0+			
				SF// 0.5					
^{257}No		257.09688	24.5 s	α/	8.222/83	(7/2+)			0.0770/100
				SF// <1.5	8.323/17				0.1018/57
					8.19/<4				0.1241/78
^{258}No		258.0982	~ 1.2 ms	SF		0+			
^{259}No		259.1010	58. m	α/78 /7.794	7.52	(9/2+)			
				EC/22/0.5	7.55				
				SF// <9.7					
^{260}No		260.1026	0.11 s	SF		0+			
^{261}No									
^{262}No		262.1073	~ 8. ms	SF		0+			
^{264}No									
$_{103}$**Lr**									
^{251}Lr		251.0944	39 m	SF					
^{252}Lr		252.0954	0.3 s	A	9.02/73				
				SF// <1	8.97/27				
^{253m}Lr			0.7 s	A	8.79				
				SF// 1.3					
^{253}Lr		253.0952	1.5 s	α/	8.72				
				SF// 8					
^{254}Lr		254.0965	18. s	α// 72	8.45				(0.042–0.306)
				EC/5.2	(8.36–8.53)				
				SF// <0.1					
^{255m2}Lr			1.4 ms						
^{255m}Lr			2.5 s	A	8.46				
^{255g}Lr		255.09669	31. s	α/	8.37/67	1/2-			
				EC/3.2	8.43/<3.6				
				SF// < 0.1	8.29/1.2				
^{256}Lr		256.0986	27. s	α/99.7 /8.554	8.43/				
				EC/4.2	8.39				

Elem. or Isot.	Natural Abundance (Atom %)	Atomic Mass or Weight	Half–life/ Resonance Width (MeV)	Decay Mode/ Energy (/MeV)	Particle Energy/ Intensity (MeV/%)	Spin ($h/2\pi$)	Nuclear Magnetic Mom. (nm)	Elect. Quadr. Mom. (b)	γ-Energy/ Intensity (MeV/%)
				SF// < 0.03					
^{257}Lr		257.0996	0.65 s	α/	8.80	7/2+			
				EC/2.5					
				SF// < 0.03					
^{258}Lr		258.1018	3.9 s	α/	8.60/46				
				EC/3.4	8.62/25				
				SF// < 5	8.56/20				
					8.65/9				
^{259}Lr		259.1029	6.1 s	α/80	8.44(1)				
				SF// 20					
^{260}Lr		260.1055	3. m	α/	8.03				
^{261}Lr		261.1069	40. m	SF					
^{262}Lr		262.1096	3.6 h	EC/2.					
				SF// <10					
^{263}Lr									
^{264}Lr									
$_{104}$**Rf**									
^{253}Rf		253.1007	~ 48. μs	SF					
				α/<10					
^{254}Rf		254.1002	23. μs	SF// >98.5		0+			
				α/<1.5					
^{255}Rf		255.1013	1.6 s	α// ~ 52	8.716/92				0.2036/49
				SF// ~ 48	8.678/3				0.1433/51
					8.906/2.5				
					8.646/1.5				
					8.575/1				
^{256}Rf		256.10117	6.2 ms	SF// 99.68		0+			
				α/0.32	8.81				
^{257m}Rf			0.14 ms						
^{257}Rf		257.1030	4.7 s	α/9.22	8.77				0.117
				EC/11	9.01				
				SF// <1.4	8.95				
					8.62				
^{258}Rf		258.1035	12. ms	SF// 87		0+			
				α/13	9.05				
^{259}Rf		259.1056	2.5 s	α/9.09/93	8.77(2)/				
				SF// 7	8.86/				
				EC// 15					
^{260}Rf		260.1064	21. ms	SF		0+			
				α// <35					
^{261m}Rf			1.2 m	α	8.29				
					8.34				
^{261}Rf		261.10877	3. s	α/60, SF/40	8.52/				
^{262}Rf		262.1099	2.1 s	SF// >99.2		0+			
^{263}Rf		263.1126	11. m	SF, α					
^{265}Rf		265.1167	~ 2. m	α					
^{267}Rf		267.122	~ 1. h	SF					
$_{105}$**Db**									
^{255}Db		255.1074	~ 1.5 s	α,					
				SF// ~ 20					
^{256}Db		256.1081	1.6 s	α/64	9.02/67				
				EC/35	8.89/11				
				SF// 0.05	9.08/11				
					9.12/11				

Elem. or Isot.	Natural Abundance (Atom %)	Atomic Mass or Weight	Half–life/ Resonance Width (MeV)	Decay Mode/ Energy (/MeV)	Particle Energy/ Intensity (MeV/%)	Spin ($h/2\pi$)	Nuclear Magnetic Mom. (nm)	Elect. Quadr. Mom. (b)	γ-Energy/ Intensity (MeV/%)
257mDb			0.7 s	α	9.17				
				SF// <13					
257Db		257.1077	2. s	α/	8.96/33				0.1022
				SF// <6	9.06/38				
					9.12/5.5				
					8.94/9				
					9.02/9				
					8.89/5.5				
258mDb			~ 1.9 s	α	9.20				
258Db		258.1092	4.2 s	α/	9.30/				0.1568
				EC/5.3/39	9.17/				0.2215
				SF// <33	9.08/				
259Db		259.1096	~ 0.51 s	SF					
				α/	9.47/				
260mDb			0.3 m						
260Db		260.1113	1.5 s	α/	9.05/				
				SF// <9.6	9.08/				
					9.13/				
261Db		261.1121	1.8 s	α/	8.93/				
				SF// ~73					
262Db		262.1141	0.5 m	SF// <33					
				α/	8.45/				
					8.53/				
					8.67/				
263Db		263.1150	~ 0.45 m	SF// 57	8.36/				
				α// 41	8.41/				
				EC/3					
266Db			~ 0.4 h	EC, SF					
267Db		267.1224	1.2 h	SF					
268Db		268.125	1.2 d	SF, EC					
270Db			1.0 d						
106 Sg									
258Sg		258.1132	2.6 ms	SF		0+			
				α// <38					
259Sg		259.1145	0.4 s	α/	9.61				
				SF// <8.6	9.35				
				EC// <13	9.03				
260Sg		260.11442	5.0 ms	α// 29	9.748	0+			
				SF// 71	9.72				
					9.81				
261mSg			~ 9 μs						
261Sg		261.1161	0.2 s	α	9.56				0.107
				EC// 1.3	9.42				
				SF// 0.6	9.41				
262Sg		262.1164	15. ms	SF		0+			
				α/<16					
263mSg			0.3 s	α	9.2				
263Sg		263.1183	0.8 s	α	9.06				
				SF// ~13	9.25				
264Sg			~ 0.04 s	α// <36					
265mSg			~ 9. s						
265Sg		265.1211	16. s	α// >65	8.84/46				
				SF// <35	8.76/23				
					8.94/23				

Elem. or Isot.	Natural Abundance (Atom %)	Atomic Mass or Weight	Half–life/ Resonance Width (MeV)	Decay Mode/ Energy (/MeV)	Particle Energy/ Intensity (MeV/%)	Spin ($h/2\pi$)	Nuclear Magnetic Mom. (nm)	Elect. Quadr. Mom. (b)	γ-Energy/ Intensity (MeV/%)
					8.69/8				
^{266}Sg		266.1221	~ 0.4. s	SF		0+			
^{267}Sg			~ 1.3 m	α					
^{269}Sg			2.1 m	A					
^{271}Sg		271.133	2. m	α/70	~ 8.54				
				SF// 30					
$_{107}$Bh									
^{260}Bh		260.122	0.04 s	α// 70	10.16				
				SF// < 18					
				EC// < 18					
^{261}Bh		261.1217	12. ms	α/, SF <10	10.40				
					10.10				
					10.03				
^{262m}Bh			22. ms	α/	10.37				0.1024
				SF// <24	10.24				
^{262}Bh		262.1229	~ 0.08 s	α/	10.1				0.0389
				SF// <11	9.9				0.1565
					9.7				
^{264}Bh		264.1246	0.44 s	α/	9.3–9.8				
				SF					
^{265}Bh		265.1252	0.9 s	α	9.24				
^{266}Bh		266.1269	~ 2 s	α	9.08				
^{267}Bh		267.1277	~ 17 s	α	8.83				
^{270}Bh			~ 1. m	α	9.0				
^{272}Bh			~ 10. s	α	8.9				
$_{108}$Hs									
^{263}Hs		263.1286	0.7 ms	α	10.6				
				SF// < 8.4	10.8				
					10.9				
^{264}Hs		264.12839	0.45 ms	α/, SF// ~ 50	11.0	0+			
^{265m}Hs			0.3 ms	α	10.57/63				
					10.73				
					10.52				
					10.34				
^{265}Hs		265.1301	1.9 ms	α/	10.30/90				
				SF// <1	10.43				
					10.37				
					10.25				
^{266}Hs		266.1301	~ 2.3 ms	α	10.2	0+			
^{267m}Hs			~ 0.8 s						
^{267}Hs		267.1318	0.05 s	α// >88	9.88				
					9.83				
					9.75				
^{269}Hs		269.1341	~ 10. s	α	9.07				
					8.92				
^{270m}Hs			0.3 m	α	8.88				
^{270}Hs		270.1347	~ 3.6 s	α	9.16	0+			
^{271}Hs			~ 4. s	α	9.13				
					9.30				
^{273}Hs			0.2 s	α	9.6				
^{275}Hs		275.146	~ 0.19 s	α	9.3				
^{277}Hs		277.150	~ 3. ms	SF					
$_{109}$Mt									
^{266m}Mt			~ 1.2 ms	α	10.46–10.81				

Elem. or Isot.	Natural Abundance (Atom %)	Atomic Mass or Weight	Half–life/ Resonance Width (MeV)	Decay Mode/ Energy (/MeV)	Particle Energy/ Intensity (MeV/%)	Spin ($h/2\pi$)	Nuclear Magnetic Mom. (nm)	Elect. Quadr. Mom. (b)	γ-Energy/ Intensity (MeV/%)
^{266}Mt		266.1373	~ 3. ms	α	10.48–11.31				
				SF// 0.25					
^{267}Mt		267.137	19 ms	α					
^{268}Mt		268.1387	~ 0.02 s	α// >68	10.3–10.8				
^{270}Mt		270.141	~ 0.8 s	α	10.0				
^{274}Mt			~ 0.4 s	α	9.8				
^{275}Mt		275.149	~ 9.7 ms	α	10.3				
^{276}Mt		276.151	~ 0.7 s	α	~ 9.71				
^{278}Mt			7.6 s						
$_{110}$**Ds**									
^{267}Ds		267.1443	~ 3 μs	α// >32	11.6				
^{269}Ds		269.1451	0.17 ms	α// >75	11.11				
^{270m}Ds			~ 6 ms	α	10.95				
					11.15				
					12.15				
^{270}Ds		270.1447	0.1 ms	α	11.03	0+			
^{271m}Ds			0.07 s	α	9.9				
^{271}Ds		271.1461	1.6 ms	α	10.8				
^{273m}Ds			0.17 ms	α	11.8				
^{273}Ds		273.1489	118 ms	α	9.73				
^{277}Ds			6. ms	α/	10.6				
^{279}Ds		279.159	0.18 s	SF// 90					
				α// 10	~ 9.70				
^{280}Ds		280.160	~ 7.6 s	SF		0+			
^{281}Ds		281.162	13. s	SF					
^{282}Ds			0.5 ms	SF					
$_{111}$**Rg**									
^{272}Rg		272.1536	~ 3.8 ms	α// >68	10.82				
^{274}Rg		274.156	~ 6. ms	α	11.2				
^{278}Rg			~ 4. ms	α	10.7				
^{279}Rg		279.162	~ 0.17 s	α	10.4				
^{280}Rg		280.164	~ 3.6 s	α	~ 9.75				
^{281}Rg			2.6 s	SF					
^{282}Rg			0.51 s	α					
$_{112}$**Cn**									
^{277}Cn		277.1639	~ 0.7 ms	α	11.45				
					11.65				
^{281}Cn			0.10 s	α					
^{282}Cn			0.8 ms	SF					
^{283}Cn		283.172	4. s	SF// < 10					
				α// ~ 100	9.52				
^{284}Cn		284.172	~ 0.10 s	SF					
^{285}Cn		285.174	~ 29. s	α	9.15				
$_{113}$**113**									
^{278}113			~ 0.24 ms	α	11.7				
^{282}113			~ 0.07 s	α	10.6				
^{283}113		283.176	~ 0.1 s	α	10.1				
^{284}113		284.178	~ 0.48 s	α	10.0				
^{285}113			~ 5.5 s	α	9.74				
					9.48				
^{286}113			20. s	α	9.6				
$_{114}$**114**									
^{285}114			0.13 s	α					
^{286}114		286.184	0.13 s	α/50	10.2				

Elem. or Isot.	Natural Abundance (Atom %)	Atomic Mass or Weight	Half–life/ Resonance Width (MeV)	Decay Mode/ Energy (/MeV)	Particle Energy/ Intensity (MeV/%)	Spin ($h/2\pi$)	Nuclear Magnetic Mom. (nm)	Elect. Quadr. Mom. (b)	γ-Energy/ Intensity (MeV/%)
				SF// 50					
[287]114		287.186	0.48 s	α	10.0				
[288]114		288.186	0.7 s	α	9.94				
[289]114		289.187	~ 2.1 s	α	9.82				
[115]**115**									
[287]115		287.191	~ 0.03 s	α	10.6				
[288]115		288.192	~ 0.09 s	α	10.5				
[289]115			0.22 s	α	10.3				
[290]115			16 ms	α	9.95				
[116]**116**									
[290]116		290.199	7. ms	α	10.8				
[291]116		291.200	0.02 s	α	~ 10.74				
[292]116		292.200	~ 18. ms	α	~ 10.66				
[293]116			0.06 s	α	10.5				
[117]**117**									
[293]117			~ 14 ms	α	11.0				
[294]117			~ 0.08 s	α	10.8				
[118]**118**									
[294]118			0.9 ms	α	11.7				

NEUTRON SCATTERING AND ABSORPTION PROPERTIES

Norman E. Holden

This table presents an evaluated set of values for experimental quantities that characterize the properties for scattering and absorption of neutrons. The neutron cross section is given for room temperature neutrons, 20.43 °C, corresponding to a thermal neutron energy of 0.0253 electron volts (eV) or a neutron velocity of 2200 meters/second. The neutron resonance integral is defined over the energy range from 0.5 eV to 0.1×10^6 eV, or 0.1 MeV.

Bound neutron scattering lengths and neutron cross sections averaged over a Maxwellian spectrum at 30 keV for astrophysical applications are also presented. A list of the major references used is given below. The literature cutoff date is January 2003. Uncertainties are given in parentheses. Parentheses with two or more numbers indicate values to the excited state(s) and to the ground state of the product nucleus.

Table Layout

Column Number	Column Title	Description
1	Isotope/Element	For elements, atomic number and chemical symbol are listed. For nuclides, mass number and chemical symbol are listed. Isomers are indicated by the addition of m, m1, or m2.
2	Isotopic Abundance	In atom percent.
3	Half-life	Half-life in decimal notation. μs = microsecond; ms = millisecond; s = second; m = minute; h = hour; d = day; y = year.
4	Thermal Neutron Cross Sections	Cross sections for neutron capture reactions in units of barns (10^{-24} cm^2) or millibarns (mb). Proton, alpha production and fission reactions are designated by σ_p, σ_a, σ_f, respectively. Separate values are listed for isomeric production.
5	Neutron Resonance Integrals	Resonance integrals for neutron capture reactions in barns (10^{-24} cm^2) or millibarns (mb). Proton, alpha production and fission reactions are designated by $R.I._p$, $R.I._a$, $R.I._f$, respectively. Separate values are listed for isomeric production.
6	Neutron Scattering Lengths	Bound coherent scattering lengths for neutron scattering reactions in units of femtometers (fm), which is equal to fermis (10^{-13} cm).
7	Maxwellian Averaged Cross Section	Astrophysical cross sections, averaged over a stellar neutron maxwellian spectrum characterized by a thermal energy of 30 keV, expressed in barns (10^{-24} cm^2), millibarns (mb) or microbarns (μb).

General Nuclear Data References

The following references represent the major sources of the nuclear data presented:

Mughabghab, S.F., Divadeenam, M., Holden, N.E.; Neutron Cross Sections, Vol. 1 *Neutron Resonance Parameters and Thermal Cross Sections*, Part A, Z = 1–60. Academic Press Inc., New York, New York (1981); Mughabghab, S.F.; Part B, Z = 61–100. Academic Press Inc., Orlando, FL (1984).

Holden, N.E.; *Fifty Years with Nuclear Fission* Conference, Wash., D.C., Gaithersburg, Md. April 26–29, 1989, p. 946. American Nuclear Society, LaGrange Park, IL (1989).

Tuli, J.K.; *Nuclear Wallet Cards*, Brookhaven National Laboratory (Jan. 2000).

Holden, N.E.; Half-lives of Selected Nuclides, *Pure & Applied Chemistry* 62, 941 (1990).

Holden, N.E., Hoffman, D.C.; Spontaneous Fission Half-lives for Ground State Nuclides, *Pure & Applied Chemistry* 72, 1525 (2000).

Koester, L., Rauch, H., Seymann, E.; Neutron Scattering Lengths: A Survey of Experimental Data and Methods, *Atomic Data Nuclear Data Tables* 49, 65 (1991).

Sears, V.F.; Neutron Scattering Lengths and Cross Sections, *Neutron News* 3, (3), 26 (1992).

Bao, Z.Y., Beer, H., Käpeler, F., Voss, F., Wisshak, K., Raucher, T.; Neutron Cross Sections for Nucleo-synthesis Studies, *Atomic Data Nuclear Data Tables* 76, 70 (2000).

Elem. or Isot.	Natural Abundance (%)	Half-Life	Thermal Neut. Cross Section (barns)	Resonance Integral (barns)	Coh. Scat. Length (fm)	σ (30 keV) Maxw. Avg. (barns)
$_1$H			0.332(2)	0.149(1)	−3.739(1)	
^{1}H	99.9885(70)	$>2.8 \times 10^{23}$ y	0.332(2)	0.149(1)	− 3.741(1)	0.25(2) mb*
^{2}H	0.0115(70)		0.51(1)mb	0.23(2) mb	6.671(4)	2.1(4) μb
^{3}H		12.33 y	< 6. μb		4.79(3)	
$_2$He			< 0.05		3.26(3)	
^{3}He	0.000134(3)		$\sigma_p = 5.33(1) \times 10^3$	$RI_p = 2.39(1) \times 10^3$	5.74(7)	
			0.05(1) mb			8.(1) μb*
^{4}He	99.999867(3)				3.26(3)	
$_3$Li			71.(2)	32.(1)	− 1.90(2)	

* Extrapolated value.

Elem. or Isot.	Natural Abundance (%)	Half-Life	Thermal Neut. Cross Section (barns)	Resonance Integral (barns)	Coh. Scat. Length (fm)	σ (30 keV) Maxw. Avg. (barns)
^{6}Li	7.59(4)		$\sigma_t = 9.4(1)\times10^2$	$RI_t = 422.(4)$	2.0(1)	$\sigma_t \approx 1.$
			39.(5) mb	17.(2) mb		0.06(1) mb*
^{7}Li	92.41(4)		45.(5) mb	20.(2) mb	$-2.22(2)$	42.(3) μb
^{8}Li		0.84 s				$< \approx 5.5$ μb
$_4$Be			8.8(4) mb	3.9(2) mb	7.79(1)	
^{7}Be		53.28 d	$\sigma_p = 3.9(1)\times10^4$	$RI_p = 1.75(5)\times10^4$		$\sigma_p = 16(4)^*$
			$\sigma_\alpha \approx 0.1$			
^{9}Be	100.		8.8(4) mb	3.9(2) mb	7.79(1)	
^{10}Be		1.52×10^6 y	<1. mb			
$_5$B			$7.6(1)\times10^2$	$3.4(1)\times10^2$	5.30(4)	
^{10}B	19.9(7)		$\sigma_\alpha = 38.4(1)\times10^2$	$RI_\alpha = 17.3(1)\times10^2$	$-0.1(3)$	
			0.3(1)	0.13(4)		
			$\sigma_p = 7.(1)$ mb			
			$\sigma_t = 8.(2)$ mb			
^{11}B	80.1(7)		5.(3) mb	2.(1) mb	6.65(4)	
$_6$C			3.5(1) mb	1.6(1) mb	6.646(1)	
^{12}C	98.93(8)		3.5(1) mb	1.6(1) mb	6.651(2)	16.(1) μb*
^{13}C	1.07(8)		1.4(1) mb	1.7(2) mb	6.19(9)	0.021(4) mb
^{14}C		5715. y	<1.4 μb			3.(1) μb*
$_7$N			2.00(6)	0.90(3)	9.36(2)	
^{14}N	99.636(20)		$\sigma_p = 1.93(5)$	$RI_p = 0.87(3)$	9.37(2)	$\sigma_p = 1.8(2)$ mb*
			0.080(1)	0.034(1)		0.04(1) mb
^{15}N	0.364(20)		0.04(1) mb	0.11(3) mb	6.44(3)	6.(1) μb*
$_8$O			0.29(1) mb	0.40(4) mb	5.805(4)	
^{16}O	99.757(16)		0.19(1) mb	0.36(4) mb	5.805(5)	34.(4) μb
^{17}O	0.038(1)		$\sigma_\alpha = 0.257(10)$	0.11(1)	5.8(2)	$\sigma_\alpha = 3.9(5)$ mb*
			0.54(7) mb	0.39(5) mb		
^{18}O	0.205(14)		0.16(1) mb	0.81(4) mb	5.84(7)	9.(1) μb*
$_9$F			9.5(1) mb	21.(3) mb	5.65(1)	6.(1) mb
^{19}F	100.		9.5(1) mb	21.(3) mb	5.65(1)	6.(1) mb
$_{10}$Ne			42.(5) mb	19.(3) mb	4.566(6)	
^{20}Ne	90.48(3)		39.(5) mb	18.(3) mb	4.631(6)	0.12(1) mb
^{21}Ne	0.27(1)		0.7(1)	0.31(5)	6.7(2)	≈ 1.5 mb
			$\sigma_\alpha = 0.18(9)$ mb			
^{22}Ne	9.25(3)		51.(5) mb	23.(3) mb	3.87(1)	58.(4) μb*
$_{11}$Na			0.53(2)	0.32(2)	3.63(2)	2.1(2) mb
^{22}Na		2.605 y	$\sigma_p = 2.8(3)\times10^4$	$RI_p < 2.\times10^5$		
			$\sigma_\alpha = 2.6(4)\times10^2$	$RI_a = 1.2(2)\times10^2$		
^{23}Na	100.		$\sigma_m = 0.43(3)$	$RI_m = 0.30(6)$	3.63(2)	2.1(2) mb
$_{12}$Mg			66.(6) mb	38.(5) mb	5.375(4)	
^{24}Mg	78.99(4)		0.053(6)	32.(4) mb	5.7(2)	3.3(4) mb
^{25}Mg	10.00(1)		0.20(1)	98.(15) mb	3.6(2)	6.4(4) mb
^{26}Mg	11.01(3)		0.038(1)	25.(2) mb	4.9(2)	0.13(1) mb*
^{27}Mg		9.45 m	0.07(2)	0.03(1)		
$_{13}$Al			0.230(2)	0.17(1)	3.45(1)	
^{26}Al		7.1×10^5 y	$\sigma_p = 1.97(10)$			0.14(2)
			$\sigma_\alpha = 0.34(1)$			
^{27}Al	100.		0.230(2)	0.17(1)	3.45(1)	2.9(3) mb
$_{14}$Si			0.166(9)	0.12(2)	4.15(1)	
^{28}Si	92.223(19)		0.17(1)	0.11(2)	4.11(1)	2.9(3) mb
^{29}Si	4.685(8)		0.12(1)	0.08(2)	4.7(1)	7.9(9) mb
^{30}Si	3.092(11)		0.107(3)	0.62(6)	4.61(1)	3.2(3) mb*
^{31}Si		2.62 h	73.(6) mb	33.(3) mb		
^{32}Si		1.6×10^2 y	< 0.5			
$_{15}$P			0.17(1)	0.08(1)	5.13(1)	
^{31}P	100.		0.17(1)	0.08(1)	5.13(1)	1.7(1) mb
$_{16}$S			0.54(2)	0.24(2)	2.847(1)	

* Extrapolated value.

Elem. or Isot.	Natural Abundance (%)	Half-Life	Thermal Neut. Cross Section (barns)	Resonance Integral (barns)	Coh. Scat. Length (fm)	σ (30 keV) Maxw. Avg. (barns)
$_{16}$S						
^{32}S	94.93(31)		0.55(5)	0.25(2)	2.804(2)	4.1(2) mb
			$\sigma_\alpha < 0.5$ mb			
^{33}S	0.76(2)		0.46(3)	0.21(2)	4.7(2)	7.4(15) mb
			$\sigma_\alpha = 0.12(1)$	$RI_\alpha = 0.05(1)$		$\sigma_\alpha = 0.18(1)$
			$\sigma_p = 2.$ mb			
^{34}S	4.29(28)		0.25(1)	0.13	3.48(3)	0.23(1) mb
^{36}S	0.02(1)		0.24(2)	0.26(3)		0.17(1) mb*
$_{17}$Cl			33.6(3)	15.(2)	9.58(1)	
^{35}Cl	75.78(4)		43.7(4)	20.(2)	11.7(1)	9.4(3) mb
			$\sigma_p = 0.44(1)$	$RI_p = 0.2$		$\sigma_p = 1.7(2)$ mb*
			$\sigma_\alpha \approx 0.08$ mb			
^{36}Cl		3.01×10^5 y	$\sigma_p = 46.(2)$ mb	$RI_p = 0.02$		$\sigma_p = 91.(8)$ mb
			<10.			
			$\sigma_\alpha = 0.59(7)$ mb			$\sigma_\alpha = 0.9(2)$ mb
^{37}Cl	24.22(4)		(0.05 + 0.38)	(0.04+0.26)	3.1(1)	2.0(2) mb
$_{18}$Ar			0.66(3)	0.42(5)	1.91(1)	
^{36}Ar	0.3365(30)		5.(1)	2.(1)	24.9(1)	
			$\sigma_\alpha = 5.4(3)$ mb			
			$\sigma_p < 1.5$ mb			
^{37}Ar		35.0 d	$\sigma_\alpha = 1.08(8)\times10^3$	$RI_\alpha = 900.$		$\sigma_\alpha \approx 1.3$
			$\sigma_p = 37.(4)$	$RI_p = 31.$		$\sigma_p \approx 0.04$
^{38}Ar	0.0632(5)		0.8(2)	0.4(1)	3.5(35)	
^{39}Ar		268. y	$6.(2)\times10^2$			
			$\sigma_\alpha <0.29$			
^{40}Ar	99.6003(30)		0.64(3)	0.41(5)	1.83(1)	2.5(3) mb
^{41}Ar		1.82 h	0.5(1)	0.2(1)		
$_{19}$K			2.1(1)	1.0(1)	3.67(2)	
^{39}K	93.2581(44)		2.1(2)	0.9(1)	3.74(2)	11.8(4) mb
			$\sigma_\alpha = 4.3(5)$ mb			
			$\sigma_p < 0.05$ mb			
^{40}K	0.0117(1)	1.26×10^9 y	30.(8)	13.(4)		$\sigma_p = 7.(1)$ mb
			$\sigma_p = 4.4(4)$	2.0(2)		$\sigma_\alpha = 40.(6)$ mb
			$\sigma_\alpha = 0.42(8)$			
^{41}K	6.7302(44)		1.46(3)	1.4(2)	2.69(8)	22.(1) mb
$_{20}$Ca			0.43(2)	0.23(2)	4.70(2)	
^{40}Ca	96.941(156)		0.41(3)	0.22(4)	4.80(2)	6.7(7) mb
			$\sigma_\alpha = 0.13(4)$ mb			
^{41}Ca		1.02×10^5 y	$\approx 4.$			
			$\sigma_\alpha = 0.18(3)$			
			$\sigma_p = 7.(2)$ mb			
^{42}Ca	0.647(23)		0.65(10)	0.39(4)	3.4(1)	16.(2) mb
^{43}Ca	0.135(10)		6.(1)	3.9(2)	− 1.56(9)	51.(6) mb
^{44}Ca	2.086(110)		0.8(2)	0.56(1)	1.42(6)	9.(1) mb
^{45}Ca		162.7 d	$\approx 15.$			
^{46}Ca	0.004(3)	$>4\times10^{15}$ y	0.70(3)	0.9(1)	3.6(2)	5.3(5) mb*
^{48}Ca	0.187(21)	4.3×10^{19} y	1.0(1)	0.5(1)	0.39(9)	0.8(1) mb*
$_{21}$Sc			27.2(2)	12.(1)	12.3(1)	
^{45}Sc	100.		(10.+17.)	(5.6+6.4)	12.3(1)	69.(5) mb
^{46}Sc		83.81 d	8.(1)	3.6(5)		
$_{22}$Ti			6.1(1)	2.8(2)	− 3.438(2)	
^{44}Ti		60 y	1.1(2)			
			$\sigma_p < 0.2$			
^{46}Ti	8.25(3)		0.6(2)	0.4(1)	4.93(6)	27.(3) mb
^{47}Ti	7.44(2)		1.6(2)	1.6(2)	3.63(1)	64.(8) mb
^{48}Ti	73.72(3)		7.9(9)	3.6(2)	− 6.09(2)	32.(5) mb
^{49}Ti	5.41(2)		1.9(5)	1.2(2)	1.04(5)	22.(2) mb
^{50}Ti	5.18(2)		0.179(3)	0.12(2)	6.18(8)	3.6(4) mb
$_{23}$V			5.0(2)	2.8(1)	− 0.382(1)	

* Extrapolated value.

Elem. or Isot.	Natural Abundance (%)	Half-Life	Thermal Neut. Cross Section (barns)	Resonance Integral (barns)	Coh. Scat. Length (fm)	σ (30 keV) Maxw. Avg. (barns)
^{50}V	0.250(4)	1.4×10^{17} y	21.(4)	50.(20)	7.6(6)	
			$\sigma_p = 0.7(4)$ mb			
^{51}V	99.750(4)		4.9(1)	2.7(2)	$-0.402(2)$	38.(4) mb
$_{24}$Cr			3.0(2)	1.7(1)	3.635(7)	
^{50}Cr	4.345(13)	$>1.8\times10^{17}$ y	15.(1)	8.(1)	$-4.5(1)$	0.05(1)
^{51}Cr		27.70 d	< 10.			
^{52}Cr	83.789(18)		0.8(1)	0.6(2)	4.91(2)	8.8(4) mb
^{53}Cr	9.501(17)		18.(2)	9.(1)	$-4.2(1)$	0.06(1)
^{54}Cr	2.365(7)		0.36(4)	0.25(5)	4.6(1)	7.(2) mb
$_{25}$Mn			13.3(1)	14.0(3)	$-3.75(2)$	
^{53}Mn		3.7×10^6 y	70.(10)	32.(5)		
^{54}Mn		312.1 d	< 10.			
^{55}Mn	100.		13.3(1)	14.0(3)	$-3.75(2)$	40.(3) mb
$_{26}$Fe			2.7(1)	1.4(2)	9.45(2)	
^{54}Fe	5.845(35)		2.3(2)	1.3(2)	4.2(1)	29.(2) mb
			$\sigma_\alpha = 10.$ μb	$RI_\alpha = 1.1(1)$ mb		
^{55}Fe		2.73 y	13.(2)	6.(1)		
			$\sigma_\alpha = 0.01$			
^{56}Fe	91.754(36)		2.8(3)	1.4(2)	9.93(3)	11.7(5) mb
^{57}Fe	2.119(10)		1.4(2)	0.8(4)	2.3(1)	40.(4) mb
^{58}Fe	0.282(4)		1.3(1)	1.3(2)	15.(7)	12.(1) mb
^{59}Fe		44.51 d	13.(3)	6.(1)		
$_{27}$Co			37.19(8)	74.(2)	2.49(2)	
^{58m}Co		9.1 h	$1.4(1)\times10^5$	$2.5(10)\times10^5$		
^{58}Co		70.88 d	$1.9(2)\times10^3$	$7.(1)\times10^3$		
^{59}Co	100.		(20.7+16.5)	(39.+35.)	2.49(2)	38.(4) mb
^{60m}Co		10.47 m	58.(3)	230.(50)		
^{60}Co		5.271 y	2.0(2)	4.3(10)		
$_{28}$Ni			4.5(2)	2.3(2)	10.3(1)	
^{58}Ni	68.0769(89)	$>4\times10^{19}$ y	4.6(4)	2.3(2)	14.4(1)	41.(2) mb
			$\sigma_\alpha < 0.03$ mb			
^{59}Ni		$\approx 7.6\times10^4$ y	$\sigma_{abs} = 92.(4)$	$RI_{abs} = 1.4(1)\times10^2$		
			$\sigma_\alpha = 14.(2)$			
			$\sigma_p = 2.(1)$			
^{60}Ni	26.2231(77)		2.9(3)	1.5(2)	2.8(1)	25.(1) mb
^{61}Ni	1.1399(6)		2.5(5)	1.5(4)	7.60(6)	82.(8) mb
			$\sigma_\alpha = 0.03$ mb			
^{62}Ni	3.6345(17)		15.(1)	6.8(3)	$-8.7(2)$	13.(4) mb
^{63}Ni		100. y	20.(5)	9.(2)		
^{64}Ni	0.9256(9)		1.6(1)	1.2(2)	$-0.37(7)$	9.(1) mb
^{65}Ni		2.517 h	22.(2)	10.(1)		
$_{29}$Cu			3.8(1)	4.1(4)	7.718(4)	
^{63}Cu	69.15(15)		4.5(2)	5.(1)	6.43(15)	0.09(1)
^{64}Cu		12.701 h	$\approx 270.$			
^{65}Cu	30.85(15)		2.17(3)	2.2(1)	10.61(19)	41.(5) mb
^{66}Cu		5.09 m	$1.4(1)\times10^2$	60.(20)		
$_{30}$Zn			1.1(2)	2.8(4)	5.680(5)	
^{64}Zn	48.27(32)	$>2.3\times10^{18}$ y	0.74(5)	1.4(3)	5.23(4)	59.(5) mb
			$\sigma_p < 12.$ μb			
			$\sigma_\alpha = 11.(3)$ μb			
^{65}Zn		243.8 d	66.(8)	30.(4)		
			$\sigma_\alpha = 2.0(2)$			
^{66}Zn	27.977(77)		0.9(3)	1.8(2)	5.98(5)	35.(3) mb
			$\sigma_\alpha < 0.02$ mb			
^{67}Zn	4.102(21)		6.9(1.4)	25.(5)	7.58(8)	0.15(2)
			$\sigma_\alpha = 0.4$ mb			
^{68}Zn	19.02(12)		(0.072 + 0.8)	(0.2 + 2.9)	6.04(3)	19.(2) mb
			$\sigma_\alpha < 0.02$ mb			$\sigma_m = 3.(1)$ mb

* Extrapolated value.

Elem. or Isot.	Natural Abundance (%)	Half-Life	Thermal Neut. Cross Section (barns)	Resonance Integral (barns)	Coh. Scat. Length (fm)	σ (30 keV) Maxw. Avg. (barns)
^{70}Zn	0.631(9)		(8.1+83.) mb	0.9(2)		0.02(1)
$_{31}$Ga			2.9(1)	22.(3)	7.288(2)	
^{69}Ga	60.108(9)		1.68(7)	16.(2)	7.88(4)	0.14(1)
^{71}Ga	39.892(9)	>2.4×10^{26} y	4.7(2)	31.(3)	6.40(3)	0.12(1)
			$\sigma_m = 0.15(5)$			
$_{32}$Ge			2.2(1)	6.(2)	8.19(2)	
^{68}Ge		270.8 d	1.0(5)			
^{70}Ge	20.370(89)		(0.3 + 2.7)	2.3(1)	10.0(1)	88.(5) mb
^{72}Ge	27.380(60)		0.9(2)	0.8(3)	8.5(1)	0.07(2)
^{73}Ge	7.759(78)	>1.8×10^{23} y	15.(1)	66.(20)	5.02(4)	0.3(1)
^{74}Ge	36.656(80)		(0.14 + 0.28)	(0.4+0.5)	7.6(1)	53.(7) mb
^{76}Ge	7.835(81)	1.6×10^{21} y	(0.09 + 0.06)	(1.3+0.6)	8.2(15)	0.03(2)
$_{33}$As			4.0(4)	61.(5)	6.58(1)	
^{75}As	100.		4.0(4)	61.(5)	6.58(1)	0.57(4)
$_{34}$Se			12.(1)	14.(3)	7.970(9)	
^{74}Se	0.89(4)		50.(2)	520(50)	0.8(3)	0.2(1)
^{75}Se		119.78 d	3.3(10)×10^2			
^{76}Se	9.37(29)		(22. + 63.)	(9.+31.)	12.2(1)	0.16(1)
^{77}Se	7.63(16)		42.(4)	30.(5)	8.25(8)	0.3(1)
			$\sigma_\alpha = 0.97(3)$ μb			
^{78}Se	23.77(28)		$\sigma_m = 0.38(2)$	RI$_m$ = 4.3(4)	8.24(9)	0.1
^{80}Se	49.61(41)		(0.05+0.54)	(0.15+0.85)	7.48(3)	42.(3) mb
^{82}Se	8.73(22)	≈ 1×10^{20} y	(39.+ 5.2) mb	39.(4) mb	6.34(8)	0.04(2)
$_{35}$Br			6.8(2)	92.(8)	6.79(2)	
^{76}Br		16.0 h	224.(42)			
^{79}Br	50.69(7)		(2.5+8.3)	(36.+96.)	6.79(7)	0.63(4)
						$\sigma_m = 0.08(1)$
^{81}Br	49.31(7)		(2.4+0.24)	51.(5)	6.78(7)	0.31(2)
$_{36}$Kr			24.(1)	39.(6)	7.81(2)	
^{78}Kr	0.353(3)	>2.3×10^{20} y	(0.17+6.)	20.(1)		(0.11+0.19)
^{80}Kr	2.286(10)		(4.6+7.)	57.(6)		(0.09+0.18)
^{82}Kr	11.593(3)		(14.+7.)	130.(13)		90.(6) mb
^{83}Kr	11.500(19)		183.(30)	183.(20)		0.24(2)
^{84}Kr	56.987(15)		$(\sigma_m + \sigma_g) = 0.11$	2.4(3)		(16.+33.) mb
			$\sigma_m = 0.09$			
^{85}Kr		10.73 y	1.7(2)	1.8(10)		0.07(2)
^{86}Kr	17.279(41)		3.(2) mb	≈ 1. mb	8.1(3)	3.2(4) mb
$_{37}$Rb			0.39(4)	6.(3)	7.08(2)	
^{84}Rb		32.9 d	$\sigma_p = 12.(2)$			
^{85}Rb	72.17(2)		(0.06+0.38)	(0.7+7.)	7.0(1)	0.24(1)
^{86}Rb		18.65 d	<20.			
^{87}Rb	27.83(2)	4.88×10^{10} y	0.10(1)	2.3(4)	7.3(1)	16.(1) mb
^{88}Rb		17.7 m	1.2(3)	0.5(1)		
$_{38}$Sr			1.2(1)	10.(1)	7.02(2)	
^{84}Sr	0.56(1)		(0.6+0.2)	(9.+1.)		0.4(1)
^{86}Sr	9.86(1)		$\sigma_m = 0.81(4)$	RI$_m$ = 4.(1)	5.68(5)	(48.+22.) mb
^{87}Sr	7.00(1)		16.(3)	118.(30)	7.41(7)	97.(5) mb
^{88}Sr	82.58(1)		5.8(4) mb	0.07(3)	7.16(6)	6.0(2) mb
^{89}Sr		50.52 d	0.42(4)	0.2		
^{90}Sr		29.1 y	10.(1) mb	0.10(2)		
$_{39}$Y			1.25(5)	1.0(1)	7.75(2)	
^{89}Y	100.		(0.001+1.25)	(0.006+1.0)	7.75(2)	19.(1) mb
^{90}Y		2.67 d	<6.5			
^{91}Y		58.5 d	1.4(3)	0.6(1)		
$_{40}$Zr			0.19(1)	0.95(9)	7.16(3)	
			$\sigma_\alpha < 0.1$ mb			
^{90}Zr	51.45(40)		≈ 0.014	0.2(1)	6.4(1)	21.(2) mb
^{91}Zr	11.22(5)		1.2(3)	5.(2)	8.8(1)	60.(8) mb

* Extrapolated value.

Elem. or Isot.	Natural Abundance (%)	Half-Life	Thermal Neut. Cross Section (barns)	Resonance Integral (barns)	Coh. Scat. Length (fm)	σ (30 keV) Maxw. Avg. (barns)
^{92}Zr	17.15(8)		0.2(1)	0.6(2)	7.5(2)	33.(4) mb
^{93}Zr		1.5×10^6 y	<4.	16.(5)		0.10(1)
^{94}Zr	17.38(28)	$>10^{17}$ y	0.049(6)	0.25(3)	8.3(2)	26.(1) mb
^{96}Zr	2.80(9)	$>1.7 \times 10^{18}$ y	0.020(3)	5.0(5)	5.5(1)	11.(1) mb
$_{41}$Nb			1.11(1)	8.5(6)	7.14(3)	
			$\sigma_\alpha < 0.1$ mb			
^{93}Nb	100.		1.1	(6.3+2.2)	7.14(3)	266.(5) mb
			$\sigma_m = 0.86$			
^{94}Nb		2.4×10^4 y	$(\sigma_m + \sigma_g) = 15.(1)$	126.(13)		
			$\sigma_m = 0.6(1)$			
^{95}Nb		34.97 d	<7.	<200.		
$_{42}$Mo			2.5(1)	26.(5)	6.72(2)	
			$\sigma_\alpha < 0.1$ mb			
^{92}Mo	14.77(31)	$>3 \times 10^{17}$ y	0.06	≈ 0.8	6.93(8)	0.07(1)
			$\sigma_m = 0.2$ µb			
^{94}Mo	9.226(99)		0.02	≈ 0.8	6.82(7)	0.10(2)
^{95}Mo	15.900(85)		13.4(3)	109.(5)	6.93(6)	0.29(1)
			$\sigma_\alpha = 30.(4)$ µb			
^{96}Mo	16.674(12)		0.5	17.(3)	6.22(6)	0.11(1)
^{97}Mo	9.560(50)		2.5(2)	14.(3)	7.26(8)	0.34(1)
			$s_a = 0.4(2)$ mb			
^{98}Mo	24.20(25)		0.14(1)	7.2(7)	6.60(7)	0.10(1)
^{100}Mo	9.67(20)	$\approx 1 \times 10^{19}$ y	0.19(1)	3.6(3)	6.75(7)	0.11(1)
$_{43}$Tc						
^{98}Tc		$\approx 6.6 \times 10^6$ y	$\sigma_m = 0.9(2)$			
^{99}Tc		2.13×10^5 y	23.(2)	$4.0(4) \times 10^2$	6.8(3)	0.93(5)
$_{44}$Ru			2.6(1)	48.(5)	7.03(3)	
^{96}Ru	5.54(14)	$>3.1 \times 10^{16}$ y	0.23(4)	7.(2)		0.21(1)
^{98}Ru	1.87(3)		< 8.			0.3(1)
^{99}Ru	12.76(14)		4.(1)	195.(20)		1.2(3)
^{100}Ru	12.60(7)		5.8(6)	11.(2)		0.21(1)
^{101}Ru	17.06(2)		5.(1)	$1.1(3) \times 10^2$		1.00(4)
			$\sigma_\alpha < 0.15$ µb			
^{102}Ru	31.55(14)		1.2(1)	4.3(5)		0.15(1)
^{103}Ru		39.27 d	<20.	≈ 30.		
^{104}Ru	18.62(27)		0.49(2)	6.(2)		0.15(1)
^{105}Ru		4.44 h	0.29(3)	0.13(1)		
^{106}Ru		1.020 y	0.15(4)	2.0(6)		
$_{45}$Rh			145.(2)	$1.2(1) \times 10^3$	5.88(4)	
^{103}Rh	100.		(11.+ 134.)	$(0.08+1.1) \times 10^3$	5.88(4)	0.81(1)
^{104m}Rh		4.36 m	800.(100)			
^{104}Rh		42.3 s	40.(30)			
^{105}Rh		35.4 h	$1.1(3) \times 10^4$	$1.7(4) \times 10^4$		
$_{46}$Pd			7.(1)	82.(8)	5.91(6)	
^{102}Pd	1.02(1)		3.2(10)	10.(2)		0.3(1)
^{104}Pd	11.14(8)			16.(2)		0.29(3)
^{105}Pd	22.33(8)		22.(2)	60.(20)	5.5(3)	1.20(6)
			$\sigma_\alpha = 0.5(2)$ µb			
^{106}Pd	27.33(3)		(0.013+0.28)	(0.2+5.5)	6.4(4)	0.25(3)
^{107}Pd		6.5×10^6 y	1.8(2)	108.(4)		1.34(6)
^{108}Pd	26.46(9)		(0.19+8.5)	(2.+240.)	4.1(3)	0.20(2)
^{110}Pd	11.72(9)		(0.033+0.7)	(0.7+8.)		0.15(2)
$_{47}$Ag			62.(1)	767.(60)	5.922(7)	
^{107}Ag	51.839(8)		(1.+35.)	(3.+105.)	7.56(1)	0.80(3)
^{109}Ag	48.161(8)		(4.1+ 87.)	$(0.7+14.1) \times 10^2$	4.17(1)	0.79(3)
^{110m}Ag		249.8 d	82.(11)	20.(4)		
^{111}Ag		7.47 d	3.(2)	105.(20)		
$_{48}$Cd			$2.52(5) \times 10^3$	73.(8)	4.87(5)	

* Extrapolated value.

Elem. or Isot.	Natural Abundance (%)	Half-Life	Thermal Neut. Cross Section (barns)	Resonance Integral (barns)	Coh. Scat. Length (fm)	σ (30 keV) Maxw. Avg. (barns)
^{106}Cd	1.25(6)	>2.6×10^{17} y	0.20(3)	4.(1)		0.30(2)
^{108}Cd	0.89(3)	>4.1×10^{17} y	1.	14.(3)	5.4(1)	0.20(1)
^{109}Cd		462.0 d	≈ 180.	6.7(12)×10^3		
			σ$_\alpha$ <0.05			
^{110}Cd	12.49(18)		(0.06+11.)	(6.+34.)	5.9(1)	(0.01+0.22)
^{111}Cd	12.80(12)		3.5(20)	51.(6)	6.5(1)	0.75(1)
^{112}Cd	24.13(21)		(0.012+2.2)	15.	6.4(1)	0.19(1)
^{113}Cd	12.22(12)	7.7×10^{15} y	2.06(4)×10^4	390.(40)	− 8.0(2)	0.67(1)
			s$_a$ <1. mb			
^{114}Cd	28.73(42)		(0.04+0.29)	16.(7)	7.5(1)	(0.01+0.12)
^{116}Cd	7.49(18)	3.8×10^{19} y	(26.+52.) mb	1.2	6.3(1)	(12.+47.) mb
$_{49}$In			197.(4)	3.3(2)×10^3	4.07(2)	
^{113}In	4.29(5)		(3.1+5.0+3.9)	(220.+90.)	5.39(6)	(0.48+0.31)
^{115}In	95.71(5)	4.4×10^{14} y	(88.+73.+44.)	(1.5+1.2+0.7)×10^3	4.01(2)	(0.69+0.02)
$_{50}$Sn			0.61(3)	8.(2)	6.225(2)	
^{112}Sn	0.97(1)		(0.15+0.40)	(8.+19.)		0.21(1)
^{113}Sn		115.1 d	≈ 9.	210.(50)		
^{114}Sn	0.66(1)		≈ 0.12	5.(1)	6.2(3)	134.(3) mb
^{115}Sn	0.34(1)		σ$_\alpha$ = 0.06 mb	29.(6)		0.34(1)
^{116}Sn	14.54(9)		(0.006+0.14)	(0.5+11.)	5.93(5)	91.(2) mb
^{117}Sn	7.68(7)		1.1(1)	16.(5)	6.48(5)	319.(7) mb
^{118}Sn	24.22(9)		σ$_m$ = 4. mb	4.7(5)	6.07(5)	62.(1) mb
^{119}Sn	8.59(4)		2.(1)	2.9(5)	6.12(5)	0.18(1)
^{120}Sn	32.58(9)		(0.001+0.13)	1.2(3)	6.49(5)	(0.5+36.) mb
^{122}Sn	4.63(3)		(0.15+0.001)	0.81(4)	5.74(5)	(18.+4.) mb
^{124}Sn	5.79(5)	>2.2×10^{18} y	(0.13+0.004)	(8.0+0.08)	5.97(5)	12.(2) mb
$_{51}$Sb			5.2(2)	169.(20)	5.57(3)	
^{121}Sb	57.21(5)		(0.4+5.8)	(13.+192.)	5.71(6)	0.53(2)
^{123}Sb	42.79(5)		(0.02+0.04+4.0)	(1.+119.)	5.38(7)	0.30(1)
^{124}Sb		60.20 d	17.(3)	≈ 8.		
$_{52}$Te			4.2(1)	47.(3)	5.80(3)	
^{120}Te	0.09(1)		(1.+5.)	≈ 1.	5.3(5)	0.4(1)
^{122}Te	2.55(12)		(0.4+3.)	(5.+75.)	3.8(2)	295.(3) mb
^{123}Te	0.89(3)	>5.3×10^{16} y	370.(40)	4.5(3)×10^3	− 0.05	0.83(1)
			σ$_\alpha$ = 0.05 mb			
^{124}Te	4.74(14)		(1.+6.)	(1.4+4.)	8.0(1)	155.(2) mb
^{125}Te	7.07(15)		1.1(2)	21.(4)	5.02(8)	431.(4) mb
^{126}Te	18.84(25)		(0.12+0.8)	(0.6+7.4)	5.56(7)	(28.+53.) mb
^{128}Te	31.74(8)	2.2×10^{24} y	(0.03+0.2)	(0.2+1.6)	5.89(7)	(3.+41.) mb
^{130}Te	34.08(62)	8.×10^{20} y	(0.01+0.19)	(0.03+0.3)	6.02(7)	(4.+11.) mb
$_{53}$I			6.2(1)	1.5(1)×10^2	5.28(2)	
^{125}I		59.4 d	900.(100)	1.4(2)×10^4		
^{127}I	100.		6.2(1)	1.5(1)×10^2	5.28(2)	0.64(3)
^{128}I		25.00 m	22.(4)	≈ 10.		
^{129}I		1.7×10^7 y	(20.7+10.3)	36.(4)		0.44(2)
^{130}I		12.36 h	18.(3)	≈ 8.		
^{131}I		8.021 d	≈ 0.7	8.(4)		
$_{54}$Xe			25.(1)	263.(50)	4.92(3)	
^{124}Xe	0.0953(27)	>10^{17} y	(28.+137.)	(0.6+3.0)×10^3		(0.13+0.51)
^{125}Xe		17.1 h	σ$_\alpha$ < 0.03			
^{126}Xe	0.0890(14)		(0.45+3.)	(8.+52.)		(0.04+0.32)
^{127}Xe		36.34 d	σ$_\alpha$ ≈ 0.01			
^{128}Xe	1.910(22)		σ$_m$ = 0.48	RI$_m$ = 38.(10)		0.26(1)
^{129}Xe	26.40(18)		22.(5)	250.(50)		0.62(2)
^{130}Xe	4.071(53)		σ$_m$ = 0.45	RI$_m$ = 16.(4)		0.132(3)
^{131}Xe	21.233(62)		90.(10)	9.(1)×10^2		0.45(8)
^{132}Xe	26.9087(680)		(0.05+0.4)	(0.9+3.7)		(5.+60.) mb
^{133}Xe		5.243 d	190.(90)			

* Extrapolated value.

Elem. or Isot.	Natural Abundance (%)	Half-Life	Thermal Neut. Cross Section (barns)	Resonance Integral (barns)	Coh. Scat. Length (fm)	σ (30 keV) Maxw. Avg. (barns)
^{134}Xe	10.436(29)	>1.1×10^{16} y	(0.003 + 0.26)	0.40(4)		20.(2) mb
^{135}Xe		9.10 h	2.65(11)×10^6	7.6(5)×10^3		
^{136}Xe	8.858(33)	>8×10^{20} y	0.26(2)	0.7(2)		0.9(1) mb
$_{55}$Cs			30.4(8)	422.(50)	5.42(2)	
^{132}Cs		6.48 d	σ$_α$ < 0.15			
^{133}Cs	100.		(2.7+27.3)	(32.+360.)	5.42(2)	(0.04+0.47)
^{134}Cs		2.065 y	140.(10)	54.(9)		
^{135}Cs		2.3×10^6 y	8.3(3)	38.(3)		
^{137}Cs		30.2 y	(0.20+0.07)	0.36(7)		
$_{56}$Ba			1.3(2)	10.(2)	5.07(3)	
^{130}Ba	0.106(1)	2.2×10^{21} y	(1.+8.)	(25.+200.)	− 3.6(6)	0.76(11)
^{132}Ba	0.101(1)	1.3×10^{21} y	(0.84+9.7)	(4.7+24.)	7.8(3)	0.6(1)
^{133}Ba		10.53 y	4.(1)	85.(30)		
^{134}Ba	2.417(18)		(0.1+1.3)	(5.6+18.)	5.7(1)	0.18(1)
^{135}Ba	6.592(12)		(0.014+5.8)	(0.47+131.)	4.7(1)	0.46(2)
^{136}Ba	7.854(24)		(0.010+0.44)	(0.1+1.5)	4.91(8)	61.(2) mb
^{137}Ba	11.232(24)		5.(1)	4.(1)	6.8(1)	76.(3) mb
^{138}Ba	71.698(42)		0.41(2)	0.4(1)	4.84(8)	4.0(2) mb
^{139}Ba		1.396 h	5.(1)	2.2(5)		
^{140}Ba		12.75 d	1.6(3)	14.(1)		
$_{57}$La			9.2(2)	12.(1)	8.24(4)	
^{138}La	0.090(1)	1.06×10^{11} y	57.(6)	4.1(9)×10^2		
^{139}La	99.910(1)		9.2(2)	12.(1)	8.24(4)	38.(3) mb
^{140}La		1.678 d	2.7(3)	69.(4)		
$_{58}$Ce			0.64(4)	0.71(6)	4.84(2)	
^{136}Ce	0.185(2)		(1.0+6.5)	58.(12)	5.80(9)	(0.028+0.3)
^{138}Ce	0.251(2)		(0.025+1.0)	(1.5+5.2)	6.70(9)	179.(5) mb
^{140}Ce	88.450(51)		0.58(4)	0.50(5)	4.84(9)	11.0(4) mb
^{141}Ce		32.50 d	29.(3)	13.(2)		
^{142}Ce	11.114(51)	>1.6×10^{17} y	0.97(3)	1.3(3)	4.75(9)	28.(1) mb
^{143}Ce		1.38 d	6.1(7)	2.7(3)		
^{144}Ce		284.6 d	1.0(1)	2.6(3)		
$_{59}$Pr			11.5(4)	14.(3)	4.58(5)	
^{141}Pr	100.		(4.+7.5)	14.(3)	4.58(5)	111.(2) mb
^{142}Pr		19.12 h	20.(3)	9.(1)		
^{143}Pr		13.57 d	90.(10)	190.(25)		
$_{60}$Nd			51.(2)	49.(5)	7.69(5)	
^{142}Nd	27.2(5)		19.(1)	34.(11)	7.7(3)	35.(1) mb
^{143}Nd	12.2(2)		330.(10) σ$_α$ = 17. mb	128.(30)		0.24(1)
^{144}Nd	23.8(3)	2.1×10^{15} y	3.6(3)	3.9(5)	2.8(3)	81.(2) mb
^{145}Nd	8.3(1)		47.(6) σ$_α$ = 12. μb	260.(40)		0.42(1)
^{146}Nd	17.2(3)		1.5(2)	3.0(4)	8.7(2)	91.(1) mb
^{147}Nd		10.98 d	440.(150)	200.		
^{148}Nd	5.7(1)		2.4(1)	13.(2)	5.7(3)	147.(2) mb
^{150}Nd	5.6(2)	≈ 1×10^{19} y	1.0(1)	14.(2)	5.3(2)	0.16(1)
$_{61}$Pm						
^{146}Pm		5.53 y	8.4(1.7)×10^3			
^{147}Pm		2.623 y	(84.+96.)	(1000.+1280.)	12.6(4)	2.(1)
^{148m}Pm		41.3 d	10600.(800)			
^{148}Pm		5.37 d	≈ 10^3	2.6(2.4)×10^3		
^{149}Pm		2.212 d	1400.(200)			
^{151}Pm		1.183 d	≈ 150.			
$_{62}$Sm			5.6(1)×10^3	1.4(2)×10^3		
^{144}Sm	3.07(7)		1.6(1)	2.4(3)		92.(6) mb
^{145}Sm		340. d	280.(20)	600.(90)		
^{147}Sm	14.99(18)	1.06×10^{11} y	56.(4), σ$_α$ = 0.6 mb	710.(50)	14.(3)	0.97(1)

* Extrapolated value.

Elem. or Isot.	Natural Abundance (%)	Half-Life	Thermal Neut. Cross Section (barns)	Resonance Integral (barns)	Coh. Scat. Length (fm)	σ (30 keV) Maxw. Avg. (barns)
^{148}Sm	11.24(10)	7×10^{15} y	2.4(6)	27.(14)		241.(2) mb
^{149}Sm	13.82(7)	10^{16} y	4.01(6)×10^4, σ$_\alpha$ = 31. mb	3.1(5)×10^3		1.82(2)
^{150}Sm	7.38(1)		102.(5)	290.(30)	14.(3)	422.(4) mb
^{151}Sm		90. y	1.52(3)×10^4	3520.(60)		2.(1)
^{152}Sm	26.75(16)		206.(15)	3.0(3)×10^3	− 5.0(6)	473.(4) mb
^{153}Sm		1.929 d	420.(180)			
^{154}Sm	22.75(29)		7.5(3)	32.(6)	9.(1)	0.21(1)
$_{63}$Eu			4570.(100)	3.8(5)×10^3	5.3(3)	
^{151}Eu	47.81(6)		(4.+3150.+6000.)	(2.+4.)×10^3		(1.6+2.2)
			σ$_\alpha$ = 8.7(3) μb			
^{152m1}Eu		9.30 h	6.8(15)×10^4	< 10^5		
^{152}Eu		13.5 y	1.1(2)×10^4	1.6(2)×10^3		5.(2)
^{153}Eu	52.19(6)		300.(20), σ$_\alpha$ <1. mb	1.8(4)×10^3	8.2(1)	2.8(1)
^{154}Eu		8.59 y	1.5(3)×10^3	1.6(2)×10^3		4.4(7)
^{155}Eu		4.76 y	3.9(2)×10^3	1.6(2)×10^4		1.3(1)
$_{64}$Gd			48.8(6)×10^3	400.(10)	9.5(2)	
^{148}Gd		75. y	1.40(14)×10^4			
^{152}Gd	0.20(1)	1.1×10^{14} y	700.(200), σ$_\alpha$ <7. mb	700.(200)		1.05(2)
^{153}Gd		240. d	2.(1)×10^4, σ$_\alpha$ = 0.03			
^{154}Gd	2.18(3)		(0.035+60.)	230.(50)		1.03(1)
^{155}Gd	14.80(12)		61.(1)×10^3, σ$_\alpha$ = .08 mb	1540.(100)		2.65(3)
^{156}Gd	20.47(9)		≈ 2.0	104.(15)	6.3(4)	615.(5) mb
^{157}Gd	15.65(2)		2.54(3)×10^5, σ$_\alpha$ <0.05	800.(100)		1.37(2)
^{158}Gd	24.84(7)		2.3(3)	73.(7)	9.(2)	324.(3) mb
^{160}Gd	21.86(19)	>1.9×10^{19} y	1.5(7)	6.(1)	9.15(5)	0.15(2)
^{161}Gd		3.66 m	2.0(6)×10^4			
$_{65}$Tb			23.2(5)	420.(50)	7.34(2)	
^{159}Tb	100.		23.2(5)	420.(50)	7.34(2)	1.6(2)
^{160}Tb		72.3 d	570.(110)			
$_{66}$Dy			9.5(2)×10^2	1.5(2)×10^3	16.9(3)	
^{156}Dy	0.056(3)		33.(3), σ$_\alpha$ < 9. mb	1000.(100)		1.6(2)
^{158}Dy	0.095(3)		43.(6), σ$_\alpha$ < 6. mb	120.(10)	6.1(5)	0.8(2)
^{159}Dy		144. d	8.(2)×10^3			
^{160}Dy	2.39(18)		60.(10), σ$_\alpha$ < 0.3 mb	1100.(200)	6.7(4)	0.89(1)
^{161}Dy	18.889(42)		600.(50), σ$_\alpha$ < 1. μb	1100.(100)	10.3(4)	1.96(2)
^{162}Dy	25.475(36)		170.(20)	2755.(300)	− 1.4(5)	446.(4) mb
^{163}Dy	24.896(42)		120.(10), σ$_\alpha$ < 20. μb	1600.(400)	5.0(4)	1.11(1)
^{164}Dy	28.260(54)		(1.7+1.0)×10^3	(4.+2.)×10^2	49.4(2)	212.(3) mb
^{165m}Dy		1.26 m	2.0(6)×10^3			
^{165}Dy		2.33 h	3.5(3)×10^3	2.2(3)×10^4		
$_{67}$Ho			61.(2)	670.(40)	8.01(8)	
^{163}Ho		4.57×10^3 y				(0.4+1.7)
^{165}Ho	100.		(3.1+58.), σ$_\alpha$ < 20. μb	(?+670.)	8.01(8)	(0.8+0.5)
^{166m}Ho		1.2×10^3 y	3.1(8)×10^3	10.(3)×10^3		
$_{68}$Er			1.5(2)×10^2	730.(10)	7.79(2)	
^{162}Er	0.139(5)		19.(3), σ$_\alpha$ < 11. mb	480.(50)	8.8(2)	1.6(1)
^{164}Er	1.601(3)		13.(3), σ$_\alpha$ < 1.2 mb	105.(10)	8.2(2)	1.08(5)
^{166}Er	33.503(36)		(3.+14.), σ$_\alpha$ < 70. μb	96.(12)	10.6(2)	0.56(6)
^{167}Er	22.869(9)		6.5(8)×10^2, σ$_\alpha$ = 3. μb	2970.(70)	3.0(3)	1.4(2)
^{168}Er	26.978(18)		2.3(3), σ$_\alpha$ = 0.09 mb	37.(5)	7.4(4)	0.34(4)
^{170}Er	14.910(36)		8.(2)	26.(4)	9.6(5)	0.17(1)
^{171}Er		7.52 h	370.(40)	170.(20)		
$_{69}$Tm			108.(4)	1.5(2)×10^3	7.07(3)	
^{169}Tm	100		(8.+100.)	1.5(2)×10^3	7.07(3)	1.13(6)
^{170}Tm		128.6 d	100.(20)	460.(50)		
^{171}Tm		1.92 y	≈ 160.	118.(6)		
$_{70}$Yb			52.(10)	1.7(2)×10^2	12.43(3)	
^{168}Yb	0.13(1)		2.4(2)×10^3, σ$_\alpha$ < 0.1 mb	2.0(5)×10^4	−4.07(2)	0.7(4)

* Extrapolated value.

Elem. or Isot.	Natural Abundance (%)	Half-Life	Thermal Neut. Cross Section (barns)	Resonance Integral (barns)	Coh. Scat. Length (fm)	σ (30 keV) Maxw. Avg. (barns)
^{169}Yb		32.02 d	$3.6(3) \times 10^3$	5200.(500)		
^{170}Yb	3.04(15)		12.(2), $\sigma_\alpha < 10.$ μb	320.(30)	6.8(1)	0.77(1)
^{171}Yb	14.28(57)		53.(5), $\sigma_\alpha < 1.5$ μb	315.(30)	9.7(1)	1.21(1)
^{172}Yb	21.83(67)		≈ 1.3, $\sigma_\alpha < 1.$ μb	25.(3)	9.4(1)	0.34(1)
^{173}Yb	16.13(27)		16.(2), $\sigma_\alpha < 1.$ mb	380.(30)	9.56(7)	0.75(1)
^{174}Yb	31.83(92)		(46.+17.), $\sigma_\alpha < 0.02$ μb	(13.+16.)	19.3(1)	151.(2) mb
^{176}Yb	12.76(41)		3.1(2), $\sigma_\alpha < 1.$ μb	8.(2)	8.7(1)	116.(2) mb
$_{71}$Lu			78.(7)	$8.3(7) \times 10^2$	7.21(3)	
^{175}Lu	97.41(2)		(16.+8.)	(550.+270.)	7.24(3)	(1.04+0.11)
^{176}Lu	2.59(2)	3.73×10^{10} y	(2.+2100.)	(3.+930.)	6.1(2)	1.53(7)
^{177m}Lu		160.7 d	3.2(3)	1.4(2)		
^{177}Lu		6.65 d	1000.(300)			
$_{72}$Hf			106.(3)	$19.7(5) \times 10^2$	7.8(1)	
^{174}Hf	0.16(1)	2.0×10^{15} y	600.(50)	400.(50)	11.(1)	0.8(2)
^{176}Hf	5.26(7)		23.(4)	700.(100)	6.6(2)	0.46(2)
^{177}Hf	18.60(9)		(1.+375.), $\sigma_\alpha < 20.$ μb	7170.(200)		1.5(1)
^{178m2}Hf		31. y	$\sigma_{m2} = 45.(5)$	$RI_{m2} = 8(1) \times 10^2$		
^{178}Hf	27.28(7)		(54.+32.)	$(0.9+1.0) \times 10^3$	5.9(2)	0.31(1)
^{179}Hf	13.62(2)		(0.43+46.)	(6.8+620.)	7.5(2)	(0.01+0.95)
^{180}Hf	35.08(16)		13.0(5), $\sigma_\alpha < 13.$ μb	32.(1)	13.2(3)	179.(5) mb
^{181}Hf		42.4 d	30.(25)			
$_{73}$Ta			20.(1)	650(20.)	6.91(7)	
^{179}Ta		1.8 y	$9.3(6) \times 10^2$	$1.22(7) \times 10^3$		
^{180m}Ta	0.012(2)	$> 1.2 \times 10^{15}$ y	≈ 560.	1350.(100)		
^{181}Ta	99.988(2)		(0.012 + 20.), $\sigma_\alpha < 1.$ μb	(0.4+650.)	6.91(7)	0.77(2)
^{182}Ta		114.43 d	8200.(600)	900.(90)		
$_{74}$W			18.(1)	$3.6(3) \times 10^2$	4.86(2)	
^{180}W	0.12(1)	7.4×10^{16} y	≈ 4.	210.(30)		0.54(6)
^{182}W	26.50(16)	8.3×10^{18} y	20.(1)	600.(90)	6.97(4)	274.(8) mb
^{183}W	14.31(4)	1.9×10^{18} y	10.5(3)	340.(50)	6.53(4)	0.52(2)
^{184}W	30.64(2)	4.0×10^{18} y	(0.002 + 2.0)	15.(2)	7.48(6)	0.22(1)
^{185}W		74.8 d	≈ 3.3	300.(50)		
^{186}W	28.43(19)	6.5×10^{18} y	37.(2)	510.(50)	− 0.72(4)	176.(5) mb
^{187}W		23.9 h	70.(10)	2760.(550)		
^{188}W		69.78 h	12.(1)			
$_{75}$Re			90.(4)	$8.4(2) \times 10^2$	9.2(3)	
^{185}Re	37.40(2)		(0.33+110.)	1700.(50)	9.0(3)	1.54(6)
^{187}Re	62.60(2)	4.2×10^{10} y	(2.+72.)	(9.+310.)	9.3(3)	1.16(6)
$_{76}$Os			17.(1)	$1.5(1) \times 10^2$	10.7(2)	
^{184}Os	0.02(1)	$>5.6 \times 10^{13}$ y	$3.3(3) \times 10^3$, $\sigma_\alpha < 10.$ mb	$1.4(1) \times 10^3$		0.4(2)
^{186}Os	1.59(3)	$2. \times 10^{15}$ y	≈ 80., $\sigma_\alpha < 0.1$ mb	$3.8(9) \times 10^2$	12(2)	0.42(2)
^{187}Os	1.96(2)		$2.(1) \times 10^2$, $\sigma_\alpha < 0.1$ mb	$5.0(7) \times 10^2$		0.90(3)
^{188}Os	13.24(8)		≈ 5., $\sigma_\alpha < 30.$ μb	$1.5(2) \times 10^2$	7.6(3)	0.40(2)
^{189}Os	16.15(5)		(0.00026+40.), $\sigma_\alpha < 10.$ μb	(0.013+670.)	10.7(3)	1.17(5)
^{190}Os	26.26(2)		(9.+4.), $\sigma_\alpha < 20.$ μb	(22.+8.)	11.0(3)	0.30(5)
^{191}Os		15.4 d	$3.8(6) \times 10^2$	$1.7(3) \times 10^2$		
^{192}Os	40.78(19)		3.(1), $\sigma_\alpha < 10.$ μb	7.(1)	11.5(4)	0.31(5)
^{193}Os		30.5 h	$2.5(5) \times 10^2$	$1.1(2) \times 10^2$		
$_{77}$Ir			$4.2(1) \times 10^2$	$2.8(4) \times 10^3$		10.6(3)
^{191}Ir	37.3(2)		(0.14+660.+260.)	$(1.0+4.2) \times 10^3$		1.35(4)
^{192}Ir		73.83 d	$1.4(3) \times 10^3$	$4.8(7) \times 10^3$		
^{193}Ir	62.7(2)		(0.04+6.+109.)	$1.4(2) \times 10^3$		0.99(7)
^{194}Ir		19.3 h	$1.6(3) \times 10^3$	$7.(2) \times 10^2$		
$_{78}$Pt			10.(1)	$1.3(1) \times 10^2$	9.60(1)	
^{190}Pt	0.014(1)	4.5×10^{11} y	$1.5(1) \times 10^2$, $\sigma_\alpha < 8.$ mb	70.(10)	9.(1)	0.7(2)
^{192}Pt	0.782(7)		(2.0+6.), $\sigma_\alpha < 0.2$ mb	115.(20)	9.9(5)	0.6(1)
^{194}Pt	32.967(99)		(0.1+1.1), $\sigma_\alpha < 5.$ μb	(4.+?)	10.55(8)	(0.03+0.34)
^{195}Pt	33.832(10)		28.(1), $\sigma_\alpha < 5.$ μb	365.(50)	8.8(1)	0.9(2)

* Extrapolated value.

Elem. or Isot.	Natural Abundance (%)	Half-Life	Thermal Neut. Cross Section (barns)	Resonance Integral (barns)	Coh. Scat. Length (fm)	σ (30 keV) Maxw. Avg. (barns)
^{196}Pt	25.242(41)		(0.045+0.55)	7.(2)	9.89(8)	(0.01+0.19)
^{198}Pt	7.163(55)		(0.3+3.1)	(5.+53.)	7.8(1)	(3.+79.) mb
^{199}Pt		30.8 m	$\approx 15.$	$\approx 7.$		
$_{79}$Au			98.7(1)	$1.55(3)\times10^3$	7.63(6)	
^{197}Au	100.		$\sigma_{m+g} = 98.7(1)$ $\sigma_m = 8.(2)$ mb	$RI_{m+g} = 1.55(3)\times10^3$ $RI_m = 0.06(2)$	7.63(6)	582.(9) mb
^{198}Au		2.695 d	$26.5(15)\times10^3$	$\approx 4.\times10^4$		
^{199}Au		3.14 d	$\approx 30.$			
$_{80}$Hg			$3.7(1)\times10^2$	87.(5)	12.69(2)	
^{196}Hg	0.15(1)	$>2.5\times10^{18}$ y	(105.+3000.)	(53.+410.)	30.(1)	0.4(2)
^{198}Hg	9.97(8)		(0.017+2.)	(1.7+70.)		0.17(2)
^{199}Hg	16.87(10)		$2.1(2)\times10^3$	435(20)	16.9(4)	0.37(2)
^{200}Hg	23.10(16)		$\approx 1.$	2.1(5)		0.12(1)
^{201}Hg	13.18(8)		$\approx 8.$	30.(3)		0.26(1)
^{202}Hg	29.86(20)		4.9(5)	4.5(2)	11.(1)	74.(6) mb
^{204}Hg	6.87(4)		0.4(1)	0.8(2)		42.(4) mb
$_{81}$Tl			3.3(1)	12.5(8)	8.776(5)	
^{203}Tl	29.524(14)		11.(1), $\sigma_\alpha < 0.3$ mb	41.(2)	7.0(2)	124.(8) mb
^{204}Tl		3.78 y	22.(2)	90.(20)		0.14(5)
^{205}Tl	70.476(14)		0.11(2)	0.6(2)	9.52(7)	54.(4) mb
$_{82}$Pb			0.172(2)	0.14(4)	9.402(2)	
^{204}Pb	1.4(1)		0.68(7)	2.0(2)	10.9(1)	90.(6) mb
^{205}Pb		1.51×10^7 y	$\approx 5.$	$\approx 2.$		0.06(1)
^{206}Pb	24.1(1)		0.027(1)	0.10(1)	9.23(5)	16.(1) mb
^{207}Pb	22.1(1)		0.61(3)	0.38(1)	9.28(2)	10.(1) mb
^{208}Pb	52.4(1)	$>2\times10^{19}$ y	0.23(1) mb, $\sigma_\alpha < 8.$ μb	2.0(2) mb	9.50(3)	0.36(4) mb
^{210}Pb		22.6 y	< 0.5			
$_{83}$Bi			0.034(1)	0.19(2)	8.532(2)	
^{209}Bi	100.		(11.+23.) mb, $\sigma_\alpha < 0.3$ μb	0.19(2)	8.532(2)	2.7(5) mb
^{210m}Bi		3.0×10^6 y	54.(4) mb	0.20(3)		
$_{84}$Po						
^{210}Po		138.4 d	$\sigma_m < 0.5$ mb, $\sigma_\alpha < 2.$ mb $\sigma_g < 30.$ mb, $\sigma_f < 0.1$			
$_{85}$At						
$_{86}$Rn						
^{220}Rn		55.6 s	<0.2			
^{222}Rn		3.823 d	0.74(5)			
$_{88}$Ra						
^{223}Ra		11.43 d	$1.3(2)\times10^2$, $\sigma_f < 0.7$			
^{224}Ra		3.66 d	12.0(5)			
^{226}Ra		1599. y	$\approx 13.$, $\sigma_f < 7.$ μb	280.(50)		10.(1)
^{228}Ra		5.76 y	36.(5), $\sigma_f < 2.$			
$_{89}$Ac						
^{227}Ac		21.77 y	$8.8(7)\times10^2$, $\sigma_f < 0.35$ mb	$1.5(4)\times10^3$		
$_{90}$Th			7.4	85.(3)		10.31(3)
^{227}Th		18.72 d	$\sigma_f = 2.0(2)\times10^2$			
^{228}Th		1.913 y	$1.2(2)\times10^2$, $\sigma_f < 0.3$	1014.(400)		
^{229}Th		7.9×10^3 y	$\approx 60.$ $\sigma_f = 30.(3)$	$1.0(2)\times10^3$ $RI_f = 466.(75)$		
^{230}Th		7.54×10^4 y	23.4(5) $\sigma_f < 0.5$ mb	$1.0(1)\times10^3$		
^{232}Th	100.	1.40×10^{10} y	7.37(4) $\sigma_f = 3.(1)$ μb $\sigma_\alpha < 1.$ μb	85.(3)		10.31(3)
^{233}Th		22.3 m	$1.5(1)\times10^3$ $\sigma_f = 15.(2)$	$4.(1)\times10^2$		

* Extrapolated value.

Elem. or Isot.	Natural Abundance (%)	Half-Life	Thermal Neut. Cross Section (barns)	Resonance Integral (barns)	Coh. Scat. Length (fm)	σ (30 keV) Maxw. Avg. (barns)
^{234}Th		24.10 d	1.8(5)			
			$\sigma_f < 0.01$			
$_{91}$Pa						
^{230}Pa		17.4 d	$1.5(3)\times10^3$			
^{231}Pa		3.25×10^4 y	$2.0(1)\times10^2$	750.(80)		9.1(3)
			$\sigma_f = 20.(1)$ mb	$RI_f = 0.05(1)$		
^{232}Pa		1.31 d	$4.6(10)\times10^2$	300.(70)		
			$\sigma_f = 1.5(5)\times10^3$	$RI_f = 1.0(1)\times10^3$		
^{233}Pa		27.0 d	39.(2)	(460.+440.)		
			$\sigma_m = 20.(4)$			
			$\sigma_g = 19.(3)$			
			$\sigma_f < 0.1$			
$_{92}$U			3.4(3); $\sigma_f = 4.2(1)$	280.(20),$RI_f = 2.0$		8.417(5)
^{230}U		20.8 d	$\sigma_f \approx 25.$			
^{231}U		4.2 d	$\sigma_f \approx 250.$			
^{232}U		70. y	73.(2)	280.(15)		
			$\sigma_f = 74.(8)$	$RI_f = 350.(30)$		
^{233}U		1.592×10^5 y	47.(2)	137.(6)		10.1(2)
			$\sigma_f = 5.3(1)\times10^2$	$RI_f = 760.(17)$		
			$\sigma_\alpha < 0.2$ mb			
^{234}U	0.0054(5)	2.455×10^5 y	96.(2)	660.(70)		12.(4)
			$\sigma_f = 0.07(2)$	$RI_f = 6.5$		
^{235}U	0.7204(6)	7.04×10^8 y	95.(5)	144.(6)		10.47(4)
			$\sigma_f = 586.(2)$	$RI_f = 275(5)$		
			$\sigma_\alpha < 0.1$ mb			
^{236}U		2.342×10^7 y	5.1(3)	360.(15)		
			$\sigma_f < 1.3$ mb	$RI_f = 4.38(50)$		
^{237}U		6.75 d	$\approx 10^2$	1200.(200)		
			$\sigma_f < 0.35$			
^{238}U	99.2742(10)	4.47×10^9 y	2.7(1)	277.(3)		8.402(5)
			$\sigma_f \approx 3.$ μb	1.54(15) mb		
			$\sigma_\alpha = 1.4(5)$ μb			
^{239}U		23.5 m	22.(2)			
			$\sigma_f = 15.(3)$			
$_{93}$Np						
^{234}Np		4.4 d	$\sigma_f = 9.(3)\times10^2$			
^{235}Np		1.085 y	$1.6(1)\times10^2$			
^{236m}Np		22.5 h	$\sigma_f = 2.7(2)\times10^3$	$7.(4)\times10^2$		
^{236}Np		1.55×10^5 y	$\sigma_f = 3.0(2)\times10^3$	$1.35(30)\times10^3$		
^{237}Np		2.14×10^6 y	$1.7(1)\times10^2$	$6.5(3)\times10^2$		10.6(1)
			$\sigma_f = 20.(1)$ mb	$RI_f = 4.7$		
^{238}Np		2.117 d	$\sigma_f = 2.6(3)\times10^3$	$1.4(3)\times10^3$		
^{239}Np		2.355 d	(32.+19.)			
			$\sigma_f < 1.$			
$_{94}$Pu						
^{236}Pu		2.87 y	$\sigma_f = 1.6(3)\times10^2$	1000.(60)		
^{237}Pu		45.7 d	$\sigma_f = 2.3(3)\times10^3$			
^{238}Pu		87.7 y	$5.1(2)\times10^2$	$1.6(2)\times10^2$		14.1(5)
			$\sigma_f = 17.(1)$	$RI_f = 26.(2)$		
^{239}Pu		2.410×10^4 y	$2.7(1)\times10^2$	$2.0(2)\times10^2$		7.7(1)
			$\sigma_f = 752.(3)$	$3.0(1)\times10^2$		
			$\sigma_\alpha \approx 0.3$ mb			
^{240}Pu		6.56×10^3 y	$2.9(1)\times10^2$	$8.4(3)\times10^3$		3.5(1)
			$\sigma_f \approx 59.$ mb	$RI_f = 3.2$		
^{241}Pu		14.4 y	$3.7(1)\times10^2$, $\sigma_\alpha < 0.2$ mb	$1.6(1)\times10^2$		
			$\sigma_f = 1.01(1)\times10^3$	$5.7(4)\times10^2$		
^{242}Pu		3.75×10^5 y	19.(1)	$1.1(1)\times10^3$		8.1(1)
			$\sigma_f < 0.2$	$RI_f = 0.23$		
^{243}Pu		4.956 h	<100.			

* Extrapolated value.

Elem. or Isot.	Natural Abundance (%)	Half-Life	Thermal Neut. Cross Section (barns)	Resonance Integral (barns)	Coh. Scat. Length (fm)	σ (30 keV) Maxw. Avg. (barns)
			$\sigma_f = 2.0(2) \times 10^2$			
^{244}Pu		8.00×10^7 y	1.7(1)	41.(3)		
^{245}Pu		10.5 h	$1.5(3) \times 10^2$	220.(40)		
$_{95}$Am						
^{241}Am		432.7 y	$(0.6+6.4) \times 10^2$	$(1.+14.) \times 10^2$		
			$\sigma_f = 3.15(10)$	14.(1)		
^{242m}Am		141. y	$1.7(4) \times 10^3$	≈ 200.		
			$\sigma_f = 5.9(3) \times 10^3$	$RI_f = 1.8(1) \times 10^3$		
^{242}Am		16.02 h	$\sigma_f = 2.1(2) \times 10^3$	$RI_f = <\ 300.$		
			$3.3(5) \times 10^2$	≈ 1.5×10^2		
^{243}Am		7.37×10^3 y	(75.+5.)	$(17.1+1.0) \times 10^2$	8.3(2)	
			$\sigma_f = 79.(2)$ mb	$RI_f = 0.056$		
^{244m}Am		≈ 26. m	$\sigma_f = 1.6(3) \times 10^3$			
^{244}Am		10.1 h	$\sigma_f = 2.2(3) \times 10^3$			
$_{96}$Cm						
^{242}Cm		162.8 d	≈ 20.	120.(50)		
			$\sigma_f ≈ 5.$			
^{243}Cm		29.1 y	$1.3(1) \times 10^2$	214.(20)		
			$\sigma_f = 6.2(2) \times 10^2$	$RI_f = 1.6(1) \times 10^3$		
^{244}Cm		18.1 y	15.(1)	640.(50)		9.5(3)
			$\sigma_f = 1.1(2)$	$RI_f = 10.8(8)$		
^{245}Cm		8.48×10^3 y	$3.5(2) \times 10^2$	110.(10)		
			$\sigma_f = 2.1(1) \times 10^3$	$RI_f = 8.(1) \times 10^2$		
^{246}Cm		4.76×10^3 y	1.2(2)	120.(10)		9.3(2)
			$\sigma_f = 0.16(7)$	13.(2)		
^{247}Cm		1.56×10^7 y	60.(30)	$5.(1) \times 10^2$		
			$\sigma_f = 82.(5)$	$7.3(7) \times 10^2$		
^{248}Cm		3.48×10^5 y	2.6(3)	270.(30)		7.7(2)
			$\sigma_f = 0.36(7)$	13.(2)		
^{249}Cm		64.15 m	≈ 1.6			
^{250}Cm		≈ 9.7×10^3 y	≈ 80.			
$_{97}$Bk						
^{249}Bk		320. d	$7.(1) \times 10^2$	$9.(1) \times 10^2$		
			$\sigma_f ≈ 0.1$			
^{250}Bk		3.217 h	$\sigma_f = 1.0(2) \times 10^3$			
$_{98}$Cf						
^{249}Cf		351. y	$5.0(3) \times 10^2$	$7.7(4) \times 10^2$		
			$\sigma_f = 1.7(1) \times 10^3$	$RI_f = 2.1(3) \times 10^3$		
^{250}Cf		13.1 y	$2.0(2) \times 10^3$	$12.(2) \times 10^3$		
			$\sigma_f = 110.(90)$	$RI_f = 160.(40)$		
^{251}Cf		9.0×10^2 y	$2.9(2) \times 10^3$	$1.6(1) \times 10^3$		
			$\sigma_f = 4.5(5) \times 10^3$	$RI_f = 5.5(3) \times 10^3$		
^{252}Cf		2.65 y	20.(2)	43.(3)		
			$\sigma_f = 32.(4)$	$RI_f = 1.1(3) \times 10^2$		
^{253}Cf		17.8 d	18.(2)	8.(1)		
			$\sigma_f = 1.3(2) \times 10^3$			
^{254}Cf		60.5 d	4.5(10)	2.		
$_{99}$Es						
^{253}Es		20.47 d	(180.+5.8)	$(37.5+1.1) \times 10^2$		
^{254m}Es		1.64 d	$\sigma_f = 1.8(1) \times 10^3$			
^{254}Es		276. d	28.(3)	18.(2)		
			$\sigma_f = 1.8(2) \times 10^3$	$RI_f = 1.2(1) \times 10^3$		
^{255}Es		40. d	≈ 55.			
$_{100}$Fm						
^{255}Fm		20.1 h	26.(3)	14.(2)		
			$\sigma_f = 3.3(2) \times 10^3$			
^{257}Fm		100.5 d	$\sigma_f = 3.0(2) \times 10^3$			

* Extrapolated value.

COSMIC RADIATION

A.G. Gregory and R.W. Clay

The Nature of Cosmic Rays

Primary cosmic radiation, in the form of high energy nuclear particles, electrons and photons from outside the solar system and from the Sun, continually bombards our atmosphere. Secondary radiation, resulting from the interaction of the primary cosmic rays with atmospheric gas, is present at sea-level and throughout the atmosphere.

Secondary radiation is collimated by absorption and scattering in the atmosphere and consists of a number of components associated with different particle species. High energy primary particles can produce large numbers of secondary particles forming an extensive air shower. Thus, a number of particles may then be detected simultaneously at sea-level.

Primary particle energies accessible in the vicinity of the Earth range from $\sim 10^8$ eV to $\sim 10^{20}$ eV. At the lower energies, the limit is determined by the inability of charged particles to traverse the heliosphere to us through the outward-moving solar wind. The upper energy limit is set by the practicality of building detectors to record particles with the extremely low fluxes found at those energies (O.C. Allkofer, 1975a; J.G. Wilson, 1976).

Primary Cosmic Rays

Primary Particle Energy Spectrum

Figure 1 shows the spectrum of primary particle energies. This includes all particle species. In differential form it is roughly a power law of intensity versus energy with an index of ~ -3. There appears to be a knee (a steepening) at a little above 10^{15} eV and an ankle (a flattening) above $\sim 10^{18}$ eV. Figure 2 emphasizes the features in the spectrum at the highest energies through multiplying the flux with a strongly rising power law of energy. This figure should be used with caution as errors for the two axes are not now independent.

Data on the high energy cosmic ray spectrum are uncertain largely because of limited event statistics due to the very low flux which might best be measured in particles per square kilometer per century. The highest energy event recorded to 1995 had an energy of 3×10^{20} eV (D.J. Bird et al., 1993).

It is expected that the highest energy cosmic rays will interact with the 2.7 K cosmic microwave background through photoproduction or photodisintegration. These interactions will appreciably reduce the observed flux of cosmic rays with energies above 5×10^{19} eV if they travel further than ~ 150 million light years. This process is known as the Greisen-Zatsepin-Kuz'min (GZK) cutoff (P. Sokolsky, 1989).

At energies below $\sim 10^{13}$ eV, solar system magnetic fields and plasma can modulate the primary component and Figure 3 shows the extent of this modulation between solar maximum and minimum (E. Juliusson, 1975; J. Linsley, 1981).

Primary Particle Energy Density

If the above spectrum is corrected for solar effects, the energy density above a particle energy of 10^9 eV outside the solar system is found to be $\sim 5 \times 10^5$ eV m^{-3}. As the threshold energy is increased, the energy density decreases rapidly, being 2×10^4 eV m^{-3} above 10^{12} eV and 10^2 eV m^{-3} above 10^{15} eV. The energy density at lower energies outside the heliosphere is unknown but may be substantially greater if the particle rest mass energy is included together with the kinetic energy (A. W. Wolfendale, 1979).

Primary Particle Isotropy

This is measured as an anisotropy $(I_{max} - I_{min})/(I_{max} + I_{min}) \times 100\%$, where I, the intensity (m^{-2}s^{-1}sr^{-1}), is usually measured with an angular resolution of a few degrees.

The measured anisotropy is small and energy dependent. It is roughly constant in amplitude at between 0.05 and 0.1% (with a phase of 0 to 6 hours in right ascension) for energies between 10^{11} eV and 10^{14} eV and appears to increase at higher energies roughly as $0.4 \times (Energy(eV)/10^{16})^{0.5}$ % up to $\sim 10^{18}$ eV. The latter rise may well be an artifact of the progressively more limited statistics as the flux drops rapidly with energy. It appears possible that a real anisotropy has been observed at the highest energies (above a few times 10^{19} eV) with a directional preference for the supergalactic plane (this plane reflects the directions of galaxies within about 100 million light years) (A.W. Wolfendale, 1979; R.W. Clay, 1987; T. Stanev et al., 1995).

Primary Particle Composition

The composition of low energy cosmic rays is close to universal nuclear abundances except where propagation effects are present. For example, Li, Be, and B which are spallation products, are overabundant by about six orders of magnitude.

Composition at 10^{11} eV per nucleus

Charge	1	2	(3–5)	(6–8)	(10–14)	(16–24)	(26–28)	≥30
% Composition (10% uncertainty)	50	25	1	12	7	4	4	0.1

Measurements at higher energies indicate that there is an increase in the relative abundances of nuclei with charge greater than 6 at energies above 50 TeV/nucleus (K. Asakimori et al., 1993) (1 TeV = 10^{12} eV).

Cosmic ray composition at low energies is often quoted at a fixed energy per nucleon. When presented in this way, protons constitute roughly 90% of the flux, helium nuclei about 10% and the remainder sum to a total of about 1%.

Certain radioactive isotopic ratios show lifetime effects. The ratio of Be^{10}/B^9 abundances is used to measure an "age" of cosmic rays since Be^{10} is unstable with a half life of about 1.6×10^6 years. A ratio of 0.6 is expected in the absence of Be^{10} decay and a ratio of about 0.2 is found experimentally (E. Juliusson, 1975; P. Meyer, 1981).

At higher energies, composition determinations are indirect and are rather contradictory and controversial. Experiments aim to differentiate between broad composition models. The measurement technique is based on studies of cosmic ray shower development. A rather direct technique for such studies is to use fluorescence observations of the shower development to determine the atmospheric depth of maximum development of the shower. Such observations suggest a heavy composition (large atomic number) at energies $\sim 10^{17}$ eV which changes with increasing energy to a light composition (perhaps protonic) above $\sim 10^{19}$ eV (T. K. Gaisser et al., 1993).

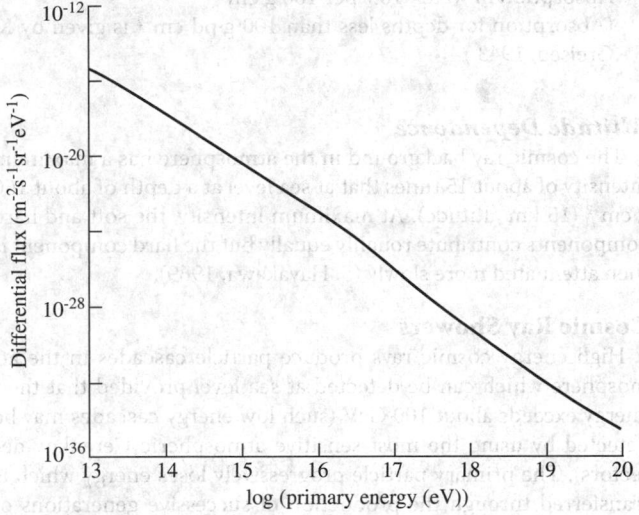

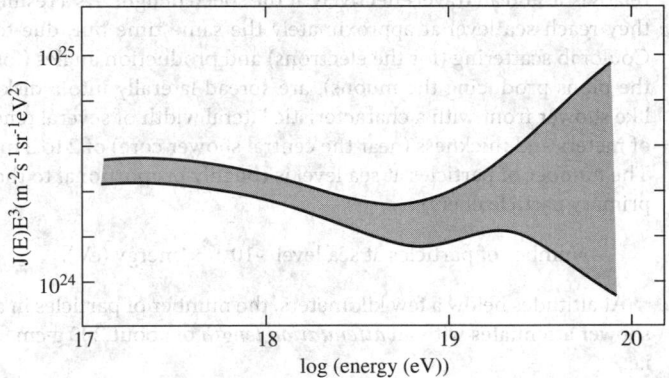

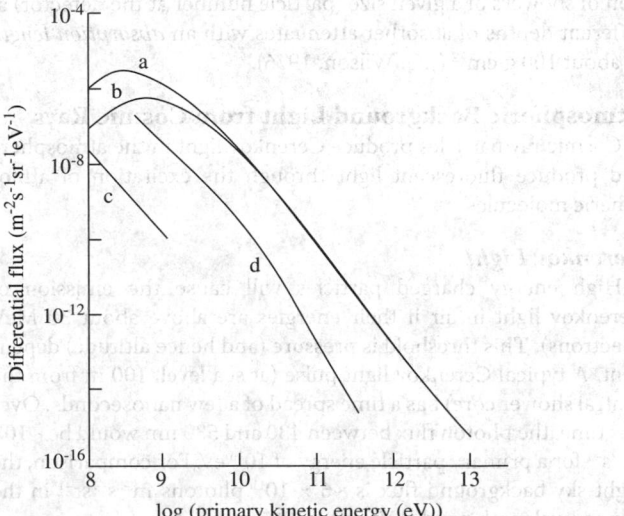

FIGURE 1. The energy spectrum of cosmic ray particles. This spectrum is of a differential form and can be converted to an integral spectrum by integration over all energies above a required threshold (E). Insofar as the spectrum approximates a power law of index −3, a simple conversion to the integral at an energy E/1.8 is obtained by multiplying the differential flux by the energy and dividing by 0.62.

FIGURE 2. Energy spectrum at the highest energies. This spectrum (after Yoshida et al., 1995) has the differential spectrum multiplied by energy cubed. It is from a compilation of a number of measurements and indicates the good general agreement at the lower energies and a spread due to inadequate statistics at the highest energies.

FIGURE 3. Energy spectrum of particles at lower energies. (a) Solar minimum proton energy spectrum. (b) Solar maximum proton energy spectrum. (c) Gamma-ray energy spectrum. (d) Local interstellar electron spectrum.

Primary Electrons

Primary electrons constitute about 1% of the cosmic ray beam. The positron to negative electron ratio is about 10% (J. M. Clem et al., 1995).

Antimatter in the Primary Beam

The ratio of antiprotons to protons in the primary cosmic ray beam (at about 400 MeV) is about 10^{-5}. At about 10 GeV the ratio is about 10^{-3}. At the highest measured energies (10 TeV), the upper limit to the ratio is about 20% (M. Amenomori et al., 1995; S. Orito et al., 1995).

Primary Gamma-Rays

The flux of primary gamma-rays is low at high energies. At 1 GeV the ratio of gamma-rays to protons is about 10^{-6}. The arrival directions of these gamma-rays are strongly concentrated in the plane of the Milky Way although there is a diffuse, near isotropic background flux and some point sources have been detected.

Since the absorption cross section for gamma-rays above 100 MeV is approximately 20 mbarn/electron, less than 10% of gamma-rays reach mountain altitudes (A. W. Wolfendale, 1979; P. F. Michelson, 1994).

Sea-Level Cosmic Radiation

The sea-level cosmic ray dose is 300 millirad·yr^{-1} and the sea-level ionization is 2.2×10^{6} ion pairs $m^{-3}s^{-1}$. The sea-level flux has a soft component, which can be absorbed in about 100 mm of lead (about 100 g·cm^{-2} of absorber) and a more penetrating (largely muon) hard component. The sea-level radiation is largely produced in the atmosphere and is a secondary component from interactions of the primary particles. The steep primary energy spectrum means that most secondaries at sea level are from rather low energy primaries. Thus the secondary flux is dependent on the solar cycle and the geomagnetic latitude of the observer.

Absolute Flux of the Hard Component

Vertical Integral Intensity I(0) ~100 $m^{-2}s^{-1}sr^{-1}$
Angular dependence $I(\theta) \sim I(0) \cos^2(\theta)$
Integrated Intensity ~200 $m^{-2}s^{-1}$
(O.C. Allkofer, 1975b).

Flux of the Soft Component

In free air, the soft component comprises about one third of the total cosmic ray flux.

Latitude Effect

The geomagnetic field influences the trajectories of lower energy cosmic rays approaching the Earth. As a result, the background flux is reduced by about 7% at the geomagnetic equator. The effect decreases toward the poles and is negligible at latitudes above about 40°.

Flux of Protons

The proton component is strongly attenuated by the atmosphere with an attenuation length (reduction by a factor of e) of about 120 g·cm^{-2}. It constitutes about 1% of the total vertical sea level flux.

Absorption

The soft component is absorbed in about 100 g·cm^{-2} of matter. The hard component is absorbed much more slowly:

Absorption in lead, 6% per 100 g·cm^{-2}
Absorption in rock, 8.5% per 100 g·cm^{-2}

Absorption in water, 10% per 100 g·cm^{-2}
(Absorption for depths less than 100 g·pd cm^{-2} is given by K. Greisen, 1943.)

Altitude Dependence

The cosmic ray background in the atmosphere has a maximum intensity of about 15 times that at sea level at a depth of about 150 g·cm^{-2} (15 km altitude). At maximum intensity, the soft and hard components contribute roughly equally but the hard component is then attenuated more slowly (S. Hayakawa, 1969).

Cosmic Ray Showers

High energy cosmic rays produce particle cascades in the atmosphere which can be detected at sea level provided that their energy exceeds about 100 GeV (such low energy cascades may be detected by using the most sensitive atmospheric Cerenkov detectors). The primary particle progressively loses energy which is transferred through the production of successive generations of secondary particles to a cascade of hadrons, an electromagnetic shower component (both positively and negatively charged electrons and gamma-rays) and muons. The secondary particles are relativistic and all travel effectively at the speed of light. As a result, they reach sea level at approximately the same time but, due to Coulomb scattering (for the electrons) and production angles (for the pions producing the muons), are spread laterally into a disk-like shower front with a characteristic lateral width of several tens of meters and thickness (near the central shower core) of 2 to 3 m. The number of particles at sea level is roughly proportional to the primary particle energy:

Number of particles at sea level ~$10^{-10} \times$ energy (eV).

At altitudes below a few kilometers, the number of particles in a shower attenuates with an *attenuation length* of about 200 g·cm^{-2}, i.e.,

particle number = original number × exp(−(depth increase)/200)

The above applies to an individual shower. The rate of observation of showers of a given size (particle number at the detector) at different depths of absorber attenuates with an *absorption length* of about 100 g·cm^{-2} (J.G. Wilson, 1976).

Atmospheric Background Light from Cosmic Rays

Cosmic ray particles produce Cerenkov light in the atmosphere and produce fluorescent light through the excitation of atmospheric molecules.

Cerenkov Light

High energy charged particles will cause the emission of Cerenkov light in air if their energies are above about 30 MeV (electrons). This threshold is pressure (and hence altitude) dependent. A typical Cerenkov light pulse (at sea level, 100 m from the central shower core) has a time spread of a few nanoseconds. Over this time, the photon flux between 430 and 530 nm would be ~10^{14} $m^{-2}s^{-1}$ for a primary particle energy of 10^{16} eV. For comparison, the night sky background flux is ~6×10^{11} photons $m^{-2}s^{-1}sr^{-1}$ in the same wavelength band (J.V. Jelley, 1967).

Fluorescence Light

Cosmic ray particles in the atmosphere excite atmospheric molecules which then emit fluorescence light. This is weak compared to the highly collimated Cerenkov component when viewed in the direction of the incident cosmic ray particle but is emitted isotropically. Typical pulse widths are longer than 50 ns and may be up to several microseconds for the total pulse from distant large showers (R.M. Baltrusaitis et al., 1985).

Effects of Cosmic Rays

Cerenkov Effects in Transparent Media

Background cosmic ray particles will produce Cerenkov light in transparent material with a photon yield between wavelengths λ_1 and λ_2

$$\sim (2\pi / 137)\sin^2(\theta_c) \int_{\lambda_1}^{\lambda_2} d\lambda / \lambda^2 \text{ photons(unit length)}^{-1}$$

where θ_c (the Cerenkov angle) = $\cos^{-1}$ (1/refractive index).

This background light is known to affect light detectors, e.g., photomultipliers, and can be a major source of background noise (R.W. Clay and A.G. Gregory, 1977).

Effects on Electronic Components

If background cosmic ray particles pass through electronic components, they may deposit sufficient energy to affect the state of, e.g., a transistor flip-flop. This effect may be significant where reliability is of great importance or the background flux is high. For instance, it has been estimated that, in communication satellite operation, an error rate of about 2×10^{-3} per transistor per year may be found. Permanent damage may also result. A significant error rate may be found even at sea level in large electronic memories. This error rate is dependent on the sensitivity of the component devices to the deposition of electrons in their sensitive volumes (J.F. Ziegler, 1981).

Biophysical Significance

When cosmic rays interact with living tissue, they produce radiation damage. The amount of the damage depends on the total dose of radiation. At sea level, this dose is small compared with doses from other sources but both the quantity and quality of the radiation change rapidly with altitude. Approximate dose rates under various conditions are:

Dose rates (mrem·yr^{-1})
Sea level cosmic rays, 30
Cosmic rays at 10 km (subsonic jets), 2000
Cosmic rays at 18 km (supersonic transports), 10,000
(c.f., mean total sea level dose, 300)

Astronauts would be subject to radiation from galactic (0.05 rads per day) and solar (a few hundred rads per solar flare) cosmic rays as well as large fluxes of low energy radiation when passing through the Van Allen belts (about 0.3 rads per traverse).

Both astronauts and SST travellers would be subject to a small flux of low energy heavy nuclei stopping in the body. Such particles are capable of destroying cell nuclei and could be particularly harmful in the early stages of the development of an embryo. The rates of heavy nuclei stopping in tissue in supersonic transports and spacecraft are approximately as follows:

Stopping nuclei ((cm^3 tissue)$^{-1}$ hr^{-1})
Supersonic transport (16 km), 0.0005
Supersonic transport (20 km), 0.005
Spacecraft, 0.15
(O. C. Allkofer, 1975a; O. C. Allkofer et al., 1974).

Carbon Dating

Radiocarbon is produced in the atmosphere due to the action of cosmic ray slow neutrons. Solar cycle modulation of the very low energy cosmic rays causes an anticorrelation of the atmospheric ^{14}C activity with sunspot number with a mean amplitude of about 0.5%. In the long term, modulation of cosmic rays by a varying magnetic field may be important (A.A. Burchuladze et al., 1979).

Practical Uses of Cosmic Rays

There are few direct practical uses of cosmic rays. Their attenuation in water and snow have, however, enabled automatic monitors of water and snow depth to be constructed. A search for hidden cavities in pyramids has been carried out using a muon "telescope."

Other Effects

Stellar X-rays have been observed to affect the transmission times of radio signals between distant stations by altering the depth of the ionospheric reflecting layer. It has also been suggested that variations in ionization of the atmosphere due to solar modulation may have observable effects on climatic conditions.

References

O.C. Allkofer (1975a) *Introduction to Cosmic Radiation*, Verlag Karl Thiemig, Munchen, Germany.

O.O. Allkofer (1975b) *J. Phys. G: Nucl. Phys.*, 1, L51.

O.C. Allkofer and W. Heinrich (1974) *Health Phys.*, 27, 543.

M. Amenomori et al. (1995) *Proc. 24th Int. Cosmic Ray Conf. Rome*, 3, 85. Universita La Sapienza, Roma.

K. Asakimori et al. (1993) *Proc. 23rd Int. Cosmic Ray Conf. Calgary*, 2, 25, University of Calgary, Canada.

R.M. Baltrusaitis et al. (1985) *Nucl. Inst. Meth.*, A420, 410.

D.J. Bird et al. (1993) *Phys. Rev. Lett.*, 71, 3401.

A.A. Burchuladze, S.V. Pagava, P. Povinec, G. I. Togonidze, S. Usacev (1979) *Proc. 16th Int. Cosmic Ray Conf. Kyoto*, 3, 201, Univ. of Tokyo, Japan.

R.W. Clay (1987) *Aust. J. Phys.*, 40, 423.

R.W. Clay and A.G. Gregory (1977) *J. Phys. A: Math. Gen.*, 10, 135.

J.M. Clem et al. (1995) *Proc. 24th Int. Cosmic Ray Conf. Rome*, 3, 5, Universita La Sapienza, Roma.

T.K. Gaisser et al. (1993) *Phys. Rev. D*, 47, 1919.

K. Greisen (1943) *Phys. Rev.*, 63, 323.

S. Hayakawa (1969) *Cosmic Ray Physics*, Wiley-Interscience, New York.

J.V. Jelley (1967) *Prog. in Elementary Particle and Cosmic Ray Physics*, 9, 41.

E. Juliusson (1975) Proc. 14th Int. Cosmic Ray Conf. Munich, 8, 2689, Max Planck Institute fur Extraterrestriche Physik, Munchen, Germany.

J. Linsley (1981) *Origin of Cosmic Rays*, I.A.U. Symposium 94, 53, D. Reidel Publishing Co. Dordrecht, Holland.

P. Meyer (1981) *Origin of Cosmic Rays*, I.A.U. Symposium 94, 7, D. Reidel Publishing Co. Dordrecht, Holland.

P.F. Michelson (1994) in *Towards a Major Atmospheric Cerenkov Detector III*, 257, Ed. T. Kifune, Universal Academy Press Inc., Tokyo, Japan.

P. Sokolsky (1989) *Introduction to Ultrahigh Energy Cosmic Ray Physics*, Addison Wesley Publishing Company.

T. Stanev et al. (1995) *Phys. Rev. Lett.*, 75, 3056.

S. Orito et al. (1995) *Proc. 24th Int. Cosmic Ray Conf. Rome*, 3, 76. Universita La Sapienza, Roma.

J.G. Wilson (1976) *Cosmic Rays*, Wykeham Pub. (London) Lt., U.K.

A.W. Wolfendale (1979) *Pramana*, 12, 631.

S. Yoshida et al. (1995) *Astroparticle Phys.*, 3, 105.

J.F. Ziegler, (1981) *IEEE Trans. Electron Devices*, ED-28, 560.

Section 12
Properties of Solids

TECHNIQUES FOR MATERIALS CHARACTERIZATION
Experimental Techniques Used to Determine the Composition, Structure, and Energy States of Solids and Liquids

H. P. R. Frederikse

The many experimental methods, originally designed to study the chemical and physical behavior of solids and liquids, have grown into a new field known as Materials Characterization (or Materials Analysis). During the past 30 years a host of techniques aimed at the study of surfaces and thin films has been added to the many tools for the analysis of bulk samples. The field has benefitted particularly from the development of computers and microprocessors, which have vastly increased the speed and accuracy of the measuring devices and the recording of their output. Materials characterization was and is a very important tool in the search for new physical and chemical phenomena. It plays an essential role in new applications of solids and liquids in industry, communications, and medicine. Many of its techniques are used in quality control, in safety regulations, and in the fight against pollution.

In most Materials Characterization experiments the sample is subjected to some kind of radiation: electromagnetic, acoustic, thermal, or particles (electrons, ions, neutrons, etc.). The surface analysis techniques usually require a high vacuum. As a result of interactions between the solid (or liquid) and the incoming radiation a beam of a similar (or a different) nature will emerge from the sample. Measurement of the physical and/or chemical attributes of this emerging radiation will yield qualitative, and often quantitative, information about the composition and the properties of the material being probed.

The modern tendency of describing practically everything in this world by a combination of a few letters (acronyms) has also penetrated the field of Materials Characterization. The table below gives the meaning of the acronym for every technique listed, the form and size of the required sample (bulk, surface, film, liquid, powder, etc.), the nature of the incoming and of the emerging radiation, the depth and the lateral spatial resolution that can be probed, and the information obtained from the experiment. The last column lists one or two major references to the technique described.

Technique	Sample	In	Out	Depth	Lateral resolution	Information obtained	Ref.
Optical and Mass Spectroscopies for Chemical Analysis							
1. AAS Atomic Absorption Spectroscopy	Atomize (flame, electro, thermal, etc.)	Light, e.g., glow discharge	Absorption spectrum	–	–	Concentration of atomic species (quantitative, using standards)	1,2
2. ICP-AES Induct. Coupled Plasma – Atomic Emission Spectroscopy	Atomize (flame, electro, thermal, ICP, etc.)	–	Emission spectrum	–	–	Concentration of atomic species (quantitative, using standards)	3
3. Dynamic SIMS Dynamic Secondary Ion Mass Spectroscopy	Surface	Ion beam (1–20 keV)	Secondary ions; analysis with mass spectrometer	2 nm–1 µm (or deeper: ion milling)	0.50 nm	Elemental and isotopic analysis; depth profile (all elements); detection limits: ppb-ppm	4
4. Static SIMS Static Secondary Ion Mass Spectroscopy	Surface	Ion beam (0.5–20 keV)	Secondary ions, analysis with mass spectrometer	0.1–0.5 nm	10 µm	Elemental analysis of surface layers; molecular analysis; detection limits: ppb-ppm	4
5. SNMS Sputtered Neutral Mass Spectroscopy	Surface, bulk	Plasma discharge; noble gases: 0.5–20 keV	Sputtered atoms ionized by atoms or electrons; then mass analyzed	0.1–0.5 nm (or deeper: ion milling)	1 cm	Elemental analysis Z ≥ 3; depth profile; detection limit: ppm	4,6
6. SALI Surface Analysis by Laser Ionization	Surface	e-beam, ion-beam, or laser for sputtering	Sputtered atoms ionized by laser; then mass analyzed	0.1–0.5 nm up to 3 µm in milling mode	60 nm	Surface analysis; depth profiling	7
7. LIMS Laser Ionization Mass Spectroscopy	Surface, bulk	u.v. laser (ns pulses)	Ionized species; analyzed with mass spectrometer	50–150 nm	5 µm–1 mm	Elemental (micro)analysis; detection limits: 1–100 ppm	8
8. SSMS Spark Source Mass Spectroscopy	Sample in the form of two electrodes	High voltage R.F. spark produces ions	Ions – analyzed in mass spectrometer	1–5 µm	–	Survey of trace elements; detection limit: 0.01–0.05 ppm	9
9. GDMS Glow Discharge Mass Spectroscopy	Sample forms the cathode for a D.C. glow discharge	Sputtered atoms ionized in plasma	Ions – analyzed in mass spectrometer	0.1–100 µm	3–4 mm	(Bulk) trace element analysis; detection limit: sub-ppb	9,10
10. ICPMS Induct. Coupled Plasma Mass Spectroscopy	Liquid-dissolved sample carried by gas stream into R.F. induction coil	Ions produced in argon plasma	Ions – analyzed in quadrupole mass spectrometer	–	–	High sensitivity analysis of trace elements	11
Photons — Absorption, Reflection and Electron Emission							
11. IRS Infrared Spectroscopy	Thin crystal, glass, liquid	I.R. light (W-filament, globar, Hg-arc)	I.R. spectrum	–	–	Electronic transitions (mainly in semiconductors and superconductors); vibrational modes (in crystals and molecules)	12,13,14
12. FTIR Fourier Transform I.R. Spectroscopy	Solid, liquid; transmission or reflection	White light (all frequencies)	Fourier Transform of spectrum (interferometer)	–	–	Spectra obtained at higher speed and resolution	15
13. ATR Attenuated Total Reflection	Surface or thin crystal	–	–	µm's	–	Atomic or molecular spectra of surfaces and films	16
14. (µ)-RS (Micro-) Raman Spectroscopy	Solid, liquid (1 µm–1 cm)	Laser beam, e.g., Ar-line, YAG-line	Raman spectra	0.5 µm	0.5 µm	Molecular and crystal vibrations	12,14,17

	Technique	Sample	In	Out	Depth	Lateral resolution	Information obtained	Ref.
15.	CARS Coherent Anti-Stokes Raman Spectroscopy	Solid, liquid (50 μm–3 cm)	Pump beam (ω_0)+ probe beam (ω_s)	Anti-Stokes spectrum	–	–	High resolution Raman spectra	14
16.	Ellipsometry	Transparent films, crystals, adsorbed layers	Polarized light	Change in polarization	0.05 nm–5 μm	25 μm (or sample thickness)	Refractive index and absorption	18,19
17.	UPS Ultraviolet Photo-electron Spectroscopy	Surfaces, adsorbed layers	u.v. light, 10–100 eV; 200 eV (synchrotron)	Electrons	0.2–10 nm	0.1–10 nm	Energies of electronic states of surfaces and free molecules	20,21
18.	PSD Photon Stimulated Desorption	Surfaces with adsorbed species	Far u.v. light E > 10 eV	Ions – analyzed with mass spectrometer	0.1–2 nm	–	Structure and desorption kinetics of adsorbed atoms and molecules	22

X-Rays

	Technique	Sample	In	Out	Depth	Lateral resolution	Information obtained	Ref.
19.	XRD X-Ray Diffraction	Single crystals, powders films	X-rays: λ = 0.05–0.2 nm (6–17 keV)	Diffracted X-ray beam	1–1000 μm	0.1–10 mm	Identification of crystallographic structures; all elements (low Z difficult)	23,24
20.	XRF/EDS X-Ray Fluorescence/Energy Dispersive Spectroscopy	Thin films, single layer	Prim. X-ray beam λ = 0.02–0.1 nm 12–80 keV	Fluorescent X-rays	1–100 μm	10 mm	Elemental analysis; all elements except H, He, Li – (EDS also used in XRD, SEM, TEM and EPMA)	25,26
21.	EXAFS Extended X-Ray Absorption Fine Structure	Films, foils	High intensity X-rays (synchrotron)	Spectrum near absorption edge	nm–μm	–	Local atomic structure: order/disorder in vicinity of absorbing atom	27
22.	XPS/ESCA X-Ray Photo-electron Spectroscopy/Electron Spect. for Chemical Analysis	Surfaces, thin films (≈20 atomic layers)	Soft X-rays (1–20 keV)	Core electrons; valence electrons	0.5–10 nm	5 nm–50 μm	(Quantitative) identification of all elements in surface layer or film	28,29

Electrons

	Technique	Sample	In	Out	Depth	Lateral resolution	Information obtained	Ref.
23.	CL Cathode Luminescence	Insulators, semiconductors	Electrons 5–50 keV	Photons 0.1–5 eV	1 nm–2 μn	1 or 2 μm	Energy levels of impurities and point defects	30
24.	APS Appearance Potential Spectroscopy	Surface (≈20 atomic layers)	Electrons (energy scan) 50–2000 eV	X-rays to pinpoint electron energy threshold	–	–	Identification of surface species	21, see also C
25.	AES Auger Electron Spectroscopy	Thin films, surfaces	Electrons 3–10 keV	Auger electrons 20–2000 eV	0.3–3 nm	≈30 nm	Elemental composition of surface (except H, He); detection limit 0.1–1%	28,29
26.	EELS Electron Energy Loss Spectroscopy	Very thin samples (<200 nm)	Electrons (100–400 keV)	(Retarded) electrons; minus 1–1000 eV	<200 nm	1–100 nm	Local elemental concentration; electronic structure, chem. bonding; interatomic distances	31
27.	EXELFS Extended Electron Energy Loss Fine Structure	Thin films	Electrons (100–400 keV)	Electrons energies 0–30 eV above edge	<200 nm	1–100 nm	Density of states of valence electrons (above Fermi level)	27,32
28.	ESD Electron Stimulated Desorption	Adsorbed species	Electrons E > 10 eV	Ions – analyzed with mass spectrometer	–	–	Structure and desorption properties of adsorbed atoms and molecules	22
29.	ESDIAD ESD-Ion Angular Distribution	(See ESD)	(See ESD)	Directional dependence of emitted ions	–	–	Geometries of adsorbed species (atoms or molecules)	22
30.	EPMA Electron Probe (X-Ray) Micro Analysis	Solid conductors and insulators <1 cm thick	Electrons 5–30 keV	Characteristic X-ray 0.1–15 keV	100 nm–5 μm	1 μm	Elemental analysis, $Z \leq 4$, major, minor and trace amounts	33,34
31.	LEED Low Energy Electron Diffraction	Surface	Mono-energetic electron beam 10–1000 eV	Diffracted electrons	0.4–2 nm	<5 μm	Crystallographic structure of surface; resolution: 0.01 nm	35
32.	RHEED Reflection High Energy Electron Diffraction	Surface	Electron beam at grazing angle 5–50 keV	Reflected electrons	0.2–10 nm	<5 μm	Surface symmetry	36,37
33.	SEM Scanning Electron Microscopy	Bulk, films (conducting)	High energy electrons usually ~30 keV	Secondary and backscattered electrons	1 nm–5 μm	1–20 nm	Surface image, defect structure; resolution 5–15 nm; magnification 300,000×	33,34
34.	(S)TEM (Scanning) Transmission Electron Microscopy	Thin specimen – <200 nm	High energy electrons typically 300 keV	Transmitted and diffracted electrons	(Sample thickness)	2–20 nm	(Defect) structure of cryst. solids; microchemistry; high resol.: 0.2 nm	33
35.	FEM Field Emission Microscopy	Metals, alloys (sharp point)	–	Electron emission (with appl. electric field – 50 kV)	≈0.5 nm	10–100 nm	Surface image, crystallographic structure	34
36.	STM Scanning Tunneling Microscopy	Polished or cleaved surface (conducting)	Tunneling current controls distance between sample and very sharp tip		1–5 nm	2–10 nm	Atomic-scale relief map of surface; resolution: vert. 0.002 nm, hor. 0.2 nm	39
37.	SPM Scanned Probe Microscopy	Very flat surface	Any field: e.g. mechan. vibration recorded with laser probe; same with magnetic, electric or thermal field		1–100 nm	1–100 nm	Surface-magnetic field, surface-thermal conductivity, etc.	39a
38.	AFM Atomic Force Microscopy	Very flat surface	Similar to STM; force measured with cantilever spring		0.5–5 nm	0.2–130 nm	Surface topography with atomic resolution; interatomic force	40

Technique	Sample	In	Out	Depth	Lateral resolution	Information obtained	Ref.
Ions and Neutrons							
39. ISS (or LEIS) Ion Scattering Spectroscopy (Low Energy Ion Scattering)	Surface	Ion beam He$^+$ or Ne$^+$ <3 keV	Sputtered ions (energy analysis)	0.1–0.5 nm	1–100 μm	Elemental analysis (better for low Z) detection limits: 0.01–1%	41
40. FIM Field Ion Microscopy	Surface: metals, alloys; very sharp tip	(He gas above sample)	He ions + high electric field produce image	≈0.1 nm	0.1–2 nm	Atomic structure of surface	34,42
41. RBS Rutherford Back Scattering	Solids, thin films	Mono-energetic ions (H$^+$ or He^{++}) 0.5–3 MeV	Backscattered ions	10 nm–1 μm	1 mm	Element identification (Li to U) detection limit: 0.01–1%	46
42. NRA Nuclear Reaction Analysis	Solids, thin films	Mono-energetic ions (Li, Be, B, etc.) 200 keV–6 MeV	Protons, deuterons ^{3}He, α-particles, γ-rays	0.1–5 μm	10 μm–10 mm	Element identification (all) detection limit: 10^{-12}–10^{-2}	47
43. PIXE Particle Induced X-ray Emission	Thin films, surface layers	High energy ions (H$^+$ or He^{++})	Characteristic X-rays	<10 μm	1 μm–2 mm	Trace impurities: Z>3 detection limit: 0.1–100 ppm (depending on sample thickness)	48
44. INS Ion Neutralization Spectroscopy	Surface	He-ions (≈5 eV)	Electrons	–	–	Energies of valence electrons	49
45. NAA Neutron Activation Analysis	Bulk, >0.5 g	Thermal neutrons	Characteristic γ-rays, (≈1 MeV)	Bulk	–	Trace concentrations (of isotopes) of elements: trans. metals, Pt-group; detection limit: 10^8–10^{14} atoms/cm^3	43
46. N(P)D Neutron (Powder) Diffraction	Crystalline solids	Thermal neutrons E ≈0.0025 eV	Diffracted neutrons	Bulk	–	Crystallographic structure; porosity, particle size	44
47. SANS Small Angle Neutron Scattering	Inhomogeneous solids; powders; porous samples	Thermal neutrons 2 θ = 10^{-2}–10^{-4}	Scattered neutrons	1–25 mm	–	Average size of inhomogeneities; range: 1 nm–1 mm	45
Acoustic							
48. SLAM Scanning Laser Acoustic Microscopy	Bulk, film	Acoustic wave produced by laser 1 MHz–1 GHz	Reflected acoustic wave	μm–cm	0.1–20 mm	Defect structure; thickness measurement	50
Thermal							
49. DTA Differential Thermal Analysis	Specimen and reference sample	Uniform heating	Temperature difference	Bulk	–	Phase transitions, crystallization	51
50. DSC Differential Scanning Calorimetry	Specimen and ref. sample	Controlled heating	Measure heat required for equal temperature	Bulk	–	Phase transitions, crystallization; activation energies	51
51. TGA Thermo Gravimetric Analysis	Bulk, 1–100 g	Controlled heating	Weight as function of temperature (and time)	Bulk	–	Decomposition, non-stoichiometry, kinetics of reaction	52
Resonance							
52. EPR (ESR) Electron Paramagnetic (Spin) Resonance	Paramagnetic solids or liquids	Microwave radiation in magnetic field 3–300 GHz; 1–100 kG	Microwave absorption (at resonance)	Bulk		Local environment of paramagnetic ion; concentration of paramagnetic, species; detection limit: 10^{11} spins/cm^3	53,54
53. ECR Electron Cyclotron Resonance	Semiconductors, metals; free electrons (low temperature)	Microwave radiation in magnetic field 10–30 GHz; 5–10 kG	Microwave absorption (at resonance)	Bulk		Electronic energy bands, effective masses	55
54. Mössbauer Effect	Source and absorber	Mono-energetic γ-rays: 5–100 keV	Mössbauer spectrum (Doppler shifted (lines)	50 m	1 cm	Interaction between nucleus and its environment (local electric, magnetic fields; bonds; valency; diffusion, etc.)	56
55. NMR (MRI) Nuclear Magnetic Resonance (Magnetic Resonance Imaging)	Solids, liquids	R.F. radiation + magnetic field; e.g. for protons: 60 MHz, 14 kG	R.F. absorption	<1 cm	1 cm	Quant. analysis; local magnetic environment; diffusion; imaging	58
56. ENDOR Electron Nuclear Double Resonance	Solids, liquids	R.F. + microwave radiation in magn. field.	Microwave absorption	–	–	Hyperfine interaction → local atomic structure	54
57. NQR Nuclear Quadrupole Resonance	Solids	R.F. radiation 0.5–1000 MHz	R.F. absorption	–	–	Asymmetry of the charge distribution at the nucleus	55,59
Other							
58. BET Brunauer-Emmett-Teller	(Large) surface area 1–20 m^2/g	Adsorbed gas (e.g., N$_2$ at low temp.) as function of pressure (monolayer coverage)		–	–	Surface area measurement	60

References

General References

A. Wachtman, J. B., *Characterization of Materials*, Butterworth-Heinemann, Boston, 1993.

B. Brundle, C. R., Evans, C. A., and Wilson, S., Eds., *Encyclopedia of Materials*, Butterworth-Heinemann, Boston, 1992.

C. Woodruff, D. P. and Delchar, T. A., *Modern Techniques of Surface Science*, Cambridge University Press, Cambridge, 1986.

D. *Metals Handbook*, 9th Edition, Vol. 10, Materials Characterization, Whan, R. E., Coordinator, American Society for Metals, Metals Park, OH, 1986.

Specific References

1. Slavin, M., *Atomic Absorption Spectroscopy*, 2nd Edition, John Wiley & Sons, New York, 1978.

2. Schrenk, W. G., *Analytical Atomic Spectroscopy*, Plenum Press, New York, 1975.

3. Dean, J. A. and Rains, T. E., *Flame Emission and Atomic Absorption Spectroscopy*, Vols. 1–3, Marcel Dekker, New York, 1969.

4. Benninghoven, A., Rudenauer, F. G., and Werner, H. W., *Secondary Ion Mass Spectroscopy*, John Wiley & Sons, New York, 1987.

5. Bird, J. R. and Williams, J. S., Eds., in *Ion Beams for Materials Analysis*, Academic Press, New York, 1989, pp. 515–537.

6. Smith, G. C., *Quantitative Surface Analysis for Materials Science*, The Institute of Metals, London, 1991.

7. Becker, E. H., in *Ion Spectroscopies for Surface Analysis*, Czanderna, A. W. and Hercules, D. M., Eds., Plenum Press, New York, 1991, p. 273.

8. Simons, D. S., *Int. J. Mass Spectrometry and Ion Processes*, 55, 15, 1983.

9. White, F. A. and Wood, G. M., *Mass Spectrometry: Applications in Science and Engineering*, John Wiley & Sons, New York, 1986.

10. Harrison, W. W. and Bentz, B. L., *Prog. Anal. Spectrometry*, 11, 53, 1988.

11. Bowmans, P. W. J. M., *Inductively Coupled Plasma Emission Spectroscopy*, Parts I and II, John Wiley & Sons, New York, 1987.

12. Brame, Jr., E. G. and Grasselli, J., *Infrared and Raman Spectroscopy*, Practical Spectroscopy Series, Vol. I, Marcel Dekker, New York, 1976.

13. Hollas, J. M., *Modern Spectroscopy*, John Wiley & Sons, New York, 1987.

14. Turrell, G., *Infrared and Raman Spectroscopy of Crystals*, Academic Press, New York and London, 1972.

15. Griffith, P. R. and Haseth, J. A., *Fourier Transform Infrared Spectroscopy*, John Wiley & Sons, New York, 1986.

16. Barnowski, M. K., *Fundamentals of Optical Fiber Communications*, Academic Press, New York, 1976.

17. Long, D. A., *Raman Spectroscopy*, McGraw-Hill, New York, 1977.

18. Azzam, R. M. A., *Ellipsometry and Polarized Light*, Elsevier-North Holland, Amsterdam, 1977.

19. Hecht, E., *Optics*, 2nd Edition, Addison-Wesley, Reading MA, 1987.

20. Brundle, C. R., in *Molecular Spectroscopy*, West, A. R., Ed., Heyden, London, 1976.

21. Park, R. L., in *Experimental Methods in Catalytic Research*, Vol. III, Anderson, R. B. and Dawson, P. T., Academic Press, New York, 1976, pp. 1–39.

22. Madey, T. E. and Stockbauer, R., in *Solid State Physics: Surfaces*, Vol. 22 of Methods of Experimental Physics, Park, R.L. and Lagally, M. G., Eds., Academic Press, New York, 1985.

23. Cullity, B. D., *Elements of X-Ray Diffraction*, 2nd Edition, Addison-Wesley, Reading, MA, 1978.

24. Schwartz, L. H. and Cohen, J. B., *Diffraction from Materials*, Springer Verlag, Berlin, 1987.

25. deBoer, D. K. G., in *Advances in X-Ray Analysis*, Vol. 34, Barrett, C. S. et. al., Eds., Plenum Press, New York, 1991.

26. Birks, L. S., *X-Ray Spectrochemical Analysis*, 2nd Edition, John Wiley & Sons, New York, 1969.

27. Bonnelle, C. and Mande, C., *Advances in X-Ray Spectroscopy*, Pergamon Press, Oxford, 1982.

28. *Practical Surface Analysis by Auger and X-Ray Photo-Electric Spectroscopy*, Briggs, D. and Seah, M. P., Eds., John Wiley & Sons, New York, 1983.

29. Powell, C. J. and Seah, M. P., *J. Vac. Sci. Technol. A*, Vol. 8, 735, 1990.

30. Yacobi, G. G. and Holt, D. B., *Cathodeluminescence Microscopy of Inorganic Solids*, Plenum Press, New York, 1990.

31. Egerton, R. F., *Electron Energy Loss Spectroscopy in the Electron Microscope*, Plenum Press, New York, 1986.

32. Disko, M. M., Krivanek, O. L., and Rez, P., *Phys. Rev.*, B25, 4252, 1982.

33. Goldstein, J. I., et. al., *Scanning Electron Microscopy and X-Ray Microanalysis*, 2nd Edition, Plenum Press, New York, 1986.

34. Murr, L. E., *Electron and Ion Microscopy and Microanalysis*, Marcel Dekker, New York, 1982.

35. Armstrong, R. A., in *Experimental Methods in Catalytic Research*, Vol. III, Anderson, R. B., and Dawson, P. T., Eds., Academic Press, New York, 1976.

36. Dobson, P. J. et. al., *Vacuum*, 33, 593, 1983.

37. Rymer, T. B., *Electron Diffraction*, Methuen, London, 1970.

38. Reimer, L., *Transmission Electron Microscopy*, Springer-Verlag, Berlin, 1984.

39. *Scanning Tunneling Microscopy and Related Methods*, Behm, R. J., Garcia, N., and Rohrer, H., Eds., Kluwer Academic Publishers, Norwell, MA, 1990.

39a. Wikramasinghe, H.K., *Scientific American*, Vol. 261, No. 4, pp. 98–105, Oct. 1989.

40. Rugar, D. and Hansma, P., *Physics Today*, 43(10), pp. 23–30, 1990.

41. Feldman, C. C. and Mayer, J. W., *Fundamentals of Surface and Thin Film Analysis*, North-Holland, Amsterdam, 1986.

42. Muller, E. W. and Tsong, T. T., *Field Ion Microscopy*, Elsevier, Amsterdam, 1969.

43. Amiel, S., *Nondestructive Activation Analysis*, Elsevier, Amsterdam, 1981.

44. Bacon, G. E., *Neutron Diffraction*, 3rd Edition, Clarendon Press, Oxford, 1975.

45. Neutron Scattering, Part A., in *Methods of Experimental Physics*, Vol. 23, Skold, K. and Price, D. L., Eds., Academic Press, New York, 1986.

46. Chu, W. K., Mayer, J. W., and Nicolet, M. A., *Backscattering Spectroscopy*, Academic Press, New York, 1987.

47. Rickey, F. A., in *High Energy and Heavy Ion Beams in Materials Analysis*, Tesmer, J. R., et. al., Eds., MRS, 1990, pp. 3–26.

48. Johansson, S. A. E. and Campbell, J. L., *PIXE: A Novel Technique for Elemental Analysis*, John Wiley & Sons, New York, 1988.

49. Hagstrum, H. D., in *Inelastic Ion-Surface Collisions*, Tolk, N. H. et. al., Eds., Academic Press, New York, 1977, pp. 1–46.

50. Nikoonahad, M., in *Research Techniques in Nondestructive Testing*, Vol. VI, Sharpe, R.S., Ed., Academic Press, New York, 1984, pp. 217–257.

51. Gallagher, P. K., *Characterization of Materials by Thermoanalytical Techniques*, MRS - Bulletin, Vol. 13, No. 7, pp. 23–27, 1988.

52. Earnest, C. M., *Compositional Analysis by Thermogravimetry*, ASTM Special Technical Publication 997, 1988.

53. Poole, C. P., *Electron Spin Resonance – A Comprehensive Treatise on Experimental Techniques*, 2nd Edition, John Wiley & Sons, New York, 1983.

54. Atherton, N. M., *Principles of Electron Spin Resonance*, Ellis Horwood Ltd., Chichester, U.K., 1993.

55. Kittel, C., *Introduction to Solid State Physics*, 6th Edition, John Wiley & Sons, New York, 1986, p. 196.

56. Gibb, T. C., *Principles of Mössbauer Spectroscopy*, Chapman & Hall, London, 1976.

57. Slichter, C. P., *Principles of Magnetic Resonance*, 3rd Edition, Springer-Verlag, Berlin, 1990.

58. *NMR Spectroscopy Techniques*, Dybrowski, C. and Lichter, R. L., Eds., Marcel Dekker, New York, 1987.

59. Das, T. P. and Hahn, E. L., *Nuclear Quadrupole Resonance Spectroscopy*, Academic Press, New York, 1958.

60. Somorjai, G. A., *Principles of Surface Chemistry*, Prentice-Hall, Englewood Cliffs, NJ, 1972, p. 216

SYMMETRY OF CRYSTALS

L. I. Berger

The ability of a body to coincide with itself in its different positions regarding a coordinate system is called its symmetry. This property reveals itself in iteration of the parts of the body in space. The iteration may be done by reflection in mirror planes, rotation about certain axes, inversions and translations. These actions are called the symmetry operations. The planes, axes, points, etc., are known as symmetry elements. Essentially, mirror reflection is the only truly primitive symmetry operation. All other operations may be done by a sequence of reflections in certain mirror planes. Hence, the mirror plane is the only true basic symmetry element. But for clarity, it is convenient to use the other symmetry operations, and accordingly, the other aforementioned symmetry elements. The symmetry elements and operations are presented in Table 1.

The entire set of symmetry elements of a body is called its symmetry class. There are thirty-two symmetry classes that describe all crystals that have ever been noted in mineralogy or been synthesized (more than 150,000). The denominations and symbols of the symmetry classes are presented in Table 2.

There are several known approaches to classification of individual crystals in accordance with their symmetry and crystallochemistry. The particles that form a crystal are distributed in certain points in space. These points are separated by certain distances (translations) equal to each other in any chosen direction in the crystal. Crystal lattice is a diagram that describes the location of particles (individual or groups) in a crystal. The lattice parameters are three non-coplanar translations that form the crystal lattice. Three basic translations form the unit cell of a crystal. August Bravais (1848) has shown that all possible crystal lattice structures belong to one or another of fourteen lattice types (Bravais lattices). The Bravais lattices, both primitive and non-primitive, are the contents of Table 3.

Among the three-dimensional figures, there is a group of polyhedrons that are called regular, which have all faces of the same shape and all edges of the same size (regular polygons). It has been shown that there are only five regular polyhedrons. Because of their importance in crystallography and solid state physics, a brief description of these polyhedrons is included in Table 4.

The systematic description of crystal structures is presented primarily in the well-known *Structurbericht*. The classification of crystals by the Structurbericht does not reflect their crystal class, the Bravais lattice, but is based on the crystallochemical type. This makes it inconvenient to use the Structurbericht categories for comparison of some individual crystals. Thus, there have been several attempts to provide a more convenient classification of crystals. Table 5 presents a compilation of different classifications which allows the reader to correlate the Structurbericht type with the international and Schoenflies point and space groups and with Pearson's symbols, based on the Bravais lattice and chemical composition of the class prototype. The information included in Table 5 has been chosen as an introduction to a more detailed crystallophysical and crystallochemical description of solids.

TABLE 1. Symmetry Operations and Elements

Symmetry operation	Name	Symmetry element International (Hermann-Mauguin)	Schoenflies	Presentation on the stereographic projection Parallel	Perpendicular
Reflection in a plane	Plane	m	C_s		
Rotation by angle $\alpha = 360°/n$ about an axis	Axis	n = 1, 2, 3, 4 or 6	C_n		
		n = 2	C_2		
		n = 3	C_3		
		n = 4	C_4		
		n = 6	C_6		
Rotation about an axis and inversion in a symmetry center lying on the axis	Inversion (improper) axis	$\bar{n} = \bar{3}, \bar{4}, \bar{6}$	C_{ni}		
		$\bar{n} = \bar{3}$	C_{3i}		
		$\bar{n} = \bar{4}$	C_{4i}		

TABLE 1. Symmetry Operations and Elements

Symmetry operation	Name	International (Hermann-Mauguin)	Schoenflies	Parallel	Perpendicular
		Symbol		Presentation on the stereographic projection	
			$\bar{n} = \bar{6}$	C_{6i}	
Inversion in a point	Center	$\bar{1}$	C_i	● ○ ✕	
Parallel translation	Translation vector $\vec{a}, \vec{b}, \vec{c}$				
Reflection in a plane and translation parallel to the plane	Glide–plane	a, b, c, n, d			
Rotation about an axis and translation parallel to the axis	Screw axis	n_m (m = 1, 2, .., n − 1)			
Rotation about an axis and reflection in a plane perpendicular to the axis	Rotatory-reflection axis	$\tilde{n}$ $\tilde{n} = \tilde{1}, \tilde{2}, \tilde{3}, \tilde{4}, \tilde{6}$	S_n		

TABLE 2. The Thirty-Two Symmetry Classes

Class name[a] and its symbol – International (Int) and Schoenflies (Sch)

Crystal symbol	Primitive Int	Primitive Sch	Central Int	Central Sch	Planal Int	Planal Sch	Axial Int	Axial Sch	Plane-axial Int	Plane-axial Sch	Inversion primitive Int	Inversion primitive Sch	Inversion-planal Int	Inversion-planal Sch
Triclinic	1	C_1	$\bar{1}$	C_i										
Monoclinic					m	C_s	2	C_2	2/m	C_{2h}				
Ortho-rhombic					mm2	C_{2v}	222	D_2	mmm	D_{2h}				
Trigonal	3	C_3	$\bar{3}$	C_{3i}	3m	C_{3v}	32	D_3	$\bar{3}m$	C_{3d}				
Tetragonal	4	C_4	4/m	C_{4h}	4mm	C_{4v}	422	D_4	4/mmm	D_{4h}	$\bar{4}$	S_4	$\bar{4}2m$	D_{2d}
Hexagonal	6	C_6	6/m	C_{6h}	6mm	C_{6v}	622	D_6	6/mmm	D_{6h}	$\bar{6}$	C_{3h}	$\bar{6}m2$	D_{3h}
Cubic	23	T	m3	T_h	$\bar{4}3m$	T_d	432	O	m3m	O_h				

[a] Per Fedorov Institute of Crystallography, Russian Academy of Sciences, nomenclature.

TABLE 3. The Fourteen Possible Space Lattices (Bravais Lattices)

Crystal system	Metric category of the system	No. of different lattices in the system	Lattice type[a] (marked by +) P C I F R	No. of identi-points per unit cell	Characteristic parameters (marked by +) a b c α β γ	Description of characteristic parameters a⊂X, b⊂Y, c⊂Z α≡(b,c), β≡(a,c), γ≡(a,b)	Symmetry of the lattice Int	Symmetry of the lattice Sch
Triclinic	Trimetric	1	+	1	+ + + + + +	a ≠ b ≠ c, α ≠ β ≠ γ	1	C
Monoclinic	Trimetric	2	+ +	1 or 2	+ + + +	a ≠ b ≠ c, α = γ = 90° ≠ β	2/m	C_{2h}
Orthorhombic	Trimetric	4	+ + + +	1, 2 or 4	+ + +	a ≠ b ≠ c, α = β = γ = 90°	mmm	D_{2h}
Trigonal (rhombohedral)	Dimetric	1	+	1	+ +	a = b = c, 120° > α = β = γ ≠ 90°	3m	D_{3d}
Tetragonal	Dimetric	2	+ +	1 or 2	+ +	a = b ≠ c, α = β = γ = 90°	4/mmm	D_{4h}
Hexagonal	Dimetric	1	+	1	+ +	a = b ≠ c, α = β = 90°, γ = 120°	6/mmm	D_{6h}
Isometric (cubic)	Monometric	3	+ + +	1, 2 or 4	+	a = b = c, α = β = γ = 90°	m3m	O_h

[a] Designations of the space-lattice types: P – primitive, C – side-centered (base-centered), I – body-centered, F – face-centered, R – rhombohedral.

TABLE 4. The Five Possible Regular Polyhedrons

Polyhedron	Symmetry (Schoenflies) Class	Symmetry (Schoenflies) Elements	Form of faces	Number of[a] Faces (F)	Number of[a] Edges (E)	Number of[a] Vertices (V)
Tetrahedron	T	$4C_3 3C_2$	Equilateral triangle	4	6	4

Cube (hexahedron)	O	$3C_44C_36C_2$	Square	6	12	8	
Octahedron	O	$3C_44C_36C_2$	Equilateral triangle	8	12	6	
Pentagonal dodecahedron	J	$6C_510C_315C_2$	Regular pentagon	12	30	20	
Icosahedron	J	$6C_510C_315C_2$	Equilateral triangle	20	30	12	

[a] Per formula by Leonhard Euler: $F + V - E = 2$

TABLE 5. Classification of Crystals

Strukturbericht symbol	Structure name	Symmetry group		Pearson symbol[a]	Standard ASTM E157-82a symbol[b]
		International	Schoenflies		
1	**2**	**3**	**4**	**5**	**6**
A1	Cu	$Fm3m$	O_h^4	cF4	F
A2	W	$Im3m$	O_h^9	cI2	B
A3	Mg	$P6_3/mmc$	D_{6h}^4	hP2	H
A4	C	$Fd3m$	O_h^7	cF8	F
A5	Sn	If_1/amd	D_{4h}^{19}	tI4	U
A6	In	$I4/mmm$	D_{4h}^{17}	tI2	U
A7	As	$R\bar{3}m$	D_{3d}^5	hR2	R
A8	Se	$P3_121$ or $P3_221$	D_3^4 (D_3^6)	hP3	H
A10	Hg	$R\bar{3}m$	D_{3d}^5	hR1	R
A11	Ga	$Cmca$	D_{2h}^{18}	oC8	Q
A12	α-Mn	$I\bar{4}3m$	T_d^3	cI58	B
A13	β-Mn	$P4_132$	O^7	cP20	C
A15	OW_3	$Pm3n$	O_h^3	cP8	C
A20	α-U	$Cmcm$	D_{2h}^{17}	oC4	Q
B1	ClNa	$Fm3m$	O_h^5	cF8	F
B2	ClCs	$Pm3m$	O_h^1	cP2	C
B3	SZn	$F\bar{4}3m$	T_d^2	cF8	F
B4	SZn	$P6_3mc$	C_{6v}^4	hP4	H
B8$_1$	AsNi	$P6_3/mmc$	D_{6h}^4	hP4	H
B8$_2$	InNi$_2$	$P6_3/mmc$	D_{6h}^4	hP6	H
B9	HgS	$P3_121$ or $P3_221$	D_3^4 or D_3^6	hP6	H
B10	OPb	$P4/nmm$	D_{4h}^7	tP4	T
B11	γ-CuTi	$P4/nmm$	D_{4h}^7	tP4	T
B13	NiS	$R\bar{3}m$	D_{3d}^5	hR6	R
B16	GeS	$Pnma$	D_{2h}^{16}	oP8	O
B17	PtS	$P4_2/mmc$	D_{4h}^9	tP4	T
B18	CuS	$P6_3/mmc$	D_{6h}^4	hP12	H
B19	AuCd	$Pmma$	D_{2h}^5	oP4	O
B20	FeSi	$P2_13$	T^4	cP8	C
B27	BFe	$Pnma$	D_{2h}^{16}	oP8	O
B31	MnP	$Pnma$	D_{2h}^{16}	oP8	O
B32	NaTl	$Fd3m$	O_h^7	cF16	F
B34	Pds	$P4_2/m$	C_{4h}^2	tP16	T
B35	CoSn	$P6/mmm$	D_{6h}^1	hP6	H
B37	SeTl	$I4/mcm$	D_{4h}^{18}	tI16	U
B$_e$	CdSb	$Pbca$	D_{2h}^{15}	oP16	O
B$_f$ (B33)	ξ-BCr	$Cmcm$	D_{2h}^{17}	oC8	Q
B$_g$	BMo	$I4_1/amd$	D_{4h}^{19}	tI4	U
B$_h$	CW	$P6m2$	D_{3h}^1	hP2	H
B$_i$	$\gamma'CMo$ (AsTi)	$P6_3/mmc$	D_{6h}^4	hP8	H
C1	CaF$_2$	$Fm\bar{3}m$	O_h^5	cF12	F
C1$_b$	AgAsMg	$F\bar{4}3m$	T_d^2	cF12	F
C2	FeS$_2$	$Pa3$	T_h^6	cP12	C
C3	Cu$_2$O	$Pn3m$	O_h^4	cP6	C
C4	O$_2$Ti	$P4_2/mnm$	D_{4h}^{14}	tP6	T
C6	CdI$_2$	$P3m1$	D_{3d}^3	hP3	H
C7	MoS$_2$	$P6_3/mmc$	D_{6h}^4	hP6	H
C11$_a$	C$_2$Ca	$I4/mmm$	D_{4h}^{17}	tI6	U
C11$_b$	MoSi$_2$	$I4/mmm$	D_{4h}^{17}	tI6	U

TABLE 5. Classification of Crystals

Strukturbericht symbol	Structure name	Symmetry group		Pearson symbol[a]	Standard ASTM E157-82a symbol[b]
		International	Schoenflies		
1	2	3	4	5	6
C12	$CaSi_2$	$R\bar{3}m$	D_{3d}^5	hR6	R
C14	$MgZn_2$	$P6_3/mmc$	D_{6h}^4	hP12	H
C15	Cu_2Mg	$Fd3m$	O_h^7	cF24	F
C15$_b$	$AuBe_5$	$F\bar{4}3m$ or F23	T_d^2 or T^2	cF24	F
C16	Al_2Cu	$I4/mcm$	D_{4h}^{18}	tI12	U
C18	FeS_2	$Pnnm$	D_{2h}^{12}	oP6	O
C19	$CdCl_2$	$R\bar{3}m$	D_{3d}^5	hR3	R
C22	Fe_2P	$P\bar{2}6m$	D_{3h}^1	hP9	H
C23	Cl_2Pb	$Pnma$	D_{2h}^{16}	oP12	O
C32	AlB_2	$P6/mmm$	D_{6h}^1	hP3	H
C33	Bi_2STe_2	$R\bar{3}m$	D_{3d}^5	hR5	R
C34	$AuTe_2$	$C2/m$ ($P2/m$)	C_{2h}^3 (C_{2h}^1)	mC6	N
C36	$MgNi_2$	$P6_3/mmc$	D_{6h}^4	hP24	H
C38	Cu_2Sb	$P4/nmm$	D_{4h}^7	tP6	T
C40	$CrSi_2$	$P6_222$	D_6^4	hP9	H
C42	SiS_2	$Ibam$	D_{2h}^{26}	oI12	P
C44	GeS_2	$Fdd2$	C_{2v}^{19}	oF72	S
C46	$AuTe_2$	$Pma2$	C_{2v}^4	oP24	O
C49	Si_2Zr	$Cmcm$	D_{2h}^{17}	oC12	Q
C54	Si_2Ti	$Fddd$	D_{2h}^{24}	oF24	S
C$_c$	Si_2Th	$I4_1/amd$	D_{4h}^{19}	tI12	U
C$_e$	$CoGe_2$	$Aba2$	C_{2v}^{17}	oC23	Q
DO$_2$	As_3Co	$Im3$	T_h^5	cI32	B
DO$_3$	BiF_3	$Fm3m$	O_h^5	cF16	F
DO$_9$	O_3Re	$Pm3m$	O_h^1	cP4	C
DO$_{11}$	CFe_3	$Pnma$	D_{2h}^{16}	oP16	O
DO$_{18}$	$AsNa_3$	$P6_3/mmc$	D_{6h}^4	hP8	H
DO$_{19}$	Ni_3Sn	$P6_3/mmc$	D_{6h}^4	hP8	H
DO$_{20}$	Al_3Ni	$Pnma$	D_{2h}^{16}	oP16	O
DO$_{21}$	Cu_3P	$P\bar{3}c1$	D_{3d}^4	hP24	H
DO$_{22}$	Cu_3P	$I4/mmm$	D_{4h}^{17}	tI8	U
DO$_{23}$	Al_3Zr	$I4/mmm$	D_{4h}^{17}	tI16	U
DO$_{24}$	Ni_3Ti	$P6_3/mmc$	D_{6h}^4	hP16	H
DO$_c$	SiU_3	$I4/mcm$	D_{4h}^{18}	tI16	U
DO$_e$	Ni_3P	$I\bar{4}$	S_4^2	tI32	U
D1$_3$	Al_4Ba	$I4/mmm$	D_{4h}^{17}	tI10	U
D1$_a$	$MoNi_4$	$I4/m$	C_{4h}^5	tI10	U
D1$_b$	Al_4U	$Imma$	D_{2h}^{28}	oI20	P
D1$_c$	$PtSn_4$	$Aba2$	C_{2v}^{17}	oC20	Q
D1$_e$	B_4Th	$P4/mbm$	D_{4h}^5	tP20	T
D1$_f$	BMn_4	$Fddd$	D_{2h}^{24}	oF40	S
D2$_1$	B_6Ca	$Pm3m$	O_h^1	cP7	C
D2$_3$	$NaZn_{13}$	$Fm3m$	O_h^5	cF112	F
D2$_b$	$Mn_{12}Th$	$I4/mmm$	D_{4h}^{17}	tI26	U
D2$_c$	MnU_6	$I4/mcm$	D_{4h}^{18}	tI28	U
D2$_d$	$CaCu_5$	$P6/mmm$	D_{6h}^1	hP6	H
D2$_f$	$B_{12}U$	$Fm3m$	O_h^5	cF52	F
D2$_h$	Al_6Mn	$Cmcm$	D_{2h}^{17}	oC28	Q
D5$_1$	$\alpha\text{-}Al_2O_3$	$R3c$	D_{3d}^6	hR10	R
D5$_2$	La_2O_3	$P\bar{3}m1$	D_{3d}^3	hP5	H
D5$_3$	Mn_2O_3	$Ia3$	T_h^7	cI80	B
D5$_8$	S_3Sb_2	$Pnma$	D_{2h}^{16}	oP20	O
D5$_9$	P_2Zn_3	$P4_2/mmc$	D_{4h}^9	tP40	T
D5$_{10}$	C_2C_3	$Pnma$	D_{2h}^{16}	oP20	O
D5$_{13}$	Al_3Ni_2	$P\bar{3}m1$	D_{3d}^3	hP5	H
D5$_a$	Si_2U_3	$P4/mbm$	D_{4h}^5	tP10	T
D5$_c$	C_3Pu_2	$I\bar{4}3d$	T_d^6	cI40	B

TABLE 5. Classification of Crystals

Strukturbericht symbol	Structure name	Symmetry group		Pearson symbol[a]	Standard ASTM E157-82a symbol[b]
1	2	International 3	Schoenflies 4	5	6
$D7_1$	Al_4C_3	$R\bar{3}m$	D^5_{3d}	hR7	R
$D7_3$	P_4Th_3	$I\bar{4}3d$	T^6_d	cI28	B
$D7_b$	B_4Ta_3	Immm	D^{25}_{2h}	oI14	P
$D8_1$	Fe_3Zn_{10}	Im3m	O^9_h	cI52	B
$D8_2$	Cu_5Zn_8	$I\bar{4}3m$	T^3_d	cI52	B
$D8_3$	Al_4Cu_9	P43m	T^1_d	cP52	C
$D8_4$	C_6Cr23	Fm3m	O^5_h	cF116	F
$D8_5$	Fe_7W_6	$R\bar{3}m$	D^5_{3d}	hR13	R
$D8_6$	$Cu_{15}Si_4$	$I\bar{4}3m$	T^3_d	cI76	B
$D8_8$	Mn_5Si_3	$P6_3/mcm$	D^3_{6h}	hP16	H
$D8_9$	Co_9S_8	Fm3m	O^5_h	cF68	F
$D8_{10}$	Al_8Cr_5	R3m	C^5_{3v}	hR26	R
$D8_{11}$	Al_5Co_2	$P6_3/mcm$	D^3_{6h}	hP28	H
$D8_a$	$Mn_{23}Th_6$	Fm3m	O^5_h	cF116	F
$D8_b$	σ-phase of Cr-Fe	$p\bar{4}_2/mnm$	D^{14}_{4h}	tP30	T
$D8_e$	$(Al,Zn)_{49}Mg_{32}$	Im3	T^5_h	cI162	B
$D8_f$	Ge_7Ir_3	Im3m	O^9_h	cI40	B
$D8_h$	B_5W_2	$P6_3/mmc$	D^4_{6h}	hP14	H
$D8_i$	B_5Mo_2	$R\bar{3}m$	D^5_{3d}	hR7	R
$D8_l$	B_3Cr_5	I4/mcm	D^{18}_{4h}	tI32	U
$D8_m$	Si_3W_5	I4/mcm	D^{18}_{4h}	tI32	U
$D10_1$	C_3Cr_7	P31c	C^4_{3v}	hP80	H
$D10_2$	Fe_3Th_7	$P6_3mc$	C^4_{6v}	hP20	H
$E0_1$	ClFPb	P4/nmm	D^7_{4h}	tP6	T
$E1_1$	$CuFeS_2$	$I\bar{4}2d$	D^{12}_{2d}	tI16	U
$E2_1$	CaO_3Ti	Pm3m	O^1_h	cP5	C
$E2_4$	S_3Sn_2	Pnma	D^{16}_{2h}	oP20	O
$E3$	Al_2CdS_4	$I\bar{4}$	S^2_4	tI14	U
$E9_3$	$SiFe_3W_3$	Fd3m	O^7_h	cF112	F
$E9_a$	Al_7Cu_2Fe	P4/mnc	D^6_{4h}	tP40	T
$E9_b$	$AlLi_3N_2$	Ia3	T^7_h	cI96	B
$F0_1$	NiSSb	$P2_13$	T^4	cP12	C
$F5_1$	$CrNaS_2$	R3m or R32	D^5_{3d} or D^7_3	hR4	R
$F5_6$	CuS_2Sb	Pnma	D^{16}_{2h}	oP16	O
$H1_1$	Al_2MgO_4	Fd3m	O^7_h	cF56	F
$H2_4$	Cu_3S_4V	P43m	T^1_d	cP8	C
$H2_5$	$AsCu_3S_4$	$Pmn2_1$	C^7_{2v}	oP16	O
$L1_0$	AuCu	P4/mmm	D^1_{4h}	tP4	T
$L1_2$	$AlCu_3$	Pm3m	O^1_h	cP4	C
$L2_1$	$AlCu_2Mn$	Fm3m	O^5_h	cF16	F
$L2_2$	Sb_2Tl_7	Im3m	O^9_h	cI54	B
$L'2_b$	H_2Th	I4/mmm	D^{17}_{4h}	tI6	U
$L'3$	Fe_2N	$P6_3/mmc$	D^4_{6h}	hP3	H
$L6_0$	$CuTi_3$	P4/mmm	D^1_{4h}	tP4	T

[a] The first letter denotes the crystal system: triclinic (a), monoclinic (m), orthorhombic (o), tetragonal (t), hexagonal (h) and cubic (c). Trigonal (rhombohedral) system is denoted by combination hR. The second letter of Pearson's symbol denotes lattice type: primitive (P), edge-(base-) centered (C), body-centered (I) or face-centered (F). The following number denotes number of atoms in the crystal unit cell.

[b] Standard ASTM E157-82a has the Bravais lattices designations as following: C – primitive cubic; B – body-centered cubic; F – face-centered cubic; T – primitive tetragonal; U – body-centered tetragonal; R – rhombohedral; H – hexagonal; O – primitive orthorhombic; P – body-centered orthorhombic; Q – base-centered orthorhombic; S – face-centered orthorhombic; M – primitive monoclinic; N – centered monoclinic; A – triclinic.

References

1. A. Schoenflies, *Kristallsysteme und Kristallstructur*, Leipzig, 1891.
2. E. S. Fedorow, Zusammenstellung der kristallographischen Resultate, *Zs. Krist.*, 20, 1892.

3. P. Groth, *Elemente der physikalischen und chemischen Krystallographie*, R. Oldenbourg, München/Berlin, 1921.

4. N. V. Belov, Class Method of Deriving Space Groups of Symmetry, *Trudy Instituta Kristallodraffi imeni Fedorova (Transactions of the Fedorov Inst. of Crystallography)*, 5, 25, 1951, in Russian.

5. W. B. Pearson, *Handbook of Lattice Spacings and Structures of Metals and Alloys*, Vol. 1, Pergamon Press, 1958; Vol. 2, 1967.

6. Ch. Kittel, *Introduction to Solid State Physics*, John Wiley & Sons, 1956.

7. G. S. Zhdanov, *Fizika Tverdogo Tela (Solid State Physics)*, Moscow University Press, 1962, in Russian.

8. M. J. Buerger, *Elementary Crystallography*, John Wiley & Sons, 1963.

9. F. D. Bloss, *Crystallography & Crystal Chemistry*, Holt, Rinehart & Winston, 1971.

10. T. Janssen, *Crystallographic Groups*, North-Holland/American Elsevier, 1973.

11. M. P. Shaskolskaya, *Kristallografiya (Crystallography)*, Vysshaya Shkola, Moscow, 1976, in Russian.

12. T. Hahn, Ed., Internat. *Tables for Crystallography*, Vol. A, D. Reidel Publishing, Boston, 1983.

13. Crystal Data. Determinative Tables, Volumes 1–6, 1966–1983, JCPDS-Intern Centre for Diffraction Data and U.S. Dept. of Commerce.

14. R. W. G. Wyckoff, *Crystal Structures*, 2nd ed., Volumes 1–6, Interscience, New York, 1963.

15. C. J. Bradley and A. P. Cracknell, *The Mathematical Theory of Symmetry in Solids*, Clarendon Press, Oxford, 1972.

16. International Tables for Crystallography. Volume A, *Space–Group Symmetry*, T. Hahn, Ed., 1989; Volume B, *Reciprocal Space*, U. Schmueli, Ed.; Volume C, *Mathematical, Physical and Chemical Tables*, A. J. C. Wilson, Ed., Kluwer Academic Publishers, Dordrecht, 1989.

17. G. R. Desiraju, *Crystal Engineering: The Design of Organic Solids*, Elsevier, Amsterdam, 1989.

18. M. Senechal, *Crystalline Symmetries: An Informal Mathematical Introduction*, Adam Hilger Publ., Bristol, 1990.

19. C. Hammond, *Introduction to Crystallography*, Oxford University Press, 1990.

20. N.W. Alcock, *Bonding and Structure: Structural Principles in Inorganic and Organic Chemistry*, Ellis Norwood Publ., 1990.

21. T. C. W. Mak and G. D. Zhou. *Crystallography in Modern Chemistry: A Resource Book of Crystal Structures*, Wiley–Interscience, New York, 1992.

22. S. C. Abrahams, K. Mirsky, and R. M. Nielson, *Acta Cryst*, B52, 806 (1996); B52, 1057 (1996).

23. C. Marcos, A. Panalague, D. B. Morciras, S. Garcia-Granda and M. R. Dias. *Acta Cryst*, B52, 899 (1996).

24. A. C. Larson, *Crystallographic Computing*, Manksgaard, Copenhagen, 1970.

25. G. M. Sheldrick, SHELXS86. Crystallographic Computing 3, Clarendon Press, Oxford, 1986; SHELXL93. Program for the Refinement of Crystal Structures, University of Göttingen Press, 1993.

26. Inorganic Crystal Structure Database, CD–ROM. Sci. Inf. Service. E-mail: SISI@Delphi.com.

IONIC RADII IN CRYSTALS

Ionic radii are a useful tool for predicting and visualizing crystal structures. This table lists a set of ionic radii R_i in Å units for the most common coordination numbers CN of positive and negative ions. The values are based on experimental crystal structure determinations, supplemented by empirical relationships, and theoretical calculations. The notation sq after the coordination number indicates a square configuration, while py indicates pyramidal.

The advice of Howard T. Evans and Marvin J. Weber in preparing this table is appreciated.

References

1. Shannon, R. D., *Acta Crystallogr.* A32, 751, 1976.
2. Jia, Y. Q., *J. Solid State Chem.* 95, 184, 1991.

Ion	CN	R_i/Å	Ion	CN	R_i/Å	Ion	CN	R_i/Å
Anions				8	1.12	Eu^{+3}	6	0.95
F^{-1}	6	1.33		10	1.23		8	1.07
Cl^{-1}	6	1.81		12	1.34	F^{+7}	6	0.08
Br^{-1}	6	1.96	Cd^{+2}	4	0.78	Fe^{+2}	4	0.63
I^{-1}	6	2.20		6	0.95		6	0.61
OH^{-1}	4	1.35		8	1.10		8	0.92
	6	1.37		12	1.31	Fe^{+3}	4	0.49
O^{-2}	2	1.21	Ce^{+3}	6	1.01		6	0.55
	6	1.40		8	1.14		8	0.78
	8	1.42		10	1.25	Fr^{+1}	6	1.80
S^{-2}	6	1.84		12	1.34	Ga^{+3}	4	0.47
Se^{-2}	6	1.98	Ce^{+4}	6	0.87		6	0.62
Te^{-2}	6	2.21		8	0.97	Gd^{+3}	6	0.94
Cations				10	1.07		8	1.05
Ac^{+3}	6	1.12		12	1.14	Ge^{+2}	6	0.73
Ag^{+1}	4	1.00	Cf^{+3}	6	0.95	Ge^{+4}	4	0.39
	6	1.15	Cf^{+4}	6	0.82		6	0.53
	8	1.28		8	0.92	Hf^{+4}	4	0.58
Ag^{+2}	4sq	0.79	Cl^{+5}	3py	0.12		6	0.71
	6	0.94	Cl^{+7}	4	0.08		8	0.83
Al^{+3}	4	0.39	Cm^{+3}	6	0.97	Hg^{+1}	6	1.19
	5	0.48	Cm^{+4}	6	0.85	Hg^{+2}	2	0.69
	6	0.54		8	0.95		4	0.96
Am^{+3}	6	0.98	Co^{+2}	4	0.56		6	1.02
	8	1.09		6	0.65		8	1.14
Am^{+4}	6	0.85		8	0.90	I^{+5}	3py	0.44
	8	0.95	Co^{+3}	6	0.55		6	0.95
As^{+3}	6	0.58	Cr^{+2}	6	0.73	I^{+7}	4	0.42
As^{+5}	4	0.34	Cr^{+3}	6	0.62		6	0.53
	6	0.46	Cr^{+4}	4	0.41	In^{+3}	4	0.62
Au^{+1}	6	1.37		6	0.55		6	0.80
Au^{+3}	4sq	0.64	Cr^{+6}	4	0.26	Ir^{+3}	6	0.68
	6	0.85		6	0.44	Ir^{+4}	6	0.63
Ba^{+2}	6	1.35	Cs^{+1}	6	1.67	Ir^{+5}	6	0.57
	8	1.42		8	1.74	K^{+1}	4	1.37
	12	1.61		10	1.81		6	1.38
Be^{+2}	4	0.27		12	1.88		8	1.51
	6	0.45	Cu^{+1}	2	0.46		12	1.64
Bi^{+3}	5	0.96		4	0.60	La^{+3}	6	1.03
	6	1.03		6	0.77		8	1.16
	8	1.17	Cu^{+2}	4sq	0.57		10	1.27
Bi^{+5}	6	0.76		6	0.73		12	1.36
Bk^{+3}	6	0.96	Dy^{+2}	6	1.07	Li^{+1}	4	0.59
Bk^{+4}	6	0.83		8	1.19		6	0.76
	8	0.93	Dy^{+3}	6	0.91		8	0.92
Br^{+5}	3py	0.31		8	1.03	Lu^{+3}	6	0.86
Br^{+7}	4	0.25	Er^{+3}	6	0.89		8	0.97
	6	0.39		8	1.00	Mg^{+2}	4	0.57
C^{+4}	4	0.15	Eu^{+2}	6	1.17		6	0.72
	6	0.16		8	1.25		8	0.89
Ca^{+2}	6	1.00		10	1.35	Mn^{+2}	4	0.66

Ion	CN	R_i/Å	Ion	CN	R_i/Å	Ion	CN	R_i/Å
	6	0.83	Pr+3	6	0.99	Tc+4	6	0.65
	8	0.96		8	1.13	Te+4	4	0.66
Mn+3	6	0.58	Pr+4	6	0.85		6	0.97
Mn+4	4	0.39		8	0.96	Te+6	4	0.43
	6	0.53	Pt+2	4sq	0.60		6	0.56
Mn+5	4	0.33		6	0.80	Th+4	6	0.94
Mn+6	4	0.26	Pt+4	6	0.63		8	1.05
Mn+7	4	0.25	Pu+3	6	1.00		10	1.13
Mo+3	6	0.69	Pu+4	6	0.86		12	1.21
Mo+4	6	0.65	Pu+5	6	0.74	Ti+2	6	0.86
Mo+5	4	0.46	Pu+6	6	0.71	Ti+3	6	0.67
	6	0.61	Ra+2	8	1.48	Ti+4	4	0.42
Mo+6	4	0.41		12	1.70		6	0.61
	6	0.59	Rb+1	6	1.52		8	0.74
	7	0.73		8	1.61	Tl+1	6	1.50
N+3	6	0.16		10	1.66		8	1.59
N+5	6	0.13		12	1.72		12	1.70
Na+1	4	0.99	Re+4	6	0.63	Tl+3	4	0.75
	6	1.02	Re+5	6	0.58		6	0.89
	8	1.18	Re+6	6	0.55		8	0.98
	9	1.24	Re+7	4	0.38	Tm+2	6	1.01
	12	1.39		6	0.53		7	1.09
Nb+3	6	0.72	Rh+3	6	0.67	Tm+3	6	0.88
	8	0.79	Rh+4	6	0.60		8	0.99
Nb+4	6	0.68	Rh+5	6	0.55	U+3	6	1.03
Nb+5	4	0.48	Ru+3	6	0.68	U+4	6	0.89
	6	0.64	Ru+4	6	0.62		8	1.00
	8	0.74	Ru+5	6	0.57		12	1.17
Nd+3	6	0.98	Ru+7	4	0.38	U+5	6	0.76
	8	1.12	Ru+8	4	0.36	U+6	2	0.45
	9	1.16	S+4	6	0.37		4	0.52
	12	1.27	S+6	4	0.12		6	0.73
Ni+2	4sq	0.49		6	0.29		8	0.86
	6	0.69	Sb+3	4py	0.76	V+2	6	0.79
Ni+3	6	0.56		6	0.76	V+3	6	0.64
Np+3	6	1.01	Sb+5	6	0.60	V+4	5	0.53
Np+4	6	0.87	Sc+3	6	0.75		6	0.58
Np+5	6	0.75		8	0.87		8	0.72
Np+6	6	0.72	Se+4	6	0.50	V+5	4	0.36
Os+4	6	0.63	Se+6	4	0.28		5	0.46
Os+5	6	0.58		6	0.42		6	0.54
Os+6	6	0.55	Si+4	4	0.26	W+4	6	0.66
Os+8	4	0.39		6	0.40	W+5	6	0.62
P+5	4	0.17	Sm+2	6	1.19	W+6	4	0.42
	6	0.38		8	1.27		5	0.51
Pa+3	6	1.04	Sm+3	6	0.96		6	0.60
Pa+4	6	0.90		8	1.08	Y+3	6	0.90
Pa+5	6	0.78		12	1.24		8	1.02
Pb+2	6	1.19	Sn+4	4	0.55		9	1.08
	8	1.29		6	0.69	Yb+2	6	1.02
	10	1.40		8	0.81		8	1.14
	12	1.49	Sr+2	6	1.18	Yb+3	6	0.99
Pb+4	4	0.65		8	1.26		9	1.04
	6	0.78		10	1.36	Zn+2	4	0.60
	8	0.94		12	1.44		6	0.74
Pd+2	4sq	0.64	Ta+3	6	0.72		8	0.90
	6	0.86	Ta+4	6	0.68	Zr+4	4	0.59
Pd+3	6	0.76	Ta+5	6	0.64		6	0.72
Pd+4	6	0.62	Tb+3	6	0.92		8	0.84
Pm+3	6	0.97		8	1.04		9	0.89
	8	1.09	Tb+4	6	0.76			
Po+4	6	0.97		8	0.88			

POLARIZABILITIES OF ATOMS AND IONS IN SOLIDS

H. P. R. Frederikse

The polarization of a solid dielectric medium, P, is defined as the dipole moment per unit volume averaged over the volume of a crystal cell. A component of P can be expanded as a function of the electric field E:

$$P_i = \sum_j a_j E_j + \sum_{jk} b_{jk} E_j E_k$$

For relatively small electric fields in isotropic substances $P = \chi_e E$, where χ_e is the electric susceptibility. If the medium is made up of N atoms (or ions) per unit volume, the polarization is $P = N p_m$ where p_m is the average dipole moment per atom. The polarizability α can be defined as $p_m = \alpha E_0$, where E_0 is the local field at the position of the atom. Using the Lorentz method to calculate the local field one finds:

$$P = N\alpha(E + 4\pi P) = \chi_e E$$

Together with the definition of the dielectric constant (relative permittivity), $\varepsilon = 1 + 4\pi\chi_e$, this leads to:

$$\alpha = \frac{3}{4\pi N}\left(\frac{\varepsilon - 1}{\varepsilon + 2}\right)$$

This expression is known as the Clausius-Mossotti equation.

The total polarization associated with atoms, ions, or molecules is due to three different sources:

1. Electronic polarization arises because the center of the local electronic charge cloud around the nucleus is displaced under the action of the field: $P_e = N\alpha_e E_0$ where α_e is the *electronic polarizability*.

2. Ionic polarization occurs in ionic materials because the electric field displaces cations and anions in opposite directions: $P_i = N\alpha_i E_0$, where α_i is the *ionic polarizability*.

3. Orientational polarization can occur in substances composed of molecules that have permanent electric dipoles. The alignment of these dipoles depends on temperature and leads to an *orientational polarizability* per molecule: $\alpha_{or} = p^2/3kT$, where p is the permanent dipole moment per molecule, k is the Boltzmann constant, and T is the temperature.

Because of the different nature of these three polarization processes the response of a dielectric solid to an applied electric field will strongly depend on the frequency of the field. The resonance of the electronic excitation in insulators (dielectrics) takes place in the ultraviolet part of the spectrum; the characteristic frequency of the lattice vibrations is located in the infrared, while the orientation of dipoles requires fields of much lower frequencies (below 10^{10} Hz). This response to electric fields of different frequencies is shown in Figure 1. Values of the electronic polarizabilities for selected atoms and ions are given in Table 1.

References

1. Kittel, C., *Introduction to Solid State Physics*, Fourth Edition, John Wiley & Sons, New York, 1971.
2. Lerner, R.G., and Trigg, G.L., Eds., *Encyclopedia of Physics, Second Edition*, VCH Publishers, New York, 1990.
3. Ralls, K.M., Courtney, T.H., and Wulff, J., *An Introduction to Materials Science and Engineering*, John Wiley & Sons, New York, 1976.

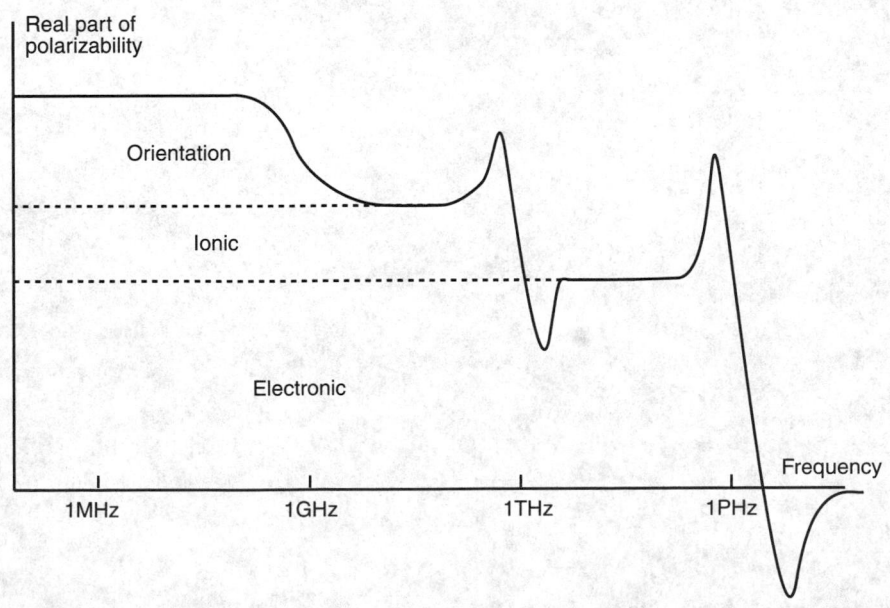

FIGURE 1. Schematic graph of the frequency dependence of the different contributions to polarizability.

TABLE 1. Electronic Polarizabilities in Units of 10^{-24} cm³

| | | | | | | He |
| | | | | | | 0.201 |

Li⁺	Be²⁺	B³⁺	C⁴⁺	O²⁻	F⁻	Ne
0.029	0.008	0.003	0.0013	3.88	1.04	0.39
Na⁺	Mg²⁺	Al³⁺	Si⁴⁺	S²⁻	Cl⁻	Ar
0.179	0.094	0.052	0.0165	10.2	3.66	1.62
K⁺	Ca²⁺	Sc³⁺	Ti⁴⁺	Se²⁻	Br⁻	Kr
0.83	0.47	0.286	0.185	10.5	4.77	2.46
Rb⁺	Sr²⁺	Y³⁺	Zr⁴⁺	Te²⁻	I⁻	Xe
1.40	0.86	0.55	0.37	14.0	7.1	3.99
Cs⁺	Ba²⁺	La³⁺	Ce⁴⁺			
2.42	1.55	1.04	0.73			

Data from Pauling, L., *Proc. R. Soc.* London, A114, 181, 1927. See also Jaswal, S.S. and Sharma, T.P., *J. Phys. Chem. Solids*, 34, 509, 1973. Values are appropriate for cgs units. To convert to SI, use the relation $\alpha(\text{SI})/\text{C m}^2\text{V}^{-1} = 1.11265 \cdot 10^{-16} \, \alpha(\text{cgs})/\text{cm}^3$.

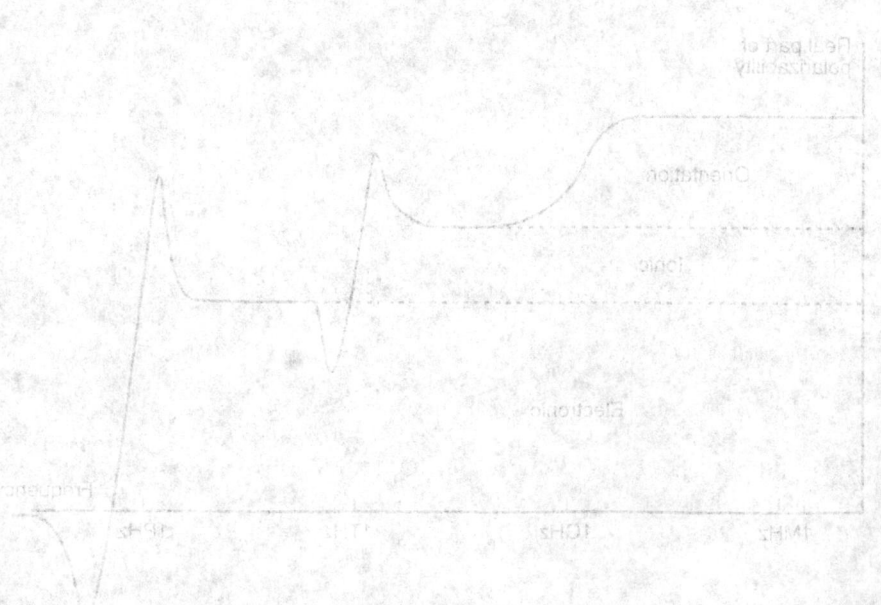

CRYSTAL STRUCTURES AND LATTICE PARAMETERS
OF ALLOTROPES OF THE ELEMENTS

H. W. King

The crystal structures of the allotropic forms of the elements are presented in terms of the Pearson symbol, the Strukturbericht designation, and the prototype of the structure. The temperatures of the phase transformations are listed in degrees Celsius and the pressures are in GPa. A consistent nomenclature is used, whereby all allotropes are labeled by Greek letters. The lattice parameters of the units cells are given in nanometers (nm) and are considered to be accurate to ±2 in the last reported digit.

This compilation is restricted to changes in crystal structures that occur as a result of a change in temperature or pressure. Low-temperature structures are included for the diatomic and rare gases, which show many similarities with respect to the metallic elements. The elements identified with an asterisk (*) have polymorphic structures based on different molecular configurations. The crystal data given for these elements refer to the most stable structure at room temperature.

Reprinted with the permission of ASM International from T.B. Massalski, Ed., *Binary Alloy Phase Diagrams*, ASM International, Metals Park, Ohio, 1986; certain data on rare earth elements were provided by K.A. Gschneidner.

Element	Temperature, °C	Pressure, GPa	Pearson symbol	Space group	Strukturbericht designation	Prototype	Lattice parameters, nm			Comment, c/a or α or β
							a	b	c	
Ac	25	atm	cF4	Fm3m	A1	Cu	0.5311	...	...	...
Ag	25	atm	cF4	Fm3m	A1	Cu	0.40857	...	...	...
αAl	25	atm	cF4	Fm3m	A1	Cu	0.40496	...	...	...
βAl	25	>20.5	hP2	P6₃/mmc	A3	Mg	0.2693	...	0.4398	1.6331
α′Am	25	atm	hP4	P6₃/mmc	A3′	αLa	0.34681	...	1.1241	2*1.621
αAm	>769	atm	cF4	Fm3m	A1	Cu	0.4894	...	...	...
βAm	>1074	atm	cI2	Im3m	A2	W	?	...	...	...
γAm	25	>15	oC4	Cmcm	A20	αU	0.3063	0.5968	0.5169	...
αAr	<−189.35	atm	cF4	Fm3m	A1	Cu	0.5316	...	...	...
(βAr)	<−189.40	atm	hP2	P6₃/mmc	A3	Mg	0.3760	...	0.6141	1.633
αAs	25	atm	hR2	R3m	A7	αAs	0.41319	...	......	α = 54.12°
εAs	>448	atm	oC8	Cmca	...	P(black)	0.362	1.085	0.448	...
Au	25	atm	cF4	Fm3m	A1	Cu	0.40782	...	...	...
βB	25	atm	hR105	R3m	...	βB	1.017	...	...	α = 65.12°
αBa	25	atm	cI2	Im3m	A2	W	0.50227	...	...	...
βBa	25	>5.33	hP2	P6₃/mmc	A3	Mg	0.3901	...	0.6154	1.5775
γBa	25	>23	?	?	...	...	...	...	...	...
αBe	25	atm	hP2	P6₃/mmc	A3	Mg	0.22859	...	0.35845	1.5681
βBe	>1270	atm	cI2	Im3m	A2	W	0.25515	...	...	...
γBe	25	>9.3	?	...	...	...	...	...	...	...
αBi	25	atm	hR2	r3m	A7	αAs	0.47460	...	...	α = 57.23°
βBi	25	>2.6	mC4	C2/m	...	βBi	0.6674	0.6117	0.3304	β = 110.33°
γBi	25	>3.0	mP3	?	...	...	0.605	0.42	0.465	β = 85.33°
σBi	25	>4.3	?	?	...	...	...	...	...	...
εBi	25	>6.5	?	?	...	...	...	...	...	...
ζBi	25	>9.0	cI2	Im3m	A2	W	0.3800	...	...	...
αBk	25	atm	hP4	P6₃/mmc	A3′	αLa	0.3416	...	1.1069	2*1.620
βBk	>977	atm	cF4	Fm3m	A1	Cu	0.4997	...	...	...
Br	<7.25	atm	oC8	Cmca	...	Cl	0.668	0.449	0.874	...
C (graphite)	25	atm	hP4	P6₃/mmc	A9	C (graphite)	0.24612	...	0.6709	2.7258
C (diamond)	25	>60	cF8	Fd3m	A4	C (diamond)	0.35669	...	...	...
C (hd)	25	HP	hP4	P6₃/mmc	...	C (hd)	0.2522	...	0.4119	1.633
αCa	25	atm	cF4	Fm3m	A1	Cu	0.55884	...	...	...
βCa	>443	atm	cI2	Im3m	A2	W	0.4480	...	...	...
γCa	25	>1.5	?	...	...	...	...	...	...	...
Cd	25	atm	hP2	P6₃/mmc	A3	Mg	0.29793	...	0.56196	1.8862
αCe	<−177	atm	cF4	Fm3m	A1	Cu	0.485	...	...	...
βCe	25	atm	hP4	P6₃/mmc	A3′	αLa	0.36810	...	1.1857	2*1.611
γCe	25	atm	cF4	Fm3m	A1	Cu	0.51610	...	...	...
δ-Ce	>726	atm	cI2	Im3m	A2	W	0.412	...	...	...

Element	Temperature, °C	Pressure, GPa	Pearson symbol	Space group	Strukturbericht designation	Prototype	Lattice parameters, nm			Comment, c/a or α or β
							a	b	c	
α′Ce	25	>5.4	oC4	Cmcm	A20	αU	0.3049	0.5998	0.5215	...
αCf	25	atm	hP4	P6₃/mmc	A3′	αLa	0.339	...	1.1015	2*1.625
βCf	>590	atm	cF4	Fm3m	A1	Cu	?	...	...	...
Cl	<-102	atm	oC8	Cmca	...	Cl	0.624	0.448	0.826	...
αCm	25	atm	hP4	P6₃/mmc	A3′	αLa	0.3496	...	1.1331	2*1.621
βCm	>1277	atm	cF4	Fm3m	A1	Cu	0.4382	...	...	...
εCo	25	atm	hP2	P6₃/mmc	A3	Mg	0.25071	...	0.40686	1.6228
αCo	>422	atm	cF4	Fm3m	A1	Cu	0.35447	...	...	...
αCr	25	atm	cI2	Im3m	A2	W	0.28848	...	...	...
α′Cr	25	HP	tI2	I4/mmm	...	α′Cr	0.2882	...	0.2887	1.002
aCs	25	atm	cI2	Im3m	A2	W	0.6141	...	...	...
βCs	25	>2.37	cF4	Fm3m	A1	Cu	0.6465	...	...	...
β′Cs	25	>4.22	cF4	Fm3m	A1	Cu	0.5800	...	...	...
γCs	25	>4.27	?	...	...	...	...	...	...	...
Cu	25	atm	cF4	Fm3m	A1	Cu	0.36146	...	...	...
α′Dy	<-187	atm	oC4	Cmcm	...	α′Dy	0.3595	0.6184	0.5678	...
αDy	25	atm	hP2	P6₃/mmc	A3	Mg	0.35915	...	0.56501	1.5732
βDy	>1381	atm	cI2	Im3m	A2	W	0.403	...	...	...
γDy	25	>7.5	hR3	R3m	...	αSm	0.3436	...	2.483	4.5*1.606
Er	25	atm	hP2	P6₃/mmc	A3	Mg	0.35592	...	0.55850	1.5692
αEs	25	atm	hP4	P6₃/mmc	A3′	αLa	?	...	...	...
βEs	?	atm	cF4	Fm3m	A1	Cu	?	...	...	...
Eu	25	atm	cI2	Im3m	A2	W	0.45827	...	...	...
αF	<-227.6	atm	mC8	C2/c	...	αF	0.550	0.338	0.728	β = 102.17°
βF	<-219.67	atm	cP16	Pm3n	...	γO	0.667	...	...	...
αFe	25	atm	cI2	Im3m	A2	W	0.28665	...	...	...
γFe	>912	atm	cF4	Fm3m	A1	Cu	0.36467	...	...	...
σFe	>1394	atm	cI2	Im3m	A2	W	0.29315	...	...	...
εFe	25	>13	hP2	P6₃/mmc	A3	Mg	0.2468	...	0.396	1.603
αGa	25	atm	oC8	Cmca	A11	αGa	0.45186	0.76570	0.45258	...
βGa	25	>1.2	tI2	I4/mmm	A6	In	0.2808	...	0.4458	1.588
γGa	-53	>3.0	oC40	Cmcm	...	γGa	1.0593	1.3523	0.5203	...
αGd	25	atm	hP2	P6₃/mmc	A3	Mg	0.36336	...	0.57810	0.5910
βGd	>1235	atm	cI2	Im3m	A2	W	0.406	...	...	...
γGd	25	>3.0	hR3	R3m	...	αSm	0.361	...	2.603	4*1.60
αGe	25	atm	cF8	Fd3m	A4	C (diamond)	0.56574	...	...	...
βGe	25	>12	tI4	I4₁/amd	A5	βSn	0.4884	...	0.2692	0.551
γGe	25	>12→atm	tP12	P4₃2₁2	...	σGe	0.593	...	0.698	1.18
σGe	LT	>12	cI16	Im3m	...	γSi	0.692	...	...	...
αH	<-271.9	atm	cF4	Fm3m	A1	Cu	0.5338	...	...	...
βH	<-259.34	atm	hP2	P6₃/mmc	A3	Mg	0.3776	...	0.6162	1.632
αHe	<-268.94	atm	hP2	P6₃/mmc	A3	Mg	0.3555	...	0.5798	1.631
βHe	>-258	0.125	cF4	Fm3m	A1	Cu	0.4240	...	...	...
γHe	<-271.47	0.03	cI2	Im3m	A2	W	0.4110	...	...	...
αHf	25	atm	hP2	P6₃/mmc	A3	Mg	0.31946	...	0.50510	1.5811
βHf	>1995	atm	cI2	Im3m	A2	W	0.3610	...	...	...
αHg	<-38.84	atm	hR1	R3m	A10	αHg	0.3005	...	...	α = 70.53°
βHg	<-194	HP	tI2	I4/mmm	...	βHg	0.3995	...	0.2825	0.707
γHg	<-194	c.w.	hR1	?	...	...	...	...	...	...
αHo	25	atm	hP2	P6₃/mmc	A3	Mg	0.35778	...	0.56178	1.5702
βHo	25	>7.5	hR3	R3m	...	αSm	0.334	...	2.45	4.5*1.63
I	25	atm	oC8	Cmca	...	Cl	0.72697	0.47903	0.97942	...
In	25	atm	tI2	I4/mmm	A6	In	0.3253	...	0.49470	1.5210
Ir	25	atm	cF4	Fm3m	A1	Cu	0.38392	...	...	...
K	25	atm	cI2	Im3m	A2	W	0.5321	...	...	...
Kr	<-157.39	atm	cF4	Fm3m	A1	Cu	0.5810	...	...	...
αLa	25	atm	hP4	P6₃/mmc	A3′	αLa	0.37740	...	1.2171	2*1.6125
βLa	>310	atm	cF4	Fm3m	A1	Cu	0.5303	...	...	...
γLa	>865	atm	cI2	Im3m	A2	W	0.426	...	...	...
β′La	25	>2.0	cF4	Fm3m	A1	Cu	0.517	...	...	...

Element	Temperature, °C	Pressure, GPa	Pearson symbol	Space group	Strukturbericht designation	Prototype	Lattice parameters, nm			Comment, c/a or α or β
							a	b	c	
αLi	<-193	atm	hP2	P6₃/mmc	A3	Mg	0.3111	...	0.5093	1.637
βLi	25	atm	cI2	Im3m	A2	W	0.35093	...	...	...
γLi	<-201	c.w.	cF4	Fm3m	A1	Cu	0.4388	...	...	...
Lu	25	atm	hP2	P6₃/mmc	A3	Mg	0.35052	...	0.55494	1.5832
Mg	25	atm	hP2	P6₃/mmc	A3	Mg	0.32094	...	0.52107	1.6236
αMn	25	atm	cI58	I43m	A12	αMn	0.89126	...	...	...
βMn	>710	atm	cP20	P4₁32	A13	βMn	0.63152	...	...	...
γMn	>1079	atm	cF4	Fm3m	A1	Cu	0.3860	...	...	...
σMn	>1143	atm	cI2	Im3m	A2	W	0.3080	...	...	...
Mo	25	atm	cI2	Im3m	A2	W	0.31470	...	...	...
αN	<-237.6	atm	cP8	Pa3	...	αN	0.5661	...	...	...
βN	<-210.00	atm	hP4	P6₃/mmc	...	βN	0.4050	...	0.6604	1.631
γN	<-253	>3.3	tP4	P4₂/mnm	...	γN	0.3957	...	0.5109	1.291
αNa	<-233	atm	hP2	P6₃/mmc	A3	Mg	0.3767	...	0.6154	1.634
βNa	25	atm	cI2	Im3m	A2	W	0.42906	...	...	...
Nb	25	atm	cI2	Im3m	A2	W	0.33004	...	...	...
αNd	25	atm	hP4	P6₃/mmc	A3′	αLa	0.36582	...	1.17966	2*1.6124
βNd	>863	atm	cI2	Im3m	A2	W	0.413	...	...	...
γNd	25	>5.0	cF4	Fm3m	A1	Cu	0.480	...	...	...
Ne	<-243.59	atm	cF4	Fm3m	A1	Cu	0.4462	...	...	...
Ni	25	atm	cF4	Fm3m	A1	Cu	0.35240	...	...	...
αNp	25	atm	oP8	Pnma	A_c	αNp	0.6663	0.4723	0.4887	...
βNp	>280	atm	tP4	P42₁2	A_d	βNp	0.4883	...	0.3389	0.694
γNp	>576	atm	cI2	Im3m	A2	W	0.352	...	...	...
αO	<-243.3	atm	mC4	C2m	...	αO	0.5403	0.3429	0.5086	β = 132.53°
βO	<-229.6	atm	hR2	R3m	...	βO	0.4210	...	...	α = 46.27°
γO	<-218.79	atm	cP16	Pm3n	...	γO	0.683	...	...	...
Os	25	atm	hP2	P6₃/mmc	A3	Mg	0.27341	...	0.43918	1.6063
P (black)	25	atm	oC8	Cmca	...	P (black)	0.33136	1.0478	0.43763	...
αPa	25	atm	tI2	I4/mmm	A_a	αPa	0.3921	...	0.3235	0.825
βPa	>1170	atm	cI2	Im3m	A2	W	0.381	...	...	...
αPb	25	atm	cF4	Fm3m	A1	Cu	0.49502	...	...	...
βPb	25	>10.3	hP2	P6₃/mmc	A3	Mg	0.3265	...	0.5387	1.650
Pd	25	atm	cF4	Fm3m	A1	Cu	0.38903	...	...	...
αPm	25	atm	hP4	P6₃/mmc	A3′	αLa	0.365	...	1.165	2*1.60
βPm	>890	atm	cI2	Im3m	A2	W	(0.410)	...	...	...
αPo	25	atm	cP1	Pm3m	A_h	αPo	0.3366	...	...	...
βPo	>54	atm	hR1	R3m	...	βPo	0.3373	...	...	α = 98.08°
αPr	25	atm	hP4	P6₃/mmc	A3′	αLa	0.36721	...	1.18326	2*1.6111
βPr	>795	atm	cI2	Im3m	A2	W	0.413	...	...	...
γPr	25	>4.0	cF4	Fm3m	A1	Cu	0.488	...	...	...
Pt	25	atm	cF4	Fm3m	A1	Cu	0.39236	...	...	...
αPu	25	atm	mP16	P2₁/m	...	αPu	0.6183	0.4822	1.0963	β = 101.97°
βPu	>125	atm	mI34	I2/m	...	βPu	0.9284	1.0463	0.7859	β = 92.13°
γPu	>215	atm	oF8	Fddd	...	γPu	0.31587	0.57682	1.0162	...
σPu	>320	atm	cF4	Fm3m	A1	Cu	0.46371	...	...	...
σ′Pu	>463	atm	tI2	I4/mmm	A6	In	0.33261	...	0.44630	1.3418
εPu	>483	atm	cI2	Im3m	A2	W	0.36343	...	...	...
Ra	25	atm	cI2	Im3m	A2	W	0.5148	...	...	...
αRb	25	atm	cI2	Im3m	A2	W	0.5705	...	...	...
βRb	25	>1.08	?	...	...	...	...	...	...	...
γRb	25	>2.05	?	...	...	...	...	...	...	...
Re	25	atm	hP2	P6₃/mmc	A3	Mg	0.27609	...	0.4458	1.6145
Rh	25	atm	cF4	Fm3m	A1	Cu	0.38032	...	...	...
Ru	25	atm	hP2	P6₂/mmc	A3	Mg	0.27058	...	0.42816	1.5824
αS	25	atm	oF128	Fddd	A16	αS	1.0464	1.28660	2.44860	...
αSb	25	atm	hR2	R3m	A7	αAs	0.45067	...	...	α = 57.11°
βSb	25	>5.0	cP1	Pm3m	A_h	αPo	0.2992	...	...	...
γSb	25	>7.5	hP2	P6₃/mmc	A3	Mg	0.3376	...	0.5341	1.582
σSb	25	>14.0	mP3	?	...	...	0.556	0.404	0.422	β = 86.0°

Crystal Structures and Lattice Parameters of Allotropes of the Elements

Element	Temperature, °C	Pressure, GPa	Pearson symbol	Space group	Strukturbericht designation	Prototype	Lattice parameters, nm			Comment, c/a or α or β
							a	b	c	
αSc	25	atm	hP2	$P6_3/mm$	A3	Mg	0.33088	...	0.52680	1.5921
βSc	>1337	atm	cI2	Im3m	A2	W	(0.373)	...	...	...
γSe	25	atm	hP3	$P3_121$	A8	γSe	0.43659	...	0.49537	1.1346
αSi	25	atm	cF8	Fd3m	A4	C (diamond)	0.54306	...	...	...
βSi	25	>9.5	tI4	$I4_1/amd$	A5	βSn	0.4686	...	0.2585	0.552
γSi	25	>16.0	cI16	Im3m	...	γSi	0.6636	...	...	...
σSi	25	>16→atm	hP4	$P6_3/mmc$	A3′	αLa	0.380	...	0.628	1.653
αSm	25	atm	hR3	R3m	...	αSm	0.36290	...	2.6207	4*1.6048
βSm	>734	atm	hP2	$P6_3/mmc$	A3	Mg	0.36630	...	0.58448	1.5956
γ′Sm	>922	atm	cI2	Im3m	A2	W	(0.410)	...	...	...
σSm	25	>4.0	hP4	$P6_3/mmc$	A3′	αLa	0.3618	...	1.166	2*1.611
αSn	<13	atm	cF8	Fd3m	A4	C (diamond)	0.64892	...	...	...
βSn	25	atm	tI4	$I4_1/amd$	A5	βSn	0.58318	...	0.31818	0.5456
γSn	25	>9.0	tI2	?	...	γSn	0.370	...	0.337	0.91
αSr	25	atm	cF4	Fm3m	A1	Cu	0.6084	...	...	...
βSr	>547	atm	cI2	Im3m	A2	W	0.487	...	...	...
β′Sr	25	>3.5	cI2	Im3m	A2	W	0.4437	...	...	...
Ta	25	atm	cI2	Im3m	A2	W	0.33030	...	...	...
α′Tb	<-53	atm	oC4	Cmcm	...	α′Dy	0.3605	0.6244	0.5706	...
aTb	25	atm	hP2	$P6_3/mmc$	A3	Mg	0.36055	...	0.56966	1.5800
βTb	>1289	atm	cI2	Im3m	A2	W	(0.407)	...	...	...
γTb	25	>6.0	hR3	R3m	...	αSm	0.341	...	2.45	4*1.60
Tc	25	atm	hP2	$P6_3/mmc$	A3	Mg	0.2738	...	0.4393	1.604
αTe	25	atm	hP3	$P3_121$	A8	γSe	0.44566	...	0.59264	1.3298
βTe	25	>2.0	hR2	R3m	A7	αAs	0.469	...	...	α = 53.30°
γTe	25	>7.0	hR1	R3m	...	βPo	0.3002	...	...	α = 103.3°
αTh	25	atm	cF4	Fm3m	A1	Cu	0.50842	...	...	...
βTh	>1360	atm	cI2	Im3m	A2	W	0.411	...	...	...
αTi	25	atm	hP2	$P6_3/mmc$	A3	Mg	0.29506	...	0.46835	1.59873
βTi	>882	atm	cI2	Im3m	A2	W	0.33065	...	...	...
ωTi	25	HP→atm	hP3	P6/mmm	...	ωTi	0.4625	...	0.2813	0.6082
αTl	25	atm	hP2	$P6_3/mmc$	A3	Mg	0.34566	...	0.55248	1.5983
βTl	>230	atm	cI2	Im3m	A2	W	0.3879	...	...	...
γTl	25	HP	cF4	Fm3m	A1	Cu	?	...	...	...
Tm	25	atm	hP2	$P6_3/mmc$	A3	Mg	0.35375	...	0.55540	1.5700
αU	25	atm	oC4	Cmcm	A20	αU	0.28537	0.58695	0.49548	...
βU	>668	atm	tP30	$P4_2/mnm$	A_b	βU	1.0759	...	0.5656	0.526
γU	>776	atm	cI2	Im3m	A2	W	0.3524	...	...	...
V	25	atm	cI2	Im3m	A2	W	0.30240	...	...	...
W	25	atm	cI2	Im3m	A2	W	0.31652	...	...	...
Xe	<-111.76	atm	cF4	Fm3m	A1	Cu	0.6350	...	...	...
αY	25	atm	hP2	$P6_3/mmc$	A3	Mg	0.36482	...	0.57318	1.5711
βY	>1478	atm	cI2	Im3m	A2	W	(0.410)	...	...	...
αYb	<-3	atm	hP2	$P6_3/mmc$	A3	Mg	0.38799	...	0.63859	1.6459
βYb	25	atm	cF4	Fm3m	A1	Cu	0.54848	...	...	...
γYb	>795	atm	cI2	Im3m	A2	W	0.444	...	...	...
Zn	25	atm	hP2	$P6_3/mmc$	A3	Mg	0.26650	...	0.49470	1.8563
αZr	25	atm	hP2	$P6_3/mmc$	A3	Mg	0.32316	...	0.51475	1.5929
βZr	>863	atm	cI2	Im3m	A2	W	0.36090	...	...	...
ωZr	25	HP→atm	hP2	P6/mmm	...	ωTi	0.5036	...	0.3109	0.617

PHASE TRANSITIONS IN THE SOLID ELEMENTS AT ATMOSPHERIC PRESSURE

This table gives the phase transition temperatures for the elements that can exist in two or more crystalline forms (allotropes). The crystal phases are labeled by Greek letters in the most common conventions, although some variation is found. All data refer to normal atmospheric pressure.

References

1. Massalski, T. B., Ed., *Binary Alloy Phase Diagrams, Second Edition*, ASM International, Metals Park, OH, 1990.
2. Cordfunke, E. H. P., and Konings, R. J. M., Eds., *Thermochemical Data for Reactor Materials and Fission Products*, North-Holland, Amsterdam, 1990.
3. Greenwood, N. N., and Earnshaw, A., *Chemistry of the Elements, Second Edition*, Butterworth-Heinemann, Oxford, 1997.
4. Rhyne, J. J., "Magnetic Phase Transitions of the Elements", *Bull. Alloy Phase Diag.* 3, 402, 1982.

Element	Symbol	Transition	$t/°C$	Comments
Americium	Am	$\alpha \to \beta$	769	
	Am	$\beta \to \gamma$	1077	
	Am	$\gamma \to$ liq	1176	
Beryllium	Be	$\alpha \to \beta$	1270	
	Be	$\beta \to$ liq	1287	
Boron	B	$\alpha \to \beta$	1100	
	B	$\beta \to \gamma$	1500	
	B	$\gamma \to$ liq	2075	
Calcium	Ca	$\alpha \to \beta$	443	
	Ca	$\beta \to$ liq	842	
Californium	Cf	$\alpha \to \beta$	590	
	Cf	$\beta \to$ liq	900	
Cerium	Ce	$\alpha \to \beta$	−177	
	Ce	$\beta \to \gamma$	61	β-Ce and γ-Ce are magnetic
	Ce	$\gamma \to \delta$	726	
	Ce	$\delta \to$ liq	799	
Cobalt	Co	$\varepsilon \to \alpha$	422	magnetic transition at 1115 °C
	Co	$\alpha \to$ liq	1495	
Curium	Cm	$\alpha \to \beta$	1277	magnetic transition at −221 °C
	Cm	$\beta \to$ liq	1345	
Dysprosium	Dy	$\alpha' \to \alpha$	−187	
	Dy	$\alpha \to \beta$	1381	magnetic transitions in α-Dy at −184 °C and −94 °C
	Dy	$\beta \to$ liq	1412	
Fluorine	F_2	$\alpha \to \beta$	−227.60	
	F_2	$\beta \to$ liq	−219.67	
Gadolinium	Gd	$\alpha \to \beta$	1235	
	Gd	$\beta \to$ liq	1313	
Hafnium	Hf	$\alpha \to \beta$	1743	
	Hf	$\beta \to$ liq	2233	
Iron	Fe	$\alpha \to \gamma$	912	magnetic transition in α-Fe at 771 °C
	Fe	$\gamma \to \delta$	1394	
	Fe	$\delta \to$ liq	1538	
Lanthanum	La	$\alpha \to \beta$	277	
	La	$\beta \to \gamma$	860	
	La	$\gamma \to$ liq	920	
Lithium	Li	$\alpha \to \beta$	−193	
	Li	$\beta \to$ liq	180.50	
Manganese	Mn	$\alpha \to \beta$	727	magnetic transition in α-Mn at −100 °C
	Mn	$\beta \to \gamma$	1100	
	Mn	$\gamma \to \delta$	1138	
	Mn	$\delta \to$ liq	1246	
Neodymium	Nd	$\alpha \to \beta$	855	magnetic transition in α-Nd at −253 °C
	Nd	$\beta \to$ liq	1016	
Neptunium	Np	$\alpha \to \beta$	280	
	Np	$\beta \to \gamma$	576	
	Np	$\gamma \to$ liq	644	

Element	Symbol	Transition	$t/°C$	Comments
Nitrogen	N_2	α→β	−237.54	
	N_2	β→liq	−210.0	
Oxygen	O_2	α→β	−249.29	
	O_2	β→γ	−229.35	
	O_2	γ→liq	−218.79	
Phosphorus	P	brown→β-white	−190	several amorphous phases (red, black, gray) exist (Ref. 3)
	P	β-white→α-white	−76.9	
	P	α-white→liq	44.15	
Plutonium	Pu	α→β	124.5	
	Pu	β→γ	214.8	
	Pu	γ→δ	320.0	
	Pu	δ→δ'	462.9	
	Pu	δ'→ε	482.6	
	Pu	ε→liq	640	
Polonium	Po	α→β	54	
	Po	β→liq	254	
Praseodymium	Pr	α→β	795	
	Pr	β→liq	931	
Promethium	Pm	α→β	890	magnetic transition in α-Pm at −175 °C
	Pm	β→liq	1042	
Protactinium	Pa	α→β	1170	
	Pa	β→liq	1572	
Samarium	Sm	α→β	734	magnetic transition in α-Sm at −167 °C
	Sm	β→γ	922	
	Sm	γ→liq	1072	
Scandium	Sc	α→β	1337	
	Sc	β→liq	1541	
Selenium	Se	α-red→gray	180	many allotropes exist (Ref. 3)
	Se	gray→liq	220.8	
Sodium	Na	α→β	−233	
	Na	β→liq	97.794	
Strontium	Sr	α→β	547	
	Sr	β→liq	777	
Sulfur	S	α→β	95.3	many allotropes exist (Ref. 3)
	S	β→liq	115.21	
Terbium	Tb	α'→α	−53	
	Tb	α→β	1289	magnetic transition in α-Tb at −230 °C
	Tb	β→liq	1359	
Thallium	Tl	α→β	230	
	Tl	β→liq	304	
Thorium	Th	α→β	1360	
	Th	β→liq	1750	
Tin	Sn	α (gray)→β (white)	13.2	
	Sn	β (white)→liq	231.928	defining fixed point on ITS-90
Titanium	Ti	α→β	882	
	Ti	β→liq	1668	
Uranium	U	α→β	669	
	U	β→γ	776	
	U	γ→liq	1135	
Ytterbium	Yb	α→β	3	
	Yb	β→γ	795	
	Yb	γ→liq	824	
Yttrium	Y	α→β	1478	
	Y	β→liq	1522	
Zirconium	Zr	α→β	866	
	Zr	β→liq	1854.7	

LATTICE ENERGIES

H. D. B. Jenkins and H. K. Roobottom

THERMOCHEMICAL CYCLE AND CALCULATED VALUES

Table 1 contains calculated values of the lattice energies (total lattice potential energies), U_{POT}, of crystalline salts, M_aX_b. U_{POT} is expressed in units of kilojoules per mole, kJ mol^{-1}. M and X can be either simple or complex ions. Substances are arranged by chemical class.

Also listed in the table is the lattice energy, $U_{POT}{}^{BFHC}$, obtained from the application of the Born–Fajans–Haber cycle (BHFC) described below, using the "Standard Thermochemical Properties of Chemical Substances" table in Section 5 of this *Handbook*, References 1 through 4, and certain other data that are given in Table 3 below.

The lattice enthalpy, ΔH_L, is given by the cycle:

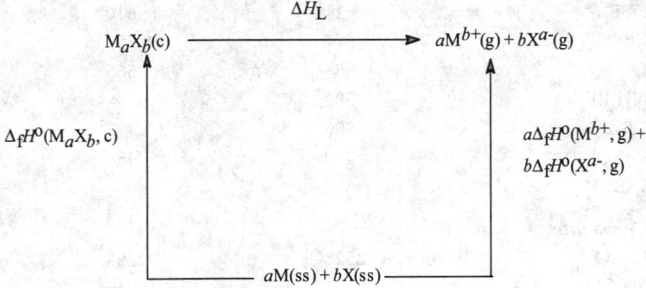

where (ss) is the standard state of the element concerned.

The lattice enthalpy, ΔH_L, is obtained using the equation:

$$\Delta H_L = a\Delta_f H°(M^{b+}, g) + b\Delta_f H°(X^{a-}, g) - \Delta_f H°(M_aX_b, c)$$

and is further related to the total lattice potential energy, U_{POT}, by the relationship:

$$\Delta H_L = U_{POT} + \left[a\left(\frac{n_M}{2} - 2\right) + b\left(\frac{n_X}{2} - 2\right)\right] RT$$

where n_M and n_X equal 3 for monatomic ions, 5 for linear polyatomic ions and 6 for polyatomic non-linear ions.

METHOD OF ESTIMATION OF VALUES NOT TABULATED

In cases where the lattice energy is not tabulated and we want to furnish an estimate, then the Kapustinskii equation (Ref. 5) can be used to obtain a value (in kJ mol^{-1}):

$$U_{POT} = \frac{121.4 z_a z_b \nu}{(r_a + r_b)}\left(1 - \frac{0.0345}{(r_a + r_b)}\right)$$

where z_a and z_b are the moduli of the charges on the ν ions in the lattice and r_a and r_b (in nm) are the thermochemical radii given in Table 2. The r_a for metal ions is taken to be the Goldschmidt radius (Ref. 6).

To cite an example, if we wish to estimate the lattice energy of the salt $[NH_4^+][HF_2^-]$ using the above procedure, we see that Table 2 gives the thermochemical radius (r_a) for NH_4^+ to be 0.136 nm and that for HF_2^- (r_b) to be 0.172 nm. The lattice potential energy is then estimated to be 700 kJ mol^{-1} compared with the calculated value of 705 kJ mol^{-1} and the Born–Fajans–Haber cycle value of 658 kJ mol^{-1}.

References

1. Wagman, D. D., Evans, W. H., Parker, V. B., Schumm, R. H., Halow, I., Bailey, S. M., Churney, K. L., and Nuttall, R. L., The NBS Tables of Chemical Thermodynamic Properties, *J. Phys. Chem. Ref. Data*, Vol. 11, Suppl. 2, 1982.
2. Chase, M. W., Davies, C. A., Downey, J. R., Frurip, D. J., McDonald, R. A., and Syverud, A. N., JANAF Thermochemical Tables, Third Edition, *J. Phys. Chem. Ref. Data*, Vol. 14, Suppl. 1, 1985.
3. Lias, S. G., Bartmess, J. E., Liebman, J. F., Holmes, J. L., Levin, R. D., and Mallard, W. G., Gas-Phase Ion and Neutral Thermochemistry, *J. Phys. Chem. Ref. Data*, Vol. 17, Suppl. 1, 1988.
4. Jenkins, H. D. B., and Pratt, K. F., *Adv. Inorg. Chem. Radiochem.* 22, 1, 1978.
5. Kapustinskii, A. F., *Quart. Rev.* 10, 283, 1956.
6. Goldschmidt, V. M., *Skrifter Norske Videnskaps-Akad.* Oslo, I, Mat.-Naturn. Kl, 1926. See also Dasent, W. E., *Inorganic Energetics*, 2nd ed., Cambridge University Press, 1982.

Table 1. Lattice Energies (kJ mol⁻¹)

Substance	Calc. U_{POT}	U_{POT}^{BHFC}
Acetates		
Li(CH₃COO)	–	843
Na(CH₃COO)	828	807
K(CH₃COO)	749	726
Rb(CH₃COO)	715	
Cs(CH₃COO)	682	
Acetylides		
CaC₂	2911	2902
SrC₂	2788	2782
BaC₂	2647	2652
Azides		
LiN₃	861	875
NaN₃	770	784
KN₃	697	
RbN₃	674	691
CsN₃	665	674
AgN₃	854	910
TlN₃	689	742
Ca(N₃)₂	2186	2316
Sr(N₃)₂	2056	2187
Ba(N₃)₂	2021	–
Mn(N₃)₂	2408	2348
Cu(N₃)₂	2730	2738
Zn(N₃)₂	2840	2970
Cd(N₃)₂	2446	2576
Pb(N₃)₂		2300
Bihalide Salts		
LiHF₂	821	847
NaHF₂	755	748
KHF₂	657	660
RbHF₂	627	631
CsHF₂	607	
NH₄HF₂	705	658
CsHCl₂	601	–
Me₄NHCl₂	427	–
Et₄NHCl₂	346	–
Bu₄NHCl₂	290	–
Bicarbonates		
NaHCO₃	820	656
KHCO₃	741	573
RbHCO₃	707	522
CsHCO₃	678	520
NH₄HCO₃	–	577
Borides		
CaB₆	5146	–
SrB₆	5104	–
BaB₆	5021	
YB₆	7447	
LaB₆	7406	
CeB₆	10083	
PrB₆	7447	
NdB₆	7447	
PmB₆	7406	
SmB₆	7447	
EuB₆	5104	
GdB₆	7489	
TbB₆	7489	
DyB₆	7489	

Substance	Calc. U_{POT}	U_{POT}^{BHFC}
HoB₆	7489	
ErB₆	7489	
TmB₆	7489	
YbB₆	5146	
LuB₆	7489	
ThB₆	10167	
Borohydrides		
LiBH₄	778	
NaBH₄	703	694
KBH₄	655	638
RbBH₄	648	
CsBH₄	628	
Borohalides		
LiBF₄	699	749
NaBF₄	657	674
KBF₄	611	616
RbBF₄	577	590
CsBF₄	556	565
NH₄BF₄	582	
KBCl₄	506	497
RbBCl₄	489	486
CsBCl₄	473	
Carbonates		
Li₂CO₃	2523	2254
Na₂CO₃	2301	2016
K₂CO₃	2084	1846
Rb₂CO₃	2000	1783
Cs₂CO₃	1920	1722
MgCO₃	3138	3122
CaCO₃	2804	2811
SrCO₃	2720	2688
BaCO₃	2615	2554
MnCO₃	3046	3092
FeCO₃	3121	3169
CoCO₃	3443	3235
CuCO₃	3494	–
ZnCO₃	3121	3273
CdCO₃	2929	3052
SnCO₃	2904	
PbCO₃	2728	2750
Cyanates		
LiNCO	849	–
NaNCO	807	816
KNCO	726	734
RbNCO	692	
CsNCO	661	
NH₄NCO	724	
Cyanides		
LiCN	874	
NaCN	766	759
KCN	692	686
RbCN	638	–
CsCN	601	–
Ca(CN)₂	2268	2240
Sr(CN)₂	2138	
Ba(CN)₂	2001	2009
NH₄CN	617	691
AgCN	(741)	935

Substance	Calc. U_{POT}	U_{POT}^{BHFC}	Substance	Calc. U_{POT}	U_{POT}^{BHFC}
$Zn(CN)_2$	2809	2817	AgI	881	892
$Cd(CN)_2$	2583	2591	AuCl	1013	1066
Formates			AuBr	1029	1059
$Li(HCO_2)$	865	–	AuI	1027	1070
$Na(HCO_2)$	791	804	InCl	–	764
$K(HCO_2)$	713	722	InBr	–	767
$Rb(HCO_2)$	685	–	InI	–	733
$Cs(HCO_2)$	651	–	TlF	–	850
$NH_4(HCO_2)$	715	–	TlCl	738	751
			TlBr	720	734
Germanates			TlI	692	710
Mg_2GeO_4	7991	–	Me_4NCl	566	–
Ca_2GeO_4	7301	7306	Me_4NBr	553	–
Sr_2GeO_4	6987	–	Me_4NI	544	–
Ba_2GeO_4	6653	6643	PH_4Br	616	–
			PH_4I	590	–
Halates			BeF_2	3464	3526
$LiBrO_3$	883	880	$BeCl_2$	3004	3033
$NaBrO_3$	803	791	$BeBr_2$	2950	2914
$KBrO_3$	740	722	BeI_2	2780	2813
$RbBrO_3$	720	705	MgF_2	2926	2978
$CsBrO_3$	694	681	$MgCl_2$	2477	2540
$NaClO_3$	770	785	$MgBr_2$	2406	2451
$KClO_3$	711	721	MgI_2	2293	2340
$RbClO_3$	690	703	CaF_2	2640	2651
$CsClO_3$	–	679	$CaCl_2$	2268	2271
$LiIO_3$	975	974	$CaBr_2$	2132	–
$NaIO_3$	883	876	CaI_2	1971	2087
KIO_3	820	780	SrF_2	2476	2513
$RbIO_3$	791	–	$SrCl_2$	2142	2170
$CsIO_3$	761	–	SrI_2	1984	1976
			BaF_2	2347	2373
Halides			$BaCl_2$	2046	2069
LiF	1030	1049	$BaBr_2$	1971	1995
LiCl	834	864	BaI_2	1862	1890
LiBr	788	820	RaF_2	2284	–
LiI	730	764	$RaCl_2$	2004	–
NaF	910	930	$RaBr_2$	1929	–
NaCl	769	790	RaI_2	1803	–
NaBr	732	754	$ScCl_2$	2380	–
NaI	682	705	$ScBr_2$	2291	–
KF	808	829	ScI_2	2201	–
KCl	701	720	TiF_2	2724	–
KBr	671	691	$TiCl_2$	2439	2514
KI	632	650	$TiBr_2$	2360	2430
RbF	774	795	TiI_2	2259	2342
RbCl	680	695	VCl_2	2607	2593
RbBr	651	668	VBr_2	–	2534
RbI	617	632	VI_2	–	2470
CsF	744	759	CrF_2	2778	2939
CsCl	657	670	$CrCl_2$	2540	2601
CsBr	632	647	$CrBr_2$	2377	2536
CsI	600	613	CrI_2	2269	2440
FrF	715	–	$MoCl_2$	2737	2746
FrCl	632	–	$MoBr_2$	2742	2753
FrBr	611	–	MoI_2	2630	–
FrI	582	–	MnF_2	2644	–
CuCl	992	996	$MnCl_2$	2510	2551
CuBr	969	978	$MnBr_2$	2448	2482
CuI	948	966	MnI_2	2212	–
AgF	953	974	FeF_2	2849	2967
AgCl	910	918			
AgBr	897	905			

Substance	Calc. U_{POT}	U_{POT}^{BHFC}	Substance	Calc. U_{POT}	U_{POT}^{BHFC}
$FeCl_2$	2569	2641	$CrCl_3$	5518	5529
$FeBr_2$	2515	2577	$CrBr_3$	5355	–
FeI_2	2439	2491	CrI_3	5275	5294
CoF_2	3004	3042	MoF_3	6459	–
$CoCl_2$	2707	2706	$MoCl_3$	5246	5253
$CoBr_2$	2640	2643	$MoBr_3$	5156	–
CoI_2	2569	2561	MoI_3	5073	–
NiF_2	3098	3089	MnF_3	6017	–
$NiCl_2$	2753	2786	$MnCl_3$	5544	–
$NiBr_2$	2729	2721	$MnBr_3$	5448	–
NiI_2	2607	2637	MnI_3	5330	–
$PdCl_2$	2778	2818	$TcCl_3$	5270	–
$PdBr_2$	2741	2751	$TcBr_3$	5215	–
PdI_2	2748	2760	TcI_3	5188	–
CuF_2	3046	3102	FeF_3	5870	–
$CuCl_2$	2774	2824	$FeCl_3$	5364	5436
$CuBr_2$	2715	2774	$FeBr_3$	5333	5347
CuI_2	2640	–	FeI_3	5117	–
AgF_2	2942	2967	$RuCl_3$	5245	5257
ZnF_2	3021	3053	$RuBr_3$	5223	5232
$ZnCl_2$	2703	2748	RuI_3	5222	5235
$ZnBr_2$	2648	2689	CoF_3	5991	–
ZnI_2	2581	2619	$RhCl_3$	5641	5665
CdF_2	2809	2830	IrF_3	(6112)	–
$CdCl_2$	2552	2565	$IrBr_3$	(4794)	–
$CdBr_2$	2507	2517	NiF_3	(6111)	–
CdI_2	2441	2455	AuF_3	(5777)	–
HgF_2	2757	–	$AuCl_3$	(4605)	–
$HgCl_2$	2657	2664	$ZnCl_3$	5832	–
$HgBr_2$	2628	2639	$ZnBr_3$	5732	–
HgI_2	2628	2624	ZnI_3	5636	–
SnF_2	2551	–	AlF_3	5924	6252
$SnCl_2$	2297	2310	$AlCl_3$	5376	5513
$SnBr_2$	2251	2256	$AlBr_3$	5247	5360
SnI_2	2193	2206	AlI_3	5070	5227
PbF_2	2535	2543	GaF_3	5829	6238
$PbCl_2$	2270	2282	$GaCl_3$	5217	5665
$PbBr_2$	2219	2230	$GaBr_3$	4966	5569
PbI_2	2163	2177	GaI_3	4611	5496
ScF_3	5492	5540	$InCl_3$	4736	5183
$ScCl_3$	4874	4901	$InBr_3$	4535	5117
$ScBr_3$	4729	4761	InI_3	4234	5001
ScI_3	4640	–	TlF_3	5493	–
YF_3	4983	–	$TlCl_3$	5258	5278
YCl_3	4506	4524	$TlBr_3$	5171	–
YI_3	4240	4258	TlI_3	5088	–
TiF_3	5644	–	$AsBr_3$	5497	5365
$TiCl_3$	5134	5153	AsI_3	4824	5295
$TiBr_3$	5012	5023	SbF_3	5295	5324
TiI_3	4845	–	$SbCl_3$	5032	4857
$ZrCl_3$	–	4791	$SbBr_3$	4954	4776
$ZrBr_3$	–	4758	SbI_3	4867	4692
ZrI_3	–	4591	$BiCl_3$	4689	4707
VF_3	5895	–	BiI_3	3774	–
VCl_3	5322	5329	LaF_3	4682	–
VBr_3	5214	5224	$LaCl_3$	4263	4242
VI_3	5121	5136	$LaBr_3$	4209	–
$NbCl_3$	5062	–	LaI_3	3916	3986
$NbBr_3$	4980	–	$CeCl_3$	4394	4348
NbI_3	4860	–	CeI_3	–	4061
CrF_3	6033	6065	$PrCl_3$	4322	4387

Substance	Calc. U_{POT}	U_{POT}^{BHFC}	Substance	Calc. U_{POT}	U_{POT}^{BHFC}
PrI_3	–	4101	RbH	686	684
$NdCl_3$	4343	4415	CsH	648	653
$SmCl_3$	4376	4450	VH	1184	(1344)
$EuCl_3$	4393	4490	NbH	1163	(1633)
$GdCl_3$	4406	4495	PdH	979	1368
$DyCl_3$	4481	4529	CuH	828	1254
$HoCl_3$	4501	4572	TiH	996	1407
$ErCl_3$	4527	4591	ZrH	916	1590
$TmCl_3$	4548	4608	HfH	904	–
TmI_3	–	4340	LaH	828	–
$YbCl_3$	–	4651	TaH	1021	–
$AcCl_3$	4096	–	CrH	1050	–
UCl_3	4243	–	NiH	929	–
$NpCl_3$	4268	–	PtH	937	–
$PuCl_3$	4289	–	AgH	941	–
$PuBr_3$	(3959)	–	AuH	1033	1108
$AmCl_3$	4293	–	TlH	745	–
TiF_4	10012	9908	GeH	950	–
$TiCl_4$	9431	–	PbH	778	–
$TiBr_4$	9288	9059	BeH_2	3205	3306
TiI_4	9108	8918	MgH_2	2791	2718
ZrF_4	8853	8971	CaH_2	2410	2406
$ZrCl_4$	8021	8144	SrH_2	2250	2265
$ZrBr_4$	7661	7984	BaH_2	2121	2133
ZrI_4	7155	7801	ScH_2	2711	2744
MoF_4	8795	–	YH_2	(2598)	2733
$MoCl_4$	8556	9603	LaH_2	2380	2522
$MoBr_4$	8510	9500	CeH_2	2414	2509
MoI_4	8427	–	PrH_2	2448	2405
$SnCl_4$	8355	8930	NdH_2	2464	2394
$SnBr_4$	7970	8852	PmH_2	2519	–
PbF_4	9519	–	SmH_2	2510	2389
CrF_2Cl	5795	–	GdH_2	2494	2651
CrF_2Br	5753	–	AcH_2	2372	–
CrF_2I	5669	–	ThH_2	2711	2738
$CrCl_2Br$	5448	–	PuH_2	2519	–
$CrCl_2I$	5381	5429	AmH_2	2544	–
$CrBr_2I$	5330	5370	TiH_2	2866	2864
CuFCl	2891	–	ZrH_2	2711	2999
CuFBr	2853	–	CuH_2	2941	–
CuFI	2803	–	ZnH_2	2870	–
CuClBr	2753	–	HgH_2	2707	–
CuClI	2694	–	AlH_3	5924	5969
CuBrI	2669	–	FeH_3	5724	–
FeF_2Cl	5711	–	ScH_3	5439	–
FeF_2Br	5653	–	YH_3	5063	4910
FeF_2I	5569	–	LaH_3	4895	4493
$FeCl_2Br$	5339	–	FeH_3	5724	–
$FeCl_2I$	5272	–	GaH_3	5690	–
$FeBr_2I$	5209	–	InH_3	5092	–
$LiIO_2F_2$	845	–	TlH_3	5092	–
$NaIO_2F_2$	766	756			
KIO_2F_2	699	689	*Hydroselenides*		
$RbIO_2F_2$	674	–	NaHSe	703	732
$CsIO_2F_2$	636	–	KHSe	644	712
$NH_4IO_2F_2$	678	–	RbHSe	623	689
$AgIO_2F_2$	736	685	CsHse	598	669
			Hydrosulphides		
Hydrides			LiHS	768	862
LiH	916	918	NaHS	723	771
NaH	807	807	RbHS	655	682
KH	711	713			

Substance	Calc. U_{POT}	U_{POT}^{BHFC}	Substance	Calc. U_{POT}	U_{POT}^{BHFC}
CsHS	628	657	CsNO$_3$	648	650
NH$_4$HS	661	718	AgNO$_3$	820	832
Ca(HS)$_2$	2184	(2171)	TlNO$_3$	690	707
Sr(HS)$_2$	2063	–	Mg(NO$_3$)$_2$	2481	2521
Ba(HS)$_2$	1979	(1956)	Ca(NO$_3$)$_2$	2268	2247
			Sr(NO$_3$)$_2$	2176	2151
Hydroxides			Ba(NO$_3$)$_2$	2062	2035
LiOH	1021	1028	Mn(NO$_3$)$_2$	2318	2478
NaOH	887	892	Fe(NO$_3$)$_2$	–	(2580)
KOH	789	796	Co(NO$_3$)$_2$	2560	2647
RbOH	766	765	Ni(NO$_3$)$_2$	–	2729
CsOH	721	732	Cu(NO$_3$)$_2$	–	2739
Be(OH)$_2$	3477	3620	Zn(NO$_3$)$_2$	2376	2649
Mg(OH)$_2$	2870	2998	Cd(NO$_3$)$_2$	2238	2462
Ca(OH)$_2$	2506	2637	Sn(NO$_3$)$_2$	2155	2254
Sr(OH)$_2$	2330	2474	Pb(NO$_3$)$_2$	2067	2208
Ba(OH)$_2$	2142	2330			
Ti(OH)$_2$	–	2953	*Nitrides*		
Mn(OH)$_2$	2909	3008	ScN	7547	7506
Fe(OH)$_2$	2653	3044	LaN	6876	6793
Co(OH)$_2$	2786	3109	TiN	8130	8033
Ni(OH)$_2$	2832	3186	ZrN	7633	7723
Pd(OH)$_2$	–	3189	VN	8283	8233
Cu(OH)$_2$	2870	3229	NbN	7939	8022
CuOH	1006	–	CrN	8269	8358
AgOH	918	845			
AuOH	1033	–	*Nitrites*		
TlOH	705	874	NaNO$_2$	774	772
Zn(OH)$_2$	2795	3151	KNO$_2$	699	687
Cd(OH)$_2$	2607	2909	RbNO$_2$	724	765
Hg(OH)$_2$	2669	–	CsNO$_2$	690	–
Sn(OH)$_2$	2489	2721			
Pb(OH)$_2$	2376	–	*Oxides*		
Sc(OH)$_3$	5063	5602	Li$_2$O	2799	2814
Y(OH)$_3$	4707	–	Na$_2$O	2481	2478
La(OH)$_3$	4443	–	K$_2$O	2238	2232
Cr(OH)$_3$	5556	6299	Rb$_2$O	2163	2161
Mn(OH)$_3$	6213	–	Cs$_2$O	2131	2063
Al(OH)$_3$	5627	–	Cu$_2$O	3273	3189
Ga(OH)$_3$	5732	6368	Ag$_2$O	3002	2910
In(OH)$_3$	5280	–	Tl$_2$O	2659	2575
Tl(OH)$_3$	5314	–	LiO$_2$	(878)	(872)
Ti(OH)$_4$	9456	–	NaO$_2$	799	821
Zr(OH)$_4$	8619	–	KO$_2$	741	751
Mn(OH)$_4$	10933	–	RbO$_2$	706	721
Sn(OH)$_4$	9188	9879	CsO$_2$	679	696
			Li$_2$O$_2$	2592	2557
Imides			Na$_2$O$_2$	2309	22717
CaNH	3293	–	K$_2$O$_2$	2114	2064
SrNH	3146	–	Rb$_2$O$_2$	2025	1994
BaNH	2975	–	Cs$_2$O$_2$	1948	1512
			MgO$_2$	3356	3526
Metavanadates			CaO$_2$	3144	3132
Li$_3$VO$_4$	3945	–	SrO$_2$	3037	2977
Na$_3$VO$_4$	3766	–	KO$_3$	697	707
K$_3$VO$_4$	3376	–	BeO	4514	4443
Rb$_3$VO$_4$	3243	–	MgO	3795	3791
Cs$_3$VO$_4$	3137	–	CaO	3414	3401
			SrO	3217	3223
Nitrates			BaO	3029	3054
LiNO$_3$	848	854	TiO	3832	3811
NaNO$_3$	755	763	VO	3932	3863
KNO$_3$	685	694	MnO	3724	3745
RbNO$_3$	662	671			

Substance	Calc. U_{POT}	U_{POT}^{BHFC}	Substance	Calc. U_{POT}	U_{POT}^{BHFC}
FeO	3795	3865	$Ca(ClO_4)_2$	1958	1971
CoO	3837	3910	$Sr(ClO_4)_2$	1862	1862
NiO	3908	4010	$Ba(ClO_4)_2$	1795	1769
PdO	3736	–			
CuO	4135	4050	**Permanganates**		
ZnO	4142	3971	$NaMnO_4$	661	–
CdO	3806	–	$KMnO_4$	607	–
HgO	3907	–	$RbMnO_4$	586	–
GeO	3919	–	$CsMnO_4$	565	–
SnO	3652	–	$Ca(MnO_4)_2$	1937	–
PbO	3520	–	$Sr(MnO_4)_2$	1845	–
Sc_2O_3	13557	13708	$Ba(MnO_4)_2$	1778	–
Y_2O_3	12705	–			
La_2O_3	12452	–	**Phosphates**		
Ce_2O_3	12661	–	$Mg_3(PO_4)_2$	11632	11407
Pr_2O_3	12703	–	$Ca_3(PO_4)_2$	10602	10479
Nd_2O_3	12736	–	$Sr_3(PO_4)_2$	10125	10075
Pm_2O_3	12811	–	$Ba_3(PO_4)_2$	9652	9654
Sm_2O_3	12878	–	$MnPO_4$	7397	–
Eu_2O_3	12945	–	$FePO_4$	7251	7300
Gd_2O_3	12996	–	BPO_4	8201	–
Tb_2O_3	13071	–	$AlPO_4$	7427	7507
Dy_2O_3	13138	–	$GaPO_4$	7381	–
Ho_2O_3	13180	–			
Er_2O_3	13263	–	**Selenides**		
Tm_2O_3	13322	–	Li_2Se	2364	–
Yb_2O_3	13380	–	Na_2Se	2130	–
Lu_2O_3	13665	–	K_2Se	1933	–
Ac_2O_3	12573	–	Rb_2Se	1837	–
Ti_2O_3	–	14149	Cs_2Se	1745	–
V_2O_3	15096	14520	Ag_2Se	2686	–
Cr_2O_3	15276	14957	Tl_2Se	2209	–
Mn_2O_3	15146	15035	BeSe	3431	–
Fe_2O_3	14309	14774	MgSe	3071	–
Al_2O_3	15916	–	CaSe	2858	2862
Ga_2O_3	15590	15220	SrSe	2736	–
In_2O_3	13928	–	BaSe	2611	–
Pb_2O_3	(14841)	–	MnSe	3176	–
CeO_2	9627	–			
ThO_2	10397	–	**Selenites**		
PaO_2	10573	–	Li_2SeO_3	2171	–
$VO_2(g)$	10644	–	Na_2SeO_3	1950	1916
NpO_2	10707	–	K_2SeO_3	1774	1749
PuO_2	10786	–	Rb_2SeO_3	1715	1675
AmO_2	10799	–	Cs_2SeO_3	1640	–
CmO_2	10832	–	Tl_2SeO_3	1879	–
TiO_2	12150	–	Ag_2SeO_3	2113	2148
ZrO_2	11188	–	$BeSeO_3$	3322	–
MoO_2	11648	–	$MgSeO_3$	3012	2998
MnO_2	12970	–	$CaSeO_3$	2732	–
SiO_2	13125	–	$SrSeO_3$	2586	2588
GeO_2	12828	–	$BaSeO_3$	2460	2451
SnO_2	11807	–			
PbO_2	11217	–	**Selenates**		
			Li_2SeO_4	2054	–
Perchlorates			Na_2SeO_4	1879	–
$LiClO_4$	709	715	K_2SeO_4	1732	–
$NaClO_4$	643	641	Rb_2SeO_4	1686	–
$KClO_4$	599	595	Cs_2SeO_4	1615	–
$RbClO_4$	564	576	Cu_2SeO_4	2201	–
$CsClO_4$	636	550	Ag_2SeO_4	2033	–
NH_4ClO_4	583	580	Tl_2SeO_4	1766	–
			Hg_2SeO_4	2163	–
			$BeSeO_4$	3448	–

Substance	Calc. U_{POT}	U_{POT}^{BHFC}	Substance	Calc. U_{POT}	U_{POT}^{BHFC}
$MgSeO_4$	2895		Cs_2PdCl_6	1426	–
$CaSeO_4$	2632	–	Rb_2PbCl_6	1343	1343
$SrSeO_4$	2489	–	Cs_2PbCl_6	1344	–
			$(NH_4)_2PbCl_6$	1355	–
Sulphates			K_2PtCl_6	1468	1471
Li_2SO_4	2229	2142	Rb_2PtCl_6	1464	–
Na_2SO_4	1827	1938	Cs_2PtCl_6	1444	–
K_2SO_4	1700	1796	$(NH_4)_2PtCl_6$	1468	–
Rb_2SO_4	1636	1748	Tl_2PtCl_6	1546	–
Cs_2SO_4	1596	1658	Ag_2PtCl_6	1773	1881
$(NH_4)_2SO_4$	1766	1777	$BaPtCl_6$	2047	2070
Cu_2SO_4	2276	2166	K_2PtBr_6	1423	1392
Ag_2SO_4	2104	1989	Ag_2PtBr_6	1791	2276
Tl_2SO_4	1828	1722	K_2PtI_6	1421	–
Hg_2SO_4	–	2127	K_2ReCl_6	1416	1442
$CaSO_4$	2489	2480	Rb_2ReCl_6	1414	–
$SrSO_4$	2577	2484	Cs_2ReCl_6	1398	–
$BaSO_4$	2469	2374	K_2ReBr_6	1375	1375
$MnSO_4$	2920	2825	K_2SiF_6	1670	1765
			Rb_2SiF_6	1639	1673
Sulphides			Cs_2SiF_6	1604	1498
Li_2S	2464	2472	Tl_2SiF_6	1675	–
Na_2S	2192	2203	K_2SnCl_6	1363	1390
K_2S	1979	(2052)	Rb_2SnCl_6	1361	1363
Rb_2S	1929	1949	Cs_2SnCl_6	1358	–
Cs_2S	1892	1850	Tl_2SnCl_6	1437	–
$(NH_4)_2S$	2008	(2026)	$(NH_4)_2SnCl_6$	1370	1344
Cu_2S	2786	2865	Rb_2SnBr_6	1309	–
Ag_2S	2606	2677	Cs_2SnBr_6	1306	–
Au_2S	2908	–	Rb_2SnI_6	1226	–
Tl_2S	2298	2258	Cs_2SnBr_6	1243	–
			K_2TeCl_6	1318	1320
Ternary Salts			Rb_2TeCl_6	1321	–
Cs_2CuCl_4	1393	–	Cs_2TeCl_6	1323	–
Rb_2ZnCl_4	1529	–	Tl_2TeCl_6	1392	–
Cs_2ZnCl_4	1492	–	$(NH_4)_2TeCl_6$	1318	–
Rb_2ZnBr_4	1498	–	K_2RuCl_6	1451	–
Cs_2ZnBr_4	1454	–	Rb_2CoF_6	1688	–
Cs_2ZnI_4	1386	–	Cs_2CoF_6	1632	–
$CsGaCl_4$	494	–	K_2NiF_6	1721	–
$NaAlCl_4$	556	–	Rb_2NiF_6	1688	–
$CsAlCl_4$	486	–	Rb_2SbCl_6	1357	–
$NaFeCl_4$	492	–	Rb_2SeCl_6	1409	–
Rb_2CoCl_4	1447	–	Cs_2SeCl_6	1397	–
Cs_2CoCl_4	1391	–	$(NH_4)_2SeCl_6$	1420	–
K_2PtCl_4	1574	1550	$(NH_4)_2PoCl_6$	1338	–
Cs_2GeF_6	1573	–	Cs_2PoBr_6	1286	–
$(NH_4)_2GeF_6$	1657	–	Cs_2CrF_6	1603	–
Cs_2GeCl_6	1404	1419	Rb_2MnF_6	1688	–
K_2HfCl_6	1345	1461	Cs_2MnF_6	1620	–
K_2IrCl_6	1442	1440	K_2MnCl_6	1462	–
Na_2MoCl_6	1526	1504	Rb_2MnCl_6	1451	–
K_2MoCl_6	1418	1412	$(NH_4)_2MnCl_6$	1464	–
Rb_2MoCl_6	1399	1399	Cs_2TeBr_6	1306	–
Cs_2MoCl_6	1347	1347	Cs_2TeI_6	1246	–
K_2NbCl_6	1375	1398	K_2TiCl_6	1412	1447
Rb_2NbCl_6	1371	1385	Rb_2TiCl_6	1415	1416
Cs_2NbCl_6	1381	1344	Cs_2TiCl_6	1402	1384
K_2OsCl_6	1447	1447	Tl_2TiCl_6	1560	1553
Cs_2OsCl_6	1409	–	K_2TiBr_6	1379	1379
K_2OsBr_6	1396	–	Rb_2TiBr_6	1341	1331
K_2PdCl_6	1481	1493			
Rb_2PdCl_6	1449	–			

Substance	Calc. U_{POT}	U_{POT}^{BHFC}
Cs_2TiBr_6	1339	1306
Na_2UBr_6	1504	–
K_2UBr_6	1484	–
Rb_2UBr_6	1473	–
Cs_2UBr_6	1459	–
K_2WCl_6	1398	1423
Rb_2WCl_6	1397	1434
Cs_2WCl_6	1392	1366
K_2WBr_6	1408	1408
Rb_2WBr_6	1361	1391
Cs_2WBr_6	1362	1332
K_2ZrCl_6	1339	1371
Rb_2ZrCl_6	1341	–
Cs_2ZrCl_6	1339	1307
Tellurides		
Li_2Te	2212	–
Na_2Te	1997	2095
K_2Te	1830	–
Rb_2Te	1837	–
Cs_2Te	1745	–
Cu_2Te	2706	2683
Ag_2Te	2607	2600
Tl_2Te	2084	2172
$BeTe$	3319	–
$MgTe$	2878	3081
$CaTe$	2721	–

Substance	Calc. U_{POT}	U_{POT}^{BHFC}
Thiocyanates		
$LiCNS$	764	(765)
$NaCNS$	682	682
$KCNS$	623	616
$RbCNS$	623	619
$CsCNS$	623	568
$NH4CNS$	605	611
$Ca(CNS)2$	2184	2118
$Sr(CNS)2$	2063	1957
$Ba(CNS)2$	1979	1852
$Mn(CNS)2$	2280	2351
$Zn(CNS)2$	2335	2560
$Cd(CNS)2$	2201	2374
$Hg(CNS)2$	2146	2492
$Sn(CNS)2$	2117	2142
$Pb(CNS)2$	2058	–
Vanadates		
$LiVO3$	810	–
$NaVO3$	761	–
$KVO3$	686	–
$RbVO3$	657	–
$CsVO3$	628	–

TABLE 2. Thermochemical Radii (nm)

Ion	Radius		Ion	Radius	
Singly Charged Anions			$GaCl_4^-$	0.328	± 0.019
			H^-	0.148	± 0.019
AgF_4^-	0.231	± 0.019	$H_2AsO_4^-$	0.227	± 0.019
$AlBr_4^-$	0.321	± 0.023	$H_2PO_4^-$	0.213	± 0.019
$AlCl_4^-$	0.317	± 0.019	HCO_2^-	0.200	± 0.019
AlF_4^-	0.214	± 0.023	HCO_3^-	0.207	± 0.019
AlH_4^-	0.226	± 0.019	HF_2^-	0.172	± 0.019
AlI_4^-	0.374	± 0.019	HSO_4^-	0.221	± 0.019
AsF_6^-	0.243	± 0.019	I^-	0.211	± 0.019
AsO_2^-	0.211	± 0.019	I_2Br^-	0.261	± 0.019
$Au(CN)_2^-$	0.266	± 0.019	I_3^-	0.272	± 0.019
$AuCl_4^-$	0.288	± 0.019	I_4^-	0.300	± 0.019
AuF_4^-	0.240	± 0.019	IBr_2^-	0.251	± 0.019
AuF_6^-	0.235	± 0.038	ICl_2^-	0.235	± 0.019
$B(OH)_4^-$	0.229	± 0.019	ICl_4^-	0.307	± 0.019
BF_4^-	0.205	± 0.019	$IO_2F_2^-$	0.233	± 0.019
BH_4^-	0.205	± 0.019	IO_3^-	0.218	± 0.019
Br^-	0.190	± 0.019	IO_4^-	0.231	± 0.019
BrF_4^-	0.231	± 0.019	IrF_6^-	0.242	± 0.019
BrO_3^-	0.214	± 0.019	MnO_4^-	0.220	± 0.019
$CF_3SO_3^-$	0.230	± 0.049	MoF_6^-	0.241	± 0.019
$CH_3CO_2^-$	0.194	± 0.019	$MoOF_5^-$	0.241	± 0.019
Cl^-	0.168	± 0.019	N_3^-	0.180	± 0.019
ClO_2^-	0.195	± 0.019	NCO^-	0.193	± 0.019
ClO_3^-	0.208	± 0.019	$NbCl_6^-$	0.338	± 0.049
ClO_4^-	0.225	± 0.019	NbF_6^-	0.254	± 0.019
$ClS_2O_6^-$	0.260	± 0.049	$Nb_2F_{11}^-$	0.311	± 0.038
CN^-	0.187	± 0.023	NbO_3^-	0.194	± 0.019
$Cr_3O_8^-$	0.276	± 0.019	NH_2^-	0.168	± 0.019
$CuBr_4^-$	0.315	± 0.019	$NH_2CH_2COO^-$	0.252	± 0.019
F^-	0.126	± 0.019	NO_2^-	0.187	± 0.019
$FeCl_4^-$	0.317	± 0.019			

Ion	Radius	
NO_3^-	0.200	± 0.019
O_2^-	0.165	± 0.019
O_3^-	0.199	± 0.034
OH^-	0.152	± 0.019
OsF_6^-	0.252	± 0.020
PaF_6^-	0.249	± 0.019
PdF_6^-	0.252	± 0.019
PF_6^-	0.242	± 0.019
PO_3^-	0.204	± 0.019
PtF_6^-	0.247	± 0.019
PuF_5^-	0.239	± 0.019
ReF_6^-	0.240	± 0.019
ReO_4^-	0.227	± 0.019
RuF_6^-	0.242	± 0.019
S_6^-	0.305	± 0.019
SCN^-	0.209	± 0.019
$SbCl_6^-$	0.320	± 0.019
SbF_6^-	0.252	± 0.019
$Sb_2F_{11}^-$	0.312	± 0.038
$Sb_3F_{14}^-$	0.374	± 0.038
$SeCl_5^-$	0.258	± 0.038
$SeCN^-$	0.230	± 0.019
SeH^-	0.195	± 0.019
SH^-	0.191	± 0.019
SO_3F^-	0.214	± 0.019
$S_3N_3^-$	0.231	± 0.038
$S_3N_3O_4^-$	0.252	± 0.038
$TaCl_6^-$	0.352	± 0.019
TaF_6^-	0.250	± 0.019
TaO_3^-	0.192	± 0.019
UF_6^-	0.301	± 0.019
VF_6^-	0.235	± 0.019
VO_3^-	0.201	± 0.019
WCl_6^-	0.337	± 0.019
WF_6^-	0.246	± 0.019
WOF_5^-	0.241	± 0.019

Doubly Charged Anions

Ion	Radius	
AmF_6^{2-}	0.255	± 0.019
$Bi_2Br_8^{2-}$	0.392	± 0.055
$Bi_6Cl_{20}^{2-}$	0.501	± 0.073
$CdCl_4^{2-}$	0.307	± 0.019
$CeCl_6^{2-}$	0.352	± 0.019
CeF_6^{2-}	0.249	± 0.019
CO_3^{2-}	0.189	± 0.019
$CoCl_4^{2-}$	0.306	± 0.019
CoF_4^{2-}	0.209	± 0.019
CoF_6^{2-}	0.256	± 0.019
$Cr_2O_7^{2-}$	0.292	± 0.019
CrF_6^{2-}	0.253	± 0.019
CrO_4^{2-}	0.229	± 0.019
$CuCl_4^{2-}$	0.304	± 0.019
CuF_4^{2-}	0.213	± 0.019
$GeCl_6^{2-}$	0.335	± 0.019
GeF_6^{2-}	0.244	± 0.019
HfF_6^{2-}	0.248	± 0.019
HgI_4^{2-}	0.377	± 0.019
$IrCl_6^{2-}$	0.332	± 0.019
$MnCl_6^{2-}$	0.314	± 0.031
MnF_4^{2-}	0.219	± 0.019
MnF_6^{2-}	0.241	± 0.019
$MoBr_6^{2-}$	0.364	± 0.019

Ion	Radius	
$MoCl_6^{2-}$	0.338	± 0.019
MoF_6^{2-}	0.274	± 0.019
MoO_4^{2-}	0.231	± 0.019
$NbCl_6^{2-}$	0.343	± 0.019
NH^{2-}	0.128	± 0.019
$Ni(CN)_4^{2-}$	0.322	± 0.019
NiF_4^{2-}	0.211	± 0.019
NiF_6^{2-}	0.249	± 0.019
O^{2-}	0.141	± 0.019
O_2^{2-}	0.167	± 0.019
$OsBr_6^{2-}$	0.365	± 0.019
$OsCl_6^{2-}$	0.336	± 0.019
OsF_6^{2-}	0.276	± 0.019
$PbCl_4^{2-}$	0.279	± 0.019
$PbCl_6^{2-}$	0.347	± 0.019
PbF_6^{2-}	0.268	± 0.019
$PdBr_6^{2-}$	0.354	± 0.019
$PdCl_4^{2-}$	0.313	± 0.019
$PdCl_6^{2-}$	0.333	± 0.019
PdF_6^{2-}	0.252	± 0.019
$PoBr_6^{2-}$	0.380	± 0.019
PoI_6^{2-}	0.428	± 0.019
$Pt(NO_2)_3Cl_3^{2-}$	0.364	± 0.019
$Pt(NO_2)_4Cl_2^{2-}$	0.383	± 0.019
$Pt(OH)_2^{2-}$	0.333	± 0.019
$Pt(SCN)_6^{2-}$	0.451	± 0.019
$PtBr_4^{2-}$	0.324	± 0.019
$PtBr_6^{2-}$	0.363	± 0.019
$PtCl_4^{2-}$	0.307	± 0.019
$PtCl_6^{2-}$	0.333	± 0.019
PtF_6^{2-}	0.245	± 0.019
$PuCl_6^{2-}$	0.349	± 0.019
$ReBr_6^{2-}$	0.371	± 0.019
$ReCl_6^{2-}$	0.337	± 0.019
ReF_6^{2-}	0.256	± 0.019
ReF_8^{2-}	0.276	± 0.019
ReH_9^{2-}	0.257	± 0.019
ReI_6^{2-}	0.421	± 0.026
RhF_6^{2-}	0.240	± 0.019
$RuCl_6^{2-}$	0.336	± 0.019
RuF_6^{2-}	0.248	± 0.019
S^{2-}	0.189	± 0.019
$S_2O_3^{2-}$	0.251	± 0.019
$S_2O_4^{2-}$	0.262	± 0.019
$S_2O_5^{2-}$	0.270	± 0.019
$S_2O_6^{2-}$	0.283	± 0.019
$S_2O_7^{2-}$	0.275	± 0.019
$S_2O_8^{2-}$	0.291	± 0.019
$S_3O_6^{2-}$	0.302	± 0.019
$S_4O_6^{2-}$	0.325	± 0.019
$S_6O_6^{2-}$	0.382	± 0.019
ScF_6^{2-}	0.276	± 0.019
Se^{2-}	0.181	± 0.019
$SeBr_6^{2-}$	0.363	± 0.019
$SeCl_6^{2-}$	0.336	± 0.019
SeO_4^{2-}	0.229	± 0.019
SiF_6^{2-}	0.248	± 0.019
SiO_3^{2-}	0.195	± 0.019
SmF_6^{2-}	0.218	± 0.019
$Sn(OH)_6^{2-}$	0.279	± 0.020
$SnBr_6^{2-}$	0.374	± 0.019

Ion	Radius	
$SnCl_6^{2-}$	0.345	± 0.019
SnF_6^{2-}	0.265	± 0.019
SnI_6^{2-}	0.427	± 0.019
SO_3^{2-}	0.204	± 0.019
SO_4^{2-}	0.218	± 0.019
$TcBr_6^{2-}$	0.363	± 0.019
$TcCl_6^{2-}$	0.337	± 0.019
TcF_6^{2-}	0.244	± 0.019
TcH_9^{2-}	0.260	± 0.019
TcI_6^{2-}	0.419	± 0.019
Te^{2-}	0.220	± 0.019
$TeBr_6^{2-}$	0.383	± 0.019
$TeCl_6^{2-}$	0.353	± 0.019
TeI_6^{2-}	0.430	± 0.019
TeO_4^{2-}	0.238	± 0.019
$Th(NO_3)_6^{2-}$	0.424	± 0.019
$ThCl_6^{2-}$	0.360	± 0.019
ThF_6^{2-}	0.263	± 0.019
$TiBr_6^{2-}$	0.356	± 0.019
$TiCl_6^{2-}$	0.335	± 0.019
TiF_6^{2-}	0.252	± 0.019
UCl_6^{2-}	0.354	± 0.019
UF_6^{2-}	0.256	± 0.019
VO_3^{2-}	0.204	± 0.019
WBr_6^{2-}	0.363	± 0.019
WCl_6^{2-}	0.339	± 0.019
WO_4^{2-}	0.237	± 0.019
$WOCl_5^{2-}$	0.334	± 0.019
$ZnBr_4^{2-}$	0.335	± 0.019
$ZnCl_4^{2-}$	0.306	± 0.019
ZnF_4^{2-}	0.219	± 0.019
ZnI_4^{2-}	0.384	± 0.019
$ZrBr_4^{2-}$	0.334	± 0.019
$ZrCl_4^{2-}$	0.306	± 0.019
$ZrCl_6^{2-}$	0.348	± 0.019
ZrF_6^{2-}	0.258	± 0.019

Multi-Charged Anions

Ion	Radius	
AlH_6^{3-}	0.256	± 0.042
AsO_4^{3-}	0.237	± 0.042
$CdBr_6^{4-}$	0.374	± 0.038
$CdCl_6^{4-}$	0.352	± 0.038
CeF_6^{3-}	0.278	± 0.038
CeF_7^{3-}	0.282	± 0.038
$Co(CN)_6^{3-}$	0.349	± 0.038
$Co(NO_2)_6^{3-}$	0.343	± 0.038
$CoCl_5^{3-}$	0.320	± 0.038
CoF_6^{3-}	0.258	± 0.042
$Cr(CN)_6^{3-}$	0.351	± 0.038
CrF_6^{3-}	0.254	± 0.042
$Cu(CN)_4^{3-}$	0.312	± 0.038
$Fe(CN)_6^{3-}$	0.347	± 0.038
FeF_6^{3-}	0.298	± 0.042
HfF_7^{3-}	0.277	± 0.042
InF_6^{3-}	0.268	± 0.038
$Ir(CN)_6^{3-}$	0.347	± 0.038
$Ir(NO_2)_6^{3-}$	0.338	± 0.038
$Mn(CN)_6^{3-}$	0.350	± 0.038
$Mn(CN)_6^{5-}$	0.401	± 0.042
$MnCl_6^{4-}$	0.349	± 0.038
N^{3-}	0.180	± 0.042
$Ni(NO_2)_6^{3-}$	0.342	± 0.038

Ion	Radius	
$Ni(NO_2)_6^{4-}$	0.383	± 0.038
NiF_6^{3-}	0.250	± 0.042
O^{3-}	0.288	± 0.038
P^{3-}	0.224	± 0.042
PaF_8^{3-}	0.299	± 0.042
PO_4^{3-}	0.230	± 0.042
PrF_6^{3-}	0.281	± 0.038
$Rh(NO_2)_6^{3-}$	0.345	± 0.038
$Rh(SCN)_6^{3-}$	0.428	± 0.042
TaF_8^{3-}	0.284	± 0.042
TbF_7^{3-}	0.290	± 0.038
$Tc(CN)_6^{5-}$	0.410	± 0.042
ThF_7^{3-}	0.282	± 0.042
$TiBr_6^{3-}$	0.315	± 0.038
TlF_6^{3-}	0.271	± 0.038
UF_7^{3-}	0.285	± 0.042
YF_6^{3-}	0.275	± 0.038
ZrF_7^{3-}	0.273	± 0.038

Singly Charged Cations

Ion	Radius	
$N(CH_3)_4^+$	0.234	± 0.019
$N_2H_5^+$	0.158	± 0.019
$N_2H_6^{2+}$	0.158	± 0.029
$NH(C_2H_5)_3^+$	0.274	± 0.019
$NH_3C_2H_5^+$	0.193	± 0.019
$NH_3C_3H_7^+$	0.225	± 0.019
$NH_3CH_3^+$	0.177	± 0.019
NH_3OH^+	0.147	± 0.019
NH_4^+	0.136	± 0.019
$NH_3C_2H_4OH^+$	0.203	± 0.019
$As_3S_4^+$	0.244	± 0.027
$As_3Se_4^+$	0.253	± 0.027
$AsCl_4^+$	0.221	± 0.027
Br_2^+	0.155	± 0.027
Br_3^+	0.204	± 0.027
Br_3^-	0.238	± 0.027
Br_5^+	0.229	± 0.027
$BrClCNH_2^+$	0.175	± 0.027
BrF_2^+	0.183	± 0.027
BrF_4^+	0.172	± 0.027
$C_{10}F_8^+$	0.265	± 0.027
$C_6F_6^+$	0.228	± 0.027
$Cl(SNSCN)_2^+$	0.347	± 0.027
$Cl_2C=NH_2^+$	0.173	± 0.027
Cl_2F^+	0.165	± 0.027
Cl_3^+	0.182	± 0.027
ClF_2^+	0.147	± 0.027
ClO_2^+	0.118	± 0.027
$GaBr_4^-$	0.317	± 0.038
I_2^+	0.185	± 0.027
I_3^+	0.225	± 0.027
I_5^+	0.263	± 0.027
IBr_2^+	0.196	± 0.027
ICl_2^+	0.175	± 0.036
IF_6^+	0.209	± 0.027
$N(S_3N_2)_2^+$	0.258	± 0.027
$N(SCl)_2^+$	0.186	± 0.027
$N(SeCl)_2^+$	0.246	± 0.027
$N(SF_2)_2^+$	0.214	± 0.027
N_2F^+	0.156	± 0.027
NO^+	0.145	± 0.027
NO_2^+	0.153	± 0.027

Ion	Radius	
O_2^+	0.140	± 0.027
$O_2(SCCF_3Cl)_2^+$	0.275	± 0.027
$ONCH_3CF_3^+$	0.200	± 0.027
$OsOF_5^-$	0.246	± 0.038
$P(CH_3)_3Cl^+$	0.197	± 0.027
$P(CH_3)_3D^+$	0.196	± 0.027
PCl_4^+	0.235	± 0.027
$ReOF_5^-$	0.245	± 0.038
$S(CH_3)_2Cl^+$	0.207	± 0.027
$S(N(C_2H_5)_3)_3^+$	0.439	± 0.027
$S_2(CH_3)_2Cl^+$	0.265	± 0.027
$S_2(CH_3)_2CN^+$	0.223	± 0.027
$S_2(CH_3)_3^+$	0.233	± 0.027
$S_2Br_5^+$	0.267	± 0.027
S_2N^+	0.159	± 0.034
$S_2N_2C_2H_4^+$	0.211	± 0.027
$S_2NC_2(PhCH_3)_2^+$	0.310	± 0.027
$S_2NC_3H_4^+$	0.218	± 0.027
$S_2NC_4H_8^+$	0.225	± 0.027
$S_3(CH_3)_3^+$	0.239	± 0.027
$S_3Br_3^+$	0.245	± 0.027
$S_3C_3H_7^+$	0.199	± 0.027
$S_3C_4F_6^+$	0.261	± 0.027
$S_3CF_3CN^+$	0.263	± 0.027
$S_3Cl_3^+$	0.233	± 0.027
$S_3N_2^+$	0.201	± 0.027
$S_3N_2Cl^+$	0.232	± 0.027
$S_4N_3^+$	0.231	± 0.027
$S_4N_3(Ph)_2^+$	0.316	± 0.027
$S_4N_4H^+$	0.178	± 0.027
$S_5N_5^+$	0.257	± 0.027
S_7I^+	0.262	± 0.027
$Sb(NPPh_3)_4^+$	0.518	± 0.027
SBr_3^+	0.220	± 0.027
$SCH_3O_2^+$	0.183	± 0.027
$SCH_3P(CH_3)_3^+$	0.248	± 0.027
$SCH_3PCH_3Cl_2^+$	0.205	± 0.027
$SCl(C_2H_5)_2^+$	0.207	± 0.027
$SCl_2CF_3^+$	0.207	± 0.027
$SCl_2CH_3^+$	0.204	± 0.027
SCl_3^+	0.185	± 0.027
$Se_3Br_3^+$	0.253	± 0.027
$Se_3Cl_3^+$	0.245	± 0.027
$Se_3N_2^+$	0.288	± 0.042
$Se_3NC_{12}^+$	0.163	± 0.027
Se_6I^+	0.260	± 0.027
$SeBr_3^+$	0.182	± 0.027
$SeCl_3^+$	0.192	± 0.027
SeF_3^+	0.179	± 0.027
SeI_3^+	0.238	± 0.027
SeN_2Cl^+	0.196	± 0.027
$SeNCl_2^+$	0.157	± 0.027
$(SeNMe_3)_3^+$	0.406	± 0.027
$SeS_2N_2^+$	0.282	± 0.042
$SF(C_6F_5)_2^+$	0.294	± 0.027
$SF_2CF_3^+$	0.198	± 0.027
$SF_2N(CH_3)_2^+$	0.210	± 0.027
SF_3^+	0.172	± 0.027
$SFS(C(CF_3)_2)_2^+$	0.275	± 0.027
$SH_2C_3H_7^+$	0.210	± 0.027
SN^+	0.158	± 0.027
$SNCl_5(CH_3CN)^-$	0.290	± 0.038

Ion	Radius	
$(SNPMe_3)_3^+$	0.308	± 0.027
$SNSC(CH_3)N^+$	0.225	± 0.027
$SNSC(CN)CH^+$	0.209	± 0.027
$SNSC(Ph)N^+$	0.251	± 0.027
$SNSC(Ph)NS_3N_2^+$	0.327	± 0.027
$SNSC(PhCH_3)N^+$	0.264	± 0.027
$(Te(N(SiMe_3)_2)_2^+$	0.371	± 0.027
$Te(N_3)_3^+$	0.226	± 0.027
$Te_4Nb_3OTe_2I_6^+$	0.407	± 0.027
$TeBr_3^+$	0.235	± 0.027
$TeCl_3^+$	0.216	± 0.027
$TeCl_3(15\text{-crown-5})^+$	0.282	± 0.027
TeI_3^+	0.243	± 0.027
$Xe_2F_{11}^+$	0.266	± 0.027
$Xe_2F_3^+$	0.221	± 0.027
XeF^+	0.174	± 0.027
XeF_3^+	0.183	± 0.027
XeF_5^+	0.186	± 0.027
$XeOF_3^+$	0.186	± 0.027
Doubly Charged Cations		
$Co_2S_2(CO)_6^{2+}$	0.263	± 0.035
$FeW(Se)_2(CO)^{2+}$	0.260	± 0.035
I_4^{2+}	0.207	± 0.035
$Mo(Te_3)(CO)_4^{2+}$	0.234	± 0.035
S_{19}^{2+}	0.292	± 0.035
$S_2(S(CH_3)_2)_2^{2+}$	0.230	± 0.035
$S_2I_4^{2+}$	0.231	± 0.035
$S_3N_2^{2+}$	0.184	± 0.035
$S_3NCCNS_3^{2+}$	0.220	± 0.035
S_3Se^{2+}	0.326	± 0.035
$S_4N_4^{2+}$	0.186	± 0.035
$S_6N_4^{2+}$	0.232	± 0.035
S_8^{2+}	0.182	± 0.035
Se_{10}^{2+}	0.253	± 0.035
Se_{17}^{2+}	0.236	± 0.035
Se_{19}^{2+}	0.296	± 0.035
$Se_2I_4^{2+}$	0.218	± 0.035
$Se_3N_2^{2+}$	0.182	± 0.035
Se_4^{2+}	0.152	± 0.035
$Se_4S_2N_4^{2+}$	0.224	± 0.035
Se_8^{2+}	0.186	± 0.035
$SeN_2S_2^{2+}$	0.182	± 0.035
$(SNP(C_2H_5)_3)_2^{2+}$	0.312	± 0.035
$TaBr_6^-$	0.351	± 0.049
$Te(trtu)_4^{2+}$	0.328	± 0.035
$Te(tu)_4^{2+}$	0.296	± 0.035
$Te_2(esu)_4Br_2^{2+}$	0.356	± 0.035
$Te_2(esu)_4Cl_2^{2+}$	0.361	± 0.035
$Te_2(esu)_4I_2^{2+}$	0.342	± 0.035
$Te_2Se_2^{2+}$	0.192	± 0.035
$Te_2Se_4^{2+}$	0.222	± 0.035
$Te_2Se_8^{2+}$	0.252	± 0.035
$Te_3S_3^{2+}$	0.217	± 0.035
Te_3Se^{2+}	0.193	± 0.035
Te_4^{2+}	0.169	± 0.035
Te_8^{2+}	0.187	± 0.035
$W(CO)_4(h3\text{-}Te)^{2+}$	0.234	± 0.035
$W_2(CO)_{10}Se_4^{2+}$	0.290	± 0.035
Multi-Charged Cations		
I_{15}^{3+}	0.442	± 0.051
$Te_2(su)_6^{4+}$	0.453	± 0.034

Ligand abbreviations: su = selenourea; esu = ethyleneselenourea; tu = thiourea; ph = phenyl.

TABLE 3. Ancillary Thermochemical Data
(kJ mol⁻¹)

Species	State	$\Delta_f H°$
AsO_4^{3-}	g	(289)
BrO_3^-	g	−145
ClO_4^-	g	−344
CN^-	g	66
CO_3^{2-}	g	−321
$Fe(NO_3)_2$	c	(−448)
HF_2^-	g	−774
$HfCl_6^{2-}$	g	−1640
$IO_2F_2^-$	g	−693
IO_3^-	g	−208
$IrCl_6^{2-}$	g	−785
$LiCH_3O_2$	c	(−745)
$NbCl_6^{2-}$	g	−1224
$NH_2CH_2CO_2^-$	g	−564
O_2^{2-}	g	553
$PdCl_6^{2-}$	g	−749
PO_4^{3-}	g	291
$PtCl_6^{2-}$	g	−774
$ReBr_6^{2-}$	g	−689
$ReCl_6^{2-}$	g	−919
$Ti(OH)_2$	c	−778

THE MADELUNG CONSTANT AND CRYSTAL LATTICE ENERGY

If U is the crystal lattice energy and M is the Madelung constant, then[a]

$$U = \frac{NMz_iz_je^2}{r}(1-1/n)$$

Substance	Ion type	Crystal form[b]	M
Sodium chloride, NaCl	M^+, X^-	FCC	1.74756
Cesium chloride, CsCl	M^+, X^-	BCC	1.76267
Calcium chloride, $CaCl_2$	$M^{++}, 2X^-$	Cubic	2.365
Calcium fluoride (fluorite), CaF_2	$M^{++}, 2X^-$	Cubic	2.51939
Cadmium chloride, $CdCl_2$	$M^{++}, 2X^-$	Hexagonal	2.244[c]
Cadmium iodide (α), CdI_2	$M^{++}, 2X^-$	Hexagonal	2.355[c]
Magnesium fluoride, MgF_2	$M^{++}, 2X^-$	Tetragonal	2.381[c]
Cuprous oxide (cuprite), Cu_2O	$2M^+, X^{--}$	Cubic	2.22124
Zinc oxide, ZnO	M^{++}, X^{--}	Hexagonal	1.4985[c]
Sphalerite (zinc blende), ZnS	M^{++}, X^{--}	FCC	1.63806
Wurtzite, ZnS	M^{++}, X^{--}	Hexagonal	1.64132[c]
Titanium dioxide (anatase), TiO_2	$M^{4+}, 2X^{--}$	Tetragonal	2.400[c]
Titanium dioxide (rutile), TiO_2	$M^{4+}, 2X^{--}$	Tetragonal	2.408[c]
β-Quartz, SiO_2	$M^{4+}, 2X^{--}$	Hexagonal	2.2197[c]
Corundum, Al_2O_3	$2M^{3+}, 3X^{--}$	Rhombohedral	4.1719

[a] N is Avogadro's number, z_i and z_j are the integral charges on the ions (in units of e), and e is the charge on the electron in electrostatic units ($e = 4.803 \times 10^{-10}$ esu). r is the shortest distance between cation–anion pairs in centimeters. Then U is in ergs (1 erg = 10^{-7} J).

[b] FCC = face centered cubic; BCC = body centered cubic.

[c] For tetragonal and hexagonal crystals the value of M depends on the details of the lattice parameters.

The Born Exponent, n is:

Ion type	n
He, Li^+	5
Ne, Na^+, F^-	7
Ar, K^+, Cu^+, Cl^-	9
Kr, Rb^+, Ag^+, Br^-	10
Xe, Cs^+, Au^+, I^-	12

For a crystal with a mixed-ion type, an average of the values of n in this table is to be used (6 for LiF, for example).

ELASTIC CONSTANTS OF SINGLE CRYSTALS

H. P. R. Frederikse

This table gives selected values of elastic constants for single crystals. The values believed most reliable were selected from the original literature. The substances are arranged by crystal system and, within each system, alphabetically by name. A reference to the original literature is given for each value; a useful compilation of published values from many sources may be found in Reference 1.

Data are given for the single-crystal density and for the elastic constants c_{ij}, in units of 10^{11} N/m^2, which is equivalent to 10^{12} dyn/cm^2.

General References

1. Simmons, G., and Wang, H., *Single Crystal Elastic Constants and Calculated Aggregate Properties: A Handbook, Second Edition*, The MIT Press, Cambridge, MA, 1971.
2. Gray, D. E., Ed., *American Institute of Physics Handbook, Third Edition*, McGraw-Hill, New York, 1972.

CUBIC CRYSTALS

Name	Formula	ρ/g cm^{-3}	T/K	Ref.	C_{11}	C_{12}	C_{44}
Aluminum	Al	2.6970	298	1	1.0675	0.6041	0.2834
Aluminum antimonide	AlSb	4.3600	300	2	0.8939	0.4427	0.4155
Ammonium bromide	NH$_4$Br	2.4314	300	3	0.3414	0.0782	0.0722
Ammonium chloride	NH$_4$Cl	1.5279	290	4	0.3814	0.0866	0.0903
Argon	Ar	1.7710	4.2	5	0.0529	0.0135	0.0159
Barium fluoride	BaF$_2$	4.8860	298	6	0.9199	0.4157	0.2568
Barium nitrate	Ba(NO$_3$)$_2$	3.2560	293	7	0.2925	0.2065	0.1277
Calcium fluoride	CaF$_2$	3.810	298	8	1.6420	0.4398	0.8406
Calcium telluride	CaTe	5.8544	298	9	0.5351	0.3681	0.1994
Cesium	Cs	1.9800	78	10	0.0247	0.0206	0.0148
Cesium bromide	CsBr	4.4560	298	11	0.3063	0.0807	0.0750
Cesium chloride	CsCl	3.9880	298	11	0.3644	0.0882	0.0804
Cesium iodide	CsI	4.5250	298	11	0.2446	0.0661	0.0629
Chromite	FeCr$_2$O$_4$	4.4500	RT	12	3.2250	1.4370	1.1670
Chromium	Cr	7.20	298	13	3.398	0.586	0.990
Cobalt oxide	CoO	6.44	298	14	2.6123	1.4699	0.8300
Cobalt zinc ferrite	CoZnFeO$_2$	5.43	303	12	2.660	1.530	0.780
Copper	Cu	8.932	298	15	1.683	1.221	0.757
Gallium antimonide	GaSb	5.6137	298	16	0.8839	0.4033	0.4316
Gallium arsenide	GaAs	5.3169	298	17	1.1877	0.5372	0.5944
Gallium phosphide	GaP	4.1297	300	18	1.4120	0.6253	0.7047
Garnet (yttrium-iron)	Y$_3$Fe$_2$(FeO$_4$)$_3$	5.17	298	19	2.680	1.106	0.766
Germanium	Ge	5.313	298	20	1.2835	0.4823	0.6666
Gold	Au	19.283	296.5	21	1.9244	1.6298	0.4200
Indium antimonide	InSb	5.7890	298	22	0.6720	0.3670	0.3020
Indium arsenide	InAs	5.6720	293	23	0.8329	0.4526	0.3959
Indium phosphide	InP	4.78	RT	24	1.0220	0.5760	0.4600
Iridium	Ir	22.52	300	25	5.80	2.42	2.56
Iron	Fe	7.8672	298	26	2.26	1.40	1.16
Lead	Pb	11.34	296	27	0.4966	0.4231	0.1498
Lead fluoride	PbF$_2$	7.79	300	28	0.8880	0.4720	0.2454
Lead nitrate	Pb(NO$_3$)$_2$	4.547	293	29	0.3729	0.2765	0.1347
Lead telluride	PbTe	8.2379	303.2	30	1.0795	0.0764	0.1343
Lithium	Li	0.5326	298	31	0.1350	0.1144	0.0878
Lithium bromide	LiBr	3.47	RT	32	0.3940	0.1880	0.1910
Lithium chloride	LiCl	2.068	295	33	0.4927	0.2310	0.2495
Lithium fluoride	LiF	2.638	RT	34	1.1397	0.4767	0.6364
Lithium iodide	LiI	4.061	RT	32	0.2850	0.1400	0.1350
Magnesium oxide	MgO	3.579	298	20	2.9708	0.9536	1.5613
Magnetite	Fe$_3$O$_4$	5.18	RT	32	2.730	1.060	0.971
Manganese oxide	MnO	5.39	298	35	2.23	1.20	0.79
Mercury telluride	HgTe	8.079	290	36	0.548	0.381	0.204
Molybdenum	Mo	10.2284	273	37	4.637	1.578	1.092
Nickel	Ni	8.91	298	15	2.481	1.549	1.242
Niobium	Nb	8.578	300	38	2.4650	1.3450	0.2873

CUBIC CRYSTALS

Name	Formula	$\rho/\text{g cm}^{-3}$	T/K	Ref.	C_{11}	C_{12}	C_{44}
Palladium	Pd	12.038	300	39	2.2710	1.7604	0.7173
Platinum	Pt	21.50	300	40	3.4670	2.5070	0.7650
Potassium	K	0.851	295	41	0.0370	0.0314	0.0188
Potassium bromide	KBr	2.740	298	11	0.3468	0.0580	0.0507
Potassium chloride	KCl	1.984	298	11	0.4069	0.0711	0.0631
Potassium cyanide	KCN	1.553	RT	32	0.1940	0.1180	0.0150
Potassium fluoride	KF	2.480	295	33	0.6490	0.1520	0.1232
Potassium iodide	KI	3.128	300	42	0.2710	0.0450	0.0364
Pyrite	FeS_2	5.016	RT	43	3.818	0.310	1.094
Rubidium	Rb	1.58	170	44	0.0296	0.0250	0.0171
Rubidium bromide	RbBr	3.350	300	45	0.3152	0.0500	0.0380
Rubidium chloride	RbCl	2.797	300	45	0.3624	0.0612	0.0468
Rubidium iodide	RbI	3.551	300	45	0.2556	0.0382	0.0278
Silicon	Si	2.331	298	46	1.6578	0.6394	0.7962
Silver	Ag	10.50	300	47	1.2399	0.9367	0.4612
Silver bromide	AgBr	5.585	300	48	0.5920	0.3640	0.0616
Sodium	Na	0.971	299	49	0.0739	0.0622	0.0419
Sodium bromate	$NaBrO_3$	3.339	RT	32	0.5450	0.1910	0.1500
Sodium bromide	NaBr	3.202	300	33	0.3970	0.1001	0.0998
Sodium chlorate	$NaClO_3$	2.485	RT	50	0.4920	0.1420	0.1160
Sodium chloride	NaCl	2.163	298	11	0.4947	0.1288	0.1287
Sodium fluoride	NaF	2.804	300	51	0.9700	0.2380	0.2822
Sodium iodide	NaI	3.6689	300	52	0.3007	0.0912	0.0733
Spinel	$MgAl_2O_4$	3.6193	298	53	2.9857	1.5372	1.5758
Strontium fluoride	SrF_2	4.277	300	54	1.2350	0.4305	0.3128
Strontium nitrate	$Sr(NO_3)_2$	2.989	293	29	0.4255	0.2921	0.1590
Strontium oxide	SrO	4.99	300	55	1.601	0.435	0.590
Strontium titanate	$SrTiO_3$	5.123	RT	56	3.4817	1.0064	4.5455
Tantalum	Ta	16.626	298	57	2.6023	1.5446	0.8255
Tantalum carbide	TaC	14.65	RT	58	5.05	0.73	0.79
Thallium bromide	TlBr	7.4529	298	59	0.3760	0.1458	0.0757
Thorium	Th	11.694	300	60	0.7530	0.4890	0.4780
Thorium oxide	ThO_2	9.991	298	61	3.670	1.060	0.797
Tin telluride	SnTe	6.445	300	62	1.1250	0.0750	0.1172
Titanium carbide	TiC	4.940	RT	107	5.00	1.13	1.75
Tungsten	W	19.257	297	64	5.2239	2.0437	1.6083
Uranium carbide	UC	13.63	300	65	3.200	0.850	0.647
Uranium dioxide	UO_2	10.97	298	66	3.960	1.210	0.641
Vanadium	V	6.022	300	67	2.287	1.190	0.432
Zinc selenide	ZnSe	5.262	298	68	0.8096	0.4881	0.4405
Zinc sulfide	ZnS	4.088	298	68	1.0462	0.6534	0.4613
Zinc telluride	ZnTe	5.636	298	68	0.7134	0.4078	0.3115
Zirconium carbide	ZrC	6.606	298	63	4.720	0.987	1.593

TETRAGONAL CRYSTALS

Name	Formula	ρ/g cm^{-3}	T/K	Ref.	C_{11}	C_{12}	C_{13}	C_{16}	C_{33}	C_{44}	C_{66}
Ammonium dihydrogen arsenate (ADA)	$NH_4H_2AsO_4$	2.3110	298	69	0.6747	−0.106	0.1652		0.3022	0.0685	0.0639
Ammonium dihydrogen phosphate (ADP)	$NH_4H_2PO_4$	1.8030	293	69	0.6200	−0.050	0.1400		0.3000	0.0910	0.0610
Barium titanate	$BaTiO_3$	5.9988	298	70	2.7512	1.7897	1.5156		1.6486	0.5435	1.1312
Calcium molybdate	$CaMoO_4$	4.255	298	79	1.447	0.664	0.466	0.134	1.265	0.369	0.451
Indium	In	7.300	RT	71	0.4450	0.3950	0.4050		0.4440	0.0655	0.1220
Magnesium fluoride	MgF_2	3.177	RT	72	1.237	0.732	0.536		1.770	0.552	0.978
Nickel sulfate hexahydrate	$NiSO_4 \cdot 6H_2O$	2.070	RT	73	0.3209	0.2315	0.0209		0.2931	0.1156	0.1779
Potassium dihydrogen arsenate (KDA)	KH_2AsO_4	2.867	RT	12	0.530	−0.060	−0.020		0.370	0.120	0.070
Potassium dihydrogen phosphate (KDP)	KH_2PO_4	2.388	RT	71	0.7140	−0.049	0.1290		0.5620	0.1270	0.0628
Rubidium dihydrogen phosphate (RDP)	RbH_2PO_4	2.800	298	74	0.5562	−0.064	0.0279		0.4398	0.1142	0.0350
Rutile	TiO_2	4.260	298	75	2.7143	1.7796	1.4957		4.8395	1.2443	1.9477
Tellurium oxide	TeO_2	5.99	RT	76	0.5320	0.4860	0.2120		1.0850	0.2440	0.5520
Tin (white)	Sn	7.29	288	77	0.7529	0.6156	0.4400		0.9552	0.2193	0.2336
Zircon	$ZrSiO_4$	4.70	RT	78	2.585	1.791	1.542		3.805	0.733	1.113

ORTHORHOMBIC CRYSTALS

Name	Formula	ρ/g cm^{-3}	T/K	Ref.	C_{11}	C_{12}	C_{13}	C_{22}	C_{23}	C_{33}	C_{44}	C_{55}	C_{66}
Acenaphthene	$C_{12}H_{10}$	1.220	293	80	0.1380	0.0210	0.0410	0.1262	0.0460	0.1117	0.0265	0.0290	0.0185
Ammonium sulfate	$(NH_4)_2SO_4$	1.774	293	81	0.3607	0.1651	0.1580	0.2981	0.1456	0.3534	0.1025	0.0717	0.0974
Aragonite	$CaCO_3$	2.93	RT	82	1.5958	0.3663	0.0197	0.8697	0.1597	0.8503	0.4132	0.2564	0.4274
Barite	$BaSO_4$	4.40	RT	82	0.8941	0.4614	0.2691	0.7842	0.2676	1.0548	0.1190	0.2874	0.2778
Benzene	C_6H_6	1.061	250	83	0.0614	0.0352	0.0401	0.0656	0.0390	0.0583	0.0197	0.0378	0.0153
Benzophenone	$(C_6H_5)_2CO$	1.219	RT	32	0.1070	0.0550	0.0169	0.1000	0.0321	0.0710	0.0203	0.0155	0.0353
Bronzite	$(MgFe)SiO_3$	3.38	RT	78	1.876	0.686	0.605	1.578	0.561	2.085	0.700	0.592	0.544
Calcium sulfate	$CaSO_4$	2.962	RT	84	0.9382	0.1650	0.1520	1.845	0.3173	1.1180	0.3247	0.2653	0.0926
Celestite	$SrSO_3$	3.96	RT	12	1.044	0.773	0.605	1.061	0.619	1.286	0.135	0.279	0.266
Cesium sulfate	Cs_2SO_4	4.243	293	81	0.4490	0.1958	0.1815	0.4283	0.1800	0.3785	0.1326	0.1319	0.1323
Fosterite	Mg_2SiO_4	3.224	298	85	3.2848	0.6390	0.6880	1.9980	0.7380	2.3530	0.6515	0.8120	0.8088
Iodic acid	HIO_3	4.630	RT	73	0.3030	0.1194	0.1169	0.5448	0.0548	0.4359	0.1835	0.2193	0.1736
Lithium ammonium tartrate	$LiNH_4C_4H_4O_6 \cdot 4H_2O$	1.71	RT	12	0.3864	0.1655	0.0875	0.5393	0.2007	0.3624	0.1190	0.0667	0.2326
Magnesium sulfate heptahydrate	$MgSO_4 \cdot 7H_2O$	1.68	RT	86	0.325	0.174	0.182	0.288	0.182	0.315	0.078	0.156	0.090
Natrolite	$(Na,Al)SiO_3$	2.25	RT	78	0.716	0.261	0.297	0.632	0.297	1.378	0.196	0.248	0.423
Nickel sulfate heptahydrate	$NiSO_4 \cdot 7H_2O$	1.948	RT	86	0.353	0.198	0.201	0.311	0.201	0.335	0.091	0.172	0.099
Olivine	$(MgFe)SiO_4$	3.324	RT	87	3.240	0.590	0.790	1.980	0.780	2.490	0.667	0.810	0.793
Potassium pentaborate	$KB_5O_8 \cdot 4H_2O$	1.74	RT	71	0.582	0.229	0.174	0.359	0.231	0.255	0.164	0.046	0.057
Potassium sulfate	K_2SO_4	2.665	293	81	0.5357	0.1999	0.2095	0.5653	0.1990	0.5523	0.195	0.1879	0.1424
Rochelle salt	$NaK(C_4H_4O_6) \cdot 4H_2O$	1.79	RT	71	0.255	0.141	0.116	0.381	0.146	0.371	0.134	0.032	0.098
Rubidium sulfate	Rb_2SO_4	3.621	293	81	0.5029	0.1965	0.1999	0.5098	0.1925	0.4761	0.1626	0.1589	0.1407
Sodium ammonium tartrate	$NaNH_4C_4H_4O_6 \cdot 4H_2O$	1.587	RT	12	0.3685	0.2725	0.3083	0.5092	0.3472	0.5541	0.1058	0.0303	0.0870
Sodium tartrate	$Na_2C_4H_4O_6 \cdot 2H_2O$	1.794	RT	12	0.461	0.286	0.320	0.547	0.352	0.665	0.124	0.031	0.098
Strontium formate dihydrate	$Sr(CHO_2)_2 \cdot 2H_2O$	2.25	RT	12	0.4391	0.1037	−0.149	0.3484	−0.014	0.3746	0.1538	0.1075	0.1724
Sulfur	S	2.07	RT	12	0.240	0.133	0.171	0.205	0.159	0.483	0.043	0.087	0.076
Thallium sulfate	$TlSO_4$	6.776	293	81	0.4106	0.2573	0.2288	0.3885	0.2174	0.4268	0.1125	0.1068	0.0751
Topaz	$Al_2SiO_3(OH,F)_2$	3.52	RT	82	2.8136	1.2582	0.8464	3.8495	0.8815	2.9452	1.0811	1.3298	1.3089
Uranium (alpha)	U	19.0453	293	88	2.1486	0.4622	0.2176	1.9983	1.0764	2.6763	1.2479	0.7379	0.7454
Zinc sulfate heptahydrate	$ZnSO_4 \cdot 7H_2O$	1.970	RT	86	0.3320	0.1720	0.2000	0.2930	0.1980	0.3200	0.0780	0.1530	0.0830

MONOCLINIC CRYSTALS

Name	Formula	ρ/g cm^{-3}	T/K	Ref.	C_{11}	C_{12}	C_{13}	C_{15}	C_{22}
Aegirine	(NaFe)Si$_2$O$_6$	3.50	RT	89	1.858	0.685	0.707	0.098	1.813
Anthracene	C$_{14}$H$_{10}$	1.258	RT	90	0.0852	0.0672	0.0590	−0.0192	0.1170
Cobalt sulfate heptahydrate	CoSO$_4$·7H$_2$O	1.948	RT	86	0.335	0.205	0.158	0.016	0.378
Diopside	(CaMg)Si$_2$O$_6$	3.31	RT	91	2.040	0.884	0.0883	−0.193	1.750
Dipotassium tartrate	KHC$_4$H$_4$O$_6$	1.97	RT	12	0.4294	0.1399	0.3129	−0.0105	0.3460
Feldspar (microceine)	KAlSi$_3$O$_8$	2.56	RT	92	0.664	0.438	0.259	−0.033	1.710
Ferrous sulfate heptahydrate	FeSO$_4$·7H$_2$O	1.898	RT	86	0.349	0.208	0.174	−0.020	0.376
Lithium sulfate monohydrate	Li$_2$SO$_4$·H$_2$O	2.221	RT	32	0.5250	0.1715	0.1730	−0.0196	0.5060
Naphthalene	C$_{10}$H$_8$	1.127	RT	93	0.0780	0.0445	0.0340	−0.006	0.0990
Potassium tartrate	K$_2$C$_4$H$_4$O$_6$	1.987	RT	32	0.3110	0.1720	0.1690	0.0287	0.3900
Sodium thiosulfate	Na$_2$S$_2$O$_3$	1.7499	RT	12	0.3323	0.1814	0.1875	0.0225	0.2953
Stilbene	(C$_6$H$_5$CH)$_2$	1.60	RT	94	0.0930	0.0570	0.0670	−0.003	0.0920
Triglycine sulfate (TGS)	(NH$_2$CH$_2$COOH)$_3$·H$_2$SO$_4$	1.68	RT	32	0.4550	0.1720	0.1980	−0.030	0.3210

Name	C_{23}	C_{25}	C_{33}	C_{35}	C_{44}	C_{46}	C_{55}	C_{66}
Aegirine	0.626	0.094	2.344	0.214	0.692	0.077	0.510	0.474
Anthracene	0.0375	−0.0170	0.1522	−0.0187	0.0272	0.0138	0.0242	0.0399
Cobalt sulfate heptahydrate	0.158	−0.018	0.371	−0.047	0.060	0.016	0.058	0.101
Diopside	0.482	−0.196	2.380	−0.336	0.675	−0.113	0.588	0.705
Dipotassium tartrate	0.1173	0.0176	0.6816	0.0294	0.0961	−0.0044	0.1270	0.0841
Feldspar (microceine)	0.192	−0.148	1.215	−0.131	0.143	−0.015	0.238	0.361
Ferrous sulfate heptahydrate	0.172	−0.019	0.360	−0.014	0.064	0.001	0.056	0.096
Lithium sulfate monohydrate	0.0368	0.0571	0.5400	−0.0254	0.1400	−0.0054	0.1565	0.2770
Naphthalene	0.0230	−0.0270	0.1190	0.0290	0.0330	−0.0050	0.0210	0.0415
Potassium tartrate	0.1330	0.0182	0.5540	0.0710	0.0870	0.0072	0.1040	0.0826
Sodium thiosulfate	0.1713	0.0983	0.4590	−0.0678	0.0569	−0.0268	0.1070	0.0598
Stilbene	0.0485	−0.005	0.0790	−0.005	0.0325	0.0050	0.0640	0.0245
Triglycine sulfate (TGS)	0.2080	−0.0036	0.2630	−0.0500	0.0950	−0.0026	0.1110	0.0620

HEXAGONAL CRYSTALS

Name	Formula	ρ/g cm^{-3}	T/K	Ref.	C_{11}	C_{12}	C_{13}	C_{33}	C_{55}
Apatite	$Ca_5(PO_4)_3(OH,F,Cl)$	3.218	RT	12	1.667	0.131	0.655	1.396	0.663
Beryl	$Be_3Al_2Si_6O_{18}$	2.68	RT	12	2.800	0.990	0.670	2.480	0.658
Beryllium	Be	1.8477	300	95	2.923	0.267	0.140	3.364	1.625
Beryllium oxide	BeO	3.01	RT	96	4.70	1.68	1.19	4.94	1.53
Cadmium	Cd	8.652	300	97	1.1450	0.3950	0.3990	0.5085	0.1985
Cadmium selenide	CdSe	5.655	298	68	0.7046	0.4516	0.3930	0.8355	0.1317
Cadmium sulfide	CdS	4.824	298	98	0.8431	0.5208	0.4567	0.9183	0.1458
Cobalt	Co	8.836	298	99	3.071	1.650	1.027	3.581	0.755
Dysprosium	Dy	8.560	298	100	0.7466	0.2616	0.2233	0.7871	0.2427
Erbium	Er	9.064	298	100	0.8634	0.3050	0.2270	0.8554	0.2809
Gadolinium	Gd	7.888	298	101	0.6667	0.2499	0.2132	0.7191	0.2089
Hafnium	Hf	12.727	298	102	1.881	0.772	0.661	1.969	0.557
Ice	H_2O(solid)	0.920	250	103	0.1410	0.0660	0.0624	0.1515	0.0288
Indium	In	7.2788	300	104	0.4535	0.4006	0.4151	0.4515	0.0651
Magnesium	Mg	1.7364	298	105	0.5950	0.2612	0.2180	0.6155	0.1635
Rhenium	Re	21.024	298	100	6.1820	2.7530	2.0780	6.8350	1.6060
Ruthenium	Ru	12.3615	298	100	5.6260	1.8780	1.6820	6.2420	1.8060
Thallium	Tl	11.560	300	106	0.4080	0.3540	0.2900	0.5280	0.0726
Titanium	Ti	4.5063	298	102	1.6240	0.9200	0.6900	1.8070	0.4670
Titanium diboride	TiB_2	4.95	RT	107	6.90	4.10	3.20	4.40	2.50
Yttrium	Y	4.472	300	108	0.7790	0.2850	0.2100	0.7690	0.2431
Zinc	Zn	7.134	295	109	1.6368	0.3640	0.5300	0.6347	0.3879
Zinc oxide	ZnO	5.6760	298	110	2.0970	1.2110	1.0510	2.1090	0.4247
Zinc sulfide	ZnS	4.089	298	96	1.2420	0.6015	0.4554	1.4000	0.2864
Zirconium	Zr	6.505	298	102	1.434	0.728	0.653	1.648	0.320

TRIGONAL CRYSTALS

Name	Formula	ρ/g cm^{-3}	T/K	Ref.	C_{11}	C_{12}	C_{13}	C_{14}	C_{33}	C_{44}
Aluminum oxide	Al_2O_3	3.986	300	111	4.9735	1.6397	1.1220	−0.2358	4.9911	1.4739
Aluminum phosphate	$AlPO_4$	2.556	RT	73	1.0503	0.2934	0.6927	−0.1271	1.3353	0.2314
Antimony	Sb	6.70	295	112	1.0130	0.3450	0.2920	0.2090	0.4500	0.3930
Bismuth	Bi	9.80	295	112	0.6370	0.2490	0.2470	0.0717	0.3820	0.1123
Calcite	$CaCO_3$	2.712	300	113	1.4806	0.5578	0.5464	−0.2058	0.8557	0.3269
Hematite	Fe_2O_3	5.240	RT	82	2.4243	0.5464	0.1542	−0.1247	2.2734	0.8569
Lithium niobate	$LiNbO_3$	4.70	RT	114	2.030	0.530	0.750	0.090	2.450	0.600
Lithium tantalate	$LiTaO_3$	7.45	RT	114	2.330	0.470	0.800	−0.110	2.750	0.940
Quartz	SiO_2	2.6485	298	115	0.8680	0.0704	0.1191	−0.1804	1.0575	0.5820
Selenium	Se	4.838	300	116	0.1870	0.0710	0.2620	0.0620	0.7410	0.1490
Sodium nitrate	$NaNO_3$	2.27	RT	12	0.8670	0.1630	0.1600	0.0820	0.3740	0.2130
Tourmaline		3.05	RT	82	2.7066	0.6927	0.0872	−0.0774	1.6070	0.6682

References

1. Thomas, J. F., *Phys. Rev.*, 175, 955–962, 1968.
2. Bolef, D. I. and M. Menes, *J. Appl. Phys.*, 31, 1426–1427, 1960.
3. Garland, C. W. and C. F. Yarnell, *J. Chem. Phys.*, 44, 1112–1120, 1966.
4. Garland, C. W. and R. Renard, *J. Chem. Phys.*, 44, 1130–1139, 1966.
5. Gsänger, M., H. Egger and E. Lüscher, *Phys. Letters*, 27A, 695–696, 1968.
6. Wong, C. and D. E. Schuele, *J. Phys. Chem. Solids*, 29, 1309–1330, 1968.
7. Haussühl, S., *Phys. Stat. Sol.*, 3, 1072–1076, 1963.
8. Wong, C. and D. E. Schuele, *J. Phys. Chem. Solids*, 28, 1225–1231, 1967.
9. McSkimin, H. J. and D. G. Thomas, *J. Appl. Phys.*, 33, 56–59, 1962.
10. Kollarits, F. J. and J. Trivisonno, *J. Phys. Chem. Solids*, 29, 2133–2139, 1968.
11. Slagle, D. D. and H. A. McKinstry, *J. Appl. Phys.*, 38, 446–458, 1967.
12. Hearmon, R. F. S., *Adv. Phys.*, 5, 323–382, 1956.
13. Sumer, A. and J. F. Smith, *J. Appl. Phys.*, 34, 2691–2694, 1963.
14. Alexandrov, K. S. et al., *Sov. Phys. Sol. State*, 10, 1316–1321, 1968.
15. Epstein, S. G. and O. N. Carlson, *Acta Metal.*, 13, 487–491, 1965.
16. McSkimin, H. J., et al., *J. Appl. Phys.*, 39, 4127–4128, 1968.
17. McSkimin, H. J., et al., *J. Appl. Phys.*, 38, 2362–2364, 1967.
18. Weil, R. and W. O. Groves, *J. Appl. Phys.*, 39, 4049–4051, 1968.
19. Bateman, T. B., *J. Appl. Phys.*, 37, 2194–2195, 1966.
20. Bogardus, E. H., *J. Appl. Phys.*, 36, 2504–2513, 1965.
21. Golding, B., S. C. Moss and B. L. Averbach, *Phys. Rev.*, 158, 637–645, 1967.
22. Bateman, T. B., H. J. McSkimin and J. M. Whelan, *J. Appl. Phys.*, 30, 544–545, 1959.
23. Gerlich, D., *J. Appl. Phys.*, 35, 3062, 1964.
24. Hickernell, F. S. and W. R. Gayton, *J. Appl. Phys.*, 37, 462, 1966.
25. MacFarlane, R. E., et al., *Phys. Letters*, 20, 234–235, 1966.
26. Leese, J. and A. E. Lord Jr., *J. Appl. Phys.*, 39, 3986–3988, 1968.
27. Miller, R. A. and D. E. Schuele, *J. Phys. Chem. Solids*, 30, 589–600, 1969.
28. Wasilik, J. H. and M. L. Wheat, *J. Appl. Phys.*, 36, 791–793, 1965.
29. Haussühl, S., *Phys. Stat. Sol.*, 3, 1072–1076, 1963.
30. Houston, B., et al., *J. Appl. Phys.*, 39, 3913–3916, 1968.
31. Trivisonno, J. and C. S. Smith, *Acta Metal.*, 9, 1064–1071, 1961.
32. Alexandrov, K. S. and T. V. Ryzhova, *Sov. Phys. Cryst.*, 6, 228–252, 1961.
33. Lewis, J. T., A. Lehoczky and C. V. Briscoe, *Phys. Rev.*, 161, 877–887, 1967.
34. Drabble, J. R. and R. E. B. Strathen, *Proc. Phys. Soc.*, 92, 1090–1995, 1967.
35. Oliver, D. W., *J. Appl. Phys.*, 40, 893, 1969.
36. Alper, T., and G. A. Saunders, *J. Phys. Chem. Solids*, 28, 1637–1642, 1967.
37. Dickinson, J. M. and P. E. Armstrong, *J. Appl. Phys.*, 38, 602–606, 1967.
38. Bolef, D. I., *J. Appl. Phys.*, 32, 100–105, 1961.
39. Rayne, J. A., *Phys. Rev.*, 112, 1125–1130, 1958.
40. MacFarlane, R. E., et al., *Phys. Letters*, 18, 91–92, 1965.
41. Smith, P. A. and C. S. Smith, *J. Phys. Chem. Solids*, 26, 279–289, 1965.
42. Norwood, M. H. and C. V. Briscoe, *Phys. Rev.*, 112, 45–48, 1958.
43. Simmons, G. and F. Birch, *J. Appl. Phys.*, 34, 2736–2738, 1963.
44. Gutman, E. J. and J. Trivisonno, *J. Appl. Phys. Sol.*, 28, 805–809, 1967.
45. Ghafelehbashi, M., et al., *J. Appl. Phys.*, 41, 652–666, 1970.
46. McSkimin, H. J. and P. Andreatch, Jr., *J. Appl. Phys.*, 35, 2161–2165, 1964.
47. Neighbours, J. R. and G. A. Alers, *Phys. Rev.*, 111, 707–712, 1958.
48. Hidshaw, W., J. T. Lewis, and C. V. Briscoe, *Phys. Rev.*, 163, 876–881, 1967.
49. Daniels, W. B., *Phys. Rev.*, 119, 1246–1252, 1960.
50. Viswanathan, R., *J. Appl. Phys.*, 37, 884–886, 1966.
51. Miller, R. A. and C. S. Smith, *J. Phys. Chem. Sol.*, 25, 1279–1292, 1964.
52. Claytor, R. N. and B. J. Marshall, *Phys. Rev.*, 120, 332–334, 1960.
53. Schreiber, E., *J. Appl. Phys.*, 38, 2508–2511, 1967.
54. Gerlich, D., *Phys. Rev.*, 136, A1366–A1368, 1964.
55. Johnston, D. L., P. H. Thrasher and R. J. Kearney, *J. Appl. Phys.*, 41, 427–428, 1970.
56. Poindexter, E. and A. A. Giardini, *Phys. Rev.*, 110, 1069, 1958.
57. Soga, N., *J. Appl. Phys.*, 37, 3416–3420, 1966.
58. Bartlett, R. W. and C. W. Smith, *J. Appl. Phys.*, 38, 5428–5429, 1967.
59. Morse, G. E. and A. W. Lawson, *J. Phys. Chem. Sol.*, 28, 939–950, 1967.
60. Armstrong, P. E., O. N. Carlson and J. F. Smith, *J. Appl. Phys.*, 30, 36–41, 1959.
61. Macedo, P. M., W. Capps and J. B. Wachtman, *J. Am. Cer. Soc.*, 47, 651, 1964.
62. Beattie, A. G., *J. Appl. Phys.*, 40, 4818–4821, 1969.
63. Chang, R. and L. J. Graham, *J. Appl Phys.*, 37, 3778–3783, 1966.
64. Lowrie, R. and A. M. Gonas, *J. Appl. Phys.*, 38, 4505–4509, 1967.
65. Graham, L. J., H. Nadler and R. Chang, *J. Appl. Phys.*, 34, 1572–1573, 1963.
66. Wachtman, J. B., Jr., et al., *J. Nucl. Mat.*, 16, 39–41, 1965.
67. Bolef, D. I., *J. Appl. Phys.*, 32, 100–105, 1961.
68. Berlincourt, D., H. Jaffe and L. R. Shiozawa, *Phys. Rev.*, 129, 1009–1017, 1963.
69. Adhav. R. S. J. *Acoust. Soc. Am.*, 43, 835–838, 1968.
70. Berlincourt, D. and H. Jaffe, *Phys. Rev.*, 111, 143–148, 1958.
71. Huntington, H. B., in *Solid State Pysics, Vol. 7*, Seitz, F., and Turnbull, D., Ed., pp. 213–285, Academic Press, New York 1958.
72. Cutler, H. R., J. J. Gibson and K. A. McCarthy, *Sol. State Comm.*, 6, 431–433, 1968.
73. Mason, W. P., *Piezoelectric Crystals and Their Application to Ultrasonics*, D. Van Nostrand Co., Inc., New York, 1950.
74. Adhav, R. S., *J. Appl. Phys.*, 40, 2725–2727, 1969.
75. Manghnani, M. H., *J. Geophys. Res.*, 74, 4317–4328, 1969.
76. Uchida, N. and Y. Ohmachi, *J. Appl Phys.*, 40, 4692–4695, 1969.
77. House, D. G. and E. Y. Vernon, *Br. J. Appl. Phys.*, 11, 254–259, 1960.
78. Ryzhova, T. V., et al., *Bull. Acad. Sci. USSR, Earth Phys. Ser.*, English Transl., no. 2, 111–113, 1966.
79. Alton, W. J. and A. J. Barlow, *J. Appl. Phys.*, 38, 3817–3820, 1967.
80. Michard, F., et al., *C. R. Acad. Sci., Paris*, 265, 565–567, 1967.
81. Haussühl, S., *Acta Cryst.*, 18, 839–842, 1965.
82. Hearmon, R. F. S., *Rev. Mod. Phys.*, 18, 409–440, 1946.
83. Heseltine, J. C. W., D. W. Elliott and O. B. Wilson, *J. Chem. Phys.*, 40, 2584–2587, 1964.
84. Schwerdtner, W. M., et al., *Canad. J. Earth Sci.*, 2, 673–683, 1965.
85. Kumazawa, M. and O. L. Anderson, *J. Geophys. Res.*, 74, 5961–5972, 1969.
86. Alexandrov, K. S., et al., *Sov. Phys. Cryst.*, 7, 753–755, 1963.
87. Verma, R. K., *J. Geophys. Soc.*, 65, 757–766, 1960.
88. McSkimin, H. J. and E. S. Fisher, *J. Appl. Phys.*, 31, 1627–1639, 1960.
89. Alexandrov, K. S. and T.V. Ryzhova, *Bull. Acad. Sci. USSR, Geophys. Ser.*, English Transl., no. 8, 871–875, 1961.
90. Afanaseva, G. K., et al, *Phys. Stat. Sol.*, 24, K61–K63, 1967.
91. Alexandrov, K. S., et al., *Sov. Phys. Cryst.*, 8, 589–591, 1964.
92. Alexandrov, K. S. and T. V Ryzhova, *Bull Acad. Sci. USSR, Geophys. Ser.*, English Transl., no. 2, 129–131, 1962.
93. Alexandrov, K. S., et al., *Sov. Phys. Cryst.*, 8, 164–166, 1963.
94. Teslenko, V. F., et al., *Sov. Phys. Cryst.*, 10, 744–747, 1966.
95. Smith, J. F. and C. L. Arbogast, *J. Appl. Phys.*, 31, 99–102, 1960.
96. Cline, C. F., H. L. Dunegan and G. M. Henderson, *J. Appl. Phys.*, 38, 1944–1948, 1967.
97. Chang, Y. A. and L. Himmel, *J. Appl. Phys.*, 37, 3787–3790, 1966.
98. Gerlich, D., *J. Phys. Chem. Solids*, 28, 2575–2579, 1967.
99. McSkimin, H. J., *J. Appl. Phys.*, 26, 406–409, 1955.
100. Fisher, E. S. and D. Dever, *Trans. Met. Soc. AIME*, 239, 48–57, 1967.
101. Fisher, E. S. and D. Dever, *Proc. Conf. Rare Earth Res.*, 6th, Gatlinburg, TN, 522–533, 1967.
102. Fisher, E. S. and C. J. Renken, *Phys. Rev.*, 135, A482–A494, 1964.
103. Proctor, T. M., Jr., *J. Acoust. Soc. Am.*, 39, 972–977, 1966.
104. Chandrasekhar, B. S. and J. A. Rayne, *Phys. Rev.*, 124, 1011–1041, 1961.
105. Wazzan, A. R. and L. B. Robinson, *Phys. Rev.*, 155, 586–594, 1967.
106. Ferris, R. W., et al., *J. Appl. Phys.*, 34, 768–770, 1963.
107. Gilman, J. J. and B. W. Roberts, *J. Appl. Phys.*, 32, 1405, 1961.
108. Smith, J. F. and J. A. Gjevre, *J. Appl. Phys.*, 31, 645–647, 1960.
109. Alers, G. A. and J. R. Neighbours, *J. Phys. Chem. Solids*, 7, 58–64, 1908.
110. Bateman, T. B., *J. Appl. Phys.*, 33, 3309–3312, 1962.
111. Tefft, W. E., *J. Res. Natl. Bur. Stand.*, 70A, 277–280, 1966.
112. DeBretteville, Jr., A. et al., *Phys. Rev.*, 148, 575–579, 1966.
113. Dandekar, D. P. and A. L. Ruoff, *J. Appl. Phys.*, 39, 6004–6009, 1968.
114. Warner, A. W., M. Onoe and G. A. Coquin, *J. Acoust. Soc. Am.*, 42, 1223–1231, 1967.
115. McSkimin, H. J., P. Andreatch and R. N. Thurston, *J. Appl. Phys.*, 36, 1624–1632, 1965.
116. Mort, J., *J. Appl. Phys.*, 38, 3414–3415, 1967.

ELECTRICAL RESISTIVITY OF PURE METALS

The first part of this table gives the electrical resistivity, in units of 10^{-8} Ω m, for 28 common metallic elements as a function of temperature. The data refer to polycrystalline samples. The number of significant figures indicates the accuracy of the values. However, at low temperatures (especially below 50 K) the electrical resistivity is extremely sensitive to sample purity. Thus the low-temperature values refer to samples of specified purity and treatment. The references should be consulted for further information on this point, as well as for values at additional temperatures.

The second part of the table gives resistivity values in the neighborhood of room temperature for other metallic elements that have not been studied over an extended temperature range.

References

1. C. Y. Ho, et al., *J. Phys. Chem. Ref. Data*, 12, 183–322, 1983; 13, 1069–1096, 1984; 13, 1097–1130, 1984, 13, 1131–1172, 1984.
2. R. A. Matula, *J. Phys Chem. Ref. Data*, 8, 1147–1298, 1979.
3. T. C. Chi, *J. Phys. Chem. Ref. Data*, 8, 339–438, 1979; 8, 439–498, 1979.
4. K. H. Hellwege, Ed., *Landolt-Börnstein Numerical Data and Functional Relationships in Science and Technology*, Group III, Vol. 15, Subvolume a, Springer-Verlag, Heidelberg, 1982.
5. L. A. Hall, *Survey of Electrical Resistivity Measurements on 16 Pure Metals in the Temperature Range 0 to 273 K*, NBS Technical Note 365, U.S. Superintendent of Documents, 1968.

Electrical Resistivity in 10^{-8} Ω m

T/K	Aluminum	Barium	Beryllium	Calcium	Cesium	Chromium	Copper
1	0.000100	0.081	0.0332	0.045	0.0026		0.00200
10	0.000193	0.189	0.0332	0.047	0.243		0.00202
20	0.000755	0.94	0.0336	0.060	0.86		0.00280
40	0.0181	2.91	0.0367	0.175	1.99		0.0239
60	0.0959	4.86	0.067	0.40	3.07		0.0971
80	0.245	6.83	0.075	0.65	4.16		0.215
100	0.442	8.85	0.133	0.91	5.28	1.6	0.348
150	1.006	14.3	0.510	1.56	8.43	4.5	0.699
200	1.587	20.2	1.29	2.19	12.2	7.7	1.046
273	2.417	30.2	3.02	3.11	18.7	11.8	1.543
293	2.650	33.2	3.56	3.36	20.5	12.5	1.678
298	2.709	34.0	3.70	3.42	20.8	12.6	1.712
300	2.733	34.3	3.76	3.45	21.0	12.7	1.725
400	3.87	51.4	6.76	4.7		15.8	2.402
500	4.99	72.4	9.9	6.0		20.1	3.090
600	6.13	98.2	13.2	7.3		24.7	3.792
700	7.35	130	16.5	8.7		29.5	4.514
800	8.70	168	20.0	10.0		34.6	5.262
900	10.18	216	23.7	11.4		39.9	6.041

T/K	Gold	Hafnium	Iron	Lead	Lithium	Magnesium	Manganese
1	0.0220	1.00	0.0225		0.007	0.0062	7.02
10	0.0226	1.00	0.0238		0.008	0.0069	18.9
20	0.035	1.11	0.0287		0.012	0.0123	54
40	0.141	2.52	0.0758		0.074	0.074	116
60	0.308	4.53	0.271		0.345	0.261	131
80	0.481	6.75	0.693	4.9	1.00	0.557	132
100	0.650	9.12	1.28	6.4	1.73	0.91	132
150	1.061	15.0	3.15	9.9	3.72	1.84	136
200	1.462	21.0	5.20	13.6	5.71	2.75	139
273	2.051	30.4	8.57	19.2	8.53	4.05	143
293	2.214	33.1	9.61	20.8	9.28	4.39	144
298	2.255	33.7	9.87	21.1	9.47	4.48	144
300	2.271	34.0	9.98	21.3	9.55	4.51	144
400	3.107	48.1	16.1	29.6	13.4	6.19	147
500	3.97	63.1	23.7	38.3		7.86	149
600	4.87	78.5	32.9			9.52	151
700	5.82		44.0			11.2	152
800	6.81		57.1			12.8	
900	7.86					14.4	

T/K	Molybdenum	Nickel	Palladium	Platinum	Potassium	Rubidium	Silver
1	0.00070	0.0032	0.0200	0.002	0.0008	0.0131	0.00100
10	0.00089	0.0057	0.0242	0.0154	0.0160	0.109	0.00115
20	0.00261	0.0140	0.0563	0.0484	0.117	0.444	0.0042
40	0.0457	0.068	0.334	0.409	0.480	1.21	0.0539
60	0.206	0.242	0.938	1.107	0.90	1.94	0.162
80	0.482	0.545	1.75	1.922	1.34	2.65	0.289
100	0.858	0.96	2.62	2.755	1.79	3.36	0.418
150	1.99	2.21	4.80	4.76	2.99	5.27	0.726
200	3.13	3.67	6.88	6.77	4.26	7.49	1.029
273	4.85	6.16	9.78	9.6	6.49	11.5	1.467
293	5.34	6.93	10.54	10.5	7.20	12.8	1.587
298	5.47	7.12	10.73	10.7	7.39	13.1	1.617
300	5.52	7.20	10.80	10.8	7.47	13.3	1.629
400	8.02	11.8	14.48	14.6			2.241
500	10.6	17.7	17.94	18.3			2.87
600	13.1	25.5	21.2	21.9			3.53
700	15.8	32.1	24.2	25.4			4.21
800	18.4	35.5	27.1	28.7			4.91
900	21.2	38.6	29.4	32.0			5.64

T/K	Sodium	Strontium	Tantalum	Tungsten	Vanadium	Zinc	Zirconium
1	0.0009	0.80	0.10	0.000016		0.0100	0.250
10	0.0015	0.80	0.102	0.000137	0.0145	0.0112	0.253
20	0.016	0.92	0.146	0.00196	0.039	0.0387	0.357
40	0.172	1.70	0.751	0.0544	0.304	0.306	1.44
60	0.447	2.68	1.65	0.266	1.11	0.715	3.75
80	0.80	3.64	2.62	0.606	2.41	1.15	6.64
100	1.16	4.58	3.64	1.02	4.01	1.60	9.79
150	2.03	6.84	6.19	2.09	8.2	2.71	17.8
200	2.89	9.04	8.66	3.18	12.4	3.83	26.3
273	4.33	12.3	12.2	4.82	18.1	5.46	38.8
293	4.77	13.2	13.1	5.28	19.7	5.90	42.1
298	4.88	13.4	13.4	5.39	20.1	6.01	42.9
300	4.93	13.5	13.5	5.44	20.2	6.06	43.3
400		17.8	18.2	7.83	28.0	8.37	60.3
500		22.2	22.9	10.3	34.8	10.82	76.5
600		26.7	27.4	13.0	41.1	13.49	91.5
700		31.2	31.8	15.7	47.2		104.2
800		35.6	35.9	18.6	53.1		114.9
900			40.1	21.5	58.7		123.1

Element	T/K	Electrical resistivity 10^{-8} Ω m	Element	T/K	Electrical resistivity 10^{-8} Ω m
Antimony	273	39	Osmium	273	8.1
Bismuth	273	107	Polonium	273	40
Cadmium	273	6.8	Praseodymium	290–300	70.0
Cerium (β, hex)	290–300	82.8	Promethium	290–300	75 est.
Cerium (γ, cub)	298	74.4	Protactinium	273	17.7
Cobalt	273	5.6	Rhenium	273	17.2
Dysprosium	290–300	92.6	Rhodium	273	4.3
Erbium	290–300	86.0	Ruthenium	273	7.1
Europium	290–300	90.0	Samarium	290–300	94.0
Gadolinium	290–300	131	Scandium	290–300	56.2
Gallium	273	13.6	Terbium	290–300	115
Holmium	290–300	81.4	Thallium	273	15
Indium	273	8.0	Thorium	273	14.7
Iridium	273	4.7	Thulium	290–300	67.6
Lanthanum	290–300	61.5	Tin	273	11.5
Lutetium	290–300	58.2	Titanium	273	39
Mercury	298	96.1	Uranium	273	28
Neodymium	290–300	64.3	Ytterbium	290–300	25.0
Niobium	273	15.2	Yttrium	290–300	59.6

ELECTRICAL RESISTIVITY OF SELECTED ALLOYS

These values were obtained by fitting all available measurements to a theoretical formulation describing the temperature and composition dependence of the electrical resistivity of metals. Some of the values listed here fall in regions of temperature and composition where no actual measurements exist. Details of the procedure may be found in the reference.

Values of the resistivity are given in units of 10^{-8} Ω m. General comments in the preceding table for pure metals also apply here.

Reference

C. Y. Ho, et al., *J. Phys. Chem. Ref. Data*, 12, 183–322, 1983.

Aluminum-Copper

Wt % Al	100 K	273 K	293 K	300 K	350 K	400 K
99[a]	0.531	2.51	2.74	2.82	3.38	3.95
95[a]	0.895	2.88	3.10	3.18	3.75	4.33
90[b]	1.38	3.36	3.59	3.67	4.25	4.86
85[b]	1.88	3.87	4.10	4.19	4.79	5.42
80[b]	2.34	4.33	4.58	4.67	5.31	5.99
70[b]	3.02	5.03	5.31	5.41	6.16	6.94
60[b]	3.49	5.56	5.88	5.99	6.77	7.63
50[b]	4.00	6.22	6.55	6.67	7.55	8.52
40[c]		7.57	7.96	8.10	9.12	10.2
30[c]		11.2	11.8	12.0	13.5	15.2
25[f]		16.3	17.2	17.6	19.8	22.2
15[h]			12.3			
10[g]	8.71	10.8	11.0	11.1	11.7	12.3
5[c]	7.92	9.43	9.61	9.68	10.2	10.7
1[b]	3.22	4.46	4.60	4.65	5.00	5.37

Aluminum-Magnesium

Wt % Al	100 K	273 K	293 K	300 K	350 K	400 K
99[c]	0.958	2.96	3.18	3.26	3.82	4.39
95[c]	3.01	5.05	5.28	5.36	5.93	6.51
90[c]	5.42	7.52	7.76	7.85	8.43	9.02
10[b]	14.0	17.1	17.4	17.6	18.4	19.2
5[b]	9.93	13.1	13.4	13.5	14.3	15.2
1[a]	2.78	5.92	6.25	6.37	7.20	8.03

Copper-Gold

Wt % Cu	100 K	273 K	293 K	300 K	350 K	400 K
99[c]	0.520	1.73	1.86	1.91	2.24	2.58
95[c]	1.21	2.41	2.54	2.59	2.92	3.26
90[c]	2.11	3.29	4.42	3.46	3.79	4.12
85[c]	3.01	4.20	4.33	4.38	4.71	5.05
80[c]	3.95	5.15	5.28	5.32	5.65	5.99
70[c]	5.91	7.12	7.25	7.30	7.64	7.99
60[c]	8.04	9.18	9.13	9.36	9.70	10.05
50[c]	9.88	11.07	11.20	11.25	11.60	11.94
40[c]	11.44	12.70	12.85	12.90	13.27	13.65
30[c]	12.43	13.77	13.93	13.99	14.38	14.78
25[c]	12.59	13.93	14.09	14.14	14.54	14.94
15[c]	11.38	12.75	12.91	12.96	13.36	13.77
10[c]	9.33	10.70	10.86	10.91	11.31	11.72
5[c]	5.91	7.25	7.41	7.46	7.87	8.28
1[c]	2.00	3.40	3.57	3.62	4.03	4.45

Copper-Nickel

Wt % Cu	100 K	273 K	293 K	300 K	350 K	400 K
99[c]	1.45	2.71	2.85	2.91	3.27	3.62
95[c]	6.19	7.60	7.71	7.82	8.22	8.62
90[c]	12.08	13.69	13.89	13.96	14.40	14.81
85[c]	18.01	19.63	19.83	19.90	20.32	20.70
80[c]	23.89	25.46	25.66	25.72	26.12	26.44
70[i]	35.73	36.67	36.72	36.76	36.85	36.89
60[i]	45.76	45.43	45.38	43.35	45.20	45.01
50[i]	50.22	50.19	50.05	50.01	49.73	49.50
40[c]	36.77	47.42	47.73	47.82	48.28	48.49
30[i]	26.73	40.19	41.79	42.34	44.51	45.40
25[c]	22.22	33.46	35.11	35.69	39.67	42.81
15[c]	13.49	22.00	23.35	23.85	27.60	31.38
10[c]	9.28	16.65	17.82	18.26	21.51	25.19
5[c]	5.20	11.49	12.50	12.90	15.69	18.78
1[c]	1.81	7.23	8.08	8.37	10.63	13.18

Copper-Palladium

Wt % Cu	100 K	273 K	293 K	300 K	350 K	400 K
99[c]	0.91	2.10	2.23	2.27	2.59	2.92
95[c]	2.99	4.21	4.35	4.40	4.74	5.08
90[c]	5.69	6.89	7.03	7.08	7.41	7.74
85[c]	8.30	9.48	9.61	9.66	10.01	10.36
80[c]	10.74	11.99	12.12	12.16	12.51	12.87
70[c]	15.67	16.87	17.01	17.06	17.41	17.78
60[c]	20.45	21.73	21.87	21.92	22.30	22.69
50[c]	26.07	27.62	27.79	27.86	28.25	28.64
40[c]	33.53	35.31	35.51	35.57	36.03	36.47
30[c]	45.03	46.50	46.66	46.71	47.11	47.47
25[c]	44.12	46.25	46.45	46.52	46.99	47.43
15[c]	31.79	36.52	36.99	37.16	38.28	39.35
10[c]	23.00	28.90	29.51	29.73	31.19	32.56
5[c]	13.09	20.00	20.75	21.02	22.84	24.54
1[c]	8.97	11.90	12.67	12.93	14.82	16.68

Copper-Zinc

Wt % Cu	100 K	273 K	293 K	300 K	350 K	400 K
99[b]	0.671	1.84	1.97	2.02	2.36	2.71
95[b]	1.54	2.78	2.92	2.97	3.33	3.69
90[b]	2.33	3.66	3.81	3.86	4.25	4.63
85[b]	2.93	4.37	4.54	4.60	5.02	5.44
80[b]	3.44	5.01	5.19	5.26	5.71	6.17
70[b]	4.08	5.87	6.08	6.15	6.67	7.19

Gold-Palladium

Wt % Au	100 K	273 K	293 K	300 K	350 K	400 K
99[c]	1.31	2.69	2.86	2.91	3.32	3.73
95[c]	3.88	5.21	5.35	5.41	5.79	6.17
90[i]	6.70	8.01	8.17	8.22	8.56	8.93
85[b]	9.14	10.50	10.66	10.72	11.10	11.48
80[b]	11.23	12.75	12.93	12.99	13.45	13.93
70[c]	16.44	18.23	18.46	18.54	19.10	19.67
60[b]	24.64	26.70	26.94	27.02	27.63	28.23
50[a]	23.09	27.23	27.63	27.76	28.64	29.42
40[a]	19.40	24.65	25.23	25.42	26.74	27.95
30[b]	14.94	20.82	21.49	21.72	23.35	24.92
25[b]	12.72	18.86	19.53	19.77	21.51	23.19
15[a]	8.54	15.08	15.77	16.01	17.80	19.61
10[a]	6.54	13.25	13.95	14.20	16.00	17.81
5[a]	4.58	11.49	12.21	12.46	14.26	16.07
1[a]	3.01	10.07	10.85	11.12	12.99	14.80

Gold-Silver

Wt % Au	100 K	273 K	293 K	300 K	350 K	400 K
99[b]	1.20	2.58	2.75	2.80	3.22	3.63
95[a]	3.16	4.58	4.74	4.79	5.19	5.59
90[j]	5.16	6.57	6.73	6.78	7.19	7.58
85[j]	6.75	8.14	8.30	8.36	8.75	9.15
80[j]	7.96	9.34	9.50	9.55	9.94	10.33
70[j]	9.36	10.70	10.86	10.91	11.29	11.68
60[j]	9.61	10.92	11.07	11.12	11.50	11.87
50[j]	8.96	10.23	10.37	10.42	10.78	11.14
40[j]	7.69	8.92	9.06	9.11	9.46	9.81
30[a]	6.15	7.34	7.47	7.52	7.85	8.19
25[a]	5.29	6.46	6.59	6.63	6.96	7.30
15[a]	3.42	4.55	4.67	4.72	5.03	5.34
10[a]	2.44	3.54	3.66	3.71	4.00	4.31
5[i]	1.44	2.52	2.64	2.68	2.96	3.25
1[b]	0.627	1.69	1.80	1.84	2.12	2.42

Iron-Nickel

Wt % Fe	100 K	273 K	293 K	300 K	400 K
99[a]	3.32	10.9	12.0	12.4	18.7
95[c]	10.0	18.7	19.9	20.2	26.8
90[c]	14.5	24.2	25.5	25.9	33.2
85[c]	17.5	27.8	29.2	29.7	37.3
80[c]	19.3	30.1	31.6	32.2	40.0
70[b]	20.9	32.3	33.9	34.4	42.4
60[c]	28.6	53.8	57.1	58.2	73.9
50[d]	12.3	28.4	30.6	31.4	43.7
40[d]	7.73	19.6	21.6	22.5	34.0
30[c]	5.97	15.3	17.1	17.7	27.4
25[b]	5.62	14.3	15.9	16.4	25.1
15[c]	4.97	12.6	13.8	14.2	21.1
10[c]	4.20	11.4	12.5	12.9	18.9
5[c]	3.34	9.66	10.6	10.9	16.1
1[b]	1.66	7.17	7.94	8.12	12.8

Silver-Palladium

Wt % Ag	100 K	273 K	293 K	300 K	350 K	400 K
99[b]	0.839	1.891	2.007	2.049	2.35	2.66
95[b]	2.528	3.58	3.70	3.74	4.04	4.34
90[b]	4.72	5.82	5.94	5.98	6.28	6.59
85[k]	6.82	7.92	8.04	8.08	8.38	8.68
80[k]	8.91	10.01	10.13	10.17	10.47	10.78
70[k]	13.43	14.53	14.65	14.69	14.99	15.30
60[i]	19.4	20.9	21.1	21.2	21.6	22.0
50[k]	29.3	31.2	31.4	31.5	32.0	32.4
40[m]	40.8	42.2	42.2	42.2	42.3	42.3
30[b]	37.1	40.4	40.6	40.7	41.3	41.7
25[k]	32.4	36.67	37.06	37.19	38.1	38.8
15[i]	21.0	27.08	26.68	27.89	29.3	30.6
10[i]	14.95	21.69	22.39	22.63	24.3	25.9
5[b]	8.91	15.98	16.72	16.98	18.8	20.5
1[a]	3.97	11.06	11.82	12.08	13.92	15.70

[a] Uncertainty in resistivity is ± 2%.
[b] Uncertainty in resistivity is ± 3%.
[c] Uncertainty in resistivity is ± 5%.
[d] Uncertainty in resistivity is ± 7% below 300 K and ± 5% at 300 and 400 K.
[e] Uncertainty in resistivity is ± 7%.
[f] Uncertainty in resistivity is ± 8%.
[g] Uncertainty in resistivity is ± 10%.
[h] Uncertainty in resistivity is ± 12%.
[i] Uncertainty in resistivity is ± 4%.
[j] Uncertainty in resistivity is ± 1%.
[k] Uncertainty in resistivity is ± 3% up to 300 K and ± 4% above 300 K.
[m] Uncertainty in resistivity is ± 2% up to 300 K and ± 4% above 300 K.

ELECTRICAL RESISTIVITY OF GRAPHITE MATERIALS

L. I. Berger

At normal conditions, the only stable crystallographic modification of carbon is graphite. The quasi-stable diamond turns into graphite starting from about 1000 °C in air. In industry, a graphitic material is commonly called either *carbon*, if it consists of small and low-oriented crystallites, or *graphite*, the material with highly ordered structure. In the 1970s, the first carbon filaments of about 7 nm in diameter were grown by Morinobu Endo at the University of Orleans, France, by the vapor-growth technique. In 1985, Sir Harold Walter Kroto of Sussex University, UK, and Richard E. Smalley and co-workers at Rice University discovered spherical carbon molecules, C_{60} (or C60), consisting of combinations of carbon atoms organized into hexagons and pentagons, named *buckminsterfullerenes* or *fullerenes* and possessing very promising mechanical and electrical properties. In 1991, Sumio Iijima, NEC Labs, Japan, and David S. Bethune, IBM Almaden Labs, observed the carbon atomic groups in the form of tubes capped by halves of the fullerene molecules and formed on the cathodes of carbon arc devices. The length of the tubes could be up to tens of micrometers and the diameter, naturally, is equal to that of the fullerene molecule. These tubes, called *nanotubes*, may be single wall (SWNT) or consist of several concentric tubes with a common axis (multi-walled nanotubes, MWNT). Two-dimensional *graphene* is another crystallographic modification of graphite (Saroj Nayak, Rensselaer U., 2004) that is a flat hexagonal network of carbon atoms with a thickness equal to the carbon atom size. The nanotube may be considered as formed by strips of graphenes turned into a cylinder. The character of the electrical conductivity (metallic or semiconductive) of a SWNT depends on orientation of the carbon hexagons of the nanotube surface regarding its axis (the chiral angle [Ref. 1]). The following table contains some typical data on electrical and electronic properties of graphite materials.

References

1. M. S. Dresselhaus, G. Dresselhaus, and Ph. Avouris (Eds.), *Carbon Nanotubes. Synthesis, Structure, Properties, and Applications,* Springer-Verlag, 2001.
2. ESPI Metals Catalog, 2007.
3. SPI Supplies Catalog, 2007.
4. F. L. Vogel, *J. Mater. Sci.* 12, 982–986, 1977.
5. K. S. Novoselov et al., *Nature* 438, 197–200, 2005.
6. Y. Zhang et al., *Nature* 438, 201–204, 2005.
7. N. Tombros et al., *Nature* 448, 571–574, 2007.
8. H. Dai, in Ref. 1, pp. 29–53.
9. CTI Carbon Nanotube Cat., 2007.
10. L. Matija et al., *Sci. Forum* 413, 49–52, 2003.

Material	Electrical resistivity ρ at R. T. mΩ cm [µΩ inch]	Energy gap at R. T. eV	Electron mobility cm²/V s	$(1/\rho)d\rho/dt$ near R. T. 10^{-4}°C^{-1}	Ref.
Bulk graphite					
Electromet graphite	1.90 [750]			−5	2
Electro graphite	1.60 [630]			−5	2
Aeromet graphite	1.47 [580]			−5	2
ESPI Superconductive	1.75 [690]			−5	2
Radioelectronics data	30 [11,800]			−5.6	3
Highly ordered pyrolytic graphite	Parallel 0.04 [15.7]				3
	Across 150 [59000]				
Single crystal graphite, normal to *c*-axis	$1 \cdot 10^{-6}$				4
Graphenes					
n-Graphene		≈5 (M); ≈10 (Γ)[c]	10^6		5,6
p-Graphene			10^4		7
Carbon nanotubes					
Metallic SWNT	12 kΩ[a]				1
Semiconducting SWNT		0.7 – 0.9[b]	128[d]		1
MWNT	10^2				9
Carbon fullerenes					
Fullerene (C_{60})	10^{12}	1.95			10

[a] Minimum resistance of individual nanotubes [Ref. 8]
[b] Est. from Ref. 1, p. 47
[c] Est. from Ref. 1, p. 116
[d] Est. from Ref. 1, p. 179.

PERMITTIVITY (DIELECTRIC CONSTANT) OF INORGANIC SOLIDS

H. P. R. Frederikse

This table lists the permittivity ε, frequently called the dielectric constant, of a number of inorganic solids. When the material is not isotropic, the individual components of the permittivity are given. A superscript S indicates a measurement made under constant strain ("clamped" dielectric constant). If the constraint is removed, the measurement yields ε^T, the "unclamped" or free dielectric constant.

The temperature of the measurement is given when available; the symbol r.t. indicates a value at nominal room temperature. The frequency of the measurement is given in the last column (i.r. indicates a measurement in the infrared).

Substances are listed in alphabetical order by chemical formula.

Reference

Young, K. F., and Frederikse, H. P. R., *J. Phys. Chem. Ref. Data* 2, 313, 1973.

Formula	Name	ε_{ijk}	T/K	v/Hz
Ag_3AsS_3	Silver thioarsenate (Proustite)	$\varepsilon_{11}^T = 16.5, \varepsilon_{11}^S = 14.5$	r.t.	2×10^7
		$\varepsilon_{33}^T = 20.0, \varepsilon_{11}^S = 18.0$	r.t.	2×10^7
AgBr	Silver bromide	12.50	r.t.	
AgCN	Silver cyanide	5.6	r.t.	10^6
AgCl	Silver chloride	11.15	r.t.	
$AgNO_3$	Silver nitrate	9.0	293	5×10^5
$AgNa(NO_2)_2$	Silver sodium nitrite	4.5 ± 0.5	r.t.	9.4×10^9
Ag_2O	Silver oxide	8.8	r.t.	
$(AlF)_2SiO_4$	Aluminum fluosilicate (topaz)	$\varepsilon_{11} = 6.62$	297	7×10^3
		$\varepsilon_{22} = 6.58$	297	7×10^3
		$\varepsilon_{33} = 6.95$	297	7×10^3
Al_2O_3	Aluminum oxide (alumina)	$\varepsilon_{11} = \varepsilon_{22} = 9.34$	298	$10^2 - 8 \times 10^9$
		$\varepsilon_{33} = 11.54$	298	$10^2 - 8 \times 10^9$
$AlPO_4$	Aluminum phosphate	$\varepsilon_{11}^T = 6.05$	r.t.	10^3
AlSb	Aluminum antimonide	11.21	300	i.r.
AsF_3	Arsenic trifluoride	5.7	r.t.	
BN	Boron nitride	7.1	r.t.	i.r.
$BaCO_3$	Barium carbonate	8.53	291	2×10^5
$Ba(COOH)_2$	Barium formate	$\varepsilon_{11} = 7.9$	r.t.	10^3
		$\varepsilon_{22} = 5.9$	r.t.	10^3
		$\varepsilon_{33} = 7.5$	r.t.	10^3
$BaCl_2$	Barium chloride	9.81	r.t.	
$BaCl_2 \cdot 2H_2O$	Barium chloride dihydrate	9.00	r.t.	10^3
BaF_2	Barium fluoride	7.32	292	$5 \times 10^2 - 10^{11}$
$Ba(NO_3)_2$	Barium nitrate	4.95	292	2×10^5
$Ba_2NaNb_5O_{15}$	Barium sodium niobate ("Bananas")	$\varepsilon_{11}^S = 222, \varepsilon_{11}^T = 235$	296	10^4
		$\varepsilon_{22}^S = 227, \varepsilon_{22}^T = 247$	296	
		$\varepsilon_{33}^S = 32, \varepsilon_{33}^T = 51$	296	
BaO	Barium oxide (baria)	34 ± 1	248, 333	60×10^7
BaO_2	Barium peroxide	10.7	r.t.	2×10^6
BaS	Barium sulfide	19.23	r.t.	7.25×10^6
$BaSO_4$	Barium sulfate	11.4	288	10^8
$BaSnO_3$	Barium stannate	18	298	25×10^5
$BaTiO_3$	Barium titanate	$\varepsilon_{11}^T = 3600$	298	10^5
		$\varepsilon_{11}^S = 2300$	298	2.5×10^8
		$\varepsilon_{33}^T = 150$	298	10^5
		$\varepsilon_{33}^S = 80$	298	2.5×10^8
$Ba_6Ti_2Nb_8O_{30}$	Barium titanium niobate	$\varepsilon_{11} = \varepsilon_{22} \approx 190$	298	
		$\varepsilon_{33} \approx 220$	298	
$BaWO_4$	Barium tungstate	$\varepsilon_{11} = \varepsilon_{22} = 35.5 \pm 0.2$	297.5	1.6×10^3
		$\varepsilon_{33} = 37.2 \pm 0.2$	297.5	1.6×10^3
$BaZrO_3$	Barium zirconate	43	r.t.	

Formula	Name	ε_{ijk}	T/K	v/Hz
$Be_3Al_2Si_6O_{18}$	Beryllium aluminum silicate (Beryl)	$\varepsilon_{33} = 5.95$	297	7×10^3
		$\varepsilon_{11} = \varepsilon_{22} = 6.86$	297	7×10^3
$BeCO_3$	Beryllium carbonate	9.7	291	2×10^5
BeO	Beryllium oxide (beryllia)	7.35 ± 0.2	293	2×10^6
$BiFeO_3$	Bismuth iron oxide	40 ± 3	300	9.4×10^9
$Bi_{12}GeO_{20}$	Bismuth germanite	$\varepsilon_{11}^S = 38$	r.t.	
$Bi(GeO_4)_3$	Bismuth germanate	16	293	
Bi_2O_3	Bismuth sesquioxide	18.2	r.t.	2×10^6
$Bi_4Ti_3O_{12}$	Bismuth titanate	112	r.t.	10^3
C	Diamond			
	Type I	5.87 ± 0.19	300	10^3
	Type IIa	5.66 ± 0.04	300	10^3
$C_4H_4O_6$	Tartaric acid	$\varepsilon_{11} = \varepsilon_{22} = 4.3$	298	
		$\varepsilon_{33} = 4.5$	298	
		$\varepsilon_{13} = 0.55$	298	
$C_6H_{14}N_2O_6$	Ethylene diamine tartrate (EDT)	$\varepsilon_{11}^T = 5.0$	293	
		$\varepsilon_{22}^T = 8.3$	293	
		$\varepsilon_{33}^T = 6.0$	293	
		$\varepsilon_{13}^T = 0.7$	293	
$C_6H_{12}O_6NaBr$	Dextrose sodium bromide	$\varepsilon_{11}^T = 4.0$	r.t.	10^3
$(CH_3NH_3)Al(SO_4)_2 \cdot 2H_2O$	Methyl ammonium alum (MASD)	19	197	
$Ca_2B_6O_{11} \cdot 5H_2O$	Colemanite	$\varepsilon_{11} = 20$	293	10^3
		$\varepsilon_{33} = 25$	293	10^3
$CaCO_3$	Calcium carbonate	$\varepsilon_{11} = 8.67$	r.t.	9.4×10^{10}
		$\varepsilon_{22} = 8.69$	r.t.	9.4×10^{10}
		$\varepsilon_{33} = 8.31$	r.t.	9.4×10^{10}
$CaCeO_3$	Calcium cerate	21	r.t.	
CaF_2	Calcium fluoride	6.81	300	$5 \times 10^2 - 10^{11}$
$CaMoO_4$	Calcium molybdate	$\varepsilon_{11} = \varepsilon_{22} = 24.0 \pm 0.2$	297.5	<10
		$\varepsilon_{33} = 20.0 \pm 0.2$	297.5	<10
$Ca(NO_3)_2$	Calcium nitrate	6.54	292	2×10^5
$CaNb_2O_6$	Calcium niobate	$\varepsilon_{11} = 22.8 \pm 1.9$	r.t.	$(5-500) \times 10^3$
$Ca_2Nb_2O_7$	Calcium pyroniobate	~45	r.t.	5×10^7
CaO	Calcium oxide	11.8 ± 0.3	283	2×10^6
CaS	Calcium sulfide	6.699	r.t.	7.25×10^6
$CaSO_4 \cdot 2H_2O$	Calcium sulfate dihydrate	$\varepsilon_{11} = 5.10$	r.t.	
		$\varepsilon_{22} = 5.24$	r.t.	
		$\varepsilon_{33} = 10.30$	r.t.	
$CaTiO_3$	Calcium titanate	165	r.t.	
$CaWO_4$	Calcium tungstate	$\varepsilon_{11} = \varepsilon_{22} = 11.7 \pm 0.1$	297.5	1.59×10^3
		$\varepsilon_{33} = 9.5 \pm 0.2$	297.5	1.59×10^3
Cd_3As_2	Cadmium arsenide	$\varepsilon_{33} = 18.5$	4	
$CdBr_2$	Cadmium bromide	8.6	293	5×10^5
CdF_2	Cadmium fluoride	8.33 ± 0.08	300	$10^5 - 10^7$
CdS	Cadmium sulfide	$\varepsilon_{11} = \varepsilon_{22} = 8.7$	300	i.r.
		$\varepsilon_{33} = 9.25$	300	i.r.
		$\varepsilon_{11} = \varepsilon_{22} = 8.37$	8	i.r.
		$\varepsilon_{33} = 9.00$	8	i.r.
		$\varepsilon_{11}^T = 8.48$	77	10^4
		$\varepsilon_{33}^T = 9.48$	77	10^4
		$\varepsilon_{11}^S = 9.02, \varepsilon_{11}^T = 9.35$	298	10^4
		$\varepsilon_{33}^S = 9.53, \varepsilon_{33}^T = 10.33$	298	10^4
$CdSe$	Cadmium selenide	$\varepsilon_{11}^S = 9.53, \varepsilon_{11}^T = 9.70$	298	10^4
		$\varepsilon_{33}^S = 10.2, \varepsilon_{33}^T = 10.65$	298	10^4
$CdTe$	Cadmium telluride	$\varepsilon_{11} = \varepsilon_{22} = 10.60 \pm 0.15$	297	i.r.

Formula	Name	ε_{ijk}	T/K	ν/Hz
		$\varepsilon_{33} = 7.05 \pm 0.05$	297	i.r.
$Cd_2Nb_2O_7$	Cadmium pyroniobate	500–580	293	10^3
CeO_2	Cerium oxide	7.0	r.t.	2×10^6
$CoNb_2O_6$	Cobalt niobate	$\varepsilon_{11} = 18.4 \pm 1.1$	r.t.	$(5–500) \times 10^3$
		$\varepsilon_{22} = 21.4 \pm 1.1$	r.t.	$(5–500) \times 10^3$
		$\varepsilon_{33} = 33.0 \pm 0.7$	r.t.	$(5–500) \times 10^3$
CoO	Cobalt oxide	12.9	298	10^2– 10^{10}
Cr_2O_3	Chromic sesquioxide	$\varepsilon_{11} = \varepsilon_{22} = 13.3$	298.5	10^3
		$\varepsilon_{33} = 11.9$	298.5	10^3
		8	315 (T_N)	6×10^{10}
$CsAl(SO_4)_2 \cdot 12H_2O$	Cesium alum	5.0	r.t.	$20–20 \times 10^3$
$CsBr$	Cesium bromide	6.38	298	1.6×10^3
Cs_2CO_3	Cesium carbonate	6.53	291	2×10^5
$CsCl$	Cesium chloride	7.2	298	
CsH_2AsO_4	Cesium dihydrogen arsenate (CDA)	4.8	273	9.5×10^9
CsH_2PO_4	Cesium dihydrogen phosphate (CDP)	6.15	285	9.5×10^9
$CsH_3(SeO_3)_2$	Cesium trihydrogen selenite	$\varepsilon_{11} = 80$	273	10^5
		$\varepsilon_{22} = 63$	273	10^5
		$\varepsilon_{33} = 12$	273	10^5
CsI	Cesium iodide	6.31	298	1.6×10^3
$CsNO_3$	Cesium nitrate	$\varepsilon_{11} = \varepsilon_{22} = 9.4$	r.t.	5×10^5
		$\varepsilon_{33} = 8.3$	r.t.	5×10^5
$CsPbCl_3$	Cesium lead chloride	14.37	300	10^5–10^6
$CuBr$	Cuprous bromide	8.0	293	5×10^5
$CuCl$	Cuprous chloride	9.8 ± 0.5	r.t.	10^3
CuO	Cupric oxide	18.1	r.t.	2×10^6
Cu_2O	Cuprous oxide (Cuprite)	7.60 ± 0.06	r.t.	10^5
$CuSO_4 \cdot 5H_2O$	Cupric sulfate pentahydrate	6.60	r.t.	
EuF_2	Europium fluoride	7.7 ± 0.2	298	$(1–300) \times 10^3$
$Eu_2(MoO_4)_3$	Europium molybdate	9.5	298	
EuS	Europium sulfide	13.10 ± 0.04	80	5×10^2–10^5
FeO	Ferrous oxide	14.2	r.t.	2×10^6
Fe_2O_3	Ferric sesquioxide	4.5	r.t.	10^5–10^7
Fe_2O_3-α	Ferric sesquioxide (hematite)	12		6×10^{10}
Fe_3O_4	Ferrosoferric oxide (magnetite)	20	r.t.	10^5–10^7
$GaAs$	Gallium arsenide	13.13	300	
		12.90	4	i.r.
GaP	Gallium phosphide	11.1	r.t.	
		10.75 ± 0.1	1.6	i.r.
$GaSb$	Gallium antimonide	15.69	r.t.	
		15.7	4	i.r.
$Gd_2(MoO_4)_3$	Gadolinium molybdate	$\varepsilon^T = 10$	298	
		$\varepsilon^S = 9.5$	298	10^3
Ge	Germanium	16.0 ± 0.3	4	9.2×10^9
		15.8 ± 0.2	r.t.	$500–3 \times 10^{10}$
GeO_2	Germanium dioxide	$\varepsilon_{11} = \varepsilon_{22} = 7.44$	r.t.	i.r.
HIO_3	Iodic acid	$\varepsilon_{11} = 7.5$	r.t.	10^3
		$\varepsilon_{22} = 12.4$	r.t.	10^3
		$\varepsilon_{33} = 8.1$	r.t.	10^3
$HNH_4(ClCH_2COO)_2$	Hydrogen ammonium dichloroacetate	$\varepsilon_{[102]} = 5.9$	r.t.	10^5
H_2O	Ice I (P = 0 kbar)	99	243	
	Ice III (P = 3 kbar)	117	243	
	Ice V (P = 5 kbar)	114	243	
	Ice VI (P = 8 kbar)	193	243	
$HgCl$	Mercurous chloride (Calumel)	$\varepsilon_{11} = \varepsilon_{22} = 14.0$	r.t.	10^{12}
$HgCl_2$	Mercuric chloride	6.5	r.t.	10^{12}
HgS	Mercurous sulfide (Cinnabar)	$\varepsilon_{11} = \varepsilon_{22} = 18.0$	r.t.	i.r.
		$\varepsilon_{33} = 32.5$	r.t.	i.r.
$HgSe$	Mercurous selenide	25.6	r.t.	10^4–10^6
I_2	Iodine	$\varepsilon_{11} = 6$	r.t.	5×10^4–10^7
		$\varepsilon_{22} = 3$	r.t.	5×10^4–10^7

Formula	Name	ε_{ijk}	T/K	ν/Hz
		$\varepsilon_{33} = 40$	r.t.	$5 \times 10^4 - 10^7$
InAs	Indium arsenide	14.55 ± 0.3	r.t.	i.r.
		15.15	4	i.r.
InP	Indium phosphide	12.61	r.t.	i.r.
InSb	Indium antimonide	17.88	4	i.r.
$KAl(SO_4)_2 \cdot 12H_2O$	Potassium alum	6.5	r.t.	$20-20 \times 10^3$
KBr	Potassium bromide	4.88	300	
		4.53	4.2	
$KBrO_3$	Potassium bromate	7.3	r.t.	2×10^6
KCN	Potassium cyanide	6.15	r.t.	2×10^6
K_2CO_3	Potassium carbonate	4.96	291	2×10^5
$K_2C_4H_4O_6 \cdot 1/2 H_2O$	Dipotassium tartrate (DKT)	$\varepsilon_{11} = 6.44$	r.t.	
		$\varepsilon_{22} = 5.80$	r.t.	
		$\varepsilon_{33} = 6.49$	r.t.	
		$\varepsilon_{13} = 0.005$	r.t.	
KCl	Potassium chloride	4.86 ± 0.02	r.t.	5×10^3
		4.50	4.2	
$KClO_3$	Potassium chlorate	5.1	r.t.	2×10^6
$KClO_4$	Potassium perchlorate	5.9	r.t.	2×10^6
K_2CrO_4	Potassium chromate	7.3	r.t.	6×10^7
$KCr(SO_4)_2 \cdot 12H_2O$	Potassium chrome alum	6.5	100–240	175×10^3
KD_2AsO_4	Potassium dideuterium arsenate (KDDA)	$\varepsilon_{11} = 70$	298	
		$\varepsilon_{33} = 31$	298	
KD_2PO_4	Potassium dideuterium phosphate (KDDP)	50 ± 2	297	10^3
KF	Potassium fluoride	6.05		2×10^6
KH_2AsO_4	Potassium dihydrogen arsenate (KDA)	$\varepsilon_{11} = 60$	298	
		$\varepsilon_{33} = 24$	298	
KH_2PO_4	Potassium dihydrogen phosphate (KDP)	46	298	10^3
		$\varepsilon_{11} = 42$	r.t.	
		$\varepsilon_{33} = 21$	r.t.	
K_2HPO_4	Dipotassium monohydrogen orthophosphate	9.05	r.t.	2×10^6
KI	Potassium iodide	5.00	r.t.	9.4×10^{10}
KIO_3	Potassium iodate	170	255	10^5
		10	293	10^5
		$\varepsilon_{[101]} \approx 40,70$	r.t.	10^5
		16.85	r.t.	2×10^6
$(K,H)Al_3(SiO_4)_3$	Mica (muscovite)	5.4	299	$10^2 - 3 \times 10^9$
$(K,H)Mg_3Al(SiO_4)_3$	Mica (Canadian)	$\varepsilon_{11} = \varepsilon_{22} = 6.9$	298	$10^2 - 10^4$
		$\varepsilon_{33} = 7.3$	298	10^4
KNO_2	Potassium nitrite	25	305	
KNO_3	Potassium nitrate	4.37	293	2×10^5
$KNbO_3$	Potassium niobate	700	r.t.	
K_3PO_4	Potassium orthophosphate	7.75	r.t.	2×10^6
KSCN	Potassium thiocyanate	7.9	r.t.	2×10^6
K_2SO_4	Potassium sulfate	6.4	r.t.	2×10^6
$K_2S_3O_6$	Potassium trithionate	5.7	293	1.8×10^6
$K_2S_4O_6$	Potassium tetrathionate	5.5	293	1.8×10^6
$K_2S_5O_6 \cdot H_2O$	Potassium pentathionate	7.8	293	1.8×10^6
$K_2S_6O_6$	Potassium hexathionate	7.8	293	1.8×10^6
K_2SeO_4	Potassium selenate	$\varepsilon_{11} = 5.9$	r.t.	10^3
		$\varepsilon_{22} = 7.7$	r.t.	10^3
$KSr_2Nb_5O_{15}$	Potassium strontium niobate	$\varepsilon_{11} = \varepsilon_{11} \approx 1200$	298	
		$\varepsilon_{33} \approx 800$	298	
$KTaNbO_3$	Potassium tantalate niobate (KTN)	34,000	273	10^4
		6,000	293	10^4
$KTaO_3$	Potassium tantalate	242	298	2×10^5
$LaScO_3$	Lanthanum scandate	30	r.t.	
LiBr	Lithium bromide	12.1	r.t.	2×10^6
Li_2CO_3	Lithium carbonate	4.9	291	2×10^5
LiCl	Lithium chloride	11.05	r.t.	2×10^6

Formula	Name	ε_{ijk}	T/K	ν/Hz
LiD	Lithium deuteride	14.0 ± 0.5	r.t.	i.r.
LiF	Lithium fluoride	9.00	298	10^2-10^7
		9.11	353	10^2-10^7
$LiGaO_2$	Lithium metagallate	$\varepsilon_{11}^T = 7.0, \varepsilon_{22}^T = 6.0$	r.t.	
		$\varepsilon_{33}^T = 9.5$	r.t.	
		$\varepsilon_{11}^S = 6.8, \varepsilon_{22}^S = 5.8$	r.t.	
Li^6H	Lithium-6 hydride	13.2 ± 0.5	r.t.	
Li^7H	Lithium-7 hydride	12.9 ± 0.5	r.t.	
$LiH_3(SeO_3)_2$	Lithium trihydrogen selenite	29	298	10^4
		$\varepsilon_{11} = 13.0$	r.t.	
		$\varepsilon_{22} = 12.9$	r.t.	
		$\varepsilon_{33} = 46$	r.t.	
LiI	Lithium iodide	11.03	r.t.	2×10^6
$LiIO_3$	Lithium iodate	$\varepsilon_{11} = \varepsilon_{22} = 65$	294.5	10^3
		$\varepsilon_{33} = 554$	298	
$LiNH_4C_4O_6 \cdot H_2O$	Lithium ammonium tartrate (LAT)	$\varepsilon_{11}^T = 7.2$	298	
		$\varepsilon_{22}^T = 8.0$	298	
		$\varepsilon_{33}^T = 6.9$	298	
$LiNa_3CrO_4 \cdot 6H_2O$	Lithium trisodium chromate	8.0	r.t.	10^3
$LiNa_3MoO_4 \cdot 6H_2O$	Lithium trisodium molybdate	$\varepsilon_{11} = 6.7$	r.t.	10^3
		$\varepsilon_{33} = 5.3$	r.t.	10^3
$LiNbO_3$	Lithium niobate	$\varepsilon_{11} = \varepsilon_{22} = 82$	298	10^5
		$\varepsilon_{33} = 30$	298	10^5
$Li_2SO_4 \cdot H_2O$	Lithium sulfate monohydrate	$\varepsilon_{11} = 5.6$	298	
		$\varepsilon_{22} = 10.3$	298	
		$\varepsilon_{33} = 6.5$	298	
		$\varepsilon_{13} = 0.07$	298	
$LiTaO_3$	Lithium tantalate	$\varepsilon_{11} = \varepsilon_{22} = 53$	r.t.	10^5
		$\varepsilon_{33} = 46$	r.t.	10^5
		$\varepsilon_{11}^S = \varepsilon_{22}^S = 41$	r.t.	
		$\varepsilon_{33}^S = 43$	r.t.	
		$\varepsilon_{11}^T = \varepsilon_{22}^T = 51$	r.t.	
		$\varepsilon_{33}^T = 45$	r.t.	
$LiTlC_4O_6 \cdot H_2O$	Lithium thallium tartrate (LTT)	$\varepsilon_{11} \approx 20$	80	
$Mg_3B_7O_{13}Cl$	Magnesium borate monochloride (boracite)	$\varepsilon_{11} = 14.1$	r.t.	5×10^5
$MgCO_3$	Magnesium carbonate	8.1	291	2×10^5
$MgNb_2O_6$	Magnesium niobate	$\varepsilon_{11} = 16.4 \pm 0.5$	r.t.	$(5-500) \times 10^3$
		$\varepsilon_{22} = 20.9 \pm 0.5$	r.t.	$(5-500) \times 10^3$
		$\varepsilon_{33} = 32.4 \pm 0.5$	r.t.	$(5-500) \times 10^3$
MgO	Magnesium oxide (Periclase)	9.65	298	10^2-10^8
$(MgO)_xAl_2O_3$	Spinel	8.6	r.t.	
$MgSO_4$	Magnesium sulfate	8.2	r.t.	
$MgSO_4 \cdot 7H_2O$	Magnesium sulfate septahydrate	5.46	r.t.	
$MgTiO_3$	Magnesium titanate	13.5	r.t.	
$MgWO_4$	Magnesium tungstate	$\varepsilon_{11} = 18.0 \pm 1$	r.t.	$(5-500) \times 10^3$
		$\varepsilon_{22} = 18.0 \pm 1$	r.t.	$(5-500) \times 10^3$
$MnNb_2O_6$	Manganese niobate	$\varepsilon_{11} = 17.4 \pm 2$	r.t.	$(5-500) \times 10^3$
		$\varepsilon_{22} = 16.1 \pm 0.5$	r.t.	$(5-500) \times 10^3$
		$\varepsilon_{33} = 30.7 \pm 1$	r.t.	$(5-500) \times 10^3$
MnO	Manganese oxide (Pyrolusite)	12.8	r.t.	6×10^{10}
MnO_2	Manganese dioxide	$\sim 10^4$	298	10^4
Mn_2O_3	Manganese sesquioxide	8	r.t.	6×10^{10}
$MnWO_4$	Manganese tungstate	$\varepsilon_{11} = 19.3 \pm 1.3$	r.t.	$(5-500) \times 10^3$
		$\varepsilon_{22} = 14.3 \pm 0.5$	r.t.	$(5-500) \times 10^3$
		$\varepsilon_{33} = 16.5 \pm 1.1$	r.t.	$(5-500) \times 10^3$
$N(CH_3)_4HgBr_3$	Tetramethylammonium tribromomercurate (TTM)	~ 10	233-373	

Formula	Name	ε_{ijk}	T/K	ν/Hz
$N(CH_3)_4HgI_3$	Tetramethylammonium triiodo mercurate (TTM)	~10	233–373	
$N_4(CH_2)_6$	Hexamethylene tetramine (HMTA)	2.6 ± 0.2	r.t.	$10^9 – 10^{10}$
$(ND_4)_2BeF_4$	Deuteroammonium fluoberyllate	$\varepsilon_{11} = 10$	r.t.	
		$\varepsilon_{22} = 9$	r.t.	
		$\varepsilon_{33} = 9$	r.t.	
$(ND_4)_2SO_4$	Deuteroammonium sulfate	$\varepsilon_{11} = 9$	r.t.	
		$\varepsilon_{22} = 10$	r.t.	
		$\varepsilon_{33} = 9$	r.t.	
$(NH_2 \cdot CH_2COOH)_3 \cdot H_2SO_4$	Triglycine sulfate (TGS)	$\varepsilon_{11} = 9$	273	10^4
		$\varepsilon_{22} = 30$	273	10^4
		$\varepsilon_{33} = 6.5$	273	10^4
$(NH_2 \cdot CH_2COOH)_3 \cdot H_2SeO_4$	Triglycine selenate (TGSe)	200	293	1.6×10^3
$(NH_2 \cdot CH_2COOH)_3 \cdot H_2BeF_4$	Triglycine fluorberyllate (TGFB)	$\varepsilon_{22} = 12$	273	10^4
$NH_4Al(SO_4)_2 \cdot 12H_2O$	Ammonium alum	6	r.t.	10^{12}
$(NH_4)_2BeF_4$	Ammonium fluorberyllate	$\varepsilon_{11} = \varepsilon_{22} = 7.8$	123	10^5
		$\varepsilon_{33} = 7.1$	123	10^5
		$\varepsilon_{11} = \varepsilon_{22} = 8.8$	293	10^5
		$\varepsilon_{33} = 9.2$	293	10^5
NH_4Br	Ammonium bromide	7.1	r.t.	7×10^5
NH_4I	Ammonium iodide	9.8	r.t.	
$(NH_4)_2C_4H_6O_6$	Ammonium tartrate	$\varepsilon_{11} = 6.45$	r.t.	10^3
		$\varepsilon_{22} = 6.8$	r.t.	10^3
		$\varepsilon_{33} = 6.0$	r.t.	10^3
$(NH_4)_2Cd_2(SO_4)_3$	Ammonium cadmium sulfate	10.0	r.t.	10^4
NH_4Cl	Ammonium chloride	6.9	r.t.	7×10^5
$NH_4(ClCH_2COO)$	Ammonium monochloroacetate	5	r.t.	2×10^6
$NH_4Cr(SO_4)_2 \cdot 12H_2O$	Ammonium chrome alum	6.5	r.t.	175×10^3
NH_4HSO_4	Ammonium bisulfate	165	273	5×10^4
$NH_4H_2AsO_4$	Ammonium dihydrogen arsenate (ADA)	5.1	265	9.5×10^9
		$\varepsilon_{11} = \varepsilon_{22} = 85$	298	10^3
		$\varepsilon_{33} = 22$	298	
$NH_4H_2PO_4$	Ammonium dihydrogen phosphate (ADP)	$\varepsilon_{11} = \varepsilon_{22} = 57.1 \pm 0.6$	294.5	$10^5 – 35 \times 10^9$
		$\varepsilon_{33} = 14.0 \pm 0.3$	294	$10^5 – 36 \times 10^9$
$ND_4D_2PO_4$	Ammonium dideuterium phosphate (ADDP)	$\varepsilon_{11} = \varepsilon_{22} = 74, \varepsilon_{33} = 24$	300	
NH_4NO_3	Ammonium nitrate	10.7	322	$(5–50) \times 10^3$
$(NH_4)_2SO_4$	Ammonium sulfate	$\varepsilon_{11} = \varepsilon_{22} = 8.0$	123	10^5
		$\varepsilon_{33} = 6.3$	123	10^5
		$\varepsilon_{11} = \varepsilon_{22} = 10.0$	293	10^5
		$\varepsilon_{33} = 9.3$	293	10^5
$(NH_4)_2UO_2(C_2O_4)_2$	Ammonium uranyl oxalate	8.03	r.t.	$10^4 – 3.3 \times 10^9$
$(NH_4)_2UO_2(C_2O_4)_2 \cdot 3H_2O$	Ammonium uranyl oxalate trihydrate	6.06	r.t.	$10^4 – 3.3 \times 10^9$
$NaBr$	Sodium bromide	6.44	298	1.6×10^3
$NaBrO_3$	Sodium bromate	$\varepsilon_{11}^T = 5.70$	298	10^3
$NaCN$	Sodium cyanide	7.55	293	10^5
$NaCO_3$	Sodium carbonate	8.75	291	2×10^5
$NaCO_3 \cdot 10H_2O$	Sodium carbonate decahydrate	5.3	r.t.	6×10^7
$NaCl$	Sodium chloride	5.9	298	$10^2 – 10^7$
		5.45	4.2	
$NaClO_3$	Sodium chlorate	$\varepsilon_{11}^T = 5.76$	301	10^3
		5.28	r.t.	10^3
$NaClO_4$	Sodium perchlorate	5.76	r.t.	10^3
NaF	Sodium fluoride	5.08 ± 0.02	r.t.	5×10^3
$NaH_3(SeO_3)_2$	Sodium trihydrogen selenite	$\varepsilon_{11} \approx 75$	273	2×10^5
$NaD_3(SeO_3)_2$	Sodium trideuterium selenite	$\varepsilon_{11} \approx 220$	273	2×10^5
NaI	Sodium iodide	7.28 ± 0.03	r.t.	

Formula	Name	ε_{ijk}	T/K	ν/Hz
$NaK(C_4H_2D_2O_6) \cdot 4D_2O$	Sodium potassium tartrate tetradeutrate (double deuterated Rochelle salt)	$\varepsilon_{11} = 70$	273	10^3
		$\varepsilon_{22} = 8.9$	273	10^3
$NaK(C_4H_4O_6) \cdot 4H_2O$	Sodium potassium tartrate tetrahydrate (Rochelle salt)	$\varepsilon_{11} = 170$	273	10^3
		$\varepsilon_{22} = 9.1$	273	10^3
$NaNH_4(C_4H_4O_6) \cdot 4H_2O$	Sodium ammonium tartrate (Ammonium Rochelle salt)	$\varepsilon_{11} = 8.4$	298	
		$\varepsilon_{22} = 9.2$	298	
		$\varepsilon_{33} = 9.5$	298	
$NaNbO_3$	Sodium niobate	$\varepsilon_{33} = 670 \pm 13$	r.t.	
		$\varepsilon_{11} = \varepsilon_{22} = 76 \pm 2$	r.t.	
$NaNO_2$	Sodium nitrite	$\varepsilon_{11} = 7.4$	r.t.	5×10^5
		$\varepsilon_{22} = 5.5$	r.t.	5×10^5
		$\varepsilon_{33} = 5.0$	r.t.	5×10^5
$NaNO_3$	Sodium nitrate	6.85	292	2×10^5
$NaSO_4$	Sodium sulfate	7.90	r.t.	
$NaSO_4 \cdot 10H_2O$	Sodium sulfate decahydrate	5.0	r.t.	
$Na_2SO_4 \cdot 5H_2O$	Sodium sulfate pentahydrate	7	250–290	$300–10^4$
$Na_2UO_2(C_2O_4)_2$	Sodium uranyl oxalate	5.18	r.t.	
$NdAlO_3$	Neodymium aluminate	17.5	r.t.	
$NdScO_3$	Neodymium scandate	27	r.t.	
$Ni_3B_7O_{13}I$	Nickel iodine boracite	$\varepsilon_{11} = 14$	260	
$NiNb_2O_6$	Nickel niobate	$\varepsilon_{11} = 16.0 \pm 0.5$	r.t.	$(5–500) \times 10^3$
		$\varepsilon_{22} = 23.8 \pm 1.8$	r.t.	$(5–500) \times 10^3$
		$\varepsilon_{33} = 31.3 \pm 2.5$	r.t.	$(5–500) \times 10^3$
NiO	Nickel oxide	11.9	298	10^5
$NiSO_4 \cdot 6H_2O$	Nickel sulfate hexahydrate	$\varepsilon_{11} = 6.2$	r.t.	
		$\varepsilon_{33} = 6.8$	r.t.	
$NiWO_4$	Nickel tungstate	$\varepsilon_{11} = 17.4 \pm 2.4$	r.t.	$(5–500) \times 10^3$
		$\varepsilon_{22} = 13.6 \pm 1.0$	r.t.	$(5–500) \times 10^3$
		$\varepsilon_{33} = 19.7 \pm 0.6$	r.t.	$(5–500) \times 10^3$
P	Phosphorous (red)	4.1	r.t.	10^8
	Phosphorous (yellow)	3.6	r.t.	10^8
$[P(CH_3)_4]HgBr_3$	Tetramethylphosphonium tribromomercurate (TTM)	~10	233–373	
$PbBr_2$	Lead bromide	>30	293	$(0.5–3) \times 10^6$
$PbCO_3$	Lead carbonate	18.6	288	10.8
$Pb(C_2H_3O_2)_2$	Lead acetate	2.6	290–295	10^6
$PbCl_2$	Lead chloride	33.5	273	$(0.5–3) \times 10^6$
Pb_2CoWO_6	Lead cobalt tungstate	~250	r.t.	
PbF_2	Lead fluoride	26.3	r.t.	
$PbHfO_3$	Lead hafnate	390	300	10^5
		185	400	
PbI_2	Lead iodide	20.8	293	$(0.5–3) \times 10^6$
$Pb_3MgNb_2O_9$	Lead magnesium niobate	10,000	297	
$PbMoO_4$	Lead molybdate	$\varepsilon_{11} = 34.0 \pm 0.4$	297.5	1.6×10^3
		$\varepsilon_{33} = 40.6 \pm 0.2$	297.5	1.6×10^3
$Pb(NO_3)_2$	Lead nitrate	16.8	r.t.	$(0.5–3) \times 10^6$
$PbNb_2O_6$	Lead niobate	$\varepsilon_{33}^T = 180$	298	
PbO	Lead oxide	25.9	r.t.	2×10^6
PbS	Lead sulfide (Galena)	190	77	i.r.
		200 ± 35	r.t.	i.r.
$PbSO_4$	Lead sulfate	14.3	290—295	10^6
$PbSe$	Lead selenide	280	r.t.	i.r.
$PbTa_2O_6$	Lead metatantalate	$\varepsilon_{11} = \varepsilon_{22} \approx 300$	r.t.	10^4
		$\varepsilon_{33} = 150$	r.t.	10^4
$PbTe$	Lead telluride	450	r.t.	i.r.
		40	77	$10^4–15 \times 10^4$
		430	4.2	$10^4–15 \times 10^4$
$PbTiO_3$	Lead titanate	~200	r.t.	10^3
$PbWO_4$	Lead tungstate	$\varepsilon_{11} = \varepsilon_{22} = 23.6 \pm 0.3$	297.5	1.59×10^3
		$\varepsilon_{33} = 31.0 \pm 0.4$	297.5	1.59×10^3
$Pb(Zn_{1/3}Nb_{2/3})O_3$	Lead zinc niobate	7	300	$10^3, 300 \times 10^3$

Formula	Name	ε_{ijk}	T/K	v/Hz
$PbZrO_3$	Lead zirconate	200	400	
$RbAl(SO_4)_2 \cdot 12H_2O$	Rubidium alum	5.1	r.t.	10^{12}
RbBr	Rubidium bromide	4.83	300	
Rb_2CO_3	Rubidium carbonate	4.87 ± 0.02	r.t.	5×10^3
RbCl	Rubidium chloride	4.91 ± 0.02	r.t.	5×10^3
$RbCr(SO_4)_2 \cdot 12H_2O$	Rubidium chrome alum	5.0	r.t.	10^{12}
RbF	Rubidium fluoride	5.91	r.t.	2×10^6
$RbHSO_4$	Rubidium bisulfate	$\varepsilon_{11} = 7$	r.t.	10^5
		$\varepsilon_{22} = 8$	r.t.	10^5
		$\varepsilon_{33} = 10$	r.t.	10^5
RbH_2AsO_4	Rubidium dihydrogen arsenate (RDA)	3.90	273	9.5×10^9
RbH_2PO_4	Rubidium dihydrogen phosphate (RDP)	6.15	285	9.5×10^9
RbI	Rubidium iodide	4.94 ± 0.02	r.t.	5×10^3
$RbInSO_4$	Rubidium indium sulfate	6.85	r.t.	
$RbNO_3$	Rubidium nitrate	20—380	433–488	10^6
		30	488–538	10^6
S	Sulfur	$\varepsilon_{11} = 3.75$	298	10^2–10^3
		$\varepsilon_{22} = 3.95$	298	10^2–10^3
		$\varepsilon_{33} = 4.44$	298	10^2–10^3
	(sublimed)	3.69	298	10^2–10^3
$SC(NH_2)_2$	Thiourea	$\varepsilon_{11} = \varepsilon_{22} \approx 3$	77–300	10^3
		$\varepsilon_{22} = 35$	300	10^3
Sb_2O_3	Antimonous sesquioxide	12.8	r.t.	$(1.5–2) \times 10^3$
Sb_2S_3	Antimonous sulfide (stibnite)	$\varepsilon_{11} = \varepsilon_{33} = 15$	r.t.	10^3
		$\varepsilon_{33} = 180$	r.t.	10^3
Sb_2Se_3	Antimonous selenide	~110	r.t.	$(10–16.5) \times 10^9$
SbSI	Antimonous sulfide iodide	2000	273	10^5
		$\varepsilon_{11} = \varepsilon_{22} \approx 25$	r.t.	10^3–10^5
		$\varepsilon_{33} \approx 5 \times 10^4$	295	10^3–10^5
Se	Selenium (monocrystal)	$\varepsilon_{11} = \varepsilon_{22} = 11$	300	24×10^9
		$\varepsilon_{33} = 21$	300	24×10^9
	(amorphous)	6.0	298	10^2–10^{10}
Si	Silicon	12.1	4.2	10^7–10^9
SiC	Silicon carbide			
	cubic	9.72	r.t.	i.r.
	6H	$\varepsilon_{11} = \varepsilon_{22} = 9.66$	r.t.	i.r.
		$\varepsilon_{33} = 10.03$	r.t.	i.r.
		9.7 ± 0.1	1.8	i.r.
Si_3N_4	Silicon nitride	4.2 (film)	r.t.	10^3
SiO	Silicon monoxide	5.8	r.t.	10^3
SiO_2	Silicon dioxide	$\varepsilon_{11} = 4.42$	r.t.	9.4×10^{10}
		$\varepsilon_{22} = 4.41$	r.t.	9.4×10^{10}
		$\varepsilon_{33} = 4.60$	r.t.	9.4×10^{10}
$Sm_2(MoO_4)_3$	Samarium molybdate	12	298	
SnO_2	Stannic dioxide	$\varepsilon_{11} = \varepsilon_{22} = 14 \pm 2$	r.t.	10^4–10^{10}
		$\varepsilon_{33} = 9.0 \pm 0.5$	r.t.	10^4–10^{10}
SnSb	Tin antimonide	147	r.t.	10^4–10^6
SnTe	Tin telluride	1770 ± 300	r.t.	i.r.
$Sr(COOH)_2 \cdot 2H_2O$	Strontium formate dihydrate	6.1	r.t.	10^3
$SrCO_3$	Strontium carbonate	8.85	298	2×10^5
$SrCl_2$	Strontium chloride	9.19	r.t.	
$SrCl_2 \cdot 6H_2O$	Strontium chloride hexahydrate	8.52	r.t.	
SrF_2	Strontium fluoride	6.50	300	5×10^2–10^{11}
$SrMoO_4$	Strontium molybdate	$\varepsilon_{11} = \varepsilon_{22} = 31.7 \pm 0.2$	297.5	1.59×10^3
		$\varepsilon_{33} = 41.7 \pm 0.2$	297.5	1.59×10^3
$Sr(NO_3)_2$	Strontium nitrate	5.33	292	2×10^5
$Sr_2Nb_2O_7$	Strontium niobate	$\varepsilon_{11} = 75$	r.t.	10^3
		$\varepsilon_{22} = 46$	r.t.	10^3
		$\varepsilon_{33} = 43$	r.t.	10^3
SrO	Strontium oxide	13.3 ± 0.3	273	2×10^6
SrS	Strontium sulfide	11.3	r.t.	7.25×10^6

Formula	Name	ε_{ijk}	T/K	ν/Hz
$SrSO_4$	Strontium sulfate	11.5	r.t.	
$SrTiO_3$	Strontium titanate	332	298	10^3
		2080	78	10^3
$SrWO_4$	Strontium tungstate	$\varepsilon_{11} = \varepsilon_{22} = 25.7 \pm 0.2$	297.5	1.6×10^3
		$\varepsilon_{33} = 34.1 \pm 0.2$	297.5	1.6×10^3
Ta_2O_5	Tantalum pentoxide (tantala)			
	α phase	$\varepsilon_{11} = \varepsilon_{22} = 30$	77	10^3
		$\varepsilon_{33} = 65$	77	10^3
	β phase	24	292	10^3
$Tb(MoO_4)_3$	Terbium molybdate	11	298	
		$\varepsilon_{11} = \varepsilon_{22} = 33$	100–200	9.4×10^9
		$\varepsilon_{33} = 53$	100–200	9.4×10^9
Te	Tellurium	$\varepsilon_{11} = \varepsilon_{22} = 33$	r.t.	
		$\varepsilon_{33} = 54$	r.t.	
	polycrystalline	27.5	r.t.	i.r.
	monocrystalline	28.0	r.t.	i.r.
ThO_2	Thorium dioxide	18.9 ± 0.4	r.t.	3×10^5
TiO_2	Titanium dioxide (rutile)	$\varepsilon_{11} = \varepsilon_{22} = 86$	300	10^4–10^6
		$\varepsilon_{33} = 170$	300	10^4–10^6
Ti_2O_3	Titanium sesquioxide	30	77	6×10^{10}
TlBr	Thallium bromide	30	293	10^3–10^7
TlCl	Thallous chloride	32.2 ± 0.2	293	10^3–10^5
TlI	Thallous iodide (orthorhombic)	20.7 ± 0.2	293	10^4
		37.3	193	10^7
$TlNO_3$	Thallous nitrate	16.5	293	5×10^5
$TlSO_4$	Thallous sulfate	25.5	293	5×10^5
UO_2	Uranium dioxide	24	r.t.	3×10^5
WO_3	Tungsten trioxide	300		
$YMnO_3$	Yttrium manganate	20	r.t.	2×10^7
Y_2O_3	Yttrium sesquioxide	10	r.t.	10^6
$YbMnO_3$	Ytterbium manganate	20	r.t.	2×10^7
Yb_2O_3	Ytterbium sesquioxide	5.0 (film)	r.t.	10^3
ZnO	Zinc monoxide	$\varepsilon_{11}^S = 8.33$	r.t.	
		$\varepsilon_{33}^S = 8.84$	r.t.	
		$\varepsilon_{11}^T = 9.26$	r.t.	
		$\varepsilon_{33}^T = 11.0$	r.t.	
		$\varepsilon_{11} = 9.26$	r.t.	
		$\varepsilon_{33} = 8.2$	r.t.	
		8.15	r.t.	i.r.
ZnS	Zinc sulfide	$\varepsilon_{11}^S = 8.08 \pm 2\%$	77	10^4
		$\varepsilon_{11}^S = 8.32 \pm 2\%$	298	10^4
		$\varepsilon_{11}^T = 8.14 \pm 2\%$	77	10^4
		$\varepsilon_{11}^T = 8.37 \pm 2\%$	298	10^4
ZnSe	Zinc selenide	$\varepsilon_{11}^T = \varepsilon_{11}^S = 9.12 \pm 2\%$	298	10^4
ZnTe	Zinc telluride	$\varepsilon_{11}^T = \varepsilon_{11}^S = 10.10 \pm 2\%$	r.t.	
$ZnWO_4$	Zinc tungstate	$\varepsilon_{22} = 16.1 \pm 0.5$	r.t.	$(5–500) \times 10^3$
ZrO_2	Zirconium dioxide (zirconia)	12.5	r.t.	2×10^6

CURIE TEMPERATURE OF SELECTED FERROELECTRIC CRYSTALS

H. P. R. Frederikse

The following table lists the major ferroelectric crystals and their Curie temperatures, T_C.

Reference

Young, K. F. and Frederikse, H. P. R., *J. Phys. Chem. Ref. Data*, 2, 313, 1973.

Name or acronym	Formula	T_C/K
Potassium dihydrogen phosphate group		
KDP	KH_2PO_4	123
KDA	KH_2AsO_4	97
KDDP	KD_2PO_4	213
KDDA	KD_2AsO_4	162
RDP	RbH_2PO_4	146
RDA	RbH_2AsO_4	111
RDDP	RbD_2PO_4	218
RDDA	RbD_2AsO_4	178
CDP	CsH_2PO_4	159
CDA	CsH_2AsO_4	143
CDDA	CsD_2AsO_4	212
Rochelle salt group		
Rochelle salt	$NaKC_4H_4O_6 \cdot 4H_2O$	255–297
Deuterated Rochelle salt	$NaKC_4H_2D_2O_6 \cdot 4H_2O$	251–308
Ammonium Rochelle salt	$NaNH_4C_4H_4O_6 \cdot 4H_2O$	109
LAT	$LiNH_4C_4H_4O_6 \cdot H_2O$	106
Triglycine sulfate group		
TGS	$(NH_2CH_2COOH)_3 \cdot H_2SO_4$	322
TGSe	$(NH_2CH_2COOH)_3 \cdot H_2SeO_4$	295
TGFB	$(NH_2CH_2COOH)_3 \cdot H_2BeF_4$	346
AFB	$(NH_4)_2BeF_4$	176
HADA	$HNH_4(ClCH_2COO)_2$	128
Perovskites and related compounds		
Barium titanate	$BaTiO_3$	406, 278, 193
Lead titanate	$PbTiO_3$	765
Potassium niobate	$KNbO_3$	712
Potassium tantalate niobate	$KTa_{2/3}Nb_{1/3}O_3$	241, 220, 170
Lithium niobate	$LiNBO_3$	1483
Lithium tantalate	$LiTaO_3$	891
Barium titanium niobate	$Ba_6Ti_2Nb_8O_{30}$	521
Ba-Na niobate ("Bananas")	$Ba_2NaNb_5O_{15}$	833
Potassium iodate	KIO_3	485, 343, 257–263, 83
Lithium iodate	$LiIO_3$	529
Potassium nitrate	KNO_3	397
Sodium nitrate	$NaNO_3$	548
Rubidium nitrate	$RbNO_3$	437–487
Miscellaneous compounds		
Cesium trihydrogen selenite	$CsH_3(SeO_3)_2$	143
Lithium trihydrogen selenite	$LiH_3(SeO_3)_2$	$T_C > T_{mp}$
Potassium selenate	K_2SeO_4	93
Methyl ammonium alum (MASD)	$CH_3NH_3Al(SO_4)_2 \cdot 12H_2O$	177
Ammonium cadmium sulfate	$(NH_4)_2Cd_2(SO_4)_3$	95
Ammonium bisulfate	$(NH_4)HSO_4$	271
Ammonium sulfate	$(NH_4)_2SO_4$	224
Ammonium nitrate	NH_4NO_3	398, 357, 305, 255
Colemanite	$CaB_3O_4(OH)_3 \cdot H_2O$	266
Cadmium pyroniobite	$Cd_2Nb_2O_7$	185
Gadolinium molybdate	$Gd_2(MoO_4)_3$	432

PROPERTIES OF ANTIFERROELECTRIC CRYSTALS

H. P. R. Frederikse

Some important antiferroelectric crystals are listed here with their Curie temperatures T_C. The last column gives the constant T_0 which appears in the Curie–Weiss law describing the dielectric constant of these materials above the Curie temperature:

$$\varepsilon = \text{const.}/(T - T_0)$$

Name or acronym	Formula	T_C/K	T_0/K
ADP	$NH_4H_2PO_4$	148	
ADA	$NH_4H_2AsO_4$	216	
ADDP	$NH_4D_2PO_4$	242, 245	
ADDA	$NH_4D_2AsO_4$	299	
A_dDDP	$ND_4D_2PO_4$	243	
A_dDDA	$ND_4D_2AsO_4$	304	
Sodium niobate	$NaNbO_3$	911, 793	
Lead hafnate	$PbHfO_3$	476	378
Lead zirconate	$PbZrO_3$	503	475
Lead metaniobate	$PbNb_2O_6$	843	530
Lead metatantalate	$PbTa_2O_6$	543	533
Tungsten trioxide	WO_3	1010	
Potassium strontium niobate	$KSr_2Nb_5O_{15}$	427	413
Sodium nitrite	$NaNO_2$	437	437
Sodium trihydrogen selenite	$NaH_3(SeO_3)_2$	193	192
Sodium trideuterium selenite	$NaD_3(SeO_3)_2$	271	245
Ammonium trihydrogen periodate	$(NH_4)_2H_3IO_6$	245	

DIELECTRIC CONSTANTS OF GLASSES

Type	Dielectric constant at 100 MHz (20 °C)	Volume resistivity (in MΩ cm at 350 °C)	Loss factor[a]
Corning 0010	6.32	10	0.015
Corning 0080	6.75	0.13	0.058
Corning 0120	6.65	100	0.012
Pyrex 1710	6.00	2,500	0.025
Pyrex 3320	4.71	–	0.019
Pyrex 7040	4.65	80	0.013
Pyrex 7050	4.77	16	0.017
Pyrex 7052	5.07	25	0.019
Pyrex 7060	4.70	13	0.018
Pyrex 7070	4.00	1,300	0.0048
Vycor 7230	3.83	–	0.0061
Pyrex 7720	4.50	16	0.014
Pyrex 7740	5.00	4	0.040
Pyrex 7750	4.28	50	0.011
Pyrex 7760	4.50	50	0.0081
Vycor 7900	3.9	130	0.0023
Vycor 7910	3.8	1,600	0.00091
Vycor 7911	3.8	4,000	0.00072
Corning 8870	9.5	5,000	0.0085
G. E. Clear (silica glass)	3.81	4,000–30,000	0.00038
Quartz (fused)	3.75 (4.1 at 1 MHz)	–	0.0002 (1 MHz)

[a] Power factor × dielectric constant equals loss factor.

PROPERTIES OF SUPERCONDUCTORS

L. I. Berger and B. W. Roberts

The following tables include superconductive properties of selected elements, compounds, and alloys. Individual tables are given for thin films, elements at high pressures, superconductors with high critical magnetic fields, and high critical temperature superconductors.

The historically first observed and most distinctive property of a superconductive body is the near total loss of resistance at a critical temperature (T_c) that is characteristic of each material. Figure 1(a) below illustrates schematically two types of possible transitions. The sharp vertical discontinuity in resistance is indicative of that found for a single crystal of a very pure element or one of a few well-annealed alloy compositions. The broad transition, illustrated by broken lines, suggests the transition shape seen for materials that are not homogeneous and contain unusual strain distributions. Careful testing of the resistivity limit for superconductors shows that it is less than 4×10^{-23} ohm cm, while the lowest resistivity observed in metals is of the order of 10^{-13} ohm cm. If one compares the resistivity of a superconductive body to that of copper at room temperature, the superconductive body is at least 10^{17} times less resistive.

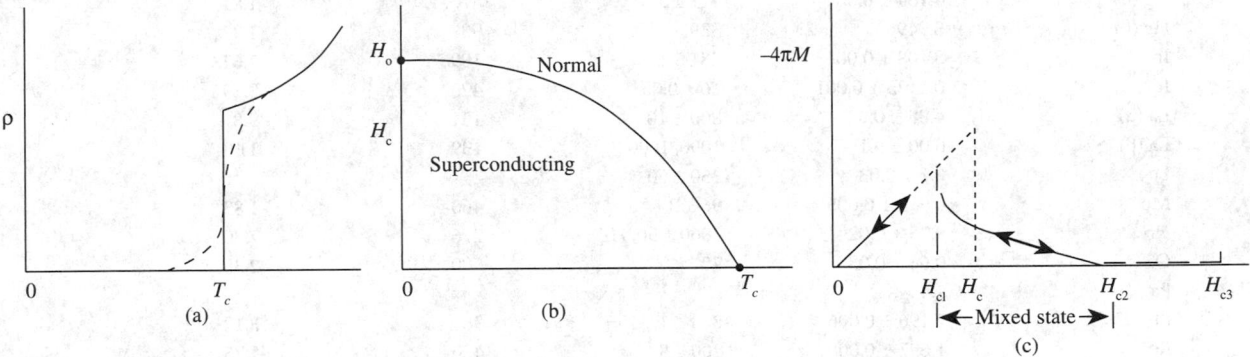

FIGURE 1. Physical properties of superconductors. (a) Resistivity vs. temperature for a pure and perfect lattice (solid line); impure and/or imperfect lattice (broken line). (b) Magnetic-field temperature dependence for Type-I or "soft" superconductors. (c) Schematic magnetization curve for "hard" or Type-II superconductors.

The temperature interval ΔT_c, over which the transition between the normal and superconductive states takes place, may be of the order of as little as 2×10^{-5} K or several K in width, depending on the material state. The narrow transition width was attained in 99.9999% pure gallium single crystals.

A Type-I superconductor below T_c, as exemplified by a pure metal, exhibits perfect diamagnetism and excludes a magnetic field up to some critical field H_c, whereupon it reverts to the normal state as shown in the H-T diagram of Figure 1(b).

The magnetization of a typical high-field superconductor is shown in Figure 1(c). The discovery of the large current-carrying capability of Nb_3Sn and other similar alloys has led to an extensive study of the physical properties of these alloys. In brief, a high-field superconductor, or Type-II superconductor, passes from the perfect diamagnetic state at low magnetic fields to a mixed state and finally to a sheathed state before attaining the normal resistive state of the metal. The magnetic field values separating the four stages are given as H_{c1}, H_{c2}, and H_{c3}. The superconductive state below H_{c1} is perfectly diamagnetic, identical to the state of most pure metals of the "soft" or Type-I superconductor. Between H_{c1} and H_{c2} a "mixed superconductive state" is found in which fluxons (a minimal unit of magnetic flux) create lines of normal flux in a superconductive matrix. The volume of the normal state is proportional to $-4\pi M$ in the "mixed state" region. Thus at H_{c2} the fluxon density has become so great as to drive the interior volume of the superconductive body completely normal. Between H_{c2} and H_{c3} the superconductor has a sheath of current-carrying superconductive material at the body surface, and above H_{c3} the normal state exists. With several types of careful measurement, it is possible to determine H_{c1}, H_{c2}, and H_{c3}. Table 6 contains some of the available data on high-field superconductive materials.

High-field superconductive phenomena are also related to specimen dimension and configuration. For example, the Type-I superconductor, Hg, has entirely different magnetization behavior in high magnetic fields when contained in the very fine sets of filamentary tunnels found in an unprocessed Vycor glass. The great majority of superconductive materials are Type-II. The elements in very pure form and a very few precisely stoichiometric and well annealed compounds are Type I with the possible exceptions of vanadium and niobium.

Metallurgical Aspects. The sensitivity of superconductive properties to the material state is most pronounced and has been used in a reverse sense to study and specify the detailed state of alloys. The mechanical state, the homogeneity, and the presence of impurity atoms and other electron-scattering centers are all capable of controlling the critical temperature and the current-carrying capabilities in high-magnetic fields. Well-annealed specimens tend to show sharper transitions than those that are strained or inhomogeneous. This sensitivity to mechanical state underlines a general problem in the tabulation of properties for superconductive materials. The occasional divergent values of the critical temperature and of the critical fields quoted for a Type-II superconductor may lie in the variation in sample preparation. Critical temperatures of materials studied early in the history of superconductivity must be evaluated in light of the probable metallurgical state of the material, as well as the availability of less pure starting elements. It has been noted that recent work has given extended consideration to the metallurgical aspects of sample preparation.

Symbols in tables: T_c: Critical temperature; H_o: Critical magnetic field in the $T = 0$ limit; θ_D: Debye temperature; and γ: Electronic specific heat.

TABLE 1. Selective Properties of Superconductive Elements

Element	T_c(K)	H_o(oersted)	θ_D(K)	γ(mJ mol^{-1}K^{-1})
Al	1.175 ± 0.002	104.9 ± 0.3	420	1.35
Am* (α,?)	0.6			
Am* (β,?)	1.0			
Be	0.026			0.21
Cd	0.517 ± 0.002	28 ± 1	209	0.69
Ga	1.083 ± 0.001	58.3 ± 0.2	325	0.60
Ga (β)	5.9, 6.2	560		
Ga (γ)	7	950, HF[a]		
Ga (Δ)	7.85	815, HF		
Hf	0.128	12.7		2.21
Hg (α)	4.154 ± 0.001	411 ± 2	87, 71.9	1.81
Hg (β)	3.949	339	93	1.37
In	3.408 ± 0.001	281.5 ± 2	109	1.672
Ir	0.1125 ± 0.001	16 ± 0.05	425	3.19
La (α)	4.88 ± 0.02	800 ± 10	151	9.8
La (β)	6.00 ± 0.1	1096, 1600	139	11.3
Lu	0.1 ± 0.03	350 ± 50		
Mo	0.915 ± 0.005	96 ± 3	460	1.83
Nb	9.25 ± 0.02	2060 ± 50, HF	276	7.80
Os	0.66 ± 0.03	70	500	2.35
Pa	1.4			
Pb	7.196 ± 0.006	803 ± 1	96	3.1
Re	1.697 ± 0.006	200 ± 5	4.5	2.35
Ru	0.49 ± 0.015	69 ± 2	580	2.8
Sn	3.722 ± 0.001	305 ± 2	195	1.78
Ta	4.47 ± 0.04	829 ± 6	258	6.15
Tc	7.8 ± 0.1	1410, HF	411	6.28
Th	1.38 ± 0.02	1.60 ± 3	165	4.32
Ti	0.40 ± 0.04	56	415	3.3
Tl	2.38 ± 0.02	178 ± 2	78.5	1.47
U	0.2			
V	5.40 ± 0.05	1408	383	9.82
W	0.0154 ± 0.0005	1.15 ± 0.03	383	0.90
Zn	0.85 ± 0.01	54 ± 0.3	310	0.66
Zr	0.61 ± 0.15	47	290	2.77
Zr (ω)	0.65, 0.95			

TABLE 2. Range of Critical Temperatures Observed for Superconductive Elements in Thin Films Condensed Usually at Low Temperatures

Element	T_c Range (K)	Comments	Element	T_c Range (K)	Comments
Al	1.15–5.7	HF[a]	Nb	2.0–10.1	
Be	5–9.75	HF	Pb	1.8–7.5	
Bi	6.17–6.6		Re	1.7–7	
Cd			Sn	3.5–6	
(Disordered)	0.79–0.91		Ta	<1.7–4.51	HF[a]
(Ordered)	0.53–0.59		Tc	4.6–7.7	
Ga	2.5–8.5	HF	Ti	1.3 Max	
Hg	3.87–4.5		Tl	2.33–2.96	
In	2.2–5.6	HF	V	1.8–6.02	
La	3.55–6.74		W	<1.0–4.1	
Mo	3.3–8.0		Zn	0.77–1.9	

[a] HF denotes high magnetic field superconductive properties.

TABLE 3. Elements Exhibiting Superconductivity Under or After Application of High Pressure

Element	T_c Range (K)	Pressure (kbar)	Element	T_c Range (K)	Pressure (kbar)
Al	1.98–0.075	0–62	Pb II	3.55	160
As	0.31–0.5	220–140	Re II	2.3 Max.	"Plastic"
	0.2–0.25	140–100			compression
Ba II	1–1.8	55–85	Sb (prepared 120	2.6–2.7	
III	1.8–5	85–144	kbar, held below		
IV	4.5–5.4	144–190	77K)		
Bi II	3.9	25–27	Sb II	3.55–3.40	85–150
III	6.55–7.25	28–38	Se II	6.75, 6.95	130
IV	7.0, 8.7–6.0	43, 43–62	Si	6.7–7.1	120–130
V	6.7, 8.3	48–80	Sn II	5.2–4.85	125–160
VI	8.55	90, 92–101	III	5.30	113
VII(?)	8.2	30	Te II	2.4–5.1	38–55
Ce (α)	0.020–0.045	20–35		4.1–4.2	53–62
Ce (α′)	1.9–1.3	45–125	IV	4.72–4	63–80
Cs V	1.5	>125	( )	3.3–2.8	100–260
Ga II	6.38	≥35	Tl (cubic form)	1.45	35
II′	7.5	≥35 then P	(hexagonal form)	1.95	35
		removed	U	2.4–0.4	10–85
Ge	5.35	115	Y	1.7–2.5	110–160
Lu	0.022–1.0	45–190	Zr (omega form, metastable)	1–1.7	60–130
P	5.8	170			

TABLE 4. Superconductive Compounds and Alloys

All compositions are denoted on an atomic basis, i.e., AB, AB_2, or AB_3 for compounds, unless noted. Solid solutions or odd compositions may be denoted as A_zB_{1-z} or A_zB. A series of three or more alloys is indicated as A_xB_{1-x} or by actual indication of the atomic fraction range, such as $A_{0-0.6}B_{1-0.4}$. The critical temperature of such a series of alloys is denoted by a range of values or possibly the maximum value.

The selection of the critical temperature from a transition in the effective permeability, or the change in resistance, or possibly the incremental changes in frequency observed by certain techniques is not often obvious from the literature. Most authors choose the mid-point of such curves as the probable critical temperature of the idealized material, while others will choose the highest temperature at which a deviation from the normal state property is observed. In view of the previous discussion concerning the variability of the superconductive properties as a function of purity and other metallurgical aspects, it is recommended that appropriate literature be checked to determine the most probable critical temperature or critical field of a given alloy.

A very limited amount of data on critical fields, H_o, is available for these compounds and alloys; these values are given at the end of the table.

A. Superconductors with $T_c < 10$ K

Substance	T_c, K	Crystal structure type	Substance	T_c, K	Crystal structure type
$Ag_{3.3}Al$	0.34	A12-cI58 (Mn)	Ag_7NO_{11}	1.04	Cubic
$Ag_xAl_yZn_{1-x-y}$	0.15	Cubic	Ag_xPb_{1-x}	7.2 max.	
$AgBi_2$	2.87–3.0		Ag_4Sn	0.1	h**
$Ag_7F_{0.25}N_{0.75}O_{10.25}$	0.85–0.90		Ag_xSn_{1-x}	1.5–3.7	
Ag_2F	0.0.066		Ag_xSn_{1-x} (film)	2.0–3.8	
Ag_7FO_8	0.3	Cubic	$AgTe_3$	2.6	Cubic
$Ag_{0.8-0.3}Ga_{0.2-0.7}$	6.5–8		AgTh	2.2	C16-tI12 (Al_2Cu)
Ag_4Ge	0.85	Hex., c.p.	$AgTh_2$	2.26	C16
$Ag_{0.438}Hg_{0.562}$	0.64	$D8_2$	$Ag_{0.03}Tl_{0.97}$	2.67	
$AgIn_2$	~2.4	C16	$Ag_{0.94}Tl_{0.06}$	2.32	
$Ag_{0.1}In_{0.9}Te$ ($n = 1.4 \times 10^{22}$)*	1.2–1.89	B1	AgY	0.33	B2-cP2 (CsCl)
$Ag_{0.2}In_{0.8}Te$ ($n = 1.07 \times 10^{22}$)	0.77–1.00	B1	Ag_xZn_{1-x}	0.5–0.845	
AgLa	0.94	B2-cP2 (CsCl)	$AlAu_4$	0.4–0.7	Like A13
AgLa (9.5 kbar)	1.2	B2	Al_2Au	0.1	C1-cF12 (CaF_2)
AgLu	0.33	B2-cP2	Al_2CMo_3	9.8–10.2	A13+trace 2nd. phase
$AgMo_4S_5$	9.1	hR15 (Mo_6PbS_8)	Al_2CaSi	5.8	
$Ag_{1.2}Mo_6Se_8$	5.9	Same	$Al_{0.131}Cr_{0.088}V_{0.781}$	1.46	Cubic
			$AlGe_2$	1.75	

Substance	T_c, K	Crystal structure type	Substance	T_c, K	Crystal structure type
Al_2Ge_2U	1.6	LI_2-cP4 (Cu_3Au)	$AuPb_2$	3.15	
$AlLa_3$	5.57	DO_{19}	$AuPb_2$ (film)	4.3	
Al_2La	3.23	C15	$AuPb_3$	4.40	
Al_2Lu	1.02	C15-cF24 (Cu_2Mg)	$AuPb_3$ (film)	4.25	
Al_3Mg_2	0.84	F.C.C.	Au_2Pb	1.18; 6–7	C15
$AlMo_3$	0.58	A15	$AuSb_2$	0.58	C2
$AlMo_6Pd$	2.1		$AuSn$	1.25	$B8_1$
AlN	1.55	B4	Au_xSn_{1-x} (film)	2.0–3.8	
Al_2NNb_3	1.3	A13	Au_5Sn	0.7–1.1	A3
Al_3Nb	0.64	tI8 (Al_3Ti)	$AuTa_{4.3}$	0.55	A15-cP8 (Cr_3Si)
$AlOs$	0.39	B2	Au_3Te_5	1.62	Cubic
Al_3Os	5.90		$AuTh_2$	3.08	C16
$AlPb$ (film)	1.2–7		$AuTl$	1.92	
Al_2Pt	0.48–0.55	C1	AuV_3	0.74	A15
Al_5Re_{24}	3.35	A12	Au_xZn_{1-x}	0.50–0.845	
$AlSb$	2.8	B4-tI4 (Sn)	$AuZn_3$	1.21	Cubic
Al_2Sc	1.02	C15-cF24 (Cu_2Mg)	Au_xZr_y	1.7–2.8	A3
Al_2Si_2U	1.34	LI_2-cP4 (Cu_3Au)	$AuZr_3$	0.92	A15
$AlTh_2$	0.1	C16-tI12 (Al_2Cu)	$B_2Ba_{0.67}Pt_3$	5.60	hP12 (B_2BaPt_3)
Al_3Th	0.75	DO_{19}	$BCMo_2$	5.4	Orthorhombic
$Al_xTi_yV_{1-x-y}$	2.05–3.62	Cubic	$BCMo_2$	5.3–7.0	Same
$Al_{0.108}V_{0.892}$	1.82	Cubic	$B_2Ca_{0.67}Pt_3$	1.57	hP12
Al_2Y	0.35	C15-cF24 (Cu_2Mg)	B_4ErIr_4	2.1	tP18 (B_4CeCo_4)
Al_3Yb	0.94	LI_2-cP4 (Cu_3Au)	B_4ErRh_4	4.3	oC108 (B_4LuRh_4)
Al_xZn_{1-x}	0.5–0.845		B_4ErRh_4	8.7	tP18 (B_4CeCo_4)
$AlZr_3$	0.73	LI_2	BHf	3.1	Cubic
$AsBiPb$	9.0		B_4HoIr_4	2.0	tP18
$AsBiPbSb$	9.0		B_4HoRh_4	1.4	oC108
$AsHfOs$	3.2	C22-hP9 (Fe_2P)	B_2Ir_3La	1.65	hP6 $(CaCu_5)$
$AsHfRu$	4.9	same	B_2Ir_3Th	2.09	Same
$As_{0.33}InTe_{0.67}$ (n = 1.24×10^{22})	0.85–1.15	B1	B_4Ir_4Tm	1.6	tP18
$As_{0.5}InTe_{0.5}$ (n = 0.97×10^{22})	0.44–0.62	B1	B_6La	5.7	
As_4La_3	0.6	cI28 (Th_3P_4)	B_2LaRh_3	2.82	hP6
$AsNb_3$	0.3	$L1_2$-tP32	$B_{12}Lu$	0.48	
$As_{0.50}Ni_{0.06}Pd_{0.44}$	1.39	C2	B_2LuOs	2.66	oP16 (B_2LuRu)
$AsNi_{0.25}Pd_{0.75}$	1.6	$B8_1$-hP4 (NiAs)	B_2LuOs_3	4.62	hP6
$AsOsZr$	8.0	C22-hP9 (Fe_2P)	B_4LuRh_4	6.2	oC108
$AsPb$	8.4		B_2LuRu	9.86	oP16
$AsPd_2$ (low-temp. phase)	0.60	Hexagonal	B_4LuRu_4	2.0	tI72 (B_4LuRu_4)
$AsPd_2$ (high-temp. phase)	1.70	C22	BMo	0.5	
$AsPd_5$	0.46	Complex		(extrapol.)	
As_3Pd_5	1.9		BMo_2	4.74	C16
$AsRh$	0.58	B31	BNb	8.25	B_f
$AsRh_{1.4–1.6}$	< 0.03–0.56	Hexagonal	B_4NdRh_4	5.3	tP18
$AsSn$	4.10		B_2OsSc	1.34	oP16
$AsSn$ (n = 2.14×10^{22})	3.41–3.65	B1	B_2OsY	2.22	oP16
	3.5–3.6;		$B_2Pt_3Sr_{0.67}$	2.78	hP12 (B_2BaPt_3)
$As_{-2}Sn_{-3}$	1.21–1.17		BRe_2	2.80; 4.6	
As_3Sn_4 (n = 0.56×10^{22})	1.16–1.19	Rhombohedral	$B_4Rh_{3.4}Ru_{0.6}$	8.38	tI72
AsV_3	0.20	A15-cP8 (Cr_3Si)	B_4Rh_4Sm	2.7	tP18
Au_5Ba	0.4–0.7	$D2_d$	B_4Rh_4Th	4.3	Same
$AuBe$	2.64	B20	B_4Rh_4Tm	9.8	Same
Au_2Bi	1.80	C15	B_4Rh_4Tm	5.4	oC108
Au_5Ca	0.34–0.38	$C15_b$	$B_{0.3}Ru_{0.7}$	2.58	$D10_2$
$AuGa_2$	1.6	C1-cF12 (CaF_2)	B_2Ru_4Sc	7.2	tI72
$AuGa$	1.2	B31	B_2Ru_3Th	1.79	hP6
$Au_{0.40–0.92}Ge_{0.60–0.08}$	<0.32–1.63	Complex	B_2Ru_3Y	2.85	Same
$AuIn_2$	0.2	C1-cF12	B_2RuY	7.80	oP16
$AuIn$	0.4–0.6	Complex	B_4Ru_4Y	1.4	tI72
$AuLu$	<0.35	B2	$B_{12}Sc$	0.39	
$AuNb_3$	1.2	A2	BTa	4.0	B_f

Substance	T_c, K	Crystal structure type	Substance	T_c, K	Crystal structure type
BTa_2	3.12	C16-tI12 (Al_2Cu)	Bi_2Pd	1.70	Monoclinic, α-phase
B_6Th	0.74		Bi_2Pd	4.25	Tetragonal, β-phase
BW_2	3.1	C16	$BiPd_{0.45}Pt_{0.55}$	3.7	$B8_1$-hP4 (NiAs)
B_6Y	6.5–7.1		$BiPdSe$	1.0	C2
$B_{12}Y$	4.7		$BiPdTe$	1.2	C2
BZr	3.4	Cubic	$BiPt$	1.21	$B8_1$
$B_{12}Zr$	5.82		$Bi_{0.1}PtSb_{0.9}$	2.05; 1.5	$B8_1$-hP4 (NiAs)
$BaBi_3$	5.69	Tetragonal	$BiPtSe$	1.45	C2
$Ba_2Mo_{15}Se_{19}$	2.75	hP15 (Mo_6PbS_8)	$BiPtTe$	1.15	C2
$Ba_xO_3Sr_{1-x}Ti$ ($n = 4.2 \times 10^{19}$)	<0.1–0.55		Bi_2Pt	0.155	Hexagonal
$Ba_{0.13}O_3W$	1.9	Tetragonal	Bi_2Rb	4.25	C15
$Ba_{0.14}O_3W$	<1.25–2.2	Hexagonal	$BiRe_2$	1.9–2.2	
$BaRh_2$	6.0	C15	$BiRh$	2.06	$B8_1$
$Be_{22}Mo$	2.51	Cubic ($Be_{22}Re$)	Bi_3Rh	3.2	Orthorhombic (NiB_3)
$Be_8Nb_5Zr_2$	5.2		Bi_4Rh	2.7	Hexagonal
$Be_{0.98-0.92}Re_{0.02-0.08}$ (quenched)	9.5–9.75	Cubic	$BiRu$	5.7	m**
$Be_{0.957}Re_{0.043}$	9.62	Cubic ($Be_{22}Re$)	Bi_3Sn	3.6–3.8	
$BeTc$	5.21	Cubic	$BiSn$	3.8	
$Be_{22}W$	4.12	Cubic ($Be_{22}Re$)	Bi_xSn_y	3.85–4.18	
$Be_{13}W$	4.1	Tetragonal	Bi_3Sr	5.62	$L1_2$
Bi_3Ca	2.0		Bi_3Te	0.75–1.0	
$Bi_{0.5}Cd_{0.13}Pb_{0.25}Sn_{0.12}$ (weight fractions)	8.2		Bi_5Tl_3	6.4	
$BiCo$	0.42–0.49		$Bi_{0.26}Tl_{0.74}$	4.4	Cubic, disordered
Bi_2Cs	4.75	C15	$Bi_{0.26}Tl_{0.74}$	4.15	$L1_2$, ordered (?)
Bi_xCu_{1-x} (electrodeposited)	2.2		Bi_2Y_3	2.25	
$BiCu$	1.33–1.40		Bi_3Zn	0.8–0.9	
Bi_3Fe	1.0	m**	$Bi_{0.3}Zr_{0.7}$	1.51	
$Bi_{0.019}In_{0.981}$	3.86		$BiZr_3$	2.4–2.8	
$Bi_{0.05}In_{0.95}$	4.65	α-phase	$BrMo_6Se_7$	7.1	hP15 (Mo_6PbS_8)
$Bi_{0.10}In_{0.90}$	5.05	Same	$Br_3Mo_6Se_5$	7.1	Same
$Bi_{0.15-0.30}In_{0.85-0.70}$	5.3–5.4	α- and β-phases	CCs_x	0.020–0.135	Hexagonal
$Bi_{0.34-0.48}In_{0.66-0.52}$	4.0–4.1		CFe_3	1.30	DO_{11}-oP16 (Fe_3C)
Bi_3In_5	4.1		$CGaMo_2$	3.7–4.1	Hexagonal
$BiIn_2$	5.65	β-phase	$CHf_{0.5}Mo_{0.5}$	3.4	B1
Bi_2Ir	1.7–2.3		$CHf_{0.3}Mo_{0.7}$	5.5	B1
Bi_2Ir (quenched)	3.0–3.96		$CHf_{0.25}Mo_{0.75}$	6.6	B1
BiK	3.6		$CHf_{0.7}Nb_{0.3}$	6.1	B1
Bi_2K	3.58	C15	$CHf_{0.6}Nb_{0.4}$	4.5	B1
$BiLi$	2.47	$L1_0$, α-phase	$CHf_{0.5}Nb_{0.5}$	4.8	B1
$Bi_{4-9}Mg$	0.7–~1.0		$CHf_{0.4}Nb_{0.6}$	5.6	B1
Bi_3Mo	3–3.7		$CHf_{0.25}Nb_{0.75}$	7.0	B1
$BiNa$	2.25	$L1_0$	$CHf_{0.2}Nb_{0.8}$	7.8	B1
$BiNb_3$	4.5	A15-cP8 (Cr_3Si)	$CHf_{0.9-0.1}Ta_{0.1-0.9}$	5.0–9.0	B1
$BiNb_3$ (high pressure and temperature)	3.05	A15	CK (excess K)	0.55	Hexagonal
$BiNi$	4.25	$B8_1$	C_8K	0.39	Hexagonal
Bi_3Ni	4.06	Orthorhombic	C_2La	1.66	tI6 (CaC_2)
$BiNi_{0.5}Rh_{0.5}$	3.0	$B8_1$-hP4 (AsNi)	C_2Lu	3.33	Same
$Bi_{0.5}NiSb_{0.5}$	2.0	Same	$C_{0.40-0.44}Mo_{0.60-0.56}$	9–13	
$Bi_{1.0}Pb_{0.1}$	7.26–9.14		C_3MoRe	3.8	B1-cF8
$Bi_{1.0}Pb_{0.1}$ (film)	7.25–8.67		$C_{0.6}Mo_{4.8}Si_3$	7.6	$D8_8$
$Bi_{0.05-0.40}Pb_{0.95-0.60}$	7.35–8.4	H.C.P. to ϵ-phase	$CMo_{0.2}Ta_{0.8}$	7.5	B1
Bi_2Pb	4.25	t**	$CMo_{0.5}Ta_{0.5}$	7.7	B1
$BiPbSb$	8.9		$CMo_{0.75}Ta_{0.25}$	8.5	B1
$Bi_{0.5}Pb_{0.31}Sn_{0.19}$ (weight fractions)	8.5		$CMo_{0.8}Ta_{0.2}$	8.7	B1
$Bi_{0.5}Pb_{0.25}Sn_{0.25}$	8.5		$CMo_{0.85}Ta_{0.15}$	8.9	B1
$BiPd_2$	4.0		CMo_xV_{1-x}	2.9–9.3	B1
$Bi_{0.4}Pd_{0.6}$	3.7–4	Hexagonal, ordered	CMo_xZr_{1-x}	9.8	B1
$BiPd$	3.7	Orthorhombic	$C_{0.984}Nb$	9.8	B1
			CNb_2	9.1	
			CNb_xTi_{1-x}	<4.2–8.8	B1
			$CNb_{0.1-0.9}Zr_{0.9-0.1}$	4.2–8.4	B1

Substance	T_c, K	Crystal structure type
CRb_x (Au)	0.023–0.151	Hexagonal
$CRe_{0.06}W$	5.0	
CRu	2.00	hP2 (CW)
$C_{0.98}7Ta$	9.7	
$C_{0.848-0.987}$	2.04–9.7	
CTa (film)	5.09	B1
CTa_2	3.26	L'_3
$CTa_{0.4}Ti_{0.6}$	4.8	B1
$Cta_{1-0.4}W_{0-0.6}$	8.5–10.5	B1
$CTa_{0.2-0.9}Zr_{0.8-0.1}$	4.6–8.3	B1
CTc (excess C)	3.85	Cubic
$CTi_{0.5-0.7}W_{0.5-0.3}$	6.7–2.1	B1
CW	1.0	
CW_2	2.74	L'_3
CW_2	5.2	F.C.C.
C_2Y	3.88	tI6 (CaC_2)
$Ca_3Co_4Sn_{13}$	5.9	cP40 ($Pr_3Rh_2Sn_{13}$)
$Ca_3Ge_{13}Rh_4$	2.1	Same
CaHg	1.6	B2-cP2 (CsCl)
$CaHg_3$	1.6	hP8 (Ni_3Sn)
$CaIr_2$	6.15	C15
$Ca_3Ir_4Sn_{13}$	7.1	cP40
$Ca_xO_3Sr_{1-x}$·Ti (n = 3.7–11 × 10^{19})	< 0.1–0.55	
$Ca_{0.1}O_3W$	1.4–3.4	Hexagonal
CaPb	7.0	
$CaRh_2$	6.40	C15
$CaRh_{1.2}Sn_{4.5}$	8.7	cP40
$CaTl_3$	2.0	B2-cP2
$Cd_{0.3-0.5}Hg_{0.7-0.5}$	1.70–1.92	
CdHg	1.77; 2.15	Tetragonal
$Cd_{0.0075-0.05}In_{0.9925-0.95}$	3.24–3.36	Tetragonal
$Cd_{0.97}Pb_{0.03}$	4.2	
CdSn	3.65	
$Cd_{0.17}Tl_{0.83}$	2.3	
$Cd_{0.18}Tl_{0.82}$	2.54	
$CeCo_2$	0.84	C15
$CeCo_{1.67}Ni_{0.33}$	0.46	C15
$CeCo_{1.67}Rh_{0.33}$	0.47	C15
$Ce_xGd_{1-x}Ru_2$	3.2–5.2	C15
$CeIr_3$	3.34	
$CeIr_5$	1.82	
$Ce_{0.005}La_{0.995}$	4.6	
Ce_xLa_{1-x}	1.3–6.3	
$Ce_xPr_{1-x}Ru_2$	1.4–5.3	C15
Ce_xPt_{1-x}	0.7–1.55	
$CeRu_2$	6.0	C15
$Ce_3Mo_6Se_5$	5.7	hR15 (Mo_6PbS_8)
$Ce_2Mo_6Te_6$	1.7	Same
$Co_xFe_{1-x}Si_2$	1.4 (max.)	C1
$CoHf_2$	0.56	$E9_3$
$CoLa_3$	4.28	
$Co_4La_3Sn_{13}$	2.8	cP40
$CoLu_3$	~0.35	
Co_xLuSn_y	1.5	cP40
$Co_{0-0.01}Mo_{0.8}Re_{0.2}$	2–10	
$Co_{0.02-0.10}Nb_3Rh_{0.98-0.90}$	2.28–1.90	A15
$Co_xNi_{1-x}Si_2$	1.4 (max.)	C1
$Co_{0.5}Rh_{0.5}Si_2$	2.5	
$Co_xRh_{1-x}Si_2$	3.65 (max.)	
$Co_{\sim0.3}So_{\sim0.7}$	~0.35	

Substance	T_c, K	Crystal structure type
$Co_4Sc_5Si_{10}$	5.0	tP38 ($Co_4Sc_5Si_{10}$)
$CoSi_2$	1.40; 1.22	C1
Co_xSn_yYb	2.5	cP40
Co_3Th_7	1.83	$D10_2$
Co_xTi_{1-x}	2.8 (max.)	Co in α-Ti
Co_xTi_{1-x}	3.8 (max.)	Co in β-Ti
$CoTi_2$	3.44	$E9_3$
CoTi	0.71	A2
CoU	1.7	B2, distorted
CoU_6	2.29	$D2_c$
$Co_{0.28}Y_{0.72}$	0.34	
CoY_3	<0.34	
$CoZr_2$	6.3	C16
$Co_{0.1}Zr_{0.9}$	3.9	A3
$Cr_{0.6}Ir_{0.4}$	0.4	H.C.P.
$Cr_{0.65}Ir_{0.35}$	0.59	H.C.P.
$Cr_{0.7}Ir_{0.3}$	0.76	H.C.P.
$Cr_{0.72}Ir_{0.28}$	0.83	
Cr_3Ir	0.45	A15
$Cr_{0-0.1}Nb_{1-0.9}$	4.6–9.2	A2
$Cr_{0.80}Os_{0.20}$	2.5	Cubic
Cr_3Os	4.68	A15-cP8 (Cr_3Si)
Cr_xRe	1.2–5.2	
$Cr_{0.4}Re_{0.6}$	2.15	$D8_b$
$Cr_{0.8-0.6}Rh_{0.2-0.4}$	0.5–1.10	A3
Cr_3Rh	0.3	A15-cP8
Cr_3Ru (annealed)	3.3	A15
Cr_2Ru	2.02	$D8_b$
Cr_3Ru_2	2.10	$D8_b$-tP30 (CrFe)
$Cr_{0.1-0.5}Ru_{0.9-0.5}$	0.34–1.65	A3
Cr_xTi_{1-x}	3.6 (max.)	Cr in α-Ti
Cr_xTi_{1-x}	4.2 (max.)	Cr in β-Ti
$Cr_{0.1}Ti_{0.3}V_{0.6}$	5.6	
$Cr_{0.0175}U_{0.9825}$	0.75	β-phase
$Cs_{0.32}O_3W$	1.12	Hexagonal
$Cu_{0.15}In_{0.85}$ (film)	3.75	
$Cu_{0.04-0.08}In_{0.94-0.92}$	4.4	
CuLa	5.85	
$Cu_2Mo_6O_2S_6$	9	hR15 (Mo_6PbS_8)
$Cu_2Mo_6Se_8$	5.9	Same
Cu_xPb_{1-x}	5.7–7.7	
CuS	1.62	B18
CuS_2	1.48–1.53	C18
CuSSe	1.5–2.0	C18
$CuSe_2$	2.3–2.43	C18
CuSeTe	1.6–2.0	C18
Cu_xSn_{1-x}	3.2–3.7	
Cu_xSn_{1-x} (film, made at 10K)	3.6–7	
Cu_xSn_{1-x} (film, made at 300K)	2.8–3.7	
$CuTe_2$	<1.25–1.3	C18
$CuTh_2$	3.49	C16
$Cu_{0-0.027}V$	3.9–5.3	A2
CuY	0.33	B2-cP2 (CsCl)
Cu_xZn_{1-x}	0.5–0.845	
$DyMo_6S_8$	2.1	hR15
Er_xLa_{1-x}	1.4–6.3	
$ErMo_6S_8$	2.2	hR15
$ErMo_6Se_8$	6.2	hR15
$Fe_3Lu_2Si_5$	6.1	tP40 ($Fe_3Sc_2Si_5$)
$Fe_{0-0.04}Mo_{0.8}Re_{0.2}$	1–10	
$Fe_{0.05}Ni_{0.05}Zr_{0.90}$	~3.9	

Substance	T_c, K	Crystal structure type	Substance	T_c, K	Crystal structure type
Fe_3Re_2	6.55	$D8_b$-tP30 (FeCr)	GeV_3	6.01	A15
$Fe_3Sc_2Si_5$	4.52	tP40	Ge_2Y	3.80	C_c
Fe_3Si_5Tm	1.3	Same	$Ge_{1.62}Y$	2.4	
$Fe_3Si_5Y_2$	2.4	Same	Ge_2Zr	0.30	oC12 ($ZrSi_2$)
Fe_3Th_7	1.86	D10	$GeZr_3$	0.4	$L1_2$-tP32 (Ti_3P)
Fe_xTi_{1-x}	3.2 (max.)	Fe in α-Ti	$H_{0.33}Nb_{0.67}$	7.28	B.C.C.
Fe_xTi_{1-x}	3.7 (max.)	Fe in β-Ti	$H_{0.1}Nb_{0.9}$	7.38	Same
$Fe_xTi_{0.6}V_{1-x}$	6.8 (max.)		$H_{0.05}Nb_{0.95}$	7.83	Same
FeU_6	3.86	$D2_c$	$H_{0.12}Ta_{0.88}$	2.81	B.C.C.
$Fe_{0.1}Zr_{0.9}$	1.0	A3	$H_{0.08}Ta_{0.92}$	3.26	Same
$Ga_{0.5}Ge_{0.5}Nb_3$	7.3	A15	$H_{0.04}Ta_{0.96}$	3.62	Same
Ga_2Ge_2U	0.87	B2-cP2	HfIrSi	3.50	C37-cP12 (Co_2Si)
$GaHf_2$	0.21	C16-tI12 (Al_2Cu)	$HfMo_2$	0.05	hP24 (Ni_2Mn)
$GaLa_3$	5.84		$HfN_{0.989}$	6.6	B1
Ga_3Lu	2.3	B2-cP2	$Hf_{0-0.5}Nb_{1-0.5}$	8.3–9.5	A2
Ga_2Mo	9.5		$Hf_{0.75}Nb_{0.25}$	> 4.2	
$GaMo_3$	0.76	A15	$HfOs_2$	2.69	C14
GaN (black)	5.85	B4	HfOsP	6.1	C22-hP9 (Fe_2P)
$Ga_{0.7}Pt_{0.3}$	2.9	C1	HfPRu	9.9	Same
GaPt	1.74	B20	$HfRe_2$	4.80	C14
GaSb (120kbar, 77K, annealed)	4.24	A5	$Hf_{0.14}Re_{0.86}$	5.86	A12
GaSb (unannealed)	~5.9		$Hf_{0.99-0.96}Rh_{0.01-0.04}$	0.85–1.51	
$Ga_{0-1}Sn_{1-0}$ (quenched)	3.47–4.18		$Hf_{0-0.55}Ta_{1-0.45}$	4.4–6.5	A2
$Ga_{0-1}Sn_{1-0}$ (annealed)	2.6–3.85		HfV_2	8.9–9.6	C15
GaTe	0.17	mC24 (GaTe)	Hg_xIn_{1-x}	3.14–4.55	
Ga_5V_2	3.55	Tetragonal (Mn_2Hg_5)	HgIn	3.81	
$GaV_{4.5}$	9.15		Hg_2K	1.20	Orthorhombic
Ga_3Zr	1.38		Hg_3K	3.18	
Ga_3Zr_5	3.8	$D8_b$-hP16 (Mn_5Si_3)	Hg_4K	3.27	
Gd_xLa_{1-x}	< 1.0–5.5		Hg_8K	3.42	
$GdMo_6S_8$	3.5	hR15	Hg_3Li	1.7	Hexagonal
$GdMo_6Se_8$	5.6	hR15	$HgMg_3$	0.17	hP8 (Na_3As)
$Gd_xOs_2Y_{1-x}$	1.4–4.7		Hg_2Mg	4.0	tI6 ($MoSi_2$)
$Gd_xRu_2Th_{1-x}$	3.6 (max.)	C15	Hg_3Mg_5	0.48	$D8_b$-hP16 (Mn_5Si_3)
$Ge_{10}As_4Y_5$	9.06	tP38 ($C0_4Sc_5Si_{10}$)	Hg_2Na	1.62	Hexagonal
GeIr	4.7	B31	Hg_4Na	3.05	
GeIrLa	1.64	tI12 (LaPtSi)	Hg_xPb_{1-x}	4.14–7.26	
$Ge_{10}Ir_4Lu_5$	2.60	tP38	HgSn	4.2	
$Ge_{10}Ir_4Y_5$	2.62	tP38	Hg_xTl_{1-x}	2.30–4.19	
Ge_2La	1.49; 2.2	Orthorhombic, distorted (Mn_2Hg_5)	Hg_5Tl_2	3.86	
			Ho_xLa_{1-x}	1.3–6.3	
GeLaPt	3.53	tI12	$Ho_{1.2}Mo_6Se_8$	6.1	$D10_2$-hR12 (Be_3Nb)
$Ge_{13}Lu_3Os_4$	3.6	cP40 ($Pr_3Rh_2Sn_{13}$)	$In_{1-0.86}Mg_{0-0.14}$	3.395–3.363	
$Ge_{10}Lu_5Rh_4$	2.79	tP38	$In_2Mo_6Te_6$	2.6	hR15 (Mo_6PbS_8)
$Ge_{13}Lu_3Ru_4$	2.3	cP40	$InNb_3$ (high pressure and temp.)	4–8; 9.2	A15
$GeMo_3$	1.43	A15			
$GeNb_2$	1.9		$In_{0.5}Nb_3Zr_{0.5}$	6.4	
$Ge_{0.29}Nb_{0.71}$	6	A15	$In_{0.11}O_3W$	< 1.25–2.8	Hexagonal
GePt	0.40	B31	$In_{0.95-0.85}Pb_{0.05-0.15}$	3.6–5.05	
Ge_3Rh_5	2.12	Orthorhombic, related to $InNi_2$	$In_{0.98-0.91}Pb_{0.02-0.09}$	3.45–4.2	
			InPb	6.65	
GeRh	0.96	B31-oP8 (MnP)	InPd	0.7	B2
$Ge_{13}Rh_4Sc_3$	1.9	c P40	InSb (quenched from 170 kbar into liquid N_2)	4.8	Like A5
$Ge_{10}Rh_4Y_5$	1.35	tP38			
$Ge_{13}Ru_4Y_3$	1.7	cP40	InSb	2.1	B4
Ge_2So	1.3		$(InSb)_{0.95-0.10}Sn_{0.05-0.90}$ (various heat treatments)	3.8–5.1	
$GeTa_3$	8.0	A15-cP8 (Cr_3Si)			
Ge_3Te_4 ($n = 1.06 \times 10^{22}$)	1.55–1.80	Rhombohedral	$(InSb)_{0-0.07}Sn_{1-0.93}$	3.67–3.74	
Ge_xTe_{1-x} ($n = 8.5$–64×10^{20})	0.07–0.41	R1	In_3Sn	~5.5	
			In_xSn_{1-x}	3.4–7.3	

Substance	T_c, K	Crystal structure type	Substance	T_c, K	Crystal structure type
$In_{0.82-1}Te$ (n = 0.83–1.71 × 10^{22})	1.02–3.45	B1	Ir_2Y_3	1.61	
$In_{1.000}Te_{1.002}$	3.5–3.7	B1	Ir_3Y	3.50	$D10_2$-hR13 (Be_3Nb)
In_3Te_4 (n = 4.7 × 10^{21})	1.15–1.25	Rhombohedral	Ir_xY_{1-x}	0.3–3.7	
In_xTl_{1-x}	2.7–3.374		Ir_2Zr	4.10	C15
$In_{0.8}Tl_{0.2}$	3.223		$Ir_{0.1}Zr_{0.9}$	5.5	A3
$In_{0.62}Tl_{0.38}$	2.760		$K_2Mo_{15}S_{19}$	3.32	hR15
$In_{0.78-0.69}Tl_{0.22-0.31}$	3.18–3.32	Tetragonal	$K_{0.27-0.31}O_3W$	0.50	Hexagonal
$In_{0.69-0.62}Tl_{0.31-0.38}$	2.98–3.3	F.C.C.	$K_{0.40-0.57}O_3W$	1.5	Tetragonal
Ir_2La	0.48	C15	$La_{0.55}Lu_{0.45}$	2.2	Hexagonal, La type
Ir_3La	2.32	$D10_2$	$La_{0.8}Lu_{0.2}$	3.4	Same
Ir_3La_7	2.24	$D10_2$	$LaMg_2$	1.05	C15
Ir_5La	2.13		$LaMo_6S_8$	7.1	hR15
$IrLaSi_2$	2.03	oC16 ($CeNiSi_2$)	LaN	1.35	
$IrLaSi_3$	2.7	tI10 ($BaNiSn_3$)	$LaOs_2$	6.5	C15
Ir_2Lu	2.47	C15	$LaPt_2$	0.46	C15
Ir_3Lu	2.89	C15	$La_{0.28}Pt_{0.72}$	0.54	C15
$Ir_4Lu_5Si_{10}$	3.9	tP38 ($Co_4Sc_5Si_{10}$)	$LaPtSi$	3.48	tI12
$IrMo$	< 1.0	A3	$LaRh_3$	2.60	
$IrMo_3$	9.6	A15	$LaRh_5$	1.62	
$IrMo_3$	6.8	$D8_b$	La_7Rh_3	2.58	$D10_2$
$IrNb_3$	1.9	A15	$LaRhSi_2$	3.42	oC16 ($CeNiSi_2$)
$Ir_{0.4}Nb_{0.6}$	9.8	$D8_b$	$La_2Rh_3Si_5$	4.45	oI40 ($Co_3Si_U_2$)
$Ir_{0.37}Nb_{0.63}$	2.32	$D8_b$	$LaRhSi_3$	2.7	tI10 ($BaNiSn_3$)
$IrNb$	7.9	$D8_b$	$LaRh_2Si_2$	3.90	tI10 (Al_4Ba)
$Ir_{1.15}Nb_{0.85}$	4.6	oP12 (IrTa)	$LaRu_2$	1.63	C15
$Ir_{0.02}Nb_3Rh_{0.98}$	2.43	A15	La_3S_4	6.5	$D7_3$
$Ir_{0.05}Nb_3Rh_{0.95}$	2.38	A15	La_3Se_4	8.6	$D7_3$
$Ir_{0.287}O_{0.14}Ti_{0.573}$	5.5	$E9_3$	$LaSi_2$	2.3	C_c
$Ir_{0.265}O_{0.035}Ti_{0.65}$	2.30	$E9_3$	La_xY_{1-x}	1.7–5.4	
Ir_xOs_{1-x}	0.3–0.98		$LaZn$	1.04	B2
$Ir_{1.5}Os_{0.5}$	2.4	C14	$Li_2Mo_6S_8$	4.2	hR15
$IrOsY$	2.6	C15	$LiPb$	7.2	
$IrSiY$	2.70	C37-oP12 (Co_2Si)	$LuOs_2$	3.49	C14
$IrSiZr$	2.04	Same	$Lu_{0.275}Rh_{0.725}$	1.27	C15
Ir_2Sc	2.07	C15	$LuRh_5$	0.49	
$Ir_{2.5}Sc$	2.46	C15	$Lu_5Rh_4Si_{10}$	3.95	tP38 ($Co_4So_5Si_{10}$)
$Ir_4Sc_5Si_{10}$	8.46	tP38	$LuRu_2$	0.86	C14
Ir_2Si_2Th	2.14	tI10	$Mg_{1.14}Mo_{6.6}S_8$	3.5	hR15
$IrSi_3Th$	1.75	tI10	$Mg2Nb$	5.6	
$IrSiTh$	6.50	tI12 (LaPtSi)	$Mg_{~0.47}Tl_{~0.53}$	2.75	B2
Ir_2Si_2Y	2.60	tI10 ($Al4Ba$)	$MgZn$	0.9	A3-oP4 (AuCd)
$Ir_4Si_{10}Y_5$	3.10	tP38	Mn_xTi_{1-x}	2.3 (max.)	Mn in -Ti
$Ir_3Si_5Y_2$	2.83	oI40	Mn_xTi_{1-x}	1.1–3.0	Mn in -Ti
$IrSn_2$	0.65–0.78	C1	MnU_6	2.32	$D2_c$
Ir_2Sr	5.70	C15	Mo_2N	5.0	F.C.C.
Ir_7Ta_{13}	1.2	$D8_b$-tP30 (FeCr)	$Mo_6Na_2S_8$	8.6	hR15
$Ir_{0.5}Te_{0.5}$	~3		Mo_xNb_{1-x}	0.016–9.2	
$IrTe_3$	1.18	C2	$Mo_{5.25}Nb_{0.75}Se_8$	6.2	hR15
$IrTh$	< 0.37	B_f	Mo_6NdSa_8	8.2	hR15
Ir_2Th	6.50	C15	Mo_3Os	7.2	A15
Ir_3Th	4.71		$Mo_{0.62}Cs_{0.38}$	5.65	$D8_b$
Ir_3Th_7	1.52	$D10_2$	Mo_3P	5.31	DO_e
Ir_5Th	3.93	$D2_d$	$Mo_6Pb_{1.2}Se_8$	6.75	hR15
$IrTi_3$	5.40	A15	$Mo_{0.5}Pd_{0.5}$	3.52	A3
IrV_2	1.39	A15	Mo_6PrSe_8	9.2	hR15
IrW_3	3.82		$MoRe$	7.8	$D8_b$-tP30
$Ir_{0.28}W_{0.72}$	4.49		$MoRe_3$	9.25; 9.89	A12
Ir_2Y	2.18; 1.38	C15	Mo_xRe_{1-x}	1.2–12.2	
$Ir_{0.69}Y_{0.31}$	1.98; 1.44	C15	$Mo_{0.42}Re_{0.58}$	6.35	$D8_b$
$Ir_{0.70}Y_{0.30}$	2.16	C15	$MoRh$	1.97	A3
			Mo_xRh_{1-x}	1.5–8.2	B.C.C.

Substance	T_c, K	Crystal structure type	Substance	T_c, K	Crystal structure type
MoRu	9.5–10.5	A3	Nb_3Rh	2.64	A15
$Mo_{0.61}Ru_{0.39}$	7.18	$D8_b$	$Nb_{0.6}Rh_{0.40}$	4.21	$D8_b$ plus other
$Mo_{0.2}Ru_{0.8}$	1.66	A3	$Nb_{0.9}Rh_{1.1}$	3.07	A3-oP4 (AuCd)
Mo_3Ru_2	7.0	$D8_b$-tP30	$Nb_3Rh_{0.98-0.90}Ru_{0.02-0.10}$	2.42–2.44	A15
$Mo_4Ru_2Te_8$	1.7	hR15	Nb_xRu_{1-x}	1.2–4.8	
Mo_6S_8	1.85	hR15	NbRuSi	2.65	oI36
Mo_6S_8Sc	3.6	hR15	NbS_2	6.1–6.3	Hexagonal, $NbSe_2$ type
$Mo_6S_8Sm_{1.2}$	2.9	hR15	NbS_2	5.0–5.5	Hexagonal, three-layer type
Mo_6S_8Tb	2.0	hR15	Nb_3Sb	0.2	$L1_2$-tP32 (Ti_3P)
Mo_6S_8Tl	8.7	hR15	$Nb_3Sb_{0-0.7}Sn_{1-0.3}$	6.8–18	A15
$Mo_6S_8Tm_{1.2}$	2.1	hR15	$NbSe_2$	5.15–5.62	Hexagonal
$Mo_6S_8Y_{1.2}$	3.0	hR15	$Nb_{1-1.05}Se_2$	2.2–7.0	Same
Mo_6S_8Yb	9.2	hR15	Nb_3Se_4	2.0	hP14
$Mo_{6.6}S_8Zn_{11}$	3.6	hR15	Nb_3Si	1.5	$L1_2$
Mo_3Sb_4	2.1		Nb_3SiSnV_3	4.0	
Mo_6Se_8	6.3	hR15	$NbSn_2$	2.60	Orthorhombic
$Mo_6Se_8Sm_{1.2}$	6.8	hR15	Nb_6Sn_5	2.8	oI44 (Sn_5Ti_6)
$Mo_6Se_8Sn_{1.2}$	6.8	hR15	NbSnTaV	6.2	A15
Mo_6Se_8Tb	5.7	hR15	$NbSnV_2$	5.5	A15
Mo_3Se_3Tl	4.0	hP14	Nb_2SnV	9.8	A15
$Mo_6Se_8Tm_{1.2}$	6.3	hR15	Nb_xTa_{1-x}	4.4–9.2	A2
Mo_6Se_8Yb	6.2	hR15	Nb_3Te_4	1.8	hP14
Mo_3Si	1.30	A15	Nb_xTi_{1-x}	0.6–9.8	
$MoSi_{0.7}$	1.34		$Nb_{0.6}Ti_{0.4}$	9.8	
Mo_xSiV_{3-x}	4.54–16.0	A15	Nb_xU_{1-x}	1.95 (max.)	
$Mo_{5.25}Ta_{0.75}Te_8$	1.7	hR15	$Nb_{0.88}V_{0.12}$	5.7	A2
Mo_6Te_8	1.7	hR15	$Nb_{0.5}V_{1.5}Zr$	4.3	C15-hP12 ($MgZn_2$)
$Mo_{0.16}Ti_{0.84}$	4.18; 4.25		$Ni_{0.3}Th_{0.7}$	1.98	$D10_2$
$Mo_{0.913}Ti_{0.087}$	2.95		$NiZr_2$	1.52	
$Mo_{0.04}Ti_{0.96}$	2.0	Cubic	$Ni_{0.1}Zr_{0.9}$	1.5	A3
$Mo_{0.025}Ti_{0.975}$	1.8		$O_3Rb_{0.27-0.29}W$	1.98	Hexagonal
Mo_xU_{1-x}	0.7–2.1		OSn	3.81	tP4 (PbO)
Mo_xV_{1-x}	0–~5.3		O_3SrTi ($n = 1.7$–12.0×10^{19})	0.12–0.37	
Mo_2Zr	4.25–4.75	C15	O_3SrTi ($n = 10^{18}$–10^{21})	0.05–0.47	
NNb (film)	6–9	B1	O_3SrTi ($n = 10^{20}$)	0.47	
$N_xO_yTi_z$	2.9–5.6	Cubic	$O_3Sr_{0.08}W$	2–4	Hexagonal
$N_xO_yV_z$	5.8–8.2	Cubic	OTi	0.58	
$N_{0.34}Re$	4–5	F.C.C.	$O_3Tl_{0.30}W$	2.0–2.14	Hexagonal
NTa (film)	4.84	B1	OV_3Zr_3	7.5	$E9_3$
$N_{0.6-0.987}Ti$	< 1.17–5.8	B1	OW_3 (film)	3.35; 1.1	A15
$N_{0.82-0.99}V$	2.9–7.9	B1	OsPti	1.2	C22-hP9 (Fe_2P)
NZr	9.8	B1	OsPZr	7.4	Same
$N_{0.906-0.984}Zr$	3.0–9.5	B1	OsReY	2.0	C14
$Na_{0.28-0.35}O_3W$	0.56	Tetragonal	Os_2Sc	4.6	C14
$Na_{0.28}Pb_{0.72}$	7.2		OsTa	1.95	A12
NbO	1.25		Os_3Th_7	1.51	$D10_2$
$NbOs_2$	2.52	A12	Os_xW_{1-x}	0.9–4.1	
Nb_3Os	1.05	A15	OsW_3	~3	
$Nb_{0.6}Os_{0.4}$	1.89; 1.78	$D8_b$	Os_2Y	4.7	C14
$Nb_3Os_{0.02-0.10}Rh_{0.98-0.90}$	2.42–2.30	A15	Os_2Zr	3.0	C14
Nb_3P	1.8	$L1_2$tP32 (Ti_3P)	Os_xZr_{1-x}	1.5–5.6	
NbPRh	4.08	C37-oP12 (Co_2Si)	PPb	7.8	
$Nb_{0.6}Pd_{0.4}$	1.60	$D8_f$ plus cubic	OsW_2	3.81	$D8_b$-tP30 (FeCr)
$Nb_3Pd_{0.02-0.10}Rh_{0.92-0.90}$	2.49–2.55	A15	$PPd_{3.0-3.2}$	<0.35–0.7	DO_{11}
$Nb_{0.62}Pt_{0.38}$	4.21	$D8_b$	P_3Pd_7 (high temperature)	1.0	Rhombohedral
Nb_5Pt_3	3.73	$D8_b$	P_3Pd_7 (low temperature)	0.70	Complex
$Nb_3Pt_{0.02-0.98}Rh_{0.98-0.02}$	2.52–9.6	A15	PRh	1.22	
$NbRe_3$	5.27	$D8_b$-tP30 (FeCr)	PRh_2	1.3	C1
$Nb_{0.38-0.18}Re_{0.62-0.82}$	2.43–9.70	A15	P_4Rh_5	1.22	oP28 ($CaFe_2O_4$)
NbRe	3.8	$D8_b$-tP30	PRhTa	4.41	C37-oP12 (Co_2Si)
NbReSi	5.1	oI36 (FeTiSi)			

Substance	T_c, K	Crystal structure type	Substance	T_c, K	Crystal structure type
PRhZr	1.55	Same	$PtV_{3.5}$	1.26	A15
PRuTi	1.3	C22-hP9 (Fe_2P)	$Pt_{0.5}W_{0.5}$	1.45	A1
PRuZr	3.46	C37-oP12	Pt_xW_{1-x}	0.4–2.7	
PW_3	2.26	DO_e	Pt_2Y_3	0.90	
Pb_2Pd	2.95	C16	Pt_2Y	1.57; 1.70	C15
Pb_4Pt	2.80	Related to C16	Pt_3Y_7	0.82	$D10_2$
Pb_2Rh	2.66	C16	PtZr	3.0	A3
PbSb	6.6		Re_2Sc	4.2	C15-hP12 ($MgZn_2$)
PbTe (plus 0.1 w/o Pb)	5.19		$Re_{24}Sc_5$	2.2	A12-cI58 (Mg)
PbTe (plus 0.1 w/o Te)	5.24–5.27		ReSiTa	4.4	oI36 (FeTiSi)
$PbTl_{0.27}$	6.43		$Re_3Si_3Y_2$	1.76	tP40 ($Fe_3Sc_2Si_5$)
$PbTl_{0.17}$	6.73		Re_3Ta_2	1.4	$D8_b$-tP30 (FeCr)
$PbTl_{0.12}$	6.88		$Re_{0.64}Ta_{0.36}$	1.46	A12
$PbTl_{0.075}$	6.98		Re_3Ta	6.78	A12-cI58 (Mn)
$PbTl_{0.04}$	7.06		$Re_{24}Ti_5$	6.60	A12
$Pb_{1-0.26}Tl_{0-0.74}$	7.20–3.68		Re_xTi_{1-x}	6.6 (max.)	
$PbTl_2$	3.75–4.1		$Re_{0.76}V_{0.24}$	4.52	$D8_b$
Pb_3Zr_5	4.60	$D8_8$	Re_3V	6.26	$D8_b$-tP30
$PbZr_3$	0.76	A15	$Re_{0.92}V_{0.08}$	6.8	A3
$Pd_{0.9}Pt_{0.1}Te_2$	1.65	C6	$Re_{0.6}W_{0.4}$	6.0	
$Pd_{0.05}Ru_{0.05}Zr_{0.9}$	~9		$Re_{0.5}W_{0.5}$	5.12	$D8_b$
$Pd_{2.2}S$ (quenched)	1.63	Cubic	$Re_{13}W_{12}$	5.2	$D8_b$-tP30
$PdSb_2$	1.25	C2	Re_3W	9.0	A12-cI58
PdSb	1.5	$B8_1$	Re_2Y	1.83	C14
PdSbSe	1.0	C2	Re_2Zr	5.9	C14
PdSbTe	1.2	C2	Re_3Zr	7.40	A12-cI58
Pd_4Se	0.42	Tetragonal	Re_6Zr	7.40	Same
$Pd_{6-7}Se$	0.66	Like Pd_4Te	$Rh_{17}S_{15}$	5.8	Cubic
$Pd_{2.8}Se$	2.3		$Rh_{-0.24}Sc_{-0.76}$	0.88; 0.92	
Pd_xSe_{1-x}	2.5 (max.)		$Rh_4Sc_5Si_{10}$	8.54	tP38
PdSi	0.93	B31	$Rh_4Sc_3Sn_{13}$	4.5	cP40
PdSn	0.41	B31	Rh_xSe_{1-x}	6.0 (max.)	
$PdSn_2$	3.34		$RhSi_3Th$	1.76	tI10
Pd_2Sn	0.41	C37	$Rh_{0.86}Sc_{1.04}Th$	6.45	tI12
Pd_3Sn	0.47–0.64	$B8_2$	Rh_2Si_2Y	3.11	tI10
Pd_2SnTm	1.77	DO_3-cF16 (BiF_3)	$Rh_3Si_5Y_2$	2.70	oI40
Pd_2SnY	4.92	Same	$Rh_4Sn_{13}Sr_3$	4.3	cP40
Pd_2SnYb	1.79	Same	Rh_xSn_yTh	1.9	cI2 (W)
PdTe	2.3; 3.85	$B8_1$	Rh_xSn_yTm	2.3	cP40
$PdTe_{1.02-1.08}$	2.56–1.88	$B8_1$	$Rh_4Sn_{13}Y_3$	3.2	cP40
$PdTe_2$	1.69	C6	Rh_2Sr	6.2	C15
$PdTe_{2.1}$	1.89	C6	$Rh_{0.4}Ta_{0.6}$	2.35	$D8_b$
$PdTe_{2.3}$	1.85	C6	$RhTe_2$	1.51	C2
$Pd_{1.1}Te$	4.07	$B8_1$	$Rh_{0.67}Te_{0.33}$	0.49	
Pd_3Te	0.76	cI2 (W)	Rh_xTe_{1-x}	1.51 (max.)	
$PdTh_2$	0.85	C16	RhTh	0.36	B_f
$Pd_{0.1}Zr_{0.9}$	7.5	A3	Rh_3Th_7	2.15	$D10_2$
PtSb	2.1	$B8_1$	Rh_5Th	1.07	
PtSi	0.88	B31	Rh_xTi_{1-x}	2.25–3.95	
PtSn	0.37	$B8_1$	$Rh_{0.02}U_{0.98}$	0.96	
$PtSn_4$	2.38	C16-oC20 ($PdSn_4$)	RhV_3	0.38	A15
Pt_3Ta_7	1.5	$D8_b$-tP30	RhW	~3.4	A3
$PtTa_3$	0.4	A15-cP8 (Cr_3Si)	RhY_3	0.65	
PtTe	0.59	Orthorhombic	Rh_2Y_3	1.48	
PtTh	0.44	B_f	Rh_3Y	1.07	C15
Pt_3Th_7	0.98	$D10_2$	Rh_5Y	0.56	
Pt_5Th	3.13		Rh_3Y_7	0.32	hP20 (Fe_3Th_7)
$PtTi_3$	0.58	A15	$Rh_{0.005}Zr$ (annealed)	5.8	
$Pt_{0.02}U_{0.98}$	0.87	β-phase	$Rh_{0-0.45}Zr_{1-0.55}$	2.1–10.8	
$PtV_{2.5}$	1.36	A15	$Rh_{0.1}Zr_{0.9}$	9.0	H.C.P.
PtV_3	2.87–3.20	A15	Ru_2Sc	1.67	C14
			RuSiTa	3.15	oI36

Substance	T_c, K	Crystal structure type
Ru_3Si_2Th	3.98	hP12
Ru_3Si_2Y	3.51	hP12
$Ru_{1.1}Sn_{3.1}Y$	1.3	cP40
Ru_2Th	3.56	C15
$RuTi$	1.07	B2
$Ru_{0.05}Ti_{0.95}$	2.5	
$Ru_{0.1}Ti_{0.9}$	3.5	
$Ru_xTi_{0.6}V_y$	6.6 (max.)	
Ru_3U	0.15	$L1_2$-cP4
$Ru_{0.45}V_{0.55}$	4.0	B2
RuW	7.5	A3
Ru_2Y	1.52	C14
Ru_2Zr	1.84	C14
$Ru_{0.1}Zr_{0.9}$	5.7	A3
STh	0.5	B1-cF8 (NaCl)
$SbSn$	1.30–1.42	B1 or distorted
$SbTa_3$	0.72	A15-cP8 (Cr_3Si)
$SbTi_3$	5.8	Same
Sb_2Ti_7	5.2	
$Sb_{0.01-0.03}V_{0.99-0.97}$	3.76–2.63	A2
SbV_3	0.80	A15
$SeTh$	1.7	B1-cF8
$SiMo_3$	1.4	A15-cP8
Si_2Th	3.2	C_c, α-phase
Si_2Th	2.4	C32, β-phase
$SiV_{2.7}Ru_{0.3}$	2.9	A15
Si_2W_3	2.8; 2.84	$L1_2$-tP32 (Ti_3P)
$SiZr_3$	0.5	$L1_2$-tP32 (Ti_3P)
$Sn_{0.174-0.104}Ta_{0.826-0.896}$	6.5–< 4.2	A15
$SnTa_3$	8.35	A15, highly ordered
$SnTa_3$	6.2	A15, partially ordered
$SnTaV_2$	2.8	A15
$SnTa_2V$	3.7	A15
Sn_xTe_{1-x} (n = $10.5-20 \times 10^{20}$)	0.07–0.22	B1
Sn_3Th	3.33	$L1_2$-cP4
$SnTi_3$	5.80	A15-cP8
Sn_xTl_{1-x}	2.37–5.2	
SnV_3	3.8	A15
$Sn_{0.02-0.057}V_{0.98-0.943}$	2.87–~1.6	A2
$SnZr_3$	0.92	A15-cP8
$Ta_{0.025}Ti_{0.975}$	1.3	Hexagonal
$Ta_{0.05}Ti_{0.95}$	2.9	Hexagonal
$Ta_{0.05-0.75}V_{0.95-0.25}$	4.30–2.65	A2
$Ta_{0.8-1}W_{0.2-0}$	1.2–4.4	A2
$Tc_{0.1-0.4}W_{0.9-0.6}$	1.25–7.18	Cubic
$Tc_{0.50}W_{0.50}$	7.52	α plus
$Tc_{0.60}W_{0.40}$	7.88	plus α
Tc_6Zr	9.7	A12
TeY	1.02	B1-cF8
$ThTl_3$	0.87	$L1_2$-cP4
$Th_{0-0.55}Y_{1-0.45}$	1.2–1.8	
$Ti_{0.70}V_{0.30}$	6.14	Cubic
Ti_xV_{1-x}	0.2–7.5	
$Ti_{0.5}Zr_{0.5}$ (annealed)	1.23	
$Ti_{0.5}Zr_{0.5}$ (quenched)	2.0	
Tl_3Y	1.52	$L1_2$-cP4
V_2Zr	8.80	C15
$V_{0.26}Zr_{0.74}$	5.9	
W_2Zr	2.16	C15
YZn	0.33	B2-cP2 (CsCl)

* n denotes current carriers concentration in cm^{-3}.

B. Superconductors with $T_c > 10K$

Substance	T_c,K	Crystal structure type	
Al_2CMo_3	10.0	A13	
$Al_{0.5}Ge_{0.5}Nb$	12.6	A15	
$Al_{-0.8}Ge_{-0.2}Nb_3$	20.7	A15	
$AlNb_3$	18.0	A15	(Cr_3Si)
$AlNb_3$	12.0	A15	(FeCr)
Al_xNb_{1-x}	<4.2–13.5	$D8_b$	
Al_xNb_{1-x}	12–17.5	A15	
$Al_{0.27}Nb_{0.73-0.48}V_{0-0.25}$	14.5–17.5	A15	
$Al\ Nb_xV_{1-x}$	4.4–13.5		
$Al_{0.1}Si_{0.9}V_3$	14.05		
AlV_3	11.8	A15	(Cr_3Si)
$AuNb_3$	11.5	A15	
$Au_{0-0.3}Nb_{1-0.7}$	1.1–11.0		
$Au_{0.02-0.98}Nb_3Rh_{0.98-0.02}$	2.53–10.9	A15	
$AuNb_{3(1-x)}V_{3x}$	1.5–11.0	A15	
$B_{0.03}C_{0.51}Mo_{0.47}$	12.5		
B_4LuRh_4	11.7		(B_4CeCo_4)
B_2LuRu	10		
B_4Rh_4Y	11.3		(B_4CeCo_4)
$B_{0.1}Si_{0.9}V_3$	15.8	A15	
$BaBi_{0.2}O_3Pb_{0.8}$	13.2		
$Ba_2CaCu_2O_8Tl_2$	120		
$Ba_2Cu_3LaO_6$	80		
$Ba_2Cu_3O_7Tm$	101		
$Ba_2Cu_3O_7Y$	90		
$(Ba,La)_2CuO_4$	36	A15	(K_2NiF_4)
$Bi_2CaCu_2O_8Sr_2$	110		
$Br_2Mo_6S_6$	13.8		(Mo_6PbS_8)
C_3La	11.0		(C_3Pu_2)
CMo	14.3	B1	(NaC1)
CMo_2	12.2	o**	
$C_{0.5}Mo_2Nb_{1-x}$	10.8–12.5	B1	
CMo_xTi_{1-x}	10.2(max)	B1	
$CMo_{0.83}Ti_{0.17}$	10.2	B1	
$C_{0-0.38}N_{1-0.62}Ta$	10.0–11.3		
CNb (whiskers)	7.5–10.5		
CNb	11.5	B1	
$C_{0.7-1.0}Nb_{0.3-0}$	6–11	B1	
CNb_xTa_{1-x}	8.2–13.9		
$CNb_{0.6-0.9}W_{0.4-0.1}$	12.5–11.6	B1	
$C_{0.1}Si_{0.9}V_3$	16.4	A15	
CTa	10.3	B1	
$CTa_{1-0.4}W_{0-0.6}$	8.5–10.5	B1	
$C_{0.66}Th_{0.13}Y_{0.21}$	17		(C_3Pu_2)
C_3Y_2	11.5		(C_3Pu_2)
CW	10	B1	
$(Ca,La)_2CuO_4$	18		(K_2NiF_4)
$Cu(La,Sr)_2O_4$	39		
$Cu_{1.8}Mo_6S_8$	10.8		(Mo_6PbS_8)
$Cr_{0.3}SiV_{2.7}$	11.3	A15	
$GaNb_3$	14.5	A15	(Cr_3Si)
$Ga_xNb_3Sn_{1-x}$	14–18.37	A15	
GaV_3	16.8	A15	
$GaV_{2.1-3.5}$	6.3–14.45	A15	
$GeNb_3$	23.2	A15	
$GeNb_3$ (quenched)	6–17	A15	
$Ge_xNb_3Sn_{1-x}$	17.6–18.0	A15	
$Ge_{0.5}Nb_3Sn_{0.5}$	11.3	A15	
$Ge_{0.1}Si_{0.9}V_3$	14.0	A15	
GeV_3	11	A15	
$InLa_3$	9.83; 10.4	LI_2	($AuCu_3$)

Substance	T_c,K	Crystal structure type	Substance	T_c,K	Crystal structure type
$In_{0-0.3}Nb_3Sn_{1-0.7}$	18.0–18.19	A15	$N_{100-42w/o}Nb_{0-58w/o}Ti$	15–16.8	
InV_3	13.9	A15	$N_{100-75w/o}Nb_{0-25w/o}Zr$	12.5–16.35	
$Ir_{0.4}Nb_{0.6}$	10	(FeCr)	NNb_xZr_{1-x}	9.8–13.8	B1
$LaMo_6Se_8$	11.4	(Mo_6PbS_8)	$N_{0.93}Nb_{0.85}Zr_{0.15}$	13.8	B1
LiO_4Ti_2	13.7	(Al_2MgO_4)	NTa	12–14	B1
MgB_2	39.0±0.5	C32	NZr	10.7	B1
MoN	12; 14.8	h*	Nb_3Pt	10.9	A15
Mo_3Os	12.7	A15	$Nb_{0.18}Re_{0.82}$	10	(Mn)
$Mo_6Pb_{0.9}S_{7.5}$	15.2	(Mo_6PbS_8)	Nb_3Si	19	A15
Mo_3Re	10.0; 15	A15	$Nb_{0.3}SiV_{2.7}$	12.8	A15
Mo_xRe_{1-x}	1.2–12.2		Nb_3Sn	18.05	A15
$Mo_{0.52}Re_{0.48}$	11.1		$Nb_{0.8}Sn_{0.2}$	18.18; 18.5	A15
$Mo_{0.57}Re_{0.43}$	14.0		Nb_3Sn_{1-x} (film)	2.6–18.5	o*
$Mo_{\sim0.60}Re_{0.395}$	10.6		Nb_3Sn_2	16.6	t*
$MoRu$	9.5–10.5	A3	$NbSnTa_2$	10.8	A15
Mo_3Ru	10.6	A15	Nb_2SnTa	16.4	A15
Mo_6Se_8Tl	12.2	(Mo_6PbS_8)	$Nb_{2.5}SnTa_{0.5}$	17.6	A15
$Mo_{0.3}SiV_{2.7}$	11.7	A15	$Nb_{2.75}SnTa_{0.25}$	17.8	A15
Mn_3Si	12.5	A15	$Nb_{3x}SnTa_{3(1-x)}$	6.0–18.0	
Mo_3Tc	15	A15	$Nb_2SnTa_{0.5}V_{0.5}$	12.2	A15
$Mo_{0.3}Tc_{0.7}$	12.0	A15	$NbTc_3$	10.5	A12
Mo_xTc_{1-x}	10.8–15.8		$Nb_{0.75}Zr_{0.25}$	10.8	
$MoTc_3$	15.8		$Nb_{0.66}Zr_{0.33}$	10.8	
NNb (whiskers)	10–14.5		$PbTa_3$	17	A15
NNb (diffusion wires)	16.10		$RhTa_3$	10	A15
$N_{0.988}Nb$	14.9; 17.3	B1	$RhZr_2$	10.8; 11.3	C16 (Al_2Cu)
$N_{0.824-0.988}Nb$	14.4–15.3	B1	$Rh_{0-0.45}Zr_{1-0.55}$	2.1–10.8	
$N_{0.7-0.795}Nb$	11.3–12.9		$SiTi_{0.3}V_{2.7}$	10.9	A15
NNb_xO_y	13.5–17.0	B1	SiV_3	17.1	A15
NNb_xO_y	6.0–11		$SiV_{2.7}Zr_{0.3}$	13.2	A15

TABLE 5. Critical Field Data

Substance	H_o (oersteds)	Substance	H_o (oersteds)
Ag_2F	2.5	$InSb$	1100
Ag_7NO_{11}	57	In_xTl_{1-x}	252–284
Al_2CMo_3	1700	$In_{0.8}Tl_{0.2}$	252
$BaBi_3$	740	$Mg_{0.47}Tl_{0.53}$	220
Bi_2Pt	10	$Mo_{0.16}Ti_{0.84}$	<985
Bi_3Sr	530	$NbSn_2$	620
Bi_5Tl_3	>400	$PbTl_{0.27}$	756
$CdSn$	>266	$PbTl_{0.17}$	796
$CoSi_2$	105	$PbTl_{0.12}$	849
$Cr_{0.1}Ti_{0.3}V_{0.6}$	1360	$PbTl_{0.075}$	880
$In_{1-0.86}Mg_{0-0.14}$	272.4–259.2	$PbTl_{0.04}$	864

TABLE 6. High Critical Magnetic-Field Superconductive Compounds and Alloys

Substance	T_c, K	H_{c1}, kOe	H_{c2}, kOe	H_{c3}, kOe	T_{obs}, K[a]
Al_2CMo_3	9.8–10.2	0.091	156		1.2
$AlNb_3$		0.375			
$Ba_xO_3Sr_{1-x}Ti$	<0.1–0.55	0.0039 max.			
$Bi_{0.5}Cd_{0.1}Pb_{0.27}Sn_{0.13}$			>24		3.06
Bi_xPb_{1-x}	7.35–8.4	0.122 max.	30 max.		4.2
$Bi_{0.56}Pb_{0.44}$	8.8		15		4.2
$Bi_{7.5w/o}Pb_{92.5w/o}$[b]			2.32		
$Bi_{0.099}Pb_{0.901}$		0.29	2.8		
$Bi_{0.02}Pb_{0.98}$		0.46	0.73		
$Bi_{0.53}Pb_{0.32}Sn_{0.16}$			>25		3.06
$Bi_{1-0.93}Sn_{0-0.07}$			0–0.032		3.7
Bi_5Tl_3	6.4		>5.6		3.35
C_8K (excess K)	0.55		0.160 (H⊥c)		0.32
			0.730 (H|c)		0.32
C_8K	0.39		0.025 (H⊥c)		0.32
			0.250 (H|c)		0.32
$C_{0.44}Mo_{0.56}$	12.5–13.5	0.087	98.5		1.2
CNb	8–10	0.12	16.9		4.2
$CNb_{0.4}Ta_{0.6}$	10–13.6	0.19	14.1		1.2
CTa	9–11.4	0.22	4.6		1.2
$Ca_xO_3Sr_{1-x}Ti$	<0.1–0.55	0.002–0.004			
$Cd_{0.1}Hg_{0.9}$ (by weight)		0.23	0.34		2.04
$Cd_{0.05}Hg_{0.95}$		0.28	0.31		2.16
$Cr_{0.10}Ti_{0.30}V_{0.60}$	5.6	0.071	84.4		0
GaN	5.85	0.725			4.2
Ga_xNb_{1-x}			>28		4.2
$GaSb$ (annealed)	4.24		2.64		3.5
$GaV_{1.95}$	5.3		73[e]		
$GaV_{2.1-3.5}$	6.3–14.45		230–300[d]		0
GaV_3		0.4	350[e]		0
			500[d]		
$GaV_{4.5}$	9.15		121[c]		0
Hf_xNb_y			>52–>102		1.2
Hf_xTa_y			>28–>86		1.2
$Hg_{0.05}Pb_{0.95}$		0.235	2.3		
$Hg_{0.101}Pb_{0.899}$		0.23	4.3		4.2
$Hg_{0.15}Pb_{0.85}$	6.75		>13		2.93
$In_{0.98}Pb_{0.02}$	3.45	0.1		0.12	2.76
$In_{0.96}Pb_{0.04}$	3.68	0.1	0.12	0.25	2.94
$In_{0.94}Pb_{0.06}$	3.90	0.095	0.18	0.35	3.12
$In_{0.913}Pb_{0.087}$	4.2	~10.17	0.55	2.65	
$In_{0.316}Pb_{0.684}$		0.155	3.7		4.2
$In_{0.17}Pb_{0.83}$			2.8	5.5	4.2
$In_{1.000}Te_{1.002}$	3.5–3.7		1.2[c]		0
$In_{0.95}Tl_{0.05}$		0.263	0.263		3.3
$In_{0.90}Tl_{0.10}$		0.257	0.257		3.25
$In_{0.83}Tl_{0.17}$		0.242	0.39		3.21
$In_{0.75}Tl_{0.25}$		0.216	0.50		3.16
LaN	1.35	0.45			0.76
La_3S_4	6.5	≈0.15	>25		1.3
La_3Se_4	8.6	≈0.2	>25		1.25
$Mo_{0.52}Re_{0.48}$	11.1		14–21	22–33	4.2
			18–28	37–43	1.3
$Mo_{0.6}Re_{0.395}$	10.6		14–20	20–37	4.2
			19–26	26–37	1.3
$Mo_{0.5}Ti_{0.5}$			75[c]		0
$Mo_{0.16}Ti_{0.84}$	4.18	0.028	98.7[c]		0
			36–38		3.0
$Mo_{0.913}Ti_{0.087}$	2.95	0.060	15		4.2
$Mo_{0.1-0.3}U_{0.9-0.7}$	1.85–2.06		>25		
$Mo_{0.17}Zr_{0.83}$			30		

Substance	T_c, K	H_{c1}, kOe	H_{c2}, kOe	H_{c3}, kOe	T_{obs}, K[a]
$N_{(12.8 \, w/o)}Nb$	15.2		>9.5		13.2
NNb (wires)	16.1		153[c]		0
			132		4.2
			95		8
			53		12
NNb_xO_{1-x}	13.5–17.0		38		
NNb_xZr_{1-x}	9.8–13.8		4- >130		4.2
$N_{0.93}Nb_{0.85}Zr_{0.15}$	13.8		>130		4.2
$Na_{0.086}Pb_{0.914}$		0.19	6.0		
$Na_{0.016}Pb_{0.984}$		0.28	2.05		
Nb	9.15		2.020		1.4
			1.710		4.2
Nb		0.4–1.1	3–5.5		4.2
Nb (unstrained)		1.1–1.8	3.40	6–9.1	4.2
Nb (strained)		1.25–1.92	3.44	6.0–8.7	4.2
Nb (cold-drawn wire)		2.48	4.10	≈10	4.2
Nb (film)			>25		4.2
$NbSc$			>30		
Nb_3Sn		0.170	221		4.2
			70		14.15
			54		15
			34		16
			17		17
$Nb_{0.1}Ta_{0.9}$		0.084	0.154		4.195
$Nb_{0.2}Ta_{0.8}$			10		4.2
$Nb_{0.65-0.73}Ta_{0.02-0.10}Zr_{0.25}$			>70->90		4.2
Nb_xTi_{1-x}			148 max.		1.2
			120 max.		4.2
$Nb_{0.222}U_{0.778}$		1.98	23		1.2
Nb_xZr_{1-x}			127 max.		1.2
			94 max.		4.2
O_3SrTi	0.43	0.0049[c]	0.504[c]		0
O_3SrTi	0.33	0.00195[c]	0.420[c]		0
$PbSb_{1 \, w/o}$ (quenched)			>1.5		4.2
$PbSb_{1 \, w/o}$ (annealed)			>0.7		4.2
$PbSb_{2.8 \, w/o}$ (quenched)			>2.3		4.2
$PbSb_{2.8 \, w/o}$ (annealed)			>0.7		4.2
$Pb_{0.871}Sn_{0.129}$		0.45	1.1		
$Pb_{0.965}Sn_{0.035}$		0.53	0.56		
$Pb_{1-0.26}Tl_{0-0.74}$	7.20–3.68		2–6.9[c]		0
$PbTl_{0.17}$	6.73		4.5[c]		0
$Re_{0.26}W_{0.74}$			>30		
$Sb_{0.93}Sn_{0.07}$			0.12		3.7
SiV_3	17.0	0.55	156[e]		
Sn_xTe_{1-x}		0.00043–0.00236	0.005–0.0775		0.012–0.079
Ta (99.95%)		0.425	1.850		1.3
		0.325	1.425		2.27
		0.275	1.175		2.66
		0.090	0.375		3.72
$Ta_{0.5}Nb_{0.5}$			3.55		4.2
$Ta_{0.65-0}Ti_{0.35-1}$	4.4–7.8		>14–138		1.2
$Ta_{0.5}Ti_{0.5}$			138		1.2
Te	3.3	0.25[c]			0
Tc_xW_{1-x}	5.75–7.88		8–44		4.2
Ti				2.7	4.2
$Ti_{0.75}V_{0.25}$	5.3	0.029[c]	199[c]		0
$Ti_{0.775}V_{0.225}$	4.7	0.024[c]	172[c]		0
$Ti_{0.615}V_{0.385}$	7.07	0.050	34		4.2
$Ti_{0.516}V_{0.484}$	7.20	0.062	28		4.2
$Ti_{0.415}V_{0.585}$	7.49	0.078	25		4.2
$Ti_{0.12}V_{0.88}$			17.3	28.1	4.2
$Ti_{0.09}V_{0.91}$			14.3	16.4	4.2
$Ti_{0.06}V_{0.94}$			8.2	12.7	4.2

Substance	T_c, K	H_{c1}, kOe	H_{c2}, kOe	H_{c3}, kOe	T_{obs}, K[a]
$Ti_{0.03}V_{0.97}$			3.8	6.8	4.2
Ti_xV_{1-x}			108 max.		1.2
V	5.31	0.8	3.4		1.79
		0.75	3.15		2
		0.45	2.2		3
		0.30	1.2		4
$V_{0.26}Zr_{0.74}$	≈5.9	0.238			1.05
		0.227			1.78
		0.185			3.04
		0.165			3.5
W (film)	1.7–4.1		>34		1

[a] Temperature of critical field measurement.
[b] w/o denotes weight percent.
[c] Extrapolated.
[d] Linear extrapolation.
[e] Parabolic extrapolation.

References

1. B. W. Roberts, in *Superconductive Materials and Some of Their Properties. Progress in Cryogenics*, Vol. IV, 1964, pp. 160–231.

2. B. W. Roberts, Superconductive Materials and Some of Their Properties, NBS Technical Notes 408 and 482, U.S. Government Printing Office, 1966 and 1969; B. W. Roberts, *J. Phys. Chem. Ref. Data*, 5, 581, 1976.

3. B. W. Roberts, Properties of Selected Superconductive Materials, 1978 Supplement, NBS Technical Note 983, 1978.

4. T. Claeson, *Phys. Rev.*, 147, 340, 1966.

5. C. J. Raub, W. H. Zachariasen, T. H. Geballe, and B. T. Matthias, *J. Phys. Chem. Solids*, 24, 1093, 1963.

6. T. H. Geballe, B. T. Matthias, V. B. Compton, E. Corenzwit, G. W. Hull, Jr., and L. D. Longinotti, *Phys. Rev.*, 1A, 119, 1965.

7. C. J. Raub, V. B. Compton, T. H. Geballe, B. T. Matthias, J. P. Maita, and G. W. Hull, Jr., *J. Phys. Chem. Solids*, 26, 2051, 1965.

8. R. D. Blaugher, J. K. Hulm, and P. N. Yocom, *J. Phys. Chem. Solids*, 26, 2037, 1965.

9. T. Claeson and H. L. Luo, *J. Phys. Chem. Solids*, 27, 1081, 1966.

10. S. C. Ng and B. N. Brockhouse, *Solid State Comm.*, 5, 79, 1967.

11. O. I. Shulishova and I. A. Shcherbak, *Izv. AN SSSR, Neorg. Materials*, 3, 1495, 1967.

12. T. F. Smith and H. L. Luo, *J. Phys. Chem. Solids*, 28, 569, 1967.

13. A. C. Lawson, *J. Less-Common Metals*, 23, 103, 1971.

14. R. Chevrel, M. Sergent, and J. Prigent, *J. Solid State Chem.*, 3, 515, 1971.

15. M. Marezio, P. D. Dernier, J. P. Remeika, and B. T. Matthias, *Mat. Res. Bull.*, 8, 657, 1973.

16. J. K. Hulm and R. D. Blaugher in *Superconductivity in d- and f-Band Metals*, D. H. Douglass, Ed., American Institute of Physics, 4, 1, 1972.

17. R. N. Shelton, A. C. Lawson, and D. C. Johnston, *Mat. Res. Bull.*, 10, 297, 1975.

18. H. D. Wiesinger, *Phys. Status Sol.*, 41A, 465, 1977.

19. O. Fisher, *Applied Phys.*, 16, 1, 1978.

20. D. C. Johnston, *Solid State Comm.*, 24, 699, 1977.

21. H. C. Ku and R. H. Shelton, *Mat. Res. Bull.*, 15, 1441, 1980.

22. H. Barz, *Mat. Res. Bull.*, 15, 1489, 1980.

23. G. P. Espinosa, A. S. Cooper, H. Barz, and J. P. Remeika, *Mat. Res. Bull.*, 15, 1635, 1980.

24. E. M. Savitskii, V. V. Baron, Yu. V. Efimov, M. I. Bychkova, and L. F. Myzenkova, in *Superconducting Materials*, Plenum Press, 1981, p. 107.

25. R. Fluckiger and R. Baillif, in Topics in *Current Physics*, O. Fischer and M. B. Maple, Eds., Springer Verlag, 34, 113, 1982.

26. R. N. Shelton, in *Superconductivity in d- and f-Band Metals*, W. Buckel and W. Weber, Eds., Kernforschungszentrum, Karlsruhe, 1982, p. 123.

27. D. C. Johnston and H. F. Braun, *Topics in Current Phys.*, 32, 11, 1982.

28. R. Chevrel and M. Sergent, *Topics in Current Phys.*, 32, 25, 1982.

29. G. P. Espinosa, A. S. Cooper, and H. Barz, *Mat. Res. Bull.*, 17, 963, 1982.

30. R. Muller, R. N. Shelton, J. W. Richardson, Jr., and R. A. Jacobson, *J. Less-Comm. Met.*, 92, 177, 1983.

31. You-Xian Zhao and Shou-An He, in *High Pressure in Science and Technology*, North Holland, 22, 51, 1983.

32. You-Xian Zhao and Shou-An He, *Solid State Comm.*, 24, 699, 1983.

33. G. P. Meisner and H. C. Ku, *Appl. Phys.*, A31, 201, 1983.

34. R. J. Cava, D. W. Murphy, and S. M. Zahurak, *J. Electrochem. Soc.*, 130, 2345, 1983.

35. R. N. Shelton, *J. Less-Comm. Met.*, 94, 69, 1983.

36. B. Chevalier, P. Lejay, B. Lloret, Wang Xian–Zhong, J. Etourneau, and P. Hagenmuller, *Annales de Chemie*, 9, 191, 1984.

37. G. Venturini, M. Meot-Meyer, E. McRae, J. F. Mareche, and B. Rogues, *Mat. Res. Bull.*, 19, 1647, 1984.

38. J. M. Tarascon, F. G. DiSalvo, D. W. Murphy, G. Hull, and J. V. Waszczak, *Phys. Rev.*, 29B, 172, 1984.

39. G. V. Subba and G. Balakrishnan, *Bull. Mat. Sci.*, 6, 283, 1984.

40. B. Batlog, *Physica*, 126B, 275, 1984.

41. M. J. Johnson, Ames Lab (USA) Report IS-T-1140, 1984.

42. I. M. Chapnik, *J. Mat. Sci. Lett.*, 4, 370, 1985.

43. W. Rong-Yao, L. Q-Guang, and Z. Xiao, *Phys. Status Sol.*, 90A, 763, 1985.

44. W. Xian-Zhong, B. Chevalier, J. Etourneau, and P. Hagenmuller, *Mat. Res. Bull.*, 20, 517, 1985.

45. H. R. Ott, F. Hulliger, H. Rudigier, and Z. Fisk, *Phys. Rev.*, 31B, 1329, 1985.

46. P. Villars and L. D. Calver, *Pearson's Handbook of Crystallographic Data for Intermetallic Phases*, Vol. 1–3, ASM, 1985.

47. G. V. Subba Rao, K. Wagner, G. Balakhrishnan, J. Jakani, W. Paulus, and R. Scollhorn, *Bull. Mat. Sci.*, 7, 215, 1985.

48. J. G. Bednorz and K. A. Muller, *Zs. Physik*, B64, 189, 1986.

49. W. Rong-Yao, *Phys. Status Sol.*, 94A, 445, 1986.

50. H. D. Yang, R. N. Shelton, and H. F. Braun, *Phys. Rev.*, 33B, 5062, 1986.

51. G. Venturini, M. Kanta, E. McRae, J. F. Mareche, B. Malaman, and B. Roques, *Mat. Res. Bull.*, 21, 1203, 1986.

52. W. Rong-Yao, *J. Mat. Sci. Lett.*, 5, 87, 1986.

53. M. K. Wu, J. R. Ashburn, C. J. Torng, P. H. Hor, R. L. Meng, L. Gao, Z. J. Huang, Y. Q. Wang, and C. W. Chu, *Phys. Rev. Lett.*, 58, 908, 1987.

54. R. J. Cava, B. B. Van Dover, B. Batlog, and E. A. Rietman, *Phys. Rev. Lett.*, 58, 408, 1987.

55. L. C. Porter, T. J. Thorn, U. Geiser, A. Umezawa, H. H. Wang, W. K. Kwok, H-C. I. Kao, M. R. Monaghan, G. W. Crabtree, K. D. Carlson, and J. M. Williams, *Inorg. Chem.*, 26, 1645, 1987.

56. A. M. Kini, U. Geiser, H-C. I. Kao, K. D. Carlson, H. H. Wang, M. R. Monaghan, and K. M. Williams, *Inorg. Chem.*, 26, 1834, 1987.

57. T. Penney, S. von Molnar, D. Kaiser, F. Holtzberg, and A. W. Kleinsasser, *Phys. Rev.*, B38, 2918, 1988.

58. Y. K. Tao, J. S. Swinnea, A. Manthiram, J. S. Kim, J. B. Goodenoug, and H. Steinfink, *J. Mat. Res.*, 3, 248, 1988.

59. G. G. Peterson, B. R. Weinberger, L. Lynds, and H. A. Krasinski, *J. Mat. Res.*, 3, 605, 1988.

60. J. B. Torrance, Y. Tokura, A. Nazzai, and S. S. P. Parkin, *Phys. Rev. Lett.*, 60, 542, 1988.

61. K. Kourtakis, M. Robbins, P. K. Gallagher, and T. Teifel, *J. Mat. Res.*, 4, 1289, 1989.

62. J. C. Phillips, *Physics of High-T_c Superconductors*, Academic Press, 1989, p. 336.

63. Shui Wai Lin and L. I. Berger, *Rev. Sci. Instrum.*, 60, 507, 1989.

64. M. Tinkham, *Introduction to Superconductivity*, McGraw–Hill, New York, 1975.

65. O. Fischer and M.B. Maple, Eds., *Topics in Current Physics*, Volume 32: Superconductivity in Ternary Compounds I; Volume 34: Superconductivity in Ternary Compounds II, Springer–Verlag, Berlin, 1982.

66. K. J. Dunn and F. P. Bundy, *Phys. Rev.*, B25, 194, 1982.

67. A. Barone and G. Paterno, *Physics and Applications of the Josephson Effect*, Wiley, New York, 1982.

68. D. H. Douglass, Ed., *Superconductivity in d- and f-Band Metals*, Plenum Press, New York, 1976.

69. D. M. Ginsberg, Ed., *Physical Properties of High Temperature Superconductors*, (Volume II, 1990; Volume III, 1992; Volume V, 1996), World Scientific, Singapore.

70. T. Ishiguro and K. Yamji, *Organic Superconductors*, Springer-Verlag, Berlin, 1990.

71. Sh. Okada, K. Shimizu, T. C. Kobayashi, K. Amaya, and Sh. Endo., *J. Phys. Soc. Jpn.*, 65, 1924, 1996.

72. A. Bourdillon and N. X. Tan Bourdillon, *High Temperature Superconductors: Processing and Science*, Academic Press, 1994.

73. J. M. Williams, J. R. Ferraro, R. J. Thorn, K. Carlson, U. Geiser, H. H. Wang, A. M. Kini, and M.-H. Whangbo, *Organic Superconductors (Including Fullerenes): Synthesis, Structure, Properties, and Theory*, Prentice–Hall, 1992.

74. J. Nagamatsu, N. Nakagawa, T. Muranaka, Y. Zenitani, and J. Akimitsu, *Nature (London)*, 410, 63, 2001.

75. Y. Boguslavsky, G. K. Perkins, X. Qi, L. F. Cohen, and A. D. Caplin, *Nature (London)*, 410, 563, 2001.

76. B. Q. Fu, Y. Feng, G. Yan, Y. Zhao, A. K. Pradhan, C. H. Cheng, P. Ji, X. H. Liu, C. F. Liu, L. Zhou, and K. F. Yau, *J. Appl. Phys.*, 92, 7341, 2002.

HIGH-TEMPERATURE SUPERCONDUCTORS

C. N. R. Rao and A. K. Raychaudhuri

The following tables give properties of a number of high-temperature superconductors. Table 1 lists the crystal structure (space group and lattice constants) and the critical transition temperature T_c for the more important high-temperature superconductors studied so far. Table 2 gives the energy gap, critical current density, and penetration depth in the superconducting state. Table 3 gives electrical and thermal properties of some of these materials in the normal state. The tables were prepared in November 1992 and updated in November 1994.

References

1. Ginsburg, D. M., Ed., *Physical Properties of High-Temperature Superconductors*, Vols. I–III, World Scientific, Singapore, 1989–1992.

2. Rao, C. N .R., Ed., *Chemistry of High-Temperature Superconductors*, World Scientific, Singapore, 1991.

3. Shackelford, J. F., *The CRC Materials Science and Engineering Handbook*, CRC Press, Boca Raton, 1992, 98–99 and 122–123.

4. Kaldis, E., Ed., *Materials and Crystallographic Aspects of HT_c-Superconductivity*, Kluwer Academic Publ., Dordrecht, The Netherlands, 1992.

5. Malik, S. K. and Shah, S. S., Ed., *Physical and Material Properties of High Temperature Superconductors*, Nova Science Publ., Commack, N.Y., 1994.

6. Chmaissem, O. et al., *Physica*, C230, 231–238, 1994.

7. Antipov, E. V. et al., *Physica*, C215, 1–10, 1993.

TABLE 1. Structural Parameters and Approximate T_c Values of High-Temperature Superconductors

Material	Structure	T_c/K (maximum value)
$La_2CuO_{4+\delta}$	Bmab; $a = 5.355$, $b = 5.401$, $c = 13.15$ Å	39
$La_{2-x}Sr_x(Ba_x)CuO_4$	I4/mmm; $a = 3.779$, $c = 13.23$ Å	35
$La_2Ca_{1-x}Sr_xCu_2O_6$	I4/mmm; $a = 3.825$, $c = 19.42$ Å	60
$YBa_2Cu_3O_7$	Pmmm; $a = 3.821$, $b = 3.885$, $c = 11.676$ Å	93
$YBa_2Cu_4O_8$	Ammm; $a = 3.84$, $b = 3.87$, $c = 27.24$ Å	80
$Y_2Ba_4Cu_7O_{15}$	Ammm; $a = 3.851$, $b = 3.869$, $c = 50.29$ Å	93
$Bi_2Sr_2CuO_6$	Amaa; $a = 5.362$, $b = 5.374$, $c = 24.622$ Å	10
$Bi_2CaSr_2Cu_2O_8$	A_2aa; $a = 5.409$, $b = 5.420$, $c = 30.93$ Å	92
$Bi_2Ca_2Sr_2Cu_3O_{10}$	A_2aa; $a = 5.39$, $b = 5.40$, $c = 37$ Å	110
$Bi_2Sr_2(Ln_{1-x}Ce_x)_2Cu_2O_{10}$	P4/mmm; $a = 3.888$, $c = 17.28$ Å	25
$Tl_2Ba_2CuO_6$	A_2aa; $a = 5.468$, $b = 5.472$, $c = 23.238$ Å; I4/mmm; $a = 3.866$, $c = 23.239$ Å	92
$Tl_2CaBa_2Cu_2O_8$	I4/mmm; $a = 3.855$, $c = 29.318$ Å	119
$Tl_2Ca_2Ba_2Cu_3O_{10}$	I4/mmm; $a = 3.85$, $c = 35.9$ Å	128
$Tl(BaLa)CuO_5$	P4/mmm; $a = 3.83$, $c = 9.55$ Å	40
$Tl(SrLa)CuO_5$	P4/mmm; $a = 3.7$, $c = 9$ Å	40
$(Tl_{0.5}Pb_{0.5})Sr_2CuO_5$	P4/mmm; $a = 3.738$, $c = 9.01$ Å	40
$TlCaBa_2Cu_2O_7$	P4/mmm; $a = 3.856$, $c = 12.754$ Å	103
$(Tl_{0.5}Pb_{0.5})CaSr_2Cu_2O_7$	P4/mmm; $a = 3.80$, $c = 12.05$ Å	90
$TlSr_2Y_{0.5}Ca_{0.5}Cu_2O_7$	P4/mmm; $a = 3.80$, $c = 12.10$ Å	90
$TlCa_2Ba_2Cu_3O_8$	P4/mmm; $a = 3.853$, $c = 15.913$ Å	110
$(Tl_{0.5}Pb_{0.5})Sr_2Ca_2Cu_3O_9$	P4/mmm; $a = 3.81$, $c = 15.23$ Å	120
$TlBa_2(La_{1-x}Ce_x)_2Cu_2O_9$	I4/mmm; $a = 3.8$, $c = 29.5$ Å	40
$Pb_2Sr_2La_{0.5}Ca_{0.5}Cu_3O_8$	Cmmm; $a = 5.435$, $b = 5.463$, $c = 15.817$ Å	70
$Pb_2(Sr,La)_2Cu_2O_6$	$P2_12_12$; $a = 5.333$, $b = 5.421$, $c = 12.609$ Å	32
$(Pb,Cu)Sr_2(La,Ca)Cu_2O_7$	P4/mmm; $a = 3.820$, $c = 11.826$ Å	50
$(Pb,Cu)(Sr,Eu)(Eu,Ce)Cu_2O_x$	I4/mmm; $a = 3.837$, $c = 29.01$ Å	25
$Nd_{2-x}Ce_xCuO_4$	I4/mmm; $a = 3.95$, $c = 12.07$ Å	30
$Ca_{1-x}Sr_xCuO_2$	P4/mmm; $a = 3.902$, $c = 3.35$ Å	110
$Sr_{1-x}Nd_xCuO_2$	P4/mmm; $a = 3.942$, $c = 3.393$ Å	40
$Ba_{0.6}K_{0.4}BiO_3$	Pm3m; $a = 4.287$ Å	31
Rb_2CsC_{60}	$a = 14.493$ Å	31
$NdBa_2Cu_3O_7$	Pmmm; $a = 3.878$, $b = 3.913$, $c = 11.753$	58
$SmBaSrCu_3O_7$	I4/mmm; $a = 3.854$, $c = 11.62$	84
$EuBaSrCu_3O_7$	I4/mmm; $a = 3.845$, $c = 11.59$	88
$GdBaSrCu_3O_7$	I4/mmm; $a = 3.849$, $c = 11.53$	86
$DyBaSrCu_3O_7$	Pmmm; $a = 3.802$, $b = 3.850$, $c = 11.56$	90
$HoBaSrCu_3O_7$	Pmmm; $a = 3.794$, $b = 3.849$, $c = 11.55$	87
$ErBaSrCu_3O_7$ (multiphase)	Pmmm; $a = 3.787$, $b = 3.846$, $c = 11.54$	82
$TmBaSrCu_3O_7$ (multiphase)	Pmmm; $a = 3.784$, $b = 3.849$, $c = 11.55$	88
$YBaSrCu_3O_7$	Pmmm; $a = 3.803$, $b = 3.842$, $c = 11.54$	84
$HgBa_2CuO_4$	I4/mmm; $a = 3.878$, $c = 9.507$	94
$HgBa_2CaCu_2O_6$ (annealed in O_2)	I4/mmm; $a = 3.862$, $c = 12.705$	127
$HgBa_2Ca_2Cu_3O_8$	Pmmm; $a = 3.85$, $c = 15.85$	133
$HgBa2Ca3Cu4O10$	Pmmm; a = 3.854, c = 19.008	126

TABLE 2. Superconducting Properties

J_c (0): Critical current density extrapolated to 0 K
λ_{ab}: Penetration depth in a-b plane
k_B: Boltzmann constant

Material	Form	Energy gap (Δ)		$10^{-6} \times J_c$ (0)/A cm^{-2}	λ_{ab}/Å
		$2\Delta_{pp}/k_B T_c$*	$2\Delta_{fit}/k_B T_c$†		
Y Ba$_2$Cu$_3$O$_7$	Single Crystal	5–6	4–5	30 (film)	1400
Bi$_2$Sr$_2$CaCu$_2$O$_8$	Single Crystal	8–9	5.5–6.5	2	2700
Tl$_2$Ba$_3$CaCu$_2$O$_8$	Ceramic	6–7	4–6	10 (film, 80 K)	2000
La$_{2-x}$Sr$_x$CuO$_4$, x = 0.15	Ceramic	7–9	4–6		
Nd$_{2-x}$Ce$_x$CuO$_4$	Ceramic	8	4–5	0.2 (film)	

* Obtained from peak to peak value.
† Obtained from fit to BCS-type relation.

TABLE 3. Normal State Properties

ρ_{ab}:	Resistivity in the a-b plane
ρ_c:	Resistivity along the c axis
+ve:	ρ_c has positive temperature coefficient of resistivity
−ve:	ρ_c has negative temperature coefficient of resistivity
n_H:	Hall density
k:	Thermal conductivity
in plane:	Along a-b plane
out of plane:	Perpendicular to a-b plane

Material	Form	ρ_{ab}/μΩ cm		ρ_c/mΩ cm		$10^{-21} \times n_H$/cm^{-3}		k/(mW/cm K) at 300 K	
		300 K	100 K	300 K	$d\rho_c/dT$	300 K	100 K	in plane	out of plane
YBa$_2$Cu$_3$O$_7$	Single								
	crystal	110	35	5	+ve	11–16	4–6	120	3
	Film	200–300	60–100			5–9	2–3		
YBa$_2$Cu$_4$O$_8$	Single								
	crystal	75	20	10	−ve	14			
	Film	100–200	20–50			22	17		
Bi$_2$Sr$_2$CuO$_6$	Single								
	crystal	300	150	5000	−ve	6	5		
Bi$_2$Sr$_2$CaCu$_2$O$_8$	Single								
	crystal	150	50	>1000	−ve	4	3	60	8
Tl$_2$Ba$_2$CuO$_6$	Single								
	crystal	300–400	50–75	200–300	+ve	3.1	2.5		
Tl$_2$Ba$_2$Ca$_2$Cu$_3$O$_{10}$	Ceramic	***	**				≈ 2*		
La$_{2-x}$Sr$_x$CuO$_4$, x = 0.12	Single				+ve for				
	crystal	900	350	200	T >225 K	2.5			
La$_{2-x}$Sr$_x$CuO$_4$, x = 0.20	Single				+ve for				
	crystal	400	200	80	T >150 K	10		50 (for x = 0.04)	20
	Film	400	160			8.4	6.3		
Nd$_{2-x}$Ce$_x$CuO$_4$, x = 0.17	Single								
	crystal	500	275			53	17		
x = 0.15	Film	140–180	35			32	11	250 (for x = 0.15)	

* At 200 K
** ρ ~0.4 mΩ cm at 120 K
*** ρ ~1.5 mΩ cm at 300 K

ORGANIC SUPERCONDUCTORS

H. P. R. Frederikse

Although the vast majority of organic compounds are insulators, a small number of organic solids show considerable electrical conductivity. Some of these materials appear to be superconductors. The superconducting organics fall primarily into two groups: those containing fulvalenes (pentagonal rings containing sulfur or selenium) and those based on fullerenes, involving the nearly spherical cluster C_{60}.

The transition temperatures T_c of the fulvalene derivatives are shown in Table 1. The abbreviations of the various molecular groups are listed in Table 2 and their chemical structures are depicted in Figure 1. Most of the T_c's are between 1 and 12 K. Several of the compounds only show superconductivity under pressure.

The fullerenes are A_3C_{60} compounds, where A represents a single or a combination of alkali atoms. The C_{60} cluster is shown in Figure 2a, while Figure 2b illustrates how the alkali atoms fit into the A_3C_{60} molecule to form the A15 crystallographic structure. Their superconducting transition temperatures range from 8 to 31.3 K (see Table 3).

References

1. Ishigura, T. and Yamaji, K., *Organic Superconductors*, Springer-Verlag, Berlin, 1990.
2. Williams, J. M. et al., *Organic Superconductors (Including Fullerenes)*, Prentice Hall, Englewood Cliffs, N.J., 1992.
3. *The Fullerenes*, Ed.: Krato, H. W., Fisher, J. E., and Cox, D. E., Pergammon Press, Oxford, 1993.
4. Schluter, M. et al., in *The Fullerenes* (Ref. 3), p. 303.

TABLE 1. Critical Pressure and Maximum Critical Temperature of Organic Superconductors

Material	P_c/kbar	T_c/K	Material	P_c/kbar	T_c/K
(TMTSF)$_2$PF$_6$	6.5	1.2	β-(ET)$_2$IBr$_2$	0	2.8
(TMTSF)$_2$AsF$_6$	9	1.3	β-(ET)$_2$AuI$_2$	0	4.8
(TMTSF)$_2$SbF$_6$	11	0.4	(ET)$_4$Hg$_{2.89}$Cl$_8$	0	4.2
(TMTSF)$_2$TaF$_6$	12	1.4	(ET)$_4$Hg$_{2.89}$Br$_8$	12	1.8
(TMTSF)$_2$ClO$_4$	0	1.4	(ET)$_3$Cl$_2$(H$_2$O)$_2$	16	2
(TMTSF)$_2$ReO$_4$	9.5	1.3	κ-(ET)$_2$Cu(NCS)$_2$	0	10.4
(TMTSF)$_2$FSO$_3$	5	3	κ-(d-ET)$_2$Cu(NCS)$_2$	0	11.4
(ET)$_4$(ReO$_4$)$_2$	4.5	2	(DMET)$_2$Au(CN)$_2$	1.5	0.9
β$_L$-(ET)$_2$I$_3$	0	1.4	(DMET)$_2$AuI$_2$	5	0.6
β$_H$-(ET)$_2$I$_3$	0	8.1	(DMET)$_2$AuBr$_2$	0	1.9
γ-(ET)$_3$I$_{2.5}$	0	2.5	(DMET)$_2$AuCl$_2$	0	0.9
ε-(ET)$_2$I$_3$(I$_8$)$_{0.5}$	0	2.5	(DMET)$_2$I$_3$	0	0.6
α-(ET)$_2$I$_3$I$_2$-doped	0	3.3	(DMET)$_2$IBr$_2$	0	0.7
α$_t$-(ET)$_2$I$_3$	0	8	(MDT-TTF)$_2$AuI$_2$	0	3.5
ε→β-(ET)$_2$I$_3$[a]	0	6	TTF[Ni(dmit)$_2$]$_2$	2	1.6[b]
θ-(ET)$_2$I$_3$	0	3.6	TTF[Pd(dmit)$_2$]$_2$	20	6.5
κ-(ET)$_2$I$_3$	0	3.6	(CH$_3$)$_4$N[Ni(dmit)$_2$]$_2$	7	5

[a] Converted form ε-type to β-type by thermal treatment.
[b] For 7 kbar.

From Ishigura, T. and Yamaji, K., *Organic Superconductors*, Springer-Verlag, Berlin, 1990. With permission.

TABLE 2. List of Symbols and Abbreviations

TTF	tetrathiafulvalene
TMTSF	tetramethyltetraselenafulvalene
BEDT-TTF or "ET"	bis(ethylenedithio)tetrathiafulvalene
MDT-TTF	methylenedithiotetrathiafulvalene
DMET	[dimethyl(ethylenedithio)diselenadithiafulvalene]
dmit	4,5-dimercapto-1,3-dithiole-2-thione
T_c	transition temperature to superconducting state
P_c	minimum pressure required for superconducting transition

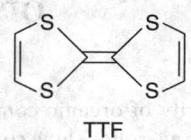

TMTSF

Tetramethyltetraselenafulvalene

TTF

Tetrathiafulvalene

BEDT – TTF or ET

Bis(ethylenedithio)tetrathiafulvalene

DMET

Dimethyl(ethylenedithio)diselenadithiafulvalene

MDT – TTF

Methylenedithiotetrathiafulvalene

M=Ni, Pd, Pt
M(dmit)$_2^{2-}$

Ligand is 4,5-dimercapto-1.3-dithiole-2-thione

FIGURE 1. Structures of various donor molecules and acceptor species.

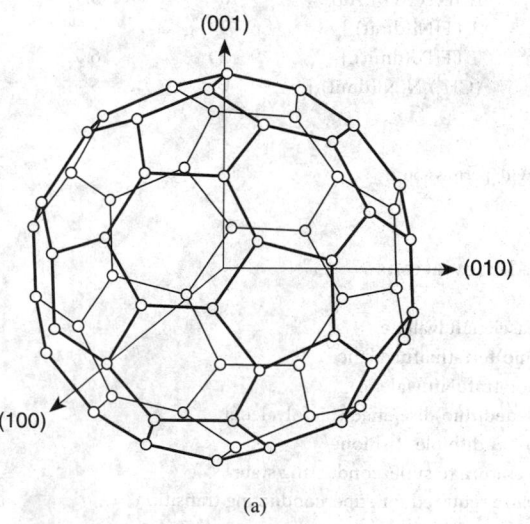

(a)

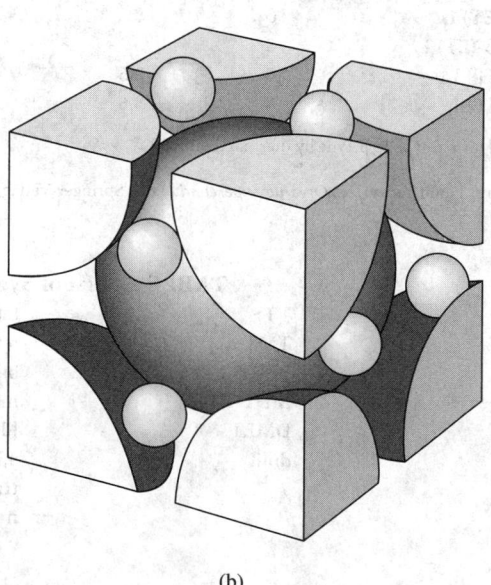

(b)

FIGURE 2. (a) C$_{60}$ cluster placed in a fcc lattice. Each crystal axis crosses a double bond shared by two hexagons. (b) A hypothetical A$_3$C$_{60}$ with the A15 structure. The structure can be seen to be an ordered defect structure of A$_6$C$_{60}$.

TABLE 3. Unit Cell and T_c for FCC-A$_3$C$_{60}$

	Lattice parameter(s) (Å)	T_c/K
Na$_2$Rb$_{0.5}$Cs$_{0.5}$C$_{60}$	14.148(3)	8.0
Na$_2$CsC$_{60}$ No. 1[a]	14.132(2)	10.5
Na$_2$CsC$_{60}$ No. 2[a]	14.176(9)	14.0
K$_3$C$_{60}$	14.253(3)	19.3
K$_2$RbC$_{60}$	14.299(2)	21.8
Rb$_2$KC$_{60}$ No. 1[a]	14.336(1)	24.4
Rb$_2$KC$_{60}$ No. 2[a]	14.364(5)	26.4
Rb$_3$C$_{60}$	14.436(2)	29.4
Rb$_2$CsC$_{60}$	14.493(2)	31.3

[a] Samples labeled No. 1 and No. 2 have the same nominal composition.

From Schluter, M et al., *The Fullerenes*, Ed.: Krato, H.W., Fisher, J.E., and Cox, D.E., Pergamon Press, Oxford, 1993. With permission.

PROPERTIES OF SEMICONDUCTORS

L. I. Berger

The term *semiconductor* is applied to a material in which electric current is carried by electrons or holes and whose electrical conductivity, when extremely pure, rises exponentially with temperature and may be increased from its low "intrinsic" value by many orders of magnitude by "doping" with electrically active impurities.

Semiconductors are characterized by an energy gap in the allowed energies of electrons in the material that separates the normally filled energy levels of the *valence band* (where "missing" electrons behave like positively charged current carriers "holes") and the *conduction band* (where electrons behave rather like a gas of free negatively charged carriers with an effective mass dependent on the material and the direction of the electrons' motion). This energy gap depends on the nature of the material and varies with direction in anisotropic crystals. It is slightly dependent on temperature and pressure, and this dependence is usually almost linear at normal temperatures and pressures.

Data are presented in five tables. Table 1 lists the main crystallographic and semiconducting properties of a large number of semiconducting materials in three main categories: "Tetrahedral Semiconductors" in which every atom is tetrahedrally coordinated to four nearest neighbor atoms (or atomic sites) as for example in the diamond structure; "Octahedral Semiconductors" in which every atom is octahedrally coordinated to six nearest neighbor atoms—as for example the halite structure; and "Other Semiconductors."

Table 2 gives electrical, magnetic, and optical properties, while Tables 3 and 4 give more details on the semiconducting properties and band structures of the most common semiconductors. Table 5 lists semiconducting minerals with typical resistivity ranges.

TABLE 1. Physico-Chemical Properties of Semiconductors (Listed by Crystal Structure)

Substance	Molecular weight	Average atomic weight	Lattice parameters (Å, room temp.)	Density (g/cm³)	Melting point (K)	Microhardness, N/mm² (M-Mohs Scale)	Specific heat, J/kg·K (300 K)	Debye temp. (K)	Coefficient of thermal linear expansion [10^{-6} K^{-1} (300K)]	Thermal conductivity [mW/cm·K (300K)]
1.1. Tetrahedral (Adamantine) Semiconductors										
1.1.1. Diamond Structure Elements (Strukturbericht symbol A4, Space Group Fd3m-O$_h^7$)										
C (Diamond)	12.01	3.56683		3.513	≈4713 (12.4 GPa) Transition to graphite > 980	10 (M)	471.5	2340	1.18	9900(I) 23200(IIA) 13600(IIB)
Si	28.09	5.43072		2.329	1687	11270	702	645	2.6	1240
Ge	72.64	5.65754		5.323	1211.35	7644	321.9	374	5.8	640
α-Sn	118.71	6.4912		5.769	505.1 (Tr. 286.4)		213	230	5.4 (220 K)	

1.1.2. Sphalerite (Zinc Blende) Structure Compounds (Strukturbericht symbol B3 Space Group F$\bar{4}$3m-T$_d^2$)

I-VII Compounds

Substance	Molecular weight	Average atomic weight	Lattice parameters (Å, room temp.)	Density (g/cm³)	Melting point (K)	Microhardness, N/mm² (M-Mohs Scale)	Specific heat, J/kg·K (300 K)	Debye temp. (K)	Coefficient of thermal linear expansion [10^{-6} K^{-1} (300K)]	Thermal conductivity [mW/cm·K (300K)]
CuF	82.54	41.27	4.255		1181					
CuCl	98.99	49.49	5.4057	3.53	695	2.3 (M)	490	240	12.1	8.4
CuBr	143.45	71.73	5.6905	4.98	770	2.5 (M)	381	207	15.4	12.5
CuI	190.45	95.23	6.60427	5.63	878	192	276	181	19.2	16.8
AgBr	187.77	93.89		6.473	>1570 (Tr. 410)	2.5 (M)	270			
AgI	234.77	117.39	6.502	5.67	831	2.5 (M)	232	134	−2.5	4.2

II-VI Compounds

Substance	Molecular weight	Average atomic weight	Lattice parameters (Å, room temp.)	Density (g/cm³)	Melting point (K)	Microhardness, N/mm² (M-Mohs Scale)	Specific heat, J/kg·K (300 K)	Debye temp. (K)	Coefficient of thermal linear expansion [10^{-6} K^{-1} (300K)]	Thermal conductivity [mW/cm·K (300K)]
BeS	41.08	20.54	4.865	2.36	dec.					
BeSe	87.97	43.99	5.139	4.315						
BeTe	136.61	68.31	5.626	5.090						
BePo	(2318)	(109)	5.838	7.3						
ZnO	81.39	40.69	4.63	5.675	2248	5.0 (M)	494	416	2.9	234
ZnS	97.46	48.72	5.4093	4.079	2100 (Tr. 1295)	1780	472	530	6.36	251
ZnSe	144.34	72.17	5.6676	5.42	1790	1350	339	400	7.2	140
ZnTe	192.99	96.5	6.101	6.34	1568	900	264	223	8.19	108
ZnPo	(274)	(137)	6.309							

Substance	Molecular weight	Average atomic weight	Lattice parameters (Å, room temp.)	Density (g/cm³)	Melting point (K)	Microhardness, N/mm² (M-Mohs Scale)	Specific heat, J/kg·K (300 K)	Debye temp. (K)	Coefficient of thermal linear expansion [10⁻⁶ K⁻¹ (300K)]	Thermal conductivity [mW/cm·K (300K)]
CdS	144.48	72.24	5.832		4.826	1750	1250	330	219	4.7	200
CdSe	191.37	95.68	6.05		5.674	1512	1300	255	181	3.8	90
CdTe	240.01	120.00	6.477		5.86	1365	600	205	200	4.9	58.5
CdPo	(321)	(161)	6.665								
HgS	232.66	116.33	5.8517		7.73	1820	3 (M)	210			
HgSe	279.55	139.78	6.084		8.25	1070	2.5 (M)	178	151	5.46	10
HgTe	328.19	164.10	6.4623		8.17	943	300	164	242	4.6	20

III-V Compounds

Substance	Molecular weight	Average atomic weight	Lattice parameters	Density	Melting point	Microhardness	Specific heat	Debye temp.	Coefficient	Thermal conductivity
BN	24.82	12.41	3.615	3.49	3239	10 (M)	793	≈1900		200
BP(L.T.)	41.78	20.87	4.538	2.9	1398 (dec)	37000		≈980		
BAs	85.73	42.87	4.777		≈2300	19000		≈625		
AlP	57.95	28.98	5.451	2.42	≈2100	5.5 (M)		588		920
AlAs	101.90	50.95	5.6622	3.81	2013	5000		417	3.5	840
AlSb	148.74	74.37	6.1355	4.218	1330	4000		292	4.2	600
GaP	100.70	50.35	5.4905	4.13	1750	9450		446	5.3	752
GaAs	144.64	72.32	5.65315	5.316	1510	7500		344	5.4	560
GaSb	191.48	95.74	6.0954	5.619	980	4480	320	265	6.1	270
InP	145.79	72.90	5.86875	4.787	1330	4100		321	4.6	800
InAs	189.74	94.87	6.05838	5.66	1215	3300	268	249	4.7	290
InSb	236.58	118.29	6.47877	5.775	798	2200	144	202	4.7	160

Other Sphalerite Structure Compounds

Substance	Molecular weight	Average atomic weight	Lattice parameters	Density	Melting point	Microhardness	Specific heat	Debye temp.	Coefficient	Thermal conductivity
MnS	87.00	43.5	5.011							
MnSe	133.90	66.95	5.82							
β-SiC (3-C SiC)	40.10	20.1	4.348	3.21	3070				2.9	4.9
Ga₂Se₃	376.32	75.26	5.429	4.92	1020	3160			8.9	50
Ga₂Te₃	522.24	104.45	5.899	5.75	1063	2370				47
In₂Te₃(H.T.)	608.44	121.7	6.173	5.8	940	1660				69
MgGeP₂	158.84	39.71	5.652							
ZnSnP₂	246.00	61.5	5.65		1200					
ZnSnAs₂(H.T.)	333.90	82.38	5.851	5.53	1050					76
ZnSnSb₂	427.56	106.89	6.281	5.67	870	2500				76

1.1.3. Wurtzite (Zincite) Structure Compounds (Strukturbericht symbol B4, Space Group P 6₃mc-C₆ᵥ⁴)

I-VII Compounds

Substance	Molecular weight	Average atomic weight	Lattice parameters		Density	Melting point
CuCl	99.0	49.5	3.91	6.42		703
CuBr	143.45	71.73	4.06	6.66		770
CuI	190.45	95.23	4.31	7.09		
AgI	234.77	117.40	4.580	7.494		

II-VI Compounds

Substance	Molecular weight	Average atomic weight	Lattice parameters		Density	Melting point					Thermal conductivity
BeO	25.01	12.51	2.698	4.380		2800					
MgTe	151.9	76.0	4.54	7.39	3.85	≈2800					
ZnO	81.37	40.69	3.24950	5.2069	5.66	2250					600
ZnS	97.43	48.72	3.8140	6.2576	4.1	2100					460
ZnTe	192.99	46.50	4.27	6.99		1568					
CdS	144.48	72.23	4.1348	6.7490	4.82	1748					401
CdSe	191.37	95.68	4.299	7.010	5.66	1512					316
CdTe	240.01	120.00	4.57	7.47							

III-V Compounds

Substance	Molecular weight	Average atomic weight	Lattice parameters		Density	Melting point					Thermal conductivity
BP(H.T.)	41.79	20.90	3.562	5.900							
AlN	40.99	20.50	3.111	4.978	3.26	≈2500					823
GaN	83.73	41.87	3.190	5.189	6.10	1500					656

Substance	Molecular weight	Average atomic weight	Lattice parameters (Å, room temp.)		Density (g/cm³)	Melting point (K)	Microhard-ness, N/mm² (M-Mohs Scale)	Specific heat, J/kg·K (300 K)	Debye temp. (K)	Coefficient of thermal linear expansion [10⁻⁶ K⁻¹ (300K)]	Thermal conductivity [mW/cm·K (300K)]
InN	128.83	64.42	3.533	5.693	6.88	1200					556

Other Wurtzite Structure Compounds

Substance	Molecular weight	Average atomic weight	Lattice parameters		Density	Melting point					
MnS	87.00	43.5	3.985	6.45	3.248						
MnSe	133.90	66.95	4.12	6.72							
SiC	40.10	20.1	3.076	5.048							
MnTe	182.54	91.27	4.078	6.701							
Al_2S_3	150.14	30.03	3.579	5.829	2.55	1400					
Al_2Se_3	290.84	58.17	3.890	6.30	3.91	1250					

1.1.4. Chalcopyrite Structure Compounds (Strukturbericht symbol E1₁, Space Group I $\overline{4}$ 2d-D_{24}^{12})

I-III-VI₂ Compounds

Substance	Molecular weight	Average atomic weight	Lattice parameters		Density	Melting point	Microhardness	Specific heat	Debye temp.	Coeff. exp.	Thermal cond.
$CuAlS_2$	154.65	38.66	5.323	10.44	3.47	2500					
$CuAlSe_2$	248.45	62.11	5.617	10.92	4.70	2260					
$CuAlTe_2$	345.73	86.43	5.976	11.80	5.50	2550					
$CuGaS_2$	197.39	49.53	5.360	10.49	4.35	2300					
$CuGaSe_2$	291.19	72.80	5.618	11.01	5.56	1970	4200		275	5.4	42
$CuGaTe_2$	388.47	97.12	6.013	11.93	5.99	2400	3500			6.9	27
$CuInS_2$	242.49	60.62	5.528	11.08	4.75	1400	2550				
$CuInSe_2$	336.29	84.07	5.785	11.56	5.77	1600	2050			6.6	37
$CuInTe_2$	433.57	108.39	6.179	12.365	6.10	1660	400		195	7.1	49
$CuTlS_2$	322.05	83.01	5.580	11.17	6.32						
$CuTlSe_2$(L.T.)	425.85	106.46	5.844	11.65	7.11	900					
$CuFeS_2$	183.51	45.88	5.29	10.32	4.088	1135					
$CuFeSe_2$	277.31	69.33				850					
$CuLaS_2$	266.58	66.65	5.65	10.86							
$AgAlS_2$	198.97	49.74	5.707	10.28	3.94						
$AgAlSe_2$	292.77	73.19	5.968	10.77	5.07	1220					
$AgAlTe_2$	390.05	97.51	6.309	11.85	6.18	1000					
$AgGaS_2$	241.71	60.43	5.755	10.28	4.72						
$AgGaSe_2$	335.51	83.88	5.985	10.90	5.84	1120	4400				
$AgGaTe_2$	432.79	108.2	6.301	11.96	6.05	990	1800		212		10
$AgInS_2$(L.T.)	286.87	71.70	5.828	11.19	5.00		2250				
$AgInSe_2$	380.61	95.15	6.102	11.69	5.81	1053	1850				30
$AgInTe_2$	477.89	119.47	6.42	12.59	6.12	965				9.49, 0.69	
$AgFeS_2$	227.83	56.96	5.66	10.30	4.53						

II-IV-V₂ Compounds

Substance	Molecular weight	Average atomic weight	Lattice parameters		Density	Melting point	Microhardness	Specific heat	Debye temp.	Coeff. exp.	Thermal cond.
$ZnSiP_2$	155.40	38.85	5.400	10.441	3.39	1640	1100				
$ZnGeP_2$	199.90	49.98	5.465	10.771	4.17	1295	8100				180
$ZnSnP_2$	246.00	61.5					6500				
$CdSiP_2$	202.43	50.61	5.678	10.431	4.00	≈1470	10500		282		
$CdGeP_2$	246.94	61.74	5.741	10.775	4.48	1049	5650				110
$CdSnP_2$	243.03	73.26	5.900	11.518		840	5000		195		140
$ZnSiAs_2$	242.20	60.55	5.61	10.88	4.70	1311	9200				
$ZnGeAs_2$	287.80	71.95	5.672	11.153	5.32	1150	6800		263		110
$ZnSnAs_2$	333.90	83.48	5.8515	11.704	5.53	1048	4550		271		150
$CdSiAs_2$	290.34	72.58	5.884	10.882		>1120	6850				
$CdGeAs_2$	334.83	83.71	5.9427	11.2172	5.60	938	4700				48
$CdSnAs_2$	380.93	95.23	6.0944	11.9182	5.72	880	3450				40

1.1.5. Other Ternary Semiconductors with Tetrahedral Coordination

I₂-IV-VI₃ Compounds

Substance	Molecular weight	Average atomic weight	Lattice parameters		Density	Melting point					Thermal cond.
Cu_2SiS_3(H.T.)	251.36	41.89	3.684	6.004	3.81	1200					23

Substance	Molecular weight	Average atomic weight	Lattice parameters (Å, room temp.)		Density (g/cm³)	Melting point (K)	Microhardness, N/mm² (M-Mohs Scale)	Specific heat, J/kg·K (300 K)	Debye temp. (K)	Coefficient of thermal linear expansion [10⁻⁶ K⁻¹ (300K)]	Thermal conductivity [mW/cm·K (300K)]
Cu_2SiS_3(L.T.)			5.290	10.156	3.63						
Cu_2SiTe_3	537.98	89.66	5.93		5.47						
Cu_2GeS_3(H.T.)	295.88	49.31	5.317		4.45	1210	4550	510	254	7.2	12
Cu_2GeS_3(L.T.)			5.327	5.215	4.46						
Cu_2GeSe_3	436.56	72.76	5.589	5.485	5.57	1030	3840	340	168	8.4	24
Cu_2GeTe_3	582.51	97.09	5.958	5.935	5.92		2890				130
Cu_2SnS_3	341.98	57.00	5.436		5.02	1110	2770	440	214	7.8	28
$CuSnSe_3$	482.66	80.44	5.687		5.94	960	2510	310	148	8.9	35
Cu_2SnTe_3	628.61	104.77	6.048		6.51	680	1970				144
Ag_2GeSe_3	525.21	87.54				810					
Ag_2SnSe_3	571.31	95.22									
Ag_2GeTe_3	671.13	111.86				600					
Ag_2SnTe_3	717.23	119.54									

I_3-V-VI_4-Compounds

Substance	Molecular weight	Average atomic weight	Lattice parameters (Å, room temp.)		Density (g/cm³)	Melting point (K)	Microhardness, N/mm² (M-Mohs Scale)	Specific heat, J/kg·K (300 K)	Debye temp. (K)	Coefficient of thermal linear expansion [10⁻⁶ K⁻¹ (300K)]	Thermal conductivity [mW/cm·K (300K)]
Cu_3PS_4	349.85	40.73	7.44	6.19							
Cu_3AsS_4	393.79	49.22	6.43	6.14	4.37	931				3.2	30.2
Cu_3AsSe_4	581.37	72.67	5.570	10.957	5.61	733			169	9.5	19
Cu_3SbS_4	440.64	55.08	5.38	16.76	4.90	830					
Cu_3SbSe_4	628.22	78.53	5.654	11.256	6.0	700			131	12.4	14.6

I-IV_2-V_3 Compounds

Substance	Molecular weight	Average atomic weight	Lattice parameters (Å, room temp.)		Density (g/cm³)	Melting point (K)	Microhardness, N/mm² (M-Mohs Scale)	Specific heat, J/kg·K (300 K)	Debye temp. (K)	Coefficient of thermal linear expansion [10⁻⁶ K⁻¹ (300K)]	Thermal conductivity [mW/cm·K (300K)]
$CuSi_2P_3$	212.64	35.44	5.25								
$CuGe_2P_3$	301.65	50.28	5.375		4.318	1113	8500	429		8.21	37.6
$AgGe_2P_3$	345.97	57.66				1015	6150				

1.1.6. "Defect Chalcopyrite" Structure Compounds (Strukturbericht Symbol E3, Space Group I $\overline{4}$-S_4^2)

Substance	Molecular weight	Average atomic weight	Lattice parameters (Å, room temp.)		Density (g/cm³)	Melting point (K)	Microhardness, N/mm² (M-Mohs Scale)	Specific heat, J/kg·K (300 K)	Debye temp. (K)	Coefficient of thermal linear expansion [10⁻⁶ K⁻¹ (300K)]	Thermal conductivity [mW/cm·K (300K)]
$ZnAl_2Se_4$	435.18	62.17	5.503	10.90	4.37						
$ZnAl_2Te_4$(?)	629.74	84.96	5.904	12.05	4.95						
$ZnGa_2S_4$(?)	333.06	47.58	5.274	10.44	3.80						
$ZnGa_2Se_4$(?)	520.66	74.38	5.496	10.99	5.21						
$ZnGa_2Te_4$(?)	715.22	102.17	5.937	11.87	5.67						
$ZnIn_2Se_4$	610.86	87.27	5.711	11.42	5.44	1250					
$ZnIn_2Te_4$	805.42	115.06	6.122	12.24	5.83	1075					
$CdAl_2S_4$	294.61	42.09	5.564	10.32	3.06						
$CdAl_2Se_4$	482.21	68.89	5.747	10.68	4.54						
$CdAl_2Te_4$(?)	676.77	97.68	6.011	12.21	5.10						
$CdGa_2S_4$	380.09	54.30	5.577	10.08	4.03						
$CdGa_2Se_4$	567.69	81.10	5.743	10.73	5.32						
$CdGa_2Te_4$	762.25	108.89	6.093	11.81	5.77						
$CdIn_2Te_4$	852.45	121.78	6.205	12.41	5.9	1060					
$HgAl_2S_4$	382.79	54.68	5.488	10.26	4.11						
$HgAl_2Se_4$	570.39	82.48	5.708	10.74	5.05						
$HgAl_2Te_4$(?)	764.48	109.28	6.004	12.11	5.81						
$HgGa_2S_4$	468.27	66.90	5.507	10.23	5.00						
$HgGa_2Se_4$	655.87	93.70	5.715	10.78	6.18						
$HgIn_2Se_4$	746.07	106.58	5.764	11.80	6.3	1100					
$HgIn_2Te_4$(?)	940.63	134.38	6.186	12.37	6.3	980					

1.1.7. Other Adamantine Compounds

Substance	Molecular weight	Average atomic weight	Lattice parameters (Å, room temp.)		Density (g/cm³)	Melting point (K)	Microhardness, N/mm² (M-Mohs Scale)	Specific heat, J/kg·K (300 K)	Debye temp. (K)	Coefficient of thermal linear expansion [10⁻⁶ K⁻¹ (300K)]	Thermal conductivity [mW/cm·K (300K)]
α–SiC	40.10	20.10	3.0817	15.12	3.21	3070					
$Hg_5Ga_2Te_8$	2163.19	144.21	6.235								
$Hg_5In_2Te_8$	2253.39	150.23	6.328								
$CdIn_2Se_4$	657.89	93.98	a = c = 5.823								

Substance	Molecular weight	Average atomic weight	Lattice parameters (Å, room temp.)	Density (g/cm³)	Melting point (K)	Microhardness, N/mm² (M-Mohs Scale)	Specific heat, J/kg·K (300 K)	Debye temp. (K)	Coefficient of thermal linear expansion [10⁻⁶ K⁻¹ (300K)]	Thermal conductivity [mW/cm·K (300K)]
1.2. Octahedral Semiconductors										
1.2.1. Halite Structure Semiconductors (Strukturbericht Symbol B1, Space Group $Fm3m\text{-}O_h^5$)										
GeTe	200.21	100.10	5.98	6.14						
SnSe	197.67	98.83	6.020		1133					
SnTe	246.31	123.15	6.313	6.45	1080 (max)					91
PbS	239.3	119.63	5.9362	7.61	1390					23
PbSe	286.2	143.08	6.1243	8.15	1340					17
PbTe	334.8	167.4	6.454	8.16	1180					23
1.2.2. Selected Other Binary Halites										
BiSe	287.94	143.97	5.99	7.98	880					
BiTe	336.58	168.29	6.47							
EuSe	230.92	115.46	6.191		2300					2.4
GdSe	236.21	118.11	5.771		2400					
NiO	74.69	37.35	4.1684	6.6	2260					
CdO	128.41	64.21	4.6953		1700					7
SrS	119.69	59.84	6.0199	3.643	3000					
1.3. Other Semiconductors										
1.3.1. Antifluorite Structure Compounds ($Fm3m\text{-}O_h^5$)										
Mg_2Si	76.70	25.57	6.338	1.88	1375				11.5	
Mg_2Ge	121.22	40.4	6.380	3.08	1388				15.0	
Mg_2Sn	167.32	55.77	6.765	3.53	1051				9.9	92
Mg_2Pb	225.81	85.27	6.836	5.1	823				10.0	
1.3.2. Tetradymite Structure Compounds ($\bar{R}3m\text{-}D_{3d}^5$)										
Sb_2Te_3	626.3	125.26	4.25 30.3	6.44	895					
Bi_2Se_3	654.84	130.97	4.14 28.7	7.51	979	167				24
Bi_2Te_3	800.76	160.15	4.38 30.45	7.73	858	155	16			30
1.3.3. Skutterudite Structure Compounds ($Im3\text{-}T_h^5$)										
CoP_3	151.85	37.96	7.7073		>1270					
$CoAs_3$	286.70	71.65	8.2060	6.73	1230					
$CoSb_3$	424.18	106.05	9.0385		1123			307		50
$NiAs_3$	283.45	70.86	8.330	6.43						
RhP_3	195.83	48.96	7.9951		>1470					
$RhAs_3$	327.67	81.92	8.4427		>1270					100
$RhSb_3$	468.16	117.04	9.2322		1170					
IrP_3	285.14	71.29	8.0151	7.36	>1470					
$IrAs_3$	416.98	104.25	8.4673	9.12	>1470					90
$IrSb_3$	557.47	139.37	9.2533	9.35	1170			303		
1.3.4. Selected Multinary Compounds										
$AgSbSe_2$	387.54	96.88	5.786	6.60	910					10.5
$AgSbTe_2$ (or $Ag_{19}Sb_{29}Te_{52}$)	484.82	121.2	6.078	7.12	830					86
$AgBiS_2$(H.T.)	380.97	95.24	5.648							
$AgBiSe_2$(H.T.)	474.77	118.69	5.82							
$AgBiTe_2$(H.T.)	572.05	143.01	6.155							
Cu_2CdSnS_4	486.43	60.80	5.586	10.83						
1.3.5. Some Elemental Semiconductors										
B		10.81	4.91 12.6	2.34	2348	9.5 (M)	1277	1370	8.3	600

Substance	Molecular weight	Average atomic weight	Lattice parameters (Å, room temp.)		Density (g/cm³)	Melting point (K)	Microhard-ness, N/mm² (M-Mohs Scale)	Specific heat, J/kg·K (300 K)	Debye temp. (K)	Coefficient of thermal linear expansion [10⁻⁶ K⁻¹ (300K)]	Thermal conductivity [mW/cm·K (300K)]
Se(gray)		78.96	4.36	4.95	4.81	493	350	292.6		(‖C) 17.89	(‖C) 45.2
										(⊥C) 74.09	(⊥C) 13.1
Te		127.60	4.45	5.91	6.23	723		196.5		16.8	(‖C) 33.8
											(⊥C) 19.7

TABLE 2. Basic Thermodynamic, Electrical, and Magnetic Properties of Semiconductors (Listed by Crystal Structure)

Substance	Heat of formation [kJ/mol] (300K)]	Volume compressibility (10⁻¹⁰m²/N)	Static dielectric constant	Atomic magnetic susceptibility (10⁻⁶ cgs)	Index of refraction	Minimum room temperature energy gap (eV)	Mobility (room temp.) (cm²/V·s)		Optical transition	Breakdown voltage kV/mm	Remarks
							Electrons	Holes			

2.1. Adamantine Semiconductors

2.1.1. Diamond Structure Elements (Strukturbericht symbol A4, Space Group Fd 3m-O$_h^7$)

C	714.4	18	5.7	−5.88	2.419 (589 nm)	5.4	1800	1400	i*	500	
Si	324	0.306	11.9	−3.9	3.49 (589 nm)	1.12	1900	500	i	30	
Ge	291	0.768	16	−0.12	3.99 (589 nm)	0.67	3800	1820	i		
α-Sn	267.5		24		2.75 (589 nm)	0.0; 0.8	2500	2400			

2.1.2. Sphalerite (Zinc Blende) Structure Compounds (Strukturbericht symbol B3 Space Group F $\overline{4}$3m-T$_d^2$)

I-VII Compounds

CuF											
CuCl	481	0.26	7.9		1.93	3.17			d		Nantokite
CuBr	481	0.26	7.9		2.12	2.91			d		
CuI	439	0.27	6.5		2.346	2.95			d		Marshite
AgBr	486		12.4		2.253	2.50	4000		i		Bromirite
AgI	389	0.41	10		2.22	2.22	30		d		Miersite

II-VI Compounds

BeS					4.17				i		
BeSe					3.61				i		
BeTe					1.45		20		d		
BePo											
ZnO											See 2.1.3.
ZnS	477		8.9	−9.9	2.356	3.54	180	5(400°C)	d		See also 2.1.3.
ZnSe	422		9.2		2.89	2.58	540	28	d		
ZnTe	376		10.4		3.56	2.26	340	100	d		
ZnP											
CdS											See 2.1.3.
CdSe											See 2.1.3.
CdTe	339		7.2		2.50	1.44	1200	50	d		
CdPo											
HgS					2.85		250		d		Metacinna-barite
HgSe	247				2.10 (α)		20000	≈1.5	s		Tiemannite
HgTe	242					−0.06	25000	350	s		Coloradoite

III-V Compounds

BN	815					4.6					Borazone

Substance	Heat of formation [kJ/mol (300K)]	Volume compressibility $(10^{-10}m^2/N)$	Static dielectric constant	Atomic magnetic susceptibility $(10^{-6}$ cgs$)$	Index of refraction	Minimum room temperature energy gap (eV)	Mobility (room temp.) $(cm^2/V{\cdot}s)$ Electrons	Holes	Optical transition	Breakdown voltage kV/mm	Remarks
BP(L.T.)						≈2.1	500	70			Ignites 470K
BAs						≈1.5					
AlP						2.45	80		i		
AlAs	627		10.9			2.16	1200	420	i		
AlSb	585	0.571	11		3.2	1.60	200–400	550	i		
GaP	635	0.110	11.1	−13.8	3.2	2.24	300	150	i		
GaAs	535	0.771	13.2	−16.2	3.30	1.35	8800	400	d		
GaSb	493	0.457	15.7	−14.2	3.8	0.67	4000	1400	d		
InP	560	0.735	12.4	−22.8	3.1	1.27	4600	150	d		
InAs	477	0.549	14.6	−27.7	3.5	0.36	33000	460	d		
InSb	447	0.442	17.7	−32.9	3.96	0.163	78000	750	d		

* i = indirect, d = direct, s = semimetal.

Other Sphalerite Structure Compounds

Substance	Heat of formation [kJ/mol (300K)]	Volume compressibility $(10^{-10}m^2/N)$	Static dielectric constant	Atomic magnetic susceptibility $(10^{-6}$ cgs$)$	Index of refraction	Minimum room temperature energy gap (eV)	Mobility Electrons	Holes	Optical transition	Breakdown voltage kV/mm	Remarks
MnS											See also 2.1.3.
MnSe											See also 2.1.3.
β-SiC					2.697	2.3	4000				
Ga_2Te_3	271			−13.5		1.35	50				
In_2Te_3 (H.T.)	198			−13.6		1.04	50				
$MgGeP_2$											El–T^{d12}
$ZnSnP_2$						2.1					Same
$ZnSnAs_2$ (H.T.)						≈0.7					Same
$ZnSnSb_2$						0.4					Same

2.1.3. Wurtzite (Zincite) Structure Compounds (Strukturbericht symbol B4, Space Group $P6_3mc\text{-}C_{6v}^4$)

I-VII Compounds

Substance	Heat of formation [kJ/mol (300K)]	Volume compressibility	Static dielectric constant	Atomic magnetic susceptibility	Index of refraction	Minimum room temperature energy gap (eV)	Mobility Electrons	Holes	Optical transition	Breakdown voltage	Remarks
CuCl											
CuBr											
CuI											
AgI						2.63					Iodargirite

II-VI Compounds

Substance	Heat of formation		Static dielectric constant		Index of refraction	Minimum room temperature energy gap (eV)	Mobility Electrons	Holes	Optical transition		Remarks
BeO											
MgTe											
ZnO	−350					3.2	180				
ZnS	−206					3.67					
ZnTe	−163										
CdS			8.45; 9.12		2.32	2.42	350	40	d		Greenockide
CdSe						1.74	900	50	d		Cadmoselite
CdTe						1.50	650				

III-V Compounds

Substance						Minimum room temperature energy gap (eV)					Remarks
BP(H.T.)											
AlN						6.02					
GaN						3.34					
InN						2.0					

Other Wurtzite Structure Compounds

Substance					Index of refraction						
MnS											
MnSe											
SiC					2.654						

Substance	Heat of formation [kJ/mol (300K)]	Volume compressibility $(10^{-10} m^2/N)$	Static dielectric constant	Atomic magnetic susceptibility $(10^{-6}$ cgs)	Index of refraction	Minimum room temperature energy gap (eV)	Mobility (room temp.) $(cm^2/V{\cdot}s)$ Electrons	Holes	Optical transition	Breakdown voltage kV/mm	Remarks
MnTe						≈1.0					
Al_2S_3	426					4.1					
Al_2Se_3	367					3.1					

2.1.4. Chalcopyrite Structure Compounds (Strukturbericht symbol E11, Space Group $I\bar{4}2d\text{-}D_{2d}^{12}$)

I-III-VI$_2$ Compounds

Substance	Heat of formation [kJ/mol (300K)]	Volume compressibility $(10^{-10} m^2/N)$	Static dielectric constant	Atomic magnetic susceptibility $(10^{-6}$ cgs)	Index of refraction	Minimum room temperature energy gap (eV)	Mobility (room temp.) $(cm^2/V{\cdot}s)$ Electrons	Holes	Optical transition	Breakdown voltage kV/mm	Remarks
$CuAlS_2$		0.106				2.5					
$CuAlSe_2$						2.67					
$CuAlTe_2$						0.88					
$CuCaS_2$		0.106				2.38					
$CuGaSe_2$		0.141				0.96, 1.63					
$CuGaTe_2$		0.227				0.82, 1.0					
$CuInS_2$		0.141				1.2					
$CuInSe_2$		0.187				0.86, 0.92					
$CuInTe_2$		0.278				0.95					
$CuTlS_2$											
$CuTlSe_2$ (L.T.)						1.07					
$CuFeS_2$						0.53					Chalcopyrite
$CuFeSe_2$						0.16					
$CuLaS_2$											
$AgAlS_2$											
$AgAlSe_2$						0.7					
$AgAlTe_2$						0.56					
$AgGaS_2$		0.150				1.66					
$AgGaSe_2$		0.182				1.1					
$AgGaTe_2$		0.280				1.9					
$AgInS_2$ (L.T.)		0.185				1.18					
$AgInSe_2$		0.238				0.96, 0.52					
$AgInTe_2$		0.338									
$AgFeS_2$											

II-IV-V$_2$ Compounds

Substance	Heat of formation [kJ/mol (300K)]	Volume compressibility $(10^{-10} m^2/N)$	Static dielectric constant	Atomic magnetic susceptibility $(10^{-6}$ cgs)	Index of refraction	Minimum room temperature energy gap (eV)	Mobility (room temp.) $(cm^2/V{\cdot}s)$ Electrons	Holes	Optical transition	Breakdown voltage kV/mm	Remarks
$ZnSiP_2$	312					2.3	1000				
$ZnGeP_2$	293					2.2					
$ZnSnP_2$	275					1.45					
$CdSiP_2$		0.103				2.2	1000				
$CdGeP_2$	289					1.8					
$CdSnP_2$	270					1.5					
$ZnSiAs_2$	290					1.7		50			
$ZnGeAs_2$	271			−14.4		0.85					
$ZnSnAs_2$	252			−18.4		0.65		300			Disorders at 910 K
$CdSiAs_2$		0.143				1.6					
$CdGeAs_2$	266			−23.4		0.53	70	25			Disorders at 903 K
$CdSnAs_2$	247		13.7	−21.5		0.26	22000	250			

2.1.5. Other Ternary Semiconductors with Tetrahedral Coordination

II$_2$-IV-VI$_3$ Compounds

Substance	Heat of formation [kJ/mol (300K)]	Volume compressibility $(10^{-10} m^2/N)$	Static dielectric constant	Atomic magnetic susceptibility $(10^{-6}$ cgs)	Index of refraction	Minimum room temperature energy gap (eV)	Mobility (room temp.) $(cm^2/V{\cdot}s)$ Electrons	Holes	Optical transition	Breakdown voltage kV/mm	Remarks
Cu_2SiS_3 (H.T.)											Wurtzite

Substance	Heat of formation [kJ/mol (300K)]	Volume compressibility (10⁻¹⁰m²/N)]	Static dielectric constant	Atomic magnetic susceptibility (10⁻⁶ cgs)	Index of refraction	Minimum room temperature energy gap (eV)	Mobility (room temp.) (cm²/V·s)		Optical transition	Breakdown voltage kV/mm	Remarks
							Electrons	Holes			
Cu_2SiS_3 (L.T.)											Tetragonal
Cu_2SiTe_3											Cubic
Cu_2GeS_3 (H.T.)				−18.7							Cubic
Cu_2GeS_3 (L.T.)							360				Tetragonal
Cu_2GeSe_3	211.5			−21.3		0.94	238				Same
Cu_2GeTe_3	190.2			−23.4							Same
Cu_2SnS_3				−18.2		0.91	405				Cubic
$CuSnSe_3$				−21.0		0.66	870				Cubic
Cu_2SnTe_3				−28.4							Cubic
Ag_2GeSe_3				−29.6		0.91 (77K)					
Ag_2SnSe_3				−29.5		0.81					
Ag_2GeTe_3				−31.4		0.25					
Ag_2SnTe_3				−31.0		0.08					

II_3-V-VI_4 Compounds

Substance	Heat			Atomic		Energy gap					Remarks
Cu_3PS_4											Enargite
Cu_3AsS_4	269.6			−15.8		1.24					
Cu_3AsSe_4	161.3			−13.1		0.88					Famatinite
Cu_3SbS_4				−8.3		0.74					Famatinite
Cu_3SbSe_4	127.1			−20.5		0.31					

II-IV_2-V_3 Compounds

Substance		Volume				Energy gap					Remarks
$CuSi_2P_3$											El
$CuGe_2P_3$		0.12				0.90					El
$AgGe_2P_3$											

2.1.6. "Defect Chalcopyrite" Structure Compounds (Strukturbericht symbol E3, Space Group $I\bar{4}$-S_4^2)

Substance	Heat					Energy gap	Electrons				Remarks
$ZnAl_2Se_4$											
$ZnAl_2Te_4$ (?)											
$ZnGa_2S_4$ (?)						≈3.4					
$ZnGa_2Se_4$ (?)						≈2.2					
$ZnGa_2Te_4$ (?)						1.35					
$ZnIn_2Se_4$	206					1.82	35				
$ZnIn_2Te_4$	198					1.2					
$CdAl_2S_4$											
$CdAl_2Se_4$											
$CdAl_2Te_4$ (?)											
$CdGa_2S_4$	256					3.44	60				
$CdGa_2Se_4$	216					2.43	33				
$CdGa_2Te_4$											
$CdIn_2Te_4$	195					(1.26 or 0.9)	4000				
$HgAl_2S_4$											
$HgAl_2Se_4$											
$HgAl_2Te_4$ (?)											
$HgGa_2S_4$	249					2.84					
$HgGa_2Se_4$	204					1.95	400				

Substance	Heat of formation [kJ/mol (300K)]	Volume compressibility $(10^{-10}m^2/N)$	Static dielectric constant	Atomic magnetic susceptibility $(10^{-6}$ cgs)	Index of refraction	Minimum room temperature energy gap (eV)	Mobility (room temp.) $(cm^2/V \cdot s)$		Optical transition	Breakdown voltage kV/mm	Remarks
							Electrons	Holes			
$HgIn_2Se_4$	196					0.6	290				
$HgIn_2Te_4$ (?)	188					0.86	200				

2.1.7. Other Adamantine Compounds

Substance	Heat of formation	Volume compressibility	Static dielectric constant	Atomic magnetic susceptibility	Index of refraction	energy gap	Electrons	Holes			Remarks
$\alpha-SiC$		10.2		−6.4	2.67	2.86	400				6H structure
$Hg_5Ga_2Te_8$											B3 with superlattice
$Hg_5In_2Te_8$						0.7	2000				B3 with superlattice
$CdIn_2Se_4$						1.55					

2.2. Octahedral Semiconductors

2.2.1. Halite Structure Semiconductors (Strukturbericht symbol B1, Space Group $Fm3m$-O_h^5)

Substance	Heat	Vol	Static	Atomic	Index	energy gap	Electrons	Holes			Remarks
GeTe											
SnSe											
SnTe											
PbS	435					0.5	600	600			
PbSe	393	161				0.37	1000	900			
PbTe	393	280				0.26	1600	600			Altaite
		360				0.25					

2.2.2. Selected Other Binary Halites

Substance	Heat	Vol	Static	Atomic	Index	energy gap	Electrons	Holes			Remarks
BiSe											
BiTe						0.4					
EuSe											
GdSe						1.8	4				
NiO						2.0 or 3.7	100				
CdO	531					2.5					
SrS						4.1					

2.3. Other Semiconductors

2.3.1. Antifluorite Structure Compounds ($Fm3m$-O_h^5)

Substance	Heat	Vol	Static	Atomic	Index	energy gap	Electrons	Holes			Remarks
Mg_2Si	79.08					0.77	405	70			
Mg_2Ge						0.74	520	110			
Mg_2Sn	76.57					0.36	320	260			
Mg_2Pb	52.72					0.1					

2.3.2. Tetradymite Structure Compounds ($R\bar{3}$-D_{3d}^5)

Substance	Heat	Vol	Static	Atomic	Index	energy gap	Electrons	Holes			Remarks
Sb_2Te_3						0.3		360			
Bi_2Se_3						0.35	600				
Bi_2Te_3						0.21	1140	680			R3m (166)

2.3.3. Skutterudite Structure Compounds ($Im3$-T_h^5)

Substance	Heat	Vol	Static	Atomic	Index	energy gap	Electrons	Holes			Remarks
CoP_3						0.43					
$CoAs_3$						0.69		~4000			
$CoSb_3$						0.63	70	~3000			
RhP_3								700			
$RhAs_3$						0.85		~3000			
$RhSb_3$						0.80		~7000			
$IrSb_3$						1.18		1500			

2.3.4. Selected Multinary Compounds

Substance	Heat	Vol	Static	Atomic	Index	energy gap	Electrons	Holes			Remarks
$AgSbSe_2$						0.58					

Substance	Heat of formation [kJ/mol (300K)]	Volume compressibility $(10^{-10} m^2/N)$	Static dielectric constant	Atomic magnetic susceptibility $(10^{-6} cgs)$	Index of refraction	Minimum room temperature energy gap (eV)	Mobility (room temp.) $(cm^2/V{\cdot}s)$ Electrons	Holes	Optical transition	Breakdown voltage kV/mm	Remarks
$AgSbTe_2$ (or Ag_{19} $Sb_{29}Te_{52}$)						0.7, 0.27					
$AgBiS_2$ (H.T.)											
$AgBiSe_2$ (H.T.)											
$AgBiTe_2$ (H.T.)											
Cu_2CdSnS_4						1.16	<2				

2.3.5. Some Elemental Semiconductors

Substance	Heat of formation [kJ/mol (300K)]	Volume compressibility $(10^{-10} m^2/N)$	Static dielectric constant	Atomic magnetic susceptibility $(10^{-6} cgs)$	Index of refraction	Minimum room temperature energy gap (eV)	Electrons	Holes	Optical transition	Breakdown voltage kV/mm	Remarks
B	397.1			−6.7	3.4	1.55	10				
Se (gray)		6.6 (0.1 GHz)		−22.1	2.5	1.5			5		$P3_121(152)$
Te				−39.5	3.3	0.33	1700	1200			Same

TABLE 3. Semiconducting Properties of Selected Materials

Substance	Minimum energy gap (eV) R.T.	0 K	$10^4\, dE/dT$ eV K^{-1}	dE/dT meV kbar^{-1}	Electron effective mass $m_{da}\,(m_o)$	Electron mobility and temperature dependence μ_a (cm^2 V^{-1} s^{-1})	$-x$	Hole effective mass $m_{dp}\,(m_a)$	Hole mobility and temperature dependence μ_p (cm^2 V^{-1} s^{-1})	$-x$
				Elements						
Si	1.110	1.169	−2.8	−1.41	1.026	1900	2.6	0.056	500	2.3
Ge	0.664	0.744	−3.7	5.1	0.0823	3800	1.66	0.0438	1820	2.33
α-Sn	0.08	0.094	−0.5		0.0236	2500	1.65	0.195	z	2.0
Se	2.11	2.48								
Te	0.335				0.08	1100		0.19	560	
				III-V Compounds						
AlAs	2.2	2.3				1200			420	
AlSb	1.6	1.7	−3.5	−1.6	0.09	200	1.5	0.4	500	1.8
GaP	2.272	2.350	−3.7	10.5	0.35	300	1.5	0.5	150	1.5
GaAs	1.441	1.579	−3.9	11.3	0.068	9000	1.0	0.5	500	2.1
GaSb	0.70	0.812	−3.7	14.5	0.050	5000	2.0	0.23	1400	0.9
InP	1.34	1.4236	−2.9	9.1	0.067	5000	2.0		200	2.4
InAs	0.356	0.418	−3.4	10.0	0.022	33 000	1.2	0.41	460	2.3
InSb	0.180	0.235	−2.8	15.7	0.014	78 000	1.6	0.4	750	2.1
				II-VI Compounds						
ZnO	3.2	3.4376	−9.5	0.6	0.38	180	1.5			
ZnS	3.80	3.91	−4.7	−5.8		180			5(400 °C)	
ZnSe	2.713	2.820	−4.5	0.7		540			28	
ZnTe	2.26	2.391	−5.2	8.3		340			100	
CdO	1.20		−6		0.1	120				
CdS	2.485	2.585	−4.1	4.5	0.165	400		0.8		
CdSe	1.751	1.841	−3.6	5.0	0.13	650	1.0	0.6		
CdTe	1.43	1.606	−5.4	8	0.14	1200		0.35	50	
HgSe	−0.061				0.030	20 000	2.0			
HgTe	−0.141	−0.3025			0.017	25 000		0.5	350	
				Halite Structure Compounds						
PbS	0.41	0.286	4		0.16	800		0.1	1000	2.2
PbSe	0.278	0.145	4		0.3	1500		0.34	1500	2.2

Substance	Minimum energy gap (eV)		$10^4 \, dE/dT$ eV K^{-1}	dE/dT meV kbar^{-1}	Electron effective mass m_{da} (m_o)	Electron mobility and temperature dependence		Hole effective mass m_{dp} (m_a)	Hole mobility and temperature dependence	
	R.T.	0 K				μ_a (cm^2 V^{-1} s^{-1})	$-x$		μ_p (cm^2 V^{-1} s^{-1})	$-x$
PbTe	0.310	0.187	4	−7	0.21	1600		0.14	750	2.2
					Others					
ZnSb	0.50	0.56			0.15	10				1.5
CdSb	0.459	0.57	−5.4		0.15	300			2000	1.5
Bi$_2$S$_3$	1.3	1.45				200			1100	
Bi$_2$Se$_3$	0.160					600			675	
Bi$_2$Te$_3$	0.13		−0.95		0.58	1200	1.68	1.07	510	1.95
Mg$_2$Si		0.77	−6.4		0.46	400	2.5		70	
Mg$_2$Ge	0.54	0.74	−9			280	2		110	
Mg$_2$Sn	0.18	0.36	−3.5		0.37	320			260	
Mg$_2$Sb$_2$		0.32				20			82	
Zn$_3$As$_2$	0.93					10	1.1		10	
Cd$_3$As$_2$	0.55				0.046	100 000	0.88			
GaSe	2.021	2.1275	−3.8						20	
GaTe	1.694	1.799	−3.6			14				
InSe	1.172	1.32				900				
TlSe	0.745		−3.9		0.3	30		0.6	20	1.5
CdSnAs$_2$	0.26				0.05	25 000	1.7			
Ga$_2$Te$_2$	1.22	1.55	−4.8							
α-In$_2$Te$_2$	0.92	1.2			0.7				50	1.1
β-In$_2$Te$_2$	1.0								5	
Hg$_5$In$_2$Te$_8$	0.5								11 000	
SnO$_2$	3.47	3.596							78	

TABLE 4. Band Properties of Semiconductors

Substance	Band curvature effective mass (Expressed as fraction of free electron mass)			Energy separation of "split-off" band (eV)	Measured (light) hole mobility (cm^2/V·s)
	Heavy holes	Light holes	"Split-off" band holes		

4.1. Data on Valence Bands of Semiconductors (Room Temperature)

4.1.1. Semiconductors with Valence Band Maximum at the Center of the Brillouin Zone ("F")

Substance	Heavy holes	Light holes	"Split-off" band holes	Energy separation "split-off" band (eV)	Measured (light) hole mobility (cm^2/V·s)
Si	0.52	0.16	0.25	0.044	500
Ge	0.34	0.043	0.08	0.3	1820
Sn	0.3				2400
AlAs					
AlSb	0.4			0.7	550
GaP				0.13	100
GaAs	0.8	0.12	0.20	0.34	400
GaSb	0.23	0.06		0.7	1400
InP				0.21	150
InAs	0.41	0.025	0.083	0.43	460
InSb	0.4	0.015		0.85	750
CdTe	0.35				50
HgTe	0.5				350

Substance	Number of equivalent valleys and direction	Band curvature effective masses		Anistrophy K = m_L/m_T	Measured (light) hole mobility (cm^2/V·s)
		Longitudinal m_L	Transverse m_T		

4.1.2. Semiconductors with Multiple Band Maxima

Substance	Number of equivalent valleys and direction	Longitudinal m_L	Transverse m_T	Anistrophy K = m_L/m_T	Measured (light) hole mobility (cm^2/V·s)
PbSe	4 "L" [111]	0.095	0.047	2.0	1500
PbTe	4 "L" [111]	0.27	0.02	10	750
Bi$_2$Te$_3$	6	0.207	~0.045	4.5	515

Substance	Energy gap (eV)	Effective mass (m_o)	Mobility (cm²/V·s)	Comments

4.2. Data on Conduction Bands of Semiconductors (Room Temperature Data)

4.2.1. Single Valley Semiconductors

Substance	Energy gap (eV)	Effective mass (m_o)	Mobility (cm²/V·s)	Comments
GaAs	1.35	0.067	8500	3(or 6?) equivalent [100] valleys 0.36 eV above this maximum with a mobility of ~50.
InP	1.27	0.067	5000	3(or 6?) equivalent [100] valleys 0.4 eV above this minimum.
InAs	0.36	0.022	33,000	Equivalent valleys ~1.0 eV above this minimum.
InSb	0.165	0.014	78,000	
CdTe	1.44	0.11	1000	4(or 8?) equivalent [111] valleys 0.51 eV above this minimum.

Substance	Energy gap	Number of equivalent valleys and direction	Band curvature effective mass		Anisotropy
			Longitudinal m_L	Transverse m_T	$K = m_L/m_T$

4.2.2. Multivalley Semiconductors

Substance	Energy gap	Number of equivalent valleys and direction	Longitudinal m_L	Transverse m_T	$K = m_L/m_T$
Si	1.107	6 in [100] "Δ"	0.00	0.192	4.7
Ge	0.67	4 in [111] at "L"	1.588	0.0815	19.5
GaSb	0.67	as Ge (?)	~1.0	~0.2	~5
PbSe	0.26	4 in [111] at "L"	0.085	0.05	1.7
PbTe	0.25	4 in [111] at "L"	0.21	0.029	5.5
Bi_2Te_3	0.13	6			~0.05

TABLE 5. Resistivity of Semiconducting Minerals

Mineral	ρ (ohm · m)	Mineral	ρ (ohm · m)
Diamond (C)	2.7	Pentlandite, $(Fe, Ni)_4S_4$	1 to 11 × 10⁻⁶
Sulfides		Pyrrhotite, Fe_7S_4	2 to 160 × 10⁻⁶
Argentite, Ag_2S	1.5 to 2.0 × 10⁻³	Pyrite, FeS_2	1.2 to 600 × 10⁻³
Bismuthinite, Bi_2S_3	3 to 570	Sphalerite, ZnS	2.7 × 10⁻³ to 1.2 × 10⁴
Bornite, $Fe_2S_3 \, nCu_2S$	1.6 to 6000 × 10⁻⁶	Antimony-sulfur compounds	
Chalcocite, Cu_2S	80 to 100 × 10⁻⁶	Berthierite, $FeSb_2S_4$	0.0083 to 2.0
Chalcopyrite, $Fe_2S_3 \, Cu_2S$	150 to 9000 × 10⁻⁶	Boulangerite, $Pb_5Sb_3S_{11}$	2 × 10³ to 4 × 10⁴
Covellite, CuS	0.30 to 83 × 10⁻⁶	Cylindrite, $Pb_3Sn_4Sb_2S_{14}$	2.5 to 60
Galena, PbS	6.8 × 10⁻⁶ to 9.0 × 10⁻²	Franckeite, $Pb_5Sn_3Sb_2S_{14}$	1.2 to 4
Haverite, MnS_2	10 to 20	Hauchecornite, $Ni_4(Bi, Sb)_2S_{14}$	1 to 83 × 10⁻⁶
Marcasite, FeS_2	1 to 150 × 10⁻³	Jamesonite, $Pb_4FeSb_6S_{14}$	0.020 to 0.15
Metacinnabarite, HgS	2 × 10⁻⁶ to 1 × 10⁻³	Tetrahedrite, Cu_3SbS_3	0.30 to 30,000
Millerite, NiS	2 to 4 × 10⁻⁷	Arsenic-sulfur compounds	
Mineral	ρ (ohm m)	Mineral	ρ (ohm m)
Molybdenite, MoS_2	0.12 to 7.5	Arsenopyrite, FeAsS	20 to 300 × 10⁻⁶
Cobaltite, CoAsS	6.5 to 130 × 10⁻³	Hessite, Ag_2Te	4 to 100 × 10⁻⁶
Enargite, Cu_3AsS_4	0.2 to 40 × 10⁻³	Nagyagite, $Pb_6Au(S,Te)_{14}$	20 to 80 × 10⁻⁶
Gersdorfite, NiAsS	1 to 160 × 10⁻⁶	Sylvanite, $AgAuTe_4$	4 to 20 × 10⁻⁶
Glaucodote, (Co, Fe)AsS	5 to 100 × 10⁻⁶	Oxides	
Antimonide		Braunite, Mn_2O_3	0.16 to 1.0
Dyscrasite, Ag_3Sb	0.12 to 1.2 × 10⁻⁶	Cassiterite, SnO_2	4.5 × 10⁻⁴ to 10,000
Arsenides		Cuprite, Cu_2O	10 to 50
Allemonite, $SbAs_3$	70 to 60,000	Hollandite, $(Ba, Na, K) Mn_8O_{16}$	2 to 100 × 10⁻³
Lollingite, $FeAs_2$	2 to 270 × 10⁻⁶	Ilmenite, $FeTiO_3$	0.001 to 4
Nicollite, NiAs	0.1 to 2 × 10⁻⁶	Magnetite, Fe_3O_4	52 × 10⁻⁶
Skutterudite, $CoAs_3$	1 to 400 × 10⁻⁶	Manganite, MnO OH	0.018 to 0.5
Smaltite, $CoAs_2$	1 to 12 × 10⁻⁶	Melaconite, CuO	6000
Tellurides		Psilomelane, $BaMn_9O_{18} \, 2H_2O$	0.04 to 6000
Altaite, PbTe	20 to 200 × 10⁻⁶	Pyrolusite, MnO_2	0.007 to 30
Calavarite, $AuTe_2$	6 to 12 × 10⁻⁶	Rutile, TiO_2	29 to 910
Coloradoite, HgTe	4 to 100 × 10⁻⁶	Uraninite, UO_2	1.5 to 200

References

1. Beer, A. C., *Galvanomagnetic Effects in Semiconductors*, Academic Press, New York, 1963.

2. Goryunova, N. A., *The Chemistry of Diamond-Like Semiconductors*, The MIT Press, Cambridge, MA, 1965.

3. Abrikosov, N. Kh., Bankina, V. F., Poretskaya, L. E., Shelimova, L. E., and Skudnova, E.V., *Semiconducting II-VI, IV-VI, and V-VI Compounds*, Plenum Press, New York, 1969.

4. Berger, L. I. and Prochukhan, V. D., *Ternary Diamond-Like Semiconductors*, Cons. Bureau/Plenum Press, New York, 1969.

5. Shay, J. L. and Wernick, J. H., *Ternary Chalcopyrite Semiconductors: Growth, Electronic Properties, and Applications*, Pergammon Press, 1975.

6. Bergman, R., *Thermal Conductivity in Solids*, Clarendon, Oxford, 1976.

7. Handbook of Semiconductors, Vol. 1, Moss, T. S. and Paul, W., Eds., *Band Theory and Transport Properties*; Vol. 2, Moss, T. S. and Balkanski, M., Eds., *Optical Properties of Solids*; Vol. 3, Moss, T. S. and Keller, S. P., Eds., *Materials Properties and Preparation*, North Holland Publ. Co., Amsterdam, 1980.

8. Böer, K. W., *Survey of Semiconductor Physics*, Van Nostrand Reinhold, 1990.

9. Rowe, D. M., Ed., *CRC Handbook of Thermoelectrics*, CRC Press, Boca Raton, FL, 1995.

10. Berger, L. I., *Semiconductor Materials*, CRC Press, Boca Raton, FL, 1997.

11. Glazov, V. M., Chizhevskaya, S.N., and Glagoleva, N.N., *Liquid Semiconductors, Plenum Press*, New York, 1969.

12. Phillips, J. C., *Bonds and Bands in Semiconductors*, Academic Press, New York, 1973.

13. Harrison, W. A., *Electronic Structure and the Properties of Solids*, Freeman Publ. House, San Francisco, 1980.

14. Balkanski, M., Ed., *Optical Properties of Solids*, North-Holland, Amsterdam, 1980.

15. *Landolt-Börnstein. Numerical Data and Functional Relationships in Science and Technology, New Series, Group III: Crystal and Solid State Physics*, Hellwege, K.-H. and Madelung, O., Eds., Volumes 17 and 22, Springer Verlag, Berlin, 1984 (and further).

16. Shklovskii, B. L. and Efros, A. L., *Electronic Processes in Doped Semiconductors*, Springer Verlag, Berlin, 1984.

17. Cohen, M. L. and Chelikowsky, J. R., *Electronic Structure and Optical Properties of Semiconductors*, Springer Verlag, New York, 1988.

18. Glass, J. T., Messier, R. F., and Fujimori, N., Eds., *Diamond, Silicon Carbide, and Related Wide Bandgap Semiconductors*, MRS Symposia Proc. 1652, Mater. Res. Soc., Pittsburgh, 1990.

19. Palik, E., Ed., *Handbook of Optical Constants of Solids II*, Academic Press, New York, 1991.

20. Reed, M., Ed., *Semiconductors and Semimetals*, Volume 35, Academic Press, Boston, 1992.

21. Haug, H. and Koch, S. W., *Quantum Theory of the Optical and Electronic Properties of Semiconductors*, 2nd Edition, World Scientific, Singapore, 1993.

22. Lockwood, D. J., Ed., *Proc. 22nd Intl. Conf. on the Physics of Semiconductors, Vancouver, 1994*, World Scientific, Singapore, 1994.

23. Morelli, D. T., Caillat, T., Fleurial, J.-P., Borschchevsky, A., Vandersande, J., Chen, B., and Uher, C., *Phys. Rev.*, B51, 9622, 1995.

24. Caillat, T., Borshchevsky, A., and Fleurial, J.-P., *J. Appl. Phys.*, 80, 4442, 1996.

25. Fleurial, J.-P., Caillat, T., and Borshchevsky, A., *Proc. XVI Intl. Conf. Thermoelectrics*, Dresden, Germany, August 26–29, 1997 (in print).

26. Borshchevsky, A. et al., U.S. Patents 5,610,366 (March 1977) and 5,831,286 (March 1998).

27. Jarrendahl, K. and Davis, R. F., *Semiconductors and Semimetals*, Vol. 52, Y.S. Park, Ed., 1998, pp. 1–20.

28. Bettini, M., *Solid State Comm.*, 13, 599, 1973.

29. Chen, A. and Sher, A., *Semiconductor Alloys, Physics and Material Engineering*, Plenum Press, New York, 1995.

30. Holloway, P. H. and McGuire, G. E., Eds., *Handbook of Compound Semiconductors*, Noyes Publ., Park Ridge, NJ, 1995.

31. Madelung, O., Ed., *Semiconductors: Group IV Elements and III-V Compounds (Data in Science and Technology)*, Springer-Verlag, Berlin, Heidelberg, 1991.

32. Madelung, O., Ed., *Semiconductors: Other than Group IV Elements and III-V Compounds (Data in Science and Technology)*, Springer-Verlag, Berlin, Heidelberg, 1992.

33. Levinshtein, M., Rumyantsev, S., and Shur, M., Eds., Handbook Series on Semiconductor Parameters. Vol. 1, *Si, Ge, C (Diamond), GaAs, GaP, GaSb, InAs, InP, InSb*, World Scientific, Singapore, 1996.

34. Levinshtein, M., Rumyantsev, S., and Shur, M., Eds., Handbook Series on Semiconductor Parameters. Vol. 2, *Ternary and Quaternary III-V Compounds*, World Scientific, Singapore, 1996.

35. *Physical Properties of Semiconductors*, NSM Archive:http://www.ioffe.ru/SVA/NSM/Semicond/

SELECTED PROPERTIES OF SEMICONDUCTOR SOLID SOLUTIONS

L. I. Berger

Alloy system	Limits of solubility	Energy gap in eV (300 K)	Remarks, references
Adamantine Semiconductors IV-IV			
		$0.8941+0.0421x+0.1691x^2$	Transition Γ - X [Ref.1]
Si_xGe_{1-x}	$0 \leq x \leq 1$	$0.7596+1.0860x+0.3306x^2$	Trans. Γ - L [Ref. 1]
Adamantine Semiconductors III-V/III-V			
Common Anion			
$Al_xGa_{1-x}N$	$0 \leq x \leq 1$		Wurtzite Structure [Ref. 2 & 3]
$Al_xGa_{1-x}P$	$0 \leq x \leq 0.5$	$2.28+0.16x$	[Ref. 2]
$Al_xIn_{1-x}P$	$0 \leq x \leq 0.44$	at Γ: $134+2.23x$; at X: $2.24+0.18x$	[Ref. 2]
$Al_xGa_{1-x}As$	$0 \leq x \leq 0.5$	$1.42=0.75x$ [Ref.3]; $1.424+1.429x-0.14x^2$ [Ref.4]	
$Al_xIn_{1-x}As$	$0 \leq x \leq 1$	at Γ: $0.37+1.91x+0.74x^2$; at X: $1.8+0.4x$	[Ref. 2 and 6]
$Al_xGa_{1-x}Sb$	$0 \leq x < 1$	$0.73+1.10x+0.47x^2$	Trans. Γ_{8v}- Γ_{6c} [Ref. 2]
$Al_xIn_{1-x}Sb$	$0 \leq x \leq 1$		[Ref. 6]
$Ga_xIn_{1-x}N$	$0 \leq x \leq 1$	$1.950+1.487x-1.000x(1-x)$	Wurtzite [Ref. 8 and 10]
$Ga_xIn_{1-x}P$	$0 \leq x \leq 1$		[Ref. 2]
$Ga_xIn_{1-x}As$	$0 \leq x \leq 1$	$0.360+0.629x+0.436x^2$	[Ref. 5]
$Ga_xIn_{1-x}Sb$	$0 < x < 1$	$0.235+0.1653x+0.413x^2$	[Ref. 2, see also Ref. 9]
Common Cation			
GaN_xAs_{1-x}	$0 \leq x \leq 0.05$	$1.42-9.9x$	[Ref. 2]
GaP_xAs_{1-x}	$0 < x < 1$	$2.270-0.846x$	[Ref. 2]
GaP_xAs_{1-x}	$0 \leq x \leq 0.05$	$1.515+1.172x+0.186x^2$	(at 2K, $\Gamma-\Gamma$) [Ref. 7]
		$1.9715+0.144x+0.211x^2$	[Ref. 2]
$GaAs_xSb_{1-x}$	$0 \leq x \leq 0.45$, $0.6 \leq x \leq 1$	$1.43-1.9x+1.2x^2$	[Ref. 5]
InP_xAs_{1-x}	$0 < x < 1$	$0.356+0.675x+0.32x^2$	[Ref. 2]
Adamantine Binary Semiconductors II-VI/II-VI [Ref. 3 and 6]			
Common Anion			
$Zn_xCd_{1-x}S$	$0 \leq x \leq 1$		Wurtzite Structure
$Zn_xHg_{1-x}S$	$0 \leq x \leq 1$		
$Cd_xHg_{1-x}S$	$0 \leq x \leq 1$		Wurtzite Structure at $x<0.6$
$Zn_xCd_{1-x}Se$	$0.7 \leq x \leq 1$		
$Zn_xHg_{1-x}Se$	$0 \leq x \leq 1$		
$Cd_xHg_{1-x}Se$	$0 \leq x \leq 0.7$ and $0.75 \leq x^* \leq 1$		x^*- Wurtzite Structure
$Zn_xCd_{1-x}Te$	$0 \leq x \leq 1$		
$Zn_xHg_{1-x}Te$	$0 \leq x \leq 1$		
$Cd_xHg_{1-x}Te$	$0 \leq x \leq 1$		
Common Cation			
ZnS_xSe_{1-x}	$0 \leq x \leq 1$		
ZnS_xTe_{1-x}	$0 \leq x \leq 0.1$ and $0.9 \leq x^* \leq 1$		x^*- Wurtzite Structure
$ZnSe_xTe_{1-x}$	$0 \leq x \leq 1$		
CdS_xSe_{1-x}	$0 \leq x \leq 1$		Wurtzite Structure
CdS_xTe_{1-x}	$0 \leq x \leq 0.25$ and $0.8 \leq x^* \leq 1$		x^*- Wurtzite Structure
$CdSe_xTe_{1-x}$	$0 \leq x \leq 0.4$ and $0.6 \leq x^* \leq 1$		x^*- Wurtzite Structure
HgS_xSe_{1-x}	$0 \leq x \leq 1$		
HgS_xTe_{1-x}	$0 \leq x \leq 1$		
$HgSe_xTe_{1-x}$	$0 \leq x \leq 1$		
Quaternary Adamantine Semiconductors II-VI/III-V [Ref. 6]			
$(ZnS)_x(AlP)_{1-x}$	$0.99 \leq x \leq 1$		
$(ZnSe)_x(GaAs)_{1-x}$	$0 \leq x < 1$		
$(CdTe)_x(InAs)_{1-x}$	$0 < x \leq 0.2$ and $0.7 \leq x \leq 1$		
$(CdTe)_x(AlSb)_{1-x}$	$0 \leq x \leq 1$		
$(HgTe)_x(InAs)_{1-x}$	$0 \leq x \leq 1$		

Alloy system	Limits of solubility	Energy gap in eV (300 K)	Remarks, references
Quaternary Adamantine Semiconductors III$_x$-III$_{1-x}$-V$_y$ - V$_{1-y}$			
Ga$_x$In$_{1-x}$As$_y$P$_{1-y}$	$0 \leq x \leq 1$, $0 \leq x \leq 1$	$1.35 + 0.668x - 1.068y + 0.758x^2 + 0.078y^2 - 0.069xy - 0.322x^2y + 0.03xy^2$	[Ref. 2 and 6]
Quaternary Adamantine Semiconductors III$_{1-x-y}$-III$_x$-III$_y$-V			
Al$_x$Ga$_y$In$_{1-x-y}$Sb	$0 \leq x \leq 1$, $0 \leq y \leq 1$	$0.095 + 1.76x + 0.28y + 0.345(x^2+y^2) + 0.085(1-x-y)^2 + xy\,(1-x-y)(23-28y)$	[Ref. 2 and 6]

References

1. Krishnamurti, S., Sher, A., and Chen, A. *Appl. Phys. Lett.* 47, 160, 1985.
2. Madelung, O., Ed., *Semiconductors Group IV Elements and III-V Compounds*, Springer, 1991; *Semiconductors Other than Group IV Elements and III-V Compounds*, Springer, 1992.
3. Goryunova, N.A., Kesamanly, F.P., and Nasledov, D.N., *Semiconductors and Semimetals*, Vol. 4, 1968, p. 413.
4. El Allali, M., Sorensen, C. B., Veje, E., and Tideman-Peterson, P., *Phys. Rev. B*, 48, 4398, 1993.
5. Nahorny, R. E., Pollack, M. A., Johnson, W. D., and Barns, R. L. *Appl. Phys. Lett.* 33, 695, 1978.
6. Goryunova, N. A. *Multicomponent Diamond-Like Semiconductors*, Sov. Radio, Moscow, 1968 (in Russian).
7. Capizzi, M., Modesti, S., Martelli, F., and Frova, A., *Solid State Comm.* 39, 333, 1981.
8. Nakamura, S., Pearton, S., and Fasol, G., *The Blue Laser Diode*, 2nd ed., Springer, 2000.
9. Roth, A. P., Keeler, W. J., and Fortin, E. *Canad. J. Phys.* 58, 560, 1980.
10. Wu, J., Walukiewicz, Yu, K. M., Ager, J. W., Haller, E. E., Lu, H., and Schaff, W. J., *Appl. Phys. Lett.* 80, 4741, 2002.

PROPERTIES OF ORGANIC SEMICONDUCTORS

L. I. Berger

Substance	Energy Gap, E, (in $E/2\,kT$) eV	Room Temperature Electrical Resistivity, ohm · cm	Mobility, μ, cm²/V · s	Sign of Majority Carriers	Temperature Range, °C	Ref.
Metal-Free Molecular Crystals						
3-Acetylamino-*N*-methylphthalimide	3.46				67 to 100	1
3-Acetylamino-*N*-phenylphthalimide	3.50				54 to 124	1
4-Amino-*N*-cyclomethylphthalimide	2.90				73 to 100	1
4-Aminophthalimide	2.78				123 to 151	1
Acridine	3.90					1
Anthanthrene	1.67	$1.5 \cdot 10^{19}$			40 to 105	1
Anthanthrene	0.84	$1.5 \cdot 10^{19}$ (15°C)				2
Anthanthrone	1.70	$7.7 \cdot 10^{18}$			20 to 150	1
Anthracene	0.83	$1.3 \cdot 10^{14}$ (15°C)				2
Anthracene	2.50	$1.5 \cdot 10^{11}$	2.3	+	20 to 130	1
Anthracene	3.88 to 4.1	$>10^{15}$	1.74(n), 2.07(p)	+ & −		4
1,2-Benzanthracene	1.04	10^{16} (30°C)				2
Benzanthrone	3.12	$1.6 \cdot 10^{16}$				1
Benzene (liquid)	0.41					2
Benzene (amorphous)	0.84	10^{15}			−14 to 5	1
Benzene (cryst.)	7		2	−	−23	4
Benzimidazole	3.0 to 4.0	$5 \cdot 10^{13}$			84 to 144	1
Benzophenone	3.34				−23 to 14	1
Benzo[*f*]quinoline	2.77				30 to 50	1
Benzo[*h*]quinoline	2.72					1
3-Benzoylamino-*N*-methylphthalimide	3.28				84 to 112	1
Benzpentacene	1.72				0 to 150	1
Biphenyl	1.46	$1.7 \cdot 10^{15}$ (50°C)				2
Biphenyl	1.45				20 to 70	1
o-Chloranil	3.0	10^{15}				1
p-Chloranil	0.61					1
Chlorpromazine	2.1	10^{12} (32°C)		+ & −	32 to 80	1
Chrysene	1.1	$4 \cdot 10^{19}$ (15°C)				2
Chrysene	2.20	$4 \cdot 10^{19}$			25 to 90	1
Circumanthracene	1.8	$6 \cdot 10^{12}$				1
Coronene	1.7	$1.7 \cdot 10^{17}$			60 to 80	1
Coronene	0.85	$1.7 \cdot 10^{17}$ (15°C)				2
Cyananthrone	0.20	$1.2 \cdot 10^{7}$			30 to 125	1
1,6-Diaminopyrene	0.6	10^{8}				
Dibenzpentacene	1.50				0 to 150	1
Dinaphthopyrene	1.60				25 to 90	1
1,8-Diphenyl-1,3,5,7-octatetraene	1.7				72 to 191	1
Diphenylpentacene	1.60				0 to 150	1
4,4′-Diphenylstilbene	1.56				160 to 280	1
4,4′-Diphenylstilbene	0.80					2
Ferrocene	1.22	10^{14}		+		1
Flavanthrone	0.70	$1.4 \cdot 10^{11}$				1
Fluorene	2.7					2
Fluorene	1.4					2
Fluoridine	1.6	$6 \cdot 10^{13}$		+	20 to 140	1
Hexacene	0.57	$3.8 \cdot 10^{10}$ (50°C)				2
Hexacene	1.3					1
Hexamethylbenzene	1.86			+	20 to 140	1
3-Hydroxy-*N*-methylphthalimide	3.80				60 to 91	1
Imidazole	2.6	10^{11}			28 to 68	1

Substance	Energy Gap, E, (in E/2 kT) eV	Room Temperature Electrical Resistivity, ohm · cm	Mobility, μ, cm²/V · s	Sign of Majority Carriers	Temperature Range, °C	Ref.
Indanthrazine	0.66	$1.4 \cdot 10^{15}$			30 to 125	1
Indanthrone	0.64	$7.5 \cdot 10^{14}$			30 to 125	1
Indanthrone (black)	0.56	$2.5 \cdot 10^{8}$			30 to 125	1
Mesitylene (liquid)	0.19					2
Mesonaphthodianthracene	0.6	$4.0 \cdot 10^{18}$ (15°C)				2
Mesonaphthodianthrene	1.48				45 to 250	1
Mesonaphthodianthrone	0.86				5 to 110	1
3-Methoxy-N-methyl-phthalimide	3.18				54 to 78	1
Naphthacene	1.7	$1 \cdot 10^{15}$				1
Naphthalene	3.5	10^{14}			27 to 47	1
Naphthalene	1.15	$2.8 \cdot 10^{14}$ (50°C)				2
Naphthalene	4.9 to 5.1		0.64(n), 1.50(p)	+ & −		4
m-Naphthodianthrene	1.20	$4 \cdot 10^{18}$			40 to 150	1
m-Naphthodianthrone	1.30	$1.5 \cdot 10^{18}$			40 to 150	1
β-Naphthol	2.36	$2 \cdot 10^{5}$			60 to 110	1
β-Naphthoquinoline	2.77					1
1-Naphthylamine	2.2				25 to 42	1
1-Naphthylamine picrate	2.7				28 to 98	1
2-Naphthylphenyl sulphone	3.5				67 to 102	1
1-Nitronaphthalene	2.5				25 to 44	1
Ovalene	1.13	$2.3 \cdot 10^{15}$				1
Pentacene	0.58	$2.4 \cdot 10^{9}$ (50°C)				2
Pentacene	1.5	$6 \cdot 10^{13}$			20 to 140	1
Perylene	2.1	$4.1 \cdot 10^{13}$			40 to 100	1
Perylene	3.10		5.53(n), 87.4(p)	+ & −	−213	4
Phenanthrene	1.15	$1.3 \cdot 10^{14}$			12 to 72	1
Phenanthrene	0.65					2
1,10-Phenanthroline	2.73				50 to 90	1
Phenazine	2.1	$7 \cdot 10^{14}$ (100°C)			98 to 143	1
Phenazine	1.1			−		4
Phenothiazine	1.6	10^{11}			50 to 150	1
Phenothiazine			2.45(n), 0.02(p)	+ & −		4
Phenylanthranilic acid	3.30				87 to 119	1
4-Phenylstilbene	1.74				140 to 220	1
4-Phenylstilbene	0.86					2
Phosphonitrilic chloride trimer	1.68	10^{15}				1
Phthalocyanine, PcH₂	1.66	10^{13}	0.1 to 0.4	+	26 to 350	1
Phthalocyanine, PcH₂	2	10^{7}	1.2(n), 1.1(p)	+ & −	100	4
Pyranthrene	1.11	$1 \cdot 10^{15}$				1
Pyranthrene	0.51	$4.5 \cdot 10^{16}$ (15°C)				2
Pyranthrone	1.06	$3.9 \cdot 10^{15}$			40 to 150	1
Pyrene	2.02	$5 \cdot 10^{17}$		−		1
Pyrene			0.50	+		4
5,6-N-Pyridine-1,9-benzanthrone	1.60					2
p-Quaterphenyl	0.89	$1.0 \cdot 10^{15}$ (50°C)				2
Quaterrylene	0.6	10^{5}		−		1
p-Quinquiphenyl	0.91	$2.0 \cdot 10^{15}$ (50°C)				2
α-Resorcin	2.10	$2 \cdot 10^{16}$			30 to 94	1
β-Resorcin	3.27	$2 \cdot 10^{18}$			30 to 94	1
Salanil	4.1	10^{4}			20 to 40	1
p-Sexiphenyl	0.91	$7.0 \cdot 10^{14}$ (50°C)				2
cis-Stilbene	2.4			+	at 20	1
trans-Stilbene	1.80		2.4	+	70 to 120	1
trans-Stilbene	0.91					2
trans-Stilbene	1.4			+		4
o-Terphenyl		$3 \cdot 10^{-5}$		+		1
m-Terphenyl			10^{-5}	+		1
p-Terphenyl	0.6	10^{14} (25°C)				2
p-Terphenyl	1.2		0.025	+		1

Substance	Energy Gap, E, (in $E/2 kT$) eV	Room Temperature Electrical Resistivity, ohm·cm	Mobility, μ, cm²/V·s	Sign of Majority Carriers	Temperature Range, °C	Ref.
p-Terphenyl			1.2(n), 0.80(p)	+ & −		4
Tetracene	0.66	$3.2 \cdot 10^{12}$ (50°C)				2
Tetracene	1.7					1
Tetracene	3.4		0.85	+		4
1,1,10,10-Tetracyanodecapentaene	2.24	10^{13}			>68	1
1,1,6,6-Tetracyanohexatriene	1.54	10^{14}		−		1
Tetracyanoethylene			0.26(max)	+		4
7,7,8,8-Tetracyanoquinodimethane			0.65	−		4
1,1,8,8-Tetracyanooctatetraene	1.42	10^{12}		−		1
Tetraphenylpentacene	1.62				0 to 150	1
Tetrathiotetracene	0.46	10^{4}				1
Triphenodioxazine	1.65	$5 \cdot 10^{14}$		−	20 to 140	1
Triphenyldiamine			$2 \cdot 10^{-2}$		at 20	1
Violanthrene	0.85	$2.1 \cdot 10^{14}$			40 to 105	1
Isoviolanthrene	0.82	$8.4 \cdot 10^{13}$			40 to 150	1
Violanthrone	0.78	$2.3 \cdot 10^{10}$			40 to 150	1
Isoviolanthrone	0.76	$5.7 \cdot 10^{9}$			40 to 150	1
o-Xylene (liquid)	0.45					2
m-Xylene (liquid)	0.41					2
p-Xylene (liquid)	0.41					2

Long-Chain Compounds and Polymers

Substance	Energy Gap, E, (in $E/2 kT$) eV	Room Temperature Electrical Resistivity, ohm·cm	Mobility, μ, cm²/V·s	Sign of Majority Carriers	Temperature Range, °C	Ref.
Acrylic acid-amylproparylaniline copolymers		10^{9}–10^{10}				3
Acrylic acid-methylproparylaniline copolymers		10^{9}–10^{10}				3
Acrylic acid-octylproparylaniline copolymers		10^{9}–10^{10}				3
Anthrone polymers		0.28 ≥100 at 1.8 kbar				3
		≥2 at 33 kbar				
$[CH(AsF_5)_{0.1}]_x$		0.0005		+		3
$[CH \cdot I_{0.22}]_x$	1.9	$trans$ 10^{5}, cis 10^{9}				3
1,6-Diacetylenes (cyclopolymerized)		10^{10}–10^{14}				3
Ionene elastomers		$2.7 \cdot 10^{7}$ to $2.2 \cdot 10^{8}$			−80 to 60	3
1,3,4-Oxydiazole polymers	0.81	$3 \cdot 10^{12}$			20 to 140	3
Oxypyrrole polymer films		0.125				3
Phenylformaldehyde polymeric pyrolysates						3
a) Pyrolysis Temperature 600°C		27	0.0014	−		
b) 1200°C		0.0044	7.84	+		
Phenylthiocyanate polymers	0.5 to 0.8	10^{5}–10^{8}				3
Polyacetylene (undoped)		10^{10}				3
Polyacetylene (I$_2$ doped)		0.04				3
Polyacetylene (cis-rich, undoped)		10^{7}				3
$trans$-Polyacetylene (I$_2$ doped, 0.22 mole %)				+		3
Polyacrylonitrile (heat treated 700°C)			0.01	−	−100 to 100	3
Poly-5,5′-biisatyl	air 0.84	air $2.6 \cdot 10^{9}$		+	20 to 140	3
thiophene-indophene	vacuum 1.0	vacuum $3.1 \cdot 10^{9}$				
Poly bis(amino)-phosphazenes	1.75	$1.8 \cdot 10^{11}$			20 to 180	3
Poly-5,5′-diisatylmetane-thiophene-indophenine	0.45	$7.3 \cdot 10^{4}$		+	20 to 140	3
Polyethylene	2.74				20 to 70	3
Polyethylene (low density)	0.17	$4 \cdot 10^{9}$			above T_g	3
Polyimide	2.84					3
Polymalonitrile	1.72			−		3
Poly(metalphthalocyanines) :Cu	0.12	$7 \cdot 10^{6}$				3
:Fe	0.15	$1.1 \cdot 10^{6}$				3
:Ni	0.46	100				3
:Sb		$3.1 \cdot 10^{6}$				3
:Zn	0.12	$5.3 \cdot 10^{3}$				3
Poly-N-methylpyrrole		$2 \cdot 10^{6}$				3
Polyoxypyrrole (black)	0.044	1790			−173 to 27	3
Polyphthalocyanines	0.01	7 to 58				3
Polypyrrole	0.01				−193 to 250	3

Substance	Energy Gap, E, (in $E/2\,kT$) eV	Room Temperature Electrical Resistivity, ohm · cm	Mobility, μ, cm²/V · s	Sign of Majority Carriers	Temperature Range, °C	Ref.
Polypyrroline II	1.74					3
Polyselenomethylene	0.7 to 2.62	$>10^{13}$			20 to 120	3
Poly(2-vinylpyridine):I_2 (1:2)	0.12	1000			−73 to 27	3
PVC (commercial)	2.84–3.04				$T<T_g$	3
PVC (commercial)	1.24–1.96				$T>T_g$	3
PVC (pure)	1.0±0.1				0 to 30	3
Salicylal-N-alkyliminate-Cu	1.62	$1.7 \cdot 10^{14}$				4
TTF-acetylacetonate polymers		$1.6 \cdot 10^4$				3
TTF-metal polymers		$1.6 \cdot 10^4$				3

References

1. F. Gutman and L. E. Lyons, *Organic Semiconductors*, John Wiley & Sons, New York, 1967.
2. Y. Okamoto and W. Brenner, *Organic Semiconductors*, Reinhold Publ. Corp., New York, 1964.
3. F. Gutman, H. Keyzer, L. E. Lyons, and R. B. Somoano, *Organic Semiconductors, Part B*, R. E. Krieger Publ. Co., Melbourne, FL, 1983.
4.. L. I. Berger, *Semiconductor Materials*, CRC Press, Boca Raton, FL, 1997.

DIFFUSION DATA FOR SEMICONDUCTORS

B. L. Sharma

The diffusion coefficient D in many semiconductors may be expressed by an Arrhenius-type relation

$$D = D_o \exp(-Q/kT)$$

where D_o is a frequency factor, Q is the activation energy for diffusion, k is the Boltzmann constant, and T is the absolute temperature. This table lists D_o and Q for various diffusants in common semiconductors.

Abbreviations used in the table are

AES – Auger Electron Spectroscopy
DLTS – Deep Level Transient Spectroscopy
SEM – Scanning Electron Microscopy
SIMS – Secondary Ion Mass Spectrometry
$D(c)$ – Concentration Dependent Diffusion Coefficient
D_{max} – Maximum Diffusion Coefficient
(f) – Fast Diffusion Component
(i) – Interstitial Diffusion Component
(s) – Slow Diffusion Component
($\parallel$) – Parallel to c Direction
($\perp$) – Perpendicular to c Direction

Semiconductor	Diffusant	Frequency factor, D_o (cm²/s)	Activation energy, Q(eV)	Temperature range (°C)	Method of measurement	Ref.
Si	H	6×10^{-1}	1.03	120–1207	Electrical and SIMS	1
	Li	2.5×10^{-3}	0.65	25–1350	Electrical	2
	Na	1.65×10^{-3}	0.72	530–800	Electrical and flame photometry	3
	K	1.1×10^{-3}	0.76	740–800	Electrical and flame photometry	3
	Cu	4×10^{-2}	1.0	800–1100	Radioactive	4
		4.7×10^{-3}	0.43 (i)	300–700	Radioactive	5
	Ag	2×10^{-3}	1.6	1100–1350	Radioactive	6
	Au	2.4×10^{-4}	0.39 (i)	700–1300	Radioactive	7
		2.75×10^{-3}	2.05 (s)			
	Be	$(D \sim 10^{-7})$	–	1050	Electrical	8
	Ca	$(D \sim 6 \times 10^{-14})$	–	1100	Electrical and SIMS	1
	Zn	1×10^{-1}	1.4	980–1270	Electrical	9
	B	2.46	3.59	1100–1250	Electrical	10
		2.4×10^{1}	3.87	840–1250	Electrical	11
	Al	1.38	3.41	1119–1390	Electrical	12
		1.8	3.2	1025–1175	Electrical	13
	Ga	3.74×10^{-1}	3.39	1143–1393	Electrical	12
		6×10^{1}	3.89	900–1050	Radioactive	14
	In	7.85×10^{-1}	3.63	1180–1389	Electrical	12
		1.94×10^{1}	3.86	1150–1242	Radioactive	15
	Tl	1.37	3.7	1244–1338	Electrical	12
		1.65×10^{1}	3.9	1105–1360	Electrical	16
	Sc	8×10^{-2}	3.2	1100–1250	Radioactive	1
	Ce	$(D \sim 3.9 \times 10^{-13})$	–	1050	SIMS	1
	Pr	2.5×10^{-7}	1.74	1100–1280	Electrical	1
	Pm	7.5×10^{-9}	1.2 (s)	730–1270	Radioactive	1
		4.2×10^{-12}	0.13 (f)			
	Er	2×10^{-3}	2.9	1100–1250	Radioactive	1
	Tm	8×10^{-3}	3.0	1100–1280	Radioactive	1
	Yb	2.8×10^{-5}	0.95	947–1097	Neutron activation	1
	Ti	1.45×10^{-2}	1.79	950–1200	DLTS	17
	C	3.3×10^{-1}	2.92	1070–1400	Radioactive	18
	Si (self)	1.54×10^{2}	4.65	855–1175	SIMS	19
		1.6×10^{3}	4.77	1200–1400	Radioactive	20
	Ge	3.5×10^{-1}	3.92	855–1000	Radioactive	21
		2.5×10^{3}	4.97	1030–1302	Radioactive	21
		7.55×10^{3}	5.08	1100–1300	SIMS	22
	Sn	3.2×10^{1}	4.25	1050–1294	Neutron activation	23
	N	2.7×10^{-3}	2.8	800–1200	Out Diffusion; SIMS	1

Semiconductor	Diffusant	Frequency factor, D_o (cm²/s)	Activation energy, Q(eV)	Temperature range (°C)	Method of measurement	Ref.
	P	2.02×10^1	3.87	1100–1250	Electrical	10
		1.1	3.4	900–1200	Radioactive	24
		7.4×10^{-2}	3.3	1130–1405	Electrical	25
	As	6.0×10^1	4.2	950–1350	Radioactive	26
		6.55×10^{-2}	3.44	1167–1394	Electrical	27
		2.29×10^1	4.1	900–1250	Electrical	28
	Sb	1.29×10^1	3.98	1190–1398	Radioactive	29
		2.14×10^{-1}	3.65	1190–1405	Electrical	27
	Bi	1.03×10^3	4.64	1220–1380	Electrical	16
		1.08	3.85	1190–1394	Electrical	27
	Cr	1×10^{-2}	1	1100–1250	Radioactive	30
	Mo	$(D \sim 2 \times 10^{-10})$	–	1000	DLTS	1
	W	$(D \sim 10^{-12})$	–	1100	DLTS	1
	O	7×10^{-2}	2.44	700–1250	SIMS	31
		1.4×10^{-1}	2.53	700–1160	SIMS	32
	S	5.95×10^{-3}	1.83	975–1200	Radioactive	33
	Se	9.5×10^{-1}	2.6	1050–1250	Electrical	34
	Te	5×10^{-1}	3.34	900–1250	SIMS	1
	Mn	6.9×10^{-4}	0.63	900–1200	Radioactive	35
	Fe	1.3×10^{-3}	0.68	30–1250	Radioactive	36
	Co	2×10^{-3}	0.69	700–1300	Radioactive	37
	Ni	2×10^{-3}	0.47	800–1300	Radioactive	38
	Ru	$(D \sim 5 \times 10^{-7}$ $- 5 \times 10^{-6})$	–	1000–1280	Electrical	1
	Rh	$(D \sim 10^{-6}$–$10^{-4})$	–	1000–1200	Electrical	39
	Pd	2.95×10^{-4}	0.22 (i)	702–1320	Nuclear Activation	1
	Pt	1.5×10^2	2.22	800–1000	Electrical	1
	Os	$(D \sim 2 \times 10^{-6})$	–	1280	Electrical	40
	Ir	4.2×10^{-2}	1.3	950–1250	Electrical	41
Ge	Li	1.3×10^{-3}	0.46	350–800	Electrical	42
		9.1×10^{-3}	0.57	800–500	Electrical	43
	Na	3.95×10^{-1}	2.03	700–850	Radioactive	44
	Cu	1.9×10^{-4}	0.18 (i)	750–900	Radioactive	45
		4×10^{-2}	0.99 (s)	600–700		
		4×10^{-3}	0.33 (i)	350–750	Radioactive	5
	Ag	4.4×10^{-2}	1.0 (i)	700–900	Radioactive	46, 47
		4×10^{-2}	2.23 (s)	800–900	Radioactive	48
	Au	2.25×10^2	2.5	600–900	Radioactive	49
	Be	5×10^{-1}	2.5	720–900	Electrical	50
	Mg	$(D \sim 8 \times 10^{-9})$	–	900	Electrical	1
	Zn	5	2.7	600–900	Radioactive and electrical	51
	Cd	1.75×10^9	4.4	760–915	Radioactive	52
	B	1.8×10^9	4.55	600–900	Electrical	51
	Al	1.0×10^3	3.45	554–905	SIMS	53
		$\sim 1.6 \times 10^2$	~ 3.24	750–850	Electrical	54
	Ga	1.4×10^2	3.35	554–916	SIMS	55
		3.4×10^1	3.1	600–900	Electrical	51
	In	1.8×10^4	3.67	554–919	SIMS	56
		3.3×10^1	3.02	700–855	Radioactive	57
	Tl	1.7×10^3	3.4	800–930	Radioactive	58
	Si	2.4×10^{-1}	2.9	650–900	(γ) resonance	59
	Ge (self)	2.48×10^1	3.14	549–891	Radioactive	60
		7.8	2.95	766–928	Radioactive	61
	Sn	1.7×10^{-2}	1.9	–	Radioactive	45
	P	3.3	2.5	600–900	Electrical	51
	As	2.1	2.39	700–900	Electrical	62
	Sb	3.2	2.41	700–855	Radioactive	57
		1.0×10^1	2.5	600–900	Radioactive and electrical	51

Semiconductor	Diffusant	Frequency factor, D_o (cm²/s)	Activation energy, Q (eV)	Temperature range (°C)	Method of measurement	Ref.
	Bi	3.3	2.57	650–850	–	63
	O	4×10^{-1}	2.08	–	Optical	64
	S	($D \sim 10^{-9}$)	–	920	–	65
	Se	($D \sim 10^{-10}$)	–	920	–	65
	Te	5.6	2.43	750–900	Radioactive	66
	Fe	1.3×10^{-1}	1.08	750–900	Radioactive	67
	Co	1.6×10^{-1}	1.12	750–850	Radioactive	47
	Ni	8×10^{-1}	0.9	670–900	Electrical	68
GaAs	Li	5.3×10^{-1}	1.0	250–500	Electrical and chemical	69
	Cu	3×10^{-2}	0.53	100–500	Radioactive	69
		6×10^{-2}	0.98	450–750	Ultrasonic	69
		1.5×10^{-3}	0.6	800–1000	Radioactive	69
	Ag	4×10^{-4}	0.8	500–1150	Radioactive	69
	Au	1×10^{-3}	1.0	740–1025	Radioactive	69
	Be	7.3×10^{-6}	1.2	800–990	Electrical	69
	Mg	4×10^{-5}	1.22	800–1200	Electrical	69
	Zn	1.5×10^1	2.49	600–980	Radioactive	69
		2.5×10^{-1}	3.0	750–1000	Radioactive	69
	Cd	1.3×10^{-3}	2.2	800–1100	Radioactive	69
		5×10^{-2}	2.43	868–1149	Radioactive	69
	Hg	($D \sim 5 \times 10^{-14}$)	–	1100	Radioactive	69
	Al	($D \sim 4 \times 10^{-18}$–10^{-14})	4.3	850–1100	AES	70
	Ga (self)	4×10^{-5}	2.6	1025–1100	Radioactive	69
		1×10^7	5.6	1125–1230	Radioactive	69
	In	($D \sim 7 \times 10^{-11}$)	–	1000	Radioactive	69
	C	($D \sim 1.04 \times 10^{-16}$)	–	825	SIMS	69
	Si	1.1×10^{-1}	2.5	850–1050	SIMS	69
	Ge	1.6×10^{-5}	2.06	650–850	SIMS	69
	Sn	6×10^{-4}	2.5	1060–1200	Radioactive	69
		1×10^{-5}	2	800–1000	Radioactive	69
	P	($D \sim 10^{-12}$–10^{-10})	2.9	800–1150	Reflectance measurements	69
	As (self)	7×10^{-1}	3.2	–	Radioactive	69
	Cr	2.04×10^{-6}	0.83 (f)	750–1000	SIMS	69
			1.7 (s)	700–900		
		7.9×10^{-3}	2.2	800–1100	Chemical analysis	69
	O	2×10^{-3}	1.1	700–900	Mass spectroscopy	69
	S	1.85×10^{-2}	2.6	1000–1300	Radioactive	69
		1.1×10^1	2.95	750–900	Electrical	69
	Se	3×10^3	4.16	1025–1200	Radioactive	69
	Te	1.5×10^{-1}	3.5	1000–1150	Radioactive	69
	Mn	6.5×10^{-1}	2.49	850–1100	Radioactive	69
	Fe	4.2×10^{-2}	1.8	850–1150	Radioactive	69
		2.2×10^{-3}	2.32	750–1050	Radioactive	69
	Co	5×10^2	2.5	800–1000	Radioactive	69
		1.2×10^{-1}	2.64	750–1050	Radioactive	69
	Tm	2.3×10^{-16}	1.0	800–1000	Radioactive	69
GaSb	Li	2.3×10^{-4}	1.9 (s)	527–657	Electrical and flame photometry	69
		1.2×10^{-1}	0.7 (f)	277–657		
	Cu	4.7×10^{-3}	0.9	470–650	Radioactive	69
	Zn	($D \sim 2 \times 10^{-13}$ – 1×10^{-11})	2	510–600	Radioactive	69
	Cd	1.5×10^{-6}	0.72	640–800	Electrical	69
	Ga (self)	3.2×10^3	3.15	658–700	Radioactive	69
	In	1.2×10^{-7}	0.53	320–650	Radioactive	69
	Sn	2.4×10^{-5}	0.8	320–650	Radioactive	69
		1.3×10^{-5}	1.1	500–650	Radioactive	69
	Sb (self)	3.4×10^4	3.45	658–700	Radioactive	69
	Se	($D \sim 2.4 \times 10^{-13}$ – 1.37×10^{-11})	–	400–500	Radioactive	69

Semiconductor	Diffusant	Frequency factor, D_o (cm²/s)	Activation energy, Q(eV)	Temperature range (°C)	Method of measurement	Ref.
	Te	3.8×10^{-4}	1.20	320–650	Radioactive	69
	Fe	5×10^{-2}	1.9 (I)	500–650	Radioactive	69
		5×10^2	2.3 (II)	500–650		
GaP	Ag	–	–	1000–1300	Radioactive	69
	Au	8	2.5 (I)	1050–1250	Radioactive	69
		20	2.4 (II)	1100–1250	Diffusion (I) A face and (II) B face	
	Be	$(D_{max} \sim 2.4 \times 10^{-9} - 8.5 \times 10^{-8})$	–	900–1000	Atomic absorption analysis	69
	Mg	5×10^{-5}	1.4	700–1050	Electrical	69
	Zn	1.0	2.1	700–1300	Radioactive	69
	Ge	–	–	900–1000	Radioactive	69
	Cr	6.2×10^{-4}	1.2	900–1130	Radioactive; ESR	69
	S	3.2×10^3	4.7	1120–1305	Radioactive	69
	Mn	2.1×10^9	4.7	T < 950	Radioactive; ESR	69
		1.1×10^{-6}	0.9	950–1130		
	Fe	1.6×10^{-1}	2.3	980–1180	Radioactive	69
	Co	2.8×10^{-3}	2.9	850–1100	Radioactive	69
InP	Cu	3.8×10^{-3}	0.69	600–900	Radioactive	69
	Ag	3.6×10^{-4}	0.59	500–900	Radioactive	69
	Au	1.32×10^{-5}	0.48	600–820	Radioactive	69
		1.37×10^{-4}	0.73	600–900	Radioactive	69
	Zn	1.6×10^{-8}	0.3	750–900	Electrical	69
		$(D \sim 2 \times 10^{-9} - 4 \times 10^{-8})$	–	700–900	Radioactive	69
	Cd	1.8	1.9	700–900	Radioactive	69
		1.1×10^{-7}	0.72	700–900	Electrical	69
		$(D \sim 7 \times 10^{-13} - 2 \times 10^{-10})$	–	450–650	Electrical	69
	In (self)	1×10^5	3.85	830–990	Radioactive	69
	Sn	$(D \sim 3 \times 10^{-8})$	–	550	Etching and cathodoluminescence	69
	P (self)	7×10^{10}	5.65	900–1000	Radioactive	69
	Cr	–	–	600–900	Radioactive	69
	S	3.6×10^{-4}	1.94	585–708	Electrical	69
	Se	$(D \sim 2 \times 10^{-8})$	–	550	Cathodoluminescence	69
	Mn	–	2.9	650–750	SIMS	69
	Fe	3	2	600–950	Radioactive	69
		6.8×10^5	3.4	600–700	SIMS	69
	Co	9×10^{-1}	1.8	600–950	Radioactive	69
InAs	Cu	3.6×10^{-3}	0.52	342–875	Radioactive	69
		2.2×10^{-2}	0.54	525–890	Radioactive	69
	Ag	7.3×10^{-4}	0.26	450–900	Radioactive	69
	Au	5.8×10^{-3}	0.65	600–900	Radioactive	69
	Mg	1.98×10^{-6}	1.17	600–900	Electrical	69
	Zn	4.2×10^{-3}	0.96	600–900	Radioactive	69
		3.11×10^{-3}	1.17	600–900	Electrical	69
	Cd	7.4×10^{-4}	1.15	650–900	Radioactive	69
	Hg	1.45×10^{-5}	1.32	650–850	Radioactive	69
	In (self)	6×10^5	4.0	740–900	Radioactive	69
	Ge	3.74×10^{-6}	1.17	600–900	Electrical	69
	Sn	1.49×10^{-6}	1.17	600–900	Electrical	69
	As (self)	3×10^7	4.45	740–900	Radioactive	69
	S	6.78	2.2	600–900	Electrical	69
	Se	12.6	2.2	600–900	Electrical	69
	Te	3.43×10^{-5}	1.28	600–900	Electrical	69
InSb	Li	7×10^{-4}	0.28	0–210	Electrical	69
	Cu	9×10^{-4}	1.08	200–500	Radioactive	69
		3×10^{-5}	0.37	230–490	Radioactive	69
	Ag	1×10^{-7}	0.25	440–510	Radioactive	69
	Au	7×10^{-4}	0.32	140–510	Radioactive	69
	Zn	5×10^{-1}	1.35	362–508	Radioactive	69

Semiconductor	Diffusant	Frequency factor, D_o (cm²/s)	Activation energy, Q(eV)	Temperature range (°C)	Method of measurement	Ref.
		–	1.5	355–455	SIMS	69
	Cd	1×10^{-5}	1.1	250–500	Radioactive	69
		1.3×10^{-4}	1.2	360–500	Electrical	69
	Hg	4×10^{-6}	1.17	425–500	Radioactive	69
	In (self)	6×10^{-7}	1.45	400–500	Radioactive	69
		1.8×10^{13}	4.3	475–517	Radioactive	69
	Sn	5.5×10^{-8}	0.75	390–512	Radioactive	69
	Pb	($D \sim 2.7 \times 10^{-15}$)	–	500	Radioactive	71
	Sb (self)	5.35×10^{-4}	1.91	400–500	Radioactive	69
		3.1×10^{13}	4.3	475–517	Radioactive	69
	S	9×10^{-2}	1.4	360–500	Electrical	69
	Se	1.6	1.87	380–500	Electrical	69
	Te	1.7×10^{-7}	0.57	300–500	Radioactive	69
	Fe	1×10^{-7}	0.25	440–510	Radioactive	69
	Co	2.7×10^{-11}	0.39	420–500	Radioactive	69
AlAs	Ga	($D \sim 2 \times 10^{-18} - 10^{-15}$)	3.6	850–1100	AES	70
	Zn	($D \sim 9 \times 10^{-11}$)	–	557	SEM	69
AlSb	Cu	3.5×10^{-3}	0.36	150–500	Radioactive	69
	Zn	3.3×10^{-1}	1.93	660–860	Radioactive	69
	Cd	$D(c) \sim 4 \times 10^{-12} - 3 \times 10^{-10}$	–	900	Radioactive	69
	Al (self)	2	1.88	570–620	X-ray	69
	Sb (self)	1	1.7	570–620	X-ray	69
ZnS	Cu	2.6×10^{-3}	0.79	470–750	Radioactive	69
		4.3×10^{-4}	0.64	250–1200	Electroluminescence	69
		9.75×10^{-3}	1.04	400–800	Luminescence	69
	Au	1.75×10^{-4}	1.16	500–800	Radioactive	69
	Zn (self)	3×10^{-4}	1.5	925<T<940	Radioactive	69
		1.5×10^4	3.26	940<T<1030		
		1×10^{16}	6.5	1030<T<1075		
	Cd	($D \sim 10^{-10}$)	–	1100	Luminescence	72
	Al	5.69×10^{-4}	1.28	800–1000	Luminescence	69
	In	3×10^1	2.2	750–1000	Radioactive	69
	S (self)	2.16×10^4	3.15	600–800	Radioactive	69
		8×10^{-5}	2.2	740–1100	Radioactive	69
	Se	($D \sim 5 \times 10^{-13}$)	–	1070	X-ray microprobe	69
	Mn	2.3×10^3	2.46	500–800	Radioactive	69
ZnSe	Li	2.66×10^{-6}	0.49	950–980	Electrical	69
	Cu	1×10^{-4}	0.66	400–800	Luminescence	69
		1.7×10^{-5}	0.56	200–570	Radioactive	69
	Ag	2.2×10^{-2}	1.18	400–800	Luminescence	69
	Zn (self)	9.8	3.0	760–1150	Radioactive	69
	Cd	6.39×10^{-4}	1.87	700–950	Photoluminescence	69
	Al	2.3×10^{-2}	1.8	800–1100	Luminescence	69
	Ga	1.81×10^2	3.0	900–1100	Luminescence	69
		–	1.3	700–850	Electron probe	69
	In	($D \sim 2 \times 10^{-12}$)	–	940	–	69
	S	($D \sim 8 \times 10^{-12}$)	–	1060	X-ray microprobe	69
	Se (self)	1.3×10^1	2.5	860–1020	Radioactive	69
		2.3×10^{-1}	2.7	1000–1050	Radioactive	69
	Ni	($D \sim 1.5 \times 10^{-8} - 1.7 \times 10^{-7}$)	–	740–910	Luminescence	69
ZnTe	Li	2.9×10^{-2}	1.22 (s)	400–700	Nuclear and chemical analysis	69
		1.7×10^{-4}	0.78 (f)			
	Zn (self)	2.34	2.56	760–860	Radioactive	69
		1.4×10^1	2.69	667–1077	Radioactive	69
	Al	–	2.0	700–1000	Electrical and optical	69
	In	4	1.96	1100–1300	Radioactive	69
	Te (self)	2×10^4	3.8	727–977	Radioactive	69
CdS	Li	3×10^{-6}	0.68	610–960	Microhardness	69
	Na	($D \sim 3 \times 10^{-7}$)	–	800	Radioactive	69

Semiconductor	Diffusant	Frequency factor, D_o (cm²/s)	Activation energy, Q(eV)	Temperature range (°C)	Method of measurement	Ref.
	Cu	1.5×10^{-3}	0.76	400–700	Radioactive	69
		1.2×10^{-2}	1.05	300–700	Ultrasonic	69
		8×10^{-5}	0.72	20–200	Electrical	69
	Ag	2.5×10^{1}	1.2 (s)	300–500	Radioactive	69
		2.4×10^{-1}	0.8 (f)			
	Au	2×10^{2}	1.8	500–800	Radioactive	69
	Zn	1.27×10^{-9}	0.86 (s)	720–1000	Radioactive	69
		1.22×10^{-8}	0.66 (f)			
	Cd (self)	3.4	2.0	700–1100	Radioactive	69
	Ga	–	–	667–967	Optical and microprobe	69
	In	6×10^{1}	2.3 (‖)	650–930	Radioactive, optical and microprobe	69
		1×10^{1}	2.03 (⊥)			
	P	6.5×10^{-4}	1.6	800–1100	Radioactive	69
	S (self)	1.6×10^{-2}	2.05	800–900	Radioactive	69
		–	2.4	750–1050	Radioactive	69
	Se	$(D \sim 1.2 \times 10^{-9})$	–	900	Radioactive	69
	Te	1.3×10^{-7}	10.4	700–1000	Radioactive	69
	Cl	$(D \sim 3 \times 10^{-10})$	–	800	Electrical	69
	I	$(D \sim 5 \times 10^{-12})$	–	1000	Radioactive	69
	Ni	6.75×10^{-3}	10.9	570–900	Luminescence	69
	Yb	$(D \sim 1.3 \times 10^{-9})$	–	960	Photoluminescence	69
CdSe	Ag	2×10^{-4}	0.53	22–400	Ultrasonic	69
	Cd (self)	1.6×10^{-3}	1.5	700–1000	Radioactive	69
		6.3×10^{-2}	1.25 (I)	600–900	Radioactive;	69
		4.12×10^{-2}	2.18 (II)	600–900	(I) saturated Cd and (II) saturated Se pressure	
	P	$(D \sim 5.3 \times 10^{-12} - 6 \times 10^{-11})$	–	900–1000	Radioactive	69
	Se (self)	2.6×10^{3}	1.55	700–1000	Radioactive; saturated Se pressure	69
CdTe	Li	$(D \sim 1.5 \times 10^{-10})$	–	300	Ion microprobe	69
	Cu	3.7×10^{-4}	0.67	97–300	Radioactive	69
		8.2×10^{-8}	0.64	290–350	Ion backscattering	69
	Ag	–	–	700–800	Electrical and photoluminescence	69
	Au	6.7×10^{1}	2.0	600–1000	Radioactive	69
	Cd (self)	1.26	2.07	700–1000	Radioactive	69
		3.26×10^{2}	2.67 (I)	650–900	Radioactive;	69
		1.58×10^{1}	2.44 (II)		(I) saturated Cd and (II) saturated Te pressure	
	In	8×10^{-2}	1.61	650–1000	Radioactive	69
		1.17×10^{2}	2.21 (I)	500–850	Radioactive; (I) saturated Cd and (II) saturated Te pressure	69
		6.48×10^{-4}	1.15 (II)			
	Sn	8.3×10^{-2}	2.2	700–925	Radioactive	69
	P	$(D \sim 1.2 \times 10^{-10})$	–	900	Radioactive	69
	As	–	–	850	–	69
	O	5.6×10^{-9}	1.22	200–650	Mass spectrometry	69
		6.0×10^{-10}	0.29	650–900		
	Se	1.7×10^{-4}	1.35	700–1000	Radioactive	69
	Te (self)	8.54×10^{-7}	1.42 (I)	600–900	Radioactive; (I) saturated Cd and (II) saturated Te pressure	69
		1.66×10^{-4}	1.38 (II)	500–800		
	Cl	7.1×10^{-2}	1.6	520–800	Radioactive	69
	Fe	$(D \sim 4 \times 10^{-8})$	0.77	900	Radioactive	69

Semiconductor	Diffusant	Frequency factor, D_o (cm²/s)	Activation energy, Q(eV)	Temperature range (°C)	Method of measurement	Ref.
HgSe	Sb	6.3×10^{-5}	0.85	540–630	Radioactive	69
	Se (self)	–	–	200–400	Radioactive	69
HgTe	Ag	6×10^{-4}	0.8	250–350	Radioactive	69
	Zn	5×10^{-8}	0.6	250–350	Radioactive	69
	Cd	3.1×10^{-4}	0.66	250–350	Radioactive	69
	Hg (self)	2×10^{-8}	0.6	200–350	Radioactive	69
	In	6×10^{-6}	0.9	200–300	Radioactive	69
	Sn	1.72×10^{-6}	0.66 (s)	200–300	Radioactive	69
		1.8×10^{-3}	0.80 (f)			
	Te (self)	10^{-6}	1.4	200–400	Radioactive	69
	Mn	1.5×10^{-4}	1.3	250–350	Radioactive	69
PbS	Cu	4.6×10^{-4}	0.36	150–450	Electrical	69
		5×10^{-3}	0.31	100–400	Electrical	69
	Pb (self)	8.6×10^{-5}	1.52	500–800	Radioactive	69
	S (self)	6.8×10^{-5}	1.38	500–750	Radioactive	69
	Ni	1.78×10^{1}	0.95	200–500	Electrical	69
PbSe	Na	1.5×10^{1}	1.74 (s)	400–850	Radioactive	69
		5.6×10^{-6}	0.4 (f)			
	Cu	2×10^{-5}	0.31	93–520	Radioactive	69
	Ag	7.4×10^{-4}	0.35	400–850	Radioactive	69
	Pb (self)	4.98×10^{-6}	0.83	400–800	Radioactive	69
	Sb	3.4×10^{-1}	2.0	650–850	Radioactive	69
	Se (self)	2.1×10^{-5}	1.2	650–850	Radioactive	69
	Cl	1.6×10^{-8}	0.45	400–850	Radioactive	69
	Ni	($D \sim 1 \times 10^{-10}$)	–	700	Radioactive	69
PbTe	Na	1.7×10^{-1}	1.91	600–850	Radioactive	69
	Sn	3.1×10^{-2}	1.56	500–800	Radioactive	69
	Pb (self)	2.9×10^{-5}	0.6	250–500	Radioactive	69
	Sb	4.9×10^{-2}	1.54	500–800	Radioactive	69
	Te	2.7×10^{-6}	0.75	500–800	Radioactive	69
	Cl	($D > 2.3 \times 10^{-10}$)	–	700	Radioactive	69
	Ni	($D > 1 \times 10^{-6}$)	–	700	Radioactive	69

References

1. N. A. Stolwijk and H. Bracht, in *Diffusion in Semiconductors and Non-Metallic Solids*, D. L. Beke, Ed., Springer-Verlag, Berlin, 1998, 2-1.
2. E. M. Pell, *Phys. Rev.*, 119, 1960; 119, 1014, 1960.
3. L. Svob, *Solid State Electron*, 10, 991, 1967.
4. B. I. Boltaks and I. I. Sosinov, *Zh. Tekh. Fiz.*, 28, 3, 1958.
5. R. N. Hall and J. N. Racette, *J. Appl. Phys.*, 35, 379, 1964.
6. B. I. Boltaks and Hsueh Shih-Yin, *Sov. Phys. Solid State*, 2, 2383, 1961.
7. W. R. Wilcox and T. J. LaChapelle, *J. Appl. Phys.*, 35, 240, 1964.
8. E. A. Taft and R. O. Carlson, *J. Electrochem. Soc.*, 117, 711, 1970.
9. R. Sh. Malkovich and N. A. Alimbarashvili, *Sov. Phys. Solid State*, 4, 1725, 1963.
10. R. N. Ghoshtagore, *Solid State Electron*, 15, 1113, 1972.
11. C. Hill, *Semiconductor Silicon 1981*, H. R. Huff, R. J. Kreiger, and Y. Takeishi, Eds., p. 988, *Electrochem. Soc.*, 1981.
12. R. N. Ghoshtagore, *Phys. Rev. B*, 3, 2507, 1971.
13. W. Rosnowski, *J. Electrochem. Soc.*, 125, 957, 1978.
14. J. S. Makris and B. J. Masters, *J. Appl. Phys.*, 42, 3750, 1971.
15. M. F. Millea, *J. Phys. Chem. Solids*, 27, 315, 1965 (refer Reference 2).
16. C. S. Fuller and J. A. Ditzenberger, *J. Appl. Phys.*, 27, 544, 1956.
17. S. Hocine and D. Mathiot, *Appl. Phys. Lett.*, 53, 1269, 1988.
18. R. C. Newman and J. Wakefield, *J. Phys. Chem. Solids*, 19, 230, 1961.
19. L. Kalinowski and R. Seguin, *Appl. Phys. Lett.*, 35, 211, 1979; *Appl. Phys. Lett.*, 36, 171, 1980.
20. R. F. Peart, *Phys. Stat. Sol.*, 15, K 119, 1966.
21. G. Hettich, H. Mehrer and K. Maler, *Inst. Phys. Conf. Ser.*, 46, 500, 1979.
22. M. Ogina, Y. Oana and M. Watanabe, *Phys. Stat. Sol. (a)*, 72, 535, 1982.
23. T. H. Yeh, S. M. Hu, and R. H. Kastl, *Appl. Phys.*, 39, 4266, 1968.
24. I. Franz and W. Langheinrich, *Solid State Electron*, 14, 835, 1971.
25. R. N. Ghoshtagore, *Hys. Rev. B*, 3, 389, 1971.
26. B. J. Masters and J. M. Fairfield, *J. Appl. Phys.*, 40, 2390, 1969.
27. R. N. Goshtagore, *Phys. Rev. B*, 3, 397, 1971.
28. R. S. Fair and J. C. C. Tsai, *J. Electrochem. Soc.*, 122, 1689, 1975.
29. J. J. Rohan, N. E. Pickering, and J. Kennedy, *J. Electrochem. Soc.*, 106, 705, 1969.
30. W. Wuerker, K. Roy, and J. Hesse, *Matsr. Res. Bull.*, 9, 971, 1974.
31. J. C. Mikkelsen, Jr., *Appl. Phys. Lett.*, 40, 336, 1982.
32. S. Tang Lee and D. Nicols, *Appl. Phys. Lett.*, 47, 1001, 1985.
33. P. L. Gruzin, S. V. Zemskii, A. D. Bullkin, and N. M. Makarov, *Sov. Phys. Sem.*, 7, 1241, 1974.
34. N. S. Zhdanovich and Yu. I. Kozlov, *Svoistva Legir, Poluprovodn.*, V. S. Zemskov, Ed., Nauka, Moscow, 1977, 115–120; *Fiz Tekh. Poluprovod.*, 9, 1594, 1975.
35. D. Gilles, W. Bergholze, and W. Schroeter, *J. Appl. Phys.*, 59, 3590, 1986.
36. E. R. Weber, *Appl. Phys. A*, 30, 1, 1983.
37. E. R. Weber, Properties of Silicon, EMIS Datareviews Ser. No. 4, INSPEC Publications, 1988, 409–451.
38. M. K. Bakhadyrkhanov, S. Zainabidinov, and A. Khamidov, *Sov. Phys. Sem.*, 14, 243, 1980.
39. S. A. Azimov, M. S. Yunosov, F. K. Khatamkulov, and G. Nasyrov, *Poluprovod.*, N. Kh. Abrikosov and V. S. Zemskov, Eds., Nauka, Moscow, 1975, 21–23.

40. S. A. Azimov, M. S. Yunosov, G. Nurkuziev, and F. R. Karimov, *Sov. Phys. Sem.*, 12, 981, 1978.

41. S. A. Azimov, B. V. Umarov, and M. S. Yunusov, *Sov. Phys. Sem.*, 10, 842, 1976.

42. C. S. Fuller and J. A. Ditzenberger, *Phys. Rev.*, 91, 193, 1953.

43. B. Pratt and F. Friedman, *J. Appl. Phys.*, 37, 1893, 1966.

44. M. Stojic, V. Spiric, and D. Kostoski, *Inst. Phys. Conf. Ser.*, 31, 304, 1976.

45. B. I. Boltaks, *Diffusion in Semiconductors*, Inforsearch, London, 1963, 162.

46. A. A. Bugai, V. E. Kosenko, and E. G. Miselyuk, *Zh. Tekh. Fiz.*, 27, 67, 1957.

47. L. Y. Wei, *J. Phys. Chem. Solids*, 18, 162, 1961.

48. V. E. Kosenko, *Sov. Phys. Solid State*, 4, 42, 1962.

49. W. C. Dunlap, Jr., *Phys. Rev.*, 97, 614, 1955

50. Yu. I. Belyaev and V. A. Zhidkov, *Sov. Phys. Solid State*, 3, 133, 1961.

51. W. C. Dunlap, Jr. *Phys. Rev.*, 94, 1531, 1954.

52. V. E. Kosenko, *Sov. Phys. Solid State*, 1, 1481, 1960.

53. P. Dorner, W. Gust, A. Lodding, H. Odelius, B. Predel, and U. Roll, *Acta Metall.*, 30, 941, 1982.

54. W. Meer and D. Pommerrening, *Z. Agnew. Phys.*, 23, 369, 1967.

55. U. Sodervall, H. Odelius, A. Lodding, U. Roll, B. Predel, W. Gust, and P. Dorner, *Phil. Mag. A*, 54, 539, 1986.

56. P. Dorner, W. Gust, A. Lodding, H. Odelius, B. Predel, and U. Roll, *Z. Metalkd.*, 73, 325, 1982.

57. P. V. Pavlov, *Sov. Phys. Solid State*, 8, 2377, 1967.

58. V. I. Tagirov and A. A. Kuliev, *Sov. Phys. Solid State*, 4, 196, 1962.

59. J. Raisanen, J. Hirvonen, and A. Anttila, *Solid State Electron.*, 24, 333, 1981.

60. C. Vogel, G. Hettich, and H. Mehrer, *J. Phys. C.*, 16, 6197, 1983.

61. H. Letaw, Jr., W. M. Portnoy, and L. Slifkin, *Phys. Rev.*, 102, 363, 1956.

62. W. Bosenberg, *Z. Naturforsch.*, 10a, 285, 1955.

63. V. M. Glazov and V. S. Zemskov, Physicochemical Principles of Semiconductor Doping, Israel Program for Scientific Translation, Jerusalem, 1968.

64. J. W. Corbett, R. S. McDonald, and G. D. Watkins, *J. Phys. Chem. Solids*, 25, 873, 1964.

65. W. W. Tyler, *J. Phys. Chem. Solids*, 8, 59, 1959.

66. V. D. Ignatkov and V. E. Kosenko, *Sov. Phys. Solid State*, 4, 1193, 1962.

67. A. A. Bugal, V. E. Kosenko, and E. G. Miseluk, *Zh. Tekh. Fiz.*, 27, 210, 1957.

68. F. van der Maesen and J. A. Brenkman, *Phillips Res. Rep.*, 9, 255, 1954.

69. M. B. Dutt and B. L. Sharma, in *Diffusion in Semiconductors and Non-Metallic Solids*, D. L. Beke, Ed., Springer-Verlag, Berlin, 1998, 3-1.

70. L. L. Chang and A. Koma, *Appl. Physics Lett.*, 29, 138, 1976.

71. D. L. Kendall, *Semiconductors and Semimetals*, Vol. 4, R. K. Willardson and A. C. Beer, Eds., Academic, 1968, 255.

72. H. J. Biter and F. Williams, *J. Luminescence*, 3, 395, 1971.

PROPERTIES OF MAGNETIC MATERIALS

H. P. R. Frederikse

Glossary of Symbols

Quantity	Symbol	Units	
		SI	emu
Magnetic field	H	A m^{-1}	Oe (oersted)
Magnetic induction	B	T (tesla)	G (gauss)
Magnetization	M	A m^{-1}	emu cm^{-3}
Spontaneous magnetization	M_s	A m^{-1}	emu cm^{-3}
Saturation magnetization	M_0	A m^{-1}	emu cm^{-3}
Magnetic flux	Φ	Wb (weber)	maxwell
Magnetic moment	m, μ	A m^2	erg/G
Coercive field	H_c	A m^{-1}	Oe
Remanence	B_r	T	G
Saturation magnetic polarization	J_s	T	G
Magnetic susceptibility	χ		
Magnetic permeability	μ	H m^{-1} (henry/meter)	
Magnetic permeability of free space	μ_0	H m^{-1}	
Saturation magnetostriction	$\lambda \, (\Delta l/l)$		
Curie temperature	T_C	K	K
Néel temperature	T_N	K	K

Magnetic moment $\mu = \gamma \hbar J = g \, \mu_B J$
where

γ = gyromagnetic ratio; J = angular momentum; g = spectroscopic splitting factor (~2)
μ_B = Bohr magneton = $9.2741 \cdot 10^{-24}$ J/T = $9.2741 \cdot 10^{-21}$ erg/G

Earth's magnetic field $H = 56$ A m^{-1} = 0.7 Oe
For iron: $M_0 = 1.7 \cdot 10^6$ A m^{-1}; $B_r = 0.8 \cdot 10^6$ A m^{-1}
1 Oe = $(1000/4\pi)$ A m^{-1}; 1 G = 10^{-4} T; 1 emu cm^{-3} = 10^3 A m^{-1}
1 Maxwell = 10^{-8} Wb
$\mu_0 = 4\pi \, 10^{-7}$ H m^{-1}

Relation between Magnetic Induction and Magnetic Field

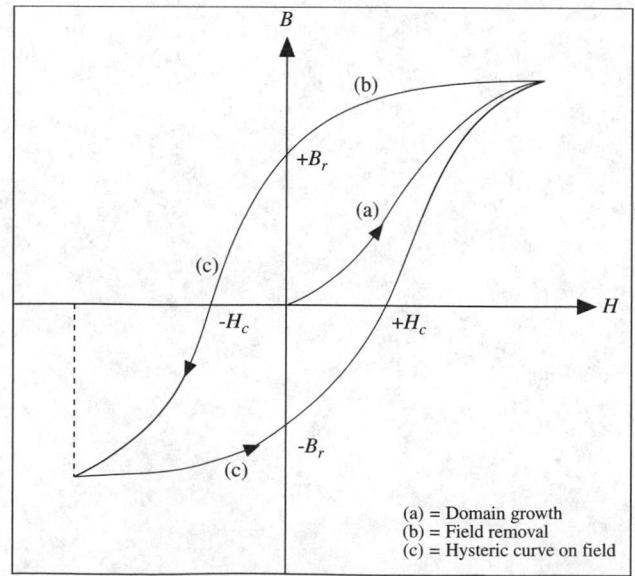

FIGURE 1. Typical curve representing the dependence of magnetic induction B on magnetic field H for a ferromagnetic material. When H is first applied, B follows curve (a) as the favorably oriented magnetic domains grow. This curve flattens as saturation is approached. When H is then reduced, B follows curve (b), but retains a finite value (the remanence B_r) at $H = 0$. In order to demagnetize the material, a negative field $-H_c$ (where H_c is called the coercive field or coercivity) must be applied. As H is further decreased and then increased to complete the cycle (curve c), a hysteresis loop is obtained. The area within this loop is a measure of the energy loss per cycle for a unit volume of the material.

(a) = Domain growth
(b) = Field removal
(c) = Hysteric curve on field

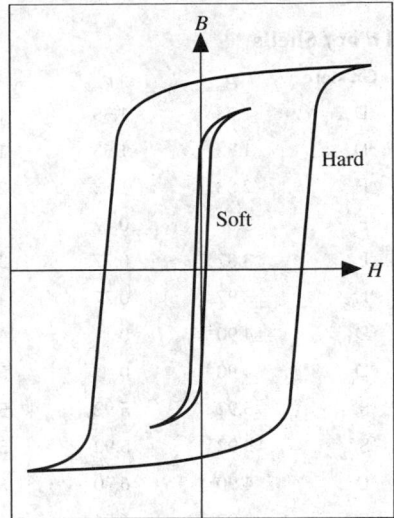

FIGURE 2. Schematic curve illustrating the *B* vs. *H* dependence for hard and soft magnetic materials. Hard materials have a larger remanence and coercive field, and a correspondingly large hysteresis loss.

Reference

Ralls, K. M., Courtney, T. H., and Wulff, J., *Introduction to Materials Science and Engineering*, J. Wiley & Sons, New York, 1976, p. 577, 582. With permission.

Magnetic Susceptibility of the Elements

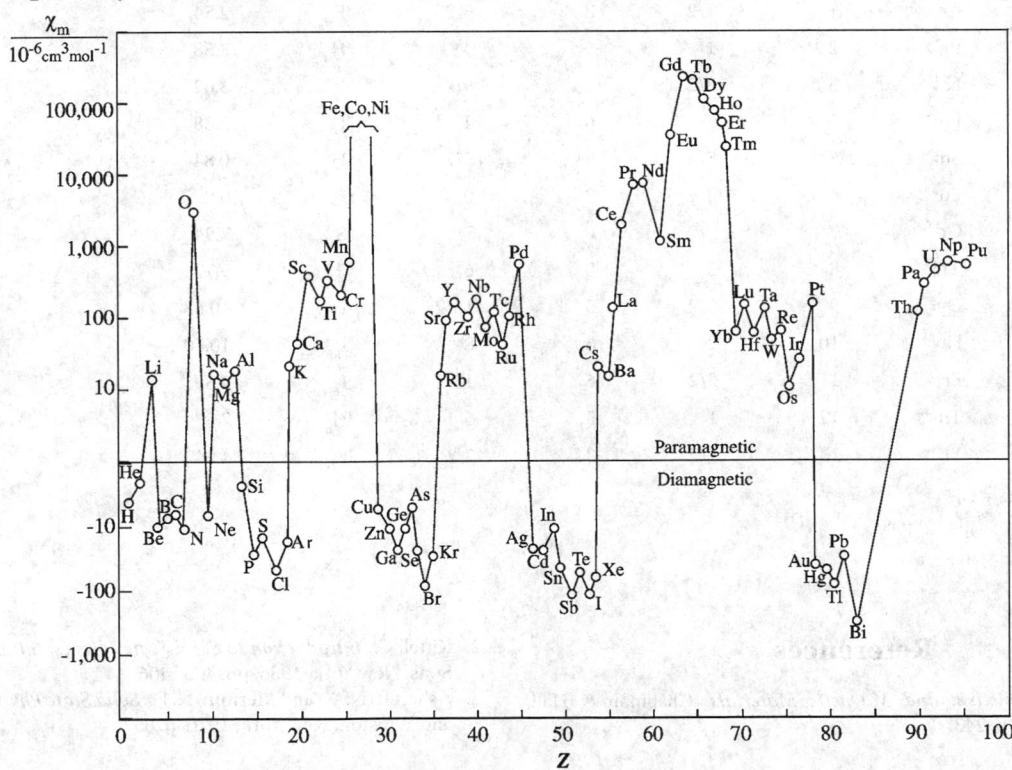

FIGURE 3. Molar susceptibility of the elements at room temperature (cgs units of 10^{-6} cm³/mol). Values are not available for Z = 9, 61, and 84–89; Fe, Co, and Ni (Z = 26–28) are ferromagnetic. Data taken from the table "Magnetic Susceptibility of the Elements and Inorganic Compounds" in Section 4.

Reference

Gray, D. E., Ed., *American Institute of Physics Handbook, Third Edition*, McGraw Hill, New York, 1972, p. 5–224. With permission.

Ground State of Ions with Partly Filled d or f Shells

Z	Element	n	S	L	J	Gr. state	p_{calc} [a]	p_{calc} [b]	p_{meas}
22	Ti^{3+}	1	1/2	2	3/2	$^2D_{3/2}$	1.73	1.55	1.8
23	V^{4+}	1	1/2	2	3/2	$^2D_{3/2}$	1.73	1.55	1.8
23	V^{3+}	2	1	3	2	$^3F_2$	2.83	1.63	2.8
23	V^{2+}	3	3/2	3	3/2	$^4F_{3/2}$	3.87	0.77	3.8
24	Cr^{3+}	3	3/2	3	3/2	$^4F_{3/2}$	3.87	0.77	3.7
25	Mn^{4+}	3	3/2	3	3/2	$^4F_{3/2}$	3.87	0.77	4.0
24	Cr^{2+}	4	2	2	0	$^5D_0$	4.90	0	4.9
25	Mn^{3+}	4	2	2	0	$^5D_0$	4.90	0	5.0
25	Mn^{2+}	5	5/2	0	5/2	$^6S_{5/2}$	5.92	5.92	5.9
26	Fe^{3+}	5	5/2	0	5/2	$^6S_{5/2}$	5.92	5.92	5.9
26	Fe^{2+}	6	2	2	4	$^5D_4$	4.90	6.70	5.4
27	Co^{2+}	7	3/2	3	9/2	$^4F_{9/2}$	3.87	6.54	4.8
28	Ni^{2+}	8	1	3	4	$^3F_4$	2.83	5.59	3.2
29	Cu^{2+}	9	1/2	2	5/2	$^2D_{5/2}$	1.73	3.55	1.9

Z	Element	n	S	L	J	Gr. state	p_{calc} [c]		p_{meas}
58	Ce^{3+}	1	1/2	3	5/2	$^2F_{5/2}$	2.54		2.4
59	Pr^{3+}	2	1	5	4	$^3H_4$	3.58		3.5
60	Nd^{3+}	3	3/2	6	9/2	$^4I_{9/2}$	3.62		3.5
61	Pm^{3+}	4	2	6	4	$^5I_4$	2.68		
62	Sm^{3+}	5	5/2	5	5/2	$^6H_{5/2}$	0.84		1.5
63	Eu^{3+}	6	3	3	0	$^7F_0$	0.0		3.4
64	Gd^{3+}	7	7/2	0	7/2	$^8S_{7/2}$	7.94		8.0
65	Tb^{3+}	8	3	3	6	$^7F_6$	9.72		9.5
66	Dy^{3+}	9	5/2	5	15/2	$^6H_{15/2}$	10.63		10.6
67	Ho^{3+}	10	2	6	8	$^5I_8$	10.60		10.4
68	Er^{3+}	11	3/2	6	15/2	$^4I_{15/2}$	9.59		9.5
69	Tm^{3+}	12	1	5	6	$^3H_6$	7.57		7.3
70	Yb^{3+}	13	1/2	3	7/2	$^2F_{7/2}$	4.54		4.5

[a] $p_{calc} = 2[S(S+1)]^{1/2}$

[b] $p_{calc} = 2[J(J+1)]^{1/2}$

[c] $p_{calc} = g[J(J+1)]^{1/2}$

References

1. Jiles, D., *Magnetism and Magnetic Materials*, Chapman & Hall, London, 1991, p. 243.

2. Kittel, C., *Introduction to Solid State Physics, 6th Edition*, J. Wiley & Sons, New York, 1986, pp. 405–406.

3. Ashcroft, N. W. and Mermin, N. D., *Solid State Physics*, Holt, Rinehart, and Winston, New York, 1976, p. 652.

Ferro- and Antiferromagnetic Elements

M_0 is the saturation magnetization at $T = 0$ K T_C is the Curie temperature
n_B is the number of Bohr magnetons per atom T_N is the Néel temperature

	M_0/gauss	n_B	T_C/K	T_N/K	Comments
Fe	22020	2.22	1043		
Co	18170	1.72	1388		
Ni	6410	0.62	627		
Cr				311	
Mn				100	
Ce				12.5	c-Axis antiferromagnetic
Nd				19.2	Basal plane modulation on hexagonal sites
				7.8	Cubic sites order (periodicity different from high-T phase)
Sm				106	Ordering on hexagonal sites
				13.8	Cubic site order
Eu				90.5	Spiral along cube axis
Gd	24880	7	293		
Tb		9	220		Basal plane ferromagnet
				230.2	Basal plane spiral
Dy		10	87		Basal plane ferromagnet
				176	Basal plane spiral
Ho		10	20		Bunched cone structure
				133	Basal plane spiral
Er		9	32		c-Axis ferrimagnetic cone structure
				80	c-Axis modulated structure
Tm		7	32		c-Axis ferrimagnetic cone structure
				56	c-Axis modulated structure

References

1. Ashcroft, N. W., and Mermin, N. D., *Solid State Physics*, Holt, Rinehart, and Winston, New York, 1976, p.652.

2. Gschneidner, K. A., and Eyring, L., *Handbook on the Physics and Chemistry of Rare Earths*, North Holland Publishing Co., Amsterdam, 1978.

Selected Ferromagnetic Compounds

M_0 is the saturation magnetization at $T = 293$ K T_C is the Curie temperature

Compound	M_0/gauss	T_C/K	Crystal system
MnB	152	578	orthorh(FeB)
MnAs	670	318	hex(FeB)
MnBi	620	630	hex(FeB)
MnSb	710	587	hex(FeB)
Mn_4N	183	743	
MnSi		34	cub(FeSi)
CrTe	247	339	hex(NiAs)
$CrBr_3$	270	37	hex(BiI_3)
CrI_3		68	hex(BiI_3)
CrO_2	515	386	tetr(TiO_2)
EuO	1910*	77	cub
EuS	1184*	16.5	cub
$GdCl_3$	550*	2.2	orthorh
FeB		598	orthorh
Fe_2B		1043	tetr ($CuAl_2$)
$FeBe_5$		75	cub($MgCu_2$)
Fe_3C		483	orthorh
FeP		215	orthorh (MnP)

* At $T = 0$ K

References

1. Kittel, C., *Introduction to Solid State Physics, 6th Edition*, J. Wiley & Sons, New York, 1986.

2. Ashcroft, N. W., and Mermin, N. D., *Solid State Physics*, Holt, Rinehart, and Winston, New York, 1976.

Magnetic Properties of High-Permeability Metals and Alloys (Soft)

μ_i is the initial permeability
μ_m is the maximum permeability
H_c is the coercive force

J_s is the saturation polarization
W_H is the hysteresis loss per cycle
T_C is the Curie temperature

Material	Composition (mass %)	μ_i/μ_0	μ_m/μ_0	H_c/A m^{-1}	J_s/T	W_H/J m^{-3}	T_C/K
Iron	Commercial 99Fe	200	6000	70	2.16	500	1043
Iron	Pure 99.9Fe	25000	350000	0.8	2.16	60	1043
Silicon-iron	96Fe-4Si	500	7000	40	1.95	50–150	1008
Silicon-iron (110) [001]	97Fe-3Si	9000	40000	12	2.01	35–140	1015
Silicon-iron {100} <100>	97Fe-3Si		100000	6	2.01		1015
Mild steel	Fe-0.1C-0.1Si-0.4Mn	800	1100	200			
Hypernik	50Fe-50Ni	4000	70000	4	1.60	22	753
Deltamax {100} <100>	50Fe-50Ni	500	200000	16	1.55		773
Isoperm {100} <100>	50Fe-50Ni	90	100	480	1.60		
78 Permalloy	78Ni-22Fe	4000	100000	4	1.05	50	651
Supermalloy	79Ni-16Fe-5Mo	100000	1000000	0.15	0.79	2	673
Mumetal	77Ni-16Fe-5Cu-2Cr	20000	100000	4	0.75	20	673
Hyperco	64Fe-35Co-0.5Cr	650	10000	80	2.42	300	1243
Permendur	50Fe-50Co	500	6000	160	2.46	1200	1253
2V-Permendur	49Fe-49Co-2V	800	4000	160	2.45	600	1253
Supermendur	49Fe-49Co-2V		60000	16	2.40	1150	1253
25Perminvar	45Ni-30Fe-25Co	400	2000	100	1.55		
7Perminvar	70Ni-23Fe-7Co	850	4000	50	1.25		
Perminvar (magnet. annealed)	43Ni-34Fe-23Co		400000	2.4	1.50		
Alfenol (or Alperm)	84Fe-16Al	3000	55000	3.2	0.8		723
Alfer	87Fe-13Al	700	3700	53	1.20		673
Aluminum-Iron	96.5Fe-3.5Al	500	19000	24	1.90		
Sendust	85Fe-10Si-5Al	36000	120000	1.6	0.89		753

References

1. McCurrie, R. A., *Structure and Properties of Ferromagnetic Materials*, Academic Press, London, 1994, p. 42.

2. Gray, D. E., Ed., *American Institute of Physics Handbook, Third Edition*, McGraw Hill, New York, 1972, p. 5–224.

Applications of High-Permeability Materials

Applications	Requirements
Power applications	
Distribution and power transformers	Low core losses, high permeability, high saturation magnetic polarization
High-quality motors and generators, stators and armatures, switched-mode power supplies	
Instrument transformers	
Audiofrequency transformers	Low core losses, high permeability, high magnetic polarization
Pulse transformers	High permeability
Cores for inductor coils	
Audiofrequency	Low hysteresis, high permeability
Carrier frequency	Very low hysteresis and eddy current loss
Radiofrequency	High permeability at low fields
Miscellaneous	
Relays, switches } Earth leakage circuit }	High permeability, low remanence, low coercivity
Magnetic shielding	Low core loss for AC applications

Applications of High-Permeability Materials

Applications	Requirements
Magnetic recording heads	High initial permeability, low or zero remanence
Magnetic amplifiers Saturable reactors Saturable transformers Transformer cores	Rectangular hysteresis loops, low hysteresis loss
Magnetic shunts for temperature compensation in magnetic circuits	Low Curie temperature, appropriate decrease in permeability with increase in temperature
Electromagnets in indicating instruments, fire detection, quartz watches, electromechanical devices	High permeability, high saturation magnetic polarization
Magnetic yokes in permanent magnet devices, such as lifting and holding magnets, loudspeakers	High permeability, high saturation magnetic polarization

Reference

McCurrie, R. A., *Structure and Properties of Ferromagnetic Materials*, Academic Press, London, 1994. With permission.

Saturation Magnetostriction of Selected Materials

The tabulated parameter λ_s is related to the fractional change in length $\Delta l/l$ by $\Delta l/l = (3/2)\lambda_s(\cos^2\theta - 1/3)$, where θ is the angle of rotation.

Material	$\lambda_s \times 10^6$
Iron	−7
Fe - 3.2% Si	+9
Nickel	−33
Cobalt	−62
45 Permalloy, 45% Ni - 55% Fe	+27
Permalloy, 82% Ni - 18% Fe	0
Permendur, 49% Co - 49% Fe - 2% V	+70
Alfer, 87% Fe - 13% Al	+30
Magnetite, Fe_3O_4	+40
Cobalt ferrite, $CoFe_2O_4$	−110
$SmFe_2$	−1560
$TbFe_2$	+1753
$Tb_{0.3}Dy_{0.7}Fe_{1.93}$ (Terfenol D)	+2000
$Fe_{66}Co_{18}B_{15}Si$ (amorphous)	+35
$Co_{72}Fe_3B_6A_{13}$ (amorphous)	0

Reference

McCurrie, R.A., *Structure and Properties of Ferromagnetic Materials*, Academic Press, London, 1994, p. 91; additional data provided by A.E. Clark, Adelphi, MD.

Properties of Various Permanent Magnetic Materials (Hard)

B_r is the remanence
$_BH_c$ is the flux coercivity
$_iH_c$ is the intrinsic coercivity

$(BH)_{max}$ is the maximum energy product
T_C is the Curie temperature
T_{max} is the maximum operating temperature

Composition	B_r/T	$_BH_c/10^3$ A m^{-1}	$_iH_c/10^3$ A m^{-1}	$(BH)_{max}$/kJ m^{-3}	T_C/°C	T_{max}/°C
Alnico1 20Ni;12Al;5Co	0.72		35	25		
Alnico2 17Ni;10Al;12.5Co;6Cu	0.72		40–50	13–14		
Alnico3 24-30Ni;12-14Al;0-3Cu	0.5–0.6		40–54	10		
Alnico4 21-28Ni;11-13Al;3-5Co;2-4Cu	0.55–0.75		36–56	11–12		
Alnico5 14Ni;8Al;24Co;3Cu	1.25	53	54	40	850	520
Alnico6 16Ni;8Al;24Co;3Cu;2Ti	1.05		75	52		
Alnico8 15Ni;7Al;35Co;4Cu;5Ti	0.83	1.6	160	45		
Alnico9 15Ni;7Al;35Co;4Cu;5Ti	1.10	1.45	1.45	75	850	520
Alnico12 13.5Ni;8Al;24.5Co;2Nb	1.20		64	76.8		

Composition	B_r/T	$_B H_c$/10^3 A m^{-1}	$_J H_c$/10^3 A m^{-1}	$(BH)_{max}$/kJ m^{-3}	T_C/°C	T_{max}/°C
BaFe$_{12}$O$_{19}$ (Ferroxdur)	0.4	1.6	192	29	450	400
SrFe$_{12}$O$_{19}$	0.4	2.95	3.3	30	450	400
LaCo$_5$	0.91			164	567	
CeCo$_5$	0.77			117	380	
PrCo$_5$	1.20			286	620	
NdCo$_5$	1.22			295	637	
SmCo$_5$	1.00	7.9	696	196	700	250
Sm(Co$_{0.76}$Fe$_{0.10}$Cu$_{0.14}$)$_{6.8}$	1.04	4.8	5	212	800	300
Sm(Co$_{0.65}$Fe$_{0.28}$Cu$_{0.05}$Zr$_{0.02}$)$_{7.7}$	1.2	10	16	264	800	300
Nd$_2$Fe$_{14}$B sintered	1.22	8.4	1120	280	300	100
Fe;52Co;14V (Vicalloy II)	1.0	42		28	700	500
Fe;24Cr;15Co;3Mo (anisotropic)	1.54	67		76	630	500
Fe;28Cr;10.5Co (Chromindur II)	0.98	32		16	630	500
Fe;23Cr;15Co;3V;2Ti	1.35	4		44	630	500
Cu;20Ni;20Fe (Cunife)	0.55	4		12	410	350
Cu;21Ni;29Fe (Cunico)	0.34	0.5		8		
Pt;23Co	0.64	4		76	480	350
Mn;29.5Al;0.5C (anisotropic)	0.61	2.16	2.4	56	300	120

References

1. McCurrie, R. A., *Structure and Properties of Ferromagnetic Materials*, Academic Press, London, 1994, p. 204.

2. Gray, D. E., Ed., *American Institute of Physics Handbook, Third Edition*, McGraw Hill, New York, 1972, p. 5–165.

3. Jiles, D., *Magnetism and Magnetic Materials*, Chapman & Hall, London, 1991.

Selected Ferrites

J_s is the saturation magnetic polarization
T_C is the Curie temperature
ΔH is the line width

Material	J_s/T	T_C/°C	ΔH/kA m^{-1}	Applications
Spinels				
γ-Fe$_2$O$_3$	0.52	575		
Fe$_3$O$_4$	0.60	585		
NiFe$_2$O$_4$	0.34	575	350	Microwave devices
MgFe$_2$O$_4$	0.14	440	70	
NiZnFe$_2$O$_4$	0.50	375	120	Transformer cores
MnFe$_2$O$_4$	0.50	300	50	Microwave devices
NiCoFe$_2$O$_4$	0.31	590	140	Microwave devices
NiCoAlFe$_2$O$_4$	0.15	450	330	Microwave devices
NiAl$_{0.35}$Fe$_{1.65}$O$_4$	0.12	430	67	Microwave devices
NiAlFe$_2$O$_4$	0.05	1860	32	Microwave devices
Mg$_{0.9}$Mn$_{0.1}$Fe$_2$O$_4$	0.25	290	56	Microwave devices
Ni$_{0.5}$Zn$_{0.5}$Al$_{0.8}$Fe$_{1.2}$O$_4$	0.14		17	Microwave devices
CuFe$_2$O$_4$	0.17	455		Electromechanical transducers
CoFe$_2$O$_4$	0.53	520		
LiFe$_5$O$_8$	0.39	670		Microwave devices
Garnets				
Y$_3$Fe$_5$O$_{12}$	0.178	280	55	Microwave devices
Y$_3$Fe$_5$O$_{12}$ (single crys.)	0.178	292	0.5	Microwave devices
(Y,Al)$_3$Fe$_5$O$_{12}$	0.12	250	80	Microwave devices
(Y,Gd)$_3$Fe$_5$O$_{12}$	0.06	250	150	Microwave devices
Sm$_3$Fe$_5$O$_{12}$	0.170	305		Microwave devices
Eu$_3$Fe$_5$O$_{12}$	0.116	293		Microwave devices
GdFe$_5$O$_{12}$	0.017	291		Microwave devices
Hexagonal crystals				
BaFe$_{12}$O$_{19}$	0.45	430	1.5	Permanent magnets
Ba$_3$Co$_2$Fe$_{24}$O$_{41}$	0.34	470	12	Microwave devices
Ba$_2$Zn$_2$Fe$_{12}$O$_{22}$	0.28	130	25	Microwave devices
Ba$_3$Co$_{1.35}$Zn$_{0.65}$Fe$_{24}$O$_{41}$		390	16	Microwave devices
Ba$_2$Ni$_2$Fe$_{12}$O$_{22}$	0.16	500	8	Microwave devices
SrFe$_{12}$O$_{19}$	0.4	450		Permanent magnets

Reference

McCurrie, R. A., *Structure and Properties of Ferromagnetic Materials*, Academic Press, London, 1994.

Spinel Structure (AB$_2$O$_4$)

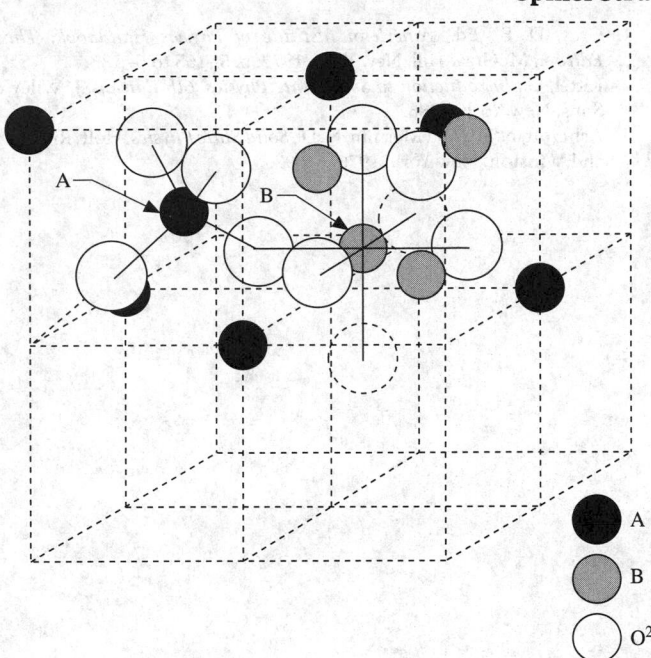

FIGURE 4. Arrangement of metal ions in the two octants A and B, showing tetrahedrally (A) and octahedrally (B) coordinated sites. (Reprinted from McCurrie, R.A., *Ferromagnetic Materials*, Academic Press, London, 1994. With permission.)

- ● A
- ◍ B
- ○ O^{2-}

Selected Antiferromagnetic Solids

T_N is the Néel temperature

Material	Structure	T_N/K	Material	Structure	T_N/K
			ZnCr$_2$O$_4$	cub	15
			ZnFe$_2$O$_4$	cub	9
Binary oxides			GeFe$_2$O$_4$	cub	10
MnO	cub(fcc)	122	MgV$_2$O$_4$	cub	45
FeO	cub(fcc)	198	MnGa$_2$O$_4$	cub	33
CoO	cub(fcc)	291			
NiO	cub(fcc)	525	*NiAs and related structures*		
α-Mn$_2$O$_3$	cub	90	CrAs	orth	300
CuO	monocl	230	CrSb	hex	705–723
UO$_2$	cub	30.8	CrSe	hex	300
Er$_2$O$_3$	cub	3.4	MnTe	hex	320–323
Gd$_2$O$_3$	cub	1.6	NiS	hex	263
			CrS	monocl	460
Perovskites					
LaCrO$_3$	orth	282	*Rutile and related structures*		
LaMnO$_3$	orth	100	CoF$_2$	tetr	38
LaFeO$_3$	orth	750	CrF$_2$	monocl	53
NdCrO$_3$	orth	224	FeF$_2$	tetr	79
NdFeO$_3$	orth	760	MnF$_2$	tetr	67
YbCrO$_3$	orth	118	NiF$_2$	tetr	83
CaMnO$_3$	cub	110	CrCl$_2$	orth	20
EuTiO$_3$	cub	5.3	MnO$_2$	tetr	84
YCrO$_3$	orth	141	FeOF	tetr	315
BiFeO$_3$	cub*	673			
KCoF$_3$	cub	125	*Corundum and related structures*		
KMnF$_3$	cub*	88.3	Cr$_2$O$_3$	rhomb	318
KFeF$_3$	cub	115	α-Fe$_2$O$_3$	rhomb	948
KNiF$_3$	cub	275	FeTiO$_3$	rhomb	68
NaMnF$_3$	cub*	60	MnTiO$_3$	rhomb	41
NaNiF$_3$	orth	149	CoTiO$_3$	rhomb	38
RbMnF$_3$	cub	82			
Spinels			*VF$_3$ and related structures*		
Co$_3$O$_4$	cub	40	CoF$_3$	rhomb	460
NiCr$_2$O$_4$	tetr	65	CrF$_3$	rhomb	80

Material	Structure	T_N/K
FeF_3	rhomb	394
MnF_3	monocl	43
MoF_3	rhomb	185
Miscellaneous		
K_2NiF_4	tetr	97
MnI_2	hex	3.4
$CoUO_4$	orth	12
$CaMn_2O_4$	orth	225
CrN	cub*	273
CeC_2	tetr	33
FeSn	hex	373
Mn_2P	hex	103

* Distorted.

References

1. Gray, D. E., Ed., *American Institute of Physics Handbook, Third Edition*, McGraw Hill, New York, 1972, p. 5–168 to 5–183.
2. Kittel, C., *Introduction to Solid State Physics, 6th Edition*, J. Wiley & Sons, New York, 1986.
3. Ashcroft, N. W., and Mermin, N. D., *Solid State Physics*, Holt, Rinehart, and Winston, New York, 1976, p. 697.

ORGANIC MAGNETS

J.S. Miller

Magnetic ordering, e.g., ferromagnetism, like superconductivity, is a property of a solid, not of an individual molecule or ion, and very rarely occurs for organic compounds. In contrast to superconductivity, where all electron spins pair to form a perfect diamagnetic material, magnetic ordering requires unpaired electron spins; hence, superconductivity and ferromagnetism are mutually exclusive.

The vast majority of organic compounds are diamagnetic (i.e., all electron spins are paired), and a relative few possess unpaired electrons (designated by an arrow, $\uparrow$) and are paramagnetic (PM), i.e., they are oriented in random directions. A few organic solids, however, exhibit strong magnetic behavior and magnetically order as ferromagnets (FO) with all spins aligned in the same direction. In some cases the spins align in the opposite direction and compensate to form an antiferromagnet (AF). In some cases these spins are not opposed to each other and do not compensate and lead to a canted antiferromagnet or weak ferromagnet (WF). If the number of spins that align in one direction differs from the number of spins that align in the opposite direction, the spins cannot compensate and a ferrimagnet (FI) results. Metamagnets (MM) are antiferromagnets in which all the spins become aligned like a ferromagnet in an applied magnetic field. Above the ordering or critical temperature, T_c, all magnets are paramagnets (PM). Organic magnets all possess electron spins in p-orbitals, but these may be in conjunction with metal ion-based spins.

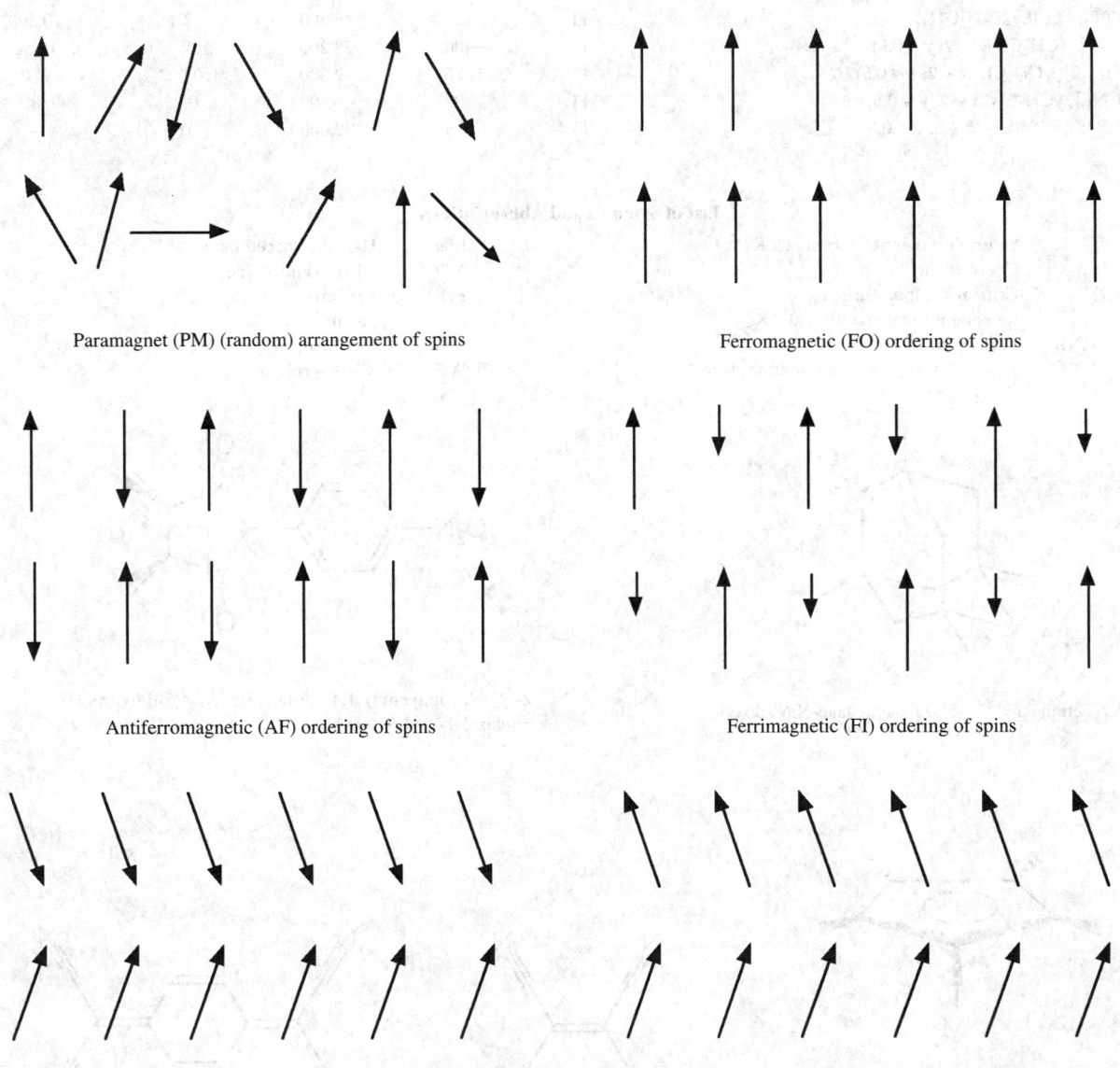

Paramagnet (PM) (random) arrangement of spins

Ferromagnetic (FO) ordering of spins

Antiferromagnetic (AF) ordering of spins

Ferrimagnetic (FI) ordering of spins

Canted antiferromagnet or weak ferromagnet (WF) ordering of spins

FIGURE 1.   Schematic illustration of the different types of magnetic behavior.

Summary of the Critical Temperature, T_c, Saturation Magnetization, M_s, Coercive Field, H_{cr}, and Remanent Magnetization, M_r, for Selected Organic-Based Magnets

Magnet	Type	T_c/K	M_s/A m^{-1}	H_{cr}/T	M_r/A m^{-1}
α-1,3,5,7-Tetramethyl-2,6-diazaadamantane-N,N'-doxyl	FO	1.48	48,300	<0.00001	—
β-2-(4'-Nitrophenyl)-4,4,5,5-tetramethyl-4,5-dihydro-1H-imidazol-1-oxyl-3-N-oxide	FO	0.6	22,300	0.00008	<200
{Fe^{III}[C$_5$(CH$_3$)$_5$]$_2$}[TCNE]	FO	4.8	37,600	0.10	2,300
{Mn^{III}[C$_5$(CH$_3$)$_5$]$_2$}[TCNE]	FO	8.8	58,200	0.12	3,700
{Cr^{III}[C$_5$(CH$_3$)$_5$]$_2$}[TCNE]	FO	3.65	46,300	—	—
α-{Fe^{III}[C$_5$(CH$_3$)$_5$]$_2$}[TCNQ]	MM	2.55	34,200	—	—
β-{Fe^{III}[C$_5$(CH$_3$)$_5$]$_2$}[TCNQ]	FO	3.0	21,600	—	—
Tanol subarate	MM	0.38	20,700	—	—
NCC$_6$F$_4$CN$_2$S$_2$	WF	35.5	45	0.00009	—
Mn^{II}(hfac)$_2$NITC$_2$H$_5$	FI	7.8	39,400	0.03	27,600
Mn^{II}(hfac)$_2$NIT(i-C$_3$H$_8$)	FI	7.6	42,400	<0.0005	<420
[Mn(hfac)$_2$]$_3$[{ON[C$_6$H$_3$(t-C(CH$_3$)$_3$)$_2$NO]$_2$}	FI	46	24,400	—	—
[MnTPP][TCNE]·2C$_6$H$_5$CH$_3$	FI	13	18,400	2.4	10,300
V[TCNE]$_x$·yCH$_2$Cl$_2$ ($x \sim 2; y \sim 0.5$)	FI	~400	28,200	0.0015 - 0.006	1,650
Mn[TCNE]$_x$·yCH$_2$Cl$_2$ ($x \sim 2; y \sim 0.5$)	FI	75	52,000	0.002	270
Fe[TCNE]$_x$·yCH$_2$Cl$_2$ ($x \sim 2; y \sim 0.5$)	FI	97	46,300	0.23	3
Co[TCNE]$_x$·yCH$_2$Cl$_2$ ($x \sim 2; y \sim 0.5$)	FI	44	22,000	0.65	—

List of Symbols and Abbreviations

M_s	Saturation magnetization at 2 K	hfac	Hexafluoroacetonate
H_{cr}	Coercive Field	NIT	Nitronyl nitroxide
T_c	Critical Temperature	FO	Ferromagnet
M_r	Remanent magnetization at 2 K	FI	Ferrimagnet
TCNE	Tetracyanoethylene	MM	Metamagnet
TCNQ	7,7,8,8-Tetracyano-p-quinodimethane	WF	Weak ferromagnet

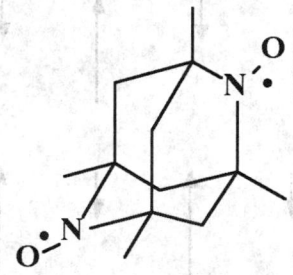

1,3,5,7-Tetramethyl-2,6-diazaadamantane-N,N'-doxyl

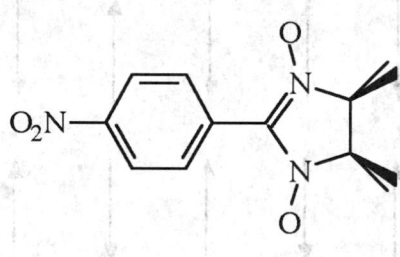

2-(4'-Nitrophyenyl)-4,4,5,5-tetramethyl-4,5-dihydro-1H-imidazol-1-oxyl-3-N-oxide

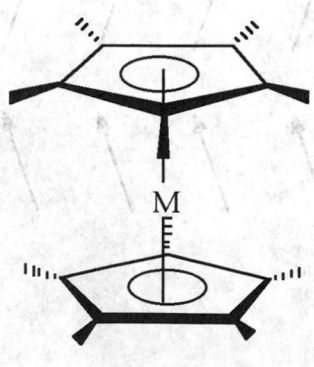

M[C$_5$(CH$_3$)$_5$]$_2$(M = Cr, Mn, Fe)

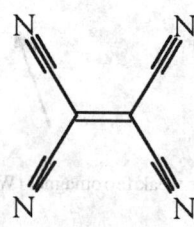

TCNE

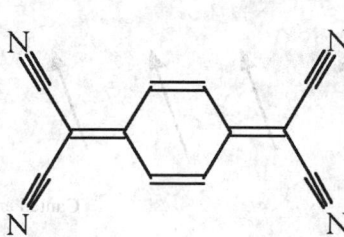

TCNQ

Mn(hfac)$_2$

Tanol subarate

NITR (R = C$_2$H$_5$, i-C$_3$H$_8$, n-C$_3$H$_8$)

{ON[C$_6$H$_3$(t-C(CH$_3$)$_3$)$_2$NO]$_2$}

MnTPP

NCC$_6$F$_4$CN$_2$S$_2$

References

1. Miller, J. S. and Epstein, A. J., *Angew. Chem. Internat. Ed.*, 33, 385, 1994.
2. Chiarelli, R., Rassat, A., Dromzee, Y., Jeannin, Y., Novak, M. A., and Tholence, J. L., *Phys. Scrip.*, T49, 706, 1993.
3. Kinoshita, M., *Jap. J. Appl. Phys.*, 33, 5718, 1994.
4. Gatteschi, D., *Adv. Mat.*, 6, 635, 1994.
5. Miller, J. S. and Epstein, A. J., *J. Chem. Soc., Chem. Commun.*, 1319, 1998.
6. Broderick, W. E., Eichorn, D. M., Lu, X., Toscano, P. J., Owens, S. M. and Hoffman, B. M., *J. Am. Chem. Soc.*, 117, 3641, 1995.
7. Banister, A. J., Bricklebank, N., Lavander, I., Rawson, J., Gregory, C. I., Tanner, B. K., Clegg, W. J., Elsegood, M. R., and Palacio, F., *Angew. Chem. Internat. Ed.*, 35, 2533, 1996.

ELECTRON INELASTIC MEAN FREE PATHS

Cedric J. Powell

The inelastic mean free path (IMFP) of electrons impinging on a solid surface is defined as the average of distances, measured along the trajectories, that electrons with a given energy travel between inelastic collisions in the substance. It is an important parameter in analyzing results from surface characterization techniques such as Auger electron spectroscopy, x-ray photoelectron spectroscopy, low-energy electron diffraction, and others. IMFPs can be measured by the elastic-peak electron spectroscopy technique and other methods, and they can be calculated from optical data. A detailed analysis of the experimental and theoretical considerations in obtaining reliable IMPF values can be found in References 4 and 5.

Table 1 below gives recommended IMFP values for 41 elemental solids in the energy range 50 eV to 30 000 eV. All values in Table 1 are taken from Reference 1. Table 2 gives IMFP values for several inorganic compounds and organic materials in the range 50 eV to

2 000 eV. The entries in Table 1are listed by atomic number, while substances in Table 2 are listed in alphabetical order by name, with inorganic compounds preceding the organic materials. IMFP values are given in Ångström units (1 Å = 10^{-10} m).

References

1. Tanuma, S., Powell, C. J., and Penn, D. R., *Surf. Interface Anal.* (in press); doi 10.1002/sia.3522.
2. Tanuma, S., Powell, C. J., and Penn, D. R., *Surf. Interface Anal.* 17, 927, 1991.
3. Tanuma, S., Powell, C. J., and Penn, D. R., *Surf. Interface Anal.* 21, 165, 1994.
4. Powell, C. J., and Jablonski, A., *J. Phys. Chem. Ref. Data* 28, 19, 1999.
5. Powell, C. J., and Jablonski, A., *Nucl. Instr. Meth. Phys. Res. A* 601, 54, 2009.

TABLE 1. Electron Inelastic Mean Free Paths of Elemental Solids in Å (10^{-10} m)

Element	Electron Energy in eV															
	50	100	250	500	750	1000	1250	1500	1750	2000	3000	5000	10000	15000	20000	30000
Li	4.6	7.0	13.3	22.7	31.4	39.7	47.8	55.7	63.4	71.0	100.4	156.2	286.6	410.1	529.6	760.6
Be	3.5	4.3	7.2	11.7	15.9	19.9	23.8	27.6	31.2	34.9	48.8	75.3	136.7	194.7	250.6	358.4
C-graphite	4.8	4.4	6.8	10.8	14.6	18.2	21.7	25.0	28.3	31.6	44.0	67.6	122.2	173.7	223.3	318.9
C-diamond	6.9	4.7	6.4	9.9	13.2	16.4	19.4	22.4	25.3	28.1	39.0	59.6	107.3	152.1	195.3	278.4
C-glassy	5.8	6.3	10.0	16.0	21.7	27.0	32.1	37.1	42.0	46.8	65.4	100.4	181.9	258.6	332.7	475.4
Na	5.0	7.6	13.8	23.1	31.8	40.0	48.0	55.8	63.5	71.0	100.1	155.1	283.4	404.7	522.0	748.5
Mg	4.0	5.4	9.4	15.3	20.7	25.9	30.9	35.8	40.6	45.3	63.5	97.9	177.7	253.0	325.7	465.9
Al	3.5	4.6	7.9	12.7	17.1	21.3	25.4	29.4	33.3	37.2	52.0	79.9	144.8	206.0	265.1	378.8
Si	4.1	5.3	9.0	14.6	19.7	24.5	29.2	33.7	38.2	42.6	59.5	91.5	165.7	235.7	303.1	433.2
K	7.1	10.0	18.3	31.4	43.5	54.8	65.7	76.4	86.8	97.1	136.8	212.0	387.5	553.5	714.0	1024.0
Sc	4.6	4.9	7.4	11.9	16.2	20.2	24.1	27.9	31.5	35.1	49.0	75.2	135.9	193.1	248.2	354.3
Ti	4.2	4.4	6.6	10.4	14.1	17.5	20.9	24.1	27.3	30.4	42.3	64.7	116.8	165.8	213.0	303.8
V	4.6	4.7	6.7	10.3	13.8	17.1	20.3	23.4	26.4	29.4	40.7	62.1	111.7	158.3	203.1	289.3
Cr	4.5	4.4	6.4	9.9	13.2	16.3	19.4	22.3	25.2	28.0	38.8	59.2	106.4	150.7	193.4	275.6
Fe	4.3	4.6	6.7	10.2	13.5	16.7	19.7	22.7	25.7	28.5	39.5	60.1	107.8	152.7	195.8	278.8
Co	5.1	4.6	6.1	9.0	11.8	14.5	17.2	19.8	22.3	24.8	34.2	51.9	92.9	131.4	168.3	239.5
Ni	4.9	4.6	6.2	9.2	12.1	14.9	17.6	20.2	22.8	25.3	35.0	53.1	95.0	134.3	172.0	244.7
Cu	5.0	5.0	7.0	10.4	13.6	16.7	19.6	22.6	25.4	28.2	39.0	59.2	105.7	149.5	191.5	272.4
Ge	4.0	5.0	8.2	12.9	17.2	21.2	25.2	29.0	32.8	36.5	50.7	77.5	139.5	197.9	254.2	362.5
Y	5.1	5.5	8.8	14.4	19.4	24.1	28.6	33.0	37.3	41.5	57.8	88.6	159.9	227.0	291.7	416.2
Nb	5.7	5.3	7.3	11.4	15.2	18.8	22.2	25.5	28.8	31.9	44.1	67.1	120.4	170.3	218.3	310.7
Mo	5.0	4.4	6.1	9.5	12.7	15.6	18.4	21.2	23.8	26.5	36.6	55.6	99.6	141.0	180.7	257.1
Ru	4.9	4.1	5.4	8.3	11.1	13.7	16.1	18.5	20.8	23.1	31.9	48.5	86.7	122.6	157.1	223.5
Rh	4.8	4.0	5.2	7.9	10.6	13.0	15.4	17.7	19.9	22.1	30.5	46.2	82.7	117.0	149.9	213.2
Pd	4.8	4.0	5.3	8.1	10.8	13.4	15.8	18.2	20.5	22.7	31.3	47.6	85.2	120.4	154.3	219.5
Ag	6.1	4.9	5.9	8.7	11.6	14.2	16.8	19.2	21.6	24.0	33.0	50.0	89.3	126.2	161.6	229.7
In	4.7	5.4	8.0	12.3	16.5	20.6	24.4	28.1	31.8	35.3	48.9	74.5	133.4	189.1	243.1	346.3
Sn	5.8	6.4	9.2	13.8	18.4	22.8	27.0	31.1	35.1	38.9	53.8	81.8	146.5	207.3	265.8	378.2
Cs	6.3	9.3	17.4	29.3	40.3	50.8	61.0	71.0	80.7	90.2	126.8	196.1	357.6	510.4	657.9	942.8
Gd	4.1	4.6	7.4	11.8	15.7	19.5	23.2	26.8	30.3	33.7	46.9	71.6	128.8	182.6	234.4	334.1
Tb	4.1	4.3	6.5	10.2	13.7	16.9	20.1	23.2	26.2	29.2	40.6	61.9	111.3	157.8	202.5	288.6
Dy	4.5	4.6	6.8	10.6	14.1	17.5	20.8	23.9	27.0	30.0	41.7	63.5	114.0	161.5	207.2	295.1
Hf	5.5	5.5	7.5	11.3	14.9	18.2	21.5	24.6	27.7	30.7	42.3	64.2	114.4	161.5	206.7	293.8
Ta	4.9	4.5	6.1	9.2	12.1	14.8	17.4	19.9	22.4	24.8	34.2	51.7	92.2	130.0	166.5	236.4
W	5.4	4.4	5.8	8.7	11.3	13.9	16.3	18.6	20.9	23.2	31.8	48.1	85.5	120.5	154.1	218.8

Element	50	100	250	500	750	1000	1250	1500	1750	2000	3000	5000	10000	15000	20000	30000
								Electron Energy in eV								
Re	5.4	4.2	5.4	8.0	10.5	12.8	15.0	17.2	19.3	21.3	29.3	44.3	78.7	111.0	142.0	201.5
Os	5.6	4.4	5.6	8.4	10.9	13.3	15.6	17.8	20.0	22.1	30.3	45.7	81.1	114.3	146.2	207.3
Ir	5.3	4.2	5.4	8.0	10.4	12.7	15.0	17.1	19.2	21.2	29.1	43.9	78.1	110.1	140.8	199.7
Pt	5.0	4.2	5.5	8.2	10.7	13.1	15.4	17.6	19.8	21.9	30.0	45.4	80.7	113.8	145.5	206.6
Au	5.1	4.3	5.6	8.4	11.0	13.5	15.9	18.2	20.5	22.7	31.1	47.1	83.9	118.4	151.5	215.1
Bi	4.9	5.5	8.0	12.3	16.4	20.2	23.9	27.5	31.0	34.4	47.5	72.1	129.3	182.8	234.4	333.6

TABLE 2. Electron Inelastic Mean Free Paths of Other Materials in Å (10^{-10} m)

Substance	Formula	50	100	150	200	300	400	600	800	1000	1200	1400	1600	1800	2000	Ref.
						Electron Energy in eV										
Gallium phosphide	GaP	5.6	6.5	7.8	9.0	11.4	13.7	18.1	22.3	26.3	30.2	34.1	37.8	41.5	45.2	2
Indium phosphide	InP	4.8	4.9	5.6	6.4	8.1	9.7	12.8	15.7	18.7	21.4	24.2	26.8	29.4	32.0	2
Lead(II) sulfide	PbS	4.8	5.6	6.7	7.8	10.0	12.1	16.1	19.8	23.6	27.1	30.6	33.9	37.2	40.5	2
Lead(II) telluride	PbTe	4.3	5.5	6.6	7.7	9.8	11.9	15.8	19.6	23.4	26.9	30.3	33.7	37.0	40.2	2
Potassium chloride	KCl	7.5	7.8	9.3	10.9	14.2	17.3	23.2	28.7	34.0	39.2	44.2	49.1	54.0	58.8	2
Silicon carbide	SiC	4.7	4.9	5.8	6.8	8.7	10.5	13.9	17.1	20.3	23.3	26.3	29.2	32.1	35.0	2
Silicon dioxide (vitreous)	SiO$_2$	8.0	7.7	8.8	10.0	12.6	15.2	20.0	24.7	29.3	33.7	38.0	42.2	46.4	50.5	2
Zinc sulfide	ZnS	5.8	6.5	7.7	8.9	11.3	13.6	17.9	22.0	26.0	29.8	33.6	37.3	40.9	44.5	2
Adenine	C$_5$H$_5$N$_5$	6.4	6.6	7.8	9.2	11.8	14.4	19.2	24.1	28.6	33.1	37.4	41.6	45.8	49.9	3
Bovine plasma albumin		7.3	7.2	8.5	9.9	12.7	15.4	20.7	25.8	30.8	35.6	40.2	44.8	49.4	53.8	3
β-Carotene	C$_{40}$H$_{56}$	6.4	7.0	8.5	10.0	13.0	15.9	21.4	26.9	32.0	37.0	41.9	46.6	51.3	56.0	3
Deoxyribonucleic acid (DNA)		7.3	7.3	8.5	9.8	12.6	15.4	20.7	25.9	30.8	35.6	40.3	44.9	49.4	53.8	3
1,6-Diphenyl-1,3,5-hexatriene	C$_{18}$H$_{16}$	6.4	7.0	8.4	9.9	12.9	15.8	21.3	26.7	31.8	36.8	41.7	46.4	51.1	55.7	3
Guanine	C$_5$H$_5$N$_5$O	6.2	6.2	7.2	8.4	10.8	13.1	17.5	21.8	25.9	29.9	33.8	37.6	41.4	45.1	3
Hexacosane	C$_{26}$H$_{54}$	7.0	7.6	9.2	10.9	14.1	17.2	23.2	29.2	34.7	40.1	45.4	50.6	55.7	60.7	3
Kapton		7.0	6.8	7.9	9.2	11.7	14.2	19.0	23.7	28.2	32.5	36.7	40.9	44.9	49.0	3
Polyacetylene		5.3	5.7	6.8	7.9	10.2	12.5	16.9	21.1	25.1	29.0	32.8	36.5	40.2	43.8	3
Poly(butene-1-sulfone)		7.1	7.2	8.5	9.9	12.7	15.4	20.7	25.8	30.6	35.3	39.9	44.4	48.8	53.2	3
Polyethylene		6.9	7.2	8.6	10.1	13.0	15.9	21.4	26.8	31.8	36.8	41.6	46.3	51.0	55.6	3
Poly(methyl methacrylate)		7.8	7.9	9.3	10.8	13.9	16.9	22.7	28.3	33.7	38.8	43.9	48.9	53.8	58.6	3
Polystyrene		6.9	7.3	8.7	10.2	13.2	16.1	21.6	27.1	32.2	37.2	42.1	46.9	51.6	56.2	3
Poly(2-vinylpyridine)		6.9	7.3	8.7	10.3	13.3	16.2	21.8	27.3	32.5	37.5	42.4	47.3	52.0	56.7	3

ELECTRON STOPPING POWERS

Cedric J. Powell

Numerical data are given for collision electron stopping powers in 22 elemental solids for energies between 100 eV and 30 keV. These stopping powers were determined with an algorithm that utilizes experimental optical data for each solid.

The stopping power for electrons and other charged particles in matter is often needed in calculations of electron transport in a medium, particularly in radiation physics and in descriptions of signal generation in analytical techniques such as electron-probe microanalysis and Auger electron spectroscopy. The stopping power is defined as the average rate at which the charged particles lose energy at any point along their trajectories. For electrons, it is customary to separate the total stopping power into two components, the collision stopping power due to inelastic-scattering events of the electrons in a medium and the radiative stopping power due to the emission of bremsstrahlung in the electric field of the atomic nucleus and atomic electrons (Ref. 1). For electron energies less than 30 keV, the radiative stopping power is less than 1% of the collision stopping power (Ref. 1), and is neglected in the numerical data given here.

Numerical data for collision and radiative stopping powers at electron energies between 10 keV and 1 GeV have been published for materials of interest in radiation physics and dosimetry (Ref. 1). Similar data can also be obtained from a web site of the National Institute of Standards and Technology (Ref. 2). The collision stopping powers were calculated from the theory of Bethe (Refs. 3 and 4) and recommended values of the one material-dependent parameter, the mean excitation energy (Ref. 1). While the Bethe theory is expected to be valid for electron energies much larger than the largest K-shell binding energy of atoms in the particular material, the Bethe stopping-power equation is frequently utilized to calculate stopping powers for energies of 10 keV and above (Refs. 1 and 2). Detailed analyses of the Bethe stopping-power theory have been published (Refs. 1 and 4).

The table gives collision stopping powers for 22 elemental solids at energies between 100 eV and 30 keV (Refs. 5 and 6). These stopping powers were determined by interpolation with a clamped cubic spline from the published data (Refs. 5 and 6) which had been calculated with an algorithm that utilizes experimental optical data for each solid. Comparisons with stopping powers from the Bethe stopping-power equation at 30 keV showed a root-mean-square difference of about 10%. This level of agreement was considered satisfactory on account of uncertainties of the algorithm and optical data used for the calculations as well as uncertainties of the mean excitation energies used with the Bethe equation.

The stopping powers in the table are given in units of eV/Å (1 Å $= 10^{-10}$ m) for a range of electron energies. The elemental solids are listed in order of atomic number. For some applications, the mass collision stopping power is desired and these values can be obtained by dividing the tabulated values by the density of the solid.

References

1. *Stopping Powers for Electrons and Positrons*, ICRU Report 37 (International Commission on Radiation Units and Measurements, Bethesda, 1984).
2. Berger, M. J., Coursey, J. S., Zucker, M. A., and Chang, J., *Stopping-Power and Range Tables for Electrons, Positrons, and Helium Ions*, Version 1.2.3, (http://physics.nist.gov/Star.html), (http://physics.nist.gov/PhysRefData/Star/Text/contents.html), 2005.
3. Bethe, H., *Ann. Physik* 5, 325, 1930.
4. Inokuti, M., *Rev. Mod. Phys.* 43, 297, 1971.
5. Tanuma, S., Powell, C. J., and Penn, D. R., *Surf. Interface Anal.* 37, 978, 2005.
6. Tanuma, S., Powell, C. J., and Penn, D. R., *J. Appl. Phys.* 103, 063707, 2008.

TABLE 1. Values of collision electron stopping powers in eV/Å for the indicated elemental solids and electron energies (Refs. 5 and 6)

	Electron energy in eV														
	100	200	300	400	500	1000	1500	2000	3000	4000	5000	10000	15000	20000	30000
Li	1.71	1.49	1.31	1.15	1.02	0.657	0.493	0.399	0.292	0.233	0.195	0.110	0.0787	0.0617	0.0437
Be	6.43	4.69	3.96	3.52	3.17	2.13	1.63	1.33	0.982	0.787	0.660	0.377	0.270	0.212	0.150
C[a]	4.84	4.09	3.39	2.90	2.54	1.73	1.39	1.16	0.881	0.715	0.606	0.353	0.255	0.201	0.143
C[b]	8.14	6.80	5.60	4.76	4.15	2.72	2.11	1.73	1.30	1.05	0.883	0.509	0.366	0.288	0.205
C[c]	10.02	9.02	7.57	6.49	5.69	3.80	2.97	2.46	1.85	1.50	1.27	0.734	0.528	0.416	0.296
Na	1.57	1.67	1.59	1.47	1.35	0.936	0.721	0.591	0.441	0.359	0.306	0.181	0.131	0.104	0.0747
Mg	3.40	3.23	3.07	2.82	2.61	1.84	1.44	1.18	0.887	0.718	0.610	0.362	0.263	0.209	0.150
Al	4.99	3.86	3.59	3.36	3.16	2.35	1.86	1.55	1.17	0.947	0.804	0.481	0.351	0.280	0.201
Si	4.66	3.34	3.02	2.87	2.72	2.06	1.64	1.37	1.04	0.845	0.716	0.426	0.312	0.248	0.178
K	1.79	1.37	1.11	0.944	0.825	0.572	0.472	0.404	0.316	0.261	0.223	0.133	0.0973	0.0778	0.0564
Sc	7.19	6.28	5.23	4.49	3.94	2.54	2.02	1.70	1.31	1.08	0.922	0.547	0.397	0.316	0.228
Ni	7.04	8.73	8.71	8.26	7.74	5.70	4.51	3.75	2.92	2.45	2.13	1.31	0.969	0.776	0.562
Cu	6.27	7.23	7.27	7.05	6.73	5.15	4.14	3.48	2.71	2.28	1.99	1.24	0.920	0.738	0.536
Ge	4.78	4.29	4.07	3.90	3.75	2.99	2.46	2.09	1.61	1.34	1.17	0.732	0.544	0.437	0.318
Zr	6.51	5.61	4.68	4.02	3.58	2.65	2.23	1.94	1.55	1.30	1.13	0.701	0.528	0.428	0.315
Ag	7.43	11.45	10.53	9.32	8.34	5.84	4.85	4.19	3.37	2.80	2.41	1.47	1.08	0.872	0.638
In	4.60	6.19	5.91	5.26	4.71	3.19	2.56	2.23	1.82	1.55	1.35	0.849	0.632	0.513	0.378

							Electron energy in eV							
	100	200	300	400	500	1000	1500	2000	3000	4000	5000	10000	15000	20000	30000
Sn	3.94	5.71	5.70	5.14	4.63	3.18	2.52	2.19	1.79	1.53	1.34	0.840	0.626	0.507	0.374
Cs	1.50	1.19	1.27	1.16	1.05	0.718	0.562	0.472	0.383	0.329	0.290	0.187	0.140	0.113	0.0839
Gd	6.28	5.58	4.85	4.62	4.32	3.20	2.58	2.19	1.70	1.43	1.26	0.829	0.629	0.512	0.378
Tb	7.87	7.25	6.24	5.73	5.28	3.82	3.06	2.58	1.99	1.65	1.44	0.926	0.697	0.563	0.413
Dy	7.43	7.12	6.25	5.79	5.38	3.97	3.20	2.71	2.10	1.74	1.52	0.978	0.736	0.596	0.437
Pt	9.35	10.81	9.88	8.98	8.33	6.54	5.60	4.95	4.06	3.45	3.02	1.95	1.50	1.23	0.920
Bi	4.79	6.16	5.81	5.21	4.70	3.41	2.85	2.50	2.05	1.75	1.53	0.968	0.744	0.605	0.450

[a] Glassy carbon
[b] Graphite
[c] Diamond

ELECTRON WORK FUNCTION OF THE ELEMENTS

The electron work function Φ is a measure of the minimum energy required to extract an electron from the surface of a solid. It is defined more precisely as the energy difference between the state in which an electron has been removed to a distance from the surface of a single crystal face that is large enough that the image force is negligible but small compared to the distance to any other face (typically about 10^{-4} cm) and the state in which the electron is in the bulk solid. In general, Φ differs for each face of a monocrystalline sample.

Since Φ is dependent on the cleanliness of the surface, measurements reported in the literature often cover a considerable range. This table contains selected values for the electron work function of the elements which may be regarded as typical values for a reasonably clean surface. The method of measurement is indicated for each value. The following abbreviations appear:

TE – Thermionic emission
PE – Photoelectric effect
FE – Field emission
CPD – Contact potential difference
polycr – Polycrystalline sample
amorp – Amorphous sample

Values in parentheses are only approximate.

References

1. Hölzl, J., and Schulte, F. K., Work Functions of Metals, in *Solid Surface Physics*, Höhler, G., Ed., Springer-Verlag, Berlin, 1979.
2. Riviere, J. C., Work Function: Measurements and Results, in *Solid State Surface Science, Vol.1*, Green, M., Ed., Decker, New York, 1969.
3. Michaelson, H. B., *J. Appl. Phys.*, 48, 4729, 1977.

Element	Plane	Φ/eV	Method
Ag	100	4.64	PE
	110	4.52	PE
	111	4.74	PE
Al	100	4.20	PE
	110	4.06	PE
	111	4.26	PE
As	polycr	(3.75)	PE
Au	100	5.47	PE
	110	5.37	PE
	111	5.31	PE
B	polycr	(4.45)	TH
Ba	polycr	2.52	TH
Be	polycr	4.98	PE
Bi	polycr	4.34	PE
C	polycr	(5.0)	CPD
Ca	polycr	2.87	PE
Cd	polycr	4.08	CPD
Ce	polycr	2.9	PE
Co	polycr	5.0	PE
Cr	polycr	4.5	PE
Cs	polycr	1.95	PE
Cu	100	5.10	FE
	110	4.48	PE
	111	4.94	PE
	112	4.53	PE
Eu	polycr	2.5	PE
Fe	100	4.67	PE
	111	4.81	PE
Ga	polycr	4.32	PE
Gd	polycr	2.90	CPD
Ge	polycr	5.0	CPD
Hf	polycr	3.9	PE
Hg	liquid	4.475	PE
In	polycr	4.09	PE
Ir	100	5.67	PE
	110	5.42	PE
	111	5.76	PE

Element	Plane	Φ/eV	Method
	210	5.00	PE
K	polycr	2.29	PE
La	polycr	3.5	PE
Li	polycr	2.93	FE
Lu	polycr	(3.3)	CPD
Mg	polycr	3.66	PE
Mn	polycr	4.1	PE
Mo	100	4.53	PE
	110	4.95	PE
	111	4.55	PE
	112	4.36	PE
	114	4.50	PE
	332	4.55	PE
Na	polycr	2.36	PE
Nb	001	4.02	TH
	110	4.87	TH
	111	4.36	TH
	112	4.63	TH
	113	4.29	TH
	116	3.95	TH
	310	4.18	TH
Nd	polycr	3.2	PE
Ni	100	5.22	PE
	110	5.04	PE
	111	5.35	PE
Os	polycr	5.93	PE
Pb	polycr	4.25	PE
Pd	polycr	5.22	PE
	111	5.6	PE
Pt	polycr	5.64	PE
	110	5.84	FE
	111	5.93	FE
	320	5.22	FE
	331	5.12	FE
Rb	polycr	2.261	PE
Re	polycr	4.72	TE
Rh	polycr	4.98	PE

Element	Plane	Φ/eV	Method
Ru	polycr	4.71	PE
Sb	amorp	4.55	
	100	4.7	
Sc	polycr	3.5	PE
Se	polycr	5.9	PE
Si	n	4.85	CPD
	p 100	(4.91)	CPD
	p 111	4.60	PE
Sm	polycr	2.7	PE
Sn	polycr	4.42	CPD
Sr	polycr	(2.59)	TH
Ta	polycr	4.25	TH
	100	4.15	TH
	110	4.80	TH
	111	4.00	TH
Tb	polycr	3.0	PE
Te	polycr	4.95	PE
Th	polycr	3.4	TH
Ti	polycr	4.33	PE
Tl	polycr	(3.84)	CPD
U	polycr	3.63	PE
	100	3.73	PE
	110	3.90	PE
	113	3.67	PE
V	polycr	4.3	PE
W	polycr	4.55	CPD
	100	4.63	FE
	110	5.22	FE
	111	4.45	FE
	113	4.46	FE
	116	4.32	TH
Y	polycr	3.1	PE
Zn	polycr	3.63	PE
	polycr	(4.9)	CPD
Zr	polycr	4.05	PE

SECONDARY ELECTRON EMISSION

The secondary emission yield, or secondary emission ratio, δ, is the average number of secondary electrons emitted from a bombarded material for every incident primary electron. It is a function of the primary electron energy E_p. The maximum yield δ_{max} corresponds to a primary electron energy E_{pmax} (see figure). The two primary electron energies corresponding to a yield of unity are denoted by the first and second crossovers (E_I and E_{II}). An insulating target, or a conducting target that is electrically floating, will charge positively or negatively depending on the primary electron energy. For $E_I < E_p < E_{II}$, $\delta > 1$ and the surface charges positively provided there is a collector present that is positive with respect to the target. For $E_p < E_I$ or $E_p > E_{II}$, $\delta < 1$, and the surface charges negatively with respect to the potential of the source of primary electrons.

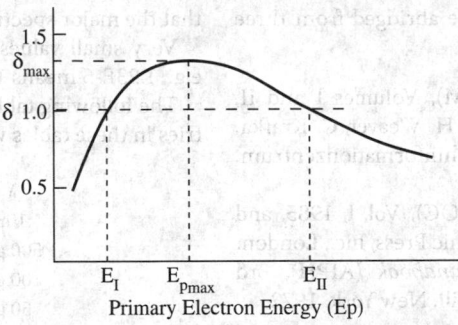

Primary Electron Energy (Ep)

Element	δ_{max}	E_{pmax} (eV)	E_I (eV)	E_{II} (eV)
Ag	1.5	800	200	>2000
Al	1.0	300	300	300
Au	1.4	800	150	>2000
B	1.2	150	50	600
Ba	0.8	400	None	None
Bi	1.2	550	None	None
Be	0.5	200	None	None
C (diamond)	2.8	750	None	>5000
C (graphite)	1.0	300	300	300
C (soot)	0.45	500	None	None
Cd	1.1	450	300	700
Co	1.2	600	200	None
Cs	0.7	400	None	None
Cu	1.3	600	200	1500
Fe	1.3	400	120	1400
Ga	1.55	500	75	None
Ge	1.15	500	150	900
Hg	1.3	600	350	>1200
K	0.7	200	None	None

Element	δ_{max}	E_{pmax} (eV)	E_I (eV)	E_{II} (eV)
Li	0.5	85	None	None
Mg	0.95	300	None	None
Mo	1.25	375	150	1200
Na	0.82	300	None	None
Nb	1.2	375	150	1050
Ni	1.3	550	150	>1500
Pb	1.1	500	250	1000
Pd	>1.3	>250	120	None
Pt	1.8	700	350	3000
Rb	0.9	350	None	None
Sb	1.3	600	250	2000
Si	1.1	250	125	500
Sn	1.35	500	None	None
Ta	1.3	600	250	>2000
Th	1.1	800	None	None
Ti	0.9	280	None	None
Tl	1.7	650	70	>1500
W	1.4	650	250	>1500
Zr	1.1	350	None	None

Compound	δ_{max}	E_{pmax} (eV)
Alkali halides		
CsCl	6.5	
KBr (crystal)	14	1800
KCl (crystal)	12	1600
KCl (layer)	7.5	1200
KI (crystal)	10	1600
KI (layer)	5.6	
LiF (crystal)	8.5	
LiF (layer)	5.6	700
NaBr (crystal)	24	1800
NaBr (layer)	6.3	
NaCl (crystal)	14	1200
NaCl (layer)	6.8	600
NaF (crystal)	14	1200
NaF (layer)	5.7	
NaI (crystal)	19	1300
NaI (layer)	5.5	
RbCl (layer)	5.8	
Oxides		
Ag_2O	1.0	
Al_2O_3 (layer)	2–9	
BaO (layer)	2.3–4.8	400
BeO	3.4	2000

Compound	δ_{max}	E_{pmax} (eV)
CaO	2.2	500
Cu_2O	1.2	400
MgO (crystal)	20–25	1500
MgO (layer)	3–15	400–1500
MoO_2	1.2	
SiO_2 (quartz)	2.1–4	400
SnO_2	3.2	640
Sulfides		
MoS_2	1.1	
PbS	1.2	500
WS_2	1.0	
ZnS	1.8	350
Others		
BaF_2 (layer)	4.5	
CaF_2 (layer)	3.2	
$BiCs_3$	6	1000
BiCs	1.9	1000
GeCs	7	700
Rb_3Sb	7.1	450
$SbCs_3$	6	700
Mica	2.4	350
Glasses	2–3	300–450

OPTICAL PROPERTIES OF SELECTED ELEMENTS

J. H. Weaver and H. P. R. Frederikse

These tables list the index of refraction n, the extinction coefficient k, and the normal incidence reflection R ($\phi = 0$) as a function of photon energy E, which is expressed in electron volts (eV). To convert the energy in eV to the wavelength in µm, use $\lambda = 1.2398/E$. To compute the dielectric function $\tilde{\varepsilon} = \varepsilon_1 + i\varepsilon_2$ from the complex index of refraction $\tilde{N} = n + ik$, use $\varepsilon_1 = n^2 - k^2$ and $\varepsilon_2 = 2nk$.

The optical constants in these tables are abridged from three more extensive tabulations:

- *Optical Properties of Metals* (OPM), Volumes I and II, *Physics Data, Nr.* 18-1 and 18-2, J. H. Weaver, C. Krafka, D. W. Lynch, and E. E. Koch, Fachinformationzentrum, Karlsruhe, Germany.
- *Handbook of Optical Constants* (HOC), Vol. I, 1985, and Vol. II, 1991. E. D. Palik, Ed., Academic Press, Inc., London.
- *American Institute of Physics Handbook* (AIPH), 3rd Edition, D. E. Gray, Ed., McGraw-Hill, New York, 1972.

The first two of these major sources provide detailed comparisons of all optical data available in the literature at the time of the compilation. For critical applications the reader should refer to the original work. References for individual metals and semiconductors are listed at the end of the tables. Generally, tabulated values for the optical properties are accurate to better than 10%. Data in parentheses are extrapolated or interpolated values. For most elements the spectral range covered is from the far infrared (0.010 or 0.10 eV) to the far ultraviolet (10, 30 or 300 eV). The intervals between successive energies in the tables are chosen in such a way that the major spectral features are preserved.

Very small values of k are expressed in exponential notation, e.g., 1.23E-5 means 1.23×10^{-5}.

The following table is convenient for associating the energy entries in these tables with the corresponding wavelengths:

λ	E/eV	λ	E/eV
1 mm	0.00124	6000 Å	2.066
500 µm	0.00248	5000 Å	2.480
100 µm	0.01240	4000 Å	3.100
50 µm	0.02480	3000 Å	4.133
10 µm	0.12398	2000 Å	6.199
5 µm	0.24797	1000 Å	12.398
1 µm	1.240	400 Å	30.996

Energy (eV)	n	k	$R(\phi = 0)$	Energy (eV)	n	k	$R(\phi = 0)$	Energy (eV)	n	k	$R(\phi = 0)$
Aluminum[1]				2.200	1.018	6.846	0.9200	14.400	0.058	0.327	0.8102
0.040	98.595	203.701	0.9923	2.400	0.826	6.283	0.9228	14.600	0.067	0.273	0.7802
0.050	74.997	172.199	0.9915	2.600	0.695	5.800	0.9238	14.800	0.086	0.211	0.7202
0.060	62.852	150.799	0.9906	2.800	0.598	5.385	0.9242	15.000	0.125	0.153	0.6119
0.070	53.790	135.500	0.9899	3.000	0.523	5.024	0.9241	15.200	0.178	0.108	0.4903
0.080	45.784	123.734	0.9895	3.200	0.460	4.708	0.9243	15.400	0.234	0.184	0.3881
0.090	39.651	114.102	0.9892	3.400	0.407	4.426	0.9245	15.600	0.280	0.073	0.3182
0.100	34.464	105.600	0.9889	3.600	0.363	4.174	0.9246	15.800	0.318	0.065	0.2694
0.125	24.965	89.250	0.9884	3.800	0.326	3.946	0.9247	16.000	0.351	0.060	0.2326
0.150	18.572	76.960	0.9882	4.000	0.294	3.740	0.9248	16.200	0.380	0.055	0.2031
0.175	14.274	66.930	0.9879	4.200	0.267	3.552	0.9248	16.400	0.407	0.050	0.1789
0.200	11.733	59.370	0.9873	4.400	0.244	3.380	0.9249	16.750	0.448	0.045	0.1460
0.250	8.586	48.235	0.9858	4.600	0.223	3.222	0.9249	17.000	0.474	0.042	0.1278
0.300	6.759	40.960	0.9844	4.800	0.205	3.076	0.9249	17.250	0.498	0.040	0.1129
0.350	5.438	35.599	0.9834	5.000	0.190	2.942	0.9244	17.500	0.520	0.038	0.1005
0.400	4.454	31.485	0.9826	6.000	0.130	2.391	0.9257	17.750	0.540	0.036	0.0899
0.500	3.072	25.581	0.9817	6.500	0.110	2.173	0.9260	18.000	0.558	0.035	0.0809
0.600	2.273	21.403	0.9806	7.000	0.095	1.983	0.9262	18.500	0.591	0.032	0.0664
0.700	1.770	18.328	0.9794	7.500	0.082	1.814	0.9265	19.000	0.620	0.030	0.0554
0.800	1.444	15.955	0.9778	8.000	0.072	1.663	0.9269	19.500	0.646	0.028	0.0467
0.900	1.264	14.021	0.9749	8.500	0.063	1.527	0.9272	20.000	0.668	0.027	0.0398
1.000	1.212	12.464	0.9697	9.000	0.056	1.402	0.9277	20.500	0.689	0.025	0.0342
1.100	1.201	11.181	0.9630	9.500	0.049	1.286	0.9282	21.000	0.707	0.024	0.0296
1.200	1.260	10.010	0.9521	10.000	0.044	1.178	0.9286	21.500	0.724	0.023	0.0258
1.300	1.468	8.949	0.9318	10.500	0.040	1.076	0.9293	22.000	0.739	0.022	0.0226
1.400	2.237	8.212	0.8852	11.000	0.036	0.979	0.9298	22.500	0.753	0.021	0.0199
1.500	2.745	8.309	0.8678	11.500	0.033	0.883	0.9283	23.000	0.766	0.021	0.0177
1.600	2.625	8.597	0.8794	12.000	0.033	0.791	0.9224	23.500	0.778	0.020	0.0157
1.700	2.143	8.573	0.8972	12.500	0.034	0.700	0.9118	24.000	0.789	0.019	0.0140
1.800	1.741	8.205	0.9069	13.000	0.038	0.609	0.8960	24.500	0.799	0.018	0.0126
1.900	1.488	7.821	0.9116	13.500	0.041	0.517	0.8789	25.000	0.809	0.018	0.0113
2.000	1.304	7.479	0.9148	14.000	0.048	0.417	0.8486	25.500	0.817	0.017	0.0102
				14.200	0.053	0.373	0.8312	26.000	0.826	0.016	0.0092

Energy (eV)	n	k	R(φ = 0)
27.000	0.840	0.015	0.0076
28.000	0.854	0.014	0.0063
29.000	0.865	0.014	0.0053
30.000	0.876	0.013	0.0044
35.000	0.915	0.010	0.0020
40.000	0.940	0.008	0.0010
45.000	0.957	0.007	0.0005
50.000	0.969	0.006	0.0003
55.000	0.979	0.005	0.0001
60.000	0.987	0.004	0.0000
65.000	0.995	0.004	0.0000
70.000	1.006	0.004	0.0000
72.500	1.025	0.004	0.0002
75.000	1.011	0.024	0.0002
77.500	1.008	0.025	0.0002
80.000	1.007	0.024	0.0002
85.000	1.007	0.028	0.0002
90.000	1.005	0.031	0.0002
95.000	0.999	0.036	0.0003
100.000	0.991	0.030	0.0002
110.000	0.994	0.025	0.0002
120.000	0.991	0.024	0.0002
130.000	0.987	0.021	0.0001
140.000	0.989	0.016	0.0001
150.000	0.990	0.015	0.0001
160.000	0.989	0.014	0.0001
170.000	0.989	0.011	0.0001
180.000	0.990	0.010	0.0000
190.000	0.990	0.009	0.0000
200.000	0.991	0.007	0.0000
220.000	0.992	0.006	0.0000
240.000	0.993	0.005	0.0000
260.000	0.993	0.004	0.0000
280.000	0.994	0.003	0.0000
300.000	0.995	0.002	0.0000

Carbon (diamond)[2]

Energy (eV)	n	k	R(φ = 0)
0.06199	2.3741		0.166
0.06888	2.3741		0.166
0.07749	2.3745		0.166
0.08856	2.3750		0.166
0.1033	2.3757		0.166
0.1240	2.3765		0.166
0.1550	2.3772		0.166
0.1907		3.1 E-05	
0.2066	2.3779	5.7 E-05	0.166
0.22		1.21E-04	
0.23		2.36E-04	
0.24		3.82E-04	
0.25		5.21E-04	
0.26		2.96E-04	
0.27		4.39E-04	
0.28		2.75E-04	
0.29		7.82E-05	
0.30		1.32E-04	
0.31	2.3787	1.30E-04	0.167
0.32		1.11E-04	
0.33		2.99E-05	
0.34		1.89E-05	
0.35		2.11E-05	

Energy (eV)	n	k	R(φ = 0)
0.36		2.47E-05	
0.37		2.80E-05	
0.38		3.11E-05	
0.39		3.67E-05	
0.40		3.58E-05	
0.41		3.25E-05	
0.4133	2.3795		0.167
0.42		2.94E-05	
0.43		2.87E-05	
0.44		3.14E-05	
0.45		3.62E-05	
0.46		3.22E-05	
0.47		1.57E-05	
0.48		6.17E-06	
0.4959	2.3801		0.167
0.6199	2.3813		0.167
0.8266	2.3837		0.167
1.240	2.3905		0.168
1.378	2.3934		0.169
1.459	2.3953		0.169
1.550	2.3975		0.169
1.653	2.4003		0.170
1.771	2.4036		0.170
1.889	2.4073		0.171
1.926	2.4084		0.171
2.066	2.4133		0.171
2.105	2.4147		0.172
2.271	2.4210		0.173
2.480	2.4299		0.174
2.650	2.4380		0.175
2.845		3.82E-07	
3.100	2.4627		0.178
3.434	2.4849		0.182
3.576	2.4955		0.183
3.961		8.97E-07	
4.160	2.5465		0.190
4.511		1.29E-06	
4.8187	2.6205	1.47E-06	0.200
5.00	2.6383		0.203
5.30		2.98E-06	
5.35		6.45E-06	
5.40		1.04E-05	
5.50		3.41E-05	
5.55		5.48E-04	
5.60	2.740	1.48E-03	0.216
5.80	2.780	5.02E-03	0.222
6.00	2.826	7.99E-03	0.228
6.10	2.852	8.62E-03	0.231
6.20	2.879	9.30E-03	0.235
6.30	2.910	9.74E-03	0.239
6.40	2.944	9.87E-03	0.243
6.50	2.985	1.10E-02	0.248
6.60	3.031	1.47E-02	0.254
6.70	3.085	2.20E-02	0.261
6.80	3.146	3.44E-02	0.268
6.90	3.220	5.24E-02	0.277
7.00	3.322	9.35E-02	0.289
7.10	3.444	0.210	0.304
7.15	3.464	0.307	0.308
7.20	3.437	0.388	0.307

Energy (eV)	n	k	R(φ = 0)
7.30	3.376	0.473	0.303
7.40	3.335	0.515	0.300
7.50	3.321	0.533	0.299
7.60	3.306	0.592	0.300
7.80	3.276	0.659	0.300
8.00	3.251	0.712	0.300
8.25	3.232	0.765	0.301
8.50	3.228	0.806	0.303
8.75	3.247	0.855	0.308
9.00	3.272	0.910	0.314
9.25	3.308	0.978	0.322
9.50	3.348	1.055	0.331
9.75	3.398	1.147	0.342
10.00	3.453	1.258	0.355
10.25	3.514	1.403	0.371
10.50	3.565	1.581	0.389
10.75	3.600	1.813	0.411
11.00	3.582	2.078	0.434
11.25	3.507	2.380	0.460
11.50	3.346	2.693	0.488
11.75	3.090	2.986	0.518
12.00	2.736	3.228	0.551
12.20	2.383	3.354	0.580
12.40	1.983	3.382	0.610
12.60	1.532	3.265	0.641
12.80	1.312	2.953	0.627
13.00	1.223	2.722	0.604
13.50	1.129	2.379	0.557
14.00	1.070	2.178	0.526
14.50	1.018	2.034	0.504
15.00	0.972	1.929	0.489
15.50	0.917	1.845	0.482
16.00	0.861	1.767	0.477
16.50	0.805	1.692	0.474
17.00	0.753	1.619	0.471
17.50	0.707	1.546	0.467
18.00	0.665	1.476	0.463
18.50	0.626	1.408	0.459
19.00	0.589	1.341	0.455
19.50	0.557	1.273	0.449
20.00	0.527	1.203	0.442
21.00	0.487	1.052	0.413
22.00	0.518	0.888	0.330
23.00	0.597	0.850	0.270
24.00	0.586	0.829	0.268
25.00	0.562	0.787	0.265
26.00	0.538	0.736	0.260
27.00	0.516	0.679	0.252
28.00	0.501	0.616	0.239
29.00	0.494	0.552	0.221
30.00	0.493	0.490	0.201

Cesium (evaporated)[3]

Energy (eV)	n	k	R(φ = 0)
2.145	0.264	1.123	0.631
2.271	0.278	0.950	0.561
2.845	0.425	0.438	0.235
3.064	0.540	0.320	0.127
3.397	0.671	0.233	0.057
3.966	0.827	0.174	0.018
4.889	0.916	0.143	0.007

Chromium[4]

Energy (eV)	n	k	R(φ = 0)
0.06	21.19	42.00	0.962
0.10	11.81	29.76	0.955
0.14	15.31	26.36	0.936
0.18	8.73	25.37	0.53
0.22	5.30	20.62	0.954
0.26	3.91	17.12	0.951
0.30	3.15	14.28	0.943
0.42	3.47	8.97	0.862
0.54	3.92	7.06	0.788
0.66	3.96	5.95	0.736
0.78	4.13	5.03	0.680
0.90	4.43	4.60	0.650
1.00	4.47	4.43	0.639
1.12	4.53	4.31	0.631
1.24	4.50	4.28	0.629
1.36	4.42	4.30	0.631
1.46	4.31	4.32	0.632
1.77	3.84	4.37	0.639
2.00	3.48	4.36	0.644
2.20	3.18	4.41	0.656
2.40	2.75	4.46	0.677
2.60	2.22	4.36	0.698
2.80	1.80	4.06	0.703
3.00	1.54	3.71	0.695
3.20	1.44	3.40	0.670
3.40	1.39	3.24	0.657
3.60	1.26	3.12	0.661
3.80	1.12	2.95	0.660
4.00	1.02	2.76	0.651
4.20	0.94	2.58	0.639
4.40	0.90	2.42	0.620
4.50	0.89	2.35	0.607
4.60	0.88	2.28	0.598
4.70	0.86	2.21	0.586
4.80	0.86	2.13	0.572
4.90	0.86	2.07	0.557
5.00	0.85	2.01	0.542
5.10	0.86	1.94	0.523
5.20	0.87	1.87	0.503
5.40	0.93	1.80	0.466
5.60	0.95	1.74	0.443
5.80	0.97	1.74	0.437
6.00	0.94	1.73	0.444
6.20	0.89	1.69	0.446
6.40	0.85	1.66	0.447
6.60	0.80	1.59	0.444
6.80	0.75	1.51	0.439
7.00	0.74	1.45	0.425
7.20	0.71	1.39	0.414
7.40	0.69	1.33	0.404
7.60	0.66	1.23	0.378
7.80	0.67	1.15	0.347
8.00	0.68	1.07	0.315
8.20	0.71	1.00	0.278
8.50	0.74	0.92	0.235
9.0	0.83	0.81	0.170
9.50	0.92	0.74	0.132
10.00	0.98	0.73	0.120
10.50	1.01	0.72	0.112
11.00	1.05	0.69	0.103
11.50	1.09	0.69	0.100
12.00	1.13	0.70	0.101
12.50	1.15	0.73	0.108
13.00	1.15	0.77	0.119
13.50	1.12	0.80	0.128
14.00	1.09	0.82	0.135
14.50	1.03	0.82	0.142
15.00	1.00	0.82	0.143
15.50	0.96	0.80	0.141
16.00	0.92	0.77	0.139
16.50	0.31	0.75	0.134
17.00	0.90	0.73	0.132
17.50	0.88	0.72	0.130
18.00	0.87	0.70	0.129
18.50	0.84	0.69	0.130
19.00	0.82	0.68	0.131
20.00	0.77	0.64	0.130
20.5	0.76	0.63	0.129
21.0	0.74	0.58	0.121
21.5	0.72	0.55	0.116
22.0	0.71	0.52	0.112
22.5	0.70	0.50	0.109
23.0	0.69	0.48	0.105
23.5	0.68	0.45	0.101
24.0	0.68	0.43	0.096
24.5	0.67	0.39	0.089
25.0	0.68	0.36	0.080
25.5	0.68	0.33	0.072
26.0	0.70	0.31	0.063
26.5	0.71	0.28	0.055
27.0	0.72	0.26	0.048
27.5	0.73	0.25	0.043
28.0	0.75	0.23	0.037
29.0	0.77	0.22	0.032
30.0	0.78	0.21	0.030

Cobalt, single crystal, $\vec{E} \parallel \hat{c}^5$

Energy (eV)	n	k	R(φ = 0)
0.10	6.71	37.87	0.982
0.15	4.66	25.47	0.973
0.20	3.55	18.78	0.962
0.25	3.98	14.59	0.933
0.30	4.04	12.16	0.907
0.40	4.24	9.13	0.847
0.50	4.41	7.19	0.782
0.60	4.91	6.13	0.729
0.70	5.24	5.85	0.713
0.80	5.17	5.89	0.716
0.90	4.94	5.95	0.720
1.00	4.46	5.86	0.722
1.10	4.07	5.61	0.715
1.20	3.81	5.36	0.706
1.30	3.60	5.20	0.701
1.40	3.37	5.09	0.701
1.50	3.10	4.96	0.701
1.60	2.84	4.77	0.697
1.70	2.66	4.57	0.690
1.80	2.45	4.41	0.687
1.90	2.31	4.18	0.675
2.00	2.21	4.00	0.664
2.10	2.13	3.85	0.654

Cobalt, single crystal, $\vec{E} \perp \hat{c}^5$

Energy (eV)	n	k	R(φ = 0)
2.20	2.07	3.70	0.642
2.30	2.01	3.59	0.634
2.40	1.95	3.49	0.627
2.50	1.88	3.40	0.622
2.60	1.81	3.32	0.618
2.70	1.73	3.24	0.615
2.80	1.66	3.13	0.607
2.90	1.61	3.05	0.600
3.00	1.55	2.96	0.594
3.20	1.46	2.80	0.579
3.40	1.38	2.64	0.563
3.60	1.31	2.48	0.544
3.80	1.28	2.33	0.519
4.00	1.26	2.20	0.495
4.20	1.25	2.10	0.471
4.40	1.24	2.01	0.452
4.60	1.24	1.94	0.435
4.80	1.23	1.88	0.423
5.00	1.22	1.83	0.411
5.20	1.21	1.79	0.403
5.40	1.19	1.77	0.399
5.60	1.16	1.75	0.400
5.80	1.10	1.73	0.406
6.00	1.03	1.68	0.407
6.20	0.97	1.62	0.401
6.40	0.94	1.53	0.386
6.60	0.91	1.46	0.368
6.80	0.91	1.38	0.345
7.00	0.91	1.32	0.326
7.00	0.91	1.26	0.305
7.40	0.92	1.21	0.286
7.60	0.93	1.17	0.269
7.80	0.94	1.13	0.253
8.00	0.95	1.09	0.239

Cobalt, single crystal, $\vec{E} \perp \hat{c}^5$

Energy (eV)	n	k	R(φ = 0)
0.10	5.83	32.36	0.979
0.15	4.24	21.37	0.965
0.20	3.87	15.53	0.042
0.30	4.34	10.01	0.865
0.40	4.66	7.39	0.785
0.50	5.17	5.75	0.709
0.60	5.77	5.17	0.682
0.70	6.15	5.20	0.685
0.80	6.08	5.61	0.702
0.90	5.57	5.93	0.715
1.00	4.83	5.94	0.721
1.10	4.31	5.60	0.711
1.20	4.02	5.34	0.701
1.30	3.78	5.16	0.694
1.40	3.55	5.05	0.692
1.50	3.26	4.93	0.692
1.60	3.03	4.74	0.687
1.70	2.83	4.60	0.684
1.80	2.61	4.45	0.683
1.90	2.41	4.27	0.677
2.00	2.25	4.09	0.670
2.10	2.13	3.89	0.659
2.20	2.04	3.72	0.646
2.30	1.99	3.56	0.632
2.40	1.95	3.44	0.620

Energy (eV)	n	k	R(φ = 0)
2.50	1.90	3.34	0.611
2.60	1.86	3.26	0.605
2.70	1.79	3.19	0.602
2.80	1.72	3.11	0.596
2.90	1.66	3.03	0.591
3.00	1.60	2.94	0.586
3.20	1.50	2.78	0.571
3.40	1.42	2.62	0.553
3.60	1.36	2.47	0.533
3.80	1.33	2.33	0.511
4.00	1.31	2.21	0.488
4.20	1.28	2.12	0.471
4.40	1.27	2.03	0.452
4.60	1.26	1.95	0.435
4.80	1.25	1.90	0.423
5.00	1.24	1.84	0.411
5.20	1.22	1.80	0.403
5.40	1.21	1.78	0.399
5.60	1.17	1.76	0.400
5.80	1.11	1.74	0.406
6.00	1.04	1.69	0.407
6.20	0.98	1.62	0.401
6.40	0.94	1.54	0.386
6.60	0.92	1.46	0.368
6.80	0.91	1.38	0.345
7.00	0.91	1.32	0.326
7.20	0.91	1.26	0.305
7.40	0.92	1.21	0.285
7.60	0.93	1.17	0.269
7.80	0.94	1.13	0.253

Copper[6]

Energy (eV)	n	k	R(φ = 0)
0.10	29.69	71.57	0.980
0.50	1.71	17.63	0.979
1.00	0.44	8.48	0.976
1.50	0.26	5.26	0.965
1.70	0.22	4.43	0.958
1.75	0.21	4.25	0.956
1.80	0.21	4.04	0.952
1.85	0.22	3.85	0.947
1.90	0.21	3.67	0.943
2.00	0.27	3.24	0.910
2.10	0.47	2.81	0.814
2.20	0.83	2.60	0.673
2.30	1.04	2.59	0.618
2.40	1.12	2.60	0.602
2.60	1.15	2.50	0.577
2.80	1.17	2.36	0.545
3.00	1.18	2.21	0.509
3.20	1.23	2.07	0.468
3.40	1.27	1.95	0.434
3.60	1.31	1.87	0.407
3.80	1.34	1.81	0.387
4.00	1.34	1.72	0.364
4.20	1.42	1.64	0.336
4.40	1.49	1.64	0.329
4.60	1.52	1.67	0.334
4.80	1.53	1.71	0.345
5.00	1.47	1.78	0.366
5.20	1.38	1.80	0.380
5.40	1.28	1.78	0.389

Energy (eV)	n	k	R(φ = 0)
5.60	1.18	1.74	0.391
5.80	1.10	1.67	0.389
6.00	1.04	1.59	0.380
6.50	0.96	1.37	0.329
7.00	0.97	1.20	0.271
7.50	1.00	1.09	0.230
8.00	1.03	1.03	0.206
8.50	1.03	0.98	0.189
9.00	1.03	0.92	0.171
9.50	1.03	0.87	0.154
10.00	1.04	0.82	0.139
11.00	1.07	0.75	0.118
12.00	1.09	0.73	0.111
13.00	1.08	0.72	0.109
14.00	1.06	0.72	0.111
14.50	1.03	0.72	0.111
15.00	1.01	0.71	0.111
15.50	0.98	0.69	0.109
16.00	0.95	0.67	0.106
17.00	0.91	0.62	0.097
18.00	0.89	0.56	0.084
19.00	0.88	0.51	0.071
20.00	0.88	0.45	0.059
21.00	0.90	0.41	0.048
22.00	0.92	0.38	0.040
23.00	0.94	0.37	0.035
24.00	0.96	0.37	0.035
25.00	0.96	0.40	0.040
26.00	0.92	0.40	0.044
27.00	0.88	0.38	0.043
28.00	0.86	0.35	0.039
29.00	0.85	0.30	0.032
30.00	0.86	0.26	0.025
31.00	0.88	0.24	0.020
32.00	0.89	0.22	0.017
33.00	0.90	0.21	0.015
34.00	0.91	0.20	0.014
35.00	0.92	0.20	0.013
36.00	0.92	0.19	0.012
37.00	0.92	0.19	0.011
38.00	0.93	0.18	0.010
39.00	0.93	0.17	0.009
40.00	0.93	0.17	0.009
41.00	0.94	0.16	0.008
42.00	0.94	0.16	0.007
43.00	0.94	0.15	0.007
44.00	0.95	0.15	0.007
45.00	0.95	0.15	0.006
46.00	0.95	0.15	0.006
47.00	0.95	0.14	0.006
48.00	0.95	0.14	0.006
49.00	0.95	0.14	0.005
50.00	0.95	0.13	0.005
51.00	0.95	0.13	0.005
52.00	0.95	0.13	0.005
53.00	0.96	0.12	0.004
54.00	0.96	0.12	0.004
55.00	0.96	0.12	0.004
56.00	0.96	0.11	0.004
57.00	0.96	0.11	0.004

Energy (eV)	n	k	R(φ = 0)
58.00	0.96	0.11	0.004
59.00	0.97	0.11	0.003
60.00	0.97	0.11	0.003
61.00	0.97	0.11	0.003
62.00	0.97	0.11	0.003
63.00	0.96	0.10	0.003
64.00	0.96	0.10	0.003
65.00	0.97	0.10	0.003
66.00	0.97	0.10	0.003
67.00	0.97	0.09	0.003
68.00	0.97	0.09	0.002
69.00	0.97	0.09	0.002
70.00	0.97	0.09	0.002
75.00	0.98	0.09	0.002
80.00	0.98	0.09	0.002
85.00	0.97	0.09	0.002
90.00	0.96	0.08	0.002

Gallium (liquid)[7]

Energy (eV)	n	k	R(φ = 0)
1.425	2.40	9.20	0.900
1.550	2.09	8.50	0.898
1.771	1.65	7.60	0.898
2.066	1.25	6.60	0.897
2.480	0.89	5.60	0.898
3.100	0.59	4.50	0.896

Germanium, single crystal[8]

Energy (eV)	n	k	R(φ = 0)
0.01240	(4.0065)	3.00E-03	0.361
0.01364	4.0063	2.40E-03	0.361
0.01488	(4.0060)	1.70E-03	0.361
0.01612	(4.0060)	1.55E-03	0.361
0.01736	(4.0060)	1.50E-03	0.361
0.01860		1.50E-03	
0.01984		1.60E-03	
0.02108		1.60E-03	
0.02232		1.55E-03	
0.02356		1.53E-03	
0.02480		1.50E-03	
0.02604		1.25E-03	
0.02728		8.50E-04	
0.02852		6.50E-04	
0.02976		7.00E-04	
0.03100	3.9827	8.50E-04	0.358
0.03224		1.55E-03	
0.03348		2.75E-03	
0.03472		3.55E-03	
0.03596	(3.9900)	3.05E-03	0.359
0.03720		2.75E-03	
0.03844		2.70E-03	
0.03968	(3.9930)	2.90E-03	0.359
0.04092		2.95E-03	
0.04215		3.20E-03	
0.04339		6.30E-03	
0.04463		3.40E-03	
0.04587	(3.9955)	2.50E-03	0.360
0.04711		2.10E-03	
0.04835		2.00E-03	
0.04959		8.00E-04	
0.05083		1.40E-03	
0.05207		1.35E-03	
0.05331		1.10E-03	
0.05455		8.00E-04	

Energy (eV)	n	k	$R(\phi = 0)$
0.05579		6.00E-04	
0.05703		9.0 E-04	
0.05827		6.5 E-04	
0.05951		4.6 E-04	
0.06075		4.0 E-04	
0.06199	3.9992	3.98E-04	0.360
0.06323		4.0 E-04	
0.06447		4.3 E-04	
0.06571		4.4 E-04	
0.06695	(4.0000)	4.3 E-04	0.360
0.06819		3.1 E-04	
0.06943		3.3 E-04	
0.07067		3.8 E-04	
0.07191		3.3 E-04	
0.07315		2.5 E-04	
0.07439		1.9 E-04	
0.07514		1.58E-04	
0.07749	4.0009	9.55E-05	0.360
0.07999	4.0011	1.71E-04	0.360
0.08266	4.0013	9.78E-05	0.360
0.08551	4.0015	5.77E-05	0.360
0.08920		3.98E-05	
0.09460		4.59E-05	
0.09840		3.51E-05	
0.1	4.0063	3.70E-05	0.361
0.2	4.0108		0.361
0.3	4.0246		0.362
0.4	4.0429		0.364
0.5	(4.074)		0.367
0.6	(4.104)	6.58E-07	0.370
0.7	4.180	1.27E-04	0.377
0.8	4.275	5.67E-03	0.385
0.9	4.285	7.45E-02	0.386
1.0	4.325	8.09E-02	0.390
1.1	4.385	0.103	0.395
1.2	4.420	0.123	0.398
1.3	4.495	0.167	0.405
1.4	4.560	0.190	0.411
1.5	4.635	0.298	0.418
1.6	4.763	0.345	0.428
1.7	4.897	0.401	0.439
1.8	5.067	0.500	0.453
1.9	5.380	0.540	0.475
2.0	5.588	0.933	0.495
2.1	5.748	1.634	0.523
2.2	5.283	2.049	0.516
2.3	5.062	2.318	0.519
2.4	4.610	2.455	0.508
2.5	4.340	2.384	0.492
2.6	4.180	2.309	0.480
2.7	4.082	2.240	0.471
2.8	4.035	2.181	0.464
2.9	4.037	2.140	0.461
3.0	4.082	2.145	0.463
3.1	4.141	2.215	0.471
3.2	4.157	2.340	0.482
3.3	4.128	2.469	0.490
3.4	4.070	2.579	0.497
3.5	4.020	2.667	0.502
3.6	3.985	2.759	0.509

Energy (eV)	n	k	$R(\phi = 0)$
3.7	3.958	2.863	0.517
3.8	3.936	2.986	0.527
3.9	3.920	3.137	0.539
4.0	3.905	3.336	0.556
4.1	3.869	3.614	0.579
4.2	3.745	4.009	0.612
4.3	3.338	4.507	0.659
4.4	2.516	4.669	0.705
4.5	1.953	4.297	0.713
4.6	1.720	3.960	0.702
4.7	1.586	3.709	0.690
4.8	1.498	3.509	0.677
4.9	1.435	3.342	0.664
5.0	1.394	3.197	0.650
5.1	1.370	3.073	0.636
5.2	1.364	2.973	0.622
5.3	1.371	2.897	0.609
5.4	1.383	2.854	0.600
5.5	1.380	2.842	0.598
5.6	1.360	2.846	0.602
5.7	1.293	2.163	0.479
5.8	1.209	2.873	0.632
5.9	1.108	2.813	0.641
6.0	1.30	2.34	0.517
6.5	1.10	2.05	0.489
7.0	1.00	1.80	0.448
7.5		1.60	
8.0	0.92	1.40	0.348
8.5	0.92	1.20	0.282
9.0	0.92	1.14	0.262
9.5		1.00	
10.0	0.93	0.86	0.167
20.0		0.237	
22.0		0.179	
24.0		0.144	
26.0		0.110	
28.0		0.0747	
30.0		0.1020	
32.0		0.0999	
34.0		0.0856	
36.0		0.0740	
38.0		0.0651	
40.0		0.0604	

Gold, electropolished, Au (110)[9]

Energy (eV)	n	k	$R(\phi = 0)$
0.10	8.17	82.83	0.995
0.20	2.13	41.73	0.995
0.30	0.99	27.82	0.995
0.40	0.59	20.83	0.995
0.50	0.39	16.61	0.994
0.60	0.28	13.78	0.994
0.70	0.22	11.75	0.994
0.80	0.18	10.21	0.993
0.90	0.15	9.01	0.993
1.00	0.13	8.03	0.992
1.20	0.10	6.54	0.991
1.40	0.08	5.44	0.989
1.60	0.08	4.56	0.986
1.80	0.09	3.82	0.979
2.00	0.13	3.16	0.953
2.10	0.18	2.84	0.925

Energy (eV)	n	k	$R(\phi = 0)$
2.20	0.24	2.54	0.880
2.40	0.50	1.86	0.647
2.50	0.82	1.59	0.438
2.60	1.24	1.54	0.331
2.70	1.43	1.72	0.356
2.80	1.46	1.77	0.368
2.90	1.50	1.79	0.368
3.00	1.54	1.80	0.369
3.10	1.54	1.81	0.371
3.20	1.54	1.80	0.368
3.30	1.55	1.78	0.362
3.40	1.56	1.76	0.356
3.50	1.58	1.73	0.349
3.60	1.62	1.73	0.346
3.70	1.64	1.75	0.351
3.80	1.63	1.79	0.360
3.90	1.59	1.81	0.366
4.00	1.55	1.81	0.369
4.10	1.51	1.79	0.368
4.20	1.48	1.78	0.367
4.30	1.45	1.77	0.368
4.40	1.41	1.76	0.370
4.50	1.35	1.74	0.370
4.60	1.30	1.69	0.364
4.70	1.27	1.64	0.354
4.80	1.25	1.59	0.344
4.90	1.23	1.54	0.332
5.00	1.22	1.49	0.319
5.20	1.21	1.40	0.295
5.40	1.21	1.33	0.275
5.60	1.21	1.27	0.256
5.80	1.21	1.20	0.236
6.00	1.22	1.14	0.218
6.20	1.24	1.09	0.203
6.40	1.25	1.05	0.190
6.60	1.27	1.01	0.177
6.80	1.30	0.97	0.167
7.00	1.34	0.95	0.162
7.20	1.36	0.95	0.161
7.40	1.38	0.96	0.164
7.60	1.38	0.98	0.169
7.80	1.35	0.99	0.171
8.00	1.31	0.96	0.165
8.20	1.30	0.92	0.155
8.40	1.30	0.89	0.147
8.60	1.31	0.88	0.144
8.80	1.31	0.86	0.140
9.00	1.30	0.83	0.133
9.20	1.31	0.81	0.126
9.40	1.33	0.78	0.122
9.60	1.36	0.78	0.121
9.80	1.37	0.79	0.124
10.00	1.37	0.80	0.126
10.20	1.36	0.80	0.127
10.40	1.35	0.80	0.125
10.60	1.34	0.79	0.123
10.80	1.34	0.77	0.120
11.00	1.34	0.76	0.116
11.20	1.34	0.74	0.113
11.40	1.35	0.73	0.111

Energy (eV)	n	k	$R(\phi = 0)$
11.60	1.36	0.72	0.109
11.80	1.38	0.71	0.108
12.00	1.39	0.71	0.109
12.40	1.44	0.73	0.115
12.80	1.45	0.79	0.127
13.20	1.42	0.84	0.137
13.60	1.37	0.86	0.140
14.00	1.33	0.86	0.140
14.40	1.29	0.86	0.139
14.80	1.26	0.84	0.135
15.20	1.24	0.83	0.132
15.60	1.22	0.81	0.127
16.00	1.21	0.79	0.123
16.40	1.20	0.78	0.119
16.80	1.19	0.76	0.116
17.20	1.19	0.75	0.114
17.60	1.19	0.74	0.111
18.00	1.19	0.74	0.109
18.40	1.19	0.73	0.109
18.80	1.20	0.74	0.110
19.20	1.21	0.76	0.116
19.60	1.21	0.80	0.125
20.00	1.18	0.83	0.133
20.40	1.14	0.85	0.141
20.80	1.10	0.87	0.149
21.20	1.05	0.88	0.156
21.60	1.00	0.88	0.162
22.00	0.94	0.86	0.164
22.40	0.89	0.83	0.163
22.80	0.85	0.79	0.157
23.20	0.82	0.75	0.149
23.60	0.80	0.70	0.138
24.00	0.80	0.66	0.125
24.40	0.80	0.62	0.113
24.80	0.80	0.58	0.101
25.20	0.82	0.56	0.090
25.60	0.83	0.54	0.084
26.00	0.84	0.52	0.079
26.40	0.85	0.51	0.074
26.80	0.85	0.50	0.071
27.20	0.86	0.49	0.068
27.60	0.86	0.49	0.065
28.00	0.87	0.48	0.063
28.40	0.88	0.48	0.062
28.80	0.88	0.48	0.062
29.20	0.88	0.48	0.062
29.60	0.87	0.48	0.064
30.00	0.86	0.48	0.064

Hafnium, single crystal, $\vec{E} \parallel \hat{c}^{10}$

Energy (eV)	n	k	$R(\phi = 0)$
0.52	1.48	4.11	0.747
0.56	1.84	3.29	0.615
0.60	2.34	2.62	0.486
0.66	3.21	2.13	0.428
0.70	3.70	2.03	0.441
0.76	4.31	2.10	0.476
0.80	4.61	2.31	0.504
0.86	4.71	2.70	0.533
0.90	4.64	2.85	0.541
0.95	4.54	2.96	0.545
1.00	4.45	3.00	0.545

Energy (eV)	n	k	$R(\phi = 0)$
1.10	4.28	3.08	0.547
1.20	4.08	3.10	0.544
1.30	3.87	3.04	0.536
1.40	3.72	2.95	0.525
1.50	3.60	2.85	0.514
1.60	3.52	2.73	0.500
1.70	3.52	2.61	0.488
1.80	3.57	2.56	0.485
1.90	3.63	2.59	0.489
2.00	3.65	2.67	0.498
2.10	3.64	2.81	0.511
2.20	3.53	2.99	0.526
2.30	3.34	3.09	0.534
2.40	3.15	3.11	0.537
2.50	2.99	3.13	0.540
2.60	2.83	3.12	0.542
2.70	2.68	3.10	0.542
2.80	2.54	3.08	0.543
2.90	2.40	3.04	0.544
3.00	2.27	3.00	0.544
3.10	2.14	2.95	0.544
3.20	2.00	2.89	0.544
3.30	1.87	2.79	0.538
3.40	1.78	2.68	0.528
3.50	1.71	2.58	0.517
3.60	1.66	2.48	0.503
3.70	1.63	2.40	0.491
3.80	1.60	2.33	0.481
3.90	1.56	2.27	0.473
4.00	1.52	2.21	0.466
4.10	1.48	2.14	0.455
4.20	1.45	2.07	0.442
4.30	1.43	2.01	0.431
4.40	1.41	1.95	0.420
4.50	1.39	1.89	0.407
4.60	1.39	1.83	0.394
4.70	1.39	1.79	0.382
4.80	1.38	1.75	0.373
4.90	1.38	1.71	0.364
5.00	1.37	1.68	0.356
5.20	1.36	1.61	0.341
5.40	1.35	1.55	0.324
5.60	1.35	1.51	0.314
5.80	1.32	1.48	0.308
6.00	1.28	1.41	0.295
6.20	1.26	1.35	0.278
6.40	1.26	1.28	0.258
6.60	1.27	1.22	0.240
6.80	1.28	1.16	0.224
7.00	1.31	1.13	0.212
7.20	1.33	1.10	0.204
7.40	1.34	1.07	0.197
7.60	1.36	1.05	0.191
7.80	1.37	1.02	0.183
8.00	1.40	1.01	0.179
8.20	1.43	1.01	0.178
8.40	1.45	1.01	0.180
8.60	1.47	1.02	0.183
8.80	1.48	1.04	0.186
9.00	1.49	1.07	0.193

Energy (eV)	n	k	$R(\phi = 0)$
9.20	1.50	1.10	0.201
9.40	1.48	1.14	0.211
9.60	1.46	1.18	0.222
9.80	1.41	1.21	0.230
10.00	1.36	1.22	0.235
10.20	1.32	1.22	0.238
10.40	1.28	1.22	0.240
10.60	1.24	1.21	0.241
10.80	1.20	1.20	0.242
11.00	1.16	1.19	0.242
11.20	1.13	1.17	0.241
11.40	1.10	1.16	0.241
11.60	1.07	1.14	0.239
11.80	1.04	1.12	0.238
12.00	1.02	1.10	0.236
12.40	0.96	1.06	0.232
12.80	0.92	1.01	0.225
13.20	0.88	0.96	0.218
13.60	0.84	0.90	0.205
14.00	0.83	0.83	0.186
14.40	0.83	0.80	0.172
14.80	0.81	0.76	0.167
15.20	0.79	0.70	0.153
15.60	0.79	0.64	0.132
16.00	0.83	0.60	0.111
16.40	0.81	0.60	0.114
16.80	0.79	0.55	0.105
17.20	0.79	0.50	0.089
17.60	0.80	0.46	0.077
18.00	0.81	0.42	0.064
18.40	0.84	0.38	0.051
18.80	0.87	0.34	0.040
19.00	0.89	0.33	0.036
19.60	0.93	0.32	0.030
20.00	0.94	0.31	0.027
20.60	0.97	0.30	0.023
21.00	0.99	0.29	0.022
21.60	1.01	0.28	0.020
22.00	1.03	0.28	0.020
22.60	1.06	0.28	0.020
23.00	1.07	0.28	0.021
23.60	1.09	0.29	0.022
24.00	1.09	0.30	0.023
24.60	1.10	0.31	0.024

Hafnium, single crystal, $\vec{E} \perp \hat{c}^{10}$

Energy (eV)	n	k	$R(\phi = 0)$
0.52	2.25	4.65	0.723
0.56	2.34	3.66	0.623
0.60	2.84	2.89	0.512
0.66	3.71	2.35	0.469
0.70	4.26	2.21	0.482
0.76	4.97	2.33	0.521
0.80	5.41	2.62	0.554
0.86	5.46	3.36	0.593
0.90	5.22	3.62	0.601
0.95	4.95	3.72	0.602
1.00	4.76	3.76	0.602
1.10	4.43	3.80	0.601
1.20	4.07	3.74	0.594
1.30	3.79	3.55	0.578
1.40	3.61	3.36	0.561

Energy (eV)	n	k	R(φ = 0)
1.50	3.55	3.13	0.540
1.60	3.58	3.01	0.529
1.70	3.63	2.98	0.526
1.80	3.66	3.02	0.530
1.90	3.63	3.14	0.541
2.00	3.51	3.26	0.551
2.10	3.35	3.33	0.558
2.20	3.18	3.36	0.563
2.30	2.99	3.39	0.568
2.40	2.78	3.35	0.569
2.50	2.65	3.26	0.562
2.60	2.54	3.22	0.560
2.70	2.42	3.17	0.559
2.80	2.31	3.13	0.558
2.90	2.20	3.08	0.558
3.00	2.08	3.05	0.561
3.10	1.94	2.98	0.560
3.20	1.83	2.88	0.555
3.30	1.74	2.78	0.547
3.40	1.68	2.69	0.538
3.50	1.62	2.61	0.529
3.60	1.57	2.52	0.519
3.70	1.53	2.45	0.510
3.80	1.49	2.38	0.501
3.90	1.45	2.32	0.493
4.00	1.41	2.25	0.484
4.10	1.38	2.18	0.474
4.20	1.35	2.11	0.462
4.30	1.33	2.05	0.451
4.40	1.31	1.99	0.438
4.50	1.30	1.93	0.427
4.60	1.29	1.88	0.415
4.70	1.28	1.82	0.402
4.80	1.28	1.77	0.389
4.90	1.27	1.73	0.379
5.00	1.27	1.69	0.367
5.20	1.27	1.62	0.349
5.40	1.27	1.57	0.335
5.60	1.26	1.52	0.322
5.80	1.24	1.48	0.313
6.00	1.21	1.42	0.302
6.20	1.19	1.36	0.285
6.40	1.18	1.29	0.265
6.60	1.19	1.22	0.244
6.80	1.21	1.18	0.230
7.00	1.22	1.14	0.217
7.20	1.23	1.10	0.206
7.40	1.26	1.06	0.194
7.60	1.28	1.04	0.187
7.80	1.30	1.02	0.180
8.00	1.33	1.00	0.174
8.20	1.35	0.99	0.173
8.40	1.38	0.99	0.173
8.60	1.40	1.00	0.174
8.80	1.42	1.02	0.178
9.00	1.43	1.04	0.184
9.20	1.45	1.08	0.193
9.40	1.43	1.12	0.204
9.60	1.40	1.16	0.214
9.80	1.37	1.19	0.223

Energy (eV)	n	k	R(φ = 0)
10.00	1.32	1.21	0.230
10.20	1.27	1.21	0.234
10.40	1.23	1.20	0.235
10.60	1.19	1.20	0.237
10.80	1.15	1.19	0.237
11.00	1.12	1.17	0.237
11.20	1.08	1.16	0.237
11.40	1.05	1.14	0.236
11.60	1.03	1.12	0.235
11.80	1.00	1.10	0.233
12.00	0.97	1.08	0.231
12.40	0.92	1.04	0.226
12.80	0.88	0.99	0.219
13.20	0.83	0.94	0.211
13.60	0.80	0.88	0.196
14.00	0.79	0.81	0.177
14.40	0.80	0.77	0.160
14.80	0.77	0.73	0.154
15.20	0.76	0.68	0.140
15.60	0.76	0.61	0.119
16.00	0.81	0.58	0.099
16.40	0.78	0.57	0.102
16.80	0.77	0.53	0.092
17.20	0.77	0.48	0.077
17.60	0.79	0.44	0.065
18.00	0.80	0.39	0.053
18.40	0.82	0.36	0.041
18.80	0.86	0.33	0.032
19.00	0.88	0.32	0.030
19.60	0.91	0.31	0.025
20.00	0.93	0.30	0.023
20.60	0.96	0.29	0.021
21.00	0.97	0.29	0.020
21.60	1.00	0.28	0.019
22.00	1.01	0.28	0.019
22.60	1.03	0.27	0.018
23.00	1.05	0.28	0.019
23.60	1.06	0.28	0.020
24.00	1.07	0.29	0.021
24.60	1.09	0.30	0.022

Iridium[11]

Energy (eV)	n	k	R(φ = 0)
0.10	28.49	60.62	0.975
0.15	15.32	45.15	0.973
0.20	9.69	35.34	0.972
0.25	6.86	28.84	0.969
0.30	5.16	24.25	0.967
0.35	4.11	20.79	0.964
0.40	3.42	18.06	0.960
0.45	3.05	15.82	0.954
0.50	2.98	14.06	0.944
0.60	2.79	11.58	0.925
0.70	2.93	9.78	0.895
0.80	3.14	8.61	0.862
0.90	3.19	7.88	0.840
1.00	3.15	7.31	0.822
1.10	3.04	6.84	0.808
1.20	2.96	6.41	0.791
1.30	2.85	6.07	0.779
1.40	2.72	5.74	0.767
1.50	2.65	5.39	0.750

Energy (eV)	n	k	R(φ = 0)
1.60	2.68	5.08	0.728
1.70	2.69	4.92	0.716
1.80	2.64	4.81	0.710
1.90	2.57	4.68	0.704
2.00	2.50	4.57	0.699
2.10	2.40	4.48	0.697
2.20	2.29	4.38	0.695
2.30	2.18	4.26	0.692
2.40	2.07	4.14	0.689
2.50	1.98	4.00	0.682
2.60	1.91	3.86	0.673
2.70	1.85	3.73	0.665
2.80	1.81	3.61	0.655
2.90	1.77	3.51	0.646
3.00	1.73	3.43	0.640
3.20	1.62	3.26	0.629
3.40	1.53	3.05	0.610
3.60	1.52	2.81	0.573
3.80	1.61	2.69	0.541
4.00	1.64	2.68	0.535
4.20	1.58	2.71	0.549
4.40	1.45	2.68	0.561
4.60	1.31	2.60	0.567
4.80	1.18	2.49	0.570
5.00	1.10	2.35	0.559
5.20	1.04	2.22	0.543
5.40	1.00	2.09	0.522
5.60	0.98	1.98	0.499
5.80	0.96	1.86	0.474
6.00	0.95	1.78	0.454
6.20	0.94	1.68	0.427
6.40	0.94	1.59	0.401
6.60	0.94	1.50	0.375
6.80	0.95	1.42	0.345
7.00	0.97	1.34	0.318
7.20	0.99	1.27	0.290
7.40	1.02	1.20	0.262
7.60	1.03	1.14	0.241
7.80	1.08	1.06	0.208
8.00	1.13	1.03	0.191
8.20	1.18	1.00	0.179
8.40	1.22	0.98	0.171
8.60	1.26	0.96	0.164
8.80	1.29	0.95	0.160
9.00	1.33	0.94	0.157
9.20	1.36	0.95	0.159
9.40	1.39	0.95	0.161
9.60	1.42	0.97	0.163
9.80	1.44	0.99	0.169
10.00	1.45	1.01	0.175
10.20	1.45	1.04	0.182
10.40	1.44	1.07	0.187
10.60	1.43	1.09	0.193
10.80	1.41	1.12	0.200
11.00	1.38	1.13	0.206
11.20	1.34	1.14	0.208
11.40	1.31	1.13	0.208
11.60	1.28	1.12	0.206
11.80	1.25	1.10	0.203
12.00	1.24	1.08	0.199

Energy (eV)	n	k	R(φ = 0)	Energy (eV)	n	k	R(φ = 0)	Energy (eV)	n	k	R(φ = 0)
12.40	1.21	1.05	0.191	1.20	3.24	4.26	0.641	9.50	0.90	1.02	0.226
12.80	1.19	1.01	0.181	1.30	3.16	4.07	0.626	9.67	0.90	1.00	0.221
13.20	1.18	0.98	0.173	1.40	3.12	3.87	0.609	9.83	0.89	0.99	0.218
13.60	1.17	0.95	0.165	1.50	3.05	3.77	0.601	10.00	0.88	0.97	0.213
14.00	1.16	0.91	0.155	1.60	3.00	3.60	0.585	10.17	0.87	0.94	0.203
14.40	1.17	0.88	0.147	1.70	2.98	3.52	0.577	10.33	0.87	0.91	0.196
14.80	1.18	0.87	0.142	1.80	2.92	3.46	0.573	10.50	0.87	0.89	0.189
15.20	1.19	0.84	0.136	1.90	2.89	3.37	0.563	10.67	0.88	0.87	0.179
15.60	1.20	0.83	0.133	2.00	2.85	3.36	0.563	10.83	0.89	0.85	0.170
16.00	1.21	0.83	0.131	2.10	2.80	3.34	0.562	11.00	0.91	0.83	0.162
16.40	1.23	0.82	0.129	2.20	2.74	3.33	0.563	11.17	0.92	0.83	0.159
16.80	1.25	0.82	0.127	2.30	2.65	3.34	0.567	11.33	0.93	0.84	0.159
17.20	1.28	0.83	0.131	2.40	2.56	3.31	0.567	11.50	0.93	0.84	0.160
17.60	1.30	0.87	0.140	2.50	2.46	3.31	0.570	11.67	0.93	0.84	0.162
18.00	1.30	0.93	0.154	2.60	2.34	3.30	0.576	11.83	0.92	0.84	0.163
18.40	1.27	0.97	0.166	2.70	2.23	3.25	0.575	12.00	0.91	0.84	0.163
18.80	1.24	1.00	0.176	2.80	2.12	3.23	0.580	12.17	0.90	0.84	0.165
19.20	1.20	1.03	0.187	2.90	2.01	3.17	0.580	12.33	0.89	0.83	0.164
19.60	1.15	1.05	0.197	3.00	1.88	3.12	0.583	12.50	0.88	0.83	0.165
20.00	1.10	1.06	0.205	3.10	1.78	3.04	0.580	12.67	0.87	0.82	0.166
20.50	1.04	1.05	0.210	3.20	1.70	2.96	0.576	12.83	0.86	0.81	0.166
21.00	0.99	1.04	0.215	3.30	1.62	2.87	0.572	13.00	0.85	0.80	0.162
21.50	0.94	1.02	0.220	3.40	1.55	2.79	0.565	13.17	0.84	0.79	0.161
22.00	0.89	1.00	0.222	3.50	1.50	2.70	0.556	13.33	0.84	0.78	0.160
22.50	0.84	0.99	0.228	3.60	1.47	2.63	0.548	13.50	0.83	0.77	0.159
23.00	0.79	0.96	0.232	3.70	1.43	2.56	0.542	13.67	0.82	0.76	0.157
23.50	0.76	0.92	0.228	3.83	1.38	2.49	0.534	13.83	0.81	0.75	0.154
24.00	0.73	0.87	0.223	4.00	1.30	2.39	0.527	14.00	0.81	0.73	0.151
24.50	0.70	0.83	0.218	4.17	1.26	2.27	0.510	14.17	0.80	0.72	0.149
25.00	0.69	0.79	0.209	4.33	1.23	2.18	0.494	14.33	0.80	0.71	0.146
25.50	0.68	0.76	0.200	4.50	1.20	2.10	0.482	14.50	0.79	0.69	0.144
26.00	0.67	0.72	0.192	4.67	1.16	2.02	0.470	14.67	0.79	0.69	0.141
26.50	0.67	0.69	0.181	4.83	1.14	1.93	0.451	14.83	0.78	0.67	0.138
27.00	0.66	0.66	0.174	5.00	1.14	1.87	0.435	15.00	0.78	0.66	0.135
27.50	0.66	0.63	0.166	5.17	1.12	1.81	0.425	15.17	0.78	0.65	0.131
28.00	0.66	0.61	0.158	5.33	1.11	1.75	0.408	15.33	0.78	0.64	0.238
28.50	0.66	0.59	0.151	5.50	1.09	1.71	0.401	15.50	0.77	0.63	0.126
29.00	0.65	0.57	0.148	5.67	1.09	1.65	0.383	15.67	0.77	0.62	0.123
29.50	0.64	0.55	0.145	5.83	1.10	1.61	0.373	15.83	0.77	0.61	0.119
30.00	0.64	0.53	0.140	6.00	1.09	1.59	0.366	16.00	0.77	0.60	0.116
32.00	0.62	0.44	0.119	6.17	1.08	1.57	0.365	16.17	0.78	0.58	0.112
34.00	0.64	0.35	0.091	6.33	1.04	1.55	0.365	16.33	0.78	0.58	0.110
36.00	0.69	0.27	0.059	6.50	1.02	1.51	0.358	16.50	0.78	0.57	0.107
38.00	0.73	0.24	0.044	6.67	1.00	1.47	0.351	16.67	0.77	0.56	0.106
40.00	0.76	0.22	0.034	6.83	0.97	1.43	0.346	16.83	0.78	0.55	0.103
				7.00	0.96	1.39	0.333	17.00	0.78	0.55	0.102
Iron[5]				7.17	0.94	1.35	0.327	17.17	0.78	0.54	0.100
0.10	6.41	33.07	0.978	7.33	0.94	1.30	0.311	17.33	0.78	0.54	0.098
0.15	6.26	22.82	0.956	7.50	0.94	1.26	0.298	17.50	0.77	0.53	0.097
0.20	3.68	18.23	0.958	7.67	0.94	1.23	0.288	17.67	0.77	0.52	0.095
0.26	4.98	13.68	0.911	7.83	0.94	1.21	0.279	17.83	0.78	0.51	0.092
0.30	4.87	12.05	0.892	8.00	0.94	1.18	0.272	18.00	0.78	0.51	0.091
0.36	4.68	10.44	0.867	8.17	0.94	1.16	0.265	18.17	0.78	0.51	0.090
0.40	4.42	9.75	0.858	8.33	0.94	1.14	0.258	18.33	0.78	0.50	0.089
0.50	4.14	8.02	0.817	8.50	0.94	1.12	0.251	18.50	0.77	0.50	0.089
0.60	3.93	6.95	0.783	8.67	0.94	1.10	0.246	18.67	0.77	0.50	0.088
0.70	3.78	6.17	0.752	8.83	0.92	1.08	0.240	18.83	0.77	0.49	0.087
0.80	3.65	5.60	0.725	9.00	0.93	1.07	0.236	19.00	0.77	0.49	0.087
0.90	3.52	5.16	0.700	9.17	0.92	1.06	0.233	19.17	0.76	0.49	0.088
1.00	3.43	4.79	0.678	9.33	0.91	1.04	0.231	19.33	0.76	0.48	0.087
1.10	3.33	4.52	0.660								

Energy (eV)	n	k	R(φ = 0)
19.50	0.75	0.47	0.086
19.67	0.75	0.47	0.085
19.83	0.75	0.46	0.084
20.00	0.74	0.45	0.083
20.17	0.74	0.44	0.081
20.33	0.74	0.44	0.081
20.50	0.74	0.42	0.080
20.67	0.73	0.43	0.079
20.83	0.73	0.42	0.078
21.00	0.73	0.41	0.077
21.17	0.72	0.40	0.076
21.33	0.72	0.39	0.074
21.50	0.72	0.38	0.073
21.67	0.72	0.38	0.071
21.83	0.72	0.37	0.070
22.00	0.72	0.36	0.068
22.17	0.71	0.35	0.067
22.33	0.72	0.34	0.064
22.50	0.72	0.34	0.063
22.67	0.72	0.33	0.062
22.83	0.72	0.32	0.059
23.00	0.72	0.31	0.058
23.17	0.72	0.30	0.056
23.33	0.72	0.29	0.054
23.50	0.73	0.28	0.050
23.67	0.73	0.28	0.049
23.83	0.74	0.27	0.047
24.00	0.74	0.27	0.045
24.17	0.74	0.26	0.044
24.33	0.74	0.26	0.043
24.50	0.74	0.25	0.042
24.67	0.75	0.25	0.040
24.83	0.75	0.24	0.039
25.00	0.75	0.24	0.038
26.00	0.76	0.21	0.031
27.00	0.78	0.18	0.026
28.00	0.79	0.16	0.021
29.00	0.81	0.14	0.017
30.00	0.82	0.13	0.014

Lithium[12]

Energy (eV)	n	k	R(φ = 0)
0.14	0.659	38.0	0.998
0.54	0.661	12.6	0.984
0.75	0.561	7.68	0.963
1.05	0.448	5.58	0.946
1.35	0.338	4.36	0.935
1.65	0.265	3.55	0.925
1.95	0.221	2.94	0.913
2.25	0.206	2.48	0.892
2.55	0.217	2.11	0.854
2.85	0.247	1.82	0.797
3.15	0.304	1.60	0.715
3.45	0.334	1.45	0.656
3.75	0.345	1.32	0.611
4.05	0.346	1.21	0.578
4.35	0.333	1.11	0.557
4.65	0.317	1.01	0.540
4.95	0.302	0.906	0.520
5.25	0.299	0.795	0.484
5.55	0.310	0.688	0.434
5.85	0.342	0.594	0.365

Energy (eV)	n	k	R(φ = 0)
6.15	0.376	0.522	0.306
6.45	0.408	0.460	0.256
6.75	0.440	0.407	0.214
7.05	0.466	0.364	0.183
7.35	0.492	0.320	0.155
7.65	0.517	0.282	0.131
7.95	0.545	0.246	0.109
8.25	0.572	0.214	0.091
8.55	0.601	0.189	0.075
8.85	0.624	0.163	0.063
9.15	0.657	0.144	0.050
9.45	0.680	0.130	0.042
9.75	0.708	0.119	0.034
10.1	0.726	0.108	0.029
10.4	0.743	0.102	0.025
10.6	0.753	0.080	0.022

Magnesium (evaporated)[13]

Energy (eV)	n	k	R(φ = 0)
2.145	0.48	3.71	0.880
2.270	0.57	3.47	0.843
2.522	0.53	2.92	0.805
2.845	0.52	2.65	0.777
3.064	0.52	2.05	0.681
5.167	0.10	1.60	0.894
5.636	0.15	1.50	0.832
6.200	0.20	1.40	0.765
6.889	0.25	1.30	0.693
7.750	0.20	1.20	0.722
8.857	0.15	0.95	0.730
10.335	0.25	0.40	0.419

Manganese[14]

Energy (eV)	n	k	R(φ = 0)
0.64	3.89	5.95	0.738
0.77	3.78	5.41	0.710
0.89	3.65	5.02	0.688
1.02	3.48	4.74	0.673
1.14	3.30	4.53	0.662
1.26	3.10	4.35	0.653
1.39	2.97	4.18	0.643
1.51	2.83	4.03	0.634
1.64	2.70	3.91	0.627
1.76	2.62	3.78	0.617
1.88	2.56	3.65	0.606
2.01	2.51	3.54	0.596
2.13	2.47	3.43	0.585
2.26	2.39	3.33	0.577
2.38	2.32	3.23	0.567
2.50	2.25	3.14	0.559
2.63	2.19	3.06	0.552
2.75	2.11	2.98	0.545
2.88	2.06	2.90	0.536
3.00	2.00	2.82	0.528
3.12	1.96	2.74	0.518
3.25	1.92	2.67	0.509
3.37	1.89	2.59	0.498
3.50	1.89	2.51	0.484
3.62	1.87	2.45	0.475
3.74	1.86	2.38	0.463
3.87	1.86	2.32	0.451
3.99	1.86	2.25	0.438
4.12	1.86	2.19	0.427

Energy (eV)	n	k	R(φ = 0)
4.24	1.85	2.14	0.417
4.36	1.85	2.08	0.406
4.49	1.86	2.03	0.395
4.61	1.85	1.99	0.388
4.74	1.84	1.94	0.378
4.86	1.83	1.91	0.372
4.98	1.82	1.86	0.362
5.11	1.82	1.82	0.354
5.23	1.81	1.79	0.348
5.36	1.78	1.76	0.342
5.48	1.74	1.73	0.337
5.60	1.73	1.70	0.331
5.73	1.72	1.67	0.325
5.85	1.70	1.64	0.319
5.98	1.67	1.61	0.313
6.10	1.63	1.58	0.307
6.22	1.62	1.55	0.301
6.35	1.59	1.52	0.295
6.47	1.55	1.50	0.292
6.60	1.48	1.47	0.288

Mercury (liquid)[15]

Energy (eV)	n	k	R(φ = 0)
0.2	13.99	14.27	0.869
0.3	11.37	11.95	0.846
0.4	9.741	10.65	0.830
0.5	8.528	9.805	0.818
0.6	7.574	9.195	0.808
0.8	6.086	8.312	0.796
1.0	4.962	7.643	0.789
1.2	4.050	7.082	0.786
1.4	3.324	6.558	0.785
1.6	2.746	6.054	0.783
1.8	2.284	5.582	0.782
2.0	1.910	5.150	0.782
2.2	1.620	4.751	0.780
2.4	1.384	4.407	0.779
2.6	1.186	4.090	0.779
2.8	1.027	3.802	0.779
3.0	0.898	3.538	0.777
3.2	0.798	3.294	0.773
3.4	0.713	3.074	0.770
3.6	0.644	2.860	0.763
3.8	0.589	2.665	0.755
4.0	0.542	2.502	0.749
4.2	0.507	2.341	0.738
4.4	0.477	2.195	0.727
4.6	0.452	2.058	0.715
4.8	0.431	1.929	0.701
5.0	0.414	1.806	0.685
5.2	0.401	1.687	0.666
5.4	0.394	1.569	0.642
5.6	0.386	1.454	0.617
5.7	0.386	1.396	0.601
5.8	0.386	1.341	0.585
5.9	0.385	1.287	0.569
6.0	0.386	1.232	0.551
6.1	0.388	1.176	0.531
6.2	0.390	1.118	0.510
6.3	0.399	1.058	0.481
6.4	0.412	1.002	0.450
6.5	0.428	0.949	0.418

Energy (eV)	n	k	R(φ = 0)	Energy (eV)	n	k	R(φ = 0)	Energy (eV)	n	k	R(φ = 0)
6.6	0.436	0.898	0.392	0.70	1.48	8.99	0.932	8.80	0.65	1.41	0.450
6.7	0.438	0.836	0.367	0.74	1.51	8.38	0.921	9.00	0.65	1.33	0.420
6.8	0.459	0.756	0.320	0.78	1.60	7.83	0.906	9.20	0.67	1.25	0.385
6.9	0.510	0.676	0.255	0.82	1.64	7.35	0.892	9.40	0.69	1.19	0.355
7.0	0.585	0.617	0.191	0.86	1.70	6.89	0.876	9.60	0.71	1.12	0.320
7.1	0.663	0.589	0.148	9.90	1.74	6.48	0.859	9.80	0.74	1.05	0.285
7.2	0.717	0.584	0.128	1.00	1.94	5.58	0.805	10.00	0.77	0.99	0.250
7.3	0.769	0.575	0.111	1.10	2.15	4.85	0.743	10.20	0.81	0.93	0.217
7.4	0.817	0.574	0.100	1.20	2.44	4.22	0.671	10.40	0.86	0.88	0.188
7.5	0.860	0.580	0.094	1.30	2.77	3.74	0.608	10.60	0.91	0.83	0.162
7.6	0.893	0.597	0.093	1.40	3.15	3.40	0.562	10.80	0.98	0.79	0.138
7.8	0.929	0.623	0.096	1.50	3.53	3.30	0.550	11.00	1.05	0.77	0.125
8.0	0.946	0.639	0.098	1.60	3.77	3.41	0.562	11.20	1.12	0.78	0.123
8.2	0.952	0.645	0.099	1.70	3.84	3.51	0.570	11.40	1.18	0.80	0.125
8.4	0.953	0.638	0.097	1.80	3.81	3.58	0.576	11.60	1.23	0.85	0.135
8.6	0.956	0.624	0.093	1.90	3.74	3.58	0.576	11.80	1.25	0.89	0.145
8.8	0.965	0.607	0.087	2.00	3.68	3.52	0.571	12.00	1.26	0.92	0.154
9.0	0.975	0.588	0.082	2.10	3.68	3.45	0.565	12.40	1.25	0.98	0.168
9.2	0.988	0.568	0.076	2.20	3.76	3.41	0.562	12.80	1.23	1.00	0.178
9.4	1.009	0.548	0.069	2.30	3.79	3.61	0.578	13.20	1.20	1.02	0.185
9.6	1.044	0.541	0.066	2.40	3.59	3.78	0.594	13.60	1.17	1.02	0.187
9.8	1.061	0.557	0.069	2.50	3.36	3.73	0.591	14.00	1.15	1.01	0.185
10.0	1.062	0.567	0.071	2.60	3.22	3.61	0.582	14.40	1.13	1.00	0.182
10.2	1.054	0.569	0.072	2.70	3.13	3.51	0.573	14.80	1.13	0.99	0.179
10.4	1.045	0.561	0.070	2.80	3.08	3.42	0.565	15.00	1.14	0.99	0.179
10.6	1.041	0.550	0.068	2.90	3.05	3.33	0.566	15.60	1.15	1.01	0.184
10.8	1.039	0.537	0.065	3.00	3.04	3.27	0.550	16.00	1.14	1.04	0.194
11.0	1.039	0.523	0.062	3.10	3.03	3.21	0.544	16.60	1.10	1.10	0.216
11.5	1.050	0.491	0.055	3.20	3.05	3.18	0.540	17.00	1.04	1.12	0.233
12.0	1.064	0.467	0.050	3.30	3.06	3.18	0.540	17.60	0.94	1.14	0.257
12.5	1.078	0.445	0.045	3.40	3.06	3.19	0.541	18.00	0.87	1.12	0.270
13.0	1.092	0.430	0.042	3.50	3.06	3.21	0.543	18.60	0.77	1.08	0.283
13.5	1.104	0.416	0.040	3.60	3.05	3.23	0.546	19.00	0.71	1.02	0.284
14.0	1.115	0.404	0.038	3.70	3.04	3.27	0.550	19.60	0.66	0.94	0.275
14.5	1.125	0.394	0.037	3.80	3.04	3.31	0.554	20.00	0.64	0.89	0.264
15.0	1.135	0.383	0.035	3.90	3.04	3.40	0.564	20.60	0.62	0.81	0.245
15.5	1.146	0.374	0.034	4.00	3.01	3.51	0.576	21.00	0.61	0.77	0.234
16.0	1.159	0.368	0.034	4.20	2.77	3.77	0.610	21.60	0.61	0.71	0.215
16.5	1.170	0.367	0.034	4.40	2.39	3.88	0.640	22.00	0.60	0.69	0.207
17.0	1.177	0.367	0.034	4.60	2.06	3.84	0.658	22.60	0.59	0.63	0.195
17.5	1.184	0.366	0.034	4.80	1.75	3.76	0.678	23.00	0.58	0.60	0.185
18.0	1.191	0.367	0.035	5.00	1.46	3.62	0.695	23.60	0.58	0.53	0.166
18.5	1.195	0.367	0.035	5.20	1.22	3.42	0.706	24.00	0.58	0.49	0.151
19.0	1.200	0.366	0.035	5.40	1.07	3.20	0.706	24.60	0.60	0.43	0.124
19.5	1.208	0.364	0.035	5.60	0.96	2.99	0.700	25.00	0.62	0.39	0.106
				5.80	0.89	2.80	0.688	25.60	0.66	0.35	0.085
Molybdenum[16]				6.00	0.85	2.64	0.674	26.00	0.68	0.33	0.072
0.10	18.53	68.51	0.985	6.20	0.81	2.50	0.660	26.50	0.71	0.31	0.060
0.15	8.78	47.54	0.985	6.40	0.79	2.36	0.641	27.00	0.73	0.29	0.050
0.20	5.10	35.99	0.985	6.60	0.78	2.24	0.619	27.50	0.76	0.28	0.041
0.25	3.36	28.75	0.984	6.80	0.78	2.13	0.592	28.00	0.79	0.27	0.036
0.30	2.44	23.80	0.983	7.00	0.80	2.04	0.568	28.50	0.81	0.26	0.031
0.34	2.00	20.84	0.982	7.20	0.81	1.98	0.548	29.00	0.83	0.26	0.028
0.38	1.70	18.44	0.980	7.40	0.81	1.95	0.542	29.50	0.86	0.26	0.025
0.42	1.57	16.50	0.978	7.60	0.75	1.90	0.552	30.00	0.88	0.26	0.023
0.46	1.46	14.91	0.975	7.80	0.71	1.81	0.542	31.00	0.92	0.29	0.024
0.50	1.37	13.55	0.971	8.00	0.69	1.73	0.530	32.00	0.92	0.32	0.030
0.54	1.35	12.36	0.966	8.20	0.67	1.65	0.512	33.00	0.90	0.33	0.032
0.58	1.34	11.34	0.960	8.40	0.66	1.57	0.495	34.00	0.91	0.34	0.034
0.62	1.38	10.44	0.952	8.60	0.65	1.49	0.475	35.00	0.87	0.37	0.043
0.66	1.43	9.67	0.942								

Energy (eV)	n	k	R(φ = 0)	Energy (eV)	n	k	R(φ = 0)	Energy (eV)	n	k	R(φ = 0)
36.00	0.82	0.34	0.043	6.60	1.01	1.40	0.325	25.00	0.89	0.42	0.050
37.00	0.81	0.30	0.038	6.80	1.02	1.35	0.308	26.00	0.88	0.39	0.046
38.00	0.81	0.27	0.033	7.00	1.03	1.30	0.291	27.00	0.87	0.37	0.042
39.00	0.82	0.25	0.029	7.20	1.03	1.27	0.282	28.00	0.87	0.35	0.040
40.00	0.83	0.23	0.025	7.40	1.03	1.24	0.273	29.00	0.86	0.34	0.037
				7.60	1.02	1.22	0.265	30.00	0.86	0.32	0.034
Nickel[17]				7.80	1.01	1.18	0.256	35.00	0.86	0.24	0.022
0.10	9.54	45.82	0.983	8.00	1.01	1.15	0.248	40.00	0.87	0.18	0.014
0.15	5.45	30.56	0.978	8.20	1.00	1.13	0.242	45.00	0.88	0.13	0.008
0.20	4.12	22.48	0.969	8.40	0.99	1.11	0.235	50.00	0.92	0.10	0.004
0.25	4.25	17.68	0.950	8.60	0.98	1.08	0.228	60.00	0.96	0.08	0.002
0.30	4.19	15.05	0.934	8.80	0.97	1.05	0.220	65.00	0.98	0.09	0.002
0.35	4.03	13.05	0.918	9.00	0.97	1.01	0.211	68.00	0.96	0.12	0.004
0.40	3.84	11.43	0.900	9.20	0.96	0.99	0.203	70.00	0.94	0.11	0.004
0.50	4.03	9.64	0.864	9.40	0.95	0.96	0.194	75.00	0.94	0.09	0.003
0.60	3.84	8.35	0.835	9.60	0.95	0.93	0.185	80.00	0.94	0.07	0.002
0.70	3.59	7.48	0.813	9.80	0.95	0.89	0.175	90.00	0.94	0.06	0.002
0.80	3.38	6.82	0.794	10.00	0.95	0.87	0.166				
0.90	3.18	6.23	0.774	10.20	0.95	0.83	0.155	*Niobium*[18]			
1.00	3.06	5.74	0.753	10.40	0.95	0.80	0.145	0.12	15.99	53.20	0.979
1.10	2.97	5.38	0.734	10.60	0.97	0.76	0.129	0.20	7.25	34.14	0.976
1.20	2.85	5.10	0.721	10.80	0.99	0.75	0.123	0.24	5.47	28.88	0.975
1.30	2.74	4.85	0.708	11.00	1.01	0.73	0.115	0.28	4.26	24.95	0.974
1.40	2.65	4.63	0.695	11.25	1.04	0.72	0.111	0.35	3.11	20.03	0.970
1.50	2.53	4.47	0.688	11.50	1.05	0.71	0.109	0.45	2.28	15.58	0.964
1.60	2.43	4.31	0.679	11.75	1.07	0.71	0.108	0.55	1.83	12.67	0.956
1.70	2.28	4.18	0.677	12.00	1.07	0.71	0.108	0.65	1.57	10.59	0.947
1.80	2.14	4.01	0.670	12.25	1.07	0.71	0.107	0.75	1.41	9.00	0.935
1.90	2.02	3.82	0.659	12.50	1.08	0.71	0.106	0.85	1.35	7.74	0.918
2.00	1.92	3.65	0.649	12.75	1.08	0.71	0.106	0.95	1.35	6.70	0.893
2.10	1.85	3.48	0.634	13.00	1.08	0.71	0.105	1.05	1.44	5.86	0.857
2.20	1.80	3.33	0.620	13.25	1.08	0.71	0.105	1.15	1.55	5.18	0.814
2.30	1.75	3.19	0.605	13.50	1.07	0.70	0.105	1.25	1.65	4.63	0.768
2.40	1.71	3.06	0.590	13.75	1.07	0.70	0.105	1.35	1.76	4.13	0.715
2.50	1.67	2.93	0.575	14.00	1.07	0.71	0.106	1.45	1.95	3.68	0.650
2.60	1.65	2.81	0.557	14.25	1.06	0.70	0.106	1.55	2.15	3.37	0.595
2.70	1.64	2.71	0.542	14.50	1.05	0.70	0.106	1.65	2.36	3.13	0.552
2.80	1.63	2.61	0.525	14.75	1.04	0.70	0.107	1.75	2.54	2.99	0.527
2.90	1.62	2.52	0.509	15.00	1.03	0.70	0.107	1.85	2.69	2.89	0.510
3.00	1.61	2.44	0.495	15.25	1.02	0.69	0.106	1.95	2.82	2.86	0.505
3.10	1.61	2.36	0.480	15.50	1.01	0.69	0.105	2.05	2.89	2.87	0.505
3.20	1.61	2.30	0.467	15.75	1.00	0.68	0.104	2.15	2.92	2.87	0.505
3.30	1.61	2.23	0.454	16.00	0.99	0.67	0.103	2.25	2.93	2.87	0.505
3.40	1.62	2.17	0.441	16.50	0.98	0.66	0.101	2.35	2.92	2.88	0.506
3.50	1.63	2.11	0.428	17.00	0.96	0.64	0.098	2.45	2.89	2.90	0.509
3.60	1.64	2.07	0.416	17.50	0.94	0.63	0.096	2.55	2.83	2.92	0.512
3.70	1.66	2.02	0.405	18.00	0.92	0.61	0.092	2.65	2.74	2.90	0.511
3.80	1.69	1.99	0.397	18.50	0.91	0.58	0.087	2.75	2.66	2.86	0.507
3.90	1.72	1.98	0.393	19.00	0.90	0.56	0.082	2.85	2.58	2.80	0.500
4.00	1.73	1.98	0.392	19.50	0.90	0.54	0.077	3.00	2.51	2.68	0.485
4.20	1.74	2.01	0.396	20.00	0.89	0.51	0.071	3.10	2.48	2.60	0.475
4.40	1.71	2.06	0.409	20.50	0.89	0.49	0.066	3.20	2.45	2.53	0.465
4.60	1.63	2.09	0.421	21.00	0.90	0.47	0.061	3.30	2.44	2.45	0.453
4.80	1.53	2.11	0.435	21.50	0.91	0.46	0.057	3.40	2.46	2.38	0.442
5.00	1.40	2.10	0.449	22.00	0.91	0.45	0.055	3.50	2.48	2.33	0.435
5.20	1.27	2.04	0.454	22.50	0.91	0.44	0.053	3.60	2.52	2.29	0.428
5.40	1.16	1.94	0.449	23.00	0.92	0.44	0.051	3.70	2.56	2.27	0.426
5.60	1.09	1.83	0.435	23.50	0.91	0.44	0.052	3.80	2.59	2.28	0.427
5.80	1.04	1.73	0.417	24.00	0.90	0.43	0.051	3.90	2.62	2.29	0.429
6.20	1.00	1.54	0.371	24.50	0.90	0.43	0.051	4.00	2.64	2.33	0.434
6.40	1.01	1.46	0.345					4.20	2.64	2.42	0.447

Energy (eV)	n	k	R(φ = 0)
4.40	2.53	2.56	0.467
4.60	2.39	2.56	0.470
4.80	2.32	2.52	0.465
5.00	2.26	2.57	0.475
5.20	2.16	2.62	0.487
5.40	2.00	2.68	0.505
5.60	1.81	2.67	0.518
5.80	1.63	2.60	0.522
6.00	1.49	2.49	0.520
6.20	1.38	2.38	0.512
6.40	1.31	2.25	0.496
6.60	1.26	2.14	0.480
6.80	1.24	2.04	0.460
7.00	1.23	1.96	0.441
7.20	1.22	1.91	0.430
7.40	1.20	1.88	0.427
7.60	1.14	1.85	0.430
7.80	1.07	1.78	0.428
8.00	1.02	1.69	0.412
8.20	1.00	1.60	0.390
8.40	0.99	1.51	0.365
8.60	0.99	1.43	0.340
8.70	0.99	1.39	0.328
8.80	1.00	1.36	0.315
9.00	1.01	1.29	0.290
9.20	1.04	1.22	0.265
9.40	1.07	1.18	0.245
9.60	1.10	1.13	0.227
9.80	1.13	1.09	0.209
10.00	1.18	1.05	0.194
10.20	1.23	1.04	0.187
10.40	1.27	1.04	0.185
10.60	1.30	1.06	0.190
10.80	1.32	1.08	0.195
11.00	1.32	1.10	0.200
11.20	1.31	1.12	0.204
11.40	1.30	1.13	0.207
11.60	1.28	1.13	0.209
11.80	1.27	1.13	0.210
12.00	1.25	1.12	0.209
12.40	1.24	1.10	0.204
12.80	1.24	1.09	0.200
13.20	1.24	1.09	0.201
13.60	1.23	1.12	0.208
14.00	1.20	1.13	0.216
14.40	1.16	1.15	0.225
14.80	1.11	1.16	0.234
15.00	1.08	1.16	0.238
15.60	0.99	1.14	0.247
16.00	0.92	1.11	0.250
16.60	0.85	1.04	0.245
17.00	0.80	0.99	0.240
17.20	0.79	0.96	0.236
17.40	0.77	0.93	0.230
17.80	0.75	0.87	0.217
18.00	0.74	0.85	0.209
18.60	0.73	0.77	0.185
19.00	0.72	0.72	0.170
19.60	0.72	0.66	0.150
20.00	0.72	0.62	0.137

Energy (eV)	n	k	R(φ = 0)
20.60	0.71	0.55	0.119
21.00	0.72	0.50	0.100
21.60	0.75	0.43	0.075
22.00	0.78	0.40	0.063
22.60	0.82	0.35	0.045
23.00	0.85	0.33	0.038
23.60	0.88	0.30	0.029
24.00	0.91	0.29	0.025
24.60	0.94	0.28	0.022
25.00	0.96	0.27	0.020
25.60	0.99	0.26	0.018
26.00	1.00	0.26	0.017
26.60	1.03	0.25	0.016
27.00	1.04	0.25	0.015
27.60	1.06	0.25	0.015
28.00	1.08	0.24	0.015
28.60	1.11	0.24	0.016
29.00	1.13	0.25	0.017
29.60	1.16	0.26	0.020
30.00	1.18	0.28	0.023
31.00	1.18	0.31	0.026
32.00	1.20	0.34	0.031
33.00	1.21	0.38	0.038
34.00	1.20	0.42	0.044
35.20	1.17	0.47	0.051
36.00	1.15	0.50	0.056
37.50	1.07	0.53	0.064
39.50	0.95	0.50	0.063
40.50	0.92	0.47	0.059

Osmium (Polycrystalline)[9]

Energy (eV)	n	k	R(φ = 0)
0.10	4.08	50.23	0.994
0.15	2.90	33.60	0.990
0.20	2.44	25.11	0.985
0.25	2.35	19.99	0.977
0.30	2.23	16.54	0.969
0.35	2.33	14.06	0.955
0.40	2.45	12.32	0.940
0.45	2.43	11.02	0.927
0.50	2.41	9.97	0.913
0.55	2.33	9.12	0.901
0.60	2.21	8.37	0.890
0.65	2.11	7.68	0.877
0.70	2.02	7.04	0.862
0.75	2.00	6.46	0.842
0.80	2.00	5.95	0.820
0.85	2.01	5.51	0.796
0.90	2.03	5.10	0.769
0.95	2.05	4.74	0.742
1.00	2.09	4.41	0.712
1.10	2.15	3.84	0.651
1.20	2.16	3.35	0.592
1.30	2.25	2.77	0.506
1.40	2.49	2.23	0.419
1.50	2.84	1.80	0.369
1.60	3.36	1.62	0.379
1.70	3.70	1.75	0.411
1.80	3.78	1.83	0.423
1.90	3.81	1.75	0.418
2.00	3.98	1.60	0.418
2.10	4.26	1.54	0.432

Energy (eV)	n	k	R(φ = 0)
2.20	4.58	1.62	0.457
2.30	4.84	1.76	0.479
2.40	5.10	2.01	0.506
2.50	5.28	2.38	0.532
2.60	5.36	2.82	0.557
2.70	5.30	3.29	0.580
2.80	5.07	3.78	0.603
2.90	4.65	4.18	0.624
3.00	4.05	4.40	0.639
3.20	3.29	3.96	0.614
3.40	2.93	3.79	0.607
3.60	2.75	3.45	0.577
3.80	2.73	3.32	0.562
4.00	2.71	3.34	0.565
4.20	2.53	3.44	0.584
4.40	2.24	3.44	0.599
4.60	2.01	3.31	0.598
4.80	1.88	3.19	0.592
5.00	1.74	3.12	0.596
5.20	1.58	3.00	0.597
5.40	1.46	2.88	0.593
5.60	1.36	2.77	0.589
5.80	1.27	2.65	0.582
6.00	1.20	2.54	0.575
6.20	1.13	2.44	0.571
6.40	1.06	2.33	0.562
6.60	1.01	2.21	0.548
6.80	0.97	2.11	0.532
7.00	0.95	2.00	0.514
7.20	0.92	1.91	0.497
7.40	0.91	1.81	0.476
7.60	0.90	1.72	0.451
7.80	0.90	1.63	0.426
8.00	0.91	1.55	0.400
8.20	0.91	1.48	0.375
8.40	0.94	1.40	0.344
8.60	0.96	1.34	0.319
8.80	0.98	1.29	0.296
9.00	1.01	1.24	0.274
9.20	1.04	1.19	0.255
9.40	1.08	1.16	0.238
9.60	1.10	1.14	0.229
9.80	1.13	1.11	0.217
10.00	1.16	1.10	0.209
10.20	1.19	1.08	0.203
10.30	1.20	1.08	0.201
10.40	1.22	1.08	0.200
10.50	1.23	1.09	0.201
10.60	1.24	1.10	0.203
10.80	1.25	1.11	0.206
11.00	1.24	1.13	0.213
11.20	1.23	1.14	0.217
11.40	1.19	1.15	0.223
11.60	1.17	1.12	0.216
11.80	1.16	1.10	0.211
12.00	1.15	1.08	0.205
12.40	1.14	1.03	0.191
12.80	1.15	1.01	0.183
13.20	1.16	0.98	0.174
13.60	1.17	0.97	0.170

Energy (eV)	n	k	R(φ = 0)
14.00	1.17	0.96	0.169
14.40	1.16	0.94	0.165
14.80	1.16	0.91	0.156
15.20	1.17	0.89	0.148
15.60	1.20	0.86	0.140
16.00	1.25	0.87	0.140
16.40	1.28	0.90	0.147
16.80	1.28	0.94	0.157
17.20	1.27	0.97	0.167
17.60	1.26	1.01	0.178
18.00	1.23	1.04	0.189
18.40	1.19	1.08	0.200
18.80	1.14	1.10	0.210
19.20	1.10	1.10	0.219
19.60	1.05	1.11	0.227
20.00	0.96	1.10	0.239
20.40	0.93	1.09	0.240
20.80	0.89	1.05	0.240
21.20	0.86	1.02	0.237
21.60	0.83	0.99	0.235
22.00	0.80	0.96	0.230
22.40	0.78	0.93	0.226
22.80	0.77	0.90	0.220
23.20	0.75	0.88	0.217
23.60	0.75	0.86	0.211
24.00	0.73	0.84	0.209
24.40	0.72	0.82	0.207
24.80	0.70	0.80	0.205
25.20	0.69	0.77	0.202
25.60	0.67	0.75	0.199
26.00	0.66	0.72	0.195
26.40	0.65	0.69	0.189
26.80	0.63	0.66	0.183
27.20	0.65	0.62	0.165
28.00	0.64	0.59	0.156
28.40	0.64	0.57	0.148
28.80	0.65	0.55	0.140
29.20	0.65	0.53	0.134
29.60	0.65	0.51	0.128
30.00	0.65	0.49	0.121
31.00	0.65	0.45	0.111
32.00	0.66	0.41	0.095
33.00	0.68	0.37	0.079
34.00	0.70	0.34	0.068
35.00	0.72	0.31	0.057
36.00	0.74	0.29	0.048
37.00	0.77	0.27	0.040
38.00	0.79	0.26	0.035
39.00	0.81	0.26	0.031
40.00	0.84	0.26	0.026

Palladium[19]

Energy (eV)	n	k	R(φ = 0)
0.10	4.13	54.15	0.994
0.15	3.13	35.82	0.990
0.20	3.07	26.59	0.983
0.26	3.11	20.15	0.971
0.30	3.56	17.27	0.955
0.36	3.98	14.41	0.932
0.40	4.27	13.27	0.916
0.46	4.27	12.11	0.902
0.50	4.10	11.44	0.896
0.56	3.92	10.49	0.883
0.60	3.80	9.96	0.876
0.72	3.51	8.70	0.854
0.80	3.35	8.06	0.840
1.00	2.99	6.89	0.811
1.10	2.81	6.46	0.800
1.20	2.65	6.10	0.790
1.30	2.50	5.78	0.781
1.40	2.34	5.50	0.774
1.50	2.17	5.22	0.767
1.60	2.08	4.95	0.755
1.70	2.00	4.72	0.745
1.80	1.92	4.54	0.737
1.90	1.82	4.35	0.729
2.00	1.75	4.18	0.721
2.10	1.67	4.03	0.714
2.20	1.60	3.88	0.707
2.30	1.53	3.75	0.700
2.40	1.47	3.61	0.693
2.50	1.41	3.48	0.685
2.60	1.37	3.36	0.676
2.70	1.32	3.25	0.668
2.80	1.29	3.13	0.658
2.90	1.26	3.03	0.648
3.00	1.23	2.94	0.639
3.10	1.20	2.85	0.630
3.20	1.17	2.77	0.622
3.30	1.14	2.68	0.613
3.40	1.12	2.60	0.602
3.50	1.10	2.52	0.591
3.60	1.08	2.45	0.581
3.70	1.07	2.38	0.570
3.80	1.06	2.31	0.558
3.90	1.05	2.25	0.547
4.00	1.03	2.19	0.537
4.20	1.04	2.09	0.510
4.40	1.03	2.01	0.493
4.60	1.03	1.94	0.476
4.80	1.01	1.90	0.470
5.00	0.96	1.86	0.472
5.20	0.90	1.79	0.474
5.40	0.85	1.70	0.463
5.60	0.81	1.62	0.449
5.80	0.78	1.54	0.437
6.00	0.76	1.45	0.418
6.20	0.74	1.37	0.397
6.40	0.73	1.29	0.375
6.60	0.72	1.21	0.350
6.80	0.73	1.13	0.316
7.00	0.73	1.05	0.287
7.20	0.75	0.98	0.255
7.40	0.77	0.91	0.223
7.60	0.79	0.85	0.195
7.80	0.83	0.78	0.163
8.00	0.88	0.73	0.133
8.20	0.94	0.70	0.117
8.40	0.96	0.70	0.114
8.60	1.00	0.65	0.097
8.80	1.04	0.65	0.094
9.00	1.07	0.64	0.090
9.50	1.12	0.65	0.089
10.00	1.14	0.65	0.088
10.50	1.16	0.65	0.087
11.00	1.18	0.64	0.086
11.50	1.19	0.65	0.087
12.00	1.20	0.66	0.089
12.50	1.19	0.67	0.091
13.00	1.18	0.67	0.091
13.50	1.18	0.67	0.092
14.00	1.17	0.67	0.093
14.50	1.15	0.68	0.095
15.00	1.13	0.69	0.098
15.50	1.10	0.68	0.096
16.00	1.08	0.66	0.092
16.50	1.06	0.63	0.086
17.00	1.07	0.61	0.081
17.50	1.06	0.61	0.080
18.00	1.07	0.59	0.077
18.50	1.07	0.59	0.077
19.00	1.08	0.59	0.077
19.50	1.08	0.61	0.080
20.00	1.07	0.65	0.090
20.50	1.03	0.67	0.098
21.00	0.99	0.67	0.103
21.50	0.95	0.66	0.103
22.00	0.91	0.64	0.103
22.50	0.88	0.62	0.101
23.00	0.86	0.59	0.097
23.50	0.85	0.56	0.091
24.00	0.84	0.54	0.086
25.00	0.81	0.51	0.084
26.40	0.80	0.43	0.066
27.80	0.81	0.38	0.052
29.20	0.82	0.35	0.046

Platinum[20]

Energy (eV)	n	k	R(φ = 0)
0.10	13.21	44.72	0.976
0.15	8.18	31.16	0.969
0.20	5.90	23.95	0.962
0.25	4.70	19.40	0.954
0.30	3.92	16.16	0.945
0.35	3.28	13.66	0.936
0.40	2.81	11.38	0.922
0.45	3.03	9.31	0.882
0.50	3.91	7.71	0.813
0.55	4.58	7.14	0.777
0.60	5.13	6.75	0.753
0.65	5.52	6.66	0.746
0.70	5.71	6.83	0.751
0.75	5.57	7.02	0.759
0.80	5.31	7.04	0.762
0.85	5.05	6.98	0.763
0.90	4.77	6.91	0.765
0.95	4.50	6.77	0.763
1.00	4.25	6.62	0.762
1.10	3.86	6.24	0.753
1.20	3.55	5.92	0.746
1.30	3.29	5.61	0.736
1.40	3.10	5.32	0.725
1.50	2.92	5.07	0.716
1.60	2.76	4.84	0.706

Energy (eV)	n	k	R(φ = 0)
1.70	2.63	4.64	0.697
1.80	2.51	4.43	0.686
1.90	2.38	4.26	0.678
2.00	2.30	4.07	0.664
2.10	2.23	3.92	0.654
2.20	2.17	3.77	0.642
2.30	2.10	3.67	0.636
2.40	2.03	3.54	0.626
2.50	1.96	3.42	0.616
2.60	1.91	3.30	0.605
2.70	1.87	3.20	0.595
2.80	1.83	3.10	0.585
2.90	1.79	3.01	0.575
3.00	1.75	2.92	0.565
3.20	1.68	2.76	0.546
3.40	1.63	2.62	0.527
3.60	1.58	2.48	0.507
3.80	1.53	2.37	0.491
4.00	1.49	2.25	0.472
4.20	1.45	2.14	0.452
4.40	1.43	2.04	0.432
4.60	1.39	1.95	0.415
4.80	1.38	1.85	0.392
5.00	1.36	1.76	0.372
5.20	1.36	1.67	0.350
5.40	1.36	1.61	0.332
5.60	1.36	1.54	0.315
5.80	1.36	1.47	0.295
6.00	1.38	1.40	0.276
6.20	1.39	1.35	0.261
6.40	1.42	1.29	0.246
6.60	1.45	1.26	0.236
6.80	1.48	1.24	0.231
7.00	1.50	1.24	0.230
7.20	1.50	1.25	0.231
7.40	1.49	1.23	0.228
7.60	1.48	1.22	0.225
7.80	1.48	1.20	0.221
8.00	1.47	1.18	0.216
8.20	1.47	1.17	0.212
8.40	1.47	1.15	0.209
8.60	1.47	1.14	0.205
8.80	1.47	1.13	0.202
9.00	1.48	1.12	0.200
9.20	1.49	1.11	0.198
9.40	1.49	1.12	0.200
9.60	1.49	1.13	0.203
9.80	1.48	1.15	0.207
10.00	1.46	1.15	0.209
10.20	1.43	1.16	0.211
10.40	1.40	1.15	0.210
10.60	1.37	1.14	0.207
10.80	1.35	1.12	0.203
11.00	1.33	1.10	0.199
11.20	1.31	1.08	0.194
11.40	1.30	1.06	0.188
11.60	1.29	1.04	0.183
11.80	1.29	1.01	0.177
12.00	1.29	1.00	0.173
12.40	1.29	0.97	0.165

Energy (eV)	n	k	R(φ = 0)
12.80	1.29	0.94	0.158
13.20	1.31	0.93	0.155
13.60	1.31	0.93	0.155
14.00	1.31	0.93	0.155
14.40	1.30	0.93	0.156
14.80	1.27	0.93	0.157
15.20	1.27	0.93	0.155
15.60	1.25	0.92	0.151
16.00	1.24	0.89	0.146
16.50	1.24	0.87	0.142
17.00	1.25	0.86	0.138
17.50	1.27	0.85	0.135
18.00	1.31	0.88	0.142
18.50	1.30	0.94	0.157
19.00	1.28	0.99	0.171
19.50	1.23	1.03	0.184
20.00	1.18	1.06	0.197
20.50	1.11	1.09	0.212
21.00	1.03	1.10	0.226
21.50	0.94	1.08	0.238
22.00	0.87	1.04	0.240
22.50	0.81	0.98	0.235
23.00	0.77	0.92	0.226
23.50	0.75	0.87	0.213
24.00	0.74	0.82	0.201
24.50	0.73	0.77	0.187
25.00	0.73	0.73	0.174
25.50	0.73	0.70	0.162
26.00	0.74	0.67	0.150
26.50	0.74	0.65	0.142
27.00	0.74	0.63	0.136
27.50	0.74	0.62	0.130
28.00	0.75	0.60	0.125
28.50	0.75	0.59	0.121
29.00	0.75	0.58	0.118
29.50	0.74	0.58	0.120
30.00	0.73	0.58	0.124

Potassium[21]

Energy (eV)	n	k	R(φ = 0)
0.55	0.139	7.10	0.989
0.58	0.119	6.72	0.990
0.63	0.106	6.32	0.990
0.67	0.091	5.79	0.990
0.73	0.079	5.30	0.989
0.81	0.066	4.75	0.989
0.92	0.056	4.19	0.988
1.05	0.044	3.58	0.987
1.23	0.040	3.04	0.985
1.44	0.040	2.56	0.979
1.65	0.044	2.19	0.970
1.87	0.050	1.84	0.955
2.07	0.053	1.62	0.943
2.27	0.049	1.43	0.938
2.45	0.046	1.28	0.933
2.64	0.043	1.14	0.928
2.82	0.043	1.02	0.919
2.95	0.041	0.898	0.913
3.06	0.041	0.799	0.905
3.40	0.052	0.549	0.852
3.71	0.089	0.288	0.719
3.97	0.287	0.091	0.310

Energy (eV)	n	k	R(φ = 0)
4.00	0.34	0.08	0.245
4.065	0.38	0.07	0.204
4.133	0.41	0.07	0.177
4.203	0.45	0.06	0.145
4.275	0.48	0.06	0.125
4.350	0.52	0.05	0.101
4.428	0.55	0.05	0.085
4.509	0.58	0.05	0.072
4.592	0.61	0.05	0.060
4.679	0.64	0.04	0.049
4.769	0.66	0.04	0.043
4.862	0.68	0.04	0.037
4.959	0.70	0.04	0.032
5.061	0.72	0.04	0.027
5.166	0.74	0.04	0.023
5.276	0.76	0.04	0.019
5.391	0.78	0.04	0.016
5.510	0.79	0.05	0.015
5.637	0.81	0.05	0.012
5.767	0.83	0.05	0.009
6.048	0.85	0.05	0.007
6.199	0.87	0.05	0.006
6.358	0.88	0.05	0.005
6.526	0.90	0.06	0.004
6.702	0.91	0.06	0.003
6.888	0.92	0.06	0.003
7.085	0.92	0.06	0.003
7.293	0.93	0.06	0.002
7.514	0.93	0.06	0.002
7.749	0.94	0.06	0.002
7.999	0.94	0.06	0.002
8.260	0.94	0.06	0.002
8.551	0.94	0.06	0.002
8.856	0.94	0.05	0.002
9.184	0.94	0.05	0.002
9.537	0.94	0.04	0.001
9.919	0.94	0.04	0.001
10.33	0.94	0.03	0.001
11.0		0.03	
12.0		0.028	

Rhenium, single crystal, $\vec{E} \parallel \vec{c}$[9]

Energy (eV)	n	k	R(φ = 0)
0.10	6.06	51.03	0.991
0.15	4.66	33.96	0.984
0.20	4.16	25.36	0.975
0.25	4.03	20.10	0.962
0.30	4.37	16.69	0.943
0.35	4.50	14.53	0.925
0.40	4.53	12.96	0.909
0.45	4.53	11.78	0.893
0.50	4.53	10.88	0.878
0.55	4.50	10.26	0.867
0.60	4.29	9.75	0.861
0.65	4.07	9.35	0.856
0.70	3.80	8.94	0.853
0.75	3.48	8.55	0.850
0.80	3.21	8.10	0.846
0.85	2.96	7.68	0.841
0.90	2.73	7.24	0.835
0.95	2.56	6.79	0.826
1.00	2.45	6.36	0.813

Energy (eV)	n	k	R(φ = 0)	Energy (eV)	n	k	R(φ = 0)	Energy (eV)	n	k	R(φ = 0)
1.10	2.38	5.61	0.778	11.40	1.28	1.28	0.252	50.00	0.80	0.30	0.038
1.20	2.35	5.02	0.742	11.60	1.26	1.28	0.252	52.00	0.78	0.30	0.044
1.30	2.39	4.54	0.702	11.80	1.24	1.26	0.249	54.00	0.72	0.30	0.055
1.40	2.44	4.13	0.662	12.00	1.23	1.24	0.244	56.00	0.66	0.24	0.061
1.50	2.50	3.79	0.624	12.40	1.22	1.21	0.237	58.00	0.65	0.16	0.055
1.60	2.59	3.49	0.587	12.80	1.21	1.18	0.230				
1.70	2.70	3.27	0.557	13.20	1.22	1.16	0.222	*Rhenium, single crystal, $\vec{E} \perp \vec{c}$[9]*			
1.80	2.82	3.10	0.535	13.60	1.22	1.13	0.215	0.10	4.25	42.83	0.991
1.90	2.90	3.00	0.520	14.00	1.24	1.12	0.209	0.15	3.28	28.08	0.984
2.00	2.97	2.91	0.510	14.40	1.27	1.11	0.204	0.20	3.28	20.66	0.971
2.10	3.03	2.86	0.504	14.80	1.29	1.15	0.213	0.25	3.47	16.27	0.951
2.20	3.06	2.84	0.501	15.20	1.29	1.19	0.225	0.30	3.73	13.44	0.926
2.30	3.07	2.82	0.499	15.60	1.26	1.22	0.236	0.35	3.93	11.54	0.900
2.40	3.06	2.81	0.498	16.00	1.23	1.25	0.248	0.40	3.99	10.15	0.875
2.50	3.02	2.80	0.497	16.40	1.19	1.27	0.259	0.45	4.17	9.03	0.846
2.60	2.96	2.77	0.493	16.80	1.14	1.29	0.269	0.50	4.34	8.26	0.821
2.70	2.89	2.68	0.482	17.00	1.12	1.30	0.275	0.55	4.45	7.73	0.801
2.80	2.89	2.57	0.468	17.40	1.07	1.30	0.286	0.60	4.53	7.40	0.788
2.90	2.99	2.47	0.457	18.00	0.99	1.30	0.300	0.65	4.44	7.26	0.784
3.00	3.11	2.57	0.470	18.40	0.93	1.29	0.311	0.70	4.13	7.09	0.784
3.20	2.90	2.68	0.482	18.80	0.87	1.28	0.321	0.75	3.77	6.75	0.779
3.40	2.83	2.50	0.459	19.20	0.81	1.25	0.330	0.80	3.55	6.32	0.766
3.60	2.93	2.48	0.457	19.60	0.77	1.21	0.332	0.85	3.39	5.95	0.752
3.80	2.86	2.56	0.467	20.00	0.73	1.18	0.333	0.90	3.26	5.61	0.737
4.00	2.81	2.51	0.460	20.40	0.70	1.14	0.332	0.95	3.17	5.27	0.719
4.20	2.86	2.55	0.466	20.80	0.67	1.11	0.332	1.00	3.09	4.96	0.701
4.40	2.81	2.74	0.489	21.20	0.64	1.08	0.334	1.10	3.05	4.39	0.658
4.60	2.56	2.83	0.504	21.60	0.61	1.04	0.335	1.20	3.08	3.89	0.613
4.80	2.41	2.71	0.493	22.00	0.58	1.01	0.340	1.30	3.20	3.56	0.578
5.00	2.39	2.68	0.488	22.40	0.55	0.97	0.341	1.40	3.23	3.38	0.559
5.20	2.34	2.75	0.500	22.80	0.53	0.93	0.338	1.50	3.23	3.12	0.532
5.40	2.20	2.81	0.515	23.20	0.51	0.89	0.334	1.60	3.29	2.88	0.507
5.60	2.02	2.84	0.530	23.60	0.50	0.85	0.329	1.70	3.38	2.72	0.491
5.80	1.83	2.80	0.538	24.00	0.48	0.80	0.319	1.80	3.47	2.59	0.480
6.00	1.65	2.71	0.541	24.40	0.48	0.76	0.207	1.90	3.54	2.50	0.473
6.20	1.54	2.59	0.532	24.80	0.47	0.72	0.296	2.00	3.63	2.43	0.469
6.40	1.45	2.50	0.526	25.20	0.47	0.68	0.282	2.10	3.74	2.40	0.470
6.80	1.32	2.31	0.508	25.60	0.47	0.65	0.270	2.20	3.83	2.38	0.472
7.00	1.26	2.23	0.500	26.00	0.47	0.61	0.255	2.30	3.93	2.44	0.481
7.20	1.20	2.15	0.493	26.40	0.48	0.57	0.240	2.40	4.00	2.55	0.492
7.40	1.16	2.06	0.480	26.80	0.48	0.54	0.225	2.50	4.01	2.70	0.505
7.60	1.12	1.99	0.470	27.20	0.49	0.51	0.208	2.60	3.90	2.84	0.514
7.80	1.08	1.89	0.454	27.60	0.50	0.48	0.193	2.70	3.74	2.92	0.517
8.00	1.05	1.80	0.435	28.00	0.51	0.45	0.176	2.80	3.57	2.88	0.511
8.20	1.05	1.71	0.411	29.00	0.54	0.39	0.145	2.90	3.49	2.75	0.497
8.40	1.05	1.62	0.386	30.00	0.57	0.33	0.114	3.00	3.53	2.71	0.493
8.60	1.06	1.55	0.360	31.00	0.62	0.29	0.086	3.20	3.55	2.84	0.506
8.80	1.09	1.48	0.336	32.00	0.66	0.26	0.065	3.40	3.34	2.88	0.508
9.00	1.11	1.43	0.317	33.00	0.68	0.24	0.054	3.60	3.25	2.83	0.501
9.20	1.13	1.39	0.301	34.00	0.72	0.21	0.041	3.80	3.24	2.84	0.502
9.40	1.16	1.34	0.281	35.00	0.76	0.20	0.031	4.00	3.19	2.94	0.513
9.60	1.18	1.32	0.274	36.00	0.79	0.20	0.025	4.20	3.05	3.06	0.526
9.80	1.20	1.29	0.264	37.00	0.82	0.19	0.021	4.40	2.88	3.15	0.539
10.00	1.23	1.26	0.252	38.00	0.85	0.20	0.018	4.60	2.67	3.18	0.548
10.20	1.25	1.25	0.246	39.00	0.89	0.21	0.016	4.80	2.44	3.17	0.554
10.40	1.28	1.25	0.242	40.00	0.88	0.26	0.022	5.00	2.25	3.12	0.556
10.60	1.29	1.25	0.242	42.00	0.88	0.26	0.022	5.20	2.10	3.04	0.555
10.80	1.30	1.26	0.244	44.00	0.89	0.29	0.026	5.40	1.96	2.96	0.553
11.00	1.30	1.27	0.247	46.00	0.85	0.32	0.035	5.60	1.84	2.88	0.551
11.20	1.29	1.28	0.249	48.00	0.82	0.30	0.036	5.80	1.73	2.81	0.549
								6.00	1.61	2.74	0.549

Energy (eV)	n	k	R(φ = 0)	Energy (eV)	n	k	R(φ = 0)	Energy (eV)	n	k	R(φ = 0)
6.20	1.51	2.64	0.545	24.80	0.48	0.75	0.303	3.10	1.41	4.20	0.760
6.40	1.42	2.56	0.541	25.20	0.47	0.72	0.295	3.20	1.30	4.09	0.764
6.80	1.28	2.37	0.526	25.60	0.47	0.68	0.286	3.30	1.20	3.97	0.767
7.00	1.22	2.28	0.517	26.00	0.46	0.64	0.276	3.40	1.11	3.84	0.769
7.20	1.16	2.19	0.508	26.40	0.46	0.61	0.263	3.50	1.04	3.71	0.768
7.40	1.12	2.08	0.493	26.80	0.46	0.57	0.249	3.60	0.99	3.58	0.764
7.60	1.12	1.98	0.468	27.20	0.47	0.53	0.231	3.70	0.95	3.45	0.759
7.80	1.08	1.93	0.463	27.60	0.48	0.50	0.216	3.80	0.91	3.34	0.753
8.00	1.05	1.83	0.443	28.00	0.49	0.47	0.198	3.90	0.88	3.23	0.747
8.20	1.05	1.74	0.418	29.00	0.51	0.41	0.164	4.00	0.86	3.12	0.739
8.40	1.05	1.66	0.397	30.00	0.55	0.34	0.129	4.20	0.83	2.94	0.722
8.60	1.06	1.58	0.372	31.00	0.59	0.29	0.097	4.40	0.80	2.76	0.706
8.80	1.07	1.52	0.351	32.00	0.64	0.26	0.072	4.60	0.78	2.60	0.684
9.00	1.09	1.46	0.327	33.00	0.67	0.24	0.060	4.80	0.79	2.46	0.659
9.20	1.11	1.41	0.309	34.00	0.70	0.22	0.047	5.00	0.79	2.34	0.635
9.40	1.14	1.36	0.290	35.00	0.74	0.20	0.036	5.20	0.79	2.23	0.613
9.60	1.17	1.31	0.273	36.00	0.77	0.19	0.029	5.40	0.80	2.14	0.591
9.80	1.20	1.27	0.258	37.00	0.80	0.19	0.023	5.60	0.80	2.06	0.573
10.00	1.24	1.24	0.244	38.00	0.84	0.19	0.018	5.80	0.79	2.00	0.561
10.20	1.29	1.22	0.234	39.00	0.88	0.21	0.016	6.00	0.76	1.93	0.556
10.40	1.33	1.23	0.233	40.00	0.87	0.25	0.023	6.20	0.73	1.85	0.544
10.60	1.36	1.25	0.238	42.00	0.87	0.25	0.023	6.40	0.70	1.77	0.534
10.80	1.38	1.28	0.245	44.00	0.88	0.28	0.026	6.60	0.68	1.69	0.518
11.00	1.37	1.31	0.253	46.00	0.84	0.31	0.035	6.80	0.67	1.60	0.498
11.20	1.36	1.33	0.259	48.00	0.82	0.30	0.036	7.00	0.66	1.52	0.476
11.40	1.33	1.34	0.264	50.00	0.80	0.30	0.039	7.20	0.66	1.43	0.452
11.60	1.31	1.34	0.266	52.00	0.77	0.30	0.044	7.40	0.66	1.35	0.423
11.80	1.28	1.33	0.266	54.00	0.71	0.29	0.055	7.60	0.67	1.27	0.394
12.00	1.26	1.32	0.264	56.00	0.66	0.23	0.061	7.80	0.68	1.20	0.363
12.40	1.23	1.29	0.257	58.00	0.64	0.16	0.055	8.00	0.69	1.12	0.329
12.80	1.22	1.26	0.251					8.20	0.71	1.04	0.288
13.20	1.20	1.23	0.245	**Rhodium**[11]				8.40	0.74	0.97	0.252
13.60	1.19	1.20	0.236	0.10	18.48	69.43	0.986	8.60	0.78	0.89	0.212
14.00	1.20	1.16	0.225	0.20	8.66	37.46	0.977	8.80	0.83	0.83	0.179
14.40	1.22	1.13	0.214	0.30	5.85	25.94	0.967	9.00	0.88	0.77	0.148
14.80	1.27	1.12	0.207	0.40	4.74	19.80	0.955	9.20	0.95	0.73	0.125
15.20	1.31	1.17	0.218	0.50	4.20	16.07	0.941	9.40	1.01	0.71	0.110
15.60	1.31	1.23	0.234	0.60	3.87	13.51	0.925	9.60	1.07	0.69	0.102
16.00	1.28	1.28	0.251	0.70	3.67	11.72	0.908	9.80	1.12	0.69	0.098
16.40	1.24	1.33	0.270	0.80	3.63	10.34	0.887	10.00	1.17	0.69	0.098
16.80	1.17	1.37	0.288	0.90	3.62	9.36	0.867	10.60	1.26	0.73	0.106
17.00	1.14	1.38	0.297	1.00	3.71	8.67	0.848	11.00	1.29	0.76	0.113
17.40	1.06	1.39	0.314	1.10	3.67	8.26	0.837	11.60	1.32	0.80	0.124
18.00	0.95	1.38	0.334	1.20	3.51	7.94	0.832	12.00	1.32	0.82	0.127
18.40	0.88	1.36	0.346	1.30	3.26	7.63	0.829	12.60	1.32	0.82	0.129
18.80	0.82	1.33	0.355	1.40	3.01	7.31	0.827	13.00	1.32	0.83	0.131
19.20	0.76	1.29	0.360	1.50	2.78	6.97	0.823	13.60	1.32	0.85	0.134
19.60	0.72	1.25	0.363	1.60	2.60	6.64	0.818	14.00	1.32	0.86	0.138
20.00	0.67	1.21	0.369	1.70	2.42	6.33	0.813	14.60	1.30	0.89	0.144
20.40	0.64	1.15	0.364	1.80	2.30	6.02	0.805	15.00	1.28	0.90	0.147
20.80	0.61	1.10	0.357	1.90	2.20	5.76	0.798	15.60	1.25	0.90	0.147
21.20	0.60	1.06	0.349	2.00	2.12	5.51	0.789	16.00	1.24	0.89	0.147
21.60	0.58	1.02	0.342	2.10	2.05	5.30	0.780	16.50	1.23	0.88	0.145
22.00	0.57	0.98	0.336	2.20	2.00	5.11	0.772	17.00	1.22	0.88	0.144
22.40	0.56	0.95	0.328	2.30	1.94	4.94	0.765	17.50	1.22	0.87	0.143
22.80	0.55	0.92	0.325	2.40	1.90	4.78	0.756	18.00	1.23	0.88	0.145
23.20	0.53	0.89	0.322	2.50	1.88	4.65	0.748	18.50	1.25	0.92	0.155
23.60	0.52	0.85	0.317	2.60	1.85	4.55	0.743	19.00	1.24	0.98	0.172
24.00	0.50	0.82	0.314	2.70	1.80	4.49	0.742	19.50	1.18	1.05	0.193
24.40	0.49	0.79	0.309	2.90	1.63	4.36	0.748	20.00	1.10	1.09	0.213
				3.00	1.53	4.29	0.753				

Energy (eV)	n	k	R(φ = 0)	Energy (eV)	n	k	R(φ = 0)	Energy (eV)	n	k	R(φ = 0)
20.50	1.00	1.09	0.230	3.30	2.00	3.91	0.671	17.50	1.32	0.93	0.155
21.00	0.91	1.05	0.234	3.40	1.87	3.83	0.673	18.00	1.26	0.99	0.173
21.50	0.86	1.00	0.228	3.50	1.76	3.74	0.674	18.50	1.18	1.02	0.185
22.00	0.83	0.95	0.219	3.60	1.66	3.65	0.675	19.00	1.11	1.02	0.192
22.50	0.81	0.92	0.214	3.70	1.57	3.55	0.673	19.50	1.05	1.02	0.199
23.00	0.79	0.90	0.213	3.80	1.49	3.45	0.672	20.00	0.99	1.02	0.208
23.50	0.75	0.87	0.214	3.90	1.42	3.35	0.668	20.50	0.92	0.99	0.212
24.00	0.73	0.84	0.210	4.00	1.37	3.24	0.661	21.00	0.86	0.94	0.209
24.50	0.70	0.81	0.208	4.20	1.29	3.08	0.649	21.50	0.83	0.90	0.203
25.00	0.69	0.77	0.202	4.40	1.22	2.93	0.639	22.00	0.81	0.86	0.193
25.50	0.67	0.74	0.195	4.60	1.16	2.79	0.628	23.00	0.77	0.79	0.182
26.00	0.66	0.70	0.188	4.80	1.11	2.67	0.617	24.00	0.74	0.74	0.171
26.50	0.65	0.66	0.176	5.00	1.06	2.56	0.607	25.00	0.71	0.69	0.163
27.00	0.65	0.64	0.168	5.20	1.01	2.46	0.600	26.00	0.68	0.63	0.154
27.50	0.65	0.61	0.159	5.40	0.95	2.35	0.593	27.00	0.67	0.57	0.140
28.00	0.65	0.59	0.152	5.60	0.92	2.23	0.576	28.00	0.66	0.51	0.124
29.00	0.65	0.54	0.137	5.80	0.90	2.14	0.559	29.00	0.67	0.46	0.107
30.00	0.66	0.51	0.127	6.00	0.88	2.05	0.545	30.00	0.67	0.43	0.097
31.00	0.64	0.49	0.127	6.20	0.87	1.98	0.531	31.00	0.67	0.37	0.084
32.00	0.61	0.44	0.126	6.40	0.84	1.91	0.521	32.00	0.69	0.33	0.070
33.00	0.60	0.37	0.110	6.60	0.82	1.84	0.510	33.00	0.71	0.30	0.058
34.00	0.65	0.30	0.074	6.80	0.79	1.77	0.500	34.00	0.73	0.27	0.048
35.00	0.69	0.28	0.058	7.00	0.76	1.69	0.489	35.00	0.75	0.25	0.039
36.00	0.73	0.27	0.049	7.20	0.75	1.61	0.472	36.00	0.77	0.24	0.035
37.00	0.74	0.28	0.047	7.40	0.73	1.54	0.455	37.00	0.79	0.23	0.039
38.00	0.74	0.27	0.045	7.60	0.73	1.46	0.433	38.00	0.80	0.22	0.027
39.00	0.75	0.25	0.041	7.80	0.73	1.39	0.411	39.00	0.82	0.22	0.024
				8.00	0.72	1.33	0.391	40.00	0.83	0.22	0.022

Ruthenium, single crystal, $\vec{E} \parallel \hat{c}^9$

Energy (eV)	n	k	R(φ = 0)
0.10	11.50	51.38	0.984
0.20	5.93	27.14	0.970
0.30	4.33	18.50	0.953
0.40	3.60	13.97	0.933
0.50	3.18	11.04	0.909
0.60	3.28	8.89	0.865
0.70	3.62	7.73	0.822
0.80	3.42	7.02	0.801
0.90	3.25	6.12	0.766
1.00	3.39	5.33	0.715
1.10	3.66	4.83	0.675
1.20	3.84	4.57	0.654
1.30	3.94	4.38	0.638
1.40	4.02	4.19	0.624
1.50	4.16	4.07	0.614
1.60	4.33	4.08	0.615
1.70	4.42	4.21	0.624
1.80	4.40	4.38	0.636
1.90	4.29	4.61	0.651
2.00	4.04	4.81	0.667
2.10	3.69	4.90	0.679
2.20	3.35	4.82	0.683
2.30	3.09	4.70	0.681
2.40	2.89	4.55	0.677
2.50	2.74	4.40	0.671
2.60	2.64	4.25	0.663
2.70	2.58	4.14	0.656
2.80	2.54	4.05	0.650
2.90	2.48	4.03	0.650
3.00	2.38	4.03	0.656
3.10	2.26	4.00	0.661
3.20	2.13	3.96	0.666

Energy (eV)	n	k	R(φ = 0)
8.20	0.72	1.26	0.366
8.40	0.73	1.20	0.342
8.60	0.74	1.14	0.318
8.80	0.74	1.08	0.295
9.00	0.75	1.02	0.267
9.20	0.77	0.97	0.243
9.40	0.79	0.91	0.217
9.60	0.82	0.86	0.190
9.80	0.85	0.81	0.167
10.00	0.88	0.76	0.144
10.20	0.92	0.72	0.125
10.40	0.96	0.69	0.110
10.60	1.01	0.67	0.100
10.80	1.05	0.66	0.094
11.00	1.09	0.65	0.090
11.20	1.12	0.65	0.088
11.40	1.15	0.65	0.087
11.60	1.18	0.65	0.088
11.80	1.21	0.66	0.090
12.00	1.23	0.67	0.092
12.40	1.26	0.69	0.098
12.80	1.27	0.72	0.104
13.20	1.28	0.74	0.108
13.60	1.28	0.75	0.111
14.00	1.28	0.76	0.114
14.40	1.27	0.76	0.114
14.80	1.27	0.76	0.114
15.00	1.27	0.76	0.114
15.60	1.28	0.77	0.115
16.00	1.30	0.78	0.118
16.50	1.32	0.80	0.123
17.00	1.34	0.85	0.136

Ruthenium, single crystal, $\vec{E} \perp \hat{c}^5$

Energy (eV)	n	k	R(φ = 0)
0.10	11.85	50.81	0.983
0.20	6.68	27.18	0.966
0.30	4.94	18.92	0.950
0.40	3.90	14.51	0.933
0.50	3.27	11.63	0.915
0.60	2.98	9.54	0.888
0.70	2.82	7.99	0.856
0.80	2.73	6.71	0.815
0.90	2.82	5.54	0.751
1.00	3.17	4.59	0.670
1.10	3.69	3.91	0.604
1.20	4.28	3.66	0.585
1.30	4.66	3.72	0.593
1.40	4.86	3.79	0.601
1.50	4.99	3.89	0.609
1.60	5.08	4.03	0.618
1.70	5.12	4.22	0.629
1.80	5.10	4.45	0.642
1.90	4.96	4.78	0.660
2.00	4.61	5.06	0.677
2.10	4.21	5.09	0.682
2.20	3.94	5.00	0.681
2.30	3.69	4.97	0.684
2.40	3.44	4.88	0.684
2.50	3.27	4.77	0.681
2.60	3.14	4.66	0.677
2.70	3.06	4.59	0.674
2.80	2.99	4.59	0.676
2.90	2.87	4.64	0.686
3.00	2.64	4.69	0.701
3.10	2.40	4.64	0.710

Energy (eV)	n	k	$R(\phi = 0)$
3.20	2.18	4.55	0.717
3.30	2.00	4.43	0.721
3.40	1.84	4.30	0.723
3.50	1.71	4.16	0.723
3.60	1.60	4.03	0.722
3.70	1.50	3.90	0.721
3.80	1.41	3.77	0.718
3.90	1.35	3.64	0.713
4.00	1.29	3.53	0.707
4.20	1.21	3.31	0.694
4.40	1.16	3.13	0.679
4.60	1.13	2.97	0.662
4.80	1.09	2.86	0.652
5.00	1.03	2.75	0.648
5.20	0.97	2.64	0.643
5.40	0.91	2.52	0.635
5.60	0.88	2.40	0.622
5.80	0.86	2.29	0.605
6.00	0.84	2.20	0.591
6.20	0.82	2.11	0.576
6.40	0.81	2.04	0.564
6.60	0.78	1.97	0.556
6.80	0.76	1.89	0.545
7.00	0.73	1.82	0.538
7.20	0.70	1.75	0.527
7.40	0.68	1.67	0.513
7.60	0.67	1.59	0.496
7.80	0.66	1.51	0.476
8.00	0.66	1.44	0.454
8.20	0.65	1.36	0.430
8.40	0.66	1.29	0.403
8.60	0.66	1.22	0.378
8.80	0.68	1.15	0.346
9.00	0.69	1.09	0.317
9.20	0.70	1.02	0.286
9.40	0.73	0.95	0.251
9.60	0.77	0.89	0.216
9.80	0.82	0.84	0.185
10.00	0.86	0.81	0.163
10.20	0.90	0.77	0.143
10.40	0.94	0.74	0.127
10.60	0.99	0.72	0.115
10.80	1.04	0.71	0.108
11.00	1.08	0.70	0.104
11.20	1.11	0.70	0.102
11.40	1.14	0.70	0.101
11.60	1.17	0.71	0.102
11.80	1.20	0.72	0.104
12.00	1.22	0.73	0.107
12.40	1.25	0.76	0.113
12.80	1.26	0.78	0.118
13.20	1.27	0.81	0.124
13.60	1.27	0.83	0.129
14.00	1.26	0.84	0.132
14.40	1.25	0.84	0.132
14.80	1.25	0.84	0.133
15.00	1.25	0.84	0.133
15.60	1.25	0.85	0.134
16.00	1.27	0.85	0.134
16.50	1.28	0.89	0.145
17.00	1.28	0.94	0.158
17.50	1.25	1.00	0.175
18.00	1.19	1.04	0.190
18.50	1.12	1.05	0.200
19.00	1.07	1.05	0.205
19.50	1.02	1.04	0.212
20.00	0.97	1.04	0.219
20.50	0.91	1.03	0.228
21.00	0.85	1.01	0.234
21.50	0.80	0.97	0.234
22.00	0.77	0.94	0.233
23.00	0.71	0.87	0.229
24.00	0.67	0.79	0.218
25.00	0.64	0.73	0.205
26.00	0.61	0.66	0.194
27.00	0.60	0.59	0.177
28.00	0.60	0.53	0.155
29.00	0.61	0.48	0.134
30.00	0.62	0.45	0.123
31.00	0.61	0.40	0.114
32.00	0.63	0.34	0.093
33.00	0.65	0.31	0.077
34.00	0.67	0.28	0.065
35.00	0.70	0.26	0.054
36.00	0.72	0.25	0.047
37.00	0.73	0.23	0.041
38.00	0.75	0.22	0.035
39.00	0.77	0.22	0.031
40.00	0.79	0.22	0.028

Selenium, single crystal, $\vec{E} \parallel \hat{c}$[22]

Energy (eV)	n	k	$R(\phi = 0)$
0.01364	2.914	0.248	0.242
0.01488	3.175	9.95E-02	0.272
0.01612	3.263	2.13E-03	0.282
0.01736	3.306	3.81E-02	0.287
0.01860	3.330	7.04E-03	0.290
0.01984	3.346	4.23E-02	0.291
0.02108	3.358	3.40E-03	0.293
0.02232	3.366	5.31E-02	0.294
0.02356	3.372	1.96E-03	0.294
0.02480	3.377	2.39E-02	0.295
0.02604	3.380		0.295
0.02728		1.16E-02	
0.02976		7.96E-03	
0.03224		8.57E-03	
0.03472		2.70E-02	
0.03720	3.397	1.72E-02	0.297
0.04463		1.13E-02	
0.04959	3.403	2.79E-03	0.298
0.05703		1.56E-03	
0.06199	3.405	1.35E-03	0.298
0.06819		5.79E-04	
0.07439	3.407	4.44E-04	0.298
0.08059		4.41E-04	
0.08679	3.408	4.32E-04	0.298
0.09299		2.44E-04	
0.09919	3.409	3.23E-04	0.299
0.1116	3.409	2.87E-04	0.299
0.1240	3.410	2.71E-04	0.299
0.2480	3.417	2.67E-04	0.299
0.3720	3.427	1.90E-04	0.301
0.4959	3.442	1.41E-04	0.302
0.6199	3.462	1.12E-04	0.304
0.7439	3.486	9.42E-05	0.307
0.8679	3.516	8.07E-05	0.310
0.9919	3.551	7.11E-05	0.314
1.116	3.592	6.37E-05	0.319
1.240	3.640	5.81E-05	0.324
1.50		1.33E-04	
1.60		1.59E-04	
1.70		6.27E-04	
1.80	4.46	2.20E-02	0.402
2.0	4.79	0.76	0.438
2.2	4.49	1.19	0.431
2.4	4.28	1.21	0.417
2.6	4.40	1.32	0.430
2.8	4.59	1.70	0.462
3.0	4.44	2.29	0.490
3.2	3.92	2.59	0.493
3.4	3.69	2.76	0.502
3.6	3.39	3.01	0.521
3.8	(3.00)		
4.0	(2.65)		
4.2	(2.30)		
4.5	1.92	2.78	0.528
5.0	1.50	2.31	0.482
6.0	1.57	1.49	0.288
7.0	1.84	1.45	0.276
8.0	1.35	1.68	0.353
9.0	1.35	1.64	0.342
10.0	0.92	1.07	0.238
12.0	1.00	1.10	0.232
14.0	0.81	0.91	0.211
16.0	0.65	0.61	0.160
18.0	0.65	0.48	0.120
20.0	0.69	0.36	0.076
22.0	0.81	0.25	0.030
24.0	0.91	0.18	0.011
26.0	0.86	0.15	0.012
28.0	0.85	0.13	0.011
30.0	0.87	0.11	0.008

Selenium, single crystal, $\vec{E} \perp \hat{c}$[22]

Energy (eV)	n	k	$R(\phi = 0)$
0.01364	2.854	0.0239	0.231
0.01488	2.932	0.0325	0.241
0.01612	3.140	0.1750	0.269
0.01736	2.959	1.3300	0.321
0.01860	2.111	0.2550	0.133
0.01984	2.356	0.0746	0.164
0.02108	2.462	0.0276	0.178
0.02232	2.502	0.0442	0.184
0.02356	2.543	0.0097	0.190
0.02480	2.550	0.0239	0.191
0.02604	2.582		0.195
0.02728	2.600	0.0101	0.198
0.02976	2.576	9.95E-03	0.194
0.03224	2.598	1.16E-02	0.197
0.03472	2.607	1.68E-02	0.199
0.03720	2.613	1.54E-02	0.199
0.04463		1.17E-02	
0.04959	2.627	3.58E-03	0.201
0.05703		8.65E-04	

Energy (eV)	n	k	R(φ = 0)
0.06199	2.632	2.07E-03	0.202
0.06819		2.89E-04	
0.07439	2.635	1.59E-04	0.202
0.08059		1.35E-04	
0.08679	2.636	1.42E-04	0.202
0.09299		1.04E-04	
0.09919	2.637	8.95E-05	0.203
0.1116	2.638	8.84E-05	0.203
0.1240	2.639	8.51E-05	0.203
0.2480	2.645	5.97E-05	0.204
0.3720	2.652	5.44E-05	0.205
0.4959	2.654	4.58E-05	0.205
0.6199	2.675	3.82E-05	0.208
0.7439	2.692	3.32E-05	0.210
0.8679	2.713	2.96E-05	0.213
0.9919	2.739	2.69E-05	0.216
1.116	2.772	2.48E-05	0.221
1.240	2.816	2.31E-05	0.226
1.50		7.37E-05	
1.60		8.63E-05	
1.70		3.60E-04	
1.80	3.32	0.11	0.289
2.00	3.38	0.65	0.310
2.20	3.07	0.73	0.282
2.40	2.93	0.61	0.259
2.60	3.00	0.53	0.263
2.80	3.12	0.58	0.279
3.00	3.30	0.70	0.305
3.20	3.35	1.01	0.328
3.40	3.22	1.24	0.334
3.60	3.06	1.47	0.344
3.80	2.84	1.66	0.351
4.00	2.51	1.81	0.356
4.20	2.18	1.83	0.352
4.50	1.75	1.94	0.382
5.00	1.25	1.50	0.316
6.00	1.32	0.73	0.107
7.00	1.62	0.61	0.105
8.00	1.81	0.69	0.135
9.00	1.66	1.02	0.182
10.00	1.72	0.95	0.171
12.00	1.25	1.02	0.181
14.00	0.98	0.92	0.178
16.00	0.68	0.96	0.274
18.00	0.61	0.65	0.191
20.00	0.73	0.48	0.094
22.00	0.78	0.39	0.060
24.00	0.78	0.32	0.046
26.00	0.78	0.26	0.036
28.00	0.80	0.19	0.023
30.00	0.79	0.14	0.020

Silicon, single crystal[23]

Energy (eV)	n	k	R(φ = 0)
0.01240	3.4185	2.90E-04	0.300
0.01488	3.4190	2.30E-04	0.300
0.01736	3.4192	1.90E-04	0.300
0.01984	3.4195	1.70E-04	0.300
0.02480	3.4197		0.300
0.03100	3.4199		0.300
0.04092	3.4200		0.300
0.04463		1.08E-04	

Energy (eV)	n	k	R(φ = 0)
0.04959	3.4201	9.15E-05	0.300
0.05703		1.56E-04	
0.06199	3.4204	2.86E-04	0.300
0.06943		3.84E-04	
0.07439		7.16E-04	
0.08059	(3.4207)	1.52E-04	0.300
0.08679		1.02E-04	
0.09299		2.59E-04	
0.09919		1.77E-04	
0.1054		1.53E-04	
0.1116		2.02E-04	
0.1178		1.22E-04	
0.1240	3.4215	6.76E-05	0.300
0.1364		5.49E-05	
0.1488		2.41E-05	
0.1612		2.49E-05	
0.1736	(3.4230)	1.68E-05	0.300
0.1798		2.45E-05	
0.1860		2.66E-06	
0.1922		1.74E-06	
0.1984		8.46E-07	
0.2046		5.64E-07	
0.2108	(3.4244)	4.17E-07	0.300
0.2170		4.05E-07	
0.2232		3.94E-07	
0.2294		3.26E-07	
0.2356		2.97E-07	
0.2418		2.82E-07	
0.2480	3.4261	1.99E-07	0.300
0.3100	3.4294		0.301
0.3626	3.4327		0.301
0.4568	3.4393	2.50E-09	0.302
0.6199	3.4490		0.303
0.8093	3.4784		0.306
1.033	3.5193		0.311
1.1	(3.5341)	1.30E-05	0.312
1.2		1.80E-04	
1.3		2.26E-03	
1.4		7.75E-03	
1.5	3.673	5.00E-03	0.327
1.6	3.714	8.00E-03	0.331
1.7	3.752	1.00E-02	0.335
1.8	3.796	0.013	0.340
1.9	3.847	0.016	0.345
2.0	3.906	0.022	0.351
2.1	3.969	0.030	0.357
2.2	4.042	0.032	0.364
2.3	4.123	0.048	0.372
2.4	4.215	0.060	0.380
2.5	4.320	0.073	0.390
2.6	4.442	0.090	0.400
2.7	4.583	0.130	0.412
2.8	4.753	0.163	0.426
2.9	4.961	0.203	0.442
3.0	5.222	0.269	0.461
3.1	5.570	0.387	0.486
3.2	6.062	0.630	0.518
3.3	6.709	1.321	0.561
3.4	6.522	2.705	0.592
3.5	5.610	3.014	0.575

Energy (eV)	n	k	R(φ = 0)
3.6	5.296	2.987	0.564
3.7	5.156	3.058	0.563
3.8	5.065	3.182	0.568
3.9	5.016	3.346	0.577
4.0	5.010	3.587	0.591
4.1	5.020	3.979	0.614
4.2	4.888	4.639	0.652
4.3	4.086	5.395	0.703
4.4	3.120	5.344	0.726
4.5	2.451	5.082	0.740
4.6	1.988	4.678	0.742
4.7	1.764	4.278	0.728
4.8	1.658	3.979	0.710
4.9	1.597	3.749	0.693
5.0	1.570	3.565	0.675
5.1	1.571	3.429	0.658
5.2	1.589	3.354	0.646
5.3	1.579	3.353	0.647
5.4	1.471	3.366	0.663
5.5	1.340	3.302	0.673
5.6	1.247	3.206	0.675
5.7	1.180	3.112	0.673
5.8	1.133	3.045	0.672
5.9	1.083	2.982	0.673
6.0	1.010	2.909	0.677
6.5	0.847	2.73	0.688
7.0	0.682	2.45	0.691
7.5	0.563	2.21	0.693
8.0	0.478	2.00	0.691
8.5	0.414	1.82	0.688
9.0	0.367	1.66	0.683
9.5	0.332	1.51	0.672
10.0	0.306	1.38	0.661
12.0	0.257	0.963	0.590
14.0	0.275	0.641	0.460
16.0	0.345	0.394	0.297
18.0	0.455	0.219	0.159
20.0	0.567	0.0835	0.079
22.14	0.675	0.0405	0.038
24.31	0.752	0.0243	0.020
26.38	0.803	0.0178	0.012
28.18	0.834	0.0152	0.008
30.24	0.860	0.0138	0.006
31.79	0.877	0.0132	0.004
34.44	0.899	0.0121	0.003
36.47	0.913	0.0113	0.002
38.75	0.925	0.0104	0.002
40.00	0.930	0.0100	0.001

Silver[6]

Energy (eV)	n	k	R(φ = 0)
0.10	9.91	90.27	0.995
0.20	2.84	45.70	0.995
0.30	1.41	30.51	0.994
0.40	0.91	22.89	0.993
0.50	0.67	18.32	0.992
1.00	0.28	9.03	0.987
1.50	0.27	5.79	0.969
2.00	0.27	4.18	0.944
2.50	0.24	3.09	0.914
3.00	0.23	2.27	0.864
3.25	0.23	1.86	0.816

Energy (eV)	n	k	$R(\phi = 0)$
3.50	0.21	1.42	0.756
3.60	0.23	1.13	0.671
3.70	0.30	0.77	0.475
3.77	0.53	0.40	0.154
3.80	0.73	0.30	0.053
3.90	1.30	0.36	0.040
4.00	1.61	0.60	0.103
4.10	1.73	0.85	0.153
4.20	1.75	1.06	0.194
4.30	1.73	1.13	0.208
4.50	1.69	1.28	0.238
4.75	1.61	1.34	0.252
5.00	1.55	1.36	0.257
5.50	1.45	1.34	0.257
6.00	1.34	1.28	0.246
6.50	1.25	1.18	0.225
7.00	1.18	1.06	0.196
7.50	1.14	0.91	0.157
8.00	1.16	0.75	0.114
9.00	1.33	0.56	0.074
10.00	1.46	0.56	0.082
11.00	1.52	0.56	0.088
12.00	1.61	0.59	0.100
13.00	1.66	0.64	0.112
14.00	1.72	0.78	0.141
14.50	1.64	0.88	0.152
15.00	1.56	0.92	0.156
16.00	1.42	0.91	0.151
17.00	1.33	0.86	0.139
18.00	1.28	0.80	0.124
19.00	1.27	0.75	0.111
20.00	1.29	0.71	0.103
21.00	1.35	0.75	0.112
21.50	1.37	0.80	0.124
22.00	1.34	0.87	0.141
22.50	1.26	0.93	0.157
23.00	1.17	0.94	0.163
23.50	1.10	0.93	0.165
24.00	1.04	0.90	0.165
24.50	0.99	0.87	0.160
25.00	0.95	0.83	0.154
25.50	0.91	0.78	0.144
26.00	0.90	0.74	0.133
26.50	0.89	0.69	0.121
27.00	0.89	0.65	0.109
27.50	0.89	0.62	0.099
28.00	0.90	0.59	0.090
28.50	0.91	0.57	0.084
29.00	0.92	0.56	0.079
30.00	0.93	0.54	0.074
31.00	0.93	0.53	0.072
32.00	0.92	0.53	0.072
33.00	0.90	0.51	0.071
34.00	0.88	0.49	0.067
35.00	0.86	0.45	0.061
36.00	0.89	0.44	0.055
38.00	0.89	0.39	0.043
40.00	0.90	0.37	0.039
42.00	0.90	0.35	0.036
44.00	0.90	0.33	0.033
46.00	0.90	0.32	0.031
48.00	0.89	0.31	0.030
50.00	0.88	0.29	0.027
52.00	0.89	0.28	0.024
54.00	0.88	0.17	0.024
56.00	0.87	0.26	0.024
58.00	0.87	0.24	0.021
60.00	0.87	0.22	0.018
62.00	0.88	0.21	0.016
64.00	0.88	0.21	0.016
66.00	0.88	0.21	0.016
68.00	0.87	0.21	0.017
70.00	0.83	0.20	0.021
72.00	0.85	0.18	0.016
74.00	0.85	0.17	0.014
76.00	0.85	0.16	0.013
78.00	0.85	0.15	0.013
80.00	0.85	0.14	0.012
85.00	0.85	0.11	0.011
90.00	0.85	0.08	0.009
95.00	0.86	0.06	0.007
100.00	0.87	0.04	0.005

Sodium[24]

Energy (eV)	n	k	$R(\phi = 0)$
0.55	0.262	9.97	0.990
0.58	0.241	9.45	0.989
0.63	0.207	8.80	0.990
0.67	0.175	8.09	0.990
0.73	0.147	7.42	0.990
0.81	0.123	6.67	0.989
0.92	0.099	5.82	0.989
1.05	0.078	5.11	0.989
1.23	0.064	4.35	0.987
1.44	0.053	3.72	0.986
1.65	0.050	3.22	0.983
1.87	0.049	2.76	0.978
2.07	0.053	2.48	0.971
2.27	0.059	2.23	0.961
2.45	0.063	2.07	0.953
2.64	0.066	1.88	0.943
2.82	0.068	1.76	0.936
2.95	0.068	1.63	0.928
3.06	0.069	1.54	0.921
3.20	0.065	1.47	0.921
3.40	0.061	1.33	0.916
3.71	0.055	1.13	0.908
3.97	0.049	1.01	0.908
6.199	0.390		0.193
6.358	0.454		0.141
6.526	0.485		0.120
6.702	0.533		0.093
6.888	0.574		0.073
7.130	0.616		0.056
7.328	0.641		0.048
7.583	0.674		0.038
7.847	0.700		0.031
8.015	0.710		0.029
8.634	0.762		0.018
9.143	0.800		0.012
9.709	0.819		0.010
10.20	0.843		0.007

Energy (eV)	n	k	$R(\phi = 0)$
11.08	0.870		0.005
11.83	0.887		0.004
12.73	0.907		0.002
13.05	0.913		0.002
13.42	0.914		0.002
13.73	0.917		0.002
14.07	0.922		0.002
14.83	0.934		0.001
15.05	0.936		0.001
15.46	0.942		0.001
16.21	0.948		0.001
18.10	0.964		0.000
21.12	0.979		0.000
25.51	0.993		0.000
26.95	1.00		0.000
27.68	1.01		0.000
28.37	1.01		0.000
29.52	1.02		0.000

Tantalum[16]

Energy (eV)	n	k	$R(\phi = 0)$
0.10	10.14	66.39	0.984
0.15	9.45	46.41	0.9834
0.20	5.77	35.46	0.982
0.26	3.67	27.53	0.981
0.30	2.87	23.90	0.980
0.38	2.03	18.87	0.978
0.50	1.37	14.26	0.974
0.58	1.15	12.19	0.970
0.70	0.96	9.92	0.962
0.78	0.89	8.77	0.956
0.90	0.84	7.38	0.942
1.00	0.89	6.47	0.992
1.10	0.93	5.75	0.899
1.20	0.98	5.14	0.872
1.30	1.00	4.62	0.842
1.40	1.04	4.15	0.805
1.50	1.09	3.73	0.762
1.60	1.15	3.33	0.707
1.70	1.24	2.95	0.640
1.80	1.35	2.60	0.560
1.90	1.57	2.24	0.460
2.00	1.83	1.99	0.388
2.10	2.10	1.84	0.354
2.20	2.36	1.81	0.351
2.30	2.56	1.86	0.365
2.40	2.68	1.92	0.378
2.50	2.75	1.98	0.388
2.60	2.80	2.02	0.395
2.70	2.84	2.08	0.405
2.80	2.85	2.14	0.412
2.90	2.84	2.20	0.420
3.00	2.81	2.24	0.425
3.20	2.73	2.31	0.432
3.40	2.61	2.33	0.435
3.60	2.49	2.30	0.430
3.80	2.40	2.22	0.418
4.00	2.36	2.14	0.406
4.20	2.35	2.06	0.392
4.40	2.39	2.01	0.384
4.60	2.45	2.00	0.384
4.80	2.53	2.06	0.394

Energy (eV)	n	k	$R(\phi = 0)$	Energy (eV)	n	k	$R(\phi = 0)$	Energy (eV)	n	k	$R(\phi = 0)$
5.00	2.58	2.20	0.416	23.00	0.73	0.24	0.043	0.7	7.00	0.24	0.563
5.20	2.52	2.44	0.450	23.60	0.80	0.26	0.033	0.8	7.23	0.48	0.574
5.40	2.31	2.61	0.480	24.00	0.80	0.26	0.034	0.9	7.48	0.94	0.589
5.60	2.06	2.67	0.501	24.60	0.82	0.25	0.029	1.0	7.70	1.56	0.606
5.80	1.83	2.63	0.510	25.00	0.83	0.25	0.026	1.2	6.99	2.22	0.593
6.00	1.63	2.56	0.515	25.60	0.86	0.24	0.022	1.4	7.11	2.46	0.604
6.20	1.48	2.45	0.512	26.00	0.88	0.25	0.022	1.6	6.75	2.91	0.606
6.40	1.37	2.33	0.504	26.60	0.87	0.26	0.023	1.8	6.89	3.70	0.637
6.60	1.29	2.22	0.492	27.00	0.87	0.25	0.022	2.0	4.67	4.67	0.654
6.80	1.23	2.11	0.478	27.60	0.89	0.23	0.019	2.2	4.94	5.16	0.681
7.00	1.18	2.01	0.462	28.00	0.90	0.23	0.017	2.4	3.94	5.08	0.686
7.20	1.15	1.91	0.445	28.60	0.91	0.22	0.015	2.6	3.25	4.77	0.681
7.40	1.13	1.82	0.425	29.00	0.92	0.22	0.014	2.8	2.73	4.42	0.674
7.60	1.12	1.75	0.406	29.60	0.94	0.22	0.014	3.0	2.30	4.16	0.674
7.80	1.11	1.68	0.390	30.00	0.95	0.22	0.014	3.5	1.69	3.44	0.646
8.00	1.11	1.61	0.370	31.00	0.97	0.23	0.014	4.0	1.33	2.64	0.571
8.20	1.12	1.55	0.350	32.00	0.98	0.24	0.015	4.5	1.32	1.96	0.428
8.40	1.13	1.50	0.332	33.00	0.98	0.25	0.015	5.0	1.63	1.60	0.312
8.60	1.14	1.45	0.317	34.00	0.99	0.25	0.016	5.5	1.72	1.57	0.302
8.80	1.17	1.41	0.301	35.00	0.99	0.26	0.017	6.0	1.73	1.45	0.276
9.00	1.19	1.40	0.294	36.00	0.99	0.27	0.018	6.5	1.78	1.36	0.257
9.20	1.21	1.38	0.289	37.00	0.99	0.28	0.019	7.0	1.83	1.36	0.257
9.40	1.21	1.38	0.287	38.00	0.98	0.28	0.021	7.5	1.72	1.51	0.289
9.60	1.21	1.38	0.285	39.00	0.97	0.29	0.022	8.0	1.54	1.37	0.260
9.80	1.21	1.37	0.285	40.00	0.95	0.29	0.023	8.5	1.55	1.23	0.226
10.00	1.20	1.37	0.286					9.0	0.99	0.93	0.179
10.20	1.19	1.37	0.286	**Tellurium, $\vec{E} \parallel \mathit{c}$[25]**				9.5	1.47	1.25	0.233
10.40	1.18	1.37	0.287	0.01364	4.82	0.118	0.431	10.0	0.86	0.86	0.181
10.60	1.16	1.36	0.288	0.01488	5.26	0.0505	0.463	11.0	0.80	0.77	0.165
10.80	1.15	1.36	0.289	0.01612	5.47	0.0278	0.477	12.0	0.79	0.76	0.164
11.00	1.13	1.35	0.290	0.01736	5.59	0.0174	0.485	14.0	0.67	0.59	0.146
11.20	1.11	1.35	0.292	0.01860		0.0796		16.0	0.59	0.49	0.147
11.40	1.09	1.34	0.293	0.01984		0.0696		18.0	0.48	0.31	0.160
11.60	1.07	1.33	0.294	0.02108		0.0749		20.0	0.74	0.20	0.035
11.80	1.05	1.32	0.295	0.02232		0.1900		22.0	0.83	0.18	0.018
12.00	1.02	1.31	0.296	0.02356		0.2220		24.0	0.85	0.15	0.013
12.20	1.00	1.29	0.295	0.02480		0.0716		26.0	0.87	0.12	0.009
12.40	0.98	1.28	0.294	0.02604		0.0682		28.0	0.89	0.090	0.006
12.60	0.96	1.26	0.292	0.02728		0.0832		30.0	0.90	0.045	0.003
12.80	0.94	1.24	0.289	0.02976		0.0149					
13.00	0.93	1.22	0.286	0.03224		2.14E-03		**Tellurium, $\vec{E} \perp \mathit{c}$[25]**			
13.60	0.91	1.16	0.272	0.03472		1.71E-02		0.01364	2.61	0.2980	0.204
14.00	0.90	1.15	0.272	0.03720	5.94	3.71E-03	0.507	0.01488	3.65	0.0894	0.325
14.60	0.85	1.15	0.285	0.03968		2.44E-03		0.01612	4.10	0.0535	0.370
15.00	0.80	1.13	0.293	0.04339	5.96	1.59E-03	0.508	0.01736	4.63	0.4990	0.420
15.60	0.72	1.08	0.301	0.04711		7.85E-04		0.01860		0.1170	
16.00	0.68	1.04	0.304	0.05083		7.38E-04		0.01984		0.0343	
16.60	0.63	0.97	0.301	0.05579		3.89E-04		0.02108	(4.42)	0.0421	0.398
17.00	0.60	0.92	0.296	0.06199	5.98	3.09E-04	0.509	0.02232		0.1060	
17.60	0.60	0.92	0.296	0.07439		2.52E-04		0.02356		0.0880	
18.00	0.55	0.79	0.274	0.08679		2.96E-04		0.02480		0.0458	
18.60	0.53	0.71	0.254	0.09919		3.68E-04		0.02604		0.0928	
19.00	0.53	0.65	0.236	0.12400	6.246	3.34E-04	0.524	0.02728		0.0886	
19.60	0.53	0.57	0.207	0.15500	6.253		0.525	0.02976		0.0232	
20.00	0.54	0.52	0.185	0.20660	6.286		0.526	0.03224		3.06E-03	
20.60	0.55	0.44	0.153	0.24800	6.316	7.48E-05	0.528	0.03472		1.25E-02	
21.00	0.57	0.39	0.127	0.31	6.372	1.18E-05	0.531	0.03720	4.71	2.65E-03	0.422
21.60	0.64	0.34	0.089	0.35		4.93E-04		0.03968		1.89E-03	
22.00	0.64	0.32	0.081	0.41		6.74E-03		0.04339	4.74	1.41E-03	0.425
22.60	0.69	0.27	0.058	0.5	6.53	2.30E-02	0.539	0.04711		8.38E-04	
				0.6	6.71	7.50E-02	0.549	0.05083		6.79E-04	

Energy (eV)	n	k	R(φ = 0)
0.05579		1.59E-04	
0.06199	4.77	1.16E-04	0.427
0.07439		7.23E-05	
0.08679		5.34E-05	
0.09919		4.28E-05	
0.1240	4.796	3.18E-05	0.429
0.1550	4.809		0.430
0.2066	4.838		0.432
0.2480	4.864	2.19E-05	0.434
0.31	4.929	3.18E-05	0.439
0.35		7.89E-02	
0.41		0.149	
0.5	4.90		0.437
0.6	4.93		0.439
0.7	4.95	0.11	0.441
0.8	5.10	0.13	0.452
0.9	5.22	0.22	0.461
1.0	5.35	0.45	0.472
1.2	5.17	0.63	0.462
1.4	5.56	0.63	0.488
1.6	5.88	1.15	0.517
1.8	6.10	1.80	0.545
2.0	5.94	2.69	0.571
2.2	5.10	3.61	0.594
2.4	4.24	3.77	0.593
2.6	3.57	3.75	0.591
2.8	3.03	3.63	0.588
3.0	2.51	3.39	0.578
3.5	1.72	2.70	0.532
4.0	1.32	2.01	0.440
4.5	1.28	1.28	0.251
5.0	1.47	0.82	0.132
5.5	1.74	0.51	0.104
6.0	1.94	0.39	0.118
6.5	2.19	0.32	0.148
7.0	2.48	0.40	0.192
7.5	2.60	0.69	0.226
8.0	2.59	0.91	0.245
8.5	2.39	1.00	0.235
9.0	1.11	1.24	0.259
9.5	2.08	1.11	0.224
10.0	0.99	1.04	0.215
11.0	0.84	1.01	0.237
12.0	0.87	0.87	0.182
14.0	0.59	0.87	0.282
16.0	0.64	0.55	0.144
18.0	0.52	0.41	0.161
20.0	0.50	0.38	0.165
22.0	0.56	0.29	0.110
24.0	0.54	0.25	0.113
26.0	0.50	0.20	0.127
28.0	0.48	0.17	0.135
30.0	0.46	0.088	0.140

Titanium (Polycrystalline)[14]

Energy (eV)	n	k	R(φ = 0)
0.10	5.03	23.38	0.965
0.15	3.00	15.72	0.954
0.20	2.12	11.34	0.939
0.25	2.05	8.10	0.890
0.30	6.39	9.94	0.833
0.35	2.74	6.21	0.792

Energy (eV)	n	k	R(φ = 0)
0.40	2.49	4.68	0.708
0.45	3.35	3.25	0.545
0.50	4.43	3.22	0.555
0.60	4.71	3.77	0.597
0.70	4.38	3.89	0.603
0.80	4.04	3.82	0.596
0.90	3.80	3.65	0.582
1.00	3.62	3.52	0.570
1.10	3.47	3.40	0.560
1.20	3.35	3.30	0.550
1.30	3.28	3.25	0.546
1.40	3.17	3.28	0.549
1.50	2.98	3.32	0.557
1.60	2.74	3.30	0.559
1.70	2.54	3.23	0.557
1.80	2.36	3.11	0.550
1.90	2.22	2.99	0.540
2.00	2.11	2.88	0.530
2.10	2.01	2.77	0.520
2.20	1.92	2.67	0.509
2.30	1.86	2.56	0.495
2.40	1.81	2.47	0.483
2.50	1.78	2.39	0.471
2.60	1.75	2.34	0.462
2.70	1.71	2.29	0.456
2.80	1.68	2.25	0.451
2.90	1.63	2.21	0.447
3.00	1.59	2.17	0.444
3.10	1.55	2.15	0.442
3.20	1.50	2.12	0.442
3.30	1.44	2.09	0.442
3.40	1.37	2.06	0.443
3.50	1.30	2.01	0.443
3.60	1.24	1.96	0.441
3.70	1.17	1.90	0.436
3.80	1.11	1.83	0.430
3.85	1.08	1.78	0.423
3.90	1.06	1.73	0.413
4.00	1.04	1.62	0.389
4.20	1.05	1.45	0.333
4.40	1.13	1.33	0.284
4.60	1.17	1.29	0.265
4.80	1.21	1.23	0.244
5.00	1.24	1.21	0.236
5.20	1.27	1.20	0.228
5.40	1.17	1.16	0.228
5.60	1.24	1.21	0.234
5.80	1.21	1.22	0.241
6.00	1.15	1.21	0.244
6.20	1.11	1.18	0.240
6.40	1.08	1.14	0.232
6.60	1.04	1.06	0.212
6.80	1.05	1.02	0.198
7.00	1.06	0.97	0.182
7.20	1.07	0.95	0.175
7.40	1.11	0.94	0.167
7.60	1.09	0.92	0.165
7.80	1.11	0.93	0.165
8.00	1.10	0.94	0.169
8.20	1.10	0.95	0.171

Energy (eV)	n	k	R(φ = 0)
8.40	1.08	0.95	0.175
8.60	1.04	0.96	0.181
8.80	1.02	0.95	0.181
9.00	1.00	0.94	0.182
9.20	0.97	0.93	0.182
9.40	0.95	0.91	0.181
9.60	0.94	0.90	0.179
9.80	0.91	0.88	0.179
10.00	0.89	0.88	0.180
10.20	0.86	0.85	0.178
10.40	0.85	0.83	0.175
10.60	0.81	0.79	0.167
10.80	0.80	0.76	0.162
11.00	0.79	0.72	0.152
11.20	0.81	0.69	0.139
11.40	0.81	0.69	0.139
11.60	0.79	0.68	0.139
11.80	0.78	0.67	0.137
12.00	0.77	0.65	0.132
12.80	0.76	0.55	0.106
13.20	0.76	0.52	0.097
13.60	0.76	0.48	0.087
14.00	0.77	0.45	0.077
14.40	0.77	0.42	0.069
14.80	0.79	0.38	0.058
15.20	0.79	0.36	0.052
15.60	0.79	0.32	0.045
16.00	0.83	0.31	0.037
16.40	0.84	0.28	0.030
16.80	0.87	0.27	0.025
17.20	0.90	0.25	0.020
17.60	0.93	0.25	0.017
18.00	0.94	0.24	0.165
18.40	0.94	0.23	0.017
18.80	0.95	0.24	0.016
19.20	0.96	0.25	0.016
19.60	0.97	0.25	0.017
20.00	0.98	0.27	0.018
20.40	0.98	0.27	0.019
20.60	1.00	0.29	0.020
21.20	0.99	0.31	0.023
21.60	0.99	0.31	0.024
22.00	0.98	0.32	0.025
22.40	0.98	0.33	0.027
22.80	0.97	0.33	0.028
23.20	0.96	0.34	0.030
23.60	0.95	0.35	0.031
24.00	0.92	0.35	0.033
24.5	0.91	0.34	0.032
25.0	0.91	0.33	0.032
25.5	0.89	0.33	0.032
26.0	0.89	0.33	0.032
26.5	0.88	0.32	0.032
27.0	0.86	0.31	0.032
27.5	0.85	0.30	0.033
28.0	0.84	0.29	0.033
28.5	0.82	0.26	0.029
29.0	0.83	0.25	0.027
30.0	0.84	0.22	0.022

Energy (eV)	n	k	R(φ = 0)	Energy (eV)	n	k	R(φ = 0)	Energy (eV)	n	k	R(φ = 0)
				5.40	2.92	3.58	0.586	22.80	0.49	0.69	0.272
				5.60	2.43	3.70	0.618	23.20	0.49	0.66	0.263
Tungsten[27]				5.80	2.00	3.61	0.637	23.60	0.48	0.62	0.252
0.10	14.06	54.71	0.983	6.00	1.70	3.42	0.643	24.00	0.49	0.57	0.234
0.20	3.87	28.30	0.981	6.20	1.47	3.24	0.646	24.40	0.50	0.53	0.213
0.25	2.56	22.44	0.980	6.40	1.32	3.04	0.640	24.80	0.51	0.49	0.191
0.30	1.83	18.32	0.979	6.60	1.21	2.87	0.631	25.20	0.53	0.46	0.171
0.34	1.71	15.71	0.973	6.80	1.12	2.70	0.619	25.60	0.55	0.43	0.150
0.38	1.86	13.88	0.963	7.00	1.06	2.56	0.607	26.00	0.57	0.40	0.132
0.42	1.92	12.63	0.954	7.20	1.01	2.43	0.593	26.40	0.59	0.38	0.117
0.46	1.69	11.59	0.952	7.40	0.98	2.30	0.573	26.80	0.61	0.37	0.105
0.50	1.40	10.52	0.952	7.60	0.95	2.18	0.556	27.00	0.62	0.36	0.099
0.54	1.23	9.45	0.948	7.80	0.93	2.06	0.533	27.50	0.64	0.34	0.085
0.58	1.17	8.44	0.938	8.00	0.94	1.95	0.505	28.00	0.67	0.32	0.073
0.62	1.28	7.52	0.917	8.20	0.94	1.86	0.481	28.50	0.69	0.31	0.065
0.66	1.45	6.78	0.888	8.40	0.96	1.76	0.449	29.00	0.71	0.30	0.057
0.70	1.59	6.13	0.856	8.60	0.99	1.70	0.422	29.50	0.73	0.30	0.052
0.74	1.83	5.52	0.810	8.80	1.01	1.65	0.401	30.00	0.75	0.29	0.047
0.78	2.12	5.00	0.759	9.00	1.01	1.60	0.388	31.00	0.78	0.29	0.042
0.82	2.36	4.61	0.710	9.20	1.02	1.55	0.369	32.00	0.79	0.29	0.040
0.86	2.92	4.37	0.661	9.40	1.03	1.50	0.352	33.00	0.82	0.28	0.033
0.90	3.11	4.44	0.660	9.60	1.05	1.44	0.329	34.00	0.84	0.29	0.032
0.94	3.15	4.43	0.658	9.80	1.09	1.38	0.307	35.00	0.85	0.31	0.033
0.98	3.15	4.36	0.653	10.00	1.13	1.34	0.287	36.00	0.85	0.32	0.036
1.00	3.14	4.32	0.649	10.20	1.19	1.33	0.274	37.00	0.84	0.33	0.039
1.10	3.05	4.04	0.627	10.40	1.24	1.34	0.270	38.00	0.83	0.33	0.040
1.20	3.00	3.64	0.590	10.60	1.27	1.36	0.274	39.00	0.81	0.33	0.042
1.30	3.12	3.24	0.545	10.80	1.29	1.39	0.282	40.00	0.80	0.33	0.045
1.40	3.29	2.96	0.515	11.00	1.28	1.42	0.290				
1.50	3.48	2.79	0.500	11.20	1.27	1.44	0.297	*Vanadium*[9]			
1.60	3.67	2.68	0.494	11.40	1.25	1.46	0.305	0.10	12.83	45.89	0.978
1.70	3.84	2.79	0.507	11.60	1.22	1.48	0.313	0.20	3.90	24.30	0.975
1.80	3.82	2.91	0.518	11.80	1.20	1.48	0.318	0.28	2.13	17.35	0.973
1.90	3.70	2.94	0.518	12.00	1.16	1.48	0.323	0.36	1.54	13.32	0.966
2.00	3.60	2.89	0.512	12.40	1.10	1.47	0.329	0.44	1.28	10.74	0.957
2.10	3.54	2.84	0.506	12.80	1.04	1.44	0.333	0.52	1.16	8.93	0.945
2.20	3.49	2.76	0.497	13.20	0.98	1.40	0.332	0.60	1.10	7.59	0.929
2.30	3.49	2.72	0.494	13.60	0.94	1.35	0.325	0.68	1.07	6.54	0.909
2.40	3.45	2.72	0.493	14.00	0.91	1.28	0.312	0.76	1.08	5.67	0.882
2.50	3.38	2.68	0.487	14.40	0.90	1.23	0.296	0.80	1.10	5.30	0.864
2.60	3.34	2.62	0.480	14.80	0.90	1.17	0.276	0.90	1.18	4.50	0.811
2.70	3.31	2.55	0.472	15.20	0.93	1.13	0.255	1.00	1.34	3.80	0.730
2.80	3.31	2.49	0.466	15.60	0.97	1.12	0.246	1.10	1.60	3.26	0.632
2.90	3.32	2.45	0.461	16.00	0.98	1.14	0.249	1.20	1.93	2.88	0.543
3.00	3.35	2.42	0.459	16.40	0.97	1.17	0.260	1.30	2.25	2.71	0.498
3.10	3.39	2.41	0.460	16.80	0.94	1.19	0.273	1.40	2.48	2.72	0.491
3.20	3.43	2.45	0.465	17.20	0.90	1.21	0.289	1.50	2.57	2.79	0.499
3.30	3.45	2.55	0.476	17.60	0.85	1.21	0.304	1.60	2.57	2.84	0.507
3.40	3.39	2.66	0.485	18.00	0.80	1.20	0.317	1.70	2.52	2.88	0.512
3.50	3.24	2.70	0.488	18.40	0.74	1.18	0.330	1.80	2.45	2.88	0.515
3.60	3.13	2.67	0.482	18.80	0.69	1.15	0.340	1.90	2.36	2.85	0.514
3.70	3.05	2.62	0.476	19.20	0.64	1.11	0.347	2.00	2.34	2.81	0.509
3.80	2.99	2.56	0.468	19.60	0.60	1.07	0.353	2.10	2.31	2.78	0.506
3.90	2.96	2.50	0.460	20.00	0.56	1.02	0.354	2.20	2.28	2.80	0.510
4.00	2.95	2.43	0.451	20.40	0.54	0.97	0.350	2.30	2.23	2.83	0.516
4.20	3.02	2.33	0.440	20.80	0.52	0.92	0.342	2.40	2.15	2.88	0.528
4.40	3.13	2.32	0.442	21.20	0.50	0.87	0.331	2.50	2.02	2.91	0.540
4.60	3.24	2.41	0.455	21.60	0.50	0.82	0.318	2.60	1.89	2.92	0.552
4.80	3.33	2.57	0.475	22.00	0.49	0.77	0.303	2.70	1.74	2.89	0.561
5.00	3.40	2.85	0.505	22.40	0.49	0.73	0.287	2.80	1.61	2.85	0.569
5.20	3.27	3.27	0.548					2.90	1.48	2.80	0.577

Energy (eV)	n	k	$R(\phi = 0)$	Energy (eV)	n	k	$R(\phi = 0)$	Energy (eV)	n	k	$R(\phi = 0)$
3.00	1.36	2.73	0.582	27.00	0.84	0.16	0.015	1.305	3.3991	2.7684	0.497
3.20	1.16	2.55	0.585	27.50	0.85	0.16	0.014	1.377	3.1807	3.4709	0.569
3.40	0.99	2.37	0.586	28.00	0.85	0.15	0.013	1.459	3.5064	4.1994	0.630
3.60	0.87	2.17	0.575	28.50	0.86	0.14	0.012	1.550	4.1241	4.7768	0.664
3.80	0.80	1.96	0.547	29.00	0.86	0.14	0.011	1.653	4.0269	4.8027	0.667
4.00	0.78	1.76	0.503	29.50	0.86	0.13	0.010	1.722	3.9369	4.6356	0.657
4.20	0.80	1.60	0.449	30.00	0.87	0.13	0.009	1.823	3.7549	4.3042	0.635
4.40	0.83	1.47	0.400	31.00	0.88	0.12	0.008	1.937	3.4512	4.1942	0.631
4.60	0.87	1.38	0.355	32.00	0.90	0.11	0.007	1.984	3.2515	4.2980	0.644
4.80	0.90	1.31	0.326	33.00	0.90	0.10	0.005	2.066	2.0802	4.7231	0.738
5.00	0.91	1.26	0.304	34.00	0.91	0.10	0.005	2.094	1.7084	4.7923	0.774
5.25	0.93	1.18	0.271	35.00	0.92	0.09	0.004	2.119	1.3329	4.4751	0.791
5.50	0.94	1.14	0.258	36.00	0.94	0.10	0.004	2.275	0.9725	4.2879	0.825
5.75	0.96	1.09	0.235	37.00	0.94	0.10	0.004	2.455	0.7568	3.7627	0.824
6.00	0.98	1.06	0.223	38.00	0.95	0.11	0.004	2.666	0.5470	3.4277	0.845
6.25	0.97	1.02	0.212	39.00	0.95	0.12	0.004	2.917	0.4774	3.0476	0.834
6.50	0.97	0.98	0.199	40.00	0.95	0.13	0.005	3.220	0.3911	2.7463	0.835
6.75	0.97	0.94	0.185					3.594	0.3147	2.3041	0.821
7.00	0.98	0.91	0.175	**Zinc, $\vec{E} \parallel \hat{c}$[28]**				4.065	0.3013	2.0077	0.789
7.33	0.97	0.89	0.170	0.7514	1.9241	7.5619	0.883	4.678	0.2806	1.7997	0.770
7.66	0.98	0.87	0.162	0.827	1.7921	6.9973	0.874				
8.00	0.98	0.85	0.155	0.866	1.5571	6.7753	0.881	**Zirconium (Polycrystalline)[28]**			
8.33	0.98	0.81	0.146	0.952	1.4824	6.2296	0.868	0.10	6.18	1.76	0.300
8.66	0.98	0.81	0.145	0.992	1.5762	5.8843	0.847	0.15	3.37	1.30	0.123
9.00	0.96	0.79	0.142	1.033	1.5407	5.3192	0.823	0.20	2.34	1.08	0.058
9.50	0.94	0.77	0.136	1.078	1.5853	4.9013	0.793	0.26	2.24	1.06	0.052
10.00	0.91	0.74	0.133	1.127	1.7768	4.5307	0.748	0.30	2.59	1.14	0.073
10.50	0.89	0.71	0.126	1.181	1.9808	4.2004	0.701	0.36	3.17	1.26	0.110
11.00	0.87	0.65	0.112	1.240	2.8821	3.4766	0.575	0.40	3.09	1.24	0.105
11.50	0.88	0.58	0.091	1.305	3.2039	3.0042	0.520	0.46	3.36	1.30	0.123
12.00	0.90	0.58	0.089	1.377	2.9459	3.5761	0.584	0.50	4.13	1.44	0.175
12.50	0.89	0.57	0.086	1.459	3.2523	4.2447	0.640	0.56	5.01	1.58	0.231
13.00	0.88	0.55	0.082	1.550	3.8086	4.6212	0.657	0.60	5.18	1.61	0.242
13.50	0.87	0.53	0.079	1.653	3.7577	4.6239	0.659	0.70	4.54	1.51	0.202
14.00	0.86	0.51	0.075	1.722	3.5908	4.4614	0.650	0.80	4.03	1.42	0.168
14.50	0.86	0.49	0.070	1.823	3.4234	4.3232	0.642	0.90	3.74	1.37	0.149
15.00	0.86	0.47	0.065	1.937	3.0132	3.9974	0.624	0.96	3.69	1.36	0.145
15.50	0.86	0.46	0.062	1.984	1.8562	3.9706	0.690	1.00	3.66	1.35	0.143
16.00	0.85	0.45	0.061	2.066	1.4856	4.0555	0.737	1.10	3.65	1.35	0.142
16.50	0.84	0.43	0.059	2.094	1.2525	3.9961	0.762	1.20	3.53	1.33	0.134
17.00	0.84	0.41	0.056	2.119	1.0017	3.8683	0.789	1.30	3.25	1.27	0.116
17.50	0.83	0.40	0.054	2.275	0.7737	3.9129	0.832	1.40	3.10	1.25	0.106
18.00	0.82	0.38	0.051	2.445	0.6395	3.4013	0.821	1.50	3.02	1.23	0.100
18.50	0.82	0.37	0.048	2.666	0.4430	3.1379	0.851	1.60	2.88	1.20	0.091
19.00	0.82	0.35	0.045	2.917	0.3589	2.8140	0.853	1.70	2.68	1.16	0.078
19.50	0.82	0.34	0.043	3.220	0.3069	2.5088	0.847	1.80	2.49	1.12	0.067
20.00	0.81	0.32	0.041	3.594	0.2737	2.1737	0.828	2.00	2.14	1.03	0.047
20.50	0.81	0.31	0.038	4.065	0.2510	1.8528	0.799	2.10	1.99	1.00	0.040
21.00	0.81	0.29	0.036	4.678	0.2354	1.6357	0.776	2.20	1.87	0.97	0.034
21.50	0.81	0.28	0.033					2.30	1.78	0.94	0.030
22.00	0.81	0.27	0.032	**Zinc, $\vec{E} \perp \hat{c}$[28]**				2.40	1.71	0.92	0.027
22.50	0.81	0.25	0.029	0.751	1.4469	7.4158	0.905	2.50	1.62	0.90	0.024
23.00	0.82	0.24	0.027	0.827	1.4744	6.9688	0.892	2.60	1.54	0.88	0.022
23.50	0.82	0.23	0.025	0.866	1.3628	6.6886	0.892	2.70	1.46	0.86	0.019
24.00	0.82	0.22	0.024	0.952	1.3165	6.2212	0.881	2.80	1.40	0.84	0.018
24.50	0.83	0.21	0.022	0.992	1.3835	5.8910	0.863	2.90	1.34	0.82	0.016
25.00	0.83	0.20	0.020	1.033	1.2889	5.4001	0.850	3.00	1.30	0.81	0.016
25.50	0.83	0.19	0.019	1.078	1.3095	4.9025	0.822	3.10	1.26	0.80	0.015
26.00	0.83	0.18	0.018	1.127	1.6897	4.4062	0.746	3.30	1.19	0.77	0.014
26.50	0.84	0.17	0.016	1.181	1.9701	4.0176	0.684	3.40	1.16	0.76	0.013
				1.240	2.8717	3.2873	0.555	3.50	1.13	0.75	0.013

Energy (eV)	n	k	R(φ = 0)	Energy (eV)	n	k	R(φ = 0)	Energy (eV)	n	k	R(φ = 0)
3.60	1.10	0.74	0.013	9.00	1.65	0.91	0.025	17.20	1.09	0.74	0.013
3.70	1.07	0.73	0.013	9.20	1.63	0.90	0.025	17.60	1.13	0.75	0.013
3.80	1.04	0.72	0.012	9.40	1.60	0.89	0.024	18.00	1.17	0.76	0.014
3.90	1.01	0.71	0.012	9.60	1.57	0.89	0.023	18.40	1.21	0.78	0.014
4.00	0.98	0.70	0.012	9.80	1.52	0.87	0.021	18.80	1.24	0.79	0.014
4.20	0.94	0.68	0.013	10.00	1.47	0.86	0.020	19.20	1.27	0.80	0.015
4.40	0.89	0.67	0.013	10.20	1.42	0.84	0.018	19.60	1.29	0.80	0.015
4.60	0.85	0.65	0.014	10.40	1.35	0.82	0.016	20.00	1.30	0.81	0.015
4.80	0.81	0.64	0.014	10.50	1.32	0.81	0.016	20.60	1.29	0.80	0.015
5.00	0.78	0.63	0.015	10.60	1.28	0.80	0.015	21.00	1.27	0.80	0.015
5.20	0.77	0.62	0.016	10.80	1.23	0.78	0.014	21.60	1.23	0.78	0.014
5.40	0.77	0.62	0.016	11.00	1.19	0.77	0.014	22.00	1.20	0.77	0.014
5.60	0.80	0.63	0.014	11.20	1.16	0.76	0.013	22.60	1.15	0.76	0.013
5.80	0.87	0.66	0.013	11.40	1.13	0.75	0.013	23.00	1.12	0.75	0.013
6.00	1.00	0.71	0.012	11.60	1.11	0.74	0.013	23.60	1.08	0.73	0.013
6.20	1.11	0.75	0.013	11.80	1.09	0.74	0.013	24.00	1.05	0.73	0.013
6.40	1.23	0.78	0.014	12.00	1.08	0.73	0.012	24.60	1.02	0.71	0.012
6.60	1.33	0.81	0.016	12.40	1.05	0.72	0.012	25.00	1.00	0.71	0.012
6.80	1.42	0.84	0.018	12.80	1.01	0.71	0.012	25.60	0.97	0.69	0.012
7.00	1.49	0.86	0.020	13.20	0.98	0.70	0.012	26.00	0.95	0.69	0.013
7.20	1.54	0.88	0.022	13.60	0.95	0.69	0.013	26.60	0.91	0.67	0.013
7.40	1.58	0.89	0.023	14.00	0.92	0.68	0.013	27.00	0.88	0.66	0.013
7.60	1.61	0.90	0.024	14.40	0.89	0.67	0.013	27.60	0.84	0.65	0.014
7.80	1.63	0.90	0.025	14.80	0.90	0.67	0.013	28.00	0.83	0.64	0.014
8.00	1.66	0.91	0.026	15.20	0.92	0.68	0.013	28.60	0.82	0.64	0.014
8.20	1.67	0.91	0.026	15.60	0.95	0.69	0.013	29.00	0.81	0.64	0.014
8.40	1.68	0.92	0.026	16.00	0.98	0.70	0.012	29.60	0.82	0.64	0.014
8.60	1.68	0.92	0.026	16.40	1.01	0.71	0.012	30.00	0.82	0.64	0.014
8.80	1.66	0.91	0.026	16.80	1.04	0.72	0.012				

References

1. Shiles, E., Sasaki, T., Inokuti, M., and Smith, D. Y., *Phys. Rev. Sect. B*, 22, 1612, 1980.
2. Edwards, D. F., and Philipp, H. R., in *HOC-I*, p. 665.
3. Ives, H. E., and Briggs, N. B., *J. Opt. Soc. Am.*, 27, 395, 1937.
4. Bos, L. W., and Lynch, D. W., *Phys. Rev. Sect. B*, 2, 4567, 1970.
5. Weaver, J. H., Colavita, E., Lynch, D. W., and Rosei, R., *Phys. Rev. Sect. B*, 19, 3850, 1979.
6. Hagemann, H. J., Gudat, W., and Kunz, C., *J. Opt. Soc. Am.*, 65, 742, 1975.
7. Schulz, L. G., *J. Opt. Soc. Am.*, 47, 64, 1957.
8. Potter, R. F., in *HOC-I*, p. 465.
9. Olson, C. G., Lynch, D. W., and Weaver, J. H., unpublished.
10. Lynch, D. W., Olson, C. G., and Weaver, J. H., unpublished.
11. Weaver, J. H., Olson, C. G., and Lynch, D. W., *Phys. Rev. Sect. B*, 15, 4115, 1977.
12. Lynch, D. W., and Hunter, W. R., in *HOC-II*, p. 345.
13. Priol, M. A., Daudé, A., and Robin, S., *Compt. Rend.*, 264, 935, 1967.
14. Johnson, P. B., and Christy, R. W., *Phys. Rev. Sect. B*, 9, 5056, 1974.
15. Arakawn, E. T., and Inagaki, T., in *HOC-II*, p. 461.
16. Weaver, J. H., Lynch, D. W., and Olson, D. G., *Phys. Rev. Sect. B*, 10, 501, 1973.
17. Lynch, D. W., Rosei, R., and Weaver, J. H., *Solid State Commun.*, 9, 2195, 1971.
18. Weaver, J. H., Lynch, D. W., and Olson, C. G., *Phys. Rev. Sect. B*, 7, 4311, 1973.
19. Weaver, J. H., and Benbow, R. L., *Phys. Rev. Sect. B*, 12, 3509, 1975.
20. Weaver, J. H., *Phys. Rev., Sect. B*, 11, 1416, 1975.
21. Lynch, D. W., and Hunter, W. R., in *HOC-II*, p. 364.
22. Palik, E. D., in *HOC-II*, p. 691.
23. Edwards, D. F., in *HOC-I*, p. 547.
24. Lynch, D. W., and Hunter, W. R., in *HOC-II*, p. 354.
25. Palik, E. D., in *HOC-II*, p. 709.
26. Lynch, D. W., Olson, C. G., and Weaver, J. H., *Phys. Rev. Sect. B*, 11, 3671, 1975.
27. Weaver, J. H., Lynch, D. W., and Olson, C. G., *Phys. Rev. Sect. B*, 12, 1293, 1975.
28. Lanham, A. P., and Terherne, D. M., *Proc. Phys. Soc.*, 83, 1059, 1964.

OPTICAL PROPERTIES OF SELECTED INORGANIC AND ORGANIC SOLIDS

L. I. Berger

Optical properties of materials are closely related to their dielectric properties. The complex dielectric function (relative permittivity) of a material is equal to

$$\varepsilon(\omega) = \varepsilon'(\omega) - j\varepsilon''(\omega),$$

where $\varepsilon'(\omega)$ and $\varepsilon''(\omega)$ are its real and imaginary parts, respectively, and ω is the angular frequency of the applied electric field. For a non-absorbing medium, the index of refraction is $n = (\varepsilon\mu)^{1/2}$, where μ is the relative magnetic permeability of the medium (material); in the majority of dielectrics, $\mu \cong 1$.

For many applications, the most important optical properties of materials are the index of refraction, the extinction coefficient, k, and the reflectivity, R. The common index of refraction of a material is equal to the ratio of the phase velocity of propagation of an electromagnetic wave of a given frequency in vacuum to that in the material. Hence, $n \geq 1$. The optical properties of highly conductive materials like metals and semiconductors (at photon energy range above the energy gap) differ from those of optically transparent media. Free electrons absorb the incident electromagnetic wave in a thin surface layer (a few hundred nanometers thick) and then release the absorbed energy in the form of secondary waves reflected from the surface. Thus, the light reflection becomes very strong; for example, highly conductive sodium reflects 99.8% of the incident wave (at 589 nm). Introduction of the effective index of refraction, $n_{eff} = (\varepsilon')^{1/2} = n - jk$, where $\varepsilon' = \varepsilon - j\delta/\omega\,\varepsilon_o$, δ is the electrical conductivity of the material in S/m, and $\varepsilon_o = 8.8542 \cdot 10^{-12}$ F/m is the permittivity of vacuum, allows one to apply the expressions of the optics of transparent media to the conductive materials. It is clear that the effective index of refraction may be smaller than 1. For example, $n = 0.05$ for pure sodium and $n = 0.18$ for pure silver (at 589.3 nm). At very high photon energies, the quantum effects, such as the internal photoeffect, start playing a greater role, and the optical properties of these materials become similar to those of insulators (low reflectance, existence of Brewster's angle, etc.).

The extinction coefficient characterizes absorption of the electromagnetic wave energy in the process of propagation of a wave through a material. The wave intensity, I, after it passes a distance x in an isotropic medium is equal to

$$I = I_0 \exp(-\alpha x),$$

where I_0 is the intensity at $x = 0$ and α is called the absorption coefficient. For many applications, the extinction coefficient, k, which is equal to

$$k = \alpha \frac{\lambda}{4\pi},$$

where λ is the wavelength of the wave in the medium, is more commonly used for characterization of the electromagnetic losses in materials.

Reflection of an electromagnetic wave from the interface between two media depends on the media indices of refraction and on the angle of incidence. It is characterized by the reflectivity, which is equal to the ratio of the intensity of the wave reflected back into the first medium to the intensity of the wave approaching the interface. For polarized light and two non-absorbing media,

$$R = \frac{(N_1 - N_2)^2}{(N_1 + N_2)^2},$$

where $N_1 = n_1/\cos\theta_1$ and $N_2 = n_2/\cos\theta_2$ for the wave polarized in the plane of incidence, and $N_1 = n_1\cos\theta_1$ and $N_2 = n_2\cos\theta_2$ for the wave polarized normal to the plane of incidence; θ_1 and θ_2 are the angles between the normal to the interface in the point of incidence and the directions of the beams in the first and second medium, respectively. The reflectivity at normal incidence in this case is

$$R = [(n_1 - n_2)/(n_1 + n_2)]^2$$

For any two opaque (absorbing) media, the normal incidence reflectivity is

$$R = \frac{(n_1 - n_2)^2 + k_2^2}{(n_1 + n_2)^2 + k_2^2}.$$

In the majority of experiments, the first medium is air ($n \approx 1$), and hence,

$$R = \frac{(1 - n)^2 + k^2}{(1 + n)^2 + k^2}.$$

The data on n and k in the following table are abridged from the sources listed in the references. The reflectivity at normal incidence, R, has been calculated from the last equation. For convenience, the energy E, wavenumber $\bar{\nu}$, and wavelength λ are given for the incidence radiation.

E/eV	$\bar{\nu}$/cm^{-1}	λ/μm	n	n_a	n_c	k	k_a	k_c	R	R_a	R_c
				Crystalline Arsenic Selenide (As$_2$Se$_3$) [Ref. 1]*								
2.194	17700	0.565				0.30					
2.168	17480	0.572				0.25					
2.141	17270	0.579				0.20					
2.123	17120	0.584				0.17					
2.098	16920	0.591				0.13					
2.094	16890	0.592						0.26			
2.091	16860	0.593						0.26			
2.073	16720	0.598				0.10		0.23			
2.060	16610	0.602						0.20			
2.049	16530	0.605					0.079	0.17			

E/eV	$\bar{\nu}$/cm^{-1}	λ/μm	n	n_a	n_c	k	k_a	k_c	R	R_a	R_c
2.036	16420	0.609						0.15			
2.023	16310	0.613						0.12			
2.013	16230	0.616					0.050				
2.009	16210	0.617						0.097			
2.000	16130	0.620						0.082			
1.987	16030	0.624						0.063			
1.977	15940	0.627					0.031				
1.974	15920	0.628						0.051			
1.962	15820	0.632						0.038			
1.953	15750	0.635						0.030			
1.949	15720	0.636					0.020				
1.937	15630	0.640						0.022			
1.925	15530	0.644						0.017			
1.922	15500	0.645					0.012				
1.905	15360	0.651					$8.6 \cdot 10^{-3}$				
1.893	15270	0.655					6.4				
1.881	15170	0.659					5.2				
1.859	14990	0.667					3.1				
1.848	14900	0.671						$1.7 \cdot 10^{-3}$			
1.845	14880	0.672					2.0				
1.842	14860	0.673						$1.2 \cdot 10^{-3}$			
1.831	14770	0.677					$1.3 \cdot 10^{-3}$	$9.0 \cdot 10^{-4}$			
1.826	14730	0.679					6.4				
1.821	14680	0.681					4.7				
1.818	14660	0.682					$8.6 \cdot 10^{-4}$				
1.815	14640	0.683					3.4				
1.807	14580	0.686					5.5				
1.802	14530	0.688					4.1				
0.06199	500.0	20.0		3.2	2.9		$1.7 \cdot 10^{-3}$	$1.8 \cdot 10^{-3}$		0.27	0.24
0.05904	476.2	21.0		3.1	2.9		$2.1 \cdot 10^{-3}$	$2.2 \cdot 10^{-3}$		0.26	0.24
0.05636	454.5	22.0		3.1	2.9		$2.5 \cdot 10^{-3}$	$2.6 \cdot 10^{-3}$		0.26	0.24
0.05391	434.8	23.0		3.1	2.9		$3.0 \cdot 10^{-3}$	$3.1 \cdot 10^{-3}$			
0.04592	370.4	27.0		3.0	2.8		$6.3 \cdot 10^{-3}$	$6.4 \cdot 10^{-3}$		0.25	0.22
0.04428	357.1	28.0		3.0	2.8		$7.6 \cdot 10^{-3}$	$7.7 \cdot 10^{-3}$		0.25	0.22
0.04275	344.8	29.0		3.0	2.8		0.0092	0.0093		0.25	0.22
0.04133	333.3	30.0		3.0	2.7		0.011	0.011		0.25	0.21
0.03542	285.7	35.0		2.7	2.5		0.037	0.034		0.21	0.18
0.03100	250.0	40.0		1.9	1.7		0.38	1.0		0.19	0.18
0.03061	247.0	40.5		2.0	2.6		0.33	0.95		0.12	0.25
0.03024	244.0	41.0		1.7	2.4		0.41	0.46		0.088	0.18
0.02883	232.6	43.0		1.2	1.3		2.2	0.94		0.50	0.16
0.02850	229.9	43.5		1.6	1.2		2.8	1.4		0.56	0.29
0.02818	227.3	44.0		2.3	1.2		3.3	2.0		0.58	0.48
0.02755	222.2	45.0		4.2	2.0		2.5	3.3		0.50	0.60
0.02480	200.0	50.0		6.5	4.0		3.6	0.26		0.62	0.36
0.02254	181.8	55.0		4.5	3.5		0.17	0.10		0.40	0.31
0.02066	166.7	60.0		4.0	3.2		0.089	0.10		0.36	0.27
0.01907	153.8	65.0		3.8	3.1		0.097	0.16		0.34	0.26
0.01771	142.9	70.0		3.6	3.0		0.19	0.30		0.32	0.25
0.01653	133.3	75.0		3.7	3.0		0.41	0.44		0.34	0.26
0.01550	125.0	80.0		3.8	3.1		0.29	0.40		0.34	0.27
0.01459	117.6	85.0		3.6	2.9		0.20	0.34		0.32	0.24
0.01378	111.1	90.0		3.2	2.6		0.43	0.49		0.28	0.21
0.01305	105.3	95.0		4.7	3.0		1.5	1.5		0.46	0.34
0.01240	100.0	100.0		4.4	2.7		0.22	0.81		0.40	0.25
0.01181	95.24	105.0		4.2	3.0		0.094	3.9		0.38	0.62
0.01127	90.91	110.0		4.1	5.3		0.059	0.70		0.37	0.47
0.01033	83.33	120.0		3.9	4.2		0.034	0.13		0.35	0.38
0.009537	76.92	130.0		3.9	4.0		0.024	0.069		0.35	0.36
0.008856	71.43	140.0		3.9	3.8		0.019	0.048		0.35	0.34
0.007749	63.50	160.0		3.8	3.7		0.014	0.032		0.34	0.33

E/eV	$\bar{\nu}$/cm^{-1}	λ/µm	n	n_a	n_c	k	k_a	k_c	R	R_a	R_c
0.006888	55.55	180.0		3.8	3.7		0.011	0.024		0.34	0.33
0.006199	50.0	200.0		3.8	3.6		0.0091	0.019		0.34	0.32

* Indices a and c relate to the radiation electric field parallel to the a and c axes of the crystal, respectively.

Vitreous Arsenic Selenide (As$_2$Se$_3$) [Ref. 1]

E/eV	$\bar{\nu}$/cm^{-1}	λ/µm	n	k	R
2.056	16580	0.603		0.12	
2.026	16340	0.612		0.11	
2.006	16180	0.618		0.099	
1.990	16050	0.623		9.0	
1.925	15530	0.644		5.6	
1.826	14730	0.679		1.4	
1.810	14600	0.685		0.012	
1.794	14470	0.691		0.0089	
1.771	14290	0.700		6.2	
1.715	13830	0.723		2.6	
1.701	13720	0.729		0.0022	
1.647	13280	0.753		0.00046	
1.629	13140	0.761	3.07	4.0	0.62
1.596	12870	0.777	3.06	2.7	0.49
1.579	12740	0.785	3.05	1.9	0.39
1.562	12590	0.794	3.05	0.00013	0.26
1.544	12450	0.803	3.04	0.000094	0.25
1.529	12330	0.811	3.03	6.3	0.78
1.512	12200	0.820	3.03	4.2	0.64
1.494	12050	0.830	3.02	2.8	0.50
1.476	11910	0.840	3.01	1.8	0.38
1.378	11110	0.90	2.98		
1.240	10000	1.00	2.93		
1.127	9091	1.10	2.90		
1.051	8475	1.18	2.89		
1.033	8333	1.20	2.88		
0.2555	1980	5.05		$1.6 \cdot 10^{-7}$	
0.2380	1919	5.21		$9.9 \cdot 10^{-8}$	
0.2344	1890	5.29		$1.1 \cdot 10^{-7}$	
0.1345	1085	9.22		4.4	
0.1339	1080	9.26		3.7	
0.1333	1075	9.30		4.4	
0.1308	1055	9.48		4.5	
0.1215	980	10.20		8.9	
0.1203	970	10.31		$9.9 \cdot 10^{-7}$	
0.1196	965	10.36		$1.0 \cdot 10^{-6}$	
0.1178	950	10.53		1.1	
0.1116	900	11.11		1.8	
0.1004	810	12.35		4.9	
0.09919	800	12.50		$7.0 \cdot 10^{-6}$	
0.09795	790	12.66		$1.0 \cdot 10^{-5}$	
0.09671	780	12.82		1.5	
0.09299	750	13.33		3.7	
0.08555	690	14.49		6.9	
0.08431	680	14.71		5.9	
0.08059	650	15.38		6.1	
0.07811	630	15.87		6.3	
0.07687	620	16.13		7.7	
0.07563	610	16.39		7.8	
0.07439	600	16.67		$9.3 \cdot 10^{-5}$	
0.07315	590	16.95	2.8	$1.2 \cdot 10^{-4}$	0.22
0.07191	580	17.24	2.8	1.4	0.32
0.07067	570	17.54	2.8	1.8	0.37
0.06943	560	17.86	2.8	2.8	0.50
0.06633	535	18.69	2.8	5.2	0.73
0.06571	530	18.87	2.8	$7.2 \cdot 10^{-4}$	0.22

E/eV	$\bar{\nu}$/cm^{-1}	λ/µm	n	n_a	n_c	k	k_a	k_c	R	R_a	R_c
0.06509	525	19.05	2.8			$1.2 \cdot 10^{-3}$			0.22		
0.06447	520	19.23	2.8			1.7			0.35		
0.06075	490	20.41	2.7			4.9			0.71		
0.06024	485.9	20.58	2.7			5.2			0.73		
0.05331	430	23.26	2.7			1.4			0.31		
0.05269	425	23.53	2.7			$1.1 \cdot 10^{-3}$			0.21		
0.05207	420	23.81	2.7			$8.5 \cdot 10^{-4}$			0.21		
0.05145	415	24.10	2.7			7.3			0.84		
0.05083	410	24.39	2.7			8.3			0.87		
0.05021	405	24.69	2.7			$9.4 \cdot 10^{-4}$			0.21		
0.04959	400	25.0	2.7			$1.2 \cdot 10^{-3}$			0.21		
0.04862	392.2	25.5	2.6			1.6			0.33		
0.04679	377.4	26.5	2.6			5.0			0.73		
0.04592	370.4	27.0	2.6			$8.0 \cdot 10^{-3}$			0.20		
0.04509	363.6	27.5	2.6			$1.2 \cdot 10^{-2}$			0.20		
0.04428	357.1	28.0	2.6			1.7			0.34		
0.03875	312.5	32.0	2.5			8.2			0.87		
0.03815	307.7	32.5	2.5			$9.3 \cdot 10^{-3}$			0.18		
0.03757	303.0	33.0	2.4			0.11			0.17		
0.02988	241.0	41.5	2.2			0.89			0.20		
0.02952	238.1	42.0	2.2			1.0			0.22		
0.02725	219.8	45.5	3.2			1.8			0.39		
0.02362	190.5	52.5	3.6			0.30			0.32		
0.01937	156.2	64.0	3.2			0.10			0.27		
0.01922	155.0	64.5	3.2			$9.6 \cdot 10^{-2}$			0.27		
0.01907	153.8	65.0	3.2			9.4			0.88		
0.01734	139.9	71.5	3.1			8.7			0.87		
0.01653	133.3	75.0	3.1			9.4			0.88		
0.01642	132.5	75.5	3.1			0.096			0.26		
0.01494	120.5	83.0	3.0			0.15			0.25		
0.01246	100.5	99.5	3.2			0.60			0.26		
0.007606	61.35	163.0	3.3			0.12			0.29		
0.006199	50.00	200.0	3.2								
0.004592	37.04	270.0	3.1			0.072			0.26		
0.002799	22.57	443.0	3.0			4.5			0.67		
0.001826	14.73	679.0	3.0			2.8			0.50		
0.001273	10.27	974.0	3.0			2.1			0.41		
0.0006491	5.236	1910.0	3.0			$1.1 \cdot 10^{-2}$			0.25		
0.0004376	3.530	2833.0	3.0			$7.5 \cdot 10^{-3}$			0.25		
0.0002903	2.341	4271.0	3.0			5.0			0.71		
0.0001716	1.384	7224.0	3.0			3.1			0.53		
0.00009047	0.7297	13704	3.0			$1.6 \cdot 10^{-3}$			0.25		
0.00005621	0.4534	22056	3.0			$9.9 \cdot 10^{-4}$			0.25		
0.00002774	0.2237	44699	3.0			5.2			0.72		
0.00001439	0.1161	86153	3.0			2.6			0.47		

Vitreous Arsenic Sulfide (As$_2$S$_3$) - [Ref. 2]

E/eV	$\bar{\nu}$/cm^{-1}	λ/µm	n	n_a	n_c	k	k_a	k_c	R	R_a	R_c
4.959	40000	0.2500	2.48			1.21			0.27		
3.100	25000	0.40	3.09			0.34			0.27		
2.48	20000	0.4999	2.83			0.013			0.23		
1.879	15150	0.66	2.59			$1.7 \cdot 10^{-6}$			0.20		
1.240	10000	1.0	2.48			$2.4 \cdot 10^{-7}$			0.18		
0.6199	5000	2.0	2.43						0.17		
0.3100	2500	4.0	2.41						0.17		
0.2480	2000	5.0	2.41						0.17		
0.1736	1400	7.143	2.40			$7.4 \cdot 10^{-7}$			0.17		
0.1240	1000	10.00	2.38			$1.3 \cdot 10^{-4}$			0.17		
0.09299	750	13.33	2.35			$3.0 \cdot 10^{-3}$			0.16		
0.07439	600	16.67	2.31			$4.6 \cdot 10^{-4}$			0.16		
0.04959	400.0	25.0	1.79			0.2			0.085		
0.03757	303.0	33.0	3.59			1.4			0.38		
0.03100	250.0	40.0	2.98			0.15			0.25		

E/eV	$\bar{\nu}$/cm^{-1}	λ/μm	n	n_a	n_c	k	k_a	k_c	R	R_a	R_c
0.02480	200.0	50	2.66			0.11			0.21		
0.02066	166.7	60	2.64			0.57			0.22		
0.01771	142.9	70	2.99			0.17			0.25		
0.01550	125.0	80	2.89			0.14			0.24		
0.01378	111.1	90	2.84			0.12			0.23		
0.01240	100	100	2.81			0.10			0.23		
0.008183	66	152	2.76			0.072			0.22		
0.004029	32.5	308	2.74			0.044			0.22		
0.002418	19.5	513	2.74			0.031			0.22		
0.001984	16	625	2.74			0.025			0.22		
0.001048	8.45	1180	2.73			$8.8 \cdot 10^{-3}$			0.22		
0.0001033	0.833	12000	2.73			$1.3 \cdot 10^{-3}$			0.22		
$4.129 \cdot 10^{-12}$	$3.33\ 10^{-8}$	$3 \cdot 10^{11}$	2.73						0.22		

Cadmium Telluride (CdTe) - [Ref. 3]

E/eV	$\bar{\nu}$/cm^{-1}	λ/μm	n	n_a	n_c	k	k_a	k_c	R	R_a	R_c
4.9	39520	0.2530	2.48			2.04			0.39		
4.1	33070	0.3024	2.33			1.59			0.32		
3.9	31460	0.3179	2.57			1.90			0.37		
3.5	28230	0.3542	2.89			1.52			0.34		
3.1	25000	0.4000	3.43			1.02			0.34		
3.0	24200	0.4133	3.37			0.861			0.32		
2.755	22220	0.45	3.080			0.485			0.27		
2.75	22180	0.4509	3.23			0.636			0.29		
2.610	21050	0.475	3.045								
2.5	20160	0.4959	3.14			0.525			0.28		
2.25	18150	0.5510	3.05			0.411			0.26		
1.771	14290	0.70	2.861			0.210			0.23		
1.512	12200	0.82	2.880			0.040			0.23		
1.50	12100	0.8266	2.98			0.319			0.25		
1.475	11900	0.840	2.905			0.00134			0.24		
1.47	11860	0.8434				0.000671					
1.465	11820	0.8463				3.37					
1.46	11780	0.8492				1.89					
1.459	11760	0.850	2.948						0.24		
1.455	11740	0.8521				$1.08 \cdot 10^{-4}$					
1.45	11690	0.8551	2.9565			$5.10 \cdot 10^{-5}$			0.24		
1.445	11650	0.8580				2.73					
1.442	11630	0.860	2.952						0.24		
1.44	11610	0.8610	2.9479			1.37			0.32		
1.43	11530	0.8670	2.9402						0.24		
1.30	10490	0.9537	2.8720						0.23		
1.24	10000	1.0	2.840						0.23		
1.20	9679	1.033	2.8353						0.23		
1.10	8872	1.127	2.8050						0.23		
1.00	8065	1.240	2.7793						0.22		
0.90	7259	1.378	2.7537						0.22		
0.80	6452	1.550	2.7384						0.22		
0.70	5646	1.771	2.7223						0.21		
0.60	4839	2.066	2.7086						0.21		
0.50	4033	2.480	2.6972						0.21		
0.40	3226	3.100	2.6878						0.21		
0.30	2420	4.133	2.6800						0.21		
0.20	1613	6.199	2.6722						0.21		
0.10	806.5	12.40	2.6535						0.20		
0.09	725.9	13.78	2.6482						0.20		
0.06819	550	18.18	2.623						0.20		
0.0573	462	21.6				$3.8 \cdot 10^{-6}$					
0.05	403.3	24.80	2.5801						0.19		
0.0469	378	26.5				$8.0 \cdot 10^{-5}$					
0.04592	370.3	27				$9.88 \cdot 10^{-5}$					
0.04133	333.3	30	2.55916			$2.86 \cdot 10^{-4}$			0.19		
0.04092	330	30.30	2.531			3.34			0.57		

E/eV	$\bar{\nu}$/cm⁻¹	λ/μm	n	n_a	n_c	k	k_a	k_c	R	R_a	R_c
0.03720	300	33.33	2.494			4.97			0.73		
0.03647	294.1	34.00				8.93					
0.03596	290	34.48	2.478			5.77·10⁻³			0.18		
0.03493	281.7	35.5				7.91					
0.03472	280	35.71	2.459			6.76			0.83		
0.03100	250	40	2.378			1.18·10⁻²			0.17		
0.02917	235.3	42.5				6.93					
0.02852	230	43.48	2.289			1.87			0.36		
0.02728	220	45.45	2.224			2.47·10⁻²			0.14		
0.02604	210	47.62	2.137			3.4·10⁻²			0.13		
0.02480	200	50.00	2.013			4.97·10⁻²			0.11		
0.02384	192.3	52.0				6.21					
0.01798	145	68.97	1.8			5.2			0.79		
0.01736	140	71.43	6.778			4.50			0.66		
0.01550	125	80.0	4.598			0.294			0.41		
0.01364	110	90.91	3.868			9.47·10⁻²			0.35		
0.01240	100	100	3.649			5.68·10⁻²			0.32		
0.009919	80	125	3.415			0.0262			0.30		
0.008679	70	142.9	3.348			0.0189			0.29		
0.007439	60	166.7	3.299			1.39			0.35		
0.006199	50	200	3.263			1.03			0.32		
0.004959	40	250	3.236			7.52·10⁻³			0.28		
0.003720	30	333.3	3.217						0.28		
0.023015	18.563		538.71			3.2096			0.28		
0.001550	12.50	800				6.18					

Gallium Arsenide (GaAs) - [Ref. 4]

E/eV	$\bar{\nu}$/cm⁻¹	λ/μm	n	n_a	n_c	k	k_a	k_c	R	R_a	R_c
155		0.007999				0.0181					
145		0.008551				0.0203					
130		0.009537				0.0224					
110		0.01127				0.0278					
90		0.01378				0.0323					
70		0.01771				0.0376					
40		0.03100				0.0426					
23		0.05391	1.037			0.228					
7.0		0.1771	1.063			1.838					
6.0	48390	0.2066	1.264			2.472			0.61		
5.00	40330	0.2480	2.273			4.084			0.67		
4.00	32260	0.3100	3.601			1.920			0.42		
3.00	24200	0.4133	4.509			1.948			0.47		
2.50	20160	0.4959	4.333			0.441			0.39		
2.00	16130	0.6199	3.878			0.211			0.35		
1.80	14520	0.8888	3.785			0.151			0.34		
1.60	12900	0.7749	3.700			0.091			0.33		
1.50	12100	0.8266	3.666			0.080			0.33		
1.40	11290	0.8856	3.6140			1.69·10⁻³			0.32		
1.20	9679	1.033	3.4920						0.31		
1.00	8065	1.240	3.4232						0.30		
0.80	6452	1.550	3.3737						0.29		
0.50	4033	2.480	3.3240						0.29		
0.25	2016	4.959	3.2978						0.29		
0.15	1210	8.266	3.2831						0.28		
0.100	806.5	12.40	3.2597			4.93·10⁻⁶			0.28		
0.090	725.9	13.78	3.2493			1.64·10⁻⁵			0.28		
0.070	564.6	17.71	3.2081			2.32·10⁻⁴			0.28		
0.060	483.9	20.66	3.1609			3.45·10⁻³			0.27		
0.0495	399.2	25.05	3.058			2.07·10⁻³			0.26		
0.03968	320	31.25	2.495			2.43·10⁻²			0.18		
0.03496	282	35.46	0.307			294·10⁻²					
0.02976	240	41.67	4.57			4.26·10⁻²			0.41		
0.02066	166.7	60	3.77			3.89·10⁻³			0.34		
0.01550	125	80	3.681			1.84·10⁻³			0.33		

E/eV	$\bar{\nu}$/cm^{-1}	λ/µm	n	n_a	n_c	k	k_a	k_c	R	R_a	R_c
0.008266	66.67	150	3.62			$2.14 \cdot 10^{-3}$			0.32		
0.002480	20	500	3.607			$1.3 \cdot 10^{-3}$			0.32		
0.001240	10	1000	3.606						0.32		

Gallium Phosphide (GaP) - [Ref. 5]

E/eV	$\bar{\nu}$/cm^{-1}	λ/µm	n	k	R
154.0		0.00805		$1.7 \cdot 10^{-2}$	
110.0		0.0113		$2.15 \cdot 10^{-2}$	
100.0		0.0124		$215 \cdot 10^{-2}$	
80.0		0.0155		$3.0 \cdot 10^{-2}$	
50.0		0.0248		$4.7 \cdot 10^{-2}$	
27.0		0.0459		$9.3 \cdot 10^{-2}$	
25.0		0.0496		0.122	
20.0		0.0620		0.180	
15.0		0.0826	0.748	0.628	
5.5	44360	0.2254	1.543	3.556	0.68
4.68	37750	0.2649	4.181	2.634	0.50
3.50	28230	0.3542	5.050	0.819	0.46
3.00	24200	0.4133	4.081	0.224	0.37
2.78	22420	0.4460	3.904	0.103	0.35
2.621	21140	0.473	3.73	$6.37 \cdot 10^{-3}$	0.33
2.480	20000	0.500	3.590	$2.47 \cdot 10^{-3}$	0.32
2.18	17580	0.5687	3.411	$2.8 \cdot 10^{-7}$	0.30
2.000	16130	0.62	3.3254		0.29
1.6	12900	0.7749	3.209		0.28
1.240	10000	1.0	3.1192		0.26
0.6888	5556	1.8	3.0439		0.26
0.4769	3846	2.6	3.0271		0.25
0.1907	1538	6.5	2.995	$4.29 \cdot 10^{-4}$	0.25
0.1550	1250	8.0	2.984		0.25
0.1240	1000	10	2.964		0.25
0.06199	500	20	2.615	$7.16 \cdot 10^{-3}$	0.20
0.03100	250	40	3.594	$1.81 \cdot 10^{-2}$	0.32
0.02480	200	50	3.461	$5.77 \cdot 10^{-3}$	0.30
0.01727	139.27	71.80	3.3922	$4.34 \cdot 10^{-3}$	0.30
0.01168	94.21	106.1	3.3621	$4.26 \cdot 10^{-3}$	0.29
0.006199	50.00	200	3.3447	$1.3 \cdot 10^{-4}$	0.29
0.004133	33.33	300	3.3413		0.29
0.001240	10.00	1000	3.3319		0.29

Indium Antimonide (InSb) - [Ref. 6]

E/eV	$\bar{\nu}$/cm^{-1}	λ/µm	n	k	R
155		0.007999		$4.77 \cdot 10^{-3}$	
60		0.02066		$7.30 \cdot 10^{-2}$	
25		0.04959	1.15	.015	
24		0.05166	1.15	0.18	
15		0.08266	0.97	0.230	
10		0.1240	0.74	0.88	
5.00	40330	0.2480	1.307	2.441	0.53
4.50	36290	0.2755	1.443	2.894	0.60
4.00	32260	0.3100	2.632	3.694	0.61
3.34	26940	0.3712	3.528	2.280	0.45
2.84	22910	0.4366	3.340	2.021	0.45
1.80	14520	0.6888	4.909	1.396	0.47
1.50	12100	0.8266	4.418	0.643	0.41
0.6	4839	2.066	4.03		0.36
0.2480	2000	5.0	4.14	$9.1 \cdot 10^{-2}$	0.37
0.1907	1538	6.5	4.30	$6.3 \cdot 10^{-2}$	0.39
0.1653	1333	7.5	4.18	$2.7 \cdot 10^{-2}$	0.38
0.06199	500	20.00	3.869	$2.0 \cdot 10^{-3}$	0.35
0.03100	250	40.00	2.98	$2.6 \cdot 10^{-3}$	0.25
0.02480	200	50.00	2.22	0.165	0.14
0.02244	181	55.25	3.05	7.59	0.84
0.02207	178	56.18	9.61	4.20	0.70

E/eV	v̄/cm⁻¹	λ/μm	n	nₐ	n_c	k	kₐ	k_c	R	Rₐ	R_c
0.02033	164	60.98	4.94			0.140			0.44		
0.01054	85	117.6	2.12			0.423			0.14		
0.005579	45	222.2	1.02			5.59			0.88		
0.001860	15	666.7	6.03			17.9			0.93		
0.001240	10	1000	10.7			24.0			0.94		

Indium Arsenide (InAs) - [Ref. 7]

E/eV	v̄/cm⁻¹	λ/μm	n	nₐ	n_c	k	kₐ	k_c	R	Rₐ	R_c
25		0.04959				1.139			0.168		
20		0.06199				1.125			0.225		
15		0.08266				0.894			0.336		
10		0.1240				0.835			1.071		
6	48390	0.2066	1.434			2.112			0.45		
5.0	40330	0.2480	1.524			2.871			0.58		
4.0	32260	0.3100	3.313			1.799			0.39		
3.5	28230	0.3542	3.008			1.754			0.37		
3.0	24200	0.4133	3.197			2.034			0.41		
2.5	20160	0.4959	4.364			1.786			0.45		
2.44	19680	0.5081	4.489			1.446			0.44		
1.86	15000	0.6666	3.889			0.554			0.36		
1.8	14520	0.6888	3.851			0.530			0.35		
1.7	13710	0.7293	3.798			0.493			0.35		
1.6	12900	0.7749	3.755			0.463			0.34		
1.5	12100	0.8266	3.714			0.432			0.34		
1.2	9679	1.033	3.613						0.32		
1.0	8065	1.240	3.548						0.31		
0.6	4839	2.066				0.161					
0.35	2823	3.542	3.608			9.58·10⁻³			0.32		
0.32	2581	3.875	3.512			1.23·10⁻⁴			0.31		
0.20	1613	6.199	3.427						0.30		
0.1240	1000	10.00	3.402						0.30		
0.06199	500	20.00	3.334						0.29		
0.04959	400	25.00	3.264						0.28		
0.04339	350	28.57	3.182			5.46·10⁻³			0.27		
0.03720	300	33.33	2.988						0.25		
0.03100	250	40.00	1.970			6.37·10⁻²			0.11		
0.02765	222	44.84	5.90			6.53			0.74		
0.02480	200	50.00	6.91			0.30			0.56		
0.01984	160	62.50	5.27			0.41			0.47		
0.01860	150	66.67	5.27			0.51			0.47		
0.01736	140	71.43	3.99			1.1·10⁻²			0.36		
0.01488	120	83.33	3.91			6.6·10⁻³			0.35		
0.01240	100	100.0	3.85			4.3·10⁻³			0.35		
0.009919	80	125.0	3.817						0.34		
0.007439	60	166.7	3.793						0.34		
0.004959	40	250.0	3.778						0.34		
0.002480	20	500	3.769						0.37		
0.001240	10	1000	3.766						0.34		

Indium Phosphide (InP) - [Ref. 8]

E/eV	v̄/cm⁻¹	λ/μm	n	nₐ	n_c	k	kₐ	k_c	R	Rₐ	R_c
20		0.06199	0.793			0.494					
15		0.08266	0.695			0.574					
10		0.1240	0.806			1.154					
5.5	44360	0.2254	1.426			2.562			0.79		
5.0	40330	0.2480	2.131			3.495			0.61		
4.0	32260	0.3100	3.141			1.730			0.38		
3.0	24200	0.4133	4.395			1.247			0.43		
2.0	16130	0.6199	3.549			0.317			0.32		
1.5	12100	0.8266	3.456			0.203			0.31		
1.25	10085	0.9915	3.324						0.29		
1.00	8068	1.239	3.220						0.28		
0.50	4034	2.479	3.114						0.26		
0.30	2420	4.131	3.089						0.26		

E/eV	$\bar{\nu}$/cm^{-1}	λ/μm	n	n_a	n_c	k	k_a	k_c	R	R_a	R_c
0.10	806.8	12.39	3.012						0.25		
0.075	605.1	16.53	2.932						0.24		
0.060	484.1	20.66	2.780			$1.46 \cdot 10^{-2}$			0.22		
0.050	403.4	24.79	2.429			$3.35 \cdot 10^{-2}$			0.17		
0.03992	322	31.06	0.307			3.57					
0.03496	282	35.46	3.89			0.282			0.35		
0.03100	250	40.00	4.27			$3.0 \cdot 10^{-2}$			0.39		
0.02728	220	45.45	3.93			$1.3 \cdot 10^{-2}$			0.35		
0.02480	200	50.0	3.81			$8.7 \cdot 10^{-3}$			0.34		
0.02418	195	51.28	3.19						0.27		
0.02232	180	55.56	3.19						0.27		
0.01860	150	66.67	3.65						0.32		
0.01240	100	100	3.57						0.32		
0.009919	80	125.0	3.551						0.31		
0.007439	60	166.7	3.538						0.31		
0.004959	40	250.0	3.529						0.31		
0.002480	20	500	3.523						0.31		
0.001240	10	1000.0	3.522						0.31		

Lead Selenide (PbSe) - [Ref. 9]

E/eV	$\bar{\nu}$/cm^{-1}	λ/μm	n	n_a	n_c	k	k_a	k_c	R	R_a	R_c
14.5		0.08551	0.72			0.20					
10		0.1240	0.68			0.50					
5	40330	0.2480	0.54			1.2					
2.0	16130	0.6199	3.65			2.9			0.51		
1.65	13310	0.7514	4.51			1.73			0.46		
1.5	12100	0.8266	4.64			2.64			0.52		
1.0	8065	1.240	4.65			1.1			0.44		
0.75	6049	1.653				0.269					
0.62	5001	2.000	4.59			0.770			0.42		
0.48	3871	2.583	4.90						0.44		
0.40	3226	3.100	4.91						0.44		
0.32	2581	3.875	4.98			0.173			0.44		
0.20	1613	6.199	4.82						0.43		
0.1190	960	10.42	4.74			$1.20 \cdot 10^{-3}$			0.42		
0.09919	800	12.50	4.72			$2.09 \cdot 10^{-3}$			0.42		
0.07935	640	15.63	4.68			$4.12 \cdot 10^{-3}$			0.42		
0.05951	480	20.83	4.59			$1.00 \cdot 10^{-2}$			0.41		
0.04959	400	25.00	4.49			$1.77 \cdot 10^{-2}$			0.40		
0.03968	320	31.25	4.31			$3.62 \cdot 10^{-2}$			0.39		
0.02976	240	41.67	3.89			$9.61 \cdot 10^{-2}$			0.24		
0.01984	160	62.50	2.34			0.56			0.18		
0.009919	80	125.0	1.73			7.38			0.88		
0.007935	64	156.3	2.91			10.1			0.90		
0.004959	40	250.0	11.2			14.6			0.88		
0.002480	20	500.0	12.6			12.2					
0.001736	14	714.3	14.1			16.6					
0.001240	10	1000	17.4			21.1					

Lead Sulfide (PbS) - [Ref. 10]

E/eV	$\bar{\nu}$/cm^{-1}	λ/μm	n	n_a	n_c	k	k_a	k_c	R	R_a	R_c
150		0.008266				$3.86 \cdot 10^{-3}$					
125		0.009919				$5.59 \cdot 10^{-3}$					
100		0.01240				$1.54 \cdot 10^{-2}$					
80		0.01550				$2.88 \cdot 10^{-2}$					
60		0.02066				$6.17 \cdot 10^{-2}$					
25		0.04959	0.845			0.171					
18.0		0.06888	0.846			0.294					
14.0		0.08856	0.651			0.665					
10.0		0.1240	0.879			1.050					
4.95	39920	0.2505	1.52			2.10			0.43		
4.0	32260	0.3100	1.73			2.83			0.55		
3.00	24200	0.4133	3.88			3.00			0.53		
2.90	23390	0.4275	4.12			2.70			0.51		

E/eV	$\bar{\nu}$/cm⁻¹	λ/μm	n	n_a	n_c	k	k_a	k_c	R	R_a	R_c
2.75	22180	0.4509	4.25			2.33			0.48		
2.55	20570	0.4862	4.35			2.00			0.47		
2.00	16130	0.6199	4.29			1.48			0.43		
1.60	12910	0.7749	4.62			0.94			0.43		
1.24	10000	1.00	4.43			0.597			0.41		
1.03	8333	1.2	4.30			0.458			0.39		
0.650	5263	1.9	4.24			0.318			0.39		
0.496	4000	2.5	4.30			0.235			0.39		
0.400	3226	3.1	4.30			2.27·10⁻²			0.39		
0.3100	2500	4.0	4.16			6.38·10⁻⁴			0.38		
0.2480	2000	5	4.115			9.25·10⁻⁴			0.37		
0.1240	1000	10	4.01			6.32·10⁻³			0.36		
0.1033	833.3	12	3.90			1.14·10⁻²			0.35		
0.08059	650	15.38	3.90						0.35		
0.06819	550	18.18	3.81						0.34		
0.04959	400	25.00	3.53						0.31		
0.03720	300	33.33	2.99						0.25		
0.02480	200.0	50	0.514			1.59					
0.01378	111.1	90	1.175			8.48			0.94		
0.01240	100.0	100	1.79			10.51			0.94		
0.008856	71.43	140	17.41			17.94			0.89		
0.006199	50.0	200	16.27			2.20			0.79		
0.003100	25.00	400	12.96			0.495			0.73		
0.001653	13.33	750	12.44			0.228			0.72		
0.001240	10.00	1000	12.35			0.167			0.72		
0.0006199	5.000	2000	12.27			0.0815			0.72		

Lead Telluride (PbTe) - [Ref. 11]

E/eV	$\bar{\nu}$/cm⁻¹	λ/μm	n	n_a	n_c	k	k_a	k_c	R	R_a	R_c
150		0.008266				2.37·10⁻³					
125		0.009919				9.71·10⁻³					
100		0.01240				4.39·10⁻²					
75		0.01653				6.43·10⁻²					
50		0.02480				6.87·10⁻²					
30		0.04133				7.77·10⁻²					
15		0.08266	0.72			0.17					
10		0.1240	0.66			0.60					
7.5		0.1653	0.8			0.92					
5.0	40330	0.2480	0.72			1.0					
3.0	24200	0.4133	1.0			2.2					
2.5	20160	0.4959	1.35			2.86			0.61		
1.5	12100	0.8266	3.8			3.1			0.53		
1.0	8065	1.240	4.55			2.2			0.49		
0.80	6452	1.550	6.25			0.71			0.53		
0.60	4839	2.066	6.10			0.521			0.52		
0.40	3226	3.100	6.075			0.331			0.52		
0.30	2420	4.133	5.95			3.55·10⁻²			0.51		
0.20	1613	6.199	5.77						0.50		
0.15	1210	8.266	5.76						0.50		
0.1017	820	12.20	5.47			9.16·10⁻³			0.48		
0.08927	720	13.89	5.38			1.37·10⁻²			0.47		
0.06943	560	17.86	5.13			3.06·10⁻²			0.45		
0.04959	400	25.00	4.50			9.6·10⁻²			0.40		
0.03968	320	31.25	3.58			0.23			0.32		
0.02976	240	41.67	1.01			1.9					
0.009919	80	125.0	2.95			16.6			0.96		
0.007439	60	166.7	4.9			22.5			0.96		
0.006199	50	200.0	6.9			27.2			0.97		
0.004959	40	250.0	11.6			34.8			0.97		
0.003720	30	333.3	27.7			35.7			0.95		
0.002480	20	500.0	27.6			39.1			0.95		
0.001240	10	1000	45.1			57.8			0.97		

E/eV	$\bar{v}$/cm^{-1}	λ/μm	n	n_a	n_c	k	k_a	k_c	R	R_a	R_c
					Lithium Fluoride (LiF) - [Ref. 12]						
2000		6.199·10^{-4}	0.9999347			4.33·10^{-6}					
1496		8.287·10^{-4}	0.999883			1.28·10^{-5}					
1016		1.220·10^{-3}	0.999757			5.18·10^{-5}					
725		1.710·10^{-3}	0.999643			1.62·10^{-4}					
504		2.460·10^{-3}	0.999162			4.96·10^{-5}					
303		4.092·10^{-3}	0.99752			3.12·10^{-4}					
250		4.959·10^{-3}	0.99632			6.17·10^{-5}					
200		6.199·10^{-3}				2.12·10^{-3}					
150		8.265·10^{-3}	0.9899			3.54·10^{-3}					
100		1.240·10^{-2}	0.9801			1.32·10^{-2}					
75		1.653·10^{-2}				2.63·10^{-2}					
50		2.480·10^{-2}				7.89·10^{-2}					
25		4.959·10^{-2}	0.558			0.521					
20		6.199·10^{-2}	1.20			0.58			0.10		
15.1		8.211·10^{-2}	1.08			0.68			0.10		
13		9.537·10^{-2}	1.04			1.64					
12.0		0.1033	2.28			0.11			0.15		
11.0		0.1127	1.77			8.07·10^{-7}			0.08		
10.00		0.12398	1.606			7.70·10^{-7}			0.05		
9		0.1375	1.53						0.04		
7		0.1771	1.46								
4.959	40000	0.250	1.4189						0.03		
4.000	32260	0.31	1.4073						0.03		
2.952	23810	0.42	1.3978						0.03		
2.000	16130	0.62	1.3915						0.03		
0.9919	8000	1.25	1.3851								
0.7999	6452	1.55	1.3858						0.03		
0.4959	4000	2.5	1.3731						0.02		
0.4000	3226	3.1	1.3650								
0.3100	2500	4.0	1.3493								
0.2480	2000	5.0	1.3266			1.8·10^{-6}			0.02		
0.2000	1613	6.2	1.2912								
0.1698	1370	7.3	1.2499								
0.1494	1205	8.3	1.2036								
0.1240	1000	10.0	1.1005			2.6·10^{-3}					
0.1127	909.1	11.0	1.0208			8.0·10^{-3}					
0.1033	833.3	12.0				1.9·10^{-2}					
0.09537	769.2	13.0				3.7·10^{-2}					
0.08679	700	14.29	0.508			7.74·10^{-2}					
0.07439	600	16.67	0.124			0.804					
0.06199	500	20.00	0.306			1.47			0.68		
0.05579	450	22.22	0.191			1.88			0.85		
0.04959	400	25.00	0.208			2.71			0.91		
0.03720	300	33.33	8.76			3.91			0.68		
0.03100	250	40.00	4.64			0.287			0.42		
0.02480	200	50.00	3.69			0.102			0.33		
0.01240	100.0	100	3.067			0.106			0.26		
0.06199	50.0	200	3.067			4.0·10^{-2}			0.26		
0.04959	40.00	250	3.067			2.2·10^{-2}			0.26		
0.02480	20.00	500	3.067			6.3·10^{-3}					
0.01378	11.11	900				3.1·10^{-3}					
4.798 10^{-4}	3.870	2584	3.023			1.19·10^{-3}			0.25		
1.464 10^{-4}	1.181	8469	3.023			6.20·10^{-4}			0.25		
4.053 10^{-5}	0.3269	30590	3.023			2.63·10^{-4}			0.25		
1.861 10^{-7}	1.501·10^{-3}	6.662·10^6	3.018			1.6·10^{-5}					
3.718 10^{-8}	2.999·10^{-4}	3.335·10^7	3.018			1.6·10^{-5}					
					Potassium Chloride (KCl) - [Ref. 13]						
2860.3		4.3347·10^{-4}				3.93·10^{-6}					
2855.3		4.3423·10^{-4}				3.39·10^{-6}					

E/eV	$\overline{\nu}$/cm^{-1}	λ/μm	n	n_a	n_c	k	k_a	k_c	R	R_a	R_c
2849.3		4.3514·10^{-4}				4.61·10^{-6}					
2835.8		4.3721·10^{-4}				5.85·10^{-6}					
2832.3		4.3775·10^{-4}				5.85·10^{-6}					
2829.8		4.3814·10^{-4}				1.57·10^{-6}					
2828.3		4.3837·10^{-4}				4.19·10^{-7}					
219		5.661·10^{-3}				1.82·10^{-3}					
215		5.767·10^{-3}				1.84·10^{-3}					
212.5		5.834·10^{-3}				2.19·10^{-3}					
211		5.876·10^{-3}				1.82·10^{-3}					
185.1		6.7·10^{-3}	0.99874						1.01 10^{-3}		
109.7		1.13·10^{-2}	0.99578						4.22 10^{-3}		
43		0.02883	0.96			3.0·10^{-2}					
40		0.03179	0.925			1.8·10^{-2}					
29.9		0.04147	0.756			0.145					
20.1		0.06168	0.910			0.495					
15.1		0.08211	0.965			0.344					
10.0		0.1240	1.16			0.38			0.035		
9.0		0.1378	1.99			0.50			0.13		
8.0		0.1550	1.15			0.46			0.048		
7.0		0.1771	2.0			8.46·10^{-7}			0.11		
6.199	50000	0.20	1.71739						0.070		
4.959	40000	0.25	1.58972								
3.999	32260	0.31	1.54005								
2.952	23810	0.42	1.50701								
2.695	21740	0.46	1.50115						0.040		
2.616	21100	0.474				7.6·10^{-11}					
2.384	19230	0.52	1.49501								
2.066	16670	0.60	1.48969						0.039		
1.550	12500	0.80	1.48291						0.038		
1.033	8333	1.2	1.47813						0.037		
0.5166	4167	2.4	1.47464						0.037		
0.2480	2000	5.0	1.47048						0.036		
0.2000	1.613	6.2	1.46796						0.036		
0.1512	1220	8.2	1.46260						0.035		
0.09999	806.5	12.4	1.44611						0.033		
0.07560	609.8	16.4	1.42295						0.030		
0.04959	400.0	25.0	1.34059			6.57·10^{-4}			0.021		
0.03999	322.6	31.0	1.2431						0.012		
0.02976	240	41.67	0.85			0.16					
0.02728	220	45.45	0.53			0.35					
0.02232	180	55.56	0.31			1.05					
0.01860	150	66.67	0.44			4.0					
0.01612	130	76.92	4.1			0.32			0.37		
0.01240	100	100.0	2.7			0.11			0.21		
0.008679	70	142.9	2.4			9.2·10^{-2}			0.17		
0.006199	50	200.0	2.2						0.14		
0.001240	10.00	1000				9.0·10^{-3}					
0.0006199	5.000	2000				3.7·10^{-3}					
0.0004133	3.333	3000				2.0·10^{-3}					

Silicon Dioxide (Glass) - [Ref. 14]

E/eV	$\overline{\nu}$/cm^{-1}	λ/μm	n	n_a	n_c	k	k_a	k_c	R	R_a	R_c
2000		6.199·10^{-4}	0.99993			1.503·10^{-5}					
1860		6.665·10^{-4}	0.99991			1.936·10^{-5}					
1609		7.705·10^{-4}	0.99989			9.941·10^{-6}					
1496		8.287·10^{-4}	0.99987			1.308·10^{-5}					
1204		1.030·10^{-3}	0.99980			2.916·10^{-5}					
1093		1.134·10^{-3}	0.99975			4.155·10^{-5}					
1016		1.220·10^{-3}	0.99971			5.423·10^{-5}					
798		1.554·10^{-3}	0.99954			1.289·10^{-4}					
597		2.077·10^{-3}	0.99917			3.560·10^{-4}					
396		3.131·10^{-3}	0.99812			4.04·10^{-4}					
303		4.092·10^{-3}	0.99678			9.91·10^{-4}					

E/eV	$\bar{\nu}$/cm^{-1}	λ/μm	n	n_a	n_c	k	k_a	k_c	R	R_a	R_c
201		$6.168 \cdot 10^{-3}$	0.99269			$3.63 \cdot 10^{-3}$					
151.2		$8.2 \cdot 10^{-3}$	0.9871			$7.3 \cdot 10^{-3}$					
99.99		$1.24 \cdot 10^{-2}$	0.9813			$7.0 \cdot 10^{-3}$					
49.59		$2.50 \cdot 10^{-2}$	0.9164			$6.5 \cdot 10^{-2}$					
40.00		$3.10 \cdot 10^{-2}$	0.907			$9.2 \cdot 10^{-2}$					
31.00		$4.00 \cdot 10^{-2}$	0.851			0.156					
25.00		0.04959	0.733			0.325					
20.00		0.06199	0.859			0.585					
15.00		0.08266	1.168			0.711			0.10		
13.00		0.09537	1.368			0.747			0.11		
11.00		0.1127	1.739			0.569			0.11		
10.00		0.1240	2.330			0.323			0.17		
9.00		0.1378	1.904			$1.89 \cdot 10^{-2}$			0.097		
7.00		0.1771	1.600						0.053		
6.00	48390	0.2066	1.543						0.046		
4.9939	40278.4	0.248272	1.50841						0.041		
4.1034	33096.1	0.302150	1.48719						0.038		
3.0640	24712.3	0.404656	1.46961						0.036		
2.5504	20570.5	0.486133	1.46313						0.035		
2.4379	19662.5	0.508582	1.46187						0.035		
2.2705	18312.5	0.546074	1.46008						0.035		
2.1489	17332.3	0.576959	1.45885						0.035		
2.1411	17269.2	0.579065	1.45877						0.035		
2.1102	17019.5	0.587561	1.45847						0.035		
2.1041	16970.4	0.589262	1.45841						0.035		
1.9257	15531.6	0.643847	1.45671						0.035		
1.8892	15237.6	0.656272	1.45637						0.035		
1.8566	14974.2	0.667815	1.45608						0.034		
1.7549	14153.9	0.706519	1.45515						0.034		
1.4550	11735.6	0.852111	1.45248						0.034		
1.0985	8860.06	1.12866	1.44888						0.034		
0.60243	4858.9	2.0581	1.43722						0.032		
0.35354	2851.4	3.5070	1.40568						0.028		
0.2976	2400	4.176	1.383			$1.07 \cdot 10^{-4}$			0.026		
0.2728	2200	4.545	1.365			$2.56 \cdot 10^{-4}$			0.024		
0.2480	2000	5.000	1.342			$3.98 \cdot 10^{-3}$			0.021		
0.2232	1800	5.556	1.306			$5.63 \cdot 10^{-3}$					
0.1984	1600	6.250	1.239			$6.52 \cdot 10^{-3}$					
0.1736	1400	7.143	1.053			$1.06 \cdot 10^{-2}$					
0.1674	1350	7.407	0.9488			$1.48 \cdot 10^{-2}$					
0.1612	1300	7.692	0.7719			$3.72 \cdot 10^{-2}$					
0.1500	1210	8.265	0.4530			0.704			0.30		
0.1401	1130	8.850	0.3563			1.53			0.66		
0.1302	1050	9.524	2.760			1.65			0.35		
0.1209	975	10.26	2.448			0.231			0.18		
0.1091	880	11.36	1.784			$7.75 \cdot 10^{-2}$			0.079		
0.09919	800	12.50	1.753			0.343			0.089		
0.08989	725	13.79	1.698			0.175			0.071		
0.06943	560	17.86	1.337			0.298			0.036		
0.06199	500	20.00	0.6616						0.882		
0.04959	400	25.0	2.739			0.397			0.23		
0.03720	300	33.33	2.210			$6.7 \cdot 10^{-2}$			0.14		
0.01240	100	100.0	1.967			$1.59 \cdot 10^{-2}$			0.11		
0.007439	60	166.7	1.959			$8.62 \cdot 10^{-3}$			0.11		
0.002480	20	500.0	1.955			$7.96 \cdot 10^{-3}$			0.10		

Silicon Monoxide (Noncrystalline) - [Ref. 15]

E/eV	$\bar{\nu}$/cm^{-1}	λ/μm	n	n_a	n_c	k	k_a	k_c	R	R_a	R_c
25		0.04959	0.8690			0.2717					
20		0.06199	0.8853			0.4919					
17.5		0.07085	0.9825			0.5961					
15		0.08266	1.132			0.6651			0.092		
12.5		0.09919	1.283			0.6523			0.090		

E/eV	$\bar{\nu}$/cm^{-1}	λ/μm	n	n_a	n_c	k	k_a	k_c	R	R_a	R_c
10		0.1240	1.378			0.6843			0.10		
7.5		0.1653	1.593			0.7473			0.12		
5	40330	0.2480	2.001			0.6052			0.15		
4	32260	0.3100	2.141			0.4006			0.15		
3	24200	0.4133	2.116			0.1211			0.13		
2.8	22580	0.4428	2.085			0.08374			0.12		
2.6	20970	0.4769	2.053			0.05544			0.12		
2.4	19360	0.5166	2.021			0.03533			0.11		
2.2	17740	0.5636	1.994			0.02153			0.11		
2	16130	0.6199	1.969			0.01175			0.11		
1.8	14520	0.6888	1.948			0.00523			0.10		
1.6	12900	0.7749	1.929			0.00151			0.10		
1.240	10000	1.000	1.87						0.092		
0.6199	5000	2.000	1.84						0.087		
0.3100	2500	4.000	1.80						0.082		
0.2480	2000	5.000	1.75						0.074		
0.2066	1667	6.000	1.70						0.067		
0.1771	1492	7.000	1.60						0.053		
0.1653	1333	7.500	1.42								
0.1459	1176	8.500	0.90			0.18					
0.1305	1053	9.500	1.20			1.20			0.024		
0.1240	1000	10.00	2.00			1.38			0.27		
0.1181	952.4	10.50	2.85			0.90			0.27		
0.1153	930.2	10.75	2.86			0.58			0.25		
0.1127	909.1	11.00	2.82			0.40			0.24		
0.1078	869.6	11.50	2.50			0.20			0.19		
0.1033	833.3	12.00	2.13			0.14			0.13		
0.09537	769.2	13.00	2.04			0.20			0.12		
0.08856	714.3	14.00	2.01			0.30			0.12		

Noncrystalline Silicon Nitride (Si$_3$N$_4$) - [Ref. 16]

E/eV	$\bar{\nu}$/cm^{-1}	λ/μm	n	n_a	n_c	k	k_a	k_c	R	R_a	R_c
24		0.05166	0.655			0.420			0.28		
23		0.05391	0.625			0.481			0.22		
22		0.05636	0.611			0.560			0.16		
21		0.05904	0.617			0.647			0.19		
20		0.06199	0.635			0.743			0.21		
19		0.06526	0.676			0.841			0.23		
18		0.06888	0.735			0.936			0.26		
17		0.07293	0.810			1.03			0.25		
16		0.07749	0.902			1.11			0.26		
15		0.08266	1.001			1.18			0.26		
14		0.08856	1.111			1.26			0.26		
13		0.09537	1.247			1.35			0.27		
12	96790	0.1033	1.417			1.43			0.28		
11	88720	0.1127	1.657			1.52			0.29		
10.5	84690	0.1181	1.827			1.53			0.29		
10	80650	0.1240	2.000			1.49			0.29		
9.5	76620	0.1305	2.162			1.44			0.28		
9	72590	0.1378	2.326			1.32			0.27		
8	64520	0.1550	2.651			0.962			0.26		
7	56460	0.1771	2.752			0.493			0.23		
6	48390	0.2066	2.541			0.102			0.19		
5	40330	0.2480	2.278			$4.9 \cdot 10^{-3}$			0.15		
4.75	38310	0.2610	2.234			$1.2 \cdot 10^{-3}$			0.15		
4.5	36290	0.2755	2.198			$2.2 \cdot 10^{-4}$			0.14		
4	32260	0.3100	2.141						0.13		
3.5	28230	0.3542	2.099						0.13		
3	24200	0.4133	2.066						0.12		
2.5	20160	0.4959	2.041						0.12		
2	16130	0.6199	2.022						0.11		
1.5	12100	0.8266	2.008						0.11		
1	8065	1.240	1.998						0.11		

E/eV	$\bar{\nu}$/cm^{-1}	λ/µm	n	n_a	n_c	k	k_a	k_c	R	R_a	R_c
					Sodium Chloride (NaCl) - [Ref. 17]						
209.5		5.918·10^{-3}				2.54·10^{-3}					
206		6.019·10^{-3}				2.62·10^{-3}					
203		6.107·10^{-3}				2.08·10^{-3}					
200		6.199·10^{-3}				1.92·10^{-3}					
26.0		0.04769	0.83			0.15			0.015		
25.0		0.04959	0.83			0.18			0.018		
22.0		0.05636	0.83			0.31			0.057		
20.0		0.06199	0.88			0.34			0.036		
18.0		0.06888	0.89			0.33			0.033		
16.1		0.07700	0.74			0.45			0.084		
14.0		0.08856	0.98			0.89			0.17		
12.0		0.1033	1.22			0.79			0.12		
10.0		0.1240	1.55			0.71			0.12		
8.00		0.1550	1.38			1.10			0.20		
6.00	48390	0.2066	1.75						0.074		
5.00	40330	0.2480	1.65						0.060		
2.952	23810	0.42	1.56324						0.048		
2.480	20000	0.50	1.55157						0.047		
2.214	17860	0.56	1.54613						0.046		
2.000	16130	0.62	1.54228						0.045		
1.771	14290	0.70	1.53865						0.045		
1.675	13510	0.74	1.53728						0.045		
1.550	12500	0.80	1.53560						0.045		
1.240	10000	1.00	1.53200						0.044		
1.033	8333	1.2	1.53000						0.044		
0.6888	5556	1.8	1.52712						0.043		
0.4959	4000	2.5	1.52531						0.043		
0.4000	3226	3.1	1.52395						0.043		
0.3263	2632	3.8	1.52226			$(1.8\pm0.2)\cdot10^{-9}$			0.043		
0.2952	2381	4.2	1.52121						0.043		
0.2755	2222	4.5	1.52036						0.043		
0.2480	2000	5.0	1.51883						0.042		
0.1240	1000	10.0	1.49473						0.039		
0.1033	833.3	12.0	1.48000						0.037		
0.08856	714.3	14.0	1.46188						0.035		
0.07749	625.0	16.0	1.4399						0.033		
0.06888	555.5	18.0	1.41364						0.029		
0.06199	500.0	20.0	1.3822						0.026		
0.04959	400	25.0	1.27			3.5·10^{-3}			0.014		
0.04215	340	29.41	1.12			1.7·10^{-2}			0.0032		
0.03720	300	33.33	0.85			0.85			0.18		
0.03410	275	36.36	0.59			0.22			0.084		
0.03286	265	37.74	0.42			0.50			0.26		
0.03224	260	38.46	0.45			0.45			0.22		
0.02480	200	50.00	0.14			1.99			0.89		
0.02108	170	58.82	1.35			6.03			0.87		
0.01984	160	62.50	6.92			2.14			0.59		
0.01922	155	64.52	5.50			0.87			0.49		
0.01860	150	66.67	4.52			0.380			0.41		
0.01736	140	71.43	3.72			0.219			0.33		
0.01612	130	76.92	3.31			0.135			0.29		
0.01488	120	83.33	3.02			0.110			0.25		
0.01240	100	100.0	2.74			0.087			0.22		
0.009919	80	125.0	2.57			0.077			0.19		
0.07439	60	166.7	2.48			0.055			0.18		
0.04959	40	250.00	2.44			0.041			0.18		
0.002480	20	500.0	2.43			0.024			0.17		
0.001240	10	1000	2.43			0.006			0.17		
0.001033	8.333	1200				8.8·10^{-3}					
0.0006888	5.556	1800				5.4·10^{-3}					

E/eV	$\bar{\nu}$/cm⁻¹	λ/μm	n	n_a	n_c	k	k_a	k_c	R	R_a	R_c
0.0006199	5.000	2000	2.43						0.17		
0.0004959	4.000	2500				4.4·10⁻³					
0.0004797	3.869	2584	2.43			2.1·10⁻³			0.17		
0.0003875	3.125	3200				3.3·10⁻³					
0.0001464	1.181	8469	2.43			5.8·10⁻⁴			0.17		
0.00004053	0.3269	30590	2.43			2.5·10⁻⁴					

Cubic Zinc Sulfide (ZnS) - [Ref. 18]

E/eV	$\bar{\nu}$/cm⁻¹	λ/μm	n	n_a	n_c	k	k_a	k_c	R	R_a	R_c
2000		6.199·10⁻⁴	0.999904			1.76·10⁻⁵					
1204		1.030·10⁻³	0.999777			1.00·10⁻⁴					
1016		1.220·10⁻³	0.999838			3.61·10⁻⁵					
901		1.376·10⁻³	0.999647			5.42·10⁻⁵					
798		1.554·10⁻³	0.999520			8.28·10⁻⁵					
707		1.754·10⁻³	0.999372			1.25·10⁻⁴					
597		2.077·10⁻³	0.999160			2.19·10⁻⁴					
377		9.50·10⁻³	0.99789			9.50·10⁻⁴					
201		6.168·10⁻³	0.99553			4.82·10⁻³					
100		1.240·10⁻²	0.99061			1.17·10⁻²					
61.99		2.000·10⁻²	0.964			3.32·10⁻²			6.2 10⁻⁴		
41.33		3.000·10⁻²	0.941			5.10·10⁻²					
31.00		4.000·10⁻²	0.847			9.95·10⁻²					
24.80		5.000·10⁻²	0.796			0.171			2.2 10⁻²		
17.71		7.000·10⁻²	0.747			0.431			7.7 10⁻²		
13.78		9.000·10⁻²	0.758			0.824			0.20		
12.40		0.1000	0.862			0.876			0.19		
9.919		0.125	1.02			1.36			0.31		
8.266		0.150	1.41			1.47			0.29		
6.199		0.200	2.32			1.62			0.32		
6.00	48390	0.2066	2.24			1.65			0.59		
4.00	32260	0.3100	2.70			0.44			0.22		
3.00	24200	0.4133	2.54			4·10⁻²			0.19		
2.50	20160	0.4959	2.42			3·10⁻²			0.17		
2.30	18550	0.5391	2.3950						0.17		
2.00	16130	0.6199	2.3576						0.16		
1.75	14110	0.7085	2.3319						0.16		
1.55	12500	0.7999	2.3146			3.50·10⁻⁶			0.16		
1.40	11290	0.8856	2.3033						0.16		
1.240	10000	1.000	2.2907			3.02·10⁻⁶			0.15		
1.00	8065	1.240	2.2795						0.15		
0.80	6452	1.550	2.2706						0.15		
0.6199	5000	2.000	2.2631			6.2·10⁻⁶			0.15		
0.45	3629	2.755	2.2587						0.15		
0.30	2420	4.133	2.2529						0.15		
0.20	1613	6.199	2.2443						0.15		
0.1550	1250	8.0	2.2213			4.5·10⁻⁶			0.14		
0.1240	1000	10.00	2.1986			8.8·10⁻⁶			0.14		
0.100	806.5	12.4	2.1969						0.14		
0.09	725.9	13.78	2.1793						0.14		
0.07999	645.2	15.5	2.1518			3.82·10⁻³			0.14		
0.07	564.6	17.71	2.1040						0.13		
0.06075	490	20.41	2.03			8.0·10⁻³			0.12		
0.05	403.3	24.80	1.6866						0.065		
0.03546	286	34.97	3.29			8.3·10⁻²			0.28		
0.03472	280	35.71	9.54			5.2·10⁻²			0.66		
0.02480	200	50.00	3.48			3.1·10⁻²			0.31		
0.01240	100	100.0	3.06			5.8·10⁻³			0.26		
0.004955	40	250.0	2.903			6.2·10⁻³			0.24		
0.004339	35	285.7	2.899			7.0·10⁻³			0.24		
0.003720	30	333.3	2.896						0.24		
0.003100	25	400.0	2.894						0.24		
0.002480	20	500.0	2.892						0.24		
0.001860	15	666.7	2.890						0.24		

E/eV	$\bar{\nu}$/cm^{-1}	λ/μm	n	n_a	n_c	k	k_a	k_c	R	R_a	R_c
					Polytetrafluoroethylene (Teflon) - [Ref. 19]						
4.960	40000	0.250							0.970		
4.769	38462	0.260							0.972		
4.593	37037	0.270							0.975		
4.426	35714	0.280							0.978		
4.276	34483	0.290							0.980		
4.133	33333	0.300							0.983		
4.000	32258	0.310							0.986		
3.875	31250	0.320							0.988		
3.758	30303	0.330							0.990		
3.647	29412	0.340							0.991		
3.543	28571	0.350							0.992		
3.444	27778	0.360							0.992		
3.351	27027	0.370							0.993		
2.255	18182	0.550							0.993		
2.067	16667	0.600							0.992		
1.378	11111	0.900							0.992		
1.305	10526	0.950							0.991		
1.078	8696	1.150							0.991		
1.033	8333	1.200							0.990		
0.9920	8000	1.250							0.990		
0.9538	7692	1.300							0.989		
0.9185	7407	1.350							0.988		
0.8857	7143	1.400							0.988		
0.8552	6897	1.450							0.989		
0.8267	6667	1.500							0.989		
0.8000	6452	1.550							0.988		
0.7750	6250	1.600							0.988		
0.7515	6061	1.650							0.987		
0.7294	5882	1.700							0.986		
0.7086	5714	1.750							0.986		
0.6889	5556	1.800							0.985		
0.6703	5405	1.850							0.980		
0.6526	5263	1.900							0.978		
0.6359	51282	1.950							0.978		
0.6200	5000	2.000							0.970		
0.6049	4878	2.050							0.959		
0.5905	4762	2.100							0.951		
0.5767	4651	2.150							0.946		
0.5636	4545	2.200							0.966		
0.5511	44444	2.250							0.965		
0.5487	44247	2.260							0.964		
0.5439	4386	2.280							0.963		
0.5415	4367	2.290							0.961		
0.5368	4329	2.310							0.959		
0.5345	4310	2.320							0.957		
0.5322	4292	2.330							0.956		
0.5299	4274	2.340							0.954		
0.5277	4255	2.350							0.951		
0.5232	4219	2.370							0.950		
0.5188	4184	2.390							0.949		
0.5167	4167	2.400							0.947		
0.5061	4082	2.450							0.946		
0.4960	4000	2.500							0.945		

References

1. Arsenic Selenide
   D. J. Treacy in *Handbook of Optical Constants of Solids*, E. D. Palik, Editor, Academic Press, 1985, p. 623. (Hereafter abbreviated as *HOCS*.)
   R. Zallen, R. E. Drews, R. L. Emerald, and M. L. Slade, *Phys. Rev. Lett.* 26, 1564 (1971).
   R. Zallen, M. L. Slade, and A. T. Ward, *Phys. Rev.* B 3, 4257 (1971).
   U. Strom and P. C. Taylor, *Phys. Rev.* B 16, 5512 (1977).
   G. Lucovsky, *Phys. Rev.* B 6, 1480 (1972).
   C. T. Moynihan, P. B. Macedo, M. S. Maklad, R. K. Mohr, and R. E. Howard, *J. Non-Cryst. Solids*, 17, 369 (1975).
   Y. Ohmachi, *J. Opt. Soc. Am.* 63, 630 (1973).

2. Arsenic Sulfide
   D. J. Treacy in *HOCS*, 1985, p. 641.
   P. A. Young, *J. Phys.* C 4, 93 (1971).
   W. S. Rodny, I. H. Malitson, and T. A. King, *J. Opt. Soc. Am.* 48, 633 (1958).
   R. Zallen, R.E. Drew, R. L. Emerald, and M.L. Slade, *Phys. Rev. Lett.* 26, 1564 (1971).
   M. S. Maklad, R. K. Mohr, R. E. Howard, P. B. Macedo, and C. T. Moynihan, *Solid State Commun.* 15, 855 (1974).
   P. B. Klein, P. C. Taylor, and D. J. Treacy, *Phys. Rev.* B16, 4511 (1977).
   G. Lucovsky, *Phys. Rev.* B 6, 1480 (1972).

3. Cadmium Telluride
   E. D. Palik in *HOCS*, 1985, p. 409.
   D. T. F. Marple and H. Ehrenreich, *Phys. Lett.* 8, 87 (1962).
   T. H. Myers, S. W. Edwards, and J. F. Schetzina, *J. Appl. Phys.* 52, 4231 (1981).
   D. T. F. Marple, *Phys. Rev.* 150, 728 (1966).
   A. N. Pikhtin and A. D. Yas'kov, *Sov. Phys. Semicond.* 12, 622 (1978).
   L. S. Ladd, *Infrared Phys.* 6, 145 (1966).
   J. E. Harvey and W. L. Wolfe, *J. Opt. Soc. Am.* 65, 1267 (1975).
   A. Manabe, A. Mitsuishi, and H. Yoshinaga, *Jpn. J. Appl. Phys.* 6, 593 (1967).
   A. Manabe, A. Mitsuishi, H. Oshinaga, Y. Ueda, and H. Sei, *Technol. Rep. Osaka Univ. Jpn.* 17, 263 (1967).
   J. R. Birch and D. K. Murrey, *Infrared Phys.* 18, 283 (1978).

4. Gallium Arsenide
   E. D. Palik in *HOCS*, 1985, p. 429.
   M. Cardona, W. Gudat, B. Sonntag, and P. Y. Yu, in *Proc. Intl. Conf. Phys. Semicond.*, 10th. Cambridge, 1970, p. 208. US Atom. Energy Commission, Oak Ridge, TN, 1970.
   H. R. Philipp and H. Ehrenreich, *Phys. Rev.* 129, 1550 (1963).
   J. B. Theeten, D. E. Aspnes, and R. P. H. Chang, *J. Appl. Phys.* 49, 6097 (1978).
   H. C. Casey, D. D. Sell, and K. W. Wecht, *J. Appl. Phys.* 46, 250 (1975).
   A. H. Kachare, W. G. Spitzer, F. K. Euler, and A. Kahan, *J. Appl. Phys.* 45, 2938 (1974).
   R. T. Holm, J. W. Gibson, and E. D. Palik, *J. Appl. Phys.* 48, 212 (1977).
   W. Cochran, S. J. Fray, F. A. Johnson, J. E. Quarrington, and N. Williams, *J. Appl. Phys. Suppl.* 32, 2102 (1961).
   C. P. Christensen, R. Joiner, S. K. T. Nieh, and W. H. Steier, *J. Appl. Phys.* 45, 4957 (1974).
   R. H. Stolen, *Phys. Rev.* B 11, 767 (1975); *Appl. Phys. Lett.* 15, 74 (1969).

5. Gallium Phosphide
   A. Borghesi and G. Guizzetti in *HOCS*, 1985, p. 445.
   M. Cardona, W. Gudat, B. Sonntag, and P. Y. Yu, *Proc. Intl. Conf. Phys. Semicond.* Cambridge, 1970, p. 208. US Atom. Energy Commission, Oak Ridge, TN, 1970.
   M. Cardona, W. Gudat, E. E. Koch, M. Skibowski, B. Sonntag, and P. Yu. *Phys. Rev. Lett.* 25, 659 (1970).
   S. E. Stokowski and D. D. Sell, *Phys. Rev.* B 5, 1636 (1972).
   S. A. Abagyan, G. A. Ivanov, Y. E. Shanurin, and V. I. Amosov, *Sov. Phys. Semicond.* 5, 889 (1971).
   P. G. Dean, G. Kaminsky, and R. B. Zetterstorm, *J. Appl. Phys.* 38, 3551 (1967).
   D. E. Aspnes and A. A. Studna, *Phys. Rev.* B 27, 985 (1983).

6. Indium Antimonide
   R. T. Holm in *HOCS*, 1985, p. 491.
   M. Cardona, W. Gudat, B. Sonntag, and P. Y. Yu, *Proc. Int. Conf. Phys. Semicond.*, 10th. Cambridge, 1970, p. 208. US Atom. Comm., Oak Ridge, TN, 1970.
   H. R. Philipp and H. Ehrenreich, *Phys. Rev.* 129, 1550 (1963).
   D. E. Aspnes and A. A. Studna, *Phys. Rev.* B 27, 985 (1983).
   T. S. Moss, S. D. Smith, and T. D. F. Hawkins, *Proc. Phys. Soc. London* 70B, 776 (1957).
   H. Yoshinaga and R. A. Oetjen, *Phys. Rev.* 101, 526 (1956).
   R. B. Sanderson, *J. Phys. Chem. Solids* 26, 803 (1965).

7. Indium Arsenide
   E. D. Palick and R. T. Holm in *HOCS*, 1985, p. 479.
   H. R. Philipp and H. Ehrenreich, *Phys. Rev.* 129, 1550 (1963).
   B. O. Seraphin and H. E. Bennett in *Semiconductors and Semimetals* (R. K. Willardson and A. C. Beer, Eds.), vol. 3, Academic, 1967, p. 499.
   D. E. Aspnes and A. A. Studna, *Phys. Rev.* B 27, 985 (1983).
   J. R. Dixon and J. M. Ellis, *Phys. Rev.* 123, 1560 (1961).
   A. Memon, T. J. Parker, and J. R. Birch, *Proc. SPIE*, 289, 20 (1981).

8. Indium Phosphide
   O. J. Glembocki and H. Piller in *HOCS*, 1985, p. 503.
   M. Cardona, *J. Appl. Phys.* 32, 958 (1961); 36, 2181 (1965).
   D. E. Aspnes and A. A. Studna, *Phys. Rev.* B 27, 985 (1983).
   G. D. Pettit and W. J. Turner, *J. Appl. Phys.* 36, 2081 (1965).
   R. Newman, *Phys. Rev.* 111, 1518 (1958).
   W. N. Reynolds, M. T. Lilburne, and R. M. Dell, *Proc. Phys. Soc. London* 71, 416 (1958).
   H. Jamshidi and T. J. Parker, Int. Meet. Infrared Mm. Waves, 7th., Marseilles, 1983.

9. Lead Selenide
   G. Bauer and H. Krenn in *HOCS*, 1985, p. 517.
   M. Cardona and D. L. Greenaway, *Phys. Rev.* A 133, 1685 (1964).
   T. S. Moss, *Optical Properties of Semiconductors*, Butterworth, 1959, p. 189.
   J. N. Zemel, J. D. Jensen, and R. B. Schoolar, *Phys. Rev.* A 140, 330 (1965).
   W. W. Scanlon, *J. Phys. Chem. Solids*, 8, 423 (1959).
   K. V. Vyatkin and A. P. Shotov, *Sov. Phys. Semicond.* 14, 785 (1980); Fiz. Tekh. Poluprovodn. 14, 1331 (1980).

10. Lead Sulfide
    G. Guizzetti and A. Borghesi in *HOCS*, 1985, p. 525.
    M. Cardona and R. Haensel, *Phys. Rev.* B 1, 2605 (1970).
    M. Cardona and D. L. Greenaway, *Phys. Rev.* A 133, 1685 (1964).
    M. Cardona, C. M. Penchina, E. E. Koch, and P. Y. Yu, *Phys. Status Solidi* B 53, 327 (1972).
    P. R. Wessel, *Phys. Rev.* 153, 836 (1967).
    C. E. Rossi and W. Paul, *J. Appl. Phys.* 38, 1803 (1967).
    J. N. Zemel, J. D. Jensen, and R. B. Schoolar, *Phys. Rev.* A 140, 330 (1965).

11. Lead Telluride
    G. Bauer and H. Krenn in *HOCS*, 1985, p. 535.
    M. Cardona and R. Haensel, *Phys. Rev.* B 1, 2605 (1970).
    M. Cardona and D. L. Greenaway, *Phys. Rev.* 133, A1685 (1964).
    D. M. Korn and R. Braunstein, *Phys. Rev.* B 5, 4837 (1972).
    W. W. Scanlon, *J. Phys. Chem. Solids* 8, 423 (1959).
    J. N. Zemel, J. D. Jensen, and R. B. Schoolar, *Phys. Rev.* 140, A330 (1965).

12. Lithium Fluoride
    E. D. Palik and W. R. Hunter in *HOCS*, 1985, p. 675.
    B. L. Henke, P. Lee, T. J. Tanaka, R. L. Shimabukuro, and B. K. Fujikawa, *Low Energy X-ray Diagnostics-1981* (D. T. Attwood and B. L. Henke, Eds.), AIP Conf. Proc. No. 75, 1981.
    A. P. Lukirskii, E. P. Savinov, O. A. Ershov, and Y. F. Shepelev, *Opt. Spektrosk.* 16, 168 (1964); 16, 310 (1964).
    F. C. Brown, C. Gahwiller, A. B. Kunz, and N. O. Lipari, *Phys. Rev. Lett.* 25, 927 (1970).
    A. Milgram and M. P. Givens, *Phys. Rev.* 125, 1506 (1962).
    T. Tomiki and T. Miyata, *J. Phys. Soc. Jpn.* 27, 658 (1969).
    A. Kachare, G. Andermann, and L. R. Brantley, *J. Phys. Chem. Solids* 33, 467 (1972).

13. Potassium Chloride
    E. D. Palik in *HOCS*, 1985, p. 703.
    O. Aita, I. Nagakura, and T. Sagawa, *J. Phys. Soc. Jpn.* 30, 1414 (1971).

A. P. Lukirskii, E. P. Savinov, O. A. Ershov, and Y. F. Shepelev, *Opt. Spectrosc.* 16, 168 (1964); *Opt. Spektrosk.* 16, 310 (1964).

T. Tomika, *J. Phys. Soc. Jpn.* 22, 463 (1967).

M. Antinori, A. Balzarotti, and M. Piacentini, *Phys. Rev.* B 7, 1541 (1973).

H. H. Li, *J. Phys. Chem. Ref. Data* 5, 329 (1976).

S. D. Allen and J. A. Harrington, *Appl. Opt.* 17, 1679 (1978).

K. W. Johnson and E. E. Bell, *Phys. Rev.* 139A, 1295 (1965).

14. Silicon Dioxide

H. R. Philipp in *HOCS*, 1985, p. 749.

J. Rife and J. Osantowski, *J. Opt. Soc. Am.* 70, 1513 (1980).

B. L. Henke, P. Lee, T. J. Tanaka, R. L. Shimabukuro, and B. K. Fujikawa, *Low Energy X-ray Diagnostics-1981* (D. T. Attwood and B. L. Henke, Eds.), AIP Conf. Proc. No. 75, 1981.

H. R. Philipp, *Solid State Commun.* 4, 73 (1966); *J. Phys. Chem. Solids*, 32, 1935 (1971).

P. L. Lamy, *Appl. Opt.* 16, 2212 (1977).

H. R. Philipp, *J. Appl. Phys.* 50 1053 (1979).

D. G. Drummond, *Proc. Roy. Soc. London*, 153, 328 (1935).

15. Silicon Monoxide

H. R. Philipp in *HOCS*, 1985, p. 765.

H. R. Philipp, *J. Phys. Chem. Solids*, 32, 1935 (1971).

G. Hass and C. D. Salzberg, *J. Opt. Soc. Am.* 44, 181 (1954).

E. Cremer, T. Kraus, and E. Ritter, *Zs. Electrochem.* 62, 939 (1958).

A. P. Bradford, G. Hass, M. McFarland, and E. Ritter, *Appl. Opt.* 4, 971 (1965).

16. Silicon Nitride

H. R. Philipp in *HOCS*, 1985, p. 771.

H. R. Philipp, *J. Electrochem. Soc.* 120, 295 (1973).

J. B. Theeten, D. E. Aspnes, F. Simondet, M. Errman, and P. C. Mürau, *J. Appl. Phys.* 52, 6788 (1981).

J. Bauer, *Phys. Status Solidi*, A 39, 411 (1977).

17. Sodium Chloride

J. E. Eldridge and E. D. Palik in *HOCS*, p. 775.

J. A. Harrington, C. J. Duthler, F. W. Patten, and M. Hass, *Solid State Commun.* 18, 1043 (1976).

T. Miyata and T. Tomiki, *J. Phys. Soc. Jpn.* 24, 1286 (1968); ibid., 22, 209 (1967).

D. M. Roessler and W. C. Walker, *J. Opt. Soc. Am.* 58, 279 (1968).

D. M. Roessler and W. C. Walker, *Phys. Rev.* 166, 599 (1968).

S. Allen and J. A. Harrington, *Appl. Opt.* 17, 1679 (1978).

O. Aita, I. Nagakura, and T. Sagawa, *J. Phys. Soc. Jpn.* 30, 1414 (1971).

18. Zinc Sulfide

E. D. Palik and A. Addamiano in *HOCS*, 1985, p. 597.

B. L. Henke, P. L. Lee, T. J. Tanaka, R. L. Shimabukuro, and B. F. Fujikawa, *Low Energy X-ray Diagnostics-1981* (D. T. Attwood and B. L. Henke, Eds.), AIP Conf. Proc. No. 75, 1981.

M. Cardona and G. Harbeke, *Phys. Rev.* 137, A1467 (1965).

Eastman Kodak, Publ. No. U-72, Rochester, New York (1981).

C. A. Klein and R. N. Donadio, *J. Appl. Phys.* 51, 797 (1980).

T. Deutsch, *Proc. Int. Conf. Phys. Semicond.*, 6th Exeter 1962, p. 505. The Inst. of Physics and the Physical Soc., London, 1962.

A. Manabe, A. Mitsuishi, and H. Yoshinaga, *Jpn. J. Appl. Phys.* 6, 593 (1967).

W. W. Piper, D. T. F. Marple, and P. D. Johnson, *Phys. Rev.* 110, 323 (1958).

19. Polytetrafluoroethylene

J. W. L. Thomas (NIST), Private communication.

NIST Certificate, STM 2044.

P. Y. Barnes, E. A. Early, and A. C. Parr, *NIST Special Publ. 250-48*, NIST Measurement Services: Spectral Reflectance.

Diffuse Reflectance Coatings and Materials Sections, Labsphere Catalog, 1996.

A. Arecchi and C. Ryder (Labsphere, North Sutten, NJ), private communication.

ELASTO-OPTIC, ELECTRO-OPTIC, AND MAGNETO-OPTIC CONSTANTS

When a crystal is subjected to a stress field, an electric field, or a magnetic field, the resulting optical effects are in general dependent on the orientation of these fields with respect to the crystal axes. It is useful, therefore, to express the optical properties in terms of the refractive index ellipsoid (or indicatrix):

$$\frac{x^2}{n_x^2} + \frac{y^2}{n_y^2} + \frac{z^2}{n_z^2} = 1$$

or

$$\sum_{ij} B_{ij} x_i y_j = 1 \, (i, j = 1, 2, 3)$$

where

$$B_{ij} = \left[\frac{1}{\varepsilon}\right]_{ij} = \left[\frac{1}{n^2}\right]_{ij}$$

ε is the dielectric constant or permeability; the quantity B_{ij} is called impermeability.

A crystal exposed to a *stress* **S** will show a change of its impermeability. The photo-elastic (or elasto-optic) constants, P_{ijkl}, are defined by

$$\Delta\left[\frac{1}{\varepsilon}\right]_{ij} = \Delta\left[\frac{1}{n^2}\right]_{ij} = \sum_{kl} P_{ijkl} S_{kl}$$

where n is the refractive index and S_{kl} are the strain tensor elements; the P_{ijkl} are the elements of a 4th rank tensor.

When a crystal is subjected to an *electric field* **E**, two possible changes of the refractive index may occur depending on the symmetry of the crystal.

1. All materials, including isotropic solids and polar liquids, show an electro-optic birefringence (Kerr effect) which is proportional to the square of the electric field, *E*:

$$\Delta\left[\frac{1}{n^2}\right]_{ij} = \sum_k K_{ijkl} E_k E_l = \sum_{k,l=1,2,3} g_{ijkl} p_k p_l$$

where E_k and E_l are the components of the electric field and P_k and P_l the electric polarizations. The coefficients, K_{ijkl}, are the quadratic electro-optic coefficients, while the constants g_{ijkl} are known as the Kerr constants.

2. The other electro-optic effect only occurs in the 20 piezoelectric crystal classes (no center of symmetry). This effect is known as the Pockels effect. The optical impermeability changes linearly with the static field

$$\Delta\left[\frac{1}{n^2}\right]_{ij} = \sum_k r_{ij,k} E_k$$

The coefficients $r_{ij,k}$ have the name (linear) electro-optic coefficients.

The values of the electro-optic coefficients depend on the boundary conditions. If the superscripts T and S denote, respectively, the conditions of zero stress (free) and zero strain (clamped) one finds:

$$r_{ij}^{T} = r_{ij}^{S} + q_{ik}^{E} e_{jk} = r_{ij}^{S} + P_{ik}^{E} d_{jk}$$

where $e_{jk} = (\partial T_k / \partial E_j)_S$ and $d_{jk} = (\partial S_k / \partial E_j)_T$ are the appropriate piezoelectric coefficients.

The interaction between a *magnetic field* and a light wave propagating in a solid or in a liquid gives rise to a rotation of the plane of polarization. This effect is known as *Faraday rotation*. It results from a difference in propagation velocity for left and right circular polarized light.

The Faraday rotation, θ_F, is linearly proportional to the magnetic field *H*:

$$\theta_F = VlH$$

where *l* is the light path length and *V* is the *Verdet* constant (minutes/oersted·cm).

For ferromagnetic, ferrimagnetic, and antiferromagnetic materials the magnetic field in the above expression is replaced by the magnetization *M* and the magneto-optic coefficient in this case is known as the Kund constant *K*:

$$\text{Specific Faraday rotation } F = KM$$

In the tables below the *Faraday rotation* is listed at the saturation magnetization per unit length, together with the absorption coefficient α, the temperature *T*, the critical temperature T_C (or T_N), and the wavelength of the measurement.

In the tables that follow, the properties are presented in groups:

- Elasto-optic coefficients (photoelastic constants)
- Linear electro-optic coefficients (Pockels constants)
- Quadratic electro-optic coefficients (Kerr constants)
- Magneto-optic coefficients:
  - Verdet constants
  - Faraday rotation parameters

Within each group, materials are classified by crystal system or physical state. References are given at the end of each group of tables.

ELASTO-OPTIC COEFFICIENTS (PHOTOELASTIC CONSTANTS)

Name Cubic (43m, 432, m3m)	Formula	$\lambda/\mu m$	p_{11}	p_{12}	p_{44}	$p_{11}-p_{12}$	Ref.
Sodium fluoride	NaF	0.633	0.08	0.20	−0.03	−0.12	1
Sodium chloride	NaCl	0.589	0.115	0.159	−0.011	−0.042	2
Sodium bromide	NaBr	0.589	0.148	0.184	−0.0036	−0.035	1
Sodium iodide	NaI	0.589	–	–	0.0048	−0.0141	3
Potassium fluoride	KF	0.546	0.26	0.20	−0.029	0.06	1
Potassium chloride	KCl	0.633	0.22	0.16	−0.025	0.06	4
Potassium bromide	KBr	0.589	0.212	0.165	−0.022	0.047	5
Potassium iodide	KI	0.590	0.212	0.171	–	0.041	6
Rubidium chloride	RbCl	0.589	0.288	0.172	−0.041	0.116	7,8
Rubidium bromide	RbBr	0.589	0.293	0.185	−0.034	0.108	7,8
Rubidium iodide	RbI	0.589	0.262	0.167	−0.023	0.095	7,8
Lithium fluoride	LiF	0.589	0.02	0.13	−0.045	−0.11	5
Lithium chloride	LiCl	0.589	–	–	−0.0177	−0.0407	3
Ammonium chloride	NH$_4$Cl	0.589	0.142	0.245	0.042	−0.103	9
Cadmium telluride	CdTe	1.06	−0.152	−0.017	−0.057	−0.135	10
Calcium fluoride	CaF$_2$	0.55–0.65	0.038	0.226	0.0254	−0.183	11
Copper chloride	CuCl	0.633	0.120	0.250	−0.082	−0.130	12
Copper bromide	CuBr	0.633	0.072	0.195	−0.083	−0.123	12
Copper iodide	CuI	0.633	0.032	0.151	−0.068	−0.119	12
Diamond	C	0.540–0.589	−0.278	0.123	−0.161	−0.385	13
Germanium	Ge	3.39	−0.151	−0.128	−0.072	−0.023	14
Gallium arsenide	GaAs	1.15	−0.165	−0.140	−0.072	−0.025	15
Gallium phosphide	GaP	0.633	−0.151	−0.082	−0.074	−0.069	15
Strontium fluoride	SrF$_2$	0.633	0.080	0.269	0.0185	−0.189	16
Strontium titanate	SrTiO$_3$	0.633	0.15	0.095	0.072	–	17
KRS-5	Tl(Br,I)	0.633	−0.140	0.149	−0.0725	−0.289	18,20
KRS-6	Tl(Br,Cl)	0.633	−0.451	−0.337	−0.164	−0.114	19,20
Zinc sulfide	ZnS	0.633	0.091	−0.01	0.075	0.101	15

Rare Gases	Formula	$\lambda/\mu m$	p_{11}	p_{12}	p_{44}	$p_{11}-p_{12}$	Ref.
Neon (T = 24.3 K)	Ne	0.488	0.157	0.168	0.004	−0.011	21
Argon (T = 82.3 K)	Ar	0.488	0.256	0.302	0.015	−0.046	22
Krypton (T = 115.6 K)	Kr	0.488	0.34	0.34	0.037	0	21
Xenon (T = 160.5 K)	Xe	0.488	0.284	0.370	0.029	−0.086	22

Garnets	Formula	$\lambda/\mu m$	p_{11}	p_{12}	p_{44}	$p_{11}-p_{12}$	Ref.
GGG	Gd$_3$Ga$_5$O$_{12}$	0.514	−0.086	−0.027	−0.078	−0.059	23
YIG	Y$_3$Fe$_5$O$_{12}$	1.15	0.025	0.073	0.041	–	15
YGG	Y$_3$Ga$_5$O$_{12}$	0.633	0.091	0.019	0.079	–	17
YAG	Y$_3$Al$_5$O$_{12}$	0.633	−0.029	0.0091	−0.0615	−0.038	15

Cubic (23, m3)	Formula	$\lambda/\mu m$	p_{11}	p_{12}	p_{44}	p_{13}	Ref.
Barium nitrate	Ba(NO$_3$)$_2$	0.589	–	$p_{11}-p_{22} =$ 0.992	−0.0205	$p_{11}-p_{13} =$ 0.713	13
Lead nitrate	Pb(NO$_3$)$_2$	0.589	0.162	0.24	−0.0198	0.20	24,25
Sodium bromate	NaBrO$_3$	0.589	0.185	0.218	−0.0139	0.213	26
Sodium chlorate	NaClO$_3$	0.589	0.162	0.24	−0.0198	0.20	26
Strontium nitrate	Sr(NO$_3$)$_2$	0.41	0.178	0.362	−0.014	0.316	27

Hexagonal (mmc, 6mm)	Formula	$\lambda/\mu m$	p_{11}	p_{12}	p_{13}	p_{31}	p_{33}	p_{44}	Ref.
Beryl	Be$_3$Al$_2$Si$_6$O$_{18}$	0.589	0.0099	0.175	0.191	0.313	0.023	−0.152	28
Cadmium sulfide	CdS	0.633	−0.142	−0.066	−0.057	−0.041	−0.20	−0.099	15,2
Zinc oxide	ZnO	0.633	±0.222	±0.099	−0.111	±0.088	−0.235	0.0585	30
Zinc sulfide	ZnS	0.633	−0.115	0.017	0.025	0.0271	−0.13	−0.0627	31

Trigonal (3m, 32, 3̄m)

Trigonal (3m, 32, 3̄m)	Formula	λ/μm	p_{11}	p_{12}	p_{13}	p_{14}	p_{31}	p_{33}	p_{41}	p_{44}	Ref.
Sapphire	Al_2O_3	0.644	−0.23	−0.03	0.02	0.00	−0.04	−0.20	0.01	−0.10	15,32
Calcite	$CaCO_3$	0.514	0.062	0.147	0.186	−0.011	0.241	0.139	−0.036	−0.058	33
Lithium niobate	$LiNbO_3$	0.633	±0.034	±0.072	±0.139	±0.066	±0.178	±0.060	±0.154	±0.300	15,34
Lithium tantalate	$LiTaO_3$	0.633	−0.081	0.081	0.093	−0.026	0.089	−0.044	−0.085	0.028	15,35
Cinnabar	HgS	0.633			±0.445			±0.115	−	−	36
Quartz	SiO_2	0.589	0.16	0.27	0.27	−0.030	0.29	0.10	−0.047	−0.079	37
Proustite	Ag_3AsS_3	0.633	±0.10	±0.19	±0.22		±0.24	±0.20	−	−	38
Sodium nitrite	$NaNO_2$	0.633		±0.21	±0.215	±0.027	±0.25		0.055	−0.06	39
Tellurium	Te	10.6	0.155	0.130	−	−	−	−	−	−	15

Tetragonal (4/mmm, 4̄2m, 422)

Tetragonal (4/mmm, 4̄2m, 422)	Formula	λ/μm	p_{11}	p_{12}	p_{13}	p_{31}	p_{33}	p_{44}	p_{66}	Ref.
Ammonium dihydrogen phosphate	ADP	0.589	0.319	0.277	0.169	0.197	0.167	−0.058	−0.091	40
Barium titanate	$BaTiO_3$	0.633	0.425	−	−	−	−	−	−	41
Cesium dihydrogen arsenate	CDA	0.633	0.267	0.225	0.200	0.195	0.227	−	−	42
Magnesium fluoride	MgF_2	0.546	−	−	−	−	−	±0.0776	±0.0488	43
Calomel	Hg_2Cl_2	0.633	±0.551	±0.440	±0.256	±0.137	±0.010	−	±0.047	44
Potassium dihydrogen phosphate	KDP	0.589	0.287	0.282	0.174	0.241	0.122	−0.019	−0.064	45
Rubidium dihydrogen arsenate	RDA	0.633	0.227	0.239	0.200	0.205	0.182	−	−	41
Rubidium dihydrogen phosphate	RDP	0.633	0.273	0.240	0.218	0.210	0.208	−	−	41
Strontium barium niobate	$Sr_{0.75}Ba_{0.25}Nb_2O_6$	0.633	0.16	0.10	0.08	0.11	0.47	−	−	46
Strontium barium niobate	$Sr_{0.5}Ba_{0.5}Nb_2O_6$	0.633	0.06	0.08	0.17	0.09	0.23	−	−	46
Tellurium oxide	TeO_2	0.633	0.0074	0.187	0.340	0.090	0.240	−0.17	−0.046	47
Rutile	TiO_2	0.633	0.017	0.143	−0.139	−0.080	−0.057	−0.009	−0.060	48

Tetragonal (4, 4̄, 4/m)

Tetragonal (4, 4̄, 4/m)	Formula	λ/μm	p_{11}	p_{12}	p_{13}	p_{16}	p_{31}	p_{33}	p_{44}	p_{45}	p_{61}	p_{66}	Ref.
Cadmium molybdate	$CdMoO_4$	0.633	0.12	0.10	0.13	−	0.11	0.18	−	−	−	−	49
Lead molybdate	$PbMoO_4$	0.633	0.24	0.24	0.255	0.017	0.175	0.300	0.067	−0.01	0.013	0.05	52
Sodium bismuth molybdate	$NaBi(MoO_4)_2$	0.633	0.243	0.205	0.25	−	0.21	0.29	−	−	−	−	

Orthorhombic (222, m22, mmm)

Orthorhombic (222, m22, mmm)	Formula	λ/μm	p_{11}	p_{12}	p_{13}	p_{21}	p_{22}	p_{23}	p_{31}	p_{32}	p_{33}	p_{44}	p_{55}	p_{66}	Ref.
Ammonium chlorate	NH_4ClO_3	0.633	−	0.24	0.18	0.23	−	0.20	0.19	0.18	±0.02	<±0.02	−	±0.04	51
Ammonium sulfate	$(NH_4)_2SO_4$	0.633	0.26	0.19	±0.260	±0.230	±0.27	±0.254	0.20	±0.26	0.26	0.015	±0.0015	0.012	52
Rochelle salt	$NaKC_4H_4O_6$	0.589	0.35	0.41	0.42	0.37	0.28	0.34	0.36	0.35	0.36	−0.030	0.0046	−0.025	53
Iodic acid (α)	HIO_3	0.633	0.302	0.496	0.339	0.263	0.412	0.304	0.251	0.345	0.336	0.084	−0.030	0.098	54
Sulfur (α)	S	0.633	0.324	0.307	0.268	0.272	0.301	0.310	0.203	0.232	0.270	0.143	0.019	0.118	54
Barite	$BaSO_4$	0.589	0.21	0.25	0.16	0.34	0.24	0.19	0.28	0.22	0.31	0.002	−0.012	0.037	55
Topaz	$Al_2SiO_4(OH,F)_2$	−	−0.085	0.069	0.052	0.095	−0.120	0.065	0.095	0.085	−0.083	−0.095	−0.031	0.098	28

Monoclinic (2, m, 2/m)

Monoclinic (2, m, 2/m)	Formula	λ/μm
Taurine	$C_2H_7NO_3S$	0.589

$p_{11} = 0.313$ $\quad p_{25} = -0.0025$ $\quad p_{51} = -0.014$
$p_{12} = 0.251$ $\quad p_{31} = 0.362$ $\quad p_{52} = 0.006$
$p_{13} = 0.270$ $\quad p_{32} = 0.275$ $\quad p_{53} = 0.0048$
$p_{15} = -0.10$ $\quad p_{33} = 0.308$ $\quad p_{55} = 0.047$
$p_{21} = 0.281$ $\quad p_{35} = -0.003$ $\quad p_{64} = 0.0024$
$p_{22} = 0.252$ $\quad p_{44} = 0.0025$ $\quad p_{66} = 0.0028$
$p_{23} = 0.272$ $\quad p_{46} = -0.0056$

Isotropic	Formula	λ/μm	p_{11}	p_{12}	p_{44}	Ref.
Fused silica	SiO$_2$	0.633	0.121	0.270	−0.075	15
Water	H$_2$O	0.633	±0.31	±0.31		15
Polystyrene		0.633	±0.30	±0.31		25
Lucite		0.633	±0.30	0.28		25
Orpiment	As$_2$S$_3$-glass	1.15	0.308	0.299	0.0045	15
Tellurium oxide	TeO$_2$-glass	0.633	0.257	0.241	0.0079	56
Laser glasses	LGS-247-2	0.488	±0.168	±0.230		57
	LGS-250-3		±0.135	±0.198		
	LGS-1		±0.214	±0.250		
	KGSS-1621		±0.205	±0.239		
Dense flint glasses	LaSF	0.633	0.088	0.147	−0.030	58
(examples)	SF$_4$		0.215	0.243	−0.014	
	U10502		0.172	0.179	−0.004	
	TaFd$_7$		0.099	0.138	−0.020	

References

A. Narasimhamurty, T. S., *Photoelastic and Electro-Optic Properties of Crystals*, Plenum Press, New York, 1981, pp. 290–293.

B. Weber, M. J., Ed., *CRC Handbook of Laser Science and Technology*, Volume IV, Part 2, CRC Press, Boca Raton, FL, 1986, pp. 324–331.

1. Petterson, H. E., *J. Opt. Soc. Am.*, 63, 1243, 1973.
2. Burstein, E. and Smith, P. L., *Phys. Rev.*, 74, 229, 1948.
3. Pakhnev, A. V., et al., *Sov. Phys. J. (transl.)*, 18, 1662, 1975.
4. Feldman, A., Horovitz, D., and Waxler, R. M., *Appl. Opt.*, 16, 2925, 1977.
5. Iyengar, K. S., *Nature (London)*, 176, 1119, 1955.
6. Bansigir, K. G. and Iyengar, K. S., *Acta Crystallogr.*, 14, 727, 1961.
7. Pakhev, A. V., et al., *Sov. Phys. J. (transl.)*, 20, 648, 1975.
8. Bansigir, K. G., *Acta Crystallogr.*, 23, 505, 1967.
9. Krishna Rao, K. V. and Krishna Murty, V. G., *Ind. J. Phys.*, 41, 150, 1967.
10. Weil, R. and Sun, M. J., *Proc. Int. Symp. CdTe (Detectors)*, Strasbourg Centre de Rech. Nucl., 1971, XIX-1 to 6, 1972.
11. Schmidt, E. D. D. and Vedam, K., *J. Phys. Chem. Solids*, 27, 1563, 1966.
12. Biegelsen, D. K., et al., *Phys. Rev. B*, 14, 3578, 1976.
13. Helwege, K. H., *Landolt-Börnstein, New Series Group III*, Vol. II, Springer-Verlag, Berlin, 1979.
14. Feldman, A., Waxler, R. M., and Horovitz, D., *J. Appl. Phys.*, 49, 2589, 1978.
15. Dixon, R. W., *J. Appl. Phys.*, 38, 5149, 1967.
16. Shabin, O. V., et al., *Sov. Phys. Sol. State (transl.)*, 13, 3141, 1972.
17. Reintjes, J. and Schultz, M. B., *J. Appl. Phys.*, 39, 5254, 1968.
18. Rivoallan, L. and Favre, F., *Opt. Commun.*, 8, 404, 1973.
19. Rivoallan, L. and Favre, F., *Opt. Commun.*, 11, 296, 1974.
20. Afanasev, I. I., et al., *Sov. J. Opt. Technol.*, 46, 663, 1979.
21. Rand, S. C., et al., *Phys. Rev. B*, 19, 4205, 1979.
22. Sipe, J. E., *Can J. Phys.*, 56, 199, 1978.
23. Christyi, I. L., et al., *Sov. Phys. Sol. State (transl.)*, 17, 922, 1975.
24. Narasimhamurty, T. S., *Curr. Sci. (India)*, 23, 149, 1954.
25. Smith, T. M. and Korpel, A., *IEEE J. Quant. Electron.*, QE-1, 283, 1965.

26. Narasimhamurty, T. S., *Proc. Ind. Acad. Sci.*, A40, 164, 1954.
27. Rabman, A., *Bhagarantam Commem. Vol.*, Bangalore Print. and Publ., 173, 1969.
28. Eppendahl, R., *Ann. Phys. (IV)*, 61, 591, 1920.
29. Laurenti, J. P. and Rouzeyre, M., *J. Appl. Phys.*, 52, 6484, 1981.
30. Sasaki, H., et al., *J. Appl. Phys.*, 47, 2046, 1976.
31. Uchida, N. and Saito, S., *J. Appl. Phys.*, 43, 971, 1972.
32. Waxler, R. M. and Farabaugh, E. M., *J. Res. Natl. Bur. Stand.*, A74, 215, 1970.
33. Nelson, D. F., Lazay, P. D., and Lax, M., *Phys. Rev.*, B6, 3109, 1972.
34. O'Brien, R. J., Rosasco, G. J., and Weber, A., *J. Opt. Soc. Am.*, 60, 716, 1970.
35. Avakyants, L. P., et al., *Sov. Phys.*, 18, 1242, 1976.
36. Sapriel, J., *Appl. Phys. Lett.*, 19, 533, 1971.
37. Narasimhamurty, T. S., *J. Opt. Soc. Am.*, 59, 682, 1969.
38. Zubrinov, I. I., et al., *Sov. Phys. Sol. State (transl.)*, 15, 1921, 1974.
39. Kachalov, O. V. and Shpilko, I. O., *Sov. Phys. JETP (transl.)*, 35, 957, 1972.
40. Narasimhamurty, T. S., et al., *J. Mater. Sci.*, 8, 577, 1973.
41. Tada, K. and Kikuchi, K., *Jpn. J. Appl. Phys.*, 19, 1311, 1980.
42. Aleksandrov, K. S., et al., *Sov. Phys. Sol. State (transl.)*, 19, 1090, 1977.
43. Afanasev, I. I., et al., *Sov. Phys. Sol. State (transl.)*, 17, 2006, 1975.
44. Silvestrova, I. M., et al., *Sov. Phys. Cryst. (transl.)*, 20, 649, 1975.
45. Veerabhadra Rao, K. and Narasimhamurty, T. S., *J. Mater. Sci.*, 10, 1019, 1975.
46. Venturini, E. L., et al., *J. Appl. Phys.*, 40, 1622, 1969.
47. Vehida, N. and Ohmachi, Y., *J. Appl. Phys.*, 40, 4692, 1969.
48. Grimsditch, M. H. and Ramdus, A. K., *Phys. Rev. B*, 22, 4094, 1980.
49. Schinke, D. P. and Viehman, W., unpublished data.
50. Coquin, G. A., et al., *J. Appl. Phys.*, 42, 2162, 1971.
51. Vasquez, F., et al., *J. Phys. Chem. Solids*, 37, 451, 1976.
52. Luspin, Y. and Hauret, G., *C.R.Ac. Sci. Paris*, B274, 995, 1972.
53. Narasimhamurty, T. S., *Phys. Rev.*, 186, 945, 1969.
54. Haussühl, S. and Weber, H. J., *Z. Kristall.*, 132, 266, 1970.
55. Vedam, K., *Proc. Ind. Ac. Sci.*, A34, 161, 1951.
56. Yano, T., Fukumoto, A., and Watanabe, A., *J. Appl. Phys.*, 42, 3674, 1971.
57. Manenkov, A. A. and Ritus, A. I., *Sov. J. Quant. Electr.*, 8, 78, 1978.
58. Eschler, H. and Weidinger, F., *J. Appl. Phys.*, 46, 65, 1975.

LINEAR ELECTRO-OPTIC COEFFICIENTS

Name Cubic ($\bar{4}3m$)	Formula	$\lambda/\mu m$	r_{41} pm/V
Cuprous bromide	CuBr	0.525	0.85
Cuprous chloride	CuCl	0.633	3.6
Cuprous iodide	CuI	0.55	−5.0
Eulytite (BSO)	$Bi_4Si_3O_{12}$	0.63	0.54
Germanium eulytite (BGO)	$Bi_4Ge_3O_{12}$	0.63	1.0
Gallium arsenide	GaAs	10.6	1.6
Gallium phosphide	GaP	0.56	−1.07
Hexamethylenetetramine	$C_6H_{12}N_4$	0.633	0.78
Sphalerite	ZnS	0.65	2.1
Zinc selenide	ZnSe	0.546	2.0
Zinc telluride	ZnTe	3.41	4.2
Cadmium telluride	CdTe	3.39	6.8

Cubic (23)	Formula	$\lambda/\mu m$	r_{41} pm/V
Ammonium chloride (77 K)	NH_4Cl	−	1.5
Ammonium cadmium langbeinite	$(NH_4)_2Cd_2(SO_4)_3$	0.546	0.70
Ammonium manganese langbeinite	$(NH_4)_2Mn_2(SO_4)_3$	0.546	0.53
Thallium cadmium langbeinite	$Tl_2Cd_2(SO_4)_3$	0.546	0.37
Potassium magnesium langbeinite	$K_2Mg_2(SO_4)_3$	0.546	0.40
Bismuth monogermanate	$Bi_{12}GeO_{20}$	−	3.3
Bismuth monosilicate	$Bi_{12}SiO_{20}$	−	3.3
Sodium chlorate	$NaClO_3$	0.589	0.4
Sodium uranyl acetate	$NaUO_2(CH_3COO)_3$	0.546	0.87
Trenhydrobromide	$N(CH_2CH_2NH_2)_3 \cdot 3HBr$	−	1.5
Trenhydrochloride	$N(CH_2CH_2NH_2)_3 \cdot 3HCl$	−	1.7

Tetragonal ($\bar{4}2m$)	Formula	T_{tran} K	r_{41} pm/V	r_{63} pm/V
Ammonium dihydrogen phosphate (ADP)	$NH_4H_2PO_4$	148	24.5	−8.5
Ammonium dideuterium phosphate (AD*P)	$NH_4D_2PO_4$	242	−	11.9
Ammonium dihydrogen arsenate (ADA)	$NH_4H_2AsO_4$	−	−	9.2
Cesium dihydrogen arsenate (CsDA)	CsH_2AsO_4	143	−	18.6
Cesium dideuterium arsenate (CsD*A)	CsD_2AsO_4	212	−	36.6
Potassium dihydrogen phosphate (KDP)	KH_2PO_4	123	8.6	−10.5
Potassium dideuterium phosphate (KD*P)	KD_2PO_4	222	8.8	23.8
Potassium dihydrogen arsenate (KDA)	KH_2AsO_4	97	12.5	10.9
Potassium dideuterium arsenate (KD*A)	KD_2AsO_4	162	−	18.2
Rubidium dihydrogen phosphate (RDP)	RbH_2PO_4	147	−	15.5
Rubidium dihydrogen arsenate (RDA)	RbH_2AsO_4	110	−	13.0
Rubidium dideuterium arsenate (RD*A)	RbD_2AsO_4	178	−	21.4

Tetragonal (4mm)	Formula	T_{tran} K	r_{13} pm/V	r_{33} pm/V	r_{51} pm/V
Barium titanate	$BaTiO_3$	406	8	28	−
Potassium lithium niobate	$K_3Li_2Nb_5O_{15}$	693	8.9	5.9	−
Lead titanate	$PbTiO_3$	765	13.8	5.9	−
Strontium barium niobate (SBN75)	$Sr_{0.75}Ba_{0.25}Nb_2O_6$	330	6.7	1340	42
Strontium barium niobate (SBN46)	$Sr_{0.46}Ba_{0.54}Nb_2O_6$	602	~180	35	−

Hexagonal (6mm)	Formula	r_{13} pm/V	r_{33} pm/V	r_{42} pm/V	r_{51} pm/V
Greenockite	CdS	3.1	2.9	2.0	3.7
Greenockite (const. strain)	CdS	1.1	2.4	–	–
Wurzite	ZnS	0.9	1.8	–	–
Zincite	ZnO	–1.4	+2.6	–	–

Hexagonal (6)	Formula	r_{13} pm/V	r_{33} pm/V	r_{42} pm/V	r_{51} pm/V
Lithium iodate	$LiIO_3$	4.1	6.4	1.4	3.3
Lithium potassium sulfate	$LiKSO_4$	$r_{13}-r_{33} = 1.6$	–	–	–

Trigonal (3m)	Formula	T_{tran} K	r_{13} pm/V	r_{22} pm/V	r_{33} pm/V	r_{42} pm/V
Cesium nitrate	$CsNO_3$	425	–	0.43	–	–
Lithium niobate	$LiNbO_3$	1483	8.6	7.0	30.8	28
Lithium tantalate	$LiTaO_3$	890	8.4	–	30.5	–
Lithium sodium sulfate	$LiNaSO_4$	–	–	<0.02	–	–
Tourmaline	–	–	–	0.3	–	–

Trigonal (32)	Formula	T_{tran} K	r_{11} pm/V	r_{41} pm/V
Cesium tartrate	$Cs_2C_4H_4O_6$	–	1.0	–
Cinnabar	HgS	659	3.1	1.5
Potassium dithionate	$K_2S_2O_6$	–	0.26	–
Strontium dithionate	$SrS_2O_6·4H_2O$	–	0.1	–
Quartz	SiO_2	1140	–0.47	0.2
Selenium	Se	398	2.5	

Orthorhombic (222)	Formula	T_{tran} K	r_{41} pm/V	r_{52} pm/V	r_{63} pm/V
Ammonium oxalate	$(NH_4)_2C_2O_4·4H_2O$	–	230	330	250
Rochelle salt	$KNaC_4H_4O_6·4H_2O$	$T_u = 297$ $T_1 = 255$	–2.0	–1.7	+0.32

Orthorhombic (mm2)	Formula	T_{trans} K	r_{13} pm/V	r_{23} pm/V	r_{33} pm/V	r_{42} pm/V	r_{51} pm/V
Barium sodium niobate (BSN)	Ba_2NaNbO_{15}	833	15	13	48	92	90
Potassium niobate	$KNbO_3$	476	28	1.3	64	380	105

Monoclinic (2)	Formula	T_{trans} K	r_{22} pm/V	r_{32} pm/V
Calcium pyroniobate	$Ca_2Nb_2O_7$	–	0.33	13.7
Triglycine sulfate (TGS)	$(NH_2CH_2COOH)_3·H_2SO_4$	322	7.2	13.6

References

1. Narasimhamurty, T. S., *Photoelastic and Electro-Optic Properties of Crystals*, Plenum Press, New York, 1981, pp. 405–407.
2. Weber, M. J., Ed., *CRC Handbook of Laser Science and Technology*, Vol. IV, CRC Press, Boca Raton, FL, 1986, pp. 258–278.

QUADRATIC ELECTRO-OPTIC COEFFICIENTS

Kerr Constants of Ferroelectric Crystals[1,2]

Name	Formula	T_{tran} K	λ μm	g_{11} 10^{10} esu	g_{12} 10^{10} esu	$g_{11}-g_{12}$ 10^{10} esu	g_{44} 10^{10} esu
Barium titanate	$BaTiO_3$	406	0.633	1.33	−0.11	1.44	
Strontium titanate	$SrTiO_3$	–	0.633	–	–	1.56	–
Potassium tantalate niobate	$KTa_{0.65}Nb_{0.35}O_3$	330	0.633	1.50	−0.42	1.92	1.63
Potassium tantalate	$KTaO_3$	13	0.633	–	–	1.77	1.33
Lithium niobate	$LiNbO_3$	1483	–	0.94	0.25	0.7	0.6
Lithium tantalate	$LiTaO_3$	938	–	1.0	0.17	0.8	0.7
Barium sodium niobate (BSN)	$Ba_{0.8}Na_{0.4}Nb_2O_6$	833	–	1.55	0.44	1.11	

Kerr Constants of Selected Liquids[2]

K is the Kerr constant at a wavelength of 589 nm and at room temperature; ε is the static dielectric constant; t_m is the melting point; and t_b is the normal boiling point

Name	Molecular formula	K 10^{-7} esu	ε	t_m °C	t_b °C
Carbon disulfide	CS_2	+3.23	2.63	−111.5	+46.3
Acetone	C_3H_6O	+16.3	21.0	−94.8	+56.1
Methyl ethyl ketone	C_4H_8O	+13.6	18.56	−86.67	+79.6
Pyridine	C_5H_5N	+20.4	13.26	−42	+115.23
Ethyl cyanoacetate	$C_5H_7NO_2$	+38.8	31.6	−22.5	205
o-Dichlorobenzene	$C_6H_4Cl_2$	+42.6	10.12	−16.7	180
Benzenesulfonyl chloride	$C_6H_5ClO_2S$	+89.9	28.90	+14.5	247
Nitrobenzene	$C_6H_5NO_2$	+326	35.6	+5.7	210.8
Ethyl 3-aminocrotonate	$C_6H_{11}NO_2$	+31.0	–	+33.9	210
Paraldehyde	$C_6H_{12}O_3$	−23.0	14.7 12.0[a]	+12.6	124
Benzaldehyde	C_7H_6O	+80.8	17.85 14.1[a]	−26	179.05
p-Chlorotoluene	C_7H_7Cl	+23.0	6.25	+7.5	162.4
o-Nitrotoluene	$C_7H_7NO_2$	+174	26.26	−10	222.3
m-Nitrotoluene	$C_7H_7NO_2$	+177	24.95	+15.5	232
p-Nitrotoluene	$C_7H_7NO_2$	+222	22.2	+51.6	238.3
Benzyl alcohol	C_7H_8O	−15.4	11.92 10.8[a]	−15.3	205.8
m-Cresol	C_7H_8O	+21.2	12.44 5.0[a]	+11.8	202.27
m-Chloroacetophenone	C_8H_7ClO	+69.1			
Acetophenone	C_8H_8O	+66.6	17.44 15.8[a]	+19.7	202.3
Quinoline	C_9H_7N	+15.0	9.16	−14.78	237.16
Ethyl salicylate	$C_9H_{10}O_3$	+19.6	8.48	+1.3	231.5
Carvone	$C_{10}H_{14}O$	+23.6	11.2	<0	230
Ethyl benzoylacetate	$C_{11}H_{12}O_3$	+16.0	13.50	<0	270
Water	H_2O	+4.0	80.10	0.00	100.0

[a] Dielectric constant at radio frequencies (108–109 Hz).

References

1. Narasimhamurty, T. S., *Photoelastic and Electro-Optic Properties of Crystals*, Plenum Press, New York, 1981, p. 408.
2. Gray, D. E., Ed., *AIP Handbook of Physics*, McGraw Hill, New York, 1972, p. 6–241.

MAGNETO–OPTIC CONSTANTS

Verdet Constants of Non-Magnetic Crystals[1]

V is the Verdet constant; n is the refractive index; and λ is the wavelength. "min" is minutes of angle.

Material	T K	λ nm	n	V min/Oe cm
Al_2O_3	300	546.1	1.771	0.0240
	300	589.3	1.768	0.0210
$BaTaO_3$	403	427		0.95
	403	496		0.38
	403	620		0.18
	403	826		0.072
$Bi_4Ge_3O_{12}$	300	442	2.077	0.289
	300	632.8	2.048	0.099
	300	1064	2.031	0.026
C (diamond)	300	589.3	2.417	0.0233
$CaCO_3$	300	589.3	1.658	0.019
CaF_2	300	589.3	1.434	0.0088
$Cd_{0.55}Mn_{0.45}Te$	300	632.8		6.87
$CuCl$	300	546.1	1.93	0.20
$GaSe$	298	632.8		0.80
$KAl(SO_4)_2 \cdot 12H_2O$	300	589.3	1.456	0.0124
KBr	300	546.1	1.564	0.0500
	300	589.3	1.560	0.0425
KCl	300	589.3	1.490	0.0275
KI	300	546.1	1.673	0.083
	300	589.3	1.666	0.070
$KTaO_3$	296	352		0.44
	296	413		0.19
	296	496		0.096
	296	620		0.051
	296	826		0.022
LaF_3	300	325	1.639	0.054
(H‖c)	300	442	1.615	0.028
	300	632.8	1.601	0.012
	300	1064	1.592	0.006
$MgAl_2O_4$	300	589.3	1.718	0.021
$NH_4AlSO_4 \cdot 12H_2O$	300	589.3	1.459	0.0128
NH_4Br	300	589.3	1.711	0.0504
NH_4Cl	300	546.1		0.0410
	300	589.3	1.643	0.0362
$NaBr$	300	546.1		0.0621
$NaCl$	300	546.1		0.0410
	300	589.3	1.544	0.0345
$NaClO_3$	300	546.1		0.0105
	300	589.3	1.515	0.0081
$NiSO_4 \cdot 6H_2O$	297	546.1		0.0256
	297	589.3	1.511	0.0221
SiO_2	300	546.1	1.546	0.0195
	300	589.3	1.544	0.0166
$SrTiO_3$	298	413	2.627	0.78
	298	496		0.31
	298	620		0.14
	298	826		0.066
ZnS	300	546.1		0.287
	300	589.3	2.368	0.226
$ZnSe$	300	476	2.826	1.50
	300	496	2.759	1.04
	300	514	2.721	0.839
	300	587	2.627	0.529
	300	632.8	2.592	0.406

Verdet Constants of Rare-Earth Aluminum Garnets at Various Wavelengths[1]

The absorption coefficient α for these materials ranges from 0.2 to 0.6 cm^{-1} at 300 K.

Material	T/K	λ = 405 nm	450 nm	480 nm	520 nm	546 nm	578 nm	635 nm	670 nm
$Tb_2Al_5O_{12}$	300	−2.266	−1.565	−1.290	−1.039	−0.912	−0.787	−0.620	−0.542
	77		−102.16	−83.45	−3.425	−3.051	−2.603	−2.008	−1.815
	4.2				−64.80	−58.35	−53.77	48.39	−45.15
	1.45		−200.95	−172.52	−139.28	−125.07	−111.27	97.47	−93.42
$Dy_3Al_5O_{12}$	300	−1.241	−0.942	−0.803	−0.667	−0.592	−0.518	−0.411	−0.359
$Ho_3Al_5O_{12}$	300	−0.709	−0.320	−0.260	−0.335	−0.304	−0.299		−0.206
$Er_3Al_5O_{12}$	300	−0.189	−0.240	−0.154	−0.162	−0.157	−0.145	−0.105	−0.089
$Tm_3Al_5O_{12}$	300	+0.151	+0.103	+0.093	0.076	0.069	+0.059	+0.048	
$Yb_3Al_5O_{12}$	298	0.287	0.215	0.186	0.140	0.133	0.116	0.094	
	77	0.718	0.540	0.481	0.393	0.342	0.302	0.239	

(V in min/Oe cm)

Verdet Constants for KDP-Type Crystals[1]

Measurements refer to T = 298 K and λ = 632.8 nm, with $k \parallel$ [001].

Material	V min/Oe cm
KH_2PO_4 (KDP)	0.0124
$KH_{0.3}D_{1.7}PO_4$ (KD*P)	0.145
$NH_4H_2PO_4$ (ADP)	0.138
KH_2AsO_4 (KDA)	0.238
$KH_{0.1}D_{1.9}AsO_4$ (KD*A)	0.245
$NH_4H_2AsO_4$ (ADH)	0.244

Verdet Constants of Gases[2]

Values refer to T = 0 °C and P = 101.325 kPa (760 mmHg); n_D is the refractive index at a wavelength of 589 nm.

Gas	$(n_D - 1) \times 10^3$	$10^6 \times V$ min/Oe cm
He	0.036	+0.40
Ar	2.81	+9.36
H_2		+6.29
N_2	0.297	+6.46
O_2	0.272	+5.69
Air	0.293	+6.27
Cl_2	0.773	+31.9
HCl	0.447	+21.5
H_2S	0.63	+41.5
NH_3	0.376	+19.0
CO	0.34	+11.0
CO_2	0.45	+9.39
NO	0.297	−58
CH4	0.444	+17.4
n-C_4H_{10}		+44.0

Verdet Constants of Liquids[2]

n_D is the refractive index at a wavelength of 589 nm and a temperature of 20 °C, unless otherwise indicated. V is the Verdet constant.

Liquid	λ/nm	t/°C	$10^2 \times V$ min/Oe cm	n_D
P	589	33	+13.3	
S	589	114	+8.1	1.929 (110 °C)
H_2O	589	20	+1.309	1.3328
D_2O	589	19.7	+1.257	1.3384
H_3PO_4	578	97.4	+1.35	
CS_2	589	20	+4.255	1.6255
CCl_4	578–589	25.1	+1.60	1.463 (15 °C)
$SbCl_5$	578	18	+7.45	1.601 (14 °C)
$TiCl_4$	578	17	−1.65	1.61
$TiBr_4$	578	46	−5.3	
Methanol	589	18.7	+0.958	1.3289
Acetone	578–589	20.0	+1.116	1.3585
Toluene	578–589	15.0	+2.71	1.4950
Benzene	578–589	15.0	+3.00	1.5005
Chlorobenzene	589	15	+2.92	1.5246
Nitrobenzene	589	15	+2.17	1.5523
Bromoform	589	17.9	+3.13	1.5960

Verdet Constants of Rare-Earth Paramagnetic Crystals[1]

n is the refractive index, and V is the Verdet constant at the wavelength and temperature indicated.

Rare Earth	Host	T/K	λ/nm	n	V min/Oe cm
Ce³⁺(30%)	CaF₂	300	325	1.516	−0.956
		300	442	1.502	−0.297
		300	633	1.494	−0.111
		300	1064	1.489	−0.035
Ce³⁺	CeF₃	300	442	1.613	−1.05
		300	633	1.598	−0.406
		77	633		−1.418
		300	1064	1.589	−0.113
Pr³⁺(5%)	CaF₂	300	266	1.471	−0.172
		300	325	1.461	−0.0818
		300	442	1.451	−0.0089
		300	633	1.445	−0.0168
		300	1064	1.441	−0.0045
Nd³⁺(2.9%)	CaF₂	4.2	426		−0.19
Nd³⁺	NdF₃	300	442	1.60	−0.553
		290	633	1.59	−0.209
		77	633		−0.755
		300	1064	1.58	−0.097
Eu³⁺(3%)	CaF₂	4.2	430		29
		4.2	440		22
Eu²⁺	EuF₂	300	450		−4.5
		300	500		−2.6
		300	550		−1.6
		300	600		−1.1
		300	650		−0.8
		300	1064		−0.19
Tb³⁺	KTb₃F₁₀	300	325	1.531	−2.174
		300	442	1.518	−0.933
		300	633	1.510	−0.386
		77	633		−1.94
		300	1064	1.505	−0.114
Tb³⁺	LiTbF₄	300	325	1.493	−1.9
		300	442	1.481	−0.98
		300	633	1.473	−0.44
		300	1064	1.469	−0.13
Tb³⁺	Tb₃Ga₅O₁₂	300	500	1.989	−0.749
		300	570	1.981	−0.581
		300	633	1.976	−0.461
		300	830	1.967	−0.21
		300	1060	1.954	−0.12

Verdet Constants of Paramagnetic Glasses[1]

The Verdet constant V is given at room temperature for the wavelengths indicated.

Rare-earth phosphate glasses of composition $R_2O_3 \cdot xP_2O_5$, where x is given in the second column

		$\lambda = 405$ nm	$\lambda = 436$ nm	$\lambda = 480$ nm	$\lambda = 500$ nm	$\lambda = 520$ nm	$\lambda = 546$ nm	$\lambda = 578$ nm	$\lambda = 600$ nm	$\lambda = 635$ nm	$\lambda = 670$ nm
R	x										
La		0.037	0.030	0.024	0.022	0.020	0.018	0.015	−0.014	0.013	–
Ce	2.67	−0.672	0.510	−0.366	−0.326	−0.287	−0.253	−0.217	−0.197	−0.173	−0.150
Pr	3.09	−0.447	−0.332	−0.283	−0.261	−0.236	−0.208	−0.182	−0.170	−0.150	−0.132
Nd	2.92	−0.250	−0.209	−0.167	−0.155	−0.136	−0.134	−0.094	−0.080	−0.080	−0.071
Sm	2.87	0.026	0.024	0.020	0.020	0.017	0.015	0.014	0.012	0.011	0.010
Eu	2.93	−0.025	−0.017	−0.010	−0.006	−0.006	−0.005	−0.004	−0.003	−0.002	−0.002
Gd	3.01	0.018	0.015	0.014	0.012	0.012	0.011	0.011	0.010	0.009	0.009
Tb	2.94	−0.560	−0.458	−0.357	−0.323	−0.295	−0.261	−0.226	−0.206	−0.190	−0.164
Dy	2.51	−0.540	−0.453	−0.359	−0.331	−0.301	0.268	−0.237	−0.217	−0.197	−0.173
Ho	2.94	−0.299	−0.313	−0.156	−0.153	−0.138	−0.138	−0.119	−0.110	−0.098	−0.084
Er	3.01	−0.139	−0.121	−0.100	−0.111	−0.095	−0.062	−0.060	−0.057	−0.051	−0.044
Tm	2.79	0.019	0.013	0.012	0.009	0.008	0.006	0.005	0.004	0.004	0.007
Yb	3.01	0.087	0.072	0.056	0.050	0.045	0.041	0.036	0.032	0.029	0.024

The following are rare-earth borate glasses with composition:

for La and Pr: $R_2O_3 \cdot xP_2O_5$; for Tb–Pr and Dy–Pr: $R_2O_3 \cdot xB_2O_3$; and for other elements: $R_2O_3 \cdot 0.85La_2O_3 \cdot xB_2O_3$.

		$\lambda = 405$ nm	$\lambda = 436$ nm	$\lambda = 480$ nm	$\lambda = 500$ nm	$\lambda = 520$ nm	$\lambda = 546$ nm	$\lambda = 578$ nm	$\lambda = 600$ nm	$\lambda = 635$ nm	$\lambda = 670$ nm
La	3.04	0.043	0.036	0.029	0.026	0.023	0.022	0.019	0.018	0.016	0.014
Pr-La	5.44	−0.380	−0.307	−0.230	−0.220	−0.201	−0.178	−0.153	−0.146	−0.128	−0.110
Nd-La	5.41	−0.180	−0.147	−0.120	−0.111	−0.096	−0.094	−0.100	−0.059	−0.056	−0.046
Sm-La	4.97	0.032	0.030	0.025	0.024	0.022	0.019	0.017	0.016	0.014	0.012
Eu-La	4.69	−0.081	−0.060	−0.038	−0.033	−0.029	−0.024	0.019	−0.016	0.014	−0.012
Gd-La	4.71	0.032	0.026	0.024	0.022	0.021	0.020	0.018	0.017	0.015	0.013
Tb-La	4.73	−0.512	−0.419	−0.319	−0.288	−0.262	−0.234	−0.205	−0.186	−0.167	−0.142
Dy-La	4.88	−0.436	−0.361	−0.299	−0.273	−0.246	−0.220	−0.193	−0.177	−0.159	−0.138
Ho-La	4.36	−0.269	−0.252	−0.123	−0.131	−0.112	−0.128	−0.104	−0.096	–	−0.074
Er-La	4.50	−0.093	−0.078	−0.068	−0.082	–	−0.045	−0.042	−0.040	−0.035	−0.034
Tm-La	4.75	0.060	0.046	0.039	0.034	0.031	0.026	0.023	0.021	0.018	0.016
Yb-La	8.58	0.115	0.094	0.073	0.066	0.060	0.054	0.046	0.043	0.037	0.033
Tb-Pr	4.99	−0.940	−0.786	−0.560	−0.536	−0.489	−0.436	−0.380	−0.348	−0.306	−0.265
Dy-Pr	4.63	−0.850	–	–	−0.497	−0.465	−0.413	−0.358	−0.332	−0.290	−0.252
Pr	2.56	−0.843	−0.646	−0.471	−0.480	−0.432	−0.390	−0.334	−0.317	−0.271	−0.243

Verdet Constants of Diamagnetic Glasses[1]

The Verdet constant V is given at room temperature for the wavelengths indicated.

Glass type	Composition (wt. %)	Verdet constant V in min/Oe cm			
		$\lambda = 325$ nm	$\lambda = 442$ nm	$\lambda = 633$ nm	$\lambda = 1064$ nm
SiO_2	100% SiO_2			0.013	
B_2O_3	100% B_2O_3			0.010	
CdO	47.5% CdO, 52.5% P_2O_5	0.079	0.033	0.022	
ZnO	36.4% ZnO, 63.6% P_2O_5	0.072	0.044	0.020	
TeO_2	88.9% TeO_2, 11.1% P_2O_5		0.196	0.076	0.022
ZrF_4	63.1% ZrF_4, 14.9% BaF_2, 7.2% LaF_3, 1.9% AlF_3, 9.1% PbF_2, 3.8% LiF			0.011	

		λ = 700 nm	λ = 853 nm	λ = 1060 nm
Bi_2O_3	95% Bi2O3, 5% B2O3	0.086	0.051	0.033
PbO	95% PbO, 5% B2O3	0.093	0.061	0.031
	82% PbO, 18% SiO2	0.077	0.045	0.027
	50% PbO, 15% K2O, 35% SiO2	0.032	0.020	0.011
Tl_2O	95% Tl2O, 5% B2O3	0.092	0.061	0.032
	82% Tl2O, 18% SiO2	0.100	0.067	0.043
	50% Tl2O, 15% K2O, 35% SiO2	0.036	0.022	0.012
SnO	76% SnO, 13% B2O3, 11% SiO2	0.071	0.046	0.026
TeO_3	75% TeO2, 25% Sb2O3	0.076	0.052	0.032
	80% TeO2, 20% ZnCl2	0.073	0.046	0.025
	84% TeO2, 16% BaO	0.056	0.041	0.029
	70% TeO2, 30% WO3	0.052	0.035	0.022
	20% TeO2, 80% PbO	0.128	0.075	0.048
Sb_2O_3	25% Sb2O3, 75% TeO2	0.076	0.050	0.032
	75% Sb2O3, 75% Cs2O, 5% Al2O3	0.074	0.044	0.025
	75% Sb2O3, 10% Cs2O, 10% Rb2O, 5% Al2O3	0.078	0.052	0.030

Verdet Constants of Commercial Glasses[1]

This table gives the density, ρ, refractive index at 589 nm, n_D, and Verdet constant, V, for the wavelengths indicated; the data refer to room temperature.

Glass type	ρ g/cm³	n_D	V in min/Oe cm				
			λ = 365.0 nm	λ = 404.7 nm	λ = 435.8 nm	λ = 546.1 nm	λ = 578.0 nm
BSC	2.49	1.5096	0.0499	0.0392	0.0333	0.02034	0.01798
HC	2.53	1.5189	0.0561	0.0440	0.0372	0.0225	0.01995
LBC	2.87	1.5406	0.0609	0.0477	0.0403	0.0245	0.0216
LF	3.23	1.5785	0.1143	0.0850	0.0693	0.0394	0.0344
BLF	3.48	1.6047	0.1112	0.0832	0.0685	0.0393	0.0344
DBC	3.56	1.6122	0.0662	0.0517	0.0435	0.0261	0.0231
DF	3.63	1.6203	0.1473	0.1076	0.0872	0.0485	0.0423
EDF	3.9	1.6533	0.1725	0.1248	0.1007	0.0556	0.0483

The composition of the glasses in weight percent is:

Glass type	SiO_2	B_2O_3	K_2O	CaO	Al_2O_3	As_2O_3	Na_2O	BaO	ZnO	PbO
BSC	69.6	6.7	20.5	2.9	0.3	0.1	–	–	–	–
HC	72.0	–	10.1	11.4	0.3	0.2	6.1	–	–	–
LBC	57.1	1.8	13.7	0.3	0.2	0.1	–	26.9	–	–
LF	52.5	–	9.5	0.3	0.2	0.1	–	–	–	37.6
BLF	45.2	–	7.8	–	–	0.4	–	16.0	8.3	22.2
DBC	36.2	7.7	0.2	0.2	3.5	0.7	–	44.6	6.7	–
DF	46.3	–	1.1	0.3	0.2	0.1	5.0	–	–	47.0
EDF	40.6	–	7.5	0.2	0.2	0.2	0.1	–	–	51.5

References

1. Weber, M. J., *CRC Handbook of Laser Science and Technology*, Vol. IV, Part 2, CRC Press, Boca Raton, FL, 1988, pp. 299–310.
2. Gray, D. E., Ed., *American Institute of Physics Handbook*, Third edition, McGraw Hill, New York, 1972, p. 6–230.

FARADAY ROTATION

Ferro-, Ferri-, and Antiferromagnetic Solids

Material	T_c K	$4\pi M_s$ gauss	F deg/cm	α cm^{-1}	$2F/\alpha$ deg	T K	λ nm
Fe	1043	21,800	4.4×10^5	6.5×10^5	1.4	300	500
			6.5×10^5	5.0×10^5	2.6	300	1000
			7×10^5	4.2×10^5	3.3	300	1500
			7×10^5	3.5×10^5	4.0	300	2000
Co	1390	18,200	2.9×10^5	–	–	300	500
			5.5×10^5	6.1×10^5	1.8	300	1000
			5.5×10^5	4.5×10^5	2.4	300	1500
			5.5×10^5	3.6×10^5	2.7	300	2000
Ni	633	6,400	0.8×10^5	–	–	300	500
			2.6×10^5	5.8×10^5	0.9	300	1000
			1.5×10^5	4.8×10^5	0.6	300	1500
			1×10^5	4.1×10^5	0.25	300	2000
Permalloy (Ni/Fe = 82/18)	803	10,700	1.2×10^5	6×10^5	0.4	300	500
Ni/Fe = 100/0		6,000	1.2×10^5	7.05×10^5	0.34	300	632.8
Ni/Fe = 80/20		10,800	2.2×10^5	7.10×10^5	0.62	300	632.8
Ni/Fe = 60/40		14,900	2.9×10^5	7.54×10^5	0.77	300	632.8
Ni/Fe = 40/60		14,400	2.2×10^5	8.17×10^5	0.54	300	632.8
Ni/Fe = 20/80		19,400	3.3×10^5	8.10×10^5	0.81	300	632.8
Ni/Fe = 0/100	639	21,600	3.5×10^5	8.13×10^5	0.86	300	632.8
MnBi		7,700	4.2×10^5	6.1×10^5	1.4	300	450
			7.5×10^5	4.2×10^5	3.6	300	900
MnAs	313	–	0.44×10^5	5.0×10^5	0.174	300	500
			0.62×10^5	4.4×10^5	0.28	300	900
CrTe	334	1015	0.5×10^5	2.0×10^5	0.5	300	550
			0.4×10^5	1.2×10^5	0.7	300	900
FeRh	333	–	0.9×10^5	3.3×10^5	0.56	348	700
$Y_3Fe_5O_{12}$ (YIG)	560	2500	2400	1500	3.2	300	555
			1250	1400	1.8	300	625
			750	450	3.3	300	770
			175	<0.06	$>3 \times 10^3$	300	5000 to 1500
$Gd_3Fe_5O_{12}$ (GdIG)	$T_n = 564$ $T = 286$	7300	−2000	6000	0.6	300	500
			−1050	900	2.3	300	600
			−300	100	6.0	300	800
			−80	70	2.3	300	1000
$NiFe_2O_4$	858	3350	2.0×10^4	5.9×10^4	0.7	300	286
			-1.0×10^4	10×10^4	0.2	300	500
			−120	38	6	300	1500
			+75	15	10	300	3000
			+110	32	7	300	5000
$CoFe_2O_4$	793	4930	2.75×10^4	12×10^4	0.5	300	286
			3.6×10^4	17×10^4	0.4	300	400
			-2.5×10^4	6×10^4	0.8	300	660
$MgFe_2O_4$	593–713[e]	1450[e]	−60	100	1	300	2500
			0	12	0	300	4000
			+35	6	11	300	6000
$Li_{0.5}Fe_{2.5}O_4$	863–953[e]	3240[e] to 3900	−440	150	6	300	1500
			+10	85	0.2	300	3000
			+110	44	5	300	5000
			+135	80	3	300	7000
$BaFe_{12}O_{19}$	723	–	−50	−38	3	300	2000
			+75	20	7.5	300	3000
			+150	20	15	300	5000
			+165	22	15	300	7000
$Ba_2Zn_2Fe_{12}O_{19}$	–	–	90	120	1.5	300	5000

Material	T_c K	$4\pi M_s$ gauss	F deg/cm	α cm^{-1}	$2F/\alpha$ deg	T K	λ nm
			75	65	2.0	300	7000
RbNiF$_3$	220	1250	360	35	20	77	450[a]
			70	10	14	77	600[a]
			310	70	9	77	800[a]
			75	25	6	77	1000[a]
RbNi$_{0.75}$Co$_{0.25}$F$_3$	109	–	180	9	40	77	600[b]
RbFeF$_3$	102	–	3400	7	900	82	300[c]
			1600	3	1100	82	400[c]
			620	1.5	830	82	600[c]
			300	2.5	240	82	800[c]
FeF$_3$	365	40 at 300 K	670	14	95	300	349[d]
			180	4.4	82	300	522.5[d]
CrCl$_3$	16.8	3880	2000	200	20	1.5	410
			–500	300		1.5	450
			–1000	70	30	1.5	590
CrBr$_3$	32.5	3390	3×10^5	3×10^3	200	1.5	478
			1.6×10^5	1.4×10^4	23	1.5	500
CrI$_3$	68	2690	1.1×10^5	6.3×10^3	35	1.5	970
			0.8×10^5	3×10^3	53	1.5	1000
FeBO$_3$	348	115 at 300 K	3200	140	45	300	500
			450	38	24	300	700
EuO	69	23700	-1.0×10^5	0.5×10^4	40	5	1100
			5×10^5	9.7×10^4	10	5	700
			0.5×10^5	7.8×10^4	1.3	5	500
			3×10^4	>0.5	~105	20	2500
			660	>1.0	1300	20	10600
EuS	16.3	–	-1.6×10^5	0	–	6	825
			-9.6×10^5	3.3×10^4	58	6	690
			$+5.5 \times 10^5$	1.2×10^5	9.2	6	563
EuSe	7.0	13,200	1.45×10^5	80	3600	4.2	750
			0.95×10^5	60	3170	4.2	800

[a] Measured along the C-axis (magnetic hard axis).

[b] Measured along the C-axis (magnetic easy axis).

[c] Measured along the C-axis ([100]-direction at room temperature).

[d] Strong natural birefringence interferes with the Faraday effect.

[e] Depends on heat treatment.

Reference

1. Weber, M. J., Ed., *CRC Handbook of Laser Science and Technology*, Vol. IV, Part 2, CRC Press, Boca Raton, FL, 1988, pp. 288–296.

NONLINEAR OPTICAL CONSTANTS

H. P. R. Frederikse

The relation between the polarization density P of a dielectric medium and the electric field E is linear when E is small, but becomes nonlinear as E acquires values comparable with interatomic electric fields (10^5 to 10^8 V/cm). Under these conditions the relation between P and E can be expanded in a Taylor's series

$$P = \varepsilon_0 \chi^{(1)} E + 2\chi^{(2)} E^2 + 4\chi^{(3)} E^3 + \cdots \cdots \quad (1)$$

where ε_0 is the permittivity of free space, while $\chi^{(1)}$ is the linear and $\chi^{(2)}$, $\chi^{(3)}$ etc. the nonlinear optical susceptibilities.

If we consider two optical fields, the first $E_j^{\omega_1}$ (along the j-direction at frequency ω_1) and the second $E_k^{\omega_2}$ (along the k-direction at frequency ω_2) one can write the second term of the Taylor's series as follows

$$P_i(\omega_1\omega_2) = 2\chi_{ijk}^{\omega_3 = \omega_1 \pm \omega_2} E_j^{\omega_1} E_k^{\omega_2}$$

When $\omega_1 \neq \omega_2$ the (parametric) mixing of the two fields gives rise to two new polarizations at the frequencies $\omega_3 = \omega_1 + \omega_2$ and $\omega_3' = \omega_1 - \omega_2$. When the two frequencies are equal, $\omega_1 = \omega_2 = \omega$, the result is Second Harmonic Generation (SHG): $\chi_{ijk}(2\omega, \omega, \omega)$, while equal and opposite frequencies, $\omega_1 = \omega$ and $\omega_2 = -\omega$ lead to Optical Rectification (OR): $\chi_{ijk}(0, \omega, -\omega)$. In the SHG case the following convention is adopted: the second order nonlinear coefficient d is equal to one half of the second order nonlinear susceptibility

$$d_{ijk} = 1/2 \chi^{(2)}$$

Because of the symmetry of the indices j and k one can replace these two by a single index (subscript) m. Consequently the notation for the SHG nonlinear coefficient in reduced form is d_{im} where m takes the values 1 to 6. Only noncentrosymmetric crystals can possess a nonvanishing d_{ijk} tensor (third rank). The unit of the SHG coefficients is m/V (in the MKSQ/SI system).

In centrosymmetric media the dominant nonlinearity is of the third order. This effect is represented by the third term in the Taylor's series (Equation 1); it is the result of the interaction of a number of optical fields (one to three) producing a new frequency $\omega_4 = \omega_1 + \omega_2 + \omega_3$. The third order polarization is given by

$$P_j(\omega_1\omega_2\omega_3) = g_4 \chi_{jklm} E_k^{\omega_1} E_1^{\omega_2} E_m^{\omega_3}$$

Third Harmonic Generation (THG) is achieved when $\omega_1 = \omega_2 = \omega_3 = \omega$. In this case the constant $g_4 = 1/4$. The third order nonlinear coefficient C is related to the third order susceptibility as follows:

$$C_{jklm} = 1/4 \chi_{jklm}$$

This coefficient is a fourth rank tensor. In the THG case the matrices must be invariant under permutation of the indices k, l, and m; as a result the notation for the third order nonlinear coefficient can be simplified to C_{jn}. The unit of C_{jn} is $m^2 \cdot V^{-2}$ (in the MKSQ/SI system).

Applications of second order nonlinear optical materials include the generation of higher (up to sixth) optical harmonics, the mixing of monochromatic waves to generate sum or difference frequencies (frequency conversion), the use of two monochromatic waves to amplify a third wave (parametric amplification) and the addition of feedback to such an amplifier to create an oscillation (parametric oscillation).

Third order nonlinear optical materials are used for THG, self-focusing, four wave mixing, optical amplification, and optical conjugation. Many of these effects — as well as the variation and modulation of optical propagation caused by mechanical, electric, and magnetic fields (see the preceding table on "Elasto-Optic, Electro-Optic, and Magneto-Optic Constants") are used in the areas of optical communication, optical computing, and optical imaging.

References

1. *Handbook of Laser Science and Technology*, Vol. 111, Part 1; Weber, M. J. Ed., CRC Press, Boca Raton, FL, 1986.
2. Dmitriev, V.G., Gurzadyan, G.G., and Nikogosyan, D., *Handbook of Nonlinear Optical Crystals*, Springer-Verlag, Berlin, 1991.
3. Shen, Y.R., *The Principles of Nonlinear Optics*, John Wiley, New York, 1984.
4. Yariv, A., *Quantum Electronics*, 3rd edition, John Wiley, New York, 1988.
5. Bloembergen, N., *Nonlinear Optics*, W.A. Benjamin, New York, 1965.
6. Zernike F. and Midwinter, J.E., *Applied Nonlinear Optics*, John Wiley, New York, 1973.
7. Hopf, F.A. and Stegeman, G.I., *Applied Classical Electrodynamics*, Volume 2: Nonlinear Optics, John Wiley, New York, 1986.
8. *Nonlinear Optical Properties of Organic Molecules and Crystals*, Chemla, D. S., and Zyss, J., Eds., Academic Press, Orlando, FL, 1987.
9. *Optical Phase Conjugation*, Fisher, R. A., Ed., Academic Press, New York, 1983.
10. Zyss, J., *Molecular Nonlinear Optics: Materials, Devices and Physics*, Academic Press, Boston, 1994.
11. Nonlinear Optics, 5 articles in *Physics Today, (Am. Inst. of Phys.)*, Vol. 47, No. 5, May, 1994.

Selected SHG Coefficients of NLO Crystals*

Material	Symmetry class	$d_{im} \times 10^{12}$ m/V	λ μm
GaAs	$\bar{4}3m$	$d_{14} = 134.1 \pm 42$	10.6
GaP	$\bar{4}3m$	$d_{14} = 71.8 \pm 12.3$	1.058
InAs	$\bar{4}3m$	$d_{14} = 364 \pm 47$	1.058
		$d_{14} = 210$	10.6
ZnSe	$\bar{4}3m$	$d_{14} = 78.4 \pm 29.3$	10.6
		$d_{36} = 26.6 \pm 1.7$	1.058
β-ZnS	$\bar{4}3m$	$d_{14} = 30.6 \pm 8.4$	10.6
		$d_{36} = 20.7 \pm 1.3$	1.058
ZnTe	$\bar{4}3m$	$d_{14} = 92.2 \pm 33.5$	10.6
		$d_{14} = 83.2 \pm 8.4$	1.058
		$d_{36} = 89.6 \pm 5.7$	1.058
CdTe	$\bar{4}3m$	$d_{14} = 167.6 \pm 63$	10.6
Bi_4GeO_{12}	$\bar{4}3m$	$d_{14} = 1.28$	1.064
$N_4(CH_2)_6$ (hexamine)	$\bar{4}3m$	$d_{14} = 4.1$	1.06
$LiIO_3$	6	$d_{33} = -7.02$	1.06
		$d_{31} = -5.53 \pm 0.3$	1.064
ZnO	6 mm	$d_{33} = -5.86 \pm 0.16$	1.058
		$d_{31} = 1.76 \pm 0.16$	1.058
		$d_{15} = 1.93 \pm 0.16$	1.058
α-ZnS	6 mm	$d_{33} = 11.37 \pm 0.07$	1.058
		$d_{33} = 37.3 \pm 12.6$	10.6
		$d_{31} = -18.9 \pm 6.3$	10.6
		$d_{15} = 21.37 \pm 8.4$	10.6
CdS	6 mm	$d_{33} = 25.8 \pm 1.6$	1.058
		$d_{31} = -13.1 \pm 0.8$	1.058
		$d_{15} = 14.4 \pm 0.8$	1.058
CdSe	6 mm	$d_{33} = 54.5 \pm 12.6$	10.6
		$d_{31} = -26.8 \pm 2.7$	10.6
$BaTiO_3$	4 mm	$d_{33} = 6.8 \pm 1.0$	1.064
		$d_{31} = 15.7 \pm 1.8$	1.064
		$d_{15} = 17.0 \pm 1.8$	1.064
$PbTiO_3$	4 mm	$d_{33} = 7.5 \pm 1.2$	1.064
		$d_{31} = 37.6 \pm 5.6$	1.064
		$d_{15} = 33.3 \pm 5$	1.064
$K_3Li_2Nb_5O_{15}$	4 mm	$d_{33} = 11.2 \pm 1.6$	1.064
		$d_{31} = 6.18 \pm 1.28$	1.064
		$d_{15} = 5.45 \pm 0.54$	1.064
$K_{0.8}Na_{0.2}Ba_2Nb_5O_{15}$	4 mm	$d_{31} = 13.6 \pm 1.6$	1.064
$SrBaNb_5O_{15}$	4 mm	$d_{33} = 11.3 \pm 3.3$	1.064
		$d_{31} = 4.31 \pm 1.32$	1.064
		$d_{15} = 5.98 \pm 2$	1.064
$NH_4H_2PO_4$ (ADP)	$\bar{4}2m$	$d_{36} = 0.53$	1.064
		$d_{36} = 0.85$	0.694
KH_2PO_4 (KDP)	$\bar{4}2m$	$d_{36} = 0.44$	1.064
		$d_{36} = 0.47 \pm 0.07$	0.694
KD_2PO_4 (KD*P)	$\bar{4}2m$	$d_{36} = 0.38 \pm 0.016$	1.058
		$d_{36} = 0.34 \pm 0.06$	0.694
		$d_{14} = 0.37$	1.058
KH_2AsO_4 (KDA)	$\bar{4}2m$	$d_{36} = 0.43 \pm 0.025$	1.06
		$d_{36} = 0.39 \pm 0.4$	0.694
$CdGeAs_2$	$\bar{4}2m$	$d_{36} = 351 \pm 105$	10.6
$AgGaS_2$	$\bar{4}2m$	$d_{36} = 18 \pm 2.7$	10.6

Material	Symmetry class	$d_{im} \times 10^{12}$ m/V	λ μm
$AgGaSe_2$	$\bar{4}2m$	$d_{36} = 37.4 \pm 6.0$	10.6
$(NH_2)_2CO$ (urea)	$\bar{4}2m$	$d_{36} = 1.3$	1.06
$AlPO_4$	32	$d_{11} = 0.35 \pm 0.03$	1.058
Se	32	$d_{11} = 97 \pm 25$	10.6
Te	32	$d_{11} = 650 \pm 30$	10.6
SiO_2 (quartz)	32	$d_{11} = 0.335$	1.064
HgS	32	$d_{11} = 50.3 \pm 17$	10.6
$(C_6H_5CO)_2$ [benzil]	32	$d_{11} = 3.6 \pm 0.5$	1.064
β-BaB_2O_4 [BBO]	3 m	$d_{22} = 2.22 \pm 0.09$	1.06
		$d_{31} = 0.16 \pm 0.08$	1.06
$LiNbO_3$	3 m	$d_{33} = 34.4$	1.06
		$d_{31} = -5.95$	1.06
		$d_{22} = 2.76$	1.06
$LiTaO_3$	3 m	$d_{33} = -16.4 \pm 2$	1.058
		$d_{31} = -1.07 \pm 0.2$	1.058
		$d_{22} = +1.76 \pm 0.2$	1.058
Ag_3AsS_3 [proustite]	3 m	$d_{31} = 11.3 \pm 2.5$	10.6
		$d_{22} = 18.0 \pm 2.5$	10.6
Ag_3SbS_3 [pyrargerite]	3m	$d_{31} = 12.6 \pm 4$	10.6
		$d_{22} = 13.4 \pm 4$	10.6
α-HIO_3	222	$d_{36} = 5.15 \pm 0.16$	1.064
$NO_2 \cdot CH_3NOC_5H_4 \cdot$ (POM)	222	$d_{36} = 6.4 \pm 1.0$	1.064
$Ba_2NaNb_5O_{15}$ [Banana]	mm 2	$d_{33} = -17.6 \pm 1.28$	1.064
		$d_{31} = -12.8 \pm 1.28$	1.064
$C_6H_4(NO_2)_2$ [MDB]	mm 2	$d_{33} = 0.74$	1.064
		$d_{32} = 2.7$	1.064
		$d_{31} = 1.78$	1.064
$Gd_2(MoO_4)_3$	mm 2	$d_{33} = -0.044 \pm 0.008$	1.064
		$d_{32} = +2.42 \pm 0.36$	1.064
		$d_{31} = -2.49 \pm 0.37$	1.064
$KNbO_3$	mm 2	$d_{33} = -19.58 \pm 1.03$	1.064
		$d_{32} = +11.34 \pm 1.03$	1.064
		$d_{31} = -12.88 \pm 1.03$	1.064
$KTiOPO_4$ [KTP]	mm 2	$d_{33} = 13.7$	1.06
		$d_{32} = \pm 5.0$	1.06
		$d_{31} = \pm 6.5$	1.06
$NO_2C_6H_4 \cdot NH_2$ [mNA]	mm 2	$d_{33} = 13.12 \pm 1.28$	1.064
		$d_{32} = 1.02 \pm 0.22$	1.064
		$d_{31} = 12.48 \pm 1.28$	1.064
$C_{10}H_{12}N_3O_6$ [MAP]	2	$d_{23} = 10.67 \pm 1.3$	1.064
		$d_{22} = 11.7 \pm 1.3$	1.064
		$d_{21} = 2.35 \pm 0.5$	1.064
		$d_{25} = -0.35 \pm 0.3$	1.064
$(NH_2CH_2COOH)_3H_2SO_4$ [TGS]	2	$d_{23} = 0.32$	0.694

* These data are taken from References 1 and 2.

Selected THG Coefficients of Some NLO Materials*

Material	NLO process	$C_{jn} \times 10^{20}$ m^2/V^{-2}	λ μm
NH$_4$H$_2$PO$_4$ [ADP]	$(-3\omega,\omega,\omega,\omega)$	$C_{11} = 0.0104$	1.06
		$C_{18} = 0.0098$	1.06
C$_6$H$_6$ [benzene]	$(-3\omega,\omega,\omega,\omega)$	$C_{11} = 0.0184 \pm 0.0042$	1.89
CdGeAs$_2$	$(-3\omega,\omega,\omega,\omega)$	$C_{11} = 182 \pm 84$	10.6
p-type: 5×10^{16} cm^{-3}		$C_{16} = 175$	10.6
		$C_{18} = -35$	10.6
C$_{40}$H$_{56}$ [β-carotene]	$(-3\omega,\omega,\omega,\omega)$	$C_{11}\ 0.263 \pm 0.08$	1.89
GaAs high-resistivity	$(-3\omega,\omega,\omega,-\omega)$	$C_{11} = 62 \pm 31$	1.06
Ge	$(-3\omega,\omega,\omega,-\omega)$	$C_{11} = 23.5 \pm 12$	1.06
LiIO$_3$	$(-3\omega,\omega,\omega,-\omega)$	$C_{12} = 0.2285$	1.06
		$C_{35} = 6.66 \pm 1$	1.06
KBr	$(-3\omega,\omega,\omega,-\omega)$	$C_{11} = 0.0392$	1.06
		$C_{18}/C_{11} = 0.3667$	1.06
KCl	$(-3\omega,\omega,\omega,-\omega)$	$C_{11} = 0.0168$	1.06
		$C_{18}/C_{11} = 0.28$	1.06
KH$_2$PO$_4$ [KDP]	$(-3\omega,\omega,\omega,-\omega)$	$C_{11}-3C_{18} = 0.04$	1.06
Si p-type: 10^{14} cm^{-3}	$(-3\omega,\omega,\omega,-\omega)$	$C_{11} = 82.8 \pm 25$	1.06
NaCl	$(-3,\omega,\omega,\omega,-\omega)$	$C_{11} = 0.0168$	1.06
		$C_{18}/C_{11} = 0.4133$	1.06
NaF	$(-3\omega,\omega,\omega,-\omega)$	$C_{11} = 0.0035$	1.06

* These data are taken from Reference 1.

PHASE DIAGRAMS

H. P. R. Frederikse

A phase is a structurally homogeneous portion of matter. Regardless of the number of chemical constituents of a gas, there is only one vapor phase. This is true also for the liquid form of a pure substance, although a mixture of several liquid substances may exist as one or several phases, depending on the interactions among the substances. On the other hand a pure solid may exist in several phases at different temperatures and pressures because of differences in crystal structure (Reference 1). At the phase transition temperature, T_{tr}, the chemical composition of the solid remains the same, but a change in the physical properties often will take place. Such changes are found in ferroelectric crystals (example BaTiO$_3$) that develop a spontaneous polarization below T_{tr}, in superconductors (example Pb) that lose all electrical resistance below the transition point, and in many other classes of solids.

In quite a few cases it is difficult to bring about the phase transition, and the high- (or low-) temperature phase persists in its metastable form. Many liquids remain in the liquid state for shorter or longer periods of time when cooled below the melting point (supercooling). However, often the slightest disturbance will cause solidification. Persistence of the high temperature phase in solid–solid transitions is usually of much longer duration. An example of this behavior is found in white tin; although gray tin is the thermodynamically stable form below T_{tr} (286.4 K), the metal remains in its undercooled, white tin state all the way to $T = 0$ K, and crystals of gray tin are very difficult to produce.

A *phase diagram* is a map that indicates the areas of stability of the various phases as a function of external conditions (temperature and pressure). Pure materials, such as mercury, helium, water, and methyl alcohol are considered one-component systems and they have *unary* phase diagrams. The equilibrium phases in two-component systems are presented in *binary* phase diagrams. Because many important materials consist of three, four, and more components, many attempts have been made to deduce their multicomponent phase diagrams. However, the vast majority of systems with three or more components are very complex, and no overall maps of the phase relationships have been worked out.

It has been shown during the last 20 to 25 years that very useful partial phase diagrams of complex systems can be obtained by means of thermodynamic modeling (References 2, 3). Especially for complicated, multicomponent alloy systems the CALPHAD method has proved to be a successful approach for producing valuable portions of very intricate phase diagrams (Reference 4). With this method thermodynamic descriptions of the free energy functions of various phases are obtained that are consistent with existing (binary) phase diagram information and other thermodynamic data. Extrapolation methods are then used to extend the thermodynamic functions into a ternary system. Comparison of the results of this procedure with available experimental data is then used to fine-tune the phase diagram and add ternary interaction functions if necessary. In principle this approximation strategy can be extended to four, five, and more component systems.

The nearly two dozen phase diagrams shown below present the reader with examples of some important types of single and multicomponent systems, especially for ceramics and metal alloys. This makes it possible to draw attention to certain features like the kinetic aspects of phase transitions (see Figure 22, which presents a time–temperature–transformation, or TTT, diagram for the precipitation of α-phase particles from the β-phase in a Ti-Mo alloy; Reference 1, pp. 358–360). The general references listed below and the references to individual figures contain phase diagrams for many additional systems.

General References

1. Ralls, K. M., Courtney, T. H., and Wulff, J., *Introduction to Materials Science and Engineering*, Chapters 16 and 17, John Wiley & Sons, New York, 1976.
2. Kaufman, L., and Bernstein, H., *Computer Calculation of Phase Diagrams*, Academic Press, New York, 1970.
3. Kattner, U. R., Boettinger, W. J. B., and Coriell, S. R., *Z. Metallkd.*, 87, 9, 1996.
4. Dinsdale, A. T., Ed., *CALPHAD*, Vol. 1–20, Pergamon Press, Oxford, 1977–1996 and continuing.
5. Baker, H., Ed., *ASM Handbook, Volume 3: Alloy Phase Diagrams*, ASM International, Materials Park, OH, 1992.
6. Massalski, T. B., Ed., *Binary Alloy Phase Diagrams, Second Edition*, ASM International, Materials Park, OH, 1990.
7. Roth. R. S., Ed., *Phase Diagrams for Ceramists*, Vol. I (1964) to Volume XI (1995), American Ceramic Society, Waterville, OH.

References to Individual Phase Diagrams

Figure 1.   Carbon: Reference 7, Vol. X (1994), Figure 8930. Reprinted with permission.

Figure 2.   Si-Ge : Ref. 5, p. 2.231. Reprinted with permission.

Figure 3.   H$_2$O (ice): See figure.

Figure 4.   SiO$_2$: Reference 7, Vol. XI (1995), Figure 9174. Reprinted with permission.

Figure 5.   Fe-O: Darken, L.S., and Gurry, R.W., *J. Am. Chem. Soc.*, 68, 798, 1946. Reprinted with permission.

Figure 6.   Ti-O: Reference 5, p. 2.324. Reprinted with permission.

Figure 7.   BaO-TiO$_2$: Reference 7, Vol. III (1975), Figure 4302. Reprinted with permission.

Figure 8.   MgO-Al$_2$O$_3$: Reference 7, Vol. XI (1995), Figure 9239. Reprinted with permission.

Figure 9.   Y$_2$O$_3$-ZrO$_2$: Reference 7, Vol. XI (1995), Figure 9348. Reprinted with permission.

Figure 10. Si-N-Al-O (Sialon): Reference 7, Vol. X (1994), Figure 8759. Reprinted with permission.

Figure 11. PbO-ZrO$_2$-TiO$_2$ (PZT): Reference 7, Vol. III (1975), Figure 4587. Reprinted with permission.

Figure 12. Al-Si-Ca-O: Reference 7 (1964), Vol. I, Figure 630. Reprinted with permission.

Figure 13. Y-Ba-Cu-O: Whitler, J.D., and Roth, R.S., *Phase Diagrams for High T$_c$ Superconductors*, Figure S-082, American Ceramic Society, Waterville, OH, 1990. Reprinted with permission.

Figure 14. Al-Cu: Reference 5, p. 2.44. Reprinted with permission.

Figure 15. Fe-C: Ralls, K.M., Courtney, T.H., and Wulff, J., *Introduction to Materials Science and Engineering*, Figure 16.13, John Wiley & Sons, New York, 1976. Reprinted with permission.

Figure 16. Fe-Cr: Reference 5, p. 2.152. Reprinted with permission.

Figure 17. Cu-Sn: Reference 5, p. 2.178. Reprinted with permission.

Figure 18. Cu-Ni: Reference 5, p. 2.173. Reprinted with permission.

Figure 19. Pb-Sn (solder): Reference 5, p. 2.335. Reprinted with permission.

Figure 20. Cu-Zn (brass): Subramanian, P.R., Chakrabarti, D.J., and Laughlin, D.E., Eds., *Phase Diagrams of Binary Copper Alloys*, p. 487, ASM International, Materials Park, OH, 1994. Reprinted with permission.

Figure 21. Co-Sm: Reference 5, p. 2.148. Reprinted with permission.

Figure 22. Ti-Mo: Reference 5, p. 2.296; Reference 1, p. 359. Reprinted with permission.

Figure 23. Fe-Cr-Ni: Reference 5, Figure 48. Reprinted with permission.

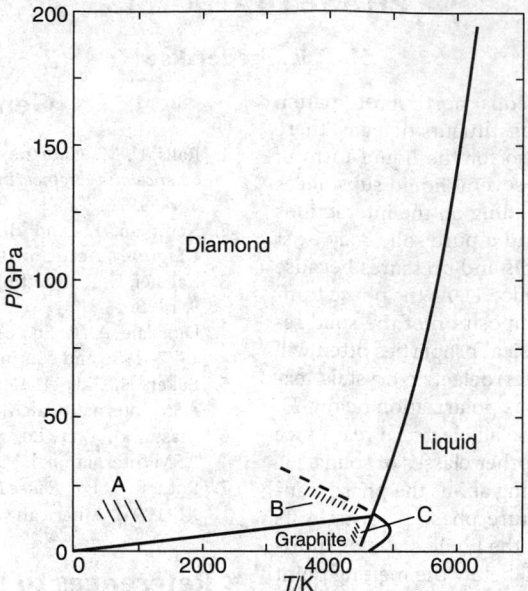

FIGURE 1. Phase diagram of carbon. (A) Martensitic transition: hex graphite → hex diamond. (B) Fast graphite-to-diamond transition. (C) Fast diamond-to-graphite transition.

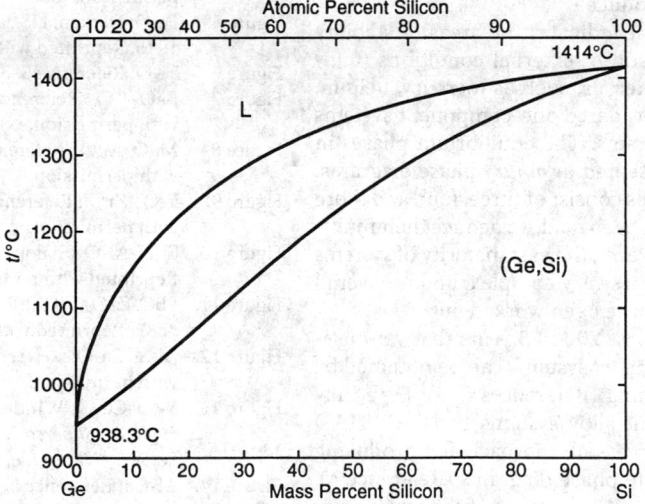

FIGURE 2. Si-Ge system.

Phase	Composition, mass % Si	Pearson symbol	Space group
(Ge,Si)	0 to 100	$cF8$	$Fd\bar{3}m$
High-pressure phases			
GeII	–	$tI4$	$I4_1/amd$
SiII	–	$tI4$	$I4_1/amd$

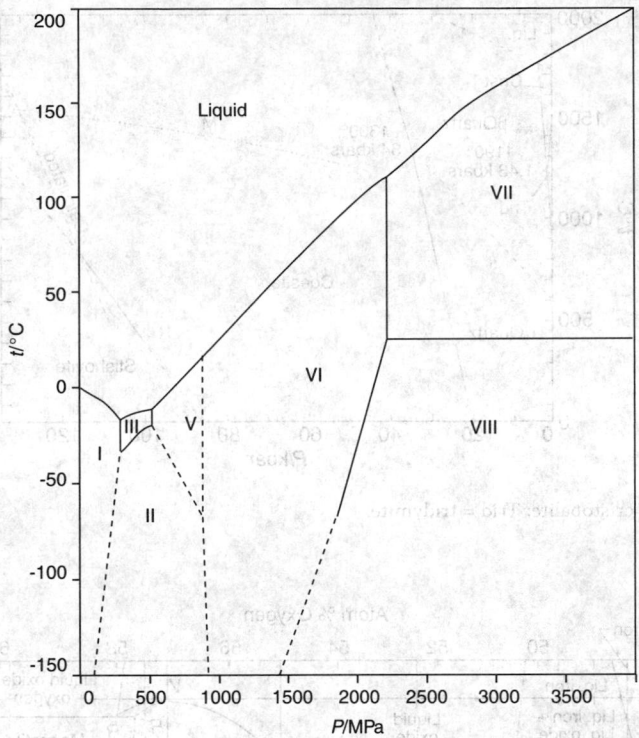

FIGURE 3. Diagram of the principal phases of ice. Solid lines are measured boundaries between stable phases; dotted lines are extrapolated. Ice IV is a metastable phase that exists in the region of ice V. Ice IX exists in the region below −100 °C and pressures in the range 200–400 MPa. Ice X exists at pressures above 44 GPa. See Table 1 for the coordinates of the triple points, where liquid water is in equilibrium with two adjacent solid phases.

TABLE 1. Crystal Structure, Density, and Transition Temperatures for the Phases of Ice

Phase	Crystal system	Cell parameters	Z	n	ρ/g cm^{-3}	Triple points
Ih	Hexagonal	$a = 4.513$; $c = 7352$	4	4	0.93	I-III: −21.99 °C, 209.9 MPa
Ic	Cubic	$a = 6.35$	8	4	0.94	
II	Rhombohedral	$a = 7.78$; $\alpha = 113.1°$	12	4	1.18	
III	Tetragonal	$a = 6.73$; $c = 6.83$	12	4	1.15	III-V: −16.99 °C, 350.1 MPa
IV	Rhombohedral	$a = 7.60$; $\alpha = 70.1°$	16	4	1.27	
V	Monoclinic	$a = 9.22$; $b = 7.54$,	28	4	1.24	V-VI: 0.16 °C, 632.4 MPa
		$c = 10.35$; $\beta = 109.2°$				
VI	Tetragonal	$a = 6.27$; $c = 5.79$	10	4	1.31	VI-VII: 82 °C, 2216 MPa
VII	Cubic	$a = 3.41$	2	8	1.56	
VIII	Tetragonal	$a = 4.80$; $c = 6.99$	8	8	1.56	
IX	Tetragonal	$a = 6.73$; $c = 6.83$	12	4	1.16	
X	Cubic	$a = 2.83$	2	8	2.51	

References

1. Wagner, W., Saul, A., and Pruss, A., *J. Phys. Chem. Ref. Data*, 23, 515, 1994.
2. Lerner, R.G. and Trigg, G.L., Eds., *Encyclopedia of Physics*, VCH Publishers, New York, 1990.
3. Donnay, J.D.H. and Ondik, H.M, *Crystal Data Determinative Tables, Third Edition, Volume 2, Inorganic Compounds*, Joint Committee on Powder Diffraction Standards, Swarthmore, PA, 1973.
4. Hobbs, P.V., *Ice Physics*, Oxford University Press, Oxford, 1974.
5. Glasser, L., *J. Chem. Edu.*, 81, 414, 2004.

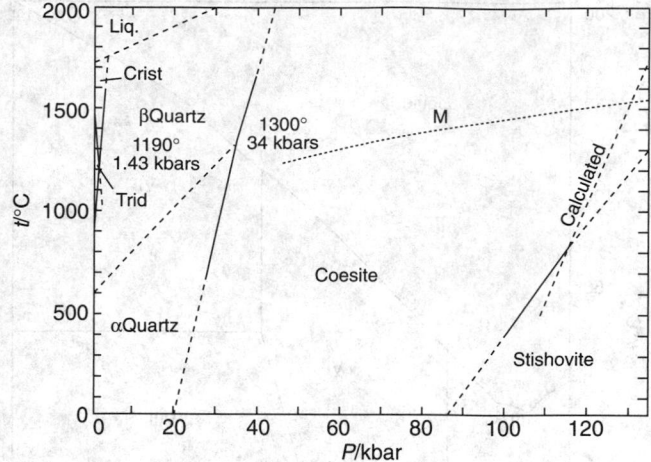

FIGURE 4.   SiO₂ system. Crist = cristobalite; Trid = tridymite.

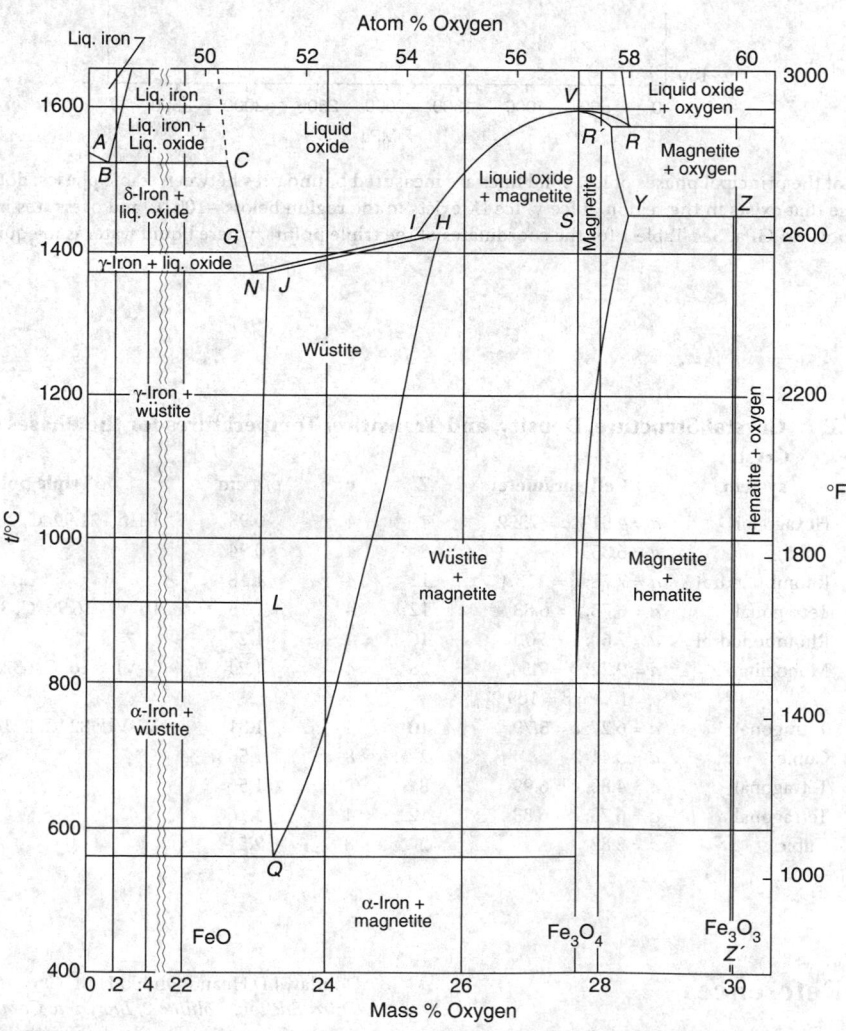

FIGURE 5.   Fe-O system.

Point	$t/°C$	% O	p_{CO_2}/p_{CO}	Point	$t/°C$	% O	p_{CO_2}/p_{CO}	p_{O_2}/atm
A	1539			Q	560	23.26	1.05	
B	1528	0.16	0.209	R	1583	28.30		1
C	1528	22.60	0.209	R′	1583	28.07		1
G	1400ª	22.84	0.263	S	1424	27.64	16.2	
H	1424	25.60	16.2	V	1597	27.64		0.0575
I	1424	25.31	16.2	Y	1457	28.36		1
J	1371	23.16	0.282	Z	1457	30.04		1
L	911ª	23.10	0.447	Z′		30.6		
N	1371	22.91	0.282					

ª Values for pure iron.

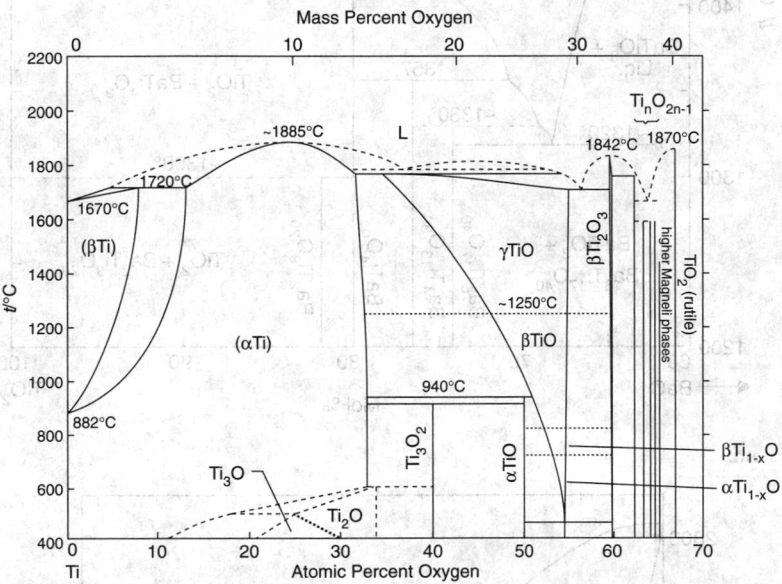

FIGURE 6. Ti–O system.

Phase	Composition, mass % O	Pearson symbol	Space group
(βTi)	0 to 3	cI2	$Im\bar{3}m$
(αTi)	0 to 13.5	hP2	$P6_3/mmc$
Ti_3O	~8 to ~13	hP~16	$P\bar{3}c$
Ti_2O	~10 to 14.4	hP3	$P\bar{3}m1$
γTiO	15.2 to 29.4	cF8	$Fm\bar{3}m$
Ti_3O_2	~18	hP~5	$P6/mmm$
βTiO	~24 to ~29.4	c**	–
αTiO	~25.0	mC16	$A2/m$ or $B*/*$
$βTi_{1-x}O$	~29.5	oI12	$I222$
$αTi_{1-x}O$	~29.5	tI18	$I4/m$
$βTi_2O_3$	33.2 to 33.6	hR30	$R\bar{3}c$
$αTi_2O_3$	33.2 to 33.6	hR30	$R\bar{3}c$
$βTi_3O_5$	35.8	m**	–
$αTi_3O_5$	35.8	mC32	$C2/m$
$α′Ti_3O_5$	35.8	mC32	Cc
$γTi_4O_7$	36.9	aP44	$P\bar{1}$
$βTi_4O_7$	36.9	aP44	$P\bar{1}$
$αTi4O_7$	36.9	aP44	$P\bar{1}$
$γTi_5O_9$	37.6	aP28	$P\bar{1}$
$βTi_6O_{11}$	38.0	aC68	$A\bar{1}$
Ti_7O_{13}	38.3	aP40	$P\bar{1}$
Ti_8O_{15}	38.5	aC92	$A\bar{1}$
Ti_9O_{17}	38.7	aP52	$P\bar{1}$
Rutile TiO_2	40.1	tP6	$P4_2/mnm$
Metastable phases			
Anatase	–	tI12	$I4_1/amd$
Brookite	–	oP24	$Pbca$
High-pressure phases			
TiO_2-II	–	oP12	$Pbcn$
TiO_2-III	–	hP~48	–

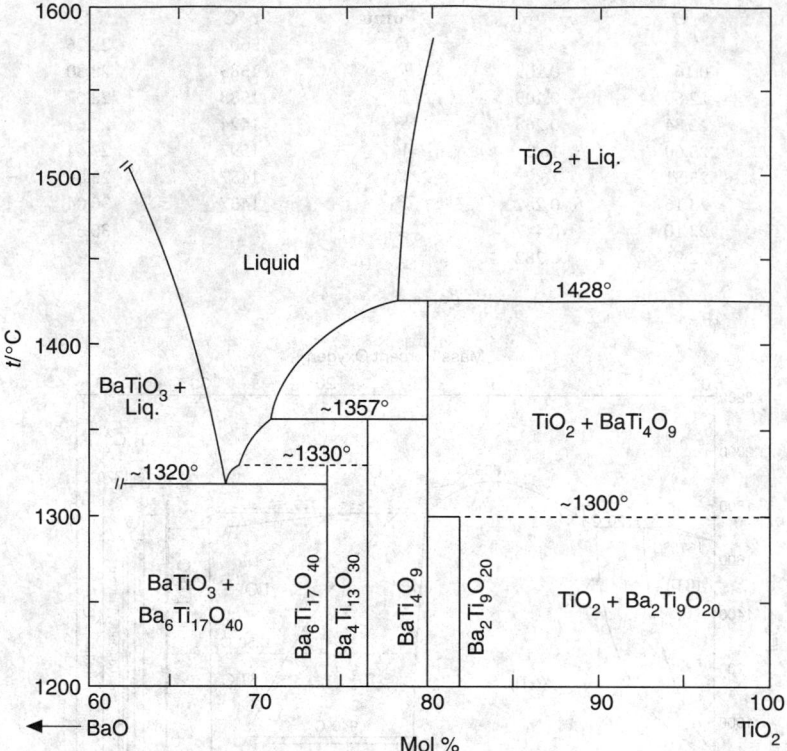

FIGURE 7. BaO-TiO$_2$ system.

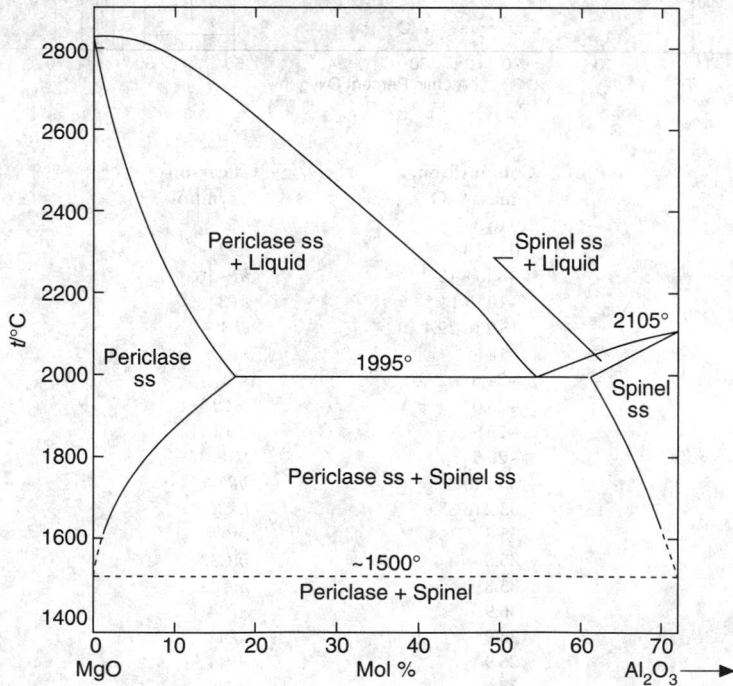

FIGURE 8. MgO-Al$_2$O$_3$ system.

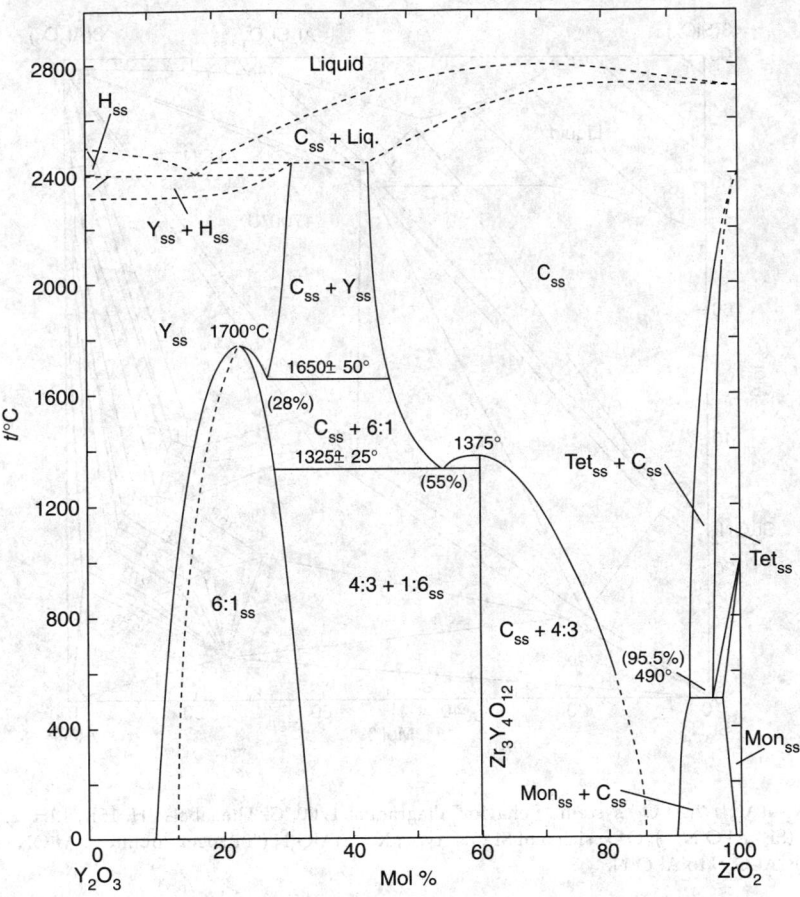

FIGURE 9. Y_2O_3-ZrO_2 system. C_{ss} = cubic ZrO_2 ss (fluorite-type ss); Y_{ss} = cubic Y_2O_3 ss; Tet_{ss} = tetragonal ZrO_2 ss; Mon_{ss} = monoclinic ZrO_2 ss; H_{ss} = hexagonal Y_2O_3 ss; 3:4 = $Zr_3Y_4O_{12}$; 1:6 = ZrY_6O_{11} ss.

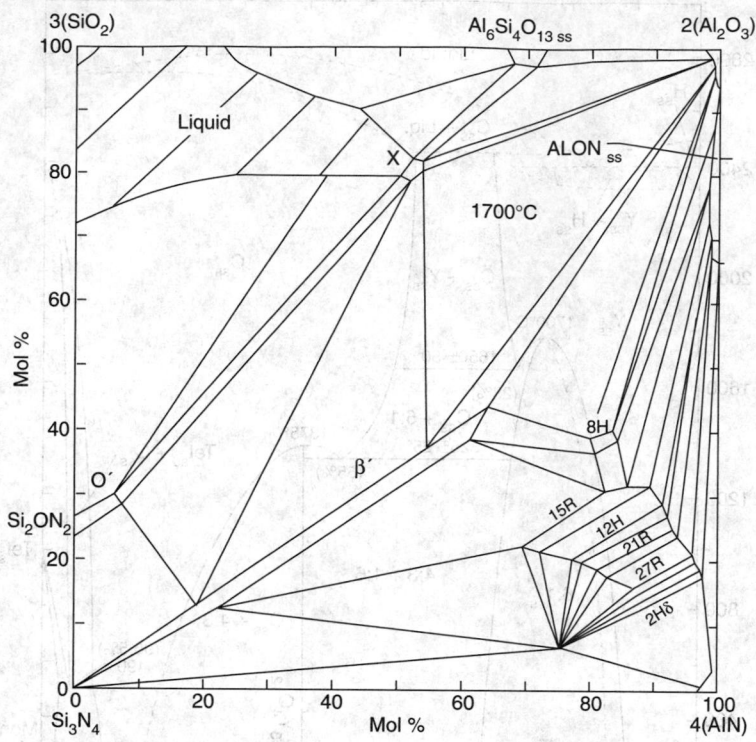

FIGURE 10. 3(SiO$_2$)-Si$_3$N$_4$-4(AlN)-2(Al$_2$O$_3$) system. "Behavior" diagram at 1700 °C. The labels 8H, 15R, 12H, 21R, 27R, 2H$^\delta$ indicate defect AlN polytypes. β′ = 3-sialon (Si$_{6-x}$Al$_x$O$_x$N$_{8-x}$); O′ = sialon of Si$_2$ON$_2$ type; X = SiAlO$_2$N ("nitrogen mullite"). ALON ss = aluminum oxynitride ss extending from approximately Al$_7$O$_9$N to Al$_3$O$_3$N.

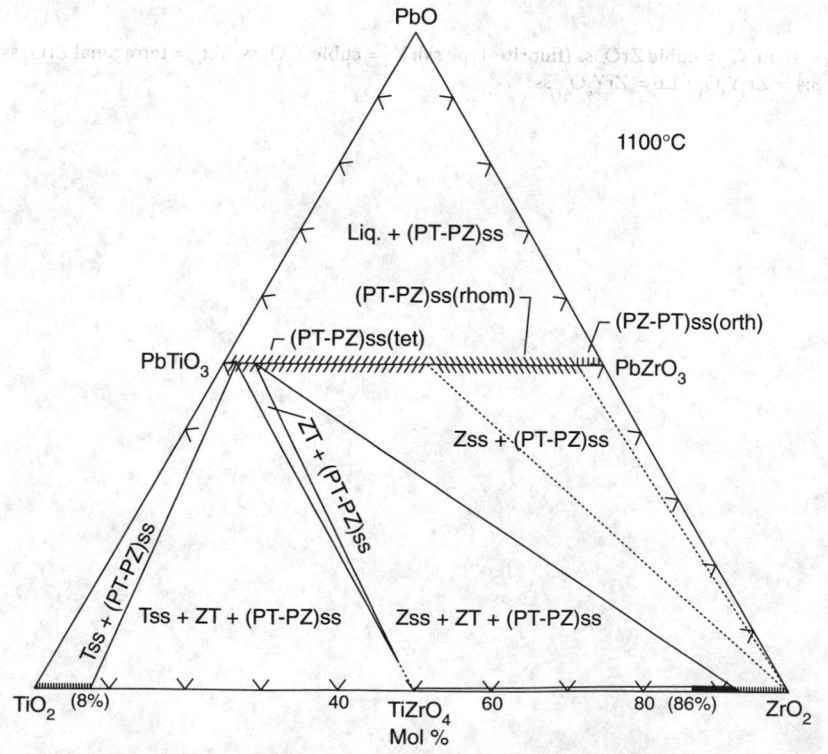

FIGURE 11. PbO-ZrO$_2$-TiO$_2$ (PZT) system, subsolidus at 1100 °C. P = PbO; T = TiO$_2$; Z = ZrO$_2$.

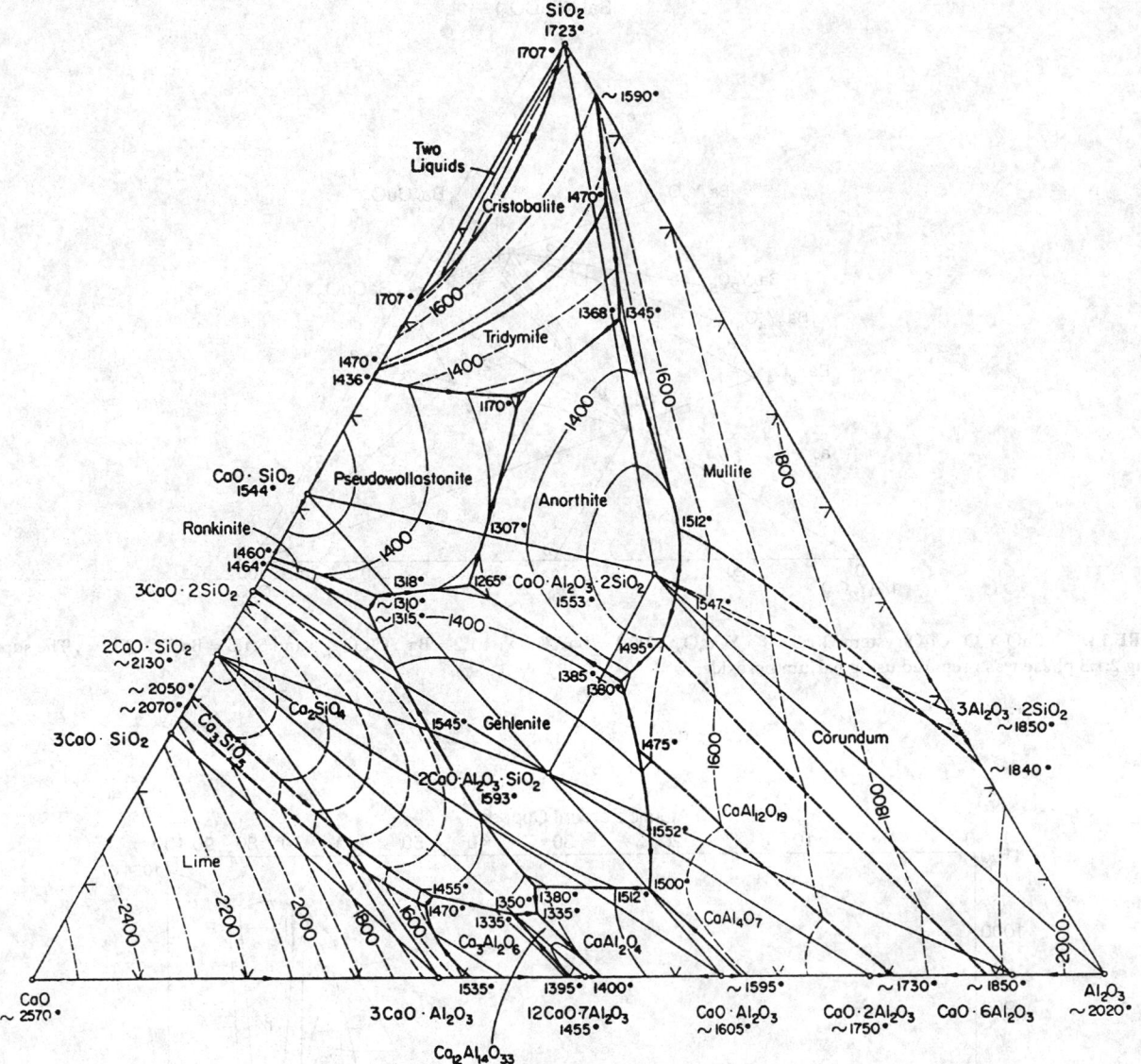

FIGURE 12. CaO-Al$_2$O$_3$-SiO$_2$ system (temperatures in °C).

Crystalline Phases

Notation	Oxide formula
Cristobalite }	SiO$_2$
Tridymite	
Pseudowollastonite	CaO·SiO$_2$
Rankinite	3CaO·2SiO$_2$
Lime	CaO
Corundum	Al$_2$O$_3$
Mullite	3Al$_2$O$_3$·2SiO$_2$
Anorthite	CaO·Al$_2$O$_3$·2SiO$_2$
Gehlenite	2CaO·Al$_2$O$_3$·SiO$_2$

Temperatures up to approximately 1550 °C are on the Geophysical Laboratory Scale; those above 1550 °C are on the 1948 International Scale.

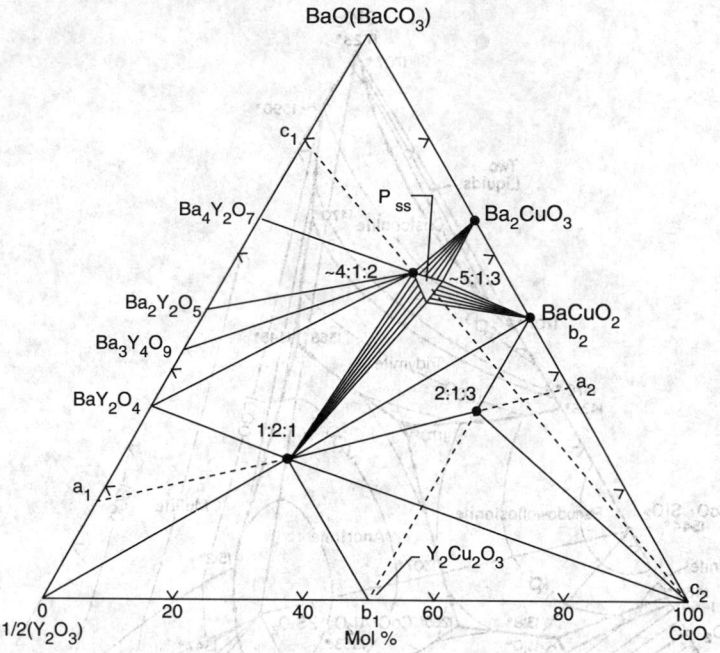

FIGURE 13. BaO-Y₂O₃-CuO system. 2:1:3 = Ba₂YCu₃O₇₋ₓ; 1:2:1 = BaY₂CuO₅; 4:1:2 = Ba₄YCu₂O₇.₅₊ₓ; and 5:1:3 = Ba₅YCu₃O₉.₅₊ₓ. The superconducting 2:1:3 phase was prepared using barium peroxide.

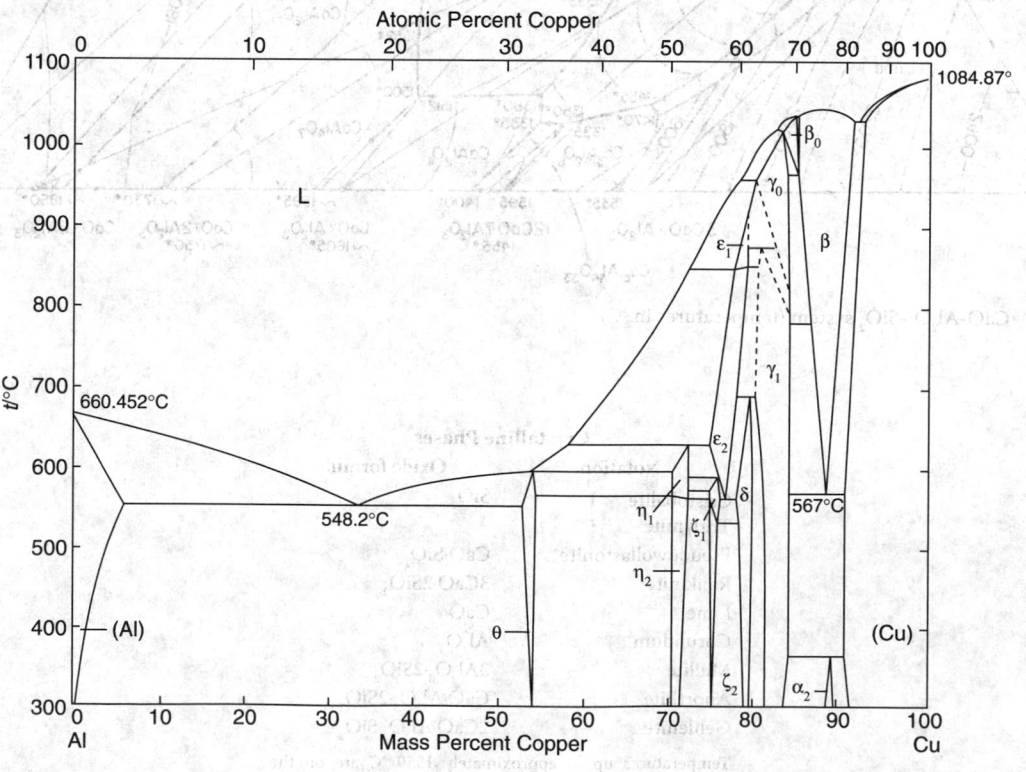

FIGURE 14. Al-Cu system.

Phase	Composition, wt % Cu	Pearson symbol	Space group
(Al)	0 to 5.65	cF4	Fm$\bar{3}$m
θ	52.5 to 53.7	tI12	I4/mcm
η$_1$	70.0 to 72.2	oP16 or oC16	Pban or Cmmm
η$_2$	70.0 to 72.1	mC20	C2/m
ζ$_1$	74.4 to 77.8	hP42	P6/mmm
ζ$_2$	74.4 to 75.2	(a)	–
ε$_1$	77.5 to 79.4	(b)	–
ε$_2$	72.2 to 78.7	hP4	P63/mmc
δ	77.4 to 78.3	(c)	R$\bar{3}$m
γ$_0$	77.8 to 84	(d)	–
γ$_1$	79.7 to 84	cP52	P$\bar{4}$3m
β$_0$	83.1 to 84.7	(d)	–
β	85.0 to 91.5	cI2	Im$\bar{3}$m
α$_2$	88.5 to 89	(e)	–
(Cu)	90.6 to 100	cF4	Fm$\bar{3}$m
Metastable phases			
θ′	–	tP6	–
β′	–	cF16	Fm$\bar{3}$m
Al$_3$Cu$_2$	61 to 70	hp5	P$\bar{3}$m1

(a) Monoclinic? (b) Cubic? (c) Rhombohedral. (d) Unknown. (e) D0$_{22}$-type long-period superlattice.

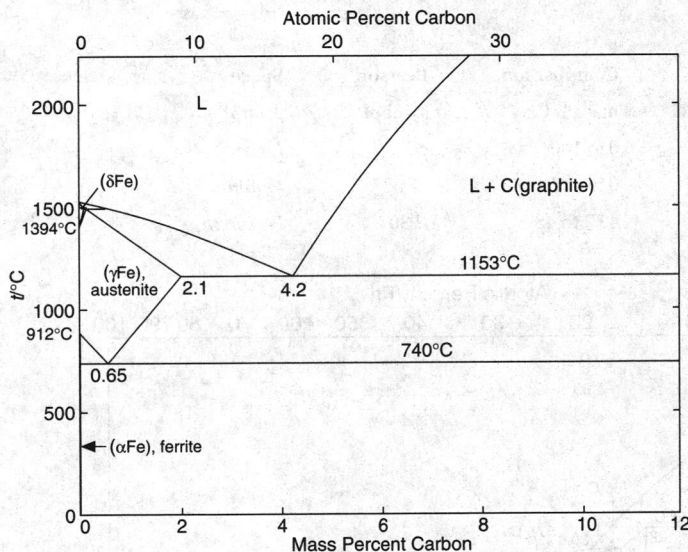

FIGURE 15.   Fe-C system.

Phase	Composition, mass % C	Pearson symbol	Space group
(δFe)	0 to 0.09	cI2	Im$\bar{3}$m
(γFe)	0 to 2.1	cF4	Fm$\bar{3}$m
(αFe)	0 to 0.021	cI2	Im$\bar{3}$m
(C)	100	hP4	P6$_3$/mmc
Metastable/high-pressure phases			
(εFe)	0	hP2	P6$_3$/mmc
Martensite	< 2.1	tI4	I4/mmm
Fe$_4$C	5.1	cP5	P$\bar{4}$3m
Fe$_3$C (θ)	6.7	oP16	Pnma
Fe$_5$C$_2$ (χ)	7.9	mC28	C2/c
Fe$_7$C$_3$	8.4	hP20	P6$_3$mc
Fe$_7$C$_3$	8.4	oP40	Pnma
Fe$_2$C (η)	9.7	oP6	Pnnm
Fe$_2$C (ε)	9.7	hP*	P6$_3$22
Fe$_2$C	9.7	hP*	P$\bar{3}$m1
(C)	100	cF8	Fd$\bar{3}$m

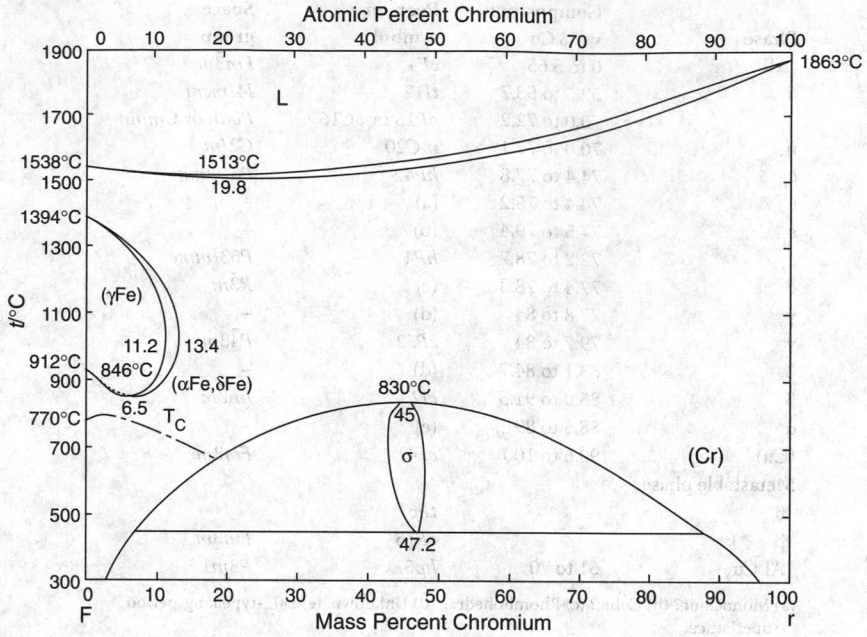

FIGURE 16. Fe-Cr system.

Phase	Composition, mass % Cr	Pearson symbol	Space group
(aFe, Cr)	0 to 100	cI2	Im3̄m
(γFe)	0 to 11.2	cF4	Fm3̄m
σ	42.7 to 48.2	tP30	P4₂/mnm

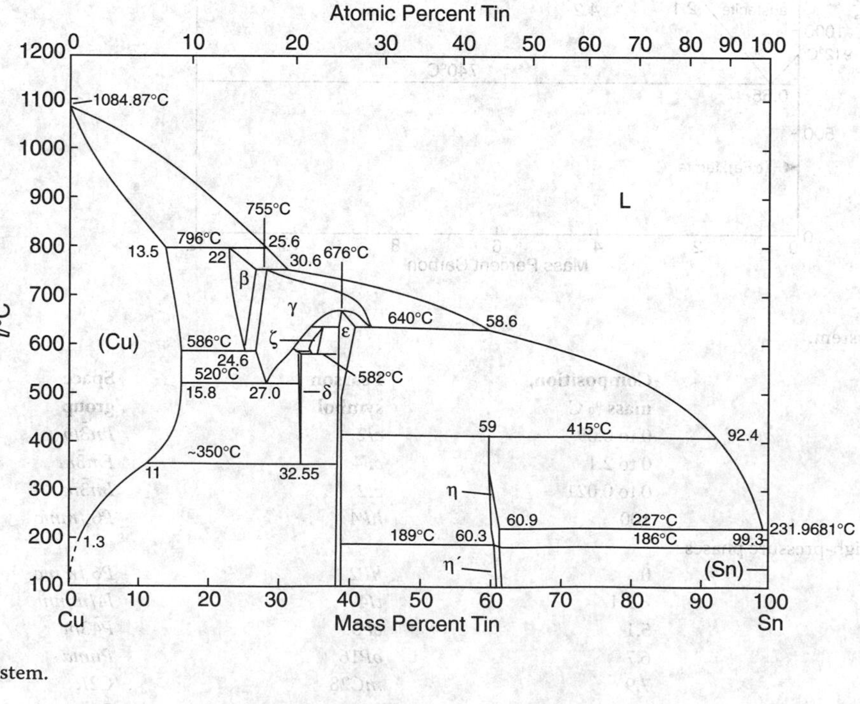

FIGURE 17. Cu-Sn system.

Phase	Composition, mass % Sn	Pearson symbol	Space group
α	0 to 15.8	cF4	$Fm\bar{3}m$
β	22.0 to 27.0	cI2	$Im\bar{3}m$
γ	25.5 to 41.5	cF16	$Fm\bar{3}m$
δ	32 to 33	cF416	$F\bar{4}3m$
ζ	32.2 to 35.2	hP26	$P6_3$
ε	27.7 to 39.5	oC80	$Cmcm$
η	59.0 to 60.9	hP4	$P6_3/mmc$
η´	44.8 to 60.9	(a)	–
(βSn)	~100	tI4	$I4_1/amd$
(αSn)	100	cF8	$Fd\bar{3}m$

(a) Hexagonal; superlattice based on NiAs-type structure.

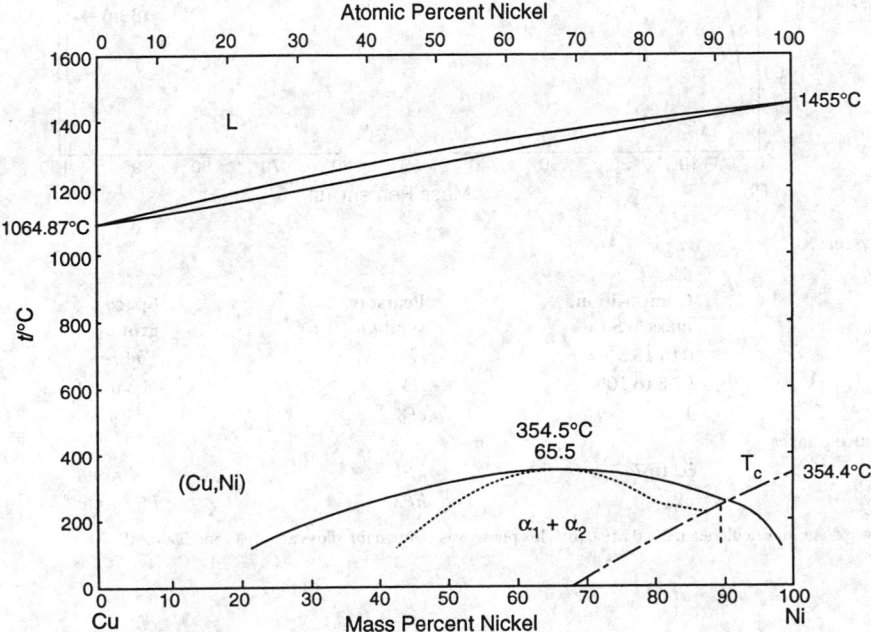

FIGURE 18. Cu-Ni system.

Phase	Composition, mass % Ni	Pearson symbol	Space group
(Cu, Ni) (above 354.5 °C)	0 to 100	cF4	$Fm\bar{3}\,m$

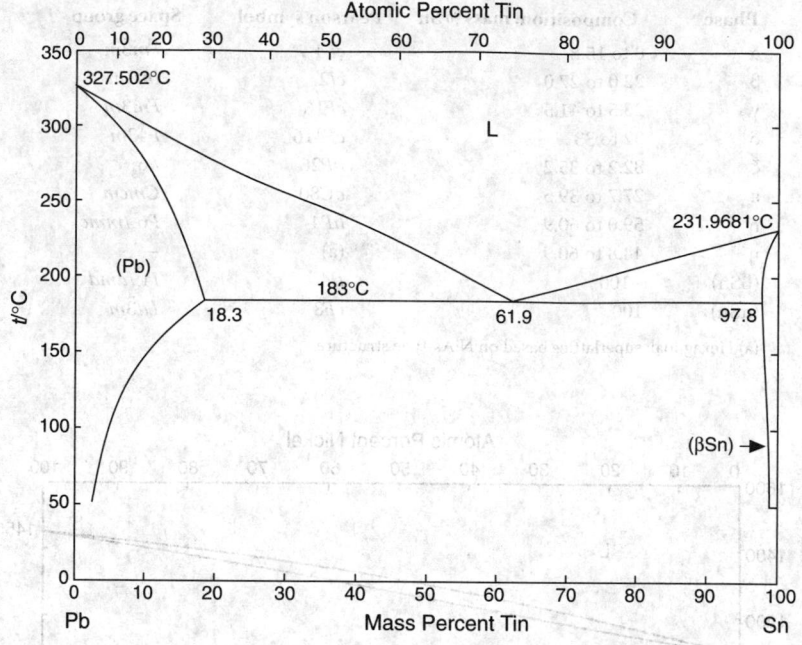

FIGURE 19.   Pb-Sn system.

Phase	Composition, mass % Sn	Pearson symbol	Space group
(Pb)	0 to 18.3	cF4	$Fm\bar{3}m$
(βSn)	97.8 to 100	tI4	$I4_1/amd$
(αSn)	100	cF8	$Fd\bar{3}m$
High-pressure phases			
ε(a)	52 to 74	hP1	$P6/mmm$
ε´(b)	52	hP2	$P6_3/mmc$

(a) From phase diagram calculated at 2500 MPa. (b) This phase was claimed for alloys at 350 °C and 5500 MPa.

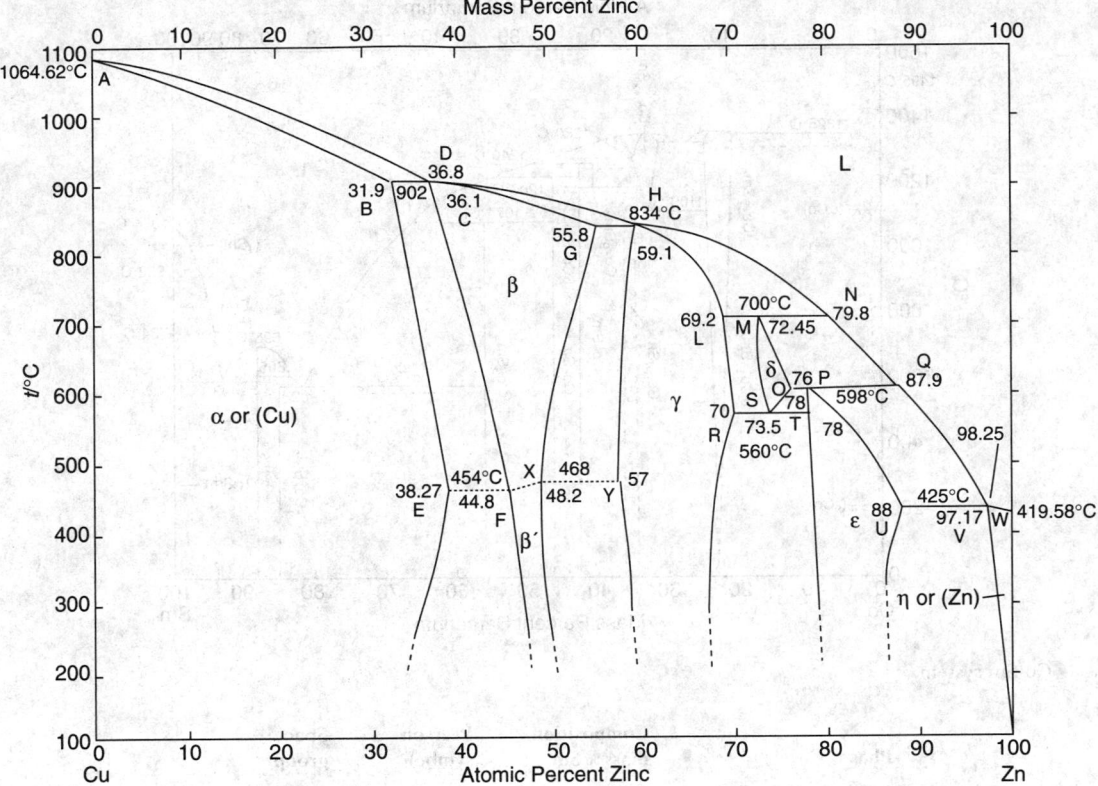

FIGURE 20. Cu-Zn system.

Phase	Composition, mass % Zn	Pearson symbol	Space group
α or (Cu)	0 to 38.95	cF4	Fm$\bar{3}$m
β	36.8 to 56.5	cI2	Im$\bar{3}$m
β′	45.5 to 50.7	cP2	Pm$\bar{3}$m
γ	57.7 to 70.6	cI52	I$\bar{4}$3m
δ	73.02 to 76.5	hP3	P$\bar{6}$
ε	78.5 to 88.3	hP2	P6$_3$/mmc
η or (Zn)	97.25 to 100	hP2	P6$_3$/mmc

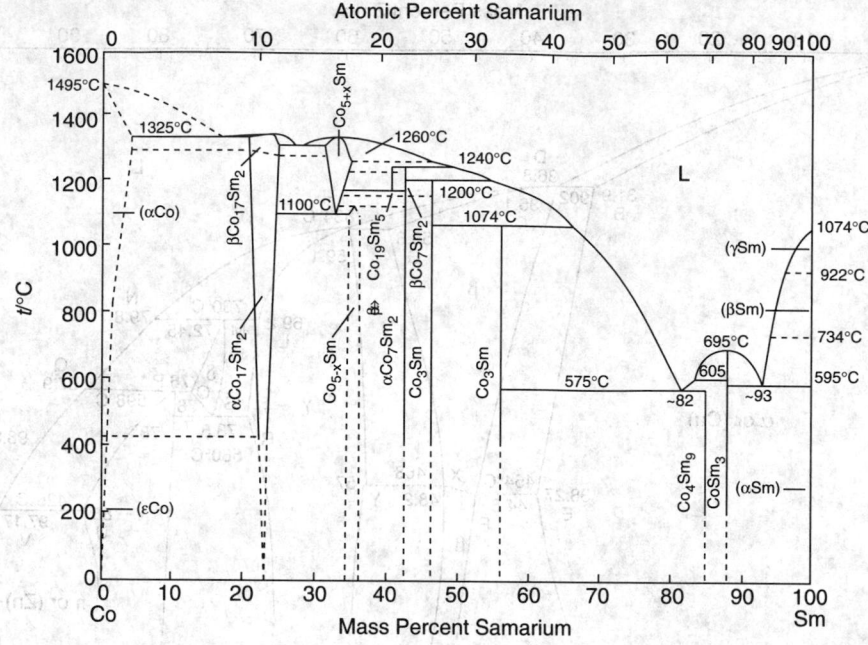

FIGURE 21.   Co-Sm system.

Phase	Composition, mass % Sm	Pearson symbol	Space group
(αCo)	0 to ~3.7	cF4	Fm$\bar{3}$m
(εCo)	~0	hP2	P6$_3$/mmc
βCo$_{17}$Sm$_2$	~23.0	hP38	P6$_3$/mmc
αCo$_{17}$Sm$_2$	~23.0	hR19	R$\bar{3}$m
		hP8	P6/mmm
Co$_{5+x}$Sm	~33 to 34	–	–
Co$_{5-x}$Sm	~34 to 35	–	–
Co$_{19}$Sm$_5$	~40.1	hR24	R$\bar{3}$m
		hP48	P6$_3$/mmc
αCo$_7$Sm$_2$	~42.1	hR18	R$\bar{3}$m
βCo$_7$Sm$_2$	~42.1	hP36	P6$_3$/mmc
Co$_3$Sm	46	hR12	R$\bar{3}$m
Co$_2$Sm	56.0	hR4	R$\bar{3}$m
		cF24	Fd$\bar{3}$m
Co$_4$Sm$_9$	~85.1	o**	–
CoSm$_3$	88	oP16	Pnma
(γSm)	~100	cI2	Im$\bar{3}$m
(βSm)	~100	hP2	P6$_3$/mmc
(αSm)	~100	hR3	R$\bar{3}$m
Other reported phases			
Co$_5$Sm	~33.8	hP6	P6/mmm
Co$_2$Sm$_5$	~86.4	mC28	C2/c

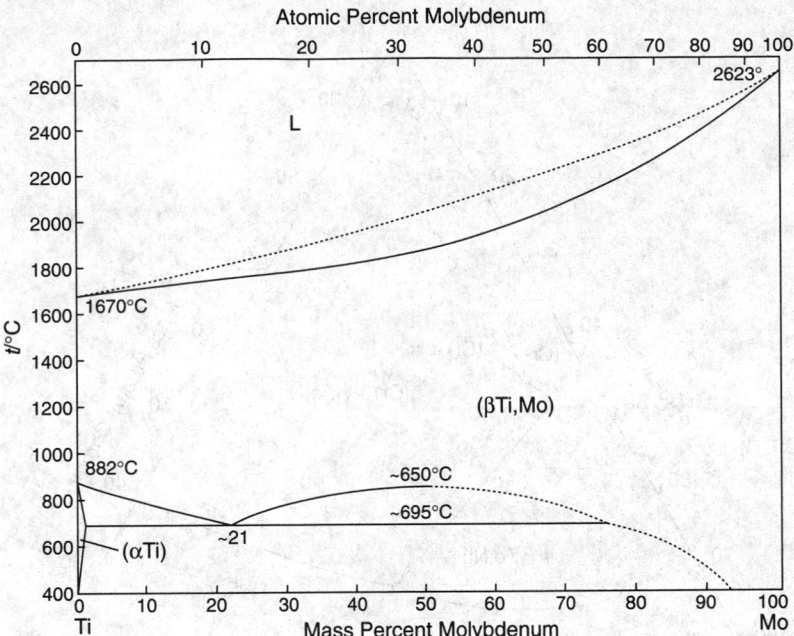

FIGURE 22.   Ti-Mo system.

Phase	Composition, mass % Mo	Pearson symbol	Space group
(βTi, Mo)	0 to 100	cI2	Im$\bar{3}$m
(αTi)	0 to 0.8	hP2	P6$_3$/mmc
α′	(a)	hP2	P6$_3$/mmc
α″	(a)	oC4	Cmcm
ω	(a)	hP3	P6/mmm

(a) Metastable.

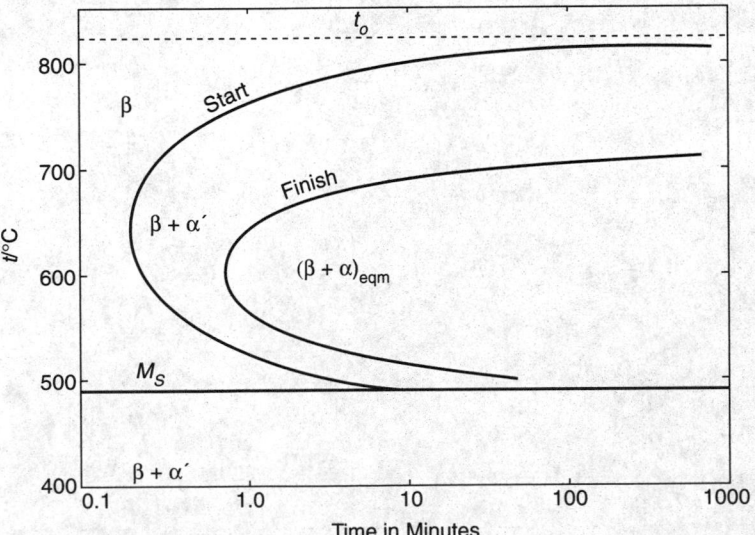

Experimental time–temperature–transformation (TTT) diagram for Ti-Mo. The start and finish times of the isothermal precipitation reaction vary with temperature as a result of the temperature dependence of the nucleation and growth processes. Precipitation is complete, at any temperature, when the equilibrium fraction of α is established in accordance with the lever rule. The solid horizontal line represents the athermal (or nonthermally activated) martensitic transformation that occurs when the β phase is quenched.

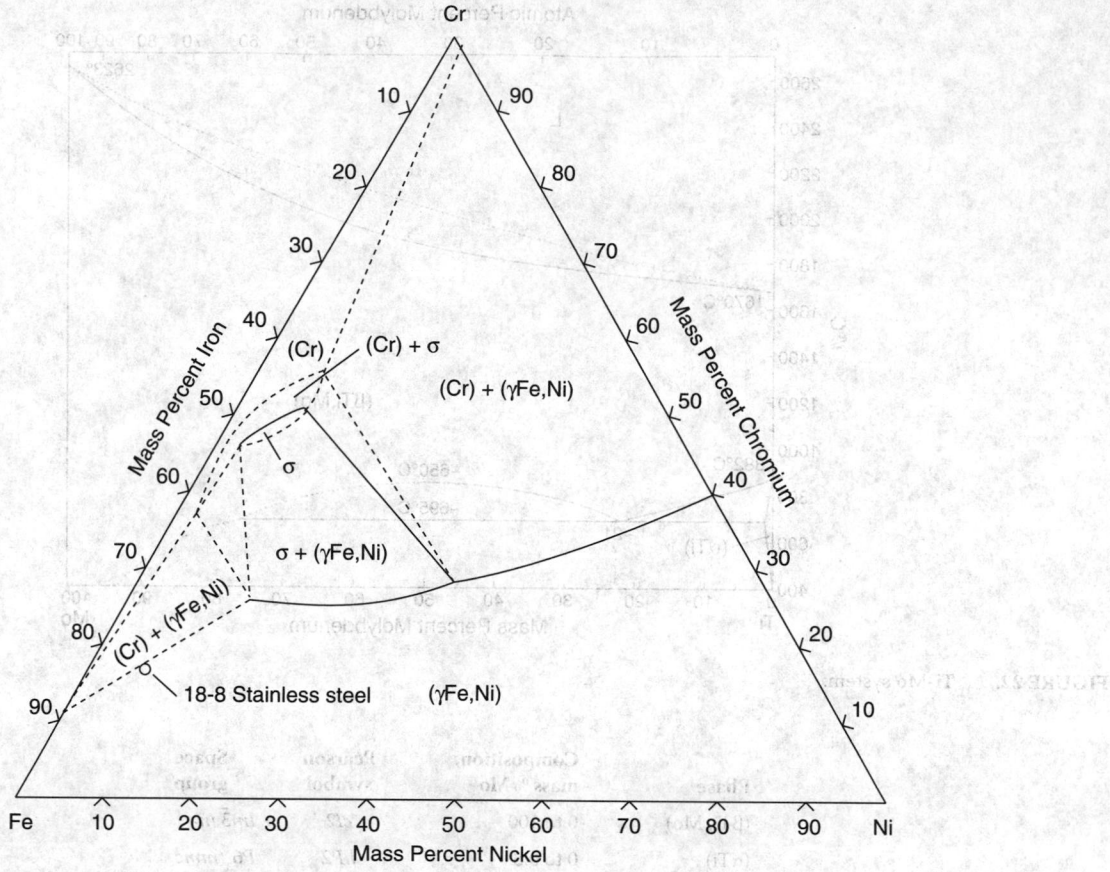

FIGURE 23. The isothermal section at 900 °C (1652 °F) of the iron-chromium-nickel ternary phase diagram, showing the nominal composition of 18-8 stainless steel.

HEAT CAPACITY OF SELECTED SOLIDS

This table gives the molar heat capacity at constant pressure of representative metals, semiconductors, and other crystalline solids as a function of temperature in the range 200 to 600 K.

References

1. Chase, M. W., et al., *JANAF Thermochemical Tables, 3rd ed., J. Phys. Chem. Ref. Data,* 14, Suppl. 1, 1985.

2. Garvin, D., Parker, V. B., and White, H. J., *CODATA Thermodynamic Tables,* Hemisphere Press, New York, 1987.

3. DIPPR Database of Pure Compound Properties, Design Institute for Physical Properties Data, American Institute of Chemical Engineers, New York, 1987.

Name	C_p in J/mol K						
	200 K	250 K	300 K	350 K	400 K	500 K	600 K
Aluminum	21.33	23.08	24.25	25.11	25.78	26.84	27.89
Aluminum oxide	51.12	67.05	79.45	88.91	96.14	106.17	112.55
Anthracene	138.6	173.9	210.7	248.8	288.4		
Benzoic acid	102.7	123.5	147.4	172.0			
Beryllium	9.98	13.58	16.46	18.53	19.95	21.94	23.34
Biphenyl	131.0	162.5	197.2				
Boron	5.99	8.82	11.40	13.65	15.69	18.72	20.78
Calcium	24.54	25.41	25.94	26.32	26.87	28.49	30.38
Calcium carbonate	66.50	75.66	83.82	91.51	96.97	104.52	109.86
Calcium oxide	33.64	38.59	42.18	45.07	46.98	49.33	50.72
Cesium chloride	50.13	51.34	52.48	53.58	54.68	56.90	59.10
Chromium	19.86	22.30	23.47	24.39	25.23	26.63	27.72
Cobalt	22.23	23.98	24.83	25.68	26.53	28.20	29.66
Copper	22.63	23.77	24.48	24.95	25.33	25.91	26.48
Copper oxide	34.80		42.41	44.95	46.78	49.19	50.83
Copper sulfate	77.01	89.25	99.25	107.65	114.93	127.19	136.31
Germanium			23.25	23.85	24.31	24.96	25.45
Gold			25.41	25.37	25.51	26.06	26.65
Graphite	5.01	6.82	8.58	10.24	11.81	14.62	16.84
Hexachlorobenzene	162.7	183.6	202.4				
Iodine	51.57	53.24	54.51	58.60			
Iron	21.59	23.74	25.15	26.28	27.39	29.70	32.05
Lead	25.87	26.36	26.85	27.30	27.72	28.55	29.40
Lithium	21.57	23.42	24.64	25.96	27.60	29.28	
Lithium chloride	43.35	46.08	48.10	49.66	50.97	53.34	55.59
Magnesium	22.72	24.02	24.90	25.57	26.14	27.17	28.18
Magnesium oxide			37.38	40.59	42.77	45.56	47.30
Manganese	23.05	24.95	26.35	27.52	28.53	30.29	31.90
Naphthalene	105.8	134.1	167.8	204.1			
Potassium	27.00	28.01	29.60				
Potassium chloride	48.44	50.10	51.37	52.31	53.08	54.71	56.35
Silicon	15.64	18.22	20.04	21.28	22.14	23.33	24.15
Silicon dioxide	32.64	39.21	44.77	49.47	53.43	59.64	64.42
Silver			25.36	25.55	25.79	26.36	26.99
Sodium	22.45	27.01	28.20	30.14			
Sodium chloride	46.89	48.85	50.21	51.25	52.14	53.96	55.81
Tantalum	24.08	24.86	25.31	25.60	25.84	26.35	26.84
Titanium	22.37	24.07	25.28	26.17	26.86	27.88	28.60
Tungsten	22.49	23.69	24.30	24.65	24.92	25.36	25.79
Vanadium	21.88	23.70	24.93	25.68	26.23	26.94	27.49
Zinc	24.05	25.02	25.45	25.88	26.35	27.39	28.59
Zirconium	23.87	24.69	25.22	25.61	25.93	26.56	27.28

THERMAL AND PHYSICAL PROPERTIES OF PURE METALS

This table gives the following properties for the metallic elements:

t_m: Melting point in °C

t_b: Normal boiling point in °C, at a pressure of 101.325 kPa (760 Torr)

$\Delta_{fus}H$: Enthalpy of fusion at the melting point in J g^{-1}

ρ: Density at 25 °C in g cm^{-3}

α: Coefficient of linear expansion at 25 °C in K^{-1} (the quantity listed is $10^6 \times \alpha$)

c_p: Specific heat capacity at constant pressure at 25 °C in J g^{-1} K^{-1}

λ: Thermal conductivity at 27 °C in W m^{-1} K^{-1}

References

1. Dinsdale, A. T., *CALPHAD* 15, 317, 1991 [melting point, enthalpy of fusion].
2. Touloukian, Y. S., *Thermophysical Properties of Matter*, Vol. 12, Thermal Expansion, IFI/Plenum, New York, 1975 [coefficient of expansion, density].
3. Ho, C. Y., Powell, R. W., and Liley, P. E., *J. Phys. Chem. Ref. Data* 3, Suppl. 1, 1974 [thermal conductivity].
4. Cox, J. D., Wagman, D. D., and Medvedev, V. A., *CODATA Key Values for Thermodynamics*, Hemisphere Publishing Corp., New York, 1989 [heat capacity].
5. Glushko, V. P., Ed., *Thermal Constants of Substances*, VINITI, Moscow, 1965-1981 [enthalpy of fusion, heat capacity].
6. Wagman, D. D., Evans, W. H., Parker, V. B., Schumm, R. H., Halow, I., Bailey, S. M., Churney, K. L., and Nuttall, R. L., *The NBS Tables of Chemical Thermodynamic Properties*, J. Phys. Chem. Ref. Data 11, Suppl. 2, 1982 [heat capacity].
7. Gschneidner, K. A., *Bull. Alloy Phase Diagrams* 11, 216–224, 1990 [various properties of the rare earth metals].
8. Hellwege, K. H., Ed., *Landolt Börnstein, Numerical Values and Functions in Physics, Chemistry, Astronomy, Geophysics, and Technology*, Vol. 2, Part 1, Mechanical-Thermal Properties of State, 1971 [density].
9. *Physical Encyclopedic Dictionary*, Vol. 1–5, Encyclopedy Publishing House, Moscow, 1960–66.

Metal (symbol)	Atomic weight	t_m °C	t_b °C	$\Delta_{fus}H$ J g^{-1}	ρ g cm^{-3}	$\alpha \times 10^6$ K^{-1}	c_p J g^{-1} K^{-1}	λ W m^{-1} K^{-1}
Actinium (Ac)	227	1050	3198	52.9	10		0.12	
Aluminum (Al)	26.98	660.323	2519	397.0	2.70	23.1	0.897	237
Antimony (Sb)	121.76	630.628	1587	162.5	6.68	11.0	0.207	24.3
Barium (Ba)	137.33	727	1845	51.8	3.62	20.6	0.204	18.4
Beryllium (Be)	9.01	1287	2468	876.1	1.85	11.3	1.825	200
Bismuth (Bi)	208.98	271.406	1564	53.1	9.79	13.4	0.122	7.87
Cadmium (Cd)	112.41	321.069	767	55.2	8.69	30.8	0.231	96.8
Calcium (Ca)	40.08	842	1484	213.1	1.54	22.3	0.647	200
Cerium (Ce)	140.12	799	3443	39.0	6.770	6.3	0.192	11.3
Cesium (Cs)	132.91	28.5	671	15.7	1.93	97	0.242	35.9
Chromium (Cr)	52.00	1907	2671	403.9	7.15	4.9	0.449	93.7
Cobalt (Co)	58.93	1495	2927	274.9	8.86	13.0	0.421	100
Copper (Cu)	63.55	1084.62	2560	208.7	8.96	16.5	0.385	401
Dysprosium (Dy)	162.50	1412	2567	69.8	8.55	9.9	0.173	10.7
Erbium (Er)	167.26	1529	2868	119	9.07	12.2	0.168	14.5
Europium (Eu)	151.96	822	1529	60.6	5.24	35.0	0.182	13.9
Gadolinium (Gd)	157.25	1313	3273	61.5	7.90	9.4	0.235	10.5
Gallium (Ga)	69.72	29.7646	2229	80.1	5.91	18	0.373	40.6
Gold (Au)	196.97	1064.18	2836	63.7	19.3	14.2	0.129	317
Hafnium (Hf)	178.49	2233	4600	152.4	13.3	5.9	0.144	23.0
Holmium (Ho)	164.93	1472	2700	71.3	8.80	11.2	0.165	16.2
Indium (In)	114.82	156.60	2027	28.7	7.31	32.1	0.233	81.6
Iridium (Ir)	192.22	2446	4428	213.9	22.56	6.4	0.131	147
Iron (Fe)	55.85	1538	2861	247.3	7.87	11.8	0.449	80.2
Lanthanum (La)	138.91	920	3464	44.6	6.15	12.1	0.195	13.4
Lead (Pb)	207.2	327.462	1749	23.0	11.3	28.9	0.130	35.3
Lithium (Li)	6.94	180.50	1342	432	0.534	46	3.582	84.7
Lutetium (Lu)	174.97	1663	3402	106.6	9.84	9.9	0.154	16.4
Magnesium (Mg)	24.31	650	1090	348.9	1.74	24.8	1.023	156
Manganese (Mn)	54.94	1246	2061	235.0	7.3	21.7	0.479	7.82
Mercury (Hg)	200.59	−38.829	356.619	11.4	13.5336	60.4	0.140	8.34
Molybdenum (Mo)	95.96	2622	4639	390.6	10.2	4.8	0.251	138

Metal (symbol)	Atomic weight	t_m °C	t_b °C	$\Delta_{fus}H$ $J\,g^{-1}$	ρ $g\,cm^{-3}$	$\alpha \times 10^6$ K^{-1}	c_p $J\,g^{-1}\,K^{-1}$	λ $W\,m^{-1}\,K^{-1}$
Neodymium (Nd)	144.24	1016	3074	49.5	7.01	9.6	0.190	16.5
Neptunium (Np)	237	644		13.5	20.2			6.3
Nickel (Ni)	58.69	1455	2913	297.8	8.90	13.4	0.444	90.7
Niobium (Nb)	92.91	2477	4741	323	8.57	7.3	0.265	53.7
Osmium (Os)	190.23	3033	5008	304.1	22.59	5.1	0.130	87.6
Palladium (Pd)	106.42	1554.8	2963	157.3	12.0	11.8	0.244	71.8
Platinum (Pt)	195.08	1768.2	3825	113.7	21.5	8.8	0.133	71.6
Plutonium (Pu)	244	640	3228	11.6	19.7	46.7		6.74
Polonium (Po)	209	254	962	48	9.20	23.5		20
Potassium (K)	39.10	63.5	759	59.7	0.89	83.3	0.757	102.4
Praseodymium (Pr)	140.91	931	3520	48.9	6.77	6.7	0.193	12.5
Promethium (Pm)	145	1042	3000		7.26	11	0.19	15
Protactinium (Pa)	231.04	1572		53.4	15.4			
Radium (Ra)	226	696		34	5			
Rhenium (Re)	186.21	3185	5590	183.0	20.8	6.2	0.137	47.9
Rhodium (Rh)	102.91	1963	3695	258.4	12.4	8.2	0.243	150
Rubidium (Rb)	85.47	39.30	688	25.6	1.53		0.363	58.2
Ruthenium (Ru)	101.07	2333	4147	381.8	12.1	6.4	0.238	117
Samarium (Sm)	150.36	1072	1794	57.3	7.52	12.7	0.196	13.3
Scandium (Sc)	44.96	1541	2836	313.6	2.99	10.2	0.568	15.8
Silver (Ag)	107.87	961.78	2162	104.8	10.5	18.9	0.235	429
Sodium (Na)	22.99	97.794	882.940	113.1	0.97	71	1.228	141
Strontium (Sr)	87.62	777	1377	84.8	2.64	22.5	0.306	35.3
Tantalum (Ta)	180.95	3017	5455	202.1	16.4	6.3	0.140	57.5
Technetium (Tc)	98	2157	4262	339.7	11			50.6
Terbium (Tb)	158.93	1359	3230	63.9	8.23	10.3	0.182	11.1
Thallium (Tl)	204.38	304	1473	20.3	11.8	29.9	0.129	46.1
Thorium (Th)	232.04	1750	4785	59.5	11.7	11.0	0.118	54.0
Thulium (Tm)	168.93	1545	1950	99.7	9.32	13.3	0.160	16.9
Tin (Sn)	118.71	231.928	2586	60.2	7.265	22.0	0.227	66.6
Titanium (Ti)	47.87	1670	3287	295.6	4.506	8.6	0.524	21.9
Tungsten (W)	183.84	3414	5555	284.5	19.3	4.5	0.132	174
Uranium (U)	238.03	1135	4131	38.4	19.1	13.9	0.116	27.6
Vanadium (V)	50.94	1910	3407	422	6.0	8.4	0.489	30.7
Ytterbium (Yb)	173.05	824	1196	44.3	6.90	26.3	0.155	38.5
Yttrium (Y)	88.91	1522	3345	128.1	4.47	10.6	0.298	17.2
Zinc (Zn)	65.38	419.527	907	108.1	7.14	30.2	0.388	116
Zirconium (Zr)	91.22	1854	4406	230.2	6.52	5.7	0.278	22.7

THERMOPHYSICAL PROPERTIES OF STAINLESS STEEL 310

Stainless steel is used in a wide variety of applications, especially at high temperatures. This table gives properties of a reference standard Stainless Steel 310 certified by the U. K. National Physical Laboratory. The properties are:

a: thermal diffusivity in $\mathrm{mm^2\,s^{-1}}$
c_p: specific heat capacity at constant pressure in $\mathrm{J\,g^{-1}\,K^{-1}}$
ρ: density in $\mathrm{g\,cm^{-3}}$
k: thermal conductivity in $\mathrm{W\,m^{-1}\,K^{-1}}$

With kind permission from Springer Science+Business Media: International Journal of Thermophysics, 28, 674, 2007, Table II.

Reference

Blumm, J., Lindemann, A. Niedrig, B., and Campbell, R., *Int. J. Thermophys.* 28, 674, 2007.

$t/°C$	$a/\mathrm{mm^2\,s^{-1}}$	$c_p/\mathrm{J\,g^{-1}\,K^{-1}}$	$\rho/\mathrm{g\,cm^{-3}}$	$k/\mathrm{W\,m^{-1}\,K^{-1}}$
−125	3.170	0.376	7.878	9.39
−100	3.130	0.411	7.871	10.12
−75	3.145	0.435	7.863	10.76
−50	3.170	0.451	7.855	11.23
−25	3.210	0.464	7.846	11.69
0	3.256	0.475	7.838	12.12
25	3.352	0.483	7.829	12.67
50	3.439	0.490	7.820	13.18
101	3.611	0.501	7.801	14.11
150	3.763	0.512	7.782	14.99
200	3.917	0.518	7.762	15.75
250	4.075	0.525	7.742	16.56
300	4.205	0.533	7.722	17.31
350	4.331	0.541	7.701	18.04
400	4.455	0.548	7.681	18.75
450	4.571	0.555	7.660	19.43
500	4.686	0.562	7.639	20.12
550	4.806	0.570	7.618	20.86
600	4.920	0.595	7.596	22.24
651	5.058	0.601	7.574	23.02
701	5.179	0.607	7.551	23.74
750	5.207	0.611	7.529	23.95
800	5.288	0.617	7.506	24.49
850	5.404	0.624	7.483	25.23
901	5.506	0.633	7.460	26.00
950	5.618	0.645	7.436	26.94
1000	5.707	0.655	7.411	27.70

THERMAL CONDUCTIVITY OF METALS AND SEMICONDUCTORS AS A FUNCTION OF TEMPERATURE

This table gives the temperature dependence of the thermal conductivity of several metals and of carbon, germanium, and silicon. For graphite, separate entries are given for the thermal conductivity parallel (∥) and perpendicular (⊥) to the layer planes. The thermal conductivity of all these materials is very sensitive to impurities at low temperatures, especially below 100 K. Therefore, the values given here should be regarded as typical values for a highly purified specimen; the thermal conductivity of different specimens can vary by more than an order of magnitude in the low-temperature range. See Reference 2 for details.

References

1. Ho, C. Y., Powell, R. W., and Liley, P. E., *J. Phys. Chem. Ref. Data*, 1, 279, 1972.
2. White, G. K., and Minges, M. L., *Thermophysical Properties of Some Key Solids*, CODATA Bulletin No. 59, 1985.

Thermal Conductivity in W/cm K

				Carbon (C)						
				Diamond (type)			Pyrolytic graphite			
T/K	Ag	Al	Au	I	IIa	IIb	∥	⊥	Cr	Cu
1	39.4	41.1	5.46						0.402*	42.2
2	78.3	81.8	10.9	0.0138*	0.033*	0.0200*			0.803	84.0
3	115	121	16.1	0.0461	0.111	0.0676			1.20	125
4	147	157	20.9	0.108	0.261	0.160			1.60	162
5	172	188	25.2	0.206	0.494	0.307			2.00	195
6	187	213	28.5	0.344	0.820	0.510			2.39	222
7	193	229	30.9	0.523	1.24	0.778			2.27	239
8	190	237	32.3	0.762	1.77	1.12			3.14	248
9	181	239	32.7	1.05	2.41	1.53			3.50	249
10	168	235	32.4	1.40	3.17	2.03	0.811	0.0116	3.85	243
15	96.0	176	24.6	3.96	8.65	5.66			5.24	171
20	51.0	117	15.8	7.87	16.8	11.2	4.20	0.0397	5.93	108
30	19.3	49.5	7.55	18.8	38.9	26.5	9.86	0.0786	5.49	44.5
40	10.5	24.0	5.15	29.4	65.9	44.0	16.4	0.120	4.25	21.7
50	7.0	13.5	4.21	35.3	92.1	59.1	23.1	0.152	3.17	12.5
60	5.5	8.5	3.74	37.4	112	67.5	29.8	0.173	2.48	8.29
70	4.97	5.85	3.48	36.9	119	69.1	36.6	0.181	2.07	6.47
80	4.71	4.32	3.32	35.1	117	65.7	42.8	0.181	1.84	5.57
90	4.60	3.42	3.28	32.7	109	60.0	47.5	0.176	1.69	5.08
100	4.50	3.02	3.27	30.0	100	54.2	49.7	0.168	1.59	4.82
150	4.32	2.48	3.25	19.5	60.2	32.5	45.1	0.125	1.29	4.29
200	4.30	2.37	3.23	14.1	40.3	22.6	32.3	0.0923	1.11	4.13
250	4.29	2.35	3.21	11.0	29.7	17.0	24.4	0.0711	1.00	4.06
300	4.29	2.37	3.17	8.95	23.0	13.5	19.5	0.0570	0.937	4.01
350	4.27	2.40	3.14	7.55*	18.5*	11.1*	16.2	0.0477	0.929	3.96
400	4.25	2.40	3.11	6.5*	15.4*	9.32*	13.9	0.0409	0.909	3.93
500	4.19	2.36	3.04				10.8	0.0322	0.860	3.86
600	4.12	2.31	2.98				8.92	0.0268	0.807	3.79
800	3.96	2.18	2.84				6.67	0.0201	0.713	3.66
1000	3.79		2.70				5.34	0.0160	0.654	3.52
1200	3.61*		2.55				4.48	0.0134	0.619	3.39
1400							3.84	0.0116	0.588	
1600							3.33	0.0100	0.556	
1800							2.93	0.00895	0.526*	
2000							2.62	0.00807	0.494*	

T/K	Fe	Geᵃ	Mg	Ni	Pb	Pt	Siᵃ	Sn	Ti	W
1	1.71	0.274	9.86	2.17	27.9	2.31	0.0693*	183	0.0144*	14.4
2	3.42	2.06	19.6	4.34	44.6	4.60	0.454	323	0.0288*	28.7
3	5.11	5.35	29.0	6.49	35.8	6.79	1.38	297	0.0432	42.8
4	6.77	8.77	37.6	8.59	22.2	8.8	2.97	181	0.0575	56.3
5	8.39	11.6	45.0	10.6	13.8	10.5	5.27	117	0.0719	68.7
6	9.93	13.9	50.8	12.5	8.10	11.8	8.23	76	0.0863	79.5
7	11.4	15.5	54.7	14.2	4.86	12.6	11.7	52	0.101	88.0
8	12.7	16.6	56.7	15.8	3.20	12.9	15.5	36	0.115	93.8
9	13.9	17.3	57.0	17.1	2.30	12.8	19.5	26	0.129	96.8
10	14.8	17.7	55.8	18.1	1.78	12.3	23.3	19.3	0.143	97.1
15	17.0	17.3	41.1	19.5	0.845	8.41	41.6	6.3	0.212	72.0
20	15.4	14.9	27.2	16.5	0.591	4.95	49.8	3.2	0.275	40.5
30	10.0	10.8	12.9	9.56	0.477	2.15	48.1	1.79	0.365	14.4
40	6.23	7.98	7.19	5.82	0.451	1.39	35.3	1.33	0.390	6.92
50	4.05	6.15	4.65	4.00	0.436	1.09	26.8	1.15	0.374	4.27
60	2.85	4.87	3.27	3.08	0.425	0.947	21.1	1.04	0.355	3.14
70	2.16	3.93	2.49	2.50	0.416	0.862	16.8	0.96	0.340	2.58
80	1.75	3.25	2.02	2.10	0.409	0.815	13.4	0.915	0.326	2.29
90	1.50	2.70	1.78	1.83	0.403	0.789	10.8	0.880	0.315	2.17
100	1.34	2.32	1.69	1.64	0.397	0.775	8.84	0.853	0.305	2.08
150	1.04	1.32	1.61	1.22	0.379	0.740	4.09	0.779	0.270	1.92
200	0.94	0.968	1.59	1.07	0.367	0.726	2.64	0.733	0.245	1.85
250	0.865	0.749	1.57	0.975	0.360	0.718	1.91	0.696	0.229	1.80
300	0.802	0.599	1.56	0.907	0.353	0.716	1.48	0.666	0.219	1.74
350	0.744	0.495	1.55	0.850	0.347	0.717	1.19	0.642	0.210	1.67
400	0.695	0.432	1.53	0.802	0.340	0.718	0.989	0.622	0.204	1.59
500	0.613	0.338	1.51	0.722	0.328	0.723	0.762	0.596	0.197	1.46
600	0.547	0.273	1.49	0.656	0.314	0.732	0.619		0.194	1.37
800	0.433	0.198	1.46*	0.676		0.756	0.422		0.197	1.25
1000	0.323	0.174		0.718		0.787	0.312		0.207	1.18
1200	0.283	0.174		0.762		0.826	0.257		0.220	1.12
1400	0.312			0.804		0.871	0.235		0.236	1.08
1600	0.330					0.919	0.221		0.253	1.04
1800	0.345*					0.961			0.270*	1.01
2000						0.994*				0.98

ᵃ Values below 300 K are typical values.

* Extrapolated.

THERMAL CONDUCTIVITY OF ALLOYS AS A FUNCTION OF TEMPERATURE

This table lists the thermal conductivity of selected alloys at various temperatures. The indicated compositions refer to weight percent. Since the thermal conductivity is sensitive to exact composition and processing history, especially at low temperatures, these values should be considered approximate.

References

1. Powell, R. L., and Childs, G. E., in *American Institute of Physics Handbook, 3rd Edition*, Gray, D. E., Ed., McGraw-Hill, New York, 1972.
2. Ho, C. Y., et al., *J. Phys. Chem. Ref. Data*, 7, 959, 1978.

Thermal conductivity in W/m K

Alloy		4 K	20 K	77 K	194 K	273 K	373 K	573 K	973 K
Aluminum:	1100	50	240	270	220	220			
	2024	3.2	17	56	95	130			
	3003	11	58	140	150	160			
	5052	4.8	25	77	120	140			
	5083, 5086	3	17	55	95	120			
	Duralumin	5.5	30	91	140	160	180		
Bismuth:	Rose metal		5.5	8.3	14	16			
	Wood's metal	4	17	23					
Copper:	electrolytic tough pitch	330	1300	550	400	390	380	370	350
	free cutting, leaded	200	800	460	380	380			
	phosphorus, deoxidized	7.5	42	120	190	220			
	brass, leaded	2.3	12	39	70	120			
	bronze, 68% Cu; 32% Zn	2.3	16	48	92	110			
	beryllium	2	17	36	70	90	113	172	
	german silver	0.75	7.5	17	20	23	25	30	40
	silicon bronze A		3.4	11	23	30			
	manganin	0.48	3.2	14	17	22			
	constantan	0.9	8.6	17	19	22			
Ferrous:	commercial pure iron	15	72	106	82	76	66	54	34
	plain carbon steel(AISI 1020)	13	20	58	65	65			
	plain carbon steel(AISI 1095)		8.5	31	41	45			
	3% Ni; 0.7% Cr; 0.6% Mo		6	22		33	35	36	30
	4% Si					20	24	28	26
	stainless steel	0.3	2	8	13	14	16	19	25
	27% Ni; 15% Cr		1.7	55		11	12	16	21
Gold:	colbalt thermocouple	1.2	8.6	20					
	65% Au; 35% Ag		12	24		61	89		
Indium:	85.5% In; 14.5% Pb	1.9	7.8	24	41				
Lead:	60% Pb; 40% Sn (soft solder)		28	44					
	64.35% Pb; 35.65% In	0.8	3.26	9.1		20.2			
Nickel:	80% Ni; 20% Cr					12	14	17	23
	contracid	0.2	2	7.3	9.5	13			
	inconel	0.5	4.2	12.5	13	15	16	19	26
	monel	0.9	7.1	15	20	21	24	30	43
Platinum:	90% Pt; 10% Ir					31	31.4		
	90% Pt; 10% Rh					30.1	30.5		
Silver:	silver solder		12	34	58				
	normal Ag thermocouple	48	230	310					
Tin:	60% Sn; 40% Pb	16	55	51					
Titanium:	5.5% Al; 2.5% Sn;0.2% Fe		1.8	4.3	6.4	7.8	8.4	10.8	
	4.7% Mn; 3.99% Al; 0.14% C		1.7	4.5	6.5	8.5			

THERMAL CONDUCTIVITY OF CRYSTALLINE DIELECTRICS

This table lists the thermal conductivity of a number of crystalline dielectrics, including some that find use as optical materials. Values are given at temperatures for which data are available.

Reference

Powell, R. L., and Childs, G. E., in *American Institute of Physics Handbook, 3rd Edition*, Gray, D. E., Ed., McGraw-Hill, New York, 1972.

Material	T/K	Ther. cond. W/m K	Material	T/K	Ther. cond. W/m K
AgCl	223	1.3		20	16
	273	1.2		77	270
	323	1.1		373	210
	373	1.1		573	120
Al,B silicate (tourmaline)	398	2.9		1273	29
∥ to c axis	540	3.2	Bi_2Te_3	80	6.4
	723	3.5		204	2.8
Al,Be silicate (beryl)	315	6.4		303	3.6
Al,F silicate (topaz)	315	17.7		370	4.6
∥ to c axis	358	15.6	C (diamond)	4.2	13
	417	13.3	type I	20	800
Al,Fe silicate (garnet)	315	35.8		77	3550
	358	35.4		194	1450
	377	35.6		273	1000
Al_2O_3 (sapphire):			$CaCO_3$		
36° to c axis	4.2	110	∥ to c axis	83	25
	20	3500		273	5.5
	35	6000	⊥ to c axis	83	17
	77	1100		194	6.5
⊥ to c axis	373	2.6		273	4.6
	523	3.9		373	3.6
	773	5.8	CaF_2	83	39
Al_2O_3 (sintered)	4.2	0.5		223	18
	20	23		273	10
	77	150		323	9.2
	194	48		373	9
	273	35	$CaWO_4$ (scheelite)	422	11.3
	373	26	CdTe	160	7.0
	973	8		297	3.6
Ar	8	6.0		422	2.9
	10	3.7	CsBr	223	1.2
	20	1.4		273	0.94
	77	0.31		323	0.81
As_2S_3 (glass)	283	0.16		373	0.77
	323	0.21	CsI	223	1.4
	373	0.27		273	1.2
BN	1047	36.2		323	1
	1475	22.7		373	0.95
	1928	21.9	Cu_2O (cuprite)	102	3.74
	2111	18.5		163	7.76
BaF_2	225	20		299	5.58
	260	13.4		360	4.86
	305	10.9	Fe_3O_4 (magnetite)	4.5	27.4
	370	10.5		20.5	293.0
$BaTiO_3$	5	4.2		126.5	7.4
	30	24.0		304	7.0
	40	25.0	Glass:		
	100	12.0	phoenix	4.2	0.095
	250	4.8		20	0.13
	300	6.2		77	0.37
BeO	4.2	0.3	plastic perspex	4.2	0.058

Material	T/K	Ther. cond. W/m K	Material	T/K	Ther. cond. W/m K
pyrex	20	0.074	NaCl	4.2	440
	77	0.44		20	300
	194	0.88		77	30
	273	1		273	6.4
H_2 (para + 0.5% ortho)	2.5	100		323	5.6
	3	150		373	5.4
	4	200	NaF	5	1100
	6	30		50	250
	10	3		100	90
H_2O (ice)	173	3.5	Ne	2	3.0
	223	2.8		3	4.6
	273	2.2		4.2	4.2
He^3 (high pressure)	0.6	25		10	0.8
	1	2		20	0.3
	1.5	0.57	NH_4Cl	77	17
	2	0.21		194	23
He^4 (high pressure)	0.5	42		230	38
	0.8	120		273	27
	1	24	$NH_4H_2PO_4$		
	2	0.18	∥ to optic axis	315	0.71
I_2	300	0.45		339	0.71
	325	0.42	⊥ to optic axis	313	1.26
	350	0.4		342	1.34
KBr	2	150	NiO	4.2	5.9
	4.2	360		40	400
	100	12		194	82
	273	5	SiO_2 (quartz)		
	323	4.8	∥ to c axis	20	720
	373	4.8		194	20
KCl	4.2	500		273	12
	25	140	⊥ to c axis	20	370
	80	35		194	10
	194	10		273	6.8
	273	7.0	SiO_2 (fused silica)	4.2	0.25
	323	6.5		20	0.7
	373	6.3		77	0.8
KI	4.2	700		194	1.2
	80	13		273	1.4
	194	4.6		373	1.6
	273	3.1		673	1.8
Kr	4.2	0.48	$SrTiO_3$	5	2.4
	10	1.7		30	21.0
	20	1.2		40	19.2
	77	0.36		100	18.5
LaF_3	78	7.8		250	12.5
	197	5.0		300	11.2
	274	5.4	TlBr	316	0.59
LiF	4.2	620	TlCl	311	0.75
	20	1800	TiO_2 (rutile)		
	77	150	∥ to optic axis	4.2	200
$MgO·Al_2O_3$ (spinel)	373	13		20	1000
	773	8.5		273	13
MnO	4.2	0.25	⊥ to optic axis	4.2	160
	40	55		20	690
	120	8		273	9
	573	3.5			

THERMAL CONDUCTIVITY OF CERAMICS AND OTHER INSULATING MATERIALS

Thermal conductivity values for ceramics, refractory oxides, and miscellaneous insulating materials are given here. The thermal conductivity refers to samples with the density indicated in the second column. Since most of these materials are highly variable, the values should only be considered as a rough guide.

References

1. Powell, R. L., and Childs, G. E., in *American Institute of Physics Handbook, 3rd Edition*, Gray, D. E., Ed., McGraw-Hill, New York, 1972.
2. Perry, R. H., and Green, D., *Perry's Chemical Engineers' Handbook, Sixth Edition*, McGraw-Hill, New York, 1984.

Material	Dens. g/cm³	t °C	Ther. cond. W/m K
Alumina (Al₂O₃)	3.8	100	30
		400	13
		1300	6
		1800	7.4
	3.5	100	17
		800	7.6
Al₂O₃ + MgO		100	15
		400	10
		1000	5.6
Asbestos	0.4	−100	0.07
		0	0.09
		100	0.10
Asbestos + 85% MgO	0.3	30	0.08
Asphalt	2.1	20	0.06
Beryllia (BeO)	2.8	100	210
		400	90
		1000	20
		1800	15
	1.85	50	64
		200	40
		600	23
Brick, dry	1.54	0	0.04
Brick, refractory:			
alosite		1000	1.3
aluminous	1.99	400	1.2
		1000	1.3
diatomaceous	0.77	100	0.2
		500	0.24
	0.4	100	0.08
		500	0.1
fireclay	2	400	1
		1000	1.2
silicon carbide	2	200	2
		600	2.4
vermiculite	0.77	200	0.26
		600	0.31
Calcium oxide		100	16
		400	9
		1000	7.5
Cement mortar	2	90	0.55
Charcoal	0.2	20	0.055
Coal	1.35	20	0.26
Concrete	1.6	0	0.8
Cork	0.05	0	0.03
		100	0.04
	0.35	0	0.06
		100	0.08
Cotton wool	0.08	30	0.04
Diatomite	0.2	0	0.05

Material	Dens. g/cm³	t °C	Ther. cond. W/m K
		400	0.09
	0.5	0	0.09
		400	0.16
Ebonite	1.2	0	0.16
Felt, flax	0.2	30	0.05
	0.3	30	0.04
Fuller's earth	0.53	30	0.1
Glass wool	0.2	−200 to 20	0.005
		50	0.04
		100	0.05
		300	0.08
Graphite			
100 mesh	0.48	40	0.18
20-40 mesh	0.7	40	1.29
Linoleum cork	0.54	20	0.08
Magnesia (MgO)		100	36
		400	18
		1200	5.8
		1700	9.2
MgO + SiO₂		100	5.3
		400	3.5
		1500	2.3
Mica:			
muscovite		100	0.72
		300	0.65
		600	0.69
phlogopite		100	0.66
Canadian		300	0.19
		600	0.2
Micanite		30	0.3
Mineral wool	0.15	30	0.04
Perlite, expanded	0.1	−200 to 20	0.002
Plastics:			
bakelite	1.3	20	1.4
celluloid	1.4	30	0.02
polystyrene foam	0.05	−200 to 20	0.033
mylar foil	0.05	−200 to 20	0.0001
nylon		−253	0.10
		−193	0.23
		25	0.30
polytetrafluoroethylene		−253	0.13
		−193	0.16
		25	0.26
		230	2.5
urethane foam	0.07	20	0.06
Porcelain		90	1
Rock:			
basalt		20	2
chalk		20	0.92

Material	Dens. g/cm³	t °C	Ther. cond. W/m K	Material	Dens. g/cm³	t °C	Ther. cond. W/m K
granite	2.8	20	2.2	Uranium dioxide		100	9.8
limestone	2	20	1			400	5.5
sandstone	2.2	20	1.3			1000	3.4
slate, ⊥		95	1.4	Wood:			
slate, ∥		95	2.5	balsa, ⊥	0.11	30	0.04
Rubber:				fir, ⊥	0.54	20	0.14
sponge	0.2	20	0.05	fir, ∥	0.54	20	0.35
92 percent		25	0.16	oak		20	0.16
Sand, dry	1.5	20	0.33	plywood		20	0.11
Sawdust	0.2	30	0.06	pine, ⊥	0.45	60	0.11
Shellac		20	0.23	pine, ∥	0.45	60	0.26
Silica aerogel	0.1	−200 to 20	0.003	walnut, ⊥	0.65	20	0.14
Snow	0.25	0	0.16	Wool	0.09	30	0.04
Steel wool	0.1	55	0.09	Zinc oxide		200	17
Thoria (ThO₂)		100	10			800	5.3
		400	5.8	Zirconia (ZrO₂)		100	2
		1500	2.4			400	2
Titanium dioxide		100	6.5			1500	2.5
		400	3.8	Zirconia + silica		200	5.6
		1200	3.3			600	4.6
						1500	3.7

THERMAL CONDUCTIVITY OF GLASSES

This table gives the composition of various types of glasses and the thermal conductivity k as a function of temperature. Because of the variability of glasses, the data should be regarded as only approximate.

Type of glass	SiO$_2$ (wt%)	Other oxides (wt%)		t °C	k W/m K
Vitreous silica	100			−150	0.85
				−100	1.05
				−50	1.20
				0	1.30
				50	1.40
				100	1.50
Vycor glass	96	B$_2$O$_3$	3	−100	1.00
				0	1.25
				100	1.40
Pyrex type chemically resistant borosilicate glasses	80–81	B$_2$O$_3$	12–13	−100	0.90
		Na$_2$O	4	0	1.10
		Al	2	100	1.25
Borosilicate crown glasses	60–65	B$_2$O$_3$	15–20	−100	0.65–0.75
				0	0.90–0.95
				100	1.00–1.05
	65–70	B$_2$O$_3$	10–15	−100	0.75–0.80
				0	0.95–1.00
				100	1.05–1.15
	70–75	B$_3$O$_3$	5–10	−100	0.80–0.85
				0	1.05–1.10
				100	1.15–1.20
Zinc crown glasses (i)	55–65	ZnO Remainder: B$_2$O$_3$, Al$_2$O$_3$	5–15	−100	0.88–0.92
				0	1.10–1.15
				100	1.15–1.25
		ZnO Remainder: Na$_2$O, K$_2$O	5–15	−100	0.60–0.70
				0	0.70–0.90
				100	0.85–0.95
		ZnO Remainder: B$_2$O$_3$, Al$_2$O$_3$	15–25	−100	0.88–0.92
				0	1.10–1.15
				100	1.15–1.20
		ZnO Remainder: Na$_2$O, K$_2$O	15–25	−100	0.65–0.80
				0	0.85–0.95
				100	0.90–1.05
Zinc crown glasses (ii)	65–75	ZnO Remainder: B$_2$O$_3$, Al$_2$O$_3$	5–15	−100	0.88–0.92
				0	1.15–1.15
				100	1.20–1.30
		ZnO Remainder: Na$_2$O, K$_2$O	5–15	−100	0.70–0.85
				0	0.90–1.05
				100	1.00–1.15
		ZnO	15–25	−100	0.90–0.95

Type of glass	SiO$_2$ (wt%)	Other oxides (wt%)		t °C	k W/m K
		Remainder: B$_2$O$_3$, Al$_2$O$_3$		0	1.15–1.15
				100	1.20–1.25
		ZnO	15–25	−100	0.65–0.85
		Remainder: Na$_2$O, K$_2$O		0	0.85–1.00
				100	1.05–1.20
Barium crown glasses	31	B$_2$O$_3$	12	−100	0.55
		Al$_2$O$_3$	8	0	0.70
		BaO	48	100	0.80
	41	B$_2$O$_3$	6	−100	0.60
		Al$_2$O$_3$	2	0	0.75
		ZnO	8	100	0.85
		BaO	43		
	47	B$_2$O$_3$	4	−100	0.65
		Na$_2$O	1	0	0.75
		K$_2$O	7	100	0.90
		ZnO	8		
		BaO	32		
	65	B$_2$O$_3$	2	−100	0.70
		Na$_2$O	5	0	0.90
		K$_2$O	15	100	1.00
		ZnO	2		
		BaO	10		
Borate glasses					
Borate flint glass	9	B$_2$O$_3$	36	−100	0.55
		Na$_2$O	1	0	0.65
		K$_2$O	2	100	0.80
		PbO	36		
		Al$_2$O$_3$	10		
		ZnO	6		
Borate flint glass	0	B$_2$O$_3$	56	−100	0.50
		Al$_2$O$_3$	12	0	0.65
		PbO	32	100	0.85
Borate flint glass	0	B$_2$O$_3$	43	−100	0.40
		Al$_2$O$_3$	5	0	0.55
		PbO	52	100	0.70
Borate glass	4	B$_2$O$_3$	55	−100	0.65
		Al$_2$O$_3$	14	0	0.80
		PbO	11	100	0.90
		K$_2$O	4		
		ZnO	12		
Borate crown glass	0	B$_2$O$_3$	64	−100	0.50
		Na$_2$O	8	0	0.65
		K$_2$O	3	100	0.85
		BaO	4		
		PbO	3		
		Al$_2$O$_3$	18		
Light borate crown glass	0	B$_2$O$_3$	69	−100	0.55
		Na$_2$O	8	0	0.70

Type of glass	Composition: SiO_2 (wt%)	Other oxides (wt%)		t °C	k W/m K
		BaO	5	100	0.90
		Al_2O_3	18		
Zinc borate glass	0	B_2O_3	40	−100	0.65
		ZnO	60	0	0.75
				100	0.85
Phosphate crown glasses					
Potash phosphate glass	0	P_2O_5	70	0	0.75
		B_2O_3	3	100	0.85
		K_2O	12		
		Al_2O_3	10		
		MgO	4		
Baryta phosphate glass	0	P_2O_5	60	45	0.75
		B_2O_3	3		
		Al_2O_3	8		
		BaO	28		
Soda-lime glasses	75	Na_2O	17	−100	0.75
		CaO	8	0	0.95
				100	1.10
	75	Na_2O	12	−100	0.90
		CaO	13	0	1.10
				100	1.15
	72	Na_2O	15	−100	0.80
		CaO	11	0	1.00
		Al_2O_3	2	100	1.15
	65	Na_2O	25	−100	0.65
		CaO	10	0	0.85
				100	0.95
	65	Na_2O	15	−100	0.85
		CaO	20	0	1.00
				100	1.10
	60	Na_2O	20	−100	0.75
		CaO	20	0	0.90
				100	1.00
Other crown glasses					
Crown glass	75	Na_2O	9	−100	0.80
		K_2O	11	0	1.00
		CaO	5	100	1.10
High dispersion crown glass	68	Na_2O	16	−100	0.65
		ZnO	3	0	0.85
		PbO	13	100	1.00
Miscellaneous flint glasses					
(i) Silicate flint glasses					
Light flint glasses	65	PbO	25	−100	0.65–0.70
		Others	10	0	0.88–0.92
				100	1.00–1.05
	55	PbO	35	−100	0.60–0.65

Type of glass	SiO₂ (wt%)	Other oxides (wt%)		t °C	k W/m K
		Others	10	0	0.75–0.85
				100	0.88–0.92
Ordinary flint glass	45	PbO	45	–100	0.50–0.60
		Others	10	0	0.65–0.75
				100	0.80–0.85
Heavy flint glass	35	PbO	60	–100	0.45–0.50
		Others	5	0	0.60–0.65
				100	0.70–0.75
Very heavy flint glasses	25	PbO	73	–100	0.40–0.45
		Others	2	0	0.55–0.60
				100	0.63–0.67
	20	PbO	80	–100	0.40
				0	0.50
				100	0.60
(ii) Borosilicate flint glass	33	B₂O₃	31	–100	0.65
		PbO	25	0	0.85
		Al₂O₃	7	100	0.95
		K₂O	3		
		Na₂O	1		
(iii) Barium flint glass	50	BaO	24	–100	0.60
		PbO	6	0	0.70
		K₂O	8	100	0.85
		Na₂O	3		
		ZnO	8		
		Sb₂O₃	1		
Other glasses					
Potassium glass	59	K₂O	33	50	0.88–0.92
		CaO	8		
Iron glasses	63	Fe₂O₃	10	–100	0.80
		Na₂O	17	0	0.95
		MgO	4	100	1.05
		CaO	3		
		Al₂O₃	2		
	67	Fe₂O₃	15	0	0.88–0.92
		Na₂O₃	18	100	1.00–1.05
	62	Fe₂O₃	20	0	0.85–0.90
		Na₂O	18	100	0.95–1.00
Rock glasses					
Obsidian				0	1.35
Artificial diabase				100	1.25

The following table heading applies:

		Composition			
Type of glass	SiO₂ (wt%)	Other oxides (wt%) Others	10	t °C	k W/m K

THERMOELECTRIC PROPERTIES OF METALS AND SEMICONDUCTORS

Lev I. Berger

There are three thermoelectric phenomena that result from correlation between propagation of heat through a conductor and displacement of the current carriers in the conductor. The Seebeck effect (Ref. 1) consists of formation of an electric current in an electrical circuit formed by two dissimilar conductors if the contacts between the conductors are held at different temperatures. A reverse phenomenon, the Peltier effect (Ref. 2), consists of formation of a temperature difference between the contacts in a circuit of this type if an electric current is created in the circuit by an external current source to which the circuit is connected. W. Thomson (Lord Kelvin), who explained both effects (Refs. 3,4), predicted and experimentally confirmed the existence of another thermoelectric phenomenon, named the Thomson effect, which consists of absorption or release of heat in a uniform conductor with a current passing through it when a temperature gradient (positive or negative) is present along the current direction.

The electromotive force, ΔU, which creates the Seebeck current in the circuit, is the algebraic sum of the emf's created in each of the conductors, and is proportional to the temperature difference, ΔT, between the electrical contact points: $\Delta U = \Delta U_1 + \Delta U_2 = \alpha_1 \Delta T + \alpha_2 \Delta T$. The coefficient of proportionality, α, called the Seebeck coefficient or thermoelectric power or thermal electromotive force (thermal emf), of each of the two materials depends on the electrical properties and temperature of the material. The Peltier effect is measured by the amount of heat, ΔQ, released or absorbed in a unit of time (in addition to the Joule heat) at a contact of two dissimilar conductors with electric current ΔI passing through the contact: $\Delta Q = \Pi \cdot \Delta I$. Thomson showed that $\Pi = \alpha T$. The Thomson effect's heat, dQ, released or absorbed in a unit of time along a part of a conductor of length dx is proportional to the current magnitude I, the temperature gradient along the conductor $\partial T/\partial x$, and the increment dx: $dQ = \tau I(\partial T/\partial x)dx$. Thomson showed that the magnitude of the coefficient of proportionality, τ, later named the Thomson coefficient, depends on only the properties of the conductor and the ambient temperature and correlates with the other thermoelectric parameters of a material through the equation $\tau = T(\partial \alpha/\partial T)$.

Another thermoelectric phenomenon, called the Bridgman effect or the internal Peltier effect (Ref. 5), occurs when an electric current passes through an anisotropic crystal, resulting in absorption or liberation of heat because of non-uniformity in current distribution.

In view of the correlations between α, Π, and τ, we need only to present data for one of these parameters, namely, thermal emf α and its dependence on temperature. These values are presented below, first for metals and then for semiconductors. In accordance with modern theory of solids, thermal emf in semiconductors is up to three or even four orders of magnitude higher than that in metals (Ref. 9).

References

1. Seebeck, T. J., *Abhand. Deut. Akad. Wiss. Berlin*, 265–373, 1822.
2. Peltier, J. C. A, *Ann. Chem.*, LVI, 371–387, 1834.
3. Thomson, W., *Proc. Roy. Soc. Edinburgh*, 91–98, 1851.
4. Thomson, W., *Math. and Phys. Papers, Cambridge*, 1, 558, 1882; 2, 306, 1882.
5. Bridgman, P. W., *Proc. Natl. Acad. Sci. USA*, 13(2), 46–50, 1927; *Phys. Rev.* 30, 911–921 (1927).
6. Blatt, F. J., *Thermoelectric Power of Metals*. Plenum Press, NY 1976.
7. Foiles, C. L., Thermopower of Pure Metals and Dilute Alloys, in *Landolt–Bornstein. Numerical Data and Functional Relationships in Science and Technology. New Series.* Group III, v. 15, Metals. Springer-Verlag, NY, 1985.
8. Burkov, A. T., and Vedernikov, M. V., in *CRC Handbook of Thermoelectrics*, D. M. Rowe, Ed., CRC Press, Boca Raton, FL, 1995, pp. 387–399.
9. Ioffe, A. F., *Semiconductor Thermoelements and Thermoelectric Cooling*, Infosearch Ltd., 1957.
10. Berger, L. I., and Prochuchan, V. D., *Ternary Diamond-Like Semiconductors*, Cons. Bureau, Plenum Press, New York, 1969.
11. Rowe, D. M., Ed., *Thermoelectrics Handbook Macro to Nano*, Taylor & Francis, Boca Raton, 2006.
12. Berger, L. I., *Semiconductor Materials*, CRC Press, Boca Raton, FL, 1996.
13. Glazov, V. M., Tshizhevskaya, S. N., and Glagoleva, N. N., *Liquid Semiconductors*, Nauka Publ. House, Moscow, 1967.
14. Shay, J. L., and Wernick, J. H., *Ternary Chalcopyrite Semiconductors: Growth, Electronic Properties and Applications*, Pergamon Press, New York, 1975.
15. Heikes, R. R., and Ure, R. W., *Thermoelectricity: Science and Engineering*, Interscience Publ., New York, 1961.
16. Goland, A. N., and Ewald, A. W., *Phys. Rev.* 104, 948 (1956).
17. Tauc J., *Photo and Thermoelectric Effects in Semiconductors*, Pergamon, New York, 1962.
18. Dugdale, J. S., *The Electrical Properties of Metals and Alloys*, Edward Arnold, London, 1977.
19. Rowe, D. M., Ed., *CRC Handbook of Thermoelectrics*, CRC Press, Boca Raton, FL, 1994.

Thermoelectric Properties of Elemental Metals

	Thermal emf $\alpha(T)$ in µV/K at Temperature T				
	100 K	300 K	500 K	1000 K	1500 K
Ag	0.73	1.51	2.82	7.95	
Al	−2.2	−1.66	1.96		
Au	0.82	1.94	2.86	3.85	
Ba	−4	12.1	28.5		
Be	−2.5	1.7	2.7	7.9	
Ca	1.05	10.3	17.1		
Cd	−0.05	2.55			
Ce	13.6	6.2	5.2	−4.8	
Co	−8.43	−30.8	−44.8	−35.9	−7.8
Cr	5	21.8	16.6	17.9	5.7

	Thermal emf $\alpha(T)$ in µV/K at Temperature T				
	100 K	300 K	500 K	1000 K	1500 K
Cs		−0.9			
Cu	1.19	1.83	2.83	5.36	
Dy	−4.1	−1.8	0.9	2.3	
Er	−3.8	−0.1	1.9	4.2	
Eu	5.3	24.5	46		
Fe	11.6	15	3	0.4	
Ga	0.5				
Gd	−4.6	−1.6	−0.5	−0.8	
Hf	0	5.5	5.7	−0.5	
Ho	−6.7	−1.6	1.4	2.8	

	Thermal emf $\alpha(T)$ in μV/K at Temperature T				
	100 K	300 K	500 K	1000 K	1500 K
In	0.56	1.68			
Ir	1.42	0.86	−0.1	−2.7	−5.7
K	−5.2	−13.7			
La	0.1	1.7	2	−1.7	
Li	4.3				
Lu	−6.9	−4.3	−2.6	0	
Mg	−2.1	−1.46			
Mn	−2.5	−9.8	−8.4	−1.5	
Mo	0.1	5.6	11.4	17.4	13.7
Na	−2.6	−6.3			
Nb	1.05	−0.44	−1.1	0.45	3.2
Nd	−4	−2.3	0	−1.2	
Ni	−8.5	−19.5	−25.8	−29.9	
Np	8.9	−3.1			
Os	−3.2	−4.4	−4.7	−6.3	−8.5
Pb	−0.58	−1.05	−1.5		
Pd	1.1	−10.7	−16.3	−32.3	−46.4
Pt	−4.1	−5.3	−7.9	−8.2	
Pu	12				
Rb	−3.6	−10			

	Thermal emf $\alpha(T)$ in μV/K at Temperature T				
	100 K	300 K	500 K	1000 K	1500 K
Re	−1.4	−5.9	−5.9	−1.9	1.8
Rh	0.8	0.6	0.5	−1.5	
Ru	0.3	−1.4	−1.8	−4.2	−7.5
Sc	−14.3	−19	−17.5	−5.4	10.2
Sm	0.7	1.2	0.6	−3	
Sn	−0.04	−1			
Sr	−3	1.1	4.2		
Ta	0.7	−1.9	−2.3	1.6	7.2
Tb	−1.6	−1	0.3	0.6	
Th	0.6	−3.2	−9.2	−14.3	−10.4
Ti	−2	9.1	5.3	−3.1	−0.5
Tl	0.6	0.3	−1.5		
Tm	−1.3	1.9	2.7	2.2	
U	3	7.1	11	16.7	
V	2.9	0.23	1.1	4.6	
W	−4.4	0.9	9	19.8	21.3
Y	−5.1	−0.7	0.3	2.9	6.6
Yb	5.1	30	20.3	12.3	
Zn	0.7	2.4			
Zr	4.4	8.9	4.6	−3	1.1

Thermoelectric Properties of Selected Semiconductors; Values Near Room Temperature unless Otherwise Indicated

Material	α/μV K^{-1}	Material	α/μV K^{-1}	Material	α/μV K^{-1}
		Elemental Semiconductors			
B	600 (500 K)	n-Si	300	p-Si	−500
n-Ge	600	p-Ge	−830	α-Sn	−40 (250 K)
		I-VI Compounds			
Cu_2S	327	Cu_2Se	135	Cu_2Te	40
Ag_2Te	120				
		II-VI Compounds			
ZnO	300	CdS	700	ZnSe	55
CdSe	200				
		III-V Compounds			
GaN	70	GaP	1200	InP	−400
AlAs	70	n-GaAs	380	p-GaAs	−310
InAs	200	AlSb	500	n-GaSb	250
p-GaSb	−55	n-InSb	240	p-InSb	200
		V-VI Compounds			
Sb_2Te_3	110	$n-Bi_2Te_3$	224	$p-Bi_2Te_3$	−227
		I-III-VI Compounds			
$CuAlS_2$	50	$AgInSe_2$	−370	$CuTlTe_2$	80
$CuGaSe_2$	40	$AgTlSe_2$	800	$AgAlTe_2$	321
$CuInSe_2$	340	$CuGaTe_2$	340	$AgGaTe_2$	950
$CuTlSe_2$	−5	$CuInTe_2$	260	$AgInTe_2$	298
$AgGaSe_2$	90	$CuTlTe_2$	80		
		I-IV-VI Compounds			
Cu_2GeS_3	300	Cu_2GeSe_3	100	Cu_2GeTe_3	10
Cu_2SnS_3	600	Cu_2SnSe_3	250	Cu_2SnTe_3	30
		I-V-VI Compounds			
Cu_3AsS_4	130	Cu_3AsSe_4	120	Cu_3SbSe_4	200
		II-IV-V Compounds			
$ZnGeP_2$	1200	$ZnSiAs_2$	1100	$CdGeAs_2$	190
$CdSnAs_2$	600				

FERMI ENERGY AND RELATED PROPERTIES OF METALS

Lev I. Berger

In the classical Drude theory of metals, the Maxwell-Boltzmann velocity distribution of electrons is used. It states that the number of electrons per unit volume with velocities in the range of $d\bar{v}$ about any magnitude $\bar{v}$ at temperature T is

$$f_B(\bar{v})d\bar{v} = n\left(\frac{m}{2\pi k_B T}\right)\exp\left(-\frac{mv^2}{2k_B T}\right)d\bar{v}$$

where n is the total number of conduction electrons in a unit volume of a metal, m is the free electron mass, and k_B is the Boltzmann constant. In an attempt to explain a substantial discrepancy between the experimental data on the specific heat of metals and the values calculated on the basis of the Drude model, Sommerfeld suggested a model of the metal in which the Pauli exclusion principle is applied to free electrons. In this case, the Maxwell-Boltzmann distribution is replaced by the Fermi-Dirac distribution:

$$f(\bar{v})d\bar{v} = 2\left(\frac{m}{h}\right)^3 d\bar{v}\left\{\exp\left[\left(\frac{mv^2}{2} - k_B T_0\right)\Big/k_B T\right]+1\right\}^{-1}$$

Here h is the Planck constant and T_0 is a characteristic temperature which is determined by the normalization condition

$$n = \int d\bar{v}\cdot f(\bar{v})$$

The magnitude of T_0 is quite high; usually, $T_0 > 10^4$ K. So, at common temperatures ($T < 10^3$ K), the free electron density of a metal is much smaller than in the case of the Maxwell-Boltzmann distribution. This allows us to explain why the experimental data on specific heat for metals are close to those for insulators.

The maximum kinetic energy the electrons of a metal may possess at $T = 0$ K is called the Fermi energy, e.g.,

$$E_F = \frac{\hbar^2 k_F^2}{2m} = \left(\frac{e^2}{2k_B}\right)(k_F r_B)^2$$

where k_F is the Fermi momentum or the Fermi wave vector

$$k_F = (3\pi^2 n)^{1/3}$$

e is the electron charge, and r_B is the Bohr radius

$$r_B = \hbar^2/me^2 = 0.529\cdot 10^{-10} \text{ m}$$

Another, more common expression for the Fermi energy is

$$E_F = \tfrac{1}{2}mv_F^2$$

where $v_F = \hbar k_F/m$ is the Fermi velocity which can be expressed using the concept of the electron radius, r_s. It is equal to radius of a sphere occupied by one free electron. If the total volume of a metal sample is V and the number of conduction electrons in this volume is N, then the volume per electron is equal to

$$\frac{V}{N} = \frac{1}{n} = \frac{4}{3}\pi r_s^3$$

and

$$r_s = \left(\frac{3}{4\pi n}\right)^{1/3}$$

The following table contains information pertinent to the Sommerfeld model for some metals. The magnitudes of T_0 are calculated using the expression

$$T_0 = \frac{E_F}{k_B} = \frac{58.2\cdot 10^4}{(r_s/r_B)^2} \text{ K}$$

Ground State Properties of the Electron Gas in Some Metals

Metal	Valency	$n/10^{28}$ m^{-3}	r_s/pm	r_s/r_B	E_F/eV	$T_0/10^4$ K	$k_F/10^{10}$ m^{-1}	$v_F/10^6$ m s^{-1}
Li[a]	1	4.70	172	3.25	4.74	5.51	1.12	1.29
Na[b]	1	2.65	208	3.93	3.24	3.77	0.92	1.07
K[b]	1	1.40	257	4.86	2.12	2.46	0.75	0.86
Rb[b]	1	1.15	275	5.20	1.85	2.15	0.70	0.81
Cs[b]	1	0.91	298	5.62	1.59	1.84	0.65	0.75
Cu	1	8.47	141	2.67	7.00	8.16	1.36	1.57
Ag	1	5.86	160	3.02	5.49	6.38	1.20	1.39
Au	1	5.90	159	3.01	5.53	6.42	1.21	1.40
Be	2	24.7	99	1.87	14.3	16.6	1.94	2.25
Mg	2	8.61	141	2.66	7.08	8.23	1.36	1.58
Ca	2	4.61	173	3.27	4.69	5.44	1.11	1.28
Sr	2	3.55	189	3.57	3.93	4.57	1.02	1.18
Ba	2	3.15	196	3.71	3.64	4.23	0.98	1.13
Nb	1	5.56	163	3.07	5.32	6.18	1.18	1.37
Fe	2	17.0	112	2.12	11.1	13.0	1.71	1.98
Mn[c]	2	16.5	113	2.14	10.9	12.7	1.70	1.96
Zn	2	13.2	122	2.30	9.47	11.0	1.58	1.83

Metal	Valency	$n/10^{28}$ m^{-3}	r_S/pm	r_S/r_B	E_F/eV	$T_0/10^4$ K	$k_F/10^{10}$ m^{-1}	$v_F/10^6$ m s^{-1}
Cd	2	9.27	137	2.59	7.47	8.68	1.40	1.62
Hg[a]	2	8.65	140	2.65	7.13	8.29	1.37	1.58
Al	3	18.1	110	2.07	11.7	13.6	1.75	2.03
Ga	3	15.4	116	2.19	10.4	12.1	1.66	1.92
In	3	11.5	127	2.41	8.63	10.0	1.51	1.74
Tl	3	10.5	131	2.48	8.15	9.46	1.46	1.69
Sn	4	14.8	117	2.22	10.2	11.8	1.64	1.90
Pb	4	13.2	122	2.30	9.47	11.0	1.58	1.83
Bi	5	14.1	119	2.25	9.90	11.5	1.61	1.87
Sb	5	16.5	113	2.14	10.9	12.7	1.70	1.96

[a] At 78 K.
[b] At 5 K.
[c] α-phase.
The data in the table are for atmospheric pressure and room temperature unless otherwise noted.

References

1. Drude, P., *Ann. Physik*, 1, 566, 1900; ibid., 3, 369, 1900.
2. Sommerfeld, A. and Bethe, H., *Handbuch der Physik*, Chapter 3, Springer, 1933.
3. Wyckoff, R. W. G., *Crystal Structures*, 2nd. ed., Interscience, 1963.
4. Ashcroft, N. W. and Mermin, N. D., *Solid State Physics*, Holt, Rinehart and Winston, 1976.

PROPERTIES OF COMMERCIAL METALS AND ALLOYS

This table gives typical values of mechanical, thermal, and electrical properties of several common commercial metals and alloys. Values refer to ambient temperature (0 to 25 °C). All values should be regarded as typical, since these properties are dependent on the particular type of alloy, heat treatment, and other factors. Values for individual specimens can vary widely.

References

1. *ASM Metals Reference Book, Second Edition*, American Society for Metals, Metals Park, OH, 1983.
2. Lynch, C. T., *CRC Practical Handbook of Materials Science*, CRC Press, Boca Raton, FL, 1989.
3. Shackelford, J. F., and Alexander, W., *CRC Materials Science and Engineering Handbook*, CRC Press, Boca Raton, FL, 1991.

Common name	Thermal conductivity W/cm K	Density g/cm³	Coeff. of linear expansion 10^{-6}/°C	Electrical resistivity μΩ cm	Modulus of elasticity GPa	Tensile strength MPa	Approx. melting point °C
Ingot iron	0.7	7.86	11.7	9.7	205	–	1540
Plain carbon steel AISI-SAE 1020	0.52	7.86	11.7	18	205	450	1515
Stainless steel type 304	0.15	7.9	17.3	72	195	550	1425
Cast gray iron	0.47	7.2	10.5	67	90	180	1175
Malleable iron		7.3	12	30	170	345	1230
Hastelloy C	0.12	8.94	11.3	125	200	780	1350
Inconel	0.15	8.25	11.5	103	200	800	1370
Aluminum alloy 3003, rolled	1.9	2.73	23.2	3.7	70	110	650
Aluminum alloy 2014, annealed	1.9	2.8	23.0	3.4	70	185	650
Aluminum alloy 360	1.5	2.64	21.0	7.5	70	325	565
Copper, electrolytic (ETP)	3.9	8.94	16.5	1.7	120	300	1080
Yellow brass (high brass)	1.2	8.47	20.3	6.4	100	300-800	930
Aluminum bronze	0.7	7.8	16.4	12	120	400-600	1050
Beryllium copper 25	0.8	8.23	17.8	7	130	500-1400	925
Cupronickel 30%	0.3	8.94	16.2		150	400-600	1200
Red brass, 85%	1.6	8.75	18.7	11	90	300-700	1000
Chemical lead	0.35	11.34	29.3	21	13	17	327
Antimonial lead (hard lead)	0.3	10.9	26.5	23	20	47	290
Solder 50-50	0.5	8.89	23.4	15	–	42	215
Magnesium alloy AZ31B	1.0	1.77	26	9	45	260	620
Monel	0.3	8.84	14.0	58	180	545	1330
Nickel (commercial)	0.9	8.89	13.3	10	200	460	1440
Cupronickel 55-45 (constantan)	0.2	8.9	18.8	49	160	–	1260
Titanium (commercial)	1.8	4.5	8.5	43	110	330-500	1670
Zinc (commercial)	1.1	7.14	32.5	6	–	130	419
Zirconium (commercial)	0.2	6.5	5.85	41	95	450	1855

HARDNESS OF MINERALS AND CERAMICS

There are several hardness scales for describing the resistance of a material to indentation or scratching. This table lists a number of common materials in order of increasing hardness. Values are given, when available, on three different hardness scales: the original Mohs Scale (range 1 to 10); the modified Mohs Scale (range 1 to 15), and the Knoop Hardness Scale. In the last case, a load of 100 g is assumed.

Reference

Shackelford, J. F. and Alexander, W., *CRC Materials Science and Engineering Handbook*, CRC Press, Boca Raton, FL, 1991.

Material	Formula	Mohs	Modified mohs	Knoop
Graphite	C	0.5		
Talc	$3MgO \cdot 4SiO_2 \cdot H_2O$	1	1	
Alabaster	$CaSO_4 \cdot 2H_2O$	1.7		
Gypsum	$CaSO_4 \cdot 2H_2O$	2	2	32
Halite (rock salt)	NaCl	2		
Stibnite (antimonite)	Sb_2S_3	2.0		
Galena	PbS	2.5		
Mica		2.8		
Calcite	$CaCO_3$	3	3	135
Barite	$BaSO_4$	3.3		
Marble		3.5		
Aragonite	$CaCO_3$	3.5		
Dolomite	$CaMg(CO_3)_2$	3.5		
Fluorite	CaF_2	4	4	163
Magnesia	MgO	5		370
Apatite	$CaF_2 \cdot 3Ca_3(PO_4)_2$	5	5	430
Opal		5		
Feldspar (orthoclase)	$K_2O \cdot Al_2O \cdot 6SiO_2$	6	6	560
Augite		6		
Hematite	Fe_2O_3	6		750
Magnetite	Fe_3O_4	6		
Rutile	TiO_2	6.2		
Pyrite	FeS_2	6.3		
Agate	SiO_2	6.5		
Uranium dioxide	UO_2	6.7		600
Silica (fused)	SiO_2		7	
Quartz	SiO_2	7	8	820
Flint		7		
Silicon	Si	7		
Andalusite	Al_2OSiO_4	7.5		
Zircon	$ZrSiO_4$	7.5		
Zirconia	ZrO_2			1200
Aluminum nitride	AlN			1225
Beryl	$Be_3Al_2Si_6O_{18}$	7.8		
Beryllia	BeO			1300
Topaz	$Al_2SiO_4(OH,F)_2$	8	9	1340
Garnet	$Al_2O_3 \cdot 3FeO \cdot 3SiO_2$		10	1360
Emery	Al_2O_3 (impure)	8		
Zirconium nitride	ZrN	8+		1510
Zirconium boride	ZrB_2			1560
Titanium nitride	TiN	9		1770
Zirconia (fused)	ZrO_2		11	
Tantalum carbide	TaC			1800
Tungsten carbide	WC			1880
Corundum (alumina)	Al_2O_3	9		2025
Zirconium carbide	ZrC			2150
Alumina (fused)	Al_2O_3		12	
Beryllium carbide	Be_2C			2400
Titanium carbide	TiC			2470
Carborundum (silicon carbide)	SiC	9.3	13	2500
Aluminum boride	AlB			2500
Tantalum boride	TaB_2			2600
Boron carbide	B_4C		14	2800
Boron	B	9.5		
Titanium boride	TiB_2			2850
Diamond	C	10	15	7000

Section 13
Polymer Properties

ABBREVIATIONS USED IN POLYMER SCIENCE AND TECHNOLOGY

ABA	triblock copolymers; acrylonitrile-butadiene acrylate	G	molar attraction constant
ABS	copolymer of acrylonitrile, butadiene, and styrene	GF	glass reinforced
ACS	acrylonitrile-chlorinated polyethylene styrene terpolymer	GRS	poly(butadiene-co-styrene)
		HDPE	high-density polyethylene
AIBN	2,2'-azobisisobutyronitrile	HIPS	high-impact polystyrene
AMA	acrylate maleic anhydride terpolymer	HMC	high strength molding compound
AMMA	acrylate-methyl methacrylate copolymer	HMWHDPE	high-molecular-weight high-density polyethylene
AN	acrylonitrile	I	ionomer
AP	ethylene-propylene copolymers	IIR	butyl rubber
APO	amorphous polyolefin	IPN	interpenetrating polymer network
AS	acrylonitrile styrene copolymer	K	constant in Mark-Houwink equation
ASA	acrylonitrile-styrene-acrylonitrile block	LC	liquid crystal
ATR	attenuated total reflectance spectroscopy	LCP	liquid crystal polymer
AU	polyurethane	LDPE	low-density polyethylene
BMC	bulk molding compound	LLDPE	linear low-density polyethylene
BMI	bis maleimide	LPE	linear polyethylene
BPO	benzoyl peroxide	MA	maleic anhydride
CA	cellulose acetate	MABS	methyl methacrylate ABS copolymer
CAB	cellulose acetate butyrate	MBS	methyl methacrylate butadiene styrene terpolymer
CAP	cellulose acetate proprionate		
CAR	carbon fiber	MDPE	medium density polyethylene
CED	cohesive energy density	MDI	methylene diphenylisocyanate
CFRP	carbon reinforced plastics	MF	melamine-formaldehyde resin
CMC	carboxymethylcellulose	MP	melamine phenolic
CN	cellulose nitrate	MWD	molecular weight distribution
COC	cycloolefin copolymer	M_n	number-average molecular weight
COP	copolyester thermoplastic elastomer	M_v	viscosity-average molecular weight
CPE	chloronated polyethylene	M_w	weight-average molecular weight
CPVC	chlorinated poly(vinyl chloride)	M_z	Z-average molecular weight
CR	neoprene	NBR	poly(butadiene-co-acrylonitrile); nitrile butadiene rubber
CTA	cellulose triacetate		
CTFE	chlorotrifluoroethylene	NR	natural rubber
C_s	chain transfer constant	OSA	olefin modified styrene acrylonitrile
DAIP	diallyl isophthalate plasticizer	P	phenolic
DAP	dially phthalate plasticizer	PA	polyamide; nylon
DNA	deoxyribonucleic acid	PA6	polyamide 6, nylon6
DP	degree of polymerization	PA11	polyamide 11, nylon11
DRS	dynamic reflectance spectroscopy	PA12	polyamide 12, nylon12
DS	degree of substitution	PA46	polyamide 46, nylon46
EAA	ethylene acrylic acid copolymer	PA66	polyamide 66, nylon6-6
EC	ethyl cellulose	PA66/6T	polyamide 66/6T
ECTFE	ethylene-chlorotrifluoroethylene copolymer	PA610	polyamide 610, nylon6-10
EEA	ethylene-ethyl acetate copolymer	PA612	polyamide 612, nylon6-12
EGG	Einstein-Guth-Gold equation	PA666	polyamide 666
EMAC	ethylene-methyl acrylate copolymer	PAA	poly(acrylic acid)
EnBA	ethylene n-butyl acetate	PAEK	polyaryletherketone
EP	epoxy resin	PAI	polyamide-imide
EPDM	poly(ethylene-co-propylene) crosslinked	PAK	polyester alkyd
EPM	ethylene-propylene copolymer	PAL	polyanaline
EPR	ethylene propylene rubber	PAN	polyacrylonitrile
EPS	expanded polystyrene	PARA	polyaryl amide
ET	thiokol	PAS	polyarylsulfone
ETFE	ethylene tetrafluoroethylene polymer	PB	polybutylene
EU	polyether polyurethane	PBAN	polybutylene-acrylonitrile copolymer
EVA	ethylene-vinyl acetate copolymer	PBD	polybutadine
EVOH	ethylene-vinyl alcohol copolymer	PBI	polybenzimidazole
FEP	fluorinated ethylene propylene	PBN	poly(butylene napthalate)
FRP	fibrous glass reinforced polyester; fiber reinforced plastic	PBS	polybutadiene-styrene copolymer
		PBT	poly(butylene terephthalate)
		PC	polycarbonate

PC/ABS	polycarbonate/acrylonitrile butadiene styrene blend
PCB	polychlorinated biphenyl
PCCE	poly(cyclohexylene dimethylene cyclohexanedicarboxylate), glycol and acid comonomer
PCL	polycaprolactone
PCT	poly(cyclohexylene terephthalate)
PCTA	poly(cyclohexylene dimethylene terephthalate) copolyester
PCTFE	polychlorotrifluoroethylene
PCTG	poly(cyclohexylene dimethylene terephthalate) copolyester
PCT-G	glycol modified polycyclohexyl terephthallate
PE	polyethylene
PEBA	polyether block amide or polyester block amide
PEEK	poly(ether ether ketone)
PEG	poly(ethylene glycol)
PEI	polyetherimide
PEK	polyetherketone
PEKEKK	polyetherketone etherketone ketone
PEKK	polyetherketoneketone
PEN	poly(ethylene napthalene)
PEO	poly(ethylene oxide)
PES	polyethersulfone
PET	poly(ethylene terephthalate)
PET-G	glycol modified poly(ethylene terephthalate)
PEX	crosslinked polyethylene
PF	phenol-formaldehyde resin
PFA	perfluoroalkoxy
PI	polyimide, polyisoprene
PIB	polyisobutylene
PIR	polyisocyanurate
PK	polyketone
PLGA	poly(lactic-co-glycolic acid)
PMAN	polymethactylonitrile
PMMA	poly(methyl methacrylate)
PMP	polymethylpentene
PMS	polymethylstyrene
PNF	poly(phosphonitrilic fluorides)
PO	polyolefin
POM	polyoxymethylene, polyformaldehyde, acetals
PP	polypropylene
PPA	polyphthalamide
PPC	chlorinated polypropylene, polyphthalate carbonate
PPE	poly(phenylene ether)
PPI	polymeric polyisocyanate
PPO	poly(phenylene oxide)
PPOX	poly(propylene oxide)
PPS	poly(phenylene sulfide)
PPSU	poly(phenylene sulfone)
PPT	poly(propylene terephthalate)
PS	polystyrene
PS-b-PI	polystyrene/polyisoprene block copolymer
PSO, PSU	polysulfone
PTFE	polytetrafluoroethylene, Teflon
PTME	poly(tetramethylene terephthalate)
PTMT	poly(tetramethylene terephthalate)
PU	polyurethane
PUR	polyurethane rubber

PVA	poly(vinyl alcohol); sometimes poly(vinyl acetate)
PVAc	poly(vinyl acetate)
PVB	poly(vinyl butyral)
PVC	poly(vinyl chloride)
PVCA	copolymer of vinyl chloride and vinyl acetate
PVDA	polyvinylidene acetate
PVDC	poly(vinylidene chloride)
PVDF	poly(vinylidene fluoride)
PVF	poly(vinyl fluoride)
PVK	poly(vinyl carbazole)
PVOH	poly(vinyl alcohol)
PVP	poly(vinyl pyrrolidone)
RIM	reaction injection molding
RNA	ribonucleic acid
ROMP	ring opening metathesis polymerization
ROP	ring opening polymerization
S	radius of gyration
SAN	poly(styrene-co-acrylonitrile)
SB	styrene butadiene copolymer
SBR	poly(butadiene-co-styrene) elastomer
SBS	styrene butadiene styrene block copolymer
SEBS	styrene ethylene butylene styrene block copolymer
SI	silicon
SIS	styrene isoprene styrene block copolymer
SMA	poly(styrene-co-maleic anhydride)
SMC	sheet molding compound
SMMA	styrene methyl methacrylate copolymer
SMS	styrene/a-methyl styrene
SN	sulfur nitride
SR	synthetic rubber
SRP	styrene-rubber plastics
SVA	styrene vinyl acrylonitrile
TDI	toluenediisocyanate
TEO	thermoplastic elastic olefin
TGA, TG	thermal gravimetric analysis
TMC	thick molding compound
TMMV	threshold molecular weight value
TPA	polyamide thermoplastic elastomer
TPC	copolyester thermoplastic elastomer
TPE	thermoplastic elastomer
TPE-O, TPO	thermoplastic elastomer - olefinic
TPE-S, TPS	thermoplastic elastomer - styrenic
TPU	thermoplastic urethane
TPX	poly-4-methylpentene
TVO	thermoplastic vulcanites
T_c	ceiling temperature; cloud-point temperature
T_g	glass transition temperature
T_m	melting point temperature
UF	urea-formaldehyde resin
UHMWPE	ultrahigh molecular weight polyethylene
ULDPE	ultra low-density polyethylene
ULPE	ultra linear polyethylene
UP, UPE	unsaturated polyester (thermoset)
VA	vinyl acetate
VAE	vinyl acetate ethylene
VLDPE	very low-density polyethylene
WLF	Williams-Landel-Ferry equation
WS	polyurethane
XLPE	crosslinked polyethylene
XPS	expandable polystyrene

PHYSICAL PROPERTIES OF SELECTED POLYMERS

The physical properties of polymers are important parameters in determining their behavior and performance in a wide range of applications. This table lists some examples of general representative physical properties (including mechanical properties) of representative polymeric compounds. For glass transition temperatures of selected polymers, see pages 13-10 through 13-16 in this section. Some of the properties in this table are defined as follows:

The **heat deflection temperature** (HDT), or heat distortion temperature, is the temperature at which a polymer or plastic sample deforms under a specified load (normally either 0.455 MPa or 1.82 MPa).

The **crystalline melting point** is the temperature (or temperature range) at which a crystalline solid changes its state from solid to liquid. Although the phrase would suggest a specific temperature, most crystalline compounds actually melt over a range of a few degrees or less.

The **coefficient of linear thermal expansion** is the fractional change in length per °C change in temperature at constant pressure.

The **compressive strength** of a material is the maximum uniaxial compressive stress (compressive force per unit area) reached when the material fails completely on being subjected to a load that pushes it together.

The **tensile strength** is a measure of the ability of a material to withstand pulling stresses. It is defined as the stress (stretching force per unit area) required to break a specimen. Polymers are approximately 20 % stronger in compression than in tension.

The **flexural strength**, or cross-breaking strength, of a material is a measure of the bending strength or stiffness of a specimen expressed as the stress required to break a specimen by exerting a torque on it.

The **impact strength** is a measure of the energy needed to break a sample. The term toughness is sometimes used to describe the impact strength of a material. The notched izod impact test is a single point test that measures the resistance of a material to impact from a swinging pendulum. Izod impact is defined as the kinetic energy needed to initiate fracture and continue the fracture until the specimen is broken. Izod specimens are notched to prevent deformation of the specimen upon impact. This test can be used as a quick and easy quality control check to determine if a material meets specific impact properties or to compare materials for general toughness.

The **ultimate elongation** is a measure of how far a material will stretch before breaking, expressed as a percentage of its original length.

The properties of the following polymers are presented in this table:

PET	poly(ethylene terephthalate)
PBT	poly(butylene terephthalate)
PC	polycarbonate
Nylon 6,6	poly(iminoadipoyliminohexamethylene)
Nylon 6	poly[imino(1-oxohexamethylene)]
PPO	poly(phenylene ether)
POM	polyoxymethylene
LDPE	low-density polyethylene
HDPE	high-density polyethylene
UHMWPE	ultrahigh molecular weight polyethylene
iPP	isotactic polypropylene
ABS	copolymer of acrylonitrile, butadiene, and styrene (extrusion grade)
PTFE	polytetrafluoroethylene, Teflon
PCTFE	polymonochlorotrifluoroethylene
PVDF	poly(vinylidene fluoride)
PVF	poly(vinyl fluoride)
PVC (rigid)	poly(vinyl chloride)
PVC (plasterized)	poly(vinyl chloride)
PMMA	poly(methyl methacrylate)

The assistance of Charles E. Carraher, Jr. in providing these data is gratefully acknowledged.

Reference

Carraher, Jr., C.E., *Seymour/Carraher's Polymer Chemistry,* 7th Edition, CRC Press, Taylor & Francis Group, Boca Raton, FL, 2008.

	PET	PBT	PC	Nylon 6,6	Nylon 6	PPO
Heat deflection temperature at 1820 kPa (°C)	100	65	130	75	80	100
Maximum resistance to continuous heat (°C)	100	60	115	120	125	80
Crystalline melting point (°C)	—	—	225	265	225	215
Coefficient of linear expansion (10^{-5}/°C)	6.5	7.0	6.8	8.0	8.0	5.0
Compressive strength (kPa)	8.6×10^4	7.5×10^4	8.6×10^4	1×10^5	9.7×10^4	9.6×10^4
Flexural strength (kPa)	1.1×10^5	9.6×10^4	9.3×10^4	1×10^5	9.7×10^4	8.9×10^4
Impact strength (Izod: cm N/cm of notch)	26	53	530	80	160	270
Tensile strength (kPa)	6.2×10^4	5.5×10^4	7.2×10^4	8.3×10^4	6.2×10^4	5.5×10^4
Ultimate elongation (%)	100	100	110	30	—	50
Density (g cm^{-3})	1.35	1.35	1.2	1.2	1.15	1.1

	POM	LDPE	HDPE	UHMWPE	iPP	ABS
Heat deflection temperature at 1820 kPa ($^\circ$C)	125	40	50	85	55	90
Maximum resistance to continuous heat ($^\circ$C)	100	40	80	80	100	90
Crystalline melting point ($^\circ$C)	180	—	—	—	—	—
Coefficient of linear expansion (10^{-5}/$^\circ$C)	10.0	10	12	12	9	9.5
Compressive strength (kPa)	1.1×10^5	—	3×10^4	—	—	4.8×10^4
Flexural strength (kPa)	9.7×10^4	—	—	—	5×10^4	6.2×10^4
Impact strength (Izod: cm N/cm of notch)	80	No break	30	No break	27	320
Tensile strength (kPa)	6.9×10^4	5×10^3	2×10^4	6×10^4	3.5×10^4	3.4×10^4
Ultimate elongation (%)	30	—	—	—	100	60
Density (g cm^{-3})	1.4	0.91	0.96	0.93	0.90	1.0

	PTFE	PCTFE	PVDF	PVF	Rigid PVC	Plasticized PVC	PMMA
Heat deflection temperature at 1820 kPa ($^\circ$C)	100	100	80	90	75	—	95
Maximum resistance to continuous heat ($^\circ$C)	250	200	150	125	60	35	75
Crystalline melting point ($^\circ$C)	—	—	—	—	170		—
Coefficient of linear expansion (10^{-5}/$^\circ$C)	10	14	8.5	10	6	12	7.0
Compressive strength (kPa)	2.7×10^4	3.8×10^4	—	—	6.8×10^4	6×10^3	1×10^5
Flexural strength (kPa)	—	6×10^4	—	—	9×10^4	—	9.6×10^4
Impact strength (Izod: cm N/cm of notch)	160	130	—	—	27	—	21
Tensile strength (kPa)	2.4×10^4	3.4×10^4	5.5×10^4	—	4.4×10^4	1×10^4	6.5×10^4
Ultimate elongation (%)	200	100	200	—	50	200	4
Density (g cm^{-3})	2.16	2.1	1.76	1.4	1.4	1.3	1.2

NOMENCLATURE FOR ORGANIC POLYMERS

Robert B. Fox and Edward S. Wilks

Organic polymers have traditionally been named on the basis of the monomer used, a hypothetical monomer or a semi-systematic structure. Alternatively, they may be named in the same way as organic compounds, i.e., on the basis of a structure as drawn. The former method, often called "source-based nomenclature" or "monomer-based nomenclature," sometimes results in ambiguity and multiple names for a single material. The latter method, termed "structure-based nomenclature," generates a sometimes cumbersome unique name for a given polymer, independent of its source. Within their limitations, both types of names are acceptable and well-documented.[1] The use of stereochemical descriptors with both types of polymer nomenclature has been published.[2]

Traditional Polymer Names

Monomer-Based Names

"Polystyrene" is the name of a homopolymer made from the single monomer styrene. When the name of a monomer comprises two or more words, the name should be enclosed in parentheses, as in "poly(methyl methacrylate)" or "poly(4-bromostyrene)" to identify the monomer more clearly. This method can result in several names for a given polymer: thus, "poly(ethylene glycol)," "poly(ethylene oxide)," and "poly(oxirane)" describe the same polymer. Sometimes, the name of a hypothetical monomer is used, as in "poly(vinyl alcohol)." Even though a name like "polyethylene" covers a multitude of materials, the system does provide understandable names when a single monomer is involved in the synthesis of a single polymer. When one monomer can yield more than one polymer, e.g., 1,3-butadiene or acrolein, some sort of structural notation must be used to identify the product, and one is not far from a formal structure-based name.

Copolymers, Block Polymers, and Graft Polymers. When more than one monomer is involved, monomer-based names are more complex. Some common polymers have been given names based on an apparent structure, as with "poly(ethylene terephthalate)." A better system has been approved by the IUPAC.[1] With this method, the arrangement of the monomeric units is introduced through use of an italicized connective placed between the names of the monomers. For monomer names represented by A, B, and C, the various types of arrangements are shown in Table 1.

Table 2 contains examples of common or semi-systematic names of copolymers. The systematic names of comonomers may also be used; thus, the polyacrylonitrile-*block*-polybutadiene-*block*-polystyrene polymer in Table 2 may also be named poly(prop-2-enenitrile)-*block*-polybuta-1,3-diene-*block*-poly(ethylbenzene). IUPAC does not require alphabetized names of comonomers within a polymer name; many names are thus possible for some copolymers.

These connectives may be used in combination and with small, non-repeating (i.e. non-polymeric) junction units; see, for example, Table 2, line 8. A long dash may be used in place of the connective -*block*-; thus, in Table 2, the polymers of lines 7 and 8 may also be written as shown on lines 9 and 10.

IUPAC also recommends an alternative scheme for naming copolymers that comprises use of "copoly" as a prefix followed by the names of the comonomers, a solidus (an oblique stroke) to separate comonomer names, and the addition before "copoly" of any applicable connectives listed in Table 2 except -*co*-.

Table 3 gives the same examples shown in Table 2 but with the alternative format. Comonomer names need not be parenthesized.

TABLE 1. IUPAC Source-Based Copolymer Classification

No.	Copolymer type	Connective	Example
1	Unspecified or unknown	-*co*-	poly(A-*co*-B)
2	Random (obeys Bernoullian distribution)	-*ran*-	poly(A-*ran*-B)
3	Statistical (obeys known statistical laws)	-*stat*-	poly(A-*stat*-B)
4	Alternating (for two monomeric units)	-*alt*-	poly(A-*alt*-B)
5	Periodic (ordered sequence for 2 or more monomeric units)	-*per*-	poly(A-*per*-B-per-C)
6	Block (linear block arrangement)	-*block*-	polyA-*block*-polyB
7	Graft (side chains connected to main chains)	-*graft*-	polyA-*graft*-polyB

TABLE 2. Examples of Source-Based Copolymer Nomenclature

No.	Copolymer name
1	poly(propene-*co*-methacrylonitrile)
2	poly[(acrylic acid)-*ran*-(ethyl acrylate)]
3	poly(butene-*stat*-ethylene-*stat*-styrene)
4	poly[(sebacic acid)-*alt*-butanediol]
5	poly[(ethylene oxide)-*per*-(ethylene oxide)-*per*-tetrahydrofuran]
6	polyisoprene-*graft*-poly(methacrylic acid)
7	polyacrylonitrile-*block*-polybutadiene-*block*-polystyrene
8	polystyrene-*block*-dimethylsilylene-*block*-polybutadiene
9	polyacrylonitrile—polybutadiene—polystyrene
10	polystyrene—dimethylsilylene—polybutadiene

TABLE 3. Examples of Source-Based Copolymer Nomenclature (Alternative Format)

No.	Polymer name
1	copoly(propene/methacrylonitrile)
2	*ran*-copoly(acrylic acid/ethyl acrylate)
3	*stat*-copoly(butene/ethylene/styrene)
4	*alt*-copoly(sebacic acid/butanediol)
5	*block*-copoly(acrylonitrile/butadiene/styrene)
6	*per*-copoly(ethylene oxide/ethylene oxide/tetrahydrofuran)
7	*graft*-copoly(isoprene/methacrylic acid)

Source-based nomenclature for non-linear macromolecules and macromolecular assemblies is covered by a 1997 IUPAC document.[11] The types of polymers in these classes, together with their connectives, are given in Table 4; the terms shown may be used as connectives, prefixes, or both to designate the features present.

TABLE 4. Connectives for Non-Linear Macromolecules and Macromolecular Assemblies

No.	Type	Connective
1	Branched (type unspecified)	branch
2	Branched with branch point of functionality f	f-branch
3	Comb	comb
4	Cross-link	ι (Greek iota)
5	Cyclic	cyclo
6	Interpenetrating polymer network	ipn
7	Long-chain branched	l-branch
8	Network	net
9	Polymer blend	blend
10	Polymer-polymer complex	compl
11	Semi-interpenetrating polymer network	sipn
12	Short-chain branched	sh-branch
13	Star	star
14	Star with f arms	f-star

Non-linear polymers are named by using the italicized connective as a *prefix* to the source-based name of the polymer component or components to which the prefix applies; some examples are listed in Table 5.

TABLE 5. Non-Linear Macromolecules

No.	Polymer name	Polymer structural features
1	poly(methacrylic acid)-*comb*-polyacrylonitrile	Comb polymer with a poly(methacrylic acid) backbone and polyacrylonitrile side chains
2	*comb*-poly[ethylene-*stat*-(vinyl chloride)]	Comb polymer with unspecified backbone composition and statistical ethylene/vinyl chloride copolymer side chains
3	polybutadiene-*comb*-(polyethylene; polypropene)	Comb polymer with butadiene backbone and side chains of polyethylene and polypropene
4	*star*-(polyA; polyB; polyC; polyD; polyE)	Star polymer with arms derived from monomers A, B, C, D, and E, respectively
5	*star*-(polyA-*block*-polyB-*block*-polyC)	Star polymer with every arm comprising a tri-block segment derived from comonomers A, B, and C
6	*star*-poly(propylene oxide)	A star polymer prepared from propylene oxide
7	5-*star*-poly(propylene oxide)	A 5-arm star polymer prepared from propylene oxide
8	*star*-(polyacrylonitrile; polypropylene) (M_r 10000: 25000)	A star polymer containing polyacrylonitrile arms of MW 10000 and polypropylene arms of MW 25000

Macromolecular assemblies held together by forces other than covalent bonds are named by inserting the appropriate italicized connective between names of individual components; Table 6 gives examples.

TABLE 6. Examples of Polymer Blends and Nets

No.	Polymer name
1	polyethylene-*blend*-polypropene
2	poly(methacrylic acid)-*blend*-poly(ethyl acrylate)
3	*net*-poly(4-methylstyrene-ι-divinylbenzene)
4	*net*-poly[styrene-*alt*-(maleic anhydride)]-ι-(polyethylene glycol; polypropylene glycol)
5	*net*-poly(ethyl methacrylate)-*sipn*-polyethylene
6	[*net*-poly(butadiene-*stat*-styrene)]-*ipn*-[*net*-poly(4-methylstyrene-ι-divinylbenzene)]

Structure-Based Polymer Nomenclature

Regular Single-Strand Polymers

Structure-based nomenclature has been approved by the IUPAC[4] and is currently being updated; it is used by *Chemical Abstracts*.[5] Monomer names are not used. To the extent that a polymer chain can be described by a repeating unit in the chain, it can be named "poly(repeating unit)." For regular single-strand polymers, "repeating unit" is a bivalent group; for regular double-strand (ladder and spiro) polymers, "repeating unit" is usually a tetravalent group.[9]

Since there are usually many possible repeating units in a given chain, it is necessary to select one, called the "constitutional repeating unit" (CRU) to provide a unique and unambiguous name, "poly(CRU)," where "CRU" is a recitation of the names of successive units as one proceeds through the CRU from left to right. For this purpose, a portion of the main chain structure that includes at least two repeating sequences is written out. These sequences will typically be composed of bivalent subunits such as -CH_2-, -O-, and groups from ring systems, each of which can be named by the usual nomenclature rules.[6,7]

Where a chain is simply one long sequence comprising repetition of a single subunit, that subunit is itself the CRU, as in "poly(methylene)" or "poly(1,4-phenylene)." In chains having more than one kind of subunit, a seniority system is used to determine the beginning of the CRU and the direction in which to move along the main chain atoms (following the shortest path in rings) to complete the CRU. Determination of the first, most senior, subunit is based on a descending order of seniority: (1) heterocyclic rings, (2) hetero atoms, (3) carbocyclic rings, and lowest, (4) acyclic carbon chains.

Within each of these classes, there is a further order of seniority that follows the usual rules of nomenclature.

Heterocycles: A nitrogen-containing ring system is senior to a ring system not containing nitrogen.[4,9] Further descending order of seniority is determined by:

   (i) the highest number of rings in the ring system
   (ii) the largest individual ring in the ring system
   (iii) the largest number of hetero atoms
   (iv) the greatest variety of hetero atoms

Hetero atoms: The senior bivalent subunit is the one nearest the top right-hand corner of the Periodic Table; the order of seniority is: O, S, Se, Te, N, P, As, Sb, Bi, Si, Ge, Sn, Pb, B, Hg.

Carbocycles: Seniority[4] is determined by:

   (i) the highest number of rings in the ring system
   (ii) the largest individual ring in the ring system
   (iii) degree of ring saturation; an unsaturated ring is senior to a saturated ring of the same size

Carbon chains: Descending order of seniority is determined by:

   (i) chain length (longer is senior to shorter)
   (ii) highest degree of unsaturation
   (iii) number of substituents (higher number is senior to lower number)
   (iv) ascending order of locants
   (v) alphabetical order of names of substituent groups

Among equivalent ring systems, preference is given to the one having lowest locants for the free valences in the subunit, and among otherwise identical ring systems, the one having least hydrogenation is senior. Lowest locants in unsaturated chains are also given preference. Lowest locants for substituents are the final determinant of seniority.

Direction within the repeating unit depends upon the shortest path, which is determined by counting main chain atoms, both cyclic and acyclic, from the most senior subunit to another subunit of the same kind or to a subunit next lower in seniority. When identification and orientation of the CRU have been accomplished, the CRU is named by writing, in sequence, the names of the largest possible subunits within the CRU from left to right. For example, the main chain of the polymer traditionally named "poly(ethylene terephthalate)" has the structure shown in Figure 1.

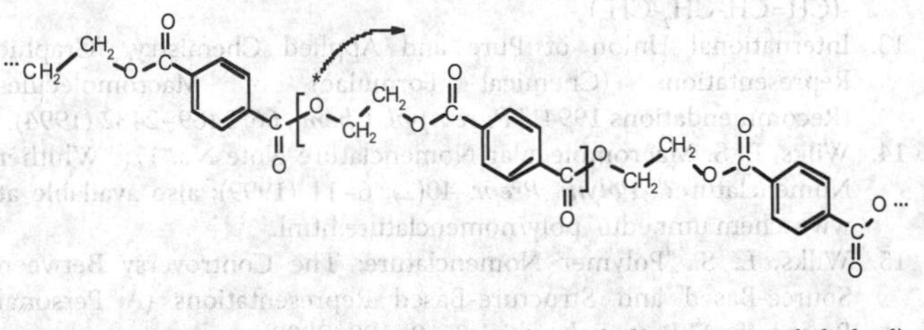

Figure 1. Structure-based name: poly(oxyethyleneoxyterephthaloyl); traditional name: poly(ethylene terephthalate).

The CRU in Figure 1 is enclosed in brackets and read from left to right. It is selected because (1) either backbone oxygen atom qualifies as the "most senior subunit," (2) the shortest path length from either -O- to the other -O- is via the ethylene subunit. Orientation of the CRU is thus defined by (1) beginning at the -O- marked with an asterisk, and (2) reading in the direction of the arrow. The structure-based name of this polymer is therefore "poly(oxyethyleneoxyterephthaloyl)," not much longer than the traditional name and much more adaptable to the complexities of substitution. As organic nomenclature evolves, more systematic names may be used for subunits, e.g., "ethane-1,2-diyl" instead of "ethylene." IUPAC still prefers "ethylene" for the $-CH_2-CH_2-$ unit, however, but also accepts "ethane-1,2-diyl."

Structure-based nomenclature can also be used when the CRU backbone has no carbon atoms. An example is the polymer traditionally named "poly(dimethylsiloxane)," which on the basis of structure would be named "poly(oxydimethylsilylene)" or "poly(oxydimethylsilanediyl)." This nomenclature method has also been applied to inorganic and coordination polymers[8] and to double-strand (ladder and spiro) organic polymers.[9]

Irregular Single-Strand Polymers

Polymers that cannot be described by the repetition of a single CRU or comprise units not all connected identically in a directional sense can also be named on a structure basis.[10] These include copolymers, block and graft polymers, and star polymers. They are given names of the type "poly(A/B/C...)," where A, B, C, etc. are the names of the component constitutional units, the number of which are minimized. The constitutional units may include regular or irregular blocks as well as atoms or atomic groupings, and each is named by the method described above or by the rules of organic nomenclature.

The solidus denotes an unspecified arrangement of the units within the main chain.[10] For example, a statistical copolymer derived from styrene and vinyl chloride with the monomeric units joined head-to-tail is named "poly(l-chloroethylene/l-phenylethylene)." A polymer obtained by 1,4-polymerization and both head-to-head and head-to-tail 1,2-polymerization of 1,3-butadi-

ene would be named "poly(but-1-ene-l,4-diyl/l-vinylethylene/2-vinylethylene)."[12] In graphic representations of these polymers, shown in Figure 2, the hyphens or dashes at each end of each CRU depiction are shown *completely within* the enclosing parentheses; this indicates that they are not necessarily the terminal bonds of the macromolecule.

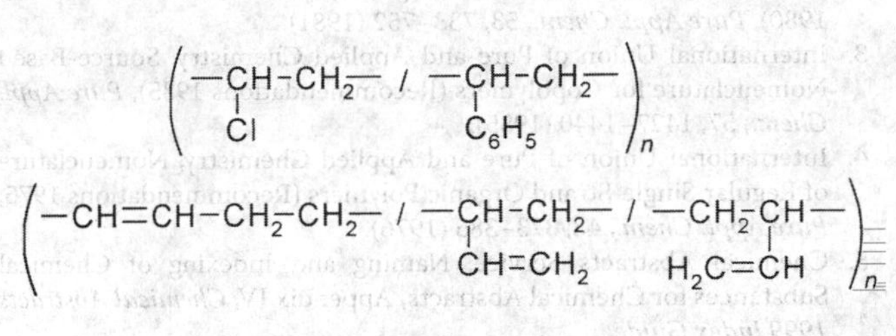

Figure 2. Graphic representations of copolymers.

A long hyphen is used to separate components in names of block polymers, as in "poly(A)—poly(B)—poly(C)," or "poly(A)—X—poly(B)" in which X is a non-polymeric junction unit, e.g., dimethylsilylene.

In graphic representations of these polymers, the blocks are shown connected when the bonding is known (Figure 3, for example); when the bonding between the blocks is unknown, the blocks are separated by solidi and are shown *completely within* the outer set of enclosing parentheses (Figure 4, for example).[10,13]

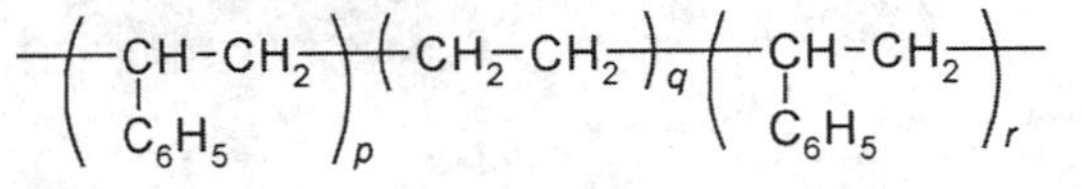

Figure 3. polystyrene—polyethylene—polystyrene.

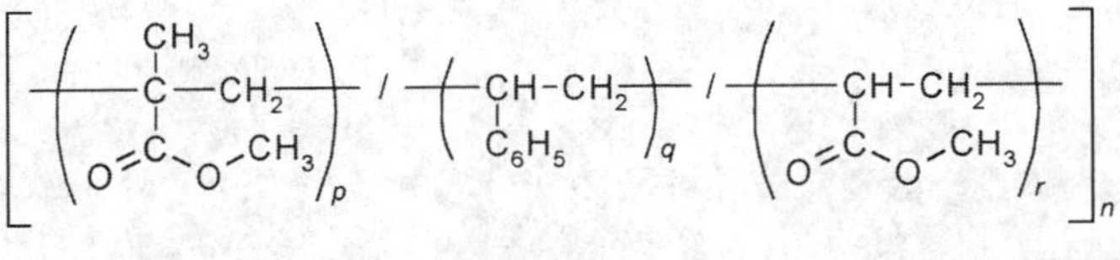

Figure 4. poly[poly(methyl methacrylate)—polystyrene—poly(methyl acrylate)].

Graft polymers are named in the same way as a substituted polymer but without the ending "yl" for the grafted chain; the name of a regular polymer, comprising Z units in which some have grafts of "poly(A)," is "poly[Z/poly(A)Z]." Star polymers are treated as a central unit with substituent blocks, as in "tetrakis(polymethylene) silane."[10,13]

Other Nomenclature Articles and Publications

In addition to the *Chemical Abstracts* and IUPAC documents cited above and listed below, other articles on polymer nomenclature are available. A 1999 article lists significant documents on polymer nomenclature published during the last 50 years in books, encyclopedias, and journals by *Chemical Abstracts*, IUPAC, and individual authors.[14] A comprehensive review of source-based and structure-based nomenclature for all of the major classes of polymers,[15] and a short tutorial on the correct identification, orientation, and naming of most commonly encountered constitutional repeating units were both published in 2000.[16]

References and Notes

1. International Union of Pure and Applied Chemistry, *Compendium of Macromolecular Nomenclature*, Blackwell Scientific Publications, Oxford, 1991.

2. International Union of Pure and Applied Chemistry, Stereochemical Definitions and Notations Relating to Polymers (Recommendations 1980), *Pure Appl. Chem.*, 53, 733–752 (1981).

3. International Union of Pure and Applied Chemistry, Source-Based Nomenclature for Copolymers (Recommendations 1985), *Pure Appl. Chem.*, 57, 1427–1440 (1985).

4. International Union of Pure and Applied Chemistry, Nomenclature of Regular Single-Strand Organic Polymers (Recommendations 1975, *Pure Appl. Chem.*, 48, 373–385 (1976).

5. Chemical Abstracts Service, Naming and Indexing of Chemical Substances for Chemical Abstracts, Appendix IV, *Chemical Abstracts 1999 Index Guide.*

6. International Union of Pure and Applied Chemistry, *A Guide to IUPAC Nomenclature of Organic Compounds* (1993), Blackwell Scientific Publications, Oxford, 1993.

7. International Union of Pure and Applied Chemistry, *Nomenclature of Organic Chemistry, Sections A, B, C, D, E, F, and H*, Pergamon Press, Oxford, 1979.

8. International Union of Pure and Applied Chemistry, Nomenclature of Regular Double-Strand and Quasi-Single-Strand Inorganic and Coordination Polymers (Recommendations 1984), *Pure Appl. Chem.*, 57, 149–168 (1985).

9. International Union of Pure and Applied Chemistry, Nomenclature of Regular Double-Strand (Ladder and Spiro) Organic Polymers (Recommendations 1993), *Pure Appl. Chem.*, 65, 1561–1580 (1993).

10. International Union of Pure and Applied Chemistry, Structure-Based Nomenclature for Irregular Single-Strand Organic Polymers (Recommendations 1994), *Pure Appl. Chem.*, 66, 873–889 (1994).

11. International Union of Pure and Applied Chemistry, "Source-Based Nomenclature for Non-Linear Macromolecules and Macromolecular Assemblies (Recommendations 1997)." *Pure Appl. Chem.*, 69, 2511–2521 (1997).

12. Poly(1,3-butadiene) obtained by polymerization of 1,3-butadiene in the so-called 1,4- mode is frequently drawn incorrectly in publications as $-(CH_2-CH=CH-CH_2)_n-$; the double bond should be assigned the lowest locant possible, i.e., the structure should be drawn as $-(CH=CH-CH_2-CH_2)_n-$.

13. International Union of Pure and Applied Chemistry, "Graphic Representations (Chemical Formulae) of Macromolecules (Recommendations 1994)." *Pure Appl. Chem.*, 66, 2469–2482 (1994).

14. Wilks, E. S. Macromolecular Nomenclature Note No. 17: "Whither Nomenclature?" *Polym. Prepr.* 40(2), 6–11 (1999); also available at www.chem.umr.edu/~poly/nomenclature.html.

15. Wilks, E. S. "Polymer Nomenclature: The Controversy Between Source-Based and Structure-Based Representations (A Personal Perspective)." *Prog. Polym. Sci.* 25, 9–100 (2000).

16. Wilks, E. S. Macromolecular Nomenclature Note No. 18: "SRUs: Using the Rules." *Polym. Prepr.* 41(1), 6a–11a (2000); also available at www.chem.umr.edu/~poly/nomenclature.html; a .pdf format version is also available.

SOLVENTS FOR COMMON POLYMERS

Abbreviations:
  HC: hydrocarbons
  MEK: methyl ethyl ketone

THF: tetrahydrofuran
DMF: dimethylformamide
DMSO: dimethylsulfoxide

Polyethylene (HDPE)	HC and halogenated HC
Polypropylene (atactic)	HC and halogenated HC
Polybutadiene	HC, THF, ketones
Polystyrene	ethylbenzene, CHCl$_3$, CCl$_4$, THF, MEK
Polyacrylates	aromatic HC, chlorinated HC, THF, esters, ketones
Polymethacrylates	aromatic HC, chlorinated HC, THF, esters, MEK
Polyacrylamide	water
Poly(vinyl ethers)	halogenated HC, MEK, butanol
Poly(vinyl alcohol)	glycols (hot), DMF
Poly(vinyl acetate)	aromatic HC, chlorinated HC, THF, esters, DMF
Poly(vinyl chloride)	THF, DMF, DMSO
Poly(vinylidene chloride)	THF (hot), dioxane, DMF
Poly(vinyl fluoride)	DMF, DMSO (hot)
Polyacrylonitrile	DMF, DMSO
Poly(oxyethylene)	aromatic HC, CHCl$_3$, alcohols, esters, DMF
Poly(2,6-dimethylphenylene oxide)	aromatic HC, halogenated HC
Poly(ethylene terephthalate)	phenol, DMSO (hot)
Polyurethanes (linear)	aromatic HC, THF, DMF
Polyureas	phenol, formic acid
Polysiloxanes	HC, THF, DMF
Poly[bis(2,2,2-trifluoroethoxy)-phosphazene]	THF, ketones, ethyl acetate

GLASS TRANSITION TEMPERATURE FOR SELECTED POLYMERS

Robert B. Fox

Polymer names are based on the IUPAC structure-based nomenclature system described in the table "Naming Organic Polymers." Within each category, names are listed in alphabetical order. Source-based and trivial names are also given (in italics) for the most common polymers. The table does not include polymers for which T_g is not clearly defined because of variability of structure or because of reactions taking place near the glass transition.

All values of T_g cited in this table have been determined by differential scanning calorimetry (DSC) except those values indicated by:

(D)	dynamic method
(Dil)	dilatometry
(M)	mechanical method

Polymer name	Glass transition temperature (T_g/K)
ACYCLIC CARBON CHAINS	
Polyalkadienes	
Poly(alkenylene) *Polyalkadiene* –[CH=CHCH₂CH₂]–	
Poly(*cis*-1-butenylene)	171
cis-1,3-polybutadiene [PBD]	
Poly(*trans*-1-butenylene)	215
trans-1,3-polybutadiene [PBD]	
Poly(1-chloro-*cis*-1-butenylene)	253
cis-1,3-polychloroprene	
Poly(1-chloro-*trans*-1-butenylene)	233
trans-1,3-polychloroprene	
Poly(1-methyl-*cis*-1-butenylene)	200
cis-1,3-polyisoprene	
Poly(1-methyl-*trans*-1-butenylene)	207
trans-1,3-polyisoprene	
Poly(1,4,4-trifluoro-1-butenylene)	238
Polyalkenes	
Poly(alkylethylene) *Poly(alkylethylene)* -[RCHCH₂]-	
Poly(1-benzylethylene)	333
Poly(1-butylethylene)	223
Poly(1-cyclohexylethylene) (atactic)	393
Poly(1-cyclohexylethylene) (isotactic)	406 (D)
Poly(1,1-dimethylethylene)	200
Polyisobutylene [PIB]	
Poly(ethylene)	148
Poly(methylene)	155
Poly(1-phenethylethylene)	283
Poly(propylene) (isotactic)	272
Poly(propylene) (syndiotactic)	ca. 265
Poly[1-(2-pyridyl)ethylene]	377
Poly[1-(4-pyridyl)ethylene]	415
Poly(1-vinylethylene)	273
Polyacrylics	
Poly[1-(alkoxycarbonyl)ethylene] *Poly(alkyl acrylate)* –[(ROCO)CHCH₂]–	
Poly[1-(benzyloxycarbonyl)ethylene]	279
Poly[1-(butoxycarbonyl)ethylene]	219 (M)
Poly(butyl acrylate) [PBA]	
Poly[1-(*sec*-butoxycarbonyl)ethylene]	251
Poly[1-(butoxycarbonyl)-1-cyanoethylene]	358
Poly[1-(butylcarbamoyl)ethylene]	319 (M)
Poly(1-carbamoylethylene)	438
Polyacrylamide [PAM]	
Poly(1-carboxyethylene)	379

Polymer name	Glass transition temperature (T_g/K)
Poly(acrylic acid) [PAA]	
Poly[1-(2-chlorophenoxycarbonyl)ethylene]	326
Poly[1-(4-chlorophenoxycarbonyl)ethylene]	331
Poly[1-(4-cyanobenzyloxycarbonyl)ethylene]	317
Poly[1-(2-cyanoethoxycarbonyl)ethylene]	277
Poly[1-(cyanomethoxycarbonyl)ethylene)]	433 Dil
Poly[1-(4-cyanophenoxycarbonyl)ethylene]	363
Poly[1-(cyclohexyloxycarbonyl)ethylene]	292
Poly[1-(2,4-dichlorophenoxycarbonyl)ethylene]	333
Poly[1-(dimethylcarbamoyl)ethylene]	362
Poly[1-(ethoxycarbonyl)ethylene]	249
Poly(ethyl acrylate) [PEA]	
Poly[1-(ethoxycarbonyl)-1-fluoroethylene]	316
Poly[1-(2-ethoxycarbonylphenoxycarbonyl)ethylene]	303
Poly[1-(3-ethoxycarbonylphenoxycarbonyl)ethylene]	297
Poly[1-(4-ethoxycarbonylphenoxycarbonyl)ethylene]	310
Poly[1-(2-ethoxyethoxycarbonyl)ethylene]	223
Poly[1-(3-ethoxypropoxycarbonyl)ethylene]	218
Poly[1-(isopropoxycarbonyl)ethylene]	267–270
Poly[1-(methoxycarbonyl)ethylene]	283
Poly(methyl acrylate) [PMA]	
Poly[1-(2-methoxycarbonylphenoxycarbonyl)ethylene]	319
Poly[1-(3-methoxycarbonylphenoxycarbonyl)ethylene]	311
Poly[1-(4-methoxycarbonylphenoxycarbonyl)ethylene]	340
Poly[1-(2-methoxyethoxycarbonyl)ethylene]	223
Poly[1-(4-methoxyphenoxycarbonyl)ethylene]	324
Poly[1-(3-methoxypropoxycarbonyl)ethylene]	198
Poly[1-(2-naphthyloxycarbonyl)ethylene]	358
Poly[1-(pentachlorophenoxycarbonyl)ethylene]	420
Poly[1-(phenethoxycarbonyl)ethylene]	270
Poly[1-(phenoxycarbonyl)ethylene]	330
Poly[1-(*m*-tolyloxycarbonyl)ethylene]	298
Poly[1-(*o*-tolyloxycarbonyl)ethylene]	325
Poly[1-(*p*-tolyloxycarbonyl)ethylene]	316
Poly[1-(2,2,2-trifluorethoxycarbonyl)ethylene]	263
Polymethacrylics	
Poly[1-(alkoxycarbonyl)-1-methylethylene] *Poly(alkyl methacrylate)* –[(ROCO)(Me)CCH$_2$]–	
Poly[1-(benzyloxycarbonyl)-1-methylethylene]	327
Poly[1-(2-bromoethoxycarbonyl)-1-methylethylene]	325
Poly[(1-(butoxycarbonyl)-1-methylethylene]	293
Poly(butyl methacrylate) [PBMA]	
Poly[1-(*sec*-butoxycarbonyl)-1-methylethylene]	333
Poly[1-(*tert*-butoxycarbonyl)-1-methylethylene)]	391
Poly[1-(2-chloroethoxycarbonyl)-1-methylethylene]	ca 315
Poly[1-(2-cyanoethoxycarbonyl)-1-methylethylene]	364
Poly[1-(4-cyanophenoxycarbonyl)-1-methylethylene]	428
Poly[1-(cyclohexyloxycarbonyl)-1-methylethylene] (atactic)	356
Poly[1-(cyclohexyloxycarbonyl)-1-methylethylene)] (isotactic)	324
Poly[1-(dimethylaminoethoxycarbonyl)-1-methylethylene]	292
Poly[1-(ethoxycarbonyl)-1-ethylethylene]	300
Poly[1-(ethoxycarbonyl)-1-methylethylene] (atactic) *Poly(ethyl methacrylate)* [PEMA]	338
Poly[1-(ethoxycarbonyl)-1-methylethylene] (isotactic)	285
Poly[1-(ethoxycarbonyl)-1-methylethylene)] (syndiotactic)	339
Poly[1-(hexyloxycarbonyl)-1-methylethylene]	268
Poly[1-(isobutoxycarbonyl)-1-methylethylene]	326
Poly[1-(isopropoxycarbonyl)-1-methylethylene]	354
Poly[1-(methoxycarbonyl)-1-methylethylene] (atactic) *Poly(methyl methacrylate)* [PMMA]	378
Poly[1-(methoxycarbonyl)-1-methylethylene)] (isotactic)	311
Poly[1-(methoxycarbonyl)-1-methylethylene)] (syndiotactic)	378
Poly[1-(4-methoxycarbonylphenoxy)-1-methylethylene]	379

Polymer name	Glass transition temperature (T_g/K)
Poly[1-(methoxycarbonyl)-1-phenylethylene)] (atactic)	391
Poly[1-(methoxycarbonyl)-1-phenylethylene)] (isotactic)	397
Poly[1-methyl-1-(phenethoxycarbonyl)ethylene]	299
Poly[1-methyl-1-(phenoxycarbonyl)ethylene]	383

Polyvinyl ethers, alcohols, and ketones

    Poly(1-alkoxyethylene) *Poly(alkyl vinyl ether)* –[ROCHCH$_2$]–
    Poly(1-hydroxyethylene) *Poly(vinyl alcohol)* –[HOCHCH$_2$]–
    Poly(1-alkanoylethylene) *Poly(alkyl vinyl ketone)* –[RCOCHCH$_2$]–

Poly(1-butoxyethylene)	218
Poly(1-*sec*-butoxyethylene)	253
Poly(1-*tert*-butoxyethylene)	361
Poly[1-(butylthio)ethylene]	253
Poly(1-ethoxyethylene)	230
Poly[1-(4-ethylbenzoyl)ethylene]	325
Poly(1-hydroxyethylene)	358 (D)
Poly(vinyl alcohol) [PVA]	
Poly(hydroxymethylene)	407
Poly(1-isopropoxyethylene)	270
Poly[1-(4-methoxybenzoyl)ethylene]	319 (M)
Poly(1-methoxyethylene)	242
Poly(methyl vinyl ether) [PMVE]	
Poly[1-(methylthio)ethylene]	272
Poly(1-propoxyethylene)	224
Poly[1-(trifluoromethoxy)trifluoroethylene]	268

Polyvinyl halides and nitriles

    Poly(1-haloethylene) *Poly(vinyl halide)* –[XCHCH$_2$]–
    Poly(1-cyanoethylene) *Poly(acrylonitrile)* –[NCCHCH$_2$]–

Poly(1-chloroethylene)	354
Poly(vinyl chloride) [PVC]	
Poly(chlorotrifluoroethylene)	373
Poly(1-cyanoethylene)	370
Polyacrylonitrile [PAN]	
Poly(1-cyano-1-methylethylene)	393
Polymethacrylonitrile	
Poly(1,1-dichloroethylene)	255
Poly(vinylidene chloride)	
Poly(1,1-difluoroethylene)	ca 233
Poly(vinylidene fluoride)	
Poly(1-fluoroethylene)	314 (M)
Poly(vinyl fluoride)	
Poly(1-hexafluoropropylene)	425
Poly[1-(2-iodoethyl)ethylene]	343
Poly(tetrafluoroethylene)	(160)
Poly[1-(trifluoromethyl)ethylene]	300

Polyvinyl esters

    Poly[1-(alkanoyloxy)ethylene] *Poly(vinyl alkanoate)* –[RCOOCHCH$_2$]–

Poly(1-acetoxyethylene)	305
Poly(vinyl acetate) [PVAc]	
Poly[1-(benzoyloxy)ethylene]	344
Poly[1-(4-bromobenzoyloxy)ethylene]	365
Poly[1-(2-chlorobenzoyloxy)ethylene]	335
Poly[1-(3-chlorobenzoyloxy)ethylene]	338
Poly[1-(4-chlorobenzoyloxy)ethylene]	357
Poly[1-(cyclohexanoyloxy)ethylene]	349 (M)
Poly[1-(4-ethoxybenzoyloxy)ethylene]	343
Poly[1-(4-ethylbenzoyloxy)ethylene]	326

Polymer name	Glass transition temperature (T_g/K)
Poly[1-(4-isopropylbenzoyloxy)ethylene]	342
Poly[1-(2-methoxybenzoyloxy)ethylene]	338
Poly[1-(3-methoxybenzoyloxy)ethylene]	ca 317
Poly[1-(4-methoxybenzoyloxy)ethylene]	360
Poly[1-(4-methylbenzoyloxy)ethylene]	343
Poly[1-(4-nitrobenzoyloxy)ethylene]	395
Poly[1-(propionoyloxy)ethylene]	283 (M)

Polystyrenes

Poly(1-phenylethylene) *Polystyrene* $-[C_6H_5CHCH_2]-$

Poly[1-(4-acetylphenyl)ethylene]	389 (M)
Poly[1-(4-benzoylphenyl)ethylene]	371 (M)
Poly[1-(4-bromophenyl)ethylene]	391
Poly[1-(4-butoxyphenyl)ethylene]	ca 320 (M)
Poly[1-(4-butoxycarbonylphenyl)ethylene]	349 (M)
Pol[(1-(4-butylphenyl)ethylene]	279
Poly[1-(4-carboxyphenyl)ethylene]	386 (M)
Poly[1-(2-chlorophenyl)ethylene]	392
Poly[1-(3-chlorophenyl)ethylene]	363
Poly[1-(4-chlorophenyl)ethylene]	383
Poly[1-(2,4-dichlorophenyl)ethylene]	406
Poly[1-(2,5-dichlorophenyl)ethylene]	379
Poly[1-(2,6-dichlorophenyl)ethylene]	440
Poly[1-(3,4-dichlorophenyl)ethylene]	401
Poly[1-(2,4-dimethylphenyl)ethylene]	385
Poly[1-(4-(dimethylamino)phenyl)ethylene]	398 (M)
Poly[1-(4-ethoxyphenyl)ethylene]	ca 359 (M)
Poly[1-(4-ethoxycarbonylphenyl)ethylene]	367 (M)
Poly[1-(4-fluorophenyl)ethylene]	368
Poly[1-(4-iodophenyl)ethylene]	429
Poly[1-(4-methoxyphenyl)ethylene]	386
Poly[1-(4-methoxycarbonylphenyl)ethylene]	386 (M)
Poly(1-methyl-1-phenylethylene)	373
Poly(α-methylstyrene)	
Poly[1-(2-(methylamino)phenyl)ethylene]	462 (M)
Poly(1-phenylethylene)	373
Polystyrene [PS]	
Poly[1-(4-propoxyphenyl)ethylene]	343 (M)
Poly[1-(4-propoxycarbonylphenyl)ethylene]	365 (M)
Poly(1-*o*-tolylethylene)	409

CHAINS WITH CARBOCYCLIC UNITS

Poly(arylenealkylene) $-[-Ar-(CH_2)_n]-$

Poly[1-(2-bromo-1,4-phenylene)ethylene]	353 (M)
Poly[1-(2-chloro-1,4-phenylene)ethylene]	343 (M)
Poly[1-(2-cyano-1,4-phenylene)ethylene]	363 (M)
Poly[1-(2,5-dimethyl-1,4-phenylene)ethylene]	373 (M)
Poly[1-(2-ethyl-1,4-phenylene)ethylene]	298 (M)
Poly[1-(1,4-naphthylene)ethylene]	433 (M)
Poly[1-(1,4-phenylene)ethylene]	ca 353 (M)

CHAINS WITH HETEROATOM UNITS

Main chain oxide units

Poly(oxyalkylene) *Poly(alkylene oxide)* $-[O(CH_2)_n]-$

Poly[oxy(1,1-bis(chloromethyl)trimethylene)]	265
Poly[oxy(1-(bromomethyl)ethylene)]	259
Poly[oxy(1-(butoxymethyl)ethylene)]	194
Poly[oxy(1-butylethylene)]	203
Poly[oxy(1-*tert*-butylethylene)]	308
Poly[oxy(1-(chloromethyl)ethylene)]	251

Polymer name	Glass transition temperature (T_g/K)
Poly(epichlorohydrin)	
Poly[oxy(2,6-dimethoxy-1,4-phenylene)]	440
Poly[oxy(1,1-dimethylethylene)]	264
Poly[oxy(2,6-dimethyl-1,4-phenylene)]	482
Poly[oxy(2,6-diphenyl-1,4-phenylene)]	493
Poly[oxy(1-ethylethylene)]	203
Poly(oxyethylidene)	243
Polyacetaldehyde	
Poly[oxy(1-(methoxymethyl)ethylene)]	211
Poly[oxy(2-methyl-6-phenyl-1,4-phenylene)]	428
Poly[oxy(1-methyltrimethylene)]	223 (D)
Poly[oxy(2-methyltrimethylene)]	218
Poly(oxy-1,4-phenylene)	358
Poly(phenylene oxide) [PPO]	
Poly[oxy(1-phenylethylene)]	313
Poly(oxytetramethylene)	189
Poly(tetrahydrofuran) [PTMO]	
Poly(oxytrimethylene)	195

Main-chain ester or anhydride units

Poly(oxyalkyleneoxyalkanedioyl) *Poly(alkylene alkanedioate)*--[O(CH$_2$)$_m$OCO(CH$_2$)$_n$CO]--

Polymer name	Glass transition temperature (T_g/K)
Poly(oxyadipoyloxydecamethylene)	217
Poly(oxyadipoyloxy-1,4-phenyleneisopropylidene-1,4-phenylene)	341
Poly(oxycarbonyloxy-1,4-phenylene-isopropylidene-1,4-phenylene)	422
Bisphenol A polycarbonate	
Poly(oxycarbonylpentamethylene)	213
Poly(oxycarbonyl-1,4-phenylenemethylene-1,4-phenylene)	395
Poly(oxycarbonyl-1,4-phenyleneisopropylidene-1,4-phenylene)	333
Poly[oxy(2,6-dimethyl-1,4-phenyleneisopropylidene-3,5-dimethyl-1,4-phenylene)oxysebacoyl]	318
Poly(oxyethylenecarbonyl-1,4-cyclohexylenecarbonyl) (trans)	291
Poly(oxyethyleneoxycarbonyl-1,4-naphthylenecarbonyl)	337
Poly(oxyethyleneoxycarbonyl-1,5-naphthylenecarbonyl)	344
Poly(oxyethyleneoxycarbonyl-2,6-naphthylenecarbonyl)	386
Poly(oxyethyleneoxycarbonyl-2,7-naphthylenecarbonyl)	392
Poly(oxyethyleneoxyterephthaloyl)	342
Poly(ethylene terephthalate) [PET]	
Poly(oxyisophthaloyl)	403 (D)
Poly(oxy(1-oxo-2,2-dimethyltrimethylene))	263
Poly(pivalolactone)	
Poly(oxy-1,4-phenyleneisopropylidene-1,4-phenyleneoxysebacoyl)	280
Poly(oxy-1,4-phenyleneoxy-1,4-phenyleneoxy-carbonyl-1-phenylene) [PEEK]	416
Poly(oxypropyleneoxyterephthaloyl)	341
Poly[oxyterephthaloyloxy(2,6-dimethyl-1,4-phenyleneisopropylidene-3,5-dimethyl-1,4-(D)phenylene)]	498
Poly(oxyterephthaloyloxyoctamethylene)	318 (D)
Poly(oxyterephthaloyloxy-1,4-phenyleneisopropylidene-1,4-phenylene)	478
Poly(bisphenol A terephthalate)	
Poly(oxytetramethyleneoxyterephthaloyl)	323
Poly(butylene terephthalate) [PBT]	

Main-chain amide units

Poly(iminoalkyleneiminoalkanedioyl) *Poly(alkylene alkanediamide)*–[NH(CH$_2$)$_m$NHCO(CH$_2$)$_n$CO]–

Polymer name	Glass transition temperature (T_g/K)
Poly(iminoadipoyliminodecamethylene)	313
Nylon 10,6	
Poly(iminoadipoyliminohexamethylene)	ca 323
Nylon 6,6	
Poly(iminoadipoyliminooctamethylene)	318
Nylon 8,6	
Poly[iminoadipoyliminotrimethylene(methylimino)trimethylene]	278
Poly(iminocarbonyl-1,4-cyclohexylenemethylene)	466
Poly[iminocarbonyl-1,4-phenylene(2-oxoethylene)iminohexamethylene]	377
Poly(iminoethylene-1,4-phenyleneethyleneiminosebacoyl)	378 (D)

Polymer name	Glass transition temperature (T_g/K)
Poly(iminohexamethyleneiminoazelaoyl)	331
Nylon 6,9	
Poly(iminohexamethyleneiminododecanedioyl)	319
Nylon 6, 12	
Poly(iminohexamethyleneiminopimeloyl)	331
Nylon 6,7	
Poly(iminohexamethyleneiminosebacoyl)	323
Nylon 6,10	
Poly(iminohexamethyleneiminosuberoyl)	330
Nylon 6,8	
Poly(iminoisophthaloylimino-4,4'-biphenylylene)	558
Poly(iminoisophthaloyliminohexamethylene)	390
Poly(iminoisophthaloyliminomethylene-1,4-cyclohexylenemethylene)	481
Poly(iminoisophthaloyliminomethylene-1,3-phenylenemethylene)	438 (M)
Poly[iminomethylene(2,5-dimethyl-1,4-phenylene)methyleneiminosuberoyl]	351
Poly(imino-1,5-naphthyleneiminoisophthaloyl)	598
Poly(imino-1,5-naphthyleneiminoterephthaloyl)	578
Poly(iminooctamethyleneiminodecanedioyl)	333
Nylon 8,10	
Poly(iminooxalyliminohexamethylene)	430
Nylon 6,2	
Poly[imino(1-oxohexamethylene)]	326
Nylon 6	
Poly[imino(1-oxodecamethylene)]	315
Nylon 10	
Poly[imino(1-oxoheptamethylene)]	325
Nylon 7	
Poly[imino(1-oxo-3-methyltrimethylene]	369
Poly[imino(1-oxononamethylene)]	319
Nylon 9	
Poly[imino(1-oxooctamethylene)]	323
Nylon 8	
Poly[imino(1-oxotrimethylene)]	384
Nylon 3	
Poly(iminopentamethyleneiminoadipoyl)	318
Nylon 5,6	
Poly[iminopentamethyleneiminocarbonyl-1,4-phenylene(2-oxoethylene)]	376
Poly(imino-1,3-phenyleneiminoisophthaloyl)	553 (M)
Poly(imino-1,4-phenyleneiminoterephthaloyl)	618
Poly(iminopimeloyliminoheptamethylene)	328
Nylon 7,7	
Poly(iminoterephthaloylimino-4,4'-biphenylylene)	613
Poly(iminotetramethyleneiminoadipoyl)	316
Nylon 4,6	
Poly[iminotetramethyleneiminocarbonyl-1,4-phenylene(2-oxoethylene)]	357
Poly(iminotrimethyleneiminoadipoyliminotrimethylene)	307
Poly[iminotrimethyleneiminocarbonyl-1,4-phenylene(2-oxoethylene)]	382
Poly(oxy-1,4-phenyleneiminoterephthaloyl-imino-1,4-phenylene)	613
Poly(sulfonylimino-1,4-phenyleneiminoadipoylimino-1,4-phenylene)	467

Main-chain urethane units

    Poly(oxyalkyleneoxycarbonyliminoalkyleneiminocarbonyl)–[O(CH$_2$)$_m$OCONH(CH$_2$)$_n$NHCO]–

Polymer name	Glass transition temperature (T_g/K)
Poly(oxyethyleneoxycarbonyliminohexamethyleneiminocarbonyl)	329
Poly[oxyethyleneoxycarbonylimino(6-methyl-1,3-phenylene)iminocarbonyl]	325
Poly(oxyethyleneoxycarbonylimino-1,4-phenylenemethylene-1,4-phenyleneiminocarbonyl)	412
Poly(oxyhexamethyleneoxycarbonyliminohexamethyleneiminocarbonyl)	332
Poly[oxyhexamethyleneoxycarbonylimino(6-methyl-1,3-phenylene)iminocarbonyl]	305
Poly(oxyhexamethyleneoxycarbonylimino-1,4-phenylenemethylene-1,4-phenyleneiminocarbonyl)	364
Poly(oxyoctamethyleneoxycarbonyliminohexamethyleneiminocarbonyl)	331
Poly[oxyoctamethyleneoxycarbonylimino(6-methyl-1,3-phenylene)iminocarbonyl]	337
Poly(oxyoctamethyleneoxycarbonylimino-1,4-phenylenemethylene-1,4-phenyleneiminocarbonyl)	352

Polymer name	Glass transition temperature (T_g/K)
Poly(oxytetramethyleneoxycarbonyliminohexamethyleneiminocarbonyl)	332
Poly[oxytetramethyleneoxycarbonylimino(6-methyl-1,3-phenylene)iminocarbonyl]	315
Poly(oxytetramethyleneoxycarbonylimino-1,4-phenylenemethylene-1,4-phenyleneiminocarbonyl)	382

Main-chain siloxanes

Poly[oxy(dialkylsilylene)] *Poly(dialkylsiloxane)* –[O(R₂Si)]–	
Poly[oxy(dimethylsilylene)]	148
Poly(dimethylsiloxane) [PDMS]	
Poly[oxy(dimethylsilylene)oxy-1,4-phenylene]	363 (M)
Poly[oxy(dimethylsilylene)oxy-1,4-phenyleneisopropylidene-1,4-phenylene]	318 (M)
Poly[oxy(diphenylsilylene)]	238
Poly(diphenylsiloxane)	
Poly[oxy(diphenylsilylene)-1,3-phenylene]	ca 331
Poly[oxy((methyl)phenylsilylene)]	187
Poly[oxy((methyl)-3,3,3-trifluoropropylsilylene)]	<193

Main-chain sulfur-containing units

Poly(dithioethylene)	223
Poly(dithiomethylene-1,4-phenylenemethylene)	296
Poly(oxy-4,4′-biphenylylene-1,4-phenylenesulfonyl-1,4-phenylene)	503 (M)
Poly(oxycarbonyloxy-1,4-phenylenethio-1,4-phenylene)	ca 383
Poly(oxyethylenedithioethylene)	220 (M)
Poly[oxy(2-hydroxytrimethylene)oxy-1,4-phenylenesulfonyl-1,4-phenylene]	428
Poly(oxymethyleneoxyethylenedithioethylene)	214
Poly(oxy-1,4-phenylenesulfinyl-1,4-phenyleneoxy-1,4-phenylenecarbonyl-1,4-phenylene)	478 (M)
Poly(oxy-1,4-phenylenesulfinyl-1,4-phenyleneoxy-1,4-phenyleneisopropylidene-1,4-phenylene)	438 (M)
Poly(oxy-1,4-phenylenesulfonyl-1,4-phenylene)	487
Poly(oxy-1,4-phenylenesulfonyl-4,4′-biphenylylenesulfonyl-1,4-phenylene)	533
Poly[oxy-1,4-phenylenesulfonyl-1,4-phenyleneoxy(2,6-dimethyl-1,4-phenylene)isopropylidene (3,5-dimethyl-1,4-phenylene)]	508 (M)
Poly[oxy-1,4-phenylenesulfonyl-1,4-phenyleneoxy-1,4-phenylenecarbonyl-1,4-phenylene]	478 (M)
Poly[oxy-1,4-phenylenesulfonyl-1,4-phenyleneoxy-1,4-phenylene(hexafluoroisopropylidene)1,4-phenylene]	478 (M)
Poly(oxy-1,4-phenylenesulfonyl-1,4-phenyleneoxy-1,4-phenyleneisopropylidene-1,4-phenylene)	449
Poly(oxy-1,4-phenylenesulfonyl-1,4-phenyleneoxy-1.4-phenylenemethylene-1,4-phenylene)	453 (M)
Poly(oxy-1,4-phenylenesulfonyl-1,4-phenyleneoxy-1.4-phenylenethio-1,4-phenylene)	448 (M)
Poly(oxy-1,4-phenylenesulfonyl-1,4-phenyleneoxyterephthaloyl)	522
Poly(oxytetramethylenedithiotetramethylene)	197
Poly(sulfonyl-1,2-cyclohexylene)	401
Poly(sulfonyl-1,3-cyclohexylene)	381
Poly(sulfonyl-1,4-phenylenemethylene-1,4-phenylene)	497
Poly(thio-1,3-cyclohexylene)	221
Poly[thio(difluoromethylene)]	155
Poly(thioethylene)	223
Poly[thio(1-ethylethylene]	218
Poly[thio(1-methyl-3-oxotrimethylene)]	285
Poly[thio(1-methyltrimethylene)]	214
Pol[(thio(1-oxohexamethylene)]	292
Poly(thio-1,4-phenylene)	370
Poly(thiopropylene)	226

Main-chain heterocyclic units

Poly(1,3-dioxa-4,6-cyclohexylenemethylene)	378
Poly(vinyl formal)	
Poly[(2,6-dioxopiperidine-1,4-diyl)trimethylene]	363
Poly[(2-methyl-1,3-dioxa-4,6-cyclohexylene)methylene]	355
Poly(vinyl acetal)	
Poly(1,4-piperazinediylcarbonyloxyethyleneoxycarbonyl)	333
Poly(1,4-piperazinediylisophthaloyl)	465 (M)
Poly[(2-propyl-1,3-dioxa-4,6-cyclohexylene)methylene]	322
Poly(vinyl butyral)	
Poly(3,6-pyridazinediyloxy-1,4-phenyleneisopropylidene-1,4-phenyleneoxy)	453 (M)
Poly(2,5-pyridinediylcarbonyliminohexamethyleneiminocarbonyl)	322

DIELECTRIC CONSTANT OF SELECTED POLYMERS

This table lists typical values of the dielectric constant (more properly called relative permittivity) of some important polymers. Values are given for frequencies of 1 kHz, 1 MHz, and 1 GHz; in most cases the dielectric constant at frequencies below 1 kHz does not differ significantly from the value at 1 kHz. Since the dielectric constant of a polymeric material can vary with density, degree of crystallinity, and other details of a particular sample, the values given here should be regarded as only typical or average values.

References

1. Gray, D. E., Ed., *American Institute of Physics Handbook, Third Edition*, p. 5-132, McGraw Hill, New York, 1972.
2. Anderson, H. L., Ed., *A Physicist's Desk Reference*, American Institute of Physics, New York, 1989.
3. Brandrup, J., and Immergut, E. H., *Polymer Handbook, Third Edition*, John Wiley & Sons, New York, 1989.

Name	$t/°C$	1 kHz	1 MHz	1 GHz
Polyacrylonitrile	25	5.5	4.2	
Polyamides (nylons)	25	3.50	3.14	2.8
	84	11	4.4	2.8
Polybutadiene	25	2.5		
Polycarbonate	23	2.92	2.8	
Polychloroprene (neoprene)	25	6.6	6.3	4.2
Polychlorotrifluoroethylene	23	2.65	2.46	2.39
Polyethylene	23	2.3		
Poly(ethylene terephthalate) (Mylar)	23	3.25	3.0	2.8
Polyisoprene (natural rubber)	27	2.6	2.5	2.4
Poly(methyl methacrylate)	27	3.12	2.76	2.6
	80	3.80	2.7	2.6
Polyoxymethylene (polyformaldehyde)	25	3.8		
Poly(phenylene oxide)	23	2.59	2.59	
Polypropylene	25	2.3	2.3	2.3
Polystyrene	25	2.6	2.6	2.6
Polysulfones	25	3.13	2.10	
Polytetrafluoroethylene (teflon)	25	2.1	2.1	2.1
Poly(vinyl acetate)	50		3.5	
	150		8.3	
Poly(vinyl chloride)	25	3.39	2.9	2.8
	100	5.3	3.3	2.7
Poly(vinylidene chloride)	23	4.6	3.2	2.7
Poly(vinylidene fluoride)	23	12.2	8.9	4.7

SECOND VIRIAL COEFFICIENTS OF POLYMER SOLUTIONS

Christian Wohlfarth

Second virial coefficients characterize the thermodynamic behavior of a dilute polymer solution. They are usually defined via the concentration dependence of the osmotic pressure, π, of a polymer solution:

$$\frac{\pi}{c_B} = RT\left[\frac{1}{M_n} + A_2 c_B + A_3 c_B^2 + \ldots\right] \quad (1)$$

where:

A_2, A_3	second, third osmotic virial coefficient
c_B	(g cm^{-3}) concentration of polymer (B)
M_n	number-average relative molar mass of the polymer
R	gas constant
T	(measuring) temperature

However, most experimental data are measured by light scattering. Scattering methods enable the determination of A_2 via the common relation:

$$\frac{Kc_B}{R(q)} = \frac{1}{M_w P_z(q)} + 2A_2 Q(q)c_B + \ldots \quad (2)$$

where:

K	a constant that summarizes the optical parameters of a scattering experiment
M_w	mass-average relative molar mass of the polymer
$P_z(q)$	z-average of the scattering function
q	scattering vector $q = \dfrac{4\pi}{\lambda}\sin\dfrac{\theta}{2}$
$Q(q)$	function for the q-dependence of A_2
$R(q)$	excess intensity of the scattered beam at the value q
λ	wavelength
θ	scattering angle

In the dilute concentration region, the virial equation is usually truncated after the second virial coefficient which leads to a linear relationship. A linearized relation over a wider concentration range can be constructed if the Stockmayer-Casassa relation between A_2 and A_3 is applied:

$$A_3 M_n = \left(\frac{A_2 M_n}{2}\right)^2 \quad (3)$$

The values of second virial coefficients depend on the chosen polymer-solvent pair, on temperature and pressure, and on molar mass. In good solvents, a scaling relation can be applied for the molar mass dependence:

$$A_2 = \alpha M_w^{-\beta} \quad (4)$$

The constants α and β for a number of polymer-solvent pairs are given in the table below. They were newly fitted to experimental data by a non-linear least-squares fitting procedure. Experimental data were taken from a recent table of experimental A_2-values in Ref. 1. The unit of A_2 (used to determine the constants α and β) is cm^3 mol g^{-2}. The used unit for the molar mass is g mol^{-1}.

Reference

1. Wohlfarth, C., *Thermodynamic Properties of Polymer Solutions*, in Landolt-Börnstein, New Series, Group VIII, Volume 6D, Lechner, M.D. (ed.), Springer Verlag, Berlin, Heidelberg, 2010.

Polymer	Solvent	T/K	α	β
Amylose	dimethylsulfoxide	298.15	0.0086	0.235
Amylose tris(N-phenyl carbamate)	2-propanone	290.15	0.0157	0.417
Cellulose tris(N-phenyl carbamate)	2-propanone	290.15	0.0023	0.177
Cellulose tris(N-phenyl carbamate)	tetrahydrofuran	298.15	0.0021	0.116
Dextran	water	298.15	0.0353	0.394
Dextran (acid-hydrolyzed)	water	298.15	0.0516	0.361
2-(Diethylamino)ethyl dextran	water	293.15	0.129	0.503
Hydroxypropyl cellulose	ethanol	298.15	0.108	0.371
Poly(N-acryloylmorpholine)	N,N-dimethylformamide	298.15	0.0084	0.295
Polybutadiene	cyclohexane	298.15	0.0112	0.204
Polybutadiene	tetrahydrofuran	296.15	0.0235	0.257
Poly(butyl methacrylate)	2-butanone	298.15	0.0099	0.285
Poly(4-tert-butylstyrene)	cyclohexane	298.15	0.0377	0.362
Poly(4-tert-butylstyrene)	toluene	310.15	0.0129	0.293
Polycarbonate-bisphenol-A	1,2-dichloroethane	298.15	0.0372	0.319
Polycarbonate-bisphenol-A	dichloromethane	298.15	0.0124	0.205
Polycarbonate-bisphenol-A	tetrahydrofuran	298.15	0.0426	0.333
Polycarbonate-bisphenol-A	trichloromethane	298.15	0.016	0.227
Poly(chloroprene)	butyl acetate	298.15	0.0041	0.229
Poly(chloroprene)	tetrachloromethane	298.15	0.0083	0.237
Poly(2-chlorostyrene)	toluene	298.15	0.0036	0.219
Poly(2-chlorostyrene)	toluene	308.15	0.0065	0.246
Poly(3-chlorostyrene)	toluene	308.15	0.0232	0.319
Poly(4-chlorostyrene)	2-butanone	294.15	0.0041	0.262

Polymer	Solvent	T/K	α	β
Poly(4-chlorostyrene)	toluene	303.15	0.0018	0.241
Poly(cyclobutyl methacrylate)	tetrahydrofuran	297.15	0.0114	0.271
Poly(cyclododecyl methacrylate)	cyclohexane	298.15	0.0029	0.242
Poly(cyclohexyl methacrylate)	cyclohexane	298.15	0.0071	0.291
Poly(cyclooctyl methacrylate)	cyclohexane	298.15	0.0056	0.271
Poly(cyclopentyl methacrylate)	tetrahydrofuran	297.15	0.0046	0.210
Poly(decyl methacrylate)	toluene	310.15	0.0389	0.398
Poly(2,5-dichlorostyrene)	1,4-dioxane	293.15	0.0044	0.295
Poly(dihexylsilylene)	tetrahydrofuran	298.15	0.0025	0.212
Poly(diisopropyl fumarate)	tetrahydrofuran	303.15	0.0044	0.197
Poly(N,N-dimethylacrylamide)	methanol	298.15	0.0168	0.266
Poly(N,N-dimethylacrylamide)	water	298.15	0.0298	0.365
Poly(3,5-dimethyl-1-methacryloylpyrazole)	trichloromethane	293.15	0.0228	0.378
Poly[N-(1,1-dimethyl-3-oxobutyl)acrylamide]	2-butanone	298.15	0.0134	0.310
Poly(2,6-dimethyl-1,4-phenylene ether)	toluene	298.15	0.0116	0.205
Poly(dimethylsiloxane)	toluene	298.15	0.0231	0.333
Poly(2,2-diphenylethyl methacrylate)	tetrahydrofuran	298.15	0.0052	0.275
Poly(diphenylmethyl methacrylate)	toluene	296.15	0.0027	0.256
Poly(docosyl methacrylate)	toluene	310.15	0.0445	0.423
Poly(dodecyl acrylate)	butyl acetate	296.15	0.0040	0.289
Poly(2-ethoxyethyl methacrylate)	2-butanone	298.15	0.0328	0.349
Poly(ethyl acrylate)	tetrahydrofuran	298.15	0.0320	0.328
Polyethylene (LDPE)	1-chloronaphthalene	398.15	0.109	0.419
Polyethylene (LDPE)	1,4-dimethylbenzene	354.15	0.0677	0.361
Polyethylene (LDPE)	tetrahydronaphthalene	354.65	0.113	0.443
Poly(ethylene-alt-propylene)	toluene	310.15	0.0451	0.333
Polyethylenimine	water	308.15	0.0813	0.490
Poly(ethyl methacrylate)	2-butanone	296.15	0.0575	0.392
Poly(ethyl methacrylate)	ethyl acetate	308.15	0.0069	0.265
Poly(3-fluorostyrene)	2-butanone	298.15	0.0542	0.397
Poly(4-fluorostyrene)	2-butanone	298.15	0.0164	0.280
Poly(hexadecyl methacrylate)	n-heptane	298.15	0.0196	0.328
Poly(hexyl acrylate)	tetrahydrofuran	298.15	0.0111	0.272
Poly(hexyl methacrylate)	2-butanone	296.15	0.0046	0.236
Poly(1-indanyl methacrylate)	tetrahydrofuran	298.15	0.0028	0.173
Poly(isobornyl methacrylate)	tetrahydrofuran	298.15	0.0042	0.244
Poly(isobutylene)	cyclohexane	298.15	0.0095	0.224
Poly(isobutylene)	n-heptane	298.15	0.0145	0.296
Poly(isobutylene)	n-hexane	297.15	0.0018	0.155
Poly(isobutylene)	trichloromethane	298.15	0.0034	0.204
Poly(isobutylene)	2,2,4-trimethylpentane	298.15	0.0051	0.219
Poly(isoprene) (70% cis, 23% trans)	cyclohexane	296.15	0.0175	0.260
Poly(isoprene) (trans)	cyclohexane	298.15	0.0064	0.179
Poly(isoprene) (3-arm star)	cyclohexane	296.15	0.0234	0.279
Poly(isoprene) (4-arm star)	cyclohexane	296.15	0.0172	0.249
Poly(isoprene) (8-arm star)	cyclohexane	296.15	0.0198	0.285
Poly(isoprene) (12-arm star)	cyclohexane	296.15	0.0108	0.257
Poly(isoprene) (18-arm star)	cyclohexane	296.15	0.0247	0.330
Poly(isoprene) (hydrogenated)	toluene	310.15	0.0455	0.333
Poly(N-isopropylacrylamide)	methanol	298.15	0.0085	0.250
Poly(N-isopropylacrylamide)	water	293.15	0.0028	0.195
Poly(isopropyl acrylate) (atactic)	bromobenzene	333.15	0.0172	0.285
Poly(isopropyl acrylate) (isotactic)	bromobenzene	333.15	0.0118	0.263
Poly(N-isopropylmethacrylamide)	water	293.15	0.0235	0.355
Poly(2-methoxyethyl methacrylate)	2-butanone	298.15	0.0525	0.397
Poly(methyl 2-butylacrylate)	2-butanone	303.15	0.0036	0.223
Poly(methyl 2-ethylacrylate)	2-butanone	303.15	0.0068	0.251
Poly(methyl methacrylate)	2-butanone	298.15	0.0052	0.254
Poly(methyl methacrylate)	1,4-dioxane	298.15	0.0164	0.295
Poly(methyl methacrylate)	nitroethane	298.15	0.0072	0.263

Polymer	Solvent	T/K	α	β
Poly(methyl methacrylate)	2-propanone	298.15	0.0035	0.235
Poly(methyl methacrylate)	tetrahydrofuran	298.15	0.0119	0.281
Poly(methyl methacrylate)	trichloromethane	298.15	0.0129	0.246
Poly(methylphenylsiloxane)	cyclohexane	298.15	0.0023	0.216
Poly(α-methylstyrene)	4-tert-butyltoluene	298.15	0.0027	0.239
Poly(α-methylstyrene)	1-chorobutane	298.15	0.0025	0.250
Poly(α-methylstyrene)	toluene	298.15	0.0124	0.298
Poly(2-methylstyrene)	toluene	303.15	0.0268	0.349
Poly(4-methylstyrene)	toluene	303.15	0.0145	0.289
Poly(2-methyl-5-vinylpyridine)	2-butanone	298.15	0.0151	0.335
Poly(2-methyl-5-vinyltetrazole)	N,N-dimethylformamide	293.15	0.0062	0.251
Poly(octadecyl methacrylate)	toluene	310.15	0.0106	0.312
Poly(1-octene)	bromobenzene	298.15	0.0051	0.222
Poly(1-octene)	n-heptane	293.15	0.0473	0.330
1,4-trans-Poly(1,3-pentadiene)	toluene	300.75	0.0377	0.405
trans-1,5-Polypentenamer	n-hexane	298.15	0.0048	0.233
Poly(1-pentene)	toluene	303.15	0.0554	0.387
Poly(phenyl methacrylate)	2-butanone	298.15	0.0105	0.328
Poly(1-phenyl-1-propyne)	toluene	298.15	0.025	0.325
Poly(phenylsilesquioxane)	toluene	310.15	0.0046	0.288
Poly(propyl acrylate)	tetrahydrofuran	298.15	0.0094	0.251
Polypropylene (atactic)	benzene	298.15	0.0108	0.311
Polypropylene (isotactic)	1-chloronaphthalene	418.15	0.0025	0.132
Polypropylene (isotactic)	tetrahydronaphthalene	408.15	0.0045	0.118
Poly(propylene sulfide)	benzene	293.15	0.0814	0.493
Polystyrene	benzene	298.15	0.0113	0.259
Polystyrene (4-arm star)	benzene	298.15	0.0092	0.258
Polystyrene (6-arm star)	benzene	298.15	0.0093	0.268
Polystyrene	bromobenzene	293.15	0.0142	0.282
Polystyrene	2-butanone	298.15	0.0028	0.233
Polystyrene (cyclic)	2-butanone	298.15	0.0207	0.395
Polystyrene	4-tert-butyltoluene	323.15	0.0035	0.221
Polystyrene	1,2-dichloroethane	308.15	0.0086	0.262
Polystyrene	ethylbenzene	298.15	0.0139	0.289
Polystyrene	tetrachloromethane	298.15	0.0084	0.246
Polystyrene	tetrahydrofuran	298.15	0.0129	0.268
Polystyrene	toluene	298.15	0.0118	0.264
Polystyrene	toluene	303.15	0.0076	0.241
Polystyrene	toluene	308.15	0.0102	0.267
Polystyrene (cyclic)	toluene	298.15	0.0144	0.292
Polystyrene (3-arm star)	toluene	293.15	0.0042	0.196
Polystyrene (12-arm star)	toluene	293.15	0.0282	0.369
Poly(tetrahydro-2H-pyran-2-yl methacrylate)	toluene	310.15	0.0182	0.362
Poly[4-(1,1,3,3-tetramethylbutyl)-phenyl methacrylate]	toluene	298.15	0.054	0.420
Poly(tridecyl methacrylate)	toluene	310.15	0.0112	0.325
Poly(vinyl acetate)	2-butanone	303.15	0.0117	0.269
Poly(vinyl acetate)	2-propanone	303.15	0.0132	0.268
Poly(N-vinylcarbazole)	benzene	298.15	0.0012	0.197
Poly(N-vinylcarbazole)	benzene	303.15	0.0014	0.193
Poly(N-vinylcarbazole)	benzene	308.15	0.0019	0.200
Poly(N-vinylcarbazole)	benzene	313.15	0.0025	0.208
Poly(N-vinylcarbazole)	1,4-dioxane	310.15	0.0038	0.261
Poly(vinyl chloride)	tetrahydrofuran	298.15	0.0096	0.157
Poly(vinyl phenylcarbamate)	1,4-dioxane	293.15	0.0130	0.307
Poly(2-vinylpyridine-1-oxide)	2-propanol	298.15	0.0332	0.365
Poly(1-vinyl-2-pyrrolidinone)	ethanol	298.15	0.0482	0.392
Poly(1-vinyl-2-pyrrolidinone)	methanol	298.15	0.0384	0.369
Poly(1-vinyl-2-pyrrolidinone)	1-propanol	298.15	0.0409	0.385
Pullulan	water	298.15	0.0092	0.291

PRESSURE–VOLUME–TEMPERATURE RELATIONSHIPS FOR POLYMER MELTS

Christian Wohlfarth

Numerous theoretical equations of state for polymer liquids have been developed. These, at the minimum, have to provide accurate fitting functions to experimental data. However, for the purpose of this table, the empirical Tait equation along with a polynomial expression for the zero pressure isobar is used. This equation is able to represent the experimental data for the melt state within the limits of experimental errors, i.e., the maximum deviations between measured and calculated specific volumes are about 0.001-0.002 cm³/g.

The general form of the Tait equation is:

$$V(P,T) = V(0,T)\{1 - C \ln[1 + P/B(T)]\} \qquad (1)$$

where the coefficient C is usually taken to be a universal constant equal to 0.0894. T is the absolute temperature in K and P the pressure in MPa. The volume V is the specific volume in cm³/g. The Tait parameter $B(T)$ has the very simple meaning that it is inversely proportional to the compressibility κ at constant temperature and zero pressure:

$$\kappa(0,T) = -[1/V(0,T)](dV/dP) = C/B(T) \qquad (2)$$

The $B(T)$ function is usually given by:

$$B(T) = B_0 \exp[-B_1(T\text{-}273.15)] \qquad (3)$$

but, sometimes a polynomial expression is used:

$$B(T) = b_0 + b_1(T\text{-}273.15) + b_2(T\text{-}273.15)^2 \qquad (4)$$

The zero-pressure isobar $V(0,T)$ is usually given by:

$$V(0,T) = A_0 + A_1(T\text{-}273.15) + A_2(T\text{-}273.15)^2 \qquad (5)$$

where A_0, A_1, A_2 are specific constants for a given polymer (the expression T-273.15 is used because fitting to the zero-pressure isobar is usually done in terms of Celsius temperature). Other forms for $V(0,T)$ are also found in the literature, such as

$$V(0,T) = A_3 \exp[A_4(T\text{-}273.15)] \qquad (6)$$

or

$$V(0,T) = A_5 \exp(A_6 T^{1.5}) \qquad (7)$$

where A_3 and A_4 or A_5 and A_6 are again specific constants for a given polymer.

The Tait equation is particularly useful to calculate derivative quantities, such as the isothermal compressibility and the thermal expansivity coefficients. The isothermal compressibility $\kappa(P,T)$ is derived from equation (1) as:

$$\kappa(P,T) = -(1/V)(dV/dP) = 1/\{[P + B(T)][1/C - \ln(1 + P/B(T))]\} \quad (8)$$

and the thermal expansivity $\alpha(P,T)$ as:

$$\alpha(P,T) = (1/V)(dV/dT) = \alpha(0,T) - PB_1\kappa(P,T) \qquad (9)$$

where $\alpha(0,T)$ represents the thermal expansivity at zero (atmospheric) pressure and is calculated from any suitable fit for the zero-pressure volume, such as equations (5) through (7) above.

Because polymer melt PVT-behavior depends only slightly on polymer molar mass above the oligomeric region, usually no information is given in the original literature for the average molar mass of the polymers.

Table 1 summarizes the polymers or copolymers considered here and the experimental ranges of pressure and temperature over which data are available. In Table 2 the Tait-equation functions, with parameters obtained from the fit, are given for 90 polymer or copolymer melts.

References

1. Zoller, P., *J. Appl. Polym. Sci.*, 23, 1051–1056, 1979.
2. Starkweather, H. W., Jones, G. A., and Zoller, P., *J. Polym. Sci., Pt. B Polym. Phys.*, 26, 257–266, 1988.
3. Fakhreddine, Y. A., and Zoller, P., *J. Polym. Sci., Pt. B Polym. Phys.*, 29, 1141–1146, 1991.
4. Rodgers, P. A., *J. Appl. Polym. Sci.*, 48, 1061–1080, 1993.
5. Rodgers, P. A., *J. Appl. Polym. Sci.*, 48, 2075–2083, 1993.
6. Yi, Y. X., and Zoller, P., *J. Polym. Sci., Pt. B Polym. Phys.*, 31, 779–788, 1993.
7. Callaghan, T. A., and Paul, D. R., *Macromolecules*, 26, 2439–2450, 1993.
8. Wang, Y. Z., Hsieh, K. H., Chen, L. W., and Tseng, H. C., *J. Appl. Polym. Sci.*, 53, 1191–1201, 1994.
9. Privalko, V. P., Arbuzova, A. P., Korskanov, V. V., and Zagdanskaya, N. E., *Polym. Intern.*, 35, 161–169, 1994.
10. Sachdev, V. K., Yashi, U., and Jain, R. K., *J. Polym. Sci., Pt. B Polym. Phys.*, 36, 841–850, 1998.

TABLE 1. Names of the Polymers, Abbreviation Used, and Range of Experimental Data Applied in the Determination of the Equation Constants

Polymer	Symbol	T/K	P/MPa	Ref.
Ethylene/propylene copolymer (50 wt%)	EP50	413–523	0.1–63	4
Ethylene/vinyl acetate copolymer				
18 wt% vinyl acetate	EVA18	385–491	0.1–177	4
25 wt% vinyl acetate	EVA25	367–506	0.1–177	4
28 wt% vinyl acetate	EVA28	367–508	0.1–177	4
40 wt% vinyl acetate	EVA40	348–508	0.1–177	4
Polyamide-6	PA6	509–569	0.1–196	4
Polyamide-11	PA11	478–542	0.1–200	5
Polyamide-66	PA66	519–571	0.1–196	4
cis-1,4-Polybutadiene	cPBD	277–328	0.1–284	4
Polybutadiene, 8% 1,2-content	PBD-8	298–473	0.1–200	6
Polybutadiene, 24% 1,2-content	PBD-24	298–473	0.1–200	6
Polybutadiene, 40% 1,2-content	PBD-40	298–473	0.1–200	6
Polybutadiene, 50% 1,2-content	PBD-50	298–473	0.1–200	6
Polybutadiene, 87% 1,2-content	PBD-87	298–473	0.1–200	6
Poly(1-butene), isotactic	iPB	406–519	0.1–196	4
Poly(butyl methacrylate)	PnBMA	307–473	0.1–200	4
Poly(butylene terephthalate)	PBT	508–576	0.1–200	3
Poly(ε-caprolactone)	PCL	373–421	0.1–200	4
Polycarbonate-bisphenol-A	PC	424–613	0.1–177	4
Polycarbonate-bisphenol-chloral	BCPC	428–557	0.1–200	4
Polycarbonate-hexafluorobisphenol-A	HFPC	432–553	0.1–200	4
Polycarbonate-tetramethylbisphenol-A	TMPC	491–563	0.1–160	4
Poly(cyclohexyl methacrylate)	PcHMA	396–471	0.1–200	4
Poly(2,5-dimethylphenylene oxide)	PPO	473–593	0.1–177	4
Poly(dimethyl siloxane)	PDMS	298–343	0.1–100	4
Poly(dimethyl siloxane) M_n = 1000	PDMS-10	304–420	0.1–250	10
Poly(dimethyl siloxane) M_n = 4000	PDMS-40	298–418	0.1–250	10
Poly(dimethyl siloxane) M_n = 6000	PDMS-60	291–423	0.1–250	10
Poly(epichlorohydrin)	PECH	333–413	0.1–200	4
Poly(ether ether ketone)	PEEK	619–671	0.1–200	4
Poly(ethyl acrylate)	PEA	310–490	0.1–196	4
Poly(ethyl methacrylate)	PEMA	386–434	0.1–196	4
Polyethylene, high density	HDPE	413–476	0.1–196	4
Polyethylene, linear	LPE	415–473	0.1–200	4
Polyethylene, linear, high MW	HMLPE	410–473	0.1–200	4
Polyethylene, branched	BPE	398–471	0.1–200	4
Polyethylene, low density	LDPE	394–448	0.1–196	4
Polyethylene, low density, type A	LDPE-A	385–498	0.1–196	1
Polyethylene, low density, type B	LDPE-B	385–498	0.1–196	1
Polyethylene, low density, type C	LDPE-C	385–498	0.1–196	1
Poly(ethylene oxide)	PEO	361–497	0.1–68	4
Poly(ethylene terephthalate)	PET	547–615	0.1–196	4
Poly(4-hexylstyrene)	P4HS	303–403	30–100	4
Polyisobutylene	PIB	326–383	0.1–100	4
Polyisoprene, 8% 3,4-content	PI-8	298–473	0.1–200	6
Polyisoprene, 14% 3,4-content	PI-14	298–473	0.1–200	6
Polyisoprene, 41% 3,4-content	PI-41	298–473	0.1–200	6
Polyisoprene, 56% 3,4-content	PI-56	298–473	0.1–200	6
Poly(methyl acrylate)	PMA	310–493	0.1–196	4
Poly(methyl methacrylate)	PMMA	387–432	0.1–200	4
Poly(4-methyl-1-pentene)	P4MP	514–592	0.1–196	4
Poly(α-methylstyrene)	PαMS	473–533	0.1–170	7
Poly(o-methylstyrene)	PoMS	412–471	0.1–180	4
Polyoxymethylene	POM	463–493	0.1–196	2
Phenoxy[a]	PH	341–573	0.1–177	4
Polysulfone[b]	PSF	475–644	0.1–196	4
Polyarylate[c]	PAr	450–583	0.1–177	4
Polypropylene, atactic	aPP	353–393	0.1–100	4

Polymer	Symbol	T/K	P/MPa	Ref.
Polypropylene, isotactic	iPP	443-570	0.1-196	4
Polystyrene	PS	388-469	0.1-200	4
Poly(tetrafluoroethylene)	PTFE	603-645	0.1- 39	4
Poly(tetrahydrofuran)	PTHF	335-439	0.1- 78	4
Poly(vinyl acetate)	PVAc	308-373	0.1- 80	4
Poly(vinyl chloride)	PVC	373-423	0.1-200	4
Poly(vinyl methyl ether)	PVME	303-471	0.1-200	4
Poly(vinylidene fluoride)	PVdF	451-521	0.1-200	5
Styrene/acrylonitrile copolymer				
2.7 wt% acrylonitrile	SAN3	378-539	0.1-200	4
5.7 wt% acrylonitrile	SAN6	370-540	0.1-200	4
15.3 wt% acrylonitrile	SAN15	405-531	0.1-200	4
18.0 wt% acrylonitrile	SAN18	377-528	0.1-200	4
40 wt% acrylonitrile	SAN40	373-543	0.1-200	4
70 wt% acrylonitrile	SAN70	373-544	0.1-200	4
Styrene/butadiene copolymer				
10 wt% styrene	SBR10	393-533	0.1-196	8
23.5 wt% styrene	SBR23	393-533	0.1-196	8
60 wt% styrene	SBR60	393-533	0.1-196	8
85 wt% styrene	SBR85	393-533	0.1-196	8
Styrene/methyl methacrylate copolymer				
20 wt% methyl methacrylate	SMMA20	383-543	0.1-200	4
60 wt% methyl methacrylate	SMMA60	383-543	0.1-200	4
N-Vinylcarbazole/4-ethylstyrene copolymer				
50 mol% ethylstyrene	VCES50	393-443	30-100	9
N-Vinylcarbazole/4-hexylstyrene copolymer				
80 mol% hexylstyrene	VCHS80	313-423	30-100	9
67 mol% hexylstyrene	VCHS67	333-423	30-100	9
60 mol% hexylstyrene	VCHS60	383-453	30-100	9
50 mol% hexylstyrene	VCHS50	373-443	30-100	9
40 mol% hexylstyrene	VCHS40	423-493	30-100	9
33 mol% hexylstyrene	VCHS33	463-523	30-100	9
20 mol% hexylstyrene	VCHS20	473-523	30-100	9
N-Vinylcarbazole/4-octylstyrene copolymer				
50 mol% octylstyrene	VCOS50	403-453	30-100	9
N-Vinylcarbazole/4-pentylstyrene copolymer				
50 mol% pentylstyrene	VCPS50	383-443	30-100	9

[a] Phenoxy = Poly(oxy-2-hydroxytrimethyleneoxy-1,4-phenyleneisopropylidene-1,4-phenylene)

[b] Polysulfone = Poly(oxy-1,4-phenylenesulfonyl-1,4-phenyleneoxy-1,4-phenyleneisopropylidene-1,4-phenylene)

[c] Polyarylate = Poly(oxyterephthaloyl/isophthaloyl T/I=50/50)oxy-1,4-phenyleneisopropylidene-1,4-phenylene

TABLE 2. Tait Equation Parameter Functions for Polymer Melts

Polymer	$V(0,T)/\text{cm}^3\text{g}^{-1}$	$B(T)/\text{MPa}$
EP50	$1.2291 + 5.799 \cdot 10^{-5}(T-273.15) + 1.964 \cdot 10^{-6}(T-273.15)^2$	$487.0 \exp[-8.103 \cdot 10^{-3}(T-273.15)]$
EVA18	$1.02391 \exp(2.173 \cdot 10^{-5}T^{1.5})$	$188.2 \exp[-4.537 \cdot 10^{-3}(T-273.15)]$
EVA25	$1.00416 \exp(2.244 \cdot 10^{-5}T^{1.5})$	$184.4 \exp[-4.734 \cdot 10^{-3}(T-273.15)]$
EVA28	$1.00832 \exp(2.241 \cdot 10^{-5}T^{1.5})$	$183.5 \exp[-4.457 \cdot 10^{-3}(T-273.15)]$
EVA40	$1.06332 \exp(2.288 \cdot 10^{-5}T^{1.5})$	$205.1 \exp[-4.989 \cdot 10^{-3}(T-273.15)]$
PA6	$0.7597 \exp[4.701 \cdot 10^{-4}(T-273.15)]$	$376.7 \exp[-4.660 \cdot 10^{-3}(T-273.15)]$
PA11	$0.9581 \exp[6.664 \cdot 10^{-4}(T-273.15)]$	$254.7 \exp[-4.178 \cdot 10^{-3}(T-273.15)]$
PA66	$0.7657 \exp[6.600 \cdot 10^{-4}(T-273.15)]$	$316.4 \exp[-5.040 \cdot 10^{-3}(T-273.15)]$
cPBD	$1.0970 \exp[6.600 \cdot 10^{-4}(T-273.15)]$	$177.7 \exp[-3.593 \cdot 10^{-3}(T-273.15)]$
PBD-8	$1.1004 + 6.718 \cdot 10^{-4}(T-273.15) + 6.584 \cdot 10^{-7}(T-273.15)^2$	$200.0 \exp[-4.606 \cdot 10^{-3}(T-273.15)]$
PBD-24	$1.1049 + 6.489 \cdot 10^{-4}(T-273.15) + 7.099 \cdot 10^{-7}(T-273.15)^2$	$193.0 \exp[-4.519 \cdot 10^{-3}(T-273.15)]$
PBD-40	$1.1013 + 6.593 \cdot 10^{-4}(T-273.15) + 5.776 \cdot 10^{-7}(T-273.15)^2$	$188.0 \exp[-4.437 \cdot 10^{-3}(T-273.15)]$
PBD-50	$1.1037 + 5.955 \cdot 10^{-4}(T-273.15) + 7.789 \cdot 10^{-7}(T-273.15)^2$	$183.0 \exp[-4.425 \cdot 10^{-3}(T-273.15)]$
PBD-87	$1.1094 + 6.729 \cdot 10^{-4}(T-273.15) + 4.470 \cdot 10^{-7}(T-273.15)^2$	$175.0 \exp[-4.538 \cdot 10^{-3}(T-273.15)]$
iPB	$1.1417 \exp[6.751 \cdot 10^{-4}(T-273.15)]$	$167.5 \exp[-4.533 \cdot 10^{-3}(T-273.15)]$
PnBMA	$0.9341 + 5.5254 \cdot 10^{-4}(T-273.15) + 6.5803 \cdot 10^{-6}(T-273.15)^2 + 1.5691 \cdot 10^{-10}(T-273.15)^3$	$226.7 \exp[-5.344 \cdot 10^{-3}(T-273.15)]$
PBT	$0.9640 - 1.017 \cdot 10^{-3}(T-273.15) + 3.065 \cdot 10^{-6}(T-273.15)^2$	$263.0 \exp[-3.444 \cdot 10^{-3}(T-273.15)]$
PCL	$0.9049 \exp[6.392 \cdot 10^{-4}(T-273.15)]$	$189.0 \exp[-3.931 \cdot 10^{-3}(T-273.15)]$
PC	$0.73565 \exp(1.859 \cdot 10^{-5}T^{1.5})$	$310.0 \exp[-4.078 \cdot 10^{-3}(T-273.15)]$
BCPC	$0.6737 + 3.634 \cdot 10^{-4}(T-273.15) + 2.370 \cdot 10^{-7}(T-273.15)^2$	$363.4 \exp[-4.921 \cdot 10^{-3}(T-273.15)]$
HFPC	$0.6111 + 4.898 \cdot 10^{-4}(T-273.15) + 1.730 \cdot 10^{-7}(T-273.15)^2$	$236.6 \exp[-5.156 \cdot 10^{-3}(T-273.15)]$
TMPC	$0.8497 + 5.073 \cdot 10^{-4}(T-273.15) + 3.832 \cdot 10^{-7}(T-273.15)^2$	$231.4 \exp[-4.242 \cdot 10^{-3}(T-273.15)]$
PcHMA	$0.8793 + 4.0504 \cdot 10^{-4}(T-273.15) + 7.774 \cdot 10^{-7}(T-273.15)^2 - 7.7534 \cdot 10^{-10}(T-273.15)^3$	$295.2 \exp[-5.220 \cdot 10^{-3}(T-273.15)]$
PPO	$0.78075 \exp(2.151 \cdot 10^{-5}T^{1.5})$	$227.8 \exp[-4.290 \cdot 10^{-3}(T-273.15)]$
PDMS	$1.0079 \exp[9.121 \cdot 10^{-4}(T-273.15)]$	$89.4 \exp[-5.701 \cdot 10^{-3}(T-273.15)]$
PDMS-10	$0.8343 + 5.991 \cdot 10^{-4}(T-273.15) + 5.734 \cdot 10^{-7}(T-273.15)^2$	$542.63 \exp[-6.69 \cdot 10^{-3}(T-273.15)]$
PDMS-40	$0.8018 + 7.072 \cdot 10^{-4}(T-273.15) + 3.635 \cdot 10^{-7}(T-273.15)^2$	$482.73 \exp[-6.09 \cdot 10^{-3}(T-273.15)]$
PDMS-60	$0.8146 + 5.578 \cdot 10^{-4}(T-273.15) + 5.774 \cdot 10^{-7}(T-273.15)^2$	$482.73 \exp[-6.09 \cdot 10^{-3}(T-273.15)]$
PECH	$0.7216 \exp[5.825 \cdot 10^{-4}(T-273.15)]$	$238.3 \exp[-4.171 \cdot 10^{-3}(T-273.15)]$
PEEK	$0.7158 \exp[6.690 \cdot 10^{-4}(T-273.15)]$	$388.0 \exp[-4.124 \cdot 10^{-3}(T-273.15)]$
PEA	$0.8756 \exp[7.241 \cdot 10^{-4}(T-273.15)]$	$193.2 \exp[-4.839 \cdot 10^{-3}(T-273.15)]$
PEMA	$0.8614 \exp[7.468 \cdot 10^{-4}(T-273.15)]$	$260.9 \exp[-5.356 \cdot 10^{-3}(T-273.15)]$
HDPE	$1.1595 + 8.0394 \cdot 10^{-4}(T-273.15)$	$179.9 \exp[-4.739 \cdot 10^{-3}(T-273.15)]$
LPE	$0.9172 \exp[7.806 \cdot 10^{-4}(T-273.15)]$	$176.7 \exp[-4.661 \cdot 10^{-3}(T-273.15)]$
HMLPE	$0.8992 \exp[8.502 \cdot 10^{-4}(T-273.15)]$	$168.3 \exp[-4.292 \cdot 10^{-3}(T-273.15)]$
BPE	$0.9399 \exp[7.341 \cdot 10^{-4}(T-273.15)]$	$177.1 \exp[-4.699 \cdot 10^{-3}(T-273.15)]$
LDPE	$1.1944 + 2.841 \cdot 10^{-4}(T-273.15) + 1.872 \cdot 10^{-6}(T-273.15)^2$	$202.2 \exp[-5.243 \cdot 10^{-3}(T-273.15)]$
LDPE-A	$1.1484 \exp[6.950 \cdot 10^{-4}(T-273.15)]$	$192.9 \exp[-4.701 \cdot 10^{-3}(T-273.15)]$
LDPE-B	$1.1524 \exp[6.700 \cdot 10^{-4}(T-273.15)]$	$196.6 \exp[-4.601 \cdot 10^{-3}(T-273.15)]$
LDPE-C	$1.1516 \exp[6.730 \cdot 10^{-4}(T-273.15)]$	$186.7 \exp[-4.391 \cdot 10^{-3}(T-273.15)]$
PEO	$0.8766 \exp[7.087 \cdot 10^{-4}(T-273.15)]$	$207.7 \exp[-3.947 \cdot 10^{-3}(T-273.15)]$
PET	$0.6883 + 5.90 \cdot 10^{-4}(T-273.15)$	$369.7 \exp[-4.150 \cdot 10^{-3}(T-273.15)]$
P4HS	$0.8251 + 6.77 \cdot 10^{-4}T$	$103.1 \exp[-2.417 \cdot 10^{-3}(T-273.15)]$
PIB	$1.0750 \exp[5.651 \cdot 10^{-4}(T-273.15)]$	$200.3 \exp[-4.329 \cdot 10^{-3}(T-273.15)]$
PI-8	$1.1030 + 6.488 \cdot 10^{-4}(T-273.15) + 5.125 \cdot 10^{-7}(T-273.15)^2$	$188.0 \exp[-4.541 \cdot 10^{-3}(T-273.15)]$
PI-14	$1.0943 + 6.293 \cdot 10^{-4}(T-273.15) + 6.231 \cdot 10^{-7}(T-273.15)^2$	$202.0 \exp[-4.653 \cdot 10^{-3}(T-273.15)]$
PI-41	$1.0951 + 6.188 \cdot 10^{-4}(T-273.15) + 6.629 \cdot 10^{-7}(T-273.15)^2$	$199.0 \exp[-4.622 \cdot 10^{-3}(T-273.15)]$
PI-56	$1.0957 + 6.655 \cdot 10^{-4}(T-273.15) + 5.661 \cdot 10^{-7}(T-273.15)^2$	$200.0 \exp[-4.644 \cdot 10^{-3}(T-273.15)]$
PMA	$0.8365 \exp[6.795 \cdot 10^{-4}(T-273.15)]$	$235.8 \exp[-4.493 \cdot 10^{-3}(T-273.15)]$
PMMA	$0.8254 + 2.8383 \cdot 10^{-4}(T-273.15) + 7.792 \cdot 10^{-7}(T-273.15)^2$	$287.5 \exp[-4.146 \cdot 10^{-3}(T-273.15)]$
P4MP	$1.4075 - 9.095 \cdot 10^{-4}(T-273.15) + 3.497 \cdot 10^{-6}(T-273.15)^2$	$37.67 + 0.2134(T-273.15) - 7.0445 \cdot 10^{-4}(T-273.15)^2$
PαMS	$0.89365 + 3.4864 \cdot 10^{-4}(T-273.15) + 5.0184 \cdot 10^{-7}(T-273.15)^2$	$297.7 \exp[-4.074 \cdot 10^{-3}(T-273.15)]$
PoMS	$0.9396 \exp[5.306 \cdot 10^{-4}(T-273.15)]$	$261.9 \exp[-4.114 \cdot 10^{-3}(T-273.15)]$
POM	$0.7484 \exp[6.770 \cdot 10^{-4}(T-273.15)]$	$305.6 \exp[-4.326 \cdot 10^{-3}(T-273.15)]$
PH	$0.76644 \exp(1.921 \cdot 10^{-5}T^{1.5})$	$359.9 \exp[-4.378 \cdot 10^{-3}(T-273.15)]$
PSF	$0.7644 + 3.419 \cdot 10^{-4}(T-273.15) + 3.126 \cdot 10^{-7}(T-273.15)^2$	$365.9 \exp[-3.757 \cdot 10^{-3}(T-273.15)]$
PAr	$0.73381 \exp(1.626 \cdot 10^{-5}T^{1.5})$	$296.9 \exp[-3.375 \cdot 10^{-3}(T-273.15)]$
aPP	$1.1841 - 1.091 \cdot 10^{-4}(T-273.15) + 5.286 \cdot 10^{-6}(T-273.15)^2$	$162.1 \exp[-6.604 \cdot 10^{-3}(T-273.15)]$
iPP	$1.1606 \exp[6.700 \cdot 10^{-4}(T-273.15)]$	$149.1 \exp[-4.177 \cdot 10^{-3}(T-273.15)]$

Polymer	$V(0,T)/\text{cm}^3\text{g}^{-1}$	$B(T)/\text{MPa}$
PS	$0.9287 \exp[5.131 \cdot 10^{-4}(T{-}273.15)]$	$216.9 \exp[-3.319 \cdot 10^{-3}(T{-}273.15)]$
PTFE	$0.3200 + 9.5862 \cdot 10^{-4}(T{-}273.15)$	$425.2 \exp[-9.380 \cdot 10^{-3}(T{-}273.15)]$
PTHF	$1.0043 \exp[6.691 \cdot 10^{-4}(T{-}273.15)]$	$178.6 \exp[-4.223 \cdot 10^{-3}(T{-}273.15)]$
PVAc	$0.82496 + 5.820 \cdot 10^{-4}(T{-}273.15) + 2.940 \cdot 10^{-7}(T{-}273.15)^2$	$204.9 \exp[-4.346 \cdot 10^{-3}(T{-}273.15)]$
PVC	$0.7196 + 5.581 \cdot 10^{-5}(T{-}273.15) + 1.468 \cdot 10^{-6}(T{-}273.15)^2$	$294.2 \exp[-5.321 \cdot 10^{-3}(T{-}273.15)]$
PVME	$0.9585 \exp[6.653 \cdot 10^{-4}(T{-}273.15)]$	$215.8 \exp[-4.588 \cdot 10^{-3}(T{-}273.15)]$
PVdF	$0.5790 \exp[8.051 \cdot 10^{-4}(T{-}273.15)]$	$244.0 \exp[-5.210 \cdot 10^{-3}(T{-}273.15)]$
SAN3	$0.9233 + 3.936 \cdot 10^{-4}(T{-}273.15) + 5.685 \cdot 10^{-7}(T{-}273.15)^2$	$239.8 \exp[-4.376 \cdot 10^{-3}(T{-}273.15)]$
SAN6	$0.9211 + 4.370 \cdot 10^{-4}(T{-}273.15) + 5.846 \cdot 10^{-7}(T{-}273.15)^2$	$226.9 \exp[-4.286 \cdot 10^{-3}(T{-}273.15)]$
SAN15	$0.9044 + 4.207 \cdot 10^{-4}(T{-}273.15) + 4.077 \cdot 10^{-7}(T{-}273.15)^2$	$238.4 \exp[-3.943 \cdot 10^{-3}(T{-}273.15)]$
SAN18	$0.9016 + 4.036 \cdot 10^{-4}(T{-}273.15) + 4.206 \cdot 10^{-7}(T{-}273.15)^2$	$240.4 \exp[-3.858 \cdot 10^{-3}(T{-}273.15)]$
SAN40	$0.8871 + 3.406 \cdot 10^{-4}(T{-}273.15) + 4.938 \cdot 10^{-7}(T{-}273.15)^2$	$289.3 \exp[-4.431 \cdot 10^{-3}(T{-}273.15)]$
SAN70	$0.8528 + 3.616 \cdot 10^{-4}(T{-}273.15) + 2.634 \cdot 10^{-7}(T{-}273.15)^2$	$335.4 \exp[-3.923 \cdot 10^{-3}(T{-}273.15)]$
SBR10	$0.9053 \exp(2.437 \cdot 10^{-5}T^{1.5})$	$530.3 \exp[-3.99 \cdot 10^{-3}(T{-}273.15)]$
SBR23	$0.8986 \exp(2.317 \cdot 10^{-5}T^{1.5})$	$551.6 \exp[-4.17 \cdot 10^{-3}(T{-}273.15)]$
SBR60	$0.8812 \exp(2.031 \cdot 10^{-5}T^{1.5})$	$486.0 \exp[-4.34 \cdot 10^{-3}(T{-}273.15)]$
SBR85	$0.8704 \exp(1.846 \cdot 10^{-5}T^{1.5})$	$356.7 \exp[-4.24 \cdot 10^{-3}(T{-}273.15)]$
SMMA20	$0.9063 + 3.570 \cdot 10^{-4}(T{-}273.15) + 6.532 \cdot 10^{-7}(T{-}273.15)^2$	$232.0 \exp[-4.143 \cdot 10^{-3}(T{-}273.15)]$
SMMA60	$0.8610 + 3.350 \cdot 10^{-4}(T{-}273.15) + 6.980 \cdot 10^{-7}(T{-}273.15)^2$	$261.0 \exp[-4.611 \cdot 10^{-3}(T{-}273.15)]$
VCES50	$0.6676 + 6.63 \cdot 10^{-4}T$	$5281.7 \exp[-9.264 \cdot 10^{-3}(T{-}273.15)]$
VCHS80	$0.7753 + 6.17 \cdot 10^{-4}T$	$247.6 \exp[-2.604 \cdot 10^{-3}(T{-}273.15)]$
VCHS67	$0.8028 + 6.50 \cdot 10^{-4}T$	$581.7 \exp[-4.553 \cdot 10^{-3}(T{-}273.15)]$
VCHS60	$0.8213 + 6.23 \cdot 10^{-4}T$	$229.1 \exp[-2.133 \cdot 10^{-3}(T{-}273.15)]$
VCHS50	$0.7827 + 5.05 \cdot 10^{-4}T$	$136.0 \exp[-1.083 \cdot 10^{-3}(T{-}273.15)]$
VCHS40	$0.7805 + 4.92 \cdot 10^{-4}T$	$155.0 \exp[-1.605 \cdot 10^{-3}(T{-}273.15)]$
VCHS33	$0.7710 + 4.86 \cdot 10^{-4}T$	$460.4 \exp[-3.453 \cdot 10^{-3}(T{-}273.15)]$
VCHS20	$0.6416 + 5.42 \cdot 10^{-4}T$	$489.8 \exp[-3.193 \cdot 10^{-3}(T{-}273.15)]$
VCOS50	$0.7081 + 7.40 \cdot 10^{-4}T$	$666.5 \exp[-4.503 \cdot 10^{-3}(T{-}273.15)]$
VCPS50	$0.7814 + 4.36 \cdot 10^{-4}T$	$880.1 \exp[-4.393 \cdot 10^{-3}(T{-}273.15)]$

UPPER CRITICAL (UCST) AND LOWER CRITICAL (LCST) SOLUTION TEMPERATURES OF BINARY POLYMER SOLUTIONS

Christian Wohlfarth

Liquid–liquid demixing in solutions of polymers in low molar mass solvents is not a rare phenomenon. Demixing depends on concentration, temperature, pressure, molar mass and molar mass distribution function of the polymer, chain branching and end groups of the polymer, the chemical nature of the solvent, isotope substitution in solvents or polymers, chemical composition of copolymers and its distributions, and other variables. Phase diagrams of polymer solutions can therefore show quite complicated behavior when they have to be considered in detail (see Ref. 1a).

Polymer solutions can undergo demixing when cooling a homogeneous solution as well as when heating such a solution. The corresponding cloud-point curves show a maximum (UCST behavior) or a minimum (LCST behavior). For common polymer solutions, the LCST region is at higher temperatures (in many cases near the critical temperature of the solvent) than the UCST region. The temperature range between both extrema provides the essential information where the one-phase region of a polymer solution can be found. In the case of monodisperse polymers the extrema are equal to the critical points. However, in the case of polydisperse polymers with distribution functions, these extrema are threshold temperatures whereas the critical point shifts to higher concentrations on the shoulder of the cloud-point curve. Usually, the critical concentration is much more strongly influenced than the critical temperature. Thus, the table below does not distinguish between threshold and critical temperatures.

UCST and LCST values depend somewhat on pressure. LCST values in the table are usually given at the vapor pressure of the solvent at this temperature. UCST values are measured in most cases at normal pressure; data at higher pressures are neglected here. The interested reader can find such information, for example, in Refs. 76, 84, 104, 157, 165, 177, 185–187, or 192.

However, UCST and LCST values of a given polymer/solvent pair depend strongly on the molar mass of the polymer. In the case of monodisperse polymers, this dependency can be described in good approximation by the so-called Shultz-Flory plot (see Refs. 6 and 8):

$$\frac{1}{T_{\text{crit}}} = \frac{1}{\theta}\left[1 + const.\cdot\left(\frac{1}{\sqrt{r}} - \frac{1}{2r}\right)\right] \tag{1}$$

where r denotes the number of segments of a polymer (being proportional to the degree of polymerization or to the molar mass or molar volume of the polymer). Extrapolation to $r \to \infty$, i.e., to infinite molar mass, leads to the value of the θ-temperature. This θ-temperature is the highest temperature for UCST behavior or the lowest temperature for LCST behavior and a given polymer/solvent pair. In the case of polydisperse polymers, the segment number in equation (1) is to be replaced by its weight average, r_w (related to M_w). The constant in equation (1) reflects further thermodynamic properties of the given polymer/solvent pair, but should not depend on molar mass. A detailed discussion can be found in Ref. 1b.

The printed table in the *Handbook* provides only one data line for a given polymer/solvent pair and does not show the molar mass dependence of UCST or LCST data. The entire table with all data at different molar masses for many of the systems is given in the electronic version, however. Nevertheless, the necessary molar mass information for a system is always provided in the table by the corresponding number average, M_n, mass average, M_w, or viscosity average, M_h, values of the polymer as given in the original sources.

Polymer	M_n/g mol⁻¹	M_w/g mol⁻¹	M_h/g mol⁻¹	Solvent	UCST/K	LCST/K	Ref.
Acrylonitrile/butadiene copolymer							
(18% Acrylonitrile)			840000	Ethyl acetate		427	220
(26% Acrylonitrile)			1000000	Ethyl acetate		412	220
Butadiene/α-methylstyrene copolymer (10% α-Methylstyrene)			100000	Ethyl acetate	387	393	220
Carbon monoxide/ethylene copolymer (1:1, alternating)		1000000		1,1,1,3,3,3-Hexafluoro-2-propanol		453	159
Cellulose diacetate			120000	Benzyl alcohol	372		86
	59900	75500		2-Butanone	279.7	471.5	111
	59300			2-Propanone	216.2	438.2	42
Cellulose diacetate/styrene graft copolymer (77.4 wt% grafted polystyrene)		750000		N,N-Dimethylformamide	262	399	106
		750000		Tetrahydrofuran		363	106
Cellulose nitrate (13.3 wt% N)	unknown			2-Propanone	328	182	148
Cellulose triacetate			20000	Benzyl alcohol	322		86
		100500		2-Propanone	290.0	472.0	42
Cellulose tricaprylate	infinite			N,N-Dimethylformamide	413		5
	infinite			3-Phenyl-1-propanol	321		5
Decamethyltetrasiloxane	310.69			Tetradecafluorohexane	332.59		195
N,N-Dimethylacrylamide/2-butoxyethyl acrylate copolymer (50 wt% 2-butoxyethyl acrylate)				Water	<273.2		164
N,N-Dimethylacrylamide/butyl acrylate copolymer							
(15 wt% Butyl acrylate)				Water		346.2	164
(20 wt% Butyl acrylate)				Water		323.2	164

Polymer	M_n/g mol^{-1}	M_w/g mol^{-1}	M_h/g mol^{-1}	Solvent	UCST/K	LCST/K	Ref.
(30 wt% Butyl acrylate)				Water		294.2	164
(35 wt% Butyl acrylate)				Water		281.2	164
N,N-Dimethylacrylamide/2-ethoxyethyl acrylate copolymer							
(50 wt% 2-Ethoxyethyl acrylate)				Water		319.2	164
(75 wt% 2-Ethoxyethyl acrylate)				Water		285.2	164
N,N-Dimethylacrylamide/ethyl acrylate copolymer							
(25 wt% Ethyl acrylate)				Water		347.2	164
(30 wt% Ethyl acrylate)				Water		334.2	164
(50 wt% Ethyl acrylate)				Water		287.2	164
(55 wt% Ethyl acrylate)				Water		<273.2	164
N,N-Dimethylacrylamide/2-methoxyethyl acrylate copolymer							
(38 mol% 2-Methoxyethyl acrylate)				Water		353	184
(45 mol% 2-Methoxyethyl acrylate)				Water		333	184
(55 mol% 2-Methoxyethyl acrylate)				Water		315	184
(68 mol% 2-Methoxyethyl acrylate)				Water		305	184
(82 mol% 2-Methoxyethyl acrylate)				Water		288	184
(92 mol% 2-Methoxyethyl acrylate)				Water		283	184
N,N-Dimethylacrylamide/methyl acrylate copolymer							
(30 wt% Methyl acrylate)				Water		371.2	164
(40 wt% Methyl acrylate)				Water		338.2	164
(50 wt% Methyl acrylate)				Water		314.2	164
(55 wt% Methyl acrylate)				Water		294.2	164
(60 wt% Methyl acrylate)				Water		279.2	164
(70 wt% Methyl acrylate)				Water		<273.2	164
N,N-Dimethylacrylamide/propyl acrylate copolymer							
(20 wt% Propyl acrylate)				Water		353.2	164
(30 wt% Propyl acrylate)				Water		337.2	164
(40 wt% Propyl acrylate)				Water		294.2	164
(50 wt% Propyl acrylate)				Water		281.2	164
Dimethylsiloxane/methylphenylsiloxane	9100	41200		Anisole	291.45		198
copolymer (15 wt% methylphenylsiloxane)	9100	41200		2-Propanone	282.45		198
Ethylene/propylene copolymer			145000	Cyclohexane		534	101
(33 mol% ethylene)			145000	Cyclopentane		490	101
			145000	2,2-Dimethylbutane		428	101
			145000	2,3-Dimethylbutane		452	101
			145000	3,4-Dimethylhexane		541	101
			145000	2,2-Dimethylpentane		472	101
			145000	2,3-Dimethylpentane		500	101
			145000	2,4-Dimethylpentane		464	101
			145000	3-Ethylpentane		511	101
			145000	Heptane		502	101
			145000	Hexane		455	101
			145000	2-Methylbutane		396	101
			145000	Methylcyclohexane		558	101
			145000	Methylcyclopentane		512	101
			145000	2-Methylhexane		486	101
			145000	Nonane		558	101
			145000	Octane		528	101
			145000	Pentane		409	101
			145000	2,2,4,4-Tetramethylpentane		539	101
			145000	2,2,3-Trimethylbutane		500	101
			145000	2,2,4-Trimethylpentane		503	101
Ethylene/propylene copolymer	70000	140000		Hexane		436	127
(43 mol% ethylene)	70000	140000		2-Methylpentane		474	127
	70000	140000		Pentane		441	127
Ethylene/propylene copolymer			154000	2,2-Dimethylbutane		407	101
(53 mol% ethylene)			154000	2,3-Dimethylbutane		437	101

Polymer	M_n/g mol⁻¹	M_w/g mol⁻¹	M_h/g mol⁻¹	Solvent	UCST/K	LCST/K	Ref.
			154000	2,2-Dimethylpentane		453	101
			154000	2,3-Dimethylpentane		488	101
			154000	2,4-Dimethylpentane		445	101
			154000	3-Ethylpentane		500	101
			154000	Heptane		493	101
			154000	Hexane		443	101
			154000	Pentane		395	101
			154000	2,2,3-Trimethylbutane		488	101
			154000	2,3,4-Trimethylhexane		565	101
			154000	2,2,4-Trimethylpentane		484	101
Ethylene/propylene copolymer (63 mol% ethylene)			236000	Cyclohexane		526	101
			236000	Cyclopentane		481	101
			236000	2,3-Dimethylbutane		429	101
			236000	3,4-Dimethylhexane		530	101
			236000	2,2-Dimethylpentane		444	101
			236000	2,3-Dimethylpentane		482	101
			236000	2,4-Dimethylpentane		434	101
			236000	3-Ethylpentane		492	101
			236000	Heptane		485	101
			236000	Hexane		436	101
			236000	2-Methylbutane		348	101
			236000	Methylcyclopentane		498	101
			236000	Nonane		547	101
			236000	Octane		512	101
			236000	Pentane		387	101
			236000	2,2,4,4-Tetramethylpentane		528	101
			236000	2,2,3-Trimethylbutane		479	101
			236000	2,2,4-Trimethylpentane		479	101
Ethylene/propylene copolymer (75 mol% ethylene)			109000	2,2-Dimethylpentane		431	101
			109000	2,4-Dimethylpentane		425	101
			109000	Heptane		475	101
			109000	Hexane		427	101
			109000	Nonane		542	101
			109000	Octane		509	101
			109000	Pentane		378	101
			109000	2,2,4,4-Tetramethylpentane		523	101
			109000	2,2,4-Trimethylpentane		469	101
Ethylene/propylene copolymer (81 mol% ethylene)			195000	Cyclohexane		522	101
			195000	Cyclopentane		474	101
			195000	2,2-Dimethylbutane		381	101
			195000	2,3-Dimethylbutane		413	101
			195000	2,4-Dimethylhexane		478	101
			195000	2,5-Dimethylhexane		466	101
			195000	3,4-Dimethylhexane		522	101
			195000	2,2-Dimethylpentane		425	101
			195000	2,3-Dimethylpentane		471	101
			195000	2,4-Dimethylpentane		420	101
			195000	3-Ethylpentane		478	101
			195000	Heptane		468	101
			195000	Hexane		425	101
			195000	2-Methylbutane		327	101
			195000	Methylcyclohexane		541	101
			195000	Methylcyclopentane		493	101
			195000	2-Methylhexane		453	101
			195000	3-Methylhexane		459	101
			195000	Nonane		540	101
			195000	Octane		506	101
			195000	Pentane		370	101
			195000	2,2,4,4-Tetramethylpentane		519	101
			195000	2,2,3-Trimethylbutane		461	101
			195000	2,2,4-Trimethylpentane		460	101

Polymer	M_n/g mol⁻¹	M_w/g mol⁻¹	M_h/g mol⁻¹	Solvent	UCST/K	LCST/K	Ref.
Ethylene/vinyl acetate copolymer							
(2.3 wt% vinyl acetate)	52000	465000		Diphenyl ether	404.2		143
(4.0 wt% vinyl acetate)	47000	280000		Diphenyl ether	392.5		143
(7.1 wt% Vinyl acetate)	34000	460000		Diphenyl ether	378.2		143
(9.5 wt% Vinyl acetate)	53000	350000		Diphenyl ether	367.3		143
(9.7 wt% Vinyl acetate)	55000	490000		Diphenyl ether	370.8		143
(12.1 wt% Vinyl acetate)	66000	300000		Diphenyl ether	360.4		143
(42.6 mol% Vinyl acetate)	14800	41500		Methyl acetate	307.0		130
Ethylene/vinyl alcohol copolymer							
(87.2 mol% Vinyl alcohol)			infinite	Water	463.55	285.65	44
(88.9 mol% Vinyl alcohol)			infinite	Water	449.15	290.75	44
(91.0 mol% Vinyl alcohol)			infinite	Water	428.45	302.95	44
(94.1 mol% Vinyl alcohol)			infinite	Water	389.25	324.45	44
Ethylene oxide/propylene oxide copolymer							
(20.0 mol% Ethylene oxide)	3400			Water		303	211
(27.0 mol% Ethylene oxide)	3000			Water		309	210
(30.0 mol% Ethylene oxide)	5400			Water		313	211
(38.5 mol% Ethylene oxide)	5000			Water		309	210
(50.0 mol% Ethylene oxide)	3900			Water		323	211
(58.8 mol% Ethylene oxide)	3000			Water		326.65	210
(72.4 mol% Ethylene oxide)		36000		Water		333	153
(79.5 mol% Ethylene oxide)		30800		Water		345	153
(86.6 mol% Ethylene oxide)		30100		Water		355.5	153
Gutta Percha			194000	Propyl acetate	318.95		7
Hydroxypropylcellulose		75000		Water		318.45	43
		300000		Water		331.25	43
N-Isopropylacrylamide/acrylamide copolymer							
(15 mol% Acrylamide)		3100000		Water		315.15	172
(30 mol% Acrylamide)		4500000		Water		326.15	172
(45 mol% Acrylamide)		3900000		Water		347.15	172
N-Isopropylacrylamide/1-deoxy-1-methacrylamido-D-glucitol							
(12.9 mol% Glucitol)	78000	170000		Water		311.3	218
(13.7 mol% Glucitol)	51600	110000		Water		314.9	218
(14.0 mol% Glucitol)	145000	432000		Water		307.5	218
N-Isopropylacrylamide/N-isopropylmethacrylamide copolymer							
(10.56 mol% N-Isopropylmethacrylamide)	55300	177000		Water		307.15	212
(30.00 mol% N-Isopropylmethacrylamide)	28800	92000		Water		309.75	212
(39.99 mol% N-Isopropylmethacrylamide)	23100	74000		Water		311.05	212
(59.89 mol% N-Isopropylmethacrylamide)	23100	74000		Water		314.65	212
(79.81 mol% N-Isopropylmethacrylamide)	16600	53000		Water		317.35	212
(89.99 mol% N-Isopropylmethacrylamide)	14700	47000		Water		318.75	212
Methylcellulose (about 30 mol% methyl substitution)			70000	Water		324.75	47
Methylcellulose/hydroxypropylcellulose copolymer (25 mol% methyl, 8 mol% hydroxypropyl substitution)			80000	Water		340.15	63
Natural rubber		300000		Pentane		403	10
			74500	2-Pentanone	274.45		7
Phenol-formaldehyde resin (acetylated)				2-Ethoxyethanol	378.2		200
Poly(acrylic acid)		120000		Tetrahydrofuran		268.3	189
Poly[bis(2,3-dimethoxypropanoxy)phosphazene]	1070000	1500000		Water		317.15	183
Poly[bis(2-(2′-methoxyethoxy)ethoxy)phosphazene]	667000	1000000		Water		338.15	183
Poly[bis(2,3-bis(2-methoxyethoxy)propanoxy)phosphazene]	714000	1000000		Water		311.15	183
Poly[bis(2,3-bis(2-(2′-methoxyethoxy)ethoxy)propanoxy)phosphazene]	1420000	1700000		Water		322.65	183
Poly[bis(2,3-bis(2-(2′-(2″-dimethoxyethoxy)ethoxy)ethoxy)propanoxy)phosphazene]	857000	1200000		Water		334.65	183
Poly(1-butene) (atactic)	infinite			Anisole	359.4		11

Polymer	M_n/g mol^{-1}	M_w/g mol^{-1}	M_h/g mol^{-1}	Solvent	UCST/K	LCST/K	Ref.
	infinite			Toluene	356.2		28
Poly(1-butene) (isotactic)	infinite			Anisole	362.3		11
			530000	Cyclopentane		498	102
			530000	2,2-Dimethylbutane		444	102
			530000	2,5-Dimethylhexane		519	102
			530000	3,4-Dimethylhexane		559	102
			530000	2,3-Dimethylpentane		517	102
			530000	2,4-Dimethylpentane		480	102
			530000	3-Ethylpentane		523	102
			530000	Heptane		509	102
	infinite			Hexane		464	102
			530000	2-Methylbutane		416	102
	infinite			Nonane		564	102
			530000	Octane		540	102
	infinite			Pentane		421	102
			530000	2,2,3-Trimethylbutane		507	102
Poly(butyl methacrylate)	278000	470000		1-Butanol	287.15		132
	278000	470000		Decane	357.25		132
	278000	470000		Ethanol	315.25		132
	278000	470000		Heptane	342.55		132
	278000	470000		Octane	345.80		132
	278000	470000		1-Pentanol	286.30		132
	278000	470000		2-Propanol	294.90		132
	278000	470000		2,2,4-Trimethylpentane	347.50		132
Poly(2-chlorostyrene)	infinite			Benzene		298	40
Poly(4-chlorostyrene)	infinite			Benzene	274.0		22
	infinite			2-(Butoxyethoxy)ethanol		323.25	46
	infinite			Butyl acetate		502.4	22
	infinite			tert-Butyl acetate		338.55	46
	infinite			Chlorobenzene	128.8		22
	infinite			2-(Ethoxyethoxy)ethanol		300.95	46
	infinite			Ethyl acetate		613.2	22
	infinite			Ethylbenzene	283.2		22
	infinite			Ethylbenzene	258.45		46
	infinite			Ethyl chloroacetate	271.35		46
	infinite			Isopropyl acetate		348.65	46
	infinite			Isopropylbenzene	332.15		46
	infinite			Isopropyl chloroacetate	264.95		46
	infinite			Methyl chloroacetate	337.75		46
	infinite			Propyl acetate		908.7	22
	infinite			Tetrachloroethene	317.55		46
	infinite			Tetrachloromethane	323.85		46
	infinite			Toluene	236.8		22
Poly(decyl methacrylate)	390000	468000		1-Butanol	304.85		113
	390000	468000		1-Pentanol	278.40		113
	220000	252000		2-Propanol	346.85		132
Polydimethylsiloxane (cyclic)	9810	10300		2,2-Dimethylpropane		433	133
	9810	10300		Tetramethylsilane		448	133, 171
Polydimethylsiloxane			626000	Butane		392.95	53
	infinite			Decane		603	30
	14750	16370		2,2-Dimethylpropane		428	133
	infinite			Dodecane		643	30
			626000	Ethane		259.65	53
			100000	Ethoxybenzene	341.99		108
	infinite			Heptane		528	30
	infinite			Hexadecane		708	30
	infinite			Hexane		493	30
	infinite			Octane		553	30
	infinite			Pentane		453	30
			203000	Propane		340.15	53
	14750	16370		Tetramethylsilane		443	133, 171

Polymer	M_n/g mol^{-1}	M_w/g mol^{-1}	M_h/g mol^{-1}	Solvent	UCST/K	LCST/K	Ref.
Poly(ethyl acrylate)			48000	1-Butanol	310.05		27
			48000	Ethanol	301.15		27
			380000	Methanol	287.25		27
			48000	1-Propanol	305.15		27
Polyethylene (branched)	8400	32000		Diphenyl ether	384.7		95, 98
	24000	123000		Diphenyl ether	396.7		95, 98
	65000	425000		Diphenyl ether	415.3		95, 98
Polyethylene (linear)			20000	Anisole	368.15		24
			20000	Benzyl acetate	459.65		24
			20000	Benzyl phenyl ether	437.15		24
			20000	Benzyl propionate	436.15		24
			50900	Biphenyl	383.55		25
			61100	Butyl acetate	448	497	70
			20000	4-tert-Butylphenol	466.15		24
			134000	Cyclohexane		518	101
			20000	Cyclohexanone	389.65		24
			134000	Cyclopentane		472	101
	36700	49300		Decane		563.75	91
			20000	1-Decanol	400.15		24
			20000	Dibenzyl ether	448.65		24
			134000	3,4-Dimethylhexane		515	101
			134000	2,2-Dimethylpentane		399	101
			134000	2,3-Dimethylpentane		463	101
			134000	2,4-Dimethylpentane		395	101
	12000	150000		Diphenyl ether	416.2		95, 98
			97200	Diphenylmethane	400.25		25
	60400	82600		Dodecane		610.85	91
		218000		1-Dodecanol	405.15		141
			134000	3-Ethylpentane		471	101
	36700	49300		Heptane		464.70	91
			20000	1-Heptanol	440.15		24
	36700	49300		Hexane		414.65	91
	7900	92000		1-Hexanol	458.15		154
			20000	2-Methoxynaphthalene	427.65		24
			20000	3-Methylbutyl acetate	407.15		24
			134000	Methylcyclohexane		537	101
			134000	Methylcyclopentane		488	101
	60400	82600		Nonane		531.90	91
			20000	1-Nonanol	431.15		24
			20000	4-Nonylphenol	410.15		24
	36700	49300		Octane		502.40	91
	7900	92000		1-Octanol	426.65		154
			20000	4-Octylphenol	424.65		24
			134000	Pentane		353	101
			20000	1-Pentanol	445.15		24
			175000	Pentyl acetate	421	528	70
			20000	4-tert-Pentylphenol	443.65		24
			20000	Phenetole	366.65		24
			134000	2,2,4,4-Tetramethylpentane		513	101
	60400	82600		Tridecane		639.30	91
			134000	2,2,3-Trimethylbutane		444	101
			134000	2,3,4-Trimethylhexane		545	101
			134000	2,2,4-Trimethylpentane		495	101
	97700	135900		Undecane		583.95	91
Poly(ethylene glycol)			8000	tert-Butyl acetate	321.2	464.2	83
			21200	tert-Butyl acetate	353.2	431.2	83
	6100	6200		Water		404.79	185
	10457	11615		Water		394.33	205
	40800	151000		Water		378.25	205

Polymer	M_n/g mol^{-1}	M_w/g mol^{-1}	M_h/g mol^{-1}	Solvent	UCST/K	LCST/K	Ref.
Poly(ethylene oxide)-b-poly[bis(methoxyethoxyethoxy)-phosphazene] block copolymer (about 67 mol% Ethylene oxide)	22000	31500		Water		338	222
Poly(ethylene oxide)-b-poly(propylene oxide)-b-poly(ethylene oxide) triblock copolymer (about 30 mol% Ethylene oxide)	4400			Water		286.65	209
Polyethylethylene	48000	52000		Diphenyl ether	411.2		95, 98
Poly(p-hexylstyrene)	infinite			2-Butanone	302.6		135
Poly(2-hydroxyethyl methacrylate)			77400	1-Butanol	337.25		35
			233600	2-Butanol	287		35
			233600	2-Metyl-1-propanol	342		35
			77400	1,2,3-Propanetriol	345		35
			77400	1-Propanol	311		35
Polyisobutylene	infinite			Anisole	377		3
			72000	Benzene		540.5	39
			703000	Butane		264.75	53
	infinite			Cycloheptane		572	34
			1500000	Cyclohexane		412	10
	infinite			Cyclooctane		637	34
			1500000	Cyclopentane		344	10
	infinite			Decane		535	34
			1500000	2,2-Dimethylbutane		376	10
			1500000	2,3-Dimethylbutane		404	10
	infinite			2,2-Dimethylhexane		454	34
	infinite			2,4-Dimethylhexane		458	34
	infinite			2,5-Dimethylhexane		446	34
	infinite			3,4-Dimethylhexane		497	34
	infinite			2,2-Dimethylpentane		404	34
	infinite			2,3-Dimethylpentane		451	34
	infinite			2,4-Dimethylpentane		403	34
	infinite			3,3-Dimethylpentane		451	34
	infinite			Diphenyl ether	306		3
	infinite			Decane		585	30
	infinite			Dodecane		582	34
	infinite			Ethylbenzene	249		3
	infinite			Ethylcyclopentane		524	34
	infinite			Ethyl heptanoate	306		3
	infinite			Ethyl hexanoate	330		3
	infinite			3-Ethylpentane		458	34
	infinite			Heptane		442	34
			72000	Hexane		428.5	39
			6030	2-Methylbutane		357.85	53
	infinite			Methylcyclohexane		526	34
	infinite			Methylcyclopentane		478	34
	infinite			2-Methylheptane		466	34
	infinite			3-Methylheptane		478	34
	infinite			2-Methylhexane		426	34
	infinite			3-Methylhexane		446	34
	infinite			2-Methylpentane		376	34
	infinite			3-Methylpentane		405	34
			470	2-Methylpropane		387	10
			72000	Octane		506.0	39
			6030	Pentane		403.55	53
			72000	Pentane		373.5	39
	infinite			Phenetole	357		3
			470	Propane		358	10
	infinite			Propylcyclopentane		547	34
	infinite			Toluene	260		3
	infinite			2,2,3-Trimethylbutane		445	34
	infinite			2,2,4-Trimethylpentane		435	34
1,4-cis-Polyisoprene			780000	2,5-Dimethylhexane		474.15	140

Polymer	M_n/g mol^{-1}	M_w/g mol^{-1}	M_η/g mol^{-1}	Solvent	UCST/K	LCST/K	Ref.
			780000	3,4-Dimethylhexane		520.15	140
			780000	2,2-Dimethylpentane		445.15	140
			780000	2,3-Dimethylpentane		484.15	140
			780000	2,4-Dimethylpentane		442.15	140
			780000	3-Methylpentane		483.15	140
			780000	Heptane		488.15	140
			780000	Hexane		434.15	140
			780000	Nonane		541.15	140
			780000	Octane		509.15	140
			780000	2,2,4,4-Tetramethylpentane		518.15	140
			780000	2,3,4-Trimethylhexane		548.15	140
			780000	2,2,4-Trimethylpentane		471.15	140
1,4-*trans*-Polyisoprene			180000	2,5-Dimethylhexane		451.15	140
			180000	3,4-Dimethylhexane		521.15	140
			180000	2,2-Dimethylpentane		405.15	140
			180000	2,3-Dimethylpentane		460.15	140
			180000	2,4-Dimethylpentane		404.15	140
			180000	3-Methylpentane		473.15	140
			180000	Heptane		467.15	140
			180000	Hexane		407.15	140
			180000	Nonane		540.15	140
			180000	Octane		503.15	140
			180000	2,2,4,4-Tetramethylpentane		519.15	140
			180000	2,3,4-Trimethylhexane		548.15	140
Poly(*N*-isopropylacrylamide)	5400	14000		Water		307.45	146
	146000	530000		Water		305.85	146
Poly(*N*-isopropylacrylamide)-poly[(*N*-acetylimino)ethylene] block copolymer (80 wt% *N*-Isopropylacrylamide)	5500			Water		306.2	223
Poly(*N*-isopropylacrylamide)-poly[(*N*-acetylimino)ethylene] graft copolymer (75 wt% *N*-Isopropylacrylamide)	6030			Water		306.2	223
Poly(*N*-isopropylmethacrylamide)	6250	20000		Water		319.95	212
Poly(methyl methacrylate)			127000	Acetonitrile	267.15		16
			970000	Acetonitrile	303.15		16
			50000	1-Butanol	353.25		2
	infinite			2-Butanone		482	80
	infinite			1-Chlorobutane	320	463	80
			970000	2,2-Dimethyl-3-pentanone	301.55		16
			127000	2,4-Dimethyl-3-pentanone	280.15		16
	200000	264000		2-Ethoxyethanol	312.15		196
		77000		Ethyl acetate	290	533	190
			127000	2-Ethylbutanal	264.65		16
	infinite			3-Heptanone	307.7		126
			970000	4-Heptanone	299.95		16
	infinite			3-Hexanone		522	80
	infinite			Methyl acetate		451	80
			50000	1-Methyl-4-isopropylbenzene	400.15		2
			1400000	2-Octanone	321.15		16
	572400	595300		3-Octanone	329.88		166
	infinite			3-Pentanone		506	80
			50000	1-Propanol	349.95		2
	infinite			2-Propanone		439	80
	200000	264000		Tetra(ethylene glycol)	390.15		196
			400000	Toluene	225.35		2
			50000	Trichloromethane	231.15		2
	200000	264000		Tri(ethylene glycol)	407.15		196
Poly(methyl methacrylate) (isotactic)	infinite			Acetonitrile	301	461	80
	infinite			2-Butanone		464	80
	infinite			1-Chlorobutane	309	454	80
	infinite			4-Heptanone	319	522	80

Polymer	M_n/g mol^{-1}	M_w/g mol^{-1}	M_h/g mol^{-1}	Solvent	UCST/K	LCST/K	Ref.
			infinite	3-Hexanone	279	511	80
			infinite	Methyl acetate		441	80
			infinite	3-Pentanone		497	80
			infinite	2-Propanone		428	80
Poly(4-methyl-1-pentene) (isotactic)			152000	Butane		388	102
			152000	Cyclopentane		505	102
			152000	2,2-Dimethylbutane		462	102
			152000	2,2-Dimethylpentane		499	102
			152000	2,4-Dimethylpentane		499	102
			infinite	Diphenyl	467.8		62
			infinite	Diphenyl ether	483.2		62
			infinite	Diphenylmethane	449.8		62
			152000	3-Ethylpentane		532	102
			152000	Heptane		522	102
			152000	Hexane		487	102
			152000	2-Methylbutane		431	102
			152000	Nonane		579	102
			152000	Octane		553	102
			152000	Pentane		441	102
			152000	2,2,3-Trimethylbutane		521	102
Poly(α-methylstyrene)	58500	61400		Butyl acetate	262.05	457.15	181
	99100	113000		Cyclohexane	293.55		152
	26000	31200		Cyclopentane	276.7	435.95	181
		289000		trans-Decahydronaphthalene	273		181
	69500	76500		Hexyl acetate	285.05	508.15	181
	72000	75600		Methylcyclohexane	328.9		203
	58500	61400		Pentyl acetate	287.1	484.6	181
Poly(2-methyl-5-vinylpyridine)			600000	Butyl acetate	287.95		20
			263000	Ethyl butyrate	319.05		20
			335000	Ethyl propionate	293.55		20
			275000	3-Methylbutyl acetate	314.75		20
			335000	4-Methyl-2-pentanone	299.95		20
			170000	2-Methylpropyl acetate	312.35		20
			165000	Pentyl acetate	316.95		20
			284000	Propionitrile	262.35		20
			152000	Propyl acetate	282.65		20
			181000	Propyl propionate	312.15		20
			233000	Tetrahydronaphthalene	316.95		20
Poly(1-pentene) (isotactic)			4500000	Cyclopentane		502	102
			4500000	2,2-Dimethylbutane		457	102
			4500000	3,4-Dimethylhexane		>569	102
			4500000	2,2-Dimethylpentane		502	102
			4500000	2,3-Dimethylpentane		529	102
			4500000	2,4-Dimethylpentane		493	102
			4500000	3-Ethylpentane		537	102
			4500000	Heptane		522	102
			4500000	Hexane		482	102
			4500000	2-Methylbutane		422	102
			4500000	Octane		556	102
			4500000	Pentane		433	102
			4500000	2,2,4-Trimethylpentane		527	102
Polypropylene (atactic)			infinite	Diphenyl ether	426.5		9
			infinite	Diethyl ether		383	68
			242000	Heptane		511	101
			infinite	Hexane		441	68
			242000	2-Methylbutane		413	101
			242000	Methylcyclohexane		564	101
			infinite	Pentane		397	68
Polypropylene (isotactic)			28000	Benzyl phenyl ether	429.2		31
			28000	Benzyl propionate	405.2		31
			28000	1-Butanol	395.2		31

Polymer	M_n/g mol^{-1}	M_w/g mol^{-1}	M_h/g mol^{-1}	Solvent	UCST/K	LCST/K	Ref.
			28000	4-*tert*-Butylphenol	413.2		31
			242000	Cyclohexane		540	101
			242000	Cyclopentane		495	101
			28000	Dibenzyl ether	433.2		31
			242000	2,2-Dimethylbutane		441	101
			242000	2,3-Dimethylbutane		465	101
			242000	3,4-Dimethylhexane		553	101
			242000	2,2-Dimethylpentane		489	101
			242000	2,3-Dimethylpentane		513	101
			242000	2,4-Dimethylpentane		481	101
			28000	Diphenyl	388.2		31
			28000	Diphenyl ether	395.2		31
			28000	Diphenylmethane	389.7		31
			242000	3-Ethylpentane		520	101
			28000	4-Ethylphenol	457.2		31
			242000	Heptane		511	101
			242000	Hexane		470	101
			242000	2-Methylbutane		413	101
			28000	3-Methylbutyl benzyl ether	384.2		31
			242000	Methylcyclohexane		564	101
			242000	Methylcyclopentane		518	101
			28000	4-Methylphenol	479.2		31
			28000	2-Methyl-1-propanol	395.2		31
			242000	Nonane		571	101
			242000	Octane		542	101
			28000	4-Octylphenol	379.2		31
			28000	4-Isooctylphenol	383.2		31
			242000	Pentane		422	101
			242000	2,2,4,4-Tetramethylpentane		548	101
			242000	2,2,3-Trimethylbutane		511	101
			242000	2,3,4-Trimethylhexane		585	101
			242000	2,2,4-Trimethylpentane		510	101
Poly(propylene glycol)	1000			Hexane	288.15		88
	575			Water		318.2	65
Polystyrene	34900	37000		Benzene		538.7	61
			62600	Butanedioic acid dimethyl ester	335.15		2
	3700	4000		1-Butanol	383.45		154
	91700	97200		2-Butanone		448.8	61
	545500	600000		Butyl acetate		489	181
	104000	110000		*tert*-Butyl acetate	250.0	417.9	74
			62600	Butyl stearate	387.15		2
	18400	19200		1-Chlorododecane	274.65		154
	18400	19200		1-Chlorohexadecane	337.05		154
	18400	19200		1-Chlorooctadecane	365.55		154
	18400	19200		1-Chlorotetradecane	309.35		154
	46400	51000		Cyclodecane	278.9		128
	46400	51000		Cycloheptane	276.2		128
	34900	37000		Cyclohexane	285.6	510.9	60
			236000	Cyclohexanol	353.5		8
	46400	51000		Cyclooctane	275.2		128
	91700	97200		Cyclopentane	275.2	445.5	61
	91500	97000		*trans*-Decahydronaphthalene	281.95		81
		4800		Decane	360.95		154
	3700	4000		1-Decanol	375.15		154
			570000	Decyl acetate		650	64
	18700	19800		Diethyl ether	235.6	314.5	51
	187000	200000		Diethyl malonate	285.8	589.6	74
	47200	50000		Diethyl oxalate	280.05		131
	151000	160000		Dimethoxymethane		401.2	51
		240000		1,4-Dimethylcyclohexane	387	482	116

Polymer	M_n/g mol^{-1}	M_w/g mol^{-1}	M_h/g mol^{-1}	Solvent	UCST/K	LCST/K	Ref.
			62600	Dimethyl malonate	409.15		2
			62600	Dimethyl oxalate	453.15		2
	116000	123000		Dodecadeuterocyclohexane	298.10		224
		25000		Dodecadeuteromethyl-cyclopentane	310.07		180
		4800		Dodecane	368.65		154
	3700	4000		1-Dodecanol	379.75		154
	infinite			Dodecyl acetate	285.2		206
	104000	110000		Ethyl acetate	213.9	435.4	72
	104000	110000		Ethyl butanoate		490.8	74
	221000	239000		Ethylcyclohexane	330.52		18
	9440	10000		Ethyl formate	272	451	74
		900000		Bis(2-ethylhexyl) phthalate	283.05		136
	4530	4800		Heptane	359	477	112
	3700	4000		1-Dexadecanol	386.25		154
	5500	5770		1,1,1,3,3,3-Hexadeutero-2-propanone	270	436	157
	1920	2030		Hexane	318	470	112
			62600	Hexanoic acid	448.15		2
	3700	4000		1-Hexanol	372.15		154
			62600	3-Hexanol	396.65		2
			90000	Hexyl acetate		578	64
	104000	110000		Methyl acetate	284.2	415.7	72
	104000	110000		3-Methyl-1-butyl acetate	210.1	510.1	72
	91700	97200		Methylcyclohexane	321.8	505.9	60
	10750	11500		Methylcyclopentane	295	480	157
	104000	110000		2-Methyl-1-propyl acetate	210.4	468.5	72
		48000		Nitroethane	303.1		151
		4800		Octadecane	403.55		154
	3700	4000		1-Octadecanol	390.55		154
	4530	4800		Octane	353	527	112
	3700	4000		1-Octanol	372.35		154
			62600	1-Octene	355.15		2
		4800		Pentadecane	385.25		154
		1100		Pentane	292		137
	3700	4000		1-Pentanol	375.05		154
	219800	233000		Pentyl acetate		519	181
		100000		1-Phenyldecane	283.60		105
	5500	5770		2-Propanone	251	452	157
	12750	13500		Propionitrile	312		187
	104000	110000		Propyl acetate	183.7	469.0	72
	104000	110000		2-Propyl acetate	220.9	414.2	72
	3700	4000		1-Tetradecanol	383.25		154
	34900	37000		Toluene		567.2	60
			62600	Vinyl acetate	384.15		2
Polystyrene (three-arm star)		230000		Cyclohexane	297.1	496.8	93
Polystyrene (four-arm star)		155000		Cyclohexane	294.13		199
Poly(trimethylene oxide)		infinite		Cyclohexane	300		79
Poly(vinyl alcohol)		40000		Water		514	45
Poly(N-vinyl caprolactam)		150000		Water		306.45	217
Poly(vinyl chloride)	55000			Dibutyl phthalate	353		114
	55000			Tricresyl phosphate	383		114
			85000	Dimethyl phthalate	355		219
Poly(N-vinylisobutyramide)	66000	105600		Water		313.25	208
Poly(vinyl methyl ether)	46500	98600		Deuterium oxide		307.2	173
	83000	155000		Water		306.95	146
Poly(N-vinyl-N-propylacetamide)			30000	Water		313.5	176
Styrene/acrylonitrile copolymer							
(21.1 wt% acrylonitrile)	infinite			Toluene	325.4		52
(23.2 wt% Acrylonitrile)	infinite			Toluene	355.1		52
(25.0 wt% Acrylonitrile)	90000	147000		Toluene	313.15		198
(51.0 wt% Acrylonitrile)		347000		Ethyl acetate		344.15	107

Polymer	M_n/g mol^{-1}	M_w/g mol^{-1}	M_h/g mol^{-1}	Solvent	UCST/K	LCST/K	Ref.
Styrene/methyl methacrylate copolymer (52.0 mol% Styrene)		infinite		Cyclohexanol	334.65		38
Styrene/α-methylstyrene copolymer (20.0 mol% Styrene)	100000	114000		Butyl acetate	288.85	453.05	181
	100000	114000		Cyclohexane	285.85	484.85	181
	100000	114000		Cyclopentane	290.95	421.05	181
	100000	114000		trans-Decahydronaphthalene	264.15		181
	100000	114000		Hexyl acetate	288.55	514.15	181
	100000	114000		Pentyl acetate	303.15	480.65	181
Trifluoronitrosomethane/tetrafluoroethylene copolymer (1:1) alternating		infinite		1,1,2-Trichloro-1,2,2-trifluoroethane	301.6		12
N-Vinylacetamide/vinyl acetate copolymer							
(58 mol% Vinyl acetate)	30000	57000		Water		340.15	225
(63 mol% Vinyl acetate)	27000	48600		Water		323.15	225
(78 mol% Vinyl acetate)	26000	46800		Water		282.15	225
Vinyl alcohol/vinyl butyrate copolymer (7.5 mol% Butyralized PVA)	infinite			Water	408.0	298.25	121
N-Vinylcaprolactam/N-vinylamine copolymer (3.8 mol% Vinyl amine)			160000	Water		308.8	176
N-Vinylformamide/vinyl acetate copolymer							
(60 mol% Vinyl acetate)	24000	45600		Water		310.15	225
(66 mol% Vinyl acetate)	25000	47500		Water		291.15	225
(73 mol% Vinyl acetate)	23000	50600		Water		277.15	225

References

1a. Koningsveld, R., Stockmayer, W.H., and Nies, E., *Polymer Phase Diagrams*, Oxford University Press, Oxford, 2001.

1b. Kamide, K., *Thermodynamics of Polymer Solutions*, Elsevier, Amsterdam, 1990.

2. Jenckel, E. and Gorke, K., *Z. Naturforsch.*, 5a, 317, 556, 1950.

3. Fox, T.G. and Flory, P.J., *J. Amer. Chem. Soc.*, 73, 1909, 1951.

4. Fox, T.G. and Flory, P.J., *J. Amer. Chem. Soc.*, 73, 1915, 1951.

5. Mandelkern, L. and Flory, P.J., *J. Amer. Chem. Soc.*, 74, 2517, 1952.

6. Shultz, A.R. and Flory, P.J., *J. Amer. Chem. Soc.*, 74, 4760, 1952.

7. Wagner, H.L. and Flory, P.J., *J. Amer. Chem. Soc.*, 74, 195, 1952.

8. Shultz, A.R.and Flory, P.J., *J. Amer. Chem. Soc.*, 75, 3888, 1953.

9. Kinsinger, J.B. and Wessling, R.A., *J. Amer. Chem. Soc.*, 81, 2908, 1959.

10. Freeman, P.I. and Rowlinson, J.S., *Polymer*, 1, 20, 1960.

11. Krigbaum, W.R., Kurz, J.E., and Smith, P., *J. Phys. Chem.*, 65, 1984, 1961.

12. Morneau, G.A., Roth, P.I., and Shultz, A.R., *J. Polym. Sci.*, 55, 609, 1961.

13. Debye, P., Coll, H., and Woermann, D., *J. Chem. Phys.*, 32, 939, 1960.

14. Debye, P., Coll, H., and Woermann, D., *J. Chem. Phys.*, 33, 1746, 1960.

15. Debye, P., Chu, B., and Woermann, D., *J. Chem. Phys.*, 36, 1803, 1962.

16. Fox, T.G., *Polymer*, 3, 111, 1962.

17. Ham, J.S., Bolen, M.C., and Hughes, J.K., *J. Polym. Sci.*, 57, 25, 1962.

18. Debye, P., Woermann, D., and Chu, B., *J. Polym. Sci.: Part A*, 1, 255, 1963.

19. Allen, G. and Baker, C.H., *Polymer*, 6, 181, 1965.

20. Gechele, G.B., Crescentini, L., *J. Polym. Sci.: Part A*, 3, 3599, 1965.

21. Myrat, C.D. and Rowlinson, J.S., *Polymer*, 6, 645, 1965.

22. Kubo, K. and Ogino, K., *Sci. Pap. Coll. Art. Sci. Univ. Tokyo*, 16, 193, 1966.

23. Rehage, G., Moeller, D., and Ernst, O., *Makromol.Chem.*, 88, 232, 1965.

24. Nakajima, A., Fujiwara, H., and Hamada, F., *J. Polym. Sci.: Part A-2*, 4, 507, 1966.

25. Nakajima, A., Hamada, F., and Hayashi, S., *J. Polym. Sci.: Part C*, 15, 285, 1966.

26. Koningsveld, R., *Proefschrift Univ. Leiden*, Heerlen, 1967.

27. Llopis, J., Albert, A., and Usobinaga P., *Eur. Polym. J.*, 3, 259, 1967.

28. Moraglio, G., Gianotti, G., and Danusso, F., *Eur. Polym. J.*, 3, 251, 1967.

29. Orwoll, R.A. and Flory, P.J., *J. Amer. Chem. Soc.*, 89, 6822, 1967.

30. Patterson, D., Delmas, G., and Somcynsky, T., *Polymer*, 8, 503, 1967.

31. Nakajima, A. and Fujiwara, H., *J. Polym. Sci.: Part A-2*, 6, 723, 1968.

32. Rehage, G. and Koningsveld, R., *J. Polym. Sci.: Polym. Lett.*, 6, 421, 1968.

33. Andreeva, V. M., et al., *Vysokomol. Soedin., Ser. B*, 11, 555, 1969.

34. Bardin, J.-M. and Patterson, D., *Polymer*, 10, 247, 1969.

35. Dusek, K., *Coll. Czech. Chem. Commun.*, 34, 3309, 1969.

36. Delmas, G. and Patterson, D., *J. Polym. Sci.: Part C*, 30, 1, 1970.

37. Koningsveld, R., Kleintjens, L.A., and Shultz, A.R., *J. Polym. Sci.: Part A-2*, 8, 1261, 1970.

38. Kotaka, T., et al., *Polym. J.*, 1, 245, 1970.

39. Liddell, A.H. and Swinton, F.L., *Discuss. Faraday Soc.*, 49, 115, 1970.

40. Matsumura, K., *Polym. J.*, 1, 322, 1970.

41. Nakayama, H., *Bull. Chem. Soc. Japan*, 43, 1683, 1970.

42. Cowie, J.M.G., Maconnachie, A., and Ranson, R.J., *Macromolecules*, 4, 57, 1971.

43. Kagemoto, A. and Baba, Y., *Kobunshi Kagaku*, 28, 784, 1971.

44. Shibatani, K. and Oyanagi, Y., *Kobunshi Kagaku*, 28 (1971) 361-367.

45. Tager, A.A., et al., *Vysokomol. Soedin., Ser. A*, 13, 659, 1971.

46. Izumi, Y. and Miyake, Y., *Polym. J.*, 3, 647, 1972.

47. Kagemoto, A., Baba, Y., and Fujishiro, R., *Makromol. Chem.*, 154, 105, 1972.

48. Kennedy, J.W., Gordon, M., and Koningsveld, R., *J. Polym. Sci.: Part C*, 39, 43, 1972.

49. Lirova, B.I., et al., *Vysokomol. Soedin., Ser. B*, 14,265, 1972.

50. Nakayama, H., *Bull. Chem. Soc. Japan*, 45, 1371, 1972.

51. Siow, K.S., Delmas, G., and Patterson, D., *Macromolecules*, 5, 29, 1972.

52. Teramachi, S. and Fujikawa, T., *J. Macromol. Sci.-Chem. A*, 6, 1393, 1972.

53. Zeman, L., Biros, J., Delmas, G., and Patterson, D., *J. Phys. Chem.*, 76, 1206, 1972.

54. Zeman, L. and Patterson, D., *J. Phys. Chem.*, 76, 1214, 1972.

55. Baba, Y., Fujita, Y., and Kagemoto, A., *Makromol. Chem.*, 164, 349, 1973.

56. Candau, F., Strazielle, C., and Benoit, H., *Makromol. Chem.*, 170, 165, 1973.

57. Hamada, F., Fujisawa, K., and Nakajima, A., *Polym. J.*, 4, 316, 1973.

58. Kuwahara, N., Nakata, M., and Kaneko, M., *Polymer*, 14, 415, 1973.

59. Kuwahara, N., Kojima, J., and Kaneko, M., *J. Polym. Sci.: Polym. Phys. Ed.*, 11, 2307, 1973.

60. Saeki, S., Kuwahara, N., Konno, S., and Kaneko, M., *Macromolecules*, 6, 246, 1973.
61. Saeki, S., Kuwahara, N., Konno, S., and Kaneko, M., *Macromolecules*, 6, 589, 1973.
62. Tani, S., Hamada, F., and Nakajima, A., *Polym. J.*, 5, 86, 1973.
63. Baba, Y. and Kagemoto, A., *Kobunshi Ronbunshu*, 31, 446, 1974.
64. Bataille, P., *J. Chem. Eng. Data*, 19, 224, 1974.
65. Bessonov, Yu.S. and Tager, A.A., *Trud. Khim. Khim. Tekhnol.*, 1, 150, 1974.
66. Cowie, J.M.G. and McEwen, L.J., *J. Chem. Soc., Faraday Trans. I*, 70, 171, 1974.
67. Cowie, J.M.G. and McEwen, I.J., *Macromolecules*, 7, 291, 1974.
68. Cowie, J.M.G. and McEwen, I.J., *J. Polym. Sci.: Polym. Phys. Ed.*, 12, 441, 1974.
69. Derham, K.W., Goldsbrough, J., and Gordon, M., *Pure Appl. Chem.*, 38, 97, 1974.
70. Kuwahara, N., Saeki, S., Chiba, T., and Kaneko, M., *Polymer*, 15, 777, 1974.
71. Nakajima, A., et al., *Makromol. Chem.*, 175, 197, 1974.
72. Saeki, S., Konno, S., Kuwahara, N., Nakata, M., and Kaneko, M., *Macromolecules*, 7, 521, 1974.
73. Ver Strate, G. and Philippoff, W., *J. Polym. Sci.: Polym. Lett. Ed.*, 12, 267, 1974.
74. Konno, S., et al., *Macromolecules*, 8, 799, 1975.
75. Nakata, M., Kuwahara, N., and Kaneko, M., *J. Chem. Phys.*, 62, 4278, 1975.
76. Saeki, S., Kuwahara, N., Nakata, M., and Kaneko, M., *Polymer*, 16, 445, 1975.
77. Strazielle, C. and Benoit, H., *Macromolecules*, 8, 203, 1975.
78. Tager, A. A., et al., *Vysokomol. Soedin., Ser. B*, 17, 61, 1975.
79. Chiu, D.S., Takahashi, Y., and Mark, J.E., *Polymer*, 17, 670, 1976.
80. Cowie, J.M.G. and McEwen, I.J., *J. Chem. Soc., Faraday Trans. I*, 72, 526, 1976.
81. Nakata, M., et al., *J. Chem. Phys.*, 64, 1022, 1976.
82. Nose, T. and Tan, T.V., *J. Polym. Sci.: Polym. Lett. Ed.*, 14, 705, 1976.
83. Saeki, S., Kuwahara, N., Nakata, M., and Kaneko, M., *Polymer*, 17, 685, 1976.
84. Saeki, S., Kuwahara, N., and Kaneko, M., *Macromolecules*, 101, 1976.
85. Slagowski, E., Tsai, B., and McIntyre, D., *Macromolecules*, 9, 687, 1976.
86. Panina, N.I., Lozgacheva, V.P., and Aver'yanova, V.M., *Vysokomol. Soedin., Ser. B*, 19, 786,. 1977.
87. Rigler, J.K., Wolf, B.A., and Breitenbach, J.W., *Angew. Makromol. Chem.*, 57, 15, 1977.
88. Vshivkov, S.A., et al., *Prots. Studneobras. Polimern. Sistem.*, (2), 3, 1977.
89. Wolf, B.A. and Jend, R., *Makromol. Chem.*, 178, 1811, 1977.
90. Wolf, B.A. and Sezen, M.C., *Macromolecules*, 10, 1010, 1977.
91. Kodama, Y. and Swinton, F.L., *Brit. Polym. J.*, 10, 191, 1978.
92. Nakata, M., Dobashi, T., Kuwahara, N., Kaneko, M., and Chu, B., *Phys. Rev. A*, 18, 2683, 1978.
93. Cowie, J.M.G., Horta, A., McEwen, I.J., and Prochazka, K., *Polym. Bull.*, 1, 329, 1979.
94. Hamano, K., Kuwahara, N., and Kaneko, M., *Phys. Rev. A*, 20, 1135, 1979.
95. Kleintjens, L.A.L., *Ph.D. Thesis*, Univ. Essex, U.K., 1979.
96. Dobashi, T., Nakata, M., and Kaneko, M., *J. Chem. Phys.*, 72, 6685, 1980.
97. Irvine, P. and Gordon, M., *Macromolecules*, 13, 761, 1980.
98. Kleintjens, L.A., Koningsveld, R., and Gordon, M., *Macromolecules*, 13, 303, 1980.
99. Lang, J.C. and Morgan, R.D., *J. Chem. Phys.*, 73, 5849, 1980.
100. Richards, R.W., *Polymer*, 21, 715, 1980.
101. Charlet, G. and Delmas, G., *Polymer*, 22, 1181, 1981.
102. Charlet, G., Ducasse, R., and Delmas, G., *Polymer*, 22, 1190, 1981.
103. Hashizume, J., Teramoto, A., and Fujita, H., *J. Polym. Sci.: Polym. Phys. Ed.*, 19, 1405, 1981.
104. Wolf, B.A. and Geerissen, H., *Colloid Polym. Sci.*, 259, 1214, 1981.
105. Geerissen, H. and Wolf, B.A., *Makromol. Chem., Rapid Commun.*, 3, 17, 1982.
106. Goloborod'ko, V.I., Valatin, S.M., und Tashmukhamedov, I.P., *Uzb. Khim. Zh.*, (3), 33, 1982.
107. Mangalam, P. V. and Kalpagam, V., *J. Polym. Sci.: Polym. Phys. Ed.*, 20, 773, 1982.
108. Shinozaki, K., Abe, M., and Nose, T., *Polymer*, 23, 722, 1982.
109. Shinozaki, K., Van Tan, T., Saito, Y., and Nose, T., *Polymer*, 23, 728, 1982.
110. Suzuki, H., Kamide, K., and Saitoh, M., *Eur. Polym. J.*, 18, 123, 1982.
111. Suzuki, H., Muraoka, Y., Saitoh, M., and Kamide, K., *Brit. Polym. J.*, 14, 23, 1982.
112. Cowie, J.M.G. and McEwen, I.J., *Polymer*, 24, 1445, 1983.
113. Herold, F.K., Schulz, G.V., and Wolf, B.A., *Materials Chem. Phys.*, 8, 243, 1983.
114. Tager, A.A., et al., *Vysokomol. Soedin., Ser. A*, 25, 1444, 1983.
115. Corti, M., Minero, C., and Degiorgio, V., *J. Phys. Chem.*, 88, 309, 1984.
116. Cowie, J.M.G. and McEwen, I.J., *Polymer*, 25, 1107, 1984.
117. Dobashi, T., Nakata, M., and Kaneko, M., *J. Chem. Phys.*, 80, 948, 1984.
118. Florin, E., Kjellander, R., and Eriksson, J.C., *J. Chem. Soc., Faraday Trans. I*, 80, 2889, 1984.
119. Gilluck, M., *Dissertation*, TH Leuna-Merseburg, 1984.
120. Rangel-Nafaile, C., Metzner, A.B., and Wissbrun, K.F., *Macromolecules*, 17, 1187, 1984.
121. Shiomi, T., et al., *J. Polym. Sci.: Polym. Phys. Ed.*, 22, 1305, 1984.
122. Tsuyumoto, M., Einaga, Y., and Fujita, H., *Polym. J.*, 16, 229, 1984.
123. Varennes, S., Charlet, G., and Delmas, G., *Polym. Eng. Sci.*, 24, 98, 1984.
124. Hamano, K., Kuwahara, N., Koyama, T., and Harada, S., *Phys. Rev. A*, 32, 3168, 1985.
125. Kraemer, H. and Wolf, B.A., *Makromol. Chem., Rapid Commun.*, 6, 21, 1985.
126. Herold, F.K. and Wolf, B.A., *Mater. Chem. Phys.*, 14, 311, 1986.
127. Irani, C.A. and Cozewith, C., *J. Appl. Polym. Sci.*, 31, 1879, 1986.
128. Cowie, J.M.G. and McEwen, I.J., *Brit. Polym. J.*, 18, 387, 1986.
129. Krüger, B., *Dissertation*, TH Leuna-Merseburg, 1986.
130. Rätzsch, M.T., et al., *J. Macromol. Sci.-Chem. A*, 23, 1349, 1986.
131. Saeki, S., et al., *Macromolecules*, 19, 2353, 1986.
132. Sander, U. and Wolf, B.A., *Angew. Makromol. Chem.*, 139, 149, 1986.
133. Barbarin-Castillo, J.-M., et al., *Polym.Commun.*, 28, 212, 1987.
134. Gruner, K. and Greer, S.C., *Macromolecules*, 20, 2238, 1987.
135. Magarik, S.Ya., Filippov, A.P., and D'yakonova, N.V., *Vysokomol. Soedin., Ser. A*, 29, 698, 1987.
136. Rangel-Nafaile, C. and Munoz-Lara, J.J., *Chem. Eng. Commun.*, 53, 177, 1987.
137. Kiepen, F. and Borchard, W., *Macromolecules*, 21, 1784, 1988.
138. Schuster, R., *Diploma Paper*, TH Leuna-Merseburg, 1988.
139. Tveekrem, J.L., Greer, S.C., and Jacobs, D.T., *Macromolecules*, 21, 147, 1988.
140. Bohossian, T., Charlet, G., and Delmas, G., *Polymer*, 30, 1695, 1989.
141. Chiu, G. and Mandelkern, L., *Macromolecules*, 23, 5356, 1990.
142. Goedel, W.A., et al., *Ber. Bunsenges. Phys. Chem.*, 94, 17, 1990.
143. Van der Haegen, R. and Van Opstal, L., *Makromol. Chem.*, 191, 1871, 1990.
144. Iwai, Y., et al., *Sekiyu Gakkaishi*, 33, 117, 1990.
145. Raetzsch, M.T., Krueger, B., and Kehlen, H., *J. Macromol. Sci.-Chem. A*, 27, 683, 1990.
146. Schild, H.G. and Tirrell, D.A., *J. Phys. Chem.*, 94, 4352, 1990.
147. Stafford, S.G., Ploplis, A.C., and Jacobs, D.T., *Macromolecules*, 23, 470, 1990.
148. Akhmadeev, I.R., et al., *Vysokomol. Soedin., Ser. B*, 33, 543, 1991.
149. Bae, Y.C., Lambert, S.M., Soane, D.S., and Prausnitz, J.M., *Macromolecules*, 24, 4403, 1991.
150. Chu, B., Linliu, K., Xie, P., Ying, Q., Wang, Z., and Shook, J.W., *Rev. Sci. Instr.*, 62, 2252, 1991.
151. Kawate, K., Imagawa, I., and Nakata, M., *Polym. J.*, 23, 233, 1991.
152. Lee, K.D. and Lee, D.C., *Pollimo*, 15, 274, 1991.
153. Louai, A., Sarazin, D., Pollet, G., Francois, J., and Moreaux, F., *Polymer*, 32, 703, 1991.
154. Van Opstal, L., Koningsveld, R., and Kleintjens, L.A., *Macromolecules*, 24, 161, 1991.
155. Schubert, K.-V., Strey, R., and Kahlweit, M., *J. Colloid Interface Sci.*, 141, 21, 1991.
156. Shen, W., Smith, G.R., Knobler, C.M., and Scott, R.L., *J. Phys. Chem.*, 95, 3376, 1991.

157. Szydlowski, J. and Van Hook, W.A., *Macromolecules*, 24, 4883, 1991.

158. Tager, A.A., et al., *Vysokomol. Soedin., Ser. B*, 33, 572, 1991.

159. Wakker, A., *Polymer*, 32, 279, 1991.

160. Yokoyama, H., Takano, A., Okada, M., and Nose, T., *Polymer*, 32, 3218, 1991.

161. Heinrich, M. and Wolf, B.A., *Polymer*, 33, 1926, 1992.

162. Heinrich, M. and Wolf, B.A., *Macromolecules*, 25, 3817, 1992.

163. Lecointe, J.P., Pascault, J.P., Suspene, L., and Yang, Y.S., *Polymer*, 33, 3226, 1992.

164. Mueller, K.F., *Polymer*, 33, 3470, 1992.

165. Szydlowski, J., Rebelo, L., and Van HooK, W.A., *Rev. Sci. Instrum.*, 63, 1717, 1992.

166. Xia, K.-Q., Franck, C., and Widom, B., *J. Chem. Phys.*, 97, 1446, 1992.

167. Arnauts, J., Berghmans, H., and Koningsveld, R., *Makromol. Chem.*, 194, 77, 1993.

168. Iwai, Y., Shigematsu, Y., Furuya, T., Fukuda, H., Arai, Y., *Polym. Eng. Sci.*, 33, 480, 1993.

169. Wakker, A., Van Dijk, F., and Van Dijk, M.A., *Macromolecules*, 26, 5088, 1993.

170. Wells, P.A., de Loos, Th.W., and Kleintjens, L.A., *Fluid Phase Equil.*, 83, 383, 1993.

171. Barbarin-Castillo, J.-M. and McLure, I.A., *Polymer*, 35, 3075, 1994.

172. Mumick, P.S. and McCormick, C.L., *Polym. Eng. Sci.*, 34, 1419, 1994.

173. Okano, K., Takada, M., Kurita, K., and Furusaka, M., *Polymer*, 35, 2284, 1994.

174. Sato, H., Kuwahara, N., and Kubota, K., *Phys. Rev. E*, 50, 1752, 1994.

175. Song, S.-W. and Torkelson, J.M., *Macromolecules*, 27, 6389, 1994.

176. Tager, A.A., et al., *Colloid Polym. Sci.*, 272, 1234, 1994.

177. Vanhee, S., et al., *Makromol. Chem. Phys.*, 195, 759, 1994.

178. Haas, C.K. and Torkelson, J.M., *Phys. Rev. Lett.*, 75, 3134, 1995.

179. Ikier, C. and Klein, H., *Macromolecules*, 28, 1003, 1995.

180. Luszczyk, M., Rebelo, L.P.N., and Van Hook, W.A., *Macromolecules*, 28, 745, 1995.

181. Pfohl, O., Hino, T., and Prausnitz, J.M., *Polymer*, 36, 2065, 1995.

182. Vshivkov, S.A. and Safronov, A.P., *Vysokomol. Soedin., Ser. B*, 37, 1779, 1995.

183. Allcock, H.R. and Dudley, G.K., *Macromolecules*, 29, 1313, 1996.

184. El-Ejmi, A.A.S. and Huglin, M.B., *Polym. Int.* 39, 113, 1996.

185. Fischer, V., Borchard, W., and Karas, M., *J. Phys. Chem.*, 100, 15992, 1996.

186. Imre, A. and Van Hook, W.A., *J. Polym. Sci.: Part B: Polym. Sci.*, 34, 751, 1996.

187. Luszczyk, M. and Van Hook, W.A., *Macromolecules*, 29, 6612, 1996.

188. Rong, Z., Wang, H., Ying, X., and Hu, Y., *J. East China Univ. Sci. Technol.*, 22, 754, 1996.

189. Safronov, A.P., Tager, A.A., and Koroleva, E.V., *Vysokomol. Soedin., Ser. B*, 38, 900, 1996.

190. Vshivkov, S.A. and Rusinova, E.V., *Vysokomol. Soedin., Ser. A*, 38, 1746, 1996.

191. Xia, K.-Q., An, X.-Q., and Shen, W.-G., *J. Chem.Phys.*, 105, 6018, 1996.

192. Imre, A. and Van Hook, W.A., *J. Polym. Sci.: Part B: Polym. Phys.*, 35, 1251, 1997.

193. Kita, R., Dobashi, T., Yamamoto, T., Nakata, M., and Kamide, K., *Phys. Rev. E*, 55, 3159, 1997.

194. Li, M., Zhu, Z.-Q., and Mei, L.-H., *Biotechnol. Progr.*, 13, 105, 1997.

195. McLure, I.A., Mokhtari, A., and Bowers, J., *J. Chem. Soc., Faraday Trans.*, 93, 249, 1997.

196. Chalykh, A.E., Dement'eva, O.V., and Gerasimov, V.K., *Vysokomol. Soedin., Ser. A*, 40, 815, 1998.

197. Kubota, K., Kita, R., and Dobashi, T., *J. Chem. Phys.*, 109, 711, 1998.

198. Schneider, A., *Dissertation*, Johannes Gutenberg Universität Mainz. 1998.

199. Terao, K., et al., *Macromolecules*, 31, 6885, 1998.

200. Yamagishi, T.-A., et al., *Macromol. Chem. Phys.*, 199, 423, 1998.

201. Lau, A.C.W. and Wu, C., *Macromolecules*, 32, 581, 1999.

202. Nakata, M., Dobashi, T., Inakuma, Y.-I., and Yamamura, K., *J. Chem. Phys.*, 111, 6617, 1999.

203. Pruessner, M.D., Retzer, M.E., and Greer, S.C., *J. Chem. Eng. Data*, 44, 1419, 1999.

204. Shimofure, S., Kubota, K., Kita, R., and Dobashi, T., *J. Chem.Phys.*, 111, 4199, 1999.

205. Fischer, V. and Borchard, W., *J. Phys. Chem. B*, 104, 4463, 2000.

206. Imre, A., and Van Hook, W.A., *Macromolecules*, 33, 5308, 2000.

207. Koizumi, J., et al., *J. Phys. Soc. Japan*, 69, 2543, 2000.

208. Kunugi, S., Tada, T., Yamazaki, Y., Yamamoto, K., and Akashi, M., *Langmuir*, 16, 2042, 2000.

209. La Mesa, C., *J. Therm. Anal. Calorim.*, 61, 493, 2000.

210. Persson, J., et al., *Bioseparation*, 9, 105, 2000.

211. Persson, J., Kaul, A., and Tjerneld, F., *J. Chromatogr. B*, 743, 115, 2000.

212. Djokpe, E. and Vogt, W., *Macromol. Chem. Phys.*, 202, 750, 2001.

213. Kujawa, P. and Winnik, F.M., *Macromolecules*, 34, 4130, 2001.

214. Pendyala, K.S., Greer, S.C., and Jacobs, D.T., *J. Chem. Phys.*, 115, 9995, 2001.

215. Berlinova, I. V., Nedelcheva, A. N., Samchikov, V., and Ivanov, Ya., *Polymer*, 43, 7243, 2002.

216. Freitag, R. and Garret-Flaudy, F., *Langmuir*, 18, 3434, 2002.

217. Maeda, Y., Nakamura, T., and Ikeda, I., *Macromolecules*, 35, 217, 2002.

218. Rebelo, L.P.N., et al., J., *Macromolecules*, 35, 1887, 2002.

219. Safronov, A.P. and Somova, T.V., *Vysokomol. Soedin., Ser. A*, 44, 2014, 2002.

220. Vshivkov, S.A., Rusinova, E.V., and Gur'ev, A.A., *Vysokomol. Soedin., Ser. B*, 44, 504, 2002.

221. Zhou, C.-S., An, X.-Q., Xia, K.-Q., Yin, X.-L., and Shen, W.-G., *J. Chem. Phys.*, 117, 4557, 2002.

222. Chang, Y., Powell, E.S., Allcock, H.R., Park, S.M., and Kim, C., *Macromolecules*, 36, 2568, 2003.

223. David, G., et al., *Eur. Polym. J.*, 39, 1209, 2003.

224. Siporska, A., Szydlowski, J., and Rebelo, L.P.N., *Phys. Chem. Chem. Phys.*, 5, 2996, 2003.

225. Yamamoto, K., Serizawa, T., and Akashi, M., *Macromol. Chem. Phys.*, 204, 1027, 2003.

VAPOR PRESSURES (SOLVENT ACTIVITIES) FOR BINARY POLYMER SOLUTIONS

Christian Wohlfarth

The vapor pressure of a binary polymer solution is given by the activity of the solvent A, a_A. Solvent activities in polymer solutions are measured either by the isopiestic method applying a reference system whose solvent activity is precisely known or by determining the solvent partial pressure, P_A, and calculating the activity of the solvent by equation (1):

$$a_A = \left(P_A / P_A^s\right) \exp\left[\frac{\left(B_{AA} - V_A^L\right)\left(P - P_A^s\right)}{RT}\right] \quad (1)$$

where B_{AA} is the second virial coefficient, P_A^s is the saturation vapor pressure, and V_A^L is the molar volume of the pure solvent A at the measuring temperature T. The exponential term is neglected in quite a lot of original papers, however, and only the reduced vapor pressures are given (such data are indicated by an asterisk in the table below). Vapor pressures of polymer solutions have been measured since the 1940s, but the amount of experimental data for polymer solutions is still relatively small in comparison to low-molecular mixtures and solutions. The data scatter with respect to temperature, concentration, molar mass, and other polymer characterization variables. Furthermore, the concentration range for measuring vapor pressures in good thermodynamic quality is often limited to the polymer mass fraction range between 0.4 and 0.85. A recent review on methods for the measurement of vapor pressures/solvent activities of polymer solutions and on related problems is given in Ref. [1]. Experimental data have been collected in several books [2-6].

The table in this *Handbook* provides data for a number of polymer solutions as smoothed values over the complete range of solvent activities between 0 (polymer mass fraction = 1) and 1 (polymer mass fraction = 0). For this purpose, the data were selected from data books [4–6] as well as from a number of original sources [7–22] which are not included in these books. The appropriate data were smoothed. The final table provides then the polymer mass fractions at given fixed solvent activities between 0.1 and 0.9. Of course, the user must keep in mind that the activity vs. concentration range of the experimental data is sometimes smaller than the below given complete range, thus the smoothed data should be used with sufficient care.

Generally, vapor pressures or solvent activities of binary polymer solutions depend on molar mass. However, for high molecular weight polymers (well above the oligomer region), this molar-mass dependence can be neglected in many cases. Therefore, the table below presents only data for polymer solutions where the number average molar mass, M_n, is in the order of 10^5 g/mol or even higher, therefore, the molar mass is not specified. The temperature is

stated, even though the temperature dependence of a_A is relatively small for the temperature ranges where most of the experimental data exist.

References

1. Wohlfarth, C., Methods for the measurement of solvent activity of polymer solutions, in *Handbook of Solvents*, Wypych, G., Ed., ChemTec Publishing, Toronto, 2000, 146.
2. Wen, H., Elbro, H. S., and Alessi, P., *Polymer Solution Data Collection.* I. Vapor-liquid equilibrium; II. Solvent activity coefficients at infinite dilution; III. Liquid-liquid equilibrium, Chemistry Data Series, Vol. 15, DECHEMA, Frankfurt am Main, 1992.
3. Danner, R. P. and High, M. S., *Handbook of Polymer Solution Thermodynamics*, American Institute of Chemical Engineers, New York, 1993.
4. Wohlfarth, C., *Vapour-Liquid Equilibrium Data of Binary Polymer Solutions: Physical Science Data*, 44, Elsevier, Amsterdam, 1994.
5. Wohlfarth, C., *CRC Handbook of Thermodynamic Data of Copolymer Solutions*, CRC Press, Boca Raton, FL, 2001.
6. Wohlfarth, C., *CRC Handbook of Thermodynamic Data of Aqueous Polymer Solutions*, CRC Press, Boca Raton, FL, 2003.
7. Wang, K., Hu, Y., and Wu, D. T., *J. Chem. Eng. Data*, 39, 916, 1994.
8. Choi, J. S., Tochigi, K., and Kojima, K., *Fluid Phase Equil.*, 111, 143, 1995.
9. Tochigi, K., Kurita, S., Ohashi, M., and Kojima, K., *Kagaku Kogaku Ronbunshu*, 23, 720, 1997.
10. Wong, H. C., Campbell, S. W., and Bhethanabotla, V. R., *Fluid Phase Equil.*, 139, 371, 1997.
11. Kim, J., Joung, K. C., Hwang, S., Huh, W., Lee, C. S., and Yoo, K.-P., *Korean J. Chem. Eng.*, 15, 199, 1998.
12. Kim, N. H., Kim, S.J., Won, Y. S., and Choi, J. S., *Korean J. Chem. Eng.*, 15, 141, 1998.
13. Feng, W., Wang, W., and Feng, Z., *J. Chem. Ind. Eng. (China)*, 49, 271, 1998.
14. French, R. N. and Koplos, G. J., *Fluid Phase Equil.*, 160, 879, 1999.
15. Striolo, A. and Praunsitz, J. M., *Polymer*, 41, 1109, 2000.
16. Fornasiero, F., Halim, M., and Prausnitz, J. M., *Macromolecules*, 33, 8435, 2000.
17. Wong, H. C., Campbell, S. W., and Bhethanabotla, V. R., *Fluid Phase Equil.*, 179, 181, 2001.
18. Wibawa, G., Takahashi, M., Sato, Y., Takishima, S., and Masuoka, H., *J. Chem. Eng. Data*, 47, 518, 2002.
19. Wibawa, G., Hatano, R., Sato, Y., Takishima, S., and Masuoka, H., *J. Chem. Eng. Data*, 47, 1022, 2002.
20. Pfohl, O., Riebesell, C., and Dohrn, R., *Fluid Phase Equil.*, 202, 289, 2002.
21. Jung, J. K., Joung, S. N., Shin, H.Y., Kim, S. Y., Yoo, K.-P., Huh, W., Lee, C. S., *Korean J. Chem. Eng.*, 19, 296, 2002.
22. Kang, S., Huang, Y., Fu, J., Liu, H., and Hu, Y., *J. Chem. Eng. Data*, 47, 788, 2002.

Solvent Activity a_A as Function of Temperature and Mass Fraction

Polymer/ solvents	a_A: T/K	0.1	0.2	0.3	0.4	0.5	0.6	0.7	0.8	0.9
					Mass Fraction of the Polymer					
Acrylonitrile/Styrene Copolymer (28 wt% Acrylonitrile)										
Benzene*)	343.15	0.982	0.962	0.940	0.915	0.886	0.851	0.809	0.753	0.670
1,2-Dimethylbenzene*)	398.15	0.983	0.964	0.942	0.918	0.890	0.857	0.817	0.764	0.685
1,3-Dimethylbenzene*)	398.15	0.983	0.965	0.944	0.921	0.893	0.861	0.821	0.769	0.690
1,4-Dimethylbenzene*)	398.15	0.983	0.964	0.942	0.918	0.890	0.857	0.817	0.763	0.684
Propylbenzene*)	398.15	0.987	0.972	0.955	0.935	0.913	0.885	0.851	0.804	0.732
Toluene*)	343.15	0.982	0.962	0.940	0.915	0.886	0.851	0.809	0.753	0.669

Polymer/ solvents	a_A: T/K	0.1	0.2	0.3	0.4	0.5	0.6	0.7	0.8	0.9
						Mass Fraction of the Polymer				
Butadiene/Styrene Copolymer (41 wt% Styrene)										
Benzene*)	343.15	0.968	0.934	0.896	0.853	0.805	0.748	0.680	0.591	0.461
Cyclohexane*)	343.15	0.978	0.953	0.925	0.893	0.856	0.811	0.754	0.678	0.556
Ethylbenzene*)	398.15	0.974	0.945	0.912	0.875	0.831	0.779	0.713	0.625	0.491
Mesitylene*)	398.15	0.977	0.950	0.921	0.887	0.847	0.799	0.738	0.656	0.526
Toluene*)	343.15	0.970	0.936	0.899	0.857	0.808	0.751	0.682	0.591	0.456
Cellulose Triacetate										
Dichloromethane	298.15	0.979	0.956	0.930	0.899	0.863	0.819	0.762	0.683	0.554
Trichloromethane	298.15	0.978	0.953	0.924	0.892	0.853	0.806	0.747	0.665	0.533
Dextran										
Water	313.15	0.988	0.975	0.960	0.942	0.921	0.894	0.860	0.810	0.725
Hydroxyethylcellulose										
Water	368.15	0.988	0.974	0.958	0.939	0.915	0.884	0.841	0.775	0.650
Hydroxypropylstarch										
Water	293.15	0.989	0.977	0.963	0.947	0.927	0.903	0.872	0.827	0.749
Nitrocellulose										
Ethyl acetate	293.15	0.938	0.885	0.835	0.786	0.737	0.685	0.627	0.560	0.471
Ethyl formate	293.15	0.958	0.916	0.873	0.828	0.780	0.728	0.668	0.595	0.494
Ethyl propionate	293.15	0.941	0.889	0.839	0.789	0.739	0.685	0.625	0.555	0.460
Methyl acetate	293.15	0.890	0.820	0.763	0.711	0.660	0.609	0.554	0.490	0.406
2-Propanone	293.15	0.922	0.861	0.807	0.756	0.706	0.653	0.596	0.530	0.443
Propyl acetate	293.15	0.937	0.881	0.827	0.775	0.722	0.665	0.602	0.528	0.426
Polybutadiene (random cis-trans-vinyl)										
Benzene	298.15	0.964	0.925	0.884	0.839	0.788	0.731	0.663	0.578	0.455
Cyclohexane	298.15	0.974	0.945	0.913	0.876	0.833	0.782	0.719	0.635	0.507
Dichloromethane	298.15	0.951	0.902	0.852	0.800	0.745	0.684	0.616	0.532	0.415
Hexane	298.15	0.984	0.965	0.943	0.916	0.881	0.837	0.775	0.683	0.534
Tetrachloromethane	298.15	0.932	0.865	0.799	0.731	0.660	0.585	0.503	0.409	0.288
Toluene	298.15	0.969	0.935	0.898	0.856	0.809	0.754	0.688	0.603	0.476
Trichloromethane	298.15	0.925	0.855	0.788	0.720	0.650	0.578	0.498	0.406	0.289
1,4-cis-Polybutadiene										
Benzene	298.15	0.966	0.930	0.890	0.846	0.796	0.738	0.668	0.580	0.450
Cyclohexane	298.15	0.977	0.951	0.922	0.888	0.849	0.803	0.747	0.677	0.581
Dichloromethane	298.15	0.948	0.898	0.848	0.796	0.742	0.683	0.616	0.536	0.424
Hexane	298.15	0.983	0.963	0.941	0.916	0.886	0.850	0.804	0.741	0.639
Tetrachloromethane	298.15	0.936	0.871	0.805	0.736	0.665	0.588	0.505	0.409	0.287
Toluene	298.15	0.969	0.936	0.900	0.860	0.815	0.763	0.701	0.622	0.506
Trichloromethane	298.15	0.915	0.840	0.770	0.702	0.634	0.562	0.485	0.396	0.283
Poly(butyl acrylate)										
Benzene	298.15	0.964	0.926	0.887	0.845	0.799	0.749	0.691	0.619	0.519
Dichloromethane	298.15	0.868	0.801	0.744	0.690	0.636	0.577	0.511	0.430	0.318
Tetrachloromethane	298.15	0.932	0.868	0.805	0.742	0.677	0.607	0.529	0.438	0.317
Toluene	298.15	0.967	0.932	0.893	0.849	0.801	0.744	0.676	0.590	0.463
Trichloromethane	298.15	0.901	0.811	0.733	0.662	0.595	0.529	0.459	0.381	0.282
Poly(butyl methacrylate)										
Benzene	313.15	0.971	0.939	0.902	0.861	0.813	0.756	0.685	0.592	0.453
1-Butanol	313.15	0.991	0.980	0.968	0.953	0.936	0.914	0.885	0.842	0.762
2-Butanol	313.15	0.992	0.982	0.969	0.953	0.933	0.906	0.869	0.815	0.719
2-Butanone	313.15	0.982	0.963	0.940	0.914	0.884	0.846	0.799	0.732	0.623
Butyl acetate*)	308.15	0.982	0.961	0.936	0.908	0.875	0.836	0.789	0.730	0.652
Cyclohexane	313.15	0.985	0.968	0.948	0.925	0.899	0.866	0.823	0.764	0.666
Cyclopentane	313.15	0.984	0.965	0.944	0.918	0.886	0.846	0.792	0.714	0.579
Diethyl ether*)	298.15	0.987	0.973	0.956	0.937	0.914	0.885	0.848	0.795	0.703
1,4-Dimethylbenzene	333.15	0.971	0.940	0.905	0.866	0.822	0.770	0.706	0.622	0.497
Ethylbenzene	333.15	0.969	0.935	0.899	0.859	0.815	0.764	0.704	0.627	0.517
Methyl acetate	313.15	0.984	0.965	0.944	0.920	0.891	0.856	0.811	0.748	0.645

Polymer/ solvents	a_A: T/K	0.1	0.2	0.3	0.4	0.5	0.6	0.7	0.8	0.9
						Mass Fraction of the Polymer				
2-Methyl-1-propanol	333.15	0.988	0.974	0.958	0.940	0.919	0.893	0.860	0.815	0.744
Octane	313.15	0.988	0.974	0.959	0.942	0.921	0.896	0.865	0.823	0.758
1-Propanol	333.15	0.990	0.980	0.967	0.952	0.934	0.911	0.881	0.834	0.746
2-Propanol	313.15	0.991	0.981	0.970	0.956	0.939	0.918	0.889	0.845	0.755
2-Propanone	313.15	0.989	0.976	0.961	0.944	0.921	0.892	0.850	0.783	0.647
Propyl acetate	313.15	0.980	0.957	0.932	0.903	0.870	0.830	0.780	0.714	0.612
Toluene	313.15	0.971	0.939	0.903	0.863	0.818	0.764	0.698	0.613	0.485
Poly(ε-caprolacton)										
Tetrachloromethane[*]	338.15	0.956	0.910	0.864	0.815	0.762	0.704	0.637	0.554	0.438
Poly(dimethylsiloxane)										
Chlorodifluoromethane	298.15	0.976	0.950	0.921	0.888	0.850	0.805	0.750	0.677	0.565
Cyclohexane	303.15	0.979	0.955	0.928	0.898	0.863	0.822	0.770	0.702	0.596
Hexane	303.15	0.982	0.962	0.939	0.912	0.880	0.842	0.793	0.724	0.611
Pentane	308.15	0.982	0.962	0.940	0.913	0.881	0.842	0.791	0.720	0.600
Pentane	423.15	0.984	0.966	0.946	0.922	0.893	0.858	0.813	0.749	0.641
Poly(ethyl acrylate)										
Benzene	298.15	0.970	0.939	0.904	0.866	0.823	0.774	0.716	0.641	0.533
Dichloromethane	298.15	0.900	0.830	0.768	0.709	0.648	0.584	0.512	0.427	0.313
Tetrachloromethane	298.15	0.950	0.900	0.848	0.794	0.736	0.672	0.598	0.509	0.385
Toluene	298.15	0.972	0.942	0.910	0.874	0.833	0.786	0.730	0.659	0.555
Trichloromethane	298.15	0.866	0.776	0.701	0.632	0.566	0.499	0.428	0.349	0.248
Poly(ethylene oxide)										
Benzene	323.15	0.972	0.942	0.908	0.869	0.824	0.771	0.706	0.620	0.490
2-Butanone	353.15	0.981	0.959	0.934	0.902	0.863	0.813	0.746	0.651	0.503
Cyclohexane	353.15	0.989	0.976	0.960	0.943	0.921	0.893	0.855	0.798	0.688
Methanol	303.15	0.964	0.927	0.887	0.844	0.797	0.744	0.682	0.604	0.494
2-Propanone	353.15	0.979	0.947	0.896	0.815	0.719	0.625	0.532	0.434	0.315
Water	293.15	0.977	0.951	0.923	0.890	0.852	0.806	0.748	0.671	0.550
Poly(ethylenimine)										
Water	353.15	0.975	0.947	0.917	0.883	0.845	0.801	0.748	0.680	0.581
Poly(ethyl methacrylate)										
Benzene	298.15	0.970	0.938	0.903	0.864	0.821	0.771	0.712	0.637	0.529
Dichloromethane	298.15	0.912	0.838	0.769	0.703	0.636	0.567	0.491	0.404	0.292
Tetrachloromethane	298.15	0.935	0.873	0.812	0.750	0.686	0.616	0.540	0.449	0.328
Toluene	298.15	0.974	0.945	0.913	0.877	0.836	0.787	0.727	0.647	0.527
Trichloromethane	298.15	0.859	0.760	0.678	0.604	0.533	0.464	0.392	0.313	0.217
Polyisobutylene										
Benzene	313.15	0.984	0.965	0.945	0.921	0.892	0.858	0.813	0.751	0.645
Cyclohexane	313.15	0.976	0.950	0.921	0.888	0.850	0.805	0.749	0.676	0.563
Cyclopentane	313.15	0.977	0.952	0.924	0.892	0.855	0.812	0.758	0.687	0.579
1,4-Dimethylbenzene	313.15	0.979	0.955	0.929	0.899	0.863	0.821	0.767	0.694	0.579
2,2-Dimethylbutane	298.15	0.983	0.964	0.942	0.917	0.887	0.852	0.806	0.743	0.640
Ethylbenzene	313.15	0.979	0.955	0.927	0.895	0.857	0.810	0.750	0.668	0.535
Heptane	298.15	0.983	0.964	0.942	0.917	0.887	0.851	0.804	0.741	0.637
Hexane	298.15	0.980	0.959	0.934	0.906	0.873	0.834	0.784	0.715	0.606
Octane	298.15	0.983	0.963	0.940	0.914	0.883	0.845	0.797	0.729	0.617
Tetrachloromethane	298.15	0.962	0.921	0.877	0.829	0.776	0.715	0.643	0.552	0.423
Toluene	313.15	0.984	0.966	0.944	0.918	0.884	0.840	0.779	0.688	0.537
Trichloromethane	298.15	0.969	0.935	0.899	0.858	0.813	0.761	0.698	0.619	0.503
2,4,4-Trimethylpentane	298.15	0.981	0.961	0.937	0.911	0.879	0.842	0.794	0.730	0.628
1,4-cis-Polyisoprene										
Benzene	313.15	0.982	0.962	0.937	0.908	0.873	0.827	0.766	0.679	0.537
2-Butanone	353.15	0.986	0.970	0.953	0.933	0.910	0.883	0.850	0.808	0.746
Cyclohexane	313.15	0.978	0.954	0.928	0.899	0.865	0.825	0.778	0.716	0.625
Dichloromethane	298.15	0.969	0.935	0.898	0.857	0.811	0.757	0.693	0.610	0.488
1,4-Dimethylbenzene	313.15	0.977	0.951	0.923	0.892	0.857	0.816	0.767	0.704	0.613

Polymer/ solvents	a_A: T/K	0.1	0.2	0.3	0.4	0.5	0.6	0.7	0.8	0.9
						Mass Fraction of the Polymer				
Ethylbenzene	313.15	0.978	0.954	0.928	0.898	0.864	0.823	0.774	0.709	0.612
Methyl acetate	313.15	0.968	0.935	0.900	0.862	0.820	0.773	0.717	0.649	0.554
Octane	313.15	0.984	0.967	0.948	0.926	0.901	0.871	0.834	0.785	0.711
Propyl acetate	333.15	0.983	0.964	0.942	0.916	0.886	0.850	0.803	0.738	0.633
Tetrachloromethane	298.15	0.929	0.864	0.800	0.737	0.672	0.602	0.526	0.435	0.316
Toluene	313.15	0.978	0.954	0.927	0.898	0.865	0.827	0.782	0.725	0.645
Trichloromethane	298.15	0.930	0.867	0.807	0.747	0.685	0.620	0.547	0.462	0.346
Poly(methyl acrylate)										
Benzene	298.15	0.979	0.956	0.930	0.901	0.867	0.826	0.776	0.710	0.608
Dichloromethane	298.15	0.917	0.851	0.791	0.732	0.671	0.605	0.532	0.444	0.326
Tetrachloromethane	298.15	0.963	0.924	0.882	0.838	0.788	0.733	0.668	0.586	0.470
Toluene	298.15	0.981	0.960	0.936	0.909	0.878	0.840	0.792	0.727	0.626
Trichloromethane	298.15	0.912	0.830	0.753	0.678	0.603	0.527	0.446	0.357	0.248
Poly(methyl methacrylate)										
Benzene	298.15	0.982	0.961	0.938	0.912	0.881	0.843	0.795	0.729	0.622
2-Butanone[*)	308.15	0.989	0.976	0.961	0.945	0.925	0.900	0.869	0.825	0.751
Cyclohexanone[*)	323.15	0.978	0.954	0.928	0.899	0.866	0.827	0.781	0.723	0.640
Dichloromethane	298.15	0.939	0.882	0.825	0.766	0.704	0.637	0.560	0.468	0.343
Ethyl acetate[*)	308.15	0.986	0.969	0.950	0.928	0.902	0.869	0.826	0.763	0.649
Toluene	298.15	0.981	0.959	0.935	0.908	0.877	0.841	0.795	0.736	0.646
Trichloromethane	298.15	0.924	0.848	0.771	0.694	0.616	0.536	0.451	0.358	0.246
Poly(α-methylstyrene)										
Cumene	338.15	0.984	0.965	0.944	0.918	0.887	0.848	0.796	0.721	0.593
α-Methylstyrene	338.15	0.978	0.954	0.927	0.896	0.859	0.816	0.761	0.687	0.570
Poly(propylene oxide)										
Benzene	333.15	0.967	0.932	0.893	0.850	0.801	0.744	0.675	0.588	0.460
Metvhanol	298.15	0.992	0.982	0.970	0.955	0.936	0.910	0.872	0.812	0.689
Polystyrene										
Benzene	333.15	0.978	0.953	0.924	0.891	0.852	0.804	0.742	0.657	0.521
2-Butanone[*)	298.15	0.986	0.971	0.954	0.935	0.912	0.885	0.851	0.804	0.724
Cyclohexane	313.15	0.990	0.978	0.965	0.949	0.931	0.908	0.877	0.833	0.754
Cyclohexanone[*)	313.15	0.970	0.937	0.900	0.858	0.810	0.753	0.684	0.593	0.459
Dichloromethane	298.15	0.949	0.899	0.849	0.797	0.743	0.684	0.617	0.536	0.423
1,3-Dimethylbenzene[*)	323.15	0.980	0.956	0.926	0.891	0.846	0.791	0.723	0.638	0.524
1,4-Dimethylbenzene	423.15	0.974	0.944	0.911	0.872	0.826	0.770	0.698	0.601	0.452
Ethyl acetate[*)	313.15	0.976	0.948	0.918	0.882	0.841	0.791	0.728	0.642	0.507
Hexane	423.15	0.980	0.958	0.933	0.904	0.869	0.827	0.772	0.697	0.574
2-Propanone	323.15	0.991	0.980	0.969	0.955	0.938	0.918	0.892	0.854	0.788
Propyl acetate	343.15	0.983	0.965	0.943	0.919	0.891	0.858	0.815	0.758	0.667
Tetrachloromethane	298.15	0.961	0.917	0.869	0.814	0.751	0.678	0.592	0.486	0.344
Toluene	313.15	0.981	0.959	0.933	0.901	0.861	0.809	0.738	0.638	0.481
Trichloromethane	298.15	0.949	0.898	0.847	0.793	0.736	0.675	0.604	0.519	0.400
Poly(tetramethylene glycol)										
Methanol	303.15	0.981	0.961	0.938	0.913	0.883	0.849	0.806	0.751	0.671
Poly(vinyl acetate)										
Benzene	313.15	0.985	0.967	0.945	0.919	0.886	0.844	0.784	0.696	0.548
1-Butanol	313.15	0.992	0.982	0.971	0.958	0.942	0.923	0.896	0.856	0.779
2-Butanol	313.15	0.987	0.972	0.956	0.937	0.915	0.889	0.856	0.813	0.747
2-Butanone	313.15	0.980	0.958	0.934	0.906	0.873	0.835	0.787	0.724	0.626
1,2-Dichloroethane[*)	300.15	0.955	0.906	0.851	0.790	0.722	0.644	0.556	0.450	0.315
1,4-Dimethylbenzene	313.15	0.990	0.978	0.964	0.948	0.928	0.903	0.868	0.814	0.705
Ethylbenzene	313.15	0.990	0.979	0.966	0.950	0.932	0.910	0.880	0.836	0.759
Methanol	333.15	0.990	0.978	0.965	0.949	0.931	0.908	0.877	0.834	0.757
Methyl acetate	313.15	0.976	0.949	0.919	0.886	0.849	0.805	0.752	0.684	0.583
2-Methyl-1-propanol	353.15	0.984	0.966	0.946	0.924	0.899	0.868	0.832	0.784	0.715
1-Propanol	353.15	0.987	0.972	0.955	0.936	0.914	0.888	0.856	0.815	0.753
2-Propanol	353.15	0.988	0.974	0.958	0.940	0.919	0.894	0.863	0.820	0.754

Polymer/ solvents	a_A: T/K	0.1	0.2	0.3	0.4	0.5	0.6	0.7	0.8	0.9
					Mass Fraction of the Polymer					
2-Propanone	333.15	0.983	0.963	0.940	0.913	0.880	0.838	0.784	0.707	0.578
Propyl acetate	333.15	0.979	0.955	0.930	0.901	0.869	0.831	0.786	0.728	0.645
Tetrahydrofuran	323.15	0.973	0.943	0.911	0.874	0.831	0.781	0.720	0.640	0.519
Toluene	333.15	0.983	0.965	0.944	0.920	0.891	0.857	0.815	0.756	0.664
Poly(vinyl chloride)										
2-Butanone*)	313.15	0.976	0.949	0.920	0.887	0.849	0.804	0.749	0.676	0.566
Cyclohexanone*)	333.15	0.971	0.934	0.889	0.839	0.781	0.714	0.635	0.536	0.397
Poly(vinyl methyl ether)										
Benzene*)	298.15	0.969	0.935	0.897	0.855	0.807	0.751	0.683	0.596	0.466
Chlorobenzene*)	343.15	0.972	0.941	0.906	0.867	0.822	0.769	0.705	0.620	0.494
1,2-Dimethylbenzene*)	363.15	0.973	0.943	0.910	0.871	0.826	0.772	0.705	0.616	0.478
Ethylbenzene*)	343.15	0.978	0.954	0.927	0.895	0.857	0.811	0.753	0.672	0.542
Propylbenzene*)	373.15	0.977	0.951	0.923	0.890	0.852	0.808	0.752	0.678	0.563
Poly(4-vinylpyridine)										
Methanol	343.15	0.986	0.971	0.953	0.931	0.905	0.871	0.825	0.756	0.627
2-Propanol	343.15	0.989	0.977	0.964	0.948	0.928	0.904	0.872	0.826	0.743
Poly(1-vinyl-2-pyrrolidinone)										
Water	368.15	0.984	0.966	0.946	0.924	0.899	0.870	0.835	0.790	0.727
Starch (amorphous)										
Water	383.15	0.991	0.981	0.970	0.956	0.939	0.918	0.889	0.845	0.754
Styrene/Methyl methacrylate Copolymer (41.45 wt% Styrene)										
Benzene*)	308.15	0.982	0.963	0.940	0.913	0.881	0.841	0.789	0.716	0.590
Vinyl acetate/Vinyl chloride Copolymer (12 wt% Vinyl acetate)										
Benzene	398.15	0.976	0.949	0.918	0.883	0.841	0.791	0.728	0.643	0.509
Chlorobenzene	398.15	0.984	0.965	0.944	0.920	0.891	0.856	0.810	0.746	0.638
1,4-Dimethylbenzene	398.15	0.989	0.977	0.963	0.946	0.926	0.899	0.863	0.807	0.692
Ethylbenzene	398.15	0.989	0.976	0.961	0.944	0.924	0.899	0.866	0.818	0.735
Octane	398.15	0.992	0.982	0.971	0.958	0.942	0.922	0.893	0.847	0.739

*) $a_A = P_A / P_A^S$

SPECIFIC ENTHALPIES OF SOLUTION OF POLYMERS AND COPOLYMERS

Christian Wohlfarth

Enthalpies of solution or mixing, expressed as the enthalpy change per unit mass of polymer, are given in the table at infinite dilution, i.e., a very small amount of polymer and a large excess of solvent were mixed isothermally to form a homogeneous solution. By thermodynamics, $\Delta_{sol} H_B^\infty$ or $\Delta_M H_B^\infty$ are obtained from the following derivatives:

$$\Delta_{sol} H_B^\infty = \lim_{m_B \to 0} \left(\partial \Delta_{sol} h / \partial m_B \right)_{P,T,m_{j \neq B}} \qquad (1)$$

$$\Delta_M H_B^\infty = \lim_{m_B \to 0} \left(\partial \Delta_M h / \partial m_B \right)_{P,T,m_{j \neq B}} \qquad (2)$$

with a unit of J/g. Thus, they are the partial specific enthalpies of solution or mixing of the polymer B at infinite dilution where $\Delta_{sol} h$ or $\Delta_M h$ is the extensive enthalpy of the solution or mixing process.

The state of the polymer before dissolution can significantly affect the enthalpy of solution. The dissolving of a semicrystalline polymer requires an additional amount of heat associated with the disordering of crystalline regions. Consequently, its enthalpy of solution is usually positive and depends on the degree of crystallinity of the given polymer sample. An amorphous polymer below its glass transition temperature, T_g (see the T_g-table of this Section), often dissolves with the release of heat. The enthalpy of solution of a glassy polymer is additionally dependent to some extent on the thermal history of the glass-forming process. An amorphous polymer above T_g can show endothermic or exothermic dissolution behavior depending on the nature of the solvent and the interaction energies involved as is the case for any enthalpy of mixing. This enthalpy of mixing is then independent of any crystalline or glassy aspects of the polymer. It can be obtained without difficulties for liquid/molten polymers mixed with a solvent. Therefore, the enthalpies given in the table are either enthalpies of solution or enthalpies of mixing, depending on the state of the polymer.

The enthalpies depend on temperature and molar mass. The necessary molar mass information for a system is provided in the table (if available) by the corresponding number average, M_n, mass average, M_w, or viscosity average, M_η, values of the polymer as given in the original sources. Outside the oligomer range, specific enthalpies of solution or mixing do not remarkably depend on molar mass, however. More enthalpy data of polymer-solvent systems can be found in Ref. 106.

Polymer	M_n/ g/mol	M_w/ g/mol	M_η/ g/mol	Solvent	T/K	ΔH_B^∞/ J/g	Ref.
Acrylonitrile/butadiene copolymer							
(18 wt% Acrylonitrile)				Benzene	298.15	0.0	18
(26 wt% Acrylonitrile)				Benzene	298.15	−1.9	18
(40 wt% Acrylonitrile)				Benzene	298.15	−2.9	18
Acrylonitrile/isoprene copolymer							
(15 mol% Isoprene)				N,N-Dimethylformamide	323.15	−32	66
Acrylonitrile/vinyl chloride copolymer							
(13 wt% Acrylonitrile)				N,N-Dimethylformamide	293.15	−38	35
(13 wt% Acrylonitrile)				N,N-Dimethylformamide	308.15	−22	35
(13 wt% Acrylonitrile)				N,N-Dimethylformamide	323.15	−18	35
(13 wt% Acrylonitrile)				N,N-Dimethylformamide	338.15	−15	35
(13 wt% Acrylonitrile)				N,N-Dimethylformamide	353.15	−12	35
(29 wt% Acrylonitrile)				N,N-Dimethylformamide	295.15	−42	35
(29 wt% Acrylonitrile)				N,N-Dimethylformamide	308.15	−27	35
(29 wt% Acrylonitrile)				N,N-Dimethylformamide	323.15	−21	35
(29 wt% Acrylonitrile)				N,N-Dimethylformamide	338.15	−19	35
(29 wt% Acrylonitrile)				N,N-Dimethylformamide	353.15	−16	35
(40 wt% Acrylonitrile)				N,N-Dimethylformamide	295.15	−47	35
(40 wt% Acrylonitrile)				N,N-Dimethylformamide	308.15	−30	35
(40 wt% Acrylonitrile)				N,N-Dimethylformamide	323.15	−28	35
(40 wt% Acrylonitrile)				N,N-Dimethylformamide	338.15	−18	35
(40 wt% Acrylonitrile)				N,N-Dimethylformamide	353.15	−17	35
Benzylcellulose							
				Benzene	298.15	−11	8
				Cyclohexanone	298.15	−15	25
				Trichloromethane	298.15	−38	25

Polymer	$M_n/$ g/mol	$M_w/$ g/mol	$M_\eta/$ g/mol	Solvent	T/K	$\Delta H_B^\infty/$ J/g	Ref.
Bisphenol A-isophthaloyl chloride/terephthaloyl chloride							
(50/50 Iso/terephtaloyl chloride)				N,N-Dimethylacetamide	298.15	−56	69
(50/50 Iso/terephtaloyl chloride)				1,1,2,2-Tetrachloroethane	298.15	+72	69
Butadiene/styrene copolymer							
(10 wt% Styrene)				Benzene	293.65	4.9	17
(30 wt% Styrene)				Benzene	293.65	3.0	17
(30 wt% Styrene)				Benzene	298.15	3.0	25
(50 wt% Styrene)				Benzene	293.65	1.8	17
(60 wt% Styrene)				Benzene	293.65	0.0	17
(70 wt% Styrene)				Benzene	293.65	0.0	17
(75 wt% Styrene)				Benzene	298.15	1.5	7
(80 wt% Styrene)				Benzene	293.65	−0.6	17
(90 wt% Styrene)				Benzene	293.65	−4.9	17
Butyl methacrylate/isobutyl methacrylate copolymer (50 wt%/50 wt%)							
Glass		150000		Cyclohexanone	303.15	5.9	98
Liquid		150000		Cyclohexanone	303.15	14.0	98
Butyl methacrylate/methyl methacrylate copolymer (45 wt%/55 wt%)							
Glass	107000	250000		Cyclohexanone	304.15	−5.4	99
Liquid	107000	250000		Cyclohexanone	304.15	+9.1	99
Cellulose acetate							
(52.2 wt% Acetate)				Formic acid	298.15	−30	10
(55.8 wt% Acetate)				Formic acid	298.15	−44	10
(52.5 wt% Acetate)				Methyl acetate	298.15	−80	1
(48 wt% Acetate)				2-Propanone	298.15	−35	25
(52.2 wt% Acetate)				2-Propanone	298.15	−30	10
(55.8 wt% Acetate)				2-Propanone	298.15	−26	10
(56 wt% Acetate)				2-Propanone	298.15	−45	25
(56 wt% Acetate)				2-Propanone	298.15	−30	4
Cellulose triacetate							
				2-Propanone	298.15	−29	4
				Trichloromethane	298.15	−47	4
Dextran							
	8200	10400		Dimethylsulfoxide	298.15	−185	75
	75900	101000		Dimethylsulfoxide	298.15	−187	70
	75900	101000		1,2-Ethanediol	298.15	−98	70
	75900	101000		Formamide	298.15	−228	70
	8200	10400		Water	298.15	−140	75
	75900	101000		Water	298.15	−150	75
(amorph)				Water	298.15	−123	65
Ethylene/propylene copolymer							
(33 mol% Ethylene)				Cyclohexane	298.15	1.4	74
(63 mol% Ethylene)				Cyclohexane	298.15	8.1	74
(75 mol% Ethylene)				Cyclohexane	298.15	11.8	74
(33 mol% Ethylene)				Cyclooctane	298.15	1.2	74

Polymer	$M_n/$ g/mol	$M_w/$ g/mol	$M_\eta/$ g/mol	Solvent	T/K	$\Delta H_B^\infty/$ J/g	Ref.
(63 mol% Ethylene)				Cyclooctane	298.15	6.9	74
(75 mol% Ethylene)				Cyclooctane	298.15	8.6	74
(33 mol% Ethylene)				Cyclopentane	298.15	−3.5	74
(63 mol% Ethylene)				Cyclopentane	298.15	1.1	74
(33 mol% Ethylene)				cis-Decahydronaphthalene	298.15	−2.4	74
(63 mol% Ethylene)				cis-Decahydronaphthalene	298.15	2.4	74
(75 mol% Ethylene)				cis-Decahydronaphthalene	298.15	3.9	74
(33 mol% Ethylene)				trans-Decahydronaphthalene	298.15	−4.8	74
(63 mol% Ethylene)				trans-Decahydronaphthalene	298.15	−1.3	74
(75 mol% Ethylene)				trans-Decahydronaphthalene	298.15	−0.3	74
(63 mol% Ethylene)				3,3-Diethylpentane	298.15	−1.4	74
(75 mol% Ethylene)				3,3-Diethylpentane	298.15	<0.1	74
(63 mol% Ethylene)				2,2-Dimethylpentane	298.15	5.3	74
(75 mol% Ethylene)				2,2-Dimethylpentane	298.15	2.3	74
(63 mol% Ethylene)				2,3-Dimethylpentane	298.15	0.7	74
(75 mol% Ethylene)				2,3-Dimethylpentane	298.15	0.4	74
(33 mol% Ethylene)				2,4-Dimethylpentane	298.15	−1.2	74
(63 mol% Ethylene)				2,4-Dimethylpentane	298.15	3.0	74
(75 mol% Ethylene)				2,4-Dimethylpentane	298.15	0.2	74
(33 mol% Ethylene)				3,3-Dimethylpentane	298.15	−2.7	74
(63 mol% Ethylene)				3,3-Dimethylpentane	298.15	0.3	74
(33 mol% Ethylene)				Dodecane	298.15	−0.1	73
(63 mol% Ethylene)				Dodecane	298.15	0.8	73
(75 mol% Ethylene)				Dodecane	298.15	−4.0	73
(63 mol% Ethylene)				3-Ethylpentane	298.15	2.6	74
(75 mol% Ethylene)				3-Ethylpentane	298.15	−0.6	74
(33 mol% Ethylene)				2,2,4,4,6,8,8-Heptamethylnonane	298.15	−0.5	73
(63 mol% Ethylene)				2,2,4,4,6,8,8-Heptamethylnonane	298.15	2.2	73
(75 mol% Ethylene)				2,2,4,4,6,8,8-Heptamethylnonane	298.15	−0.9	73
(33 mol% Ethylene)				Hexadecane	298.15	0.7	73
(63 mol% Ethylene)				Hexadecane	298.15	−1.1	73
(75 mol% Ethylene)				Hexadecane	298.15	−4.6	73
(63 mol% Ethylene)				3-Methylhexane	298.15	0.7	74
(75 mol% Ethylene)				3-Methylhexane	298.15	1.7	74
(33 mol% Ethylene)				Octane	298.15	−1.6	73
(63 mol% Ethylene)				Octane	298.15	3.6	73
(75 mol% Ethylene)				Octane	298.15	0.3	73
(33 mol% Ethylene)				2,2,4,6,6-Pentamethylheptane	298.15	−0.3	73
(63 mol% Ethylene)				2,2,4,6,6-Pentamethylheptane	298.15	3.6	73
(75 mol% Ethylene)				2,2,4,6,6-Pentamethylheptane	298.15	0.0	73
(63 mol% Ethylene)				2,2,4,4-Tetramethylpentane	298.15	2.7	74
(75 mol% Ethylene)				2,2,4,4-Tetramethylpentane	298.15	3.1	74
(33 mol% Ethylene)				2,2,4-Trimethylpentane	298.15	−0.2	73
(63 mol% Ethylene)				2,2,4-Trimethylpentane	298.15	1.9	73
(75 mol% Ethylene)				2,2,4-Trimethylpentane	298.15	3.5	73

Polymer	$M_n/$ g/mol	$M_w/$ g/mol	$M_\eta/$ g/mol	Solvent	T/K	$\Delta H_B^\infty/$ J/g	Ref.
Ethylene/vinylacetate copolymer							
(85 wt% Vinyl acetate)				Cyclopentanone	298.15	−0.5	104
(70 wt% Vinyl acetate)		220000		Tetrahydrofuran	304.65	−1.3	93
Gelatine							
				Water	293.15	−92	29
				Water	323.15	−63	29
Guttapercha							
				Trichloromethane	303.15	47	22
Isobutyl methacrylate/methyl methacrylate copolymer (51 wt%/49 wt%)							
Glass	150000			Cyclohexanone	303.15	−11	98
Liquid	150000			Cyclohexanone	303.15	15	98
Natural rubber							
				Benzene	298.15	10	25
				Benzene	298.15	12	20
Nitrocellulose							
			16600	2-Butanone	298.15	−80	4
			23000	2-Butanone	298.15	−81	4
			40000	2-Butanone	298.15	−81	4
				Butyl acetate	293.15	−75	23
				Butyl acetate	298.15	−75	23
				Butyl acetate	298.15	−73	26
				Butyl acetate	303.15	−75	23
				Butyl acetate	308.15	−71	23
				Butyl acetate	313.15	−65	23
				Butyl acetate	313.15	−67	26
				Butyl acetate	318.15	−59	23
				Butyl acetate	323.15	−54	23
				Butyl acetate	328.15	−50	23
				Butyl acetate	333.15	−59	26
				Butyl acetate	343.15	−55	26
				Butyl acetate	353.15	−47	26
				Diethyl ether	295.15	−62	3
				Dibutyl phthalate	273.15	−45	26
				Dibutyl phthalate	298.15	−46	26
				Dibutyl phthalate	313.15	−46	26
				Dibutyl phthalate	333.15	−42	26
				Ethanol	295.15	−46	3
				Ethyl acetate	293.15	−76	23
				Ethyl acetate	298.15	−75	23
				Ethyl acetate	303.15	−69	23
				Ethyl acetate	308.15	−61	23
				Ethyl acetate	313.15	−54	23
				Ethyl acetate	318.15	−50	23
				Ethyl acetate	323.15	−50	23
				Ethyl acetate	328.15	−50	23
				Methanol	293.15	−69	23

Polymer	$M_n/$ g/mol	$M_w/$ g/mol	$M_\eta/$ g/mol	Solvent	T/K	$\Delta H_B^\infty/$ J/g	Ref.
				Methanol	298.15	−56	23
				Methanol	303.15	−50	23
				Methanol	308.15	−50	23
				Methanol	313.15	−50	23
				Methanol	318.15	−50	23
				Methanol	323.15	−50	23
				Methanol	328.15	−50	23
				2,4-Pentanedione	298.15	−74	4
				2-Pentanone	298.15	−64	4
				2-Propanone	273.15	−75	26
				2-Propanone	293.15	−75	23
				2-Propanone	298.15	−83	2
				2-Propanone	298.15	−68	4
				2-Propanone	298.15	−71	8
				2-Propanone	298.15	−74	23
				2-Propanone	298.15	−79	25
				2-Propanone	298.15	−75	26
				2-Propanone	303.15	−60	23
				2-Propanone	308.15	−51	23
				2-Propanone	313.15	−50	23
				2-Propanone	313.15	−65	26
				2-Propanone	318.15	−50	23
				2-Propanone	323.15	−50	23
				2-Propanone	323.15	−50	26
				2-Propanone	328.15	−50	23
				Pyridine	298.15	−106	2
				Tri(4-methylphenyl) phosphate	298.15	−16	26
				Tri(4-methylphenyl) phosphate	313.15	−28	26
				Tri(4-methylphenyl) phosphate	333.15	−41	26
				Tri(4-methylphenyl) phosphate	343.15	−44	26
				Tri(4-methylphenyl) phosphate	353.15	−47	26
Nylon-6 (unoriented)							
				Formic acid	295.15	−53	24
				Tricresol	323.55	−66	22
				Tricresol	345.55	−66	22
Poly(acrylonitrile)							
				Benzene	298.15	0.0	18
				N,N-Dimethylformamide	295.15	−23	35
				N,N-Dimethylformamide	298.15	−21	18
				N,N-Dimethylformamide	298.15	−43	42
				N,N-Dimethylformamide	308.15	−17	35
				N,N-Dimethylformamide	323.15	−13	35

Polymer	$M_n/$ g/mol	$M_w/$ g/mol	$M_\eta/$ g/mol	Solvent	T/K	$\Delta H_B^\infty/$ J/g	Ref.
				N,N-Dimethylformamide	323.15	−15	66
				N,N-Dimethylformamide	338.15	−10	35
				Dimethylsulfoxide	298.15	−70	42
Poly(γ-benzyl-ʟ-glutamate)							
		160000		Dichloroacetic acid	303.15	−35	46
		160000		1,2-Dichloroethane	303.15	−1.6	46
Polybutadiene							
				Benzene	298.15	6.1	7
				Benzene	298.15	7.1	25
				Benzene	298.15	10.5	32
				2,2,4-Trimethylpentane	298.15	1.1	32
1,4-cis-Polybutadiene							
		low		Cyclohexane	298.15	5.4	74
		low		Cyclooctane	298.15	5.8	74
		low		Cyclopentane	298.15	<0.1	74
		low		cis-Decahydronaphthalene	298.15	4.2	74
		low		trans-Decahydronaphthalene	298.15	2.6	74
		low		3,3-Diethylpentane	298.15	5.2	74
		low		2,2-Dimethylpentane	298.15	4.1	74
		low		2,3-Dimethylpentane	298.15	4.5	74
		low		2,4-Dimethylpentane	298.15	3.2	74
		low		3,3-Dimethylpentane	298.15	3.2	74
		low		Dodecane	298.15	4.2	74
		low		3-Ethylpentane	298.15	3.7	74
		low		2,2,4,4,6,8,8-Heptamethylnonane	298.15	4.8	74
		low		Hexadecane	298.15	4.9	74
		low		3-Methylhexane	298.15	3.6	74
		low		Octane	298.15	4.3	74
		low		2,2,4,6,6-Pentamethylheptane	298.15	5.0	74
		low		2,2,4,4-Tetramethylpentane	298.15	5.8	74
		low		2,3,3,4-Tetramethylpentane	298.15	5.1	74
Poly(1-butene)							
			20000	Cyclohexane	298.15	1.0	74
			20000	Cyclooctane	298.15	1.8	74
			20000	Cyclopentane	298.15	−2.9	74
			20000	cis-Decahydronaphthalene	298.15	<0.1	74
			20000	trans-Decahydronaphthalene	298.15	−2.0	74
			20000	Decane	298.15	1.2	62
			20000	3,3-Diethylpentane	298.15	−2.6	74
			20000	2,2-Dimethylpentane	298.15	−4.0	74
			20000	2,3-Dimethylpentane	298.15	−2.8	74
			20000	2,4-Dimethylpentane	298.15	−2.3	74
			20000	3,3-Dimethylpentane	298.15	−2.2	74
			20000	Dodecane	298.15	2.1	74

Polymer	$M_n/$ g/mol	$M_w/$ g/mol	$M_\eta/$ g/mol	Solvent	T/K	$\Delta H_B^\infty/$ J/g	Ref.
			20000	3-Ethylpentane	298.15	−2.8	74
			20000	Heptane	298.15	0.0	73
			20000	Hexadecane	298.15	3.0	62
			20000	Hexane	298.15	−1.2	62
			20000	3-Methylhexane	298.15	−2.1	74
			20000	Nonane	298.15	0.9	73
			20000	Octane	298.15	0.4	73
			20000	2,2,4,6,6-Pentamethylheptane	298.15	0.6	73
			20000	Pentane	298.15	−2.6	62
			20000	Tetradecane	298.15	2.7	62
			20000	2,2,4,4-Tetramethylpentane	298.15	−1.4	74
			20000	2,3,3,4-Tetramethylpentane	298.15	−2.2	74
			20000	2,2,4-Trimethylpentane	298.15	−0.5	73
Poly(butyl acrylate)							
				2-Propanone	298.15	0.8	25
Poly(butyl methacrylate)							
Glass	91300	210000		Cyclohexanone	304.15	7.7	99
Liquid	91300	210000		Cyclohexanone	304.15	8.2	99
				2-Propanone	298.15	19.5	25
Polychloroprene							
				Benzene	298.15	0.5	7
Poly(2,6-dimethyl phenylene oxide)							
	17000	46400		1,2-Dichlorobenzene	303.05	55	89
Poly(dimethylsiloxane)							
	13000			Benzene	298.15	11.2	50
			20000	Benzene	298.15	13.5	61
			100000	Benzene	298.15	14.2	40
			170000	Bromocyclohexane	303.15	10.2	51
			80000	2-Butanone	293.15	14.4	77
	30900			2-Butanone	303.15	14.2	44
			170000	2-Butanone	303.15	14.7	51
			80000	2-Butanone	308.15	14.3	77
			80000	2-Butanone	323.15	14.3	77
			80000	Butyl acetate	298.15	6.1	41
			80000	Butyl propanoate	298.15	4.9	41
			100000	Chlorobenzene	298.15	7.5	40
	13000			Cyclohexane	298.15	3.0	50
			20000	Cyclohexane	298.15	5.2	74
			100000	Cyclohexane	298.15	5.2	40
			20000	Cyclooctane	298.15	6.8	74
			20000	Cyclopentane	298.15	1.0	74
			20000	cis-Decahydronapthalene	298.15	7.1	74
			20000	trans-Decahydronapthalene	298.15	4.3	74
			20000	Decamethyltetrasiloxane	297.65	0.45	37
			20000	Decane	298.15	3.8	37
			20000	Decane	298.15	3.9	61

Polymer	$M_n/$ g/mol	$M_w/$ g/mol	$M_\eta/$ g/mol	Solvent	T/K	$\Delta H_B^\infty/$ J/g	Ref.
			80000	Decane	298.15	3.8	41
			80000	Decyl acetate	298.15	4.5	41
			80000	Dibutyl ether	298.15	0.6	41
			80000	Diethoxymethane	298.15	1.8	41
			80000	Diethyl ether	298.15	−1.3	41
			20000	3,3-Diethylpentane	298.15	1.9	74
			80000	Dihexyl ether	298.15	3.0	41
			80000	1,2-Dimethoxyethane	298.15	12.2	41
			80000	Dimethoxymethane	298.15	7.4	41
		13000		1,2-Dimethylbenzene	298.15	4.3	50
		13000		1,3-Dimethylbenzene	298.15	3.0	50
		13000		1,4-Dimethylbenzene	298.15	3.2	50
			20000	1,4-Dimethylbenzene	298.15	4.2	61
			80000	2,6-Dimethyl-4-heptanone	293.15	6.1	77
			20000	2,2-Dimethylpentane	298.15	0.8	74
			20000	2,3-Dimethylpentane	298.15	1.4	74
			20000	2,4-Dimethylpentane	298.15	1.6	74
			20000	3,3-Dimethylpentane	298.15	0.5	74
			80000	Dipentyl ether	298.15	2.1	41
			80000	Dipropyl ether	298.15	−1.2	41
			20000	Dodecamethylpentasiloxane	297.65	−0.3	37
			20000	Dodecane	297.65	4.5	37
			20000	Dodecane	298.15	4.4	73
			80000	Dodecane	298.15	4.5	41
			80000	Ethyl acetate	298.15	12.7	41
			170000	Ethyl acetate	303.15	13.7	51
		13000		Ethylbenzene	298.15	6.4	50
			20000	Ethylbenzene	298.15	6.2	61
			80000	Ethyl butanoate	298.15	6.0	41
			80000	Ethyl decanoate	298.15	3.8	41
			80000	Ethyl dodecanoate	298.15	3.8	41
			80000	Ethyl heptanoate	298.15	4.1	41
			80000	Ethyl hexanoate	298.15	4.3	41
			80000	Ethyl nonanoate	298.15	3.7	41
			80000	Ethyl octanoate	298.15	3.8	41
			20000	3-Ethylpentane	298.15	0.6	74
			80000	Ethyl propanoate	298.15	8.0	41
			20000	2,2,4,4,6,8,8-Heptamethylnonane	298.15	3.5	74
		13000		Heptane	298.15	1.8	50
			20000	Heptane	298.15	1.9	73
			20000	Heptane	297.65	2.0	37
			20000	Heptane	298.15	2.0	61
			80000	Heptane	298.15	2.0	41
			100000	Heptane	298.15	2.1	40
			170000	Heptane	303.15	1.9	51
			80000	3-Heptanone	308.15	8.8	77
			80000	3-Heptanone	323.15	8.8	77

Polymer	$M_n/$ g/mol	$M_w/$ g/mol	$M_\eta/$ g/mol	Solvent	T/K	$\Delta H_B^\infty/$ J/g	Ref.
			20000	Hexadecane	297.65	5.5	37
			20000	Hexadecane	298.15	5.5	73
			20000	Hexamethyldisiloxane	298.15	−1.2	37
			20000	Hexamethyldisiloxane	298.15	−1.6	61
			170000	Hexamethyldisiloxane	303.15	−1.5	51
			20000	Hexane	297.65	0.7	37
			20000	Hexane	298.15	0.7	61
			80000	Hexane	298.15	0.7	41
			170000	Hexane	303.15	0.3	51
			80000	Hexyl acetate	298.15	5.0	41
		13000		Isopropylbenzene	298.15	4.1	50
			80000	Methyl butanoate	298.15	8.6	41
		13000		Methylcyclohexane	298.15	2.9	50
			100000	Methylcyclohexane	298.15	1.9	40
			80000	Methyl decanoate	298.15	4.8	41
			20000	3-Methylhexane	298.15	1.3	74
			80000	Methyl hexanoate	298.15	5.3	41
			80000	Methyl octanoate	298.15	5.0	41
			80000	4-Methyl-2-pentanone	293.15	9.9	77
			80000	4-Methyl-2-pentanone	308.15	9.0	77
			80000	Methyl propanoate	298.15	12.1	41
			20000	Nonane	297.65	3.4	37
			20000	Nonane	298.15	3.3	73
			80000	Nonane	298.15	3.4	41
			80000	Octamethylcyclotetrasiloxane	293.15	−0.4	78
			20000	Octamethyltrisiloxane	297.65	−0.6	37
			20000	Octamethyltrisiloxane	298.15	−0.8	61
			170000	Octamethyltrisiloxane	303.15	−1.0	51
			20000	Octane	297.65	2.6	37
			20000	Octane	298.15	2.4	73
			20000	Octane	298.15	2.6	61
			80000	Octane	298.15	2.6	41
			20000	2,2,4,6,6-Pentamethylheptane	298.15	2.7	73
			20000	Pentane	298.15	−0.9	37
			20000	Pentane	298.15	−0.9	61
			80000	Pentane	298.15	−1.0	41
			80000	Pentyl acetate	298.15	5.8	41
			80000	Pentyl propanoate	298.15	3.8	41
			80000	Propyl acetate	298.15	8.6	41
			170000	Propyl acetate	303.15	9.9	51
			80000	Propyl propanoate	298.15	5.6	41
			100000	Tetrachloromethane	298.15	2.4	40
			20000	Tetradecane	297.65	5.1	37
			20000	Tetradecane	298.15	5.1	61
			80000	Tetradecane	298.15	5.1	41
			20000	2,2,4,4-Tetramethylpentane	298.15	2.1	73
			20000	2,2,4,4-Tetramethylpentane	298.15	2.3	74

Polymer	$M_n/$ g/mol	$M_w/$ g/mol	$M_\eta/$ g/mol	Solvent	T/K	$\Delta H_B^\infty/$ J/g	Ref.
			20000	2,3,3,4-Tetramethylpentane	298.15	1.9	74
	13000			Toluene	298.15	5.5	50
			20000	Toluene	298.15	6.7	61
			20000	Tridecane	297.65	4.8	37
			80000	Tridecane	298.15	4.8	41
	13000			1,3,5-Trimethylbenzene	298.15	3.7	50
			20000	2,2,4-Trimethylpentane	298.15	1.4	73
			20000	Undecane	297.65	4.2	37
			80000	Undecane	298.15	4.3	41
Polyethylene							
Semicrystalline		65000		1-Chloronaphthalene	373.15	780	47
Semicrystalline		65000		1-Chloronaphthalene	383.15	980	47
Semicrystalline		65000		1-Chloronaphthalene	393.15	800	47
Liquid		65000		1-Chloronaphthalene	403.15	49	47
Semicrystalline		144000		1-Chloronaphthalene	383.15	920	47
Semicrystalline		144000		1-Chloronaphthalene	393.15	990	47
Semicrystalline		144000		1-Chloronaphthalene	403.15	690	47
Liquid		144000		1-Chloronaphthalene	413.15	67	47
Liquid		144000		1-Chloronaphthalene	423.15	85	47
Semicrystalline		670000		1-Chloronaphthalene	363.15	380	47
Semicrystalline		670000		1-Chloronaphthalene	373.15	430	47
Semicrystalline		670000		1-Chloronaphthalene	383.15	165	47
Liquid		670000		1-Chloronaphthalene	393.15	39	47
Liquid		670000		1-Chloronaphthalene	403.15	36	47
Semicrystalline		900000		1-Chloronaphthalene	391.80	245	105
Semicrystalline		900000		Cyclohexane	379.50	205	105
Semicrystalline		900000		Cyclopentane	380.00	190	105
Alkathene				Decahydronaphthalene	349.85	142	56
Rigidex-3				Decahydronaphthalene	366.65	180	56
Rigidex-50				Decahydronaphthalene	374.05	233	56
Semicrystalline		900000		Decahydronaphthalene	384.00	260	105
Semicrystalline				1,2-Dichloroethane	333.15	30	27
Semicrystalline				1,2-Dichloroethane	338.15	38	27
Semicrystalline				1,2-Dichloroethane	343.15	54	27
Semicrystalline				1,2-Dichloroethane	348.15	65	27
Semicrystalline			10000	1,4-Dimethylbenzene	354.15	139	5
Semicrystalline			11800	1,4-Dimethylbenzene	352.15	139	5
Semicrystalline			15600	1,4-Dimethylbenzene	353.65	154	5
Semicrystalline			15600	1,4-Dimethylbenzene	363.65	113	5
Semicrystalline			15600	1,4-Dimethylbenzene	368.65	104	5
Semicrystalline		900000		2,4-Dimethylpentane	393.00	230	105
Semicrystalline		900000		2,2,4,4,6,8,8-Heptamethylnonane	399.50	170	105
Semicrystalline		900000		Hexadecane	399.50	262	105
Semicrystalline		900000		2-Methylbutane	394.20	165	105
Semicrystalline		65000		1,2,3,4-Tetrahydronaphthalene	373.15	940	47

Polymer	$M_n/$ g/mol	$M_w/$ g/mol	$M_\eta/$ g/mol	Solvent	T/K	$\Delta H_B^\infty/$ J/g	Ref.
Semicrystalline		65000		1,2,3,4-Tetrahydronaphthalene	383.15	990	47
Semicrystalline		65000		1,2,3,4-Tetrahydronaphthalene	393.15	790	47
Liquid		65000		1,2,3,4-Tetrahydronaphthalene	403.15	58	47
Semicrystalline		84000		1,2,3,4-Tetrahydronaphthalene	373.15	835	47
Semicrystalline		130000		1,2,3,4-Tetrahydronaphthalene	353.15	630	47
Semicrystalline		130000		1,2,3,4-Tetrahydronaphthalene	373.15	520	47
Liquid		130000		1,2,3,4-Tetrahydronaphthalene	393.15	69	47
Semicrystalline		144000		1,2,3,4-Tetrahydronaphthalene	373.15	610	47
Semicrystalline		144000		1,2,3,4-Tetrahydronaphthalene	383.15	1200	47
Semicrystalline		144000		1,2,3,4-Tetrahydronaphthalene	393.15	1130	47
Semicrystalline		144000		1,2,3,4-Tetrahydronaphthalene	403.15	800	47
Liquid		144000		1,2,3,4-Tetrahydronaphthalene	413.15	136	47
Liquid		144000		1,2,3,4-Tetrahydronaphthalene	423.15	88	47
Semicrystalline		310000		1,2,3,4-Tetrahydronaphthalene	343.15	485	47
Semicrystalline		670000		1,2,3,4-Tetrahydronaphthalene	353.15	560	47
Semicrystalline		670000		1,2,3,4-Tetrahydronaphthalene	363.15	560	47
Semicrystalline		670000		1,2,3,4-Tetrahydronaphthalene	373.15	460	47
Semicrystalline		670000		1,2,3,4-Tetrahydronaphthalene	383.15	155	47
Liquid		670000		1,2,3,4-Tetrahydronaphthalene	393.15	67	47
Liquid		670000		1,2,3,4-Tetrahydronaphthalene	403.15	39	47
Semicrystalline			16000	Toluene	353.15	110	22
Semicrystalline			22000	Toluene	358.35	118	22
Semicrystalline			22000	Toluene	367.35	106	22
Semicrystalline		900000		1,2,4-Trichlorobenzene	386.50	255	105
Poly(ethylene glycol)							
	180			Benzene	303.15	110	71
	385			Benzene	303.15	60	71
	560			Benzene	303.15	40	71
	1050			Benzene	303.15	90	71
	1610			Benzene	303.15	140	71
	1940			Benzene	303.15	215	71
	3200			Benzene	303.15	195	71
	4330			Benzene	303.15	195	71

Polymer	M_n/ g/mol	M_w/ g/mol	M_η/ g/mol	Solvent	T/K	ΔH_B^∞/ J/g	Ref.
	5850			Benzene	303.15	190	71
	9950			Benzene	303.15	195	71
			43400	Benzene	303.15	190	71
	400	420		Trichloromethane	303.15	−79	95
	590	615		Trichloromethane	303.15	−88	95
	180			Water	303.15	−136	71
	200			Water	321.35	−125	83
	355			Water	303.15	−159	71
	400			Water	321.35	−150	83
	560			Water	303.15	−150	71
	990			Water	321.35	−101	83
	1050			Water	303.15	−106	71
	1460			Water	321.35	−137	83
	1610			Water	303.15	−6	71
	1940			Water	303.15	57	71
	3200			Water	303.15	58	71
	4330			Water	303.15	28	71
	5850			Water	303.15	39	71
	9950			Water	303.15	30	71
			14000	Water	303.15	7	55
			14000	Water	313.15	27	55
			20300	Water	303.15	45	71
			34500	Water	303.15	34	71
			43300	Water	303.15	+40	71
Poly(ethylene glycol) dimethyl ether							
	250			Tetrachloromethane	303.15	−12	95
	250			Tetrachloromethane	318.15	−7.6	95
	400			Tetrachloromethane	303.15	−12	95
	520	550		Tetrachloromethane	303.15	−12	95
	520	550		Tetrachloromethane	303.15	−7.6	95
	250			Trichloromethane	303.15	−184	95
	520	550		Trichloromethane	303.15	−135	95
Poly(ethylene glycol) monododecyl ether							
	230			Dodecane	302.15	42	68
	274			Dodecane	302.15	23	68
	318			Dodecane	302.15	34	68
	362			Dodecane	302.15	37	68
	406			Dodecane	302.15	42	68
Poly(ethylene glycol) monomethyl ether							
	353	377		Trichloromethane	303.15	−125	95
	550	580		Trichloromethane	303.15	−117	95
Poly(ethylene oxide)							
Semicrystalline	6000			Dichloromethane	303.15	+84	58
Liquid	6000			Dichloromethane	303.15	−160	58

Polymer	M_n/ g/mol	M_w/ g/mol	M_η/ g/mol	Solvent	T/K	ΔH_B^∞/ J/g	Ref.
Semicrystalline	6000			Trichloromethane	303.15	+52	58
Liquid	6000			Trichloromethane	303.15	−186	58
Quenched	1520	1720		Water	293.15	−403	64
Annealed	1520	1720		Water	293.15	−392	64
Quenched	1520	1720		Water	298.15	−180	64
Annealed	1520	1720		Water	298.15	−150	64
Quenched	1520	1720		Water	303.15	+68	64
Annealed	1520	1720		Water	303.15	+109	64
Liquid	6000			Water	303.15	−50	58
Quenched	6840	7525		Water	293.15	−28	64
Annealed	6840	7525		Water	293.15	+209	64
Quenched	6840	7525		Water	298.15	+241	64
Annealed	6840	7525		Water	298.15	+540	64
Quenched	16600	19600		Water	293.15	−160	64
Annealed	16600	19600		Water	293.15	−143	64
Quenched	16600	19600		Water	298.15	+59	64
Annealed	16600	19600		Water	298.15	+155	64
Quenched	16600	19600		Water	303.15	+353	64
Annealed	16600	19600		Water	303.15	+490	64
Semicrystalline		20000		Water	298.15	+10	84
Polyindene							
	765	1023		Anisole	299.15	2.1	102
	765	1023		Benzene	299.15	−0.04	102
	765	1023		Benzonitrile	299.15	−4.4	102
	765	1023		Bromobenzene	299.15	−3.9	102
	765	1023		2-Butanone	299.15	1.9	102
	765	1023		Chlorobenzene	299.15	−3.9	102
	765	1023		1-Chlorobutane	299.15	−4.0	102
	765	1023		1-Chloroheptane	299.15	1.5	102
	765	1023		Cyclohexane	299.15	15	102
	765	1023		N,N-dimethylaniline	299.15	−8.2	102
	765	1023		Ethyl acetate	299.15	4.2	102
	765	1023		Ethylbenzene	299.15	−1.4	102
	765	1023		Ethyl benzoate	299.15	−0.7	102
	765	1023		Nitrobenzene	299.15	4.6	102
	765	1023		1-Nitropropane	299.15	8.4	102
	765	1023		Pyridine	299.15	−6.7	102
	765	1023		1,1,2,2-Tetrachloroethane	299.15	−19	102
	765	1023		Tetrachloromethane	299.15	−2.5	102
	765	1023		1,1,1-Trichloroethane	299.15	−1.8	102
	765	1023		Trichloromethane	299.15	−20	102
Polyisobutylene							
	360		700	Benzene	298.15	30	67
	1000		2000	Benzene	298.15	25	67
	1300		2500	Benzene	298.15	23	67
			30000	Benzene	297.65	19	38
			30000	Benzene	298.15	19	40

Polymer	$M_n/$ g/mol	$M_w/$ g/mol	$M_\eta/$ g/mol	Solvent	T/K	$\Delta H_B^\infty/$ J/g	Ref.
	44700			Benzene	303.15	16	44
			48000	Benzene	303.4	19	72
			50000	Benzene	303.15	16	45
			72000	Benzene	300.15	19	53
			72000	Benzene	323.15	18	53
			72000	Benzene	343.15	16	53
			72000	Benzene	375.15	13	53
			72000	Benzene	394.15	9.2	53
			72000	Benzene	423.15	3.5	53
			72000	Benzene	437.15	−0.5	53
			72000	Benzene	453.15	−4.7	53
			90000	Benzene	298.15	6.7	32
			160000	Benzene	303.15	16	51
			560000	Benzene	298.15	18	34
				Benzene	298.15	6.8	7
				Benzene	298.15	6.8	10
			30000	Chlorobenzene	297.65	12	38
			30000	Chlorobenzene	298.15	13	40
			50000	Chlorobenzene	303.15	12	45
			160000	Chlorobenzene	303.15	12	51
			560000	Chlorobenzene	298.15	12	34
	360		700	Cyclohexane	298.15	3.8	67
	1000		2000	Cyclohexane	298.15	1.2	67
	1300		2500	Cyclohexane	298.15	1.1	67
			4500	Cyclohexane	298.15	−0.6	74
			30000	Cyclohexane	297.65	−0.7	38
			30000	Cyclohexane	298.15	−0.6	40
			50000	Cyclohexane	303.15	−0.7	45
			160000	Cyclohexane	303.15	−0.6	51
			1990000	Cyclohexane	298.15	−0.7	39
			4500	Cyclooctane	298.15	+0.3	74
			4500	Cyclopentane	298.15	−5.9	74
			4500	cis-Decahydronaphthalene	298.15	0.2	74
			4500	trans-Decahydronaphthalene	298.15	−0.8	74
			30000	Decane	297.65	−0.5	38
			50000	Decane	303.15	−0.5	45
			30000	Dibutyl ether	297.65	1.2	37
			30000	Diethyl ether	297.65	2.8	37
			30000	Diethyl ether	297.65	2.8	38
			4500	3,3-Diethylpentane	298.15	−1.4	74
			30000	Dihexyl ether	297.65	0.9	37
			4500	2,2-Dimethylpentane	298.15	−1.1	74
			4500	2,3-Dimethylpentane	298.15	−1.9	74
			4500	2,4-Dimethylpentane	298.15	−1.1	74
			4500	3,3-Dimethylpentane	298.15	−1.7	74
			30000	Dipentyl ether	297.65	1.0	37
			30000	Dipropyl ether	297.65	1.8	37

Polymer	M_n/ g/mol	M_w/ g/mol	M_η/ g/mol	Solvent	T/K	ΔH_B^∞/ J/g	Ref.
	360		700	Dodecane	298.15	1.9	67
	1000		2000	Dodecane	298.15	0.7	67
	1300		2500	Dodecane	298.15	0.5	67
			4500	Dodecane	298.15	0.2	73
			30000	Dodecane	297.65	−0.1	38
			30000	Dodecane	298.15	−0.1	40
			48000	Ethylbenzene	291.15	9.5	72
			48000	Ethylbenzene	343.15	3.6	72
			30000	Ethyl decanoate	297.65	3.0	37
			30000	Ethyl heptanoate	297.65	5.6	37
			30000	Ethyl hexadecanoate	297.65	1.3	37
			30000	Ethyl hexanoate	297.65	6.7	37
			30000	Ethyl nonanoate	297.65	3.7	37
			30000	Ethyl octanoate	297.65	4.6	37
			4500	3-Ethylpentane	298.15	−2.0	74
			30000	Ethyl tetradecanoate	297.65	1.8	37
			4500	2,2,4,4,6,8,8-Heptamethylnonane	298.15	−0.5	74
	360		700	Heptane	298.15	−0.5	67
	1000		2000	Heptane	298.15	−1.0	67
	1300		2500	Heptane	298.15	−1.4	67
			4500	Heptane	298.15	−1.7	73
			30000	Heptane	297.65	−1.8	38
			30000	Heptane	298.15	−2.0	40
			50000	Heptane	303.15	−1.8	45
			160000	Heptane	303.15	−1.6	51
				Heptane	298.15	−1.4	7
				Heptane	298.15	−1.4	10
	360		700	Hexadecane	298.15	4.5	67
	1000		2000	Hexadecane	298.15	2.1	67
	1300		2500	Hexadecane	298.15	1.0	67
			4500	Hexadecane	298.15	0.9	73
			30000	Hexadecane	297.65	0.04	38
			30000	Hexane	297.65	−2.5	38
			30000	Hexane	298.15	−2.6	40
			50000	Hexane	303.15	−2.5	45
			72000	Hexane	303.15	−1.8	53
			72000	Hexane	324.15	−2.3	53
			72000	Hexane	348.15	−2.9	53
			72000	Hexane	373.15	−3.7	53
			72000	Hexane	393.15	−5.3	53
			72000	Hexane	408.15	−6.7	53
			72000	Hexane	423.15	−9.0	53
			72000	Hexane	433.15	−9.9	53
			160000	Hexane	303.15	−2.5	51
			30000	2-Methylbutane	297.65	−3.1	38
			30000	Methylcyclohexane	297.65	−1.2	38
			50000	Methylcyclohexane	303.15	−1.2	45

Polymer	$M_n/$ g/mol	$M_w/$ g/mol	$M_\eta/$ g/mol	Solvent	T/K	$\Delta H_B^\infty/$ J/g	Ref.
			160000	Methylcyclohexane	303.15	−1.2	51
			4500	3-Methylhexane	298.15	−1.0	74
			30000	3-Methylpentane	297.65	−2.8	38
			30000	Nonane	297.65	−0.8	38
			4500	Nonane	298.15	−0.8	73
			4500	Octane	298.15	−1.1	73
			30000	Octane	297.65	−1.2	38
			72000	Octane	303.15	−0.3	53
			72000	Octane	324.15	−0.8	53
			72000	Octane	348.15	−0.9	53
			72000	Octane	373.15	−1.1	53
			72000	Octane	393.15	−1.3	53
			72000	Octane	423.15	−3.6	53
			4500	2,2,4,6,6-Pentamethylheptane	298.15	−0.1	73
	360		700	Pentane	298.15	−1.9	67
	1000		2000	Pentane	298.15	−2.9	67
	1300		2500	Pentane	298.15	−3.2	67
			30000	Pentane	297.65	−3.6	38
			72000	Pentane	303.15	−2.8	53
			72000	Pentane	333.15	−3.4	53
			72000	Pentane	352.15	−4.5	53
			72000	Pentane	365.15	−5.5	53
	360		700	Tetrachloromethane	298.15	5.9	67
	1000		2000	Tetrachloromethane	298.15	5.8	67
	1300		2500	Tetrachloromethane	298.15	5.0	67
			1990000	Tetrachloromethane	298.15	4.1	39
			30000	Tetradecane	297.65	0.0	38
			4500	2,2,4,4-Tetramethylpentane	298.15	−0.6	74
			4500	2,3,3,4-Tetramethylpentane	298.15	−2.3	74
			50000	Toluene	303.15	7.4	45
			160000	Toluene	303.15	7.4	51
			1990000	Toluene	298.15	8.8	39
				Toluene	298.15	1.8	7
				Toluene	298.15	1.8	10
			30000	Tridecane	297.65	−0.04	38
			4500	2,2,4-Trimethylpentane	298.15	−0.4	73
			30000	2,2,4-Trimethylpentane	297.65	−0.6	38
			1990000	2,2,4-Trimethylpentane	298.15	0.0	39
				2,2,4-Trimethylpentane	298.15	0.0	7
				2,2,4-Trimethylpentane	298.15	0.0	10
			30000	Undecane	297.65	−0.4	38
Poly(isobutyl methacrylate)							
Glass		260000		Cyclohexanone	303.15	−5.2	98
Liquid		260000		Cyclohexanone	303.15	13	98
Poly(methyl acrylate)							
				2-Propanone	298.15	0.0	25

Polymer	M_n/ g/mol	M_w/ g/mol	M_η/ g/mol	Solvent	T/K	ΔH_B^∞/ J/g	Ref.
Poly(methyl methacrylate)							
Glass	73900	170000		Cyclohexanone	304.15	−14	98
Liquid	73900	170000		Cyclohexanone	304.15	17	98
	1930			1,2-Dichloroethane	298.15	−20	32
	240000			1,2-Dichloroethane	298.15	−27	32
			53000	Ethylbenzene	298.15	−31	28
			180000	Ethylbenzene	298.15	−29	28
	28900	35900		4-Methyl-2-pentanone	303.15	−21	76
	93940	101000		4-Methyl-2-pentanone	303.15	−24	76
	137000	215000		4-Methyl-2-pentanone	303.15	−28	76
				2-Propanone	298.15	−30	25
	93940	101000		Toluene	303.15	−22	76
	689000	782000		Toluene	303.15	−24	76
			12000	Trichloromethane	298.15	−65	52
			54000	Trichloromethane	298.15	−80	52
			80000	Trichloromethane	298.15	−81	52
			100000	Trichloromethane	298.15	−84	52
			320000	Trichloromethane	298.15	−83	52
	93940	101000		Trichloromethane	303.15	−71	76
	689000	782000		Trichloromethane	303.15	−72	76
		2320000		Trichloromethane	303.15	−73	76
Poly(4-methyl-1-pentene)							
Semicrystalline			350000	Cyclohexane	303.15	30	79
Poly(α-methylstyrene)							
	1030	1180		Toluene	298.15	−7.1	43
		1430		Toluene	298.15	−30	43
	1820	2230		Toluene	298.15	−34	43
	1920			Toluene	298.15	−37	43
	2700	3300		Toluene	298.15	−39	43
	3280			Toluene	298.15	−46	43
	5260			Toluene	298.15	−46	43
	8600			Toluene	298.15	−45	43
	12200			Toluene	298.15	−46	43
		10500		Toluene	310.15	−8.4	92
		53000		Toluene	310.15	−13	92
		55000		Toluene	333.15	−16	96
		87000		Toluene	298.15	−17	90
		87000		Toluene	310.15	−16	92
		87000		Toluene	333.15	−11	90
Poly(2-methyl-5-vinyltetrazole)							
				Acetic acid	298.15	47	100
				Acetonitrile	298.15	14	100
				1,2-Dichloroethane	298.15	17	100
				N,N-Diethylacetamide	298.15	17	100
				N,N-Dimethylformamide	298.15	33	100
				Dimethylsulfoxide	298.15	10	100

Polymer	M_n/ g/mol	M_w/ g/mol	M_η/ g/mol	Solvent	T/K	ΔH_B^∞/ J/g	Ref.
				Formamide	298.15	12	100
				Formic acid	298.15	110	100
				Nitromethane	298.15	10	100
				Pyridine	298.15	16	100
Poly(octamethylene oxide)							
	7000			Benzene	298.15	20	64
	7000			Benzene	303.15	22	64
	7000			Benzene	308.15	25	64
Polypentenamer							
			50000	Cyclohexane	298.15	4.6	74
			50000	Cyclooctane	298.15	5.1	74
			50000	Cyclopentane	298.15	−2.3	74
			50000	*cis*-Decahydronaphthalene	298.15	2.6	74
			50000	*trans*-Decahydronaphthalene	298.15	<0.1	74
			50000	3,3-Diethylpentane	298.15	2.4	74
			50000	2,2-Dimethylpentane	298.15	3.3	74
			50000	2,3-Dimethylpentane	298.15	2.2	74
			50000	2,4-Dimethylpentane	298.15	3.3	74
			50000	3,3-Dimethylpentane	298.15	2.7	74
			50000	Dodecane	298.15	2.9	74
			50000	3-Ethylpentane	298.15	2.1	74
			50000	2,2,4,4,6,8,8-Heptamethylnonane	298.15	3.2	74
			50000	Hexadecane	298.15	2.6	74
			50000	3-Methylhexane	298.15	2.4	74
			50000	Octane	298.15	2.2	74
			50000	2,2,4,6,6-Pentamethylheptane	298.15	3.8	74
			50000	2,2,4,4-Tetramethylpentane	298.15	4.5	74
			50000	2,3,3,4-Tetramethylpentane	298.15	2.4	74
			50000	2,2,4-Trimethylpentane	298.15	4.3	74
Poly(m-phenyleneisophthalamide)							
Glass				*N,N*-Dimethylacetamide	298.15	−171	60
Semicrystalline				*N,N*-Dimethylacetamide	298.15	−128	60
Glass				*N,N*-Dimethylformamide	298.15	−149	60
Semicrystalline				*N,N*-Dimethylformamide	298.15	−125	60
Glass				1-Methyl-2-pyrrolidone	298.15	−177	60
Semicrystalline				1-Methyl-2-pyrrolidone	298.15	−118	60
Polypropylene (atactic)							
	18000			Benzene	298.15	31	80
	6000			Cyclohexane	298.15	2.3	74
	18000			Cyclohexane	298.15	3.9	80
	6000			Cyclooctane	298.15	3.0	74
	6000			Cyclopentane	298.15	−2.3	74
	6000			*cis*-Decahydronaphthalene	298.15	0.5	74

Polymer	$M_n/$ g/mol	$M_w/$ g/mol	$M_\eta/$ g/mol	Solvent	T/K	$\Delta H_B^\infty/$ J/g	Ref.
			6000	*trans*-Decahydronaphthalene	298.15	−2.4	74
			18000	Decane	298.15	3.1	80
			6000	3,3-Diethylpentane	298.15	−3.9	74
			18000	1,2-Dimethylbenzene	298.15	13	80
			18000	1,3-Dimethylbenzene	298.15	12	80
			18000	1,4-Dimethylbenzene	298.15	10	80
			6000	2,2-Dimethylpentane	298.15	−2.2	74
			6000	2,3-Dimethylpentane	298.15	−2.5	74
			6000	2,4-Dimethylpentane	298.15	−1.8	74
			6000	3,3-Dimethylpentane	298.15	−3.0	74
			6000	Dodecane	298.15	1.7	73
			18000	Ethylbenzene	298.15	14	80
			6000	3-Ethylpentane	298.15	−2.5	74
			6000	2,2,4,4,6,8,8-Heptamethylnonane	298.15	−0.7	73
			6000	Heptane	298.15	−1.6	73
			18000	Heptane	298.15	0.5	80
			6000	Hexadecane	298.15	2.3	73
			18000	Hexane	298.15	−1.4	80
			6000	3-Methylhexane	298.15	−1.8	74
			6000	Nonane	298.15	0.8	73
			18000	Nonane	298.15	2.4	80
			6000	Octane	298.15	−1.2	73
			18000	Octane	298.15	1.0	80
			6000	2,2,4,6,6-Pentamethylheptane	298.15	−0.2	73
			18000	Pentane	298.15	−4.7	80
			18000	Tetrachloromethane	298.15	6.6	80
			6000	2,2,4,4-Tetramethylpentane	298.15	−0.8	74
			6000	2,3,3,4-Tetramethylpentane	298.15	−3.1	74
			18000	Toluene	298.15	17	80
			18000	Trichloromethane	298.15	17	80
			6000	2,2,4-Trimethylpentane	298.15	−1.0	73
Polypropylene (isotactic)							
				1-Chloronaphthalene	383.15	26	47
				1-Chloronaphthalene	393.15	170	47
				1-Chloronaphthalene	403.15	245	47
				1-Chloronaphthalene	423.15	275	47
				1,2,3,4-Tetrahydronaphthalene	373.15	140	47
				1,2,3,4-Tetrahydronaphthalene	383.15	215	47
				1,2,3,4-Tetrahydronaphthalene	393.15	330	47
				1,2,3,4-Tetrahydronaphthalene	403.15	330	47
				1,2,3,4-Tetrahydronaphthalene	413.15	335	47
				1,2,3,4-Tetrahydronaphthalene	423.15	290	47

Polymer	M_n/ g/mol	M_w/ g/mol	M_η/ g/mol	Solvent	T/K	ΔH_B^∞/ J/g	Ref.
Poly(propylene glycol)							
	150			Benzene	321.35	200	103
	425			Benzene	321.35	80	103
	2025			Benzene	321.35	45	103
	150			Ethanol	321.35	40	103
	425			Ethanol	321.35	60	103
	2025			Ethanol	321.35	65	103
	396	412		Tetrachloromethane	303.15	4.7	95
	396	412		Tetrachloromethane	318.15	5.2	95
	1900			Tetrachloromethane	303.15	−8.2	95
	1900			Tetrachloromethane	318.15	11	95
	1900			Trichloromethane	303.15	−81	95
		400		Water	298.15	−165	97
	150			Water	321.35	−90	103
	425			Water	321.35	−95	103
Polystyrene							
		600		Benzene	298.15	−1.3	54
		600		Benzene	313.15	−2.5	54
		900		Benzene	291.15	−10	54
		900		Benzene	318.15	−5.8	54
		2000		Benzene	291.15	−16	54
		2000		Benzene	318.15	−6.8	54
		5000		Benzene	291.15	−23	54
		5000		Benzene	318.15	−12	54
		10300		Benzene	291.15	−26	54
		10300		Benzene	318.15	−18	54
			18000	Benzene	298.15	−4.1	19
		20000		Benzene	296.15	−15	9
			29000	Benzene	298.15	−5.0	19
			30000	Benzene	298.15	−7.5	19
			59000	Benzene	298.15	−13	19
			91000	Benzene	298.15	−15	19
		97200		Benzene	318.15	−21	54
			142000	Benzene	298.15	−17	19
			190000	Benzene	303.15	−18	51
				Benzene	300.15	−16	85
			216000	Benzene	298.15	−18	19
			272000	Benzene	298.15	−21	19
			300000	Benzene	298.15	−21	12
				Benzene	298.15	−27	25
				Benzene	298.15	−10	7
		20000		Butyl acetate	296.15	−13	9
		20000		2-Butanone	296.15	−15	9
			142000	2-Butanone	296.15	−17	30
			190000	Butylbenzene	303.15	−14	51
		150000		Chlorobenzene	293.15	−32	49
			266000	Chlorobenzene	298.15	5.4	34

Polymer	$M_n/$ g/mol	$M_w/$ g/mol	$M_\eta/$ g/mol	Solvent	T/K	$\Delta H_B^\infty/$ J/g	Ref.
				Chlorobenzene	293.15	−39	21
	1260			Cyclohexane	298.15	10	16
	1910			Cyclohexane	298.15	5.4	16
	3160			Cyclohexane	298.15	−5.4	16
	3980			Cyclohexane	298.15	−6.9	16
	5630			Cyclohexane	298.15	−9.3	16
	9070			Cyclohexane	298.15	−11	16
	20000			Cyclohexane	296.15	2.5	9
			190000	Cyclohexane	303.15	−2.1	51
				Cyclohexane	293.15	−14	21
	22400			Cyclohexanone	298.15	−29	25
	20000			Cyclohexene	296.15	−9.4	9
			190000	Decahydronaphthalene	303.15	3.8	51
	110000	115000		1,2-Dichlorobenzene	303.05	26	89
	20000			1,2-Dimethylbenzene	296.15	−13	9
	20000			1,3-Dimethylbenzene	296.15	−12	9
			190000	1,3-Dimethylbenzene	303.15	−15	51
			190000	1,4-Dioxane	303.15	−12	51
	20000			Ethyl acetate	296.15	−11	9
			142000	Ethyl acetate	296.15	−13	30
			785	Ethylbenzene	298.15	0.0	14
			18000	Ethylbenzene	298.15	−3.8	14
			18000	Ethylbenzene	298.15	−3.9	19
			30000	Ethylbenzene	298.15	−5.7	19
			35000	Ethylbenzene	298.15	−6.5	19
			45000	Ethylbenzene	298.15	−8.4	19
			91000	Ethylbenzene	298.15	−11	19
			142000	Ethylbenzene	298.15	−13	19
			216000	Ethylbenzene	298.15	−17	19
	60000			Ethylbenzene	303.15	−22	57
	113000	122000		Ethylbenzene	306.65	−24	63
	113000	122000		Ethylbenzene	317.15	−19	63
	113000	122000		Ethylbenzene	337.15	−11	63
	113000	122000		Ethylbenzene	347.15	−6.4	63
	113000	122000		Ethylbenzene	350.65	−4.9	63
	113000	122000		Ethylbenzene	366.65	−2.3	63
	113000	122000		Ethylbenzene	367.15	−2.1	63
	113000	122000		Ethylbenzene	368.65	−2.6	63
	113000	122000		Ethylbenzene	372.15	−4.2	63
	113000	122000		Ethylbenzene	378.15	−4.6	63
	113000	122000		Ethylbenzene	385.15	−5.7	63
	150000			Ethylbenzene	293.15	−34	49
			190000	Ethylbenzene	303.15	−15	51
			272000	Ethylbenzene	298.15	−18	14
			272000	Ethylbenzene	298.15	−18	19
			413000	Ethylbenzene	298.15	−24	39
				Ethylbenzene	293.15	−34	21

Polymer	M_n/ g/mol	M_w/ g/mol	M_η/ g/mol	Solvent	T/K	ΔH_B^∞/ J/g	Ref.
				Ethylbenzene	298.15	−17	11
				Ethylbenzene	298.15	−30	28
	1260			2-Propanone	298.15	−0.6	16
	1910			2-Propanone	298.15	−7.7	16
	3160			2-Propanone	298.15	−16	16
	3980			2-Propanone	298.15	−17	16
	5630			2-Propanone	298.15	−19	16
	9070			2-Propanone	298.15	−21	16
	20000			2-Propanone	296.15	−11	9
			190000	Propylbenzene	303.15	−14	51
	20000			Styrene	296.15	−18	9
				Styrene	296.15	−35	6
			413000	Tetrachloromethane	298.15	−22	39
	600			Toluene	296.15	−2.1	16
	600			Toluene	309.15	−1.8	16
	600			Toluene	318.15	−1.5	16
	1260			Toluene	296.15	−11	16
	1260			Toluene	303.15	−8.0	16
	1260			Toluene	309.15	−5.9	16
	1260			Toluene	318.15	−3.4	16
	1260			Toluene	328.15	−2.3	16
	1260			Toluene	338.15	−1.9	16
	1260			Toluene	346.65	−1.3	16
	1910			Toluene	298.15	−16	16
	1910			Toluene	318.15	−6.7	16
	1910			Toluene	338.15	−3.4	16
	1910			Toluene	348.15	−2.5	16
	3160			Toluene	298.15	−23	16
	3980			Toluene	298.15	−24	16
	3980			Toluene	318.15	−17	16
	3980			Toluene	338.15	−10	16
	5630			Toluene	298.15	−26	16
	9070			Toluene	298.15	−28	16
	270000			Toluene	298.15	−33	16
		600		Toluene	298.15	−1.4	54
		600		Toluene	313.15	−3.2	54
		900		Toluene	291.15	−7.3	54
		900		Toluene	318.15	−6.6	54
		2000		Toluene	291.15	−11	54
		2000		Toluene	318.15	−7.2	54
		5000		Toluene	291.15	−21	54
		5000		Toluene	318.15	−11	54
		10300		Toluene	291.15	−24	54
		10300		Toluene	318.15	−15	54
		97200		Toluene	318.15	−17	54
		9000		Toluene	310.15	−9.2	92
	20000			Toluene	296.15	−17	9

Polymer	$M_n/$ g/mol	$M_w/$ g/mol	$M_\eta/$ g/mol	Solvent	T/K	$\Delta H_B^\infty/$ J/g	Ref.
	20400			Toluene	298.15	−8.2	90
	20400			Toluene	310.15	−8.4	92
	20400			Toluene	333.15	−6.4	90
	47000			Toluene	310.15	−5.0	92
	50000			Toluene	333.15	−6.8	96
	60000			Toluene	303.15	−21	57
	113000	122000		Toluene	304.15	−29	63
	113000	122000		Toluene	306.15	−27	63
	113000	122000		Toluene	306.65	−26	63
	113000	122000		Toluene	316.15	−23	63
	113000	122000		Toluene	333.15	−15	63
	113000	122000		Toluene	337.15	−12	63
	113000	122000		Toluene	346.15	−8.3	63
	113000	122000		Toluene	347.15	−7.8	63
	113000	122000		Toluene	348.15	−8.2	63
	113000	122000		Toluene	350.65	−6.5	63
	113000	122000		Toluene	359.15	−4.3	63
	113000	122000		Toluene	362.15	−2.7	63
	113000	122000		Toluene	369.15	−3.3	63
	113000	122000		Toluene	372.15	−2.8	63
	115000			Toluene	310.15	−5.0	92
	150000			Toluene	293.15	−34	49
			190000	Toluene	303.15	−18	51
	214000			Toluene	300.15	−19	85
			250000	Toluene	303.15	−18	51
				Toluene	293.15	−34	21
				Toluene	298.65	−39	27
				Toluene	308.15	−34	27
				Toluene	318.15	−30	27
				Toluene	333.15	−23	27
				Toluene	343.15	−13	27
				Toluene	353.15	−13	27
		600		Trichloromethane	298.15	−13	54
		600		Trichloromethane	313.15	−9.9	54
		900		Trichloromethane	291.15	−22	54
		900		Trichloromethane	313.15	−15	54
		2000		Trichloromethane	291.15	−28	54
		2000		Trichloromethane	313.15	−16	54
		5000		Trichloromethane	291.15	−30	54
		5000		Trichloromethane	313.15	−18	54
		10300		Trichloromethane	291.15	−33	54
		10300		Trichloromethane	313.15	−23	54
	22400			Trichloromethane	298.15	−17	25
		97200		Trichloromethane	313.15	−25	54
	20000			1,3,5-Trimethylbenzene	296.15	−11	9
			190000	1,3,5-Trimethylbenzene	303.15	−13	51

Polymer	$M_n/$ g/mol	$M_w/$ g/mol	$M_\eta/$ g/mol	Solvent	T/K	$\Delta H_B^\infty/$ J/g	Ref.
Poly(tetramethylene oxide)							
	650			Benzene	313.15	4.0	86
	1000			Benzene	313.15	2.0	86
	2000			Benzene	313.15	1.1	86
	650			1,2-Dichloroethane	313.35	3.1	88
	2000			1,2-Dichloroethane	313.35	0.3	88
	650			1,2-Dimethylbenzene	313.15	5.9	87
	1000			1,2-Dimethylbenzene	313.15	1.8	87
	2000			1,2-Dimethylbenzene	313.15	0.9	87
	650			1,3-Dimethylbenzene	313.15	6.4	87
	1000			1,3-Dimethylbenzene	313.15	0.6	87
	2000			1,3-Dimethylbenzene	313.15	0.8	87
	650			1,4-Dimethylbenzene	313.15	4.3	87
	1000			1,4-Dimethylbenzene	313.15	1.8	87
	2000			1,4-Dimethylbenzene	313.15	0.7	87
	650			1,4-Dioxane	321.35	4.0	81
	1000			1,4-Dioxane	321.35	2.3	81
	2000			1,4-Dioxane	321.35	1.0	81
	650			Ethylbenzene	313.15	6.9	86
	1000			Ethylbenzene	313.15	3.4	86
	2000			Ethylbenzene	313.15	−0.05	86
	650			Propylbenzene	313.15	5.8	86
	1000			Propylbenzene	313.15	1.3	86
	2000			Propylbenzene	313.15	0.9	86
	650			Tetrachloromethane	313.15	3.3	88
	1000			Tetrachloromethane	313.15	1.4	88
	2000			Tetrachloromethane	321.35	0.7	82
	650			Toluene	313.15	4.3	86
	1000			Toluene	313.15	2.0	86
	2000			Toluene	313.15	0.9	86
	650			1,3,5-Trimethylbenzene	313.15	6.1	87
	1000			1,3,5-Trimethylbenzene	313.15	2.7	87
	2000			1,3,5-Trimethylbenzene	313.15	0.6	87
Poly(vinyl acetate)							
			140000	Benzene	298.15	2.3	13
			350000	2-Butanone	303.15	−1.7	51
			350000	Butyl acetate	303.15	1.0	51
			135000	Chlorobenzene	298.15	5.0	34
			350000	Ethyl acetate	303.15	−6.7	51
				Ethyl acetate	303.15	0.0	11
			26000	3-Heptanone	303.15	7.0	44
			350000	3-Heptanone	303.15	4.9	51
			140000	Methanol	298.15	−45	13
			350000	Methyl acetate	303.15	−9.7	51
			350000	2-Pentanone	303.15	0.0	51
			93000	2-Propanone	303.15	−0.4	32

Polymer	$M_n/$ g/mol	$M_w/$ g/mol	$M_\eta/$ g/mol	Solvent	T/K	$\Delta H_B^\infty/$ J/g	Ref.
			350000	2-Propanone	303.15	−3.9	51
				2-Propanone	303.15	−2.9	25
			350000	Propyl acetate	303.15	−2.7	51
		150000		Tetrahydrofuran	304.65	4.5	93
			140000	Trichloromethane	298.15	28	13
Poly(vinyl alcohol)							
				Ethanol	298.15	3.8	11
				Ethanol	298.15	9.6	31
	7260			Water	303.15	−34	15
	17000			Water	303.15	−18	32
	61600			Water	303.15	−41	15
				Water	303.15	−8.4	11
Poly(vinyl chloride)							
				Chlorobenzene	298.15	−17	36
Glass	23200			Cyclohexanone	303.15	−27	59
Liquid	23200			Cyclohexanone	303.15	−7.5	59
Glass	38700			Cyclohexanone	303.15	−29	59
Liquid	38700			Cyclohexanone	303.15	−6.6	59
Glass	53500			Cyclohexanone	303.15	−28	59
Liquid	53500			Cyclohexanone	303.15	−6.3	59
Glass	66700			Cyclohexanone	303.15	−29	59
Liquid	66700			Cyclohexanone	303.15	−6.1	59
Glass	136000			Cyclohexanone	303.15	−31	59
Liquid	136000			Cyclohexanone	303.15	−5.8	59
Glass	155400			Cyclohexanone	303.15	−32	59
Liquid	155400			Cyclohexanone	303.15	−5.8	59
	48000			Cyclopentanone	298.15	−28	104
				1,2-Dichloroethane	323.65	24	27
				1,2-Dichloroethane	328.15	34	27
				1,2-Dichloroethane	333.15	38	27
				1,2-Dichloroethane	368.15	44	27
				1,2-Dichloroethane	373.15	46	27
				1,2-Dichloroethane	378.15	46	27
				N,N-Dimethylformamide	293.15	−28	35
				N,N-Dimethylformamide	308.15	−19	35
				N,N-Dimethylformamide	323.15	−14	35
				N,N-Dimethylformamide	338.15	−7.5	35
				N,N-Dimethylformamide	353.15	2.4	35
Glass	23200			Tetrahydrofuran	303.15	−34	59
Liquid	23200			Tetrahydrofuran	303.15	−14	59
Glass	38700			Tetrahydrofuran	303.15	−35	59
Liquid	38700			Tetrahydrofuran	303.15	−14	59
Glass	53500			Tetrahydrofuran	303.15	−39	59
Liquid	53500			Tetrahydrofuran	303.15	−14	59
Glass	66700			Tetrahydrofuran	303.15	−36	59
Liquid	66700			Tetrahydrofuran	303.15	−14	59
Glass	136000			Tetrahydrofuran	303.15	−39	59

Polymer	M_n/ g/mol	M_w/ g/mol	M_η/ g/mol	Solvent	T/K	ΔH_B^∞/ J/g	Ref.
Liquid	136000			Tetrahydrofuran	303.15	−14	59
Glass	155400			Tetrahydrofuran	303.15	−39	59
Liquid	155400			Tetrahydrofuran	303.15	−14	59
Poly(1-vinyl-3,5-dimethyl-1,2,4-triazole)							
				N,N-Dimethylformamide	298.15	−28	94
				Water	298.15	−139	94
Poly(1-vinylimidazole)							
	20700			Acetic acid	298.15	−393	94
	20700			Butanoic acid	298.15	−322	94
	20700			N,N-Dimethylacetamide	298.15	−48	94
	20700			N,N-Dimethylformamide	298.15	−48	94
	20700			1-Methyl-2-pyrrolidinone	298.15	−54	91
	20700			Pentanoic acid	298.15	−325	94
	20700			Propanoic acid	298.15	−278	94
	20700			Water	298.15	−119	91
Poly(1-vinylpyrazole)							
	18900			Acetic acid	298.15	−88	94
	18900			Butanoic acid	298.15	−60	94
	18900			N,N-Dimethylacetamide	298.15	−26	94
	18900			N,N-Dimethylformamide	298.15	−28	94
	18900			Pentanoic acid	298.15	−52	94
	18900			Propanoic acid	298.15	−36	94
Poly(1-vinyl-2-pyrrolidone)							
			32000	Trichloromethane	298.15	−75	48
			32000	Water	298.15	−150	48
Poly(1-vinyl-1,2,4-triazole)							
	69500			Acetic acid	298.15	−85	94
	69500			Butanoic acid	298.15	−74	94
	69500			N,N-Dimethylacetamide	298.15	−49	94
	69500			N,N-Dimethylformamide	298.15	−47	94
	69500			1-Methyl-2-pyrrolidinone	298.15	−55	91
	69500			Pentanoic acid	298.15	−75	94
	69500			Propanoic acid	298.15	−72	94
	69500			Water	298.15	−68	91
Vinyl acetate/vinyl alcohol copolymer							
(9 wt% Vinyl acetate)				2-Propanone	298.15	6.3	33
(44 wt% Vinyl acetate)				2-Propanone	298.15	4.6	33
(57 wt% Vinyl acetate)				2-Propanone	298.15	0.0	33
(67 wt% Vinyl acetate)				2-Propanone	298.15	−1.3	33
(4.2 mol% Vinyl acetate) 7560				Water	303.15	−41	15
(4.3 mol% Vinyl acetate) 64300				Water	303.15	−49	15
(9.0 mol% Vinyl acetate) 66900				Water	303.15	−55	15
(10.3 mol% Vinyl acetate) 7970				Water	303.15	−41	15

Polymer	M_n/ g/mol	M_w/ g/mol	M_η/ g/mol	Solvent	T/K	ΔH_B^∞/ J/g	Ref.
(15.3 mol% Vinyl acetate) 8300				Water	303.15	−60	15
(15.4 mol% Vinyl acetate) 70700				Water	303.15	−65	15
(19.5 mol% Vinyl acetate) 73100				Water	303.15	−66	15
(22.1 mol% Vinyl acetate) 8800				Water	303.15	−60	15
(26.2 mol% Vinyl acetate) 77000				Water	303.15	−64	15
(30.6 mol% Vinyl acetate) 9370				Water	303.15	−53	15
(34.0 mol% Vinyl acetate) 81600				Water	303.15	−60	15
(34.7 mol% Vinyl acetate) 9670				Water	303.15	−44	15
Vinyl acetate/vinyl chloride copolymer (90 wt% Vinyl chloride)							
Glass	12400	26000		Cyclohexanone	304.15	−37	101
Liquid	12400	26000		Cyclohexanone	304.15	−16	101
Vinyl chloride/vinylidene chloride copolymer							
				Trichloromethane	297.15	−17	2

References

1. Liepatoff, S. and Preobagenskaja, S., *Kolloid Z. Z. Polym.*, 68, 324, 1934.
2. Kargin, V. and Papkov, S., *Acta Physicochim. URSS*, 3, 839, 1935.
3. Papkov, S. and Kargin, V., *Acta Physicochim. URSS*, 7, 667, 1937.
4. Tager, A. and Kargin, V., *Acta Physicochim. URSS*, 14, 713, 1941.
5. Raine, H.C., Richards, R.B., and Ryder, H., *Trans Faraday Soc.*, 41, 56, 1945.
6. Roberts, D.E., Walton, W.W., and Jessup, R.S., *J. Polym. Sci.*, 2, 420, 1947.
7. Tager, A. and Sanatina, V., *Kolloidn. Zhur.*, 12, 474, 1950.
8. Glikman, S.A. and Root, L.A., *Zh. Obshch. Khim.*, 21, 58, 1951.
9. Hellfritz, H., *Makromol. Chem.*, 7, 191, 1951.
10. Tager, A. and Vershkain, R., *Kolloidn. Zhur.*, 13, 123, 1951.
11. Tager, A.A., and Kargin, V.A., *Kolloidn. Zhur.*, 14, 367, 1952.
12. Tager, A.A. and Dombek, Zh.S., *Kolloidn. Zhur.*, 15, 69, 1953.
13. Daoust, H. and Rinfret, M., *Can. J. Chem.*, 32, 492, 1954.
14. Gatovskaya, T.V., Kargin, V.A., and Tager, A.A., *Zh. Fiz. Khim.*, 29, 883, 1955.
15. Oya, S., *Chem. High Polym. Japan*, 12, 122, 1955.
16. Schulz, G.V., Guenner, K. von, and Gerrens, H., *Z. Phys. Chem., N. F.*, 4, 192, 1955.
17. Tager, A.A., Kosova, L.K., Karlinskaya, D.Yu., and Yurina, I.A., *Kolloid. Zhur.*, 17, 315, 1955.
18. Tager, A.A. and Kosova, L.K., *Kolloid. Zhur.*, 17, 391, 1955.
19. Tager, A.A., Krivokorytova, R.V., and Khodorov, P.M., *Dokl. Akad. Nauk SSSR*, 100, 741, 1955.
20. Glikman, S.A. and Root, L.A., *Kolloidn. Zhur.*, 18, 523, 1956.
21. Jenckel, E. and Gorke, K., *Z. Elektrochem.*, 60, 579, 1956.
22. Lipatov, Yu.S., Kargin, V.A., and Slonimskii, G.L., *Zh. Fiz. Khim.*, 30, 1202, 1956.
23. Meerson, S.I. and Lipatov, S.M., *Kolloidn., Zh.*, 18, 447, 1956.
24. Mikhailov, N.V. and Fainberg, E.Z., *Kolloidn. Zhur.*, 18, 44, 1956.
25. Struminskii, G.V. and Slonimskii, G.L., *Zh. Fiz. Khim.*, 30, 1941, 1956.
26. Gal'perin, D.I. and Moseev, L.I., *Kolloidn. Zhur.*, 19, 167, 1957.
27. Akhmedov, K.S., *Uzb. Khim. Zh.* (1), 19, 1958.
28. Kargin, V.A. and Lipatov, Yu.S., *Zh. Fiz. Khim.*, 32, 326, 1958.
29. Meerson, S.I. and Lipatov, S.M., *Kolloidn. Zhur.*, 20, 353, 1958.
30. Tager, A.A. and Galkina, L.A., *Nauchn. Dokl. Vyssh. Shkol., Khim. Khim. Tekhnol.*, (2), 357, 1958.
31. Tager, A. A. and Kargin, V. A., *Zh. Fiz. Khim.*, 32, 1362, 1958.
32. Tager, A. A. and Kargin, V. A., *Zh. Fiz. Khim.*, 32, 2694, 1958.
33. Tager, A.A. and Iovleva, M., *Zh. Fiz. Khim.*, 32, 1774, 1958.
34. Horth, A., Patterson, D., and Rinfret, M., *J. Polym. Sci.*, 39. 189, 1959.
35. Zelikman, S.G. and Mikhailov, N.V., *Vysokomol. Soedin.*, 1, 1077, 1959.
36. Mueller, F.H. and Engelter, A., *Kolloid Z.*, 171, 152, 1960.
37. Delmas, G., Patterson, D., and Boehme, A., *Trans. Faraday Soc.*, 58, 2116, 1962.
38. Delmas, G., Patterson, D., and Somcynsky, T., *J. Polym. Sci.*, 57, 79, 1962.
39. Tager, A.A. and Podlesnyak, A.I., *Vysokomol. Soedin., Ser. A*, 5, 87, 1963.
40. Delmas, G., Patterson, D., and Bhattacharyya, S.N., *J. Phys. Chem.*, 68, 1468, 1964.
41. Patterson, D., *J. Polym. Sci.: Part A*, 2, 5177, 1964.
42. Zverev, M.P., Barash, A.N., and Zubov, P.I., *Vysokomol. Soedin., Ser. A*, 6, 1012, 1964.
43. Cottam, B.J., Cowie, J.M.G., and Bywater, S., *Makromol. Chem.*, 86, 116, 1965.
44. Bianchi, U., Pedemonte, E., and Rossi, C., *Makromol. Chem.*, 92, 114, 1966.
45. Cuniberti, C. and Bianchi, U., *Polymer*, 7, 151, 1966.
46. Giacommeti, G. and Turolla, A., *Z. Phys. Chem., N.F.*, 51, 108, 1966.
47. Schreiber, H.P. and Waldman, M.H., *J. Polym. Sci.: Part A-2*, 5, 555, 1967.
48. Goldfarb, J. and Rodriguez, S., *Makromol. Chem.*, 116, 96, 1968.
49. Maron, S.H. and Daniels, C.A., *J. Macromol. Sci.-Phys. B*, 2, 769, 1968.
50. Morimoto, S., *J. Polym. Sci.: Part A-1*, 6, 1547, 1968.
51. Bianchi, U., Cuniberti, C., Pedemonte, E., and Rossi, C., *J. Polym. Sci.: Part A-2*, 7, 855, 1969.
52. Gerth, Ch. and Mueller, F.H., *Kolloid-Z. Z. Polym.*, 241, 1071, 1970.
53. Liddell, A.H. and Swinton, F.L., *Discuss. Faraday Soc.*, 49, 115, 1970.
54. Morimoto, S., *Nippon Kagaku Zasshi*, 91, 31, 1970.
55. Nakayama, H., *Bull. Chem. Soc. Japan*, 43, 1683, 1970.

56. Blackadder, D.A. and Roberts, T.L., *Angew. Makromol. Chem.*, 27, 165, 1972.
57. Maron, S.H. and Filisko, F.E., *J. Macromol. Sci.-Phys. B*, 6, 57, 1972.
58. Maron, S.H. and Filisko, F.E., *J. Macromol. Sci.-Phys. B*, 6, 79, 1972.
59. Maron, S.H. and Filisko, F.E., *J. Macromol. Sci.-Phys. B*, 6, 413, 1972.
60. Sokolova, D.F., Sokolov, L.B., and Gerasimov, V.D., *Vysokomol. Soedin., Ser. B*, 14, 580, 1972.
61. Chahal, R.S., Kao, W.-P., and Patterson, D., *J. Chem. Soc., Faraday Trans. I*, 69, 1834, 1973.
62. Delmas, G. and Tancrede, P., *Eur. Polym. J.*, 9, 199, 1973.
63. Filisko, F.E., Raghava, R.S., and Yeh, G.S.Y., *J. Macromol. Sci.-Phys. B*, 10, 371, 1974.
64. Ikeda, M., Suga, H., and Seki, S., *Polymer*, 16, 634, 1975.
65. Kiselev, V.P., Shakhova, E.M., Fainberg, E.Z., Virnik, A.D, and Rogovin, Z.A., *Vysokomol. Soedin., Ser. B*, 18, 847, 1976.
66. Petrosyan, V.A., Gabrielyan, G.A., and Rogovin, Z.A., *Arm. Khim. Zhur.*, 29, 516, 1976.
67. Deshpande, D.D. and Prabhu, C.S., *Macromolecules*, 10, 433, 1977.
68. Miura, T. and Nakamura, M., *Bull. Chem. Soc. Japan*, 50, 2528, 1977.
69. Sokolova, D.F., Kudim, T.V., Sokolov, L.B., Zhegalova, N.I., and Zhuravlev, N.D., *Vysokomol. Soedin., Ser. B*, 20, 596, 1978.
70. Basedow, A.M. and Ebert, K.H., *J. Polym. Sci.: Polym. Symp.*, 66, 101, 1979.
71. Koller, J., *Dissertation*, TU München, 1979.
72. Lee, J.-O., Ono, M., Hamada, F., and Nakajima, A., *Polym. Bull.*, 1, 763, 1979.
73. Phuong-Nguyen, H. and Delmas, G., *Macromolecules*, 12, 740, 1979.
74. Phuong-Nguyen, H. and Delmas, G., *Macromolecules*, 12, 746, 1979.
75. Basedow, A.M., Ebert, K.H., and Feigenbutz, W., *Makromol. Chem.*, 181, 1071, 1980.
76. Graun, K., *Dissertation*, TU München, 1980.
77. Shiomi, T., Izumi, Z., Hamada, F., and Nakajima, A., *Macromolecules*, 13, 1149, 1980.
78. Shiomi, T., Kohra, Y., Hamada, F., and Nakajima, A., *Macromolecules*, 13, 1154, 1980.
79. Aharoni, S.M., Charlet, G., and Delmas, G., *Macromolecules*, 14, 1390, 1981.
80. Ochiai, H., Ohashi, T., Tadokoro, Y., and Murakami, I., *Polym. J.*, 14, 457, 1982.
81. Sharma, S.C., Mahajan, R., Sharma, V.K., and Lakhanpal, M.L., *Indian J. Chem.*, 21A, 682, 1982.
82. Sharma, S.C., Mahajan, R., Sharma, V.K., and Lakhanpal, M.L., *Indian J. Chem.*, 21A, 685, 1982.
83. Lakhanpal, M.L. and Parashar, R.N., *Indian J. Chem.*, 22A, 48, 1983.
84. Daoust, H. and St-Cyr, D., *Macromolecules*, 17, 596, 1984.
85. Aeleni, N., *Mater. Plast. (Bucharest)*, 22, 92, 1985.
86. Sharma, S.C. and Sharma, V.K., *Indian J. Chem.*, 24A, 292, 1985.
87. Sharma, S.C., Bhalla, S., and Sharma, V.K., *Indian J. Chem.*, 25A, 131, 1986.
88. Sharma, S.C., Syngal, M., and Sharma, V.K., *Indian J. Chem.*, 26A, 285, 1987.
89. Aukett, P.N. and Brown, C.S., *J. Therm. Anal.*, 33, 1079, 1988.
90. Lanzavecchia, L. and Pedemonte, E., *Thermochim. Acta*, 137, 123, 1988.
91. Tager, A.A., Safronov, A.P., Voit, V.V., Lopyrev, V.A., Ermakova, T.G., Tatarova, L.A., and Shagelaeva, N.S., *Vysokomol. Soedin., Ser. A*, 30, 2360, 1988.
92. Pedemonte, E. and Lanzavecchia, L., *Thermochim. Acta*, 162, 223, 1990.
93. Shiomi, T., Ishimatsu, H., Eguchi, T., and Imai, K., *Macromolecules*, 23, 4970, 1990.
94. Tager, A.A. and Safronov, A.P., *Vysokomol. Soedin., Ser. A*, 33, 67, 1991.
95. Zellner, H., *Dissertation*, TU München, 1993.
96. Brunacci, A., Pedemonte, E., Cowie, J.M.G., and McEwen, I. J., *Polymer*, 35, 2893, 1994.
97. Carlsson, M., Hallen, D., and Linse, P., *J. Chem. Soc. Faraday Trans.*, 91, 2081, 1995.
98. Sato, T., Tohyama, M., Suzuki, M., Shiomi, T., and Imai, K., *Macromolecules*, 29, 8231, 1996.
99. Shiomi, T., Tohyama, M., Endo, M., Sato, T., and Imai, K., *J. Polym. Sci.: Part B: Polym. Phys.*, 34, 2599, 1996.
100. Kizhnyaev, V.N., Gorkovenko, O.P., Bazhenov, D.N., and Smirnov, A.I., *Vysokomol. Soedin., Ser. A*, 39, 856, 1997.
101. Sato, T., Suzuki, M., Tohyama, M., Endo, M., Shiomi, T., and Imai, K., *Polym. J.*, 29, 417, 1997.
102. Vanderryn, J. and Zettlemoyer, A.C., *Ind. Eng. Chem., Chem. Eng. Data Ser.*, 2, 56, 1957.
103. Parashar, R. and Sharma, S.C., *Indian J. Chem.*, 27A, 1092, 1988.
104. Righetti, M.C., Cardelli, C., Scalari, M., Tombari, E., and Conti, G., *Polymer*, 43, 5035, 2002.
105. Phuong-Nguyen, H. and Delmas, G., *J. Solution Chem.*, 23, 249, 1994.
106. Wohlfarth, C., *CRC Handbook of Enthalpy Data of Polymer-Solvent Systems*, CRC Press, Boca Raton, 2006.

SOLUBILITY PARAMETERS OF SELECTED POLYMERS

Christian Wohlfarth

The concept of cohesive energy density and solubility parameter was introduced by Hildebrand:

$$\delta^2 = \frac{\Delta U_m^{LV}}{V_m} = \frac{\Delta H_m^{LV} - RT}{V_m} \qquad (1)$$

V_m is the molar volume, ΔU_m^{LV} is the molar energy of vaporization, and ΔH_m^{LV} is the molar enthalpy of vaporization. Units for the solubility parameter are $(MPa)^{1/2} = (J/cm^3)^{1/2} = 0.4887(cal/cm^3)^{1/2}$. The energy of vaporization is not accessible for polymers, but cohesive energy density of polymers can be determined from *PVT*-data. However, common ways for determining polymer solubility parameters use thermodynamic properties of polymer solutions and their relations to excess enthalpy or excess Gibbs energy per unit volume. These excess quantities are related to the (square) difference between the solubility parameters of solvents and polymers, i.e. $(\delta_1 - \delta_2)^2$.

$$\frac{H^E}{V} \sim (\delta_1 - \delta_2)^2 \quad \text{or} \quad \frac{G^E}{V} \sim (\delta_1 - \delta_2)^2 \qquad (2)$$

Often, the Flory-Huggins solvent-polymer interaction parameter is applied instead of H^E or G^E. There are some books (Refs. 1–3) giving details for such procedures as well as extensive tables of polymer solubility parameters from which the table below is extracted. Methods for calculating solubility parameters can be found in Refs. 4–7.

References

1. Barton, A.F.M., *CRC Handbook of Polymer-Liquid Interaction Parameters and Solubility Parameters*, CRC Press, Boca Raton, 1991.
2. Brandrup, J., Immergut, E.H., Grulke, E.A. (eds.), *Polymer Handbook*, 4th ed., J. Wiley & Sons, New York, 1999.
3. Wohlfarth, C., *Thermodynamic Properties of Polymer Solutions*, in Landolt-Börnstein, New Series, Group VIII, Vol. 6D, Lechner, M.D., (ed.), Springer Verlag, Berlin, Heidelberg, 2010.
4. Hildebrand, J.H., Prausnitz, J.M., Scott, R.L., *Regular and Related Solutions*, Van Nostrand Reinhold Co., New York, 1970.
5. [Van] Krevelen, D.W., *Properties of Polymers*, 3rd ed., Elsevier, Amsterdam, 1990.
6. Bicerano, J., *Prediction of Polymer Properties*, 3rd ed., CRC Press, Boca Raton, 2002.
7. Hansen, C.M., *Hansen Solubility Parameters: A User's Handbook*, 2nd ed., CRC Press, Boca Raton, 2007.

Polymer	T/K	$\delta/(J/cm^3)^{1/2}$
Benzyl cellulose	298	25.2
Butadiene/acrylonitrile copolymer		
(25 wt % acrylonitrile)	298	19.4
(30 wt % acrylonitrile)	298	19.2
(34 wt % acrylonitrile)	298	20.4
(39 wt % acrylonitrile)	298	21.35
Butadiene/styrene copolymer		
(6 wt % styrene)	298	16.5
(12.5 wt % styrene)	298	16.5
(15 wt% styrene)	298	17.5
(25 wt% styrene)	298	17.6
(40 wt% styrene)	298	17.8
Cellulose	298	26.0
Cellulose acetate	298	25.1
Cellulose diacetate	298	22.3
Cellulose nitrate	298	22.0
Cellulose triacetate	298	19.0
Ethyl cellulose	298	21.1
Ethylene/1-octene copolymer		
(2.0 wt% 1-octene)	473	17.2
(7.5 wt% 1-octene)	473	16.7
(12.0 wt% 1-octene)	473	16.5
(25.0 wt% 1-octene)	473	16.5
(39.4 wt% 1-octene)	473	16.4
(55.0 wt% 1-octene)	473	16.3
(64.0 wt% 1-octene)	473	16.3
Ethylene/vinyl acetate copolymer		

Polymer	T/K	$\delta/(J/cm^3)^{1/2}$
(11 wt% vinyl acetate)	323	16.6
(16 wt% vinyl acetate)	323	17.1
(24 wt% vinyl acetate)	323	16.8
(37 wt% vinyl acetate)	323	17.0
(42 wt% vinyl acetate)	323	17.2
Gelatine	298	24.6
Hydroxypropyl cellulose	298	26.8
Natural rubber	298	16.6
Poly(acrylonitrile)	298	26.0
Polyamide 4	298	24.0
Polyamide 6	298	21.7
Polyamide 66	298	22.9
Polyamide 7	298	24.1
Polyamide 8	298	20.3
Polyamide 9	298	22.6
Polyamide 10	298	19.4
Polyamide 11	298	22.9
Polyamide 12	298	20.8
Poly(*p*-benzamide)	298	23.0
Poly(benzyl methacrylate)	298	15.3
Polybutadiene	298	16.6
1,2-Polybutadiene	298	16.5
1,4-*cis*-Polybutadiene	298	16.5
Poly(1-butene), *isotactic*	298	16.0
Poly(butyl acrylate)	298	18.0
Poly(butyl methacrylate)	298	17.9
Poly(2-butyl methacrylate)	413	14.7

Polymer	T/K	$\delta/(\text{J/cm}^3)^{1/2}$
Poly(ε-caprolactone)	298	19.9
Polycarbonate bisphenol-A	298	20.0
Polycarbonate hexafluorobisphenol-A	298	20.1
Poly(chloroprene)	298	17.6
Poly(4-chlorostyrene)	298	19.3
Poly(cyclohexyl methacrylate)	298	19.8
Poly(3,3-diethyloxetane)	298	16.2
Poly(3,3-dimethyloxetane)	298	16.2
Poly(2,6-dimethyl-1,4-phenylene ether)	298	18.1
Poly(dimethylsiloxane)	298	15.4
Poly(1,3-dioxepane)	298	18.8
Poly(1,3-dioxolane)	298	20.7
Poly(dodecyl methacrylate)	298	16.8
Poly(epichlorohydrin)	298	16.2
Poly(ethoxyethyl methacrylate)	298	18.4
Poly(ethyl acrylate)	298	19.2
Polyethylene, branched	298	16.2
Polyethylene, linear	298	16.2
Poly(ethylene adipate)	298	19.8
Poly(ethylene glycol)	298	23.7
Poly(ethylene oxide)	298	20.5
Poly(ethylene terephthalate)	298	21.9
Poly(ethyl methacrylate)	298	18.3
Poly(hexyl methacrylate)	298	17.7
Poly(4-hydroxystyrene)	298	24.0
Poly(isobornyl acrylate)	298	16.8
Poly(isobornyl methacrylate)	298	17.0
Poly(isobutylene)	298	16.2
Poly(isobutyl methacrylate)	413	14.6
1,4-cis-Poly(isoprene)	298	16.5
Poly(N-isopropylacrylamide)	298	23.5
Poly(DL-lactic acid)	298	20.5
Poly(L-lactide)	298	19.3
Poly(methacrylonitrile)	298	21.0
Poly(methyl acrylate)	298	20.5
Poly(methyl methacrylate)	298	19.3
Poly(4-methyl-1-pentene)	298	15.3
Poly(2-methylpropene)	298	17.7
Poly(2-methylstyrene)	298	18.4
Poly(4-methylstyrene)	298	19.3

Polymer	T/K	$\delta/(\text{J/cm}^3)^{1/2}$
Poly(methylvinylsiloxane)	298	15.65
Polynorbornene	298	14.0
Poly(1-octene)	298	16.6
Poly(octyl methacrylate)	298	18.0
Poly(propyl acrylate)	298	18.4
Poly(propyl methacrylate)	298	16.0
Polypropylene, atactic	298	15.5
Polypropylene, isotactic	298	17.5
Polypropylene, syndiotactic	298	17.6
Poly(propylene glycol)	298	19.5
Poly(propylene oxide)	298	18.5
Poly(propyl methacrylate)	298	20.0
Polystyrene	298	19.0
Polysulfone	298	19.9
Poly(tetrafluoroethylene)	298	19.6
Poly(tetramethylene oxide)	298	16.8
Poly(thioethylene)	298	18.8
Poly(trimethylene sulfide)	298	23.0
Poly(vinyl acetate)	298	20.6
Poly(vinyl alcohol)	298	22.0
Poly(vinyl bromide)	298	19.4
Poly(vinyl butyl ether)	298	19.2
Poly(N-vinylcarbazole)	298	19.2
Poly(vinyl chloride)	298	19.6
Poly(vinyl ethyl ether)	298	19.5
Poly(vinylidene fluoride)	298	23.2
Poly(vinyl methyl ether)	298	21.0
Poly(vinyl phenyl ether)	298	20.2
Poly(vinyl propionate)	298	18.1
Poly(vinyl propyl ether)	298	19.3
Poly(1-vinyl-2-pyrrolidinone)	298	25.6
Vinyl acetate/vinyl alcohol copolymer		
(43.4 mol% vinyl acetate)	298	21.8
(60.9 mol% vinyl acetate)	298	21.4
(74.4 mol% vinyl acetate)	298	20.9
(94.8 mol% vinyl acetate)	298	20.2
Vinyl acetate/vinyl chloride copolymer		
(3 wt% vinyl acetate)	298	18.8
(10 wt% vinyl acetate)	298	17.3
(17 wt% vinyl acetate)	298	19.1

Section 14
Geophysics, Astronomy, and Acoustics

ASTRONOMICAL CONSTANTS

Victor Abalakin

The constants in this table are based originally on the set of constants adopted by the International Astronomical Union (IAU) in 1976. Updates have been made when new data were available. All values are given in SI Units; thus masses are expressed in kilograms and distances in meters.

The astronomical unit of time is a time interval of one day (1 d) equal to 86400 s. An interval of 36525 d is one Julian century (1 cy).

References

1. Seidelmann, P. K., *Explanatory Supplement to the Astronomical Almanac*, University Science Books, Mill Valley, CA, 1990.
2. Lang, K. R., *Astrophysical Data: Planets and Stars*, Springer-Verlag, New York, 1992.
3. *The Astronomical Almanac for the Year 2007*, U.S. Government Printing Office, Washington, and Her Majesty's Stationary Office, London (2005).

Defining constants

Gaussian gravitational constant $\qquad$ $k = 0.01720209895 \ \mathrm{m^3 \ kg^{-1} \ s^{-2}}$

Speed of light $\qquad$ $c = 299792458 \ \mathrm{m \ s^{-1}}$

Primary constants

Light-time for unit distance (1 ua) $\qquad$ $\tau_A = 499.004786 \ \mathrm{s}$

Equatorial radius of Earth $\qquad$ $a_e = 6378140 \ \mathrm{m}$

Equatorial radius of Earth (IUGG value) $\qquad$ $a_e = 6378136 \ \mathrm{m}$

Dynamical form-factor for Earth $\qquad$ $J_2 = 0.001082636$

Geocentric gravitational constant $\qquad$ $GE = 3.986004 \times 10^{14} \ \mathrm{m^3 \ s^{-2}}$

Constant of gravitation $\qquad$ $G = 6.67428 \times 10^{-11} \ \mathrm{m^3 \ kg^{-1} \ s^{-2}}$

Ratio of mass of moon to that of Earth $\qquad$ $\mu = 0.01230002$

$\qquad$ $1/\mu = 81.300587$

General precession in longitude, per Julian century, at standard epoch J2000 $\qquad$ $\rho = 5028''.796$

Obliquity of the ecliptic at standard epoch J2000 $\qquad$ $\varepsilon = 23°26'21''.448$

Derived constants

Constant of nutation at standard epoch J2000 $\qquad$ $N = 9''.2025$

Unit distance (ua $= c\tau_A$) $\qquad$ $\mathrm{ua} = 1.49597871464 \times 10^{11} \ \mathrm{m}$

Solar parallax ($\pi_0 = \arcsin(a_e/\mathrm{ua})$) $\qquad$ $\pi_0 = 8''.794143$

Constant of aberration for standard epoch J2000 $\qquad$ $\kappa = 20''.49552$

Flattening factor for the Earth $\qquad$ $f = 1/298.256 = 0.00335282$

Heliocentric gravitational constant ($GS = A^3 k^2/D^2$) $\qquad$ $GS = 1.32712438 \times 10^{20} \ \mathrm{m^3 \ s^{-2}}$

Ratio of mass of sun to that of the Earth ($S/E = (GS)/(GE)$) $\qquad$ $S/E = 332946.0$

Ratio of mass of sun to that of Earth + moon $\qquad$ $(S/E)/(1 + \mu) = 328900.56$

Mass of the sun ($S = (GS)/G$) $\qquad$ $S = 1.98844 \times 10^{30} \ \mathrm{kg}$

Ratios of mass of sun to masses of the planets

Mercury	6023600
Venus	408523.7
Earth + moon	328900.56
Mars	3098708
Jupiter	1047.349
Saturn	3497.898
Uranus	22902.98
Neptune	19412.24

PROPERTIES OF THE SOLAR SYSTEM

The following tables give various properties of the planets and characteristics of their orbits in the solar system. Certain properties of the sun and of the earth's moon are also included.

Explanations of the column headings:

- *Mass*: mass of the planet in units of 10^{24} kg
- *Radius*: radius at the equator in km
- *Density*: mean density in g/cm^3
- *Flattening*: degree of oblateness, defined as $(r_e - r_p)/r_e$, where r_e and r_p are the equatorial and polar radii, respectively
- *Potential coefficients*: coefficients in the spherical harmonic representation of the gravitational potential U by the equation

$$U(r, \phi) = (GM/r) [1 - \Sigma J_n (a/r)^n P_n (\sin \phi)],$$

where G is the gravitational constant, r the distance from the center of the planet, a the radius of the planet, M the mass, ϕ the latitude, and P_n the Legendre polynomial of degree n.

- *Gravity*: acceleration due to gravity at the surface
- *Escape vel.*: velocity needed at the surface of the planet to escape the gravitational pull
- *Dist. to sun*: semi-major axis of the elliptical orbit in astronomical units (1 ua ≈ $1.496 \cdot 10^8$ km)
- *ε*: eccentricity of the orbit
- *Ecliptic angle*: angle between the planetary orbit and the plane of the earth's orbit around the sun
- *Inclin.*: angle between the equatorial plane of the planet and the plane of the planetary orbit
- *Orbit period*: period of revolution around the sun measured in years
- *Rotation period*: period of rotation of the planet measured in hours. A negative value indicates retrograde rotation.
- *Albedo*: ratio of the light reflected from the planet to the light incident on it
- *No. of satellites*: Number of confirmed satellites; this includes satellites that have not been named.
- T_{sur}: mean temperature at the surface
- P_{sur}: pressure of the atmosphere at the surface

The last four entries in the table are *dwarf planets* as defined by the International Astronomical Union. These are bodies in orbit around the sun that are massive enough to adopt a near-spherical shape as a result of their self-gravity, but are appreciably smaller than the major planets. *Plutoids* form a subset of the dwarf plan-

ets; their orbits are larger than that of Neptune (see Ref. 9). As of 2008, the IAU has recognized the names for three plutoids: Pluto, Eris, and Makemake.

The following general information on the solar system is of interest:

Mass of the earth = M_e = $5.9736 \cdot 10^{24}$ kg
Total mass of planetary system = $2.669 \cdot 10^{27}$ kg = 447 M_e
Total angular momentum of planetary system = $3.148 \cdot 10^{43}$ kg m^2 s^{-1}
Total kinetic energy of the planets = $1.99 \cdot 10^{35}$ J
Total rotational energy of planets = $0.7 \cdot 10^{35}$ J

Properties of the sun:

Mass = $1.9884 \cdot 10^{30}$ kg = 332943 M_e
Radius = $6.9551 \cdot 10^8$ m
Surface area = $6.079 \cdot 10^{18}$ m^2
Volume = $1.409 \cdot 10^{27}$ m^3
Mean density = 1.411 g/cm^3
Gravity at surface = 27398 cm/s^2
Escape velocity at surface = $6.177 \cdot 10^5$ m/s
Effective temperature = 5780 K
Total radiant power emitted (luminosity) = $3.8427 \cdot 10^{26}$ W
Surface flux of radiant energy = $6.322 \cdot 10^7$ W/m^2
Flux of radiant energy at the earth (Solar Constant) = 1366.4 W/m^2 (Ref. 8)

References

1. *Planetary Fact Sheet*, NASA Goddard Space Flight Center, <nssdc.gsfc.nasa.gov/planetary/factsheet>, November 2007.
2. The Planetary Society, <www.planetary.org/explore/topics/groups/our_solar_system/>.
3. Arnet, B., *The Nine Planets*, <www.nineplanets.org>.
4. Onasch, B., *Our Solar System*, <www.onasch.de/astro/>.
5. *The Astronomical Almanac for the year 2007*, U.S. Government Printing Office, Washington, 2005; available online at <asa.usno.navy.mil/>.
6. Lang, K. R., *Astrophysical Data: Planets and Stars*, Springer-Verlag, New York, 1992.
7. Cox, A. N., *Allen's Astrophysical Quantities, Fourth Edition*, Springer-Verlag, New York, 2000; this is a revision of Allen, C. W., *Astrophysical Quantities, Third Edition*, 1983.
8. Amsler, C., et al, *Physics Letters* B667, 1, 2008; section on astrophysical constants available at <pdg.lbl.gov/2008/reviews/astrorpp.pdf>.
9. IAU Press Release, <www.iau.org/public_press/news/release/iau0804>, 11 June 2008.
10. Seidelmann, P. K., Editor, *Explanatory Supplement to the Astronomical Almanac*, University Science Books, Mill Valley, CA, 1992.

Planet	Mass 10^{24} kg	Radius km	Density g/cm^3	Flattening	Potential coefficients $10^3 J_2$	$10^6 J_3$	$10^6 J_4$	Gravity cm/s^2	Escape vel. km/s
Mercury	0.33022	2439.7	5.43	0.0000				370	4.25
Venus	4.8685	6051.8	5.24	0.000	0.027			887	10.36
Earth	5.9736	6378.14	5.52	0.00335364	1.08263	−2.54	−1.61	980	11.18
(Moon)	0.07349	1738.1	3.35	0.0012	0.2027			162	2.38
Mars	0.64185	3396.2	3.93	0.0064763	1.964	36		371	5.03
Jupiter	1898.6	71492	1.33	0.0648744	14.75	−580		2312	59.54
Saturn	568.46	60268	0.69	0.09796	16.45	−1000		896	35.5
Uranus	86.832	25559	1.27	0.02293	3.343			869	21.29
Neptune	102.43	24764	1.64	0.01708	3.41			1100	23.5

Dwarf planets:

Planet	Mass 10^{24} kg	Radius km	Density g/cm^3	Flattening	Potential coefficients $10^3 J_2$	$10^6 J_3$	$10^6 J_4$	Gravity cm/s^2	Escape vel. km/s
Pluto	0.0125	1195	1.75	0				58	1.2
Eris	0.0166	1225						≈80	
Makemake	≈0.004	750	≈2					≈47	≈0.84
Ceres	0.000943	471	2.08					27	0.51

Planet	Dist. to Sun ua	ε	Ecliptic angle	Inclin.	Orbit period (yr)	Rotation period (hr)	Albedo	No. of satellites
Mercury	0.38710	0.2056	7.00°	0.01°	0.2408467	1407.6	0.106	0
Venus	0.72333	0.0068	3.39°	177.36°	0.61519726	−5832.5	0.65	0
Earth	1.00000	0.0167	0	23.45°	1.0000174	23.9345	0.367	1
Mars	1.52366	0.0934	1.85°	25.19°	1.8808476	24.6229	0.150	2
Jupiter	5.20336	0.0484	1.305°	3.12°	11.862615	9.9250	0.52	63
Saturn	9.53707	0.0542	2.48°	26.73°	29.447498	10.656	0.47	60
Uranus	19.19126	0.0472	0.77°	97.86°	84.016846	−17.24	0.51	27
Neptune	30.06896	0.0086	1.77°	28.32°	164.79132	16.11	0.41	13

Dwarf planets:

Planet	Dist. to Sun ua	ε	Ecliptic angle	Inclin.	Orbit period (yr)	Rotation period (hr)	Albedo	No. of satellites
Pluto	39.48168	0.2488	17.14°	118°	247.92065	−153.29	0.6	3
Eris	67.7	0.44		44.19°	557		0.8	1
Makemake	45.79	0.159		28.96°	309.88			
Ceres	2.77	0.08		10.59°	4.599	9.074	0.090	

Planet	T_{sur} K	P_{sur} bar	Atmospheric composition CO_2	N_2	O_2	H_2O	H_2	He	A	Ne	CO	CH_4
Mercury	440	$1\cdot10^{-15}$										
Venus	737	90	96.5%	3.5%	69 ppm	20 ppm		12 ppm	70 ppm	7 ppm	17 ppm	
Earth	288	1.014	0.038%	78.084%	20.946%	0 to 3%	0.55 ppm	5.24 ppm	0.934%	18.18 ppm	1 ppm	1.7 ppm
Mars	210	0.007	95.32%	2.7%	0.13%	0.021%			1.6%	2.5 ppm	0.08%	
Jupiter	165	>>1000				4 ppm	89.8%	10.2%				0.30%
Saturn	134	>>1000					96.3%	3.25%				0.45%
Uranus	76	>>1000					82.5%	15.2%				2.3%
Neptune	56						80%	19%				1%
Pluto	50	$3\cdot10^{-6}$										

SATELLITES OF THE PLANETS

This table gives characteristics of the known satellites of the planets. The parameters covered are:

- Orbital period in units of earth days. An R following the value indicates a retrograde motion.
- Distance from the planet, as measured by the semi-major axis of the orbit
- Eccentricity of the orbit
- Inclination of the satellite orbit with respect to the equator of the planet
- Mass of the satellite in kilograms
- Radius of the satellite in kilometers
- Geometric albedo, which is a measure of the fraction of incident sunlight reflected by the satellite.

Since this is a very active field of research, the Internet sites listed below should be consulted for the most recent data.

References

1. *Solar System Dynamics*, Jet Propulsion Laboratory, California Institute of Technology, <ssd.jpl.nasa.gov/?phys_data>, June 2008.
2. The Planetary Society, <www.planetary.org/explore/topics/groups/our_solar_system/>.
3. Arnet, B., *The Nine Planets*, <www.nineplanets.org>.
4. Onasch, B., *Our Solar System*, <www.onasch.de/astro/>.
5. Sheppard, S. S., *The Giant Satellite and Moon Page*, <www.dtm.ciw.edu/sheppard/satellites/>.
6. *Gazetteer of Planetary Nomenclature*, U. S. Geological Survey, <planetarynames.wr.usgs.gov/append7.html>.
7. Seidelmann, P. K., Editor, *Explanatory Supplement to the Astronomical Almanac*, University Science Books, Mill Valley, CA, 1992.
8. Lang, K. R., *Astrophysical Data: Planets and Stars*, Springer-Verlag, New York, 1992.
9. Allen, C. W., *Astrophysical Quantities, Second Edition*, Athlone Press, London, 1955.

Planet		Satellite	Orb. period Earth days	Distance 10^3 km	Eccentricity	Inclination	Mass kg	Radius km	Albedo
Earth		Moon	27.321661	384.400	0.054900489	18.28–28.58°	$7.3483 \cdot 10^{22}$	1737.5	0.12
Mars	I	Phobos	0.31891023	9.378	0.0151	1.0°	$1.06 \cdot 10^{16}$	13.5×10.8×9.4	0.07
	II	Deimos	1.2624407	23.460	0.0005	0.9–2.7°	$2.4 \cdot 10^{15}$	7.5×6.1×5.5	0.07
Jupiter	I	Io	1.769137786	421.8	0.0041	0.04°	$8.932 \cdot 10^{22}$	1821.6	0.63
	II	Europa	3.551181041	671.1	0.0101	0.47°	$4.8 \cdot 10^{22}$	1560.8	0.67
	III	Ganymede	7.15455296	1070.4	0.0015	0.21°	$1.4819 \cdot 10^{23}$	2631.2	0.43
	IV	Callisto	16.6890184	1882.7	0.007	0.51°	$1.0759 \cdot 10^{23}$	2410.3	0.17
	V	Amalthea	0.49817905	181.4	0.003	0.40°	$7.17 \cdot 10^{18}$	131×73×67	0.09
	VI	Himalia	250.5662	11460	0.162	27.63°	$9.56 \cdot 10^{18}$	85	0.04
	VII	Elara	259.6528	11737	0.217	24.77°	$7.77 \cdot 10^{17}$	40	0.04
	VIII	Pasiphae	743.63 R	23620	0.409	145°	$1.91 \cdot 10^{17}$	18	0.04
	IX	Sinope	758.90 R	23940	0.250	153°	$7.77 \cdot 10^{16}$	14	0.04
	X	Lysithea	259.20	11720	0.112	29.02°	$7.77 \cdot 10^{16}$	12	0.04
	XI	Carme	734.17 R	23400	0.253	164°	$9.56 \cdot 10^{16}$	15	0.04
	XII	Ananke	629.77 R	21280	0.244	147°	$3.82 \cdot 10^{16}$	10	0.04
	XIII	Leda	240.92	11170	0.164	26.07°	$5.68 \cdot 10^{15}$	5	0.04
	XIV	Thebe	0.6745	221.9	0.018	0.8°	$7.77 \cdot 10^{17}$	55×45	0.05
	XV	Adrastea	0.29826	129	0.0015		$1.91 \cdot 10^{16}$	13×10×8	0.10
	XVI	Metis	0.29478	128	0.0002		$9.56 \cdot 10^{16}$	20	0.06
	XVII	Callirrhoe	758.77	24100	0.283		$8.7 \cdot 10^{14}$	4	0.04
	XVIII	Themisto	130.02	7507	0.242		$6.9 \cdot 10^{14}$	4	0.04
	XIX	Megaclite	752.86	23810	0.425		$2.1 \cdot 10^{14}$	2.7	0.04
	XX	Taygete	732.41	23360	0.251		$1.6 \cdot 10^{14}$	2.5	0.04
	XXI	Chaldene	723.72	23180	0.238		$7.5 \cdot 10^{13}$	1.9	0.04
	XXII	Harpalyke	623.32	21110	0.227		$1.2 \cdot 10^{14}$	2.2	0.04
	XXIII	Kalyke	742.06	23580	0.243		$1.9 \cdot 10^{14}$	2.6	0.04
	XXIV	Iocaste	631.60	21270	0.218		$1.9 \cdot 10^{14}$	2.6	0.04
	XXV	Erinome	728.46	23280	0.270		$4.5 \cdot 10^{13}$	1.6	0.04
	XXVI	Isonoe	726.63	23220	0.261		$7.5 \cdot 10^{13}$	1.9	0.04
	XVII	Praxidike	625.39	21150	0.220		$4.3 \cdot 10^{14}$	3.4	0.04
	XXVIII	Autonoe	760.95	23039	0.334		$9.0 \cdot 10^{13}$	2.0	0.04
	XXIX	Thyone	627.21	20940	0.229		$9.0 \cdot 10^{13}$	2.0	0.04
	XXX	Hermippe	633.90	21131	0.210		$9.0 \cdot 10^{13}$	2.0	0.04
	XXXI	Aitne	730.18	23231	0.264		$4.5 \cdot 10^{13}$	1.5	0.04
	XXXII	Eurydome	717.33	22685	0.276		$4.5 \cdot 10^{13}$	1.5	0.04
	XXXIII	Euanthe	620.49	20721	0.232		$4.5 \cdot 10^{13}$	1.5	0.04
	XXXIV	Euporie	550.74	19302	0.144		$1.5 \cdot 10^{13}$	1	0.04
	XXXV	Orthosie	622.56	20721	0.281		$1.5 \cdot 10^{13}$	1	0.04
	XXXVI	Sponde	748.34	23487	0.312		$1.5 \cdot 10^{13}$	1	0.04

Planet		Satellite	Orb. period Earth days	Distance 10^3 km	Eccentricity	Inclination	Mass kg	Radius km	Albedo
	XXXVII	Kale	729.47	23217	0.260		$1.5 \cdot 10^{13}$	1	0.04
	XXXVIII	Pasithee	719.44	23096	0.267		$1.5 \cdot 10^{13}$	1	0.04
	XXXIX	Hegemone	739.6	23947	0.328		$4.5 \cdot 10^{13}$	1.5	
	XL	Mneme	620.0	21069	0.227		$1.5 \cdot 10^{13}$	1	
	XLI	Aoede	761.5	23981	0.432		$9.0 \cdot 10^{13}$	2.0	
	XLII	Thelxinoe	628.1	21162	0.221		$1.5 \cdot 10^{13}$	1	
	XLIII	Arche	723.9	22931	0.259		$4.5 \cdot 10^{13}$	1.5	
	XLIV	Kallichore	764.7	24043	0.264		$1.5 \cdot 10^{13}$	1	
	XLV	Helike	634.8	21263	0.156		$9.0 \cdot 10^{13}$	2.0	
	XLVI	Carpo	456.1	16989	0.430		$4.5 \cdot 10^{13}$	1.5	
	XLVII	Eukelade	746.4	23661	0.272		$9.0 \cdot 10^{13}$	2.0	
	XLVIII	Cyllene	737.8	24349	0.319		$1.5 \cdot 10^{13}$	1	
	XLIX	Kore	779.2	24543	0.325			1	
Saturn	I	Mimas	0.942421813	185.52	0.0202	1.53°	$3.75 \cdot 10^{19}$	196	0.5
	II	Enceladus	1.370217855	238.02	0.00452	1.86°	$6.50 \cdot 10^{19}$	250	1.0
	III	Tethys	1.887802160	294.66	0.00000	1.86°	$6.27 \cdot 10^{20}$	530	0.9
	IV	Dione	2.736914742	377.40	0.002230	0.02°	$1.10 \cdot 10^{21}$	560	0.7
	V	Rhea	4.517500436	527.04	0.00100	0.35°	$2.31 \cdot 10^{21}$	765	0.7
	VI	Titan	15.94542068	1221.83	0.029192	0.33°	$1.3455 \cdot 10^{23}$	2575	0.21
	VII	Hyperion	21.2766088	1481.1	0.104	0.43°	$1.59 \cdot 10^{19}$	205×130×110	0.3
	VIII	Iapetus	79.3301825	3561.3	0.02828	14.72°	$1.59 \cdot 10^{21}$	730	0.6
	IX	Phoebe	550.31 R	12952	0.16326	177°	$7.2 \cdot 10^{18}$	110	0.08
	X	Janus	0.6945	151.472	0.007	0.14°	$1.92 \cdot 10^{18}$	110×100×80	0.6
	XI	Epimetheus	0.6942	151.422	0.009	0.34°	$5.4 \cdot 10^{17}$	70×60×50	0.5
	XII	Helene	2.7369	377.40	0.005	0.0°	$2.5 \cdot 10^{16}$	18×16×15	0.6
	XIII	Telesto	1.8878	294.66			$7.2 \cdot 10^{15}$	17×14×13	1.0
	XIV	Calypso	1.8878	294.66			$3.6 \cdot 10^{15}$	17×11×11	0.7
	XV	Atlas	0.6019	137.670		0.3°	$1.1 \cdot 10^{16}$	20×10	0.4
	XVI	Prometheus	0.6130	139.353	0.0024	0.0°	$3.3 \cdot 10^{17}$	70×50×40	0.6
	XVII	Pandora	0.6285	141.70	0.0042	0.0°	$1.9 \cdot 10^{17}$	55×45×35	0.5
	XVIII	Pan	0.5750	133.583			$2.7 \cdot 10^{15}$	10	0.5
	XIX	Ymir	1315.14	23096	0.470		$4.9 \cdot 10^{15}$	8	0.06
	XX	Paaliaq	686.95	15199	0.364		$8.2 \cdot 10^{15}$	9.5	0.06
	XXI	Tarvos	926.23	18247	0.536		$2.7 \cdot 10^{15}$	6.5	0.06
	XXII	Ijiraq	451.42	11440	0.322		$1.2 \cdot 10^{15}$	5	0.06
	XXIII	Suttungr	1016.67	19463	0.114		$2.1 \cdot 10^{14}$	2.8	0.06
	XXIV	Kiviuq	449.22	11365	0.334		$3.3 \cdot 10^{15}$	7	0.06
	XXV	Mundilfari	952.77	18709	0.208		$2.1 \cdot 10^{14}$	2.8	0.06
	XXVI	Albiorix	783.45	16404	0.478		$2.1 \cdot 10^{16}$	13	0.06
	XXVII	Skathi	728.20	15647	0.270		$3.1 \cdot 10^{14}$	3.2	0.06
	XXVIII	Erriapus	871.19	17616	0.474		$7.6 \cdot 10^{14}$	4.3	0.06
	XXIX	Siarnaq	895.53	18160	0.295		$3.9 \cdot 10^{16}$	16	0.06
	XXX	Thrymr	1094.11	20382	0.470		$2.1 \cdot 10^{14}$	2.8	0.06
	XXXI	Narvi	1003.86	19007	0.431		$4.9 \cdot 10^{15}$	3.3	0.04
	XXXII	Methone	1.010	194			$1.65 \cdot 10^{13}$	1.5	
	XXXIII	Pallene	1.154	211			$3.92 \cdot 10^{13}$	2	
	XXXIV	Polydeuces	2.737	377.4				4	
	XXXV	Daphnis	0.594	136.5				3.5	
	XXXVI	Aegir	1117.52	20735				3.5	
	XXXVII	Bebhionn	834.84	17119				3	
	XXXVIII	Bergelmir	1005.74	19338				3	
	XXXIX	Bestla	1088.72	20129				3.5	
	XL	Farbauti	1085.55	20390				2.5	
	XLI	Fenrir	1260.35	22453				2	
	XLII	Fornjot	1494.20	25108				3	
	XLIII	Hati	1038.61	19856				3	
	XLIV	Hyrrokkin	931.86	18437				4	
	XLV	Kari	1230.97	22118				3.5	
	XLVI	Loge	1311.36	23065				3	
	XLVII	Skoll	878.29	17665				3	

Planet		Satellite	Orb. period Earth days	Distance 10^3 km	Eccentricity	Inclination	Mass kg	Radius km	Albedo
	XLVIII	Surtur	1297.36	22707				3	
	XLIX	Anthe		197.7				1	
	L	Jarnsaxa		18600				3	
	LI	Greip		18105				3	
	LII	Tarqeq		19720				3.5	
Uranus	I	Ariel	2.52037935	191.02	0.0034	0.3°	$1.35 \cdot 10^{21}$	579	0.39
	II	Umbriel	4.1441772	266.30	0.0050	0.36°	$1.17 \cdot 10^{21}$	584.7	0.21
	III	Titania	8.7058717	435.91	0.0022	0.14°	$3.52 \cdot 10^{21}$	788.9	0.27
	IV	Oberon	13.4632389	583.52	0.0008	0.10°	$3.01 \cdot 10^{21}$	761.4	0.23
	V	Miranda	1.41347925	129.39	0.0027	4.2°	$6.59 \cdot 10^{19}$	236	0.32
	VI	Cordelia	0.335033	49.77	0.0003	0.1°	$5.4 \cdot 10^{16}$	20.1	0.07
	VII	Ophelia	0.376409	53.79	0.0099	0.1°	$5.4 \cdot 10^{16}$	21.4	0.07
	VIII	Bianca	0.434577	59.17	0.0009	0.2°	$9.3 \cdot 10^{16}$	25.7	0.07
	IX	Cressida	0.463570	61.78	0.0004	0.0°	$3.4 \cdot 10^{17}$	39.8	0.07
	X	Desdemona	0.473651	62.68	0.0001	0.2°	$1.8 \cdot 10^{17}$	32.0	0.07
	XI	Juliet	0.493066	64.35	0.0007	0.1°	$5.6 \cdot 10^{17}$	46.8	0.07
	XII	Portia	0.513196	66.09	0.0001	0.1°	$1.7 \cdot 10^{18}$	67.6	0.07
	XIII	Rosalind	0.558459	69.94	0.0001	0.3°	$2.6 \cdot 10^{17}$	36.0	0.07
	XIV	Belinda	0.623525	75.26	0.0001	0.0°	$3.6 \cdot 10^{17}$	40.3	0.07
	XV	Puck	0.761832	86.01	0.0001	0.31°	$2.9 \cdot 10^{18}$	81.0	0.07
	XVI	Caliban	579.73	7231	0.1587			49	0.07
	XVII	Sycorax	1288.30	12179	0.5224		$5.4 \cdot 10^{18}$	95	0.07
	XVIII	Prospero	1978.29	16256	0.4448		$2.1 \cdot 10^{16}$	15	0.07
	XIX	Setebos	2225.21	17418	0.5914		$2.1 \cdot 10^{16}$	15	0.07
	XX	Stephano	677.36	8004	0.2292		$6.0 \cdot 10^{15}$	10	0.07
	XXI	Trinculo	749.24	8504	0.2200		$7.5 \cdot 10^{14}$	5	0.04
	XXII	Francisco	266.56	4276	0.146		$1.3 \cdot 10^{15}$	11	
	XXIII	Margaret	1687.01	14345	0.661		$1.0 \cdot 10^{15}$	5.5	
	XXIV	Ferdinand	2887.21	20901	0.368		$1.3 \cdot 10^{15}$	6	
	XXV	Perdita	0.638	76.42	0.0		$4.0 \cdot 10^{17}$	40	
	XXVI	Mab	0.923	97.73	0.0		$4.0 \cdot 10^{15}$	8	
	XXVII	Cupid	0.613	74.8	0.0		$1.2 \cdot 10^{15}$	6	
Neptune	I	Triton	5.8768541 R	354.76	0.000016	157.345°	$2.147 \cdot 10^{22}$	1353.4	0.76
	II	Nereid	360.13619	5513.4	0.7512	27.6°	$3.1 \cdot 10^{19}$	170	0.15
	III	Naiad	0.294396	48.227	0.0003	4.74°	$1.3 \cdot 10^{17}$	33	0.07
	IV	Thalassa	0.311485	50.075	0.0002	0.21°	$3.5 \cdot 10^{17}$	41	0.09
	V	Despina	0.334655	52.526	0.0001	0.07°	$2.3 \cdot 10^{18}$	75	0.09
	VI	Galatea	0.428745	61.953	0.0001	0.05°	$2.7 \cdot 10^{18}$	88	0.08
	VII	Larissa	0.554654	73.548	0.0014	0.20°	$4.8 \cdot 10^{18}$	104×89	0.09
	VIII	Proteus	1.122315	117.647	0.0004	0.55°	$4.9 \cdot 10^{19}$	218×208×201	0.10
	IX	Halimede	1879.08	16611	0.2646				
	X	Psamathe	9074.30	48096	0.3809		$1.5 \cdot 10^{16}$	14	
	XI	Sao	2912.72	22228	0.1365				
	XII	Laomedeia	3171.33	23567	0.3969				
	XIII	Neso	9740.73	49285	0.5714				
Plutoids*									
Pluto	I	Charon	6.387	17.536	0.0022	99°	$1.6 \cdot 10^{21}$	593	0.37
	II	Nix	24.86	48.708	0.0030		$5 \cdot 10^{16}$	22–65	
	III	Hydra	38.20	64.749	0.0051		$5 \cdot 10^{16}$	22–65	
Eris	I	Dysnomia		30				100–200	

* In June 2008 the International Astronomical Union decided on the name *plutoid* for the category of transneptunian dwarf planets. Plutoids are celestial bodies in orbit around the sun at a semimajor axis greater than that of Neptune and sufficiently massive to adopt a near-spherical shape. See <www.iau.org/public_press/news/release/iau0804/>.

INTERSTELLAR MOLECULES

Frank J. Lovas and Lewis E. Snyder

A number of molecules have been detected in the interstellar medium, in circumstellar envelopes around evolved stars, and comae and tails of comets through observation of their microwave, infrared, or optical spectra. The following list gives the molecules and the particular isotopic species that have been reported so far. Molecules are listed by molecular formula in the Hill order. All species not footnoted otherwise are observed in interstellar clouds, while some are also found in comets and circumstellar clouds. The list was last updated in October 2008 and lists 162 molecules (298 isotopic forms).

References

1. Lovas, F. J., Recommended Rest Frequencies for Observed Interstellar Molecule Microwave Transitions — 2002 Revision, *J. Phys. Chem. Ref. Data* 33, 177–355 (2004); and update appearing at http://physics.nist.gov/PhysRefData/micro/html/contents.html
2. Snyder, L. E., Cometary Molecules, Internat. Astron. Union Symposium No. 150, *Astrochemistry of Cosmic Phenomena*, ed. P. D. Singh, Kluwer Academic Publishers, Dordrecht, The Netherlands, pp. 427–434 (1992).

Molecular formula	Name	Isotopic species
AlCl	Aluminum monochloride	$Al^{35}Cl$[a]
		$Al^{37}Cl$[a]
AlF	Aluminum monofluoride	AlF[a]
CAlN	Aluminum isocyanide	$AlNC$[a]
CF^+	Fluoromethylidynium ion	CF^+
CH	Methylidyne	CH
CH^+	Methyliumylidene	CH^+
CHN	Hydrogen cyanide	HCN
		$H^{13}CN$
		$HC^{15}N$
		DCN
CHN	Hydrogen isocyanide	HNC
		$H^{15}NC$
		$HN^{13}C$
		DNC
		$D^{15}NC$
CHNO	Isocyanic acid	HNCO
		DNCO
CHNO	Hydroxyl cyanide	HOCN
CHNS	Isothiocyanic acid	HNCS
CHO	Oxomethyl	HCO
CHO^+	Oxomethylium	HCO^+
		$H^{13}CO^+$
		$HC^{17}O^+$
		$HC^{18}O^+$
		DCO^+
		$D^{13}CO^+$
CHO^+	Hydroxymethylidyne	HOC^+
CHO_2^+	Hydroxyoxomethylium	$HOCO^+$
CHP	Phosphaethyne	HCP[a]
CHS^+	Thiooxomethylium	HCS^+
CH_2	Methylene	CH_2
CH_2N^+	Iminomethylium	$HCNH^+$
CH_2N	Methylene amidogen	CH_2N
CH_2N_2	Cyanamide	NH_2CN
CH_2O	Formaldehyde	H_2CO
		$H_2^{13}CO$
		$H_2C^{18}O$
		HDCO
		D_2CO
CH_2O_2	Formic acid	HCOOH
		$H^{13}COOH$
		HCOOD
		DCOOH
CH_2S	Thioformaldehyde	H_2CS
		$H_2^{13}CS$
		$H_2C^{34}S$
		HDCS
		D_2CS
CH_3	Methyl	CH_3[a]
CH_3N	Methanimine	CH_2NH
		$^{13}CH_2NH$
CH_3NO	Formamide	NH_2CHO
		$NH_2^{13}CHO$
CH_3O^+	Hydroxymethylium ion	H_2COH^+
CH_4	Methane	CH_4
CH_4O	Methanol	CH_3OH
		$^{13}CH_3OH$
		$CH_3^{18}OH$
		CH_2DOH
		CH_3OD
		CHD_2OH
		CD_3OH
CH_4S	Methanethiol	CH_3SH
CH_5N	Methylamine	CH_3NH_2
CMgN	Magnesium cyanide	$MgCN$[a]
CMgN	Magnesium isocyanide	$^{24}MgNC$[a]
		$^{25}MgNC$[a]
		$^{26}MgNC$[a]
CN	Cyanide radical	CN
		^{13}CN
		$C^{15}N$
CN^+	Cyanide radical ion	CN^+[b]
CNNa	Sodium cyanide	$NaCN$[a]
CNSi	Silicon cyanide	$SiCN$[a]
CNSi	Silicon isocyanide	$SiNC$[a]
CN_2	Cyanoimidogen	NCN[b]
CO	Carbon monoxide	CO
		^{13}CO
		$C^{17}O$
		$C^{18}O$
		$^{13}C^{17}O$
		$^{13}C^{18}O$
		^{14}CO

Molecular formula	Name	Isotopic species
CO^+	Carbon monoxide ion	CO^+
COS	Carbon oxysulfide	OCS
		$OC^{34}S$
		$O^{13}CS$
		^{18}OCS
CO_2	Carbon dioxide	CO_2
CO_2^+	Carbon dioxide ion	$CO_2^{+\,b}$
CP	Carbon phosphide	CP^a
CS	Carbon monosulfide	CS
		$C^{33}S$
		$C^{34}S$
		$C^{36}S$
		^{13}CS
		$^{13}C^{34}S$
CSi	Silicon carbide	SiC^a
C_2	Dicarbon	C_2
C_2H	Ethynyl	C_2H
		^{13}CCH
		$C^{13}CH$
		C_2D
C_2HN	Cyanomethylene	HCCN
C_2HNO	Cyanoformaldehyde	CNCHO
C_2H_2	Acetylene	HCCH
C_2H_2N	Cyanomethyl	CH_2CN
C_2H_2O	Ketene	H_2CCO
C_2H_3N	Acetonitrile	CH_3CN
		$^{13}CH_3CN$
		$CH_3^{13}CN$
		$CH_3C^{15}N$
		CH_2DCN
C_2H_3N	Isocyanomethane	CH_3NC
C_2H_3N	Keteneimine	CH_2CNH
C_2H_4	Ethylene	H_2CCH_2
$C_2H_4N_2$	Aminoacetonitrile	NH_2CH_2CN
C_2H_4O	Acetaldehyde	CH_3CHO
C_2H_4O	Ethylene oxide	$c-C_2H_4O$
C_2H_4O	*anti*-Ethenol	$a-CH_2CHOH$
C_2H_4O	*syn*-Ethenol	$s-CH_2CHOH$
$C_2H_4O_2$	Methyl formate	CH_3OCHO
$C_2H_4O_2$	Acetic acid	CH_3COOH
$C_2H_4O_2$	Glycolaldehyde	CH_2OHCHO
C_2H_5NO	Acetamide	CH_3CONH_2
C_2H_6	Ethane	$CH_3CH_3^b$
C_2H_6O	*trans*-Ethanol	$t-CH_3CH_2OH$
C_2H_6O	*gauche*-Ethanol	$g-CH_3CH_2OH$
C_2H_6O	Dimethyl ether	CH_3OCH_3
$C_2H_6O_2$	Ethylene glycol	$HOCH_2CH_2OH$
C_2O	Oxoethenylidene	CCO
C_2P	Phosphaethenylidene	CCP^a
C_2S	Thioxoethenylidene	CCS
		$CC^{34}S$
		^{13}CCS
		$C^{13}CS$
C_2Si	Silicon dicarbide	$c-SiC_2$
		$c-^{29}SiC_2$
		$c-^{30}SiC_2$
		$c-Si^{13}CC$

Molecular formula	Name	Isotopic species
C_3	Tricarbon	C_3
C_3H	Cyclopropenylidyne	$c-C_3H$
		$c-CC^{13}CH$
C_3H	Propenylidyne	$l-C_3H$
C_3HN	Cyanoacetylene	HCCCN
		$H^{13}CCCN$
		$HC^{13}CCN$
		$HCC^{13}CN$
		$HCCC^{15}N$
		DCCCN
C_3HN	Isocyanoacetylene	HCCNC
C_3HN	3-Imino-1,2-propadienylidene	HNCCC
C_3H_2	Cyclopropenylidene	$c-C_3H_2$
		$c-H^{13}CCCH$
		$c-HC^{13}CCH$
		$c-C_3HD$
C_3H_2	Propadienylidene	$l-H_2CCC$
$C_3H_2N^+$	Protonated cyanoacetylene	$HCCCNH^+$
C_3H_2O	2-Propynal	HCCCHO
C_3H_2O	Cyclopropenone	$c-C_3H_2O$
C_3H_3N	Acrylonitrile (vinyl cyanide)	CH_2CHCN
		$^{13}CH_2CHCN$
		$CH_2^{13}CHCN$
C_3H_4	Propyne	CH_3CCH
		$CH_3C^{13}CH$
		$^{13}CH_3CCH$
		CH_2DCCH
		CH_3CCD
C_3H_4O	Propenal	CH_2CHCHO
C_3H_5N	Propanenitrile (ethyl cyanide)	CH_3CH_2CN
		$^{13}CH_3CH_2CN$
		$CH_3^{13}CH_2CN$
		$CH_3CH_2^{13}CN$
C_3H_6	Propylene	CH_2CHCH_3
C_3H_6O	Acetone	$(CH_3)_2CO$
C_3H_6O	Propanal	CH_3CH_2CHO
C_3N	Cyanoethynyl	CCCN
		$^{13}CCCN$
		$C^{13}CCN$
		$CC^{13}CN$
C_3N^-	Cyanoethynyl anion	$CCCN^-$
C_3O	1,2-Propadienylidene, 3-oxo	CCCO
C_3S	1,2-Propadienylidene, 3-thioxo	CCCS
		$CCC^{34}S$
		$C^{13}CCS$
C_3Si	Silicon tricarbon	SiC_3
C_4H	1,3-Butadiynyl radical	HCCCC
		$H^{13}CCCC$
		$HC^{13}CCC$
		$HCC^{13}CC$
		$HCCC^{13}C$
		DCCCC
C_4H^-	1,3-Butadiynyl anion	$HCCCC^-$
C_4HN	3-Cyano-1,2-propadienylidene	HCCCCN
C_4H_2	Butatrienylidene	H_2CCCC
C_4H_2	1,3-Butadiyne	$HCCCCH^a$
C_4H_3N	2-Butynenitrile	CH_3CCCN

Molecular formula	Name	Isotopic species	Molecular formula	Name	Isotopic species
C_4H_3N	Cyanoallene	CH_2CCHCN	H_2O	Water	H_2O
C_4Si	Silicon tetracarbide	SiC_4^a			$H_2^{18}O$
		$SiCCC^{13}C$			HDO
C_5	Pentacarbon	C_5^a			D_2O
C_5H	2,4-Pentadiynylidyne	$HCCCCC$	H_2O^+	Oxoniumyl	$H_2O^{+\,b}$
C_5HN	2,4-Pentadiynenitrile	$HCCCCCN$	H_2S	Hydrogen sulfide	H_2S
		$H^{13}CCCCCN$			$H_2^{34}S$
		$HC^{13}CCCCN$			HDS
		$HCC^{13}CCCN$			D_2S
		$HCCC^{13}CCN$	H_3^+	Trihydrogen ion	H_3^+
		$HCCCC^{13}CN$			H_2D^+
		$DCCCCCN$			D_2H^+
C_5H_4	1,3-Pentadiyne	CH_3C_4H	H_3N	Ammonia	NH_3
C_5N	1,3-Butadiynylium, 4-cyano	C_5N			$^{15}NH_3$
C_6H	1,3,5-Hexatriynyl	$HCCCCCC$			NH_2D
C_6H^-	1,3,5-Hexatriynyl anion	$HCCCCCC^-$			NHD_2
C_6H_2	1,3,5-Hexatriyne	$HCCCCCCH^a$			ND_3
C_6H_2	1,2,3,4,5-Hexapentaenylidene	$H_2CCCCCC$	H_3O^+	Oxonium hydride	H_3O^+
C_6H_3N	Methylcyanodiacetylene	CH_3C_4CN	H_4Si	Silane	SiH_4^a
C_6H_6	Benzene	C_6H_6	NO	Nitric oxide	NO
C_7H	2,4,6-Heptatriynylidyne	$HCCCCCCC$	NP	Phosphorous nitride	PN
C_7HN	2,4,6-Heptatriynenitrile	HC_7N	NS	Nitrogen sulfide	NS
C_7H_4	Methyltriacetylene	CH_3C_6H			$N^{34}S$
C_8H	1,3,5,7-Octatraynyl	HC_8	NSi	Silicon nitride	SiN
C_8H^-	1,3,5,7-Octatraynyl anion	HC_8^-	N_2	Nitrogen	N_2
C_9HN	2,4,6,8-Nonatetraynenitrile	HC_9N	N_2^+	Nitrogen ion	$N_2^{+\,b}$
$C_{11}HN$	2,4,6,8,10-Undecapentaynenitrile	$HC_{11}N$	N_2O	Nitrous oxide	N_2O
ClH	Hydrogen chloride	$H^{35}Cl$	OP	Phosphorus monoxide	PO^a
		$H^{37}Cl$	OS	Sulfur monoxide	SO
ClK	Potassium chloride	$K^{35}Cl^a$			^{34}SO
		$K^{37}Cl^a$			^{33}SO
$ClNa$	Sodium chloride	$Na^{35}Cl^a$			$S^{18}O$
		$Na^{37}Cl^a$	OS^+	Sulfur monoxide ion	SO^+
FH	Hydrogen fluoride	HF	OSi	Silicon monoxide	SiO
FeO	Iron monoxide	FeO			$Si^{18}O$
HLi	Lithium hydride	^7LiH			^{29}SiO
HN	Imidogen	HN			^{30}SiO
HNO	Nitrosyl hydride	HNO	O_2	Oxygen	O_2
HN_2^+	Hydrodinitrogen(1+)	N_2H^+	O_2S	Sulfur dioxide	SO_2
		$^{15}NNH^+$			$^{33}SO_2$
		$N^{15}NH^+$			$^{34}SO_2$
		N_2D^+			$OS^{17}O$
HO	Hydroxyl	OH			$OS^{18}O$
		^{17}OH	SSi	Silicon monosulfide	SiS
		^{18}OH			$Si^{33}S$
HO^+	Oxoniumylidene	$OH^{+\,b}$			$Si^{34}S$
HS	Mercapto	SH			^{29}SiS
H_2	Hydrogen	H_2			$^{29}Si^{34}S$
H_2N	Amidogen	NH_2			^{30}SiS
					$^{30}Si^{34}S$
			S_2	Sulfur	S_2^b

l- before the isotopic species indicates a linear configuration, while *c-* indicates a cyclic molecule.

[a] Reported only in circumstellar clouds.

[b] Reported only in comets.

MASS, DIMENSIONS, AND OTHER PARAMETERS OF THE EARTH

This table is a collection of data on various properties of the Earth. Most of the values are given in SI units. Note that 1 ua (astronomical unit) = 149,597,870 km.

References

1. Seidelmann, P. K., Ed., *Explanatory Supplement to the Astronomical Almanac*, University Science Books, Mill Valley, CA, 1992.
2. Lang, K. R., *Astrophysical Data: Planets and Stars*, Springer-Verlag, New York, 1992.

Quantity	Symbol	Value	Unit
Mass	M	$5.9723 \cdot 10^{27}$	g
Major orbital semi-axis	a_{orb}	1.000000	ua
		$1.4959787 \cdot 10^8$	km
Distance from Sun at perihelion	r_π	0.9833	ua
Distance from Sun at aphelion	r_α	1.0167	ua
Moment of perihelion passage	T_π	Jan. 2, 4 h 52 min	
Moment of aphelion passage	T_α	July 4, 5 h 05 min	
Siderial rotation period around Sun	P_{orb}	$31.5581 \cdot 10^6$	s
		365.25636	d
Mean rotational velocity	U_{orb}	29.78	km/s
Mean equatorial radius	$\bar{a}$	6378.140	km
Mean polar compression (flattening factor)	α	1/298.257	
Difference in equatorial and polar semi-axes	$a - c$	21.385	km
Compression of meridian of major equatorial axis	α_a	1/295.2	
Compression of meridian of minor equatorial axis	α_b	1/298.0	
Equatorial compression	ε	1/30 000	
Difference in equatorial semi-axes	$a - b$	213	m
Difference in polar semi-axes	$c_N - c_S$	~70	m
Polar asymmetry	η	$\sim 1 \cdot 10^{-5}$	
Mean acceleration of gravity at equator	g_e	9.78036	m/s^2
Mean acceleration of gravity at poles	g_p	9.83208	m/s^2
Difference in acceleration of gravity at pole and at equator	$g_p - g_e$	5.172	cm/s^2
Mean acceleration of gravity for entire surface of terrestrial ellipsoid	g	9.7978	m/s^2
Mean radius	R	6371.0	km
Area of surface	S	$5.10 \cdot 10^8$	km^2
Volume	V	$1.0832 \cdot 10^{12}$	km^3
Mean density	ρ	5.515	g/cm^3
Siderial rotational period	P	86,164.09	s
Rotational angular velocity	ω	$7.292116 \cdot 10^{-5}$	rad/s
Mean equatorial rotational velocity	v	0.46512	km/s
Rotational angular momentum	L	$5.861 \cdot 10^{33}$	J s
Rotational energy	E	$2.137 \cdot 10^{29}$	J
Ratio of centrifugal force to force of gravity at equator	q_c	0.0034677 = 1/288	
Moment of inertia	I	$8.070 \cdot 10^{37}$	kg m^2
Relative braking of Earth's rotation due to tidal friction	$\Delta\omega_e/\omega$	$-4.2 \cdot 10^{-8}$	century^{-1}
Relative secular acceleration of Earth's rotation	$\Delta\omega_i/\omega$	$+1.4 \cdot 10^{-8}$	century^{-1}
Not secular braking of Earth's rotation	$\Delta\omega/\omega$	$-2.8 \cdot 10^{-8}$	century^{-1}
Probable value of total energy of tectonic deformation of Earth	E_t	$\sim 1 \cdot 10^{23}$	J/century
Secular loss of heat of Earth through radiation into space	$\Delta'E_k$	$1 \cdot 10^{23}$	J/century
Portion of Earth's kinetic energy transformed into heat as a result of lunar and solar tides in the hydrosphere	$\Delta''E_k$	$1.3 \cdot 10^{23}$	J/century
Differences in duration of days in March and August	ΔP	0.0025 (March-August)	s
Corresponding relative annual variation in Earth's rotational velocity	$\Delta^*\omega/\omega$	$2.9 \cdot 10^{-8}$ (Aug.-March)	
Presumed variation in Earth's radius between August and March	Δ^*R	−9.2 (Aug.-March)	cm
Annual variation in level of world ocean	Δh_o	~10 (Sept.-March)	cm
Area of continents	S_c	$1.49 \cdot 10^8$	km^2
		29.2	% of surface
Area of world ocean	S_o	$3.61 \cdot 10^8$	km^2
		70.8	% of surface

Quantity	Symbol	Value	Unit
Mean height of continents above sea level	h_c	875	m
Mean depth of world ocean	h_o	3794	m
Mean thickness of lithosphere within the limits of the continents	$h_{c.l.}$	35	km
Mean thickness of lithosphere within the limits of the ocean	$h_{o.l.}$	4.7	km
Mean rate of thickening of continental lithosphere	$\Delta h/\Delta t$	10 – 40	m/10^6 y
Mean rate of horizontal extension of continental lithosphere	$\Delta l/\Delta t$	0.75 – 20	km/10^6 y
Mass of crust	m_1	$2.36 \cdot 10^{22}$	kg
Mass of mantle		$4.05 \cdot 10^{24}$	kg
Amount of water released from the mantle and core in the course of geological time		$3.40 \cdot 10^{21}$	kg
Total reserve of water in the mantle		$2 \cdot 10^{23}$	kg
Present content of free and bound water in the Earth's lithosphere		$2.4 \cdot 10^{21}$	kg
Mass of hydrosphere	m_h	$1.664 \cdot 10^{21}$	kg
Amount of oxygen bound in the Earth's crust		$1.300 \cdot 10^{21}$	kg
Amount of free oxygen		$1.5 \cdot 10^{18}$	kg
Mass of atmosphere	m_a	$5.136 \cdot 10^{18}$	kg
Mass of biosphere	m_b	$1.148 \cdot 10^{16}$	kg
Mass of living matter in the biosphere		$3.6 \cdot 10^{14}$	kg
Density of living matter on dry land		0.1	g/cm^2
Density of living matter in ocean		$15 \cdot 10^{-8}$	g/cm^3
Age of the Earth		$4.55 \cdot 10^9$	y
Age of oldest rocks		$4.0 \cdot 10^9$	y
Age of most ancient fossils		$3.4 \cdot 10^9$	y

GEOLOGICAL TIME SCALE

Period or epoch	Beginning and end, in 10^6 years	Key events
Cenozoic era		
Quaternary		
Holocene	0–0.0115	
Pleistocene*	0.0115–1.81	Homo Erectus breakout
Tertiary		
Pliocene*	1.81–5.3	Ape man fossils
Miocene	5.3–23	Origin of grass
Oligocene	23–34	Rise of cats, dogs, pigs
Eocene	34–56	Debut of hoofed mammals
Paleocene	56–65	Earliest primates
Mesozoic era		
Cretaceous	65–145	Demise of dinosaurs
Jurassic	145–200	First birds
Triassic	200–251	Appearance of dinosaurs
Paleozoic era		
Permian	251–299	Flowers, insect pollination
Carboniferous	299–359	First conifers
Devonian	359–416	First vertebrates ashore
Silurian	416–444	Spore-bearing plants
Ordovician	444–488	First animals ashore
Cambrian	488–542	Vertebrates appear
Pre-Cambrian		
Pre-Cambrian III (Proterozoic)	542–2500	First plants, jellyfish
Pre-Cambrian II (Archean)	2500–3850	Photosynthetic bacteria
Pre-Cambrian I (Hadean)	3850–4600	Earth formed 4600 million years ago

* Some authorities place the boundary between the Pleistocene and Pliocene at $2.6 \cdot 10^6$ years.

References

1. U.S. Geological Survey Geologic Names Committee, 2007, Divisions of geologic time—Major chronostratigraphic and geochronologic units: U.S. Geological Survey Fact Sheet 2007-3015. Available on the Internet at <pubs.usgs.gov/fs/2007/3015/index.html>.
2. Walker, J. D., and Geissman, J. W., compilers, 2009, Geologic Time Scale: Geological Society of America, <www.geosociety.org/science/timescale/timescl.pdf>.
3. Calder, N., *Timescale - An Atlas of the Fourth Dimension*, Viking Press, New York, 1983.

ACCELERATION DUE TO GRAVITY

The acceleration due to gravity is tabulated here as a function of latitude and height above the Earth's surface. Values were calculated from the expression

$$g/(m/s^2) = 9.780356 (1 + 0.0052885 \sin^2 \phi - 0.0000059 \sin^2 2\phi) - 0.003086 H$$

where ϕ is the latitude and H is the height in kilometers.

Reference

Jursa, A. S., Ed., *Handbook of Geophysics and the Space Environment*, 4th ed., Air Force Geophysics Laboratory, 1985, p. 14–17.

ϕ	$H = 0$	$H = 1$ km	$H = 5$ km	$H = 10$ km
0	9.78036	9.77727	9.76493	9.74950
5	9.78075	9.77766	9.76532	9.74989
10	9.78191	9.77882	9.76648	9.75105
15	9.78381	9.78072	9.76838	9.75295
20	9.78638	9.78330	9.77095	9.75552
25	9.78956	9.78647	9.77413	9.75870
30	9.79324	9.79016	9.77781	9.76238
35	9.79732	9.79424	9.78189	9.76646
40	9.80167	9.79858	9.78624	9.77081
45	9.80616	9.80307	9.79073	9.77530
50	9.81065	9.80757	9.79522	9.77979
55	9.81501	9.81193	9.79958	9.78415
60	9.81911	9.81602	9.80368	9.78825
65	9.82281	9.81972	9.80738	9.79195
70	9.82601	9.82292	9.81058	9.79515
75	9.82860	9.82551	9.81317	9.79774
80	9.83051	9.82743	9.81508	9.79965
85	9.83168	9.82860	9.81625	9.80082
90	9.83208	9.82899	9.81665	9.80122

DENSITY, PRESSURE, AND GRAVITY AS A FUNCTION OF DEPTH WITHIN THE EARTH

This table gives the density ρ, pressure *p*, and acceleration due to gravity *g* as a function of depth below the Earth's surface, as calculated from the model of the structure of the Earth in Reference 1. The model assumes a radius of 6371 km for the Earth. The boundary between the crust and mantle (the Mohorovicic discontinuity) is taken as 21 km, while in reality it varies considerable with location.

References

1. Anderson, D. L., and Hart, R. S., *J. Geophys. Res.*, 81, 1461, 1976.
2. Carmichael, R. S., *CRC Practical Handbook of Physical Properties of Rocks and Minerals*, p. 467, CRC Press, Boca Raton, FL, 1989.

Depth km	ρ g/cm³	*p* kbar	*g* cm/s²	Depth km	ρ g/cm³	*p* kbar	*g* cm/s²
Crust				1771	4.96	752	994
0	1.02	0	981	2071	5.12	903	1002
3	1.02	3	982	2371	5.31	1061	1017
3	2.80	3	982	2671	5.45	1227	1042
21	2.80	5	983	2886	5.53	1352	1069
Mantle (solid)				**Outer core (liquid)**			
21	3.49	5	983	2886	9.96	1352	1069
41	3.51	12	983	2971	10.09	1442	1050
61	3.52	19	984	3371	10.63	1858	953
81	3.48	26	984	3671	11.00	2154	874
101	3.44	33	984	4071	11.36	2520	760
121	3.40	39	985	4471	11.69	2844	641
171	3.37	56	987	4871	11.99	3116	517
221	3.34	73	989	5156	12.12	3281	427
271	3.37	89	991	**Inner core (solid)**			
321	3.47	106	993	5156	12.30	3281	427
371	3.59	124	994	5371	12.48	3385	355
571	3.95	199	999	5771	12.52	3529	218
871	4.54	328	997	6071	12.53	3592	122
1171	4.67	466	992	6371	12.58	3617	0
1471	4.81	607	991				

OCEAN PRESSURE AS A FUNCTION OF DEPTH AND LATITUDE

The following table is based upon an ocean model which takes into account the equation of state of standard seawater and the dependence on latitude of the acceleration of gravity. The tabulated pressure value is the excess pressure over the ambient atmospheric pressure at the surface.

References

1. *International Oceanographic Tables, Volume 4*, Unesco Technical Papers in Marine Science No. 40, Unesco, Paris, 1987.
2. Saunders, P. M., and Fofonoff, N. P., *Deep-Sea Res.* 23, 109–111, 1976.

Pressure in MPa at the Specified Latitude

Depth (meters)	0°	15°	30°	45°	60°	75°	90°
0	0.0000	0.0000	0.0000	0.0000	0.0000	0.0000	0.0000
500	5.0338	5.0355	5.0404	5.0471	5.0537	5.0586	5.0605
1000	10.0796	10.0832	10.0930	10.1064	10.1198	10.1296	10.1333
1500	15.1376	15.1431	15.1577	15.1778	15.1980	15.2127	15.2182
2000	20.2076	20.2148	20.2344	20.2613	20.2882	20.3080	20.3153
2500	25.2895	25.2985	25.3231	25.3568	25.3905	25.4153	25.4244
3000	30.3831	30.3940	30.4236	30.4641	30.5047	30.5345	30.5453
3500	35.4886	35.5012	35.5358	35.5832	35.6307	35.6654	35.6782
4000	40.6056	40.6201	40.6598	40.7140	40.7683	40.8082	40.8229
4500	45.7342	45.7505	45.7952	45.8564	45.9176	45.9626	45.9791
5000	50.8742	50.8924	50.9421	51.0102	51.0785	51.1285	51.1469
5500	56.0255	56.0456	56.1004	56.1755	56.2508	56.3059	56.3262
6000	61.1882	61.2100	61.2700	61.3521	61.4344	61.4947	61.5168
6500	66.3619	66.3857	66.4508	66.5399	66.6292	66.6947	66.7187
7000	71.5467	71.5724	71.6427	71.7388	71.8352	71.9059	71.9318
7500	76.7426	76.7701	76.8456	76.9488	77.0523	77.1282	77.1560
8000	81.9493	81.9788	82.0594	82.1697	82.2804	82.3614	82.3911
8500	87.1669	87.1983	87.2841	87.4016	87.5193	87.6057	87.6373
9000	92.3950	92.4284	92.5194	92.6440	92.7689	92.8606	92.8941
9500	97.6346	97.6698	97.7661	97.8978	98.0300	98.1269	98.1624
10000	102.8800	102.9170	103.0185	103.1572	103.2961	103.3981	103.4355

PROPERTIES OF SEAWATER

In addition to the dependence on temperature and pressure, the physical properties of seawater vary with the concentration of the dissolved constituents. A convenient parameter for describing the composition is the salinity, S, which is defined in terms of the electrical conductivity of the seawater sample. The defining equation for the practical salinity is:

$$S = a_0 + a_1 K^{1/2} + a_2 K + a_3 K^{3/2} + a_4 K^2 + a_5 K^{5/2},$$

where K is the ratio of the conductivity of the seawater sample at 15 °C and atmospheric pressure to the conductivity of a potassium chloride solution in which the mass fraction of KCl is 0.0324356, at the same temperature and pressure. The values of the coefficients are:

$$a_0 = 0.0080 \qquad a_3 = 14.0941$$
$$a_1 = -0.1692 \qquad a_4 = -7.0261$$
$$a_2 = 25.3851 \qquad a_5 = 2.7081$$
$$\Sigma\, a_i = 35.0000$$

Thus when $K = 1$, $S = 35$ exactly (S is normally quoted in units of ‰, i.e., parts per thousand). The value of S can be roughly equated with the mass of dissolved material in grams per kilogram of seawater. Salinity values in the open oceans at midlatitudes typically fall between 34 and 36.

It is customary in oceanography to define the pressure at a given point as the pressure due to the column of water between that point and the surface. Thus by convention $P = 0$ at the sea surface. To a good approximation the pressure in decibars (dbar) can be equated to the depth in meters. Thus at 45° latitude the pressure is 5000 dbar at 4902 m, 10000 dbar at 9700 m.

The first table below gives several properties of seawater as a function of temperature for a salinity of 35. The second and third give density and electrical conductivity as a function of salinity at several temperatures, and the fourth lists typical concentrations of the main constituents of seawater as a function of salinity. The final table gives the freezing point as a function of salinity and pressure.

References

1. *The Practical Salinity Scale 1978 and the International Equation of State of Seawater 1980*, Unesco Technical Papers in Marine Science No. 36, Unesco, Paris, 1981; sections No. 37, 38, 39, and 40 in this series give background papers and detailed tables.
2. Kennish, M. J., *CRC Practical Handbook of Marine Science*, CRC Press, Boca Raton, FL, 1989.
3. Poisson, A. *IEEE J. Ocean. Eng.* OE-5, 50, 1981.
4. Webster, F., in *AIP Physics Desk Reference*, E. R. Cohen, D. R. Lide and G. L. Trigg, Eds., Springer-Verlag, New York, 2003.

Properties of Seawater as a Function of Temperature at Salinity $S = 35$ and Normal Atmospheric Pressure

ρ = density in g/cm^3

$\beta = (1/\rho)\,(d\rho/dS)$ = fractional change in density per unit change in salinity

$\alpha = -(1/\rho)\,(d\rho/dt)$ = fractional change in density per unit change in temperature (°C^{-1})

κ = electrical conductivity in S/cm

η = viscosity in mPa s (equal to cP)

c_p = specific heat in J/kg °C

v = speed of sound in m/s

t/°C	ρ/g cm^{-3}	$10^7\beta$	$10^7\,\alpha$/°C^{-1}	κ/S cm^{-1}	η/mPa s	c_p/J kg^{-1}°C^{-1}	v/m s^{-1}
0	1.028106	7854	526	0.029048	1.892	3986.5	1449.1
5	1.027675	7717	1136	0.033468	1.610		
10	1.026952	7606	1668	0.038103	1.388	3986.3	1489.8
15	1.025973	7516	2141	0.042933	1.221		
20	1.024763	7444	2572	0.047934	1.085	3993.9	1521.5
25	1.023343	7385	2970	0.053088	0.966		
30	1.021729	7338	3341	0.058373	0.871	4000.7	1545.6
35	1.019934	7300	3687				
40		7270	4004			4003.5	1563.2

Density of Surface Seawater in g/cm^3 as a Function of Temperature and Salinity

t/°C	$S = 0$	$S = 5$	$S = 10$	$S = 15$	$S = 20$	$S = 25$	$S = 30$	$S = 35$	$S = 40$
0	0.999843	1.003913	1.007955	1.011986	1.016014	1.020041	1.024072	1.028106	1.032147
5	0.999967	1.003949	1.007907	1.011858	1.015807	1.019758	1.023714	1.027675	1.031645
10	0.999702	1.003612	1.007501	1.011385	1.015269	1.019157	1.023051	1.026952	1.030862
15	0.999102	1.002952	1.006784	1.010613	1.014443	1.018279	1.022122	1.025973	1.029834
20	0.998206	1.002008	1.005793	1.009576	1.013362	1.017154	1.020954	1.024763	1.028583
25	0.997048	1.000809	1.004556	1.008301	1.012050	1.015806	1.019569	1.023343	1.027128
30	0.995651	0.999380	1.003095	1.006809	1.010527	1.014252	1.017985	1.021729	1.025483
35	0.994036	0.997740	1.001429	1.005118	1.008810	1.012509	1.016217	1.019934	1.023662
40	0.992220	0.995906	0.999575	1.003244	1.006915	1.010593	1.014278	1.017973	1.021679

Electrical Conductivity of Seawater in S/cm as a Function of Temperature and Salinity

$t/°C$	$S = 5$	$S = 10$	$S = 15$	$S = 20$	$S = 25$	$S = 30$	$S = 35$	$S = 40$
0	0.004808	0.009171	0.013357	0.017421	0.021385	0.025257	0.029048	0.032775
5	0.005570	0.010616	0.015441	0.020118	0.024674	0.029120	0.033468	0.037734
10	0.006370	0.012131	0.017627	0.022947	0.028123	0.033171	0.038103	0.042935
15	0.007204	0.013709	0.019905	0.025894	0.031716	0.037391	0.042933	0.048355
20	0.008068	0.015346	0.022267	0.028948	0.035438	0.041762	0.047934	0.053968
25	0.008960	0.017035	0.024703	0.032097	0.039276	0.046267	0.053088	0.059751
30	0.009877	0.018771	0.027204	0.035330	0.043213	0.050888	0.058373	0.065683

Composition of Seawater and Ionic Strength at Various Salinities (Ref. 2)

	Expressed as molality			As grams per kilogram of seawater		
Constituent	$S = 30$	$S = 35$	$S = 40$	$S = 30$	$S = 35$	$S = 40$
Cl^-	0.482	0.562	0.650	16.58	19.33	22.36
Br^-	0.00074	0.00087	0.00100	0.057	0.067	0.078
F^-		0.00007			0.001	
SO_4^{2-}	0.0104	0.0114	0.0122	0.97	1.06	1.14
HCO_3^-	0.00131	0.00143	0.00100	0.078	0.085	0.059
$NaSO_4^-$	0.0085	0.0108	0.0139	0.98	1.25	1.60
KSO_4^-	0.00010	0.00012	0.00015	0.013	0.016	0.020
Na^+	0.405	0.472	0.544	9.03	10.53	12.13
K^+	0.00892	0.01039	0.01200	0.338	0.394	0.455
Mg^{2+}	0.0413	0.0483	0.0561	0.974	1.139	1.323
Ca^{2+}	0.00131	0.00143	0.00154	0.051	0.056	0.060
Sr^{2+}	0.00008	0.00009	0.00011	0.007	0.008	0.009
$MgHCO_3^+$	0.00028	0.00036	0.00045	0.023	0.030	0.037
$MgSO_4$	0.00498	0.00561	0.00614	0.582	0.655	0.717
$CaSO_4$	0.00102	0.00115	0.00126	0.135	0.152	0.166
$NaHCO_3$	0.00015	0.00020	0.00024	0.012	0.016	0.020
H_3BO_3	0.00032	0.00037	0.00042	0.019	0.022	0.025
Ionic strength	0.5736	0.6675	0.7701			

Freezing Point of Seawater in °C as a Function of Salinity and Pressure

$P/dbar$	$S = 0$	5	10	15	20	25	30	35	40
0	0.000	−0.274	−0.542	−0.812	−1.083	−1.358	−1.638	−1.922	−2.212
50	−0.038	−0.311	−0.580	−0.849	−1.121	−1.396	−1.676	−1.960	−2.250
100	−0.075	−0.349	−0.618	−0.887	−1.159	−1.434	−1.713	−1.998	−2.287
500	−0.377	−0.650	−0.919	−1.188	−1.460	−1.735	−2.014	−2.299	−2.589

ABUNDANCE OF ELEMENTS IN THE EARTH'S CRUST AND IN THE SEA

This table gives the estimated abundance of the elements in the continental crust (in mg/kg, equivalent to parts per million by mass) and in seawater near the surface (in mg/L). Values represent the median of reported measurements. The concentrations of the less abundant elements may vary with location by several orders of magnitude.

References

1. Carmichael, R. S., Ed., *CRC Practical Handbook of Physical Properties of Rocks and Minerals*, CRC Press, Boca Raton, FL, 1989.
2. Bodek, I., et al., *Environmental Inorganic Chemistry*, Pergamon Press, New York, 1988.
3. Ronov, A. B., and Yaroshevsky, A. A., "Earth's Crust Geochemistry," in *Encyclopedia of Geochemistry and Environmental Sciences*, Fairbridge, R. W., Ed., Van Nostrand, New York, 1969.

Element	Abundance Crust mg/kg	Abundance Sea mg/L
Ac	5.5×10^{-10}	
Ag	7.5×10^{-2}	4×10^{-5}
Al	8.23×10^{4}	2×10^{-3}
Ar	3.5	4.5×10^{-1}
As	1.8	3.7×10^{-3}
Au	4×10^{-3}	4×10^{-6}
B	1.0×10^{1}	4.44
Ba	4.25×10^{2}	1.3×10^{-2}
Be	2.8	5.6×10^{-6}
Bi	8.5×10^{-3}	2×10^{-5}
Br	2.4	6.73×10^{1}
C	2.00×10^{2}	2.8×10^{1}
Ca	4.15×10^{4}	4.12×10^{2}
Cd	1.5×10^{-1}	1.1×10^{-4}
Ce	6.65×10^{1}	1.2×10^{-6}
Cl	1.45×10^{2}	1.94×10^{4}
Co	2.5×10^{1}	2×10^{-5}
Cr	1.02×10^{2}	3×10^{-4}
Cs	3	3×10^{-4}
Cu	6.0×10^{1}	2.5×10^{-4}
Dy	5.2	9.1×10^{-7}
Er	3.5	8.7×10^{-7}
Eu	2.0	1.3×10^{-7}
F	5.85×10^{2}	1.3
Fe	5.63×10^{4}	2×10^{-3}
Ga	1.9×10^{1}	3×10^{-5}
Gd	6.2	7×10^{-7}
Ge	1.5	5×10^{-5}
H	1.40×10^{3}	1.08×10^{5}
He	8×10^{-3}	7×10^{-6}
Hf	3.0	7×10^{-6}
Hg	8.5×10^{-2}	3×10^{-5}
Ho	1.3	2.2×10^{-7}
I	4.5×10^{-1}	6×10^{-2}
In	2.5×10^{-1}	2×10^{-2}
Ir	1×10^{-3}	
K	2.09×10^{4}	3.99×10^{2}
Kr	1×10^{-4}	2.1×10^{-4}
La	3.9×10^{1}	3.4×10^{-6}
Li	2.0×10^{1}	1.8×10^{-1}
Lu	8×10^{-1}	1.5×10^{-7}
Mg	2.33×10^{4}	1.29×10^{3}
Mn	9.50×10^{2}	2×10^{-4}
Mo	1.2	1×10^{-2}

Element	Abundance Crust mg/kg	Abundance Sea mg/L
N	1.9×10^{1}	5×10^{-1}
Na	2.36×10^{4}	1.08×10^{4}
Nb	2.0×10^{1}	1×10^{-5}
Nd	4.15×10^{1}	2.8×10^{-6}
Ne	5×10^{-3}	1.2×10^{-4}
Ni	8.4×10^{1}	5.6×10^{-4}
O	4.61×10^{5}	8.57×10^{5}
Os	1.5×10^{-3}	
P	1.05×10^{3}	6×10^{-2}
Pa	1.4×10^{-6}	5×10^{-11}
Pb	1.4×10^{1}	3×10^{-5}
Pd	1.5×10^{-2}	
Po	2×10^{-10}	1.5×10^{-14}
Pr	9.2	6.4×10^{-7}
Pt	5×10^{-3}	
Ra	9×10^{-7}	8.9×10^{-11}
Rb	9.0×10^{1}	1.2×10^{-1}
Re	7×10^{-4}	4×10^{-6}
Rh	1×10^{-3}	
Rn	4×10^{-13}	6×10^{-16}
Ru	1×10^{-3}	7×10^{-7}
S	3.50×10^{2}	9.05×10^{2}
Sb	2×10^{-1}	2.4×10^{-4}
Sc	2.2×10^{1}	6×10^{-7}
Se	5×10^{-2}	2×10^{-4}
Si	2.82×10^{5}	2.2
Sm	7.05	4.5×10^{-7}
Sn	2.3	4×10^{-6}
Sr	3.70×10^{2}	7.9
Ta	2.0	2×10^{-6}
Tb	1.2	1.4×10^{-7}
Te	1×10^{-3}	
Th	9.6	1×10^{-6}
Ti	5.65×10^{3}	1×10^{-3}
Tl	8.5×10^{-1}	1.9×10^{-5}
Tm	5.2×10^{-1}	1.7×10^{-7}
U	2.7	3.2×10^{-3}
V	1.20×10^{2}	2.5×10^{-3}
W	1.25	1×10^{-4}
Xe	3×10^{-5}	5×10^{-5}
Y	3.3×10^{1}	1.3×10^{-5}
Yb	3.2	8.2×10^{-7}
Zn	7.0×10^{1}	4.9×10^{-3}
Zr	1.65×10^{2}	3×10^{-5}

SOLAR IRRADIANCE AT THE EARTH

The solar luminosity (total radiant power emitted by the Sun) is $3.86 \cdot 10^{26}$ W, of which about 1366 W m^{-2} (the solar irradiance or "solar constant") reaches the top of the Earth's atmosphere. To a zeroth approximation, the sun can be considered a black body with an effective temperature of 5780 K, which implies a peak in the radiation at around 0.520 μm (5200 Å). The actual solar spectral emission is more complex, especially at ultraviolet and shorter wavelengths. The graph in Fig. 1, which was taken from Ref. 1, summarizes the solar irradiance at the top of the atmosphere in the range 0.3 μm to 10 μm.

While the solar irradiance has been known for some time to undergo both long-term and short-term variations, accurate measurements have become possible only recently. Figure 2, which is taken from Ref. 4, shows the variation over the last 35 years.

References

1. Jursa, A. S., Ed., *Handbook of Geophysics and the Space Environment*, Air Force Geophysics Laboratory, 1985.
2. Pierce, A. K., and Allen, R. G., "The Solar Spectrum between 0.3 and 10 μm", in *The Solar Output and Its Variation*, White, O. R., Ed., Colorado Associated University Press, Boulder, CO, 1977.
3. Lang, K. R., *Astrophysical Data. Planets and Stars*, Springer-Verlag, New York, 1992.
4. Hansen, J. E., and Sato, M., <www.columbia.edu/~mhs119/Solar>, 2011.
5. Frohlich, C., and Lean, J., *Astron. Astrophys. Rev.* 12, 273, 2004.

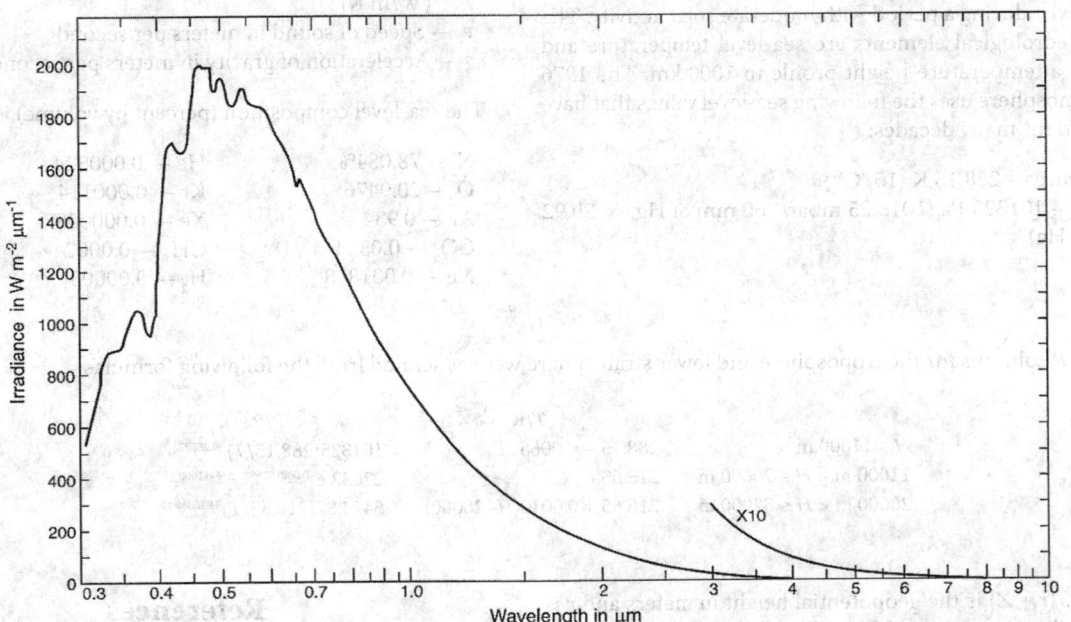

FIGURE 1. Wavelength dependence of solar irradiance.

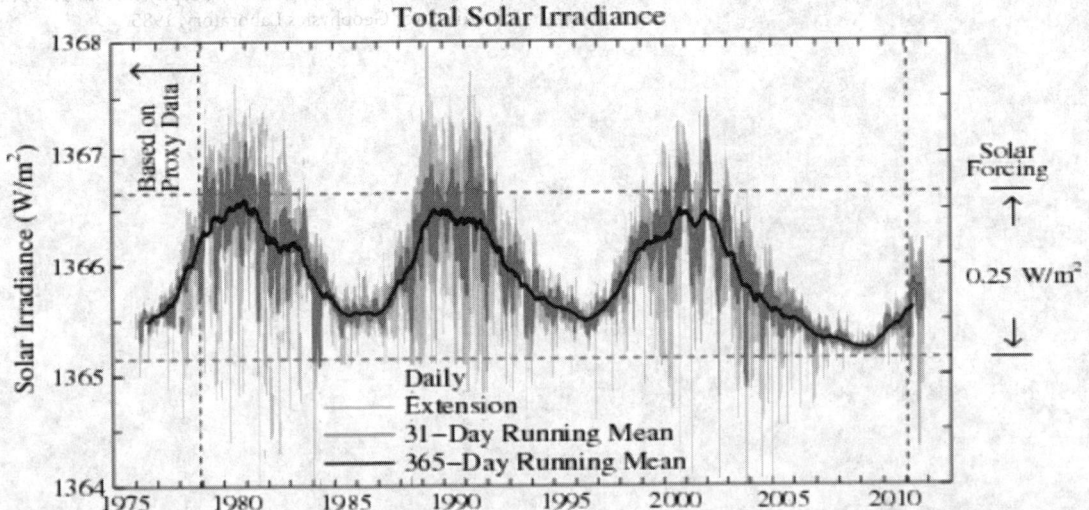

FIGURE 2. Variation of solar irradiance in the period 1976–2010.

U.S. STANDARD ATMOSPHERE (1976)

A Standard Atmosphere is a hypothetical vertical distribution of atmospheric temperature, pressure, and density that is roughly representative of year-round, midlatitude conditions. Typical uses are to serve as a basis for pressure altimeter calibrations, aircraft performance calculations, aircraft and rocket design, ballistic tables, meteorological diagrams, and various types of atmospheric modeling. The air is assumed to be dry and to obey the perfect gas law and the hydrostatic equation which, taken together, relate temperature, pressure, and density with vertical position. The atmosphere is considered to rotate with the Earth and to be an average over the diurnal cycle, the semiannual variation, and the range from active to quiet geomagnetic and sunspot conditions.

The U.S. Standard Atmosphere (1976) is an idealized, steady-state representation of mean annual conditions of the Earth's atmosphere from the surface to 1000 km at latitude 45° N, as it is assumed to exist during a period with moderate solar activity. The defining meteorological elements are sea-level temperature and pressure and a temperature-height profile to 1000 km. The 1976 Standard Atmosphere uses the following sea-level values that have been standard for many decades:

Temperature — 288.15 K (15 °C)
Pressure — 101325 Pa (1013.25 mbar, 760 mm of Hg, or 29.92 in. of Hg)

Density — 1225 g/m^3 (1.225 g/L)
Mean molar mass — 28.964 g/mol

The parameters included in this condensed version of the U.S. Standard Atmosphere are:

Z — Height (geometric) above mean sea level in meters
T — Temperature in kelvins
P — Pressure in pascals (1 Pa = 0.01 millibars)
ρ — Density in kilograms per cubic meter (1 kg/m^3 = 1 g/L)
n — Number density in molecules per cubic meter
v — Mean collision frequency in collisions per second
l — Mean free path in meters
η — Absolute viscosity in pascal seconds (1 Pa s = 1000 cP)
k — Thermal conductivity in joules per meter second kelvin (W/m K)
v_s — Speed of sound in meters per second
g — Acceleration of gravity in meters per second square

The sea-level composition (percent by volume) is taken to be:

N_2 — 78.084%	He — 0.000524
O_2 — 20.9476	Kr — 0.000114
Ar — 0.934	Xe — 0.0000087
CO_2 — 0.0314	CH_4 — 0.0002
Ne — 0.001818	H_2 — 0.00005

The T and P columns for the troposphere and lower stratosphere were generated from the following formulas:

	T/K	P/Pa
$H \leq 11000$ m	$288.15 - 0.0065\,H$	$101325(288.15/T)^{-5.25577}$
11000 m $< H \leq 20000$ m	216.65	$22632\,e^{-0.00015768832(H-11000)}$
20000 m $< H \leq 32000$ m	$216.65 + 0.0010(H-20000)$	$5474.87(216.65/T)^{34.16319}$

where $H = rZ/(r + Z)$ is the geopotential height in meters and r is the mean Earth radius at 45° N latitude, taken as 6356766 m. For altitudes up to 32 km, $\rho = 0.003483677(P/T)$ in the units used here. Formulas for the other quantities may be found in the references.

References

1. COESA, *U.S. Standard Atmosphere*, 1976, U.S. Government Printing Office, Washington, D.C., 1976.
2. Jursa, A. S., Ed., *Handbook of Geophysics and the Space Environment*, Air Force Geophysics Laboratory, 1985.

Z/m	T/K	P/Pa	ρ/kg m⁻³	n/m⁻³	ν/s⁻¹	l/m	η/Pa s	W m⁻¹ K⁻¹	v_s/m s⁻¹	g/m s⁻²
−5000	320.68	1.778E+05	1.931	4.015E+25	1.151E+10	4.208E−08	1.942E−05	0.02788	359.0	9.822
−4500	317.42	1.685E+05	1.849	3.845E+25	1.096E+10	4.395E−08	1.927E−05	0.02763	357.2	9.830
−4000	314.17	1.596E+05	1.770	3.680E+25	1.044E+10	4.592E−08	1.912E−05	0.02738	355.3	9.819
−3500	310.91	1.511E+05	1.693	3.520E+25	9.933E+09	4.800E−08	1.897E−05	0.02713	353.5	9.818
−3000	307.66	1.430E+05	1.619	3.366E+25	9.448E+09	5.019E−08	1.882E−05	0.02688	351.6	9.816
−2500	304.41	1.352E+05	1.547	3.217E+25	8.982E+09	5.252E−08	1.867E−05	0.02663	349.8	9.814
−2000	301.15	1.278E+05	1.478	3.102E+25	8.623E+09	5.447E−08	1.852E−05	0.02638	347.9	9.813
−1500	297.90	1.207E+05	1.411	2.935E+25	8.106E+09	5.757E−08	1.836E−05	0.02613	346.0	9.811
−1000	294.65	1.139E+05	1.347	2.801E+25	7.693E+09	6.032E−08	1.821E−05	0.02587	344.1	9.810
−500	291.40	1.075E+05	1.285	2.672E+25	7.298E+09	6.324E−08	1.805E−05	0.02562	342.2	9.808
0	288.15	1.013E+05	1.225	2.547E+25	6.919E+09	6.633E−08	1.789E−05	0.02533	340.3	9.807
500	284.90	9.546E+04	1.167	2.427E+25	6.556E+09	6.961E−08	1.774E−05	0.02511	338.4	9.805
1000	281.65	8.988E+04	1.112	2.311E+25	6.208E+09	7.310E−08	1.758E−05	0.02485	336.4	9.804
1500	278.40	8.456E+04	1.058	2.200E+25	5.874E+09	7.680E−08	1.742E−05	0.02459	334.5	9.802
2000	275.15	7.950E+04	1.007	2.093E+25	5.555E+09	8.073E−08	1.726E−05	0.02433	332.5	9.801
2500	271.91	7.469E+04	0.957	1.990E+25	5.250E+09	8.491E−08	1.710E−05	0.02407	330.6	9.799
3000	268.66	7.012E+04	0.909	1.891E+25	4.959E+09	8.937E−08	1.694E−05	0.02381	328.6	9.797
3500	265.41	6.579E+04	0.863	1.795E+25	4.680E+09	9.411E−08	1.678E−05	0.02355	326.6	9.796
4000	262.17	6.166E+04	0.819	1.704E+25	4.414E+09	9.917E−08	1.661E−05	0.02329	324.6	9.794
4500	258.92	5.775E+04	0.777	1.616E+25	4.160E+09	1.046E−07	1.645E−05	0.02303	322.6	9.793
5000	255.68	5.405E+04	0.736	1.531E+25	3.918E+09	1.103E−07	1.628E−05	0.02277	320.6	9.791
5500	252.43	5.054E+04	0.697	1.450E+25	3.687E+09	1.165E−07	1.612E−05	0.02250	318.5	9.790
6000	249.19	4.722E+04	0.660	1.373E+25	3.467E+09	1.231E−07	1.595E−05	0.02224	316.5	9.788
6500	245.94	4.408E+04	0.664	1.299E+25	3.258E+09	1.302E−07	1.578E−05	0.02197	314.4	9.787
7000	242.70	4.111E+04	0.590	1.227E+25	3.058E+09	1.377E−07	1.561E−05	0.02170	312.3	9.785
7500	239.46	3.830E+04	0.557	1.159E+25	2.869E+09	1.458E−07	1.544E−05	0.02144	310.2	9.784
8000	236.22	3.565E+04	0.526	1.093E+25	2.689E+09	1.545E−07	1.527E−05	0.02117	308.1	9.782
8500	232.97	3.315E+04	0.496	1.031E+25	2.518E+09	1.639E−07	1.510E−05	0.02090	306.0	9.781
9000	229.73	3.080E+04	0.467	9.711E+24	2.356E+09	1.740E−07	1.493E−05	0.02063	303.9	9.779
9500	226.49	2.858E+04	0.440	9.141E+24	2.202E+09	1.848E−07	1.475E−05	0.02036	301.7	9.777
10000	223.25	2.650E+04	0.414	8.598E+24	2.056E+09	1.965E−07	1.458E−05	0.02009	299.5	9.776
10500	220.01	2.454E+04	0.389	8.079E+24	1.918E+09	2.091E−07	1.440E−05	0.01982	297.4	9.774
11000	216.77	2.270E+04	0.365	7.585E+24	1.787E+09	2.227E−07	1.422E−05	0.01954	295.2	9.773
11500	216.65	2.098E+04	0.337	7.016E+24	1.653E+09	2.408E−07	1.422E−05	0.01953	295.1	9.771
12000	216.65	1.940E+04	0.312	6.486E+24	1.528E+09	2.605E−07	1.422E−05	0.01953	295.1	9.770
12500	216.65	1.793E+04	0.288	5.996E+24	1.412E+09	2.818E−07	1.422E−05	0.01953	295.1	9.768
13000	216.65	1.658E+04	0.267	5.543E+24	1.306E+09	3.048E−07	1.422E−05	0.01953	295.1	9.767
13500	216.65	1.533E+04	0.246	5.124E+24	1.207E+09	3.297E−07	1.422E−05	0.01953	295.1	9.765
14000	216.65	1.417E+04	0.228	4.738E+24	1.116E+09	3.566E−07	1.422E−05	0.01953	295.1	9.764
14500	216.65	1.310E+04	0.211	4.380E+24	1.032E+09	3.857E−07	1.422E−05	0.01953	295.1	9.762
15000	216.65	1.211E+04	0.195	4.049E+24	9.538E+08	4.172E−07	1.422E−05	0.01953	295.1	9.761
16000	216.65	1.035E+04	0.166	3.461E+24	8.153E+08	4.881E−07	1.422E−05	0.01953	295.1	9.758
17000	216.65	8.850E+03	0.142	2.959E+24	6.969E+08	5.710E−07	1.422E−05	0.01953	295.1	9.754
18000	216.65	7.565E+03	0.122	2.529E+24	5.958E+08	6.680E−07	1.422E−05	0.01953	295.1	9.751
19000	216.65	6.467E+03	0.104	2.162E+24	5.093E+08	7.814E−07	1.422E−05	0.01953	295.1	9.748
20000	216.65	5.529E+03	8.891E−02	1.849E+24	4.354E+08	9.139E−07	1.422E−05	0.01953	295.1	9.745
21000	217.58	4.729E+03	7.572E−02	1.574E+24	3.716E+08	1.073E−06	1.427E−05	0.01961	295.1	9.742
22000	218.57	4.048E+03	6.451E−02	1.341E+24	3.173E+08	1.260E−06	1.432E−05	0.01970	296.4	9.739
23000	219.57	3.467E+03	5.501E−02	1.144E+24	2.712E+08	1.477E−06	1.438E−05	0.01978	297.1	9.736
24000	220.56	2.972E+03	4.694E−02	9.759E+23	2.319E+08	1.731E−06	1.443E−05	0.01986	297.7	9.733
25000	221.55	2.549E+03	4.008E−02	8.334E+23	1.985E+08	2.027E−06	1.448E−05	0.01995	298.4	9.730
26000	222.54	2.188E+03	3.426E−02	7.123E+23	1.700E+08	2.372E−06	1.454E−05	0.02003	299.1	9.727
27000	223.54	1.880E+03	2.930E−02	6.092E+23	1.458E+08	2.773E−06	1.459E−05	0.02011	299.7	9.724
28000	224.53	1.610E+03	2.508E−02	5.214E+23	1.250E+08	3.240E−06	1.465E−05	0.02020	300.4	9.721
29000	225.52	1.390E+03	2.148E−02	4.466E+23	1.073E+08	3.783E−06	1.470E−05	0.02028	301.1	9.718
30000	226.51	1.197E+03	1.841E−02	3.828E+23	9.219E+07	4.414E−06	1.475E−05	0.02036	301.7	9.715
31000	227.50	1.031E+03	1.579E−02	3.283E+23	7.925E+07	5.146E−06	1.481E−05	0.02044	302.4	9.712
32000	228.49	8.891E+02	1.356E−02	2.813E+23	6.818E+07	5.995E−06	1.486E−05	0.02053	303.0	9.709
33000	230.97	7.673E+02	1.157E−02	2.406E+23	5.852E+07	7.021E−06	1.499E−05	0.02073	304.7	9.706
34000	233.74	6.634E+02	9.887E−03	2.056E+23	5.030E+07	8.218E−06	1.514E−05	0.02096	306.5	9.703
35000	236.51	5.746E+02	8.463E−03	1.760E+23	4.331E+07	9.601E−06	1.529E−05	0.02119	308.3	9.700

Z/m	T/K	P/Pa	ρ/kg m⁻³	n/m⁻³	ν/s⁻¹	l/m	η/Pa s	W m⁻¹ K⁻¹	v_s/m s⁻¹	g/m s⁻²
36000	239.28	4.985E+02	7.258E−03	1.509E+23	3.736E+07	1.120E−05	1.543E−05	0.02142	310.1	9.697
38000	244.82	3.771E+02	5.367E−03	1.116E+23	2.794E+07	1.514E−05	1.572E−05	0.02188	313.7	9.690
40000	250.35	2.871E+02	3.996E−03	8.308E+22	2.104E+07	2.034E−05	1.601E−05	0.02233	317.2	9.684
42000	255.88	2.200E+02	2.995E−03	6.227E+22	1.594E+07	2.713E−05	1.629E−05	0.02278	320.7	9.678
44000	261.40	1.695E+02	2.259E−03	4.697E+22	1.215E+07	3.597E−05	1.657E−05	0.02323	324.1	9.672
46000	266.93	1.313E+02	1.714E−03	3.564E+22	9.318E+06	4.740E−05	1.685E−05	0.02376	327.5	9.666
48000	270.65	1.023E+02	1.317E−03	2.738E+22	7.208E+06	6.171E−05	1.704E−05	0.02397	329.8	9.660
50000	270.65	7.978E+01	1.027E−03	2.135E+22	5.620E+06	7.913E−05	1.703E−05	0.02397	329.8	9.654
52000	269.03	6.221E+01	8.056E−04	1.675E+22	4.397E+06	1.009E−04	1.696E−05	0.02384	328.8	9.648
54000	263.52	4.834E+01	6.390E−04	1.329E+22	3.452E+06	1.272E−04	1.660E−05	0.02340	325.4	9.642
56000	258.02	3.736E+01	5.045E−04	1.049E+22	2.696E+06	1.611E−04	1.640E−05	0.02296	322.0	9.636
58000	252.52	2.872E+01	3.963E−04	8.239E+21	2.095E+06	2.051E−04	1.612E−05	0.02251	318.6	9.632
60000	247.02	2.196E+01	3.097E−04	6.439E+21	1.620E+06	2.624E−04	1.584E−05	0.02206	315.1	9.624
65000	233.29	1.093E+01	1.632E−04	3.393E+21	8.294E+05	4.979E−04	1.512E−05	0.02093	306.2	9.609
70000	219.59	5.221	8.283E−05	1.722E+21	4.084E+05	9.810E−04	1.438E−05	0.01978	297.1	9.594
75000	208.40	2.388	3.992E−05	8.300E+20	1.918E+05	2.035E−03	1.376E−05	0.01883	289.4	9.579
80000	198.64	1.052	1.846E−05	3.838E+20	8.656E+04	4.402E−03	1.321E−05	0.01800	282.5	9.564
85000	188.89	4.457E−01	8.220E−06	1.709E+20	3.766E+04	9.886E−03	1.265E−05	0.01716	275.5	9.550
90000	186.87	1.836E−01	3.416E−06	7.116E+19	1.560E+04	2.370E−02				9.535
95000	188.42	7.597E−02	1.393E−06	2.920E+19	6.440E+03	5.790E−02				9.520
100000	195.08	3.201E−02	5.604E−07	1.189E+19	2.680E+03	1.420E−01				9.505
110000	240.00	7.104E−03	9.708E−08	2.144E+18	5.480E+02	7.880E−01				9.476
120000	360.00	2.538E−03	2.222E−08	5.107E+17	1.630E+02	3.310				9.447
130000	469.27	1.251E−03	8.152E−09	1.930E+17	7.100E+01	8.800				9.418
140000	559.63	7.203E−04	3.831E−09	9.322E+16	3.800E+01	1.800E+01				9.389
150000	634.39	4.542E−04	2.076E−09	5.186E+16	2.300E+01	3.300E+01				9.360
160000	696.29	3.040E−04	1.233E−09	3.162E+16	1.500E+01	5.300E+01				9.331
170000	747.57	2.121E−04	7.815E−10	2.055E+16	1.000E+01	8.200E+01				9.302
180000	790.07	1.527E−04	5.194E−10	1.400E+16	7.200	1.200E+02				9.274
190000	825.16	1.127E−04	3.581E−10	9.887E+15	5.200	1.700E+02				9.246
200000	854.56	8.474E−05	2.541E−10	7.182E+15	3.900	2.400E+02				9.218
220000	899.01	5.015E−05	1.367E−10	4.040E+15	2.300	4.200E+02				9.162
240000	929.73	3.106E−05	7.858E−11	2.420E+15	1.400	7.000E+02				9.106
260000	950.99	1.989E−05	4.742E−11	1.515E+15	9.300E−01	1.100E+03				9.051
280000	965.75	1.308E−05	2.971E−11	9.807E+14	6.100E−01	1.700E+03				8.997
300000	976.01	8.770E−06	1.916E−11	6.509E+14	4.200E−01	2.600E+03				8.943
320000	983.16	5.980E−06	1.264E−11	4.405E+14	2.900E−01	3.800E+03				8.889
340000	988.15	4.132E−06	8.503E−12	3.029E+14	2.000E−01	5.600E+03				8.836
360000	991.65	2.888E−06	5.805E−12	2.109E+14	1.400E−01	8.000E+03				8.784
380000	994.10	2.038E−06	4.013E−12	1.485E+14	1.000E−01	1.100E+04				8.732
400000	995.83	1.452E−06	2.803E−12	1.056E+14	7.200E−02	1.600E+04				8.680
450000	998.22	6.447E−07	1.184E−12	4.678E+13	3.300E−02	3.600E+04				8.553
500000	999.24	3.024E−07	5.215E−13	2.192E+13	1.600E−02	7.700E+04				8.429
550000	999.67	1.514E−07	2.384E−13	1.097E+13	8.400E−03	1.500E+05				8.307
600000	999.85	8.213E−08	1.137E−13	5.950E+12	4.800E−03	2.800E+05				8.188
650000	999.93	4.887E−08	5.712E−14	3.540E+12	3.100E−03	4.800E+05				8.072
700000	999.97	3.191E−08	3.070E−14	2.311E+12	2.200E−03	7.300E+05				7.958
750000	999.98	2.260E−08	1.788E−14	1.637E+12	1.700E−03	1.000E+06				7.846
800000	999.99	1.704E−08	1.136E−14	1.234E+12	1.400E−03	1.400E+06				7.737
850000	1000.00	1.342E−08	7.824E−15	9.717E+11	1.200E−03	1.700E+06				7.630
900000	1000.00	1.087E−08	5.759E−15	7.876E+11	1.000E−03	2.100E+06				7.525
950000	1000.00	8.982E−09	4.453E−15	6.505E+11	8.700E−04	2.600E+06				7.422
1000000	1000.00	7.514E−09	3.561E−15	5.442E+11	7.500E−04	3.100E+06				7.322

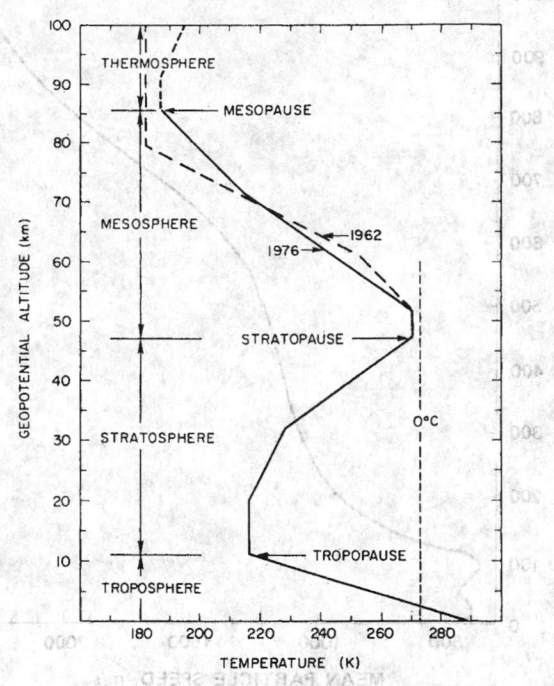

FIGURE 1. Temperature-height profile for U.S. Standard Atmosphere.

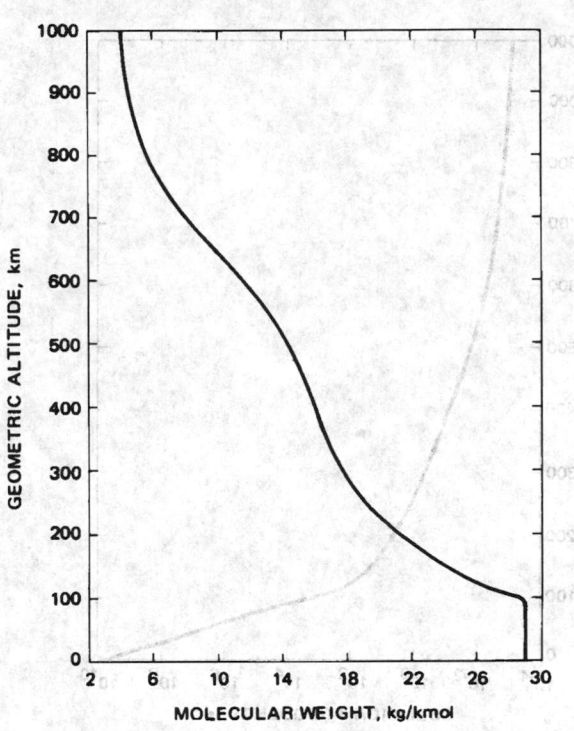

FIGURE 3. Mean molecular weight as a function of geometric altitude.

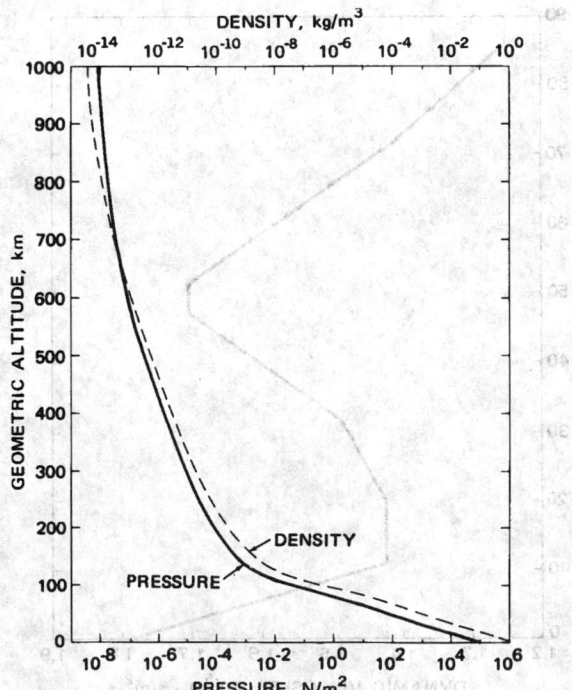

FIGURE 2. Total pressure and mass density as a function of geometric altitude.

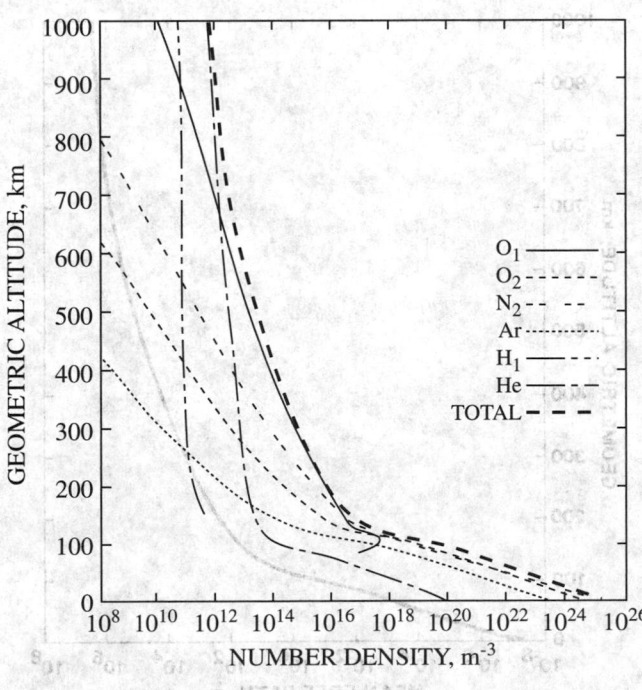

FIGURE 4. Number density of individual species and total number density as a function of geometric altitude.

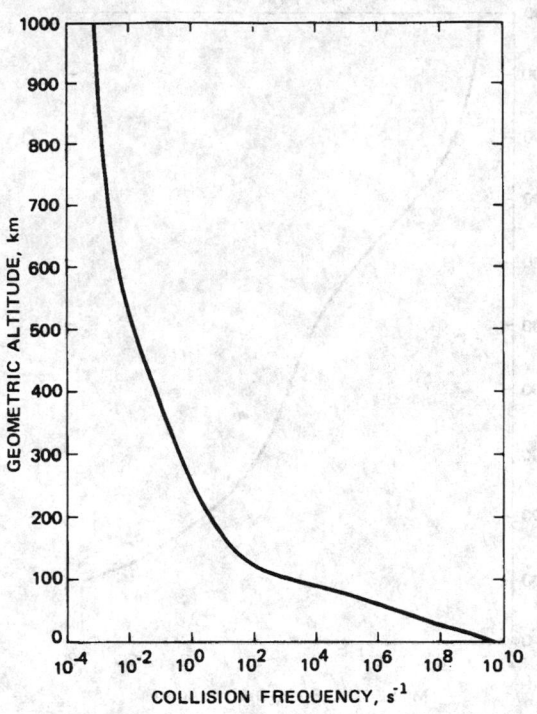

FIGURE 5. Collision frequency as a function of geometric altitude.

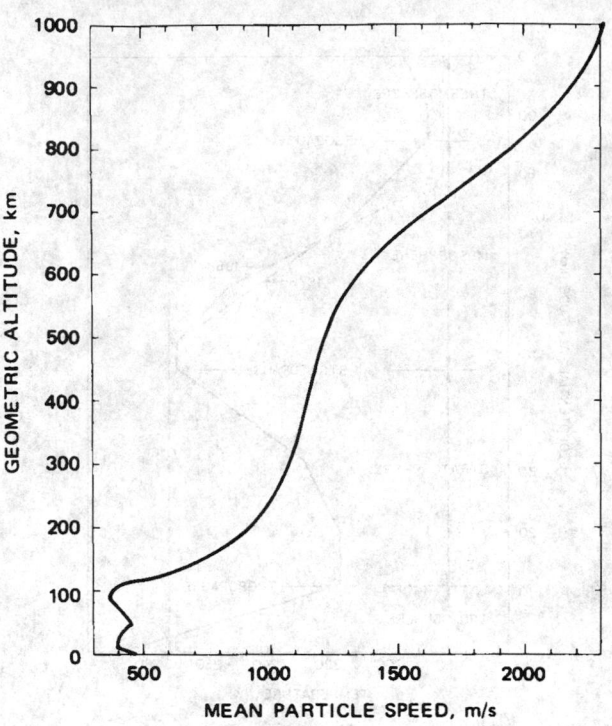

FIGURE 7. Mean air-particle speed as a function of geometric altitude.

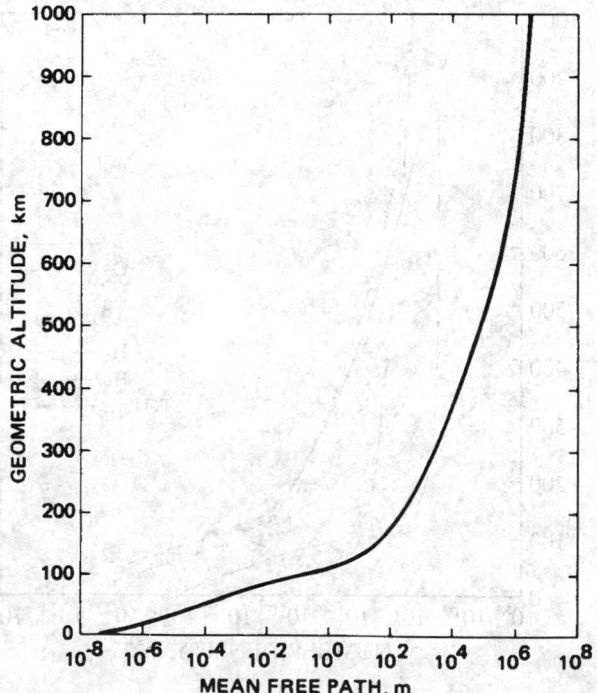

FIGURE 6. Mean free path as a function of geometric altitude.

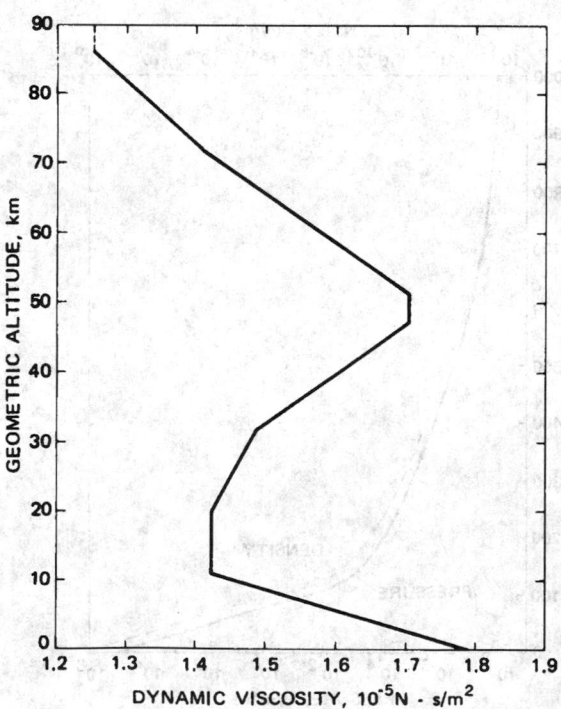

FIGURE 8. Dynamic viscosity as a function of geometric altitude.

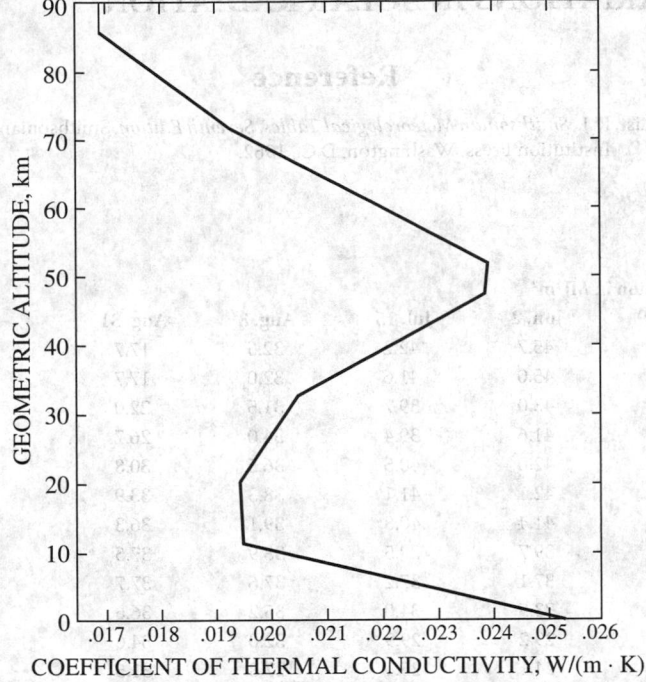

FIGURE 9. Coefficient of thermal conductivity as a function of geometric altitude.

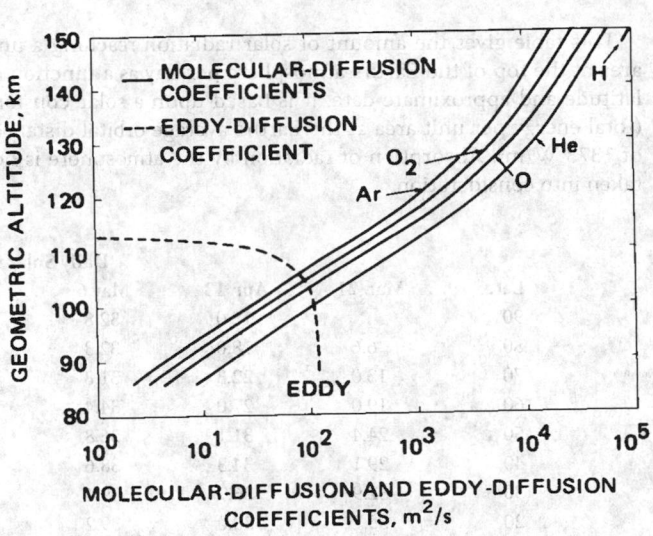

FIGURE 11. Molecular-diffusion and eddy-diffusion coefficients as a function of geometric altitude.

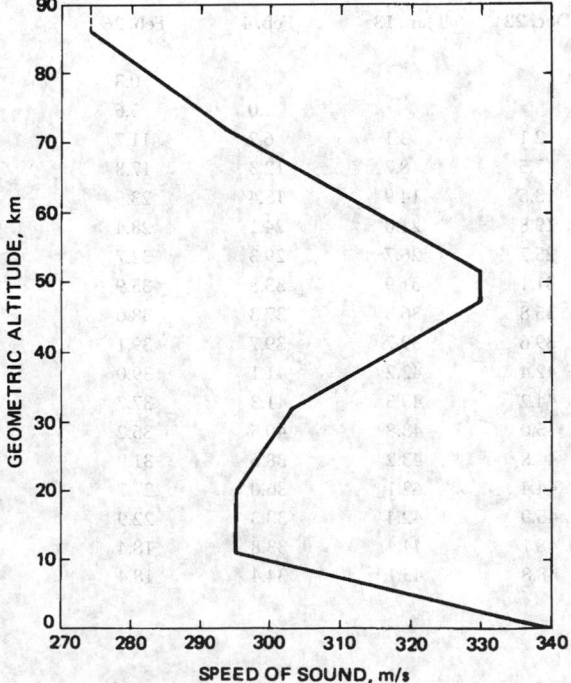

FIGURE 10. Speed of sound as a function of geometric altitude.

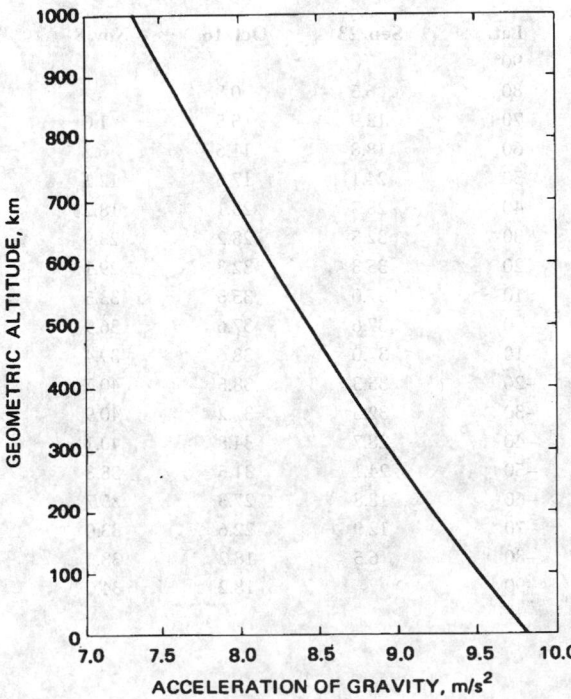

FIGURE 12. Acceleration of gravity as a function of geometric altitude.

GEOGRAPHICAL AND SEASONAL VARIATIONS IN SOLAR RADIATION

This table gives the amount of solar radiation reaching a unit area at the top of the Earth's atmosphere per day as a function of latitude and approximate date. It is based upon a solar constant (total energy per unit area at the Earth's average orbital distance) of 1373 W/m². Absorption of radiation by the atmosphere is not taken into consideration.

Reference

List, R. J., *Smithsonian Meteorological Tables, Seventh Edition*, Smithsonian Institution Press, Washington, D.C., 1962.

Daily Solar Radiation in MJ/m²

Lat.	Mar. 21	Apr. 13	May 6	May 29	Jun. 2	Jul. 15	Aug. 8	Aug. 31
90°		18.0	32.8	42.4	45.7	42.2	32.5	17.7
80	6.6	18.0	32.3	41.8	45.0	41.6	32.0	17.7
70	13.0	22.3	31.8	39.9	43.0	39.7	31.5	22.0
60	19.0	27.0	34.4	39.7	41.6	39.4	34.0	26.7
50	24.4	31.1	36.8	40.7	42.0	40.5	36.5	30.8
40	29.1	34.3	38.6	41.3	42.1	41.1	38.3	33.9
30	32.9	36.7	39.4	41.1	41.4	40.8	39.1	36.3
20	35.7	38.0	39.2	39.7	39.7	39.5	38.9	37.5
10	37.4	38.1	37.9	37.4	37.1	37.2	37.6	37.7
0	38.0	37.1	35.5	34.1	33.5	34.0	35.2	36.6
−10	37.4	35.0	32.3	30.0	29.2	29.9	32.0	34.6
−20	35.7	31.8	28.0	25.2	24.1	25.1	27.8	31.5
−30	32.9	27.8	23.1	19.7	18.5	19.7	22.8	27.4
−40	29.1	22.8	17.5	14.0	12.6	13.9	17.4	22.6
−50	24.4	17.3	11.7	8.2	7.0	8.2	11.6	17.2
−60	19.0	11.4	5.9	2.9	2.0	2.9	5.9	11.3
−70	13.0	5.4	1.0				1.0	5.3
−80	6.6	0.3						0.3
−90								

Lat.	Sep. 23	Oct. 16	Nov. 8	Nov. 30	Dec. 22	Jan. 13	Feb. 4	Feb. 26
90°								
80	6.5	0.3						0.3
70	12.9	5.5	1.0				1.0	5.6
60	18.8	11.6	6.2	3.1	2.1	3.1	6.2	11.7
50	24.1	17.6	12.1	8.7	7.5	8.7	12.3	17.8
40	28.7	23.1	18.2	14.8	13.5	14.9	18.4	23.5
30	32.5	28.2	23.9	20.9	19.8	21.0	24.1	28.4
20	35.3	32.3	29.1	26.6	25.7	26.7	29.3	32.7
10	37.0	35.5	33.5	31.8	31.1	31.9	33.8	35.9
0	37.6	37.6	36.9	36.1	35.8	36.3	37.3	38.0
−10	37.0	38.6	39.4	39.5	39.6	39.7	39.7	39.1
−20	35.3	38.5	40.7	42.0	42.4	42.2	41.1	39.0
−30	32.5	37.2	40.9	43.3	44.2	43.5	41.3	37.7
−40	28.7	34.8	40.1	43.6	45.0	43.8	40.5	35.2
−50	24.1	31.5	38.3	43.1	44.8	43.2	38.6	31.9
−60	18.8	27.3	35.7	41.9	44.4	42.1	36.0	27.7
−70	12.9	22.6	33.0	42.2	45.9	42.4	33.3	22.9
−80	6.5	18.2	33.5	44.2	48.1	44.4	33.8	18.4
−90		18.2	34.0	44.8	48.8	45.1	34.4	18.4

MAJOR WORLD EARTHQUAKES

The United States Geological Survey maintains a database of historic earthquakes throughout the world (Reference 1). The table below is extracted from that database; it includes about 300 major earthquakes, based upon the magnitude and the degree of destruction. All recorded earthquakes of magnitude 7.5 or greater are listed, even if the fatalities are unknown or small. The death toll is often a rough estimate; in many cases the true toll could be much greater. More details on the exact location and degree of destruction can be found in References 2 and 3.

The magnitude is given on the Richter scale, which was developed in 1935 by Charles F. Richter of the California Institute of Technology as a mathematical device to compare the size of earthquakes. The magnitude of an earthquake is measured by the logarithm of the amplitude of waves recorded by seismographs. Adjustments are included for the variation in the distance between the various seismographs and the epicenter of the earthquake. On the Richter Scale, magnitude is expressed in whole numbers and decimal fractions, e.g. 6.3. Because of the logarithmic basis of the scale, each whole number increase in magnitude represents a tenfold increase in measured amplitude; as an estimate of energy, each whole number step in the magnitude scale corresponds to the release of about 31 times more energy than the amount associated with the preceding whole number value.

References

1. Historic Worldwide Earthquakes, http://earthquake.usgs.gov/regional/world/historical.php.
2. Most Destructive Known Earthquakes on Record in the World, http://earthquake.usgs.gov/regional/world/most_destructive.php.
3. Earthquakes with 1,000 or More Deaths since 1900, http://earthquake.usgs.gov/regional/world/world_deaths.php.

Date	Location	Magnitude	Fatalities
856/12/22	Damghan, Iran		200,000
893/03/23	Ardabil, Iran		150,000
1138/08/09	Aleppo, Syria		230,000
1268	Silicia, Asia Minor		60,000
1290/09/27	Chihli, China		100,000
1556/01/23	Shensi, China	8.0	830,000
1619/02/14	Trujillo, Peru	7.7	350
1667/11	Shemakha, Caucasia		80,000
1668/08/17	Anatolia, Turkey	8.0	
1687/10/20	Lima, Peru	8.5	
1693/01/11	Sicily, Italy	7.5	60,000
1700/01/26	Cascadia Subduction Zone (Oregon to British Columbia)	9.0	
1727/11/18	Tabriz, Iran		77,000
1730/07/08	Valparasio, Chile	8.7	
1755/11/01	Lisbon, Portugal	8.7	70,000
1783/02/04	Calabria, Italy		50,000
1787/05/02	Puerto Rico	8.0	
1811/12/16	New Madrid Region, Missouri	8.1	
1812/02/07	New Madrid Region, Missouri	8.0	
1812/03/26	Caracas, Venezuela	7.7	
1812/12/08	Southwest of San Bernardino County, California	6.9	40
1812/12/21	West of Ventura, California	7.1	1
1821/07/10	Camana, Peru	8.2	162
1835/02/20	Concepcion, Chile	8.2	500
1843/02/08	Leeward Islands	8.3	
1855/01/23	Wellington, New Zealand	8.0	4
1857/01/09	Fort Tejon, California	7.9	1
1868/04/03	Ka'u District, Island of Hawaii	7.9	77
1868/08/13	Arica, Peru (now Chile)	9.0	400
1872/03/26	Owens Valley, California	7.4	27
1877/05/10	Offshore Tarapaca, Chile	8.3	34
1886/09/01	Charleston, South Carolina	7.3	60
1887/05/03	Northern Sonora, Mexico	7.4	51
1891/10/27	Mino-Owari, Japan	8.0	
1892/02/24	Imperial Valley, California	7.8	
1896/06/15	Sanriku, Japan	8.5	28,000
1897/06/12	Assam, India	8.3	1,500
1899/09/04	Cape Yakataga, Alaska	7.9	
1899/09/10	Yakutat Bay, Alaska	8.0	
1900/10/09	Kodiak Island, Alaska	7.7	
1902/04/19	Quezaltenango and San Marcos, Guatemala	7.5	2,000
1902/12/16	Eastern Uzbekistan (Turkestan)	6.4	4,700
1903/04/28	Malazgirt, Turkey	7.0	3,500
1903/05/28	Gole, Turkey (Ottoman Empire)	5.8	1,000
1903/08/11	Southern Greece	8.3	
1905/04/04	Kangra, India	7.5	19,000
1905/07/09	Mongolia	8.4	
1905/09/08	Calabria, Italy	7.9	557
1906/01/31	Off the Coast of Esmeraldas, Ecuador	8.8	1,000
1906/03/16	Chia-i, Taiwan	6.8	1,250
1906/04/18	San Francisco, California	7.8	3,000
1906/08/17	Valparaiso, Chile	8.2	20,000
1907/01/14	Kingston, Jamaica	6.5	800-1,000
1907/04/15	Guerrero, Mexico	7.7	
1907/10/21	Qaratog, Tajikistan, Russia	8.0	12,000
1908/12/12	Off the Coast of Central Peru	8.2	
1908/12/28	Messina, Italy	7.2	72,000
1909/01/23	Silakhor, Iran (Persia)	7.3	5,000-6,000
1910/04/12	Taiwan region	7.6	
1911/01/03	Chong-Kemin, Kyrgyzstan	7.8	450
1911/02/18	Sarez, Tajikistan	7.4	90
1911/06/07	Off Guerrero, Mexico	7.7	45
1912/08/09	Murefte, Turkey (Ottoman Empire)	7.8	2,800

Date	Location	Magnitude	Fatalities
1914/10/03	Burdur, Turkey (Ottoman Empire)	7.0	4,000
1915/01/13	Avezzano, Italy	7.0	32,610
1917/01/20	Bali, Indonesia		1,500
1917/07/30	Daguan, Yunnan, China	7.5	1,800
1918/02/13	Nan'ao, Guangdong, (Kwangtung), China	7.3	1,000
1918/10/11	Mona Passage	7.5	116
1920/06/05	Taiwan region	8.0	
1920/12/16	Haiyuan, Ningxia, China	7.8	200,000
1922/11/11	Chile-Argentina Border	8.5	100
1923/02/03	Kamchatka Peninsula	8.5	
1923/03/24	Near Luhuo, Sichuan, China	7.3	3,500
1923/05/25	Torbat-e Heydariyeh, Iran	5.7	2,200
1923/09/01	Kanto (Kwanto), Japan	7.9	142,800
1925/03/16	Yunnan, China	7.1	5,800
1927/03/07	Tango, Japan	7.6	3,020
1927/05/22	Tsinghai (Kansu), China	7.6	40,900
1928/12/01	Talca, Chile	7.6	225
1929/03/07	Fox Islands, Aleutian Islands, Alaska	7.8	
1929/05/01	Koppeh Dagh, Iran (Persia)	7.4	3,800
1930/05/06	Salmas, Iran (Persia)	7.2	2,500
1930/07/23	Irpinia, Italy	6.5	1,400
1931/01/15	Oaxaca, Mexico	7.8	114
1931/02/02	Hawke's Bay, New Zealand	7.9	256
1931/03/31	Managua, Nicaragua	6.0	2,500
1931/04/27	Zangezur Mountains, Armenia — Azerbaijan border	5.7	2,800
1931/08/10	Xinjiang, China	8.0	10,000
1932/06/03	Jalisco, Mexico	8.1	45
1932/06/18	Colima, Mexico	7.8	
1932/12/25	Gansu, China	7.6	275
1933/03/02	Sanriku, Japan	8.4	3,000
1933/03/11	Long Beach, California	6.4	115
1933/08/25	Sichuan, China	7.4	9,300
1934/01/15	Bihar, India — Nepal	8.1	10,700
1935/04/20	Taiwan (Formosa)	7.1	3,270
1935/05/30	Quetta, Pakistan	7.5	30,000
1935/07/16	Taiwan (Formosa)	6.5	2,740
1938/02/01	Banda Sea, Indonesia	8.5	
1938/11/10	Shumagin Islands, Alaska	8.2	
1939/01/25	Chillan, Chile	7.8	28,000
1939/12/26	Erzincan, Turkey	7.8	32,700
1940/05/19	Imperial Valley, California	7.1	9
1940/05/24	Callao, Peru	8.2	249
1940/11/10	Vrancea, Romania	7.3	1,000
1942/08/06	Guatemala	7.9	38
1942/08/24	Off the coast of central Peru	8.2	30
1942/11/26	Turkey	7.6	
1942/12/20	Erbaa, Turkey	7.3	1,100
1943/04/06	Illapel — Salamanca, Chile	8.2	25
1943/09/10	Tottori, Japan	7.4	1,190
1943/11/26	Ladik, Turkey	7.6	4,000
1944/01/15	San Juan, Argentina	7.4	8,000
1944/02/01	Gerede, Turkey	7.4	2,790
1944/12/07	Tonankai, Japan	8.1	998
1945/01/12	Mikawa, Japan	7.1	1,961
1945/11/27	Makran Coast, Pakistan	8.0	4,000
1946/04/01	Unimak Island, Alaska	8.1	165
1946/05/31	Ustukran, Turkey	5.9	840-1,300
1946/08/04	Samana, Dominican Republic	8.0	100
1946/11/10	Ancash, Peru	7.3	1,400
1946/12/20	Nankaido, Japan	8.1	1,362
1947/11/01	Satipo, Peru	7.3	233
1948/05/11	Moquegua, Peru	7.4	70
1948/05/25	Sichuan, China	7.3	800
1948/06/28	Fukui, Japan	7.3	3,769
1948/10/05	Ashgabat, Turkmenistan	7.3	110,000
1949/04/13	Puget Sound, Washington	7.1	8
1949/07/10	Khait, Tajikistan	7.5	12,000
1949/08/05	Ambato, Ecuador	6.8	5,050
1949/08/22	Queen Charlotte Islands, British Columbia, Canada	8.1	
1950/08/15	Assam — Tibet	8.6	1,526
1951/08/02	Cosiguina, Nicaragua	5.8	1,000
1952/07/21	Kern County, California	7.3	12
1952/11/04	Kamchatka Peninsula	9.0	
1953/02/12	Torud, Iran	6.5	970
1953/03/18	Yenice-Gonen, Turkey	7.3	1,070
1953/08/12	Kefallinia, Greece	7.1	455
1953/12/12	Tumbes, Peru	7.4	7
1954/03/29	Spain	7.9	
1954/04/30	Greece	7.1	31
1954/09/09	Orleansville, Algeria	6.8	1,250
1957/03/09	Andreanof Islands, Alaska	8.6	
1957/04/25	Fethiye, Turkey	7.1	15
1957/05/26	Bolu Province, Turkey	7.1	66
1957/06/27	Stanovoy Mountains, Russia (USSR)	7.6	1,200
1957/07/02	Mazandaran, Iran	7.1	1,200
1957/07/28	Guerrero, Mexico	7.9	68
1957/12/04	Gobi-Altay, Mongolia	8.1	30
1957/12/13	Sahneh, Iran	7.1	1,130
1958/01/15	Arequipa, Peru	7.3	26
1958/07/10	Lituya Bay, Alaska	7.7	5
1958/11/06	Kuril Islands	8.3	
1959/04/26	Taiwan region	7.5	2
1959/08/18	Hebgen Lake, Montana	7.3	28
1960/01/13	Arequipa, Peru	7.5	57
1960/02/29	Agadir, Morocco	5.7	12,000-15,000
1960/05/21	Arauco Peninsula, Chile	7.9	
1960/05/22	Chile (off coast)	9.5	1,655
1962/05/11	Guerrero, Mexico	7.0	4
1962/05/19	Guerrero, Mexico	7.1	3
1962/09/01	Qazvin, Iran	7.1	12,225
1963/07/26	Skopje, Macedonia	6.0	1,100

Date	Location	Magnitude	Fatalities	Date	Location	Magnitude	Fatalities
1963/10/13	Kuril Islands	8.5		1981/02/24	Greece	6.8	3,000
1964/03/28	Prince William Sound, Alaska	9.2	128	1981/06/11	Southern Iran	6.9	3,000
1964/06/16	Niigata, Japan	7.5	26	1981/07/28	Southern Iran	7.3	1,500
1964/10/06	Western Turkey	7.0	36	1982/12/13	Yemen	6.0	2,800
1965/01/24	Sanana, Indonesia (Ceram Sea)	7.6	71	1983/10/30	Erzurum Province, Turkey	6.9	1,342
1965/02/04	Rat Islands, Alaska	8.7		1985/03/03	Offshore Valparaiso, Chile	7.8	177
1965/03/14	Hindu Kush, Afghanistan	7.8		1985/09/19	Michoacan, Mexico	8.0	9,500
1965/03/28	La Ligua, Chile	7.4	400	1986/05/07	Andreanof Islands, Alaska	7.9	
1965/03/31	Central Greece	7.1	6	1986/10/10	El Salvador	5.5	1,000
1965/08/23	Oaxaca, Mexico	7.3	6	1987/03/06	Colombia-Ecuador	7.0	1,000
1966/03/07	Hebei, China	7.0	1,000	1987/11/30	Gulf of Alaska	7.8	
1966/03/22	Hebei, China	6.9	1,000	1988/03/06	Gulf of Alaska	7.7	
1966/08/19	Varto, Turkey	6.8	2,529	1988/08/20	Nepal-India border region	6.8	1,000
1966/10/17	Near the Coast of Peru	8.1	125	1988/12/07	Spitak, Armenia	6.8	25,000
1967/07/22	Mudurnu Valley, Turkey	7.3	173	1989/10/18	Loma Prieta, California	6.9	63
1968/05/23	Inangahua, New Zealand	7.1	2	1990/06/20	Western Iran	7.4	50,000
1968/08/02	Oaxaca, Mexico	7.1	18	1990/07/16	Luzon, Philippine Islands	7.7	1,621
1968/08/31	Dasht-e Bayaz, Iran	7.3	12,000	1991/04/22	Costa Rica	7.6	47
1969/02/28	Portugal-Morocco area	7.8	13	1991/10/19	Northern India	6.8	2,000
1969/07/25	Guangdong, China	5.9	3,000	1992/09/02	Nicaragua	7.6	116
1970/01/04	Yunnan Province, China	7.5	10,000	1992/12/12	Flores Region, Indonesia	7.8	2,500
1970/03/28	Gediz, Turkey	6.9	1,086	1993/08/08	South of the Mariana Islands	7.8	
1970/05/31	Chimbote, Peru	7.9	70,000	1993/09/29	Latur-Killari, India	6.2	9,748
1970/07/31	Colombia	8.0	1	1994/01/17	Northridge, California	6.7	60
1971/02/09	San Fernando, California	6.6	65	1994/06/09	Bolivia	8.2	10
1971/05/22	Eastern Turkey	6.9	1,000	1995/01/16	Kobe, Japan	6.9	5,502
1971/07/09	Valparaiso region, Chile	7.5	90	1995/05/27	Sakhalin Island	7.1	1,989
1972/01/25	Taiwan region	7.5		1996/06/10	Andreanof Islands, Alaska	7.9	
1972/04/10	Southern Iran	7.1	5,054	1997/05/10	Northern Iran	7.3	1,567
1972/04/24	Taiwan region	7.2	4	1997/10/14	South of Fiji Islands	7.8	
1972/07/30	Sitka, Alaska	7.6		1997/12/05	Near East Coast of Kamchatka	7.8	
1972/12/23	Nicaragua	6.2	5,000	1998/01/04	Loyalty Islands Region	7.5	
1974/05/10	Near Zhaotong, China	6.8	20,000	1998/02/04	Afghanistan-Tajikistan Border Region	5.9	2,323
1974/07/13	Panama-Colombia border region	7.3	11	1998/03/25	Balleny Islands Region (off Antarctica)	8.1	
1974/10/03	Near the Coast of Central Peru	8.1	78	1998/05/03	Southeast of Taiwan	7.5	
1974/10/08	Leeward Islands	7.5		1998/05/30	Afghanistan-Tajikistan Border Region	6.6	4,000
1974/12/28	Northern Pakistan	6.2	5,300	1998/07/17	Near North Coast of New Guinea, Papua New Guinea	7.0	2,183
1975/02/02	Near Islands, Alaska	7.6		1999/01/25	Colombia	6.1	1,185
1975/02/04	Haicheng, China	7.0	2,000	1999/08/17	Izmit, Turkey	7.6	17,118
1975/09/06	Diyarbakir Province, Turkey	6.7	2,300	1999/09/20	Taiwan	7.6	2,400
1975/11/29	Kalapana, Hawaii	7.2	2	1999/09/30	Oaxaca, Mexico	7.5	
1976/02/04	Guatemala	7.5	23,000	1999/11/12	Duzce, Turkey	7.2	894
1976/05/06	Northeastern Italy	6.5	1,000	2000/06/04	Southern Sumatera, Indonesia	7.9	103
1976/06/25	Papua, Indonesia	7.1	422	2000/06/18	South Indian Ocean	7.9	
1976/07/27	Tangshan, China	7.5	255,000	2000/11/16	New Ireland Region, Papua New Guinea	8.0	2
1976/08/16	Mindanao, Philippines	7.9	8,000	2001/01/01	Mindanao, Philippines	7.5	
1976/11/24	Turkey-Iran border region	7.3	5,000	2001/01/13	El Salvador	7.7	852
1977/03/04	Romania	7.2	1,500	2001/01/26	Gujarat, India	7.6	20,085
1978/09/16	Iran	7.8	15,000	2001/02/13	El Salvador	6.6	315
1979/02/28	Mt. St. Elias, Alaska	7.5					
1980/10/10	El Asnam (formerly Orleansville), Algeria	7.7	5,000				

Date	Location	Magnitude	Fatalities
2001/06/23	Near the Coast of Peru	8.4	138
2002/03/03	Hindu Kush Region, Afghanistan	7.4	166
2002/03/05	Mindanao, Philippines	7.5	15
2002/03/25	Hindu Kush Region, Afghanistan	6.1	1,000
2002/03/31	Taiwan region	7.1	5
2002/08/19	Fiji Islands	7.7	
2002/09/08	New Guinea, Papua New Guinea	7.6	
2002/10/10	Irian Jaya, Indonesia	7.6	8
2002/11/02	Northern Sumatera, Indonesia	7.4	3
2002/11/03	Denali Fault, Alaska	7.9	
2003/01/22	Offshore Colima, Mexico	7.6	29
2003/05/21	Northern Algeria	6.8	2,226
2003/05/26	Halmahera, Indonesia	7.0	1
2003/07/15	Carlsberg Ridge	7.6	
2003/08/04	Scotia Sea	7.6	
2003/09/25	Hokkaido, Japan Region	8.3	
2003/09/27	Southwestern Siberia, Russia	7.3	3
2003/11/17	Rat Islands, Aleutian Islands, Alaska	7.8	
2003/12/26	Southeastern Iran	6.6	31,000
2004/02/05	Irian Jaya, Indonesia	7.0	37
2004/11/11	Kepulauan Alor, Indonesia	7.5	34
2004/11/26	Papua, Indonesia	7.1	32
2004/12/23	North of Macquarie Island, New Zealand	8.1	
2004/12/26	Sumatra-Andaman Islands	9.1	227,898
2005/03/28	Northern Sumatra, Indonesia	8.6	1,313
2005/06/13	Tarapaca, Chile	7.8	11
2005/09/09	New Ireland Region, Papua New Guinea	7.6	
2005/09/26	Northern Peru	7.5	5
2005/10/08	Pakistan	7.6	86,000
2006/01/27	Banda Sea	7.6	
2006/02/22	Mozambique	7.0	4
2006/04/20	Koryakia, Russia	7.6	
2006/05/03	Tonga	8.0	
2006/05/26	Java, Indonesia	6.3	5,749
2006/07/17	South of Java, Indonesia	7.7	730
2006/11/15	Kuril Islands	8.3	
2006/12/26	Taiwan Region	7.1	2
2007/01/13	East of the Kuril Islands	8.1	
2007/01/21	Molucca Sea	7.5	
2007/04/01	Solomon Islands	8.1	40
2007/08/08	Java, Indonesia	7.5	
2007/08/15	Near the Coast of Central Peru	8.0	514
2007/09/12	Southern Sumatra, Indonesia	8.5	25
2007/09/12	Kepulauan Mentawai region, Indonesia	7.9	
2007/09/28	Mariana Islands region	7.5	
2007/11/14	Antofagasta, Chile	7.7	2
2007/12/09	South of the Fiji Islands	7.8	
2008/05/12	Eastern Sichuan, China	7.9	87,652
2008/07/05	Sea of Okhotsk	7.7	
2009/01/03	Near the North Coast of Papua, Indonesia	7.7	5
2009/03/19	Tonga region	7.6	
2009/04/06	Central Italy	6.3	295
2009/07/15	Off West Coast of the South Island, New Zealand	7.8	
2009/08/10	Andaman Islands, India region	7.5	
2009/09/29	Samoa Islands region	8.1	192
2009/09/30	Southern Sumatra, Indonesia	7.5	1117
2009/10/07	Santa Cruz Islands	7.8	
2009/10/07	Vanuatu, Coral Sea	7.7	
2010/01/12	Haiti region	7.0	316,000
2010/02/27	Offshore Maule, Chile	8.8	577
2010/03/08	Eastern Turkey	6.1	51
2010/04/04	Baja California, Mexico	7.2	2
2010/04/13	Southern Qinghai, China	6.9	2,968
2010/07/23	Moro Gulf, Mindanao, Philippines	7.6	
2011/02/21	South Island of New Zealand	6.1	181
2011/03/11	Near the East Coast of Honshu, Japan	9.0	20,352

WEATHER-RELATED SCALES

Saffir-Simpson Hurricane Scale

- **Tropical Storm**
  Winds 39–73 mph
- **Category 1 Hurricane** — winds 74–95 mph (64–82 knots); pressure greater than 980 mbar; storm surge 3–5 ft (1.0–1.7 m)
  No real damage to buildings. Damage to unanchored mobile homes. Some damage to poorly constructed signs. Also, some coastal flooding and minor pier damage.
  — Examples: Irene 1999 and Allison 1995
- **Category 2 Hurricane** — winds 96–110 mph (83–95 knots); pressure 979–965 mbar; storm surge 6–8 ft (1.8–2.6 m)
  Some damage to building roofs, doors and windows. Considerable damage to mobile homes. Flooding damages piers and small craft in unprotected moorings may break their moorings. Some trees blown down.
  — Examples: Bonnie 1998, Georges (FL & LA) 1998 and Gloria 1985
- **Category 3 Hurricane** — winds 111–130 mph (96–113 knots); pressure 964–945 mbar; storm surge 9–12 ft (2.7–3.8 m)
  Some structural damage to small residences and utility buildings. Large trees blown down. Mobile homes and poorly built signs destroyed. Flooding near the coast destroys smaller structures with larger structures damaged by floating debris. Terrain may be flooded well inland.
  — Examples: Keith 2000, Fran 1996, Opal 1995, Alicia 1983 and Betsy 1965

- **Category 4 Hurricane** — winds 131–155 mph (114–135 knots); pressure 944–920 mbar; storm surge 13–18 ft (3.9–5.6 m)
  More extensive curtainwall failures with some complete roof structure failure on small residences. Major erosion of beach areas. Terrain may be flooded well inland.
  — Examples: Hugo 1989 and Donna 1960
- **Category 5 Hurricane** — winds 156 mph and up (135+ knots); pressure less than 920 mbar; storm surge 19+ ft (5.7+ m)
  Complete roof failure on many residences and industrial buildings. Some complete building failures with small utility buildings blown over or away. Flooding causes major damage to lower floors of all structures near the shoreline. Massive evacuation of residential areas may be required.
  — Examples: Andrew (FL) 1992, Camille 1969 and Labor Day 1935

Fujita Tornado Damage Scale

The original Fujita Scale was modified by NOAA in February 2007 and is now called the Enhanced Fujita Scale (EF). It is an operational scale based on the estimated speed of three-second wind gusts, as indicated by typical damage levels. The table below describes the damage levels according to the original scale. In the enhanced scale, the damage is measured by a more elaborate set of criteria (see http://www.spc.noaa.gov/efscale/ef-scale.html).

EF Number	3 s Gusts (mph)	Typical damage (according to the original Fujita Scale)
0	65–85	**Light damage.** Some damage to chimneys; branches broken off trees; shallow-rooted trees pushed over; sign boards damaged.
1	86–110	**Moderate damage.** Peels surface off roofs; mobile homes pushed off foundations or overturned; moving autos blown off roads.
2	111–135	**Considerable damage.** Roofs torn off frame houses; mobile homes demolished; boxcars overturned; large trees snapped or uprooted; light-object missiles generated; cars lifted off ground.
3	136–165	**Severe damage.** Roofs and some walls torn off well-constructed houses; trains overturned; most trees in forest uprooted; heavy cars lifted off the ground and thrown.
4	166–200	**Devastating damage.** Well-constructed houses leveled; structures with weak foundations blown away some distance; cars thrown and large missiles generated.
5	Over 200	**Incredible damage.** Strong frame houses leveled off foundations and swept away; automobile-sized missiles fly through the air in excess of 100 meters (109 yd); trees debarked; incredible phenomena will occur.

Beaufort Wind Scale

The Beaufort Wind Scale was devised by British Rear-Admiral Sir Francis Beaufort in 1805 based on observations of the effects of the wind.

Force	Wind (knots)	WMO classification	Appearance of wind effects On the water	On land
0	< 1	Calm	Sea surface smooth and mirror-like	Calm, smoke rises vertically
1	1–3	Light Air	Scaly ripples, no foam crests	Smoke drift indicates wind direction, still wind vanes
2	4–6	Light Breeze	Small wavelets, crests glassy, no breaking	Wind felt on face, leaves rustle, vanes begin to move
3	7–10	Gentle Breeze	Large wavelets, crests begin to break, scattered whitecaps	Leaves and small twigs constantly moving, light flags extended

Force	Wind (knots)	WMO classification	Appearance of wind effects	
			On the water	On land
4	11–16	Moderate Breeze	Small waves 1–4 ft. becoming longer, numerous whitecaps	Dust, leaves, and loose paper lifted, small tree branches move
5	17–21	Fresh Breeze	Moderate waves 4–8 ft taking longer form, many whitecaps, some spray	Small trees in leaf begin to sway
6	22–27	Strong Breeze	Larger waves 8–13 ft, whitecaps common, more spray	Larger tree branches moving, whistling in wires
7	28–33	Near Gale	Sea heaps up, waves 13–20 ft, white foam streaks off breakers	Whole trees moving, resistance felt walking against wind
8	34–40	Gale	Moderately high (13–20 ft) waves of greater length, edges of crests begin to break into spindrift, foam blown in streaks	Whole trees in motion, resistance felt walking against wind
9	41–47	Strong Gale	High waves (20 ft), sea begins to roll, dense streaks of foam, spray may reduce visibility	Slight structural damage occurs, slate blows off roofs
10	48–55	Storm	Very high waves (20–30 ft) with overhanging crests, sea white with densely blown foam, heavy rolling, lowered visibility	Seldom experienced on land, trees broken or uprooted, "considerable structural damage"
11	56–63	Violent Storm	Exceptionally high (30–45 ft) waves, foam patches cover sea, visibility more reduced	
12	64+	Hurricane	Air filled with foam, waves over 45 ft, sea completely white with driving spray, visibility greatly reduced	

Wind Chill

The following chart prepared by the U. S. National Weather Service gives the temperature perceived by an average person as a function of the real air temperature and the wind speed. The current scale was adopted in 2001.

Wind Chill (°F) = 35.74 + 0.6215T - 35.75(V^0.16) + 0.4275T(V^0.16)
Where, T = Air Temperature (°F) V = Wind Speed (mph)
Effective 11/01/01

Reference

National Oceanic and Atmospheric Administration, http://www.noaa.gov

INFRARED ABSORPTION BY THE EARTH'S ATMOSPHERE

Several constituents of the Earth's atmosphere absorb infrared radiation. At ground level the strongest absorbers are H_2O and CO_2, but 30 to 40 other compounds can make significant contributions. The centers of the most important absorption bands are listed below:

Molecule	Vibrational mode	Band center in cm^{-1}
H_2O	Bend	1595
H_2O	Symmetric O-H stretch	3657
H_2O	Antisymmetric O-H stretch	3756
CO_2	Bend	667
CO_2	Antisymmetric C-O stretch	2349
O_3	Bend	701
O_3	Antisymmetric O-O stretch	1042
O_3	Symmetric O-O stretch	1103
N_2O	Bend	589
N_2O	N-O stretch	1285
N_2O	N-N stretch	2224
CO	C-O stretch	2143
CH_4	Degenerate deformation	1306
CH_4	Degenerate stretch	3019

The HITRAN Molecular Spectroscopy Database (References 1 and 2) is a compilation of wavenumbers and intensities of more than 1.7 million spectral lines of atmospheric constituents. It is a valuable resource for calculating transmission of the atmosphere, radiative energy transfer, and other phenomena. The graph below, which was supplied by Walter J. Lafferty (Reference 3), gives the transmittance of the atmosphere for one set of conditions.

References

1. Rothman, L. S., et al., *J. Quant. Spectros. Radiat. Transfer* 82, 5, 2003; *ibid.*, to be published, 2005.
2. HITRAN Molecular Spectroscopy Database, <http://cfa-www. Harvard.edu/HITRAN/hitrandata04/>.
3. Lafferty, W. J., Some Aspects of High Resolution Molecular Spectroscopy, in *Lectures on Molecular Physics*, Institute for the Structure of Matter, Centro de Fisica Miguel A. Catalan, Madrid, 1997.

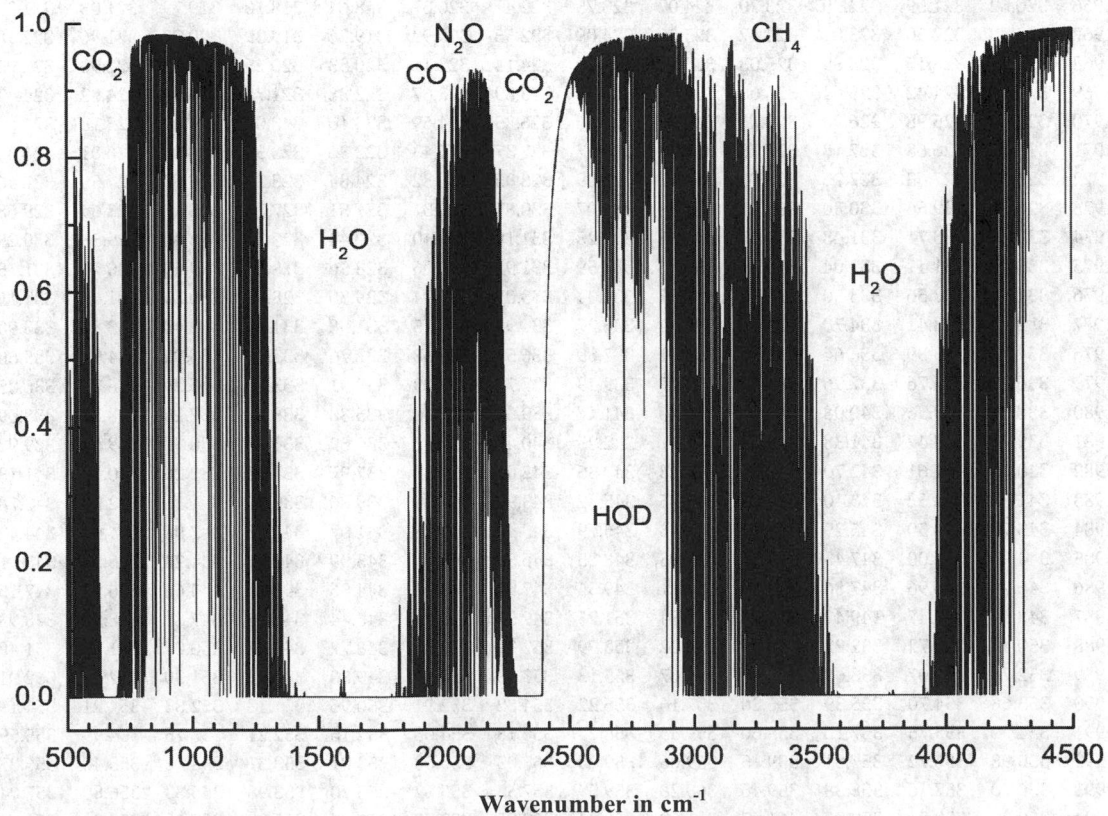

Transmittance of U.S. Standard Atmosphere at Ground Level for a Path of 1 km at 296 K

ATMOSPHERIC CONCENTRATION OF CARBON DIOXIDE, 1958–2008

The data in this table were taken at the Mauna Loa Observatory in Hawaii and represent averages adjusted to the 15th of each month. The last column gives the average over the year. The concentration of CO_2 is given in parts per million by volume. Data from other measurement sites may be found in Ref. 1.

The first graph illustrates the seasonal variation of CO_2 concentration and the steady increase over the last 50 years. The second graph summarizes the growth in global emissions of CO_2 into the atmosphere as a result of burning of fossil fuels (Ref. 2).

References

1. Keeling, C. D., Piper, S. C., Bollenbacher, A. F., and Walker, S. J., Atmospheric carbon dioxide record from Mauna Loa, Carbon Dioxide Information Analysis Center, Oak Ridge National Laboratory, U.S. Department of Energy, Oak Ridge, TN, <cdiac.ornl.gov/trends/co2/sio-mlo.html>, February 2009.
2. Boden, T. A., Marland, G., and Andres, R. J., Global, Regional, and National Fossil-Fuel CO_2 Emissions, Carbon Dioxide Information Analysis Center, Oak Ridge National Laboratory, U.S. Department of Energy, Oak Ridge, TN, <cdiac.ornl.gov/trends/emis/tre_glob.html>, 2009.

CO_2 Concentration in ppm at Mauna Loa

Year	Jan.	Feb.	March	April	May	June	July	Aug.	Sept.	Oct.	Nov.	Dec.	Annual
1958			315.71	317.45	317.50		315.85	314.93	313.19		313.34	314.67	
1959	315.58	316.47	316.65	317.72	318.29	318.16	316.55	314.80	313.84	313.34	314.82	315.59	315.98
1960	316.43	316.97	317.58	319.03	320.03	319.59	318.18	315.91	314.16	313.84	315.00	316.19	316.91
1961	316.89	317.70	318.54	319.48	320.58	319.77	318.58	316.79	314.99	315.31	316.10	317.01	317.65
1962	317.94	318.56	319.69	320.58	321.01	320.61	319.61	317.40	316.26	315.42	316.69	317.69	318.45
1963	318.74	319.08	319.86	321.39	322.24	321.47	319.74	317.77	316.21	315.99	317.06	318.36	318.99
1964	319.57				322.24	321.89	320.44	318.70	316.70	316.87	317.68	318.71	
1965	319.44	320.44	320.89	322.13	322.16	321.87	321.21	318.87	317.81	317.30	318.87	319.42	320.03
1966	320.62	321.59	322.39	323.70	324.07	323.75	322.41	320.37	318.64	318.10	319.79	321.03	321.37
1967	322.33	322.50	323.04	324.42	325.00	324.09	322.55	320.92	319.26	319.39	320.72	321.96	322.18
1968	322.57	323.15	323.89	325.03	325.57	325.36	324.14	322.11	320.33	320.25	321.33	322.90	323.05
1969	324.00	324.42	325.64	326.66	327.38	326.70	325.89	323.67	322.38	321.78	322.85	324.12	324.62
1970	325.06	325.98	326.93	328.14	328.07	327.66	326.35	324.69	323.10	323.07	324.01	325.13	325.68
1971	326.17	326.68	327.18	327.78	328.92	328.57	327.37	325.43	323.36	323.57	324.80	326.01	326.32
1972	326.77	327.63	327.75	329.72	330.07	329.09	328.05	326.32	324.84	325.20	326.50	327.55	327.46
1973	328.54	329.56	330.30	331.50	332.48	332.07	330.87	329.31	327.51	327.18	328.16	328.64	329.68
1974	329.35	330.71	331.48	332.65	333.08	332.25	331.18	329.40	327.44	327.37	328.46	329.58	330.25
1975	330.40	331.41	332.04	333.31	333.96	333.59	331.91	330.06	328.56	328.34	329.49	330.76	331.15
1976	331.74	332.56	333.50	334.58	334.87	334.34	333.05	330.94	329.30	328.94	330.31	331.68	332.15
1977	332.92	333.41	334.70	336.07	336.74	336.27	334.93	332.75	331.58	331.16	332.40	333.85	333.90
1978	334.97	335.39	336.64	337.76	338.01	337.89	336.54	334.68	332.76	332.54	333.92	334.95	335.50
1979	336.23	336.76	337.96	338.89	339.47	339.29	337.73	336.09	333.91	333.86	335.29	336.73	336.85
1980	338.01	338.36	340.08	340.77	341.46	341.17	339.56	337.60	335.88	336.02	337.10	338.21	338.69
1981	339.23	340.47	341.38	342.51	342.91	342.25	340.49	338.43	336.69	336.85	338.36	339.61	339.93
1982	340.75	341.61	342.70	343.57	344.13	343.35	342.06	339.82	337.97	337.86	339.26	340.49	341.13
1983	341.37	342.52	343.10	344.94	345.75	345.32	343.99	342.39	339.86	339.99	341.16	342.99	342.78
1984	343.70	344.50	345.29	347.08	347.43	346.79	345.40	343.28	341.07	341.35	342.98	344.22	344.42
1985	344.97	346.00	347.43	348.35	348.93	348.25	346.56	344.69	343.09	342.80	344.24	345.56	345.91
1986	346.29	346.96	347.86	349.55	350.21	349.54	347.94	345.91	344.86	344.17	345.66	346.90	347.15
1987	348.02	348.47	349.42	350.99	351.84	351.25	349.52	348.11	346.44	346.36	347.81	348.96	348.93
1988	350.43	351.72	352.22	353.59	354.22	353.79	352.39	350.44	348.72	348.88	350.07	351.34	351.48
1989	352.76	353.07	353.68	355.42	355.67	355.13	353.90	351.67	349.80	349.99	351.30	352.53	352.91
1990	353.66	354.70	355.39	356.20	357.16	356.22	354.82	352.91	350.96	351.18	352.83	354.21	354.19
1991	354.72	355.75	357.16	358.60	359.33	358.24	356.18	354.03	352.16	352.21	353.75	354.99	355.59
1992	355.98	356.72	357.81	359.15	359.66	359.25	357.03	355.00	353.01	353.31	354.16	355.40	356.37
1993	356.70	357.16	358.38	359.46	360.28	359.59	357.58	355.52	353.70	353.98	355.33	356.80	357.04
1994	358.36	358.91	359.97	361.27	361.68	360.94	359.55	357.49	355.84	355.99	357.58	359.04	358.89
1995	359.96	361.00	361.64	363.45	363.79	363.26	361.90	359.46	358.06	357.75	359.56	360.70	360.88
1996	362.05	363.25	364.03	364.72	365.41	364.97	363.65	361.49	359.46	359.60	360.76	362.33	362.64
1997	363.18	364.00	364.57	366.35	366.80	365.62	364.47	362.51	360.19	360.77	362.43	364.28	363.76
1998	365.32	366.15	367.31	368.61	369.29	368.87	367.64	365.77	363.90	364.23	365.46	366.97	366.63
1999	368.15	368.87	369.59	371.14	371.00	370.35	369.27	366.94	364.63	365.12	366.67	368.01	368.31
2000	369.14	369.46	370.52	371.66	371.82	371.70	370.12	368.12	366.62	366.73	368.29	369.53	369.48

Year	Jan.	Feb.	March	April	May	June	July	Aug.	Sept.	Oct.	Nov.	Dec.	Annual
2001	370.28	371.50	372.12	372.87	374.02	373.30	371.62	369.55	367.96	368.09	369.68	371.24	371.02
2002	372.43	373.09	373.52	374.86	375.55	375.40	374.02	371.49	370.71	370.24	372.08	373.78	373.10
2003	374.68	375.63	376.11	377.65	378.35	378.13	376.62	374.50	372.99	373.00	374.35	375.70	375.64
2004	376.79	377.37	378.41	380.52	380.63	379.57	377.79	375.86	374.06	374.24	375.86	377.48	377.38
2005	378.37	379.69	380.41	382.10	382.28	382.13	380.66	378.71	376.42	376.88	378.32	380.04	379.67
2006	381.38	382.03	382.64	384.62	384.95	384.06	382.29	380.47	378.67	379.06	380.14	381.74	381.84
2007	382.45	383.68	384.23	386.26	386.39	385.87	384.39	381.78	380.73	380.81	382.33	383.69	383.55
2008	385.07	385.72	385.85	386.71	388.45	387.64	386.10	383.95	382.91	382.73	383.96	385.02	385.34

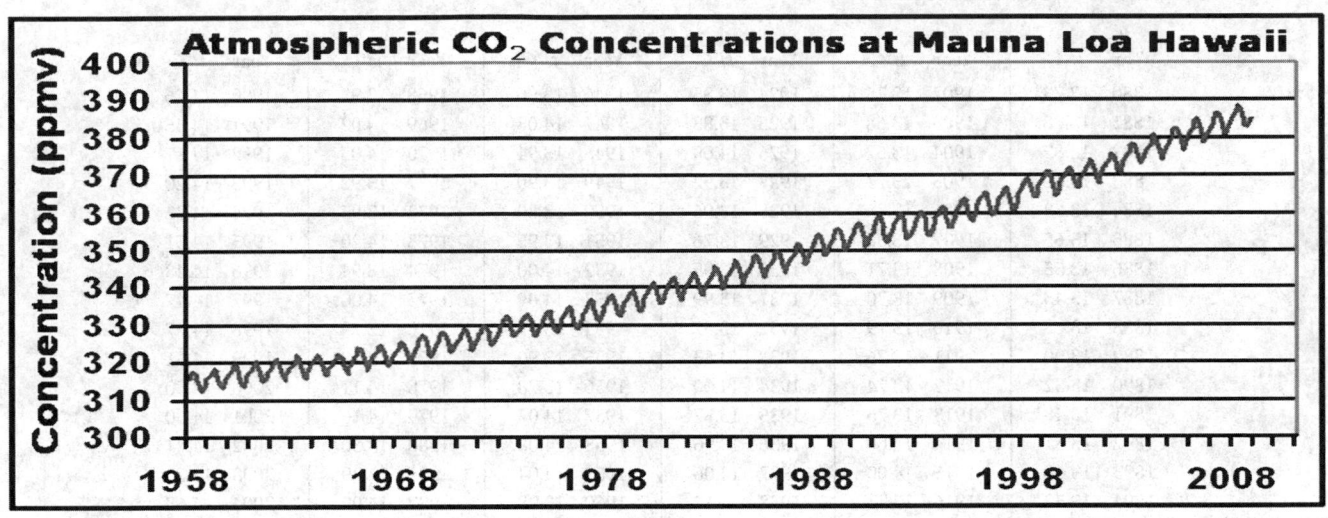

CO$_2$ Emissions from Burning of Fossil Fuels, 1750–2008

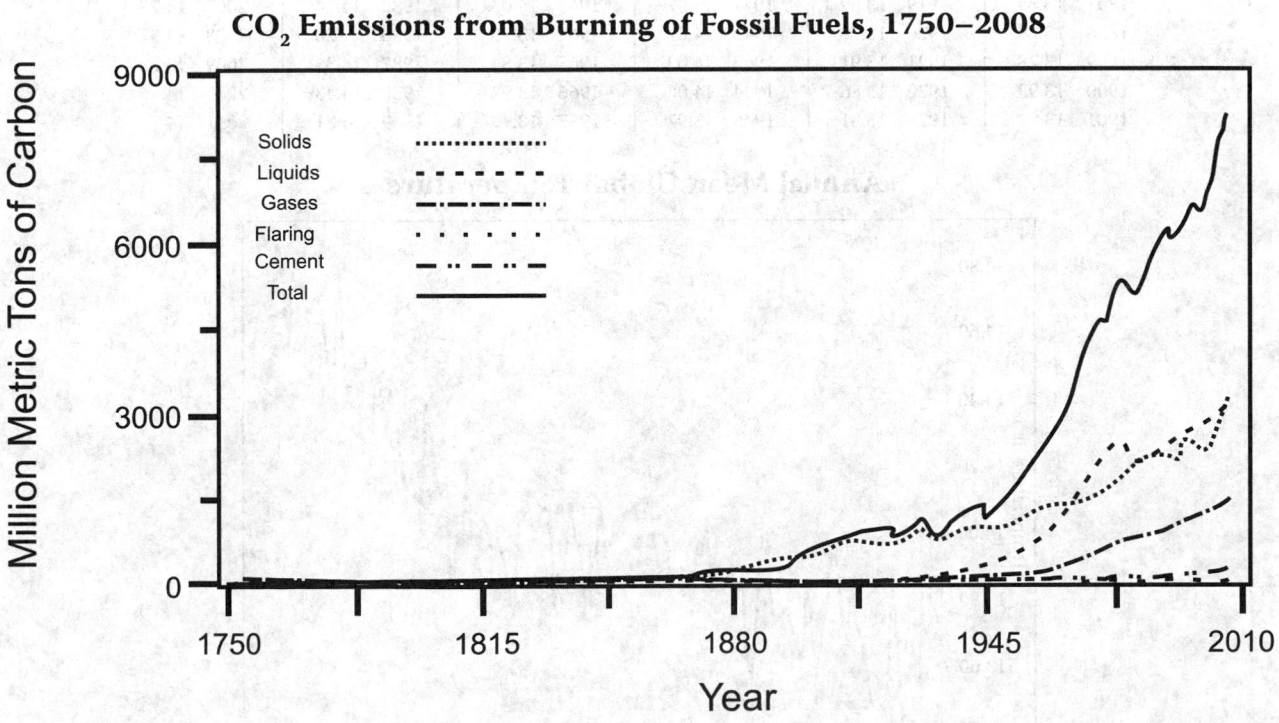

GLOBAL TEMPERATURE TREND, 1880–2011

This table and graph summarize the trend in annual mean global surface temperature from 1880 to 2011. The values were calculated from the global mean temperature anomalies in Ref. 2 by assuming an absolute global mean of 14.00 °C, which is the best estimate for the 1951–1980 period. The 95% confidence interval for comparing the annual mean temperature values for recent years is 0.05 °C.

Reference 1 gives links to more extensive data sets that cover individual months, seasons, and zonal regions.

References

1. NASA Goddard Institute for Space Studies, <data.giss.nasa.gov/gistemp>.
2. Annual mean Temperature Anomalies in 0.01 degrees Celsius: selected zonal means, <data.giss.nasa.gov/gistemp/tabledata_v3/ZonAnn.Ts.txt>, January 2012.

Year	t/°C	Year	t/°C	Year	t/°C	Year	t/°C	Year	t/°C	Year	t/°C
1880	13.73	1902	13.75	1924	13.85	1946	13.96	1968	13.93	1990	14.46
1881	13.78	1903	13.68	1925	13.83	1947	14.08	1969	14.02	1991	14.45
1882	13.82	1904	13.57	1926	14.05	1948	13.94	1970	14.05	1992	14.17
1883	13.82	1905	13.74	1927	13.92	1949	13.90	1971	13.92	1993	14.20
1884	13.53	1906	13.86	1928	13.96	1950	13.80	1972	13.95	1994	14.32
1885	13.65	1907	13.56	1929	13.76	1951	13.95	1973	14.20	1995	14.51
1886	13.55	1908	13.71	1930	13.95	1952	14.00	1974	13.94	1996	14.43
1887	13.43	1909	13.70	1931	13.99	1953	14.09	1975	14.00	1997	14.47
1888	13.68	1910	13.79	1932	13.97	1954	13.89	1976	13.79	1998	14.77
1889	13.90	1911	13.76	1933	13.83	1955	13.90	1977	14.18	1999	14.51
1890	13.52	1912	13.74	1934	14.00	1956	13.78	1978	14.11	2000	14.50
1891	13.48	1913	13.76	1935	13.87	1957	14.07	1979	14.16	2001	14.60
1892	13.51	1914	14.00	1936	13.96	1958	14.08	1980	14.30	2002	14.71
1893	13.48	1915	14.00	1937	14.06	1959	14.04	1981	14.39	2003	14.70
1894	13.59	1916	13.77	1938	14.12	1960	13.98	1982	14.09	2004	14.63
1895	13.68	1917	13.56	1939	13.94	1961	14.09	1983	14.34	2005	14.80
1896	13.73	1918	13.64	1940	14.09	1962	14.05	1984	14.16	2006	14.70
1897	13.84	1919	13.89	1941	14.07	1963	14.03	1985	14.14	2007	14.78
1898	13.69	1920	13.79	1942	14.07	1964	13.75	1986	14.20	2008	14.57
1899	13.78	1921	13.91	1943	14.02	1965	13.84	1987	14.35	2009	14.72
1900	13.93	1922	13.86	1944	14.06	1966	13.93	1988	14.43	2010	14.86
1901	13.94	1923	13.81	1945	13.97	1967	13.99	1989	14.31	2011	14.72

Annual Mean Global Temperature

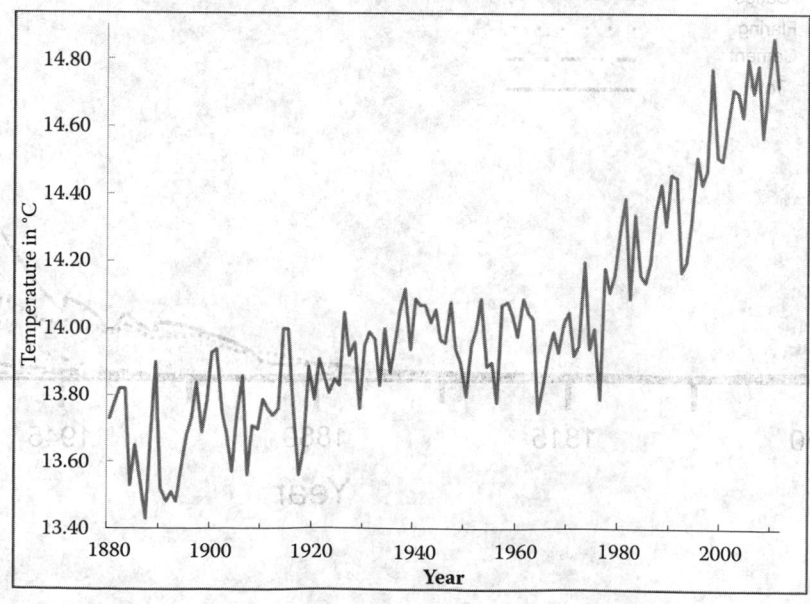

GLOBAL WARMING POTENTIAL OF GREENHOUSE GASES

The Global Warming Potential (GWP) of a gas is a measure of the degree, relative to carbon dioxide, to which the presence of that gas in the atmosphere will contribute to a long-term increase in global temperature. The calculation of the GWP for a given gas takes into account the efficiency of the gas in absorbing solar radiation (primarily determined by the infrared spectrum of the compound) and the time the compound will remain in the atmosphere before it is removed by natural processes. Thus if a pulse of 1 kg of the gas is emitted to the atmosphere at the same time as a pulse of 1 kg of CO_2, the GWP compares the warming effect of the gas relative to CO_2 over various time horizons.

This table, which is taken from the 2007 report of the Intergovernmental Panel of Climate Change (IPCC), gives the lifetime in years and the radiative efficiency in watts per square meter for a concentration of one part per billion for the major compounds identified in the Kyoto Protocol as contributing to global climate change. Radiative efficiency is a measure of the radiative forcing that influences the energy balance in the Earth–atmosphere system. The last four columns of the table give the Global Warming Potential, first as estimated in 1995 for a 100 year time horizon, and then as estimated with improved data in 2007 for 20-,

100-, and 500-year horizons. The calculation of a GWP involves a number of assumptions, and other measures have been proposed (see Reference).

The list of compounds includes those identified in the Montreal Protocol as contributing to ozone depletion, since these compounds also contribute to global warming. It also includes compounds used or proposed as replacements for the ozone-depleting compounds but which still have global warming potential.

Reference

Forster, P., V. Ramaswamy, P. Artaxo, T. Berntsen, R. Betts, D. W. Fahey, J. Haywood, J. Lean, D.C. Lowe, G. Myhre, J. Nganga, R. Prinn, G. Raga, M. Schulz, and R. Van Dorland, 2007: Changes in Atmospheric Constituents and in Radiative Forcing. In: *Climate Change 2007: The Physical Science Basis. Contribution of Working Group I to the Fourth Assessment Report of the Intergovernmental Panel on Climate Change* [Solomon, S., D. Sin, M. Manning, Z. Chen, M. Marquis, K. B. Averyt, M. Tignor and H.L. Miller (eds.)], Cambridge University Press, Cambridge, United Kingdom and New York, NY, USA. Available on the Internet at < http://ipcc-wg1.ucar. edu/wg1/wg1-report.html>.

Compound	Synonym/Code	Formula	Lifetime (years)	Rad. eff. $W\,m^{-2}ppb^{-1}$	1995 100 yr	GWP for given time horizon Current estimate		
						20 yr	100 yr	500 yr
Natural atmospheric constituents								
Carbon dioxide		CO_2		1.4×10^{-5}	1	1	1	1
Methane		CH_4	12	3.7×10^{-4}	21	72	25	7.6
Nitrous oxide		N_2O	114	3.03×10^{-3}	310	289	298	153
Substances controlled by the Montreal Protocol								
Trichlorofluoromethane	CFC-11	CCl_3F	45	0.25	3800	6730	4750	1620
Dichlorodifluoromethane	CFC-12	CCl_2F_2	100	0.32	8100	11000	10900	5200
Chlorotrifluoromethane	CFC-13	CCl_3F	640	0.25		10800	14400	16400
1,1,2-Trichloro-1,2,2-trifluoroethane	CFC-113	CCl_2FCClF_2	85	0.3	4800	6540	6130	2700
1,2-Dichloro-1,1,2,2-tetrafluoroethane	CFC-114	$CClF_2CClF_2$	300	0.31		8040	10000	8730
Chloropentafluoroethane	CFC-115	$CClF_2CF_3$	1700	0.18		5310	7370	9990
Bromotrifluoromethane	Halon-1301	$CBrF_3$	65	0.32	5400	8480	7140	2760
Bromochlorodifluoromethane	Halon-1211	$CBrClF_2$	16	0.3		4750	1890	575
1,2-Dibromotetrafluoroethane	Halon-2402	$CBrF_2CBrF_2$	20	0.33		3680	1640	503
Tetrachloromethane	Carbon tetrachloride	CCl_4	26	0.13	1400	2700	1400	435
Bromomethane	Methyl bromide	CH_3Br	0.7	0.01		17	5	1
1,1,1-Trichloroethane	Methyl chloroform	CH_3CCl_3	5	0.06		506	146	45
Chlorodifluoromethane	HCFC-22	$CHClF_2$	12	0.2	1500	5160	1810	549
2,2-Dichloro-1,1,1-trifluoroethane	HCFC-123	$CHCl_2CF_3$	1.3	0.14	90	273	77	24
1-Chloro-1,2,2,2-tetrafluoroethane	HCFC-124	$CHClFCF_3$	5.8	0.22	470	2070	609	185
1,1-Dichloro-1-fluoroethane	HCFC-141b	CH_3CCl_2F	9.3	0.14		2250	725	220
1-Chloro-1,1-difluoroethane	HCFC-142b	CH_3CClF_2	17.9	0.2	1800	5490	2310	705
3,3-Dichloro-1,1,1,2,2-pentafluoropropane	HCFC-225ca	$CHCl_2CF_2CF_3$	1.9	0.2		429	122	37
1,3-Dichloro-1,1,2,2,3-pentafluoropropane	HCFC-225cb	$CHClFCF_2CClF_2$	5.8	0.32		2030	595	181
Hydrofluorocarbons								
Trifluoromethane	HFC-23	CHF_3	270	0.19	11700	12000	14800	12200
Difluoromethane	HFC-32	CH_2F_2	4.9	0.11	650	2330	675	205
Pentafluoroethane	HFC-125	CHF_2CF_3	29	0.23	2800	6350	3500	1100

Compound	Synonym/Code	Formula	Lifetime (years)	Rad. eff. $W\,m^{-2}ppb^{-1}$	1995 100 yr	Current estimate 20 yr	100 yr	500 yr
1,1,1,2-Tetrafluoroethane	HFC-134a	CH_2FCF_3	14	0.16	1300	3830	1430	435
1,1,1-Trifluoroethane	HFC-143a	CH_3CF_3	52	0.13	3800	5890	4470	1590
1,1-Difluoroethane	HFC-152a	CH_3CHF_2	1.4	0.09	140	437	124	38
1,1,1,2,3,3,3-Heptafluoropropane	HFC-227ea	CF_3CHFCF_3	34.2	0.26	2900	5310	3220	1040
1,1,1,3,3,3-Hexafluoropropane	HFC-236fa	$CF_3CH_2CF_3$	240	0.28	6300	8100	9810	7660
1,1,1,3,3-Pentafluoropropane	HFC-245fa	$CHF_2CH_2CF_3$	7.6	0.28		3380	1030	314
1,1,1,3,3-Pentafluorobutane	HFC-365mfc	$CH_3CF_2CH_2CF_3$	8.6	0.21		2520	794	241
1,1,1,2,3,4,4,5,5,5-Decafluoropentane	HFC-43-10mee	$CF_3CHFCHFCF_2CF_3$	15.9	0.4	1300	4140	1640	500

Perfluorinated compounds

Compound	Synonym/Code	Formula	Lifetime (years)	Rad. eff.	1995 100 yr	20 yr	100 yr	500 yr
Sulfur hexafluoride		SF_6	3200	0.52	23900	16300	22800	32600
Nitrogen trifluoride		NF_3	740	0.21		12300	17200	20700
Tetrafluoromethane	PFC-14	CF_4	50000	0.10	6500	5210	7390	11200
Hexafluoroethane	PFC-116	C_2F_6	10000	0.26	9200	8630	12200	18200
Perfluoropropane	PFC-218	C_3F_8	2600	0.26	7000	6310	8830	12500
Perfluorocyclobutane	PFC-318	$c\text{-}C_4F_8$	3200	0.32	8700	7310	10300	14700
Perfluorobutane	PFC-3-1-10	C_4F_{10}	2600	0.33	7000	6330	8860	12500
Perfluoropentane	PFC-4-1-12	C_5F_{12}	4100	0.41		6510	9160	13300
Perfluorohexane	PFC-5-1-14	C_6F_{14}	3200	0.49	7400	6600	9300	13300
Perfluorodecalin	PFC-9-1-18	$C_{10}F_{18}$	>1,000	0.56		>5500	>7500	>9500
(Trifluoromethyl)sulfur pentafluoride		SF_5CF_3	800	0.57		13200	17700	21200

Fluorinated ethers

Compound	Synonym/Code	Formula	Lifetime (years)	Rad. eff.	1995 100 yr	20 yr	100 yr	500 yr
Trifluoromethyl difluoromethyl ether	HFE-125	CHF_2OCF_3	136	0.44		13800	14900	8490
Bis(difluoromethyl) ether	HFE-134	CHF_2OCHF_2	26	0.45		12200	6320	1960
Methyl trifluoromethyl ether	HFE-143a	CH_3OCF_3	4.3	0.27		2630	756	230
2-Chloro-2-(difluoromethoxy)-1,1,1-trifluoroethane	HCFE-235da2	$CHF_2OCHClCF_3$	2.6	0.38		1230	350	106
Methyl 1,1,2,2-tetrafluoroethyl ether	HFE-245cb2	$CH_3OCF_2CHF_2$	5.1	0.32		2440	708	215
2-(Difluoromethoxy)-1,1,1-trifluoroethane	HFE-245fa2	$CHF_2OCH_2CF_3$	4.9	0.31		2280	659	200
Methyl pentafluoroethyl ether	HFE-254cb2	$CH_3OCF_2CHF_2$	2.6	0.28		1260	359	109
Perfluoropropyl methyl ether	HFE-347mcc3	$CH_3OCF_2CF_2CF_3$	5.2	0.34		1980	575	175
1,1,2,2-Tetrafluoroethyl 1,1,1-trifluoroethyl ether	HFE-347pcf2	$CHF_2CF_2OCH_2CF_3$	7.1	0.25		1900	580	175
1-Methoxy-1,1,2,2,3,3-hexafluoropropane	HFE-356pcc3	$CH_3OCF_2CF_2CHF_2$	0.33	0.93		386	110	33
Methyl nonafluorobutyl ether	HFE-449sl (HFE-7100)	$C_4F_9OCH_3$	3.8	0.31		1040	297	90
1-Ethoxy-1,1,2,2,3,3,4,4,4-nonafluorobutane	HFE-569sf2 (HFE-7200)	$C_4F_9OC_2H_5$	0.77	0.3		207	59	18
1-(Difluoromethoxy)-2-[(difluoromethoxy)difluoromethoxy]-1,1,2,2-tetrafluoroethane	HFE-43-10pccc124 (H-Galden 1040x)	$CHF_2OCF_2OC_2F_4OCHF_2$	6.3	1.37		6320	1870	569
Bis(difluoromethoxy)difluoromethane	HFE-236ca12 (HG-10)	$CHF_2OCF_2OCHF_2$	12.1	0.66		8000	2800	860
1,2-Bis(difluoromethoxy)-1,1,2,2-tetrafluoroethane	HFE-338pcc13 (HG-01)	$CHF_2OCF_2CF_2OCHF_2$	6.2	0.87		5100	1500	460
Perfluoropolymethylisopropyl ether	PFPMIE	$CF_3OCF(CF_3)CF_2OCF_2OCF_3$	800	0.65		7620	10300	12400

Other compounds - Direct effects

Compound	Synonym/Code	Formula	Lifetime (years)	Rad. eff.	1995 100 yr	20 yr	100 yr	500 yr
Dimethyl ether	Methyl ether	CH_3OCH_3	0.015	0.02		1	1	<<1
Dichloromethane	Methylene chloride	CH_2Cl_2	0.38	0.03		31	8.7	2.7
Chloromethane	Methyl chloride	CH_3Cl	1	0.01		45	13	4

ATMOSPHERIC ELECTRICITY

Hans Dolezalek, Hannes Tammet, John Latham, and Martin A. Uman

I. SURVEY AND GLOBAL CIRCUIT

Hans Dolezalek

The science of atmospheric electricity originated in 1752 by an experimental proof of a related earlier hypothesis (that lightning is an electrical event). In spite of a large effort, in part by such eminent physicists as Coulomb, Lord Kelvin, and many others, an overall, proven theory able to generate models with sufficient resolution is not yet available. Generally accepted and encompassing textbooks are now more than 20 years old. The voluminous proceedings of the, so far, nine international atmospheric electricity conferences (1954 to 1992) give much valuable detail and demonstrate impressive progress, as do a number of less comprehensive textbooks published in the last 20 years, but a general theory as indicated above is not yet created. Only now, are certain related measuring techniques and mathematical possibilities emerging.

Applications to practical purposes do exist in the field of lightning research (including the electromagnetic radiation emanating from lightning) by the establishment of lightning-location networks and by the now developing possibility of detecting electrified clouds that pose hazards to aircraft. Application of atmospheric electricity to other parts of meteorology seems to be promising but so far has seldom been instituted. Because some atmospheric electric signals propagate around the Earth and because of the existence of a global circuit, applications for the monitoring of global change processes and conditions are now being proposed. Significant secular changes in the global circuit would indicate a change in the global climate; the availability of many old data (about a span of 100 years) could help detect a long-term trend.

The concept of the "global circuit" is based on the theory of the global spherical capacitor: both the solid (and liquid) Earth as one electrode, and the high atmospheric layers (about the ionosphere) as the other, are by orders of magnitude more electrically conductive than the atmosphere between them. According to the "classical picture of atmospheric electricity," this capacitor is continuously charged by the common action of all thunderstorms to a d.c. voltage difference of several hundred kilovolts, the Earth being negative. The much smaller but still existing conductivity of the atmosphere allows a current flowing from the ionosphere to the ground, integrated for all sink areas of the whole Earth, of the order of 1.5 kA. In this way, a global circuit is created with many generators and sink-areas both interspaced and distributed over the whole globe, all connected to two nodes: ionosphere and ground. Within the scope of the global circuit, for each location, the current density (order of several pA/m^2) is determined by the voltage difference between ionosphere and ground (which is the same for all locations but varying in time) and the columnar resistance reaching from the ground up to the ionosphere (in the order of 10^{17} Ωm^2).

Natural processes, especially meteorological processes and some human activity, which produce or move electric charges ("space charges") or affect the ion distribution, constitute local generators and thereby "local circuits," horziontally and/or in parallel or antiparallel to the local part of the global circuit. In many cases, the local currents are much stronger than the global ones, making the measurement of the global current at a given location and/or during a period of time very difficult or, often, impossible. The strongest local circuits usually occur with certain weather conditions (precipitation, fog, high wind, blown-up dust or snow, heavy cloudiness) that make measurement of the global circuit impossible everywhere; but even in their absence local generators exist in varying magnitudes and of different characters. The separation of the local and global shares in the measured values of current density is a central problem of the science of atmospheric electricity. Aerological measurements are of high value in this regard.

The above description is within the "classical picture" of atmospheric electricity, a group of hypotheses to explain the electrification of the atmosphere. It is probably fundamentally correct but certainly not complete; it has not yet been confirmed by systems of measurements resulting in no inner contradictions. In particular, extraterrestrial influences must be permitted; their general significance is still under debate.

Within this "classical picture" a kind of electric standard atmosphere may be constructed as shown in Table 1.

Values with a star, *, are rough average values from measurement. A star in parentheses, (*), points to a typical value from one or a few measurements. All other values have been calculated from starred values, under the assumption that at 2 km 50% and at 12 km 90% of the columnar resistance is reached. Voltage drop along one of the partial columns can be calculated by subtracting the value for the lower column from that of the upper one. Columnar resistances, conductances, and capacitances are valid for that particular part of the column which is indicated at the left. Capacitances are calculated with the formula for plate capacitors, and this fact must be considered also for the time constants for columns.

According to measurements, U, the potential difference between 0 m and 65 km may vary by a factor of approximately 2. The total columnar resistance, R_c, is estimated to vary up to a factor of 3, the variation being due to either reduction of conductivity in the exchange layer (about lowest 2 km of this table) or to the presence of high mountains; in both cases the variation is caused in the troposphere. Smaller variations in the stratosphere and mesosphere are being discussed because of aerosols there. The air–Earth current density in fair weather varies by a factor of 3 to 6 accordingly. Conductivity near the ground varies by a factor of about 3 but only decreasing; increase of conductivity due to extraordinary radioactivity is a singular event. The field strength near the ground varies as a consequence of variations of air–Earth current density and conductivity from about 1/3 to about 10 times of the value quoted in the table. Conductivity near the ground shows a diurnal and an annual variation which depends strongly on the locality: air–Earth current density shows a diurnal and annual variation because the Earth–ionosphere potential difference undergoes such variations, and also because the columnar resistance is supposed to have a diurnal and probably an annual variation.

Conductivities and air–earth current densities on high mountains are greater than at sea level by factors of up to 10. Conductivity decreases when atmospheric humidity increases. Values for space charges are not quoted because measurements are too few to allow calculation of average values. Values of parameters over the oceans are still rather uncertain.

Theoretically, in fair weather conditions, Ohm's law must be fulfilled for the electric field, the conduction current density, and the electrical conductivity of the atmosphere. Deviations point to shortcomings in the applied measuring techniques. Data which are representative for a large area (in the extreme, "globally representative data," i.e., data on the global circuit), can be obtained on the ground only by stations on an open plane and only if local generators are either small or constant or are independently measured. Certain measurements with instrumented aircraft provide globally representative information valid for the period of the actual measurement.

TABLE 1. Electrical Parameters of the Clear (Fair Weather) Atmosphere, Pertinent to the Classical Picture of Atm. Electricity (Electric Standard Atmosphere)

Part of atmosphere for which the values are calculated (elements are in free, cloudless atmosphere)	Currents, I, in A; and current densities, i, in A/m²	Potential differences, U, in V; field strength E in V/m; $U = 0$ at sea level	Resistances, R, in Ω; columnar resistances, R_c, in Ω m² and resistivities, ρ, in Ω m	Conductances, G, in Ω⁻¹; columnar conductances G_c, in Ω⁻¹m²; total conductivities, γ, in Ω⁻¹m⁻¹	Capacitances, C, in F; columnar capacitances, C_c, in F m⁻² and capacitivities, ε, in F m⁻¹	Time constants τ, in seconds
Volume element at about sea level, 1 m³	$i = 3 \times 10^{-12*}$	$E_0 = 1.2 \times 10^{2*}$	$\rho_0 = 4 \times 10^{13}$	$\gamma_0 = 2.5 \times 10^{-14}$	$\varepsilon_0 = 8.9 \times 10^{-12*}$	$\tau_0 = 3.6 \times 10^2$
Lower column of 1 m² cross section from sea level to 2 km height	$i = 3 \times 10^{-12}$	At upper end: $U_1 = 1.8 \times 10^5$	$R_{c1} = 6 \times 10^{16}$	$G_{c1} = 1.7 \times 10^{-17}$	$C_{c1} = 4.4 \times 10^{-15}$	$\tau_{c1} = 2.6 \times 10^2$
Volume element at about 2 km height, 1 m³	$i = 3 \times 10^{-12}$	$E_2 = 6.6 \times 10^1$	$\rho_2 = 2.2 \times 10^{13(*)}$	$\gamma_2 = 4.5 \times 10^{-14}$	$\varepsilon_2 = 8.9 \times 10^{-12*}$	$\tau_2 = 2 \times 10^2$
Center column of 1 m² cross section from 2 to 12 km	$i = 3 \times 10^{-12}$	At upper end: $U_m = 3.15 \times 10^5$	$R_{cm} = 4.5 \times 10^{16}$	$G_{cm} = 5 \times 10^{-17}$	$C_{cm} = 8.8 \times 10^{-16}$	$\tau_{cm} = 1.8 \times 10^1$
Volume element at about 12 km height, 1 m³	$i = 3 \times 10^{-12}$	$E_{12} = 3.9 \times 10^0$	$\rho_{12} = 1.3 \times 10^{12(*)}$	$\gamma_{12} = 7.7 \times 10^{-13}$	$\varepsilon_{12} = 8.9 \times 10^{-12}$	$\tau_{12} = 1.2 \times 10^1$
Upper column of 1 m² cross section from 12 to 65 km height	$i = 3 \times 10^{-12}$	At upper end: $U_u = 3.5 \times 10^5$	$R_{cu} = 1.5 \times 10^{16}$	$G_{cu} = 2.5 \times 10^{-17}$	$C_{cu} = 1.67 \times 10^{-16}$	$\tau_{cm} = 6.7 \times 10^0$
Whole column of 1 m² cross section from 0 to 65 km height	$i = 3 \times 10^{-12}$	At upper end: $U = 3.5 \times 10^5$	$R_c = 1.2 \times 10^{17}$	$G_c = 8.3 \times 10^{-18}$	$C_c = 1.36 \times 10^{-16}$	$\tau_c = 1.64 \times 10^1$
Total spherical capacitor area: 5×10^{14} m²	$i = 1.5 \times 10^3$	$U = 3.5 \times 10^{5*}$	$R = 2.4 \times 10^2$	$G = 4.2 \times 10^{-3}$	$C = 6.8 \times 10^{-2}$	$\tau = 1.64 \times 10^1$

Note: All currents and fields listed are part of the global circuit, i.e., circuits of local generators are not included. Values are subject to variations due to latitude and altitude of the point of observation above sea level, locality with respect to sources of disturbances, meteorological and climatological factors, and man-made changes. For more explanations, see text.

II. AIR IONS

Hannes Tammet

The term "air ions" signifies all airborne particles that are the carriers of the electrical current in the air and have drift velocities determined by the electric field.

The probability of electrical dissociation of molecules in the atmospheric air under thermodynamic equilibrium is near to zero. The average ionization at the ground level over the ocean is $2 \cdot 10^6$ ion pairs m⁻³s⁻¹. This ionization is produced mainly by cosmic rays. Over the continents the ionizing radiation from soil and from radioactive substances in the air each add about $4 \cdot 10^6$ m⁻³s⁻¹. The total average ionization rate of 10^7 m³s⁻¹ is equivalent to 17 μR/h, which is a customary expression of the background level of the ionizing radiations. The ionization rate over the ground varies in space due to the radioactivity of soil, and in time depending on the exchange of air between the atmosphere and radon-containing soil. Radioactive pollution increases the ionization rate. A temporary increase of about 10 times was registered in Sweden after the Chernobyl accident in 1986. The emission of Kr⁸⁵ from nuclear power plants can noticeably increase the global ionization rate in the 21st century. The ionization rate decreases with altitude near the ground and increases at higher altitudes up to 15 km, where it has a maximum of about $5 \cdot 10^7$ m⁻³s⁻¹. Solar X-ray and extreme UV radiation cause a new increase at altitudes over 60 km.

Local sources of air ions are point discharges in strong electric fields, fluidization of charged drops from waves, etc.

The enhanced chemical activity of an ion results in a chain of ion-molecule reactions with the colliding neutrals, and, in the first microsecond of the life of an air ion, a charged molecular cluster called the *cluster ion* is formed. According to theoretical calcula-tions in the air free from exotic trace gases the following cluster ions should be dominant:

$NO_3^- \cdot (HNO_3) \cdot H_2O$, $NO_2^- \cdot (H_2O)_2$, $NO_3^- \cdot H_2O$, $O_2^- \cdot (H_2O)_4$, $O_2^- \cdot (H_2O)_5$,

$H_3O^+ \cdot (H_2O)_6$, $NH_4^+ \cdot (H_2O)_2$, $NH_4^+ \cdot (H_2O)$, $H_3O^+ \cdot (H_2O)_5$, $NH_4^+ \cdot NH_3$

A measurable parameter of air ions is the electrical mobility k, characterizing the drift velocity in the unit electric field. The mobility is inversely proportional to the density of air, and the re-sults of measurements are as a rule reduced to normal conditions. According to mobility the air ions are called fast or small or light ions with mobility $k > 5 \cdot 10^{-5}$ m²V⁻¹s¹, intermediate ions, and slow or large or heavy ions with mobility $k > 10^{-6}$ m²V⁻¹s¹. The boundary between intermediate and slow ions is conventional.

Cluster ions are fast ions. The masses of cluster ions may be measured with mass spectrometers, but the possible ion-molecule reactions during the passage of the air through nozzles to the vac-uum chamber complicate the measurement. Mass and mobility of cluster ions are highly correlated. The experimental results[5] can be expressed by the empirical formula

$$m \approx \frac{850 \text{ u}}{[0.3 + k / (10^{-4} \text{ m}^2 \text{V}^{-1}\text{s}^{-1})]^3}$$

where u is the unified atomic mass unit.

The value of the transport cross-section of a cluster ion is needed to calculate its mobility according to the kinetic theory of Chapman and Enskog. The theoretical estimation of transport cross-sections is rough and cannot be used to identify the chemical structures of cluster ions. Mass spectrometry is the main technique of identification of cluster ions.[2]

Märk and Castleman[4] presented an overview of over 1000 publications on the experimental studies of cluster ions. Most of them present information about ions of millisecond age range. The low concentration makes it difficult to get detailed information about masses and mobilities of the natural atmospheric ions at ground level. The results of a 1-year continuous measurement[6] are as follows:

	+ ions	– ions	unit
Average mobility	1.36	1.56	$10^{-4}\,m^2V^{-1}s^{-1}$
The corresponding mass	190	130	u
The corresponding diameter	0.69	0.61	nm
The average concentration	400	360	$10^6\,m^{-3}$
The corresponding conductivity	8.7	9.0	fS

The distribution of tropospheric cluster ions according to the mobility and estimated mass is depicted in Figure 1.

The problems and results of direct mass spectrometry of natural cluster ions are analyzed by Eisele[2] for ground level and by Meyerott, Reagan and Joiner[5] for stratospheric measurements. Air ions in the high atmosphere are a subject of ionospheric physics.

During its lifetime (about 1 min), a cluster ion at ground level collides with nearly 10^{12} molecules. Thus the cluster ions are able to concentrate trace gases of very low concentration if they have an extra high electron or proton affinity. For example, Eisele[2] demonstrated that a considerable fraction of positive atmospheric cluster ions in the unpolluted atmosphere at ground level probably consists of a molecule derived from pyridine. The concentration of these constituents is estimated to be about 10^{-12}. Therefore, air-ion mass and mobility spectrometry is considered as a promising technique for trace analysis in the air. Mass and mobility spectrometry of millisecond-age air ions has been developed as a technique of chemical analysis known as "plasma chromatography."[1] The sensitivity of the detection grows with the age of the cluster ions measured.

The mechanisms of annihilation of cluster ions are ion-ion recombination (on the average 3%) and sedimentation on aerosol particles (on the average 97% of cluster ions at ground level). The result of the combination of a cluster ion and neutral particle is a charged particle called an *aerosol ion*. In conditions of detailed thermodynamic equilibrium the probability that a spherical particle of diameter d carries q elementary charges is calculated from the Boltzmann distribution:

$$p_q(d) = (2\pi d/d_0)^{1/2}\exp(-q^2 d_0/2d)$$

where d_0 = 115 nm (at 18 °C). The supposition about the detailed equilibrium is an approximation and the formula is not valid for particles less than d_0. On the basis of numerical calculations by Hoppel and Frick[3] the following charge probabilities can be derived:

d	3	10	30	100	300	1000	3000	nm
P_0	98	90	70	42	24	14	8	%
$p_{-1}+p_1$	2	10	30	48	41	25	15	%
$p_{-2}+p_2$	0	0	0	10	23	21	14	%
$P_{q>2}$	0	0	0	0	12	40	63	%
k_1	15000	1900	250	28	5.1	1.11	0.33	$10^{-9}\,m^2V^{-1}s^{-1}$

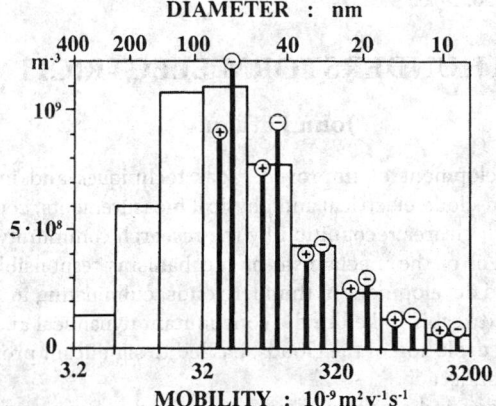

FIGURE 1. Average mobility and mass spectra of natural tropospheric cluster ions. Concentrations of the mobility fractions were measured in a rural site every 5 min over 1 year.[6] Ion mass is estimated according to the above empirical formula.

FIGURE 2. Mobility and size spectra of tropospheric aerosol ions.[6] The wide bars mark the fraction concentrations theoretically estimated on the basis of the standard size distribution of tropospheric aerosol. The pin bars with head + and – mark average values of positive and negative aerosol ion fraction concentrations measured in a rural site every 5 min during 4 months.

The last line of the table presents the mobility of a particle carrying one elementary charge. The distribution of the atmospheric aerosol ions over mobility is demonstrated in Figure 2.

Although the concentration of aerosol in continental air at ground level is an order of magnitude higher than the concentration of cluster ions, the mobilities of aerosol ions are so small that their percentage in air conductivity is less than 1%.

A specific class of aerosol ions is condensed aerosol ions produced as a result of the condensation of gaseous matter on the cluster ions. In aerosol physics the process is called ion-induced nucleation; it is considered as one of the processes of gas-to-particle conversion. The condensed aerosol ions have an inherent charge. Their sizes and mobilities are between the sizes and mobilities of cluster ions and of ordinary aerosol ions. Water and standard constituents of atmospheric air are not able to condense on the cluster ions in the real atmosphere. Thus the concentration of condensed aerosol ions depends on the trace constituents in the air and is very low in unpolluted air. Knowledge about condensed aerosol ions is poor because of measurement difficulties.

References

1. Carr, T. W., Ed., *Plasma Chromatography*, Plenum Press, New York and London, XII + 259 pp., 1984.
2. Eisele, F. L., Identification of Tropospheric Ions, *J. Geophys. Res.*, vol. 91, no. D7, pp. 7897–7906, 1986.
3. Hoppel, W. A., and Frick, G. M., The Nonequilibrium Character of the Aerosol Charge Distributions Produced by Neutralizers, *Aerosol Sci. Technol.*, vol. 12, no. 3, pp. 471–496, 1990.
4. Mark, T. D., and Castleman, A. W., Experimental Studies on Cluster Ions, in *Advances in Atomic and Molecular Physics*, vol. 20, pp. 65–172, Academic Press, 1985.
5. Meyerott, R. E., Reagan, J. B., and Joiner, R. G., The Mobility and Concentration of Ions and the Ionic Conductivity in the Lower Stratosphere, *J. Geophys. Res.*, vol. 85, no. A3, pp. 1273–1278, 1980.
6. Salm, J., Tammet, H., Iher, H., and Hörrak, U., Atmospheric Electrical Measurements in Tahkuse, Estonia (in Russian), in *Voprosy Atmosfernogo Elektrichestva*, pp. 168–175, Gidrometeoizdat, Leningrad, 1990.

III. THUNDERSTORM ELECTRICITY

John Latham

The development of improved radar techniques and instruments for in-cloud electrical and physical measurements, coupled with a much clearer recognition by the research community that establishment of the mechanism or mechanisms responsible for electric field development in thunderclouds, culminating in lightning, is inextricably linked to the concomitant dynamical and microphysical evolution of the clouds, has led to significant progress over the past decade.

Field studies indicate that in most thunderclouds the electrical development is associated with the process of glaciation, which can occur in a variety of incompletely understood ways. In the absence of ice, field growth is slow, individual hydrometeor charges are low, and lightning is produced only rarely. Precipitation — in the solid form, as graupel — also appears to be a necessary ingredient for significant electrification, as does significant convective activity and mixing between the clouds and their environments, via entrainment.

Increasingly, the view is being accepted that charge transfer leading to field-growth is largely a consequence of rebounding collisions between graupel pellets and smaller vapor-grown ice crystals, followed by the separation under gravity of these two types

of hydrometeor. These collisions occur predominantly within the temperature range −15 to −30 °C, and for significant charge transfer need to occur in the presence of supercooled cloud droplets.

The field evidence is inconsistent with an inductive mechanism, and extensive laboratory studies indicate that the principal charging mechanism is non-inductive and associated — in ways yet to be identified — with differences in surface characteristics of the interacting hydrometeors.

Laboratory studies indicate that the two most favored sites for corona emission leading to the lightning discharge are the tips of ephemeral liquid filaments, produced during the glancing collisions of supercooled raindrops, and protuberances on large ice crystals or graupel pellets. The relative importance of these alternatives will depend on the hydrometeor characteristics and the temperature in the regions of strongest fields; these features are themselves dependent on air-mass characteristics and climatological considerations.

A recently identified but unresolved question is why, in continental Northern Hemisphere thunderclouds at least, the sign of the charge brought to ground by lightning is predominantly negative in summer but more evenly balanced in winter.

IV. LIGHTNING

Martin A. Uman

From both ground-based weather-station data and satellite measurements, it has been estimated that there are about 100 lightning discharges, both cloud and ground flashes, over the whole Earth each second; representing an average global lightning flash density of about 6 $km^{-2}yr^{-1}$. Most of this lightning occurs over the Earth's land masses. For example, in central Florida, where thunderstorms occur about 90 days/yr, the flash density for discharges to earth is about 15 $km^{-2}yr^{-1}$. Some tropical areas of the Earth have thunderstorms up to 300 days/yr.

Lightning can be defined as a transient, high-current electric discharge whose path length is measured in kilometers and whose most common source is the electric charge separated in the ordinary thunderstorm or cumulonimbus cloud. Well over half of all lightning discharges occur totally within individual thunderstorm clouds and are referred to as intracloud discharges. Cloud-to-ground lightning, however, has been studied more extensively than any other lightning form because of its visibility and its more practical interest. Cloud-to-cloud and cloud-to-air discharges are less common than intracloud or cloud-to-ground lightning.

Lightning between the cloud and Earth can be categorized in terms of the direction of motion, upward or downward, and the sign of the charge, positive or negative, of the developing discharge (called a *leader*) that initiates the overall event. Over 90% of the worldwide cloud-to-ground discharges is initiated in the thundercloud by downward-moving negatively charged leaders and subsequently results in the lowering of negative charge to Earth. Cloud-to-ground lightning can also be initiated by downward-moving positive leaders, less than 10% of the worldwide cloud-to-ground lightning being of this type although the exact percentage is a function of season and latitude. Lightning between cloud and ground can also be initiated by leaders that develop upward from the Earth. These upward-initiated discharges are relatively rare, may be of either polarity, and generally occur from mountaintops and tall man-made structures.

We discuss next the most common type of cloud-to-ground lightning. A negative cloud-to-ground discharge or *flash* has an overall duration of some tenths of a second and is made up of various components, among which are typically three or four high-current pulses called *strokes*. Each stroke lasts about a millisecond, the separation time between strokes being typically several tens of milliseconds. Such lightning often appears to "flicker" because the human eye can just resolve the individual light pulse associated with each stroke. A drawing of the components of a negative cloud-to-ground flash is found in Figure 3. Some values for salient parameters are found in Table 1. The negatively charged *stepped leader* initiates the first stroke in a flash by propagating from cloud to ground through virgin air in a series of discrete steps. Photographically observed leader steps in clear air are typically 1 μs in duration and tens of meters in length, with a pause time between steps of about 50 μs. A fully developed stepped leader lowers up to 10 or more coulombs of negative cloud charge toward ground in tens of milliseconds with an average downward speed of about 2×10^5 m/s. The average leader current is in the 100 to 1000 A range. The steps have pulse currents of at least 1 kA. Associated with these currents are electric- and magnetic-field pulses with widths of about 1 μs or less and risetimes of about 0.1 μs or less. The stepped leader, during its trip toward ground, branches in a downward direction, resulting in the characteristic downward-branched geometrical structure commonly observed. The electric potential of the bottom of the negatively charged leader channel with respect to ground has a magnitude in excess of 10^7 V. As the leader tip nears ground, the electric field at sharp objects on the ground or at irregularities of the ground itself exceeds the breakdown value of air, and one or more upward-moving discharges (often called upward leaders) are initiated from those points, thus beginning the *attachment process*. An understanding of the physics of the attachment process is central to an understanding of the operation of lightning protection of ground-based objects and the effects of lightning on humans and animals, since it is the attachment process that determines where the lightning connects to objects on the ground and the value of the early currents which flow. When one of the upward-moving discharges from the ground (or from a lightning rod or an individual) contacts the tip of the downward-moving stepped leader, typically some tens of meters above the ground, the leader tip is effectively connected to ground potential. The negatively charged leader channel is then discharged to earth when a ground potential wave, referred to as the first *return stroke*, propagates continuously up the leader path. The upward speed of a return stroke near the ground is typically near one third the speed of light, and the speed decreases with height. The first return stroke produces a peak current near ground of typically 30 kA, with a time from zero to peak of a few microseconds. Currents measured at the ground fall to half of the peak value in about 50 μs, and currents of the order of hundreds of amperes may flow for times of a few milliseconds up to several hundred milliseconds. The longer-lasting currents are known as *continuing currents*. The rapid release of return stroke energy heats the leader channel to a temperature near 30,000 K and creates a high-pressure channel which expands and generates the shock waves that eventually become thunder, as further discussed later. The return stroke effectively lowers to ground the charge originally deposited onto the stepped-leader channel and additionally initiates the lowering of other charges that may be available to the top of its channel. First return-stroke electric fields exhibit a microsecond scale rise to peak with a typical peak value of 5 V/m, normalized to a distance of 100 km by an inverse distance relationship. Roughly half of the field rise to peak, the so-called "fast transition," takes place in tenths of a microsecond, an observation that can only be made if the field propagation is over a highly conducting surface such as salt water.

After the first return-stroke current has ceased to flow, the flash, including charge motion in the cloud, may end. The lightning is then called a single-stroke flash. On the other hand, if additional charge is made available to the top of the channel, a continuous or *dart leader* may propagate down the residual first-stroke channel at a typical speed of about 1×10^7 m/s. The dart leader lowers a charge of the order of 1 C by virtue of a current of about 1 kA. The dart leader then initiates the second (or any subsequent) return stroke. Subsequent return-stroke currents generally have faster zero-to-peak rise times than do first-stroke currents, but similar maximum rates of change, about 100 kA/μs. Some leaders begin as dart leaders, but toward the end of their trip toward ground become stepped leaders. These leaders are known as *dart-stepped leaders* and may have different ground termination points (and separate upward leaders) from the first stroke. Most often the dart-stepped leaders are associated with the second stroke of the flash. Nearly half of all flashes exhibit more than one termination point on ground with the distance between separate terminations being up to several kilometers. Subsequent return-stroke radiated electric and magnetic fields are similar to, but usually a factor of two or so smaller than first return-stroke fields. About one third of all multiple-stroke flashes has at least one subsequent stroke that is larger than the first stroke.

Cloud-to-ground flashes that lower positive charge, though not common, are of considerable practical interest because their peak currents and total charge transfer can be much larger than for the more common negative ground flash. The largest recorded peak currents, those in the 200- to 300-kA range, are due to the return strokes of positive lightning. Such positive flashes to ground are initiated by downward-moving leaders that do not exhibit the distinct steps of their negative counterparts. Rather, they show a luminosity that is more or less continuous but modulated in intensity. Positive flashes are generally composed of a single stroke followed by a period of continuing current. Positive flashes are probably initiated from the upper positive charge in the thundercloud charge dipole when that cloud charge is horizontally separated from the negative charge beneath it, the source of the usual negative cloud-to-ground lightning. Positive flashes are relatively common in winter thunderstorms (snow storms), which produce few flashes overall, and are relatively uncommon in summer thunderstorms. The fraction of positive lightning in summer thunderstorms apparently increases with increasing latitude and with increasing height of the ground above sea level.

Distant lightning return stroke fields are often referred to as sferics (called "atmospherics" in the older literature). The peak in the sferics frequency spectrum is near 5 kHz due to the bipolar or ringing nature of the distant return-stroke electromagnetic signal and to the effects of propagation.

Thunder, the acoustic radiation associated with lightning, is sometimes divided into the categories "audible," sounds that one can hear, and "infrasonic," below a few tens of hertz, a frequency range that is inaudible. This division is made because it is thought that the mechanisms that produce audible and infrasonic thunder are different. Audible thunder is thought to be due to the expansion of a rapidly heated return stroke channel, as noted earlier, whereas infrasonic thunder is thought to be associated with the conversion to sound of the energy stored in the electrostatic field of the thundercloud when lightning rapidly reduces that cloud field.

The technology of artificially initiating lightning by firing upward small rocket trailing grounded wire of a few hundred meters length has been well-developed during the past decade. Such "triggered" flashes are similar to natural upward-initiated discharges from tall structure. They often contain subsequent strokes which,

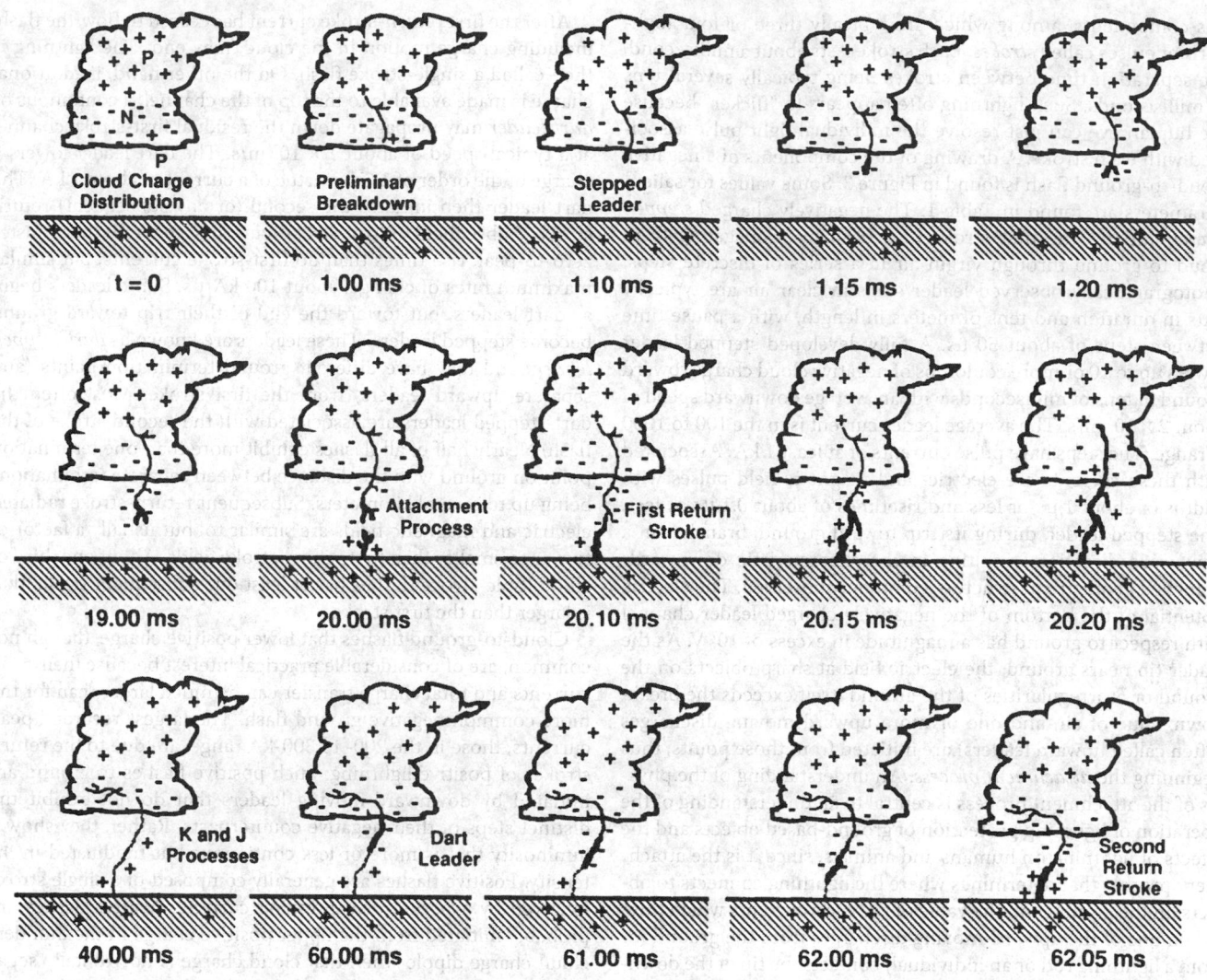

FIGURE 3. Sequence of steps in cloud-to-ground lightning.

when they occur, are similar to the subsequent strokes in natural lightning. These triggered subsequent strokes have been the subject of considerable recent research.

Also in the past 10 years or so sophisticated lightning locating equipment has been installed throughout the world. For example, all ground flashes in the U.S. are now centrally monitored for research, for better overall weather prediction, and for hazard warning for aviation, electric utilities and other lightning-sensitive facilities.

Information on lightning physics can be found in M. A. Uman, *The Lightning Discharge*, Academic Press, San Diego, 1987; on lightning death and injury in *Medical Aspects of Lightning Injury*, C. Andrews, M. A. Cooper, M. Darveniza, and D. Mackerras, Eds.,

CRC Press, Boca Raton, FL, 1992. Ground flash location information for the U.S., in real time or archived, is available from Geomet Data Service of Tucson, AZ, which is also a source of the names of providers of those data in other countries.

Table 2 has data for cloud-to-ground lightning discharges bringing negative charge to earth. The values listed are intended to convey a rough feeling for the various physical parameters of lightning. No great accuracy is claimed since the results of different investigators are often not in good agreement. These values may, in fact, depend on the particular environment in which the lightning discharge is generated. The choice of some of the entries in the table is arbitrary.

TABLE 2. Data for Cloud-to-Ground Lightning Discharges

	Minimum[a]	Representative values	Maximum[a]
Stepped leader			
Length of step, m	3	50	200
Time interval between steps, μs	30	50	125
Average speed of propagation of stepped leader, m/s[b]	1.0×10^5	2.0×10^5	3.0×10^6
Charged deposited on stepped-leader channel, coulombs	3	5	20
Dart leader			
Speed of propagation, m/s[b]	1.0×10^6	1.0×10^7	2.4×10^7
Charged deposited on dart-leader channel, coulombs	0.2	1	6
Return stroke[c]			
Speed of propagation, m/s[b]	2.0×10^7	1.0×10^8	2.0×10^8
Maximum current rate of increase, kA/μs	<1	100	400
Time to peak current, μs	<1	2	30
Peak current, kA	2	30	200
Time to half of peak current, μs	10	50	250
Charge transferred excluding continuing current, coulombs	0.02	3	20
Channel length, km	2	5	15
Lightning flash			
Number of strokes per flash	1	4	26
Time interval between strokes in absence of continuing current, ms	3	60	100
Time duration of flash, s	10^{-2}	0.5	2
Charge transferred including continuing current, coulombs	3	30	200

[a] The words maximum and minimum are used in the sense that most measured values fall between these limits.

[b] Speeds of propagation are generally determined from photographic data and are "two-dimensional." Since many lightning flashes are not vertical, values stated are probably slight underestimates of actual values.

[c] First return strokes have longer times to current peak and generally larger charge transfer than do subsequent return strokes

Adapted from Uman, M.A., *Lightning*, Dover Paperbook, New York, 1986, and Uman, M.A., *The Lightning Discharge*, Academic Press, San Diego, 1987.

SPEED OF SOUND IN VARIOUS MEDIA

The speed of sound in various solids, liquids, and gases is given in these tables. While only a single parameter v is needed for liquids and gases, sound propagation in isotropic solids is characterized by three velocity parameters. For a solid of infinite extent (or of finite extent if all dimensions are much larger than a wavelength), there are two relevant quantities,

v_l: velocity of longitudinal waves
v_s: velocity of shear waves.

For a cylindrical rod with diameter much smaller than a wavelength,

v_{ext}: velocity of extensional waves along the rod. (Torsional waves in the rod are propagated at the same speed as sheer waves in an infinite solid.)

Table 1 lists values for a variety of solid materials. Table 2 covers liquids and gases; values for cryogenic liquids are given at the normal boiling point. Table 3 gives the speed of sound in pure water and in seawater of salinity $S = 3.5\%$ as a function of temperature.

All values are in meters per second and are given for normal atmospheric pressure.

References

1. Gray, D. E., Ed., *American Institute of Physics Handbook, Third Edition*, McGraw Hill, New York, 1972.
2. Anderson, H.L., Ed., *A Physicist's Desk Reference*, American Institute of Physics, New York, 1989.
3. Younglove, B. A., Thermophysical Proeprties of Fluids. Part I, *J. Phys. Chem. Ref. Data*, 11, Suppl. 1, 1982.
4. Younglove, B. A., and Ely, J. F., Thermophysical Properties of Fluids. Part II, *J. Phys. Chem. Ref. Data*, 16, 577, 1987.
5. Harvey, A. H., Peskin, A. P., and Klein, S. A., NIST Standard Reference Database 10: NIST/ASME Steam Properties, Version 2.22, National Institute of Standards and Technology, Standard Reference Data Program, Gaithersburg, Maryland, 2008 (www.nist.gov/srd/nist10.cfm).
6. Mason, W. P., *Physical Acoustics and the Properties of Solids*, D. Van Nostrand Co., Princeton, N.J., 1958.
7. *Landolt-Börnstein, Numerical Data and Functional Relationships in Science and Technology*, New Series, II/5, Molecular Acoustics, Springer-Verlag, Heidelberg, 1967.

TABLE 1. Speed of Sound in Solids at Room Temperature

Name	v_l/m s^{-1}	v_s/m s^{-1}	v_{ext}/m s^{-1}	Name	v_l/m s^{-1}	v_s/m s^{-1}	v_{ext}/m s^{-1}
Metals				Steel, 347 Stainless	5790	3100	5000
Aluminum, rolled	6420	3040	5000	Steel, K9	5940	3250	5250
Beryllium	12890	8880	12870	Tin, rolled	3320	1670	2730
Brass (70 Cu, 30 Zn)	4700	2110	3480	Titanium	6070	3125	5090
Constantan	5177	2625	4270	Tungsten, annealed	5220	2890	4620
Copper, annealed	4760	2325	3810	Tungsten, drawn	5410	2640	4320
Copper, rolled	5010	2270	3750	Zinc, rolled	4210	2440	3850
Duralumin 17S	6320	3130	5150				
Gold, hard-drawn	3240	1200	2030	**Other materials**			
Iron, cast	4994	2809	4480	Fused silica	5968	3764	5760
Iron, electrolytic	5950	3240	5120	Glass, heavy silicate flint	3980	2380	3720
Iron, Armco	5960	3240	5200	Glass, light borate crown	5100	2840	4540
Lead, annealed	2160	700	1190	Glass, pyrex	5640	3280	5170
Lead, rolled	1960	690	1210	Lucite	2680	1100	1840
Magnesium, annealed	5770	3050	4940	Nylon 6-6	2620	1070	1800
Molybdenum	6250	3350	5400	Polyethylene	1950	540	920
Monel metal	5350	2720	4400	Polystyrene	2350	1120	1840
Nickel	6040	3000	4900	Rubber, butyl	1830		
Platinum	3260	1730	2800	Rubber, gum	1550		
Silver	3650	1610	2680	Rubber, neoprene	1600		
Steel (1% C)	5940	3220	5180	Tungsten carbide	6655	3980	6220

TABLE 2. Speed of Sound in Liquids and Gases

Name	$t/°C$	$v/m\ s^{-1}$	Name	$t/°C$	$v/m\ s^{-1}$
			1-Pentadecene	20	1351
Liquids			Pentane	20	1008
Acetone	20	1203	Propane	−42.1	1158
Argon	−185.9	813	1-Propanol	20	1223
Benzene	25	1310	Tetrachloromethane	25	930
Bromobenzene	20	1169	Trichloromethane	25	987
Butane	−0.5	1034	1-Undecene	20	1275
1-Butanol	20	1258	Water	25	1497
Carbon disulfide	25	1140	Water (sea, S = 3.5%)	25	1534
Chlorobenzene	20	1311			
Cyclohexane	19	1280	*Gases at 1 atm*		
1-Decene	20	1250	Air, dry	25	346
Diethyl ether	25	976	Ammonia	0	415
Ethane	−88.6	1326	Argon	27	323
Ethanol	20	1162	Carbon monoxide	0	338
Ethylene	−103.8	1309	Carbon dioxide	0	259
Ethylene glycol	25	1658	Chlorine	0	206
Fluorobenzene	20	1183	Deuterium	0	890
Glycerol	25	1904	Ethane	27	312
Helium	−268.9	180	Ethylene	27	331
Heptane	20	1162	Helium	0	965
1-Heptene	20	1128	Hydrogen	27	1310
Hexane	20	1083	Hydrogen bromide	0	200
Hydrogen	−252.9	1101	Hydrogen chloride	0	296
Iodobenzene	20	1114	Hydrogen iodide	0	157
Mercury	25	1450	Hydrogen sulfide	0	289
Methane	−161.5	1337	Methane	27	450
Methanol	20	1121	Neon	0	435
Nitrobenzene	25	1463	Nitric oxide	10	325
Nitrogen	−195.8	939	Nitrogen	27	353
1-Nonene	20	1218	Nitrous oxide	0	263
Octane	20	1197	Oxygen	27	330
1-Octene	20	1184	Sulfur dioxide	0	213
Oxygen	−183.0	906	Water (steam)	100	472

TABLE 3. Speed of Sound in Water and Seawater (S = 3.5%) at Different Temperatures

$t/°C$	$v/m\ s^{-1}$ Water	Seawater
0	1402.4	1449.1
10	1447.3	1489.8
20	1483.3	1521.5
25	1496.7	1534.4
30	1509.2	1545.6
40	1528.9	1563.2
50	1542.6	
60	1551.0	
70	1554.7	
80	1554.4	
90	1550.4	
100	1543.5	

ATTENUATION AND SPEED OF SOUND IN AIR AS A FUNCTION OF HUMIDITY AND FREQUENCY

This table gives the attenuation and speed of sound as a function of frequency at various values of relative humidity. All values refer to still air at 20 °C.

References

1. Tables of Absorption and Velocity of Sound in Still Air at 68 °F (20 °C), AD-738576, National Technical Information Service, Springfield, VA.
2. Evans, L. B., Bass, H. E., and Sutherland, L. C., *J. Acoust. Soc. Am.*, 51, 1565, 1972.

Frequency (Hz)	Attenuation (dB/km)	Speed (m/s)
Relative humidity 0%		
20	0.51	343.477
40	1.07	343.514
50	1.26	343.525
63	1.43	343.536
100	1.67	343.550
200	1.84	343.559
400	1.96	343.561
630	2.11	343.562
800	2.27	343.562
1250	2.82	343.562
2000	4.14	343.562
4000	8.84	343.564
6300	14.89	343.565
10000	26.28	343.566
12500	35.81	343.566
16000	52.15	343.567
20000	75.37	343.567
40000	267.01	343.567
63000	644.66	343.567
80000	1032.14	343.567
Relative humidity 30%		
20	0.03	343.807
40	0.11	343.808
50	0.17	343.810
63	0.25	343.810
100	0.50	343.814
200	1.01	343.821
400	1.59	343.826
630	2.24	343.827
800	2.85	343.828
1250	5.09	343.828
2000	10.93	343.829
4000	38.89	343.831
6300	90.61	343.836
10000	204.98	343.846
12500	294.08	343.854
16000	422.51	343.865
20000	563.66	343.877
40000	1110.97	343.911
63000	1639.47	343.924
80000	2083.08	343.929

Frequency (Hz)	Attenuation (dB/km)	Speed (m/s)
Relative humidity 60%		
20	0.02	344.182
40	0.06	344.183
50	0.09	344.183
63	0.15	344.184
100	0.34	344.185
200	0.99	344.190
400	1.94	344.197
630	2.57	344.200
800	2.94	344.201
1250	4.01	344.202
2000	6.55	344.203
4000	18.73	344.204
6300	42.51	344.204
10000	101.84	344.206
12500	155.67	344.208
16000	247.78	344.211
20000	373.78	344.215
40000	1195.37	344.238
63000	2220.64	344.262
80000	2951.71	344.274
Relative humidity 100%		
20	0.01	344.685
40	0.04	344.685
50	0.06	344.685
63	0.09	344.685
100	0.22	344.686
200	0.77	344.689
400	2.02	344.695
630	3.05	344.699
800	3.57	344.701
1250	4.59	344.704
2000	6.29	344.705
4000	13.58	344.706
6300	27.72	344.706
10000	63.49	344.706
12500	96.63	344.707
16000	154.90	344.708
20000	237.93	344.709
40000	884.28	344.718
63000	1973.62	344.731
80000	2913.01	344.742

SPEED OF SOUND IN DRY AIR

Eric W. Lemmon

These values were calculated from the equation of state for dry air (average molecular weight 28.96) treated as a real gas. Values refer to standard atmospheric pressure. The speed of sound varies only slightly with pressure; at two atmospheres and −100 °C the value decreases by 0.16%, while at two atmospheres and 80 °C the speed increases by 0.05%. For additional values, see the table in Section 6 labeled "Thermophysical Properties of Air."

Reference

Lemmon, E.W., Jacobsen, R.T, Penoncello, S.G., and Friend, D.G., Thermodynamic Properties of Air and Mixtures of Nitrogen, Argon, and Oxygen from 60 to 2000 K at Pressures to 2000 MPa, *J. Phys. Chem. Ref. Data* 29, 331, 2000.

$t/°C$	$v_s/\text{m s}^{-1}$	$t/°C$	$v_s/\text{m s}^{-1}$	$t/°C$	$v_s/\text{m s}^{-1}$
−100	263.5	−30	312.7	40	354.9
−95	267.3	−25	315.9	45	357.7
−90	271.1	−20	319.1	50	360.4
−85	274.8	−15	322.2	55	363.2
−80	278.5	−10	325.4	60	365.9
−75	282.1	−5	328.4	65	368.7
−70	285.7	0	331.5	70	371.3
−65	289.2	5	334.5	75	374.0
−60	292.7	10	337.5	80	376.7
−55	296.1	15	340.5	85	379.3
−50	299.5	20	343.4	90	381.9
−45	302.9	25	346.3	95	384.5
−40	306.2	30	349.2	100	387.0
−35	309.5	35	352.0		

MUSICAL SCALES

Equal Tempered Chromatic Scale
$A_4 = 440$ Hz
International Concert Pitch

Note	Frequency	Note	Frequency	Note	Frequency	Note	Frequency
C_0	16.35	C_2	65.41	C_4	261.63	C_6	1046.50
$C\#_0$	17.32	$C\#_2$	69.30	$C\#_4$	277.18	$C\#_6$	1108.73
D_0	18.35	D_2	73.42	D_4	293.66	D_6	1174.66
$D\#_0$	19.45	$D\#_2$	77.78	$D\#_4$	311.13	$D\#_6$	1244.51
E_0	20.60	E_2	82.41	E_4	329.63	E_6	1318.51
F_0	21.83	F_2	87.31	F_4	349.23	F_6	1396.91
$F\#_0$	23.12	$F\#_2$	92.50	$F\#_4$	369.99	$F\#_6$	1479.98
G_0	24.50	G_2	98.00	G_4	392.00	G_6	1567.98
$G\#_0$	25.96	$G\#_2$	103.83	$G\#_4$	415.30	$G\#_6$	1661.22
A_0	27.50	A_2	110.00	A_4	440.00	A_6	1760.00
$A\#_0$	29.14	$A\#_2$	116.54	$A\#_4$	466.16	$A\#_6$	1864.66
B_0	30.87	B_2	123.47	B_4	493.88	B_6	1975.53
C_1	32.70	C_3	130.81	C_5	523.25	C_7	2093.00
$C\#_1$	34.65	$C\#_3$	138.59	$C\#_5$	554.37	$C\#_7$	2217.46
D_1	36.71	D_3	146.83	D_5	587.33	D_7	2349.32
$D\#_1$	38.89	$D\#_3$	155.56	$D\#_5$	622.25	$D\#_7$	2489.02
E_1	41.20	E_3	164.81	E_5	659.26	E_7	2637.02
F_1	43.65	F_3	174.61	F_5	698.46	F_7	2793.83
$F\#_1$	46.25	$F\#_3$	185.00	$F\#_5$	739.99	$F\#_7$	2959.96
G_1	49.00	G_3	196.00	G_5	783.99	G_7	3135.96
$G\#_1$	51.91	$G\#_3$	207.65	$G\#_5$	830.61	$G\#_7$	3322.44
A_1	55.00	A_3	220.00	A_5	880.00	A_7	3520.00
$A\#_1$	58.27	$A\#_3$	233.08	$A\#_5$	932.33	$A\#_7$	3729.31
B_1	61.74	B_3	246.94	B_5	987.77	B_7	3951.07
						C_8	4186.01

Scientific or Just Scale
$C_4 = 256$ Hz

Note	Frequency	Note	Frequency	Note	Frequency	Note	Frequency
C_0	16	C_2	64	C_4	256	C_6	1024
D_0	18	D_2	72	D_4	288	D_6	1152
E_0	20	E_2	80	E_4	320	E_6	1280
F_0	21.33	F_2	85.33	F_4	341.33	F_6	1365.33
G_0	24	G_2	96	G_4	384	G_6	1536
A_0	26.67	A_2	106.67	A_4	426.67	A_6	1706.67
B_0	30	B_2	120	B_4	480	B_6	1920
C_1	32	C_3	128	C_5	512	C_7	2048
D_1	36	D_3	144	D_5	576	D_7	2304
E_1	40	E_3	160	E_5	640	E_7	2560
F_1	42.67	F_3	170.67	F_5	682.67	F_7	2730.67
G_1	48	G_3	192	G_5	768	G_7	3072
A_1	53.33	A_3	213.33	A_5	853.33	A_7	3413.33
B_1	60	B_3	240	B_5	960	B_7	3840
						C_8	4096

CHARACTERISTICS OF HUMAN HEARING

The human ear is sensitive to sound waves with frequencies in the range from a few hertz to almost 20 kHz. Auditory response is usually expressed in terms of the *loudness level* of a sound, which is a measure of the sound pressure. The reference level, which is given in the unit *phon*, is a pure tone of frequency 1000 Hz with sound pressure of 20 μPa (in cgs units, $2 \cdot 10^{-4}$ dyn/cm^2); loudness level is usually expressed in decibels (dB) relative to this reference level. If a normal observer perceives an arbitrary sound to be equally loud as this reference sound, the sound is said to have the loudness level of the reference. The sensitivity of the typical human ear ranges from about 0 dB, the threshold loudness level, to about 140 dB, the level at which pain sets in. The minimum detectable level thus represents a sound wave of pressure 20 μPa and intensity (power density) 10^{-16} W/cm^2.

The following figure illustrates the frequency dependence of the threshold for an average young adult.

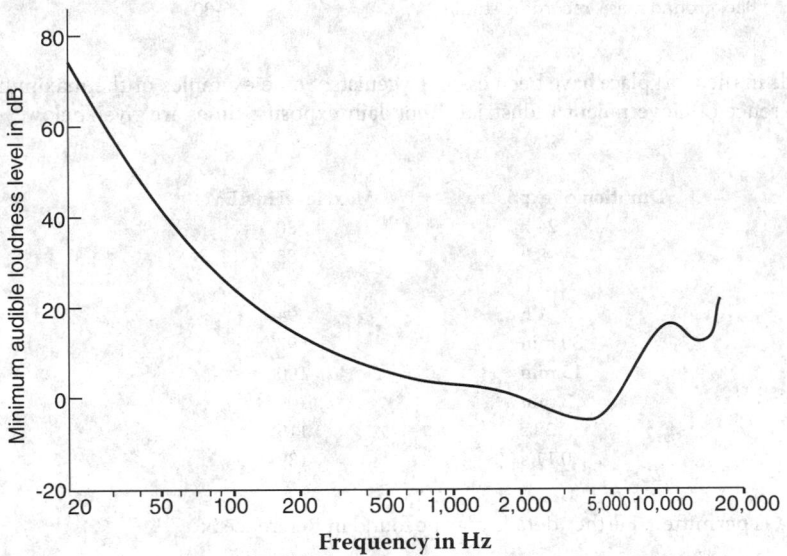

The relation between loudness level and frequency for a typical person is expressed by the following table:

Sound pressure level in dB relative to 20 μPa	Frequency in Hz					
	125	500	1000	4000	8000	10000
10			10	18		
20		16	20	28	11	
30	4	27	30	37	21	17
40	17	39	40	45	30	26
50	34	52	50	54	38	35
60	52	65	60	64	47	44
70	70	76	70	73	56	54
80	86	86	80	83	66	64
90	98	96	90	94	77	74
100	108	105	100	106	88	86

Thus, a 10,000 Hz tone at a pressure level of 50 dB seems equally loud as a 1000 Hz tone at a pressure of 35 dB.

The term *noise* refers to any unwanted sound, either a pure tone or a mixture of frequencies. Since the sensitivity of the ear is frequency dependent, as illustrated by the above table, noise level is expressed in a frequency-weighted scale, known as A-weighting. Decibel readings on this scale are designated as dBa. Typical noise levels from various sources are illustrated in this table:

Source	Noise level in dBa
Rocket engine	200
Jet aircraft engine	160
Light aircraft, cruising	140
Tractor, 150 hp	115
Electric motor, 100 hp at 2600 rpm	105
Pneumatic drill	100
Subway train	90
Vacuum cleaner	85
Heavy automobile traffic	75
Conversational speech	65
Whispered speech	40
Background noise, recording studio	25-30

Recommended noise thresholds in the workplace have been established by the American Conference of Government Industrial Hygenists. Some examples of the maximum safe levels for different daily exposure times are given below.

Duration of exposure	Max. level in dBa
24 h	80
8 h	85
4 h	88
1 h	94
30 min	97
15 min	100
2 min	109
28 s	115
0.11 s	139

No exposure greater than 140 dBa is permitted. Further details may be found in Reference 3.

References

1. Anderson, H. L., Ed., *A Physicist's Desk Reference*, American Institute of Physics, New York, 1989, chap. 2.
2. Gray, D. E., Ed., *American Institute of Physics Handbook, Third Edition*, McGraw Hill, New York, 1972, chap. 3.
3. *Threshold Limit Values for Chemical Substances and Physical Agents; Biological Exposure Indices*, 2008 Edition, American Conference of Governmental Industrial Hygienists, 1330 Kemper Meadow Drive, Cincinnati, OH 45240-1634; <www.acgih.org>.

Section 15
Practical Laboratory Data

STANDARD ITS-90 THERMOCOUPLE TABLES

The Instrument Society of America (ISA) has assigned standard letter designations to a number of thermocouple types having specified emf-temperature relations. These designations and the approximate metal compositions that meet the required relations, as well as the useful temperature ranges, are given below:

Type B	(Pt + 30% Rh) vs. (Pt + 6% Rh)	0 to 1820 °C
Type E	(Ni + 10% Cr) vs. (Cu + 43% Ni)	−270 to 1000 °C
Type J	Fe vs. (Cu + 43% Ni)	−210 to 1200 °C
Type K	(Ni + 10% Cr) vs. (Ni + 2% Al + 2% Mn + 1% Si)	−270 to 1372 °C
Type N	(Ni + 14% Cr + 1.5% Si) vs. (Ni + 4.5% Si + 0.1% Mg)	−270 to 1300 °C
Type R	(Pt + 13% Rh) vs. Pt	−50 to 1768 °C
Type S	(Pt + 10% Rh) vs. Pt	−50 to 1768 °C
Type T	Cu vs. (Cu + 43% Ni)	−270 to 400 °C

The compositions are given in weight percent, and the positive leg is listed first. It should be emphasized that the standard letter designations do not imply a precise composition but rather that the specified emf-temperature relation is satisfied.

The first set of tables below lists, for each thermocouple type, the emf as a function of temperature on the International Temperature Scale of 1990 (ITS-90). The coefficients in the equation used to generate the table are also given. The second set of tables gives the inverse relationships, i.e., the coefficients in the polynomial equation that expresses the temperature as a function of thermocouple emf. The accuracy of these equations is also stated.

Further details and tables at closer intervals may be found in Reference 1.

References

1. Burns, G. W., Seroger, M. G., Strouse, G. F., Croarkin, M. C., and Guthrie, W. F., *Temperature-Electromotive Force Reference Functions and Tables for the Letter-Designated Thermocouple Types Based on the ITS-90*, Natl. Inst. Stand. Tech. (U.S.) Monogr. 175, 1993.
2. Schooley, J. F., *Thermometry*, CRC Press, Boca Raton, FL, 1986.

Type B Thermocouples: emf-Temperature (°C) Reference Table and Equations
Thermocouple emf as a Function of Temperature in Degrees Celsius (ITS-90)

°C	0	10	20	30	40	50	60	70	80	90	100
0	0.000	−0.002	−0.003	−0.002	−0.000	0.002	0.006	0.011	0.017	0.025	0.033
100	0.033	0.043	0.053	0.065	0.078	0.092	0.107	0.123	0.141	0.159	0.178
200	0.178	0.199	0.220	0.243	0.267	0.291	0.317	0.344	0.372	0.401	0.431
300	0.431	0.462	0.494	0.527	0.561	0.596	0.632	0.669	0.707	0.746	0.787
400	0.787	0.828	0.870	0.913	0.957	1.002	1.048	1.095	1.143	1.192	1.242
500	1.242	1.293	1.344	1.397	1.451	1.505	1.561	1.617	1.675	1.733	1.792
600	1.792	1.852	1.913	1.975	2.037	2.101	2.165	2.230	2.296	2.363	2.431
700	2.431	2.499	2.569	2.639	2.710	2.782	2.854	2.928	3.002	3.078	3.154
800	3.154	3.230	3.308	3.386	3.466	3.546	3.626	3.708	3.790	3.873	3.957
900	3.957	4.041	4.127	4.213	4.299	4.387	4.475	4.564	4.653	4.743	4.834
1000	4.834	4.926	5.018	5.111	5.205	5.299	5.394	5.489	5.585	5.682	5.780
1100	5.780	5.878	5.976	6.075	6.175	6.276	6.377	6.478	6.580	6.683	6.786
1200	6.786	6.890	6.995	7.100	7.205	7.311	7.417	7.524	7.632	7.740	7.848
1300	7.848	7.957	8.066	8.176	8.286	8.397	8.508	8.620	8.731	8.844	8.956
1400	8.956	9.069	9.182	9.296	9.410	9.524	9.639	9.753	9.868	9.984	10.099
1500	10.099	10.215	10.331	10.447	10.563	10.679	10.796	10.913	11.029	11.146	11.263
1600	11.263	11.380	11.497	11.614	11.731	11.848	11.965	12.082	12.199	12.316	12.433
1700	12.433	12.549	12.666	12.782	12.898	13.014	13.130	13.246	13.361	13.476	13.591
1800	13.591	13.706	13.820								

NOTE—Temperature ranges and coefficients of equations used to compute the above table: The equations are of the form: $E = c_0 + c_1 t + c_2 t^2 + c_3 t^3 + \ldots c_n t^n$, where E is the emf in millivolts, t is the temperature in degrees Celsius (ITS-90), and c_0, c_1, c_2, c_3, etc. are the coefficients. These coefficients are extracted from Reference 1.

		0 °C to 630.615 °C	**630.615 °C to 1820 °C**
c_0	=	0.000 000 000 0	$-3.893\ 816\ 862\ 1\ \dots$
c_1	=	$-2.465\ 081\ 834\ 6 \times 10^{-4}$	$2.857\ 174\ 747\ 0 \times 10^{-2}$
c_2	=	$5.904\ 042\ 117\ 1 \times 10^{-6}$	$-8.488\ 510\ 478\ 5 \times 10^{-5}$
c_3	=	$-1.325\ 793\ 163\ 6 \times 10^{-9}$	$1.578\ 528\ 016\ 4 \times 10^{-7}$
c_4	=	$1.566\ 829\ 190\ 1 \times 10^{-12}$	$-1.683\ 534\ 486\ 4 \times 10^{-10}$
c_5	=	$-1.694\ 452\ 924\ 0 \times 10^{-15}$	$1.110\ 979\ 401\ 3 \times 10^{-13}$
c_6	=	$6.299\ 034\ 709\ 4 \times 10^{-19}$	$-4.451\ 543\ 103\ 3 \times 10^{-17}$
c_7	=	....	$9.897\ 564\ 082\ 1 \times 10^{-21}$
c_8	=	....	$-9.379\ 133\ 028\ 9 \times 10^{-25}$

Type E Thermocouples: emf-Temperature (°C) Reference Table and Equations
Thermocouple emf as a Function of Temperature in Degrees Celsius (ITS-90)

emf in Millivolts   Reference junctions at 0 °C

°C	0	-10	-20	-30	-40	-50	-60	-70	-80	-90	-100
-200	-8.825	-9.063	-9.274	-9.455	-9.604	-9.718	-9.797	-9.835			
-100	-5.237	-5.681	-6.107	-6.516	-6.907	-7.279	-7.632	-7.963	-8.273	-8.561	-8.825
0	0.000	-0.582	-1.152	-1.709	-2.255	-2.787	-3.306	-3.811	-4.302	-4.777	-5.237

°C	0	10	20	30	40	50	60	70	80	90	100
0	0.000	0.591	1.192	1.801	2.420	3.048	3.685	4.330	4.985	5.648	6.319
100	6.319	6.998	7.685	8.379	9.081	9.789	10.503	11.224	11.951	12.684	13.421
200	13.421	14.164	14.912	15.664	16.420	17.181	17.945	18.713	19.484	20.259	21.036
300	21.036	21.817	22.600	23.386	24.174	24.964	25.757	26.552	27.348	28.146	28.946
400	28.946	29.747	30.550	31.354	32.159	32.965	33.772	34.579	35.387	36.196	37.005
500	37.005	37.815	38.624	39.434	40.243	41.053	41.862	42.671	43.479	44.286	45.093
600	45.093	45.900	46.705	47.509	48.313	49.116	49.917	50.718	51.517	52.315	53.112
700	53.112	53.908	54.703	55.497	56.289	57.080	57.870	58.659	59.446	60.232	61.017
800	61.017	61.801	62.583	63.364	64.144	64.922	65.698	66.473	67.246	68.017	68.787
900	68.787	69.554	70.319	71.082	71.844	72.603	73.360	74.115	74.869	75.621	76.373
1000	76.373										

NOTE—Temperature ranges and coefficients of equations used to compute the above table: The equations are of the form: $E = c_0 + c_1 t + c_2 t^2 + c_3 t^3 + \dots c_n t^n$, where E is the emf in millivolts, t is the temperature in degrees Celsius (ITS-90), and c_0, c_1, c_2, c_3, etc. are the coefficients. These coefficients are extracted from Reference 1.

		−270 to 0 °C	**0 °C to 1000 °C**
c_0	=	0.000 000 000 0 ...	0.000 000 000 0 ...
c_1	=	$5.866\ 550\ 870\ 8 \times 10^{-2}$	$5.866\ 550\ 871\ 0 \times 10^{-2}$
c_2	=	$4.541\ 097\ 712\ 4 \times 10^{-5}$	$4.503\ 227\ 558\ 2 \times 10^{-5}$
c_3	=	$-7.799\ 804\ 868\ 6 \times 10^{-7}$	$2.890\ 840\ 721\ 2 \times 10^{-8}$
c_4	=	$-2.580\ 016\ 084\ 3 \times 10^{-8}$	$-3.305\ 689\ 665\ 2 \times 10^{-10}$
c_5	=	$-5.945\ 258\ 305\ 7 \times 10^{-10}$	$6.502\ 440\ 327\ 0 \times 10^{-13}$
c_6	=	$-9.321\ 405\ 866\ 7 \times 10^{-12}$	$-1.919\ 749\ 550\ 4 \times 10^{-16}$
c_7	=	$-1.028\ 760\ 553\ 4 \times 10^{-13}$	$-1.253\ 660\ 049\ 7 \times 10^{-18}$
c_8	=	$-8.037\ 012\ 362\ 1 \times 10^{-16}$	$2.148\ 921\ 756\ 9 \times 10^{-21}$
c_9	=	$-4.397\ 949\ 739\ 1 \times 10^{-18}$	$-1.438\ 804\ 178\ 2 \times 10^{-24}$
c_{10}	=	$-1.641\ 477\ 635\ 5 \times 10^{-20}$	$3.596\ 089\ 948\ 1 \times 10^{-28}$
c_{11}	=	$-3.967\ 361\ 951\ 6 \times 10^{-23}$	....
c_{12}	=	$-5.582\ 732\ 872\ 1 \times 10^{-26}$	....
c_{13}	=	$-3.465\ 784\ 201\ 3 \times 10^{-29}$	....

Type J Thermocouples: emf-Temperature (°C) Reference Table and Equations
Thermocouple emf as a Function of Temperature in Degrees Celsius (ITS-90)

emf in Millivolts   Reference junctions at 0 °C

°C	0	-10	-20	-30	-40	-50	-60	-70	-80	-90	-100
-200	-7.890	-8.095									
-100	-4.633	-5.037	-5.426	-5.801	-6.159	-6.500	-6.821	-7.123	-7.403	-7.659	-7.890
0	0.000	-0.501	-0.995	-1.482	-1.961	-2.431	-2.893	-3.344	-3.786	-4.215	-4.633

emf in Millivolts — Reference junctions at 0 °C

°C	0	10	20	30	40	50	60	70	80	90	100
0	0.000	0.507	1.019	1.537	2.059	2.585	3.116	3.650	4.187	4.726	5.269
100	5.269	5.814	6.360	6.909	7.459	8.010	8.562	9.115	9.669	10.224	10.779
200	10.779	11.334	11.889	12.445	13.000	13.555	14.110	14.665	15.219	15.773	16.327
300	16.327	16.881	17.434	17.986	18.538	19.090	19.642	20.194	20.745	21.297	21.848
400	21.848	22.400	22.952	23.504	24.057	24.610	25.164	25.720	26.276	26.834	27.393
500	27.393	27.953	28.516	29.080	29.647	30.216	30.788	31.362	31.939	32.519	33.102
600	33.102	33.689	34.279	34.873	35.470	36.071	36.675	37.284	37.896	38.512	39.132
700	39.132	39.755	40.382	41.012	41.645	42.281	42.919	43.559	44.203	44.848	45.494
800	45.494	46.141	46.786	47.431	48.074	48.715	49.353	49.989	50.622	51.251	51.877
900	51.877	52.500	53.119	53.735	54.347	54.956	55.561	56.164	56.763	57.360	57.953
1000	57.953	58.545	59.134	59.721	60.307	60.890	61.473	62.054	62.634	63.214	63.792
1100	63.792	64.370	64.948	65.525	66.102	66.679	67.255	67.831	68.406	68.980	69.553
1200	69.553										

NOTE—Temperature ranges and coefficients of equations used to compute the above table: The equations are of the form: $E = c_0 + c_1 t + c_2 t^2 + c_3 t^3 + \ldots c_n t^n$, where E is the emf in millivolts, t is the temperature in degrees Celsius (ITS-90), and c_0, c_1, c_2, c_3, etc. are the coefficients. These coefficients are extracted from Reference 1.

		−260 °C to 760 °C	760 °C to 1200 °C
c_0	=	$0.000\ 000\ 000\ 0\ \ldots$	$2.964\ 562\ 568\ 1 \times 10^2$
c_1	=	$5.038\ 118\ 781\ 5 \times 10^{-2}$	$-1.497\ 612\ 778\ 6\ \ldots$
c_2	=	$3.047\ 583\ 693\ 0 \times 10^{-5}$	$3.178\ 710\ 392\ 4 \times 10^{-3}$
c_3	=	$-8.568\ 106\ 572\ 0 \times 10^{-8}$	$-3.184\ 768\ 670\ 1 \times 10^{-6}$
c_4	=	$1.322\ 819\ 529\ 5 \times 10^{-10}$	$1.572\ 081\ 900\ 4 \times 10^{-9}$
c_5	=	$-1.705\ 295\ 833\ 7 \times 10^{-13}$	$-3.069\ 136\ 905\ 6 \times 10^{-13}$
c_6	=	$2.094\ 809\ 069\ 7 \times 10^{-16}$	....
c_7	=	$-1.253\ 839\ 533\ 6 \times 10^{-19}$	....
c_8	=	$1.563\ 172\ 569\ 7 \times 10^{-23}$	....

Type K Thermocouples: emf-Temperature (°C) Reference Table and Equations

Thermocouple emf as a Function of Temperature in Degrees Celsius (ITS-90)

emf in Millivolts — Reference junctions at 0 °C

°C	0	-10	-20	-30	-40	-50	-60	-70	-80	-90	-100
-200	-5.891	-6.035	-6.158	-6.262	-6.344	-6.404	-6.441	-6.458			
-100	-3.554	-3.852	-4.138	-4.411	-4.669	-4.913	-5.141	-5.354	-5.550	-5.730	-5.891
0	0.000	-0.392	-0.778	-1.156	-1.527	-1.889	-2.243	-2.587	-2.920	-3.243	-3.554

°C	0	10	20	30	40	50	60	70	80	90	100
0	0.000	0.397	0.798	1.203	1.612	2.023	2.436	2.851	3.267	3.682	4.096
100	4.096	4.509	4.920	5.328	5.735	6.138	6.540	6.941	7.340	7.739	8.138
200	8.138	8.539	8.940	9.343	9.747	10.153	10.561	10.971	11.382	11.795	12.209
300	12.209	12.624	13.040	13.457	13.874	14.293	14.713	15.133	15.554	15.975	16.397
400	16.397	16.820	17.243	17.667	18.091	18.516	18.941	19.366	19.792	20.218	20.644
500	20.644	21.071	21.497	21.924	22.350	22.776	23.203	23.629	24.055	24.480	24.905
600	24.905	25.330	25.755	26.179	26.602	27.025	27.447	27.869	28.289	28.710	29.129
700	29.129	29.548	29.965	30.382	30.798	31.213	31.628	32.041	32.453	32.865	33.275
800	33.275	33.685	34.093	34.501	34.908	35.313	35.718	36.121	36.524	36.925	37.326
900	37.326	37.725	38.124	38.522	38.918	39.314	39.708	40.101	40.494	40.885	41.276
1000	41.276	41.665	42.053	42.440	42.826	43.211	43.595	43.978	44.359	44.740	45.119
1100	45.119	45.497	45.873	46.249	46.623	46.995	47.367	47.737	48.105	48.473	48.838
1200	48.838	49.202	49.565	49.926	50.286	50.644	51.000	51.355	51.708	52.060	52.410
1300	52.410	52.759	53.106	53.451	53.795	54.138	54.479	54.819			

NOTE—Temperature ranges and coefficients of equations used to compute the above table: The equations are of the form $E = c_0 + c_1 t + c_2 t^2 + c_3 t^3 + \ldots c_n t^n$, where E is the emf in millivolts, t is the temperature in degrees Celsius (ITS-90), and c_0, c_1, c_2, c_3, etc. are the coefficients. In the 0 °C to 1372 °C range there is also an exponential term that must be evaluated and added to the equation. The exponential term is of the form $c_0 e^{c_1 (t - 126.9686)^2}$, where t is the temperature in °C, e is the natural logarithm base, and c_0 and c_1 are the coefficients. These coefficients are extracted from Reference 1.

		-270 °C to 0 °C	0 °C to 1372 °C	0 °C to 1372 °C (Exponential term)
c_0	=	0.000 000 000 0	$-1.760\ 041\ 368\ 6 \times 10^{-2}$	$1.185\ 976 \times 10^{-1}$
c_1	=	$3.945\ 012\ 802\ 5 \times 10^{-2}$	$3.892\ 120\ 497\ 5 \times 10^{-2}$	$-1.183\ 432 \times 10^{-4}$
c_2	=	$2.362\ 237\ 359\ 8 \times 10^{-5}$	$1.855\ 877\ 003\ 2 \times 10^{-5}$	....
c_3	=	$-3.285\ 890\ 678\ 4 \times 10^{-7}$	$-9.945\ 759\ 287\ 4 \times 10^{-8}$	....
c_4	=	$-4.990\ 482\ 877\ 7 \times 10^{-9}$	$3.184\ 094\ 571\ 9 \times 10^{-10}$	....
c_5	=	$-6.750\ 905\ 917\ 3 \times 10^{-11}$	$-5.607\ 284\ 488\ 9 \times 10^{-13}$	....
c_6	=	$-5.741\ 032\ 742\ 8 \times 10^{-13}$	$5.607\ 505\ 905\ 9 \times 10^{-16}$	....
c_7	=	$-3.108\ 887\ 289\ 4 \times 10^{-15}$	$-3.202\ 072\ 000\ 3 \times 10^{-19}$	....
c_8	=	$-1.045\ 160\ 936\ 5 \times 10^{-17}$	$9.715\ 114\ 715\ 2 \times 10^{-23}$	....
c_9	=	$-1.988\ 926\ 687\ 8 \times 10^{-20}$	$-1.210\ 472\ 127\ 5 \times 10^{-26}$	....
c_{10}	=	$-1.632\ 269\ 748\ 6 \times 10^{-23}$	......	....

Type N Thermocouples: emf-Temperature (°C) Reference Table and Equations
Thermocouple emf as a Function of Temperature in Degrees Celsius (ITS-90)

emf in Millivolts   Reference junctions at 0 °C

°C	0	-10	-20	-30	-40	-50	-60	-70	-80	-90	-100
-200	3.990	-4.083	-4.162	-4.226	-4.277	-4.313	-4.336	-4.345			
-100	-2.407	-2.612	2.808	2.994	3.171	3.336	-3.491	3.634	-3.766	-3.884	-3.990
0	0.000	-0.260	-0.518	-0.772	-1.023	-1.269	-1.509	1.744	-1.972	-2.193	-2.407

°C	0	10	20	30	40	50	60	70	80	90	100
0	0.000	0.261	0.525	0.793	1.065	1.340	1.619	1.902	2.189	2.480	2.774
100	2.774	3.072	3.374	3.680	3.989	4.302	4.618	4.937	5.259	5.585	5.913
200	5.913	6.245	6.579	6.916	7.255	7.597	7.941	8.288	8.637	8.988	9.341
300	9.341	9.696	10.054	10.413	10.774	11.136	11.501	11.867	12.234	12.603	12.974
400	12.974	13.346	13.719	14.094	14.469	14.846	15.225	15.604	15.984	16.366	16.748
500	16.748	17.131	17.515	17.900	18.286	18.672	19.059	19.447	19.835	20.224	20.613
600	20.613	21.003	21.393	21.784	22.175	22.566	22.958	23.350	23.742	24.134	24.527
700	24.527	24.919	25.312	25.705	26.098	26.491	26.883	27.276	27.669	28.062	28.455
800	28.455	28.847	29.239	29.632	30.024	30.416	30.807	31.199	31.590	31.981	32.371
900	32.371	32.761	33.151	33.541	33.930	34.319	34.707	35.095	35.482	35.869	36.256
1000	36.256	36.641	37.027	37.411	37.795	38.179	38.562	38.944	39.326	39.706	40.087
1100	40.087	40.466	40.845	41.223	41.600	41.976	42.352	42.727	43.101	43.474	43.846
1200	43.846	44.218	44.588	44.958	45.326	45.694	46.060	46.425	46.789	47.152	47.513
1300	47.513										

NOTE—Temperature ranges and coefficients of equations used to compute the above table: The equations are of the form $E = c_0 + c_1 t + c_2 t^2 + c_3 t^3 + \ldots c_n t^n$, where E is the emf in millivolts, t is the temperature in degrees Celsius (ITS-90), and c_0, c_1, c_2, c_3, etc. are the coefficients. These coefficients are extracted from Reference 1.

		-270 °C to 0 °C	0 °C to 1300 °C
c_0	=	0.000 000 000 0 ...	0.000 000 000 0...
c_1	=	$2.615\ 910\ 596\ 2 \times 10^{-2}$	$2.592\ 939\ 460\ 1 \times 10^{-2}$
c_2	=	$1.095\ 748\ 422\ 8 \times 10^{-5}$	$1.571\ 014\ 188\ 0 \times 10^{-5}$
c_3	=	$-9.384\ 111\ 155\ 4 \times 10^{-8}$	$4.382\ 562\ 723\ 7 \times 10^{-8}$
c_4	=	$-4.641\ 203\ 975\ 9 \times 10^{-11}$	$-2.526\ 116\ 979\ 4 \times 10^{-10}$
c_5	=	$-2.630\ 335\ 771\ 6 \times 10^{-12}$	$6.431\ 181\ 933\ 9 \times 10^{-13}$
c_6	=	$-2.265\ 343\ 800\ 3 \times 10^{-14}$	$-1.006\ 347\ 151\ 9 \times 10^{-15}$
c_7	=	$-7.608\ 930\ 079\ 1 \times 10^{-17}$	$9.974\ 533\ 899\ 2 \times 10^{-19}$
c_8	=	$-9.341\ 966\ 783\ 5 \times 10^{-20}$	$-6.086\ 324\ 560\ 7 \times 10^{-22}$
c_9	=	....	$2.084\ 922\ 933\ 9 \times 10^{-25}$
c_{10}	=	....	$-3.068\ 219\ 615\ 1 \times 10^{-29}$

Type R Thermocouples: emf-Temperature (°C) Reference Table and Equations
Thermocouple emf as a Function of Temperature in Degrees Celsius (ITS-90)

emf in Millivolts								Reference junctions at 0 °C		
°C	0	-10	-20	-30	-40	-50	-60	-70	-80	-90	-100
0	0.000	-0.051	-0.100	-0.145	-0.188	-0.226					

°C	0	10	20	30	40	50	60	70	80	90	100
0	0.000	0.054	0.111	0.171	0.232	0.296	0.363	0.431	0.501	0.573	0.647
100	0.647	0.723	0.800	0.879	0.959	1.041	1.124	1.208	1.294	1.381	1.469
200	1.469	1.558	1.648	1.739	1.831	1.923	2.017	2.112	2.207	2.304	2.401
300	2.401	2.498	2.597	2.696	2.796	2.896	2.997	3.099	3.201	3.304	3.408
400	3.408	3.512	3.616	3.721	3.827	3.933	4.040	4.147	4.255	4.363	4.471
500	4.471	4.580	4.690	4.800	4.910	5.021	5.133	5.245	5.357	5.470	5.583
600	5.583	5.697	5.812	5.926	6.041	6.157	6.273	6.390	6.507	6.625	6.743
700	6.743	6.861	6.980	7.100	7.220	7.340	7.461	7.583	7.705	7.827	7.950
800	7.950	8.073	8.197	8.321	8.446	8.571	8.697	8.823	8.950	9.077	9.205
900	9.205	9.333	9.461	9.590	9.720	9.850	9.980	10.111	10.242	10.374	10.506
1000	10.506	10.638	10.771	10.905	11.039	11.173	11.307	11.442	11.578	11.714	11.850
1100	11.850	11.986	12.123	12.260	12.397	12.535	12.673	12.812	12.950	13.089	13.228
1200	13.228	13.367	13.507	13.646	13.786	13.926	14.066	14.207	14.347	14.488	14.629
1300	14.629	14.770	14.911	15.052	15.193	15.334	15.475	15.616	15.758	15.899	16.040
1400	16.040	16.181	16.323	16.464	16.605	16.746	16.887	17.028	17.169	17.310	17.451
1500	17.451	17.591	17.732	17.872	18.012	18.152	18.292	18.431	18.571	18.710	18.849
1600	18.849	18.988	19.126	19.264	19.402	19.540	19.677	19.814	19.951	20.087	20.222
1700	20.222	20.356	20.488	20.620	20.749	20.877	21.003				

NOTE—Temperature ranges and coefficients of equations used to compute the above table: The equations are of the form $E = c_0 + c_1 t + c_2 t^2 + c_3 t^3 + ... c_n t^n$, where E is the emf in millivolts, t is the temperature in degrees Celsius (ITS-90), and c_0, c_1, c_2, c_3, etc. are the coefficients. These coefficients are extracted from Reference 1.

		–50 °C to 1064.18 °C	1064.18 °C to 1664.5 °C	1664.5 °C to 1768.1 °C
c_0	=	0.000 000 000 00 ...	2.951 579 253 16 ...	$1.522\ 321\ 182\ 09 \times 10^2$
c_1	=	$5.289\ 617\ 297\ 65 \times 10^{-3}$	$-2.520\ 612\ 513\ 32 \times 10^{-3}$	$-2.688\ 198\ 885\ 45 \times 10^{-1}$
c_2	=	$1.391\ 665\ 897\ 82 \times 10^{-5}$	$1.595\ 645\ 018\ 65 \times 10^{-5}$	$1.712\ 802\ 804\ 71 \times 10^{-4}$
c_3	=	$-2.388\ 556\ 930\ 17 \times 10^{-8}$	$-7.640\ 859\ 475\ 76 \times 10^{-9}$	$-3.458\ 957\ 064\ 53 \times 10^{-8}$
c_4	=	$3.569\ 160\ 010\ 63 \times 10^{-11}$	$2.053\ 052\ 910\ 24 \times 10^{-12}$	$-9.346\ 339\ 710\ 46 \times 10^{-15}$
c_5	=	$-4.623\ 476\ 662\ 98 \times 10^{-14}$	$-2.933\ 596\ 681\ 73 \times 10^{-16}$	....
c_6	=	$5.007\ 774\ 410\ 34 \times 10^{-17}$	....	....
c_7	=	$-3.731\ 058\ 861\ 91 \times 10^{-20}$	....	....
c_8	=	$1.577\ 164\ 823\ 67 \times 10^{-23}$	....	....
c_9	=	$-2.810\ 386\ 252\ 51 \times 10^{-27}$	....	....

Type S Thermocouples: emf-Temperature (°C) Reference Table and Equations.
Thermocouple emf as a Function of Temperature in Degrees Celsius (ITS-90)

emf in Millivolts								Reference junctions at 0 °C		
°C	0	-10	-20	-30	-40	-50	-60	-70	-80	-90	-100
0	0.000	-0.053	-0.103	-0.150	-0.194	-0.236					

°C	0	10	20	30	40	50	60	70	80	90	100
0	0.000	0.055	0.113	0.173	0.235	0.299	0.365	0.433	0.502	0.573	0.646
100	0.646	0.720	0.795	0.872	0.950	1.029	1.110	1.191	1.273	1.357	1.441
200	1.441	1.526	1.612	1.698	1.786	1.874	1.962	2.052	2.141	2.232	2.323
300	2.323	2.415	2.507	2.599	2.692	2.786	2.880	2.974	3.069	3.164	3.259
400	3.259	3.355	3.451	3.548	3.645	3.742	3.840	3.938	4.036	4.134	4.233
500	4.233	4.332	4.432	4.532	4.632	4.732	4.833	4.934	5.035	5.137	5.239
600	5.239	5.341	5.443	5.546	5.649	5.753	5.857	5.961	6.065	6.170	6.275
700	6.275	6.381	6.486	6.593	6.699	6.806	6.913	7.020	7.128	7.236	7.345
800	7.345	7.454	7.563	7.673	7.783	7.893	8.003	8.114	8.226	8.337	8.449
900	8.449	8.562	8.674	8.787	8.900	9.014	9.128	9.242	9.357	9.472	9.587
1000	9.587	9.703	9.819	9.935	10.051	10.168	10.285	10.403	10.520	10.638	10.757
1100	10.757	10.875	10.994	11.113	11.232	11.351	11.471	11.590	11.710	11.830	11.951
1200	11.951	12.071	12.191	12.312	12.433	12.554	12.675	12.796	12.917	13.038	13.159

Type S Thermocouples: emf-Temperature (°C) Reference Table and Equations.
Thermocouple emf as a Function of Temperature in Degrees Celsius (ITS-90)

emf in Millivolts   Reference junctions at 0 °C

°C	0	-10	-20	-30	-40-	-50	-60	-70	-80	-90	-100
1300	13.159	13.280	13.402	13.523	13.644	13.766	13.887	14.009	14.130	14.251	14.373
1400	14.373	14.494	14.615	14.736	14.857	14.978	15.099	15.220	15.341	15.461	15.582
1500	15.582	15.702	15.822	15.942	16.062	16.182	16.301	16.420	16.539	16.658	16.777
1600	16.777	16.895	17.013	17.131	17.249	17.366	17.483	17.600	17.717	17.832	17.947
1700	17.947	18.061	18.174	18.285	18.395	18.503	18.609				

NOTE—Temperature ranges and coefficients of equations used to compute the above table: The equations are of the form $E = c_0 + c_1 t + c_2 t^2 + c_3 t^3 + \dots c_n t^n$, where E is the emf in millivolts, t is the temperature in degrees Celsius (ITS-90), and c_0, c_1, c_2, c_3, etc. are the coefficients. These coefficients are extracted from Reference 1.

		−50 °C to 1064.18 °C	1064.18 °C to 1664.5 °C	1664.5 °C to 1768.1 °C
c_0	=	0.000 000 000 00 …	1.329 004 440 85 …	$1.466\ 282\ 326\ 36 \times 10^{2}$
c_1	=	$5.403\ 133\ 086\ 31 \times 10^{-3}$	$3.345\ 093\ 113\ 44 \times 10^{-3}$	$-2.584\ 305\ 167\ 52 \times 10^{-1}$
c_2	=	$1.259\ 342\ 897\ 40 \times 10^{-5}$	$6.548\ 051\ 928\ 18 \times 10^{-6}$	$1.636\ 935\ 746\ 41 \times 10^{-4}$
c_3	=	$-2.324\ 779\ 686\ 89 \times 10^{-8}$	$-1.648\ 562\ 592\ 09 \times 10^{-9}$	$-3.304\ 390\ 469\ 87 \times 10^{-8}$
c_4	=	$3.220\ 288\ 230\ 36 \times 10^{-11}$	$1.299\ 896\ 051\ 74 \times 10^{-14}$	$-9.432\ 236\ 906\ 12 \times 10^{-15}$
c_5	=	$-3.314\ 651\ 963\ 89 \times 10^{-14}$	…………	…………
c_6	=	$2.557\ 442\ 517\ 86 \times 10^{-17}$	…………	…………
c_7	=	$-1.250\ 688\ 713\ 93 \times 10^{-20}$	…………	…………
c_8	=	$2.714\ 431\ 761\ 45 \times 10^{-24}$	…………	…………

Type T Thermocouples: emf-Temperature (°C) Reference Table and Equations.
Thermocouple emf as a Function of Temperature in Degrees Celsius (ITS-90)

emf in Millivolts   Reference junctions at 0 °C

°C	0	-10	-20	-30	-40	-50	-60	-70	-80	-90	-100
-200	-5.603	-5.753	-5.888	-6.007	-6.105	-6.180	-6.232	-6.258			
-100	-3.379	-3.657	-3.923	-4.177	-4.419	-4.648	-4.865	-5.070	-5.261	-5.439	-5.603
0	0.000	-0.383	-0.757	-1.121	-1.475	-1.819	-2.153	-2.476	-2.788	-3.089	-3.379

°C	0	10	20	30	40	50	60	70	80	90	100
0	0.000	0.391	0.790	1.196	1.612	2.036	2.468	2.909	3.358	3.814	4.279
100	4.279	4.750	5.228	5.714	6.206	6.704	7.209	7.720	8.237	8.759	9.288
200	9.288	9.822	10.362	10.907	11.458	12.013	12.574	13.139	13.709	14.283	14.862
300	14.862	15.445	16.032	16.624	17.219	17.819	18.422	19.030	19.641	20.255	20.872
400	20.872										

NOTE—Temperature ranges and coefficients of equations used to compute the above table: The equations are of the form $E = c_0 + c_1 t + c_2 t^2 + c_3 t^3 + \dots c_n t^n$, where E is the emf in millivolts, t is the temperature in degrees Celsius (ITS-90), and c_0, c_1, c_2, c_3, etc. are the coefficients. These coefficients are extracted from Reference 1.

		270 °C to 0 °C	0 °C to 400 °C
c_0	=	0.000 000 000 0 …	0.000 000 000 0 …
c_1	=	$3.874\ 810\ 636\ 4 \times 10^{-2}$	$3.874\ 810\ 636\ 4 \times 10^{-2}$
c_2	=	$4.419\ 443\ 434\ 7 \times 10^{-5}$	$3.329\ 222\ 788\ 0 \times 10^{-5}$
c_3	=	$1.184\ 432\ 310\ 5 \times 10^{-7}$	$2.061\ 824\ 340\ 4 \times 10^{-7}$
c_4	=	$2.003\ 297\ 355\ 4 \times 10^{-8}$	$-2.188\ 225\ 684\ 6 \times 10^{-9}$
c_5	=	$9.013\ 801\ 955\ 9 \times 10^{-10}$	$1.099\ 688\ 092\ 8 \times 10^{-11}$
c_6	=	$2.265\ 115\ 659\ 3 \times 10^{-11}$	$-3.081\ 575\ 877\ 2 \times 10^{-14}$
c_7	=	$3.607\ 115\ 420\ 5 \times 10^{-13}$	$4.547\ 913\ 529\ 0 \times 10^{-17}$
c_8	=	$3.849\ 393\ 988\ 3 \times 10^{-15}$	$-2.751\ 290\ 167\ 3 \times 10^{-20}$
c_9	=	$2.821\ 352\ 192\ 5 \times 10^{-17}$	…………
c_{10}	=	$1.425\ 159\ 477\ 9 \times 10^{-19}$	…………
c_{11}	=	$4.876\ 866\ 228\ 6 \times 10^{-22}$	…………
c_{12}	=	$1.079\ 553\ 927\ 0 \times 10^{-24}$	…………
c_{13}	=	$1.394\ 502\ 706\ 2 \times 10^{-27}$	…………
c_{14}	=	$7.979\ 515\ 392\ 7 \times 10^{-31}$	…………

Type B Thermocouples: Coefficients (c_i) of Polynomials for the Computation of Temperatures in °C as a Function of the Thermocouple emf in Various Temperature and emf Ranges

Temperature range:	250 °C to 700 °C	700 °C to 1820 °C
emf range:	0.291 mV to 2.431 mV	2.431 mV to 13.820 mV
$c_0 =$	$9.842\,332\,1 \times 10^1$	$2.131\,507\,1 \times 10^2$
$c_1 =$	$6.997\,150\,0 \times 10^2$	$2.851\,050\,4 \times 10^2$
$c_2 =$	$-8.476\,530\,4 \times 10^2$	$-5.274\,288\,7 \times 10^1$
$c_3 =$	$1.005\,264\,4 \times 10^3$	$9.916\,080\,4 \dots$
$c_4 =$	$-8.334\,595\,2 \times 10^2$	$-1.296\,530\,3 \dots$
$c_5 =$	$4.550\,854\,2 \times 10^2$	$1.119\,587\,0 \times 10^{-1}$
$c_6 =$	$-1.552\,303\,7 \times 10^2$	$-6.062\,519\,9 \times 10^{-3}$
$c_7 =$	$2.988\,675\,0 \times 10^1$	$1.866\,169\,6 \times 10^{-4}$
$c_8 =$	$-2.474\,286\,0 \dots$	$-2.487\,858\,5 \times 10^{-6}$

NOTE— The above coefficients are extracted from Reference 1 and are for an expression of the form shown in Section 10.3.2. They yield approximate values of temperature that agree within ±0.03 °C with the values given in Table 10.2.

Type E Thermocouples: Coefficients (c_i) of Polynomials for the Computation of Temperatures in °C as a Function of the Thermocouple emf in Various Temperature and emf Ranges

Temperature range:	-200 °C TO 0 °C	0 °C to 1000 °C
emf range:	-8.825 mV to 0.0 mV	0.0 mV to 76.373 mV
$c_0 =$	$0.000\,000\,0 \dots$	$0.000\,000\,0 \dots$
$c_1 =$	$1.697\,728\,8 \times 10^1$	$1.705\,703\,5 \times 10^1$
$c_2 =$	$-4.351\,497\,0 \times 10^{-1}$	$-2.330\,175\,9 \times 10^{-1}$
$c_3 =$	$-1.585\,969\,7 \times 10^{-1}$	$6.543\,558\,5 \times 10^{-3}$
$c_4 =$	$-9.250\,287\,1 \times 10^{-2}$	$-7.356\,274\,9 \times 10^{-5}$
$c_5 =$	$-2.608\,431\,4 \times 10^{-2}$	$-1.789\,600 \times 10^{-6}$
$c_6 =$	$-4.136\,019\,9 \times 10^{-3}$	$8.403\,616\,5 \times 10^{-8}$
$c_7 =$	$-3.403\,403\,0 \times 10^{-4}$	$-1.373\,587\,9 \times 10^{-9}$
$c_8 =$	$-1.156\,489\,0 \times 10^{-5}$	$1.062\,982\,3 \times 10^{-11}$
$c_9 =$	$\dots$	$-3.244\,708\,7 \times 10^{-14}$

NOTE— The above coefficients are extracted from Reference 1 and are for an expression of the form shown in Section 10.3.2. They yield approximate values of temperature that agree within ±0.02 °C with the values given in Table 10.4

Type J Thermocouples: Coefficients (c_i) of Polynomials for the Computation of Temperatures in °C as a Function of the Thermocouple emf in Various Temperature and emf Ranges

Temperature range:	-210 °C to 0 °C	0 °C to 760 °C	760 °C to 1200 °C
emf Range:	-8.095 mV to 0.0 mV	0.0 mV to 42.919 mV	42.919 mV to 69.553 mV
$c_0 =$	$0.000\,000\,0 \dots$	$0.000\,000 \dots$	$-3.113\,581\,87 \times 10^3$
$c_1 =$	$1.952\,826\,8 \times 10^1$	$1.978\,425 \times 10^1$	$3.005\,436\,84 \times 10^2$
$c_2 =$	$-1.228\,618\,5 \dots$	$-2.001\,204 \times 10^{-1}$	$-9.947\,732\,30 \dots$
$c_3 =$	$-1.075\,217\,8 \dots$	$1.036\,969 \times 10^{-2}$	$1.702\,766\,30 \times 10^{-1}$
$c_4 =$	$-5.908\,693\,3 \times 10^{-1}$	$-2.549\,687 \times 10^{-4}$	$1.430\,334\,68 \times 10^{-3}$
$c_5 =$	$-1.725\,671\,3 \times 10^{-1}$	$3.585\,153 \times 10^{-6}$	$4.438\,860\,84 \times 10^{-6}$
$c_6 =$	$-2.813\,151\,3 \times 10^{-2}$	$-5.344\,285 \times 10^{-8}$	$\dots$
$c_7 =$	$-2.396\,337\,0 \times 10^{-3}$	$5.099\,890 \times 10^{-10}$	$\dots$
$c_8 =$	$-8.382\,332\,1 \times 10^{-5}$	$\dots$	$\dots$

NOTE— The above coefficients are extracted from Reference 1 and are for an expression of the form shown in Section 10.3.2. They yield approximate values of temperature that agree within ± 0.5 °C with the values given in Table 10.6.

Type K Thermocouples: Coefficients (c_i) of Polynomials for the Computation of Temperatures in °C as a Function of the Thermocouple emf in Various Temperature and emf Ranges

Temperature range:	−200 °C to 0 °C	0 °C to 500 °C	500 °C to 1372 °C
emf Range:	−5.891 mV to 0.0 mV	0.0 mV to 20.644 mV	20.644 mV to 54.886 mV
$c_0 =$	0.000 000 0 …	0.000 000 …	$-1.318\ 058 \times 10^2$
$c_1 =$	$2.517\ 346\ 2 \times 10^1$	$2.508\ 355 \times 10^1$	$4.830\ 222 \times 10^1$
$c_2 =$	−1.166 287 8 …	$7.860\ 106 \times 10^{-2}$	−1.646 031 …
$c_3 =$	−1.083 363 8 …	$-2.503\ 131 \times 10^{-1}$	$5.464\ 731 \times 10^{-2}$
$c_4 =$	$-8.977\ 354\ 0 \times 10^{-1}$	$8.315\ 270 \times 10^{-2}$	$-9.650\ 715 \times 10^{-4}$
$c_5 =$	$-3.734\ 237\ 7 \times 10^{-1}$	$-1.228\ 034 \times 10^{-2}$	$8.802\ 193 \times 10^{-6}$
$c_6 =$	$-8.663\ 264\ 3 \times 10^{-2}$	$9.804\ 036 \times 10^{-4}$	$3.110\ 810 \times 10^{-8}$
$c_7 =$	$-1.045\ 059\ 8 \times 10^{-2}$	$-4.413\ 030 \times 10^{-5}$	…….
$c_8 =$	$-5.192\ 057\ 7 \times 10^{-4}$	$1.057\ 734 \times 10^{-6}$	…….
$c_9 =$	…….	$-1.052\ 755 \times 10^{-8}$	…….

NOTE—The above coefficients are extracted from Reference 1 and are for an expression of the form shown in Section 10.3.2. They yield approximate values of temperature that agree within ±0.05 °C with the values given in Table 10.8.

Type N Thermocouples: Coefficients (c_i) of Polynomials for the Computation of Temperatures in °C as a Function of the Thermocouple emf in Various Temperature and emf Ranges

Temperature range:	−200 °C to 0 °C	0 °C to 600 °C	600 °C to 1300 °C
emf Range:	−3.990 mV to 0.0 mV	0.0 mV to 20.613 mV	20.613 mV to 47.513 mV
$c_0 =$	0.000 000 0 …	0.000 00 …	$1.972\ 485 \times 10^1$
$c_1 =$	$3.843\ 684\ 7 \times 10^1$	$3.868\ 96 \times 10^1$	$3.300\ 943 \times 10^1$
$c_2 =$	1.101 048 5 …	−1.082 67 …	$-3.915\ 159 \times 10^{-1}$
$c_3 =$	5.222 931 2 …	$4.702\ 05 \times 10^{-2}$	$9.855\ 391 \times 10^{-3}$
$c_4 =$	7.206 052 5 …	$-2.121\ 69 \times 10^{-6}$	$-1.274\ 371 \times 10^{-4}$
$c_5 =$	5.848 858 6 …	$-1.172\ 72 \times 10^{-4}$	$7.767\ 022 \times 10^{-7}$
$c_6 =$	2.775 491 6 …	$5.392\ 80 \times 10^{-6}$	………
$c_7 =$	$7.707\ 516\ 6 \times 10^{-1}$	$-7.981\ 56 \times 10^{-8}$	………
$c_8 =$	$1.158\ 266\ 5 \times 10^{-1}$	………	………
$c_9 =$	$7.313\ 886\ 8 \times 10^{-3}$	………	………

NOTE—The above coefficients are extracted from Reference 1 and are for an expression of the form shown in Section 10.3.2. They yield approximate values of temperature that agree within ± 0.04 °C with the values given in Table 10.10.

Type R Thermocouples: Coefficients (c_i) of Polynomials for the Computation of Temperatures in °C as a Function of the Thermocouple emf in Various Temperature and emf Ranges

Temperature range:	−50 °C to 250 °C	250 °C to 1200 °C	1064 °C to 1664.5 °C	1664.5 °C to 1768.1 °C
emf Range:	−0.226 mV to 1.923 mV	1.923 mV to 13.228 mV	11.361 mV to 19.739 mV	19.739 mV to 21.103 mV
$c_0 =$	0.000 000 0 …	$1.334\ 584\ 505 \times 10^1$	$-8.199\ 599\ 416 \times 10^1$	$3.406\ 177\ 836 \times 10^4$
$c_1 =$	$1.889\ 138\ 0 \times 10^2$	$1.472\ 644\ 573 \times 10^2$	$1.553\ 962\ 042 \times 10^2$	$-7.023\ 729\ 171 \times 10^3$
$c_2 =$	$-9.383\ 529\ 0 \times 10^1$	$-1.844\ 024\ 844 \times 10^1$	$-8.342\ 197\ 663$	$5.582\ 903\ 813 \times 10^2$
$c_3 =$	$1.306\ 861\ 9 \times 10^2$	4.031 129 726 …	$4.279\ 433\ 549 \times 10^{-1}$	$-1.952\ 394\ 635 \times 10^1$
$c_4 =$	$-2.270\ 358\ 0 \times 10^2$	$-6.249\ 428\ 360 \times 10^{-1}$	$-1.191\ 577\ 910 \times 10^{-2}$	$2.560\ 740\ 231 \times 10^{-1}$
$c_5 =$	$3.514\ 565\ 9 \times 10^2$	$6.468\ 412\ 046 \times 10^{-2}$	$1.492\ 290\ 091 \times 10^{-4}$	………
$c_6 =$	$-3.895\ 390\ 0 \times 10^2$	$-4.458\ 750\ 426 \times 10^{-3}$	………	………
$c_7 =$	$2.823\ 947\ 1 \times 10^2$	$1.994\ 710\ 149 \times 10^{-4}$	………	………
$c_8 =$	$-1.260\ 728\ 1 \times 10^2$	$-5.313\ 401\ 790 \times 10^{-6}$	………	………
$c_9 =$	$3.135\ 361\ 1 \times 10^1$	$6.481\ 976\ 217 \times 10^{-8}$	………	………
$c_{10} =$	−3.318 776 9 …	………	………	………

NOTE—The above coefficients are extracted from Reference 1 and are for an expression of the form shown in Section 10.3.2. They yield approximate values of temperature that agree within ±0.02 °C with the values given in Table 10.12.

Type S Thermocouples: Coefficients (c_i) of Polynomials for the Computation of Temperatures in °C as a Function of the Thermocouple emf in Various Temperature and emf Ranges

Temperature range:	−50 °C to 250 °C	250 °C to 1200 °C	1064 °C to 1664.5 °C	1664.5 °C to 1768.1 °C
emf Range:	−0.235 mV to 1.874 mV	1.874 mV to 11.950 mV	10.332 mV to 17.536 mV	17.536 mV to 18.693 mV
$c_0 =$	0.000 000 00 …	$1.291\ 507\ 177 \times 10^1$	$-8.087\ 801\ 117 \times 10^1$	$5.333\ 875\ 126 \times 10^4$
$c_1 =$	$1.849\ 494\ 60 \times 10^2$	$1.466\ 298\ 863 \times 10^2$	$1.621\ 573\ 104 \times 10^2$	$-1.235\ 892\ 298 \times 10^4$
$c_2 =$	$-8.005\ 040\ 62 \times 10^1$	$-1.534\ 713\ 402 \times 10^1$	$-8.536\ 869\ 453 \ …$	$1.092\ 657\ 613 \times 10^3$
$c_3 =$	$1.022\ 374\ 30 \times 10^2$	$3.145\ 945\ 973 \ …$	$4.719\ 686\ 976 \times 10^{-1}$	$-4.265\ 693\ 686 \times 10^1$
$c_4 =$	$-1.522\ 485\ 92 \times 10^2$	$-4.163\ 257\ 839 \times 10^{-1}$	$-1.441\ 693\ 666 \times 10^{-2}$	$6.247\ 205\ 420 \times 10^{-1}$
$c_5 =$	$1.888\ 213\ 43 \times 10^2$	$3.187\ 963\ 771 \times 10^{-2}$	$2.081\ 618\ 890 \times 10^{-4}$	....
$c_6 =$	$-1.590\ 859\ 41 \times 10^2$	$-1.291\ 637\ 500 \times 10^{-3}$	....	....
$c_7 =$	$8.230\ 278\ 80 \times 10^1$	$2.183\ 475\ 087 \times 10^{-5}$	....	....
$c_8 =$	$-2.341\ 819\ 44 \times 10^1$	$-1.447\ 379\ 511 \times 10^{-7}$	....	....
$c_9 =$	$2.797\ 862\ 60 \ …$	$8.211\ 272\ 125 \times 10^{-9}$	....	....

NOTE—The above coefficients are extracted from Reference 1 and are for an expression of the form shown in Section 10.3.2. They yield approximate values of temperature that agree within ± 0.02 °C with the values given in Table 10.14.

Type T Thermocouples: Coefficients (c_i) of Polynomials for the Computation of temperatures in °C as a Function of the Thermocouple emf in Various Temperature and emf Ranges

Temperature range:	−200 °C to 0 °C	0 °C to 400 °C
emf Range:	−5.603 mV to 0.0 mV	0.0 mV to 20.872 mV
$c_0 =$	0.000 000 0 …	0.000 000 …
$c_1 =$	$2.594\ 919\ 2 \times 10^1$	$2.592\ 800 \times 10^1$
$c_2 =$	$-2.131\ 696\ 7 \times 10^{-1}$	$-7.602\ 961 \times 10^{-1}$
$c_3 =$	$7.901\ 869\ 2 \times 10^{-1}$	$4.637\ 791 \times 10^{-2}$
$c_4 =$	$4.252\ 777\ 7 \times 10^{-1}$	$-2.165\ 394 \times 10^{-3}$
$c_5 =$	$1.330\ 447\ 3 \times 10^{-1}$	$6.048\ 144 \times 10^{-5}$
$c_6 =$	$2.024\ 144\ 6 \times 10^{-2}$	$-7.293\ 422 \times 10^{-7}$
$c_7 =$	$1.266\ 817\ 1 \times 10^{-3}$	....

NOTE—The above coefficients are extracted from Reference 1 and are for an expression of the form shown in Section 10.3.2. They yield approximate values of temperature that agree within ± 0.04 °C with the values given in Table 10.16.

SECONDARY REFERENCE POINTS ON THE ITS-90 TEMPERATURE SCALE

The International Temperature Scale of 1990 is described in Section 1 of this *Handbook*, where the defining fixed points are listed. The Consultative Committee on Thermometry (CCT) of the International Committee on Weights and Measures (CIPM), which oversees the temperature scale, has recommended a number of secondary reference points whose values have been accurately determined with respect to the primary fixed points. The most accurate of these, referred to as "first quality points," satisfy several criteria involving purity of the material, reproducibility, and documentation of the measurements. The CCT also lists "second quality points" that do not yet satisfy all the criteria but are still useful. Taken together,

these secondary reference points, help fill in the gaps between the primary fixed points.

The table below describes these secondary reference points. The best values resulting from the CCT evaluation are listed on both the Kelvin and Celsius scales, along with an estimate of uncertainty. Full details are given in the reference.

The entries within each quality group are listed in order of increasing temperature.

Reference

Bedford, R. E., Bonnier, G., Maas, H., and Pavese, F., *Metrologia* 33, 133, 1996.

Substance	Type of Transition	T_{90}/K	t_{90}/°C	Uncert.
First Quality Points				
Zinc	Superconductive transition	0.8500	−272.300	0.0030
Aluminum	Superconductive transition	1.1810	−271.9690	0.0025
Helium (^{4}He)	Superfluid transition	2.1768	−270.9732	0.0001
Indium	Superconductive transition	3.4145	−269.7355	0.0025
Lead	Superconductive transition	7.1997	−265.9503	0.0025
Niobium	Superconductive transition	9.2880	−263.8620	0.0025
Deuterium (^{2}H$_2$)	Triple point (equilibrium D$_2$)	18.689	−254.461	0.001
Deuterium (^{2}H$_2$)	Triple point (normal D$_2$)	18.724	−254.426	0.001
Neon (^{20}Ne)	Triple point	24.541	−248.609	0.001
Neon	Boiling point	27.097	−246.053	0.001
Nitrogen	Triple point	63.151	−209.999	0.001
Nitrogen	Boiling point	77.352	−195.798	0.002
Argon	Boiling point	87.303	−185.847	0.001
Oxygen	Condensation point	90.197	−182.953	0.001
Methane	Triple point	90.694	−182.456	0.001
Xenon	Triple point	161.405	−111.745	0.001
Carbon dioxide	Triple point	216.592	−56.558	0.001
Mercury	Freezing point	234.3210	−38.8290	0.0005
Water	Ice point	273.15	0	
Gallium	Triple point	302.9166	29.7666	0.0001
Water	Boiling point	373.124	99.974	0.001
Indium	Triple point	429.7436	156.5936	0.0002
Bismuth	Freezing point	544.552	271.402	0.001
Cadmium	Freezing point	594.219	321.069	0.001
Lead	Freezing point	600.612	327.462	0.001
Antimony	Freezing point	903.778	630.628	0.001
Copper/71.9% silver	Eutectic melting point	1052.78	779.63	0.05
Palladium	Freezing point	1828.0	1554.8	0.1
Platinum	Freezing point	2041.3	1768.2	0.4
Rhodium	Freezing point	2236	1963	3
Iridium	Freezing point	2719	2446	6
Molybdenum	Melting point	2895	2622	4
Tungsten	Melting point	3687	3414	7
Second Quality Points				
Hydrogen	Triple point (normal H$_2$)	13.952	−259.198	0.002
Hydrogen	Boiling point (normal H$_2$)	20.388	−252.762	0.002
Oxygen	α-β transition	23.868	−249.282	0.005
Nitrogen	α-β transition	35.614	−237.536	0.006
Oxygen	β-γ transition	43.796	−229.354	0.001

Substance	Type of Transition	T_{90}/K	t_{90}/°C	Uncert.
Krypton	Triple point	115.775	−157.375	0.001
Carbon dioxide	Sublimation point	194.686	−78.464	0.003
Sulfur hexafluoride	Triple point	223.554	−49.596	0.005
Gallium/20% indium	Eutectic melting point	288.800	15.650	0.001
Gallium/8% tin	Eutectic melting point	293.626	20.476	0.002
Diphenyl ether	Triple point	300.014	26.864	0.001
Ethylene carbonate	Triple point	309.465	36.315	0.001
Succinonitrile	Triple point	331.215	58.065	0.002
Sodium	Freezing point	370.944	97.794	0.005
Benzoic acid	Triple point	395.486	122.336	0.002
Benzoic acid	Freezing point	395.502	122.352	0.007
Mercury	Boiling point	629.769	356.619	0.004
Sulfur	Boiling point	717.764	444.614	0.002
Copper/66.9% aluminum	Eutectic melting point	840.957	567.807	0.010
Silver/30% aluminum	Eutectic melting point	840.957	567.807	0.002
Sodium chloride	Freezing point	1075.168	802.018	0.011
Sodium	Boiling point	1156.090	882.940	0.005
Nickel	Freezing point	1728	1455	1
Cobalt	Freezing point	1768	1495	3
Iron	Freezing point	1811	1538	3
Titanium	Melting point	1943	1670	2
Zirconium	Melting point	2127	1854	8
Aluminum oxide	Melting point	2326	2053	2
Ruthenium	Melting point	2606	2333	10

RELATIVE SENSITIVITY OF BAYARD-ALPERT IONIZATION GAUGES TO VARIOUS GASES

Paul Redhead

The ion current I_+ in a hot-cathode ionization gauge is given by $I_+ = KI_eP$. The gauge constant is $K = (I_+/I_e)(1/P)$, where I_e is the electron current, and P the pressure. The sensitivity is given by $S = KI_e = I_+/P$. The constant K is independent of pressure below about 10^{-3} Pa.

Relative sensitivities for different Bayard-Alpert ionization gauges may differ by as much as ± 15% as a result of differences in applied voltages, electron current, and electrode structure. The table below presents the average of the measurements of 12 experimenters on Bayard-Alpert ionization gauges in various gases. The sensitivity relative to nitrogen is tabulated.

Gas		Relative sensitivity $S/S(N_2)$
Helium	He	0.18
Neon	Ne	0.31
Argon	Ar	1.4
Krypton	Kr	1.9
Xenon	Xe	2.7
Nitrogen	N_2	1.00
Hydrogen	H_2	0.43
Oxygen	O_2	0.96
Carbon monoxide	CO	1.0
Carbon dioxide	CO_2	1.4
Water	H_2O	0.93
Sulfur hexafluoride	SF_6	2.3
Mercury	Hg	3.5
Methane	CH_4	1.6
Ethane	C_2H_6	2.6
Propane	C_3H_8	3.5
Butane	C_4H_{10}	4.3
Ethene	C_2H_4	1.3
Propene	C_3H_6	1.8
Acetylene	C_2H_2	0.61
Allene	C_3H_4	1.3
1-Propyne (Methyl acetylene)	C_3H_4	1.4
Benzene	C_6H_6	3.8

References

1. Hollanda, R., *J. Vac. Sci. Technol.*, **10**, 1133, 1973.
2. Nakayama, K., and Hojo, H., *Jap. J. Appl. Phys.*, Suppl. 2, part 1, p. 113, 1974.
3. Tilford, C. R., *J. Vac. Sci. Technol. A*, **1**, 152, 1983.
4. Tilford, C. R., in *Physical Methods of Chemistry, vol.6, Determination of Thermodynamic Properties*, B. W. Rossiter and R. C. Baetzoid, Eds., pp. 101-173, John Wiley, New York, 1992.

LABORATORY SOLVENTS AND OTHER LIQUID REAGENTS

This table summarizes the properties of 575 liquids that are commonly used in the laboratory as solvents or chemical reagents.

The properties tabulated are:

M_r: Molecular weight

t_m: Melting point in °C

t_b: Normal boiling point in °C

ρ: Density in g/mL at the temperature in °C indicated by the superscript

η: Viscosity in mPa s (1 mPa s = 1 centipoise) at 25 °C

ε: Dielectric constant at ambient temperature (15 to 30 °C)

μ: Dipole moment in D

c_p: Specific heat capacity of the liquid at constant pressure at 25 °C in J/g K

vp: Vapor pressure at 25 °C in kPa (1 kPa = 7.50 mmHg)

FP: Flash point in °C

Fl. lim.:   Flammable (explosive) limit in air in percent by volume

IT: Autoignition temperature in °C

TLV: Threshold limit for allowable airborne concentration in parts per million by volume at 25 °C and atmospheric pressure

Data on the temperature dependence of viscosity, dielectric constant, and vapor pressure can be found in the pertinent tables in this *Handbook*.

References

1. Lide, D. R., *Handbook of Organic Solvents*, CRC Press, Boca Raton, FL, 1994.
2. Lide, D. R., and Kehiaian, H. V., *Handbook of Thermophysical and Thermochemical Data*, CRC Press, Boca Raton, FL, 1994.
3. Riddick, J. A., Bunger, W. B., and Sakano, T. K., *Organic Solvents, Fourth Edition*, John Wiley & Sons, New York, 1986.
4. *Fire Protection Guide to Hazardous Materials, 11th Edition*, National Fire Protection Association, Quincy, MA, 1994.
5. Urben, P. G., Ed., *Bretherick's Handbook of Reactive Chemical Hazards, 5th Edition*, Butterworth-Heinemann, Oxford, 1995.
6. *2004 TLV's and BEI's*, American Conference of Governmental Industrial Hygienists, Cincinnati, OH, 2004.

Name	Mol. form.	M_r	t_m/°C	t_b/°C	ρ/g mL^{-1}	η/mPa s	ε	μ/D	c_p/J g^{-1}K^{-1}	vp/kPa	FP/°C	Fl. lim.	IT/°C	TLV/ppm
Acetaldehyde	C_2H_4O	44.052	-123.37	20.1	0.7834[18]		21.0	2.750	2.020	120	-39	4-60%	175	25
Acetic acid	$C_2H_4O_2$	60.052	16.64	117.9	1.0446[25]	1.056	6.20	1.70	2.053	2.07	39	4-20%	463	10
Acetic anhydride	$C_4H_6O_3$	102.089	-74.1	139.5	1.082[20]	0.843	22.45	≈ 2.8	1.648	0.680	49	2.7-10.3%	316	5
Acetone	C_3H_6O	58.079	-94.7	56.05	0.7845[25]	0.306	21.01	2.88	2.175	30.8	-20	3-13%	465	500
Acetone cyanohydrin	C_4H_7NO	85.105	-19	95	0.932[19]						74	2.2-12%	688	4.6
Acetonitrile	C_2H_3N	41.052	-43.82	81.65	0.7857[20]	0.369	36.64	3.92	2.229	11.9	6	3-16%	524	20
Acetophenone	C_8H_8O	120.149	20.5	202	1.0281[20]	1.681	17.44	3.02	1.703	0.049	77		570	10
Acetyl bromide	C_2H_3BrO	122.948	-96	76	1.6625[16]					16.2				
Acetyl chloride	C_2H_3ClO	78.497	-112.8	50.7	1.1051[20]	0.368	15.8	2.72	1.491	38.4	4		390	
Acrolein	C_3H_4O	56.063	-87.7	52.6	0.840[20]			3.1		36.2	-26	2.8-31%	220	0.1
Acrylic acid	$C_3H_4O_2$	72.063	12.5	141	1.0511[20]				2.022	0.53	50	2.4-8%	438	2
Acrylonitrile	C_3H_3N	53.063	-83.48	77.3	0.8007[25]		33.0	3.87	2.05	14.1	0	3-17%	481	2
Allyl alcohol	C_3H_6O	58.079	-129	97.0	0.8540[20]	1.218	19.7	1.60	2.392	3.14	21	3-18%	378	0.5
Allylamine	C_3H_7N	57.095	-88.2	53.3	0.758[20]			1.2		33.1	-29	2-22%	374	
2-Amino-2-methyl-1-propanol	$C_4H_{11}NO$	89.136	25.5	165.5	0.934[20]						67			
3-Amino-1-propanol	C_3H_9NO	75.109	12.4	187.5	0.9824[26]						80			
Aniline	C_6H_7N	93.127	-6.02	184.17	1.0217[20]	3.85	7.06	1.13	2.061	0.090	70	1.3-11%	615	2
Anisole	C_7H_8O	108.138	-37.13	153.7	0.9940[20]	1.056	4.30	1.38	1.840	0.472	52		475	
Antimony(V) chloride	Cl_5Sb	299.024	4	140 dec	2.34		3.222							
Antimony(V) fluoride	F_5Sb	216.752	8.3	141	3.10									
Arsenic(III) chloride	$AsCl_3$	181.280	-16	130	2.150			1.59		5.38				
Benzaldehyde	C_7H_6O	106.122	-57.1	178.8	1.0401[25]		17.85	3.0	1.621	0.169	63		192	
Benzene	C_6H_6	78.112	5.49	80.09	0.8765[20]	0.604	2.2825	0	1.741	12.7	-11	1-8%	498	0.5
Benzeneacetonitrile	C_8H_7N	117.149	-23.8	233.5	1.0205[15]		17.87	3.5		0.012	113			
Benzeneethanamine	$C_8H_{11}N$	121.180	<0	195	0.9640[25]									
Benzeneethanol	$C_8H_{10}O$	122.164	-27	218.2	1.0202[20]		12.31		2.068	0.01	96			
Benzenemethanethiol	C_7H_8S	124.204	-30	194.5	1.058[20]		4.705							
Benzenesulfonyl chloride	$C_6H_5ClO_2S$	176.621	14.5	251 dec	1.3470[15]		28.90			0.008				
Benzenethiol	C_6H_6S	110.177	-14.93	169.1	1.0775[20]		4.26	1.23	1.572	0.26				0.1
Benzonitrile	C_7H_5N	103.122	-13.99	191.1	1.0093[15]	1.267	25.9	4.18	1.602	0.11				
Benzoyl chloride	C_7H_5ClO	140.567	-0.4	197.2	1.2120[20]		23.0			0.084	72			0.5
Benzyl acetate	$C_9H_{10}O_2$	150.174	-51.3	213	1.0550[20]		5.34	1.22	0.989	0.022	90		460	10
Benzyl alcohol	C_7H_8O	108.138	-15.4	205.31	1.0419[24]	5.47	11.916	1.71	2.015	0.015	93		436	
Benzylamine	C_7H_9N	107.153		185	0.9813[20]	1.624	5.18			0.096				
2,2'-Bioxirane	$C_4H_6O_2$	86.090	2.0	144	1.113[20]									

Name	Mol. form.	M_r	t_m/°C	t_b/°C	ρ/g mL⁻¹	η/mPa s	ε	μ/D	c_p/J g⁻¹K⁻¹	vp/kPa	FP/°C	Fl. lim.	IT/°C	TLV/ppm
Bis(2-aminoethyl)amine	C₄H₁₃N₃	103.166	-39	207	0.9569[20]		12.62	1.9	2.462	0.03	98	2-7%	358	1
N,N'-Bis(2-aminoethyl)-1,2-ethanediamine	C₆H₁₈N₄	146.234	12	266.5			10.76							
Bis(2-chloroethyl) ether	C₄H₈Cl₂O	143.012	-51.9	178.5	1.22[20]		21.20	2.6	1.545	0.143	55	3%-	369	5
Bis(chloromethyl) ether	C₂H₄Cl₂O	114.958	-41.5	106	1.323[15]		3.51							0.001
Bis(2-ethylhexyl) phthalate	C₂₄H₃₈O₄	390.557	-55	384	0.981[25]		5.3	2.84	1.804	0.00000005	218			0.3
Bis(2-hydroxyethyl) sulfide	C₄H₁₀O₂S	122.186	-10.2	282	1.1793[25]		28.61			0.08	160		298	
Boron tribromide	BBr₃	250.523	-45	91	2.6			0						1
Boron trichloride	BCl₃	117.169	-107	12.65				0	0.911	156				
Bromine	Br₂	159.808	-7.2	58.8	3.1028	0.944	3.1484	0	0.474	28.2				0.1
Bromobenzene	C₆H₅Br	157.008	-30.72	156.06	1.4950[20]	1.074	5.45	1.70	0.983	0.556	51		565	
1-Bromobutane	C₄H₉Br	137.018	-112.6	101.6	1.2758[20]	0.606	7.315	2.08	0.798	5.26	18	2.6-6.6%	265	
2-Bromobutane, (±)-	C₄H₉Br	137.018	-112.65	91.3	1.2585[20]		8.64	2.23		9.32	21			
Bromochloromethane	CH₂BrCl	129.384	-87.9	68.0	1.9344[20]			1.7	0.41	19.5				200
Bromodichloromethane	CHBrCl₂	163.829	-57	90	1.980[20]									
Bromoethane	C₂H₅Br	108.965	-118.6	38.5	1.4604[20]	0.374	9.01	2.03	0.925	62.5		7-8%	511	5
Bromoethene	C₂H₃Br	106.949	-139.54	15.8	1.4933[20]		5.63	1.42	1.007	141		9-15%	530	0.5
2-Bromo-2-methylpropane	C₄H₉Br	137.018	-16.2	73.3	1.4278[20]		10.98	2.17	1.102	17.7				
1-Bromopentane	C₅H₁₁Br	151.045	-88.0	129.8	1.2182[20]		6.31	2.20	0.875	1.68	32			
1-Bromopropane	C₃H₇Br	122.992	-110.3	71.1	1.3537[20]	0.489	8.09	2.18	0.702	18.6			490	
2-Bromopropane	C₃H₇Br	122.992	-89.0	59.5	1.3140[20]	0.458	9.46	2.21	1.075	28.9				
3-Bromopropene	C₃H₅Br	120.976	-119	70.1	1.398[20]	0.471	7.0	≈ 1.9		18.6	-1	4.4-7.3%	295	
2-Bromotoluene	C₇H₇Br	171.035	-27.8	181.7	1.4232[20]		4.641			0.17	79			
Bromotrichloromethane	CBrCl₃	198.274	-5.65	105	2.012[25]		2.405			5.35				
Butanal	C₄H₈O	72.106	-96.86	74.8	0.8016[20]		13.45	2.72	2.270	15.7	-22	2-12.5%	218	
1,3-Butanediol	C₄H₁₀O₂	90.121	-77	207.5	1.0053[20]		28.8		2.521	0.008	121		395	
1,4-Butanediol	C₄H₁₀O₂	90.121	20.4	235	1.0171[20]		31.9	2.58	2.220	0.002	121			
2,3-Butanediol	C₄H₁₀O₂	90.121	7.6	182.5	1.0033[20]				2.363	0.02			402	
2,3-Butanedione	C₄H₆O₂	86.090	-1.2	88	0.9808[18]		4.04			7.45	27			
Butanenitrile	C₄H₇N	69.106	-111.9	117.6	0.7936[20]	0.553	24.83	3.9	2.301	2.55	24	>1.6%	501	
1-Butanethiol	C₄H₁₀S	90.187	-115.7	98.5	0.8416[20]		5.204	1.53	1.898	6.07	2			0.5
2-Butanethiol	C₄H₁₀S	90.187	-165	85.0	0.8295[20]		5.645			10.8	-23			
Butanoic acid	C₄H₈O₂	88.106	-5.1	163.75	0.9528[25]	1.426	2.98	1.65	2.027	0.221	72	2-10%	443	
Butanoic anhydride	C₈H₁₄O₃	158.195	-75	200	0.9668[20]		12.8		1.793	0.07	54	0.9-5.8%	279	
1-Butanol	C₄H₁₀O	74.121	-88.6	117.73	0.8095[20]	2.54	17.84	1.66	2.391	0.86	37	1-11%	343	20
2-Butanol	C₄H₁₀O	74.121	-88.5	99.51	0.8063[20]	3.10	17.26	1.8	2.656	2.32	24	2-10%	405	100
2-Butanone	C₄H₈O	72.106	-86.64	79.59	0.7999[25]	0.405	18.56	2.78	2.201	12.6	-9	1-11%	404	200
trans-2-Butenal	C₄H₆O	70.090	-76	102.2	0.8516[20]			3.67	1.361	4.92	13	2.1-15.5%	232	0.3
cis-2-Butenoic acid	C₄H₆O₂	86.090	15	169	1.0267[20]					0.06				
2-Butoxyethanol	C₆H₁₄O₂	118.174	-74.8	168.4	0.9015[20]		9.30	2.1	2.378	0.15	69	4-13%	238	20
Butyl acetate	C₆H₁₂O₂	116.158	-78	126.1	0.8825[20]	0.685	5.07	1.9	1.961	1.66	22	2-8%	425	150
sec-Butyl acetate	C₆H₁₂O₂	116.158	-98.9	112	0.8748[20]		5.135	1.87			31	1.7-9.8%		200
Butyl acrylate	C₇H₁₂O₂	128.169	-64.6	145	0.8898[20]		5.25		1.958	0.731	29	1.7-9.9%	292	2
Butylamine	C₄H₁₁N	73.137	-49.1	77.00	0.7414[20]	0.574	4.71	1.0	2.450	12.2	-12	2-10%	312	5
sec-Butylamine	C₄H₁₁N	73.137	<-72	62.73	0.7246[20]			1.28			-9			
tert-Butylamine	C₄H₁₁N	73.137	-66.94	44.04	0.6958[20]		58.5	1.3	2.627	48.4	-9	2-9%	380	
Butylbenzene	C₁₀H₁₄	134.218	-87.85	183.31	0.8601[20]	0.950	2.359	≈ 0	1.813	0.150	71	0.8-5.8%	410	
tert-Butylbenzene	C₁₀H₁₄	134.218	-57.8	169.1	0.8665[20]		2.359	≈ 0.83	1.773	0.280	60	0.7-5.7%	450	
Butyl benzoate	C₁₁H₁₄O₂	178.228	-22.4	250.3	1.000[20]		5.52			0.005	107			
tert-Butyl ethyl ether	C₆H₁₄O	102.174	-94	72.6	0.736[25]				2.13	16.5				5
tert-Butyl hydroperoxide	C₄H₁₀O₂	90.121	6	89 dec	0.8960[20]						27			
1-tert-Butyl-4-methylbenzene	C₁₁H₁₆	148.245	-52	190	0.8612[20]			≈ 0		0.09	68			1
Butyl vinyl ether	C₆H₁₂O	100.158	-92	94	0.7888[20]			1.25	2.316	6.65	-9		255	
γ-Butyrolactone	C₄H₆O₂	86.090	-43.61	204	1.1296[20]		39.0	4.27	1.642	0.43	98			
Carbon disulfide	CS₂	76.141	-112.1	46	1.2632[20]	0.352	2.6320	0	1.003	48.2	-30	1-50%	90	10
Chloroacetaldehyde	C₂H₃ClO	78.497	-16.3	85.5	1.19									1
Chloroacetone	C₃H₅ClO	92.524	-44.5	119	1.15[20]					2				1
Chloroacetyl chloride	C₂H₂Cl₂O	112.942	-22	106	1.4202[20]			2.23		3.33				0.05
2-Chloroaniline	C₆H₆ClN	127.572	-1.9	208.8		3.32	13.40	1.77		0.034				
3-Chloroaniline	C₆H₆ClN	127.572	-10.28	230.5	1.2161[20]		13.3		1.558	0.0156			705	
Chlorobenzene	C₆H₅Cl	112.557	-45.31	131.72	1.1058[20]	0.753	5.6895	1.69	1.334	1.6	28	1-10%	593	10
2-Chloro-1,3-butadiene	C₄H₅Cl	88.536	-130	59.4	0.956[20]		4.914			29.5	-20	4-20%		10
1-Chlorobutane	C₄H₉Cl	92.567	-123.1	78.4	0.8857[20]	0.422	7.276	2.05	1.891	13.7	-12	2-10%	240	
2-Chlorobutane	C₄H₉Cl	92.567	-131.3	68.2	0.8732[20]		8.564	2.04		21.0	-10			
Chlorocyclohexane	C₆H₁₁Cl	118.604	-43.81	142	1.000[20]		7.9505	2.1		1.0	32			
Chlorodibromomethane	CHBr₂Cl	208.280	-20	120	2.451[20]									
Chloroethane	C₂H₅Cl	64.514	-138.4	12.3	0.9239[0]		9.45	2.05	1.617	160	-50	4-15%	519	100

Name	Mol. form.	M_r	t_m/°C	t_b/°C	ρ/g mL⁻¹	η/mPa s	ε	μ/D	c_p/J g⁻¹K⁻¹	vp/kPa	FP/°C	Fl. lim.	IT/°C	TLV/ppm
2-Chloroethanol	C₂H₅ClO	80.513	-67.5	128.6	1.2019²⁰		25.80	1.78		1.2	60	5-16%	425	1
2-Chloroethyl vinyl ether	C₄H₇ClO	106.551	-70	108	1.0495²⁰						27			
(Chloromethyl)benzene	C₇H₇Cl	126.584	-45	179	1.1004²⁰		6.854	1.8	1.44	0.164	67	1%-	585	1
Chloromethyl methyl ether	C₂H₅ClO	80.513	-103.5	59.5	1.063¹⁰					24.9				
1-Chloro-2-methylpropane	C₄H₉Cl	92.567	-130.3	68.5	0.8773²⁰		7.027	2.00	1.713	19.9	-6	2-8.7%		
2-Chloro-2-methylpropane	C₄H₉Cl	92.567	-25.60	50.9	0.8420²⁰		9.663	2.13	1.867	42.7	0			
1-Chloronaphthalene	C₁₀H₇Cl	162.616	-2.5	259	1.1880²⁵		5.04	1.57	1.307	0.003	121		>558	
1-Chlorooctane	C₈H₁₇Cl	148.674	-57.8	183.5	0.8734²⁰		5.05	2.00	1.335	0.11	70			
1-Chloropentane	C₅H₁₁Cl	106.594	-99.0	108.4	0.8820²⁰		6.654	2.16		4.36	13	1.6-8.6%	260	
2-Chlorophenol	C₆H₅ClO	128.556	9.4	174.9	1.2634²⁰	3.59	7.40		1.468	0.308	64			
1-Chloropropane	C₃H₇Cl	78.541	-122.9	46.5	0.8899²⁰	0.334	8.588	2.05	1.683	45.8	<-18	2.6-11%	520	
2-Chloropropane	C₃H₇Cl	78.541	-117.18	35.7	0.8617²⁰	0.303		2.17		68.9	-32	2.8-11%	593	
3-Chloro-1,2-propanediol	C₃H₇ClO₂	110.540		213 dec	1.325¹⁸		31.0							
3-Chloropropanenitrile	C₃H₄ClN	89.524	-51	175.5	1.1573²⁰						76			
2-Chloropropene	C₃H₅Cl	76.525	-137.4	22.6	0.9017²⁰		8.92	1.647		110	-37	4.5-16%		
3-Chloropropene	C₃H₅Cl	76.525	-134.5	45.1	0.9376²⁰	0.314	8.2	1.94	1.635	48.9	-32	2.9-11%	485	1
Chlorosulfonic acid	ClHO₃S	116.525	-80	152	1.75					0.42				
2-Chlorotoluene	C₇H₇Cl	126.584	-35.8	159.0	1.0825²⁰	0.964	4.721	1.56	1.318	0.482				50
4-Chlorotoluene	C₇H₇Cl	126.584	7.5	162.4	1.0697²⁰	0.837	6.25	2.21		0.4				
Chromyl chloride	Cl₂CrO₂	154.900	-96.5	117	1.91									0.025
trans-Cinnamaldehyde	C₉H₈O	132.159	-7.5	246	1.0497²⁰		17.72			0.005				
o-Cresol	C₇H₈O	108.138	31.03	191.04	1.0327³⁵		6.76	1.45	2.160	0.041	81	>1.4%	599	5
m-Cresol	C₇H₈O	108.138	12.24	202.27	1.0339²⁰	12.91	12.44	1.48	2.080	0.019	86	>1.1%	558	5
p-Cresol	C₇H₈O	108.138	34.77	201.98	1.0185⁴⁰		13.05	1.48	2.044	0.017	86	>1.1%	558	5
Cyanogen chloride	CClN	61.471	-6.5	13	1.186²⁰			2.8331						0.3
Cyclobutane	C₄H₈	56.107	-90.7	12.6	0.7038⁰			0		157	<10	>1.8%		
Cyclohexane	C₆H₁₂	84.159	6.59	80.73	0.7739²⁵	0.894	2.0243	≈ 0	1.841	13.0	-20	1-8%	245	100
Cyclohexanol	C₆H₁₂O	100.158	25.93	160.84	0.9624²⁰	57.5	16.40		2.079	0.10	68	1-9%	300	50
Cyclohexanone	C₆H₁₀O	98.142	-27.9	155.43	0.9478²⁰	2.02	16.1	2.87	1.856	0.53	44	1-9%	420	20
Cyclohexene	C₆H₁₀	82.143	-103.5	82.98	0.8110²⁰	0.625	2.2176	0.33	1.805	11.8	-12	>1.2%	310	300
Cyclohexylamine	C₆H₁₃N	99.174	-17.8	134	0.8191²⁰	1.944	4.547	1.3		1.20	31	1-9%	293	10
1,3-Cyclopentadiene	C₅H₆	66.102	-85	41	0.8021²⁰			0.419		58.5				75
Cyclopentane	C₅H₁₀	70.133	-93.4	49.3	0.7457²⁰	0.413	1.9687	≈ 0	1.837	42.3	-25	2%-	361	600
Cyclopentanol	C₅H₁₀O	86.132	-17.5	140.42	0.9488²⁰		18.5		2.119	0.294	51			
Cyclopentanone	C₅H₈O	84.117	-51.90	130.57	0.9487²⁰		13.58	3.3	1.84	1.55	26			
cis-Decahydronaphthalene	C₁₀H₁₈	138.250	-42.9	195.8	0.8965²⁰	3.04	2.219	≈ 0	1.678	0.10				
trans-Decahydronaphthalene	C₁₀H₁₈	138.250	-30.4	187.3	0.8659²⁵	1.948	2.184	≈ 0	1.653	0.164	54	1-5%	255	
Decamethylcyclopenta-siloxane	C₁₀H₃₀O₅Si₅	370.770	-38	210	0.9593²⁰		2.50			0.02				
Decanal	C₁₀H₂₀O	156.265	-4.0	208.5	0.830¹⁵					0.02				
Decane	C₁₀H₂₂	142.282	-29.6	174.15	0.7266²⁵	0.838	1.9853	≈ 0	2.210	0.170	51	0.8-5.4%	210	
Decanoic acid	C₁₀H₂₀O₂	172.265	31.4	268.7	0.8858⁴⁰				2.761					
1-Decanol	C₁₀H₂₂O	158.281	6.9	231.1	0.8297²⁰	10.91	7.93		2.341	0.009	82		288	
1-Decene	C₁₀H₂₀	140.266	-66.3	170.5	0.7408²⁰	0.756	2.136	≈ 0	2.144	0.210	<55		235	
Diacetone alcohol	C₆H₁₂O₂	116.158	-44	167.9	0.9387²⁰	2.80	18.2	3.2	1.905	0.224	58	2-7%	643	50
Dibenzyl ether	C₁₄H₁₄O	198.260	1.8	298	1.0428²⁰		3.821				135			
Dibromodifluoromethane	CBr₂F₂	209.816	-110.1	22.76				0.66		110				100
1,2-Dibromoethane	C₂H₄Br₂	187.861	9.84	131.6	2.1683²⁵	1.595	4.9612	1.2	0.724	1.55				
Dibromomethane	CH₂Br₂	173.835	-52.5	97	2.4969²⁰	0.980	7.77	1.43	0.61	6.12				
1,2-Dibromotetrafluoroethane	C₂Br₂F₄	259.823	-110.32	47.35	2.149²⁵		2.34		0.69	43.4				
Dibutylamine	C₈H₁₉N	129.244	-62	159.6	0.7670²⁰	0.918	2.765	1.0	2.266	0.34	47	1-6%		
Dibutyl ether	C₈H₁₈O	130.228	-95.2	140.28	0.7684²⁰	0.637	3.0830	1.17	2.136	0.898	25	1.5-7.6%	194	
Di-tert-butyl peroxide	C₈H₁₈O₂	146.228	-40	111	0.704²⁰					3.43	18			
Dibutyl phthalate	C₁₆H₂₂O₄	278.344	-35	340	1.0465²⁰	16.63	6.58	2.82	1.789		157	>0.5%	402	0.4
Dibutyl sebacate	C₁₈H₃₄O₄	314.461	-10	344.5	0.9405¹⁵		4.54	2.48	1.968		178	>0.4%	365	
Dibutyl sulfide	C₈H₁₈S	146.294	-79.7	185	0.8386²⁰		4.29	1.61	1.943	0.09	76			
Dichloroacetic acid	C₂H₂Cl₂O₂	128.942	10	194	1.5634²⁰		8.33			0.03				
o-Dichlorobenzene	C₆H₄Cl₂	147.002	-17.0	180	1.3059²⁰	1.324	10.12	2.50	1.105	0.18	66	2-9%	648	25
m-Dichlorobenzene	C₆H₄Cl₂	147.002	-24.8	173	1.2884²⁰	1.044	5.02	1.72	1.163	0.252	72			
trans-1,4-Dichloro-2-butene	C₄H₆Cl₂	124.997	1.0	155.4	1.183²⁵									0.005
Dichlorodimethylsilane	C₂H₆Cl₂Si	129.061	-16	70.3	1.064²⁵					18.9	<21	3.4-9.5%		
1,1-Dichloroethane	C₂H₄Cl₂	98.959	-96.9	57.3	1.1757²⁰	0.464	10.10	2.06	1.276	30.5	-17	5-11%	458	100
1,2-Dichloroethane	C₂H₄Cl₂	98.959	-35.7	83.5	1.2454²⁵	0.779	10.42	1.8	1.298	10.6	13	6-16%	413	10
1,1-Dichloroethene	C₂H₂Cl₂	96.943	-122.56	31.6	1.213²⁰		4.60	1.34	1.148	80.0	-28	7-16%	570	5
cis-1,2-Dichloroethene	C₂H₂Cl₂	96.943	-80.0	60.1	1.2837²⁰	0.445	9.20	1.90	1.201	26.8	6	3-15%	460	200
trans-1,2-Dichloroethene	C₂H₂Cl₂	96.943	-49.8	48.7	1.2565²⁰	0.317	2.14	0	1.205	44.2	2	6-13%	460	200
Dichloromethane	CH₂Cl₂	84.933	-97.2	40	1.3266²⁰	0.413	8.93	1.60	1.192	58.2		13-23%	556	50

Name	Mol. form.	M_r	t_m/°C	t_b/°C	ρ/g mL^{-1}	η/mPa s	ε	μ/D	c_p/J g^{-1}K^{-1}	vp/kPa	FP/°C	Fl. lim.	IT/°C	TLV/ppm
(Dichloromethyl)benzene	$C_7H_6Cl_2$	161.029	-17	205	1.26[25]		6.9	2.1		0.06				
1,1-Dichloropropane	$C_3H_6Cl_2$	112.986		88.1	1.1321[20]					9.09				
1,2-Dichloropropane, (±)-	$C_3H_6Cl_2$	112.986	-100.53	96.4	1.1560[20]		8.37	1.8	1.320	6.62	21	3-15%	557	75
1,3-Dichloropropane	$C_3H_6Cl_2$	112.986	-99.5	120.9	1.1785[25]		10.27	2.08		2.44				
2,3-Dichloropropene	$C_3H_4Cl_2$	110.970	10	94	1.211[20]						15	2.6-7.8%		
2,4-Dichlorotoluene	$C_7H_6Cl_2$	161.029	-13.5	201	1.2476[20]		5.68	1.70		0.055				
Dicyclohexylamine	$C_{12}H_{23}N$	181.318	-0.1	256 dec	0.9123[20]					0.003	>99			
Diethanolamine	$C_4H_{11}NO_2$	105.136	28	268.8	1.0966[20]		25.75	2.8	2.22	<0.01	172	2-13%	662	0.5
1,1-Diethoxyethane	$C_6H_{14}O_2$	118.174	-100	102.25	0.8254[20]		3.80	1.4	2.01	3.68	-21	2-10%	230	
1,2-Diethoxyethane	$C_6H_{14}O_2$	118.174	-74.0	121.2	0.8351[25]		3.90		2.195	4.33	27		205	
Diethylamine	$C_4H_{11}N$	73.137	-49.8	55.5	0.7056[20]	0.319	3.680	0.92	2.313	30.1	-23	2-10%	312	5
N,N-Diethylaniline	$C_{10}H_{15}N$	149.233	-38.8	216.3	0.9307[20]		5.15			0.025	85		630	
o-Diethylbenzene	$C_{10}H_{14}$	134.218	-31.2	184	0.8800[20]		2.594			0.13	57		395	
m-Diethylbenzene	$C_{10}H_{14}$	134.218	-83.9	181.1	0.8602[20]		2.369			0.14	56		450	
p-Diethylbenzene	$C_{10}H_{14}$	134.218	-42.83	183.7	0.8620[20]		2.259			0.13	55	0.7-6%	430	
Diethyl carbonate	$C_5H_{10}O_3$	118.131	-43	126	0.9692[25]		2.820	1.10	1.80	1.63	25			
Diethylene glycol	$C_4H_{10}O_3$	106.120	-10.4	245.8	1.1197[15]	30.2	31.82	2.3	2.307	0.001	124	2-17%	224	
Diethylene glycol diethyl ether	$C_8H_{18}O_3$	162.227	-45	188	0.9063[20]		5.70		2.104	0.10	82			
Diethylene glycol dimethyl ether	$C_6H_{14}O_3$	134.173	-68	162	0.9434[20]	0.989	7.23	2.0	2.043	0.315	67			
Diethylene glycol monobutyl ether	$C_8H_{18}O_3$	162.227	-68	231	0.9553[20]				2.188	0.0032				
Diethylene glycol monoethyl ether	$C_6H_{14}O_3$	134.173		196	0.9885[20]			1.6	2.243	0.017	96			
Diethylene glycol monoethyl ether acetate	$C_8H_{16}O_4$	176.211	-25	218.5	1.0096[20]			1.8		0.029	110		425	
Diethylene glycol monomethyl ether	$C_5H_{12}O_3$	120.147		193	1.035[20]			1.6	2.256	0.024	96	1-23%	240	
Diethyl ether	$C_4H_{10}O$	74.121	-116.2	34.5	0.7138[20]	0.224	4.2666	1.15	2.369	71.7	-45	2-36%	180	400
Diethyl maleate	$C_8H_{12}O_4$	172.179	-8.8	223	1.0662[20]		7.560			0.015	121		350	
Diethyl malonate	$C_7H_{12}O_4$	160.168	-50	200	1.0551[20]		7.550	2.54	1.779	0.048	93			
Diethyl oxalate	$C_6H_{10}O_4$	146.141	-40.6	185.7	1.0785[20]		8.266	2.49	1.784	0.030	76			
Diethyl phthalate	$C_{12}H_{14}O_4$	222.237	-40.5	295	1.232[14]		7.86		1.647	0.002	161	>0.7%	457	0.6
Diethyl succinate	$C_8H_{14}O_4$	174.195	-21	217.7	1.0402[20]		6.098			0.15	90			
Diethyl sulfate	$C_4H_{10}O_4S$	154.185	-24	208	1.172[25]		29.2			0.05	104		436	
Diethyl sulfide	$C_4H_{10}S$	90.187	-103.91	92.1	0.8362[20]	0.422	5.723	1.54	1.900	7.78				
Diiodomethane	CH_2I_2	267.836	6.1	182	3.3211[20]		5.32	1.08	0.500	0.172				
Diiodosilane	H_2I_2Si	283.911	-1	150										
Diisobutylamine	$C_8H_{19}N$	129.244	-73.5	139.6		0.723				0.972	29			
Diisopentyl ether	$C_{10}H_{22}O$	158.281		172.5	0.7777[20]		2.817	1.23	2.394	0.210				
Diisopropylamine	$C_6H_{15}N$	101.190	-61	83.9	0.7153[20]	0.393		1.15		10.7	-1	1.1-7.1%	316	5
Diisopropyl ether	$C_6H_{14}O$	102.174	-85.4	68.4	0.7192[25]	0.379	3.805	1.13	2.122	19.9	-28	1-8%	443	250
1,2-Dimethoxyethane	$C_4H_{10}O_2$	90.121	-69.20	84.5	0.8637[25]	0.455	7.30		2.145	9.93	-2		202	
Dimethoxymethane	$C_3H_8O_2$	76.095	-105.1	42	0.8593[20]		2.644	0.7	2.129	53.1	-32	2-14%	237	1000
Dimethylacetal	$C_4H_{10}O_2$	90.121	-113.2	64.5	0.8501[20]					22.9				
N,N-Dimethylacetamide	C_4H_9NO	87.120	-18.59	165	0.9372[25]	1.927	38.85	3.7	2.016	0.075	70	2-12%	490	10
2,3-Dimethylaniline	$C_8H_{11}N$	121.180	<-15	221.5	0.9931[20]						97	>1%		
2,6-Dimethylaniline	$C_8H_{11}N$	121.180	11.2	215	0.9842[20]			1.63	1.971	0.45	96			
N,N-Dimethylaniline	$C_8H_{11}N$	121.180	2.42	194.15	0.9557[20]	1.300	4.90	1.68	1.771	0.107	63		371	5
2,2-Dimethylbutane	C_6H_{14}	86.175	-98.8	49.73	0.6444[25]	0.351	1.869	≈ 0	2.227	42.5	-48	1.2-7%	405	500
2,3-Dimethylbutane	C_6H_{14}	86.175	-128.10	57.93	0.6616[20]	0.361	1.889	≈ 0	2.201	31.3	-29	1.2-7%	405	500
3,3-Dimethyl-2-butanone	$C_6H_{12}O$	100.158	-52.5	106.1	0.7229[25]		12.73			4.27				
Dimethylcarbamic chloride	C_3H_6ClNO	107.539	-33	167	1.168[25]									
Dimethyl disulfide	$C_2H_6S_2$	94.199	-84.67	109.74	1.0625[20]		9.6	1.8	1.551	3.82	24			
N,N-Dimethylethanolamine	$C_4H_{11}NO$	89.136	-59	134	0.8866[20]					0.9				
N,N-Dimethylformamide	C_3H_7NO	73.094	-60.48	153	0.9445[25]	0.794	38.25	3.82	2.060	0.439	58	2-15%	445	10
2,6-Dimethyl-4-heptanone	$C_9H_{18}O$	142.238	-41.5	169.4	0.8062[20]		9.91	2.7	2.090	0.23	49	1-7%	396	25
1,1-Dimethylhydrazine	$C_2H_8N_2$	60.098	-57.20	63.9	0.791[22]				2.731	20.9	-15	2-95%	249	0.01
Dimethyl phthalate	$C_{10}H_{10}O_4$	194.184	5.5	283.7	1.1905[20]	14.36	8.66		1.561	0.001	146	>0.9%	490	0.6
2,6-Dimethylpyridine	C_7H_9N	107.153	-6.1	144.01	0.9226[20]		7.33	1.7	1.728	0.746				
Dimethyl sulfate	$C_2H_6O_4S$	126.132	-31.7	188 dec	1.3322[20]		55.0			0.13	83		188	0.1
Dimethyl sulfide	C_2H_6S	62.134	-98.24	37.33	0.8483[20]	0.284	6.70	1.554	1.901	64.4	-37	2.2-20%	206	10
Dimethyl sulfoxide	C_2H_6OS	78.133	17.89	189	1.1010[25]	1.987	47.24	3.96	1.958	0.084	95	3-42%	215	
1,4-Dioxane	$C_4H_8O_2$	88.106	11.85	101.5	1.0337[20]	1.177	2.2189	0	1.726	4.95	12	2-22%	180	20
1,3-Dioxolane	$C_3H_6O_2$	74.079	-97.22	78	1.060[20]			1.19	1.593	14.6	2			20
Dipentyl ether	$C_{10}H_{22}O$	158.281	-69	190	0.7833[20]		2.798	1.20	1.579	0.13	57		170	
Dipropylamine	$C_6H_{15}N$	101.190	-63	109.3	0.7400[20]	0.517	2.923	1.03	2.500	3.21	17		299	

Name	Mol. form.	M_r	t_m/°C	t_b/°C	ρ/g mL⁻¹	η/mPa s	ε	μ/D	c_p/J g⁻¹K⁻¹	vp/kPa	FP/°C	Fl. lim.	IT/°C	TLV/ppm
Dipropylene glycol monomethyl ether	C₇H₁₆O₃	148.200	-80	188.3	0.95									
Dipropyl ether	C₆H₁₄O	102.174	-114.8	90.08	0.7466²⁰	0.396	3.38	1.21	2.169	8.35	21	1.3-7%	188	
Dodecane	C₁₂H₂₆	170.334	-9.57	216.32	0.7495²⁰	1.383	2.0120	≈0	2.206	0.016	74	>0.6%	203	
1-Dodecanol	C₁₂H₂₆O	186.333	23.9	260	0.8309²⁴		5.82		2.351	0.000016	127		275	
1-Dodecene	C₁₂H₂₄	168.319	-35.2	213.8	0.7584²⁰	1.20	2.152	≈0	2.143	0.019	79			
Epichlorohydrin	C₃H₅ClO	92.524	-26	118	1.1812²⁰	1.073	22.6	1.8	1.422	2.2	31	4-21%	411	0.5
1,2-Epoxybutane	C₄H₈O	72.106	-150	63.4	0.8297²⁰			1.891	2.039	31.7	-22	1.7-19%	439	
1,2-Epoxy-4-(epoxyethyl)cyclohexane	C₈H₁₂O₂	140.180	<-55	227	1.0966²⁰									0.1
1,2-Ethanediamine	C₂H₈N₂	60.098	11.14	117	0.8979²⁰		13.82	1.99	2.872	1.62	40	3-12%	385	10
1,2-Ethanediol	C₂H₆O₂	62.068	-12.69	197.3	1.1135²⁰	16.06	41.4	2.28	2.394	0.01	111	3-22%	398	40
1,2-Ethanediol, diacetate	C₆H₁₀O₄	146.141	-31	190	1.1043²⁰		7.7	2.34	2.121	0.030	88	1.6-8.4%	482	
1,2-Ethanediol, dinitrate	C₂H₄N₂O₆	152.062	-22.3	198.5	1.4918²⁰		28.26			0.009				0.05
1,2-Ethanedithiol	C₂H₆S₂	94.199	-41.2	146.1	1.234²⁰		7.26	2.03						
Ethanethiol	C₂H₆S	62.134	-147.88	35.0	0.8315²⁵	0.287	6.667	1.60	1.898	70.3	-17	2.8-18%	300	0.5
Ethanol	C₂H₆O	46.068	-114.14	78.29	0.7893²⁰	1.074	25.3	1.69	2.438	7.87	13	3-19%	363	1000
Ethanolamine	C₂H₇NO	61.083	10.5	171	1.0180²⁰	21.1	31.94	2.3	3.201	0.05	86	3-24%	410	3
4-Ethoxyaniline	C₈H₁₁NO	137.179	4.6	254	1.0652¹⁶		7.43			0.0007	116			
Ethoxybenzene	C₈H₁₀O	122.164	-29.43	169.81	0.9651²⁰	1.197	4.216	1.45	1.870	0.204	63			
2-Ethoxyethanol	C₄H₁₀O₂	90.121	-70	135	0.9253²⁵		13.38	2.1	2.339	0.71	43	3-18%	235	5
2-Ethoxyethyl acetate	C₆H₁₂O₃	132.157	-61.7	156.4	0.9740²⁰		7.567	2.2	2.845	0.24	56	2-8%	379	5
Ethyl acetate	C₄H₈O₂	88.106	-83.8	77.11	0.9003²⁰	0.423	6.0814	1.78	1.937	12.6	-4	2-12%	426	400
Ethyl acetoacetate	C₆H₁₀O₃	130.141	-45	180.8	1.0368¹⁰		14.0		1.906	0.095	57	1-10%	295	
Ethyl acrylate	C₅H₈O₂	100.117	-71.2	99.4	0.9234²⁰		6.05	1.96		5.14	10	1.4-14%	372	5
Ethylamine	C₂H₇N	45.084	-80.5	16.5	0.689¹⁵		8.7	1.22	2.884	141	-16	4-14%	385	5
N-Ethylaniline	C₈H₁₁N	121.180	-63.5	203.0	0.9625²⁰	2.05	5.87			0.039	85			
Ethylbenzene	C₈H₁₀	106.165	-94.96	136.19	0.8626²⁵	0.631	2.4463	0.59	1.726	1.28	21	1-7%	432	100
Ethyl benzoate	C₉H₁₀O₂	150.174	-34	212	1.0415²⁵		6.20	2.00	1.638	0.04	88		490	
Ethyl butanoate	C₆H₁₂O₂	116.158	-98	121.3	0.8735²⁵	0.639	5.18	1.74	1.963	2.01	24		463	
2-Ethyl-1-butanol	C₆H₁₄O	102.174	<-15	147	0.8326²⁰		6.19			0.206	57			
Ethyl chloroacetate	C₄H₇ClO₂	122.551	-21	144.3	1.1585²⁰					0.640	64			
Ethyl chloroformate	C₃H₅ClO₂	108.524	-80.6	95	1.1352²⁰		9.736				16		500	
Ethyl cyanoacetate	C₅H₇NO₂	113.116	-22.5	205	1.0654²⁰		31.62	2.17	1.947	0.003	110			
Ethyleneimine	C₂H₅N	43.068	-77.9	56	0.832²⁵		18.3	1.90		28.9	-11	3.3-55%	320	0.5
Ethyl formate	C₃H₆O₂	74.079	-79.6	54.4	0.9208²⁰	0.380	8.57	1.9	2.015	32.3	-20	3-16%	455	100
2-Ethylhexanal	C₈H₁₆O	128.212	<-100	163	0.8540²⁰						44	0.9-7.2%	190	
2-Ethyl-1,3-hexanediol	C₈H₁₈O₂	146.228	-40	244	0.9325²²		18.73				127		360	
2-Ethyl-1-hexanol	C₈H₁₈O	130.228	-70	184.6	0.8319²⁵	6.27	7.58	1.74	2.438	0.019	73	0.8-9.7%	231	
2-Ethylhexyl acetate	C₁₀H₂₀O₂	172.265	-80	199	0.8718²⁰			1.8		0.09	71	1-8%	268	
Ethyl lactate	C₅H₁₀O₃	118.131	-26	154.5	1.0328²⁰		15.4	2.4	2.150		46	>1.5%	400	
Ethyl 3-methylbutanoate	C₇H₁₄O₂	130.185	-99.3	135.0	0.8656²⁰		4.71			1.07				
Ethyl 2-methylpropanoate	C₆H₁₂O₂	116.158	-88.2	110.1	0.868²⁰					3.25	13			
Ethyl nitrite	C₂H₅NO₂	75.067		18	0.899¹⁵					135	-35	4-50%	90	
Ethyl propanoate	C₅H₁₀O₂	102.132	-73.9	99.1	0.8843²⁵	0.501	5.76	1.74	1.920	4.97	12	1.9-11%	440	
Ethyl silicate	C₈H₂₀O₄Si	208.329	-82.5	168.8	0.9320²⁰		2.50		1.749	1.17	52			10
Eucalyptol	C₁₀H₁₈O	154.249	0.8	176.4	0.9267²⁰		4.57			0.260	48			
Fluorobenzene	C₆H₅F	96.102	-42.18	84.73	1.0225²⁰	0.550	5.465	1.60	1.523	10.4	-15			
Fluorosulfonic acid	FHO₃S	100.070	-89	163	1.726					0.08				
Formamide	CH₃NO	45.041	2.49	220	1.1334²⁰	3.34	111.0	3.73	2.389	0.01	154			10
Formic acid	CH₂O₂	46.026	8.3	101	1.220²⁰	1.607	51.1	1.425	2.151	5.75	50	18-57%	434	5
Furan	C₄H₄O	68.074	-85.61	31.5	0.9514²⁰	0.361	2.94	0.66	1.686	80.0	-36	2-14%		
Furfural	C₅H₄O₂	96.085	-38.1	161.7	1.1594²⁰	1.587	42.1	3.5	1.698	0.29	60	2-19%	316	2
Furfuryl alcohol	C₅H₆O₂	98.101	-14.6	171	1.1296²⁰		16.85	1.9	2.079	0.097	75	2-16%	491	10
Germanium(IV) chloride	Cl₄Ge	214.42	-51.50	86.55	1.88			0						
Glycerol	C₃H₈O₃	92.094	18.1	290	1.2613²⁰	934	46.53	2.6	2.377	<0.01	199	3-19%	370	2.7
Glycerol triacetate	C₉H₁₄O₆	218.203	-78	259	1.1583²⁰		7.11		1.763	<0.01	138	1%-	433	
Glycerol trioleate	C₅₇H₁₀₄O₆	885.432	-4		0.915¹⁵		3.109							
Heptanal	C₇H₁₄O	114.185	-43.4	152.8	0.8132²⁵		9.07		2.015	0.46				
Heptane	C₇H₁₆	100.202	-90.55	98.4	0.6795²⁵	0.387	1.9209	≈0	2.242	6.09	-4	1-7%	204	400
Heptanoic acid	C₇H₁₄O₂	130.185	-7.17	222.2	0.9124²⁵	3.84	3.04		2.039	0.001			275	
1-Heptanol	C₇H₁₆O	116.201	-33.2	176.45	0.8219²⁰	5.81	11.75		2.342	0.0044				
2-Heptanone	C₇H₁₄O	114.185	-35	151.05	0.8111²⁰	0.714	11.95	2.6	2.037	0.49	39	1-8%	393	50
3-Heptanone	C₇H₁₄O	114.185	-39	147	0.8183²⁰		12.7	2.78		0.5	46			50
4-Heptanone	C₇H₁₄O	114.185	-33	144	0.8174²⁰		12.60			0.164	49			50
1-Heptene	C₇H₁₄	98.186	-118.9	93.64	0.6970²⁰	0.340	2.092	≈0	2.157	7.52	-1		260	
Hexachloro-1,3-butadiene	C₄Cl₆	260.761	-21	215	1.556²⁵		2.55			0.13			610	0.02

Name	Mol. form.	M_r	t_m/°C	t_b/°C	ρ/g mL⁻¹	η/mPa s	ε	μ/D	c_p/J g⁻¹K⁻¹	vp/kPa	FP/°C	Fl. lim.	IT/°C	TLV/ppm
Hexachloro-1,3-cyclopentadiene	C_5Cl_6	272.772	-9	239	1.7019[25]									0.01
Hexafluorobenzene	C_6F_6	186.054	5.03	80.26	1.6184[20]	2.79	2.029	0	1.191	11.3				
Hexamethyldisiloxane	$C_6H_{18}OSi_2$	162.377	-66	99	0.7638[20]		2.179		1.918	5.57				
Hexamethylphosphoric triamide	$C_6H_{18}N_3OP$	179.200	7.2	232.5	1.03[20]		31.3	5.5	1.791					
Hexanal	$C_6H_{12}O$	100.158	-56	131	0.8335[20]				2.101	1.48	32			
Hexane	C_6H_{14}	86.175	-95.35	68.73	0.6606[25]	0.300	1.8865	≈ 0	2.270	20.2	-22	1-8%	225	50
Hexanedinitrile	$C_6H_8N_2$	108.141	1	295	0.9676[20]				1.190	<0.01	93	2-5%	550	2
Hexanoic acid	$C_6H_{12}O_2$	116.158	-3	205.2	0.9212[25]		2.600	1.13	1.937	0.005	102		380	
1-Hexanol	$C_6H_{14}O$	102.174	-47.4	157.6	0.8136[20]	4.58	13.03		2.353	0.11	63		290	
2-Hexanone	$C_6H_{12}O$	100.158	-55.5	127.6	0.8113[20]	0.583	14.56	2.7	2.130	1.54	25	1-8%	423	5
1-Hexene	C_6H_{12}	84.159	-139.76	63.48	0.6685[25]	0.252	2.077	≈ 0	2.178	24.8	-26	1.2-6.9%	253	50
Hexyl acetate	$C_8H_{16}O_2$	144.212	-80.9	171.5	0.8779[15]		4.42		1.961	0.185	45			
Hydrazine	H_4N_2	32.045	1.4	113.55	1.0036	0.876	51.7	1.75	3.086	1.91	38	5-100%		0.01
Hydrazoic acid	HN_3	43.028	-80	35.7				1.70		68.2				0.11
Hydrogen cyanide	CHN	27.026	-13.29	26	0.6876[20]	0.183	114.9	2.985	2.612	98.8	-18	6-40%	538	4.7
Hydrogen peroxide	H_2O_2	34.015	-0.43	150.2	1.44		74.6	1.573	2.619	0.26				1
3-Hydroxypropanenitrile	C_3H_5NO	71.078	-46	221	1.0404[25]			3.2		0.010	129			
Indan	C_9H_{10}	118.175	-51.38	177.97	0.9639[20]	1.357			1.609	0.2				
Indene	C_9H_8	116.160	-1.5	182	0.9960[25]				1.609	0.220				10
Iodine bromide	BrI	206.808	40	116 dec	4.3			0.726						
Iodine chloride	ClI	162.357	27.39	100 dec	3.24			1.24		3.59				
Iodobenzene	C_6H_5I	204.008	-31.3	188.4	1.8308[20]	1.554	4.59	1.70	0.778	0.133				
1-Iodobutane	C_4H_9I	184.018	-103	130.5	1.6154[20]		6.27	1.93		1.85				
Iodoethane	C_2H_5I	155.965	-111.1	72.3	1.9357[20]	0.556	7.82	1.976	0.738	18.2				
Iodomethane	CH_3I	141.939	-66.4	42.43	2.2789[20]	0.469	6.97	1.62	0.888	53.9				2
1-Iodopropane	C_3H_7I	169.992	-101.3	102.5	1.7489[20]	0.703	7.07	2.04	0.746	5.75				
2-Iodopropane	C_3H_7I	169.992	-90	89.5	1.7042[20]	0.653	8.19	1.95	0.535	9.36				
Iron pentacarbonyl	C_5FeO_5	195.896	-20	103	1.5[20]		2.602		1.228	4				0.1
Isobutanal	C_4H_8O	72.106	-65.9	64.5	0.7891[20]			2.75		23.0	-18	1.6-10.6%	196	
Isobutyl acetate	$C_6H_{12}O_2$	116.158	-98.8	116.5	0.8712[20]	0.676	5.068	1.9	2.013	2.39	18	1-11%	421	150
Isobutyl acrylate	$C_7H_{12}O_2$	128.169	-61	132	0.8896[20]						30		427	
Isobutylamine	$C_4H_{11}N$	73.137	-86.7	67.75	0.724[25]	0.571	4.43	1.3	2.505	19.0	-9	2-12%	378	
Isobutylbenzene	$C_{10}H_{14}$	134.218	-51.4	172.79	0.8532[20]		2.318	≈ 0	1.793	0.257	55	0.8-6%	427	
Isobutyl formate	$C_5H_{10}O_2$	102.132	-95.8	98.2	0.8776[20]		6.41	1.88		5.34	5	2-9%	320	
Isobutyl isobutanoate	$C_8H_{16}O_2$	144.212	-80.7	148.6	0.8542[20]			1.9		0.552	38	1-8%	432	
Isopentane	C_5H_{12}	72.149	-159.77	27.88	0.6201[20]	0.214	1.845	0.13	2.284	91.7	-51	1.4-7.6%	420	600
Isopentyl acetate	$C_7H_{14}O_2$	130.185	-78.5	142.5	0.876[15]		4.72	1.9	1.909	0.728	25	1-8%	360	50
Isophorone	$C_9H_{14}O$	138.206	-8.1	215.2	0.9255[20]	2.33			1.834	0.06	84	1-4%	460	5
Isopropenyl acetate	$C_5H_8O_2$	100.117	-92.9	94	0.9090[20]					6.02	26		432	
Isopropenylbenzene	C_9H_{10}	118.175	-23.2	165.4	0.9106[20]		2.28		1.711	0.40	54	1.9-6.1%	574	50
Isopropyl acetate	$C_5H_{10}O_2$	102.132	-73.4	88.6	0.8718[20]				1.952	7.88	2	2-8%	460	100
Isopropylamine	C_3H_9N	59.110	-95.13	31.76	0.6891[20]	0.325	5.6268	1.19	2.771	78.0	-37		402	5
Isopropylbenzene	C_9H_{12}	120.191	-96.02	152.41	0.8640[25]	0.737	2.381	0.79	1.753	0.61	36	1-7%	424	50
Isopropylbenzene hydroperoxide	$C_9H_{12}O_2$	152.190		153	1.03[20]					0.004				
1-Isopropyl-2-methylbenzene	$C_{10}H_{14}$	134.218	-71.5	178.1	0.8766[20]					0.2				
1-Isopropyl-3-methylbenzene	$C_{10}H_{14}$	134.218	-63.7	175.1	0.8610[20]					0.22				
1-Isopropyl-4-methylbenzene	$C_{10}H_{14}$	134.218	-67.94	177.1	0.8573[20]		2.2322	≈0	1.761	0.19	47	1-6%	436	
Isoquinoline	C_9H_7N	129.159	26.47	243.22	1.0910[30]		11.0	2.73	1.519	0.007				
d-Limonene	$C_{10}H_{16}$	136.234	-74.0	178	0.8411[20]	1.47	2.3746		1.828	0.277	45	0.7-6.1%	237	
l-Limonene	$C_{10}H_{16}$	136.234		178	0.843[20]		2.3738			0.254				
Mesityl oxide	$C_6H_{10}O$	98.142	-59	130	0.8653[20]	0.602	15.6	2.8	2.165	1.47	31	1-7%	344	15
Methacrylic acid	$C_4H_6O_2$	86.090	16	162.5	1.0153[20]			1.65	1.871	0.12	77	1.6-8.8%	68	20
Methanol	CH_4O	32.042	-97.53	64.6	0.7914[20]	0.544	33.0	1.70	2.531	16.9	11	6-36%	464	200
2-Methoxyaniline	C_7H_9NO	123.152	6.2	224	1.0923[20]		5.230			0.013	118			0.1
4-Methoxybenzaldehyde	$C_8H_8O_2$	136.149	0	248	1.119[15]		22.0			0.004				
2-Methoxyethanol	$C_3H_8O_2$	76.095	-85.1	124.1	0.9647[20]		17.2	2.36	2.249	1.31	39	2-14%	285	5
2-Methoxyethyl acetate	$C_5H_{10}O_3$	118.131	-70	143	1.0074[19]		8.25	2.1	2.624	0.67	49	2-12%	392	5
Methyl acetate	$C_3H_6O_2$	74.079	-98.25	56.87	0.9342[20]	0.364	7.07	1.72	1.916	28.8	-10	3-16%	454	200
Methyl acrylate	$C_4H_6O_2$	86.090	<-75	80.7	0.9535[20]		7.03	1.77	1.845	11.0	-3	2.8-25%	468	2
2-Methylacrylonitrile	C_4H_5N	67.090	-35.8	90.3	0.8001[20]			3.69	1.883	8.26	1	2-6.8%		1
2-Methylaniline	C_7H_9N	107.153	-14.41	200.3	0.9984[20]	3.82	6.138	1.6	1.96	0.043	85		482	2
3-Methylaniline	C_7H_9N	107.153	-31.3	203.3	0.9889[20]	3.31	5.816	1.45	2.118	0.036				2
N-Methylaniline	C_7H_9N	107.153	-57	196.2	0.9891[20]	2.04	5.96		1.933	0.05				0.5
Methyl benzoate	$C_8H_8O_2$	136.149	-12.4	199	1.0837[25]	1.857	6.642	1.9	1.625	0.052	83			

Name	Mol. form.	M_r	t_m/°C	t_b/°C	ρ/g mL⁻¹	η/mPa s	ε	μ/D	c_p/J g⁻¹K⁻¹	vp/kPa	FP/°C	Fl. lim.	IT/°C	TLV/ppm
2-Methyl-1,3-butadiene	C_5H_8	68.118	-145.9	34.0	0.679[20]		2.098	0.25	2.240	73.4	-54	1.5-8.9%	395	
Methyl butanoate	$C_5H_{10}O_2$	102.132	-85.8	102.8	0.8984[20]	0.541	5.48		1.941	4.30	14			
3-Methylbutanoic acid	$C_5H_{10}O_2$	102.132	-29.3	176.5	0.931[20]			0.63	1.930	0.067			416	
3-Methyl-1-butanol	$C_5H_{12}O$	88.148	-117.2	131.1	0.8104[20]	3.69	15.63		2.382	0.315	43	1.2-9%	350	100
2-Methyl-2-butanol	$C_5H_{12}O$	88.148	-9.1	102.4	0.8096[20]	3.55	5.78	1.82	2.803	2.19	19	1.2-9%	437	
3-Methyl-2-butanol, (±)-	$C_5H_{12}O$	88.148		112.9	0.8180[20]		12.1			1.20	38			
3-Methyl-2-butanone	$C_5H_{10}O$	86.132	-93.1	94.33	0.8051[20]		10.37		2.089	6.99				200
2-Methyl-1-butene	C_5H_{10}	70.133	-137.53	31.2	0.6504[20]		2.180		2.241	81.4	-20			
2-Methyl-2-butene	C_5H_{10}	70.133	-133.72	38.56	0.6623[20]	0.203	1.979		2.179	62.1	-20			
Methyl tert-butyl ether	$C_5H_{12}O$	88.148	-108.6	55.0	0.7353[25]				2.127	33.6				50
Methyl chloroacetate	$C_3H_5ClO_2$	108.524	-32.1	129.5	1.236[20]		12.0			1.0	57	7.5-18.5%		
Methylcyclohexane	C_7H_{14}	98.186	-126.6	100.93	0.7694[20]	0.679	2.024	≈0	1.882	6.18	-4	1-7%	250	400
Methylcyclopentane	C_6H_{12}	84.159	-142.42	71.8	0.7486[20]	0.479	1.9853	≈0	1.886	18.3	-29	1-8%	258	
N-Methylformamide	C_2H_5NO	59.067	-3.8	199.51	1.011[19]	1.678	189.0	3.83	2.096	0.03				
Methyl formate	$C_2H_4O_2$	60.052	-99	31.7	0.9713[20]	0.325	9.20	1.77	1.983	78.1	-19	5-23%	449	100
5-Methyl-2-hexanone	$C_7H_{14}O$	114.185		144	0.888[20]		13.53			0.691	36	1-8%	191	50
Methylhydrazine	CH_6N_2	46.072	-52.36	87.5					2.928	6.61	-8	2.5-92%	194	0.01
Methyl isocyanate	C_2H_3NO	57.051	-45	39.5	0.9230[27]		21.75	≈2.8		57.7	-7	5.3-26%	534	0.02
Methyl lactate, (±)-	$C_4H_8O_3$	104.105		144.8	1.0928[20]					0.62	49	>2.2%	385	
Methyl methacrylate	$C_5H_8O_2$	100.117	-47.55	100.5	0.9377[25]		6.32	1.67	1.910	5.10	10	1.7-8.2%		50
1-Methylnaphthalene	$C_{11}H_{10}$	142.197	-30.43	244.7	1.0202[20]		2.915	≈0	1.578	0.009			529	
Methyloxirane	C_3H_6O	58.079	-111.9	35	0.859[0]			2.01	2.073	71.7	-37	3.1-27.5%	449	2
2-Methylpentane	C_6H_{14}	86.175	-153.6	60.26	0.650[25]	0.286	1.886	≈0	2.248	28.2	<-29	1-7%	264	500
3-Methylpentane	C_6H_{14}	86.175	-162.90	63.27	0.6598[25]	0.306	1.886	≈0	2.213	25.3	-7	1.2-7%	278	500
2-Methyl-2,4-pentanediol	$C_6H_{14}O_2$	118.174	-50	197.1	0.923[15]		23.4	2.9	2.843	<0.01	102	1-9%	306	25
2-Methyl-1-pentanol	$C_6H_{14}O$	102.174		149	0.8263[20]				2.427	0.236	54	1.1-9.65%	310	
4-Methyl-2-pentanol	$C_6H_{14}O$	102.174	-90	131.6	0.8075[20]	4.07			2.672	0.698	41	1-6%		25
4-Methyl-2-pentanone	$C_6H_{12}O$	100.158	-84	116.5	0.7965[25]	0.545	13.11		2.130	2.64	18	1-8%	448	50
2-Methylpropanenitrile	C_4H_7N	69.106	-71.5	103.9	0.7704[20]		24.42	4.29			8		482	
2-Methyl-2-propanethiol	$C_4H_{10}S$	90.187	-0.5	64.2	0.7943[25]		5.475	1.66		24.2	<-29			
Methyl propanoate	$C_4H_8O_2$	88.106	-87.5	79.8	0.9150[20]	0.431	6.200		1.943	11.5	-2	2.5-13%	469	
2-Methylpropanoic acid	$C_4H_8O_2$	88.106	-46	154.45	0.9681[20]	1.226	2.58	1.08	1.964	0.17	56	2-9.2%	481	
2-Methyl-1-propanol	$C_4H_{10}O$	74.121	-101.9	107.89	0.8018[20]	3.33	17.93	1.64	2.449	1.39	28	2-11%	415	50
2-Methyl-2-propanol	$C_4H_{10}O$	74.121	25.69	82.4	0.7887[20]	4.31	12.47	1.7	2.949	5.52	11	2-8%	478	100
2-Methylpyridine	C_6H_7N	93.127	-66.68	129.38	0.9443[20]		10.18	1.85	1.703	1.5	39		538	
3-Methylpyridine	C_6H_7N	93.127	-18.14	144.14	0.9566[20]		11.10	2.40	1.704	0.795				
4-Methylpyridine	C_6H_7N	93.127	3.67	145.36	0.9548[20]		12.2	2.70	1.707	0.759	57			
N-Methyl-2-pyrrolidone	C_5H_9NO	99.131	-23.09	202	1.0230[25]		32.55	4.1	3.105	0.04	96	1-10%	346	
Methyl salicylate	$C_8H_8O_3$	152.148	-8	222.9	1.181[25]		8.80	2.47	1.637	0.015	96		454	
4-Methylstyrene	C_9H_{10}	118.175	-34.1	172.8	0.9173[25]					0.245	53	0.8-11%	538	50
Morpholine	C_4H_9NO	87.120	-4.8	128	1.0005[20]	2.02	7.42	1.55	1.892	1.34	37	1-11%	290	20
β-Myrcene	$C_{10}H_{16}$	136.234		167	0.8013[15]		2.3			0.280				
Nickel carbonyl	C_4NiO_4	170.734	-19.3	43 (exp 60)	1.31[25]				1.198					0.05
L-Nicotine	$C_{10}H_{14}N_2$	162.231	-79	247	1.0097[20]		8.937							0.1
Nitric acid	HNO_3	63.013	-41.6	83	1.55			2.17	1.744	8.34				2
2-Nitroanisole	$C_7H_7NO_3$	153.136	10.5	272	1.2540[20]		45.75	5.0		0.002				
Nitrobenzene	$C_6H_5NO_2$	123.110	5.7	210.8	1.2037[20]	1.863	35.6	4.22	1.509	0.03	88	2-9%	482	1
Nitroethane	$C_2H_5NO_2$	75.067	-89.5	114.0	1.0448[25]	0.688	29.11	3.23	1.790	2.79	28	3-17%	414	100
Nitromethane	CH_3NO_2	61.041	-28.38	101.19	1.1371[20]	0.630	37.27	3.46	1.746	4.79	35	7-22%	418	20
1-Nitropropane	$C_3H_7NO_2$	89.094	-108	131.1	0.9961[25]	0.798	24.70	3.66	1.97	1.36	36	2%-	421	25
2-Nitropropane	$C_3H_7NO_2$	89.094	-91.3	120.2	0.9821[25]		26.74	3.73	1.911	2.3	24	3-11%	428	10
N-Nitrosodiethylamine	$C_4H_{10}N_2O$	102.134		176.9	0.9422[20]									
N-Nitrosodimethylamine	$C_2H_6N_2O$	74.081		152	1.0048[20]					0.73				
2-Nitrotoluene	$C_7H_7NO_2$	137.137	-10.4	222	1.1611[19]		26.26		1.474	0.0014	106			2
3-Nitrotoluene	$C_7H_7NO_2$	137.137	15.5	232	1.1581[20]		24.95		1.474	0.03	106			2
Nonane	C_9H_{20}	128.255	-53.46	150.82	0.7192[20]	0.665	1.9722	≈0	2.217	0.570	31	0.8-2.9%	205	200
Nonanoic acid	$C_9H_{18}O_2$	158.238	12.4	254.5	0.9052[20]	7.01	2.475	0.79	2.290	0.00005				
1-Nonanol	$C_9H_{20}O$	144.254	-5	213.37	0.8280[20]	9.12	8.83		2.470	0.00050			260	
1-Nonene	C_9H_{18}	126.239	-81.3	146.9	0.7253[25]	0.586	2.180	≈0	2.142	0.714	26			
4-Nonylphenol	$C_{15}H_{24}O$	220.351	42	≈295	0.950[20]									
cis,cis-9,12-Octadecadienoic acid	$C_{18}H_{32}O_2$	280.446	-7		0.9022[20]		2.754							
cis-9-Octadecenoic acid	$C_{18}H_{34}O_2$	282.462	13.4	360	0.8935[20]		2.336	1.18	2.043	0.000001	189		363	
Octane	C_8H_{18}	114.229	-56.82	125.67	0.6986[25]	0.508	1.948	≈0	2.229	1.86	13	1-7%	206	300
Octanoic acid	$C_8H_{16}O_2$	144.212	16.5	239	0.9073[25]	5.02	2.85	1.15	2.066	0.0002				
1-Octanol	$C_8H_{18}O$	130.228	-14.8	195.16	0.8262[25]	7.29	10.30	1.8	2.344	0.01	81		270	
2-Octanol	$C_8H_{18}O$	130.228	-31.6	179.3	0.8193[20]	6.49	8.13	1.71	2.535		88		265	

Name	Mol. form.	M_r	t_m/°C	t_b/°C	ρ/g mL⁻¹	η/mPa s	ε	μ/D	c_p/J g⁻¹K⁻¹	vp/kPa	FP/°C	Fl. lim.	IT/°C	TLV/ppm
2-Octanone	$C_8H_{16}O$	128.212	-16	172.5	0.820[20]		9.51	2.7	2.132	0.12	52			
1-Octene	C_8H_{16}	112.213	-101.7	121.29	0.7149[20]	0.447	2.113	≈ 0	2.148	2.30	21		230	
Oxetane	C_3H_6O	58.079	-97	47.6	0.8930[25]			1.94						
2-Oxetanone	$C_3H_4O_2$	72.063	-33.4	162	1.1460[20]			4.18	1.694	0.3	74	>2.9%		0.5
Oxirane	C_2H_4O	44.052	-112.5	10.6	0.8821[10]		12.42	1.89	1.998	175	-20	3-100%	429	1
Oxiranemethanol, (±)-	$C_3H_6O_2$	74.079	-45	167 dec	1.1143[25]									2
Paraldehyde	$C_6H_{12}O_3$	132.157	12.6	124.3	0.9943[20]	1.079		1.43		1.6	36	>1.3%	238	
Parathion	$C_{10}H_{14}NO_5PS$	291.261	6.1	375	1.2681[20]									0.01
Pentachloroethane	C_2HCl_5	202.294	-28.78	162.0	1.6796[20]	2.25	3.716	0.92	0.859	0.478				
cis-1,3-Pentadiene	C_5H_8	68.118	-140.8	44.1	0.6910[20]		2.319	0.500		50.6				
trans-1,3-Pentadiene	C_5H_8	68.118	-87.4	42	0.6710[25]			0.585		54.7				
Pentanal	$C_5H_{10}O$	86.132	-91.5	103	0.8095[20]		10.00			4.58	12		222	50
Pentane	C_5H_{12}	72.149	-129.67	36.06	0.6262[20]	0.224	1.8371	≈ 0	2.317	68.3	-40	2-8%	260	600
Pentanedial	$C_5H_8O_2$	100.117	-14	188 dec										0.05
1,5-Pentanediol	$C_5H_{12}O_2$	104.148	-18	239	0.9914[20]		26.2	2.5	3.08	0.001	129		335	
2,4-Pentanedione	$C_5H_8O_2$	100.117	-23	138	0.9721[25]		26.524	2.8	2.08	1.02	34		340	
1-Pentanethiol	$C_5H_{12}S$	104.214	-75.65	126.6	0.850[20]		4.847			1.83	18			
Pentanoic acid	$C_5H_{10}O_2$	102.132	-33.6	186.1	0.9339[25]		2.661	1.61	2.059	0.024	96		400	
1-Pentanol	$C_5H_{12}O$	88.148	-77.6	137.98	0.8144[20]	3.62	15.13	1.7	2.361	0.259	33	1-10%	300	
2-Pentanol	$C_5H_{12}O$	88.148	-73	119.3	0.8094[20]	3.47	13.71	1.66	2.716	0.804	34	1.2-9%	343	
3-Pentanol	$C_5H_{12}O$	88.148	-69	116.25	0.8203[20]	4.15	13.35	1.64	2.719	1.10	41	1.2-9%	435	
2-Pentanone	$C_5H_{10}O$	86.132	-76.8	102.26	0.809[20]	0.470	15.45	2.7	2.137	4.97	7	2-8%	452	200
3-Pentanone	$C_5H_{10}O$	86.132	-39	101.7	0.8098[25]	0.444	17.00	2.82	2.216	4.72	13	>1.6%	450	200
1-Pentene	C_5H_{10}	70.133	-165.12	29.96	0.6405[20]	0.195	2.011	≈ 0.5	2.196	85.0	-18	1.5-8.7%	275	
cis-2-Pentene	C_5H_{10}	70.133	-151.36	36.93	0.6556[20]			≈ 0	2.163	66.0	<-20			
trans-2-Pentene	C_5H_{10}	70.133	-140.21	36.34	0.6431[25]			≈ 0	2.239	67.4	<-20			
Pentyl acetate	$C_7H_{14}O_2$	130.185	-70.8	149.2	0.8756[20]		4.79	1.75	2.005	0.60	16	1-8%	360	50
Pentylamine	$C_5H_{13}N$	87.164	-55	104.3	0.7544[20]	0.702	4.27		2.501	4.00	-1	2.2-22%		
Perchloric acid	$ClHO_4$	100.459	-112	≈ 90 dec	1.77									
Peroxyacetic acid	$C_2H_4O_3$	76.051	-0.2	110	1.226[15]					1.93	41			
Phenol	C_6H_6O	94.111	40.89	181.87	1.0545[45]		12.40	1.224	2.123	0.055	79	1.8-8.6%	715	5
2-Phenoxyethanol	$C_8H_{10}O_2$	138.164	14	245	1.102[22]					0.001	121			
Phenylhydrazine	$C_6H_8N_2$	108.141	20.6	243.5	1.0986[20]	13.03	7.15		2.007	0.003	88			0.1
1-Phenyl-2-propylamine, (±)-	$C_9H_{13}N$	135.206		203	0.9306[25]					0.06	<100			
Phosphinic acid	H_3O_2P	65.997	26.5	130	1.49									
Phosphoric acid	H_3O_4P	97.995	42.4	407					1.480					0.25
Phosphorothioc trichloride	Cl_3PS	169.398	-36.2	125	1.635		4.94							
Phosphorus(III) bromide	Br_3P	270.686	-41.5	173.2	2.8					0.38				
Phosphorus(III) chloride	Cl_3P	137.332	-93.6	76.1	1.574	0.529	3.498	0.56		16.1				0.2
Phosphoryl chloride	Cl_3OP	153.331	1.18	105.5	1.645		14.1	2.54	0.905	4.97				0.1
α-Pinene	$C_{10}H_{16}$	136.234	-64	156.2	0.8539[25]		2.1787			0.64	33		255	
β-Pinene	$C_{10}H_{16}$	136.234	-61.5	166	0.860[25]		2.4970			0.61	38		275	
Piperidine	$C_5H_{11}N$	85.148	-11.02	106.22	0.8606[20]	1.573	4.33	1.2	2.113	4.28	16	1-10%		
Propanal	C_3H_6O	58.079	-80	48	0.8657[25]	0.321	18.5	2.72	2.362	42.2	-30	2.6-17%	207	20
1,2-Propanediol	$C_3H_8O_2$	76.095	-60	187.6	1.0361[20]	40.4	27.5	2.2	2.507	0.02	99	3-13%	371	
1,3-Propanediol	$C_3H_8O_2$	76.095	-27.7	214.4	1.0538[20]		35.1	2.5		0.007			400	
Propanenitrile	C_3H_5N	55.079	-92.78	97.14	0.7818[20]	0.294	29.7	4.05	2.166	6.14	2	3-14%	512	
Propanoic acid	$C_3H_6O_2$	74.079	-20.5	141.15	0.9882[25]	1.030	3.44	1.75	2.063	0.553	52	2.9-12.1%	465	10
Propanoic anhydride	$C_6H_{10}O_3$	130.141	-45	170	1.0110[20]		18.30		1.806	0.45	63	1.3-9.5%	285	
1-Propanol	C_3H_8O	60.095	-124.39	97.2	0.7997[25]	1.945	20.8	1.55	2.395	2.76	23	2-14%	412	200
2-Propanol	C_3H_8O	60.095	-87.9	82.3	0.7809[25]	2.04	20.18	1.56	2.604	6.02	12	2-13%	399	200
Propargyl alcohol	C_3H_4O	56.063	-51.8	113.6	0.9478[20]		20.8	1.13			36			1
Propyl acetate	$C_5H_{10}O_2$	102.132	-93	101.54	0.8878[20]	0.544	5.62	1.8	1.921	4.49	13	2-8%	450	200
Propylamine	C_3H_9N	59.110	-84.75	47.22	0.7173[20]	0.376	5.08	1.17	2.776	42.1	-37	2-10%	318	
Propylbenzene	C_9H_{12}	120.191	-99.6	159.24	0.8593[25]		2.370	≈ 0	1.786	0.45	30	1-6%	450	
Propyl butanoate	$C_7H_{14}O_2$	130.185	-95.2	143.0	0.8730[20]		4.3			0.618	37			
Propylene carbonate	$C_4H_6O_3$	102.089	-48.8	242	1.2047[20]		66.14	4.9	2.141	0.05	135			
Propyl formate	$C_4H_8O_2$	88.106	-92.9	80.9	0.9073[20]	0.485	6.92	1.89	1.945	10.9	-3		455	
Propyl propanoate	$C_6H_{12}O_2$	116.158	-75.9	122.5	0.8809[20]		5.249			1.88	79			
Pyridine	C_5H_5N	79.101	-41.70	115.23	0.9819[20]	0.879	13.260	2.21	1.678	2.76	20	2-12%	482	1
Pyrrole	C_4H_5N	67.090	-23.39	129.79	0.9698[20]	1.225	8.00	1.74	1.903	1.10	39			
Pyrrolidine	C_4H_9N	71.121	-57.79	86.56	0.8586[20]	0.704	8.30	1.6	2.202	8.40	3			
2-Pyrrolidone	C_4H_7NO	85.105	25	251	1.120[20]		28.18	3.5	1.99		129			
Quinoline	C_9H_7N	129.159	-14.78	237.16	1.0977[15]	3.34	9.16	2.29	1.51	0.011			480	
Safrole	$C_{10}H_{10}O_2$	162.185	11.2	234.5	1.1000[25]					0.01	100			
Salicylaldehyde	$C_7H_6O_2$	122.122	-7	197	1.1674[20]		18.35	2.86	1.818	0.075	78			

Name	Mol. form.	M_r	t_m/°C	t_b/°C	ρ/g mL⁻¹	η/mPa s	ε	μ/D	c_p/J g⁻¹K⁻¹	vp/kPa	FP/°C	Fl. lim.	IT/°C	TLV/ppm
Selenium chloride	Cl₂Se₂	228.83	-85	130 dec	2.774									
Selenium oxychloride	Cl₂OSe	165.86	8.5	177	2.44		46.2			0.02				
Selenium oxyfluoride	F₂OSe	132.96	15	125	2.8					0.56				
Styrene	C₈H₈	104.150	-30.65	145	0.9016²⁵	0.695	2.4737	0.123	1.747	0.81	31	1-7%	490	20
Sulfolane	C₄H₈O₂S	120.171	27.6	287.3	1.2723¹⁸		43.26	4.8	1.498	<0.01	177			
Sulfur chloride	Cl₂S₂	135.037	-77	137	1.69		4.79			1.27				1
Sulfur dichloride	Cl₂S	102.971	-122	59.6	1.62		2.915	0.36		17.9				
Sulfuric acid	H₂O₄S	98.080	10.31	337	1.8				1.416					0.05
Sulfuryl chloride	Cl₂O₂S	134.970	-51	69.4	1.680		9.1	1.81	0.993	18.7				
α-Terpinene	C₁₀H₁₆	136.234		174	0.8375¹⁹		2.4526							
1,1,2,2-Tetrabromoethane	C₂H₂Br₄	345.653	0	243.5	2.9655²⁰		6.72	1.38	0.479	0.003			335	1
Tetrabromosilane	Br₄Si	347.702	5.39	154	2.8			0						
1,1,2,2-Tetrachloro-1,2-difluoroethane	C₂Cl₄F₂	203.830	24.8	92.8	1.5951⁵⁰		2.52		0.852	7.51				500
1,1,1,2-Tetrachloroethane	C₂H₂Cl₄	167.849	-70.2	130.2	1.5406²⁰	1.437			0.92	1.6	47	5-12%		
1,1,2,2-Tetrachloroethane	C₂H₂Cl₄	167.849	-42.4	145.2	1.5953²⁰		8.50	1.32	0.967	0.622	62	20-54%		1
Tetrachloroethene	C₂Cl₄	165.833	-22.3	121.3	1.6230²⁰	0.844	2.268	0	0.865	2.42	45			25
Tetrachloromethane	CCl₄	153.823	-22.62	76.8	1.5940²⁰	0.908	2.2379	0	0.850	15.2				5
Tetrachlorosilane	Cl₄Si	169.897	-68.74	57.65	1.5	99.4		0	0.855	31.3				
Tetradecane	C₁₄H₃₀	198.388	5.82	253.58	0.7596²⁰	2.13	2.0343	≈0		0.002	112	>0.5%	200	
Tetraethylene glycol	C₈H₁₈O₅	194.226	-6.2	328	1.1285¹⁵		20.44		2.208	0.000001	182			
Tetrafluoroboric acid	BF₄H	87.813		130 dec	~1.8									
Tetrahydrofuran	C₄H₈O	72.106	-108.44	65	0.8833²⁵	0.456	7.52	1.75	1.720	21.6	-14	2-12%	321	200
Tetrahydrofurfuryl alcohol	C₅H₁₀O₂	102.132	<-80	178	1.0524²⁰		13.48	2.1	1.774	0.100	75	1.5-9.7%	282	
1,2,3,4-Tetrahydronaphthalene	C₁₀H₁₂	132.202	-35.7	207.6	0.9645²⁵	2.14	2.771	≈0	1.645	0.05	71	1-5%	385	
Tetrahydropyran	C₅H₁₀O	86.132	-49.1	88	0.8814²⁰		5.66	1.74	1.82	9.54	-20			
Tetrahydrothiophene	C₄H₈S	88.172	-96.2	121.1	0.9987²⁰	0.973		1.90		2.45				
Tetramethylsilane	C₄H₁₂Si	88.224	-99.06	26.6	0.648¹⁹		1.921	0	2.313	94.2				
Tetramethylurea	C₅H₁₂N₂O	116.161	-0.6	176.5	0.9687²⁰		23.10	3.5		0.138	77			
Tetranitromethane	CN₄O₈	196.033	13.8	126.1	1.6380²⁰		2.317	0		1.13				0.005
Thionyl bromide	Br₂OS	207.873	-50	140			9.06			0.84				
Thionyl chloride	Cl₂OS	118.970	-101	75.6	1.631		8.675	1.45	1.017	16.0				1
Thiophene	C₄H₄S	84.140	-38.21	84.0	1.0649²⁰		2.739	0.55	1.471	10.6	-1			
Tin(IV) chloride	Cl₄Sn	260.521	-34.07	114.15	2.234			0	0.634					
Titanium(IV) chloride	Cl₄Ti	189.678	-24.12	136.45	1.73				0.766					
Toluene	C₇H₈	92.139	-94.95	110.63	0.8668²⁰	0.560	2.379	0.37	1.707	3.79	4	1-7%	480	50
Toluene-2,4-diisocyanate	C₉H₆N₂O₂	174.156	20.5	251	1.2244²⁰		8.433		1.653	0.003	127	0.9-9.5%		0.005
Tribromomethane	CHBr₃	252.731	8.69	149.1	2.8788²⁵	1.857	4.404	0.99	0.517	0.726	83			0.5
Tributylamine	C₁₂H₂₇N	185.349	-70	216.5	0.7770²⁰		2.340	0.8		0.01	63	1-5%		
Tributyl borate	C₁₂H₂₇BO₃	230.151	<-70	234	0.8567²⁰		2.23	0.77			93			
Tributyrin	C₁₅H₂₆O₆	302.363	-75	307.5	1.0350²⁰		5.72		1.837		180	>0.5%	407	
Trichloroacetaldehyde	C₂HCl₃O	147.387	-57.5	97.8	1.512²⁰		6.8		1.025	6.66				
1,2,4-Trichlorobenzene	C₆H₃Cl₃	181.447	16.92	213.5	1.459²⁵					0.057	105	2.5-6.6%	571	5
1,1,1-Trichloroethane	C₂H₃Cl₃	133.404	-30.01	74.09	1.3390²⁰	0.793	7.243	1.76	1.082	16.5	-1	8-13%	500	350
1,1,2-Trichloroethane	C₂H₃Cl₃	133.404	-36.3	113.8	1.4397²⁰		7.1937	1.4	1.131	3.1	32	6-28%	460	10
Trichloroethene	C₂HCl₃	131.388	-84.7	87.21	1.4642²⁰	0.545	3.390	0.8	0.947	9.91	32	8-11%	420	50
Trichloroethylsilane	C₂H₅Cl₃Si	163.506	-105.6	100.5	1.2373²⁰			2.04		6.29	22			
Trichlorofluoromethane	CCl₃F	137.368	-110.44	23.7	1.4879²⁰	0.421	3.00	0.46	0.885	106				1000
Trichloromethane	CHCl₃	119.378	-63.41	61.17	1.4788²⁵	0.537	4.8069	1.04	0.957	26.2				10
(Trichloromethyl)benzene	C₇H₅Cl₃	195.474	-4.42	221	1.3723²⁰		6.9	2.03		0.35	127		211	0.1
Trichloromethylsilane	CH₃Cl₃Si	149.480	-90	65.6	1.273²⁰			1.91	1.091	22.5	-9	7.6->20%	>404	
Trichloronitromethane	CCl₃NO₂	164.376	-64	112	1.6558²⁰		7.319			3.18				0.1
1,2,3-Trichloropropane	C₃H₅Cl₃	147.431	-14.7	157	1.3889²⁰		7.5		1.245	0.492	71	3.2-12.6%		10
Trichlorosilane	Cl₃HSi	135.452	-128.2	33	1.331	0.326		0.86			-50		104	
1,1,2-Trichloro-1,2,2-trifluoroethane	C₂Cl₃F₃	187.375	-36.22	47.7	1.5635²⁵	0.656	2.41		0.908	44.8				1000
Tri-o-cresyl phosphate	C₂₁H₂₁O₄P	368.363	11	410	1.1955²⁰		6.7	2.87	1.57	0.0000002	225		385	0.01
Tridecane	C₁₃H₂₈	184.361	-5.4	235.47	0.7564²⁰	1.724	2.0213	≈0	2.206	0.005	79			
1-Tridecene	C₁₃H₂₆	182.345	-13	232.8	0.7658²⁰	1.50	2.139	≈0	2.149	0.0047	79			
Triethanolamine	C₆H₁₅NO₃	149.188	20.5	335.4	1.1242²⁰	609	29.36	3.6	2.61	<0.01	179	1-10%		0.8
Triethylamine	C₆H₁₅N	101.190	-114.7	89	0.7275²⁰	0.347	2.418	0.66	2.173	7.70	-7	1-8%	249	1
Triethylene glycol	C₆H₁₄O₄	150.173	-7	285	1.1274¹⁵		23.69		2.18	0.0002	177	1-9%	371	
Triethylene glycol dimethyl ether	C₈H₁₈O₄	178.227	-45	216	0.986²⁰		7.62				111			
Triethyl phosphate	C₆H₁₅O₄P	182.154	-56.4	215.5	1.0695²⁰		13.20	3.1			115		454	
Trifluoroacetic acid	C₂HF₃O₂	114.023	-15.2	73	1.5351²⁵	0.808	8.42	2.28		15.1				
(Trifluoromethyl)benzene	C₇H₅F₃	146.110	-28.95	102.1	1.1884²⁰		9.22	2.86	1.289	5.14	12			

Name	Mol. form.	M_r	t_m/°C	t_b/°C	ρ/g mL⁻¹	η/mPa s	ε	μ/D	c_p/J g⁻¹K⁻¹	vp/kPa	FP/°C	Fl. lim.	IT/°C	TLV/ppm
1,2,3-Trimethylbenzene	C₉H₁₂	120.191	-25.4	176.12	0.8944[20]		2.656	≈ 0	1.800	0.20	44	0.8-6.6%	470	25
1,2,4-Trimethylbenzene	C₉H₁₂	120.191	-43.77	169.38	0.8758[20]		2.377	≈ 0	1.789	0.30	44	1-6%	500	25
1,3,5-Trimethylbenzene	C₉H₁₂	120.191	-44.72	164.74	0.8615[25]		2.279	0	1.741	0.33	50	1-5%	559	25
Trimethyl borate	C₃H₉BO₃	103.912	-29.3	67.5	0.915[25]		2.2762		1.828	17.2	-8			
Trimethylchlorosilane	C₃H₉ClSi	108.642	-40	60	0.856[25]					30.7	-28		395	
2,2,4-Trimethylpentane	C₈H₁₈	114.229	-107.3	99.22	0.6878[25]		1.943	≈ 0	2.093	6.50	-12		418	300
2,3,3-Trimethylpentane	C₈H₁₈	114.229	-100.9	114.8	0.7262[20]		1.9780	≈ 0	2.150	3.60	<21		425	300
Trimethyl phosphate	C₃H₉O₄P	140.074	-46	197.2	1.2144[20]		20.6	3.2		0.11	107			
2,4,6-Trimethylpyridine	C₈H₁₁N	121.180	-46	170.6	0.9166[22]		7.807	2.05		4.1				
Trinitroglycerol	C₃H₅N₃O₉	227.087	13.5	exp 218	1.5931[20]		19.25			0.00005			270	0.05
Undecane	C₁₁H₂₄	156.309	-25.5	195.9	0.7402[20]	1.098	1.9972	≈ 0	2.207	0.05	69			
Vanadium(IV) chloride	Cl₄V	192.753	-25.7	148	1.816		3.05							
Vanadyl trichloride	Cl₃OV	173.299	-79	127	1.829		3.4							
Vinyl acetate	C₄H₆O₂	86.090	-93.2	72.8	0.9256[25]			1.79	1.969	15.4	-8	2.6-13.4%	402	10
4-Vinylcyclohexene	C₈H₁₂	108.181	-108.9	128	0.8299[20]					1.87	16		269	0.1
Water	H₂O	18.015	0.00	100.0	0.9970	0.890	80.100	1.8546	4.180	3.17				
o-Xylene	C₈H₁₀	106.165	-25.2	144.5	0.8802[10]	0.760	2.562	0.64	1.753	0.88	32	1-7%	463	100
m-Xylene	C₈H₁₀	106.165	-47.8	139.12	0.8596[25]	0.581	2.359	≈ 0	1.724	1.13	27	1-7%	527	100
p-Xylene	C₈H₁₀	106.165	13.25	138.37	0.8566[25]	0.603	2.2735	0	1.710	1.19	27	1-7%	528	100
2,4-Xylenol	C₈H₁₀O	122.164	24.5	210.98	0.9650[20]		5.060	1.4		0.022				

The chart below gives qualitative information on the miscibility of pairs of organic liquids. Two liquids are considered miscible (indicated by **M** in the chart) if mixing equal volumes produces a single liquid phase. If two phases separate, they are considered immiscible (**I**). An entry of **P** indicates two phases whose volumes differ appreciably, suggesting a partial miscibility of the components. The symbol **R** indicates a reaction between the components. All data refer to room temperature.

The codes for the columns are:

A Acetone	**J** Diethyl ether	**S** Methyl isopropyl ketone
B Benzaldehyde	**K** *N,N*-Dimethylaniline	**T** Nitromethane
C Benzene	**L** Dipentylamine	**U** 1-Octanol
D Butyl acetate	**M** Ethyl alcohol	**V** 1,3-Propanediol
E Butyl alcohol	**N** Ethylene glycol	**W** Pyridine
F Carbon tetrachloride	**O** Ethylene glycol monoethyl ether	**X** Triethylenetetramine
G 2-Chloroethanol	**P** Formamide	**Y** Triethyl phosphate
H Chloroform	**Q** Furfuryl alcohol	
I *o*-Cresol	**R** Glycerol	

References

1. Drury, J. S., *Ind. Eng. Chem.* 44, 2744, 1959.
2. Jackson, W. M., and Drury, J. S., *Ind. Eng. Chem.* 51, 1491, 1959.

	A	B	C	D	E	F	G	H	I	J	K	L	M	N	O	P	Q	R	S	T	U	V	W	X	Y
Acetone	-	M	M	M	M	M	M	M		M	M	M	M	M	M	M	M	I	M	M	M	M	M	M	M
Adiponitrile	M		M		M	I				I	M		M	I		M	M	I							
2-Amino-2-methyl-1- propanol	M	M	M	M	M	M	M				M	M		M	M		M	M	M			M	M	M	M
p-Anisaldehyde							M						I	M			I				I				M
Benzaldehyde	M	-	M	M	M	M				M	M		M	P		M	M	P		M	M	M	M		
Benzene	M	M	-	M	M	M	M	M		M	M	M	M	I	M	I	M	I	M	I	M	I	M	M	M
Benzonitrile	M		M	M	M					M	M		M	I		I	M	I	M			I	M		
Benzothiazole	M		M	M	M					M	M		M	M		I	M	I				M	M		
Benzyl alcohol	M	M	M	M	M	M				M	M		M	M		M	M	M			M	M	M	M	
Benzyl mercaptan	M		M	M	M					M	M		M	I		I	M	I				I	M		
2-Bromoethyl acetate	M		P				M				R	M					M							R	
1,3-Butanediol	M		I				M	M	P		M	M					M					M	M	
2,3-Butanediol	M		P				M	M	M		M	M					M					M	M	
Butyl acetate	M	M	M	-	M	M				M	M		M	P		I	M	I			M	M	P	M	
Butyl alcohol	M	M	M	M	-	M				M	M		M	M		M	M	M			M	M	M	M	
Carbon tetrachloride	M	M	M	M	M	-				M	M		M	I		I	M	I			M	M	I	M	
2-Chloroethanol	M		M				-			M	M		M					M				M	M	M	
Chloroform	M		M			M		-		M	M		M	P	M			I	M			M	M	M	
3-Chloro-1,2-propanediol	M		I				M	M			R	M					M					M	R	
Cinnamaldehyde	M		M				M	M			M	M		I	M			I	M			I		R	M
o-Cresol			M		-					M	M		M				M				M			M
Diacetone alcohol	M	M	P	M	M	P				M	M		M	M		M	M	I	M		M	M	M	
Dibenzyl ether	M	M			M	M				M			I	M			I	M				M	M	
Dibutylamine					R						M	M				P						M		
Dibutyl carbonate	M	M			M		M			M	M						M						I	
Dibutyl ether	M	M	M	M	M	M	M	M		M	M		I			I	M	I	M		I	M	I	M	
Diethanolamine	M	I	I	I	I	M	I			I	P		M	M		M	M	M			I	M	M	M
Diethylacetic acid	M		M			M	M				R	M	M				I	M					R	M
Diethylene glycol dibutyl ether	M	M			M	M	M				R	M					M							M
Diethylene glycol diethyl ether	M	M			M	M	M				M	M					M						M	M
Diethylene glycol monobutyl ether	M	M			M	M	M				M	M					M						M	M
Diethylene glycol monoethyl ether	M	M			M	M	M				M	M					M						M	M
Diethylene glycol monomethyl ether	M	M			M	M	M				M	M					M						M	M
Diethylenetriamine	M	M			R		I			I	M	M					M	R					M	M
Diethyl ether	M	M	M	M	M	M	M	M		M	-	M	M	M	I	M	I	M	I	M	M	M	M	I	M	M	M
Diethylformamide	M	M			M	M				R	M	M					M	M				M		R	M
Dihexyl ether	M	M			M	M				M	M	I	M				I	M				I		I	

	A	B	C	D	E	F	G	H	I	J	K	L	M	N	O	P	Q	R	S	T	U	V	W	X	Y
Diisobutyl ketone	M		M			M	M				M	M	I	M			I	M			I			M	M
Diisopropylamine	M		M				R	M				M	M	M	M			M	M			M		M	M
N,N-Dimethylaniline	M	M	M	M	M	M				M	-		M	I		I	M	I		M	M	I	M		
Dipentylamine	M		M				M		M		-	M	P	M			P	M			M			I	M
N,N-Dipropylaniline	M		M		M	M	M	M			M	M	M	M	I	M	I	M	I	M			I	M	M
Dipropylene glycol	M		M			M	M	M	M			M						M						M	M
Ethyl alcohol	M	M	M	M	M	M		M			M	M	-	M	M	M	M	M	M	M	M	M	M	M	M
Ethyl benzoate	M	M	M	M	M	M		M			M	M	M	I		I	M	I	M	M	M	P		M	M
Ethyl chloroacetate	M		M			M	M				M	M	I	M			I	M			I			R	M
Ethyl cinnamate	M		M				M				M	M	I	M			I	M			I			M	M
Ethylene glycol	M	P	I	P	M	I		P	M	I	I	P	M	-			M	M	M	I	I	M	M	M	M
Ethylene glycol monobutyl ether	M		M			M	M	M		M	M	M						M						M	M
Ethylene glycol monoethyl ether	M		M			M	M	M		M	M	M	-					M						M	M
Ethylene glycol monomethyl ether	M		M			M	M	M		M	M	M												M	M
2-Ethyl-1-hexanol	M	M	M	M	M			M	M		M	M					I	M	I		I	M	M	M	
Ethyl phenylacetate	M		M			M	M				M		I	M			I	M			I			M	M
Ethyl thiocyanate	M		M			M	M			M	M		I	M		I	M	I			I		M		
Formamide	M	M	I	I	M	I				I	I		M	M		-	M	M		M	I	M	M		
Furfuryl alcohol	M	M	M	M	M			M	M		M	M		M	-	M		M	M	M	M				
Glycerol	I	P	I	I	M	I		I	M	I	I	P	M	M		M	M	-	I	I	I	M	M	M	
1-Heptadecanol	M		M			M					M	M						M							M
3-Heptanol	M		M			M	M				M	M	M	M			I				M			M	M
Heptyl acetate	M		M			M	M				M		I	M			I	M			I			R	M
Hexanenitrile	M		M			M	M				M	M	I	M			I	M			I			M	M
Isobutyl mercaptan	M		M		M	M			M	M		M	I		I	M	I				R	M			
Isopentyl acetate	M		M			M	M				M	M	I	M			I	M			I			M	M
Isopentyl alcohol	M	M	M	M	M	M			M	M		M	M		M	M	I		M	M	M	M			
Isopentyl sulfide	M		M		M	M			M	M		M	I		I		I	I			I	M			
Methyl disulfide	M		M		M	M			M	M		M	I		I	M	I				R	M			
Methyl isobutyl ketone	M	M	M	M	M	M			M	M		M	I			P	M	I			M	M	I	M	
Methyl isopropyl ketone	M		M			M	M		M		M	M	M	I	M		I	-			M			R	M
4-Methylpentanoic acid	M		M				M				M	M	M	M			I	M			M			R	M
Nitromethane	M	M	I	M	M	M			M	M		M	I		M	M	I		-		P	I	M		
1-Octanol	M	M	M	M	M	M			M	M		M	M		I	M	I		P	-	M	M			
o-Phenetidine	M		M			M	M				M	M	M	M				M				M			
1,2-Propanediol	M	I				M	M	P			M	M						M						M	M
1,3-Propanediol	M	M	I	P	M	I			M	M	I	M	M		M	M	M	M	I	M	-		M	M	
Pyridine	M	M	M	M	M	M	M		M	M		M	M	M	M	M	M		M	M	M	-			M
Tetradecanol	M		M			M	M				M	M	I	M			I	M			P			M	M
Tributyl phosphate	M		M			M	M				M	M	P	M			I	M			M			M	M
Triethylene glycol	M	P				M	M	I		P	M							M					M	M	
Triethylenetetramine	M		M			M	M		M		I	M	M				M	R			M		-		M
Triethyl phosphate	M		M			M	M	M			M	M						M					M	M	-
2,6,8-Trimethyl-4-nonanone	M		M			M	M				M	M	I	M			I	M			I		I		M

DENSITY OF SOLVENTS AS A FUNCTION OF TEMPERATURE

The table below lists the density of several common solvents in the temperature range from 0 °C to 100 °C. The values have been calculated from the Rackett Equation using parameters in the reference. Density values refer to the liquid at its saturation vapor pressure; thus entries for temperatures above the normal boiling point are for pressures greater than atmospheric.

Reference

Lide, D. R., and Kehiaian, H. V., *Handbook of Thermophysical and Thermochemical Data*, CRC Press, Boca Raton, FL, 1994.

					Density in g/mL						
Solvent	0 °C	10 °C	20 °C	30 °C	40 °C	50 °C	60 °C	70 °C	80 °C	90 °C	100 °C
Acetic acid			1.051	1.038	1.025	1.012	0.9993	0.9861	0.9728	0.9592	0.9454
Acetone	0.8129	0.8016	0.7902	0.7785	0.7666	0.7545	0.7421	0.7293	0.7163	0.7029	0.6890
Acetonitrile			0.7825	0.7707	0.7591	0.7473	0.7353	0.7231	0.7106	0.6980	0.6851
Aniline	1.041	1.033	1.025	1.016	1.008	1.000	0.9909	0.9823	0.9735	0.9646	0.9557
Benzene		0.8884	0.8786	0.8686	0.8584	0.8481	0.8376	0.8269	0.8160	0.8049	0.7935
1-Butanol	0.8293	0.8200	0.8105	0.8009	0.7912	0.7812	0.7712	0.7609	0.7504	0.7398	0.7289
Butylamine	0.7606	0.7512	0.7417	0.7320	0.7221	0.7120	0.7017	0.6911	0.6803	0.6693	0.6579
Carbon disulfide	1.290	1.277	1.263	1.248	1.234						
Chlorobenzene	1.127	1.116	1.106	1.096	1.085	1.074	1.064	1.053	1.042	1.030	1.019
Cyclohexane		0.7872	0.7784	0.7694	0.7602	0.7509	0.7414	0.7317	0.7218	0.7117	0.7013
Decane	0.7447	0.7374	0.7301	0.7226	0.7151	0.7074	0.6997	0.6919	0.6839	0.6758	0.6676
1-Decanol			0.8294	0.8229	0.8162	0.8093	0.8024	0.7955	0.7884	0.7813	0.7740
Dichloromethane	1.362	1.344	1.326	1.307	1.289	1.269	1.250	1.229	1.208	1.187	1.165
Diethyl ether	0.7368	0.7254	0.7137	0.7018	0.6896	0.6770	0.6639	0.6505	0.6366	0.6220	0.6068
N,N-Dimethylaniline		0.9638	0.9562	0.9483	0.9401	0.9318	0.9234	0.9150	0.9064	0.8978	0.8890
Ethanol	0.8121	0.8014	0.7905	0.7793	0.7680	0.7564	0.7446	0.7324	0.7200	0.7073	0.6942
Ethyl acetate	0.9245	0.9126	0.9006	0.8884	0.8759	0.8632	0.8503	0.8370	0.8234	0.8095	0.7952
Ethylbenzene	0.8836	0.8753	0.8668	0.8582	0.8495	0.8407	0.8318	0.8228	0.8136	0.8043	0.7948
Ethyl formate	0.9472	0.9346	0.9218	0.9087	0.8954	0.8818	0.8678	0.8535	0.8389	0.8238	0.8082
Ethyl propanoate	0.9113	0.9005	0.8895	0.8784	0.8671	0.8556	0.8439	0.8319	0.8197	0.8072	0.7944
Heptane	0.7004	0.6921	0.6837	0.6751	0.6664	0.6575	0.6485	0.6393	0.6298	0.6202	0.6102
Hexane	0.6774	0.6685	0.6594	0.6502	0.6407	0.6311	0.6212	0.6111	0.6006	0.5899	0.5789
1-Hexanol	0.8359	0.8278	0.8195	0.8111	0.8027	0.7941	0.7854	0.7766	0.7676	0.7585	0.7492
Isopropylbenzene	0.8769	0.8696	0.8615	0.8533	0.8450	0.8366	0.8280	0.8194	0.8106	0.8017	0.7927
Methanol	0.8157	0.8042	0.7925	0.7807	0.7685	0.7562	0.7435	0.7306	0.7174	0.7038	0.6898
Methyl acetate	0.9606	0.9478	0.9346	0.9211	0.9074	0.8933	0.8790	0.8643	0.8491	0.8336	0.8176
N-Methylaniline	1.0010	0.9933	0.9859	0.9785	0.9709	0.9633	0.9556	0.9478	0.9399	0.9319	0.9239
Methylcyclohexane	0.7858	0.7776	0.7693	0.7608	0.7522	0.7435	0.7346	0.7255	0.7163	0.7069	0.6973
Methyl formate	1.003	0.9887	0.9739	0.9588	0.9433	0.9275	0.9112	0.8945	0.8772	0.8594	0.8409
Methyl propanoate	0.9383	0.9268	0.9150	0.9030	0.8907	0.8783	0.8656	0.8526	0.8393	0.8257	0.8117
Nitromethane			1.139	1.125	1.111	1.097	1.083	1.069	1.055	1.040	1.026
Nonane	0.7327	0.7252	0.7176	0.7099	0.7021	0.6941	0.6861	0.6779	0.6696	0.6611	0.6525
Octane	0.7185	0.7106	0.7027	0.6945	0.6863	0.6779	0.6694	0.6608	0.6520	0.6430	0.6338
Pentanoic acid	0.9563	0.9476	0.9389	0.9301	0.9211	0.9121	0.9029	0.8937	0.8843	0.8748	0.8652
1-Propanol	0.8252	0.8151	0.8048	0.7943	0.7837	0.7729	0.7619	0.7506	0.7391	0.7273	0.7152
2-Propanol	0.8092	0.7982	0.7869	0.7755	0.7638	0.7519	0.7397	0.7272	0.7143	0.7011	0.6876
Propyl acetate	0.9101	0.8994	0.8885	0.8775	0.8662	0.8548	0.8432	0.8313	0.8192	0.8069	0.7942
Propylbenzene	0.8779	0.8700	0.8619	0.8538	0.8456	0.8373	0.8289	0.8204	0.8117	0.8030	0.7943
Propyl formate	0.9275	0.9166	0.9053	0.8938	0.8821	0.8702	0.8581	0.8457	0.8330	0.8201	0.8068
Tetrachloromethane	1.629	1.611	1.593	1.575	1.557	1.538	1.518	1.499	1.479	1.458	1.437
Toluene	0.8846	0.8757	0.8667	0.8576	0.8483	0.8389	0.8294	0.8197	0.8098	0.7998	0.7896
Trichloromethane	1.524	1.507	1.489	1.471	1.452	1.433	1.414	1.394			
2,2,4-Trimethylpentane			0.6921	0.6836	0.6750	0.6663	0.6574	0.6484	0.6391	0.6296	0.6199
o-Xylene			0.8801	0.8717	0.8633	0.8547	0.8460	0.8372	0.8282	0.8191	0.8099
m-Xylene	0.8813	0.8729	0.8644	0.8558	0.8470	0.8382	0.8292	0.8201	0.8109	0.8015	0.7920
p-Xylene			0.8609	0.8523	0.8436	0.8347	0.8258	0.8167	0.8075	0.7981	0.7886

DEPENDENCE OF BOILING POINT ON PRESSURE

The normal boiling point of a liquid is defined as the temperature at which the vapor pressure reaches standard atmospheric pressure, 101.325 kPa. The change in boiling point with pressure may be calculated from the representation of the vapor pressure by the Antoine Equation,

$$\ln p = A_1 - A_2/(T + A_3)$$

where p is the vapor pressure, T the absolute temperature, and A_1, A_2, and A_3 are constants. This table, which has been calculated using the Antoine constants in Reference 1, gives values of $\Delta t/\Delta p$ for a number of liquids, in units of both °C/kPa and °C/mmHg. The correction to the boiling point is generally accurate to 0.1 to 0.2 °C as long as the pressure is within 10% of standard atmospheric pressure.

A slightly less accurate estimate of $\Delta t/\Delta p$ may be obtained from the Clausius-Clapeyron equation, with the assumption that the change in volume upon vaporization equals the ideal-gas volume of the vapor. This leads to the equation

$$\Delta t/\Delta p = RT_b^2/p_0\Delta_{vap}H(T_b)$$

where R is the molar gas constant, p_0 is 101.325 kPa, T_b is the normal boiling point temperature (absolute), and $\Delta_{vap}H(T_b)$ is the molar enthalpy of vaporization at the normal boiling point. Values of the last quantity may be obtained from the table "Enthalpy of Vaporization" in Section 6.

Reference

1. Lide, D. R., and Kehiaian, H. V., *CRC Handbook of Thermophysical and Thermochemical Data*, CRC Press, Boca Raton, FL, 1994, pp. 49-59.

Compound	t_b °C	$\Delta t/\Delta p$ °C/kPa	$\Delta t/\Delta p$ °C/mmHg
Acetaldehyde	20.1	0.261	0.0348
Acetic acid	117.9	0.324	0.0432
Acetone	56.0	0.289	0.0385
Acetonitrile	81.6	0.316	0.0421
Ammonia	-33.33	0.198	0.0264
Aniline	184.1	0.378	0.0504
Anisole	153.7	0.367	0.0489
Benzaldehyde	179.0	0.392	0.0523
Benzene	80.0	0.321	0.0428
Bromine	58.8	0.300	0.0400
Butane	-0.5	0.267	0.0356
1-Butanol	117.7	0.278	0.0371
Carbon disulfide	46.2	0.304	0.0405
Chlorine	-34.04	0.224	0.0299
Chlorobenzene	131.7	0.365	0.0487
1-Chlorobutane	78.6	0.321	0.0428
Chloroethane	12.3	0.262	0.0349
Chloroethylene	-13.3	0.241	0.0321
Cyclohexane	80.7	0.328	0.0437
Cyclohexanol	160.8	0.344	0.0459
Cyclohexanone	155.4	0.382	0.0509
Decane	174.1	0.388	0.0517
Dibutyl ether	140.2	0.363	0.0484
Dichloromethane	39.6	0.276	0.0368
Diethyl ether	34.5	0.278	0.0371
Dimethyl sulfoxide	189.0	0.379	0.0505
1,4-Dioxane	101.5	0.321	0.0428
Dipropyl ether	90.0	0.326	0.0435
Ethanol	78.2	0.249	0.0332
Ethyl acetate	77.1	0.300	0.0400
Ethylene glycol	197.3	0.331	0.0441
Heptane	98.5	0.336	0.0448
Hexafluorobenzene	80.2	0.305	0.0407
Hexane	68.7	0.314	0.0419
1-Hexanol	157.6	0.318	0.0424
Hydrogen fluoride	20.1	0.276	0.0368
Iodomethane	42.5	0.291	0.0388
Isobutane	-11.7	0.254	0.0339
Methanol	64.6	0.251	0.0335
Methyl acetate	56.8	0.282	0.0376
Methyl formate	31.7	0.582	0.0776
N-Methylaniline	196.2	0.396	0.0528
N-Methylformamide	199.5	0.371	0.0495
Nitrobenzene	210.8	0.418	0.0557
Nitromethane	101.1	0.320	0.0427
1-Octanol	195.1	0.360	0.0480
Pentane	36.0	0.289	0.0385
1-Pentanol	137.9	0.296	0.0395
Phenol	181.8	0.349	0.0465
Propane	-42.1	0.224	0.0299
1-Propanol	97.2	0.261	0.0348
2-Propanol	82.3	0.247	0.0329
Pyridine	115.2	0.340	0.0453
Pyrrole	129.7	0.330	0.0440
Pyrrolidine	86.5	0.309	0.0412
Styrene	145.1	0.369	0.0492
Sulfur dioxide	-10.05	0.221	0.0295
Tetrachloroethylene	121.3	0.354	0.0472
Tetrachloromethane	76.8	0.325	0.0433
Toluene	110.6	0.353	0.0471
Trichloroethylene	87.2	0.330	0.0440
Trichloromethane	61.1	0.302	0.0403
Trimethylamine	2.8	0.248	0.0331
Water	100.0	0.276	0.0368
o-Xylene	144.5	0.373	0.0497
m-Xylene	139.1	0.368	0.0491
p-Xylene	138.3	0.369	0.0492

EBULLIOSCOPIC CONSTANTS FOR CALCULATION OF BOILING POINT ELEVATION

The boiling point T_b of a dilute solution of a non-volatile, non-dissociating solute is elevated relative to that of the pure solvent. If the solution is ideal (i.e., follows Raoult's Law), the amount of elevation depends only on the number of particles of solute present. Hence the change in boiling point ΔT_b can be expressed as

$$\Delta T_b = E_b \, m_2$$

where m_2 is the molality (moles of solute per kilogram of solvent) and E_b is the Ebullioscopic Constant, a characteristic property of the solvent. The Ebullioscopic Constant may be calculated from the relation

$$E_b = R \, T_b^2 \, M/\Delta_{vap}H$$

where R is the molar gas constant, T_b is the normal boiling point temperature (absolute) of the solvent, M the molar mass of the solvent, and $\Delta_{vap}H$ the molar enthalpy (heat) of vaporization of the solvent at its normal boiling point.

This table lists E_b values for some common solvents, as calculated from data in the table "Enthalpy of Vaporization" in Section 6.

Compound	E_b/K kg mol^{-1}	Compound	E_b/K kg mol^{-1}
Acetic acid	3.22	Hexane	2.90
Acetone	1.80	Iodomethane	4.31
Acetonitrile	1.44	Methanol	0.86
Aniline	3.82	Methyl acetate	2.21
Anisole	4.20	N-Methylaniline	4.3
Benzaldehyde	4.24	N-Methylformamide	2.2
Benzene	2.64	Nitrobenzene	5.2
1-Butanol	2.17	Nitromethane	2.09
Carbon disulfide	2.42	1-Octanol	5.06
Chlorobenzene	4.36	Phenol	3.54
1-Chlorobutane	3.13	1-Propanol	1.66
Cyclohexane	2.92	2-Propanol	1.58
Cyclohexanol	3.5	Pyridine	2.83
Decane	6.10	Pyrrole	2.33
Dichloromethane	2.42	Pyrrolidine	2.32
Diethyl ether	2.20	Tetrachloroethylene	6.18
Dimethyl sulfoxide	3.22	Tetrachloromethane	5.26
1,4-Dioxane	3.01	Toluene	3.40
Ethanol	1.23	Trichloroethylene	4.52
Ethyl acetate	2.82	Trichloromethane	3.80
Ethylene glycol	2.26	Water	0.513
Heptane	3.62	o-Xylene	4.25

CRYOSCOPIC CONSTANTS FOR CALCULATION OF FREEZING POINT DEPRESSION

The freezing point T_f of a dilute solution of a non-volatile, non-dissociating solute is depressed relative to that of the pure solvent. If the solution is ideal (i.e., follows Raoult's Law), this lowering is a function only of the number of particles of solute present. Thus the absolute value of the lowering of freezing point ΔT_f can be expressed as

$$\Delta T_f = E_f m_2$$

where m_2 is the molality (moles of solute per kilogram of solvent) and E_f is the Cryoscopic Constant, a characteristic property of the solvent. The Cryoscopic Constant may be calculated from the relation

$$E_f = R\, T_f^2\, M / \Delta_{fus} H$$

where R is the molar gas constant, T_f is the freezing point temperature (absolute) of the solvent, M the molar mass of the solvent, and $\Delta_{fus} H$ the molar enthalpy (heat) of fusion of the solvent.

This table lists cryscopic constants for selected substances, as calculated from data in the table "Enthalpy of Fusion" in Section 6.

Compound	E_f/K kg mol^{-1}
Acetamide	3.92
Acetic acid	3.63
Acetophenone	5.16
Aniline	5.23
Benzene	5.07
Benzonitrile	5.35
Benzophenone	8.58
(+)-Camphor	37.8
1-Chloronaphthalene	7.68
o-Cresol	5.92
m-Cresol	7.76
p-Cresol	7.20
Cyclohexane	20.8
Cyclohexanol	42.2
cis-Decahydronaphthalene	6.42
trans-Decahydronaphthalene	4.70
Dibenzyl ether	6.17
p-Dichlorobenzene	7.57
Diethanolamine	3.16
Dimethyl sulfoxide	3.85

Compound	E_f/K kg mol^{-1}
1,4-Dioxane	4.63
Diphenylamine	8.38
Ethylene glycol	3.11
Formamide	4.25
Formic acid	2.38
Glycerol	3.56
Methylcyclohexane	2.60
Naphthalene	7.45
Nitrobenzene	6.87
Phenol	6.84
Pyridine	4.26
Quinoline	6.73
Succinonitrile	19.3
1,1,2,2-Tetrabromoethane	21.4
1,1,2,2-Tetrachloro-1,2-difluoroethane	41.0
Toluene	3.55
p-Toluidine	4.91
Tribromomethane	15.0
Water	1.86
p-Xylene	4.31

FREEZING POINT LOWERING BY ELECTROLYTES IN AQUEOUS SOLUTION

Reference

Forsythe, W. E., *Smithsonian Physical Tables, Ninth Edition*, Smithsonian Institution, Washington, D.C., 1956.

Compound	Lowering of freezing point of water (in °C) as function of molality (mol/kg)									
	0.05	0.10	0.25	0.50	0.75	1.00	1.50	2.00	2.50	3.00
CaCl$_2$	0.25	0.49	1.27	2.66	4.28	6.35	10.78	15.27	20.42	28.08
CuSO$_4$	0.13	0.23	0.47	0.96						
HCl	0.18	0.36	0.90	1.86	2.90	4.02	6.63	9.94		
HNO$_3$	0.18	0.35	0.88	1.80	2.78	3.80	5.98	8.34	10.95	13.92
H$_2$SO$_4$	0.20	0.39	0.96	1.95	3.04	4.28	7.35	11.35	16.32	
KBr	0.18	0.36	0.92	1.78						
KCl	0.17	0.35	0.86	1.68	2.49	3.29	4.88	6.50	8.14	9.77
KNO$_3$	0.17	0.33	0.78	1.47	2.11	2.66				
K$_2$SO$_4$	0.23	0.43	1.01	1.87						
LiCl	0.18	0.35	0.88	1.80	2.78					
MgSO$_4$	0.13	0.24	0.55	1.01	1.50	2.08	3.41			
NH$_4$Cl	0.17	0.34	0.85	1.70	2.55					
NaCl	0.18	0.35	0.85	1.68	2.60					
NaNO$_3$	0.18	0.36	0.80	1.62	2.63	3.10				

CORRECTION OF BAROMETER READINGS TO 0 °C TEMPERATURE

The following corrections are used to reduce the reading of a mercury barometer with a brass scale to 0 °C. The number in the table should be subtracted from the observed height of the mercury column to give the true pressure in mmHg (1mmHg = 133.322 Pa). The table is calculated from the formula

$$\Delta h = -0.0001634 \, ht/(1+0.0001818 \, t),$$

where h is the observed column height in mm and t the Celsius temperature. This relation is based on thermal expansion coefficients of $181.8 \cdot 10^{-6}$ °C^{-1} for mercury and $18.4 \cdot 10^{-6}$ °C^{-1} for brass.

t/°C	620	630	640	650	660	670	680	690	700	710	720	730	740	750	760	770	780	790	800
									Observed Height in mm										
0	0.00	0.00	0.00	0.00	0.00	0.00	0.00	0.00	0.00	0.00	0.00	0.00	0.00	0.00	0.00	0.00	0.00	0.00	0.00
1	0.10	0.10	0.10	0.11	0.11	0.11	0.11	0.11	0.11	0.12	0.12	0.12	0.12	0.12	0.12	0.13	0.13	0.13	0.13
2	0.20	0.21	0.21	0.21	0.22	0.22	0.22	0.23	0.23	0.23	0.24	0.24	0.24	0.25	0.25	0.25	0.25	0.26	0.26
3	0.30	0.31	0.31	0.32	0.32	0.33	0.33	0.34	0.34	0.35	0.35	0.36	0.36	0.37	0.37	0.38	0.38	0.39	0.39
4	0.40	0.41	0.42	0.42	0.43	0.44	0.44	0.45	0.46	0.46	0.47	0.48	0.48	0.49	0.50	0.50	0.51	0.52	0.52
5	0.51	0.51	0.52	0.53	0.54	0.55	0.56	0.56	0.57	0.58	0.59	0.60	0.60	0.61	0.62	0.63	0.64	0.64	0.65
6	0.61	0.62	0.63	0.64	0.65	0.66	0.67	0.68	0.69	0.70	0.71	0.71	0.72	0.73	0.74	0.75	0.76	0.77	0.78
7	0.71	0.72	0.73	0.74	0.75	0.77	0.78	0.79	0.80	0.81	0.82	0.83	0.85	0.86	0.87	0.88	0.89	0.90	0.91
8	0.81	0.82	0.84	0.85	0.86	0.87	0.89	0.90	0.91	0.93	0.94	0.95	0.97	0.98	0.99	1.01	1.02	1.03	1.04
9	0.91	0.92	0.94	0.95	0.97	0.98	1.00	1.01	1.03	1.04	1.06	1.07	1.09	1.10	1.12	1.13	1.15	1.16	1.17
10	1.01	1.03	1.04	1.06	1.08	1.09	1.11	1.13	1.14	1.16	1.17	1.19	1.21	1.22	1.24	1.26	1.27	1.29	1.30
11	1.11	1.13	1.15	1.17	1.18	1.20	1.22	1.24	1.26	1.27	1.29	1.31	1.33	1.35	1.36	1.38	1.40	1.42	1.44
12	1.21	1.23	1.25	1.27	1.29	1.31	1.33	1.35	1.37	1.39	1.41	1.43	1.45	1.47	1.49	1.51	1.53	1.55	1.57
13	1.31	1.34	1.36	1.38	1.40	1.42	1.44	1.46	1.48	1.50	1.53	1.55	1.57	1.59	1.61	1.63	1.65	1.67	1.70
14	1.41	1.44	1.46	1.48	1.51	1.53	1.55	1.57	1.60	1.62	1.64	1.67	1.69	1.71	1.73	1.76	1.78	1.80	1.83
15	1.52	1.54	1.56	1.59	1.61	1.64	1.66	1.69	1.71	1.74	1.76	1.78	1.81	1.83	1.86	1.88	1.91	1.93	1.96
16	1.62	1.64	1.67	1.69	1.72	1.75	1.77	1.80	1.82	1.85	1.88	1.90	1.93	1.96	1.98	2.01	2.03	2.06	2.09
17	1.72	1.74	1.77	1.80	1.83	1.86	1.88	1.91	1.94	1.97	1.99	2.02	2.05	2.08	2.10	2.13	2.16	2.19	2.22
18	1.82	1.85	1.88	1.91	1.93	1.96	1.99	2.02	2.05	2.08	2.11	2.14	2.17	2.20	2.23	2.26	2.29	2.32	2.35
19	1.92	1.95	1.98	2.01	2.04	2.07	2.10	2.13	2.17	2.20	2.23	2.26	2.29	2.32	2.35	2.38	2.41	2.44	2.48
20	2.02	2.05	2.08	2.12	2.15	2.18	2.21	2.25	2.28	2.31	2.34	2.38	2.41	2.44	2.47	2.51	2.54	2.57	2.60
21	2.12	2.15	2.19	2.22	2.26	2.29	2.32	2.36	2.39	2.43	2.46	2.50	2.53	2.56	2.60	2.63	2.67	2.70	2.73
22	2.22	2.26	2.29	2.33	2.36	2.40	2.43	2.47	2.51	2.54	2.58	2.61	2.65	2.69	2.72	2.76	2.79	2.83	2.86
23	2.32	2.36	2.40	2.43	2.47	2.51	2.54	2.58	2.62	2.66	2.69	2.73	2.77	2.81	2.84	2.88	2.92	2.96	2.99
24	2.42	2.46	2.50	2.54	2.58	2.62	2.66	2.69	2.73	2.77	2.81	2.85	2.89	2.93	2.97	3.01	3.05	3.08	3.12
25	2.52	2.56	2.60	2.64	2.68	2.72	2.77	2.81	2.85	2.89	2.93	2.97	3.01	3.05	3.09	3.13	3.17	3.21	3.25
26	2.62	2.66	2.71	2.75	2.79	2.83	2.88	2.92	2.96	3.00	3.04	3.09	3.13	3.17	3.21	3.26	3.30	3.34	3.38
27	2.72	2.77	2.81	2.85	2.90	2.94	2.99	3.03	3.07	3.12	3.16	3.20	3.25	3.29	3.34	3.38	3.42	3.47	3.51
28	2.82	2.87	2.91	2.96	3.00	3.05	3.10	3.14	3.19	3.23	3.28	3.32	3.37	3.41	3.46	3.51	3.55	3.60	3.64
29	2.92	2.97	3.02	3.06	3.11	3.16	3.21	3.25	3.30	3.35	3.39	3.44	3.49	3.54	3.58	3.63	3.68	3.72	3.77
30	3.02	3.07	3.12	3.17	3.22	3.27	3.32	3.36	3.41	3.46	3.51	3.56	3.61	3.66	3.71	3.75	3.80	3.85	3.90
31	3.12	3.17	3.22	3.27	3.32	3.37	3.43	3.48	3.53	3.58	3.63	3.68	3.73	3.78	3.83	3.88	3.93	3.98	4.03
32	3.22	3.28	3.33	3.38	3.43	3.48	3.54	3.59	3.64	3.69	3.74	3.79	3.85	3.90	3.95	4.00	4.05	4.11	4.16
33	3.32	3.38	3.43	3.48	3.54	3.59	3.64	3.70	3.75	3.81	3.86	3.91	3.97	4.02	4.07	4.13	4.18	4.23	4.29
34	3.42	3.48	3.53	3.59	3.64	3.70	3.75	3.81	3.87	3.92	3.98	4.03	4.09	4.14	4.20	4.25	4.31	4.36	4.42
35	3.52	3.58	3.64	3.69	3.75	3.81	3.86	3.92	3.98	4.03	4.09	4.15	4.21	4.26	4.32	4.38	4.43	4.49	4.55
36	3.62	3.68	3.74	3.80	3.86	3.92	3.97	4.03	4.09	4.15	4.21	4.27	4.32	4.38	4.44	4.50	4.56	4.62	4.68
37	3.72	3.78	3.84	3.90	3.96	4.02	4.08	4.14	4.20	4.26	4.32	4.38	4.44	4.50	4.56	4.62	4.68	4.74	4.80
38	3.82	3.88	3.95	4.01	4.07	4.13	4.19	4.25	4.32	4.38	4.44	4.50	4.56	4.62	4.69	4.75	4.81	4.87	4.93
39	3.92	3.99	4.05	4.11	4.18	4.24	4.30	4.37	4.43	4.49	4.56	4.62	4.68	4.75	4.81	4.87	4.94	5.00	5.06
40	4.02	4.09	4.15	4.22	4.28	4.35	4.41	4.48	4.54	4.61	4.67	4.74	4.80	4.87	4.93	5.00	5.06	5.13	5.19

DETERMINATION OF RELATIVE HUMIDITY FROM DEW POINT

The relative humidity of a water vapor–air mixture is defined as 100 times the partial pressure of water divided by the saturation vapor pressure of water at the same temperature. The relative humidity may be determined from the dew point t_{dew}, which is the temperature at which liquid water first condenses when the mixture is cooled from an initial temperature t. This table gives relative humidity as a function of the dew point depression $t - t_{dew}$ for several values of the dew point. Values are calculated from the vapor pressure table in Section 6.

$t - t_{dew}$	t_{dew}/°C				
	−10	0	10	20	30
0.0	100	100	100	100	100
0.2	99	99	99	99	99
0.4	97	97	97	98	98
0.6	95	96	96	96	97
0.9	94	94	95	95	96
1.0	92	93	94	94	94
1.2	91	92	92	93	93
1.4	90	90	91	92	92
1.6	88	89	90	91	91
1.8	87	88	89	90	90
2.0	86	87	88	88	89
2.2	84	85	86	87	89
2.4	83	84	85	86	87
2.6	82	83	84	85	86
2.8	80	82	83	84	85
3.0	79	81	82	83	84
3.2	78	80	81	82	83
3.4	77	79	80	81	82
3.6	76	77	79	80	82
3.8	75	76	78	79	81
4.0	73	75	77	78	80
4.2	72	74	76	77	79
4.4	71	73	75	77	78
4.6	70	72	74	76	77
4.8	69	71	73	75	76
5.0	69	70	72	74	75
5.2	67	69	71	73	75
5.4	66	68	70	72	74
5.6	65	67	69	71	73
5.9	64	66	69	70	72
6.0	63	66	68	70	71
6.2	62	65	67	69	71
6.4	61	64	66	68	70
6.6	60	63	65	67	69
6.8	60	62	64	66	68
7.0	59	61	63	66	68
7.2	58	60	63	65	67
7.4	57	60	62	64	66
7.6	56	59	61	63	65
7.8	55	58	60	63	65
8.0	54	57	60	62	64

$t - t_{dew}$	t_{dew}/°C				
	−10	0	10	20	30
8.2	54	56	59	61	63
8.4	53	56	58	60	63
8.6	52	55	57	60	62
8.8	51	54	57	59	61
9.0	51	53	56	58	61
9.2	50	53	55	58	60
9.4	49	52	55	57	59
9.6	48	51	54	56	59
9.8	48	51	53	56	58
10.0	47	50	53	55	57
10.5	45	48	51	54	56
11.0	44	47	49	52	55
11.5	42	45	48	51	53
12.0	41	44	47	49	52
12.5	39	42	45	48	50
13.0	38	41	44	46	49
13.5	37	40	43	45	48
14.0	35	38	41	44	47
14.5	34	37	40	43	45
15.0	33	36	39	42	44
15.5	32	35	38	40	
16.0	31	34	37	39	
16.5	30	33	36	38	
17.0	29	32	35	37	
17.5	28	31	34	36	
18.0	27	30	33	35	
18.5	26	29	32	34	
19.0	25	28	31	33	
19.5	24	27	30	33	
20.0	24	26	29	32	
21.0	22	25	27	30	
22.0	21	23	26	29	
23.0	19	22	24	27	
24.0	18	21	23	26	
25.0	17	19	22	24	
26.0	16	18	21	23	
27.0	15	17	20	22	
28.0	14	16	19	21	
29.0	13	15	18	20	
30.0	12	14	17	19	

DETERMINATION OF RELATIVE HUMIDITY FROM WET AND DRY BULB TEMPERATURES

Relative humidity may be determined by comparing temperature readings of wet and dry bulb thermometers. The following table, extracted from more extensive U.S. National Weather Service tables, gives the relative humidity as a function of air temperature t_d (dry bulb) and the difference $t_d - t_w$ between dry and wet bulb temperatures. The data assume a pressure near normal atmospheric pressure and an instrumental configuration with forced ventilation.

t_d/°C	$(t_d - t_w)$/°C											
	0.5	1.0	1.5	2.0	2.5	3.0	3.5	4.0	4.5	5.0	5.5	6.0
−10	83	67	51	35	19							
−8	86	71	57	43	29	15						
−6	88	74	61	49	37	25	8					
−4	89	77	66	55	44	33	23	12				
−2	90	79	69	60	50	40	31	22	12			
0	91	81	72	64	55	46	38	29	21	13	5	
2	91	84	76	68	60	52	44	37	29	22	14	7
4	92	85	78	71	63	57	49	43	36	29	22	16
6	93	86	79	73	66	60	54	48	41	35	29	24
8	93	87	81	75	69	63	57	51	46	40	35	29
10	94	88	82	77	71	66	60	55	50	44	39	34
12	94	89	83	78	73	68	63	58	53	48	43	39
14	95	90	85	79	75	70	65	60	56	51	47	42
16	95	90	85	81	76	71	67	63	58	54	50	46
18	95	91	86	82	77	73	69	65	61	57	53	49
20	96	91	87	83	78	74	70	66	63	59	55	51
22	96	92	87	83	80	76	72	68	64	61	57	54
24	96	92	88	84	80	77	73	69	66	62	59	56
26	96	92	88	85	81	78	74	71	67	64	61	58
28	96	93	89	85	82	78	75	72	69	65	62	59
30	96	93	89	86	83	79	76	73	70	67	64	61
35	97	94	90	87	84	81	78	75	72	69	67	64
40	97	94	91	88	85	82	80	77	74	72	69	67

t_d/°C	$(t_d - t_w)$/°C											
	6.5	7.0	7.5	8.0	8.5	9.0	10.0	11.0	12.0	13.0	14.0	15.0
4	9											
6	17	11	5									
8	24	19	14	8								
10	29	24	20	15	10	6						
12	34	29	25	21	16	12	5					
14	38	34	30	26	22	18	10					
16	42	38	34	30	26	23	15	8				
18	45	41	38	34	30	27	20	14	7			
20	48	44	41	37	34	31	24	18	12	6		
22	50	47	44	40	37	34	28	22	17	11	6	
24	53	49	46	43	40	37	31	26	20	15	10	5
26	54	51	49	46	43	40	34	29	24	19	14	10
28	56	53	51	48	45	42	37	32	27	22	18	13
30	58	55	52	50	47	44	39	35	30	25	21	17
32	60	57	54	51	49	46	41	37	32	28	24	20
34	61	58	56	53	51	48	43	39	35	30	26	23
36	62	59	57	54	52	50	45	41	37	33	29	25
38	63	61	58	56	54	51	47	43	39	35	31	27
40	64	62	59	57	54	53	48	44	40	36	33	29

CONSTANT HUMIDITY SOLUTIONS

Anthony Wexler

An excess of a water soluble salt in contact with its saturated solution and contained within an enclosed space produces a constant relative humidity and water vapor pressure according to

$$RH = A \exp(B/T)$$

where RH is the percent relative humidity (generally accurate to ±2 %), T is the temperature in kelvin, and the constants A and B and the range of valid temperatures are given in the table below. The vapor pressure, p, can be calculated from

$$p = (RH/100) \times p_0$$

where p_0 is the vapor pressure of pure water at temperature T as given in the table in Section 6 titled "Vapor Pressure of Water from 0 to 370 °C".

References

1. Wexler, A. S. and Seinfeld, J. H., *Atmospheric Environment*, 25A, 2731, 1991.
2. Greenspan, L., *J. Res. National Bureau of Standards*, 81A, 89, 1977.
3. Broul, et al., *Solubility of Inorganic Two-Component Systems*, Elsevier, New York, 1981.
4. Wagman, D. D. et al., *J. Phys. Chem. Ref. Data, Vol. 11*, Suppl. 2, 1982.

Compound	Temperature range (°C)	RH (25 °C)	A	B
$NaOH \cdot H_2O$	15–60	6	5.48	27
$LiBr \cdot 2H_2O$	10–30	6	0.23	996
$ZnBr_2 \cdot 2H_2O$	5–30	8	1.69	455
$KOH \cdot 2H_2O$	5–30	9	0.014	1924
$LiCl \cdot H_2O$	20–65	11	14.53	−75
$CaBr_2 \cdot 6H_2O$	11–22	16	0.17	1360
$LiI \cdot 3H_2O$	15–65	18	0.15	1424
$CaCl_2 \cdot 6H_2O$	15–25	29	0.11	1653
$MgCl_2 \cdot 6H_2O$	5–45	33	29.26	34
$NaI \cdot 2H_2O$	5–45	38	3.62	702
$Ca(NO_3)_2 \cdot 4H_2O$	10–30	51	1.89	981
$Mg(NO_3)_2 \cdot 6H_2O$	5–35	53	25.28	220
$NaBr \cdot 2H_2O$	0–35	58	20.49	308
NH_4NO_3	10–40	62	3.54	853
KI	5–30	69	29.35	254
$SrCl_2 \cdot 6H_2O$	5–30	71	31.58	241
$NaNO_3$	10–40	74	26.94	302
$NaCl$	10–40	75	69.20	25
NH_4Cl	10–40	79	35.67	235
KBr	5–25	81	40.98	203
$(NH_4)_2SO_4$	10–40	81	62.06	79
KCl	5–25	84	49.38	159
$Sr(NO_3)_2 \cdot 4H_2O$	5–25	85	28.34	328
$BaCl_2 \cdot 2H_2O$	5–25	90	69.99	75
CsI	5–25	91	70.77	75
KNO_3	0–50	92	43.22	225
K_2SO_4	10–50	97	86.75	34

STANDARD SALT SOLUTIONS FOR HUMIDITY CALIBRATION

Saturated aqueous solutions of inorganic salts are convenient secondary standards for calibration of instruments for measurement of relative humidity. The International Union of Pure and Applied Chemistry has recommended salt solutions for calibrations in the range of 10% to 90% relative humidity, and the American Society for Testing and Materials has published similar standards. The data in this table are taken from the IUPAC recommendations, except for K_2CO_3 and K_2SO_4, which are ASTM recommendations.

Details on the preparation and use of these standards may be found in References 1 and 2. Data for other salts are given in Reference 3.

References

1. Marsh, K. N., Editor, *Recommended Reference Materials for the Realization of Physicochemical Properties*, Blackwell Scientific Publications, Oxford, 1987, pp. 157–162.
2. *Standard Practice for Maintaining Constant Relative Humidity by Means of Aqueous Solutions*, ASTM Standard E 104-85, Reapproved 1991.
3. Greenspan, L., *J. Res. Nat. Bur. Stand.*, 81A, 89, 1977.

t/°C	LiCl	$MgCl_2$	K_2CO_3	$Mg(NO_3)_2$	NaCl	KCl	K_2SO_4
				Relative Humidity in %			
0		33.66 ± 0.33	43.1 ± 0.7	60.35 ± 0.55	75.51 ± 0.34	88.61 ± 0.53	98.8 ± 2.1
5		33.60 ± 0.28	43.1 ± 0.5	58.86 ± 0.43	75.65 ± 0.27	87.67 ± 0.45	98.5 ± 0.9
10		33.47 ± 0.24	43.1 ± 0.4	57.36 ± 0.33	75.67 ± 0.22	86.77 ± 0.39	98.2 ± 0.8
15		33.30 ± 0.21	43.2 ± 0.3	55.87 ± 0.27	75.61 ± 0.18	85.92 ± 0.33	97.9 ± 0.6
20	11.31 ± 0.31	33.07 ± 0.18	43.2 ± 0.3	54.38 ± 0.23	75.47 ± 0.14	85.11 ± 0.29	97.6 ± 0.5
25	11.30 ± 0.27	32.78 ± 0.16	43.2 ± 0.4	52.89 ± 0.22	75.29 ± 0.12	84.34 ± 0.26	97.3 ± 0.5
30	11.28 ± 0.24	32.44 ± 0.14	43.2 ± 0.5	51.40 ± 0.24	75.09 ± 0.11	83.62 ± 0.25	97.0 ± 0.4
35	11.25 ± 0.22	32.05 ± 0.13		49.91 ± 0.29	74.87 ± 0.12	82.95 ± 0.25	96.7 ± 0.4
40	11.21 ± 0.21	31.60 ± 0.13		48.42 ± 0.37		82.32 ± 0.25	96.4 ± 0.4
45	11.16 ± 0.21	31.10 ± 0.13		46.93 ± 0.47		81.74 ± 0.28	96.1 ± 0.4
50	11.10 ± 0.22	30.54 ± 0.14		45.44 ± 0.60		81.20 ± 0.31	95.8 ± 0.5
55	11.03 ± 0.23	29.93 ± 0.16				80.70 ± 0.35	
60	10.95 ± 0.26	29.26 ± 0.18				80.25 ± 0.41	
65	10.86 ± 0.29	28.54 ± 0.21				79.85 ± 0.48	
70	10.75 ± 0.33	27.77 ± 0.25				79.49 ± 0.57	
75	10.64 ± 0.38	26.94 ± 0.29				79.17 ± 0.66	
80	10.51 ± 0.44	26.05 ± 0.34				78.90 ± 0.77	

LOW-TEMPERATURE BATHS FOR MAINTAINING CONSTANT TEMPERATURE

A liquid–solid slurry is a convenient means of maintaining a constant temperature environment below room temperature. The following is a list of readily available organic liquids suitable for this purpose, arranged in order of their melting (freezing) points t_m. The normal boiling points t_b are also given.

Compound	t_m/°C	t_b/°C
Isopentane (2-Methylbutane)	−159.9	27.8
Methylcyclopentane	−142.5	71.8
3-Chloropropene (Allyl chloride)	−134.5	45.1
Pentane	−129.7	36.0
Allyl alcohol	−129	97.0
Ethanol	−114.1	78.2
Carbon disulfide	−111.5	46
Isobutyl alcohol	−108	107.8
Toluene	−94.9	110.6
Acetone	−94.8	56.0
Ethyl acetate	−83.6	77.1
Dry ice + acetone	−78	
p-Cymene	−68.9	177.1
Trichloromethane (Chloroform)	−63.6	61.1
N-Methylaniline	−57	196.2
Chlorobenzene	−45.2	131.7
Anisole	−37.5	153.7
Bromobenzene	−30.6	156.0
Tetrachloromethane (Carbon tetrachloride)	−23	76.8
Benzonitrile	−12.7	191.1

METALS AND ALLOYS WITH LOW MELTING TEMPERATURE

L. I. Berger

Metal or alloy system	Composition, % * Weight	Composition, % * Atomic	Melting temperature (°C)	Comments	Ref.
Hg	100	100	−38.829		
Cs–K	77.0–23.0	50.0–50.0	−37.5	Eutectic (?)	1
Cs–Na	94.5–5.5	75.0–25.0	−30.0	Eutectic	2
K–Na	76.7–23.3	65.9–34.1	−12.65	Eutectic	3
Na–Rb	8.0–92.0	24.4–75.6	−5	Eutectic	4
Ga–In–Sn	62.5–21.5–16.0	73.6–15.3–11.1	11	Eutectic	5
Ga–Sn–Zn	82.0–12.0–6.0	86.0–7.3–6.7	17	Eutectic	5
Cs	100	100	28.44		
Ga	100	100	29.7646		
K–Rb	32.0–68.0	50–50	33	Eutectic	4
Bi–Cd–In–Pb–Sn	44.7–5.3–19.1–22.6–8.3	35.1–8.2–27.3–17.9–11.5	46.7	Eutectic	6
Bi–In–Pb–Sn	49.5–21.3–17.6–11.6	39.2–30.7–14.0–16.2	58.2	Eutectic	6
Bi–In–Sn	32.5–51.0–16.5	21.1–60.1–18.8	60.5	Eutectic	7
K	100	100	63.38		
Bi–Cd–Pb–Sn	50.0–12.5–25.0–12.5	41.5–19.3–21.0–18.2	70	Wood's alloy	6
Bi–In	33.0–67.0	21.3–78.7	72	Eutectic	8
Bi–Cd–Pb	51.6–8.2–40.2	48.1–14.2–37.7	91.5	Eutectic	6
Bi–Pb–Sn	52.5–32.0–15.5	46.8–28.7–24.5	95	Eutectic	6
Na	100	100	97.8		
Bi–Cd–Sn	54.0–20.0–26.0	39.4–27.2–33.4	102.5	Eutectic	6
In–Sn	51.8–48.2	52.6–47.4	119	Eutectic	9
Cd–In	25.3–74.7	25.7–74.3	120	Eutectic	10
Bi–Pb	55.5–44.5	55.3–44.7	124	Eutectic	11
Bi–Sn–Zn	56.0–40.0–4.0	40.2–50.6–9.2	130	Eutectic	6, 7
Bi–Sn	70–30	57.0–43.0	138.5	Eutectic	6, 12
Bi–Cd	60.3–39.7	45.0–55.0	145.5	Eutectic	13, 14
In	100	100	156.6		
Li	100	100	180.5		
Pb–Sn	38.1–61.9	26.1–73.9	183	Eutectic	6,15
Bi–Tl	48.0–52.0	47.5–52.5	185	Eutectic	13
Sn–Zn	91.0–9.0	85.0–15.0	198	Eutectic	14
Sb–Sn	8.0–92.0	7.8–92.2	199	White Metal	16
Au–Pb	14.6–85.4	15.2–84.8	212	Eutectic	17
Ag–Sn	3.5–96.5	3.8–96.2	221	Eutectic	13,18
Bi–Pb–Sb–Sn	48.0–28.5–9.0–14.5	40.8–24.5–13.1–21.6	226	Matrix Alloy	6
Cu–Sn	0.75–99.25	1.3–98.7	227	Eutectic	13, 19
Sn	100	100	231.928		

* The useful expressions for correlations between the atomic and weight concentrations of an alloy components are:

$$f(a, A_k) = \frac{f(w, A_k)}{M_k \sum_{i=1}^{N} \frac{f(w, A_i)}{M_i}} \quad \text{and} \quad f(w, A_k) = \frac{M_k \cdot f(a, A_k)}{\sum_{i=1}^{N} M_i \cdot f(a, A_i)} \quad (i = 1, \ldots, k, \ldots, N)$$

where $f(a, A_i)$ and $f(w, A_i)$ are the atomic and weight concentrations of component A_i, respectively, and M_i is the atomic weight of this component.

References

1. Zintle, E. and Hauke, W., *Z. Electrochem.*, 44, 104, 1938.
2. Rinck, E., *Compt. Rend.*, 199, 1217, 1934.
3. Krier, C. A., Craign, R. S., and Wallace, W. E., *J. Phys. Chem.*, 61, 522, 1957.
4. Goria, C., *Gazz. Chim. Ital.*, 65, 865, 1935.
5. Baker, H., Ed., *ASM Handbook, Volume 3: Alloy Phase Diagrams*, ASM Intl., Materials Park, OH, 1992.
6. Sedlacek, V., *Non-Ferrous Metals and Alloys*, Elsevier, 1986.
7. Villars, P., Prince, A., Okamoto, H., Eds., *Handbook of Ternary Alloy Phase Diagrams*, ASM Intl., 1994.
8. Palatnik, L. S., Kosevich, V. M., and Tyrina, L. V., *Phys. Metals Metallog. (USSR)*, 11, 75, 1961.
9. Neumann, T. and Alpout, O., *J. Less-Common Metals*, 6, 108, 1964.
10. Neumann, T. and Predel, B., *Z. Metallk.*, 50, 309, 1959.
11. Roy, P., Orr, R. L., and Hultgren, R., *J. Phys. Chem.*, 64, 1034, 1960.
12. Dobovicek, B. and Smajic, N., *Rudarsko–Met. Zbornik*, 4, 353, 1962.
13. Massalski, T. B., Okamoto, H., Subramanian, P. R., and Kacprzak, L., Eds., *Binary Alloy Phase Diagrams*, 2nd ed., ASM Intl., 1990.
14. Dobovicek, B. and Straus, B., *Rudarsko–Met. Zbornik*, 3, 273, 1960.
15. Schurmann, E. and Gilhaus, F. J., *Arch. Eisenhuettenw.*, 32, 867, 1961.
16. Rosenblatt, G. M. and Birchenall, C. E., *Trans. AIME*, 224, 481, 1962.
17. Evans, D. S. and Prince, A., in *Alloy Phase Diagrams*, MRS Simposia Proc., Vol. 19, North Holland, 1983, p. 383.
18. Umanskiy, M. M., *Zh. Fiz. Khim.*, 14, 846, 1940.
19. Homer, C. E. and Plummer, H., *J. Inst. Met.*, 64, 169, 1939.

WIRE TABLES

The resistance per unit length of wires of various metals is tabulated here. Values were calculated from resistivity values in the tables "Electrical Resistivity of Pure Metals" and "Electrical Resistivity of Selected Alloys", which appear in Section 12. In practice, resistance may vary because of differing heat treatments and metal composition. The values in the table refer to 20 °C, but values at other temperatures may be calculated from the following resistivity data:

| Metal | Resistivity in 10^{-8} Ω m at temperature | | | |
	0 °C	20 °C	25 °C	100 °C
Aluminum	2.417	2.650	2.709	3.56
Brass (70% Cu, 30% Zn)	5.87	6.08	6.13	6.91
Constantan (60% Cu, 40% Ni)	45.43	45.38	45.35	45.11
Copper	1.543	1.678	1.712	2.22
Nichrome (79% Ni, 21% Cr)	107.3	107.5	107.6	108.3
Platinum	9.6	10.5	10.7	13.6
Silver	1.467	1.587	1.617	2.07
Tungsten	4.82	5.28	5.39	7.18

Resistance per unit length at 20 °C in Ω/m

AWG Gauge[a]	Diameter (mm)	Aluminum	Brass	Constantan	Copper	Nichrome	Platinum	Silver	Tungsten
0	8.252	0.000495	0.00114	0.00848	0.000314	0.0201	0.00196	0.000297	0.00099
2	6.543	0.000788	0.00181	0.0135	0.000499	0.0320	0.00312	0.000472	0.00157
4	5.189	0.00125	0.00287	0.0214	0.000793	0.0508	0.00496	0.000750	0.00250
6	4.115	0.00199	0.00457	0.0341	0.00126	0.0808	0.00789	0.00119	0.00397
8	3.264	0.00317	0.00727	0.0542	0.00200	0.128	0.0125	0.00190	0.00631
10	2.588	0.00504	0.0115	0.0863	0.00319	0.204	0.0200	0.00302	0.0100
12	2.053	0.00800	0.0184	0.137	0.00507	0.325	0.0317	0.00479	0.0159
14	1.628	0.0127	0.0292	0.218	0.00806	0.516	0.0504	0.00762	0.0254
16	1.291	0.0202	0.0464	0.347	0.0128	0.821	0.0802	0.0121	0.0403
18	1.024	0.0322	0.0738	0.551	0.0204	1.30	0.127	0.0193	0.0641
20	0.8118	0.0512	0.117	0.877	0.0324	2.08	0.203	0.0307	0.102
22	0.6439	0.0814	0.187	1.39	0.0515	3.30	0.322	0.0487	0.162
24	0.5105	0.129	0.297	2.22	0.0820	5.25	0.513	0.0775	0.258
26	0.4049	0.206	0.472	3.52	0.130	8.35	0.815	0.123	0.410
28	0.3211	0.327	0.751	5.60	0.207	13.3	1.30	0.196	0.652
30	0.2548	0.520	1.19	8.90	0.329	21.1	2.06	0.311	1.03
32	0.2019	0.828	1.90	14.2	0.524	33.6	3.28	0.496	1.65
34	0.1601	1.32	3.02	22.5	0.833	53.4	5.22	0.788	2.62
36	0.1270	2.09	4.80	35.8	1.32	84.9	8.29	1.25	4.17
38	0.1007	3.33	7.63	57.0	2.11	135	13.2	1.99	6.63
40	0.07988	5.29	12.1	90.5	3.35	214	20.9	3.17	10.5

[a] Often called Brown & Sharpe Gauge.

CHARACTERISTICS OF PARTICLES AND PARTICLE DISPERSOIDS

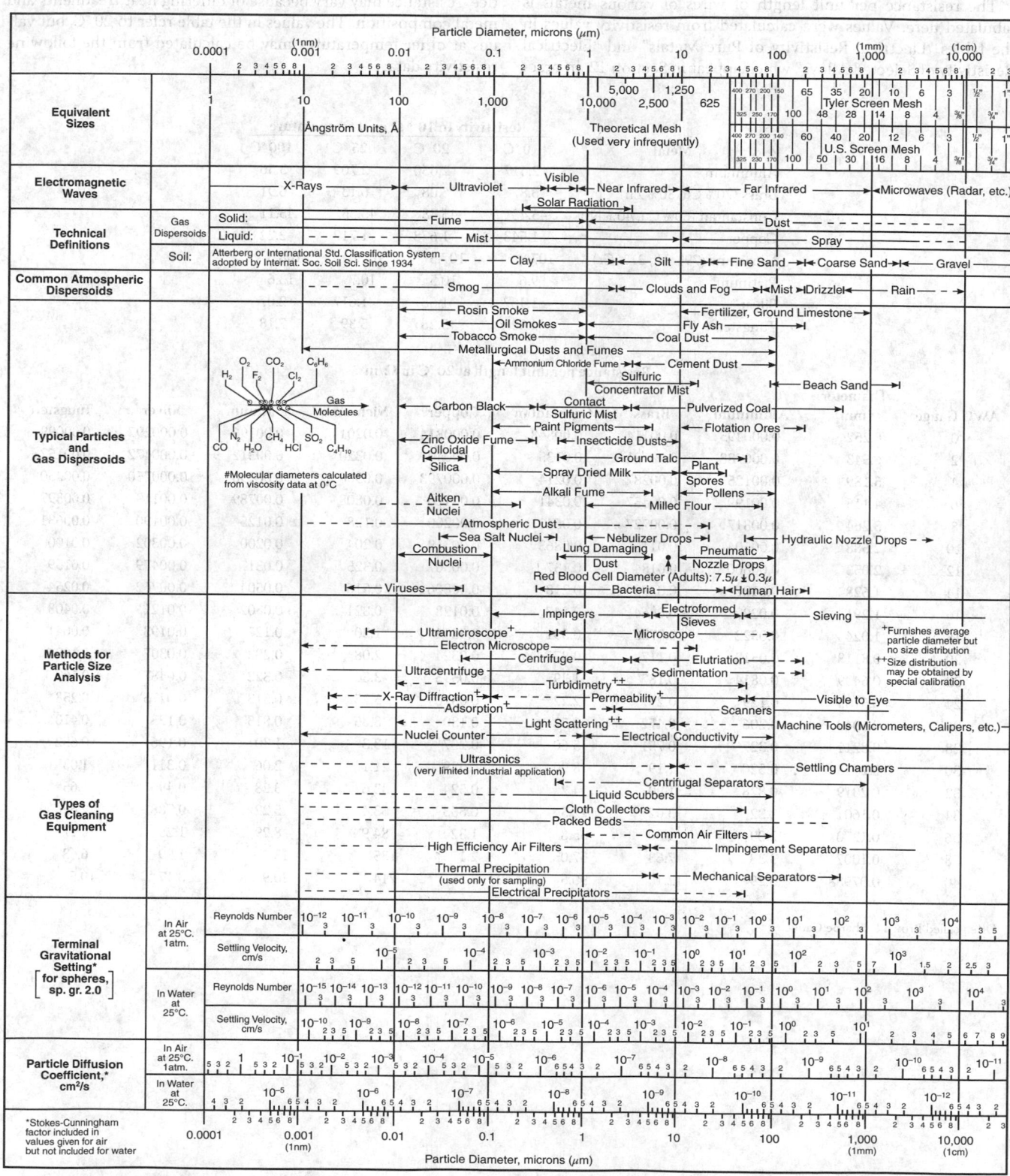

From C.E. Lapple, *Stanford Research Institute Journal*, Vol. 5, p. 95 (Third Quarter) 1961.

DENSITY OF VARIOUS SOLIDS

This table gives the range of density for miscellaneous solid materials whose characteristics depend on the source or method of preparation.

References

1. Forsythe, W. E., *Smithsonian Physical Tables, Ninth Edition*, Smithsonian Institution, Washington, D.C., 1956.
2. Kaye, G. W. C., and Laby, T. H., *Tables of Physical and Chemical Constants, 16th Edition*, Longman, London, 1995.
3. Brandrup, J., and Immergut, E. H., *Polymer Handbook, Third Edition*, John Wiley & Sons, New York, 1989.

Material	$\rho/\ g\ cm^{-3}$	Material	$\rho/\ g\ cm^{-3}$	Material	$\rho/\ g\ cm^{-3}$
Agate	2.5-2.7	Pyrex	2.23	Solder	8.7-9.4
Alabaster,		Granite	2.64-2.76	Starch	1.53
carbonate	2.69-2.78	Graphite	2.30-2.72	Steel, stainless	7.8
sulfate	2.26-2.32	Gum arabic	1.3-1.4	Sugar	1.59
Albite	2.62-2.65	Gypsum	2.31-2.33	Talc	2.7-2.8
Amber	1.06-1.11	Hematite	4.9-5.3	Tallow, beef	0.94
Amphiboles	2.9-3.2	Hornblende	3.0	Tar	1.02
Anorthite	2.74-2.76	Ice	0.917	Topaz	3.5-3.6
Asbestos	2.0-2.8	Iron, cast	7.0-7.4	Tourmaline	3.0-3.2
Asbestos slate	1.8	Ivory	1.83-1.92	Tungsten carbide	14.0-15.0
Asphalt	1.1-1.5	Kaolin	2.6	Wax, sealing	1.8
Basalt	2.4-3.1	Leather, dry	0.86	Wood (seasoned)	
Beeswax	0.96-0.97	Lime, slaked	1.3-1.4	alder	0.42-0.68
Beryl	2.69-2.70	Limestone	2.68-2.76	apple	0.66-0.84
Biotite	2.7-3.1	Linoleum	1.18	ash	0.65-0.85
Bone	1.7-2.0	Magnetite	4.9-5.2	balsa	0.11-0.14
Brasses	8.44-8.75	Malachite	3.7-4.1	bamboo	0.31-0.40
Brick	1.4-2.2	Marble	2.6-2.84	basswood	0.32-0.59
Bronzes	8.74-8.89	Meerschaum	0.99-1.28	beech	0.70-0.90
Butter	0.86-0.87	Mica	2.6-3.2	birch	0.51-0.77
Calamine	4.1-4.5	Muscovite	2.76-3.00	blue gum	1.00
Calcspar	2.6-2.8	Ochre	3.5	box	0.95-1.16
Camphor	0.99	Opal	2.2	butternut	0.38
Cardboard	0.69	Paper	0.7-1.15	cedar	0.49-0.57
Celluloid	1.4	Paraffin	0.87-0.91	cherry	0.70-0.90
Cement, set	2.7-3.0	Peat blocks	0.84	dogwood	0.76
Chalk	1.9-2.8	Pitch	1.07	ebony	1.11-1.33
Charcoal,		Polyamides	1.15-1.25	elm	0.54-0.60
oak	0.57	Polyethylene	0.92-0.97	hickory	0.60-0.93
pine	0.28-0.44	Poly(methyl methacrylate)	1.19	holly	0.76
Cinnabar	8.12	Polypropylene	0.91-0.94	juniper	0.56
Clay	1.8-2.6	Polystyrene	1.06-1.12	larch	0.50-0.56
Coal,		Polytetrafluoroethylene	2.28-2.30	locust	0.67-0.71
anthracite	1.4-1.8	Poly(vinyl acetate)	1.19	logwood	0.91
bituminous	1.2-1.5	Poly(vinyl chloride)	1.39-1.42	mahogany	0.66-0.85
Coke	1.0-1.7	Porcelain	2.3-2.5	maple	0.62-0.75
Copal	1.04-1.14	Porphyry	2.6-2.9	oak	0.60-0.90
Cork	0.22-0.26	Pyrite	4.95-5.10	pear	0.61-0.73
Corundum	3.9-4.0	Quartz (α)	2.65	pine, pitch	0.83-0.85
Diamond	3.51	Resin	1.07	white	0.35-0.50
Dolomite	2.84	Rock salt	2.18	yellow	0.37-0.60
Ebonite	1.15	Rubber,		plum	0.66-0.78
Emery	4.0	hard	1.19	poplar	0.35-0.50
Epidote	3.25-3.50	soft	1.1	satinwood	0.95
Feldspar	2.55-2.75	pure gum	0.91-0.93	spruce	0.48-0.70
Flint	2.63	Neoprene	1.23-1.25	sycamore	0.40-0.60
Fluorite	3.18	Sandstone	2.14-2.36	teak, Indian	0.66-0.98
Galena	7.3-7.6	Serpentine	2.50-2.65	walnut	0.64-0.70
Garnet	3.15-4.3	Silica, fused,	2.21	water gum	1.00
Gelatin	1.27	Silicon carbide	3.16	willow	0.40-0.60
Glass,		Slag	2.0-3.9	Wood's metal	9.70
common	2.4-2.8	Slate	2.6-3.3		
lead	3-4	Soapstone	2.6-2.8		

DENSITY OF SULFURIC ACID

This table gives the density of aqueous sulfuric acid solutions as a function of concentration (in mass percent of H_2SO_4) and temperature.

Reference

Washburn, E. W., Ed., *International Critical Tables of Numerical Data of Physics, Chemistry, and Technology*, Vol. 3, p. 56, McGraw-Hill, New York, 1926–1932.

| Mass % | \multicolumn{11}{c}{Density in g/mL} |
	0 °C	10 °C	15 °C	20 °C	25 °C	30 °C	40 °C	50 °C	60 °C	80 °C	100 °C
1	1.0074	1.0068	1.0060	1.0051	1.0038	1.0022	0.9986	0.9944	0.9895	0.9779	0.9645
2	1.0147	1.0138	1.0129	1.0118	1.0104	1.0087	1.0050	1.0006	0.9956	0.9839	0.9705
3	1.0219	1.0206	1.0197	1.0184	1.0169	1.0152	1.0113	1.0067	1.0017	0.9900	0.9766
4	1.0291	1.0275	1.0264	1.0250	1.0234	1.0216	1.0176	1.0129	1.0078	0.9961	0.9827
5	1.0364	1.0344	1.0332	1.0317	1.0300	1.0281	1.0240	1.0192	1.0140	1.0022	0.9888
6	1.0437	1.0414	1.0400	1.0385	1.0367	1.0347	1.0305	1.0256	1.0203	1.0084	0.9950
7	1.0511	1.0485	1.0469	1.0453	1.0434	1.0414	1.0371	1.0321	1.0266	1.0146	1.0013
8	1.0585	1.0556	1.0539	1.0522	1.0502	1.0481	1.0437	1.0386	1.0330	1.0209	1.0076
9	1.0660	1.0628	1.0610	1.0591	1.0571	1.0549	1.0503	1.0451	1.0395	1.0273	1.0140
10	1.0735	1.0700	1.0681	1.0661	1.0640	1.0617	1.0570	1.0517	1.0460	1.0338	1.0204
12	1.0886	1.0846	1.0825	1.0802	1.0780	1.0756	1.0705	1.0651	1.0593	1.0469	1.0335
14	1.1039	1.0994	1.0971	1.0947	1.0922	1.0897	1.0844	1.0788	1.0729	1.0603	1.0469
16	1.1194	1.1145	1.1120	1.1094	1.1067	1.1040	1.0985	1.0927	1.0868	1.0740	1.0605
18	1.1351	1.1298	1.1271	1.1243	1.1215	1.1187	1.1129	1.1070	1.1009	1.0879	1.0744
20	1.1510	1.1453	1.1424	1.1394	1.1365	1.1335	1.1275	1.1215	1.1153	1.1021	1.0885
22	1.1670	1.1609	1.1579	1.1548	1.1517	1.1486	1.1424	1.1362	1.1299	1.1166	1.1029
24	1.1832	1.1768	1.1736	1.1704	1.1672	1.1640	1.1576	1.1512	1.1448	1.1313	1.1176
26	1.1996	1.1929	1.1896	1.1862	1.1829	1.1796	1.1730	1.1665	1.1599	1.1463	1.1325
28	1.2160	1.2091	1.2057	1.2023	1.1989	1.1955	1.1887	1.1820	1.1753	1.1616	1.1476
30	1.2326	1.2255	1.2220	1.2185	1.2150	1.2115	1.2046	1.1977	1.1909	1.1771	1.1630
32	1.2493	1.2421	1.2385	1.2349	1.2314	1.2278	1.2207	1.2137	1.2068	1.1928	1.1787
34	1.2661	1.2588	1.2552	1.2515	1.2479	1.2443	1.2371	1.2300	1.2229	1.2088	1.1946
36	1.2831	1.2757	1.2720	1.2684	1.2647	1.2610	1.2538	1.2466	1.2394	1.2251	1.2109
38	1.3004	1.2929	1.2891	1.2855	1.2818	1.2780	1.2707	1.2635	1.2561	1.2418	1.2276
40	1.3179	1.3103	1.3065	1.3028	1.2991	1.2953	1.2880	1.2806	1.2732	1.2589	1.2446
42	1.3357	1.3280	1.3242	1.3205	1.3167	1.3129	1.3055	1.2981	1.2907	1.2762	1.2619
44	1.3538	1.3461	1.3423	1.3384	1.3346	1.3308	1.3234	1.3160	1.3086	1.2939	1.2796
46	1.3724	1.3646	1.3608	1.3569	1.3530	1.3492	1.3417	1.3343	1.3269	1.3120	1.2976
48	1.3915	1.3835	1.3797	1.3758	1.3719	1.3680	1.3604	1.3528	1.3455	1.3305	1.3159
50	1.4110	1.4029	1.3990	1.3951	1.3911	1.3872	1.3795	1.3719	1.3644	1.3494	1.3348
52	1.4310	1.4228	1.4188	1.4148	1.4109	1.4069	1.3991	1.3914	1.3837	1.3687	1.3540
54	1.4515	1.4431	1.4391	1.4350	1.4310	1.4270	1.4191	1.4113	1.4036	1.3884	1.3735
56	1.4724	1.4640	1.4598	1.4557	1.4516	1.4475	1.4396	1.4317	1.4239	1.4085	1.3934
58	1.4937	1.4852	1.4809	1.4768	1.4726	1.4685	1.4604	1.4524	1.4446	1.4290	1.4137
60	1.5154	1.5067	1.5024	1.4983	1.4940	1.4898	1.4816	1.4735	1.4656	1.4497	1.4344
62	1.5375	1.5287	1.5243	1.5200	1.5157	1.5115	1.5031	1.4950	1.4869	1.4708	1.4554
64	1.5600	1.5510	1.5465	1.5421	1.5378	1.5335	1.5250	1.5167	1.5086	1.4923	1.4766
66	1.5828	1.5736	1.5691	1.5646	1.5602	1.5558	1.5472	1.5388	1.5305	1.5140	1.4981
68	1.6059	1.5965	1.5920	1.5874	1.5829	1.5785	1.5697	1.5611	1.5528	1.5359	1.5198
70	1.6293	1.6198	1.6151	1.6105	1.6059	1.6014	1.5925	1.5838	1.5753	1.5582	1.5417
72	1.6529	1.6433	1.6385	1.6338	1.6292	1.6246	1.6155	1.6067	1.5981	1.5806	1.5637
74	1.6768	1.6670	1.6622	1.6574	1.6526	1.6480	1.6387	1.6297	1.6209	1.6031	1.5857
76	1.7008	1.6908	1.6858	1.6810	1.6761	1.6713	1.6619	1.6526	1.6435	1.6252	1.6074
78	1.7247	1.7144	1.7093	1.7043	1.6994	1.6944	1.6847	1.6751	1.6657	1.6469	1.6286
80	1.7482	1.7376	1.7323	1.7272	1.7221	1.7170	1.7069	1.6971	1.6873	1.6680	1.6493
82	1.7709	1.7599	1.7544	1.7491	1.7437	1.7385	1.7281	1.7180	1.7080	1.6882	1.6692
84	1.7916	1.7804	1.7748	1.7693	1.7639	1.7585	1.7479	1.7375	1.7274	1.7072	1.6878
86	1.8095	1.7983	1.7927	1.7872	1.7818	1.7763	1.7657	1.7552	1.7449	1.7245	1.7050
88	1.8243	1.8132	1.8077	1.8022	1.7968	1.7914	1.7809	1.7705	1.7602	1.7397	1.7202
90	1.8361	1.8252	1.8198	1.8144	1.8091	1.8038	1.7933	1.7829	1.7729	1.7525	1.7331
91	1.8410	1.8302	1.8248	1.8195	1.8142	1.8090	1.7986	1.7883	1.7783	1.7581	1.7388
92	1.8453	1.8346	1.8293	1.8240	1.8188	1.8136	1.8033	1.7932	1.7832	1.7633	1.7439
93	1.8490	1.8384	1.8331	1.8279	1.8227	1.8176	1.8074	1.7974	1.7876	1.7681	1.7485
94	1.8520	1.8415	1.8363	1.8312	1.8260	1.8210	1.8109	1.8011	1.7914		

					Density in g/mL						
Mass %	0 °C	10 °C	15 °C	20 °C	25 °C	30 °C	40 °C	50 °C	60 °C	80 °C	100 °C
95	1.8544	1.8439	1.8388	1.8337	1.8286	1.8236	1.8137	1.8040	1.7944		
96	1.8560	1.8457	1.8406	1.8355	1.8305	1.8255	1.8157	1.8060	1.7965		
97	1.8569	1.8466	1.8414	1.8364	1.8314	1.8264	1.8166	1.8071	1.7977		
98	1.8567	1.8463	1.8411	1.8361	1.8310	1.8261	1.8163	1.8068	1.7976		
99	1.8551	1.8445	1.8393	1.8342	1.8292	1.8242	1.8145	1.8050	1.7958		
100	1.8517	1.8409	1.8357	1.8305	1.8255	1.8205	1.8107	1.8013	1.7922		

DENSITY OF ETHANOL–WATER MIXTURES

This table gives the density of mixtures of ethanol and water as a function of composition and temperature. The composition is specified in weight percent of ethanol, i.e., mass of ethanol per 100 g of solution. Values from the reference have been converted to true densities.

Reference

Washburn, E. W., Ed., *International Critical Tables of Numerical Data of Physics, Chemistry, and Technology*, Vol. 3, McGraw-Hill, New York, 1926–1932.

Weight % Ethanol	Density in g/cm³						
	10 °C	15 °C	20 °C	25 °C	30 °C	35 °C	40 °C
0	0.99970	0.99910	0.99820	0.99705	0.99565	0.99403	0.99222
5	0.99095	0.99029	0.98935	0.98814	0.98667	0.98498	0.98308
10	0.98390	0.98301	0.98184	0.98040	0.97872	0.97682	0.97472
15	0.97797	0.97666	0.97511	0.97331	0.97130	0.96908	0.96667
20	0.97249	0.97065	0.96861	0.96636	0.96392	0.96131	0.95853
25	0.96662	0.96421	0.96165	0.95892	0.95604	0.95303	0.94988
30	0.95974	0.95683	0.95379	0.95064	0.94738	0.94400	0.94052
35	0.95159	0.94829	0.94491	0.94143	0.93787	0.93422	0.93048
40	0.94235	0.93879	0.93515	0.93145	0.92767	0.92382	0.91989
45	0.93223	0.92849	0.92469	0.92082	0.91689	0.91288	0.90881
50	0.92159	0.91773	0.91381	0.90982	0.90577	0.90165	0.89747
55	0.91052	0.90656	0.90255	0.89847	0.89434	0.89013	0.88586
60	0.89924	0.89520	0.89110	0.88696	0.88275	0.87848	0.87414
65	0.88771	0.88361	0.87945	0.87524	0.87097	0.86664	0.86224
70	0.87599	0.87184	0.86763	0.86337	0.85905	0.85467	0.85022
75	0.86405	0.85985	0.85561	0.85131	0.84695	0.84254	0.83806
80	0.85194	0.84769	0.84341	0.83908	0.83470	0.83027	0.82576
85	0.83948	0.83522	0.83093	0.82658	0.82218	0.81772	0.81320
90	0.82652	0.82225	0.81795	0.81360	0.80920	0.80476	0.80026
95	0.81276	0.80850	0.80422	0.79989	0.79553	0.79112	0.78668
100	0.79782	0.79358	0.78932	0.78504	0.78073	0.77639	0.77201

DIELECTRIC STRENGTH OF INSULATING MATERIALS

L. I. Berger

The loss of the dielectric properties by a sample of a gaseous, liquid, or solid insulator as a result of application to the sample of an electric field* greater than a certain critical magnitude is called *dielectric breakdown*. The critical magnitude of electric field at which the breakdown of a material takes place is called the *dielectric strength* of the material (or *breakdown voltage*). The dielectric strength of a material depends on the specimen thickness (as a rule, thin films have greater dielectric strength than that of thicker samples of a material), the electrode shape**, the rate of the applied voltage increase, the shape of the voltage vs. time curve, and the medium surrounding the sample, e.g., air or other gas (or a liquid — for solid materials only).

Breakdown in Gases

The current carriers in gases are free electrons and ions generated by external radiation. The equilibrium concentration of these particles at normal pressure is about 10^3 cm^{-3}, and hence the electrical conductivity is very small, of the order of $10^{-16} - 10^{-15}$ S/cm. But in a strong electric field, these particles acquire kinetic energy along their free path, large enough to ionize the gas molecules. The new charged particles ionize more molecules; this avalanche-like process leads to formation between the electrodes of channels of conducting plasma (streamers), and the electrical resistance of the space between the electrodes decreases virtually to zero.

Because the dielectric strength (breakdown voltage) of gases strongly depends on the electrode geometry and surface condition and the gas pressure, it is generally accepted to present the data for a particular gas as a fraction of the dielectric strength of either nitrogen or sulfur hexafluoride measured at the same conditions. In Table 1, the data are presented in comparison with the dielectric strength of nitrogen, which is considered equal to 1.00. For convenience to the reader, a few average magnitudes of the dielectric strength of some gases are expressed in kilovolts per millimeter. The data in the table relate to the standard conditions, unless indicated otherwise.

Breakdown in Liquids

If a liquid is pure, the breakdown mechanism in it is similar to that in gases. If a liquid contains liquid impurities in the form of small drops with greater dielectric constant than that of the main liquid, the breakdown is the result of formation of ellipsoids from these drops by the electric field. In a strong enough electric field, these ellipsoids merge and form a high-conductivity channel between the electrodes. The current increases the temperature in the channel, liquid boils, and the current along the steam canal leads to breakdown. Formation of a conductive channel (bridge) between the electrodes is observed also in liquids with solid impurities. If a liquid contains gas impurities in the form of small bubbles, breakdown is the result of heating of the liquid in strong electric fields. In the locations with the highest current density, the liquid boils, the size of the gas bubbles increases, they merge and form gaseous channels between the electrodes, and the breakdown medium is again the gas plasma.

Breakdown in Solids

It is known that the current in solid insulators does not obey Ohm's law in strong electric fields. The current density increases almost exponentially with the electric field, and at a certain field magnitude it jumps to very high magnitudes at which a specimen of a material is destroyed. The two known kinds of electric breakdown are thermal and electrical breakdowns. The former is the result of material heating by the electric current. Destruction of a sample of a material happens when, at a certain voltage, the amount of heat produced by the current exceeds the heat release through the sample surface; the breakdown voltage in this case is proportional to the square root of the ratio of the thermal conductivity and electrical conductivity of the material. A semi-empirical expression for dependence of the breakdown voltage, V_B, on the physical properties and geometry of a sample of a solid material for the one-dimensional case is

$$V_B = [A\rho\kappa / a\varphi(d)]^{1/2}$$

where A is a numerical constant related to the system of units used, ρ and κ are the volume resistivity and thermal conductivity of the sample material, a is a constant related to the chemical bond nature and crystal structure of the sample material, and $\varphi(d)$ is a function of the sample geometry, first of all, thickness, d (see, e.g., Ref. R6). In the majority of materials, $\varphi(d)$ increases with d, hence, the magnitude of V_B is greater in the thinner samples of a particular material.

The electrical breakdown results from the tunneling of the charge carriers from electrodes or from the valence band or from the impurity levels into the conduction band, or by the impact ionization. The tunnel effect breakdown happens mainly in thin layers, e.g., in thin p-n junctions. Otherwise, the impact ionization mechanism dominates. For this mechanism, the dielectric strength of an insulator can be estimated using Boltzmann's kinetic equation for electrons in a crystal.

In the following tables, the dielectric strength values are for room temperature and normal atmospheric pressure, unless indicated otherwise.

---

* The unit of electric field in the SI system is newton per coulomb or volt per meter.
** For example, the U.S. standard ASTM D149 is based on use of symmetrical electrodes, while per U.K. standard BS2918 one electrode is a plane and the other is a rod with the axis normal to the plane.

TABLE 1. Dielectric Strength of Gases

Material	Dielectric* strength	Ref.	Material	Dielectric* strength	Ref.
Nitrogen, N_2	1.00		Trichlorofluoromethane, CCl_3F	3.50	1
Hydrogen, H_2	0.50	1,2		4.53	2
Helium, He	0.15	1	Trichloromethane, $CHCl_3$	4.2	1
Oxygen, O_2	0.92	2		4.39	2
Air	0.97	6	Methylamine, CH_3NH_2	0.81	1
Air (flat electrodes), kV/mm	3.0	3	Difluoromethane, CH_2F_2	0.79	2
Air, kV/mm	0.4-0.7	4	Trifluoromethane, CHF_3	0.71	2
Air, kV/mm	1.40	5	Bromochlorodifluoromethane, CF_2ClBr	3.84	2
Neon, Ne	0.25	1	Chlorodifluoromethane, $CHClF_2$	1.40	1
	0.16	2		1.11	2
Argon, Ar	0.18	2	Dichlorofluoromethane, $CHCl_2F$	1.33	1
Chlorine, Cl_2	1.55			2.61	2
Carbon monoxide, CO	1.02	1	Chlorofluoromethane, CH_2ClF	1.03	1
	1.05	2	Hexafluoroethane, C_2F_6	1.82	1
Carbon dioxide, CO_2	0.88	1		2.55	2
	0.82	2	Ethyne (Acetylene), C_2H_2	1.10	1
	0.84	6		1.11	2
Nitrous oxide, N_2O	1.24	2	Chloropentafluoroethane, C_2ClF_5	2.3	1
Sulfur dioxide, SO_2	2.63	2		3.0	6
	2.68	6	Dichlorotetrafluoroethane, $C_2Cl_2F_4$	2.52	2
Sulfur monochloride, S_2Cl_2	1.02	1	Chlorotrifluoroethylene, C_2ClF_3	1.82	2
(at 12.5 Torr)			1,1,1-Trichloro-2,2,2-trifluoroethane	6.55	1
Thionyl fluoride, SOF_2	2.50	1	1,1,2-Trichloro-1,2,2-trifluoroethane	6.05	1
Sulfur hexafluoride, SF_6	2.50	1	Chloroethane, C_2H_5Cl	1.00	1
	2.63	2	1,1-Dichloroethane	2.66	2
Sulfur hexafluoride, SF_6, kV/mm	8.50	7	Trifluoroacetonitrile, CF_3CN	3.5	1
	9.8	8	Acetonitrile, CH_3CN	2.11	2
Perchloryl fluoride, ClO_3F	2.73	1	Dimethylamine, $(CH_3)_2NH$	1.04	1
Tetrachloromethane, CCl_4	6.33	1	Ethylamine, $C_2H_5NH_2$	1.01	1
	6.21	2	Ethylene oxide (oxirane), CH_3CHO	1.01	1
Tetrafluoromethane, CF_4	1.01	1	Perfluoropropene, C_3F_6	2.55	2
Methane, CH_4	1.00	1	Octafluoropropane, C_3F_8	2.19	1
	1.13	2		2.47	2
Bromotrifluoromethane, CF_3Br	1.35	1	3,3,3-Trifluoro-1-propene, CH_2CHCF_3	2.11	2
	1.97	2	Pentafluoroisocyanoethane, C_2F_5NC	4.5	1
Bromomethane, CH_3Br	0.71	2	1,1,1,4,4,4-Hexafluoro-2-butyne, CF_3CCCF_3	5.84	2
Chloromethane, CH_3Cl	1.29	2	Octafluorocyclobutane, C_4F_8	3.34	2
Iodomethane, CH_3I	3.02	1	1,1,1,2,3,4,4,4-Octafluoro-2-butene	2.8	1
Iodomethane, CH_3I, at 370 Torr	2.20	7	Decafluorobutane, C_4F_{10}	3.08	1
Dichloromethane, CH_2Cl_2	1.92	2	Perfluorobutanenitrile, C_3F_7CN	5.5	1
Dichlorodifluoromethane, CCl_2F_2	2.42	1	Perfluoro-2-methyl-1,3-butadiene, C_5F_8	5.5	1
	2.63	2,6	Hexafluorobenzene, C_6F_6	2.11	2
Chlorotrifluoromethane, $CClF_3$	1.43	1	Perfluorocyclohexane, C_6F_{12}, (saturated vapor)	6.18	2
	1.53	2			

* Relative to nitrogen, unless units of kV/mm are indicated.

TABLE 2. Dielectric Strength of Liquids

Material	Dielectric strength kV/mm	Ref.	Material	Dielectric strength kV/mm	Ref.
Helium, He, liquid, 4.2 K	10	9	Octane, C_8H_{18}	16.6	14
Static	10	11		20.4	15
Dynamic	5	11		179	17,18
	23	12	Ethylbenzene, C_8H_{10}	226	17,18
Nitrogen, N_2, liquid, 77K			Propylbenzene, C_9H_{12}	250	17,18
Coaxial cylinder electrodes	20	10	Isopropylbenzene, C_9H_{12}	238	17,18
Sphere to plane electrodes	60	10	Decane, $C_{10}H_{22}$	192	17,18
Water, H_2O, distilled	65-70	13	Synthetic Paraffin Mixture		
Carbon tetrachloride, CCl_4	5.5	14	Synfluid 2cSt PAO	29.5	37
	16.0	15	Butylbenzene, $C_{10}H_{14}$	275	17,18
Hexane, C_6H_{14}	42.0	16	Isobutylbenzene, $C_{10}H_{14}$	222	17,18
Two 2.54 cm diameter spherical			Silicone oils—polydimethylsiloxanes,		
electrodes, 50.8 µm space	156	17,18	$(CH_3)_3Si\text{-}O\text{-}[Si(CH_3)_2]_x\text{-}O\text{-}Si(CH_3)_3$		
Cyclohexane, C_6H_{12}	42-48	16	Polydimethylsiloxane silicone fluid	15.4	20
2-Methylpentane, C_6H_{14}	149	17,18	Dimethyl silicone	24.0	21,22
2,2-Dimethylbutane, C_6H_{14}	133	17,18	Phenylmethyl silicone	23.2	22
2,3-Dimethylbutane, C_6H_{14}	138	17,18	Silicone oil, Basilone M50	10-15	23
Benzene, C_6H_6	163	17,18	Mineral insulating oils	11.8	6
Chlorobenzene, C_6H_5Cl	7.1	14	Polybutene oil for capacitors	13.8	6
	18.8	15	Transformer dielectric liquid	28-30	6
2,2,4-Trimethylpentane, C_8H_{18}	140	17,18	Isopropylbiphenyl capacitor oil	23.6	6
Phenylxylylethane	23.6	19	Transformer oil	110.7	24
Heptane, C_7H_{16}	166	17,18	Transformer oil Agip ITE 360	9-12.6	23
2,4-Dimethylpentane, C_7H_{16}	133	17,18	Perfluorinated hydrocarbons		
Toluene, $C_6H_5CH_3$	199	17,18	Fluorinert FC 6001	8.0	23
	46	16	Fluorinert FC 77	10.7	23
	12.0	14	Perfluorinated polyethers		
	20.4	15	Galden XAD (Mol. wt. 800)	10.5	23
			Galden D40 (Mol. wt. 2000)	10.2	23
			Castor oil	65	25

TABLE 3. Dielectric Strength of Solids

Material	Dielectric strength kV/mm	Ref	Material	Dielectric strength kV/mm	Ref
Sodium chloride, NaCl, crystalline	150	26	Phlogopite, amber, natural	118	6
Potassium bromide, KBr, crystalline	80	26	Fluorophlogopite, synthetic	118	6
Ceramics			Glass-bonded mica	14.0-15.7	6
Alumina (99.9% Al_2O_3)	13.4	6,27a	Thermoplastic Polymers		
Aluminum silicate, Al_2SiO_5	5.9	6	Polypropylene	23.6	6
Berillia (99% BeO)	13.8	6,27b	Amide polymer nylon 6/6, dry	23.6	6
Boron nitride, BN	37.4	6	Polyamide-imide copolymer	22.8	6
Cordierite, $Mg_2Al_4Si_5O_{18}$	7.9	6,27c	Modified polyphenylene oxide	21.7	6
Forsterite, Mg_2SiO_4	9.8	28	Polystyrene	19.7	6
Porcelain	35-160	26	Polymethyl methacrylate	19.7	6
Steatite, $Mg_3Si_4O_{11}\cdot H_2O$	9.1-15.4	6	Polyetherimide	18.9	6
Titanates of Mg, Ca, Sr, Ba, and Pb	20-120	3	Amide polymer nylon 11(dry)	16.7	6
Barium titanate, glass bonded	>30	36	Polysulfone	16.7	6
Zirconia, ZrO_2	11.4	29	Styrene-acrylonitrile copolymer	16.7	6
Glasses			Acrylonitrile-butadiene-styrene	16.7	6
Fused silica, SiO_2	470-670	26	Polyethersulfone	15.7	6
Alkali-silicate glass	200	26	Polybutylene terephthalate	15.7	6
Standard window glass	9.8-13.8	28	Polystyrene-butadiene copolymer	15.7	6
Micas			Acetal homopolymer	15.0	6
Muscovite, ruby, natural	118	6	Acetal copolymer	15.0	6
			Polyphenylene sulfide	15.0	6

Material	Dielectric strength kV/mm	Ref
Polycarbonate	15.0	6
Acetal homopolymer resin (molding resin)	15.0	6
Acetal copolymer resin	15.0	6
Thermosetting Molding Compounds		
Glass-filled allyl	15.7	6
(Type GDI-30 per MIL-M-14G)		
Glass-filled epoxy, electrical grade	15.4	6
Glass-filled phenolic	15.0	6
(Type GPI-100 per MIL-M-14G)		
Glass-filled alkyd/polyester	14.8	6
(Type MAI-60 per MIL-M-14G)		
Glass-filled melamine	13.4	6
(Type MMI-30 per MIL-M-14G)		
Extrusion Compounds for High-Temperature Insulation		
Polytetrafluoroethylene	19.7	6
Perfluoroalkoxy polymer	21.7	6
Fluorinated ethylene-propylene copolymer	19.7	6
Ethylene-tetrafluoroethylene copolymer	15.7	6
Polyvinylidene fluoride	10.2	6
Ethylene-chlorotrifluoroethylene copolymer	19.3	6
Polychlorotrifluoroethylene	19.7	6
Extrusion Compounds for Low-Temperature Insulation		
Polyvinyl chloride		
Flexible	11.8-15.7	30
Rigid	13.8-19.7	30
Polyethylene	18.9	28
Polyethylene, low-density	21.7	6
	300	31
Polyethylene, high-density	19.7	6
Polypropylene/polyethylene copolymer	23.6	6
Embedding Compounds		
Basic epoxy resin: bisphenol-A/epichlorohydrin polycondensate	19.7	6
Cycloaliphatic epoxy: alicyclic diepoxy carboxylate	19.7	6
Polyetherketone	18.9	30
Polyurethanes		
Two-component, polyol-cured	25.4	6
Two-part solventless, polybutylene-based	24.0	6
Silicones		
Clear two-part heat curing eletrical grade silicone embedding resin	21.7	6
Red insulating enamel (MIL-E-22118)		
Dry	47.2	6
Wet	11.8	6
Enamels		
Red enamel, fast cure		
Standard conditions	78.7	6
Immersion conditions	47.2	6
Black enamel		
Standard conditions	70.9	6
Immersion conditions	47.2	6
Varnishes		
Vacuum-pressure impregnated baking type solventless polyester varnish		

Material	Dielectric strength kV/mm	Ref
Rigid, two-part	70.9	6
Semiflexible high-bond thixotropic	78.7	6
Rigid high-bond high-flash freon-resistant	68.9	6
Baking type epoxy varnish		
Solventless, rigid, low viscosity, one-part	90.6	6
Solventless, semiflexible, one-part	82.7	6
Solventless, semirigid, chemical resistant, low dielectric constant	106.3	6
Solvable, for hermetic electric motors	181.1	6
Polyurethane coating		
Clear conformal, fast cure		
Standard conditions	78.7	6
Immersion conditions	47.2	6
Insulating Films and Tapes		
Low-density polyethylene film (40 μm thick)	300	31
Poly-p-xylylene film	410-590	32
Aromatic polymer films		
Kapton H (Du Pont)	389-430	33
Ultem (GE Plastic and Roem AG)	437-565	33
Hostaphan (Hoechst AG)	338-447	33
Amorphous Stabar K2000 (ICI film)	404-422	33
Stabar S100 (ICI film)	353-452	33
Polyetherimide film (26 μm)	486	34
Parylene N/D (poly-p-xylylene/poly-dichloro-p-xylylene) 25 μm film	275	6
Cellulose acetate film	157	6
Cellulose triacetate film	157	6
Polytetrafluoroethylene film	87-173	6
Perfluoroalkoxy film	157-197	6
Fluorinated ethylene-propylene copolymer film	197	6
Ethylene-tetrafluoroethylene film	197	6
Ethylene-chlorotrifluoroethylene copolymer film	197	6
Polychlorotrifluoroethylene film	118-153.5	6
High-voltage rubber insulating tape	28	6
Composites		
Isophthalic polyester (vinyl toluene monomer) filled with		
Calcium carbonate, $CaCO_3$	15.0	38
Gypsum, $CaSO_4$	14.4	38
Alumina trihydrate	15.4	38
Clay	14.4	38
BPA fumarate polyester (vinyl toluene monomer) filled with		
Calcium carbonate	6.1	38
Gypsum	5.9	38
Alumina trihydrate	11.8	38
Clay	12.6	38
Polysulfone resin—30% glass fiber	16.5-18.7	38
Polyamid resin (Nylon 66)— 30% carbon fiber	13.0	38
Polyimide thermoset resin, glass reinforced	12.0	39
Polyester resin (thermoplastic)—		

Material	Dielectric strength kV/mm	Ref	Material	Dielectric strength kV/mm	Ref
40% glass fiber	20.0	38	Room-temperature vulcanized silicone rubber	9.2-10.9	35
Epoxy resin (diglycidyl ether of bisphenol A), glass reinforced	16.0	40	Ureas (from carbamide to tetraphenylurea)	11.8-15.7	28
Various Insulators			Dielectric papers		
Rubber, natural	100-215	26	Aramid paper, calendered	28.7	6
Butyl rubber	23.6	6	Aramid paper, uncalendered	12.2	6
Neoprene	15.7-27.6	6	Aramid with Mica	39.4	6
Silicone rubber	26-36	6			

References

1. Vijh, A. K., *IEEE Trans.*, EI-12, 313, 1997.
2. Brand, K. P., *IEEE Trans.*, EI-17, 451, 1982.
3. *Encyclopedic Dictionary in Physics*, Vedensky, B. A. and Vul, B. M., Eds., Vol. 4, Soviet Encyclopedia Publishing House, Moscow, 1965.
4. Kubuki, M., Yoshimoto, R., Yoshizumi, K., Tsuru, S., and Hara, M., *IEEE Trans.*, DEI-4, 92, 1997.
5. Al-Arainy, A. A., Malik, N. H., and Qureshi, M. I., *IEEE Trans.*, DEI-1, 305, 1994.
6. Shugg, W. T., *Handbook of Electrical and Electronic Insulating Materials*, Van Nostrand Reinhold, New York, 1986.
7. Devins, J. C., *IEEE Trans.*, EI-15, 81, 1980.
8. Xu, X., Jayaram, S., and Boggs, S. A., *IEEE Trans.*, DEI-3, 836, 1996.
9. Okubo, H., Wakita, M., Chigusa, S., Nayakawa, N., and Hikita, M., *IEEE Trans.*, DEI-4, 120, 1997.
10. Hayakawa, H., Sakakibara, H., Goshima, H., Hikita, M., and Okubo, H., *IEEE Trans.*, DEI-4, 127, 1997.
11. Okubo, H., Wakita, M., Chigusa, S., Hayakawa, N., and Hikita, M., *IEEE Trans.*, DEI-4, 220, 1997.
12. Von Hippel, A. R., *Dielectric Materials and Applications*, MIT Press, Cambridge, MA, 1954.
13. Jones, H. M. and Kunhards, E. E., *IEEE Trans.*, DEI-1, 1016, 1994.
14. Nitta, Y. and Ayhara, Y., *IEEE Trans.*, EI-11, 91, 1976.
15. Gallagher, T. J., *IEEE Trans.*, EI-12, 249, 1977.
16. Wong, P. P. and Forster, E. O., in *Dielectric Materials. Measurements and Applications*, IEE Conf. Publ. 177, 1, 1979.
17. Kao, K. C. *IEEE Trans.*, EI-11, 121, 1976.
18. Sharbaugh, A. H., Crowe, R. W., and Cox, E. B., *J. Appl. Phys.*, 27, 806, 1956.
19. Miller, R. L., Mandelcorn, L., and Mercier, G. E., in *Proc. Intl. Conf. on Properties and Applications of Dielectric Materials*, Xian, China, June 24-28, 1985; cited in Ref. 6, p. 492.
20. Hakim, R. M., Oliver, R. G., and St-Onge, H., *IEEE Trans.*, EI-12, 360, 1977.
21. Hosticka, C., *IEEE Trans.*, 389, 1977.
22. Yasufuku, S., Umemura, T., and Ishioka, Y., *IEEE Trans.*, EI-12, 402, 1977.
23. Forster, E. O., Yamashita, H., Mazzetti, C., Pompini, M., Caroli, L., and Patrissi, S., *IEEE Trans.*, DEI-1, 440, 1994.
24. Bell, W. R., *IEEE Trans.*, 281, 1977.
25. Ramu, T. C. and Narayana Rao, Y., in *Dielectric Materials. Measurements and Applications*, IEE Conf. Publ. 177, 37.
26. Skanavi, G. I., *Fizika Dielektrikov; Oblast Silnykh Polei* (Physics of Dielectrics; Strong Fields). Gos. Izd. Fiz. Mat. Nauk (State Publ. House for Phys. and Math. Scis.), Moscow, 1958.
27. Kleiner, R. N., in *Practical Handbook of Materials Science*, Lynch, C. T., Ed., CRC Press, 1989; 27a: p. 304; 27b: p.300; 27c: p. 316.
28. *Materials Selector Guide. Materials and Methods*, Reinhold Publ., New York, 1973.
29. Flinn, R. A. and Trojan, P. K., *Engineering Materials and Their Applications*, 2nd ed., Houghton Mifflin, 1981, p. 614.
30. Lynch, C. T., Ed., *Practical Handbook of Materials Science*, CRC Press, Boca Raton, FL, 1989.
31. Suzuki, H., Mukai, S., Ohki, Y., Nakamichi, Y., and Ajiki, K., *IEEE Trans.*, DEI-4, 238, 1997.
32. Mori, T., Matsuoka, T., and Muzitani, T., *IEEE Trans.*, DEI-1, 71, 1994.
33. Bjellheim, P. and Helgee, B., *IEEE Trans.*, DEI-1, 89, 1994.
34. Zheng, J. P., Cygan, P. J., and Jow, T. R., *IEEE Trans.*, DEI-3, 144, 1996.
35. Danukas, M. G., *IEEE Trans.*, DEI-1, 1196, 1994.
36. Burn, I. and Smithe, D. H., *J. Mater. Sci.*, 7, 339, 1972.
37. Hope, K.D., Chevron Chemical, Private Communication.
38. *Engineering Materials Handbook*, Vol. 1, Composites, C.A. Dostal, Ed., ASM Intl., 1987.
39. 1985 Materials Selector, *Mater. Eng.*, (12) 1984.
40. *Modern Plastics Encyclopedia*, McGraw-Hill, v. 62 (No. 10A) 1985–1986.

Review Literature on the Subject

R1. Kuffel, E. and Zaengl, W. S., *HV Engineering Fundamentals*, Pergamon, 1989.
R2. Kok, J. A., *Electrical Breakdown of Insulating Liquids*, Phillips Tech. Library, Cleaver-Hum, London, 1961.
R3. Gallagher, T. J., *Simple Dielectric Liquids*, Clarendon, Oxford, 1975.
R4. Meek, J. M. and Craggs, J. D., Eds., *Electric Breakdown in Gases*, John Wiley & Sons, 1976.
R5. Von Hippel, A. R., *Dielectric Materials and Applications*, MIT Press, Cambridge, MA, 1954.
R6. O'Dwyer, J. J. *The Theory of Dielectric Breakdown of Solids*, Clarendon Press, 1964.

COEFFICIENT OF FRICTION

The coefficient of friction between two surfaces is the ratio of the force required to move one over the other to the force pressing the two together. Thus if F is the minimum force needed to move one surface over the other, and W is the force pressing the surfaces together, the coefficient of friction μ is given by $\mu = F/W$. A greater force is generally needed to initiate movement from rest than to continue the motion once sliding has started. Thus the static coefficient of friction μ (static) is usually larger that the sliding or kinetic coefficient μ (sliding).

This table gives characteristic values of both the static and sliding coefficients of friction for a number of material combinations. In each case Material 1 is moving over the surface of Material 2.

The type of lubrication or any other special condition is indicated in the third column. All values refer to room temperature unless otherwise indicated. It should be emphasized that the coefficient of friction is very sensitive to the condition of the surface, so that these values represent only a rough guide.

References

1. Minshall, H., in *CRC Handbook of Chemistry and Physics, 73rd Edition*, Lide, D. R., Ed., CRC Press, Boca Raton, FL, 1992.
2. Fuller, D. D., in *American Institute of Physics Handbook, 3rd Edition*, Gray, D. E., Ed., McGraw-Hill, New York, 1972.

Material 1	Material 2	Conditions	μ (static)	μ (sliding)
Metals				
Hard steel	Hard steel	Dry	0.78	0.42
		Castor oil	0.15	0.081
		Steric acid	0.005	0.029
		Lard	0.11	0.084
		Light mineral oil	0.23	
		Graphite		0.058
Hard steel	Graphite	Dry	0.21	
Mild steel	Mild steel	Dry	0.74	0.57
		Oleic acid		0.09
Mild steel	Phosphor bronze	Dry		0.34
Mild steel	Cast iron	Dry		0.23
Mild steel	Lead	Dry	0.95	0.95
		Mineral oil	0.5	0.3
Mild steel	Brass	Dry	0.35	
Cast iron	Cast iron	Dry	1.10	0.15
Aluminum	Aluminum	Dry	1.05	1.4
Aluminum	Mild steel	Dry	0.61	0.47
Brass	Mild steel	Dry	0.51	0.44
		Castor oil	0.11	
Brass	Cast iron	Dry		0.30
Bronze	Cast iron	Dry		0.22
Cadmium	Mild steel	Dry		0.46
Copper	Copper	Dry	1.6	
Copper	Mild steel	Dry	0.53	0.36
		Oleic acid		0.18
Copper	Cast iron	Dry	1.05	0.29
Copper	Glass	Dry	0.68	0.53
Lead	Cast iron	Dry		0.43
Magnesium	Magnesium	Dry	0.6	
Magnesium	Mild steel	Dry		0.42
Magnesium	Cast iron	Dry		0.25
Nickel	Nickel	Dry	1.10	0.53
Nickel	Mild steel	Dry		0.64
Tin	Cast iron	Dry		0.32
Zinc	Cast iron	Dry	0.85	0.21
Nonmetals				
Diamond	Diamond	Dry	0.1	
Diamond	Metals	Dry	0.12	
Garnet	Mild steel	Dry		0.39

Material 1	Material 2	Conditions	μ (static)	μ (sliding)
Glass	Glass	Dry	0.94	0.4
Glass	Nickel	Dry	0.78	0.56
Graphite	Graphite	Dry	0.1	
Mica	Mica	Freshly cleaved	1.0	
Nylon	Nylon	Dry	0.2	
Nylon	Steel	Dry	0.40	
Polyethylene	Polyethylene	Dry	0.2	
Polyethylene	Steel	Dry	0.2	
Polystyrene	Polystyrene	Dry	0.5	
Polystyrene	Steel	Dry	0.3	
Sapphire	Sapphire	Dry	0.2	
Teflon	Teflon	Dry	0.04	0.04
Teflon	Steel	Dry	0.04	0.04
Tungsten carbide	Tungsten carbide	Dry, room temp.	0.17	
		Dry, 1000 °C	0.45	
		Dry, 1600 °C	1.8	
		Oleic acid	0.12	
Tungsten carbide	Graphite	Dry	0.15	
Tungsten carbide	Steel	Dry	0.5	
		Oleic acid	0.08	

Miscellaneous Materials

Material 1	Material 2	Conditions	μ (static)	μ (sliding)
Cotton	Cotton	Threads	0.3	
Leather	Cast iron	Dry	0.6	0.56
Leather	Oak	Parallel to grain	0.61	0.52
Oak	Oak	Parallel to grain	0.62	0.48
		Perpendicular to grain	0.54	0.32
Silk	Silk	Clean	0.25	
Wood	Wood	Dry	0.35	
		Wet	0.2	
Wood	Brick	Dry	0.6	
Wood	Leather	Dry	0.35	

Various Materials on Ice and Snow

Material 1	Material 2	Conditions	μ (static)	μ (sliding)
Ice	Ice	Clean, 0 °C	0.1	0.02
		Clean, −12 °C	0.3	0.035
		Clean, −80 °C	0.5	0.09
Aluminum	Snow	Wet, 0 °C	0.4	
		Dry, 0 °C	0.35	
Brass	Ice	Clean, 0 °C		0.02
		Clean, −80 °C		0.15
Nylon	Snow	Wet, 0 °C	0.4	
		Dry, −10 °C	0.3	
Teflon	Snow	Wet, 0 °C	0.05	
		Dry, 0 °C	0.02	
Wax, ski	Snow	Wet, 0 °C	0.1	
		Dry, 0 °C	0.04	
		Dry, −10 °C	0.2	

FLAME TEMPERATURES

This table gives the adiabatic flame temperatures for stoichemetric mixtures of various fuels and oxidizers. The temperatures are calculated from thermodynamic and transport properties under ideal adiabatic conditions, using methods described in the reference.

Reference

Fristrom, R. M., *Flame Structures and Processes*, Oxford University Press, New York, 1995.

Adiabatic Flame Temperature in K for Various Fuel-Oxidizer Combinations

Fuel	Oxidizer					
	Air	O_2	F_2	Cl_2	N_2O	NO
Organic liquids and gases						
Acetaldehyde	2288					
Acetone	2253					
Acetylene	2607					
Benzene	2363					
Butane	2248					
Carbon disulfide	2257					
Cyanogen	2596	4855				
Cyclohexane	2250					
Cyclopropane	2370					
Decane	2286					
Ethane	2244					
Ethanol	2238					
Ethylene	2375					
Hexane	2238					
Methane	2236					
Methanol	2222					
Oxirane	2177					
Pentane	2250					
Propane	2250					
Toluene	2344					
Solids						
Aluminum		4005				
Lithium		2711				
Phosphorus (white)		3242				
Zirconium		4278				
Other						
Ammonia		2845				
Carbon monoxide	1388					
Diborane		3350				
Hydrazine		3037				
Hydrogen	2169	3000	4006	2493	2965	3127
Hydrogen sulfide	2091	3414				
Phosphine		3139				
Silane		3043				

ALLOCATION OF FREQUENCIES IN THE RADIO SPECTRUM

In the United States the National Telecommunications and Information Administration (NTIA) has responsibility for assigning each portion of the radio spectrum (9 kHz to 300 GHz) for different uses. These assignments must be compatible with the rules of the International Telecommunications Union (ITU), to which the United States is bound by treaty. The current assignments are given in a wall chart (Reference 1) and may also be found on the NTIA web site (Reference 2). The list below summarizes the broad features of the spectrum allocation, with particular attention to those sections of scientific interest. The references should be consulted for details of the allocations in the frequency bands listed here, which in some cases are quite complex.

References

1. *United States Frequency Allocations*, 1996 Spectrum Wall Chart, Stock No. 003-000-00652-2, U. S. Government Printing Office, P. O. Box 371954, Pittsburgh, PA 15250-7954.
2. http://www.ntia.doc.gov/osmhome/allochrt.html

Frequency range	Allocation
9–19.95 kHz	Maritime communication, navigation
19.95–20.05 kHz	Standard frequency and time signal (also at 60 kHz and 2.5, 5, 10, 15, 20, 25 MHz)
20.05–535 kHz	Maritime and aeronautical communication, navigation
535–1605 kHz	AM radio broadcasting
1605–3500 kHz	Mobile communication and navigation, amateur radio (1800–1900 kHz)
3.5–4.0 MHz	Amateur radio
4.0–5.95 MHz	Mobile communication
5.95–13.36 MHz	Mobile communication, amateur, short-wave broadcasting
13.36–13.41 MHz	Radioastronomy
13.41–25.55 MHz	Mobile communication, amateur, short-wave broadcasting
25.55–25.67 MHz	Radioastronomy
25.67–37.5 MHz	Mobile communication, amateur, short-wave broadcasting
37.5–38.25 MHz	Radioastronomy
38.25–50.0 MHz	Mobile communication
50.0–54.0 MHz	Amateur
54.0–72.0 MHz	TV channels 2–4
72.0–73.0 MHz	Mobile communication
73.0–74.6 MHz	Radioastronomy
74.6–76.0 MHz	Mobile communication
76.0–88.0 MHz	TV channels 5–6
88.0–108.0 MHz	FM radio broadcasting
108.0–118.0 MHz	Aeronautical navigation
118.0–174.0 MHz	Mobile communication, space research, meteorological satellites
174.0–216.0 MHz	TV channels 7–13
216.0–400.05 MHz	Mobile communication
400.05–400.15 MHz	Standard frequency and time satellite (also 20 and 25 GHz)
400.15–406.1 MHz	Meteorological aids (radiosonde)
406.1–410.0 MHz	Radioastronomy
410.0–470.0 MHz	Mobile communication, amateur
470.0–512.0 MHz	TV channels 14–20
512.0–608.0 MHz	TV channels 21–36
608.0–614.0 MHz	Radioastronomy
614.0–806.0 MHz	TV channels 38–69
806–1400 MHz	Mobile communication, navigation
1400–1427 MHz	Radioastronomy, space research
1427–1660 MHz	Various navigation and satellite applications
1660–1710 MHz	Radioastronomy, space research, meteorology
1710–2655 MHz	Various navigation and satellite applications
2655–2700 MHz	Radioastronomy, space research
2.7–4.99 GHz	Various navigation and satellite applications
4.99–5.0 GHz	Radioastronomy, space research
5.0–10.6 GHz	Various navigation and satellite applications
10.6–10.7 GHz	Radioastronomy, space research
10.7–15.35 GHz	Various navigation and satellite applications
15.35–15.4 GHz	Radioastronomy, space research
15.4–22.21 GHz	Various navigation and satellite applications

Frequency range	Allocation
22.21–22.5 GHz	Radioastronomy, space research
22.25–23.6 GHz	Various navigation and satellite applications
23.6–24.0 GHz	Radioastronomy, space research
24.0–31.3 GHz	Various navigation and satellite applications
31.3–31.8 GHz	Radioastronomy, space research
31.8–42.5 GHz	Various navigation and satellite applications
42.5–43.5 GHz	Radioastronomy
43.5–51.4 GHz	Various navigation and satellite applications
51.4–54.25 GHz	Radioastronomy, space research
54.25–58.2 GHz	Space research
58.2–59.0 GHz	Radioastronomy, space research
59.0–64.0 GHz	Satellite applications
64.0–65.0 GHz	Radioastronomy, space research
65.0–72.77 GHz	Various navigation and satellite applications
72.77–72.91 GHz	Radioastronomy, space research
72.91–86.0 GHz	Various navigation and satellite applications
86.0–92.0 GHz	Radioastronomy, space research
92.0–105.0 GHz	Various navigation and satellite applications
105.0–116.0 GHz	Radioastronomy, space research
116.0–164.0 GHz	Various navigation and satellite applications
164.0–168.0 GHz	Radioastronomy, space research
168.0–182.0 GHz	Various navigation and satellite applications
182.0–185.0 GHz	Radioastronomy, space research
185.0–217.0 GHz	Various navigation and satellite applications
217.0–231.0 GHz	Radioastronomy, space research
231.0–265.0 GHz	Various navigation and satellite applications
265.0–275.0 GHz	Radioastronomy
275.0–300.0 GHz	Mobile communications

Section 16
Health and Safety Information

HANDLING AND DISPOSAL OF CHEMICALS IN LABORATORIES

Robert Joyce and Blaine C. McKusick

The following material has been extracted from two books prepared under the auspices of the Committee on Hazardous Substances in the Laboratory of the National Academy of Sciences – National Research Council. Readers are referred to these books for full details:

Prudent Practices for Handling Hazardous Chemicals in Laboratories, National Academy Press, Washington, D.C., 1981.

Prudent Practices for Disposal of Chemicals from Laboratories, National Academy Press, Washington, D.C., 1983.

The permission of the National Academy Press to use these extracts is gratefully acknowledged.

INCOMPATIBLE CHEMICALS

The term "incompatible chemicals" refers to chemicals that can react with each other

- Violently
- With evolution of substantial heat
- To produce flammable products
- To produce toxic products

Good laboratory safety practice requires that incompatible chemicals be stored, transported, and disposed of in ways that will prevent their coming together in the event of an accident. Tables 1 and 2 give some basic guidelines for the safe handling of acids, bases, reactive metals, and other chemicals. Neither of these tables is exhaustive, and additional information on incompatible chemicals can be found in the following references.

1. Urben, P. G., Ed., *Bretherick's Handbook of Reactive Chemical Hazards*, 5th ed., Butterworth-Heinemann, Oxford, 1995.
2. Luxon, S. G., Ed., *Hazards in the Chemical Laboratory*, 5th ed., Royal Society of Chemistry, Cambridge, 1992.
3. *Fire Protection Guide to Hazardous Materials*, 11th ed., National Fire Protection Association, Quincy, MA, 1994.

TABLE 1. General Classes of Incompatible Chemicals

A	B
Acids	Bases, reactive metals
Oxidizing agents[a]	Reducing agents[a]
Chlorates	Ammonia, anhydrous and aqueous
Chromates	Carbon
Chromium trioxide	Metals
Dichromates	Metal hydrides
Halogens	Nitrites
Halogenating agents	Organic compounds
Hydrogen peroxide	Phosphorus
Nitric acid	Silicon
Nitrates	Sulfur
Perchlorates	
Peroxides	
Permanganates	
Persulfates	

[a] The examples of oxidizing and reducing agents are illustrative of common laboratory chemicals; they are not intended to be exhaustive.

TABLE 2. Examples of Incompatible Chemicals

Chemical	Is incompatible with
Acetic acid	Chromic acid, nitric acid, hydroxyl compounds, ethylene glycol, perchloric acid, peroxides, permanaganates
Acetylene	Chlorine, bromine, copper, fluorine, silver, mercury
Acetone	Concentrated nitric and sulfuric acid mixtures
Alkali and alkaline earth metals (such as powdered aluminum or magnesium, calcium, lithium, sodium, potassium)	Water, carbon tetrachloride or other chlorinated hydrocarbons, carbon dioxide, halogens
Ammonia (anhydrous)	Mercury (in manometers, for example), chlorine, calcium hypochlorite, iodine, bromine, hydrofluoric acid (anhydrous)
Ammonium nitrate	Acids, powdered metals, flammable liquids, chlorates, nitrites, sulfur, finely divided organic or combustible materials
Aniline	Nitric acid, hydrogen peroxide
Arsenical materials	Any reducing agent
Azides	Acids

Chemical	Is incompatible with
Bromine	See Chlorine
Calcium oxide	Water
Carbon (activated)	Calcium hypochlorite, all oxidizing agents
Carbon tetrachloride	Sodium
Chlorates	Ammonium salts, acids, powdered metals, sulfur, finely divided organic or combustible materials
Chromic acid and chromium troixide	Acetic acid, naphthalene, camphor, glycerol, alcohol, flammable liquids in general
Chlorine	Ammonia, acetylene, butadiene, butane, methane, propane (or other petroleum gases), hydrogen, sodium carbide, benzene, finely divided metals, turpentine
Chlorine dioxide	Ammonia, methane, phosphine, hydrogen sulfide
Copper	Acetylene, hydrogen peroxide
Cumene hydroperoxide	Acids (organic or inorganic)
Cyanides	Acids
Flammable liquids	Ammonium nitrate, chromic acid, hydrogen peroxide, nitric acid, sodium peroxide, halogens
Fluorine	Everything
Hydrocarbons (such as butane, propane, benzene)	Fluorine, chlorine, bromine, chromic acid, sodium peroxide
Hydrocyanic acid	Nitric acid, alkali
Hydrofluoric acid (anhydrous)	Ammonia (aqueous or anhydrous)
Hydrogen peroxide	Copper, chromium, iron, most metals or their salts, alcohols, acetone, organic materials, aniline, nitro-methane, combustible materials
Hydrogen sulfide	Fuming nitric acid, oxidizing gases
Hypochlorites	Acids, activated carbon
Iodine	Acetylene, ammonia (aqueous or anhydrous), hydrogen
Mercury	Acetylene, fulminic acid, ammonia
Nitrates	Sulfuric acid
Nitric acid (concentrated)	Acetic acid, aniline, chromic acid, hydrocyanic acid, hydrogen sulfide, flammable liquids, flammable gases, copper, brass, any heavy metals
Nitrites	Acids
Nitroparaffins	Inorganic bases, amines
Oxalic acid	Silver, mercury
Oxygen	Oils, grease, hydrogen, flammable liquids, solids, or gases
Perchloric acid	Acetic anhydride, bismuth and its alloys, alcohol, paper, wood, grease, oils
Peroxides, organic	Acids (organic or mineral), avoid friction, store cold
Phosphorus (white)	Air, oxygen, alkalis, reducing agents
Potassium	Carbon tetrachloride, carbon dioxide, water
Potassium chlorate	Sulfuric and other acids
Potassium perchlorate (see also chlorates)	Sulfuric and other acids
Potassium permanganate	Glycerol, ethylene glycol, benzaldehyde, surfuric acid
Selenides	Reducing agents
Silver	Acetylene, oxalic acid, tartartic acid, ammonium compounds, fulminic acid
Sodium	Carbon tetrachloride, carbon dioxide, water
Sodium nitrite	Ammonium nitrate and other ammonium salts
Sodium peroxide	Ethyl or methyl alcohol, glacial acetic acid, acetic anhydride, benzaldehyde, carbon disulfide, glycerin, ethylene glycol, ethyl acetate, methyl acetate, furfural
Sulfides	Acids
Sulfuric acid	Potassium chlorate, potassium perchlorate, potassium permanganate (similar compounds of light metals, such as sodium, lithium)
Tellurides	Reducing agents

EXPLOSION HAZARDS

Table 3 lists some common classes of laboratory chemicals that have potential for producing a violent explosion when subjected to shock or friction. These chemicals should never be disposed of as such, but should be handled by procedures given in *Prudent Practices for Disposal of Chemicals from Laboratories*, National Academy Press, 1983, chapters 6 and 7. Additional information on these, as well as on some less common classes of explosives, can be found in L. Bretherick, *Handbook of Reactive Chemical Hazards*, 5th ed., Butterworth-Heinemann, Oxford, 1995.

Table 4 lists some illustrative combinations of common laboratory reagents that can produce explosions when they are brought together or that form reaction products that can explode without any apparent external initiating action. This list is not exhaustive, and additional information on potentially explosive reagent combinations can be found in *Manual of Hazardous Chemical Reactions, A Compilation of Chemical Reactions Reported to be Potentially Hazardous*, National Fire Protection Association, NFPA 491M, 1991, NFPA, Quincy, MA.

WATER-REACTIVE CHEMICALS

Table 5 lists some common laboratory chemicals that react violently with water and that should always be stored and handled so that they do not come into contact with liquid water or water vapor.

Procedures for decomposing laboratory quantities are given in *Prudent Practices for Disposal of Chemicals from Laboratories*, chapter 6; the pertinent section of that chapter is given in parentheses.

PYROPHORIC CHEMICALS

Many members of the classes of readily oxidized, common laboratory chemicals listed in Table 6 ignite spontaneously in air. A more extensive list can be found in L. Bretherick, *Handbook of Reactive Chemical Hazards*, 3rd ed., Butterworths, London-Boston, 1985. Pyrophoric chemicals should be stored in tightly closed containers under an inert atmosphere (or, for some, an inert liquid),

and all transfers and manipulations of them must be carried out under an inert atmosphere or liquid. Suggested procedures for decomposing them are given in *Prudent Practices for Disposal of Chemicals from Laboratories*, chapter 6; the pertinent section of that chapter is given in parentheses.

TABLE 3. Shock-Sensitive Compounds

Acetylenic compounds, especially polyacetylenes, haloacetylenes, and heavy metal salts of acetylenes (copper, silver, and mercury salts are particularly sensitive)

Acyl nitrates

Alkyl nitrates, particularly polyol nitrates such as nitrocellulose and nitroglycerine

Alkyl and acyl nitrites

Alkyl perchlorates

Amminemetal oxosalts: metal compounds with coordinated ammonia, hydrazine, or similar nitrogenous donors and ionic perchlorate, nitrate, permanganate, or other oxidizing group

Azides, including metal, nonmetal, and organic azides

Chlorite salts of metals, such as $AgClO_2$ and $Hg(ClO_2)_2$

Diazo compounds such as CH_2N_2

Diazonium slats, when dry

Fulminates (silver fulminate, AgCNO, can form in the reaction mixture from the Tollens' test for aldehydes if it is allowed to stand for some time; this can be prevented by adding dilute nitric acid to the test mixture as soon as the test has been completed)

Hydrogen peroxide becomes increasingly treacherous as the concentration rises above 30%, forming explosive mixtures with organic materials and decomposing violently in the presence of traces of transition metals

N–Halogen compounds such as difluoroamino compounds and halogen azides

N–Nitro compounds such as N–nitromethylamine, nitrourea, nitroguanidine, and nitric amide

Oxo salts of nitrogenous bases: perchlorates, dichromates, nitrates, iodates, chlorites, chlorates, and permanganates of ammonia, amines, hydroxylamine, guanidine, etc.

Perchlorate salts. Most metal, nonmetal, and amine perchlorates can be detonated and may undergo violent reaction in contact with combustible materials

Peroxides and hydroperoxides, organic (see Chapter 6, Section II.P)

Peroxides (solid) that crystallize from or are left from evaporation of peroxidizable solvents (see Chapter 6 and Appendix I)

Peroxides, transition–metal salts

Picrates, especially salts of transition and heavy metals, such as Ni, Pb, Hg, Cu, and Zn; picric acid is explosive but is less sensitive to shock or friction than its metal salts and is relatively safe as a water–wet paste (see Chapter 7)

Polynitroalkyl compounds such as tetranitromethane and dinitroacetonitrile

Polynitroaromatic compounds, especially polynitro hydrocarbons, phenols, and amines

TABLE 4. Potentially Explosive Combinations of Some Common Reagents

Acetone + chloroform in the presence of base

Acetylene + copper, silver, mercury, or their salts

Ammonia (including aqueous solutions) + Cl_2, Br_2, or I_2

Carbon disulfide + sodium azide

Chlorine + an alcohol

Chloroform or carbon tetrachloride + powdered Al or Mg

Decolorizing carbon + an oxidizing agent

Diethyl ether + chlorine (including a chlorine atmosphere)

Dimethyl sulfoxide + an acyl halide, $SOCl_2$ or $POCl_3$

Dimethyl sulfoxide + CrO_3

Ethanol + calcium hypochlorite

Ethanol + silver nitrate

Nitric acid + acetic anhydride or acetic acid

Picric acid + a heavy–metal salt, such as of Pb, Hg, or Ag

Silver oxide + ammonia + ethanol

Sodium + a chlorinated hydrocarbon

Sodium hypochlorite + an amine

TABLE 5. Water-Reactive Chemicals

Alkali metals (III.D)

Alkali metal hydrides (III.C.2)

Alkali metal amides (III.C.7)

Metal alkyls, such as lithium alkyls and aluminum alkyls (IV.A)

Grignard reagents (IV.A)

Halides of nonmetals, such as BCl_3, BF_3, PCl_3, PCl_5, $SiCl_4$, S_2Cl_2 (III.F)

Inorganic acid halides, such as $POCl_3$, $SOCl_2$, SO_2Cl_2 (III.F)

Anhydrous metal halides, such as $AlCl_3$, $TiCl_4$, $ZrCl_4$, $SnCl_4$ (III.E)

Phosphorus pentoxide (III.I)

Calcium carbide (IV.E)

Organic acid halides and anhydrides of low molecular weight (II.J)

TABLE 6. Classes of Pyrophoric Chemicals

Grignard reagents, RMgX (IV.A)

Metal alkyls and aryls, such as RLi, RNa, R_3Al, R_2Zn (IV.A)

Metal carbonyls, such as Ni $(CO)_4$, $Fe(CO)_5$, $Co_2(CO)_8$ (IV.B)

Alkali metals such as Na, K (III.D.1)

Metal powders, such as Al, Co, Fe, Mg, Mn, Pd, Pt, Ti, Sn, Zn, Zr (III.D.2)

Metal hydrides, such as NaH, $LiAlH_4$ (IV.C.2)

Nonmetal hydrides, such as B_2H_6 and other boranes, PH_3, AsH_3 (III.G)

Nonmetal alkyls, such as R_3B, R_3P, R_3As (IV.C)

Phosphorus (white) (III.H)

HAZARDS FROM PEROXIDE FORMATION

Many common laboratory chemicals can form peroxides when allowed access to air over a period of time. A single opening of a container to remove some of the contents can introduce enough air for peroxide formation to occur. Some types of compounds form peroxides that are treacherously and violently explosive in concentrated solution or as solids. Accordingly, peroxide-containing liquids should never be evaporated near to or to dryness. Peroxide formation can also occur in many polymerizable unsaturated compounds, and these peroxides can initiate a runaway, sometimes explosive, polymerization reaction. Procedures for testing for peroxides and for removing small amounts from laboratory chemicals are given in *Prudent Practices for Disposal of Chemicals from Laboratories*, chapter 6, Section II.P.

Table 7 provides a list of structural characteristics in organic compounds that can peroxidize. These structures are listed in approximate order of decreasing hazard. Reports of serious incidents involving the last five structural types are extremely rare, but these structures are listed because laboratory workers should be aware that they can form peroxides that can influence the course of experiments in which they are used.

Table 8 gives examples of common laboratory chemicals that are prone to form peroxides on exposure to air. The lists are not exhaustive, and analogous organic compounds that have any of the structural features given in Table 7 should be tested for peroxides before being used as solvents or reagents, or before being distilled. The recommended retention times begin with the date of synthesis or of opening the original container.

DISPOSAL OF TOXIC CHEMICALS

It is often desirable to precipitate toxic cations or hazardous anions from solution to facilitate recovery or disposal. Table 9 lists precipitants for many common cations, and Table 10 gives precipitants for some hazardous anions. Many cations can be precipitated as sulfides by adding sodium sulfide solution (preferable to the highly toxic hydrogen sulfide) to a neutral solution of the cation (Table 11). Control of pH is important because some sulfides will redissolve in excess sulfide ion. After precipitation, excess sulfide can be destroyed by addition of hypochlorite.

Most metal cations are precipitated as hydroxides or oxides at high pH. Since many of these precipitates will redissolve in excess base, it is often necessary to control pH. Table 12 shows the recommended pH range for precipitating many cations in their most common oxidation state. The notation "1 N" in the right-hand column indicates that the precipitate will not dissolve in 1 N sodium hydroxide (pH 14).

The distinctions between high and low toxicity or hazard are based on toxicological and other data, and are relative. There is no implication of a sharp distinction between high and low, or that any cations or anions are totally without hazard.

TABLE 7. Types of Chemicals That Are Prone to Form Peroxides

A. Organic structures (in approximate order of decreasing hazard)

1.	$\overset{H}{\underset{}{\underset{\diagdown}{C}}}-O-$	Ethers and acetals with α hydrogen atoms
2.	$C=C-\overset{H}{\underset{}{C}}$	Olefins with allylic hydrogen atoms
3.	$C=C-\overset{X}{\underset{}{}}$	Chloroolefins and fluoroolefins
4.	$CH_2=C$	Vinyl halides, esters, and ethers
5.	$C=C-C=C$	Dienes
6.	$\overset{H}{\underset{}{C}}=C-C\equiv CH$	Vinylacetylenes with α hydrogen atoms
7.	$\overset{H}{\underset{}{C}}-C\equiv CH$	Alkylacetylenes with α hydrogen atoms
8.	$\overset{H}{\underset{}{C}}-Ar$	Alkylarenes that contain tertiary hydrogen atoms
9.	$-\overset{}{\underset{}{C}}-H$	Alkanes and cycloalkanes that contain tertiary hydrogen atoms

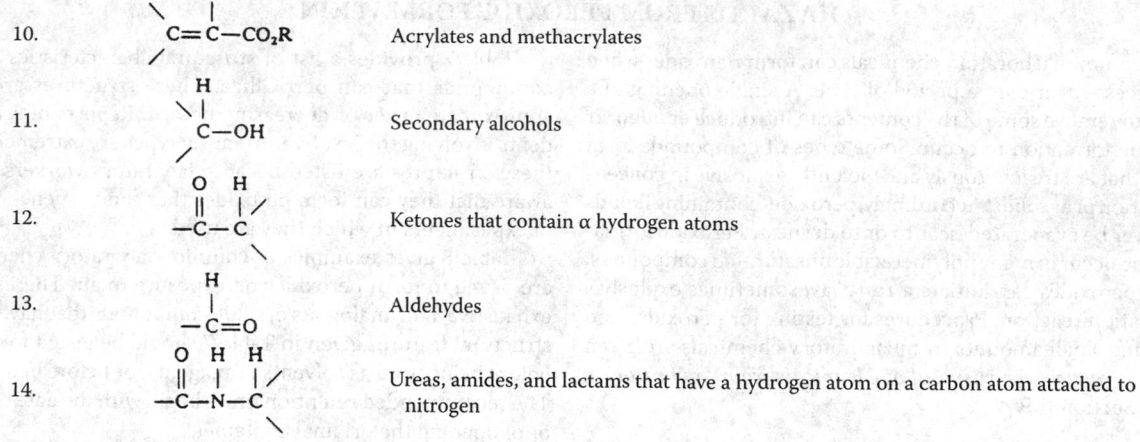

10. Acrylates and methacrylates

11. Secondary alcohols

12. Ketones that contain α hydrogen atoms

13. Aldehydes

14. Ureas, amides, and lactams that have a hydrogen atom on a carbon atom attached to nitrogen

B. Inorganic substances

1. Alkali metals, especially potassium, rubidium, and cesium (see Chapter 6, Section III.D)
2. Metal amides (see Chapter 6, Section III.C.7)
3. Organometallic compounds with a metal atom bonded to carbon (see Chapter 6, Section IV)
4. Metal alkoxides

TABLE 8. Common Peroxide-Forming Chemicals
LIST A
Severe Peroxide Hazard on Storage with Exposure to Air

Discard within 3 months

- Diisopropyl ether (isopropyl ether)
- Divinylacetylene (DVA)[a]
- Potassium metal
- Potassium amide
- Sodium amide (sodamide)
- Vinylidene chloride (1,1-dichloroethylene)[a]

LIST B
Peroxide Hazard on Concentration; Do Not Distill or Evaporate without First Testing for the Presence of Peroxides

Discard or Test for Peroxides after 6 Months

- Acetaldehyde diethyl acetal (acetal)
- Cumene (isopropylbenzene)
- Cyclohexene
- Cyclopentene
- Decalin (decahydronaphthalene)
- Diacetylene
- Dicyclopentadiene
- Diethyl ether (ether)
- Diethylene glycol dimethyl ether (diglyme)
- Dioxane
- Ethylene glycol dimethyl ether (glyme)
- Ethylene glycol ether acetates
- Ethylene glycol monoethers (cellosolves)
- Furan
- Methylacetylene
- Methylcyclopentane
- Methyl isobutyl ketone
- Tetrahydrofuran (THF)
- Tetralin (tetrahydronaphthalene)
- Vinyl ethers[a]

LIST C
Hazard of Rapid Polymerization Initiated by Internally Formed Peroxides[a]

a. Normal Liquids; discard or test for peroxides after 6 months[b]

- Chloroprene (2-chloro-1,3-butadiene)[c]
- Styrene
- Vinyl acetate
- Vinylpyridine

b. Normal Gases; discard after 12 months[d]

- Butadiene[c]
- Tetrafluoroethylene (TFE)[c]
- Vinylacetylene (MVA)[c]
- Vinyl chloride

---

[a] Polymerizable monomers should be stored with a polymerization inhibitor from which the monomer can be separated by distillation just before use.

[b] Although common acrylic monomers such as acrylonitrile, acrylic acid, ethyl acrylate, and methyl methacrylate can form peroxides, they have not been reported to develop hazardous levels in normal use and storage.

[c] The hazard from peroxides in these compounds is substantially greater when they are stored in the liquid phase, and if so stored without an inhibitor they should be considered as in LIST A.

[d] Although air will not enter a gas cylinder in which gases are stored under pressure, these gases are sometimes transferred from the original cylinder to another in the laboratory, and it is difficult to be sure that there is no residual air in the receiving cylinder. An inhibitor should be put into any such secondary cylinder before one of these gases is transferred into it; the supplier can suggest inhibitors to be used. The hazard posed by these gases is much greater if there is a liquid phase in such a secondary container, and even inhibited gases that have been put into a secondary container under conditions that create a liquid phase should be discarded within 12 months.

Note: Laboratory workers should label all containers of peroxidizable solvents or reagents with one of the following:

[LIST A]

Peroxidizable compound

Received   Opened

Date

Discard 3 months after opening

[LISTS B AND C]

Peroxidizable compound

Received   Opened

Date

Discard or test for peroxides
6 months after opening

TABLE 9. Relative Toxicity of Cations

High toxic hazard	Precipitant[a]	Low toxic hazard	Precipitant[a]
Antimony	OH^-, S^{2-}	Aluminum	OH^-
Arsenic	S^{2-}	Bismuth	OH^-, S^{2-}
Barium	SO_4^{2-}, CO_3^{2-}	Calcium	SO_4^{2-}, CO_3^{2-}
Beryllium	OH^-	Cerium	OH^-
Cadmium	OH^-, S^{2-}	Cesium	
Chromium (III)[b]	OH^-	Copper[c]	OH^-, S^{2-}
Cobalt (II)[b]	OH^-, S^{2-}	Gold	OH^-, S^{2-}
Gallium	OH^-	Iron[c]	OH^-, S^{2-}
Germanium	OH^-, S^{2-}	Lanthanides	OH^-
Hafnium	OH^-	Lithium	
Indium	OH^-, S^{2-}	Magnesium	OH^-
Iridium	OH^-, S^{2-}	Molybdenum (VI)[b,d]	
Lead	OH^-, S^{2-}	Niobium (V)	OH^-
Manganese (II)[b]	OH^-, S^{2-}	Palladium	OH^-, S^{2-}
Mercury	OH^-, S^{2-}	Potassium	
Nickel	OH^-, S^{2-}	Rubidium	
Osmium (IV)[b,e]	OH^-, S^{2-}	Scandium	OH^-
Platinum (II)[b]	OH^-, S^{2-}	Sodium	
Rhenium (VII)[b]	S^{2-}	Strontium	SO_4^{2-}, CO_3^{2-}
Rhodium (III)[b]	OH^-, S^{2-}	Tantalum	OH^-
Ruthenium (III)[b]	OH^-, S^{2-}	Tin	OH^-, S^{2-}
Selenium	S^{2-}	Titanium	OH^-
Silver	Cl^-, OH^-, S^{2-}	Yttrium	OH^-
Tellurium	S^{2-}	Zinc[c]	OH^-, S^{2-}
Thallium	OH^-, S^{2-}	Zirconium	OH^-
Tungsten (VI)[b,d]			
Vanadium	OH^-, S^{2-}		

[a] Precipitants are listed in order of preference:
   OH^- = base (sodium hydroxide or sodium carbonate)
   S^{2-} = sulfide
   Cl^- = chloride
   SO_4^{2-} = sulfate
   CO_3^{2-} = carbonate
[b] The precipitant is for the indicated valence state.
[c] Maximum tolerance levels have been set for these low-toxicity ions by the U.S. Public Health Service, and large amounts should not be put into public sewer systems. The small amounts typically used in laboratories will not normally affect water supplies.
[d] These ions are best precipitated as calcium molybdate or calcium tungstate.
[e] CAUTION: OsO_4, a volatile, extremely poisonous substance, is formed from almost any osmium compound under acid conditions in the presence of air.

TABLE 10. Relative Hazard of Anions

High-hazard anions			Low-hazard anions
Ion	**Hazard type[a]**	**Precipitant**	**Low-hazard anions**
Aluminum hydride, AlH_4^-	F	—	Bisulfite, HSO_3^-
Amide, NH_2^-	F,E[b]	—	Borate, BO_3^{3-}, $B_4O_7^{2-}$
Arsenate, AsO_3^-, AsO_4^{3-}	T	Cu^{2+}, Fe^{2+}	Bromide, Br^-
Arsenite, AsO_2^-, AsO_3^{3-}	T	Pb^{2+}	Carbonate, CO_3^{2-}
Azide, N_3^-	E, T	—	Chloride, Cl^-
Borohydride, BH_4^-	F	—	Cyanate, OCN^-
Bromate, BrO_3^-	O, E	—	Hydroxide, OH^-
Chlorate, ClO_3^-	O, E	—	Iodide, I^-
Chromate, CrO_4^{2-}, $Cr_2O_7^{2-}$	T, O	c	Oxide, O^{2-}
Cyanide, CN^-	T	—	Phosphate, PO_4^{3-}
Ferricyanide, $Fe(CN)_6^{3-}$	T	Fe^{2+}	Sulfate, SO_4^2
Ferrocyanide, $Fe(CN)_6^{4-}$	T	Fe^{3+}	Sulfite, SO_3^{2-}
Fluoride, F^-	T	Ca^{2+}	Thiocyanate, SCN^-
Hydride, H^-	F	—	
Hydroperoxide, O_2H^-	O, E	—	
Hydrosulfide, SH^-	T	—	
Hypochlorite, OCl^-	O	—	
Iodate, IO_3^-	O, E	—	
Nitrate, NO_3^-	O	—	
Nitrite, NO_2^-	T, O	—	
Perchlorate, ClO_4^-	O, E	—	
Permanganate, MnO_4^-	T, O	d	
Peroxide, O_2^{2-}	O, E	—	
Persulfate, $S_2O_8^{2-}$	O	—	
Selenate, SeO_4^{2-}	T	Pb^{2+}	
Selenide, Se^{2-}	T	Cu^{2+}	
Sulfide, S^{2-}	T	e	

[a] Toxic, T: oxidant, O; flammable, F; explosive, E.
[b] Metal amides readily form explosive peroxides on exposure to air.
[c] Reduce and precipitate as Cr(III); see Table 9.
[d] Reduce and precipitate as Mn(II); see Table 9.
[e] See Table 11.

TABLE 11. Precipitation of Sulfides

Precipitated at pH 7	Not precipitated at low pH	Forms a soluble complex at high pH
Ag^+		
As^{3+a}		X
Au^{+a}		X
Bi^{3+}		
Cd^{2+}		
Co^{2+}	X	
Cr^{3+a}		
Cu^{2+}		
Fe^{2+a}	X	
Ge^{2+}		X
Hg^{2+}		X
In^{3+}	X	
Ir^{4+}		X
Mn^{2+a}	X	
Mo^{3+}		X
Ni^{2+}	X	
Os^{4+}		
Pb^{2+}		
Pd^{2+a}		
Pt^{2+a}		X
Re^{4+}		
Rh^{2+a}		
Ru^{4+}		

TABLE 11. Precipitation of Sulfides

Precipitated at pH 7	Not precipitated at low pH	Forms a soluble complex at high pH
Sb^{3+a}		X
Se^{2+}		X
Sn^{2+}		X
Te^{4+}		X
Tl^{+a}	X	
V^{4+a}		
Zn^{2+}	X	

[a] Higher oxidation states of this ion are reduced by sulfide ion and precipitated as this sulfide.

TABLE 12. pH Range for Precipitation of Metal Hydroxides and Oxides

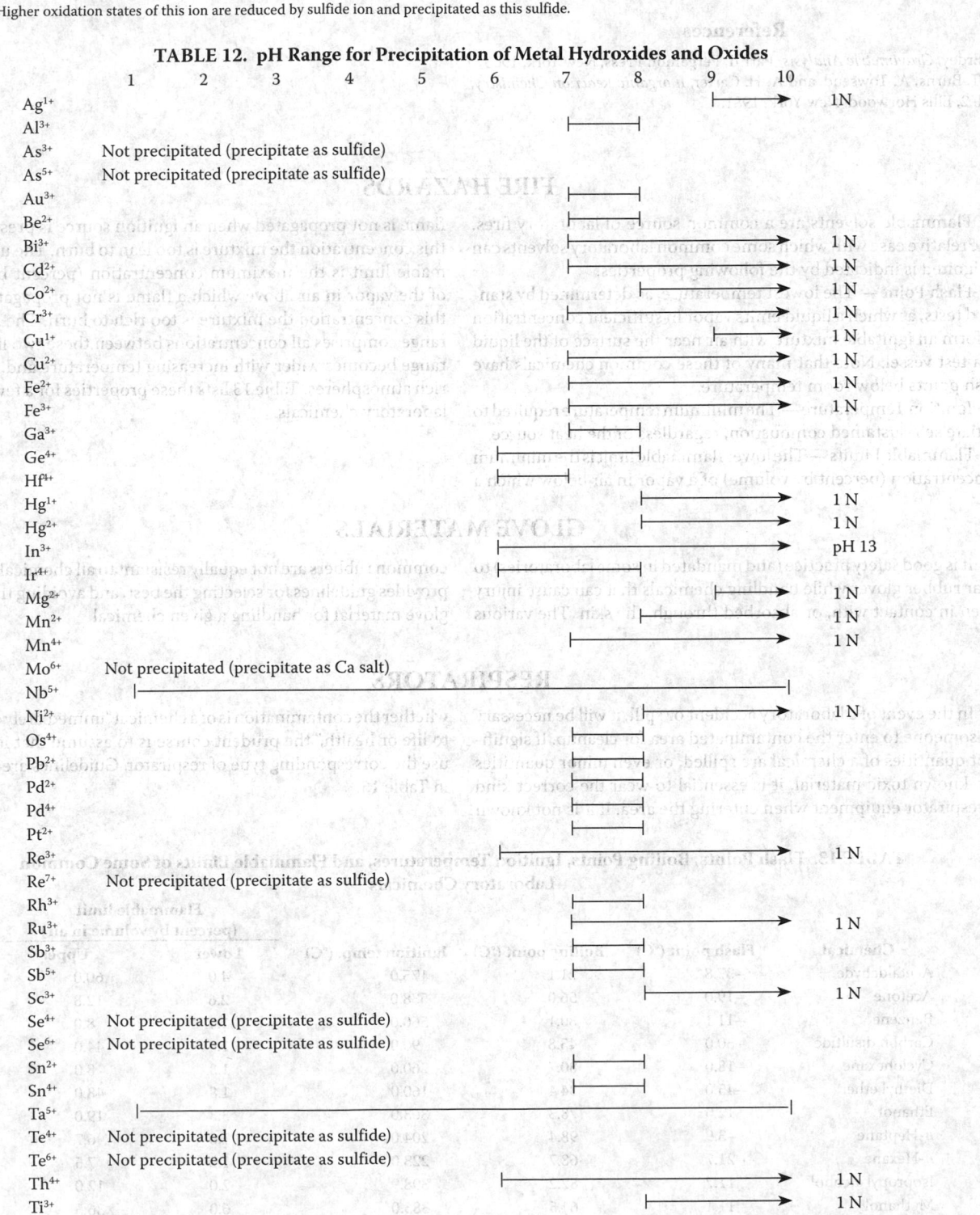

TABLE 12. pH Range for Precipitation of Metal Hydroxides and Oxides

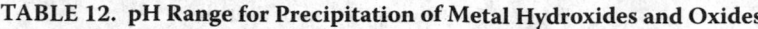
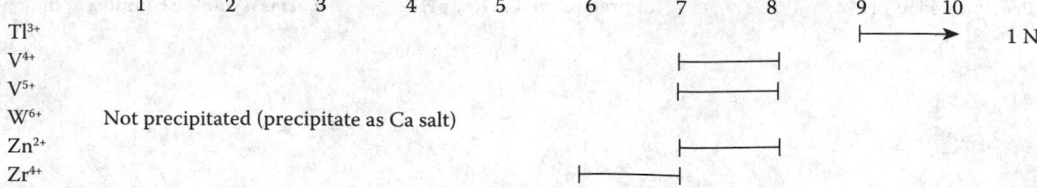

	1	2	3	4	5	6	7	8	9	10	
Tl^{3+}											1 N
V^{4+}											
V^{5+}											
W^{6+}	Not precipitated (precipitate as Ca salt)										
Zn^{2+}											
Zr^{4+}											

References

L. Erdey, *Gravimetric Analysis*, Part II, Pergamon Press, New York, 1965.
D. T. Burns, A. Towsend, and A. H. Carter, *Inorganic Reaction Chemistry*,
Vol. 2, Ellis Horwood, New York, 1981.

FIRE HAZARDS

Flammable solvents are a common source of laboratory fires. The relative ease with which some common laboratory solvents can be ignited is indicated by the following properties.

Flash Point — The lowest temperature, as determined by standard tests, at which a liquid emits vapor in sufficient concentration to form an ignitable mixture with air near the surface of the liquid in a test vessel. Note that many of these common chemicals have flash points below room temperature.

Ignition Temperature — The minimum temperature required to initiate self-sustained combustion, regardless of the heat source.

Flammable Limits — The lower flammable limit is the minimum concentration (percent by volume) of a vapor in air below which a flame is not propagated when an ignition source is present. Below this concentration the mixture is too lean to burn. The upper flammable limit is the maximum concentration (percent by volume) of the vapor in air above which a flame is not propagated. Above this concentration the mixture is too rich to burn. The flammable range comprises all concentrations between these two limits. This range becomes wider with increasing temperature and in oxygen-rich atmospheres. Table 13 lists these properties for a few common laboratory chemicals.

GLOVE MATERIALS

It is good safety practice (and mandated in some laboratories) to wear rubber gloves while handling chemicals that can cause injury when in contact with, or absorbed through, the skin. The various common rubbers are not equally resistant to all chemicals. Table 14 provides guidelines for selecting the best, and avoiding the poorest, glove material for handling a given chemical.

RESPIRATORS

In the event of a laboratory accident or spill, it will be necessary for someone to enter the contaminated area for cleanup. If significant quantities of a chemical are spilled, or even minor quantities of a known toxic material, it is essential to wear the correct kind of respirator equipment when entering the area. If it is not known whether the contamination is of a chemical "immediately dangerous to life or health," the prudent course is to assume that it is, and to use the corresponding type of respirator. Guidelines are presented in Table 15.

TABLE 13. Flash Points, Boiling Points, Ignition Temperatures, and Flammable Limits of Some Common Laboratory Chemicals

Chemical	Flash point (°C)	Boiling point (°C)	Ignition temp. (°C)	Flammable limit (percent by volume in air)	
				Lower	Upper
Acetaldehyde	−37.8	21.1	175.0	4.0	60.0
Acetone	−19.0	56.0	538.0	2.6	12.8
Benzene	−11.1	80.1	560.0	1.4	8.0
Carbon disulfide	−30.0	45.8	90.0	1.0	44.0
Cyclohexane	−18.0	80.7	260.0	1.3	8.0
Diethyl ether	−45.0	34.4	160.0	1.8	48.0
Ethanol	12.0	78.3	363.0	3.3	19.0
n-Heptane	−3.9	98.4	204.0	1.0	6.7
n-Hexane	−21.7	68.7	223.0	1.2	7.5
Isopropyl alcohol	11.7	82.2	398.9	2.0	12.0
Methanol	11.1	64.5	385.0	6.0	36.5
Methyl ethyl ketone	−6.1	79.6	515.6	1.9	11.0

TABLE 13. Flash Points, Boiling Points, Ignition Temperatures, and Flammable Limits of Some Common Laboratory Chemicals

Chemical	Flash point (°C)	Boiling point (°C)	Ignition temp. (°C)	Flammable limit (percent by volume in air)	
				Lower	Upper
Pentane	−40.0	36.1	260.0	1.4	7.8
Styrene	31.0	145.0	490.0	1.1	6.1
Toluene	4.4	110.6	530.0	1.3	7.0
p-Xylene	25.0	132.4	529.0	1.1	7.0

Note: For a more extensive listing, see the table "Properties of Common Solvents" in Section 15.

TABLE 14. Resistance to Chemicals of Common Glove Materials (E = Excellent, G = Good, F = Fair, P = Poor)

Chemical	Natural rubber	Neoprene	Nitrile	Vinyl
Acetaldehyde	G	G	E	G
Acetic acid	E	E	E	E
Acetone	G	G	G	F
Acrylonitrile	P	G	—	F
Ammonium hydroxide (sat)	G	E	E	E
Aniline	F	G	E	G
Benzaldehyde	F	F	E	G
Benzene[a]	P	F	G	F
Benzyl chloride[a]	F	P	G	P
Bromine	G	G	—	G
Butane	P	E	—	P
Butyraldehyde	P	G	—	G
Calcium hypochlorite	P	G	G	G
Carbon disulfide	P	P	G	F
Carbon tetrachloride[a]	P	F	G	F
Chlorine	G	G	—	G
Chloroacetone	F	E	—	P
Chloroform[a]	P	F	G	P
Chromic acid	P	F	F	E
Cyclohexane	F	E	—	P
Dibenzyl ether	F	G	—	P
Dibutyl phthalate	F	G	—	E
Diethanolamine	F	E	—	E
Diethyl ether	F	G	E	P
Dimethyl sulfoxide[b]	—	—	—	—
Ethyl acetate	F	G	G	F
Ethylene dichloride[a]	P	F	G	P
Ethylene glycol	G	G	E	E
Ethylene trichloride[a]	P	P	—	P
Fluorine	G	G	—	G
Formaldehyde	G	E	E	E
Formic acid	G	E	E	E
Glycerol	G	G	E	E
Hexane	G	E	—	P
Hydrobromic acid (40%)	G	E	—	E
Hydrochloric acid (conc)	G	G	G	E
Hydrofluoric acid (30%)	G	G	G	E
Hydrogen peroxide	G	G	G	E
Iodine	G	G	—	E
Methylamine	G	G	E	E
Methyl cellosolve	F	E	—	P
Methyl chloride[a]	P	E	—	P
Methyl ethyl ketone	F	G	G	P
Methylene chloride[a]	F	F	G	F
Monoethanolamine	F	E	—	E
Morpholine	F	E	—	G
Naphthalene[a]	G	G	E	G
Nitric acid (conc)	P	P	P	G
Perchloric acid	F	G	F	E

TABLE 14. Resistance to Chemicals of Common Glove Materials (E = Excellent, G = Good, F = Fair, P = Poor)

Chemical	Natural rubber	Neoprene	Nitrile	Vinyl
Phenol	G	E	—	E
Phosphoric acid	G	E	—	E
Potassium hydroxide (sat)	G	G	G	E
Propylene dichloride[a]	P	F	—	P
Sodium hydroxide	G	G	G	E
Sodium hypochlorite	G	P	F	G
Sulfuric acid (conc)	G	G	F	G
Toluene[a]	P	F	G	F
Trichloroethylene[a]	P	F	G	F
Tricresyl phosphate	P	F	—	F
Triethanolamine	F	E	E	E
Trinitrotoluene	P	E	—	P

[a] Aromatic and halogenated hydrocarbons will attack all types of natural and synthetic glove materials. Should swelling occur, the user should change to fresh gloves and allow the swollen gloves to dry and return to normal.

[b] No data on the resistance to dimethyl sulfoxide of natural rubber, neoprene, nitrile rubber, or vinyl materials are available; the manufacturer of the substance recommends the use of butyl rubber gloves.

TABLE 15. Guide for Selection of Respirators

Type of hazard	Type of respirator
Oxygen deficiency	Self-contained breathing apparatus
	Hose mask with blower
	Combination of air-line respirator and auxiliary self-contained air supply or air-storage receiver with alarm
Gas and vapor contaminants	
Immediately dangerous to life or health	Self-contained breathing apparatus
	Hose mask with blower
	Air-purifying full-facepiece respirator with chemical canister (gas mask)
	Self-rescue mouthpiece respirator (for escape only)
	Combination of air-line respirator and auxiliary self-contained air supply or air-storage receiver with alarm
Not immediately dangerous to life or health	Air-line respirator
	Hose mask with blower
	Air-purifying half-mask or mouthpiece respirator with chemical cartridge
Particulate contaminants	
Immediately dangerous to life or health	Self-contained breathing apparatus
	Hose mask with blower
	Air-purifying full-facepiece respirator with appropriate filter
	Self-rescue mouthpiece respirator (for escape only)
	Combination of air-line respirator and auxiliary self-contained air supply or air-storage receiver with alarm
Not immediately dangerous to life or health	Air-purifying half-mask or mouthpiece respirator with filter pad or cartridge
	Air-line respirator
	Air-line abrasive-blasting respirator
	Hose mask with blower
Combination of gas, vapor, and particulate contaminants	
Immediately dangerous to life or health	Self-contained breathing apparatus
	Hose mask with blower
	Air-purifying full-facepiece respirator with chemical canister and appropriate filter (gas mask with filter)
	Self-rescue mouthpiece respirator (for escape only)
	Combination of air-line respirator and auxiliary self-contained air supply or air-storage receiver with alarm
Not immediately dangerous to life or health	Air-line respirator
	Hose mask without blower
	Air-purifying half-mask or mouthpiece respirator with chemical cartridge and appropriate filter

Source: ANSI Standard Z88.2 (1969).

FLAMMABILITY OF CHEMICAL SUBSTANCES

This table gives properties related to the flammability of about 900 chemical substances. The properties listed are:

t_B: Normal boiling point in °C (at 101.325 kPa pressure).

FP: Flash point, which is the minimum temperature at which the vapor pressure of a liquid is sufficient to form an ignitable mixture with air near the surface of the liquid. Flash point is not an intrinsic physical property but depends on the conditions of measurement (see Reference 1).

Fl. limits: Flammable limits (often called explosive limits), which specify the range of concentration of the vapor in air (in percent by volume) for which a flame can propagate. Below the lower flammable limit, the gas mixture is too lean to burn; above the upper flammable limit, the mixture is too rich. Values refer to ambient temperature and pressure and are dependent on the precise test conditions. A ? indicates that one of the limits is not known.

IT: Ignition temperature (sometimes called autoignition temperature), which is the minimum temperature required for self-sustained combustion in the absence of an external ignition source. As in the case of flash point, the value depends on specified test conditions.

Even in cases where very careful measurements of flash point have been replicated in several laboratories, observed values can differ by 3 to 6 °C (Reference 4). For more typical measurements, larger uncertainties should be assumed in both flash points and autoignition temperatures. The absence of a flash point entry in this table does not mean that the substance is nonflammable, but only that no reliable value is available.

Compounds are listed by molecular formula following the Hill convention. Substances not containing carbon are listed first, followed by those that contain carbon. To locate an organic compound by name or CAS Registry Number when the molecular formula is not known, use the table "Physical Constants of Organic Compounds" in Section 3 and its indexes to determine the molecular formula.

References

1. *Fire Protection Guide to Hazardous Materials, 11th Edition*, National Fire Protection Association, Quincy, MA, 1994.
2. Urben, P. G., Ed., *Bretherick's Handbook of Reactive Chemical Hazards, 5th Edition*, Butterworth-Heinemann, Oxford, 1995.
3. Daubert, T. E., Danner, R. P., Sibul, H. M., and Stebbins, C. C., *Physical and Thermodynamic Properties of Pure Compounds: Data Compilation*, extant 1994 (core with 4 supplements), Taylor & Francis, Bristol, PA.
4. *Report of Investigation: Flash Point Reference Materials*, National Institute of Standards and Technology, Standard Reference Materials Program, Gaithersburg, MD, 1995.

Mol. form.	Name	t_B/°C	FP/°C	Fl. limits	IT/°C
		Compounds not containing carbon			
B_2H_6	Diborane	−92.4	−90	1–98%	≈40
B_5H_9	Pentaborane(9)	60	30	0.4–?	35
BrH_3Si	Bromosilane	1.9	<0		≈20
Br_3HSi	Tribromosilane	109			≈20
Cl_2H_2Si	Dichlorosilane	8.3		4.1–99%	36
Cl_3HSi	Trichlorosilane	33	−50		104
GeH_4	Germane	−88.1			≈20
Ge_2H_6	Digermane	29			≈50
H_2	Hydrogen	−252.8		4–74%	
H_2S	Hydrogen sulfide	−59.55		4–44%	260
H_2S_2	Hydrogen disulfide	70.7	<22		
H_2Te	Hydrogen telluride	−2			−50
H_3N	Ammonia	−33.33		16–25%	
H_3P	Phosphine	−87.75		1.8–?	
H_4N_2	Hydrazine	113.55	38	5–100%	
H_4P_2	Diphosphine	63.5			≈20
H_4Si	Silane	−111.9	−112	1.4–?	≈20
H_6Si_2	Disilane	−14.3	−14		≈20
H_8Si_3	Trisilane	52.9	<0		≈20
P	Phosphorus (white)	280.5			38
		Compounds containing carbon			
CHN	Hydrogen cyanide	26	−18	6–40%	538
CH_2Cl_2	Dichloromethane	40		13–23%	556
CH_2N_2	Cyanamide		141		
CH_2O	Formaldehyde	−19.1	85	7.0–73%	424
$(CH_2O)_x$	Paraformaldehyde		70	7.0–73%	300
CH_2O_2	Formic acid	101	50	18–57%	434
CH_3Br	Bromomethane	3.5		10–16%	537
CH_3Cl	Chloromethane	−24.0		8.1–17.4%	632

Mol. form.	Name	t_B/°C	FP/°C	Fl. limits	IT/°C
CH₃Cl₃Si	Methyltrichlorosilane	65.6	−9	7.6−>20%	>404
CH₃NO	Formamide	220	154		
CH₃NO₂	Nitromethane	101.1	35	7.3−?	418
CH₄	Methane	−161.5		5.0−15.0%	537
CH₄Cl₂Si	Dichloromethylsilane	41	−9	6.0−55%	316
CH₄O	Methanol	64.6	11	6.0−36%	464
CH₄S	Methanethiol	5.9	−18	3.9−21.8%	
CH₅N	Methylamine	−6.3	0	4.9−20.7%	430
CH₆N₂	Methylhydrazine	87.5	−8	2.5−92%	194
CO	Carbon monoxide	−191.5		12.5−74%	609
COS	Carbon oxysulfide	−50		12−29%	
CS₂	Carbon disulfide	46	−30	1.3−50.0%	90
C₂ClF₃	Chlorotrifluoroethylene	−27.8		8.4−16.0%	
C₂F₄	Tetrafluoroethylene	−75.9		10.0−50.0%	200
C₂HCl₃	Trichloroethylene	87.2		8−10.5%	420
C₂HCl₃O	Dichloroacetyl chloride	108	66		
C₂H₂	Acetylene	−84.7		2.5−100%	305
C₂H₂Cl₂	1,1-Dichloroethylene	31.6	−28	6.5−15.5%	570
C₂H₂Cl₂	cis-1,2-Dichloroethylene	60.1	6	3−15%	460
C₂H₂Cl₂	trans-1,2-Dichloroethylene	48.7	2	6−13%	460
C₂H₂F₂	1,1-Difluoroethylene	−85.7		5.5−21.3%	
C₂H₃Br	Bromoethylene	15.8		9−15%	530
C₂H₃Cl	Chloroethylene	−13.3	−78	3.6−33.0%	472
C₂H₃ClF₂	1-Chloro-1,1-difluoroethane	−9.7		6−18%	632
C₂H₃ClO	Acetyl chloride	50.7	4		390
C₂H₃Cl₂NO₂	1,1-Dichloro-1-nitroethane	123.5	76		
C₂H₃Cl₃	1,1,1-Trichloroethane	74.0		8−10.5%	500
C₂H₃Cl₃	1,1,2-Trichloroethane	113.8	32	6−28%	460
C₂H₃Cl₃Si	Trichlorovinylsilane	91.5	21		
C₂H₃F	Fluoroethylene	−72		2.6−21.7%	
C₂H₃N	Acetonitrile	81.6	6	3.0−16.0%	524
C₂H₃NO	Methyl isocyanate	39.5	−7	5.3−26%	534
C₂H₄	Ethylene	−103.7		2.7−36%	450
C₂H₄ClNO₂	1-Chloro-1-nitroethane	124.5	56		
C₂H₄Cl₂	1,1-Dichloroethane	57.4	−17	5.4−11.4%	458
C₂H₄Cl₂	1,2-Dichloroethane	83.5	13	6.2−16%	413
C₂H₄O	Acetaldehyde	20.1	−39	4.0−60%	175
C₂H₄O	Ethylene oxide	10.6	−20	3.0−100%	429
C₂H₄O₂	Acetic acid	117.9	39	4.0−19.9%	463
C₂H₄O₂	Methyl formate	31.7	−19	4.5−23%	449
C₂H₄O₃	Ethaneperoxoic acid	110	41		
C₂H₅Br	Bromoethane	38.5		6.8−8.0%	511
C₂H₅Cl	Chloroethane	12.3	−50	3.8−15.4%	519
C₂H₅ClO	Ethylene chlorohydrin	128.6	60	4.9−15.9%	425
C₂H₅Cl₃Si	Trichloroethylsilane	100.5	22		
C₂H₅N	Ethyleneimine	56	−11	3.3−54.8%	320
C₂H₅NO₂	Nitroethane	114.0	28	3.4−17%	414
C₂H₅NO₂	Ethyl nitrite	18	−35	4.0−50%	90
C₂H₅NO₃	Ethyl nitrate	87.2	10	4−?	
C₂H₆	Ethane	−88.6		3.0−12.5%	472
C₂H₆Cl₂Si	Dichlorodimethylsilane	70.3	<21	3.4−9.5%	
C₂H₆O	Ethanol	78.2	13	3.3−19%	363
C₂H₆O	Dimethyl ether	−24.8	−41	3.4−27.0%	350
C₂H₆OS	2-Mercaptoethanol	158	74		
C₂H₆OS	Dimethyl sulfoxide	189	95	2.6−42%	215
C₂H₆O₂	Ethylene glycol	197.3	111	3.2−22%	398
C₂H₆O₄S	Dimethyl sulfate		83		188
C₂H₆S	Ethanethiol	35.1	−17	2.8−18.0%	300
C₂H₆S	Dimethyl sulfide	37.3	−37	2.2−19.7%	206
C₂H₆S₂	Dimethyl disulfide	109.8	24		
C₂H₇N	Ethylamine	16.5	−16	3.5−14%	385

Mol. form.	Name	t_B/°C	FP/°C	Fl. limits	IT/°C
C₂H₇N	Dimethylamine	6.8	20	2.8–14.4%	400
C₂H₇NO	Ethanolamine	171	86	3.0–23.5%	410
C₂H₈N₂	1,2-Ethanediamine	117	40	2.5–12.0%	385
C₂H₈N₂	1,1-Dimethylhydrazine	63.9	−15	2–95%	249
C₂N₂	Cyanogen	−21.1		6.6–32%	
C₃H₃Br	3-Bromo-1-propyne	89	10	3.0–?	324
C₃H₃N	2-Propenenitrile	77.3	0	3.0–17.0%	481
C₃H₄	Propyne	−23.2		2.1–12.5%	
C₃H₄ClN	3-Chloropropanenitrile	175.5	76		
C₃H₄Cl₂	2,3-Dichloropropene	94	15	2.6–7.8%	
C₃H₄O	Propargyl alcohol	113.6	36		
C₃H₄O	Acrolein	52.6	−26	2.8–31%	220
C₃H₄O₂	Propenoic acid	141	50	2.4–8.0%	438
C₃H₄O₂	2-Oxetanone	162	74	2.9–?	
C₃H₄O₃	Ethylene carbonate	248	143		
C₃H₅Br	3-Bromopropene	70.1	−1	4.4–7.3%	295
C₃H₅Cl	2-Chloropropene	22.6	−37	4.5–16%	
C₃H₅Cl	3-Chloropropene	45.1	−32	2.9–11.1%	485
C₃H₅ClO	Epichlorohydrin	118	31	3.8–21.0%	411
C₃H₅ClO	Propanoyl chloride	80	12		
C₃H₅ClO₂	2-Chloropropanoic acid	185	107		500
C₃H₅ClO₂	Ethyl chloroformate	95	16		500
C₃H₅ClO₂	Methyl chloroacetate	129.5	57	7.5–18.5%	
C₃H₅Cl₂NO₂	1,1-Dichloro-1-nitropropane	145	66		
C₃H₅Cl₃	1,2,3-Trichloropropane	157	71	3.2–12.6%	
C₃H₅Cl₃Si	Trichloro-2-propenylsilane	117.5	35		
C₃H₅N	Propanenitrile	97.1	2	3.1–14%	512
C₃H₅NO	3-Hydroxypropanenitrile	221	129		
C₃H₅N₃O₉	Trinitroglycerol				270
C₃H₆	Propene	−47.6		2.0–11.1%	455
C₃H₆	Cyclopropane	−32.8		2.4–10.4%	498
C₃H₆ClNO₂	1-Chloro-1-nitropropane	142	62		
C₃H₆ClNO₂	2-Chloro-2-nitropropane		57		
C₃H₆Cl₂	1,2-Dichloropropane	96.4	21	3.4–14.5%	557
C₃H₆Cl₂O	1,3-Dichloro-2-propanol	176	74		
C₃H₆N₂	Dimethylcyanamide	163.5	71		
C₃H₆O	Allyl alcohol	97.0	21	2.5–18.0%	378
C₃H₆O	Methyl vinyl ether	5.5			287
C₃H₆O	Propanal	48	−30	2.6–17%	207
C₃H₆O	Acetone	56.0	−20	2.5–12.8%	465
C₃H₆O	Methyloxirane	35	−37	3.1–27.5%	449
C₃H₆O₂	Propanoic acid	141.1	52	2.9–12.1%	465
C₃H₆O₂	Ethyl formate	54.4	−20	2.8–16.0%	455
C₃H₆O₂	Methyl acetate	56.8	−10	3.1–16%	454
C₃H₆O₂	1,3-Dioxolane	78	2		
C₃H₆O₃	Dimethyl carbonate	90.5	19		
C₃H₆O₃	1,3,5-Trioxane	114.5	45	3.6–29%	414
C₃H₇Br	1-Bromopropane	71.1			490
C₃H₇Cl	1-Chloropropane	46.5	<−18	2.6–11.1%	520
C₃H₇Cl	2-Chloropropane	35.7	−32	2.8–10.7%	593
C₃H₇ClO	2-Chloro-1-propanol	133.5	52		
C₃H₇ClO	1-Chloro-2-propanol	127	52		
C₃H₇Cl₃Si	Trichloropropylsilane	123.5	37		
C₃H₇N	Allylamine	53.3	−29	2.2–22%	374
C₃H₇NO	N,N-Dimethylformamide	153	58	2.2–15.2%	445
C₃H₇NO₂	1-Nitropropane	131.1	36	2.2–?	421
C₃H₇NO₂	2-Nitropropane	120.2	24	2.6–11.0%	428
C₃H₇NO₃	Propyl nitrate	110	20	2–100%	175
C₃H₈	Propane	−42.1	−104	2.1–9.5%	450
C₃H₈O	1-Propanol	97.2	23	2.2–13.7%	412
C₃H₈O	2-Propanol	82.3	12	2.0–12.7%	399

Mol. form.	Name	t_B/°C	FP/°C	Fl. limits	IT/°C
C₃H₈O	Ethyl methyl ether	7.4	−37	2.0–10.1%	190
C₃H₈O₂	1,2-Propylene glycol	187.6	99	2.6–12.5%	371
C₃H₈O₂	1,3-Propylene glycol	214.4			400
C₃H₈O₂	Ethylene glycol monomethyl ether	124.1	39	1.8–14%	285
C₃H₈O₂	Dimethoxymethane	42	−32	2.2–13.8%	237
C₃H₈O₃	Glycerol	290	199	3–19%	370
C₃H₉BO₃	Trimethyl borate	67.5	−8		
C₃H₉ClSi	Trimethylchlorosilane	60	−28		395
C₃H₉N	Propylamine	47.2	−37	2.0–10.4%	318
C₃H₉N	Isopropylamine	31.7	−37		402
C₃H₉N	Trimethylamine	2.8	−5	2.0–11.6%	190
C₃H₉NO	3-Amino-1-propanol	187.5	80		
C₃H₉NO	1-Amino-2-propanol	159.4	77		374
C₃H₉NO	*N*-Methyl-2-ethanolamine	158	74		
C₃H₉O₃P	Trimethyl phosphite	111.5	54		
C₃H₉O₄P	Trimethyl phosphate	197.2	107		
C₃H₁₀N₂	1,3-Propanediamine	139.8	24		
C₄Cl₆	Hexachloro-1,3-butadiene	215			610
C₄H₂O₃	Maleic anhydride	202	102	1.4–7.1%	477
C₄H₄	1-Buten-3-yne	5.1		21–100%	
C₄H₄N₂	Succinonitrile	266	132		
C₄H₄O	Furan	31.5	−36	2.3–14.3%	
C₄H₄O₂	Diketene	126.1	34		
C₄H₄S	Thiophene	84.0	−1		
C₄H₅Cl	2-Chloro-1,3-butadiene	59.4	−20	4.0–20.0%	
C₄H₅N	2-Butenenitrile	120.5	16		
C₄H₅N	Methylacrylonitrile	90.3	1	2–6.8%	
C₄H₅N	Pyrrole	129.7	39		
C₄H₆	1,3-Butadiene	−4.4		2.0–12.0%	420
C₄H₆	2-Butyne	26.9	−31	1.4–?	
C₄H₆O	Divinyl ether	28.3	<−30	1.7–27%	360
C₄H₆O	Ethoxyacetylene	50	<−7		
C₄H₆O	*trans*-2-Butenal	102.2	13	2.1–15.5%	232
C₄H₆O	3-Buten-2-one	81.4	−7	2.1–15.6%	491
C₄H₆O	Vinyloxirane	68	<−50		
C₄H₆O₂	Methacrylic acid	162.5	77	1.6–8.8%	68
C₄H₆O₂	Vinyl acetate	72.5	−8	2.6–13.4%	402
C₄H₆O₂	Methyl acrylate	80.7	−3	2.8–25%	468
C₄H₆O₂	2,3-Butanedione	88	27		
C₄H₆O₂	γ-Butyrolactone	204	98		
C₄H₆O₃	Acetic anhydride	139.5	49	2.7–10.3%	316
C₄H₆O₃	Propylene carbonate	242	135		
C₄H₆O₆	*L*-Tartaric acid		210		425
C₄H₇Br	1-Bromo-2-butene	104.5		4.6–12.0%	
C₄H₇BrO₂	Ethyl bromoacetate	168.5	48		
C₄H₇Cl	2-Chloro-1-butene	58.5	−19	2.3–9.3%	
C₄H₇Cl	3-Chloro-2-methylpropene	71.5	−12	3.2–8.1%	
C₄H₇ClO	2-Chloroethyl vinyl ether	108	27		
C₄H₇ClO₂	Ethyl chloroacetate	144.3	64		
C₄H₇N	Butanenitrile	117.6	24	1.6–?	501
C₄H₇N	2-Methylpropanenitrile	103.9	8		482
C₄H₇NO	Acetone cyanohydrin		74	2.2–12.0%	688
C₄H₇NO	2-Pyrrolidone	251	129		
C₄H₈	1-Butene	−6.2		1.6–10.0%	385
C₄H₈	*cis*-2-Butene	3.7		1.7–9.0%	325
C₄H₈	*trans*-2-Butene	0.8		1.8–9.7%	324
C₄H₈	Isobutene	−6.9		1.8–9.6%	465
C₄H₈	Cyclobutane	12.6	<10	1.8–?	
C₄H₈Cl₂	1,2-Dichlorobutane	124.1			275
C₄H₈Cl₂	1,4-Dichlorobutane	161	52		
C₄H₈Cl₂O	Bis(2-chloroethyl) ether	178.5	55	2.7–?	369

Mol. form.	Name	$t_B/°C$	FP/°C	Fl. limits	IT/°C
C_4H_8O	2-Buten-1-ol	121.5	27	4.2–35.3%	349
C_4H_8O	2-Methyl-2-propenol	114.5	33		
C_4H_8O	Ethyl vinyl ether	35.5	<−46	1.7–28%	202
C_4H_8O	1,2-Epoxybutane	63.4	−22	1.7–19%	439
C_4H_8O	Butanal	74.8	−22	1.9–12.5%	218
C_4H_8O	Isobutanal	64.5	−18	1.6–10.6%	196
C_4H_8O	2-Butanone	79.5	−9	1.4–11.4%	404
C_4H_8O	Tetrahydrofuran	65	−14	2–11.8%	321
C_4H_8OS	1,4-Oxathiane	147	42		
$C_4H_8O_2$	Butanoic acid	163.7	72	2.0–10.0%	443
$C_4H_8O_2$	2-Methylpropanoic acid	154.4	56	2.0–9.2%	481
$C_4H_8O_2$	Propyl formate	80.9	−3		455
$C_4H_8O_2$	Isopropyl formate	68.2	−6		485
$C_4H_8O_2$	Ethyl acetate	77.1	−4	2.0–11.5%	426
$C_4H_8O_2$	Methyl propanoate	79.8	−2	2.5–13%	469
$C_4H_8O_2$	3-Hydroxybutanal		66		250
$C_4H_8O_2$	1,4-Dioxane	101.5	12	2.0–22%	180
$C_4H_8O_2S$	Sulfolane	287.3	177		
$C_4H_8O_3$	Methyl lactate	144.8	49	2.2–?	385
$C_4H_8O_3$	Ethylene glycol monoacetate	188	102		
C_4H_9Br	1-Bromobutane	101.6	18	2.6–6.6%	265
C_4H_9Br	2-Bromobutane	91.2	21		
C_4H_9Cl	1-Chlorobutane	78.6	−12	1.9–10.1%	240
C_4H_9Cl	2-Chlorobutane	68.2	−10		
C_4H_9Cl	1-Chloro-2-methylpropane	68.5	−6	2.0–8.7%	
C_4H_9Cl	2-Chloro-2-methylpropane	50.9	0		
$C_4H_9Cl_3Si$	Butyltrichlorosilane	148.5	54		
C_4H_9N	Pyrrolidine	86.5	3		
C_4H_9NO	N-Ethylacetamide	205	110		
C_4H_9NO	N,N-Dimethylacetamide	165	70	1.8–11.5%	490
C_4H_9NO	Butanal oxime	154	58		
C_4H_9NO	2-Butanone oxime	152.5	≈70		
C_4H_9NO	Morpholine	128	37	1.4–11.2%	290
$C_4H_9NO_2$	N-Acetylethanolamine		179		460
$C_4H_9NO_3$	Butyl nitrate	133	36		
C_4H_{10}	Butane	−0.5	−60	1.9–8.5%	287
C_4H_{10}	Isobutane	−11.7	−87	1.8–8.4%	460
$C_4H_{10}N_2$	Piperazine	146	81		
$C_4H_{10}O$	1-Butanol	117.7	37	1.4–11.2%	343
$C_4H_{10}O$	2-Butanol	99.5	24	1.7–9.8%	405
$C_4H_{10}O$	2-Methyl-1-propanol	107.8	28	1.7–10.6%	415
$C_4H_{10}O$	2-Methyl-2-propanol	82.4	11	2.4–8.0%	478
$C_4H_{10}O$	Diethyl ether	34.5	−45	1.9–36.0%	180
$C_4H_{10}O$	Methyl propyl ether	39.1	−20	2.0–14.8%	
$C_4H_{10}O_2$	1,2-Butanediol	190.5	40		
$C_4H_{10}O_2$	1,3-Butanediol	207.5	121		395
$C_4H_{10}O_2$	1,4-Butanediol	235	121		
$C_4H_{10}O_2$	2,3-Butanediol	182.5			402
$C_4H_{10}O_2$	Ethylene glycol monoethyl ether	135	43	3–18%	235
$C_4H_{10}O_2$	Ethylene glycol dimethyl ether	85	−2		202
$C_4H_{10}O_2$	tert-Butyl hydroperoxide		27		
$C_4H_{10}O_2S$	2,2′-Thiodiethanol	282	160		298
$C_4H_{10}O_3$	Diethylene glycol	245.8	124	2–17%	224
$C_4H_{10}O_4S$	Diethyl sulfate	208	104		436
$C_4H_{10}S$	1-Butanethiol	98.5	2		
$C_4H_{10}S$	2-Butanethiol	85	−23		
$C_4H_{10}S$	2-Methyl-1-propanethiol	88.5	2		
$C_4H_{10}S$	2-Methyl-2-propanethiol	64.3	<−29		
$C_4H_{10}Se$	Diethyl selenide	108		2.5–?	
$C_4H_{11}N$	Butylamine	77.0	−12	1.7–9.8%	312
$C_4H_{11}N$	sec-Butylamine	63.5	−9		

Mol. form.	Name	t_B/°C	FP/°C	Fl. limits	IT/°C
C$_4$H$_{11}$N	*tert*-Butylamine	44.0	−9	1.7–8.9%	380
C$_4$H$_{11}$N	Isobutylamine	67.7	−9	2–12%	378
C$_4$H$_{11}$N	Diethylamine	55.5	−23	1.8–10.1%	312
C$_4$H$_{11}$NO	2-Amino-1-butanol	178	74		
C$_4$H$_{11}$NO	2-Amino-2-methyl-1-propanol	165.5	67		
C$_4$H$_{11}$NO$_2$	Diethanolamine	268.8	172	2–13%	662
C$_4$H$_{12}$Sn	Tetramethylstannane	78	−12	1.9–?	
C$_4$H$_{13}$N$_3$	Diethylenetriamine	207	98	2–6.7%	358
C$_5$H$_4$O$_2$	Furfural	161.7	60	2.1–19.3%	316
C$_5$H$_5$N	Pyridine	115.2	20	1.8–12.4%	482
C$_5$H$_6$	2-Methyl-1-buten-3-yne	32	<−7		
C$_5$H$_6$N$_2$	2-Methylpyrazine	137	50		
C$_5$H$_6$O	3-Methylfuran	66	−30		
C$_5$H$_6$O$_2$	Furfuryl alcohol	171	75	1.8–16.3%	491
C$_5$H$_7$N	1-Methylpyrrole	115	16		
C$_5$H$_7$NO	2-Furanmethanamine	145.5	37		
C$_5$H$_7$NO$_2$	Ethyl cyanoacetate	205	110		
C$_5$H$_8$	2-Methyl-1,3-butadiene	34.0	−54	1.5–8.9%	395
C$_5$H$_8$	1-Pentyne	40.1	<−20		
C$_5$H$_8$	Cyclopentene	44.2	−29		395
C$_5$H$_8$O	3-Methyl-3-buten-2-one	98		1.8–9.0%	
C$_5$H$_8$O	Cyclopentanone	130.5	26		
C$_5$H$_8$O	3,4-Dihydro-2H-pyran	86	−18		
C$_5$H$_8$O$_2$	Allyl acetate	103.5	22		374
C$_5$H$_8$O$_2$	Isopropenyl acetate	94	26		432
C$_5$H$_8$O$_2$	Vinyl propanoate	91.2	1		
C$_5$H$_8$O$_2$	Ethyl acrylate	99.4	10	1.4–14%	372
C$_5$H$_8$O$_2$	Methyl methacrylate	100.5	10	1.7–8.2%	
C$_5$H$_8$O$_2$	2,4-Pentanedione	138	34		340
C$_5$H$_8$O$_3$	Methyl acetoacetate	171.7	77		280
C$_5$H$_9$NO	*N*-Methyl-2-pyrrolidone	202	96	1–10%	346
C$_5$H$_{10}$	1-Pentene	29.9	−18	1.5–8.7%	275
C$_5$H$_{10}$	*cis*-2-Pentene	36.9	<−20		
C$_5$H$_{10}$	*trans*-2-Pentene	36.3	<−20		
C$_5$H$_{10}$	2-Methyl-1-butene	31.2	−20		
C$_5$H$_{10}$	3-Methyl-1-butene	20.1	−7	1.5–9.1%	365
C$_5$H$_{10}$	2-Methyl-2-butene	38.5	−20		
C$_5$H$_{10}$	Cyclopentane	49.3	−25	1.5–?	361
C$_5$H$_{10}$Cl$_2$	1,5-Dichloropentane	179	>27		
C$_5$H$_{10}$N$_2$	3-(Dimethylamino)propanenitrile	173	65		
C$_5$H$_{10}$O	Cyclopentanol	140.4	51		
C$_5$H$_{10}$O	Pentanal	103	12		222
C$_5$H$_{10}$O	2-Pentanone	102.2	7	1.5–8.2%	452
C$_5$H$_{10}$O	3-Pentanone	101.9	13	1.6–?	450
C$_5$H$_{10}$O	Tetrahydropyran	88	−20		
C$_5$H$_{10}$O	2-Methyltetrahydrofuran	78	−11		
C$_5$H$_{10}$O$_2$	Pentanoic acid	186.1	96		400
C$_5$H$_{10}$O$_2$	3-Methylbutanoic acid	176.5			416
C$_5$H$_{10}$O$_2$	Butyl formate	106.1	18	1.7–8.2%	322
C$_5$H$_{10}$O$_2$	Isobutyl formate	98.2	5	2–9%	320
C$_5$H$_{10}$O$_2$	Propyl acetate	101.5	13	1.7–8%	450
C$_5$H$_{10}$O$_2$	Isopropyl acetate	88.6	2	1.8–8%	460
C$_5$H$_{10}$O$_2$	Ethyl propanoate	99.1	12	1.9–11%	440
C$_5$H$_{10}$O$_2$	Methyl butanoate	102.8	14		
C$_5$H$_{10}$O$_2$	3-Ethoxypropanal	135.2	38		
C$_5$H$_{10}$O$_2$	Tetrahydrofurfuryl alcohol	178	75	1.5–9.7%	282
C$_5$H$_{10}$O$_3$	Diethyl carbonate	126	25		
C$_5$H$_{10}$O$_3$	Ethylene glycol monomethyl ether acetate	143	49	1.5–12.3%	392
C$_5$H$_{10}$O$_3$	Ethyl lactate	154.5	46	1.5–?	400
C$_5$H$_{11}$Br	1-Bromopentane	129.8	32		

Mol. form.	Name	t_B/°C	FP/°C	Fl. limits	IT/°C
C₅H₁₁Cl	1-Chloropentane	107.8	13	1.6–8.6%	260
C₅H₁₁Cl	2-Chloro-2-methylbutane	85.6		1.5–7.4%	345
C₅H₁₁Cl	1-Chloro-3-methylbutane	98.9	<21	1.5–7.4%	
C₅H₁₁Cl₃Si	Trichloropentylsilane	172	63		
C₅H₁₁N	Piperidine	106.2	16		
C₅H₁₁N	N-Methylpyrrolidine	81	−14		
C₅H₁₁NO	4-Methylmorpholine	116	24		
C₅H₁₁NO₂	Isopentyl nitrite	99.2			210
C₅H₁₂	Pentane	36.0	−40	1.4–8.0%	260
C₅H₁₂	Isopentane	27.8	−51	1.4–7.6%	420
C₅H₁₂	Neopentane	9.4	−65	1.4–7.5%	450
C₅H₁₂N₂	1-Methylpiperazine	138	42		
C₅H₁₂N₂O	Tetramethylurea	176.5	77		
C₅H₁₂O	1-Pentanol	137.9	33	1.2–10.0%	300
C₅H₁₂O	2-Pentanol	119.3	34	1.2–9.0%	343
C₅H₁₂O	3-Pentanol	116.2	41	1.2–9.0%	435
C₅H₁₂O	2-Methyl-1-butanol	128	50		385
C₅H₁₂O	3-Methyl-1-butanol	131.1	43	1.2–9.0%	350
C₅H₁₂O	2-Methyl-2-butanol	102.4	19	1.2–9.0%	437
C₅H₁₂O	3-Methyl-2-butanol	112.9	38		
C₅H₁₂O	2,2-Dimethyl-1-propanol	113.5	37		
C₅H₁₂O	Ethyl propyl ether	63.2	<−20	1.7–9.0%	
C₅H₁₂O₂	1,5-Pentanediol	239	129		335
C₅H₁₂O₂	2-Isopropoxyethanol	145	33		
C₅H₁₂O₂	2,2-Dimethyl-1,3-propanediol	208	129		399
C₅H₁₂O₃	Diethylene glycol monomethyl ether	193	96	1.38–22.7%	240
C₅H₁₂S	1-Pentanethiol	126.6	18		
C₅H₁₂S	3-Methyl-2-butanethiol		3		
C₅H₁₃N	Pentylamine	104.3	−1	2.2–22%	
C₅H₁₃N	Butylmethylamine	91	13		
C₆H₂Cl₄	1,2,4,5-Tetrachlorobenzene	244.5	155		
C₆H₃ClN₂O₄	1-Chloro-2,4-dinitrobenzene	315	194	2.0–22%	
C₆H₃Cl₃	1,2,4-Trichlorobenzene	213.5	105	2.5–6.6%	571
C₆H₄ClNO₂	1-Chloro-4-nitrobenzene	242	127		
C₆H₄Cl₂	o-Dichlorobenzene	180	66	2.2–9.2%	648
C₆H₄Cl₂	m-Dichlorobenzene	173	72		
C₆H₄Cl₂	p-Dichlorobenzene	174	66		
C₆H₄Cl₂O	2,4-Dichlorophenol	210	114		
C₆H₅Br	Bromobenzene	156.0	51		565
C₆H₅Cl	Chlorobenzene	131.7	28	1.3–9.6%	593
C₆H₅ClO	o-Chlorophenol	174.9	64		
C₆H₅ClO	p-Chlorophenol	220	121		
C₆H₅Cl₂N	3,4-Dichloroaniline	272	166		
C₆H₅Cl₃Si	Trichlorophenylsilane	201	91		
C₆H₅F	Fluorobenzene	84.7	−15		
C₆H₅NO₂	Nitrobenzene	210.8	88	1.8–?	482
C₆H₅N₃O₄	2,4-Dinitroaniline	332	224		
C₆H₆	1,5-Hexadien-3-yne	85	<−20	1.5–?	
C₆H₆	Benzene	80.0	−11	1.2–7.8%	498
C₆H₆N₂O₂	p-Nitroaniline	332	199		
C₆H₆O	Phenol	181.8	79	1.8–8.6%	715
C₆H₆O₂	1,2-Benzenediol	245	127		
C₆H₆O₂	Resorcinol		127	1.4–?	608
C₆H₆O₂	p-Hydroquinone	287	165		516
C₆H₇N	Aniline	184.1	70	1.3–11%	615
C₆H₇N	2-Methylpyridine	129.3	39		538
C₆H₇N	4-Methylpyridine	145.3	57		
C₆H₈ClN	Aniline, hydrochloride		193		
C₆H₈Cl₂O₂	Hexanedioyl dichloride		72		
C₆H₈N₂	Adiponitrile	295	93	1.0–?	550
C₆H₈N₂	o-Phenylenediamine	257	156	1.5–?	

Mol. form.	Name	t_B/°C	FP/°C	Fl. limits	IT/°C
C$_6$H$_8$N$_2$	Phenylhydrazine	243.5	88		
C$_6$H$_8$N$_2$	2,5-Dimethylpyrazine	155	64		
C$_6$H$_8$O	2,5-Dimethylfuran	93.5	7		
C$_6$H$_8$O$_4$	Dimethyl maleate	202	113		
C$_6$H$_{10}$	1,4-Hexadiene	65	−21	2.0–6.1%	
C$_6$H$_{10}$	2-Methyl-1,3-pentadiene	75.8	−12		
C$_6$H$_{10}$	4-Methyl-1,3-pentadiene	76.5	−34		
C$_6$H$_{10}$	2-Hexyne	84.5	−10		
C$_6$H$_{10}$	Cyclohexene	82.9	−12	1.2–?	310
C$_6$H$_{10}$O	Diallyl ether	94	−7		
C$_6$H$_{10}$O	Cyclohexanone	155.4	44	1.1–9.4%	420
C$_6$H$_{10}$O	Mesityl oxide	130	31	1.4–7.2%	344
C$_6$H$_{10}$O$_2$	Vinyl butanoate	116.7	20	1.4–8.8%	
C$_6$H$_{10}$O$_2$	Ethyl 2-butenoate	136.5	2		
C$_6$H$_{10}$O$_2$	Ethyl methacrylate	117	20		
C$_6$H$_{10}$O$_2$	2,5-Hexanedione	194	79		499
C$_6$H$_{10}$O$_3$	Ethyl acetoacetate	180.8	57	1.4–9.5%	295
C$_6$H$_{10}$O$_3$	Propanoic anhydride	170	63	1.3–9.5%	285
C$_6$H$_{10}$O$_4$	Adipic acid	337.5	196		420
C$_6$H$_{10}$O$_4$	Diethyl oxalate	185.7	76		
C$_6$H$_{10}$O$_4$	Ethylene glycol diacetate	190	88	1.6–8.4%	482
C$_6$H$_{11}$Cl	Chlorocyclohexane	142	32		
C$_6$H$_{11}$NO	Caprolactam	270	125		
C$_6$H$_{11}$NO$_2$	Nitrocyclohexane	205	88		
C$_6$H$_{11}$NO$_2$	4-Acetylmorpholine		113		
C$_6$H$_{12}$	1-Hexene	63.4	−26	1.2–6.9%	253
C$_6$H$_{12}$	cis-2-Hexene	68.8	−21		
C$_6$H$_{12}$	2-Methyl-1-pentene	62.1	−28		300
C$_6$H$_{12}$	4-Methyl-1-pentene	53.9	−7		300
C$_6$H$_{12}$	4-Methyl-cis-2-pentene	56.3	−32		
C$_6$H$_{12}$	4-Methyl-trans-2-pentene	58.6	−29		
C$_6$H$_{12}$	2-Ethyl-1-butene	64.7	<−20		315
C$_6$H$_{12}$	2,3-Dimethyl-1-butene	55.6	<−20		360
C$_6$H$_{12}$	2,3-Dimethyl-2-butene	73.3	<−20		401
C$_6$H$_{12}$	Cyclohexane	80.7	−20	1.3–8%	245
C$_6$H$_{12}$	Methylcyclopentane	71.8	−29	1.0–8.35%	258
C$_6$H$_{12}$	Ethylcyclobutane	70.8	−15	1.2–7.7%	210
C$_6$H$_{12}$	2-Methyl-2-pentene	67.3	<−7		
C$_6$H$_{12}$Cl$_2$O$_2$	1,2-Bis(2-chloroethoxy)ethane	232	121		
C$_6$H$_{12}$O	cis-3-Hexen-1-ol	156.5	54		
C$_6$H$_{12}$O	Butyl vinyl ether	94	−9		255
C$_6$H$_{12}$O	Isobutyl vinyl ether	83	−9		
C$_6$H$_{12}$O	Hexanal	131	32		
C$_6$H$_{12}$O	2-Ethylbutanal		21	1.2–7.7%	
C$_6$H$_{12}$O	2-Methylpentanal	117	17		199
C$_6$H$_{12}$O	2-Hexanone	127.6	25	1–8%	423
C$_6$H$_{12}$O	3-Hexanone	123.5	35	1–8%	
C$_6$H$_{12}$O	4-Methyl-2-pentanone	116.5	18	1.2–8.0%	448
C$_6$H$_{12}$O	Cyclohexanol	160.8	68	1–9%	300
C$_6$H$_{12}$O$_2$	Hexanoic acid	205.2	102		380
C$_6$H$_{12}$O$_2$	2-Methylpentanoic acid	195.6	107		378
C$_6$H$_{12}$O$_2$	Diethylacetic acid	194	99		400
C$_6$H$_{12}$O$_2$	Pentyl formate	130.4	26		
C$_6$H$_{12}$O$_2$	Butyl acetate	126.1	22	1.7–7.6%	425
C$_6$H$_{12}$O$_2$	sec-Butyl acetate	112	31	1.7–9.8%	
C$_6$H$_{12}$O$_2$	Isobutyl acetate	116.5	18	1.3–10.5%	421
C$_6$H$_{12}$O$_2$	Propyl propanoate	122.5	79		
C$_6$H$_{12}$O$_2$	Ethyl butanoate	121.5	24		463
C$_6$H$_{12}$O$_2$	Ethyl 2-methylpropanoate	110.1	13		
C$_6$H$_{12}$O$_2$	Diacetone alcohol	167.9	58	1.8–6.9%	643

Mol. form.	Name	t_B/°C	FP/°C	Fl. limits	IT/°C
$C_6H_{12}O_3$	Ethylene glycol monoethyl ether acetate	156.4	56	2–8%	379
$C_6H_{12}O_3$	Paraldehyde	124.3	36	1.3–?	238
$C_6H_{12}S$	Cyclohexanethiol	158.9	43		
$C_6H_{13}Cl$	1-Chlorohexane	135	35		
$C_6H_{13}N$	Cyclohexylamine	134	31	1.9–9.4%	293
$C_6H_{13}NO$	N-Butylacetamide	229	116		
$C_6H_{13}NO$	2,6-Dimethylmorpholine	146.6	44		
$C_6H_{13}NO$	N-Ethylmorpholine	138.5	32		
$C_6H_{13}NO_2$	4-Morpholineethanol	227	99		
C_6H_{14}	Hexane	68.7	−22	1.1–7.5%	225
C_6H_{14}	2-Methylpentane	60.2	<−29	1.0–7.0%	264
C_6H_{14}	3-Methylpentane	63.2	−7	1.2–7.0%	278
C_6H_{14}	2,2-Dimethylbutane	49.7	−48	1.2–7.0%	405
C_6H_{14}	2,3-Dimethylbutane	57.9	−29	1.2–7.0%	405
$C_6H_{14}N_2O$	1-Piperazineethanol	246	124		
$C_6H_{14}O$	1-Hexanol	157.6	63		
$C_6H_{14}O$	2-Methyl-1-pentanol	149	54	1.1–9.65%	310
$C_6H_{14}O$	4-Methyl-2-pentanol	131.6	41	1.0–5.5%	
$C_6H_{14}O$	2-Ethyl-1-butanol	147	57		
$C_6H_{14}O$	Dipropyl ether	90.0	21	1.3–7.0%	188
$C_6H_{14}O$	Diisopropyl ether	68.5	−28	1.4–7.9%	443
$C_6H_{14}O$	Butyl ethyl ether	92.3	4		
$C_6H_{14}O_2$	2,5-Hexanediol	218	110		
$C_6H_{14}O_2$	2-Methyl-2,4-pentanediol	197.1	102	1–9%	306
$C_6H_{14}O_2$	Ethylene glycol monobutyl ether	168.4	69	4–13%	238
$C_6H_{14}O_2$	1,1-Diethoxyethane	102.2	−21	1.6–10.4%	230
$C_6H_{14}O_2$	Ethylene glycol diethyl ether	119.4	27		205
$C_6H_{14}O_3$	1,2,6-Hexanetriol		191		
$C_6H_{14}O_3$	Diethylene glycol monoethyl ether	196	96		
$C_6H_{14}O_3$	Diethylene glycol dimethyl ether	162	67		
$C_6H_{14}O_3$	Trimethylolpropane		149		
$C_6H_{14}O_4$	Triethylene glycol	285	177	0.9–9.2%	371
$C_6H_{15}N$	Hexylamine	132.8	29		
$C_6H_{15}N$	Butylethylamine	107.5	18		
$C_6H_{15}N$	Dipropylamine	109.3	17		299
$C_6H_{15}N$	Diisopropylamine	83.9	−1	1.1–7.1%	316
$C_6H_{15}N$	Triethylamine	89	−7	1.2–8.0%	249
$C_6H_{15}NO_2$	Diisopropanolamine	250	127		374
$C_6H_{15}NO_3$	Triethanolamine	335.4	179	1–10%	
$C_6H_{15}N_3$	1-Piperazineethanamine	220	93		
$C_6H_{15}O_4P$	Triethyl phosphate	215.5	115		454
$C_6H_{16}N_2$	N,N-Diethylethylenediamine	144	46		
$C_7H_3ClF_3NO_2$	1-Chloro-4-nitro-2-(trifluoromethyl)benzene	232	135		
$C_7H_4ClF_3$	1-Chloro-2-(trifluoromethyl)benzene	152.2	59		
$C_7H_4F_3NO_2$	1-Nitro-3-(trifluoromethyl)benzene	202.8	103		
C_7H_5ClO	Benzoyl chloride	197.2	72		
C_7H_5ClO	4-Chlorobenzaldehyde	213.5	88		
$C_7H_5Cl_3$	(Trichloromethyl)benzene	221	127		211
$C_7H_5F_3$	(Trifluoromethyl)benzene	102.1	12		
$C_7H_6N_2O_4$	1-Methyl-2,4-dinitrobenzene		207		
C_7H_6O	Benzaldehyde	179.0	63		192
$C_7H_6O_2$	Benzoic acid	249.2	121		570
$C_7H_6O_2$	Salicylaldehyde	197	78		
$C_7H_6O_3$	Salicylic acid		157	1.1–?	540
C_7H_7Br	o-Bromotoluene	181.7	79		
C_7H_7Br	p-Bromotoluene	184.3	85		
C_7H_7Cl	(Chloromethyl)benzene	179	67	1.1–?	585
$C_7H_7NO_2$	o-Nitrotoluene	222	106		
$C_7H_7NO_2$	m-Nitrotoluene	232	106		
$C_7H_7NO_2$	p-Nitrotoluene	238.3	106		

Mol. form.	Name	t_B/°C	FP/°C	Fl. limits	IT/°C
C₇H₈	Toluene	110.6	4	1.1–7.1%	480
C₇H₈	Bicyclo[2.2.1]hepta-2,5-diene	89.5	−21		
C₇H₈O	o-Cresol	191.0	81	1.4–?	599
C₇H₈O	m-Cresol	202.2	86	1.1–?	558
C₇H₈O	p-Cresol	201.9	86	1.1–?	558
C₇H₈O	Benzyl alcohol	205.3	93		436
C₇H₈O	Anisole	153.7	52		475
C₇H₈O₂	4-Methoxyphenol	243	132		421
C₇H₈O₃S	p-Toluenesulfonic acid		184		
C₇H₉N	o-Methylaniline	200.3	85		482
C₇H₉N	p-Methylaniline	200.4	87		482
C₇H₉NO	o-Anisidine	224	118		
C₇H₁₀O	3-Cyclohexene-1-carboxaldehyde	105	57		
C₇H₁₀O₄	3,3-Diacetoxy-1-propene	180	82		
C₇H₁₂	4-Methylcyclohexene	102.7	−1		
C₇H₁₂O₂	Butyl acrylate	145	29	1.7–9.9%	292
C₇H₁₂O₂	Isobutyl acrylate	132	30		427
C₇H₁₂O₂	Cyclohexyl formate	162	51		
C₇H₁₂O₄	Diethyl malonate	200	93		
C₇H₁₄	1-Heptene	93.6	−1		260
C₇H₁₄	trans-2-Heptene	98	<0		
C₇H₁₄	Cycloheptane	118.4	<21	1.1–6.7%	
C₇H₁₄	Methylcyclohexane	100.9	−4	1.2–6.7%	250
C₇H₁₄	Ethylcyclopentane	103.5	<21	1.1–6.7%	260
C₇H₁₄O	2-Heptanone	151.0	39	1.1–7.9%	393
C₇H₁₄O	3-Heptanone	147	46		
C₇H₁₄O	4-Heptanone	144	49		
C₇H₁₄O	5-Methyl-2-hexanone	144	36	1.0–8.2%	191
C₇H₁₄O	cis-2-Methylcyclohexanol	165	65		296
C₇H₁₄O	trans-2-Methylcyclohexanol	167.5	65		296
C₇H₁₄O	cis-3-Methylcyclohexanol	174.5	70		295
C₇H₁₄O	trans-3-Methylcyclohexanol	174.5	70		295
C₇H₁₄O	cis-4-Methylcyclohexanol	173	70		295
C₇H₁₄O	trans-4-Methylcyclohexanol	174	70		295
C₇H₁₄O₂	Pentyl acetate	149.2	16	1.1–7.5%	360
C₇H₁₄O₂	Isopentyl acetate	142.5	25	1.0–7.5%	360
C₇H₁₄O₂	sec-Pentyl acetate	130.5	32		
C₇H₁₄O₂	Butyl propanoate	146.8	32		426
C₇H₁₄O₂	Propyl butanoate	143.0	37		
C₇H₁₅NO₂	Ethyl N-butylcarbamate	202	92		
C₇H₁₆	Heptane	98.5	−4	1.05–6.7%	204
C₇H₁₆	2-Methylhexane	90.0	−1	1.0–6.0%	280
C₇H₁₆	3-Methylhexane	92	−4		280
C₇H₁₆	2,3-Dimethylpentane	89.7	−56	1.1–6.7%	335
C₇H₁₆	2,4-Dimethylpentane	80.4	−12		
C₇H₁₆	2,2,3-Trimethylbutane	80.8	<0		412
C₇H₁₆N₂O	4-Morpholinepropanamine	220	104		
C₇H₁₆O	2-Heptanol	159	71		
C₇H₁₆O	3-Heptanol	157	60		
C₇H₁₆O	2,4-Dimethyl-3-pentanol	138.7	49		
C₇H₁₆O	2,3,3-Trimethyl-2-butanol	131	<0		375
C₇H₁₇N	Heptylamine	156	54		
C₇H₁₈N₂	N,N-Diethyl-1,3-propanediamine	168.5	59		
C₈H₄O₃	Phthalic anhydride	295	152	1.7–10.5%	570
C₈H₆O₄	Phthalic acid		168		
C₈H₆O₄	Terephthalic acid		260		496
C₈H₇ClO	α-Chloroacetophenone	247	118		
C₈H₇N	Benzeneacetonitrile	233.5	113		
C₈H₈	Styrene	145	31	0.9–6.8%	490
C₈H₈O	Phenyloxirane	194.1	74		498
C₈H₈O	Benzeneacetaldehyde	195	71		

Mol. form.	Name	$t_B/°C$	FP/°C	Fl. limits	IT/°C
C_8H_8O	Acetophenone	202	77		570
$C_8H_8O_2$	Benzeneacetic acid	265.5	>100		
$C_8H_8O_2$	Phenyl acetate	196	80		
$C_8H_8O_2$	Methyl benzoate	199	83		
$C_8H_8O_2$	2-Methoxybenzaldehyde	243.5	118		
$C_8H_8O_3$	Methyl salicylate	222.9	96		454
C_8H_9Cl	1-Chloro-4-ethylbenzene	184.4	64		
C_8H_9NO	Acetanilide	304	169		530
$C_8H_9NO_2$	Methyl 2-aminobenzoate	256	>100		
C_8H_{10}	Ethylbenzene	136.1	21	0.8–6.7%	432
C_8H_{10}	o-Xylene	144.5	32	0.9–6.7%	463
C_8H_{10}	m-Xylene	139.1	27	1.1–7.0%	527
C_8H_{10}	p-Xylene	138.3	27	1.1–7.0%	528
$C_8H_{10}O$	p-Ethylphenol	217.9	104		
$C_8H_{10}O$	Benzeneethanol	218.2	96		
$C_8H_{10}O$	α-Methylbenzyl alcohol	205	93		
$C_8H_{10}O$	Phenetole	169.8	63		
$C_8H_{10}O$	Benzyl methyl ether	170	135		
$C_8H_{10}O$	4-Methylanisole	175.5	60		
$C_8H_{10}O_2$	2-Phenoxyethanol	245	121		
$C_8H_{11}N$	N-Ethylaniline	203.0	85		
$C_8H_{11}N$	N,N-Dimethylaniline	194.1	63		371
$C_8H_{11}N$	2,3-Xylidine	221.5	97	1.0–?	
$C_8H_{11}N$	2,6-Xylidine	215	96		
$C_8H_{11}N$	α-Methylbenzylamine	187	79		
$C_8H_{11}N$	5-Ethyl-2-picoline	178.3	68	1.1–6.6%	
$C_8H_{11}NO$	N-Phenylethanolamine	279.5	152		
$C_8H_{11}NO$	o-Phenetidine	232.5	115		
$C_8H_{11}NO$	p-Phenetidine	254	116		
C_8H_{12}	1,5-Cyclooctadiene	150.8	35		
C_8H_{12}	4-Vinylcyclohexene	128	16		269
$C_8H_{12}O_4$	Diethyl maleate	223	121		350
$C_8H_{12}O_4$	Diethyl fumarate	214	104		
$C_8H_{14}O_2$	Cyclohexyl acetate	173	58		335
$C_8H_{14}O_2$	Butyl methacrylate	160	52		
$C_8H_{14}O_3$	Butanoic anhydride	200	54	0.9–5.8%	279
$C_8H_{14}O_3$	2-Methylpropanoic anhydride	183	59	1.0–6.2%	329
$C_8H_{14}O_3$	Butyl acetoacetate		85		
$C_8H_{14}O_4$	Ethyl succinate	217.7	90		
$C_8H_{14}O_5$	Diethylene glycol diacetate	200	135		
$C_8H_{14}O_6$	Diethyl tartrate	281	93		
$C_8H_{15}ClO$	Octanoyl chloride	195.6	82		
C_8H_{16}	1-Octene	121.2	21		230
C_8H_{16}	2,4,4-Trimethyl-1-pentene	101.4	−5	0.8–4.8%	391
C_8H_{16}	2,4,4-Trimethyl-2-pentene	104.9	2		305
C_8H_{16}	Ethylcyclohexane	131.9	35	0.9–6.6%	238
C_8H_{16}	cis-1,2-Dimethylcyclohexane	129.8	16		304
C_8H_{16}	trans-1,2-Dimethylcyclohexane	123.5	11		304
C_8H_{16}	cis-1,4-Dimethylcyclohexane	124.4	16		
C_8H_{16}	Propylcyclopentane	131			269
$C_8H_{16}O$	Octanal	171	52		
$C_8H_{16}O$	2-Ethylhexanal	163	44	0.85–7.2%	190
$C_8H_{16}O$	2-Octanone	172.5	52		
$C_8H_{16}O_2$	Hexyl acetate	171.5	45		
$C_8H_{16}O_2$	sec-Hexyl acetate	147.5	45		
$C_8H_{16}O_2$	2-Ethylbutyl acetate	162.5	54		
$C_8H_{16}O_2$	Pentyl propanoate	168.6	41		378
$C_8H_{16}O_2$	Butyl butanoate	166	53		
$C_8H_{16}O_2$	Isobutyl butanoate	156.9	50		
$C_8H_{16}O_2$	Isobutyl isobutanoate	148.6	38	0.96–7.59%	432
$C_8H_{16}O_2$	Ethyl hexanoate	167	49		

Mol. form.	Name	t_B/°C	FP/°C	Fl. limits	IT/°C
$C_8H_{16}O_2$	1,4-Cyclohexanedimethanol	283	167		316
$C_8H_{16}O_3$	Pentyl lactate		79		
$C_8H_{16}O_4$	Diethylene glycol monoethyl ether acetate	218.5	110		425
$C_8H_{17}Cl$	1-Chlorooctane	181.5	70		
$C_8H_{17}Cl$	3-(Chloromethyl)heptane	172	60		
C_8H_{18}	Octane	125.6	13	1.0–6.5%	206
C_8H_{18}	2,3-Dimethylhexane	115.6	7		438
C_8H_{18}	2,4-Dimethylhexane	109.5	10		
C_8H_{18}	3-Ethyl-2-methylpentane	115.6	<21		460
C_8H_{18}	2,2,3-Trimethylpentane	110	<21		346
C_8H_{18}	2,2,4-Trimethylpentane	99.2	−12		418
C_8H_{18}	2,3,3-Trimethylpentane	114.8	<21		425
$C_8H_{18}O$	1-Octanol	195.1	81		
$C_8H_{18}O$	2-Octanol	180	88		
$C_8H_{18}O$	2-Ethyl-1-hexanol	184.6	73	0.88–9.7%	231
$C_8H_{18}O$	Dibutyl ether	140.2	25	1.5–7.6%	194
$C_8H_{18}O_2$	2-Ethyl-1,3-hexanediol	244	127		360
$C_8H_{18}O_2$	2,2,4-Trimethyl-1,3-pentanediol	235	113		346
$C_8H_{18}O_2$	Di-tert-butyl peroxide	111	18		
$C_8H_{18}O_3$	Diethylene glycol diethyl ether	188	82		
$C_8H_{18}O_4$	2,5,8,11-Tetraoxadodecane	216	111		
$C_8H_{18}O_5$	Tetraethylene glycol	328	182		
$C_8H_{18}S$	1-Octanethiol	199.1	69		
$C_8H_{18}S$	Dibutyl sulfide	185	76		
$C_8H_{19}N$	Octylamine	179.6	60		
$C_8H_{19}N$	Dibutylamine	159.6	47	1.1–6%	
$C_8H_{19}N$	Diisobutylamine	139.6	29		
$C_8H_{19}N$	2-Ethylhexylamine	169.2	60		
$C_8H_{20}O_4Si$	Ethyl silicate	168.8	52		
$C_8H_{23}N_5$	Tetraethylenepentamine	341.5	163		321
$C_9H_6N_2O_2$	Toluene-2,4-diisocyanate	251	127	0.9–9.5%	
C_9H_7N	Quinoline	237.1			480
C_9H_{10}	o-Methylstyrene	169.8	53	0.8–11.0%	538
C_9H_{10}	m-Methylstyrene	164	53	0.8–11.0%	538
C_9H_{10}	p-Methylstyrene	172.8	53	0.8–11.0%	538
C_9H_{10}	Isopropenylbenzene	165.4	54	1.9–6.1%	574
$C_9H_{10}O$	1-Phenyl-1-propanone	217.5	99		
$C_9H_{10}O$	4-Methylacetophenone	226	96		
$C_9H_{10}O_2$	Ethyl benzoate	212	88		490
$C_9H_{10}O_2$	Benzyl acetate	213	90		460
$C_9H_{10}O_2$	Methyl 2-phenylacetate	216.5	91		
$C_9H_{11}NO$	4-Methylacetanilide	307	168		
C_9H_{12}	Propylbenzene	159.2	30	0.8–6.0%	450
C_9H_{12}	Isopropylbenzene	152.4	36	0.9–6.5%	424
C_9H_{12}	o-Ethyltoluene	165.2			440
C_9H_{12}	m-Ethyltoluene	161.3			480
C_9H_{12}	p-Ethyltoluene	162			475
C_9H_{12}	1,2,3-Trimethylbenzene	176.1	44	0.8–6.6%	470
C_9H_{12}	1,2,4-Trimethylbenzene	169.3	44	0.9–6.4%	500
C_9H_{12}	1,3,5-Trimethylbenzene	164.7	50	1–5%	559
$C_9H_{12}O$	α−Ethylbenzyl alcohol	219	100		
$C_9H_{12}O_2$	Ethylene glycol monobenzyl ether	256	129		352
$C_9H_{12}O_3S$	Ethyl p-toluenesulfonate		158		
$C_9H_{13}N$	Amphetamine	203	<100		
$C_9H_{14}O$	Phorone	197.5	85		
$C_9H_{14}O$	Isophorone	215.2	84	0.8–3.8%	460
$C_9H_{14}O_6$	Triacetin	259	138	1.0–?	433
C_9H_{16}	Octahydroindene	167			296
$C_9H_{16}O_2$	Allyl hexanoate	186	66		
C_9H_{18}	1-Nonene	146.9	26		

Mol. form.	Name	$t_B/°C$	FP/°C	Fl. limits	IT/°C
C₉H₁₈	Propylcyclohexane	156.7			248
C₉H₁₈	Isopropylcyclohexane	154.8			283
C₉H₁₈	Butylcyclopentane	156.6			250
C₉H₁₈O	2-Nonanone	195.3	60	0.9–5.9%	360
C₉H₁₈O	Diisobutyl ketone	169.4	49	0.8–7.1%	396
C₉H₁₈O₂	Pentyl butanoate	186.4	57		
C₉H₁₈O₂	Isopentyl butanoate	179	59		
C₉H₁₈O₂	Butyl 3-methylbutanoate		53		
C₉H₂₀	Nonane	150.8	31	0.8–2.9%	205
C₉H₂₀	3-Ethyl-4-methylhexane	140	24		
C₉H₂₀	4-Ethyl-2-methylhexane	133.8	<21	0.7–?	280
C₉H₂₀	2,2,5-Trimethylhexane	124.0	13		
C₉H₂₀	3,3-Diethylpentane	146.3		0.7–5.7%	290
C₉H₂₀	3-Ethyl-2,4-dimethylpentane	136.7	390		
C₉H₂₀	2,2,3,3-Tetramethylpentane	140.2	<21	0.8–4.9%	430
C₉H₂₀	2,2,3,4-Tetramethylpentane	133.0	<21		
C₉H₂₁BO₃	Triisopropyl borate	140	28		
C₉H₂₁N	Tripropylamine	156	41		
C₉H₂₁NO₃	Triisopropanolamine		160		320
C₁₀H₇Cl	1-Chloronaphthalene	259	121		>558
C₁₀H₈	Naphthalene	217.9	79	0.9–5.9%	526
C₁₀H₈O	2-Naphthol	285	153		
C₁₀H₉N	1-Naphthalenamine	300.8	157		
C₁₀H₁₀O₂	Safrole	234.5	100		
C₁₀H₁₀O₄	Dimethyl phthalate	283.7	146	0.9–?	490
C₁₀H₁₀O₄	Dimethyl isophthalate	282	138		
C₁₀H₁₀O₄	Dimethyl terephthalate	288	153		518
C₁₀H₁₁NO₂	Acetoacetanilide		185		
C₁₀H₁₂	1,2,3,4-Tetrahydronaphthalene	207.6	71	0.8–5.0%	385
C₁₀H₁₂O₂	Isopropyl benzoate	216	99		
C₁₀H₁₂O₂	Ethyl phenylacetate	227	99		
C₁₀H₁₄	Butylbenzene	183.3	71	0.8–5.8%	410
C₁₀H₁₄	sec-Butylbenzene	173.3	52	0.8–6.9%	418
C₁₀H₁₄	tert-Butylbenzene	169.1	60	0.7–5.7%	450
C₁₀H₁₄	Isobutylbenzene	172.7	55	0.8–6.0%	427
C₁₀H₁₄	p-Cymene	177.1	47	0.7–5.6%	436
C₁₀H₁₄	1,2,3,4-Tetramethylbenzene	205	74		427
C₁₀H₁₄	1,2,3,5-Tetramethylbenzene	198	71		427
C₁₀H₁₄	1,2,4,5-Tetramethylbenzene	196.8	54		
C₁₀H₁₄	o-Diethylbenzene	184	57		395
C₁₀H₁₄	m-Diethylbenzene	181.1	56		450
C₁₀H₁₄	p-Diethylbenzene	183.7	55	0.7–6.0%	430
C₁₀H₁₄O	Butyl phenyl ether	210	82		
C₁₀H₁₄O₂	4-tert-Butyl-1,2-benzenediol	285	130		
C₁₀H₁₅N	N-Butylaniline	243.5	107		
C₁₀H₁₅N	N,N-Diethylaniline	216.3	85		630
C₁₀H₁₅NO₂	N-Phenyl-N,N-diethanolamine		196	0.7–?	387
C₁₀H₁₆	Dipentene	178	45		237
C₁₀H₁₆	d-Limonene	178	45	0.7–6.1%	237
C₁₀H₁₆	α-Pinene	156.2	33		255
C₁₀H₁₆	β-Pinene	166	38		275
C₁₀H₁₆	β-Phellandrene	171.5	49		
C₁₀H₁₆O	Camphor	207.4	66	0.6–3.5%	466
C₁₀H₁₈	trans-Decahydronaphthalene	187.3	54	0.7–5.4%	255
C₁₀H₁₈O	Borneol		66		
C₁₀H₁₈O	Linalol	198	71		
C₁₀H₁₈O	α-Terpineol	220	90		
C₁₀H₁₈O	Cineole	176.4	48		
C₁₀H₁₈O	trans-Geraniol	230	>100		
C₁₀H₁₈O₄	Dibutyl oxalate	241	104		
C₁₀H₁₉NO₂	N-tert-Butylaminoethyl methacrylate		96		

Mol. form.	Name	t_B/°C	FP/°C	Fl. limits	IT/°C
$C_{10}H_{20}$	1-Decene	170.5	<55		235
$C_{10}H_{20}$	Butylcyclohexane	180.9			246
$C_{10}H_{20}$	Isobutylcyclohexane	171.3			274
$C_{10}H_{20}$	*tert*-Butylcyclohexane	171.5			342
$C_{10}H_{20}O$	Citronellol	224	96		
$C_{10}H_{20}O_2$	2-Ethylhexyl acetate	199	71	0.76–8.14%	268
$C_{10}H_{20}O_2$	Ethyl octanoate	208.5	79		
$C_{10}H_{21}N$	*N*-Butylcyclohexanamine		93		
$C_{10}H_{22}$	Decane	174.1	51	0.8–5.4%	210
$C_{10}H_{22}$	2-Methylnonane	167.1			210
$C_{10}H_{22}$	3-Ethyloctane	166.5			230
$C_{10}H_{22}$	4-Ethyloctane	163.7			229
$C_{10}H_{22}O$	1-Decanol	231.1	82		288
$C_{10}H_{22}O$	Dipentyl ether	190	57		170
$C_{10}H_{22}O_2$	Ethylene glycol dibutyl ether	203.3	85		
$C_{10}H_{22}O_5$	Tetraethylene glycol dimethyl ether	275.3	141		
$C_{10}H_{22}S$	Dipentyl sulfide		85		
$C_{10}H_{23}N$	Decylamine	220.5	99		
$C_{10}H_{23}N$	Dipentylamine	202.5	51		
$C_{11}H_{10}$	1-Methylnaphthalene	244.7			529
$C_{11}H_{12}O_3$	Ethyl benzoylacetate		141		
$C_{11}H_{14}O_2$	Butyl benzoate	250.3	107		
$C_{11}H_{16}$	*p-tert*-Butyltoluene	190	68		
$C_{11}H_{16}$	Pentylbenzene	205.4	66		
$C_{11}H_{16}$	1,3-Diethyl-5-methylbenzene	205			455
$C_{11}H_{16}$	Pentamethylbenzene	232	93		427
$C_{11}H_{16}O$	4-*tert*-Butyl-2-methylphenol	237	118		
$C_{11}H_{17}N$	*p-tert*-Pentylaniline	260.5	102		
$C_{11}H_{20}O_2$	2-Ethylhexyl acrylate		82		252
$C_{11}H_{22}$	Pentylcyclohexane	203.7			239
$C_{11}H_{22}O$	2-Undecanone	231.5	89		
$C_{11}H_{22}O_2$	Nonyl acetate	210	68		
$C_{11}H_{24}$	Undecane	195.9	69		
$C_{11}H_{24}$	2-Methyldecane	189.3			225
$C_{11}H_{24}O$	2-Undecanol	228	113		
$C_{12}H_9Br$	4-Bromo-1,1'-biphenyl	310	144		
$C_{12}H_{10}$	Biphenyl	256.1	113	0.6–5.8%	540
$C_{12}H_{10}Cl_2Si$	Dichlorodiphenylsilane	305	142		
$C_{12}H_{10}O$	*o*-Phenylphenol	286	124		530
$C_{12}H_{10}O$	Diphenyl ether	258.0	112	0.8–1.5%	618
$C_{12}H_{11}N$	2-Aminobiphenyl	299			450
$C_{12}H_{11}N$	Diphenylamine	302	153		634
$C_{12}H_{12}$	1-Ethylnaphthalene	258.6			480
$C_{12}H_{14}O_4$	Diethyl phthalate	295	161	0.7–?	457
$C_{12}H_{14}O_4$	Diethyl terephthalate	302	117		
$C_{12}H_{16}$	Cyclohexylbenzene	240.1	99		
$C_{12}H_{16}O_3$	Pentyl salicylate	270	132		
$C_{12}H_{17}NO$	*N*-Butyl-*N*-phenylacetamide	281	141		
$C_{12}H_{18}$	1,5,9-Cyclododecatriene	240	71		
$C_{12}H_{20}O_4$	Dibutyl maleate	280	141		
$C_{12}H_{22}O_4$	Dimethyl sebacate		145		
$C_{12}H_{22}O_6$	Dibutyl tartrate	320	91		284
$C_{12}H_{23}N$	Dicyclohexylamine		>99		
$C_{12}H_{24}$	1-Dodecene	213.8	79		
$C_{12}H_{24}O_2$	Ethyl decanoate	241.5	>100		
$C_{12}H_{25}Br$	1-Bromododecane	276	144		
$C_{12}H_{26}$	Dodecane	216.3	74	0.6–?	203
$C_{12}H_{26}O$	1-Dodecanol	259	127		275
$C_{12}H_{26}O$	2-Butyl-1-octanol	246.5	110		
$C_{12}H_{26}O_3$	Diethylene glycol dibutyl ether	256	118		310
$C_{12}H_{26}S$	1-Dodecanethiol	277	128		

Mol. form.	Name	t_B/°C	FP/°C	Fl. limits	IT/°C
$C_{12}H_{27}BO_3$	Tributyl borate	234	93		
$C_{12}H_{27}N$	Tributylamine	216.5	63		
$C_{12}H_{27}O_4P$	Tributyl phosphate	289	146		
$C_{13}H_{12}$	2-Methylbiphenyl	255.5	137		502
$C_{13}H_{12}$	Diphenylmethane	265.0	130		485
$C_{13}H_{14}N_2$	p,p′-Diaminodiphenylmethane	398	220		
$C_{13}H_{26}$	1-Tridecene	232.8	79		
$C_{13}H_{26}O$	2-Tridecanone	263	107		
$C_{13}H_{28}$	Tridecane	235.4	79		
$C_{13}H_{28}O$	1-Tridecanol		121		
$C_{14}H_8O_2$	9,10-Anthracenedione	377	185		
$C_{14}H_{10}$	Anthracene	339.9	121	0.6–?	540
$C_{14}H_{10}$	Phenanthrene	340	171		
$C_{14}H_{12}O_2$	Benzyl benzoate	323.5	148		480
$C_{14}H_{12}O_3$	Benzyl salicylate	320	>100		
$C_{14}H_{14}$	1,1-Diphenylethane	272.6	>100		440
$C_{14}H_{14}O$	Dibenzyl ether	298	135		
$C_{14}H_{16}$	1-Butylnaphthalene	289.3	360		
$C_{14}H_{16}N_2O_2$	o-Dianisidine		206		
$C_{14}H_{23}N$	N,N-Dibutylaniline	274.8	110		
$C_{14}H_{28}$	1-Tetradecene	233	110		235
$C_{14}H_{30}$	Tetradecane	253.5	112	0.5–?	200
$C_{14}H_{30}O$	1-Tetradecanol	289	141		
$C_{15}H_{18}$	1-Pentylnaphthalene	307	124		
$C_{15}H_{24}$	Nonylbenzene	280.5	99		
$C_{15}H_{24}O$	2,6-Di-tert-butyl-4-methylphenol	265	127		
$C_{15}H_{26}O_6$	Tributyrin	307.5	180	0.5–?	407
$C_{15}H_{33}N$	Tripentylamine	242.5	102		
$C_{16}H_{14}O$	1,3-Diphenyl-2-buten-1-one	342.5	177		
$C_{16}H_{18}$	2-Butyl-1,1′-biphenyl		>100		430
$C_{16}H_{22}O_4$	Dibutyl phthalate	340	157	0.5–?	402
$C_{16}H_{26}$	Decylbenzene	298	107		
$C_{16}H_{34}$	Hexadecane	286.8	136		202
$C_{16}H_{34}O$	Dioctyl ether	283	>100		205
$C_{16}H_{35}N$	Bis(2-ethylhexyl)amine		132		
$C_{17}H_{20}N_2O$	N,N′-Diethylcarbanilide		150		
$C_{17}H_{34}O$	2-Heptadecanone	320	120		
$C_{17}H_{36}O$	1-Heptadecanol	333	154		
$C_{18}H_{14}$	o-Terphenyl	332	163		
$C_{18}H_{14}$	m-Terphenyl	363	191		
$C_{18}H_{15}O_3P$	Triphenyl phosphite	360	218		
$C_{18}H_{15}O_4P$	Triphenyl phosphate		220		
$C_{18}H_{15}P$	Triphenylphosphine		180		
$C_{18}H_{30}$	Dodecylbenzene	328	140		
$C_{18}H_{32}O_7$	Butyl citrate		157		368
$C_{18}H_{34}O_2$	Oleic acid	360	189		363
$C_{18}H_{34}O_4$	Dibutyl sebacate	344.5	178	0.4–?	365
$C_{18}H_{36}O_2$	Stearic acid		196		395
$C_{18}H_{37}Cl_3Si$	Trichlorooctadecylsilane		89		
$C_{18}H_{38}$	Octadecane	316.3	>100		227
$C_{18}H_{38}O$	1-Octadecanol				450
$C_{19}H_{16}$	Triphenylmethane	359	>100		
$C_{19}H_{38}O$	2-Nonadecanone		124		
$C_{19}H_{38}O_2$	Methyl stearate	443	153		
$C_{19}H_{40}$	Nonadecane	329.9	>100		230
$C_{20}H_{14}O_4$	Diphenyl phthalate		224		
$C_{20}H_{28}$	1-Decylnaphthalene	379	177		
$C_{20}H_{42}$	Eicosane	343	>100		232
$C_{21}H_{21}O_4P$	Tri-o-cresyl phosphate	410	225		385
$C_{21}H_{26}O_3$	4-Octylphenyl salicylate		216		416
$C_{21}H_{32}O_2$	Methyl abietate		180		

Mol. form.	Name	t_B/°C	FP/°C	Fl. limits	IT/°C
C$_{22}$H$_{42}$O$_2$	Butyl oleate		180		
C$_{22}$H$_{42}$O$_4$	Bis(2-ethylhexyl) adipate		206	0.4–?	377
C$_{22}$H$_{44}$O$_2$	Butyl stearate	343	160		355
C$_{23}$H$_{46}$O$_2$	Pentyl stearate		185		
C$_{24}$H$_{20}$Sn	Tetraphenylstannane	420	232		
C$_{24}$H$_{38}$O$_4$	Bis(2-ethylhexyl) phthalate	384	218		
C$_{25}$H$_{48}$O$_4$	Bis(2-ethylhexyl) azelate		227	0.3–?	374

THRESHOLD LIMITS FOR AIRBORNE CONTAMINANTS

Several organizations recommend limits of exposure to airborne contaminants in the workplace. These include the Occupational Safety and Health Administration (OSHA), the National Institute for Occupational Safety and Health (NIOSH), and the non-governmental organization, American Conference of Governmental Industrial Hygienists (ACGIH). The threshold limit value (TLV) for a substance (also known as persissible exposure limit, PEL) is defined as the concentration level under which the majority of workers may be repeatedly exposed, day after day, without adverse effects. The TLV recommendations are given in two forms:

- Time-weighted average (TWA) concentration for a normal 8 h workday and 40 h workweek.
- Short-term exposure limit (STEL), which should not be exceeded for more than 15 min.

Both kinds of limits are specified for some substances.

The following table gives threshold limit values for a number of substances that may be encountered in the atmosphere of a chemical laboratory or industrial facility. All values refer to the concentration in air at 25 °C and normal atmospheric pressure. Data for gases are given in parts per million by volume (ppm). Values for liquids refer to mists or aerosols, and those for solids to dusts or fumes; both are stated in mass concentration (mg/m³). A "C" following a STEL value indicates a ceiling limit, which should not be exceeded even for very brief periods because of acute toxic effects of the substance. The notation "levels as low as possible" in the Comments column indicates such a high degree of hazard that no safe limit can be recommended.

Substances are listed alphabetically by systematic name. The Formula column gives the molecular formula in the Hill convention for organic compounds and the customary line formula for inorganic compounds. The Comments column gives further information on the form of the substance and the basis to which the limit is referred. The TWA and STEL limits appear in the last two columns.

References

1. *2010 TLV's and BEI's*, American Conference of Governmental Industrial Hygienists, 1330 Kemper Meadow Drive, Cincinnati, OH 45240-1634, 2010 (www.acgih.org).
2. *NIOSH Pocket Guide to Chemical Hazards*, U.S. Department of Health and Human Services, National Institute for Occupational Health and Safety, DHHS (NIOSH) Publication No. 2005-149, September 2007 (www.cdc.gov/niosh/npg/pdfs/2005-149.pdf).
3. OSHA Standard Number 1910.1000, TABLE Z-1, Limits for Air Contaminants, U.S. Department of Labor, Occupational Safety and Health Administration (www.osha.gov/pls/oshaweb/owadisp.show_document?p_table=STANDARDS&p_id=9992).

Name	Synonym	Formula	CAS Reg. No.	Comments	Time-weighted average	Short-term exposure limit
Abate	Temephos	$C_{16}H_{20}O_6P_2S_3$	3383-96-8		10 mg/m³	
Acetaldehyde	Ethanal	CH_3CHO	75-07-0			25 ppm C
Acetic acid	Ethanoic acid	CH_3COOH	64-19-7		10 ppm	15 ppm
Acetic anhydride	Acetyl acetate	$C_4H_6O_3$	108-24-7		5 ppm	
Acetone	2-Propanone	$(CH_3)_2CO$	67-64-1		500 ppm	750 ppm
Acetone cyanohydrin		C_4H_7NO	75-86-5	as CN		5 mg/m³ C
Acetonitrile	Methyl cyanide	CH_3CN	75-05-8		20 ppm	
Acetophenone	Methyl phenyl ketone	C_8H_8O	98-86-2		10 ppm	
2-(Acetyloxy)benzoic acid	Acetylsalicylic acid (Aspirin)	$C_9H_8O_4$	50-78-2		5 mg/m³	
Acrolein	2-Propenal	$CH_2{=}CHCHO$	107-02-8			0.1 ppm C
Acrylamide	2-Propenamide	C_3H_5NO	79-06-1		0.03 mg/m³	
Acrylic acid	2-Propenoic acid	$C_3H_4O_2$	79-10-7		2 ppm	
Acrylonitrile	Propenenitrile	$CH_2{=}CHCN$	107-13-1		2 ppm	
Alachlor	Acetamide, 2-chloro-N-(2,6-diethylphenyl)-N-(methoxymethyl)-	$C_{14}H_{20}ClNO_2$	15972-60-8		1 mg/m³	
Aldrin		$C_{12}H_8Cl_6$	309-00-2		0.05 mg/m³	
Allyl alcohol	2-Propen-1-ol	C_3H_6O	107-18-6		0.5 ppm	
Allyl glycidyl ether	AGE	$C_6H_{10}O_2$	106-92-3		1 ppm	
Allyl propyl disulfide		$C_6H_{12}S_2$	2179-59-1		0.5 ppm	
Aluminum		Al	7429-90-5	metal dust and insoluble compounds	1 mg/m³	
4-Aminobiphenyl	p-Biphenylamine	$C_{12}H_{11}N$	92-67-1	levels as low as possible		
4-Amino-3,5,6-trichloro-2-pyridinecarboxlic acid	Picloram	$C_6H_3Cl_3N_2O_2$	1918-02-1		10 mg/m³	
Ammonia		NH_3	7664-41-7		25 ppm	35 ppm
Ammonium chloride	Sal ammoniac	NH_4Cl	12125-02-9	fume	10 mg/m³	20 mg/m³
Ammonium perfluorooctanoate		$C_8H_4F_{15}NO_2$	3825-26-1		0.01 mg/m³	
Ammonium sulfamate		$NH_4NH_2SO_3$	7773-06-0		10 mg/m³	
Aniline	Benzenamine	$C_6H_5NH_2$	62-53-3		2 ppm	
Antimony	Stibium	Sb	7440-36-0	and compounds, as Sb	0.5 mg/m³	
Antimony(III) oxide (senarmontite)	Senarmontite	Sb_2O_3	1309-64-4	levels as low as possible		
Arsenic		As	7440-38-2	and inorganic compounds, as As	0.01 mg/m³	
Arsine	Arsenic hydride	AsH_3	7784-42-1		0.005 ppm	

Name	Synonym	Formula	CAS Reg. No.	Comments	Time-weighted average	Short-term exposure limit
Asbestos			1332-21-4	all forms; 0.1 fibers/mL		
Asphalt			8052-42-4	fume, as aerosol	0.5 mg/m³	
Atrazine	6-Chloro-N-ethyl-N'-(1-methylethyl)-1,3,5-triazine-2,4-diamine	$C_8H_{14}ClN_5$	1912-24-9		5 mg/m³	
Azinphos-methyl		$C_{10}H_{12}N_3O_3PS_2$	86-50-0	inhalable fraction & vapor	0.2 mg/m³	
Barium		Ba	7440-39-3	and soluble compounds, as Ba	0.5 mg/m³	
Barium sulfate	Barite	$BaSO_4$	7727-43-7		10 mg/m³	
Benomyl		$C_{14}H_{18}N_4O_3$	17804-35-2		1 mg/m³	
Benz[a]anthracene	1,2-Benzanthracene	$C_{18}H_{12}$	56-55-3	levels as low as possible		
Benzene	[6]Annulene	C_6H_6	71-43-2		0.5 ppm	2.5 ppm
1,2-Benzenediamine	o-Phenylenediamine	$C_6H_8N_2$	95-54-5		0.1 mg/m³	
1,3-Benzenediamine	m-Phenylenediamine	$C_6H_8N_2$	108-45-2		0.1 mg/m³	
1,4-Benzenediamine	p-Phenylenediamine	$C_6H_8N_2$	106-50-3		0.1 mg/m³	
1,3-Benzenedimethanamine	m-Xylene diamine	$C_8H_{12}N_2$	1477-55-0			0.1 mg/m³ C
Benzenethiol	Phenyl mercaptan	C_6H_5SH	108-98-5		0.1 ppm	
p-Benzidine	[1,1'-Biphenyl]-4,4'-diamine	$C_{12}H_{12}N_2$	92-87-5	levels as low as possible		
Benzo[b]fluoranthene	Benz[e]acephenanthrylene	$C_{20}H_{12}$	205-99-2	levels as low as possible		
Benzo[a]pyrene	2,3-Benzopyrene	$C_{20}H_{12}$	50-32-8	levels as low as possible		
p-Benzoquinone	2,5-Cyclohexadiene-1,4-dione	$C_6H_4O_2$	106-51-4		0.1 ppm	
Benzoyl chloride	Benzoic acid, chloride	C_6H_5COCl	98-88-4			0.5 ppm C
Benzoyl peroxide		$C_{14}H_{10}O_4$	94-36-0		5 mg/m³	
Benzyl acetate	(Acetoxymethyl)benzene	$C_9H_{10}O_2$	140-11-4		10 ppm	
Beryllium	Glucinium	Be	7440-41-7	and compounds, as Be	0.00005 mg/m³	
Biphenyl	Diphenyl	$C_{12}H_{10}$	92-52-4		0.2 ppm	
Bis(2-aminoethyl)amine	Diethylenetriamine	$C_4H_{13}N_3$	111-40-0		1 ppm	
Bis(2-chloroethyl) ether	Dichloroethyl ether	$C_4H_8Cl_2O$	111-44-4		5 ppm	10 ppm
Bis(chloromethyl) ether	Chloromethyl ether	$C_2H_4Cl_2O$	542-88-1		0.001 ppm	
Bis(2-dimethylaminoethyl) ether	2,2'-Oxybis[N,N-dimethylethanamine] (DMAEE)	$C_8H_{20}N_2O$	3033-62-3		0.05 ppm	0.15 ppm
Bis(2-ethylhexyl) phthalate	Di-sec-octyl phthalate (DEHP)	$C_{24}H_{38}O_4$	117-81-7		5 mg/m³	
Bismuth telluride	Tetradymite	Bi_2Te_3	1304-82-1		10 mg/m³	
Boric acid	Orthoboric acid	H_3BO_3	10043-35-3	and inorganic borate compounds	2 mg/m³	6 mg/m³
Boron oxide	Boric oxide	B_2O_3	1303-86-2		10 mg/m³	
Boron tribromide	Tribromoborane	BBr_3	10294-33-4			1 ppm C
Boron trifluoride	Trifluoroborane	BF_3	7637-07-2			1 ppm C
Bromacil	5-Bromo-3-sec-butyl-6-methyluracil	$C_9H_{13}BrN_2O_2$	314-40-9		10 mg/m³	
Bromine		Br_2	7726-95-6		0.1 ppm	0.2 ppm
Bromine pentafluoride		BrF_5	7789-30-2		0.1 ppm	
Bromochloromethane	Halon 1011	CH_2BrCl	74-97-5		200 ppm	
2-Bromo-2-chloro-1,1,1-trifluoroethane	Halothane	$C_2HBrClF_3$	151-67-7		50 ppm	
Bromoethane	Ethyl bromide	C_2H_5Br	74-96-4		5 ppm	
Bromoethene	Vinyl bromide	$CH_2=CHBr$	593-60-2		0.5 ppm	
Bromomethane	Methyl bromide	CH_3Br	74-83-9		1 ppm	
1-Bromopropane	Propyl bromide	C_3H_7Br	106-94-5		10 ppm	
Bromotrifluoromethane	Halon-1301	CF_3Br	75-63-8		1000 ppm	
1,3-Butadiene	Divinyl	C_4H_6	106-99-0		2 ppm	
Butane		C_4H_{10}	106-97-8	both isomers	1000 ppm	
1-Butanethiol	Butyl mercaptan	$C_4H_{10}S$	109-79-5		0.5 ppm	
1-Butanol	Butyl alcohol	$C_4H_{10}O$	71-36-3		20 ppm	
2-Butanol	sec-Butyl alcohol	$C_4H_{10}O$	78-92-2		100 ppm	
2-Butanone	Methyl ethyl ketone (MEK)	C_4H_8O	78-93-3		200 ppm	300 ppm
2-Butanone peroxide	Methyl ethyl ketone peroxide	$C_8H_{16}O_4$	1338-23-4			0.2 ppm C
trans-2-Butenal	trans-Crotonaldehyde	C_4H_6O	123-73-9			0.3 ppm C
1-Butene	1-Butylene	C_4H_8	106-98-9		250 ppm	
cis-2-Butene		C_4H_8	590-18-1		250 ppm	
trans-2-Butene		C_4H_8	624-64-6		250 ppm	
3-Buten-2-one	Methyl vinyl ketone	C_4H_6O	78-94-4			0.2 ppm C
2-Butoxyethanol	Ethylene glycol monobutyl ether (EGBE)	$C_6H_{14}O_2$	111-76-2		20 ppm	
2-Butoxyethyl acetate	Ethylene glycol monobutyl ether acetate (EGBEA)	$C_8H_{16}O_3$	112-07-2		20 ppm	
Butyl acetate		$C_6H_{12}O_2$	123-86-4		150 ppm	200 ppm
sec-Butyl acetate	1-Methylpropyl acetate	$C_6H_{12}O_2$	105-46-4		200 ppm	
tert-Butyl acetate		$C_6H_{12}O_2$	540-88-5		200 ppm	
Butyl acrylate	Butyl 2-propenoate	$C_7H_{12}O_2$	141-32-2		2 ppm	
Butylamine	1-Butanamine	$C_4H_{11}N$	109-73-9			5 ppm C
tert-Butyl chromate		$C_8H_{18}CrO_4$	1189-85-1	as CrO_3		0.1 mg/m³ C
tert-Butyl ethyl ether	Ethyl tert-butyl ether (ETBE)	$C_6H_{14}O$	637-92-3		5 ppm	
Butyl glycidyl ether	BGE	$C_7H_{14}O_2$	2426-08-6		3 ppm	

Name	Synonym	Formula	CAS Reg. No.	Comments	Time-weighted average	Short-term exposure limit
Butyl lactate		$C_7H_{14}O_3$	34451-18-8		5 ppm	
1-*tert*-Butyl-4-methylbenzene	4-*tert*-Butyltoluene	$C_{11}H_{16}$	98-51-1		1 ppm	
2-*sec*-Butylphenol		$C_{10}H_{14}O$	89-72-5		5 ppm	
Cadmium		Cd	7440-43-9	metal	0.01 mg/m³	
Cadmium		Cd	7440-43-9	compounds, as Cd	0.002 mg/m³	
Calcium chromate		$CaCrO_4$	13765-19-0	as Cr	0.001 mg/m³	
Calcium cyanamide	Calcium carbimide	$CaCN_2$	156-62-7		0.5 mg/m³	
Calcium hydroxide	Portlandite	$Ca(OH)_2$	1305-62-0		5 mg/m³	
Calcium metasilicate	Parawollastonite	$CaSiO_3$	1344-95-2	synthetic, nonfibrous	10 mg/m³	
Calcium oxide	Lime	CaO	1305-78-8		2 mg/m³	
Calcium sulfate	Anhydrite	$CaSO_4$	7778-18-9		10 mg/m³	
Camphor, (+)	1,7,7-Trimethylbicyclo[2.2.1]heptan-2-one, (1*R*)-	$C_{10}H_{16}O$	464-49-3		2 ppm	3 ppm
Caprolactam	6-Hexanelactam	$C_6H_{11}NO$	105-60-2		5 mg/m³	
Captafol		$C_{10}H_9Cl_4NO_2S$	2425-06-1		0.1 mg/m³	
Captan		$C_9H_8Cl_3NO_2S$	133-06-2		5 mg/m³	
Carbaryl		$C_{12}H_{11}NO_2$	63-25-2		0.5 mg/m³	
Carbofuran	7-Benzofuranol, 2,3-dihydro-2,2-dimethyl-, methylcarbamate	$C_{12}H_{15}NO_3$	1563-66-2		0.1 mg/m³	
Carbon black	Carbon (amorphous)	C	1333-86-4		3.5 mg/m³	
Carbon dioxide	Carbonic anhydride	CO_2	124-38-9		5000 ppm	30,000 ppm
Carbon disulfide	Carbon bisulfide	CS_2	75-15-0		1 ppm	
Carbon (graphite)	Graphite	C	7782-42-5	except fibers	2 mg/m³	
Carbon monoxide	Carbon oxide	CO	630-08-0		25 ppm	
Carbonyl chloride	Phosgene	$COCl_2$	75-44-5		0.1 ppm	
Carbonyl fluoride		COF_2	353-50-4		2 ppm	5 ppm
Cellulose			9004-34-6		10 mg/m³	
Cesium hydroxide		CsOH	21351-79-1		2 mg/m³	
Chlordane	1,2,4,5,6,7,8,8-Octachloro-2,3,3a,4,7,7a-hexahydro-4,7-methano-1*H*-indene	$C_{10}H_6Cl_8$	57-74-9		0.5 mg/m³	
o-Chlorinated diphenyl oxide			31242-93-0		0.5 mg/m³	
Chlorine		Cl_2	7782-50-5		0.5 ppm	1 ppm
Chlorine dioxide		ClO_2	10049-04-4		0.1 ppm	0.3 ppm
Chlorine trifluoride		ClF_3	7790-91-2			0.1 ppm C
Chloroacetaldehyde	2-Chloro-1-ethanal	C_2H_3ClO	107-20-0			1 ppm C
Chloroacetic acid		$CH_2ClCOOH$	79-11-8		0.5 ppm	
Chloroacetone		C_3H_5ClO	78-95-5			1 ppm C
α-Chloroacetophenone	ω-Chloroacetophenone	C_8H_7ClO	532-27-4		0.05 ppm	
Chloroacetyl chloride		$C_2H_2Cl_2O$	79-04-9		0.05 ppm	0.15 ppm
Chlorobenzene	Phenyl chloride	C_6H_5Cl	108-90-7		10 ppm	
o-Chlorobenzylidene malononitrile		$C_{10}H_5ClN_2$	2698-41-1			0.05 ppm C
2-Chloro-1,3-butadiene	Chloroprene	C_4H_5Cl	126-99-8		10 ppm	
Chlorodifluoromethane	HCFC-22	CF_2HCl	75-45-6		1000 ppm	
Chloroethane	Ethyl chloride	C_2H_5Cl	75-00-3		100 ppm	
2-Chloroethanol	Ethylene chlorohydrin	C_2H_5ClO	107-07-3			1 ppm C
Chloroethene	Vinyl chloride	$CH_2{=}CHCl$	75-01-4		1 ppm	
Chloromethane	Methyl chloride	CH_3Cl	74-87-3		50 ppm	100 ppm
(Chloromethyl)benzene	Benzyl chloride	C_7H_7Cl	100-44-7		1 ppm	
Chloromethyl methyl ether		C_2H_5ClO	107-30-2	levels as low as possible		
1-Chloro-4-nitrobenzene	*p*-Chloronitrobenzene	$C_6H_4ClNO_2$	100-00-5		0.1 ppm	
1-Chloro-1-nitropropane		$C_3H_6ClNO_2$	600-25-9		2 ppm	
Chloropentafluoroethane	CFC-115	CF_3CF_2Cl	76-15-3		1000 ppm	
2-Chloropropanoic acid	2-Chloropropionic acid	$C_3H_5ClO_2$	598-78-7		0.1 ppm	
2-Chloro-1-propanol	Propylene chlorohydrin	C_3H_7ClO	78-89-7		1 ppm	
1-Chloro-2-propanol	*sec*-Propylene chlorohydrin	C_3H_7ClO	127-00-4		1 ppm	
3-Chloropropene	Allyl chloride	C_3H_5Cl	107-05-1		1 ppm	2 ppm
2-Chlorostyrene		C_8H_7Cl	2039-87-4		50 ppm	75 ppm
2-Chlorotoluene	1-Chloro-2-methylbenzene	C_7H_7Cl	95-49-8		50 ppm	
Chlorpyrifos	Phosphorothioic acid, *O,O*-diethyl *O*-(3,5,6-trichloro-2-pyridinyl) ester	$C_9H_{11}Cl_3NO_3PS$	2921-88-2		0.1 mg/m³	
Chromium		Cr	7440-47-3	metal	0.5 mg/m³	
Chromium		Cr	7440-47-3	Cr(III) compounds, as Cr	0.5 mg/m³	
Chromium		Cr	7440-47-3	soluble Cr(VI) compounds, as Cr	0.05 mg/m³	
Chromium		Cr	7440-47-3	insoluble Cr(VI) compounds, as Cr	0.01 mg/m³	
Chromium(VI) dichloride dioxide	Chromyl chloride	CrO_2Cl_2	14977-61-8		0.025 ppm	
Chrysene	Benzo[a]phenanthrene	$C_{18}H_{12}$	218-01-9	levels as low as possible		

Name	Synonym	Formula	CAS Reg. No.	Comments	Time-weighted average	Short-term exposure limit
Clopidol		$C_7H_7Cl_2NO$	2971-90-6		10 mg/m³	
Coal				dust, anthracite	0.4 mg/m³	
Coal				dust, bituminous	0.9 mg/m³	
Coal tar			65966-93-2	volatiles, as aerosol	0.2 mg/m³	
Cobalt		Co	7440-48-4	metal and inorganic compounds, as Co	0.02 mg/m³	
Cobalt carbonyl	Dicobalt octacarbonyl	$Co_2(CO)_8$	10210-68-1	as Co	0.1 mg/m³	
Cobalt hydrocarbonyl	Tetracarbonylhydrocobalt	C_4HCoO_4	16842-03-8	as Co	0.1 mg/m³	
Copper		Cu	7440-50-8	fume	0.2 mg/m³	
Copper		Cu	7440-50-8	dusts & mists, as Cu	1 mg/m³	
Cotton				dust	0.1 mg/m³	
Coumaphos		$C_{14}H_{16}ClO_5PS$	56-72-4		0.05 mg/m³	
o-Cresol	2-Methylphenol	C_7H_8O	95-48-7	inhalable fraction & vapor	20 mg/m³	
m-Cresol	3-Methylphenol	C_7H_8O	108-39-4	inhalable fraction & vapor	20 mg/m³	
p-Cresol	4-Methylphenol	C_7H_8O	106-44-5	inhalable fraction & vapor	20 mg/m³	
Crufomate		$C_{12}H_{19}ClNO_3P$	299-86-5		5 mg/m³	
Cyanamide	Cyanogenamide	H_2NCN	420-04-2		2 mg/m³	
Cyanide ion [CN⁻]		CN⁻	57-12-5	cyanide salts, as CN		5 mg/m³ C
Cyanogen		C_2N_2	460-19-5		10 ppm	
Cyanogen chloride	Chlorine cyanide	CICN	506-77-4			0.3 ppm C
Cyclohexane	Hexahydrobenzene	C_6H_{12}	110-82-7		100 ppm	
Cyclohexanol	Cyclohexyl alcohol	$C_6H_{12}O$	108-93-0		50 ppm	
Cyclohexanone	Pimelic ketone	$C_6H_{10}O$	108-94-1		20 ppm	50 ppm
Cyclohexene	Tetrahydrobenzene	C_6H_{10}	110-83-8		300 ppm	
Cyclohexylamine	Cyclohexanamine	$C_6H_{13}N$	108-91-8		10 ppm	
1,3-Cyclopentadiene	Pyropentylene	C_5H_6	542-92-7		75 ppm	
Cyclopentane	Pentamethylene	C_5H_{10}	287-92-3		600 ppm	
Cyhexatin	Tricyclohexylhydroxystannane	$C_{18}H_{34}OSn$	13121-70-5		5 mg/m³	
Decaborane(14)		$B_{10}H_{14}$	17702-41-9		0.05 ppm	0.15 ppm
Demeton	Systox	$C_8H_{19}O_3PS_2$	8065-48-3		0.05 mg/m³	
Demeton-S-methyl	Phosphorothioic acid, S-[2-(ethylthio)ethyl] O,O-dimethyl ester	$C_6H_{15}O_3PS_2$	919-86-8		0.05 mg/m³	
Diacetone alcohol	4-Hydroxy-4-methyl-2-pentanone	$C_6H_{12}O_2$	123-42-2		50 ppm	
4,4'-Diaminodiphenylmethane	4,4'-Methylenedianiline	$C_{13}H_{14}N_2$	101-77-9		0.1 ppm	
Diazinon		$C_{12}H_{21}N_2O_3PS$	333-41-5		0.01 mg/m³	
Diazomethane		CH_2-NN	334-88-3		0.2 ppm	
Diborane		B_2H_6	19287-45-7		0.1 ppm	
Dibromodifluoromethane		CBr_2F_2	75-61-6		100 ppm	
2-Dibutylaminoethanol		$C_{10}H_{23}NO$	102-81-8		0.5 ppm	
2,6-Di-tert-butyl-4-methylphenol	Butylated hydroxytoluene (BHT)	$C_{15}H_{24}O$	128-37-0		2 mg/m³	
Dibutylphenyl phosphate		$C_{14}H_{23}O_4P$	2528-36-1		0.3 ppm	
Dibutyl phosphate		$C_8H_{19}O_4P$	107-66-4	inhalable fraction & vapor	5 mg/m³	
Dibutyl phthalate	Butyl phthalate	$C_{16}H_{22}O_4$	84-74-2		5 mg/m³	
Dichloroacetic acid		$CHCl_2COOH$	79-43-6		0.5 ppm	
Dichloroacetylene		C_2Cl_2	7572-29-4			0.1 ppm C
o-Dichlorobenzene	1,2-Dichlorobenzene	$C_6H_4Cl_2$	95-50-1		25 ppm	50 ppm
p-Dichlorobenzene	1,4-Dichlorobenzene	$C_6H_4Cl_2$	106-46-7		10 ppm	
3,3'-Dichloro-p-benzidine	3,3'-Dichloro[1,1'-biphenyl]-4,4'-diamine	$C_{12}H_{10}Cl_2N_2$	91-94-1	levels as low as possible		
cis-1,4-Dichloro-2-butene		$C_4H_6Cl_2$	1476-11-5		0.005 ppm	
trans-1,4-Dichloro-2-butene		$C_4H_6Cl_2$	110-57-6		0.005 ppm	
Dichlorodifluoromethane	CFC-12	CF_2Cl_2	75-71-8		1000 ppm	
1,3-Dichloro-5,5-dimethyl hydantoin		$C_5H_6Cl_2N_2O_2$	118-52-5		0.2 mg/m³	0.4 mg/m³
1,1-Dichloroethane	Ethylidene dichloride	CH_3CHCl_2	75-34-3		100 ppm	
1,2-Dichloroethane	Ethylene dichloride	$C_2H_4Cl_2$	107-06-2		10 ppm	
1,1-Dichloroethene	Vinylidene chloride	$CH_2=CCl_2$	75-35-4		5 ppm	
cis-1,2-Dichloroethene	cis-1,2-Dichloroethylene	$C_2H_2Cl_2$	156-59-2		200 ppm	
trans-1,2-Dichloroethene	trans-1,2-Dichloroethylene	$C_2H_2Cl_2$	156-60-5		200 ppm	
Dichlorofluoromethane	Refrigerant 21	$CHCl_2F$	75-43-4		10 ppm	
Dichloromethane	Methylene chloride	CH_2Cl_2	75-09-2		50 ppm	
1,1-Dichloro-1-nitroethane	Ethide	$C_2H_3Cl_2NO_2$	594-72-9		2 ppm	
(2,4-Dichlorophenoxy)acetic acid	2,4-D	$C_8H_6Cl_2O_3$	94-75-7		10 mg/m³	
1,2-Dichloropropane, (±)-	Propylene dichloride	$C_3H_6Cl_2$	26198-63-0		10 ppm	
2,2-Dichloropropanoic acid	2,2-Dichloropropionic acid	$C_3H_4Cl_2O_2$	75-99-0		5 mg/m³	
cis-1,3-Dichloropropene	cis-1,3-Dichloropropylene	$C_3H_4Cl_2$	10061-01-5		1 ppm	
trans-1,3-Dichloropropene	trans-1,3-Dichloropropylene	$C_3H_4Cl_2$	10061-02-6		1 ppm	
1,2-Dichloro-1,1,2,2-tetrafluoroethane	CFC-114	$C_2Cl_2F_4$	76-14-2		1000 ppm	

Name	Synonym	Formula	CAS Reg. No.	Comments	Time-weighted average	Short-term exposure limit
Dichlorvos	Phosphoric acid, 2,2-dichloroethenyl dimethyl ester	$C_4H_7Cl_2O_4P$	62-73-7		0.1 mg/m³	
Dicrotophos		$C_8H_{16}NO_5P$	141-66-2		0.05 mg/m³	
m-Dicyanobenzene	m-Phthalodinitrile	$C_8H_4N_2$	626-17-5	inhalable fraction & vapor	5 mg/m³	
Dicyclopentadiene		$C_{10}H_{12}$	1755-01-7		5 ppm	
Dieldrin		$C_{12}H_8Cl_6O$	60-57-1	inhalable fraction & vapor	0.1 mg/m³	
Diesel fuel			68334-30-5	as total hydrocarbons	100 mg/m³	
Diethanolamine	Bis(2-hydroxyethyl)amine	$C_4H_{11}NO_2$	111-42-2	inhalable fraction & vapor	1 mg/m³	
Diethylamine	N-Ethylethanamine	$(C_2H_5)_2NH$	109-89-7		5 ppm	15 ppm
2-Diethylaminoethanol		$C_6H_{15}NO$	100-37-8		2 ppm	
Diethyl ether	Ethyl ether	$(C_2H_5)_2O$	60-29-7		400 ppm	500 ppm
3,3-Diethylpentane	Tetraethylmethane	C_9H_{20}	1067-20-5		200 ppm	
Diethyl phthalate		$C_{12}H_{14}O_4$	84-66-2		5 mg/m³	
1,1-Difluoroethene	Vinylidene fluoride	$CH_2{=}CF_2$	75-38-7		500 ppm	
Diglycidyl ether	Bis(2,3-epoxypropyl) ether (DGE)	$C_6H_{10}O_3$	2238-07-5		0.1 ppm	
Diisopropylamine	N-Isopropyl-2-propanamine	$C_6H_{15}N$	108-18-9		5 ppm	
Diisopropyl ether	Isopropyl ether	$C_6H_{14}O$	108-20-3		250 ppm	310 ppm
Dimethoxymethane	Methylal	$C_3H_8O_2$	109-87-5		1000 ppm	
N,N-Dimethylacetamide	N,N-Dimethylethanamide	C_4H_9NO	127-19-5		10 ppm	
Dimethylamine	N-Methylmethanamine	$(CH_3)_2NH$	124-40-3		5 ppm	15 ppm
Dimethylaniline (unspecified isomer)	Xylidine (unspecified isomer)	$C_8H_{11}N$	1300-73-8	all isomers	0.5 ppm	
2,3-Dimethylaniline	2,3-Xylidine	$C_8H_{11}N$	87-59-2		0.5 ppm	
2,4-Dimethylaniline	2,4-Xylidine	$C_8H_{11}N$	95-68-1		0.5 ppm	
2,5-Dimethylaniline	2,5-Xylidine	$C_8H_{11}N$	95-78-3		0.5 ppm	
2,6-Dimethylaniline	2,6-Xylidine	$C_8H_{11}N$	87-62-7		0.5 ppm	
3,4-Dimethylaniline	3,4-Xylidine	$C_8H_{11}N$	95-64-7		0.5 ppm	
3,5-Dimethylaniline	3,5-Xylidine	$C_8H_{11}N$	108-69-0		0.5 ppm	
N,N-Dimethylaniline	N,N-Dimethylbenzenamine	$C_8H_{11}N$	121-69-7		5 ppm	10 ppm
2,2-Dimethylbutane	Neohexane	C_6H_{14}	75-83-2		500 ppm	1000 ppm
2,3-Dimethylbutane	Diisopropyl	C_6H_{14}	79-29-8		500 ppm	1000 ppm
Dimethylcarbamic chloride	Dimethylcarbamoyl chloride	C_3H_6ClNO	79-44-7		0.005 mg/m³	
Dimethyl disulfide	Methyl disulfide	$C_2H_6S_2$	624-92-0		0.5 mg/m³	
N,N-Dimethylformamide	DMF	C_3H_7NO	68-12-2		10 ppm	
2,2-Dimethylheptane		C_9H_{20}	1071-26-7		200 ppm	
2,3-Dimethylheptane		C_9H_{20}	3074-71-3		200 ppm	
2,4-Dimethylheptane		C_9H_{20}	2213-23-2		200 ppm	
2,5-Dimethylheptane		C_9H_{20}	2216-30-0		200 ppm	
2,6-Dimethylheptane		C_9H_{20}	1072-05-5		200 ppm	
3,3-Dimethylheptane		C_9H_{20}	4032-86-4		200 ppm	
3,4-Dimethylheptane		C_9H_{20}	922-28-1		200 ppm	
3,5-Dimethylheptane		C_9H_{20}	926-82-9		200 ppm	
4,4-Dimethylheptane		C_9H_{20}	1068-19-5		200 ppm	
2,6-Dimethyl-4-heptanone	Diisobutyl ketone	$C_9H_{18}O$	108-83-8		25 ppm	
2,2-Dimethylhexane		C_8H_{18}	590-73-8		300 ppm	
2,3-Dimethylhexane		C_8H_{18}	584-94-1		300 ppm	
2,4-Dimethylhexane		C_8H_{18}	589-43-5		300 ppm	
2,5-Dimethylhexane	Biisobutyl	C_8H_{18}	592-13-2		300 ppm	
3,3-Dimethylhexane		C_8H_{18}	563-16-6		300 ppm	
3,4-Dimethylhexane		C_8H_{18}	583-48-2		300 ppm	
1,1-Dimethylhydrazine	UDMH	$C_2H_8N_2$	57-14-7		0.01 ppm	
Dimethyl mercury	Mercury dimethyl	$Hg(CH_3)_2$	593-74-8		0.01 mg/m³	
trans-3,7-Dimethyl-2,6-octadienal	Citral	$C_{10}H_{16}O$	141-27-5	inhalable fraction & vapor	5 ppm	
Dimethyl phthalate	Methyl phthalate	$C_{10}H_{10}O_4$	131-11-3		5 mg/m³	
2,2-Dimethyl-1-propanol acetate		$C_7H_{14}O_2$	926-41-0		50 ppm	100 ppm
Dimethyl sulfate		$C_2H_6O_4S$	77-78-1		0.1 ppm	
Dimethyl sulfide	2-Thiapropane	$(CH_3)_2S$	75-18-3		10 ppm	
1,2-Dinitrobenzene	o-Dinitrobenzene	$C_6H_4N_2O_4$	528-29-0		0.15 ppm	
1,3-Dinitrobenzene	m-Dinitrobenzene	$C_6H_4N_2O_4$	99-65-0		0.15 ppm	
1,4-Dinitrobenzene	p-Dinitrobenzene	$C_6H_4N_2O_4$	100-25-4		0.15 ppm	
1,4-Dioxane	1,4-Dioxacyclohexane	$C_4H_8O_2$	123-91-1		20 ppm	
Dioxathion		$C_{12}H_{26}O_6P_2S_4$	78-34-2		0.1 mg/m³	
1,3-Dioxolane	1,3-Dioxacyclopentane	$C_3H_6O_2$	646-06-0		20 ppm	
Diphenylamine	N-Phenylbenzenamine	$(C_6H_5)_2NH$	122-39-4		10 mg/m³	
Diphenyl ether	Oxybisbenzene	$(C_6H_5)_2O$	101-84-8		1 ppm	2 ppm
4,4'-Diphenylmethane diisocyanate	Methylene diphenyl diisocyanate (MDI)	$C_{15}H_{10}N_2O_2$	101-68-8		0.005 ppm	
Dipropylene glycol monomethyl ether	1-(2-Methoxyisopropoxy)-2-propanol (DPGME)	$C_7H_{16}O_3$	34590-94-8		100 ppm	150 ppm

Name	Synonym	Formula	CAS Reg. No.	Comments	Time-weighted average	Short-term exposure limit
Diquat		$C_{12}H_{12}N_2$	2764-72-9		0.5 mg/m³	
Disulfiram		$C_{10}H_{20}N_2S_4$	97-77-8		2 mg/m³	
Disulfoton	Phosphorodithioic acid, O,O-diethyl S-[2-(ethylthio)ethyl] ester	$C_8H_{19}O_2PS_3$	298-04-4		0.05 mg/m³	
Diuron		$C_9H_{10}Cl_2N_2O$	330-54-1		10 mg/m³	
o-Divinylbenzene	1,2-Divinylbenzene	$C_{10}H_{10}$	91-14-5		10 ppm	
m-Divinylbenzene	1,3-Divinylbenzene	$C_{10}H_{10}$	108-57-6		10 ppm	
p-Divinylbenzene	1,4-Divinylbenzene	$C_{10}H_{10}$	105-06-6		10 ppm	
1-Dodecanethiol		$C_{12}H_{26}S$	112-55-0		0.1 ppm	
Endosulfan		$C_9H_6Cl_6O_3S$	115-29-7	inhalable fraction & vapor	0.1 mg/m³	
Endrin		$C_{12}H_8Cl_6O$	72-20-8		0.1 mg/m³	
Enflurane		$C_3H_2ClF_5O$	13838-16-9		75 ppm	
Epichlorohydrin	(Chloromethyl)oxirane	C_3H_5ClO	13403-37-7		0.5 ppm	
1,2-Epoxy-4-(epoxyethyl)cyclohexane	4-Vinyl-1-cyclohexene dioxide	$C_8H_{12}O_2$	106-87-6		0.1 ppm	
Ethane		C_2H_6	74-84-0		1000 ppm	
1,2-Ethanediamine	Ethylenediamine	$C_2H_8N_2$	107-15-3		10 ppm	
1,2-Ethanediol	Ethylene glycol	$(CH_2OH)_2$	107-21-1			100 mg/m³ C
1,2-Ethanediol, dinitrate	Ethylene glycol dinitrate (EGDN)	$C_2H_4N_2O_6$	628-96-6		0.05 ppm	
Ethanethiol	Ethyl mercaptan	C_2H_5SH	75-08-1		0.5 ppm	
Ethanol	Ethyl alcohol	C_2H_5OH	64-17-5			1000 ppm
Ethanolamine	Glycinol; 2-Aminoethanol	C_2H_7NO	141-43-5		3 ppm	6 ppm
Ethion	Phosphorodithioic acid, S,S-methylene O,O,O',O'-tetraethyl ester	$C_9H_{22}O_4P_2S_4$	563-12-2		0.05 mg/m³	
Ethoxydimethylsilane	Dimethylethoxysilane	$C_4H_{12}OSi$	14857-34-2		0.5 ppm	1.5 ppm
2-Ethoxyethanol	Ethylene glycol monoethyl ether (EGEE)	$C_4H_{10}O_2$	110-80-5		5 ppm	
2-Ethoxyethyl acetate	Ethylene glycol monoethyl ether acetate (EGEEA)	$C_6H_{12}O_3$	111-15-9		5 ppm	
Ethyl acetate		$C_4H_8O_2$	141-78-6		400 ppm	
Ethyl acrylate	Ethyl propenoate	$C_5H_8O_2$	140-88-5		5 ppm	15 ppm
Ethylamine	Ethanamine	$C_2H_5NH_2$	75-04-7		5 ppm	15 ppm
Ethylbenzene	Phenylethane	C_8H_{10}	100-41-4		100 ppm	125 ppm
Ethyl 2-cyanoacrylate	Ethyl 2-cyano-2-propenoate	$C_6H_7NO_2$	7085-85-0		0.2 ppm	
3-Ethyl-2,2-dimethylpentane		C_9H_{20}	16747-32-3		200 ppm	
3-Ethyl-2,3-dimethylpentane		C_9H_{20}	16747-33-4		200 ppm	
3-Ethyl-2,4-dimethylpentane		C_9H_{20}	1068-87-7		200 ppm	
Ethylene	Ethene	$CH_2=CH_2$	74-85-1		200 ppm	
Ethyleneimine	Aziridine	C_2H_5N	151-56-4		0.05 ppm	0.1 ppm
Ethyl formate		$C_3H_6O_2$	109-94-4		100 ppm	
3-Ethylheptane		C_9H_{20}	15869-80-4		200 ppm	
4-Ethylheptane		C_9H_{20}	2216-32-2		200 ppm	
3-Ethylhexane		C_8H_{18}	619-99-8		300 ppm	
2-Ethylhexanoic acid		$C_8H_{16}O_2$	149-57-5		5 mg/m³	
5-Ethylidene-2-norbornene	5-Ethylidenebicyclo[2.2.1]hept-2-ene	C_9H_{12}	16219-75-3			5 ppm C
3-Ethyl-2-methylhexane		C_9H_{20}	16789-46-1		200 ppm	
3-Ethyl-3-methylhexane		C_9H_{20}	3074-76-8		200 ppm	
3-Ethyl-4-methylhexane	2,3-Diethylpentane	C_9H_{20}	3074-77-9		200 ppm	
4-Ethyl-2-methylhexane		C_9H_{20}	3074-75-7		200 ppm	
3-Ethyl-2-methylpentane	2-Methyl-3-ethylpentane	C_8H_{18}	609-26-7		300 ppm	
3-Ethyl-3-methylpentane	3-Methyl-3-ethylpentane	C_8H_{18}	1067-08-9		300 ppm	
N-Ethylmorpholine		$C_6H_{13}NO$	100-74-3		5 ppm	
O-Ethyl O-p-nitrophenyl benzenethiophosphonate	EPN	$C_{14}H_{14}NO_4PS$	2104-64-5		0.1 mg/m³	
Ethyl silicate	Tetraethoxysilane	$Si(OC_2H_5)_4$	78-10-4		10 ppm	
Fenamiphos		$C_{13}H_{22}NO_3PS$	22224-92-6		0.05 mg/m³	
Fensulfothion	Phosphorothioic acid, O,O-diethyl O-[4-(methylsulfinyl)phenyl] ester	$C_{11}H_{17}O_4PS_2$	115-90-2		0.01 mg/m³	
Fenthion	Phosphorothioic acid, O,O-dimethyl O-[3-methyl-4-(methylthio)phenyl] ester	$C_{10}H_{15}O_3PS_2$	55-38-9		0.05 mg/m³	
Ferbam	Iron, tris(dimethylcarbamodithioato-S,S)-, (OC-6-11)-	$C_9H_{18}FeN_3S_6$	14484-64-1	inhalable fraction	5 mg/m³	
Ferrocene	Bis(cyclopentadienyl)iron	$Fe(C_5H_5)_2$	102-54-5		10 mg/m³	
Ferrovanadium			12604-58-9	dust	1 mg/m³	3 mg/m³
Flour				dust	0.5 mg/m³	
Fluoride ion [F⁻]		F^-	16984-48-8	fluoride salts, as F	2.5 mg/m³	
Fluorine		F_2	7782-41-4		1 ppm	2 ppm
Fluorine monoxide	Oxygen difluoride	F_2O	7783-41-7			0.05 ppm C
Fluoroethene	Vinyl fluoride	$CH_2=CHF$	75-02-5		1 ppm	

Name	Synonym	Formula	CAS Reg. No.	Comments	Time-weighted average	Short-term exposure limit
Fonofos	Phosphonodithioic acid, ethyl-, O-ethyl S-phenyl ester	$C_{10}H_{15}OPS_2$	944-22-9		0.01 mg/m³	
Formaldehyde	Methanal	HCHO	50-00-0			0.3 ppm C
Formamide	Methanamide	$HCONH_2$	75-12-7		10 ppm	
Formic acid	Methanoic acid	HCOOH	64-18-6		5 ppm	10 ppm
Furfural	2-Furaldehyde	$C_5H_4O_2$	98-01-1		2 ppm	
Furfuryl alcohol	2-Furanmethanol	$C_5H_6O_2$	98-00-0		10 ppm	15 ppm
Gallium arsenide		GaAs	1303-00-0		0.0003 mg/m³	
Gasoline			8006-61-9		300 ppm	500 ppm
Germane	Germanium tetrahydride	GeH_4	7782-65-2		0.2 ppm	
Glycerol	1,2,3-Propanetriol	$CH_2OHCHOHCH_2OH$	56-81-5	mist	10 mg/m³	
Glyoxal		$H_2C_2O_2$	107-22-2		0.1 mg/m³	
Grain				dust	4 mg/m³	
Hafnium		Hf	7440-58-6	metal and compounds, as Hf	0.5 mg/m³	
Heptachlor		$C_{10}H_5Cl_7$	76-44-8		0.05 mg/m³	
Heptachlor epoxide		$C_{10}H_5Cl_7O$	1024-57-3		0.05 mg/m³	
Heptane		C_7H_{16}	142-82-5	all isomers	400 ppm	500 ppm
2-Heptanone	Methyl pentyl ketone	$C_7H_{14}O$	110-43-0		50 ppm	
3-Heptanone	Ethyl butyl ketone	$C_7H_{14}O$	106-35-4		50 ppm	75 ppm
4-Heptanone	Dipropyl ketone	$C_7H_{14}O$	123-19-3		50 ppm	
Hexachlorobenzene	Perchlorobenzene	C_6Cl_6	118-74-1		0.002 mg/m³	
Hexachloro-1,3-butadiene	Perchlorobutadiene	C_4Cl_6	87-68-3		0.02 ppm	
1,2,3,4,5,6-Hexachlorocyclohexane, (1α,2α,3β,4α,5α,6β)	Lindane	$C_6H_6Cl_6$	58-89-9		0.5 mg/m³	
Hexachloro-1,3-cyclopentadiene	Perchlorocyclopentadiene	C_5Cl_6	77-47-4		0.01 ppm	
Hexachloroethane	Perchloroethane	CCl_3CCl_3	67-72-1		1 ppm	
Hexachloronaphthalene (unspecified isomer)		$C_{10}H_2Cl_6$	1335-87-1	all isomers	0.2 mg/m³	
Hexahydro-1,3-isobenzofurandione	Hexahydrophthalic anhydride	$C_8H_{10}O_3$	85-42-7			0.005 mg/m³ C
Hexahydro-1,3,5-trinitro-1,3,5-triazine	Cyclonite	$C_3H_6N_6O_6$	121-82-4		0.5 mg/m³	
Hexamethylene diisocyanate		$C_8H_{12}N_2O_2$	822-06-0		0.005 ppm	
Hexane		C_6H_{14}	110-54-3		50 ppm	
1,6-Hexanediamine	Hexamethylenediamine	$C_6H_{16}N_2$	124-09-4		0.5 ppm	
Hexanedinitrile	Adiponitrile	$C_6H_8N_2$	111-69-3		2 ppm	
1,6-Hexanedioic acid	Adipic acid	$C_6H_{10}O_4$	124-04-9		5 mg/m³	
2-Hexanone	Butyl methyl ketone	$C_6H_{12}O$	591-78-6		5 ppm	10 ppm
1-Hexene		C_6H_{12}	592-41-6		50 ppm	
sec-Hexyl acetate	4-Methyl-2-pentyl acetate	$C_8H_{16}O_2$	108-84-9		50 ppm	
Hydrazine		N_2H_4	302-01-2		0.01 ppm	
Hydrazoic acid	Hydrogen azide	HN_3	7782-79-8	vapor		0.11 ppm C
Hydrogen bromide	Hydrobromic acid	HBr	10035-10-6			2 ppm C
Hydrogen chloride	Hydrochloric acid	HCl	7647-01-0			2 ppm C
Hydrogen cyanide	Hydrocyanic acid	HCN	74-90-8			4.7 ppm C
Hydrogen fluoride	Hydrofluoric acid	HF	7664-39-3		0.5 ppm	2 ppm C
Hydrogen peroxide		H_2O_2	7722-84-1		1 ppm	
Hydrogen selenide		H_2Se	7783-07-5		0.05 ppm	
Hydrogen sulfide		H_2S	7783-06-4		1 ppm	5 ppm
p-Hydroquinone	1,4-Benzenediol	$C_6H_6O_2$	123-31-9		1 mg/m³	
2-Hydroxypropyl acrylate		$C_6H_{10}O_3$	999-61-1		0.5 ppm	
Indene	Indonaphthene	C_9H_8	95-13-6		5 ppm	
Indium		In	7440-74-6	metal and compounds, as In	0.1 mg/m³	
Iodine		I_2	7553-56-2	and volatile iodides	0.01 ppm	0.1 ppm C
Iodomethane	Methyl iodide	CH_3I	74-88-4		2 ppm	
Iron ion [Fe+2]		Fe^{+2}	15438-31-0	soluble ferrous salts, as Fe	1 mg/m³	
Iron ion [Fe+3]		Fe^{+3}	20074-52-6	soluble ferric salts, as Fe	1 mg/m³	
Iron(III) oxide	Hematite	Fe_2O_3	1309-37-1	dust and fume, as Fe	5 mg/m³	
Iron pentacarbonyl	Iron carbonyl [Fe(CO)$_5$]	$Fe(CO)_5$	13463-40-6	as Fe	0.1 ppm	0.2 ppm
Isobutane	2-Methylpropane	C_4H_{10}	75-28-5		1000 ppm	
Isobutene	2-Methyl-1-propene	$(CH_3)_2C=CH_2$	115-11-7		250 ppm	
Isobutyl acetate	2-Methylpropyl acetate	$C_6H_{12}O_2$	110-19-0		150 ppm	
Isobutyl nitrite		$C_4H_9NO_2$	542-56-3			1 ppm C
Isopentane	2-Methylbutane	C_5H_{12}	78-78-4		600 ppm	
Isopentyl acetate	Isoamyl acetate	$C_7H_{14}O_2$	123-92-2		50 ppm	100 ppm
Isophorone	3,5,5-Trimethyl-2-cyclohexen-1-one	$C_9H_{14}O$	78-59-1			5 ppm C
Isophorone diisocyanate		$C_{12}H_{18}N_2O_2$	4098-71-9		0.005 ppm	
Isopropenylbenzene	α-Methyl styrene	C_9H_{10}	98-83-9		10 ppm	
2-Isopropoxyethanol		$C_5H_{12}O_2$	109-59-1		25 ppm	
Isopropyl acetate	1-Methylethyl acetate	$C_5H_{10}O_2$	108-21-4		100 ppm	200 ppm

Name	Synonym	Formula	CAS Reg. No.	Comments	Time-weighted average	Short-term exposure limit
Isopropylamine	2-Propanamine	C_3H_9N	75-31-0		5 ppm	10 ppm
N-Isopropylaniline		$C_9H_{13}N$	768-52-5		2 ppm	
Isopropylbenzene	Cumene	C_9H_{12}	98-82-8		50 ppm	
Isopropyl glycidyl ether	(1-Methylethoxy)methyloxirane (IGE)	$C_6H_{12}O_2$	4016-14-2		50 ppm	75 ppm
Kaolin			1332-58-7		2 mg/m³	
Kerosene			8008-20-6		200 mg/m³	
Ketene		$CH_2=C=O$	463-51-4		0.5 ppm	1.5 ppm
Lead		Pb	7439-92-1	metal & inorganic compounds, as Pb	0.05 mg/m³	
Lead(II) arsenate		$Pb_3(AsO_4)_2$	3687-31-8		0.15 mg/m³	
Lead(II) chromate	Crocoite	$PbCrO_4$	7758-97-6	as Pb	0.05 mg/m³	
Lithium hydride		LiH	7580-67-8		0.025 mg/m³	
Magnesium oxide	Magnesia	MgO	1309-48-4		10 mg/m³	
Malathion		$C_{10}H_{19}O_6PS_2$	121-75-5		1 mg/m³	
Maleic anhydride		$C_4H_2O_3$	108-31-6		0.1 ppm	
Manganese		Mn	7439-96-5	metal and inorganic compounds, as Mn	0.2 mg/m³	
Manganese cyclopentadienyl tricarbonyl		$C_8H_5MnO_3$	12079-65-1	as Mn	0.1 mg/m³	
Manganese 2-methylcyclopentadienyl tricarbonyl		$C_9H_7MnO_3$	12108-13-3	as Mn	0.2 mg/m³	
Mercury	Quicksilver	Hg	7439-97-6	metal & inorganic compounds, as Hg	0.025 mg/m³	
Mercury	Quicksilver	Hg	7439-97-6	alkyl compounds, as Hg	0.01 mg/m³	0.03 mg/m³
Mercury	Quicksilver	Hg	7439-97-6	aryl compounds, as Hg	0.1 mg/m³	
Mesityl oxide	Isobutenyl methyl ketone	$C_6H_{10}O$	141-79-7		15 ppm	25 ppm
Methacrylic acid	2-Methylpropenoic acid	$C_4H_6O_2$	79-41-4		20 ppm	
Methane		CH_4	74-82-8		1000 ppm	
Methanethiol	Methyl mercaptan	CH_3SH	74-93-1		0.5 ppm	
Methanol	Methyl alcohol	CH_3OH	67-56-1		200 ppm	250 ppm
Methomyl	Acetimidic acid, N-[(methylcarbamoyl)oxy]thio-, methyl ester	$C_5H_{10}N_2O_2S$	16752-77-5		2.5 mg/m³	
2-Methoxyaniline	o-Anisidine	C_7H_9NO	90-04-0		0.5 mg/m³	
4-Methoxyaniline	p-Anisidine	C_7H_9NO	104-94-9		0.5 mg/m³	
Methoxychlor		$C_{16}H_{15}Cl_3O_2$	72-43-5		10 mg/m³	
2-Methoxyethanol	Ethylene glycol monomethyl ether (EGME)	$C_3H_8O_2$	109-86-4		0.1 ppm	
2-Methoxyethyl acetate	Ethylene glycol monomethyl ether acetate (EGMEA)	$C_5H_{10}O_3$	110-49-6		0.1 ppm	
2-Methoxy-2-methylbutane	Methyl tert-pentyl ether (TAME)	$C_6H_{14}O$	994-05-8		20 ppm	
4-Methoxyphenol		$C_7H_8O_2$	150-76-5		5 mg/m³	
1-Methoxy-2-propanol	1,2-Propylene glycol monomethyl ether (PGME)	$C_4H_{10}O_2$	107-98-2		100 ppm	150 ppm
Methyl acetate		$C_3H_6O_2$	79-20-9		200 ppm	250 ppm
Methyl acrylate	Methyl propenoate	$C_4H_6O_2$	96-33-3		2 ppm	
2-Methylacrylonitrile	2-Methylpropenenitrile	C_4H_5N	126-98-7		1 ppm	
Methylamine	Methanamine	CH_3NH_2	74-89-5		5 ppm	15 ppm
2-Methylaniline	o-Toluidine	C_7H_9N	95-53-4		2 ppm	
3-Methylaniline	m-Toluidine	C_7H_9N	108-44-1		2 ppm	
4-Methylaniline	p-Toluidine	C_7H_9N	106-49-0		2 ppm	
N-Methylaniline	N-Methylbenzenamine	C_7H_9N	100-61-8		0.5 ppm	
3-Methyl-1-butanol	Isopentyl alcohol (Isoamyl alcohol)	$C_5H_{12}O$	123-51-3		100 ppm	125 ppm
2-Methyl-1-butanol acetate		$C_7H_{14}O_2$	624-41-9		50 ppm	100 ppm
3-Methyl-2-butanol acetate		$C_7H_{14}O_2$	5343-96-4		50 ppm	100 ppm
3-Methyl-2-butanone	Methyl isopropyl ketone	$C_5H_{10}O$	563-80-4		200 ppm	
Methyl tert-butyl ether	tert-Butyl methyl ether (MTBE)	$C_5H_{12}O$	1634-04-4		50 ppm	
Methyl 2-cyanoacrylate	Mecrylate	$C_5H_5NO_2$	137-05-3		0.2 ppm	
Methylcyclohexane		C_7H_{14}	108-87-2		400 ppm	
1-Methylcyclohexanol		$C_7H_{14}O$	590-67-0		50 ppm	
cis-2-Methylcyclohexanol		$C_7H_{14}O$	615-38-3		50 ppm	
trans-2-Methylcyclohexanol, (±)-		$C_7H_{14}O$	615-39-4		50 ppm	
cis-3-Methylcyclohexanol, (±)-		$C_7H_{14}O$	5454-79-5		50 ppm	
trans-3-Methylcyclohexanol, (±)-		$C_7H_{14}O$	7443-55-2		50 ppm	
cis-4-Methylcyclohexanol		$C_7H_{14}O$	7731-28-4		50 ppm	
trans-4-Methylcyclohexanol		$C_7H_{14}O$	7731-29-5		50 ppm	
2-Methylcyclohexanone, (±)-		$C_7H_{12}O$	24965-84-2		50 ppm	75 ppm
Methyl demeton		$C_6H_{15}O_3PS_2$	8022-00-2		0.05 mg/m³	
2-Methyl-3,5-dinitrobenzamide	Dinitolmide	$C_8H_7N_3O_5$	148-01-6		1 mg/m³	
1-Methyl-2,3-dinitrobenzene	2,3-Dinitrotoluene	$C_7H_6N_2O_4$	602-01-7		0.2 mg/m³	
1-Methyl-2,4-dinitrobenzene	2,4-Dinitrotoluene	$C_7H_6N_2O_4$	121-14-2		0.2 mg/m³	

Name	Synonym	Formula	CAS Reg. No.	Comments	Time-weighted average	Short-term exposure limit
1-Methyl-3,5-dinitrobenzene	3,5-Dinitrotoluene	$C_7H_6N_2O_4$	618-85-9		0.2 mg/m^3	
2-Methyl-1,3-dinitrobenzene	2,6-Dinitrotoluene	$C_7H_6N_2O_4$	606-20-2		0.2 mg/m^3	
2-Methyl-1,4-dinitrobenzene	2,5-Dinitrotoluene	$C_7H_6N_2O_4$	619-15-8		0.2 mg/m^3	
4-Methyl-1,2-dinitrobenzene	3,4-Dinitrotoluene	$C_7H_6N_2O_4$	610-39-9		0.2 mg/m^3	
2-Methyl-4,6-dinitrophenol	4,6-Dinitro-o-cresol	$C_7H_6N_2O_5$	534-52-1		0.2 mg/m^3	
4,4'-Methylenebis[2-chloroaniline]	3,3'-Dichloro-4,4'-diaminodiphenylmethane (MBOCA)	$C_{13}H_{12}Cl_2N_2$	101-14-4		0.01 ppm	
Methylenebis(4-cyclohexylisocyanate)		$C_{15}H_{22}N_2O_2$	5124-30-1		0.005 ppm	
Methyl formate		$C_2H_4O_2$	107-31-3		100 ppm	150 ppm
2-Methylheptane		C_8H_{18}	592-27-8		300 ppm	
3-Methylheptane		C_8H_{18}	589-81-1		300 ppm	
4-Methylheptane		C_8H_{18}	589-53-7		300 ppm	
6-Methyl-1-heptanol	Isooctyl alcohol	$C_8H_{18}O$	1653-40-3		50 ppm	
5-Methyl-3-heptanone	Ethyl 2-methylbutyl ketone	$C_8H_{16}O$	541-85-5		10 ppm	
5-Methyl-2-hexanone	Methyl isopentyl ketone	$C_7H_{14}O$	110-12-3		50 ppm	
Methylhydrazine		CH_6N_2	60-34-4		0.01 ppm	
Methyl isocyanate		CH_3NCO	624-83-9		0.02 ppm	
Methyl methacrylate	Methyl 2-methyl-2-propenoate	$C_5H_8O_2$	80-62-6		50 ppm	100 ppm
1-Methylnaphthalene		$C_{11}H_{10}$	90-12-0		0.5 ppm	
2-Methylnaphthalene		$C_{11}H_{10}$	91-57-6		0.5 ppm	
2-Methyl-5-nitroaniline	5-Nitro-o-toluidine	$C_7H_8N_2O_2$	99-55-8		1 mg/m^3	
2-Methyloctane		C_9H_{20}	3221-61-2		200 ppm	
3-Methyloctane		C_9H_{20}	2216-33-3		200 ppm	
4-Methyloctane		C_9H_{20}	2216-34-4		200 ppm	
Methyloxirane	1,2-Propylene oxide	C_3H_6O	16033-71-9		2 ppm	
Methyl parathion		$C_8H_{10}NO_5PS$	298-00-0	inhalable fraction & vapor	0.02 mg/m^3	
2-Methylpentane	Isohexane	C_6H_{14}	107-83-5		500 ppm	1000 ppm
3-Methylpentane		C_6H_{14}	96-14-0		500 ppm	1000 ppm
2-Methyl-2,4-pentanediol	Hexylene glycol	$C_6H_{14}O_2$	107-41-5			25 ppm C
4-Methyl-2-pentanol	Methyl isobutyl carbinol	$C_6H_{14}O$	108-11-2		25 ppm	40 ppm
4-Methyl-2-pentanone	Isobutyl methyl ketone	$C_6H_{12}O$	108-10-1		20 ppm	75 ppm
2-Methyl-1-propanol	Isobutyl alcohol	$C_4H_{10}O$	78-83-1		50 ppm	
2-Methyl-2-propanol	tert-Butyl alcohol	$(CH_3)_3COH$	75-65-0		100 ppm	
Methyl silicate	Tetramethoxysilane	$Si(OCH_3)_4$	681-84-5		1 ppm	
2-Methylstyrene	2-Vinyl toluene	C_9H_{10}	611-15-4		50 ppm	100 ppm
3-Methylstyrene	3-Vinyl toluene	C_9H_{10}	100-80-1		50 ppm	100 ppm
4-Methylstyrene	4-Vinyl toluene	C_9H_{10}	622-97-9		50 ppm	100 ppm
N-Methyl-N,2,4,6-tetranitroaniline	Tetryl	$C_7H_5N_5O_8$	479-45-8		1.5 mg/m^3	
Metribuzin		$C_8H_{14}N_4OS$	21087-64-9		5 mg/m^3	
Mevinphos		$C_7H_{13}O_6P$	7786-34-7		0.01 ppm	
Mica			12001-26-2		3 mg/m^3	
Mineral oil				mist	5 mg/m^3	10 mg/m^3
Molybdenum		Mo	7439-98-7	metal and insoluble compounds, as Mo	10 mg/m^3	
Molybdenum		Mo	7439-98-7	soluble compounds, as Mo	0.5 mg/m^3	
Monocrotophos	trans-Dimethyl 1-methyl-3-(methylamino)-3-oxo-1-propenyl phosphate	$C_7H_{14}NO_5P$	6923-22-4		0.05 mg/m^3	
Morpholine	Tetrahydro-1,4-oxazine	C_4H_9NO	110-91-8		20 ppm	
Naled	1,2-Dibromo-2,2-dichloroethylphosphoric acid, dimethyl ester	$C_4H_7Br_2Cl_2O_4P$	300-76-5		0.1 mg/m^3	
Naphtha			8030-30-6		400 ppm	
Naphthalene		$C_{10}H_8$	91-20-3		10 ppm	15 ppm
1-Naphthalenylthiourea	ANTU	$C_{11}H_{10}N_2S$	86-88-4		0.3 mg/m^3	
2-Naphthylamine	β-Naphthylamine	$C_{10}H_9N$	91-59-8	levels as low as possible		
Neopentane	2,2-Dimethylpropane	$C(CH_3)_4$	463-82-1		600 ppm	
Nickel		Ni	7440-02-0	metal	1.5 mg/m^3	
Nickel		Ni	7440-02-0	soluble compounds, as Ni	0.1 mg/m^3	
Nickel		Ni	7440-02-0	insoluble compounds, as Ni	0.2 mg/m^3	
Nickel carbonyl [Ni[CO]$_4$]	Nickel tetracarbonyl	$Ni(CO)_4$	13463-39-3	as Ni	0.05 ppm	
Nickel subsulfide	Heazlewoodite	Ni_3S_2	12035-72-2	as Ni	0.1 mg/m^3	
L-Nicotine	3-(1-Methyl-2-pyrrolidinyl)pyridine, (S)-	$C_{10}H_{14}N_2$	54-11-5		0.5 mg/m^3	
Nitrapyrin	Pyridine, 2-chloro-6-(trichloromethyl)-	$C_6H_3Cl_4N$	1929-82-4		10 mg/m^3	
Nitric acid		HNO_3	7697-37-2		2 ppm	4 ppm
Nitric oxide		NO	10102-43-9		25 ppm	
4-Nitroaniline		$C_6H_6N_2O_2$	100-01-6		3 mg/m^3	
Nitrobenzene		$C_6H_5NO_2$	98-95-3		1 ppm	
4-Nitrobiphenyl		$C_{12}H_9NO_2$	92-93-3	levels as low as possible		

Name	Synonym	Formula	CAS Reg. No.	Comments	Time-weighted average	Short-term exposure limit
Nitroethane		$C_2H_5NO_2$	79-24-3		100 ppm	
Nitrogen dioxide		NO_2	10102-44-0		3 ppm	5 ppm
Nitrogen trifluoride		NF_3	7783-54-2		10 ppm	
Nitromethane	Nitrocarbol	CH_3NO_2	75-52-5		20 ppm	
1-Nitropropane		$C_3H_7NO_2$	108-03-2		25 ppm	
2-Nitropropane	Isonitropropane	$C_3H_7NO_2$	79-46-9		10 ppm	
N-Nitrosodimethylamine	Dimethylnitrosamine	$C_2H_6N_2O$	62-75-9	levels as low as possible		
2-Nitrotoluene	1-Methyl-2-nitrobenzene	$C_7H_7NO_2$	88-72-2		2 ppm	
3-Nitrotoluene	1-Methyl-3-nitrobenzene	$C_7H_7NO_2$	99-08-1		2 ppm	
4-Nitrotoluene	1-Methyl-4-nitrobenzene	$C_7H_7NO_2$	99-99-0		2 ppm	
Nitrous oxide		N_2O	10024-97-2		50 ppm	
3,3,4,4,5,5,6,6,-Nonafluoro-1-hexene	Perfluorobutylethene	$C_6H_3F_9$	19430-93-4		100 ppm	
Nonane		C_9H_{20}	111-84-2	all isomers	200 ppm	
Octachloronaphthalene	Perchloronaphthalene	$C_{10}Cl_8$	2234-13-1		0.1 mg/m³	0.3 mg/m³
Octane		C_8H_{18}	111-65-9	all isomers	300 ppm	375 ppm
Osmium(VIII) oxide	Osmic acid (Osmium tetroxide)	OsO_4	20816-12-0		0.0002 ppm	0.0006 ppm
Oxalic acid		$H_2C_2O_4$	144-62-7		1 mg/m³	2 mg/m³
2-Oxetanone	β-Propiolactone	$C_3H_4O_2$	57-57-8		0.5 ppm	
Oxirane	Ethylene oxide	C_2H_4O	75-21-8		1 ppm	
Oxiranemethanol, (±)-	Glycidol	$C_3H_6O_2$	61915-27-3		2 ppm	
4,4'-Oxybis(benzenesulfonyl hydrazide)		$C_{12}H_{14}N_4O_5S_2$	80-51-3		0.1 mg/m³	
Ozone		O_3	10028-15-6	depends on workload	0.1 ppm	
Paraquat		$C_{12}H_{14}N_2$	4685-14-7		0.5 mg/m³	
Parathion		$C_{10}H_{14}NO_5PS$	56-38-2		0.05 mg/m³	
Pentaborane(9)		B_5H_9	19624-22-7		0.005 ppm	0.015 ppm
Pentachloronaphthalene		$C_{10}H_3Cl_5$	1321-64-8	all isomers	0.5 mg/m³	
Pentachloronitrobenzene	Quintozene	$C_6Cl_5NO_2$	82-68-8		0.5 mg/m³	
Pentachlorophenol		C_6HCl_5O	87-86-5		0.5 mg/m³	
Pentaerythritol		$C_5H_{12}O_4$	115-77-5		10 mg/m³	
Pentanal	Valeraldehyde	$C_5H_{10}O$	110-62-3		50 ppm	
Pentane		C_5H_{12}	109-66-0	all isomers	600 ppm	750 ppm
Pentanedial	Glutaraldehyde	$C_5H_8O_2$	111-30-8			0.05 ppm C
3-Pentanol acetate		$C_7H_{14}O_2$	620-11-1		50 ppm	100 ppm
2-Pentanone	Methyl propyl ketone	$C_5H_{10}O$	107-87-9			150 ppm
3-Pentanone	Diethyl ketone	$C_5H_{10}O$	96-22-0		200 ppm	300 ppm
Pentyl acetate	Amyl acetate	$C_7H_{14}O_2$	628-63-7	all isomers	50 ppm	100 ppm
sec-Pentyl acetate, (R)-	sec-Amyl acetate, (R)-	$C_7H_{14}O_2$	54638-10-7		50 ppm	100 ppm
Perchloryl fluoride	Chlorine trioxide fluoride	ClO_3F	7616-94-6		3 ppm	6 ppm
Perfluoroacetone	Hexafluoroacetone	C_3F_6O	684-16-2		0.1 ppm	
Perfluoroisobutene	Perfluoroisobutylene	C_4F_8	382-21-8			0.01 ppm C
Perfluoropropene	Hexafluoropropene	C_3F_6	116-15-4		0.1 ppm	
Phenol	Hydroxybenzene	C_6H_5OH	108-95-2		5 ppm	
10H-Phenothiazine	Thiodiphenylamine	$C_{12}H_9NS$	92-84-2		5 mg/m³	
Phenyl glycidyl ether	PGE	$C_9H_{10}O_2$	122-60-1		0.1 ppm	
Phenylhydrazine		$C_6H_8N_2$	100-63-0		0.1 ppm	
N-Phenyl-2-naphthalenamine	N-Phenyl-β-naphthylamine	$C_{16}H_{13}N$	135-88-6	levels as low as possible		
Phenylphosphine	Monophenylphosphine	$C_6H_5PH_2$	638-21-1			0.05 ppm C
Phorate	Phosphorodithioic acid, O,O-diethyl S-[(ethylthio)methyl] ester	$C_7H_{17}O_2PS_3$	298-02-2		0.05 mg/m³	0.2 mg/m³
Phosphine	Phosphorus hydride	PH_3	7803-51-2		0.3 ppm	1 ppm
Phosphoric acid	Orthophosphoric acid	H_3PO_4	7664-38-2		1 mg/m³	3 mg/m³
Phosphorus (white)	Yellow phosphorus	P	7723-14-0		0.1 mg/m³	
Phosphorus(III) chloride	Phosphorus trichloride	PCl_3	7719-12-2		0.2 ppm	0.5 ppm
Phosphorus(V) chloride	Phosphorus pentachloride	PCl_5	10026-13-8		0.1 ppm	
Phosphorus(V) sulfide	Phosphorus pentasulfide	P_2S_5	1314-80-3		1 mg/m³	3 mg/m³
Phosphoryl chloride	Phosphorus oxychloride	$POCl_3$	10025-87-3		0.1 ppm	
Phthalic anhydride		$C_8H_4O_3$	85-44-9		1 ppm	
α-Pinene	2-Pinene	$C_{10}H_{16}$	80-56-8		20 ppm	
β-Pinene	Nopinene	$C_{10}H_{16}$	127-91-3		20 ppm	
Piperazine dihydrochloride	Diethylenediamine dihydrochloride	$C_4H_{12}Cl_2N_2$	142-64-3		5 mg/m³	
2-Pivaloyl-1,3-indandione	Pindone	$C_{14}H_{14}O_3$	83-26-1		0.1 mg/m³	
Platinum		Pt	7440-06-4	metal dust	1 mg/m³	
Platinum		Pt	7440-06-4	soluble salts, as Pt	0.002 mg/m³	
Polychlorinated biphenyls (42% chlorine)	PCBs		53469-21-9		1 mg/m³	
Polychlorinated biphenyls (54% chlorine)	PCBs		11097-69-1		0.5 mg/m³	
Poly(vinyl chloride)	PVC		9002-86-2		1 mg/m³	

Name	Synonym	Formula	CAS Reg. No.	Comments	Time-weighted average	Short-term exposure limit
Portland cement			65997-15-1	respirable fraction; no asbestos	1 mg/m³	
Potassium hydroxide		KOH	1310-58-3			2 mg/m³ C
Propanal	Propionaldehyde	C_2H_5CHO	123-38-6		20 ppm	
Propane	LPG	C_3H_8	74-98-6		1000 ppm	
1,3-Propane sultone	1,2-Oxathiolane, 2,2-dioxide	$C_3H_6O_3S$	1120-71-4	levels as low as possible		
Propanoic acid	Propionic acid	$C_3H_6O_2$	79-09-4		10 ppm	
1-Propanol	Propyl alcohol	$CH_3CH_2CH_2OH$	71-23-8		100 ppm	400 ppm
2-Propanol	Isopropyl alcohol	$CH_3CHOHCH_3$	67-63-0		200 ppm	400 ppm
Propargyl alcohol	3-Hydroxy-1-propyne (2-Propyn-1-ol)	C_3H_4O	107-19-7		1 ppm	
Propene	Propylene	$CH_3CH=CH_2$	115-07-1		500 ppm	
Propoxur	Phenol, 2-(1-methylethoxy)-, methylcarbamate	$C_{11}H_{15}NO_3$	114-26-1		0.5 mg/m³	
Propyl acetate		$C_5H_{10}O_2$	109-60-4		200 ppm	250 ppm
1,2-Propylene glycol dinitrate		$C_3H_6N_2O_6$	6423-43-4		0.05 ppm	
Propyleneimine	2-Methylaziridine	C_3H_7N	75-55-8		0.2 ppm	0.4 ppm
Propyl nitrate		$C_3H_7NO_3$	627-13-4		25 ppm	40 ppm
Propyne	Methylacetylene	CH_3CCH	74-99-7		1000 ppm	
Pyrethrin I	Pyrethrum	$C_{21}H_{28}O_3$	121-21-1		5 mg/m³	
2-Pyridinamine	2-Aminopyridine	$C_5H_6N_2$	504-29-0		0.5 ppm	
Pyridine	Azine	C_5H_5N	110-86-1		1 ppm	
Pyrocatechol	1,2-Benzenediol (Catechol)	$C_6H_6O_2$	120-80-9		5 ppm	
Resorcinol	1,3-Benzenediol	$C_6H_6O_2$	108-46-3		10 ppm	20 ppm
Rhodium		Rh	7440-16-6	metal and insoluble compounds, as Rh	1 mg/m³	
Rhodium		Rh	7440-16-6	soluble compounds, as Rh	0.01 mg/m³	
Ronnel		$C_8H_8Cl_3O_3PS$	299-84-3		5 mg/m³	
Rotenone		$C_{23}H_{22}O_6$	83-79-4		5 mg/m³	
Rubber				natural latex, as inhalable proteins	0.0001 mg/m³	
Selenium		Se	7782-49-2	element and compounds, as Se	0.2 mg/m³	
Selenium hexafluoride		SeF_6	7783-79-1		0.05 ppm	
Sesone	Sodium 2-(2,4-dichlorophenoxy)ethyl sulfate	$C_8H_7Cl_2NaO_5S$	136-78-7		10 mg/m³	
Silane	Silicon tetrahydride	SiH_4	7803-62-5		5 ppm	
Silicon carbide (hexagonal)	Moissanite	SiC	409-21-2	lower limits for fibrous SiC	10 mg/m³	
Silicon dioxide (α-quartz)	Silica	SiO_2	14808-60-7	respirable fraction	0.025 mg/m³	
Silicon dioxide (cristobalite)	Cristobalite	SiO_2	14464-46-1	respirable fraction	0.025 mg/m³	
Silver		Ag	7440-22-4	metal	0.1 mg/m³	
Silver		Ag	7440-22-4	soluble compounds, as Ag	0.01 mg/m³	
Sodium azide	Smite	NaN_3	26628-22-8			0.29 mg/m³ C
Sodium fluoroacetate		$C_2H_2FNaO_2$	62-74-8		0.05 mg/m³	
Sodium hydrogen sulfite	Sodium bisulfite	$NaHSO_3$	7631-90-5		5 mg/m³	
Sodium hydroxide	Caustic soda	NaOH	1310-73-2			2 mg/m³ C
Sodium metabisulfite	Sodium pyrosulfite	$Na_2S_2O_5$	7681-57-4		5 mg/m³	
Sodium tetraborate decahydrate	Borax	$Na_2B_4O_7 \cdot 10H_2O$	1303-96-4		2 mg/m³	6 mg/m³
Starch			9005-25-8		10 mg/m³	
Stibine	Antimony(III) hydride	SbH_3	7803-52-3		0.1 ppm	
Stoddard solvent			8052-41-3		100 ppm	
Strontium chromate		$SrCrO_4$	7789-06-2	as Cr	0.0005 mg/m³	
Strychnine		$C_{21}H_{22}N_2O_2$	57-24-9		0.15 mg/m³	
Styrene	Vinylbenzene	C_8H_8	100-42-5		20 ppm	40 ppm
Subtilisins			9014-01-1	as crystalline active enzyme	0.00006 mg/m³ C	
Sucrose		$C_{12}H_{22}O_{11}$	57-50-1		10 mg/m³	
Sulfometuron methyl		$C_{15}H_{16}N_4O_5S$	74222-97-2		5 mg/m³	
Sulfotep	Tetraethyl thiodiphosphate (TEDP)	$C_8H_{20}O_5P_2S_2$	3689-24-5		0.2 mg/m³	
Sulfur chloride [SSCl₂]	Sulfur monochloride (Disulfur dichloride)	S_2Cl_2	10025-67-9			1 ppm C
Sulfur decafluoride	Disulfur decafluoride (Sulfur pentafluoride)	S_2F_{10}	5714-22-7			0.01 ppm C
Sulfur dioxide		SO_2	7446-09-5			0.25 ppm
Sulfur hexafluoride		SF_6	2551-62-4		1000 ppm	
Sulfuric acid	Oil of vitriol	H_2SO_4	7664-93-9		0.2 mg/m³	
Sulfur tetrafluoride		SF_4	7783-60-0			0.1 ppm C
Sulfuryl fluoride		SO_2F_2	2699-79-8		5 ppm	10 ppm
Sulprofos		$C_{12}H_{19}O_2PS_3$	35400-43-2	inhalable fraction & vapor	0.1 mg/m³	
Talc		$3MgO \cdot 4SiO_2 \cdot H_2O$	14807-96-6	respirable fraction; no asbestos	2 mg/m³	
Tantalum		Ta	7440-25-7	dust	5 mg/m³	

Name	Synonym	Formula	CAS Reg. No.	Comments	Time-weighted average	Short-term exposure limit
Tantalum(V) oxide	Tantalum pentoxide	Ta_2O_5	1314-61-0	dust, as Ta	5 mg/m³	
Tellurium		Te	13494-80-9	and compounds, as Te (except H_2Te)	0.1 mg/m³	
Tellurium hexafluoride		TeF_6	7783-80-4		0.02 ppm	
Terbufos		$C_9H_{21}O_2PS_3$	13071-79-9		0.01 mg/m³	
Terephthalic acid	1,4-Benzenedicarboxylic acid	$C_8H_6O_4$	100-21-0		10 mg/m³	
o-Terphenyl		$C_{18}H_{14}$	84-15-1			5 mg/m³ C
m-Terphenyl		$C_{18}H_{14}$	92-06-8			5 mg/m³ C
p-Terphenyl		$C_{18}H_{14}$	92-94-4			5 mg/m³
1,1,2,2-Tetrabromoethane	Acetylene tetrabromide	$C_2H_2Br_4$	79-27-6		0.1 ppm	
Tetrabromomethane	Carbon tetrabromide	CBr_4	558-13-4		0.1 ppm	0.3 ppm
1,1,1,2-Tetrachloro-2,2-difluoroethane	Tetrachloro-1,1-difluoroethane	$C_2Cl_4F_2$	76-11-9		100 ppm	
1,1,2,2-Tetrachloro-1,2-difluoroethane	Tetrachloro-1,2-difluoroethane	$C_2Cl_4F_2$	76-12-0		50 ppm	
1,1,2,2-Tetrachloroethane	Acetylene tetrachloride	$C_2H_2Cl_4$	79-34-5		1 ppm	
Tetrachloroethene	Perchloroethylene	C_2Cl_4	127-18-4		25 ppm	100 ppm
Tetrachloromethane	Carbon tetrachloride	CCl_4	56-23-5		5 ppm	10 ppm
1,2,3,4-Tetrachloronaphthalene		$C_{10}H_4Cl_4$	20020-02-4		2 mg/m³	
Tetraethyl lead		$C_8H_{20}Pb$	78-00-2	as Pb	0.1 mg/m³	
Tetraethyl pyrophosphate	TEPP	$C_8H_{20}O_7P_2$	107-49-3		0.01 mg/m³	
Tetrafluoroethene	Tetrafluoroethylene	$F_2C=CF_2$	116-14-3		2 ppm	
Tetrahydrofuran	Tetramethylene oxide (Oxolane)	C_4H_8O	109-99-9		50 ppm	100 ppm
2,2,3,3-Tetramethylbutane		C_8H_{18}	594-82-1		300 ppm	
Tetramethyl lead		$C_4H_{12}Pb$	75-74-1	as Pb	0.15 mg/m³	
2,2,3,3-Tetramethylpentane		C_9H_{20}	7154-79-2		200 ppm	
2,2,3,4-Tetramethylpentane		C_9H_{20}	1186-53-4		200 ppm	
2,2,4,4-Tetramethylpentane	Di-tert-butylmethane	C_9H_{20}	1070-87-7		200 ppm	
2,3,3,4-Tetramethylpentane		C_9H_{20}	16747-38-9		200 ppm	
Tetramethylsuccinonitrile	Tetramethylbutanedinitrile	$C_8H_{12}N_2$	3333-52-6		0.5 ppm	
Tetranitromethane		$C(NO_2)_4$	509-14-8		0.005 ppm	
Thallium		Tl	7440-28-0	and compounds, as Tl	0.02 mg/m³	
4,4'-Thiobis(6-tert-butyl-m-cresol)	Bis(5-tert-butyl-4-hydroxy-2-methylphenyl) sulfide	$C_{22}H_{30}O_2S$	96-69-5		10 mg/m³	
Thioglycolic acid		$C_2H_4O_2S$	68-11-1		1 ppm	
Thionyl chloride	Sulfinyl dichloride	$SOCl_2$	7719-09-7			0.2 ppm C
Thiram		$C_6H_{12}N_2S_4$	137-26-8		0.05 mg/m³	
Tin		Sn	7440-31-5	metal	2 mg/m³	
Tin		Sn	7440-31-5	inorganic compounds, as Sn	2 mg/m³	
Tin		Sn	7440-31-5	organic compounds, as Sn	0.1 mg/m³	
Titanium(IV) oxide (rutile)	Rutile (Titanium dioxide)	TiO_2	1317-80-2		10 mg/m³	
Toluene	Methylbenzene	C_7H_8	108-88-3		20 ppm	
Toluene-2,4-diisocyanate		$C_9H_6N_2O_2$	584-84-9		0.005 mg/m³	0.02 ppm
Toluene-2,6-diisocyanate		$C_9H_6N_2O_2$	91-08-7		0.005 mg/m³	0.02 ppm
Toxaphene	Polychlorocamphene	$C_{10}H_{10}Cl_8$	8001-35-2		0.5 mg/m³	
1H-1,2,4-Triazol-3-amine	Amitrole	$C_2H_4N_4$	61-82-5		0.2 mg/m³	
Tribromomethane	Bromoform	$CHBr_3$	75-25-2		0.5 ppm	
Tributyl phosphate	Butyl phosphate	$C_{12}H_{27}O_4P$	126-73-8		0.2 ppm	
Trichlorfon	2,2,2-Trichloro-1-hydroxyethylphosphonic acid, dimethyl ester	$C_4H_8Cl_3O_4P$	52-68-6		1 mg/m³	
Trichloroacetic acid		CCl_3COOH	76-03-9		1 ppm	
1,2,4-Trichlorobenzene		$C_6H_3Cl_3$	120-82-1			5 ppm C
1,1,1-Trichloro-2,2-bis(4-chlorophenyl)ethane	Dichlorodiphenyltrichloroethane (DDT)	$C_{14}H_9Cl_5$	50-29-3		1 mg/m³	
1,1,1-Trichloroethane	Methyl chloroform	$CHCCl_3$	71-55-6		350 ppm	450 ppm
1,1,2-Trichloroethane	Vinyl trichloride	$C_2H_3Cl_3$	79-00-5		10 ppm	
Trichloroethene	Trichloroethylene	C_2HCl_3	79-01-6		10 ppm	25 ppm
Trichlorofluoromethane	CFC-11	CCl_3F	75-69-4			1000 ppm C
Trichloromethane	Chloroform	$CHCl_3$	67-66-3		10 ppm	
Trichloromethanesulfenyl chloride	Perchloromethyl mercaptan	CCl_4S	594-42-3		0.1 ppm	
(Trichloromethyl)benzene	Benzotrichloride	$C_6H_5CCl_3$	98-07-7			0.1 ppm C
Trichloronaphthalene (unspecified isomer)		$C_{10}H_5Cl_3$	1321-65-9	all isomers	5 mg/m³	
Trichloronitromethane	Chloropicrin	CCl_3NO_2	76-06-2		0.1 ppm	
2,4,5-Trichlorophenoxyacetic acid	2,4,5-T	$C_8H_5Cl_3O_3$	93-76-5		10 mg/m³	
1,2,3-Trichloropropane	Allyl trichloride	$C_3H_5Cl_3$	96-18-4		10 ppm	
1,1,2-Trichloro-1,2,2-trifluoroethane	CFC-113	$C_2Cl_3F_3$	76-13-1		1000 ppm	1250 ppm
Tri-o-cresyl phosphate	Tri-o-tolyl phosphate	$C_{21}H_{21}O_4P$	78-30-8		0.1 mg/m³	
Triethanolamine	Tris(2-hydroxyethyl)amine	$C_6H_{15}NO_3$	102-71-6		5 mg/m³	
Triethylamine	N,N-Diethylethanamine	$C_6H_{15}N$	121-44-8		1 ppm	3 ppm

Name	Synonym	Formula	CAS Reg. No.	Comments	Time-weighted average	Short-term exposure limit
1,3,5-Triglycidyl-s-triazinetrione	1,3,5-Tris(oxiranemethyl)-1,3,5-triazine-2,4,6(1H,3H,5H)-trione	$C_{12}H_{15}N_3O_6$	2451-62-9		0.05 mg/m^3	
Triiodomethane	Iodoform	CHI_3	75-47-8		0.6 ppm	
Trimellitic anhydride	1,2,4-Benzenetricarboxylic anhydride	$C_9H_4O_5$	552-30-7		0.0005 mg/m^3	0.002 mg/m^3 C
Trimethylamine	N,N-Dimethylmethanamine	$(CH_3)_3N$	75-50-3		5 ppm	15 ppm
1,2,3-Trimethylbenzene	Hemimellitene	C_9H_{12}	526-73-8		25 ppm	
1,2,4-Trimethylbenzene	Pseudocumene	C_9H_{12}	95-63-6		25 ppm	
1,3,5-Trimethylbenzene	Mesitylene	C_9H_{12}	108-67-8		25 ppm	
2,2,3-Trimethylhexane		C_9H_{20}	16747-25-4		200 ppm	
2,2,4-Trimethylhexane		C_9H_{20}	16747-26-5		200 ppm	
2,2,5-Trimethylhexane		C_9H_{20}	3522-94-9		200 ppm	
2,3,3-Trimethylhexane		C_9H_{20}	16747-28-7		200 ppm	
2,3,4-Trimethylhexane		C_9H_{20}	921-47-1		200 ppm	
2,3,5-Trimethylhexane		C_9H_{20}	1069-53-0		200 ppm	
2,4,4-Trimethylhexane		C_9H_{20}	16747-30-1		200 ppm	
3,3,4-Trimethylhexane		C_9H_{20}	16747-31-2		200 ppm	
2,2,3-Trimethylpentane	2-tert-Butylbutane	C_8H_{18}	564-02-3		300 ppm	
2,2,4-Trimethylpentane	Isooctane	C_8H_{18}	540-84-1		300 ppm	
2,3,3-Trimethylpentane		C_8H_{18}	560-21-4		300 ppm	
2,3,4-Trimethylpentane		C_8H_{18}	565-75-3		300 ppm	
Trimethyl phosphite		$C_3H_9O_3P$	121-45-9		2 ppm	
Trinitroglycerol	Nitroglycerin	$C_3H_5N_3O_9$	55-63-0		0.05 ppm	
2,4,6-Trinitrophenol	Picric acid	$C_6H_3N_3O_7$	88-89-1		0.1 mg/m^3	
2,4,6-Trinitrotoluene	2-Methyl-1,3,5-trinitrobenzene (TNT)	$C_7H_5N_3O_6$	118-96-7		0.1 mg/m^3	
Triphenyl phosphate		$C_{18}H_{15}O_4P$	115-86-6		3 mg/m^3	
Tungsten	Wolfram	W	7440-33-7	metal and insoluble compounds, as W	5 mg/m^3	10 mg/m^3
Tungsten	Wolfram	W	7440-33-7	soluble compounds, as W	1 mg/m^3	3 mg/m^3
Turpentine			8006-64-2		20 ppm	
Uranium		U	7440-61-1	metal and compounds, as U	0.2 mg/m^3	0.6 mg/m^3
Vanadium(V) oxide	Vanadium pentoxide	V_2O_5	1314-62-1	inhalable fraction, as V	0.05 mg/m^3	
Vinyl acetate		$C_4H_6O_2$	108-05-4		10 ppm	15 ppm
4-Vinylcyclohexene		C_8H_{12}	100-40-3		0.1 ppm	
1-Vinyl-2-pyrrolidinone		C_6H_9NO	88-12-0		0.05 ppm	
Warfarin	Coumadin	$C_{19}H_{16}O_4$	81-81-2		0.1 mg/m^3	
Wood				inhalable fraction (0.5 mg/m³ for western red cedar)	1 mg/m^3	
o-Xylene	1,2-Dimethylbenzene	C_8H_{10}	95-47-6		100 ppm	150 ppm
m-Xylene	1,3-Dimethylbenzene	C_8H_{10}	108-38-3		100 ppm	150 ppm
p-Xylene	1,4-Dimethylbenzene	C_8H_{10}	106-42-3		100 ppm	150 ppm
Yttrium		Y	7440-65-5	metal and compounds, as Y	1 mg/m^3	
Zinc chloride		$ZnCl_2$	7646-85-7	fume	1 mg/m^3	2 mg/m^3
Zinc chromate		$ZnCrO_4$	13530-65-9	as Cr	0.01 mg/m^3	
Zinc oxide	Zincite	ZnO	1314-13-2		2 mg/m^3	10 mg/m^3
Zirconium		Zr	7440-67-7	metal and compounds, as Zr	5 mg/m^3	10 mg/m^3

OCTANOL–WATER PARTITION COEFFICIENTS

The octanol–water partition coefficient, P, is a widely used parameter for correlating biological effects of organic substances. It is a property of the two-phase system in which water and 1-octanol are in equilibrium at a fixed temperature and the substance is distributed between the water-rich and octanol-rich phases. P is defined as the ratio of the equilibrium concentration of the substance in the octanol-rich phase to that in the water-rich phase, in the limit of zero concentration. In general, P tends to be large for compounds with extended non-polar structures (such as long chain or multi-ring hydrocarbons) and small for compounds with highly polar groups. Thus P (or, in its more common form of expression, log P) provides a measure of the lipophilic vs. hydrophilic nature of a compound, which is an important consideration in assessing the potential toxicity. A discussion of methods of measurement and accuracy considerations for log P may be found in Reference 1.

This table gives selected values of log P for about 450 organic compounds, including many of environmental importance. All values refer to a nominal temperature of 25 °C. The source of each value is indicated in the last column. These references contain data on many more compounds than are included here.

Compounds are listed by molecular formula following the Hill convention. To locate a compound by name or CAS Registry Number when the molecular formula is not known, use the table "Physical Constants of Organic Compounds" in Section 3 and its indexes to determine the molecular formula.

References

1. Sangster, J., *J. Phys. Chem. Ref. Data*, 18, 1111, 1989.
2. Mackay, D., Shiu, W. Y., and Ma, K. C., *Illustrated Handbook of Physical-Chemical Properties and Environmental Fate for Organic Chemicals*, Lewis Publishers/CRC Press, Boca Raton, FL, 1992.
3. Shiu, W. Y., and Mackay, D., *J. Phys. Chem. Ref. Data*, 15, 911, 1986.
4. Pinsuwan, S., Li, L., and Yalkowsky, S. H., *J. Chem. Eng. Data*, 40, 623, 1995.
5. *Solubility Data Series, International Union of Pure and Applied Chemistry, Vol. 20*, Pergamon Press, Oxford, 1985.
6. *Solubility Data Series, International Union of Pure and Applied Chemistry, Vol. 38*, Pergamon Press, Oxford, 1985.
7. Miller, M. M., Ghodbane, S., Wasik, S. P., Tewari, Y. B., and Martire, D. E., *J. Chem. Eng. Data*, 29, 184, 1984.

Mol. form.	Name	log P	Ref.
CCl_2F_2	Dichlorodifluoromethane	2.16	2
CCl_3F	Trichlorofluoromethane	2.53	2
CCl_4	Tetrachloromethane	2.64	2
$CHBr_3$	Tribromomethane	2.38	2
$CHCl_3$	Trichloromethane	1.97	2
CH_2BrCl	Bromochloromethane	1.41	2
CH_2Br_2	Dibromomethane	2.3	2
CH_2Cl_2	Dichloromethane	1.25	2
CH_2F_2	Difluoromethane	0.20	1
CH_2I_2	Diiodomethane	2.5	2
CH_2O	Formaldehyde	0.35	1
CH_2O_2	Formic acid	−0.54	1
CH_3Br	Bromomethane	1.19	2
CH_3Cl	Chloromethane	0.91	2
CH_3F	Fluoromethane	0.51	1
CH_3I	Iodomethane	1.5	2
CH_3NO	Formamide	−1.51	1
CH_3NO_2	Nitromethane	−0.33	1
CH_4O	Methanol	−0.74	1
CH_5N	Methylamine	−0.57	1
$C_2Cl_3F_3$	1,1,2-Trichlorotrifluoroethane	3.16	2
C_2Cl_4	Tetrachloroethylene	2.88	2
C_2Cl_6	Hexachloroethane	4.00	4
C_2HCl_3	Trichloroethylene	2.53	2
C_2HCl_5	Pentachloroethane	2.89	2
$C_2H_2Cl_2$	1,1-Dichloroethylene	2.13	2
$C_2H_2Cl_2$	*cis*-1,2-Dichloroethylene	1.86	2
$C_2H_2Cl_2$	*trans*-1,2-Dichloroethylene	1.93	2
$C_2H_2Cl_4$	1,1,2,2-Tetrachloroethane	2.39	2
C_2H_3Cl	Chloroethylene	1.38	2
$C_2H_3Cl_3$	1,1,1-Trichloroethane	2.49	2
$C_2H_3Cl_3$	1,1,2-Trichloroethane	2.38	2
C_2H_3N	Acetonitrile	−0.34	1
$C_2H_4Cl_2$	1,1-Dichloroethane	1.79	2
$C_2H_4Cl_2$	1,2-Dichloroethane	1.48	2
C_2H_4O	Acetaldehyde	0.45	1
C_2H_4O	Ethylene oxide	−0.30	1

Mol. form.	Name	log P	Ref.
$C_2H_4O_2$	Acetic acid	−0.17	1
C_2H_5Br	Bromoethane	1.6	2
C_2H_5Cl	Chloroethane	1.43	2
C_2H_5I	Iodoethane	2	2
C_2H_5NO	Acetamide	−1.26	1
$C_2H_5NO_2$	Nitroethane	0.18	1
C_2H_6O	Ethanol	−0.30	1
C_2H_6O	Dimethyl ether	0.10	1
C_2H_6OS	Dimethyl sulfoxide	−1.35	1
$C_2H_6O_2S$	Dimethyl sulfone	−1.41	1
C_2H_7N	Ethylamine	−0.13	1
C_2H_7N	Dimethylamine	−0.38	1
C_3H_3N	2-Propenenitrile	0.25	1
$C_3H_4Cl_2$	*cis*-1,3-Dichloropropene	2.03	2
C_3H_4O	Propargyl alcohol	−0.38	1
C_3H_4O	Acrolein	−0.01	1
C_3H_5Br	3-Bromopropene	1.79	1
C_3H_5ClO	Epichlorohydrin	0.30	2
$C_3H_5Cl_3$	1,2,3-Trichloropropane	2.63	2
C_3H_5N	Propanenitrile	0.16	1
C_3H_5NO	Acrylamide	−0.78	1
$C_3H_6Cl_2$	1,2-Dichloropropane	2.0	2
C_3H_6O	Allyl alcohol	0.17	1
C_3H_6O	Propanal	0.59	1
C_3H_6O	Acetone	−0.24	1
C_3H_6O	Methyloxirane	0.03	1
$C_3H_6O_2$	Propanoic acid	0.33	1
$C_3H_6O_2$	Methyl acetate	0.18	1
C_3H_7Br	1-Bromopropane	2.1	2
C_3H_7Br	2-Bromopropane	1.9	2
C_3H_7Cl	1-Chloropropane	2.04	1
C_3H_7Cl	2-Chloropropane	1.90	1
C_3H_7I	1-Iodopropane	2.5	2
C_3H_7N	Allylamine	0.03	1
C_3H_7NO	*N,N*-Dimethylformamide	−1.01	1
C_3H_7NO	*N*-Methylacetamide	−1.05	1
$C_3H_7NO_2$	1-Nitropropane	0.87	1

Mol. form.	Name	log P	Ref.	Mol. form.	Name	log P	Ref.
C_3H_8O	1-Propanol	0.25	1	$C_5H_{10}O$	2-Methyltetrahydrofuran	1.85	2
C_3H_8O	2-Propanol	0.05	1	$C_5H_{10}O_2$	Pentanoic acid	1.39	1
C_3H_8S	1-Propanethiol	1.81	1	$C_5H_{10}O_2$	Propyl acetate	1.24	1
C_3H_9N	Propylamine	0.48	1	$C_5H_{10}O_2$	Ethyl propanoate	1.21	1
C_3H_9N	Isopropylamine	0.26	1	$C_5H_{10}O_3$	Diethyl carbonate	1.21	1
C_3H_9N	Ethylmethylamine	0.15	1	$C_5H_{11}Br$	1-Bromopentane	3.37	1
C_3H_9N	Trimethylamine	0.16	1	$C_5H_{11}F$	1-Fluoropentane	2.33	1
C_4H_4O	Furan	1.34	1	$C_5H_{11}N$	Piperidine	0.84	1
C_4H_4S	Thiophene	1.81	1	$C_5H_{11}NO_2$	1-Nitropentane	2.01	1
C_4H_5N	Pyrrole	0.75	1	C_5H_{12}	Pentane	3.45	1
C_4H_6	1,3-Butadiene	1.99	1	C_5H_{12}	Neopentane	3.11	1
C_4H_6	2-Butyne	1.46	1	$C_5H_{12}O$	1-Pentanol	1.51	1
C_4H_6O	2,5-Dihydrofuran	0.46	1	$C_5H_{12}O$	2-Pentanol	1.25	1
$C_4H_6O_2$	Methacrylic acid	0.93	1	$C_5H_{12}O$	3-Pentanol	1.21	1
$C_4H_6O_2$	Vinyl acetate	0.73	1	$C_5H_{12}O$	3-Methyl-1-butanol	1.28	1
$C_4H_6O_2$	Methyl acrylate	0.80	1	$C_5H_{12}O$	2-Methyl-2-butanol	0.89	1
C_4H_7N	Butanenitrile	0.60	1	$C_5H_{12}O$	3-Methyl-2-butanol	1.28	1
C_4H_8	cis-2-Butene	2.33	1	$C_5H_{12}O$	2,2-Dimethyl-1-propanol	1.31	1
C_4H_8	trans-2-Butene	2.31	1	$C_5H_{12}O$	Methyl tert-butyl ether	0.94	1
C_4H_8	Isobutene	2.35	1	$C_5H_{13}N$	Pentylamine	1.49	1
$C_4H_8Cl_2O$	Bis(2-chloroethyl) ether	1.12	2	C_6Cl_6	Hexachlorobenzene	5.47	5
C_4H_8O	Ethyl vinyl ether	1.04	1	C_6HCl_5	Pentachlorobenzene	5.03	5
C_4H_8O	Butanal	0.88	1	C_6HCl_5O	Pentachlorophenol	5.07	4
C_4H_8O	2-Butanone	0.29	1	$C_6H_2Cl_4$	1,2,3,4-Tetrachlorobenzene	4.55	5
C_4H_8O	Tetrahydrofuran	0.46	1	$C_6H_2Cl_4$	1,2,3,5-Tetrachlorobenzene	4.65	5
$C_4H_8O_2$	Butanoic acid	0.79	1	$C_6H_2Cl_4$	1,2,4,5-Tetrachlorobenzene	4.51	5
$C_4H_8O_2$	Propyl formate	0.83	1	$C_6H_3Cl_3$	1,2,3-Trichlorobenzene	4.04	5
$C_4H_8O_2$	Ethyl acetate	0.73	1	$C_6H_3Cl_3$	1,2,4-Trichlorobenzene	3.98	5
C_4H_9Br	1-Bromobutane	2.75	1	$C_6H_3Cl_3$	1,3,5-Trichlorobenzene	4.02	5
C_4H_9Cl	1-Chlorobutane	2.64	2	$C_6H_4Cl_2$	o-Dichlorobenzene	3.38	5
C_4H_9F	1-Fluorobutane	2.58	1	$C_6H_4Cl_2$	m-Dichlorobenzene	3.48	5
C_4H_9I	1-Iodobutane	3	2	$C_6H_4Cl_2$	p-Dichlorobenzene	3.38	5
C_4H_9N	Pyrrolidine	0.46	1	$C_6H_4Cl_2O$	2,4-Dichlorophenol	3.23	4
C_4H_9NO	Butanamide	−0.21	1	C_6H_5Br	Bromobenzene	2.99	2
C_4H_9NO	N,N-Dimethylacetamide	−0.77	1	C_6H_5Cl	Chlorobenzene	2.84	1
$C_4H_9NO_2$	1-Nitrobutane	1.47	1	C_6H_5F	Fluorobenzene	2.27	2
C_4H_{10}	Isobutane	2.8	2	C_6H_5I	Iodobenzene	3.28	2
$C_4H_{10}O$	1-Butanol	0.84	1	$C_6H_5NO_2$	Nitrobenzene	1.85	1
$C_4H_{10}O$	2-Butanol	0.65	1	C_6H_6	Benzene	2.13	1
$C_4H_{10}O$	2-Methyl-1-propanol	0.76	1	C_6H_6O	Phenol	1.48	4
$C_4H_{10}O$	2-Methyl-2-propanol	0.35	1	C_6H_6S	Benzenethiol	2.52	1
$C_4H_{10}O$	Diethyl ether	0.89	1	C_6H_7N	Aniline	0.90	1
$C_4H_{10}S$	1-Butanethiol	2.28	1	C_6H_7N	2-Methylpyridine	1.11	1
$C_4H_{10}S$	Diethyl sulfide	1.95	1	C_6H_7N	3-Methylpyridine	1.20	1
$C_4H_{11}N$	Butylamine	0.86	1	C_6H_7N	4-Methylpyridine	1.22	1
$C_4H_{11}N$	tert-Butylamine	0.40	1	C_6H_8	1,4-Cyclohexadiene	2.3	2
$C_4H_{11}N$	Diethylamine	0.58	1	C_6H_8O	5-Hexyn-2-one	0.58	1
C_5H_5N	Pyridine	0.65	1	C_6H_8O	2-Cyclohexen-1-one	0.61	1
C_5H_6O	2-Methylfuran	1.85	1	C_6H_8O	2-Ethylfuran	2.40	1
C_5H_7N	1-Methylpyrrole	1.21	1	C_6H_{10}	1,5-Hexadiene	2.8	2
C_5H_8	1,4-Pentadiene	2.48	1	C_6H_{10}	1-Hexyne	2.73	2
C_5H_8	1-Pentyne	1.98	1	C_6H_{10}	Cyclohexene	2.86	1
$C_5H_8O_2$	Methyl methacrylate	1.38	1	$C_6H_{10}O$	5-Hexen-2-one	1.02	1
$C_5H_8O_2$	Ethyl acrylate	1.32	1	$C_6H_{10}O$	Cyclohexanone	0.81	1
C_5H_9N	Pentanenitrile	0.94	1	$C_6H_{10}O_2$	Ethyl methacrylate	1.94	1
C_5H_{10}	1-Pentene	2.2	2	$C_6H_{11}Br$	Bromocyclohexane	3.20	1
C_5H_{10}	Cyclopentane	3.00	1	$C_6H_{11}N$	Hexanenitrile	1.66	1
$C_5H_{10}O$	2-Pentanone	0.84	1	C_6H_{12}	1-Hexene	3.40	1
$C_5H_{10}O$	3-Pentanone	0.82	1	C_6H_{12}	4-Methyl-1-pentene	2.5	2
$C_5H_{10}O$	3-Methyl-2-butanone	0.56	1	C_6H_{12}	Cyclohexane	3.44	1
$C_5H_{10}O$	Tetrahydropyran	0.82	1	C_6H_{12}	Methylcyclopentane	3.37	2

Mol. form.	Name	log P	Ref.
$C_6H_{12}O$	Cyclohexanol	1.23	1
$C_6H_{12}O$	Hexanal	1.78	1
$C_6H_{12}O$	2-Hexanone	1.38	1
$C_6H_{12}O$	4-Methyl-2-pentanone	1.31	1
$C_6H_{12}O_2$	Hexanoic acid	1.92	1
$C_6H_{12}O_2$	Butyl acetate	1.82	1
$C_6H_{13}Br$	1-Bromohexane	3.80	1
$C_6H_{13}N$	Cyclohexylamine	1.49	1
C_6H_{14}	Hexane	4.00	1
C_6H_{14}	3-Methylpentane	3.60	2
C_6H_{14}	2,2-Dimethylbutane	3.82	1
C_6H_{14}	2,3-Dimethylbutane	3.85	2
$C_6H_{14}O$	1-Hexanol	2.03	1
$C_6H_{14}O$	2-Hexanol	1.76	1
$C_6H_{14}O$	3-Hexanol	1.65	1
$C_6H_{14}O$	3,3-Dimethyl-2-butanol	1.48	1
$C_6H_{14}O$	Dipropyl ether	2.03	1
$C_6H_{14}O$	Diisopropyl ether	1.52	1
$C_6H_{15}N$	Hexylamine	2.06	1
$C_6H_{15}N$	Dipropylamine	1.67	1
$C_6H_{15}N$	Triethylamine	1.45	1
$C_7H_5BrO_2$	2-Bromobenzoic acid	2.20	4
$C_7H_5BrO_2$	3-Bromobenzoic acid	2.87	4
$C_7H_5BrO_2$	4-Bromobenzoic acid	2.86	4
C_7H_5N	Benzonitrile	1.56	1
C_7H_6O	Benzaldehyde	1.48	1
$C_7H_6O_2$	Benzoic acid	1.88	4
$C_7H_6O_2$	Phenyl formate	1.26	1
$C_7H_6O_3$	Salicylic acid	2.20	4
C_7H_7Br	(Bromomethyl)benzene	2.92	1
C_7H_7Cl	o-Chlorotoluene	3.42	1
C_7H_7Cl	m-Chlorotoluene	3.28	1
C_7H_7Cl	p-Chlorotoluene	3.33	1
C_7H_7Cl	(Chloromethyl)benzene	2.30	1
$C_7H_7NO_2$	p-Nitrotoluene	2.42	1
C_7H_8	Toluene	2.73	1
C_7H_8	1,3,5-Cycloheptatriene	2.63	2
C_7H_8O	o-Cresol	1.98	1
C_7H_8O	m-Cresol	1.98	1
C_7H_8O	p-Cresol	1.97	1
C_7H_8O	Benzyl alcohol	1.05	1
C_7H_8O	Anisole	2.11	1
C_7H_9N	Benzylamine	1.09	1
C_7H_9N	o-Methylaniline	1.32	1
C_7H_9N	m-Methylaniline	1.40	1
C_7H_9N	p-Methylaniline	1.39	1
C_7H_9N	N-Methylaniline	1.66	1
C_7H_{14}	1-Heptene	3.99	1
C_7H_{14}	Methylcyclohexane	3.88	1
$C_7H_{14}O$	2-Heptanone	1.98	1
$C_7H_{14}O$	5-Methyl-2-hexanone	1.88	1
$C_7H_{15}Br$	1-Bromoheptane	4.36	1
$C_7H_{15}Cl$	1-Chloroheptane	4.15	1
$C_7H_{15}I$	1-Iodoheptane	4.70	1
C_7H_{16}	Heptane	4.50	1
$C_7H_{16}O$	1-Heptanol	2.62	1
$C_7H_{16}O$	2-Heptanol	2.31	1
$C_7H_{16}O$	3-Heptanol	2.24	1
$C_7H_{16}O$	4-Heptanol	2.22	1
$C_7H_{17}N$	Heptylamine	2.57	1
C_8H_6	Phenylacetylene	2.40	1
C_8H_6O	Benzofuran	2.67	1
C_8H_6S	Benzo[b]thiophene	3.12	1
C_8H_7N	Benzeneacetonitrile	1.56	1
C_8H_7N	Indole	2.14	1
C_8H_8	Styrene	3.05	1
C_8H_8O	Acetophenone	1.63	1
C_8H_8O	2-Methylbenzaldehyde	2.26	1
C_8H_8O	Benzeneacetaldehyde	1.78	1
C_8H_8O	2,3-Dihydrobenzofuran	2.14	1
C_8H_8O	Phenyloxirane	1.61	1
$C_8H_8O_2$	o-Toluic acid	2.32	4
$C_8H_8O_2$	m-Toluic acid	2.37	1
$C_8H_8O_2$	p-Toluic acid	2.34	1
$C_8H_8O_2$	Benzeneacetic acid	1.41	1
$C_8H_8O_2$	Phenyl acetate	1.49	1
$C_8H_8O_2$	Methyl benzoate	2.20	1
C_8H_{10}	Ethylbenzene	3.15	1
C_8H_{10}	o-Xylene	3.12	1
C_8H_{10}	m-Xylene	3.20	1
C_8H_{10}	p-Xylene	3.15	1
$C_8H_{10}O$	o-Ethylphenol	2.47	1
$C_8H_{10}O$	m-Ethylphenol	2.50	1
$C_8H_{10}O$	p-Ethylphenol	2.50	1
$C_8H_{10}O$	2,4-Xylenol	2.35	1
$C_8H_{10}O$	2,5-Xylenol	2.34	1
$C_8H_{10}O$	2,6-Xylenol	2.36	1
$C_8H_{10}O$	3,4-Xylenol	3.23	1
$C_8H_{10}O$	3,5-Xylenol	2.35	1
$C_8H_{10}O$	Benzeneethanol	1.36	1
$C_8H_{10}O$	α-Methylbenzyl alcohol	1.42	1
$C_8H_{10}O$	3-Methylbenzenemethanol	1.60	1
$C_8H_{10}O$	4-Methylbenzenemethanol	1.58	1
$C_8H_{10}O$	Phenetole	2.51	1
$C_8H_{10}O$	Benzyl methyl ether	1.35	1
$C_8H_{10}O$	2-Methylanisole	2.74	1
$C_8H_{10}O$	3-Methylanisole	2.66	1
$C_8H_{10}O$	4-Methylanisole	2.81	1
$C_8H_{11}N$	p-Ethylaniline	1.96	1
$C_8H_{11}N$	N,N-Dimethylaniline	2.31	1
$C_8H_{11}N$	Benzeneethanamine	1.41	1
$C_8H_{14}O_2$	Butyl methacrylate	2.88	1
$C_8H_{15}N$	Octanenitrile	2.75	1
C_8H_{16}	1-Octene	4.57	1
C_8H_{16}	Cyclooctane	4.45	2
C_8H_{16}	2-Octanone	2.37	1
$C_8H_{16}O_2$	Octanoic acid	3.05	1
$C_8H_{17}Br$	1-Bromooctane	4.89	1
C_8H_{18}	Octane	5.15	1
$C_8H_{18}O$	1-Octanol	3.07	1
$C_8H_{18}O$	2-Octanol	2.90	1
$C_8H_{18}O$	4-Octanol	2.68	1
$C_8H_{18}O$	Dibutyl ether	3.21	1
C_9H_7N	Quinoline	2.03	1
C_9H_7N	Isoquinoline	2.08	1
C_9H_8	Indene	2.92	1
$C_9H_8O_2$	trans-Cinnamic acid	2.13	1
C_9H_9N	Benzenepropanenitrile	1.72	1
C_9H_{10}	Indan	3.33	1
$C_9H_{10}O$	1-Phenyl-1-propanone	2.19	1
$C_9H_{10}O$	1-Phenyl-2-propanone	1.44	1
$C_9H_{10}O$	4-Methylacetophenone	2.19	1

Mol. form.	Name	log P	Ref.
$C_9H_{10}O_2$	2-Phenylpropanoic acid	1.80	1
$C_9H_{10}O_2$	Benzyl acetate	1.96	1
$C_9H_{10}O_2$	4-Methylphenyl acetate	2.11	1
$C_9H_{10}O_2$	Ethyl benzoate	2.64	1
C_9H_{12}	Propylbenzene	3.69	1
C_9H_{12}	Isopropylbenzene	3.66	1
C_9H_{12}	o-Ethyltoluene	3.53	1
C_9H_{12}	p-Ethyltoluene	3.63	2
C_9H_{12}	1,2,3-Trimethylbenzene	3.60	1
C_9H_{12}	1,2,4-Trimethylbenzene	3.63	1
C_9H_{12}	1,3,5-Trimethylbenzene	3.42	1
$C_9H_{12}O$	2-Propylphenol	2.93	1
$C_9H_{12}O$	4-Propylphenol	3.20	1
$C_9H_{12}O$	2,3,6-Trimethylphenol	2.67	1
$C_9H_{12}O$	2,4,6-Trimethylphenol	2.46	1
$C_9H_{12}O$	Benzenepropanol	1.88	1
$C_9H_{13}N$	N,N-Dimethylbenzylamine	1.98	1
$C_9H_{13}N$	Amphetamine	1.76	1
C_9H_{18}	1-Nonene	5.15	1
$C_9H_{18}O$	2-Nonanone	3.16	1
$C_9H_{18}O$	5-Methyl-2-octanone	2.92	1
C_9H_{20}	Nonane	5.65	1
$C_9H_{20}O$	1-Nonanol	4.02	1
$C_9H_{21}N$	Tripropylamine	2.79	1
$C_{10}H_7Cl$	1-Chloronaphthalene	3.90	1
$C_{10}H_7Cl$	2-Chloronaphthalene	3.98	1
$C_{10}H_8$	Naphthalene	3.34	4
$C_{10}H_8$	Azulene	3.22	1
$C_{10}H_8O$	1-Naphthol	2.84	1
$C_{10}H_8O$	2-Naphthol	2.70	1
$C_{10}H_{12}O_2$	Isopropyl benzoate	3.18	1
$C_{10}H_{14}$	Butylbenzene	4.26	1
$C_{10}H_{14}$	tert-Butylbenzene	4.11	1
$C_{10}H_{14}$	Isobutylbenzene	4.01	2
$C_{10}H_{14}$	p-Cymene	4.10	1
$C_{10}H_{14}$	1,2,4,5-Tetramethylbenzene	4.10	2
$C_{10}H_{14}$	1,2,3,4-Tetramethylbenzene	4.00	1
$C_{10}H_{14}$	1,2,3,5-Tetramethylbenzene	4.10	1
$C_{10}H_{14}O$	4-Butylphenol	3.65	1
$C_{10}H_{20}O$	2-Decanone	3.77	1
$C_{10}H_{20}O_2$	Decanoic acid	4.09	1
$C_{10}H_{22}$	Decane	6.25	1
$C_{10}H_{22}O$	1-Decanol	4.57	1
$C_{11}H_9N$	4-Phenylpyridine	2.59	1
$C_{11}H_{10}$	1-Methylnaphthalene	3.87	1
$C_{11}H_{10}$	2-Methylnaphthalene	4.00	1
$C_{11}H_{16}$	Pentylbenzene	4.90	1
$C_{11}H_{16}$	Pentamethylbenzene	4.56	1
$C_{11}H_{22}O$	2-Undecanone	4.09	1
$C_{11}H_{22}O_2$	Methyl decanoate	4.41	1
$C_{12}Cl_{10}$	Decachlorobiphenyl	8.26	3
$C_{12}HCl_9$	2,2',3,3',4,5,5',6,6'-Nonachlorobiphenyl	8.16	3
$C_{12}H_2Cl_8$	2,2',3,3',5,5',6,6'-Octachlorobiphenyl	7.10	3
$C_{12}H_3Cl_7$	2,2',3,3',4,4',6-Heptachlorobiphenyl	6.70	3
$C_{12}H_4Cl_6$	2,2',3,3',4,4'-Hexachlorobiphenyl	7.00	3
$C_{12}H_4Cl_6$	2,2',4,4',6,6'-Hexachlorobiphenyl	7.00	3
$C_{12}H_4Cl_6$	2,2',3,3',6,6'-Hexachlorobiphenyl	6.70	3
$C_{12}H_5Cl_5$	2,3,4,5,6-Pentachlorobiphenyl	6.30	3
$C_{12}H_5Cl_5$	2,2',4,5,5'-Pentachlorobiphenyl	6.40	3
$C_{12}H_6Cl_4$	2,3,4,5-Tetrachlorobiphenyl	5.72	3

Mol. form.	Name	log P	Ref.
$C_{12}H_6Cl_4$	2,2',4',5-Tetrachlorobiphenyl	5.73	7
$C_{12}H_7Cl_3$	2,4,5-Trichlorobiphenyl	5.60	3
$C_{12}H_7Cl_3$	2,4,6-Trichlorobiphenyl	5.47	3
$C_{12}H_8Cl_2$	2,5-Dichlorobiphenyl	5.10	3
$C_{12}H_8Cl_2$	2,6-Dichlorobiphenyl	5.00	3
$C_{12}H_8O$	Dibenzofuran	4.12	1
$C_{12}H_9Cl$	2-Chlorobiphenyl	4.52	1
$C_{12}H_9Cl$	3-Chlorobiphenyl	4.58	1
$C_{12}H_9Cl$	4-Chlorobiphenyl	4.61	1
$C_{12}H_9N$	Carbazole	3.72	1
$C_{12}H_{10}$	Acenaphthene	3.96	4
$C_{12}H_{10}$	Biphenyl	3.76	6
$C_{12}H_{10}N_2$	Azobenzene	3.82	1
$C_{12}H_{10}O$	Diphenyl ether	4.21	1
$C_{12}H_{10}S$	Diphenyl sulfide	4.45	1
$C_{12}H_{11}N$	Diphenylamine	3.44	4
$C_{12}H_{12}$	1-Ethylnaphthalene	4.40	1
$C_{12}H_{12}$	1,2-Dimethylnaphthalene	4.31	1
$C_{12}H_{12}$	1,4-Dimethylnaphthalene	4.37	1
$C_{12}H_{14}O$	4-Phenylcyclohexanone	2.45	1
$C_{12}H_{18}$	Hexylbenzene	5.52	1
$C_{12}H_{18}$	Hexamethylbenzene	4.69	4
$C_{12}H_{22}O$	Cyclododecanone	4.10	1
$C_{12}H_{24}O_2$	Dodecanoic acid	4.6	1
$C_{12}H_{26}O$	1-Dodecanol	5.13	1
$C_{13}H_8O$	9H-Fluoren-9-one	3.58	1
$C_{13}H_9N$	Acridine	3.40	1
$C_{13}H_{10}$	9H-Fluorene	4.20	4
$C_{13}H_{10}O$	Benzophenone	3.18	1
$C_{13}H_{10}O_2$	Phenyl benzoate	3.59	1
$C_{13}H_{11}NO$	N-Phenylbenzamide	2.62	1
$C_{13}H_{12}$	Diphenylmethane	4.14	1
$C_{13}H_{12}$	4-Methylbiphenyl	4.63	1
$C_{13}H_{12}O$	Diphenylmethanol	2.67	1
$C_{13}H_{12}O$	Benzyl phenyl ether	3.79	1
$C_{14}H_{10}$	Anthracene	4.56	4
$C_{14}H_{10}$	Phenanthrene	4.52	4
$C_{14}H_{12}$	trans-Stilbene	4.81	1
$C_{14}H_{12}$	1-Methylfluorene	4.97	1
$C_{14}H_{12}O$	2-Phenylacetophenone	3.18	1
$C_{14}H_{12}O_2$	Benzyl benzoate	3.97	1
$C_{14}H_{14}$	1,2-Diphenylethane	4.70	1
$C_{14}H_{14}$	4,4'-Dimethylbiphenyl	5.09	1
$C_{14}H_{22}$	Octylbenzene	6.30	1
$C_{14}H_{28}O_2$	Tetradecanoic acid	6.1	1
$C_{15}H_{12}$	2-Methylanthracene	5.15	2
$C_{15}H_{12}$	9-Methylanthracene	5.07	1
$C_{15}H_{12}$	1-Methylphenanthrene	5.14	2
$C_{16}H_{10}$	Fluoranthene	5.07	4
$C_{16}H_{10}$	Pyrene	5.08	4
$C_{16}H_{14}$	9,10-Dimethylanthracene	5.69	1
$C_{16}H_{32}O_2$	Hexadecanoic acid	7.17	1
$C_{17}H_{12}$	11H-Benzo[a]fluorene	5.40	1
$C_{17}H_{12}$	11H-Benzo[b]fluorene	5.75	1
$C_{18}H_{12}$	Benz[a]anthracene	5.91	1
$C_{18}H_{12}$	Chrysene	5.73	4
$C_{18}H_{12}$	Naphthacene	5.76	1
$C_{18}H_{12}$	Triphenylene	5.49	4
$C_{18}H_{15}N$	Triphenylamine	5.74	1
$C_{18}H_{30}O_2$	Linolenic acid	6.46	1
$C_{18}H_{32}O_2$	Linoleic acid	7.05	1

Mol. form.	Name	log P	Ref.	Mol. form.	Name	log P	Ref.
$C_{18}H_{34}O_2$	Oleic acid	7.64	1	$C_{20}H_{40}O_2$	Arachidic acid	9.29	1
$C_{18}H_{36}O_2$	Stearic acid	8.23	1	$C_{21}H_{16}$	1,2-Dihydro-3-methylbenz[j] aceanthrylene	6.75	1
$C_{19}H_{16}O$	Triphenylmethanol	3.68	1				
$C_{20}H_{12}$	Perylene	6.25	1	$C_{22}H_{12}$	Benzo[ghi]perylene	6.90	1
$C_{20}H_{12}$	Benzo[a]pyrene	6.20	4	$C_{24}H_{12}$	Coronene	6.05	4
$C_{20}H_{32}O_2$	Arachidonic acid	6.98	1				

PROTECTION AGAINST IONIZING RADIATION

The following data and rules of thumb are helpful in estimating the penetrating capability of and danger of exposure to various types of ionizing radiation. More precise data should be used for critical applications.

Alpha Particles

Alpha particles of at least 7.5 MeV are required to penetrate the epidermis, the protective layer of skin, which is about 0.07 mm thick.

Electrons

Electrons of at least 70 keV are required to penetrate the epidermis, the protective layer of skin, which is about 0.07 mm thick.

The range of electrons in g/cm^2 is approximately equal to the maximum energy (E) in MeV divided by 2.

The range of electrons in air is about 3.65 m per MeV; for example, a 3 MeV electron has a range of about 11 m in air.

A chamber wall thickness of 30 mg/cm^2 will transmit 70% of the initial fluence of 1 MeV electrons and 20% of that of 0.4 MeV electrons.

When electrons of 1 to 2 MeV pass through light materials such as water, aluminum, or glass, less than 1% of their energy is dissipated as bremsstrahlung.

The bremsstrahlung from 1 Ci of ^{32}P aqueous solution in a glass bottle is about 1 mR/h at 1 meter distance.

When electrons from a 1 Ci source of $^{90}Sr - ^{90}Y$ are absorbed, the bremsstrahlung hazard is approximately equal to that presented by the gamma radiation from 12 mg of radium. The average energy of the bremsstrahlung is about 300 keV.

Gamma Rays

The air-scattered radiation (sky-shine) from a 100 Ci ^{60}Co source placed 1 ft behind a 4 ft high shield is about 100 mrad/h at 6 ft from the outside of the shield.

Within ±20% for point source gamma emitters with energies between 0.07 and 4 MeV, the exposure rate (R/h) at 1 ft is $6C \cdot E \cdot n$ where C is the activity in curies, E is the energy in MeV, and n is the number of gammas per disintegration.

Neutrons

An approximate HVL (thickness of absorber for which the neutron flux falls to half its initial value) for 1 MeV neutrons is 3.2 cm of paraffin; that for 5 MeV neutrons is 6.9 cm of paraffin.

Miscellaneous

The activity of any radionuclide is reduced to less than 1% after 7 half-lives (i.e., $2^{-7} = 0.8\%$).

For nuclides with a half-life greater than 6 days, the change in activity in 24 hours will be less than 10%.

10 HVL (half-value layers) attenuates approximately by 10^{-3}.

There is 0.64 mm^3 of radon gas at STP in transient equilibrium with 1 Ci of radium.

The natural background from all sources in most parts of the world leads to an equivalent dose rate of about 0.04 to 4 mSv per year for the average person. About 84% of this comes from terrestrial sources, the remainder from cosmic rays. The U. S. average is about 3.6 mSv/yr but can range up to 50 mSv/yr in some areas. A passenger in a plane flying at 12,000 meters receives 5 μSv/hr from cosmic rays (as compared to about 0.03 μSv/hr at sea level).

The ICRP recommended exposure limit to man-made sources of ionizing radiation (Reference 2) is 20 mSv/yr averaged over 5 years, with the dose in any one year not to exceed 50 mSv.

A whole-body dose of about 3 Gy over a short time interval will typically lead to 50% mortality in 30 days assuming no medical treatment.

Units

The gray (Gy) is the SI unit of absorbed dose; it is a measure of the mean energy imparted to a sample of irradiated matter, divided by the mass of the sample. Gy is a special name for the SI unit J/kg.

The sievert (Sv) is the SI unit of equivalent dose, which is defined as the absorbed dose multiplied by a weighting factor that expresses the long-term biological risk from low-level chronic exposure to a specified type of radiation. The Sv is another special name for J/kg.

1 curie (Ci) = $3.7 \cdot 10^{10}$ becquerel (Bq); i.e., $3.7 \cdot 10^{10}$ disintegrations per second.

1 roentgen (R) = $2.58 \cdot 10^{-4}$ coulomb per kilogram (C/kg); a measure of the charge (positive or negative) liberated by x-ray or gamma radiation in air, divided by the mass of air.

1 rad = 0.01 Gy

1 rem = 0.01 Sv

References

1. Padikal, T. N., and Fivozinsky, S. P., *Medical Physics Data Book, National Bureau of Standards Handbook 138*, U.S. Government Printing Office, Washington, D.C., 1981.
2. *1990 Recommendations of the International Commission on Radiological Protection*, ICRP Publication 60, *Annals of the ICRP*, Pergamon Press, Oxford, 1991.
3. *Radiation: Doses, Effects, Risks*, United Nations Sales No. E.86.III.D.4, 1985.
4. Eidelman, S., et al., *Physics Letters*, B592, 1, 2004.

ANNUAL LIMITS ON INTAKES OF RADIONUCLIDES

K. F. Eckerman

The following table lists, for workers, the annual limits on oral and inhalation intakes (ALI) for selected radionuclides based on the occupational radiation protection guidance of the International Commission on Radiological Protection (References 1 and 2). An intake of one ALI corresponds to an annual whole body dose of 0.02 Sv (2 rem).

The ALI is expressed in the SI unit of activity, the becquerel (Bq), and in the conventional unit, the microcurie (μCi); 1 μCi = $3.7 \cdot 10^4$ Bq. The chemical form of inhaled radionuclides is, in most instances, stated in terms of the rate of absorption to blood from the lungs and the fractional absorption from the small intestine. Types F, M, and S denote chemical forms that are absorbed from the lungs at rates characterized as fast, moderate, and slow, respectively. The time to absorb 90% of the deposited radionuclide, in the absence of radioactive decay, corresponds to about 10 minutes,

150 days, and 7000 days for Type F, M, and S compounds, respectively. Type F compounds can be considered to be more soluble than M or S, S being the most insoluble. Chemical form consideration for ingestion is specified by the fractional absorption from the small intestine, denoted as f_1. The f_1 values range from 10^{-5} to 1. Higher fractional absorption is associated with greater solubility of the compound.

References

1. *1990 Recommendations of the International Commission on Radiological Protection, ICRP Publication 60, Annals of the ICRP 21, (1–3), Pergamon Press, Oxford, 1991.*
2. *Dose Coefficients for Intakes of Radionuclides by Workers, ICRP Publication 68, Annals of the ICRP, 24(4), Pergamon Press, Oxford, 1995.*

	Physical half-life	Chemical form type/f_1	Inhalation intakes ALI Bq	Inhalation intakes ALI μCi	Chemical form f_1	Oral intakes ALI Bq	Oral intakes ALI μCi
³H	12.3 y	HT gas	1.1E+13	3.0E+08	1.000	1.1E+13	3.0E+08
		HTO vapor	1.1E+09	3.0E+04			
¹¹C	0.340 h	CO	1.7E+10	4.5E+05	1.000	8.3E+08	2.3E+04
		CO_2	9.1E+09	2.5E+05			
		Organic compounds	6.2E+09	1.7E+05			
¹⁴C	5730 y	CO	2.5E+10	6.8E+05	1.000	3.4E+07	9.3E+02
		CO_2	3.1E+09	8.3E+04			
		Organic compounds	3.4E+07	9.3E+02			
¹⁸F	1.83 h	F 1.000	3.7E+08	1.0E+04	1.000	4.1E+08	1.1E+04
		M 1.000	2.2E+08	6.1E+03			
		S 1.000	2.2E+08	5.8E+03			
²²Na	2.60 y	F 1.000	1.0E+07	2.7E+02	1.000	6.3E+06	1.7E+02
²⁴Na	15.0 h	F 1.000	3.8E+07	1.0E+03	1.000	4.7E+07	1.3E+03
³²P	14.3 d	F 0.800	1.8E+07	4.9E+02	0.800	8.3E+06	2.3E+02
		M 0.800	6.9E+06	1.9E+02			
³⁵S	87.4 d	Inorganic compounds					
		F 0.800	2.5E+08	6.8E+03	0.800	1.4E+08	3.9E+03
		M 0.800	1.8E+07	4.9E+02	0.100	1.1E+08	2.8E+03
		Vapor	1.7E+08	4.5E+03			
		Organic compounds			1.000	2.6E+07	7.0E+02
⁴²K	12.4 h	F 1.000	1.0E+08	2.7E+03	1.000	4.7E+07	1.3E+03
⁴³K	22.6 h	F 1.000	7.7E+07	2.1E+03	1.000	8.0E+07	2.2E+03
⁴⁵Ca	163 d	M 0.300	8.7E+06	2.4E+02	0.300	2.6E+07	7.1E+02
⁴⁷Ca	4.53 d	M 0.300	9.5E+06	2.6E+02	0.300	1.3E+07	3.4E+02
⁵¹Cr	27.7 d	F 0.100	6.7E+08	1.8E+04	0.100	5.3E+08	1.4E+04
		M 0.100	5.9E+08	1.6E+04	0.010	5.4E+08	1.5E+04
		S 0.100	5.6E+08	1.5E+04			
⁵⁴Mn	312 d	F 0.100	1.8E+07	4.9E+02	0.100	2.8E+07	7.6E+02
		M 0.100	1.7E+07	4.5E+02			
⁵²Fe	8.28 h	F 0.100	2.9E+07	7.8E+02	0.100	1.4E+07	3.9E+02
		M 0.100	2.1E+07	5.7E+02			
⁵⁵Fe	2.70 y	F 0.100	2.2E+07	5.9E+02	0.100	6.1E+07	1.6E+03
		M 0.100	6.1E+07	1.6E+03			

	Physical half-life	Chemical form type/f_1	Inhalation intakes ALI Bq	µCi	Chemical form f_1	Oral intakes ALI Bq	µCi
^{59}Fe	44.5 d	F 0.100	6.7E+06	1.8E+02	0.100	1.1E+07	3.0E+02
		M 0.100	6.3E+06	1.7E+02			
^{57}Co	271 d	M 0.100	5.1E+07	1.4E+03	0.100	9.5E+07	2.6E+03
		S 0.050	3.3E+07	9.0E+02	0.050	1.1E+08	2.8E+03
^{58}Co	70.8 d	M 0.100	1.4E+07	3.9E+02	0.100	2.7E+07	7.3E+02
		S 0.050	1.2E+07	3.2E+02	0.050	2.9E+07	7.7E+02
^{60}Co	5.27 y	M 0.100	2.8E+06	7.6E+01	0.100	5.9E+06	1.6E+02
		S 0.050	1.2E+06	3.2E+01	0.050	8.0E+06	2.2E+02
^{64}Cu	12.7 h	F 0.500	2.9E+08	7.9E+03	0.500	1.7E+08	4.5E+03
		M 0.500	1.3E+08	3.6E+03			
		S 0.500	1.3E+08	3.6E+03			
^{59}Ni	75000 y	F 0.050	9.1E+07	2.5E+03	0.050	3.2E+08	8.6E+03
		M 0.050	2.1E+08	5.8E+03			
		Vapor	2.4E+07	6.5E+02			
^{63}Ni	96.0 y	F 0.050	3.8E+07	1.0E+03	0.050	1.3E+08	3.6E+03
		M 0.050	6.5E+07	1.7E+03			
		Vapor	1.0E+07	2.7E+02			
^{65}Zn	244 d	S 0.500	7.1E+06	1.9E+02	0.500	5.1E+06	1.4E+02
^{67}Ga	3.26 d	F 0.001	1.8E+08	4.9E+03	0.001	1.1E+08	2.8E+03
		M 0.001	7.1E+07	1.9E+03			
^{68}Ga	1.13 h	F 0.001	4.1E+08	1.1E+04	0.001	2.0E+08	5.4E+03
		M 0.001	2.5E+08	6.7E+03			
^{68}Ge	288 d	F 1.000	2.4E+07	6.5E+02	1.000	1.5E+07	4.2E+02
		M 1.000	2.5E+06	6.8E+01			
^{75}Se	120 d	F 0.800	1.4E+07	3.9E+02	0.800	7.7E+06	2.1E+02
		M 0.800	1.2E+07	3.2E+02	0.050	4.9E+07	1.3E+03
^{79}Se	65000 y	F 0.800	1.3E+07	3.4E+02	0.800	6.9E+06	1.9E+02
		M 0.800	6.5E+06	1.7E+02	0.050	5.1E+07	1.4E+03
^{86}Rb	18.6 d	F 1.000	1.5E+07	4.2E+02	1.000	7.1E+06	1.9E+02
^{85}Sr	64.8 d	F 0.300	3.6E+07	9.7E+02	0.300	3.6E+07	9.7E+02
		S 0.010	3.1E+07	8.4E+02	0.010	6.1E+07	1.6E+03
^{87m}Sr	2.80 h	F 0.300	9.1E+08	2.5E+04	0.300	6.7E+08	1.8E+04
		S 0.010	5.7E+08	1.5E+04	0.010	6.1E+08	1.6E+04
^{89}Sr	50.5 d	F 0.300	1.4E+07	3.9E+02	0.300	7.7E+06	2.1E+02
		S 0.010	3.6E+06	9.7E+01	0.010	8.7E+06	2.4E+02
^{90}Sr	29.1 y	F 0.300	6.7E+05	1.8E+01	0.300	7.1E+05	1.9E+01
		S 0.010	2.6E+05	7.0E+00	0.010	7.4E+06	2.0E+02
^{99}Mo	2.75 d	F 0.800	5.6E+07	1.5E+03	0.800	2.7E+07	7.3E+02
		S 0.050	1.8E+07	4.9E+02	0.050	1.7E+07	4.5E+02
^{99m}Tc	6.02 h	F 0.800	1.0E+09	2.7E+04	0.800	9.1E+08	2.5E+04
		M 0.800	6.9E+08	1.9E+04			
^{99}Tc	213000 y	F 0.800	5.0E+07	1.4E+03	0.800	2.6E+07	6.9E+02
		M 0.800	6.3E+06	1.7E+02			
^{106}Ru	1.01 y	F 0.050	2.0E+06	5.5E+01	0.050	2.9E+06	7.7E+01
		M 0.050	1.2E+06	3.2E+01			
		S 0.050	5.7E+05	1.5E+01			
^{111}In	2.83 d	F 0.020	9.1E+07	2.5E+03	0.020	6.9E+07	1.9E+03
		M 0.020	6.5E+07	1.7E+03			
^{113m}In	1.66 h	F 0.020	1.1E+09	2.8E+04	0.020	7.1E+08	1.9E+04
		M 0.020	6.3E+08	1.7E+04			
^{113}Sn	115 d	F 0.020	2.5E+07	6.8E+02	0.020	2.7E+07	7.4E+02
		M 0.020	1.1E+07	2.8E+02			
^{123}I	13.2 h	F 1.000	1.8E+08	4.9E+03	1.000	9.5E+07	2.6E+03
		Vapor	9.5E+07	2.6E+03			
^{125}I	60.1 d	F 1.000	2.7E+06	7.4E+01	1.000	1.3E+06	3.6E+01
		Vapor	1.4E+06	3.9E+01			
^{129}I	$1.57 \cdot 10^7$ y	F 1.000	3.9E+05	1.1E+01	1.000	1.8E+05	4.9E+00
		Vapor	2.1E+05	5.6E+00			

	Physical half-life	Chemical form type/f_1	Inhalation intakes		Chemical form f_1	Oral intakes	
			ALI Bq	ALI µCi		ALI Bq	ALI µCi
^{131}I	8.04 d	F 1.000	1.8E+06	4.9E+01	1.000	9.1E+05	2.5E+01
		Vapor	1.0E+06	2.7E+01			
^{129}Cs	1.34 d	F 1.000	2.5E+08	6.7E+03	1.000	3.3E+08	9.0E+03
^{134}Cs	2.06 y	F 1.000	2.1E+06	5.6E+01	1.000	1.1E+06	2.8E+01
^{136}Cs	13.1 d	F 1.000	1.1E+07	2.8E+02	1.000	6.7E+06	1.8E+02
^{137}Cs	30.0 y	F 1.000	3.0E+06	8.1E+01	1.000	1.5E+06	4.2E+01
^{141}Ce	32.5 d	M 5.0E-04	7.4E+06	2.0E+02	5.0E-04	2.8E+07	7.6E+02
		S 5.0E-04	6.5E+06	1.7E+02			
^{144}Ce	284 d	M 5.0E-04	8.7E+05	2.4E+01	5.0E-04	3.8E+06	1.0E+02
		S 5.0E-04	6.9E+05	1.9E+01			
^{133}Ba	10.7 y	F 0.100	1.1E+07	3.0E+02	0.100	2.0E+07	5.4E+02
^{140}Ba	12.7 d	F 0.100	1.3E+07	3.4E+02	0.100	8.0E+06	2.2E+02
^{169}Yb	32.0 d	M 5.0E-04	9.5E+06	2.6E+02	5.0E-04	2.8E+07	7.6E+02
		S 5.0E-04	8.3E+06	2.3E+02			
^{198}Au	2.69 d	F 0.100	5.1E+07	1.4E+03	0.100	2.0E+07	5.4E+02
		M 0.100	2.0E+07	5.5E+02			
		S 0.100	1.8E+07	4.9E+02			
^{198m}Au	2.30 d	F 0.100	3.4E+07	9.2E+02	0.100	1.5E+07	4.2E+02
		M 0.100	1.0E+07	2.7E+02			
		S 0.100	1.1E+07	2.8E+02			
^{197}Hg	2.67 d	Inorganic compounds					
		F 0.400	2.4E+08	6.4E+03	1.000	2.0E+08	5.5E+03
					0.400	1.2E+08	3.2E+03
		Vapor	4.5E+06	1.2E+02			
		Organic compounds					
		F 0.020	2.0E+08	5.4E+03	0.020	8.7E+07	2.4E+03
		M 0.020	7.1E+07	1.9E+03			
^{203}Hg	46.6 d	Inorganic compounds					
		F 0.400	2.7E+07	7.2E+02	1.000	1.1E+07	2.8E+02
					0.400	1.8E+07	4.9E+02
		Vapor	2.9E+06	7.7E+01			
		Organic compounds					
		F 0.020	3.4E+07	9.2E+02	0.020	3.7E+07	1.0E+03
		M 0.020	1.1E+07	2.8E+02			
^{201}Tl	3.04 d	F 1.000	2.6E+08	7.1E+03	1.000	2.1E+08	5.7E+03
^{210}Pb	22.3 y	F 0.200	1.8E+04	4.9E-01	0.200	2.9E+04	7.9E-01
^{207}Bi	38.0 y	F 0.050	2.4E+07	6.4E+02	0.050	1.5E+07	4.2E+02
		M 0.050	6.3E+06	1.7E+02			
^{210}Po	138 d	F 0.100	2.8E+04	7.6E-01	0.100	8.3E+04	2.3E+00
		M 0.100	9.1E+03	2.5E-01			
^{224}Ra	3.66 d	M 0.200	8.3E+03	2.3E-01	0.200	3.1E+05	8.3E+00
^{226}Ra	1600 y	M 0.200	1.7E+03	4.5E-02	0.200	7.1E+04	1.9E+00
^{228}Ra	5.75 y	M 0.200	1.2E+04	3.2E-01	0.200	3.0E+04	8.1E-01
^{228}Th	1.91 y	M 5.0E-04	8.7E+02	2.4E-02	5.0E-04	2.9E+05	7.7E+00
		S 2.0E-04	6.3E+02	1.7E-02	2.0E-04	5.7E+05	1.5E+01
^{230}Th	77000 y	M 5.0E-04	7.1E+02	1.9E-02	5.0E-04	9.5E+04	2.6E+00
		S 2.0E-04	2.8E+03	7.5E-02	2.0E-04	2.3E+05	6.2E+00
^{232}Th	1.40·10^{10} y	M 5.0E-04	6.9E+02	1.9E-02	5.0E-04	9.1E+04	2.5E+00
		S 2.0E-04	1.7E+03	4.5E-02	2.0E-04	2.2E+05	5.9E+00
^{234}U	2.44·10^5 y	F 0.020	3.1E+04	8.4E-01	0.020	4.1E+05	1.1E+01
		M 0.020	9.5E+03	2.6E-01	0.002	2.4E+06	6.5E+01
		S 0.002	2.9E+03	7.9E-02			
^{235}U	7.04·10^8 y	F 0.020	3.3E+04	9.0E-01	0.020	4.3E+05	1.2E+01
		M 0.020	1.1E+04	3.0E-01	0.002	2.4E+06	6.5E+01

		Inhalation intakes				Oral intakes		
	Physical half-life	Chemical form type/f_1	ALI		Chemical form f_1	ALI		
			Bq	μCi		Bq	μCi	
		S 0.002	3.3E+03	8.9E-02				
^{238}U	$4.47 \cdot 10^9$ y	F 0.020	3.4E+04	9.3E-01	0.020	4.5E+05	1.2E+01	
		M 0.020	1.3E+04	3.4E-01	0.002	2.6E+06	7.1E+01	
		S 0.002	3.5E+03	9.5E-02				
^{237}Np	$2.14 \cdot 10^6$ y	M 5.0E-04	1.3E+03	3.6E-02	5.0E-04	1.8E+05	4.9E+00	
^{239}Np	2.36 d	M 5.0E-04	1.8E+07	4.9E+02	5.0E-04	2.5E+07	6.8E+02	
^{238}Pu	87.7 y	M 5.0E-04	6.7E+02	1.8E-02	5.0E-04	8.7E+04	2.4E+00	
		S 1.0E-05	1.8E+03	4.9E-02	1.0E-05	2.3E+06	6.1E+01	
					1.0E-04	4.1E+05	1.1E+01	
^{239}Pu	24100 y	M 5.0E-04	6.3E+02	1.7E-02	5.0E-04	8.0E+04	2.2E+00	
		S 1.0E-05	2.4E+03	6.5E-02	1.0E-05	2.2E+06	6.0E+01	
					1.0E-04	3.8E+05	1.0E+01	
^{241}Pu	14.4 y	M 5.0E-04	3.4E+04	9.3E-01	5.0E-04	4.3E+06	1.2E+02	
		S 1.0E-05	2.4E+05	6.4E+00	1.0E-05	1.8E+08	4.9E+03	
					1.0E-04	2.1E+07	5.6E+02	
^{241}Am	432 y	M 5.0E-04	7.4E+02	2.0E-02	5.0E-04	1.0E+05	2.7E+00	
^{244}Cm	18.1 y	M 5.0E-04	1.2E+03	3.2E-02	5.0E-04	1.7E+05	4.5E+00	
^{252}Cf	2.64 y	M 5.0E-04	1.5E+03	4.2E-02	5.0E-04	2.2E+05	6.0E+00	

CHEMICAL CARCINOGENS

The following substances are listed in the *12th Report on Carcinogens, 2011*, released by the National Institute of Environmental Health Sciences (NIEHS) under the National Toxicology Program (NTP). Substances are grouped in two classes:

- **Known to be human carcinogens:** There is sufficient evidence of cancer from studies in humans showing a cause-and-effect relationship between exposure to the substance and human cancer.

- **Reasonably anticipated to be human carcinogens:** There is limited evidence of cancer from studies in humans or sufficient evidence of cancer from studies in experimental animals showing a cause-and-effect relationship between exposure to the substance and cancer. Alternatively, a substance can be listed in this category if there is evidence that it is a member of a class of substances already listed in the Report on Carcinogens or causes biological effects known to lead to the development of cancer.

The NTP report also lists many poorly defined materials such as soots, tars, mineral oils, and coke oven emissions, as well as viruses, sunlight, ionizing radiation, etc. These carcinogenic agents are not included here.

The table lists the substance names normally used in the *Handbook of Chemistry and Physics*. Explanatory comments and additional names and acronyms by which the substance is known are given in the second column. In many cases the primary name given here is different from that used in the NTP report; however, names used in the NTP report appear in the second column. The Chemical Abstracts Service Registry Number (CAS RN) is given in the last column. Extensive details on each substance are given in the reference.

Reference

National Toxicology Program, *12th Report on Carcinogens*, 2011, <ntp.niehs.nih.gov/go/roc12>.

Substance	Comments & Other Names	CAS RN
Known to be Human Carcinogens		
Aflatoxins		1402-68-2
4-Aminobiphenyl	*p*-Biphenylamine	92-67-1
Aristolochic acids		
Arsenic	and inorganic arsenic compounds	7440-38-2
Asbestos		1332-21-4
Azathioprine	6-[(1-Methyl-4-nitro-1*H*-imidazol-5-yl)thio]-1*H*-purine	446-86-6
Benzene	[6]Annulene	71-43-2
p-Benzidine	includes dyes metabolized to Benzidine; [1,1′-Biphenyl]-4,4′-diamine	92-87-5
Beryllium	and beryllium compounds	7440-41-7
Bis(2-chloroethyl) sulfide	Mustard gas	505-60-2
Bis(chloromethyl) ether	and technical grade Chloromethyl methyl ether	542-88-1
1,3-Butadiene	Divinyl	106-99-0
Cadmium	and cadmium compounds	7440-43-9
Chlorambucil		305-03-3
Chloroethene	Vinyl chloride	75-01-4
1-(2-Chloroethyl)-3-(4-methylcyclohexyl)-1-nitrosourea	Semustine; MeCCNU	13909-09-6
Chloromethyl methyl ether		107-30-2
Chromium	hexavalent compounds only	7440-47-3
Cyclophosphamide	Cyclophosphane	50-18-0
Cyclosporin A	Cyclosporine	59865-13-3
trans-Diethylstilbestrol		56-53-1
Erionite		66733-21-9
Estrogens, steroidal		
N-(4-Ethoxyphenyl)acetamide	in analgesic mixtures; Phenacetin	62-44-2
Formaldehyde	Methanal	50-00-0
Melphalan	*L*-Phenylalanine, 4-[bis(2-chloroethyl)amino]-	148-82-3
Methoxsalen	with UV-A therapy; 9-Methoxy-7*H*-furo[3,2-*g*][1]benzopyran-7-one; PUVA	298-81-7
2-Naphthylamine	β-Naphthylamine	91-59-8
Nickel compounds	both inorganic and organic	
Oxirane	Ethylene oxide	75-21-8
Radon	source of ionizing radiation	10043-92-2
Silicon dioxide (α-quartz)	respirable size; Silica	14808-60-7
Silicon dioxide (tridymite)	respirable size; Tridymite	15468-32-3
Silicon dioxide (cristobalite)	respirable size; Cristobalite	14464-46-1
Sulfuric acid	in strong acid mists	7664-93-9
Tamoxifen		10540-29-1
2,3,7,8-Tetrachlorodibenzo-*p*-dioxin	Dioxin; TCDD	1746-01-6

Substance	Comments & Other Names	CAS RN
Thorium(IV) oxide	source of ionizing radiation; Thoria; Thorium dioxide	1314-20-1
Triethylenethiophosphoramide	Thiotepa	52-24-4

Reasonably Anticipated to be Human Carcinogens

Substance	Comments & Other Names	CAS RN
Acetaldehyde	Ethanal	75-07-0
2-(Acetylamino)fluorene		53-96-3
Acrylamide	2-Propenamide	79-06-1
Acrylonitrile	Propenenitrile	107-13-1
4-Allyl-1,2-dimethoxybenzene	Methyleugenol	93-15-2
2-Amino-9,10-anthracenedione	2-Aminoanthraquinone	117-79-3
1-Amino-2,4-dibromo-9,10-anthracenedione	1-Amino-2,4-dibromoanthraquinone	81-49-2
2-Amino-3,4-dimethylimidazo[4,5-*f*]quinoline	Me-IQ	77094-11-2
2-Amino-3,8-dimethylimidazo[4,5-*f*]quinoxaline	MeIQx	77500-04-0
1-Amino-2-methyl-9,10-anthracenedione	1-Amino-2-methylanthraquinone	82-28-0
2-Amino-3-methyl-3*H*-imidazo(4,5-*f*)quinoline	IQ	76180-96-6
2-Amino-1-methyl-6-phenylimidazo[4,5-*b*]pyridine	PhIP	105650-23-5
Azacitidine	4-Amino-1-β-*D*-ribofuranosyl-1,3,5-triazine-2(1*H*)-one	320-67-2
Benz[*a*]anthracene	1,2-Benzanthracene	56-55-3
Benzo[*b*]fluoranthene	Benz[*e*]acephenanthrylene	205-99-2
Benzo[*j*]fluoranthene	Dibenzo[*a,jk*]fluorene	205-82-3
Benzo[*k*]fluoranthene	2,3,1′,8′-Binaphthylene	207-08-9
Benzo[*a*]pyrene	2,3-Benzopyrene	50-32-8
2,2′-Bioxirane	Diepoxybutane	1464-53-5
2,2-Bis(bromomethyl)-1,3-propanediol	Pentaerythritol dibromide	3296-90-0
Bis(2-chloroethyl)methylamine hydrochloride	Nitrogen mustard hydrochloride; Mechlorethamine	55-86-7
N,N′-Bis(2-chloroethyl)-*N*-nitrosourea	Carmustine; BCNU	154-93-8
Bis[4-(dimethylamino)phenyl]methane	Michler's Base	101-61-1
1,3-Bis(2,3-epoxypropoxy)benzene	Diglycidyl resorcinol ether	101-90-6
Bis(2-ethylhexyl) phthalate	DEHP; Di-*sec*-octyl phthalate	117-81-7
Bromodichloromethane		75-27-4
Bromoethene	Vinyl bromide	593-60-2
1,4-Butanediol dimethylsulfonate	Busulfan	55-98-1
tert-Butyl-4-hydroxyanisole	Butylated hydroxyanisole	25013-16-5
Captafol	Difolatan	2425-06-1
Chloramphenicol		56-75-7
Chlorendic acid	1,4,5,6,7,7-Hexachloro-5-norbornene-2,3-dicarboxylic acid	115-28-6
Chlorinated paraffins (C$_{12}$, 60% Cl)		108171-26-2
4-Chloro-1,2-benzenediamine	4-Chloro-*o*-phenylenediamine	95-83-0
2-Chloro-1,3-butadiene	Chloroprene	126-99-8
1-(2-Chloroethyl)-3-cyclohexyl-1-nitrosourea	CCNU; Lomustine; Belustine	13010-47-4
4-Chloro-2-methylaniline	also the hydrochloride; *p*-Chloro-*o*-toluidine	95-69-2
1-Chloro-2-methylpropene	Dimethylvinyl chloride	513-37-1
3-Chloro-2-methylpropene		563-47-3
Chlorozotocin	2-[[[(2-Chloroethyl)nitrosoamino]carbonyl]amino]-2-deoxy-*D*-glucose	54749-90-5
Cobalt(II) sulfate	Cobaltous sulfate	10124-43-3
Cobalt-tungsten carbide	powders and hard metal	
Cupferron		135-20-6
Dacarbazine	5-(3,3-Dimethyl-1-triazenyl)-1*H*-imidazole-4-carboxamide	4342-03-4
Decabromobiphenyl		13654-09-6
2,4-Diaminoanisole sulfate	1,3-Benzenediamine, 4-methoxy, sulfate	39156-41-7
4,4′-Diaminodiphenyl ether	4,4′-Oxydianiline	101-80-4
4,4′-Diaminodiphenylmethane	also the dihydrochloride; 4,4′-Methylenedianiline	101-77-9
4,4′-Diaminodiphenylmethane dihydrochloride	4,4′-Methylenedianiline dihydrochloride	13552-44-8
4,4′-Diaminodiphenyl sulfide	4,4′-Thiodianiline	139-65-1
cis-Diamminedichloroplatinum	Cisplatin	15663-27-1
Dibenz[*a,h*]acridine		226-36-8
Dibenz[*a,j*]acridine	7-Azadibenz[*a,j*]anthracene	224-42-0
Dibenz[*a,h*]anthracene	1,2:5,6-Dibenzanthracene	53-70-3
7*H*-Dibenzo[*c,g*]carbazole		194-59-2
Dibenzo[*a,e*]pyrene	Naphtho[1,2,3,4-*def*]chrysene	192-65-4
Dibenzo[*a,h*]pyrene	Dibenzo[*b,def*]chrysene	189-64-0

Substance	Comments & Other Names	CAS RN
Dibenzo[a,i]pyrene	Benzo[rst]pentaphene	189-55-9
Dibenzo[a,l]pyrene	Dibenzo[def,p]chrysene	191-30-0
1,2-Dibromo-3-chloropropane		96-12-8
1,2-Dibromoethane	Ethylene dibromide	106-93-4
2,3-Dibromo-1-propanol	DBP	96-13-9
2,3-Dibromo-1-propanol, phosphate (3:1)	Tris(2,3-dibromopropyl) phosphate	126-72-7
p-Dichlorobenzene	1,4-Dichlorobenzene	106-46-7
3,3′-Dichloro-p-benzidine	3,3′-Dichloro[1,1′-biphenyl]-4,4′-diamine	91-94-1
3,3′-Dichloro-p-benzidine dihydrochloride	3,3′-Dichloro-[1,1′-biphenyl]-4,4′-diamine dihydrochloride	612-83-9
1,2-Dichloroethane	Ethylene dichloride	107-06-2
Dichloromethane	Methylene chloride	75-09-2
1,3-Dichloropropene (unspecified isomer)	technical grade	542-75-6
Diethyl sulfate		64-67-5
2,3-Dihydro-6-propyl-2-thioxo-4(1H)-pyrimidinone	6-Propylthiouracil; PROP	51-52-5
1,8-Dihydroxy-9,10-anthracenedione	Danthron	117-10-2
3,3′-Dimethoxybenzidine	and dyes metabolized to 3,3′-Dimethoxybenzidine; Dianisidine	119-90-4
4-(Dimethylamino)azobenzene		60-11-7
2′,3-Dimethyl-4-aminoazobenzene	o-Aminoazotoluene; 4-o-Tolylazo-o-toluidine	97-56-3
Dimethylcarbamic chloride	Dimethylcarbamoyl chloride	79-44-7
1,1-Dimethylhydrazine	UDMH	57-14-7
Dimethyl sulfate		77-78-1
1,6-Dinitropyrene		42397-64-8
1,8-Dinitropyrene		42397-65-9
1,4-Dioxane	1,4-Dioxacyclohexane	123-91-1
1,2-Diphenylhydrazine	Hydrazobenzene	122-66-7
1,3-Diphenyl-1-triazene	Diazoaminobenzene	136-35-6
Disperse Blue No. 1	1,4,5,8-Tetraamino-9,10-anthracenedione	2475-45-8
Doxorubicin hydrochloride	Adriamycin	25316-40-9
Epichlorohydrin	(Chloromethyl)oxirane	13403-37-7
1,2-Epoxy-4-(epoxyethyl)cyclohexane	4-Vinyl-1-cyclohexene dioxide	106-87-6
Ethyl carbamate	Urethane	51-79-6
Ethyl methanesulfonate		62-50-0
N-Ethyl-N-nitrosourea	N-Nitroso-N-ethylurea; ENU	759-73-9
Fluoroethene	Vinyl fluoride	75-02-5
Fuchsin	C.I. Basic Red 9, monohydrochloride	569-61-9
Furan	Oxacyclopentadiene	110-00-9
Glass wool fibers (inhalable)	certain types	
Hexabromobiphenyl (unspecified isomer)	Firemaster FF-1	67774-32-7
Hexachlorobenzene	Perchlorobenzene	118-74-1
1,2,3,4,5,6-Hexachlorocyclohexane, (1α,2α,3β,4α,5α,6β)	Lindane	58-89-9
1,2,3,4,5,6-Hexachlorocyclohexane, (1α,2α,3β,4α,5β,6β)	α-Hexachlorocyclohexane	319-84-6
1,2,3,4,5,6-Hexachlorocyclohexane, (1α,2β,3α,4β,5α,6β)	β-Hexachlorocyclohexane	319-85-7
Hexachlorocyclohexane (unspecified isomer)	all isomers; Lindane	608-73-1
Hexachloroethane	Perchloroethane	67-72-1
Hexamethylphosphoric triamide	HMPA; Tris(dimethylamino)phosphine oxide	680-31-9
Hydrazine		302-01-2
Hydrazine sulfate		10034-93-2
2-Imidazolidinethione	Ethylene thiourea	96-45-7
Indeno[1,2,3-cd]pyrene	1,10-(1,2-Phenylene)pyrene	193-39-5
Kepone	Chlordecone	143-50-0
Lead	and lead compounds	7439-92-1
2-Methoxyaniline	also the hydrochloride; o-Anisidine	90-04-0
2-Methoxy-5-methylaniline	p-Cresidine; 5-Methyl-o-anisidine	120-71-8
2-Methylaniline	also the hydrochloride; o-Toluidine	95-53-4
2-Methylaniline, hydrochloride	o-Toluidine, hydrochloride	636-21-5
4-Methyl-1,3-benzenediamine	Toluene-2,4-diamine; 2,4-Diaminotoluene	95-80-7
2-Methyl-1,3-butadiene	Isoprene	78-79-5
5-Methylchrysene	5-MC	3697-24-3
4,4′-Methylenebis[2-chloroaniline]	MBOCA; 3,3′-Dichloro-4,4′-diaminodiphenylmethane	101-14-4
Methyl methanesulfonate		66-27-3
N-Methyl-N′-nitro-N-nitrosoguanidine		70-25-7

Substance	Comments & Other Names	CAS RN
N-Methyl-N-nitrosourea	N-Nitroso-N-methylurea	684-93-5
Methyloxirane	1,2-Propylene oxide	16033-71-9
Metronidazole	2-Methyl-5-nitro-1H-imidazole-1-ethanol	443-48-1
Mirex	Hexachloropentadiene dimer	2385-85-5
Naphthalene		91-20-3
Nickel	metallic (nickel compounds are known carcinogens)	7440-02-0
Nitrilotriacetic acid	N,N-Bis(carboxymethyl)glycine	139-13-9
2-Nitroanisole	1-Methoxy-2-nitrobenzene	91-23-6
Nitrobenzene		98-95-3
6-Nitrochrysene		7496-02-8
Nitrofen	2,4-Dichloro-1-(4-nitrophenoxy)benzene	1836-75-5
Nitromethane	Nitrocarbol	75-52-5
2-Nitropropane	Isonitropropane	79-46-9
1-Nitropyrene		5522-43-0
4-Nitropyrene		57835-92-4
N-Nitrosodibutylamine	Dibutylnitrosamine	924-16-3
N-Nitrosodiethanolamine	2,2′-(Nitrosoimino)ethanol	1116-54-7
N-Nitrosodiethylamine	DEN; Diethylnitrosamine	55-18-5
N-Nitrosodimethylamine	DMN; Dimethylnitrosamine	62-75-9
4-(N-Nitrosomethylamino)-1-(3-pyridyl)-1-butanone	NNK	64091-91-4
N-Nitroso-N-methylvinylamine	N-Methyl-N-nitrosoethenamine	4549-40-0
4-Nitrosomorpholine	N-Nitrosomorpholine	59-89-2
N-Nitrosonornicotine	N′-Nitroso-3-(2-pyrrolidinyl)pyridine	16543-55-8
N-Nitrosopiperidine	1-Nitrosopiperidine	100-75-4
N-Nitrosodipropylamine	N-Nitroso-N-propyl-1-propanamine	621-64-7
N-Nitrosopyrrolidine		930-55-2
N-Nitrososarcosine	N-Methyl-N-nitrosoglycine	13256-22-9
2-Nitrotoluene	1-Methyl-2-nitrobenzene	88-72-2
Norethisterone	19-Norpregn-4-en-20-yn-3-one, 17-hydroxy-, (17 α)-	68-22-4
Ochratoxin A		303-47-9
Octabromobiphenyl (unspecified isomer)		61288-13-9
2-Oxetanone	β-Propiolactone	57-57-8
Oxiranemethanol, (±)-	Glycidol	61915-27-3
Oxymetholone	Androstan-3-one, 17-hydroxy-2-(hydroxymethylene)-17-methyl-	434-07-1
Phenazopyridine hydrochloride	3-(Phenylazo)-2,6-pyridinediamine, monohydrochloride	136-40-3
Phenolphthalein	3,3-Bis(4-hydroxyphenyl)-1(3H)-isobenzofuranone	77-09-8
Phenoxybenzamine hydrochloride		63-92-3
Phenyloxirane	Styrene-7,8-oxide	96-09-3
Phenytoin	also Phenytoin sodium; Dilantin; 5,5-Diphenyl-2,4-imidazolidinedione	57-41-0
Polybrominated biphenyls	PBBs	
Polychlorinated biphenyls	PCBs	1336-36-3
Procarbazine	also the hydrochloride	671-16-9
Progesterone	Pregn-4-ene-3,20-dione	57-83-0
1,3-Propane sultone	1,2-Oxathiolane, 2,2-dioxide	1120-71-4
Propyleneimine	2-Methylaziridine	75-55-8
Reserpine		50-55-5
Riddelline		23246-96-0
Safrole	5-(2-Propenyl)-1,3-benzodioxole	94-59-7
Selenium monosulfide		7446-34-6
Streptozotocin	D-Glucopyranose, 2-deoxy-2-[[(methylnitrosoamino)carbonyl]amino]-	18883-66-4
Styrene	Vinylbenzene	100-42-5
Sulfallate	N,N-Diethyldithiocarbamic acid, 2-chloroallyl ester	95-06-7
Tetrachloroethene	Perchloroethylene	127-18-4
Tetrachloromethane	Carbon tetrachloride	56-23-5
Tetrafluoroethene	Tetrafluoroethylene	116-14-3
N,N,N′,N′-Tetramethyl-4,4′-diaminobenzophenone	Michler's ketone; 4,4′-Methylenebis[N,N-dimethylaniline]	90-94-8
Tetranitromethane		509-14-8
Thioacetamide	Ethanethioamide	62-55-5
Thiourea	Thiocarbamide	62-56-6
o-Tolidine	and dyes metabolized to o-Tolidine; 3,3′-Dimethylbenzidine	119-93-7
Toluene diisocyanate (unspecified isomer)	includes both 2,4- and 2,6- isomers; TDI	26471-62-5

Substance	Comments & Other Names	CAS RN
Toxaphene	Polychlorocamphene	8001-35-2
1*H*-1,2,4-Triazol-3-amine	Amitrole	61-82-5
1,1,1-Trichloro-2,2-bis(4-chlorophenyl)ethane	DDT; Dichlorodiphenyltrichloroethane	50-29-3
Trichloroethene	Trichloroethylene	79-01-6
Trichloromethane	Chloroform	67-66-3
(Trichloromethyl)benzene	Benzotrichloride	98-07-7
2,4,6-Trichlorophenol		88-06-2
1,2,3-Trichloropropane	Allyl trichloride	96-18-4

Appendix A
Mathematical Tables

MISCELLANEOUS MATHEMATICAL CONSTANTS

π CONSTANTS

$$\pi = 3.14159\ 26535\ 89793\ 23846\ 26433\ 83279\ 50288\ 41971\ 69399\ 37511$$
$$1/\pi = 0.31830\ 98861\ 83790\ 67153\ 77675\ 26745\ 02872\ 40689\ 19291\ 48091$$
$$\pi^2 = 9.86960\ 44010\ 89358\ 61883\ 44909\ 99876\ 15113\ 53136\ 99407\ 24079$$
$$\log_e \pi = 1.14472\ 98858\ 49400\ 17414\ 34273\ 51353\ 05871\ 16472\ 94812\ 91531$$
$$\log_{10} \pi = 0.49714\ 98726\ 94133\ 85435\ 12682\ 88290\ 89887\ 36516\ 78324\ 38044$$
$$\log_{10} \sqrt{2\pi} = 0.39908\ 99341\ 79057\ 52478\ 25035\ 91507\ 69595\ 02099\ 34102\ 92128$$

CONSTANTS INVOLVING e

$$e = 2.71828\ 18284\ 59045\ 23536\ 02874\ 71352\ 66249\ 77572\ 47093\ 69996$$
$$1/e = 0.36787\ 94411\ 71442\ 32159\ 55237\ 70161\ 46086\ 74458\ 11131\ 03177$$
$$e^2 = 7.38905\ 60989\ 30650\ 22723\ 04274\ 60575\ 00781\ 31803\ 15570\ 55185$$
$$M = \log_{10} e = 0.43429\ 44819\ 03251\ 82765\ 11289\ 18916\ 60508\ 22943\ 97005\ 80367$$
$$1/M = \log_e 10 = 2.30258\ 50929\ 94045\ 68401\ 79914\ 54684\ 36420\ 76011\ 01488\ 62877$$
$$\log_{10} M = 9.63778\ 43113\ 00536\ 78912\ 29674\ 98645 - 10$$

π^e AND e^π CONSTANTS

$$\pi^e = 22.45915\ 77183\ 61045\ 47342\ 71522$$
$$e^\pi = 23.14069\ 26327\ 79269\ 00572\ 90864$$
$$e^{-\pi} = 0.04321\ 39182\ 63772\ 24977\ 44177$$
$$e^{\pi/2} = 4.81047\ 73809\ 65351\ 65547\ 30357$$
$$i^i = e^{-\pi/2} = 0.20787\ 95763\ 50761\ 90854\ 69556$$

NUMERICAL CONSTANTS

$$\sqrt{2} = 1.41421\ 35623\ 73095\ 04880\ 16887\ 24209\ 69807\ 85696\ 71875\ 37695$$
$$\sqrt[3]{2} = 1.25992\ 10498\ 94873\ 16476\ 72106\ 07278\ 22835\ 05702\ 51464\ 70151$$
$$\log_e 2 = 0.69314\ 71805\ 59945\ 30941\ 72321\ 21458\ 17656\ 80755\ 00134\ 36026$$
$$\log_{10} 2 = 0.30102\ 99956\ 63981\ 19521\ 37388\ 94724\ 49302\ 67881\ 89881\ 46211$$
$$\sqrt{3} = 1.73205\ 08075\ 68877\ 29352\ 74463\ 41505\ 87236\ 69428\ 05253\ 81039$$
$$\sqrt[3]{3} = 1.44224\ 95703\ 07408\ 38232\ 16383\ 10780\ 10958\ 83918\ 69253\ 49935$$
$$\log_e 3 = 1.09861\ 22886\ 68109\ 69139\ 52452\ 36922\ 52570\ 46474\ 90557\ 82275$$
$$\log_{10} 3 = 0.47712\ 12547\ 19662\ 43729\ 50279\ 03255\ 11530\ 92001\ 28864\ 19070$$

OTHER CONSTANTS

$$\text{Euler's Constant } \gamma = 0.57721\ 56649\ 01532\ 86061$$
$$\log_e \gamma = -0.54953\ 93129\ 81644\ 82234$$
$$\text{Golden Ratio } \phi = 1.61803\ 39887\ 49894\ 84820\ 45868\ 34365\ 63811\ 77203\ 09180$$

DECIMAL EQUIVALENTS OF COMMON FRACTIONS

		1/64	0.015625			33/64	0.515625
	1/32	2/64	0.03125		17/32	34/64	0.53125
		3/64	0.046875			35/64	0.546875
1/16	2/32	4/64	0.0625	9/16	18/32	36/64	0.5625
		5/64	0.078125			37/64	0.578125
	3/32	6/64	0.09375		19/32	38/64	0.59375
		7/64	0.109375			39/64	0.609375
1/8	4/32	8/64	0.125	5/8	20/32	40/64	0.625
		9/64	0.140625			41/64	0.640625
	5/32	10/64	0.15625		21/32	42/64	0.65625
		11/64	0.171875			43/64	0.671875
3/16	6/32	12/64	0.1875	11/16	22/32	44/64	0.6875
		13/64	0.203125			45/64	0.703125
	7/32	14/64	0.21875		23/32	46/64	0.71875
		15/64	0.234375			47/64	0.734375
1/4	8/32	16/64	0.25	3/4	24/32	48/64	0.75
		17/64	0.265625			49/64	0.765625
	9/32	18/64	0.28125		25/32	50/64	0.78125
		19/64	0.296875			51/64	0.796875
5/16	10/32	20/64	0.3125	13/16	26/32	52/64	0.8125
		21/64	0.328125			53/64	0.828125
	11/32	22/64	0.34375		27/32	54/64	0.84375
		23/64	0.359375			55/64	0.859375
3/8	12/32	24/64	0.375	7/8	28/32	56/64	0.875
		25/64	0.390625			57/64	0.890625
	13/32	26/64	0.40625		29/32	58/64	0.90625
		27/64	0.421875			59/64	0.921875
7/16	14/32	28/64	0.4375	15/16	30/32	60/64	0.9375
		29/64	0.453125			61/64	0.953125
	15/32	30/64	0.46875		31/32	62/64	0.96875
		31/64	0.484375			63/64	0.984375
1/2	16/32	32/64	0.5	1/1	32/32	64/64	1

QUADRATIC FORMULA

The solutions of the equation $ax^2 + bx + c = 0$, where $a \neq 0$, are given by:

$$x = \frac{-b \pm \sqrt{b^2 - 4ac}}{2a}.$$

EXPONENTIAL AND HYPERBOLIC FUNCTIONS AND THEIR COMMON LOGARITHMS

	e^x		e^{-x}	$\sinh x$		$\cosh x$		$\tanh x$
x	Value	$\log_{10}$	Value	Value	$\log_{10}$	Value	$\log_{10}$	Value
0.00	1.0000	0.00000	1.00000	0.0000	$-\infty$	1.0000	0.00000	0.00000
0.01	1.0101	.00434	0.99005	.0100	−2.00001	1.0001	.00002	.01000
0.02	1.0202	.00869	.98020	.0200	−2.30106	1.0002	.00009	.02000
0.03	1.0305	.01303	.97045	.0300	−2.47719	1.0005	.00020	.02999
0.04	1.0408	.01737	.96079	.0400	−2.60218	1.0008	.00035	.03998
0.05	1.0513	.02171	.95123	.0500	−2.69915	1.0013	.00054	.04996
0.06	1.0618	.02606	.94176	.0600	−2.77841	1.0018	.00078	.05993
0.07	1.0725	.03040	.93239	.0701	−2.84545	1.0025	.00106	.06989
0.08	1.0833	.03474	.92312	.0801	−2.90355	1.0032	.00139	.07983
0.09	1.0942	.03909	.91393	.0901	−2.95483	1.0041	.00176	.08976
0.10	1.1052	.04343	.90484	.1002	−1.00072	1.0050	.00217	.09967
0.11	1.1163	.04777	.89583	.1102	−1.04227	1.0061	.00262	.10956
0.12	1.1275	.05212	.88692	.1203	−1.08022	1.0072	.00312	.11943
0.13	1.1388	.05646	.87809	.1304	−1.11517	1.0085	.00366	.12927
0.14	1.1503	.06080	.86936	.1405	−1.14755	1.0098	.00424	.13909
0.15	1.1618	.06514	.86071	.1506	−1.17772	1.0113	.00487	.14889
0.16	1.1735	.06949	.85214	.1607	−1.20597	1.0128	.00554	.15865
0.17	1.1853	.07383	.84366	.1708	−1.23254	1.0145	.00625	.16838
0.18	1.1972	.07817	.83527	.1810	−1.25762	1.0162	.00700	.17808
0.19	1.2092	.08252	.82696	.1911	−1.28136	1.0181	.00779	.18775
0.20	1.2214	.08686	.81873	.2013	−1.30392	1.0201	.00863	.19738
0.21	1.2337	.09120	.81058	.2115	−1.32541	1.0221	.00951	.20697
0.22	1.2461	.09554	.80252	.2218	−1.34592	1.0243	.01043	.21652
0.23	1.2586	.09989	.79453	.2320	−1.36555	1.0266	.01139	.22603
0.24	1.2712	.10423	.78663	.2423	−1.38437	1.0289	.01239	.23550
0.25	1.2840	.10857	.77880	.2526	−1.40245	1.0314	.01343	.24492
0.26	1.2969	.11292	.77105	.2629	−1.41986	1.0340	.01452	.25430
0.27	1.3100	.11726	.76338	.2733	−1.43663	1.0367	.01564	.26362
0.28	1.3231	.12160	.75578	.2837	−1.45282	1.0395	.01681	.27291
0.29	1.3364	.12595	.74826	.2941	−1.46847	1.0423	.01801	.28213
0.30	1.3499	.13029	.74082	.3045	−1.48362	1.0453	.01926	.29131
0.31	1.3634	.13463	.73345	.3150	−1.49830	1.0484	.02054	.30044
0.32	1.3771	.13897	.72615	.3255	−1.51254	1.0516	.02187	.30951
0.33	1.3910	.14332	.71892	.3360	−1.52637	1.0549	.02323	.31852
0.34	1.4049	.14766	.71177	.3466	−1.53981	1.0584	.02463	.32748
0.35	1.4191	.15200	.70469	.3572	−1.55290	1.0619	.02607	.33638
0.36	1.4333	.15635	.69768	.3678	−1.56564	1.0655	.02755	.34521
0.37	1.4477	.16069	.69073	.3785	−1.57807	1.0692	.02907	.35399
0.38	1.4623	.16503	.68386	.3892	−1.59019	1.0731	.03063	.36271
0.39	1.4770	.16937	.67706	.4000	−1.60202	1.0770	.03222	.37136
0.40	1.4918	.17372	.67032	.4108	−1.61358	1.0811	.03385	.37995
0.41	1.5063	.17806	.66365	.4216	−1.62488	1.0852	.03552	.33847
0.42	1.5220	.18240	.65705	.4325	−1.63594	1.0895	.03723	.39693
0.43	1.5373	.18675	.65051	.4434	−1.64677	1.0939	.03897	.40532
0.44	1.5527	.19109	.64404	.4543	−1.65738	1.0984	.04075	.41364
0.45	1.5683	.19543	.63763	.4653	−1.66777	1.1030	.04256	.42190
0.46	1.5841	.19978	.63128	.4764	−1.67797	1.1077	.04441	.43008
0.47	1.6000	.20412	.62500	.4875	−1.68797	1.1125	.04630	.43820
0.48	1.6161	.20846	.61878	.4986	−1.69779	1.1174	.04822	.44624
0.49	1.6323	.21280	.61263	.5098	−1.70744	1.1225	.05018	.45422
0.50	1.6487	.21715	.60653	.5211	−1.71692	1.1276	.05217	.46212
0.51	1.6653	.22149	.60050	.5324	−1.72624	1.1329	.05419	.46995
0.52	1.6820	.22583	.59452	.5438	−1.73540	1.1383	.05625	.47770
0.53	1.6989	.23018	.58860	.5552	−1.74442	1.1438	.05834	.48538

x	e^x Value	e^x $\log_{10}$	e^{-x} Value	$\sinh x$ Value	$\sinh x$ $\log_{10}$	$\cosh x$ Value	$\cosh x$ $\log_{10}$	$\tanh x$ Value
0.54	1.7160	.23452	.58275	.5666	−1.75330	1.1494	.06046	.49299
0.55	1.7333	.23886	.57695	.5782	−1.76204	1.1551	.06262	.50052
0.56	1.7507	.24320	.57121	.5897	−1.77065	1.1609	.06481	.50798
0.57	1.7683	.24755	.56553	.6014	−1.77914	1.1669	.06703	.51536
0.58	1.7860	.25189	.55990	.6131	−1.78751	1.1730	.06929	.52267
0.59	1.8040	.25623	.55433	.6248	−1.79576	1.1792	.07157	.52990
0.60	1.8221	.26058	.54881	.6367	−1.80390	1.1855	.07389	.53705
0.61	1.8404	.26492	.54335	.6485	−1.81194	1.1919	.07624	.54413
0.62	1.8589	.26926	.53794	.6605	−1.81987	1.1984	.07861	.55113
0.63	1.8776	.27361	.53259	.6725	−1.82770	1.2051	.08102	.55805
0.64	1.8965	.27795	.52729	.6846	−1.83543	1.2119	.08346	.56490
0.65	1.9155	.28229	.52205	.6967	−1.84308	1.2188	.08593	.57167
0.66	1.9348	.28664	.51685	.7090	−1.85063	1.2258	.08843	.57836
0.67	1.9542	.29098	.51171	.7213	−1.85809	1.2330	.09095	.58498
0.68	1.9739	.29532	.50662	.7336	−1.86548	1.2402	.09351	.59152
0.69	1.9937	.29966	.50158	.7461	−1.87278	1.2476	.09609	.59798
0.70	2.0138	.30401	.49659	.7586	−1.88000	1.2552	.09870	.60437
0.71	2.0340	.30835	.49164	.7712	−1.88715	1.2628	.10134	.61068
0.72	2.0544	.31269	.48675	.7838	−1.89423	1.2706	.10401	.61691
0.73	2.0751	.31703	.48191	.7966	−1.90123	1.2785	.10670	.62307
0.74	2.0959	.32138	.47711	.8094	−1.90817	1.2865	.10942	.62915
0.75	2.1170	.32572	.47237	.8223	−1.91504	1.2947	.11216	.63515
0.76	2.1383	.33006	.46767	.8353	−1.92185	1.3030	.11493	.64108
0.77	2.1598	.33441	.46301	.8484	−1.92859	1.3114	.11773	.64693
0.78	2.1815	.33875	.45841	.8615	−1.93527	1.3199	.12055	.65721
0.79	2.2034	.34309	.45384	.8748	−1.94190	1.3286	.12340	.65841
0.80	2.2255	.34744	.44933	.8881	−1.94846	1.3374	.12627	.66404
0.81	2.2479	.35178	.44486	.9015	−1.95498	1.3464	.12917	.66959
0.82	2.2705	.35612	.44043	.9150	−1.96144	1.3555	.13209	.67507
0.83	2.2933	.36046	.43605	.9286	−1.96784	1.3647	.13503	.68048
0.84	2.3164	.36481	.43171	.9423	−1.97420	1.3740	.13800	.68581
0.85	2.3396	.36915	.42741	.9561	−1.98051	1.3835	.14099	.69107
0.86	2.3632	.37349	.42316	.9700	−1.98677	1.3932	.14400	.69626
0.87	2.3869	.37784	.41895	.9840	−1.99299	1.4029	.14704	.70137
0.88	2.4100	.38218	.41478	.9981	−1.99916	1.4128	.15009	.70642
0.89	2.4351	.38652	.41066	1.0122	0.00528	1.4229	.15317	.71139
0.90	2.4596	.39087	.40657	1.0265	.01137	1.4331	.15627	.21630
0.91	2.4843	.39521	.40242	1.0409	.01741	1.4434	.15939	.72113
0.92	2.5093	.39955	.39852	1.0554	.02341	1.4539	.16254	.72590
0.93	2.5345	.40389	.39455	1.0700	.02937	1.4645	.16570	.73059
0.94	2.5600	.40824	.39063	1.0847	.03530	1.4753	.16888	.73522
0.95	2.5857	.41258	.38674	1.0995	.04119	1.4862	.17208	.73978
0.96	2.6117	.41692	.38289	1.1144	.04704	1.4973	.17531	.74428
0.97	2.6379	.42127	.37908	1.1294	.05286	1.5085	.17855	.74870
0.98	2.6645	.42561	.37531	1.1446	.05864	1.5199	.18181	.75307
0.99	2.6912	.42995	.37158	1.1598	.06439	1.5314	.18509	.75736
1.00	2.7183	.43429	.36788	1.1752	.07011	1.5431	.18839	.76159
1.10	3.0042	.47772	.33287	1.3356	.12569	1.6685	.22233	.80050
1.20	3.3201	.52115	.30119	1.5095	.17882	1.8107	.25784	.83365
1.30	3.6693	.56458	.27253	1.6984	.23004	1.9709	.29467	.86172
1.40	4.0552	.60801	.24660	1.9043	.27974	2.1509	.33262	.88535
1.50	4.4817	.65144	.22313	2.1293	.32823	2.3524	.37151	.90515
1.60	4.9530	.69487	.20190	2.3756	.37577	2.5775	.41119	.92167
1.70	5.4739	.73830	.18268	2.6456	.42253	2.8283	.45153	.93541
1.80	6.0496	.78173	.16530	2.9422	.46867	3.1075	.49241	.94681
1.90	6.6859	.82516	.14957	3.2682	.51430	3.4177	.53374	.95624
2.00	7.3891	.86859	.13534	3.6269	.55953	3.7622	.57544	.96403

	e^x		e^{-x}	$\sinh x$		$\cosh x$		$\tanh x$
x	Value	$\log_{10}$	Value	Value	$\log_{10}$	Value	$\log_{10}$	Value
2.10	8.1662	.91202	.12246	4.0219	.60443	4.1443	.61745	.97045
2.20	9.0250	.95545	.11080	4.4571	.64905	4.5679	.65972	.97574
2.30	9.9742	.99888	.10026	4.9370	.69346	5.0372	.70219	.98010
2.40	11.023	1.04231	.09072	5.4662	.73769	5.5569	.74484	.98367
2.50	12.182	1.08574	.08208	6.0502	.78177	6.1323	.78762	.98661
2.60	13.464	1.12917	.07427	6.6947	.82573	6.7690	.83052	.98903
2.70	14.880	1.17260	.06721	7.4063	.86960	7.4735	.87352	.99101
2.80	16.445	1.21602	.06081	8.1919	.91339	8.2527	.91660	.99263
2.90	18.174	1.25945	.05502	9.0596	.95711	9.1146	.95974	.99396
3.00	20.086	1.30288	.04979	10.018	1.00078	10.068	1.00293	0.99505
3.50	33.115	1.52003	.03020	16.543	1.21860	16.573	1.21940	0.99818
4.00	54.598	1.73718	.01832	27.290	1.43600	27.308	1.43629	0.99933
4.50	90.017	1.95433	.01111	45.003	1.65324	45.014	1.65335	0.99975
5.00	148.41	2.17147	.00674	74.203	1.87042	74.210	1.87046	0.99991
5.50	244.69	2.38862	.00409	122.34	2.08758	122.35	2.08760	0.99997
6.00	403.43	2.60577	.00248	201.71	2.30473	201.72	2.30474	0.99999
6.50	665.14	2.82291	.00150	332.57	2.52188	332.57	2.52189	1.00000
7.00	1096.6	3.04006	.00091	548.32	2.73904	548.32	2.73903	1.00000
7.50	1808.0	3.25721	.00055	904.02	2.95618	904.02	2.95618	1.00000
8.00	2981.0	3.47436	.00034	1490.5	3.17333	1490.5	3.17333	1.00000
8.50	4914.8	3.69150	.00020	2457.4	3.39047	2457.4	3.39047	1.00000
9.00	8103.1	3.90865	.00012	4051.5	3.60762	4051.5	3.60762	1.00000
9.50	13360.	4.12580	.00007	6679.9	3.82477	6679.9	3.82477	1.00000
10.00	22026.	4.34294	.00005	11013.	4.04191	11013.	4.04191	1.00000

NATURAL TRIGONOMETRIC FUNCTIONS TO FOUR PLACES

x radians	x degrees	$\sin x$	$\cos x$	$\tan x$	$\cot x$	$\sec x$	$\csc x$		
.0000	0° 00′	.000	1.0000	.0000	–	1.000	–	90° 00′	1.5708
.0029	10	.0029	1.0000	.0029	343.8	1.000	343.8	50	1.5679
.0058	20	.0058	1.0000	.0058	171.9	1.000	171.9	40	1.5650
.0087	30	.0087	1.0000	.0087	114.6	1.000	114.6	30	1.5621
.0116	40	.0116	.9999	.0116	85.94	1.000	85.95	20	1.5592
.0145	50	.0145	.9999	.0145	68.75	1.000	68.76	10	1.5563
.0175	1° 00′	.0175	.9998	.0175	57.29	1.000	57.30	89° 00′	1.5533
.0262	30	.0262	.9997	.0262	38.19	1.000	38.20	30	1.5446
.0349	2° 00′	.0349	.9994	.0349	28.64	1.001	28.65	88° 00′	1.5359
.0436	30	.0436	.9990	.0437	22.90	1.001	22.93	30	1.5272
.0524	3° 00′	.0523	.9986	.0524	19.08	1.001	19.11	87° 00′	1.5184
.0611	30	.0610	.9981	.0612	16.35	1.002	16.38	30	1.5097
.0698	4° 00′	.0698	.9976	.0699	14.30	1.002	14.34	86° 00′	1.5010
.0785	30	.0785	.9969	.0787	12.71	1.003	12.75	30	1.4923
.0873	5° 00′	.0872	.9962	.0875	11.43	1.004	11.47	85° 00′	1.4835
.0960	30	.0958	.9954	.0963	10.39	1.005	10.43	30	1.4748
.1047	6° 00′	.1045	.9945	.1051	9.514	1.006	9.597	84° 00′	1.4661
.1134	30	.1132	.9936	.1139	8.777	1.006	8.834	30	1.4573
.1222	7° 00′	.1219	.9925	.1228	8.144	1.008	8.206	83° 00′	1.4486
.1309	30	.1305	.9914	.1317	7.596	1.009	7.661	30	1.4399
.1396	8° 00′	.1392	.9903	.1405	7.115	1.010	7.185	82° 00′	1.4312
.1484	30	.1478	.9890	.1495	6.691	1.011	6.765	30	1.4224
.1571	9° 00′	.1564	.9877	.1584	6.314	1.012	6.392	81° 00′	1.4137
.1658	30	.1650	.9863	.1673	5.976	1.014	6.059	30	1.4050
.1745	10° 00′	.1736	.9848	.1763	5.671	1.015	5.759	80° 00′	1.3963
.1833	30	.1822	.9833	.1853	5.396	1.017	5.487	30	1.3875
.1920	11° 00′	.1908	.9816	.1944	5.145	1.019	5.241	79° 00′	1.3788
.2007	30	.1994	.9799	.2035	4.915	1.020	5.016	30	1.3701
.2094	12° 00′	.2079	.9781	.2126	4.705	1.022	4.810	78° 00′	1.3614
.2182	30	.2164	.9763	.2217	4.511	1.025	4.620	30	1.3526
.2269	13° 00′	.2250	.9744	.2309	4.331	1.026	4.445	77° 00′	1.3439
.2356	30	.2334	.9724	.2401	4.165	1.028	4.284	30	1.3352
.2443	14° 00′	.2419	.9703	.2493	4.011	1.031	4.134	76° 00′	1.3265
.2531	30	.2404	.9681	.2586	3.867	1.033	3.994	30	1.3177
.2618	15° 00′	.2588	.9659	.2679	3.732	1.035	3.864	75° 00′	1.3090
.2705	30	.2672	.9636	.2773	3.606	1.038	3.742	30	1.3003
.2793	16° 00′	.2756	.9613	.2867	3.487	1.040	3.628	74° 00′	1.2915
.2880	30	.2840	.9588	.2962	3.376	1.043	3.521	30	1.2828
.2967	17° 00′	.2924	.9563	.3057	3.271	1.046	3.420	73° 00′	1.2741
.3054	30	.3007	.9537	.3153	3.172	1.049	3.326	30	1.2654
.3142	18° 00′	.3090	.9511	.3249	3.078	1.051	3.236	72° 00′	1.2566
.3229	30	.3173	.9483	.3346	2.989	1.054	3.152	30	1.2479
.3316	19° 00′	.3256	.9455	.3443	2.904	1.058	3.072	71° 00′	1.2392
.3403	30	.3338	.9426	.3541	2.824	1.061	2.996	30	1.2305
.3491	20° 00′	.3420	.9397	.3640	2.747	1.064	2.924	70° 00′	1.2217
.3578	30	.3502	.9367	.3739	2.675	1.068	2.855	30	1.2130
.3665	21° 00′	.3584	.9336	.3839	2.605	1.071	2.790	69° 00′	1.2043
.3752	30	.3665	.9304	.3939	2.539	1.075	2.729	30	1.1956
.3840	22° 00′	.3746	.9272	.4040	2.475	1.079	2.669	68° 00′	1.1868
.3927	30	.3827	.9239	.4142	2.414	1.082	2.613	30	1.1781
.4014	23° 00′	.3907	.9205	.4245	2.356	1.086	2.559	67° 00′	1.1694
.4102	30	.3987	.9171	.4348	2.300	1.090	2.508	30	1.1606
.4189	24° 00′	.4067	.9135	.4452	2.246	1.095	2.459	66° 00′	1.1519
		$\cos y$	$\sin y$	$\cot y$	$\tan y$	$\csc y$	$\sec y$	y degrees	y radians

x radians	x degrees	$\sin x$	$\cos x$	$\tan x$	$\cot x$	$\sec x$	$\csc x$		
.4276	30	.4147	.9100	.4557	2.194	1.099	2.411	30	1.1432
.4363	25° 00′	.4226	.9063	.4663	2.145	1.103	2.366	65° 00′	1.1345
.4451	30	.4305	.9026	.4770	2.097	1.108	2.323	30	1.1257
.4538	26° 00′	.4384	.8988	.4877	2.050	1.113	2.281	64° 00′	1.1170
.4625	30	.4462	.8949	.4986	2.006	1.117	2.241	30	1.1083
.4712	27° 00′	.4540	.8910	.5095	1.963	1.122	2.203	63° 00′	1.0996
.4800	30	.4617	.8870	.5206	1.921	1.127	2.166	30	1.0908
.4887	28° 00′	.4695	.8829	.5317	1.881	1.133	2.130	62° 00′	1.0821
.4974	30	.4772	.8788	.5430	1.842	1.138	2.096	30	1.0734
.5061	29° 00′	.4848	.8746	.5543	1.804	1.143	2.063	61° 00′	1.0647
.5149	30	.4924	.8704	.5658	1.767	1.149	2.031	30	1.0559
.5236	30° 00′	.5000	.8660	.5774	1.732	1.155	2.000	60° 00′	1.0472
.5323	30	.5075	.8616	.5890	1.698	1.161	1.970	30	1.0385
.5411	31° 00′	.5150	.8572	.6009	1.664	1.167	1.942	59° 00′	1.0297
.5498	30	.5225	.8526	.6128	1.632	1.173	1.914	30	1.0210
.5585	32° 00′	.5299	.8480	.6249	1.600	1.179	1.887	58° 00′	1.0123
.5672	30	.5373	.8434	.6371	1.570	1.186	1.861	30	1.0036
.5760	33° 00′	.5446	.8397	.6494	1.540	1.192	1.836	57° 00′	.9948
.5847	30	.5519	.8339	.6619	1.511	1.199	1.812	30	.9861
.5934	34° 00′	.5592	.8290	.6745	1.483	1.206	1.788	56° 00′	.9774
.6021	30	.5664	.8241	.6873	1.455	1.213	1.766	30	.9687
.6109	35° 00′	.5736	.8192	.7002	1.428	1.221	1.743	55° 00′	.9599
.6196	30	.5807	.8141	.7133	1.402	1.228	1.722	30	.9512
.6283	36° 00′	.5878	.8090	.7265	1.376	1.236	1.701	54° 00′	.9425
.6370	30	.5948	.8039	.7400	1.351	1.244	1.681	30	.9338
.6458	37° 00′	.6018	.7986	.7536	1.327	1.252	1.662	53° 00′	.9250
.6545	30	.6088	.7934	.7673	1.303	1.260	1.643	30	.9163
.6632	38° 00′	.6157	.7880	.7813	1.280	1.269	1.624	52° 00′	.9076
.6720	30	.6225	.7826	.7954	1.257	1.278	1.606	30	.8988
.6807	39° 00′	.6293	.7771	.8098	1.235	1.287	1.589	51° 00′	.8901
.6894	30	.6361	.7716	.8243	1.213	1.296	1.572	30	.8814
.6981	40° 00′	.6428	.7660	.8391	1.192	1.305	1.556	50° 00′	.8727
.7069	30	.6494	.7604	.8541	1.171	1.315	1.540	30	.8639
.7156	41° 00′	.6561	.7547	.8693	1.150	1.325	1.524	49° 00′	.8552
.7243	30	.6626	.7490	.8847	1.130	1.335	1.509	30	.8465
.7330	42° 00′	.6691	.7431	.9004	1.111	1.346	1.494	48° 00′	.8378
.7418	30	.6756	.7373	.9163	1.091	1.356	1.480	30	.8290
.7505	43° 00′	.6820	.7314	.9325	1.072	1.367	1.466	47° 00′	.8203
.7592	30	.6884	.7254	.9490	1.054	1.379	1.453	30	.8116
.7679	44° 00′	.6947	.7193	.9657	1.036	1.390	1.440	46° 00′	.8029
.7767	30	.7009	.7133	.9827	1.018	1.402	1.427	30	.7941
.7854	45° 00′	.7071	.7071	1.0000	1.0000	1.414	1.414	45° 00′	.7854
		$\cos y$	$\sin y$	$\cot y$	$\tan y$	$\csc y$	$\sec y$	y degrees	y radians

RELATION OF ANGULAR FUNCTIONS IN TERMS OF ONE ANOTHER

Trigonometric Functions

Function	$\sin\alpha$	$\cos\alpha$	$\tan\alpha$	$\cot\alpha$	$\sec\alpha$	$\csc\alpha$
$\sin\alpha$	$\sin\alpha$	$\pm\sqrt{1-\cos^2\alpha}$	$\dfrac{\tan\alpha}{\pm\sqrt{1+\tan^2\alpha}}$	$\dfrac{1}{\pm\sqrt{1+\cot^2\alpha}}$	$\dfrac{\pm\sqrt{\sec^2\alpha-1}}{\sec\alpha}$	$\dfrac{1}{\csc\alpha}$
$\cos\alpha$	$\pm\sqrt{1-\sin^2\alpha}$	$\cos\alpha$	$\dfrac{1}{\pm\sqrt{1+\tan^2\alpha}}$	$\dfrac{\cot\alpha}{\pm\sqrt{1+\cot^2\alpha}}$	$\dfrac{1}{\sec\alpha}$	$\dfrac{\pm\sqrt{\csc^2\alpha-1}}{\csc\alpha}$
$\tan\alpha$	$\dfrac{\sin\alpha}{\pm\sqrt{1-\sin^2\alpha}}$	$\dfrac{\pm\sqrt{1-\cos^2\alpha}}{\cos\alpha}$	$\tan\alpha$	$\dfrac{1}{\cot\alpha}$	$\pm\sqrt{\sec^2\alpha-1}$	$\dfrac{1}{\pm\sqrt{\csc^2\alpha-1}}$
$\cot\alpha$	$\dfrac{\pm\sqrt{1-\sin^2\alpha}}{\sin\alpha}$	$\dfrac{\cos\alpha}{\pm\sqrt{1-\cos^2\alpha}}$	$\dfrac{1}{\tan\alpha}$	$\cot\alpha$	$\dfrac{1}{\pm\sqrt{\sec^2\alpha-1}}$	$\pm\sqrt{\csc^2\alpha-1}$
$\sec\alpha$	$\dfrac{1}{\pm\sqrt{1-\sin^2\alpha}}$	$\dfrac{1}{\cos\alpha}$	$\pm\sqrt{1+\tan^2\alpha}$	$\dfrac{\pm\sqrt{1+\cot^2\alpha}}{\cot\alpha}$	$\sec\alpha$	$\dfrac{\csc\alpha}{\pm\sqrt{\csc^2\alpha-1}}$
$\csc\alpha$	$\dfrac{1}{\sin\alpha}$	$\dfrac{1}{\pm\sqrt{1-\cos^2\alpha}}$	$\dfrac{\pm\sqrt{1+\tan^2\alpha}}{\tan\alpha}$	$\pm\sqrt{1+\cot^2\alpha}$	$\dfrac{\sec\alpha}{\pm\sqrt{\sec^2\alpha-1}}$	$\csc\alpha$

Note: The choice of sign depends upon the quadrant in which the angle terminates.

Hyperbolic Functions

Function	$\sinh x$	$\cosh x$	$\tanh x$
$\sinh x =$	$\sinh x$	$\pm\sqrt{\cosh^2 x-1}$	$\dfrac{\tanh x}{\sqrt{1-\tanh^2 x}}$
$\cosh x =$	$\sqrt{1+\sinh^2 x}$	$\cosh x$	$\dfrac{1}{\sqrt{1-\tanh^2 x}}$
$\tanh x =$	$\dfrac{\sinh x}{\sqrt{1+\sinh^2 x}}$	$\dfrac{\pm\sqrt{\cosh^2 x-1}}{\cosh x}$	$\tanh x$
$\operatorname{cosech} x =$	$\dfrac{1}{\sinh x}$	$\pm\dfrac{1}{\sqrt{\cosh^2 x-1}}$	$\dfrac{\sqrt{1-\tanh^2 x}}{\tanh x}$
$\operatorname{sech} x =$	$\dfrac{1}{\sqrt{1+\sinh^2 x}}$	$\dfrac{1}{\cosh x}$	$\sqrt{1-\tanh^2 x}$
$\coth x =$	$\dfrac{\sqrt{1+\sinh^2 x}}{\sinh x}$	$\dfrac{\pm\cosh x}{\sqrt{\cosh^2 x-1}}$	$\dfrac{1}{\tanh x}$

Function	$\operatorname{cosech} x$	$\operatorname{sech} x$	$\coth x$
$\sinh x =$	$\dfrac{1}{\operatorname{cosech} x}$	$\pm\dfrac{\sqrt{1-\operatorname{sech}^2 x}}{\operatorname{sech} x}$	$\dfrac{\pm 1}{\sqrt{\coth^2 x-1}}$
$\cosh x =$	$\pm\dfrac{\sqrt{\operatorname{cosech}^2 x+1}}{\operatorname{cosech} x}$	$\dfrac{1}{\operatorname{sech} x}$	$\pm\dfrac{\coth x}{\sqrt{\coth^2 x-1}}$
$\tanh x =$	$\dfrac{1}{\sqrt{\operatorname{cosech}^2 x+1}}$	$\pm\sqrt{1+\operatorname{sech}^2 x}$	$\dfrac{1}{\coth x}$
$\operatorname{cosech} x =$	$\operatorname{cosech} x$	$\pm\dfrac{\operatorname{sech} x}{\sqrt{1-\operatorname{sech}^2 x}}$	$\pm\dfrac{\sqrt{\coth^2 x-1}}{1}$
$\operatorname{sech} x =$	$\pm\dfrac{\operatorname{cosech} x}{\sqrt{\operatorname{cosech}^2 x+1}}$	$\operatorname{sech} x$	$\pm\dfrac{\sqrt{\coth^2 x-1}}{\coth x}$
$\coth x =$	$\sqrt{\operatorname{cosech}^2 x+1}$	$\pm\dfrac{1}{\sqrt{1-\operatorname{sech}^2 x}}$	$\coth x$

Whenever two signs are shown, choose $+$ sign if x is positive, $-$ sign if x is negative.

DERIVATIVES

In the following formulas u, v, w represent functions of x, while a, c, n represent fixed real numbers. All arguments in the trigonometric functions are measured in radians, and all inverse trigonometric and hyperbolic functions represent principal values. *Let $y = f(x)$ and $\frac{dy}{dx} = \frac{d[f(x)]}{dx} = f'(x)$ define, respectively, a function and its derivative for any value x in their common domain. The differential for the function at such a value x is accordingly defined as

$$dy = d[f(x)] = \frac{dy}{dx}dx = \frac{d[f(x)]}{dx}dx = f'(x)\,dx$$

Each derivative formula has an associated differential formula. For example, formula 6 below has the differential formula

$$d(uvw) = uv\,dw + vw\,du + uw\,dv$$

1. $\dfrac{d}{dx}(a) = 0$

2. $\dfrac{d}{dx}(x) = 1$

3. $\dfrac{d}{dx}(au) = a\dfrac{du}{dx}$

4. $\dfrac{d}{dx}(u + v - w) = \dfrac{du}{dx} + \dfrac{dv}{dx} - \dfrac{dw}{dx}$

5. $\dfrac{d}{dx}(uv) = u\dfrac{dv}{dx} + v\dfrac{du}{dx}$

6. $\dfrac{d}{dx}(uvw) = uv\dfrac{dw}{dx} + vw\dfrac{du}{dx} + uw\dfrac{dv}{dx}$

7. $\dfrac{d}{dx}\left(\dfrac{u}{v}\right) = \dfrac{v\frac{du}{dx} - u\frac{dv}{dx}}{v^2} = \dfrac{1}{v}\dfrac{du}{dx} - \dfrac{u}{v^2}\dfrac{dv}{dx}$

8. $\dfrac{d}{dx}(u^n) = nu^{n-1}\dfrac{du}{dx}$

9. $\dfrac{d}{dx}(\sqrt{u}) = \dfrac{1}{2\sqrt{u}}\dfrac{du}{dx}$

10. $\dfrac{d}{dx}\left(\dfrac{1}{u}\right) = -\dfrac{1}{u^2}\dfrac{du}{dx}$

11. $\dfrac{d}{dx}\left(\dfrac{1}{u^n}\right) = -\dfrac{n}{u^{n+1}}\dfrac{du}{dx}$

12. $\dfrac{d}{dx}\left(\dfrac{u^n}{v^m}\right) = \dfrac{u^{n-1}}{v^{m+1}}\left(nv\dfrac{du}{dx} - mu\dfrac{dv}{dx}\right)$

13. $\dfrac{d}{dx}(u^n v^m) = u^{n-1}v^{m-1}\left(nv\dfrac{du}{dx} + mu\dfrac{dv}{dx}\right)$

14. $\dfrac{d}{dx}[f(u)] = \dfrac{d}{du}[f(u)] \cdot \dfrac{du}{dx}$

15. $\dfrac{d^2}{dx^2}[f(u)] = \dfrac{df(u)}{du} \cdot \dfrac{d^2u}{dx^2} + \dfrac{d^2f(u)}{du^2} \cdot \left(\dfrac{du}{dx}\right)^2$

16. $\dfrac{d^n}{dx^n}[uv] = \binom{n}{0}v\dfrac{d^n u}{dx^n} + \binom{n}{1}\dfrac{dv}{dx}\dfrac{d^{n-1}u}{dx^{n-1}} + \binom{n}{2}\dfrac{d^2 v}{dx^2}\dfrac{d^{n-2}u}{dx^{n-2}}$

    $+ \cdots + \binom{n}{k}\dfrac{d^k v}{dx^k}\dfrac{d^{n-k}u}{dx^{n-k}} + \cdots + \binom{n}{n}u\dfrac{d^n v}{dx^n}$

    where $\binom{n}{r} = \frac{n!}{r!(n-r)!}$ is the binomial coefficient, n non-negative integer, and $\binom{n}{0} = 1$.

17. $\dfrac{du}{dx} = \dfrac{1}{\frac{dx}{du}}$ if $\dfrac{dx}{du} \neq 0$

18. $\dfrac{d}{dx}(\log_a u) = (\log_a e)\dfrac{1}{u}\dfrac{du}{dx}$

19. $\dfrac{d}{dx}(\log_e u) = \dfrac{1}{u}\dfrac{du}{dx}$

20. $\dfrac{d}{dx}(a^u) = a^u(\log_e a)\dfrac{du}{dx}$

21. $\dfrac{d}{dx}(e^u) = e^u\dfrac{du}{dx}$

22. $\dfrac{d}{dx}(u^v) = vu^{v-1}\dfrac{du}{dx} + (\log_e u)\,u^v\dfrac{dv}{dx}$

23. $\dfrac{d}{dx}(\sin u) = (\cos u)\dfrac{du}{dx}$

24. $\dfrac{d}{dx}(\cos u) = -(\sin u)\dfrac{du}{dx}$

25. $\dfrac{d}{dx}(\tan u) = (\sec^2 u)\dfrac{du}{dx}$

26. $\dfrac{d}{dx}(\cot u) = -(\csc^2 u)\dfrac{du}{dx}$

27. $\dfrac{d}{dx}(\sec u) = \sec u \cdot \tan u\dfrac{du}{dx}$

28. $\dfrac{d}{dx}(\csc u) = -\csc u \cdot \cot u\dfrac{du}{dx}$

29. $\dfrac{d}{dx}(\operatorname{vers} u) = \sin u\dfrac{du}{dx}$

30. $\dfrac{d}{dx}(\arcsin u) = \dfrac{1}{\sqrt{1 - u^2}}\dfrac{du}{dx}, \quad \left(-\dfrac{\pi}{2} \le \arcsin u \le \dfrac{\pi}{2}\right)$

31. $\dfrac{d}{dx}(\arccos u) = -\dfrac{1}{\sqrt{1 - u^2}}\dfrac{du}{dx}, \quad (0 \le \arccos u \le \pi)$

32. $\dfrac{d}{dx}(\arctan u) = \dfrac{1}{1 + u^2}\dfrac{du}{dx}, \quad \left(-\dfrac{\pi}{2} < \arctan u < \dfrac{\pi}{2}\right)$

33. $\dfrac{d}{dx}(\operatorname{arc\,cot} u) = -\dfrac{1}{1 + u^2}\dfrac{du}{dx}, \quad (0 \le \operatorname{arc\,cot} u \le \pi)$

34. $\dfrac{d}{dx}(\operatorname{arc\,sec} u) = \dfrac{1}{u\sqrt{u^2 - 1}}\dfrac{du}{dx}, \quad \left(0 \le \operatorname{arc\,sec} u < \dfrac{\pi}{2}, -\pi \le \operatorname{arc\,sec} u < -\dfrac{\pi}{2}\right)$

35. $\dfrac{d}{dx}(\operatorname{arc\,csc} u) = -\dfrac{1}{u\sqrt{u^2 - 1}}\dfrac{du}{dx}, \quad \left(0 < \operatorname{arc\,csc} u \le \dfrac{\pi}{2}, -\pi < \operatorname{arc\,csc} u \le -\dfrac{\pi}{2}\right)$

36. $\dfrac{d}{dx}(\operatorname{arc\,vers} u) = \dfrac{1}{\sqrt{2u - u^2}}\dfrac{du}{dx}, \quad (0 \le \operatorname{arc\,vers} u \le \pi)$

37. $\dfrac{d}{dx}(\sinh u) = (\cosh u)\dfrac{du}{dx}$

38. $\dfrac{d}{dx}(\cosh u) = (\sinh u)\dfrac{du}{dx}$

39. $\dfrac{d}{dx}(\tanh u) = (\operatorname{sech}^2 u)\dfrac{du}{dx}$

40. $\dfrac{d}{dx}(\coth u) = -(\operatorname{csch}^2 u)\dfrac{du}{dx}$

41. $\dfrac{d}{dx}(\operatorname{sech} u) = -(\operatorname{sech} u \cdot \tanh u)\dfrac{du}{dx}$

42. $\dfrac{d}{dx}(\operatorname{csch} u) = -(\operatorname{csch} u \cdot \coth u)\dfrac{du}{dx}$

43. $\dfrac{d}{dx}(\sinh^{-1} u) = \dfrac{d}{dx}[\log(u + \sqrt{u^2 + 1})] = \dfrac{1}{\sqrt{u^2 + 1}}\dfrac{du}{dx}$

44. $\dfrac{d}{dx}(\cosh^{-1} u) = \dfrac{d}{dx}[\log(u + \sqrt{u^2 - 1})] = \dfrac{1}{\sqrt{u^2 - 1}}\dfrac{du}{dx}, \quad (u > 1, \cosh^{-1} u > 0)$

45. $\dfrac{d}{dx}(\tanh^{-1} u) = \dfrac{d}{dx}\left[\dfrac{1}{2}\log\dfrac{1+u}{1-u}\right] = \dfrac{1}{1-u^2}\dfrac{du}{dx}, \quad (u^2 < 1)$

46. $\dfrac{d}{dx}(\coth^{-1} u) = \dfrac{d}{dx}\left[\dfrac{1}{2}\log\dfrac{u+1}{u-1}\right] = \dfrac{1}{1-u^2}\dfrac{du}{dx}, \quad (u^2 > 1)$

47. $\dfrac{d}{dx}(\operatorname{sech}^{-1} u) = \dfrac{d}{dx}\left[\log\dfrac{1+\sqrt{1-u^2}}{u}\right] = -\dfrac{1}{u\sqrt{1-u^2}}\dfrac{du}{dx}, \quad (0 < u < 1, \ \operatorname{sech}^{-1} u > 0)$

48. $\dfrac{d}{dx}(\operatorname{csch}^{-1} u) = \dfrac{d}{dx}\left[\log\dfrac{1+\sqrt{1+u^2}}{u}\right] = -\dfrac{1}{|u|\sqrt{1+u^2}}\dfrac{du}{dx}$

49. $\dfrac{d}{dq}\displaystyle\int_p^q f(x)\,dx = f(q), \quad [p\ \text{constant}]$

50. $\dfrac{d}{dp}\displaystyle\int_p^q f(x)\,dx = -f(p), \quad [q\ \text{constant}]$

51. $\dfrac{d}{da}\displaystyle\int_p^q f(x,a)\,dx = \displaystyle\int_p^q \dfrac{\partial}{\partial a}[f(x,a)]\,dx + f(q,a)\dfrac{dq}{da} - f(p,a)\dfrac{dp}{da}$

INTEGRATION

The following is a brief discussion of some integration techniques. A more complete discussion can be found in a number of good textbooks. However, the purpose of this introduction is simply to discuss a few of the important techniques which may be used, in conjunction with the integral table which follows, to integrate particular functions.

No matter how extensive the integral table, it is a fairly uncommon occurrence to find in the table the exact integral desired. Usually some form of transformation will have to be made. The simplest type of transformation, and yet the most general, is substitution. Simple forms of substitution, such as $y = ax$, are employed almost unconsciously by experienced users of integral tables. Other substitutions may require more thought. In some sections of the tables, appropriate substitutions are suggested for integrals that are similar to, but not exactly like, integrals in the table. Finding the right substitution is largely a matter of intuition and experience.

Several precautions must be observed when using substitutions:

1. Be sure to make the substitution in the dx term, as well as everywhere else in the integral.

2. Be sure that the function substituted is one-to-one and continuous. If this is not the case, the integral must be restricted in such a way as to make it true. See the example following.

3. With definite integrals, the limits should also be expressed in terms of the new dependent variable. With indefinite integrals, it is necessary to perform the reverse substitution to obtain the answer in terms of the original independent variable. This may also be done for definite integrals, but it is usually easier to change the limits.

Example:

$$\int \frac{x^4}{\sqrt{a^2 - x^2}}\,dx$$

Here we make the substitution $x = |a|\sin\theta$. Then $dx = |a|\cos\theta\,d\theta$, and

$$\sqrt{a^2 - x^2} = \sqrt{a^2 - a^2\sin^2\theta} = |a|\sqrt{1 - \sin^2\theta} = |a\cos\theta|$$

Notice the absolute value signs. It is very important to keep in mind that a square root radical always denotes the positive square root, and to assure the sign is always kept positive. Thus $\sqrt{x^2} = |x|$. Failure to observe this is a common cause of errors in integration.

Notice also that the indicated substitution is not a one-to-one function; that is, it does not have a unique inverse. Thus we must restrict the range of θ in such a way as to make the function one-to-one. Fortunately, this is easily done by solving for θ

$$\theta = \sin^{-1}\frac{x}{|a|}$$

and restricting the inverse sine to the principal values, $-\frac{\pi}{2} \le \theta \le \frac{\pi}{2}$.

Thus the integral becomes

$$\int \frac{a^4 \sin^4 \theta |a| \cos \theta \, d\theta}{|a| \, |\cos \theta|}$$

Now, however, in the range of values chosen for θ, $\cos \theta$ is always positive. Thus we may remove the absolute value signs from $\cos \theta$ in the denominator. (This is one of the reasons that the principal values of the inverse trigonometric functions are defined as they are.) Then the $\cos \theta$ terms cancel, and the integral becomes

$$a^4 \int \sin^4 \theta \, d\theta$$

By application of integral formulas 299 and 296, we integrate this to

$$-a^4 \frac{\sin^3 \theta \cos \theta}{4} - \frac{3a^4}{8} \cos \theta \sin \theta + \frac{3a^4}{8} \theta + C$$

We now must perform the inverse substitution to get the result in terms of x. We have

$$\theta = \sin^{-1} \frac{x}{|a|}$$

$$\sin \theta = \frac{x}{|a|}$$

Then

$$\cos \theta = \pm \sqrt{1 - \sin^2 \theta} = \pm \sqrt{1 - \frac{x^2}{a^2}} = \pm \frac{\sqrt{a^2 - x^2}}{|a|}.$$

Because of the previously mentioned fact that $\cos \theta$ is positive, we may omit the $\pm$ sign. The reverse substitution then produces the final answer

$$\int \frac{x^4}{\sqrt{a^2 - x^2}} \, dx = -\frac{1}{4} x^3 \sqrt{a^2 - x^2} - \frac{3}{8} a^2 x \sqrt{a^2 - x^2} + \frac{3}{8} a^4 \sin^{-1} \frac{x}{|a|} + C.$$

Any rational function of x may be integrated, if the denominator is factored into linear and irreducible quadratic factors. The function may then be broken into partial fractions, and the individual partial fractions integrated by use of the appropriate formula from the integral table. See the section on partial fractions for further information.

Many integrals may be reduced to rational functions by proper substitutions. For example,

$$z = \tan \frac{x}{2}$$

will reduce any rational function of the six trigonometric functions of x to a rational function of z. (Frequently there are other substitutions that are simpler to use, but this one will always work. See integral formula number 484.)

Any rational function of x and $\sqrt{ax + b}$ may be reduced to a rational function of z by making the substitution

$$z = \sqrt{ax + b}.$$

Other likely substitutions will be suggested by looking at the form of the integrand.

The other main method of transforming integrals is integration by parts. This involves applying formula number 5 or 6 in the accompanying integral table. The critical factor in this method is the choice of the functions u and v. In order for the method to be successful, $v = \int dv$ and $\int v \, du$ must be easier to integrate than the original integral. Again, this choice is largely a matter of intuition and experience.

Example:

$$\int x \sin x \, dx$$

Two obvious choices are $u = x$, $dv = \sin x \, dx$, or $u = \sin x$, $dv = x \, dx$. Since a preliminary mental calculation indicates that $\int v \, du$ in the second choice would be more, rather than less, complicated than the original integral (it would contain x^2), we use the first choice.

$$u = x \qquad\qquad du = dx$$
$$dv = \sin x \, dx \qquad\qquad v = -\cos x$$
$$\int x \sin x \, dx = \int u \, dv = uv - \int v \, du = -x \cos x + \int \cos x \, dx$$
$$= \sin x - x \cos x$$

Of course, this result could have been obtained directly from the integral table, but it provides a simple example of the method. In more complicated examples the choice of u and v may not be so obvious, and several different choices may have to be tried. Of course, there is no guarantee that any of them will work.

Integration by parts may be applied more than once, or combined with substitution. A fairly common case is illustrated by the following example.

Example:

$$\int e^x \sin x \, dx$$

Let

$$u = e^x \qquad \text{Then } du = e^x \, dx$$
$$dv = \sin x \, dx \qquad v = -\cos x$$
$$\int e^x \sin x \, dx = \int u \, dv = uv - \int v \, du = -e^x \cos x + \int e^x \cos x \, dx$$

In this latter integral,

$$\text{Let } u = e^x \qquad \text{Then } du = e^x \, dx$$
$$dv = \cos x \, dx \qquad v = \sin x$$

$$\int e^x \sin x \, dx = -e^x \cos x + \int e^x \cos x \, dx \;=\; -e^x \cos x + \int u \, dv$$

$$= \; -e^x \cos x + uv - \int v \, du$$

$$= \; -e^x \cos x + e^x \sin x - \int e^x \sin x \, dx$$

This looks as if a circular transformation has taken place, since we are back at the same integral we started from. However, the above equation can be solved algebraically for the required integral:

$$\int e^x \sin x \, dx = \frac{1}{2} e^x \sin x - \frac{1}{2} e^x \cos x$$

In the second integration by parts, if the parts had been chosen as $u = \cos x$, $dv = e^x \, dx$, we would indeed have made a circular transformation, and returned to the starting place.

In general, when doing repeated integration by parts, one should never choose the function u at any stage to be the same as the function v at the previous stage, or a constant times the previous v.

The following rule is called the extended rule for integration by parts. It is the result of $n+1$ successive applications of integration by parts. If

$$g_1(x) \;=\; \int g(x) \, dx, \qquad g_2(x) = \int g_1(x) \, dx,$$

$$g_3(x) \;=\; \int g_2(x) \, dx, \ldots, g_m(x) = \int g_{m-1}(x) \, dx, \ldots,$$

then

$$\int f(x) \cdot g(x) \, dx \;=\; f(x) \cdot g_1(x) - f'(x) \cdot g_2(x) + f''(x) \cdot g_3(x) - + \cdots$$

$$+ (-1)^n f^{(n)}(x) g_{n+1}(x) + (-1)^{n+1} \int f^{(n+1)}(x) g_{n+1}(x) \, dx.$$

A useful special case of the above rule is when $f(x)$ is a polynomial of degree n. Then $f^{(n+1)}(x) = 0$, and

$$\int f(x) \cdot g(x) \, dx = f(x) \cdot g_1(x) - f'(x) \cdot g_2(x) + f''(x) \cdot g_3(x) - + \cdots + (-1)^n f^{(n)}(x) g_{n+1}(x) + C.$$

Example: If $f(x) = x^2$, $g(x) = \sin x$

$$\int x^2 \sin x \, dx = -x^2 \cos x + 2x \sin x + 2 \cos x + C.$$

Another application of this formula occurs if

$$f''(x) = af(x) \qquad \text{and} \qquad g''(x) = bg(x),$$

where a and b are unequal constants. In this case, by a process similar to that used in the above example for $\int e^x \sin x \, dx$, we get the formula

$$\int f(x)g(x) \, dx = \frac{f(x) \cdot g'(x) - f'(x) \cdot g(x)}{b - a} + C.$$

This formula could have been used in the example mentioned. Here is another example.

Example: If $f(x) = e^{2x}$, $g(x) = \sin 3x$, then $a = 4$, $b = -9$, and

$$\int e^{2x} \sin 3x \, dx = \frac{3 e^{2x} \cos 3x - 2 e^{2x} \sin 3x}{-9 - 4} + C = \frac{e^{2x}}{13} (2 \sin 3x - 3 \cos 3x) + C$$

The following additional points should be observed when using this table.

1. A constant of integration is to be supplied with the answers for indefinite integrals.

2. Logarithmic expressions are to base $e = 2.71828\ldots$, unless otherwise specified, and are to be evaluated for the absolute value of the arguments involved therein.

3. All angles are measured in radians, and inverse trigonometric and hyperbolic functions represent principal values, unless otherwise indicated.

4. If the application of a formula produces either a zero denominator or the square root of a negative number in the result, there is usually available another form of the answer which avoids this difficulty. In many of the results, the excluded values are specified, but when such are omitted it is presumed that one can tell what these should be, especially when difficulties of the type herein mentioned are obtained.

5. When inverse trigonometric functions occur in the integrals, be sure that any replacements made for them are strictly in accordance with the rules for such functions. This causes little difficulty when the argument of the inverse trigonometric function is positive, since then all angles involved are in the first quadrant. However, if the argument is negative, special care must be used. Thus if $u > 0$,

$$\sin^{-1} u = \cos^{-1} \sqrt{1 - u^2} = \csc^{-1} \frac{1}{u}, \text{ etc.}$$

However, if $u < 0$,

$$\sin^{-1} u = - \cos^{-1} \sqrt{1 - u^2} = -\pi - \csc^{-1} \frac{1}{u}, \text{ etc.}$$

See the section on inverse trigonometric functions for a full treatment of the allowable substitutions.

6. In integrals 340–345 and some others, the right side includes expressions of the form

$$A \tan^{-1} [B + C \tan f(x)].$$

In these formulas, the $\tan^{-1}$ does not necessarily represent the principal value. Instead of always employing the principal branch of the inverse tangent function, one must instead use that branch of the inverse tangent function upon which $f(x)$ lies for any particular choice of x. (This is not an issue when the antiderivative is continuous.)

Example:

$$\int_0^{4\pi} \frac{dx}{2 + \sin x} = \frac{2}{\sqrt{3}} \tan^{-1} \frac{2 \tan(x/2 + 1)}{\sqrt{3}} \Big]_0^{4\pi}$$

$$= \frac{2}{\sqrt{3}} \left[\tan^{-1} \left(\frac{2 \tan 2\pi + 1}{\sqrt{3}} \right) - \tan^{-1} \left(\frac{2 \tan 0 + 1}{\sqrt{3}} \right) \right]$$

$$= \frac{2}{\sqrt{3}} \left[\frac{13\pi}{6} - \frac{\pi}{6} \right] = \frac{4\pi}{\sqrt{3}} = \frac{4\sqrt{3}\pi}{3}$$

Here

$$\tan^{-1}\frac{2\tan 2\pi + 1}{\sqrt{3}} = \tan^{-1}\frac{1}{\sqrt{3}} = \frac{13\pi}{6},$$

since $f(x) = 2\pi$; and

$$\tan^{-1}\frac{2\tan 0 + 1}{\sqrt{3}} = \tan^{-1}\frac{1}{\sqrt{3}} = \frac{\pi}{6},$$

since $f(x) = 0$.

7. B_n and E_n where used in integrals represents the Bernoulli and Euler numbers as defined in tables of Bernoulli and Euler polynomials contained in certain mathematics reference and handbooks.

INTEGRALS

ELEMENTARY FORMS

1. $\displaystyle\int a\,dx = ax$

2. $\displaystyle\int a\cdot f(x)\,dx = a\int f(x)\,dx$

3. $\displaystyle\int \phi(y)\,dx = \int \frac{\phi(y)}{y'}\,dy, \quad \text{where } y' = \frac{dy}{dx}$

4. $\displaystyle\int (u+v)\,dx = \int u\,dx + \int v\,dx, \quad \text{where } u \text{ and } v \text{ are any functions of } x$

5. $\displaystyle\int u\,dv = u\int dv - \int v\,du = uv - \int v\,du$

6. $\displaystyle\int u\frac{dv}{dx}\,dx = uv - \int v\frac{du}{dx}\,dx$

7. $\displaystyle\int x^n\,dx = \frac{x^{n+1}}{n+1}, \quad \text{except } n = -1$

8. $\displaystyle\int \frac{f'(x)\,dx}{f(x)} = \log f(x), \quad (df(x) = f'(x)\,dx)$

9. $\displaystyle\int \frac{dx}{x} = \log x$

10. $\displaystyle\int \frac{f'(x)\,dx}{2\sqrt{f(x)}} = \sqrt{f(x)}, \quad (df(x) = f'(x)\,dx)$

11. $\displaystyle\int e^x\,dx = e^x$

12. $\displaystyle\int e^{ax}\,dx = e^{ax}/a$

13. $\displaystyle\int b^{ax}\,dx = \frac{b^{ax}}{a\log b}, \quad (b > 0)$

14. $\displaystyle\int \log x\,dx = x\log x - x$

15. $\displaystyle\int a^x \log a\,dx = a^x, \quad (a > 0)$

16. $\displaystyle\int \frac{dx}{a^2 + x^2} = \frac{1}{a}\tan^{-1}\frac{x}{a}$

17. $\displaystyle\int \frac{dx}{a^2 - x^2} = \begin{cases} \frac{1}{a}\tanh^{-1}\frac{x}{a} \\ \text{or} \\ \frac{1}{2a}\log\frac{a+x}{a-x}, \quad (a^2 > x^2) \end{cases}$

18. $\displaystyle\int \frac{dx}{x^2 - a^2} = \begin{cases} -\frac{1}{a}\coth^{-1}\frac{x}{a} \\ \text{or} \\ \frac{1}{2a}\log\frac{x-a}{x+a}, \quad (x^2 > a^2) \end{cases}$

19. $\displaystyle\int \frac{dx}{\sqrt{a^2 - x^2}} = \begin{cases} \sin^{-1}\frac{x}{|a|} \\ \text{or} \\ -\cos^{-1}\frac{x}{|a|}, \quad (a^2 > x^2) \end{cases}$

20. $\displaystyle\int \frac{dx}{\sqrt{x^2 \pm a^2}} = \log(x + \sqrt{x^2 \pm a^2})$

21. $\displaystyle\int \frac{dx}{x\sqrt{x^2 - a^2}} = \frac{1}{|a|} \sec^{-1}\frac{x}{a}$

22. $\displaystyle\int \frac{dx}{x\sqrt{a^2 \pm x^2}} = -\frac{1}{a} \log\left(\frac{a + \sqrt{a^2 \pm x^2}}{x}\right)$

FORMS CONTAINING $(a + bx)$

For forms containing $a + bx$, but not listed in the table, the substitution $u = \frac{a+bx}{x}$ may prove helpful.

23. $\displaystyle\int (a + bx)^n \, dx = \frac{(a + bx)^{n+1}}{(n + 1)b}, \quad (n \neq -1)$

24. $\displaystyle\int x(a + bx)^n \, dx = \frac{1}{b^2(n+2)}(a + bx)^{n+2} - \frac{a}{b^2(n+1)}(a + bx)^{n+1}, \quad (n \neq -1, -2)$

25. $\displaystyle\int x^2(a + bx)^n \, dx = \frac{1}{b^3}\left[\frac{(a + bx)^{n+3}}{n+3} - 2a\frac{(a + bx)^{n+2}}{n+2} + a^2\frac{(a + bx)^{n+1}}{n+1}\right]$

26. $\displaystyle\int x^m(a + bx)^n \, dx = \begin{cases} \frac{x^{m+1}(a+bx)^n}{m+n+1} + \frac{an}{m+n+1}\displaystyle\int x^m(a + bx)^{n-1} \, dx \\ \text{or} \\ \frac{1}{a(n+1)}\left[-x^{m+1}(a + bx)^{n+1} + (m + n + 2)\displaystyle\int x^m(a + bx)^{n+1} \, dx\right] \\ \text{or} \\ \frac{1}{b(m+n+1)}\left[x^m(a + bx)^{n+1} - ma\displaystyle\int x^{m-1}(a + bx)^n \, dx\right] \end{cases}$

27. $\displaystyle\int \frac{dx}{a + bx} = \frac{1}{b} \log(a + bx)$

28. $\displaystyle\int \frac{dx}{(a + bx)^2} = -\frac{1}{b(a + bx)}$

29. $\displaystyle\int \frac{dx}{(a + bx)^3} = -\frac{1}{2b(a + bx)^2}$

30. $\displaystyle\int \frac{x \, dx}{a + bx} = \begin{cases} \frac{1}{b^2}[a + bx - a \log(a + bx)] \\ \text{or} \\ \frac{x}{b} - \frac{a}{b^2} \log(a + bx) \end{cases}$

31. $\displaystyle\int \frac{x \, dx}{(a + bx)^2} = \frac{1}{b^2}\left[\log(a + bx) + \frac{a}{a + bx}\right]$

32. $\displaystyle\int \frac{x \, dx}{(a + bx)^n} = \frac{1}{b^2}\left[\frac{-1}{(n-2)(a + bx)^{n-2}} + \frac{a}{(n-1)(a + bx)^{n-1}}\right], \quad n \neq 1, 2$

33. $\displaystyle\int \frac{x^2 \, dx}{a + bx} = \frac{1}{b^3}\left[\frac{1}{2}(a + bx)^2 - 2a(a + bx) + a^2 \log(a + bx)\right]$

34. $\displaystyle\int \frac{x^2 \, dx}{(a + bx)^2} = \frac{1}{b^3}\left[a + bx - 2a \log(a + bx) - \frac{a^2}{a + bx}\right]$

35. $\displaystyle\int \frac{x^2 \, dx}{(a + bx)^3} = \frac{1}{b^3}\left[\log(a + bx) + \frac{2a}{a + bx} - \frac{a^2}{2(a + bx)^2}\right]$

36. $\displaystyle\int \frac{x^2 \, dx}{(a + bx)^n} = \frac{1}{b^3}\left[\frac{-1}{(n-3)(a + bx)^{n-3}} + \frac{2a}{(n-2)(a + bx)^{n-2}} - \frac{a^2}{(n-1)(a + bx)^{n-1}}\right], \quad n \neq 1, 2, 3$

37. $\displaystyle\int \frac{dx}{x(a + bx)} = -\frac{1}{a} \log\frac{a + bx}{x}$

38. $\displaystyle\int \frac{dx}{x(a + bx)^2} = \frac{1}{a(a + bx)} - \frac{1}{a^2} \log\frac{a + bx}{x}$

39. $\displaystyle \int \frac{dx}{x(a+bx)^3} = \frac{1}{a^3}\left[\frac{1}{2}\left(\frac{2a+bx}{a+bx}\right)^2 + \log\frac{x}{a+bx}\right]$

40. $\displaystyle \int \frac{dx}{x^2(a+bx)} = -\frac{1}{ax} + \frac{b}{a^2}\log\frac{a+bx}{x}$

41. $\displaystyle \int \frac{dx}{x^3(a+bx)} = \frac{2bx-a}{2a^2x^2} + \frac{b^2}{a^3}\log\frac{x}{a+bx}$

42. $\displaystyle \int \frac{dx}{x^2(a+bx)^2} = -\frac{a+2bx}{a^2x(a+bx)} + \frac{2b}{a^3}\log\frac{a+bx}{x}$

FORMS CONTAINING $c^2 \pm x^2$ or $x^2 - c^2$

43. $\displaystyle \int \frac{dx}{c^2+x^2} = \frac{1}{c}\tan^{-1}\frac{x}{c}$

44. $\displaystyle \int \frac{dx}{c^2-x^2} = \frac{1}{2c}\log\frac{c+x}{c-x}, \quad (c^2 > x^2)$

45. $\displaystyle \int \frac{dx}{x^2-c^2} = \frac{1}{2c}\log\frac{x-c}{x+c}, \quad (x^2 > c^2)$

46. $\displaystyle \int \frac{x\,dx}{c^2\pm x^2} = \pm\frac{1}{2}\log(c^2\pm x^2)$

47. $\displaystyle \int \frac{x\,dx}{(c^2\pm x^2)^{n+1}} = \mp\frac{1}{2n(c^2\pm x^2)^n}$

48. $\displaystyle \int \frac{dx}{(c^2\pm x^2)^n} = \frac{1}{2c^2(n-1)}\left[\frac{x}{(c^2\pm x^2)^{n-1}} + (2n-3)\int\frac{dx}{(c^2\pm x^2)^{n-1}}\right]$

49. $\displaystyle \int \frac{dx}{(x^2-c^2)^n} = \frac{1}{2c^2(n-1)}\left[-\frac{x}{(x^2-c^2)^{n-1}} - (2n-3)\int\frac{dx}{(x^2-c^2)^{n-1}}\right]$

50. $\displaystyle \int \frac{x\,dx}{x^2-c^2} = \frac{1}{2}\log(x^2-c^2)$

51. $\displaystyle \int \frac{x\,dx}{(x^2-c^2)^{n+1}} = -\frac{1}{2n(x^2-c^2)^n}$

FORMS CONTAINING $a+bx$ AND $c+dx$

Define $u = a+bx$, $v = c+dx$, and $k = ad - bc$. If $k = 0$, then $v = \frac{c}{a}u$.

52. $\displaystyle \int \frac{dx}{u\cdot v} = \frac{1}{k}\cdot\log\left(\frac{v}{u}\right)$

53. $\displaystyle \int \frac{x\,dx}{u\cdot v} = \frac{1}{k}\left[\frac{a}{b}\log(u) - \frac{c}{d}\log(v)\right]$

54. $\displaystyle \int \frac{dx}{u^2\cdot v} = \frac{1}{k}\left(\frac{1}{u} + \frac{d}{k}\log\frac{v}{u}\right)$

55. $\displaystyle \int \frac{x\,dx}{u^2\cdot v} = \frac{-a}{bku} - \frac{c}{k^2}\log\frac{v}{u}$

56. $\displaystyle \int \frac{x^2\,dx}{u^2\cdot v} = \frac{a^2}{b^2ku} + \frac{1}{k^2}\left[\frac{c^2}{d}\log(v) + \frac{a(k-bc)}{b^2}\log(u)\right]$

57. $\displaystyle \int \frac{dx}{u^n\cdot v^m} = \frac{1}{k(m-1)}\left[\frac{-1}{u^{n-1}\cdot v^{m-1}} - (m+n-2)b\int\frac{dx}{u^n\cdot v^{m-1}}\right]$

58. $\displaystyle \int \frac{u}{v}dx = \frac{bx}{d} + \frac{k}{d^2}\log(v)$

59. $\displaystyle \int \frac{u^m\,dx}{v^n} = \begin{cases} \frac{-1}{k(n-1)}\left[\frac{u^{m+1}}{v^{n-1}} + b(n-m-2)\int\frac{u^m}{v^{n-1}}dx\right] \\ \text{or} \\ \frac{-1}{d(n-m-1)}\left[\frac{u^m}{v^{n-1}} + mk\int\frac{u^{m-1}}{v^n}dx\right] \\ \text{or} \\ \frac{-1}{d(n-1)}\left[\frac{u^m}{v^{n-1}} - mb\int\frac{u^{m-1}}{v^{n-1}}dx\right] \end{cases}$

FORMS CONTAINING $(a + bx^n)$

60. $\displaystyle\int \frac{dx}{a + bx^2} = \frac{1}{\sqrt{ab}}\tan^{-1}\frac{x\sqrt{ab}}{a}, \quad (ab > 0)$

61. $\displaystyle\int \frac{dx}{a + bx^2} = \begin{cases} \frac{1}{2\sqrt{-ab}}\log\frac{a + x\sqrt{-ab}}{a - x\sqrt{-ab}}, & (ab < 0) \\ \text{or} \\ \frac{1}{\sqrt{-ab}}\tanh^{-1}\frac{x\sqrt{-ab}}{a}, & (ab < 0) \end{cases}$

62. $\displaystyle\int \frac{dx}{a^2 + b^2x^2} = \frac{1}{ab}\tan^{-1}\frac{bx}{a}$

63. $\displaystyle\int \frac{x\,dx}{a + bx^2} = \frac{1}{2b}\log(a + bx^2)$

64. $\displaystyle\int \frac{x^2\,dx}{a + bx^2} = \frac{x}{b} - \frac{a}{b}\int\frac{dx}{a + bx^2}$

65. $\displaystyle\int \frac{dx}{(a + bx^2)^2} = \frac{x}{2a(a + bx^2)} + \frac{1}{2a}\int\frac{dx}{a + bx^2}$

66. $\displaystyle\int \frac{dx}{a^2 - b^2x^2} = \frac{1}{2ab}\log\frac{a + bx}{a - bx}$

67. $\displaystyle\int \frac{dx}{(a + bx^2)^{m+1}} = \begin{cases} \frac{1}{2ma}\frac{x}{(a+bx^2)^m} + \frac{2m-1}{2ma}\int\frac{dx}{(a+bx^2)^m} \\ \text{or} \\ \frac{(2m)!}{(m!)^2}\left[\frac{x}{2a}\sum_{r=1}^{m}\frac{r!(r-1)!}{(4a)^{m-r}(2r)!(a+bx^2)^r} + \frac{1}{(4a)^m}\int\frac{dx}{a+bx^2}\right] \end{cases}$

68. $\displaystyle\int \frac{x\,dx}{(a + bx^2)^{m+1}} = -\frac{1}{2bm(a + bx^2)^m}$

69. $\displaystyle\int \frac{x^2\,dx}{(a + bx^2)^{m+1}} = \frac{-x}{2mb(a + bx^2)^m} + \frac{1}{2mb}\int\frac{dx}{(a + bx^2)^m}$

70. $\displaystyle\int \frac{dx}{x(a + bx^2)} = \frac{1}{2a}\log\frac{x^2}{a + bx^2}$

71. $\displaystyle\int \frac{dx}{x^2(a + bx^2)} = -\frac{1}{ax} - \frac{b}{a}\int\frac{dx}{a + bx^2}$

72. $\displaystyle\int \frac{dx}{x(a + bx^2)^{m+1}} = \begin{cases} \frac{1}{2am(a+bx^2)^m} + \frac{1}{a}\int\frac{dx}{x(a+bx^2)^m} \\ \text{or} \\ \frac{1}{2a^{m+1}}\left[\sum_{r=1}^{m}\frac{a^r}{r(a+bx^2)^r} + \log\frac{x^2}{a+bx^2}\right] \end{cases}$

73. $\displaystyle\int \frac{dx}{x^2(a + bx^2)^{m+1}} = \frac{1}{a}\int\frac{dx}{x^2(a + bx^2)^m} - \frac{b}{a}\int\frac{dx}{(a + bx^2)^{m+1}}$

74. $\displaystyle\int \frac{dx}{a + bx^3} = \frac{k}{3a}\left[\frac{1}{2}\log\frac{(k+x)^3}{a + bx^3} + \sqrt{3}\tan^{-1}\frac{2x-k}{k\sqrt{3}}\right], \quad \left(k = \sqrt[3]{\frac{a}{b}}\right)$

75. $\displaystyle\int \frac{x\,dx}{a + bx^3} = \frac{1}{3bk}\left[\frac{1}{2}\log\frac{a + bx^3}{(k+x)^3} + \sqrt{3}\tan^{-1}\frac{2x-k}{k\sqrt{3}}\right], \quad \left(k = \sqrt[3]{\frac{a}{b}}\right)$

76. $\displaystyle\int \frac{x^2\,dx}{a + bx^3} = \frac{1}{3b}\log(a + bx^3)$

77. $\displaystyle\int \frac{dx}{a + bx^4} = \frac{k}{2a}\left[\frac{1}{2}\log\frac{x^2 + 2kx + 2k^2}{x^2 - 2kx + 2k^2} + \tan^{-1}\frac{2kx}{2k^2 - x^2}\right], \quad \left(ab > 0, k = \sqrt[4]{\frac{a}{4b}}\right)$

78. $\displaystyle\int \frac{dx}{a + bx^4} = \frac{k}{2a}\left[\frac{1}{2}\log\frac{x + k}{x - k} + \tan^{-1}\frac{x}{k}\right], \quad \left(ab < 0, k = \sqrt[4]{-\frac{a}{b}}\right)$

79. $\displaystyle\int \frac{x\,dx}{a + bx^4} = \frac{1}{2bk}\tan^{-1}\frac{x^2}{k}, \quad \left(ab > 0, k = \sqrt{\frac{a}{b}}\right)$

80. $\displaystyle\int \frac{x\,dx}{a + bx^4} = \frac{1}{4bk}\log\frac{x^2 - k}{x^2 + k}, \quad \left(ab < 0, k = \sqrt{-\frac{a}{b}}\right)$

81. $\displaystyle\int \frac{x^2\,dx}{a + bx^4} = \frac{1}{4bk}\left[\frac{1}{2}\log\frac{x^2 - 2kx + 2k^2}{x^2 + 2kx + 2k^2} + \tan^{-1}\frac{2kx}{2k^2 - x^2}\right], \quad \left(ab > 0, k = \sqrt[4]{\frac{a}{4b}}\right)$

82. $\displaystyle\int \frac{x^2\,dx}{a + bx^4} = \frac{1}{4bk}\left[\log\frac{x - k}{x + k} + 2\tan^{-1}\frac{x}{k}\right], \quad \left(ab < 0, k = \sqrt[4]{-\frac{a}{b}}\right)$

83. $\displaystyle\int \frac{x^3\,dx}{a + bx^4} = \frac{1}{4b}\log(a + bx^4)$

84. $\displaystyle\int \frac{dx}{x(a + bx^n)} = \frac{1}{an}\log\frac{x^n}{a + bx^n}$

85. $$\int \frac{dx}{(a+bx^n)^{m+1}} = \frac{1}{a} \int \frac{dx}{(a+bx^n)^m} - \frac{b}{a} \int \frac{x^n\,dx}{(a+bx^n)^{m+1}}$$

86. $$\int \frac{x^m\,dx}{(a+bx^n)^{p+1}} = \frac{1}{b} \int \frac{x^{m-n}\,dx}{(a+bx^n)^p} - \frac{a}{b} \int \frac{x^{m-n}\,dx}{(a+bx^n)^{p+1}}$$

87. $$\int \frac{dx}{x^m(a+bx^n)^{p+1}} = \frac{1}{a} \int \frac{dx}{x^m(a+bx^n)^p} - \frac{b}{a} \int \frac{dx}{x^{m-n}(a+bx^n)^{p+1}}$$

88. $$\int x^m(a+bx^n)^p\,dx = \begin{cases} \frac{1}{b(np+m+1)}\left[x^{m-n+1}(a+bx^n)^{p+1} - a(m-n+1)\int x^{m-n}(a+bx^n)^p\,dx\right] \\ \text{or} \\ \frac{1}{np+m+1}\left[x^{m+1}(a+bx^n)^p + anp\int x^m(a+bx^n)^{p-1}\,dx\right] \\ \text{or} \\ \frac{1}{a(m+1)}\left[x^{m+1}(a+bx^n)^{p+1} - (m+1+np+n)b\int x^{m+n}(a+bx^n)^p\,dx\right] \\ \text{or} \\ \frac{1}{an(p+1)}\left[-x^{m+1}(a+bx^n)^{p+1} + (m+1+np+n)\int x^m(a+bx^n)^{p+1}\,dx\right] \end{cases}$$

FORMS CONTAINING $c^3 \pm x^3$

89. $$\int \frac{dx}{c^3 \pm x^3} = \pm\frac{1}{6c^2}\log\frac{(c\pm x)^3}{c^3 \pm x^3} + \frac{1}{c^2\sqrt{3}}\tan^{-1}\frac{2x \mp c}{c\sqrt{3}}$$

90. $$\int \frac{dx}{(c^3 \pm x^3)^2} = \frac{x}{3c^3(c^3 \pm x^3)} + \frac{2}{3c^3}\int \frac{dx}{c^3 \pm x^3}$$

91. $$\int \frac{dx}{(c^3 \pm x^3)^{n+1}} = \frac{1}{3nc^3}\left[\frac{x}{(c^3 \pm x^3)^n} + (3n-1)\int \frac{dx}{(c^3 \pm x^3)^n}\right]$$

92. $$\int \frac{x\,dx}{c^3 \pm x^3} = \frac{1}{6c}\log\frac{c^3 \pm x^3}{(c\pm x)^3} \pm \frac{1}{c\sqrt{3}}\tan^{-1}\frac{2x \mp c}{c\sqrt{3}}$$

93. $$\int \frac{x\,dx}{(c^3 \pm x^3)^2} = \frac{x^2}{3c^3(c^3 \pm x^3)} + \frac{1}{3c^3}\int \frac{x\,dx}{c^3 \pm x^3}$$

94. $$\int \frac{x\,dx}{(c^3 \pm x^3)^{n+1}} = \frac{1}{3nc^3}\left[\frac{x^2}{(c^3 \pm x^3)^n} + (3n-2)\int \frac{x\,dx}{(c^3 \pm x^3)^n}\right]$$

95. $$\int \frac{x^2\,dx}{c^3 \pm x^3} = \pm\frac{1}{3}\log(c^3 \pm x^3)$$

96. $$\int \frac{x^2\,dx}{(c^3 \pm x^3)^{n+1}} = \mp\frac{1}{3n(c^3 \pm x^3)^n}$$

97. $$\int \frac{dx}{x(c^3 \pm x^3)} = \frac{1}{3c^3}\log\frac{x^3}{c^3 \pm x^3}$$

98. $$\int \frac{dx}{x(c^3 \pm x^3)^2} = \frac{1}{3c^3(c^3 \pm x^3)} + \frac{1}{3c^6}\log\frac{x^3}{c^3 \pm x^3}$$

99. $$\int \frac{dx}{x(c^3 \pm x^3)^{n+1}} = \frac{1}{3nc^3(c^3 \pm x^3)^n} + \frac{1}{c^3}\int \frac{dx}{x(c^3 \pm x^3)^n}$$

100. $$\int \frac{dx}{x^2(c^3 \pm x^3)} = -\frac{1}{c^3 x} \mp \frac{1}{c^3}\int \frac{x\,dx}{c^3 \pm x^3}$$

101. $$\int \frac{dx}{x^2(c^3 \pm x^3)^{n+1}} = \frac{1}{c^3}\int \frac{dx}{x^2(c^3 \pm x^3)^n} \mp \frac{1}{c^3}\int \frac{x\,dx}{(c^3 \pm x^3)^{n+1}}$$

FORMS CONTAINING $c^4 \pm x^4$

102. $$\int \frac{dx}{c^4 + x^4} = \frac{1}{2c^3\sqrt{2}}\left[\frac{1}{2}\log\frac{x^2 + cx\sqrt{2} + c^2}{x^2 - cx\sqrt{2} + c^2} + \tan^{-1}\frac{cx\sqrt{2}}{c^2 - x^2}\right]$$

103. $$\int \frac{dx}{c^4 - x^4} = \frac{1}{2c^3}\left[\frac{1}{2}\log\frac{c+x}{c-x} + \tan^{-1}\frac{x}{c}\right]$$

104. $$\int \frac{x\,dx}{c^4 + x^4} = \frac{1}{2c^2}\tan^{-1}\frac{x^2}{c^2}$$

105. $$\int \frac{x\,dx}{c^4 - x^4} = \frac{1}{4c^2}\log\frac{c^2 + x^2}{c^2 - x^2}$$

106. $$\int \frac{x^2\,dx}{c^4 + x^4} = \frac{1}{2c\sqrt{2}}\left[\frac{1}{2}\log\frac{x^2 - cx\sqrt{2} + c^2}{x^2 + cx\sqrt{2} + c^2} + \tan^{-1}\frac{cx\sqrt{2}}{c^2 - x^2}\right]$$

107. $\int \dfrac{x^2\,dx}{c^4 - x^4} = \dfrac{1}{2c}\left[\dfrac{1}{2}\log\dfrac{c+x}{c-x} - \tan^{-1}\dfrac{x}{c}\right]$

108. $\int \dfrac{x^3\,dx}{c^4 \pm x^4} = \pm\dfrac{1}{4}\log(c^4 \pm x^4)$

FORMS CONTAINING $(a + bx + cx^2)$

Define $X = a + bx + cx^2$ and $q = 4ac - b^2$. If $q = 0$, then $X = c\left(x + \dfrac{b}{2c}\right)^2$, and formulas starting with 23 should be used in place of these.

109. $\int \dfrac{dx}{X} = \dfrac{2}{\sqrt{q}}\tan^{-1}\dfrac{2cx + b}{\sqrt{q}}, \quad (q > 0)$

110. $\int \dfrac{dx}{X} = \begin{cases} \dfrac{-2}{\sqrt{-q}}\tanh^{-1}\dfrac{2cx+b}{\sqrt{-q}} \\ \text{or} \\ \dfrac{1}{\sqrt{-q}}\log\dfrac{2cx+b-\sqrt{-q}}{2cx+b+\sqrt{-q}}, \quad (q < 0) \end{cases}$

111. $\int \dfrac{dx}{X^2} = \dfrac{2cx + b}{qX} + \dfrac{2c}{q}\int \dfrac{dx}{X}$

112. $\int \dfrac{dx}{X^3} = \dfrac{2cx + b}{q}\left(\dfrac{1}{2X^2} + \dfrac{3c}{qX}\right) + \dfrac{6c^2}{q^2}\int \dfrac{dx}{X}$

113. $\int \dfrac{dx}{X^{n+1}} = \begin{cases} \dfrac{2cx + b}{nq\,X^n} + \dfrac{2(2n-1)c}{qn}\int\dfrac{dx}{X^n} \\ \text{or} \\ \dfrac{(2n)!}{(n!)^2}\left(\dfrac{c}{q}\right)^n\left[\dfrac{2cx+b}{q}\displaystyle\sum_{r=1}^{n}\left(\dfrac{q}{cX}\right)^r\left(\dfrac{(r-1)!\,r!}{(2r)!}\right) + \int\dfrac{dx}{X}\right] \end{cases}$

114. $\int \dfrac{x\,dx}{X} = \dfrac{1}{2c}\log X - \dfrac{b}{2c}\int \dfrac{dx}{X}$

115. $\int \dfrac{x\,dx}{X^2} = \dfrac{bx + 2a}{qX} - \dfrac{b}{q}\int \dfrac{dx}{X}$

116. $\int \dfrac{x\,dx}{X^{n+1}} = -\dfrac{2a + bx}{nq\,X^n} - \dfrac{b(2n-1)}{nq}\int \dfrac{dx}{X^n}$

117. $\int \dfrac{x^2}{X}\,dx = \dfrac{x}{c} - \dfrac{b}{2c^2}\log X + \dfrac{b^2 - 2ac}{2c^2}\int \dfrac{dx}{X}$

118. $\int \dfrac{x^2}{X^2}\,dx = \dfrac{(b^2 - 2ac)x + ab}{cq\,X} + \dfrac{2a}{q}\int \dfrac{dx}{X}$

119. $\int \dfrac{x^m\,dx}{X^{n+1}} = -\dfrac{x^{m-1}}{(2n-m+1)c\,X^n} - \dfrac{n-m+1}{2n-m+1}\cdot\dfrac{b}{c}\int \dfrac{x^{m-1}\,dx}{X^{n+1}} + \dfrac{m-1}{2n-m+1}\cdot\dfrac{a}{c}\int \dfrac{x^{m-2}\,dx}{X^{n+1}}$

120. $\int \dfrac{dx}{xX} = \dfrac{1}{2a}\log\dfrac{x^2}{X} - \dfrac{b}{2a}\int \dfrac{dx}{X}$

121. $\int \dfrac{dx}{x^2 X} = \dfrac{b}{2a^2}\log\dfrac{X}{x^2} - \dfrac{1}{ax} + \left(\dfrac{b^2}{2a^2} - \dfrac{c}{a}\right)\int \dfrac{dx}{X}$

122. $\int \dfrac{dx}{xX^n} = \dfrac{1}{2a(n-1)X^{n-1}} - \dfrac{b}{2a}\int \dfrac{dx}{X^n} + \dfrac{1}{a}\int \dfrac{dx}{xX^{n-1}}$

123. $\int \dfrac{dx}{x^m X^{n+1}} = -\dfrac{1}{(m-1)ax^{m-1}X^n} - \dfrac{n+m-1}{m-1}\cdot\dfrac{b}{a}\int \dfrac{dx}{x^{m-1}X^{n+1}} - \dfrac{2n+m-1}{m-1}\cdot\dfrac{c}{a}\int \dfrac{dx}{x^{m-2}X^{n+1}}$

FORMS CONTAINING $\sqrt{a + bx}$

124. $\int \sqrt{a + bx}\,dx = \dfrac{2}{3b}\sqrt{(a+bx)^3}$

125. $\int x\sqrt{a + bx}\,dx = -\dfrac{2(2a - 3bx)\sqrt{(a+bx)^3}}{15b^2}$

126. $\int x^2\sqrt{a + bx}\,dx = \dfrac{2(8a^2 - 12abx + 15b^2x^2)\sqrt{(a+bx)^3}}{105b^3}$

127. $\int x^m\sqrt{a + bx}\,dx = \begin{cases} \dfrac{2}{b(2m+3)}\left[x^m\sqrt{(a+bx)^3} - ma\int x^{m-1}\sqrt{a+bx}\,dx\right] \\ \text{or} \\ \dfrac{2}{b^{m+1}}\sqrt{a+bx}\displaystyle\sum_{r=0}^{m}\dfrac{m!(-a)^{m-r}}{r!(m-r)!(2r+3)}(a+bx)^{r+1} \end{cases}$

128. $\displaystyle\int \frac{\sqrt{a+bx}}{x}\,dx = 2\sqrt{a+bx} + a\int \frac{dx}{x\sqrt{a+bx}}$

129. $\displaystyle\int \frac{\sqrt{a+bx}}{x^2}\,dx = \frac{\sqrt{a+bx}}{x} + \frac{b}{2}\int \frac{dx}{x\sqrt{a+bx}}$

130. $\displaystyle\int \frac{\sqrt{a+bx}}{x^m}\,dx = -\frac{1}{(m-1)a}\left| \frac{\sqrt{(a+bx)^3}}{x^{m-1}} + \frac{(2m-5)b}{2}\int \frac{\sqrt{a+bx}}{x^{m-1}}\,dx \right|$

131. $\displaystyle\int \frac{dx}{\sqrt{a+bx}} = \frac{2\sqrt{a+bx}}{b}$

132. $\displaystyle\int \frac{x\,dx}{\sqrt{a+bx}} = -\frac{2(2a-bx)}{3b^2}\sqrt{a+bx}$

133. $\displaystyle\int \frac{x^2\,dx}{\sqrt{a+bx}} = \frac{2(8a^2 - 4abx - 3b^2x^2)}{15b^3}\sqrt{a+bx}$

134. $\displaystyle\int \frac{x^m\,dx}{\sqrt{a+bx}} = \begin{cases} \frac{2}{(2m+1)b}\left[x^m\sqrt{a+bx} - ma\int \frac{x^{m-1}\,dx}{\sqrt{a+bx}} \right] \\ \text{or} \\ \frac{2(-a)^m\sqrt{a+bx}}{b^{m+1}}\sum_{r=0}^{m} \frac{(-1)^r m!(a+bx)^r}{(2r+1)r!(m-r)!a^r} \end{cases}$

135. $\displaystyle\int \frac{dx}{x\sqrt{a+bx}} = \frac{1}{\sqrt{a}}\log\left(\frac{\sqrt{a+bx} - \sqrt{a}}{\sqrt{a+bx} + \sqrt{a}} \right), \quad (a>0)$

136. $\displaystyle\int \frac{dx}{x\sqrt{a+bx}} = \frac{2}{\sqrt{-a}}\tan^{-1}\sqrt{\frac{a+bx}{-a}}, \quad (a<0)$

137. $\displaystyle\int \frac{dx}{x^2\sqrt{a+bx}} = -\frac{\sqrt{a+bx}}{ax} - \frac{b}{2a}\int \frac{dx}{x\sqrt{a+bx}}$

138. $\displaystyle\int \frac{dx}{x^n\sqrt{a+bx}} = \begin{cases} -\frac{\sqrt{a+bx}}{(n-1)ax^{n-1}} - \frac{(2n-3)b}{(2n-2)a}\int \frac{dx}{x^{n-1}\sqrt{a+bx}} \\ \text{or} \\ \frac{(2n-2)!}{[(n-1)!]^2}\left[-\frac{\sqrt{a+bx}}{a}\sum_{r=1}^{n-1} \frac{r!(r-1)!}{x^r 2(r)!}\left(-\frac{b}{4a} \right)^{n-r-1} + \left(-\frac{b}{4a} \right)^{n-1}\int \frac{dx}{x\sqrt{a+bx}} \right] \end{cases}$

139. $\displaystyle\int (a+bx)^{\pm\frac{n}{2}}\,dx = \frac{2(a+bx)^{\frac{2\pm n}{2}}}{b(2\pm n)}$

140. $\displaystyle\int x(a+bx)^{\pm\frac{n}{2}}\,dx = \frac{2}{b^2}\left[\frac{(a+bx)^{\frac{4\pm n}{2}}}{4\pm n} - \frac{a(a+bx)^{\frac{2\pm n}{2}}}{2\pm n} \right]$

141. $\displaystyle\int \frac{dx}{x(a+bx)^{\frac{m}{2}}} = \frac{1}{a}\int \frac{dx}{x(a+bx)^{\frac{m-2}{2}}} - \frac{b}{a}\int \frac{dx}{(a+bx)^{\frac{m}{2}}}$

142. $\displaystyle\int \frac{(a+bx)^{n/2}\,dx}{x} = b\int (a+bx)^{(n-2)/2}\,dx + a\int \frac{(a+bx)^{(n-2)/2}}{x}\,dx$

143. $\displaystyle\int f(x,\sqrt{a+bx})\,dx = \frac{2}{b}\int f\left(\frac{z^2-a}{b}, z \right) z\,dz, \quad (z=\sqrt{a+bx})$

FORMS CONTAINING $\sqrt{a+bx}$ and $\sqrt{c+dx}$

Define $u = a+bx$, $v = c+dx$, and $k = ad-bc$. If $k=0$, then, $v = (\frac{c}{a})u$, and formulas starting with 124 should be used in place of these.

144. $\displaystyle\int \frac{dx}{\sqrt{uv}} = \begin{cases} \frac{2}{\sqrt{bd}}\tanh^{-1}\frac{\sqrt{bduv}}{bv}, & bd>0, \quad k<0 \\ \text{or} \\ \frac{2}{\sqrt{bd}}\tanh^{-1}\frac{\sqrt{bduv}}{du}, & bd>0, \quad k>0 \\ \text{or} \\ \frac{1}{\sqrt{bd}}\log\frac{(bv+\sqrt{bduv})^2}{v}, & (bd>0) \end{cases}$

145. $\displaystyle\int \frac{dx}{\sqrt{uv}} = \begin{cases} \frac{2}{\sqrt{-bd}}\tan^{-1}\frac{\sqrt{-bduv}}{bv} \\ \text{or} \\ -\frac{1}{\sqrt{-bd}}\sin^{-1}\left(\frac{2bdx+ad+bc}{|k|} \right), & (bd<0) \end{cases}$

146. $\displaystyle\int \sqrt{uv}\, dx = \frac{k+2bv}{4bd}\sqrt{uv} - \frac{k^2}{8bd}\int \frac{dx}{\sqrt{uv}}$

147. $\displaystyle\int \frac{dx}{v\sqrt{u}} = \begin{cases} \frac{1}{\sqrt{kd}}\log\frac{d\sqrt{u}-\sqrt{kd}}{d\sqrt{u}+\sqrt{kd}} \\ \text{or} \\ \frac{1}{\sqrt{kd}}\log\frac{(d\sqrt{u}-\sqrt{kd})^2}{v}, \quad (kd>0) \end{cases}$

148. $\displaystyle\int \frac{dx}{v\sqrt{u}} = \frac{2}{\sqrt{-kd}}\tan^{-1}\frac{d\sqrt{u}}{\sqrt{-kd}}, \quad (kd<0)$

149. $\displaystyle\int \frac{x\, dx}{\sqrt{uv}} = \frac{\sqrt{uv}}{bd} - \frac{ad+bc}{2bd}\int \frac{dx}{\sqrt{uv}}$

150. $\displaystyle\int \frac{dx}{v\sqrt{uv}} = \frac{-2\sqrt{uv}}{kv}$

151. $\displaystyle\int \frac{v\, dx}{\sqrt{uv}} = \frac{\sqrt{uv}}{b} - \frac{k}{2b}\int \frac{dx}{\sqrt{uv}}$

152. $\displaystyle\int \sqrt{\frac{v}{u}}\, dx = \frac{v}{|v|}\int \frac{v\, dx}{\sqrt{uv}}$

153. $\displaystyle\int v^m\sqrt{u}\, dx = \frac{1}{(2m+3)d}\left(2v^{m+1}\sqrt{u} + k\int \frac{v^m\, dx}{\sqrt{u}}\right)$

154. $\displaystyle\int \frac{dx}{v^m\sqrt{u}} = -\frac{1}{(m-1)k}\left(\frac{\sqrt{u}}{v^{m-1}} + \left(m-\frac{3}{2}\right)b\int \frac{dx}{v^{m-1}\sqrt{u}}\right)$

155. $\displaystyle\int \frac{v^m\, dx}{\sqrt{u}} = \begin{cases} \frac{2}{b(2m+1)}\left[v^m\sqrt{u} - mk\int \frac{v^{m-1}}{\sqrt{u}}\, dx\right] \\ \text{or} \\ \frac{2(m!)^2\sqrt{u}}{b(2m+1)!}\sum_{r=0}^m \left(-\frac{4k}{b}\right)^{m-r}\frac{(2r)!}{(r!)^2}v^r \end{cases}$

FORMS CONTAINING $\sqrt{x^2 \pm a^2}$

156. $\displaystyle\int \sqrt{x^2\pm a^2}\, dx = \frac{1}{2}\left[x\sqrt{x^2\pm a^2} \pm a^2\log\left(x+\sqrt{x^2\pm a^2}\right)\right]$

157. $\displaystyle\int \frac{dx}{\sqrt{x^2\pm a^2}} = \log\left(x+\sqrt{x^2\pm a^2}\right)$

158. $\displaystyle\int \frac{dx}{x\sqrt{x^2-a^2}} = \frac{1}{|a|}\sec^{-1}\frac{x}{a}$

159. $\displaystyle\int \frac{dx}{x\sqrt{x^2+a^2}} = -\frac{1}{a}\log\left(\frac{a+\sqrt{x^2+a^2}}{x}\right)$

160. $\displaystyle\int \frac{\sqrt{x^2+a^2}}{x}\, dx = \sqrt{x^2+a^2} - a\log\left(\frac{a+\sqrt{x^2+a^2}}{x}\right)$

161. $\displaystyle\int \frac{\sqrt{x^2-a^2}}{x}\, dx = \sqrt{x^2-a^2} - |a|\sec^{-1}\frac{x}{a}$

162. $\displaystyle\int \frac{x\, dx}{\sqrt{x^2\pm a^2}} = \sqrt{x^2\pm a^2}$

163. $\displaystyle\int x\sqrt{x^2\pm a^2}\, dx = \frac{1}{3}\sqrt{(x^2\pm a^2)^3}$

164. $\displaystyle\int \sqrt{(x^2\pm a^2)^3}\, dx = \frac{1}{4}\left[x\sqrt{(x^2\pm a^2)^3} \pm \frac{3a^2 x}{2}\sqrt{x^2\pm a^2} + \frac{3a^4}{2}\log(x+\sqrt{x^2\pm a^2})\right]$

165. $\displaystyle\int \frac{dx}{\sqrt{(x^2\pm a^2)^3}} = \frac{\pm x}{a^2\sqrt{x^2\pm a^2}}$

166. $\displaystyle\int \frac{x\, dx}{\sqrt{(x^2\pm a^2)^3}} = \frac{-1}{\sqrt{x^2\pm a^2}}$

167. $\displaystyle\int x\sqrt{(x^2\pm a^2)^3}\, dx = \frac{1}{5}\sqrt{(x^2\pm a^2)^5}$

168. $\displaystyle\int x^2\sqrt{x^2\pm a^2}\, dx = \frac{x}{4}\sqrt{(x^2\pm a^2)^3} \mp \frac{a^2}{8}x\sqrt{x^2\pm a^2} - \frac{a^4}{8}\log\left(x+\sqrt{x^2\pm a^2}\right)$

169. $\displaystyle\int x^3\sqrt{x^2+a^2}\,dx = (\tfrac{1}{5}x^2 - \tfrac{2}{15}a^2)\sqrt{(a^2+x^2)^3}$

170. $\displaystyle\int x^3\sqrt{x^2-a^2}\,dx = \tfrac{1}{5}\sqrt{(x^2-a^2)^5} + \tfrac{a^2}{3}\sqrt{(x^2-a^2)^3}$

171. $\displaystyle\int \frac{x^2\,dx}{\sqrt{x^2\pm a^2}} = \frac{x}{2}\sqrt{x^2\pm a^2} \mp \frac{a^2}{2}\log(x+\sqrt{x^2\pm a^2})$

172. $\displaystyle\int \frac{x^3\,dx}{\sqrt{x^2\pm a^2}} = \frac{1}{3}\sqrt{(x^2\pm a^2)^3} \mp a^2\sqrt{x^2\pm a^2}$

173. $\displaystyle\int \frac{dx}{x^2\sqrt{x^2\pm a^2}} = \mp\frac{\sqrt{x^2\pm a^2}}{a^2 x}$

174. $\displaystyle\int \frac{dx}{x^3\sqrt{x^2+a^2}} = \frac{\sqrt{x^2+a^2}}{2a^2 x^2} + \frac{1}{2a^3}\log\frac{a+\sqrt{x^2+a^2}}{x}$

175. $\displaystyle\int \frac{dx}{x^3\sqrt{x^2-a^2}} = \frac{\sqrt{x^2-a^2}}{2a^2 x^2} + \frac{1}{2|a^3|}\sec^{-1}\frac{x}{a}$

176. $\displaystyle\int x^2\sqrt{(x^2\pm a^2)^3}\,dx = \frac{x}{6}\sqrt{(x^2\pm a^2)^5} \mp \frac{a^2 x}{24}\sqrt{(x^2\pm a^2)^3} - \frac{a^4 x}{16}\sqrt{x^2\pm a^2} \mp \frac{a^6}{16}\log(x+\sqrt{x^2\pm a^2})$

177. $\displaystyle\int x^3\sqrt{(x^2\pm a^2)^3}\,dx = \frac{1}{7}\sqrt{(x^2\pm a^2)^7} \mp \frac{a^2}{5}\sqrt{(x^2\pm a^2)^5}$

178. $\displaystyle\int \frac{\sqrt{x^2\pm a^2}\,dx}{x^2} = -\frac{\sqrt{x^2\pm a^2}}{x} + \log(x+\sqrt{x^2\pm a^2})$

179. $\displaystyle\int \frac{\sqrt{x^2+a^2}}{x^3}\,dx = -\frac{\sqrt{x^2+a^2}}{2x^2} - \frac{1}{2a}\log\frac{a+\sqrt{x^2+a^2}}{x}$

180. $\displaystyle\int \frac{\sqrt{x^2-a^2}}{x^3}\,dx = -\frac{\sqrt{x^2-a^2}}{2x^2} + \frac{1}{2|a|}\sec^{-1}\frac{x}{a}$

181. $\displaystyle\int \frac{\sqrt{x^2\pm a^2}}{x^4}\,dx = \mp\frac{\sqrt{(x^2\pm a^2)^3}}{3a^2 x^3}$

182. $\displaystyle\int \frac{x^2\,dx}{\sqrt{(x^2\pm a^2)^3}} = \frac{-x}{\sqrt{x^2\pm a^2}} + \log(x+\sqrt{x^2\pm a^2})$

183. $\displaystyle\int \frac{x^3\,dx}{\sqrt{(x^2\pm a^2)^3}} = \sqrt{x^2\pm a^2} \pm \frac{a^2}{\sqrt{x^2\pm a^2}}$

184. $\displaystyle\int \frac{dx}{x\sqrt{(x^2+a^2)^3}} = \frac{1}{a^2\sqrt{x^2+a^2}} - \frac{1}{a^3}\log\frac{a+\sqrt{x^2+a^2}}{x}$

185. $\displaystyle\int \frac{dx}{x\sqrt{(x^2-a^2)^3}} = -\frac{1}{a^2\sqrt{x^2-a^2}} - \frac{1}{|a^3|}\sec^{-1}\frac{x}{a}$

186. $\displaystyle\int \frac{dx}{x^2\sqrt{(x^2\pm a^2)^3}} = -\frac{1}{a^4}\left[\frac{\sqrt{x^2\pm a^2}}{x} + \frac{x}{\sqrt{x^2\pm a^2}}\right]$

187. $\displaystyle\int \frac{dx}{x^3\sqrt{(x^2+a^2)^3}} = -\frac{1}{2a^2 x^2\sqrt{x^2+a^2}} - \frac{3}{2a^4\sqrt{x^2+a^2}} + \frac{3}{2a^5}\log\frac{a+\sqrt{x^2+a^2}}{x}$

188. $\displaystyle\int \frac{dx}{x^3\sqrt{(x^2-a^2)^3}} = \frac{1}{2a^2 x^2\sqrt{x^2-a^2}} - \frac{3}{2a^4\sqrt{x^2-a^2}} - \frac{3}{2|a^5|}\sec^{-1}\frac{x}{a}$

189. $\displaystyle\int \frac{x^m}{\sqrt{x^2\pm a^2}}\,dx = \frac{1}{m}x^{m-1}\sqrt{x^2\pm a^2} \mp \frac{m-1}{m}a^2\int\frac{x^{m-2}}{\sqrt{x^2\pm a^2}}\,dx$

190. $\displaystyle\int \frac{x^{2m}}{\sqrt{x^2\pm a^2}}\,dx = \frac{(2m)!}{2^{2m}(m!)^2}\left[\sqrt{x^2\pm a^2}\sum_{r=1}^{m}\frac{r!(r-1)!}{(2r)!}(\mp a^2)^{m-r}(2x)^{2r-1} + (\mp a^2)^m\log(x+\sqrt{x^2\pm a^2})\right]$

191. $\displaystyle\int \frac{x^{2m+1}}{\sqrt{x^2\pm a^2}}\,dx = \sqrt{x^2\pm a^2}\sum_{r=0}^{m}\frac{(2r)!(m!)^2}{(2m+1)!(r!)^2}(\mp 4a^2)^{m-r}x^{2r}$

192. $\displaystyle\int \frac{dx}{x^m\sqrt{x^2\pm a^2}} = \mp\frac{\sqrt{x^2\pm a^2}}{(m-1)a^2 x^{m-1}} \mp \frac{(m-2)}{(m-1)a^2}\int\frac{dx}{x^{m-2}\sqrt{x^2\pm a^2}}$

193. $\displaystyle\int \frac{dx}{x^{2m}\sqrt{x^2 \pm a^2}} = \sqrt{x^2 \pm a^2} \sum_{r=0}^{m-1} \frac{(m-1)!\,m!\,(2r)!\,2^{2m-2r-1}}{(r!)^2(2m)!(\mp a^2)^{m-r}x^{2r+1}}$

194. $\displaystyle\int \frac{dx}{x^{2m+1}\sqrt{x^2+a^2}} = \frac{(2m)!}{(m!)^2}\left[\frac{\sqrt{x^2+a^2}}{a^2} \sum_{r=1}^{m} (-1)^{m-r+1}\frac{r!(r-1)!}{2(2r)!(4a^2)^{m-r}x^{2r}} \right.$
$\displaystyle \left. + \frac{(-1)^{m+1}}{2^{2m}a^{2m+1}}\log \frac{\sqrt{x^2+a^2}+a}{x} \right]$

195. $\displaystyle\int \frac{dx}{x^{2m+1}\sqrt{x^2-a^2}} = \frac{(2m)!}{(m!)^2}\left[\frac{\sqrt{x^2-a^2}}{a^2} \sum_{r=1}^{m} \frac{r!(r-1)!}{2(2r)!(4a^2)^{m-r}x^{2r}} + \frac{1}{2^{2m}|a|^{2m+1}}\sec^{-1}\frac{x}{a} \right]$

196. $\displaystyle\int \frac{dx}{(x-a)\sqrt{x^2-a^2}} = -\frac{\sqrt{x^2-a^2}}{a(x-a)}$

197. $\displaystyle\int \frac{dx}{(x+a)\sqrt{x^2-a^2}} = \frac{\sqrt{x^2-a^2}}{a(x+a)}$

198. $\displaystyle\int f(x,\sqrt{x^2+a^2})\,dx = a\int f(a\tan u, a\sec u)\sec^2 u\,du, \quad \left(u = \tan^{-1}\frac{x}{a}, a > 0\right)$

199. $\displaystyle\int f(x,\sqrt{x^2-a^2})\,dx = a\int f(a\sec u, a\tan u)\sec u \tan u\,du, \quad \left(u = \sec^{-1}\frac{x}{a}, a > 0\right)$

FORMS CONTAINING $\sqrt{a^2-x^2}$

200. $\displaystyle\int \sqrt{a^2-x^2}\,dx = \frac{1}{2}\left[x\sqrt{a^2-x^2} + a^2 \sin^{-1}\frac{x}{|a|} \right]$

201. $\displaystyle\int \frac{dx}{\sqrt{a^2-x^2}} = \begin{cases} \sin^{-1}\frac{x}{|a|} \\ \text{or} \\ -\cos^{-1}\frac{x}{|a|} \end{cases}$

202. $\displaystyle\int \frac{dx}{x\sqrt{a^2-x^2}} = -\frac{1}{a}\log\left(\frac{a+\sqrt{a^2-x^2}}{x} \right)$

203. $\displaystyle\int \frac{\sqrt{a^2-x^2}}{x}\,dx = \sqrt{a^2-x^2} - a\log\left(\frac{a+\sqrt{a^2-x^2}}{x} \right)$

204. $\displaystyle\int \frac{x\,dx}{\sqrt{a^2-x^2}} = -\sqrt{a^2-x^2}$

205. $\displaystyle\int x\sqrt{a^2-x^2}\,dx = -\frac{1}{3}\sqrt{(a^2-x^2)^3}$

206. $\displaystyle\int \sqrt{(a^2-x^2)^3}\,dx = \frac{1}{4}\left[x\sqrt{(a^2-x^2)^3} + \frac{3a^2x}{2}\sqrt{a^2-x^2} + \frac{3a^4}{2}\sin^{-1}\frac{x}{|a|} \right]$

207. $\displaystyle\int \frac{dx}{\sqrt{(a^2-x^2)^3}} = \frac{x}{a^2\sqrt{a^2-x^2}}$

208. $\displaystyle\int \frac{x\,dx}{\sqrt{(a^2-x^2)^3}} = \frac{1}{\sqrt{a^2-x^2}}$

209. $\displaystyle\int x\sqrt{(a^2-x^2)^3}\,dx = -\frac{1}{5}\sqrt{(a^2-x^2)^5}$

210. $\displaystyle\int x^2\sqrt{a^2-x^2}\,dx = -\frac{x}{4}\sqrt{(a^2-x^2)^3} + \frac{a^2}{8}\left(x\sqrt{a^2-x^2} + a^2\sin^{-1}\frac{x}{|a|} \right)$

211. $\displaystyle\int x^3\sqrt{a^2-x^2}\,dx = \left(-\frac{1}{5}x^2 - \frac{2}{15}a^2\right)\sqrt{(a^2-x^2)^3}$

212. $\displaystyle\int x^2\sqrt{(a^2-x^2)^3}\,dx = -\frac{1}{6}x\sqrt{(a^2-x^2)^5} + \frac{a^2x}{24}\sqrt{(a^2-x^2)^3} + \frac{a^4x}{16}\sqrt{a^2-x^2} + \frac{a^6}{16}\sin^{-1}\frac{x}{|a|}$

213. $\displaystyle\int x^3\sqrt{(a^2-x^2)^3}\,dx = \frac{1}{7}\sqrt{(a^2-x^2)^7} - \frac{a^2}{5}\sqrt{(a^2-x^2)^5}$

214. $\displaystyle\int \frac{x^2\,dx}{\sqrt{a^2-x^2}} = -\frac{x}{2}\sqrt{a^2-x^2} + \frac{a^2}{2}\sin^{-1}\frac{x}{|a|}$

215. $\displaystyle\int \frac{dx}{x^2\sqrt{a^2-x^2}} = -\frac{\sqrt{a^2-x^2}}{a^2x}$

216. $\displaystyle\int \frac{\sqrt{a^2-x^2}}{x^2}\,dx = -\frac{\sqrt{a^2-x^2}}{x} - \sin^{-1}\frac{x}{|a|}$

217. $\int \dfrac{\sqrt{a^2 - x^2}}{x^3}\, dx = -\dfrac{\sqrt{a^2 - x^2}}{2x^2} + \dfrac{1}{2a} \log \dfrac{a + \sqrt{a^2 - x^2}}{x}$

218. $\int \dfrac{\sqrt{a^2 - x^2}}{x^4}\, dx = -\dfrac{\sqrt{(a^2 - x^2)^3}}{3a^2 x^3}$

219. $\int \dfrac{x^2\, dx}{\sqrt{(a^2 - x^2)^3}} = \dfrac{x}{\sqrt{a^2 - x^2}} - \sin^{-1} \dfrac{x}{|a|}$

220. $\int \dfrac{x^3\, dx}{\sqrt{a^2 - x^2}} = -\dfrac{2}{3}(a^2 - x^2)^{3/2} - x^2(a^2 - x^2)^{1/2} = -\dfrac{1}{3}\sqrt{a^2 - x^2}(x^2 + 2a^2)$

221. $\int \dfrac{x^3\, dx}{\sqrt{(a^2 - x^2)^3}} = 2(a^2 - x^2)^{1/2} + \dfrac{x^2}{(a^2 - x^2)^{1/2}} = -\dfrac{a^2}{\sqrt{a^2 - x^2}} + \sqrt{a^2 - x^2}$

222. $\int \dfrac{dx}{x^3 \sqrt{a^2 - x^2}} = -\dfrac{\sqrt{a^2 - x^2}}{2a^2 x^2} - \dfrac{1}{2a^3} \log \dfrac{a + \sqrt{a^2 - x^2}}{x}$

223. $\int \dfrac{dx}{x \sqrt{(a^2 - x^2)^3}} = \dfrac{1}{a^2 \sqrt{a^2 - x^2}} - \dfrac{1}{a^3} \log \dfrac{a + \sqrt{a^2 - x^2}}{x}$

224. $\int \dfrac{dx}{x^2 \sqrt{(a^2 - x^2)^3}} = \dfrac{1}{a^4}\left[-\dfrac{\sqrt{a^2 - x^2}}{x} + \dfrac{x}{\sqrt{a^2 - x^2}} \right]$

225. $\int \dfrac{dx}{x^3 \sqrt{(a^2 - x^2)^3}} = -\dfrac{1}{2a^2 x^2 \sqrt{a^2 - x^2}} + \dfrac{3}{2a^4 \sqrt{a^2 - x^2}} - \dfrac{3}{2a^5} \log \dfrac{a + \sqrt{a^2 - x^2}}{x}$

226. $\int \dfrac{x^m}{\sqrt{a^2 - x^2}}\, dx = -\dfrac{x^{m-1} \sqrt{a^2 - x^2}}{m} + \dfrac{(m-1)a^2}{m} \int \dfrac{x^{m-2}}{\sqrt{a^2 - x^2}}\, dx$

227. $\int \dfrac{x^{2m}}{\sqrt{a^2 - x^2}}\, dx = \dfrac{(2m)!}{(m!)^2}\left[-\sqrt{a^2 - x^2} \sum_{r=1}^{m} \dfrac{r!(r-1)!}{2^{2m-2r+1}(2r)!} a^{2m-2r} x^{2r-1} + \dfrac{a^{2m}}{2^{2m}} \sin^{-1} \dfrac{x}{|a|} \right]$

228. $\int \dfrac{x^{2m+1}}{\sqrt{a^2 - x^2}}\, dx = -\sqrt{a^2 - x^2} \sum_{r=0}^{m} \dfrac{(2r)!(m!)^2}{(2m+1)!(r!)^2}(4a^2)^{m-r} x^{2r}$

229. $\int \dfrac{dx}{x^m \sqrt{a^2 - x^2}} = -\dfrac{\sqrt{a^2 - x^2}}{(m-1)a^2 x^{m-1}} + \dfrac{m-2}{(m-1)a^2} \int \dfrac{dx}{x^{m-2} \sqrt{a^2 - x^2}}$

230. $\int \dfrac{ax}{x^{2m} \sqrt{a^2 - x^2}} = -\sqrt{a^2 - x^2} \sum_{r=0}^{m-1} \dfrac{(m-1)!m!(2r)!2^{2m-2r-1}}{(r!)^2 (2m)! a^{2m-2r} x^{2r+1}}$

231. $\int \dfrac{dx}{x^{2m+1} \sqrt{a^2 - x^2}} = \dfrac{(2m)!}{(m!)^2}\left[-\dfrac{\sqrt{a^2 - x^2}}{a^2} \sum_{r=1}^{m} \dfrac{r!(r-1)!}{2(2r)!(4a^2)^{m-r} x^{2r}} + \dfrac{1}{2^{2m}a^{2m+1}} \log \dfrac{a - \sqrt{a^2 - x^2}}{x} \right]$

232. $\int \dfrac{dx}{(b^2 - x^2)\sqrt{a^2 - x^2}} = \dfrac{1}{2b\sqrt{a^2 - b^2}} \log \dfrac{(b\sqrt{a^2 - x^2} + x\sqrt{a^2 - b^2})^2}{b^2 - x^2}, \quad (a^2 > b^2)$

233. $\int \dfrac{dx}{(b^2 - x^2)\sqrt{a^2 - x^2}} = \dfrac{1}{b\sqrt{b^2 - a^2}} \tan^{-1} \dfrac{x\sqrt{b^2 - a^2}}{b\sqrt{a^2 - x^2}}, \quad (b^2 > a^2)$

234. $\int \dfrac{dx}{(b^2 + x^2)\sqrt{a^2 - x^2}} = \dfrac{1}{b\sqrt{a^2 + b^2}} \tan^{-1} \dfrac{x\sqrt{a^2 + b^2}}{b\sqrt{a^2 - x^2}}$

235. $\int \dfrac{\sqrt{a^2 - x^2}}{b^2 + x^2}\, dx = \dfrac{\sqrt{a^2 + b^2}}{|b|} \sin^{-1} \dfrac{x\sqrt{a^2 + b^2}}{|a|\sqrt{x^2 + b^2}} - \sin^{-1} \dfrac{x}{|a|}$

236. $\int f(x, \sqrt{a^2 - x^2})\, dx = a \int f(a \sin u, a \cos u) \cos u\, du, \quad \left(u = \sin^{-1} \dfrac{x}{a}, a > 0 \right)$

FORMS CONTAINING $\sqrt{a + bx + cx^2}$

Define $X = a + bx + cx^2$, $q = 4ac - b^2$, and $k = \dfrac{4c}{q}$. If $q = 0$, then $\sqrt{X} = \sqrt{c}\left| x + \dfrac{b}{2c} \right|$.

237. $\int \dfrac{dx}{\sqrt{X}} = \begin{cases} \dfrac{1}{\sqrt{c}} \log(2\sqrt{cX} + 2cx + b) \\ \text{or} \\ \dfrac{1}{\sqrt{c}} \sinh^{-1} \dfrac{2cx+b}{\sqrt{q}}, \quad (c > 0) \end{cases}$

238. $\int \dfrac{dx}{\sqrt{X}} = -\dfrac{1}{\sqrt{-c}} \sin^{-1} \dfrac{2cx + b}{\sqrt{-q}}, \quad (c < 0)$

239. $\displaystyle\int \frac{dx}{X\sqrt{x}} = \frac{2(2cx+b)}{q\sqrt{x}}$

240. $\displaystyle\int \frac{dx}{X^2\sqrt{x}} = \frac{2(2cx+b)}{3q\sqrt{x}}\left(\frac{1}{X} + 2k\right)$

241. $\displaystyle\int \frac{dx}{X^n\sqrt{x}} = \begin{cases} \frac{2(2cx+b)\sqrt{x}}{(2n-1)qX^n} + \frac{2k(n-1)}{2n-1}\int \frac{dx}{X^{n-1}\sqrt{x}} \\ \text{or} \\ \frac{(2cx+b)(n!)(n-1)!4^nk^{n-1}}{q[(2n)!]\sqrt{x}}\sum_{r=0}^{n-1} \frac{(2r)!}{(4kX)^r(r!)^2} \end{cases}$

242. $\displaystyle\int \sqrt{x}\,dx = \frac{(2cx+b)\sqrt{x}}{4c} + \frac{1}{2k}\int \frac{dx}{\sqrt{x}}$

243. $\displaystyle\int X\sqrt{x}\,dx = \frac{(2cx+b)\sqrt{x}}{8c}\left(X + \frac{3}{2k}\right) + \frac{3}{8k^2}\int \frac{dx}{\sqrt{x}}$

244. $\displaystyle\int X^2\sqrt{x}\,dx = \frac{(2cx+b)\sqrt{x}}{12c}\left(X^2 + \frac{5X}{4k} + \frac{15}{8k^2}\right) + \frac{5}{16k^3}\int \frac{dx}{\sqrt{x}}$

245. $\displaystyle\int X^n\sqrt{x}\,dx = \begin{cases} \frac{(2cx+b)X^n\sqrt{x}}{4(n+1)c} + \frac{2n+1}{2(n+1)k}\int X^{n-1}\sqrt{x}\,dx \\ \text{or} \\ \frac{(2n+2)!}{[(n+1)!]^2(4k)^{n+1}}\left[\frac{k(2cx+b)\sqrt{x}}{c}\sum_{r=0}^{n} \frac{r!(r+1)!(4kX)^r}{(2r+2)!} + \int \frac{dx}{\sqrt{x}}\right] \end{cases}$

246. $\displaystyle\int \frac{x\,dx}{\sqrt{x}} = \frac{\sqrt{x}}{c} - \frac{b}{2c}\int \frac{dx}{\sqrt{x}}$

247. $\displaystyle\int \frac{x\,dx}{X\sqrt{x}} = -\frac{2(bx+2a)}{q\sqrt{x}}$

248. $\displaystyle\int \frac{x\,dx}{X^n\sqrt{x}} = -\frac{\sqrt{x}}{(2n-1)cX^n} - \frac{b}{2c}\int \frac{dx}{X^n\sqrt{x}}$

249. $\displaystyle\int \frac{x^2\,dx}{\sqrt{x}} = \left(\frac{x}{2c} - \frac{3b}{4c^2}\right)\sqrt{x} + \frac{3b^2-4ac}{8c^2}\int \frac{dx}{\sqrt{x}}$

250. $\displaystyle\int \frac{x^2\,dx}{X\sqrt{x}} = \frac{(2b^2-4ac)x+2ab}{cq\sqrt{x}} + \frac{1}{c}\int \frac{dx}{\sqrt{x}}$

251. $\displaystyle\int \frac{x^2\,dx}{X^n\sqrt{x}} = \frac{(2b^2-4ac)x+2ab}{(2n-1)cq\,X^{n-1}\sqrt{x}} + \frac{4ac+(2n-3)b^2}{(2n-1)cq}\int \frac{dx}{X^{n-1}\sqrt{x}}$

252. $\displaystyle\int \frac{x^3\,dx}{\sqrt{x}} = \left(\frac{x^2}{3c} - \frac{5bx}{12c^2} + \frac{5b^2}{8c^3} - \frac{2a}{3c^2}\right)\sqrt{x} + \left(\frac{3ab}{4c^2} - \frac{5b^3}{16c^3}\right)\int \frac{dx}{\sqrt{x}}$

253. $\displaystyle\int \frac{x^n\,dx}{\sqrt{x}} = \frac{1}{nc}x^{n-1}\sqrt{x} - \frac{(2n-1)b}{2nc}\int \frac{x^{n-1}\,dx}{\sqrt{x}} - \frac{(n-1)a}{nc}\int \frac{x^{n-2}\,dx}{\sqrt{x}}$

254. $\displaystyle\int x\sqrt{x}\,dx = \frac{X\sqrt{x}}{3c} - \frac{b(2cx+b)}{8c^2}\sqrt{x} - \frac{b}{4ck}\int \frac{dx}{\sqrt{x}}$

255. $\displaystyle\int xX\sqrt{x}\,dx = \frac{X^2\sqrt{x}}{5c} - \frac{b}{2c}\int X\sqrt{x}\,dx$

256. $\displaystyle\int xX^n\sqrt{x}\,dx = \frac{X^{n+1}\sqrt{x}}{(2n+3)c} - \frac{b}{2c}\int X^n\sqrt{x}\,dx$

257. $\displaystyle\int x^2\sqrt{x}\,dx = \left(x - \frac{5b}{6c}\right)\frac{X\sqrt{x}}{4c} + \frac{5b^2-4ac}{16c^2}\int \sqrt{x}\,dx$

258. $\displaystyle\int \frac{dx}{x\sqrt{x}} = -\frac{1}{\sqrt{a}}\log \frac{2\sqrt{aX}+bx+2a}{x}, \quad (a>0)$

259. $\displaystyle\int \frac{dx}{x\sqrt{x}} = \frac{1}{\sqrt{-a}}\sin^{-1}\left(\frac{bx+2a}{|x|\sqrt{-q}}\right), \quad (a<0)$

260. $\displaystyle\int \frac{dx}{x\sqrt{x}} = -\frac{2\sqrt{x}}{bx}, \quad (a=0)$

261. $\displaystyle\int \frac{dx}{x^2\sqrt{x}} = -\frac{\sqrt{x}}{ax} - \frac{b}{2a}\int \frac{dx}{x\sqrt{x}}$

262. $\displaystyle\int \frac{\sqrt{x}\,dx}{x} = \sqrt{x} + \frac{b}{2}\int \frac{dx}{\sqrt{x}} + a\int \frac{dx}{x\sqrt{x}}$

263. $\displaystyle\int \frac{\sqrt{x}\,dx}{x^2} = -\frac{\sqrt{x}}{x} + \frac{b}{2}\int \frac{dx}{x\sqrt{x}} + c\int \frac{dx}{\sqrt{x}}$

FORMS INVOLVING $\sqrt{2ax - x^2}$

264. $\displaystyle \int \sqrt{2ax - x^2}\, dx = \frac{1}{2}\left[(x - a)\sqrt{2ax - x^2} + a^2 \sin^{-1}\frac{x - a}{|a|}\right]$

265. $\displaystyle \int \frac{dx}{\sqrt{2ax - x^2}} = \begin{cases} \cos^{-1}\frac{a - x}{|a|} \\ \text{or} \\ \sin^{-1}\frac{x - a}{|a|} \end{cases}$

266. $\displaystyle \int x^n \sqrt{2ax - x^2}\, dx = \begin{cases} -\frac{x^{n-1}(2ax - x^2)^{3/2}}{n + 2} + \frac{(2n+1)a}{n+2}\int x^{n-1}\sqrt{2ax - x^2}\, dx \\ \text{or} \\ \sqrt{2ax - x^2}\left[\frac{x^{n+1}}{n+2} - \sum_{r=0}^{n} \frac{(2n+1)!(r!)^2 a^{n-r+1}}{2^{n-r}(2r+1)!(n+2)!n!}x^r\right] \\ \quad + \frac{(2n+1)!a^{n+2}}{2^n n!(n+2)!}\sin^{-1}\frac{x - a}{|a|} \end{cases}$

267. $\displaystyle \int \frac{\sqrt{2ax - x^2}}{x^n}\, dx = \frac{(2ax - x^2)^{1/2}}{(3 - 2n)ax^n} + \frac{n - 3}{(2n - 3)a}\int \frac{\sqrt{2ax - x^2}}{x^{n-1}}\, dx$

268. $\displaystyle \int \frac{x^n\, dx}{\sqrt{2ax - x^2}} = \begin{cases} \frac{-x^{n-1}\sqrt{2ax - x^2}}{n} + \frac{a(2n-1)}{n}\int \frac{x^{n-1}}{\sqrt{2ax - x^2}}\, dx \\ \text{or} \\ -\sqrt{2ax - x^2}\sum_{r=1}^{n} \frac{(2n)!r!(r-1)!a^{n-r}}{2^{n-r}(2r)!(n!)^2}x^{r-1} + \frac{(2n)!a^n}{2^n(n!)^2}\sin^{-1}\frac{x - a}{|a|} \end{cases}$

269. $\displaystyle \int \frac{dx}{x^n\sqrt{2ax - x^2}} = \begin{cases} \frac{\sqrt{2ax - x^2}}{a(1 - 2n)x^n} + \frac{n - 1}{(2n-1)a}\int \frac{dx}{x^{n-1}\sqrt{2ax - x^2}} \\ \text{or} \\ -\sqrt{2ax - x^2}\sum_{r=0}^{n-1} \frac{2^{n-r}(n-1)!n!(2r)!}{(2n)!(r!)^2 a^{n-r}x^{r+1}} \end{cases}$

270. $\displaystyle \int \frac{dx}{(2ax - x^2)^{3/2}} = \frac{x - a}{a^2\sqrt{2ax - x^2}}$

271. $\displaystyle \int \frac{x\, dx}{(2ax - x^2)^{3/2}} = \frac{x}{a\sqrt{2ax - x^2}}$

MISCELLANEOUS ALGEBRAIC FORMS

272. $\displaystyle \int \frac{dx}{\sqrt{2ax + x^2}} = \log(x + a + \sqrt{2ax + x^2})$

273. $\displaystyle \int \sqrt{ax^2 + c}\, dx = \frac{x}{2}\sqrt{ax^2 + c} + \frac{c}{2\sqrt{a}}\log\left(x\sqrt{a} + \sqrt{ax^2 + c}\right), \quad (a > 0)$

274. $\displaystyle \int \sqrt{ax^2 + c}\, dx = \frac{x}{2}\sqrt{ax^2 + c} + \frac{c}{2\sqrt{-a}}\sin^{-1}\left(x\sqrt{-\frac{a}{c}}\right), \quad (a < 0)$

275. $\displaystyle \int \sqrt{\frac{1 + x}{1 - x}}\, dx = \sin^{-1}x - \sqrt{1 - x^2}$

276. $\displaystyle \int \frac{dx}{x\sqrt{ax^n + c}} = \begin{cases} \frac{1}{n\sqrt{c}}\log\frac{\sqrt{ax^n + c} - \sqrt{c}}{\sqrt{ax^n + c} + \sqrt{c}} \\ \text{or} \\ \frac{2}{n\sqrt{c}}\log\frac{\sqrt{ax^n + c} - \sqrt{c}}{\sqrt{x^n}}, \quad (c > 0) \end{cases}$

277. $\displaystyle \int \frac{dx}{x\sqrt{ax^n + c}} = \frac{2}{n\sqrt{-c}}\sec^{-1}\sqrt{-\frac{ax^n}{c}}, \quad (c < 0)$

278. $\displaystyle \int \frac{dx}{\sqrt{ax^2 + c}} = \frac{1}{\sqrt{a}}\log(x\sqrt{a} + \sqrt{ax^2 + c}), \quad (a > 0)$

279. $\displaystyle \int \frac{dx}{\sqrt{ax^2 + c}} = \frac{1}{\sqrt{-a}}\sin^{-1}\left(x\sqrt{-\frac{a}{c}}\right), \quad (a < 0)$

280. $\displaystyle \int (ax^2 + c)^{m+1/2}\, dx = \begin{cases} \frac{x(ax^2 + c)^{m+1/2}}{2(m+1)} + \frac{(2m+1)c}{2(m+1)}\int (ax^2 + c)^{m-1/2}\, dx \\ \text{or} \\ x\sqrt{ax^2 + c}\sum_{r=0}^{m} \frac{(2m+1)!(r!)^2 c^{m-r}}{2^{2m-r+1}m!(m+1)!(2r+1)!}(ax^2 + c)^r \\ \quad + \frac{(2m+1)!c^{m+1}}{2^{2m+1}m!(m+1)!}\int \frac{dx}{\sqrt{ax^2 + c}} \end{cases}$

281. $\displaystyle \int x(ax^2 + c)^{m+\frac{1}{2}}\, dx = \frac{(ax^2 + c)^{m+\frac{3}{2}}}{(2m + 3)a}$

282. $\displaystyle\int \frac{(ax^2+c)^{m+1/2}}{x}\, dx = \begin{cases} \frac{(ax^2+c)^{m+1/2}}{2m+1} + c \displaystyle\int \frac{(ax^2+c)^{m-1/2}}{x}\, dx \\ \text{or} \\ \sqrt{ax^2+c}\ \sum_{r=0}^{m} \frac{c^{m-r}(ax^2+c)^r}{2r+1} + c^{m+1} \displaystyle\int \frac{dx}{x\sqrt{ax^2+c}} \end{cases}$

283. $\displaystyle\int \frac{dx}{(ax^2+c)^{m+1/2}} = \begin{cases} \frac{x}{(2m-1)c(ax^2+c)^{m-1/2}} + \frac{2m-2}{(2m-1)c} \displaystyle\int \frac{dx}{(ax^2+c)^{m-1/2}} \\ \text{or} \\ \frac{x}{\sqrt{ax^2+c}} \sum_{r=0}^{m-1} \frac{2^{2m-2r-1}(m-1)!\,m!\,(2r)!}{(2m)!\,(r!)^2\,c^{m-r}(ax^2+c)^r} \end{cases}$

284. $\displaystyle\int \frac{dx}{x^m\sqrt{ax^2+c}} = -\frac{\sqrt{ax^2+c}}{(m-1)cx^{m-1}} - \frac{(m-2)a}{(m-1)c} \int \frac{dx}{x^{m-2}\sqrt{ax^2+c}}$

285. $\displaystyle\int \frac{1+x^2}{(1-x^2)\sqrt{1+x^4}}\, dx = \frac{1}{\sqrt{2}} \log \frac{x\sqrt{2}+\sqrt{1+x^4}}{1-x^2}$

286. $\displaystyle\int \frac{1-x^2}{(1+x^2)\sqrt{1+x^4}}\, dx = \frac{1}{\sqrt{2}} \tan^{-1} \frac{x\sqrt{2}}{\sqrt{1+x^4}}$

287. $\displaystyle\int \frac{dx}{x\sqrt{x^n+a^2}} = -\frac{2}{na} \log \frac{a+\sqrt{x^n+a^2}}{\sqrt{x^n}}$

288. $\displaystyle\int \frac{dx}{x\sqrt{x^n-a^2}} = -\frac{2}{na} \sin^{-1} \frac{a}{\sqrt{x^n}}$

289. $\displaystyle\int \sqrt{\frac{x}{a^3-x^3}}\, dx = \frac{2}{3} \sin^{-1} \left(\frac{x}{a}\right)^{3/2}$

FORMS INVOLVING TRIGONOMETRIC FUNCTIONS

290. $\displaystyle\int (\sin ax)\, dx = -\frac{1}{a} \cos ax$

291. $\displaystyle\int (\cos ax)\, dx = \frac{1}{a} \sin ax$

292. $\displaystyle\int (\tan ax)\, dx = -\frac{1}{a} \log \cos ax = \frac{1}{a} \log \sec ax$

293. $\displaystyle\int (\cot ax)\, dx = \frac{1}{a} \log \sin ax = -\frac{1}{a} \log \csc ax$

294. $\displaystyle\int (\sec ax)\, dx = \frac{1}{a} \log(\sec ax + \tan ax) = \frac{1}{a} \log \tan \left(\frac{\pi}{4} + \frac{ax}{2}\right)$

295. $\displaystyle\int (\csc ax)\, dx = \frac{1}{a} \log(\csc ax - \cot ax) = \frac{1}{a} \log \tan \frac{ax}{2}$

296. $\displaystyle\int (\sin^2 ax)\, dx = -\frac{1}{2a} \cos ax \sin ax + \frac{1}{2}x = \frac{1}{2}x - \frac{1}{4a} \sin 2ax$

297. $\displaystyle\int (\sin^3 ax)\, dx = -\frac{1}{3a} (\cos ax)(\sin^2 ax + 2)$

298. $\displaystyle\int (\sin^4 ax)\, dx = \frac{3x}{8} - \frac{\sin 2ax}{4a} + \frac{\sin 4ax}{32a}$

299. $\displaystyle\int (\sin^n ax)\, dx = -\frac{\sin^{n-1} ax \cos ax}{na} + \frac{n-1}{n} \int (\sin^{n-2} ax)\, dx$

300. $\displaystyle\int (\sin^{2m} ax)\, dx = -\frac{\cos ax}{a} \sum_{r=0}^{m-1} \frac{(2m)!\,(r!)^2}{2^{2m-2r}(2r+1)!\,(m!)^2} \sin^{2r+1} ax + \frac{(2m)!}{2^{2m}(m!)^2} x$

301. $\displaystyle\int (\sin^{2m+1} ax)\, dx = -\frac{\cos ax}{a} \sum_{r=0}^{m} \frac{2^{2m-2r}(m!)^2(2r)!}{(2m+1)!\,(r!)^2} \sin^{2r} ax$

302. $\displaystyle\int (\cos^2 ax)\, dx = \frac{1}{2a} \sin ax \cos ax + \frac{1}{2}x = \frac{1}{2}x + \frac{1}{4a} \sin 2ax$

303. $\displaystyle\int (\cos^3 ax)\, dx = \frac{1}{3a} (\sin ax)(\cos^2 ax + 2)$

304. $\displaystyle\int (\cos^4 ax)\, dx = \frac{3x}{8} + \frac{\sin 2ax}{4a} + \frac{\sin 4ax}{32a}$

305. $\displaystyle\int (\cos^n ax)\, dx = \frac{1}{na}\cos^{n-1} ax \sin ax + \frac{n-1}{n}\int (\cos^{n-2} ax)\, dx$

306. $\displaystyle\int (\cos^{2m} ax)\, dx = \frac{\sin ax}{a}\sum_{r=0}^{m-1}\frac{(2m)!(r!)^2}{2^{2m-2r}(2r+1)!(m!)^2}\cos^{2r+1} ax + \frac{(2m)!}{2^{2m}(m!)^2}x$

307. $\displaystyle\int (\cos^{2m+1} ax)\, dx = \frac{\sin ax}{a}\sum_{r=0}^{m}\frac{2^{2m-2r}(m!)^2(2r)!}{(2m+1)!(r!)^2}\cos^{2r} ax$

308. $\displaystyle\int \frac{dx}{\sin^2 ax} = \int (\csc^2 ax)\, dx = -\frac{1}{a}\cot ax$

309. $\displaystyle\int \frac{dx}{\sin^m ax} = \int (\csc^m ax)\, dx = -\frac{1}{(m-1)a}\cdot\frac{\cos ax}{\sin^{m-1} ax} + \frac{m-2}{m-1}\int \frac{dx}{\sin^{m-2} ax}$

310. $\displaystyle\int \frac{dx}{\sin^{2m} ax} = \int (\csc^{2m} ax)\, dx = -\frac{1}{a}\cos ax \sum_{r=0}^{m-1}\frac{2^{2m-2r-1}(m-1)!m!(2r)!}{(2m)!(r!)^2 \sin^{2r+1} ax}$

311. $\displaystyle\int \frac{dx}{\sin^{2m+1} ax} = \int (\csc^{2m+1} ax)\, dx = -\frac{1}{a}\cos ax \sum_{r=0}^{m-1}\frac{(2m)!(r!)^2}{2^{2m-2r}(m!)^2(2r+1)! \sin^{2r+2} ax} + \frac{1}{a}\cdot\frac{(2m)!}{2^{2m}(m!)^2}\log\tan\frac{ax}{2}$

312. $\displaystyle\int \frac{dx}{\cos^2 ax} = \int (\sec^2 ax)\, dx = \frac{1}{a}\tan ax$

313. $\displaystyle\int \frac{dx}{\cos^n ax} = \int (\sec^n ax)\, dx = \frac{1}{(n-1)a}\cdot\frac{\sin ax}{\cos^{n-1} ax} + \frac{n-2}{n-1}\int \frac{dx}{\cos^{n-2} ax}$

314. $\displaystyle\int \frac{dx}{\cos^{2m} ax} = \int (\sec^{2m} ax)\, dx = \frac{1}{a}\sin ax \sum_{r=0}^{m-1}\frac{2^{2m-2r-1}(m-1)!m!(2r)!}{(2m)!(r!)^2 \cos^{2r+1} ax}$

315. $\displaystyle\int \frac{dx}{\cos^{2m+1} ax} = \int (\sec^{2m+1} ax)\, dx = \frac{1}{a}\sin ax \sum_{r=0}^{m-1}\frac{(2m)!(r!)^2}{2^{2m-2r}(m!)^2(2r+1)! \cos^{2r+2} ax} + \frac{1}{a}\cdot\frac{(2m)!}{2^{2m}(m!)^2}\log(\sec ax + \tan ax)$

316. $\displaystyle\int (\sin mx)(\sin nx)\, dx = \frac{\sin(m-n)x}{2(m-n)} - \frac{\sin(m+n)x}{2(m+n)}, \quad (m^2 \neq n^2)$

317. $\displaystyle\int (\cos mx)(\cos nx)\, dx = \frac{\sin(m-n)x}{2(m-n)} + \frac{\sin(m+n)x}{2(m+n)}, \quad (m^2 \neq n^2)$

318. $\displaystyle\int (\sin ax)(\cos ax)\, dx = \frac{1}{2a}\sin^2 ax$

319. $\displaystyle\int (\sin mx)(\cos nx)\, dx = -\frac{\cos(m-n)x}{2(m-n)} - \frac{\cos(m+n)x}{2(m+n)}, \quad (m^2 \neq n^2)$

320. $\displaystyle\int (\sin^2 ax)(\cos^2 ax)\, dx = -\frac{1}{32a}\sin 4ax + \frac{x}{8}$

321. $\displaystyle\int (\sin ax)(\cos^m ax)\, dx = -\frac{\cos^{m+1} ax}{(m+1)a}$

322. $\displaystyle\int (\sin^m ax)(\cos ax)\, dx = \frac{\sin^{m+1} ax}{(m+1)a}$

323. $\displaystyle\int (\cos^m ax)(\sin^n ax)\, dx = \begin{cases} \frac{\cos^{m-1} ax \sin^{n+1} ax}{(m+n)a} + \frac{m-1}{m+n}\int (\cos^{m-2} ax)(\sin^n ax)\, dx \\ \text{or} \\ -\frac{\sin^{n-1} ax \cos^{m+1} ax}{(m+n)a} + \frac{n-1}{m+n}\int (\cos^m ax)(\sin^{n-2} ax)\, dx \end{cases}$

324. $\displaystyle\int \frac{\cos^m ax}{\sin^n ax}\, dx = \begin{cases} -\frac{\cos^{m+1} ax}{(n-1)a \sin^{n-1} ax} - \frac{m-n+2}{n-1}\int \frac{\cos^m ax}{\sin^{n-2} ax}\, dx \\ \text{or} \\ \frac{\cos^{m-1} ax}{a(m-n)\sin^{n-1} ax} + \frac{m-1}{m-n}\int \frac{\cos^{m-2} ax}{\sin^n ax}\, dx \end{cases}$

325. $\displaystyle\int \frac{\sin^m ax}{\cos^n ax}\, dx = \begin{cases} \frac{\sin^{m+1} ax}{a(n-1)\cos^{n-1} ax} - \frac{m-n+2}{n-1}\int \frac{\sin^m ax}{\cos^{n-2} ax}\, dx \\ \text{or} \\ -\frac{\sin^{m-1} ax}{a(m-n)\cos^{n-1} ax} + \frac{m-1}{m-n}\int \frac{\sin^{m-2} ax}{\cos^n ax}\, dx \end{cases}$

326. $\displaystyle\int \frac{\sin ax}{\cos^2 ax}\, dx = \frac{1}{a\cos ax} = \frac{\sec ax}{a}$

327. $\int \dfrac{\sin^2 ax}{\cos ax}\, dx = -\dfrac{1}{a}\sin ax + \dfrac{1}{a}\log\tan\left(\dfrac{\pi}{4} + \dfrac{ax}{2}\right)$

328. $\int \dfrac{\cos ax}{\sin^2 ax}\, dx = -\dfrac{1}{a\sin ax} = -\dfrac{\csc ax}{a}$

329. $\int \dfrac{dx}{(\sin ax)(\cos ax)} = \dfrac{1}{a}\log\tan ax$

330. $\int \dfrac{dx}{(\sin ax)(\cos^2 ax)} = \dfrac{1}{a}\left(\sec ax + \log\tan\dfrac{ax}{2}\right)$

331. $\int \dfrac{dx}{(\sin ax)(\cos^n ax)} = \dfrac{1}{a(n-1)\cos^{n-1} ax} + \int \dfrac{dx}{(\sin ax)(\cos^{n-2} ax)}$

332. $\int \dfrac{dx}{(\sin^2 ax)(\cos ax)} = -\dfrac{1}{a}\csc ax + \dfrac{1}{a}\log\tan\left(\dfrac{\pi}{4} + \dfrac{ax}{2}\right)$

333. $\int \dfrac{dx}{(\sin^2 ax)(\cos^2 ax)} = -\dfrac{2}{a}\cot 2ax$

334. $\int \dfrac{dx}{\sin^m ax \cos^n ax} = \begin{cases} -\dfrac{1}{a(m-1)(\sin^{m-1} ax)(\cos^{n-1} ax)} \\ \quad + \dfrac{m+n-2}{m-1}\displaystyle\int \dfrac{dx}{(\sin^{m-2} ax)(\cos^n ax)} \\ \text{or} \\ \dfrac{1}{a(n-1)\sin^{m-1} ax\cos^{n-1} ax} + \dfrac{m+n-2}{n-1}\displaystyle\int \dfrac{dx}{\sin^m ax\cos^{n-2} ax} \end{cases}$

335. $\int \sin(a+bx)\, dx = -\dfrac{1}{b}\cos(a+bx)$

336. $\int \cos(a+bx)\, dx = \dfrac{1}{b}\sin(a+bx)$

337. $\int \dfrac{dx}{1 \pm \sin ax} = \mp\dfrac{1}{a}\tan\left(\dfrac{\pi}{4} \mp \dfrac{ax}{2}\right)$

338. $\int \dfrac{dx}{1 + \cos ax} = \dfrac{1}{a}\tan\dfrac{ax}{2}$

339. $\int \dfrac{dx}{1 - \cos ax} = -\dfrac{1}{a}\cot\dfrac{ax}{2}$

340. $\int \dfrac{dx}{a + b\sin x} = \begin{cases} \dfrac{2}{\sqrt{a^2-b^2}}\tan^{-1}\dfrac{a\tan\frac{x}{2}+b}{\sqrt{a^2-b^2}} \\ \text{or} \\ \dfrac{1}{\sqrt{b^2-a^2}}\log\dfrac{a\tan\frac{x}{2}+b-\sqrt{b^2-a^2}}{a\tan\frac{x}{2}+b+\sqrt{b^2-a^2}} \end{cases}$

341. $\int \dfrac{dx}{a + b\cos x} = \begin{cases} \dfrac{2}{\sqrt{a^2-b^2}}\tan^{-1}\dfrac{\sqrt{a^2-b^2}\tan\frac{x}{2}}{a+b} \\ \text{or} \\ \dfrac{1}{\sqrt{b^2-a^2}}\log\left(\dfrac{\sqrt{b^2-a^2}\tan\frac{x}{2}+a+b}{\sqrt{b^2-a^2}\tan\frac{x}{2}-a-b}\right) \end{cases}$

342. $\int \dfrac{dx}{a + b\sin x + c\cos x}$

$= \begin{cases} \dfrac{1}{\sqrt{b^2+c^2-a^2}}\log\left(\dfrac{b-\sqrt{b^2+c^2-a^2}+(a-c)\tan\frac{x}{2}}{b+\sqrt{b^2+c^2-a^2}+(a-c)\tan\frac{x}{2}}\right) & (\text{if } a^2 < b^2+c^2,\, a\neq c), \\[2mm] \dfrac{2}{\sqrt{a^2-b^2-c^2}}\tan^{-1}\left(\dfrac{b+(a-c)\tan\frac{x}{2}}{\sqrt{a^2-b^2-c^2}}\right) & (\text{if } a^2 > b^2+c^2), \\[2mm] \dfrac{1}{a}\left[\dfrac{a-(b+c)\cos x-(b-c)\sin x}{a-(b-c)\cos x+(b+c)\sin x}\right] & (\text{if } a^2 = b^2+c^2,\, a\neq c). \end{cases}$

343. $\int \dfrac{\sin^2 x\, dx}{a + b\cos^2 x} = \dfrac{1}{b}\sqrt{\dfrac{a+b}{a}}\tan^{-1}\left(\sqrt{\dfrac{a}{a+b}}\tan x\right) - \dfrac{x}{b}, \quad (ab>0, \text{ or } |a|>|b|)$

344. $\int \dfrac{dx}{a^2\cos^2 x + b^2\sin^2 x} = \dfrac{1}{ab}\tan^{-1}\left(\dfrac{b\tan x}{a}\right)$

345. $\int \dfrac{\cos^2 cx}{a^2 + b^2\sin^2 cx}\, dx = \dfrac{\sqrt{a^2+b^2}}{ab^2 c}\tan^{-1}\dfrac{\sqrt{a^2+b^2}\tan cx}{a} - \dfrac{x}{b^2}$

346. $\int \dfrac{\sin cx\cos cx}{a\cos^2 cx + b\sin^2 cx}\, dx = \dfrac{1}{2c(b-a)}\log(a\cos^2 cx + b\sin^2 cx)$

347. $\displaystyle \int \frac{\cos cx}{a\cos cx + b\sin cx}\,dx = \int \frac{dx}{a + b\tan cx}$
$$= \frac{1}{c(a^2+b^2)}[acx + b\log(a\cos cx + b\sin cx)]$$

348. $\displaystyle \int \frac{\sin cx}{a\sin cx + b\cos cx}\,dx = \int \frac{dx}{a + b\cot cx} = \frac{1}{c(a^2+b^2)}[acx - b\log(a\sin cx + b\cos cx)]$

349. $\displaystyle \int \frac{dx}{a\cos^2 x + 2b\cos x\sin x + c\sin^2 x} = \begin{cases} \dfrac{1}{2\sqrt{b^2-ac}}\log\dfrac{c\tan x + b - \sqrt{b^2-ac}}{c\tan x + b + \sqrt{b^2-ac}}, & (b^2 > ac) \\ \text{or} \\ \dfrac{1}{\sqrt{ac-b^2}}\tan^{-1}\dfrac{c\tan x + b}{\sqrt{ac-b^2}}, & (b^2 < ac) \\ \text{or} \\ -\dfrac{1}{c\tan x + b}, & (b^2 = ac) \end{cases}$

350. $\displaystyle \int \frac{\sin ax}{1 \pm \sin ax}\,dx = \pm x + \frac{1}{a}\tan\left(\frac{\pi}{4} \mp \frac{ax}{2}\right)$

351. $\displaystyle \int \frac{dx}{(\sin ax)(1 \pm \sin ax)} = \frac{1}{a}\tan\left(\frac{\pi}{4} \mp \frac{ax}{2}\right) + \frac{1}{a}\log\tan\frac{ax}{2}$

352. $\displaystyle \int \frac{dx}{(1 + \sin ax)^2} = -\frac{1}{2a}\tan\left(\frac{\pi}{4} - \frac{ax}{2}\right) - \frac{1}{6a}\tan^3\left(\frac{\pi}{4} - \frac{ax}{2}\right)$

353. $\displaystyle \int \frac{dx}{(1 - \sin ax)^2} = \frac{1}{2a}\cot\left(\frac{\pi}{4} - \frac{ax}{2}\right) + \frac{1}{6a}\cot^3\left(\frac{\pi}{4} - \frac{ax}{2}\right)$

354. $\displaystyle \int \frac{\sin ax}{(1 + \sin ax)^2}\,dx = -\frac{1}{2a}\tan\left(\frac{\pi}{4} - \frac{ax}{2}\right) + \frac{1}{6a}\tan^3\left(\frac{\pi}{4} - \frac{ax}{2}\right)$

355. $\displaystyle \int \frac{\sin ax}{(1 - \sin ax)^2}\,dx = -\frac{1}{2a}\cot\left(\frac{\pi}{4} - \frac{ax}{2}\right) + \frac{1}{6a}\cot^3\left(\frac{\pi}{4} - \frac{ax}{2}\right)$

356. $\displaystyle \int \frac{\sin x\,dx}{a + b\sin x} = \frac{x}{b} - \frac{a}{b}\int \frac{dx}{a + b\sin x}$

357. $\displaystyle \int \frac{dx}{(\sin x)(a + b\sin x)} = \frac{1}{a}\log\tan\frac{x}{2} - \frac{b}{a}\int \frac{dx}{a + b\sin x}$

358. $\displaystyle \int \frac{dx}{(a + b\sin x)^2} = \frac{b\cos x}{(a^2 - b^2)(a + b\sin x)} + \frac{a}{a^2 - b^2}\int \frac{dx}{a + b\sin x}$

359. $\displaystyle \int \frac{\sin x\,dx}{(a + b\sin x)^2} = \frac{a\cos x}{(b^2 - a^2)(a + b\sin x)} + \frac{b}{b^2 - a^2}\int \frac{dx}{a + b\sin x}$

360. $\displaystyle \int \frac{dx}{a^2 + b^2\sin^2 cx} = \frac{1}{ac\sqrt{a^2 + b^2}}\tan^{-1}\frac{\sqrt{a^2 + b^2}\tan cx}{a}$

361. $\displaystyle \int \frac{dx}{a^2 - b^2\sin^2 cx} = \begin{cases} \dfrac{1}{ac\sqrt{a^2-b^2}}\tan^{-1}\dfrac{\sqrt{a^2-b^2}\tan cx}{a}, & (a^2 > b^2) \\ \text{or} \\ \dfrac{1}{2ac\sqrt{b^2-a^2}}\log\dfrac{\sqrt{b^2-a^2}\tan cx + a}{\sqrt{b^2-a^2}\tan cx - a}, & (a^2 < b^2) \end{cases}$

362. $\displaystyle \int \frac{\cos ax}{1 + \cos ax}\,dx = x - \frac{1}{a}\tan\frac{ax}{2}$

363. $\displaystyle \int \frac{\cos ax}{1 - \cos ax}\,dx = -x - \frac{1}{a}\cot\frac{ax}{2}$

364. $\displaystyle \int \frac{dx}{(\cos ax)(1 + \cos ax)} = \frac{1}{a}\log\tan\left(\frac{\pi}{4} + \frac{ax}{2}\right) - \frac{1}{a}\tan\frac{ax}{2}$

365. $\displaystyle \int \frac{dx}{(\cos ax)(1 - \cos ax)} = \frac{1}{a}\log\tan\left(\frac{\pi}{4} + \frac{ax}{2}\right) - \frac{1}{a}\cot\frac{ax}{2}$

366. $\displaystyle \int \frac{dx}{(1 + \cos ax)^2} = \frac{1}{2a}\tan\frac{ax}{2} + \frac{1}{6a}\tan^3\frac{ax}{2}$

367. $\displaystyle \int \frac{dx}{(1 - \cos ax)^2} = -\frac{1}{2a}\cot\frac{ax}{2} - \frac{1}{6a}\cot^3\frac{ax}{2}$

368. $\displaystyle \int \frac{\cos ax}{(1 + \cos ax)^2}\,dx = \frac{1}{2a}\tan\frac{ax}{2} - \frac{1}{6a}\tan^3\frac{ax}{2}$

369. $\displaystyle \int \frac{\cos ax}{(1 - \cos ax)^2}\,dx = \frac{1}{2a}\cot\frac{ax}{2} - \frac{1}{6a}\cot^3\frac{ax}{2}$

370. $\displaystyle \int \frac{\cos x\,dx}{a + b\cos x} = \frac{x}{b} - \frac{a}{b}\int \frac{dx}{a + b\cos x}$

371. $\displaystyle \int \frac{dx}{(\cos x)(a + b\cos x)} = \frac{1}{a}\log\tan\left(\frac{x}{2} + \frac{\pi}{4}\right) - \frac{b}{a}\int \frac{dx}{a + b\cos x}$

372. $\displaystyle \int \frac{dx}{(a + b\cos x)^2} = \frac{b\sin x}{(b^2 - a^2)(a + b\cos x)} - \frac{a}{b^2 - a^2}\int \frac{dx}{a + b\cos x}$

373. $\displaystyle\int \frac{\cos x}{(a + b\cos x)^2}\,dx = \frac{a\sin x}{(a^2 - b^2)(a + b\cos x)} - \frac{b}{a^2 - b^2}\int \frac{dx}{a + b\cos x}$

374. $\displaystyle\int \frac{dx}{a^2 + b^2 - 2ab\cos cx} = \frac{2}{c(a^2 - b^2)}\tan^{-1}\left(\frac{a + b}{a - b}\tan\frac{cx}{2}\right)$

375. $\displaystyle\int \frac{dx}{a^2 + b^2\cos^2 cx} = \frac{1}{ac\sqrt{a^2 + b^2}}\tan^{-1}\frac{a\tan cx}{\sqrt{a^2 + b^2}}$

376. $\displaystyle\int \frac{dx}{a^2 - b^2\cos^2 cx} = \begin{cases} \frac{1}{ac\sqrt{a^2 - b^2}}\tan^{-1}\frac{a\tan cx}{\sqrt{a^2 - b^2}}, & (a^2 > b^2) \\ \text{or} \\ \frac{1}{2ac\sqrt{b^2 - a^2}}\log\frac{a\tan cx - \sqrt{b^2 - a^2}}{a\tan cx + \sqrt{b^2 - a^2}}, & (b^2 > a^2) \end{cases}$

377. $\displaystyle\int \frac{\sin ax}{1 \pm \cos ax}\,dx = \mp\frac{1}{a}\log(1 \pm \cos ax)$

378. $\displaystyle\int \frac{\cos ax}{1 \pm \sin ax}\,dx = \pm\frac{1}{a}\log(1 \pm \sin ax)$

379. $\displaystyle\int \frac{dx}{(\sin ax)(1 \pm \cos ax)} = \pm\frac{1}{2a(1 \pm \cos ax)} + \frac{1}{2a}\log\tan\frac{ax}{2}$

380. $\displaystyle\int \frac{dx}{(\cos ax)(1 \pm \sin ax)} = \mp\frac{1}{2a(1 \pm \sin ax)} + \frac{1}{2a}\log\tan\left(\frac{\pi}{4} + \frac{ax}{2}\right)$

381. $\displaystyle\int \frac{\sin ax}{(\cos ax)(1 \pm \cos ax)}\,dx = \frac{1}{a}\log(\sec ax \pm 1)$

382. $\displaystyle\int \frac{\cos ax}{(\sin ax)(1 \pm \sin ax)}\,dx = -\frac{1}{a}\log(\csc ax \pm 1)$

383. $\displaystyle\int \frac{\sin ax}{(\cos ax)(1 \pm \sin ax)}\,dx = \frac{1}{2a(1 \pm \sin ax)} \pm \frac{1}{2a}\log\tan\left(\frac{\pi}{4} + \frac{ax}{2}\right)$

384. $\displaystyle\int \frac{\cos ax}{(\sin ax)(1 \pm \cos ax)}\,dx = -\frac{1}{2a(1 \pm \cos ax)} \pm \frac{1}{2a}\log\tan\frac{ax}{2}$

385. $\displaystyle\int \frac{dx}{\sin ax \pm \cos ax} = \frac{1}{a\sqrt{2}}\log\tan\left(\frac{ax}{2} \pm \frac{\pi}{8}\right)$

386. $\displaystyle\int \frac{dx}{(\sin ax \pm \cos ax)^2} = \frac{1}{2a}\tan\left(ax \mp \frac{\pi}{4}\right)$

387. $\displaystyle\int \frac{dx}{1 + \cos ax \pm \sin ax} = \pm\frac{1}{a}\log\left(1 \pm \tan\frac{ax}{2}\right)$

388. $\displaystyle\int \frac{dx}{a^2\cos^2 cx - b^2\sin^2 cx} = \frac{1}{2abc}\log\frac{b\tan cx + a}{b\tan cx - a}$

389. $\displaystyle\int x(\sin ax)\,dx = \frac{1}{a^2}\sin ax - \frac{x}{a}\cos ax$

390. $\displaystyle\int x^2(\sin ax)\,dx = \frac{2x}{a^2}\sin ax - \frac{a^2x^2 - 2}{a^3}\cos ax$

391. $\displaystyle\int x^3(\sin ax)\,dx = \frac{3a^2x^2 - 6}{a^4}\sin ax - \frac{a^2x^3 - 6x}{a^3}\cos ax$

392. $\displaystyle\int x^m\sin ax\,dx = \begin{cases} -\frac{1}{a}x^m\cos ax + \frac{m}{a}\int x^{m-1}\cos ax\,dx \\ \text{or} \\ \cos ax\sum_{r=0}^{\left[\frac{m}{2}\right]}(-1)^{r+1}\frac{m!}{(m-2r)!}\cdot\frac{x^{m-2r}}{a^{2r+1}} \\ \quad + \sin ax\sum_{r=0}^{\left[\frac{m-1}{2}\right]}(-1)^r\frac{m!}{(m-2r-1)!}\cdot\frac{x^{m-2r-1}}{a^{2r+2}} \end{cases}$

Note: $[s]$ means greatest integer $\leq s$; Thus $[3.5]$ means 3; $[5] = 5$, $\left[\frac{1}{2}\right] = 0$.

393. $\displaystyle\int x(\cos ax)\,dx = \frac{1}{a^2}\cos ax + \frac{x}{a}\sin ax$

394. $\displaystyle\int x^2(\cos ax)\,dx = \frac{2x\cos ax}{a^2} + \frac{a^2x^2 - 2}{a^3}\sin ax$

395. $\displaystyle\int x^3(\cos ax)\,dx = \frac{3a^2x^2 - 6}{a^4}\cos ax + \frac{a^2x^3 - 6x}{a^3}\sin ax$

396. $\displaystyle\int x^m(\cos ax)\,dx = \begin{cases} \frac{x^m\sin ax}{a} - \frac{m}{a}\int x^{m-1}\sin ax\,dx \\ \text{or} \\ \sin ax\sum_{r=0}^{\lfloor m/2\rfloor}(-1)^r\frac{m!}{(m-2r)!}\cdot\frac{x^{m-2r}}{a^{2r+1}} \\ \quad + \cos ax\sum_{r=0}^{\lfloor(m-1)/2\rfloor}(-1)^r\frac{m!}{(m-2r-1)!}\cdot\frac{x^{m-2r-1}}{a^{2r+2}} \end{cases}$

Note: $[s]$ means greatest integer $\leq s$; Thus $[3.5]$ means 3; $[5] = 5$, $\left[\frac{1}{2}\right] = 0$.

397. $\displaystyle\int \frac{\sin ax}{x}\, dx = \sum_{n=0}^{r} (-1)^n \frac{(ax)^{2n+1}}{(2n+1)(2n+1)!}$

398. $\displaystyle\int \frac{\cos ax}{x}\, dx = \log x + \sum_{n=1}^{r} (-1)^n \frac{(ax)^{2n}}{2n(2n)!}$

399. $\displaystyle\int x(\sin^2 ax)\, dx = \frac{x^2}{4} - \frac{x\sin 2ax}{4a} - \frac{\cos 2ax}{8a^2}$

400. $\displaystyle\int x^2(\sin^2 ax)\, dx = \frac{x^3}{6} - \left(\frac{x^2}{4a} - \frac{1}{8a^3}\right)\sin 2ax - \frac{x\cos 2ax}{4a^2}$

401. $\displaystyle\int x(\sin^3 ax)\, dx = \frac{x\cos 3ax}{12a} - \frac{\sin 3ax}{36a^2} - \frac{3x\cos ax}{4a} + \frac{3\sin ax}{4a^2}$

402. $\displaystyle\int x(\cos^2 ax)\, dx = \frac{x^2}{4} + \frac{x\sin 2ax}{4a} + \frac{\cos 2ax}{8a^2}$

403. $\displaystyle\int x^2(\cos^2 ax)\, dx = \frac{x^3}{6} + \left(\frac{x^2}{4a} - \frac{1}{8a^3}\right)\sin 2ax + \frac{x\cos 2ax}{4a^2}$

404. $\displaystyle\int x(\cos^3 ax)\, dx = \frac{x\sin 3ax}{12a} + \frac{\cos 3ax}{36a^2} + \frac{3x\sin ax}{4a} + \frac{3\cos ax}{4a^2}$

405. $\displaystyle\int \frac{\sin ax}{x^m}\, dx = -\frac{\sin ax}{(m-1)x^{m-1}} + \frac{a}{m-1}\int \frac{\cos ax}{x^{m-1}}\, dx$

406. $\displaystyle\int \frac{\cos ax}{x^m}\, dx = -\frac{\cos ax}{(m-1)x^{m-1}} - \frac{a}{m-1}\int \frac{\sin ax}{x^{m-1}}\, dx$

407. $\displaystyle\int \frac{x}{1 \pm \sin ax}\, dx = \mp\frac{x\cos ax}{a(1 \pm \sin ax)} + \frac{1}{a^2}\log(1 \pm \sin ax)$

408. $\displaystyle\int \frac{x}{1 + \cos ax}\, dx = \frac{x}{a}\tan\frac{ax}{2} + \frac{2}{a^2}\log\cos\frac{ax}{2}$

409. $\displaystyle\int \frac{x}{1 - \cos ax}\, dx = -\frac{x}{a}\cot\frac{ax}{2} + \frac{2}{a^2}\log\sin\frac{ax}{2}$

410. $\displaystyle\int \frac{x + \sin x}{1 + \cos x}\, dx = x\tan\frac{x}{2}$

411. $\displaystyle\int \frac{x - \sin x}{1 - \cos x}\, dx = -x\cot\frac{x}{2}$

412. $\displaystyle\int \sqrt{1 - \cos ax}\, dx = -\frac{2\sin ax}{a\sqrt{1 - \cos ax}} = -\frac{2\sqrt{2}}{a}\cos\left(\frac{ax}{2}\right)$

413. $\displaystyle\int \sqrt{1 + \cos ax}\, dx = \frac{2\sin ax}{a\sqrt{1 + \cos ax}} = \frac{2\sqrt{2}}{a}\sin\left(\frac{ax}{2}\right)$

414. $\displaystyle\int \sqrt{1 + \sin x}\, dx = \pm 2\left(\sin\frac{x}{2} - \cos\frac{x}{2}\right),$
[use + if $(8k-1)\frac{\pi}{2} < x \le (8k+3)\frac{\pi}{2}$, otherwise − ; k an integer]

415. $\displaystyle\int \sqrt{1 - \sin x}\, dx = \pm 2\left(\sin\frac{x}{2} + \cos\frac{x}{2}\right),$
[use + if $(8k-3)\frac{\pi}{2} < x \le (8k+1)\frac{\pi}{2}$, otherwise −; k an integer]

416. $\displaystyle\int \frac{dx}{\sqrt{1 - \cos x}} = \pm\sqrt{2}\,\log\tan\frac{x}{4},$
[use + if $4k\pi < x < (4k+2)\pi$, otherwise −; k an integer]

417. $\displaystyle\int \frac{dx}{\sqrt{1 + \cos x}} = \pm\sqrt{2}\,\log\tan\left(\frac{x + \pi}{4}\right),$
[use + if $(4k-1)\pi < x < (4k+1)\pi$, otherwise −; k an integer]

418. $\displaystyle\int \frac{dx}{\sqrt{1 - \sin x}} = \pm\sqrt{2}\,\log\tan\left(\frac{x}{4} - \frac{\pi}{8}\right),$
[use + if $(8k+1)\frac{\pi}{2} < x < (8k+5)\frac{\pi}{2}$, otherwise −; k an integer]

419. $\displaystyle\int \frac{dx}{\sqrt{1 + \sin x}} = \pm\sqrt{2}\,\log\tan\left(\frac{x}{4} + \frac{\pi}{8}\right),$
[use + if $(8k-1)\frac{\pi}{2} < x < (8k+3)\frac{\pi}{2}$, otherwise −; k an integer]

420. $\displaystyle\int \tan^2(ax)\, dx = \frac{1}{a}\tan ax - x$

421. $\displaystyle\int \tan^3(ax)\, dx = \frac{1}{2a}\tan^2 ax + \frac{1}{a}\log\cos ax$

422. $\displaystyle\int \tan^4(ax)\,dx = \frac{\tan^3 ax}{3a} - \frac{1}{a}\tan ax + x$

423. $\displaystyle\int \tan^n(ax)\,dx = \frac{\tan^{n-1} ax}{a(n-1)} - \int (\tan^{n-2} ax)\,dx$

424. $\displaystyle\int \cot^2(ax)\,dx = -\frac{1}{c}\cot ax - x$

425. $\displaystyle\int \cot^3(ax)\,dx = -\frac{1}{2a}\cot^2 ax - \frac{1}{a}\log\sin ax$

426. $\displaystyle\int \cot^4(ax)\,dx = -\frac{1}{3a}\cot^3 ax + \frac{1}{a}\cot ax + x$

427. $\displaystyle\int \cot^n(ax)\,dx = -\frac{\cot^{n-1} ax}{a(n-1)} - \int (\cot^{n-2} ax)\,dx$

428. $\displaystyle\int \frac{x}{\sin^2 ax}\,dx = \int x(\csc^2 ax)\,dx = -\frac{x\cot ax}{a} + \frac{1}{a^2}\log\sin ax$

429. $\displaystyle\int \frac{x}{\sin^n ax}\,dx = \int x(\csc^n ax)\,dx = -\frac{x\cos ax}{a(n-1)\sin^{n-1} ax} - \frac{1}{a^2(n-1)(n-2)\sin^{n-2} ax} + \frac{(n-2)}{(n-1)}\int \frac{x}{\sin^{n-2} ax}\,dx$

430. $\displaystyle\int \frac{x}{\cos^2 ax}\,dx = \int x(\sec^2 ax)\,dx = \frac{1}{a}x\tan ax + \frac{1}{a^2}\log\cos ax$

431. $\displaystyle\int \frac{x}{\cos^n(ax)}\,dx = \int x(\sec^n ax)\,dx = \frac{x\sin ax}{a(n-1)\cos^{n-1} ax} - \frac{1}{a^2(n-1)(n-2)\cos^{n-2} ax} + \frac{n-2}{n-1}\int \frac{x}{\cos^{n-2} ax}\,dx$

432. $\displaystyle\int \frac{\sin ax}{\sqrt{1+b^2\sin^2 ax}}\,dx = -\frac{1}{ab}\sin^{-1}\frac{b\cos ax}{\sqrt{1+b^2}}$

433. $\displaystyle\int \frac{\sin ax}{\sqrt{1-b^2\sin^2 ax}}\,dx = -\frac{1}{ab}\log(b\cos ax + \sqrt{1-b^2\sin^2 ax})$

434. $\displaystyle\int \sin(ax)\sqrt{1+b^2\sin^2 ax}\,dx = -\frac{\cos ax}{2a}\sqrt{1+b^2\sin^2 ax} - \frac{1+b^2}{2ab}\sin^{-1}\frac{b\cos ax}{\sqrt{1+b^2}}$

435. $\displaystyle\int \sin(ax)\sqrt{1-b^2\sin^2 ax}\,dx = -\frac{\cos ax}{2a}\sqrt{1-b^2\sin^2 ax} - \frac{1-b^2}{2ab}\log(b\cos ax + \sqrt{1-b^2\sin^2 ax})$

436. $\displaystyle\int \frac{\cos ax}{\sqrt{1+b^2\sin^2 ax}}\,dx = \frac{1}{ab}\log(b\sin ax + \sqrt{1+b^2\sin^2 ax})$

437. $\displaystyle\int \frac{\cos ax}{\sqrt{1-b^2\sin^2 ax}}\,dx = \frac{1}{ab}\sin^{-1}(b\sin ax)$

438. $\displaystyle\int \cos(ax)\sqrt{1+b^2\sin^2 ax}\,dx = \frac{\sin ax}{2a}\sqrt{1+b^2\sin^2 ax} + \frac{1}{2ab}\log(b\sin ax + \sqrt{1+b^2\sin^2 ax})$

439. $\displaystyle\int \cos(ax)\sqrt{1-b^2\sin^2 ax}\,dx = \frac{\sin ax}{2a}\sqrt{1-b^2\sin^2 ax} + \frac{1}{2ab}\sin^{-1}(b\sin ax)$

440. $\displaystyle\int \frac{dx}{\sqrt{a+b\tan^2 cx}} = \frac{\pm 1}{c\sqrt{a-b}}\sin^{-1}\left(\sqrt{\frac{a-b}{a}}\sin cx\right), \quad (a > |b|)$

[use $+$ if $(2k-1)\frac{\pi}{2} < x \le (2k+1)\frac{\pi}{2}$, otherwise $-$; k an integer]

FORMS INVOLVING INVERSE TRIGONOMETRIC FUNCTIONS

441. $\displaystyle\int \sin^{-1}(ax)\,dx = x\sin^{-1} ax + \frac{\sqrt{1-a^2x^2}}{a}$

442. $\displaystyle\int \cos^{-1}(ax)\,dx = x\cos^{-1} ax - \frac{\sqrt{1-a^2x^2}}{a}$

443. $\displaystyle\int \tan^{-1}(ax)\,dx = x\tan^{-1} ax - \frac{1}{2a}\log(1+a^2x^2)$

444. $\displaystyle\int \cot^{-1}(ax)\,dx = x\cot^{-1} ax + \frac{1}{2a}\log(1+a^2x^2)$

445. $\int \sec^{-1}(ax)\, dx = x\sec^{-1} ax - \frac{1}{a}\log\left(ax + \sqrt{a^2x^2 - 1}\right)$

446. $\int \csc^{-1}(ax)\, dx = x\csc^{-1} ax + \frac{1}{a}\log\left(ax + \sqrt{a^2x^2 - 1}\right)$

447. $\int \sin^{-1}\frac{x}{a}\, dx = x\sin^{-1}\frac{x}{a} + \sqrt{a^2 - x^2}, \qquad (a > 0)$

448. $\int \cos^{-1}\frac{x}{a}\, dx = x\cos^{-1}\frac{x}{a} - \sqrt{a^2 - x^2}, \qquad (a > 0)$

449. $\int \tan^{-1}\frac{x}{a}\, dx = x\tan^{-1}\frac{x}{a} - \frac{a}{2}\log(a^2 + x^2)$

450. $\int \cot^{-1}\frac{x}{a}\, dx = x\cot^{-1}\frac{x}{a} + \frac{a}{2}\log(a^2 + x^2)$

451. $\int x\sin^{-1}(ax)\, dx = \frac{1}{4a^2}\left[(2a^2x^2 - 1)\sin^{-1}(ax) + ax\sqrt{1 - a^2x^2}\right]$

452. $\int x\cos^{-1}(ax)\, dx = \frac{1}{4a^2}\left[(2a^2x^2 - 1)\cos^{-1}(ax) - ax\sqrt{1 - a^2x^2}\right]$

453. $\int x^n\sin^{-1}(ax)\, dx = \frac{x^{n+1}}{n+1}\sin^{-1}(ax) - \frac{a}{n+1}\int \frac{x^{n+1}\, dx}{\sqrt{1 - a^2x^2}}, \; (n \neq -1)$

454. $\int x^n\cos^{-1}(ax)\, dx = \frac{x^{n+1}}{n+1}\cos^{-1}(ax) + \frac{a}{n+1}\int \frac{x^{n+1}\, dx}{\sqrt{1 - a^2x^2}}, \; (n \neq -1)$

455. $\int x\tan^{-1}(ax)\, dx = \frac{1 + a^2x^2}{2a^2}\tan^{-1} ax - \frac{x}{2a}$

456. $\int x^n\tan^{-1}(ax)\, dx = \frac{x^{n+1}}{n+1}\tan^{-1} ax - \frac{a}{n+1}\int \frac{x^{n+1}}{1 + a^2x^2}\, dx$

457. $\int x(\cot^{-1} ax)\, dx = \frac{1 + a^2x^2}{2a^2}\cot^{-1} ax + \frac{x}{2a}$

458. $\int x^n\cot^{-1}(ax)\, dx = \frac{x^{n+1}}{n+1}\cot^{-1} ax + \frac{a}{n+1}\int \frac{x^{n+1}}{1 + a^2x^2}\, dx$

459. $\int \frac{\sin^{-1}(ax)}{x^2}\, dx = a\log\left(\frac{1 - \sqrt{1 - a^2x^2}}{x}\right) - \frac{\sin^{-1}(ax)}{x}$

460. $\int \frac{\cos^{-1}(ax)\, dx}{x^2} = -\frac{1}{x}\cos^{-1}(ax) + a\log\frac{1 + \sqrt{1 - a^2x^2}}{x}$

461. $\int \frac{\tan^{-1}(ax)\, dx}{x^2} = -\frac{1}{x}\tan^{-1}(ax) - \frac{a}{2}\log\frac{1 + a^2x^2}{x^2}$

462. $\int \frac{\cot^{-1}(ax)}{x^2}\, dx = -\frac{1}{x}\cot^{-1} ax - \frac{a}{2}\log\frac{x^2}{a^2x^2 + 1}$

463. $\int \sin^{-1}(ax)^2\, dx = x(\sin^{-1} ax)^2 - 2x + \frac{2\sqrt{1 - a^2x^2}}{a}\sin^{-1} ax$

464. $\int \cos^{-1}(ax)^2\, dx = x(\cos^{-1} ax)^2 - 2x - \frac{2\sqrt{1 - a^2x^2}}{a}\cos^{-1} ax$

465. $\int (\sin^{-1} ax)^n\, dx = \begin{cases} x(\sin^{-1} ax)^n + \dfrac{n\sqrt{1 - a^2x^2}}{a}(\sin^{-1} ax)^{n-1} - n(n-1)\displaystyle\int (\sin^{-1} ax)^{n-2}\, dx \\[2mm] \text{or} \\[2mm] \displaystyle\sum_{r=0}^{[n/2]}(-1)^r\frac{n!}{(n-2r)!}x(\sin^{-1} ax)^{n-2r} + \sum_{r=0}^{[n-1/2]}(-1)^r\frac{n!\sqrt{1 - a^2x^2}}{(n - 2r - 1)!a}(\sin^{-1} ax)^{n-2r-1} \end{cases}$

Note: [s] means greatest integer $\leq s$. Thus [3.5] means 3; [5] = 5, $\left[\frac{1}{2}\right] = 0$.

466. $\int (\cos^{-1} ax)^n\, dx = \begin{cases} x(\cos^{-1} ax)^n - \dfrac{n\sqrt{1 - a^2x^2}}{a}(\cos^{-1} ax)^{n-1} - n(n-1)\displaystyle\int (\cos^{-1} ax)^{n-2}\, dx \\[2mm] \text{or} \\[2mm] \displaystyle\sum_{r=0}^{[n/2]}(-1)^r\frac{n!}{(n-2r)!}x(\cos^{-1} ax)^{n-2r} \times \sum_{r=0}^{[n-1/2]}(-1)^r\frac{n!\sqrt{1 - a^2x^2}}{(n - 2r - 1)!a}(\cos^{-1} ax)^{n-2r-1} \end{cases}$

467. $\int \frac{1}{\sqrt{1 - a^2x^2}}(\sin^{-1} ax)\, dx = \frac{1}{2a}(\sin^{-1} ax)^2$

468. $\int \frac{x^n}{\sqrt{1 - a^2x^2}}(\sin^{-1} ax)\, dx = -\frac{x^{n-1}}{na^2}\sqrt{1 - a^2x^2}\sin^{-1} ax + \frac{x^n}{n^2a} + \frac{n-1}{na^2}\int \frac{x^{n-2}}{\sqrt{1 - a^2x^2}}\sin^{-1} ax\, dx$

469. $\int \dfrac{1}{\sqrt{1-a^2x^2}}(\cos^{-1} ax)\, dx = -\dfrac{1}{2a}(\cos^{-1} ax)^2$

470. $\int \dfrac{x^n}{\sqrt{1-a^2x^2}}(\cos^{-1} ax)\, dx = -\dfrac{x^{n-1}}{na^2}\sqrt{1-a^2x^2}\cos^{-1} ax - \dfrac{x^n}{n^2 a} + \dfrac{n-1}{na^2}\int \dfrac{x^{n-2}}{\sqrt{1-a^2x^2}}\cos^{-1} ax\, dx$

471. $\int \dfrac{\tan^{-1} ax}{a^2x^2+1}\, dx = \dfrac{1}{2a}(\tan^{-1} ax)^2$

472. $\int \dfrac{\cot^{-1} ax}{a^2x^2+1}\, dx = -\dfrac{1}{2a}(\cot^{-1} ax)^2$

473. $\int x\sec^{-1} ax\, dx = \dfrac{x^2}{2}\sec^{-1} ax - \dfrac{1}{2a^2}\sqrt{a^2x^2-1}$

474. $\int x^n \sec^{-1} ax\, dx = \dfrac{x^{n+1}}{n+1}\sec^{-1} ax - \dfrac{1}{n+1}\int \dfrac{x^n\, dx}{\sqrt{a^2x^2-1}}$

475. $\int \dfrac{\sec^{-1} ax}{x^2}\, dx = -\dfrac{\sec^{-1} ax}{x} + \dfrac{\sqrt{a^2x^2-1}}{x}$

476. $\int x\csc^{-1} ax\, dx = \dfrac{x^2}{2}\csc^{-1} ax + \dfrac{1}{2a^2}\sqrt{a^2x^2-1}$

477. $\int x^n \csc^{-1} ax\, dx = \dfrac{x^{n+1}}{n+1}\csc^{-1} ax + \dfrac{1}{n+1}\int \dfrac{x^n\, dx}{\sqrt{a^2x^2-1}}$

478. $\int \dfrac{\csc^{-1} ax}{x^2}\, dx = -\dfrac{\csc^{-1} ax}{x} - \dfrac{\sqrt{a^2x^2-1}}{x}$

FORMS INVOLVING TRIGONOMETRIC SUBSTITUTIONS

479. $\int f(\sin x)\, dx = 2\int f\left(\dfrac{2z}{1+z^2}\right)\dfrac{dz}{1+z^2}, \quad \left(z = \tan \dfrac{x}{2}\right)$

480. $\int f(\cos x)\, dx = 2\int f\left(\dfrac{1-z^2}{1+z^2}\right)\dfrac{dz}{1+z^2}, \quad \left(z = \tan \dfrac{x}{2}\right)$

481. $\int f(\sin x)\, dx = \int f(u)\dfrac{du}{\sqrt{1-u^2}}, \quad (u = \sin x)$

482. $\int f(\cos x)\, dx = -\int f(u)\dfrac{du}{\sqrt{1-u^2}}, \quad (u = \cos x)$

483. $\int f(\sin x, \cos x)\, dx = \int f\left(u, \sqrt{1-u^2}\right)\dfrac{du}{\sqrt{1-u^2}}, \quad (u = \sin x)$

484. $\int f(\sin x, \cos x)\, dx = 2\int f\left(\dfrac{2z}{1+z^2}, \dfrac{1-z^2}{1+z^2}\right)\dfrac{dz}{1+z^2}, \quad \left(z = \tan \dfrac{x}{2}\right)$

LOGARITHMIC FORMS

485. $\int (\log x)\, dx = x\log x - x$

486. $\int x(\log x)\, dx = \dfrac{x^2}{2}\log x - \dfrac{x^2}{4}$

487. $\int x^2(\log x)\, dx = \dfrac{x^3}{3}\log x - \dfrac{x^3}{9}$

488. $\int x^n(\log ax)\, dx = \dfrac{x^{n+1}}{n+1}\log ax - \dfrac{x^{n+1}}{(n+1)^2}$

489. $\int (\log x)^2\, dx = x(\log x)^2 - 2x\log x + 2x$

490. $\int (\log x)^n\, dx = \begin{cases} x(\log x)^n - n\int (\log x)^{n-1}\, dx, & (n \neq -1) \\ \quad\text{or} \\ (-1)^n n!\, x\sum_{r=0}^{n}\dfrac{(-\log x)^r}{r!} \end{cases}$

491. $\int \dfrac{(\log x)^n}{x}\, dx = \dfrac{1}{n+1}(\log x)^{n+1}$

492. $\int \dfrac{dx}{\log x} = \log(\log x) + \log x + \dfrac{(\log x)^2}{2\cdot 2!} + \dfrac{(\log x)^3}{3\cdot 3!} + \cdots$

493. $\int \dfrac{dx}{x\log x} = \log(\log x)$

494. $\displaystyle\int \frac{dx}{x(\log x)^n} = -\frac{1}{(n-1)(\log x)^{n-1}}$

495. $\displaystyle\int \frac{x^m\,dx}{(\log x)^n} = -\frac{x^{m+1}}{(n-1)(\log x)^{n-1}} + \frac{m+1}{n-1}\int \frac{x^m\,dx}{(\log x)^{n-1}}$

496. $\displaystyle\int x^m(\log x)^n\,dx = \begin{cases} \dfrac{x^{m+1}(\log x)^n}{m+1} - \dfrac{n}{m+1}\displaystyle\int x^m(\log x)^{n-1}\,dx \\[2mm] \text{or} \\[2mm] (-1)^n\dfrac{n!}{m+1}x^{m+1}\displaystyle\sum_{r=0}^{n}\dfrac{(-\log x)^r}{r!(m+1)^{n-r}} \end{cases}$

497. $\displaystyle\int x^p\cos(b\ln x)\,dx = \frac{x^{p+1}}{(p+1)^2+b^2}\left[b\sin(b\ln x)+(p+1)\cos(b\ln x)\right]+c$

498. $\displaystyle\int x^p\sin(b\ln x)\,dx = \frac{x^{p+1}}{(p+1)^2+b^2}\left[(p+1)\sin(b\ln x)-b\cos(b\ln x)\right]+c$

499. $\displaystyle\int [\log(ax+b)]\,dx = \frac{ax+b}{a}\log(ax+b) - x$

500. $\displaystyle\int \frac{\log(ax+b)}{x^2}\,dx = \frac{a}{b}\log x - \frac{ax+b}{bx}\log(ax+b)$

501. $\displaystyle\int x^m[\log(ax+b)]\,dx = \frac{1}{m+1}\left[x^{m+1}-\left(-\frac{b}{a}\right)^{m+1}\right]\log(ax+b) - \frac{1}{m+1}\left(-\frac{b}{a}\right)^{m+1}\sum_{r=1}^{m+1}\frac{1}{r}\left(-\frac{ax}{b}\right)^r$

502. $\displaystyle\int \frac{\log(ax+b)}{x^m}\,dx = -\frac{1}{m-1}\frac{\log(ax+b)}{x^{m-1}} + \frac{1}{m-1}\left(-\frac{a}{b}\right)^{m-1}\log\frac{ax+b}{x} + \frac{1}{m-1}\left(-\frac{a}{b}\right)^{m-1}\sum_{r=1}^{m-2}\frac{1}{r}\left(-\frac{b}{ax}\right)^r,\ (m>2)$

503. $\displaystyle\int \left[\log\frac{x+a}{x-a}\right]dx = (x+a)\log(x+a)-(x-a)\log(x-a)$

504. $\displaystyle\int x^m\left[\log\frac{x+a}{x-a}\right]dx = \frac{x^{m+1}-(-a)^{m+1}}{m+1}\log(x+a) - \frac{x^{m+1}-a^{m+1}}{m+1}\log(x-a) + \frac{2a^{m+1}}{m+1}\sum_{r=1}^{\left[\frac{m+1}{2}\right]}\frac{1}{m-2r+2}\left(\frac{x}{a}\right)^{m-2r+2}$

Note: $[s]$ means greatest integer $\le s$; Thus $[3.5]$ means 3; $[5]=5$, $\left[\frac{1}{2}\right]=0$.

505. $\displaystyle\int \frac{1}{x^2}\left[\log\frac{x+a}{x-a}\right]dx = \frac{1}{x}\log\frac{x-a}{x+a} - \frac{1}{a}\log\frac{x^2-a^2}{x^2}$

506. $\displaystyle\int (\log X)\,dx = \begin{cases} \left(x+\dfrac{b}{2c}\right)\log X - 2x + \dfrac{\sqrt{4ac-b^2}}{c}\tan^{-1}\dfrac{2cx+b}{\sqrt{4ac-b^2}}, & (b^2-4ac<0) \\[2mm] \text{or} \\[2mm] \left(x+\dfrac{b}{2c}\right)\log X - 2x + \dfrac{\sqrt{b^2-4ac}}{c}\tanh^{-1}\dfrac{2cx+b}{\sqrt{b^2-4ac}}, & (b^2-4ac>0) \\[2mm] \text{where} \\ X = a+bx+cx^2 \end{cases}$

507. $\displaystyle\int x^n(\log(a+bx+cx^2)\,dx = \frac{x^{n+1}}{n+1}\log X - \frac{2c}{n+1}\int \frac{x^{n+2}}{X}\,dx - \frac{b}{n+1}\int \frac{x^{n+1}}{X}\,dx$

508. $\displaystyle\int \log(x^2+a^2)\,dx = x\log(x^2+a^2) - 2x + 2a\tan^{-1}\frac{x}{a}$

509. $\displaystyle\int \log(x^2-a^2)\,dx = x\log(x^2-a^2) - 2x + a\log\frac{x+a}{x-a}$

510. $\displaystyle\int x\log(x^2\pm a^2)\,dx = \frac{1}{2}(x^2\pm a^2)\log(x^2\pm a^2) - \frac{1}{2}x^2$

511. $\displaystyle\int \log(x+\sqrt{x^2\pm a^2})\,dx = x\log(x+\sqrt{x^2\pm a^2}) - \sqrt{x^2\pm a^2}$

512. $\displaystyle\int x\log(x+\sqrt{x^2\pm a^2})\,dx = \left(\frac{x^2}{2}\pm\frac{a^2}{4}\right)\log(x+\sqrt{x^2\pm a^2}) - \frac{x\sqrt{x^2\pm a^2}}{4}$

513. $\displaystyle\int x^m\log(x+\sqrt{x^2\pm a^2})\,dx = \frac{x^{m+1}}{m+1}\log(x+\sqrt{x^2\pm a^2}) - \frac{1}{m+1}\int \frac{x^{m+1}}{\sqrt{x^2\pm a^2}}\,dx$

514. $\displaystyle\int \frac{\log(x+\sqrt{x^2+a^2})}{x^2}\,dx = -\frac{\log(x+\sqrt{x^2+a^2})}{x} - \frac{1}{a}\log\frac{a+\sqrt{x^2+a^2}}{x}$

515. $\displaystyle\int \frac{\log(x+\sqrt{x^2-a^2})}{x^2}\,dx = -\frac{\log(x+\sqrt{x^2-a^2})}{x} + \frac{1}{|a|}\sec^{-1}\frac{x}{a}$

516. $\displaystyle\int x^n\log(x^2-a^2)\,dx = \frac{1}{n+1}\left[x^{n+1}\log(x^2-a^2) - a^{n+1}\log(x-a) - (-a)^{n+1}\log(x+a) - 2\sum_{r=0}^{[n/2]}\frac{a^{2r}x^{n-2r+1}}{n-2r+1}\right]$

Note: $[s]$ means greatest integer $\le s$; Thus $[3.5]$ means 3; $[5]=5$, $\left[\frac{1}{2}\right]=0$.

EXPONENTIAL FORMS

517. $\int e^x\,dx = e^x$

518. $\int e^{-x}\,dx = -e^{-x}$

519. $\int e^{ax}\,dx = \dfrac{e^{ax}}{a}$

520. $\int x\,e^{ax}\,dx = \dfrac{e^{ax}}{a^2}(ax-1)$

521. $\int x^m e^{ax}\,dx = \begin{cases} \dfrac{x^m e^{ax}}{a} - \dfrac{m}{a}\displaystyle\int x^{m-1}e^{ax}\,dx \\ \quad\text{or} \\ e^{ax}\displaystyle\sum_{r=0}^m (-1)^r \dfrac{m!\,x^{m-r}}{(m-r)!\,a^{r+1}} \end{cases}$

522. $\int \dfrac{e^{ax}\,dx}{x} = \log x + \dfrac{ax}{1!} + \dfrac{a^2 x^2}{2\cdot 2!} + \dfrac{a^3 + x^3}{3\cdot 3!} + \cdots$

523. $\int \dfrac{e^{ax}}{x^m}\,dx = -\dfrac{1}{m-1}\dfrac{e^{ax}}{x^{m-1}} + \dfrac{a}{m-1}\int \dfrac{e^{ax}}{x^{m-1}}\,dx$

524. $\int e^{ax}\log x\,dx = \dfrac{e^{ax}\log x}{a} - \dfrac{1}{a}\int \dfrac{e^{ax}}{x}\,dx$

525. $\int \dfrac{dx}{1+e^x} = x - \log(1+e^x) = \log\dfrac{e^x}{1+e^x}$

526. $\int \dfrac{dx}{a+be^{px}} = \dfrac{x}{a} - \dfrac{1}{ap}\log(a+be^{px})$

527. $\int \dfrac{dx}{ae^{mx}+be^{-mx}} = \dfrac{1}{m\sqrt{ab}}\tan^{-1}\left(e^{mx}\sqrt{\dfrac{a}{b}}\right), \quad (a>0,\,b>0)$

528. $\int \dfrac{dx}{ae^{mx}-be^{-mx}} = \begin{cases} \dfrac{1}{2m\sqrt{ab}}\log\dfrac{\sqrt{a}\,e^{mx}-\sqrt{b}}{\sqrt{a}\,e^{mx}+\sqrt{b}} \\ \quad\text{or} \\ \dfrac{-1}{m\sqrt{ab}}\tanh^{-1}\left(\sqrt{\dfrac{a}{b}}\,e^{mx}\right), \quad (a>0,\,b>0) \end{cases}$

529. $\int (a^x - a^{-x})\,dx = \dfrac{a^x + a^{-x}}{\log a}$

530. $\int \dfrac{e^{ax}}{b+ce^{ax}}\,dx = \dfrac{1}{ac}\log(b+ce^{ax})$

531. $\int \dfrac{x\,e^{ax}}{(1+ax)^2}\,dx = \dfrac{e^{ax}}{a^2(1+ax)}$

532. $\int x\,e^{-x^2}\,dx = -\dfrac{1}{2}e^{-x^2}$

533. $\int e^{ax}\sin(bx)\,dx = \dfrac{e^{ax}[a\sin(bx) - b\cos(bx)]}{a^2+b^2}$

534. $\int e^{ax}\sin(bx)\sin(cx)\,dx = \dfrac{e^{ax}[(b-c)\sin(b-c)x + a\cos(b-c)x]}{2[a^2+(b-c)^2]} - \dfrac{e^{ax}[(b+c)\sin(b+c)x + a\cos(b+c)x]}{2[a^2+(b+c)^2]}$

535. $\int e^{ax}\sin(bx)\cos(cx)\,dx = \begin{cases} \dfrac{e^{ax}[a\sin(b-c)x-(b-c)\cos(b-c)x]}{2[a^2+(b-c)^2]} + \dfrac{e^{ax}[a\sin(b+c)x-(b+c)\cos(b+c)x]}{2[a^2+(b+c)^2]} \\ \quad\text{or} \\ \dfrac{e^{ax}}{\rho}[(a\sin bx - b\cos bx)[\cos(cx-\alpha)] - c(\sin bx)\sin(cx-\alpha)] \\ \text{where} \\ \rho = \sqrt{(a^2+b^2-c^2)^2 + 4a^2 c^2}, \\ \quad \rho\cos\alpha = a^2+b^2-c^2, \quad \rho\sin\alpha = 2ac \end{cases}$

536. $\int e^{ax}\sin(bx)\sin(bx+c)\,dx = \dfrac{e^{ax}\cos c}{2a} - \dfrac{e^{ax}[a\cos(2bx+c)+2b\sin(2bx+c)]}{2(a^2+4b^2)}$

537. $\int e^{ax}\sin(bx)\cos(bx+c)\,dx = -\dfrac{e^{ax}\sin c}{2a} + \dfrac{e^{ax}[a\sin(2bx+c)-2b\cos(2bx+c)]}{2(a^2+4b^2)}$

538. $\int e^{ax}\cos(bx)\,dx = \dfrac{e^{ax}}{a^2+b^2}[a\cos(bx)+b\sin(bx)]$

539. $\int e^{ax}\cos(bx)\cos(cx)\,dx = \dfrac{e^{ax}[(b-c)\sin(b-c)x + a\cos(b-c)x]}{2[a^2+(b-c)^2]} + \dfrac{e^{ax}[(b+c)\sin(b+c)x + a\cos(b+c)x]}{2[a^2+(b+c)^2]}$

540. $\int e^{ax}\cos(bx)\cos(bx+c)\,dx = \dfrac{e^{ax}\cos c}{2a} + \dfrac{e^{ax}[a\cos(2bx+c)+2b\sin(2bx+c)]}{2(a^2+4b^2)}$

541. $\int e^{ax}\cos(bx)\sin(bx+c)\,dx = \dfrac{e^{ax}\sin c}{2a} + \dfrac{e^{ax}[a\sin(2bx+c)-2b\cos(2bx+c)]}{2(a^2+4b^2)}$

542. $\int e^{ax}\sin^n(bx)\,dx = \dfrac{1}{a^2+n^2b^2}\left[(a\sin bx - nb\cos bx)e^{ax}\sin^{n-1}bx + n(n-1)b^2\int e^{ax}[\sin^{n-2}bx]\,dx\right]$

543. $\int e^{ax}\cos^n(bx)\,dx = \dfrac{1}{a^2+n^2b^2}\left[(a\cos bx + nb\sin bx)e^{ax}\cos^{n-1}bx + n(n-1)b^2\int e^{ax}[\cos^{n-2}bx]\,dx\right]$

544. $\int x^m e^x \sin x\,dx = \dfrac{1}{2}x^m e^x(\sin x - \cos x) - \dfrac{m}{2}\int x^{m-1}e^x\sin x\,dx + \dfrac{m}{2}\int x^{m-1}e^x\cos x\,dx$

545. $\int x^m e^{ax}\sin(bx)\,dx = \begin{cases} x^m e^{ax}\dfrac{a\sin bx - b\cos bx}{a^2+b^2} - \dfrac{m}{a^2+b^2}\int x^{m-1}e^{ax}(a\sin bx - b\cos bx)\,dx \\[2mm] \text{or} \\[1mm] e^{ax}\sum_{r=0}^{m}\dfrac{(-1)^r m! x^{m-r}}{\rho^{r+1}(m-r)!}\sin[bx - (r+1)\alpha] \\[1mm] \text{where} \\[1mm] \rho = \sqrt{a^2+b^2},\quad \rho\cos\alpha = a,\quad \rho\sin\alpha = b \end{cases}$

546. $\int x^m e^x \cos x\,dx = \dfrac{1}{2}x^m e^x(\sin x + \cos x) - \dfrac{m}{2}\int x^{m-1}e^x\sin x\,dx - \dfrac{m}{2}\int x^{m-1}e^x\cos x\,dx$

547. $\int x^m e^{ax}\cos(bx)\,dx = \begin{cases} x^m e^{ax}\dfrac{a\cos bx + b\sin bx}{a^2+b^2} - \dfrac{m}{a^2+b^2}\int x^{m-1}e^{ax}(a\cos bx + b\sin bx)\,dx \\[2mm] \text{or} \\[1mm] e^{ax}\sum_{r=0}^{m}\dfrac{(-1)^r m! x^{m-r}}{\rho^{r+1}(m-r)!}\cos[bx - (r+1)\alpha] \\[1mm] \rho = \sqrt{a^2+b^2},\quad \rho\cos\alpha = a,\quad \rho\sin\alpha = b \end{cases}$

548. $\int e^{ax}(\cos^m x)(\sin^n x)\,dx = \begin{cases} \dfrac{e^{ax}\cos^{m-1}x\,\sin^n x[a\cos x + (m+n)\sin x]}{(m+n)^2+a^2} \\[2mm] -\dfrac{na}{(m+n)^2+a^2}\int e^{ax}(\cos^{m-1}x)(\sin^{n-1}x)\,dx \\[2mm] +\dfrac{(m-1)(m+n)}{(m+n)^2+a^2}\int e^{ax}(\cos^{m-2}x)(\sin^n x)\,dx \\[2mm] \text{or} \\[1mm] \dfrac{e^{ax}\cos^m x\,\sin^{n-1}x[a\sin x - (m+n)\cos x]}{(m+n)^2+a^2} \\[2mm] +\dfrac{ma}{(m+n)^2+a^2}\int e^{ax}(\cos^{m-1}x)(\sin^{n-1}x)\,dx \\[2mm] +\dfrac{(n-1)(m+n)}{(m+n)^2+a^2}\int e^{ax}(\cos^m x)(\sin^{n-2}x)\,dx \\[2mm] \text{or} \\[1mm] \dfrac{e^{ax}(\cos^{m-1}x)(\sin^{n-1}x)(a\sin x\cos x + m\sin^2 x - n\cos^2 x)}{(m+n)^2+a^2} \\[2mm] +\dfrac{m(m-1)}{(m+n)^2+a^2}\int e^{ax}(\cos^{m-2}x)(\sin^n x)\,dx \\[2mm] +\dfrac{n(n-1)}{(m+n)^2+a^2}\int e^{ax}(\cos^m x)(\sin^{n-2}x)\,dx \\[2mm] \text{or} \\[1mm] \dfrac{e^{ax}(\cos^{m-1}x)(\sin^{n-1}x)(a\cos x\sin x + m\sin^2 x - n\cos^2 x)}{(m+n)^2+a^2} \\[2mm] +\dfrac{m(m-1)}{(m+n)^2+a^2}\int e^{ax}(\cos^{m-2}x)(\sin^{n-2}x)\,dx \\[2mm] +\dfrac{(n-m)(n+m-1)}{(m+n)^2+a^2}\int e^{ax}(\cos^m x)(\sin^{n-2}x)\,dx \end{cases}$

549. $\int x e^{ax}\sin(bx)\,dx = \dfrac{xe^{ax}}{a^2+b^2}(a\sin bx - b\cos bx) - \dfrac{e^{ax}}{(a^2+b^2)^2}[(a^2-b^2)\sin bx - 2ab\cos bx]$

550. $\int x e^{ax}\cos(bx)\,dx = \dfrac{xe^{ax}}{a^2+b^2}(a\cos bx - b\sin bx) - \dfrac{e^{ax}}{(a^2+b^2)^2}[(a^2-b^2)\cos bx - 2ab\sin bx]$

551. $\int \dfrac{e^{ax}}{\sin^n x}\,dx = -\dfrac{e^{ax}[a\sin x + (n-2)\cos x]}{(n-1)(n-2)\sin^{n-1}x} + \dfrac{a^2+(n-2)^2}{(n-1)(n-2)}\int \dfrac{e^{ax}}{\sin^{n-2}x}\,dx$

552. $\int \dfrac{e^{ax}}{\cos^n x}\,dx = -\dfrac{e^{ax}[a\cos x - (n-2)\sin x]}{(n-1)(n-2)\cos^{n-1}x} + \dfrac{a^2+(n-2)^2}{(n-1)(n-2)}\int \dfrac{e^{ax}}{\cos^{n-2}x}\,dx$

553. $\int e^{ax}\tan^n x\,dx = e^{ax}\dfrac{\tan^{n-1}x}{n-1} - \dfrac{a}{n-1}\int e^{ax}\tan^{n-1}x\,dx - \int e^{ax}\tan^{n-2}x\,dx$

HYPERBOLIC FORMS

554. $\int \sinh x\,dx = \cosh x$

555. $\int \cosh x\,dx = \sinh x$

556. $\int \tanh x\,dx = \log\cosh x$

557. $\int \coth x \, dx = \log \sinh x$

558. $\int \operatorname{sech} x \, dx = \tan^{-1}(\sinh x)$

559. $\int \operatorname{csch} x \, dx = \log \tanh \left(\dfrac{x}{2}\right)$

560. $\int x \sinh x \, dx = x \cosh x - \sinh x$

561. $\int x^n \sinh x \, dx = x^n \cosh x - n \int x^{n-1}(\cosh x) \, dx$

562. $\int x \cosh x \, dx = x \sinh x - \cosh x$

563. $\int x^n \cosh x \, dx - x^n \sinh x - n \int x^{n-1}(\sinh x) \, dx$

564. $\int \operatorname{sech} x \tanh x \, dx = -\operatorname{sech} x$

565. $\int \operatorname{csch} x \coth x \, dx = -\operatorname{csch} x$

566. $\int \sinh^2 x \, dx = \dfrac{\sinh 2x}{4} - \dfrac{x}{2}$

567. $\int (\sinh^m x)(\cosh^n x) \, dx = \begin{cases} \frac{1}{m+n}(\sinh^{m+1} x)(\cosh^{n-1} x) + \frac{n-1}{m+n} \int (\sinh^m x)(\cosh^{n-2} x) \, dx \\ \quad \text{or} \\ \frac{1}{m+n} \sinh^{m-1} x \cosh^{n+1} x - \frac{m-1}{m+n} \int (\sinh^{m-2} x)(\cosh^n x) \, dx, \quad (m+n \neq 0) \end{cases}$

568. $\int \dfrac{dx}{(\sinh^m x)(\cosh^n x)} \begin{cases} -\frac{1}{(m-n)(\sinh^{m-1} x)(\cosh^{n-1} x)} - \frac{m+n-2}{m-1} \int \frac{dx}{(\sinh^{m-2} x)(\cosh^n x)}, \quad (m \neq 1) \\ \quad \text{or} \\ \frac{1}{(n-1)\sinh^{m-1} x \cosh^{n-1} x} + \frac{m+n-2}{n-1} \int \frac{dx}{(\sinh^m x)(\cosh^{n-2} x)}, \quad (n \neq 1) \end{cases}$

569. $\int \tanh^2 x \, dx = x - \tanh x$

570. $\int \tanh^n x \, dx = -\dfrac{\tanh^{n-1} x}{n-1} + \int (\tanh^{n-2} x) \, dx, \quad (n \neq 1)$

571. $\int \operatorname{sech}^2 x \, dx = \tanh x$

572. $\int \cosh^2 x \, dx = \dfrac{\sinh 2x}{4} + \dfrac{x}{2}$

573. $\int \coth^2 x \, dx = x - \coth x$

574. $\int \coth^n x \, dx = -\dfrac{\coth^{n-1} x}{n-1} + \int \coth^{n-2} x \, dx, \quad (n \neq 1)$

575. $\int \operatorname{csch}^2 x \, dx = -\operatorname{ctnh} x$

576. $\int \sinh(mx) \sinh(nx) \, dx = \dfrac{\sinh(m+n)x}{2(m+n)} - \dfrac{\sinh(m-n)x}{2(m-n)}, \quad (m^2 \neq n^2)$

577. $\int \cosh(mx) \cosh(nx) \, dx = \dfrac{\sinh(m+n)x}{2(m+n)} + \dfrac{\sinh(m-n)x}{2(m-n)}, \quad (m^2 \neq n^2)$

578. $\int \sinh(mx) \cosh(nx) \, dx = \dfrac{\cosh(m+n)x}{2(m+n)} + \dfrac{\cosh(m-n)x}{2(m-n)}, \quad (m^2 \neq n^2)$

579. $\int \sinh^{-1} \dfrac{x}{a} \, dx = x \sinh^{-1} \dfrac{x}{a} - \sqrt{x^2 + a^2}, \quad (a > 0)$

580. $\int x \sinh^{-1} \dfrac{x}{a} \, dx = \left(\dfrac{x^2}{2} + \dfrac{a^2}{4}\right) \sinh^{-1} \dfrac{x}{a} - \dfrac{x}{4}\sqrt{x^2 + a^2}, \quad (a > 0)$

581. $\int x^n \sinh^{-1} x \, dx = \left(\dfrac{x^{n+1}}{n+1}\right) \sinh^{-1} x - \dfrac{1}{n+1} \int \dfrac{x^{n+1}}{(1+x^2)^{\frac{1}{2}}} \, dx, \quad (n \neq -1)$

582. $\int \cosh^{-1} \dfrac{x}{a} \, dx = \begin{cases} x \cosh^{-1} \frac{x}{a} - \sqrt{x^2 - a^2}, \quad \left(\cosh^{-1} \frac{x}{a} > 0\right) \\ \quad \text{or} \\ x \cosh^{-1} \frac{x}{a} + \sqrt{x^2 - a^2}, \quad \left(\cosh^{-1} \frac{x}{a} < 0\right), \quad (a > 0) \end{cases}$

583. $\displaystyle \int x \cosh^{-1} \frac{x}{a} \, dx = \frac{2x^2 - a^2}{4} \cosh^{-1} \frac{x}{a} - \frac{x}{4}(x^2 - a^2)^{\frac{1}{2}}$

584. $\displaystyle \int x^n (\cosh^{-1} x) \, dx = \frac{x^{n+1}}{n+1} \cosh^{-1} x - \frac{1}{n+1} \int \frac{x^{n+1}}{(x^2-1)^{\frac{1}{2}}} \, dx, \quad (n \neq -1)$

585. $\displaystyle \int \tanh^{-1} \frac{x}{a} \, dx = x \tanh^{-1} \frac{x}{a} + \frac{a}{2} \log(a^2 - x^2), \quad \left(\left| \frac{x}{a} \right| < 1 \right)$

586. $\displaystyle \int \coth^{-1} \frac{x}{a} \, dx = x \coth^{-1} \frac{x}{a} + \frac{a}{2} \log(x^2 - a^2), \quad \left(\left| \frac{x}{a} \right| > 1 \right)$

587. $\displaystyle \int x \tanh^{-1} \frac{x}{a} \, dx = \frac{x^2 - a^2}{2} \tanh^{-1} \frac{x}{a} + \frac{ax}{2}, \quad \left(\left| \frac{x}{a} \right| < 1 \right)$

588. $\displaystyle \int x^n \tanh^{-1} x \, dx = \frac{x^{n+1}}{n+1} \tanh^{-1} x - \frac{1}{n+1} \int \frac{x^{n+1}}{1 - x^2} \, dx, \quad (n \neq -1)$

589. $\displaystyle \int x \coth^{-1} \frac{x}{a} \, dx = \frac{x^2 - a^2}{2} \coth^{-1} \frac{x}{a} + \frac{ax}{2}, \quad \left(\left| \frac{x}{a} \right| > 1 \right)$

590. $\displaystyle \int x^n \coth^{-1} x \, dx = \frac{x^{n+1}}{n+1} \coth^{-1} x + \frac{1}{n+1} \int \frac{x^{n+1}}{x^2 - 1} \, dx, \quad (n \neq -1)$

591. $\displaystyle \int \operatorname{sech}^{-1} x \, dx = x \operatorname{sech}^{-1} x + \sin^{-1} x$

592. $\displaystyle \int x \operatorname{sech}^{-1} x \, dx = \frac{x^2}{2} \operatorname{sech}^{-1} x - \frac{1}{2} \sqrt{1 - x^2}$

593. $\displaystyle \int x^n \operatorname{sech}^{-1} x \, dx = \frac{x^{n+1}}{n+1} \operatorname{sech}^{-1} x + \frac{1}{n+1} \int \frac{x^n}{\sqrt{1 - x^2}} \, dx, \quad (n \neq -1)$

594. $\displaystyle \int \operatorname{csch}^{-1} x \, dx = x \operatorname{csch}^{-1} x + \frac{x}{|x|} \sinh^{-1} x$

595. $\displaystyle \int x \operatorname{csch}^{-1} x \, dx = \frac{x^2}{2} \operatorname{csch}^{-1} x + \frac{1}{2} \frac{x}{|x|} \sqrt{1 + x^2}$

596. $\displaystyle \int x^n \operatorname{csch}^{-1} x \, dx = \frac{x^{n+1}}{n+1} \operatorname{csch}^{-1} x + \frac{1}{n+1} \frac{x}{|x|} \int \frac{x^n}{\sqrt{x^2 + 1}} \, dx, \quad (n \neq -1)$

DEFINITE INTEGRALS

597. $\displaystyle \int_0^\infty x^{n-1} e^{-x} \, dx = \int_0^1 \left(\log \frac{1}{x} \right)^{n-1} dx = \frac{1}{n} \prod_{m=1}^\infty \frac{\left(1 + \dfrac{1}{m} \right)^n}{1 + \dfrac{n}{m}} = \Gamma(n)$

for $n \neq 0, -1, -2, -3, \ldots$ (This is the Gamma function)

598. $\displaystyle \int_0^\infty t^n p^{-t} \, dt = \frac{n!}{(\log p)^{n+1}}, \quad (n = 0, 1, 2, 3, \ldots \text{ and } p > 0)$

599. $\displaystyle \int_0^\infty t^{n-1} e^{-(a+1)t} \, dt = \frac{\Gamma(n)}{(a+1)^n}, \quad (n > 0, a > -1)$

600. $\displaystyle \int_0^1 x^m \left(\log \frac{1}{x} \right)^n dx = \frac{\Gamma(n+1)}{(m+1)^{n+1}}, \quad (m > -1, n > -1)$

601. $\Gamma(n)$ is finite if $n > 0$; $\Gamma(n+1) = n\Gamma(n)$

602. $\Gamma(n) \cdot \Gamma(1-n) = \frac{\pi}{\sin n\pi}$

603. $\Gamma(n) = (n-1)!$ if $n = $ integer > 0

604. $\displaystyle \Gamma\left(\frac{1}{2}\right) = 2 \int_0^\infty e^{-t^2} \, dt = \sqrt{\pi} = 1.7724538509 \cdots = \left(-\frac{1}{2} \right)!$

605. $\Gamma\left(n + \frac{1}{2}\right) = \frac{1 \cdot 3 \cdot 5 \ldots (2n-1)}{2^n} \sqrt{\pi} \quad n = 1, 2, 3, \ldots$

606. $\Gamma\left(-n + \frac{1}{2}\right) = \frac{(-1)^n 2^n \sqrt{\pi}}{1 \cdot 3 \cdot 5 \ldots (2n-1)} \quad n = 1, 2, 3, \ldots$

607. $\displaystyle \int_0^1 x^{m-1}(1-x)^{n-1} \, dx = \int_0^\infty \frac{x^{m-1}}{(1+x)^{m+n}} \, dx = \frac{\Gamma(m)\Gamma(n)}{\Gamma(m+n)} = B(m, n)$

(This is the Beta function)

608. $B(m, n) = B(n, m) = \frac{\Gamma(m)\Gamma(n)}{\Gamma(m+n)}$, where m and n are any positive real numbers.

609. $\displaystyle \int_a^b (x-a)^m (b-x)^n \, dx = (b-a)^{m+n+1} \frac{\Gamma(m+1) \cdot \Gamma(n+1)}{\Gamma(m+n+2)}, \quad (m > -1, n > -1, b > a)$

610. $\displaystyle \int_1^\infty \frac{dx}{x^m} = \frac{1}{m-1}, \quad [m > 1]$

611. $\displaystyle \int_0^\infty \frac{dx}{(1+x)x^p} = \pi \csc p\pi, \quad [0 < p < 1]$

612. $\int_0^\infty \dfrac{dx}{(1-x)x^p} = -\pi \cot p\pi, \quad [0 < p < 1]$

613. $\int_0^\infty \dfrac{x^{p-1}\,dx}{(1+x)} = \dfrac{\pi}{\sin p\pi} = B(p, 1-p) = \Gamma(p)\Gamma(1-p), \quad [0 < p < 1]$

614. $\int_0^\infty \dfrac{x^{m-1}\,dx}{1+x^n} = \dfrac{\pi}{n\sin\frac{m\pi}{n}}, \quad [0 < m < n]$

615. $\int_0^\infty \dfrac{x^a\,dx}{(m+x^b)^c} = \dfrac{m^{\frac{a+1-bc}{b}}}{b}\left[\dfrac{\Gamma\left(\frac{a+1}{b}\right)\Gamma\left(c-\frac{a+1}{b}\right)}{\Gamma(c)}\right] \quad \left(a > -1,\, b > 0,\, m > 0,\, c > \frac{a+1}{b}\right)$

616. $\int_0^\infty \dfrac{dx}{(1+x)\sqrt{x}} = \pi$

617. $\int_0^\infty \dfrac{a\,dx}{a^2+x^2} = \begin{cases} \frac{\pi}{2} & (\text{if } a > 0), \\ 0 & (\text{if } a = 0), \\ -\frac{\pi}{2} & (\text{if } a < 0) \end{cases}$

618. $\int_0^a (a^2-x^2)^{n/2}\,dx = \dfrac{1}{2}\int_{-a}^a (a^2-x^2)^{n/2}\,dx = \dfrac{1\cdot 3\cdot 5\ldots n}{2\cdot 4\cdot 6\ldots(n+1)}\cdot\dfrac{\pi}{2}\cdot a^{n+1} \quad (n \text{ odd},\, a > 0)$

619. $\int_0^a x^m(a^2-x^2)^{n/2}\,dx = \begin{cases} \frac{1}{2}a^{m+n+1}B\left(\frac{m+1}{2}, \frac{n+2}{2}\right) & (a > 0,\, m > -1,\, n > -2) \\ \text{or} \\ \frac{1}{2}a^{m+n+1}\dfrac{\Gamma\left(\frac{m+1}{2}\right)\Gamma\left(\frac{n+2}{2}\right)}{\Gamma\left(\frac{m+n+3}{2}\right)} & (a > 0,\, m > -1,\, n > -2) \end{cases}$

620. $\int_0^{\pi/2}\sin^n x\,dx = \begin{cases} \int_0^{\pi/2}(\cos^n x)\,dx \\ \frac{1\cdot 3\cdot 5\cdot 7\ldots(n-1)}{2\cdot 4\cdot 6\cdot 8\ldots(n)}\frac{\pi}{2}, & (n \text{ an even integer},\, n \neq 0), \\ \frac{1\cdot 3\cdot 5\cdot 7\ldots(n-1)}{2\cdot 4\cdot 6\cdot 8\ldots(n)}, & (n \text{ an odd integer},\, n \neq 0), \\ \frac{\sqrt{\pi}}{2}\dfrac{\Gamma\left(\frac{n+1}{2}\right)}{\Gamma\left(\frac{n}{2}+1\right)} & (n > -1) \end{cases}$

621. $\int_0^\infty \dfrac{\sin mx\,dx}{x} = \dfrac{\pi}{2};\text{ if } m > 0;\ 0, \text{ if } m = 0;\ -\dfrac{\pi}{2}, \text{ if } m < 0$

622. $\int_0^\infty \dfrac{\cos x\,dx}{x} = \infty$

623. $\int_0^\infty \dfrac{\tan x\,dx}{x} = \dfrac{\pi}{2}$

624. $\int_0^\pi \sin ax \cdot \sin bx\,dx = \int_0^\pi \cos ax \cdot \cos bx\,dx = 0, \quad (a \neq b;\, a,\, b \text{ integers})$

625. $\int_0^{\pi/a} [\sin(ax)][\cos(ax)]\,dx = \int_0^\pi [\sin(ax)][\cos(ax)]\,dx = 0$

626. $\int_0^\pi [\sin(ax)][\cos(bx)]\,dx = \dfrac{2a}{a^2-b^2}, \text{ if } a - b \text{ is odd, or } 0 \text{ if } a - b \text{ is even}$

627. $\int_0^\infty \dfrac{\sin x \cos mx\,dx}{x} = 0, \quad \text{if } m < -1 \text{ or } m > 1;\ \dfrac{\pi}{4}, \text{ if } m = \pm 1;\ \dfrac{\pi}{2}, \text{ if } m^2 < 1$

628. $\int_0^\infty \dfrac{\sin ax \sin bx}{x^2}\,dx = \dfrac{\pi a}{2}, \quad (a \leq b)$

629. $\int_0^\pi \sin^2 mx\,dx = \int_0^\pi \cos^2 mx\,dx = \dfrac{\pi}{2} \quad (m \text{ is a non-zero integer})$

630. $\int_0^\infty \dfrac{\sin^2(px)}{x^2}\,dx = \dfrac{\pi|p|}{2}$

631. $\int_0^\infty \dfrac{\sin x}{x^p}\,dx = \dfrac{\pi}{2\Gamma(p)\sin(p\pi/2)}, \quad 0 < p < 1$

632. $\int_0^\infty \dfrac{\cos x}{x^p}\,dx = \dfrac{\pi}{2\Gamma(p)\cos(p\pi/2)}, \quad 0 < p < 1$

633. $\int_0^\infty \dfrac{1-\cos px}{x^2}\,dx = \dfrac{\pi|p|}{2}$

634. $\int_0^\infty \dfrac{\sin px \cos qx}{x}\,dx = \left\{0, q > p > 0;\ \dfrac{\pi}{2}, p > q > 0;\ \dfrac{\pi}{4}, p = q > 0\right\}$

635. $\int_0^\infty \dfrac{\cos(mx)}{x^2+a^2}\,dx = \dfrac{\pi}{2|a|}e^{-|ma|}$

636. $\displaystyle\int_0^\infty \cos(x^2)\,dx = \int_0^\infty \sin(x^2)\,dx = \frac{1}{2}\sqrt{\frac{\pi}{2}}$

637. $\displaystyle\int_0^\infty \sin ax^n\,dx = \frac{1}{na^{1/n}}\Gamma(1/n)\sin\frac{\pi}{2n},\qquad \text{if } n > 1$

638. $\displaystyle\int_0^\infty \cos ax^n\,dx = \frac{1}{na^{1/n}}\Gamma(1/n)\cos\frac{\pi}{2n},\qquad \text{if } n > 1$

639. $\displaystyle\int_0^\infty \frac{\sin x}{\sqrt{x}}\,dx = \int_0^\infty \frac{\cos x}{\sqrt{x}}\,dx = \sqrt{\frac{\pi}{2}}$

640. (a) $\int_0^\infty \frac{\sin^3 x}{x}\,dx = \frac{\pi}{4}$ $\qquad$ (b) $\int_0^\infty \frac{\sin^3 x}{x^2}\,dx = \frac{3}{4}\log 3$

641. $\displaystyle\int_0^\infty \frac{\sin^3 x}{x^3}\,dx = \frac{3\pi}{8}$

642. $\displaystyle\int_0^\infty \frac{\sin^4 x}{x^4}\,dx = \frac{\pi}{3}$

643. $\displaystyle\int_0^{\pi/2} \frac{dx}{1 + a\cos x} = \frac{\cos^{-1} a}{\sqrt{1 - a^2}},\qquad (|a| < 1)$

644. $\displaystyle\int_0^\pi \frac{dx}{a + b\cos x} = \frac{\pi}{\sqrt{a^2 - b^2}},\qquad (a > b \geq 0)$

645. $\displaystyle\int_0^{2\pi} \frac{dx}{1 + a\cos x} = \frac{2\pi}{\sqrt{1 - a^2}},\qquad (a^2 < 1)$

646. $\displaystyle\int_0^\infty \frac{\cos ax - \cos bx}{x}\,dx = \log\left|\frac{b}{a}\right|$

647. $\displaystyle\int_0^{\pi/2} \frac{dx}{a^2\sin^2 x + b^2\cos^2 x} = \frac{\pi}{2|ab|}$

648. $\displaystyle\int_0^{\pi/2} \frac{dx}{(a^2\sin^2 x + b^2\cos^2 x)^2} = \frac{\pi(a^2 + b^2)}{4a^3 b^3},\qquad (a, b > 0)$

649. $\displaystyle\int_0^{\pi/2} \sin^{n-1} x \cos^{m-1} x\,dx = \frac{1}{2}B\left(\frac{n}{2}, \frac{m}{2}\right),\qquad (\text{if } m \text{ and } n \text{ are positive integers})$

650. $\displaystyle\int_0^{\pi/2} (\sin^{2n+1}\theta)\,d\theta = \frac{2\cdot 4\cdot 6\ldots(2n)}{1\cdot 3\cdot 5\ldots(2n+1)},\qquad (n = 1, 2, 3, \ldots)$

651. $\displaystyle\int_0^{\pi/2} (\sin^{2n}\theta)\,d\theta = \frac{1\cdot 3\cdot 5\ldots(2n-1)}{2\cdot 4\ldots(2n)}\left(\frac{\pi}{2}\right),\qquad (n = 1, 2, 3, \ldots)$

652. $\displaystyle\int_0^{\pi/2} \frac{x}{\sin x}\,dx = 2\left\{\frac{1}{1^2} - \frac{1}{3^2} + \frac{1}{5^2} - \frac{1}{7^2} + \cdots\right\}$

653. $\displaystyle\int_0^{\pi/2} \frac{dx}{1 + \tan^m x} = \frac{\pi}{4}$

654. $\displaystyle\int_0^{\pi/2} \sqrt{\cos\theta}\,d\theta = \frac{(2\pi)^{\frac{3}{2}}}{\left[\Gamma(\frac{1}{4})\right]^2}$

655. $\displaystyle\int_0^{\pi/2} (\tan^h\theta)\,d\theta = \frac{\pi}{2\cos\left(\frac{h\pi}{2}\right)},\qquad (0 < h < 1)$

656. $\displaystyle\int_0^\infty \frac{\tan^{-1}(ax) - \tan^{-1}(bx)}{x}\,dx = \frac{\pi}{2}\log\frac{a}{b},\qquad (a, b > 0)$

657. The area enclosed by a curve defined through the equation $x^{\frac{b}{c}} + y^{\frac{b}{c}} = a^{\frac{b}{c}}$ where $a > 0$, c a positive odd integer and b a positive even integer is given by $\frac{\left[\Gamma\left(\frac{c}{b}\right)\right]^2}{\Gamma\left(\frac{2c}{b}\right)}\left(\frac{2ca^2}{b}\right)$

658. $I = \displaystyle\iiint_R x^{h-1} y^{m-1} z^{n-1}\,dv$, where R denotes the region of space bounded by the co-ordinate planes and that portion of the surface $\left(\frac{x}{a}\right)^p + \left(\frac{y}{b}\right)^q + \left(\frac{z}{c}\right)^k = 1$, which lies in the first octant and where $h, m, n, p, q, k, a, b, c$, denote positive real numbers is given by

$$\int_0^a x^{h-1}\,dx \int_0^b \left[1 - \left(\frac{x}{a}\right)^p\right]^{\frac{1}{e}} y^m\,dy \int_0^c \left[1 - \left(\frac{x}{a}\right)^p - \left(\frac{y}{b}\right)^q\right]^{\frac{1}{e}} z^{n-1}\,dz = \frac{a^h b^m c^n}{pqk} \frac{\Gamma\left(\frac{h}{p}\right)\Gamma\left(\frac{m}{q}\right)\Gamma\left(\frac{n}{k}\right)}{\Gamma\left(\frac{h}{p} + \frac{m}{q} + \frac{n}{k} + 1\right)}$$

659. $\displaystyle\int_0^\infty e^{-ax}\,dx = \frac{1}{a}, \quad (a > 0)$

660. $\displaystyle\int_0^\infty \frac{e^{-ax} - e^{-bx}}{x}\,dx = \log\frac{b}{a}, \quad (a, b > 0)$

661. $\displaystyle\int_0^\infty x^n e^{-ax}\,dx = \begin{cases} \frac{\Gamma(n+1)}{a^{n+1}} & \text{(if } n > -1 \text{ and } a > 0) \\ \text{or} \\ \frac{n!}{a^{n+1}} & \text{(if } a > 0 \text{ and } n \text{ is a positive integer)} \end{cases}$

662. $\displaystyle\int_0^\infty x^n \exp(-ax^p)\,dx = \frac{\Gamma(k)}{pa^k}, \quad \left(n > -1,\ p > 0,\ a > 0,\ k = \frac{n+1}{p}\right)$

663. $\displaystyle\int_0^\infty e^{-a^2 x^2}\,dx = \frac{1}{2a}\sqrt{\pi} = \frac{1}{2a}\Gamma\left(\frac{1}{2}\right), \quad (a > 0)$

664. $\displaystyle\int_0^\infty x e^{-x^2}\,dx = \frac{1}{2}$

665. $\displaystyle\int_0^\infty x^2 e^{-x^2}\,dx = \frac{\sqrt{\pi}}{4}$

666. $\displaystyle\int_0^\infty x^{2n} e^{-ax^2}\,dx = \frac{1 \cdot 3 \cdot 5 \ldots (2n-1)}{2^{n+1} a^n}\sqrt{\frac{\pi}{a}} \quad (a > 0,\ n > -\tfrac{1}{2})$

667. $\displaystyle\int_0^\infty x^{2n+1} e^{-ax^2}\,dx = \frac{n!}{2a^{n+1}}, \quad (a > 0,\ n > -1)$

668. $\displaystyle\int_0^1 x^m e^{-ax}\,dx = \frac{m!}{a^{m+1}}\left[1 - e^{-a}\sum_{r=0}^m \frac{a^r}{r!}\right]$

669. $\displaystyle\int_0^\infty e^{\left(-x^2 - \frac{a^2}{x^2}\right)}\,dx = \frac{e^{-2a}\sqrt{\pi}}{2}, \quad (a \geq 0)$

670. $\displaystyle\int_0^\infty e^{-nx}\sqrt{x}\,dx = \frac{1}{2n}\sqrt{\frac{\pi}{n}} \quad (n > 0)$

671. $\displaystyle\int_0^\infty \frac{e^{-nx}}{\sqrt{x}}\,dx = \sqrt{\frac{\pi}{n}} \quad (n > 0)$

672. $\displaystyle\int_0^\infty e^{-ax}(\cos mx)\,dx = \frac{a}{a^2 + m^2}, \quad (a > 0)$

673. $\displaystyle\int_0^\infty e^{-ax}(\sin mx)\,dx = \frac{m}{a^2 + m^2}, \quad (a > 0)$

674. $\displaystyle\int_0^\infty x e^{-ax}[\sin(bx)]\,dx = \frac{2ab}{(a^2 + b^2)^2}, \quad (a > 0)$

675. $\displaystyle\int_0^\infty x e^{-ax}[\cos(bx)]\,dx = \frac{a^2 - b^2}{(a^2 + b^2)^2}, \quad (a > 0)$

676. $\displaystyle\int_0^\infty x^n e^{-ax}[\sin(bx)]\,dx = \frac{n![(a+ib)^{n+1} - (a-ib)^{n+1}]}{2i(a^2 + b^2)^{n+1}}, \quad (i^2 = -1,\ a > 0)$

677. $\displaystyle\int_0^\infty x^n e^{-ax}[\cos(bx)]\,dx = \frac{n![(a-ib)^{n+1} + (a+ib)^{n+1}]}{2(a^2 + b^2)^{n+1}}, \quad (i^2 = -1,\ a > 0,\ n > -1)$

678. $\displaystyle\int_0^\infty \frac{e^{-ax}\sin x}{x}\,dx = \cot^{-1} a, \quad (a > 0)$

679. $\displaystyle\int_0^\infty e^{-a^2 x^2}\cos bx\,dx = \frac{\sqrt{\pi}}{2|a|}\exp\left(-\frac{b^2}{4a^2}\right), \quad (ab \neq 0)$

680. $\displaystyle\int_0^\infty e^{-t\cos\phi} t^{b-1}[\sin(t\sin\phi)]\,dt - [\Gamma(b)]\sin(b\phi), \quad \left(b > 0,\ -\frac{\pi}{2} < \phi < \frac{\pi}{2}\right)$

681. $\displaystyle\int_0^\infty e^{-t\cos\phi} t^{b-1}[\cos(t\sin\phi)]\,dt - [\Gamma(b)]\cos(b\phi), \quad \left(b > 0,\ -\frac{\pi}{2} < \phi < \frac{\pi}{2}\right)$

682. $\displaystyle\int_0^\infty t^{b-1}\cos t\,dt = [\Gamma(b)]\cos\left(\frac{b\pi}{2}\right), \quad (0 < b < 1)$

683. $\displaystyle\int_0^\infty t^{b-1}(\sin t)\,dt = [\Gamma(b)]\sin\left(\frac{b\pi}{2}\right), \quad (0 < b < 1)$

684. $\displaystyle\int_0^1 (\log x)^n\,dx = (-1)^n \cdot n! \quad (n > -1)$

685. $\displaystyle\int_0^1 \left(\log\frac{1}{x}\right)^{\frac{1}{2}}\,dx = \frac{\sqrt{\pi}}{2}$

686. $\displaystyle\int_0^1 \left(\log\frac{1}{x}\right)^{-\frac{1}{2}}\,dx = \sqrt{\pi}$

687. $\displaystyle\int_0^1 \left(\log \frac{1}{x}\right)^n dx = n!$

688. $\displaystyle\int_0^1 x \log(1-x)\, dx = -\frac{3}{4}$

689. $\displaystyle\int_0^1 x \log(1+x)\, dx = \frac{1}{4}$

690. $\displaystyle\int_0^1 x^m (\log x)^n\, dx = \frac{(-1)^n n!}{(m+1)^{n+1}}, \qquad (m > -1,\ n = 0, 1, 2, \ldots)$

   If $n \neq 0, 1, 2, \ldots$ replace $n!$ by $\Gamma(n+1)$.

691. $\displaystyle\int_0^1 \frac{\log x}{1+x}\, dx = -\frac{\pi^2}{12}$

692. $\displaystyle\int_0^1 \frac{\log x}{1-x}\, dx = -\frac{\pi^2}{6}$

693. $\displaystyle\int_0^1 \frac{\log(1+x)}{x}\, dx = \frac{\pi^2}{12}$

694. $\displaystyle\int_0^1 \frac{\log(1-x)}{x}\, dx = -\frac{\pi^2}{6}$

695. $\displaystyle\int_0^1 \log(x) \log(1+x)\, dx = 2 - 2\log 2 - \frac{\pi^2}{12}$

696. $\displaystyle\int_0^1 \log(x) \log(1-x)\, dx = 2 - \frac{\pi^2}{6}$

697. $\displaystyle\int_0^1 \frac{\log x}{1-x^2}\, dx = -\frac{\pi^2}{8}$

698. $\displaystyle\int_0^1 \log\left(\frac{1+x}{1-x}\right) \cdot \frac{dx}{x} = \frac{\pi^2}{4}$

699. $\displaystyle\int_0^1 \frac{\log x\, dx}{\sqrt{1-x^2}} = -\frac{\pi}{2} \log 2$

700. $\displaystyle\int_0^1 x^m \left[\log\left(\frac{1}{x}\right)\right]^n dx = \frac{\Gamma(n+1)}{(m+1)^{n+1}}, \qquad (\text{if } m+1 > 0 \text{ and } n+1 > 0)$

701. $\displaystyle\int_0^1 \frac{(x^p - x^q)\, dx}{\log x} = \log\left(\frac{p+1}{q+1}\right), \qquad (p+1 > 0,\ q+1 > 0)$

702. $\displaystyle\int_0^1 \frac{dx}{\sqrt{\log\left(\frac{1}{x}\right)}} = \sqrt{\pi}, \qquad (\text{same as integral 686})$

703. $\displaystyle\int_0^\infty \log\left(\frac{e^x + 1}{e^x - 1}\right) dx = \frac{\pi^2}{4}$

704. $\displaystyle\int_0^{\pi/2} \log(\sin x)\, dx = \int_0^{\pi/2} \log\cos x\, dx = -\frac{\pi}{2}\log 2$

705. $\displaystyle\int_0^{\pi/2} \log(\sec x)\, dx = \int_0^{\pi/2} \log\csc x\, dx = \frac{\pi}{2}\log 2$

706. $\displaystyle\int_0^\pi x \log(\sin x)\, dx = -\frac{\pi^2}{2}\log 2$

707. $\displaystyle\int_0^{\pi/2} \sin x \log(\sin x)\, dx = \log 2 - 1$

708. $\displaystyle\int_0^{\pi/2} \log\tan x\, dx = 0$

709. $\displaystyle\int_0^\pi \log(a \pm b\cos x)\, dx = \pi \log\left(\frac{a + \sqrt{a^2 - b^2}}{2}\right), \qquad (a \geq b)$

710. $\displaystyle\int_0^\pi \log(a^2 - 2ab\cos x + b^2)\, dx = \begin{cases} 2\pi \log a & a \geq b > 0 \\ 2\pi \log b & b \geq a > 0 \end{cases}$

711. $\displaystyle\int_0^\infty \frac{\sin ax}{\sinh bx}\, dx = \frac{\pi}{2|b|} \tanh \frac{a\pi}{2b}$

712. $\displaystyle\int_0^\infty \frac{\cos ax}{\cosh bx}\, dx = \frac{\pi}{2|b|} \operatorname{sech} \frac{a\pi}{2b}$

713. $\displaystyle\int_0^\infty \frac{dx}{\cosh ax} = \frac{\pi}{2|a|}$

714. $\displaystyle\int_0^\infty \frac{x\,dx}{\sinh ax} = \frac{\pi^2}{4a^2} \quad (a > 0)$

715. $\displaystyle\int_0^\infty e^{-ax}\cosh bx\,dx = \frac{a}{a^2 - b^2}, \quad (0 \le |b| \; < a)$

716. $\displaystyle\int_0^\infty e^{-ax}\sinh bx\,dx = \frac{b}{a^2 - b^2}, \quad (0 \le |b| \; < a)$

717. $\displaystyle\int_0^\infty \frac{\sinh ax}{e^{bx}+1}\,dx = \frac{\pi}{2b}\csc\frac{a\pi}{b} - \frac{1}{2a} \quad (b > 0)$

718. $\displaystyle\int_0^\infty \frac{\sinh ax}{e^{bx}-1}\,dx = \frac{1}{2a} - \frac{\pi}{2b}\cot\frac{a\pi}{b} \quad (b > 0)$

719. $\displaystyle\int_0^{\pi/2} \frac{dx}{\sqrt{1 - k^2\sin^2 x}} = \frac{\pi}{2}\left[1 + \left(\frac{1}{2}\right)^2 k^2 + \left(\frac{1\cdot 3}{2\cdot 4}\right)^2 k^4 + \left(\frac{1\cdot 3\cdot 5}{2\cdot 4\cdot 6}\right)^2 k^6 + \cdots\right], \quad \text{if } k^2 < 1$

720. $\displaystyle\int_0^{\pi/2} \sqrt{1 - k^2\sin^2 x}\,dx = \frac{\pi}{2}\left[1 - \left(\frac{1}{2}\right)^2 k^2 - \left(\frac{1\cdot 3}{2\cdot 4}\right)^2 \frac{k^4}{3} - \left(\frac{1\cdot 3\cdot 5}{2\cdot 4\cdot 6}\right)2\frac{k^6}{5} - \cdots\right], \quad \text{if } k^2 < 1$

721. $\displaystyle\int_0^\infty e^{-x}\log x\,dx = -\gamma = -0.5772157\ldots$

722. $\displaystyle\int_0^\infty e^{-x^2}\log x\,dx = -\frac{\sqrt{\pi}}{4}(\gamma + 2\log 2)$

723. $\displaystyle\int_0^\infty \left(\frac{1}{1-e^{-x}} - \frac{1}{x}\right)e^{-x}\,dx = \gamma = 0.5772157\ldots \quad \text{[Euler's Constant]}$

724. $\displaystyle\int_0^\infty \frac{1}{x}\left(\frac{1}{1+x} - e^{-x}\right)dx = \gamma = 0.5772157\ldots$

For n even :

725. $\displaystyle\int \cos^n x\,dx = \frac{1}{2^{n-1}}\sum_{k=0}^{n/2-1}\binom{n}{k}\frac{\sin(n-2k)x}{(n-2k)} + \frac{1}{2^n}\binom{n}{n/2}x$

726. $\displaystyle\int \sin^n x\,dx = \frac{1}{2^{n-1}}\sum_{k=0}^{n/2-1}\binom{n}{k}\frac{\sin[(n-2k)(\frac{\pi}{2}-x)]}{2k-n} + \frac{1}{2^n}\binom{n}{n/2}x$

For n odd:

727. $\displaystyle\int \cos^n x\,dx = \frac{1}{2^{n-1}}\sum_{k=0}^{(n-1)/2}\binom{n}{k}\frac{\sin(n-2k)x}{n-2k}$

728. $\displaystyle\int \sin^n x\,dx = \frac{1}{2^{n-1}}\sum_{k=0}^{(n-1)/2}\binom{n}{k}\frac{\sin\left[(n-2k)\left(\frac{\pi}{2}-x\right)\right]}{2k-n}$

DIFFERENTIAL EQUATIONS

Certain types of differential equations occur sufficiently often to justify the use of formulas for the corresponding particular solutions. The following set of Tables I to XIV covers all first, second, and nth order ordinary linear differential equations with constant coefficients for which the right members are of the form $P(x)e^{rx}\sin sx$ or $P(x)e^{rx}\cos sx$, where r and s are constants and $P(x)$ is a polynomial of degree n.

When the right member of a reducible linear partial differential equation with constant coefficients is not zero, particular solutions for certain types of right members are contained in Tables XV to XXI. In these tables both F and P are used to denote polynomials, and it is assumed that no denominator is zero. In any formula the roles of x and y may be reversed throughout, changing a formula in which x dominates to one in which y dominates. Tables XIX, XX, XXI are applicable whether the equations are reducible or not. The symbol $\binom{m}{n}$ stands for $\frac{m!}{(m-n)!n!}$ and is the $(n+1)^{\text{st}}$ coefficient in the expansion of $(a+b)^m$. Also $0! = 1$ by definition.

The tables as herewith given are those contained in the text *Differential Equations* by Ginn and Company (1955) and are published with their kind permission and that of the author, Professor Frederick H. Steen.

SOLUTION OF LINEAR DIFFERENTIAL EQUATIONS WITH CONSTANT COEFFICIENTS

Any linear differential equation with constant coefficients may be written in the form

$$p(D)y = R(x)$$

where

- D is the differential operation: $Dy = \frac{dy}{dx}$
- $p(D)$ is a polynomial in D,
- y is the dependent variable,
- x is the independent variable,
- $R(x)$ is an arbitrary function of x.

A power of D represents repeated differentiation, that is

$$D^n y = \frac{d^n y}{dx^n}$$

For such an equation, the general solution may be written in the form

$$y = y_c + y_p$$

where y_p is any particular solution, and y_c is called the *complementary function*. This complementary function is defined as the general solution of the *homogeneous equation*, which is the original differential equation with the right side replaced by zero, i.e.,

$$p(D)y = 0$$

The complementary function y_c may be determined as follows:

1. Factor the polynomial $p(D)$ into real and complex linear factors, just as if D were a variable instead of an operator.

2. For each nonrepeated linear factor of the form $(D - a)$, where a is real, write down a term of the form

$$ce^{ax}$$

   where c is an arbitrary constant.

3. For each repeated real linear factor of the form $(D - a)^n$, write down n terms of the form

$$c_1 e^{ax} + c_2 x e^{ax} + c_3 x^2 e^{ax} + \cdots + c_n x^{n-1} e^{ax}$$

   where the c_i's are arbitrary constants.

4. For each non-repeated conjugate complex pair of factors of the form $(D - a + ib)(D - a - ib)$, write down two terms of the form

$$c_1 e^{ax} \cos bx + c_2 e^{ax} \sin bx$$

5. For each repeated conjugate complex pair of factors of the form $(D - a + ib)^n (D - a - ib)^n$, write down $2n$ terms of the form

$$c_1 e^{ax} \cos bx + c_2 e^{ax} \sin bx + c_3 x e^{ax} \cos bx + c_4 x e^{ax} \sin bx$$
$$+ \cdots + c_{2n-1} x^{n-1} e^{ax} \cos bx + c_{2n} x^{n-1} e^{ax} \sin bx$$

6. The sum of all the terms thus written down is the complementary function y_c.

To find the particular solution y_p, use the following tables, as shown in the examples. For cases not shown in the tables, there are various methods of finding y_p. The most general method is called *variation of parameters*. The following example illustrates the method:

Example: Find y_p for $(D^2 - 4)y = e^x$.

This example can be solved most easily by use of equation 63 in the tables following. However, it is given here as an example of the method of variation of parameters.

The complementary function is

$$y_c = c_1 e^{2x} + c_2 e^{-2x}$$

To find y_p, replace the constants in the complementary function with unknown functions,

$$y_p = u e^{2x} + v e^{-2x}$$

We now prepare to substitute this assumed solution into the original equation. We begin by taking all the necessary derivatives:

$$y_p = ue^{2x} + ve^{-2x}$$

$$y'_p = 2ue^{2x} - 2ve^{-2x} + u'e^{2x} + v'e^{-2x}$$

For each derivative of y_p except the highest, we set the sum of all the terms containing u' and v' to 0. Thus the above equation becomes

$$u'e^{2x} + v'e^{-2x} = 0 \quad \text{and} \quad y'_p = 2ue^{2x} - 2ve^{-2x}$$

Continuing to differentiate, we have

$$y''_p = 4ue^{2x} + 4ve^{-2x} + 2u'e^{2x} - 2v'e^{-2x}$$

When we substitute into the original equation, all the terms not containing u' or v' cancel out. This is a consequence of the method by which y_p was set up.

Thus all that is necessary is to write down the terms containing u' or v' in the highest order derivative of y_p, multiply by the constant coefficient of the highest power of D in $p(D)$, and set it equal to $R(x)$. Together with the previous terms in u' and v' which were set equal to 0, this gives us as many linear equations in the first derivatives of the unknown functions as there are unknown functions. The first derivatives may then be solved for by algebra, and the unknown functions found by integration. In the present example, this becomes

$$u'e^{2x} + v'e^{-2x} = 0$$

$$2u'e^{2x} - 2v'e^{-2x} = e^x$$

We eliminate v' and u' separately, getting

$$4u'e^{2x} = e^x$$

$$4v'e^{-2x} = -e^x$$

Thus

$$u' = \tfrac{1}{4}e^{-x}$$

$$v' = -\tfrac{1}{4}e^{3x}$$

Therefore, by integrating

$$u = -\tfrac{1}{4}e^{-x}$$

$$v = -\tfrac{1}{12}e^{3x}$$

A constant of integration is not needed, since we need only one particular solution. Thus

$$y_p = ue^{2x} + ve^{-2x} = -\frac{1}{4}e^{-x}e^{2x} - \frac{1}{12}e^{3x}e^{-2x}$$

$$= -\frac{1}{4}e^x - \frac{1}{12}e^x = -\frac{1}{3}e^x$$

and the general solution is

$$y = y_c + y_p = c_1e^{2x} + c_2e^{-2x} - \frac{1}{3}e^x$$

The following samples illustrate the use of the tables.

Example 1: Solve $(D^2 - 4)y = \sin 3x$. Substitution of $q = -4$, $s = 3$ in formula 24 gives

$$y_p = \frac{\sin 3x}{-9 - 4}$$

wherefore the general solution is

$$y = c_1e^{2x} + c_2e^{-2x} - \frac{\sin 3x}{13}$$

Example 2: Obtain a particular solution of $(D^2 - 4D + 5)y = x^2 e^{3x} \sin x$.

Applying formula 40 with $a = 2$, $b = 1$, $r = 3$, $s = 1$, $P(x) = x^2$, $s + b = 2$, $s - b = 0$, $a - r = -1$, $(a - r)^2 + (s + b)^2 = 5$, $(a - r)^2 + (s - b)^2 = 1$, we have

$$
\begin{aligned}
y_p &= \frac{e^{3x} \sin x}{2} \left[\left(\frac{2}{5} - \frac{0}{1} \right) x^2 + \left(\frac{2(-1)2}{25} - \frac{2(-1)0}{1} \right) 2x + \left(\frac{3 \cdot 1 \cdot 2 - 2^3}{125} - \frac{3 \cdot 1 \cdot 0 - 0}{1} \right) 2 \right] \\
&\quad - \frac{e^{3x} \cos x}{2} \left[\left(\frac{-1}{5} - \frac{-1}{1} \right) x^2 + \left(\frac{1 - 4}{25} - \frac{1 - 0}{1} \right) 2x + \left(\frac{-1 - 3(-1)4}{125} - \frac{-1 - 3(-1)0}{1} \right) 2 \right] \\
&= \left(\frac{1}{5} x^2 - \frac{4}{25} x - \frac{2}{125} \right) e^{3x} \sin x + \left(-\frac{2}{5} x^2 + \frac{28}{25} - \frac{136}{125} \right) e^{3x} \cos x
\end{aligned}
$$

The special formulas effect a very considerable saving of time in problems of this type.

Example 3: Obtain a particular solution of $(D^2 - 4D + 5)y = x^2 e^{2x} \cos x$. (Compare with Example 2.)

Formula 40 is not applicable here since for this equation $r = a$, $s = b$, wherefore the denominator $(a - r)^2 + (s - b)^2 = 0$. We turn instead to formula 44. Substituting $a = 2$, $b = 1$, $P(x) = x^2$ and replacing sin by cos, cos by $-$ sin, we obtain

$$
\begin{aligned}
y_p &= \frac{e^{2x} \cos x}{4} \left(x^2 - \frac{2}{4} \right) + \frac{e^{2x} \sin x}{2} \int \left(x^2 - \frac{1}{2} \right) dx \\
&= \left(\frac{x^2}{4} - \frac{1}{8} \right) e^{2x} \cos x + \left(\frac{x^3}{6} - \frac{x}{4} \right) e^{2x} \sin x
\end{aligned}
$$

which is the required solution.

Example 4: Find z_p for $(D_x - 3D_y) z = \ln(y + 3x)$. Referring to Table XV we note that formula 69 (not 68) is applicable. This gives

$$z_p = x \ln(y + 3x)$$

It is easily seen that $-y/3 \ln(y + 3x)$ would serve equally well.

Example 5: Solve $(D_x + 2D_y - 4) z = y \cos(y - 2x)$.

Since R in formula 76 contains a polynomial in x, not y, we rewrite the given equation in the form $(D_y + \frac{1}{2} D_x - 2) z = \frac{1}{2} y \cos(y - 2x)$. Then

$$z_c = e^{2y} F \left(x - \frac{1}{2} y \right) = e^{2x} f(2x - y)$$

and by the formula

$$z_p = -\frac{1}{2} \cos(y - 2x) \cdot \left(\frac{y}{2} + \frac{\frac{1}{2}}{2} \right) = -\frac{1}{8} (2y + 1) \cos(y - 2x)$$

Example 6: Find z_p for $(D_x + 4D_y)^3 z = (2x - y)^2$.

Using formula 79, we obtain

$$z_p = \frac{\iiint u^2 \, du^3}{[2 + 4(-1)]^3} = \frac{u^5}{5 \cdot 4 \cdot 3 \cdot (-8)} = -\frac{(2x - y)^5}{480}$$

Example 7: Find z_p for $(D_x^3 + 5 D_x^2 D_y - 7D_x + 4)z = e^{2x+3y}$. By formula 87

$$z_p = \frac{e^{2x+3y}}{2^3 + 5 \cdot 2^2 \cdot 3 - 7 \cdot 2 + 4} = \frac{e^{2x+3y}}{58}$$

Example 8: Find z_p for

$$(D_x^4 + 6D_x^3 D_y + D_x D_y + D_y^2 + 9)z = \sin(3x + 4y)$$

Since every term in the left number is of even degree in the two operators D_x and D_y, formula 90 is applicable. It gives

$$
\begin{aligned}
z_p &= \frac{\sin(3x + 4y)}{(-9)^2 + 6(-9)(-12) + (-12) + (-16) + 9} \\
&= \frac{\sin(3x + 4y)}{710}
\end{aligned}
$$

Table I: $(D - a)y = R$

R y_p

1. e^{rx} $\dfrac{e^{rx}}{r-a}$

2. $\sin sx*$ $-\dfrac{a\sin sx + s\cos sx}{a^2+s^2} = \dfrac{1}{\sqrt{a^2+s^2}}\sin\left(sx + \tan^{-1}\dfrac{s}{a}\right)$

3. $P(x)$ $-\dfrac{1}{a}\left[P(x) + \dfrac{P'(x)}{a} + \dfrac{P''(x)}{a^2} + \cdots + \dfrac{P^{(n)}(x)}{a^n}\right]$

4. $e^{rx}\sin sx*$ Replace a by $a - r$ in formula 2 and multiply by e^{rx}.

5. $P(x)\ e^{rx}$ Replace a by $a - r$ in formula 3 and multiply by e^{rx}.

6. $P(x)\sin sx*$ $-\sin sx\left[\dfrac{a}{a^2+s^2}P(x) + \dfrac{a^2-s^2}{(a^2+s^2)^2}P'(x) + \dfrac{a^3-3as^2}{(a^2+s^2)^3}P''(x) + \cdots + \dfrac{a^k - \binom{k}{2}a^{k-2}s^2 + \binom{k}{4}a^{k-4}s^4 - \cdots}{(a^2+s^2)^k}P^{(k-1)}(x) + \cdots\right]$

                $-\cos sx\left[\dfrac{s}{a^2+s^2}P(x) + \dfrac{2as}{(a^2+s^2)^2}P'(x) + \dfrac{3a^2s-s^3}{(a^2+s^2)^3}P''(x) + \cdots + \dfrac{\binom{k}{1}a^{k-1}s - \binom{k}{3}a^{k-3}s^3 + \cdots}{(a^2+s^2)^k}P^{(k-1)}(x) + \cdots\right]$

7. $P(x)e^{rx}\sin sx*$ Replace a by $a - r$ in formula 6 and multiply by e^{rx}.

8. e^{ax} xe^{ax}

9. $e^{ax}\sin sx*$ $-\dfrac{e^{ax}\cos sx}{s}$

10. $P(x)e^{ax}$ $e^{ax}\displaystyle\int^s P(x)\,dx$

11. $P(x)e^{ax}\sin sx$ $\dfrac{e^{ax}\sin sx}{s}\left[\dfrac{P'(x)}{s^3} - \dfrac{P'''(x)}{s^3} + \dfrac{P^v(x)}{s^5} - \cdots\right] - \dfrac{e^{ax}\cos sx}{s}\left[P(x) - \dfrac{P''(x)}{s^2} + \dfrac{P^{iv}(x)}{s^4} - \cdots\right]$

       * For $\cos sx$ in R replace "sin" by "cos" and "cos" by "$-$sin" in y_p.

$$D^n = \frac{d^n}{dx^n} \qquad \binom{m}{n} = \frac{m!}{(m-n)!n!} \qquad 0! = 1$$

Table II: $(D - a)^2 y = R$

R y_p

12. e^{rx} $\dfrac{e^{rz}}{(r-a)^2}$

13. $\sin sx*$ $\dfrac{1}{(a^2+s^2)^2}[(a^2-s^2)\sin sx + 2as\cos sx] = \dfrac{1}{a^2+s^2}\sin\left(sx + \tan^{-1}\dfrac{2as}{a^2-s^2}\right)$

14. $P(x)$ $\dfrac{1}{a^2}\left[P(x) + \dfrac{2P'(x)}{a} + \dfrac{3P''(x)}{a^2} + \cdots + \dfrac{(n+1)P^{(n)}(x)}{a^n}\right]$

15. $e^{rx}\sin sx*$ Replace a by $a - r$ in formula 13 and multiply by e^{rx}.

16. $P(x)e^{rx}$ Replace a by $a - r$ in formula 14 and multiply by e^{rx}.

17. $P(x)\sin sx*$ $\sin sx\left[\dfrac{a^2-s^2}{(a^2+s^2)^2}P(x) + 2\dfrac{a^3-3as^2}{(a^2+s^2)^3}P'(x) + 3\dfrac{a^4-6a^2s^2+s^4}{(a^2+s^2)^4}P''(x) + \cdots\right.$

            $\left. +(k-1)\dfrac{a^k - \binom{k}{2}a^{k-2}s^2 + \binom{k}{4}a^{k-4}s^4 - \cdots}{(a^2+s^2)^k}P^{(k-2)}(x) + \cdots\right]$

            $+\cos sx\left[\dfrac{2as}{(a^2+s^2)^2}P(x) + 2\dfrac{3a^2s-s^3}{(a^2+s^2)^3}P'(x) + 3\dfrac{4a^3s-4as^3}{(a^2+s^2)^4}P''(x) + \cdots\right.$

            $\left. +(k-1)\dfrac{\binom{k}{1}a^{k-1}s - \binom{k}{3}a^{k-3}s^3 + \cdots}{(a^2+s^2)^k}P^{(k-2)}(x) + \cdots\right]$

18. $P(x)e^{rx}\sin sx*$ Replace a by $a - r$ in formula 17 and multiply by e^{rx}.

19. e^{ax} $\frac{1}{2}x^2 e^{ax}$

20. $e^{ax}\sin sx*$ $-\dfrac{e^{ax}\sin sx}{s^2}$

21. $P(x)e^{ax}$ $e^{ax}\displaystyle\iint P(x)\,dx\,dx$

22. $P(x)e^{ax}\sin sx*$ $-\dfrac{e^{ax}\sin sx}{s^2}\left[P(x) - \dfrac{3P''(x)}{s^2} + \dfrac{5P^{iv}(x)}{s^4} - \dfrac{7P^{vi}(x)}{s^6} + \cdots\right]$

            $-\dfrac{e^{ax}\cos sx}{s^2}\left[\dfrac{2P'(x)}{s} + \dfrac{4P'''(x)}{s^3} - \dfrac{6P^v(x)}{s^5} - \cdots\right]$

       * For $\cos sx$ in R replace "sin" by "cos" and "cos" by "$-$ sin" in y_p.

Table III: $(D^2 + q)y = R$

R	y_p
23. e^{rx}	$\frac{e^{rx}}{r^2+q}$
24. $\sin sx*$	$\frac{\sin sx}{-s^2+q}$

25. $P(x)$ $\frac{1}{q}\left[P(x) - \frac{P''(x)}{q} + \frac{P^{iv}(x)}{q^2} - \cdots + (-1)^k \frac{P^{(2k)}(x)}{qk} \cdots\right]$

26. $e^{rx}\sin sx$ $\frac{(r^2-s^2+q)e^{rx}\sin sx - 2rse^{rx}\cos sx}{(r^2-s^2+q)^2+(2rs)^2} = \frac{e^{rx}}{\sqrt{(r^2-s^2+q)^2+(2rs)^2}}\sin\left[sx - \tan^{-1}\frac{2rs}{r^2-s^2+q}\right]$

27. $P(x)e^{rx}$ $\frac{e^{rx}}{r^2+q}\left[P(x) - \frac{2r}{r^2+q}P'(x) + \frac{3r^2-q}{(r^2+q)^2}P''(x) - \frac{4r^3-4qr}{(r^2+q)^3}P'''(x) + \cdots \right.$

      $\left. + \cdots + (-1)^{k-1}\frac{\binom{k}{1}r^{k-1}-\binom{k}{3}r^{k-3}q+\binom{k}{5}r^{k-5}q^2-\cdots}{(r^2+q)^{k-1}}P^{(k-1)}(x) + \cdots\right]$

28. $P(x)\sin sx*$ $\frac{\sin sx}{(-s^2+q)}\left[P(x) - \frac{3s^2+q}{(-s^2+q)^2}P''(x) + \frac{5s^4+10s^2q+q^2}{(-s^2+q)^4}P^{iv}(x) + \cdots\right.$

      $\left. +(-1)^k\frac{\binom{2k+1}{1}s^{2k}+\binom{2k+1}{3}s^{2k-2}q+\binom{2k+1}{5}s^{2k-4}q^2+\cdots}{(-s^2+q)^{2k}}P^{(2k)}(x) + \cdots\right]$

      $-\frac{s\cos sx}{(-s^2+q)}\left[\frac{2P'(x)}{(-s^2+q)} - \frac{4s^2+4q}{(-s^2+q)^3}P'''(x) + \cdots\right.$

      $\left. +(-1)^{k+1}\frac{\binom{2k}{1}s^{2k-2}+\binom{2k}{3}s^{2k-4}q+\cdots}{(-s^2+q)^{2k-1}}P^{(2k-1)}(x) + \cdots\right]$

Table IV: $(D^2 + b^2)y = R$

R	y_p
29. $\sin bx*$	$-\frac{x\cos bx}{2b}$

30. $P(x)\sin bx*$ $\frac{\sin bx}{(2b)^2}\left[P(x) - \frac{P''(x)}{(2b)^2} + \frac{P^{iv}(x)}{(2b)^4} - \cdots\right] - \frac{\cos bx}{2b}\int\left[P(x) - \frac{P''(x)}{(2b)^2} + \cdots\right]dx$

* For $\cos sx$ in R replace "sin" by "cos" and "cos" by "$-\sin$" in y_p.

Table V: $(D^2 pD + q)y = R$

R	y_p
31. e^{rx}	$\frac{e^{rx}}{r^2+pr+q}$
32. $\sin sx*$	$\frac{(q-s^2)\sin sx - ps\cos sx}{(q-s^2)^2+(ps)^2} = \frac{1}{\sqrt{(q-s^2)^2+(ps)^2}}\sin\left(sx - \tan^{-1}\frac{ps}{q-s^2}\right)$

33. $P(x)$ $\frac{1}{q}\left[P(x) - \frac{p}{q}P'(x) + \frac{p^2-q}{q^2}P''(x) - \frac{p^3-2pq}{q^3}P'''(x) + \cdots + (-1)^n\frac{p^n-\binom{n-1}{1}p^{n-2}q+\binom{n-2}{2}p^{n-4}q^2-\cdots}{q^n}P^{(n)}(x)\right]$

34. $e^{rx}\sin sx*$ Replace p by $p+2r$, q by $q+pr+r^2$ in formula 32 and multiply by e^{rx}.

35. $P(x)e^{rx}$ Replace p by $p+2r$, q by $q+pr+r^2$ in formula 33 and multiply by e^{rx}.

Table VI: $(D - b)(D - a)y = R$

36. $P(x)\sin sx*$ $\frac{\sin sx}{b-a}\left[\left(\frac{a}{a^2+s^2} - \frac{b}{b^2+s^2}\right)P(x) + \left(\frac{a^2-s^2}{(a^2+s^2)^2} - \frac{b^2-s^2}{(b^2+s^2)^2}\right)P'(x)\right.$

      $\left. + \left(\frac{a^3-3as^2}{(a^2+s^2)^3} - \frac{b^3-3bs^2}{(b^2+s^2)^3}\right)P''(x) + \cdots\right]$

      $+ \frac{\cos sx}{b-a}\left(\frac{s}{a^2+s^2} - \frac{s}{b^2+s^2}\right)P(x) + \left(\frac{2as}{(a^2+s^2)^2} - \frac{2bs}{(b^2+s^2)^2}\right)P'(x)$

      $+ \left(\frac{3a^2s-s^2}{(a^2+s^2)^3} - \frac{3b^2s-s^3}{(b^2+s^2)^3}\right)P''(x) + \cdots\right]^{\dagger}$

37. $P(x)e^{rx}\sin sx*$ Replace a by $a-r$, b by $b-r$ in formula 36 and multiply by e^{rx}.

38. $P(x)e^{ax}$ $\frac{e^{ax}}{a-b}\left[\int P(x)\,dx + \frac{P(x)}{(b-a)} + \frac{P'(x)}{(b-a)^2} + \frac{P''(x)}{(b-a)^3} + \cdots + \frac{P^{(n)}(x)}{(b-a)^{n+1}}\right]$

* For $\cos sx$ in R replace "sin" by "cos" and "cos" by "$-\sin$" in y_p.

† For additional terms, compare with formula 6.

Table VII: $(D^2 - 2aD + a^2 + b^2)y = R$

R	y_p

39. $P(x)\sin sx^*$

$\dfrac{\sin sx}{2b}\left[\left(\dfrac{s+b}{a^2+(s+b)^2} - \dfrac{s-b}{a^2+(s-b)^2}\right)P(x) + \left(\dfrac{2a(s+b)}{[a^2+(s+b)^2]^2} - \dfrac{2a(s-b)}{[a^2+(s-b)^2]^2}\right)P'(x)\right.$

$\left. + \left(\dfrac{3a^2(s+b)-(s+b)^3}{[a^2+(s+b)^2]^3} - \dfrac{3a^2(s-b)-(s-b)^3}{[a^2+(s-b)^2]^3}\right)P''(x) + \cdots\right]$

$- \dfrac{\cos sx}{2b}\left[\left(\dfrac{a}{a^2+(s+b)^2} - \dfrac{a}{a^2+(s-b)^2}\right)P(x) + \left(\dfrac{a^2-(s+b)^2}{[a^2+(s+b)^2]^2} - \dfrac{a^2-(s-b)^2}{[a^2+(s-b)^2]^2}\right)P'(x)\right.$

$\left. + \left(\dfrac{a^2-3a(s+b)^2}{[a^2+(s+b)^2]^3} - \dfrac{a^3-3as(s-b)^2}{[a^2+(s-b)^2]^3}\right)P''(x) + \cdots\right]^\dagger$

40. $P(x)e^{rx}\sin sx*$ Replace a by $a - r$ in formula 39 and multiply by e^{rx}.

41. $P(x)e^{ax}$ $\dfrac{e^{ax}}{b^2}\left[P(x) - \dfrac{P''(x)}{b^2} + \dfrac{P^{iv}(x)}{b^4} - \cdots\right]$

42. $e^{ax}\sin sx*$ $\dfrac{e^{ax}\sin sx}{-s^2+b^2}$

43. $e^{ax}\sin bx*$ $-\dfrac{xe^{ax}\cos bx}{2b}$

44. $P(x)e^{ax}\sin bx*$ $\dfrac{e^{ax}\sin bx}{(2b)^2}\left[P(x) - \dfrac{P''(x)}{(2b)^2} + \dfrac{P^{iv}(x)}{(2b)^4} - \cdots\right]$

$\qquad - \dfrac{e^{ax}\cos bx}{2b}\int\left[P(x) - \dfrac{P''(x)}{(2b)^2} + \dfrac{P^{iv}(x)}{(2b)^4} - \cdots\right]dx$

* For $\cos sx$ in R replace "sin' by "cos' and "cos" by "$-$sin" in y_p.

† For additional terms, compare with formula 6.

Table VIII: $f(D)y = [D^n + a_{n-1}D^{n-1} + \cdots + a_1 D + a_0]y = R$

R	y_p

45. e^{rx} $\dfrac{e^{rx}}{f(r)}$

46. $\sin sx*$ $\dfrac{[a_0-a_2s^2+a_4s^4-\cdots]\sin sx-[a_1s-a_3s^3+a_5s^5+\cdots]\cos sx}{[a_0-a_2s^2+a_4s^4-\cdots]^2+[a_1s-a_3s^3+a_5s^5-\cdots]^2}$

Table IX: $f(D^2)y = R$

47. $\sin sx*$ $\dfrac{\sin sx}{f(-s^2)} = \dfrac{\sin sx}{a_0-a_2s^2+\cdots\pm s^{2n}}$

Table X: $(D-a)^n y = R$

48. e^{rx} $\dfrac{e^{rx}}{(r-a)^n}$

49. $\sin sx*$ $\dfrac{(-1)^n}{(a^2+s^2)^2}\{[a^n - \binom{n}{2}a^{n-2}s^2 + \binom{4}{n}a^{n-4}s^4 - \cdots]\sin sx$

$\qquad\qquad + [\binom{n}{1}a^{n-1}s - \binom{n}{3}a^{n-3}s^3 + \cdots]\cos sx\}$

50. $P(x)$ $\dfrac{(-1)^n}{a^n}\left[P(x) + \binom{n}{1}\dfrac{P'(x)}{a} + \binom{n+1}{2}\dfrac{P'(x)}{a^2} + \binom{n+2}{3}\dfrac{P'''(x)}{a^2} + \cdots\right]$

51. $e^{rx}\sin sx*$ Replace a by $a - r$ in formula 49 and multiply by e^{rx}.

52. $e^{rx}P(x)$ Replace a by $a - r$ in formula 50 and multiply by e^{rx}.

53. $P(x)\sin sx*$ $(-1)^n\sin sx[A_n P(x) + \binom{n}{1}A_{n+1}P'(x) + \binom{n+1}{2}A_{n+2}P''(x) + \binom{n+2}{3}A_{n+3}P'''(x) + \cdots]$

$\qquad + (-1)^n\cos sx[B_n P(x) + \binom{n}{1}B_{n+1}P'(x) + \binom{n+1}{2}B_{n+2}P''(x) + \binom{n+2}{3}B_{n+3}P'''(x) + \cdots]$

$\qquad A_1 = \dfrac{a}{a^2+s^2},\ A_2 = \dfrac{a^2-s^2}{(a^2+s^2)^2},\ \ldots,\ A_k = \dfrac{a^k - \binom{k}{2}a^{k-2}s^2 + \binom{k}{4}a^{k-4}s^4 - \cdots}{(a^2+s^2)^k}$

$\qquad B_1 = \dfrac{a}{a^2+s^2},\ B_2 = \dfrac{2as}{(a^2+s^2)^2},\ \ldots,\ B_k = \dfrac{\binom{k}{1}a^{k-1}s - \binom{k}{3}a^{k-3}s^3 + \cdots}{(a^2+s^2)^k}$

54. $e^{rx}\sin sx*$ Replace a by $a - r$ in formula 53 and multiply by e^{rx}.

55. $e^{ax}P(x)$ $e^{ax}\displaystyle\int\int\cdots\int P(x)\,dx^n$

56. $P(x)e^{ax}\sin sx*$ $\dfrac{(-1)^{\frac{n-1}{2}}e^{ax}\sin sx}{s^n}\left[\binom{n}{n-1}\dfrac{P'(x)}{s} - \binom{n+2}{n-1}\dfrac{P'''(x)}{s^3} + \binom{n+4}{n-1}\dfrac{P^v(x)}{s^5} - \cdots\right]$

$\qquad + \dfrac{(-1)^{\frac{n+1}{2}}e^{ax}\cos sx}{s^n}\left[\binom{n-1}{n-1}P(x) - \binom{n+1}{n-1}\dfrac{P''(x)}{s^2} + \binom{n+3}{n-1}\dfrac{P^{iv}(x)}{s^4} - \cdots\right]$ (n odd)

$\qquad \dfrac{(-1)^{\frac{n}{2}}e^{ax}\sin sx}{s^n}\left[\binom{n-1}{n-1}P(x) - \binom{n+1}{n-1}\dfrac{P''(x)}{s^2} + \binom{n+3}{n-1}\dfrac{P^{iv}(x)}{s^4} - \cdots\right]$

$\qquad + \dfrac{(-1)^{\frac{n}{2}}e^{ax}\cos sx}{s^n}\left[\binom{n}{n-1}\dfrac{P'(x)}{s} - \binom{n+2}{n-1}\dfrac{P'''(x)}{s^3} + \binom{n+4}{n-1}\dfrac{P^v(x)}{s^5} - \cdots\right]$ (n even)

* For $\cos sx$ in R replace "sin" by "cos" and "cos" by "$-$sin" in y_p.

Table XI: $(D - a)^n f(D)y = R$

57. e^{ax} $\dfrac{x^n}{n!} \cdot \dfrac{e^{ax}}{f(a)}$

     * For $\cos sx$ in R replace "sin" by "cos" and "cos" by "$-$sin" in y_p.

Table XII: $(D^2 + q)^n y = R$

R	y_p

58. e^{rx} $e^{rx}/(r^2 + q)^n$

59. $\sin sx*$ $\sin sx/(q - s^2)^n$

60. $P(x)$ $\dfrac{1}{q^n}\left[P(x) - \binom{n}{1}\dfrac{P''(x)}{q^2} + \binom{n+1}{2}\dfrac{P^{\mathrm{iv}}(x)}{q^2} - \binom{n+2}{3}\dfrac{P^{\mathrm{vi}}(x)}{q^3} + \cdots \right]$

61. $e^{rx}\sin sx*$ $\dfrac{e^{rx}}{(A^2+B^2)^n}\left\{ \left[A^n - \binom{n}{2}A^{n-2}B^2 + \binom{n}{4}A^{n-4}B^4 - \cdots \right]\sin sx \right.$

$$\left. - \left[\binom{n}{1}A^{n-1}B - \binom{n}{3}A^{n-3}B^3 + \cdots \right]\cos sx \right\}$$

$$A = r^2 - s^2 + q, \quad B = 2rs$$

Table XIII: $(D^2 + b^2)^n y = R$

62. $\sin bx*$ $(-1)^{n+1/2}\dfrac{x^n \cos bx}{n!(2b)^n}$ (n odd), $(-1)^{n/2}\dfrac{x^n \sin bx}{n!(2b)^n}$ (n even)

Table XIV: $(D^n - q)y = R$

63. e^{rx} $e^{rx}/(r^n - q)$

64. $P(x)$ $-\dfrac{1}{q}\left[P(x)\dfrac{P^{(n)}(x)}{q} + \dfrac{P^{(2n)}(x)}{q^2} + \cdots \right]$

65. $\sin sx*$ $-\dfrac{q\sin sx + (-1)^{\frac{n-1}{2}}s^n \cos sx}{q^2 + s^{2n}}$ (n odd), $\dfrac{\sin sx}{(-s^2)^{n/2}-q}$ (n even)

66. $e^{rx}\sin sx*$ $\dfrac{Ae^{rx}\sin sr - Be^{rx}\cos sx}{A^2+B^2} = \dfrac{e^{rx}}{\sqrt{A^2+B^2}}\sin\left(sx - \tan^{-1}\dfrac{B}{A}\right)$

$$A = \left[r^n - \binom{n}{2}r^{n-2}s^2 + \binom{n}{4}r^{n-4}s^4 - \cdots \right] - q,$$
$$B = \left[\binom{n}{1}r^{n-1}s - \binom{n}{3}r^{n-3}s^3 + \cdots \right]$$

     * For $\cos sx$ in R replace "sin" by "cos" and cos by "$-$sin" in y_p.

Table XV: $(D_x + mD_y)z = R$

R	z_p

67. e^{ax+by} $\dfrac{e^{ax+by}}{a+mb}$

68. $f(ax + by)$ $\dfrac{\int f(u)du}{a+mb}, \quad u = ax + by$

69. $f(y - mx)$ $xf(y - mx)$

70. $\phi(x, y) f(y - mx)$ $f(y - mx) \int \phi(x, a + mx)\, dx$ ($a = y - mx$ after integration)

Table XVI: $(D_x + mD_y - k)z = R$

71. e^{ax+by} $\dfrac{e^{ax+by}}{a+mb-k}$

72. $\sin(ax + by)*$ $-\dfrac{(a+bm)\cos(ax+by)+k\sin(ax+by)}{(a+bm)^2+k^2}$

73. $e^{\alpha x+\beta y}\sin(ax+by)*$ Replace k in 72 by $k - \alpha - m\beta$ and multiply by $e^{\alpha x+\beta y}$

74. $e^{xk}f(ax+by)$ $\dfrac{e^{kx}\int f(u)du}{a+mb}, u = ax + by$

75. $f(y - mx)$ $-\dfrac{f(y-mx)}{k}$

76. $p(x)f(y-mx)$ $-\dfrac{1}{k}f(y-mx)\left[p(x) + \dfrac{p'(x)}{k} + \dfrac{p''(x)}{k^2} + \cdots + \dfrac{p^{(n)}(x)}{k^n} \right]$

77. $e^{kx}f(y-mx)$ $xe^{kx}f(y-mx)$

     * For $\cos(ax+by)$ replace "sin" by "cos" and "cos" by "$-$sin" in z_p.

$$D_x = \frac{\partial}{\partial x}; \quad D_y = \frac{\partial}{\partial y}; \quad D_x{}^k D_y{}^r = \frac{\partial^{k+r}}{\partial_x{}^k \partial_y{}^r}$$

Table XVII: $(D_z + mD_y)^n z = R$

R	z_p
78. e^{ax+by}	$\frac{e^{ax+by}}{(a+mb)^n}$
79. $f(ax + by)$	$\frac{\iint \cdots \int f(u)du^n}{(a+mb)^n}, \ u = ax + by$
80. $f(y - mx)$	$\frac{x^n}{n!} f(y - mx)$
81. $\phi(x, y) f(y + mx)$	$f(y - mx) \iint \cdots \int \phi(x, a + mx) \, dx^n \ (a = y - mx \ \text{after integration})$

Table XVIII: $(D_x + mD_y - k)^n z = R$

82. e^{ax+by}	$\frac{e^{ax+by}}{(a+mb-k)^n}$
83. $f(y - mx)$	$\frac{(-1)^n f(y-mx)}{k^n}$
84. $P(x) f(y - mx)$	$\frac{(-1)^n}{k^n} f(y - mx) \left[p(x) + \binom{n}{1}\frac{p'(x)}{k} + \binom{n+1}{2}\frac{p''(x)}{k^3} + \binom{n+2}{3}\frac{p'''(x)}{k^3} + \cdots \right]$
85. $e^{kz} f(ax + by)$	$\frac{e^{kx} \int \int \cdots \int f(u)du^n}{(a+mb)^n}, \ u = ax + by$
86. $e^{kx} f(y - mx)$	$\frac{x^n}{n!} e^{kx} f(y - mx)$

Table XIX: $[D_x^n + a_1 D_x^{n-1} D_y + a_2 D_x^{n-2} D_y^2 + \cdots + a^n D_y^n]z = R$

87. e^{ax+by}	$\dfrac{e^{ax+by}}{a + a_1 a^{n-1}b + a_2 a^{n-2}b^2 + \cdots + a_n b^n}$
88. $f(ax + by)$	$\dfrac{\iint \cdots \iint f(u)du^n}{a^n + a_1 a^{n-1}b + a_2 a^{n-2}b^2 + \cdots + a^n b^n}, \ (u = ax + by)$

Table XX: $F(D_x, D_y)z = R$

89. e^{ax+by}	$\dfrac{e^{ax+by}}{F(a,b)}$

Table XXI: $F(D_x^2, D_x D_y, D_y^2)z = R$

90. $\sin(ax + by)*$	$\dfrac{\sin(ax+by)}{F(-a^2, -ab, -b^2)}$

* For $\cos(ax + by)$ replace "sin" by "cos", and "cos" by "$-\sin$" in z_p.

Differential equation	Method of solution
$yF(xy)\,dx + x\,G(xy)\,dy = 0$	$\ln x = \displaystyle\int \frac{G(v)\,dv}{v\{G(v) - F(v)\}} + c$ where $v = xy$. If $G(v) = F(v)$, then the solution is $xy = c$.
Linear, homogeneous, second order equation $\dfrac{d^2 y}{dx^2} + b\dfrac{dy}{dx} + cy = 0$ where b and c are real constants	Let m_1, m_2 be the roots of $m^2 + bm + c = 0$. Then there are 3 cases: Case 1. m_1, m_2 real and distinct: $\qquad\qquad y = c_1 e^{m_1 x} + c_2 e^{m_2 x}$ Case 2. m_1, m_2 real and equal: $\qquad\qquad y = c_1 e^{m_1 x} + c_2 x e^{m_1 x}$ Case 3. $m_1 = p + qi,\ m_2 = p - qi$: $\qquad\qquad y = e^{px}(c_1 \cos qx + c_2 \sin qx)$ where $p = -b/2$, $q = \sqrt{4c - b^2}/2$
Linear, nonhomogeneous, second order equation $\dfrac{d^2 y}{dx^2} + b\dfrac{dy}{dx} + cy = R(x)$ where b and c are real constants	There are 3 cases corresponding to those above: Case 1. $\qquad y = c_1 e^{m_1 x} + c_2 e^{m_2 x}$ $\qquad\quad + \dfrac{e^{m_1 x}}{m_1 - m_2} \displaystyle\int e^{-m_1 x} R(x)\,dx$ $\qquad\quad + \dfrac{e^{m_2 x}}{m_2 - m_1} \displaystyle\int e^{-m_2 x} R(x)\,dx$ Case 2. $\qquad y = c_1 e^{m_1 x} + c_2 x e^{m_1 x}$ $\qquad\quad + x e^{m_1 x} \displaystyle\int e^{-m_1 x} R(x)\,dx$ $\qquad\quad - e^{m_1 x} \displaystyle\int x e^{-m_1 x} R(x)\,dx$ Case 3. $\qquad y = e^{px}(c_1 \cos qx + c_2 \sin qx)$ $\qquad\quad + \dfrac{e^{px} \sin qx}{q} \displaystyle\int e^{-px} R(x) \cos qx\,dx$ $\qquad\quad - \dfrac{e^{px} \cos qx}{q} \displaystyle\int e^{-px} R(x) \sin qx\,dx$

Differential equation	Method of solution
Euler or Cauchy equation $$x^2 \frac{d^2y}{dx} + bx\frac{dy}{dx} + cy = S(x)$$	Putting $x = e^t$, the equation becomes $$\frac{d^2y}{dt^2} + (b-1)\frac{dy}{dt} + cy = S(e^t)$$ and can then be solved as a linear second order equation.
Bessel's equation $$x^2 \frac{d^2y}{dx^2} + x\frac{dy}{dx} + (\lambda^2 x^2 - n^2)y = 0$$	$y = c_1 J_n(\lambda x) + c_2 Y_n(\lambda x)$
Transformed Bessel's equation $$x^2 \frac{d^2y}{dx^2} + (2p+1)x\frac{dy}{dx} +$$ $$(\alpha^2 x^{2r} + \beta^2)\,y = 0$$	$$y = x^{-p}\left\{ c_1 J_{q/r}\left(\frac{\alpha}{r}x^r\right) + c_2 Y_{q/r}\left(\frac{\alpha}{r}x^r\right) \right\}$$ where $q = \sqrt{p^2 - \beta^2}$.
Legendre's equation $$(1 - x^2)\frac{d^2y}{dx^2} - 2x\frac{dy}{dx} + n(n+1)y = 0$$	$y = c_1 P_n(x) + c_2 Q_n(x)$
Separation of variables $$f_1(x)g_1(y)\,dx + f_2(x)g_2(y)\,dy = 0$$	$$\int \frac{f_1(x)}{f_2(x)}\,dx + \int \frac{g_2(y)}{g_1(y)}\,dy = c$$
Exact equation $$M(x, y)\,dx + N(x, y)\,dy = 0$$ where $\partial M/\partial y = \partial N/\partial x$	$\int M\partial x + \int \left(n - \frac{\partial}{\partial y}\int M\partial x \right) dy = c$ where ∂x indicates that the integration is to be performed with respect to x keeping y constant.
Linear first order equation $$\frac{dy}{dx} + P(x)y = Q(x)$$	$$ye^{\int P\,dx} = \int Qe^{\int P\,dx}\,dx + c$$
Bernoulli's equation $$\frac{dy}{dx} + P(x)y = Q(x)y^n$$	$ve^{(1-n)\int P\,dx} = \int Qe^{(1-n)\int P\,dx}\,dx + c$ where $v = y^{1-n}$. If $n = 1$, then the solution is $\ln y = \int (Q - P)\,dx + c$.
Homogeneous equation $$\frac{dy}{dx} = F\left(\frac{y}{x}\right)$$	$\ln x = \int \frac{dv}{F(v) - v} + c$ where $v = y/x$. If $F(v) = v$, then the solution is $y = cx$.
Reducible to homogeneous $(a_1 x + b_1 y + c_1)\,dx$ $+(a_2 x + b_2 y + c_2)\,dy = 0$ with $\dfrac{a_1}{a_2} \neq \dfrac{b_1}{b_2}$	Set $u = a_1 x + b_1 y + c_1$ and $v = a_2 x + b_2 y + c_2$. Then eliminate x and y and the equation becomes homogenous.
Reducible to separable $(a_1 x + b_1 y + c_1)\,dx$ $+(a_2 x + b_2 y + c_2)\,dy = 0$ with $\dfrac{a_1}{a_2} = \dfrac{b_1}{b_2}$	Set $u = a_1 x + b_1 y$. Then eliminate x or y and the equation becomes separable.

FOURIER SERIES

1. If $f(x)$ is a bounded periodic function of period 2L (i.e., $f(x + 2L) = f(x)$), and satisfies the *Dirichlet conditions*:

   (a) In any period $f(x)$ is continuous, except possibly for a finite number of jump discontinuities.

   (b) In any period $f(x)$ has only a finite number of maxima and minima.

   Then $f(x)$ may be represented by the *Fourier series*

   $$\frac{a_0}{2} + \sum_{n=1}^{\infty} \left(a_n \cos \frac{n\pi x}{L} + b_n \sin \frac{n\pi x}{L} \right)$$

   where a_n and b_n are as determined below. This series will converge to $f(x)$ at every point where $f(x)$ is continuous, and to

   $$\frac{f(x^+) + f(x^-)}{2}$$

   (i.e., the average of the left-hand and right-hand limits) at every point where $f(x)$ has a jump discontinuity.

   $$a_n = \frac{1}{L} \int_{-L}^{L} f(x) \cos \frac{n\pi x}{L} \, dx, \quad n = 0, 1, 2, 3, \ldots,$$

   $$b_n = \frac{1}{L} \int_{-L}^{L} f(x) \sin \frac{n\pi x}{L} \, dx, \quad n = 1, 2, 3, \ldots$$

   We may also write

   $$a_n = \frac{1}{L} \int_{\alpha}^{\alpha+2L} f(x) \cos \frac{n\pi x}{L} \, dx \text{ and } b_n = \frac{1}{L} \int_{\alpha}^{\alpha+2L} f(x) \sin \frac{n\pi x}{L} \, dx$$

   where α is any real number. Thus if $\alpha = 0$,

   $$a_n = \frac{1}{L} \int_{0}^{2L} f(x) \cos \frac{n\pi x}{L} \, dx, \quad n = 0, 1, 2, 3, \ldots,$$

   $$b_n = \frac{1}{L} \int_{0}^{2L} f(x) \sin \frac{n\pi x}{L} \, d, \quad n = 1, 2, 3, \ldots$$

2. If in addition to the restrictions in (1), $f(x)$ is an even function (i.e., $f(-x) = f(x)$), then the Fourier series reduces to

   $$\frac{a_0}{2} + \sum_{n=1}^{\infty} a_n \cos \frac{n\pi x}{L}$$

   That is, $b_n = 0$. In this case, a simpler formula for a_n is

   $$a_n = \frac{2}{L} \int_{0}^{L} f(x) \cos \frac{n\pi x}{L} \, dx, \quad n = 0, 1, 2, 3, \ldots$$

3. If in addition to the restrictions in (1), $f(x)$ is an odd function (i.e., $f(-x) = -f(x)$), then the Fourier series reduces to

   $$\sum_{n=1}^{\infty} b_n \sin \frac{n\pi x}{L}$$

   That is, $a_n = 0$. In this case, a simpler formula for the b_n is

   $$b_n = \frac{2}{L} \int_{0}^{L} f(x) \sin \frac{n\pi x}{L} \, dx, \quad n = 1, 2, 3, \ldots$$

4. If in addition to the restrictions in (2) above, $f(x) = -f(L - x)$, then a_n will be 0 for all even values of n, including $n = 0$. Thus in this case, the expansion reduces to

   $$\sum_{m=1}^{\infty} a_{2m-1} \cos \frac{(2m - 1)\pi x}{L}$$

5. If in addition to the restrictions in (3) above, $f(x) = f(L - x)$, then b_n will be 0 for all even values of n. Thus in this case, the expansion reduces to

$$\sum_{m=1}^{\infty} b_{2m-1} \sin \frac{(2m-1)\pi x}{L}$$

(The series in (4) and (5) are known as *odd-harmonic series*, since only the odd harmonics appear. Similar rules may be stated for even-harmonic series, but when a series appears in the even-harmonic form, it means that $2L$ has not been taken as the smallest period of $f(x)$. Since any integral multiple of a period is also a period, series obtained in this way will also work, but in general computation is simplified if $2L$ is taken to be the smallest period.)

6. If we write the Euler definitions for $\cos\theta$ and $\sin\theta$, we obtain the complex form of the Fourier series known either as the "Complex Fourier Series" or the "Exponential Fourier Series" of $f(x)$. It is represented as

$$f(x) = \frac{1}{2} \sum_{n=-\infty}^{n=+\infty} c_n e^{i\omega_n x}$$

where

$$c_n = \frac{1}{L} \int_{-L}^{L} f(x)\, e^{-i\omega_n x}\, dx, \quad n = 0, \pm 1, \pm 2, \pm 3, \ldots$$

with $\omega_n = \frac{n\pi}{L}$ for $n = 0, \pm 1, \pm 2, \ldots$ The set of coefficients c_n is often referred to as the Fourier spectrum.

7. If both sine and cosine terms are present and if $f(x)$ is of period $2L$ and expandable by a Fourier series, it can be represented as

$$f(x) = \frac{a_0}{2} + \sum_{n=1}^{\infty} c_n \sin\left(\frac{n\pi x}{L} + \phi_n\right), \quad \text{where}$$

$$a_n = c_n \sin\phi_n, \quad b_n = c_n \cos\phi_n, \quad c_n = \sqrt{a_n^2 + b_n^2}, \quad \phi_n = \arctan\left(\frac{a_n}{b_n}\right)$$

It can also be represented as

$$f(x) = \frac{a_0}{2} + \sum_{n=1}^{\infty} c_n \cos\left(\frac{n\pi x}{L} + \phi_n\right), \quad \text{where}$$

$$a_n = c_n \cos\phi_n, \quad b_n = -c_n \sin\phi_n, \quad c_n = \sqrt{a_n^2 + b_n^2}, \quad \phi_n = \arctan\left(-\frac{b_n}{a_n}\right)$$

where ϕ_n is chosen so as to make a_n, b_n, and c_n hold.

8. The following table of trigonometric identities should be helpful for developing Fourier series.

	n	n even	n odd	$n/2$ odd	$n/2$ even
$\sin n\pi$	0	0	0	0	0
$\cos n\pi$	$(-1)^n$	$+1$	-1	$+1$	$+1$
$*\sin \frac{n\pi}{2}$		0	$(-1)^{(n-1)/2}$	0	0
$*\cos \frac{n\pi}{2}$		$(-1)^{n/2}$	0	-1	$+1$
$\sin \frac{n\pi}{4}$		$\frac{\sqrt{2}}{2}(-1)^{(n^2+4n+11)/8}$	$(-1)^{(n-2)/4}$	0

*A useful formula for $\sin \frac{n\pi}{2}$ and $\cos \frac{n\pi}{2}$ is given by

$$\sin \frac{n\pi}{2} = \frac{(i)^{n+1}}{2}[(-1)^n - 1] \quad \text{and} \quad \cos \frac{n\pi}{2} = \frac{(i)^n}{2}[(-1)^n + 1], \quad \text{where} \quad i^2 = -1.$$

Auxiliary Formulas for Fourier Series

$$1 = \frac{4}{\pi}\left[\sin\frac{\pi x}{k} + \frac{1}{3}\sin\frac{3\pi x}{k} + \frac{1}{5}\sin\frac{5\pi x}{k} + \cdots\right] \quad [0 < x < k]$$

$$x = \frac{2k}{\pi}\left[\sin\frac{\pi x}{k} - \frac{1}{2}\sin\frac{2\pi x}{k} + \frac{1}{3}\sin\frac{3\pi x}{k} - \cdots\right] \quad [-k < x < k]$$

$$x = \frac{k}{2} - \frac{4k}{\pi^2}\left[\cos\frac{\pi x}{k} + \frac{1}{3^2}\cos\frac{3\pi x}{k} + \frac{1}{5^2}\cos\frac{5\pi x}{k} + \cdots\right] \quad [0 < x < k]$$

$$x^2 = \frac{2k^2}{\pi^3}\left[\left(\frac{\pi^2}{1} - \frac{4}{1}\right)\sin\frac{\pi x}{k} - \frac{\pi^2}{2}\sin\frac{2\pi x}{k} + \left(\frac{\pi^2}{3} - \frac{4}{3^3}\right)\sin\frac{3\pi x}{k}\right.$$
$$\left. - \frac{\pi^2}{4}\sin\frac{4\pi x}{k} + \left(\frac{\pi^2}{5} - \frac{4}{5^3}\right)\sin\frac{5\pi x}{k} + \cdots\right] \quad [0 < x < k]$$

$$x^2 = \frac{k^2}{3} - \frac{4k^2}{\pi^2}\left[\cos\frac{\pi x}{k} - \frac{1}{2^2}\cos\frac{2\pi x}{k} + \frac{1}{3^2}\cos\frac{3\pi x}{k} - \frac{1}{4^2}\cos\frac{4\pi x}{k} + \cdots\right]$$
$$[-k < x < k]$$

$$1 - \frac{1}{3} + \frac{1}{5} - \frac{1}{7} + \cdots = \frac{\pi}{4}$$

$$1 - \frac{1}{2^2} + \frac{1}{3^2} + \frac{1}{4^2} + \cdots = \frac{\pi^2}{6}$$

$$1 - \frac{1}{2^2} + \frac{1}{3^2} - \frac{1}{4^2} + \cdots = \frac{\pi^2}{12}$$

$$1 + \frac{1}{3^2} + \frac{1}{5^2} - \frac{1}{7^2} + \cdots = \frac{\pi^2}{8}$$

$$\frac{1}{2^2} + \frac{1}{4^2} + \frac{1}{6^2} + \frac{1}{8^2} + \cdots = \frac{\pi^2}{24}$$

FOURIER EXPANSIONS FOR BASIC PERIODIC FUNCTIONS

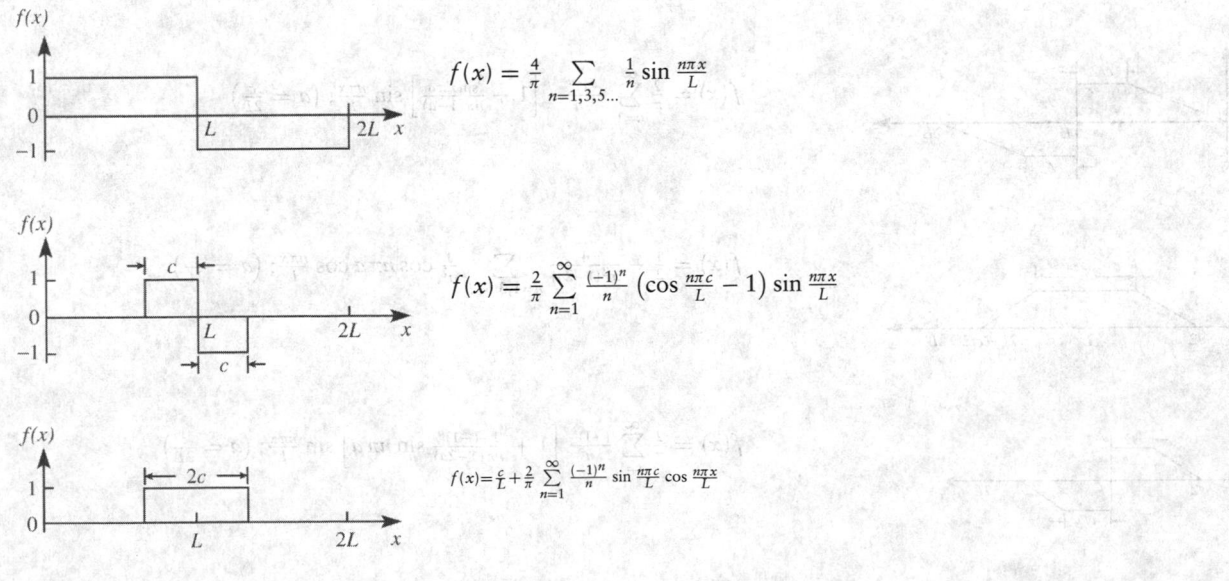

$$f(x) = \frac{4}{\pi}\sum_{n=1,3,5\ldots}\frac{1}{n}\sin\frac{n\pi x}{L}$$

$$f(x) = \frac{2}{\pi}\sum_{n=1}^{\infty}\frac{(-1)^n}{n}\left(\cos\frac{n\pi c}{L} - 1\right)\sin\frac{n\pi x}{L}$$

$$f(x) = \frac{c}{L} + \frac{2}{\pi}\sum_{n=1}^{\infty}\frac{(-1)^n}{n}\sin\frac{n\pi c}{L}\cos\frac{n\pi x}{L}$$

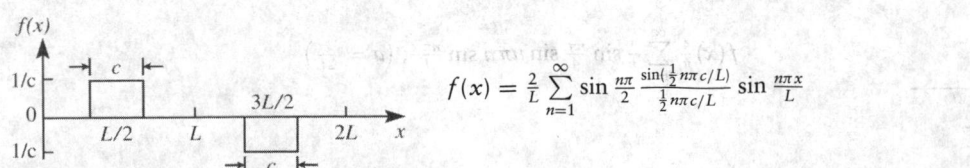

$$f(x) = \frac{2}{L}\sum_{n=1}^{\infty}\sin\frac{n\pi}{2}\frac{\sin(\frac{1}{2}n\pi c/L)}{\frac{1}{2}n\pi c/L}\sin\frac{n\pi x}{L}$$

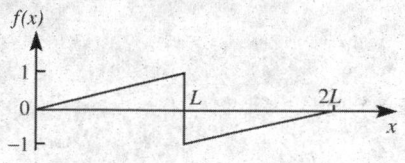

$$f(x) = \frac{2}{\pi} \sum_{n=1}^{\infty} \frac{(-1)^{n+1}}{n} \sin \frac{n\pi x}{L}$$

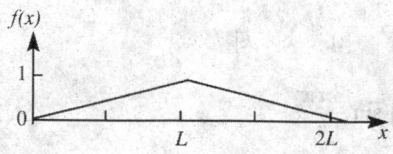

$$f(x) = \frac{1}{2} - \frac{4}{\pi^2} \sum_{n=1,3,5,\ldots} \frac{1}{n^2} \cos \frac{n\pi x}{L}$$

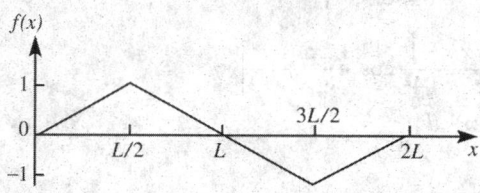

$$f(x) = \frac{8}{\pi^2} \sum_{n=1,3,5,\ldots} \frac{(-1)^{(n-1)/2}}{n^2} \sin \frac{n\pi x}{L}$$

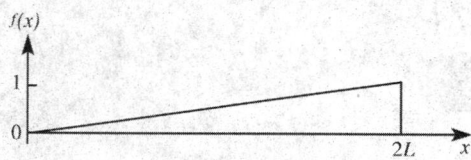

$$f(x) = \frac{1}{2} - \frac{1}{\pi} \sum_{n=1}^{\infty} \frac{1}{n} \sin \frac{n\pi x}{L}$$

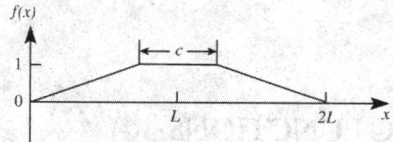

$$f(x) = \frac{1}{2}(1+a) + \frac{2}{\pi^2(1-a)} \sum_{n=1}^{\infty} \frac{1}{n^2} [(-1)^n \cos n\pi a - 1] \cos \frac{n\pi x}{L};$$
$$\left(a = \frac{c}{2L}\right)$$

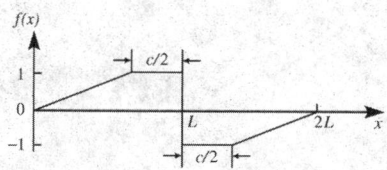

$$f(x) = \frac{2}{\pi} \sum_{n=1}^{\infty} \frac{(-1)^{n-1}}{n} \left[1 + \frac{\sin n\pi a}{n\pi(1-a)}\right] \sin \frac{n\pi x}{L}; \left(a = \frac{c}{2L}\right)$$

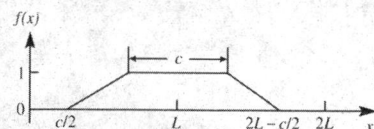

$$f(x) = \frac{1}{2} - \frac{4}{\pi^2(1-2a)} \sum_{n=1,3,5,\ldots} \frac{1}{n^2} \cos n\pi a \cos \frac{n\pi x}{L}; \left(a = \frac{c}{2L}\right)$$

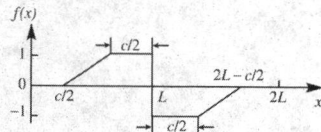

$$f(x) = \frac{2}{\pi} \sum_{n=1}^{\infty} \frac{(-1)^n}{n} \left[1 + \frac{1+(-1)^n}{n\pi(1-2a)} \sin n\pi a\right] \sin \frac{n\pi x}{L}; \left(a = \frac{c}{2L}\right)$$

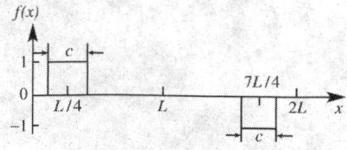

$$f(x) \frac{4}{\pi} \sum_{n=1}^{\infty} \frac{1}{n} \sin \frac{n\pi}{4} \sin n\pi a \sin \frac{n\pi x}{L}; \left(a = \frac{c}{2L}\right)$$

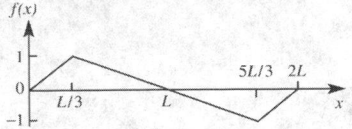

$$f(x) = \frac{9}{\pi^2} \sum_{n=1}^{\infty} \frac{1}{n^2} \sin\frac{n\pi}{3} \sin\frac{n\pi x}{L}; \ \left(a = \frac{c}{2L}\right)$$

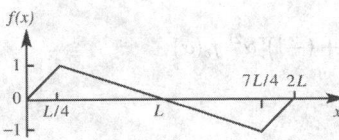

$$f(x) = \frac{32}{3\pi^2} \sum_{n=1}^{\infty} \frac{1}{n^2} \sin\frac{n\pi}{4} \sin\frac{n\pi x}{L}; \ \left(a = \frac{c}{2L}\right)$$

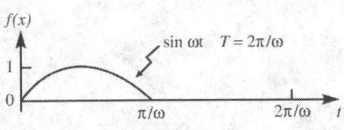

$$f(x) = \frac{1}{\pi} + \frac{1}{2} \sin\omega t - \frac{2}{\pi} \sum_{n=2,4,6,\dots} \frac{1}{n^2-1} \cos n\omega t$$

Extracted from graphs and formulas, pages 372, 373, *Differential Equations in Engineering Problems*, Salvadori and Schwarz, published by Prentice-Hall, Inc., 1954.

THE FOURIER TRANSFORMS

For a piecewise continuous function $F(x)$ over a finite interval $0 \le x \le \pi$; the *finite Fourier cosine transform* of $F(x)$ is

$$f_c(n) = \int_0^{\pi} F(x) \cos nx \, dx \quad (n = 0, 1, 2, \dots)$$

If x ranges over the interval $0 \le x \le L$, the substitution $x' = \pi x/L$ allows the use of this definition, also. The inverse transform is written.

$$\overline{F}(x) = \frac{1}{\pi} f_c(0) - \frac{2}{\pi} \sum_{n=1}^{x} f_c(n) \cos nx \quad (0 < x < \pi)$$

where $F(x) = \frac{F(x+\epsilon)+F(x-\epsilon)}{2}$. We observe that $F(x+) = F(x-) = F(x)$ at points of continuity. The formula

$$
\begin{aligned}
f_c^{(2)}(n) &= \int_0^{\pi} F''(x) \cos nx \, dx \\
&= -n^2 f_c(n) - F'(0) + (-1)^n F'(\pi)
\end{aligned}
\tag{1}
$$

makes the finite Fourier cosine transform useful in certain boundary value problems. Analogously, the *finite Fourier sine transform* of $F(x)$ is

$$f_s(n) = \int_0^{\pi} F(x) \sin nx \, dx \quad (n = 1, 2, 3, \dots)$$

and

$$\overline{F}(x) = \frac{2}{\pi} \sum_{n=1}^{\infty} f_s(n) \sin nx \quad (0 < x < \pi)$$

Corresponding to (1) we have

$$
\begin{aligned}
f_s^{(2)}(n) &= \int_0^{\pi} F''(x) \sin nx \, dx \\
&= -n^2 f_s(n) - n F(0) - n(-1)^n F(\pi)
\end{aligned}
\tag{2}
$$

If $F(x)$ is defined for $x \le 0$ and is piecewise continuous over any finite interval, and if $\int_0^x F(x)\,dx$ is absolutely convergent, then

$$f_c(\alpha) = \sqrt{\frac{2}{\pi}} \int_0^x F(x) \cos(\alpha x) \, dx$$

is the *Fourier cosine transform* of $F(x)$. Furthermore,

$$\overline{F}(x) = \sqrt{\frac{2}{\pi}} \int_0^x f_c(\alpha) \cos(\alpha x) \, d\alpha.$$

If $\lim_{x \to \infty} d^n F/dx^n = 0$, then an important property of the Fourier cosine transform is

$$f_c^{(2r)}(\alpha) = \sqrt{\frac{2}{\pi}} \int_0^x \left(\frac{d^{2r} F}{dx^{2r}} \right) \cos(\alpha x) \, dx = -\sqrt{\frac{2}{\pi}} \sum_{n=0}^{r-1} (-1)^n a_{2r-2n-1} \alpha^{2n} + (-1)^r \alpha^{2r} f_c(\alpha) \tag{3}$$

where $\lim_{x \to \infty} d^r F/dx^r = a_r$, makes it useful in the solution of many problems.

Under the same conditions.

$$f_s(\alpha) = \sqrt{\frac{2}{\pi}} \int_0^x F(x) \sin(\alpha x) \, dx$$

defines the *Fourier sine transform* of $F(x)$, and

$$\overline{F}(x) = \sqrt{\frac{2}{\pi}} \int_0^x f_s(\alpha) \sin(\alpha x) \, d\alpha$$

Corresponding to (3) we have

$$f_s^{(2r)}(\alpha) = \sqrt{\frac{2}{\pi}} \int_0^\infty \frac{d^{2r} F}{dx^{2r}} \sin(\alpha x) \, dx = -\sqrt{\frac{2}{\pi}} \sum_{n=1}^{r} (-1)^n \alpha^{2n-1} a_{2r-2n} + (-1)^{r-1} \alpha^{2r} f_s(\alpha)$$

Similarly, if $F(x)$ is defined for $-\infty < x < \infty$, and if $\int_{-\infty}^\infty F(x) \, dx$ is absolutely convergent, then

$$f(\alpha) = \frac{1}{\sqrt{2\pi}} \int_{-\infty}^\infty F(x) e^{i\alpha x} \, dx$$

is the *Fourier transform* of $F(x)$, and

$$\overline{F}(x) = \frac{1}{\sqrt{2\pi}} \int_{-\infty}^\infty f(\alpha) e^{-i\alpha x} \, d\alpha$$

Also, if

$$\lim_{|x| \to \infty} \left| \frac{d^n F}{dx^n} \right| = 0 \quad (n = 1, 2, \ldots, r-1)$$

then

$$f^{(r)}(\alpha) = \frac{1}{\sqrt{2\pi}} \int_{-\infty}^\infty F^{(r)}(x) e^{i\alpha x} \, dx = (-i\alpha)^r f(\alpha)$$

Finite Sine Transforms

$f_s(n)$	$F(x)$
1. $f_s(n) = \int_0^\pi F(x) \sin nx \, dx \ (n = 1, 2, \ldots)$	$F(x)$
2. $(-1)^{n+1} f_s(n)$	$F(\pi - x)$
3. $\frac{1}{n}$	$\frac{\pi - x}{\pi}$
4. $\frac{(-1)^{n+1}}{n}$	$\frac{x}{\pi}$
5. $\frac{1-(-1)^n}{n}$	1
6. $\frac{2}{n^2} \sin \frac{n\pi}{2}$	$\begin{cases} x & \text{when } 0 < x < \pi/2 \\ \pi - x & \text{when } \pi/2 < x < \pi \end{cases}$
7. $\frac{(-1)^{n+1}}{n^3}$	$\frac{x(\pi^2 - x^2)}{6\pi}$
8. $\frac{1-(-1)^n}{n^3}$	$\frac{x(\pi - x)}{2}$
9. $\frac{\pi^2(-1)^{n-1}}{n} - \frac{2[1-(-1)^n]}{n^3}$	x^2
10. $\pi(-1)^n \left(\frac{6}{n^3} - \frac{\pi^2}{n} \right)$	x^3
11. $\frac{n}{n^2+c^2}[1 - (-1)^n e^{c\pi}]$	e^{cx}
12. $\frac{n}{n^2+c^2}$	$\frac{\sinh c(\pi - x)}{\sinh c\pi}$

$f_s(n)$	$F(x)$
13. $\frac{n}{n^2-k^2} \quad (k \neq 0, 1, 2, \ldots)$	$\frac{\sin k(\pi - x)}{\sin k\pi}$
14. $\begin{cases} \frac{\pi}{2} & \text{when } n = m \\ 0 & \text{when } n \neq m \end{cases} \quad (m = 1, 2, \ldots)$	$\sin mx$
15. $\frac{n}{n^2 - k^2}[1 - (-1)^n \cos k\pi]$ $(k \neq 1, 2, \ldots)$	$\cos kx$
16. $\begin{cases} \frac{n}{n^2-m^2}[1 - (-1)^{n+m}] \\ \quad \text{when } n \neq m = 1, 2, \ldots \\ 0 \quad\quad \text{when } n = m \end{cases}$	$\cos mx$
17. $\frac{n}{(n^2-k^2)^2} (k \neq 0, 1, 2, \ldots)$	$\frac{\pi \sin kx}{2k \sin^2 k\pi} - \frac{x \cos k(\pi - x)}{2k \sin k\pi}$
18. $\frac{b^n}{n} (|b| \leq 1)$	$\frac{2}{\pi} \arctan \frac{b \sin x}{1 - b \cos x}$
19. $\frac{1-(-1)^n}{n} b^n \quad (|b| \leq 1)$	$\frac{2}{\pi} \arctan \frac{2b \sin x}{1 - b^2}$

Finite Cosine Transforms

$f_c(n)$	$F(x)$
1. $f_c(n) = \int_0^\pi F(x)\cos nx\,dx$ $(n = 0, 1, 2, \ldots)$	$F(x)$
2. $(-1)^n f_c(n)$	$F(\pi - x)$
3. 0 when $n = 1, 2, \cdots$; $\quad f_c(0) = \pi$	1
4. $\frac{2}{n}\sin\frac{n\pi}{2}$; $\quad f_c(0) = 0$	$\begin{cases} 1 & \text{when } 0 < x < \pi/2 \\ -1 & \text{when } \pi/2 < x < \pi \end{cases}$
5. $-\frac{1-(-1)^n}{n^2}$; $\quad f_c(0) = \frac{\pi^2}{2}$	x
6. $\frac{(-1)^n}{n^2}$; $\quad f_c(0) = \frac{\pi^2}{6}$	$\frac{x^2}{2\pi}$
7. $\frac{1}{n^2}$; $\quad f_c(0) = 0$	$\frac{(\pi-x)^2}{2\pi} - \frac{\pi}{6}$
8. $3\pi^2\frac{(-1)^n}{n^2} - 6\frac{1-(-1)^n}{n^4}$; $\quad f_c(0) = \frac{\pi^4}{4}$	x^3
9. $\frac{(-1)^n e^{c\pi}-1}{n^2+c^2}$	$\frac{1}{c}e^{cx}$
10. $\frac{1}{n^2+c^2}$	$\frac{\cosh c(\pi-x)}{c\sinh c\pi}$
11. $\frac{k}{n^2-k^2}[(-1)^n\cos\pi k - 1]$ $\quad (k \neq 0, 1, 2, \cdots)$	$\sin kx$
12. $\frac{(-1)^{n+m}-1}{n^2-m^2}$; $f_c(m) = 0$ $(m = 1, 2, \cdots)$	$\frac{1}{m}\sin mx$
13. $\frac{1}{n^2-k^2}$ $(k \neq 0, 1, 2, \ldots)$	$-\frac{\cos k(\pi-x)}{k\sin k\pi}$
14. $\begin{cases} 0 & \text{for } n = 1, 2, \cdots; \ n \neq m \\ \frac{\pi}{2} & \text{for } n = m \end{cases}$	$\cos mx \quad$ for $m = 1, 2, 3, \ldots$

Fourier Sine Transforms

$F(x)$	$f_s(\alpha)$
1. $\begin{cases} 1 & (0 < x < a) \\ 0 & (x > a) \end{cases}$	$\sqrt{\frac{2}{\pi}}\left[\frac{1-\cos\alpha}{\alpha}\right]$
2. $x^{p-1}(0 < p < 1)$	$\sqrt{\frac{2}{\pi}}\frac{\Gamma(p)}{\alpha^p}\sin\frac{p\pi}{2}$
3. $\begin{cases} \sin x & (0 < x < a) \\ 0 & (x > a) \end{cases}$	$\frac{1}{\sqrt{2\pi}}\left[\frac{\sin[a(1-\alpha)]}{1-\alpha} - \frac{\sin[a(1+\alpha)]}{1+\alpha}\right]$
4. e^{-x}	$\sqrt{\frac{2}{\pi}}\left[\frac{\alpha}{1+\alpha^2}\right]$
5. $xe^{-x^2/2}$	$\alpha e^{-\alpha^2/2}$
6. $\cos\frac{x^2}{2}$	$\sqrt{2}\left[\sin\frac{\alpha^2}{2}C\left(\frac{\alpha^2}{2}\right) - \cos\frac{\alpha^2}{2}S\left(\frac{\alpha^2}{2}\right)\right]^*$
7. $\sin\frac{x^2}{2}$	$\sqrt{2}\left[\cos\frac{\alpha^2}{2}C\left(\frac{\alpha^2}{2}\right) + \sin\frac{\alpha^2}{2}S\left(\frac{\alpha^2}{2}\right)\right]^*$

Here $C(y)$ and $S(y)$ are the Fresnel integrals:

$$C(y) = \frac{1}{\sqrt{2\pi}}\int_0^y \frac{1}{\sqrt{t}}\cos t\,dt, \qquad S(y) = \frac{1}{\sqrt{2\pi}}\int_0^y \frac{1}{\sqrt{t}}\sin t\,dt$$

*More extensive tables of the Fourier sine and cosine transforms can be found in Fritz Oberhettinger, *Tabellen zur-Fourier Transformation*, Springer, 1957.

Fourier Cosine Transforms

$F(x)$	$f_c(\alpha)$
1. $\begin{cases} 1 & (0 < x < a) \\ 0 & (x > a) \end{cases}$	$\sqrt{\frac{2}{\pi}}\frac{\sin a\alpha}{\alpha}$
2. x^{p-1} $(0 < p < 1)$	$\sqrt{\frac{2}{\pi}}\frac{\Gamma(p)}{\alpha^p}\cos\frac{p\pi}{2}$
3. $\begin{cases} \cos x & (0 < x < a) \\ 0 & (x > a) \end{cases}$	$\frac{1}{\sqrt{2\pi}}\left[\frac{\sin[a(1-\alpha)]}{1-\alpha} + \frac{\sin[a(1+\alpha)]}{1+\alpha}\right]$
4. e^{-x}	$\sqrt{\frac{2}{\pi}}\left(\frac{1}{1+\alpha^2}\right)$
5. $e^{-x^2/2}$	$e^{-\alpha^1/2}$
6. $\cos\frac{x^2}{2}$	$\cos\left(\frac{\alpha^2}{2} - \frac{\pi}{4}\right)$
7. $\sin\frac{x^2}{2}$	$\cos\left(\frac{\alpha^2}{2} + \frac{\pi}{4}\right)$

Fourier Transforms

	$F(x)$	$f(\alpha)$
1.	$\dfrac{\sin ax}{x}$	$\begin{cases} \sqrt{\dfrac{\pi}{2}} & \lvert\alpha\rvert < a \\ 0 & \lvert\alpha\rvert > a \end{cases}$
2.	$\begin{cases} e^{iwx} & (p < x < q) \\ 0 & (x < p,\; x > q) \end{cases}$	$\dfrac{i}{\sqrt{2\pi}}\dfrac{e^{ip(w+\alpha)} - e^{iq(w+\alpha)}}{(w+\alpha)}$
3.	$\begin{cases} e^{-cx+iwx} & (x > 0) \\ 0 & (x < 0) \end{cases} \quad (c > 0)$	$\dfrac{i}{\sqrt{2\pi}(w+\alpha+ic)}$
4.	$e^{-px^2}\ R(p) > 0$	$\dfrac{1}{\sqrt{2p}}e^{-\alpha^2/4p}$
5.	$\cos px^2$	$\dfrac{1}{\sqrt{2p}}\cos\left[\dfrac{\alpha^2}{4p} - \dfrac{\pi}{4}\right]$
6.	$\sin px^2$	$\dfrac{1}{\sqrt{2p}}\cos\left[\dfrac{\alpha^2}{4p} + \dfrac{\pi}{4}\right]$
7.	$\lvert x\rvert^{-p} \quad (0 < p < 1)$	$\sqrt{\dfrac{2}{\pi}}\dfrac{\Gamma(1-p)\sin\frac{p\pi}{2}}{\lvert\alpha\rvert^{(1-p)}}$
8.	$\dfrac{e^{-a\lvert x\rvert}}{\sqrt{\lvert x\rvert}}$	$\dfrac{\sqrt{\sqrt{(a^2+\alpha^2)}+a}}{\sqrt{a^2+\alpha^2}}$
9.	$\dfrac{\cosh ax}{\cosh \pi x} \quad (-\pi < a < \pi)$	$\sqrt{\dfrac{2}{\pi}}\dfrac{\cos\frac{a}{2}\cosh\frac{\alpha}{2}}{\cosh\alpha+\cos a}$
10.	$\dfrac{\sinh ax}{\sinh \pi x} \quad (-\pi < a < \pi)$	$\dfrac{1}{\sqrt{2\pi}}\dfrac{\sin a}{\cosh\alpha+\cos a}$
11.	$\begin{cases} \dfrac{1}{\sqrt{a^2-x^2}} & (\lvert x\rvert < a) \\ 0 & (\lvert x\rvert > a) \end{cases}$	$\sqrt{\dfrac{\pi}{2}}J_0(a\alpha)$
12.	$\dfrac{\sin[b\sqrt{a^2+x^2}]}{\sqrt{a^2+x^2}}$	$\begin{cases} 0 & (\lvert\alpha\rvert > b) \\ \sqrt{\dfrac{\pi}{2}}J_0(a\sqrt{b^2-\alpha^2}) & (\lvert\alpha\rvert < b) \end{cases}$
13.	$\begin{cases} p_n(x) & (\lvert x\rvert < 1) \\ 0 & (\lvert x\rvert > 1) \end{cases}$	$\dfrac{i^n}{\sqrt{\alpha}}J_{n+\frac{1}{2}}(\alpha)$
14.	$\begin{cases} \dfrac{\cos[b\sqrt{a^2-x^2}]}{\sqrt{a^2-x^2}} & (\lvert x\rvert < a) \\ 0 & (\lvert x\rvert > a) \end{cases}$	$\sqrt{\dfrac{\pi}{2}}J_0(a\sqrt{a^2+b^2})$
15.	$\begin{cases} \dfrac{\cosh[b\sqrt{a^2-x^2}]}{\sqrt{a^2-x^2}} & (\lvert x\rvert < a) \\ 0 & (\lvert x\rvert > a) \end{cases}$	$\sqrt{\dfrac{\pi}{2}}J_0(a\sqrt{\alpha^2-b^2})$

*More extensive tables of Fourier transforms can be found in W. Magnus and F. Oberhettinger, *Formulas and Theorems of the Special Functions of Mathematical Physics*. Chelsea, 1949, 116–120.

SERIES EXPANSION

The expression in parentheses following certain of the series indicates the region of convergence. If not otherwise indicated it is to be understood that the series converges for all finite values of x.

Binomial Series

$$(x + y)^n = x^n + nx^{n-1}y + \frac{n(n-1)}{2!}x^{n-2}y^2 + \frac{n(n-1)(n-2)}{3!}x^{n-3}y^3 + \cdots (y^2 < x^2)$$

$$(1 \pm x)^n = 1 \pm nx + \frac{n(n-1)x^2}{2!} \pm \frac{n(n-1)(n-2)x^3}{3!} + \cdots (x^2 < 1)$$

$$(1 \pm x)^{-n} = 1 \mp nx + \frac{n(n+1)x^2}{2!} \mp \frac{n(n+1)(n+2)x^3}{3!} + \cdots (x^2 < 1)$$

$$(1 \pm x)^{-1} = 1 \mp x + x^2 \mp x^3 + x^4 \mp x^5 + \cdots \quad (x^2 < 1)$$

$$(1 \pm x)^{-2} = 1 \mp 2x + 3x^2 \mp 4x^3 + 5x^4 \mp 6x^5 + \cdots \quad (x^2 < 1)$$

Reversion of Series

Let a series be represented by

$$y = a_1 x + a_2 x^2 + a_3 x^3 + a_4 x^4 + a_5 x^5 + a_6 x^6 + \cdots$$

with $a_1 \neq 0$. The coefficients of the series

$$x = A_1 y + A_2 y^2 + A_3 y^3 + A_4 y^4 + \cdots$$

are

$$A_1 = \frac{1}{a_1} \qquad A_2 = -\frac{a_2}{a_1^3} \qquad A_3 = \frac{1}{a_1^5}(2a_2^2 - a_1 a_3)$$

$$A_4 = \frac{1}{a_1^7}(5a_1 a_2 a_3 - a_1^2 a_4 - 5a_2^3)$$

$$A_5 = \frac{1}{a_1^9}(6a_1^2 a_2 a_4 + 3a_1^2 a_3^2 + 14a_2^4 - a_1^3 a_5 - 21a_1 a_2^2 a_3)$$

$$A_6 = \frac{1}{a_1^{11}}(7a_1^3 a_2 a_5 + 7a_1^3 a_3 a_4 + 84a_1 a_2^3 a_3 - a_1^4 a_6 - 28a_1^2 a_2^2 a_4 - 28a_1^2 a_2 a_3^2 - 42a_2^5)$$

$$A_7 = \frac{1}{a_1^{13}}(8a_1^4 a_2 a_6 + 8a_1^4 a_3 a_5 + 4a_1^4 a_4^2 + 120a_1^2 a_2^3 a_4 + 180a_1^2 a_2^2 a_3^2 + 132a_2^6 - a_1^5 a_7$$

$$-36a_1^3 a_2^2 a_5 - 72a_1^3 a_2 a_3 a_4 - 12a_1^3 a_3^3 - 330a_1 a_2^4 a_3)$$

Taylor Series

1. $$f(x) = f(a) + (x-a)f'(a) + \frac{(x-a)^2}{2!}f''(a) + \frac{(x-a)^3}{3!}f'''(a)$$

    $$+ \cdots + \frac{(x-a)^n}{n!}f^{(n)}(a) + \cdots \qquad \text{(Taylor Series)}$$

    (Increment form)

2. $$f(x+h) = f(x) + hf'(x) + \frac{h^2}{2!}f''(x) + \frac{h^3}{3!}f'''(x) + \cdots$$

    $$= f(h) + xf'(h) + \frac{x^2}{2!}f''(h) + \frac{x^3}{3!}f'''(h) + \cdots$$

3. If $f(x)$ is a function possessing derivatives of all orders throughout the interval $a \le x \le b$, then there is a value X, with $a < X < b$, such that

    $$f(b) = f(a) + (b-a)f'(a) + \frac{(b-a)^2}{2!}f''(a) + \cdots + \frac{(b-a)^{n-1}}{(n-1)!}f^{(n-1)}(a) + \frac{(b-a)^n}{n!}f^{(n)}(X)$$

    $$f(a+h) = f(a) + hf'(a) + \frac{h^2}{2!}f''(a) + \cdots + \frac{h^{n-1}}{(n-1)!}f^{(n-1)}(a) + \frac{h^n}{n!}f^{(n)}(a+\theta h)$$

    where $b = a + h$ and $0 < \theta < 1$. Or

    $$f(x) = f(a) + (x-a)f'(a) + \frac{(x-a)^2}{2!}f''(a) + \cdots + (x-a)^{n-1}\frac{f^{(n-1)}(a)}{(n-1)!} + R_n,$$

    where

    $$R_n = \frac{f^{(n)}[a + \theta \cdot (x-a)]}{n!}(x-a)^n, 0 < \theta < 1.$$

    The above forms are known as Taylor series with the remainder term.

4. Taylor series for a function of two variables

    If $\left(h\frac{\partial}{\partial x} + k\frac{\partial}{\partial y}\right) f(x,y) = h\frac{\partial f(x,y)}{\partial x} + k\frac{\partial f(x,y)}{\partial y};$

    $$\left(h\frac{\partial}{\partial x} + k\frac{\partial}{\partial y}\right)^2 f(x,y) = h^2\frac{\partial^2 f(x,y)}{\partial x^2} + 2hk\frac{\partial^2 f(x,y)}{\partial x \partial y} + k^2\frac{\partial^2 f(x,y)}{\partial y^2}$$

    etc., and if $\left(h\frac{\partial}{\partial x} + k\frac{\partial}{\partial y}\right)^n f(x,y)\Big|_{x=a}^{y=b}$ where the bar and subscripts mean that after differentiation we are to replace x by a and y by b, then

    $$f(a+h, b+k) = f(a,b) + \left(h\frac{\partial}{\partial x} + k\frac{\partial}{\partial y}\right) f(x,y)\Big|_{x=a}^{y=b} + \cdots + \frac{1}{n!}\left(h\frac{\partial}{\partial x} + k\frac{\partial}{\partial y}\right)^n f(x,y)\Big|_{x=a}^{y=b} + \cdots$$

Maclaurin Series

$$f(x) = f(0) + xf'(0) + \frac{x^2}{2!}f''(0) + \frac{x^3}{3!}f'''(0) + \cdots + x^{n-1}\frac{f^{(n-1)}(0)}{(n-1)!} + R_n,$$

where

$$R_n = \frac{x^n f^{(n)}(\theta x)}{n!}, \quad 0 < \theta < 1.$$

Exponential Series

$$e = 1 + \frac{1}{1!} + \frac{1}{2!} + \frac{1}{3!} + \frac{1}{4!} + \cdots$$

$$e^x = 1 + x + \frac{x^2}{2!} + \frac{x^3}{3!} + \frac{x^4}{4!} + \cdots$$

$$a^x = 1 + x\log_e a + \frac{(x\log_e a)^2}{2!} + \frac{(x\log_e a)^3}{3!} + \cdots$$

$$e^x = e^a\left[1 + (x-a) + \frac{(x-a)^2}{2!} + \frac{(x-a)^3}{3!} + \cdots\right]$$

Logarithmic Series

$$\log_e x = \frac{x-1}{x} + \frac{1}{2}\left(\frac{x-1}{x}\right)^2 + \frac{1}{3}\left(\frac{x-1}{x}\right)^3 + \cdots \qquad (x > \tfrac{1}{2})$$

$$\log_e x = (x-1) - \frac{1}{2}(x-1)^2 + \frac{1}{3}(x-1)^3 - \cdots \qquad (2 \geq x > 0)$$

$$\log_e x = 2\left[\frac{x-1}{x+1} + \frac{1}{3}\left(\frac{x-1}{x+1}\right)^3 + \frac{1}{5}\left(\frac{x-1}{x+1}\right)^5 + \cdots\right] \qquad (x > 0)$$

$$\log_e(1+x) = x - \frac{1}{2}x^2 + \frac{1}{3}x^3 - \frac{1}{4}x^4 + \cdots \qquad (-1 < x \leq 1)$$

$$\log_e(n+1) - \log_e(n-1) = 2\left[\frac{1}{n} + \frac{1}{3n^3} + \frac{1}{5n^5} + \cdots\right]$$

$$\log_e(a+x) = \log_e a + 2\left[\frac{x}{2a+x} + \frac{1}{3}\left(\frac{x}{2a+x}\right)^3 \right.$$
$$\left. + \frac{1}{5}\left(\frac{x}{2a+x}\right)^5 + \cdots\right] \qquad (a > 0, -a < x < +\infty)$$

$$\log_e\frac{1+x}{1-x} = 2\left[x + \frac{x^3}{3} + \frac{x^5}{5} + \cdots + \frac{x^{2n-1}}{2n-1} + \cdots\right] \qquad -1 < x < 1$$

$$\log_e x = \log_e a + \frac{(x-a)}{a} - \frac{(x-a)^2}{2a^2} + \frac{(x-a)^3}{3a^3} - + \cdots \qquad 0 < x \leq 2a$$

Trigonometric Series

$$\sin x = x - \frac{x^3}{3!} + \frac{x^5}{5!} - \frac{x^7}{7!} + \cdots \quad \text{(all real values of } x)$$

$$\cos x = 1 - \frac{x^2}{2!} + \frac{x^4}{4!} - \frac{x^6}{6!} + \cdots \quad \text{(all real values of } x)$$

$$\tan x = x + \frac{x^3}{3} + \frac{2x^5}{15} + \frac{17x^7}{315} + \frac{62x^9}{2835} + \cdots + \frac{(-1)^{n-1}2^{2n}(2^{2n}-1)B_n}{(2n)!}x^{2n-1} + \cdots,$$
$$\left[x^2 < \frac{\pi^2}{4} \quad \text{and } B_n \text{ represents the } n^{\text{th}} \text{ Bernoulli number}\right]$$

$$\cot x = \frac{1}{x} - \frac{x}{3} - \frac{x^3}{45} - \frac{2x^5}{945} - \frac{x^7}{4725} - \cdots - \frac{(-1)^{n+1}2^{2n}}{(2n)!}B_{2n}x^{2n-1} - \cdots,$$
$$\left[x^2 < \pi^2 \quad \text{and } B_n \text{ represents the } n^{\text{th}} \text{ Bernoulli number}\right]$$

$$\sec x = 1 + \frac{x^2}{2} + \frac{5}{24}x^4 + \frac{61}{720}x^6 + \frac{277}{8064}x^8 + \cdots + \frac{(-1)^n}{(2n)!}E_{2n}x^{2n} + \cdots,$$
$$\left[x^2 < \frac{\pi^2}{4} \quad \text{and } E_n \text{ represents the } n^{\text{th}} \text{ Euler number}\right]$$

$$\csc x = \frac{1}{x} + \frac{x}{6} + \frac{7}{360}x^3 + \frac{31}{15,120}x^5 + \frac{127}{604,800}x^7 + \cdots + \frac{(-1)^{n+1}2(2^{2n-1}-1)}{(2n)!}B_{2n}x^{2n-1} + \cdots,$$
$$\left[x^2 < \pi^2 \quad \text{and } B_n \text{ represents the } n^{\text{th}} \text{ Bernoulli number}\right]$$

$$\sin x = x\left(1 - \frac{x^2}{\pi^2}\right)\left(1 - \frac{x^2}{2^2\pi^2}\right)\left(1 - \frac{x^2}{3^2\pi^2}\right)\cdots \qquad (x^2 < \infty)$$

$$[2pt]\cos x = \left(1 - \frac{4x^2}{\pi^2}\right)\left(1 - \frac{4x^2}{3^2\pi^2}\right)\left(1 - \frac{4x^2}{5^2\pi^2}\right)\cdots \qquad (x^2 < \infty)$$

$$[2pt]\sin^{-1}x = x + \frac{x^3}{2\cdot 3} + \frac{1\cdot 3}{2\cdot 4\cdot 5}x^5 + \frac{1\cdot 3\cdot 5}{2\cdot 4\cdot 6\cdot 7}x^7 + \cdots \qquad \left(x^2 < 1, -\frac{\pi}{2} < \sin^{-1}x < \frac{\pi}{2}\right)$$

$$[2pt]\cos^{-1}x = \frac{\pi}{2} - \left(x + \frac{x^3}{2\cdot 3} + \frac{1\cdot 3}{2\cdot 4\cdot 5}x^5 + \frac{1\cdot 3\cdot 5}{2\cdot 4\cdot 6\cdot 7}x^7 + \cdots\right) \qquad (x^2 < 1, 0 < \cos^{-1}x < \pi)$$

$$[2pt]\tan^{-1}x = x - \frac{x^3}{3} + \frac{x^5}{5} - \frac{x^7}{7} + \cdots \qquad (x^2 < 1)$$

$$[2pt]\tan^{-1}x = \frac{\pi}{2} - \frac{1}{x} + \frac{1}{3x^3} - \frac{1}{5x^5} + \frac{1}{7x^7} - \cdots \qquad (x > 1)$$

$$[2pt]\tan^{-1}x = -\frac{\pi}{2} - \frac{1}{x} + \frac{1}{3x^3} - \frac{1}{5x^5} + \frac{1}{7x^7} - \cdots \qquad (x < -1)$$

$$[2pt]\cot^{-1}x = \frac{\pi}{2} - x + \frac{x^3}{3} - \frac{x^5}{5} + \frac{x^7}{7} - \cdots \qquad (x^2 < 1)$$

$$\log_e \sin x = \log_e x - \frac{x^2}{6} - \frac{x^4}{180} - \frac{x^6}{2835} - \cdots \qquad (x^2 < \pi^2)$$

$$\log_e \cos x = -\frac{x^2}{2} - \frac{x^4}{12} - \frac{x^6}{45} - \frac{17x^8}{2520} - \cdots \qquad \left(x^2 < \frac{\pi^2}{4}\right)$$

$$\log_e \tan x = \log_e x + \frac{x^2}{3} + \frac{7x^4}{90} + \frac{62x^6}{2835} + \cdots \qquad \left(x^2 < \frac{\pi^2}{4}\right)$$

$$e^{\sin x} = 1 + x + \frac{x^2}{2!} - \frac{3x^4}{4!} - \frac{8x^5}{5!} - \frac{3x^6}{6!} + \frac{56x^7}{7!} + \cdots$$

$$e^{\cos x} = e\left(1 - \frac{x^2}{2!} + \frac{4x^4}{4!} - \frac{31x^6}{6!} + \cdots\right)$$

$$e^{\tan x} = 1 + x + \frac{x^2}{2!} + \frac{3x^3}{3!} + \frac{9x^4}{4!} + \frac{37x^5}{5!} + \cdots \qquad \left(x^2 < \frac{\pi^2}{4}\right)$$

$$\sin x = \sin a + (x - a)\cos a - \frac{(x-a)^2}{2!}\sin a$$
$$- \frac{(x-a)^3}{3!}\cos a + \frac{(x-a)^4}{4!}\sin a + \cdots$$

VECTOR ANALYSIS

Definitions

Any quantity that is completely determined by its magnitude is called a *scalar*. Examples of such are mass, density, temperature, etc. Any quantity that is completely determined by its magnitude and direction is called a *vector*. Examples of such are velocity, acceleration, force, etc. A vector quantity is represented by a directed line segment, the length of which represents the magnitude of the vector. A vector quantity is usually represented by a boldfaced letter such as V. Two vectors V_1 and V_2 are equal to one another if they have equal magnitudes and are acting in the same directions. A negative vector, written as $-V$, is one that acts in the opposite direction to V, but is of equal magnitude to it. If we represent the magnitude of V by v, we write $|V| = v$. A vector parallel to V, but equal to the reciprocal of its magnitude is written as V^{-1} or as $1/V$.

The *unit vector* V/v (when $v \neq 0$) is that vector which has the same direction as V, but has a magnitude of unity (sometimes represented as V_0 or $\hat{v}$).

Vector Algebra

The vector sum of V_1 and V_2 is represented by $V_1 + V_2$. The vector sum of V_1 and $-V_2$, or the difference of the vector V_2 from V_1 is represented by $V_1 - V_2$.

If r is a scalar, then $rV = Vr$, and represents a vector r times the magnitude of V, in the same direction as V if r is positive, and in the opposite direction if r is negative. If r and s are scalars, V_1, V_2, V_3, vectors, then the following rules of scalars and vectors hold:

$$V_1 + V_2 = V_2 + V_1$$
$$(r + s)V_1 = rV_1 + sV_1; \quad r(V_1 + V_2) = rV_1 + rV_2$$
$$V_1 + (V_2 + V_3) = (V_1 + V_2) + V_3 = V_1 + V_2 + V_3$$

Vectors in Space

A plane is described by two distinct vectors V_1 and V_2. Should these vectors not intersect each other, then one is displaced parallel to itself until they do (Figure 1). Any other vector V lying in this plane is given by

$$V = rV_1 + sV_2$$

A *position vector* specifies the position in space of a point relative to a fixed origin. If therefore V_1 and V_2 are the position vectors of the points A and B, relative to the origin O, then any point P on the line AB has a position vector V given by

$$V = rV_1 + (1 - r)V_2$$

The scalar "r" can be taken as the metric representation of P since $r = 0$ implies $P = B$ and $r = 1$ implies $P = A$ (Figure 2). If P divides the line AB in the ratio $r:s$ then

$$V = \left(\frac{r}{r + s}\right)V_1 + \left(\frac{s}{r + s}\right)V_2$$

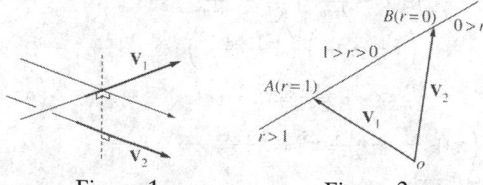

Figure 1.　　　　　Figure 2.

The vectors $V_1, V_2, V_3, \ldots, V_n$ are said to be *linearly dependent* if there exist scalars $r_1, r_2, r_3, \ldots, r_n$, not all zero, such that

$$r_1V_1 + r_2V_2 + \cdots + r_nV_n = 0$$

A vector $\mathbf{V}$ is linearly dependent upon the set of vectors $\mathbf{V}_1, \mathbf{V}_2, \mathbf{V}_3,\ldots,\mathbf{V}_n$ if

$$\mathbf{V} = r_1\mathbf{V}_1 + r_2\mathbf{V}_2 + r_3\mathbf{V}_3 + \cdots + r_n\mathbf{V}_n$$

Three vectors are linearly dependent if and only if they are co-planar.

All points in space can be uniquely determined by linear dependence upon three *base vectors*, i.e., three vectors any one of which is linearly independent of the other two. The simplest set of base vectors is the unit vectors along the coordinate Ox, Oy and Oz axes. These are usually designated by $\mathbf{i}$, $\mathbf{j}$ and $\mathbf{k}$, respectively.

If $\mathbf{V}$ is a vector in space, and a, b and c are the respective magnitudes of the projections of the vector along the axes then

$$\mathbf{V} = a\mathbf{i} + b\mathbf{j} + c\mathbf{k}$$

and

$$v = \sqrt{a^2 + b^2 + c^2}$$

and the direction cosines of $\mathbf{V}$ are

$$\cos\alpha = a/v, \quad \cos\beta = b/v, \quad \cos\gamma = c/v.$$

The law of addition yields

$$\mathbf{V}_1 + \mathbf{V}_2 = (a_1 + a_2)\mathbf{i} + (b_1 + b_2)\mathbf{j} + (c_1 + c_2)\mathbf{k}$$

The Scalar, Dot, or Inner Product of Two Vectors

This product is represented as $\mathbf{V}_1 \cdot \mathbf{V}_2$ and is defined to be equal to $v_1 v_2 \cos\theta$, where θ is the angle from $\mathbf{V}_1$ to $\mathbf{V}_2$, i.e.,

$$\mathbf{V}_1 \cdot \mathbf{V}_2 = v_1 v_2 \cos\theta$$

The following rules apply for this product:

$$\mathbf{V}_1 \cdot \mathbf{V}_2 = a_1 a_2 + b_1 b_2 + c_1 c_2 = \mathbf{V}_2 \cdot \mathbf{V}_1$$

It should be noted that this verifies that scalar multiplication is commutative.

$$(\mathbf{V}_1 + \mathbf{V}_2) \cdot \mathbf{V}_3 = \mathbf{V}_1 \cdot \mathbf{V}_3 + \mathbf{V}_2 \cdot \mathbf{V}_3$$
$$\mathbf{V}_1 \cdot (\mathbf{V}_2 + \mathbf{V}_3) = \mathbf{V}_1 \cdot \mathbf{V}_2 + \mathbf{V}_1 \cdot \mathbf{V}_3$$

If $\mathbf{V}_1$ is perpendicular to $\mathbf{V}_2$ then $\mathbf{V}_1 \cdot \mathbf{V}_2 = 0$, and if $\mathbf{V}_1$ is parallel to $\mathbf{V}_2$, then $\mathbf{V}_1 \cdot \mathbf{V}_2 = v_1 v_2 = rw_1^2$. In particular

$$\mathbf{i} \cdot \mathbf{i} = \mathbf{j} \cdot \mathbf{j} = \mathbf{k} \cdot \mathbf{k} = 1,$$

and

$$\mathbf{i} \cdot \mathbf{j} = \mathbf{j} \cdot \mathbf{k} = \mathbf{k} \cdot \mathbf{i} = 0$$

The Vector or Cross Product of Two Vectors

This product is represented as $\mathbf{V}_1 \times \mathbf{V}_2$ and is defined to be equal to $v_1 v_2(\sin\theta)\mathbf{1}$, where θ is the angle from $\mathbf{V}_1$ to $\mathbf{V}_2$ and $\mathbf{1}$ is a unit vector perpendicular to the plane of $\mathbf{V}_1$ and $\mathbf{V}_2$ and so directed that a right-handed screw driven in the direction of $\mathbf{1}$ would carry $\mathbf{V}_1$ into $\mathbf{V}_2$, i.e.,

$$\mathbf{V}_1 \times \mathbf{V}_2 = v_1 v_2(\sin\theta)\mathbf{1}$$

and $\tan\theta = \dfrac{|\mathbf{V}_1 \times \mathbf{V}_2|}{\mathbf{V}_1 \cdot \mathbf{V}_2}$

The following rules apply for vector products:

$$\mathbf{V}_1 \times \mathbf{V}_2 = -\mathbf{V}_2 \times \mathbf{V}_1$$
$$\mathbf{V}_1 \times (\mathbf{V}_2 + \mathbf{V}_3) = \mathbf{V}_1 \times \mathbf{V}_2 + \mathbf{V}_1 \times \mathbf{V}_3$$
$$(\mathbf{V}_1 + \mathbf{V}_2) \times \mathbf{V}_3 = \mathbf{V}_1 \times \mathbf{V}_3 + \mathbf{V}_2 \times \mathbf{V}_3$$
$$\mathbf{V}_1 \times (\mathbf{V}_2 \times \mathbf{V}_3) = \mathbf{V}_2(\mathbf{V}_3 \cdot \mathbf{V}_1) - \mathbf{V}_3(\mathbf{V}_1 \cdot \mathbf{V}_2)$$
$$\mathbf{i} \times \mathbf{i} = \mathbf{j} \times \mathbf{j} = \mathbf{k} \times \mathbf{k} = 0 \quad \text{(the zero vector)}$$
$$\mathbf{i} \times \mathbf{j} = \mathbf{k}, \quad \mathbf{j} \times \mathbf{k} = \mathbf{i}, \quad \mathbf{k} \times \mathbf{i} = \mathbf{j}$$

If $\mathbf{V}_1 = a_1\mathbf{i} + b_1\mathbf{j} + c_1\mathbf{k}$, $\mathbf{V}_2 = a_2\mathbf{i} + b_2\mathbf{j} + c_2\mathbf{k}$, and $\mathbf{V}_3 = a_3\mathbf{i} + b_3\mathbf{j} + c_3\mathbf{k}$, then

$$\mathbf{V}_1 \times \mathbf{V}_2 = \begin{vmatrix} \mathbf{i} & \mathbf{j} & \mathbf{k} \\ a_1 & b_1 & c_1 \\ a_2 & b_2 & c_2 \end{vmatrix} = (b_1 c_2 - b_2 c_1)\mathbf{i} + (c_1 a_2 - c_2 a_1)\mathbf{j} + (a_1 b_2 - a_2 b_1)\mathbf{k}$$

It should be noted that, since $\mathbf{V}_1 \times \mathbf{V}_2 = -\mathbf{V}_2 \times \mathbf{V}_1$, the vector product is not commutative.

Scalar Triple Product

There is only one possible interpretation of the expression $V_1 \cdot V_2 \times V_3$ and that is $V_1 \cdot (V_2 \times V_3)$ which is obviously a scalar. Further

$$V_1 \cdot (V_2 \times V_3) = (V_1 \times V_2) \cdot V_3 = V_2 \cdot (V_3 \times V_1)$$

$$= \begin{vmatrix} a_1 & b_1 & c_1 \\ a_2 & b_2 & c_2 \\ a_3 & b_3 & c_3 \end{vmatrix}$$

$$= r_1 r_2 r_3 \cos\phi \sin\theta,$$

Where θ is the angle between V_2 and V_3 and ϕ is the angle between V_1 and the normal to the plane of V_2 and V_3.

This product is called the *scalar triple product* and is written as $[V_1 V_2 V_3]$.

The determinant indicates that it can be considered as the volume of the parallelepiped whose three determining edges are V_1, V_2 and V_3.

It also follows that cyclic permutation of the subscripts does not change the value of the scalar triple product so that

$$[V_1 V_2 V_3] = [V_2 V_3 V_1] = [V_3 V_1 V_2]$$

$$\text{but} \quad [V_1 V_2 V_3] = -[V_2 V_1 V_3] \quad \text{etc.} \quad \text{and} \quad [V_1 V_1 V_2] \equiv 0 \quad \text{etc.}$$

Given three non-coplanar reference vectors V_1, V_2 and V_3, the *reciprocal system* is given by V_1^*, V_2^* and V_3^*, where

$$1 = v_1 v_1^* = v_2 v_2^* = v_3 v_3^*$$

$$0 = v_1 v_2^* = v_1 v_3^* = v_2 v_1^* \quad \text{etc.}$$

$$V_1^* = \frac{V_2 \times V_3}{[V_1 V_2 V_3]}, \quad V_2^* = \frac{V_3 \times V_1}{[V_1 V_2 V_3]}, \quad V_3^* = \frac{V_1 \times V_2}{[V_1 V_2 V_3]}$$

The system i, j, k is its own reciprocal.

Vector Triple Product

The product $V_1 \times (V_2 \times V_3)$ defines the *vector triple product*. Obviously, in this case, the brackets are vital to the definition.

$$V_1 \times (V_2 \times V_3) = (V_1 \cdot V_3)V_2 - (V_1 \cdot V_2)V_3$$

$$= \begin{vmatrix} i & j & k \\ a_1 & b_1 & c_1 \\ \begin{vmatrix} b_2 & c_2 \\ b_3 & c_3 \end{vmatrix} & \begin{vmatrix} c_2 & a_2 \\ c_3 & a_3 \end{vmatrix} & \begin{vmatrix} a_2 & b_2 \\ a_3 & b_3 \end{vmatrix} \end{vmatrix}$$

i.e., it is a vector, perpendicular to V_1, lying in the plane of V_2, V_3. Similarly

$$(V_1 \times V_2) \times V_3 = \begin{vmatrix} i & j & k \\ \begin{vmatrix} b_1 & c_1 \\ b_2 & c_2 \end{vmatrix} & \begin{vmatrix} c_1 & a_1 \\ c_2 & a_2 \end{vmatrix} & \begin{vmatrix} a_1 & b_1 \\ a_2 & b_2 \end{vmatrix} \\ a_3 & b_3 & c_3 \end{vmatrix}$$

$$V_1 \times (V_2 \times V_3) + V_2 \times (V_3 \times V_1) + V_3 \times (V_1 \times V_2) \equiv 0$$

If $V_1 \times (V_2 \times V_3) = (V_1 \times V_2) \times V_3$, then V_1, V_2, V_3 form an *orthogonal set*. Thus i, j, k form an orthogonal set.

Geometry of the Plane, Straight Line and Sphere

The position vectors of the fixed points A, B, C, D relative to O are V_1, V_2, V_3, V_4 and the position vector of the variable point P is V.

The vector form of the equation of the straight line through A parallel to V_2 is

$$V = V_1 + rV_2$$

$$\text{or} \quad (V - V_1) = rV_2$$

$$\text{or} \quad (V - V_1) \times V_2 = 0$$

while that of the plane through A perpendicular to V_2 is

$$(V - V_1) \cdot V_2 = 0$$

The equation of the line AB is

$$V = rV_1 + (1 - r)V_2$$

and those of the bisectors of the angles between V_1 and V_2 are

$$V = r\left(\frac{V_1}{v_1} \pm \frac{V_2}{v_2}\right) \quad \text{or}$$

$$V = r(\hat{v}_1 \pm \hat{v}_2)$$

The perpendicular from C to the line through A parallel to $\mathbf{V}_2$ has as its equation

$$\mathbf{V} = \mathbf{V}_1 - \mathbf{V}_3 - \hat{\mathbf{v}}_2 \cdot (\mathbf{V}_1 - \mathbf{V}_3)\hat{\mathbf{v}}_2.$$

The condition for the intersection of the two lines, $\mathbf{V} = \mathbf{V}_1 + r\mathbf{V}_3$ and $\mathbf{V} = \mathbf{V}_2 + s\mathbf{V}_4$, is

$$[(\mathbf{V}_1 - \mathbf{V}_2)\mathbf{V}_3\mathbf{V}_4] = 0.$$

The common perpendicular to the above two lines is the line of intersection of the two planes

$$[(\mathbf{V} - \mathbf{V}_1)\mathbf{V}_3(\mathbf{V}_3 \times \mathbf{V}_4)] = 0 \quad \text{and} \quad [(\mathbf{V} - \mathbf{V}_2)\mathbf{V}_4(\mathbf{V}_3 \times \mathbf{V}_4)] = 0$$

and the length of this perpendicular is

$$\frac{[(\mathbf{V}_1 - \mathbf{V}_2)\mathbf{V}_3\mathbf{V}_4]}{|\mathbf{V}_3 \times \mathbf{V}_4|}.$$

The equation of the line perpendicular to the plane ABC is

$$\mathbf{V} = \mathbf{V}_1 \times \mathbf{V}_2 + \mathbf{V}_2 \times \mathbf{V}_3 + \mathbf{V}_3 \times \mathbf{V}_1$$

and the distance of the plane from the origin is

$$\frac{[\mathbf{V}_1\mathbf{V}_2\mathbf{V}_3]}{|(\mathbf{V}_2 - \mathbf{V}_1) \times (\mathbf{V}_3 - \mathbf{V}_1)|}.$$

In general the vector equation

$$\mathbf{V} \cdot \mathbf{V}_2 = r$$

defines the plane which is perpendicular to $\mathbf{V}_2$, and the perpendicular distance from A to this plane is

$$\frac{r - \mathbf{V}_1 \cdot \mathbf{V}_2}{v_2}$$

The distance from A, measured along a line parallel to $\mathbf{V}_3$, is

$$\frac{r - \mathbf{V}_1 \cdot \mathbf{V}_2}{\mathbf{V}_2 \cdot \hat{\mathbf{v}}_3} \quad \text{or} \quad \frac{r - \mathbf{V}_1 \cdot \mathbf{V}_2}{v_2 \cos\theta}$$

where θ is the angle between $\mathbf{V}_2$ and $\mathbf{V}_3$. (If this plane contains the point C then $r = \mathbf{V}_3 \cdot \mathbf{V}_2$ and if it passes through the origin, then $r = 0$.) Given two planes

$$\mathbf{V} \cdot \mathbf{V}_1 = r$$
$$\mathbf{V} \cdot \mathbf{V}_2 = s$$

then any plane through the line of intersection of these two planes is given by

$$\mathbf{V} \cdot (\mathbf{V}_1 + \lambda\mathbf{V}_2) = r + \lambda s$$

where λ is a scalar parameter. In particular $\lambda = \pm v_1/v_2$ yields the equation of the two planes bisecting the angle between the given planes.

The plane through A parallel to the plane of $\mathbf{V}_2$, $\mathbf{V}_3$ is

$$\mathbf{V} = \mathbf{V}_1 + r\mathbf{V}_2 + s\mathbf{V}_3$$
$$\text{or} \quad (\mathbf{V} - \mathbf{V}_1) \cdot \mathbf{V}_2 \times \mathbf{V}_3 = 0$$
$$\text{or} \quad [\mathbf{V}\mathbf{V}_2\mathbf{V}_3] - [\mathbf{V}_1\mathbf{V}_2\mathbf{V}_3] = 0$$

so that the expansion in rectangular Cartesian coordinates yields (where $\mathbf{V} \equiv x\mathbf{i} + y\mathbf{j} + z\mathbf{k}$):

$$\begin{vmatrix} (x - a_1) & (y - b_1) & (z - c_1) \\ a_2 & b_2 & c_2 \\ a_3 & b_3 & c_3 \end{vmatrix} = 0$$

which is obviously the usual linear equation in x, y, and z.

The plane through AB parallel to $\mathbf{V}_3$ is given by

$$[(\mathbf{V} - \mathbf{V}_1)(\mathbf{V}_1 - \mathbf{V}_2)\mathbf{V}_3] = 0$$
$$\text{or} \quad [\mathbf{V}\mathbf{V}_2\mathbf{V}_3] - [\mathbf{V}\mathbf{V}_1\mathbf{V}_3] - [\mathbf{V}_1\mathbf{V}_2\mathbf{V}_3] = 0$$

The plane through the three points A, B and C is

$$V = V_1 + s(V_2 - V_1) + t(V_3 - V_1)$$

$$\text{or} \quad V = rV_1 + sV_2 + tV_3 \qquad (r + s + t \equiv 1)$$

$$\text{or} \quad [(V - V_1)(V_1 - V_2)(V_2 - V_3)] = 0$$

$$\text{or} \quad [VV_1V_2] + [VV_2V_3] + [VV_3V_1] - [V_1V_2V_3] = 0$$

For four points A, B, C, D to be coplanar, then

$$rV_1 + sV_2 + tV_3 + uV_4 \equiv 0 \equiv r + s + t + u$$

The following formulas relate to a sphere when the vectors are taken to lie in three-dimensional space and to a circle when the space is two dimensional. For a circle in three dimensions, take the intersection of the sphere with a plane.

The equation of a sphere with center O and radius OA is

$$V \cdot V = v_1^2 \quad (\text{not}\, V = V_1)$$

$$\text{or} \quad (V - V_1) \cdot (V + V_1) = 0$$

while that of a sphere with center B radius v_1 is

$$(V - V_2) \cdot (V - V_2) = v_1^2$$

$$\text{or} \quad V \cdot (V - 2V_2) = v_1^2 - v_2^2$$

If the above sphere passes through the origin, then

$$V \cdot (V - 2V_2) = 0$$

Note that in two-dimensional polar coordinates this is simply

$$r = 2a \cdot \cos\theta$$

while in three-dimensional Cartesian coordinates it is

$$x^2 + y^2 + z^2 - 2(a_2x + b_2y + c_2x) = 0.$$

The equation of a sphere having the points A and B as the extremities of a diameter is

$$(V - V_1) \cdot (V - V_2) = 0.$$

The square of the length of the tangent from C to the sphere with center B and radius v_1 is given by

$$(V_3 - V_2) \cdot (V_3 - V_2) = v_1^2$$

The condition that the plane $V \cdot V_3 = s$ is tangential to the sphere $(V - V_2) \cdot (V - V_2) = v_1^2$ is

$$(s - V_3 \cdot V_2) \cdot (s - V_3 \cdot V_2) = v_1^2 v_3^2.$$

The equation of the tangent plane at D, on the surface of sphere $(V - V_2) \cdot (V - V_2) = v_1^2$, is

$$(V - V_4) \cdot (V_4 - V_2) = 0$$

$$\text{or} \quad V \cdot V_4 - V_2 \cdot (V + V_4) = v_1^2 - v_2^2$$

The condition that the two circles $(V - V_2) \cdot (V - V_2) = v_1^2$ and $(V - V_4) \cdot (V - V_4) = v_3^2$ intersect orthogonally is clearly

$$(V_2 - V_4) \cdot (V_2 - V_4) = v_1^2 + v_3^2$$

The polar plane of D with respect to the circle

$$(V - V_2) \cdot (V - V_2) = v_1^2 \quad \text{is}$$
$$V \cdot V_4 - V_2 \cdot (V + V_4) = v_1^2 - v_2^2$$

Any sphere through the intersection of the two spheres $(V - V_2) \cdot (V - V_2) = v_1^2$ and $(V - V_4) \cdot (V - V_4) = v_3^2$ is given by

$$(V - V_2) \cdot (V - V_2) + \lambda(V - V_4) \cdot (V - V_4) = v_1^2 + \lambda v_3^2$$

while the radical plane of two such spheres is

$$V \cdot (V_2 - V_4) = -\frac{1}{2}(v_1^2 - v_2^2 - v_3^2 + v_4^2)$$

Differentiation of Vectors

If $V_1 = a_1\mathbf{i} + b_1\mathbf{j} + c_1\mathbf{k}$, and $V_2 = a_2\mathbf{i} + b_2\mathbf{j} + c_2\mathbf{k}$, and if V_1 and V_2 are functions of the scalar t, then

$$\frac{d}{dt}(V_1 + V_2 + \cdots) = \frac{dV_1}{dt} + \frac{dV_2}{dt} + \cdots$$

$$\frac{dV_1}{dt} = \frac{da_1}{dt}\mathbf{i} + \frac{db_1}{dt}\mathbf{j} + \frac{dc_1}{dt}\mathbf{k}, \quad \text{etc}$$

$$\frac{d}{dt}(V_1 \cdot V_2) = \frac{dV_1}{dt} \cdot V_2 + V_1 \cdot \frac{dV_2}{dt}$$

$$\frac{d}{dt}(V_1 \times V_2) = \frac{dV_1}{dt} \times V_2 + V_1 \times \frac{dV_2}{dt}$$

$$V \cdot \frac{dV}{dt} = v \cdot \frac{dv}{dt}$$

In particular, if V is a vector of constant length, then the right-hand side of the last equation is identically zero showing that V is perpendicular to its derivative.

The derivatives of the triple products are

$$\frac{d}{dt}[V_1 V_2 V_3] = \left[\left(\frac{dV_1}{dt}\right)V_2 V_3\right] + \left[V_1\left(\frac{dV_2}{dt}\right)V_3\right] + \left[V_1 V_2\left(\frac{dV_3}{dt}\right)\right] \quad \text{and}$$

$$\frac{d}{dt}\{V_1 \times (V_2 \times V_3)\} = \left(\frac{dV_1}{dt}\right) \times (V_2 \times V_3) + V_1 \times \left(\left(\frac{dV_2}{dt}\right) \times V_3\right) + V_1 \times \left(V_2 \times \left(\frac{dV_3}{dt}\right)\right)$$

Geometry of Curves in Space

s = the *length of arc*, measured from some fixed point on the curve (Figure 3).

V_1 = the position vector of the point A on the curve.

$V_1 + \delta V_1$ = the position vector of the point P in the neighborhood of A.

$\hat{\mathbf{t}}$ = the *unit tangent* to the curve at the point A, measured in the direction of s increasing.

The *normal plane* is that plane which is perpendicular to the unit tangent. The principal normal is defined as the intersection of the normal plane with the plane defined by V_1 and $V_1 + \delta V_1$ in the limit as $\delta V_1 - 0$.

$\hat{\mathbf{n}}$ = the *unit normal* (principal) at the point A. The plane defined by $\hat{\mathbf{t}}$ and $\hat{\mathbf{n}}$ is called the *osculating plane* (alternatively plane of curvature or local plane).

ρ = the radius of curvature at A.

$\delta\theta$ = the angle subtended at the origin by δV_1.

$$\kappa = \frac{d\theta}{ds} = \frac{1}{\rho}$$

$\hat{\mathbf{b}}$ = the *unit binormal* i.e., the unit vector which is parallel to $\hat{\mathbf{t}} \times \hat{\mathbf{n}}$ at the point A

λ = the *torsion* of the curve at A.

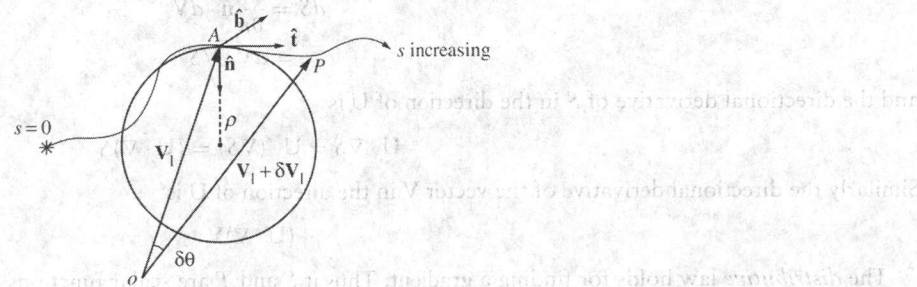

Figure 3.

Frenet's Formulas:

$$\frac{d\hat{\mathbf{t}}}{ds} = \kappa\hat{\mathbf{n}}$$

$$\frac{d\hat{\mathbf{n}}}{ds} = -\kappa\hat{\mathbf{t}} + \lambda\hat{\mathbf{b}}$$

$$\frac{d\hat{\mathbf{b}}}{ds} = -\lambda\hat{\mathbf{n}}$$

The following formulas are also applicable:

Unit tangent $\hat{\mathbf{t}} = \frac{d\mathbf{V}_1}{ds}$

Equation of the tangent $(\mathbf{V} - \mathbf{V}_1) \times \hat{\mathbf{t}} = 0$ or $\mathbf{V} = \mathbf{V}_1 + q\hat{\mathbf{t}}$

Unit normal $\hat{\mathbf{n}} = \frac{1}{\kappa}\frac{d^2\mathbf{V}_1}{ds^2}$

Equation of the normal plane $(\mathbf{V} - \mathbf{V}_1) \cdot \hat{\mathbf{t}} = 0$

Equation of the normal $(\mathbf{V} - \mathbf{V}_1) \times \hat{\mathbf{n}} = 0$ or $\mathbf{V} = \mathbf{V}_1 + r\hat{\mathbf{n}}$

Unit binormal $\hat{\mathbf{b}} = \hat{\mathbf{t}} \times \hat{\mathbf{n}}$

Equation of the binormal $(\mathbf{V} - \mathbf{V}_1) \times \hat{\mathbf{b}} = 0$

or $\mathbf{V} = \mathbf{V}_1 + u\hat{\mathbf{b}}$

or $\mathbf{V} = \mathbf{V}_1 + w\frac{d\mathbf{V}_1}{ds} \times \frac{d^2\mathbf{V}_1}{ds^2}$

Equation of the osculating plane $[(\mathbf{V} - \mathbf{V}_1)\hat{\mathbf{t}}\hat{\mathbf{n}}] = 0$

or $\left[(\mathbf{V} - \mathbf{V}_1)\left(\frac{d\mathbf{V}_1}{ds}\right)\left(\frac{d^2\mathbf{V}_1}{ds^2}\right)\right] = 0$

Differential Operators—Rectangular Coordinates

$$dS = \frac{\partial S}{\partial x} \cdot dx + \frac{\partial S}{\partial y} \cdot dy + \frac{\partial S}{\partial z} \cdot dz$$

By definition

$$\nabla \equiv \text{del} \equiv \mathbf{i}\frac{\partial}{\partial x} + \mathbf{j}\frac{\partial}{\partial y} + \mathbf{k}\frac{\partial}{\partial z}$$

$$\nabla^2 \equiv \text{Laplacian} \equiv \frac{\partial^2}{\partial x^2} + \frac{\partial^2}{\partial y^2} + \frac{\partial^2}{\partial z^2}$$

If S is a scalar function, then $\nabla S \equiv \text{grad } S \equiv \frac{\partial S}{\partial x}\mathbf{i} + \frac{\partial S}{\partial y}\mathbf{j} + \frac{\partial S}{\partial z}\mathbf{k}$.

Grad S defines both the direction and magnitude of the maximum rate of increase of S at any point. Hence the name *gradient* and also its vectorial nature. ∇S is independent of the choice of rectangular coordinates.

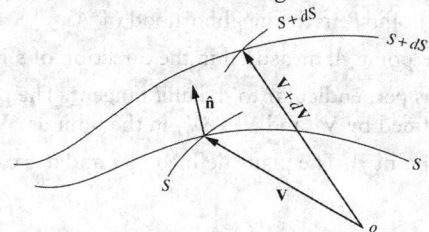

Figure 4.

$$\nabla S = \frac{\partial S}{\partial n}\hat{\mathbf{n}} \tag{4}$$

where $\hat{\mathbf{n}}$ is the unit normal to the surface $S = $ constant, in the direction of S increasing. The total derivative of S at a point having the position vector $\mathbf{V}$ is given by (Figure 4)

$$dS = \frac{\partial S}{\partial n}\hat{\mathbf{n}} \cdot d\mathbf{V}$$

$$= d\mathbf{V} \cdot \nabla S$$

and the directional derivative of S in the direction of $\mathbf{U}$ is

$$\mathbf{U} \cdot \nabla S = \mathbf{U} \cdot (\nabla S) = (\mathbf{U} \cdot \nabla)S$$

Similarly the directional derivative of the vector $\mathbf{V}$ in the direction of $\mathbf{U}$ is

$$(\mathbf{U} \cdot \nabla)\mathbf{V}$$

The *distributive* law holds for finding a gradient. Thus if S and T are scalar functions

$$\nabla(S + T) = \nabla S + \nabla T$$

The *associative* law becomes the rule for differentiating a product:

$$\nabla(ST) = S\nabla T + T\nabla S$$

If $\mathbf{V}$ is a vector function with the magnitudes of the components parallel to the three coordinate axes V_x, V_y, V_z, then

$$\nabla \cdot \mathbf{V} \equiv \text{div } \mathbf{V} \equiv \frac{\partial V_x}{\partial x} + \frac{\partial V_y}{\partial y} + \frac{\partial V_z}{\partial z}$$

The divergence obeys the distributive law. Thus, if **V** and **U** are vector functions, then

$$\nabla \cdot (\mathbf{V} + \mathbf{U}) = \nabla \cdot \mathbf{V} + \nabla \cdot \mathbf{U}$$
$$\nabla \cdot (S\mathbf{V}) = (\nabla S) \cdot \mathbf{V} + S(\nabla \cdot \mathbf{V})$$
$$\nabla \cdot (\mathbf{U} \times \mathbf{V}) = \mathbf{V} \cdot (\nabla \times \mathbf{U}) - \mathbf{U} \cdot (\nabla \times \mathbf{V})$$

As with the gradient of a scalar, the divergence of a vector is invariant under a transformation from one set of rectangular coordinates to another.

$$\nabla \times \mathbf{V} \equiv \text{curl}\,\mathbf{V} \qquad (\text{ sometimes } \nabla \wedge \mathbf{V} \text{ or } \text{rot}\,\mathbf{V})$$

$$\equiv \left(\frac{\partial V_x}{\partial y} - \frac{\partial V_y}{\partial z} \right)\mathbf{i} + \left(\frac{\partial V_x}{\partial z} - \frac{\partial V_z}{\partial x} \right)\mathbf{j} + \left(\frac{\partial V_y}{\partial x} - \frac{\partial V_x}{\partial y} \right)\mathbf{k}$$

$$= \begin{vmatrix} \mathbf{i} & \mathbf{j} & \mathbf{k} \\ \frac{\partial}{\partial x} & \frac{\partial}{\partial y} & \frac{\partial}{\partial z} \\ V_x & V_y & V_z \end{vmatrix}$$

The *curl* (or *rotation*) of a vector is a vector that is invariant under a transformation from one set of rectangular coordinates to another.

$$\nabla \times (\mathbf{U} + \mathbf{V}) = \nabla \times \mathbf{U} + \nabla \times \mathbf{V}$$
$$\nabla \times (S\mathbf{V}) = (\nabla S) \times \mathbf{V} + S(\nabla \times \mathbf{V})$$
$$\nabla \times (\mathbf{U} \times \mathbf{V}) = (\mathbf{V} \cdot \nabla)\mathbf{U} - (\mathbf{U} \cdot \nabla)\mathbf{V} + \mathbf{U}(\nabla \cdot \mathbf{V}) - \mathbf{V}(\nabla \cdot \mathbf{U})$$

If $\mathbf{V} = V_x\mathbf{i} + V_y\mathbf{j} + V_z\mathbf{k}$, then

$$\nabla \cdot \mathbf{V} = \nabla V_x \cdot \mathbf{i} + \nabla V_y \cdot \mathbf{j} + \nabla V_z \cdot \mathbf{k}$$
$$\text{and} \quad \nabla \times \mathbf{V} = \nabla V_x \times \mathbf{i} + \nabla V_y \times \mathbf{j} + \nabla V_z \times \mathbf{k}$$

The operator ∇ can be used more than once. The possibilities where ∇ is used twice are:

$$\nabla \cdot (\nabla \theta) \equiv \text{div}\,\text{grad}\,\theta$$
$$\nabla \times (\nabla \theta) \equiv \text{curl}\,\text{grad}\,\theta$$
$$\nabla(\nabla \cdot \mathbf{V}) \equiv \text{grad}\,\text{div}\,\mathbf{V}$$
$$\nabla \cdot (\nabla \times \mathbf{V}) \equiv \text{div}\,\text{curl}\,\mathbf{V}$$
$$\nabla \times (\nabla \times \mathbf{V}) \equiv \text{curl}\,\text{curl}\,\mathbf{V}$$

Thus, if S is a scalar and **V** is a vector:

$$\text{div}\,\text{grad}\,S \equiv \nabla \cdot (\nabla S) \equiv \text{Laplacian}\,S \equiv \nabla^2 S \equiv \frac{\partial^2 S}{\partial x^2} + \frac{\partial^2 S}{\partial y^2} + \frac{\partial^2 S}{\partial z^2}$$

$$\text{curl}\,\text{grad}\,S \equiv 0$$

$$\text{curl}\,\text{curl}\,\mathbf{V} \equiv \text{grad}\,\text{div}\,\mathbf{V} - \nabla^2\mathbf{V};$$

$$\text{div}\,\text{curl}\,\mathbf{V} \equiv 0$$

Taylor expansion in three dimensions can be written

$$f(\mathbf{V} + \varepsilon) = e^{\varepsilon \cdot \nabla} f(\mathbf{V}) \qquad \text{where} \quad \mathbf{V} = x\mathbf{i} + y\mathbf{j} + z\mathbf{k}$$
$$\text{and} \quad \varepsilon = h\mathbf{i} + l\mathbf{j} + m\mathbf{k}$$

ORTHOGONAL CURVILINEAR COORDINATES

If at a point P there exist three uniform point functions u, v and w so that the surfaces $u = \text{const.}$, $v = \text{const.}$, and $w = \text{const.}$, intersect in three distinct curves through P, then the surfaces are called the *coordinate surfaces* through P. The three lines of intersection are referred to as the *coordinate lines* and their tangents a, b, and c as the *coordinate axes*. When the coordinate axes form an orthogonal set the system is said to define *orthogonal curvilinear coordinates* at P.

Consider an infinitesimal volume enclosed by the surfaces u, v, w, $u + du$, $v + dv$, and $w + dw$ (Figure 5).

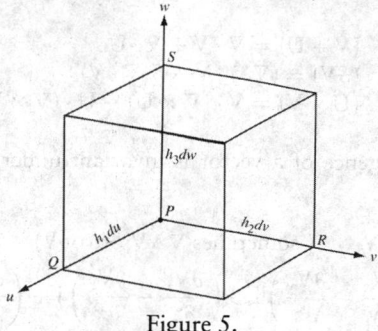

Figure 5.

The surface $PRS \equiv u = $ constant, and the face of the curvilinear figure immediately opposite this is $u + du = $ constant, etc. In terms of these surface constants

$$P = P(u, v, w)$$
$$Q = Q(u + du, v, w) \quad \text{and} \quad PQ = h_1 \, du$$
$$R = R(u, v + dv, w) \quad \text{and} \quad PR = h_2 \, dv$$
$$S = S(u, v, w + dw) \quad \text{and} \quad PS = h_3 \, dw$$

where h_1, h_2, and h_3 are functions of u, v, and w.

- In rectangular Cartesians $\mathbf{i}, \mathbf{j}, \mathbf{k}$

$$h_1 = 1, \qquad h_2 = 1, \qquad h_3 = 1.$$

$$\frac{\hat{\mathbf{a}}}{h_1} \frac{\partial}{\partial u} = \hat{\mathbf{i}} \frac{\partial}{\partial x}, \qquad \frac{\hat{\mathbf{b}}}{h_2} \frac{\partial}{\partial v} = \hat{\mathbf{j}} \frac{\partial}{\partial y}, \qquad \frac{\hat{\mathbf{c}}}{h_3} \frac{\partial}{\partial w} = \hat{\mathbf{k}} \frac{\partial}{\partial z}.$$

- In cylindrical Cartesians $\hat{\mathbf{r}}, \hat{\theta}, \hat{\mathbf{z}}$

$$h_1 = 1, \qquad h_2 = r, \qquad h_3 = 1.$$

$$\frac{\hat{\mathbf{a}}}{h_1} \frac{\partial}{\partial u} = \hat{\mathbf{r}} \frac{\partial}{\partial r}, \qquad \frac{\hat{\mathbf{b}}}{h_2} \frac{\partial}{\partial v} = \frac{\hat{\theta}}{r} \frac{\partial}{\partial \theta}, \qquad \frac{\hat{\mathbf{c}}}{h_3} \frac{\partial}{\partial w} = \hat{\mathbf{z}} \frac{\partial}{\partial z}.$$

- In spherical coordinates $\hat{\mathbf{r}}, \hat{\theta}, \hat{\psi}$

$$h_1 = 1, \qquad h_2 = r, \qquad h_3 = r \sin \theta$$

$$\frac{\hat{\mathbf{a}}}{h_1} \frac{\partial}{\partial u} = \hat{\mathbf{r}} \frac{\partial}{\partial r}, \qquad \frac{\hat{\mathbf{b}}}{h_2} \frac{\partial}{\partial v} = \frac{\hat{\theta}}{r} \frac{\partial}{\partial \theta}, \qquad \frac{\hat{\mathbf{c}}}{h_3} \frac{\partial}{\partial w} = \frac{\hat{\psi}}{r \sin \theta} \frac{\partial}{\partial \psi}$$

The general expressions for grad, div and curl together with those for ∇^2 and the directional derivative are, in orthogonal curvilinear coordinates, given by:

$$\nabla S = \frac{\hat{\mathbf{a}}}{h_1} \frac{\partial S}{\partial u} + \frac{\hat{\mathbf{b}}}{h_2} \frac{\partial S}{\partial v} + \frac{\hat{\mathbf{c}}}{h_3} \frac{\partial S}{\partial w}$$

$$(\mathbf{V} \cdot \nabla) S = \frac{V_1}{h_1} \frac{\partial S}{\partial u} + \frac{V_2}{h_2} \frac{\partial S}{\partial v} + \frac{V_3}{h_3} \frac{\partial S}{\partial w}$$

$$\nabla \cdot \mathbf{V} = \frac{1}{h_1 h_2 h_3} \left\{ \frac{\partial}{\partial u}(h_2 h_3 V_1) + \frac{\partial}{\partial v}(h_3 h_1 V_2) + \frac{\partial}{\partial w}(h_1 h_2 V_3) \right\}.$$

$$\nabla \times \mathbf{V} = \frac{\hat{\mathbf{a}}}{h_2 h_3} \left\{ \frac{\partial}{\partial v}(h_3 V_3) - \frac{\partial}{\partial w}(h_2 V_2) \right\} + \frac{\hat{\mathbf{b}}}{h_3 h_1} \left\{ \frac{\partial}{\partial w}(h_1 V_1) - \frac{\partial}{\partial u}(h_3 V_3) \right\}$$

$$+ \frac{\hat{\mathbf{c}}}{h_1 h_2} \left\{ \frac{\partial}{\partial u}(h_2 V_2) - \frac{\partial}{\partial v}(h_1 V_1) \right\}$$

$$\nabla^2 S = \frac{1}{h_1 h_2 h_3} \left\{ \frac{\partial}{\partial u} \left(\frac{h_2 h_3}{h_1} \frac{\partial S}{\partial u} \right) + \frac{\partial}{\partial v} \left(\frac{h_3 h_1}{h_2} \frac{\partial S}{\partial v} \right) + \frac{\partial}{\partial w} \left(\frac{h_1 h_2}{h_3} \frac{\partial S}{\partial w} \right) \right\}$$

Formulas of Vector Analysis

	Rectangular coordinates	Cylindrical coordinates	Spherical coordinates
Conversion to rectangular coordinates		$x = r\cos\varphi \quad y = r\sin\varphi \quad z = z$	$x = r\cos\varphi\sin\theta \quad y = r\sin\varphi\sin\theta$ $z = r\cos\theta$
Gradient...	$\nabla\phi = \frac{\partial\phi}{\partial x}\mathbf{i} + \frac{\partial\phi}{\partial y}\mathbf{j} + \frac{\partial\phi}{\partial z}\mathbf{k}$	$\nabla\phi = \frac{\partial\phi}{\partial r}\mathbf{r} + \frac{1}{r}\frac{\partial\phi}{\partial\varphi}\Phi + \frac{\partial\phi}{\partial z}\mathbf{k}$	$\nabla\phi = \frac{\partial\phi}{\partial r}\mathbf{r} + \frac{1}{r}\frac{\partial\phi}{\partial\theta}\theta + \frac{1}{r\sin\theta}\frac{\partial\phi}{\partial\varphi}\Phi$
Divergence...	$\nabla\cdot\mathbf{A} = \frac{\partial A_x}{\partial x} + \frac{\partial A_y}{\partial y} + \frac{\partial A_z}{\partial z}$	$\nabla\cdot\mathbf{A} = \frac{1}{r}\frac{\partial(r A_r)}{\partial r} + \frac{1}{r}\frac{\partial A_\varphi}{\partial\varphi}$ $+ \frac{\partial A_z}{\partial z}$	$\nabla\cdot\mathbf{A} = \frac{1}{r^2}\frac{\partial(r^2 A_r)}{\partial r} + \frac{1}{r\sin\theta}\frac{\partial(A_\theta\sin\theta)}{\partial\theta}$ $+ \frac{1}{r\sin\theta}\frac{\partial A_\varphi}{\partial\varphi}$
Curl...	$\nabla\times\mathbf{A} = \begin{vmatrix} \mathbf{i} & \mathbf{j} & \mathbf{k} \\ \frac{\partial}{\partial x} & \frac{\partial}{\partial y} & \frac{\partial}{\partial z} \\ A_x & A_y & A_z \end{vmatrix}$	$\nabla\times\mathbf{A} = \begin{vmatrix} \frac{1}{r}\mathbf{r} & \Phi & \frac{1}{r}\mathbf{k} \\ \frac{\partial}{\partial r} & \frac{\partial}{\partial\varphi} & \frac{\partial}{\partial z} \\ A_r & r A_\varphi & A_z \end{vmatrix}$	$\nabla\times\mathbf{A} = \begin{vmatrix} \frac{\mathbf{r}}{r^2\sin\theta} & \frac{\theta}{r\sin\theta} & \frac{\Phi}{r} \\ \frac{\partial}{\partial r} & \frac{\partial}{\partial\theta} & \frac{\partial}{\partial\varphi} \\ A_r & r A_\theta & r A_\varphi\sin\theta \end{vmatrix}$
Laplacian...	$\nabla^2\phi = \frac{\partial^2\phi}{\partial x^2} + \frac{\partial^2\phi}{\partial y^2} + \frac{\partial^2\phi}{\partial z^2}$	$\nabla^2\phi = \frac{1}{r}\frac{\partial}{\partial r}\left(r\frac{\partial\phi}{\partial r}\right) + \frac{1}{r^2}\frac{\partial^2\phi}{\partial\varphi^2}$ $+ \frac{\partial^2\phi}{\partial z^2}$	$\nabla^2\phi = \frac{1}{r^2}\frac{\partial}{\partial r}\left(r^2\frac{\partial\phi}{\partial r}\right) + \frac{1}{r^2\sin\theta}\frac{\partial}{\partial\theta}\left(\sin\theta\frac{\partial\phi}{\partial\theta}\right)$ $+ \frac{1}{r^2\sin^2\theta}\frac{\partial^2\phi}{\partial\varphi^2}$

TRANSFORMATION OF INTEGRALS

If

1. s is the distance along a curve "C" in space and is measured from some fixed point.
2. S is a surface area
3. V is a volume contained by a specified surface
4. $\hat{\mathbf{t}}$ = the unit tangent to C at the point
5. $P\,\hat{\mathbf{n}}$ = the unit outward pointing normal
6. F is some vector function
7. ds is the vector element of curve $(= \hat{\mathbf{t}}\,ds)$
8. dS is the vector element of surface $(= \hat{\mathbf{n}}\,dS)$

then

$$\int_{(c)} \mathbf{F}\cdot\hat{\mathbf{t}}\,ds = \int_{(c)} \mathbf{F}$$

and when $\mathbf{F} = \nabla\phi$

$$\int_C (\nabla\phi)\cdot\hat{\mathbf{t}}\,ds = \int_C d\phi$$

Gauss' Theorem

When S defines a closed region having a volume V:

$$\iiint_V (\nabla\cdot\mathbf{F})\,dV = \iint_S \mathbf{F}\cdot(\hat{\mathbf{n}})\,dS = \iint_S \mathbf{F}\cdot dS$$

also

$$\iiint_V (\nabla\phi)\,dV = \iint_S \phi\,\hat{\mathbf{n}}\,dS$$

and

$$\iiint_V (\nabla\times\mathbf{F})\,dV = \iint_S (\hat{\mathbf{n}}\times\mathbf{F})\,dS$$

Stokes' Theorem

When C is closed and bounds the open surface S:

$$\iint_S \hat{\mathbf{n}}\cdot(\nabla\times\mathbf{F})\,dS = \int_C \mathbf{F}\cdot d\mathbf{s}$$

also

$$\iint_S (\hat{\mathbf{n}}\times\nabla\phi)\,dS = \int_{(c)} \phi\,d\mathbf{s}$$

Green's Theorem

$$\iint_S (\nabla\phi\cdot\nabla\theta)\,dS = \iint_S \phi\,\hat{\mathbf{n}}\cdot(\nabla\theta)\,dS = \iiint_V \phi(\nabla^2\theta)\,dV$$

$$= \iint_S \theta\cdot\hat{\mathbf{n}}(\nabla\phi)\,dS = \iiint_V \phi(\nabla^2\theta)\,dV$$

BESSEL FUNCTIONS

1. Bessel's differential equation for a real variable x is

$$x^2 \frac{d^2 y}{dx^2} + x \frac{dy}{dx} + (x^2 - n^2) y = 0$$

2. When n is not an integer, two independent solutions of the equation are $J_n(x)$ and $J_{-n}(x)$ where

$$J_n(x) = \sum_{k=0}^{\infty} \frac{(-1)^k}{k! \Gamma(n+k+1)} \left(\frac{x}{2}\right)^{n+2k}$$

3. If n is an integer, then $J_n(x) = (-1)^n J_n(x)$, where

$$J_n(x) = \frac{x^n}{2^n n!} \left\{ 1 - \frac{x^2}{2^2 \cdot 1!(n+1)} + \frac{x^4}{2^4 \cdot 2!(n+1)(n+2)} + \frac{x^6}{2^6 \cdot 3!(n+1)(n+2)(n+3)} + \cdots \right\}$$

4. For $n = 0$ and $n = 1$, this formula becomes

$$J_0(x) = 1 - \frac{x^2}{2^2(1!)^2} + \frac{x^4}{2^4(2!)^2} - \frac{x^6}{2^6(3!)^2} + \frac{x^8}{2^8(4!)^2} - \cdots$$

$$J_1(x) = \frac{x}{2} - \frac{x^3}{2^3 \cdot 1!2!} + \frac{x^5}{2^5 \cdot 2!3!} - \frac{x^7}{2^7 \cdot 3!4!} + \frac{x^9}{2^9 \cdot 4!5!} - \cdots$$

5. When x is large and positive, the following asymptotic series may be used

$$J_0(x) = \left(\frac{2}{\pi x}\right)^{\frac{1}{2}} \left\{ P_0(x) \cos\left(x - \frac{\pi}{4}\right) - Q_0(x) \sin\left(x - \frac{\pi}{4}\right) \right\}$$

$$J_1(x) = \left(\frac{2}{\pi x}\right)^{\frac{1}{2}} \left\{ P_1(x) \cos\left(x - \frac{3\pi}{4}\right) - Q_1(x) \sin\left(x - \frac{3\pi}{4}\right) \right\}$$

where

$$P_0(x) \sim 1 - \frac{1^2 \cdot 3^2}{2!(8x)^2} + \frac{1^2 \cdot 3^2 \cdot 5^2 \cdot 7^2}{4!(8x)^4} - \frac{1^2 \cdot 3^2 \cdot 5^2 \cdot 7^2 \cdot 9^2 \cdot 11^2}{6!(8x)^6} + \cdots$$

$$Q_0(x) \sim -\frac{1^2}{1!8x} + \frac{1^2 \cdot 3^2 \cdot 5^2}{3!(8x)^3} - \frac{1^2 \cdot 3^2 \cdot 5^2 \cdot 7^2 \cdot 9^2}{5!(8x)^5} + - \cdots$$

$$P_1(x) \sim 1 + \frac{1^2 \cdot 3 \cdot 5}{2!(8x)^2} - \frac{1^2 \cdot 3^2 \cdot 5^2 \cdot 7 \cdot 9}{4!(8x)^4} + \frac{1^2 \cdot 3^2 \cdot 5^2 \cdot 7^2 \cdot 9^2 \cdot 11 \cdot 13}{6!(8x)^6} - + \cdots$$

$$Q_1(x) \sim \frac{1 \cdot 3}{1!8x} - \frac{1^2 \cdot 3^2 \cdot 5 \cdot 7}{3!(8x)^3} + \frac{1^2 \cdot 3^2 \cdot 5^2 \cdot 7^2 \cdot 9 \cdot 11}{5!(8x)^5} - \cdots$$

[In $P_1(x)$ the signs alternate from + to − after the first term]

6. The zeros of $J_0(x)$ and $J_1(x)$.

If j_{0s} and j_{1s} are the sth zeros of $J_0(x)$ and $J_1(x)$, respectively, and if $a = 4_s - 1$, $b = 4_s + 1$

$$j_{0,s} \sim \frac{1}{4}\pi a \left\{ 1 + \frac{2}{\pi^2 a^2} - \frac{62}{3\pi^4 a^4} + \frac{15,116}{15\pi^6 a^6} - \frac{12,554,474}{105\pi^8 a^8} + \frac{8,368,654,292}{315\pi^{10}a^{10}} - + \cdots \right\}$$

$$j_{1,s} \sim \frac{1}{4}\pi b \left\{ 1 - \frac{6}{\pi^2 b^2} + \frac{6}{\pi^4 b^4} - \frac{4716}{5\pi^6 b^6} + \frac{3,902,418}{35\pi^8 b^8} - \frac{895,167,324}{35\pi^{10}b^{10}} + \cdots \right\}$$

$$J_1(j_{0,s}) \sim \frac{(-1)^{s+1}2^{\frac{3}{2}}}{\pi a^{\frac{1}{2}}} \left\{ 1 - \frac{56}{3\pi^4 a^4} + \frac{9664}{5\pi^6 a^6} - \frac{7,381,280}{21\pi^8 a^8} + \cdots \right\}$$

$$J_0(j_{1,s}) \sim \frac{(-1)^s 2^{\frac{3}{2}}}{\pi b^{\frac{1}{2}}} \left\{ 1 + \frac{24}{\pi^4 b^4} - \frac{19,584}{10\pi^6 b^6} + \frac{2,466,720}{7\pi^8 b^8} - \cdots \right\}$$

7. Table of zeros for $J_0(x)$ and $J_1(x)$

Define $\{\alpha_n, \beta_n\}$ by $J_0(\alpha_n) = 0$ and $J_1(\beta_n) = 0$.

Roots α_n	$J_1(\alpha_n)$	Roots β_n	$J_0(\beta_n)$
2.4048	0.5191	0.0000	1.0000
5.5201	−0.3403	3.8317	−0.4028
8.6537	0.2715	7.0156	0.3001
11.7915	−0.2325	10.1735	−0.2497
14.9309	0.2065	13.3237	0.2184
18.0711	−0.1877	16.4706	−0.1965
21.2116	0.1733	19.6159	0.1801

8. Recurrence formulas

$$J_{n-1}(x) + J_{n+1}(x) = \frac{2n}{x} J_n(x) \qquad nJ_n(x) + xJ_n'(x) = xJ_{n-1}(x)$$
$$J_{n-1}(x) - J_{n+1}(x) = 2J_n'(x) \qquad nJ_n(x) - xJ_n'(x) = xJ_{n+1}(x)$$

9. If J_n is written for $J_n(x)$ and $J_n^{(k)}$ is written for $\frac{d^k}{dx^k}\{J_n(x)\}$, then the following derivative relationships are important

$$J_0^{(r)} = -J_1^{(r-1)}$$
$$J_0^{(2)} = -J_0 + \frac{1}{x}J_1 = \frac{1}{2}(J_2 - J_0)$$
$$J_0^{(3)} = \frac{1}{x}J_0 + \left(1 - \frac{2}{x^2}\right)J_1 = \frac{1}{4}(-J_3 + 3J_1)$$
$$J_0^{(4)} = \left(1 - \frac{3}{x^2}\right)J_0 - \left(\frac{2}{x} - \frac{6}{x^3}\right)J_1 = \frac{1}{8}(J_4 - 4J_2 + 3J_0), \text{ etc.}$$

10. Half-order Bessel functions

$$J_{\frac{1}{2}}(x) = \sqrt{\frac{2}{\pi x}} \sin x$$
$$J_{-\frac{1}{2}}(x) = \sqrt{\frac{2}{\pi x}} \cos x$$
$$J_{n+\frac{3}{2}}(x) = -x^{n+\frac{1}{2}} \frac{d}{dx}\{x^{-(n+\frac{1}{2})} J_{n+\frac{1}{2}}(x)\}$$
$$J_{n-\frac{1}{2}}(x) = x^{-(n+\frac{1}{2})} \frac{d}{dx}\{x^{n+\frac{1}{2}} J_{n+\frac{1}{2}}(x)\}$$

n	$\left(\frac{\pi x}{2}\right)^{\frac{1}{2}} J_{n+\frac{1}{2}}(x)$	$\left(\frac{\pi x}{2}\right)^{\frac{1}{2}} J_{-(n+\frac{1}{2})}(x)$
0	$\sin x$	$\cos x$
1	$\frac{\sin x}{x} - \cos x$	$-\frac{\cos x}{x} - \sin x$
2	$\left(\frac{3}{x^2} - 1\right)\sin x - \frac{3}{x}\cos x$	$\left(\frac{3}{x^2} - 1\right)\cos x + \frac{3}{x}\sin x$
3	$\left(\frac{15}{x^3} - \frac{6}{x}\right)\sin x - \left(\frac{15}{x^2} - 1\right)\cos x$ etc.	$-\left(\frac{15}{x^3} - \frac{6}{x}\right)\cos x - \left(\frac{15}{x^2} - 1\right)\sin x$

11. Additional solutions to Bessel's equation are

$Y_n(x)$ (also called Weber's function, and sometimes denoted by $N_n(x)$)

$H_n^{(1)}(x)$ and $H_n^{(2)}(x)$ (also called Hankel functions)

These solutions are defined as follows

$$Y_n(x) = \begin{cases} \dfrac{J_n(x)\cos(n\pi) - J_{-n}(x)}{\sin(n\pi)} & n \text{ not an integer} \\ \lim\limits_{\nu \to n} \dfrac{J_\nu(x)\cos(\nu\pi) - J_{-\nu}(x)}{\sin(\nu\pi)} & n \text{ an integer} \end{cases}$$

$$H_n^{(1)}(x) = J_n(x) + iY_n(x)$$
$$H_n^{(2)}(x) = J_n(x) - iY_n(x)$$

The additional properties of these functions may all be derived from the above relations and the known properties of $J_n(x)$.

12. Complete solutions to Bessel's equation may be written as

$$c_1 J_n(x) + c_2 J_{-n}(x) \qquad \text{if } n \text{ is not an integer}$$

or, for any value of n,

$$c_1 J_n(x) + c_2 Y_n(x) \qquad \text{or} \qquad c_1 H_n^{(1)} x + c_2 H_n^{(2)}(x)$$

13. The modified (or hyperbolic) Bessel's differential equation is

$$x^2 \frac{d^2 y}{dx^2} + x \frac{dy}{dx} - (x^2 + n^2)y = 0$$

14. When n is not an integer, two independent solutions of the equation are $I_n(x)$ and $I_{-n}(x)$, where

$$I_n(x) = \sum_{k=0}^{\infty} \frac{1}{k! \Gamma(n+k+1)} \left(\frac{x}{2}\right)^{n+2k}$$

15. If n is an integer,

$$I_n(x) = I_{-n}(x) = \frac{x^n}{2^n n!} \left\{ 1 + \frac{x^2}{2^2 \cdot 1!(n+1)} + \frac{x^4}{2^4 \cdot 2!(n+1)(n+2)} \right.$$
$$\left. + \frac{x^6}{2^6 \cdot 3!(n+1)(n+2)(n+3)} + \cdots \right\}$$

16. For $n = 0$ and $n = 1$, this formula becomes

$$I_0(x) = 1 + \frac{x^2}{2^2(1!)^2} + \frac{x^4}{2^4(2!)^2} + \frac{x^6}{2^6(3!)^2} + \frac{x^8}{2^8(4!)^2} + \cdots$$

$$I_1(x) = \frac{x}{2} + \frac{x^3}{2^3 \cdot 1!2!} + \frac{x^5}{2^5 \cdot 2!3!} + \frac{x^7}{2^7 \cdot 3!4!} + \frac{x^9}{2^9 \cdot 4!5!} + \cdots$$

17. Another solution to the modified Bessel's equation is

$$K_n(x) = \begin{cases} \frac{1}{2}\pi \frac{I_{-n}(x) - I_n(x)}{\sin(n\pi)} & n \text{ not an integer} \\ \lim_{\nu \to n} \frac{1}{2}\pi \frac{I_{-\nu}(x) - I_\nu(x)}{\sin(\nu\pi)} & n \text{ an integer} \end{cases}$$

This function is linearly independent of $I_n(x)$ for all values of n. Thus the complete solution to the modified Bessel's equation may be written as

$$c_1 I_n(x) + c_2 I_{-n}(x) \quad n \text{ not an integer}$$

or

$$c_1 I_n(x) + c_2 K_n(x) \quad \text{any } n$$

18. The following relations hold among the various Bessel functions:

$$I_n(z) = i^{-m} J_m(iz)$$
$$Y_n(iz) = (i)^{n+1} I_n(z) - \frac{2}{\pi} i^{-n} K_n(z)$$

Most of the properties of the modified Bessel function may be deduced from the known properties of $J_n(x)$ by use of these relations and those previously given.

19. Recurrence formulas

$$I_{n-1}(x) - I_{n+1}(x) = \frac{2n}{x} I_n(x) \quad I_{n-1}(x) + I_{n+1}(x) = 2 I_n'(x)$$
$$I_{n-1}(x) - \frac{n}{x} I_n(x) = I_n'(x) \quad I_n'(x) = I_{n+1}(x) + \frac{n}{x} I_n(z)$$

THE FACTORIAL FUNCTION

For non-negative integers n, the factorial of n, denoted $n!$, is the product of all positive integers less than or equal to n; $n! = n \cdot (n-1) \cdot (n-2) \cdots 2 \cdot 1$. If n is a negative integer ($n = -1, -2, \dots$), then $n! = \pm\infty$.

Approximations to $n!$ for large n include Stirling's formula

$$n! \approx \sqrt{2\pi e} \left(\frac{n}{e}\right)^{n+\frac{1}{2}},$$

and Burnsides's formula

$$n! \approx \sqrt{2\pi} \left(\frac{n+\frac{1}{2}}{e}\right)^{n+\frac{1}{2}}.$$

n	$n!$	$\log_{10} n!$	n	$n!$	$\log_{10} n!$
0	1	0.00000	1	1	0.00000
2	2	0.30103	3	6	0.77815
4	24	1.38021	5	120	2.07918
6	720	2.85733	7	5040	3.70243
8	40320	4.60552	9	3.6288×10^5	5.55976
10	3.6288×10^6	6.55976	11	3.9917×10^7	7.60116
12	4.7900×10^8	8.68034	13	6.2270×10^9	9.79428
14	8.7178×10^{10}	10.94041	15	1.3077×10^{12}	12.11650
16	2.0923×10^{13}	13.32062	17	3.5569×10^{14}	14.55107
18	6.4024×10^{15}	15.80634	19	1.2165×10^{17}	17.08509
20	2.4329×10^{18}	18.38612	25	1.5511×10^{25}	25.19065
30	2.6525×10^{32}	32.42366	40	8.1592×10^{47}	47.91165
50	3.0414×10^{64}	64.48307	60	8.3210×10^{81}	81.92017
70	1.1979×10^{100}	100.07841	80	7.1569×10^{118}	118.85473
90	1.4857×10^{138}	138.17194	100	9.3326×10^{157}	157.97000
110	1.5882×10^{178}	178.20092	120	6.6895×10^{198}	198.82539
130	6.4669×10^{219}	219.81069	150	5.7134×10^{262}	262.75689
500	1.2201×10^{1134}	1134.0864	1000	4.0239×10^{2567}	2567.6046

THE GAMMA FUNCTION

Definition: $\Gamma(n) = \int_0^\infty t^{n-1} e^{-t}\, dt \quad n > 0$

Recursion Formula: $\Gamma(n+1) = n\Gamma(n)$

$\Gamma(n+1) = n!$ if $n = 0, 1, 2, \ldots$ where $0! = 1$

For $n < 0$ the gamma function can be defined by using

$\Gamma(n) = \frac{\Gamma(n+1)}{n}$

Graph:

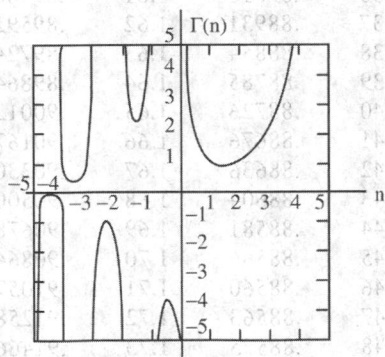

Special Values: $\Gamma(1/2) = \sqrt{\pi}$

$$\Gamma(m + 1/2) = \frac{1 \cdot 3 \cdot 5 \cdots (2m-1)}{2^m} \sqrt{\pi} \quad m = 1, 2, 3, \ldots$$

$$\Gamma(-m + 1/2) = \frac{(-1)^m 2^m \sqrt{\pi}}{1 \cdot 3 \cdot 5 \cdots (2m-1)} \quad m = 1, 2, 3, \ldots$$

Definition:

$$\Gamma(x+1) = \lim_{k \to \infty} \frac{1 \cdot 2 \cdot 3 \cdots k}{(x+1)(x+2) \cdots (x+k)} k^x$$

$$\frac{1}{\Gamma(x)} = x e^{\gamma x} \prod_{m=1}^\infty \left\{ \left(1 + \frac{x}{m}\right) e^{-x/m} \right\}$$

This is an infinite product representation for the gamma function where γ is Euler's constant.

Properties:

$$\Gamma'(1) = \int_0^\infty e^{\gamma x} \ln x \, dx = -\gamma$$

$$\frac{\Gamma'(x)}{\Gamma(x)} = -\gamma + \left(\frac{1}{1} - \frac{1}{x}\right) + \left(\frac{1}{2} - \frac{1}{x+1}\right) + \cdots + \left(\frac{1}{n} - \frac{1}{x+n-1}\right) + \cdots$$

$$\Gamma(x+1) = \sqrt{2\pi x}\, x^x e^{-x} \left\{1 + \frac{1}{12x} + \frac{1}{288x^2} - \frac{139}{51,840x^3} + \cdots\right\}$$

This is called *Stirling's asymptotic series.*

Values of $\Gamma(n) = \int_0^\infty e^{-x} x^{n-1} \, dx$; $\quad \Gamma(n+1) = n\Gamma(n)$

n	$\Gamma(n)$	n	$\Gamma(n)$	n	$\Gamma(n)$	n	$\Gamma(n)$
1.00	1.00000	1.25	.90640	1.50	.88623	1.75	.91906
1.01	.99433	1.26	.90440	1.51	.88659	1.76	.92137
1.02	.98884	1.27	.90250	1.52	.88704	1.77	.92376
1.03	.98355	1.28	.90072	1.53	.88757	1.78	.92623
1.04	.97844	1.29	.89904	1.54	.88818	1.79	.92877
1.05	.97350	1.30	.89747	1.55	.88887	1.80	.93138
1.06	.96874	1.31	.89600	1.56	.88964	1.81	.93408
1.07	.96415	1.32	.89464	1.57	.89049	1.82	.93685
1.08	.95973	1.33	.89338	1.58	.89142	1.83	.93969
1.09	.95546	1.34	.89222	1.59	.89243	1.84	.94261
1.10	.95135	1.35	.89115	1.60	.89352	1.85	.94561
1.11	.94740	1.36	.89018	1.61	.89468	1.86	.94869
1.12	.94359	1.37	.88931	1.62	.89592	1.87	.95184
1.13	.93993	1.38	.88854	1.63	.89724	1.88	.95507
1.14	.93642	1.39	.88785	1.64	.89864	1.89	.95838
1.15	.93304	1.40	.88726	1.65	.90012	1.90	.96177
1.16	.92980	1.41	.88676	1.66	.90167	1.91	.96523
1.17	.92670	1.42	.88636	1.67	.90330	1.92	.96877
1.18	.92373	1.43	.88604	1.68	.90500	1.93	.97240
1.19	.92089	1.44	.88581	1.69	.90678	1.94	.97610
1.20	.91817	1.45	.88566	1.70	.90864	1.95	.97988
1.21	.91558	1.46	.88560	1.71	.91057	1.96	.98374
1.22	.91311	1.47	.88563	1.72	.91258	1.97	.98768
1.23	.91075	1.48	.88575	1.73	.91466	1.98	.99171
1.24	.90852	1.49	.88595	1.74	.91683	1.99	.99581
						2.00	1.00000

THE BETA FUNCTION

Definition: $B(m, n) = \int_0^1 t^{m-1}(1 - t)^{n-1} dt \quad m > 0, n > 0$

Relationship with Gamma function: $B(m, n) = \dfrac{\Gamma(m)\Gamma(n)}{\Gamma(m + n)}$

Properties:

$$B(m, n) = B(n, m)$$

$$B(m, n) = 2 \int_0^{\pi/2} \sin^{2m-1} \theta \cos^{2n-1} \theta \, d\theta$$

$$B(m, n) = \int_0^\infty \frac{t^{m-1}}{(1+t)^{m+n}} dt$$

$$B(m, n) = r^n(r + 1)^m \int_0^1 \frac{t^{m-1}(1-t)^{n-1}}{(r+t)^{m+n}} dt$$

THE ERROR FUNCTION

Definition: $\operatorname{erf}(x) = \dfrac{2}{\sqrt{\pi}} \displaystyle\int_0^x e^{-t^2}\, dt$

Series: $er f(x) = \dfrac{2}{\sqrt{\pi}} \left(x - \dfrac{x^3}{3} + \dfrac{1}{2!}\dfrac{x^5}{5} - \dfrac{1}{3!}\dfrac{x^7}{7} + \cdots \right)$

Property: $\operatorname{erf}(x) = -\operatorname{erf}(-x)$

Relationship with Normal Probability Function $f(t)$: $\displaystyle\int_0^x f(t)\, dt = \dfrac{1}{2}\operatorname{erf}\left(\dfrac{x}{\sqrt{2}}\right)$ To evaluate $\operatorname{erf}(2.3)$, one proceeds as follows:

For $\frac{x}{\sqrt{2}} = 2.3$, one finds $x = (2.3)(\sqrt{2}) = 3.25$. In the normal probability function table (page A-104), one finds the entry 0.4994 opposite the value 3.25. Thus $\operatorname{erf}(2.3) = 2(0.4994) = 0.9988$.

$$\operatorname{erfc}(z) = 1 - \operatorname{erf}(z) = \dfrac{2}{\sqrt{\pi}} \int_z^\infty e^{-t^2}\, dt$$

is known as the complementary error function.

ORTHOGONAL POLYNOMIALS

I: Legendre

Name: Legendre   *Symbol:* $P_n(x)$ *Interval:* $[-1, 1]$

Differential Equation: $(1 - x^2)y'' - 2xy' + n(n+1)y = 0$

Explicit Expression: $P_n(x) = \dfrac{1}{2^n} \displaystyle\sum_{m=0}^{[n/2]} (-1)^m \binom{n}{m}\binom{2n-2m}{n} x^{n-2m}$

Recurrence Relation: $(n+1)P_{n+1}(x) = (2n+1)xP_n(x) - nP_{n-1}(x)$

Weight: 1

Standardization: $P_n(1) = 1$

Norm: $\displaystyle\int_{-1}^{+1} [P_n(x)]^2\, dx = \dfrac{2}{2n+1}$

Rodrigues' Formula: $P_n(x) = \dfrac{(-1)^n}{2^n n!} \dfrac{d^n}{dx^n}\{(1-x^2)^n\}$

Generating Function: $R^{-1} = \displaystyle\sum_{n=0}^{\infty} P_n(x)z^n; \; -1 < x < 1, \quad |z| \; < 1,$

$$R = \sqrt{1 - 2xz + z^2}$$

Inequality: $|P_n(x)| \leq 1,\; -1 \leq x \leq 1.$

II: Tschebysheff, First Kind

Name: Tschebysheff, First Kind   *Symbol:* $T_n(x)$ *Interval:* $[-1, 1]$

Differential Equation: $(1 - x^2)y - xy' + n^2 y = 0$

Explicit Expression: $\dfrac{n}{2} \displaystyle\sum_{m=0}^{[n/2]} (-1)^m \dfrac{(n-m-1)!}{m!(n-2m)!}(2x)^{n-2m} = \cos(n \arccos x) = T_n(x)$

Recurrence Relation: $T_{n+1}(x) = 2xT_n(x) - T_{n-1}(x)$

Weight: $(1 - x^2)^{-1/2}$

Standardization: $T_n(1) = 1$

Norm: $\int_{-1}^{+1} (1 - x^2)^{-1/2} [T_n(x)]^2 \, dx = \begin{cases} \pi/2, & n \neq 0 \\ \pi, & n = 0 \end{cases}$

Rodrigues' Formula: $\dfrac{(-1)^n (1 - x^2)^{1/2} \sqrt{\pi}}{2^{n+1} \Gamma(n + \frac{1}{2})} \dfrac{d^n}{dx^n} \{(1 - x^2)^{n-(1/2)}\} = T_n(x)$

Generating Function: $\dfrac{1 - xz}{1 - 2xz - z^2} = \sum_{n=0}^{\infty} T_n(x) z^n, \; -1 < x < 1, \quad |z| < 1$

Inequality: $|T_n(x)| \leq 1, \; -1 \leq x \leq 1.$

III: Tschebysheff, Second Kind

Name: Tschebysheff, Second Kind   *Symbol $U_n(x)$*   *Interval:* $[-1, 1]$
Differential Equation: $(1 - x^2) y'' - 3xy' + n(n + 2)y = 0$

Explicit Expression: $U_n(x) = \sum_{m=0}^{[n/2]} (-1)^m \dfrac{(m - n)!}{m!(n - 2m)!} (2x)^{n-2m}$

$$U_n(\cos \theta) = \frac{\sin[(n + 1)\theta]}{\sin \theta}$$

Recurrence Relation: $U_{n+1}(x) = 2xU_n(x) - U_{n-1}(x)$
Weight: $(1 - x^2)^{1/2}$ *Standardization:* $U_n(1) = n + 1$

Norm: $\displaystyle\int_{-1}^{+1} (1 - x^2)^{1/2} [U_n(x)]^2 \, dx = \frac{\pi}{2}$

Rodrigues' Formula: $U_n(x) = \dfrac{(-1)^n (n + 1) \sqrt{\pi}}{(1 - x^2)^{1/2} 2^{n+1} \Gamma(n + \frac{3}{2})} \dfrac{d^n}{dx^n} \{(1 - x^2)^{n+(1/2)}\}$

Generating Function: $\dfrac{1}{1 - 2xz + z^2} = \displaystyle\sum_{n=0}^{\infty} U_n(x) z^n, \; -1 < x < 1, \quad |z| < 1$

Inequality: $|U_n(x)| \leq n + 1, \; -1 \leq x \leq 1.$

IV: Jacobi

Name: Jacobi   *Symbol:* $P_n^{(\alpha, \beta)}(x)$ *Interval:* $[-1, 1]$
Differential Equation: $(1 - x^2) y'' + [\beta - \alpha - (\alpha + \beta + 2)x]y' + n(n + \alpha + \beta + 1)y = 0$

Explicit Expression: $P_n^{(\alpha, \beta)}(x) = \dfrac{1}{2^n} \displaystyle\sum_{m=0}^{n} \binom{n + \alpha}{m} \binom{n + \beta}{n - m} (x - 1)^{n-m} (x + 1)^m$

Recurrence Relation:

$$2(n + 1)(n + \alpha + \beta + 1)(2n + \alpha + \beta) P_{n+1}^{(\alpha, \beta)}(x)$$
$$= (2n + \alpha + \beta + 1)[(\alpha^2 - \beta^2) + (2n + \alpha + \beta + 2)$$
$$\times (2n + \alpha + \beta)x] P_n^{(\alpha, \beta)}(x)$$
$$- 2(n + \alpha)(n + \beta)(2n + \alpha + \beta + 2) P_{n-1}^{(\alpha, \beta)}(x)$$

Weight: $(1 - x)^{\alpha}(1 + x)^{\beta}; \alpha, \beta > 1$
Standardization: $P_n^{(\alpha, \beta)}(x) = \binom{n + \alpha}{n}$

Norm: $\displaystyle\int_{-1}^{+1} (1 - x)^{\alpha}(1 + x)^{\beta} [P_n^{(\alpha, \beta)}(x)]^2 \, dx = \dfrac{2^{\alpha+\beta+1} \Gamma(n + \alpha + 1) \Gamma(n + \beta + 1)}{(2n + \alpha + \beta + 1)n! \Gamma(n + \alpha + \beta + 1)}$

Rodrigues' Formula: $P_n^{(\alpha, \beta)}(x) = \dfrac{(-1)^n}{2^n n! (1 - x)^{\alpha}(1 + x)^{\beta}} \dfrac{d^n}{dx^n} \{(1 - x)^{n+\alpha}(1 + x)^{n+\beta}\}$

Generating Function: $R^{-1}(1 - z + R)^{-\alpha}(1 + z + R)^{-\beta} = \displaystyle\sum_{n=0}^{\infty} 2^{-\alpha-\beta} P_n^{(\alpha, \beta)}(x) z^n,$

$$R = \sqrt{1 - 2xz + z^2}, \quad |z| < 1$$

$$\text{Inequality: } \max_{-1 \le x \le 1} |P_n^{(\alpha,\beta)}(x)| = \begin{cases} \binom{n+q}{n} \sim n^q \text{ if } q = \max(\alpha,\beta) \ge -\frac{1}{2} \\ |P_n^{(\alpha,\beta)}(x')| \sim n^{-1/2} \text{ if } q < -\frac{1}{2} \\ x' \text{ is one of the two maximum points nearest} \\ \frac{\beta-\alpha}{\alpha+\beta+1} \end{cases}$$

V: Generalized Laguerre

Name: Generalized Laguerre   *Symbol*: $L_n^{(\alpha)}(x)$ *Interval*: $[0, \infty]$

Differential Equation: $xy'' + (\alpha + 1 - x)y' + ny = 0$

Explicit Expression: $L_n^{(\alpha)}(x) = \sum_{m=0}^{n} (-1)^m \binom{n+\alpha}{n-m} \frac{1}{m!} x^m$

Recurrence Relation: $(n+1)L_n^{(\alpha)} + 1(x) = [(2n+\alpha+1) - x]L_n^{(\alpha)}(x) - (n+\alpha)L_n^{(\alpha)} - 1(x)$

Weight: $x^\alpha e^{-x}, \alpha > -1$ *Standardization*: $L_n^{(\alpha)}(x) = \frac{(-1)^n}{n!} x^n + \cdots$

Norm: $\int_0^\infty x^\alpha e^{-x} [L_n^{(\alpha)}(x)]^2 \, dx = \frac{\Gamma(n+\alpha+1)}{n!}$

Rodrigues' Formula: $L_n^{(\alpha)}(x) = \frac{1}{n! x^\alpha e^{-x}} \frac{d^n}{dx^n} \{x^{n+\alpha} e^{-x}\}$

Generating Function: $(1-z)^{-\alpha-1} \exp\left(\frac{xz}{z-1}\right) = \sum_{n=0}^{\infty} L_n^{(\alpha)}(x) z^n$

Inequality: $|L_n^{(\alpha)}(x)| \le \frac{\Gamma(n+\alpha+1)}{n! \Gamma(\alpha+1)} e^{x/2}; \quad \begin{matrix} x \ge 0 \\ \alpha > 0 \end{matrix}$

$|L_n^{(a)}(x)| \le \left[2 - \frac{\Gamma(\alpha+n+1)}{n! \Gamma(\alpha+1)}\right] e^{x/2}; \quad \begin{matrix} x \ge 0 \\ -1 < \alpha < 0 \end{matrix}$

VI: Hermite

Name: Hermite   *Symbol*: $H_n(x)$ *Interval*: $[-\infty, \infty]$

Differential Equation: $y'' - 2xy' + 2ny = 0$

Explicit Expression: $H_n(x) = \sum_{m=0}^{[n/2]} \frac{(-1)^m n! (2x)^{n-2m}}{m!(n-2m)!}$

Recurrence Relation: $H_{n+1}(x) = 2xH_n(x) - 2nH_{n-1}(x)$

Weight: e^{-x^2} *Standardization*: $H_n(1) = 2^n x^n + \cdots$

Norm: $\int_{-\infty}^{\infty} e^{-x^2} [H_n(x)]^2 \, dx = 2^n n! \sqrt{\pi}$

Rodrigues' Formula: $H_n(x) = (-1)^n e^{x^2} \frac{d^n}{dx^n}(e^{-x^2})$

Generating Function: $e^{-x^2+2zx} = \sum_{n=0}^{\infty} H_n(x) \frac{z^n}{n!}$

Inequality: $|H_n(x)| e^{x^2/2} k 2^{n/2} \sqrt{n!} \, k \approx 1.086435$

TABLES OF ORTHOGONAL POLYNOMIALS

$H_0 = 1$

$H_1 = 2x$

$H_2 = 4x^2 - 2$

$H_3 = 8x^3 - 12x$

$H_4 = 16x^4 - 48x^2 + 12$

$H_5 = 32x^5 - 160x^3 + 120x$

$H_6 = 64x^6 - 480x^4 + 720x^2 - 120$

$H_7 = 128x^7 - 1344x^5 + 3360x^3 - 1680x$

$H_8 = 256x^8 - 3584x^6 + 13440x^4 - 13440x^2 + 1680$

$H_9 = 512x^9 - 9216x^7 + 48384x^5 - 80640x^3 + 30240x$

$H_{10} = 1024x^{10} - 23040x^8 + 161280x^6 - 403200x^4 + 302400x^2 - 30240$

$x^{10} = (30240H_0 + 75600H_2 + 25200H_4 + 2520H_6 + 90H_8 + H_{10})/1024$

$x^9 = (15120H_1 + 10080H_3 + 1512H_5 + 72H_7 + H_9)/512$

$x^8 = (1680H_0 + 3360H_2 + 840H_4 + 56H_6 + H_8)/256$

$x^7 = (840H_1 + 420H_3 + 42H_5 + H_7)/128$

$x^6 = (120H_0 + 180H_2 + 30H_4 + H_6)/64$

$x^5 = (60H_1 + 20H_3 + H_5)/32$

$x^4 = (12H_0 + 12H_2 + H_4)/16$

$x^3 = (6H_1 + H_3)/8$

$x^2 = (2H_0 + H_2)/4$

$x = (H_1)/2$

$1 = H_0$

$L_0 = 1$

$L_1 = -x + 1$

$L_2 = (x^2 - 4x + 2)/2$

$L_3 = (-x^3 + 9x^2 - 18x + 6)/6$

$L_4 = (x^4 - 16x^3 + 72x^2 - 96x + 24)/24$

$L_5 = (-x^5 + 25x^4 - 200x^3 + 600x^2 - 600x + 120)/120$

$L_6 = (x^6 - 36x^5 + 450x^4 - 2400x^3 + 5400x^2 - 4320x + 720)/720$

$x^6 = 720L_0 - 4320L_1 + 10800L_2 - 14400L_3 + 10800L_4 - 4320L_5 + 720L_6$

$x^5 = 120L_0 - 600L_1 + 1200L_2 - 1200L_3 + 600L_4 - 120L_5$

$x^4 = 24L_0 - 96L_1 + 144L_2 - 96L_3 + 24L_4$

$x^3 = 6L_0 - 18L_1 + 18L_2 - 6L_3$

$x^2 = 2L_0 - 4L_1 + 2L_2$

$x = L_0 - L_1$

$1 = L_0$

$P_0 = 1$

$P_1 = x$

$P_2 = (3x^2 - 1)/2$

$P_3 = (5x^3 - 3x)/2$

$P_4 = (35x^4 - 30x^2 + 3)/8$

$P_5 = (63x^5 - 70x^3 + 15x)/8$

$P_6 = (231x^6 - 315x^4 + 105x^2 - 5)/16$

$P_7 = (429x^7 - 693x^5 + 315x^3 - 35x)/16$

$P_8 = (6435x^8 - 12012x^6 + 6930x^4 - 1260x^2 + 35)/128$

$P_9 = (12155x^9 - 25740x^7 + 18018x^5 - 4620x^3 + 315x)/128$

$P_{10} = (46189x^{10} - 109395x^8 + 90090x^6 - 30030x^4 + 3465x^2 - 63)/256$

$x^{10} = (4199P_0 + 16150P_2 + 15504P_4 + 7904P_6 + 2176P_8 + 256P_{10})/46189$

$x^9 = (3315P_1 + 4760P_3 + 2992P_5 + 960P_7 + 128P_9)/12155$

$x^8 = (715P_0 + 2600P_2 + 2160P_4 + 832P_6 + 128P_8)/6435$

$x^7 = (143P_1 + 182P_3 + 88P_5 + 16P_7)/429$

$x^6 = (33P_0 + 110P_2 + 72P_4 + 16P_6)/231$

$x^5 = (27P_1 + 28P_3 + 8P_5)/63$

$x^4 = (7P_0 + 20P_2 + 8P_4)/35$

$x^3 = (3P_1 + 2P_3)/5$

$x^2 = (P_0 + 2P_2)/3$

$x = P_1$

$1 = P_0$

$T_0 = 1$

$T_1 = x$

$T_2 = 2x^2 - 1$

$T_3 = 4x^3 - 3x$

$T_4 = 8x^4 - 8x^2 + 1$

$T_5 = 16x^5 - 20x^3 + 5x$

$T_6 = 32x^6 - 48x^4 + 18x^2 - 1$

$T_7 = 64x^7 - 112x^5 + 56x^3 - 7x$

$T_8 = 128x^8 - 256x^6 + 160x^4 - 32x^2 + 1$

$T_9 = 256x^9 - 576x^7 + 432x^5 - 120x^3 + 9x$

$T_{10} = 512x^{10} - 1280x^8 + 1120x^6 - 400x^4 + 50x^2 - 1$

$x^{10} = (126T_0 + 210T_2 + 120T_4 + 45T_6 + 10T_8 + T_{10})/512$

$x^9 = (126T_1 + 84T_3 + 36T_5 + 9T_7 + T_9)/256$

$x^8 = (35T_0 + 56T_2 + 28T_4 + 8T_6 + T_8)/128$

$x^7 = (35T_1 + 21T_3 + 7T_5 + T_7)/64$

$x^6 = (10T_0 + 15T_2 + 6T_4 + T_6)/32$

$x^5 = (10T_1 + 5T_3 + T_5)/16$

$x^4 = (3T_0 + 4T_2 + T_4)/8$

$x^3 = (3T_1 + T_3)/4$

$x^2 = (T_0 + T_2)/2$

$x = T_1$

$1 = T_0$

$U_0 = 1$

$U_1 = 2x$

$U_2 = 4x^2 - 1$

$U_3 = 8x^3 - 4x$

$U_4 = 16x^4 - 12x^2 + 1$

$U_5 = 32x^5 - 32x^3 + 6x$

$U_6 = 64x^6 - 80x^4 + 24x^2 - 1$

$U_7 = 128x^7 - 192x^5 + 80x^3 - 8x$

$U_8 = 256x^8 - 448x^6 + 240x^4 - 40x^2 + 1$

$U_9 = 512x^9 - 1024x^7 + 672x^5 - 160x^3 + 10x$

$U_{10} = 1024x^{10} - 2304x^8 + 1792x^6 - 560x^4 + 60x^2 - 1$

$x^{10} = (42U_0 + 90U_2 + 75U_4 + 35U_6 + 9U_8 + U_{10})/1024$

$x^9 = (42U_1 + 48U_3 + 27U_5 + 8U_7 + U_9)/512$

$x^8 = (14U_0 + 28U_2 + 20U_4 + 7U_6 + U_8)/256$

$x^7 = (14U_1 + 14U_3 + 6U_5 + U_7)/128$

$x^6 = (5U_0 + 9U_2 + 5U_4 + U_6)/64$

$x^5 = (5U_1 + 4U_3 + U_5)/32$

$x^4 = (2U_0 + 3U_2 + U_4)/16$

$x^3 = (2U_1 + U_3)/8$

$x^2 = (U_0 + U_2)/4$

$x = (U_1)/2$

$1 = U_0$

CLEBSCH–GORDAN COEFFICIENTS

$$\begin{pmatrix} j_1 & j_2 \\ m_1 & m_2 \end{pmatrix} \begin{matrix} j \\ m \end{matrix} = \delta_{m, m_1 + m_2} \sqrt{\frac{(j_1 + j_2 - j)!(j + j_1 - j_2)!(j + j_2 - j_1)!(2j + 1)}{(j + j_1 + j_2 + 1)!}}$$

$$\times \sum_k \frac{(-1)^k \sqrt{(j_1 + m_1)!(j_1 - m_1)!(j_2 + m_2)!(j_2 - m_2)!(j + m)!(j - m)!}}{k!(j_1 + j_2 - j - k)!(j_1 - m_1 - k)!(j_2 + m_2 - k)!(j - j_2 + m_1 + k)!(j - j_1 - m_2 + k)!}.$$

1. Conditions:

   (a) Each of $\{j_1, j_2, j, m_1, m_2, m\}$ may be an integer, or half an integer. Additionally: $j > 0$, $j_1 > 0$, $j_2 > 0$ and $j + j_1 + j_2$ is an integer.
   (b) $j_1 + j_2 - j \geq 0$.
   (c) $j_1 - j_2 + j \geq 0$.
   (d) $-j_1 + j_2 + j \geq 0$.
   (e) $|m_1| \leq j_1$, $|m_2| \leq j_2$, $|m| \leq j$.

2. Special values:

   (a) $\begin{pmatrix} j_1 & j_2 \\ m_1 & m_2 \end{pmatrix} \begin{matrix} j \\ m \end{matrix} = 0$ if $m_1 + m_2 \neq m$.

   (b) $\begin{pmatrix} j_1 & 0 \\ m_1 & 0 \end{pmatrix} \begin{matrix} j \\ m \end{matrix} = \delta_{j_1, j} \delta_{m_1, m}$.

   (c) $\begin{pmatrix} j_1 & j_2 \\ 0 & 0 \end{pmatrix} \begin{matrix} j \\ 0 \end{matrix} = 0$ when $j_1 + j_2 + j$ is an odd integer.

   (d) $\begin{pmatrix} j_1 & j_1 \\ m_1 & m_1 \end{pmatrix} \begin{matrix} j \\ m \end{matrix} = 0$ when $2j_1 + j$ is an odd integer.

3. Symmetry relations: all of the following are equal to $\begin{pmatrix} j_1 & j_2 \\ m_1 & m_2 \end{pmatrix} \begin{matrix} j \\ m \end{matrix}$:

   (a) $\begin{pmatrix} j_2 & j_1 \\ -m_2 & -m_1 \end{pmatrix} \begin{matrix} j \\ -m \end{matrix}$,

   (b) $(-1)^{j_1 + j_2 - j} \begin{pmatrix} j_2 & j_1 \\ m_1 & m_2 \end{pmatrix} \begin{matrix} j \\ m \end{matrix}$,

   (c) $(-1)^{j_1 + j_2 - j} \begin{pmatrix} j_1 & j_2 \\ -m_1 & -m_2 \end{pmatrix} \begin{matrix} j \\ -m \end{matrix}$,

   (d) $\sqrt{\frac{2j + 1}{2j_1 + 1}} (-1)^{j_2 + m_2} \begin{pmatrix} j & j_2 \\ -m & m_2 \end{pmatrix} \begin{matrix} j_1 \\ -m_1 \end{matrix}$,

   (e) $\sqrt{\frac{2j + 1}{2j_1 + 1}} (-1)^{j_1 - m_1 + j - m} \begin{pmatrix} j & j_2 \\ m & -m_2 \end{pmatrix} \begin{matrix} j_1 \\ m_1 \end{matrix}$,

   (f) $\sqrt{\frac{2j + 1}{2j_1 + 1}} (-1)^{j - m + j_1 - m_1} \begin{pmatrix} j_2 & j \\ m_2 & -m \end{pmatrix} \begin{matrix} j_1 \\ -m_1 \end{matrix}$,

   (g) $\sqrt{\frac{2j + 1}{2j_2 + 1}} (-1)^{j_1 - m_1} \begin{pmatrix} j_1 & j \\ m_1 & -m \end{pmatrix} \begin{matrix} j_2 \\ -m_2 \end{matrix}$,

   (h) $\sqrt{\frac{2j + 1}{2j_2 + 1}} (-1)^{j_1 - m_1} \begin{pmatrix} j & j_1 \\ m & -m_1 \end{pmatrix} \begin{matrix} j_2 \\ m_2 \end{matrix}$.

By use of the symmetry relations, Clebsch–Gordan coefficients may be put in the standard form $j_1 \leq j_2 \leq j$ and $m \geq 0$.

m_2	m	j_1	j	$\begin{pmatrix} j_1 & \frac{1}{2} & j \\ m_1 & m_2 & m \end{pmatrix}$	
$-\frac{1}{2}$	0	$\frac{1}{2}$	1	$\frac{\sqrt{2}}{2}$	≈ 0.707107
0	$\frac{1}{2}$	$\frac{1}{2}$	1	$\frac{\sqrt{3}}{2}$	≈ 0.866025
$\frac{1}{2}$	0	$\frac{1}{2}$	1	$\frac{\sqrt{2}}{2}$	≈ 0.707107
$\frac{1}{2}$	$\frac{1}{2}$	$\frac{1}{2}$	1	$\frac{\sqrt{3}}{2}$	≈ 0.866025
$\frac{1}{2}$	1	$\frac{1}{2}$	1	1	≈ 1.000000

m_2	m	j_1	j	$\begin{pmatrix} j_1 & 1 & j \\ m_1 & m_2 & m \end{pmatrix}$	
-1	0	1	1	$\frac{\sqrt{2}}{2}$	≈ 0.707107
-1	0	1	2	$\frac{\sqrt{6}}{6}$	≈ 0.408248
$-\frac{1}{2}$	0	$\frac{1}{2}$	$\frac{3}{2}$	$\frac{\sqrt{2}}{2}$	≈ 0.707107
$-\frac{1}{2}$	$\frac{1}{2}$	1	1	$\frac{3}{4}$	≈ 0.750000
$-\frac{1}{2}$	$\frac{1}{2}$	1	2	$\frac{\sqrt{5}}{4}$	≈ 0.559017
0	0	1	2	$\frac{\sqrt{6}}{3}$	≈ 0.816496
0	0	$\frac{1}{2}$	$\frac{3}{2}$	$\frac{\sqrt{3}}{2}$	≈ 0.866025
0	$\frac{1}{2}$	$\frac{1}{2}$	$\frac{3}{2}$	$\frac{\sqrt{6}}{3}$	≈ 0.8164967
0	$\frac{1}{2}$	1	1	$\frac{\sqrt{2}}{4}$	≈ 0.353553
0	$\frac{1}{2}$	1	2	$\frac{\sqrt{10}}{4}$	≈ 0.790569
0	1	1	1	$\frac{\sqrt{2}}{2}$	≈ 0.707107

m_2	m	j_1	j	$\begin{pmatrix} j_1 & 1 & j \\ m_1 & m_2 & m \end{pmatrix}$	
0	1	1	2	$\frac{\sqrt{2}}{2}$	≈ 0.707107
$\frac{1}{2}$	0	$\frac{1}{2}$	$\frac{3}{2}$	$\frac{\sqrt{2}}{2}$	≈ 0.707107
$\frac{1}{2}$	$\frac{1}{2}$	1	1	$-\frac{\sqrt{2}}{4}$	≈ -0.353553
$\frac{1}{2}$	$\frac{1}{2}$	1	2	$\frac{\sqrt{10}}{4}$	≈ 0.790569
$\frac{1}{2}$	1	$\frac{1}{2}$	$\frac{3}{2}$	$\frac{\sqrt{30}}{6}$	≈ 0.912871
$\frac{1}{2}$	$\frac{3}{2}$	1	2	$\frac{\sqrt{105}}{12}$	≈ 0.853913
1	0	1	1	$-\frac{\sqrt{2}}{2}$	≈ -0.707107
1	0	1	2	$\frac{\sqrt{6}}{6}$	≈ 0.408248
1	$\frac{1}{2}$	$\frac{1}{2}$	$\frac{3}{2}$	$\frac{\sqrt{3}}{3}$	≈ 0.577350
1	$\frac{1}{2}$	1	1	$-\frac{3}{4}$	≈ -0.750000
1	$\frac{1}{2}$	1	2	$\frac{\sqrt{5}}{4}$	≈ 0.559017
1	1	$\frac{1}{2}$	$\frac{3}{2}$	$\frac{\sqrt{10}}{4}$	≈ 0.790569
1	1	1	1	$-\frac{\sqrt{2}}{2}$	≈ -0.707107
1	1	1	2	$\frac{\sqrt{2}}{2}$	≈ 0.707107
1	$\frac{3}{2}$	$\frac{1}{2}$	$\frac{3}{2}$	1	≈ 1.000000
1	$\frac{3}{2}$	1	2	$\frac{\sqrt{105}}{12}$	≈ 0.853913
1	2	1	2	1	≈ 1.000000

NORMAL PROBABILITY FUNCTION

Table of the Normal Distribution

For a standard normal random variable ($\Phi(z)$ is the area under the Standard Normal Curve from $-\infty$ to z).

Limits		Proportion of the total area	Remaining area
$\mu - \lambda\sigma$	$\mu + \lambda\sigma$	(%)	(%)
$\mu - \sigma$	$\mu + \sigma$	68.27	31.73
$\mu - 1.65\sigma$	$\mu + 1.65\sigma$	90	10
$\mu - 1.96\sigma$	$\mu + 1.96\sigma$	95	5
$\mu - 2\sigma$	$\mu + 2\sigma$	95.45	4.55
$\mu - 2.58\sigma$	$\mu + 2.58\sigma$	99.0	0.99
$\mu - 3\sigma$	$\mu + 3\sigma$	99.73	0.27
$\mu - 3.09\sigma$	$\mu + 3.09\sigma$	99.8	0.2
$\mu - 3.29\sigma$	$\mu + 3.29\sigma$	99.9	0.1

x	1.282	1.645	1.960	2.326	2.576	3.090
$\Phi(x)$	0.90	0.95	0.975	0.99	0.995	0.999
$2[1 - \Phi(x)]$	0.20	0.10	0.05	0.02	0.01	0.002

x	3.09	3.72	4.26	4.75	5.20	5.61	6.00	6.36
$1 - \Phi(x)$	10^{-3}	10^{-4}	10^{-5}	10^{-6}	10^{-7}	10^{-8}	10^{-9}	10^{-10}

Areas under the Standard Normal Curve from 0 to z

z	0	1	2	3	4	5	6	7	8	9
0.0	.0000	.0040	.0080	.0120	.0160	.0199	.0239	.0279	.0319	.0359
0.1	.0398	.0438	.0478	.0517	.0557	.0596	.0636	.0675	.0714	.0754
0.2	.0793	.0832	.0871	.0910	.0948	.0987	.1026	.1064	.1103	.1141
0.3	.1179	.1217	.1255	.1293	.1331	.1368	.1406	.1443	.1480	.1517
0.4	.1554	.1591	.1628	.1664	.1700	.1736	.1772	.1808	.1844	.1879
0.5	.1915	.1950	.1985	.2019	.2054	.2088	.2123	.2157	.2190	.2224
0.6	.2258	.2291	.2324	.2357	.2389	.2422	.2454	.2486	.2518	.2549
0.7	.2580	.2612	.2652	.2673	.2704	.2734	.2764	.2794	.2823	.2852
0.8	.2881	.2910	.2939	.2967	.2996	.3023	.3051	.3078	.3106	.3133
0.9	.3159	.3186	.3212	.3238	.3264	.3289	.3315	.3340	.3365	.3389
1.0	.3413	.3438	.3461	.3485	.3508	.3531	.3554	.3577	.3599	.3621
1.1	.3643	.3665	.3686	.3708	.3729	.3749	.3770	.3790	.3810	.3830
1.2	.3849	.3869	.3888	.3907	.3925	.3944	.3962	.3980	.3997	.4015
1.3	.4032	.4049	.4066	.4082	.4099	.4115	.4131	.4147	.4162	.4177
1.4	.4192	.4207	.4222	.4236	.4251	.4265	.4279	.4292	.4306	.4319
1.5	.4332	.4345	.4357	.4370	.4382	.4394	.4406	.4418	.4429	.4441
1.6	.4452	.4463	.4474	.4484	.4495	.4505	.4515	.4525	.4535	.4545
1.7	.4554	.4564	.4573	.4582	.4591	.4599	.4608	.4616	.4625	.4633
1.8	.4641	.4649	.4656	.4664	.4671	.4678	.4686	.4693	.4699	.4706
1.9	.4713	.4719	.4726	.4732	.4738	.4744	.4750	.4756	.4761	.4767
2.0	.4772	.4778	.4783	.4788	.4793	.4798	.4803	.4808	.4812	.4817
2.1	.4821	.4826	.4830	.4834	.4838	.4842	.4846	.4850	.4854	.4857
2.2	.4861	.4864	.4868	.4871	.4875	.4878	.4881	.4884	.4887	.4890
2.3	.4893	.4896	.4898	.4901	.4904	.4906	.4909	.4911	.4913	.4916
2.4	.4918	.4920	.4922	.4925	.4927	.4929	.4931	.4932	.4934	.4936
2.5	.4938	.4940	.4941	.4943	.4945	.4946	.4948	.4949	.4951	.4952
2.6	.4953	.4955	.4956	.4957	.4959	.4960	.4961	.4962	.4963	.4964
2.7	.4965	.4966	.4967	.4968	.4969	.4970	.4971	.4972	.4973	.4974
2.8	.4974	.4975	.4976	.4977	.4977	.4978	.4979	.4979	.4980	.4981
2.9	.4981	.4982	.4982	.4983	.4984	.4984	.4985	.4985	.4986	.4986
3.0	.4987	.4987	.4987	.4988	.4988	.4989	.4989	.4989	.4990	.4990
3.1	4990	.4991	.4991	.4991	.4992	.4992	.4992	.4992	.4993	.4993
3.2	4993	.4993	.4994	.4994	.4994	.4994	.4994	.4995	.4995	.4995
3.3	4995	.4995	.4995	.4996	.4996	.4996	.4996	.4996	.4996	.4997
3.4	4997	.4997	.4997	.4997	.4997	.4997	.4997	.4997	.4997	.4998
3.5	4998	.4998	.4998	.4998	.4998	.4998	.4998	.4998	.4998	.4998
3.6	4998	.4998	.4999	.4999	.4999	.4999	.4999	.4999	.4999	.4999
3.7	4999	.4999	.4999	.4999	.4999	.4999	.4999	.4999	.4999	.4999
3.8	4999	.4999	.4999	.4999	.4999	.4999	.4999	.4999	.4999	.4999
3.9	5000	.5000	.5000	.5000	.5000	.5000	.5000	.5000	.5000	.5000

Common sample size calculations

Parameter	Estimate	Sample size
μ	$\bar{x}$	$n = \left(\dfrac{z_{\alpha/2} \cdot \sigma}{E}\right)^2$
p	$\hat{p}$	$n = \dfrac{(z_{\alpha/2})^2 \cdot pq}{E^2}$
$\mu_2 - \mu_2$	$\bar{x}_1 - \bar{x}_2$	$n_1 = n_2 = \dfrac{(z_{\alpha/2})^2(\sigma_1^2 + \sigma_2^2)}{E^2}$
$p_1 - p_2$	$\hat{p}_1 - \hat{p}_2$	$n_1 = n_2 = \dfrac{(z_{\alpha/2})^2(p_1 q_1 + p_2 q_2)}{E^2}$

Common one-sample confidence intervals

Parameter	Assumptions	$100(1-\alpha)\%$ Confidence interval
μ	n large, σ^2 known, or normality, σ^2 known	$\bar{x} \pm z_{\alpha/2} \cdot \dfrac{\sigma}{\sqrt{n}}$
μ	normality, σ^2 unknown	$\bar{x} \pm t_{\alpha/2,n-1} \cdot \dfrac{s}{\sqrt{n}}$
σ^2	normality	$\left(\dfrac{(n-1)s^2}{\chi^2_{\alpha/2,n-1}}, \dfrac{(n-1)s^2}{\chi^2_{1-\alpha/2,n-1}}\right)$
p	binomial experiment, n large	$\hat{p} \pm z_{\alpha/2} \cdot \sqrt{\dfrac{\hat{p}(1-\hat{p})}{n}}$

Common two-sample confidence intervals

Parameter	Assumptions	$100(1-\alpha)\%$ Confidence interval
$\mu_1 - \mu_2$	normality, independence, σ_1^2, σ_2^2 known or n_1, n_2 large, independence, σ_1^2, σ_2^2 known	$(\bar{x}_1 - \bar{x}_2) \pm z_{\alpha/2} \cdot \sqrt{\dfrac{\sigma_1^2}{n_1} + \dfrac{\sigma_2^2}{n_2}}$
$\mu_1 - \mu_2$	normality, independence, $\sigma_1^2 = \sigma_2^2$ unknown	$(\bar{x}_1 - \bar{x}_2) \pm$ $t_{\frac{\alpha}{2},n_1+n_2-2} \cdot s_p \sqrt{\dfrac{1}{n_1} + \dfrac{1}{n_2}}$ $s_p^2 = \dfrac{(n_1-1)s_1^2 + (n_2-1)s_2^2}{n_1+n_2-2}$
$\mu_1 - \mu_2$	normality, independence, $\sigma_1^2 \neq \sigma_2^2$ unknown	$(\bar{x}_1 - \bar{x}_2) \pm t_{\alpha/2,\nu} \cdot \sqrt{\dfrac{s_1^2}{n_1} + \dfrac{s_2^2}{n_2}}$ $\nu \approx \dfrac{\left(\frac{s_1^2}{n_1} + \frac{s_2^2}{n_2}\right)^2}{\frac{(s_1^2/n_1)^2}{n_1-1} + \frac{(s_2^2/n_2)^2}{n_2-1}}$
$\mu_1 - \mu_2$	normality, n pairs, dependence	$\bar{d} \pm t_{\alpha/2,n-1} \cdot \dfrac{s_d}{\sqrt{n}}$
$p_1 - p_2$	binomial experiments, n_1, n_2 large, independence	$(\hat{p}_1 - \hat{p}_2) \pm$ $z_{\alpha/2} \cdot \sqrt{\dfrac{\hat{p}_1(1-\hat{p}_1)}{n_1} + \dfrac{\hat{p}_2(1-\hat{p}_2)}{n_2}}$

PERCENTAGE POINTS, STUDENT'S T-DISTRIBUTION

This table gives values of t such that

$$F(t) = \int_{-\infty}^{t} \frac{\Gamma\left(\frac{n+1}{2}\right)}{\sqrt{n\pi}\,\Gamma\left(\frac{n}{2}\right)} \left(1 + \frac{x^2}{n}\right) - \frac{n+1}{2} dx$$

for n, the number of degrees of freedom, equal to $1, 2, \ldots, 30, 40, 60, 120, \infty$; and for $F(t) = 0.60, 0.75, 0.90, 0.95, 0.975, 0.99,$ 0.995, and 0.9995. The t-distribution is symmetrical, so that $F(-t) = 1 - F(t)$

n/F	.60	.75	.90	.95	.975	.99	.995	.9995
1	.325	1.000	3.078	6.314	12.706	31.821	63.657	636.619
2	.289	.816	1.886	2.920	4.303	6.965	9.925	31.598
3	.277	.765	1.638	2.353	3.182	4.541	5.841	12.924
4	.271	.741	1.533	2.132	2.776	3.747	4.604	8.610
5	.267	.727	1.476	2.015	2.571	3.365	4.032	6.869
6	.265	.718	1.440	1.943	2.447	3.143	3.707	5.959
7	.263	.711	1.415	1.895	2.365	2.998	3.499	5.408
8	.262	.706	1.397	1.860	2.306	2.896	3.355	5.041
9	.261	.703	1.383	1.833	2.262	2.821	3.250	4.781
10	.260	.700	1.372	1.812	2.228	2.764	3.169	4.587
11	.260	.697	1.363	1.796	2.201	2.718	3.106	4.437
12	.259	.695	1.356	1.782	2.179	2.681	3.055	4.318
13	.259	.694	1.350	1.771	2.160	2.650	3.012	4.221
14	.258	.692	1.345	1.761	2.145	2.624	2.977	4.140
15	.258	.691	1.341	1.753	2.131	2.602	2.947	4.073
16	.258	.690	1.337	1.746	2.120	2.583	2.921	4.015
17	.257	.689	1.333	1.740	2.110	2.567	2.898	3.965
18	.257	.688	1.330	1.734	2.101	2.552	2.878	3.922
19	.257	.688	1.328	1.729	2.093	2.539	2.861	3.883
20	.257	.687	1.325	1.725	2.086	2.528	2.845	3.850
21	.257	.686	1.323	1.721	2.080	2.518	2.831	3.819
22	.256	.686	1.321	1.717	2.074	2.508	2.819	3.792
23	.256	.685	1.319	1.714	2.069	2.500	2.807	3.767
24	.256	.685	1.318	1.711	2.064	2.492	2.797	3.745
25	.256	.684	1.316	1.708	2.060	2.485	2.787	3.725
26	.256	.684	1.315	1.706	2.056	2.479	2.779	3.707
27	.256	.684	1.314	1.703	2.052	2.473	2.771	3.690
28	.256	.683	1.313	1.701	2.048	2.467	2.763	3.674
29	.256	.683	1.311	1.699	2.045	2.462	2.756	3.659
30	.256	.683	1.310	1.697	2.042	2.457	2.750	3.646
40	.255	.681	1.303	1.684	2.021	2.423	2.704	3.551
60	.254	.679	1.296	1.671	2.000	2.390	2.660	3.460
120	.254	.677	1.289	1.658	1.980	2.358	2.617	3.373
∞	.253	.674	1.282	1.645	1.960	2.326	2.576	3.291

*This table is abridged from the *Statistical Tables* by R. A. Fisher and Frank Yates published by Oliver & Boyd. Ltd., Edinburgh and London, 1938. It is published here with the kind permission of the authors and their publishers.

PERCENTAGE POINTS, CHI-SQUARE DISTRIBUTION

This table gives values of χ^2 such that

$$F(\chi)^2 = \int_0^{\chi^2} \frac{1}{2^{n/2}\Gamma\left(\frac{n}{2}\right)} x^{(n-2)/2} e^{-x/2}\, dx$$

for n, the number of degrees of freedom, equal to $1, 2, \ldots, 30$. For $n > 30$, a normal approximation is quite accurate. The expression $\sqrt{2x^2} - \sqrt{2n-1}$ is approximately normally distributed as the standard normal distribution. Thus χ_α^2, the α-point of the distribution, may be computed by the formula

$$\chi_\alpha^2 = \frac{1}{2}[x_\alpha + \sqrt{2n-1}]^2,$$

where x_α is the α-point of the cumulative normal distribution. For even values of n, $F(\chi^2)$ can be written as

$$1 - F(\chi^2) = \sum_{x=0}^{x'-1} \frac{e^{-\lambda}\lambda^x}{x!}$$

with $\lambda = \frac{1}{2}\chi^2$ and $x' = \frac{1}{2}n$. Thus the cumulative chi-square distribution is related to the cumulative Poisson distribution.
Another approximate formula for large n

$$\chi_\alpha^2 = n\left(1 - \frac{2}{9n} + z_\alpha\sqrt{\frac{2}{9n}}\right)^3$$

n = degrees of freedom
z_α = the normal deviate (the value of x for which $F(x)$ = the desired percentile).

x	1.282	1.645	1.960	2.326	2.576	3.090
$F(x)$	.90	.95	.975	.99	.995	.999

$\chi_{.99}^2 = 60[1 - 0.00370 + 2.326(0.06086)]^3 = 88.4$ is the 99th percentile for 60 degrees of freedom.

$$F(\chi^2) = \int_0^{\chi^2} \frac{1}{2^{n/2}\Gamma\left(\frac{n}{2}\right)} x^{n-2/2} e^{-x/2}\, dx$$

$n \backslash F$	.005	.010	.025	.050	.100	.250	.500	.750	.900	.950	.975	.990	.995
1	.0000393	.000157	.000982	.00393	.0158	.102	.455	1.32	2.71	3.84	5.02	6.63	7.88
2	.0100	.0201	.0506	.103	.211	.575	1.39	2.77	4.61	5.99	7.38	9.21	10.6
3	.0717	.115	.216	.352	.584	1.21	2.37	4.11	6.25	7.81	9.35	11.3	12.8
4	.207	.297	.484	.711	1.06	1.92	3.36	5.39	7.78	9.49	11.1	13.3	14.9
5	.412	.554	.831	1.15	1.61	2.67	4.35	6.63	9.24	11.1	12.8	15.1	16.7
6	.676	.872	1.24	1.64	2.20	3.45	5.35	7.84	10.6	12.6	14.4	16.8	18.5
7	.989	1.24	1.69	2.17	2.83	4.25	6.35	9.04	12.0	14.1	16.0	18.5	20.3
8	1.34	1.65	2.18	2.73	3.49	5.07	7.34	10.2	13.4	15.5	17.5	20.1	22.0
9	1.73	2.09	2.70	3.33	4.17	5.90	8.34	11.4	14.7	16.9	19.0	21.7	23.6
10	2.16	2.56	3.25	3.94	4.87	6.74	9.34	12.5	16.0	18.3	20.5	23.2	25.2
11	2.60	3.05	3.82	4.57	5.58	7.58	10.3	13.7	17.3	19.7	21.9	24.7	26.8
12	3.07	3.57	4.40	5.23	6.30	8.44	11.3	14.8	18.5	21.0	23.3	26.2	28.3
13	3.57	4.11	5.01	5.89	7.04	9.30	12.3	16.0	19.8	22.4	24.7	27.7	29.8
14	4.07	4.66	5.63	6.57	7.79	10.2	13.3	17.1	21.1	23.7	26.1	29.1	31.3
15	4.60	5.23	6.26	7.26	8.55	11.0	14.3	18.2	22.3	25.0	27.5	30.6	32.8
16	5.14	5.81	6.91	7.96	9.31	11.9	15.3	19.4	23.5	26.3	28.8	32.0	34.3
17	5.70	6.41	7.56	8.67	10.1	12.8	16.3	20.5	24.8	27.6	30.2	33.4	35.7
18	6.26	7.01	8.23	9.39	10.9	13.7	17.3	21.6	26.0	28.9	31.5	34.8	37.2
19	6.84	7.63	8.91	10.1	11.7	14.6	18.3	22.7	27.2	30.1	32.9	36.2	38.6
20	7.43	8.26	9.59	10.9	12.4	15.5	19.3	23.8	28.4	31.4	34.2	37.6	40.0
21	8.03	8.90	10.3	11.6	13.2	16.3	20.3	24.9	29.6	32.7	35.5	38.9	41.4
22	8.64	9.54	11.0	12.3	14.0	17.2	21.3	26.0	30.8	33.9	36.8	40.3	42.8
23	9.26	10.2	11.7	13.1	14.8	18.1	22.3	27.1	32.0	35.2	38.1	41.6	44.2
24	9.89	10.9	12.4	13.8	15.7	19.0	23.3	28.2	33.2	36.4	39.4	43.0	45.6
25	10.5	11.5	13.1	14.6	16.5	19.9	24.3	29.3	34.4	37.7	40.6	44.3	46.9
26	11.2	12.2	13.8	15.4	17.3	20.8	25.3	30.4	35.6	38.9	41.9	45.6	48.3
27	11.8	12.9	14.6	16.2	18.1	21.7	26.3	31.5	36.7	40.1	43.2	47.0	49.6
28	12.5	12.6	15.3	16.9	18.9	22.7	27.3	32.6	37.9	41.3	44.5	48.3	51.0
29	13.1	14.3	16.0	17.7	19.8	23.6	28.3	33.7	39.1	42.6	45.7	49.6	52.3
30	13.8	15.0	16.8	18.5	20.6	24.5	29.3	34.8	40.3	43.8	47.0	50.9	53.7

PERCENTAGE POINTS, *F*-DISTRIBUTION

This table gives values of F such that

$$F(F) = \int_0^F \frac{\Gamma\left(\frac{m+n}{2}\right)}{\Gamma\left(\frac{m}{2}\right)\Gamma\left(\frac{n}{2}\right)} m^{m/2} n^{n/2} x^{m-2/2} (n+mx)^{-(m+n)/2}\, dx$$

for selected values of m, the number of degrees of freedom of the numerator of F; and for selected values of n, the number of degrees freedom of the denominator of F. The table also provides values corresponding to $F(F) = .10, .05, .025, .01, .005, .001$ since $F_{1-\alpha}$ for m and n degrees of freedom is the reciprocal of F_α for n and m degrees of freedom. Thus

$$F_{.05}(4, 7) = \frac{1}{F_{.95}(7, 4)} = \frac{1}{6.09} = .164$$

$$F(F) = \int_0^F \frac{\Gamma\left(\frac{m+n}{2}\right)}{\Gamma\left(\frac{m}{2}\right)\Gamma\left(\frac{n}{2}\right)} m^{m/2} n^{n/2} x^{(m/2)-1} (n+mx)^{-(m+n)/2}\, dx = .90$$

n\m	1	2	3	4	5	6	7	8	9	10	12	15	20	24	30	40	60	120	∞
1	39.86	49.50	53.59	55.83	57.24	58.20	58.91	59.44	59.86	60.19	60.71	61.22	61.74	62.00	62.26	62.53	62.79	63.06	63.33
2	8.53	9.00	9.16	9.24	9.29	9.33	9.35	9.37	9.38	9.39	9.41	9.42	9.44	9.45	9.46	9.47	9.47	9.48	9.49
3	5.54	5.46	5.39	5.34	5.31	5.28	5.27	5.25	5.24	5.23	5.22	5.20	5.18	5.18	5.17	5.16	5.15	5.14	5.13
4	4.54	4.32	4.19	4.11	4.05	4.01	3.98	3.95	3.94	3.92	3.90	3.87	3.84	3.83	3.82	3.80	3.79	3.78	3.76
5	4.06	3.78	3.62	3.52	3.45	3.40	3.37	3.34	3.32	3.30	3.27	3.24	3.21	3.19	3.17	3.16	3.14	3.12	3.10
6	3.78	3.46	3.29	3.18	3.11	3.05	3.01	2.98	2.96	2.94	2.90	2.87	2.84	2.82	2.80	2.78	2.76	2.74	2.72
7	3.59	3.26	3.07	2.96	2.88	2.83	2.78	2.75	2.72	2.70	2.67	2.63	2.59	2.58	2.56	2.54	2.51	2.49	2.47
8	3.46	3.11	2.92	2.81	2.73	2.67	2.62	2.59	2.56	2.54	2.50	2.46	2.42	2.40	2.38	2.36	2.34	2.32	2.29
9	3.36	3.01	2.81	2.69	2.61	2.55	2.51	2.47	2.44	2.42	2.38	2.34	2.30	2.28	2.25	2.23	2.21	2.18	2.16
10	3.29	2.92	2.73	2.61	2.52	2.46	2.41	2.38	2.35	2.32	2.28	2.24	2.20	2.18	2.16	2.13	2.11	2.08	2.06
11	3.23	2.86	2.66	2.54	2.45	2.39	2.34	2.30	2.27	2.25	2.21	2.17	2.12	2.10	2.08	2.05	2.03	2.00	1.97
12	3.18	2.81	2.61	2.48	2.39	2.33	2.28	2.24	2.21	2.19	2.15	2.10	2.06	2.04	2.01	1.99	1.96	1.93	1.90
13	3.14	2.76	2.56	2.43	2.35	2.28	2.23	2.20	2.16	2.14	2.10	2.05	2.01	1.98	1.96	1.93	1.90	1.88	1.85
14	3.10	2.73	2.52	2.39	2.31	2.24	2.19	2.15	2.12	2.10	2.05	2.01	1.96	1.94	1.91	1.89	1.86	1.83	1.80
15	3.07	2.70	2.49	2.36	2.27	2.21	2.16	2.12	2.09	2.06	2.02	1.97	1.92	1.90	1.87	1.85	1.82	1.79	1.76
16	3.05	2.67	2.46	2.33	2.24	2.18	2.13	2.09	2.06	2.03	1.99	1.94	1.89	1.87	1.84	1.81	1.78	1.75	1.72
17	3.03	2.64	2.44	2.31	2.22	2.15	2.10	2.06	2.03	2.00	1.96	1.91	1.86	1.84	1.81	1.78	1.75	1.72	1.69
18	3.01	2.62	2.42	2.29	2.20	2.13	2.08	2.04	2.00	1.98	1.93	1.89	1.84	1.81	1.78	1.75	1.72	1.69	1.66
19	2.99	2.61	2.40	2.27	2.18	2.11	2.06	2.02	1.98	1.96	1.91	1.86	1.81	1.79	1.76	1.73	1.70	1.67	1.63
20	2.97	2.59	2.38	2.25	2.16	2.09	2.04	2.00	1.96	1.94	1.89	1.84	1.79	1.77	1.74	1.71	1.68	1.64	1.61
21	2.96	2.57	2.36	2.23	2.14	2.08	2.02	1.98	1.95	1.92	1.87	1.83	1.78	1.75	1.72	1.69	1.66	1.62	1.59
22	2.95	2.56	2.35	2.22	2.13	2.06	2.01	1.97	1.93	1.90	1.86	1.81	1.76	1.73	1.70	1.67	1.64	1.60	1.57
23	2.94	2.55	2.34	2.21	2.11	2.05	1.99	1.95	1.92	1.89	1.84	1.80	1.74	1.72	1.69	1.66	1.62	1.59	1.55
24	2.93	2.54	2.33	2.19	2.10	2.04	1.98	1.94	1.91	1.88	1.83	1.78	1.73	1.70	1.67	1.64	1.61	1.57	1.53
25	2.92	2.53	2.32	2.18	2.09	2.02	1.97	1.93	1.89	1.87	1.82	1.77	1.72	1.69	1.66	1.63	1.59	1.56	1.52
26	2.91	2.52	2.31	2.17	2.08	2.01	1.96	1.92	1.88	1.86	1.81	1.76	1.71	1.68	1.65	1.61	1.58	1.54	1.50
27	2.90	2.51	2.30	2.17	2.07	2.00	1.95	1.91	1.87	1.85	1.80	1.75	1.70	1.67	1.64	1.60	1.57	1.53	1.49
28	2.89	2.50	2.29	2.16	2.06	2.00	1.94	1.90	1.87	1.84	1.79	1.74	1.69	1.66	1.63	1.59	1.56	1.52	1.48
29	2.89	2.50	2.28	2.15	2.06	1.99	1.93	1.89	1.86	1.83	1.78	1.73	1.68	1.65	1.62	1.58	1.55	1.51	1.47
30	2.88	2.49	2.28	2.14	2.05	1.98	1.93	1.88	1.85	1.82	1.77	1.72	1.67	1.64	1.61	1.57	1.54	1.50	1.46
40	2.84	2.44	2.23	2.09	2.00	1.93	1.87	1.83	1.79	1.76	1.71	1.66	1.61	1.57	1.54	1.51	1.47	1.42	1.38
60	2.79	2.39	2.18	2.04	1.95	1.87	1.82	1.77	1.74	1.71	1.66	1.60	1.54	1.51	1.48	1.44	1.40	1.35	1.29
120	2.75	2.35	2.13	1.99	1.90	1.82	1.77	1.72	1.68	1.65	1.60	1.55	1.48	1.45	1.41	1.37	1.32	1.26	1.19
∞	2.71	2.30	2.08	1.94	1.85	1.77	1.72	1.67	1.63	1.60	1.55	1.49	1.42	1.38	1.34	1.30	1.24	1.17	1.00

$F = \frac{s_1^2}{s_2^2} = \frac{S_1}{m} \Big/ \frac{S_2}{n}$, where $s_1^2 = S_1/m$ and $s_2^2 = S_2/n$ are independent mean squares estimating a common variance σ^2 and based on m and n degrees of freedom, respectively.

$$F(F) = \int_0^F \frac{\Gamma\left(\frac{m+n}{2}\right)}{\Gamma\left(\frac{m}{2}\right)\Gamma\left(\frac{n}{2}\right)} m^{m/2} n^{n/2} x^{(m/2)-1}(n+mx)^{-(m+n)/2}\, dx = .95$$

n\m	1	2	3	4	5	6	7	8	9	10	12	15	20	24	30	40	60	120	∞
1	161.4	199.5	215.7	224.6	230.2	234.0	236.8	238.9	240.5	241.9	243.9	245.9	248.0	249.1	250.1	251.1	252.2	253.3	254.3
2	18.51	19.00	19.16	19.25	19.30	19.33	19.35	19.37	19.38	19.40	19.41	19.43	19.45	19.45	19.46	19.47	19.48	19.49	19.50
3	10.13	9.55	9.28	9.12	9.01	8.94	8.89	8.85	8.81	8.79	8.74	8.70	8.66	8.64	8.62	8.59	8.57	8.55	8.53
4	7.71	6.94	6.59	6.39	6.26	6.16	6.09	6.04	6.00	5.96	5.91	5.86	5.80	5.77	5.75	5.72	5.69	5.66	5.63
5	6.61	5.79	5.41	5.19	5.05	4.95	4.88	4.82	4.77	4.74	4.68	4.62	4.56	4.53	4.50	4.46	4.43	4.40	4.36
6	5.99	5.14	4.76	4.53	4.39	4.28	4.21	4.15	4.10	4.06	4.00	3.94	3.87	3.84	3.81	3.77	3.74	3.70	3.67
7	5.59	4.74	4.35	4.12	3.97	3.87	3.79	3.73	3.68	3.64	3.57	3.51	3.44	3.41	3.38	3.34	3.30	3.27	3.23
8	5.32	4.46	4.07	3.84	3.69	3.58	3.50	3.44	3.39	3.35	3.28	3.22	3.15	3.12	3.08	3.04	3.01	2.97	2.93
9	5.12	4.26	3.86	3.63	3.48	3.37	3.29	3.23	3.18	3.14	3.07	3.01	2.94	2.90	2.86	2.83	2.79	2.75	2.71
10	4.96	4.10	3.71	3.48	3.33	3.22	3.14	3.07	3.02	2.98	2.91	2.85	2.77	2.74	2.70	2.66	2.62	2.58	2.54
11	4.84	3.98	3.59	3.36	3.20	3.09	3.01	2.95	2.90	2.85	2.79	2.72	2.65	2.61	2.57	2.53	2.49	2.45	2.40
12	4.75	3.89	3.49	3.26	3.11	3.00	2.91	2.85	2.80	2.75	2.69	2.62	2.54	2.51	2.47	2.43	2.38	2.34	2.30
13	4.67	3.81	3.41	3.18	3.03	2.92	2.83	2.77	2.71	2.67	2.60	2.53	2.46	2.42	2.38	2.34	2.30	2.25	2.21
14	4.60	3.74	3.34	3.11	2.96	2.85	2.76	2.70	2.65	2.60	2.53	2.46	2.39	2.35	2.31	2.27	2.22	2.18	2.13
15	4.54	3.68	3.29	3.06	2.90	2.79	2.71	2.64	2.59	2.54	2.48	2.40	2.33	2.29	2.25	2.20	2.16	2.11	2.07
16	4.49	3.63	3.24	3.01	2.85	2.74	2.66	2.59	2.54	2.49	2.42	2.35	2.28	2.24	2.19	2.15	2.11	2.06	2.01
17	4.45	3.59	3.20	2.96	2.81	2.70	2.61	2.55	2.49	2.45	2.38	2.31	2.23	2.19	2.15	2.10	2.06	2.01	1.96
18	4.41	3.55	3.16	2.93	2.77	2.66	2.58	2.51	2.46	2.41	2.34	2.27	2.19	2.15	2.11	2.06	2.02	1.97	1.92
19	4.38	3.52	3.13	2.90	2.74	2.63	2.54	2.48	2.42	2.38	2.31	2.23	2.16	2.11	2.07	2.03	1.98	1.93	1.88
20	4.35	3.49	3.10	2.87	2.71	2.60	2.51	2.45	2.39	2.35	2.28	2.20	2.12	2.08	2.04	1.99	1.95	1.90	1.84
21	4.32	3.47	3.07	2.84	2.68	2.57	2.49	2.42	2.37	2.32	2.25	2.18	2.10	2.05	2.01	1.96	1.92	1.87	1.81
22	4.30	3.44	3.05	2.82	2.66	2.55	2.46	2.40	2.34	2.30	2.23	2.15	2.07	2.03	1.98	1.94	1.89	1.84	1.78
23	4.28	3.42	3.03	2.80	2.64	2.53	2.44	2.37	2.32	2.27	2.20	2.13	2.05	2.01	1.96	1.91	1.86	1.81	1.76
24	4.26	3.40	3.01	2.78	2.62	2.51	2.42	2.36	2.30	2.25	2.18	2.11	2.03	1.98	1.94	1.89	1.84	1.79	1.73
25	4.24	3.39	2.99	2.76	2.60	2.49	2.40	2.34	2.28	2.24	2.16	2.09	2.01	1.96	1.92	1.87	1.82	1.77	1.71
26	4.23	3.37	2.98	2.74	2.59	2.47	2.39	2.32	2.27	2.22	2.15	2.07	1.99	1.95	1.90	1.85	1.80	1.75	1.69
27	4.21	3.35	2.96	2.73	2.57	2.46	2.37	2.31	2.25	2.20	2.13	2.06	1.97	1.93	1.88	1.84	1.79	1.73	1.67
28	4.20	3.34	2.95	2.71	2.56	2.45	2.36	2.29	2.24	2.19	2.12	2.04	1.96	1.91	1.87	1.82	1.77	1.71	1.65
29	4.18	3.33	2.93	2.70	2.55	2.43	2.35	2.28	2.22	2.18	2.10	2.03	1.94	1.90	1.85	1.81	1.75	1.70	1.64
30	4.17	3.32	2.92	2.69	2.53	2.42	2.33	2.27	2.21	2.16	2.09	2.01	1.93	1.89	1.84	1.79	1.74	1.68	1.62
40	4.08	3.23	2.84	2.61	2.45	2.34	2.25	2.18	2.12	2.08	2.00	1.92	1.84	1.79	1.74	1.69	1.64	1.58	1.51
60	4.00	3.15	2.76	2.53	2.37	2.25	2.17	2.10	2.04	1.99	1.92	1.84	1.75	1.70	1.65	1.59	1.53	1.47	139
120	3.92	3.07	2.68	2.45	2.29	2.17	2.09	2.02.	1.96	1.91	1.83	1.75	1.66	1.61	1.55	1.50	1.43	1.35	1.25
∞	3.84	3.00	2.60	2.37	2.21	2.10	2.01	1.94	1.88	1.83	1.75	1.67	1.57	1.52	1.46	1.39	1.32	1.22	1.00

$F = \frac{s_1^2}{s_2^2} = \frac{S_1}{m} / \frac{S_2}{n}$, where $s_1^2 = S_1/m$ and $s_2^2 = S_2/n$ are independent mean squares estimating a common variance σ^2 and based on m and n degrees of freedom, respectively.

$$F(F) = \int_0^F \frac{\Gamma\left(\frac{m+n}{2}\right)}{\Gamma\left(\frac{m}{2}\right)\Gamma\left(\frac{n}{2}\right)} m^{m/2} n^{n/2} x^{(m/2)-1} (n+mx)^{-(m+n)/2} \, dx = .975$$

$n \backslash m$	1	2	3	4	5	6	7	8	9	10	12	15	20	24	30	40	60	120	∞
1	647.8	799.5	864.2	899.6	921.8	937.1	948.2	956.7	963.3	968.6	976.7	984.9	993.1	997.2	1001	1006	1010	1014	1018
2	38.51	39.00	39.17	39.25	39.30	39.33	39.36	39.37	39.39	39.40	39.41	39.43	39.45	39.46	39.46	39.47	39.48	39.49	39.50
3	17.44	16.04	15.44	15.10	14.88	14.73	14.62	14.54	14.47	14.42	14.34	14.25	14.17	14.12	14.08	14.04	13.99	13.95	13.90
4	12.22	10.65	9.98	9.60	9.36	9.20	9.07	8.98	8.90	8.84	8.75	8.66	8.56	8.51	8.46	8.41	8.36	8.31	8.26
5	10.01	8.43	7.76	7.39	7.15	6.98	6.85	6.76	6.68	6.62	6.52	6.43	6.33	6.28	6.23	6.18	6.12	6.07	6.02
6	8.81	7.26	6.60	6.23	5.99	5.82	5.70	5.60	5.52	5.46	5.37	5.27	5.17	5.12	5.07	5.01	4.96	4.90	4.85
7	8.07	6.54	5.89	5.52	5.29	5.12	4.99	4.90	4.82	4.76	4.67	4.57	4.47	4.42	4.36	4.31	4.25	4.20	4.14
8	7.57	6.06	5.42	5.05	4.82	4.65	4.53	4.43	4.36	4.30	4.20	4.10	4.00	3.95	3.89	3.84	3.78	3.73	3.67
9	7.21	5.71	5.08	4.72	4.48	4.32	4.20	4.10	4.03	3.96	3.87	3.77	3.67	3.61	3.56	3.51	3.45	3.39	3.33
10	6.94	5.46	4.83	4.47	4.24	4.07	3.95	3.85	3.78	3.72	3.62	3.52	3.42	3.37	3.31	3.26	3.20	3.14	3.08
11	6.72	5.26	4.63	4.28	4.04	3.88	3.76	3.66	3.59	3.53	3.43	3.33	3.23	3.17	3.12	3.06	3.00	2.94	2.88
12	6.55	5.10	4.47	4.12	3.89	3.73	3.61	3.51	3.44	3.37	3.28	3.18	3.07	3.02	2.96	2.91	2.85	2.79	2.72
13	6.41	4.97	4.35	4.00	3.77	3.60	3.48	3.39	3.31	3.25	3.15	3.05	2.95	2.89	2.84	2.78	2.72	2.66	2.60
14	6.30	4.86	4.24	3.89	3.66	3.50	3.38	3.29	3.21	3.15	3.05	2.95	2.84	2.79	2.73	2.67	2.61	2.55	2.49
15	6.20	4.77	4.15	3.80	3.58	3.41	3.29	3.20	3.12	3.06	2.96	2.86	2.76	2.70	2.64	2.59	2.52	2.46	2.40
16	6.12	4.69	4.08	3.73	3.50	3.34	3.22	3.12	3.05	2.99	2.89	2.79	2.68	2.63	2.57	2.51	2.45	2.38	2.32
17	6.04	4.62	4.01	3.66	3.44	3.28	3.16	3.06	2.98	2.92	2.82	2.72	2.62	2.56	2.50	2.44	2.38	2.32	2.25
18	5.98	4.56	3.95	3.61	3.38	3.22	3.10	3.01	2.93	2.87	2.77	2.67	2.56	2.50	2.44	2.38	2.32	2.26	2.19
19	5.92	4.51	3.90	3.56	3.33	3.17	3.05	2.96	2.88	2.82	2.72	2.62	2.51	2.45	2.39	2.33	2.27	2.20	2.13
20	5.87	4.46	3.86	3.51	3.29	3.13	3.01	2.91	2.84	2.77	2.68	2.57	2.46	2.41	2.35	2.29	2.22	2.16	2.09
21	5.83	4.42	3.82	3.48	3.25	3.09	2.97	2.87	2.80	2.73	2.64	2.53	2.42	2.37	2.31	2.25	2.18	2.11	2.04
22	5.79	4.38	3.78	3.44	3.22	3.05	2.93	2.84	2.76	2.70	2.60	2.50	2.39	2.33	2.27	2.21	2.14	2.08	2.00
23	5.75	4.35	3.75	3.41	3.18	3.02	2.90	2.81	2.73	2.67	2.57	2.47	2.36	2.30	2.24	2.18	2.11	2.04	1.97
24	5.72	4.32	3.72	3.38	3.15	2.99	2.87	2.78	2.70	2.64	2.54	2.44	2.33	2.27	2.21	2.15	2.08	2.01	1.94
25	5.69	4.29	3.69	3.35	3.13	2.97	2.85	2.75	2.68	2.61	2.51	2.41	2.30	2.24	2.18	2.12	2.05	1.98	1.91
26	5.66	4.27	3.67	3.33	3.10	2.94	2.82	2.73	2.65	2.59	2.49	2.39	2.28	2.22	2.16	2.09	2.03	1.95	1.88
27	5.63	4.24	3.65	3.31	3.08	2.92	2.80	2.71	2.63	2.57	2.47	2.36	2.25	2.19	2.13	2.03	2.00	1.93	1.85
28	5.61	4.22	3.63	3.29	3.06	2.90	2.78	2.69	2.61	2.55	2.45	2.34	2.23	2.17	2.11	2.05	1.98	1.91	1.83
29	5.59	4.20	3.61	3.27	3.04	2.88	2.76	2.67	2.59	2.53	2.43	2.32	2.21	2.15	2.09	2.03	1.96	1.89	1.81
30	5.57	4.18	3.59	3.25	3.03	2.87	2.75	2.65	2.57	2.51	2.41	2.31	2.20	2.14	2.07	2.01	1.94	1.87	1.79
40	5.42	4.05	3.46	3.13	2.90	2.74	2.62	2.53	2.45	2.39	2.29	2.18	2.07	2.01	1.94	1.88	1.80	1.72	1.64
60	5.29	3.93	3.34	3.01	2.79	2.63	2.51	2.41	2.33	2.27	2.17	2.06	1.94	1.88	1.82	1.74	1.67	1.58	1.48
120	5.15	3.80	3.23	2.89	2.67	2.52	2.39	2.30	2.22	2.16	2.05	1.94	1.82	1.76	1.69	1.61	1.53	1.43	1.31
∞	5.02	3.69	3.12	2.79	2.57	2.41	2.29	2.19	2.11	2.05	1.94	1.83	1.71	1.64	1.57	1.48	1.39	1.27	1.00

$F = \frac{s_1^2}{s_2^2} = \frac{S_1}{m} / \frac{S_2}{n}$, where $s_1^2 = S_1/m$ and $s_2^2 = S_2/n$ are independent mean squares estimating a common variance σ^2 and based on m and n degrees of freedom, respectively.

$$F(F) = \int_0^F \frac{\Gamma\left(\frac{m+n}{2}\right)}{\Gamma\left(\frac{m}{2}\right)\Gamma\left(\frac{n}{2}\right)} m^{m/2} n^{n/2} x^{(m/2)-1} (n+mx)^{-(m+n)/2}\, dx = .99$$

$n \backslash m$	1	2	3	4	5	6	7	8	9	10	12	15	20	24	30	40	60	120	∞
1	4052	4999.5	5403	5625	5764	5859	5928	5982	6022	6056	6106	6157	6209	6235	6261	6287	6313	6339	6366
2	98.50	99.00	99.17	99.25	99.30	99.33	99.36	99.37	99.39	99.40	99.42	99.43	99.45	99.46	99.47	99.47	99.48	99.49	99.50
3	34.12	30.82	29.46	28.71	28.24	27.91	27.67	27.49	27.35	27.23	27.05	26.87	26.69	26.60	26.50	26.41	26.32	26.22	26.13
4	21.20	18.00	16.69	15.98	15.52	15.21	14.98	14.80	14.66	14.55	14.37	14.20	14.02	13.93	13.84	13.75	13.65	13.56	13.46
5	16.26	13.27	12.06	11.39	10.97	10.67	10.46	10.29	10.16	10.05	9.89	9.72	9.55	9.47	9.38	9.29	9.20	9.11	9.02
6	13.75	10.92	9.78	9.15	8.75	8.47	8.26	8.10	7.98	7.87	7.72	7.56	7.40	7.31	7.23	7.14	7.06	6.97	6.88
7	12.25	9.55	8.45	7.85	7.46	7.19	6.99	6.84	6.72	6.62	6.47	6.31	6.16	6.07	5.99	5.91	5.82	5.74	5.65
8	11.26	8.65	7.59	7.01	6.63	6.37	6.18	6.03	5.91	5.81	5.67	5.52	5.36	5.28	5.20	5.12	5.03	4.95	4.86
9	10.56	8.02	6.99	6.42	6.06	5.80	5.61	5.47	5.35	5.26	5.11	4.96	4.81	4.73	4.65	4.57	4.48	4.40	4.31
10	10.04	7.56	6.55	5.99	5.64	5.39	5.20	5.06	4.94	4.85	4.71	4.56	4.41	4.33	4.25	4.17	4.08	4.00	3.91
11	9.65	7.21	6.22	5.67	5.32	5.07	4.89	4.74	4.63	4.54	4.40	4.25	4.10	4.02	3.94	3.86	3.78	3.69	3.60
12	9.33	6.93	5.95	5.41	5.06	4.82	4.64	4.50	4.39	4.30	4.16	4.01	3.86	3.78	3.70	3.62	3.54	3.45	3.36
13	9.07	6.70	5.74	5.21	4.86	4.62	4.44	4.30	4.19	4.10	3.96	3.82	3.66	3.59	3.51	3.43	3.34	3.25	3.17
14	8.86	6.51	5.56	5.04	4.69	4.46	4.28	4.14	4.03	3.94	3.80	3.66	3.51	3.43	3.35	3.27	3.18	3.09	3.00
15	8.68	6.36	5.42	4.89	4.56	4.32	4.14	4.00	3.89	3.80	3.67	3.52	3.37	3.29	3.21	3.13	3.05	2.96	2.87
16	8.53	6.23	5.29	4.77	4.44	4.20	4.03	3.89	3.78	3.69	3.55	3.41	3.26	3.18	3.10	3.02	2.93	2.84	2.75
17	8.40	6.11	5.18	4.67	4.34	4.10	3.93	3.79	3.68	3.59	3.46	3.31	3.16	3.08	3.00	2.92	2.83	2.75	2.65
18	8.29	6.01	5.09	4.58	4.25	4.01	3.84	3.71	3.60	3.51	3.37	3.23	3.08	3.00	2.92	2.84	2.75	2.66	2.57
19	8.18	5.93	5.01	4.50	4.17	3.94	3.77	3.63	3.52	3.43	3.30	3.15	3.00	2.92	2.84	2.76	2.67	2.58	2.49
20	8.10	5.85	4.94	4.43	4.10	3.87	3.70	3.56	3.46	3.37	3.23	3.09	2.94	2.86	2.78	2.69	2.61	2.52	2.42
21	8.02	5.78	4.87	4.37	4.04	3.81	3.64	3.51	3.40	3.31	3.17	3.03	2.88	2.80	2.72	2.64	2.55	2.46	2.36
22	7.95	5.72	4.82	4.31	3.99	3.76	3.59	3.45	3.35	3.26	3.12	2.98	2.83	2.75	2.67	2.58	2.50	2.40	2.31
23	7.88	5.66	4.76	4.26	3.94	3.71	3.54	3.41	3.30	3.21	3.07	2.93	2.78	2.70	2.62	2.54	2.45	2.35	2.26
24	7.82	5.61	4.72	4.22	3.90	3.67	3.50	3.36	3.26	3.17	3.03	2.89	2.74	2.66	2.58	2.49	2.40	2.31	2.21
25	7.77	5.57	4.68	4.18	3.85	3.63	3.46	3.32	3.22	3.13	2.99	2.85	2.70	2.62	2.54	2.45	2.36	2.27	2.17
26	7.72	5.53	4.64	4.14	3.82	3.59	3.42	3.29	3.18	3.09	2.96	2.81	2.66	2.58	2.50	2.42	2.33	2.23	2.13
27	7.68	5.49	4.60	4.11	3.78	3.56	3.39	3.26	3.15	3.06	2.93	2.78	2.63	2.55	2.47	2.38	2.29	2.20	2.10
28	7.64	5.45	4.57	4.07	3.75	3.53	3.36	3.23	3.12	3.03	2.90	2.75	2.60	2.52	2.44	2.35	2.26	2.17	2.06
29	7.60	5.42	4.54	4.04	3.73	3.50	3.33	3.20	3.09	3.00	2.87	2.73	2.57	2.49	2.41	2.33	2.23	2.14	2.03
30	7.56	5.39	4.51	4.02	3.70	3.47	3.30	3.17	3.07	2.98	2.84	2.70	2.55	2.47	2.39	2.30	2.21	2.11	2.01
40	7.31	5.18	4.31	3.83	3.51	3.29	3.12	2.99	2.89	2.80	2.66	2.52	2.37	2.29	2.20	2.11	2.02	1.92	1.80
60	7.08	4.98	4.13	3.65	3.34	3.12	2.95	2.82	2.72	2.63	2.50	2.35	2.20	2.12	2.03	1.94	1.84	1.73	1.60
120	6.85	4.79	3.95	3.48	3.17	2.96	2.79	2.66	2.56	2.47	2.34	2.19	2.03	1.95	1.86	1.76	1.66	1.53	1.38
∞	6.63	4.61	3.78	3.32	3.02	2.80	2.64	2.51	2.41	2.32	2.18	2.04	1.88	1.79	1.70	1.59	1.47	1.32	1.00

$F = \frac{s_1^2}{s_2^2} = \frac{S_1}{m} / \frac{S_2}{n}$, where $s_1^2 = S_1/m$ and $s_2^2 = S_2/n$ are independent mean squares estimating a common variance σ^2 and based on m and n degrees of freedom, respectively.

MOMENT OF INERTIA FOR VARIOUS BODIES OF MASS

The mass of the body is indicated by m

Body	Axis	Moment of inertia
Uniform thin rod of length l	Normal to the length, at one end	$m\frac{1}{3}l^2$
Uniform thin rod of length l	Normal to the length, at the center	$m\frac{1}{12}l^2$
Thin rectangular sheet, sides a and b	Through the center parallel to b	$m\frac{1}{12}a^2$
Thin rectangular sheet, sides a and b	Through the center perpendicular to the sheet	$m\frac{1}{12}(a^2 + b^2)$
Thin circular sheet of radius r	Normal to the plate through the center	$m\frac{1}{2}r^2$
Thin circular sheet of radius r	Along any diameter	$m\frac{1}{4}r^2$
Thin circular ring. Radii r_1 and r_2	Through center normal to plane of ring	$m\frac{1}{2}(r_1^2 + r_2^2)$
Thin circular ring. Radii r_1 and r_2	Any diameter	$m\frac{1}{4}(r_1^2 + r_2^2)$
Rectangular parallelepiped, edges a, b, and c	Through center perpendicular to face ab, (parallel to edge c)	$m\frac{1}{12}(a^2 + b^2)$
Sphere, radius r	Any diameter	$m\frac{2}{5}r^2$
Spherical shell, external radius, r_1, internal radius r_2	Any diameter	$m\frac{2}{5}\frac{(r_1^5 - r_2^5)}{(r_1^3 - r_2^3)}$
Spherical shell, very thin, mean radius, r	Any diameter	$m\frac{2}{3}r^2$
Right circular cylinder of radius r, length l	The longitudinal axis of the solid	$m\frac{1}{2}r^2$
Right circular cylinder of radius r, length l	Transverse diameter	$m\left(\frac{r^2}{4} + \frac{l^2}{12}\right)$
Hollow circular cylinder, length l, radii r_1 and r_2	The longitudinal axis of the figure	$m\frac{1}{2}(r_1^2 + r_2^2)$
Thin cylindrical shell, length l, mean radius, r	The longitudinal axis of the figure	mr^2
Hollow circular cylinder, length l, radii r_1 and r_2	Transverse diameter	$m\left(\frac{r_1^2 + r_2^2}{4} + \frac{l^2}{12}\right)$
Hollow circular cylinder, length l, very thin, mean radius r	Transverse diameter	$m\left(\frac{r^2}{2} + \frac{l^2}{12}\right)$
Elliptic cylinder, length l, transverse semiaxes a and b	Longitudinal axis	$m\frac{1}{4}(a^2 + b^2)$
Right cone, altitude h, radius of base r	Axis of the figure	$m\frac{3}{10}r^2$
Spheroid of revolution, equatorial radius r	Polar axis	$m\frac{2}{5}r^2$
Ellipsoid, axes $2a$, $2b$, $2c$	Axis $2a$	$m\frac{1}{5}(b^2 + c^2)$

Appendix B

Sources of Physical and Chemical Data

SOURCES OF PHYSICAL AND CHEMICAL DATA

In addition to the primary research journals, there are many useful sources of property data of the type contained in the *CRC Handbook of Chemistry and Physics*. A selected list of these is presented here, with emphasis on print and electronic sources whose contents have been subject to a reasonable level of quality control.

A. Data Journals

1. *Journal of Physical and Chemical Reference Data* — Published jointly by the National Institute of Standards and Technology and the American Institute of Physics, this quarterly journal contains compilations of evaluated data in chemistry, physics, and materials science. It is available in print and on the Internet. [ojps.aip.org/jpcrd/]

2. *Journal of Chemical and Engineering Data* — This bimonthly journal of the American Chemical Society publishes articles reporting original experimental measurements carried out under carefully controlled conditions. The main emphasis is on thermochemical and thermophysical properties. Review articles with evaluated data from the literature are also published. [pubs.acs.org/journals/jceaax/index.html]

3. *Journal of Chemical Thermodynamics* — This journal publishes original research papers that include highly accurate measurements of thermodynamic and thermophysical properties. [www.sciencedirect.com/science/journal/00219614]

4. *Atomic Data and Nuclear Data Tables* — This is a bimonthly journal containing compilations of data in atomic physics, nuclear physics, and related fields. [www.sciencedirect.com/science/journal/0092640X]

5. *Journal of Phase Equilibria and Diffusion* — This journal presents critically evaluated phase diagrams and related data on alloy systems. It is published by ASM International and is the successor to the previous ASM periodical *Bulletin of Alloy Phase Diagrams*. [www.asm-intl.org]

B. Data Centers

This section lists selected organizations that perform a continuing function of compiling and critically evaluating data in specific fields of science.

1. **National Institute of Standards and Technology** — Under its Standard Reference Data program, NIST supports a number of data centers in chemistry, physics, and materials science. Topics covered include thermodynamics, fluid properties, chemical kinetics, mass spectroscopy, atomic spectroscopy, fundamental physical constants, ceramics, and crystallography. Address: Office of Standard Reference Data, National Institute of Standards and Technology, Gaithersburg, MD 20899 [www.nist.gov/srd/].

2. **Thermodynamics Research Center** — Now located at the National Institute of Standards and Technology, TRC maintains an extensive archive of data covering thermodynamic, thermochemical, and transport properties of organic compounds and mixtures. Data are distributed in both print and electronic form. Address: Mail code 838.00, 325 Broadway, Boulder, CO 80305-3328 [www.trc.nist.gov].

3. **Design Institute for Physical Property Data** — Under the auspices of the American Institute of Chemical Engineers [www.aiche.org/dippr/], DIPPR offers evaluated data on industrially important chemical compounds. The largest project deals with physical, thermodynamic, and transport properties of pure compounds. Address: Brigham Young University, Provo, UT 84602 [dippr.byu.edu].

4. **Dortmund Data Bank** — Maintains extensive databases on thermodynamic and transport properties of pure compounds and mixtures of industrial interest. The data are distributed through DECHEMA, FIZ CHEMIE, and other outlets. Address: DDBST GmbH, Industriestr. 1, 26121 Oldenburg, Germany [www.ddbst.de].

5. **Cambridge Crystallographic Data Centre** — Maintains the Cambridge Structural Database of over 500,000 organic compounds. The data files and manipulation software are distributed in several ways. Address: 12 Union Rd., Cambridge CB2 1EZ, U.K. [www.ccdc.cam.ac.uk].

6. **FIZ Karlsruhe** — In addition to many bibliographic databases, FIZ Karlsruhe maintains the Inorganic Crystal Structure Database in collaboration with the National Institute of Standards and Technology. The ICSD contains the atomic coordinates and related data on over 50,000 inorganic crystals. Address: Fachinformationszentrum (FIZ) Karlsruhe, Hermann-von-Helmholtz-Platz 1, D-76344 Eggenstein-Leopoldshafen, Germany [www.fiz-karlsruhe.de].

7. **International Centre for Diffraction Data** — Maintains and distributes the Powder Diffraction File (PDF), a file of over 500,000 X-ray powder diffraction patterns used for identification of crystalline materials. The ICDD also distributes the NIST Crystal Data file, which contains lattice parameters for over 235,000 inorganic, organic, metal, and mineral crystalline materials. Address: 12 Campus Blvd., Newton Square, PA 19073-3273 [www.icdd.com].

8. **Research Collaboratory for Structural Bioinformatics** — Maintains the Protein Data Bank (PDB), a file of 3-dimensional structures of proteins and other biological macromolecules. Address: Department of Chemistry and Chemical Biology, Rutgers University, 610 Taylor Road, Piscataway, NJ 08854-8087 [www.rcsb.org].

9. **Toth Information Systems** — Maintains the Metals Crystallographic Data File (CRYSTMET). Address: 2045 Quincy Ave., Gloucester, ON, Canada K1J 6B2 [www.tothcanada.com].

10. **Atomic Mass Data Center** — Collects and evaluates high-precision data on masses of individual isotopes and maintains a comprehensive database. Address: C.S.N.S.M (IN2P3-CNRS), Batiment 108, F-91405 Orsay Campus, France [www.nndc.bnl.gov/amdc].

11. **Particle Data Group** — International center for data of high-energy physics; maintains a database of properties of fundamental particles that is published in both print and electronic form. Address: MS 50-308, Lawrence Berkeley National Laboratory, Berkeley, CA 94720 [pdg.lbl.gov].

12. **National Nuclear Data Center** — Maintains databases on nuclear structure and reactions, including neutron cross sections. The NNDC is the U.S. node in an international network of nuclear data centers. Address: Brookhaven National Laboratory, Upton, NY 11973-5000 [www.nndc.bnl.gov].

13. **International Union of Pure and Applied Chemistry** — Address: PO Box 13757, Research Triangle Park, NC 27709-3757 [www.iupac.org]. IUPAC supports a number of long-term data projects, including these examples:

a. **Solubility Data Project** — Carries out evaluation of all types of solubility data. The results are published in the Solubility Data Series, whose current outlet is the *Journal of Physical and Chemical Reference Data.* [www.iupac.org/divisions/V/cp5.html]

b. **Kinetic Data for Atmospheric Chemistry** — Maintains a comprehensive database on the kinetics of reactions important in the chemistry of the atmosphere. [www.iupac-kinetic.ch.cam.ac.uk/]

c. **International Thermodynamic Tables for the Fluid State** — Prepares definitive tables of the thermodynamic properties of industrially important fluids. Thirteen volumes have been published by IUPAC. [www.iupac.org/publications/books/seriestitles/]

d. **Stability Constants Database** — Collection of metal-ligand stability constants and associated software. [www.acadsoft.co.uk]

C. Major Multi-Volume Handbook Series

1. *CRC Chemical Dictionaries* — These originally appeared in print form as the *Dictionary of Organic Compounds, Dictionary of Natural Products,* etc. They are now published in electronic form and are available in DVD format [www.crcpress.com] and on the Internet [www.chemnetbase.com]. The consolidated version, called the *Combined Chemical Dictionary,* has data on more than 560,000 compounds spanning all branches of chemistry. The coverage includes physical properties, biological sources, hazard information, uses, and literature references.

2. *Properties of Organic Compounds* — Originally published in three editions as the *Handbook of Data on Organic Compounds,* it is now in electronic form as *Properties of Organic Compounds.* The database includes about 30,000 compounds; physical properties and spectral data (mass, infrared, Raman, ultraviolet, and NMR) are covered. It is offered as CDROM [www.crcpress.com] and by Web access [www.chemnetbase.com].

3. *Beilstein Handbook of Organic Chemistry* — The classic source of data on organic compounds, dating from the 19th century, *Beilstein* was converted to electronic form in the last decade of the 20th century. Over 8 million compounds and 10 million chemical reactions are now covered, with a broad range of physical properties as well as synthetic methods and ecological data. The database is accessed by the CrossFire software [accelrys.com/products].

4. *Gmelin Handbook of Inorganic and Organometallic Chemistry* — A subset of the information in the print series has been converted to electronic form and is now distributed in the same manner as *Beilstein.* In addition to the standard physical properties, the coverage includes a wide range of optical, magnetic, spectroscopic, thermal, and transport properties for about 1.4 million compounds [accelrys.com/products].

5. *DECHEMA Chemical Data Series* — DECHEMA distributes the DETHERM database, which emphasizes data used in process design in the chemical industry, including thermodynamic and transport properties of about 20,000 pure compounds and 90,000 mixtures. Access is available through in-house databases and via the Internet [www.dechema.de].

6. *Landolt-Börnstein Numerical Data and Functional Relationships in Science and Technology* — *Landolt-Börnstein* covers a very broad range of data in physics, chemistry, crystallography, materials science, biophysics, astronomy, and geophysics. Hard-copy volumes are no longer published, but the entire collection is available on-line [www.springermaterials.com].

D. Selected Single-Volume Handbooks

The following handbooks offer broad coverage of high-quality data in a single volume. This list is only representative; an extensive listing of handbooks in all fields of science may be found in *Handbooks and Tables in Science and Technology, Third Edition* (Russell H. Powell, ed., Oryx Press, Westport, CT, 1994).

1. *American Institute of Physics Handbook* — Although an old book, it contains much data that are still useful, especially in acoustics, mechanics, optics, and solid state physics. (Dwight E. Gray, ed., McGraw-Hill, New York, 1972)

2. *Constants of Inorganic Substances* — This book presents physical constants, thermodynamic data, solubility, reactivity, and other information on over 3000 inorganic compounds. Since it draws heavily on Russian literature, it contains a great deal of data that do not make their way into most U.S. handbooks. (R. A. Lidin, L. L. Andreeva, and V. A. Molochko, Begell House, New York, 1995)

3. *Handbook of Chemistry and Physics* — Now in the 92nd Edition, the *CRC Handbook* covers data from most branches of chemistry and physics. The annual revisions permit regular updating of the information. Also available on CDROM [www.crcpress.com] and the Web [hbcpnetbase.com]. (W. M. "Mickey" Haynes, ed., CRC Press, Boca Raton, FL, 2009)

4. *Handbook of Inorganic Compounds* — This book covers physical constants and solubility for about 3300 inorganic compounds. Also available on CDROM [www.crcpress.com]. (Dale L. Perry and Sidney L. Phillips, eds., CRC Press, Boca Raton, FL, 1995)

5. *Handbook of Physical Properties of Liquids and Gases* — This is a valuable source of data on all types of fluids, ranging from liquid and gaseous hydrocarbons to molten metals and ionized gases. Detailed tables of physical, thermodynamic, and transport properties are given for temperatures from the cryogenic region to 6000 K. Western and Russian literature is covered. (N. B. Vargaftik, Y. K. Vinogradov, and V. S. Yargin, Begell House, New York, 1996)

6. *Handbook of Physical Quantities* — The range of coverage is somewhat similar to the *CRC Handbook of Chemistry and Physics,* but with a stronger emphasis on physics than on chemistry. Solid state physics, lasers, nuclear physics, geophysics, and astronomy receive considerable attention. (Igor S. Grigoriev and Evgenii Z. Meilikhov, eds., CRC Press, Boca Raton, FL, 1997)

7. *Kaye & Laby Tables of Physical and Chemical Constants* — *Kaye & Laby* dates from 1911, and the 16th Edition was prepared in 1995 by a committee of experts. The coverage extends to almost every field of physics and chemistry; data on a limited number of representative substances or materials are given for each topic. (Longman Group Limited, Harlow, Essex, U.K., 1995)

8. *Lange's Handbook of Chemistry* — Provides broad coverage of chemical data; last updated in 2005. Also available on the Web [www.knovel.com]. (James G. Speight, ed., McGraw-Hill, New York, 2005)

9. ***Recommended Reference Materials for the Realization of Physicochemical Properties*** — This IUPAC book emphasizes highly accurate data on substances and materials that can be used as calibration standards. It covers physical, thermal, optical, and electrical properties. (K. N. Marsh, ed., Blackwell Scientific Publications, Oxford, 1987)

10. *The Merck Index* — Now in its 14th Edition (published in 2006), *The Merck Index* is a widely used source of data on over 10,000 compounds, chosen particularly for their importance in biology, medicine, and ecology. A short monograph on each compound gives information on the synthesis and uses as well as physical and toxicological properties. A CD-ROM accompanies the book. (Maryadele J. O'Neil, ed., John Wiley & Sons, Indianapolis, IN, 2006)

E. Summary of Useful Web Sites for Physical and Chemical Properties

Most of the Web sites in the following list provide direct access to factual data on physical and chemical properties. However, the list also includes portals that link to different property databases or describe the procedure for gaining access to electronic sources of property data. There are also a few chemical directory sites that are useful for obtaining formulas, synonyms, and registry numbers for substances of interest.

Web Site	Address	Comments
ACD/Labs Spectral Data	www.acdlabs.com/products/adh/	Infrared, Raman and NMR spectra collections from Coblentz Society and other sources
Advanced Chemistry Development	www.acdlabs.com	Chemical directory, with programs for estimating physical and spectral properties
Alloy Center	products.asminternational.org/alloycenter/	Physical, electrical, thermal, and mechanical properties of alloys
American Mineralogist Crystal Structure Database	www.geo.arizona.edu/AMS/amcsd.php	Lattice constants of minerals
Atomic Mass Data Center	www.nndc.bnl.gov/amdc	See B.10
Beilstein	www.accelrys.com/products/databases	See C.3
Biocatalysis/Biodegradation Database	umbbd.msl.umn.edu/	Biocatalytic reactions, biodegradation of chemical compounds
BioCyc	biocyc.org/	Metabolic pathways of microorganisms
BioInfo Bank	gibk26.bse.kyutech.ac.jp	Portal to ProTherm (protein thermodynamics), ProNit (protein–nucleic acid interactions), biomolecule structures
Biological Macromolecule Crystallization Database	xpdb.nist.gov:8060/BMCD4/index.faces	Crystal data and crystallization conditions for proteins, nucleic acids, and complexes
BRENDA	www.brenda-enzymes.info/	Enzyme nomenclature and properties
Cambridge Structural Database	www.ccdc.cam.ac.uk	See B.5
Carbon Dioxide Information Center	cdiac.esd.ornl.gov/	Data on atmospheric carbon dioxide
Ceramic Properties Databases	www.ceramics.org/knowledge-center/ceramic-resource-center/	Mechanical, thermal, and other properties of ceramic materials
ChemExper	www.chemexper.com/	Consolidated chemical catalogs from various suppliers; provides physical properties and safety data; links to molfiles and MSDS
Chemfinder	www.chemfinder.com	Chemical directory, with links to several property databases
Chemical Acronyms Database	bl-libg-doghill.ads.iu.edu/chem-web/databases/acronyms/index.php	Useful for associating chemical names and acronyms
Chemical Entities of Biological Interest (ChEBI)	www.ebi.ac.uk/chebi/	Dictionary of molecules and fragments, with identifiers and structures
ChemID*plus*	chem.sis.nlm.nih.gov/chemidplus/	Chemical directory
ChemIndustry	www.chemindustry.com/chemicals/	Chemical directory
CHEMnetBASE	www.chemnetbase.com	Portal to *CRC Chemical Dictionaries, Handbook of Chemistry and Physics, Properties of Organic Compounds*, etc.
ChemSpider	www.chemspider.com	Aggregation of chemical structures and other information from many public sources
ChemSynthesis Chemical Database	www.chemsynthesis.com	References to synthesis; limited property data
CODATA Databases	www.codata.org/resources/databases/	Thermodynamic key values and fundamental constants
Comparative Toxicogenomics Database (CTD)	ctd.mdibl.org/	Chemical – gene/protein interactions
CRC Combined Chemical Dictionary	www.chemnetbase.com/	See C.1
Crystallography Open Database (COD)	www.crystallography.net	Crystal data on 52,000 compounds

Web Site	Address	Comments
DECHEMA (DETHERM)	www.dechema.de/en/dtherm.html	See C.5
DIPPR Pure Compound Database	dippr.byu.edu	See B.3
Dortmund Data Bank	www.ddbst.de	See B.4
eMolecules	www.emolecules.com	Portal to databases of chemical suppliers
Enzyme Nomenclature Database	www.expasy.ch/enzyme/	IUBMB nomenclature for enzymes
European Bioinformatics Institute	www.ebi.ac.uk/Databases/	Nucleotide and protein sequences, protein structures, enzyme nomenclature and reactions
FDM Reference Spectra Databases	www.fdmspectra.com/	Infrared, Raman, and mass spectra
FIZ Chemie Berlin	www.fiz-chemie.de	Portal to Infotherm, Acronyms, thermophysical properties
FIZ Karlsruhe — ICSD	www.fiz-karlsruhe.de	See B.6
Fundamental Physical Constants	physics.nist.gov/cuu/constants	CODATA fundamental constants
Gmelin	www.accelrys.com/products/databases	See C.4
Handbook of Chemistry and Physics	hbcpnetbase.com	Web version of *CRC Handbook*
Hazardous Substances Data Bank	toxnet.nlm.nih.gov/cgi-bin/sis/htmlgen?HSDB	Physical and toxicological properties of chemicals of health or environmental importance
HITRAN Database	www.cfa.harvard.edu/hitran/	High resolution spectroscopic data for constituents of the atmosphere; parameters for calculating atmospheric transmission
Human Metabolome Database	hmdb.ca	Chemical and biological data on small molecule metabolites in humans
Infotherm	www.fiz-chemie.de/infotherm/servlet/ infothermsearch	Physical and thermal properties of pure compounds and mixtures
International Centre for Diffraction Data	www.icdd.com	See B.7
International Spectroscopic Data Bank	www.is-db.org	All types of spectra, deposited by users. Access is free
Ionic Liquids Database (ILThermo)	ilthermo.boulder.nist.gov/	Thermodynamic and thermophysical properties of ionic liquids and mixtures
IUBMB	www.chem.qmul.ac.uk/iubmb/	Enzyme and nucleic acid nomenclature
IUCr Data Activities	www.iucr.org/resources/data	Portal to crystallographic databases
IUPAC Home Page	www.iupac.org/resources/data	See B.13
IUPAC Kinetics Data	www.iupac-kinetic.ch.cam.ac.uk/	See B.13.b
IUPAC Nomenclature Rules	www.chem.qmul.ac.uk/iupac/	Useful site for organic and biochemical nomenclature
IUPAC-NIST Solubility Database	srdata.nist.gov/solubility/	See B.13.a
Klotho Biochemical Compounds Declarative Database	www.biocheminfo.org/klotho/	Structure diagrams of biochemical molecules
Knovel.com	www.knovel.com	Portal to *Lange's Handbook, Perry's Chemical Engineers' Handbook*, etc.
Kyoto Encyclopedia of Genes and Genomes (KEGG)	www.genome.ad.jp/kegg/	Includes data on drugs and other biochemical compounds
Landolt-Börnstein Online	www.springermaterials.com	See C.6
Lipidat	www.lipidat.tcd.ie	Structures and thermodynamic properties of lipids; crystal polymorphic transitions
MatWeb	www.matweb.com	Thermal, electrical, and mechanical properties of engineering materials
Metals Crystallographic Data File	www.tothcanada.com	See B.9
NASA Chemical Kinetics Data	jpldataeval.jpl.nasa.gov	Kinetic and photochemical data for stratospheric modeling
National Center for Biotechnology Information	www.ncbi.nlm.nih.gov	Portal to GenBank and other sequence databases
National Nuclear Data Center	www.nndc.bnl.gov	See B.12
National Toxicology Program	ntp.niehs.nih.gov	Chemical health and safety data
NIST Atomic Spectra Database	www.nist.gov/pml/data/atomicspec.cfm	Energy levels, wavelengths, and transition probabilities of atoms and atomic ions
NIST Ceramics Webbook	www.nist.gov/mml/ceramics/	See B.1
NIST Chemistry Webbook	webbook.nist.gov	Broad range of physical, thermal, and spectral properties
NIST Data Gateway	srdata.nist.gov/gateway/	Portal to all NIST data systems; see B.1
NIST Physical Reference Data	www.nist.gov/pml/data	Atomic and molecular spectra, cross sections, X-ray attenuation, and dosimetry data
NLM Gateway	www.nlm.nih.gov/databases	Portal to all National Library of Medicine databases
NMR Shift DB	www.nmrshiftdb.org	NMR data submitted by users
Nucleic Acid Database	ndbserver.rutgers.edu/	Crystal structures of nucleic acids

Web Site	Address	Comments
Particle Data Group	pdg.lbl.gov	See B.11
Polymers — A Property Database	www.polymersdatabase.com/	Properties of commercial polymers
Powder Diffraction File	www.icdd.com	See B.7
Properties of Organic Compounds	www.chemnetbase.com/scripts/pocweb.exe	See C.2
Protein Data Bank	www.rcsb.org	See B.8
PubChem	pubchem.ncbi.nlm.nih.gov/	Chemical directory with links to biological information
SABIO-Reaction Kinetics Database	sabio.villa-bosch.de/	Data on kinetics of biochemical reations
Sigma-Aldrich	www.sigmaaldrich.com/	Chemical catalogs; includes some physical property data
Spectral Database for Organic Compounds	riodb01.ibase.aist.go.jp/sdbs/cgi-bin/cre_index_cgi?lang=eng	MS, NMR, IR, Raman, and ESR spectra; 32,000 compounds measured at AIST, Japan
SpecInfo	onlinelibrary.wiley.com/book/10.1002/9780471692294	IR, NMR, and mass spectra
Spectra Online	www.ftirsearch.com/	FTIR and Raman spectra
SPRESI-web	www.spresi.de/	Structures, reactions, and some physical properties
SpringerMaterials	www.springermaterials.com	The new online version of Landolt-Börnstein Tables (C.6)
STN Easy	stneasy.cas.org	Chemical directory (and access to Chemical Abstracts databases)
STN Easy-Europe	stneasy.fiz-karlsruhe.de	European node of STN Easy
STN Easy-Japan	stneasy-japan.cas.org	Japanese node of STN Easy
Swissprot	bo.expasy.org/enzyme/	Enzyme nomenclature and related information
Thermodynamics of Enzyme-Catalyzed Reactions	xpdb.nist.gov/enzyme_thermodynamics/	Equilibrium constants of biochemical reactions
Thermodynamics Research Center	www.trc.nist.gov	See B.2
TOXNET	toxnet.nlm.nih.gov	Portal to HSDB and other databases on hazardous chemicals

Index

The most efficient way to use this index is to look for the pertinent *property* (e.g., vapor pressure, entropy), *process* (e.g., disposal of chemicals, calibration), or *general concept* (e.g., units, radiation). Most primary entries are subdivided into several secondary entries, e.g., under heat capacity there are 17 secondary entries such as air, metals, water, etc. Primary entries will be found for certain *classes of substances*, such as alloys, elements, organic compounds, refrigerants, semiconductors, etc. Primary entries are also given for the individual chemical elements and for a few compounds such as water and carbon dioxide. However, only the most important tables are listed under these substances. Therefore, the user will find in most cases that it is best to look first for the property of interest, then examine the table or tables that are referenced.

Entries in boldface type are the titles of tables as they appear in the Table of Contents.

The reference given for each index term is the inclusive pages of the pertinent table (e.g., **8**-45 to 55). The introduction to each table describes the method of ordering the substances within that table.

The editor would be grateful for comments and suggestions on this index.

INDEX

I

STANDARD ATOMIC WEIGHTS (2009)

Atomic Number	Element	Symbol	Atomic Weight
1	Hydrogen*	H	1.008 [1.00784; 1.00811]
2	Helium	He	4.002602(2)
3	Lithium*	Li	6.94 [6.938; 6.997]
4	Beryllium	Be	9.012182(3)
5	Boron*	B	10.81 [10.806; 10.821]
6	Carbon*	C	12.011 [12.0096; 12.0116]
7	Nitrogen*	N	14.007 [14.00643; 14.00728]
8	Oxygen*	O	15.999 [15.99903; 15.99977]
9	Fluorine	F	18.9984032(5)
10	Neon	Ne	20.1797(6)
11	Sodium	Na	22.98976928(2)
12	Magnesium	Mg	24.3050(6)
13	Aluminum	Al	26.9815386(8)
14	Silicon*	Si	28.085 [28.084; 28.086]
15	Phosphorus	P	30.973762(2)
16	Sulfur*	S	32.06 [32.059; 32.076]
17	Chlorine*	Cl	35.45 [35.446; 35.457]
18	Argon	Ar	39.948(1)
19	Potassium	K	39.0983(1)
20	Calcium	Ca	40.078(4)
21	Scandium	Sc	44.955912(6)
22	Titanium	Ti	47.867(1)
23	Vanadium	V	50.9415(1)
24	Chromium	Cr	51.9961(6)
25	Manganese	Mn	54.938045(5)
26	Iron	Fe	55.845(2)
27	Cobalt	Co	58.933195(5)
28	Nickel	Ni	58.6934(4)
29	Copper	Cu	63.546(3)
30	Zinc	Zn	65.38(2)
31	Gallium	Ga	69.723(1)
32	Germanium	Ge	72.63(1)
33	Arsenic	As	74.92160(2)
34	Selenium	Se	78.96(3)
35	Bromine	Br	79.904(1)
36	Krypton	Kr	83.798(2)
37	Rubidium	Rb	85.4678(3)
38	Strontium	Sr	87.62(1)
39	Yttrium	Y	88.90585(2)
40	Zirconium	Zr	91.224(2)
41	Niobium	Nb	92.90638(2)
42	Molybdenum	Mo	95.96(2)
43	Technetium**	Tc	
44	Ruthenium	Ru	101.07(2)
45	Rhodium	Rh	102.90550(2)
46	Palladium	Pd	106.42(1)
47	Silver	Ag	107.8682(2)
48	Cadmium	Cd	112.411(8)
49	Indium	In	114.818(3)
50	Tin	Sn	118.710(7)
51	Antimony	Sb	121.760(1)
52	Tellurium	Te	127.60(3)
53	Iodine	I	126.90447(3)
54	Xenon	Xe	131.293(6)
55	Cesium	Cs	132.9054519(2)
56	Barium	Ba	137.327(7)
57	Lanthanum	La	138.90547(7)
58	Cerium	Ce	140.116(1)
59	Praseodymium	Pr	140.90765(2)
60	Neodymium	Nd	144.242(3)
61	Promethium**	Pm	
62	Samarium	Sm	150.36(2)
63	Europium	Eu	151.964(1)
64	Gadolinium	Gd	157.25(3)
65	Terbium	Tb	158.92535(2)
66	Dysprosium	Dy	162.500(1)
67	Holmium	Ho	164.93032(2)
68	Erbium	Er	167.259(3)
69	Thulium	Tm	168.93421(2)
70	Ytterbium	Yb	173.054(5)
71	Lutetium	Lu	174.9668(1)
72	Hafnium	Hf	178.49(2)
73	Tantalum	Ta	180.94788(2)
74	Tungsten	W	183.84(1)
75	Rhenium	Re	186.207(1)
76	Osmium	Os	190.23(3)
77	Iridium	Ir	192.217(3)
78	Platinum	Pt	195.084(9)
79	Gold	Au	196.966569(4)
80	Mercury	Hg	200.59(2)
81	Thallium*	Tl	204.38 [204.382; 204.385]
82	Lead	Pb	207.2(1)
83	Bismuth	Bi	208.98040(1)
84	Polonium**	Po	
85	Astatine**	At	
86	Radon**	Rn	
87	Francium**	Fr	
88	Radium**	Ra	
89	Actinium**	Ac	
90	Thorium	Th	232.03806(2)
91	Protactinium	Pa	231.03588(2)
92	Uranium	U	238.02891(3)
93	Neptunium**	Np	
94	Plutonium**	Pu	
95	Americium**	Am	
96	Curium**	Cm	
97	Berkelium**	Bk	
98	Californium**	Cf	
99	Einsteinium**	Es	
100	Fermium**	Fm	
101	Mendelevium**	Md	
102	Nobelium**	No	
103	Lawrencium**	Lr	
104	Rutherfordium**	Rf	
105	Dubnium**	Db	
106	Seaborgium**	Sg	
107	Bohrium**	Bh	
108	Hassium**	Hs	
109	Meitnerium**	Mt	
110	Darmstadtium**	Ds	
111	Roentgenium**	Rg	
112	Copernicium**	Cn	
113	Ununtrium**	Uut	
114	Ununquadium**	Uuq	
115	Ununpentium**	Uup	
116	Ununhexium**	Uuh	
117	Ununseptium**	Uus	
118	Ununoctium**	Uuo	

* The first value for this element is the conventional value to be used if there is no information on the origin of the material. This is followed by the interval in which atomic weights in natural terrestrial materials are known to fall. See page 1-12 for further information.

** Since the element has no stable isotopes and no characteristic isotopic composition, an atomic weight is not tabulated.